Powders and Grains 2017 – 8th International Conference on Micromechanics on Granular Media

EPJ Web of Conferences Volume 140 (2017)

Montpellier, France
3 - 7 July 2017

Part 1 of 2

Editors:

Farhang Radjai
Saeid Nezamabadi
Stefan Luding
Jean-Yves Delenne

ISBN: 978-1-5108-4434-6

Printed from e-media with permission by:

Curran Associates, Inc.
57 Morehouse Lane
Red Hook, NY 12571

Some format issues inherent in the e-media version may also appear in this print version.

Printed by Curran Associates, Inc. (2017)

For additional information, please contact EDP Sciences – Web of Conferences
at the address below.

EDP Sciences – Web of Conferences
17, Avenue du Hoggar
Parc d'Activité de Courtabœuf
BP 112
F-91944 Les Ulis Cedex A
France

Phone: +33 (0) 1 69 18 75 75
Fax: +33 (0) 1 69 28 84 91

contact-edps@webofconferences.org

Additional copies of this publication are available from:

Curran Associates, Inc.
57 Morehouse Lane
Red Hook, NY 12571 USA
Phone: 845-758-0400
Fax: 845-758-2633
Email: curran@proceedings.com
Web: www.proceedings.com

TABLE OF CONTENTS

PART 1

PART 2

Editorial: Powders & Grains 2017

Farhang Radjai[1,2,⋆], *Saeid* Nezamabadi[1,⋆⋆], *Stefan* Luding[3,⋆⋆⋆], and *Jean-Yves* Delenne[4,⋆⋆⋆⋆]

[1]*LMGC, University of Montpellier, CNRS, 163 rue Auguste Broussonnet, 34090 Montpellier, France.*
[2]*$\langle MSE \rangle^2$, UMI 3466 CNRS-MIT, CEE, MIT, 77 Massachusetts Avenue, Cambridge 02139, USA.*
[3]*MSM, ET, MESA+, Universiteit Twente, P.O. Box 217, 7500 AE Enschede, Netherlands.*
[4]*IATE, UMR1208 INRA - CIRAD - University of Montpellier - SupAgro, 34060 Montpellier, France.*

Abstract. Granular materials challenge researchers and engineers in various fields not only because they occur with a broad variety of grain sizes, shapes, and interactions in nature and industry, but also because they show a rich panoply of mechanical states. Despite their polymorphism, all these different types of soils, powders, granules, ores, pharmaceutical products, etc., are instances of granular matter with the same least common denominator of being sandlike (psammoid in Greek), i.e., solid grains interacting via frictional contacts. With nearly 400 research papers, this volume provides a general picture of current research on granular materials across the world. The topics covered by this volume range from small deformations of particles interacting through enduring frictional-elastic contacts to rapid flows of particles that exchange momentum and energy in multiple collisions, either in single phase or particle-fluid multi-phase flows. Applications range from powder processing to chemical engineering and modeling of geomaterials. This volume also covers four emerging sectors of research: particle shape effects (with broad scopes for designing innovative behaviors), granular suspensions (bridging between the two distinct communities of non-Brownian suspensions and immersed granular flows), particle breakage (with crucial relevance to most fields of application), material instability (implying failure mechanisms under various loading conditions), and environmental granular processes (including bio-stabilization of slopes, root-soil interactions, large-scale geological processes and astryphysics of, e.g., asteroids).

1 Powders and Grains: goals and scope

The main scope of the *Powders and Grains* conference is to communicate recent developments for a better understanding of particle-scale interactions and microstructure in granular materials, and their upscaling to macroscopic scales. Understanding the behavior of granular materials from a micromechanical perspective is of fundamental importance in many areas of science and technology. On one hand, granular materials are key elements in a wide variety of industrial and natural processes. On the other hand, they provide a rich model for all non-equilibrium systems of interacting particles at

⋆e-mail: franck.radjai@umontpellier.fr
⋆⋆e-mail: saeid.nezamabadi@umontpellier.fr
⋆⋆⋆e-mail: s.luding@utwente.nl
⋆⋆⋆⋆e-mail: jean-yves.delenne@inra.fr

all scales. For this reason, the micromechanics of granular materials is a common meeting ground for applied mathematics, geology, and physics, as well as chemical, civil and mechanical engineering.

The papers published in this volume were presented during the conference *Powders and Grains 2017* held in *Montpellier, France on 3-7 July 2017.* This is the 8th International Conference on the Micromechanics of Granular Media organized under the auspices of the Association pour l'Etude de la Micromécanique des Milieux Granulaires (AEMMG). Earlier conferences were held in 1989 at Clermont-Ferrand, France; 1993 at Birmingham, UK; 1997 at Durham, North Carolina, USA; 2001 at Sendai, Japan; 2005 at Stuttgart, Germany; 2009 at Golden, Colorado, USA; 2013 at Sydney, Australia.

The proceedings of the last two conferences can be accessed following the links
PG2009: http://aip.scitation.org/toc/apc/1145/1?expanded=1145
PG2013: http://aip.scitation.org/toc/apc/1542/1?expanded=1542 .

The present PG2017 collection includes 10 review papers. The papers have been organized in the following sections:

1. Review papers
2. Granular solids
3. Granular flow
4. Granular gas
5. Particle properties
6. Particle shape effects
7. Particle breakage
8. Cohesive granular materials
9. Fluids and particles
10. Material instability
11. Continuum modeling
12. Geomaterials
13. Powder processing
14. Environmental granular processes
15. Particle simulations and particle-based methods
16. Miscellaneous

Despite this classification, meant to highlight the central research question investigated in each paper, most papers potentially belong to two or more sections. For example, several papers considered in "Particle simulations and particle-based methods" are actually on the rheology of granular flows. In the same way, the papers in "Miscellaneous" could well be distributed in other sections. The only exception is the one dealing with the statistics of publications on granular materials, showing both the extent and impact of this research direction in one particular well-known journal.

In fact, all papers deal in some way with one or more of the first three items: granular solids, flows, and gases. All other sections may therefore be considered as papers focusing on one feature of granular rheology. The item "Environmental granular processes" is new and concerns natural phenomena in which granular materials play an important role. It includes plant-soil interactions, large-scale geological processes and even astrophysical systems such as asteroids. Although there are only a few papers in this section, they reflect important research fields that may benefit from progress in understanding the micromechanics of granular materials.

The papers were reviewed by the members of the Scientific Committee, Invited Speakers, selected experts, and members of the Organizing Committee. The editors needed nearly four months to get all papers reviewed, and would like to sincerely acknowledge the help and efforts of all reviewers, some of whom reviewed more than 30 papers! Without such strong support, this volume would not have appeared in time. Our special thanks go to Heinrich Jaeger, Ferenc Kun, François Nicot, Felix Darve, Jeffrey Morris and Guillaume Ovarlez, who accepted to handle papers related to the focus sessions.

We also thank graduate students P. Schumacher and L. Amarsid from the LMGC for their help with the processing of papers.

2 Organizing Committee

2.1 AEMMG Association

The "Association pour l'Etude de la MicroMécanique des Milieux Granulaires" (AEMMG) is an association created to promote the Powders and Grains meeting every four years. It was created by Roland Gourvès in 1989 in France. The board is composed of

- Stefan Luding, President
- Antoinette Tordesillas, Secretary
- Jean-Noël Roux, Treasurer

The committee is composed of international researchers working on granular materials. See www2.msm.ctw.utwente.nl/msm_www/AEMMG. The committee meeting takes place every four years during the conference. The members of the committee act as the Scientific Committee for the conference. Here is the full list:

- Tomaso Aste
  UK, University College London
- Katalin Bagi
  Hungary, Budapest University of Technology & Economics
- Jean-Pierre Bardet
  USA, University of Miami
- Bob Behringer
  USA, Duke University
- Bernard Cambou
  France, Ecole Centrale de Lyon
- Ching S. Chang
  USA, University of Massachusetts Amherst
- Felix Darve
  France, Institut polytechnique de Grenoble
- Olivier Dauchot
  France, ESPCI
- Itai Einav
  Australia, University of Sydney
- Pierre Evesque
  France, Ecole Centrale Paris
- Mojtaba Ghadiri
  UK, University of Leeds
- Nico Gray
  UK, University of Manchester
- Gerd Gudehus
  Germany, Universität Karlsruhe
- Takahiro Hatano
  Japan, University of Tokyo
- Hisao Hayakawa
  Japan, University of Kyoto
- Hans Herrmann
  Switzerland, ETH-Zurich
- May Hou
  China, Chinese Academy of Sciences
- Heinrich M. Jaeger
  USA, University of Chicago
- Jim Jenkins
  USA, Cornell University
- Mingjing Jiang
  China, Tongji University
- Luigi La Ragione
  Italy, Politecnico di Bari

- Mario Liu
  Germany, University of Tübingen
- Stefan Luding
  Netherlands, University of Twente
- Kenichi Maeda
  Japan, Nagoya Institute of Technology
- Hernan Makse
  USA, City University of New York
- Anita Mehta
  India, Bose National Centre for Basic Sciences
- Francisco Melo
  Chile, Universidad de Santiago de Chile
- Hans Muhlhaus
  Australia, University of Queensland
- Masami Nakagawa
  USA, Colorado School of Mines
- Prabhu Nott
  India, Indian Institute of Science, Bangalore
- Catherine O'Sullivan
  UK, Imperial College London
- Luc Oger
  France, Université de Rennes 1
- Jin Ooi
  UK, University of Edinburgh
- Thorsten Pöschel
  Germany, University of Erlangen-Nuremberg
- Olivier Pouliquen
  France, CNRS, Université d'Aix-Marseille
- Farhang Radjai
  France, CNRS, Université de Montpellier
- Rasmussen Keld
  Denmark, Aarhus University
- Jean-Noël Roux
  France, Université Paris-Est
- Mark Shattuck
  USA, City University of New York
- Rodrigo Soto
  Chile, Universidad de Chile
- Colin Thornton
  UK, University of Birmingham
- Antoinette Tordesillas
  Australia, University of Melbourne
- Gioacchino Viggiani
  France, Université Joseph Fourier
- Aibing Yu
  Australia, Monash University
- Iker Zuriguel
  Spain, Universidad de Navarra

With this we also want to commemorate two of our members who recently passed away, namely Jürgen Tomas, Germany, and Antonio Castellanio, Spain.

2.2 Local Committee

The local organizing committee of the *Powders and Grains 2017* conference was mainly composed of the members of the Réseau MiDi (Milieux Divisés), a local network of researchers and laboratories in Montpellier, working on granular materials (see www.reseau-midi.org). It was extended to several scientists from other laboratories in France. Here is the list:

- Farhang Radjai (***chair***)
  LMGC, CNRS, University of Montpellier
- Jean-Yves Delenne (***co-chair***)
  IATE, INRA, University of Montpellier

- Emilien Azéma
  LMGC, University of Montpellier
- Ludovic Berthier
  L2C, CNRS, University of Montpellier
- Gael Combe
  3SR, University of Grenoble Alpes
- Carole Delenne
  Hydrosciences, University of Montpellier
- Agnès Duri
  IATE, INRA, University of Montpellier
- Thierry Fourcaud
  AMAP, CIRAD, Montpellier
- Xavier Frank
  IATE, INRA, University of Montpellier
- Claire Mayer
  IATE, INRA, University of Montpellier
- Serge Mora
  LMGC, University of Montpellier
- Saeid Nezamabadi
  LMGC, University of Montpellier
- François Nicot
  IRSTEA, Grenoble
- Luc Oger
  IPR, University Rennes 1
- Olivier Pouliquen
  IUSTI, CNRS, University Aix-Marseille
- Vincent Richefeu
  3SR, University of Grenoble Alpes
- Jean-Noël Roux
  Navier, IFSTTAR, University Paris-Est
- Thierry Ruiz
  IATE, University of Montpellier
- Lydie Staron
  Institut Jean Le Rond d'Alembert, University Pierre and Marie-Curie
- Alfredo Taboada
  Geosciences, University of Montpellier

3 Invited speakers

The invited talks of the *Powders and Grains 2017* conference were:

- Malcolm Bolton, University of Cambridge, UK
  Geo-phenomena: finding the granular mechanism
- Karen Daniels, North Carolina State University, USA
  The role of rorce Networks in granular materials
- Elisabeth Guazzelli, Aix-Marseille University, France
  Rheology of dense suspensions of non colloidal particles
- Ken Kamrin, Massachusetts Institute of Technology, USA
  A hierarchy of granular continuum models: Why flowing grains are both simple and complex?
- Jim Jenkins, Cornell University, USA
  Dense, collisional, shearing flows of compliant spheres
- Takashi Matsushima, University of Tsukuba, Japan
  The granular processes in planetary surfaces
- Vanessa Magnanimo, University of Twente, Netherlands
  Micromechanics of complex granular materials: a focus on small strain behavior

- Thorsten Pöeschel, Universität Erlangen-Nürnberg, Germany
  History and structure of granular packings

- Roland Pellenq, Massachusetts Institute of Technology, USA
  The potential of mean force concept for bridging (length and time) scales in the modeling of complex porous materials

- Matthias Schröter, Friedrich-Alexander University in Erlangen, Germany
  A local view on the role of shape and friction

- Matthieu Wyart, École Polytechnique Fédérale de Lausanne, Switzerland
  Unifying suspension and granular flows near jamming

- Runyu Yang, University of New South Wales, Australia
  Discrete modelling of compaction of non-spherical particles

- Iker Zuriguel, Universidad de Navarra, Spain
  Clogging and unclogging of many-particle systems passing through a bottleneck

Acknowledgements

We acknowledge the support of Eventime (www.eventime.fr) for the organization of this conference in Montpellier. We also gratefully acknowledge financial support for the conference and publication of this volume by the following organizations:

- Région Occitanie (www.laregion.fr)

- University of Montpellier (www.umontpellier.fr)

- LabEx Numev (www.lirmm.fr/numev)

- Laboratoire de Mécanique et Génie Civil (www.lmgc.univ-montp2.fr), University of Montpellier

- Laboratoire Charles Coulomb (www.coulomb.univ-montp2.fr), University of Montpellier

- HydroSciences (www.hydrosciences.org), University of Montpellier

- Agropolis Fondation (www.agropolis-fondation.fr)

- Laboratoire de Micromécanique et Intégrité des Structures (www.irsn.fr/FR/Larecherche/Organisation/equipes/combustible-situations-accidentelles/MIST)

Statement of Peer review

In submitting conference proceedings to EPJ Web of Conferences, the editors of the proceedings certify to the Publisher that they adhere to the **Policy on Publishing Integrity** of the journal in order to safeguard good scientific practice in publishing.

1. All articles have been subjected to peer review administered by the proceedings editors.
2. Reviews have been conducted by expert referees, who have been requested to provide unbiased and constructive comments aimed, whenever possible, at improving the work.
3. Proceedings editors have taken all reasonable steps to ensure the quality of the materials they publish and their decision to accept or reject a paper for publication has been based only on the merits of the work and the relevance to the journal.

Title, date and place of the conference

Powders and Grains 2017
8th International Conference on Micromechanics on Granular Media

3-7 July, 2017
Montpellier, France

Proceedings editor(s):

Farhang Radjai, Saeid Nezamabadi, Stefan Luding and Jean-Yves Delenne

Date and editor's signature

17 March 2017

EPJ Web of Conferences **140**, 13001 (2017) DOI: 10.1051/epjconf/201714013001
Powders & Grains 2017

Flow characteristics of metallic powder grains for additive manufacturing

Bernhard Peters[1,*] and *Gabriele* Pozzetti[1,**]

[1] *Université du Luxembourg 6, rue Coudenhove-Kalergi L-1359 Luxembourg*

Abstract. Directed energy deposition technologies for additive manufacturing such as laser selective melting (SLM) or electron beam melting (EBM) is a fast growing technique mainly due to its flexibility in product design. However, the process is a complex interaction of multi-physics on multiple length scales that are still not entirely understood. A particular challenging task are the flow characteristics of metallic powder ejected as jets from a nozzle and shielded by an inert turbulent gas flow. Therefore, the objective is to describe numerically the complex interaction between turbulent flow and powder grains. In order to include both several physical processes and length scales an Euler-Lagrange technology is applied. Within this framework powder is treated by the Discrete-Element-Method, while gas flow is described by Euler approaches as found in classical Computational Fluid Dynamics (CFD). The described method succeeded in delivering more accuracy and consistency than a standard approach based on the volume averaging technique and therefore, is suited for the solution of problems within an engineering framework.

1 Introduction

Additive manufacturing, also known as 3D printing, processes synthesise a three- dimensional object by adding successive layers of materials under computer control [1] and is sometimes referred to as the third industrial revolution. In its broad sense additive manufacturing includes processes such as binder jetting, directed energy deposition, material extrusion, material jetting, powder bed fusion, sheet lamination and vat photo-polymerisation that apply a variety of materials such as polymeric materials, biological materials, ceramic materials and metallic materials. For metal-based additive manufacturing methods, the metal is supplied in a powder form, with each particle typically having a diameter of 20 -100 μm. Selective laser melting (SLM) and electron beam melting (EBM) apply metallic powders through one or several nozzles onto the substrate's surface where the powder is fused layer by layer into the substrate through an energy input e.g. laser or electron beam. The entire environment is protected by a shielding gas due to the reactive atmosphere of the fusing particles. Thus, the powder interacts strongly with the the fluid dynamics of the shielding gas, that effects crucially the trajectory of single powder particles and hence, its location of deposition. Numerical approaches developed by He et al. [2], Peyre et al. [3], Wen et al. [4] and Qi et al. [5] describe the fusion process, however, neglect the transport of powder between nozzles and substrate. Therefore, the objective of the current contribution is to describe numerically the complex process of powder deposition on a substrate.

*e-mail: bernhard.peters@uni.lu
**e-mail: gabriele.pozzetti@uni.lu

2 Numerical Approach

The Discrete Element Method (DEM), also called a Distinct Element Method, is probably the most often applied numerical approach to track the trajectories of particles in a system. Thus, DEM is a widely accepted and effective method to address engineering problems in granular and discontinuous materials, especially in granular flows, rock mechanics, and powder mechanics. Pioneering work in this domain has been carried out by Cundall [6], Haff [7], Herrmann [8] and Walton [9]. The volume of Allen and Tildesley [10] is perceived as a standard reference for this field.

2.1 Extended Discrete Element Method

The Extended Discrete Element Method considers each particle of an ensemble as an individual entity with motion attached to it. The motion module of the Extended Discrete Element Method (XDEM) handles a sufficient number of geometric shapes that are believed to cover a large range of engineering applications. The exchange of data between continuous and discrete solutions requires careful coordination and a complex feed-back loop so that the coupled analysis converges to an accurate solution. This is performed by coupling algorithms between the Discrete Element Method to Computational Fluid Dynamics (CFD) [11]. For a more detailed review the reader is referred to Peters [12–15]. Hence, an ensemble of discrete and moving particles offers the highest potential to describe transport processes that is characterised by the motion of a rigid body through six degrees of freedom for translation along the three directions in space and rotation about the centre-of-mass.

DOI: 10.1051/epjconf/201714013001

By describing the degrees of freedom for each particle its motion is entirely determined. Newton's Second Law for conservation of linear and angular momentum describe position and orientation of a particle i as follows:

$$m_i \frac{d^2\vec{r}_i}{dt^2} = \sum_{i=1}^{N} \vec{F}_{ij}(\vec{r}_j, \vec{v}_j, \vec{\phi}_j, \vec{\omega}_j) + \vec{F}_{extern} \quad (1)$$

$$\bar{I}_i \frac{d^2\vec{\phi}_i}{dt^2} = \sum_{i=1}^{N} \vec{M}_{ij}(\vec{r}_j, \vec{v}_j, \vec{\phi}_j, \vec{\omega}_j) + \vec{M}_{extern} \quad (2)$$

where $\vec{F}_{ij}(\vec{r}_j, \vec{v}_j, \vec{\phi}_j, \vec{\omega}_j)$ and $\vec{M}_{ij}(\vec{r}_j, \vec{v}_j, \vec{\phi}_j, \vec{\omega}_j)$ are the forces and torquues acting on a particle i of mass m_i and tensor moment of inertia $\bar{I}_i$. Both forces and torques depend on position $\vec{r}_j$, velocity $\vec{v}_j$, orientation $\vec{\phi}_j$, and angular velocity $\vec{\omega}_j$ of neighbour particles j that undergo impact with particle i. The contact forces comprise all forces as a result from material contacts between a particle and its neighbours. Forces may include external forces due to moving grate bars, fluid forces and contact forces between the particles in contact with a bounding wall. This results in a system of coupled non-linear differential equations which usually cannot be solved analytically.

2.2 Computational Fluid Dynamics

Contrary to a pure continuous approach, the current concept treats powder as a discrete entity, while the flow of carrier gas is described by classical Computational Fluid Dynamics (CFD). Since the powder grains are tracked by a Lagrangian approach, their positions are known and thus, the corresponding CFD cell that contains them. Hence, powder grains present in a CFD cell occupy a certain volume meaning introducing a void fraction for the relevant transient differential conservation equations e.g. mass and momentum. Therefore, the flow is treated as a flow in porous media that introduces the porosity concept into the conservation equations. Hence, the turbulent flow ($Re \sim 2700$) is resolved by a Large Eddy Simulation (LES), that exchanges momentum with the powder grains. This coupling in conjunction with flow predictions is handled by an extended and advanced solver based on the OpenFOAM open-source framework.

3 Results

As mentioned-above transport of metallic powder from nozzles to the substrate within a shielding gas atmosphere was predicted by a software framework for advanced multi-physics simulation technology. The set-up of the geometry as depicted in figure 1 includes a central inlet for inert gas, through which also optical access for a laser is provided.

A further ring is arranged concentric to the optical access and serves as location for four nozzles. They inject the metallic powder into the gas phase, that is shielded by an inert gas. It is injected over a conical outer ring and is intended to create an inert atmosphere and to protect the

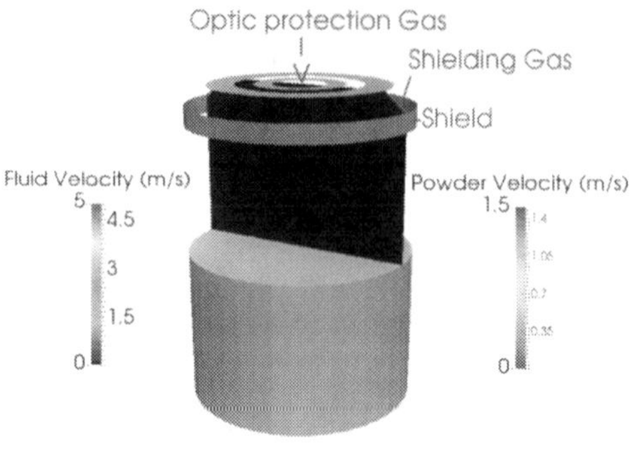

Figure 1: Geometrical set-up of the flow domain for gas and metallic powder

entire environment from entering of oxygen. The shielding gas is redirected radially and outwards due to its direction of the inlet velocity towards the centre axis of the laser. Hence, the velocity field resembles closely a torus-like structure shown in figure 2.

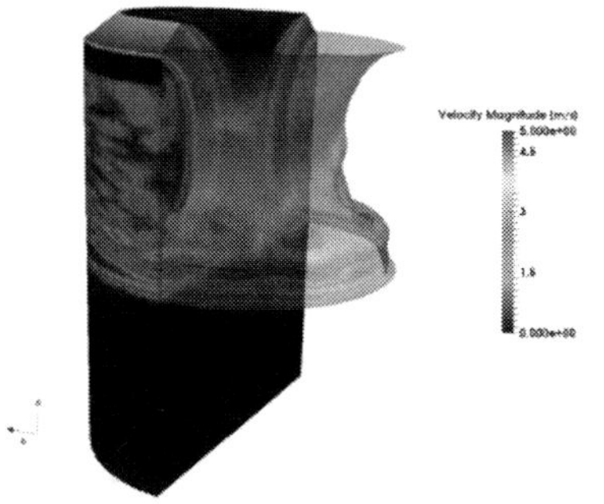

Figure 2: Centre cut through the velocity field with an iso-surface of pressure coloured with the velocity magnitude

The iso-surface in figure 2 clearly indicates the spatially varying velocity distribution that undergoes fluctuation over a velocity range of $\sim 5\ m/s$. Furthermore, the injected shielding gas directed towards the centre axis of the cylindrical arrangement is redirected radially, that requires radially dependent pressure field. It also is subject to strong spatial variations as indicated by the geometry of the iso-surface. These fluctuations in both velocity and pressure are also strongly reflected by the velocity distribution on the circumferential surface of the cylindrical geometry as depicted in figure 3.

Figure 3 highlights the typical steak-like structure of this class of flow configurations. Additionally, the velocity field shows significant temporal characteristics indicating a swirling flow about the centre axis. Similar to the velocity field in figure 2, the radial velocity undergoes equally strong fluctuations of $\sim 5\ m/s$. Both velocity and pressure field have a significant impact on the trajectories of the

DOI: 10.1051/epjconf/201714013001

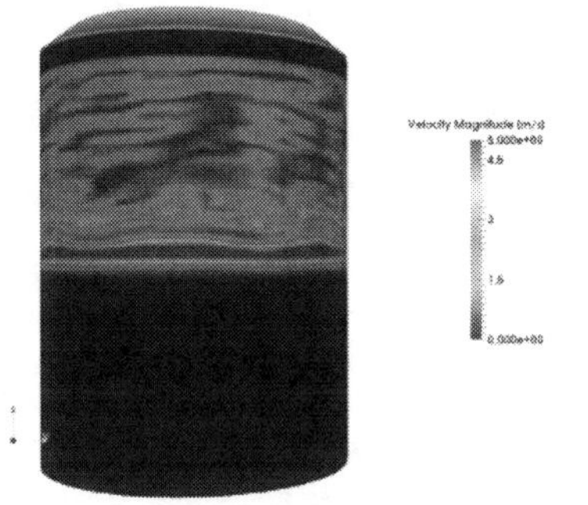

Figure 3: Velocity distribution on the cylindrical circumference

metallic powder that is shown in the sequence of figures 4a to 4c.

Metallic powder with its size distribution in the range of μm has a low inertia and therefore, is bound to follow the gas velocity profiles e.g. streamlines without mayor deviations. thus, the particles follow the velocity contours from the start of injection from the nozzles as depicted in figure 4a. Furthermore, figure 4a shows also the developing velocity field with a velocity tip penetrating the inert gas atmosphere. Particles are injected almost horizontally with a minor inclination only which requires them to be a radially accelerated to change their direction of the trajectory which can be observed in figure 4b. Thus, the direction of the metallic powder has changed from an almost horizontal injection direction to a downward flow meaning. The four jets of metallic powder very much focus to the centre area of the laser which is caused by the almost fully developed flow profile of the gas. An inclined injection through a conical area at the inlet generates also a strongly converging flow to the centre axis that the powder follows due to its small inertia as depicted in figure 4b. The following figure 5 depicts a magnified illustration of the powder grains coloured with their velocity magnitude (powder grains are enlarged for improved visibility and therefore, partially overlap) and the jet closely folowing the streamlines of the carrier gas.

However, the presence of a surface of the work piece, that cannot be penetrated by the gas, induces a diverging flow pattern as shown in figure 4c. Since the flow leaves the geometry over the circumferential area, the direction of the flow has to change again to an outward radial direction as already depicted in figure 3. Hence, after a converging flow a rapidly diverging gas flow is installed that has also a strong impact on the flow pattern of the metallic powder. As the powder follows very closely the gas velocity, the initially focussed powder jets experience likewise severe diverging flow characteristics. These characteristics generate four detached impact areas at the edge of the melt pool. Hence, no defined focal point exists where the four powder jets penetrate the melt pool that would enable controlled deposition and build-up of the substrate. Additionally, lowest temperatures are expected

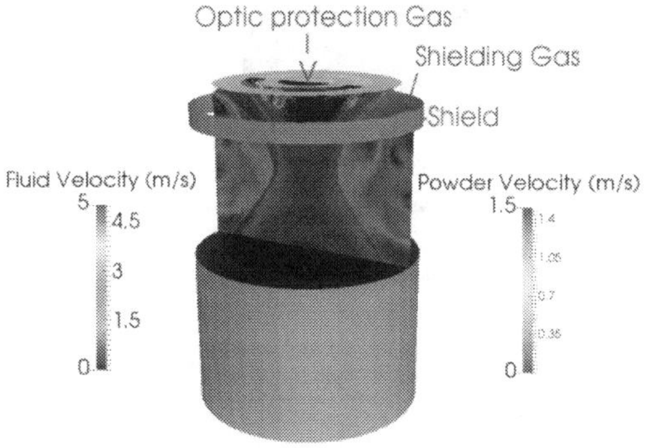

(a) at time: t=0.005 s

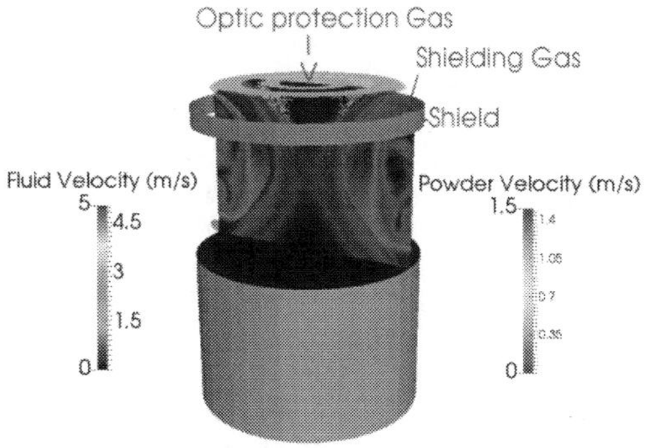

(b) at time: t=0.0085

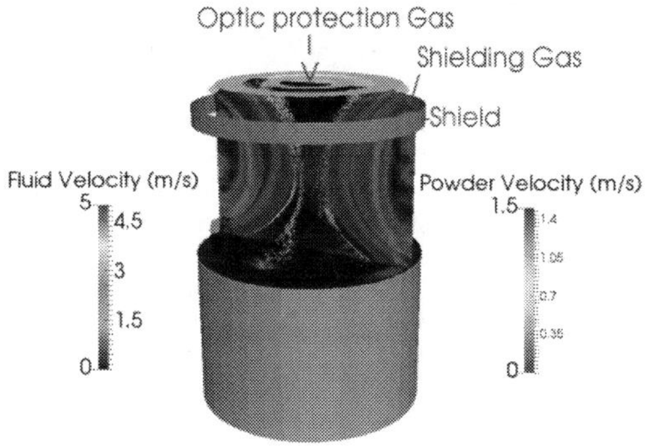

(c) at time: t=0.2 s

Figure 4: Centre slice of fluctuating velocity field with instantaneous positions of powder grains

at the edges of the melt pool so that less energy is available for the powder to melt as compared to the centre of the melt pool where the laser energy generates highest temperatures. Reduced temperatures may not be sufficient to melt the biggest powder particles entirely so that they appear as a partially melted inclusion inside the crystal structure of the solidified substrate. Consequently, no homogeneous

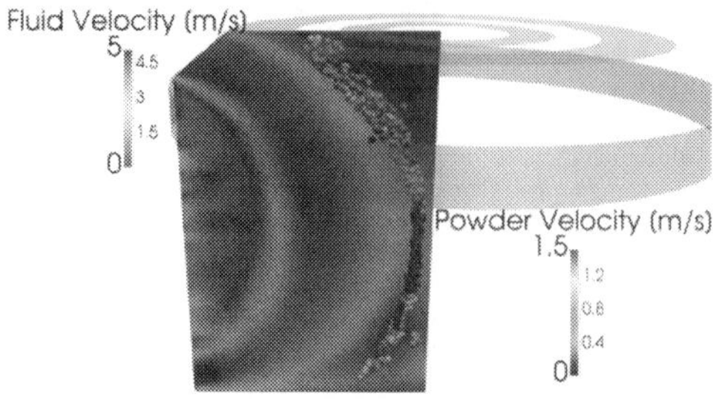

Figure 5: Close-up of a powder jet following closely the streamlines of the carrier gas

crystal structure is obtained, that in addition produces reduced mechanical strength.

4 Summary

Within the current study the flow pattern of both shielding gas and four jets of injected metallic powder were investigated numerically. For this purpose, an advanced multi-physics simulation technology was employed to obtain detailed results. The simulation technology is based on the Extended Discrete Element Method (XDEM) that couples discrete with continuous numerical techniques. Hence, the trajectory of the metallic powder grains are resolved with the Discrete Element Method (DEM), while the velocity field of the shielding gas is analysed by classical Computational Fluid Dynamics (CFD). Both phases discrete and continuous exchange momentum, that determines the trajectories of metallic powder to a large extent. Thus, position and velocity of each grain are predicted at all instances of time, that allows tracking the grains until they impact with the melt pool. Identifying the location of deposition of the metallic powder jets is of utmost importance for building up a desired geometry of the substrate. This is very difficult to achieve with experimental devices only, so that numerical techniques are complementing experiments in a favourable way.

References

[1] J. Excell, *The rise of additive manufacturing*, retrieved 2016-08-31, https://www.theengineer.co.uk/issues/24-may-2010/the-rise-of-additive-manufacturing/
[2] X. He, G. Yu, J. Mazumder, Journal of Physics D-Applied Physics **43**, 1 (2010)
[3] P. Peyre, P. Aubry, R. Fabbro, R. Neveu, A. Longuet, Journal of Physics D: Applied Physics **41**, 025403 (2008)
[4] S.Y. Wen, Y.C. Shin, Journal of Laser Applications **23**, 1 (2011)
[5] H. Qi, J. Mazumder, , H. Ki, Journal of Applied Physics **100**, 24903 (2006)
[6] P.A. Cundall, O.D.L. Strack, Geotechnique **29**, 47 (1979)
[7] P.K. Haff, B.T. Werner, Poweder Techn. **48**, 23 (1986)
[8] J.A.C. Gallas, H.J. Herrmann, S. Sokolowski, Phys. Rev. Lett. **69**, 1371 (1992)
[9] O. Walton, R. Braun, J. of Rheology **30**, 949 (1986)
[10] M.P. Allen, D.J. Tildesley, *Computer Simulation of Liquids* (Claredon Press Oxford, 1990)
[11] J.H. Ferzinger, M. Peric, *Computation Methods for Fluid Dynamics* (Springer Verlag, Heidelberg, 1996)
[12] B. Peters, *Thermal Conversion of Solid Fuels* (WIT Press, Southampton, 2003)
[13] A. Estupinan Donoso, F. Hoffmann, B. Peters, International Review of Mechanical Engineering **7**, 328 (2013)
[14] K. Samiei, B. Peters, Chemical Engineering Science **87**, 234 (2013)
[15] K. Samiei, G. Berhe, B. Peters, KONA Powder and Particle Journal (2013)

DOI: 10.1051/epjconf/201714005001

On the torsional loading of elastoplastic spheres in contact

Sadegh Nadimi[1,*], and *Joana* Fonseca[1]

[1]City, University of London, London, UK

Abstract. The mechanical interaction between two bodies involves normal loading in combination with tangential, torsional and rotational loading. This paper focuses on the torsional loading of two spherical bodies which leads to twisting moment. The theoretical approach for calculating twisting moment between two spherical bodies has been proposed by Lubkin [1]. Due to the complexity of the solution, this has been simplified by Deresiewicz for discrete element modelling [2]. Here, the application of a simplified model for elastoplastic spheres is verified using computational modelling. The single grain interaction is simulated in a combined finite discrete element domain. In this domain a grain can deform using a finite element formulation and can interact with other objects based on discrete element principles. For an elastoplastic model, the contact area is larger in comparison with the elastic model, under a given normal force. Therefore, the plastic twisting moment is stiffer. The results presented here are important for describing any granular system involving torsional loading of elastoplastic grains. In particular, recent research on the behaviour of soil has clearly shown the importance of plasticity on grain interaction and rearrangement.

1 Background

Mathematical models have been incorporated into discrete modelling of granular system describing the force-displacement relationship between two contacting rigid bodies. This has proven the reliability in simulating granular materials such as soil [3, 4]. The interaction includes normal contact with relative contact area motions such as sliding, rolling or spinning. The spinning around the axis of the contact normal creates a twisting moment (M_T). When M_T in combination with normal loading is applied to two grains in contact, the contact area will undergo some angular displacement (β). The shear forces at the contact will provide some resistance to sliding. Depending on the distribution of normal forces, the region that meets the Coulomb friction condition will experience sliding and the rest of the contact area will undergo sticking [5, 6]. The schematic of this mixed boundary problem for two identical spheres in contact where stick and slip regions coexist is shown in Fig. 1.

The relationship between the applied moment and the radius of stick region can be written as follows:

$$M_T = 2\pi \int_0^a q(r) r^2 dr \tag{1}$$

where $q(r)$ is torsional shear traction that is a function of r, the radial position from centre of the contact area. If $c \leq r \leq a$, the traction is limited to $q(r)=\mu F_N(r)$.

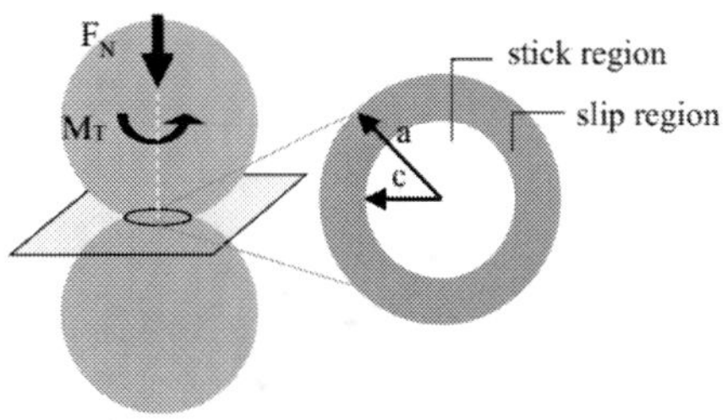

Fig. 1. Contact area including stick and slip regions for two identical spheres subjected to torsional moment and normal loading.

Lubkin (1951) delivers the solution to this problem by proposing an equation to define the shear stress at the contact surface within the stick region [1]. By combining Lubkin's solution with normal force distribution, the twisting moment can be obtained from the following expression [1, 5]:

$$M_T = \frac{\mu F_N a}{4\pi} \times \left\{ \frac{\dfrac{3\pi^2}{4} + k'k^2\left[6K(k) + (4k'^2 - 3)D\right] - 3kK(k)\sin^{-1}k' - 3k^2}{\left[K(k)\displaystyle\int_0^{\pi/2} \frac{\sin^{-1}(k'\sin\alpha)d\alpha}{(1-k'^2\sin^2\alpha)^{3/2}} - D(k)\int_0^{\pi/2} \frac{\sin^{-1}(k'\sin\alpha)d\alpha}{(1-k'^2\sin^2\alpha)^{1/2}}\right]} \right\} \tag{2}$$

---

[*] Corresponding author: sadegh.nadimi-shahraki@city.ac.uk

where $k=\sqrt{1-(c/a)^2}$, $k'=\frac{c}{a}$ and $\sin\alpha = r/c$.

$D(k)$ is the complete elliptical integral with modulus K, given by $D(k)=(K-E)/k2$ with K and E being the complete elliptical integrals of the first and second type, respectively.
Given the complexity of Eq. 1, a simplified solution was proposed by Deresiewicz (1954) between a, M_T and β based on an explicit approximation for numerical modelling which is defined as follows:

$$\frac{Ga^2\beta}{\mu F_N} = \frac{1}{8}\left[1-\sqrt{1-\frac{3}{2}\frac{M_T}{\mu F_N a}}\right]\times\left[3-\sqrt{1-\frac{3}{2}\frac{M_T}{\mu F_N a}}\right] \qquad (3)$$

where G is the shear modulus. Therefore, the torsional stiffness can be specified as:

$$k_t = \frac{dM_T}{d\beta} = \frac{16Ga^3}{3}\left[2\left(1-\frac{3}{2}\frac{M_T}{\mu F_N a}\right)^{-1/2} - 1\right]^{-1} \qquad (4)$$

It is important to note that the Eqs. 3 and 4 are only applicable for small values of twisting moment where $M_T/\mu F_N a << 1$ [5]. Fig. 2 shows a non-dimensional moment-twist profile derived from Eq. 3.

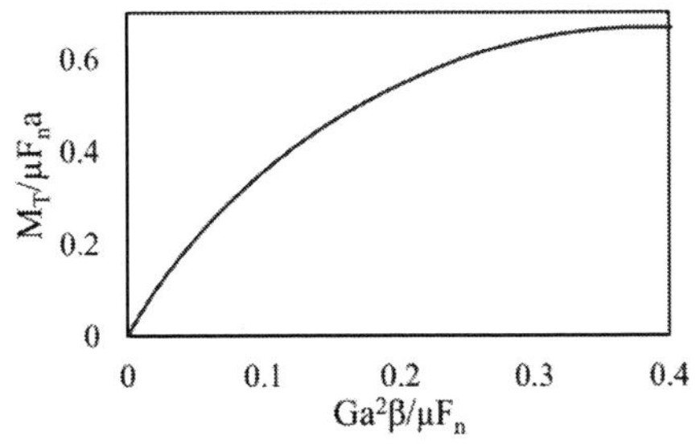

Fig. 2. Non-dimensional relationship between torque and twisting angle for spherical grains.

Despite the development of this theoretical approach for the elastic interaction of spheres in the fifties, the elastoplastic interaction under torsion has not been verified yet, as discussed in Thornton [4].

In this paper, the problem is replicated in the framework of a finite-discrete element method [7]. In this framework, a continuum body can deform using finite element formulation and can interact with other objects (rigid and deformable) based on the law of motion, *i.e.* Newton's second law, used in discrete element modelling. Firstly, the elastic behaviour of a sphere in contact is verified against theoretical equations. Then, the elastoplastic behaviour is presented. The aim of this study is to investigate the applicability of Deresiewicz's solution for elastoplastic grains in contact under twisting moment. The results have implications for describing a granular system with elastoplastic grains.

2 Elastic interaction

The interaction of two identical spheres in contact is simplified, due to symmetry, by the interaction of a sphere in contact with a rigid plate. In order to apply pure torsion to a deformable sphere, a rigid core was generated inside the sphere and was tied to the sphere (Fig. 3). Fig. 4 shows the numerical mesh of the problem in Abaqus software package [8]. The sphere has a diameter of 2.2mm and is represented by a mesh formed by 60,743 elements and 18,112 nodes. The mesh at the contact area was refined for a more accurate presentation of the problem. The material parameters used in the simulation are listed in Table 1. Explicit time discretization was employed to allow for future work on a large number of grains, since the implicit time discretization is computationally very expensive. The property of hard contact was defined between the sphere and the plate. Using 'hard contact' behaviour means that all the force is transmitted through the contact. Due to body deformability, the relation of normal force versus normal displacement with hard contact assumption follows exactly Hertzian theory [6, 9].
The simulation includes two steps: normal loading and torsional loading. In step one, controlled displacement of 10μm was applied to the sphere in the normal direction, which corresponds to 95N normal loading for this problem. In the second step, the sphere was purely rotated around the contact normal using controlled angular displacement of 0.04rad.

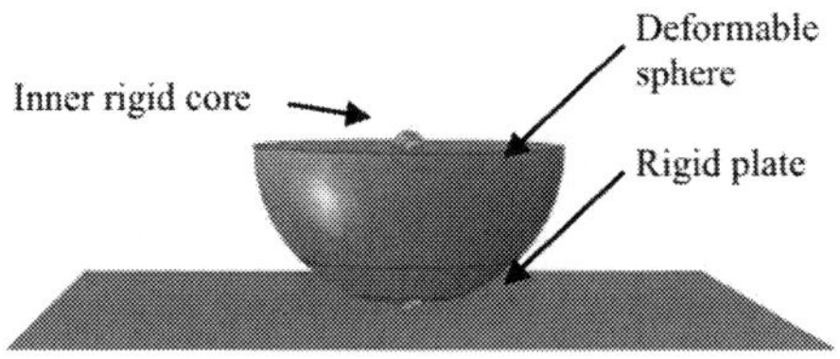

Fig. 3. Schematic showing the inner core and cut section of the deformable sphere.

Table 1. Physical and mechanical parameters of the spheres.

Elastic modulus	E (GPa)	63
Poisson ratio	υ (-)	0.3
Density	ρ (gr/mm^3)	2.5
Friction coefficient	μ (-)	0.22
Diameter	D (mm)	2.2

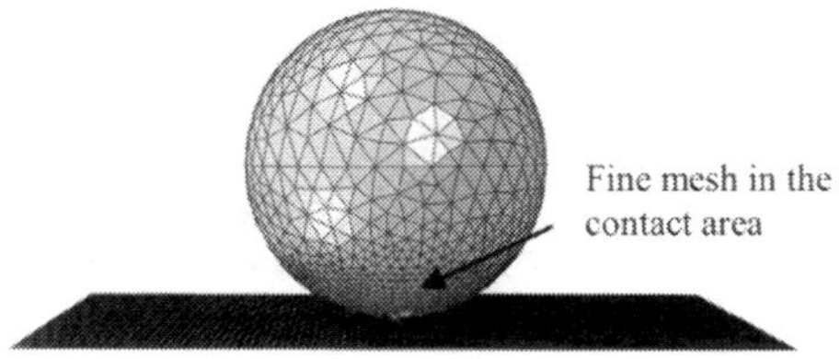

Fig. 4. Meshed sphere in contact with the rigid plate; mesh defined by smaller elements in the contact area.

The comparison between Deresiewicz theory and the numerical simulation is presented in Fig. 5. The plot shows a good agreement between the numerical model and theory. As can be seen in Fig. 5, there is a small discrepancy in the angular displacement (β) corresponding to the occurrence of pure slipping.

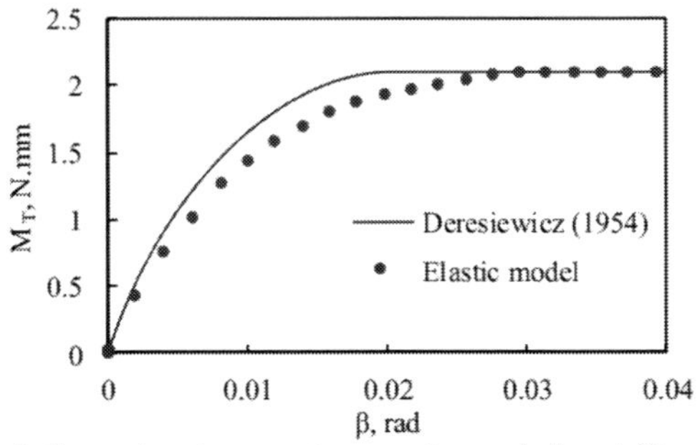

Fig. 5. Comparison between theory and numerical modelling for elastic interaction of two identical elastic spheres.

3 Elastoplastic interaction

The torsional interaction becomes plastic when the yield strength is reached in normal loading. This is particularly important for soil grains due to initial contact plasticity [4, 9]. The contact area would be larger for elastoplastic interaction in comparison with elastic interaction for a given normal load. Therefore, pure shear force shows higher tangential stiffness [4]. It is also expected to observe higher stiffness (k_t) due to the larger radius of contact area (a) in torsional loading.

To incorporate the plastic behaviour in numerical modelling, isotropic hardening constitutive laws were assigned to the sphere. It is assumed that the material yields at 100MPa and then hardens with hardening modulus of 20GPa (Table 2). The stress:strain relation assigned to the material is shown in Fig. 6.

The simulation steps were exactly the same as for the elastic model. In order to keep the normal load of 95N, a 47μm controlled displacement was applied in the normal direction which changes the 'a' value from 149μm (for elastic) to 319μm for plastic interaction. This was obtained by trial and error. In the second step, controlled angular displacement of 0.04rad was applied to the sphere around the contact normal.

Table 2. Isotropic hardening parameters used.

Yield strength	Y (MPa)	100
Hardening modulus	Et (GPa)	20

Fig. 7 shows the comparison of the numerical simulations for elastic and elastoplastic interaction in terms of twisting moment and angular displacement. It can be seen that the value of plastic twisting moment is 2.5 times the elastic twisting moment for a constant normal loading, while the contact area was nearly doubled. The normalised twisting moment, presenting in Fig. 8, shows that the plastic interaction is stiffer than the elastic interaction and pure slipping occurs at smaller values of angular displacement. Finally, the comparison of the theoretical, elastic and plastic models for normalised twisting moment versus normalised angular displacement are presented in Fig. 9. The good agreement observed, suggests the applicability of Deresiewicz solution for elastoplastic interaction.

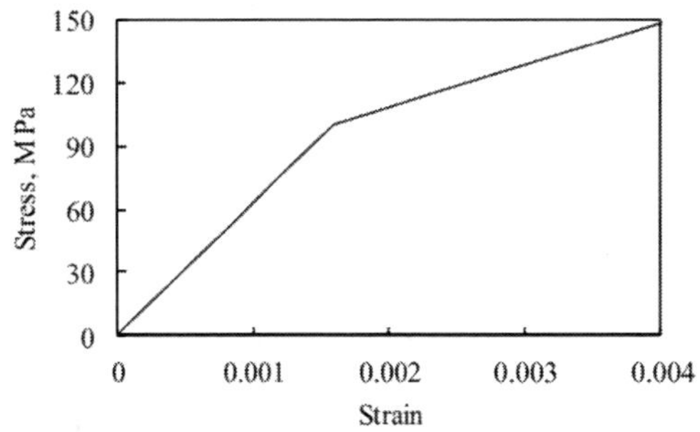

Fig. 6. Stress:strain for isotropic hardening material used in the simulation.

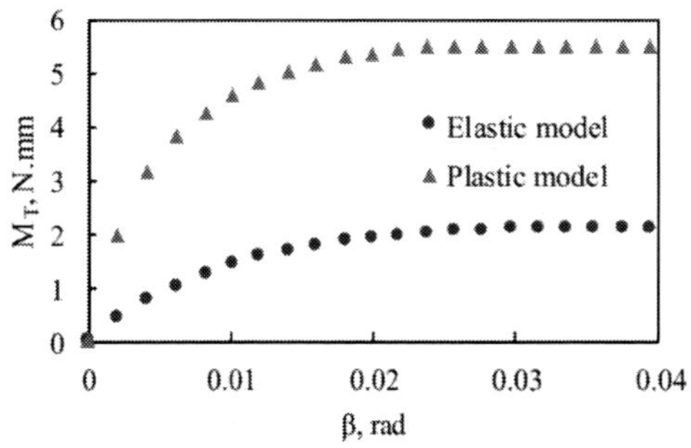

Fig. 7. Comparison between the numerical modelling of elastic and plastic interaction of two identical spheres under torsion and constant normal loading.

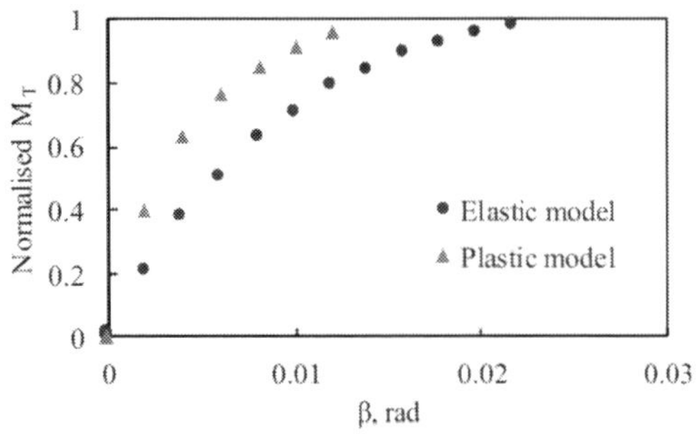

Fig. 8. Comparison between the normalised twisting moments derived from numerical modelling of elastic and plastic interaction of two identical spheres under torsion and constant normal loading.

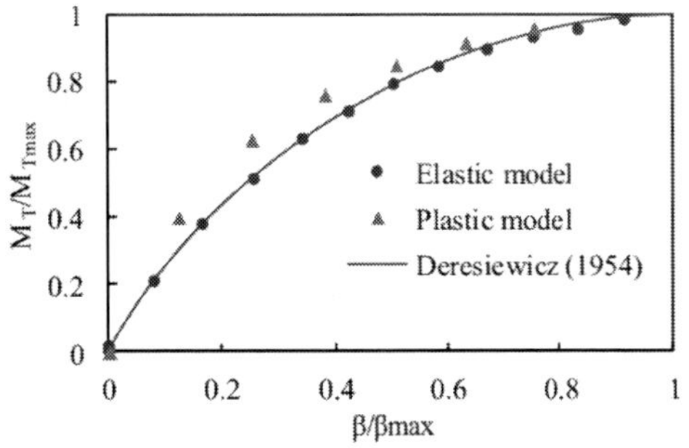

Fig. 9. Comparison between the numerical modelling of elastic and plastic interaction and Deresiewicz's solution of two identical spheres under torsion and constant normal loading.

4 Closing remarks

This study makes use of the general contact model for two identical spherical grains twisted around their contact normal as proposed by Lubkin and later simplified by Deresiewicz. The formulations proposed were verified for elastic and more importantly for elastoplastic interactions by means of a finite-discrete element method.

It was shown that the plastic torsional interaction is much stiffer than the elastic interaction. This is due to the larger contact area for plastic interaction under a given normal force. It was also presented that pure slipping occurs at smaller values of angular displacement for plastic interaction. The data presented here confirm the applicability of Deresiewicz's solution for elastoplastic torsional interaction.

The first author would like to express thanks to City, University of London for his doctoral scholarship.

References

1. J. L. Lubkin, ASME J. Appl. Mech., Vol. **73**, pp. 183-187 (1951).
2. H. Deresiewicz, J. Appl. Mech., Vol. **21**, pp. 52-56, (1954).
3. C. O'Sullivan, *Particulate Discrete Element Modelling: A Geomechanics Perspective*. Spon Press (2011).
4. C. Thornton, *Granular Dynamics, Contact Mechanics and Particle System simulations*, Springer (2015).
5. E. Dintwa, M. van Zeebroeck, E. Tijskens, H. Ramon, Granular Matter **7**: 169-179 (2005).
6. S. Nadimi, J. Fonseca, Géotechnique (under review, 2017).
7. A. Munjiza, *The Combined Finite-Discrete Element Method*, Wiley, London (2004).
8. ABAQUS User's Manual 2016, Dassault Systèmes, version 6.14.
9. S. Nadimi, J. Fonseca, R.N. Taylor, In *Deformation characteristics of geomaterials*: Proceedings of the 6th International symposium on deformation characteristics of geomaterials, Buenos Aires, Argentina (2015).
10. E. Dintwa, M. van Zeebroeck, E. Tijskens, H. Ramon, Eur. Phys. J. B **39**: 77 (2004).

A discrete element model of brittle damages generated by thermal expansion mismatch of heterogeneous media

Damien André[1,*], *Truong thi* Nguyen[2,**], *Nicolas* Tessier-doyen[2,***], and *Marc* Huger[1,****]

[1] *ENSCI, SPCTS, UMR 7315, F-87000 Limoges, France*
[2] *Univ. Limoges, SPCTS, UMR 7315, F-87000 Limoges, France*

Abstract. At the macroscopic scale, such media as rocks or ceramics can be seen as homogeneous continuum. However, at the microscopic scale these materials involve sophisticated micro-structures that mix several phases. Generally, these micro-structures are composed by a large amount of inclusions embedded in a brittle matrix that ensures the cohesion of the structure. These materials generally exhibit complex non linear mechanical behaviors that result from the interactions between the different phases. This paper proposes to study the impact of the diffuse damages that result from the thermal expansion mismatch between the phases in presence. The Discrete Element Method (DEM) that naturally take into account discontinuities is proposed to study these phenomena.

1 Introduction

Rocks, ceramics or refractories are heterogeneous materials that exhibit a multi-phase composition and involve different sizes of aggregates, bonding phases and various additives. The description and the prediction of thermo-mechanical behavior of such materials present a real difficulty due to their complex microstructure. Hashin & Shtrikman (H&S) have developed an analytical approach that predicts thermo-mechanical properties of perfectly cohesive multi-phase materials free of damages. However, most of real materials present numerous micro-cracks at room temperature resulting from thermal expansion mismatches between their phases. These damages highly influence the thermo-mechanical properties of such materials.

For instance, Young's modulus is strongly affected by the presence of micro-cracks and the measured values are often in disagreement with H&S's prediction [1]. The study proposed here proposes a numerical method in order to predict the occurrences of these damages and their influences on macroscopic properties such as Young's modulus. The Finite Element Method (FEM) is often used to study mechanical properties of materials. However, at the microscopic scale, this method is not adapted to describe discontinuities (micro-cracks) without assumptions on their localizations, their paths, their growths and their coalescences.

The DEM could be an interesting alternative to study multi-damaged materials because it takes naturally into account these discontinuities. The DEM implements a group of distinct elements (also named *discrete element*) that are in interaction though contacts or cohesive laws. This model consists of an assembly of discrete elements, deformable or not, linked up by simple mechanical laws to mimic the behavior of the material. The advantages are the description, in a natural way, of the crack initiations, the crack propagations and coalescences. Researchers have used this method to study damages in solids, such as rocks or ceramics. However, the main difficulty is to tackle quantitative results due to the necessity to find the relations between unknown microscopic laws, at the discrete element scale, and the macroscopic properties, at the structure scale, such as Young's modulus, Poisson's ratio or Coefficient of Thermal Expansion (CTE).

*e-mail: damien.andre@unilim.fr
**e-mail: truong_thi.nguyen@unilim.fr
***e-mail: nicolas.tessier-doyen@unilim.fr
****e-mail: marc.huger@unilim.fr

2 Model material

The reference material used for this work is a *model material*. A model material mimics, through a simplified framework, a given discoupled phenomenon observed with real and complex materials. In order to study the impact of thermal expansion mismatches, a two-phase model material, composed of mono-diameter spherical alumina inclusion embedded in a borosilicate glass matrix, is preferred. The thermo-mechanical parameters values for alumina and glass were chosen to produce a micro-crack network during the cooling stage of the sample preparation.

2.1 Processing of two-phase model materials

The model materials used in this study are composed of spherical monomodal alumina inclusions (average diameter equals to 500 μm) which are randomly placed in a borosilicate glass matrix. Alumina (corrundum phase) is a single oxide-based ceramic exhibiting fine grains. The main requirements of the selected glass matrix are homogeneity, isotropy, a rather chemical inertia and the capability to monitor thermal expansion coefficient. In this way, a borosilicate glass has been prepared from the melting of a vitrifiable mixture initially constituted by different raw materials containing silica, boron oxide and other secondary oxides. After grinding and sieving at $40 \mu m$, the glass powder was mixed during 3 hours with a slight proportion of organic additives used as binder and lubricant. A perfectly controlled volume fraction of alumina inclusions is incorporated in the mixture and is homogenized during 1 hour to ensure the dispersion of spherical alumina inclusions. Green specimens ($80 \times 40 \times 10$ mm^3) are shaped by uni-axial pressing (80 MPa) before debinding and sintering under uni-axial pressure (15MPa at 600°C) to remove residual porosity. Three different volume fractions of inclusion were prepared (15%, 30% and 45%). Figure 1 shows the microstructure of a final two-phase model material highlighting the micro-crack network. The main thermo-mechanical properties of both individual materials are given in the table 1. For details about the experimental processing of model materials, readers may refer to [1, 2].

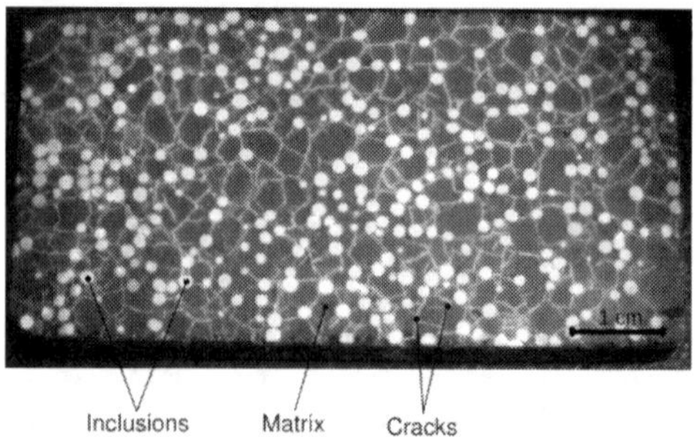

Figure 1: Damaged microstructure of the model material

Table 1: The main thermo-mechanical parameter values of the borosilicate glass and alumina

Properties	Matrix	Inclusion
Material	Glass	Alumina
Expansion coefficent α (K^{-1})	11.6×10^{-6}	7.6×10^{-6}
Young's modulus E (GPa)	72	340
Poisson's ratio ν	0.23	0.24
Bulk density ρ (kg.m^{-3})	2570	3970
Tensile stress limit (MPa)	50	380

2.2 Induced residual thermal stresses

The introduction of spherical particles in the matrix leads the occurring of thermal stresses during the cooling stage of the sample processing. Because of the CTE mismatch, the matrix is under tensile mode and the inclusions are subjected to compressive stresses. More precisely, The matrix is subjected to tensile orthoradial stresses σ_{orth} and radial compression σ_{rad}. Brittleness of the glass matrix induces orthoradial cracks that occurs and propagates in the matrix. Figure 1 shows the resulting micro-crack network with several inclusions on a model material.

3 The Hashin & Shtrikman model

Among the various theoretical approaches developed to predict elastic and thermal properties of two-phase materials, the H&S bounds are used here because this analytical approach is well suited to describe inclusional materials. The apparent Young's modulus can be predicted from the knowledge of (i) both volume fractions of the matrix and inclusional particles, (ii) bulk and shear modulus of each phase. For details about this analytical approach, reader may refer to [3].

As shown in Figure 2, the plots of the experimental Young's modulus (noted E) are lower than H&S lower bound (HS$^-$) prediction. The Young's modulus was measured through an ultrasonic technique [4]. The presented results underline the impact of the micro-cracked microstructure on the macroscopic thermo-mechanical behavior. The micro-cracks lead to a significative loss of rigidity compared to the H&S prediction. When the volume fraction of inclusions (15%, 30% and 45%) increases, the results for E values tend to enhance.

The following parts of this paper will introduce the discrete element model used to predict apparent Young's modulus that can not be treated with standard H&S models and FEM models.

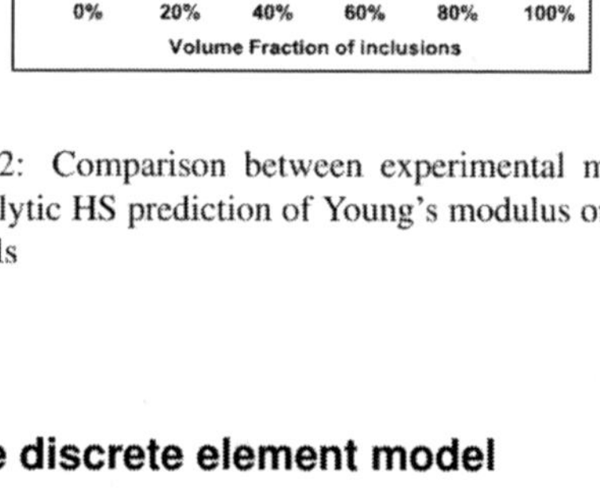

Figure 2: Comparison between experimental measures and analytic HS prediction of Young's modulus on model materials

4 The discrete element model

The discrete element approach used here is a mix between the *lattice* models and the *particle* models as it was first proposed by POTYONDY in [5]. Unlike continuous approaches, the main difficulty for DEM is to simulate quantitatively a continuum. The purpose of the free DEM software *GranOO* [1] is to face this difficulty. At this time, *GranOO* embeds some models that enable the quantitative simulation of mechanical and thermal behavior of continua with DEM [6].

4.1 Overview of the numerical procedure

Figure 3 shows the main simulation steps implemented in this study. The first step consists in calibrating the thermo-mechanical parameters of the discrete element model (for details, see the next sections). These microscopic parameters, related to the scale of the discrete elements, are denoted by the μ indice. These parameters are the microscopic Young's modulus E_μ, radius ratio R_μ, tensile strength limit σ_{μ_f} and CTE α_μ. Considering the values reported in Table 1, both borosilicate glass and alumina parameters are calibrated separately. After building the initial cubic discrete domains, the second step consists in inserting the spherical alumina inclusions. This step involves a simple geometrical algorithm. Then, the virtual sample is cooled from 450°C (the glass transition temperature) to 20°C (room temperature). This step, that involves thermo-mechanical simulations, leads to cracks initiation and propagation in the virtual samples (see Figure 4). In this study, the temperature is supposed constant within the sample. As a first approach, thermal conduction is neglected. Finally, the damaged virtual samples given by the last step are frozen. In this case, only the elastic behavior is taking into account in order to forbid crack extensions. These frozen samples are submitted to virtual tensile tests to evaluate their apparent Young's modulii E_{M}.

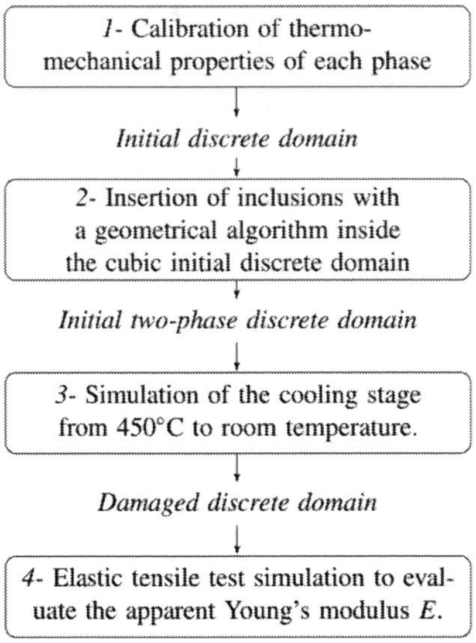

Figure 3: General view of the whole simulation steps

[1] see www.granoo.org

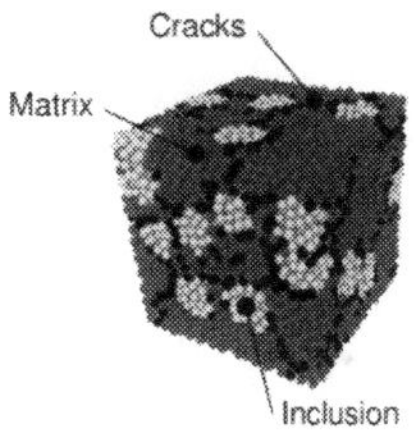

Figure 4: Virtual sample of the two-phase model material

4.2 Elastic behavior

The cohesive beam bond model [7] is used here to simulate perfectly elastic media characterized by a Young's modulus and a Poisson's ratio. The bond network given by the initial domain building is replaced by a beam network.

In such model, the discrete elements are bonded by Euler-Bernoulli beams. These beams are able to work in tensile, bending and torsion modes. The interactions resulting from the cohesive beams at the discrete element scale are denoted *microscopic*. They are symbolized by the μ suffix. In opposition, the emergent behaviors of the whole assembly network are denoted *macroscopic*. The macroscopic scale is symbolized by the *M* suffix. The macroscopic behavior corresponds to the behavior of the simulated material.

The cohesive beams are simply defined by two microscopic parameters : a Young's modulus E_μ and a radius ratio R_μ. The E_μ and the R_μ values are quantified thanks to a calibration process. The aim of this process is to reach the required values of macroscopic Young's modulus E_{M} and Poisson's ratio ν_{M} of the simulated material.

4.3 Brittle behavior

Within the discrete element approach, cracks are generally simulated by breaking the cohesive bonds if a criterion is reached. The main existing approaches are based on the computation of bond strains or stresses [5]. However, these approaches are not able to simulate complex crack paths such as the hertzian cone crack that exhibits on fused silica glass [8]. In a previous study, a new criterion, based on the computation of an equivalent Cauchy stress tensor and a maximal hydrostatic pressure value was developed. It has been shown that this criterion gives more accurate results than the standard DEM failure criteria. To compute this hydrostatic stress, an equivalent Cauchy stress tensor for each discrete element is computed as :

$$\bar{\bar{\sigma}}_i = \frac{1}{2\Omega_i}\left(\frac{1}{2}\sum_j \mathbf{r}_{ij}\otimes\mathbf{f}_{ij} + \mathbf{f}_{ij}\otimes\mathbf{r}_{ij}\right) \quad (1)$$

where : $\otimes$ is the tensor product between two vectors, $\bar{\bar{\sigma}}_i$ is the equivalent Cauchy stress tensor of the discrete element i, Ω_i is the volume of the discrete element i, $\mathbf{f}_{ij}$ is the force imposed on the discrete element i by a cohesive beam that bonds the discrete element i to another discrete element j and, $\mathbf{r}_{ij}$ is the relative position vector between the center of the two bonded discrete elements i and j. Finally, the developed criterion assumes that fracture occurs if the computed tensile hydrostatic stress is higher than a threshold value σ^f:

$$\frac{1}{3}\text{trace}(\bar{\bar{\sigma}}_i) \geq \sigma^f \quad (2)$$

4.4 Thermal expansion behavior

As the other mechanical parameters described in the previous sections, the thermal linear expansion α is considered at two levels : microscopically (α_μ) and macroscopically (α_M). The α_μ is the input parameter related to the thermal expansion coefficient of the cohesive beams and the α_M is the macroscopic thermal expansion of the discrete domain. To study the relation between the microscopic and the macroscopic thermal expansion, some simulations were carried out. These simulations conduct to the following simple result :

$$\alpha_\mu = \alpha_M \quad (3)$$

So, the thermal expansion coefficient can be introduced directly without any calibration.

5 Results and conclusion

Following the approach described in the last sections, discrete domains were subjected to cooling tests and then, to tensile tests in order to quantify their apparent Young's modulii. During the tensile tests, the micro-cracks are "frozen" and are not allowed to extend. Experimentally, the Young's modulii are measured by ultrasonic pulse echography technique that do not damage the materials. Finally, the results of numerical simulations and experimental observations are plotted in Figure 5, where : the absciss *Volume fraction* is the volumic fraction of inclusion in samples, the *Hv-* and *Hv+* curves denotes bounds of the H&S predictions, the *Numerical 3D model without cracks* curve corresponds to the undamaged DEM samples, the *Numerical 3D model with cracks* curve corresponds to the damaged DEM samples and the *Experimental* curve corresponds to the experimental observations.

These results shows that the *Numerical 3D model without cracks* match perfectly with the H&S predictions. This result confirm that the H&S predictions are not able to take into account the damages inside the material. However, the *Numerical 3D model with cracks* match with the experimental results measured by ultrasonic pulse echography technique. It allows to validate quantitatively the numerical 3D DEM model.

This work related to the simulation of thermo-elastic behaviour using DEM presents a significant improvement of the discrete element methods applied to the simulation of continuum media. This model has been applied to study the influence of the damages generated during thermal cooling of multiphase materials. These damages highly influence the rigidity of materials and is of high level of importance for common engineering applications. The proposed method seems to be adapted to predict this rigidity and allows to consider further studies to improve the understanding of more complex multiphase materials.

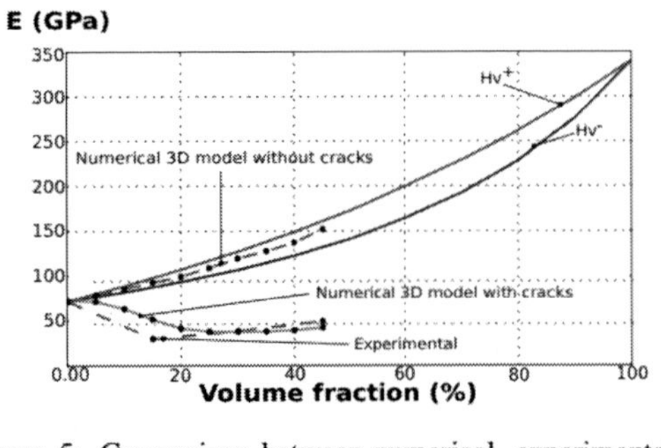

Figure 5: Comparison between numerical, experimental and H&S predictal Young's modulus

References

[1] N. Tessier-Doyen, J. Glandus, M. Huger, Journal of the European Ceramic Society **26**, 289 (2006)

[2] Y. Joliff, J. Absi, M. Huger, J. Glandus, Computational materials science **39**, 267 (2007)

[3] Z. Hashin, S. Shtrikman, Journal of the Mechanics and Physics of Solids **11**, 127 (1963)

[4] M. Huger, D. Fargeot, C. Gault, High Temperatures. High Pressures **34**, 193 (2002)

[5] D. Potyondy, P.A. Cundall, International Journal of Rock Mechanics and Mining Sciences **41**, 1329 (2004), rock Mechanics Results from the Underground Research Laboratory, Canada

[6] D. Andre, J. Charles, I. Iordanoff, *3D Discrete Element Workbench for Highly Dynamic Thermo-mechanical Analysis: GranOO* (Wiley, 2015), ISBN 9781848217720

[7] D. André, I. Iordanoff, J. luc Charles, J. Néauport, Computer Methods in Applied Mechanics and Engineering **213–216**, 113 (2012)

[8] D. André, M. Jebahi, I. Iordanoff, J. luc Charles, J. Néauport, Computer Methods in Applied Mechanics and Engineering **265**, 136 (2013)

Estimation of the velocity at the impact of high rate sliding granular masses

Francesco Federico[1,*], and *Chiara* Cesali[1]

[1]University of Rome 'Tor Vergata', Department of Civil Engineering, Via del Politecnico 1, 00133 Rome, Italy

Abstract. Two analytical *block* models to estimate the maximum velocity reached by granular flows are proposed. The first one models the speed evolution of *coarse-grained* materials flows (e.g. debris flows, avalanches); it is based on the power balance of a granular mass sliding along planar surface, written by taking into account the volume of the debris mass, an assigned interstitial pressure, the energy dissipation due to *(i)* grain inelastic collisions ('granular temperature' within a basal '*shear layer*'); *(ii)* friction along sliding surface; *(iii)* fragmentation of grains. The second model allows to simulate the speed evolution of *fine-grained* materials flows (e.g. mudflows, quick clays) by taking into account the dissipation of the excess pore water pressure due to consolidation phenomena. Finally, the comparison between the results obtained through the proposed models and field and laboratory measures is carried out.

1 Introduction

Very high stresses arise at the contact surface between an impacting fluid and a solid body that constitutes an obstacle to the fluid flow; furthermore, these stresses act only for a very short time period. Thus, structures exposed to the impact with a fluid or a fluidized granular mass may be affected by serious damages, mainly due to the shock pressures. To avoid undesirable damages to inhabited areas, technical countermeasures (e.g. check dams, barriers and walls) must be put into action; their design must be based on the values of the impact force. Several relationships to estimate the maximum impact force per unit width (F) or pressure (p_{max}) applied by a debris flow against structures have been proposed [1–4]:

$$p_{\max} = \frac{\rho v^4}{gh}; \quad F = (5 \div 12)\rho h v^2 \tag{1–2}$$

$$p_{\max} = (2 \div 5)\rho v^2; \quad F = \delta \cdot (\rho \cdot c \cdot v) \cdot h \tag{3–4}$$

ρ being the flow density; v, the impact velocity; h, the flow depth; c, the fluid celerity; g, the acceleration gravity; δ, a numerical coefficient ranged between 0.75÷0.9 [4]. For the impact of mudflows against structures, the following relationship is proposed [5]:

$$p_{\max} = c_d \rho A \frac{v^2}{2} \tag{5}$$

in which $A = bh$ represents the contact area; b, the flow width (for 2D cases, $b = 1$); c_d, the resistance coefficient evaluable as: $c_d = 0.18 \cdot F_r + 1.4$, with F_r, Froude number [5]. All previously described expressions show the dependence of the maximum pressure p_{max} or force F on the velocity (v) of the granular flow at the impact zone with rigid structures. A granular flow generally reaches the maximum velocity at the end of the '*first*' slope, where safeguarding measures (e.g. walls, barriers) are usually built. Thus, two analytical (*block*) models to estimate the maximum velocity reached by *coarse* and *fine* grained material flows are developed and proposed.

2 Main phenomena affecting coarse grained material flows

Experimental observations showed the growth of a basal "*shear layer*" [6], where initially large deformations and then dilation, *collisions* and *granular temperature* occur. By colliding each on others and with an irregular sliding surface, the particles change instantaneously their velocity (module and direction) and exhibit a sudden fluctuation (Fig. 1). To synthetically represent the grains' motion within the *shear layer*, the concept of *granular temperature* (T_g), defined as the average value of the square of the grains' velocity fluctuations with respect to their mean velocity, is introduced [7]. These velocity fluctuations nearly assume a Maxwellian distribution [8]; thus, T_g should be more properly defined by the variance of the velocity fluctuations [9]. For simplicity, T_g can be assumed proportional to the square of the shear rate [10]:

$$T_g = \frac{1}{15(1-e)}\left[1 + \frac{\pi}{12}\left(1 + \frac{5}{8 \cdot v_s g_0}\right)^2\right] \cdot \left(d_p \frac{dv}{dy}\right)^2 \tag{6}$$

* Corresponding author: fdrfnc@gmail.com

e being the restitution coefficient (ϵ [0,1]; $e < 1$ must be considered due to the inelastic nature of the collisions); $v(y)$, the grain velocity profile (y, normal to the shear direction); d_p, the average grain diameter; v_s , the solid fraction, which may assume the maximum value v_{max} (equal to 0.74 for spherical grains); g_0, the radial distribution function: $g_0(v_s) = (1 - v_s/v_{max})^{-2.5 \cdot v max}$ [11].

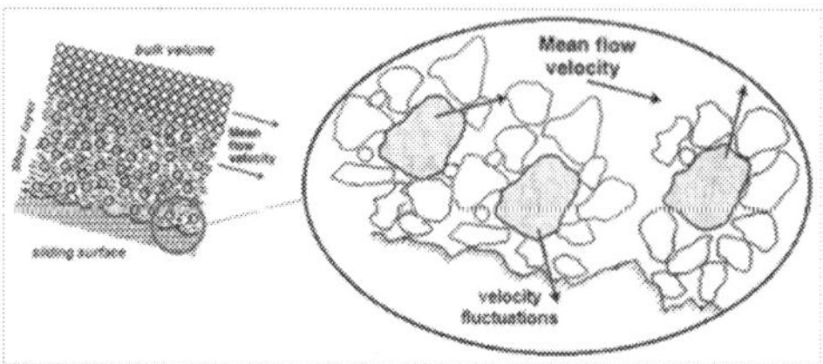

Fig. 1. Particles colliding with an irregular sliding surface, within the *shear layer*, inducing fluctuations of their velocity.

High speed relative motion and collisions between grains taking place within the *shear layer*, causing a *fluidification* effect [12] coupled with energy dissipations, require the development of complex resistance laws. The velocity field in coarse-grained material flows can be assumed to be composed of a *frictional* regime and a *collisional* regime [13]. In *frictional regime*, the shear stresses satisfy the Mohr-Coulomb resistance law: $\tau_{fr} = \sigma' \cdot \tan\varphi_b$, σ' being the effective normal stress; φ_b, the dynamic friction angle at the base of the granular mass. In *collisional regime*, the stress component normal to the boundary (*"dispersive pressure"*) and the shear stress component are expressed as follows [14]:

$$p_{disp} = a_i \rho_s \lambda^2 d_p^2 \cos\phi \left(\frac{\partial v}{\partial y}\right)^2 ; \tau_{disp} = a_i \rho_s \lambda^2 d_p^2 \sin\phi \left(\frac{\partial v}{\partial y}\right)^2 \quad (7\text{–}8)$$

a_i being the Bagnold coefficient (suggested value 0.042); ρ_s, the solid fraction mass density; λ, the linear concentration, that is a function of solid fraction v_s ($\lambda = 1/[(v_{max}/v_s)^{1/3} - 1]$); ϕ, the internal dynamic friction angle of granular bulk. Thus, the (total) shear stress τ_{max} is obtained by adding the frictional shear stress τ_{fr} to the dispersive shear stresses τ_{disp} . Resistance laws splitting these two contributions ($\tau_{max} = \tau_{fr} + \tau_{disp}$) have been therefore proposed [15, 16]; rheological models where both the *frictional* and *collisional* contributions are coupled into a single term, are also available [17].

Rapid change of pores volumes related to the continuous grains rearrangement [18] within the *shear layer* can generate excess pore water pressures at the base of coarse-grained material flows. Specific test show that the pore water pressure is maintained during the motion of sliding granular mass without dissipation [19]. Fragmentation (or grains' size reduction) can reduce the available energy and then mobility. The specific energy (kW·h/ton) required for grains' fragmentation is expressed as follows [20]:

$$W_{frag} = 10 \cdot W_{Bond}\left(d_{p,fin}^{-1/2} - d_{p,in}^{-1/2}\right) \ge 0 \quad (9)$$

only if $d_{p,fin} < d_{p,in}$. W_{Bond} is the Bond work index (kW·h/ton), depending on the type of granular material; $d_{p,in}$ and $d_{p,fin}$ are the size (in μm) of the grains before and after crushing, respectively; $d_{p,fin} = 1.04 \cdot 10^{-4} \cdot (W_{bond}/H_g)^2$, H_g being the geodetic difference in level; in this case W_{bond} is expressed in J/kg and $d_{p,fin}$ in cm.

3 Main phenomena affecting fine grained material flows

Fine - grained material flows are typically characterized by a high concentration of silts and clays. According to field observations and measurements [21], it is possible to state that the triggering mechanism and successive propagation mainly depend on excess pore water pressures followed by consolidation process. The excess pore water pressures can be generated by *(i)* shear strains induced by an earthquake; *(ii)* undrained loading following human activities; *(iii)* rapid accumulation of rainwater in soil layers with low permeability; *(iv)* seepage flow in boundary materials. The considerable shear strength reduction, following the generation of the excess pore water pressures, is often the main reason of high mobility of unstable fine-grained material volumes, even on very gentle slopes; furthermore, high pore water pressures can also induce the partial or complete liquefaction of the soil [19]. Conversely, the consolidation process of fine-grained materials [22-23], along the motion, progressively reduces the pore water pressure; the corresponding increase of the shear strength causes a reduction of the travelled distance.
For materials, characterized by $d_p < 0.02$ m, the energy dissipation due to grain collisions can be neglected [16].

4 Proposed models

Both proposed models are based on the following hypotheses (Fig. 2a): *(i)* the sliding granular mass is represented by a block (parallelepipedal shape) of thickness h, length l and width b (d_w is the depth, from the upper surface of the block, of the groundwater); *(ii)* the granular flow runs along planar surface of constant slope α_l and length L (deformations induced by the possible variations of sliding surface are neglected [24]); *(iii)* although it is well-known that the mass of a debris flow may change due to erosion or deposition processes [12, 16], for the sake of simplicity, the debris flow mass is assumed constant.

4.1 Coarse-grained material flows modeling

An *energy-based* model is developed to describe the mobility of coarse-grained material flows [16], by taking into account the effects of *collisions* and *granular temperature*. The granular sliding body is composed of two layers (Fig. 2b) of equal basal area ($\Omega = l \cdot b$) and length l, representing the *shear layer* (thickness s^s) and the overlying mass (thickness s^b, *"block"*) respectively. By imposing the equilibrium orthogonally to the sliding surfaces, the resulting force N_{tot} (= $W_b \cdot \cos\alpha_l$, W_b being the weight of the block, Fig. 2b) must be balanced by the lithostatic (total) stresses, $\sigma = \rho^b gh \cdot \cos\alpha_1$ (ρ^b being the density of the block), coupled to the dispersive pressures, p_{disp} (eq. (7)):

$$W_b \cos\alpha_1 = \bar{\sigma} \cdot r(\dot{x})\Omega + p_{disp} \cdot r(\dot{x})\Omega \quad (10)$$

r and $\bar{r}$ (=1 − r), dependent on the velocity ($v = \dot{x}$), have been suitably introduced to allow a *weighted* balance of the force N_{tot} according to the resulting lithostatic force and the resultant of the colliding forces statistically acting along the irregular sliding surface [16]:

$$r(v) = \frac{1}{\pi}\{\arctan[\eta(v - v_{cr})] - \arctan[-\eta(v + v_{cr})]\} \quad (11)$$

η is a parameter falling in the range 0.005÷0.5, which modulates the shape of r; v_{cr} is the critical speed for which the regime dominated by the inertial forces becomes a *collisional* regime [16].

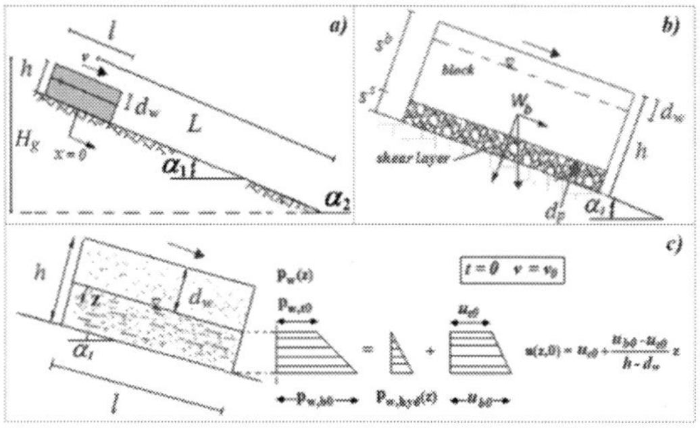

Fig. 2. *a)* Problem's setting; *b)* and *c)* assumptions concerning coarse and fine grained material flows modeling, respectively.

The eq. (10) allows to determine the thickness of the overlying block (if a linear change of velocity along y is assumed, the dispersive pressures p_{disp}, τ_{disp} and T_g become proportional to $\dot{x}^2$): $s^b(\dot{x}) = (1 - r) \cdot h + r \cdot Q_1\lambda^2 \dot{x}^2$, where $Q_1 = a_i\rho^s d_p^2 \cdot \cos\phi/(\rho^b g \cdot \cos\alpha_1)$ and ρ^s is the density of *shear layer*. Because the sum of the masses of the *shear layer* (m^s) and the overlying block (m^b) equals the total sliding mass (m): $m = m^b + m^s = (\rho^b s^b + \rho^s s^s) \cdot \Omega$, the thickness of the shear layer s^s is thus expressed as: $s^s(\dot{x}) = (m - \rho^b s^b(\dot{x})\Omega)/(\rho^s\Omega)$. If a constant value is assigned to the interstitial pressure ($p_{w,b}$) at the base of the granular mass [18] and isopiezic lines are orthogonal to the motion direction: $p_{w,b} = p_{w,b,hyd} = \gamma_w(h - d_w)\cdot\cos\alpha_1$ (γ_w being the water unit weight; $d_w < 0$ allows to simulate the excess pore water pressure). The power balance of the sliding mass is expressed as [16]:

$$\dot{E}_p(t) + \dot{E}_k(t) + \dot{E}_r(t) + \dot{E}_{coll}(t) + \dot{E}_{gt}(t) + \dot{E}_{frag}(t) = 0 \quad (12)$$

where $\dot{E}_p$ is the potential power ($=\dot{E}_p^b + \dot{E}_p^s$, $\dot{E}_p^b$ and $\dot{E}_p^s$ being the potential power of the block and *shear layer*, respectively); $\dot{E}_k$ is the kinetic power of the sliding mass ($=\dot{E}_k^b + \dot{E}_k^s$, $\dot{E}_k^b$ and $\dot{E}_k^s$ being the kinetic power of the block and *shear layer*, respectively); $\dot{E}_r$ is the power related to the energies dissipated along the sliding surface due to friction and dispersive pressures ($\dot{E}_r = T_{max} \cdot \dot{x}$, $T_{max} = T_{fr} + T_{disp}$; $T_{fr} = \tau_{fr} \cdot (1-r) \cdot \Omega$, $T_{disp} = \tau_{disp} \cdot r \cdot \Omega$); $\dot{E}_{coll}$ is the power dissipated due to grain collisions; $\dot{E}_{gt}$ is the power stored as *granular temperature*; $\dot{E}_{frag}$ is the power lost by crashing or fragmentation [16].

4.2 Fine-grained material flows modeling

It is assumed a trapezoidal initial pore water pressure distribution ($p_w(z)$) within a basal saturated layer of thickness ($h - d_w$) (Fig. 2c); the initial values $p_{w,t0}$ and $p_{w,b0}$ of pore water pressure, at the top and the base of the saturated layer, depend on the values of hydrostatic interstitial pressure ($p_{w,hyd}(z) = \gamma_w z \cdot \cos\alpha_1$) and on the excess pore water pressure $u(z,t)$ (z, normal to the sliding surface, directed downward, from the upper surface of the saturated soil layer, Fig. 2c). Thus, the resultant U of the pore water pressures at the base of the sliding mass ($p_{w,b}(t)$), is equal to the sum of the basal hydrostatic interstitial pressure ($p_{w,b,hyd}$) and the basal excess pore water pressure at time t ($u(z=h-d_w,t)=u_b(t)$): $U = p_{w,b}(t)\cdot\Omega = (p_{w,b,hyd} + u_b(t))\cdot\Omega$. If the involved material is not affected by dilatancy, the basal excess pore water pressures are dissipated during the motion due to consolidation process. Their evolution, referred to the case of impermeable horizontal base and drainage only through the upper surface of the saturated soil layer, is simply described through the following dissipation law $u_b(t)$ [25]: $u_b(t) = u_{b,0} \cdot e^{-at}$, where a is a parameter related to the variables that govern the consolidation process of the involved material ($a = \pi^2 c_v/(4 \cdot H^2)$, c_v being the 1-D consolidation coefficient; H, the drainage distance); u_{b0} is the initial basal excess pore water pressure, evaluable as: $u_{b0} = r_{0,b} \cdot u_{b,0,max}$, with $u_{b,0,max} = [(h-d_w)\gamma' + d_w \cdot \rho g] \cdot \cos\alpha_1$ (γ' is the submerged soil unit weight) and $r_{0,b} < 1$ to avoid the soil liquefaction. Under the previous assumptions, the motion's law is expressed as follows:

$$mg\sin\alpha_1 - \left[mg\cos\alpha_1 - \left(p_{w,b,hyd} + u_b(t)\right)\Omega\right]\tan\varphi_b = m\frac{dv}{dt} \quad (13)$$

5 Comparison among measured data and theoretical results

Coarse-grained materials. The Acquabona Creek's debris flow, occurred in June 1997 in the Dolomites (Eastern Alps, Italy), was directly observed. Channel cross-section measurements taken along the flow channel indicate debris flow velocity ranging from 3.1 to 9.0 m/s [26]. After reworking the measured data, the following main parameters may be selected: L = 1250 m; α_1 = 18°; l = 75 m; h = 2 m; Ω = 300 m^2; d_p = 0.03 m. The values of remaining parameters describing the mechanical behaviour, reported in Fig. 3, have been chosen according to previous parametrical analyses [16]. The proposed model allows to fit the measured maximum velocity, the speed evolution ($v(x)$) along the path and the observed maximum traveled distance. If the value of a single parameter (e.g. η) is changed, the fitting is not obtainable (curves b, c; Fig. 3).

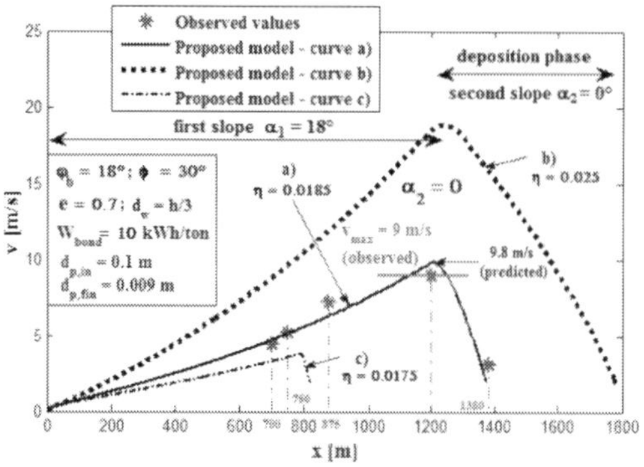

Fig. 3. Acquabona Creek's debris flow: measured and theoretical velocity (v) vs distance travel (x).

Fine-grained materials. Laboratory tests were performed with mixtures of water and sediment collected from the Rio Gadria material flow deposits (Eastern Alps, Italy) [27]. The involved material was characterized by an appreciable muddy component (d_p = 1 mm; v_s = 0.55). The experimental apparatus is shown in Fig. 4: it is composed of flume of length 2 m, width 0.15 m, height 0.40 m, slope angle 20°. During the sliding of the material, the velocity was measured in four different cross sections (P_i, Fig. 4) of the flume. By assigning the following input parameters: L = 2 m; h = 0.20 m; $l = b$ = 0.15 m; α_1 = 20° (the other parameters, reported in Fig. 4, have been determined according to parametrical analyses [22]), through the proposed model, the results shown in Fig. 4 are obtained.

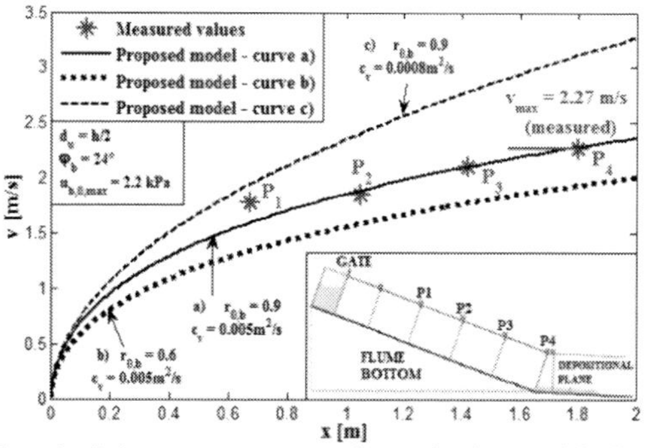

Fig. 4. Laboratory tests on fine-grained material flow: measured and theoretical velocity (v) vs distance travel (x).

Through the rational choice of input parameters, the proposed model allow to fit the measured values of velocity reached by material flow along its path. Again, the fitting is not obtainable by changing the selected parameters (e. g. c_v or $r_{0,b}$; curves b, c; Fig. 4).

6 Concluding remarks

Two analytical (block) models to estimate the velocity of *coarse* and *fine* – grained materials flows are proposed. The models are based on some simplified assumptions concerning the geometry of the granular mass (*block*) and sliding surface (*planar*), the evolution of pore water pressures (*1-D consolidation*), the energy dissipation (*friction, collisions, fragmentation*). Furthermore, they depend on some unconventional and empirical input parameters (e, η, v_{cr}, $r_{0,b}$), necessary to describe the physical and mechanical behaviour of *coarse* and *fine* – grained material flows. Their values may be obtained by the comparison between the measured and theoretical values of velocity.

References

1. Y. Fukui, M. Nakamura, H. Shiraishi, Y. Sasaki, Coastal Eng. Jpn., **6**, 67 – 82 (1963)
2. A. Daido, *XXV IAHR Congress, Technical Session B*. **3**, 211 – 213 (1993)
3. S. Zhang, Natural Hazards, **7**, 1 – 23 (1993)
4. F. Federico, A. Amoruso, IJEGE **1**, 5 – 24 (2008)
5. A. Armanini, C. Dalri, F. Della Putta, M. Larcher, L. Rampanelli, M. Righetti, *Hydraulics of Dams Structures*, Taylor & Francis Group, London, ISBN 9058096327 (2004)
6. O. Hungr, Can. Geot. J., **32** (4), 610 – 623 (1995)
7. S. Ogawa, *US Japan Seminar on Continuum Mechanical and Statistical Approaches in the Mechanics of Granular Materials*, Tokyo (1978)
8. S. Warr, G.T.H. Jacques, J.M. Huntley, Powder Technology, **81**, 41 - 56 (1994)
9. D. Gollin, E. Bowman, P. Shepley, *6th Int. Symp. On Deformation Characteristics of Geomaterials*, Buenos Aires (2015)
10. A. Armanini, *XXXII Convegno Nazionale di Idraulica e costruzioni Idrauliche*, Palermo (2010)
11. S.B. Savage, C.K.K. Lun, J. of Fluid Mech., **189**, 311–335 (1988)
12. O. Hungr, S.G. Evans, *7th International Symposium on Landslides*, Trondheim, **11**, 233–238 (1996)
13. D. Zhang, M.A. Foda, Acta Mechanica, **136** (3), 155–170 (1999)
14. R.A. Bagnold, *R. Soc. London A*, **225**, 49–63 (1954)
15. A. Armanini, J. of Hydr. Res., **51**, 111–120 (2013)
16. F. Federico, C. Cesali, Can. Geot. J., **52**,1–21 (2015)
17. P. Jop, Y. Forterre, O. Pouliquen, Nature, 441, 727-730 (2006)
18. A. Musso, F. Federico, G. Troiano, Comp. and Geotech., **31**, 209–226 (2004)
19. R.M. Iverson, American Geophysical Union (1997)
20. F.V. De Blasio, Acta Mech., **221**, 375–382 (2011)
21. J.N. Hutchinson, R.K. Bhandari, Geotechnique, **21**, 353–358 (1971)
22. J.N. Hutchinson, Can. Geot. J., **23**, 115–126 (1986)
23. M. Pastor, J.A. Quecedo, F. Merodo, M.I. Herreros, E. González, P. Mira. Numerical modelling in Geomechanics, **6** (6), 1213-1232 (2002)
24. A. Taboada, K.J. Chang, *J. Geoph. Res.*, **114** (2009)
25. J.A. Fernández-Merodo, G. Herrera, P. Mira, J. Mulas, M. Pastor, L. Noferini, D. Mecatti, G. Luzi, *Inter. Env. Model. and Soft. S.*, 1476-1483 (2008)
26. M. Berti, R. Genevois, A. Simoni, P.R. Tecca, Geomorphology, **29**, 265-274 (1999)
27. F. Bettella, G.B. Bischetti, V. D'Agostino, S.V. Marai, E. Ferrari, T. Michelini, J. of Agricultural Engineering, **46**, 4 (2015)

EPJ Web of Conferences **140**, 03002 (2017) DOI: 10.1051/epjconf/201714003002
Powders & Grains 2017

Numerical modeling of the debris flows runout

Francesco Federico[1,*], and *Chiara* Cesali[1]

[1]University of Rome 'Tor Vergata', Department of Civil and Information Engineering, Via del Politecnico 1, 00133 Rome, Italy

Abstract. Rapid debris flows are identified among the most dangerous of all landslides. Due to their destructive potential, the runout length has to be predicted to define the hazardous areas and design safeguarding measures. To this purpose, a continuum model to predict the debris flows mobility is developed. It is based on the well known depth-integrated avalanche model proposed by Savage and Hutter (*S&H model*) to simulate the dry granular materials flows. Conservation of mass and momentum equations, describing the evolving geometry and the depth averaged velocity distribution, are re-written taking into account the effects of the interstitial pressures and the possible variation of mass along the motion due to erosion/deposition processes. Furthermore, the mechanical behaviour of the debris flow is described by a recently developed rheological law, which allows to take into account the dissipative effects of the grain inelastic collisions and friction, simultaneously acting within a '*shear layer*', typically at the base of the debris flows. The governing PDEs are solved by applying the finite difference method. The analysis of a documented case is finally carried out.

1 Introduction

Savage and Hutter (1989) applied for the first time the depth averaged Saint Venant equations to analyze the propagation of sliding granular masses along an inclined planar surface [1].

By considering the x-coordinate in the downflow direction and the z-coordinate in the normal direction to the sliding surface and assuming that *(i)* the downslope velocity is constant along the depth ($v = v(x,t)$); *(ii)* the density ρ of the granular material is constant; *(iii)* the granular material is dry; *(iv)* the mass is constant; *(v)* the granular material obeys the Mohr-Coulomb resistance law within the deforming mass (with internal friction angle ϕ) and along the sliding surface (with basal friction angle δ), by referring to an element of length dx (Fig. 1), the following mass balance and x-momentum balance equations may be written [2]:

$$\frac{\partial H}{\partial t} + \frac{\partial}{\partial x}(v \cdot H) = 0 \qquad (1)$$

$$\frac{\partial v}{\partial t} + v\frac{\partial v}{\partial x} = g\sin\theta - \mathrm{sgn}(v) g \tan\delta(\cos\theta) - gK_{act,pas}\cos\theta\frac{\partial H}{\partial x} \qquad (2)$$

$H(x,t)$ is the thickness or depth of the sliding granular mass; $v(x,t)$ is the x-component of the velocity vector; g is the gravity acceleration; θ is the slope of the sliding surface; $K_{act,pas}$ is the active (if $\partial v/\partial x \geq 0$) or passive (if $\partial v/\partial x < 0$) earth pressure coefficient:

$$K_{act,\,pas} = 2 \cdot \left[\left(1 \pm \sqrt{1-\cos^2\phi(1+\tan^2\delta)}\right)\Big/\cos^2\phi\right] - 1 \qquad (3)$$

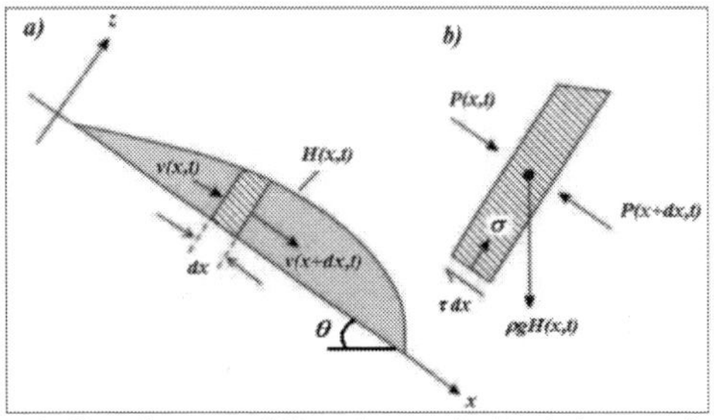

Fig. 1. Granular mass sliding along an inclined planar surface: *a)* geometry and coordinate system; *b)* forces acting on an infinitesimal element.

2 Proposed model

Equations (1) and (2) are rewritten by taking into account the effects of the pore water pressures and the variation of mass due to erosion or deposition processes.

Moreover, the curvature of the sliding surface and the typical (*frictional-collisional*) mechanical behaviour of a debris flow [3] are considered.

#

2.1 Curvature of the sliding surface

If the sliding surface is geometrically described by an arc of circumference (curvature $k \neq 0$), the centrifugal force occurs; this force modifies the normal resultant force ($\sigma\, dx$) and, then, the shear resistance force ($\tau\, dx$):

$$\tau \cdot dx = -\,\mathrm{sgn}(v) \cdot \rho H \cdot (g\cos\theta + kv^2)\tan\delta \cdot dx \qquad (4)$$

ρ being the mass density per unit volume.

* Corresponding author: fdrfnc@gmail.com

2.2 Pore water pressure

Pore water pressure (p_w) induces the reduction of the normal resultant force and modifies the lateral pressures (p_L). Furthermore, if the curvature k of the sliding surface assumes high values, p_w doesn't vary hydrostatically [4]: a change of the direction of motion of a fluid particle (curvilinear path) implies an increase of piezometric head along the direction normal to the sliding surface. Thus, the pore water pressure at the base ($p_{w,b}$) of the sliding mass is expressed as follows:

$$p_{w,b} = \gamma_w H_w \left(\cos\theta + k\frac{v^2}{g}\right) \tag{5}$$

γ_w being the specific weight of water and H_w the elevation of phreatic surface on the slope bed (Fig. 2). Additional excess pore water pressures due to changes of total stresses under undrained conditions, typically taking place within fine-grained materials as shown in centrifuge tests [5], are herein neglected. Moreover, the dissipation of excess pore water pressures due to consolidation processes is not taken into account [6].

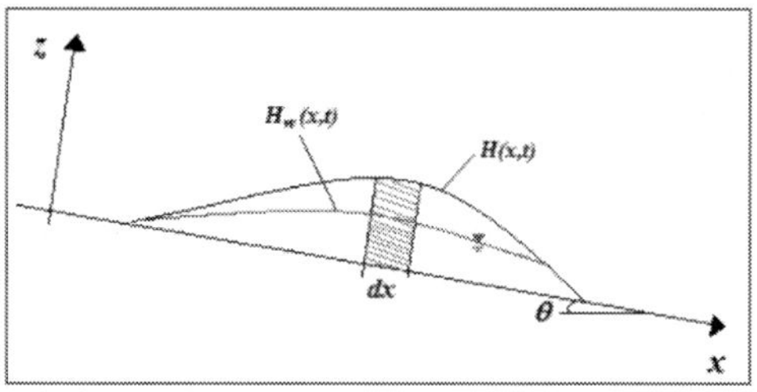

Fig. 2. Saturated or partially saturated sliding granular mass: definition of H_w.

According to (5), the shear resistance force becomes:

$$\tau \cdot dx = -\mathrm{sgn}(v)\cdot(\rho g H - H_w \gamma_w)\cdot\left(\cos\theta + k\frac{v^2}{g}\right)\tan\delta \cdot dx \tag{6}$$

and the resultant P (= $P(x,t) - P(x+dx,t)$, Fig. 1) of the lateral pressures (p_L) is expressed as follows:

$$P = -\int_0^H \frac{\partial p_L}{\partial x} dz dx = -[K_{act,pass}\cdot(\cos\theta + k\frac{v^2}{g})\cdot (\rho g H\frac{\partial H}{\partial x} - \gamma_w H_w \frac{\partial H_w}{\partial x}) + \gamma_w H_w \frac{\partial H_w}{\partial x}(\cos\theta + k\frac{v^2}{g})]dx \tag{7}$$

2.3 Mass variation

The mass of a debris flow may change along the motion due to erosion or deposition processes [7]. Typically, the erosion phenomenon mainly occurs at high elevation, induced by the high slope and the great travel speed, while the deposition, caused by the slowdown due in turn to the reduction of slope, downstream occurs. The erosion (or deposition) process is considered by introducing the erosion (or deposition) rate $\varepsilon(v)$, dependent on the velocity of the sliding mass. It is simply assumed: $\varepsilon(v) = \varepsilon_{er}$, if $v \geq v_{crit}$; $\varepsilon(v) = \varepsilon_{dep}$, if $v < v_{crit}$. ε_{er} (> 0, constant) is the erosion rate; ε_{dep} (< 0, constant) is the deposition rate; v_{crit} is the critical velocity for which the transition from erosion to deposition process occurs. Generally, v_{crit} corresponds to a limit angle of erosion, β_e, evaluable through the relation [8]:

$$\beta_e = \arctan\left\{\tan\delta\left[1 - 2\frac{\rho_w}{\rho_{bed}}\left(1 + \frac{\rho}{\rho_{bed}}\right)^{-1}\right]\right\} \tag{8}$$

where ρ_w is the density of water and ρ_{bed} is the density of the material lying on the bed of the channel (typically, 1800–2000 kg/m^3).

2.4 Frictional–collisional mechanical behaviour

Experimental observations showed the growth of a "*shear layer*" at the base of a sliding granular mass [9], where initially great deformations and then dilation and collisions between solid grains occur, differently from the top, causing a fluidification effect [10] coupled with energy dissipations. Thus, in a rapid sliding granular mass, normal and shear stresses, associated with *frictional* and *collisional* regimes, respectively, simultaneously occur [11].

Since a rapidly debris flowing is affected by a 'mixed' *frictional – collisional* mechanical behaviour, the shear resistance (resultant) force may be approximately expressed as [3, 12]:

$$\tau \cdot dx = -\mathrm{sgn}(v)\left(\bar{r}\cdot\tau_{fr} + r\cdot\tau_{disp}\right)\cdot dx \tag{9}$$

where τ_{fr} is the shear stress associated with *frictional* regime and τ_{disp} is the shear stress associated with *collisional* regime. For the normal to the boundary and shear components of the collisional (*dispersive*) stresses (pressures), the following relationships are respectively proposed [13]:

$$p_{disp} = a_i \rho_s \lambda^2 d_p^2 \cos\phi\left(\frac{\partial u}{\partial z}\right)^2 ; \tau_{disp} = a_i \rho_s \lambda^2 d_p^2 \sin\phi\left(\frac{\partial u}{\partial z}\right)^2 \tag{10-11}$$

a_i being the "Bagnold coefficient" (suggested value 0.042); ρ_s, the solid fraction mass density (2300÷2800 kg/m^3); λ, the "linear concentration", dependent on the solid fraction v_s ($\lambda = 1/[(v_{max}/v_s)^{1/3} - 1]$, v_{max} (maximum solid fraction) = 0.74 for spherical grains); d_p is the average grain diameter.
According to Mohr-Coulomb resistance law, the *frictional* shear stress τ_{fr} is thus expressed as follows:

$$\tau_{fr} = \left[(\rho H - \rho_w H_w)\cdot(g\cos\theta + kv^2) + p_{disp}\right]\tan\delta \tag{12}$$

r and $\bar{r}$ (defined as 1 - r), dependent on the velocity v, allow a "weighted" balance of the shear resultant force between the *frictional* and *dispersive* stresses, statistically acting along the irregular sliding surface. It is proposed [3]:

$$r(v) = \frac{1}{\pi}\{\arctan[\eta(v - U_{cr})] - \arctan[-\eta(v + U_{cr})]\} \quad (13)$$

η is a parameter falling in the range 0.005÷0.5 [3], which modulates the shape of r; U_{cr} is the critical speed for which the regime dominated by the inertial forces becomes a collisional regime: $U_{cr} = [(\rho_s - \rho_w)\cdot N_{sav} b_s^2 g H_0/(\rho_s \cdot d_p^2)]^{1/2}$ [14], where N_{sav} is the Savage number equal to 0.1, b_s is the average *shear layer* thickness (=10÷15·d_p) [15], H_0 is the maximum initial thickness of the granular mass. If a linear change of velocity v along z within the *shear layer* is assumed, p_{disp} and τ_{disp} become proportional to v^2.

2.5 Fundamental equations

Under the previous considerations, the mass balance and x-momentum balance equations are rewritten as follows:

$$\frac{\partial H}{\partial t} + \frac{\partial}{\partial x}(v \cdot H) - \varepsilon(v) = 0 \quad (14)$$

$$\frac{\partial v}{\partial t} + v\frac{\partial v}{\partial x} = g\sin\theta - \mathrm{sgn}(v)\cdot\{\bar{r}\cdot[(1 - R_u)\cdot(g\cos\theta + kv^2) + (p_{disp}/\rho H)]\tan\delta + r\cdot(\tau_{disp}/\rho H)\} - [(1 - R_u \cdot d_w)K_{act,pass} + R_u \cdot d_w]\cdot(g\cos\theta + kv^2)\frac{\partial H}{\partial x} \quad (15)$$

where d_w and R_u are respectively defined as $d_w = H_w/H$ and $R_u = \gamma_w H_w/(\gamma H) = (\gamma_w/\gamma)\cdot d_w$. The coefficient R_u can assume the following values: $R_u = 0$, if $d_w = 0$ (dry material); $R_u \in (0, \gamma_w/\gamma)$, if $d_w \in (0, 1)$; $R_u = \gamma_w/\gamma$, if $d_w = 1$ (saturated flow); $R_u \in (\gamma_w/\gamma, 1)$, if $d_w \in (1, \gamma/\gamma_w)$; $R_u = 1$, if $d_w = \gamma/\gamma_w$. $d_w \in (1, \gamma/\gamma_w)$ roughly allows to simulate the excess pore water pressure at the base of high speed coarse-grained materials flows, due to the rapid change of pores volume related to the continuous rearrangement of grains, localized within the *shear layer* [16]. $R_u = 1$ implies that the pore water pressure (p_w) equals the total normal stress (σ) and the frictional strength of the granular material is null. The system of governing PDEs (eqs. (14), (15)) is numerically solved.

3 Discretization of governing PDEs

The numerical solution of the system of governing PDEs is obtained by applying the *Finite Difference Method* (FDM). To this purpose, the granular mass profile is divided in N cells (Fig. 3): the index j (= 0, 1…N) corresponds to the cell boundary points. The mass balance equation (14) is discretized according to *H&S* formulation (first order approximation) [2]:

$$H_j^t = (H_j^{t-1} + \varepsilon\cdot\Delta t)\cdot\frac{(x_{j+1}^{t-1} - x_j^{t-1})}{(x_{j+1}^t - x_j^t)} \quad (16)$$

with $x_j^t = x_j^{t-1} + u_j^{t-1}\cdot\Delta t$ (Δt = integration time interval). The x-momentum balance equation (15) is discretized according to an implicit method.

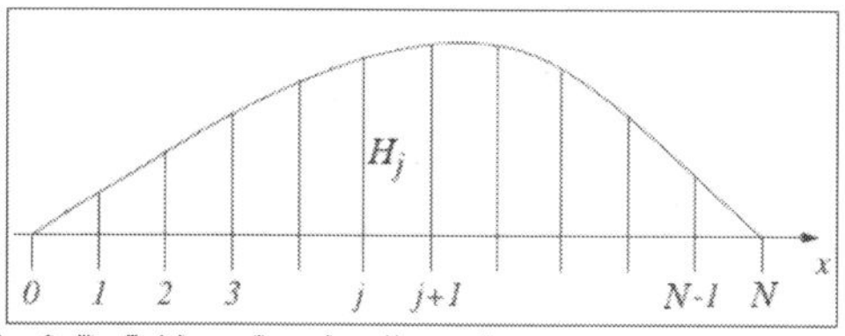

Fig. 3. Definition of mesh cell notation.

In particular, for the intermediate cells (except the first and the last two ones), $\partial H/\partial x$ (and $\partial v/\partial x$) is replaced by the following fourth order approximate expression:

$$\frac{\partial H}{\partial x} = \frac{H_{j-2}^t - 8H_{j-1}^t + 8H_{j+1}^t - H_{j+2}^t}{12\Delta x} \quad (17)$$

(Δx = integration space interval) and $\partial v/\partial t$ is replaced by its second order approximation:

$$\frac{\partial v}{\partial t} = \frac{1}{\Delta t}\cdot[v_j^t - \frac{1}{2}(v_{j+1}^{t-1} + v_{j-1}^{t-1})] \quad (18)$$

For the first and the last cells, the spatial ($\partial H/\partial x$, $\partial v/\partial x$) and temporal ($\partial v/\partial t$) derivatives are replaced by the second order *'one side'* and first order approximations, respectively:

$$\frac{\partial H}{\partial x} = \frac{\mp 3H_j^t \pm 4H_{j\pm1}^t \mp H_{j\pm2}^t}{2\Delta x}; \quad \frac{\partial v}{\partial t} = \frac{v_j^t - v_j^{t-1}}{\Delta t} \quad (19\text{-}20)$$

For the second and the second-last cells, $\partial H/\partial x$ (and $\partial v/\partial x$) is replaced by the following approximation:

$$\frac{\partial H}{\partial x} = \frac{H_{j+1}^t - H_{j-1}^t}{2\Delta x} \quad (21)$$

and $\partial v/\partial t$ is replaced by eq. (18). The numerical scheme proposed to solve the system of governing PDEs is implemented through a MATLAB iterative code.

In order to prevent the blow-up of the numerical solution, the *CFL* condition is imposed; in particular, $v_j^t \cdot \Delta t/L_j^t < 1$, where L_j^t (= x_{j+1}^t - x_j^t) is the width of the *i-th* cell. If this condition is not fulfilled, smaller values of Δt must be chosen, also taking into account the computational time for the simulation.

The iterative procedure of generation of the solution of the system of governing PDEs (H_j^t, v_j^t) definitively stops if during deceleration of the sliding mass: $v_m^t < tol$, where v_m^t is the average velocity of the cells (or debris flow) and *tol* is the speed tolerance.

Discretization variables (e.g. N, Δt) as well as the role of the additional parameters related to the *collisional behaviour* [3] have been parametrically evaluated.

4 Case study: Champlong landslide

The Champlong landslide occurred in the Val d'Aosta on October 2000. The failure surface developed within a moraine cover, inducing the detachment of a volume of approximately 10^4 m^3. The debris flow involved the deposits lying on the bed of the channel (width 40m), increasing its volume. The maximum moved volume, the traveled distance (projected on horizontal plane) and the maximum velocity are estimated at approximately $2 \cdot 10^4$ m^3, 870 m, 15 m/s, respectively [17]. To apply the proposed model, the sliding surface is schematized (Fig. 4) by two planar surfaces (θ_1 =27° and length =500 m, θ_2 =14°) linked by an arc of circumference (k = 1/1000m).

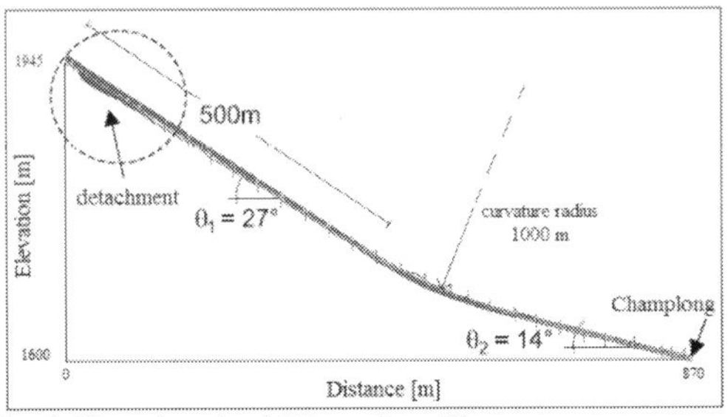

Fig. 4. Champlong landslide: path profile.

The initial (detached) mass is represented by a block (parallelepipedal shape) of thickness H_0 = *2m* and width/length L = 125m; the following values for the remaining input parameters are selected: ϕ=40°; d_w =0.7; δ =23°; N =20; Δt =0.01s; ρ_s = 2500 kg/m^3; v_s =0.6; d_p =0.04 m; η =0.005; ε_{er} =0.01 m/s; ε_{dep} = - 0.0001 m/s. The erosion/deposition rate has been chosen in order to obtain the maximum value of the involved volume, in situ measured.

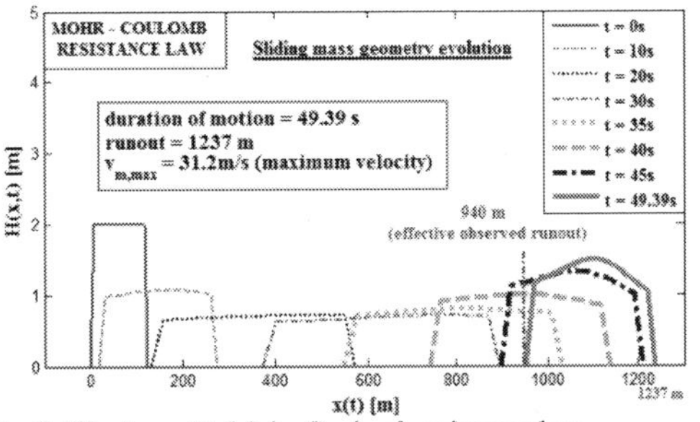

Fig. 5. $H(x,t)$ vs $x(t)$: Mohr-Coulomb resistance law.

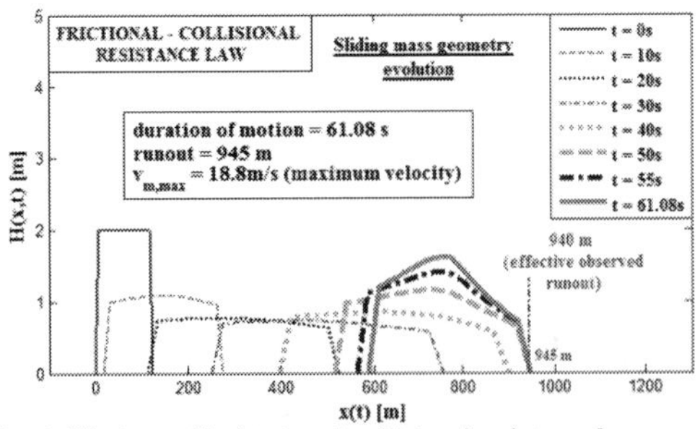

Fig. 6. $H(x,t)$ vs $x(t)$: *frictional-collisional* resistance law.

The results of simulations, in terms of sliding mass geometry evolution, computed according to Mohr-Coulomb and *frictional-collisional* resistance laws are shown in Figs. 5, 6. The maximum velocity (31.2 m/s) and runout length (1237 m) computed according to Mohr-Coulomb resistance law appear remarkably greater than the corresponding values (18.8 m/s, 945 m) obtained according to *frictional-collisional* resistance law, close to those ones *in situ* observed (15 m/s, 940 m).

5 Concluding remarks

Based on the well known *S&H model*, a continuum model to predict the debris flows mobility is proposed. It allows to take into account the effects of the interstitial pressures and the possible variation of mass along the motion due to erosion/deposition processes as well as the typical (*frictional-collisional*) mechanical behaviour of a debris flow, affected by grain inelastic collisions and friction, simultaneously acting within a basal '*shear layer*', causing complex dissipative effects.

The analysis of the Champlong landslide showed that, through a suitable choice of the input parameters, the proposed model allows to produce the *in situ* observed runout. Particularly, the comparison with the results obtained by assuming the conventional purely *frictional* resistance law (Mohr-Coulomb) pointed out the key role played by the additional input parameters related to the *collisional behaviour* (e.g. d_p, the average grain diameter; η, the modulator parameter; v_s, the solid fraction of the *shear layer*) on the final runout length.

References

1. S.B. Savage, K. Hutter, J. Fluid Mech. **199**, 177 – 215 (1989)
2. S.P. Pudasaini, K. Hutter, *Avalanche dynamics. Dynamics of rapid flows of dense granular avalanches* (Springer-Verlag Berlin, 2007)
3. F. Federico, C. Cesali, Can. Geotech. J. **52**, 1 – 21 (2015)
4. G. De Marchi, *Idraulica. Basi scientifiche e applicazioni tecniche* (Ulrico Hoepli, Milano, 1961)
5. A.F. Rauch, K. Ladwig, P.J. Amaya, *World of Coal Ash Conference* (Nasvhille, Ash Library, 2015)
6. J.N. Hutchinson, Can. Geotech. J. **23**, 115 – 126 (1986)
7. O. Hungr, *Int. W. on Occurrence and Mechanisms of Flow-Like Landslides in Natural Slopes and Earthfills* (Kluwer Publishers, Sorrento, 2004)
8. T. Takahashi, *Debris Flow* (IAHR Monograph Series, Balkema Publishers, Rotterdam, 1991)
9. O. Hungr, Can. Geotech. J. **32**, 610 – 623 (1995)
10. O. Hungr, S.G. Evans, *7th Int. Symp. on Landslides* **1**, 233 – 238 (Trondheim, 1996)
11. D. Zhang, M.A. Foda, Ac. Mech. **121**, 1 – 19 (1997)
12. A. Armanini, J. of Hydraulic Research **51** (2), 111 – 120 (2013)
13. R. A. Bagnold, P.R.S. London **225**, 49 – 63 (1954)
14. R.M. Iverson, *The physics of debris flows* (American Geophysical Union, 1997)
15. S. Straub, Geol. Rundschau **86**, 415 – 425 (1997)
16. A. Musso, F. Federico, G. Troiano, Comp. and Geotech. **31**, 209 – 226 (2004)
17. M. Rosso, C. Cavallo, R. Sesenna, E. Citera, M. Cussotto, GE.A.M. **110**, 61 – 69 (2003)

Why granular media are thermal after all

Mario Liu[1,*] and *Yimin* Jiang[2]

[1] *Theoretische Physik, Universität Tübingen,72076 Tübingen, Germany*
[2] *Central South University, Changsha 410083, China*

Abstract. Two approaches exist to account for granular behavior. The thermal one considers the total entropy, which includes microscopic degrees of freedom such as phonons; the *athermal* one (as with the Edward entropy) takes grains as elementary. Granular solid hydrodynamics (GSH) belongs to the first, DEM, granular kinetic theory and athermal statistical mechanics (ASM) to the second. A careful discussion of their conceptual differences is given here. Three noteworthy insights or results are: (1) While DEM and granular kinetic theory are well justified to take grains as elementary, any athermal entropic consideration is bound to run into trouble. (2) Many general principles are taken as invalid in granular media. Yet within the thermal approach, energy conservation and fluctuation-dissipation theorem remain valid, granular temperatures equilibrate, and phase space is well explored in a grain at rest. Hence these are abnormalities of the athermal approximation, not of granular media as such. (3) GSH is a wide-ranged continuum mechanical description of granular dynamics.

1 Introduction

Taking grains as elementary particles interacting via the Newtonian law, the discrete element method (DEM) is the tool of choice coming to terms with granular behavior. Similarly, granular kinetic theory also assumes that grains are structureless particles undergoing dissipative collisions. Both results have contributed to the wide-spread believe in the physics community that grains may *generally* be approximated as elementary. Starting from this believe, "athermal statistical mechanics" (ASM) defines a reduced entropy S_g that does not contain any microscopic degrees of freedom, only those of granular configurations, and assumes it is maximal in equilibrium [1]. One example is the Edward entropy, S_{Ed}, given by the number of possibilities grains may be stably packed [2, 3].

This is a leap of faith. The irrelevance of microscopics for DEM or the kinetic theory does not imply that the true entropy S, of all the microscopic degrees of freedom such as phonons and free electrons, is irrelevant. Brownian motion is negligible because grains are macroscopically large, each containing many, many microscopic degrees of freedom, such that $S \gg S_g$. Taking $S + S_g$ to be maximal in equilibrium implies that equilibrium holds as long as S is maximal, pretty much irrespective of S_g.

The lack of Brownian motion shows that phonons and electrons do not move grains. But this is quite insufficient for the conclusion of athermality. Crucial is the question: *Which of the two entropies, S_g or S, is being increased by dissipation off equilibrium, during collisions and relaxations, and becomes maximal in equilibrium?* Grains are athermal only if S_g is the answer. However, whenever a system loses energy and heats up measurably – by fingers or a thermometer, it is S that is being increased. This is what grains do in experiments, and – if one would trace the lost energy – also in DEM. And this is why GSH takes S to be maximal in equilibrium, with dissipative terms constructed to increase S off-equilibrium. ASM takes S_g to be maximal, and as yet shuns discussion of dissipation.

Note that since neither DEM nor the kinetic theory deal with the entropy, their obvious usefulness in accounting for experiments does not support the ASM assumption. What is more, both possess the appropriate dissipative terms to reach the proper equilibrium, with all grains at rest, given by S = max. Note also that GSH successfully describes a wide range of granular phenomena [4], including fast dense flow, elaso-plastic motion, static stress distribution, propagation of elastic waves, and compaction. And it reduces, in the respective limits, to hypoplasticity [5], Kamrin's nonlocal constitutive relation [6], the $\mu(I)$-rheology, and the hydrodynamic equations of granular gases. GSH also contains mathematical structures of striking similarity to glassy dynamics, including critical slowing down, aging and memory. Given these results, GSH appears well-positioned to present a unified macroscopic framework for coming to terms with the multitude of granular phenomena.

As yet, GSH is a qualitative theory, providing a bird's eye view of granular behavior. But we are striving to make it quantitative. Its input in basic understanding are *two-stage irreversibility* and *variable transient elasticity*. The first addresses the three length scales of any granular media – macroscopic, granular and microscopic, and the fact that energy in the macroscopic degrees of freedom typically first decays into the granular, then the microscopic

*e-mail: mliu@uni-tuebingen.de

ones. The second addresses the fact that the elastic stress relaxes when grains jiggle, and that it relaxes faster the stronger the jiggling is.

These two concepts introduce the granular temperature T_g (conjugate to S_g) and the elastic strain u_{ij} as additional state variables. Hereby, u_{ij} accounts for the coarse-grained elastic deformation of the grains, and the associated elastic energy, while T_g quantifies the quickly fluctuating elastic and kinetic energy of individual grains. Both obey relaxation equations, pushing the system toward equilibrium, S = max, and accounting for granular dynamics on their way there. This is the most ostentatious link between microscopics (phonons and electrons) and macroscopics (granular flow and stacking). In addition, GSH contains the conservation equations for mass, momentum and energy, and a balance equation for S.

2 Two models for grains

There are two models for grains, a gas of particles taken as elementary, and a block of thermal rock that has been sliced and diced. The first is athermal, the second thermal. Both work well within their respective range of validity, but one needs to know what they are. The simple example of a pendulum is helpful here.

2.1 Is a pendulum athermal?

A pendulum is, like a grain, a macroscopic object. The equation, $\ddot{\theta}+\alpha\dot{\theta}+\theta g/l = 0$, accounts for its small-amplitude motion, with no need to consider its entropy S. (θ: pendulum angle, g: gravitational constant, l length of the string, α: friction coefficient.) Still, a pendulum is not athermal, because the frictional force $\alpha\dot{\theta}$ increases S.

This force is derived by starting from the second law of thermodynamics, ie. S only increases, until it is maximal in equilibrium: The *conserved* energy E is given by the potential, kinetic and heat contributions, $E = \frac{1}{2}Mgl\theta^2 + \frac{1}{2}Ml^2\dot{\theta}^2 + \int T\mathrm{d}S$, or $\dot{E} = 0 = Mgl\theta\dot{\theta} + Ml^2\dot{\theta}\ddot{\theta} + T\dot{S}$. Inserting the pendulum equation, $\ddot{\theta} + Y + \theta g/l = 0$, with Y unknown, we find $\dot{S} = (Ml^2/T)\dot{\theta}\,Y > 0$. Since $Ml^2/T > 0$, but not necessarily $\dot{\theta}$, we set $Y = \alpha\dot{\theta}$, $\alpha > 0$. This force acts until $\theta, \dot{\theta} = 0$, implying $E = \int T\mathrm{d}S$, or S = max.

Clearly, a pendulum is both athermal and thermal – athermal when the dissipative force is known, thermal if it needs to be derived. Grains are not different. Because DEM and granular kinetic theory possess the proper dissipative forces, they may treat grains as elementary. But if one needs to derive the dissipative terms of a macroscopic theory, it is not clear how one can possibly avoid S.

2.2 Athermal gas and thermal solid

Faced with the task to setup a macroscopic, continuum-mechanical theory, it may seem natural to model grains as a gas of elementary particles, because it seems hard to imagine how phonons inside the grains could possibly influence their flow and stacking. However, see the last paragraph of the introduction. Moreover, athermal grains and gaseous molecules are far apart, mainly because energy conservation does not hold for grains, hence statistical mechanics does not work. ASM tries to ameliorate this by replacing the energy with volume or stress. Yet energy conservation, related to time translational symmetry, is a fundamental property of matter; constancy of volume or stress are experimental prescriptions.

There are further problems: Failure of granular temperatures to equilibrate, invalidity of the fluctuation-dissipation theorem, jammed grains lack the possibility to explore the phase space... All these in addition to the basic problem, $S \gg S_g$. If the gas model poses such difficulties, the question whether a solid is better suited as a starting point, to develop a macroscopic theory for granular media, naturally arises. The answer: *grains are often well modeled by a thermal rock that has been sliced and diced.*

Modelling a block of solid including the gas of quasi-particles such as phonons and free electrons, the energy dissipating into them is not lost. With the energy conserved, thermodynamics holds, and statistical mechanics works. Crucially, same is true for a stack of two blocks, or a pile of grains – as long as they are macroscopic.

The notion that grains at rest are jammed, in need of shaking for phase space exploration, is now inappropriate: Phonons roam nearly as freely in a pile of grains as in a block of rock. The fluctuation-dissipation theorem, dealing with thermal fluctuations of phonons and electrons, also holds in granular media. Finally, that granular temperatures T_g do not equilibrate is traceable to the neglect of the true temperature T: Given two systems, each with a granular and a true temperature, the behavior is completely analogous to four systems with four temperatures. There is little unique or incomprehensible about T_g's behavior.

With the microscopics reassuringly healthy, we proceed to consider the macroscopic description of granular dynamics. A block of solid is an elastic medium. Cutting the block in half, with one part on top of the other, we expect them to remain elastic under shear if they do not slip. If they do, we subtract the slipping portion from the total displacement to obtain the *deforming* one. Further slicing the block to eventually arrive at many (macroscopic) pieces, the system is still elastic – as long as we keep track of the deforming displacement. We call the associated strain field elastic, coarse-grain it, and denote it as u_{ij}. Since the elastic stress π_{ij} stems from deformed grains, it is a function of u_{ij}, and of the density, $\pi_{ij} = \pi_{ij}(\rho, u_{kl})$.

Because force equilibrium, $\nabla_j\pi_{ij} = \rho g_i$, is the condition for S = max with respect to variations of u_{ij} [7–9], any stable stacking of grains (the ones being counted in the Edward entropy), is an equilibrium configuration.

To see how misleading the reduced entropy S_g may be, consider an ensemble of pendulums in equilibrium – all hanging down, no deviations allowed. Since there is only this one state, we have $S_g = 0$ being minimal. And the reason is: A pendulum coming to rest has maximal S.

Grains are of course more complicated than non-interacting pendulums. Jamming and obstructing one another, they have $S_g \neq 0$, and give rise to a panoply of phenomena. Nevertheless, taking the total entropy, S_{tot} =

$S + S_g$, from all available degrees of freedom, to be maximal in equilibrium is always correct. With $S_g \ll S$, taking $S = \max$ is an excellent approximation. Taking $S_g = \max$ implies S does not change. Yet since any dissipation heats up the grains, increases S, this simply never happens.

2.3 Tapping: thermal and athermal

Tapping is accepted as the one experiment naturally explained by ASM [10]. We therefore scrutinize it more closely. A heap of grains has many stable configurations, many local maxima of S. The logarithm of this number, typically an increasing function of the density, is the Edward entropy, $S_{Ed}(\rho)$. Assuming it is $S_{Ed}(\rho)$ that is maximal in equilibrium implies that the density increases under tapping, simply because this increases S_{Ed}.

This consideration is justified only if S is independent of the density, if the number of microscopic energy states remains the same when the density increases – a proposition that seems unlikely to be true. We note in passing that since S_g counts all granular states, in and off equilibrium, including the much more numerous ones with grains flying and jiggling, it is much larger than S_{Ed}, or

$$S_{Ed} \ll S_g \ll S. \tag{1}$$

The alternative explanation employing GSH involves S. In a free column of grains, every layer (carrying the same load from above) has a given pressure. In GSH, with $\pi_{ij} = \pi_{ij}(\rho, u_{kl})$, the pressure $P \equiv \frac{1}{3}\pi_{\ell\ell}$ also depends on ρ, u_{kl}. Under tapping, because grains periodically lose contact with one another, their deformation is slowly lost, $u_{ij} \to 0$. As u_{ij} relaxes, ρ increases to maintain the pressure P. This is why the density increases, and it is connected to an increase of S: As $u_{ij} \to 0$, the associated elastic energy dissipates into heat, increasing S. This increase is much, much larger than any changes in S_{Ed}.

It is not hard to discriminate between both explanations experimentally, because the consideration involving $S_{Ed}(\rho)$ is independent of the pressure, while in GSH, it results directly from holding the pressure (or stress) constant. If grains are submerged in a liquid of the same density, gravitation does not produce any pressure. Starting from any stress, tapping will diminish u_{ij}, and with it the stress, but will not increase the density. Compaction happens then only if an external pressure is applied.

2.4 The concerns of the athermal community

A large fraction of the granular physics community believes granular media cannot be treated by conventional tools of theoretical physics, because general principles, including energy conservation, thermodynamics and the concept of equilibrium, are invalid. As should be obvious by now, these are consequences of the athermal model, not abnormalities of granular media as such.

Then there are those who do accept that equilibrium is given by $S = \max$, but take this as a result of equilibrium statistical mechanics alone, lacking any relevance off-equilibria. This is a partial and erroneous view. First of all, the entropy is well defined off-equilibrium, if the system is in local equilibrium, or in generalized equilibrium, in which a few slowly relaxing variables are off their equilibrium values. And it grows continually as the system moves toward global equilibrium. This evolution is accounted for by dissipative terms, which are derived by requiring that S always increases – as we did for the pendulum above, and as was done setting up GSH.

3 Granular Thermodynamics

3.1 Two versions of granular thermodynamics

The basic thermodynamic relation for a *"mechanically stable agglomerate of infinitely rigid grains at rest"* is (with V the volume) widely taken as [2, 3]

$$\mathrm{d}V = (\mathrm{d}V/\mathrm{d}S_{Ed})\mathrm{d}S_{Ed} \equiv X\mathrm{d}S_{Ed}. \tag{2}$$

This ansatz is best appreciated by taking S_g as a function of the energy E and volume V: $S_g = S_g(E, V)$, or

$$\mathrm{d}S_g = \frac{\partial S_g}{\partial E}\mathrm{d}E + \frac{\partial S_g}{\partial V}\mathrm{d}V \equiv \frac{1}{T_g}\mathrm{d}E + \frac{P}{T_g}\mathrm{d}V. \tag{3}$$

For infinitely rigid grains at rest, as there is no kinetic or deformation energy, we have $E \equiv 0$ — the energy is zero however the grains are arranged. This implies $\mathrm{d}S_g = (P/T_g)\mathrm{d}V$, or $\mathrm{d}V = (T_g/P)\mathrm{d}S_g \equiv X\mathrm{d}S_g$. Then, clearly, S_g reduces to S_{ED}.

Because grains are elastic and frequently in motion, the energy does not usually vanish. Hence Eq.(3) holds and is usually written as $\mathrm{d}E = T_g\mathrm{d}S_g - P\mathrm{d}V$. Including microscopic degrees of freedom, we have $E(S, S_g, V)$, or

$$\mathrm{d}E = T\mathrm{d}S + T_g\mathrm{d}S_g - P\mathrm{d}V. \tag{4}$$

Next, as the change of the energy with volume depends on the form change, the variable V has to be replaced by u_{ij} and ρ, and the description has to be local: Hence we take an energy density $w = w(s, s_g, \rho, u_{ij})$, depending on the entropy density s, reduced entropy density s_g, or

$$\mathrm{d}w = T\mathrm{d}s + T_g\mathrm{d}s_g + \mu\mathrm{d}\rho - \pi_{ij}\mathrm{d}u_{ij}. \tag{5}$$

This is the thermodynamic starting point for GSH [7, 8].

3.2 More on the granular temperature T_g

In any uniform medium such as water or air, there are two length scales, macro- and microscopic. All degrees of freedom may be divided into either of these two groups. A macroscopic theory takes those from the first as explicit variables, each with an equation of motion. These including mass, momentum and energy density. Those from the second group are treated summarily, with their contribution to the energy lumped together as heat, and characterized by the temperature T. Irreversibility is caused by the macroscopic energy decaying into heat.

In granular media, there is an intermediate group of degrees – momentum and deformation of individual grains.

In DEM, these are explicit variables. For a macroscopic theory, a summary inclusion again suffices, with their energy lumped into granular heat, quantified by T_g. That macroscopic energy decays into granular heat, then on to true heat, is what we termed *two-stage irreversibilty*.

We separate T_g from T, instead of lumping them into one temperature, not only because of the different length scales. Equally important is the fact that T_g is an independent state variable, on which granular dynamics critically depends: The elastic stress relaxes when grains jiggle, when $T_g \neq 0$, while it is (within limits) independent of T. Lumping both into one temperature obscures this difference. Extending this argument, we see that a third temperature is superfluous. For instance, it is not useful to introduce a temperature for those degrees characterizing the surface roughness of grains. Though the length scale is distinctly smaller, no aspect of the macroscopic granular dynamics depends on an associated temperature. These degrees are part of T_g. Similarly, it is not useful to introduce a configuration temperature independent from T_g.

Both T_g and T are genuine temperatures, as each characterizes the energy of a group of degrees of freedom. Same holds for the granular and true entropy, S_g and S: Each is the logarithm of the number of states in the associated group. Any difficulties treating T_g as a temperature arises only because T is ignored. For instance, the granular temperatures T_g of two systems in contact are typically different – seemingly a failure of T_g to equilibrate. Given two granular systems, 1 and 2, with only 1 being excited, there are, in the steady state, four generally unequal temperatures: T^1, T_g^1, T^2, T_g^2. And there are three ongoing energy fluxes: $(T_g^1 \to T^1)$, $(T_g^1 \to T_g^2)$, $(T_g^2 \to T^2)$. This is in complete analogy to four conventional thermal systems, (1, 1a, 2, 2a), with only (1a) being heated by an external source, and (1a,1), (1a, 2a), (2a, 2) in pairwise thermal contact. We expect all four temperatures to be different.

Taking the energy density as a function of the two entropy densities, $w = w(s, s_g)$, the conjugate variables are:

$$T \equiv \partial w/\partial s, \qquad T_g \equiv \partial w/\partial s_g. \tag{6}$$

Denoting $s_{tot} = s + s_g$, we may also write

$$dw = T ds + T_g ds_g = T ds_{tot} + (T_g - T) ds_g, \tag{7}$$

and identify $T ds_{tot}$ as the equilibrium energy change for changes of the total entropy, and $(T_g - T)ds_g$ as the extra energy contribution for $T_g \neq T$. Since $(T_g - T)$ characterizes the non-optimal energy distribution between the granular and microscopic degrees of freedom, the energy w has a minimum at $T_g = T$, and is, if expanded, given as: $w \propto (T_g - T)^2$. This implies $T_g - T$ relaxes until it vanishes.

Now, with $s \gg s_g$, and any granular motion at all occurs at $T_g \gg T$. Therefore, $T_g - T \approx T_g$, $s_{tot} \approx s$, and the rewriting of Eq.(7) does not change anything. And we may simply take $w \propto T_g^2$, with T_g relaxing until $T_g = 0$.

Since the granular heat $\int T_g ds_g \propto T_g^2$ denotes the elastic and kinetic energy of the grains, it reduces, in the limit of vanishing ρ, in which only the kinetic part remains, to the energy or temperature of the kinetic theory T_k, or

$$T_k \propto T_g^2. \tag{8}$$

A division into three scales works when they are well separated – though this is a problem of accuracy, not viability. Scale separation is well satisfied in large-scaled, engineering-type experiments, less so in small-scaled ones. Using glass or steel beads (typically larger) aggravates the problem. Nevertheless, when the system is too small for spatial averaging, one may still average over time and runs to get rid of the fluctuations not contained in a hydrodynamic theory.

References

[1] D. Bi, S. Henkes, K.E. Daniels, B. Chakraborty. Ann. Rev. Condensed Matter Physics 6: 63-83 (2015);

[2] S.F. Edwards, R.B.S. Oakeshott, Physica A**157**, 1080 (1989);

[3] R. Blumenfeld, J.F. Jordan, S.F. Edwards. Phys. Rev. Lett. **109**, 238001 (2012)

[4] Y.M. Jiang and M. Liu. Eur. Phys. J. E **38:** 15 (2015)

[5] D. Kolymbas. *Introduction to hypoplasticity*, Balkema, Rotterdam, 2000

[6] D.L. Henann and K. Kamrin. *Proceedings of the National Academy of Sciences*, **110**, 6730 (2012).

[7] Y.M. Jiang, M. Liu. Granular Matter, 11:139 (2009). Free copy:www.springerlink.com/content/a8016874j8868u8r/fulltext

[8] Y.M. Jiang and M. Liu. *Acta Mech.* 225, 2363–2384 (2014)

[9] Y.M. Jiang and M. Liu. *Why granular media are thermal, and quite normal, after all.* Eur. Phys. J. E (2017) 40: 10

[10] P. Richard, M. Nicodemi, R. Delannay, P. Ribiere, D. Bideau, Nature, **4**, 121 (2005)

DOI: 10.1051/epjconf/201714011001

Accounting for segregation employing granular solid hydrodynamics

Mario Liu[1,*] and *Yimin* Jiang[2]

[1] *Theoretische Physik, Universität Tübingen,72076 Tübingen, Germany*
[2] *Central South University, Changsha 410083, China*

Abstract. Granular solid hydodynamics (GSH) is generalized to account for segregation. The necessary change is minimal, and consists in introducing the concentration c as an additional variable. There is no need for the stress and velocity of every population. GSH's functionality with respect to all other granular phenomena – from elasto-plastic motion and shear band to the μ(I)-rheology – is preserved.

Segregation is a granular phenomenon of considerable practical importance. Many researchers have been interested in various aspects of this phenomenon, and have succeeded in setting up theories accounting for them. The descriptions are diverse and possess a high level of sophistication, see [1–3] and references therein. Surprising highlights, for a person entering this area, are especially the description of self-channelizing flows that spontaneously form static levees [4], or the account of grain size dynamics in a five-dimensional space, using a continuous size variable s in addition to time t and space $\vec{r}$ [5].

Though incapable of accounting for segregation, GSH is otherwise a fairly wide-ranged macroscopic theory of granular behavior. The phenomena accounted for include static stress distribution, elasto-plastic motion, and the $\mu(I)$-rheology; also angles of repose and stability, propagation of elastic waves, compaction [6–8]. Furthermore, the theory is derived employing thermodynamics and conservation laws, treating granular media as a thermal, and not-at-all abnormal, system, see [9]. (This last reference contains a minimalist version of GSH that may smoothen one's introduction to it.)

Generalizing GSH to include segregation seems a useful step, as it would clarify the thermodynamic basis of segregation, and would ascertain that the employed equations comply with all constraints. Moreover, this step becomes a necessary one if one aims to account for hybrid phenomena such as motion and segregation in a partially filled rotating cylinder. This is a phenomenon in which especially the following effects and their mutual interaction are relevant: static stress distribution, the $\mu(I)$-rheology, the angles of repose and stability, all in addition to the concentration diffusion. A simple-minded combination of the GSH equations with the advection-diffusion equation is not likely to suffice.

There are two points in the conventional approach that may merit more thoughts and some scrutiny. First, focusing on segregation, these theories and equations are tailored to yield a proper account, with less thoughts spent on other granular phenomena such as the ones mentioned above. Yet the ultimate goal of any theory is to eventually arrive at a unified account, similarly broad-ranged as the sum of our observations.

Second, most papers start their description by introducing stress, velocity, and the interaction force β for each population of grains, even though distinguishing the stress from forces often lacks uniqueness. Employing scaling laws for the stresses and forces, the given set of differential equations is typically first used to derive an advection-diffusion equation for the concentration c – usually assuming incompressibility – which is then solved. The scaling law was postulated at first, but then shown to hold through DEM simulations [2]. A classic reference for this type of approach, referred to as *mixture theory*, is [10].

On the other hand, it is known that the diffusion equation for the concentration c may be derived directly, with no ambiguities, no approximations, and all thermodynamic constraints spelt out, see eg. the well-known textbook by Landau and Lifshitz [11]. For instance, the diffusion equation holds irrespective whether the system is compressible, or whether the scaling laws is valid under all circumstances, say for large soft grains mixed with small hard ones. Also, the derivation reveals the constraint that the diffusion current J_k is necessarily proportional to $\nabla_k(\chi/\rho)$, where ρ is the total density, see Eq.(1), and χ is the derivative of the thermodynamic energy w with respect to the concentration c, or $\chi \equiv \partial w/\partial c$, that is usually referred to as the chemical potential, see the detailed discussion on the energy w below.

In the following, we shall confine ourselves to a mixture of two types of grains, say large and small, postponing the inclusion of the continuous variable s for a future work. Then we only need to introduce the concentration as a scalar state variable,

$$c \equiv \varrho^1/\varrho, \quad \varrho \equiv \varrho^1 + \varrho^2. \tag{1}$$

*e-mail: mliu@uni-tuebingen.de

For c and ρ – as for the rest of the GSH-variables, but unlike the velocity or stress of one grain species – coarse-graining is a simple and unique procedure, carried out for all particles within a given volume element.

For a hydrodynamic theory such as GSH, the most important quantity is the total, conserved, thermodynamic energy density $w + \frac{1}{2}\rho v^2$, with w the energy of the local rest frame, and v_i the barycentric velocity. For monodisperse grains, the energy of GSH depends on the entropy s, granular entropy s_g, density ρ, and the elastic strain u_{ik}, or $w = w(s, s_g, \rho, u_{ij})$. Being the total energy, w contains the energy contributions of all degrees of freedom. The entropy s accounts for the energy of microscopic ones such as electrons and phonons, which is usually referred to as heat. The granular entropy s_g accounts for the energy in the mesoscopic, granular degrees of freedom, ie. the quickly fluctuating kinetic and elastic energy of individual grains. One may call this energy contribution, connected to granular jiggling, granular heat. The macroscopic, coarse-grained degrees of freedom, given by the density ρ and the elastic strain u_{ij}, are explicitly accounted for, where u_{ij} quantifies the large-scale deformation of the grains. To consider segregation, the dependence on the coarse-grained concentration c needs to be included,

$$w = w(s, s_g, \rho, c, u_{ij}). \tag{2}$$

Much physics is encoded in the derivatives of w, usually referred to as the conjugate variables,

$$T \equiv \frac{\partial w}{\partial s}, \quad T_g \equiv \frac{\partial w}{\partial s_g}, \quad \pi_{ij} \equiv -\frac{\partial w}{\partial u_{ij}}, \quad \chi \equiv \frac{\partial w}{\partial c}, \tag{3}$$

which are, respectively, the temperature, the granular temperature, the elastic stress and the chemical potential. For instance, the total stress σ_{ij}, defined by $\partial_t(\rho v_i) + \nabla_i \sigma_{ij} = 0$ (with $\partial_t \equiv \partial/\partial t$), is given as

$$\sigma_{ij} = P_T \delta_{ij} + \pi_{ij} + \rho v_i v_j - \sigma^D_{ij}, \tag{4}$$

where the thermodynamic pressure P_T is another derivative,

$$P_T \equiv -\frac{\partial(w/\rho)}{\partial(1/\rho)}, \tag{5}$$

while σ^D_{ij} is the dissipative contribution. (For the simplest case, it is the viscous stress.) Similarly, the evolution equation for c is necessarily given as

$$\rho(\partial_t + v_i \nabla_i)c - \nabla_k J_k = 0, \tag{6}$$

$$J_k = D \nabla_k(\chi/\rho). \tag{7}$$

Only this form is compatible with the conservation laws of energy and the two masses, in addition to the second law of thermodynamics, see the derivation in appendix A. We note that although Eq.(6) is quite similar to the advection-diffusion equation employed in the literature, the basic idea is somewhat different. The latter has two counteracting terms,

$$\partial_t c + \nabla_k(c v_k) - \nabla_z[S_r F(c)] = \nabla_z(D \nabla_z c),$$

with the diffusive one $\nabla_z(D\nabla_z c)$ homogenizing the concentration, and the advective one $\nabla_z[S_r F(c)]$ segregating the two populations. In the stationary case, both terms balance. Eqs.(6,7), on the other hand, have the chemical potential χ/ρ as a single driving field that both equalize and segregates, stopping to operate only when it is uniform, though c may well be nonuniform then. This is similar to temperature diffusion: The heat current stops when T is uniform, though the entropy s need not be.

More specifically, we have $J_k = 0$ for $v_i \perp \nabla_i c$ and a steady state, $\partial_t c = 0$. (The case $J_k = \text{const} \neq 0$ is not compatible with the boundary condition $J_k = 0$.) This implies

$$\chi/\rho = \text{const}. \tag{8}$$

Being a derivative of the energy w, the chemical potential χ shares the same variables with $w = w(s_g, \rho, u_{ij}, c)$. (We ignore the entropy s and inhomogeneity of the temperature T here.) Any non-uniformity in the first three variables induces a compensating c, to keep χ/ρ = const. A stationary, non-uniform c is of course segregation.

Moreover, a steady-state T_g (which we shall see below is $\sim s_g$) is proportional to the shear rate, $T_g \propto \dot\gamma$: Quantifying granular jiggling, T_g relaxes if left alone and is excited by $\dot\gamma$, such that $T_g \propto \dot\gamma$ holds in the steady state. Since flows frequently have non-uniform shear rates and densities, they give rise to segregation.

Note we need to have the diffusion rate $D = D(T_g)$ such, that $D \to 0$ for grains at rest, $T_g \to 0$. Because whatever the chemical potential then is, concentration distribution is frozen in. Most simply, we may set $D \propto T_g$. This implies slow diffusion rates for regions of small $\dot\gamma$.

Eqs.(6,7) comprise the structural change of GSH, though we still need to specify the concentration dependence of the energy density $w(c)$, to obtain the chemical potential $\chi \equiv \partial w/\partial c$. Finding the c-dependence of the energy is a trial and error process, same as the one we went through to find the energy for mono-disperse systems, an expression that is by now well verified [6–8]. Both are constitutive choices. The energy for mono-disperse grains has the form:

$$w = w_\Delta + w_T, \tag{9}$$

$$w_\Delta = \sqrt{\Delta}[2\mathcal{B}(\rho)\Delta^2/5 + \mathcal{A}(\rho)u_s^2], \tag{10}$$

$$\Delta \equiv -u_{kk}, \; u_s^2 \equiv u^*_{ij}u^*_{ij}, \; u^*_{ij} \equiv u_{ij} + \Delta\delta_{ij}/3,$$

$$w_T = s_g^2/(2\rho b) = \rho b T_g^2/2, \tag{11}$$

$$T_g \equiv \partial w/\partial s_g = s_g/\rho b,$$

$$b = b_1/\rho + b_0\left[1 - \rho/\rho_{cp}\right]^a, \quad a \approx 0.1.$$

The contribution w_Δ is the elastic energy, a function of the elastic strain u_{ij}, which quantifies the coarse-grained deformation of the grains.The term w_T is the energy of granular jiggling, the quickly fluctuating kinetic and elastic energy of the grains. (Note $T_g^2 \propto T_k$, where T_k is the temperature used in the kinetic theory.) With the expressions $\mathcal{B}(\rho)$, $\mathcal{A}(\rho)$ given in [6, 7], w_Δ and the elastic stress π_{ij} have proven suitable for the account of various static stress distributions, propagation of elastic waves and elasto-plastic motion [8]. w_T and

$$P_T \equiv -\frac{\partial(w/\rho)}{\partial(1/\rho)} = -\frac{\partial(w_T/\rho)}{\partial(1/\rho)},$$

from granular jiggling are also well-verified. (The contribution from w_Δ is of higher order and may be neglected.) For $\rho \to 0$, the first term in b dominates, $b \approx b_1/\rho$, then P_T reduces to that found in kinetic theories. (There is then no elastic stress, $\pi_{ij} = 0$.) For densities closer to the close-packed value ρ_{cp}, the term $\propto b_0$ dominates, accounting for compaction and the $\mu(I)$-rheology.

If w is to depend on c, we need to specify its dependence such that it reverts to the above formulas for $c = 0$ and $c = 1$. Furthermore, thermodynamic stability or the convexity of energy requires

$$\frac{\partial^2 w}{\partial c^2} > 0, \quad \frac{\partial^2 w}{\partial c^2}\frac{\partial^2 w}{\partial \rho^2} - \frac{\partial^2 w}{\partial\rho\partial c} > 0. \tag{12}$$

However, a partially concave w does exist in nature, eg. in mixtures of liquids. It then implies that certain concentration values cannot be realized in this mixture. The system will demix, with a region of high concentration, $c = c_1$, another of low concentration, $c = c_2$, such that $w(c_1)$ and $w(c_2)$ are convex. Analogously, a partially concave w for a mixture of two types of grains does not have to be ill-chosen. It may appropriately reflect the observation that certain c-values is not realized in equilibrium. The system will then also demix, with a region of say only large grains, another with a c value possessing a stable, convex energy.

In specifying the c-dependence, we shall for two reasons only modify w_T, not w_Δ. First, the fluctuating part of the kinetic stress was identified as the dominant cause for segregation, see [12], and it is closely related to $P_T \propto T_g^2$.

Second, the elastic contribution of $w_\Delta(c)$ to the chemical potential is typically small, $\chi_\Delta \ll \chi_T$, hence $\chi = \chi_\Delta + \chi_T \approx \chi_T$. The reason is, in the elastic pressure, $P_\Delta \equiv \frac{1}{3}\pi_{kk} \propto \mathcal{B}\Delta^{1.5}$, with typically $\Delta \lesssim 10^{-3}$, a large $\mathcal{B}$ compensates a small $\Delta^{1.5}$. Now, since χ_T is of the order of P_T/c, and $\chi_\Delta \propto (\mathcal{B}/c)\Delta^{2.5} \propto \Delta P_\Delta$, we have $\chi_T \gg \chi_\Delta$, if P_Δ is comparable to P_T. (For T_g^2 sufficiently small, we have $P_T \propto T_g^2 \ll P_\Delta$. But this regime is probably irrelevant for segregation.)

Modifying w_T, one easy way to proceed is to replace its constants, especially

$$\rho_{cp}(c) = \rho^0_{cp} + hc(1-c), \tag{13}$$

because we know that ρ_{cp} changes in the mixture, with h positive and growing with the ratio of the two radia, because the small grains fill the void between the large ones. If we take h depending on c, say linearly, $\rho_{cp}(c)$ is not symmetric with respect to $c \to 1-c$, and its maximum is not at $c = \frac{1}{2}$. If ρ^0_{cp} is different for $c = 1$ and $c = 0$, it is to be replaced by

$$\rho^0_{cp} \to c\rho^1_{cp} + (1-c)\rho^2_{cp}. \tag{14}$$

If Eq.(13) is the only change, we have

$$\chi \approx \partial w_T/\partial c|_{s_g} = -(\rho T_g^2/2)\,\partial b/\partial c,$$
$$\frac{\partial b}{\partial c} = \frac{b_0 a\rho}{\rho^2_{cp}}\left(1 - \frac{\rho}{\rho_{cp}}\right)^{a-1}\frac{\partial\rho_{cp}}{\partial c}, \tag{15}$$
$$\partial\rho_{cp}/\partial c = h(1-2c) + c(1-c)\partial h/\partial c.$$

Note $\partial^2 w/\partial c^2 > 0$ for h = const, though it becomes negative for $c < 1/3$ if $h \propto c$. If h = const, we have

$$\frac{\chi}{\rho} = -\frac{T_g^2 h[1-2c]b_0 a\rho[2\rho^2_{cp}(c)]^{-1}}{\left[1-\rho/\rho_{cp}(c)\right]^{1-a}}. \tag{16}$$

Since the c-dependence of the denominator varies more weakly, we may frequently neglect it, replacing $\rho_{cp}(c)$ with ρ^0_{cp}. Note that the structure of this formula,

$$\chi/\rho = -T_g^2 h[1-2c]f(\rho),$$

is retained [with a modified $f(\rho)$], if Eq.(13) works also for replacing b_0 alone, though replacing b_0 and ρ_{cp} simultaneously is certainly more realistic.

Replacing b_1 does not lead to segregation: For a rarefied granular gas, we have $\frac{1}{2}b_1T_g^2 = \frac{3}{2}(\rho/M)k_BT$, and if bi-disperse $\frac{1}{2}b_1T_g^2 = \frac{3}{2}\rho k_BT[c/M_1 + (1-c)/M_2]$. This implies the substitution

$$b_1 \to b_1^0[c + (1-c)(M_1/M_2)]. \tag{17}$$

Being linear in c, this form does not lead to a c-dependence of the chemical potential χ.

A robust conclusion is, flows with a constant $\dot\gamma$ and a constant density do not give rise to segregation.

References

[1] J.M.N.T. Gray and A.R. Thorton Proc. R. Soc. **A 461**, 1447 (2005).

[2] D. R. Tunuguntla, O. Bokhove, and A. R. Thornton, Journal of Fluid Mechanics 749, 99 (2014).

[3] P. Gajjar, K. van der Vaart, A. R. Thornton, C. G. Johnson, C. Ancey and J. M. N. T. Gray. *J. Fluid Mech.*, vol. 794, pp. 460–505 (2016)

[4] S. Deboeuf, E. Lajeunesse, O. Dauchot, B. Andreotti. Phys.Rev.Lett. 97, 158303 (2006)

[5] B. Marks, P. Rognon, and I. Einav, Journal of Fluid Mechanics 690, 499 (2012).

[6] Y.M. Jiang, M. Liu. Granular Matter, 11:139(2009), free download: www.springerlink.com/content/a8016874j8868u8r/fulltext

[7] Y.M. Jiang and M. Liu. *Acta Mech.* 225, 2363–2384 (2014)

[8] Y.M. Jiang and M. Liu. *Applying GSH to a wide range of experiments in granular media.* Eur. Phys. J. E (2015) 38:15

[9] Y.M. Jiang and M. Liu. *Why granular media are thermal, and quite normal, after all.* Eur. Phys. J. E (2017) 40: 10

[10] L.W.Morland, *Flow of viscous fluid through a porous deformable matrix* Surveys in Geophysics (1992) 13:209 268.

[11] L. D. Landau and E. M. Lifshitz. *Fluid Mechanics*. Butterworth-Heinemann (1987).

[12] K.M. Hill and D.S. Tan. *Segregation in dense sheared flows: gravity, temperature gradients, and stress partit*. J. Fluid Mech., **756**, 54-88. (2014)

A Deriving the c-equation

Starting with the equations for ρ^1, ρ^2,

$$\partial_t\rho^1 + \nabla_k(\rho^1 v_k^1) = 0,$$
$$\partial_t\rho^2 + \nabla_k(\rho^2 v_k^2) = 0, \quad (18)$$

we rewrite them as

$$\partial_t\rho^1 + \nabla_k(\rho^1 v_k) - \nabla_k J_k = 0, \quad (19)$$
$$\partial_t\rho^2 + \nabla_k(\rho^2 v_k) + \nabla_k J_k = 0, \quad (20)$$
$$\text{where} \quad J_k = \rho^1\rho^2(v_k^2 - v_k^1)/\rho, \quad (21)$$
$$v_k = (\rho^1 v_k^1 + \rho^2 v_k^2)/\rho.$$

Combining both equations, we obtain

$$\partial_t\rho + \nabla_k(\rho v_k) = 0, \quad (22)$$
$$\rho(\partial_t c + v_k\nabla_k c) - \nabla_k J_k = 0, \quad (23)$$

which is the continuity equation and the equation for $c \equiv \rho^1/\rho$ of the form given in Eq.(2).

Next, we derive the expression for J_k, in terms of the variables already present, as Eq.(20), containing v_k^1, v_k^2, is not useful. For simplicity, we shall go to the rest frame, $v_k \equiv 0$, because the same consideration in a general frame is rather more complicated, and produces results we are not interested in here.

As the conserved energy density $w = w(s, \rho^1, \rho^2)$ is a function of the entropy density s, in addition to ρ^1, ρ^2, we write

$$\mathrm{d}w = T\mathrm{d}s + \mu^1\mathrm{d}\rho^1 + \mu^2\mathrm{d}\rho^2,$$

where T, μ^1, μ^2 are the temperature and the respective chemical potentials. (The dependence on the granular entropy s_g and the elastic strain u_{ij} is also neglected here for simplicity.) We rewrite this relation as

$$\mathrm{d}w = T\mathrm{d}s + \mu^2\mathrm{d}\rho + (\mu^1 - \mu^2)\mathrm{d}\rho^1, \quad (24)$$
$$= T\mathrm{d}s + \mu\mathrm{d}\rho + \chi\mathrm{d}c, \quad \text{with}$$
$$\chi = \rho(\mu^1 - \mu^2), \ \mu = (1-c)\mu^2 + c\mu^1. \quad (25)$$

Taking the temporal derivative of Eq.(24), inserting the conservation equations (for the rest frame): $\partial_t w = \nabla_k Q_k$, $\partial_t\rho = 0$, $\partial_t\rho^1 = \nabla_k J_k$, and the balance equation $T\partial_t s = R > 0$, we find

$$\nabla_k Q_k - R = (\mu^1 - \mu^2)\nabla_k J_k \quad (26)$$
$$= \nabla_k[(\mu^1 - \mu^2)J_k] - J_k\nabla_k(\mu^1 - \mu^2).$$

This equation is an identity that has to be satisfied at all time and in all of space. Taking $Q_k = (\mu^1 - \mu^2)J_k$, and Eq.(3),

$$J_k = D\nabla_k(\mu^1 - \mu^2) = D\nabla_k(\chi/\rho), \quad (27)$$

with $D > 0$, satisfies energy conservation and ensures that the entropy production R is positive-definite.

 DOI: 10.1051/epjconf/201714002002

Angle of repose revisited: When is a heap a cone?

Michael Rackl[1,*], *Florian E.* Grötsch[1], and *Willibald A.* Günthner[1]

[1] *Institute for Materials Handling, Material Flow and Logistics*
Technical University of Munich, Boltzmannstraße 15, 85748 Garching, Germany

Abstract. The angle of repose is a frequently reported value in bulk solids characterisation and its popularity is based on the simple measurement setup. It can be useful to estimate the flowability of a bulk material or to estimate the inner friction coefficient for cohesionless solids. The latter relies on Coulomb material theory, which yields that a bulk solid heap resembles a perfect cone. The aim of this study was to investigate if the cone assumption is valid for three distinct bulk materials. In order to address this question, the angle of repose was measured for each sample and the resulting ideal cone model compared with a respective 3D-scanned surface model of the physical heap. Differences were mapped onto a two-dimensional representation of the ideal cone's lateral surface. The heaps of two of the three investigated bulk solids showed considerable differences with regard to the cone assumption. Discrepancies can be divided into local and global shape deviations. The results show that measuring and reporting the angle of repose should be undertaken with care. Many publications list angle of repose values without reporting the measurement setup or the shape of the heap. However, the latter can vary significantly from Coulomb theory.

1 Introduction

The angle of repose is a frequently reported value in bulk solids characterisation. Its prevalence is mostly based on the simple and straightforward measurement setup. Furthermore, Coulomb material theory allows for estimation of the bulk solid's inner friction coefficient from the angle of repose. In doing so, the angle of repose measurement relies on the assumption that the concerning bulk solid heap resembles the geometric shape of a perfect cone, which roughly holds true for free-flowing materials. For cohesive or other non-free-flowing bulk solids the question arises, whether the angle of repose is a meaningful characteristic to measure and report. Local effects of cohesion or widely spread variations of particle shapes can result in the physical heaps' shapes to considerably differ from a cone. The aim of this study was to investigate if the cone assumption is valid for lime stone, corn grains and milk powder.

2 Materials and Methods

Three-dimensional surface scans of bulk solid heaps were compared against cones based on an ideal Coulomb material.

2.1 Angle of repose theory

An ideal bulk solid cone with the angle of repose, Φ, is shown on the left of Figure 1. If one considers an ideal

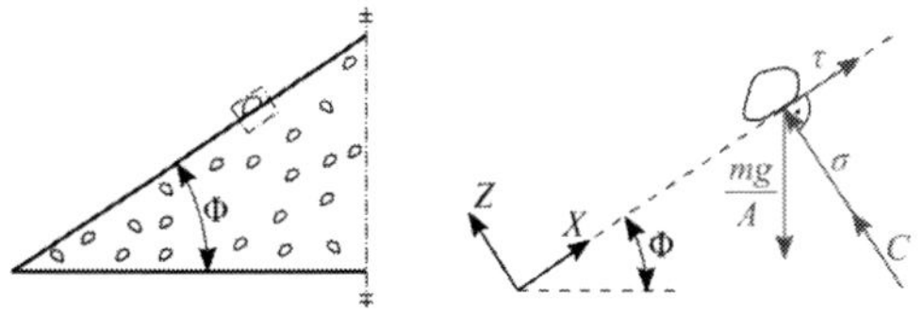

Figure 1. Ideal bulk solid cone (left) and free body diagram of one particle on the lateral surface (right).

Coulomb material [1, p. 55] then the right part of Figure 1 is the free body diagram of a single, tiny particle sitting on other particles of the heap. The equations for the static equilibrium with regard to the axes X and Z are Equations 1 and 2, where τ is shear stress, σ is normal stress, m is the particle mass, g the gravitational acceleration and A the contact surface area between particle and the heap. C is a constant representing cohesional stress between the particle and the surface.

$$X : 0 = \tau - \frac{mg}{A} \sin\Phi \quad (1)$$

$$Z : 0 = \sigma + C - \frac{mg}{A} \cos\Phi \quad (2)$$

The combination of both equations yields Equation 3. After the cohesion constant, $C_0 = C\tan\Phi$, is introduced and $\tan\Phi$ is replaced by the internal friction coefficient, μ, the Coulomb yield criterion is found in Equation 4 (cf. [1,

*e-mail: rackl@fml.mw.tum.de, mi.rackl@gmx.de

p. 55]); for cohesionless materials: $C_0 \approx 0$.

$$\tau = \sigma \tan\Phi + C \tan\Phi \quad (3)$$

$$\tau = \mu\sigma + C_0 \quad (4)$$

A particle sitting on the lateral surface of a heap will only stay in place if $\tau \geq \mu\sigma + C_0$. Otherwise particles are going to start sliding under the influence of gravity until $\tau \geq \mu\sigma + C_0$ is satisfied. Hence, for an ideal cohesionless Coulomb material the angle of repose is equal to $\arctan\mu$ and the resulting heap resembles a perfect cone.

2.2 Experimental investigation of the angle of repose

The materials investigated in this study were lime stone, corn grains and skimmed milk powder. Graded lime stones with a size range from 20 to 70 mm (d_{50} = 26 mm) were purchased from a local quarry [2]. d_{50} is the mass median diameter as obtained from the particle size distribution. The dried corn grains were obtained from an agricultural wholesale trader [3] and had d_{50} = 6 mm. The investigated skimmed milk powder [4] had a d_{50} of 0.06 mm.

Slump tests were carried out with a cylindrical container (ø425 mm) made from rolled mild steel, which was placed on a bottom plate made from the same material. The container was filled to a height of 545 mm and then lifted up vertically at a constant velocity of 142 mm/s, such that a heap of the respective bulk solid was formed under the influence of gravity (cf. Figure 2). The angle of repose of the respective heaps was measured according to an FEM standard [5]. In addition to this, the heaps were scanned by means of a KINECT v2 sensor to generate three-dimensional surface meshes. Each measurement was repeated for four heaps (N = 4).

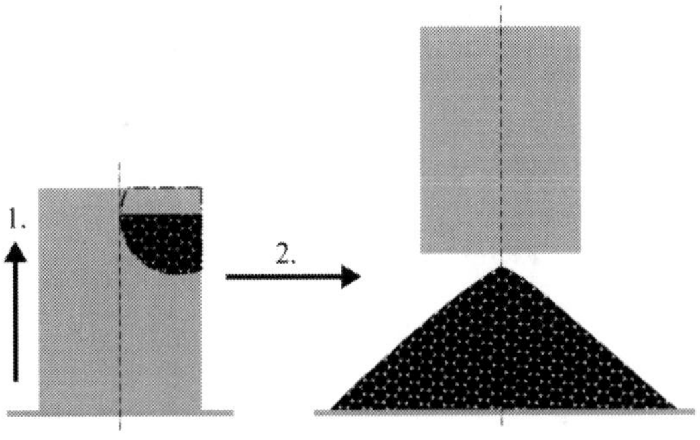

Figure 2. Experimental setup.

Each 3D-scanned surface mesh was imported into Matlab [6] and optimally aligned and positioned with regard to its ideal cone according to FEM. Figure 3 shows an exemplary image of an ideal cone and a 3D-scanned heap surface. The ideal cone was created based on the measured value of the angle of repose and heap height. After the alignment, differences between the surface geometry of the ideal cone and the 3D-scanned surface mesh were determined. Each difference was computed as the distance between a point on the scanned surface mesh and a point on the cone's lateral surface, along a vector perpendicular to the cone's surface [7]. Finally, the differences were mapped onto a two-dimensional developed view of the ideal cone's lateral surface.

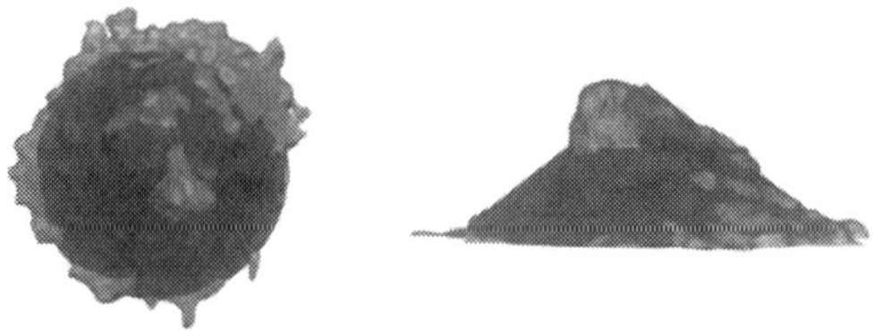

Figure 3. 3D-scanned heap surface (yellow), aligned with regard to its ideal cone counterpart (blue), in a top-town (left) as well as side view (right, not to scale).

3 Results and Discussion

Two out of three heap shapes measured by means of three-dimensional scans revealed considerable differences with respect to an ideal cone.

3.1 Angle of repose

The angle of repose results, measured according to the FEM standard [5], are listed in Table 1. Based on the angle of repose, descriptions of flowability can be assigned to the materials [1, p. 52], whereas lime stone and corn grains would be "very free flowing", while milk powder is considered "free-flowing". These descriptions are of course very rough, since they are exclusively based on the angle of repose.

In addition to this, the ratio between the heap diameter and the particle diameter, d_{50}, is listed in Table 1. The FEM standard suggests that this ratio should exceed 20 for angle of repose measurements.

Table 1. Measured angles of repose with standard deviation and heap diameter to d_{50} particle size ratio. (N = 4).

Material	Angle of repose (°)	Ratio (-)
Lime stone	26 ± 1	~38
Corn grains	16 ± 1	~192
Milk powder	32 ± 0	~17,500

3.2 Comparison of the angle of repose with the heap surfaces

Results for the mapped differences between the 3D-scanned heaps and the respective ideal cones for exemplary heaps of each material are shown in the Figures 4, 5 and 6.

Figure 4 shows good overall agreement in shape, i. e., large areas with small differences. Nonetheless, there are spots, displayed in deep blue and red respectively, where differences are as large as 40 mm. This noise originates from the coarse granularity of the lime stone particles (cf. ratio in Table 1). Since the single particles are rather large compared to the overall heap size, the surface structure of the lime stone heap is rough.

The size ratio of lime stone in this study is considerably larger than the value of 20 which the FEM standards recommends, which brings up questions regarding the significance of measurements with a low heap diameter to particle size ratio. Fluctuations and fluctuation-induced measurement errors of the angle of repose are expected to increase considerably when the size ratio is lower than in this study. The recommended minimum ratio of 20 seems much too low and 38 from this work is on the borderline, at best.

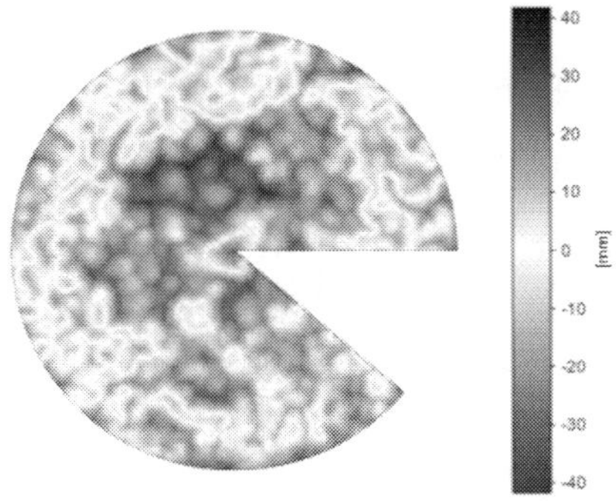

Figure 4. Developed view of the ideal cone with differences to the scanned heap for lime stone.

The developed view for corn grains is depicted in Figure 5. Its surface shows significantly less granularity than that of lime stone and hence appears smoother. Furthermore, the absolute differences between the ideal cone and the scanned heap are considerably smaller. However, a systematically different shape can be observed in a ring around the developed view's centre. From the top to the bottom of the physical heap, i. e. the inside to the outside of the circle sector, the differences increase to a maximum and then decrease until they approximately reach zero again at the bottom of the heap.

Figure 6 displays the result for a milk powder heap. The heap's surface is very smooth due to the tiny particle size of the milk powder. Nonetheless, there are large differences with regard to the ideal cone and the transition between ditches and knobs, i. e. negative and positive differences, is abrupt. In addition to this, the heap is far from rotationally symmetrical.

The causes of the asymmetrical shape of the milk powder heaps are supposedly connected to avalanching effects. When the container was lifted, the powder did not begin

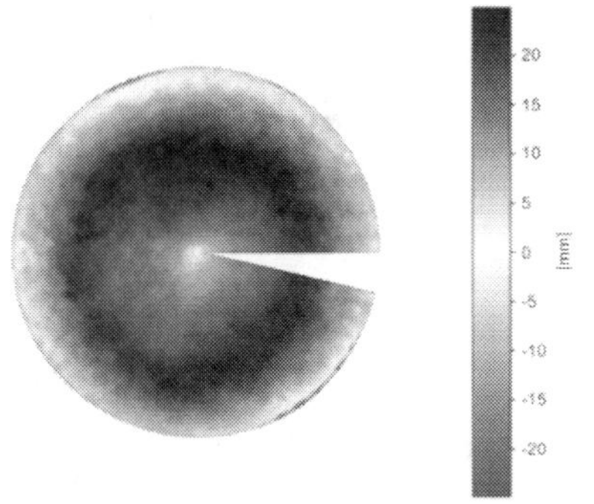

Figure 5. Developed view of the ideal cone with differences to the scanned heap for corn grains.

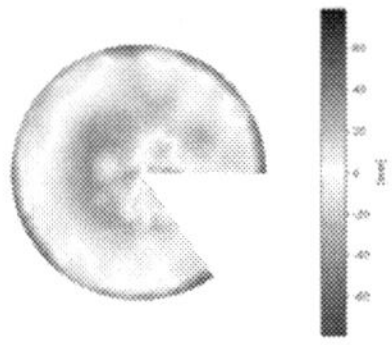

Figure 6. Developed view of the ideal cone with differences to the scanned heap for milk powder.

to flow immediately, but instead kept its cylindrical shape on the first few centimetres, until the cohesive strength was exceeded. Thereafter, larger amounts of milk powder started to pour out of the container intermittently, where each batch generated a small avalanche. Thus, the final shape is assumed to be caused by landslides, which occur when the shear strength and cohesive strength within the milk powder are locally exceeded.

3.3 Conclusion and outlook

The main findings with regard to the shape deviations for the three investigated materials are displayed in Figure 7. Lime stone came closest to the global shape of a cone. Yet, its overall shape quality was disturbed by local shape deviations originating from the coarse granularity, i. e. large particle size. The influence of large particle size on the surface smoothness can be seen in the leftmost sketch in Figure 7. The particles form gaps between them and lead to the granular characteristic, where the overall shape fits, yet positive and negative deviations can be observed. The corn grain's heap is similar to a cone and smooth in shape, however the heap has a rounded tip, as shown in the middle of Figure 7. This is in agreement with findings reported in the literature [8]. The heap shape for milk powder is depicted in the rightmost part of Figure 7. It becomes obvious that the angle of repose cannot be directly measured

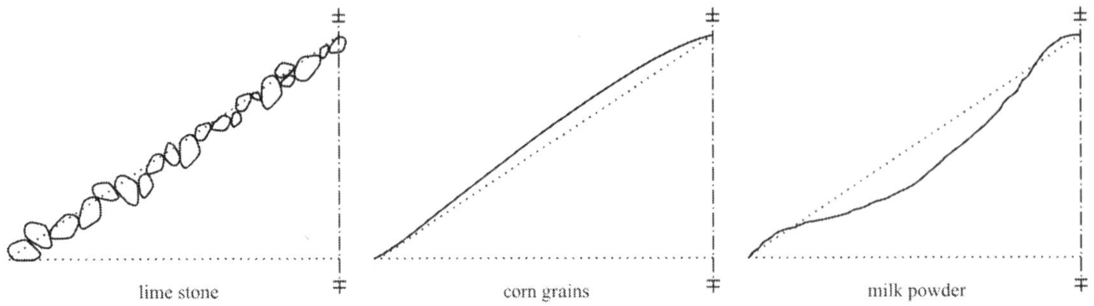

Figure 7. Approximate shape of each of the heaps (solid lines) displayed over the ideal cone (dotted lines).

from a side view, since the inclination changes considerably from the centre to the edge.

The aim of this study was to investigate if the implicit cone assumption of the angle of repose is valid for lime stone, corn grains and milk powder. It can be seen from the scope of the results of this study that this assumption is only valid for lime stone, with the exception of local shape deviations due to its granularity. Corn grains and milk powder show considerable global deviations from the ideal cone shape. Strictly speaking, milk powder is cohesive and hence its heap shape should not be compared with the cone of an ideal cohesionless material. Although cohesive bulk solids do conform to angle of repose theory, the angle of repose is widely reported for such materials in practice [9–11] and can be useful to identify cohesive materials. Corn grains can be considered "free-flowing" according to their angle of repose, however this does not mean that their heap shape resembles a perfect cone. Even though in this study the angle of repose's standard deviation was the smallest for milk powder, its heap shape showed the largest differences from a cone.

Care has to be taken when choosing the dimensions of the angle of repose measurement setup. The results form this work indicate that a heap diameter to particle size ratio of 20 (recommended from the FEM standard) and even 38 (this work) may not be large enough to form heaps which comply with angle of repose theory.

Many publications list values for the angle of repose without reporting the measurement setup or the shape of the heap. The latter however can differ significantly from theoretical considerations, as shown in this study. Future research could focus on three-dimensional scanning of a large variety of bulk materials and quantitative comparison with ideal cones, as well as among each other. These results are going to be helpful in gaining better understanding with regard to heap formation and could provide a base for the link between Coulomb material theory and physical observations of the angle of repose.

Acknowledgement

This study received funding by the AiF (no. 18371 N/1), within the programme for sponsorship by Industrial Joint Research (IGF) of the German Federal Ministry of Economic Affairs and Energy based on an enactment of the German Parliament.

References

[1] D. McGlinchey, *Characterisation of bulk solids* (Blackwell and CRC Press, Oxford and Boca Raton, FL, 2005), ISBN 1-4051-1624-2

[2] Märker Kalk GmbH, Harburg, Germany, *Märker lump lime: Technical Data Sheet*

[3] BayWa AG, Bockhorn, Germany, *Dried corn grains, batch 63, harvest 2014*

[4] EUROSÉRUM, Port-sur-Saône, France, *Skimmed milk MH PREMIUM - ADPI Extra grade*

[5] The European Federation of Materials Handling "FEM", *FEM 2.582 (11.1991): General properties of bulk materials and their symbolization*

[6] Matlab, *version R2014b* (The MathWorks Inc, Natick, Massachusetts, 2014)

[7] M. Rusch, *Studie zur 3D-gestützten Analyse von Schütthaufen (semester thesis)* (16.12.2015), Institute for Materials Handling, Material Flow and Logistics. Technical University of Munich. Germany.

[8] E. Kermani, T. Qiu, T. Li, Int. J. Geomech. **15**, 04015004 (2015)

[9] A. Sharma, A.H. Jana, R.S. Chavan, Compr. Rev. Food Sci. F. **11**, 518 (2012)

[10] G. Lumay, F. Boschini, K. Traina, S. Bontempi, J.C. Remy, R. Cloots, N. Vandewalle, Powder Technol. **224**, 19 (2012)

[11] A. Samadani, A. Kudrolli, Phys. Rev. E **64**, 051301 (2001)

Application of spherical harmonics analysis on LBS particles and LBS fragments

Budi Zhao[1,*], *Deheng* Wei[1], and *Jianfeng* Wang[1]

[1] *Department of Architecture and Civil Engineering, City University of Hong Kong, Hong Kong*

Abstract. This paper applies surface parameterization and spherical harmonics analysis to the characterization of particle shapes of Leighton Buzzard sand (LBS) particles and LBS fragments obtained from X-ray micro-tomography (μCT). The rotation, transition and scale independent spherical coefficients were obtained. The relationship between spherical coefficients and shape parameters of form, roundness and compactness was investigated. The coefficients of degree one determine the principal dimensions of an ellipsoid, which has a similar aspect ratio with the original surface. The coefficients of higher degree characterise more details by increasing the percentage of higher and lower mean curvature on the reconstructed surface. As the spherical degree increases, the reconstructed surface tend to have lower particle roundness, sphericity and convexity, and higher aspect ratio.

1 Introduction

The studies on particle shape characterization and reconstruction mainly have two objectives. The first one is to classify and quantify particle shapes. It could be used to relate shape characteristics with performance properties (e.g. mechanical behaviour of granular materials) [1] and understand the broad class of phenomena that involve shape evolution (e.g. weathering and fragmentation) [2–4]. The simplest way to obtain shape parameters (e.g. aspect ratio, roundness and compactness) is to form dimensionless indices with principal dimensions, surface area, volume or local curvature. The second objective is to randomly generate shapes for numerical simulation models. Fourier transformation and spherical harmonics method could mathematically reconstruct and randomly generate particle shapes in two-dimension and three-dimension, respectively [5, 6].

This study investigates three-dimensional (3D) particle shape characterization with spherical harmonics analysis. Leighton Buzzard sand (LBS) particles and LBS fragments were analyzed. The transition, rotation and size independent coefficients were obtained. The influence of spherical coefficients on particle form, roundness and compactness was investigated.

2 Shape description using spherical harmonics

Spherical harmonics are a complete set of orthogonal functions defined on the surface of a sphere. It has been widely used to mathematically represent 3D particle shapes. It originally used radial representation ($r(\theta, \phi)$) which is restricted to star-shaped surfaces [7]. This limitation was overcome by a surface parameterization, which defines a continuous, one-to-one mapping from the surface of the original object to a unit sphere [8].

*e-mail: bdzhao2-c@my.cityu.edu.hk

2.1 Spherical surface parameterization

Parameterization could be considered as a constrained optimization problem that embeds the object surface onto the surface of a unit sphere while minimizing the distortion of the surface net. Earlier parameterization approaches used voxel assembly of objects [8]. It has been recently extended to triangular meshes [9]. In this study, we follows the Control of Area and Length Distortions (CALD) framework proposed in [9].

In the CALD algorithm, a bijective mapping was obtained between each vertex on a surface $\mathbf{v} = (x, y, z)$ and a pair of coordinates θ and ϕ. $\theta \in [0, \pi]$ and $\phi \in [0, 2\pi)$ are the latitudinal and longitudinal coordinate, respectively. The θ and ϕ values of each vertice were determined mainly through local smoothing method and global smoothing method to reduce the area and length distortions. It is performed on each component of $\mathbf{v}$ independently, which resulted in: $\mathbf{v}(\theta, \phi) = (x(\theta, \phi), y(\theta, \phi), z(\theta, \phi))^T$. An illustration of the surface parameterization is shown in figure 1(b).

2.2 Spherical harmonics transformation

Spherical harmonics analysis is generally seen as the extension of Fourier analysis on the unit sphere. Any spherical scaler function ($f(\theta, \phi)$) could be decomposed as the

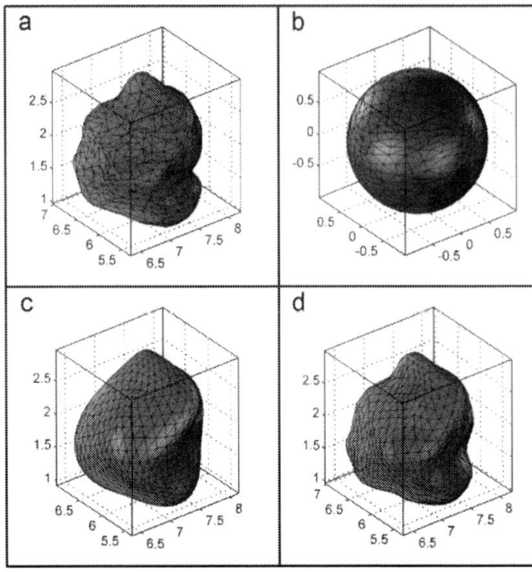

Figure 1. Surface parameterization and reconstruction using SPHARM-MAT toolbox: (a) original surface; (b) mapping on the unit-sphere; (c) reconstruction using $l = 5$; (d) reconstruction using $l = 10$

sum of spherical harmonics:

$$f(\theta,\phi) = \sum_{l=0}^{\infty}\sum_{m=-l}^{m=l} c_l^m Y_l^m(\theta,\phi) \tag{1}$$

where Y_l^m and c_l^m are called the spherical harmonic function and coefficient of degree l and order m. The set of spherical harmonics Y_l^m are called Laplace's spherical harmonics, which form an orthonormal basis of the Hilbert space of square-integrable functions:

$$Y_l^m(\theta,\phi) = \frac{(2l+1)}{4\pi}\frac{(l-m)!}{(l+m)!}P_l^m(\cos\theta e^{im\phi}) \tag{2}$$

where P_l^m is the associated Legendre polynomial.

The spherical harmonics analysis is performed on each component of $\mathbf{v}(\theta,\phi)$ independently, which resulted in harmonic coefficients with three components, $\mathbf{c}_l^m = (c_{xl}^m, c_{yl}^m, c_{zl}^m)^T$. The coefficients $\mathbf{c}_l^m$ up to a user-defined degree can be estimated by solving a set of linear equations in a least square fashion. Figure 1(c,d) shows the reconstruction with different spherical degrees.

2.3 Rotation, transition and size-independent coefficients

The coefficients ($\mathbf{c}_l^m$) obtained so far depend not only on particle shape, but also on the relative position, orientation and size of a particle. The rotational invariance was obtained by considering the contribution of the spherical harmonics of degree $l = 1$, an ellipsoid. It involves rotating the object in both parameter space and object space. The parameter space is rotated so that the north pole ($\theta = 0$) will be at one end of the shortest main axis of this first-order ellipsoid and the point where the Greenwich meridian ($\phi = 0$) crosses the equator ($\theta = \pi/2$) is at one end of the longest main axis. Then, the ellipsoid is rotated in the object space to make its main axes coincide with the coordinate axes. The resulted coefficients are invariant under rotation of the object, except mirroring.

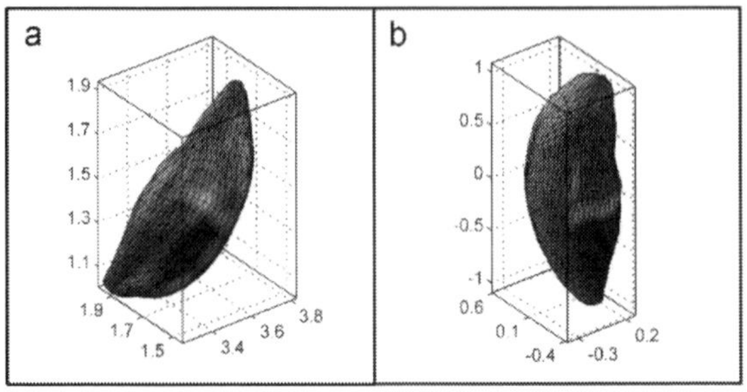

Figure 2. Surface alignment using SPHARM-MAT toolbox: (a) original surface; (b) surface after translation, rotation and scaling

The transitional invariance could be achieved by ignoring $\mathbf{c}_0^0$. Scaling invariance can be achieved by dividing all coefficients by $\sqrt[3]{V}$, where V is the volume of the particle. The scaled particles have a unit volume of 1 mm^3. The resulted coefficients ($\mathbf{c}''^m_l$) are rotation, transition and scale-independent. Figure 2 shows the original particle surface and the surface reconstructed from $\mathbf{c}''^m_l$.

3 Application to LBS particles and LBS fragments

The shapes of Leighton Buzzard sand (LBS) particles and LBS fragments generated from single particle crushing tests were analyzed. LBS particles and LBS fragments were scanned with X-ray micro-tomography (μCT) systems with a resolution of 15 μm and 3 μm, respectively. The LBS particles have particle size between 1 and 2 mm, while the LBS fragments are larger than 0.001 mm^3. The raw CT images were put through a series of image processing methods to obtain the smooth triangular surface mesh [3], based on which we performed the spherical harmonics analysis and shape quantification at form, roundness and compactness.

3.1 Extraction of the spherical harmonics coefficients

We used the SPHARM-MAT Toolbox (available at http://www.iu.edu/ shenlab/software.html) to perform the spherical parameterization and compute the spherical harmonic coefficients independent from transition, rotation and scaling.

3.2 Shape quantification on form, roundness and compactness

We followed the framework proposed in [3] to quantify the surface reconstructed with spherical harmonics at form,

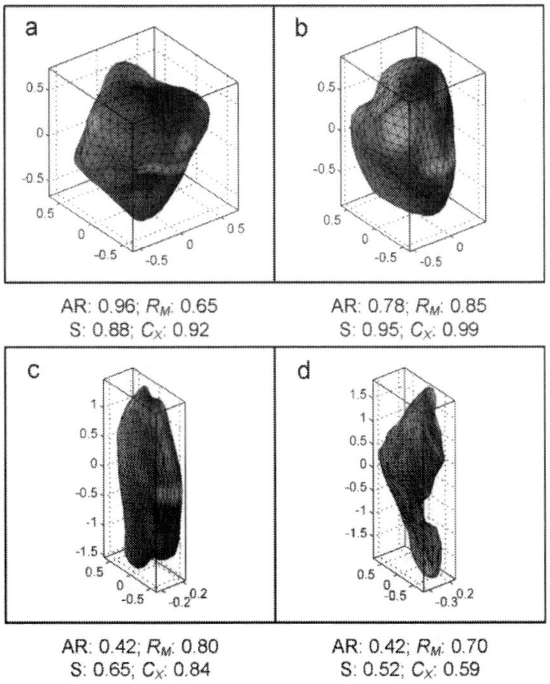

Figure 3. Reconstruction of LBS particles (a and b) and LBS fragments (c and d) from $\mathbf{c}''^m_l$ ($l \leq 15$) along with shape parameters

roundness and compactness. A 3D particle has three principal dimensions ($a \geq b \geq c$). Two aspect ratios were defined as elongation index ($EI = b/a$) and flatness index ($FI = c/b$). Then, the representative aspect ratio (AR) was defined as the average value of EI and FI. Sphericity (S) was defined as the ratio between the surface area of a particle's volume-equivalent sphere and the particle. Convexity (C_X) was defined as the ratio between the volume of the particle and the volume of its convex hull. To evaluate particle roundness, the surface was simplified with 1280 triangular meshes to represent the cut-off between roundness and roughness. Then, particle roundness index (R_M) was defined from local mean curvature values around the corners, which were defined as the part of surface that has larger local mean curvature than its area-equivalent sphere. Area-equivalent sphere instead of maximum inscribed sphere was used to define corners due to the difficulty in obtaining the maximum inscribed sphere from triangular surface mesh. The area-equivalent sphere is always larger than the maximum inscribed sphere, which leads to larger area of identified corners.

4 Results

Totally 80 LBS particles and 46 LBS fragments were analyzed in this study. Figure 3 shows the reconstruction of two LBS particles and two LBS fragments along with their shape parameters. The influence of spherical coefficients

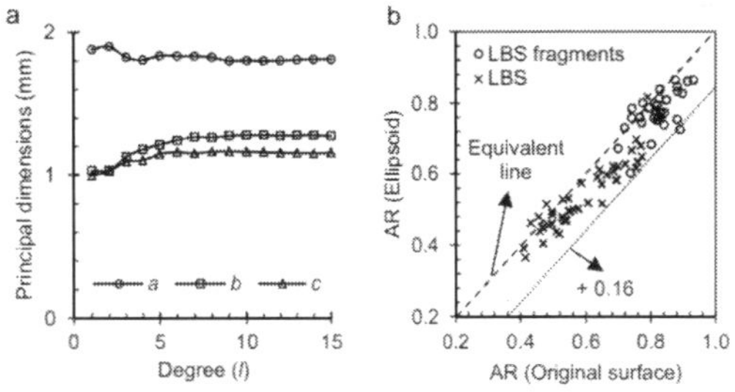

Figure 4. (a) Principal dimensions of a reconstructed LBS particle as spherical degree increases; (b) relationship between AR values of original surfaces and AR values of ellipsoids reconstructed with $\mathbf{c}''^m_1$

on the form, roundness and compactness of reconstructed surface is described as follows.

- Particle form. As shown in figure 4(a), the maximum principal dimension (a) does not change much as the degree increases, while the medium and minimum dimensions (b and c) slightly increase before degree 7. The coefficients of degree one determine an ellipsoid. The three principal dimensions of the ellipsoid could be calculated with $a = \|c''^m_{z,1}\|$, $b = \|c''^m_{y,1}\|$ and $c = \|c''^m_{x,1}\|$. Figure 4(b) shows that most particles have AR values slightly larger than the ellipsoids defined by c''^m_1.
- Particle roundness. Particle roundness characterises the variances around corners. It is a surface feature between particle form and roughness. A LBS particle and a LBS fragment were reconstructed with highest spherical degree of 1, 5, 8 and 15. The reconstructed surface was all represented by 1280 triangular meshes, which serve as the cut-off between roundness and roughness. Figure 5 shows the cumulative distributions of the mean curvatures on these surfaces. It is found that as the degree increases the variance of mean curvature increases, which characterises better the finer features of particle shape. For example, the proportion of surface with higher mean curvature increases.
- Particle compactness. Particle compactness describes the relationship between surface area, volume and volume of convex hull. It is thus an overall feature which is related to particle form, roundness and roughness. As the degree increases, the finer features will be better characterised, which will result in smaller particle compactness values.

The distributions of shape parameters for LBS particles and LBS fragments were fitted with Weibull's distributions with two parameters, i.e. the characteristic value (L_0) and Weibull modulus (m). The characteristic value and Weibull modulus describe the average value and the variance of shape parameters, respectively. The evolution of L_0 and m for LBS particles and LBS fragments reconstructed with different spherical degrees was shown in figure 6. It is found that as the degree increases, the characteristic value (L_0) of AR decreases, while that of R_M, S

 DOI: 10.1051/epjconf/201714005002

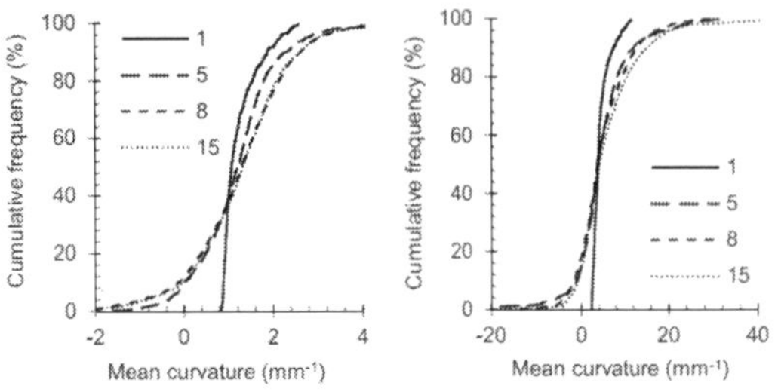

Figure 5. Distributions of mean curvature for reconstructed surface with 1,280 triangular meshes with different degrees for (a) a LBS particle and (b) a LBS fragment

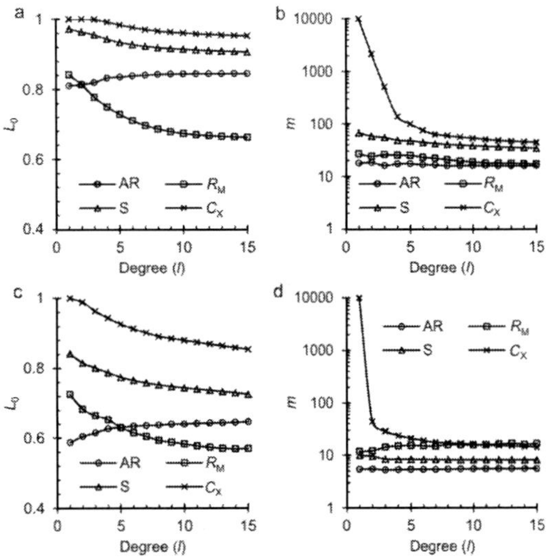

Figure 6. Variance of characteristic shape parameters (L_0) and Weibull modulus (m) for reconstructed surfaces with different spherical harmonics degree (l) for (a-b) LBS particles and (c-d) LBS fragments

and C_X decreases. It tends to converge at a higher degree. If we consider a 5% difference from the shape parameters of surface reconstructed with $\mathbf{c}''^m_l$ ($l \leq 15$) as the threshold for a good particle reconstruction, the degree needs to characterize form, roundness and compactness of both LBS particles and LBS fragments is 8. The increase of Weibull modulus (m) indicates that the variance of shape parameters increase as the degree increases, especially for convexity.

5 Conclusions

This study performed spherical harmonics analysis on LBS particles and LBS fragments. The reconstructed surface was quantified at form, roundness and compactness. It is found that particle form could be well described with an ellipsoid represented by coefficients of degree one. The coefficients of degrees larger than one characterise more detail surface features, which could be reflected by the variance of mean curvature distributions. In general, as the degree increases, particle form will increase; particle roundness, sphericity and convexity will decrease. The statistical analysis on the shape parameters of reconstructed surface indicates that a maximum degree of 8 is enough for characterizing the form, roundness and compactness of LBS particles and LBS fragments.

Acknowledgement

The study presented in this article was supported by General Research Fund No. CityU 120512 from the Research Grant Council of the Hong Kong SAR and Research Grant No. 51379180 from the National Science Foundation of China.

References

[1] Cho G C, Dodds J & Santamarina J C, Journal of geotechnical and geoenvironmental engineering **132**(5), 591-602 (2006)
[2] Zhao B D, Wang J F, Coop M R, Viggiani G & Jiang M J, Géotechnique **65**(8), 625-641 (2015)
[3] Zhao B D & Wang J F, Powder technology **291**, 262-275 (2016)
[4] Domokos G, Kun F, Sipos A Á, & Szabó T, Scientific reports **5**, 9147 (2015)
[5] Mollon G & Zhao J D, Granular matter **14**(5), 621-638 (2012)
[6] Zhou B, Wang J F & Zhao B D, Engineering geology **184**, 126-137 (2015)
[7] Ballard D H & Brown C M, *Computer Vision* (Prentice-Hall, Englewood Cliffs, NJ, 1981)
[8] Brechbühler CH, Gerig G & Kübler, O., Computer vision and image understanding **61**(2), 154-170 (1995)
[9] Shen L & Makedon F, Image and vision computing **24**(7), 743-761 (2006)
[10] Barrett P J, Sedimentology **27**(3), 291-303 (1980)
[11] Bowman E T, Soga K & Drummond W, Géotechnique, **51**(6), 545-554 (2001)

Numerical study of particle capture efficiency in fibrous filter

Jianhua Fan[1,*], *Franck* Lominé[1,**], and *Mustapha* Hellou[1,***]
[1] *INSA, LGCGM, EA3913, F-35708, Rennes, France*

Abstract. Numerical simulations are performed for transport and deposition of particles over a fixed obstacle in a fluid flow. The effect of particle size and Stokes number on the particle capture efficiency is investigated using two methods. The first one is one-way coupling combining Lattice Boltzmann (LB) method with Lagrangian point-like approach. The second one is two-way coupling based on the coupling between Lattice Boltzmann method and discrete element (DE) method, which consider the particle influence on the fluid. Then the single fiber collection efficiency characterized by Stokes number (St) are simulated by LB-DE methods. Results show that two-way coupling method is more appropriate in our case for particles larger than 8 μm. A good agreement has also been observed between our simulation results and existing correlations for single fiber collection efficiency. The numerical simulations presented in this work are useful to understand the particle transport and deposition and to predict the capture efficiency.

1 Introduction

Particle transport and deposition on various obstacles placed in a fluid flow are fundamental phenomena in many environmental and industrial applications such as transport in porous media, air purification, biological engineering and chemical processes. In early studies, many researchers employed theoretical analysis [1, 2], numerical simulations [3–5] and experimental measurements [6, 7] to examine particle deposition in flow over different obstacles such as cylinders, spheres or disks [8].

The main mechanisms for particle capture are Brownian diffusion, interception and inertial impaction [9]. It is possible to estimate which collection mechanism is dominant according to the following non-dimensional parameters, Peclet number, $Pe = UD_f/D$, where D and D_f are Brownian diffusion coefficient and fiber diameter, intercept coefficient $R = d_p/D_f$ and Stokes number $St = \rho_p d_p^2 U/(18\mu D_f)$, in which ρ_p and d_p are the density and diameter of the particle respectively. μ and U are fluid viscosity and mean fluid velocity. Capture efficiency corresponding to Brownian diffusion is related to Pe. For this case, particles diameter less than 0.01 μm with small Pe number have more opportunity to be captured. Inertial impaction depends on St number. High capture efficiency is also obtained for large particles (≥ 10 μm) with large St number dominated by inertial impaction. However, the interception relies on intercept coefficient R, which is dominant for the particle in the range of 0.1-1 μm, where these other two collection mechanisms, Brownian diffusion and inertial impaction, are relatively weak [9].

The velocity and position of the solid particle rely evidently on the fluid velocity field, thus, it is essential to acquire an accurate distribution of the velocity field for the particle trajectory analysis which is appropriately described by the Lagrangian approaches. Conventional CFD model, based on the incompressible Navier-Stokes equation, combined with Lagrangian approach is usually used to simulate the fluid-solid flows within porous media such as in fibrous filters [10]. The solid particle is viewed as mass point without volume in dilute regime (point-like approach), thus the trajectory and velocity of the particle are calculated from the equation of the motion using Lagrangian approach under consideration of the different external forces such as drag force and Brownian force [9]. It is reasonable to neglect the effect of particle on the fluid when the particle-laden flows are dilute enough [11]. Obviously, point-like approach estimates roughly the actual displacement of particle within a time step. However, particle with large size can affect significantly the flow field. In this case the volume of particle and its influence on the flow field should be considered.

For particle-laden flows, the conventional numerically methods have difficulties in adaptive unstructured grids and have to adopt moving mesh for example to assure a high-resolution near the deposit when considering the particle motion. It leads to increase the computational cost. Lattice Boltzmann method (LBM) demonstrated significant advantages due to easily implementing dynamic and complex boundary conditions. In the last decades LBM has gained increasing popularity for solving fluid-solid problems [12].

Here we focus on the study of particle with large size where inertial impaction and interception mechanisms are considered. The first aim of this work is to investigate

*e-mail: jianhua.fan@insa-rennes.fr
**e-mail: franck.lomine@insa-rennes.fr
***e-mail: mustapha.hellou@insa-rennes.fr

 DOI: 10.1051/epjconf/201714003003

the particle capture process using a self-written code to implement the coupling of Lattice Boltzmann and Discrete element methods (LB-DE methods), which consider the influence of the particle motion on the flow field. Another aim is to analyze the effects of Stokes number on single fiber collection efficiency.

2 Study of particle deposition : comparison between one-way coupling and two-way coupling methods

In the particle-laden flow, inertial impaction is one of the major mechanisms. We show in Fig.1 the schematic of the limit trajectory of the particle and deposit on the obstacle surface. The particles released from the distance δ to the lattice axis can be trapped. Thus we define the particle capture efficiency as $\eta = 2\delta/D_f$.

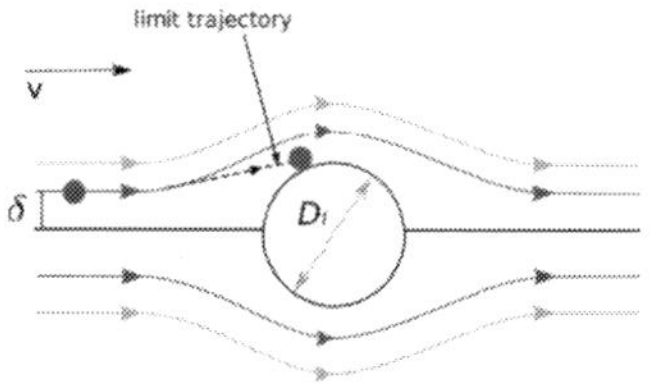

Figure 1. Particle trajectory and deposition on a single obstacle.

The inertial effect in particle capture can be quantified by the dimensionless Stokes number St [9, 10]. In the limiting case $St \rightarrow 0$, the trajectories of the particles basically correspond to the fluid streamlines and almost follow them closely. Particles can only impact the surface of the obstacle when the distance from the center of particles to the boundary of the obstacle is less than or equal to the particles radius. As St continuously increases (still for low St), both inertial impaction and interception mechanism exist and dominate under certain conditions. In this case, particles follow slightly the fluid streamlines. The reason is that the inertia of particle with low St is not intense enough and interception still plays a little role. When St is sufficiently large, inertial impaction mechanism becomes more and more stronger and dominant. Particles deviate from the corresponding streamlines as they approach the obstacle and finally deposit on the surface of the obstacle. Generally speaking, the particle with large size have a larger St number (i.e., $d_p = 10\ \mu m$ and the inertial impaction dominates). Hence in order to reduce the effect of interception on the particle capture efficiency, we mainly study the particle with large size in this work.

2.1 One-way coupling method

The whole fluid space is computed using the LB method. After that each solid particle is injected one by one at the inlet of the domain. The initial velocity along the flow direction is set to be equal to the fluid velocity at that position. For one-way coupling method, the force of particle on fluid dynamics can be neglected if the particle volume fraction is less than 10^{-6}[11]. The solid particle is viewed as a point without volume. It is assumed that the velocity flow field remains unchanged and is not affected by the injected particles. Then particles velocities and trajectories are directly calculated by the traditional point-like approach tracing individual particle in the Lagrangian frame [13, 14]. Following Newton's equation is applied to calculate the particle velocity and trajectory at any time.

$$St\frac{d\vec{u}_p^*}{dt^*} = \vec{u}^* - \vec{u}_p^* \quad (1)$$

$$\frac{d\vec{x}_p^*}{dt^*} = \vec{u}_p^* \quad (2)$$

where $\vec{u}_p^*$ and $\vec{u}^*$ are dimensionless velocities of particle and fluid. Using the mean inlet velocity as reference velocity. $\vec{x}_p^*$ and t^* are dimensionless position and time of particle. It is worth to note that solid particles are not constrained to locate only at the same lattice nodes as the fluid particles of the LBM. Velocity of fluid $\vec{u}^*$ located at position of the solid particles is calculated using bilinear interpolation among the four closest nodes of the lattice. A fourth order Runge-Kutta (RK) has been used to numerically solve the equations (1) and (2). Then the particle capture efficiency is estimated. The result is compared with the work of Araujo et al. [10], presented in Fig.2, who used the analytical solution for the biharmonic equation $\nabla^4\psi = 0$ corresponding to Stokes equation written with stream function, given by Marshall et al. [15] to obtain the velocity flow field. Then the velocities and trajectories of point-like particles are calculated with some assumptions. The influence of particle motion on the fluid field and the interaction between particles are negligible. From Fig.2 we can conclude that fluid velocity field obtained with LB method combined with bilinear interpolation is good enough to reproduce previous results presented in literature.

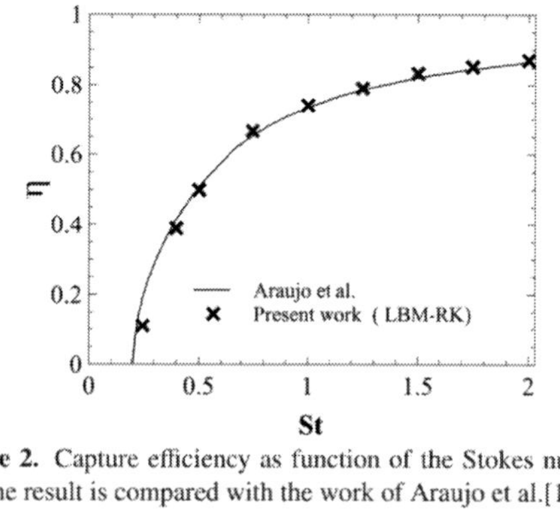

Figure 2. Capture efficiency as function of the Stokes number St. The result is compared with the work of Araujo et al.[10].

2.2 Two-way coupling method

Now we use a two-way coupling method (LB-DE methods) to examine the particle transport and deposition taking into account the effect of particle volume on the flow field. The fluid phase is computed using the LB method.

For DEM part, linear spring dashpot (LSD) model is used for the normal component of the contact force. An extension of the method proposed by Cundall-Strack for shear force[16]. The effect of particle volume on the flow field is due to the particle size. Hence with the purpose of investigating influence on the particle size to the capture efficiency, various simulations with different particle diameters (8 μm, 10 μm, 12 μm) have been performed. The result of particle capture efficiency is compared with the one with obtained one-way coupling method (Fig.3).

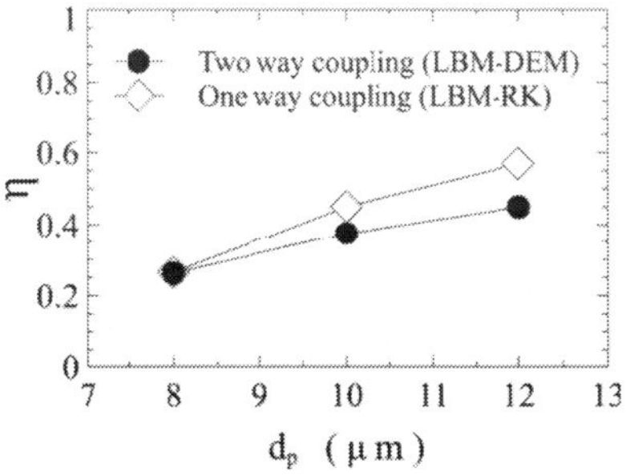

Figure 3. Capture efficiency as function of particle diameter obtained from one-way and two-way coupling method, respectively.

When the solid particle size is small (d_p < 8 μm for our model), particle volumes have little effect on capture efficiency. However as the particle size grows, the curve of capture efficiency obtained with one-way coupling deviates from the cure obtained with two-way coupling. For $d_p \geq 10$ μm the model of one-way coupling seems to be not very accurate.

3 Single fiber collection efficiency

The single fiber is a filtration unit, hence it is essential to investigate the single fiber collection efficiency. In this section we focus on the study of the single fiber collection efficiency with a quantity of solid particles injected (d_p =10 μm). The two-way coupling method (LB-DE) is used, which viewed the solid particles as objects with volume. The two-dimensional computational domain is shown in Fig 4. A circular obstacle with diameter D_f representing the fiber is placed at the center of a rectangular domain of width H and length L. A periodic boundary condition is applied on the left and right side of the domain. Both the fluid and solid particles flow out from one side will enter into the domain from other side. No-slip boundary condition is used at the surface of the fiber. The constant velocity $U = 0.67m/s$ is applied at the inlet and outlet. After the flow field achieves a steady state (velocity vector is shown in Fig 4), 500 particles are injected randomly in the projected area of fiber at the inlet of computational domain. The initial velocity of the particle along y axis is assumed to be equal to fluid velocity $u_{py} = U$, while x component is equal to zero.

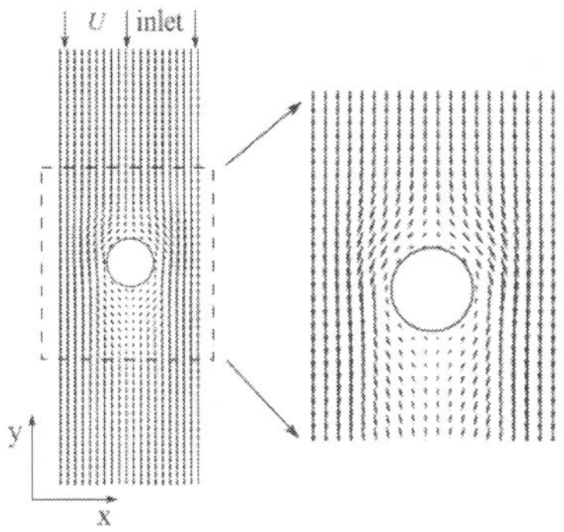

Figure 4. A steady state velocity vector of flow field around a circle calculated by LB method.

For initial stage of filter, the particles in the flow field are assumed to be captured once they collide on the surface of the circular fiber[17, 18]. The single fiber collection efficiency is thus defined as[19]:

$$\eta_s = N_{collect}/N_{project} \quad (3)$$

where the $N_{collect}$ and $N_{project}$ represent the number of solid particle collected and the number of injected particles in the projected area of fiber respectively.

Series of snapshots from one simulation of particle transport and deposition with coupled LB-DE method are shown in Fig.5. The particle volume fraction is about 10^{-3}. It is seen that after the particles injected in the upstream of fiber, the velocity field of fluid changed remarkably which, in turn, lead to the variation of the particle velocities and trajectories.

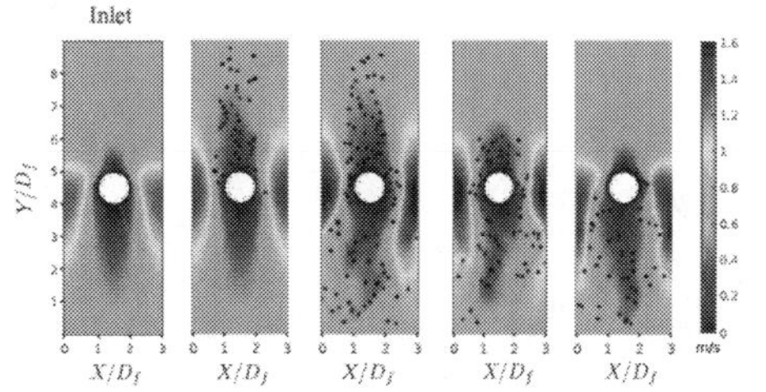

Figure 5. Velocity field of fluid and particles distribution for St=0.5

Fiber capture efficiency is influenced significantly by the inertial parameter St. Among the published literatures [20, 21], a number of impaction theories and empirical expressions have been developed to predict single fiber efficiency resulting from the inertial impaction.

Schweers et al.[7] improved the approximation of formula proposed by Suneja and Lee[22] and developed following equation for single fiber efficiency:

$$\eta_s = \left(\frac{St}{St + 0.8} - \frac{2.56 - \log_{10} Re - 3.2R}{10\sqrt{St}}\right)(1 + R) \quad (4)$$

where the validity range is for $1 < Re < 60$, $\eta_s > 0.05$, $R \leq 0.15$.

Israel and Rosner[23] investigated the aerodynamic capture efficiency of non-stokesian particles and determined the following expression for $St > 0.14$:

$$\eta_s = [1 + 1.25(St - 0.125)^{-1} - 1.4 \times 10^{-2}(St - 0.125)^{-2} + 0.508 \times 10^{-4}(St - 0.125)^{-3}]^{-1} \quad (5)$$

In order to examine the effect of St on the single fiber efficiency with considering large particle with volume, a comparison between the single fiber collection efficiency obtained with our LB-DE method and that obtained from the empirical relationships calculated according to Eqs (4) and (5) is made Figure 6. In our model the Reynolds number (Re) and intercept coefficient are $Re \approx 7$ and $R = 0.0625$, respectively. The collection efficiency is plotted as a function of the St ranging from $0.5 < St < 3$. The agreement between the simulation and empirical formula Eqs (4) is observed. However with respect to the Eqs (5), there exist a little differences with our simulation result but with the same tendency.

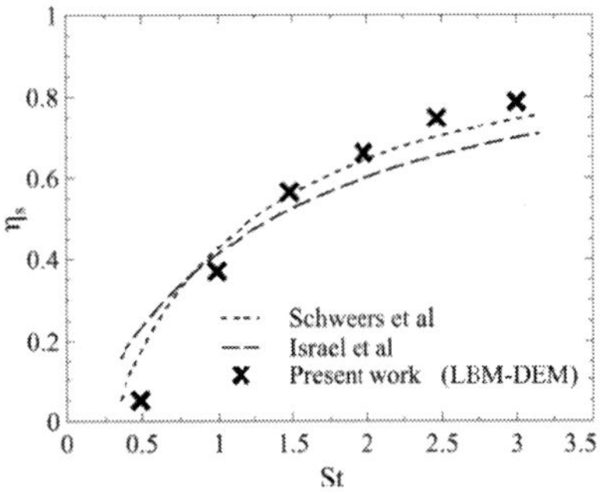

Figure 6. Comparison between single fiber collection efficiency obtained from numerically method (LB-DE) and from the empirical correlations. $Re \approx 7$, $R = 0.0625$.

4 Conclusions

In this paper we studied the effect of the particle size on the capture efficiency obtained with one-way coupling and two-way coupling respectively. For small particles, one-way coupling in our model is sufficient precise to estimate the particle transport and deposition. The capture efficiency from our simulation agrees well with the prediction of Araujo et al. However, with respect to the larger particles, two-way coupling is much more accurate owing to correct consideration of the influence of particle motion on the flow field.

The single fiber collection efficiency is investigated by LB-DE methods. The particle deposition on the fiber surface is essentially characterized by the particle Stokes number as inertial impaction plays a dominant role. It is found that the rate of single fiber collection efficiency increased from rapidly to slowly with increasing Stokes number. In dilute case, good agreement with previous study of particle capture efficiency have been observed. This coupled model can also be used to study capture efficiency in relatively dense cases. Based on these simulations, the presented model offers an appealing tool for understanding the filtration process and predicting the capture efficiency.

References

[1] J.A. Harrop, J.I. Stenhouse, Chemical Engineering Science **24**, 1475 (1969)
[2] K.W. Lee, B.Y.H. Liu, Aerosol Science and Technology **1**, 147 (1982)
[3] D. Michael, P. Norey, Journal of Fluid Mechanics **37**, 565 (1969)
[4] C.S. Wang, M. Beizaie, C. Tien, AIChE Journal **23**, 879 (1977)
[5] D.J. Brandon, S.K. Aggarwal, Aerosol Science and Technology **34**, 340 (2001)
[6] J.B. Wong, W.E. Ranz, H.F. Johnstone, Journal of Applied Physics **26**, 244 (1955)
[7] E. Schweers, H. Umhauer, F. Lbffler, Particle and Particle Systems Characterization **1**, 275 (1994)
[8] K.R. May, R. Clifford, Annals of Occupational Hygiene **10**, 83 (1967)
[9] H. Wang, H. Zhao, Z. Guo, C. Zheng, Powder Technology **227**, 111 (2012)
[10] A.D. Araújo, J.S. Andrade, H.J. Herrmann, Physical Review Letters **97**, 138001 (2006)
[11] S. Elghobashi, Applied Scientific Research **52**, 309 (1994)
[12] F. Lominé, L. Scholtès, L. Sibille, P. Poullain, International Journal for Numerical and Analytical Methods in Geomechanics **37**, 577 (2013)
[13] K. Luo, F. Wu, K. Qiu, Z. Wang, J. Fan, International Journal of Heat and Mass Transfer **83**, 372 (2015)
[14] O. Filippova, D. Hänel, Computers & Fluids **26**, 697 (1997)
[15] H. Marshall, M. Sahraoui, M. Kaviany, Physics of Fluids **6**, 507 (1994)
[16] P.A. Cundall, O. Strack, Geomechanics **29**, 47 (1979)
[17] J. Wang, D.Y.H. Pui, Journal of Nanoparticle Research **11**, 185 (2009)
[18] K.C. Lin, H. Tao, K.W. Lee, Aerosol Science and Technology **48**, 1299 (2014)
[19] S. Jafari, M. Salmanzadeh, M. Rahnama, G. Ahmadi, Journal of Aerosol Science **41**, 198 (2010)
[20] I. Stechkina, A. Kirsch, N. Fuchs, Annals of Occupational Hygiene **12**, 1 (1969)
[21] H. Landahl, R. Herrmann, Journal of colloid science **4**, 103 (1949)
[22] S. Suneja, C. Lee, Atmos. Env **8**, 1081 (1974)
[23] R. Israel, D.E. Rosner, Aerosol Science and Technology **2**, 45 (1982)

Statistical properties of gravity-driven granular discharge flow under the influence of an obstacle

Keita Endo[1] and *Hiroaki* Katsuragi[1,*]

[1] *Department of Earth and Environmental Sciences, Nagoya University, Furocho, Chikusa, Nagoya 464-8601, Japan*

Abstract. Two-dimensional granular discharge flow driven by gravity under the influence of an obstacle is experimentally investigated. A horizontal exit of width W is opened at the bottom of vertical Hele-Shaw cell filled with stainless-steel particles to start the discharge flow. In this experiment, a circular obstacle is placed in front of the exit. Thus, the distance between the exit and obstacle L is also an important parameter. During the discharge, granular-flow state is acquired by a high-speed camera. The bulk discharge-flow rate is also measured by load cell sensors. The obtained high-speed-image data are analyzed to clarify the particle-level granular-flow dynamics. Using the measured data, we find that the obstacle above the exit affects the granular-flow field. Specifically, the existence of obstacle results in large horizontal granular temperature and small packing fraction. This tendency becomes significant when L is smaller than approximately $6D_g$ when $W \simeq 4D_g$, where D_g is diameter of particles.

1 Introduction

Flow rate of gravity-driven granular discharge flow through a narrow orifice is usually constant and independent of the amount of particles above the orifice. This well-known property of granular discharge flow enables hourglass to be a good time keeper. The relation between flow rate, particle size, and orifice size has been empirically developed by Beberloo et al. [1]. This Beberloo's law is a useful relation to estimate the granular flow rate when the size of orifice is large enough. However, when the orifice size is smaller than six times of particle diameter, clogging of the flow can readily be observed [2, 3]. In such unstable flow regime, size distribution of avalanches has been measured and analyzed extensively [4]. They considered Markov process for calculating the clogging probability since the avalanche size distribution obeys exponential form. Arch formation at the exit region is a main reason for the clogging. The clogging probability can be reduced when an obstacle is placed just above the orifice [5, 6]. Similar effect can also be observed in various other phenomena relating to bottleneck clogging [6–8]. According to Ref. [5], the reduction of clogging probability due to the obstacle comes from the effective pressure decrease in the region of arch formation (just above the orifice). On the other hand, Roussel et al. reported that the packing fraction in the region of arch formation is an important parameter for discussing the clogging occurrence. To reveal the principal factor governing clogging occurrence, detail analysis of granular-flow field in the discharge flow is necessary.

*e-mail: katsurag@eps.nagoya-u.ac.jp

Statistics of granular discharge flow has been studied by several research groups [9–11]. According to these works, velocity distribution of particle motion is not Gaussian and diffusive motion can be observed in horizontal direction. Moreover, self-similarity of velocity distribution around the exit region has also been reported [12]. These features in velocity distribution show particular nature of granular discharge flow. Although they could be related to the mechanics of clogging, its detail has not yet been clarified.

As mentioned above, an obstacle has frequently been used in order to avoid the clogging. However, the detail of velocity distribution and its role for clogging-probability reduction has not yet been revealed well. Therefore, in this study, we study particle motions in granular discharge flow under the influence of an obstacle by means of particle tracking. As a first step to discuss the obstacle effect to the granular discharge flow, we will simply focus on the velocity distribution and packing fraction at the exit region just above the orifice.

2 Experiment

A transparent (acrylic) quasi-two-dimensional cell (Hele-Shaw cell) of inner dimensions 300 × 210 × 6.5 mm is vertically held and filled with stainless-steel particles of diameter D_g = 6.35 mm. Before filling the stainless particles, a circular obstacle made with stainless steel is inserted into the cell. Diameter and thickness of the obstacle are 50 and 6.0 mm, respectively. The obstacle is hung on a universal testing machine (AG-X, Shimadzu) with a thin stainless-steel rod of diameter 6 mm. After filling the

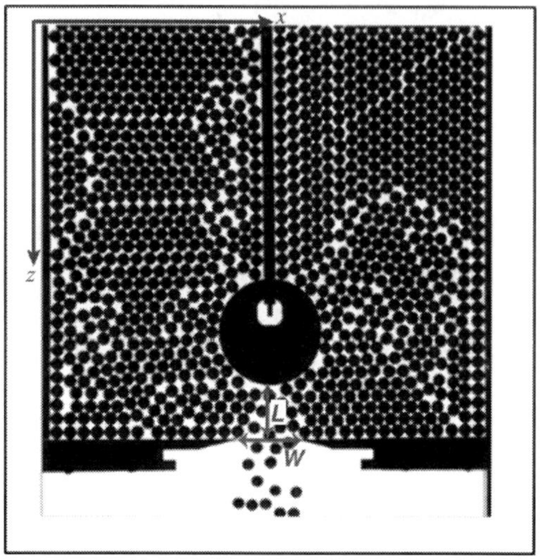

Figure 1. An example snapshot image of particles and obstacle taken by the high-speed camera. x and z directions correspond to horizontal and vertical directions, respectively. Width of the orifice W and vertical distance between the obstacle and orifice L are main control parameters. Diameter of particles is 6.35 mm.

particle, an exit orifice is opened at the center of cell bottom to start the discharge flow. During the discharge, flow rate is measured by load cell sensors (LMB-A, Kyowa) which quantify the mass of discharged particles. At the same time, particle motions are captured by a high-speed camera (FASTCAM SA5, Photron) with 500 fps. The spacial resolution of this measurement is 0.26 mm/pixel. An example snapshot of the particle flow is shown in Fig. 1. Main control parameters in this experiment are the width of orifice W and the vertical distance between obstacle and exit orifice L. As shown in Fig. 1, vertical and horizontal directions correspond to z and x directions, respectively. From the high-speed video data, all particles are tracked and their velocity distributions and packing fraction are computed.

3 Results

3.1 Flow rate

First, the measured discharge-flow rate is shown in Fig. 2. Discharged mass M is shown as a function of time t. The important character in this plot is the linearity of these data. This linear relation indicates the constant flow rates. As can be confirmed in Fig. 2, the flow rate Q is reduced by introducing the obstacle. Here, in both of with- and without-obstacle cases, W is fixed to 25 mm, and $L = 30$ mm is employed for the case of with-obstacle.

The value of Q was systematically measured by varying W and L. The measured results of $Q(L)$ for various W are plotted in Fig. 3. We can confirm that $Q = |dM/dt|$ is basically the increasing function of L in large W regime. For small W, Q is almost constant independent of L. If L is

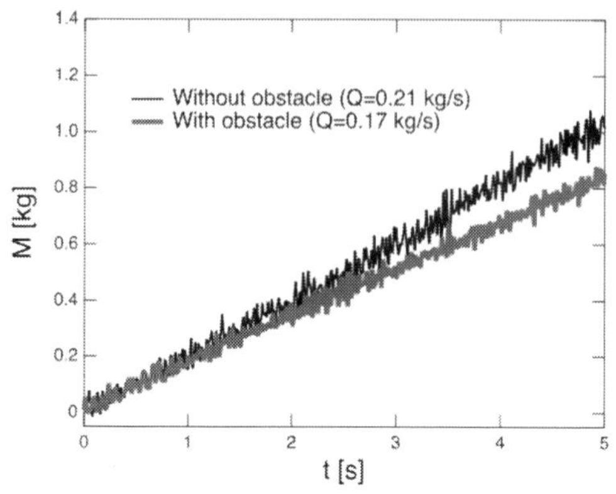

Figure 2. Discharged mass M as a function of time t. Both cases (with- and without-obstacle) show liner relation. This means that the flow rates are constant. The value of flow rate (slope in this plot) is reduced by introducing the obstacle.

fixed and W is varied, Q becomes the increasing function of W for all L. These experimental results suggest that the obstacle effectively narrows the exit region. In other words, the obstacle basically impedes discharge flow. This result contrasts with a numerical work in which the flow rate shows a peak at a certain L [13]. Although the similar flow rate peak has been reported in experiment [14], its peaks are not very clear. While the current data rather show dips at certain L, it is difficult to identify the clear peak of Q. When $W < 5D_g$ is fulfilled in two-dimensional silo, the clogging could occur [2]. Since the local arch formation is the principal reason for clogging, the narrow exit region naturally increases the risk of clogging. Although the obtained result is still preliminary, threshold value for the clogging occurrence is slightly different between horizontal (W) and vertical (L) directions. This difference might come from the anisotropy introduced by gravity. Since the current experimental system is driven by gravity, such an anisotropy might not be surprising. Furthermore, according to a recent study [15], the clogging probability could remain finite even in very wide exit situation. Thus, the threshold might only have a practical meaning.

3.2 Velocity distribution

To understand the physics governing gravity-driven discharge flow, we analyzed the particle motions by using high-speed video data. Here, we restrict ourselves to the small region just above the orifice. Specifically, a small square region of width 50 mm and height 30 mm is particularly analyzed in the following analysis. Although the fixed square width (50 mm) is used in all measurements, the height is decreased to L when L is smaller than 30 mm. In Fig. 4, probability density functions (PDF) of particle velocities are shown. The PDF of vertical component V_z and horizontal component V_x are presented in Fig. 4(a) and (b), respectively.

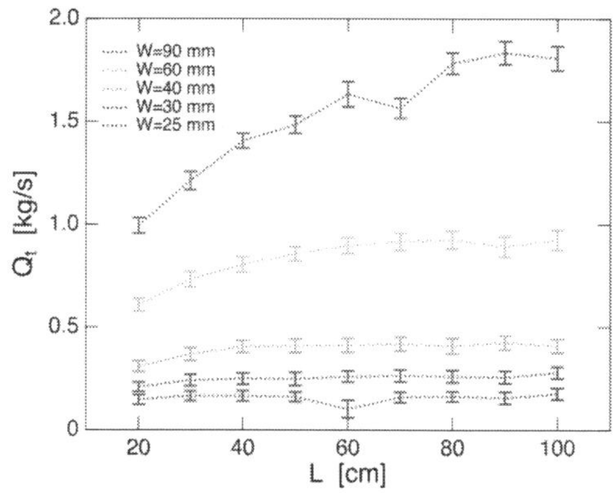

Figure 3. Obstacle distance L dependence of flow rate Q with various orifice width W. Q is the increasing function of L in large W regime.

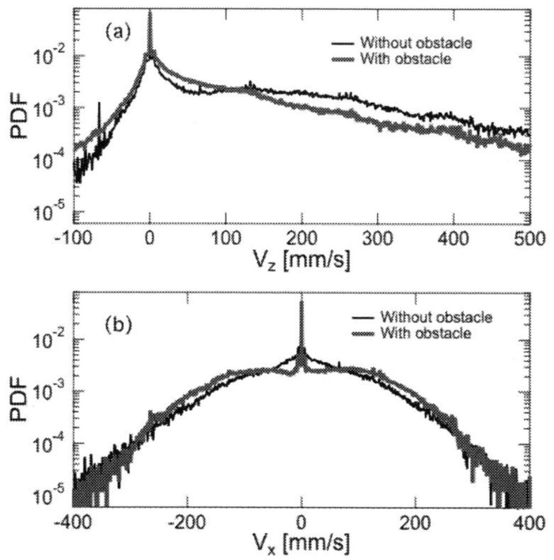

Figure 4. Probability density function (PDF) of particle velocities in (a) vertical [z] and (b) horizontal [x] directions. W is fixed to 25 mm and L = 30 mm is used for the obstacle position.

Since the flow is driven by gravity, PDF of V_z has an asymmetric form (Fig. 4(a)). The exponential-like tail (linearity in large V_x regime of Fig. 4(a)) in PDF of V_z is qualitatively consistent with the constant flow rate (Fig. 2). The difference in large V_z tail part between with- and without-obstacle cases causes the difference in Q. One can also confirm the small amount of particles exhibiting negative V_z. This negative V_z comes from the backscattering due to the particle collisions.

In the horizontal-velocity PDF (Fig. 4(b)), symmetric distribution forms can clearly be confirmed. This is natural because the gravity has nothing to do with horizontal direction. The sharp peaks around $V_x = 0$ come from almost stopping particles population. Actually, the motion of particles shows stop-and-go behavior. This behavior is easily confirmed by the direct observation of high-speed video data and the similar intermittency has been observed in a recent study [16]. In addition, the qualitative form of PDF is different between with- and without-obstacle cases. PDF of V_x with obstacle shows the splitting (or broadening) of the distribution. The broader part in PDF of V_x stems from the enhanced particle motions. This peak splitting (or broadening) due to the existence of obstacle can be understood by two possible effects. One is the detouring of particles. If there is no obstacle on the way to exit, particles can directly approach the exit. However, the obstacle prevents this direct (nearly vertical) motion of particles. The particles must detour the obstacle to approach the exit. This detouring can directly raise the variance (i.e., granular temperature) of horizontal velocity V_x distribution. The other possibility is the increase of particle collisions. As a result of detouring of particles, their collisions beneath the obstacle could be enhanced. Then, these collisions activate the particle motions. This collisional effect is also able to raise the variance (granular temperature) of V_x distribution. This is a secondary (indirect) effect of particle detouring. Note that, for V_z, gravity effect is much stronger than this effect.

4 Discussion

In order to characterize the granular-flow state under the influence of obstacle, we measured horizontal granular temperature T_{gx} and packing fraction ϕ in the considered square region just above the exit orifice. Here, we simply define T_{gx} by the variance of V_x distribution. In Fig. 5, measured T_{gx} and ϕ are displayed. In this measurement, W was fixed to 25 mm and L was varied from 20 to 100 mm. Horizontal dashed lines correspond to the levels of the without-obstacle case. As can be seen in Fig. 5, T_{gx} and ϕ become almost constant in the range of $L > 40$ mm. This critical value $L \simeq 40$ mm roughly corresponds to $6D_g = 38.1$ mm. And the values of T_{gx} and ϕ are roughly identical to those in without-obstacle case in the large $L(> 40$ mm) regime. This is the reason for the small variation of Q for narrow (small W) case observed in Fig. 3. That is, both W and L are relevant to the flow characterization. However, to clearly observe the regime in which the obstacle effect is negligible, here we plot the small W case in Fig. 5.

In the small $L(< 40$ mm) regime, T_{gx} increases as L is decreased (Fig. 5(a)). This large granular temperature causes the sparse particle configuration, i.e., the effective thermal expansion might be observed. Indeed, the effective thermal expansion (small packing fraction ϕ) in granular flow can clearly be confirmed in small $L(< 40$ mm) regime in Fig. 5(b). Perhaps, this effective thermal expansion could reduce the risk of clogging. However, the authors of Ref. [5] did not consider the low packing fraction is the principal reason for the clogging-probability reduction. Rather, they have considered that the pressure screening due to the backscattered particles' negative V_z

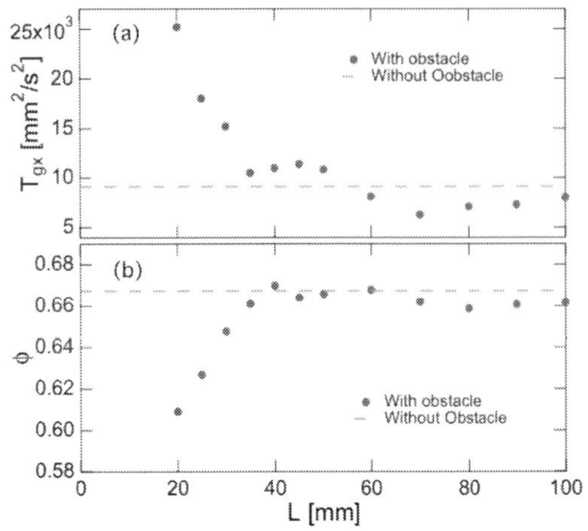

Figure 5. (a) Horizontal granular temperature T_{gx} and (b) packing fraction ϕ in the square region above the exit orifice (see text for the place and size of the square region). In both plots, W is fixed to 25 mm. The increase (or decrease) of T_{gx} (or ϕ) can be confirmed in small $L(< 40$ mm) regime.

could be the main reason for clogging-probability reduction. The increase of population of negative V_z under the influence of obstacle can surely be confirmed in this experiment, too (Fig. 4(a)). This large V_z results in the effective reduction of confining pressure in granular flow. Thus, we think that both effects (the low packing fraction due to enhanced particle collisions and the increase of negative V_z population) could result in the reduction of risk of clogging.

Perhaps, varying the shape of obstacle could be useful to investigate further details of clogging reduction due to the obstacle. In general, flow field around an obstacle can be affected by the shape of obstacle. Such a geometric effect might vary the clogging probability through the variation of granular-flow field. Actually, we are going to analyze the granular discharge flow with different obstacle shapes. The result will be presented elsewhere. The systematic examination with various W and obstacle size like Ref. [14] could also be an interesting future problem.

5 Summary

We performed simple granular discharge flow experiment. Flow rate Q, velocity distributions (V_x: horizontal and V_z: vertical), horizontal granular temperature T_{gx}, and packing fraction ϕ were measured and analyzed. Particularly, T_{gx} and ϕ were measured in the vicinity of exit orifice. From the measured and analyzed results, we confirmed that the flow rate depends on both orifice width W and obstacle distance L. Velocity distribution in vertical direction shows asymmetric form and has exponential-like tail. In addition, we found that horizontal velocity distribution shows the peak splitting (or broadening) by approaching the obstacle to the exit. To characterize this behavior, horizontal granular temperature was measured. At the same time, packing fraction was also measured. Both of them show similar increasing behaviors when the obstacle approaches as close as $L < 40$ mm $\simeq 6D_g$ when $W = 25$ mm $\simeq 4D_g$.

Acknowledgements

This work has been supported by JSPS KAKENHI No. 15H03707.

References

[1] W.A. Beberloo, H.A. Leniger, J. van de Velde, Chem. Eng. Sci. **15**, 260 (1961)
[2] K. To, P.Y. Lai, H.K. Pak, Phys. Rev. Lett. **86**, 71 (2001)
[3] I. Zuriguel, Pap. Phys. **6**, 1 (2014)
[4] K. To, Phys. Rev. E **71**, 060301 (2005)
[5] I. Zuriguel, A. Janda, A. Garcimartín, C. Lozano, R. Arévalo, D. Maza, Phys. Rev. Lett. **107**, 278001 (2011)
[6] I. Zuriguel, D.R. Parisi, R.C. Hidalgo, C. Lozano, A. Janda, P.A. Gago, J.P. Peralta, L.M. Ferrer, L.A. Pugnaloni, E. Clément et al., Sci. Rep. **4**, 7324 (2014)
[7] D. Yanagisawa, A. Kimura, A. Tomoeda, R. Nishi, Y. Suma, K. Ohtsuka, K. Nishinari, Phys. Rev. E **80**, 036110 (2009)
[8] G.A. Frank, C.O. Dorso, Physica A **390**, 2135 (2011)
[9] J. Choi, A. Kudrolli, R.R. Rosales, M.Z. Bazant, Phys. Rev. Lett. **92**, 174301 (2004)
[10] S. Moka, P.R. Nott, Phys. Rev. Lett. **95**, 068003 (2005)
[11] A.V. Orpe, A. Kudrolli, Phys. Rev. Lett. **98**, 238001 (2007)
[12] A. Janda, I. Zuriguel, D. Maza, Phys. Rev. Lett. **108**, 248001 (2012)
[13] F. Alonso-Marroquin, S.I. Azeezullah, S.A. Galindo-Torres, L.M. Olsen-Kettle, Phys. Rev. E **85**, 020301 (2012)
[14] C. Lozano, A. Janda, A. Garcimartín, R. Arévalo, D. Maza, I. Zuriguel, Phys. Rev. E **86**, 031306 (2012)
[15] C.C. Thomas, D.J. Durian, Phys. Rev. Lett. **114**, 178001 (2015)
[16] C.C. Thomas, D.J. Durian, Phys. Rev. E **94**, 022901 (2016)

Dynamical weakening of pyroclastic flows by mechanical vibrations

Jose Manuel Valverde[1,*], *Carlos* Soria-Hoyo[1], and *Olivier* Roche[2]

[1]*Faculty of Physics. University of Seville. Ada. Reina Mercedes s/n 41012 Seville, Spain*
[2]*Laboratoire Magmas et Volcans. Université Blaise Pascal-CNRS-IRD, OPGC. 6 Avenue Blaise Pascal, TSA 60026-CS 60026, F-63178 Aubière, France*

Abstract. Dynamical weakening of dense granular flows plays a critical role on diverse geological events such as seismic faulting and landslides. A common feature of these processes is the development of fluid-solid relative flows that could lead to fluidization by hydrodynamic viscous stresses. Volcanic ash landslides (pyroclastic flows) are characterized by their high mobility often attributed to fluidization of the usually fine and/or low-density particles by their interaction with the entrapped gas. However, the physical mechanism that might drive sustained fluidization of these dense granular flows over extraordinarily long runout distances is elusive. The behavior of volcanic ash in a slowly rotating drum subjected to mechanical vibrations shown in this work suggests that fluid-particle relative oscillations in dense granular flows present in volcanic eruption events can promote pore gas pressure at reduced shear rates as to sustain fluidization.

Pyroclastic flows, triggered by the collapse of lava domes and explosive volcanic eruptions, represent a serious hazard for densely populated areas around volcanoes. These dense granular flows are chiefly characterized by their high mobility. Usually, extraordinary long runout distances over 100 km and across topographic obstacles are reached [2–5]. A possible mechanism responsible for such astonishing behavior is sustained fluidization caused by the enhancement of pore gas pressure [2, 5]. Since pyroclastic flows commonly contain large amounts of fine ash and light pumice rock fragments, their flow dynamics could be indeed determined by the prevalence of the interstitial gas drag force on the particles over their weight. The role of fluidization on the extraordinary mobility of pyroclastic flows is suggested by the typical signature observed in the patterns of prehistoric ignimbrite deposits indicating negligible friction and suppressed turbulence at the depositional boundary layer [3]. It is believed that long runout landslides involving large release and deposit volumes ($> 10^6 m^3$) are associated with high flow velocities ranging from tens to hundreds of m/s [4]. At such high velocities granular flows of fine and/or light particles become fluidized by the entrapped gas at the avalanche front and acquire the behavior of a low viscosity fluid lacking any resistance to shear [6]. Recent observations of sediments from dense pre-historic pyroclastic flows showing the signatures of fluidization and traveling distances over 170 km demonstrate relatively modest and uniform velocities (<10 m/s) suggesting that high velocities are not necessary for fluidization of pyroclastic flows [5]. These observations, together with lab-scale experiments, indicate that the main factor in driving sustained fluidization is long-lived gas pore pressure presumably due to hampered diffusion by the reduced permeability of the trapped gas through large amounts of fine ash [5].

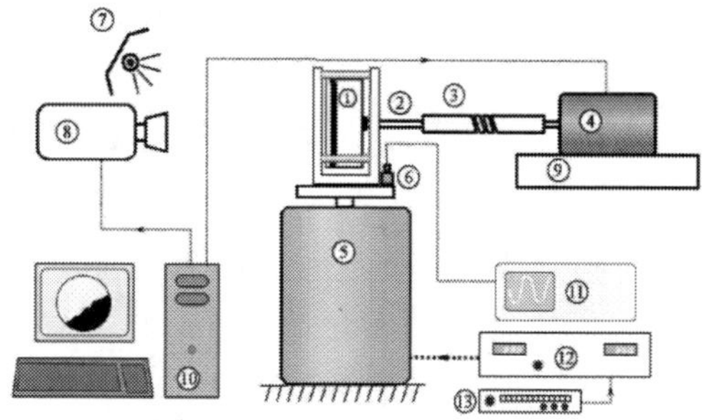

Figure 1. Experimental setup: (1) rotating drum (side view), (2) aluminium shaft, (3) flexible coupling (cardan), (4) step motor and controller, (5) vibrator (Tira TV 52120), (6) accelerometer, (7) halogen lamp, (8) CCD camera, (9) optical antivibration table, (10) desktop computer, (11) oscilloscope, (12) power amplifier, (13) function generator.

In our work we investigate the possible role of mechanical vibrations on sustained fluidization of pyroclastic flows at reduced flow velocities by means of lab-scale experiments. In a previous work [7] vibrations were shown to significantly weaken the dynamical resistance of a cornstarch fine powder and glass beads to avalanche in a slowly rotating drum. In the present paper, experimental measurements are shown using ignimbrite powder samples from prehistoric pyroclastic flows deposits. Long-lived high intensity mechanical vibrations could be generated in pyro-

*e-mail: jmillan@us.es

clastic flows by collisions of pyroclasts onto the mountain slope or other obstacles [8, 9] with a seismic energy that can be comparable with the energy released by the precursory swarm of volcano-tectonic earthquakes [10]. These intense vibrations could play a role on the enhancement of pore fluid pressure as suggested, for example, by the correlation found between the volume of pyroclastic flows during the Unzen volcano eruption (1991) and the magnitude of the generated earthquakes [10].

We have used ignimbrite powder samples from pyroclastic deposits at Clermont-Ferrand area (France), which consisted of mainly three types of particles: i) pumices as the most abundant, white, vesiculated, with typical density of 0.5-2 g/cm^3 that represent the fragmented magma; ii) rock fragments commonly dark, with typical density of 2.5-3 g/cm^3 coming from the conduit or the ground; and, by a small amount, iii) crystals of several colors mainly in size range ~250 μm - 1mm. The average solid density of the samples measured by a gas pycnometer was ρ_p =2.4 g/cm^3. As a common feature regarding their morphology, Scanning Electron Microscopy pictures show a marked angular shape of the particles with a noticeable presence of sub-micron fragments and asperities on their surface. Since the gas-solid hydrodynamic interaction is strongly dependent on particle size the powder as received was sieved at 100^{0}C to avoid ambient moisture induced cohesion in order to obtain samples in diverse ranges of particle size d_p (between <63 μm and 2 mm).

Figure 1 shows a layout of the experimental setup used in our work. The powder half-fills a cylindrical Plexiglas drum (4.5 cm internal radius and 2cm depth), which is mounted on the base of an electromagnetic vibration exciter and slowly rotated ($\Omega \simeq 4.8$ rpm) around its horizontal axis. The vibrator is driven by a signal generator that provides sinusoidal, vertical vibrations of controlled amplitude A, and frequency f (in the range 25-200 Hz). The shaft is fitted at the other end to the motor axis by means of an elastic cardan to isolate the motor from vibrations. A CCD camera interfaced to a computer for image processing records the profile of the powder as affected by rotation and vibration. In order to avoid moisture induced cohesion the samples were heated up to 100^{0}C immediately before the test, which serves also to mimic the high temperature usually found in pyroclastic flows. Peak vibration velocity $u_1 = A2\pi f$ is monitored using a piezoelectric accelerometer and image analysis.

Figure 2 illustrates the effect of adding vibrations to the slowly rotating drum. As may be seen vibrations have a relevant influence on the dynamical behavior of the ignimbrite samples regardless of their particle size. The slowly rotated bed loses progressively its frictional strength as the intensity of vibrations is increased. For sufficiently strong vibrations, the angle of the slope drops to nearly zero even for the largest particle size samples. A remarkable behavior is observed for the finest ignimbrite sample (d_p < 63μm) that resembles the behavior of relatively large glass beads (d_p > 1 mm) fluidized by liquids [11]. Instabilities in these liquid-fluidized beds are characterized by the early appearance of a succession of es-

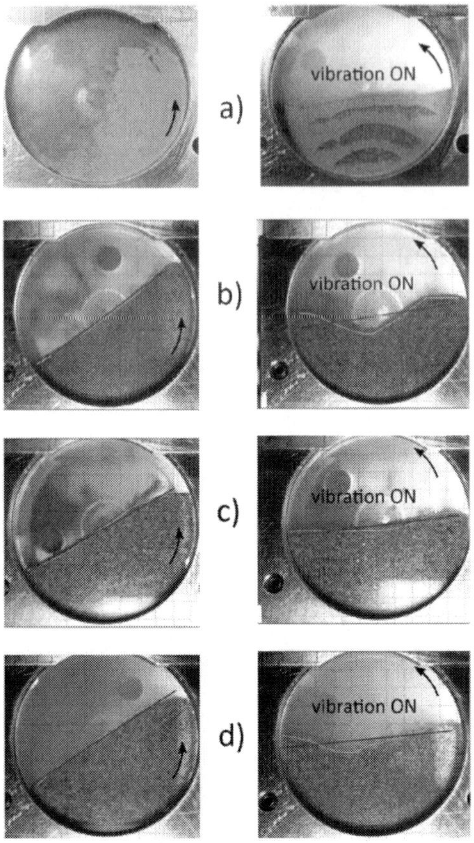

Figure 2. Snapshots of the slowly rotating drum half-filled with ignimbrite powder samples in the absence of vibrations (left) and for high intensity vibrations (f = 200 Hz, vibration velocity ~ 0.1 m/s) applied (right). a) d_p < 63μm; b) 250 - 500 μm; c) 500 μm - 1 mm; d) 1 - 2 mm.

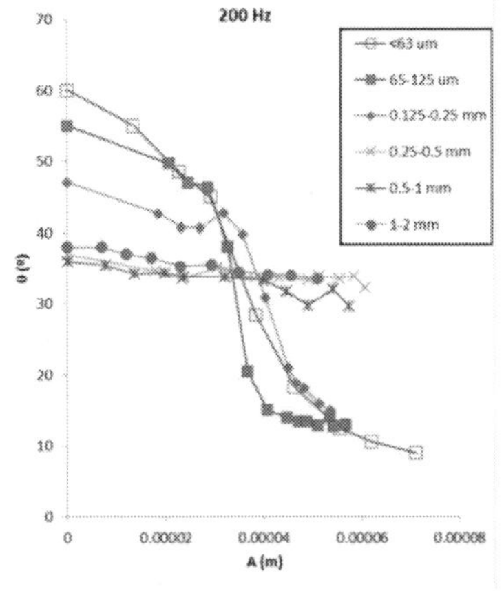

Figure 3. Average angle of avalanche in the slowly rotating drum as a function of the amplitude of vertical vibrations imposed (f = 200 Hz) for ignimbrite samples of diverse particle size ranges as indicated.

sentially one-dimensional upward traveling voidage wavefronts (1-DTWs) whose destabilization at sufficiently high liquid flow rates lead to bubble-like voidage pockets [11]. A similar flow structure evolution is observed in our rotating drum experiment for the finest ignimbrite sample as the vibration intensity is increased indicating the extraordinary relevance of gas-solid hydrodynamic interactions as promoted by vibrations. Figure 2a demonstrates the formation of these gas pockets arising for destabilization of 1-DTWs in close analogy with observations in liquid-fluidized beds. For the larger particle size ($d_p > 100\mu$m) ignimbrite samples, the flow regime in the rotating drum becomes inertial in the absence of vibrations since the gas-particle drag force is negligible as compared to inertial forces [6]. Yet, we see that the addition of vibrations leads also to a loss of resistance to shear (Figures 2b-d). A characteristic feature observed in some of the experiments is the formation of a double slope (Fig. 2b) arguably caused by boundary effects due to the small diameter of the drum used in our tests. This undesirable effect would be avoided by using drums of much larger diameter, which was prevented in our setup due to technical limitations related to lack of mechanical stability by the simultaneous imposition of rotation and vibrations but should be a subject for future work.

Figure 3 shows data on the average angle of avalanche as a function of the vibration amplitude (f = 200 Hz). The effect of vibration becomes more relevant as particle size is decreased which indicates a marked enhancement of the gas-solid hydrodynamic interaction for the small particle size samples. Thus, it may be foreseen that the effect of mechanical vibrations on the weakening of pyroclastic flows would be promoted as the amount of fine particles is increased. As seen in Fig. 3, a sudden drop of the angle occurs in the range of vibration amplitudes around $A = 0.03 - 0.05$ mm for samples with particle size below 0.25 mm, which would be caused by the transition to fluidization of most of the material.

According to a simple model on the fluid pore pressure due to relative gas-solid oscillations in a granular medium (see [7] for a detailed description), the relative decrease of the friction coefficient in the avalanching powder would be given by

$$\frac{\Delta\mu}{\mu} = \beta\Upsilon u_1' \tag{1}$$

where $\beta = \frac{6\pi\eta R\Lambda n_0}{\rho_p\phi g}$, $\Lambda \simeq \frac{E\phi}{18(1-\phi)^3}$, $\Upsilon = \left[\left(1+\frac{R}{\delta}\right)^2 + \left(\frac{R}{\delta}\right)^2\left(1+\frac{2R}{9\delta}\right)^2\right]^{1/2}$, η is the fluid dynamic viscosity, R is the particle radius (spherical particles are assumed), ϕ is the particle volume fraction, $E \simeq 180$ is the Ergun's constant, $\delta = \sqrt{\frac{\eta}{\rho\omega}}$ is the thickness of the Stokes boundary layer, ρ is the fluid density, $\omega = 2\pi f$, and u_1' is the r.m.s oscillation velocity ($u_1' = u_1/\sqrt{2}$). Data on the relative decrease of the friction coefficient (absolute value) estimated from the average angle of avalanche in our rotating drum experiments are plotted in Fig. 4 as a function of $\beta\Upsilon u_1'$. To calculate $\beta\Upsilon$ average values of particle size were taken as d_p = 63 μm, 95 μm, 188 μm, 325 μm, 750 μm, and 1.5 mm, respectively. The particle volume fraction for each sample in the drum was calculated from the volume filled and the solid density. In the case of the finest ignimbrite, the particle volume fraction is small ($\phi \simeq 0.25$) as typical for fine cohesive powders [6]. As particle size is increase, ϕ increases up to $\phi \simeq 0.4$ for the noncohesive samples. This is still a low value as compared with the packing fraction of randomly packed noncohesive hard spheres ($\phi \simeq 0.56$). The reason why ignimbrite particles pack in more porous structures is their irregular shape [12]. Ignimbrite particles are notably angular shaped, which affects also both the interparticle attractive force and the gas-solid hydrodynamic drag. The latter would be decreased as the contact surface area is reduced whereas the former would be increased as the surface to volume ratio is increased for low sphericity particles. This can been accounted introducing the sphericity Ψ of the particles in the effective Ergun's constant (E/Ψ) [13]. Sphericity of volcanic ash particles typically varies in the range $\Psi \simeq 0.6 - 0.8$ [14]. In our calculations we have used $\Psi = 0.7$.

As seen in Fig. 4 experimental data on the relative decrease of the friction coefficient follows the trend predicted despite the model simplicity albeit the predicted result deviates considerably from data for the lowest frequencies tested and the smallest particle size powders. In these limits the relative decrease of the static friction coefficient is fairly below the theoretical expectation. On the other hand, Fig. 4 shows a good agreement between the experimental data obtained for the relative decrease of the friction coefficient and the prediction by Eq. 1 for the largest particle size powders suggesting that the enhancement of pore pressure by vibrations weakens the inertial flow of these noncohesive granular materials that otherwise (under steady flows) would not be affected by the gas-solid hydrodynamic interaction [6]. Probable causes for the deviation observed in the case of fine powders may be compaction induced vibration and the prevalence of van der Waals attractive forces between the particles on their weight, which confers the powder a nonnegligible cohesiveness although the typical angular shape of the particles as well as the large presence of asperities on their surface would lead to a reduction of van der Waals forces [15]. A consequence of the relative strength of interparticle attractive forces is that the angle of avalanche depends critically on the typical size of the system [16]. Moreover, interparticle attractive forces are usually enhanced by consolidation forces [17], which would increase the frictional resistance of the powder. Adhesive forces lead also to the aggregation of particles of size typically below $\sim 50\mu$m due to their prevalence over particle weight [18]. These aggregates screen external fluid flow and may behave as low density effective grains that can be fluidized at reduced velocities and exhibit a liquid-like behavior [18].

Dynamical weakening by oscillatory flows can be expected in a variety of geological process involving intense vibrations and/or acoustic waves of very long duration and high frequencies leading to viscous fluid-solid oscillatory

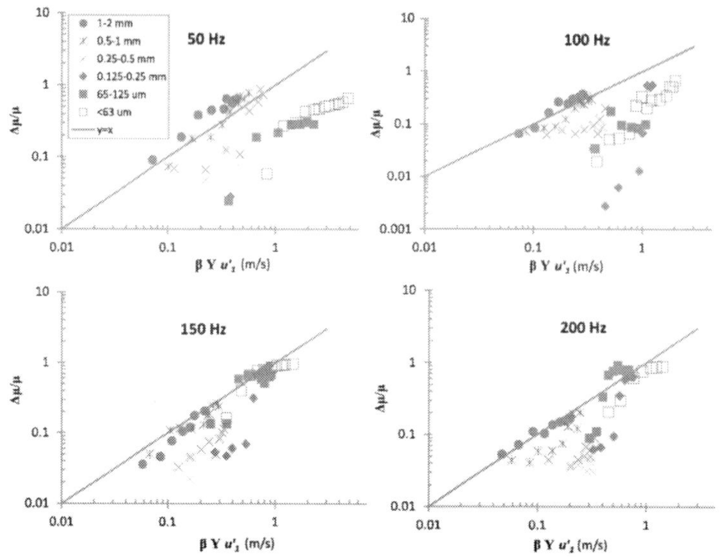

Figure 4. Relative decrease (absolute value) of the friction coefficient measured for ignimbrite samples of various particle size ranges in a slowly rotating drum subjected to vibrations of diverse frequencies as a function of $\beta\Upsilon u_1'$ (Eq. 1).

flows. This may be the case of seismic faulting, lahars, non-volcanic landslides, detachment faulting and impact crater formation. A major outstanding question for understanding the flow behavior of this variety of geological processes is what is the physical mechanism that drives dynamical weakening as a commonly observed feature in all of them. Our results suggest that the enhancement of pore fluid pressure by viscous oscillatory flows could be a strong mechanism to be given further consideration.

Acknowledgements

This work was supported by the Spanish Government Agency Ministerio de Economia y Competitividad and FEDER Funds (Contract CTQ2014-52763-C2-2-R).

References

[1] O. Roche, M.A. Gilbertson, J.C. Phillips, R.S.J. Sparks, Journal of Geophysical Research: Solid Earth **109**, 1 (2004), b10201

[2] R. Cas, H. Wright, C. Folkes, C. Lesti, M. Porreca, G. Giordano, J. Viramonte, Bulletin of Volcanology **73**, 1583 (2011)

[3] O. Korup, D. Schneider, C. Huggel, A. Dufresne, in *Treatise on Geomorphology*, edited by J.F. Shroder (Academic Press, San Diego, 2013), pp. 183 – 199, ISBN 978-0-08-088522-3

[4] O. Roche, D.C. Buesch, G.A. Valentine, Nature Communications **7**, 10890 (2016)

[5] A. Castellanos, J.M. Valverde, A.T. Perez, A. Ramos, P.K. Watson, Phys. Rev. Lett. **82**, 1156 (1999)

[6] J.M. Valverde, C. Soria-Hoyo, Journal of Geophysical Research: Solid Earth **120**, 6182 (2015), 2015JB012317

[7] K. Uhira, H. Yamasato, M. Takeo, Journal of Geophysical Research: Solid Earth **99**, 17757 (1994)

[8] S. De Angelis, V. Bass, V. Hards, G. Ryan, Natural Hazards and Earth System Science **7**, 467 (2007)

[9] V.M. Zobin, in *Introduction to Volcanic Seismology (Second Edition)*, edited by V.M. Zobin (Elsevier, Oxford, 2012), pp. 261 – 293, second edition edn., ISBN 978-0-444-56375-0

[10] S. Sundaresan, Annu. Rev. Fluid Mech. **35**, 63 (2003)

[11] R.Y. Yang, R.P. Zou, A.B. Yu, Phys. Rev. E **62**, 3900 (2000)

[12] P. Carman, Chemical Engineering Research and Design **75, Supplement**, S32 (1997)

[13] C.M. Riley, W.I. Rose, G.J.S. Bluth, Journal of Geophysical Research: Solid Earth **108**, n/a (2003), 2504

[14] A. Castellanos, Adv. Phys. **54**, 263 (2005)

[15] J.M. Valverde, A. Castellanos, A. Ramos, P.K. Watson, Phys. Rev. E **62**, 6851 (2000)

[16] M.A.S. Quintanilla, A. Castellanos, J.M. Valverde, Phys. Rev. E **64**, 031301(1) (2001)

[17] A. Castellanos, J.M. Valverde, M.A.S. Quintanilla, Phys. Rev. Lett. **94**, 075501(1) (2005)

Enhancement of CO_2 capture in limestone and dolomite granular beds by high intensity sound waves

Jose Manuel Valverde[1,*], *Jose Manuel* Perez-Ebri[1], and *Miguel Angel* Sanchez-Quintanilla[1]

[1] *Faculty of Physics*
University of Seville
Ada. Reina Mercedes s/n 41012 Seville, Spain

Abstract. The calcium looping (CaL) process, based on the calcination/carbonation of $CaCO_3$ at high temperatures, has emerged in the last years as a potentially low cost technology for CO_2 capture. In this work, we show that the application of high intensity sound waves to granular beds of limestone and dolomite in a CaL reactor enhances significantly their multicycle CO_2 capture capacity. Sound waves are applied either during the calcination stage of each CaL cycle or in the carbonation stage. The effect of sound is to intensify the transfer of heat, mass and momentum and is more marked when sound is applied during calcination by promoting CaO regeneration. The application of sound would allow reducing the calcination temperature thereby mitigating the decay of capture capacity with the number of cycles and reducing the energy penalty of the technology.

1 Introduction

The calcium looping (CaL) process is a second generation technology to capture CO_2 by means of cyclic carbonation and calcination of CaO [1–5]. In this process, the gas is passed through a bed of CaO particles at atmospheric pressure in a carbonator reactor operated at around 650 ^{0}C. At this temperature, the kinetics of carbonation is fast enough as required by industrial applications whereas the equilibrium CO_2 concentration is low (around 1% vol.), which allows for a high CO_2 capture efficiency [6]. The carbonated solids are circulated into a second gas-solid reactor (calciner) operated at temperatures above 900 ^{0}C where CaO regeneration takes place. Gas fluidized beds are operated in the fast fluidization regime (gas velocities above m/s) in the case of post-combustion CO_2 capture, which favors the gas-solids contacting efficiency [3]. On the other hand, pre-combustion capture conditions require either fixed or bubbling beds operated at low gas velocities (~ 1-10 cm/s) [7]. The usually poor thermal conductivity in a fixed bed usually leads to local heating in exothermic reactions (such as carbonation) and local cooling in endothermic reactions (such as calcination), which slow down gas-solid heterogeneous reactions at high temperature. In this work, we investigate the effect of high intensity low frequency sound on the multicycle capture performance of fixed limestone and dolomite beds operated at CaL conditions. If the inertia of a particle in a fluid subjected to a sound wave is large enough to be unmovable by the sound wave, the gas-solid heat/mass transfer can be greatly enhanced because of the development of a convective fluid flow in a boundary layer adjacent to the solid (*acoustic streaming*) [8]. It is therefore envisaged that low frequency sonoprocessing might be a useful technique to promote the gas-solid heat/mass transfer in fixed beds at CaL conditions.

2 Experimental setup and methods

Figure 1 shows the experimental setup used in this work. The reactor consists of a cylindrical quartz vessel (45 mm inner diameter) with a gas distributor made of a quartz plate fitted at its base (2mm thick, 16 to 40 μm pore size). Dry compressed air or a mixture of 15% CO_2 and 85% N_2 is supplied to the inlet port of the reactor by means of two mass flow controllers and a set of valves. For carbonation, a mixture of 15% CO_2+85% N_2 is passed through the reactor while dry compressed air is used for calcination. In either case, a total gas flow of 2000 cm^3/min in standard conditions (25 ^{o}C, 1 atm) is kept fixed by means of the mass flow controllers. The gas exiting the reactor passes through a gas analyzer and a flow meter to monitor its composition. A pressure gauge connected to the inlet port of the reactor is used to measure the total pressure drop across the reactor. An electric signal generator and amplifier are used to excite the loudspeaker to produce the low frequency sound wave that reaches the reactor trough a closed guide. A silicone elastic membrane inside the guide with good sound transmission prevents gas exchange between the reactor and the sound guide. The frequency of the sound was fixed to 130 Hz and the sound level intensity was 157 dB. The temperature is kept at 610 ^{o}C for 45 minutes during the carbonation stage and at 900 ^{o}C for 30 minutes during the calcination stage, with a 15 minutes

* e-mail: jmillan@us.es

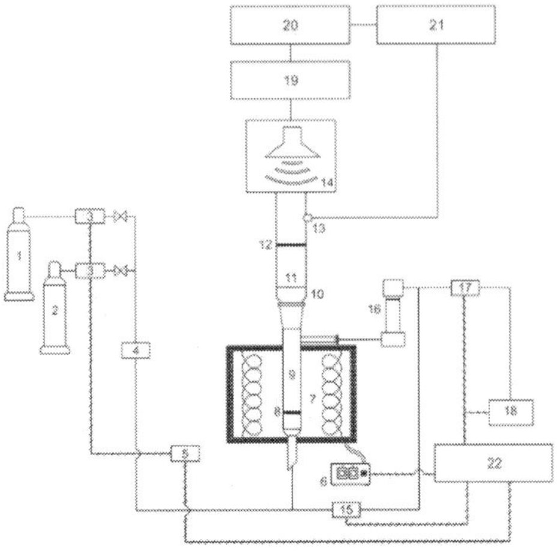

Figure 1. 1: Compressed tank of 15% CO_2 /85 % N_2 vol/vol). 2: Tank of compressed dry air. 3: Mass flow controllers. 4: Safety maximum pressure regulator. 5: Mass flow controller readout. 6: Oven temperature controller. 7: Oven. 8: Quartz fritted filter. 9: Quartz reactor. 10: Teflon adapter. 11: Acoustic waveguide. 12: Silicone elastic membrane. 13: Microphone. 14: Low frequency loudspeaker. 15: Differential pressure transducer. 16: Particle filter. 17: Mass flow meter. 18 CO_2 gas analyzer. 19: Electric amplifier. 20: Electric signal generator. 21: Oscilloscope. 22: Data acquisition and control.

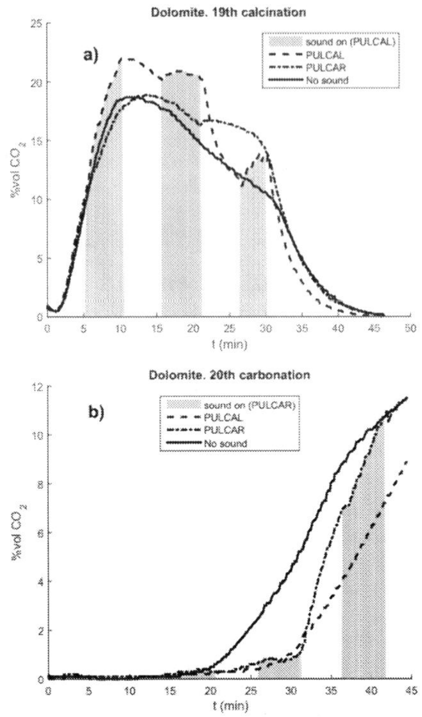

Figure 2. a): CO_2 breakthrough curve during calcination in the 19th cycle. b): CO_2 breakthrough curve during carbonation in the 20th cycle. Both plots are for dolomitic CaO under various operation modes: PULCAR: pulsed sound applied during carbonation; PULCAL pulsed sound applied during calcination.

intermediate period between the calcination and the carbonation stages in which the reactor is vented using dry air while the temperature drops from the calcination to the carbonation temperature. As CaO precursors we have used sieved natural limestone and dolomite particles supplied to us by a local quarry (Taljedi S.L., Gilena, Spain). Only the fraction of particles with mesh sizes d_p in the range 710 μm $\geq d_p \geq$ 425 μm was used. In this range of sizes, the particles have large enough inertia not to be moved by the low frequency sound wave applied. In all the runs, the fresh samples (178 g mass) were initialized by precalcination in-situ at 900 ^{o}C during 30 minutes under dry air.

3 Experimental results and discussion

Figs. 5 and 2 show the time evolution of the CO_2 %vol concentration measured in the effluent gas of the reactor (CO_2 breakthrough curve) during the calcination and the carbonation stages of the 19th and 20th cycles, respectively, for limestone (Fig. 5) and dolomite (Fig. 2). The grey bands in the figures indicate the time span of sound application for the PULCAL data (pulsed sound applied only in calcination stages) and the PULCAR data (pulsed sound applied only in carbonation stages). In the experiments pulsed sound was employed to clearly appreciate the effect of sound application and to allow for refrigeration of the loudspeaker. Thus, the sound generator was turned on for 5 minutes and turned off for another 5 minutes cyclically. As can be seen, the application of high intensity sound waves has a significant effect on enhancing both carbonation and calcination. This effect is more marked for the PULCAL operation mode, in which local maxima of the CO_2 concentration measured in the effluent gas are seen to nearly coincide with the application of sound. Moreover, the effect of sound application is further prolonged to the part of the stage when it is off. As can be seen in Figs. 5 and 2, the CO_2 breakthrough curve obtained for the PULCAL mode remains neatly below the curve obtained for the other two operation modes in the carbonation cycle during the first minutes (although sound is off in PULCAL operation). Likewise the CO_2 breakthrough measured during calcination for the PULCAR mode is well above that measured in the absence of sound (although the sound is off in PULCAR mode). The observation that PULCAL operation causes an increase in the capture rate of CO_2 during the carbonation cycle can be explained by the enhancement of decarbonation in the previous calcination stage, which yields a CaO skeleton with increased porosity and higher available surface area for carbonation. Thus, the PULCAL operation is the mode for which the capture capacity of the sorbent is mostly enhanced.

Fig. 3 and Fig. 4 show the CO_2 breakthrough curves for several carbonation cycles of experimental runs op-

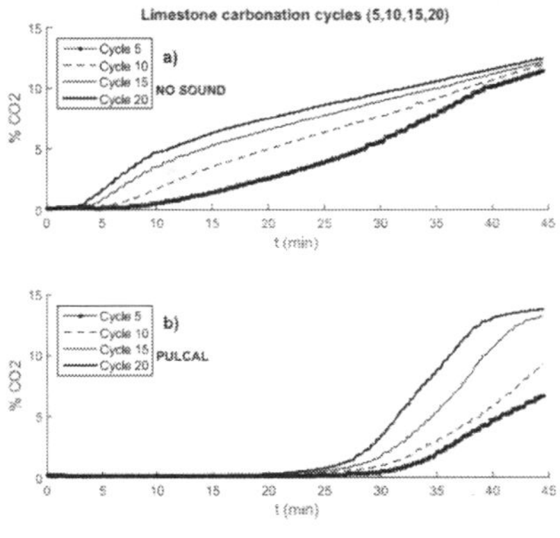

Figure 3. CO_2 breakthrough curves for carbonation cycles under: a) "no sound" and b) PULCAL operation modes for limestone derived CaO.

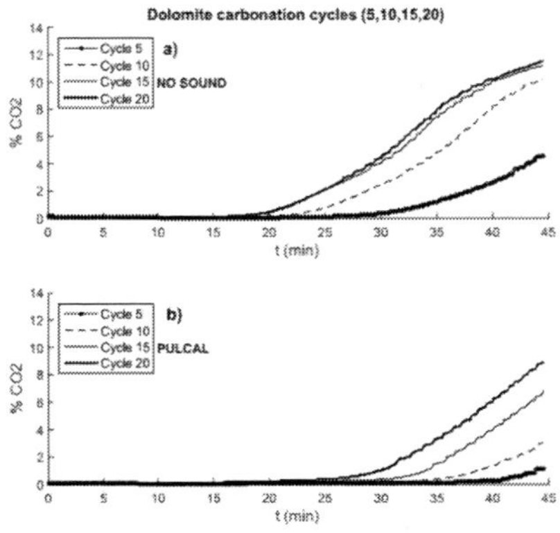

Figure 4. CO_2 breakthrough curves for carbonation cycles under: a) "no sound" and b) PULCAL operation modes for dolomite derived CaO.

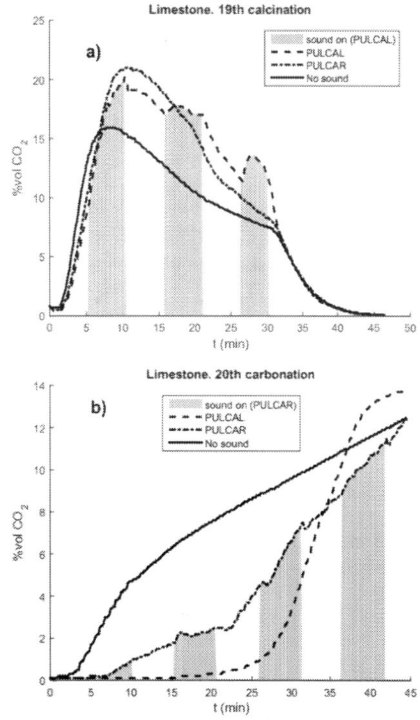

Figure 5. a): CO_2 breakthrough curve during calcination in the 19th cycle. b): CO_2 breakthrough curve during carbonation in the 20th cycle. Both plots are for limestone derived CaO under various operation modes: "No sound": no sound applied; PULCAR: pulsed sound applied during carbonation; PULCAL pulsed sound applied during calcination. Grey bands indicate the time over which high intensity sound is applied.

erated in the PULCAL and "no sound" modes. As may be seen, the effect of PULCAL operation mode on the is much more marked for limestone derived CaO (Fig. 5) than for the dolomitic CaO as might be expected since the MgO inert skeleton would serve already to promote mass/heat transfer in the case of dolomite in the absence of sound as compared to lime.

The total amount of CO_2 captured at each cycle m_{CO_2} has been obtained from the CO_2 breakthrough curves measured in the carbonation stage. It must be taken into account that the MgO, once formed, does not react with CO_2 in the range of temperatures and partial pressures of CO_2 used in our experiments [10] and therefore it remains as inert material. Thus, the maximum theoretical value of the capture capacity for lime is $C_{lim}^{max} = 0.78$ whereas for the dolomitic lime is $C_{dol}^{max} = 0.46$.

Figure 6 shows the capture capacity C as a function of the cycle number measured at the end of the 45 minutes of each carbonation stage. As can be observed, the capture capacity is for both sorbents well below the maximum uptake capacity specially for the first few cycles mainly due to incomplete decarbonation under fixed bed conditions, which severely hinder the transfer of heat and mass. The capture capacity increases in the first cycles as subsequent calcinations allow for a higher degree of decarbonation. After a few cycles, all the available CaO is regenerated during the calcination stage and the capture capacity reaches a maximum, after which it follows a gradual decline with the cycle number arguably caused by a loss of surface area of the sorbent due to sintering in the calcination stages. It is also seen in Fig 6 that the decline of the capture capacity is mitigated for the dolomite derived sorbent as might be expected since the inert MgO present in the dolomite-based sorbent acts as a skeleton that supports the active CaO surface area. Thus, although the dolomite derived sorbent has a smaller percentage of CaO than lime, its capture capacity is larger in accordance with previous results obtained from thermogravimetric tests [11].

The application of high intensity sound would cause an increase in the pressure drop across a particle bed even if the particles are not movable by the sound wave (as is

 DOI: 10.1051/epjconf/201714014001

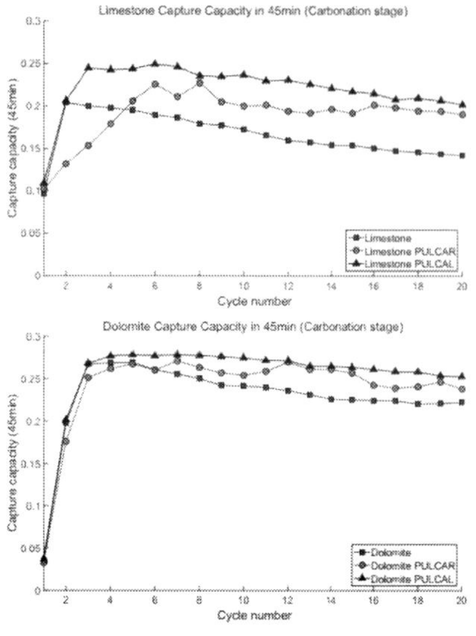

Figure 6. CO_2 capture capacity for limestone and dolomite derived sorbents in the absence of sound and when pulsed sound is applied during the carbonation stage (PULCAR) and in the calcination stage (PULCAL).

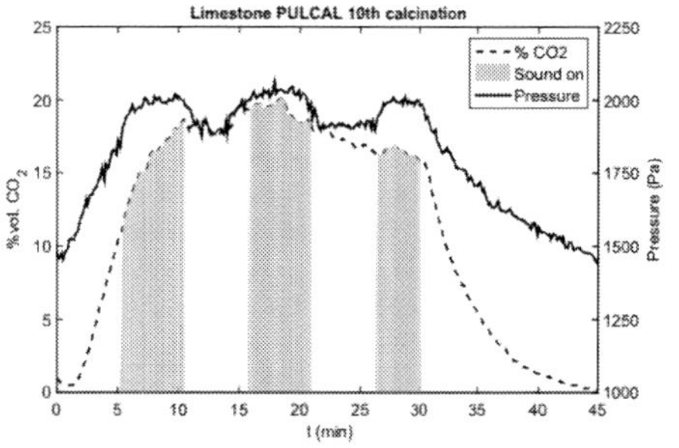

Figure 7. Total pressure drop across the reactor during the 10th calcination cycle of limestone derived CaO in PULCAL operation mode as a function of time together with the %vol CO_2 of the gas leaving the reactor. Grey bands indicate the periods of application of high intensity sound.

the case of our experiments). This increase of pressure drop as due to the relative gas-solid oscillations is caused by the enhancement of viscous dissipation due to acoustic streaming, which consists of the development of a steady rotational gas current in a boundary layer (the so-called Stokes boundary layer) of typical thickness $\delta \sim \sqrt{\nu/\omega}$, where ν is the kinematic gas viscosity and ω is the angular oscillation frequency [8]. Figure 7 shows the time evolution of the pressure drop across the reactor for the 10th calcination stage of limestone derived CaO under PULCAL operation together with the evolution of the % CO_2 in the effluent gas. Remarkably, under PULCAL operation mode, the pressure drop across the reactor follows the same trend than the CO_2 breakthrough curve as the CO_2 released by the sorbent contributes to the net gas flow through the material in the reactor. On top of this general trend, the pressure drop is seen to increase temporarily during the sound pulses.

4 Conclusions

This work sows that application of high-intensity low-frequency sound enhances the capture capacity of natural limestone and dolomite fixed beds along carbonation/calcination cycles due to acoustic streaming resulting from the frictional dissipation of energy in a boundary layer nearby the solids. Acoustic streaming promotes heat and mass transfer thus accelerating both carbonation and calcination. This effect is more marked when the pulsed sound is applied during the calcination stage arguably due to the relatively higher temperature, which promotes further the gas viscosity and therefore acoustic streaming.

Acknowledgements

This work was supported by the Spanish Government Agency Ministerio de Economia y Competitividad and FEDER Funds (Contract CTQ2014-52763-C2-2-R).

References

[1] V. Manovic, E.J. Anthony, J. Phys. Chem. A **114**, 3997 (2010)
[2] W. Wang, S. Ramkumar, D. Wong, L.S. Fan, Fuel **92**, 94 (2012)
[3] B. Arias, M.E. Diego, J.C. Abanades, M. Lorenzo, L. Diaz, D. Martínez, J. Alvarez, A. Sanchez-Biezma, Int. J. Greenh. Gas Control **18**, 237 (2013)
[4] I. Martínez, G. Grasa, R. Murillo, B. Arias, J.C. Abanades, Energy Fuels **26**, 1432 (2012)
[5] A. Perejón, L.M. Romeo, Y. Lara, P. Lisbona, A. Martínez, J.M. Valverde, Appl. Energy **162**, 787 (2016)
[6] C. Ortiz, R. Chacartegui, J. Valverde, J. Becerra, Appl. Energy **169**, 408 (2016)
[7] K. K. Johnsen, H. Ryu, J. Grace, C. Lim, Chem. Eng. Sci. **61**, 1195 (2006)
[8] J. Valverde, Eur. Phys. J. E **38**, 38 (2015)
[9] J. Readman, B. Richard, Phys. Chem. Chem. Phys. **7**, 1214 (2005)
[10] J. Valverde, P. Sanchez-Jimenez, L. Perez-Maqueda, Appl. Energy **138**, 202 (2015)

On the tensile strength of soil grains in Hertzian response

Sadegh Nadimi[1,*] and *Joana* Fonseca[1]

[1]City, University of London, London, UK

Abstract. The breakage initiation of soil grains is controlled by its tensile capacity. Despite the importance of tensile strength, it is often disregarded due to difficulties in measurement. This paper presents an experimental and numerical investigation on the effect of tensile strength on Hertzian response of a single soil grain. Hertz theory is commonly used in numerical simulation to present the contact constitutive behaviour of a purely elastic grain under normal loading. This normal force:displacement comes from stress distribution and concentration inside the grain. When the stress reaches the tensile capacity, a crack initiates. A series of numerical tests have been conducted to determine the sensitivity of Hertzian response to the selected tensile strength used as an input data. An elastic-damage constitutive model has been employed for spherical grains in a combined finite-discrete element framework. The interpretation of results was enriched by considering previous theoretical work. In addition, systematic experimental tests have been carried out on both spherical glass beads and grains of two different sands, i.e. Leighton Buzzard silica sand and coarse carbonate sand from Persian Gulf. The preliminary results suggest that lower tensile strength leads to a softer response under normal loading. The wider range of responses obtained for the carbonate sand, are believed to be related to the large variety of grain shape associated with bioclastic origin of the constituent grains.

1 Introduction

1.1 Single grain breakage

The importance of grain breakage in granular assemblies has been well documented [1-4]. This is of interest to geotechnical, mineral, chemical, food and transportation industries. Recent research advances have been obtained by focusing on the response of a single grain [5, 6]. The elastic normal force:displacement response of a grain under compression is called Hertzian response due to pioneering work by German physicist, Hertz [7]. Soil grains are brittle material and break under loading. They will ideally follow a Hertzian response up to the grain strength has been reached. The breakage of a grain is believed to be a tensile phenomenon (cracking) rather than be related to compressive strength (crushing) of the material.

Russell & Muir Wood [8] proposed an approximate expression for compressive and tensile strength of an ideal grain (sphere) under diametrically-compression. They adopted Christensen multiaxial failure criterion [9] which has two parameters for brittle materials: (1) intrinsic strength, κ and (2) microstructure factor, χ. The latter parameter (χ) which was described as the microstructural deviations from the ideal is further investigated in this paper by means of numerical and experimental investigation. The theoretical model is presented in the first part of the paper. This is followed by a numerical investigation presented in the second part and finally the experiment is described in part three and compared with the numerical investigation.

1.2 Theoretical consideration

The internal stress field propagation in a single spherical grain subjected to the contact forces was presented in [8]. This has been linked to failure criterion in [9] which states a material is not at failure as long as the following condition is compiled:

$$\frac{\chi\kappa}{\sqrt{3}} I_1 + (1+\chi)^2 J_2 < \frac{\kappa^2}{1+\chi} \tag{1}$$

$$J_2 = \left(\frac{I_1^2}{3} - I_2 \right) \tag{2}$$

where I_1 and I_2 are the first and second invariants of stress tensor. J_2 is the second invariant of the deviatoric stress tensor. Here, for unconfined compression, $I_1=\sigma_c$, $I_2=0$, $J2=(\sigma_c)^2/3$ and σ_c is the uniaxial compressive strength.

The intrinsic strength, κ and the microstructure factor, χ are defined as follows:

$$\chi = \frac{|\sigma_c|}{\sigma_t} - 1 \text{ and } \kappa = \frac{1+\chi}{\sqrt{3}} |\sigma_c| \tag{3}$$

where σ_t is the tensile strength.

* Corresponding author: sadegh.nadimi-shahraki@city.ac.uk

According to Christensen [9] κ is the strength of the material with no microstructural damage and is associated to atomic scale properties. The unconfined compressive and tensile strengths of the material are:

$$\sigma_c = -\frac{\sqrt{3}\kappa}{1+\chi} \text{ and } \sigma_t = -\frac{\sqrt{3}\kappa}{(1+\chi)^2} \quad (4)$$

The approximate expressions for compressive and tensile strength of an elastic sphere under compression are [8]:

$$\sigma_c = \frac{aF}{\pi R^2 \sin^2\theta} \text{ and } \sigma_t = \frac{-aF}{\chi\pi R^2 \sin^2\theta} \quad (5)$$

where R is radius, F is the contact force at failure, θ is the contact area in degrees and a is defined as:

$$a = \frac{3\left(\frac{3}{32}+\frac{\sqrt{2}}{24}+\left(\frac{\sqrt{2}}{12}-\frac{1}{4}\right)\mu+\left(\frac{1}{2}-\frac{\sqrt{2}}{3}\right)\mu^2\right)}{(2-\sqrt{2})(1+\mu)} \quad (6)$$

where μ is the coefficient of friction.

In Eq. (5), the contact area θ can be derived by considering the Hertz solution, Eq. (7):

$$F_n = \frac{4}{3}\frac{E}{1-\upsilon^2}\sqrt{R}\delta_n^{3/2} \quad (7)$$

where F_n is the normal force, E is the elastic modulus, υ is the Poisson ratio and δ_n is the normal displacement. Therefore, θ can be computed using Eq. (8):

$$\theta = 2\cos^{-1}\left(1-\left(\frac{3F(1-\upsilon^2)}{4ER}\right)^{2/3}\right) \quad (8)$$

To the best knowledge of the authors, the microstructural origin of κ and χ has not been supported by any atomic scale studies. In approximate solution for strength of sphere, σ_c is unaffected by the value of χ (Eq. 5). According to literature, the parameter χ varies from 10 to 170 for a range of rock types [8, 10]. Russell & Muir Wood [8] considered four values for χ = 19, 49, 99, 199 in their parametric study. In this paper, the effect of χ on the normal force:displacement behaviour of a sphere under compression is presented.

2 Numerical investigation

The problem is investigated in the framework of combined finite-discrete element framework [11, 12]. The interaction of two contacting spheres is simplified to interaction of a sphere with rigid plate, due to symmetry. The effect of χ on the normal force:displacement is investigated by means of an elastic-damage constitutive model. This model uses isotropic damage elasticity in combination with isotropic tensile and compressive plasticity. The constitutive model, numerical model and results are presented in this section.

2.1 Constitutive model

The response of the material is modelled using damage plasticity based on two failure mechanisms: (1) tensile cracking and (2) compressive crushing [13, 14]. Damage states are characterised by two hardening parameters: the equivalent plastic strain in tension, ε_t^{pl}, and in compression, ε_c^{pl}. The stress:strain relation is governed by scalar damage elasticity as defined as follows:

$$\sigma = (1-d)D_0^{el} : (\varepsilon - \varepsilon^{pl}) = D^{el} : (\varepsilon - \varepsilon^{pl}) \quad (9)$$

where d is a scalar describing the amount of isotropic damage. D_0^{el} is the initial elastic stiffness, D^{el} is the damaged elastic stiffness, ε is the total strain ε^{pl} is the plastic strain.

The states of failure and damage are determined using a yield surface in the effective stress space. The yield condition proposed by [13] and modified by [14] has been incorporated to model both tensile and compressive behaviour. The flow potential surfaces are not described here, as it is out of scope of this paper. The typical uniaxial tensile stress strain is shown in Fig. 1a. The degradation of stress depends on the formulation of d. Similarly, for the compressive stress:strain curve, the degradation is controlled by the value of d after the ultimate compressive strength (Fig. 1b). The material hardens from yield compressive strength to the ultimate strength and then softens.

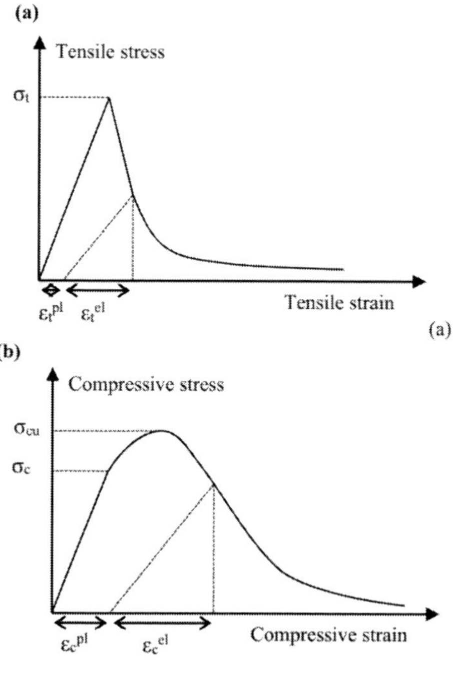

Fig. 1. Response of the material under uniaxial loading (a) in tension and (b) in compression.

This constitutive model is attractive, as it can represent a different failure mechanism for cracking and for crushing. Thus, enabling investigating the effect of χ by changing σ_t for constant σ_c.

2.2 Numerical Model

The model sphere has a diameter of 2.2mm and is represented by a mesh formed by 60,743 elements and 18,112 nodes. The material parameters used in the simulation are listed in Table 1. Explicit time discretization was employed. The property of hard contact was defined between the sphere and the plate. Using '*hard contact*' behaviour means that all the force is transmitted through the contact. The relation of normal force:displacement with hard contact assumption follows exactly the Hertzian theory for pure elasticity, due to the deformability of bodies [12]. A series of six simulations were conducted. Different values of χ were considered which is listed in Table 2 with corresponding tensile and compressive strengths.

Table 1. Physical and mechanical parameters used

Elastic modulus	E (GPa)	63
Poisson ratio	υ (-)	0.3
Density	ρ (gr/mm^3)	2.5
Diameter	D (mm)	2.2

Table 2. Compressive and tensile strengths for different value of parameter χ and the maximum normal force obtained corresponds to the onset of breakage

χ	σ_c (MPa)	σ_t (MPa)	F_N (N)
9	2000	200	942
19	2000	100	589
49	2000	40	263
99	2000	20	138
199	2000	10	65

2.3 Results

Figure 2 shows the results of the simulations. The pure elastic model is also presented to observe the deviation of the elastic-damage simulations from pure elasticity. The increase in the value of χ shows the decrease in the normal force which corresponds to the onset of breakage. The exact value of the normal force, F_N, for different χ values is listed in the last column of Table 2.

3 Experimental investigation

Single grain experiments are presented in this section on the three type of grains, including glass bead, silica sand and carbonate sand. The schematic of the set-up is shown in Fig. 3. A strain controlled machine '*Instron 5969*' was used over a range of force applications up to 2kN. The instrumentation accuracy was <1μm for displacement and <0.1N for load measurements.

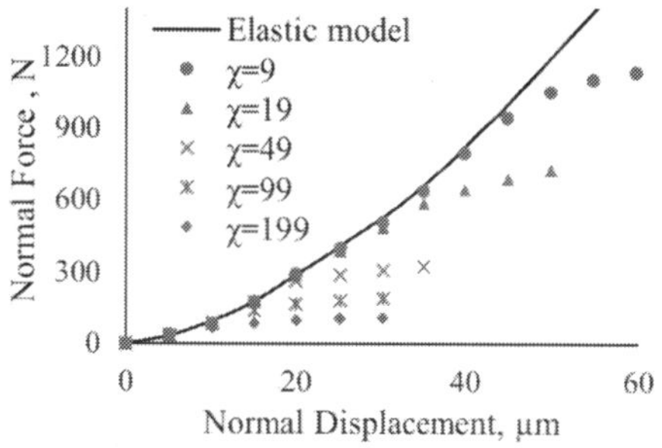

Fig. 2. Response of a spherical grain under compression for different tensile strengths.

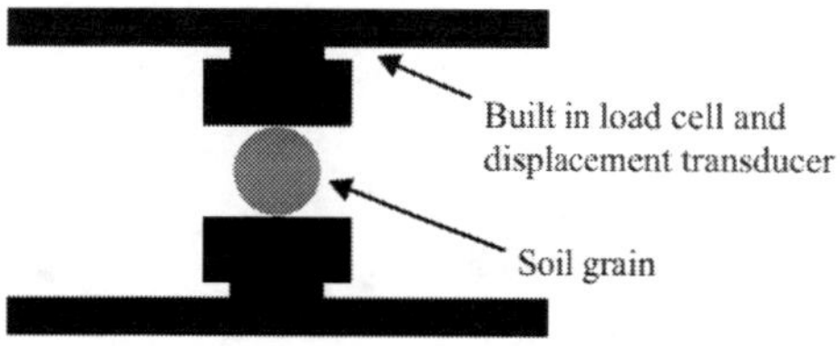

Fig. 3. Schematic of the single grain test under normal load.

3.1 Glass bead

Forty single grain tests were carried out on glass beads. They are commercially supplied by Sigmund Lindner GmbH as type S beads with specific gravity of 2.57. Three range of diameters were chosen, including 1.1 to 1.4mm, 2.0 to 2.4mm, and 3.6 to 4.1mm. In previous experimental work, the tensile strength of a grain has been defined as the maximum normal force over the squared grain diameter [1]. In order to show the variation of tensile strength, the histograms of maximum normal loads applied on the grains are presented. Figs 4.a, b and c show the variation of the maximum normal force applied to the glass beads with the different diameters mentioned above. The Standard Deviation (SD) and median values (Med) are also presented in Fig. 4 caption. McDowell & Bolton [1] proposed a relationship between tensile strength and grain diameter as follows:

$$\sigma_t \propto D^{-3/m} \tag{10}$$

They reported the value of m is in the range of 5-10 for silica sand, and limestone aggregates. Here, the value of m was obtained from power regression to be 2.89 (R^2=0.75) for glass beads. This will be further discussed in the next Section. According to Fig 3.b and numerical simulation, a low value of χ (approximately 9) is expected for glass beads. Fig. 4.d shows the comparison of a typical experimental response observed with the Hertzian response.

3.2 Silica sand

Forty grains of Leighton Buzzard sand, from England, with median diameter of 0.84mm were randomly picked.

Fig 5.a and 5.b show the typical response of silica sand grain under compression. It comprises an initial plateau associated with grain rotation and asperity damage, followed by inelastic Hertzian type of hardening and finally breakage. Fig. 6.a shows the variation of maximum normal force for silica grains with a standard deviation of 6.1N and a median value of 15.5N, excluding 8 grains with capacity from 40N to 80N, for better visualisation. Despite the similarity in mineralogy and elastic modulus between glass beads and silica sand, the tensile strength is significantly different which suggests different value for χ.

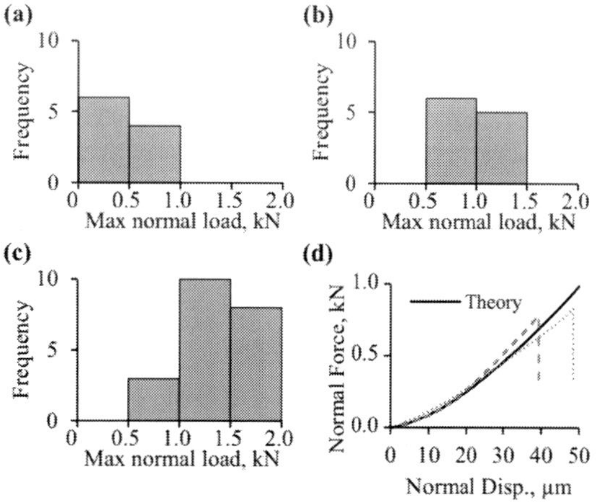

Fig. 4. Distribution of the maximum normal load applied on the glass beads of a) 1.1-1.4mm (SD=123, Med=459), b) 2.0-2.4mm (SD=136, Med=979), c) 3.6-4.0mm diameter (SD=357, Med=1290); d) comparison between the typical normal force: displacement response of glass beads and Hertzian theory.

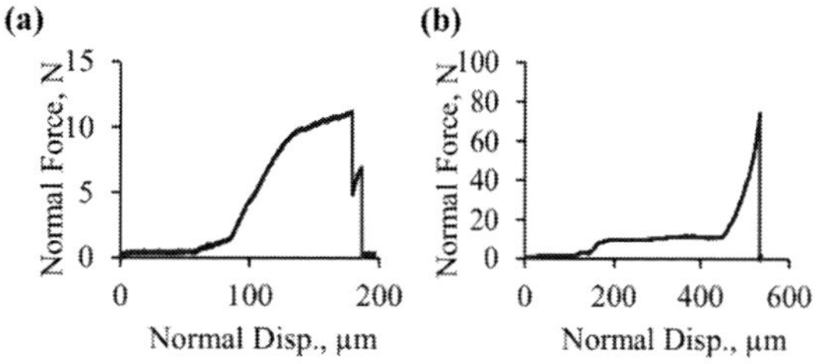

Fig. 5. Typical normal force:displacement response of silica sand.

3.3 Carbonate sand

Forty grains of a shelly carbonate sand from the Persian Gulf with median diameter of 2mm were randomly picked. Fig. 6.b shows the histogram of maximum normal load carried by carbonated sand grains with a standard deviation of 7.9N and a median value of 9.2N, excluding 10 grains with very wide capacity from 40N to 384N, for better visualisation. Although the majority of grains can bear less than 10N, this diversity in response can be attributed to the various shape found in this bioclastic material [15].

4 Closing remarks

The numerical and experimental tests presented here show that the failure of a single grain under compression is a tensile (cracking) phenomenon. We can observe that the lower the tensile capacity, the softer the Hertzian response. The low value of χ for glass beads in comparison with the high value for silica grains can be related to geological history and weathering of the sand grains. There might be a direct relation between the parameter m in [1] and χ in [8], which needs further investigation. Carbonate sand bearing capacity can be categorised by considering grain morphology and intergranular void ratio.

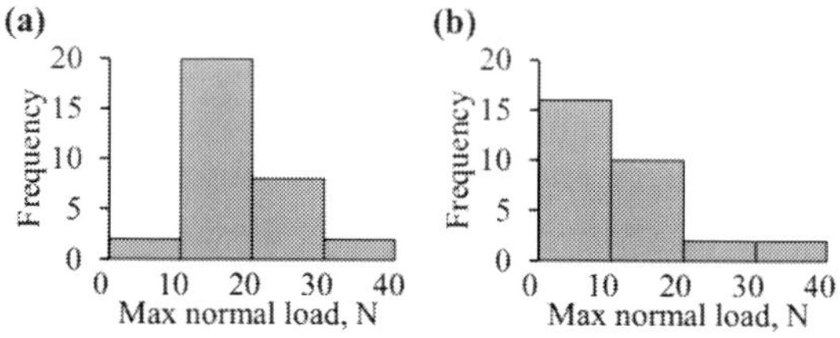

Fig. 6. Distribution of maximum normal load applied on the grains of, a) silica sand, b) carbonate sand.

The authors would like to thank City, University of London for the doctoral scholarship of the first author and EPSRC for the financial support (EP/N018168/1).

References

1. G.R. McDowell, M.D. Bolton, Géotechnique, **48** (1998)
2. Y. Nakata, A.F.L. Hyde, M. Hyodo, H. Murata, Géotechnique, **49** (1999)
3. G. Marketos, M. D. Bolton, Journal of the Mechanics and Physics of Solids, **55** (2007)
4. F.N. Altuhafi, M. R. Coop, Géotechnique, **61** (2011)
5. B. Zhao, J. Wang, M.R. Coop, G. Viggiani, M. Jiang, Géotechnique, **65** (2015)
6. W. Wang, M.R. Coop, Géotechnique http://dx.doi.org/10.1680/jgeot.15.P.247
7. H. Hertz, Journal für die reine und angewandte Mathematik, **92** (1882)
8. A.R. Russell, D. Muir Wood, International Journal of Rock Mechanics & Mining Sciences, **46** (2009)
9. R.M. Christensen, Mathematics and Mechanics of Solids, **5** (2000)
10. R.E. Goodman, *Introduction to rock mechanics*. 2nd Ed. New York: Wiley (1989)
11. A. Munjiza, *The Combined Finite-Discrete Element Method,* London: Wiley (2004)
12. S. Nadimi, J. Fonseca, Géotechnique (under review, February 2017).
13. J. Lubliner, J. Oliver, S. Oller, E. Onate, Inter. Journal of Solids and Structures, **25** (1989)
14. J. Lee, G.L. Fenves, ASCE Journal of Engineering Mechanics, **124** (1998)
15. J. Fonseca, et al. In *Deformation characteristics of geomaterials*: Buenos Aires, Argentina (2015).

 DOI: 10.1051/epjconf/201714015003

Micro-scale anisotropy of contacts and pores in granular media

Adnan Sufian[1,*], *Adrian R.* Russell[1], and *Andrew J.* Whittle[2]

[1] *Centre for Infrastructure Engineering and Safety, School of Civil and Environmental Engineering, The University of New South Wales, Sydney, NSW 2052, Australia*

[2] *Department of Civil and Environmental Engineering, Massachusetts Institute of Technology, 77 Massachusetts Avenue, Cambridge, MA 02139, United States of America*

Abstract. This paper considers the anisotropy associated with the interparticle contact network and interconnected pore space, which is quantified through a scalar quantity based on the fabric tensor. The study focused on the anisotropy associated with subsets of contacts and pores. It is shown that the stress ratio can be directly related to the geometric anisotropy of strong and non-sliding contacts, and that the anisotropy of pores is correlated with the anisotropy of contacts.

1 Introduction

Understanding the micro-scale origins of the complex macroscopic behaviour of granular media is essential to develop better predictive models. This requires the characterisation of micro-scale features including the network of interparticle contacts and the interconnected pore space. Numerous studies have suggested that anisotropy of micro-scale features plays a crucial role in the behaviour of granular materials [1–3]. While most prior studies have focused on the contact network, this paper explores the evolving anisotropy of both contacts and pores. Interparticle contacts contribute to the stress transmission process, while pores influence deformation characteristics and hence, both play distinct roles in the behaviour of granular materials. Micro-scale features are explored through numerical simulations of two-way cyclic axisymmetric tests on monodisperse assemblies of spherical particles. A novel feature of this study is that anisotropy is quantified for subsets of contacts and pores, thereby providing an indication of which micro-scale features have dominant contribution on the macroscopic behaviour.

2 Numerical Simulations

Discrete element simulations were conducted using the open-source software LIGGGHTS [4]. Details of these simulations are presented in [5]. Axisymmetric simulations mimicking a drained triaxial test were carried out for loose (CP-L) and dense (CP-D) samples. Samples were generated by randomly inserting particles within a cubical domain (with no contact between particles), and then compressing the assembly by applying isotropic pressure on the rigid frictionless walls of the domain. The current simulations were conducted using a monodisperse assembly

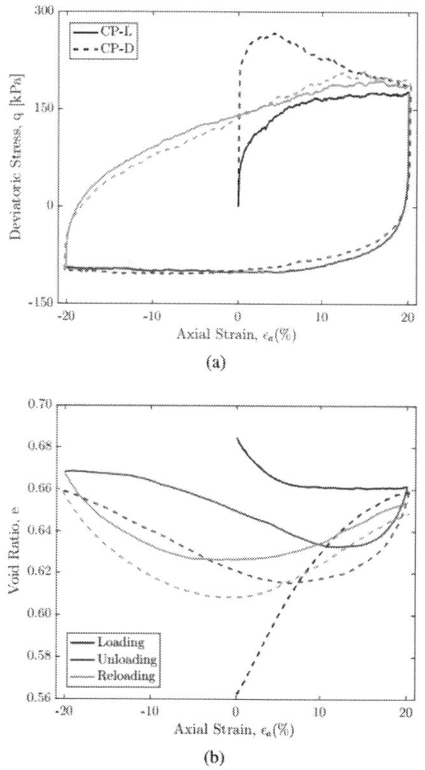

Figure 1: Stress-strain and volumetric response for numerical simulations. Note that the colour scheme for loading, unloading, and reloading, along with the line type distinguishing CP-L and CP-D samples is maintained for the rest of the paper, unless stated otherwise.

*e-mail: a.sufian@unsw.edu.au

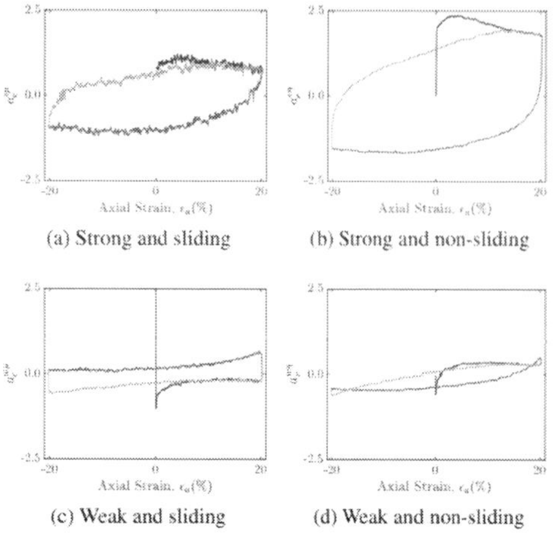

(a) Strong and sliding (b) Strong and non-sliding

(c) Weak and sliding (d) Weak and non-sliding

Figure 2: Evolution of geometric anisotropy for the interparticle contact network in the CP-D simulation.

of elastic spherical particles (diameter d_s = 1mm, elastic modulus E = 70MPa, and Poisson ratio ν = 0.25) compressed to a pressure, p = 200kPa. Particle contacts are represented using the Hertz-history model in LIGGGHTS. The interparticle friction, μ, is varied during isotropic compression to achieve samples with different pre-shear densities, but a constant value, $\mu = 0.25$, was used for all shear simulations. The numerical analyses simulate quasi-static cyclic shearing (with an inertia number of $I \approx 10^{-3}$) to a target axial strain of $|\varepsilon_a| = 20\%$ through loading, unloading and reloading phases (Fig. 1).

The macroscopic shear stress (Fig. 1a) and volumetric strain (void ratio, e, Fig. 1b) show behaviour that is qualitatively consistent with the measured behaviour of dense and loose sands. A critical state condition is approached in the loading phase, where the loose and dense sample exhibit similar stress state and void ratio at large axial strain. It should be noted that the void ratio was computed over a subset of the sample to avoid boundary irregularities associated with the rigid walls.

3 Anisotropy of Interparticle Contacts

The anisotropy of interparticle contacts can be quantified through a fabric tensor [6, 7]. This study is particularly concerned with the anisotropy associated with subsets of the interparticle contact network as forces are predominantly transmitted through a sparse subset of contacts.

In this study, a quad-partition is considered comprising four subnetworks: (i) strong and sliding contacts; (ii) strong and non-sliding contacts; (iii) weak and sliding contacts; and (iv) weak and non-sliding contacts, where the subnetworks are denoted $\{s\mu, s\eta, w\mu, w\eta\}$ respectively. Strong contacts are defined as those contacts with an interparticle normal force (f_n) greater than the mean, $f_n \geq \langle f_n \rangle$, while weak contacts are those with $f_n < \langle f_n \rangle$. Sliding contacts are at the Coulomb friction limit, that is, $|f_t| = \mu |f_n|$, where f_t is the interparticle tangential force.

The anisotropy of these subnetworks is quantified by considering the directional distribution of contact normals. This follows the stress-force-fabric relationship proposed in [5] for partitioned subnetworks, where further details can be found. If **n** is a unit vector along the contact normal direction, then the probability density function of contact normals with orientation **n** in the kth subnetwork (where k can have the values $\{s\mu, s\eta, w\mu, w\eta\}$) can be expressed by the second-order approximation:

$$E^k(\mathbf{n}) = \frac{1}{4\pi}\left(1 + D^k_{ij} n_i n_j\right) \tag{1}$$

where the subscripts $\{i, j\}$ are indices of the vector and tensorial quantities. D^k_{ij} is termed the deviatoric directional probability density tensor in the kth subnetwork which is given by:

$$D^k_{ij} = \frac{15}{2}\left(N^k_{ij} - \frac{N^k_{kk}}{3}\delta_{ij}\right) \tag{2}$$

where

$$N^k_{ij} = \frac{1}{N^k_c}\sum_{\alpha=1}^{N^k_c} n^\alpha_i n^\alpha_j \tag{3}$$

is termed the moment tensor (or fabric tensor) for interparticle contacts. δ_{ij} is the Kronecker delta and N^k_c is the number of contacts in the kth subnetwork. A scalar measure of anisotropy can be calculated by considering the second-invariant of the directional probability density tensor:

$$a^k_c = \sqrt{\frac{3}{2} D^k_{ij} D^k_{ij}} \tag{4}$$

a^k_c provides a measure of geometric anisotropy, as it is purely concerned with the arrangement of contacts rather than the forces transmitted through the contacts. Figures 2a-2d illustrate the evolution of a^k_c for each of the four subnetworks in the CP-D simulation (similar trends occur for the CP-L sample). Note that a signed measure of anisotropy is considered, where positive values imply preferential orientation in the axial direction, while negative values indicate preferential orientation in the radial direction.

Figures 2a and 2b show that contacts in the $\{s\mu, s\eta\}$ subnetworks are preferentially orientated in the direction of loading, while contacts in the $\{w\mu, w\eta\}$ subnetworks are orientated perpendicular to the direction of loading (Fig. 2c-2d). This reflects the force chain description of the contact network [8], where contacts with relatively large force magnitudes form columnar structures, which are braced laterally by contact with relatively small force magnitudes.

The subnetwork of strong and non-sliding contacts, $\{s\eta\}$, is clearly the dominant contributor to the geometric anisotropy of the contact network (Fig. 2b). What is particularly interesting about the behaviour of this subnetwork is that it virtually mimics the stress-strain response in Figure 1a. In fact, the relationship between anisotropy of the strong and non-sliding contacts and the stress ratio can

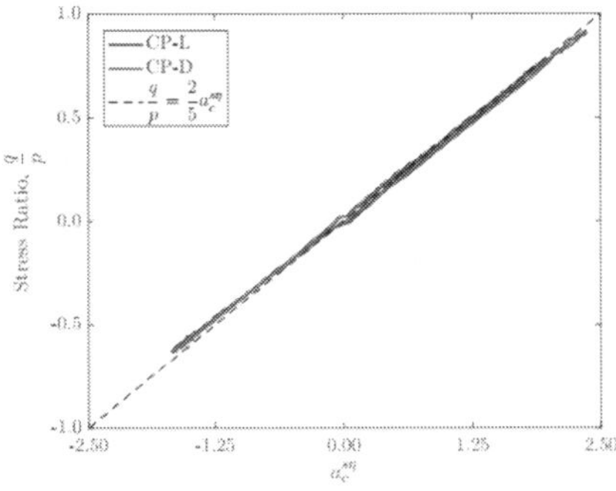

Figure 3: Unique relationship between stress ratio and the anisotropy associated with strong and non-sliding contacts.

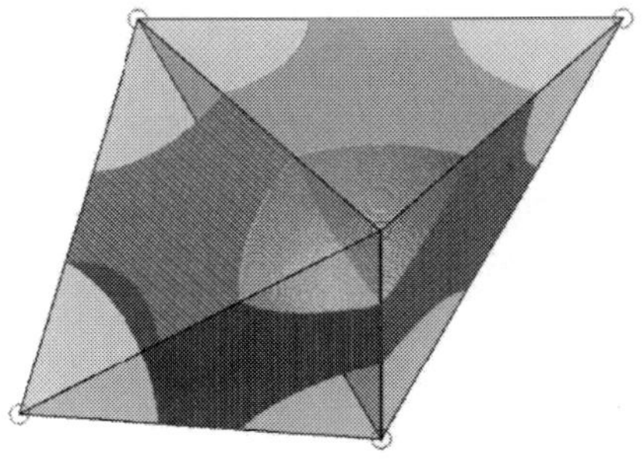

Figure 4: Example of an individual pore comprising two Delaunay cells. Note that the particle sectors have been removed leaving just the pore space.

be expressed as $\frac{q}{p} = \frac{2}{5}a_c^{s\eta}$. This is demonstrated in Figure 3, which indicates that stress ratio can be directly related to the geometric configuration of strong and non-sliding contacts throughout the two-way cyclic test. The potential implications of this relationship is discussed further in [5].

4 Anisotropy of Pores

The anisotropy of pores can be quantified in a similar manner to Section 3 but requires a clear and unambiguous identification of individual pore bodies. In this study, individual pores are identified using the modified Delaunay tessellation detailed in [9]. Initially, a classical Delaunay tessellation is performed to identify tetrahedral Delaunay cells, where the vertices of the tetrahedron are the particle centres. These Delaunay cells are merged together according to a connectivity criterion based on the inscribed sphere associated with each Delaunay cell. If the inscribed spheres of two adjacent Delaunay cells overlap, then the Delaunay cells are merged together (Fig. 4). This results in pores that are described as polyhedral cells, providing a physically intuitive delineation of the pore space that better captures local density fluctuations. This approach has been shown to describe water retention characteristics in monodisperse assemblies [10].

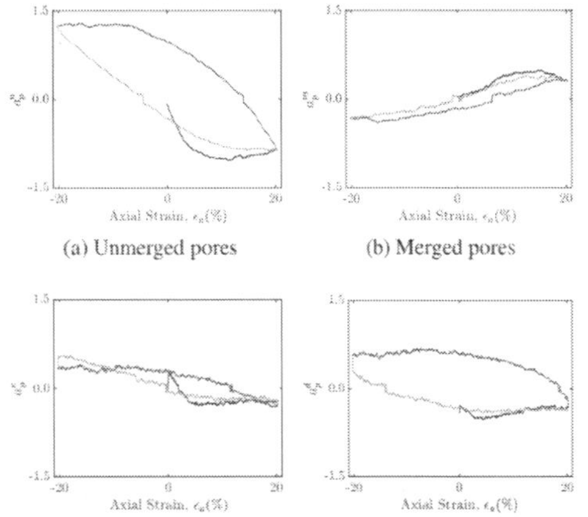

(a) Unmerged pores (b) Merged pores

(c) Contracting pores (d) Dilating pores

Figure 5: Evolution of anisotropy in pore orientation for the CP-D simulation.

In order to quantify anisotropy of pores, the orientation of individual pores must be calculated. This is achieved using the pore orientation tensor [9]:

$$P_{ij} = \frac{1}{N_s} \sum_{\alpha=1}^{N_s} a^\alpha t_i^\alpha t_j^\alpha \tag{5}$$

where $\mathbf{t}^\alpha$ and a^α are the unit normal vector and surface area of the triangular faces forming the polyhedral pore (Fig. 4) and N_s is the number of faces. The major principal orientation can be determined by an eigen-decomposition. If $\mathbf{p}$ is the unit vector along the major principal direction, then the probability density function of pores with an orientation $\mathbf{p}$ is given by:

$$E^k(\mathbf{p}) = \frac{1}{4\pi}\left(1 + C_{ij}^k p_i p_j\right) \tag{6}$$

This is analogous to the expression in Equation 1, and in a similar manner to Section 3:

$$C_{ij}^k = \frac{15}{2}\left(M_{ij}^k - \frac{M_{kk}^k}{3}\delta_{ij}\right) \tag{7}$$

where

$$M_{ij}^k = \frac{1}{N_p^k} \sum_{\alpha=1}^{N_p^k} p_i^\alpha p_j^\alpha \tag{8}$$

is the moment tensor for the pore space. N_p^k is the number of pores in the kth subset of pores. As per the contact subnetworks considered in Section 3, the anisotropy associated with distinct subsets of pores will be determined. Two dual-partitions are considered: (i) merged and unmerged pores, denoted $\{m, u\}$ respectively, and (ii) contracting and dilating pores, denoted $\{c, d\}$.

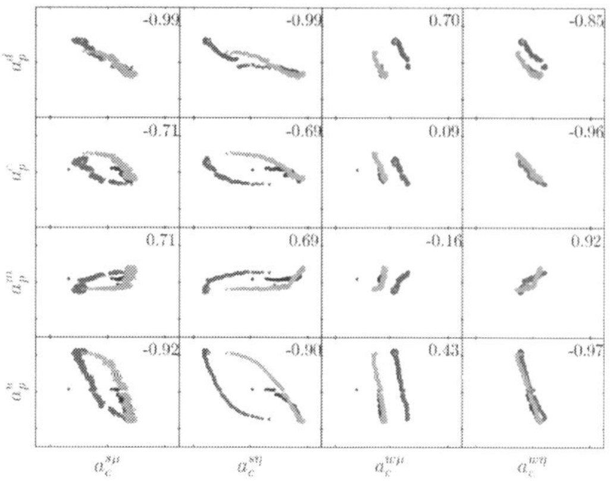

Figure 6: Correlation between anisotropy parameters for contacts and pores. The Pearson correlation coefficient is also presented.

Merged and unmerged pores are delineated through the modified Delaunay tessellation mentioned above. Unmerged pores have tetrahedral geometry, while merged pores are polyhedral (Fig. 4). This partition also reflects the contribution of pore volume, as unmerged pores have relatively small pore volume (and vice versa).

Contracting and dilating pores are delineated by considering the incremental change in pore volume. This is calculated by performing the classical Delaunay tessellation at time step $t + \Delta t$, and applying this tessellation to the previous time step t. If the incremental time step Δt is small, then the error in applying the same Delaunay tessellation to two time steps is minimal. The incremental volume change for a given pore is simply the summation of volume change associated with the corresponding Delaunay cells.

A scalar measure of anisotropy in pore orientation is defined by:

$$a_p^k = \sqrt{\frac{3}{2} C_{ij}^k C_{ij}^k} \tag{9}$$

The evolution of a_p^k for both dual-partitions of the pore space is shown in Figure 5 for the CP-D simulation. Unmerged pores have a preferential orientation perpendicular to the loading direction (Fig. 5a), while merged pores have a slight degree of anisotropy in the direction of loading (Fig. 5b). This observation can be explained by considering the particle contacts along the edges of the polyhedral pore cell. Merged pores (which have large pore volumes) can only be sustained by particle contacts with substantial interparticle normal forces. It has been observed that the average interparticle normal force increases with pore volume. Section 3 demonstrated that contacts with relatively large normal force magnitudes tend to align parallel with the loading direction and hence merged pores will align in this direction. Further, Figure 5c shows that contracting pores are approximately isotropically orientated, while dilating pores tend to align perpendicular to the loading direction (Fig. 5d). This agrees with the intuitive notion of lateral expansion under compression in a drained triaxial test.

These observations suggests that anisotropy of pores and contacts are somehow related. This is confirmed in Figure 6, which shows the correlation between anisotropy parameters of contacts and pores, along with their respective Pearson correlation coefficient. Certain parameters (for example, $a_c^{w\eta}$ vs. a_p^u) exhibit a strong linear correlation throughout the entire two-way cycle. Other sets of parameters have strong correlation for the individual loading, unloading and reloading branches (for example, $a_c^{w\mu}$ vs. a_p^u). Figure 6 suggests that links can be established between the contact and pore space networks. The anisotropy of unmerged pores, a_p^u, appears to be highly correlated with the anisotropy of weak contacts ($a_c^{w\mu}$ and $a_c^{w\eta}$). This suggests that unmerged pores predominantly comprise weak contacts, which agrees with the above discussion. Of particular interest is the observations that the anisotropy of dilating pores is correlated with strong contacts, while the anisotropy of contracting pores is somewhat correlated with weak contacts. This provides a connection between deformation characteristics (which may be controlled by the behaviour of pores) and stress transmission (which is governed by contacts). This presents an interesting angle to approach the constitutive behaviour of granular materials and is the subject of ongoing research.

5 Conclusion

This study quantifies the anisotropy of contacts and pores in a numerical simulation of a drained triaxial test. The anisotropy of subsets of contacts and pores are considered. This revealed that the stress ratio can be described uniquely as a function of geometric anisotropy of strong and non-sliding contacts and that the anisotropy of pores can be related to the anisotropy of contacts.

References

[1] Oda, M., Nemat-Nasser, S. & Konishi, J., Soils Found. **25**, 85 (1985)

[2] Wan, R. & Guo, P., Can. Geotech. J. **38**, 16 (2001).

[3] Li, X.S. & Dafalias, Y., J. Eng. Mech. **138**, 263 (2012)

[4] Kloss, C., Goniva, C., Hager, A., Amberger, S. & Pirker, S., Prog. Comput. Fluid Dy. **12**, 140 (2012)

[5] Sufian, A., Russell, A.R. & Whittle, A.J., Geotechnique (accepted, 2017)

[6] Satake, M., Deformation and Failure of Granular Materials (Balkema, Rotterdam, 1982) 63

[7] Kanatani, K.I., Int. J. Eng. Sci. **22**, 149 (1984)

[8] Peters, J.F., Muthuswamy, M., Wibowo, J., & Tordesillas, A., Phys. Rev. E **72**, 041307 (2005)

[9] Sufian, A., Russell, A.R., Whittle, A.J., & Saadatfar, M., Granul. Matter **17**, 727 (2015)

[10] Ferraro, A., Sufian, A. & Russell, A.R., Acta Geotech. (submitted, 2016)

Dry granular flows – rheological measurements of the $\mu(I)$ – Rheology

Abdoulaye Fall[1*], *Michel* Badetti[1], *Guillaume* Ovarlez[2], *François* Chevoir[1] and *Jean-Noël* Roux[1]

[1]Laboratoire NAVIER, UMR 8205 CNRS-ENPC-IFSTTAR, 2 Allée Kepler 77420 Champs sur Marne, France
[2]CNRS, LOF, UMR 5258, 33600 Pessac, France

Abstract: Granular materials do not always flow homogeneously like fluids when submitted to external stress, but often form rigid regions that are separated by narrow shear bands where the material yields and flows. This shear localization impacts their apparent rheology, which makes it difficult to infer a constitutive behavior from conventional rheometric measurements. Moreover, they present a dilatant behavior, which makes their study in classical fixed-volume geometries difficult. These features led numerous groups to perform extensive studies with inclined plane flows, which were of crucial importance for the development and the validation of the μ(*I*)-rheology. Our aim is to develop a method to characterize granular materials with rheometrical tools. Using rheometry measurements in an annular shear cell, dense granular flows of 0.5mm spherical and monodisperse beads are studied. A focus is placed on the comparison between the present results and the μ(*I*)-rheology.

1 Introduction

Granular matter shows both solid and fluid behavior [1]. These materials are very sensitive to various parameters: geometry of the flow, wall roughness, flow rate, shape and size distribution of the grains, and coupling with the interstitial fluid [2]. In the dry case, the rheology is solely governed by momentum transfer and energy dissipation occurring in direct contacts between grains and with the walls. Despite the seeming simplicity of the system, the behavior of dry granular material is very rich and extends from solid to gaseous properties depending on the flow regime. In the absence of a unified framework, granular flows are generally divided into three different regimes. (i) At low shear, particles stay in contact and interact frictionally with their neighbours over long periods of time. This "quasistatic" regime of granular flow has been classically studied using modified plasticity models based on a Coulomb friction criterion [3]. The response in terms of velocity or solid fraction profiles is independent of the shear rate. (ii) Upon increasing the deformation rate, a viscouslike [4] regime occurs and the material flows more as a liquid. In this intermediate regime, the particles experience multicontact interactions. (iii) At very high velocity, a transition occurs toward a gaseous regime, in which the particles interact through binary collisions [5]. For the modeling of dense granular flows, the concept of inertial number I has been widely used and investigated with regard to its relationship with dynamic parameters, such as velocity, stress, and friction coefficient, which leads to constitutive relations for granular flows. Thus, "dynamic dilatancy" law and "friction" law were deduced from discrete simulation of two dimensional simple shear of a granular material without gravity [4]. It was observed that both dimensionless quantities: the internal friction coefficient $\mu = \tau / \sigma_N$ and the solid fraction ϕ are functions of I [4,6]. Following general results from simulations of planar shear [4] and successful applications to inclined plane flows [7], the experiments of Jop et al. [7] were carried out to quantify, for glass beads, the $\mu(I)$ – rheology from the quasistatic to the rapid flow regime, corresponding to moderate I (from 0.01 to 0.5) as $\mu = \mu_s + (\mu_2 - \mu_1)/(1 + I_0 / I)$ in which μ_s, μ_2, and I_0 are three fitting parameters dependent on material properties. However, the asymptotic value of μ at high I was not obtained by da Cruz et al. [4] who observed an approximately linear increase of the internal friction coefficient from the static internal friction value. In 3D-simulation studies, Hatano did not either observe the asymptotic value of μ at high I: from $I = 10^{-4}$ to $I = 0.2$, he reported a law in which the friction coefficient increases as a power of I $\mu = \mu_s + \alpha I^n$ [8].

In the present work (see Fall et al. for more details [9]), we show that it is not necessary to develop specific setups (such as the inclined plane) to study dense granular flows. Indeed, we show that a simple annular shear cell can be adapted to a standard rheometer to study the rheology of granular materials under controlled confining pressure. It allows us, in particular, to obtain

* Corresponding author: abdoulaye.fall@ifsttar.fr

the dilatancy law $\phi(I)$ and also to study very accurately the quasistatic limit. Thus, from the steady state measurements of the torque and the gap during an imposed shear flow under an applied normal confining stress σ_N, we report two laws in which the internal friction coefficient and the solid fraction are functions of I. An effort is then made to compare the present results with the $\mu(I)$ – rheology described in the literature.

2 Materials and methods

To investigate the steady flows of dry granular materials and determine the $\mu(I)$ – rheology, three main features are required: (i) to avoid shear banding, (ii) to apply a confining stress in the velocity gradient direction, and (iii) to allow volume fraction variations. If one wants to use a rheometer, (i) implies that the use of a Couette cell should be avoided since it is characterized by a shear stress inhomogeneity that naturally leads to shear banding. Both the cone-and-plate and the parallel-plate geometries allow a normal force to be imposed in the velocity gradient direction; however, the analysis of the cone-and-plate flow can be performed only at a single gap value, i.e., it cannot be used to characterize a material whose the volume varies under shear. On the other hand, any gap variation in parallel-plate geometry can be accounted for in the determination of the shear rate value. Moreover, shear banding should in principle be avoided in this last geometry since the shear stress is independent of the vertical position in the gap. Nevertheless, the shear rate varies along the radial position and is equal to zero in the center. One should thus try to avoid the use of the central zone of the gap. Finally, the material should be confined by lateral walls to make sure that any gap variation actually leads to a material volume fraction variation. These requirements led us to develop a home-made annular shear cell, inspired by [10], in which pressure-imposed measurements can be performed. Annular shear cells have been extensively used to characterize the flow of pharmaceutical powders and dry granular materials [11,12]. We use a granular material made of rigid polystyrene beads (from Dynoseeds) of density $\rho_p = 1050$ kg/m^3, of diameter $d_p = 0.5$ mm (with a standard deviation of 5%). Spherical beads fill the annular box between two static concentric cylinders with, respectively, an inner and outer radii of R_i=21 mm and R_o=45 mm. The width of the annular trough is about $48 d_p$ leading to a ratio of inner to outer wall radii of 0.46. We also used another annular channel with the same width but with a larger ratio of inner to outer cylinder radii equal to 0.61 (R_i=38 mm and R_o=62 mm). We have verified that changing the ratio of inner to outer cylinder radii of the annular shear cell does not significantly affect the results. The filling height (initial gap, h_0) of the annular box is adjustable from a few grain diameters (typically $5 d_p$) to $30 d_p$. The cylinders were finished as smoothly as possible to permit the granular material to slip there as readily as possible. For that, they are made of polyoxymethylene (POM) resin which exhibits a low friction coefficient due to the flexibility of the linear molecular chains. We have measured the friction coefficient at the wall μ_w between polystyrene beads and a plane made of POM by measuring, with our rheometer, the sliding stress under different normal stresses such as in Jenike's shear tester [13]; it is found to be very small: $\mu_w = 0.05$. The experiments were performed initially on a very dense piling $\phi_0 \approx 0.625$, close to the so called random close packing of 0.637 [14] obtained by combining a rain filling and tapping the box [15] to get a reasonably uniform packing. However, we show below that the steady state obtained when the material is sheared is the same for an initially looser piling. Granular beads are then driven by the ring-shaped upper boundary made of POM which is assembled on a *Kinexus Pro rheometer by Malvern*. To avoid wall slip, both the moving upper boundary and the static lower boundary are serrated, with 0.5mm ridges which correspond to the size of grains.

In our rheometer, instead of setting the value of the gap size for a given experiment, as in previous studies [16] and generally in rheometric measurements, we impose the normal force (i.e., the confining normal stress) and then, under shear, we let the gap size vary in order to maintain the desired value of the normal force. A typical measurement is shown in Fig. 1, where we start out with a given gap $h_0 \approx 6d_p$, impose a constant shear rate $\dot{\gamma}$ and normal force F_N and measure the torque T and the gap h as a function of strain (or time). In this case, the solid fraction ϕ is not fixed but adjusts to the imposed shear. However, it remains important to notice that, in order to keep the imposed shear rate constant, the rheometer adjusts the rotation velocity Ω since the gap varies as will be discussed below. The system reaches a steady state after a certain amount of shear strain, but we have carefully compare the transient dynamics of these quantities, beginning with freshly poured grains for two preparations, "rain piling" and "rain coupled to tapping piling" (respectively) which allow us to get the looser piling and the denser piling (respectively) samples with an initial solid fraction of 0.612 and 0.625 (respectively) [9]. As a consequence, there is no need for the sample preparation for the grains we study, namely, monodisperse spheres (note that this might not be a general result). All the annular shear cell requires is that the sample is sheared at the required normal pressure until the critical state is reached. However, it is necessary to adopt a reference packing state and ensure that the piling method used allows us to achieve sufficient repeatability for a given set materials and system parameters. Hence, we prepare all samples in the same way: a very dense material, from a rain piling coupled to tapping, is used in all the experiments presented below (the initial average volume fraction is 0.625).

DOI: 10.1051/epjconf/201714003005

3 Experimental results and analyses

We have measured the driving torque and the gap as a function of shear strain for various imposed normal force F_N (between 1 and 5N) and applied shear rate $\dot{\gamma}$ (between 0.01 and 77 s^{-1}) for a given gap $h_0 \approx 6d_p$. In Figs. 1(a-b), we show those measurements for F_N=3N and various $\dot{\gamma}$. At low shear rate, the driving torque increases slowly before reaching a steady plateau within strain of order of unity. Meanwhile, the gap fluctuates around its initial value. (We show the gap size rescaled by its initial value before shearing h/h_0 called the rescaled gap in the following.)

Upon increasing the imposed shear rate, an overshoot occurs: Its amplitude increases with increasing the shear rate. In steady state, a rate dependence of the torque is observed [Fig. 1(a)]. Moreover, increasing the imposed shear rate causes an increase of the gap size [Fig. 1(b)] allowing to quantify the dynamic dilatancy of the granular material. Notice that, with increasing the applied shear rate, large fluctuations of the driving torque and also of the gap size evolution occur in steady state flows. Similarly, in experiments in which different normal forces are imposed at a given shear rate, the steady torque is observed to increase while the steady solid fraction (steady gap) decreases when the normal force is increased. Similarly, in experiments in which different normal forces are imposed at a given shear rate [Figs. 1(c,d)], the steady torque is observed to increase while the steady solid fraction (steady gap) decreases when the normal force is increased.

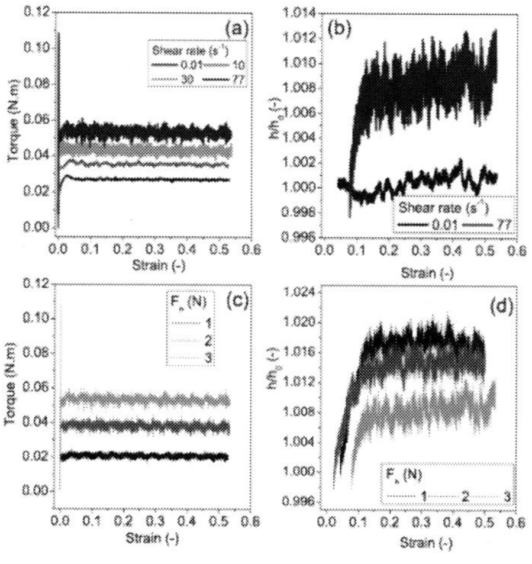

Fig. 1: Evolution as a function of the strain at 3N imposed normal force under different applied shear rates of: (a) The driving torque and (b) the rescaled gap size (only two curves are shown for clarity). Evolution as a function of the strain at 77 s-1 imposed shear rate under different applied normal forces of: (c) The driving torque and (d) the gap size rescaled by its initial value before shearing.

Once the above described experiments are combined, we can obtain the constitutive laws of the dry granular material, i.e., the dependence of the steady solid fraction ϕ and the ratio between shear and normal stresses $\mu = \tau / \sigma_N$ variation on shear rate. Indeed, from macroscopic quantities T, Ω, F_N, and h, the shear stress τ, the normal stress σ_N, shear rate $\dot{\gamma}$, and the solid fraction ϕ can be computed.

With this analysis, one can plot the shear stress in the steady state as a function of the normal stress as shown in Fig. 2(a). The first observation is that a linear relationship between the shear and normal stresses is seen for all imposed shear rates, with a slope that increases from 0.265 to 0.6 with increasing the shear rate. If an internal friction coefficient μ is defined as the ratio between shear and normal stresses, we evidence here that l is rate dependent: It increases with $\dot{\gamma}$.

The second observation is that, for all imposed shear rates, the steady value of the solid fraction decreases when one decreases the normal stress [Fig. 2(b)]. Indeed, since the grains cannot escape from the cell, one can measure unambiguously the solid fraction from the gap variation as $\phi = m / \pi\rho h(R_0^2 - R_i^2)$ in which m represents the mass of grains. Thus, $h_0 / h = \phi / \phi_0$ will simply reflect the impact of shear and confinement on dilatancy.

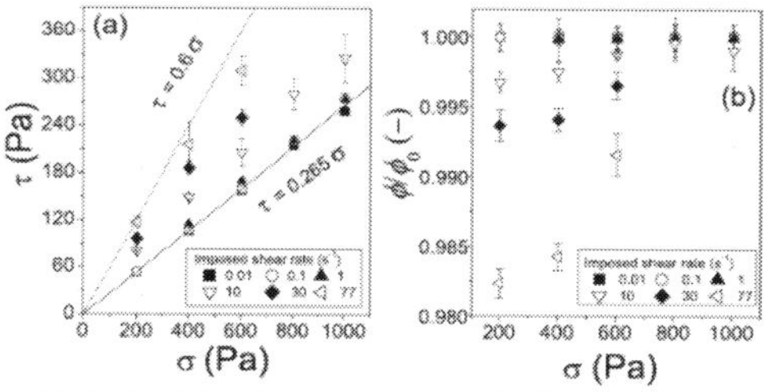

Fig. 2: Plot of the shear stress (a) and of the solid fraction rescaled by its initial value before shearing (b) as a function of the normal stress. The stress and the gap are measured in the steady state, for different imposed shear rate. The error bars come from three experiment runs.

In order to analyze the results in term of $\mu(I)$ – rheology, one has to define an inertial number such as $I = \dot{\gamma} d_p / \sqrt{\sigma_N / \rho_p}$. It varies between $10^{-7}$and 0.1 in the range of applied normal force and shear rate. This corresponds to the usual range of quasistatic to dense flow regimes. It should be noted that our annular shear geometry does not allow higher values of *I*.

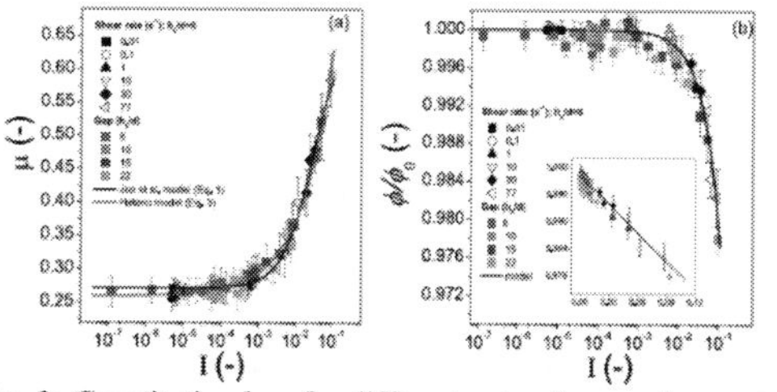

Fig. 3: Constitutive law for different sets of mean shear rates and imposed normal forces (a) friction law, i.e., effective internal friction coefficient as a function of inertial number; (b) dynamic dilatancy law, i.e., solid fraction as a function of

inertial number. The error bars come from three experiment runs.

Figure 3(a) shows how μ vary throughout the flow regimes since I characterizes the local rapidity of the flow. All the data obtained for different sets of shear rate and normal force collapse on a single curve τ / μ *vs.* I. For low inertial number, the internal friction coefficient tends to a finite value $\mu \approx$ 0.265 and increases with increasing I.

At the same time, the solid fraction variation ϕ / ϕ_0 with the inertial number I is shown in Fig. 3(b). Once again, all the data collapse on a single curve. At low I, ϕ / ϕ_0 is quasi-constant: this is the quasistatic regime. When I increases, the inertia starts influencing the flow and the system becomes rate dependent: the ratio ϕ / ϕ_0 decreases; this regime corresponds to the dense flow in which the granular material dilates. Moreover, in order to check the robustness of our results, we have varied the initial size of the gap. Here, with our annular shear geometry, the same experiments discussed above are made with different gap sizes from $6d_p$ to $22d_p$ and different sets of imposed shear rate and normal force. It is shown in Fig. 3 that changing the gap does not significantly affect these results. This suggests a total absence of shear localization at high inertial number. However, at small inertial number, the resolution of our measurements is not sufficient to dismiss the possibility that shear localization arises. It was indeed shown that in confined annular flow at small shear velocities and high confining pressures—small I—the shear may be not homogeneous and solid and fluid phases coexist [17].

4 Conclusion

From the steady state measurement of the torque and the gap, the internal friction coefficient, the solid fraction, and the inertial number I are measured. For low I, the flow goes to the quasistatic limit and the internal friction coefficient and the solid fraction profiles are independent of I. Upon increasing I, dilation occurs and the solid fraction decreases linearly when I increases while the friction coefficient increases. Comparing now these experimental results with existing models such as those of Jop et al. [7] and Hatano [8] on dry granular flows in terms of quasistatic and dense flow behaviors, our experimental data include points in range of I from 10^{-7} to 0.1 which covers quasistatic and dense flow regimes. Our measurements then show that, in this range of I, both models can describe our data. As a consequence, we bring evidence that rheometric measurements can be relevant to describe dry granular flows. However, additional experimental work should be carried out in order to measure the dependence of the boundary layer constitutive law on the state of the bulk material, so as to be able to describe properly the rheology when approaching the quasistatic limit.

References

1. J. Duran and R. P. Behringer, *Sands, Powders, and Grains: An Introduction to the Physics of Granular Materials* (2001).
2. B. Andreotti, Y. Forterre and O. Pouliquen, *Granular Media: Between Fluid and Solid* (2013).
3. A. Schofield and P. Wroth, Soil Use Manag. **25**, 128 (1968).
4. F. Da Cruz, S. Emam, M. Prochnow, J. N. Roux, and F. Chevoir, Phys. Rev. E - Stat. Nonlinear, Soft Matter Phys. **72**, (2005).
5. I. Goldhirsch, Annu. Rev. Fluid Mech. **35**, 267 (2003).
6. F. Chevoir, J. N. Roux, F. da Cruz, P. G. Rognon, and G. Koval, Powder Technol. **190**, 264 (2009).
7. P. Jop, Y. Forterre, and O. Pouliquen, Nature **441**, 727 (2006).
8. T. Hatano, Phys. Rev. E - Stat. Nonlinear, Soft Matter Phys. **75**, 1 (2007).
9. A. Fall, G. Ovarlez, D. Hautemayou, C. Mézière, J.-N. Roux, and F. Chevoir, J. Rheol. (N. Y. N. Y). **59**, 1065 (2015).
10. F. Boyer, É. Guazzelli, and O. Pouliquen, Phys. Rev. Lett. **107**, 1 (2011).
11. J. F. Carr and D. M. Walker, Powder Technol. **1**, 369 (1968).
12. S. B. Savage and M. Sayed, J. Fluid Mech. **53142**, 391 (1984).
13. J. Schwedes, Granul. Matter **5**, 1 (2003).
14. S. Torquato, T. M. Truskett, and P. G. Debenedetti, Phys. Rev. Lett. **84**, 2064 (2000).
15. P. Philippe and D. Bideau, Europhys. Lett. **60**, 677 (2002).
16. D. Schulze, Aufbereitungs-Technik **35**, 524 (1994).
17. Aharonov, E., and D. Sparks, Phys. Rev. E 65, 051302 (2002).

Discontinuous Shear Thickening in Cornstarch Suspensions

Abdoulaye Fall[1,*], *Anaël Lemaître*[1], *and Guillaume Ovarlez*[2]

[1]Laboratoire NAVIER, UMR 8205 CNRS-ENPC-IFSTTAR, 2 Allée Kepler, 77420 Champs sur Marne, France
[2]CNRS, LOF, UMR 5258, 33600 Pessac, France

Abstract. We study the emergence of discontinuous shear-thickening (DST) in cornstarch, the well know system for this phenomenon, by combining macroscopic rheometry with local Magnetic Resonance Imaging (MRI) measurements. We bring evidence that macroscopic DST is characterized in wide gap with a shear localization, part of the material close to the inner cylinder is flowing and the rest is not. The flow seperates into a low-density flowing and a high-density jammed region. Moreover, the local rheology of the flowing region does not directly reflect DST but, strikingly, is most often shear-thinning. Our data are not consistent with recent theoretical suggestions based on the presumed existence of s-shaped flow curves. Instead, they support that DST should be attributed to the existence of a shear jamming limit at volume fractions quite significantly below random close packing.

1 Introduction

Granular suspensions are seemingly simple systems composed of non-colloidal rigid particles in Newtonian fluids, where the only relevant interactions are hydrodynamic and contact forces. Yet, they present a rich rheology [1], which includes shear-thinning or shear thickening, normal stress differences, shear banding and yield stress behaviour. With increasing packing fractions, continuous shear thickening (CST) appears at lower and lower strain rates and becomes more and more abrupt. This eventually leads to the emergence of a spectacular phenomenon: discontinuous shear thickening (DST), an order-of-magnitude jump of the stationary macroscopic stress when strain rate crosses some threshold value. DST is observed in a narrow range of packing fractions near random close packing. Recent works thus suggest that it is a dynamic or shear jamming transition, which motivates a considerable attention both from experiments [2], numerical simulation [3–5] and theory [6]. Shear thickening is also observed in dense colloidal suspension, where it has been related to the formation of dense clusters of particles [7–9]. In these systems, the competition between shear induced cluster formation and Brownian motion that homogenizes the suspension naturally determine a critical shear rate for the onset of shear thickening.

The origin of shear thickening is essentially unknown. Proposed mechanisms include: the crossover to Bagnold scaling due to particle inertia [10]; the coupling between normal and shear stresses associated with material dilatancy [11,12]. In numerical simulations DST was observed in systems which are forced to remain homogeneous by periodic boundary conditions [5]. It was thus suggested that it might arise from a jump of the local response, which occurs at fixed packing fraction, and is caused by the proliferation of frictional contacts [5,6].

Examples abound, however, in granular suspension rheology where the macroscopic response is sharply different from the local behaviour: the apparent yield stress [13,14] or a transient DST behavior [10], for example, were both shown to result from the emergence of flow inhomogeneities due respectively to density differences and migration. The observation of finite size effects in DST measurements [2,11], at sizes up to 100 particles, questions the idea that DST is a purely local phenomenon. In this context, it is unclear how a phase diagram for the local response can be deduced from experimental data and compared with numerical [5], and theoretical [6] works.

This situation motivates us to study DST in cornstarch [15], the well-known system for this phenomenon, while accessing experimentally the local response. To this aim, we have designed a new velocity controlled rheometer which can support torques up to 30 N.m in a wide-gap Couette cell inserted in our Magnetic Resonance Imaging (MRI) scanner. It allows us to access local velocity and particle volume fraction profiles in the flowing sample [13], for a wide range of strain rates, which covers CST and DST. We thus bring evidence that macroscopic DST is characterized in wide gap with a shear localization, part of the material close to the inner cylinder is flowing and the rest is not. The flow separates into a low-density flowing and a high-density jammed region. Moreover, the local rheology of the flowing region does not directly reflect DST but, strikingly, is most often shear-thinning. Our data are not consistent with recent theoretical suggestions based on the

* Corresponding author: abdoulaye.fallifsttar.fr

presumed existence of s-shaped flow curves. Instead, they support that DST should be attributed to the existence of a shear jamming limit at volume fractions quite significantly below random close packing.

2 Materials

The cornstarch (Sigma) is the same as in [15]: it presents slightly polydisperse particles of size ~ 20μm with irregular shapes (Fig. 1). We have measured the random close packing fraction $\phi_{RCP} \approx 55\%$ by tapping the powder for long time in a container [16]. We have also measured the random loose packing $\phi_{RLP} \approx 35\%$ as the density achieved after sedimentation of a dilute suspension [17]. The suspensions are prepared by mixing cornstarch in demineralized water. The initial volume fractions range from 33.5% to 43.9%.

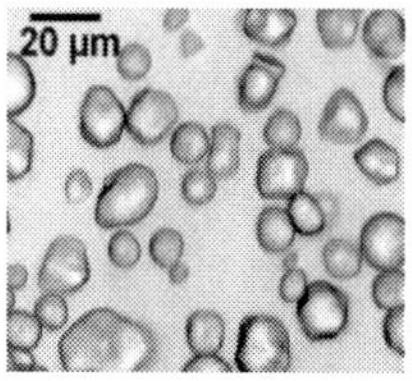

Fig.1: Micrograph of the cornstarch particles.

3 Results

Let us start with the macroscopic rheological behavior. Experiments are carried out with a Couette geometry on a *Malvern Kinexus Pro* stress-controlled rheometer. In order to avoid confinement effects, a wide gap is used [11]. The inner and outer cylinders are respectively of 25 mm and 36 mm diameter; both cylinders are roughened. The inner cylinder height is 37.5 mm. Fig. 2a shows the evolution of torque measured on the inner cylinder during a 3 min logarithmic rotation rate Ω ramp from 5.10^{-3} rad/s to 200 rad/s. near the inner cylinder, the local shear stress is $\tau(R_i;\Omega)=T(\Omega)/(2\pi R_i^2 h)$ and the local strain rate is a priori unknown, but classically estimated as $\dot\gamma(R_i;\Omega)=2\Omega R_o^2/(R_o^2-R_i^2)$. Combining the stress and the strain rate equations yields the apparent viscosity $\eta \equiv \tau/\dot\gamma$ versus $\dot\gamma$ relation reported on Fig. 2b. Note that at low viscosities, macroscopic inertial effects may arise (Taylor Couette instabilities), which limit the accessible strain rate range at low volume fractions.

At low $\dot\gamma$, the response is clearly shear-thinning: the viscosity decreases with the shear rate with the exception of our highest packing fraction data, it crosses over to CST around a characteristic strain rate $\dot\gamma_{CST}$ defined at that where the viscosity reaches a minimum. Viscosity jumps at some higher strain rate $\dot\gamma_{DST}$ which characterizes the onset of DST. This behavior is generic: cornstarch suspensions from 38% to 44% volume fractions show a DST.

Since the classical estimate of $\dot\gamma(R_i)$, which is based on the Newtonian solution, remains unsatisfactory; an exact expression exists so long as the material is homogeneous in the studied torque range:

$$\dot\gamma(R_i;T_o)=2\sum_{n=0}^{\infty}\left(T\frac{d\Omega}{dT}\right)\Bigg|_{T=T_o\left(\frac{R_i}{R_o}\right)^{2n}} \qquad \text{(Eq. 1)}$$

It cannot be used around the DST transition, however, because the apparent singularity of $d\Omega/dT$ cannot be resolved experimentally. Nevertheless we have checked that using the first two terms of (Eq. 1) gives the same qualitative behaviour as in Fig. 1b up to DST. We will show below that the flow does remain homogeneous below DST, which allows us to use (Eq. 1) to compute accurately $\dot\gamma_{CST}$ and $\dot\gamma_{DST}$ (Fig. 2c). Interestingly, (i) at any volume fraction, CST is always observed before DST; (ii) both $\dot\gamma_{CST}$ and $\dot\gamma_{DST}$ seem to vanish around the same volume fraction ϕ_c which we estimate to be ≈ 45% ≪ ϕ_{RCP} by linear extrapolation.

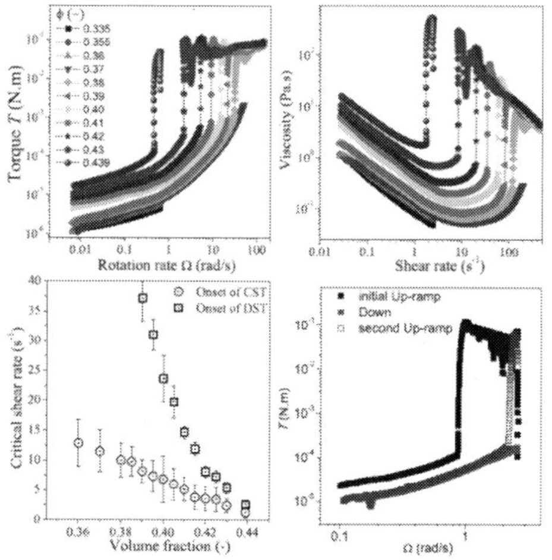

Fig.2: Macroscopic rheometry data. (a) Torque T vs rotation rate Ω during a logarithmic ramp at various packing fractions ϕ. (b) Apparent viscosity vs apparent shear rate extracted from (a). (c) Critical shear rates $\dot\gamma_{CST}$ and $\dot\gamma_{DST}$ vs ϕ. (d) Reversibility test at ϕ = 43.9%: succession of up, down, and up ramps.

To investigate further this thickening behavior, we look at the steady-state behavior when the shear-rate is increased up to a value $\dot\gamma > \dot\gamma_{DST}$ and subsequently go down in rotation rate. We find that the system has completely changed: we descend on a completely different branch that is significantly below the original one. If we increase the shear (= rotation) rate again, we remain on this second branch; we can go up and down as long as we want but the system does not change anymore: there was an irreversible change in the system when we were on the first branch (Fig. 1d).

In order to see what happens, we measure velocity and concentration profiles with the help of the velocity-controlled MRI rheometry. Our Couette cell has inner and outer radii R_i = 3cm and R_o = 5cm (resp.) and inner cylinder height h = 11cm. Both cylinders are roughened to avoid slip, which we checked from velocity profiles. All experiments discussed below are performed by preparing a homogeneous material with mean volume fraction ϕ_0 = 43.9% (experiments at 40%, 41% and 42.5% show similar features), and imposing a fixed inner cylinder rotation rate Ω with values ranging between 5 and 100 rpm, until steady state is reached.

Our MRI [13,18] provides the stationary packing fraction $\phi(r;\Omega)$ and azimuthal velocity $v(r;\Omega)$ at any radial position r. From the latter we extract the local strain rate $\dot{\gamma}(r;\Omega) = v/r - \partial v/\partial r$. We do not have access to torque measurement. But, since the local stress is $\tau(r;\Omega) = \tau(R_i;\Omega)R_i^2/r^2$ in the Couette geometry, we can estimate the local viscosity profile as:

$$\eta(r;\Omega) = \frac{\tau(R_i;\Omega)}{\dot{\gamma}(r;\Omega)}\frac{R_i^2}{r^2} \qquad \text{(Eq. 2)}$$

up to the unknow prefactor $\tau(R_i;\Omega)$.

In Fig 3, we plot the steady velocity and concentration profiles thus measured for a few Ωs. Velocity is normalized by its value at the inner cylinder.

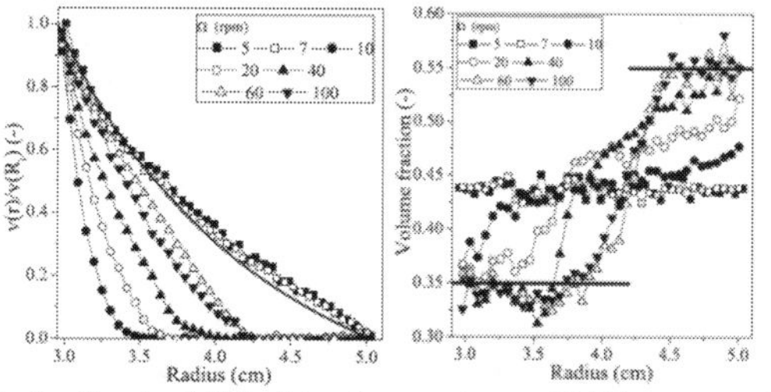

Fig.3: Steady MRI data for a ϕ_0 = 43.9% cornstarch suspension and different rotational velocities Ω. (a) Velocity profiles. (b) Density profiles; solid lines indicate ϕ_{RLP} and ϕ_{RCP}.

The results show that the system is well behaved and the particle concentration remains homogeneous when the imposed rotation rate is below a critical value around 10 rpm. However as soon as we go beyond this critical rotation rate, instead of decreasing continuously over the entire gap, the velocity profile goes to zero at some finite distance from the inner cylinder: there is shear localization. Upon increasing the rotation rate, we observe that the gap cannot be fully sheared: there still is a 8-25 mm region near the outer cylinder where the material is now not sheared, and for the highest rotation speeds, the sheared region does not occupy the whole gap of the Couette cell: the thickness of the sheared layer does not increase anymore whatever the imposed rotation rate and the concentration profile is essentially Ω-independent in steady state.

This shear localization is associated with the sudden emergence of density inhomogeneities. Namely, the volume fraction decreases in the flowing layer, while it increases in the jammed region, as required by the conservation of particle number. Indeed, the concentration profile measurements (Fig. 3b) show that this happens because a strong gradient in particle concentration develops: the system shows particle migration. The result of the migration is that the particle concentration becomes smaller near the inner cylinder and larger near the outer cylinder. The latter observation also explains the velocity profile: a region develops that is so concentrated in particles that it is jammed. Interestingly to note that the concentration of this region corresponds to ϕ_{RCP}. While the lower part of the concentration profile corresponds to ϕ_{RLP}. It is noteworthy that the density profile can achieve multiple forms depending on shear history.

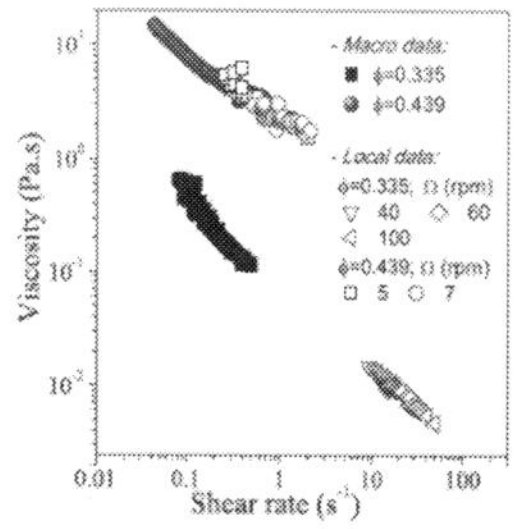

Fig.4: Comparison of local rheometry data obtained from MRI measurements (open symbols) and near the inner cylinder in macroscopic rheometry (filled symbols) in homogeneous conditions. Upper data: ϕ = 43.9%; lower data: $\phi \approx$ 33.5%.

We now examine how the local rheology can be accessed within the shear-thickened flow. Three of our datasets, for Ω = 40, 60, and 100rpm, present a finite width region of roughly uniform packing fraction $\phi_{\min} \simeq$ 33%. Using only the data collected in these regions, we can hence implement the same approach as in the fully homogeneous case, to access the local rheology $\eta(\dot{\gamma};\phi)$ at $\phi = \phi_{\min}$. For each Ω, we match the prefactor $\tau(R_i;\Omega)$ and plot in Fig. 4 the resulting $\eta(\dot{\gamma};\phi)$ vs $\dot{\gamma}(r;\Omega)$ data points. Clearly, the local $\eta(\dot{\gamma};\phi_{\min})$ rheology is shear-thinning, and extrapolates the $\phi = 33.5\%$ rheometry measurements. These data demonstrate that macroscopic DST cannot be interpreted as a reflecting DST in the local response at a fixed volume fraction. Instead, it results from the separation of the system into a low density flowing and a high density jammed regions. Somewhat paradoxically, the local response was found above to be shear-thinning in the flowing layer of the macroscopic shear-thickened state. In the flowing layer, the density plateau only develops around $\phi_{\min}$. It is hence at this packing fraction only that we can extract the local rheology $\eta(\dot{\gamma};\phi)$ from MRI data. To further qualify the local response, we now construct a phase diagram for all the local flow and jammed states. To this end, we collect the operating points ($\dot{\gamma}(r;\Omega), \phi(r;\Omega)$) obtained with the MRI for all Ω's and all r's and report them on Fig. 5, where they are compared with the curves $\dot{\gamma}_{CST}(\phi)$ and $\dot{\gamma}_{DST}(\phi)$ that delimit the flow regimes previously identified using macroscopic rheometry. Our MRI data include points obtained in the flowing layer as well as in jammed regions: in the scatter plot, they are respectively found on the left (low ϕ, with non-vanishing $\dot{\gamma}$) and right (high ϕ, $\dot{\gamma} \simeq 0$) hand sides.

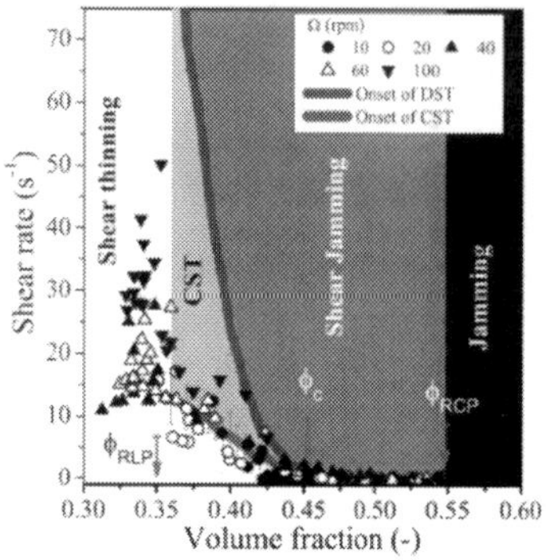

Fig.5: Phase diagram: scatter plot of the local $(\dot{\gamma}(r;\Omega), \phi(r;\Omega))$ values compared with the $\dot{\gamma}_{CST}(\phi)$ and $\dot{\gamma}_{DST}(\phi)$ lines.

All the local flow points clearly lie below the $\dot{\gamma}_{DST}(\phi)$ line, which thus appears to be an intrinsic material limit beyond which no steady flow can be sustained. No flow is found in particular at packing fractions above the value $\phi_c \simeq 45\%$ where $\dot{\gamma}_{CST}(\phi)$ and $\dot{\gamma}_{DST}(\phi)$ vanish. $\phi_c \ll \phi_{RCP} \simeq 55\%$ is quite distinct from the jamming point, and a continuum of jammed states is thus accessed beween ϕ_c and ϕ_{RCP}.

4 Conclusion

In the literature, macroscopic DST was proposed to result from the coexistence, in a range of packing fractions, of two flow curves for the local response. This bi-stability is justified by introducing an auxiliary state variable, such as the density of frictional contacts. Within this viewpoint, at low strain rate, the local response lies on a low-viscosity curve; DST occurs when the flow becomes unstable and jumps on high-viscosity curve. Macroscopic DST would thus result from an intrinsic DST behavior of the local response. Our data, however, supports a completely different scenario. Indeed, at the onset of macroscopic DST, the part of the flow that reaches locally a limit of stability, i.e. the $\dot{\gamma}_{DST}$ line, is the region near the inner cylinder, where the local shear rate is maximal. Within existing interpretations of DST based on s-shaped flow curves, it is this locally unstable region that should jump to a high viscosity state. But, in stark contrast, we find that when DST occurs, this region jumps toward a low-density, low viscosity shear-thinning state. Meanwhile jamming occurs in the region near the outer cylinder which, at the onset of DST, was well within its stability limit. DST thus occurs when a global reorganization of the flow takes place. It does not naively reflect local jumps of the local rheology.

Hence we do not find any necessity to account for a secondary flow curve associated with a change of a state variable such as a density of contacts. Instead, DST is clearly determined by two major factors: (i) the existence of a limit $\dot{\gamma}_{DST}(\phi)$ beyond which no steady flow can be accessed; (ii) the fact that as $\phi_c \ll \phi_{RCP}$, a continuum of jammed states exists. Both are direct consequences of a unique physical phenomenon: shear jamming. We thus understand that when the inner cylinder region reaches the shear-jamming limit, it can only escape the $\dot{\gamma}_{DST}(\phi)$ line by decreasing its local volume fraction, thus driving migration throughout the gap. This is possible because density can be increased significantly above ϕ_c in the growing jammed layer. Many questions, of course, remain especially about the microscopic mechanisms which lead to the emergence of a shear-jamming line at packing fraction quite separated from ϕ_{RCP}. It is quite plausible that the angularity of particles, which is significant for cornstarch, is a determining factor.

References

[1] M. M. Denn and J. F. Morris, Annu. Rev. Chem. Biomol. Eng. **5**, 203 (2014).
[2] E. Brown, H. Zhang, N. A. Forman, B. W. Maynor, D. E. Betts, J. M. DeSimone, and H. M. Jaeger, J. Rheol. (N. Y. N. Y). **54**, 1023 (2010).
[3] R. Seto, R. Mari, J. F. Morris, and M. M. Denn, Phys. Rev. Lett. **111**, (2013).
[4] C. Heussinger, Phys. Rev. E - Stat. Nonlinear, Soft Matter Phys. **88**, (2013).
[5] R. Mari, R. Seto, J. F. Morris, and M. M. Denn, J. Rheol. (N. Y. N. Y). **58**, (2014).
[6] M. Wyart and M. E. Cates, Phys. Rev. Lett. **112**, 1 (2014).
[7] J. Bender, J. Rheol. (N. Y. N. Y). **40**, 899 (1996).
[8] B. J. Maranzano and N. J. Wagner, J. Chem. Phys. **114**, 10514 (2001).
[9] N. J. Wagner and J. F. Brady, Phys. Today **62**, 27 (2009).
[10] A. Fall, A. Lemaître, F. Bertrand, D. Bonn, and G. Ovarlez, Phys. Rev. Lett. **105**, (2010).
[11] A. Fall, N. Huang, F. Bertrand, G. Ovarlez, and D. Bonn, Phys. Rev. Lett. **100**, (2008).
[12] A. Fall, F. Bertrand, G. Ovarlez, and D. Bonn, J. Rheol. (N. Y. N. Y). **56**, (2012).
[13] G. Ovarlez, F. Bertrand, and S. Rodts, J. Rheol. (N. Y. N. Y). **50**, 1 (2006).
[14] A. Fall, F. Bertrand, G. Ovarlez, and D. Bonn, Phys. Rev. Lett. **103**, (2009).
[15] F. S. Merkt, R. D. Deegan, D. I. Goldman, E. C. Rericha, and H. L. Swinney, Phys. Rev. Lett. **92**, 184501 (2004).
[16] P. Philippe and D. Bideau, Europhys. Lett. **60**, 677 (2002).
[17] M. P. Ciamarra and A. Coniglio, Phys. Rev. Lett. **101**, (2008).
[18] A. Fall, F. Bertrand, D. Hautemayou, C. Mezière, P. Moucheront, A. Lemaître, and G. Ovarlez, Phys. Rev. Lett. **114**, 1 (2015).

Influence of obstacles on bubbles rising in water-saturated sand

Raphaël Poryles[1], *Germán* Varas[2,*], and *Valérie* Vidal[1,**]

[1] *Univ. Lyon, ENS de Lyon, Univ. Claude Bernard, CNRS, Laboratoire de Physique, F-69342 Lyon, France*
[2] *Instituto de Física, Pontificia Universidad Católica de Valparaíso, Av. Universidad 330, Valparaíso, Chile*

Abstract. This work investigates the dynamics of air rising through a water-saturated sand confined in a Hele-Shaw cell in which a circular obstacle is trapped. The air is injected at constant flow rate through a single nozzle at the bottom center of the cell. Without obstacle, in a similar configuration, previous studies pointed out the existence of a fluidized zone generated by the central upward gas motion which entrains two granular convection rolls on its sides. Here, a circular obstacle which diameter is of the order of the central air channel width is trapped at the vertical of the injection nozzle. We analyze the influence of the obstacle location on the size of the fluidized zone and its impact on the morphology of the central air channel. Finally, we quantify the variations of the granular free surface. Two configurations with multiple obstacles are also considered.

1 Introduction

Gas rising in liquid-saturated sands is a phenomenon widely encountered in nature, with strong implications in natural or industrial hazard mitigation. Examples of soil liquefaction, methane venting areas [1], CO_2 sequestration [2] and air sparging for ground decontamination [3] are among the many which focus the attention of both fundamental and applied research since decades. Since the beginning of the 90's, the remediation of contaminated soils has motivated studies of the morphology of gas invasion in saturated granular media. In the particular case of a single gas injection point in highly packed beds, it has been shown that the gas rising through the immersed granular layer generates specific patterns. Cumulated in time, the gas pathways delimitate a zone of parabolic shape above the injection point [3–5]. For mobile sands (e.g. loose-packed sands), the gas and fluid recirculation generate a so-called fluidized zone, at the vertical of the injection point. Its shape is also parabolic and consists of a central air channel with two granular convection rolls on the side, while outside this region, the grains remain at rest [6–11]. Most of these studies have been performed with either model sands, consisting in spherical glass beads, or real sands sieved in order to get a peaked size distribution, and further assess the influence of the grain size on the process. However, soils are intrinsically heterogeneous, with the presence of layering or large heterogeneities.

Ground heterogeneities can have drastic consequences on the dynamics of the sand layer fluidization. Although the case of a stratified medium modeled with rectangular, horizontal obstacles has been investigated qualitatively for highly packed beds [4], the influence of heterogeneities trapped into mobile sediments has received, to our knowledge, few or no attention. In this work, we propose to investigate experimentally the air-induced fluidization of a mobile, water-saturated granular layer with a single or multiple trapped obstacle(s). The quasi-2D setup makes it possible to visualize the fluidized zone and quantify the influence of obstacles on its morphology.

2 Experimental setup

The experimental setup is a vertical Hele-Shaw cell (glass plates 40 × 30 cm, gap e = 2mm) filled with dense particles immersed in water. The particles are polydisperse spherical glass beads (Sovitec) with a typical diameter d = 318 ± 44 μm and a roughly gaussian size distribution [11]. The obstacle consists of a cylindrical magnet (HKCM Engineering e.K. S15x02Ni-N35, 1.17 T) of diameter D = 1.5 cm and thickness 2 mm, equal to the cell gap to prevent air, liquid or grains to flow on its faces. It is inserted in the empty cell by means of another magnet (cylinder of diameter 5 mm and height 28 mm, HKCM Engineering e.K., Z05x28Ep-N52, 1.43 T) located outside the glass plates. This stronger magnet ensures that the obstacle remains fixed in the cell through all the experiment. The obstacle is located at the vertical of the air injection nozzle, with its center at height h (Fig. 1a, third panel). The cell is then filled with grains and water, following a prococol described in [12] which ensures reproducible initial conditions. The initial granular bed height is fixed at $h_g \simeq 20$ cm, and the water layer above the grains free surface is $h_w \simeq 2$ cm.

Eight experiments are performed with different obstacle location h = [2.5,5,7.5,10,12.5,15] cm (Fig. 1a). A reference experiment without obstacle is also performed, as well as two additional experiments where three identical

*e-mail: german.varas@pucv.cl
**e-mail: valerie.vidal@ens-lyon.fr

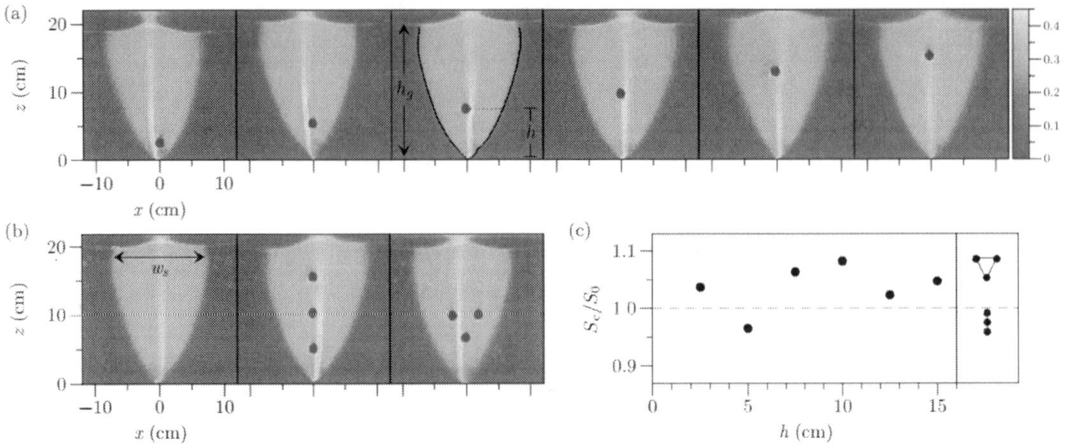

Figure 1. (a,b) Cumulated motion (matrix M, see text and Eq. 1) for the different experiments: (a) when varying the obstacle height h; (b) without obstacle (left), with three in-line obstacles (middle) and three obstacles with a triangular configuration (right). The colorbar indicates the percentage of motion in the system [threshold 1%], h_g the initial granular layer height, h the obstacle height and w_s the fluidized zone extension at z = 18 cm. The black line (a, third image) denotes the contour of the fluidized zone (see text). (c) Surface of the zone delimited by this contour up to z = 18 cm, S_c (see text), normalized by the surface of the fluidized zone without obstacle, S_0, as a function of the obstacle height [• single obstacle (a); the line and triangle on the right of the figure correspond to the in-line (b, middle) or triangular (b, right) configuration]. The red dashed line indicates the reference without obstacle (Fig. 1b, left).

obstacles are set following two geometries (Fig. 1b): (1) the obstacles are located in-line, at the vertical of the injection point, at $h \simeq 5, 10, 15$ cm; (2) the obstacles are placed in a triangular shape, with a first obstacle at the vertical of the injection point at $h = 7$ cm and two more obstacles at $h = 10.5$ cm, separated by a distance of 4 cm.

At time $t = 0$ s, air is injected from the bottom nozzle (inner diameter 1 mm) at constant flow rate $Q = 1.98$ mL/s (mass flow controller D-6311, M+W Instruments). The cell is lightened from behind by a LED panel. Images of the experiment are recorded with a camera (PixeLINK PL-B741U, 1280×800 px^2) at a frame rate of 1 image every 2 seconds during 20 hours, which makes it possible to observe the long-time dynamics of the system.

3 Results

3.1 Fluidized zone

First we focus on the fluidized zone, which is the central region where the grains move during the experiment. To delimitate this region, we compute from the raw images I_k a *cumulated motion* image M, defined as follows:

$$M = \frac{1}{N}\sum_{k=1}^{N-1} \mathcal{B}_{1\%}(|I_{k+1} - I_k|) \qquad (1)$$

where N is the total number of frames, $|I_{k+1} - I_k|$ the absolute value of the difference between two consecutive images and $\mathcal{B}_{1\%}$ a binarization function returning 1 when $|I_{k+1} - I_k| > 1\% \max(|I_{k+1} - I_k|)$, and 0 otherwise. The sum generates a matrix where each value corresponds to the number of images for which the corresponding pixel in the raw image sequence has changed its value by more than 1%. After dividing by the total number of images (Eq. 1), M thus represents the percentage of images - i.e., of time - a given spot has experienced motion (with a 1% threshold) during the whole experiment. Changing the threshold to 0.5 or 2% does not change significantly the results.

The images of cumulated motion for each experiment are displayed in Figures 1a and 1b. Without obstacle (Fig. 1b, left), we recover the parabolic shape reported in previous works [4–6, 8, 10, 13]. We compute the surface of the fluidized zone as the region where the percentage of cumulated motion is above 20% (third image, Fig. 1a). This threshold is chosen to match correctly the contour of the observed motion zone. As we are interested in the total area where the grains are affected both by the air rise and the presence of obstacles, we consider the zone defined by the fluidized zone contour including both the mobile region and the obstacle(s), up to z = 18 cm to avoid surface effects, and denote its surface S_c. Figure 1c displays S_c, normalized by the surface of the fluidized zone without obstacle, S_0, as a function of the obstacle height h. The case of multiple obstacles (in-line or triangular configuration) is also reported (Fig. 1c, right side). Except for h = 5 cm and the in-line configuration, where the height of the first obstacle is also h = 5 cm, the presence of an obstacle tends to increase the size of the fluidized zone by a few percent.

3.2 Central channel

To investigate further the impact of the obstacle on the structure of the fluidized zone, we quantify in Figure 2 the bulk properties of the cumulated motion, and the average characteristics of the central channel. To do so, we consider the intensity profiles at constant height $z = z_0$ in the cumulated motion images, and fit for different z_0 the

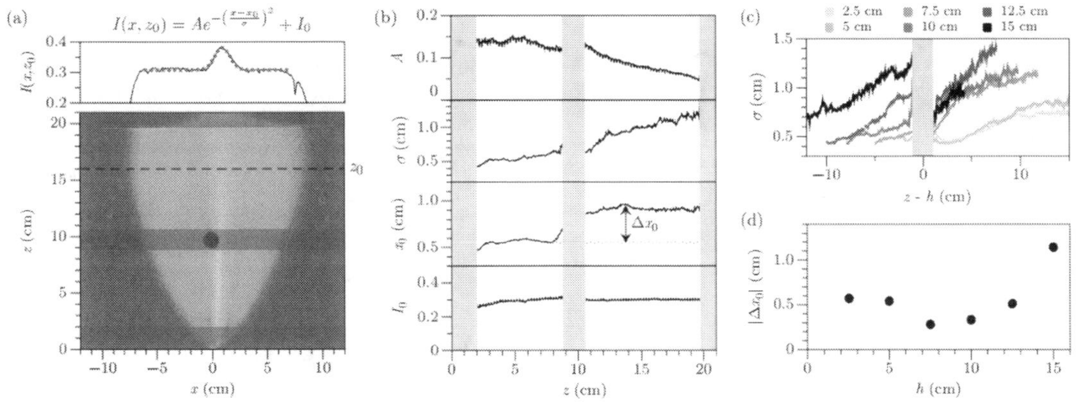

Figure 2. (a) Cumulated motion for $h = 10$ cm [bottom image]. The upper plot displays the intensity profile [solid black line] at $z = z_0$ [dashed line indicated in the image below], and the fit [red dots] corresponding to a gaussian function (the central channel) above a plateau (the granular convection). (b) Amplitude A, width σ, average horizontal shift x_0 and intensity plateau I_0 as a function of z for the gaussian fit in (a). (c) Width σ as a function of $z - h$ for different obstacle height h [gray scale, see legend above]. (d) Horizontal shift $|\Delta x_0|$ of the central air channel due to the obstacle, as a function of the obstacle height h. The gray zones in (a,b,c) correspond to regions close to the injection nozzle, the obstacle and the free surface, which are ignored in the analysis.

granular convection and the central channel by a plateau and a gaussian function (Fig. 2a, upper panel, red dots):

$$I(x, z_0) = A\, e^{-\left(\frac{x-x_0}{\sigma}\right)^2} + I_0 \tag{2}$$

The amplitude A and width σ can be interpreted as the average amplitude of motion and channel width, respectively. A decreases while σ increases along the channel height z (Fig. 2b), which can be interpreted as the result of flow conservation in the central channel. The intensity plateau I_0 corresponding to the granular convection on the sides remains roughly constant (Fig. 2b), indicating that the presence of the obstacle, in average, has no influence on the side regions. Finally, note that the horizontal shift x_0 experiences a jump $|\Delta x_0|$ when passing the obstacle. In the gray zones (Figs. 2a,b,c) the gaussian fit does not work, due either to the proximity of the injection nozzle (the air channel is pinched and the distribution strongly peaked), the free surface disturbances or the obstacle.

Figures 2c,d summarize the experimental results with a single obstacle. The higher the obstacle, the larger is the width of the central channel both before ($z - h < 0$) and after ($z - h > 0$) the obstacle (Fig. 2c). The lateral jump $|\Delta x_0|$ experiences a slight decrease, then a noticeable increase for $h = 7.5$ cm. Note that $|\Delta x_0|$ always remains smaller than the obstacle diameter, $D = 1.5$ cm, although for $h = 15$ cm, it reaches a value close to D.

3.3 Grains free surface

Lastly, we focus on the grains free surface behavior. Indeed, potential field applications are mostly based on surface observations, and the goal here is to find a possible link between the presence of obstacles and the surface morphology. To do so, we compute the mean free surface geometry by averaging in time the contour of the granular bed surface extracted from the raw images by simple contour detection. Figure 3a compiles the average free surface vertical displacement, $\langle \Delta z_s \rangle$, as a function of the horizontal coordinate x, for different obstacle height. The red curve is the free surface deformation without obstacle, and $x = 0$ cm corresponds to the injection nozzle location. Note that even without obstacle, the maximum surface vertical displacement can exhibit a shift respect to the air injection point, here of about 1 cm.

The first quantitative feature which can be extracted is the free surface extend, w_s. Because the initial height of the granular bed can slightly vary from one experiment to the other, and due to free surface fluctuations which may be difficult to quantify, we chose to estimate w_s as the fluidized zone extent at $z = 18$ cm, right below the free surface (Fig. 1b, left). Figure 3b displays w_s, normalized by the extent without obstacle, w_s^0. Interestingly, whichever the depth of the obstacle trapped in the granular bed, the extent of the fluidized zone at the surface is larger by a few percent ($w_s/w_s^0 > 1$). A noticeable jump is observed for $h \geq 5$ cm, followed by a plateau, although no explanation can be given at present. To estimate further the free surface deformation, Fig. 3c displays the skewness γ of the profiles shown in Fig. 3a. The skewness is a measurement of the profile asymmetry, defined as

$$\gamma = \frac{\frac{1}{n}\sum_{i=1}^{n}(x_i - \bar{x})^3}{\left(\sqrt{\frac{1}{n}\sum_{i=1}^{n}(x_i - \bar{x})^2}\right)^3} \tag{3}$$

where $\{x_i\}$ correspond to the nth discretized values of $\langle \Delta z_s \rangle$ and $\bar{x}$ to its average. A minimum is reported for $h = 7.5$ cm, at the same value where the plateau in w_s appears.

The right part of the figure presents the in-line and triangular configurations for the three-obstacle configuration. Interestingly, the triangular geometry is close to the case without obstacle, while the in-line geometry is close to the single obstacle for $h = 10$ cm.

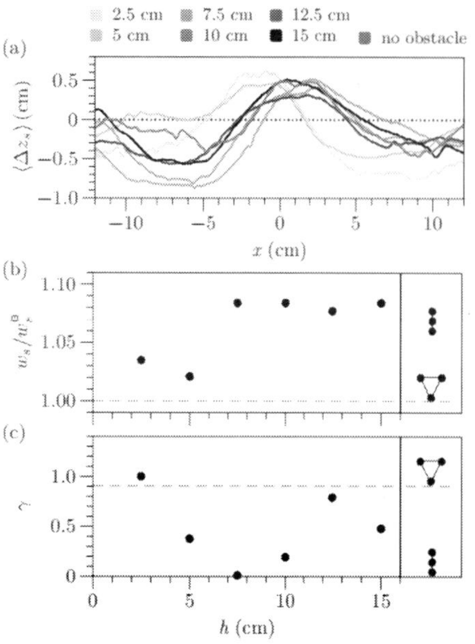

Figure 3. (a) Average vertical displacement of the grains free surface, $\langle \Delta z_s \rangle$, as a function of the horizontal coordinate, x [the gray scale indicates the obstacle height and the red curve the experiment without obstacle, see legend above]. (b) Width of the fluidized zone measured at z = 18 cm, w_s, normalized by the same width without obstacle, w_s^0, as a function of the obstacle height h. (c) Skewness γ of the average free surface deformation displayed in (a) as a function of the obstacle height h. The red line in (b,c) indicates the reference experiment without obstacle; the line and triangle symbols on the right correspond to the in-line (Fig. 1b, middle) or triangular (Fig. 1b, right) configuration].

4 Conclusion and perspectives

We quantified the influence of obstacles on air rising in water-saturated sand. The presence of a single obstacle increases both the size of the central fluidized zone and its extent close to the surface, independently of its depth as long as the obstacle is not too close from the air injection point ($h \geq 7.5$ cm). The analysis of the average free surface deformation shows that its skewness undergoes a minimum at a critical height h = 7.5 cm. No explication has been found at present for this peculiar value, although it could correspond to a transition between fingering and fracturing reported in a previous work [14]. A minimum is also observed at the same value for the horizontal shift in the central channel provoked by the obstacle.

In the field, obstacles would correspond to heterogeneities such as large grains or rocks trapped in sediments layer. This work demonstrates that even in a well-controlled experiment, the observation of the free surface is not sufficient to infer the depth of a ground heterogeneity. One can expect that the obstacle size or shape will play an important role on the surface deformation. It will be the central point of a further study which may unravel the origin of the critical height. The case of multiple obstacles is more complex, but closer to what is found in nature. Preliminary experiments considering two configurations with multiple obstacles show a discrepancy between the influence of the three obstacles on the fluidized zone size and its extent. A thorough study with multiple obstacles of different sizes, shapes and configurations is also among the perspectives of this work.

Acknowledgements

The authors acknowledge financial support from FONDECYT Project No. 11121300 and Programa de Cooperación Científica ECOS/CONICYT C14E07.

References

[1] H. Svensen, S. Planke, A.M.S. Renssen, B. Jamtveit, R. Myklebust, T.R. Eidem, S.S. Rey, Nature **429**, 542 (2004)
[2] Q. Kang, I.N. Tsimpanogiannis, D. Zhang, P.C. Lichtner, Fuel Process. Technol. **86**, 1647 (2005)
[3] K.R. Reddy, S. Kosgi, J. Zhou, Hazard. Waste Hazard. Mater. **12**, 97 (1995)
[4] W. Ji, A. Dahmani, D.P. Ahlfeld, J.D. Lin, E. Hill, Ground Water Monit. Rem. **13**, 115 (1993)
[5] J.S. Selker, M. Niemet, N.G. McDuffie, S.M. Gorelick, J.Y. Parlange, Transp. Porous Med. **68**, 107 (2007)
[6] X.Z. Kong, W. Kinzelbach, F. Stauffer, Chem. Eng. Sci. **64**, 1528 (2009)
[7] X.Z. Kong, W. Kinzelbach, F. Stauffer, Chem. Eng. Sci. **65**, 4652 (2010)
[8] G. Varas, V. Vidal, J.C. Géminard, Phys. Rev. E **83**, 011302 (2011)
[9] G. Varas, J.C. Géminard, V. Vidal, Gran. Matt. **15**, 801 (2013)
[10] G. Varas, G. Ramos, J.C. Géminard, V. Vidal, Front. Phys. **3**, 44 (2015)
[11] G. Ramos, G. Varas, J.C. Géminard, V. Vidal, Phys. Rev. E **92**, 062210 (2015)
[12] R. Poryles, V. Vidal, G. Varas, Phys. Rev. E **93**, 032904 (2016)
[13] R. Semer, J.A. Adams, K.R. Reddy, Geotech. Geol. Eng. **16**, 59 (1998)
[14] V. De Zotti, V. Vidal, in *Compte-rendus de la 17e Rencontre du Non-Linéaire*, edited by E. Falcon, M. Lefranc, F. Pétrélis, C.T. Pham (Non-Linéaire Publications, 2014), pp. 29–34

Investigations of formation of quasi-static vortex-structures in granular bodies using DEM

Jan Kozicki[1] and *Jacek* Tejchman[1,*]

[1]Gdańsk University of Technology, 80-952 Gdańsk, Poland

Abstract. The paper presents some two-dimensional simulation results of vortex-structures in cohesionless initially dense sand during quasi-static passive wall translation. The sand behaviour was simulated using the discrete element method (DEM). Sand grains were modelled by spheres with contact moments to approximately capture the irregular grain shape. In order to detect vortex-structures, the Helmholtz-Hodge decomposition of a flow displacement field from DEM calculations was used. This approach enabled us to distinguish both incompressibility and vorticity in the granular displacement field.

1 Introduction

Granular vortex-structures (swirling motion of several grains around its central point) were frequently observed in experiments on granular materials [1], [2] and in calculations using the discrete element method (DEM) [3]-[6]. They became apparent when the motion associated with uniform (affine) strain was subtracted from the actual granular deformation. A dominant mechanism responsible for the vortex formation was buckling of force chains [3], [5]. The collapse of main force chains lead to a formation of larger voids and their build-up to the formation of smaller voids [3]. The vortex-structures have been mainly observed in shear zones [4] which are the fundamental phenomenon in granular bodies [7].

The objective of this paper is to report the results of comprehensive 2D studies by DEM on quasi-static vortex-structures in sand behind a rigid wall during its passive translation by using the Helmholtz-Hodge decomposition of a displacement vector field [8], [9]. This detection method has the following advantages over other techniques [4]: 1) it operates directly on displacement vectors, 2) it is designed for both 2D and 3D flow and 3) it does not need any additional parameters for calculations. DEM analyses were carried out with spheres with contact moments [3], [4], [10] to approximately capture the irregular grain shape (the contact moment increments were calculated by means of the rolling stiffness multiplied by the angular rotational increment vectors). In order to accelerate the computation time, some simplifications were assumed in numerical studies: larger spheres, linear grain size distribution and linear normal contact. Attention was paid to a relationship between the vortex occurrence and shear localization. In addition, the predominant periods of vortex-structures were calculated.

2 DEM results of passive earth pressure model tests

In order to simulate the behaviour of real sand, the 3D spherical discrete model YADE, developed at University of Grenoble [11] was used by taking advantage of the so-called soft-particle approach (i.e. the model allows for particle deformation which is modelled as their overlap).

The DEM calculation results were described in detail in [3]. The quasi-static simulations were performed for a 2D initially dense sand body of 0.40 m length and 0.20 m height in order to compare with experiments with Karlsruhe sand (mean grain diameter d_{50}=0.5 mm) [7]. The vertical retaining wall and the bottom of the granular specimen were assumed to be stiff and very rough. Since the experiments were idealized as a 2D boundary value problem, in order to significantly accelerate simulations, the computations were performed with the specimen depth equal to the grain size (i.e. one layer of spheres was simulated along the depth only). The particles with d_{50}=1.0 mm, characterized by a linear grain size distribution, were assumed (grain size range was 0.5-1.5 mm, 62'600 particles). The initial void ratio of sand was e_o=0.62 (volumetric weight γ=25.5 kN/m^3).

Figure 1A shows the evolution of the resultant normalized horizontal earth pressure force (earth pressure coefficient) $K_p=2E_h/(\gamma h^2 d_{50})$ versus the normalized horizontal wall displacement u/h (h=0.2 m, E_h – the horizontal force acting on the wall) for d_{50}=1 mm. The specimen exhibited the initial strain hardening up to the peak (u/h=0.038), followed by some softening before a common asymptote was reached. The earth pressure coefficient was K_p^{max}=30 for d_{50}=1 mm. The earth pressure coefficient was K_p^{max}=30 for d_{50}=1 mm. It can be thus anticipated that for d_{50}=0.5 mm (real sand), K_p^{max} should be about 25-27. The value of K_p^{max}=30 for d_{50}=1 mm was slightly smaller than

* Corresponding author: tejchmk@pg.gda.pl

DOI: 10.1051/epjconf/201714003006

K_p^{max}=31 obtained by FEM (d_{50}=0.5 mm) [7] and was closer to engineering earth pressure coefficients [7].

The distribution of single sphere rotations ω during wall translation is presented in Fig.1B (red denotes the sphere rotation ω>+30° and blue ω<-30°, dark grey is related to the sphere rotation in the range 5°≤ω≤30° and light grey to the range -30°≤ω≤-5°, positive sign means clockwise rotation). Accepting such a colour convention makes shear zones clearly observable (only particles within shear zones significantly rotate). There exists a clear grain separation regarding clockwise (red) and anti-clockwise (blue) rotation. Except of two main shear zones (curved and radial shear zone), some secondary shear zone might also form [3]. The curved shear zone started to develop along the specimen bottom for the normalized wall translation u/h=0.02 (Fig.2Bb). It was fully developed for u/h=0.06. Its thickness was t_s=20 mm (20×d_{50}). The radial shear zone began to form for u/h=0.04 (Fig.1Bd) and for u/h=0.06 connected the curved shear zone. Its thickness was t_s=10 mm (10×d_{50}). The horizontal force strongly fluctuated at the residual state [12]. There was a good qualitative agreement between DEM simulation results and real corresponding experimental outcomes [7], [13].

3 Numerical results using Helmholtz-Hodge decomposition

The Helmholtz-Hodge decomposition (HHD) of 3D vector fields is one of the fundamental theorems in fluid dynamics [8], [9]. It describes a vector displacement field $\vec{\xi}$ in terms of its curl-free and divergence-free components based on potential functions:

$$\vec{\xi} = \vec{\nabla} u + \vec{\nabla} \times \vec{v} + \vec{h}, \tag{1}$$

wherein $\nabla = \left(\frac{\partial}{\partial x}, \frac{\partial}{\partial y}, \frac{\partial}{\partial z}\right)^T$ is the gradient, $\nabla \cdot = \left(\frac{\partial}{\partial x} + \frac{\partial}{\partial y} + \frac{\partial}{\partial z}\right)$ denotes the divergence operator, $\nabla \times$ is the curl operator, u denotes the scalar potential field, $\vec{v}$ is the vector potential field and $\vec{h}$ denotes the harmonic vector field. The gradient of the scalar potential function $\vec{\nabla} u$ is called the curl-free component and is related to expansion/contraction (because is irrotational) while the curl of the vector potential function $\vec{\nabla} \times \vec{v}$ is called the divergence-free component and is related to vorticity (because is incompressible). The harmonic component is related to pure translation. A discrete vector field decomposition was done according to the method described in [8], [9].

The system of linear equations was solved using the following general boundary conditions: $\vec{\nabla} \times \vec{v}$ was tangential to the domain boundary $\vec{v}|_{\partial T}$=0 and $\vec{\nabla} u$ was orthogonal to the boundary domain $u|_{\partial T}$=0. The calculations were carried out for one granular specimen.

Figure 2 presents the evolution of the vector field curl $\vec{\nabla} \times \vec{v}$ (divergence-free component related to vorticity) during normalized wall translation u/h. The green circles describe the local minima (right-handed vortices) and the red circles the local minima (left handed vortices). Vortex-structures appeared from the wall translation beginning. They were immediately concentrated in shear zones only. Right-handed and left-handed vortices alternately occurred. The right-handed vortices dominated in the curved shear zone and left-handed vortices dominated in the radial shear zone. Their distance was different.

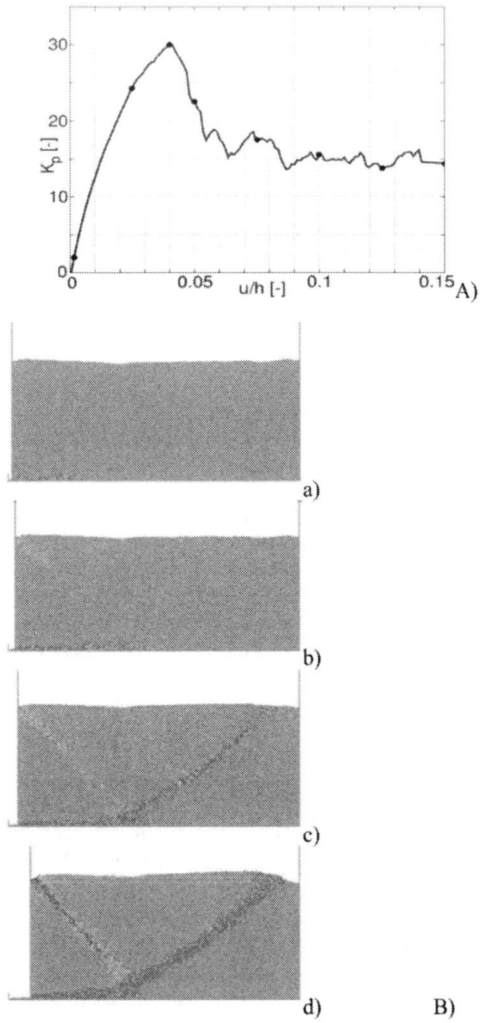

Fig. 1. DEM results (passive case, translating wall, initially medium dense sand, height h=0.2 m): A) evolution of resultant normalized horizontal earth pressure force $2E_h/(\gamma h^2 d_{50})$ versus normalized horizontal wall displacement u/h and B) deformed granular body 0.2×0.4 m² with distribution of single sphere rotations: a) u/h=0.02, b) u/h=0.04, c) u/h=0.06 and d) u/h=0.15 (red colour denotes clockwise rotation ω>+30°, blue colour denotes anticlockwise rotation ω<-30°) [3]

The maximum/minimum number of right-handed vortices in the curved shear zone was 65/15 (0.01<u/h≤0.06) or 35 (u/h>0.06).

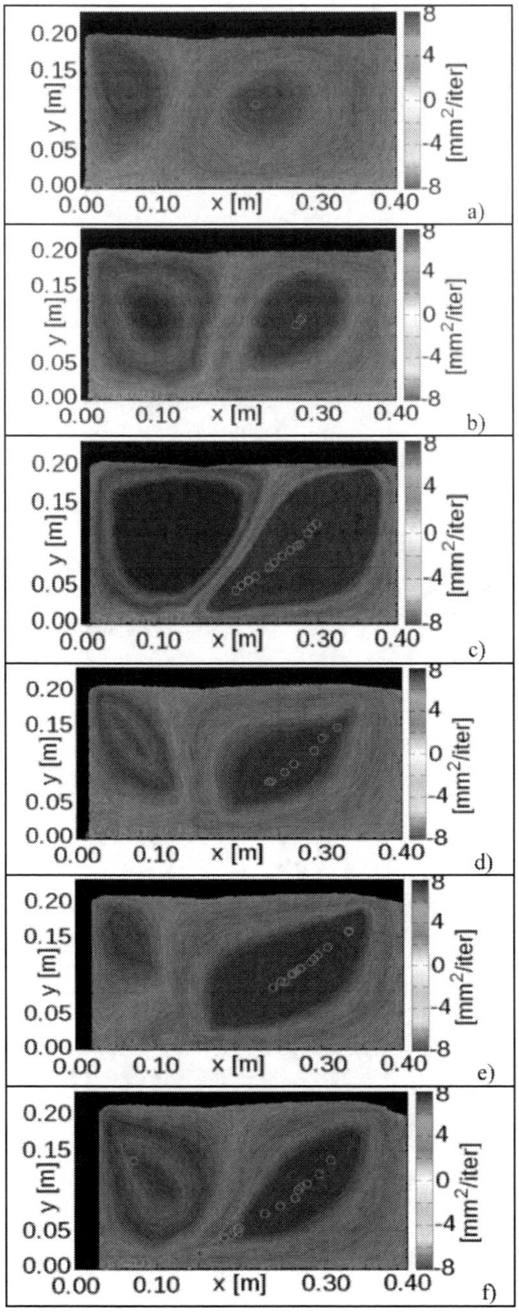

Fig. 2. Evolution of vector field curl $\vec{\nabla} \times \vec{v}$ (divergence-free component related to vorticity) in granular specimen area $x \times y$ during normalized wall translation *u/h*: a) *u/h*=0.025, b) *u/h*=0.04, c) *u/h*=0.05, d) *u/h*=0.075, e) *u/h*=0.1 and f) *u/h*=0.15 (scale denotes velocity perpendicular to specimen v_z in [mm^2/iteration]), green circles describe local minima (right-handed vortices) and red circles local minima (left handed vortices))

The maximum/minimum number of left-handed vortices in the radial shear zone was 24/1. During the entire wall translation (*u/h*=0-0.15), the predominant period of the right-handed vortices was equal to 1.5% of *u/h* (entire specimen) and to 1.5% of *u/h* (curved shear zone). The predominant period of the left-handed vortices was equal to 2% of *u/h* (entire specimen) and to 4% of *u/h* (radial shear zone). For the residual state (*u/h*=0.06-0.15), the predominant period of the right-handed vortices in the curved shear zone was 1.25% of *u/h* and of the left-handed vortices in the radial shear zone was 1% of *u/h*.

Figure 3 shows the vector displacement field $\vec{\xi}$ during the normalized horizontal wall translation *u/h*. Sphere displacement directions are marked by white arrows. Based on the displacement vector length and direction changes, a curved shear zone between the wall bottom and free upper boundary already began in the first calculation step. Later it moved to the right to reach its ultimate position (Fig.1Bd) (at the maximum resultant normalized horizontal earth pressure force of Fig.1A, *u/h*=0.04). Behind the curved shear zone the material was totally rigid. The radial shear zone was obtained for *u/h*=0.065.

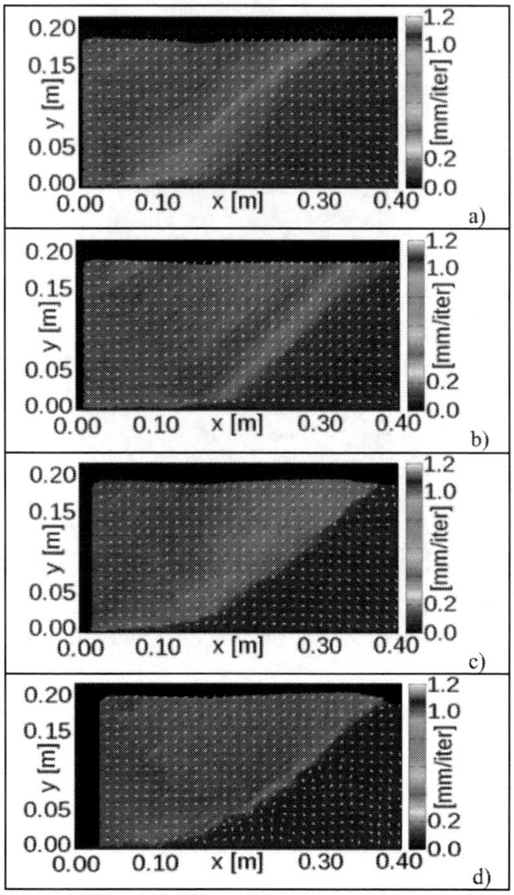

Fig. 3. Vector field $\vec{\xi}$ in granular specimen area $x \times y$ during normalized wall translation *u/h*: a) *u/h*=0.025, b) *u/h*=0.04, c) *u/h*=0.075 and d) *u/h*=0.15 (scale denotes vector length in [mm/iteration])

The evolution of the scalar field gradient $\vec{\nabla} u$ (curl-free component related to compressibility) during normalized wall translation *u/h* is described in Fig.4. The green circles describe the sources (local dilatancy minima) and red circles the sinks (local contractancy maxima). Initially global contractancy and later global dilatancy occurred in the granular specimen. The dilatancy was maximum after the peak and later it diminished. In the residual state ($u/h \geq 0.075$, Fig.1A), local regions of dilatancy and contractancy alternately happened along both shear zones (with the prevalence of dilatancy) (Fig.4).

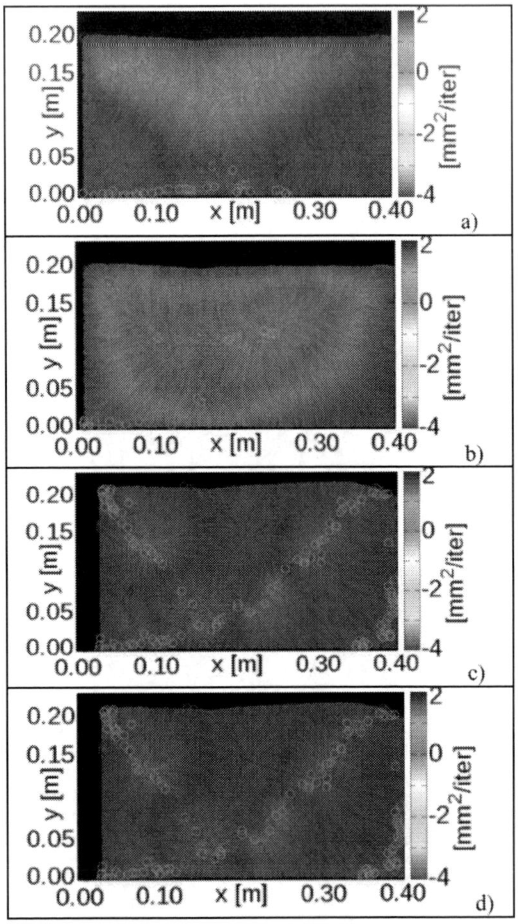

Fig. 4. Scalar field gradient $\vec{\nabla}u$ (curl-free component related to compressibility) in granular specimen area $x \times y$ during normalized wall translation *u/h*: a) u/h=0.025, b) u/h=0.04, c) u/h=0.075 and d) u/h=0.15 (scale denotes *u* in [mm^2/iteration] (sign (-) - dilatancy, sign (+) contractancy, green circles - sources (local dilatancy minima), red circles (local contractancy maxima))

4 Conclusions

The following main conclusions may be listed from DEM simulations of vortex-structures in sand during a passive earth pressure problem:

A strong relationship was found between the location of vortex-structures and shear localization. Vortex-structures were the precursor of shear localization since they clearly appeared at the place of a curved and radial shear zone from the beginning of the deformation process, i.e. significantly earlier than e.g. based on single grain rotations. They vortex-structures solely appeared in shear zones during the entire deformation process. They had a tendency to move along shear zones. Their number varied and was larger on average at the residual state. The right-handed vortices were dominant in the curved shear zone and left-handed ones were dominant in the radial shear zone.

The right-handed vortices were dominant in the curved shear zone and left-handed ones were dominant in the radial shear zone. In the curved shear zone, the predominant period of right-handed vortices was 1.5% of *u/h* during the entire wall movement (calculated based on the number of vortices versus *u/h* using FFT). In the radial shear zone, the predominant period of left-handed vortices was 4% of *u/h*. In the case of the residual state (u/h=0.06-0.15), the predominant period of right-handed vortices in the curved shear zone was 1.25% of *u/h* and of left-handed vortices in the radial shear zone was 1% of *u/h*.

In the residual state, local regions of dilatacy and contractancy alternately happened along main shear zones with a dominance of local dilatancy.

The authors would like to acknowledge the support by the grant 2011/03/B/ST8/05865 "Experimental and theoretical investigations of micro-structural phenomena inside of shear localization in granular materials" financed by the Polish National Science Centre (Poland).

References

1. S. Abedi S, A.L. Rechenmacher, A.D. Orlando, Granul. Matter **14**, 695-705 (2012)
2. V. Richefeu, G. Combe, G. Viggiani G, Geotech. Letters **2**, 113-118 (2012)
3. M. Nitka, J. Tejchman, J. Kozicki, D. Leśniewska, Granul. Matter **3**, 325-343 (2015)
4. J. Kozicki, J. Tejchman, Granul. Matter **2**, 20 (2016)
5. A. Tordesillas, S. Pucilowski, D.M. Walker, J.F. Peters, L.E. Walizer, Int. J. Numer. Anal. Meths. in Geomech. **38**, 12 (2014)
6. M.R. Kuhn, Mech. of Materials **31**, 407-442 (1999)
7. J. Tejchman, *FE modeling of shear localization in granular bodies with micro-polar hypoplasticity* (Springer Verlag, Berlin-Heidelberg 2008)
8. H. Bhatia, V. Pascucci, IEEE Transactions on Visualization and Computer Graphics **19**, 8 (2013)
9. Y. Tong, S. Lombeyda, A.N. Hirani, M. Desbrun, ACM Transactions on Graphics (TOG) **22**, 3 (2003)
10. J. Kozicki, J. Tejchman, H.B. Mühlhaus, Int. J. Num. Anal. Methods in Geomech. **38**, 1923-1952 (2014)
11. J. Kozicki, F.V. Donze, Computer Methods in Applied Mechanics and Engineering **197**, 4429-4443 (2008)
12. G. Combe, V. Richefeu, M. Stasiak, A.P. F. Atman, Phys. Rev. Lett. **115**, 238301 (2015)
13. M. Niedostatkiewicz, D. Leśniewska, J. Tejchman, Strain **47**, s2 (2011)

Dynamic Self-assembly of Non-Brownian Spheres

J. Marcos Salazar[1,*], *J. Marc* Simon[1], *J. Carlos* Ruiz-Suárez[2], *Francisco* Peñuñuri[3] and *Osvaldo* Carvente[3]

[1] *Laboratoire Interdisciplinaire Carnot de Bourgogne, UMR 6303 CNRS Université de Bourgogne Franche- Comté, 9 Avenue Alain Savary, B.P. 47870, 21078 Dijon Cedex, France*
[2] *CINVESTAV Monterrey, Autopista Nueva al Aeropuerto km 9.5, Apodaca, Nuevo León 66600, México*
[3] *Departamento de Ingeniería Física, Universidad Autónoma de Yucatán, 97310, Mérida, Yucatán, México*

Abstract. Granular self-assembly of confined non-Brownian spheres under gravity is studied by Molecular Dynamics simulations. Starting from a disordered phase, dry or cohesive spheres organize, by vibrational annealing into BCT or FCC structures, respectively. During the self-assembling process, isothermal and isodense points are observed. The existence of such points indicates that both granular temperature and packing fraction undergo an inversion process. Around the isothermal point, a sudden growth of beads having the maximum coordination number takes place. We show by a density fluctuation analysis that a transition form a disordered phase to a crystalline structure may be associated to a first-order transition.

1 Introduction

Spontaneous pattern formation by self-assembly is an appealing research subject given its potentiality as a manufacturing process. Self-assembly processes are common through nature and technology. The formation of regular cristalline lattices is a fundamental process in self-assembly and is a method for converting $\approx 100nm$ structures into photonic materials [1]. In natural systems crystal formation is widely present, e.g., in the geological production of opals and in the biological formation of proteins and virus [2]. Some years ago the methodology of self-assembly has been used for the fabrication of electronic devices and appears as a practical strategy for manufacturing nanostructure. Beyond the technological interest of self assembly it has always fascinated human curiosity for its æsthetic aspects. But, the physical processes governing self-assembly is an open question and the present manuscript address some issues that may be of interest.

Several years ago we serendipitously discovered that a collection of dry spheres subjected to vibrational annealing, forming a body centered tetragonal (BCT) lattice in a commensurable container, self-assembly instead, upon the addition of a small drop of oil, into a face centered cubic (FCC) [3]. We speculated that the observed polymorphism was due to cohesion introduced by capillary forces. In the present article, we perform Molecular Dynamics simulations (MD) with the aim to confirm our earlier speculations. Intriguingly, we found that the interrelating dynamics of the grains, subjected to a vibration that changes with time, gives rise to temperature and density inversions. This creates two peculiar points easily recognizable as isosbestic points. When the system starts to cross these points density fluctuations decrease and the number of beads having the maximum coordination number (12 for FCC and 10 for BCT) suddenly appear to finally give place to a granular crystallization. Our density fluctuation analysis shows that this transition may be associated to a first order phase transition.

The grains are indifferent to thermal fluctuations so one must provide some sort of stochastic mobility. Such mobility is simply achieved by shaking the box where the grains are contained. Thermodynamic-like macroscopic quantities can be well defined to describe the collective behavior of these constituents. In these systems, the packing fraction, pressure and granular temperature are also well defined [4–14]. However, due to the fact that granular materials are dissipative and subjected to gravity influence, they exhibit a non-uniform temperature and packing fraction when energy is injected to them. Indeed, in a vibrating box filled by spheres one can clearly observe that both quantities change with height [15] and these gradients persist as we change the frequency of the vibration.

2 Molecular Dynamics

MD simulations were performed for tracking the packing evolution of dry and cohesive ensembles composed of mono-dispersed steel beads (radius $r = 1.19 \times 10^{-3}$m and mass $m = 5.54 \times 10^{-5}$kg) inside a container subjected to a vertical vibration strength. We used ensembles having 666, 1099 and 6120 beads. Given that the results obtained for 1099 and 6120 are similar to those of 666 beads, we focus mainly on the results for 666 beads. The dimensions of the container ensures (within a small tolerance) either a BCT structure ($L_x = L_y = 2r(1 + (N_{cells=} = 4)\sqrt{3}) = 18.87 \times 10^{-3}m)$ or a FCC cohesive structure

* e-mail: jmarcos@u-bourgogne.fr

$(L_x = L_y = 2r(1 + (N_{cells} = 5)\sqrt{2}) = 19.3 \times 10^{-3}m)$. The vibrated ensemble is subjected to an iso-gamma cooling, where the dimensionless acceleration, $\Gamma = A(2\pi\nu)^2/g$, remains constant while the frequency ν (amplitude A) is increased (decreased). This iso-gamma annealing drives the system from a dilute state to a compact state [3]. It is important to remark that in the absence of annealing there is no possibility to observe self-assembling (Fig.1).

The interacting force acting between two colliding particles i (mass m_i) and j (mass m_j) is given by $\vec{F}_{ij} = \vec{F}_N + \vec{F}_T$. The former term is given by: $\vec{F}^n = k_n \xi_{ij} \vec{n} - \frac{\gamma_n}{2} m_{eff} \vec{V}_n$, where $\vec{V}_n = |\vec{v}_j - \vec{v}_i|\vec{n}$, $\vec{n} = \frac{\vec{r}_j - \vec{r}_i}{|\vec{r}_j - \vec{r}_i|}$, k_n is a material constant associated to the hardness of the beads and $\xi_{ij} = R_j + R_i - |\vec{r}_j - \vec{r}_i|$, representing the overlapping between particles (R_i, $\vec{r}_i$, $\vec{v}_i$ are respectively, the radius, position and velocity of particle i). The second term in $\vec{F}^n$ represents the energy dissipation by the collision and is proportional to the relative velocities of the particles. Where γ_n represents the damping coefficient, $m_{eff} = m_i m_j/(m_i + m_j)$ the effective mass.

The tangential forces, $\vec{F}^T$, are modeled by using the tangential relative velocities. For two colliding particles the relative velocity of the spheres at the point of contact is given by: $\vec{v}^T_{rel} = \vec{v}_{ij} - \vec{n}(\vec{n} \cdot \vec{v}_{ij})$, with $\vec{v}_{ij} = \vec{v}_i - \vec{v}_j + R_i \vec{n} \times \vec{\omega}_i + R_j \vec{n} \times \vec{\omega}_j$ with $\vec{\omega}_i$, $\vec{\omega}_j$ the angular velocities. We use the force model introduced by Cundall [16], where the static friction is described by a spring acting along the tangential direction of the contact plane. The spring elongation is given by: $\zeta(t) = \int_{t_k}^{t} v^T_{rel}(t')\,dt'$. This equation allows to determine the restoring tangential force, $|k_t\zeta|$, (k_t represents the spring stiffness) which is limited by the Coulomb's rule. Henceforth, the tangential force can be written as: $F^T = -sign(\vec{v}^T_{rel}) * \min(|k_t\zeta|, \mu|\vec{F}^n|)$ where μ is the friction coefficient of the beads and $\vec{F}^T = F^T \vec{t}$ with $\vec{t} = \vec{n}^{\perp}$. In our simulations we consider the restitution coefficient independent of the impact velocity. The parameters used in the simulations are: $k_n(N/m)$= 81293.298, $\gamma_n(s^{-1})$= 0.245, γ_t(N/m)=0.122, μ =0.5, ϵ =0.9, m_{eff}(g)=2.77 $\times$ 10^{-5}, τ_c(sec)= 5. The capillary forces of wet particles represented in our simulations are given by:

$$F_{coh}(s) = \frac{rA\pi\sigma[A\,(\tan(\beta) - \sec(\beta)) + 1]\tan(\beta)}{A\sec(\beta) - 1}, \quad (1)$$

where $A = (1 + \frac{s}{2r})$, s the separation between beads, σ the liquid surface tension and β the meniscus angle formed by the liquid between two interacting beads. The parameters used here are: $\sigma = 0.03(Nt/m)$, $1 \leq s \leq 60(\mu m)$ and $\beta = 12.5°$ (see Ref.[17] for more details).

3 Temperature and density inversion

The simulation box height is divided in bins, bin_k, of thickness $z_{k+1} - z_k$ equal to two beads diameter. The system vibrates at a fixed frequency ν_j which is active for a period of time (five seconds). In these time intervals, the local granular temperature $\delta v_k(bin_k, \nu_j)$ [13], and the local packing fraction $\rho_k(bin_k, \nu_j)$ are estimated by bin. Every 6 ms the instantaneous state of the system is sampled for calculating the mean temperature and packing fraction (Fig.1 and Fig.4). We use the Voronoi tessellation to calculate the volume variation, $V_{voro}(i)$, associated to each particle i [18] for all the frequencies ν_j. Then $\rho_k(bin_k, \nu_j) = \langle V_i/V_{voro}(i)\rangle\ \forall\ i \in bin_k$, where V_i is the volume of the spheres. The brackets represent average over particles and time.

In Fig.1 we show snapshots of the system for several frequencies of a) dry and c) wet beads, respectively. The colorbar is proportional to the granular temperature (red for hot and blue for cold). From the first figure, we can easily notice that at the beginning of the annealing there is a temperature gradient along the direction of gravity (warmer at the bottom and colder at the top). As the annealing continues, the gradient eventually disappears at a given frequency $\nu_c = 34Hz$ (temperature homogeneity represented by uniform white color). Above this frequency, a temperature inversion appears (the bottom becomes colder than the top). The mean local temperature curves for each bin_k, as a function of frequency, are given in Fig. 1(c, d). Accordingly,

$$\left.\frac{\partial(\delta v_k(bin_k, \nu_j))}{\partial bin_k}\right|_{\nu_c} = 0. \quad (2)$$

When the system crosses the region of the isothermal point established by the Eq.2, the maximum coordination number suddenly starts to increase as seen in Fig.2 where we show, for each annealing frequency, the mean number of particles having the maximum first neighbors (10 for BCT and 12 for FCC). While the BCT lattice can form only with the annealing of the system, the FCC structure requires also cohesion. Fig.3 shows how the mean Voronoï volume decreases during the annealing process and we can readily see that near 100 seconds an intriguing change of regime occurs (detailed below). Above this time the mentioned volume per bead remains unchanged.

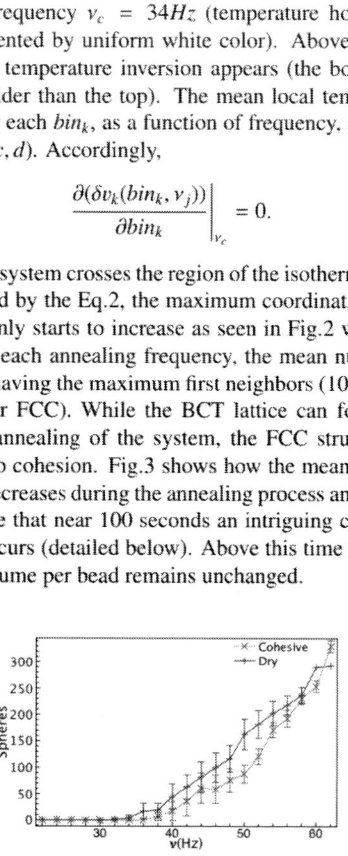

Figure 2. The mean number of particles having the maximum coordination number. The maximum coordination number sharply grows around 34-36 Hz, when the temperature vs frequency profiles cross.

Previous studies have shown how the density and the temperature of a vibrated granular system changes with height [15]. For the studied system presented here we find similar temperature profiles. Concerning the local density $\rho_k(bin_k, \nu_j)$ as a function of the container's height, Fig.4 shows distinctive crossing points. These points define a

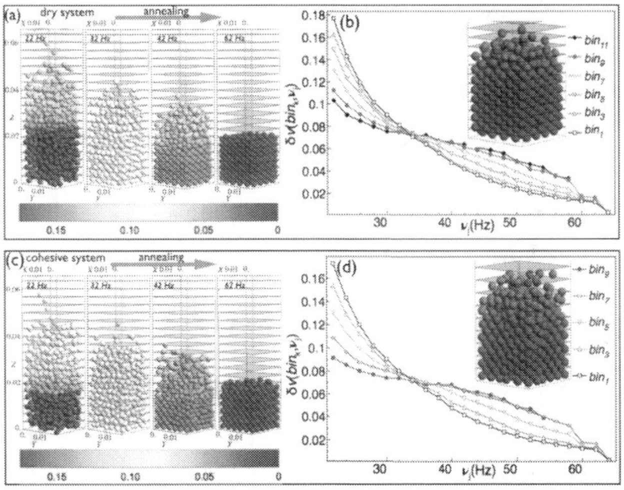

Figure 1. a) Snapshots of the dry system at 22, 32, 42, and 62 (Hz). The system crystallizes into a BCT structure; b) Granular temperature versus frequency for each bin_k. We clearly observe an inversion beyond the isothermal point. Panels c) and d) are for the cohesive system crystallizing into a FCC. The colorbar indicates that particles with the highest and lowest temperature are in red and blue respectively.

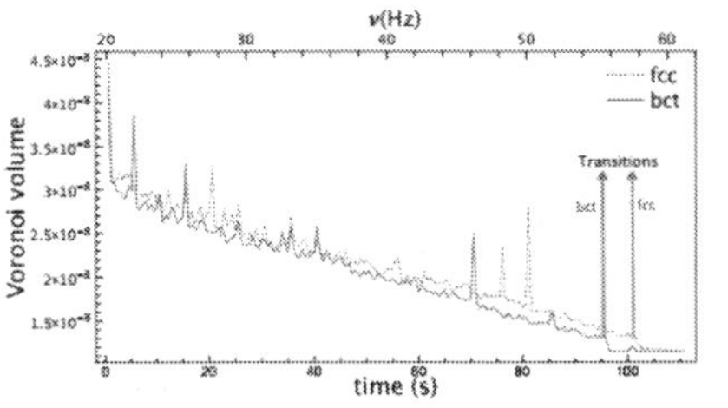

Figure 3. Time evolution of mean Voronoï volume of beads during the annealing process for a BCT and a FCC structure. It is interesting to remark that near ninety seconds the Voronoï volume decay shows a change of regime.

small region of constant packing fraction that does not depend on vibration conditions. We call it *isodense* point. Then, it is possible to establish that: $\left.\frac{\partial\langle\rho_k(bin_k,\nu_j)\rangle}{\partial\nu_j}\right|_{bin_{k^*}} = 0,$. For the dry (BCT) and cohesive (FCC) structures $k^* = 9$ or (6), respectively.

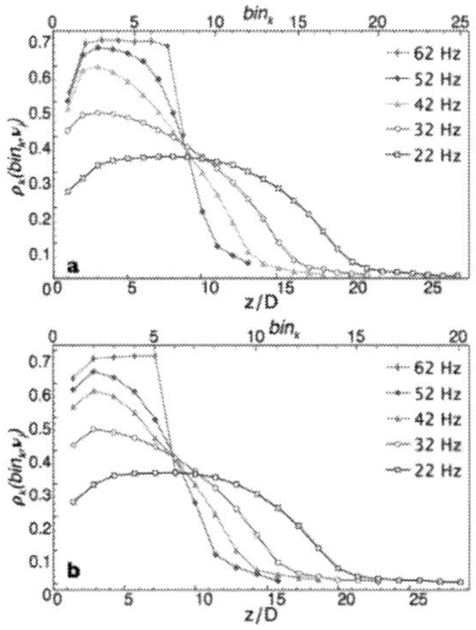

Figure 4. Local packing fraction for some frequencies as a function of height (bin_k) using Voronoi tessellation. a) BCT crystal and b) FCC crystal. Isodense points appear near the final heights for each lattices.

4 Phase transition

For characterizing the transition obtained from a disorder ensemble of steel beads to a crystalline structure (Fig.1) we use concepts from statistical mechanics. An efficient way to track a transition is to follow the time evolution of the thermal compressibility, χ_T. This quantity in the grand canonical (GC) ensemble can be computed from density fluctuations [19] according to the following expression:

$$\frac{N}{V}k_B T\chi_T = (\frac{\langle\delta N^2\rangle}{N})_{GC} = \Gamma^{-1}(V) \qquad (3)$$

where k_B is the Boltzmann constant and δN the fluctuations associated to the number of particles N inside a volume V. Actually, the thermodynamic factor Γ^{-1} gives a measure of the deviation from the ideal behavior of the system. Given that the system sizes simulated are not infinite, we compute the temporal evolution of the density fluctuations by considering (for each configuration) spherical volumes randomly located within the simulation box and by discarding the container surfaces. These spheres have a radius equal to 7.0×10^{-3} cm. Even when this choice is rather arbitrary, it has to be taken between twice the particles size and half the simulation box size [20]. We tested different sizes from 2.0×10^{-3} to 9.0×10^{-3} and we verified that the results are similar. Fig.5 shows that by starting from a disordered system Γ^{-1} starts to increase indicating the re-organization of beads into an ordered structure. Afterwards an abrupt discontinuity of Γ^{-1} appears which may be associated to a first order transition. The behavior Γ^{-1} observed for ensembles of beads leading to a BCT or FCC structure are differentiated by the time at which the transition occurs. In Fig.5 we can readily notice that the transitions to a BCT and FCC structures are accompanied by the presence of small peak around 1600 and 1700, respectively, which can be seen as precursors of the transition. Our results suggest that the transition occurs by jumps to well differentiated stable states for reaching the final state. This conjecture is under analysis and will be presented elsewhere.

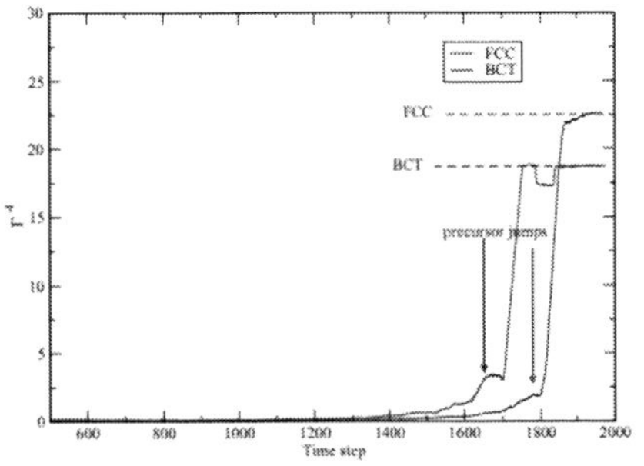

Figure 5. The Γ^{-1} time evolution shows a phase transition for a BCT crystal and a FCC crystal around 1650 and 1800 iterations steps ($\approx$ 100 sec), respectively. Notice that the transition is characterized by a grow of Γ^{-1} and the presence a small peak before the transition takes place.

Acknowledgements

This work has been supported by CONACYT, Mexico, under grants 168281 and ANR-France project MI2C 2015-2019.

References

[1] T. D. Clark, J. Tien, D. C. Duffy, .K E. Paul, and G. M. Whitesides J. Am. Chem. Soc., **123**, 7677-7682, (2001).

[2] G. M. Whitesides and B. Grzybowski, Science **295**, 2418 (2002).

[3] O. Carvente and J. C. Ruiz-Suárez, Phys. Rev. Lett. **95**, 018001 (2005).

[4] J. B. Knight, C. G. Fandrich, C. N. Lau, H. M. Jaeger, and S. R. Nagel, Phys. Rev. E. **51**, 3957 (1995).

[5] E. R. Nowak, J. B. Knight, E. Ben-Naim, H. M. Jaeger and S. R. Nagel, Phys. Rev. E. **57**, 1971 (1998).

[6] J. S. Olafsen and J. S. Urbach, Phys. Rev. E. **60** R2468 (1999).

[7] S. Warr and J. P. Hansen, Europhys. Lett. **36**, 589 (1996).

[8] F. Rouyer and N. Menon, Phys. Rev. Lett. **85**, 3676 (2000).

[9] C. R. K. Windows-Yule and D. J. Parker, Phys. Rev. E. **87**, 022211 (2013).

[10] G. D'Anna, P. Mayor, A. Barrat, V. Loreto and F. Nori, Nature **424**, 909 (2003).

[11] T. Shinbrot, Nature Physics **9**, 263 (2013).

[12] R. Ramírez and. R. Soto, Physica A. **322**, 372 (2003).

[13] I. Goldhirsch, Powder Technology, **182**, 130 (2008).

[14] A. Bodrova, A. K. Dubey, S.Puri and N. Brilliantov, Phys. Rev. Lett. **109** 178001 (2012).

[15] R. D. Wildman, J. M. Huntley, J.-P. Hansen, D. J. Parker, and D. A. Allen, Phys. Rev. E. **62**, 3826 (2000).

[16] P.A. Cundall and O.D.L Strack. *Géotechnique* **29**, No 1, 45-67 (1979).

[17] J. M. Salazar. *Granular Matter*, DOI 10.1007/s10035-014-0497-1 (2014).

[18] C. Briscoe, C. Song, P. Wang and H. A. Makse, Phys. Rev. Lett. **101**, 188001 (2008).

[19] Computer simulation of liquids, M. P. Allen and D. Tildesley, Oxford University Press, New-York (1987).

[20] S. K. Schnell, T. J. H. Vlugt, J.-M. Simon, D. Bedeaux and S. Kjelstrup Chem. Phys. Letters **504**, 199 (2011).

[21] O. Carvente, M. Salazar-Cruz, F. Peñuñuri and J.C. Ruiz-Suarez. Phys. Rev. E. **93**, 020902(R) (2016).

 DOI: 10.1051/epjconf/201714008001

Granular compaction and stretched exponentials

Experiments and a numerical stochastic model

Maxime Nicolas[1,*], *Jean-Eric* Mathonnet[2,**], *Blanche* Dalloz[1,***], and *Philippe* Sornay[2,****]

[1] *Aix Marseille Univ, CNRS, IUSTI, Marseille, France*
[2] *CEA, DEN, DEC, SFER, LCU, 13108 Saint-Paul-Lez-Durance, France*

Abstract. We present a stochastic model to investigate the compaction kinetics of a granular material submitted to vibration. The model is compared to experimental results obtained with glass beads and with a cohesive powder. We also propose a physical interpretation of the characteristic time τ and the exponent β of the stretched exponential function widely used to represent the granular compaction kinetics, and we show that the characteristic time is proportional to the number of grains to move. The exponent β is expressed as a logarithmic compaction rate.

1 Introduction

The compaction of a non-cohesive granular material (glass beads, sand, rice for example) under vibrations or vertical tapping is often well represented by a stretched exponential function [1], inspired by the Kohlrausch-Williams-Watts (KWW) relaxation model:

$$C(t) = \frac{\phi_\infty - \phi(t)}{\phi_\infty - \phi_0} = e^{-(t/\tau)^\beta} \quad (1)$$

In this expression, the packing fraction ϕ evolves between an initial value ϕ_0 and a saturated value ϕ_∞, t is the number of applied taps or vibrations, τ is a characteristic time, and β an exponent. While other empirical expressions are proposed in the literature [2–4], we will only focus on the expression (1) throughout this paper with aim of providing a physical meaning to τ and β using a stochastic model and some experiments.

It is easy to show that τ is the time where the curve $\phi(\ln t)$ exhibits an inflexion point in a log-lin space. At this particular time $t = \tau$, the packing fraction is $\phi^* = \phi_\infty - (\phi_\infty - \phi_0)/e$, and the exponent β is proportional to the slope of the $\phi(\ln t)$ curve:

$$\beta = \frac{e}{\phi_\infty - \phi_0}\left(\frac{d\phi}{d\ln t}\right)_{t=\tau} \quad (2)$$

Hence β can be interpreted as a logarithmic rate of compaction. A graphical meaning of β can also be found when plotting Y as a function of $\ln t$,

$$Y = \ln\left(\ln\frac{1}{C}\right) = \beta\ln t - \beta\ln\tau. \quad (3)$$

In these axis, the expression (1) is then a straight line of slope β.

We suggest in this paper a simple one-dimensional stochastic model of compaction to understand the physical origins of τ and β. Moreover, the « universality » of the KWW relaxation expression may be questioned when the granular material exhibits a cohesive property. Widely used in the industry, cohesive powders are often difficult to handle and transport. A cohesive powder usually presents a high angle of repose and a low bulk volume fraction. The cohesiveness of a powder may be measured through its ability to flow under gravity, and this ill-defined "flowability" is often described by the Hausner ratio I_H, the ratio of the the tapped bulk density of the powder over the freely settled bulk density [5]. A Hausner ratio greater than 1.25 is considered to be an indication of weak flowability.

It is thus interesting to trial the role of cohesion on the compaction curve, with an extension of the stochastic model, and also with some experiments on a cohesive powder under vibrations.

2 Stochastic model

2.1 Non cohesive model

We first describe here a simple stochastic model to simulate the compaction of a non-cohesive granular material. the model is a set of N unit grains shared out on a discretized one-dimensional space of size H_0 bounded with a static grain at the bottom $z = 0$ (see Figure 1a). The initial linear fraction is $\phi_{0g} = N/H_0$ and we write $\phi_{\infty g} = 1$ the maximum packing fraction. The free spaces between two consecutive grains model the pore space between physical grains and are of the same order of magnitude as the grain volume [6].

*e-mail: maxime.nicolas@univ-amu.fr
**e-mail: jean.eric.mathonnet@gmail.com
***e-mail: blanche.dalloz@univ-amu.fr
****e-mail: philippe.sornay@cea.fr

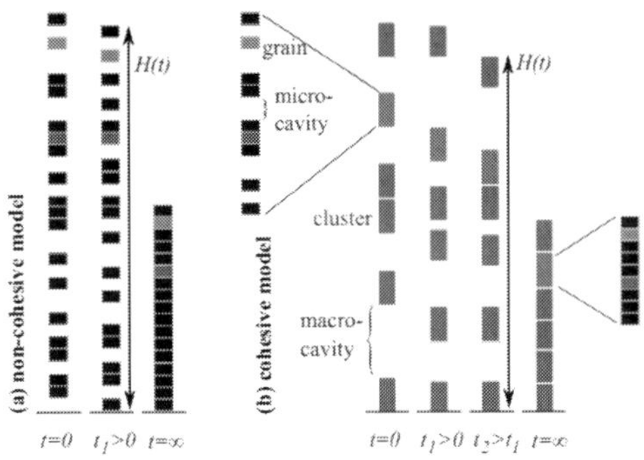

Figure 1. (a) Sketch of the non-cohesive model with individual grains only. From $t = 0$ to t_1, the green-labeled grain is allowed to move downwards whereas the red-labeled grain can not move. (b) Sketch of the cohesive model with 6 clusters containing 9 grains each. The blue-labeled cluster may move down of a distance of its size (from $t = 0$ to t_1) or of the available space (from t_1 to t_2).

At each time step, all the grains are tested in a random order. For each grain a random number r determines its ability to move: if $r \leqslant p_g$, it may move down of a space unit only if the space below is free. The grain motion probability p_g is governed by the packing fraction as

$$p_g(\phi_g) = \frac{\phi_{\infty g} - \phi_g}{\phi_{\infty g} - \phi_{0g}}, \quad (4)$$

which is the ratio of the free volume [7] at time t by the free volume at time $t = 0$. Many other expressions of this probability (also named mobility) are available in the literature [7, 8] are derived from statistical physics principles, but we prefer an expression which expresses a decrease of this probability from the initial state (ϕ_{0g}) to the final state ($\phi_{g\infty}$) in the simplest way. We checked that the order of tests of the N grains has no influence on the global dynamics. At the end of the loop on the N grains, the global packing fraction $\phi(t) = N/H(t)$ is simply computed with the height $H(t)$ of the highest grain of the set at time step t. The computation stops after a predefined number of time steps. The system obviously does not evolve anymore when the packing fraction has reached its maximum limit $\phi_{\infty g} = 1$. To avoid random fluctuations on the results, several runs were averaged before presenting the results.

A first example of result of this model is shown on Figure 2 where the height of the packing is plotted in a log-lin space (blue curve). During the process, a compaction front is uprising. This front is located at $H_c(t)$, and the grains below H_c are at the maximum packing fraction $\phi_{\infty g}$. If we assume that the packing fraction above H_c is a constant value $\phi_m = \phi_0$, the front location must be

$$H_c(t) = \frac{N}{\phi(t)} \frac{\phi(t) - \phi_0}{\phi_\infty - \phi_0} \quad (5)$$

Looking at the dashed curve in Figure 2, this assumption seems to be checked at any time.

Simulated compaction curves averaged on 20 000 runs are shown in Figure 3 for different initial packing fractions

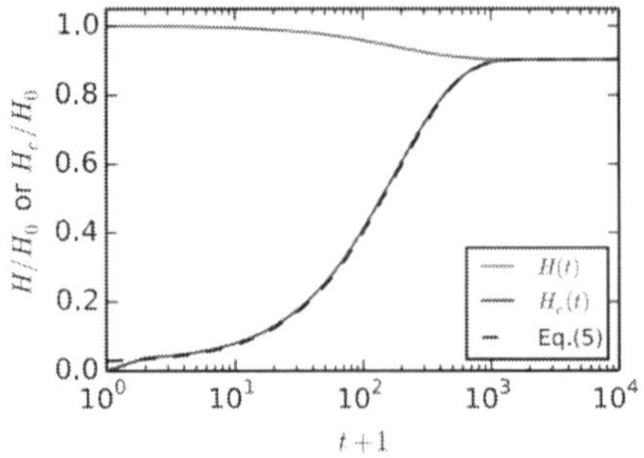

Figure 2. Height of the modeled packing $H(t)$ (red curve) normalized by the initial height H_0, height of the compacted grains $H_c(t)$ (blue curve), compared with expression 5 (black dashed line). The model parameters are $N = 250$, $\phi_0 = 0.9$ from an average of 20 000 runs.

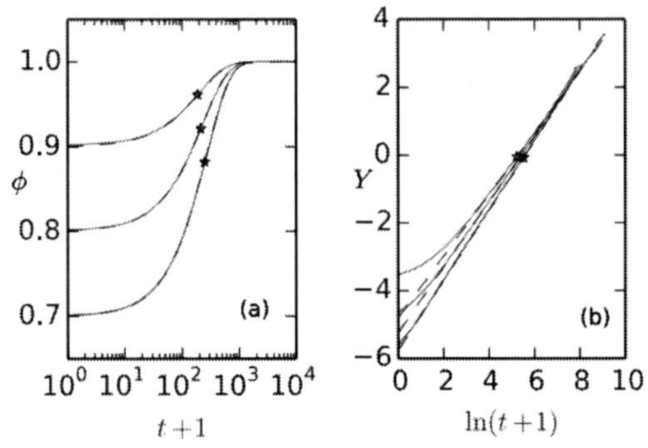

Figure 3. Simulation results (colored continuous lines) for $N = 250$ grains, averaged on 20 000 runs. Each curve is fitted by Eq. 1 (dashed black line) (a) Compaction curves for three different initial conditions. The star symbols indicate the inflexion points at time $t = T$, (b) Same data in the $(\ln t, Y)$ space.

ϕ_0 and for various representations. From these data, we can compute the characteristic time T and the logarithmic compaction rate b defined by

$$\left(\frac{d^2\phi}{d(\ln t)^2}\right)_{t=T} = 0, \qquad b = \frac{e}{\phi_\infty - \phi_0}\left(\frac{d\phi}{d\ln t}\right)_{t=T} \quad (6)$$

These compaction curves are compared with the KWW expression, and we can notice that the numerical result $(\ln t, Y)$ is not a straight line at all time, the stretched exponential expression does not completely fit the numerical data, especially at short time. However, around $t = T$ (indicated by the stars), the numerical data are fairly well approximated by the KWW expression, showing that τ is indeed a good approximate of T and $\beta \approx b$. Varying both ϕ_0 and N, this model shows that the characteristic time is expressed by

$$T = N\left(\frac{\phi_{\infty g} - \phi_{0g}}{\phi_{0g}} + A\right), \quad \phi_{0g} < \phi_{\infty g} \quad (7)$$

with a fitting parameter $A = 0.6$, and the logarithmic compaction velocity

$$b = 0.34\frac{\phi_{0g}}{\phi_{\infty g}} + A, \quad \phi_{0g} < \phi_{\infty g}. \tag{8}$$

The characteristic compaction time is thus governed by the number of grains N. The ratio $\phi_{0g}/\phi_{\infty g}$ is also a governing parameter for T and β.

2.2 Cohesive model

The previous model can be extended for a cohesive granular material. the cohesive granular system is modeled as a set of N unit grains shared out between N_c clusters, each cluster containing n grains. As previously, the grains and clusters are located on a discretized one-dimensional space of size H_0 bounded with a static grain at the bottom $z = 0$. The initial state is prepared first by placing randomly the clusters of even size n/ϕ_{0g} without overlap, with a linear fraction of clusters ϕ_{0c}. Then the grains are randomly placed inside each cluster with a linear fraction ϕ_{0g}. The initial global packing fraction is then $\phi_0 = \phi_{0c}\phi_{0g}$. At each time step, the particles and the clusters may move according to the motion probability laws

$$p_c(\phi_c) = \frac{\phi_{\infty c} - \phi_c}{\phi_{\infty c} - \phi_{0c}}, \quad p_g(\phi_g) = \frac{\phi_{\infty g} - \phi_g}{\phi_{\infty g} - \phi_{0g}} \tag{9}$$

where subscript c is for clusters, g for individual grains, $\phi_c(t)$ is the cluster linear fraction, and $\phi_g(t)$ is the linear fraction of grains inside the clusters.

A representative result of the simulation is given in Figure 4. The compaction curve shows an obvious two-stages evolution. The first stage corresponds to the fast compaction of the clusters while the second stage corresponds to the compaction of the individual grains. This approach clearly shows that two separate characteristic times T_c and T_g are present, where the subscripts c and g stand for clusters and grains. As for the non-cohesive model, the characteristic times are defined through the inflexion points of the $\phi(\log t)$ curve, and scale on the number of clusters or individual grains,

$$T_c = N_c\left(\frac{\phi_{\infty c} - \phi_{0c}}{\phi_{0c}} + A\right) \tag{10}$$

and

$$T_g = N\left[\frac{A}{2}\exp\left(\frac{\phi_{\infty g} - \phi_{0g}}{A/2}\right)\right]. \tag{11}$$

Whatever the model parameters (ϕ_{0c}, ϕ_{0g}, N, n), the numerical results can be well fitted by an extension of the stretched exponential function expression with two exponentials:

$$\phi = \phi_\infty - (\phi_p - \phi_0)e^{-(t/\tau_c)^{\beta_c}} - (\phi_\infty - \phi_p)e^{-(t/\tau_g)^{\beta_g}} \tag{12}$$

with two time-scales τ_c and τ_g, two exponents β_c and β_g and a plateau packing fraction ϕ_p (See Figure 4a) . The fitted characteristic times τ_c and τ_g are again very close to the computed times T_c and T_g, and the exponents β_c and β_g may be interpreted as logarithmic compaction rates for clusters and grains respectively.

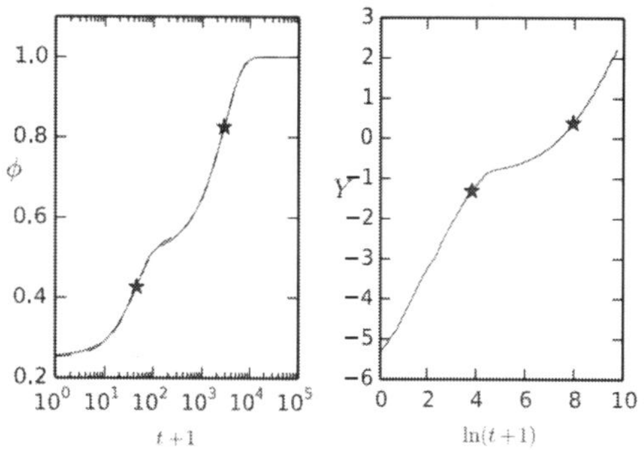

Figure 4. Simulation result for a modeled cohesive granular material. The parameters are $N = 1250$, $n = 50$, $\phi_{0c} = \phi_{0g} = 0.5$. (a) volume fraction as a function of time in a log-lin space (blue curve) and the fitted expression (12) (black dashed line). The stars indicate the T_c and T_g characteristic times. (b) Same data plotted as $Y = \ln(\ln(1/C))$ as a function of $\ln t$.

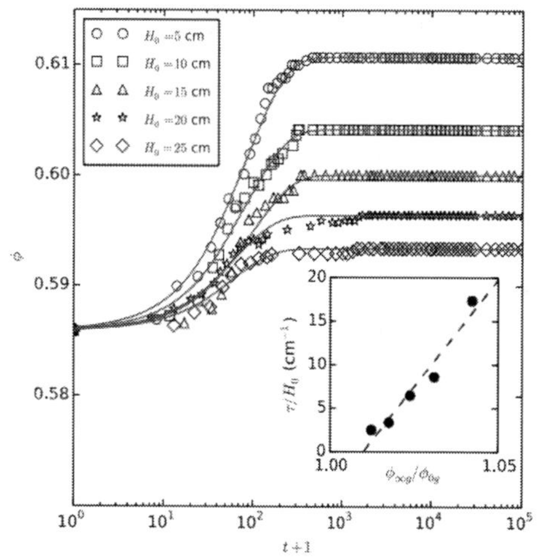

Figure 5. Experimental results of the compaction of a glass bead assembly for different initial height. The red lines are the KWW expression (1). Insert: the fitted characteristic time τ as a function of the initial volume fraction.

3 Experiments

The results from the stochastic model are compared with compaction experiments with an experimental setup based on a horizontal vibration of a vertical tank of square section 15×15 mm^2. A first set of experiments was made with glass beads of 130 μm, a non-cohesive granular material ($I_H = 1.08$). The compaction curves are shown in Figure 5 for 5 different initial heights. While it is difficult to compare the 3 dimensional experimental results with 1 dimensional numerical results, we were able to test one of the main result from the stochastic model. By varying the mass of grains in the shaken tank, hence varying the num-

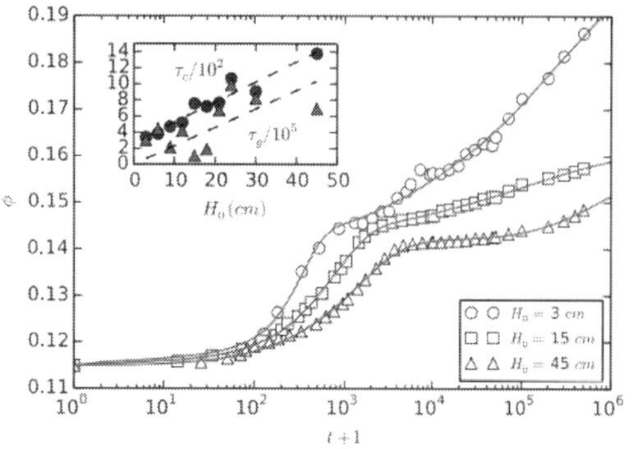

Figure 6. Experimental results of the compaction of a cohesive powder under vibrations (100 Hz, 7g). The red lines are the KWW2 expression (12). Insert: fitted characteristic times vs packing height.

ber of grains in the system, the fitted characteristic time over the initial packing height τ/H_0 varies linearly with $\phi_{\infty g}/\phi_{0g}$, as predicted by equation (7).

A second set of experiments was conducted with a cohesive UO_2 powder. This powder is made of grains of $d = 30\ \mu$m diameter with rough surfaces and a Hausner ratio $I_H = 1.53$. The figure 6 presents three sets of data for three different filling heights of the tank. Starting at a low initial volume fraction (the UO_2 grains have an intrinsic porosity), a first increase occurs at $t \approx 10^3$ cycles of vibrations. A direct observation of the system during the beginning of the vibration process shows the existence of macro-cavities rising upwards at least near the front plate of the tank, and these large void structures may exist also in the bulk (Figure 7). For a larger time, a second stage of compaction occurs at $t \approx 10^4$ cycles.

Despite a long experimental time (10^6 cycles), we did not observe a saturation of the volume fraction, the limit ϕ_∞ seems to be ill-defined in our experiment. However, the double exponential (12) fits well the experimental data, and we were able to extract the characteristic times τ_c and τ_g as a function of the number of particles in the system.

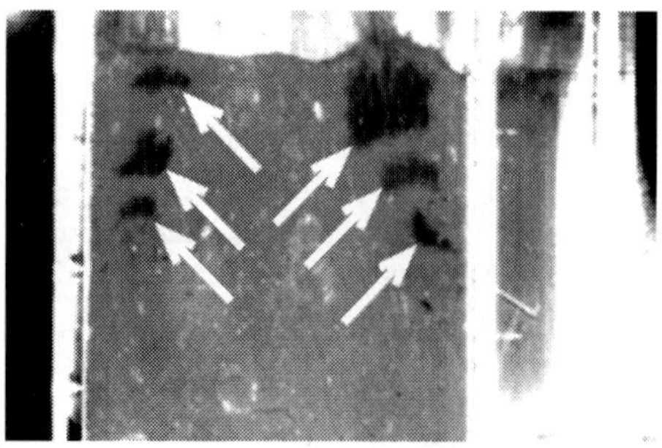

Figure 7. Image of the UO_2 powder during the vibration process. The arrows indicate macro-cavities near the front plate. Only the top of the tank is shown here.

The insert of Figure 6 seems to show a linear trend between the characteristic times and the height of the initial packing.

4 Conclusions

With this stochastic model we demonstrate that the characteristic time of compaction is proportional to the number of moveable objects. This result seems to be confirmed by experiments on two examples of granular material: a non-cohesive glass beads assembly, and a cohesive powder. Through this work we do not agree with the previous work of Hao [9, 10] where the time of compaction is related to the inverse of the mass of granular material.

A two-stages compaction evolution has already been proposed by Barker and Mehta [11] but here we associate the first stage of compaction with a collective motion of grains, and the second stage of compaction with the individual motion of the grains.

References

[1] P. Philippe and D. Bideau, *Europhys. Lett.* **60** (5), 677-683 (2002)
[2] J. B. Knight, C. G. Fandrich, C. N. Lau, H. M. Jaeger ans S. R. Nagel, *Phys. Rev. E* **51**(5) 3957-3963 (1995)
[3] E. R. Nowak, J. B. Knight, E. Ben-Naim, H. M. Jaeger and S. R. Nagel, *Phys. Rev. E* **57**(2) 1971-1982 (1998)
[4] E. Rondet, M. Delalonde Michèle and T. Ruiz, *Rec. Progr. Génie des Procédés*, **107** 4.8-1–4.8-9 (2015)
[5] R. L. Carr, *Chem. Eng.* **72**, 163-168 (1965)
[6] P. Philippe, F. Barbe, S. Bourlès, X. Thibault and D. Bideau, *Phys. Rev. E* **68**, 020301(R) (2003)
[7] T. Boutreux and P.-G. de Gennes, *Physica A* **244** 59-67 (1997)
[8] J.J. Arenzon and Y. Levin, *Physica* (Amsterdam) **325A**, 371 (2003).
[9] T. Hao, *Soft Matter* **11** 1554 (2015)
[10] T. Hao, *Soft Matter* **11** 3056 (2015)
[11] G. C. Barker and A. Mehta, *Phys. Rev. E* **47**(1) 184-188 (1993)

Experimental study of shear bands formation in a granular material

Thai Binh Nguyen[1], *Sean* McNamara[1], *Jérôme* Crassous[1], and *Axelle* Amon[1,*]

[1]*Université de Rennes 1, Institut de Physique de Rennes (UMR UR1-CNRS 6251), F-35042 Rennes cedex, France*

Abstract. We present an experimental investigation of the formation of shear bands in a granular sample submitted to a biaxial test. Our principal result is the direct observation of the bifurcation at the origin of the localization process in the material. At the bifurcation, the shear band is spatially extended: we observe a breaking of symmetry without any sudden localization of the deformation in a narrow band. Our work thus allows to clearly distinguish different phenomena: *bifurcation* which is a ponctual event which occurs before the peak, *localization* which is a process that covers a range of deformation of several percents during which the peak occurs and finally *stationary shear bands* which are well-defined permanent structures that can be observed at the end of the localization process, after the peak.

1 Introduction

Localization of the deformation and shear band formation in granular materials is an ubiquitous phenomenon which is still very far from being understood [1–7]. Theoretically the phenomenon is described as a bifurcation [2, 8, 9]. Experimental works using full-field measurements have studied the distribution of the deformation in a granular materials showing that strain heterogeneities are observed before failure [3, 10–14] but, to the best of our knowledge, the bifurcation initiating the localization process has never been reported in a quantitative study. Here, we report an experimental study where we are able to characterize objectively the distribution of plastic strain in the sample. Using image analysis tools we are able to exhibit a symmetry breaking in the strain distribution coining the bifurcation that occurs in the system.

Our quantitative measurements also allow to propose some disambiguation about different notions. We will show in the following that we are able to distinguish between three different phenomena:

- *bifurcation*: we are able to evidence a symetry breaking in the plastic strain distribution which allows us to identify the critical value of the strain at which a bifurcation takes place. This phenomenon can be qualified as sudden as it is ponctual during the loading, but it does not correspond to the emergence of a failure plane as the strain field is still diffuse at this stage. This phenomenon occurs before the peak of the loading curve.
- *localization*: starting at the bifurcation, a process of focussing in the plastic strain field is observed from a diffuse, extended field towards a narrow band. We thus underline that localization is not a sudden, catastrophic event but qualifies an evolution in the spatial extension of the strain distribution from very diffuse field to a strongly confined one.
- *stationary shear bands*: the narrowing of the spatial distribution of the plastic activity finally reaches a stationary state. At this stage well-defined, stationary shear bands are observed.

*e-mail: axelle.amon@univ-rennes1.fr

2 Experiment

The experimental setup is a biaxial test in a plane strain configuration that has been extensively described elsewhere [12].

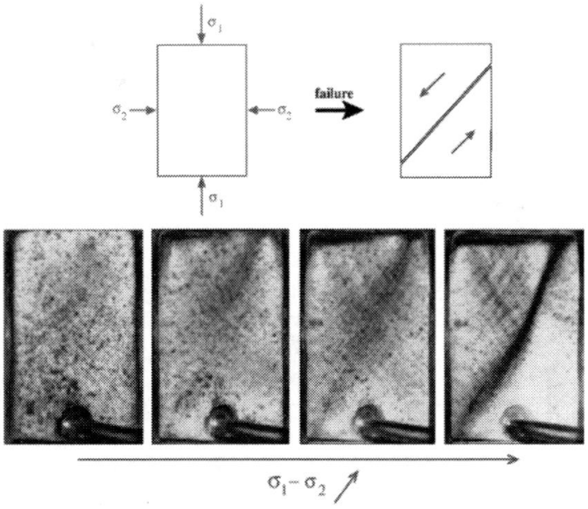

Figure 1. Top: Principle of a biaxial test. Bottom: Examples of deformation maps obtained at different stages of the loading process during the test. Colorcode: yellow, local strain $\simeq 10^{-5}$, black, $\gtrsim 10^{-3}$.

DOI: 10.1051/epjconf/201714010001

A rectangular sample of size $85 \times 55 \times 25$ mm^3 of granular material (glass beads, $d = 70 - 110$ μm, volume fraction ~ 0.60), submitted to a confining pressure of 30 kPa, is loaded by a quasi-static uniaxial compression. An original interferometric measurement method of micro-deformations based on diffusing-wave spectroscopy [15, 16] is used to obtain deformation maps of the sample (see bottom row of Fig. 1). This method gives access to deformation of the order of 10^{-5} all along the loading with a spatial resolution of a few beads diameter. Those maps are directly link to the local deformation between two successive images so that we have access to a spatially-resolved measurement of the incremental strain.

The observed behavior is very reproducible (see References [12, 17] and the movies in supplemental material of those references). Intermittent heterogeneities of the strain are observed from the very beginning of the loading [12–14]. This behavior corresponds to correlated plastic events in the amorphous material and have been analyzed and interpreted in references [13, 18]. The evolution of this pattern as the loading proceeds is complex. Finally, from an axial deformation of about 6.5%, well defined stationary shear bands are observed [12, 17] with an angle corresponding to the Mohr-Coulomb angle [19]. The angles measured directly on the maps are shown in insert Figure 3(a) and the Mohr-Coulomb values deduced independently from the loading curve are indicated on the graph. We see that we have a good correspondence between those angles for each of our experiments, showing that we indeed observe the Mohr-Coulomb angle.

Here we study more specifically the formation of the final shear bands by image analysis of the incremental deformation maps obtained during loading.

3 Image analysis method

We have developed different image analysis tools [17, 20, 21] in order to characterize quantitatively the process of formation of the shear band. We present here only the results obtained using projection analysis. Other methods allows to obtain complementary informations that confort the results presented here (see Ref. [17] for a full detailed study).

The image analysis consists in a projection of the values of the pixels on a line of variable inclination (see Figure 2 top). A projection at a given angle provides a profile of intensity. For an optimal angle of the line a maximum in the profile is detected (see Figure 2 bottom). This optimal profile allows to deduce both the inclination of the shear band (from the angle of the optimal line) and the width of the shear band from the width of the profile.

Those observables allow us to quantify the orientation and degree of localization of the strain distribution process during the loading and to characterize the bifurcation which precedes and initiates the localization process in the material. It also allows us to identify the axial strain from which stationary shear bands can be unambiguously identify.

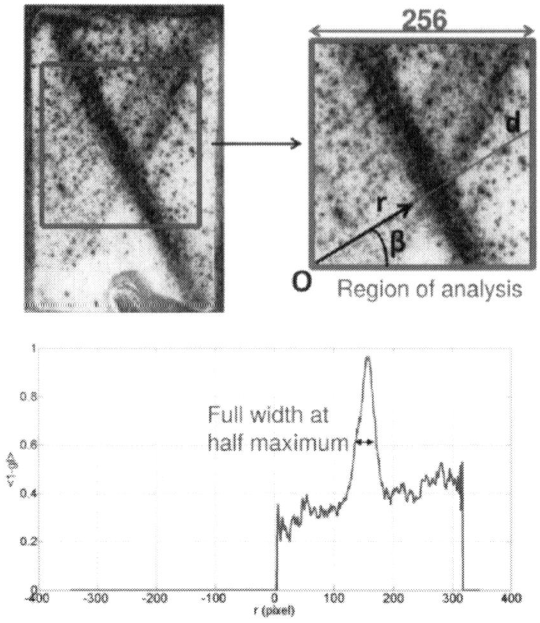

Figure 2. Principle of the image analysis: we define a region of interest for the analysis to avoid artefacts due to boundaries. We projet the grey level of the pixels on a line and search for the inclination β leading to a maximum of intensitiy in the projected profile. We obtain from the inclination of the line the orientation of the strain field and from the width of the curve at half maximum the width of the distribution of the strain.

4 Results

Figure 3 gathers our observations for three different experiments done in the same conditions and corresponding each to a color. Fig. 3(a) shows the loading curves for each test and in inset are shown the final deformation maps also for each experiment. We observe on those maps the final shear bands. The angles measured directly on the maps are in agreement with the ones deduce from the loading curves using a Mohr-Coulomb failure criterium [19].

Figure 3(b) presents the angle of inclination of the strain field obtained from image analysis (see previous section and Fig. 2) as a function of the axial strain for each experiment. Two different zones can be identified: from 0 to roughly 4.5%, the angles are widely fluctuating and the error bars on their determination are huge. This means that during the first half of the loading no definite direction can be determined in the strain field. The plastic field distribution is thus isotropic. After an axial strain of 4.5% a definite and well-resolved angle is observed for each experiments. The angles observed are not identical from one experiment to the other but all the observed angles gather in two groups of values symmetrical with respect to the loading axis. This coins a symmetry breaking and thus a bifurcation: at $\epsilon \approx 4.5\%$ an orientation emerges spontaneously in the strain field. We observe that the chosen direction is well-defined and stays roughly constant until the

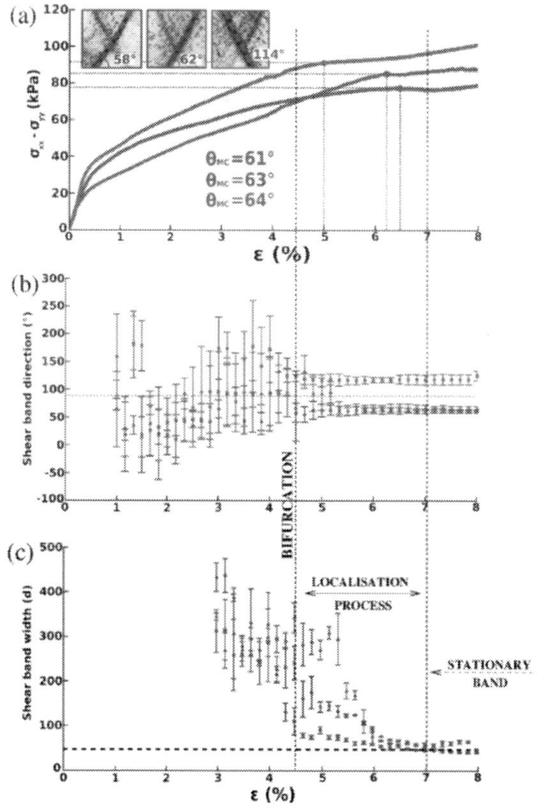

Figure 3. (a) Loading curves of the three different tests done in the same conditions. Insert: final deformation maps exhibiting stationary shear bands. The angles directly measured are in agreement with the Mohr-Coulomb angles θ_{MC} that can be deduced from the loading curves. (b) Principal orientation of the spatial distribution of the deformation field for the same three experiments. (c) Width of the distribution of the deformation field.

end of the test and thus corresponds to the Mohr-Coulomb angle. We underline that we do not observe at this stage a failure plane but only an overall orientation in a diffuse field. This direction is chosen spontenaously by the system between two conjugated symmetrical direction.

Figure 3(b) shows the evolution of the width of the projected profiles as a function of the axial strain for each test. Before the bifurcation this width has no real meaning as the image is approximatively isotropical. We thus measure width of the order of the size of the region of interest. When the bifurcation occurs, the width of the distribution becomes meaningful. We observe in Figure 3(b) that at the bifurcation the profiles are still very large, meaning that the strain field is very diffuse when the bifurcation occurs. From $\epsilon \approx 4.5\%$ to $\epsilon \approx 6.5\%$ we observe a narrowing of the spatial extension of the profiles until a stationary value is reached, identical for all our experiments. We thus observe that the phenomenon of localization is not a sudden event but on the contrary a progressive process that starts with a bifurcation and ends when the width of the bands have reached a stationary value. Interestingly, such an evolution of the shear band width has been reported in very recent theoretical and numerical works on shear band formation [22].

Our experimental results lead us to distinguish the bifurcation phenomenon from the formation of the failure plane, i.e. the stabilization of shear bands on a constant width. Those two phenomena are not concomittant in a test but on the contrary occurs at two well-separated values of the axial strain from either side of the local maxima of the loading curve. Our analysis allows to clearly identify and distinguish them. A more thorough analysis of this localization process can be found in Ref. [17].

5 Conclusion

We have presented an experimental study of the shear band formation during a biaxial test of a granular sample. Using an unmatching resolution in deformation measurement as well as efficient image analysis tools, we provide a quantitative study of the localization process and most especially of the bifurcation that initiates this localization.

Our experiment clearly shows that it is necessary to distinguish the *bifurcation* which coins a symetry breaking generating a general orientation in a diffuse, extended strain field, from the establishment of the *stationary shear bands*. Those two phenomena are not simultaneous but on the contrary are well separated during the loading and are respectively observed before and after the local maximum of the loading curve. Between those two events, a *localization* process takes place. We think that our experiment is the first to be able to characterize quantitatively all those phenomena and thus that it will certainly be a benchmark for future theoretical works. An very surprising result is the fact that the angle chosen at the bifurcation does not evolve strongly during the remaining of the loading. Consequently, the Mohr-Coulomb angle emerges in the system very soon during the loading and before a modeling of the system as two parts in frictional sliding has any relevancy.

To understand the micromechanical processes at the origin of the bifurcation, we are presently performing numerical simulations of a bidimensional biaxial test using discrete elements. Those studies should allow us to observe the influence of different parameters on the shear band formation and study the processes underlying the failure process [4–7].

Acknowledgements

We thank J. Weiss for insightful discussions.

References

[1] *Plasticity and geomechanics*, Davis, R. O. and Selvadurai, A. P. S. (Cambridge University Press, 2002).
[2] *Bifurcation analysis in geomechanics*. Vardoulakis, I. and Sulem, J. (Blackie Academic and Professional, Glasgow, England, 1995).

[3] Desrues, J. and Viggiani, G. *Int. J. Numer. Anal. Methods Geomech.*, **28**, 279–321 (2004).

[4] Wren, J. R., and Borja, R. I., *Comput. Methods Appl. Mech. Engrg.* **141**, 221 (1997).

[5] Tordesillas, A. *et al. Proc. R. Soc. A* **469**, 20120606 (2013).

[6] Gimbert, F., Amitrano, D., and Weiss, J., *EPL* **104**, 46001 (2013).

[7] La Ragione, L., Prantil V. C., and Jenkins, J. T., *J. Mech. Phys. Solids* **83**, 146 (2015).

[8] Rudnicki, J. W. and Rice, J. R. *J. Mech. Phys. Solids*, **23**, 371 (1975).

[9] Bardet, J.P. *Computers and geotechnics* **10**, 163 (1990).

[10] Rechenmacher, A. L. *J. Mech. Phys. Solids* **54** 22 (2006).

[11] Hall, S. *et al. Géotechnique* **60**, 315 (2010).

[12] Le Bouil, A. *et al. Gran. Matt.* **16**, 1 (2014).

[13] Le Bouil, A., Amon, A., McNamara, S., and Crassous, J., *Phys. Rev. Lett.* **112**, 246001 (2014).

[14] Desrues, J. and Andó, E. *C. R. Physique* **16**, 26 (2015).

[15] Erpelding, M., Amon, A., and Crassous, J. *Phys. Rev. E* **78**, 046104 (2008).

[16] Erpelding, M. *et al. Strain* **49**, 167 (2013).

[17] Nguyen, T. B., and Amon, A., *EPL* **116**, 28007 (2016).

[18] McNamara, S., Crassous, J., and A., Amon, *Phys. Rev. E* **94**, 022907 (2016).

[19] *Statics and Kinematics of Granular Materials*, R. M. Nedderman (Cambridge University Press, 1992).

[20] Lehoucq, R. *et al. Front. Phys.*, **2**, 84 (2015).

[21] Welker, P., PhD Thesis, Universität Stuttgart (2011) http://dx.doi.org/10.18419/opus-5019.

[22] Benzi, R. *et al.*, *Soft Matter* **12**, 514 (2016).

On the relevance of the meso-scale in an upscaling approach for granular materials

Sinh Khoa Nguyen[1], *Hélène* Magoariec[1,*], *Eric* Vincens[1], and *Bernard* Cambou[1]

[1] *University of Lyon, LTDS - CNRS UMR 5513, Ecole Centrale de Lyon, 36 avenue Guy de Collongue, Ecully 69134 Cedex, France*

Abstract. This paper deals with the interest of upscaling from a meso-scale understand and model the mechanical behavior of granular materials. The meso-domains associated to the meso-scale are defined as closed loops of particles in contact. Six sets of meso-domains with different local textures called "phases" have been defined. The analysis of the behavior of these phases through 2D DEM simulations of a compression test followed by an extension test has allowed the discriminant evolution of the structure and then the role of the different phases at the meso-scale to be evidenced. Constitutive modeling of granular materials can advantageously be developed from the meso-scale where the principles of continuum mechanics apply.

1 Introduction

So far, a straightforward and reliable modeling of granular materials remains an open issue. The difficulty lies in the very discrete nature of these materials composed of grains in contact. In this study, we explored a new way to perform an uspcaling approach considering a meso-scale, defined at the level of clusters of particles [1, 2]. Following [3, 4], six sets of meso-domains with different local textures called "phases" has been defined. An analysis of the behavior of these phases through 2D DEM simulations is provided. It tends to emphasize the key role of the meso-scale in an upscaling approach to model the mechanical behavior of granular materials.

The numerical biaxial compression test is achieved using the commercial software PFC2D. The sample is composed of 25,000 circular particles, whose diameter is uniformly distributed from 4.6 mm to 9.2 mm. The normal and tangential stiffnesses at contact are equal to 10^9 N/m, the intergranular friction ratio is equal to 0.57, the walls are rigid and frictionless, the numerical damping are taken equal to 0.3, and the simulations are gravity-free. The compaction process leads to an isotropic stress state of 100 kPa with a porosity of the sample equal to 0.167. From this isotropic state, the biaxial test consists of a loading stage followed by an unloading stage (Figure 5 – gray lines). Herein, the compression direction $\mathbf{1}'$ is such that:

$$\dot{E}_1^d > 0 \tag{1}$$

where $\dot{E}_1^d$ is the diagonal component of the incremental deviatoric strain tensor $\dot{E}^d$.

*e-mail: helene.magoariec@ec-lyon.fr

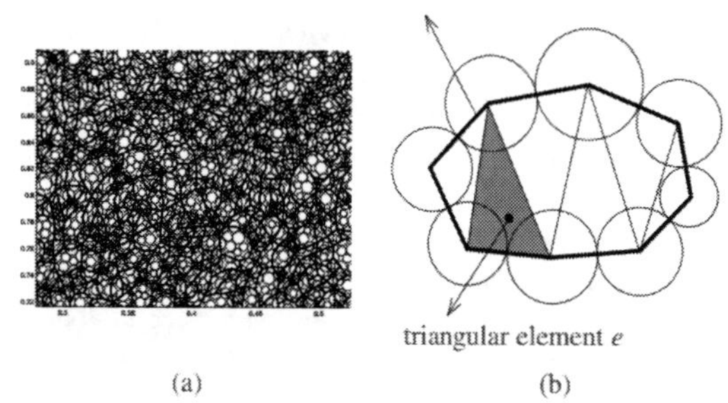

Figure 1. (*a*) Set of polygonal meso-domains and (*b*) Triangular elements, with open and closed branches, defined by the Delaunay triangulation and meso-domain (bold lines) obtained by association of the triangles sharing an open branch.

2 Definitions for the meso-scale

We propose to define an intermediate scale at which both the local strain and stress fields can be consistently defined and the local texture can be consistently described [3, 5, 6]. To this end, we rely upon methods proposed in the literature for partitioning a granular medium into local structures. It must be noted that a significant number of particles in the sample do not have any contact or have only one or two contacts because of the gravity-free assumption. These particles, called rattlers, do not participate in supporting the external loading. They are thus disregarded for all the following analysis.

The sample is first paved with triangles connecting the centers of mass of three neighboring particles using a Delaunay triangulation [7]. Each edge of this triangulation is then considered either as a closed branch if connecting two particles in contact, or as an open branch if connecting two neighboring particles without contact. Triangles sharing a common open branch are then associated to obtain local polygonal structures, here called the meso-domains (Figure 1). The meso-scale is then defined as the scale of these local structures. Upscaling from basic Delaunay cells cannot provide a correct upscaling because they do not hold a physical meaning as a local pore around which a mechanical transfer of forces is made possible. As a consequence, they cannot capture the local change of volume.

The structure at the meso-scale is characterized by two meso-variables: the meso-density and the meso-texture. The meso-density is usually defined through the porosity, which is the ratio of the void volume inside the meso-domain m, V_v^m, to its total volume, V^m. For each meso domain m, the meso-texture is characterized by the loop tensor $\boldsymbol{L}^m$, from which two scalars are defined [3, 8]: its orientation, θ^m, and its elongation degree, β^m:

$$\boldsymbol{L}^m = \sum_{k=1}^{r^m} l^k \boldsymbol{n}^k \otimes \boldsymbol{n}^k, \quad \theta^m = (\widehat{\boldsymbol{m}^m, \boldsymbol{1}'}), \quad \beta^m = \frac{\sqrt{\boldsymbol{D}^m : \boldsymbol{D}^m}}{tr(\boldsymbol{L}^m)}, \tag{2}$$

where r^m is the valence of the meso-domain m, l^k is the length of the closed branch k and $\boldsymbol{n}^k$ its outward unit vector, $\boldsymbol{D}^m$ is the deviatoric part of the loop tensor $\boldsymbol{L}^m$ and $\boldsymbol{m}^m$ its the major principal direction, and $\boldsymbol{1}'$ is the compression direction. To analyze the role of the meso-domains texture throughout a loading path, the sample is divided into sets of meso-domains sharing the same texture characteristics. This sets, here called *the phases*, are obtained considering two sorts of classes:

- 3 orientation classes: the class 1′ of meso-domains nearing the compression direction $\boldsymbol{1}'$, the class 2′ of meso-domains perpendicular to the compression direction, and the class 0′ of meso-domains oriented in an intermediate direction;
- 2 elongation classes: the class W of meso-domains weakly elongated ($\beta^m < \langle\beta^m\rangle_V$, where $\langle\cdot\rangle_V$ is the volume-weighted operator) and the class S of meso-domains strongly elongated ($\beta^m > \langle\beta^m\rangle_V$).

The six phases are then denoted by: $S1'$, $S0'$, $S2'$, $W1'$, $W0'$, $W2'$ and the the mean orientation of a given phase p is denoted by θ^p. The relevance of these phases to characterize the meso-scale and to be a framework for a classical volume average based change of scale are justified through the analyse of their spatial distribution which showed no spatial localization of any phase [4].

The definition of the strain tensor increment of a meso-domain m, $\delta\boldsymbol{\varepsilon}^m$, is based on the definition of the incremental gradient tensor for a triangular element e proposed by [9, 10] and used in [3–6, 11]:

$$\delta\boldsymbol{a}^e = \frac{2}{3}\sum_{s=1}^{3} \delta\boldsymbol{u}^s \otimes \Delta\boldsymbol{y}^s, \quad \text{with} \quad \sum_s \Delta(\boldsymbol{x}^s \otimes \Delta\boldsymbol{y}^s) = \boldsymbol{I}, \tag{3}$$

where $s \in \{1, 2, 3\}$ is the sides index of the cell e, $\delta\boldsymbol{u}^s$ is the difference of the incremental displacements between the two vertices of a side s, $\Delta\boldsymbol{x}^s$ is the vector defined by side s, and $\boldsymbol{I}$ is the identity matrix. The incremental displacement gradient tensor for a meso-domain m, $\delta\boldsymbol{a}^m$, is then calculated as the volume average of $\delta\boldsymbol{a}^e$ and the strain tensor increment of a meso–domain, $\delta\boldsymbol{\varepsilon}^m$, is obtained as the symmetric part of $\delta\boldsymbol{a}^m$.

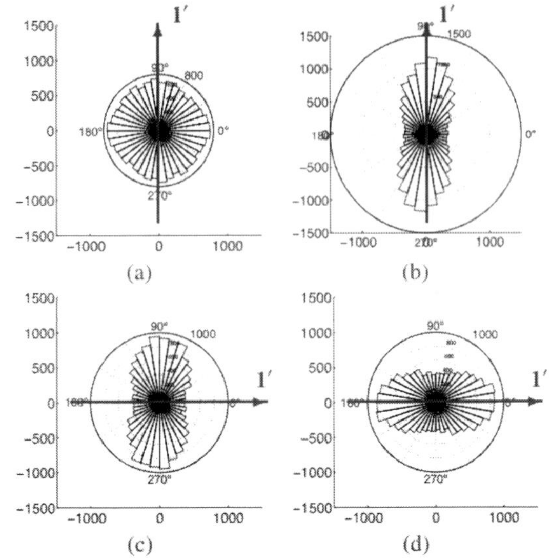

Figure 2. Polar distributions of the number of meso-domains for the reference states. Loading path: (*a*) initial state, (*b*) peak state. Unloading path: (*c*) initial state, and (*d*) critical state.

The definition used for the stress tensor in a meso-domain m is the one proposed by [5] and used in [4, 6, 10, 11], based on the external contact forces acting on the particles of the given meso-domain:

$$\hat{\boldsymbol{\sigma}}^m = \frac{1}{\hat{V}^m}\sum_{c\in K^m_{ext}} \boldsymbol{f}^c \otimes \boldsymbol{x}^c, \tag{4}$$

where K^m_{ext} is the set of external contacts for the meso-domain m, $\boldsymbol{x}^c$ is the position vector of contact c, $\boldsymbol{f}^c$ is the external contact force acting at contact c, and $\hat{V}^m$ is the volume delimited by the boundary on which the contact forces $\boldsymbol{f}^c$ applied. Then, the following assumption is done: $\boldsymbol{\sigma}^m$ in volume V^m is equal to $\hat{\boldsymbol{\sigma}}^m$ defined in volume $\hat{V}^m$. The validity of this assumption was proved in [5].

For a given phase p, the strain tensor increment, $\delta\boldsymbol{\varepsilon}^p$, and the stress tensor, $\boldsymbol{\sigma}^p$, are obtained as the volume average of $\delta\boldsymbol{\varepsilon}^m$ and $\boldsymbol{\sigma}^m$ over all the meso-domains of the considered phase, respectively:

$$\delta\boldsymbol{\varepsilon}^p = \langle\delta\boldsymbol{\varepsilon}^m\rangle_{V^p}, \quad \delta\boldsymbol{\sigma}^p = \langle\delta\boldsymbol{\sigma}^m\rangle_{V^p}. \tag{5}$$

3 Evolution of the meso-structure

Figure 2 depicts the evolution of the distribution of orientations of all meso-domains for some usual reference states: the initial state, the maximum shearing resistance state (peak state), and the critical state. At the initial state

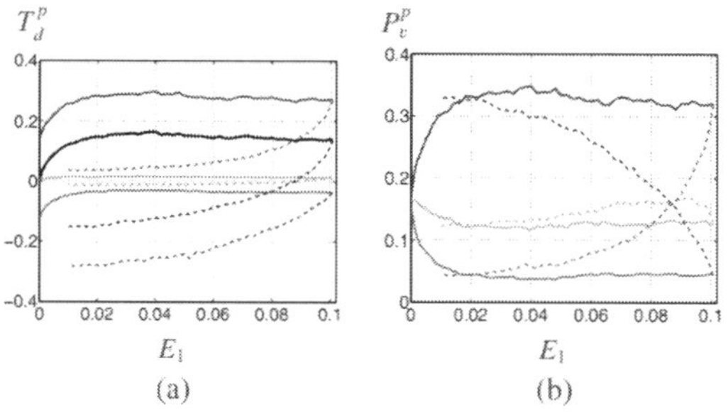

Figure 3. (a) Anisotropy T_d^p and Percentage by volume P_v^p occupied by the six phases for elongation class S and for different orientation classes. Blue lines: 1′; Red lines: 2′; Green lines: 0′. Solid lines: loading path; Dashed lines: unloading path.

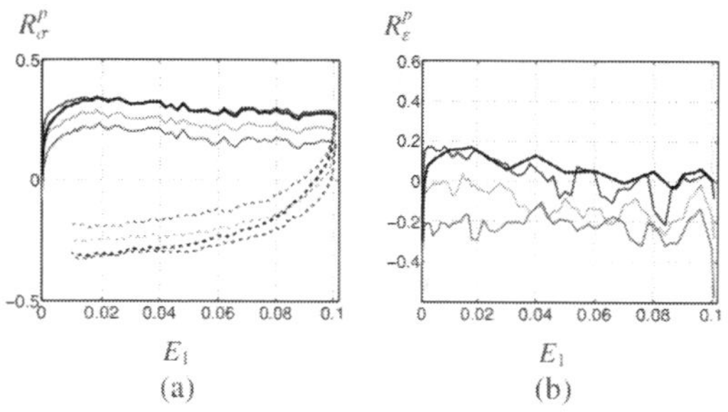

Figure 4. (*a*) Stress ratio and (*b*) Strain ratio for phases of elongation class S and for different orientation classes. Blue lines: 1′; Red lines: 2′; Green lines: 0′; Black lines: sample. Solid lines: loading path; Dashed lines: unloading path.

of the loading path, which corresponds to an isotropic state of stress for the sample, one can note that the meso-domains are not strictly isotropically oriented due to boundary effects. During loading, the meso-domains are mainly oriented in the major principal compression direction. Then, during unloading, the major principal compression direction changes from the vertical direction to the horizontal direction, and the meso-domains are subjected to a reorientation according to this new major principal compression direction. Similar results can be observed when considering the two elongation classes W and S.
To quantify the evolution of this anisotropy, the texture tensor of a given phase defined in [4] is used:

$$\boldsymbol{T}^p = \frac{1}{V^p}\sum_{1}^{N^m} V^m \boldsymbol{m}^m \otimes \boldsymbol{m}^m, \tag{6}$$

where V^p is the volume of the phase p, N^m the number of meso-domains in phase p, $\boldsymbol{m}^m$ is defined by Equation (2). The second invariant of the texture tensor is used to quantify the intensity of anisotropy. Whatever the loading direction, the phases of the orientation class 1′ develop a high level of anisotropy T_d^p, while the phases of the orientation class 2′ have a weak contribution to overall anisotropy (Figure 3(a)). On the contrary, the anisotropy of the phases 0′ hardly evolves and remains stable throughout loading and unloading. During the loading, new meso-domains are created and others are destroyed. This leads to an evolution of the volume of the six phases during the loading–unloading test.

In terms of percentage by volume of the different phases, the initial state is homogenous. Throughout loading, the volume of the meso-domains of orientation class 1′ increases, reaching a maximum and then decreases slightly, while the volume of the meso-domains of orientation class 2′ decreases quickly before stabilizing (Figure 3(b)). The volume of the meso-domains of orientation class 0′ does not evolve significantly. In fact, new contacts are preferentially created in the compression direction **1′**, which generates many meso-domains in this direction and increases their total volume. Throughout unloading, the same observations hold if we consider the new compression direction which has changed (see on Figure 2). Whatever the loading direction, the granular material mainly bases its mechanical resistance on the meso-domains of orientation class 1′.

4 Analysis of meso-stress/strain

The analysis of the behavior of the different phases is performed by means of the evolution of the stress ratio and of the incremental strain ratio respectively defined by:

$$R_\sigma^p = \frac{\sigma_{1'}^p - \sigma_{2'}^p}{\sigma_{1'}^p + \sigma_{2'}^p}, \quad R_\varepsilon^p = \frac{\delta\varepsilon_{1'}^p + \delta\varepsilon_{2'}^p}{\delta\varepsilon_{1'}^p - \delta\varepsilon_{2'}^p}, \tag{7}$$

where $\sigma_{i'}^p$ is the stress component of phase p in the principal direction i' of the macroscopic stress and where $\delta\varepsilon_{i'}^p$ is a component of the incremental strain tensor for phase p in the principal macroscopic stress direction i'. A positive (respectively negative) value of $\delta\varepsilon_{i'}^p$ corresponds to a dilating (respectively contracting) phase.
Figure 4(a-b) show the evolution of these ratios. For class S phases, the initial stress state in the six phases is not necessarily isotropic and is very different from one to another. The six stress ratios R_σ^p reach very different plateaus with values greater for phases 1′ and smaller for phases 2′. Furthermore, these values are independent of the considered stress path (corresponding to the critical state). The behavior of the phases is closely related to their orientation with respect to the loading path as well as to the distance between the current anisotropy level and the one at the critical state for the considered loading path. In Figure 4(b), at the beginning of the loading stage, the incremental strain ratio for class S take very different values for the different phase, then these values increase and reach more or less stabilized values. This ratio remains positive (dilation) for phases 1′ and negative (contractancy) for phases 2′. The conclusions that can be drawn for the unloading stage are similar to those obtained in the loading stage but with the adequate definition of orientation **1′**. These conclusions are consistent with

DOI: 10.1051/epjconf/201714012002

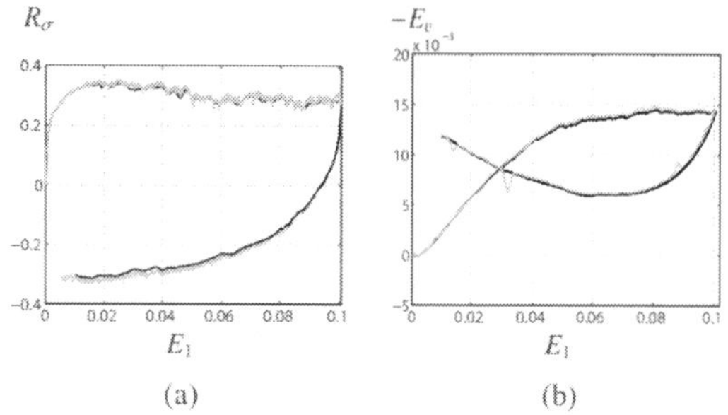

Figure 5. Comparison of the results of the DEM simulations (grey) and of the upscaling process (black) for the loading and unloading paths at the REV scale: (*a*) macroscopic stress ratio and (*b*) macroscopic strain ratio. Dashed lines: DEM. Solid lines: model.

the ones observed for the stress ratios. Phases reaching highest values of the stress ratio correspond to phases showing greatest values of dilatancy. Similar features were observed for class *W*.

5 Upscaling process

The upscaling process from the meso-scale to the macro-scale is based on the previously defined phases. The consistency of the mechanical behavior of these sets with respect to the loading direction has been here-before demonstrated. The percentage by volume of these sets evolves accordingly. So it is possible to define the evolution of the stress ratio and the volumetric strain at the sample level by considering the volumetric means of these two variables:

$$\boldsymbol{\Sigma} = \langle \boldsymbol{\sigma}^p \rangle_V , \quad \delta \boldsymbol{E} = \langle \delta \boldsymbol{\varepsilon}^p \rangle_V . \qquad (8)$$

These two volumetric means are compared to the values obtained from the boundary conditions (Figure 5). The very good agreement between these two results implies that:

- the meso-scale is an appropriated local scale at which the local mechanical behavior can be defined;
- the proposed meso-variables defined in the framework of continuum mechanics are relevant static and kinematic variables, even though the definition of the meso-stress raises some intrinsic issues;
- given the behavior of each phase, the knowledge of the percentage by volume occupied by each of them is a sufficient information to perform the change of scale.

6 Conclusion

A DEM simulation of a biaxial loading and unloading has been carried out in this work. Six sets of meso-domains called "phases" have been defined. The mechanical behavior of all the considered phases was found clearly dependent on their orientations with respect to the loading direction and on their current anisotropy level. Furthermore, the volumetric amount of phases oriented in the compression direction (respectively extension direction) increases (respectively decreases) during the loading. The very good agreement between the mechanical behavior defined from the sample boundary and the one defined from the volumetric average of the mechanical behavior of the set of phases proves that the complex behavior of granular material at a sample scale can be understood by an upscale from the meso-scale.

References

[1] N.P. Kruyt, L. Rothenburg. *Journal of Applied Mechanics*, **118**, p. 706–711 (1996)

[2] M.R. Kuhn. *Mechanics of Materials*, **31**, p. 407–429 (1999)

[3] N.S. Nguyen, H. Magoariec, B. Cambou, A. Danescu. *Int. J. Solids Struct.*, **46**, p. 63257–3271 (2009)

[4] S.-K. Nguyen, H. Magoariec, E. Vincens, B. Cambou. *Int. J. Solids Struct.*; **97–98**, p.256–274 (2016)

[5] N.S. Nguyen, H. Magoariec, B. Cambou. *Int. J. Numer. Anal. Methods Geomech.*, **36** (14), p. 1609–1635 (2011)

[6] N.S. Nguyen, H. Magoariec, B. Cambou. *C.R. Mécanique*, **342** (3), p. 156–173 (2014)

[7] L.J. Guibas, D.E. Knuth, M. Sharir. *Lecture of Notes in Computer Science*, **443**, p. 414–431 (1990)

[8] T. Tsuchikura, M. Satake. Statistical measure tensors and their application to computer simulation analysis of biaxial compression test, In *Engineering Mechanics: A Force for 21st Century*, Murakami H, Luco JE (eds). ASCE: Reston, p. 1732–1735 (1998)

[9] B. Cambou, M. Chaze, F. Dedecker. *Eur. J. Mech. A*, **19**, p. 999–1014 (2000)

[10] M. Chaze, B. Cambou. *Granular Matter*, **16** (5), p. 657–673 (2014)

[11] B. Cambou, H. Magoariec, N.S. Nguyen, *Granular Materials at Meso-scale: Towards a Change of Scale Approach*, ISTE Press and Elsevier Ltd (2016)

Assessing the representativeness of durability tests for wood pellets by DEM Simulation – Comparing conditions in a durability test with transfer chutes

Aditya Mahajan[1], *Ioannis* Dafnomilis[2], *Victoria* Hancock[2], *Gabriel* Lodewijks[2] and *Dingena* Schott[2,*]

[1] Indian Institute of Technology Delhi, Department of Mechanical Engineering, Hauz Khas, New Delhi, India – 110016
[2] TU Delft, Department of Maritime and Transport Technology, 2628 CD Delft, The Netherlands

Abstract. Dust generation when handling wood pellets is related to the durability of the product, in other words the wear rate of particles subject to forces. During transport, storage and handling wood pellets undergo different forces when interacting with different pieces of equipment.
This paper assesses the representativeness of the tumbling can test in relation to transfer chutes, by comparing forces acting on wood pellets in durability tests and in transfer chutes using DEM. The study also incorporates effects such as shape and size variations. The results showed that the tumbling can test underestimates compressive and tangential forces. Since the tested material is subject to milder conditions than in reality, it can be concluded that this test is not representative for the conditions in the supply chain of wood pellets.

1 Introduction

Over the years a standard for wood pellets (EN15210 or ISO/NP17831-1) [1] has been developed to assess the durability of materials amongst others applied to wood pellets, such as a tumbling can. However, it is unlikely that this is representative for the handling in the whole supply chain because real operational conditions can greatly differ in terms of forces from tests on lab-scale as is suggested by previous work on the topic comparing durability tests to filling and reclaiming of silos [2]. Moreover, in the biomass industry the problem of dust generation and generation of fines (crumbling off of smaller parts of the pellets) remains, despite a standard being in place.

This paper assesses the representativeness of the tumbling can test in relation to transfer chutes, by comparing forces acting on wood pellets in durability tests and in five transfer chutes using DEM. In addition the effect of the particle composition of a wood pellet is investigated.

2 Materials and Methods

2.1 DEM simulations

Discrete Element Method [3] is a numerical technique to analyse the movement and interaction of multiple small bodies in a system. In this paper EDEM 2.6.1 [4] has been used as the DEM simulation software.

2.1.1 Materials: Pellet Models

Wood pellets are cylindrical shaped particles of varying length. They are created by compressing smaller wood particles together in a pellet mill. Here 12 mm wood pellets are considered.

For the simulations 3 different pellet shapes were used. The shape variation is shown in Figure 1.

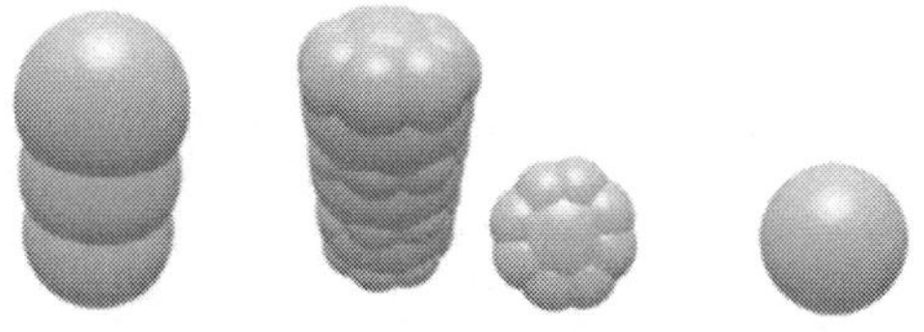

a) b) c)

Figure 1 : Pellet Models a) Simple Particle (3 spheres) b) Advanced Shape (35 spheres) c) Sphere Equivalent

The equivalent sphere diameter (D_{eq}) of shape c) was calculated using :

$$D_{eq} = 2\sqrt[3]{\frac{3}{4}\frac{V_{pellet}}{\pi}} \quad (1)$$

For shape a) and b) 3 different lengths were defined. The size distribution follows a total mass ratio of 4:3:3 for the 22mm, 30mm and 18mm pellet lengths respectively. An overview of particle shapes and size distribution is given in Table 1.

[*] Corresponding author: D.L.Schott@tudelft.nl

Table 1: Size and size distribution for for the 3 different particle shapes simulated with a tumbling can.

Para-meters	Simple Pellets	Advanced Pellets	Simple Pellets with size distribution			Sphere equivalent
D (mm)	12	12	12	12	12	15.568
L (mm)	22	22	22	30	18	-
Particle mass (g)	2.74	2.8	2.74	3.55	2.18	2.8
Number of pellets	182	179	73	42	68	179
			Total 183			

For shapes a) 3 spheres and b) 35 spheres, two different contact models were applied: Hertz-Mindlin no-slip and Hertz-Mindlin with bonding. This means that the spheres are flexibly bonded together instead of a rigid connection. This gives a total of 7 different pellet models tested.

2.2 Calibration

In order to ensure realistic behaviour of the particles inside the tumbling can and the transfer chutes, the material model will be calibrated with 3 laboratory tests; the bulk density test, the lifting can test and the ledge test.

The particle density was measured by putting a sample of 100g in a measuring cylinder filled with water. The volume was determined by the immediate increase of the water level, which indicates that minimum water is absorbed and the particle density is unaffected.

The ledge test is a box open at the top (Figure 2a). Two faces are made of wood and the rest is made of transparent acrylic. One wooden face is latched and hinged at the upper end. The inner height is 26cm, the door is 20cm and the viewing acrylic walls are 22 cm wide. First, the box is filled with pellets up to an approximate height of 14-15cm from the base. Then the door is unlatched and the material flows out until it reaches a certain angle of repose which was noted down. This was repeated 20 times.

The lifting can test is a cylinder open at both ends of 20cm height and 26cm diameter, i.e. more than 20 times the diameter of the sample pellet. The can is put on a base and then filled by pouring in pellets from the top until it is filled up to a height of 14-15cm. It is then lifted vertically and material escapes out from the bottom and spreads on the base resulting in formation of a pile (Figure 2b). The slope of this pile is measured from 5 different directions. This is repeated thrice, leading to 15 observations.

After determining the mean and standard deviation of the observations, the upper and lower limit values of the 95% confidence interval were obtained for the sample.

These tests will be simulated with DEM as well. For that, different combinations of parameters were tested and the results compared to the experimental results. After a number of iterations a calibrated model was achieved.

Since the value of the coefficient of static friction (wood-wood) in general lies between 0.2-0.5 and the rolling friction coefficient lies between 0-0.05 [5][6], the possible combinations were tested within this range.

Once calibrated, the model will then be used in simulations with the tumbling can test and the transfer chutes.

a)

b)

Figure 2: a) Ledge test, b) Pile formed after lifting can test

2.3 Tumbling can test

The tumbling can or rotating drum (Figure 3) is a test device with which the durability of wood pellets is determined under mechanical handling conditions. Other durability testers, such as the Holmen durability tester and the Ligno tester focus on the pneumatic handling of pellets. Compared to the tumbling can, both testers operate in a smaller time frame, where the pellets are exposed to higher destructive forces [7]. Nevertheless, according to Temmerman et al. [8] more repeatable and reproducible results are achievable with the tumbling can compared to the Ligno tester.

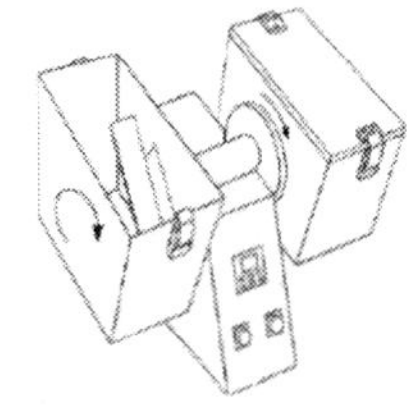

Figure 3 Tumbling can tester for pellet durability test according to EN15210-1

A single chamber of the tumbling can was modelled according to the EN15210 standard [1] as a stainless steel box of dimensions 0.3x0.3x0.125m. Inside a stainless steel baffle of 0.23x0.05m is mounted to one side. The amount of material (500g) and operating conditions (constant rotational speed of 50 rpm) were as described in the standard procedure. In practice the result is the mechanical durability derived from the measured number of fines created in the test due to breakage of particles as the result of forces acting on them. Here the forces are being used to compare to the conditions in the transfer chutes. The simulations were run for 30sec comprising 25 rotations.

2.4 Transfer chutes

Five different transfer chute designs were compared: The Rockbox, Sloped chute, Hood only, Hood & Spoon and Tralix are presented in Figure 4 and 5.

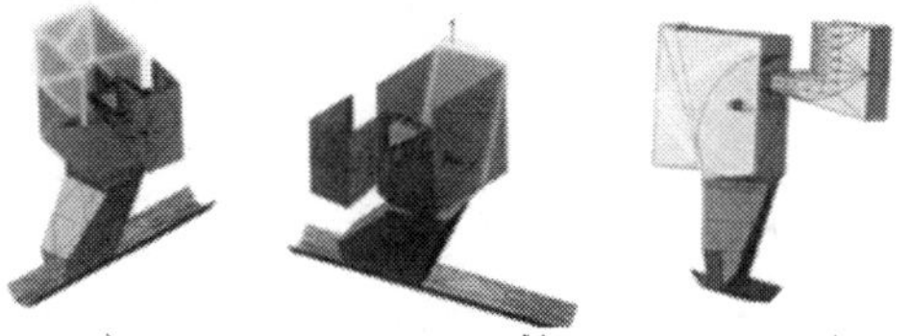

a) b) c)

Figure 4: a) Rockbox, b) Sloped Chute, c) Hood only

The Rockbox (Figure 4a) first fills to form a pile and after the pile attains a particular slope a regular material flow is generated. This layer of material then acts as a slope. As such the rockbox prevents damage of the equipment by sacrificing the biomass instead.

In a Sloped chute (Figure 4b) the material does not accumulate, instead the sloping face allows it to directly flow out to the receiving conveyor.

The Hood only model (Figure 4c) has a hood instead of a slope. The hood gives a smoother flow instead of abruptly stopping the particles.

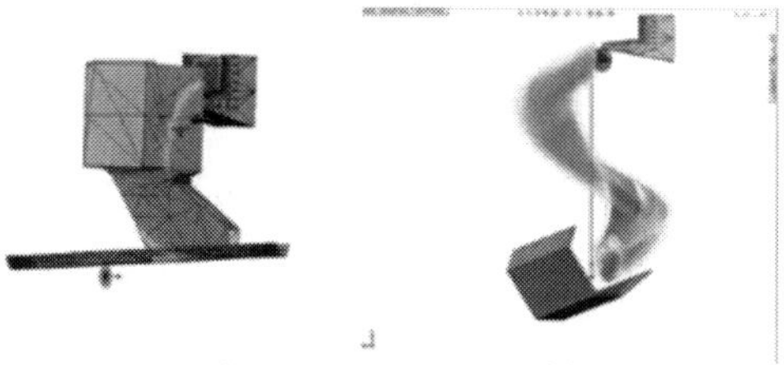

a) b)

Figure 5: a) Hood & spoon, b) Tralix (transfer helix)

The Hood & Spoon model (Figure 5a) is an upgraded version of the above model. It has an added curved (spoon) exit face to ensure smooth exit of particles to the output conveyor belt besides the smooth entry.

The Tralix (Transfer Helix) is depicted in Figure 5b and provides a helical path for particles to flow in a more streamlined and non-accumulative manner.

The modelled input and output belt speed was 4 m/s. The vertical distance between belts is 4.7m, except in helix where it is 11.2m. The mass flow is 1044 tph. The simulations were run twice with the simple pellet model (Figure 1a) and each took between 25 and 30 hours.

3 Results

3.1 Calibrated material

The angle of repose had a mean value of 41.1 degrees, standard deviation of 2.4 degrees (95% confidence interval being 39.9-42.3 degrees) for the ledge test and a mean of 24.4 degrees, standard deviation of 2.4 degrees (95% confidence interval being 23.1-25.7 degrees) for the lifting can test.

The angle of repose results obtained via simulations were matched to lie in the confidence interval of the experimental results and then ranked in order of closeness to the mean value, with the value closer to the mean having a higher rank. Then the data that qualified for both cases was checked by summing the ranks achieved in the 2 cases. The lower the sum, the closer the results are to the experiments.

From the analysis the coefficient of static and rolling friction was ascertained as 0.4 and 0.04 for 12 mm wood pellets. The full listing of DEM parameters is given in Table 2.

Table 2: DEM input parameters for wood pellets.

Parameter	Value	Parameter	Value
ρ_p [kg/m^3]	1389	ρ_w [kg/m^3]	7800
E_p [Pa]	1e08	E_w [Pa]	7e10
ν_p	0.1	ν_w	0.3
$C_{R,p}$	0.02	$C_{R,w}$	0.02
$\mu_{s,p\text{-}p}$	0.4	$\mu_{s,p\text{-}w}$	0.4
$\mu_{r,p\text{-}p}$	0.04	$\mu_{r,p\text{-}w}$	0.04
D_p [mm]	12	Δt [s]	1.73e-05
If bonded:			
Bonded disk radius	6 mm		
Normal stiffness	5e6 N/m^3	Critical normal stress	4e7 Pa
Shear stiffness	5e6 N/m^3	Critical shear stress	2e7 Pa

3.2 Tumbling can test

With the calibrated model, simulations of the tumbling can were performed for the various pellet models introduced in Section 2.1.1. The particle trajectories in the tumbling can are illustrated in Figure 6.

a) b) c)

Figure 6: Tumbling can particle trajectories at a) t=1.5 s, b) t=1.7 s and c) t=1.8 s

The forces acting on the particles during rotation of the can are analysed and the compressive force is depicted in Figure 7.

The advanced shape particle is at the higher end in all cases, i.e. when all are non-bonded models (1st three series) and when all are bonded models (next three series). This is probably due to the fact that the advanced shape model has a more realistic shape, lesser vacant voids and since it is made of more spheres, it has a higher area of contact with the surroundings. The other shapes and sizes undergo slightly smaller compressive forces.

Also the pellets built with bonded contact models experience slightly higher compressive forces compared to their respective non-bonded counterparts. However, a

DOI: 10.1051/epjconf/201714015004

significant effect of the pellet model on the average compressive forces could not be shown. From a comparison of tangential and compressive forces the average tangential force is lower by a factor of 6-7 (not shown here). It can be inferred that the tangential forces do not play a critical role compared to the compressive forces for the durability of particles.

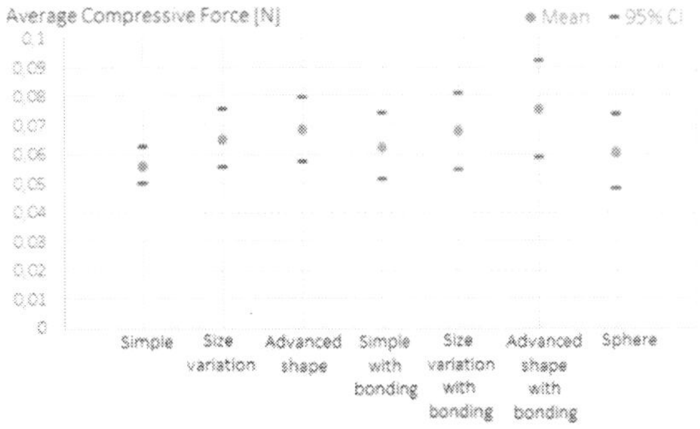

Figure 7: Average compressive force and the 95% confidence intervals in the tumbling can for different pellet models.

The angular velocity analysis reveals how size variation leads to bigger size particles providing more gaps for the smaller particles to rotate and roll in, whereas the advanced shape provides indented surface leading to more friction and difficulty in rolling.

3.3 Transfer chutes

The average particle compressive force for the five different configurations of the transfer chutes is depicted in Figure 8.

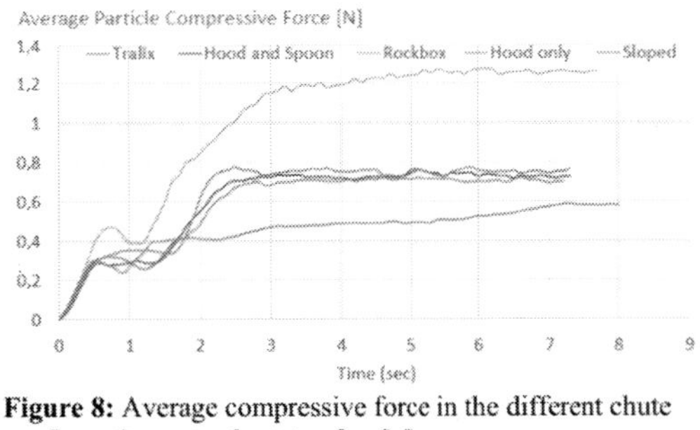

Figure 8: Average compressive force in the different chute configurations, steady state after 2.5s

The initial behaviour (up to 1.5s) is common as the particles are in the feeding conveyer and are about to enter the chutes. After that the difference can be seen among the different chutes. The hood & spoon, the hood only and the sloped models have almost identical performance. After 2.5s the particle forces in the transfer chutes reach a steady state except for the rockbox and the Tralix. For the rockbox, this is because the accumulation and initial heap formation takes more time. After the flow in the rockbox is established it can be seen that the average compressive force is almost twice the value of the other configurations. The Tralix is gentler as it carries the bulk down gradually in uncrowded flow instead of particles falling down and impacting.

The pattern for tangential forces (not shown here) is similar except that the tangential forces are 7-8 times smaller than the compressive forces. The values of the hood and spoon configuration are closer to the rockbox.

4 Conclusions

Based on the DEM simulation results in this paper the tumbling can cannot be considered as representative for the conditions in transfer chutes. From a comparison of forces acting on the particles it follows that the average compressive forces and tangential forces in the tumbling can are underestimated by a factor of 10-20. This is compared to a mass flow of 1044 tph and velocities of 4m/s which in industrial practice are often higher.

Several shapes and particle contact models of wood pellets were tested in the tumbling can. The advanced shape consists of 35 spheres and is most realistic in design due to greater contact area and sharper edges that greatly affect the particle dynamics. However, in the tumbling can simulation a significant difference in compressive and tangential forces could not be shown. In addition, a disadvantage of the advanced shape is the increase in computational times.

Five designs of transfer chutes under similar simulation conditions were compared. The Tralix has the lowest average compressive forces and the traditional rockbox is the worst in that aspect.

The durability testing mechanism should be changed to ensure a better representation of the actual supply chain conditions. The improved tests must subject the material to higher compressive and tangential forces.

References

[1] ISO/TC 238, Solid biofuels - Determination of mechanical durability of pellets and briquettes - Part 1: Pellets (2011)

[2] D.L. Schott, R. Tans, I. Dafnomilis, V. Hancock & G. Lodewijks, On the durability tests for handling of solid biomass such as wood pellets, FME Transactions **44**, 279-284 (2016)

[3] P.A. Cundall and O.D.L. Strack, A discrete numerical model for granular assemblies, Geotechnique, **29** (No. 1), 47–65, (1979)

[4] DEM Solutions, EDEM 2.6.1 (2014)

[5] R.L. Stroshine, Physical Properties of Agricultural Materials and Food Products (2004)

[6] http://www.engineeringtoolbox.com/friction-coefficients-d_778.html

[7] Kaliyan N. and R.V. Morey, Factors affecting strength and durability of densified biomass products; Biomass and Bioenergy, **33**, 337-359 (2009)

[8] Temmerman,M., F.Rabier, P.D.Jensen, H.Hartmann, Th.Böhm, Comparative study of durability test methods for pellets and briquettes; Biomass and Bioenergy **30**, 964-972 (2006)

 DOI: 10.1051/epjconf/201714012003

Comparing the effects of interparticle friction coefficient and intermediate stress ratio on critical-state DEM simulations using Delaunay triangulations

Kevin J. Hanley[1,*], *Xin* Huang[2], and *Catherine* O'Sullivan[3]

[1]Institute for Infrastructure and Environment, School of Engineering, The University of Edinburgh, Edinburgh EH9 3JL, UK
[2]Department of Geotechnical Engineering, Tongji University, 1239 Siping Road, Shanghai 200092, China
[3]Department of Civil and Environmental Engineering, Skempton Building, Imperial College London, London SW7 2AZ, UK

Abstract. Strong force chains form when any granular material is subjected to load. A prior study used Delaunay triangulations to investigate the role of interparticle friction coefficient, μ, in stabilising the strong force chains [1]. In this follow-on paper, the effects of μ and the intermediate stress ratio, b, are compared. Numerical samples were sheared triaxially until critical state was attained. The contact networks and Delaunay triangulations of the particle centroids were both obtained at the end of each simulation. As μ is increased, the numbers of contacts in the contact networks decrease consistently. The numbers of edges, faces or tetrahedra in the Delaunay triangulations all increase with increasing μ up to 0.25 and become approximately constant thereafter. Changing b has no significant effect. The percentage of faces in the triangulation comprising three contacts shows a linearly-decreasing trend with increasing angle of shearing resistance. This is because only orthogonal loads are applied. Triangular structures require larger lateral supporting forces to maintain their stability than columnar structures when subjected to an axial load; hence, σ_3' is expected to be larger relative to σ_1' when triangular motifs are more prevalent. An increased proportion of triangular structures therefore leads to a lower ϕ_{cv}'.

1 Introduction

The stress distribution is highly heterogeneous within any granular material subjected to load. Strong force chains, consisting of contacting particles aligned with the major principal stress orientation, form upon loading to transmit the majority of the load and carry the deviatoric stress during shear deformation. These strong force chains are supported by a weak orthogonal contact network. A range of complementary investigation tools including DEM simulation [2], photoelasticity experiments [3] and analytical modelling [4,5] indicate that the dominant failure mechanism in a granular material is buckling failure of the strong force chains. The stability of the strong force chains increases with increasing interparticle friction coefficient, μ [6]. A prior study used Delaunay triangulations to further investigate the role of μ in stabilising the strong force chains [1]. In this paper, Delaunay triangulations are again used, on this occasion to compare the effects of μ and the intermediate stress ratio, b, on force chain stability. The intermediate stress ratio is defined as

$$b = \frac{\sigma_2'-\sigma_3'}{\sigma_1'-\sigma_3'} \tag{1}$$

where σ_1', σ_2' and σ_3' are the major, intermediate and minor principal effective stresses, respectively.

2 DEM simulations and analysis

A total of 20,164 polydisperse spherical particles were randomly placed within a cubical periodic cell, ensuring no overlaps between particles. Fig. 1 shows the particle size distribution of the numerical sample which is representative of Toyoura sand. A simplified Hertz-Mindlin contact model was used with a particle shear modulus of 29 GPa and Poisson's ratio of 0.12. The particle density was set at 2650 kg/m^3. Gravity was inactive. The simulation code used was a version of LAMMPS [7] which included some modifications made by the authors. The starting point for all triaxial shearing simulations was the same: a dense, stable, frictionless isotropic assembly at a confining pressure of 100 kPa which was created by moving opposing pairs of periodic boundaries closer together under stress control.

Two different data sets were analysed: 'varying b' and 'varying μ'. The 'varying b' data set [8] consisted of eight individual simulations. Seven of these simulations used constant b values of 0.0, 0.2, 0.3, 0.4, 0.5, 0.6 or 0.8, while one plane strain simulation was run which had b varying between 0.3 and 0.4 (b = 0.3383 at critical state). Starting from the stable, isotropic sample described above, each of these 'varying b' triaxial shearing simulations was run by initially increasing μ from 0.0 to 0.25, and then using a stress-control algorithm to either impose plane strain conditions or else

* Corresponding author: k.hanley@ed.ac.uk

maintain b at a defined, constant value under a constant σ_3' condition. The 'varying μ' data set [6] consisted of six drained shearing simulations at constant μ values of 0.0, 0.1, 0.25, 0.5, 0.75 or 1.0. Taking the same stable, isotropic sample, each 'varying μ' simulation was run by changing μ from 0.0 to one of these six values before subjecting the sample to drained triaxial compression at b = 0.0. Comparing the 'varying b' and 'varying μ' data sets, one simulation with b = 0.0 and μ = 0.25 is common to both. Hence, thirteen simulations are analysed in this paper in total. Each shearing simulation continued to an axial strain of around 54%.

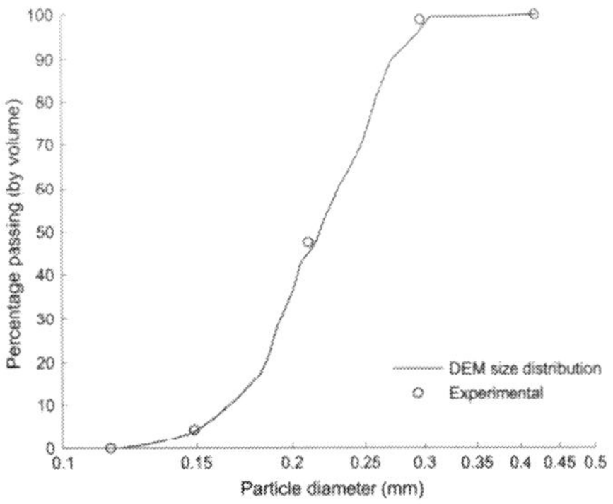

Fig. 1. Particle size distribution of numerical samples compared with that of a real Toyoura sand (adapted from [1]).

Samples were compared only at critical state, i.e., the state at which volume and stress become constant, using data obtained at the end of each simulation. The complete macro-scale responses obtained in these simulations may be seen in [6] for five of the six samples from the 'varying μ' data set (excluding the frictionless simulation). The effect of b on the macro-scale behaviour under shearing is presented in detail in [8].

By joining the centroids of the contacting particles, the contact networks were obtained. The particle centroids were also used to compute Voronoi tessellations using Voro++ [9]. Both the conventional unweighted tessellation and the radical/Laguerre tessellation which includes particle diameter in its calculation were obtained. These tessellations were converted to their Delaunay duals to enable a direct comparison between the edges of the contact networks and the edges of the associated Delaunay triangulations.

3 Results and discussion

Table 1 summarises the key macro-scale results at the critical state of each simulation. e is the void ratio (the volume of voids divided by the volume of solids), ϕ_{cv}' is the angle of shearing resistance, $p' = (\sigma_1' + \sigma_2' + \sigma_3')/3$ is the mean effective stress and $q = \sqrt{3 \cdot J_2}$ where J_2 is the second invariant of the stress deviatoric tensor $s_{ij}' = \sigma_{ij}' - p' \cdot \delta_{ij}$ and δ_{ij} represents the Kronecker parameter [8]. For triaxial compression with b = 0.0, q simplifies to the conventional definition $q = \sigma_1' - \sigma_3'$.

Table 1. Key macro-scale data at the end of all simulations after the attainment of critical state.

'varying μ' data set					
μ	b	ϕ_{cv}' (°)	q (kPa)	p' (kPa)	e
0.0	0.0	4.98	23.951	107.99	0.5344
0.1	0.0	13.54	61.885	120.63	0.5909
0.25	0.0	18.17	87.794	129.27	0.6317
0.5	0.0	19.97	99.514	133.18	0.6504
0.75	0.0	20.19	90.418	130.14	0.6527
1.0	0.0	20.98	100.49	133.50	0.6542
'varying b' data set					
μ	b	ϕ_{cv}' (°)	q (kPa)	p' (kPa)	e
0.25	0.0	18.17	87.794	129.27	0.6317
0.25	0.2	19.73	101.86	140.75	0.6316
0.25	0.3	20.07	106.91	146.34	0.6335
0.25	0.3383	20.48	101.25	147.26	0.6331
0.25	0.4	20.34	109.16	154.59	0.6353
0.25	0.5	20.18	105.69	156.38	0.6381
0.25	0.6	19.52	96.666	158.01	0.6385
0.25	0.8	19.18	106.98	147.73	0.6319

As μ increases, the number of contacts decreases from around 57,000 at μ = 0.0 to 32,000 when μ = 1.0 whereas e shows the opposite trend (Fig. 2a). The number of contacts is independent of b (Fig. 2b). There is also no obvious relationship between e and b. Analysis of a more comprehensive data set has shown a slight sensitivity might be expected [8]; nonetheless, the effect of b on e is negligibly small compared to the effect of μ on e. Results for the plane strain simulation, shown as light grey markers on Fig. 2b, are in line with those for the simulations at constant b values.

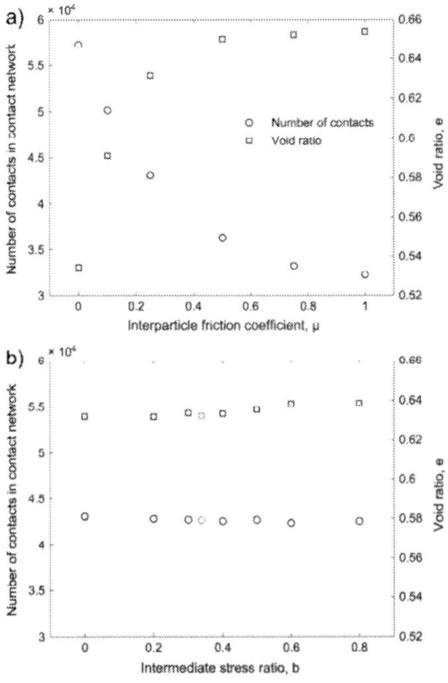

Fig. 2. The variations in the number of contacts and the void ratio at critical state with (a) μ; (b) b.

The number of contacts decreases with increasing μ; however Fig. 3a shows that the number of edges in the Delaunay triangulations increases with μ up to $\mu = 0.25$ and remains approximately constant thereafter. As for the number of contacts, the number of edges in the Delaunay triangulations is unaffected by b (Fig. 3b). Similar results are obtained for the numbers of faces or tetrahedra in the Delaunay triangulations (Fig. 4 and Fig. 5, respectively). The trends obtained for both types of triangulation are similar; the unweighted triangulations have a consistent offset from the radical equivalents which contain fewer edges (Fig. 3).

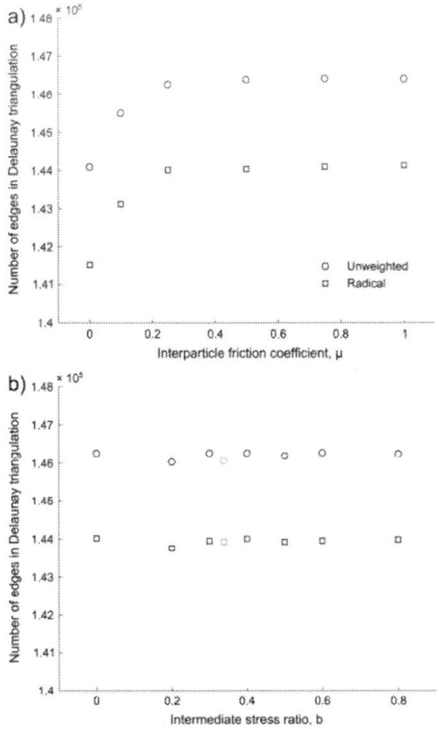

Fig. 3. The number of edges at critical state in the Delaunay triangulations against (a) μ; (b) b.

These increases in the numbers of edges or faces with increasing μ (up to $\mu = 0.25$) are implied in the literature, although usually for more idealised assemblies than were used for this paper. For mono-sized [10] or binary [11] assemblies of hard spheres, the average number of faces comprising a Delaunay tetrahedron decreases with increasing packing fraction (η). Redenbach [12] points out that this behaviour is expected as a dense packing of spheres becomes increasingly ordered as η increases. In this study, the total number of faces increases with increasing μ (Fig. 4a), with increasing e (Table 1) or equivalently with decreasing η. [13] show that the average number of edges per face is almost invariant for radical tessellations using varying ternary mixtures of spheres simulated using DEM. Hence, it may be inferred that the average number of edges per tetrahedron also decreases with increasing η. As η increases, so too does the coordination number. For an idealised 2D random tessellation, an approximate relationship is presented in [14] based on Euler's formula: increasing the coordination number, and hence η, reduces the average number of edges per tetrahedron. In 3D, a reduction in the average number of edges per tetrahedron also implies a reduction in the average number of faces per tetrahedron [14].

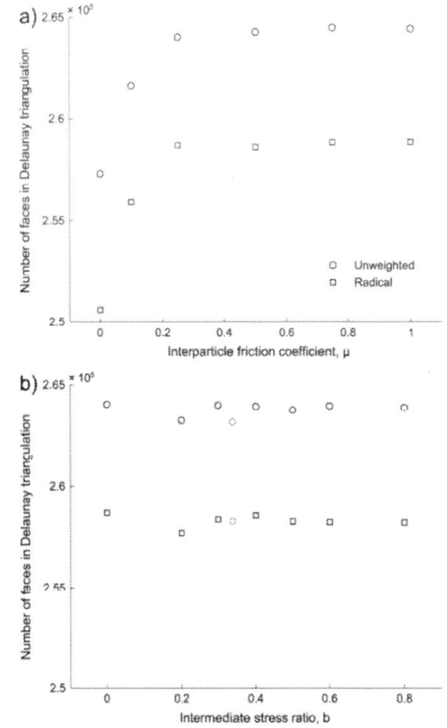

Fig. 4. The number of faces at critical state in the Delaunay triangulations against (a) μ; (b) b.

The percentage of faces in the triangulation comprising three interparticle contacts (P_{fc}) decreases monotonically as μ is increased [1]. Fig. 6 shows that increasing μ reduces P_{fc} but increases ϕ'_{cv}. The relationship between P_{fc} and ϕ'_{cv} is almost linear. The shear resistance increases when the proportion of triangular structures is reduced. A tentative explanation was proposed in [1], that the effect of μ on strength is much more significant than the proportion of triangular structures. It is likely that the observations in Fig. 6 are because only orthogonal loads are applied in these simulations. Triangular structures require larger lateral supporting forces to maintain their stability than columnar structures when subjected to an axial load; hence, σ'_3 is expected to be larger relative to σ'_1 when triangular motifs are more prevalent. Since the angle of shearing resistance is calculated as $\sin(\phi'_{cv}) = (\sigma'_1 - \sigma'_3)/(\sigma'_1 + \sigma'_3)$, an increased proportion of triangular structures leads to a lower ϕ'_{cv}. This contribution to strength due to micro-scale particle geometry reinforces the contribution to strength due to interparticle friction [6] as μ is increased.

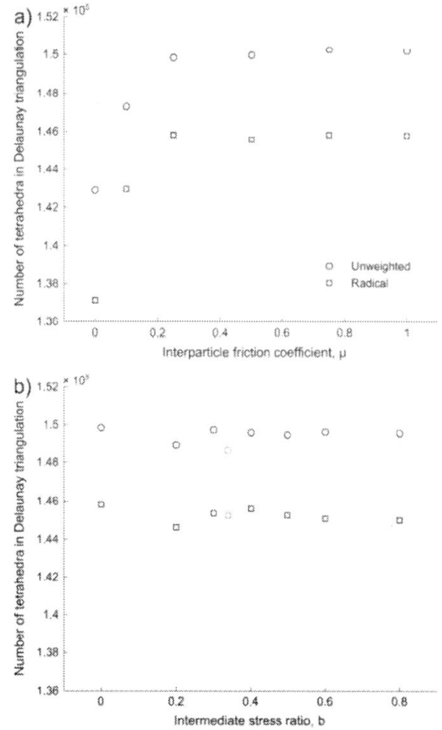

Fig. 5. The number of tetrahedra at critical state in the Delaunay triangulations against (a) μ; (b) b.

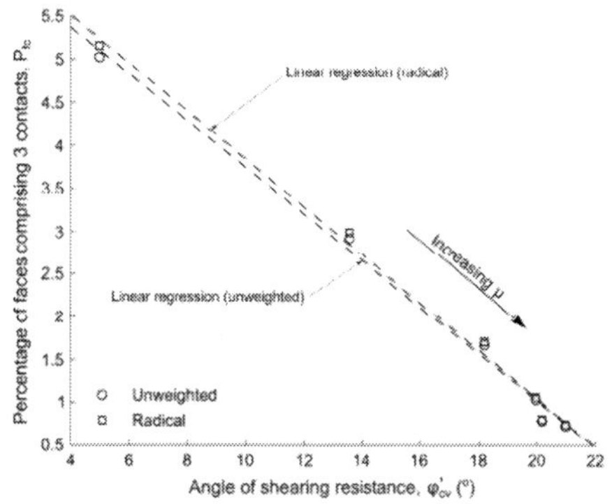

Fig. 6. Percentage of triangulation faces comprising three interparticle contacts, P_{fc}, against ϕ'_{cv} (°).

4 Conclusions

Six polydisperse numerical samples with different interparticle friction coefficients, μ, were sheared triaxially at b = 0.0 until critical state was attained for each. Eight similar samples were sheared at a constant μ of 0.25 but differing intermediate stress ratios from 0.0 to 0.8. At the end of each simulation, the particle centroids were used to obtain the contact networks and compute Delaunay tessellations. For all measures, the intermediate stress ratio had effectively no influence on behaviour, as shown in [8] for the coordination number. As μ is increased, the numbers of edges in the contact networks decrease consistently whereas the numbers of edges, faces or tetrahedra in the Delaunay triangulations all increase with increasing μ up to μ = 0.25 and become approximately constant thereafter.

The percentage of faces in the triangulation comprising three contacts (P_{fc}) decreases monotonically as μ is increased. The relationship between P_{fc} and ϕ'_{cv} is almost linear. The shear resistance increases when the proportion of triangular structures is reduced. A partial explanation was proposed in [1] based on the effect of μ on the rolling/sliding behaviour. Another reason is the contribution to strength due to micro-scale particle geometry. Columnar structures possess greater self-stability than triangular structures when subjected to orthogonal loads. Triangular structures require larger lateral supporting forces to maintain their stability than columnar structures when subjected to an axial load; hence, σ'_3 is expected to be larger relative to σ'_1 when triangular motifs are more prevalent. The definition of ϕ'_{cv} thus indicates that an increased proportion of triangular structures leads to a lower ϕ'_{cv}. This contribution to strength due to micro-scale particle geometry reinforces the contribution to strength due to interparticle friction [6] as μ is increased.

References

1. K.J. Hanley, X. Huang, C. O'Sullivan, *Proc. of 4th International Conference on Particle-Based Methods (PARTICLES 2015)*, 890–899 (2015)
2. C. Thornton, Géotechnique **50**(1), 43–53 (2000)
3. R.P. Behringer, K.E. Daniels, T.S. Majmudar, M. Sperl, Philos. Trans. A Math. Phys. Eng. Sci. **366**, 493–504 (2008)
4. A. Tordesillas, M. Muthuswamy, J. Mech. Phys. Solids **57**, 706–727 (2009)
5. C. O'Sullivan, M.A. Wadee, K.J. Hanley, D. Barreto, Géotechnique **63**(15), 1298–1309 (2013)
6. X. Huang, K.J. Hanley, C. O'Sullivan, C.Y. Kwok, Int. J. Numer. Anal. Meth. Geomech. **38**(12), 1276–1297 (2014)
7. S.J. Plimpton, Comput. Phys. **117**(1), 1–19 (1995)
8. X. Huang, K.J. Hanley, C. O'Sullivan, C.Y. Kwok, M.A. Wadee, Granul. Matter **16**(5), 641–655 (2014)
9. C.H. Rycroft, Chaos **19**, 041111 (2009)
10. P. Richard, L. Oger, J.P. Troadec, A. Gervois, Phys. Rev. E **60**(4), 4551 (1999)
11. P. Richard, L. Oger, J.P. Troadec, A. Gervois, Physica A **259**(1–2), 205–221 (1998)
12. C. Redenbach, Comp. Mater. Sci. **44**(4), 1397–1407 (2009)
13. L.Y. Yi, K.J. Dong, R.P. Zou, A.B. Yu, Powder Technol. **224**, 129–137 (2012)
14. S. Torquato, *Random Heterogeneous Materials: Microstructure and Macroscopic Properties*. 1st ed. Springer-Verlag, New York (2002)

DOI: 10.1051/epjconf/201714008002

Experimental investigation of impinging jet erosion on model cohesive granular materials

Florian Brunier-Coulin[1,*], *Jean-Loup* Sarrat[1], *Pablo* Cuéllar[1], and *Pierre* Philippe[1]

[1]IRSTEA, 3275 route de Cézanne, CS40061, 13182 Aix-en-Provence Cedex 5, France

Abstract. Erosion of soils affects both natural landscapes and engineering constructions as embankment dams or levees. Improving the safety of such earthen structures requires in particular finding out more about the elementary mechanisms involved in soil erosion. Towards this end, an experimental work was undertaken in three steps. First, several model materials were developed, made of grains (mostly glass beads) with solid bridges at particle contacts whose mechanical yield strength can be continuously varied. Furthermore, for most of them, we succeeded in obtaining a translucent system for the purpose of direct visualization. Second, these materials were tested against surface erosion by an impinging jet to determine a critical shear stress and a kinetic coefficient [2, 3]. Note that an adapted device based on optical techniques (combination of Refractive Index Matching and Planar Laser Induced Fluorescence [3]) was used specifically for the transparent media. Third, some specifically developed mechanical tests, and particularly traction tests, were implemented to estimate the mechanical strength of the solid bridges both at micro-scale (single contact) and at macro-scale (sample) and to investigate a supposed relationship with soil resistance to erosion.

1 Introduction

Hydraulic erosion of soils is a phenomenon due to the action of water on the detachment and transport of elementary soil particles. The physical process by which the fluid flow causes the dislodgment of primary particles is a key problem for many fields. In civil engineering, Foster and co-workers [1] suspected that 90% of embankment dam failures are due to erosion while, according to the International Commission Of Large Dams, an estimated proportion of 75% of dams are built in sand, clay or muddy sediments, and are consequently threatened by the risk of erosion. As a consequence, the stability and durability of earthen hydraulic structures depend on the response of soil against hydrodynamic stresses. In France there are on average approximately one failure of dams and one failure of protection dikes per year. In order to enhance the safety of these structures, it is important to comprehend better the elementary mechanisms involved during erosion processes at the surface of a cohesive material.

Different erosion laws can be found in the literature. Most of them involve two erosion parameters accounting for the erosion threshold and the kinetics of erosion. These laws are usually based on a mean hydrodynamic shear stress related to the fluid velocity and pressure gradient, both of them averaged in time and space. The erosion parameters are *a priori* inherent to the material and can be estimated experimentally by adjusting specifically an erosion law with data obtained from specific erosion tests. However, it is difficult to identify a clear correlation between these parameters and other more common soil properties.

A parametric study based on model materials is proposed here, to seek for a clearer relation between common soil properties and resistance against erosion. We will first present the model materials, based on glass beads with a cohesion made of solid bonds, and then the erosion and mechanical tests used in this study. Finally, the results of this analysis are presented and discussed in the last section.

2 Model materials

2.1 Granular sample

The model material used in the present study is made of millimetric glass beads with inter-particle solid bonds. The spherical glass beads were provided by *SiLiBeads* with two distinct types of glass. Some samples are based on silica glass beads with a bulk density ρ_s=2500kg.m^{-3}, and with diameters d ranging either between 2.85 and 3.30mm, or between 0.75 and 1mm. For purpose of index-matching as discussed later, borosilicate glass beads with a lower refractive index n=1.472, a bulk density ρ_b=2230kg.m^{-3} and a diameter d=3±0.02mm are also used.

2.2 Solid bridges

To obtain a cohesive sample, we chose to mix each granular sample with a liquid matrix so as to bond the beads with capillary bridges. With specific liquids like paraffin or resin, it is possible to obtain solid bridges after cooling or drying. In this study, different volumes of transparent liquid resin (provided by *Syntilor*) are mixed with the beads in a rectangular box of 20x20x10cm^3, before drying at 60°c during 7 days to obtain solid bonds. A typical solid bridge between two beads with d=3mm is shown on Figure 1.

The protocol used to prepare the cohesive samples enables to assume a random loose packing with, typically, a volume fraction ϕ=0,6 and a number of bonds per particle z=6 [5].

* Corresponding author: florian.brunier@irstea.fr

 DOI: 10.1051/epjconf/201714008002

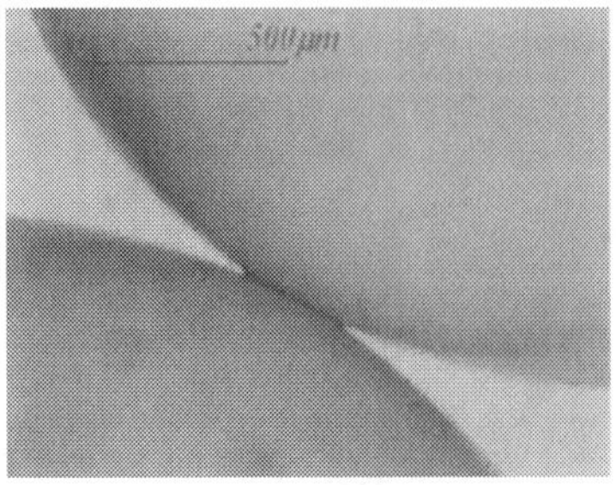

Fig. 1. Typical solid bridge of resin between two glass beads of diameter 3mm.

3 Experimental set up and protocol

3.1. Index matched round impinging jet

One of the most popular erosion test is the JET test proposed by [2], used *in situ* or *ex situ* in laboratory. As sketched in Figure 2, the set-up developed in this study is adapted from the JET test to fulfil the requirements of the optical techniques used to visualise inside the material. The cohesive sample is implemented under a cell of 40cm high and a cross section area 10x20cm^2. The cell is slowly saturated with an oil mixture made of light mineral oil supplied by *Sigma-Aldrich* (90% in mass) and immersion oil supplied by *Cargille* (10% in mass). This mixture has the same refractive index than the borosilicate glass beads with a dynamic viscosity $\mu_f \approx 0.28cP$ and a bulk density ρ_f=847kg.m^{-3}.

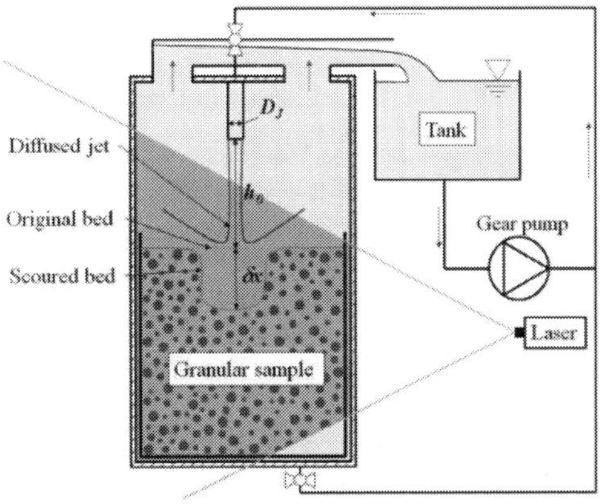

Fig. 2. Index matched erosion tests with a round impinging jet.

The upper side of the cell includes two outlets to create an overflow system, and a central inlet connected to a gear pump. The inlet is composed of a metal tube of 4cm length and 5mm of inner diameter D_J to generate a downward vertical immersed jet at a controlled flow rate supplied by the pump from a liquid reservoir tank.

To visualise the erosion process, a fluorescent dye (Nile Red supplied by *Fluka*) is added to the immersion fluid, and a planar laser of wavelength 532nm, inside the dye fluorescence spectrum, is placed in the jet axis. Images are recorded through a high pass optical filter at 580nm by a digital camera with a frame rate up to 20 images per second.

To analyse the erosion threshold and the evolution of the crater depth (δx), the jet injection velocity is slowly increased from 0 to 10m/s in almost 1min. This slow acceleration ramp enables to assume a quasi-steady fully developed jet (*i.e.* local velocity on the granular surface can be calculated directly from the injection velocity).

3.2 Mechanical traction test

To quantify the cohesive forces contributing to the resistance against erosion, two mechanical tests have been developed to measure the mechanical resistance at different scales. For the larger beads (>1mm), it is possible to extract from the cohesive sample two beads still bonded together. One bead is glued to a movable deck when the upper bead is held up with a micrometric clamp connected to a scale (see Figure 3 on the left). Then the deck is slowly moving downward and the scale measurement is recording until the failure of the bond. The resulting normal traction force of the solid bridge at failure is noted F_t^*.

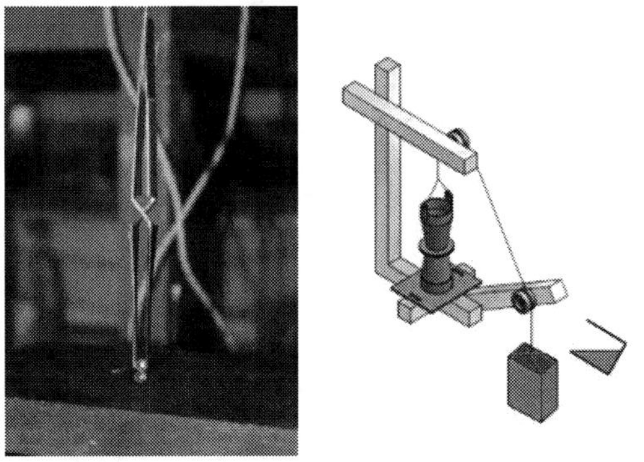

Fig. 3. Micro-scale traction test between two beads of 3mm (left) and macro-scale traction test on a representative volume of beads of 1mm (right).

For the smaller beads (≤1mm), a second device enables to measure the global de-cohesion stress on a representative volume of cohesive soil. Before drying, the beads mixed with the liquid resin are set in place inside two conical cylinders as showing in Figure 3. After solidification of the capillary bridges (see section 2.2), the load is slowly increased vertically on the upper part until ultimate separation of both parts. The resulting critical force is divided by the circular failure section to calculate a normal stress of de-bonding named τ_t^*.

4 Results and analysis

4.1. Experimental results

4.1.1. Erosion tests

To compare the impact of cohesion according to the different bead sizes, an adapted variable seems to be the mean thickness of liquid resin (e), calculated as the ratio of the volume of resin to the total bead surface. Figure 4 shows as the critical injection velocity generating first beads removal (U_J^*) increases with e for the different bead sizes.

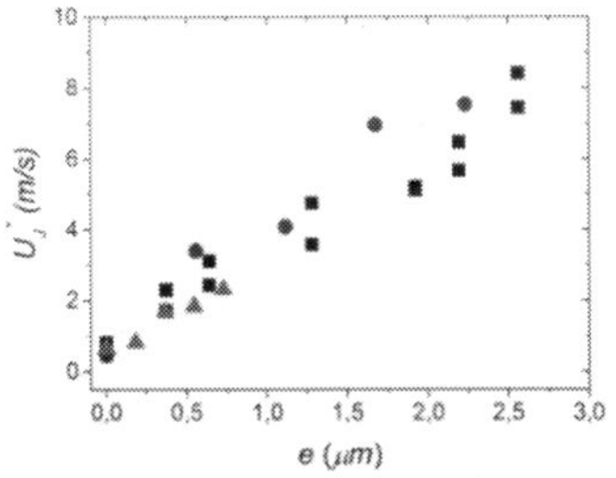

Fig. 4. Critical jet injection velocity (U_J^*) versus mean resin thickness (e). Blue circles, black squares, and red triangles stand for borosilicate glass beads of 3mm, silica glass beads of 3mm, and silica glass beads of 1mm respectively.

From the critical injection velocities, it is possible to calculate the local velocity in the jet centreline, u_0^*, at the distance $h_0$between the nozzle and the granular bed by a theoretical relation given by [6] for a self-similar laminar jet model and validated for the present jet configuration:

$$u_0^*(h_0) = U_J^* \frac{Re_J^*}{8} \frac{D_J}{(h_0+0.048Re_J^*)} \quad (1)$$

where $U_J^*$is the injection velocity and $Re_J^* = \frac{U_J^* D_J}{\nu_f}$ with ν_f the kinematic viscosity of the fluid.

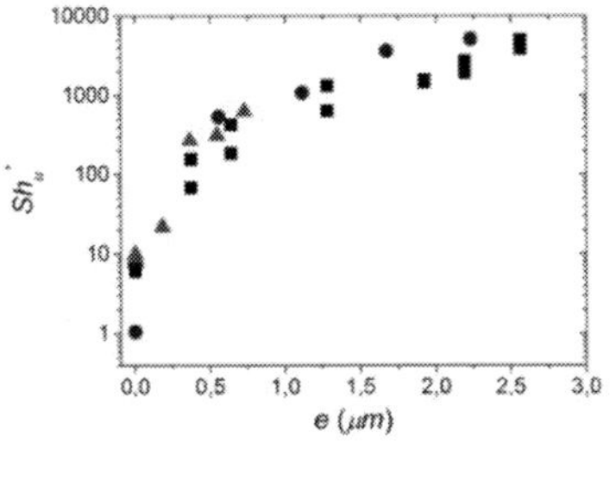

Fig. 5. Critical Shields number (Sh_{cl}^*) as a function of the mean thickness of resin (e) with the same symbols as used in Figure 4.

Then, following a previous study by [7], an inertial cohesion-less Shields number Sh_{cl} is used to account for the erosion threshold of a grain without cohesion and compares the fluid stress, expressed by $\rho_f u_0^2$, and the buoyant weight of the particle without any shape factor. Consequently, the resulting critical value reads:

$$Sh_{cl}^* = \frac{\rho_f u_0^{*2}}{\Delta\rho g d} \quad (2)$$

with $\Delta\rho$ the difference between the particle density and the fluid density.

Figure 5 shows a strong increase of the Shields number with e over more than three orders of magnitude, highlighting a great influence of adhesive or cohesive forces on the Shields number as expected [8].

4.1.2. Mechanical resistance of the bonds

The traction tests realised at different scales enable to quantify the ratio between the buoyant weight of the bead and the critical failure force of the solid bridges by a generalised Bond number. Dimensionally this number can be expressed either from the local traction force or from the global traction stress (Bo_m or Bo_M respectively) as:

$$Bo_m = \frac{\Delta\rho g d^3}{F_t^*} \quad (3)$$

$$Bo_M = \frac{\Delta\rho g d}{\tau_t^*} \quad (4)$$

The Bond numbers calculated from equations 3 and 4 for beads of 1 and 3mm almost collapse as shown in Figure 6 when plotting against e. The decrease of Bo with e (in μm) can be fitted by an inverse relation as represented in dashed line on Figure 6. This relation can be used for purpose of Bond number interpolation.

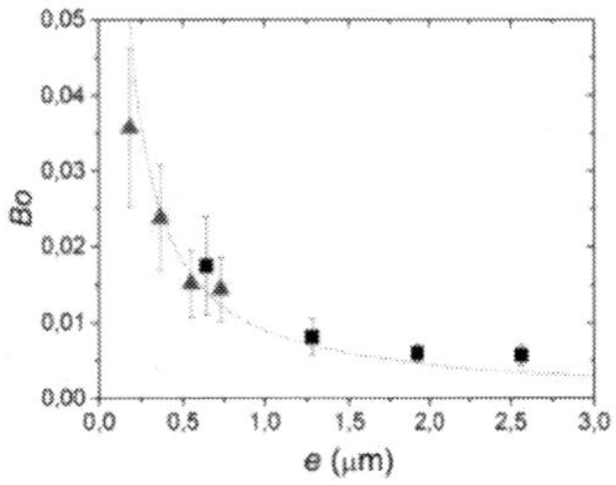

Fig.6. Evolution of the Bond number as a function of the mean thickness of resin with the same symbols as used in Figure 4.

4.2. Discussion

The traction stress measured at micro and macro-scale can be used to model the impact of the cohesion on the erosion threshold, additionally to weight and friction. A simple assumption is to simply sum the buoyant weight and the cohesive stress in the Shields formulation. The resulting generalised Shields number is then expressed via the Bond number as:

$$Sh = \frac{\tau_f}{\tau_m(1+Bo)} \qquad (5)$$

where τ_m is the mechanical traction stress and τ_f is the fluid stress.

The critical cohesion-less Shields number Sh_{cl}^* and generalised Shields number Sh^* are represented on Figure 7 as a function of the particle Reynolds number $Re_p = \frac{u_0^* d}{\nu_f}$. This Figure shows that the new formulation of the Shields number enables to rationalize the data, in spite of the dependency between the inertial Shields and the Reynolds number due to the fluid velocity.

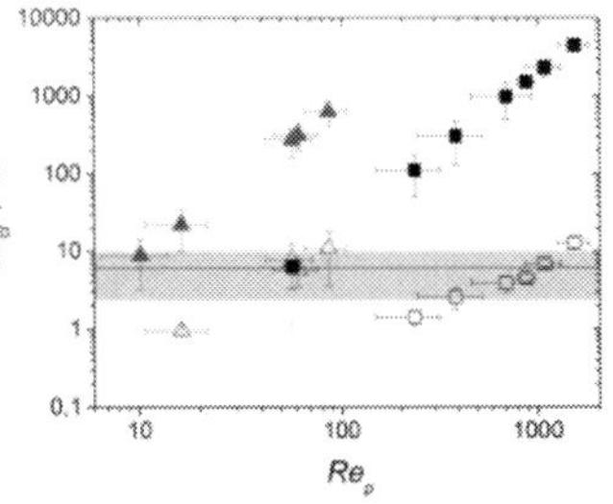

Fig.7. Cohesion-less Shields number (filled symbols) and cohesive Shields number (open symbols) as a function of the Reynolds Particular for both beads size (red triangle $d\approx 1$mm and black square $d\approx 3$mm).

The values of the generalised Shields number can be averaged to obtain a value of $Sh^* = 3.0 \pm 1.9$ (black curve and grey area in Figure 7). This value is in good agreement with those obtained for cohesion-less beads with an inertial Shields number either by us or by [7] with a planar jet.

In the case of Bond numbers far below 1 (i.e. $\tau_m \gg \Delta\rho g d$), the relation between Sh_{cl}^* and Sh^* can be expressed:

$$Sh_{cl}^* = Sh^*\left(1 + \frac{1}{Bo}\right) \sim \frac{Sh^*}{Bo} \qquad (6)$$

Accordingly, Figure 8 shows the dependency between Sh_{cl}^* and $1/Bo$. The slope equal to the averaged value of $Sh^* \sim 3$ is in approximate agreement with most of the data.

5. Conclusion

The development of model cohesive granular materials used in parallel with an index matched impinging jet and traction tests at micro and macro-scale allowed to propose a generalised expression of the Shields number for both cohesive and cohesion-less materials in good agreement with the present experimental data.

A first perspective is to better understand the link between micro-scale and macro-scale in realising local and global traction tests with the same material. For that, the size of the set-up has to be adapted to the bead diameters. Finally, this study was realised with solid bounding and a specific size of beads. The perspectives for a future work are to expand this study to a larger range of particle sizes and shapes and to different kinds of bonds like capillary bridges with a viscous fluid or a synthetic clay.

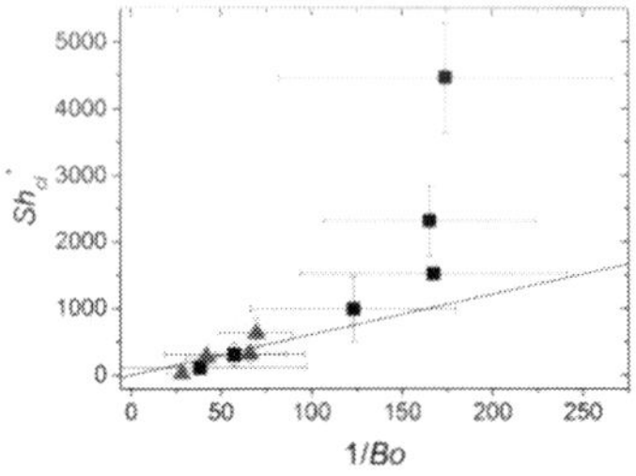

Fig. 8. Cohesion-less Shields number as a function of $1/Bo$ for silica beads of 1 and 3mm.

Acknowledgment

We acknowledge financial support from the region PACA (Provence-Alpes-Côte d'Azur).

References

1. M. Foster, R. Fell, and M. Spannagle, The statistics of embankment dam failures and accidents. Can. Geotech. J. **37**:1000-1024 (2000)
2. G. J. Hanson and K. R. Cook, Apparatus, test procedures, and analytical methods to measure soil erodibility in situ. Applied Engineering in Agriculture **20(4)**:455-462 (2004)
3. E. Partheniades, Erosion and deposition of cohesive soils. ASCE **91(HY1)**:105-139 (1965)
4. P. Philippe and M. Badiane, Localized fluidization in a granular medium. Phys. Rev. E **87**, 042206 (2013)
5. J.-Y. Delenne, F. Soulié, M. S. El Youssoufi and F. Radjaï, From liquid to solid bounding in cohesive granular media. Mechanics of Materials **43**:529-537 (2011)
6. A. Revuelta, A. L. Sánchez and A. Liñán, Confined axisymmetric laminar jets with large expansion ratios. J. Fluid Mech. **456**:319-352 (2002)
7. S. Badr, G. Gauthier and P. Gondret, Erosion threshold of a liquid immersed granular bed by an impinging plane liquid jet. Phys. Fluids **26**:023302 (2014)
8. S. E. Coleman and V. I. Nikora, A unifying framework for particle entrainment. Water Resources Research **44**:W04415 (2008)

EPJ Web of Conferences 140, 03007 (2017) DOI: 10.1051/epjconf/201714003007
Powders & Grains 2017

Non-spherical granular flows down inclined chutes

R.C. Hidalgo[1,*], *S.M. Rubio-Largo*[1], *F. Alonso-Marroquin*[2], and *T. Weinhart*[3,**]

[1] *Department of Physics and Applied Mathematics, University of Navarra, 31080 Pamplona, Navarra, Spain*
[2] *School of Civil Engineering, The University of Sydney, Sydney NSW 2006, Australia*
[3] *Multi Scale Mechanics, CTW, UTwente, 7500 AE Enschede, Netherlands*

Abstract. In this work, we numerically examine the steady-state granular flow of 3D non-spherical particles down an inclined plane. We use a hybrid CPU/GPU implementation of the discrete element method of non-spherical elongated particles. Thus, a systematic study of the system response is performed varying the particle aspect ratio and the plane inclination. Similarly to the case of spheres, we observe three well-defined regimes: arresting flows, steady uniform flows and accelerating flows. Both steady and dynamic macroscopic fields are derived from microscopic data, by time-averaging and spatial smoothing (coarse-graining), including density, velocity, as well as the kinetic and contact stress tensors. The internal morphology of the flow was quantified exploring the solid fraction profiles and the particle orientation distribution. Furthermore, the system's characteristic time and length scales are investigated in detail. Our aim is to achieve a continuum mechanical description of granular flows composed of non-spherical particles based on the micromechanical details. Thus, to evaluate the influence of particle shape on the constitutive response in granular of those systems. However, to meet the proceeding's page restrictions here we will only discuss the dependence of some terms of the continuum averaged equations on the coarse-graining scale, specifically the case of the kinetic part of the stress tensor.

1 Introduction

Granular flows [1] usually show quite complex behaviors that are also found in diverse systems, such as colloidal suspensions [2], pedestrian dynamics [3] and animal herds [4]. Unfortunately, a complete theoretical framework or rheological model explaining the constitutive response of granular systems is not available yet. However, several promising approaches have recently been published [5–7], and it is known that several modes of energy dissipation, such as mechanic friction or plastic deformations lead to non-equilibrium steady state situations.

In examining granular flows there are several experimental restrictions, due to the opaque nature of the grains. Thus, full access to the 3D- behavior of the grains is generally not feasible and, hence, there is a real need to perform numerical simulations in this framework. Discrete element modeling (DEM) is widely accepted as an effective method to address engineering problems concerning dense granular media [8]. Moreover, in typical applications the formulation of granular macroscopic fields are also necessary. The micro-mechanical details, *i.e.* velocity and position of individual particles, allow one to find the continuum field profiles using coarse-grain averaging technique [9–13]. Furthermore, with this homogenization approach, the static and dynamic parts of the stress tensor are deduced in terms of contact forces and velocity fluctuations, respectively.

Although spherical particles are a special case in nature, the majority of both experimental and numerical studies have considered packings of spherical particles so far. However, the appropriate description of systems composed of non-spherical particles is of significant importance for practical applications such as the handling of rocks, rice, wheat or tablets. Thus, there is currently and increasing interest in the behavior of non-spherical grains both experimentally [14–17] and numerically [18, 19]. The goal of this study is to extend existing theories of frictional avalanching flows of spherical particles in steady state situations, to the case of non-spherical grains. The ultimate aim is to achieve a continuum-mechanical description of granular flows composed by non-spherical grains based on micromechanics, This will help us to understand the role that particle shape plays in governing the system's mechanical response.

Our DEM simulations and the coarse-grain averaging technique [9–13] allows us to compute the mean velocity, density and stress fields in detail. However, the paper size limitation precludes us to present a complete description. Here we will only discuss the dependence of some terms of the continuum averaged equations on the size of the averaging domain, specifically the case of the kinetic part of the stress tensor, which is know to be scale-dependent [10–13] even for spherical particles.

*e-mail: raulcruz@unav.es
**e-mail: t.weinhart@utwente.nl

Figure 1. Snapshot of a system of 32768 spherocylinders with elongation of $\xi = 3$. Sketch of the interaction among two particles is also shown.

2 DEM and Coarse-Graining Formulation

We have developed a hybrid GPU-CPU discrete element algorithm to investigate granular chute flows of rod-shaped particles. In the model, the rods are considered sphero-cylinders characterized by their length l and sphero-radius r. Thus, their aspect ratio is defined by $\zeta = (l+2r)/2r$ (see figure 1). For calculating the interaction force between two particles i and j, $\vec{F}_{ij}$, we use an algorithm based on the concept of spherocylinder [20], which is defined by two vertices and the sphero-radius r; the surface of a spherocylinder is delineated by all points at distance r from the edge between the two vertices. Thus, the contact detection between to spherocylinders is reduced to find the closest point between two edges, resulting an overlap distance δ, defined by the overlap of two spheres of radius r. Fig.1 illustrates the sketch of the interaction between two spherocylinders. Thus, the force $\vec{F}_{ij}$ exerted on particle i by the particle j is defined by: $\vec{F}_{ij} = -\vec{F}_{ji}$. The force $\vec{F}_{ij}$ can be decomposed as $\vec{F}_{ij} = F^n \cdot \vec{n} + F^t \cdot \vec{t}$, where F^n is the component normal to the contact plane $\vec{n}$. Additionally, F^t acts on the tangential direction $\vec{t}$. To define the normal interaction F^n, we use a linear elastic force, which is governed by the overlap distance δ between two spherocylinders. To account for dissipation, a velocity dependent viscous damping is assumed. Hence, the total normal force reads as $F^n = -k^n\delta - \gamma^n m_r v^t_{rel}$, where k^n is the spring constant in the normal direction, $m_r = \frac{m_p}{2}$ stands for the pair's reduced mass, γ^n is the damping coefficient in the normal direction and v^n_{rel} is the normal relative velocity between i and j. The tangential force F^t also contains an elastic term and a tangential frictional term accounting for static friction between the grains. Taking into account Coulomb's friction constrain, which reads as, $F^t = \min\{-k^t\xi - \gamma^t m_r \cdot |v^t_{rel}|, \mu F^n\}$, where γ^t is the damping coefficient in tangential direction, v^T_{rel} is the tangential component of the relative contact velocity of the overlapping pair. ξ represents the elastic elongation of an imaginary spring with spring constant k^t at the contact, which increases as $\frac{d\xi(t)}{dt} = v^t_{rel}$ as long as there is an overlap between the interacting particles. The elastic tangential elongation $\vec{\xi}$ is kept orthogonal to the normal vector (truncated if necessary) [11]. μ is the friction coefficient of the particles.

Table 1. Particle's elongation ζ and number of particles used in each case. In all cases, we kept total mass $M_T \approx 1.4 \times 10^4$.

ζ	1.01	1.3	1.5	2.0	2.5	3.0
N_p	5664	8448	10752	17824	26784	37632

We use a coordinate system where x denotes the flow direction, z the in-plane vorticity direction, and y the depth direction normal to the base. The chute is inclined at an angle θ such that gravity acts in the direction $(sin(\theta), -cos(\theta), 0)$. The size of the simulation cell was $L_x = L_z = 16\,l$ and $L_y = 32\,l$ with periodic boundaries conditions on the $x-$ and $z-$ directions. The base of the system is a rough surface consisting of $N_b = 385$ fixed particles spherical particles with $R = l/2$. Note those conditions resemble one particular surface roughness of Ref.[12]. Moreover, similar to Ref.[12] we have use $k_t = (2/7)k_n$, $\gamma_t = \gamma_n$ with $e_n = 0.88$ and $\rho_p = 2500$. The microscopic friction coefficient is set to $\mu = 0.5$, the gravity to $g = 1$ and $k_n = 2 \times 10^5 m_p g/l$.

In the simulations presented here, all particles are mono-dispersed with the same length $l = 1/8$. Simulations are computed using rods of different aspect ratios, from $\xi = 1.01$ to $\xi = 3.0$, but always keeping the total mass of particles of $M_T \approx 1.4 \times 10^4$. In each case, the values of r and N_p are adjusted to the choice of ζ (see Table 1). The N_p flowing particles are introduced to the system at random non-overlapping positions well above the base. The gravity field induces the particle motion and they fall and accelerate down the plane until they reach a steady state, which is then analyzed.

In order to explore the dynamical and mechanical properties of the particle flow, a coarse graining methodology is used to analyze the results [9–13] The numerical data provide the position and velocities of every particle as well as the particle interaction forces. According to [10–12], the macroscopic mass density of a granular flow at time t is defined by $\rho(\vec{r}, t) = \sum_{i=1}^{N_p} m_i \phi(\vec{r} - \vec{r}_i(t))$ where the sum runs over all the particles within the system and the coarse-grained (CG) function, $\phi(\vec{\mathcal{R}})$. In our case, we use a truncated Gaussian coarse-graining function $\phi(\vec{\mathcal{R}}) = A_w e^{-(|\vec{\mathcal{R}}|/2w)^2}$ with cutoff $r_c = 6w$ where the value of w defines the coarse-grained scale. A_w is calculated in order to guarantee the normalization condition. Thus, the flow solid fraction can be found $\varphi(\vec{r},t) = \rho(\vec{r},t)/\rho_p$, where ρ_p is the material density. In the same way, the coarse grained momentum density function $P(\vec{r},t)$ is defined by $P(\vec{r},t) = \sum_{i=1}^{N_p} m_i \vec{v}_i \phi(\vec{r} - \vec{r}_i(t))$, where the $\vec{v}_i$ represent the velocity of particle i. The macroscopic velocity field $\vec{V}(\vec{r},t)$ is then defined as the ratio of momentum and density fields, $V(\vec{r},t) = P(\vec{r},t)/\rho(\vec{r},t)$.

On the other hand, the mean stress field $\bar{\sigma}_{\alpha\beta}$ is composed by the mean contact stress field $\sigma^c{}_{\alpha\beta}(\vec{r})$ and the mean kinetic stress field $\sigma^k{}_{\alpha\beta}(\vec{r})$. The spatial behavior of $\sigma^c{}_{\alpha\beta}(\vec{r})$ can be deduced in terms of the microscopic entities that characterize each contact, *i.e.* $\vec{F}_{ij}$ and the position of the contacting particles r_i and r_j [10–12]. In general the outcomes obtained for $\sigma^c{}_{\alpha\beta}(\vec{r})$ are independent of the coarse-graining scale.

Following Refs.[9–12], the kinetic stress reads as,

$$\sigma^k_{\alpha\beta}(\vec{r}) = \sum_i^{N_p} m_i v'_{i\alpha} v'_{i\beta} \phi(\vec{r} - \vec{r}_i(t)), \tag{1}$$

where the sum runs over all the particles, $\vec{v}'_i$ is the velocity fluctuation of particle i, respect to the mean field. $\vec{v}'_i(\vec{r},t) = \vec{v}_i(t) - \vec{V}(\vec{r},t)$. Note that the velocity fluctuation are defined with respect to the center of the averaging volume at $\vec{r}$.

Very recently, Artoni and Richard [13] has proposed to define the velocity fluctuation of the particles respect to the particle location $\vec{r}_i$. As a result they found a decomposition of the kinetic stress tensor in two terms. The first, $T^k_{\alpha\beta}(\vec{r})$, that is independent of the averaging domain size and a new term

$$T^\gamma_{\alpha\beta}(\vec{r}) = \rho(\mathbf{D} \cdot \nabla\vec{V}(\vec{r},t))(\mathbf{D} \cdot \nabla\vec{V}(\vec{r},t)) \tag{2}$$

Remarkably, the correction $T^\gamma_{\alpha\beta}(\vec{r})$ accounts for the dependence of averaging domain size, the used coarse-graining function, particle shape and polidespersity. Following their scheme the vector D read as,

$$\rho D_k(\vec{r}) = \sum_{i=1}^{N_p} m_i (x_k - x_i(t))^2 \phi(\vec{r} - \vec{r}_i(t)) \tag{3}$$

Hence, using Eqs(1) and (2) one can find

$$T^k_{\alpha\beta}(\vec{r}) = \sigma^k_{\alpha\beta}(\vec{r}) - T^\gamma_{\alpha\beta}(\vec{r}). \tag{4}$$

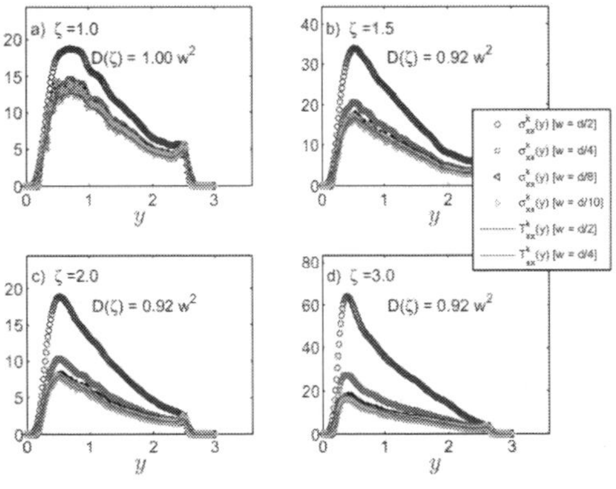

Figure 2. Vertical profiles of the xx component of the kinetic stress $\sigma^k_{xx}(y)$ calculated using Eq.(1) as a function of y. The fields are deduced using using a Gaussian coarse-graining function $\phi(\vec{\mathcal{R}})$ with different values of w. Results for several particle elongation are shown: a) $\zeta = 1.01$, b) $\zeta = 1.5$, c) $\zeta = 2.0$, d) $\zeta = 3.0$.

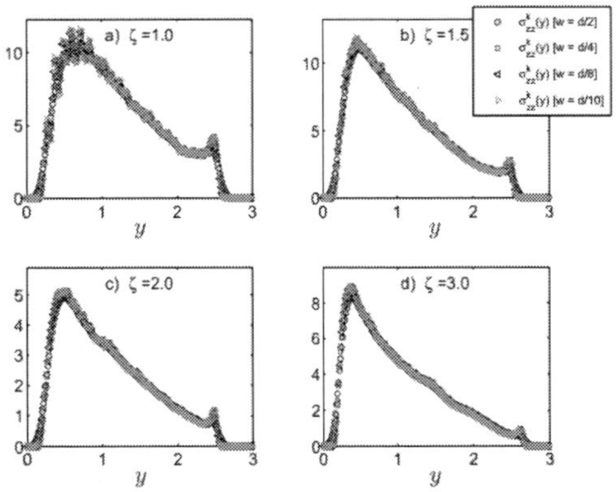

Figure 3. Vertical profiles of the zz component of the kinetic stress $\sigma^k_{zz}(y)$ calculated using Eq.(1) as a function of y. The fields are deduced using using a Gaussian coarse-graining function $\phi(\vec{\mathcal{R}})$ with different values of w. Results for several particle elongation are shown: a) $\zeta = 1.01$, b) $\zeta = 1.5$, c) $\zeta = 2.0$, d) $\zeta = 3.0$.

2.1 Results and Discussion

The particles are introduced in the system at random positions. Thus gravity induces their motion and they fall and accelerate down the plane. Here, we will only discuss results corresponding to certain domain of angles in which the system reaches a steady state of motion on the x directions. To obtain detailed information about steady flows, we used the expressions defined above and [10–12]. Since the flows are uniform in x and z, we further averaged over those directions. Thus, we obtained the time-, width- and length-averaged density $\rho(y)$, momentum density $P(y)$ and velocity $V(y) = P(y)/\rho(y)$ fields (data not shown). The spatial profiles of the coarse contact stress, $\sigma^c{}_{\alpha\beta}(\vec{r})$, was also computed in all cases (data not shown). Our outcomes are totally consistent with earlier findings [11, 12].

It is important to remark that in dense granular flows, the values of $\sigma^k{}_{\alpha\beta}(\vec{r})$ generally result several orders of magnitude smaller than $\sigma^c{}_{\alpha\beta}(\vec{r})$. However, the spatial pattern of $\sigma^k{}_{\alpha\beta}(\vec{r})$ has been successfully used to distinguish between the role of force chains and velocity fluctuations in applications [21]. That's why a precise calculation of $\sigma^k{}_{\alpha\beta}(\vec{r})$ is necessary. Next we discuss the dependence of some components of the tensor $\sigma^k{}_{\alpha\beta}(\vec{r})$ on the coarse-graining scale, as well as the role that the velocity gradients play.

In figure 2, the profiles of the xx component of the kinetic stress $\sigma^k_{xx}(\vec{r})$ are illustrated. We present systematic study, with results obtained using particles with four different elongations are shown. The fields were deduced using using a Gaussian coarse-graining function $\phi(\vec{\mathcal{R}})$ with different values of w. As it noticeable, the values of σ^k_{xx} in general depends on w. Those results correlate with the fact that $\frac{\partial V_x}{\partial y} \neq 0$, which causes to the dependency of the results on the averaging domain size. However, as it was obtained earlier [10–12], there is a length scale much less than a particle diameter d, where the kinetic stress can be accurately computed using Eq.(1). The consistency of our outcomes is proven by the values of $T^k_{xx}(y) = \sigma^k_{xx}(y) - T^\gamma_{xx}(y)$, which are also shown for comparison in figure 3. Note that computing $T^\gamma_{xx}(y)$ involves the calculation of the components of the vector $D_k(\zeta)$, which depends on the coarse graining function $\phi(\vec{\mathcal{R}})$, the coarse graining scale w^2, and the particle elongation ζ. Once the system reaches a steady state of motion on the x directions we can find $T^\gamma_{xx}(y) = D_k(\zeta)\left(\frac{\partial V_x}{\partial y}\right)^2$ and the other components of $T^\gamma(y)$ diminishes. For the case of spherical particles and a Gaussian coarse-graining function $\phi(\vec{\mathcal{R}})$ with w, result $D_k(\zeta = 1) = w^2$. The values $D_k(\zeta)$ in terms of $D_k(1)$ are shown for clarifying purposes.

In figure 3, the profiles of the zz component of the kinetic stress $\sigma^k_{zz}(y)$ are illustrated. As it is noticeable, the component $\sigma^k_{zz}(y)$ in general do not depend on w and all the curves collapse. The fact that the only non zero component of the velocity gradient is $\frac{\partial V_x}{\partial y}$, leads to $T^\gamma_{zz}(y) \approx 0$ and, accommodatingly, the outcomes of $\sigma^k_{zz}(y)$ are independent of the averaging domain size. Similar outcomes are obtained for the profiles of the yy component of the kinetic stress $\sigma^k_{yy}(y)$, which is consistent for symmetry reasons.

Summarizing, we used a hybrid CPU/GPU implementation of DEM, examining the steady-state flow of 3D non-spherical particles down an inclined plane. A systematic study of the system response was performed varying the particle aspect ratio and the inclination angle. Similarly to the case of spheres, we observed three well-defined regimes: arresting flows, steady uniform flows and accelerating flows. Both steady and dynamic macroscopic fields were derived from microscopic data, by time-averaging and spatial smoothing (coarse- graining). Here, we have discussed that the values of $\sigma^k_{xx}(y)$, in general depends on w. This results correlates with the fact that $\frac{\partial V_x}{\partial y} \neq 0$, with the definition of the velocity fluctuation with respect to the center of the averaging volume. However, we also found there is a length scale much less than a particle diameter d, where the dependence on the coarse graining scale is diminished. The consistency of our outcomes was tested comparing with the scale free kinetic stress $T^k_{xx}(y)$ (see Eq.(2) and Ref.[13]). In our case, we found that the relevant component of $D_k(\zeta)$ depends on the coarse graining scale w, and the particle elongation ζ. Our ultimate aim is to achieve a continuum mechanical description of granular flows composed by non-sherical particles based on the micromechanical details and evaluate the influence of particle shape on the constitive response of those systems.

Acknowledgements

The Spanish MINECO (Projects FIS2014-57325), the German DFG project LU 450/10 (SPP PiKo) and Australian ARC Linkage Project LP160100280 have supported this work.

References

[1] I. Aranson, L. Tsimring, Rev. Mod. Phys. **78**, 641 (2006)
[2] S.L. Sanderson, A.Y. Cheer, J.S. Goodrich, J.D. Graziano, W.T. Callan, Nature **412**, 439 (2001)
[3] J.M. Pastor, A. Garcimartín, P.A. Gago, J.P. Peralta, C. Martín-Gómez, L.M. Ferrer, D.R. Maza, D.and Parisi, L.A. Pugnaloni, I. Zuriguel, Phys. Rev. E **92**, 062817 (2015)
[4] A. Garcimartin, J.M. Pastor, L.M. Ferrer, J.J. Ramos, C. Martin-Gomez, I. Zuriguel, Phys. Rev. E **91**, 022808 (2015)
[5] P. Jop, Y. Forterre, O. Pouliquen, Nature **441**, 727 (2006)
[6] Y. Jiang, M. Liu, Granular Matter **11**, 139 (2009)
[7] K. Kamrin, G. Koval, Phys. Rev. Lett. **108**, 178301 (2012)
[8] T. Pöschel, T. Schwager, *Computational Granular Dynamics* (Springer-Verlag Berlin Heidelberg New York, 2005)
[9] M. Babic, International Journal of Engineering Science **35**, 523 (1997)
[10] I. Goldhirsch, Granular Matter **12**, 239 (2010)
[11] T. Weinhart, A.R. Thornton, S. Luding, O. Bokhove, Granular Matter **14**, 531 (2012)
[12] T. Weinhart, R. Hartkamp, A.R. Thornton, S. Luding, Physics of Fluids **25** (2013)
[13] R. Artoni, P. Richard, Phys. Rev. E **91**, 032202 (2015)
[14] T. Kanzaki, M. Acevedo, I. Zuriguel, I. Pagonabarraga, D. Maza, R.C. Hidalgo, Eur. Phys. J. E, **34**, 133 (2011)
[15] K. Harth, U. Kornek, T. Trittel, U. Strachauer, S. Höme, K. Will, R. Stannarius, Phys. Rev. Lett. **110**, 144102 (2013)
[16] M. Acevedo, R.C. Hidalgo, I. Zuriguel, D. Maza, Phys. Rev. E **87**, 012202 (2012)
[17] S. Wegner, R. Stannarius, A. Boese, G. Rose, B. Szabo, E. Somfai, T. Borzsonyi, Soft Matter **10**, 5157 (2014)
[18] F. Villemot, J. Talbot, Granul. Matter **14**, 91 (2012)
[19] T. Kanzaki, R.C. Hidalgo, D. Maza, I. Pagonabarraga, J. Stat. Mech. **2010**, P06020 (2010)
[20] F. Alonso-Marroquín, Europhys. Lett. **83**, 14001 (2008)
[21] S.M. Rubio-Largo, A. Janda, D. Maza, I. Zuriguel, R.C. Hidalgo, Phys. Rev. Lett. **114**, 238002 (2015)

Effect of the particle shape on the particle dynamics in a spheronization process

Dominik Weis[1,*], *Maria* Niesing[2], *Markus* Thommes[2], and *Sergiy* Antonyuk[1]

[1]Institute of Particle Process Engineering, University of Kaiserslautern, 67663 Kaiserslautern, Germany
[2]Chair of Solids Process Engineering, Technical University Dortmund, 44227 Dortmund, Germany

Abstract. Spherical granules with a narrow size distribution are widely used in many pharmaceutical applications. Extrusion-spheronization is a well-established process to produce such pharmaceutical pellets. The cylindrical extrudates from the extrusion step are rounded in the spheronizer. The formation mechanisms inside of a spheronizer depend strongly on the particle dynamics. To describe the complex particle flow and interactions, the Discrete Element Method can be used. In our previous works the spherical particles during the last part of the spheronization process were studied. Since the pellets have a cylindrical shape at the beginning and undergo different stages of deformation during the rounding process, the objective of this study was the description of the influence of the particle shape on the particle dynamics. To predict the interactions of the pellets, their dominant plastic behaviour was described with an appropriate contact model and the material parameters were calibrated with compression and impact tests.

1 Introduction

Spherical granules with a narrow size distribution and a homogeneous surface are highly interesting for pharmaceutical applications. To produce such pharmaceutical pellets, a combined extrusion and spheronization process is widely used [1-3]. A spheronizer consists of a stationary cylindrical wall and a rotating disk with a structured surface called friction plate (Figure 1).

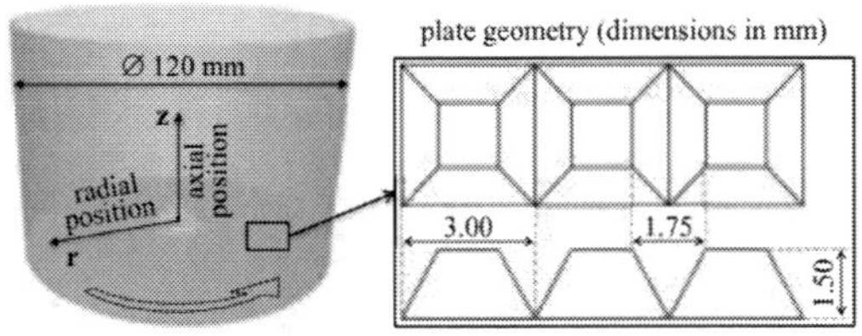

Fig. 1. Spheronizer geometry used in the simulations

The rounding of the wet cylindrical extrudates in the spheronization process is influenced by various interfering mechanisms. An important factor affecting those mechanisms is the particle dynamics in the spheronizer, which depends on the process conditions and the pellet properties. For this purpose, simulations of the particle dynamics were performed with the Discrete Element Method (DEM). In our previous work, the particle dynamics for different process conditions and material properties were studied for a late stage of the spheronizsation process, in which our particles can be approximated as spheres. [4]. Since the pellets have a cylindrical shape at the beginning of the spheronization process and undergo different stages of deformation (cylindrical – dumbbell - spherical) during the rounding process [5-7], this contribution focuses on the influence of the particle shape on the particle dynamics. The simulation results provide information about the influence of the particle shape on the particle flow regime and the mixing of the particle torus, as well as the influence on the collision rates and forces. This information obtained on the micro scale can be used as the basis to describe the overall rounding process on the macro scale for example with a population balance model as it was done in [8] for a melt granulation process. To calibrate the contact model for the simulations, the dominant plastic behaviour of the pellets was investigated in experiments. The required material parameters were derived by compression and impact tests with spheronized pellets.

2 Materials and methods

2.1 Material properties of pharmaceutical pellets

The studied pellets were produced by extrusion/ spheronization. A mixture consisting of microcrystalline cellulose (Vivapur 102, JRS Pharma, Germany) with a mass fraction $w_i = 0.2$ g/g and α-lactose monohydrate (Granulac 200, Meggle, Germany) with $w_i = 0.8$ g/g was extruded (Micro 27 GL-28D, Leistritz, Germany) with water at 300 rpm. Afterwards, 450 g of the extrudates were spheronized in a lab scale spheronizer (RM300, Schlüter, Germany) with a rotational speed of the friction plate of 750 rpm for 5 minutes. The deformation behaviour of spheronized pellets was obtained with

* Corresponding author: dominik.weis@mv.uni-kl.de

uniaxial compression tests. A single pellet was loaded by a punch with a rate of 20 µm/s while the force and displacement were continuously recorded. Moreover, loading/unloading tests were performed. The pellet was loaded up to a predefined maximal force of 0.1 N. From measured force-displacement curves, the important parameters like stiffness and static coefficient of restitution were estimated. Figure 2 shows a typical force-displacement curve of the tested pellets. It exhibits the dominant plastic material behaviour.

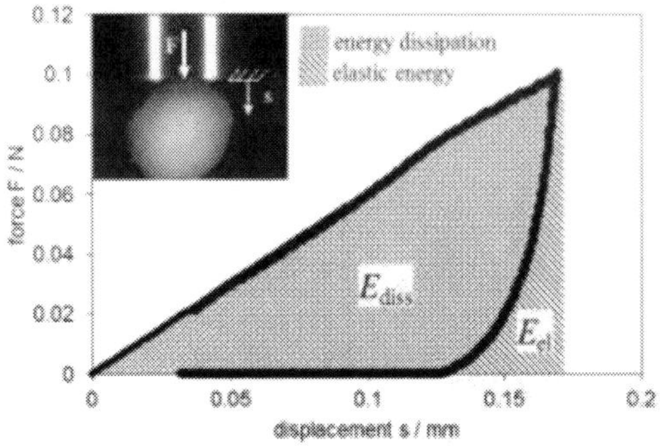

Fig. 2. Typical force-displacement curve obtained by uniaxial compression test

The energy dissipation based on plastic deformation can be expressed with a static restitution coefficient e_{stat}:

$$e_{stat} = \sqrt{\frac{E_{el}}{E_{load}}} = \sqrt{\frac{E_{el}}{E_{diss}+E_{el}}} \tag{1}$$

For dynamic impacts, the energy dissipation can be different, due to viscous effects [9]. For this reason, additional impact tests are necessary. Normal collisions with impact speeds of 0.3 - 0.6 m/s on the horizontal surface of the pyramids of the friction plate were recorded with a high speed camera (Y 8 Speedgrade 3, IDT, Tallahassee, US). The restitution coefficient was calculated as a ratio of the rebound and impact velocities. The results of the compression and impact tests are shown in Table 1. Due to the small deviation of the results obtained by compression and impact tests, viscous effects can be assumed to be negligible.

Table 1. Material parameters of pellets obtained by compression and impact experiments

parameter	unit	arithmetic mean	standard deviation
pellet diameter	mm	1.364	0.137
e_{stat} (compression)	-	0.227	0.035
e_{dyn} (impact)	-	0.218	0.056
breakage force	N	0.146	0.031
loading stiffness	N/mm	0.658	0.164

2.2 Discrete Element Method

The DEM, introduced firstly in 1979 by Cundall and Strack [10], can be used to simulate the dynamics of particulate systems. The method describes each particle in the system and its interactions with other particles or rigid walls via contact forces F_c. The particles are assumed to be spherical and indestructible. Based on the force balance for every single particle, the equations of motion (2) and (3) are numerically solved.

$$m_p \frac{dv_p}{dt} = F_g + \sum_i^k F_{c,i} \tag{2}$$

$$J_p \frac{d\omega_p}{dt} = \sum_i^k M_{p,i} \tag{3}$$

In the equations m_p and J_p are the mass of a particle and its moment of inertia while v_p and w_p are the velocity and the angular velocity. $F_{c,i}$ and $M_{p,i}$ are the force and the torque acting on a particle due to the interactions with contact partner i. These contact forces need to be modeled with an appropriate contact model. To describe the dominant plastic material behavior of the pellets used in this study, the model of Walton and Braun [11-13] can be utilized. It has a linear force-displacement relationship and can be assigned to the hysteresis models. That means the energy dissipation due to plastic deformation is modeled by assuming different stiffnesses K_1 and K_2 for loading and unloading. The force-displacement curve is shown in Figure 3. It can be seen that for a reloading, before an interaction is finished, the greater stiffness K_2 is used because of the local consolidation in the contact area. The normal force can therefore be calculated as follows:

$$F_{c,n} = -\begin{cases} K_1 \cdot s_n & \text{if } K_1 \cdot s_n < K_2 \cdot (s_n - s_0) \\ K_2 \cdot (s_n - s_0) & \text{if } s_n > s_0 \\ 0 & \text{if } s_n \leq s_0 \end{cases} \tag{4}$$

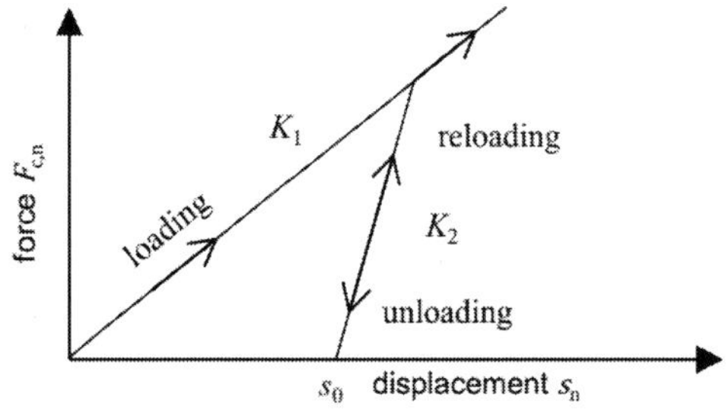

Fig. 3. Force-displacement curve of Walton and Braun model

Since the model does not include energy dissipation based on viscous damping terms, the relationship in (5) can be derived between the stiffness of the loading and unloading case. With a constant value for the coefficient of restitution the unloading stiffness can be calculated from the loading stiffness.

$$e = \sqrt{\frac{K_1}{K_2}} \tag{5}$$

In DEM there are different methods to approximate the real, non-spherical shape of particles. Generally there are two strategies to model the real particle shape: the single-particle and the clustered-particle approach [14]. In this study the clustered-particle approach, in which smaller spherical particle are connected in order to form a particle of complex shape [15, 13], is used because it is a relatively simple model, which can approximately predict the dynamic of non-spherical particles [17].

3 Results

The three different stages of the pellet rounding, which can be seen in Figure 4 (left: cylindrical, mid: dumbbell, right: spherical), were simulated in this study. For every particle shape the pellets had a diameter of 1 mm. Simulations were performed with a bed mass of m = 35 g and a rotational speed of the plate of n = 2500 rpm. Since, during spheronization, the bed mass remains constant and breakage of cylindrical particles takes places, the total number of pellets in the simulation of dumbbell and spherical stages was assumed to be higher than for the cylindrical stage.

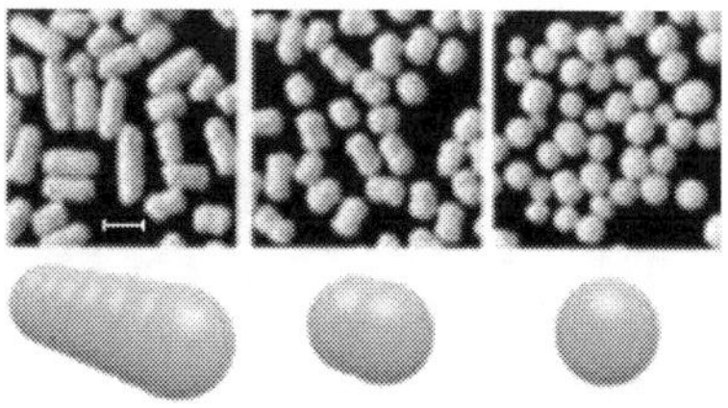

Fig. 4. Pellets after spheronization times of 20, 120, 480 s [18] modelled with the multi-sphere approach, (scale: 2.0 mm)

Figure 5 shows the residence probabilities of the different shaped pellets in a poloidal cut plane. As can be seen on the left side, a poloidal cut plane is a rz-cut through the particle torus. Due to the rotational symmetry, such a poloidal cut plane is representative for the whole particle bed. The residence probabilities, defined as the average number of particles in the different regions scaled with the total number, exhibit a decrease in the bed porosity with increasing sphericity of the particles. For every particle shape, there is a zone with very small values for the residence probability close to the friction plate. This zone shows the biggest axial expansion for the cylindrical particles and decreases with increasing sphericity. It is also noticeable that the spherical particles are regularly arranged close to the wall. Those maxima in the residence probability get less and less with decreasing sphericity. The influence of the particle shape on the rate of the collisions of particles with the friction plate is shown in Figure 6. The collision rate is defined as the number of collisions with the friction plate per particle within a second. All curves for the different particle shapes are qualitatively similar. The particles collide at the inner radial positions for the first time with the friction plate. The maximum of the collision rate is reached at about r = 59 mm for all studied particle shapes. In the middle of the torus, at 50 < z < 58, no significant difference between the spherical and the dumbbell shape can be obtained. The cylindrical shaped particles collide more often with the friction plate in this region. In the region close to the wall, the dumbbell shaped particles collide less often with the friction plate compared to the other shapes.

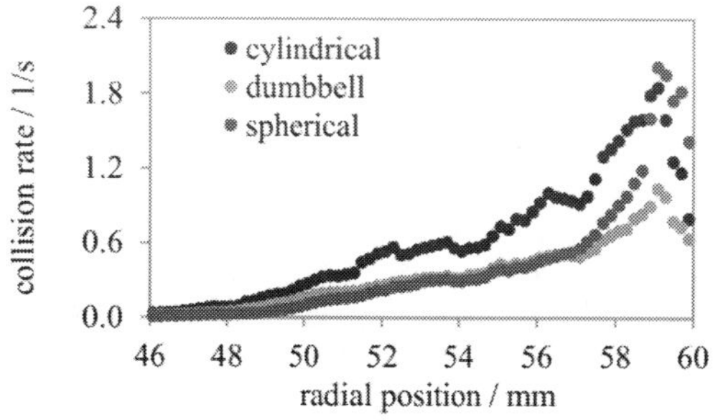

Fig. 6. Averaged rates of the collisions with the friction plate

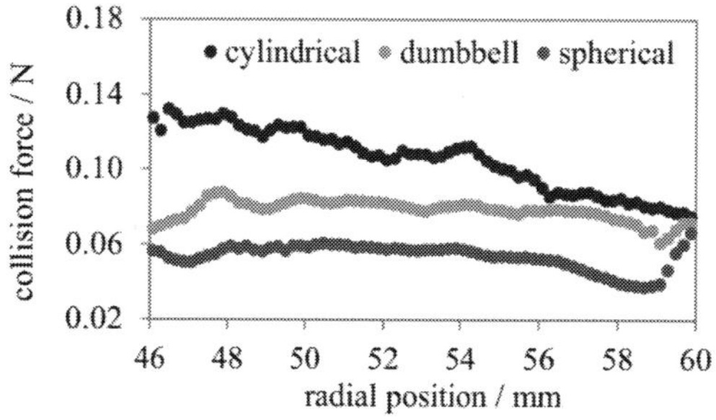

Fig. 7. Averaged forces of the collisions with the friction plate

Figure 7 shows the averaged collision force of the collisions of the different shaped particles with the friction plate. It can be seen that the biggest forces act on the cylindrical shaped particles. Both distributions, the collision rates and forces, are important functions to describe the overall rounding process in a spheronizer on the macro scale. For this reason the different rounding

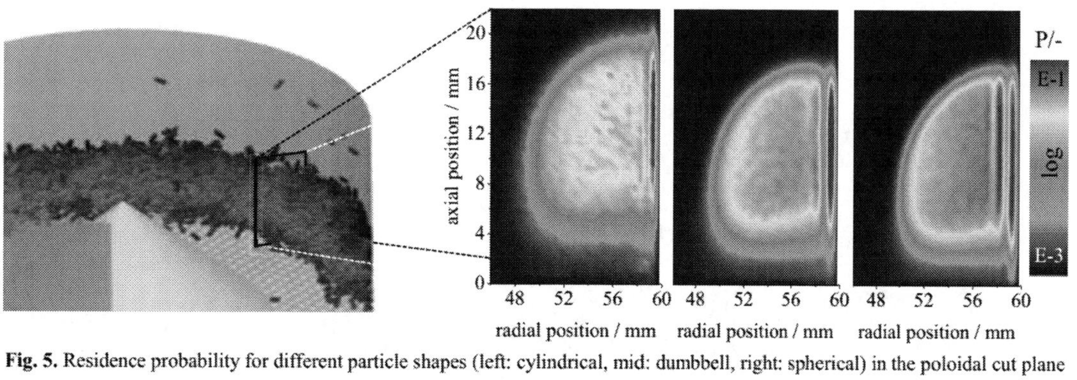

Fig. 5. Residence probability for different particle shapes (left: cylindrical, mid: dumbbell, right: spherical) in the poloidal cut plane

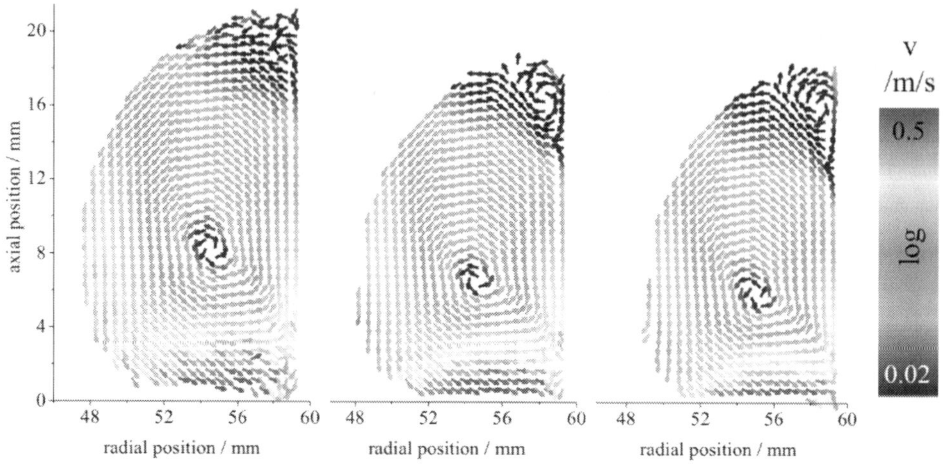

Fig. 8. Averaged poloidal particle movement for different particle shapes (left: cylindrical, mid: dumbbell, right: spherical) in the poloidal cut plane

mechanisms like deformation, attrition and breakage have to be described as a function of the collision forces and pellet shape and can then be used in a population balance. Comparing the breakage forces measured in the compression tests with spherical pellets in Table 1 for example, the simulation results indicate that no breakage will occur for the spherical shaped particles.

In Figure 8 the averaged poloidal velocity map is shown for the different shapes. The overall flow regime is quite similar. The largest poloidal velocities are reached in the region with high porosity close to the friction plate. In accordance to the axial expansion of this region for cylindrical particles, the region with high poloidal velocities also increases for cylindrical shaped particles. The primary rotational movement, which occurs for every particle shape, is superimposed by a smaller counter rotating movement at the top of the torus for the dumbbell and the spherical shape. For every particle shape, there are two zones which exhibit comparative small poloidal velocities, what has a negative effect on the mixing behaviour. The omission of the counter rotating movement at the top of the particle torus for the cylindrical particles should have positive effects on the mixing. Since we assume that the main part of the rounding happens due to the interaction with the plate, the mixing has also to be taken into account to describe the overall rounding process.

4 Conclusion

In this contribution the effect of the particle shape on the particle dynamics in a spheronization process is analysed with DEM simulations. For this reason the plastic material behaviour was described with an appropriate contact model. To calibrate the model, the required material parameters were determined by compression and impact tests with spheronized pellets. The different deformation stages were modelled using the multi-sphere approach. The simulations provide information about the residence probabilities and the flow regime of different shaped particles. Moreover the impact on the collision rates and forces of the collisions with the friction plate were examined. The parameter obtained on the micro scale are the basis to describe the overall rounding in a spheronizer on the macro scale in a population balance model.

References

1. S. Muley, T. Nandgude, S. Poddar, Asian J. Biomed. Pharm. Sci. (to be published)
2. A. Bathool, G.D. Vishakante, M.S. Khan, V.K. Gupta, J. Pharm. Res. **4**, 3282-3286 (2011)
3. N. Shinde et al., Asian J. Res. Pharm, Sci. **4**, 38-47 (2014)
4. D. Weis, S. Antonyuk, D. Thaete, M. Thommes, Proceedings of IV International Conference on Particle-Based Methods, 576-578 (2015)
5. C. Krueger, M. Thommes, P. Kleinebudde, Powder Technol. **238**, 176-187 (2013)
6. C.L.S. Lau et al., Chem. Eng. Res. Des. **92**, 2413–2424 (2014)
7. M.P. Bryan et al., Powder Technol. **270**, 163-175 (2015)
8. H.S. Tan et al., Powder Technol. **143**, 65-83 (2004)
9. S. Antonyuk et al., Granular Matter **12**, 15-47 (2010)
10. P. A.Cundall, O. D. Strack, Geotechnique **29**, 47-65 (1979)
11. O. R. Walton,R. L. Braun, Acta Mechanica **63**, 73–86 (1986)
12. O. R. Walton,R. L. Braun, J. Rheol. **30**, 949-980 (1986)
13. O. R. Walton, Particulate two-phase flow **25**, 884-907 (1993)
14. D. Höhner et al., Powder Technol. **208**, 643–656 (2011)
15. J. Favier et al., Eng. Comput. **16**, 467–480 (1999)
16. M.H. Abbaspour-Fard, Biosystems Engineering **88**, 153–161 (2004)
17. D. Höhner et al., Powder Technol. **208**, 643-656 (2011)
18. M. Koester, E. Willemsen, C. Krueger, M. Thommes, Int. J. Pharm. **431**, 84-89 (2012)

Kinetic theory of discontinuous shear thickening

Hisao Hayakawa[1,*] and *Satoshi* Takada[2,**]

[1] *Yukawa Institute for Theoretical Physics, Kyoto University, Kyoto 606-8502, Japan*
[2] *Department of Physics, Kyoto University, Kyoto University, Kyoto 606-8502, Japan*

Abstract. A simple kinetic theory to exhibit a discontinuous shear thickening (DST) is proposed. The model includes a collision integral, a drag from environment and an environmental temperature T_{ex} as well as a shear term. The viscosity of this model is proportional to $\dot{\gamma}^2$ for large shear rate $\dot{\gamma}$, while it is independent of $\dot{\gamma}$ for low sheae rate. The emergence of the DST is enhanced for lower density and lower nonzero T_{ex}.

1 Introduction

The viscosity of a Newtonian fluid is independent of the shear rate $\dot{\gamma}$. This Newtonian viscosity is no longer valid, if the fluid contains some suspensions. For instance, a dense suspension often behaves as a Newtonian fluid at rest or weakly stirred situation, but its resistance becomes large as if it is a solid, when the suspension is sheared with large $\dot{\gamma}$. This process is known as a shear thickening which can be used in various industrial applications.

Shear thickening attracts much attention even from physicists [1, 2]. In many of shear thickening fluids, the continuous shear thickening (CST) becomes steeper as the density increases, up to the point that the shear viscosity exhibits a discontinuous shear thickening (DST). The suspensions exhibiting DSTs have some common features, where (i) the normal stress difference becomes large when the DST takes place [2], and (ii) it is only observed below the jamming point [3].

To understand the DST, Brown *et al.* proposed three mechanisms, hydro-clustering, order-disorder transition and dilatancy for the shear thickening as well as their combinations [1, 4, 5]. However, none of three mechanisms is necessary for some DSTs as in Ref. [6]. Many researchers, instead, indicate that the mutual frictions between grains play important roles in the DST mainly from simulations of suspensens and granular particles [6–13]. Nevertheless, we may ask a question whether the mutual friction is always necessary for any DST.

Some researchers [14–17] proposed phenomenologies for the DST by the introduction of the order parameter which exhibits a S-shape in the relationship, i. e. the flow curve, between the stress and $\dot{\gamma}$. The most successful phenomenology has been proposed by Wyart and Cates [18] in which they have introduced the fraction of lubrication film to interpolate two divergent densities for purely frictional grains and frictionless grains. They have demontsrated that the rescaled pressure exhibit a saddle-node bifurcation at a critical density. Then, their flow curve exhibits a S-shape above the critical density, which is an explanation of the DST. In spite of their sucess we still do not know how to measure the fraction of lubrication film as well as the justification of their starting equation.

There are only a few microscopic theories on shear thickening. One of remarkable achievements is that by Santos *et al.*[19] in which they demonstrate the existence of a CST in moderately dense hard-core gases by using the revised Enskog theory. We do not know any systematic microscopic theory to describe DSTs, so far, as long as we know. Moreover, we may look for the possiblity to have a DST without the mutual friction between grains.

The purpose of this paper is to demonstrate the existence of a DST without the mutual friction between grains in terms of a simple kinetic theory. We believe that the methodology is applicable to wide class of sheared fluids, though the model contains still some phenomenological assumptions.

2 Kinetic theory: general properties

Let us consider a collection of mono-disperse smooth spheres (the diameter d and the mass m without the mutual friction between spheres) influenced by the background fluid under a uniform shear flow characterized by the macroscopic velocity field $\boldsymbol{u} = (u_x, u_y, u_z)$ as

$$u_x = \dot{\gamma} y, \quad u_y = u_z = 0. \tag{1}$$

Introducing the peculiar momentum $\boldsymbol{p}_i \equiv m(v_x - \dot{\gamma}y, v_y, v_z)$ we assume that the one-body distribution function under the simple shear satisfies

$$\left(\frac{\partial}{\partial t} + \dot{\gamma}\left\{y\frac{\partial}{\partial x} - p_y\frac{\partial}{\partial p_x}\right\}\right) f(\boldsymbol{r}, \boldsymbol{p}, t) = \zeta \frac{\partial}{\partial \boldsymbol{p}} \cdot \left(\left\{\boldsymbol{p} + mT_{ex}\frac{\partial}{\partial \boldsymbol{p}}\right\} f(\boldsymbol{r}, \boldsymbol{p}, t)\right) + J(f), \tag{2}$$

* e-mail: hisao@yukawa.kyoto-u.ac.jp
** e-mail: takada.satoshi.3s@kyoto-u.ac.jp

where we have introduced ζ and $T_{\rm ex}$ to characterize the drag from the solvent and the environmental temperature, respectively. Here, we ignore hydrodynamic interactions between spheres. We also assume that ζ is independent of $T_{\rm ex}$, and can be treated as a constant. Here, $J(f)$ is a collisional integral, which must be a functional of the two-body distribution. Although the two-body distribution depends on the three-body distribution in general situations, the two-body distribution can be represented by a functional of the one-body distribution in idealistic situations such as dilute limit. Note that the position $\boldsymbol{r}$ in $f(\boldsymbol{r},\boldsymbol{p},t)$ should be interpreted as the relative position of adjacent two particles in a uniform system. Otherwise, one can regard f as the distribution function for a tracer particle.

For simplicity, we adopt the simple Bhatnagar-Gross-Krook (BGK) model for the collision integral for moderate dense hard spheres: [19]

$$J(f)=\frac{f_{\rm eq}-f}{\tau};\quad \frac{1}{\tau}=\frac{16}{5}nd^2g_0(\varphi)\sqrt{\frac{\pi T}{m}},\qquad (3)$$

where we have introduced the density n, the volume fraction $\varphi=\pi nd^3/6$ and the first peak of the radial distribution $g_0(\varphi)=(1-\varphi/2)/(1-\varphi)^3$ [20]. Here the kinetic temperature T introduced in Eq. (3) is defined by $T\equiv{\rm tr}\overleftrightarrow{\sigma}^{(K)}/(3n)$, where $\sigma^{(K)}_{\alpha\beta}=\int d\boldsymbol{r}d\boldsymbol{p}f(\boldsymbol{r},\boldsymbol{p})p_\alpha p_\beta/m$. Note that T is the characterization of the motion of suspended particles, which differs from $T_{\rm ex}$ for the environment. For later convenience, we also introduce anisotropic temperatures $T_x\equiv\sigma^{(K)}_{xx}/n$. It should be noted that the standard BGK model for moderately dense hard spheres should be associated with the extra terms to represent the potential contribution [19]. Thus, Eq. (3) is quantitatively correct only for dilute cases.

Now, we adopt Grad's 13 moment-like expansion [21–23], i. e. the expansion in terms of currents such as the heat current and the stress current in addition to the conventional 5 collisional conserved quantities. If the system is uniform, we can assume

$$f(\boldsymbol{r},\boldsymbol{p})=f_{\rm eq}(\boldsymbol{r},\boldsymbol{p})\left\{1+\frac{1}{2T}(\mu_{\alpha\beta}-\delta_{\alpha\beta})\tilde{\sigma}_{\alpha\beta}\right\}\qquad (4)$$

where $f_{\rm eq}=\exp[\{p^2/(2m)+U_R(r)\}/T]/\mathcal{Z}$ with $\mathcal{Z}=\int d\boldsymbol{r}\int d\boldsymbol{p}\exp[-\{p^2/(2m)+U_R(r)\}/T]$ is the equilibrium distribution and

$$\tilde{\sigma}_{\alpha\beta}\equiv\frac{p_\alpha p_\beta}{m}-\frac{r_\alpha r_\beta}{r}U_R'(r)\qquad (5)$$

is the microscopic stress. Here, we adopt Einstein's rule in which duplicated Greek indices take summations over the components. Note that the macroscopic stress satisfies $\sigma_{\alpha\beta}=\int d\boldsymbol{p}d\boldsymbol{r}f(\boldsymbol{r},\boldsymbol{p})\tilde{\sigma}_{\alpha\beta}$. We also note that $U_R(r)$ is the sum of pairwise potentials, which can be replaced by the renormalized potential satisfying $U_R(r)=-T\ln g(r,\varphi)$ in the low density case, where $g(r,\varphi)$ is the radial distribution function [24]. Here, we have introduced the macroscopic stress ratio $\mu_{\alpha\beta}\equiv\sigma_{\alpha\beta}/P$ with the pressure $P\equiv{\rm tr}\overleftrightarrow{\sigma}/3$. This picture is consistent with the previous statistical mechanical treatments, where the correction term has the form $\tau\dot\gamma\tilde{\sigma}_{\alpha\beta}/T$ for $\alpha\neq\beta$ if $\tau\dot\gamma$ satisfies relationship $\tau\dot\gamma=\mu\equiv-\mu_{xy}=-\mu_{yx}$ [25–28].

Multiplying $\tilde{\sigma}_{\alpha\beta}$ on both sides of Eq.(2) with the aid of Eq.(4) and integrating over $\boldsymbol{p}$ and $\boldsymbol{r}$, we obtain

$$\frac{\partial}{\partial t}T=\frac{1}{Z(\varphi)}\left\{-\frac{2\dot\gamma Y(\varphi)\sigma_{xy}}{3n}+2\zeta(T_{\rm ex}-T)\right\},\qquad (6)$$

$$\frac{\partial}{\partial t}\Delta T=-\frac{2\dot\gamma Y(\varphi)\sigma_{xy}}{nZ(\varphi)}-\left(\nu+\frac{2\zeta}{Z(\varphi)}\right)\Delta T,\qquad (7)$$

$$\frac{\partial}{\partial t}\sigma_{xy}=\dot\gamma Y(\varphi)nZ(\varphi)\left(\frac{\Delta T}{3}-T\right)-\left(\nu+\frac{2\zeta}{Z(\varphi)}\right)\sigma_{xy},\qquad (8)$$

where we have introduced

$$Z(\varphi)=\frac{P}{nT}=\frac{1+\varphi+\varphi^2-\varphi^3}{(1-\varphi)^3};\quad \Delta T\equiv T_x-T_y\qquad (9)$$

and

$$Y(\varphi)=\frac{2}{Z(\varphi)}-1=\frac{1-7\varphi+5\varphi^2-\varphi^3}{1+\varphi+\varphi^2-\varphi^3}.\qquad (10)$$

Because $Y(\varphi_c)$ becomes negative for $\varphi>\varphi_c$ with

$$\varphi_c=\frac{1}{3}\left(5-\frac{4}{19-3\sqrt{33}}-(19-3\sqrt{33})^{1/3}\right)=0.160713\cdots,\qquad (11)$$

our analysis is only valid for $\varphi<\varphi_c$. It is remarkable that the normal stress is coupled with P and σ_{xy} in Eqs. (6)-(8), which is consistent with Ref. [2].

Thus, we obtain a closed set of equations (6) - (8). To derive Eq. (8) we have used

$$\sigma^{(K)}_{xy}=-\mu nT=\sigma_{xy}Z(\varphi)^{-1},\qquad (12)$$

where we note $\sigma^{(K)}_{\alpha\beta}=-(\mu/(m^2T))\int d\boldsymbol{p}p_\alpha^2p_\beta^2f_{\rm eq}(\boldsymbol{p})=-\mu nT$ and $\sigma_{xx}-\sigma_{yy}=Z(\varphi)n\Delta T$.

The damping frequency ν has been introduced in Eqs. (6)-(8) by the relation $\Lambda_{xy}=\int d\boldsymbol{p}d\boldsymbol{r}\tilde{\sigma}_{xy}f/\tau=\nu\sigma_{xy}$. Note that the contribution from the diagonal term is zero for the BGK model (3). Thanks to the assumption (4), it is straightforward to show

$$\nu^*=\frac{D(\varphi)}{Z(\varphi)}\sqrt{T^*}(1+C(\varphi))\qquad (13)$$

for $\nu^*\equiv\nu/\zeta$ and $T^*=T/(md^2\zeta^2)$ with $D(\varphi)\equiv 48\varphi g_0(\varphi)/5\sqrt{\pi}$. Although the precise estimation of the potential contribution $C(\varphi)$ is not easy, this term becomes zero if the system size is enough large and the stiffness of particles is finite. Therefore, we omit the contribution of $C(\varphi)$ in the later discussion.

It should be noted that Eqs. (6)-(8) cannot determine T, ΔT and σ_{xy} as functions of n, $\dot\gamma$, and ν because all of them are homogeneous equations. Instead, we obtain the relations e.g. μ, $\Delta T/T$ and $\dot\gamma$ as functions of φ and ν. We also note that the motion of all particles stops, i.e. there is no dynamic steady state in the absence of $\dot\gamma$ and $T_{\rm ex}$.

In the steady state, we immediately obtain

$$\frac{\Delta T}{T}=\frac{6(1-\theta^{-1})}{\nu^*Z(\varphi)+2}\qquad (14)$$

from Eqs.(6) and (7), where we have introduced $\theta \equiv T/T_{\rm ex}$. Substituting this into Eq.(6) we obtain

$$\sigma_{xy} = -3\frac{1-\theta^{-1}}{\dot\gamma^* Y(\varphi)}nT. \tag{15}$$

From Eqs.(8), (14) and (15) we obtain

$$\dot\gamma^* = \frac{\nu^* Z(\varphi)+2}{Y(\varphi)Z(\varphi)}\sqrt{\frac{3(\theta-1)}{\theta\nu^* Z(\varphi)+2}}. \tag{16}$$

Then, the dimensionless viscosity is given by

$$\eta^* = \frac{6\varphi Y(\varphi)Z(\varphi)^2 T_{\rm ex}^*\{\theta\nu^* Z(\varphi)+2\}}{\pi(\nu^* Z(\varphi)+2)^2}, \tag{17}$$

where $\eta^* = \eta d/(m\zeta)$ with $\eta \equiv -\sigma_{xy}/\dot\gamma$ and $T_{\rm ex}^* = T_{\rm ex}\zeta^2/(md^2)$. Substituting Eq. (16) into Eq. (15) we obtain

$$\mu = -\frac{\sigma_{xy}}{Z(\varphi)nT} = \frac{\sqrt{3(\theta-1)(\theta\nu^* Z(\varphi)+2)}}{\nu^* Z(\varphi)+2} \tag{18}$$

for $\theta > 1$. Note that there is no steady solution for $\theta \le 1$, because the system is heated by the shear.

3 Rheology at $T_{\rm ex} = 0$

Now, let us determine the expression of η^* at $T_{\rm ex}^* = 0$. Note that ν^* can be determined by Eq. (16) as $\sqrt{\nu^*}Z(\varphi)^{3/2}Y(\varphi)\dot\gamma^* = \sqrt{3}(\nu^* Z(\varphi)+2)$ with $\dot\gamma^* = \dot\gamma/\zeta$, whose solution is given by $\sqrt{\nu^*} = \dot\gamma^* Y(\varphi)\sqrt{Z(\varphi)}/(2\sqrt{3})\{1 \pm \sqrt{1-24/(Y(\varphi)Z(\varphi)\dot\gamma^*)^2}$. Therefore, we obtain

$$\nu^* = \frac{Z(Y\dot\gamma^*)^2}{6}\left\{1-\frac{12}{(YZ\dot\gamma^*)^2}\pm\sqrt{1-\frac{24}{(YZ\dot\gamma^*)^2}}\right\}. \tag{19}$$

Because ν^* must be positive, ν^* exists only if $\dot\gamma^* \ge \dot\gamma_c^* \equiv 2\sqrt{6}/(Z(\varphi)Y(\varphi))$. Equation (19) indicates the absence of any dynamic steady state for $\dot\gamma^* < \dot\gamma_c^*$ because of the existence of damping terms in kinetic equation (2) proportional to ζ. Equation (19) is also characteristics that there are two branches in ν^*. Note that Eq. (19) is independent of the constitutive equation (13).

From Eqs. (13) and (19) we obtain

$$T^* = \left(\frac{5\sqrt{\pi}Z(\varphi)}{48\varphi g_0(\varphi)}\right)^2 \nu^{*2} \approx \left(\frac{5\sqrt{\pi}(Y(\varphi)Z(\varphi)\dot\gamma^*)^2}{144\varphi g_0(\varphi)}\right)^2 \tag{20}$$

for $\dot\gamma^* \gg 1$. Then, η^* satisfies

$$\eta^* = \frac{25\pi Z(\varphi)^5\nu^{*3}}{384\varphi g_0(\varphi)^2(\nu^* Z(\varphi)+2)^2} \approx \frac{25\pi Z(\varphi)^4 Y(\varphi)^2\dot\gamma^{*2}}{1152\varphi g_0(\varphi)} \tag{21}$$

for $\dot\gamma^* \gg 1$, where we have used Eqs.(3), (13) and (20). Note that Eq.(21) recovers the numerical observation in Ref. [13] for $\dot\gamma^* \gg 1$, while smaller η^* corresponding to smaller ν^* in Eq. (19) becomes zero for large $\dot\gamma^*$.

Figure 1 is the plot of η^*/η_c^* against $\dot\gamma^*/\dot\gamma_c^*$ for various φ, where η_c^* is η^* at $\dot\gamma_c^*$ based on Eqs. (16) and (17). It is remarkable that the scaled flow curve is independent of φ. This figure exhibits the existence of two branches for η^* above $\dot\gamma_c^*$, while the nonexistence of η^* below γ_c^*. The viscosity sharply increases near γ_c^* in the upper branch corresponding to the CST but there is no DST at $T_{\rm ex} = 0$.

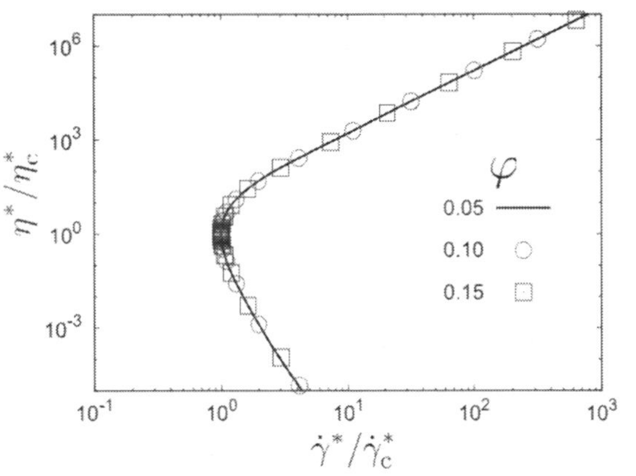

Figure 1. Plot of the dimensionless viscosity η^*/η_c^* against the shear rate $\dot\gamma^*/\dot\gamma_c^*$ for various φ, where we adopt $T_{\rm ex} = 0$.

4 Rheology at finite $T_{\rm ex}$

Now, let us discuss the behavior at finite $T_{\rm ex}$. Figure 2 is the plots of the viscosities η^*/η_c^* against $\dot\gamma^*/\dot\gamma_c^*$ for various φ.

It is remarkable that the viscosity takes a S-shape in the flow curve at $T_{\rm ex}^* = 0.1$. The appearance of S-shape is the result of gluing of two branches or the saddle-node bifurcation, as indicated by Ref. [18]. This S-shape is expected to appear if the Newtonian viscosity $\eta_0^* \equiv 6\varphi Z(\varphi)^2 T_{\rm ex}^*/\{\pi D(\varphi)(2+D(\varphi)\sqrt{T_{\rm ex}^*}\}$ is less than $\eta_c^* = 25\pi Z(\varphi)/(48\varphi g_0(\varphi)^2)$.

More precise condition of the DST is given by $\partial\eta^*/\partial\dot\gamma^* = (\partial\eta^*/\partial\theta)/(\partial\dot\gamma^*/\partial\theta) \to \infty$. Therefore, the critical condition $\partial\dot\gamma^*/\partial\theta = 0$ can be read

$$8\theta^{1/2} + B^2\theta^{3/2}(1+\theta) - 2B(2-7\theta+\theta^2) = 0, \tag{22}$$

which is a quintic equation for $\sqrt{\theta}$, where $B \equiv D(\varphi)\sqrt{T_{\rm ex}^*}$. This S-shape in the flow cure is the requirement condition to have a DST.

It is straightforward to analyze the linear stability of the steady solution. The result is counter intuitive, because all steady solutions are linearly stable. Nevertheless, when we control $\dot\gamma$, the viscosity discontinuously changes, if the flow curve has the S-shape. The amount of discontinuity in the flow curve is finite as shown in Fig. 2.

5 Discussion and Conclusion

Now, let us discuss our results. We should note that the DST in our model takes place for $\varphi < \varphi_c(T_{\rm ex}^*)$. This result is contrast to experimental observations, though our result is obtained from the fact that the Newtonian viscosity η_0^* becomes lower if φ decreases. After the completion of this work, we have confirmed that the Boltzmann equation instead of BGK model in the dilute limit exhibits a DST which perfectly agrees with the simulaion result [29] and the DST disappears as the density increases from the analysis of both the Enskog equation and the simulation [30].

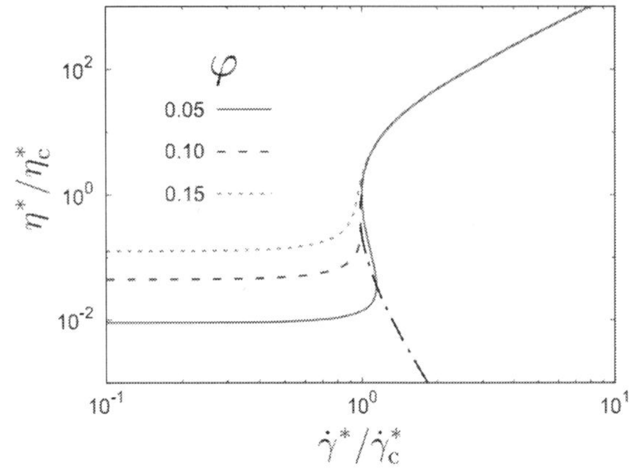

Figure 2. Plot of η^*/η_c^* against $\dot\gamma^*/\dot\gamma_c^*$ at $T_{ex}^* = 0.1$ for various φ. The back dot-dashed line satisfies $\partial\dot\gamma^*/\partial\eta^* = 0$.

One of key assumptions in our analysis is Eq. (4). Fortunately, Eq. (4) seems to be consistent with the Green-Kubo formula in nonequilibrium statistical mechanics [25–28].

We have assumed that the system is uniform to simplify the argument. This assumption cannot be used if the phase separation takes place.

In conclusion, we demonstrate that our simple kinetic model which has a constant damping ζ and an external temperature T_{ex} exhibits the DST. The emergence of the DST is the result of a saddle-node bifurcation by the gluing between Newtonian branch and the branch at $T_{ex} = 0$. We also note that the normal stress difference plays an important role in the DST, which is consistent with Ref. [2]. The proposed mechanism is new, because the DST is unrelated to neither mutual friction of grains, the order-disorder transition, boundary effects nor hydro-clustering mechanism.

Acknowledgment: We thank K. Suzuki, T. Kawasaki, M. Otsuki, K. Saitoh and R. Seto for his useful comments. This work is partially supported by the Grant-in-Aid of MEXT for Scientific Research (Grant No. 16H04025).

References

[1] E. Brown and H. M. Jeager, Rep. Prog. Phys. **77**, 040602 (2014).
[2] C. D. Cwalina and N. J. Wagner, J. Rheol. **58**, 949 (2014)
[3] E. Brown and H. M. Jeager, Phys. Rev. Lett. **103**, 080601 (2009).
[4] E. Brown, N. A. Forman, C. S. Orellana, H. Zhang, B. W. Maynor, D. E. Betts, J. M. DeShimone and H. M. Jaeger, Nature Mat. **9**, 220 (2010).
[5] S. Waitukaitis, S. and H. M. Jaeger, Nature, **487**, 205 (2012)
[6] M. Otsuki and H. Hayakawa, Phys. Rev. E **83**, 051301 (2011).
[7] D. Bi, J. Zhang, B. Chakraborty, R. P. Behringer, Nature **480**, 355 (2011).
[8] R. Seto, R. Mari, J. F. Morris, and M. M. Denn, Phys. Rev. Lett. **111**, 218301 (2013).
[9] R. Mari, R. Seto, J. F. Morris, M. M. Denn, J. Rheol. **58**, 1693 (2014).
[10] M. Pica Ciamarra, R. Pastore, M. Nicodemi, and A. Coniglio, Phys. Rev. E **84**, 041308 (2011).
[11] C. Haussinger, Phys. Rev. E **88**, 050201(2013).
[12] M. Grob, C. Heussinger, and A. Zippelius, Phys. Rev. E **89**, 050201(2014).
[13] T. Kawasaki, A. Ikeda and L. Berthier, EPL **107**, 28009 (2014).
[14] H. Nakanishi and N. Mitarai, J. Phys. Soc. Jpn. **80**, 033801 (2011).
[15] H. Nakanishi, S. Nagahiro, and N. Mitarai, Phys. Rev. E **85**, 011401 (2012).
[16] S. Nagahiro, H. Nakanishi, and N. Mitarai, Europhysics Letters **104**, 28002 (2013).
[17] M. Grob, A. Zippelius and C. Heussinger, Phys. Rev. E **93**, 030901 (2016).
[18] M. Wyart and M. E. Cates, Phys. Rev. Lett. **112**, 098302 (2014).
[19] A. Santos, J. M. Montanero, J. W. Dufty and J. J. Brey, Phys. Rev. E **57**, 1644 (1998).
[20] N. F. Carnaharn and K. E. Starling, J. Chem. Phys. **45**, 2102 (1969).
[21] H. Grad, Commun. Pure Appl. Math. **2**, 331 (1949).
[22] V. Garzó, Phys. Rev. E **66**, 021308 (2002).
[23] A. Santos, V. Garzó and J. W. Dufty, Phys. Rev. E **69**, 061303 (2004).
[24] J.-P. Hansen and I. R. McDonald, "The Theory of Simple Liquids, 2nd Edition" (Academic Press, London, 1990).
[25] D. J. Evans and G. P. Morriss, "Statistical Mechanics of Nonequilibrium Liquids, 2nd Ed." (Cambridge University Press, Cambridge, England, 2008).
[26] S.-H. Chong, M. Otsuki, and H. Hayakawa, Phys. Rev. E **81**, 041130 (2010).
[27] H. Hayakawa and M. Otsuki, Phys. Rev. E **88**, 032117 (2013).
[28] K. Suzuki and H. Hayakawa, Phys. Rev. Lett. **115**, 098001 (2015).
[29] H. Hayakawa and S. Takada, arXiv:1611.07295.
[30] H. Hayakawa, S. Takada and V. Garzó, in preparation,

Rheology of dense suspensions of non colloidal particles

Élisabeth Guazzelli[1,*]

[1] *Aix Marseille Univ, CNRS, IUSTI, Marseille, France*

Abstract. Dense suspensions are materials with broad applications both in industrial processes (e.g. waste disposal, concrete, drilling muds, metalworking chip transport, and food processing) and in natural phenomena (e.g. flows of slurries, debris, and lava). Despite its long research history and its practical relevance, the mechanics of dense suspensions remain poorly understood. The major difficulty is that the grains interact both by hydrodynamic interactions through the liquid and by mechanical contact. These systems thus belong to an intermediate regime between pure suspensions and granular flows. We show that we can unify suspension and granular rheology under a common framework by transferring the frictional approach of dry granular media to wet suspensions of spherical particles. We also discuss non-Newtonian behavior such as normal-stress differences and shear-induced migration. Beyond the classical problem of dense suspension of hard spheres which is far from being completely resolved, there are also entirely novel avenues of study concerning more complex mixtures of particles and fluids such as those involving other types of particles (e.g. fibers) or non-Newtonian fluids that we will also address.

1 Introduction

Dense suspensions are materials with broad applications both in industrial processes (e.g. waste disposal, concrete, drilling muds, metalworking chip transport, and food processing) and in natural phenomena (e.g. flows of slurries, debris, and lava). Despite its long research history and its practical relevance, the mechanics of dense suspensions remain poorly understood.

In order to understand their flowing behavior, it is desirable to know their response to imposed forces and motions at their boundary. The fundamental problem is then to determine the rheological properties of these media (considered as equivalent homogeneous materials) from a knowledge of the mechanics of the particles and the interstitial fluid. In other words, the key problem is to understand the relationship between the macroscopic or bulk properties of the medium and its microscopic structure at these large concentrations.

The major difficulty of dense particulate flows is that the grains interact both by hydrodynamic interactions through the liquid and by mechanical contact. Dense or highly concentrated particulate flows indeed belong to an intermediate regime between pure suspensions and granular flows. The complex nature of these particle interactions greatly contribute to the lack of understanding of these systems.

The present contribution aims at providing a review of recent work on the rheology of dense suspensions of non colloidal particles. These studies were done in collaboration with F. Boyer, J. E. Butler, E. Couturier, S. Dagois-Bohy, L. M. Davidson, S. Hormozi, O. Pouliquen, P. R. Nott, S. Shaikh, B. Snook, S. Strednak, F. Tapia.

*e-mail: Elisabeth.Guazzelli@univ-amu.fr

2 Suspension viscosity

Adding suspended particles to a fluid increases its viscosity. This is known for a Newtonian fluid since the seminal work of Einstein demonstrating that the viscosity of the mixture is increased above that of the suspending fluid [1]. The suspension viscosity increases with increasing particle concentration and diverges at the jamming transition where the particle concentration reaches a maximum value, see e.g. [2–5].

To be more precise, for a suspension of neutrally-buoyant solid spheres under steady shearing conditions, the linearity of the Stokes equations leads to a viscous scaling of the shear stress, $\tau = \eta_s(\phi)\,\eta_f\dot{\gamma}$, where $\dot{\gamma} = \sqrt{2\mathbf{E}:\mathbf{E}}$ is the shear rate (where $\mathbf{E}$ is the rate of strain) and η_f is the viscosity of the suspending Newtonian fluid. For monodisperse suspensions of non-Brownian rigid spheres, the dimensionless effective shear viscosity, η_s, only depends on the particle volume fraction, ϕ, and increases with increasing ϕ, diverging at maximum packing fraction, ϕ_m.

3 Non-Newtonian behavior: Normal stresses

This quasi-Newtonian behavior of the shear stress does not fully describe the rheology of suspensions, as it does not account for the possible existence of normal stress differences. These normal stress differences also scale viscously in Stokes flows and, since the normal stresses do not depend on the sign of the shear rate, are proportional to the modulus of the shear stress, $|\tau|$. These normal stress differences can be written as $N_1 = \alpha_1|\tau|$ and $N_2 = \alpha_2|\tau|$, where N_1 and N_2 are the first and second normal stress differences. The ratios of normal-stress differences to shear stress, α_1 and α_2, are called the normal stress difference coefficients. They are sole functions of ϕ and do not diverge at ϕ_m since the normal stress differences and the shear stress present the same divergence when approaching jamming at ϕ_m [6]. .

Normal stresses are difficult to measure using standard rheological tools. They were chosen to be inferred from measurements of the deflection of the free-surface in a Weissenberg, or rotating-rod, geometry and in a tilted-trough [7, 8]. The first method is well-known in polymers as the Weissenberg or rod-climbing effect, see figure 1 (*a*). For suspensions of spheres, the climbing is down instead of up the rod and measurement of the free-surface profile provides a combination of the normal stress differences. The second tilted-trough method provides the second normal stress difference in isolation, see figure 1 (*b*). Combining both methods yields a complete measurement of the two differences which are both linear in the modulus of the shear rate. These approaches have some significant advantages over using a standard rheometer as confinement effect can be reduced and sensitivity improved [7–9].

For suspensions of spheres, the second normal stress difference, N_2, is observed to be negative and to increase with increasing particle volume fraction (growing especially quickly for $\phi \gtrsim 0.20$), see e.g. [5]. The properties of the first normal stress difference are more elusive. The magnitude of N_1 is unquestionably much smaller than that of N_2, but assessing the sign is difficult. Using these non conventional rheological tools (rotating-rod and tilted-trough), the value is found too close to zero to determine whether it is negative, positive or null within experimental accuracy. Some experiments using more conventional rheometry find that N_1 is quite small and negative [10–13], while others report positive values [14]. Some recent studies point to the effect of polydispersity [15] and of confinement [16] which may explain these discrepancies.

Whereas the whole suspension, i.e. the mixture of the particles and the fluid, is incompressible, the particle phase is not. There exists a pressure coming from the dispersed particulate phase that could develop. It has been termed particle pressure (or more generally particle normal-stress) and is analogous to the osmotic pressure exerted by both colloidal particles and dissolved molecules, see e.g. [17, 18]. This particle normal stress, P, also scales viscously and is linear in the modulus of the shear rate since it must be independent of the sign of the shear rate as previously noted for the normal stresses of the whole suspension. It can be written as $P = \eta_n(\phi)\eta_f\dot{\gamma}$, where $\dot{\gamma}$ is conveniently defined in an invariant form as $\dot{\gamma} = \sqrt{2\,\mathbf{E}:\mathbf{E}}$, where $\mathbf{E}$ is the rate of strain. The dimensionless effective normal viscosity, η_n, is again a sole function of ϕ and presents the same divergence as $\eta_s(\phi)$ when approaching jamming at ϕ_m [6].

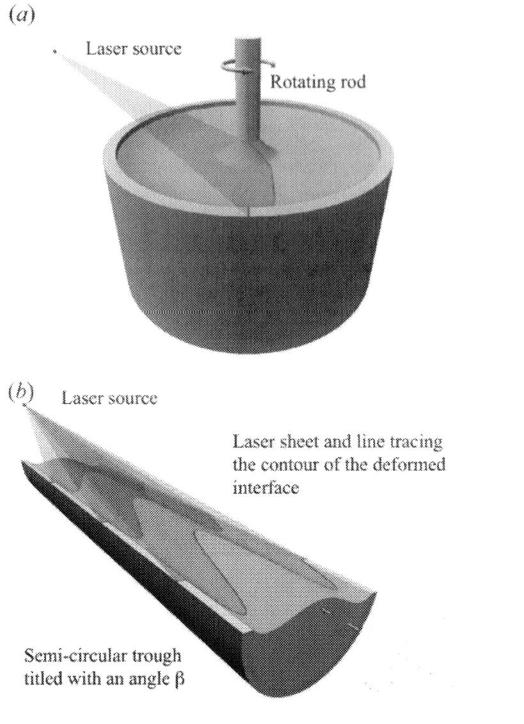

Figure 1. Sketches adapted from figure 2 of [9] of the (a) rotating-rod and (b) tilted-trough flows. Laser sheets, projected at low angles onto the surfaces of the flowing suspensions, enable quantification of the surface deformations and subsequently to the determination of the normal stress differences.

Quantitative measurements of the particle-phase stresses are not easily performed and only few experiments are reported in the literature [19–21]. The imposed-pressure approach of [20] will be discussed in section 5. The particle pressure is also of crucial importance in the understanding of shear-induced migration which is discussed in the following section 4.

4 Irreversible dynamics: Shear-induced migration

Shear-induced migration is a manifestation of irreversible dynamics in shearing flows of non-Brownian suspension. This irreversibility may be unexpected as, at low Reynolds numbers, the velocity and pressure fields are supposed to be governed by the linear and reversible Stokes equations. It is due to the combined effects of hydrodynamics interactions between the particles and non-hydrodynamic forces such as close contact interactions (short-ranged repulsive forces or small roughnesses).

Shear-induced migration drives particles from regions of high to low shear rate. This phenomenon was first clearly identified in a Couette rheometer [22] and moti-

DOI: 10.1051/epjconf/201714001001

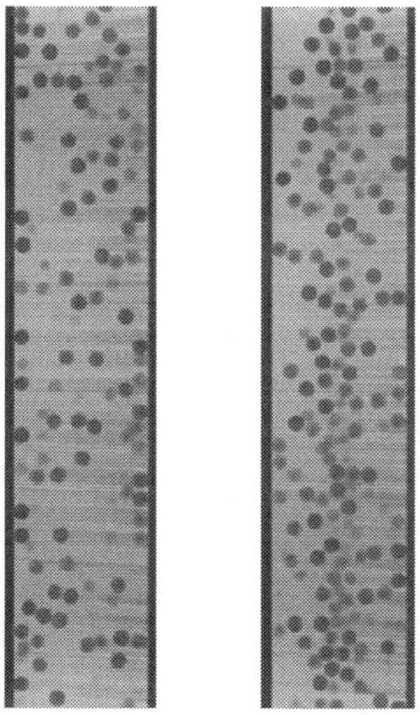

Figure 2. Images of a suspension, consisting of neutrally buoyant spheres, at a bulk volume fraction of 20% subjected to large oscillating displacements at low Reynolds number in a pipe, adapted from movie 1 of [24]. Particle volume fraction and velocity can be measured during the oscillations by matching the refractive index of the fluid to that of the particles and using fluorescence to distinguish the particle and fluid phases for imaging with a thin laser sheet. The left image corresponds to the initially mixed suspension and the right image to the suspension after 32 oscillations where fully-developed migration toward the centerline is observed.

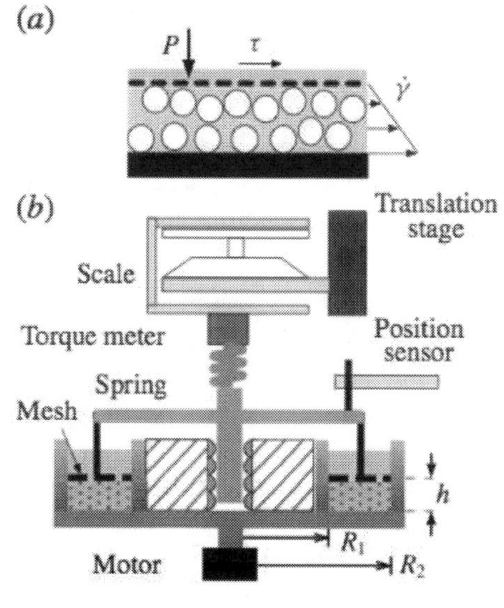

Figure 3. (a) Principle of the pressure-imposed rheometry: the porous plate exerts a pressure (P) on the grains while the suspension is sheared ($\dot{\gamma}$) and the shear stress (τ) is measured. (b) Sketch of the pressure-imposed shear cell, adapted from figure 2 of [29]. The shear cell consists of (i) a bottom annular cylinder (of radii R_1 = 43.95 mm and R_2 = 90.28 mm) that contains the suspension and rotates and (ii) a top cover plate that can be moved vertically and which is porous to enable fluid flow through it. From the measurements of the position of the top plate h, the rotation rate of the bottom annulus, and the normal force and torque exerted on the top plate, it is possible to extract ϕ, $\dot{\gamma}$, P, and τ.

vated a large body of research due to its implication on the characterization of suspension rheology, see e.g. [3–5]. In pressure-driven Poiseuille flow, the particles migrate toward the centerline, see figure 2. Again, since the first observation of such inhomogeneities in suspension pipe flows [23], several experimental studies have been performed to measure migration in pressure-driven flow in a pipe or a channel, see e.g. [24] and references therein.

Since the first observation of shear-induced migration, two types of migration models have been proposed. Early efforts used a diffusion model, in which the particle migration flux was expressed in terms of the gradients of the particle concentration and shear rate. This diffusion model is successful in predicting migration in wide-gap Couette and pressure-driven Poiseuille flows, but fails to predict the absence of migration in curvilinear torsional flows, see e.g. [6] and references therein. A more recent and rather successful model, termed the suspension balance model, relates the migration flux to the rheology of the suspension [25]. This modeling uses a two-phase approach of the suspension and finds that the particle migration flux is driven by the divergence of the normal stress of the particle phase [6, 25–27]. This model requires correlations for the particle phase stress which cannot be easily obtained experimentally. When using realistic correlations, the suspension balance model qualitatively supports the experimental observations but presents however some quantitative discrepancies [24].

Another avenue to study inhomogeneities in suspension flows is to perform discrete particle simulations. Earlier simulations of the pressure-driven flow in a two-dimensional channel of a suspension were conducted using Stokesian Dynamics [25]. Numerical capabilities are expanding rapidly and new methods such as the force coupling method [28] are providing effective information on the migration process in agreement with experiments [24].

5 A frictional approach: Unifying suspension and granular rheology

As explained in sections 2 and 3, for a suspension consisting of neutrally-buoyant spheres in a Newtonian fluid of viscosity, η_f, the scaling of the stresses is viscous, and consequently, when the suspension is sheared under a constant shear rate, $\dot{\gamma}$, at concentration, ϕ, the shear stress is $\tau = \eta_s(\phi)\,\eta_f\dot{\gamma}$ and the particle pressure is $P = \eta_n(\phi)\,\eta_f\dot{\gamma}$. The rheology thus reduces to the knowledge of the two functions $\eta_s(\phi)$ and $\eta_n(\phi)$ which are known to diverge in a similar way at maximum volume fraction, ϕ_m.

There is an equivalent approach to this classical view in terms of effective viscosities which is coming from the rheology of dry granular materials and hinges on a frictional view of the problem, see e.g. [30]. When an assembly of particles is subjected to steady shear under a confining particle pressure, P, there is only one dimensionless control parameter, a dimensionless shear rate which can be interpreted as the ratio of the time scale for particles to rearrange due to the pressure, P, to the time scale of the flow, $\dot{\gamma}^{-1}$, see figure 3 (*a*). The friction, $\mu = \tau/P$, and the volume fraction, ϕ, are sole function of this dimensionless number. In the case of immersed granular media for which viscous forces are dominant, the dimensionless number is $J = \eta_f \dot{\gamma}/P$ and is viscous in contrast to dry granular flows where it is inertial. Using an original pressure-imposed shear cell such as sketched in figure 3 (*b*), $\mu(J)$ and $\phi(J)$ has been found once again to collapse onto universal curves [20].

Describing the flow in terms of an effective friction coefficient thus works for both granular flows and suspensions, and therefore unifies suspension and granular rheology. This alternative way of looking at suspensions enabled to circumvent the divergence observed in volume-imposed rheometry and provided examination of the rheology extremely close to the jamming transition. In particular, this approach yields accurate measurements of the particle pressure, a quantity not often easily captured as mentioned earlier at the end of section 3.

6 Toward more complex particulate systems

The pressure-imposed rheometry presented in section 5 complemented by the free-surface-profilometry measurements giving access to normal stresses differences presented in section 3 provide a full description of the stress tensor of dense particulate system. These unconventional rheological tools offer a brand new perspective of analysis which can be applied to more complex particulate systems and offer new avenues of research.

Studies have been undertaken to analyze the influence of the shape of the particles (e.g. large-aspect-ratio particles such as fibers) on the rheological behaviors. In particular, measurements of normal stress differences have been reported for suspensions of rigid, non-Brownian fibers at high concentrations [9]. The second normal stress difference is found to be negative and its magnitude increases with increasing concentration and decreasing aspect ratio. The first normal stress difference is positive and its magnitude is approximately twice that of the second normal stress difference. Numerical simulations which account for both hydrodynamic and contact contributions reveal that the contact interactions are largely responsible for the observed normal stress differences [9]. The rheological behaviors observed for suspensions of fibers, which are qualitatively different from those for suspensions of spheres, have implications for flows of fibers in more general situations (e.g. in pressure-driven flows).

Systematic investigations have been also performed to examine how the macroscopic rheology close to the jamming transition is influenced by the nature of the suspending fluid. In particular, the rheology of non colloidal suspensions composed of a yield-stress fluid and of neutrally-buoyant solid spheres has been explored in the dense regime [29]. The rheological measurements are in agreement with a model based on scaling arguments and homogenization methods [31, 32]. A striking result is that all the data including a Newtonian test case show a perfect collapse of the friction, $\tau/P = \mu(\phi)$. This shows that the constitutive laws close to jamming has a form similar to that for a Newtonian suspending fluid and thus demonstrates unambiguously that the dynamics of the particles close to jamming are mainly controlled by geometrical constraints and are independent of the suspending fluid. This scaling approach has been further tested by inferring the lever function relating the local shear rate seen by the particles to the macroscopic shear rate imposed by the rheological flow and by showing that an excellent collapse is again obtained with a sole dependence in ϕ. The properties of the viscoplastic suspensions can be satisfactorily modeled as that of a Herschel-Bulkley fluid with an exponent equal to that of the suspending fluid. The dimensionless effective yield-stress and viscosity (to be more precise, consistency) are found to be sole functions of ϕ that can be deduced from the knowledge of the lever function and the dimensionless effective viscosity, $\eta_s(\phi)$, of a test suspension having a Newtonian suspending fluid. This study offers new perspective in the study of even more complex particulate systems as the knowledge of the rheological constitutive laws in the Newtonian case and the calculation of the lever function relating the local to the macroscopic shear rate seem to be the sole ingredients needed to infer the rheological properties of these suspensions in the dense regime. This may be extended to shear-thinning or shear-thickening suspending fluids along the same lines.

Acknowledgements

This work was undertaken under the auspices of the ANR projects 'Rhéologie des suspensions concentrées' (ANR-08-BLAN-0048-CSD2) and 'Dense Particulate Systems' (ANR-13-IS09-0005-01), the 'Laboratoire d'Excellence Mécanique et Complexité' (ANR-11-LABX-0092), the 'Initiative d'Excellence' A*MIDEX (ANR-11-IDEX-0001-02), and the COST Action MP1305 'Flowing Matter'. It also benefited from funding from 'Études et Productions Schlumberger'.

References

[1] A. Einstein, Annalen der Physik **17**, 549 (1905)
[2] J. Stickel, R. Powell, Annual Review of Fluid Mechanics **37**, 129 (2005)
[3] J. Morris, Rheologica Acta **48**, 909 (2009)
[4] É. Guazzelli, J.F. Morris, *A Physical Introduction to Suspension Dynamics* (Cambridge University Press, 2012)
[5] M.M. Denn, J.F. Morris, Annual Review of Chemical and Biomolecular Engineering **5**, 203 (2014)
[6] J. Morris, F. Boulay, Journal of Rheology **43**, 1213 (1999)

[7] F. Boyer, O. Pouliquen, É. Guazzelli, Journal of Fluid Mechanics **686**, 5 (2011)

[8] E. Couturier, F. Boyer, O. Pouliquen, É. Guazzelli, Journal of Fluid Mechanics **686**, 26 (2011)

[9] B. Snook, L.M. Davidson, J.E. Butler, O. Pouliquen, É. Guazzelli, Journal of Fluid Mechanics **758**, 486 (2014)

[10] I. Zarraga, D. Hill, D. Leighton Jr, Journal of Rheology **44**, 185 (2000)

[11] I. Zarraga, D. Hill, D. Leighton Jr, Journal of Rheology **45**, 1065 (2001)

[12] A. Singh, P. Nott, Journal of Fluid Mechanics **490**, 293 (2003)

[13] S.C. Dai, E. Bertevas, F. Qi, R.I. Tanner, Journal of Rheology **57**, 493 (2013)

[14] T. Dbouk, L. Lobry, E. Lemaire, Journal of Fluid Mechanics **715**, 239 (2013)

[15] C. Gamonpilas, J.F. Morris, M.M. Denn, Journal of Rheology **60**, 289 (2016)

[16] S. Gallier, E. Lemaire, L. Lobry, F. Peters, Journal of Fluid Mechanics **799**, 100 (2016)

[17] Y. Yurkovetsky, J. Morris, Journal of Rheology **52**, 141 (2008)

[18] A. Deboeuf, G. Gauthier, J. Martin, Y. Yurkovetsky, J. Morris, Physical Review Letters **102**, 108301 (2009)

[19] D. Prasad, H. Kytömaa, International Journal of Multiphase Flow **21**, 775 (1995)

[20] F. Boyer, É. Guazzelli, O. Pouliquen, Physical Review Letters **107**, 188301 (2011)

[21] S. Garland, G. Gauthier, J. Martin, J.F. Morris, Journal of Rheology **57**, 71 (2013)

[22] D. Leighton, A. Acrivos, Journal of Fluid Mechanics **181**, 415 (1987)

[23] A. Karnis, H. Goldsmith, S. Mason, Journal of Colloid and Interface Science **22**, 531 (1966)

[24] B. Snook, J.E. Butler, É. Guazzelli, Journal of Fluid Mechanics **786**, 128 (2016)

[25] P. Nott, J. Brady, Journal of Fluid Mechanics **275**, 157 (1994)

[26] D. Lhuillier, Physics of Fluids **21**, 023302 (2009)

[27] P.R. Nott, É. Guazzelli, O. Pouliquen, Physics of Fluids **23**, 043304 (2011)

[28] K. Yeo, M.R. Maxey, Journal of Fluid Mechanics **682**, 491 (2011)

[29] S. Dagois-Bohy, S. Hormozi, É. Guazzelli, O. Pouliquen, Journal of Fluid Mechanics **776**, R2 (2015)

[30] B. Andreotti, Y. Forterre, O. Pouliquen, *Granular Media: Between Fluid and Solid* (Cambridge University Press, 2013)

[31] X. Chateau, G. Ovarlez, K.L. Trung, Journal of Rheology **52**, 489 (2008)

[32] E. Lerner, G. Düring, M. Wyart, Proceedings of the National Academy of Sciences of the United States of America **109**, 4798 (2012)

EPJ Web of Conferences **140**, 08003 (2017) DOI: 10.1051/epjconf/201714008003
Powders & Grains 2017

Clustering and melting in a wet granular monolayer

Philipp Ramming[1] and *Kai* Huang[1,*]

[1] *Experimentalphysik V, Universität Bayreuth, 95440 Bayreuth, Germany*

Abstract. We investigate experimentally the collective behavior of a wet granular monolayer under vertical vibrations. The spherical particles are partially wet such that there are short-ranged attractive interactions between adjacent particles. As the vibration strength increases, clustering, reorganizing and melting regimes are identified subsequently through a characterization with the bond-orientational order parameters and the mean kinetic energy of the particles. The melting transition is found to be a continuous process starting from the defects inside the crystal.

1 Introduction

For partially wet granular materials, liquid mediated particle-particle interactions lead to dramatically different collective behavior than noncohesive dry granular materials, even if the mean liquid film thickness is only tens of atomic layers thick [1]. In our planet largely covered with water, understanding how wetting liquid plays a role in determining the transitions of granular materials between solid-like, liquid-like or gas-like states is important concerning the prediction of natural disasters such as landslide or debris flow, as well as the enhancement of the efficiency in industrial processes [2]. In quasi-two-dimensions, driven noncohesive granular monolayers have attracted continuous attentions in the past decade, because the dissipative nature of granular matter marks it as a model system to understand phase transitions far from thermodynamic equilibrium [3–5]. Particularly, a previous investigation on agitated dry granular monolayers suggests a two-stage melting scenario [6] in accordance to the Kosterlitz, Thouless, Halperin, Nelson and Young (KTHNY) theory for crystals in two-dimensions (2D) [7, 8]. For the partially wet case, the capillary bridges formed between adjacent particles, which give rise to short-ranged attractive interactions, can be considered as 'molecular bonds' in the model system. The nonequilibrium stationary states (NESS) depend on the balance between energy injection through, e.g., mechanical agitations, and dissipation through, e.g., inelastic collisions or viscous forces from wetting liquid [9, 10]. In a previous investigation, it was shown that a 2D wet granular crystal under horizontal swirling motion (i.e., horizontal vibrations in two orthogonal directions with a phase shift of $\pi/2$ in a plane perpendicular to gravity) melts from its free surface [11]. In a following investigation, the fluctuations of the internal structures quantified by the bond-orientational order parameters (BOOP) show different types of noises for different NESS, and surface melting was found to be associated with the emerging $1/f$ noise [12].

Here, we focus on the collective behavior of a wet granular monolayer subjected to vertical vibrations along the direction of gravity. From both the internal structure and the mobility of the particles, we identify three distinct regimes: Clustering of mobilized particles, reorganization of small clusters into a hexagonal crystal with defects, and a continuous melting initiated at the defects.

2 Experimental Setup and Procedure

Figure 1a shows a sketch of the experiment setup. We use a computer program to adjust vibration parameters, take images, and collect acceleration data automatically in order to explore the parameter space in a well-controlled manner. Cleaned white glass beads (SiLibeads P) with a diameter $d = 2.0 \pm 0.02$ mm and a density $\rho_p = 2.47 \pm 0.05\,\mathrm{g/cm^3}$, after being mixed with purified water (LaborStar TWF, surface tension $\sigma = 0.072\,\mathrm{N/m}$), are added into a cylindrical polytetrafluoroethylene (PTFE) container. The height $H = 2.60 \pm 0.05$ mm and inner diameter $D = 13.00$ cm of the container is controlled by the embedded spacer (see Figure1b). H is chosen to be only slightly larger than d to ensure a monolayer. The global area fraction is $Nd^2/D^2 \approx 50.5\%$ with particle number $N = 2134$. The liquid content is defined as $W \equiv V_w/V_g$, where V_w and V_g are the volume of the wetting liquid and that of the glass beads correspondingly. It is kept within 3% so that cohesion arises mainly from liquid bridges formed between adjacent particles [13]. The container is agitated vertically against gravity with an electromagnetic shaker (Tira TV50350). The frequency f and amplitude z_0 of the sinusoidal vibrations are controlled with a function generator (Agilent FG33220) and the dimensionless acceleration $\Gamma = 4\pi^2 f^2 z_0/g$ is measured with an accelerometer (Dytran 3035B2), where g is the gravitational

*e-mail: kai.huang@uni-bayreuth.de

DOI: 10.1051/epjconf/201714008003

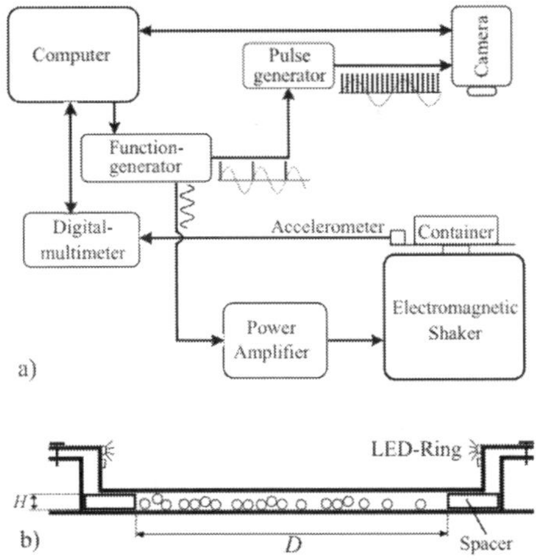

Figure 1. Schematic of the experimental setup (a) and the container (b). Shaking control, camera triggering, acceleration detection, and image capturing are carried out automatically with a computer program. In (b), the spacer is a PTFE ring used to confine the particles into a monolayer.

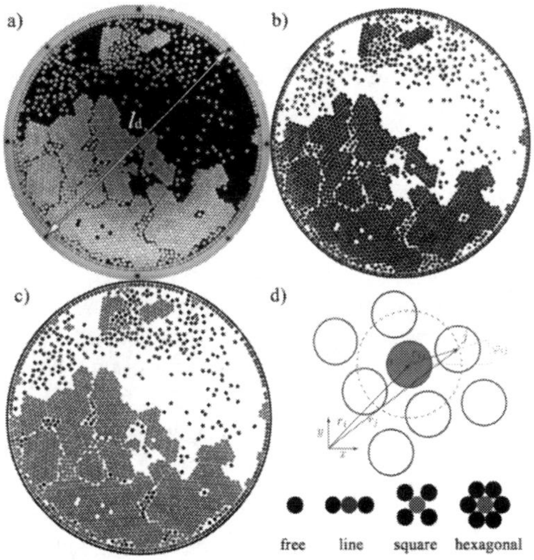

Figure 2. (a) A sample raw image from the high speed camera. Eight black spheres are embedded in the PTFE ring symmetrically in the azimuth direction for scaling purposes. (b) Negative image of (a) after the background is removed. (c) Reconstructed image of (a) based on the positions and local structures of all particles. (d) Color coding for the local structure (free, line, square, or hexagonal) identified with bond-orientational order parameters and a sketch defining the position of particle i and one of its neighbors j in the laboratory system.

acceleration. The collective behavior of the sample is captured with a high speed camera (IDT MotionScope M3) mounted above the container. The camera is triggered by a synchronized multi-pulse generator to capture images at fixed phases of each vibration cycle. More details on the set-up control can be found in Ref. [14].

At the beginning of each experimental run, the sample is driven to the gas-like state with $\Gamma \approx 40$ for 10 minutes to have a reproducible initial condition with the wetting liquid homogeneously distributed. Subsequently, Γ is quenched to ≈ 4 and increases stepwise to ≈ 32 at a fixed $f = 50\,\mathrm{Hz}$. In each step, $M = 500$ images are taken by the high speed camera with a frame rate of 500 frames per second, corresponding to $K = 10$ frames per vibration cycle.

Figure 2 shows the image analysis procedure, from raw images captured by the high speed camera (a) to the local structures of individual particles (c) identified with BOOP. Because shaking is along the viewing direction of the camera, the images captured at various phases of each vibration cycle have a slightly different scale, which leads to a shift of each particle position up to a few pixels, depending on the particle location and Γ. For particles at the rim of the container, the deviation can reach $0.15d$ at $\Gamma = 15$. In order to correct this error, we group images according to the capturing phase k and use different scales for different phases. The scales are obtained with $\epsilon(k) = \langle l_d(k)/D_l \rangle$, where $l_d(k)$ is the distance between the centers of two oppositely placed marker particles in pixels (see Figure 2a), $D_l = 135.0\,\mathrm{mm}$ is the measured length, and $\langle ... \rangle$ denotes an average over all pairs of markers and all frames at phase k.

After image enhancement (see Figure 2b), we apply a Hough algorithm to locate the positions of all particles in all frames in a Cartesian system centered at the image center with a sub-pixel resolution [15]. In order to detect all the particles for all frames captured, the threshold used in the Hough transformation is automatically adjusted such that the correct number of particles are detected. Subsequently, the position of particle i at frame number n is rescaled with $(x_i, y_i) = \epsilon(k)(x_{i0}, y_{i0})$, where phase k is the modulo of n over K, and (x_{i0}, y_{i0}) is the position of particle i in the original image.

The local structure of a particle is identified with BOOP [16, 17], which are calculated as follows: As shown in Figure 2(d), we first identify bonds connecting a particle to all its neighbors, e.g., $\mathbf{r}_{ij}$ pointing from particle i to j. The neighbors are identified with the criterion $|\mathbf{r}_{ij}| \leq r_c$, where the critical bond length $r_c = 1.05d$ is slightly larger than the particle diameter. From the azimuth (ϕ_{ij}) and polar angles ($\theta_{ij} = 0$ for the 2D case here) of all bonds originating from particle i, we calculate

$$Q_{lm}(i) = \frac{1}{N_b} \sum_{j=1}^{N_b} Y_{lm}(\phi_{ij}, \theta_{ij}), \tag{1}$$

where $N_b(i)$ is the total number of bonds for particle i, and Y_{lm} corresponds to the spherical harmonics of bond $\mathbf{r}_{ij}$. Subsequently, we obtain BOOP of weight l with

$$Q_l = \sqrt{\frac{4\pi}{2l+1} \sum_{m=-l}^{l} |Q_{lm}|^2}. \tag{2}$$

Here, Q_6 is chosen as the order parameter because of its sensitivity to the hexagonal order. Based on a comparison of Q_6 to the values obtained from perfectly hexagonal, square, line structures, as well as for free particles (see Figure 2d), the local structure of each individual particle is identified (see Figure 2c). In comparison to other local measures such as coordination number or local area fraction, the advantage of using BOOP is that the influence of particles on the edge of a cluster is minimized, which is essential for analyzing the structure of small clusters.

3 Results and discussion

Figure 3 presents the collective behavior of the wet spherical particles as Γ increases. It is quantified with two order parameters. The percentage of particles in a locally hexagonal packing (i.e., appear red in the reconstructed images), ξ_6, is used as a measure of the static structure. The rescaled kinetic energy $\tilde{E}_k(t) \equiv \langle|\mathbf{v}_i(t)-\bar{\mathbf{v}}(t)|^2\rangle/(fd)^2$ is used to characterize particle dynamics, where $\mathbf{v}_i(t)$ is the velocity of particle i at time t, $\bar{\mathbf{v}}(t) = \sum_{i=1}^{N} \mathbf{v}_i(t)/N$ corresponds to the mean velocity of all particles at time t, and $\langle ... \rangle$ denotes an average over all particles and frames. The particle velocity is obtained with $\mathbf{v}_i(t) = [\mathbf{r}_i(t + \Delta t) - \mathbf{r}_i(t)]/\Delta t$, where the time step $\Delta t = 1/(Kf)$ is determined by the frame rate of the camera. Based on visual inspections (see the top panel of Figure 3) as well as characterizations with the two order parameters, the following three regimes are identified.

In the clustering regime, the particles are gradually being mobilized, and subsequently form small clusters due to the formation of capillary bridges. At the initial $\Gamma \approx 4.0$, no movements of the particles are observed from the top view images. As Γ increases to ≈ 4.7, some particles start to move and form small clusters (see the reconstructed image at $\Gamma = 6.8$). The critical acceleration Γ_c can be understood from the force balance $\mathbf{F}_d = \mathbf{F}_b + \mathbf{G}$, where the magnitude of the capillary force $\mathbf{F}_b$, gravity $\mathbf{G}$ and driving force $\mathbf{F}_d$ can be estimated with $|\mathbf{F}_b| = \pi\sigma d \cos\alpha$ with contact angle $\alpha \approx 0$, $|\mathbf{G}| = \pi\rho_p d^3 g/6$, and $|\mathbf{F}_d| = \Gamma_c G$ respectively [18]. Consequently, we estimate $\Gamma_c \approx 4.5$, which agrees well with the experiments. Above Γ_c, the mobilized particles start to form chains as well as fractal structures. Correspondingly, the order parameter ξ_6 grows with the agglomeration process from ≈ 0.25 to more than 0.60.

As Γ increases further, visual inspections reveal that more and more compact structures (reminiscent to those observed in a previous investigation of wet granular monolayers under shear [19]) are forming along with a more significant growth of ξ_6 in the range of $6.5 \leq \Gamma \leq 7.7$. Interestingly, the mobility of particles grows rapidly in this range (see Figure 3b), suggesting that the enhanced mobility of particles facilitates the agglomeration process. The rapid growth of $\tilde{E}_k$ in this regime could be attributed to the additional freedom gained during the rupture of capillary bridges between the particles and the container bottom. Note that overcoming the capillary force at Γ_c does not necessarily lead to the rupture of capillary bridges, as the latter also requires the relative separation distance to be greater than the rupture distance of the bridge.

In the reorganizing regime starting at $\Gamma \approx 7.7$, the small clusters merge and form a large one with a hexagonal structure. Initially, ξ_6 decays slightly as Γ grows. This feature can be attributed to the enhanced mobility of clusters, because intense interactions between rigid clusters can drive more particles into the liquid-like phase and consequently lead to a lower ξ_6. Note that the transition into the liquid-like state requires the rupture of capillary bridges, and the rupture energy is $\propto \sqrt{W}$ [20], therefore ξ_6 is smaller for $W = 2\%$ in comparison to 3% for a given agitation strength. For the same reason, the mobility of particles for $W = 2\%$ is slightly larger than 3%, because less energy is required for rupturing at a given energy injection. As Γ approaches the transition to the melting regime (iii), the interactions between clusters become less important because of the merging of clusters. The fluctuations of ξ_6 suggest that the internal structure of the merged cluster reorganizes frequently. Visual inspections reveal that the frequent generation and diminishing of defects in the crystalline structure (see the reconstructed image for $\Gamma = 14.7$) account for the fluctuations.

As melting starts at $\Gamma = 15.5$, a sudden drop of ξ_6 along with a rapid growth of $\tilde{E}_k$ are observed. Different from a previous investigation on horizontally driven systems [11], melting does not predominately start from the surface of the crystal. Instead, melting through cluster breaking, preferably along the defects, is more likely to be

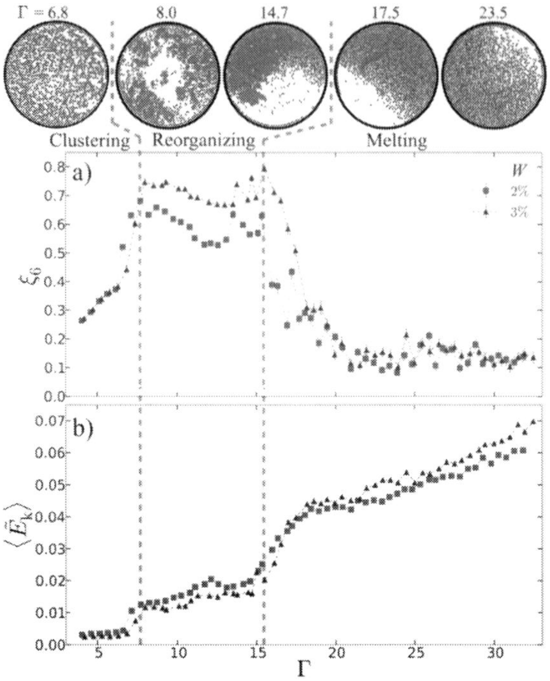

Figure 3. Upper panel shows the local structure of all particles at various Γ and $W = 3\%$. (a) Percentage of particles in a hexagonal structure, ξ_6 as a function of Γ for different W. (b) Rescaled kinetic energy, $\langle\tilde{E}_k\rangle$, averaged over all particles and $M - 1$ consequent frames as a function of Γ. Both order parameters suggest three regimes separated by the two dash lines (located at $\Gamma = 7.7$ and 15.5 respectively).

observed. For the case of horizontal driving in a plane perpendicular to gravity, the mobility of the particles are confined in a strict monolayer. For the vertical vibrations used here, neighboring particles may move out of phase due to the small gap above the granular layer. It facilitates the formation of topological defects at which melting starts. Further increase of Γ leads to a slow erosion of the crystal by the particles in the liquid-like phase, and consequently a gradual decay of ξ_6. At $\Gamma \approx 23$, melting completes for both liquid contents. Further increase of agitation strength leads to a continuous growth of $\bar{E}_k$ with Γ, because energy injection increases. Strong density fluctuations are present at high Γ, even though the container is leveled carefully before each experimental run. Occasionally, the particles are accumulated at the side of the container (see, e.g. reconstructed image at $\Gamma = 17.5$), and sometimes the whole cluster swirls collectively. The reason for this collective behavior will be a topic of further investigations.

Concerning the influence of liquid content, our results show that the amount of wetting liquid plays a role in determining the amount of particles in a liquid-like state, as soon as the particles are effectively mobilized in the reorganizing and melting regimes. However, it does not influence the transitions between different regimes, which appear to be predominately determined by the agitation strength. In the melting regime, the averaged kinetic energy for both W agrees with each other, indicating that the energy dissipation from the rupture of liquid bridges does not play a dominating role in determining the mobility of the particles.

4 Conclusions

To summarize, the collective behavior of a wet granular monolayer under vertical vibrations is investigated experimentally. The internal structures of the wet granular assemblies are obtained with the bond-orientational order parameter Q_6, and the dynamics is characterized with the averaged mobility of individual particles. From both internal structures of the particles and the mean kinetic energy, three regimes (clustering, reorganizing and melting) are identified. If the peak vibration acceleration Γ is greater than a critical value, at which the driving force is sufficient to overcome the capillary force and gravity, the particles are mobilized and form small compact clusters. In the reorganizing regime, the clusters merge into a hexagonal crystal with defects. Further increase of Γ leads to a continuous melting predominately from the defects inside the crystal, reminiscent to a second-order phase transition. The difference to surface melting observed in a horizontally driven system [11] can be attributed to the additional degree of freedom of the particles along the direction of vibrations. The defects mediated melting transition observed in this nonequilibrium system triggers the question whether this transition falls into the scenario of KTHNY theory or not.

Acknowledgements

We thank Ingo Rehberg, Simeon Völkel, and Andreas Zippelius for helpful discussions. This work is partly supported by the Deutsche Forschungsgemeinschaft through Grant No. HU1939/4-1.

References

[1] K. Huang, M. Sohaili, M. Schröter, S. Herminghaus, Phys. Rev. E **79**, 010301 (2009)
[2] K. Huang, Habilitation thesis, University of Bayreuth (2014)
[3] P.M. Reis, R.A. Ingale, M.D. Shattuck, Phys. Rev. Lett. **96**, 258001 (2006)
[4] T. Müller, D. de las Heras, I. Rehberg, K. Huang, Phys. Rev. E **91**, 062207 (2015)
[5] A. Fortini, K. Huang, Phys. Rev. E **91**, 032206 (2015)
[6] J.S. Olafsen, J.S. Urbach, Phys. Rev. Lett. **95**, 098002 (2005)
[7] J.M. Kosterlitz, D.J. Thouless, J. Phys. C: Solid State Phys. **6**, 1181 (1973)
[8] K.J. Strandburg, Rev. Mod. Phys. **60**, 161 (1988)
[9] F. Gollwitzer, I. Rehberg, C.A. Kruelle, K. Huang, Phys. Rev. E **86**, 011303 (2012)
[10] T. Müller, K. Huang, Phys. Rev. E **93**, 042904 (2016)
[11] C. May, M. Wild, I. Rehberg, K. Huang, Phys. Rev. E **88**, 062201 (2013)
[12] K. Huang, New J.Phys. **17**, 083055 (2015)
[13] M. Scheel, R. Seemann, M. Brinkmann, M. Di Michiel, A. Sheppard, B. Breidenbach, S. Herminghaus, Nat. Mater. **7**, 189 (2008)
[14] K. Huang, I. Rehberg, Phys. Rev. Lett. **107**, 028001 (2011)
[15] C. Kimme, D. Ballard, J. Sklansky, Communications of the ACM **18**, 120 (1975)
[16] P.J. Steinhardt, D.R. Nelson, M. Ronchetti, Phys. Rev. B **28**, 784 (1983)
[17] Y. Wang, S. Teitel, C. Dellago, J. Chem. Phys. **122**, 214722 (2005)
[18] L. Butzhammer, S. Völkel, I. Rehberg, K. Huang, Physical Review E **92**, 012202 (2015)
[19] K. Huang, M. Brinkmann, S. Herminghaus, Soft Matter **8**, 11939 (2012)
[20] K. Huang, K. Roeller, S. Herminghaus, Eur. Phys. J. Spec. Top. **179**, 25 (2009)

Grain Deformation Processes in Porous Quartz Sandstones – Insight from the Clusters of Cataclastic Deformation Bands

Sven Philit[1], *Roger* Soliva[1], *Gregory* Ballas[2], *Haakon* Fossen[3]

[1]Géosciences Montpellier – Université de Montpellier, France
[2]Centre de Bretagne – Ifremer (Brest), France
[3]Department of Earth Science - University of Bergen, and Instituto de Geociências – Universidade de São Paulo, Brazil

Abstract. Porous sandstones represent classical reservoirs for water or hydrocarbons. Deformation in such granular material is due to tectonic events and occurs through the process of cataclasis, implying the crushing of the grains to a diversity of smaller clasts. Cataclasis is generally accompanied by porosity and permeability decrease. Although it is known that cataclastic deformation localizes to form individual bands and clusters of bands, the parameters controlling the distribution of this deformation were not well understood until recently. We used scanline measurements to show a favoured localization and clustering of the deformation on the case of normal-fault stress regime and potentially in strike-slip fault regime. The reverse regime favours the formation of distributed networks of conjugate deformation bands. At the scale of a cluster, field data reveals that the minimum modal grain size value of the host sandstone(s) controls the band density. Finally, microscopic cathodoluminescence analysis reveals enhanced quartz cementation for high degree of cataclasis. Hence, because band clustering, high degree of cataclasis and band cementation are favoured in normal-fault stress regime, tectonic extension appears to be favourable conditions for the formation of efficient barriers to fluid-flow in porous sandstone reservoirs.

1 Introduction

Clusters of cataclastic deformation bands (CDB clusters) (Fig. 1) are natural geological objects originating from the deformation of porous sandstones, which are granular materials. They are a specific type of strain localization features since they locally accommodate the deformation (shear displacement), but do not necessarily develop discrete slip surface [1]–[3]. These structures consist of a cluster of discrete bands, each with a thickness of 0.5 - 5 mm and more densely distributed than in the adjacent sandstone. The CDB clusters are up to a few tens of centimetres wide and their lateral extent ranges from tens to hundreds of metres; they can be relaying over distances of several kilometres (e.g. near Goblin Valley State Park, Utah, USA).

The main process of deformation in porous sandstones is cataclasis. This process consists of a brittle deformation implying grain powdering by rolling, flaking and crushing due to stress concentrations at grain contacts induced by shear and compression [4]–[6]. The process of cataclasis mostly yields compactant structures and can be considered as a ductile deformation at macroscopic scale [7].

Because of porosity and permeability loss during deformation, CDB clusters are likely to affect fluid flow in reservoir settings (permeability reduction up to 6 orders of magnitudes – e.g. [8]). Hence, we think it is important to understand their internal structure and their associated deformation processes in order to understand their resulting sealing vs. draining property. In order to address these issues, we study the distribution of the cataclastic deformation and the structures of the CDB clusters with respect to the following parameters: (1) tectonic regime, (2) grain size of the host sandstone, and (3) cementation.

Figure 1. Example of CDB cluster resulting from the localization of deformation in porous sandstones.

2 Results

2.1 Distribution of the deformation

The distribution of deformation at the macroscale was measured by counting the number of bands per metres along scanlines oriented in the direction of maximum extension in the case of normal Andersonian regimes, in the direction of maximum shortening in thrust regimes, and normal to the deformation in strike-slip regimes. Band spatial distribution is represented by the cluster factor (see [2] for reference), which is a standard deviation type function allowing to quantify to what extent bands are clustered (high value) or pervasively distributed (low value). These data come from a total of 73 deformation band sets formed in various sandstones (medium sands to fine gravel), various tectonic stress regimes (normal, thrust or strike-slip), and different burial depths (0.3-2.5 km) [9].

In the normal-fault regime, the deformation consists of Shear Bands and/or Compactional Shear Bands localised as conjugate sets of CDB clusters (9-162 bands/m, Fig. 2), sometimes including a slip surface (fault). Each band typically absorbs a centimetre- to decimetre- displacement.

In the thrust-fault regime, the deformation consists of a network of conjugate Shear-Enhanced Compaction Bands and Compactional Shear Bands distributed from a few centimetres to tens of centimetres apart (6-67 bands/m, Fig. 2) and are extensively distributed at the outcrop (distributed over a zone of up to hundreds of metres). Each band typically absorbs a few millimetres to a few centimetres of shear displacement.

In the strike-slip-fault regime, based on the only 2 studied exposures, the deformation consists of Compactional Shear Bands to Shear Bands localising as CDB clusters (22 and 112 bands/m, Fig. 2).

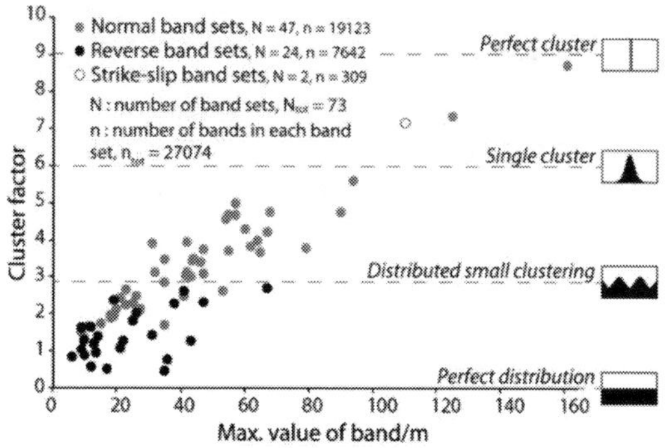

Figure 2. Graph illustrating the geometry of the cataclastic deformation as a function of tectonic regimes in terms of distribution (cluster factor) and band density (Max value of band/m), modified after [9])

As displayed in Fig. 2, normal and strike-slip cataclastic deformation have high cluster factors (1.4-8.8), implying strong clustering (0 representing a perfect even pervasive distribution, and 9, closing to a perfect CDB clusters). Thrust-fault cataclastic deformation has low cluster factors (0.4-2.8).

2.2 Intra-cluster deformation density

At the scale of the CDB clusters, we measure the deformation by counting the cumulated thickness of bands (mm precision) in the thickness of clusters from seven outcrops formed in different tectonic regimes, at different burial depths, and in host sandstones of variable granulometry (see properties in Tab. 1).

For a given cluster, we repeated the measurement 17 to 60 times on a distance N of several metres along the cluster to validate the accuracy of our data. We represent the density of deformation as the ratio (r_{band}) of cumulated thickness of bands divided by the cluster thickness. The burial of the formation at the time of deformation was obtained in the literature; the granulometry was obtained through laser diffraction analysis.

Table 1. Properties of the deformed granular sandstone and their associated deformation density

Site	*Tectonic Regime*	*Burial (m)*	*Grain Size*	*Modal values (µm)*	*Min modal value (µm)*	*Cluster r_{band}*
Cum	Normal	100-400	Medium	0.37	0.37	0.40
GV	Normal	1000-2000	Fine to Medium	0.20; 0.28	0.20	0.26
Bon	Normal	300-600	Fine to Coarse	0.21; 0.41; 0.75	0.21	0.11
BG	Thrust	500-1500	Medium	0.29	0.29	0.19
Bla	Thrust	100-800	Medium	0.37	0.37	0.41
Bed	S-S	50-200	Medium	0.29	0.29	0.35
StM	S-S	300-600	Medium to Coarse	0.37; 0.78	0.37	0.35

In the normal regime, the deformation density r_{band} ranges from 0.11 (Bon) to 0.26 (GV) to 0.40 (Cum). The measured r_{band} for the cluster formed in the thrust regime (much rarer) equals 0.19 at BG and 0.41 at Bla. The measured r_{band} for the clusters formed in strike slip (rare) equals 0.35 for both Bed and StM. In Fig. 3, we plotted the deformation density as a function of most fine-grained host sandstone.

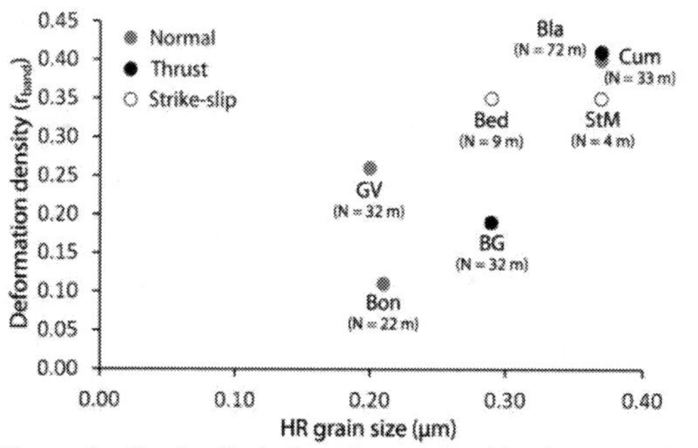

Figure 3. Graph displaying the relationship between the deformation density in the clusters and the modal grain size of the most fine-grained host sandstone, if several sandstones were displaced against each other.

2.3 Cementation in cataclastic textures

In clean sandstones, CDB clusters often show quartz cementation [10], [11] under the form of 1-10s of μm coatings around the clasts (Fig. 4). This cementation is observed into CDB clusters formed in sand series having been buried to 400 m or more. We observe particularly well cemented clusters at Bédoin, Provence (France) in clean arenite sandstone buried at a maximum depth of 800 m. The alteration profile at this site demonstrates an aquifer environment at the time of deformation.

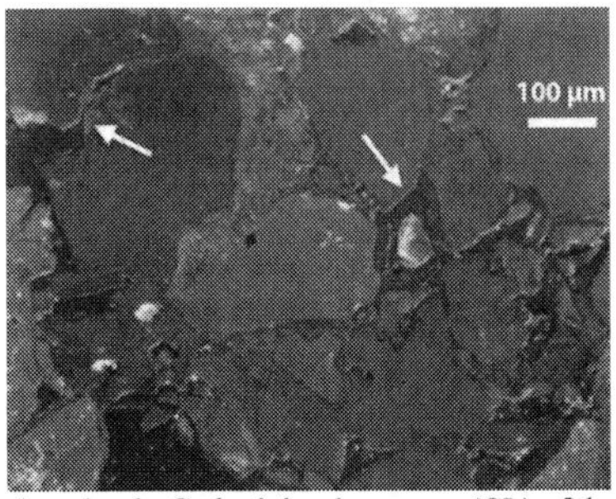

Figure 4. Imaging in Cathodoluminescence (CL) of the coating of quartz cements around the clasts (arrows) in partially cemented Crush microbreccia.

At the macroscopic scale, a census of the deformation and cementation in the thickness of the cluster and on a distance of 9 m along-strike the cluster indicates that cemented cataclastic bands account for at least 19% of the cluster thickness. These structures are variably connecting alongside the cluster depending on the distribution and orientation of the bands.

We undertake Cathodoluminescence microscopy (CL) to assess the degree of cataclasis and the relative amount of cement in several parts of an 80 cm wide CDB cluster. The cataclasis is considered as the fraction of rock occupied by clasts; we consider clasts as the fragments of sand grains whose Ferret diameter is lower than 10% of the modal grain diameter of the host sandstone. We use the classification proposed by Sibson [12] to designate different degrees of cataclasis. We use the Intergranular Volume (IGV) [13] to qualify the porosity of the rock previous to cementation.

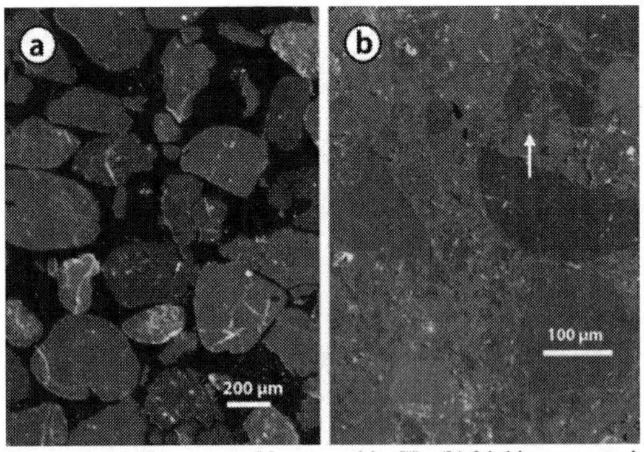

Figure 5. (a) Uncemented host sand in CL; (b) highly cemented protocataclasite in CL.

In the host sand next to the cluster, porosity equals 31% and no cement is observed (Fig. 5a). The parts of the cluster where cataclasis represents 9% of the rock correspond to *uncemented* to *partially cemented* crush microbreccia with IGV of 22% (Fig. 4). In the partially cemented crush microbreccia, the resulting porosity can be as low as 12 %. The parts of the cluster where cataclasis represents 30% of the rock correspond to *partially cemented* to *highly cemented* protocataclasite with IGV of 10% (Fig. 5b). The resulting porosity is estimated to 5% or lower. The parts of the cluster where cataclasis represents 60% of the rock correspond to *highly cemented* cataclasite with IGV ≤ 5%. The resulting porosity is estimated as lower than 4%. These observations are summarized in Fig. 6.

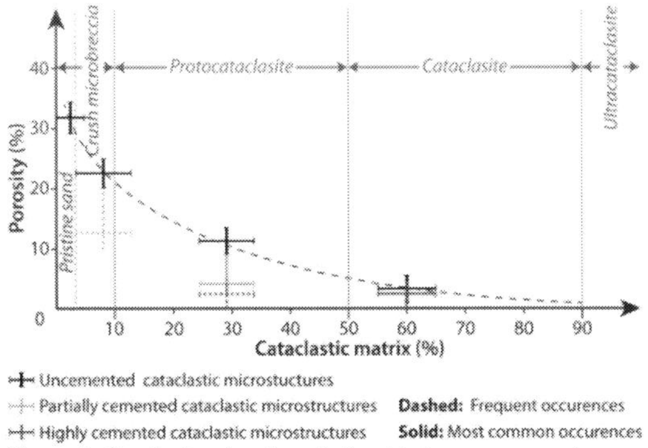

Figure 6. Graph showing the relationship between porosity, degree of cataclasis and degree of cementation.

3 Concluding discussion

The data from the scanline measurements shows the clustering of cataclastic shear band in the normal-fault tectonic regime. In the thrust-fault regime, deformation is for the most part widely distributed as conjugate network of individual cataclastic compaction bands. This difference in behaviour could be explained by the different stress paths to which the sandstone was submitted [3]: the normal-fault regime would imply low mean stress and high differential stress leading to localized shear, whereas the thrust-fault regime would induce high mean stress and moderate differential stress and cause distributed deformation. This hypothesis is consistent with laboratory mechanical tests (e.g. [14]).

At the scale of the cluster thickness, no direct relation between deformation density (r_{band}) and tectonic regime or the burial depth is observed for the studied clusters. However, our data suggest a positive correlation between deformation density and the modal grain size of the most fine-grained HR (if two different sandstones displaced against each other): the higher the sandstone grain size, the higher the deformation density. We hypothesize that, for comparable sorting, high shear strain density could be favoured by more grain contacts (more resistance to compaction) in fine sandstones than the medium to coarse sandstones [4]. This is consistent with brittle-ductile transition occurring for high mean stress in fine sandstones, then favouring strain localisation under coulomb-type shear behaviour for low confinement in such material (e.g. [14-15]). Although the grain size seems to have a significant impact on the morphology of the cluster (e.g. the density of the deformation as we just showed, or the cluster thickness [16]), other properties such as the grain sorting or the grain shape of the grains may also influence the density of deformation bands in a cluster. The analysis of the deformation density in the cluster should be undertaken on more sites to strengthen our conclusion and to better understand the main control of the most fine-grained HR.

Our study of the influence of cataclasis on the diagenesis shows that increasing cataclasis favours quartz cementation. Consequently, the low porosity of the cataclastic structures is further reduced and the permeability across the clusters potentially decreased, even at shallow burial. As argued by several authors [17–21], cementation would be promoted by the large amount of fresh-reactional quartz surfaces (nucleation discontinuities) in contact with the nourishing fluid. In particular, it is suggested that the high silica concentration responsible for this cementation may be released and reprecipitated during the cataclastic deformation, implying an immediate self-sealing of the structure [22]. We note that the nucleation of silica may be possible if no oxide (or clay) coats the reactional surface, as they may inhibit quartz precipitation [23], [24]. Nevertheless, these conclusions are important because although well-known at depths > 1.5 km [18], [25], [20], quartz diagenesis at low burial was rarely reported (e.g. [26]).

We note that further microscopic study of the deformation could be completed by micro-mechanical analysis through Discrete Element Modelling (DEM). Over the past ten years, DEM studies have shown efficient grain fragmentation modelling, using well-defined bond properties at grain contacts (contact force, friction and cohesion). In particular, the geometry, the clast properties and the dilatancy and softening behaviours of shear bands can be anticipated as functions of the loading conditions, the cementation and the sample size of the modelled sandstones (e.g. [27-29]).

Since CDB clusters extend for 10s to 100s of metres, they are potential barriers or baffles to fluid flow in reservoir settings. This work actually suggests that the link between the tectonic regime and the degree of cataclasis must be studied carefully as we showed that intense cataclasis can generate high permeability decrease both through mechanical and chemical (cementation) processes. Preliminary results seem to show that localizing normal fault-stress regime generate more cataclasis through more shearing. Also, the grain size of the host sandstone could influence the barrier efficiency of a band cluster since it seems to affect the deformation density in clusters.

References

1. H. Fossen, R. A. Schultz, Z. K. Shipton, and K. Mair, J. Geol. Soc. **164**, 4 (2007)
2. C. A. J. Wibberley, J.-P. Petit, and T. Rives, Geol. Soc. Lond. Spec. Publ. **289**, 1 (2007)
3. R. Soliva, R. A. Schultz, G. Ballas, A. Taboada, C. Wibberley, E. Saillet, A. Benedicto, J. Struct. Geol. **49**, pp. 50–63 (2013).
4. J. J. Gallagher Jr, M. Friedman, J. Handin, and G. M. Sowers, Tectonophysics **21**, 3 (1974)
5. A. Aydin and A. M. Johnson, Pure Appl. Geophys. **116**, 4–5 (1978)
6. M. Antonellini and A. Aydin, AAPG Bull. **78**, 3 (1994)
7. M. Friedman and J. M. Logan, Geol. Soc. Am. Bull. **84**, 4 (1973)
8. G. Ballas, H. Fossen, and R. Soliva, *J.* Struct. Geol. **76**, pp. 1–21 (2015)
9. R. Soliva, G. Ballas, H. Fossen, and S. Philit, Geology **44**, 6 (2016)
10. N. L. Watts, Mar. Pet. Geol. **4**, 4 (1987)
11. Q. J. Fisher and R. J. Knipe, Mar. Pet. Geol. **18**, 10 (2000)
12. R. H. Sibson, J. Geol. Soc. **133**, 3 (1977)
13. S. T. Paxton, J. O. Szabo, J. M. Ajdukiewicz, and R. E. Klimentidis, AAPG Bull. **86**, 12 (2002)
14. T. Wong and P. Baud, J. Struct. Geol. **44**, pp. 25–53, (2012)
15. E. H. Rutter and C. T. Glover, J. Struct. Geol. **44**, pp. 129–140, (2012)
16. H. Fossen, R. Soliva, G. Ballas, B. Trzaskos, C. Cavalcante, and R. A. Schultz, *J. Geol. Soc.*, (to be published)
17. M. Canals and J. D. Meunier, Geochim. Cosmochim. Acta **59**, 4 (1995)
18. Q. J. Fisher and R. J. Knipe, Geol. Soc. Lond. Spec. Publ. **147**, 1 (1998)
19. Q. J. Fisher, R. J. Knipe, and R. H. Worden, in *Quartz Cementation in Sandstones*, Blackwell Science., R. H. Worden and S. Morad, Eds. International Association of Sedimentologists - Special Publication **29**, pp. 129–146 (2000)
20. R. H. Lander, R. E. Larese, and L. M. Bonnell, AAPG Bull. **92**, 11 (2008)
21. R. T. Williams, J. R. Farver, C. M. Onasch, and D. F. Winslow, J. Struct. Geol. **74**, pp. 24–30 (2015)
22. B. T. Ngwenya, S. C. Elphick, I. G. Main, and G. B. Shimmield, Earth Planet. Sci. Lett., **183** (2000)
23. E. F. McBride, "Quartz cement in sandstones: a review," *Earth-Sci. Rev.*, **26**, 1-3 1989.
24. S. N. Ehrenberg, AAPG Bull. **77**, 7 (1993)
25. O. Walderhaug, AAPG Bull. **84**, 1–2 (2000)
26. K. L. Milliken, M. H. Reed, and S. E. Laubach, in *Faults, fluid flow, and petroleum traps*, R. Sorkhabi and Y. Tsuji, Eds. pp. 237–249 (2005)
27. Y. H. Wang and S. C. Leung, J. Geotech. Geoenvironmental Eng. **134**, 7 (2008)
28. M. Jiang, H. Chen, and G. B. Crosta, Int. J. Rock Mech. Min. Sci. **78**, pp. 175–189 (2015)
29. G. Pál, Z. Jánosi, F. Kun, and I. G. Main, Phys. Rev. E **94**, 5 (2016)

Sensitivity analysis of Immersed Boundary Method simulations of fluid flow in dense polydisperse random grain packings

Chris Knight[1,*], *Mohd Hazmil* Abdol Azis[2], *Catherine* O'Sullivan[3], *Berend* van Wachem[2], and *Daniele* Dini[2]

[1] *Department of Physics, Imperial College London, UK*
[2] *Department of Mechanical Engineering, Imperial College London, UK*
[3] *Department of Civil Engineering, Imperial College London, UK*

Abstract. Polydisperse granular materials are ubiquitous in nature and industry. Despite this, knowledge of the momentum coupling between the fluid and solid phases in dense saturated grain packings comes almost exclusively from empirical correlations [2–4, 8] with monosized media. The Immersed Boundary Method (IBM) is a Computational Fluid Dynamics (CFD) modelling technique capable of resolving pore scale fluid flow and fluid-particle interaction forces in polydisperse media at the grain scale. Validation of the IBM in the low Reynolds number, high concentration limit was performed by comparing simulations of flow through ordered arrays of spheres with the boundary integral results of Zick and Homsy [10]. Random grain packings were studied with linearly graded particle size distributions with a range of coefficient of uniformity values (C_u = 1.01, 1.50, and 2.00) at a range of concentrations ($\phi \in [0.396, 0.681]$) in order to investigate the influence of polydispersity on drag and permeability. The sensitivity of the IBM results to the choice of radius retraction parameter [1] was investigated and a comparison was made between the predicted forces and the widely used Ergun correlation [3].

1 Introduction

Multi-phase flows involving granular materials occur in many industrial processes for example in the oil and gas, construction and pharmaceutical industries as well as in natural processes such as landslides, scour and erosion, and sedimentation. Knowledge of the momentum coupling between the fluid and solid phases is therefore of great commercial and societal value yet a general treatment over a wide range of solid volume fractions, ϕ, and Reynolds number, Re, is at present not possible due to the complex nature of these systems.

For flows involving saturated soil and sand packings such as those occurring in internal erosion and liquefaction processes, two independent terms, the pressure gradient or buoyancy force, F_b, and the drag force, F_d, dominate the fluid-particle interaction. Stokes law gives the drag force acting on a sphere of diameter D_p moving through a fluid with dynamic viscosity μ in the limit $\phi \to 0, Re \to 0$,

$$F_d^S = 3\pi\mu D_p U. \tag{1}$$

In the context of grain packings, U is the superficial or apparent flow velocity which can be expressed in terms of u, the interstitial flow velocity, as $U = (1 - \phi)u$. As the concentration of solids and the Reynolds number are increased the drag force varies with a complex dependence on both ϕ and Re.

*e-mail: christopher.knight13@imperial.ac.uk

Previous experimental and numerical efforts to characterise the dependence of F_d on ϕ and Re have focussed on monodisperse systems. Broadly speaking the experimental approaches are either based on fluidisation experiments [3] or sedimentation experiments [2]. Hill et al. used the Lattice Boltzmann (LB) method to study the drag force in ordered and random arrays at low to moderate Re [4]. More recently Immersed Boundary Methods (IBM) [8] have been used to provide more accurate results, improving on these monodisperse drag correlations.

In reality granular media typically exhibit not just a single grain size but a range of grain sizes. This size polydispersity affects the packing density and pore space configuration of dense granular materials significantly with implications for the fluid-particle interaction forces. Due to the prevalence of polydisperse media in nature and industry there is a clear need for improved understanding of the relationship between grain size distributions and fluid-particle momentum coupling.

An outline of the IBM is given in section 2, followed by a description of validation work conducted for the IBM in section 3. In section 4 results of IBM simulations with random packings are presented and finally, conclusions are given in section 5.

2 The Immersed boundary method

The Immersed Boundary Method is a CFD technique which enables simulations of flow around complex geometries, such as the pore space in soils. Fluid is represented by a fixed structured grid of Eulerian cells in which the Navier-Stokes equations for mass and momentum conservation are solved locally. Solid surfaces are represented by rigid meshes of points inserted into the fluid domain at which no-slip, no-penetration (ns-np) boundary conditions are imposed through additional forces applied locally to the fluid equation system. These points can move relative to the fixed Eulerian cells adopting arbitrary positions and are therefore referred to as Lagrangian points. An illustration of the two meshes is shown in figure 1. In the following, lower and upper case symbols are used to refer to Eulerian and Lagrangian variables respectively. In the present work a *multi-direct forcing* type IBM was used to solve the Navier-Stokes equations for the flow of fluid in the pore spaces between particles and determine the fluid-particle interaction forces.

Uhlmann [9] developed a *direct forcing* type IBM applicable to structured, uniform, cubical fluid meshes in which the additional Lagrangian forces are smoothed out across the Eulerian domain by interpolation and spreading processes using a discrete Dirac delta function (DDF). Forces to be applied at the Lagrangian points on the surface of an immersed object are determined explicitly from the current velocity of the fluid in the local neighbourhood of the point by requiring that a no-slip condition is enforced. The required velocity, $\mathbf{U}^d(\mathbf{X}_i^l)$, at position $\mathbf{X}_i^l$ of Lagrangian point l on the surface of object i is determined by the rigid body motion of the object,

$$\mathbf{U}^d(\mathbf{X}_i^l) = \mathbf{v}_i + \omega_i \times (\mathbf{X}_i^l - \mathbf{x}_i^c), \tag{2}$$

where $\mathbf{x}_i^c$ and $\mathbf{v}_i$ are the position and translational velocity vectors of the centre of mass of object i respectively and ω_i is the rotational velocity vector about $\mathbf{x}_i^c$. Expressing the fluid momentum conservation equation in terms of the fluid-particle interaction force and inserting the desired velocity from Equation (2) into the transient term gives the force required to impose the no-slip condition at each Lagrangian point,

$$\mathbf{F}^N(\mathbf{X}_i^l) = \frac{1}{\Delta t}\left[\mathbf{U}^d(\mathbf{X}_i^l) - \mathbf{U}^N(\mathbf{X}_i^l)\right] - \mathbf{RHS}^N(\mathbf{X}_i^l). \tag{3}$$

Here $\mathbf{U}^N(\mathbf{X}_i^l)$ is the current value of the fluid velocity interpolated at the Lagrangian point at time level N, Δt is the fluid time step and $\mathbf{RHS}^N(\mathbf{X}_i^l)$ represents the interpolated values of the derivative fields corresponding to the pressure, viscous and convective terms in the momentum conservation equations. The Lagrangian forces are then spread across the fluid cells in the support region of the DDF of each point and summed up to give the additional forcing in each Eulerian fluid cell. The force $\mathbf{f}$ applied to Eulerian cell m with centre at $\mathbf{x}_m$ is then given by

$$\mathbf{f}(\mathbf{x}_m) = \sum_{i=1}^{N_P}\sum_{l=1}^{N_L} \mathbf{F}(\mathbf{X}_i^l)\delta_h(\mathbf{X}_i^l - \mathbf{x}_m)\Delta V_i^l, \tag{4}$$

where δ_h is the three point DDF of Roma [6] in the present work, ΔV_i^l is the Lagrangian point volume and the sums are taken over all N_P particles and all N_L Lagrangian points of each particle.

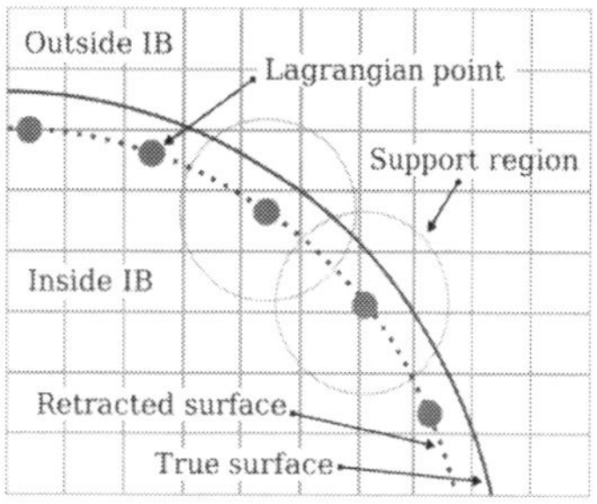

Figure 1. Illustration of Lagrangian points (red circles) immersed within an Eulerian fluid grid (blue lines). The Lagrangian points are placed a small distance r_d from the true surface of the particle (solid black line). It can be seen that some overlap of the DDF support regions occurs for points in close proximity.

As a consequence of the overlapping support regions, there is some interference of one Lagrangian point with another. This means that in practice the ns-np condition is not well enforced after a single round of forcing. Luo et al. [5] proposed the *multi-direct forcing* scheme in which forces are applied iteratively until a convergence criteria or maximum number of iterations has been reached.

Another side effect of the interpolation of fluid velocities and spreading of forces is the enhancement of particle size which occurs as fluid in the immediate vicinity of an immersed object is frozen in the diffuse cloud of forces surrounding the object. In the case of the Roma DDF [6], with a support region three cells wide, this means that the smeared out particle interface extends a distance $3\Delta x/2$ into the fluid where Δx is the Eulerian grid spacing. As F_d is proportional to the projected area of the particle this enhancement of size has the undesirable consequence of increasing F_d. Breugem [1] suggested the *radius retraction* scheme in which the Lagrangian points are placed a small distance $r_d = b\Delta x$ inside the particle interface in order that the actual particle size is better approximated after the interpolation and spreading procedures by suitable choice of the parameter b. Breugem found empirically that the choice $b = 0.3$ was optimal, giving second order convergence of the predicted drag forces with increasing grid resolution. Tang et al. [7] showed that $b = 0.3$ is only optimal in the low concentration limit previously considered by Breugem, with the optimal value becoming more sensitive to ϕ and Re at higher concentrations.

3 Validation of the IBM with ordered packings

Flow through ordered arrays of monodisperse spheres in the low Re limit has been used for validation purposes by several researchers for LB [4] and IBM simulations [1, 8]. Zick and Homsy [10] considered flow through periodic simple cubic (SC), body centred cubic (BCC) and face centred cubic arrays (FCC) in an infinite space in the limit

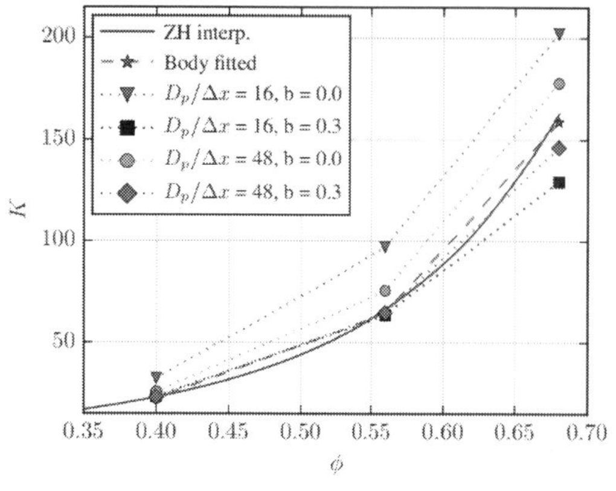

Figure 2. Drag coefficient K against ϕ for the BCC case.

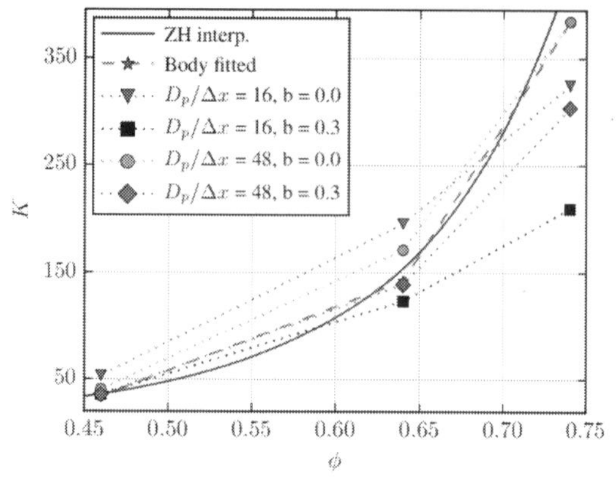

Figure 3. Drag coefficient K against ϕ for the FCC case.

$Re \rightarrow 0$ from dilute to maximum packing fraction in each case (SC: $\phi_{SC}^{max} = 0.5236$; BCC: $\phi_{BCC}^{max} = 0.6802$; FCC: $\phi_{FCC}^{max} = 0.7405$). The Navier-Stokes equations were expressed in terms of a set of two-dimensional integral equations of the hydrodynamic stress at the sphere surfaces. For each configuration and concentration considered, Zick and Homsy determined a drag coefficient $K = F_d/F_d^s$ numerically by evaluating these integral equations using increasing numbers of terms and extrapolating the results to infinity. The results obtained by Zick and Homsy were shown to compare well with other values available in the literature.

The steep dependence of K on ϕ in the densely packed limit means that any small variation in the effective particle diameter as $\phi \rightarrow \phi^{max}$ leads to large degrees of error in the predicted values of K. To establish confidence in the predicted drag forces the influence of r_d on K was investigated using the results of Zick and Homsy as benchmark values [10]. SC, BCC and FCC cases were considered. In each case an inlet boundary condition was applied along the $\hat{\mathbf{x}}$ direction with periodic boundary conditions in the $\hat{\mathbf{y}}$ and $\hat{\mathbf{z}}$ directions. Several layers of particles (4 for SC and 3 for BCC and FCC cases) were arranged along the $\hat{\mathbf{x}}$ direction with only the forces acting on internal layers considered in order to minimise boundary effects. All simulations were conducted at $Re \leq 0.1$ with the fluid time step chosen to yield a Courant number of $C \approx 0.1$. For each configuration several values of ϕ were considered by varying the particle diameter whilst the centre positions and domain dimensions remained fixed. Δx was varied to satisfy the target resolutions $D_p/\Delta x = 16, 32, 48$, and 64 for each configuration and concentration considered.

Representative results of the variation of K with ϕ for the BCC case for $b = 0.0$ and 0.3 at low resolution with $D_p/\Delta x = 16$ and high resolution with $D_p/\Delta x = 48$ are shown in Figure 2. An interpolation of the results of Zick and Homsy [10] and results from boundary fitted CFD simulations are used for comparison. Results for the BCC case up to $\phi \approx \phi_{SC}^{max}$ along with results from SC cases (not shown here) confirm that $b = 0.3$ is optimal, as suggested by Breugem [1], giving agreement to within 2% of the benchmark values at the low resolution of $D_p/\Delta x = 16$. The results at $\phi = \phi_{BCC}^{max}$ show that $b = 0.3$ leads to an underprediction of K with the optimal value appearing to lie somewhere in between $b = 0.0$ and $b = 0.3$. Figure 3 shows representative results for the variation of K with ϕ for the FCC case for $b = 0.0$ and 0.3 at $D_p/\Delta x = 16$ and 48. Results at $\phi \approx \phi_{BCC}^{max}$ confirm that the optimal value of b lies in the range $0.0 < b < 0.3$ whilst results at higher concentration suggest that $b \rightarrow 0.0$ as $\phi \rightarrow \phi_{FCC}^{max}$.

4 Drag in polydisperse packings

Random grain packings with linearly graded size distributions with coefficient of uniformity values $C_u = 1.01, 1.50$ and 2.00 were prepared using the Discrete Element Method (DEM). Keeping the minimum diameter fixed at $D_p^{min} = 0.50$ mm, ensembles of particle diameters were generated using a pseudo-random number generator with standard deviations of the errors between the generated and target size distributions for each C_u of less than 2%. A disperse cloud of particles was produced by choosing random initial positions with no particles overlapping. Isotropic compression to 1.0×10^2 kPa was then performed with periodic boundaries in all directions. The resulting particle positions and domain dimensions were then used to generate the Lagrangian points and fluid domain dimensions respectively for IBM simulations with an inlet/outlet condition along one coordinate direction and periodic boundary conditions on the Lagrangian points and fluid equations in the two orthogonal directions. A grid resolution of $D_p^{min}/\Delta x = 24$ was used and the inlet velocity, U, was chosen to satisfy $Re \approx 0.1$ in all cases, with the fluid time step, Δt, chosen to yield a Courant number of $C \approx 0.1$. In order to not over constrain the velocities and pressures in the pore spaces at the inlet and outlet, additional fluid cells were added in these regions such that all particles lay entirely inside the fluid domain along the inlet/outlet direction.

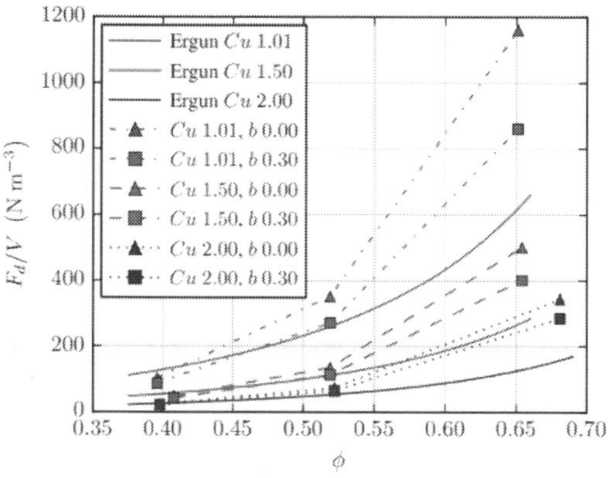

Figure 4. Drag force per unit volume F_d/V against ϕ for all values of C_u considered with $b = 0.00$ and 0.30. Solid lines show the predictions of the Ergun correlation.

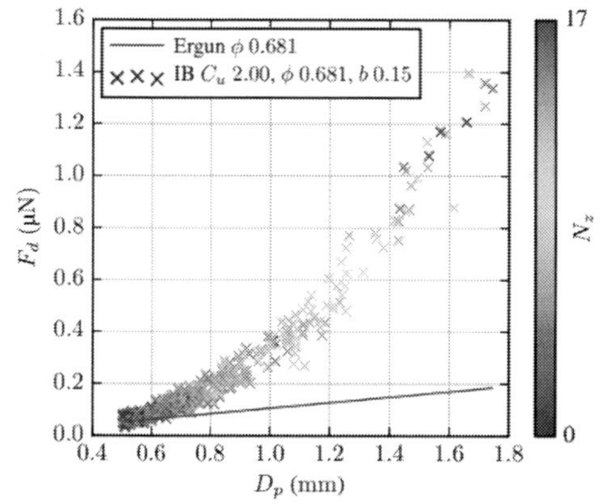

Figure 5. Individual particle drag F_d against diameter D_p at ϕ = 0.681 for 395 of 497 particles in a random configuration with C_u = 2.00. Colour of points indicates coordination number N_z.

Results for the total drag on each packing normalised by domain volume V against concentration for all C_u values with $b = 0.00$ and 0.30 are shown in figure 4. For comparison the predictions of the widely used Ergun correlation [3] are shown where the characteristic diameter is taken as the average diameter by mass for each packing. It can be seen that in the range $\phi \in [0.39, 0.52]$ the Ergun correlation and the IB simulation results are in good agreement. At the highest concentrations considered there is a large discrepancy between the Ergun correlation and the IB results. As C_u increases the surface area to volume ratio decreases and hence the total drag is seen to decrease for approximately equal concentrations. The results become less sensitive to changes in b as C_u increases which can be explained by the decrease in dF_d/dD_p as D_p increases.

Drag forces on individual particles are shown against diameter in figure 5. The Ergun correlation provides good predictions for the smallest particles, however the prediction diverges from the IB results as D_p increases. In a polydisperse packing the larger particles experience a locally higher concentration than they would in a monodisperse packing at the same ϕ as small particles can fill the gaps between larger particles. A first order indication of this increase in local ϕ is given by the number of interparticle contacts, or coordination number, N_z, which is seen to increase with increasing particle size. Monodisperse correlations do not account for this local increase in ϕ which explains the discrepancy. Particles in the inlet and outlet boundary regions where the flow is developing experience systematically less drag and are not shown in figure 5.

5 Conclusions

Results with ordered packings shown in section 3 support the findings of Breugem [1] that $b = 0.30$ is optimal for disperse systems whilst results at higher concentrations indicate that $b \to 0$ as $\phi \to 1$ confirming the findings of Tang et al. [7] that the optimal value of b is dependent on ϕ. Results for random packings in section 4 show that the Ergun correlation [3] underpredicts the drag acting on individual particles in polydisperse packings with an error which increases with particle diameter which is expected as the effect of the local increase in ϕ with D_p is not taken into account. To remove boundary effects for cases presented in section 4, alternate boundary conditions along the inlet/outlet direction will be explored.

Acknowledgements

This work was made possible by funding from EPSRC awarded through the CDT in Theory and Simulation of Materials at Imperial College London.

References

[1] Breugem W. P., *J. Comput. Phys.* **231**(13), 4469-4498 (2012)
[2] Di Felice R., *Int. J. Multiphase Flow* **20**(1), 153-159 (1994)
[3] Ergun S., *Chem. Eng. Prog.* **48**(2), 89-94 (1952)
[4] Hill R. J., Koch, D. L., and Ladd, A. J. C., *J. Fluid Mech.* **448**, 213-241 (2001)
[5] Luo K., Wang Z., Fan J., Cen K., *Phys. Rev. E* **76**(6), 1-9 (2007).
[6] Roma A. M., Peskin C. S., and Berger M. J., *J. Comput. Phys.* **153**(2), 509-534 (1999).
[7] Tang Y., Kriebitzsch S. H. L., Peters E. A. J. F., van der Hoef M. A., Kuipers J. A. M., *Int. J. Multiphase Flow* **62**, 73-86 (2014).
[8] Tenneti S., Garg R., and Subramaniam S., *Int. J. Multiphase Flow* **37**(9), 1072-1092 (2011)
[9] Uhlmann M., *J. Comput. Phys.* **209**(2), 448-476 (2005).
[10] Zick A. A., and Homsy G. M., *J. Fluid Mech.* **115**, 13-26 (1982)

EPJ Web of Conferences **140**, 13002 (2017) DOI: 10.1051/epjconf/201714013002
Powders & Grains 2017

Experimental study of tensile strength of pharmaceutical tablets: effect of the diluent nature and compression pressure

Audrey Juban[1,*], *Stéphanie* Briançon[1,**], *François* Puel[2,***], *Thierry* Hoc[3,****], and *Cécile* Nouguier-Lehon[3,†]

[1] *Univ Lyon, Universite Lyon 1, CNRS, UMR5007, LAGEP, 43 bd du 11 Novembre 1918, F-69622 LYON, France*
[2] *LGPM, CentraleSupélec, Université Paris-Saclay, Grande Voie des Vignes, 92295 Châtenay-Malabry, France*
[3] *Univ Lyon, Ecole Centrale de Lyon, LTDS UMR CNRS 5513, 36 avenue Guy de Collongue, F-69134 ECULLY cedex, France*

Abstract. In the pharmaceutical field, tablets are the most common dosage form for oral administration in the world. Among different manufacturing processes, direct compression is widely used because of its economics interest and it is a process which avoids the steps of wet granulation and drying processes. Tablets are composed of at least two ingredients: an active pharmaceutical ingredient (API) which is mixed with a diluent. The nature of the powders and the processing conditions are crucial for the properties of the blend and, consequently, strongly influence the mechanical characteristics of tablets. Moreover, tablets have to present a suitable mechanical strength to avoid crumbling or breaking when handling, while ensuring an appropriate disintegration after administration. Accordingly, this mechanical property is an essential parameter to consider. Experimental results showed that proportion of the diluent, fragmentary (DCPA) or plastic (MCC), had a large influence on the tensile strength evolution with API content as well as the compression load applied during tableting process. From these results a model was developed in order to predict the tensile strength of binary tablets by knowing the compression pressure. The validity of this model was demonstrated for the two studied systems and a comparison was made with two existing models.

1 Introduction

Pharmaceutical tablet is the most common dosage form for oral administration, mainly because it is an easier way for drug intake by patients. Other reasons for this success are the ease of large-scale production, their safety and easy handling and good stability. Among different manufacturing processes, direct compression is widely used because of its economics interest and its process which avoids the steps of wet granulation and drying processes [1–4]. Tablets are composed of at least two ingredients: an active pharmaceutical ingredient (API) which is mixed with a diluent [5]. Other ingredients could also be included, like lubricants, disintegrating agents, binders, etc. It is well known that the nature of the powders used and the processing conditions are crucial for the properties of the blend (fluidity and compressibility) and, consequently, strongly influence the mechanical characteristics of the obtained tablets [6, 7]. Tablets have to present a suitable mechanical strength to avoid crumbling or breaking when handling, while ensuring an appropriate disintegration after administration. Accordingly, the tablet tensile strength is an essential parameter to consider [8–11] and is therefore standardly tested during the manufacturing process [12]. This mechanical characteristic is calculated from the breaking force of the tablet obtained from a standard method, generally using a diametral compression test [13].

The objective of this work is first to experimentally study the effect of operating conditions: composition and compression load (formulation and process parameter, respectively). In a second step, a model allowing the prediction of the tensile strength of binary tablets from the pressure applied during direct tableting process was proposed.

2 Materials and methods

2.1 Powders

The active pharmaceutical ingredient (API) used in this work is the anhydrous caffeine form I (CFI) obtained by using the same method as [14] from commercial anhydrous caffeine Form II (Cooper). To study the effect of the diluent nature, two different powders were used: microcrystalline cellulose (MCC) Avicel®PH 102 (FMC Health and nutrition) and anhydrous dicalcium phosphate (DCPA) from JRS Pharma. CFI and MCC are known to present a plastic behavior under compression, whereas DCPA exhibits a fragmentary behavior [5, 15, 16].

* e-mail: audrey.juban.pro@gmail.com
** e-mail: stephanie.briancon@univ-lyon1.fr
*** e-mail: francois.puel@centralesupelec.fr
**** e-mail: thierry.hoc@ec-lyon.fr
† e-mail: cecile.nouguier@ec-lyon.fr

2.2 Tablet preparation

Cylindrical tablets were prepared from binary mixtures of CFI (one day after its manufacture) and MCC or DCPA as diluent. Each diluent was mixed in various proportions with caffeine: 10, 20, 30, 40, 50, 60, 78 and 90 wt% of CFI for MCC and 20, 40, 60 and 80 wt% of CFI for DCPA. The blending was performed with a tridimensional mixer (Turbula®T2F) at a rotational rate of 49 rpm for a duration of 10 minutes. Series of pure CFI and MCC tablets were also made. All tablets were produced at a constant mass (about 300 mg) by using a compaction simulator Styl'One Classic (MEDEL'PHARM, Beynost, France) and its data acquisition software (Analis, 2.03 version, MEDEL'PHARM). Standard Euro D tools with flat-faces of diameter 11.28 mm were fitted on the simulator. Refer to [14] for a complete description of this single station press. Tablets were made without lubricant using one main compression, driven by force. To study the effect of compression pressure, tablets were produced under three different compression loads corresponding to 50, 100 and 200 MPa in compression pressure, noted *P* thereafter. According to the European Pharmacopoeia [12], the tablet mass *m*, thickness *t* and diameter *D* were measured just after compression, allowing the determination of tablet density and porosity.

2.3 Diametral compression test

Diametral compression tests were performed on tablets one day after their fabrication using a tensile test device (ElectroPulsTM E10000, Instron®, Elancourt, France) and its data acquisition software (WaveMatrixTM, Instron®). A sensor of 1 kN was used. Diametral compression tests were performed at constant velocity of 0.1 mm.min^{-1} with an acquisition rate of 10 Hz. A minimum of three tablets per composition and per compression load were tested. The maximal diametral crushing force *F* was measured for each tablet and the tensile strength was calculated using the classical equation [17]:

$$\sigma_T = \frac{2F}{\pi D t} \tag{1}$$

3 Results

3.1 Tablet tensile strength

Table 1 reports the tablet tensile strength for all compositions studied. As expected, and in accordance with the literature [15, 18], for a given composition, the tablet tensile strength increased with the compression pressure. Moreover, the tensile strengths of tablets containing DCPA were lower than those obtained for CFI/MCC mixtures. For CFI/MCC mixtures, tablet tensile strength generally increased with a decreasing of the caffeine content in the formulation. By contrast, with the binary mixtures containing DCPA, the tablet tensile strength slightly increased with the caffeine content. At the lowest compaction pressure (50 MPa) the trend of tablet tensile strength with caffeine content was more erratic than for the greater compression pressure. This variability can be explained by the tablet porosity dispersion at low pressure.

Table 1. Tablet tensile strength (σ_T in MPa) obtained from diametral compression test. Average calculated from 3 tablets.

Diluent (-)	Composition (wt% CFI)	Compression pressure P (MPa) 50	100	200
	100	1.08 ± 0.11	1.84 ± 0.09	2.39 ± 0.29
MCC[a]	90	1.34 ± 0.05	2.36 ± 0.09	3.20 ± 0.10
	78	1.21 ± 0.03	2.56 ± 0.02	4.16 ± 0.11
	60	1.65 ± 0.04	3.34 ± 0.12	5.41 ± 0.29
	50	1.54 ± 0.02	3.34 ± 0.06	6.06 ± 0.17
	40	1.65 ± 0.01	3.79 ± 0.04	6.79 ± 0.09
	30	1.50 ± 0.01	3.60 ± 0.06	6.48 ± 0.07
	20	1.80 ± 0.01	4.27 ± 0.06	7.87 ± 0.12
	10	1.69 ± 0.06	4.21 ± 0.08	8.16 ± 0.12
	0	2.00 ± 0.07	5.04 ± 0.11	9.73 ± 0.24
DCPA	80	0.97 ± 0.04	1.69 ± 0.05	2.79 ± 0.09
	60	0.74 ± 0.01	1.33 ± 0.07	2.52 ± 0.06
	40	0.49 ± 0.03	1.08 ± 0.05	1.99 ± 0.06
	20	0.30 ± 0.00	0.67 ± 0.02	1.62 ± 0.03

[a] From [14]

3.2 Modeling the tablet tensile strength

As observed in table 1, the compression pressure, *P*, is the parameter having the greatest impact on the value of the tablet tensile strength, σ_T. Hence, the objective of this section is to propose a model allowing the prediction of the tensile strength of binary tablet by knowing the compression pressure applied during the direct tableting process. Moreover, since the pharmaceutical tablets are usually composed of a minimum of two components, it seems relevant to only consider the two extreme compositions of each studied mixture for the construction of the model. Consequently, a linear regression was plotted for the two extreme formulations of each studied mixture (10 and 90 wt% of CFI and 20 and 80 wt% of CFI for MCC and DCPA, respectively). The coefficient of determination R^2 was always greater than 0.94. Values of regression parameters obviously depend on each blend and hence on the percentage of CFI. Assuming a linear relationship for these parameters with the caffeine content, the tablet tensile strength can finally be predicted by:

$$\sigma_T = (A_1 \times \text{wt\%}\,\text{CFI} + A_2) \times P + (B_1 \times \text{wt\%}\,\text{CFI} + B_2) \tag{2}$$

The values of coefficients A_1, A_2, B_1 and B_2 obtained for the two studied mixtures are given in table 2.

Table 2. Coefficients required for the application of the proposed model (equation 2)

	A_1 (-)	A_2 (-)	B_1 (MPa)	B_2 (MPa)
CFI/MCC	$-3.9 \cdot 10^{-4}$	0.0464	0.0151	-0.4356
CFI/DCPA	$5.2 \cdot 10^{-5}$	0.0079	0.0099	-0.3733

DOI: 10.1051/epjconf/201714013002

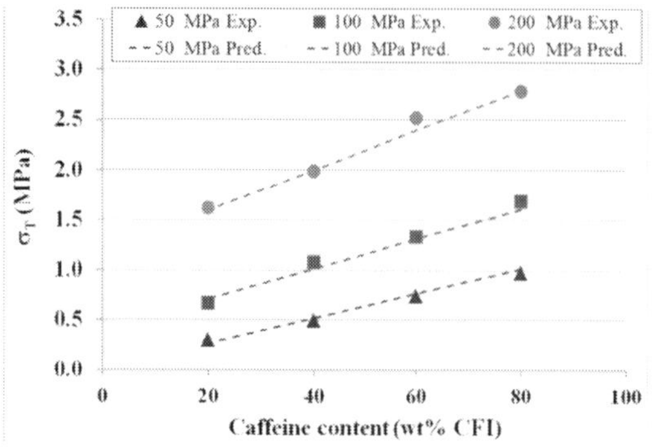

Figure 1. Evolution of experimental tensile strength σ_T obtained by diametral compression test (filled symbols) and predicted values (lines) versus the percentage of caffeine in tablets for binary mixture of CFI/DCPA

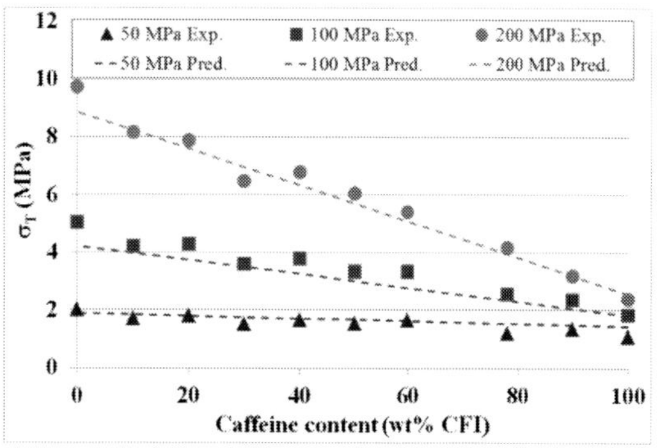

Figure 2. Evolution of experimental tensile strength σ_T obtained by diametral compression test (filled symbols) and predicted values (lines) versus the percentage of caffeine in tablets for binary mixture of CFI/MCC

4 Discussion

4.1 Validation of the proposed model

Figures 1 and 2 show the evolution of tablet tensile strength according to the percentage of CFI for binary mixtures containing DCPA and MCC, respectively. The comparison of experimental data (filled symbols) with the values predicted by equation 2 (lines) showed that the proposed model is relevant, and one can note that most of the experimental values were slightly underestimated by the model. Mean relative deviations between predicted and experimental values of tablet tensile strength are of 4.2% and 8.2% for binary mixtures of CFI/DCPA and CFI/MCC, respectively. In the case of CFI/MCC mixture, pure tablets were manufactured and the mean relative deviation was 9.1% using the data for 0 and 100 wt%, close to value obtained with 10 and 90 wt%. It is worth noting that the proposed model is efficient for binary mixtures composed of two components exhibiting plastic deformation under compression (CFI/MCC) but also for mixtures containing a plastic and a fragmentary compound (CFI/DCPA).

Table 3. Mean relative deviations between experimental tablet tensile strength and values predicted by models

Diluent	Equation (2)	Model [14]	Model [22]
MCC	8.2%	12.0%	14.6%
DCPA	4.2%	7.8%	-

4.2 Comparison with models from literature

Several authors have been interested in the study of tablet tensile strength, looking especially for a method allowing its prediction. Most of them are based on the knowledge of the relative density (or the porosity) of mixture. One can cite the studies based on the percolation theory [18–21] (which also requires several unknown parameters such as *critical relative density*) and the studies based on the Ryskewitch-Duckworth equation [22, 23]. However, recently a model between the tablet tensile strength and the surface hardness of tablet was also proposed [14]. For more details about these models refer to [14].

In order to test the efficiency of the model proposed in this study, we compared results obtained by equation 2 with values predicted by Wu et al.'s model [22] and Juban et al.'s model [14]. Data required to apply these models were determined using the methodology already described in [14]. Especially, for each composition studied, the relative density and the porosity of tablets were calculated and the hardness measurements were performed by using the same device and method. The mean relative deviations between experimental and predicted data calculated from the three predictive models were given in table 3, showing the good ability for all models to predict the tensile strength of binary tablets. Nevertheless, Wu et al.'s model could not be applied for binary mixtures composed of CFI/DCPA, because it has been impossible to manufacture pure tablets of DCPA due to important binding and sticking phenomena occurring during the direct tableting process. This problem highlighted a restriction of this model, since some pharmaceutical powders, especially API, do not have the required compression properties such as fluidity and compressibility to form pure tablets [24, 25]. Moreover, the predicted data calculated from equation 2 and from the model developed in [14] were slightly underestimated compared to the experimental ones, for the two mixtures studied. By contrast, for tablets containing MCC as diluent, the tensile strengths predicted from [22] presented an overestimation.

The main difference between these models is the data required for their application. The model proposed in this work presents the advantage of only being based on a tableting process parameter, the compression pressure, which is easily accessible since the tablets are manufactured, and by the knowledge of the tensile strength of two different compositions for a given system.

5 Conclusion

In this work, the impact of operating conditions on the mechanical strength of binary tablets was evaluated, including both process and formulation parameters.

As expected, for the two diluents used in the formulation, the tablet tensile strength increased with the compression pressure. Whatever the compression pressure, the CFI/MCC tablets exhibited a decreasing tensile strength with an increase of the caffeine content. By contrast, an opposite behavior was observed for CFI/DCPA tablets (tensile strength increasing with caffeine content). This is in accordance with the literature [18, 24] and attests that the choice of the ingredients is essential and could have many consequences on the quality of the final tablets.

Moreover, all of these results showed that a relationship between a process parameter and a mechanical characteristic of tablets can be found. So, a predictive model for the tensile strength of pharmaceutical tablets was developed for a given binary system as a function of the compression pressure. Constants used in the proposed model were determined from the tablet tensile strength of the two extreme compositions of each studied system. This model was validated with experimental results, for both studied binary systems and on a wide range of compositions and compression pressures which are quite commonly used in the pharmaceutical manufacturing. The proposed model is relevant for the two studied mixtures, whether it consists of two plastic compounds or a mixture containing a plastic and a fragmentary compound. Moreover, it does not require the fabrication of pure component tablets.

Finally, the tablet tensile strength can be connected directly to the compression load, which is easily accessible since the tablets are manufactured, and by the knowledge of the tensile strength of two compositions, for a chosen binary system. By extending the applicability domain of the developed model for tablets composed of three or more ingredients, it could be part of a "Quality by Design" approach.

Acknowledgements

The support of the Ministère de l'Enseignement Supérieur et de la Recherche is gratefully acknowledged. We would like to thank MEDEL'PHARM and the IVTV platform (ANR-10-EQPX-06-01).

References

[1] T. Martinello, T.M. Kaneko, M.V.R. Velasco, M.E.S.T. Taqueda, V.O. Consiglieri, Int. J. Pharm. **322**, 87-95 (2006)

[2] L.X. Liu, I. Marziano, A.C. Bentham, J.D. Litster, E.T. White, T. Howes, Powder Technol. **240**, 66-73 (2013)

[3] C. Hirschberg, C.C. Sun, J. Rantanen, J. Pharm. Biomed. Anal. **128**, 462-468 (2016)

[4] T. Osamura, Y. Takeuchi, R. Onodera, M. Kitamura, Y. Takahashi, K. Tahara, H. Takeuchi, Int. J. Pharm. **510**(1), 195-202 (2016)

[5] M. Jivraj, L.G. Martini, C.M. Thomson, Pharm Sci Technology Today **3**(2), 58-63 (2000)

[6] M.B. Tejedor, N. Nordgren, M. Schuleit, M.W. Rutland, Int. J. Pharm. **486**, 315-323 (2015)

[7] A.D. Rajkumar, G.K. Reynolds, D Wilson, S. Wren, M.J. Hounslow, A.D. Salman, Eur. J. Pharm. Biopharm. **106**, 88-96 (2016)

[8] C.Y. Wu, J.P.K. Seville, Powder Technol. **189**, 285-294 (2009)

[9] D.A. Pai, A.A. Hayes, M.R. Okos, Powder Technol. **239**, 441-450 (2013)

[10] C. Shang, I.C. Sinka, B. Jayaraman, J. Pan, Int. J. Pharm. **442**, 57-64 (2013)

[11] S.M. Razavi, G. Callegari, G. Drazer, A.M. Cuitino, Int. J. Pharm. **484**, 29-37 (2015)

[12] European Directorate for the Quality of Medicines and Healthcare, *European Pharmacopoeia*, 8th ed. (Strasbourg, 2014)

[13] F. Podczeck, Int. J. Pharm. **436**, 214-232 (2012)

[14] A. Juban, C. Nouguier-Lehon, S. Briancon, T. Hoc, F. Puel, Int. J. Pharm. **490**, 438-445 (2015)

[15] Y. Zhang, Y. Law, S. Chakrabarti, AAPS PharmSciTech. **4**(4) 62 (2003)

[16] J. Ilkka, P. Paronen, Int. J. Pharm. **94** 181-187 (1993)

[17] J.T. Fell, J.M. Newton, J. Pharm. Sci. **59** 688-691 (1970)

[18] A. Michrafy, M. Michrafy, M.S. Kadiri, J.A. Dodds, Int. J. Pharm. **333**,118-126 (2007)

[19] M. Kuentz, H. Leuenberger, Eur. J. Pharm. Biopharm. **49**, 151-159 (2000)

[20] N. Ramirez, L.M. Melgoza, M. Kuentz, H. Sandoval, I. Caraballo, Eur. J. Pharm. Sci. **22**, 19-23 (2004)

[21] V. Busignies, B. Leclerc, P. Porion, P. Evesque, G. Couarraze, P. Tchoreloff, Eur. J. Pharm. Biopharm. **67**, 507-514 (2005)

[22] C.Y. Wu, S.M. Best, A.C. Bentham, B.C. Hancock, W. Bonfield, Eur. J. Pharm. Sci. **25**, 331-336 (2005)

[23] W. Duckworth, J. Am. Ceram. Soc. **36** 68 (1953)

[24] S. Mangal, F. Meiser, G. Tan, T. Gengenbach, D.A.V. Morton, I. Larson, Eur. J. Pharm. Biopharm. **104**, 110-116 (2016)

[25] A. Koynov, F. Romanski, A.M. Cuitino, Powder Technol. **236** 5-11 (2013)

Characterising powder flow properties – the need for a multivariate approach

Tim Freeman[1]*, *Katrina* Brockbank[1] *and Jerome* Sabathier[2]

[1] Freeman Technology Ltd, Tewkesbury, UK
[2] ForS Instruments, Bailleau Pine, France

Abstract. Despite their widespread and well-established use, powders are challenging materials to work with, as evidenced by the common problems encountered during storage and processing, as well as in the quality and consistency of final products. The diverse range of unit operations used to handle and manipulate powders subject them to extremes of stress and flow regimes; from the high stress, static conditions present in hoppers to the dispersed, dynamic state of a fluidised bed dryer. It is therefore possible for a powder to behave a certain way in a given unit operation, but entirely differently in another. Many existing powder testing techniques don't deliver the required information as the test conditions do not represent the conditions in the process. Modern powder rheometers generate process relevant data by accurately measuring dynamic flow, bulk and shear properties. This approach enables a powder's response to aeration, consolidation, forced flow and changes in flow rate to be reliably quantified thereby simulating the conditions which a powder will be subjected to in process. This paper provides an introduction to powder rheology, including a comparison with traditional techniques, and uses case studies to demonstrate how powder rheology can be applied to optimise production processes and enhance product quality

1 Introduction

Powder and granular systems must be considered as a multiphase continuum whose behaviour is a function of both the fundamental properties of the powder and individual particles (for example, size, shape, surface texture and moisture content), and external factors such as humidity, temperature and consolidation history. As such powders and granular materials exhibit diverse and complex behaviour

Whilst manufacturers may be able to control some fundamental properties, such as particle size and crystal habit, most fundamental properties cannot not be dictated, especially for natural and biological materials. External factors such as temperature and humidly can also be regulated however this can often be extremely expensive and may only cover part of the lifecycle of the powder. As such powder behaviour is difficult to both predict and control. The manufacturer and end user must use a pro-active approach in understanding the behaviour of a given powder or granular system through accurate and repeatable quantification of the powder properties, in particular the flowability. This is further complicated by the dependence of the flow behaviour on the stress and strain rate, meaning the test conditions should ideally be representative of the process conditions to which the powder is subjected

A number of well documented, traditional methods such as Angle of Repose (AOR), Flow meters (such as Critical Orifice Diameter) and Tapped Density tests (Carr's Index / Hauser Ratio) are widely available. Whilst these methods are typically cheap, quick and simple to use they are highly operator dependent, poorly repeatable and not overly differentiating [1]. Another widely accepted methodology is Shear Cell analysis. Whilst the mathematical principles are clearly defined, this method was originally designed to determine the incipient flow of continuous, cohesive powders at higher stress levels whereas as many processes and unit operations occur at low to moderate stresses. One major advantage of Shear Cell analysis is its use in hopper design,

Arguably, there are two major disadvantages with these methodologies. Firstly, these methods cannot characterise the full range of powder flowability/ cohesion. AOR and Flow Meters require the powder or granular material to flow through a funnel and are therefore unsuitable for cohesive powders. Tapped Density tests are also unsuitable for highly cohesive materials as the tapping force is insufficient to overcome the strong interparticular cohesive bonds meaning that the powder bed will not necessarily repack / consolidate when

* Corresponding author: tim.freeman@freemantech.co.uk

tapped. A further issue arises with this test method when dealing with granular materials, as the heavier / denser particles are more likely to pack efficiently under their own mass, resulting in falsely low results.

Conversely Shear Cell analysis is typically less differentiating with respect to free-flowing powders and may not detect minor variations in the powder properties which impact on the process performance. Furthermore, due to particle size limitations, commercially available instruments are not readily available for larger granular materials. The second major disadvantage is that these methods is that they are not generally representative of the range of process conditions to which powders and grains are subjected during processing, indeed there is no standard test for measuring powder flow in all four flow regimes (plastic, inertial, fluidised and entrained flow) [2].

More recent development of powder rheometers has overcome these limitations, with these methods becoming more widely recognised and accepted, as demonstrated by the large number of journal articles available [2, 3]. Whilst this technology was primary developed to measure powders under dynamic conditions, the modular design and adaptability means that both the bulk (Compressibility and Permeability) and shear (Shear Cell and Wall Friction) properties are also easily measured. As such this technique is capable of evaluating both free-flowing and highly cohesive materials, even at larger particular sizes, not only allowing direct comparison of different materials (i.e. different raw ingredients) but also meaning that a single instrument can be used to monitor the process train from raw materials through to the finished product.

2 Dynamic principle

The rheological properties of the powder or granular systems are quantified through measurements of the resistance to flow. The FT4 Powder Rheometer (Freeman Technology, UK) reports this as a flow energy. This is obtained by means of a patented measurement principle that evaluates the resistance to the motion of a twisted blade passing along a prescribed path through a precise volume of the powder. The torque and force required to maintain movement of the blade are recorded and converted into a flow energy (figure 1). This principle has recently been modelled using DEM with the results of the computer model shown to be comparable to experimental values. [4]. Modification of the tip speed and helix angle also allow the stress and strain rates to be easily altered whilst different vessel and blade sizes allow the characterisation of powders with particle sizes ranging from a few nanometres to several millimetres.

Unlike traditional methodologies, dynamic test methods also incorporate sample conditioning to remove operator variability and consolidation history. This is achieved through the displacement of the whole sample to loosen and slightly aerate the powder with the aim of producing a standardised packing state. For ease of use the conditioning cycle(s) is normally incorporated into the test program however where consolidation/storage history is being investigated the sample can be pre -conditioned before storing the sample. Further improvements to repeatability are achieved by use of a vessel design which incorporates a splitting mechanism to ensure a consistent sample volume.

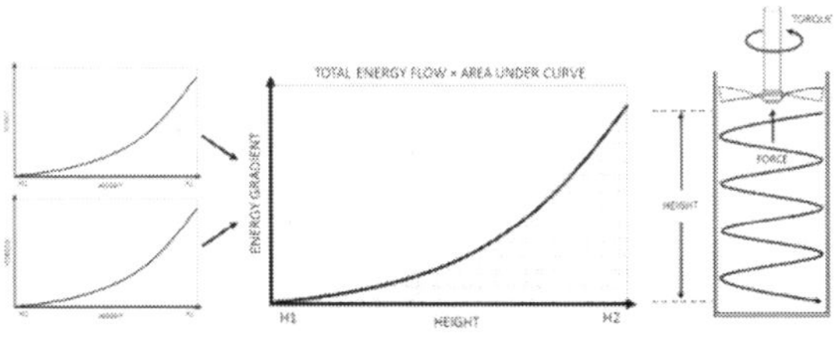

Fig. 1 Dynamic measurement principle

Other dynamic systems which measure the resistance to the passage of a blade, typically only measure axial force. This assumes that the powder bed acts as a continuous uniform body, whereas most powder beds contain air pockets and channels, especially more cohesive powders. As particles displaced by the blade fill these voids rather than colliding with and displacing surrounding particles, the energy is not fully transmitted to an axially positioned load cell [1]. Comparison of force and torque measurements demonstrates that the force is often an order of magnitude lower than torque. As such, not only are axial force measurements much less differentiating, but may also fail to correctly asses and rank more cohesive powders [1].

In addition to measuring the properties of the powder in a "poured" state, the addition of a controlled air feed allows the measurement of the flow energy as a function of air velocity. This is particularly useful for understanding the properties of powders in a low stress, aerated/fluidised state, for example during pneumatic conveying. Not all powders reach a fully fluidised state, especially cohesive or large/dense powders, this can be problematic for rheological techniques that require powders to be fluidisable.

At the opposite end of the spectrum, dynamic properties can be measured at higher stresses through application of an external force, such as compaction or vibration [5]. More recent advancements have also demonstrated the benefits of rheological measurements for quantifying caking behaviour in powders, especially for non-homogenous caking [6].

3 Additional test methodologies

Whilst not a direct measure of the powder flowability, the bulk properties of Compressibility and Permeability can have considerable influence on powder behaviour during unit operations. For example, permeability can affect flow from hoppers and silos as well as filling applications, and compressibility is relevant to storage or during compaction processes such as tabletting and roll compaction. Finally, shear properties are useful for understanding how powders transition from a high stress, static state to a dynamic state. Measurements of Wall Friction quantify the ease with which a powder will move against process equipment surfaces.

4 Case studies

In both case studies, the powders were characterised using the FT4 Powder Rheometer (Dynamic Flow, Bulk and Shear tests). The results were compared with known process behaviour in order to identify the most relevant parameters.

4.1 Spray coating

Three samples of a polymer powder were used in a spray coating application using a corona charging system. Sample A exhibited good performance in terms of flowing through the nozzle and adhering to the substrate, and Sample B showed acceptable behaviour, but Sample C was poor in both aspects; causing blockages in the nozzle and subsequently falling away from the substrate during transportation to the kiln. Both particle size analysis and shear testing failed to discriminate between the samples,

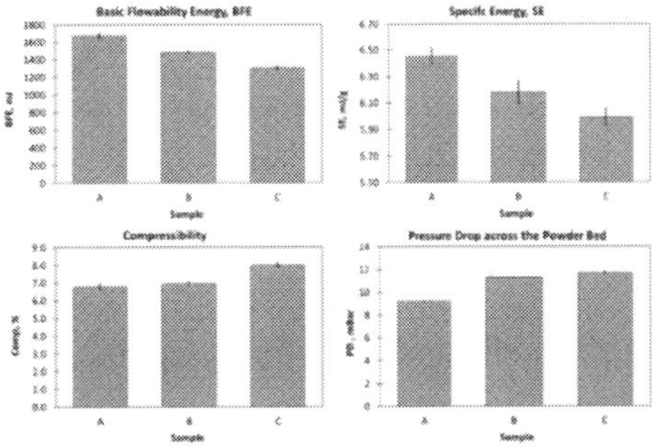

Fig. 2. Dynamic and Bulk results for polymers A to C used in a spray coating application.

Sample A generated the highest Basic Flowability Energy (BFE) and Specific Energy (SE) of the three samples (fig. 2), which together indicate greater cohesion and particle-particle interlocking. Sample C generated the lowest BFE and SE, suggesting that a degree of inter-particular cohesion is required to form a uniform coating on the substrate, and that Sample C does not meet this criterion.

Sample C was the most compressible of the samples, indicating a greater propensity to compact under forced flow conditions, such as when drawing the powder from the storage vessel into the fluidisation chamber. The greater propensity to compact will promote the formation of agglomerates, inhibiting both the spraying and charging operations in the nozzle.

Sample A generated the lowest Pressure Drop across the Powder Bed indicating that it is the most permeable. This suggests that it will be the most free-flowing under conveyance, and that once fluidised, is likely to flow more readily within an air stream. Sample C was least permeable, generating the highest Pressure Drop across the Powder Bed, which is likely to cause more erratic, pulsatile flow into the fluidisation chamber, and unstable flow of the fluidised mass.

The results suggest that a degree of cohesion is required to form a uniform coating but susceptibility to agglomeration and erratic flow is problematic to the process. Sample C, with the lowest BFE and Permeability, and the highest Compressibility, is most sensitive to compaction during conveyance to the fluidisation chamber, forming agglomerates that can block the nozzle and cause inconsistent charging. This also highlights a common misconception that the most free-flowing samples will provide the best process performance, with the least cohesive sample having the worst process performance in this example. Indeed, free-flowing powders have their own problems, such segregation and flooding.

4.2 Capsule filling

Three batches of a pharmaceutical blend were used to fill sachets for oral dosing. The three batches flowed differently from the filling shoe into the sachet, resulting in significant weight variation in the sachets. Sample A exhibited good performance and Sample B (with a wider particle size distribution) was classed as average. Sample C had the same particle size distribution as Sample A, but performed very poorly in the process, suggesting that particle size alone did not dictate performance. Samples where characterised using the FT4 Powder Rheometer.

Sample A generated the lowest BFE, requiring less energy to move the blade through the powder (fig 3). Sample C generated the highest. The low BFE is indicative of a powder that is able to flow more freely under the forced flow conditions present in a shoe feeder operation. Sample C generated the highest Pressure Drop across the Powder Bed (lowest permeability), indicating a greatest resistance to the passage of air. Low permeability means that air entrained in the powder when it enters the sachet cannot escape, which can lead to high weight variation across a manufacturing run. Sample A was less permeable than Sample B, suggesting that an extreme value for this parameter may contribute to sub-optimal performance in this specific unit operation. Other filling applications may be more sensitive to variations in the permeability and therefore necessitate different acceptance criteria[7].

Differences in the permeability may also explain the aeration behaviour. Sample A generated the highest Aerated Energy (AE). Highly permeable powders allow air to traverse the bed readily with little influence on its packing structure. In contrast, Sample C generated the lowest AE, likely as a result of its lower permeability. Furthermore, powders that are sensitive to aeration may also be more prone to segregation and dusting, which can both have a detrimental impact on content uniformity.

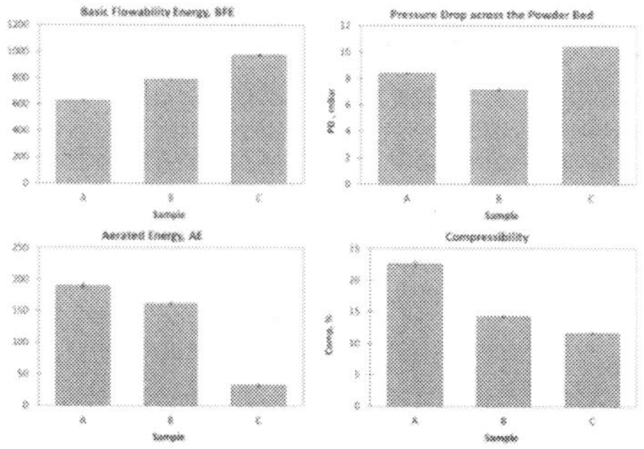

Fig. 3. Dynamic and Bulk results for pharmaceutical powders A to C used in a sachet filling application.

Sample A generated the highest Compressibility value, likely a result of the formation of agglomerates. however, this does not appear to have a negative impact on the filling operation, probably due to the low stress conditions imposed in the process demonstrating the importance of identifying test methodologies which represent process conditions.

The results show that powders with a low resistance to dynamic flow (low BFE), and a high permeability (high AE, lower Pressure Drop) perform best in this operation. In contrast, powders with a higher resistance to flow, and a greater sensitivity to aeration, perform poorly. This enables a design space to be defined, against which new formulations can be assessed to predict performance.

5 Conclusion

Both cases studies demonstrate that a single parameter may not be sufficient to fully rationalise the process performance, and that a multivariate approach is required. The case studies also demonstrate that the key properties are specific to each process; while BFE was an important parameter in understanding the process performance in both applications, the acceptance criteria were different. Expansion of the sample set would enable the upper and lower limits for each parameter to be more accurately defined, providing a more robust design window for each process or unit operation.

The spray coating application required a higher BFE, as degree of cohesion was necessary to ensure that the powders adhered to each other and formed a uniform coating, whereas higher cohesion was detrimental to the sachet filling application as it resulted in poor flow in the shoe feeder operation. A powder which works in the spray coating application is likely to perform badly in the sachet filling application. As such, it is necessary to understand the specific requirements/constraints for a given unit operation, furthermore it cannot be automatically assumed that the most free-flowing powder/granular system is the most suitable

Powder flowability is not an inherent material property, but is more about the ability of powder to flow in a desired manner in a specific piece of equipment. Successful processing demands that powders and processes are well-matched and it is not uncommon for the same powder to perform well in one process but poorly in another. This means that several characterisation methodologies are required, the results from which can be correlated with process ranking to produce a design space of parameters that correspond to acceptable process behaviour. Rather than relying on single number characterisation to describe behaviour across all processes, the multivariate approach of powder rheology simulates a range of unit operations, allowing for the direct investigation of a powder's response to various process and environmental conditions.

References

1. K. Brockbank, *Novel Approaches to the Assessment of Pharmaceutical Powder Flow Behaviour*, in *School of Life Sciences*. University of Bradford: Bradford, UK, (2011).
2. C. Ghoroi, L. Gurumurthy, D. J. Mcdaniel, L. J. Jallo, R. N. Davé, Powder Technology. **236**, 63(2013).
3. D. Am Ende, M. Birch, S. J. Brenek, M. T. Maloney, Organic Process Research & Development. **17**, 1345(2013).
4. Z. Yan, S. A. Wilkinson, E. H. Turnbull, E. H. Stitt, M. M, *Parametric Evalution for Powder Flowability using a Freeman Rheometer: A Discrete Element Method Study*, in *IV International Conference on Particle-based Methods – Fundamentals and Applications*. CIMNE: Barcelona, Spain, (2015).
5. R. Freeman, Powder Technology. **174**, 25(2007).
6. T. Freeman, K. Brockbank, B. Armstrong, Procedia Engineering. **102**, 35(2015).
7. T. Freeman, J. Clayton, B. Armstrong, X. Fu, *The Influence of Powder Flow Properties on the Filling of Dies and Capsules*, in *Particulate Systems Analysis 2011*: Edinburgh, UK, (2011).

Flexible fiber in interaction with a dense granular flow close to the jamming transition

Nicolas Algarra[1,*], *Marguerite* Leang[1], *Arnaud* Lazarus[2], *Damien* Vandembroucq[1], and *Evelyne* Kolb[1,**]

[1] PMMH, CNRS UMR 7636, ESPCI Paris, UPMC Univ Paris 06 and Univ Paris 07, 10 rue Vauquelin, 75005 Paris, France
[2] Sorbonne Universités, UPMC Univ Paris 06, CNRS, UMR 7190, Institut Jean Le Rond d'Alembert, 75005, Paris, France

Abstract. We propose a new fluid/structure interaction in the unusual case of a dense granular medium flowing against an elastic fiber acting as a flexible intruder. We study experimentally the reconfiguration and the forces exerted on the flexible fiber produced by the flow at a constant and low velocity of a two-dimensional disordered packing of grains close but below the jamming transition.

1 Introduction

The study of fluid-structure interaction has been traditionally developed in the context of mechanics and engineering for optimizing structures against wind or water flow. Recently there have been fundamental works dealing with the drag reduction and reconfiguration of flexible objects into streamlined shapes in controlled water flows [1], wind tunnels [2] or into 2D soap film flows [3]. This field has been extended to bio-physical domains, like the reconfiguration of plants and leaves under wind or of algae in marine flows [4]. In animals, the flexible nature of the body, wings or fins interacting with the external medium could be of particular interest for optimizing the thrust efficiency, like for the swimming of fishes or eels, the flying of birds and insects [5], the crawling of snakes and the diving of sandfish lizards [6] or the digging of worms in mud sediments [7]. Even at a smaller scale, the swimming of micro-organisms with flagella or even the beating of ciliary flagella [8] take advantage of the flexibility.

In the present work, we propose a new fluid/structure interaction in the unusual case of a 2D dense granular medium flowing against an elastic fiber acting as a flexible intruder: One tip of the fiber is clamped while the other is free and initially faces the granular flow, as if the fiber was penetrating the granular medium. The mechanical resistance produced by the cluster of grains contacting the fiber tends to bend or buckle the flexible intruder, which in turn modifies the granular flow.

More precisely, we study experimentally the reconfiguration and the forces exerted on an elastic mylar beam (the fiber) produced by the flow at a constant and low velocity of a two-dimensional, bidisperse, dense and disordered packing of metallic cylinders (the granular fluid). Imposing the fiber geometry like its length or thickness sets the critical buckling force the fiber is able to resist if it was not supported by lateral grains, while increasing the granular packing fraction might laterally consolidate the fiber and prevent it from buckling. But on the other side, the approach to jamming transition by increasing the granular packing fraction will be characterized by a dramatically increasing size of the cluster of connected grains forming a solid block acting against the fiber, which might promote the fiber deflection. Thus, we investigated the fiber deflexion as well as the forces experienced by the fiber and compared them with theoretical predictions from *Elastica* for particular models of loadings along the fiber.

2 Experimental setup

The granular medium is an horizontal assembly of around 6800 rigid cylinders of external diameters d_1=4 mm (in number N_1) and d_2=5 mm (in number $N_2 = \frac{4}{7}N_1$) and height h=3 mm. The grains lie on a rectangular glass plate closed by rigid walls (cell width $W \approx 54\ d_2$ and maximum cell length $D \approx 94\ d_2$) and are moved at a constant velocity V_0=5/6 mm/s corresponding to a quasi-static regime for granular flows. The initial packing fraction of grains ϕ_0 can be adjusted by slightly shifting one of the wall along the cell length, such that ϕ_0 is in the range 80%-83%, close to but below the 2D (bidimensional) jamming transition at 83.56% observed in our similar preceding experiments performed on rigid intruders [9],[10]. The fiber is a mylar beam of rectangular cross-section with a thickness $t = 350\,\mu$m, a height h=3 mm and a length L which is fixed in a range of 2 to 10 diameters of large grains d_2. One tip of the fiber is clamped and connected to a 6-axis force and torque sensor mounted slightly above the horizontal plate. The other fiber extremity is free. The fiber is initially aligned along the y-axis such that the free extremity faces the granular flow, as if the fiber was penetrating the granular medium. A CCD camera of 1600×1200 pixels

*e-mail: nicolas.algarra@espci.fr
**e-mail: evelyne.kolb@upmc.fr

placed above the set-up records images at a frequency of 2 Hz, corresponding to a plate displacement (along the y-axis) of $U_0 = 5/12$ mm $= d_2/12$ between 2 successive images (see Fig. 1). Ten similar experiments are performed for each set of parameters L and ϕ_0.

3 Results

From the series of images obtained for fiber lengths $L \geq 2.5cm = 5d_2$, we observe that the fiber experiences three successive regimes during its penetration relative to the granular medium. In a first regime (I), the fiber slightly fluctuates around its initial straight position (see Fig.1a for example for $L = 3$ cm, $\phi_0 = 80.85\%$). After a traveling distance ℓ that depends on the initial organization of grains, the fiber irremediably inclines itself towards one side, either right or left, and enters a regime of irreversible bending (Regime II) (Fig.1b), i.e. the fiber can not come back to its straight position. We also observe the formation of a cavity empty of grains behind the fiber. The area of the cavity increases while the fiber continues to bend further. Once the main part of the fiber is placed perpendicular to the arrival flux of grains, we define a third regime (III) called avalanche regime: successive events of loading occur when the fiber bends further (Fig.1c), followed by rapid unloading events where the fiber comes back (Fig.1d). These unloadings are associated with notable grains recirculations from upstream to downstream around the free end of the fiber. For fiber lengths $L < 2.5$ cm and $\phi_0 < 82.0\%$, only the first regime of small fluctuation (I) is observed and in average, the fiber stays straight over the whole penetration distance. The threshold for straight penetration is obtained for slightly smaller lengths ($L < 2.0$ cm $= 4d_2$) when $\phi_0 \geq 82.0\%$ approaches the jamming transition.

3.1 Fiber deflection

From each image we extract the fiber shape. Then we compute the lateral deflection (along the x-axis) of the fiber free end δ as a function of the penetration distance ℓ in the granular medium. As shown in Fig.2, the three regimes are clearly visible and the deflections corresponding to the images of Fig.1 are indicated with letters. In regime I, the value of δ oscillates around zero without exceeding one grain size. Then during regime II, the deflection increases with some fluctuations but without returning to zero. The limit between regime I and II is defined in the following way: On Fig. 2, we determine the location of the last passage of the deflection δ through zero, corresponding to the last time when the fiber was straight. Then we identify the maximum δ_1 of the amplitude of deflection before this point. This is done for each experiment performed with the same control parameters L and ϕ_0, which allows determining a maximum deflection in the regime I equals to $\delta_{1Max} = 0.29d_2 = 0.48L$ for $L = 3$ cm and $\phi_0 = 80.85\%$ in Fig. 2. Note that the typical deflection at the transition I-II is always of the order of one grain size. The deflection dramatically increases during regime II, which corresponds to the fiber bending aside. Then the value of δ saturates in Fig.2 for L=3 cm when the regime III is obtained. For longer lengths, there is no clear plateau as the fiber continues to bend further with a pronounced hook shape. But we observed large fluctuations corresponding to numerous loading-unloading events characteristic of the last regime. In all cases, the limit between regime II and III is defined as the average over the 10 experiments of the maximum deflection. It is obtained at $\delta = 0.90\,L$ for the Fig.2.

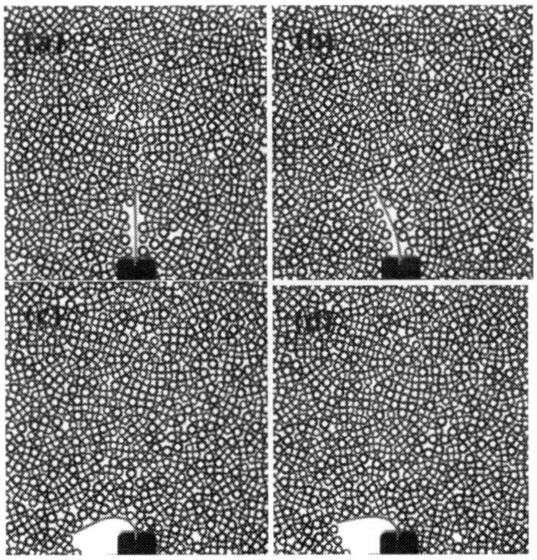

Figure 1. Images of the fiber of length $L = 3$ *cm* and thickness $t = 350\,\mu m$ inside a 2D granular material of initial packing fraction $\phi_0 = 80.85\%$. (a) After a relative displacement of $\ell = 2\,d_2$ (Regime I). (b) $\ell = 10.16\,d_2$ (Regime II). (c) Image just before an unloading event at $\ell = 34.83\,d_2$ (Regime III) (d) Following image just after the unloading (avalanche) event at $\ell = 35\,d_2$.

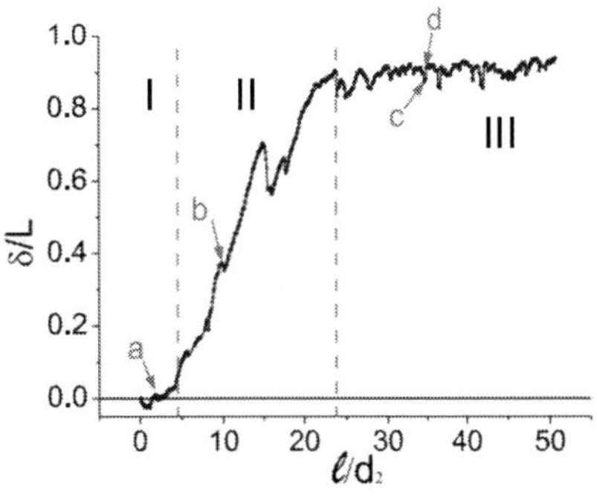

Figure 2. Lateral deflection of the fiber free end δ normalized by the fiber length L as a function of the distance of penetration ℓ (normalized by the grain diameter d_2) for the experiment shown in Fig.1 ($L = 3$ *cm*, $\phi_0 = 80.85\%$ and t=350 μm). The numbers I, II, III correspond to the three successive regimes, and the letters a, b, c et d to the corresponding labelled images of Fig.1.

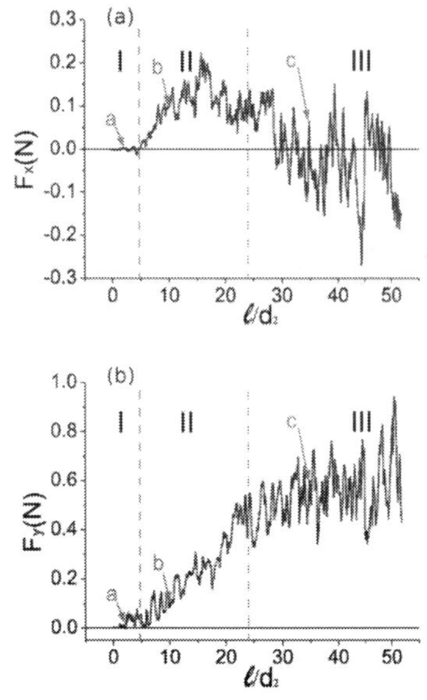

Figure 3. Evolution with the penetration distance ℓ of the components of forces detected by the force sensor at the fiber clamping point for the individual experiment of Fig.1 with $L = 3$ *cm*, $\phi_0 = 80.85\%$ et $t = 350$ μm. The zones I, II, III correspond to the ones defined in Fig.2 as well as the images a, b et c. (a) Cross-flow force F_x. (b) Drag force F_y.

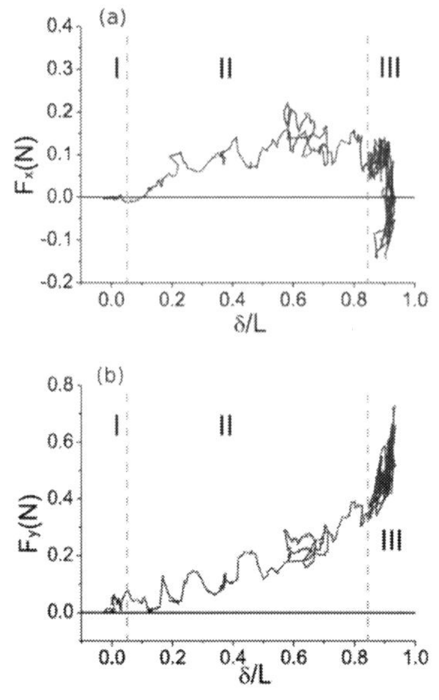

Figure 4. Forces measured at the fiber clamping point as a function of the lateral deflection δ obtained from the images and normalized by the fiber length $L = 3$ *cm* for $\phi_0 = 80.85\%$.(a) Cross-flow force F_x. (b) Drag force F_y.

3.2 Drag force

With the sensor located at the clamping point, we also record the resultant of forces acting on the fiber: the drag force F_y opposing the grains flux as well as the cross-flow force F_x. Both components are plotted as a function of the travelling distance ℓ (Fig. 3) and the limits of the three different regimes obtained in Fig. 2 are also reported on this graph. In the first regime (I), the cross-flow force F_x is near zero, as the fiber stays almost straight. The drag force F_y is small but always positive, as the fiber faces the arrival of grains. In the second regime (II), the fiber bends irreversibly on one side, leading to a rather regular increase of the drag force F_y with the penetration distance ℓ till a value of around 0.5 N. At the same time, we observe a positive component of the cross-flow force F_x related to the deflection (Fig. 1b). In the third regime (III), the drag force F_y seems to saturate although highly fluctuating, while the cross-flow force F_x becomes negative. This is due to the hook shape of the fiber in this regime: The grains in contact with the fiber near its free extremity push the fiber in the reverse direction (towards the right side) compared with the grains near the clamping point that push the fiber to the left side.

As the evolution of forces and deflections seems to behave in a similar way, we plot them together in Fig. 4. Again the distinction between the three regimes is clearly visible. Of particular interest are the loop shapes in regime III, that correspond to temporal cycles of loading and unloading.

4 Comparison with simulations

For the 10 experiments performed with the same control parameters L and ϕ_0, we obtained the forces on the fiber as well as the successive fiber shapes and corresponding lateral deflections. When plotting the cross-flow force (Fig. 5a) and drag force (Fig. 5b) as a function of the deflection, we observe a reasonable collapse of all the data provided by the 10 experiments. This suggests a generic mechanism of loading for the bending of the fiber observed in regime II. Moreover the presence of a cavity behind the fiber indicates that the fiber is loaded on its upstream side. Finally, as the grains contact the fiber along their height and as the corresponding friction is low, we propose to compare our experimental results with *Elastica* simulation of a fiber deflected by an homogenous orthogonal loading (Fig. 6). *Elastica* theory provides a way to treat large scale elastic deflections of beams [11].

Numerical results are reported in Fig. 5. They are obtained without any adjustable parameter and only depend on the mechanical and geometrical properties of the fiber (Young modulus $E = 3.8$ GPa and Poisson ratio $\nu = 0.4$). This simple model of loading gives a reasonable agreement for capturing the envelopes of Fig.5 for the experiments for $L = 3$ cm. This loading is associated to the cluster of grains that piles up on the upstream side of the fiber. For better comparison with the simulation, the same force data provided by the 10 similar experiments have been averaged inside windows of δ/L of width 0.05 (Fig.7). The error bars correspond to the standard deviation.

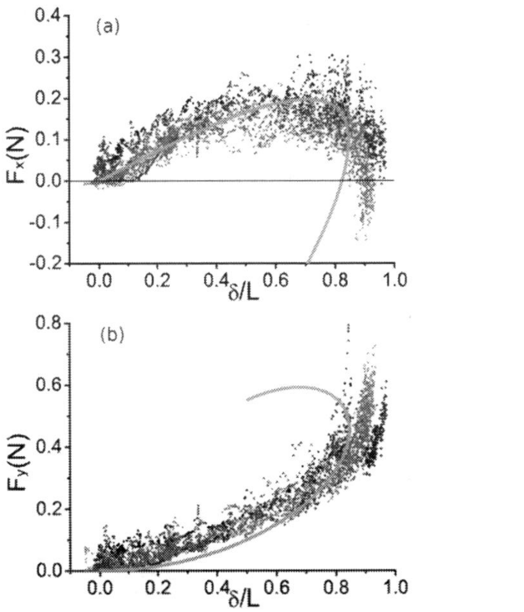

Figure 5. Superposition of the experimental forces for the 10 different experiments performed with the same control parameters ($L = 3\ cm$, $\phi_0 = 80.85\%$ and $t = 350\ \mu m$) vs the lateral fiber deflection normalized by the fiber length L. Points of different colours correspond to the different experiments. The orange line is the simulation for an homogeneous orthogonal loading of the fiber. (a) For the cross-flow force F_x (b) For the drag force F_y.

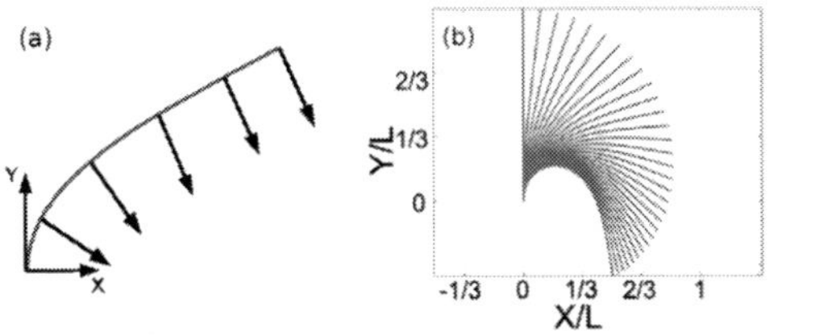

Figure 6. (a) Schematics of the homogeneous orthogonal loading of the fiber for *Elastica* simulations. (b) Superposition of the simulated fiber shapes for increasing forces.

5 Conclusions

We investigated the fluid/structure interaction between a flexible fiber and a dense granular flow close to the jamming transition. Above a critical fiber length, the fiber bends towards one side instead of oscillating around its initial straight position facing the granular flow. The critical length slightly decays with the 2D granular packing fraction in the investigated range of density ($80\% - 83\%$).

In the case when the fiber bends, we showed that a simple model of homogeneous orthogonal loading produced by the grain flow against the fiber reasonably reproduces the experimental measurements of cross-flow and drag forces for intermediate fiber lengths.

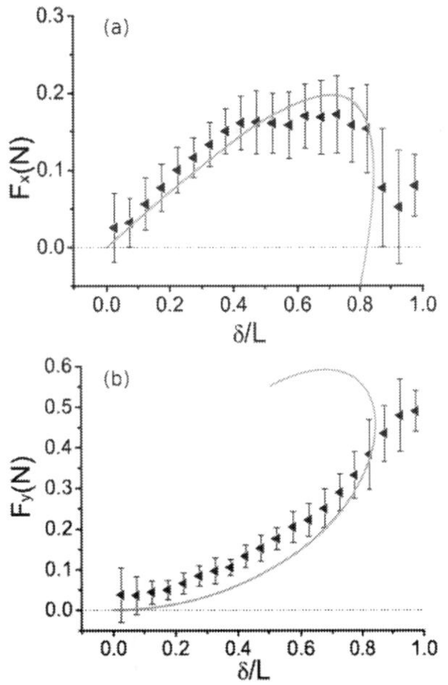

Figure 7. Comparison between the numerical results (lines) obtained for an homogeneous orthogonal loading of the fiber and the mean experimental forces (symbols). (a) Cross-flow force F_x (b) Drag force F_y.

Acknowledgements

Thanks to N. Cheng, K. Luginbuhl, D. Wendell of A. Hosoi's group (MIT-France Seed Fund Project, Flexible Objects in Granular Media), Y. Takehara and F. Postic.

References

[1] L. Schouveiler, A. Boudaoud, J. Fluid Mech. **563**, 71 (2006)
[2] F. Gosselin, E. de Langre, B.A. Machado-Almeida, J. Fluid Mech. **650**, 319 (2010)
[3] S. Alben, M. Shelley, J. Zhang, Nature **420**, 479 (2002)
[4] E. de Langre, Annu. Rev. Fluid Mech. **40**, 141 (2008)
[5] S. Ramananarivo, R. Godoy-Diana, B. Thiria, Proc. Natl. Acad. Sci. **108**, 5964 (2011)
[6] A.E. Hosoi, D.I. Goldman, Ann. Rev. Fluid Mech. **47**, 431 (2015)
[7] K.M. Dorgan, P.A. Jumars, B. Johnson, B.P. Boudreau, E. Landis, Nature **433**, 475 (2005)
[8] E. Lauga, T.R. Powers, Rep. Prog. Phys. **72** (2009)
[9] E. Kolb, P. Cixous, N. Gaudouen, T. Darnige, Phys. Rev. E **87**, 032207 (2013)
[10] E. Kolb, P. Cixous, J.C. Charmet, Granular Matter **16**, 223 (2014)
[11] A. Lazarus, C. Maurini, S. Neukirch, *Stability of discretized nonlinear elastic systems* (Springer Vienna, Vienna, 2015), pp. 1–53

 DOI: 10.1051/epjconf/201714013003

Numerical modelling of powder caking at REV scale by using DEM

Mohamed Guessasma[1,*], *Homayra* Silva Tavares[1], *Zahra* Afrassiabian[2], and *Khashayar* Saleh[2]

[1] *UPJV - Campus Universitaire Pôle Technologique - LTI-EA3899, Saint-Quentin, France*
[2] *Sorbonne Universités - Université de Technologie de Compiègne, TIMR-EA4297, Compiègne, France*

Abstract. This work deals with numerical simulation of powder caking process caused by capillary condensation phenomenon. Caking consists in unwanted agglomeration of powder particles. This process is often irreversible and not easy to predict. To reproduce mechanism involved by caking phenomenon we have used the Discrete Elements Method (DEM). In the present work, we mainly focus on the role of capillary condensation and subsequent liquid bridge formation within a granular medium exposed to fluctuations of ambient relative humidity. Such bridges cause an attractive force between particles, leading to the formation of a cake with intrinsic physicochemical and mechanical properties. By considering a Representative Elementary Volume (REV), the DEM is then performed by means of a MULTICOR-3D software tacking into account the properties of the cake (degree of saturation) in order to establish relationships between the microscopic parameters and the macroscopic behaviour (tensile strength).

1 Introduction

Powder solids are regularly faced to the problem of caking during their periods of storage and/or transport. The agglomeration of particles that compose a powder depends not only on their water content, the temperature and the applied pressure, but also on the interactions between the solid substance and water molecules present in the atmosphere, i.e., on relative humidity (R_h) at which the product is stored [1]. It is well known that ambient humidity plays an important role in most stages leading to caking: capillary condensation of water at contact points between particles, subsequent dissolution of a solid and formation of saturated solution eventually followed by precipitation of solid during evaporation of water. Almost all types of powder solids are subjected to caking mechanism, including crystalline or amorphous materials, soluble or insoluble, fine or coarse, etc. Caking is involved by attractive forces acting between particles under environmental conditions over time. As mentioned above, caking arises due to interparticle forces depending on powder state and physical aspects. For example, van der Waals forces in the case of dry caking, capillary forces for wet caking, material bridge formation for solid caking and more occasionally, electrostatic, magnetic or interlocking forces are also involved. According to many theoretical studies and experimental works carried out on this issue, capillary forces are at least 1 or 2 orders of magnitude higher than van der Waals forces. In conclusion whatever the main mechanism of caking is, the presence of water either in vapour, liquid, adsorbed or absorbed state could largely accentuate the process. In fact, capillary condensation is the first step of caking for both amorphous and crystalline powders [2–6]. This phenomenon considerably affects the amount of water which appears in a bulk media at a given R_h. The water will then act as a plasticizer in case of amorphous materials or as a solvent for crystalline powders [7]. In the present work, we are interested in simulating capillary force effects due to caking mechanism by using DEM-based approach. Indeed, in view of the nature of the powder, the DEM seems to be the best way for modelling a large number of particles taking into account physical and mechanical aspects. The state of art on DEM modelling of discrete media with capillary interactions reveals many works dealing with numerous applications fields such geomechanics and process engineering [8, 9]. With DEM, the second law of Newton is solved for each particle and the contact interactions between particles are computed by specific constitutive laws. This allows to characterise the behaviour of discrete media at a macroscopic scale starting from interactions at the binary contact scale. By considering a REV, the DEM is then performed using a numerical tool to study the influence of physical parameters on the mechanical behaviour, for example saturation degree and tensile strength. All numerical simulations discussed in this contribution have been carried out with MULTICOR-3D software [10].

2 Theoretical modelling

At relatively higher water activities, $a_w > 90\%$, water molecules become mobile enough to form liquid bridges, subsequently, the adsorbed molecules on the solid surface form a liquidlike a meniscus [2, 3]. Geometric as-

*e-mail: mohamed.guessasma@u-picardie.fr

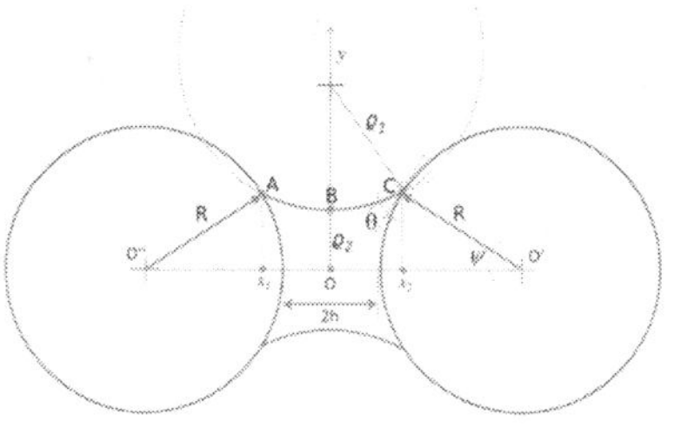

Figure 1. Schematic representation of a pendular liquid [14]

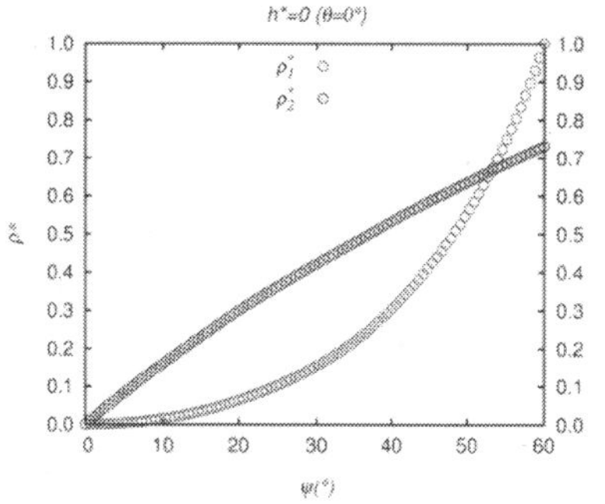

Figure 2. Dimensionless principal radii ρ_1^* and ρ_2^* as function of filling angle ψ

sumptions and analytic models to characterise the capillary bond are given below.

2.1 Fundamental equations

Usually the liquid interface is described by two principal radii, ρ_1 and ρ_2, that define the capillary radius r_c 1.

$$\frac{1}{r_c} = \frac{1}{2}\left(\frac{1}{\rho_1} - \frac{1}{\rho_2}\right) \quad (1)$$

Kelvin model is commonly used to describe the capillary condensation phenomenon at high a_w. Under mechanical equilibrium condition at the liquid interface, the Kelvin equation is given by 2.

$$ln\ a_w = \frac{-2V_l\gamma}{r_c\mathscr{R}T} \quad (2)$$

with γ the vapour-liquid surface tension, $\mathscr{R}$ the universal gas constant, T the temperature and V_l the molar volume of liquid.

Figure 1 gives a schematic representation of liquid meniscus in a pendular bond state when two identical particles ($R_1 = R_2 = R$) at a distant of $2h$ are joined by a liquid bridge. Considering the capillary torque profile [11, 12], the first and second principal radii of curvature are given as follows 3.

$$\begin{aligned} \rho_1 &= \frac{R\,(1 - cos\psi) + h}{cos(\psi + \theta)} \\ \rho_2 &= R\ sin\psi - \rho_1\left[1 - sin(\psi + \theta)\right] \end{aligned} \quad (3)$$

where ψ and θ are the filling and average liquid-solid contact angles, respectively.

The volume of liquid V_l is obtained by integrating the function expressing the arc $\widehat{ABC}$ (figure 1) with respect the variable x. The dimensionless form of liquid volume, $V^* = \frac{3V_l}{4\pi R^3}$, is commonly considered [14].

As evoked in introduction, the fluctuations of ambient R_h led to the formation of liquid bridge within a powder, which causes an attractive force between particles. According to Laplace theory for capillary forces, the total attractive force [13] between the two particles including both the capillary and the surface tension components is given by equation 4.

$$F_{cap} = 2\pi R\gamma\ sin\psi\left[sin(\psi + \theta) + \frac{R}{2}sin\psi\left(\frac{1}{\rho_1} - \frac{1}{\rho_2}\right)\right] \quad (4)$$

2.2 Parametric studies on a binary agglomerate

When a binary agglomerate is subjected to humidification the capillary condensation happens between two limiting values of the filling angle ψ. The variation of principal radii, ρ_1 and ρ_2, as a function of ψ, in the case of a perfectly wetting system ($\theta = 0$), leads to extremal values of ψ given by the intersection of curves. As shown on figure 2, the filling angle ψ, for $2h^* = 0$ ($h^* = \frac{h}{R}$) and $\theta = 0$, reaches extremal values, ψ_{min} and ψ_{max} at 0° and 53°, respectively. Beyond these values, the equivalent capillary radius r_c becomes negative which means that the capillary condensation is brought to an end. Besides, if the relative gap $2h^* > 0.3$, capillary condensation phenomenon stops [14].

The capillary force acting between two particles is very sensitive to the water activity, a_w, due to the liquid amount of the capillary bridge. The influence of the liquid volume is shown through the evolution of the capillary force as function of the separating gap for different volume of water (figure 3). Obviously, the attractive force increases when the relative gap between particles decreases. Furthermore, the attractive force varies slightly when the amount of liquid reaches its maximum (5% of particle volume) and exponentially when the amount of liquid becomes lower. The next study deals with the variation of capillary force with respect to the filling angle for all considered relative gap $2h^*$ (figure 4). The volume of liquid is kept constant. As shown on the figures a lower value of filling angle involves a capillary force close to zero, except when the separating gap $2h^*$ is nil. In such case, the dimensionless attractive force reaches a symbolic value, namely $F^* = \pi$.

3 DEM modelling

In this section we are interested on modelling of capillary forces acting into a Representative Elementary Volume

 DOI: 10.1051/epjconf/201714013003

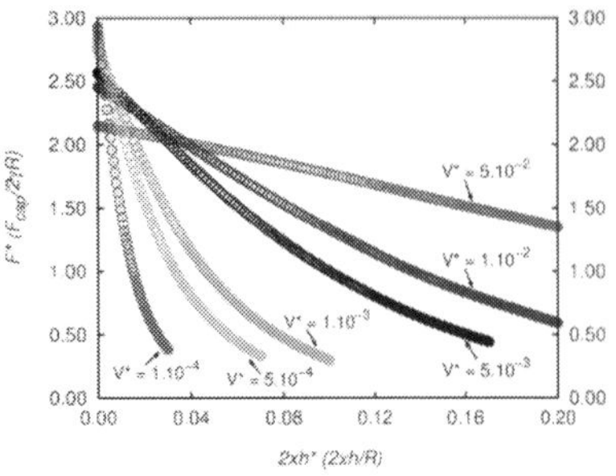

Figure 3. Dimensionless capillary force F^* as function of $2h^*$

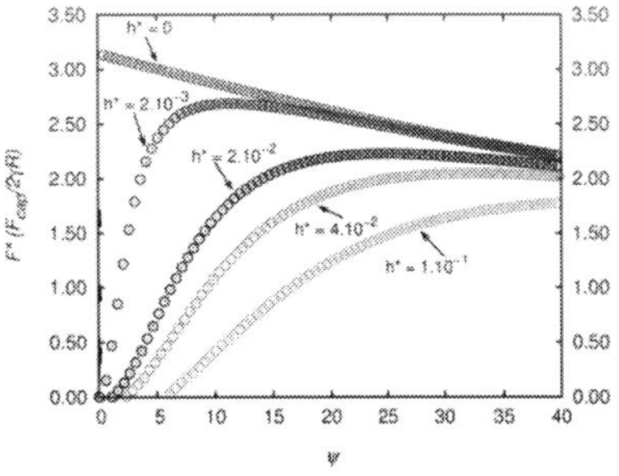

Figure 4. Dimensionless capillary force F^* as function of filling angle ψ

(REV) composed of a set of particles subjected to a given water activity. The numerical modelling is made with Discrete Element Method (DEM) approach by using the explicit formulation [15], which is also called smoothed formulation.

3.1 Contact law

The spherical particles are rigid, a local contact deformation is considered and the interactions (F_n, F_t) are governed by analogies with damped springs $(K_{n,t}, C_{n,t})$ in normal and tangential directions $(\vec{n}, \vec{t})$. Therefore, by using discrete approach, the contact forces between particles, without capillary effect, are described with a contact model depending on elastic force displacement law, Coulomb friction and viscous damping coefficient 5.

$$\begin{cases} F_n = -K_n\delta_n - C_n v_n \\ F_t = -\min(K_t\delta_t + C_t v_t, \mu|F_n|) \times \mathrm{sgn}(v_t) \\ \text{with :} \\ \delta_{n,t}, v_{n,t} : \text{relative displacements and velocities} \\ \delta_t = v_t \times \Delta t, \; \Delta t : \text{time step} \\ C_{n,t} = 2\sqrt{K_{n,t} m^*}, \; m^* : \text{reduced mass} \end{cases} \tag{5}$$

In the presence of water vapour condensation, the particles in the REV are subjected to attractive normal forces due to the formation of capillary bounds. The normal component of the contact force is, then, composed of two parts; repulsive one and attractive one 6. The viscous effect of liquid is neglected in the modelling.

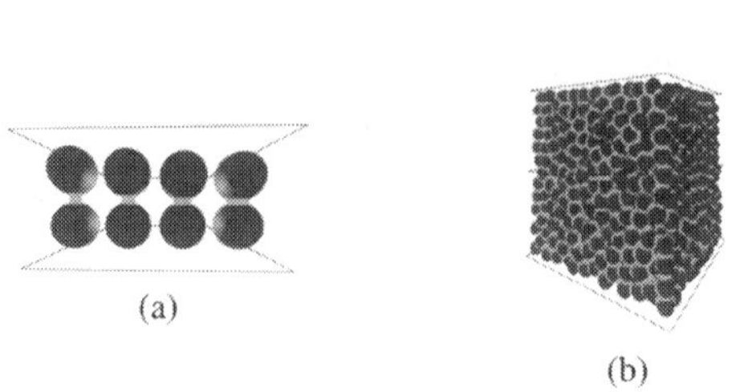

Figure 5. Capillary bridges at: binary contact scale for several values of $2h^*$ (a) - REV scale (b)

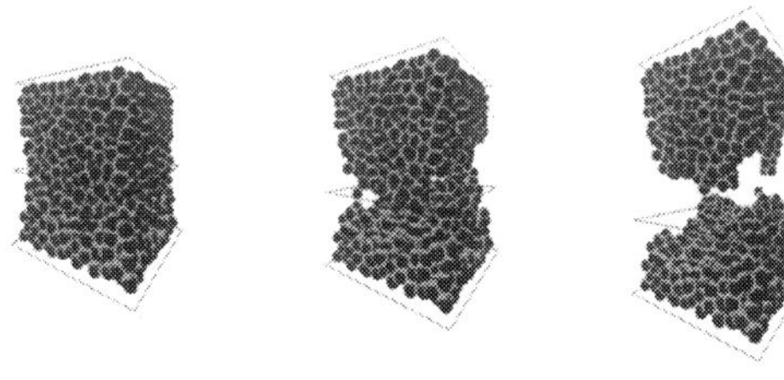

Figure 6. REV under tensile loading

$$F_n = -K_n\delta_n - C_n v_n - F_{cap} \tag{6}$$

On figure 5-a, representing a snapshot from MULTICOR-3D software, are shown four binary contacts for different separating distances $2h^*$ (0.02, 0.08, 0.2, 0.3). At each contact, the capillary bridge shape is drawn with respect their respective to geometric parameters, namely the principal radii ρ_1, ρ_2 and the filling angle ψ.

3.2 Discrete simulations of capillary effects at a REV scale

A REV of 10 000 monosized particles with radius $R = 2.285\ \mu m$ and density $\rho_v = 1500\ kg/m^3$ is generated with MULTICOR-3D software by using a radius expansion algorithm [16] until the porosity of packing reaches a desired value. In the present case, the porosity is fixed to 50%. Saturation degree, υ, of the REV is of the order of 0.64% and considered as a reference value for simulations with multiple saturation degree (υ varies in the interval [0.64% – 12.83%]). The amount of liquid is spread equally at each contact and kept constant during the simulations (figure 5-b). It should be recalled that only the pairs of particles with a relative gap $2h^* \leq 0.3$ are subjected to capillary forces. Thereafter, the attractive forces are computed numerically using an iterative algorithm.

The influence of saturation degree is investigated by considering tensile tests for several values of υ. The tensile strength of the REV is obtained numerically and compared to analytical one, according to Rumpf's model [13], for a random packing of monosized spherical

DOI: 10.1051/epjconf/201714013003

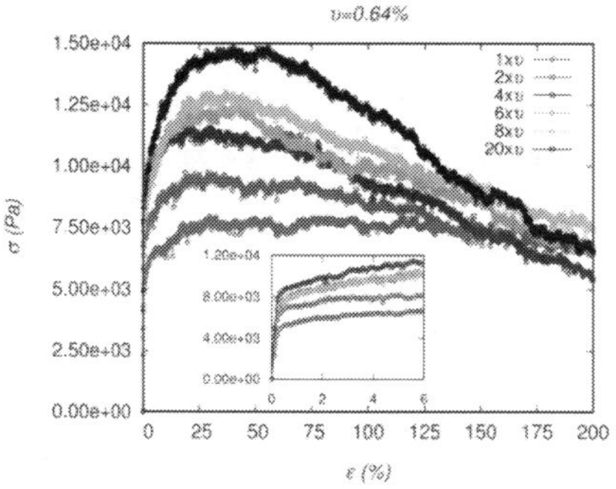

Figure 7. Tensile curves for several saturation degrees

particles and an even distribution of liquid. Regarding the tensile test, the REV is placed between two walls, at the lower and upper sides 6. Before starting the tensile test, the REV should first be in an equilibrium state. In fact, when the saturation degree is affected to the REV the particles start to move slightly as a result of the capillary forces. Once the REV reaches its equilibrium (minimum of kinetic energy) the tensile test is started. The lower side of the REV is fixed during the test while the upper one moves vertically with a constant velocity ($10^{-4}m.s^{-1}$). The particles of both lower and upper sides are subjected to the same kinematic conditions as the fixed and mobile walls. On figure 7 are plotted the tensile curves for six values of the saturation degree. The tensile stress, σ, is estimated from the resultant of capillary force acting on the cross section at the middle of the REV. The relative displacement, ϵ, is computed with respect to the particle radius. As expected, the higher the saturation degree of the REV, the higher the tensile strength is. This is due to both increasing of liquid amount at binary contact scale and coordination number of particles. Comparison made with analytical model exhibits discrepancies about 30% with numerical predictions. This is mainly due to the statistical parameters of the Rumpf's model, namely the coordination number and the mean capillary force. Nevertheless, the DEM remains more suitable to model mechanical behaviour of powder agglomerate starting from binary contact scale.

Acknowledgements

This study has been carried out under project MOTA-MORPHE (2015–2018). Thanks to the "Région des Hauts-de-France" and the "FEDER" for their financial support.

References

[1] Langlet M., Benali M., Pezron I., Saleh K., Guigon P., Metlas-Komunjer L., Chemical Engineering Science **86**, 78-86 (2013)

[2] Fisher L.R., Israelachvili J.N., Nature, **277**, 548-549 (1979)

[3] Fisher L.R., Israelachvili J.N., Colloids and surfaces, **3**, 303-319 (1981)

[4] Lian G., Thornton C., Adams M.J., Journal of Colloid and Interface Science, **161**, 138-147, 1993

[5] Feeny J., Fitzpatrick J.J., Particulate Science and Technology **29**,397-406 (2011)

[6] Gagneux G., Millet O., Transp Porous Med **105**, 117–139 (2014)

[7] Palzer S., Chemical Engineering Science **14**, 3959-3968 (2005)

[8] Delenne J.-Y., Soulié F., El Youssoufi M.S., Radjai F., Powder Technology **208**, 308-311 (2011)

[9] Kim B.S., Park S.W., Kato S., Computers and Geotechnics **42**, 52-61 (2012)

[10] Sanni I., Bellenger E., Fortin J., Coorevits P., Journal of Computational and Applied Mathematics **234**, 1161-1171, 2010

[11] Mehrotra V.P. ,Sastry K.V.S., Powder Technology **25**, 203-214 (1980)

[12] Tselishchev Y.G., Val'Tsifer V.A., colloid Journal *65*, 418-423, (2003)

[13] Rumpf H., *Agglomeration* (John Wiley, New York, 1962), 379-418

[14] Afrassiabian Z., Leturia M., BenaliM., Guessasma M., Khashayar S., Chemical Engineering Research and Design **110**, 245-254 (2016)

[15] Cundall P.A., Strack O.D.L., Geotechnique **29**, 47–65, (1979)

[16] Lubachevsky B. D., Stillinger F. H., Journal of Statistical Physics **60**, 561–583 (1990)

Dense granular Flows: a conceptual design of high-power neutron source

Lei Yang[1,*], *Sheng* Zhang[1], *Ping* Lin[1], *Guanghui* Yang[1,2], *Yuan* Tian[1], *Jiang-feng* Wan[1,2]

[1]Institute of Modern Physics, Chinese Academy of Sciences, 730000, Nanchang Road 509, Chengguan, Lanzhou, China
[2]Univeristy of Chinese Academy of Sciences, 100049, Yuquan Road 19 A, Shijingshan, Beijing, China

Abstract. A high-power neutron source system is very useful for multifunctional applications, such as material facilities for advanced nuclear power, space radiation studies, radiography and tomography. Here the idea of inclined dense granular flow is utilized and developed in a new conceptual design of a compact high-power target to produce a high-energy and high-flux neutron irradiation (the flux is up to 10^{15} n/cm^2/s or even 10^{16}). Comparing to the traditional solid and liquid heavy metal targets, this design has advantages in material choice, fluid stability, heat removal, etc. In this paper the natures of the granular flows in an inclined chute are investigated and preliminary experimental and numerical results are reported. Then the feasibility of this design is discussed.

1 Background

In many areas of physics, chemistry, biology, materials, and nuclear engineering, to study the structure and functionality of materials, it is very valuable for a very intense source of neutrons. In complement to X-ray or synchrotron sources, neutron source has several unique advantages [1]. Spallation neutron source is an accelerator-driven facility to produce high intensity neutrons based on spallation reactions. After the successful first spallation neutron sources, several high-power neutron sources are built, constructed and planned, such as SNS [2], J-PARC [3] and ESS [4].

For engineering of spallation target, there are many design options such as plate targets, rod targets, heavy metal targets or rotation targets [5]. Recently, the new concept of gravity-driven Dense Granular Target (DGT) was proposed [6]. This new concept will contribute to the design of MW spallation target for CIADS project [7] (see details in Figure 1).

Compared to currently wide used targets, the attractions for DGT include:
1) The flowing behaviors of grains in DGT are analogous to the fluids [8] and the deposited high power will be removed off-line.
2) In DGT, heat shocks induced by proton beam are dispersed since granular materials usually show an excellent buffering performance [9].
3) Grains in DGT are renewed continuously off line. There are benefits in selecting proper materials to reduce the corrosion, chemistry-toxicity and radio-toxicity.
4) There are mature technologies in industry capable for conveying, filtering and cooling grains.

Fig. 1. Engineering application of granular target was put into practice during the target research of ADS project, which is supported by the "Strategic Priority Research Program" of the Chinese Academy of Sciences. Top-left: A principle testing facility coupled with 1.2MeV electron beam was operated in 2013. Top-right: Later in 2015, a large scale facility started its construction. Bottom: Supported by these preliminary work, the next stage namely CIADS project will setup a prototype of ADS system applying granular target, in Huizhou, Guangdong.

For the nuclear and space engineering, a strong and rapid testing facility is much needed. Here a conceptual design of a high-energy and high-flux neutron source is

* Corresponding author: lyang@impcas.ac.cn

 DOI: 10.1051/epjconf/201714003009

proposed which is able to offer a neutron flux up to 10^{15} n/cm^2/s. Due to a hopper was used as the container of DGT in [6], the placement of specimens is limited. For the new inclined granular flow target, the flow can be easily controlled and modified [10], which will be shown below. The dynamics and heat transfer in this target are plotted after a series of large-scale simulations.

Table 1. Geometrical and material parameters in this paper.

Quantity(Units)	Symbol	Value
Diameter of particles (mm)	d	5
Width of the chute (d)	W	40
Length of the chute (d)	L	600
Height of the gate (d)	G	25
Inclination angle of the chute (°)	θ_c	25
Length/width of the reservoir (d)	D	40
Elastic modulus (GPa)	E	287
Poisson's ratio	v	0.032
Particle-particle and particle-wall friction coefficients	μ	0.2
Density (kg/m^3)	ρ	1850
Particle-particle and particle-wall coefficient of restitution	ε	0.9
Thermal conductivity of particles (W/m/K)	λ_p	200
Coefficient of linear thermal expansion (10^{-6}/K)	α	11.3
Heat capacity(J/K/kg)	C_p	1825
Hardness (GPa)	H	1.67
Thermal conductivity of gas (W/m/K)	λ_g	0.1517

2 Flow behaviors

DEM (Discrete Element Method), closely related to MD (Molecular Dynamics), is a widely-accepted method for simulating contact dynamics in many-body systems consisting of particles, such as powders, bubbles, grains and colloids. To simulate the granular flows in the proposed target, a DEM code containing heat-transfer module was run on GPUs (Graphic Processing Units) [11-14].

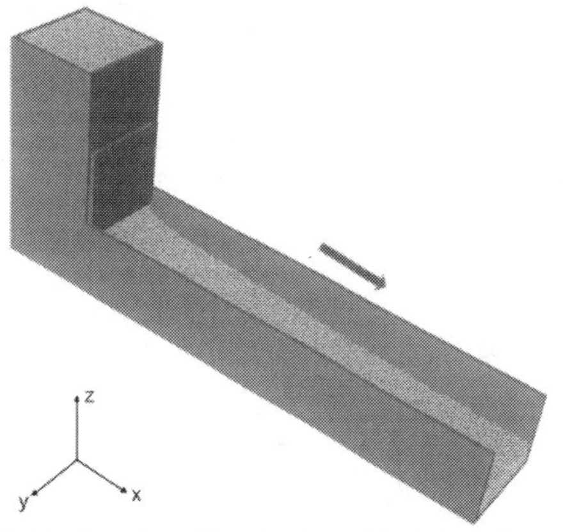

Fig. 2. Configuration of the simple model of chute flows. The red arrow denotes the flowing direction.

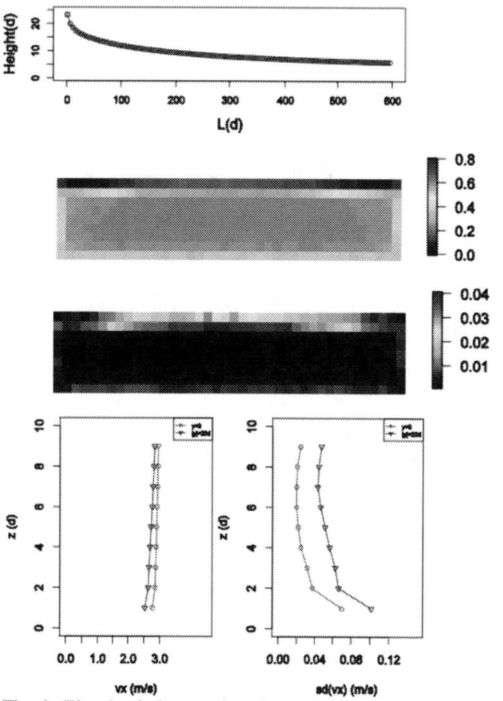

Fig. 3. The simulation results of granular flows in a chute with gate height of $25d$, chute width of $40d$ and inclination angle of 25°. Top: the variation of the shape of the flow. Middle-top: spatial profile of volume fraction in the cross section which is perpendicular to x direction at x = 1500 mm. Middle-bottom: Spatial profile of standard deviation of volume fraction in the cross section which is perpendicular to x direction at x = 1,500 mm. Bottom-left: spatial profile of v_x with different depths at x=1,500 mm. Bottom-right: spatial profile of standard deviation of v_x with different depths at x = 1,500 mm. Red circles represent v_x of the central particles and blue inverse triangles represent v_x of the particles near the sidewalls.

To simplify the model, monosized beryllium particles (diameter d is 5 mm) are simulated. The material of the reservoir and chute is the same as particles and their roughness is assumed as nil. The configuration of the 'the basic' case is schematically shown in Figure 2. The geometrical and material parameters in the basic case are shown in Table 1. The inclination angle is set as 25° to avoid potential stagnant or gaseous state in the downstream flow [15]. Initially the particles are packed in the reservoir and then are discharged through an opening gate. The time-averaged velocity and volume fraction fields along the chute flow are plotted and the stability of the flow is analyzed.

For this design, the stability of the flow is very important. The morphology of the flow is stable over time (see in Figure 3). To evaluate the stability of the flow, temporal profiles of velocity and volume fraction fields are analyzed and their standard error is calculated. Here it is showed that the standard errors of volume fraction are very small except at the free surface. For v_x, the standard error on the bottom is more than the standard error at the free surface and although both are relatively small.

3 Neutronics study

The results of deuterium beams hitting a monolithic beryllium target are calculated by using GMT (GPU-based Monte Carlo Transport program) [16] and checked by Geant4 [17]. The power density distributions along beam direction for different beam energies are plotted in Figure 4. When the beam energy rises from 10 MeV to 80 MeV, the maximum penetration depth grows from 0.1 cm to 2.0 cm and the maximum power density in the target decreases slightly. This profile is used for the following simulations of heat transfer.

A beryllium target bombarded by an 80MeV, 20mA deuterium beam is simulated and the energy spectrum of neutrons is shown in Figure 5. It is showed that this target will provide sufficient high-energy neutrons to replicate the radiation environment inside fission and fusion reactors. The higher energy neutrons can be produced to satisfy the DEMO 14 MeV peak by increasing the energy of the beam. On the other side, the low-energy neutrons can be produced by utilizing moderators.

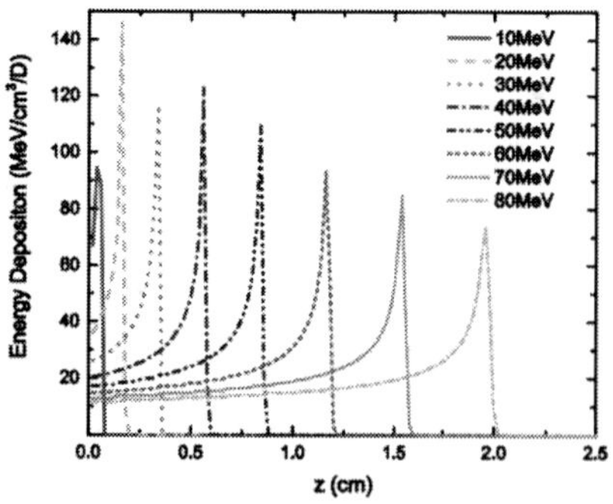

Fig. 4. The profiles of power density distributions along beam direction with different energy deuterium beams.

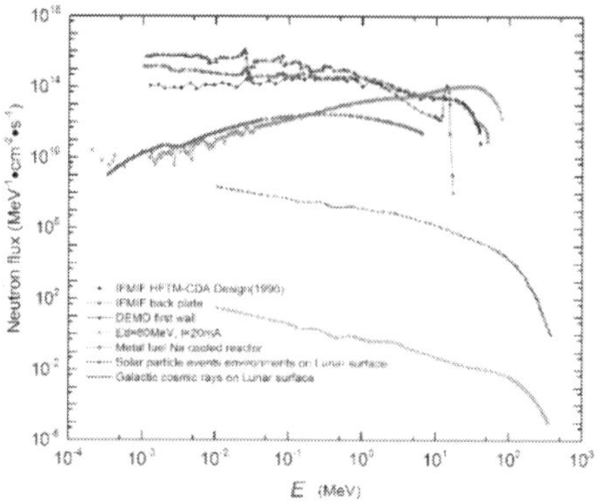

Fig. 5. The energy spectrum of neutrons in different situations. Pink inverse triangles denote the spectrum in the current design.

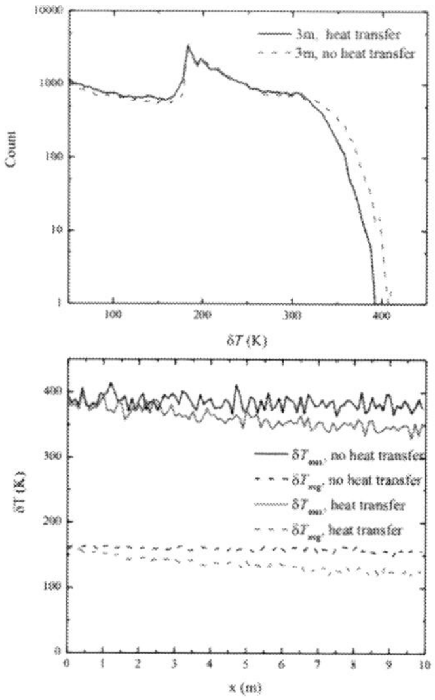

Fig. 6. Top: the distribution of temperature rises. Bottom: the distribution of temperature profile along the x direction.

4 Heat limit

The heat transfer process in chute flow with an 80 MeV, 20 mA deuterium beam (beam spot size is 20×20 mm²) is simulated. It is supposed that the system is operated in a Helium environment (1 atm). The distances between release gate and the shooting point L varies between 2,000 and 3,000 mm. The profile of energy deposition (shown in Figure 4) is used here. In this simulation, four modes of heat transfer are considered: a) thermal conduction through the particle contacts, b) thermal conduction through the stagnant gas film, c) thermal conduction through the contact gaps between two particles, d) radiant heat transfer between surface of particles [18]. The material parameters used for the heat transfer models are listed in Table 1. In this simulation the temperature of particles released from reservoir is set around 290 ℃. Quantitative results are shown in Figure 6 (where L = 3,000 mm). For the case L = 3,000 mm, the ratio of

DOI: 10.1051/epjconf/201714003009

particles whose temperature rise is more than 400℃ is small (less than 5%). The average temperature rise of particles is less than 180℃ and the maximum temperature rise is no more than 400℃. For comparison, a system without heat transfer is also simulated, where each particle is assumed to be thermally insulated. In this system, the maximum temperature rise is no more than 420℃ (can be used as a maximum estimate of maximum temperature rise).

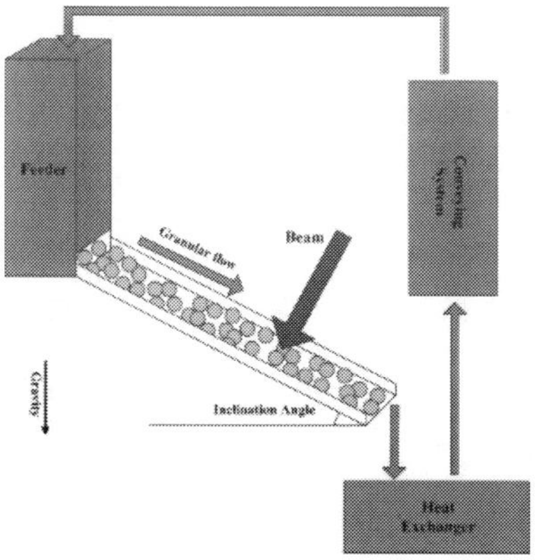

Fig. 7. Schematic diagram of the gravity-driven chute flow target system.

5 Conceptual design and Discussions

A conceptual design of the gravity-driven chute flow target is schematically shown in Figure 7. As a feeder, there is a reservoir or a hopper filled with heavy metal particles. Downstream there is an inclined chute and the particles discharging from the feeder will flow down the chute by gravity. The beam hits the flowing particles at a fixed position. At the end of the chute, the particles will flow into a heat exchanger where they are cooled. After filtering, the particles are re-injected into the feeder by a conveying system. The stability of the flows is illustrated above and can be conveniently controlled by adjusting the inclination angles of chute. The chute can be polished to avoid the surface waves mentioned above. The numerical study shows that the neutron flux can be up to 5×10^{15} n/cm^2/s and the temperature rise in the material is acceptable.

This work is supported by the National Magnetic Confinement Fusion Science Program of China (Grant No. 2014GB104002).

References

1. Wei, J., et al., China Spallation Neutron Source - an overview of application prospects. Chinese Physics C, 2009. 33(11): p. 1033-1042.
2. Gabriel, T.A., J.R. Haines, and T.J. McManamy, Overview of the Spallation Neutron Source (SNS) with emphasis on target systems. Journal of Nuclear Materials, 2003. 318: p. 1-13.
3. Oyama, Y., J-PARC and new era of science. Nuclear Instruments & Methods in Physics Research Section a-Accelerators Spectrometers Detectors and Associated Equipment, 2006. 562(2): p. 548-552.
4. Lindroos, M., et al., The European Spallation Source. Nuclear Instruments & Methods in Physics Research Section B-Beam Interactions with Materials and Atoms, 2011. 269(24): p. 3258-3260.
5. Bauer, G.S., Overview on spallation target design concepts and related materials issues. Journal of Nuclear Materials, 2010. 398(1-3): p. 19-27.
6. Yang, L. and W. Zhan, New concept for ADS spallation target: Gravity-driven dense granular flow target. Science China Technological Sciences, 2015: p. 1-7.
7. Zhan, W. ADS Programme and Key Technology R&D in China. in Proc. 2013 International Particle Accelerator Conference, Shanghai. 2013.
8. Massoudi, M. and T.X. Phuoc, Conduction and dissipation in the shearing flow of granular materials modeled as non-Newtonian fluids. Powder Technology, 2007. 175(3): p. 146-162.
9. Grujicic, M., et al., Shock-Wave Attenuation and Energy-Dissipation Potential of Granular Materials. Journal of Materials Engineering and Performance, 2012. 21(2): p. 167-179.
10. Silbert, L.E., et al., Granular flow down an inclined plane: Bagnold scaling and rheology. Physical Review E, 2001. 64(5).
11. Tian, Y., et al. A heterogeneous CPU-GPU implementation for discrete elements simulation with multiple GPUs. in Awareness Science and Technology and Ubi-Media Computing (iCAST-UMEDIA), 2013 International Joint Conference on. 2013. IEEE.
12. Qi, J., et al., GPU-accelerated DEM implementation with CUDA. International Journal of Computer Science and Engineering, Inderscience, 2015. 11(3): p. 330-337.
13. Zhang, S., et al., Investigating the influence of wall frictions on hopper flows. Granular Matter, 2014. 16(6): p. 857-866.
14. Lin, P., et al., Numerical study of free-fall arches in hopper flows. Physica a-Statistical Mechanics and Its Applications, 2015. 417: p. 29-40.
15. Borzsonyi, T. and R.E. Ecke, Rapid granular flows on a rough incline: Phase diagram, gas transition, and effects of air drag. Physical Review E, 2006. 74(6).
16. Cai, H., et al., Code development and target station design study for Chinese Accelerator-Driven System Project. Nuclear Science and Engineering, accept.
17. Agostinelli, S., et al., GEANT4-a simulation toolkit. Nuclear Instruments & Methods in Physics Research Section a-Accelerators Spectrometers Detectors and Associated Equipment, 2003. 506(3): p. 250-303.
18. Vargas-Escobar, W.L., Discrete modelling of heat conduction in granular media. 2002, University of Pittsburgh.

Relaxation processes after instantaneous shear rate reversal in a dense granular flow

Eduardo Rojas[1], *Rodrigo* Soto[2,*], *Eric* Clement[3], *Martin* Trulsson[3], and *Bruno* Andreotti[3]

[1]*Departamento de Ingeniería Mecánica, Facultad de Ingeniería, Universidad de Antofagasta, Antofagasta, Chile.*
[2]*Departamento de Física, Facultad de Ciencias Físicas y Matemáticas, Universidad de Chile, Santiago, Chile.*
[3]*École Supérieure de Physique et de Chimie Industrielles (ESPCI), Paris, France.*

Abstract. A numerical model of granular material at different packing fractions and under steady shear is submitted to a sudden shear reversal. We monitor consequences of the strong density and shear rate spatio-temporal heterogeneities, on the constitutive relations. We show that the dynamics can be decomposed into two subsequent regimes spanning a time scale inversely proportional to the shear rate. In the first regime, a non-local constitutive relation is satisfied, hence accounting for the spatial heterogeneity of the fluidity parameter. However at later time, we find that the local $\mu(I)$ constitutive relation can be applied, in spite of the fact that the fluidity parameter remains heterogeneous.

1 Introduction

The understanding of dense granular flows received an important impulse by the discovery of the local rheology laws [1–3]. Dry granular media in avalanches or planar Couette geometries are standard benchmarks where the normal pressure P and shear stress τ can be measured and analysed separately. It was found that in hard grain systems, the normalized shear stress $\mu = \frac{\tau}{P}$ –which can be associated to an effective friction coefficient– depends only on the dimensionless inertial number $I = \frac{\dot{\gamma} d}{\sqrt{P/\rho_p}}$, for a flow of particles of diameter d and mass density ρ_p, and shear rate $\dot{\gamma}$. The relation $\mu(I)$, shown in Fig. 3(a), constitutes a local rheology relation [1–3]. At steady state, it is associated with a second relation: $\phi(I)$ expressing the decrease of packing fraction with the inertial number. The limit $I \to \pm 0$ yields the maximum packing fraction value ϕ^* depending on the microscopic granular details, particularly on the friction coefficients. It is important to keep in mind that the only true limit, where a solid phase would emerge as a real rigidity transition, is obtained for zero friction and is called the jamming point, sometimes called the random close packing limit [4]. Therefore, as we will see later, nothing prevents the observation, as a transient, of local packing fractions larger than ϕ^* as long as they remain smaller than the random close packing limit.

The local rheology picture was extended to non-planar geometries by assuming that stress and shear-rate tensors remain parallel with a proportionally constant associated to μ, with I computed from the invariants of the tensor [5]. Computer simulations of a rotating drum show that those tensors are indeed not parallel but still their invariants are related via the $\mu(I)$ rheology [6]. Simulations of a flow on an inclined plane have shown also a small but finite angular deviation when an objective analysis of the tensors is performed [7]. For non-homogeneous fields, non-local effects were observed either close to jamming [1, 8, 9] or in the vicinity of a shear band [10–12]. Different theoretical visions were proposed to model or interpret these features via fluidity parameters [9, 13–16].

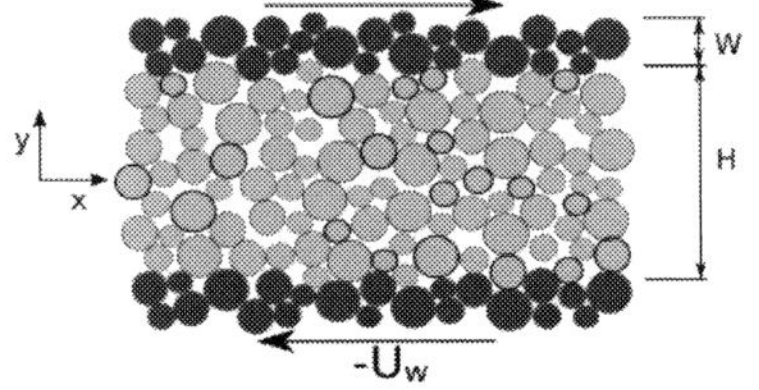

Figure 1. Set-up used in the simulations. The base width is $L = 45d$, the height is $H = 18d$, and the thickness of the confining walls is $W = 1.5d$.

In the application of the rheological laws to model flow regimes and hydrodynamic fields it is assumed that they would retain their form under time-dependent conditions and, therefore, could be used as input in the momentum conservation equations. In a recent paper, we analysed the rheological response in time-dependent planar Couette flows using molecular dynamics simulations of a frictional granular packing [17]. It was found that the system evolves in two distinct phases. In the first one the non-local rheology describes accurately the system evolution, with similar values for the non-local correction as compared to Refs. [16, 18]. In the second phase the tempo-

*e-mail: rsoto@dfi.uchile.cl

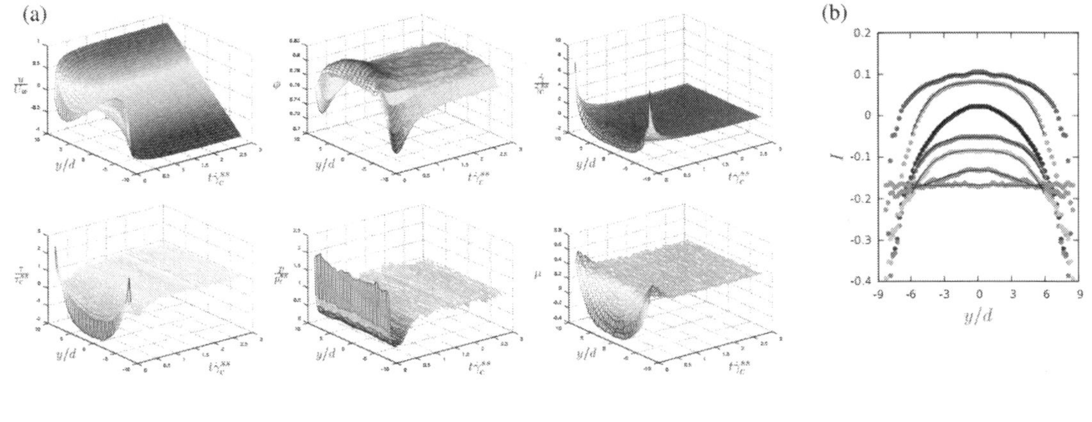

Figure 2. (a) Temporal evolution after the instantaneous velocity reversion from an initial (negative) steady shear, for $\phi_G = 0.790$ of the following fields: from left to right and top to bottom, horizontal velocity $u(x,t)$, volume fraction $\phi(y,t)$, shear rate $\dot{\gamma}(y,t)$, shear stress $\tau(y,t)$, pressure $p(y,t)$, and friction coefficient $\mu(y,t)$. Velocities are normalised to U_w and other field to the central value at the steady state. Colors, giving the field value, are used to enhance readability. (b) Vertical profile $I(y)$ for $\phi_G = 0.767$, at different instants of time (from top to bottom at $t\dot{\gamma}_c^{ss} = 0^+, 0.05, 0.25, 0.5, 1.0, 2.5, 5.0$). Solid lines are fits in the central region, used to compute $\kappa = d^2\nabla^2 I/I$.

ral evolution is characterised by the local rheology, even though the fluidity parameter does not vanish. Here, we present additional elements of this study.

2 Simulation method

The numerical set-up consists in a two dimensional dry granular medium confined between two rough walls in absence of gravity (see Fig. 1). The medium is made of a polydisperse mixture of circular grains with uniform mass density and diameters that are uniformly distributed in the range $[0.5d, 1.5d]$, where d is the average diameter and m_p is the average grain mass. The walls are made of similar particles, which will be forced to move at imposed velocities $\pm U_w$ to produce a planar Couette flow. The shear rate is $\dot{\gamma}_w = 2U_w/H$, H being the height of the flowing region. The system is periodic in the horizontal x direction, while it is limited by the walls in the vertical one. The horizontal size is $L = 45d$ and the system height is $H = 18d$. Finally, the wall thickness is $W = 1.5d$, such that all particles with their center in this region move together with the wall.

The granular model is the one used in Refs. [19] and [17]. The grains interact with visco-elastic forces in the normal and tangential directions, added to a tangential Coulomb friction, where the spring constant for the normal force is much larger than the maximal pressure reached in the simulations. Hence we remain in the hard particle limit for which the $\mu(I)$ rheology was obtained. Therefore, the microscopic dynamics is characterized by a unique time scale given by the shear rate, used to fix time units.

The initial configurations are prepared as follow. $N = 800$ particles are placed in the system between the walls. A fixed vertical pressure P_e is imposed on the walls to compress the system and a shearing motion is added until the system relaxes to the steady state. Depending on the imposed pressure, different global volume fractions ϕ^G between the walls are obtained in the range $0.767 < \phi^G < 0.805$, which is the only control parameter in our simulations. Having obtained the initial configurations, the wall separation is fixed and the following simulations are done *at constant volume*. Planar Couette flows are simulated with an imposed wall velocity U_w, until a stationary state is achieved. The stationary state is slightly inhomogeneous in the vertical direction (see Fig. 2) as a result of boundary effects, which have been shown to increase when decreasing the inertial number [20, 21]. Hence, we measure in the central region of the box (averaged over one third of the system height) the steady state volume fraction ϕ_c^{ss}, shear rate $\dot{\gamma}_c^{ss}$, pressure p_c^{ss}, and shear stress τ_c^{ss}. From these, the friction coefficient $\mu_c^{ss}$and inertial number I_c^{ss} in the centre are obtained. The friction coefficient is fitted to the standard expression $\mu_c^{ss} = \mu_1 + \frac{\mu_2-\mu_1}{1+I_0/I_c^{ss}}$ [1], where the fit parameters are $\mu_1 = 0.277$, $\mu_2 = 0.85$ and $I_0 = 0.364$, the same as those obtained in Ref. [22]. Also, the volume fraction and the inertial number in the central region are related in the usual form $\phi_c^{ss} = \phi^* - mI_c^{ss}$, with $\phi^* = 0.813$, as the extrapolated values for vanishing I_c^{ss}, and $m = 0.28$.

Transient regimes are obtained by instantly reversing the imposed velocities on the walls $\pm U_w$, where the sign alternates periodically. The alternation period is large allowing the system to reach a steady state before it is perturbed by the sudden change in shear direction. The process is repeated and averaged considering 1000 cycles. Here, we will analyse the relaxation process as a function of time, measured from the shear rate reversion.

3 System evolution

The relaxation process is strongly inhomogeneous temporally and spatially. Figure 2 displays the temporal evolution of all relevant fields, averaged over the horizontal direction x and the cycles, with time measure from the reversion instant: horizontal velocity $u(x,t)$, volume fraction $\phi(y,t)$, shear rate $\dot{\gamma}(y,t)$, shear stress $\tau(y,t)$, pres-

 DOI: 10.1051/epjconf/201714003010

sure $p(y,t)$, friction coefficient $\mu(y,t)$, and inertial number $I(y,t)$. The shear rate and inertial number are obtained by fitting locally the velocity profile to low degree polynomials, to later derive them.

Throughout all the simulations we performed, we noticed that the system evolution can be characterized by two subsequent time phases, which can be simply visualized by inspection of the relaxation profiles. First, just after the shear reversion, large shear rates are produced near the walls as u changes from a linear Couette profile to the reverse one. These large shear rates heat up the granular medium, increasing globally the pressure. Associated to the large shear rates, there is an increase of pressure near the walls that push the system towards the centre, generating a transverse mass flux as shown by the vertical velocity v. This flux increases the density at the centre, which reaches a maximum at a characteristic time τ_1. Then, a second phase begins, where the density relaxes down to its steady state value. In this phase, the pressure field is spatially homogeneous and approaches its steady state value. Thus, the density relaxation is not driven any more by pressure gradients but qualitatively, by processes more akin to a diffusional relaxation, associated to non-affine motion. At all times, the corresponding shear rate fields $\dot{\gamma}(y,t)$ are spatially heterogeneous. The characteristic time of this phase is τ_2. Finally, both τ_1 and τ_2 scale with $1/\dot{\gamma}_c^{ss}$, however, the second phase lasts about ten times longer than the first phase. Also, no critical phenomena is observed when approaching ϕ^* [17].

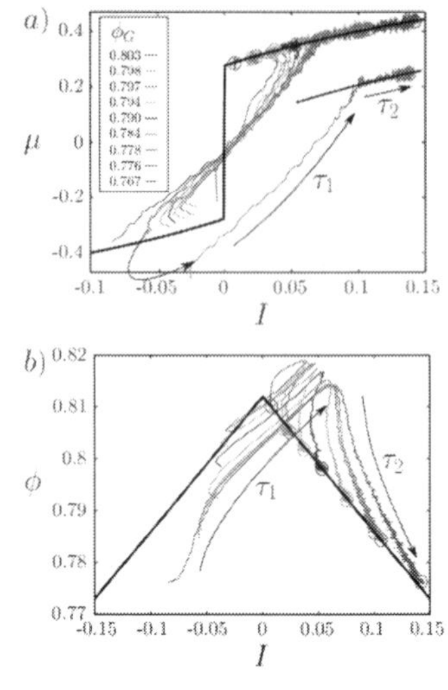

Figure 3. Temporal evolution (indicated by an arrow) of the system in the (a) μ-I, (b) ϕ-I spaces for quantities measured in the central region of the box. Different colours correspond to various global volume fractions ϕ_G. The black solid lines show the steady state relations $\mu(I)$ and $\phi(I)$ and the open circles show the final steady state values. To highlight the evolution in the two phases, for illustration purposes only, in the top figure we offset on both axis the relaxation curve for $\phi_G = 0.776$. First, during τ_1 the system does not follow the local rheology and the volume fraction increases until a maximum value is reached. In the second phase, of duration τ_2, the local rheology $\mu(I)$ is fulfilled but the volume fraction, which decreases monotonically while I increases, is not given by the steady state relation $\phi(I)$.

4 Local and non-local rheology

Now we investigate the relations linking the dynamical friction coefficient and the packing fraction, to the inertial number. For this purpose we measure the volume fraction $\phi_c(t)$, inertial number $I_c(t)$, and friction coefficient $\mu_c(t)$ in the central region of the simulation box (averaged over one third of the system height). Figure 3 presents the system evolution of those variables in the μ-I and the ϕ-I spaces, for different global volume fractions ϕ_G. As a reference, in black, we display the relations $\mu(I)$ [Fig. 3(a)] and $\phi(I)$ [Fig. 3(b)] obtained in the central region for the stationary state. After the velocity reversal, the first phase corresponds to a strong departure from the $\mu(I)$ relation. However in the second phase, the relaxation to steady-state is taking place on the $\mu(I)$ curve. This final relaxation takes place either by increase or decrease of I_c according to the value of the global packing fraction ϕ_G.

For the packing fraction evolution [see on Fig. 3(b)], in no instance of the relaxation process, the local $\phi(I)$ relation is satisfied. Interestingly, the maximal ϕ_c value reached in the centre is always higher than the steady state value ϕ^* but still, I_c remains finite. In all cases, the volume fractions are smaller than the random close packing limit.

Figure 4 presents the temporal evolution of the central values of μ, ϕ, p and τ. It is remarkable the μ reaches the local rheology much before the pressure and the stress reach their steady values. That is, during the second phase p and τ continue to evolve, but they are not independent as they are linked by the relation $\tau = \mu(I)p$. Finally, the volume fraction reaches steady state in a much longer scale.

To test the non-local rheology in the spirit of Bouzid *et al.* [16], we computed the parameter $\kappa = d^2\nabla^2 I/I$ associated with the local fluidity relative to the neighborhood. We obtain the second derivative of I in the centre, by fitting $I(y)$ by a fourth degree even polynomial (see Fig. 2). Figure 5(a) displays for $\phi_G = 0.767$ the time evolution of κ and one sees that it reaches zero after a time τ_1 but thereafter displays a rebound before reaching zero in the second part of the relaxation dynamics. Note that despite the boundary effects, which produce non-uniform shear rates [20, 21], the inertial number is almost uniform at the center in the steady state. Remarkably, when $\Delta\mu = \frac{\mu(I_c(t))-\mu_c(t)}{\mu(I_c(t))}$ is displayed as a function of κ [see Fig. 5(b)], one obtains a good collapse for the different volume fractions on a single curve $\chi(\kappa)$, hence validating the non-local relation as proposed by Bouzid *et al.* [16] of the type $\mu_c = \mu(I_c)[1-\chi(\kappa)]$. However for $t > \tau_1$, the non-local extension to the rheology is not valid. The departure in κ from zero is small but, systematic.

5 Summary and discussion

The relaxation process of a granular system in a planar Couette geometry at fixed volume is studied after the im-

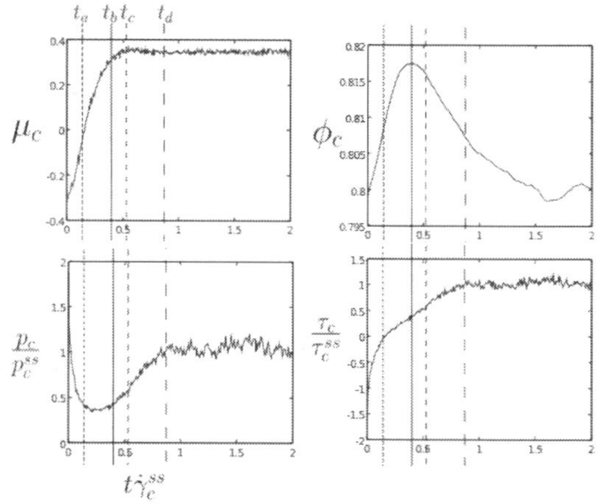

Figure 4. Temporal evolution of the central averaged value of μ, ϕ, p and τ, for $\phi_G = 0.790$. Different instants are represented by vertical lines. t_a: inertial number changes sign from negative to positive; t_b: maximum density is reached at the centre; t_c: μ_c reaches the local rheology value; and t_d: the shear stress reaches the steady value. For $t_c \leq t \leq t_d$, μ_c is not constant but evolves following the local rheology relation $\mu(I)$.

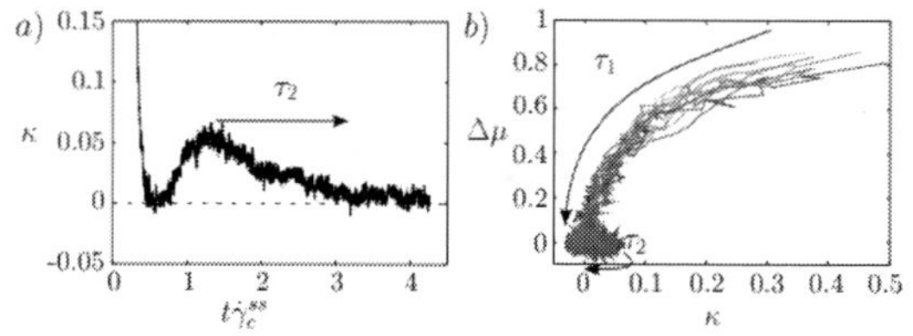

Figure 5. (a) Temporal evolution of the non-local parameter $\kappa = d^2\nabla^2 I/I$ at $\phi_G = 0.767$. Other volume fractions show similar behaviour, with smaller rebounds in κ when approaching ϕ_G^*. All quantities are evaluated at the center of the box. Time is measured from the instant of shear rate reversion. (b) $\Delta\mu$ plotted as a function of κ. Different colours correspond to the various simulated volume fractions (labels shown in Fig. 3).

posed shear rate is instantly reversed. We found that the system evolves in two phases. First, the steady local relation $\mu(I)$ is not satisfied but the measured value approaches exponentially the steady relation. In this phase the non-local rheology is satisfied with similar values for the non-local correction $\chi(\kappa)$ as compared to Refs. [16, 18]. Because of the pressure increase at the boundary there is a mass flux towards the centre of the cell such that density increases but the local constitutive relation $\phi(I)$ is not satisfied. Thereafter, there is a second phase where the system evolves following the local relation $\mu(I)$ found for steady flows. Finally, the time scales for both phases depend on the distance to the maximum packing fraction, but do not show any critical behaviour.

A salient result of this study is that κ is a good predictor for the non-local rheology observed in the first phase. However, in the second phase, where we identified smaller but systematic non-zero values for κ, the non-locality does not manifest itself in the $\mu(I)$ rheology. This observation points on the possibility of a relevant hidden variable accounting for non-locality. This variable would be similar to the fluidity parameter as defined by Bouzid *et al.* [16], but would relax faster to reach zero in the second phase.

This research is supported by Fondecyt Grant No. 1140778 and Ecos-Conicyt Grant C11E04. E.R. acknowledges the support of a Becas CONICYT. B.A. is supported by Institut Universitaire de France. This work is funded by the ANR JamVibe and a CNES research grant

References

[1] GDR MiDi, Eur. Phys. J. E **14**, 341 (2004).
[2] da Cruz F., Emam S., Prochnow M., Roux J. N., and Chevoir F., Phys. Rev. E **72**, 021309 (2005).
[3] Fortere Y. and Pouliquen O., Ann. Rev. of Fluid Mech. **40**, 1 (2008).
[4] O'Hern C. S., Langer S. A., Liu A. J., and Nagel S. R., Phys. Rev. Lett. **88**, 075507 (2002).
[5] Jop P., Forterre Y., and Pouliquen O., Nature **441**, 727 (2006).
[6] Cortet P.-P., Bonamy D., Daviaud F., Dauchot O., Dubrulle B., and Renouf M., EPL **88**, 14001 (2009).
[7] Weinhart T., Hartkamp R., Thornton A. R., and Luding S., Phys. Fluids **25**, 070605 (2013).
[8] Deboeuf S., Lajeunesse E., Dauchot O., and Andreotti B., Phys. Rev. Lett. **97**, 158303 (2006).
[9] Aranson I., Tsimring L.,t Malloggi F., and Clement E., Phys. Rev. E **72**, 031303 (2008).
[10] Komatsu T. S., Inagaki S., Nakagawa N., and Nasuno S., Phys. Rev. Lett. **86**, 1757 (2001).
[11] Nichol K., Zanin A., Bastien R., Wandersman E., and van Hecke M., Phys. Rev. Lett. **104**, 078302 (2010).
[12] Reddy K. A., Forterre Y., and Pouliquen O., Phys. Rev. Lett. **106**, 108301 (2011).
[13] Pouliquen O. and Fortere Y., Phil. Trans. R. Rol A **367**, 5091 (2009).
[14] Kamrin K. and Koval G., Phys. Rev. Lett. **108**, 178301 (2012).
[15] Henann D. L. and Kamrin K., Proc. Natl. Acad. Sci. USA **110**, 6730 (2013).
[16] Bouzid M., Trulsson M., Claudin P., Clement E., and Andreotti B., Phys. Rev. Lett. **111**, 238301 (2013).
[17] Rojas E., Trulsson M., Andreotti B., Clement E. and Soto R., EPL **109**, 64002 (2015).
[18] Bouzid M., Trulsson M., Claudin P., Clement E., and Andreotti B., EPL **109**, 24002 (2015).
[19] Luding S., Granular Matter **10**, 235 (2008).
[20] Miller T., Rognon P., Metzger B., and Einav I., Phys. Rev. Lett. **111**, 058002 (2013).
[21] Rognon P., Miller T., Metzger B., and Einav I., J. Fluid. Mech. **764**,171 (2014).
[22] Trulsson M., Andreotti B., and Claudin P., Phys. Rev. Lett. **109**, 118305 (2012).

Does modifying the particle size distribution of a granular material (i.e., material scalping) alters its shear strength?

Emilien Azéma[1,*], *Sandra* Linero[2,**], *Nicolas* Estrada[3,***], and *Arcesio* Lizcano[4,****]

[1] *LMGC, Université de Montpellier, Montpellier, France*
[2] *SRK Consulting Perth, Australia /University of Newcastle, Newcastle, Australia*
[3] *Departamento de Ingeniería Civil y Ambiental, Universidad de Los Andes, Bogota, Colombia*
[4] *SRK Consulting Vancouver, Canada*

Abstract. By means of two dimensional contact dynamics simulations, we analyzed the effect of the particle size distribution (PSD) on the shear strength of granular materials composed of un-breakable disks. We modelled PSDs with a normalized beta function, which allows for building S-shaped gradation curves, such as those that typically occur in soils. We systematically controlled and varied the size span and the shape of the PSD, and found that the shear strength is independent both characteristics. This implies that PSD modification procedures such as material scalping (i.e., removing the smallest and/or largest particles in the sample) should not affect significantly the shear strength of the material composed of unbreakable discs. In order to explore the origins of the invariance of the shear strength with PSD, we analyzed the connectivity, force transmission, and friction mobilization in terms of anisotropies, finding that the constant shear strength is due to a subtle compensation of anisotropies.

1 Introduction

The study of the mechanical behavior of mine wastes containing particles of metric order is a challenging task because commercial testing devices can only accommodate samples composed of particles a few centimeters in dimension. The construction of larger devices for laboratory testing is not economical. To overcome the equipment size limitations, testing of coarse material is conducted on samples prepared by convenient modification of the particle size distribution (PSD). Parallel scaling (i.e., reducing the particles' size while keeping the PSD shape constant) or material scalping (i.e., removing the smallest and/or largest particles) are some of the technics used to prepare samples for laboratory testing [1]. The tested sample may thus differ from the prototype in both the size of the particles and the form of the PSD.

Sample preparation procedures such as parallel scaling or material scalping are widely used in engineering practice. However, when and how these procedures affect the material's mechanical response (e.g., the shear strength) remains an open question. Important elements to be considered for tackling this question are: (i) the relation between particle size and fragmentation probability; (ii) the link between PSD, packing fraction, and connectivity; and (iii) the evolution of particle's shape and interaction's forces with particle size.

*e-mail: emilien.azema@umontpellier.fr
**e-mail: slinero@srk.com.au
***e-mail: n.estrada22@uniandes.edu.co
****e-mail: alizcano@srk.com

A possible strategy for studying this problem theoretically is to analyze simplified systems (e.g., two-dimensional systems, comprised of circular and unbreakable grains, with simple interaction laws such as Coulomb friction). This allows for singling out the effects of the PSD while leaving out the "noise" of other important factors. Discrete element methods are well suited for this type of simulations. Recently, this strategy has been used successfully to analyze the effect of the PSD on packing fraction and shear strength (see Refs. [2–6]), focusing on the comparison between bi-disperse and continuous PSDs, as well as on uniform distributions by volume fraction. Works exploring the effect of the PSD shape by means of discrete element simulations are rare (see Refs. [7, 8]).

The main objective of this work was to analyze the effect of the particle size distribution (PSD) on the shear strength of granular materials composed of unbreakable disks. We build different PSDs using a normalized beta function, which allows for building S-shaped gradation curves, such as those that typically occur in soils. We varied two parameters independently over a broad parametric space, controlling both the size span and the shape of the PSD. We found that the shear strength is invariant with both the size span and the shape of the PSD. From a micromechanical viewpoint, we show that the shear strength remains constant for different PSDs thanks to a subtle compensation of anisotropies, specifically, contact and branch length anisotropies. The results support several hypotheses proposed recently and extend their validity to a larger family of PSDs.

2 Numerical procedures

The size span of the PSDs is defined by:

$$s = (d_{max} - d_{min})/(d_{max} + d_{min}), \quad (1)$$

where d_{max} and d_{min} are the minimum and maximum particle diameters, respectively; $s = 0$ corresponds to a monodisperse packing whereas $s = 1$ corresponds to an infinitely large polydispersity.

A β-function was use to model the shape of the PSDs. It is defined by:

$$\beta(x) = \frac{1}{B(a,b)} \int_0^x t^{a-1}(1-t)^{b-1} dt, \quad (2)$$

where, $a > 0$ and $b > 0$ are the parameters of the distribution and $B(a,b) = \Gamma(a)\Gamma/\Gamma(a+b)$, where Γ is the Gamma function. The PSD can then be described using the cumulate distribution function $h(d)$ (also called grading curve in industrial contexts), given by

$$h(d) = \beta[x = d_r(d); a, b], \quad (3)$$

where $d_r(d) = (d - d_{min})/(d_{max} - d_{min})$ is the the reduced diameter. As shown by Voivret et al. in Ref. [9], this model allows for controlling both the size span and the shape of the PSD, including reference distributions such as monodisperse, power law, and S-shaped PSD curves. In our simulations, d_{max} was 0.15 m and the size span was varied in the range $s \in [0.2, 0.4, 0.6, 0.8, 0.9]$.

In order to generate a set of particle diameters following a given distribution, the range $d_{max} - d_{min}$ was divided into N_c classes. Within each class, a uniform distribution by volume (area) of grains was considered. By construction, the number of particles increased with b, thus N_c and the total number of particles N_p were chosen in order to: (i) obtain a good fit with the theoretical curve and (ii) obtain a "reasonable" sample size, in terms of the number of particles. In the simulations presented here, the parameter a was fixed to 1 and the parameter b varied from 1 (a uniform distribution by volume fraction) to 5 (S-shaped gradation with a pronounced curvature). The total number of particles increased from 10,000 for $b = 1$ to $\sim 70,000$ for $b = 5$. Figure 1 shows the theoretical PSDs, i.e., Eq.3, as well as the PSD of the simulated samples (points), for several combinations of s and b.

The simulations used the contact dynamics (CD) method [10], which assumes perfectly rigid particles interacting through mutual exclusion and Coulomb friction. We used the code LMGC90 which is a multipurpose software developed in the University of Montpellier, capable of modeling a collection of deformable or undeformable particles of various shapes by different algorithms. In each simulation, the particles were first compacted by isotropic compression inside a box. The gravity was set to 0. The friction coefficients between particles was set to $\mu = 0.4$. The friction coefficient between the particles and the walls

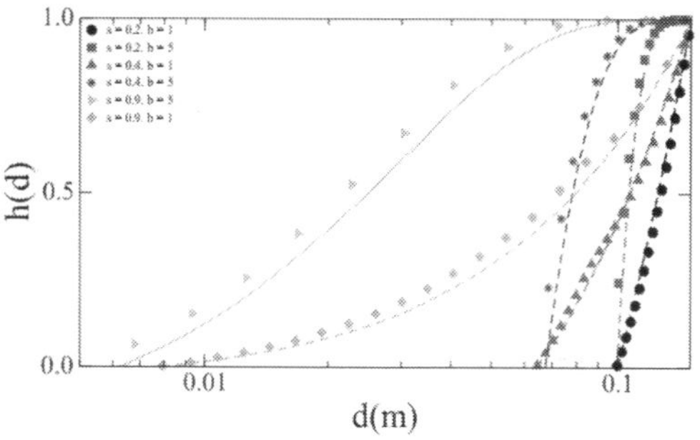

Figure 1. Theoretical and simulated particle size distributions for several combinations of parameters s and b.

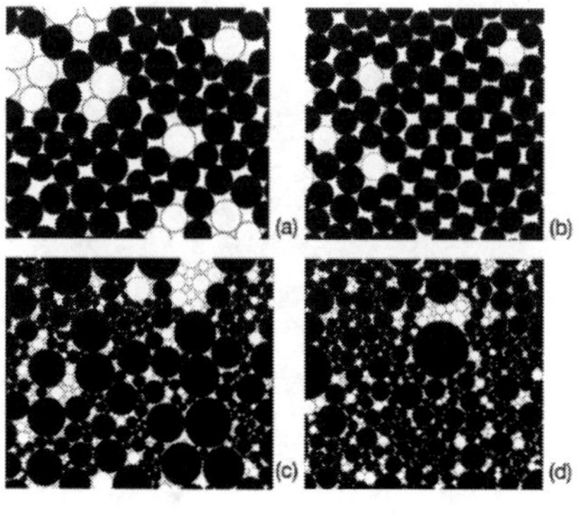

Figure 2. Zoom for ($s = 0.2, b = 1$)(a), ($s = 0.2, b = 5$)(b), ($s = 0.9, b = 1$)(c) and ($s = 0.9, b = 5$)(d) at the end of the isotropic compression. Floating particles (i.e., particles with one or no contacts) are shown in white.

was set to $\mu_w = 0$, in order to obtain homogeneous and isotropic particles packing. Figure 2 shows zooms of the packings obtained for four different values of s and b at the end of isotropic compaction. Small crystalized regions can be observed for small values of s, which happens because most of the particles have the same diameter. In contrast, disorder increases with s.

The samples were then sheared in a biaxial configuration. A downward velocity v_y was imposed on the upper wall while keeping a constant confining stress on the lateral walls. The strain rate v_y/H was low, so that the test can be considered as quasi-static.

3 Macroscopic shear strength

The stress tensor σ can be evaluated from the simulation data as an average over all the contacts of the dyadic product of the contact force $\mathbf{f}^c$ and the branch vector ℓ^c: $\sigma_{\alpha\beta} = n_c\langle f^c_\alpha \ell^c_\beta \rangle_c$ [10], where n_c is the number density of contacts c and the average $\langle ... \rangle_c$ runs over all contacts in a control volume. In 2D, we define the stress deviator as $q = (\sigma_1 - \sigma_2)/2$ and the mean stress as $p = (\sigma_1 + \sigma_2)/2$, where $\sigma_{1,2}$ are the principal stresses. The stress state is

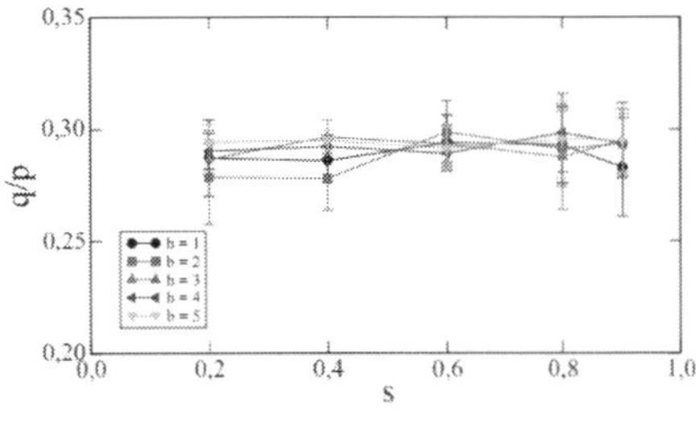

Figure 3. Shear strength $(q/p)^*$ in the steady state as function of s, for all values of b.

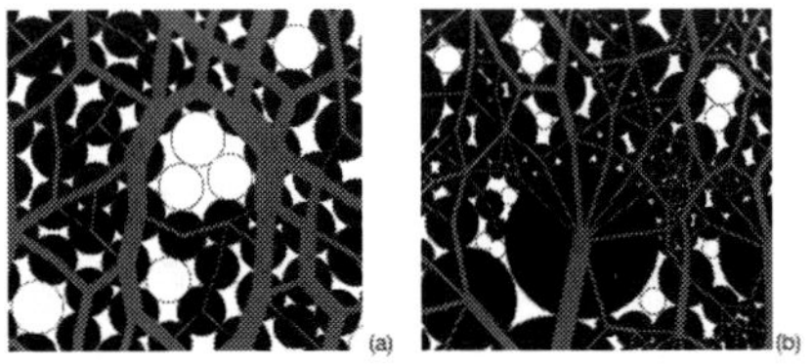

Figure 4. Maps of normal forces in a portion of the sample for ($s = 0.2, b = 5$)(a) and ($s = 0.9, b = 5$)(b) after a large shear strain. Line thickness is proportional to the normal force.

described by the mean stress p and the normalized shear stress q/p [11]. During shear, q/p increased initially to a peak value and then decreased to a nearly constant value $(q/p)^*$, which we used to characterize the material's shear strength in the steady state.

Figure 3 shows the evolution of $(q/p)^*$ as a function of s, for all values of b. It can be seen that, regardless of b, the shear strength is almost independent of s and close to 0.28. This has been previously observed for samples composed of circular [12] and polygonal [6] particles with a uniform distribution by volume fraction. Our results extend this counterintuitive observation to the family of S-shaped PSDs described by Eq. 3.

4 Anisotropies of the contact and force network

The shear strength of dry granular materials is generally attributed to the buildup of an anisotropic structure during shear due to friction between the particles and as a result of steric effects depending on particle shape [13].This is well illustrated in Fig. 4, which shows two maps of normal forces in a portion of the sample for $(s, b) = (0.2, 5)$ and $(s, b) = (0.9, 5)$ after a large shear strain. The contact forces are represented by segments joining the particle centers, whose thickness is proportional to the force magnitude. Visual inspection reveals that the connectivity of the contact network varies drastically with s in terms of both the mean number of contacts per particle (i.e., coordination number) and the contacts' orientation and length.

A way to capture this microstructure is to consider the probability distribution $P(\boldsymbol{n})$ of the contact normals $\boldsymbol{n}$, which is generically nonuniform. In two dimensions, the unit vector $\boldsymbol{n}$ is described by a single angle θ, the orientation of the contact normal. The probability density function $P(\theta)$ of contact normals provides detailed statistical information about the fabric. It is π-periodic in the absence of an intrinsic polarity for $\boldsymbol{n}$. At the lowest order, we have [14, 15]

$$P(\theta) = \{1 + a_c \cos 2(\theta - \theta_\sigma)\}/2\pi, \tag{4}$$

where θ_σ is the major principal stress direction ($\theta_\sigma = \pi/2$), and a_c is the contact anisotropy.

This description of the granular microstructure can be enriched by considering also the angular average of normal branch length and forces:

$$\begin{cases} \langle \ell_n \rangle(\theta) &= \langle \ell \rangle \{1 + a_{\ell n} \cos 2(\theta - \theta_\sigma)\} \\ \langle f_n \rangle(\theta) &= \langle f_n \rangle \{1 + a_{fn} \cos 2(\theta - \theta_\sigma)\} \\ \langle f_t \rangle(\theta) &= \langle f_n \rangle a_{ft} \sin 2(\theta - \theta_\sigma), \end{cases} \tag{5}$$

where $\langle f_n \rangle$ and $\langle \ell \rangle$ are the mean force and the mean branch length, and a_{ln}, a_{fn}, and a_{ft} the branch length, normal force and tangential forces anisotropies.

The anisotropies a_c, $a_{\ell n}$, a_{fn}, and a_{ft} are interesting descriptors of the granular microstructure and force transmission properties, because they underlie the different microscopic origins of shear strength. The general expression of the stress tensor leads to the following simple relation[14, 15]:

$$\frac{q}{p} \simeq \frac{1}{2}(a_c + a_{ln} + a_{fn} + a_{ft}). \tag{6}$$

This expression is very useful, since it allows for describing the shear strength as the addition of several microstructural anisotropies. It is worth noting that the shear strength can be understood as the material's ability to bear anisotropic stress.

Figure 5(a) shows the evolution of a_c and $a_{\ell n}$ as functions of s, for all values of b. For small values of b (i.e., $b = 1$ and 2), it can be seen that a_c decreases with s, indicating that the anisotropy of the contact network is decreasing. This happens because, as s increases, the coordination number also increases, and the particles are surrounded by an increasing number of contacts. On the other side, it can be seen that $a_{\ell n}$ increases with s, indicating that the branch lengths of the contacts that are oriented along the principal stress direction tend to be longer. These results show that the shear strength remains constant due to a compensation of anisotropies: the contact network becomes less anisotropic but the force chains along the principal stress direction tend to be captured by the larger particles. Lastly, for large values of b (i.e., $b = 4$ and 5), both a_c and $a_{\ell n}$ remain approximately constant with s and b.

Figure 5(b) shows the evolution of a_{fn} and a_{ft} as functions of s for all values of b. In contrast to contact and branch length anisotropies, it can be seen that a_{fn} and a_{ft} are approximately constant with s and b. This means that the anisotropy of force chains as well as the degree of friction mobilization are not affected by changing the PSD.

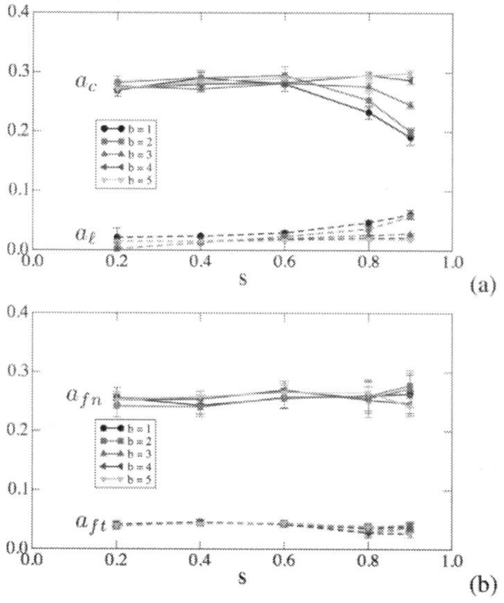

Figure 5. (a) Contact and branch length anisotropies, a_c and $a_{\ell n}$ respectively, and (b) Normal and tangential force anisotropies, a_{fn} and a_{ft} respectively, all averaged in the steady state, as functions of s, for all values of b.

5 Conclusions

We investigated the joint effect of the size span and shape of the particle size distribution (PSD) on the shear strength of packings composed of unbreakable disks, by means of contact dynamics simulations. For this, we independently varied the size span and shape of the distribution across a broad parametric space, and investigated the micromechanical origin of the shear strength for each simulated PSD.

On the first hand, we showed that the shear strength is independent of both the size span and the shape of the PSD. This has already been observed, also by means of discrete element simulations, for samples with uniform distributions by volume fraction [6, 12]. In these cases, only the PSD size span was varied in samples with uniform distributions by volume fraction. Our results extend the validity of this result by showing that the shear strength is invariant, not only with the size span, but also with the shape of the PSD, at least for those shapes that can be described by Eq. 3. These findings are in agreement with those reported recently in [8] for particle distributions described by a power law.

On the second hand, we investigated the origins of the shear strength by analyzing the contact, branch length, and force anisotropies. It was shown that the shear strength remains constant due to a compensation between the contact and branch length anisotropies, while force anisotropy and friction mobilization remain constant. As shown previously [6, 12], these results confirm that, as the size span of the PSD increases, the system is more and more connected and larger forces tend to be captured by larger particles.

The results suggest that the differences observed on the laboratory, in the mechanical response of prototype soil samples and its scaled representations, can be attributed to factors other than the PSD changes introduced with the scaling procedure. It should be noted that this investigation considers that particle characteristics (e.g., shape, strength, and interaction laws) do not depend on particle size. In real materials, several important characteristics, such as grain fragmentation probability, grain shape, and interaction forces, depend notably on particle size. As a consequence, modifying the PSD of a real soil change also the *nature* of the particles that compose the sample, which can provoke appreciable changes in the mechanical response of the granular system.

References

[1] J. Duncan, S. Wright, in *Soil Strength and Slope Stability*, edited by J. Wiley & Sons (2005), ISBN 9780471691631

[2] M. Wackenhut, H. Herrmann, Physical Review E **68**, 041303 (2003)

[3] C. Voivret, Ph.D. thesis, Université de Montpellier (2009)

[4] B. Yohannes, K.M. Hill, Physical Review E **82**, 1 (2010)

[5] D.H. Nguyen, E. Azéma, F. Radjai, P. Sornay, Physical Review E **90**, 012202 (2014)

[6] D.H. Nguyen, E. Azéma, P. Sornay, F. Radjai, Physical Review E **91**, 1 (2015)

[7] P. Stroeven, M. Stroeven, Cement and Concrete Research **29**, 1201 (1999)

[8] N. Estrada, Physical Review E **Accepted** (2016)

[9] C. Voivret, F. Radjaï, J.Y. Delenne, M. El Youssoufi, Physical Review E **76**, 1 (2007)

[10] J.J. Moreau, European Journal of Mechanics - A/Solids **13**, 93 (1994)

[11] J. Mitchell, K. Soga, *Fundamentals of Soil behavior* (John Wiley, 2005), ISBN 978-0-471-46302-3

[12] C. Voivret, F. Radjaï, J.Y. Delenne, M. El Youssoufi, Physical Review Letters **102**, 178001 (2009)

[13] F. Radjai, J.Y. Delenne, E. Azéma, S. Roux, Granular Matter **14**, 259 (2012)

[14] L. Rothenburg, R.J. Bathurst, Geotechniques **39**, 601 (1989)

[15] E. Azéma, N. Estrada, F. Radjaï, Physical Review E **86**, 041301 (2012)

Chemical analysis of extracting transition metal oxides from polymetallic ore by sulphate process

Otgon-Uul *Enkh-Uyanga*[1], Baatar *Munkhtsetseg* [1*], Urtnasan *Urangoo*[2], Enkhtur *Tserendulam*[1] and Davaadorj *Agiimaa*[1]

[1]Chemistry Department, Ulaanbaatar State University, 13343, Ulaanbaatar , Mongolia
[2]Charles Sturt University, Canberra 2602 ACT, Australia

Abstract. In this research work we attempt to improve the purity of polymetallic ores in Mongolia whilst developing practical applications of its refinement processes and this paper presents the results of chemical research of extracting transition metal titanium oxides, ferrous oxide and rare earth oxides from polymetallic ore. Thereby, chemical and mineral analysis of polymetallic ore is carried out basis of responses to the support process at various degrees of water whereas transition metal sulphates solubility differ. As a result of sulphate and resulphurization process we have extracted anatase with 62.5 percent titanium dioxide and brookite mineral with 89.6 percent of titanium dioxide as well as mineral with 83.8 percent of ferrous oxide hematite and rare earth oxides with 57.6 percent of cerium oxide. These oxides are identified under various conditions in the thermal processing. The morphology structure and chemical content compound of the mineral has been verified as a result of the XRF, XRD, SEM-EDX analysis.

Keywords: polymetallic ore, anatase, brookite, hematite, rare earth oxides

Introduction: Mongolia is a country rich in polymetallic ores. There is an increased demand for research of the processing polymetallic ores by removing impurities and forming refined minerals as transition metals have found countless applications in many industries. Therefore, this research attempts to improve the purity of such ores whilst developing practical applications of refinement process.

Polymetallic ores are complex materials containing several metals and metal compositions. As metals and metal compositions in the ore increases, technology to extract these metals become more difficult and complicated [1]. However transition metal minerals such as titanium, iron can be recovered by the methods; leaching minerals from ores using an acidic solution such as sulphuric acid [2]. Titanium dioxide can be produced from high content of titanium such as an ilmenite mineral, which is treated with the sulphuric acid (H_2SO_4), chemical reaction (1,2) follow as:

$$FeTiO_3+H_2SO_4+2(n-1)H_2O=TiOSO_4\cdot nH_2O+FeSO_4+nH_2O \quad (1)$$

$$TiOSO_4+2H_2O=TiO_2\cdot H_2O+H_2SO_4 \quad (2)$$

First of all, we determined its optimal condition to extract transition metal by sulphate method [3] and then titanium dioxide and other minerals have extracted whilst removing related elements during sulphate processes of the ore sample, which has been repeated.

Thereby, chemical and mineral analysis of polymetallic ore is carried out basis of responses to the support process at various degrees of water whereas transition metal sulphates solubility differ.

There are certain solutes, which have independently fair degree of solubility to temperature change. One of the them is cerium(III) sulphate that has lesser degree of solubility at temperature increases. Rare earth elements belonged in cerium group have very good degree of solubility in water at $+2^0C$ temperature [4].

The solubility of ferrous sulphate increases with an increase in sulphuric acid concentration and steeply decreases with an increase in temperature in pure water [5].

Experimental

Materials and methods: As a research raw material is 2306.33 hectare Orkhontuul either “Anand” polymetallic ore in the Orkhontuul soum has selected. This ore is occupied in South East of the Selenge province that is 238 km away from Ulaanbaatar city and 250 km away from center of the province.

On the basic of the laboratory tests, a scheme flow sheet for the preparation of TiO_2, ferrous oxide, and rare earth oxides from polymetallic ore sample has been proposed involving the following main process: milling, sulphuric acid leaching, calcining, hydrochloric acid leaching, water hydrolysis, concentrated sulphuric acid decomposition. The full experimental procedure is shown in Fig.1.

* Corresponding author: bmunkhtsetseg@gmail.com

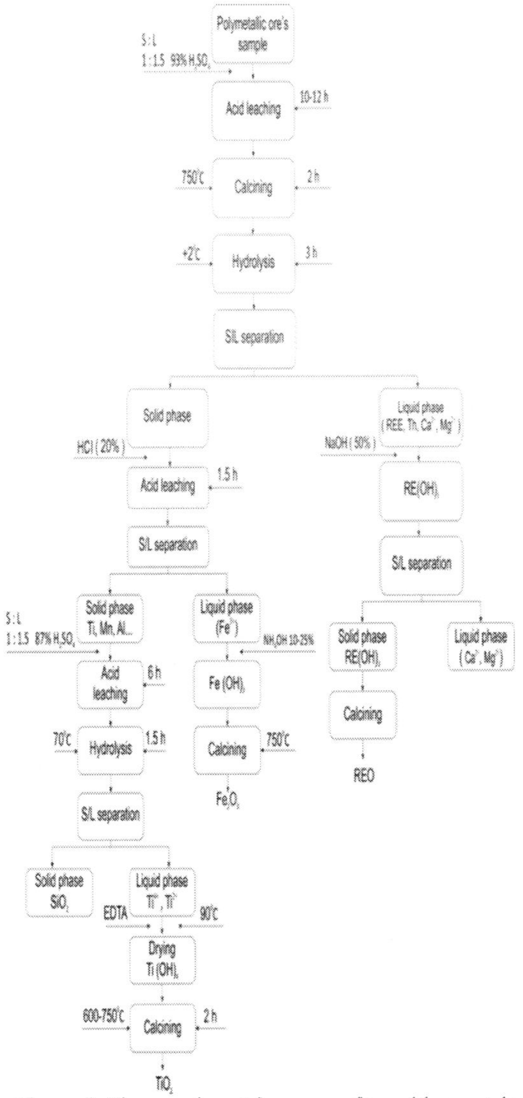

Figure 1. The experimental process of transition metals and rare earth oxides from polymetallic ore

The sample was milled before leaching process. Sulphurization process occurs more efficient when solid phase of the ore and concentrate has been milled in appropriate size. In this research, firstly polymetallic ore was milled by Jaw crusher device to get 3.35mm, and milled by LM in order to get 75 micro-sized polymetallic ore sample.

We determined optimal conditions of sulphuric acid leaching process. In the sulphate process, the milled sample is treated with a 93% concentrated sulphuric acid at 140^{0}C for 10-12 hours, where S:L ratio is 1:1.5. Once sulphate processing completed then it calcined at 750^{0}C degrees. (Part I).

After sulphurization and calcining process we are dissoluted the sample in a water with +2^{0}C temperature and then separated solid and liquid phase. At this step, soluble rare earth sulphates are transferred into liquid phase (REE, Th, Ca^{2+} and Mg^{2+}). Liquid phase contained REE, Th, Ca^{2+} and Mg^{2+} is treated with 50% concentrated sodium hydroxide then solid and liquid phases separeted in order to isolate hydroxides of rare earth metals from other metals. Separated solid phase contained rare earth metals is calcined to extract rare earth oxides. (Part II).

After rare earth metals isolated, the separated solid phase (part I) is treated with a 20% concentrated hydrochloric acid for 1.5 hours, whereas acid leaching has taken place, whereas iron ionic is in liquid phase.

Process to separating solid and liquid phases repeated again, and a separated solid phase is treated with 87% concentrated H_2SO_4 for 6 hours and then hydrolysis process has carried on at 70^{0}C for 1.5 hours in order to remove silica oxide.

At the next stage, 0,1 mol/l EDTA solution was added to titanium sulphate solutions. The solution was heated to 90^{0}C for 90 min. The precipitate was filtered, washed with deionized water and dried in air at room temperature [6]. The powder was calcined at different temperatures 600^{0}C to 750^{0}C 900^{0}C for 2 hours [7].

It is well known that EDTA plays a key role in improving the purity. Therefore, we used EDTA for improving the purity of TiO_2 particles and EDTA was introduced as an additive to precipitate the precipitation of Fe^{3+} during the hydrolysis of titanium sulphate, chemical reaction (3,4) follow as [5]. (Part III).

$$EDTA+4OH^{-}=EDTA^{4-}+4H_2O \quad (3)$$

$$[Fe(H_2O)_6]^{3+}(aq)+EDTA^{4-}(aq)=[Fe(EDTA)]^{-}(aq)+6H_2O \quad (4)$$

Sametime, separated liquid phase containing Fe^{3+} is treated with 10 to 25% concentrated NH_4OH and then calcining process is carried on at 750^{0}C to extract ferrous oxide hematite. (Part IV)

Chemical analysis of the selected polymetallic ore sample and extracted transition metal oxides are accurately studied and computed by inductively coupled plasma mass spectrometer (ICP-MS) and full reflecting X-ray flourescence (XRF) spectrometer, minerals analysis determinate by X-ray diffractometer (Enraf Nonius Delft), and mineral morphology structure by Hitachi TM-1000 SEM-EDX a device for analytical method.

Results and discussion

In the result of the chemical analysis of the polymetallic ore, macro elements such as an iron (Fe-28.12%), aluminium (Al-9.5%), titanium (Ti-0.84%), magnesium (Mg-0.61%), potassium (K-0.57%), calcium (Ca-0.71%) contents are determined by ICP-MS method (Table 1) and quartz SiO_2 of silica group, anorthite $CaAl_2Si_2O_8$, albite $NaAlSi_3O_8$, magnetite Fe_3O_4 minerals detected by X-ray phase analysis of the sample. Figure 2.

Table 1. Some macro elements chemical content of polymetallic ore sample (%)

Elements	Al	Ca	Fe	K	Mg	P	Ti
Polymetallic ore's sample (%)	9.5	0.71	28.12	0.57	0.61	0.03	0.84

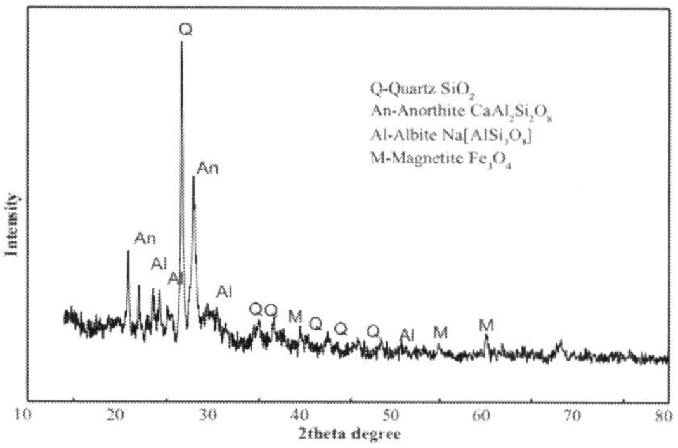

Figure 2. X-ray diffractograms of the polymetallic ore

It is difficult process to carry out if polymetallic ore contains silica group minerals and once sulphate processing takes place where metal minerals in the ore are formed sulphate forms and at sametime as soluble rate increases the process turns out to become much easy. Once sulphate process has carried on polymetallic ore which contained silica group minerals then has calcined at 750^{0}C. Meanwhile the transition metals, iron and titanium minerals formed into sulphate forms and in the calcining process minerals are eliminated into oxides of the these elements which are richened magnitite 8.16%, hematite 5.96%, rutile 4.52% that have transferred into solid phase [2].

In the result (part I) of this analysis minerals such as a quartz SiO_2, anorthite $CaAl_2Si_2O_8$, albite $NaAlSi_3O_8$ magnitite Fe_3O_4, hematite Fe_2O_3, rutile TiO_2, dickite $Al_2[Si_2O_5](OH)_4$ have detected, which shows that sulphate and calcining processing carried out succesfully figure 3.

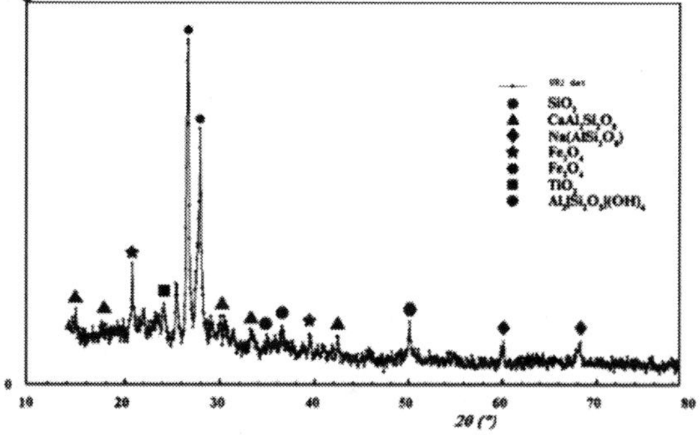

Figure 3. X-ray diffractogram of the sample after sulphate and calcining process of the polymetallic ore

The experimental section part II phase analysis result shown quartz SiO_2-56.27%, anatase TiO_2-28.77%, albite $Na[AlSi_3O8]$-7.23%, anorthite $Eu_{0.92}Al_{1.76}Si_{2.24}O_8$ - 4.84% and aeschynite $Ce(Nb,Ti)_2O_6$-2.87% have been detected by X-ray analysis of rare earth elements oxides extracted at 750^{0}C Figure 4.

Cerium oxide content in complex rare earth element oxide was 57.6% which has been determined by cerimetry.

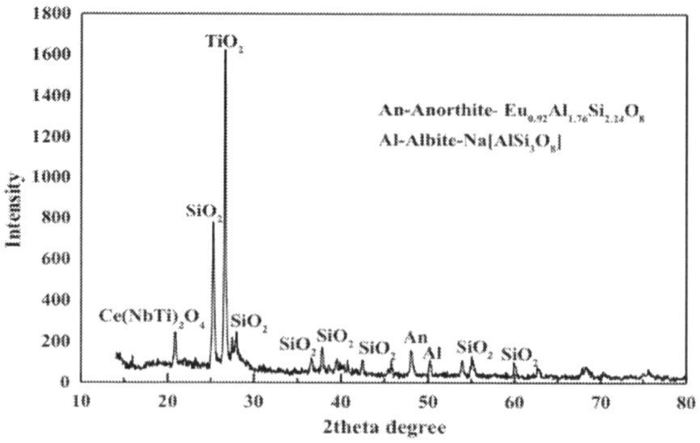

Figure 4. X-ray diffractogram of the extracted total rare earth oxides

The experimental process part III result is shown extracted titanium hydroxide is calcined at 600^{0}C, 750^{0}C, 900^{0}C for 2 hours to extract titanium oxide and X-ray phase analysis carried on to determine crystall structure. As a result of the X-ray phase analysis amorph titanium oxide at 600^{0}C, polymorph anatase mineral of the titanium oxide 750^{0}C and brookite minerals at 900^{0}C determined, respectively Figure 5.

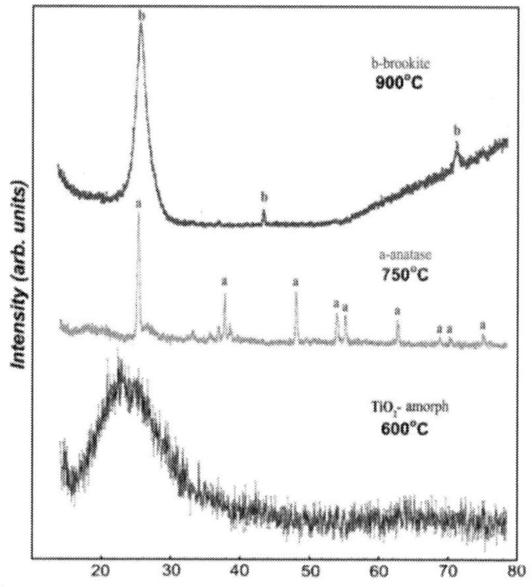

Figure 5. X-ray diffractogram of the extracted titanium dioxide TiO_2

Chemical properties of the extracted titanium oxide in the anatase minerals of the titanium oxide extracted at 750^{0}C has studied by the analytical method X-ray flourescence.

According to the result of the X-ray flourescence analytical method, macro elements content in sample is 90.9%, which contains titanium oxide 62.5%, aluminium oxide 2.27%, iron oxide 5.46%, sulfur oxide 9.89% and silicon oxide 4.42%, respectively Table 2.

According the study of the chemical properties of the brookite mineral of the titanium oxide extracted at

900^{0}C, total macro element content is 99.9% where titanium oxide content increased to 89.6%. Through the X-ray flourescence analysis has determined that the concentrate contains aluminium oxide 1.17%, iron oxide 6.7% sulphur oxide 0.96% and silicon oxide 1.92%, respective Table 3.

Table 2. Chemical contents of the extracted anatase mineral

Content %	Al_2O_3	BaO	CaO	CuO	Fe_2O_3	K_2O	MnO_2	Nb_2O_3	Ta_2O_3	ZnO	ZrO	P_2O_5	PbO	SO_2	SiO_2	**TiO_2**	**Total**
	2.27	1.94	1.35	0.13	7.8	0.22	0.14	0.13	0.011	0.02	0.05	0.02	0.01	9.89	4.42	**62.5**	**90.9**

Table 3. Chemical contents of the extracted brookite mineral

Content %	Al_2O_3	BaO	CaO	CuO	Fe_2O_3	K_2O	MnO_2	Nb_2O_3	Ta_2O_3	ZnO	ZrO	P_2O_5	PbO	SO_2	SiO_2	**TiO_2**	**Total**
	1.17	0.01	0.07	0.04	6.7	1.2	0.02	0.07	0.01	0.01	0.02	0.07	0.01	0.96	1.92	**89.6**	**99.9**

When a morphology structure of the extracted anatase mineral is studied by SEM-EDX analysis, finding presents that the titanium oxide anatase has tetragonal crystal structure and contains many polycrystals, which shown at Figure 6.

Figure 6. Morphology structure of extracted anatase mineral - SEM-EDX

As a by-product of extraction of titanium dioxide process, we have extracted ferrous oxide hematite successfully. (Part IV)

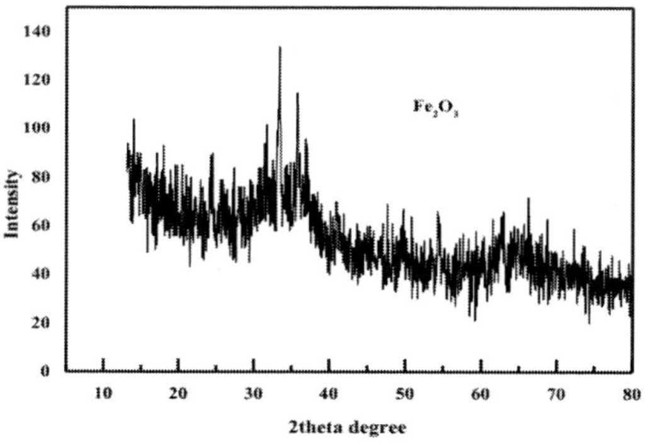

Figure 7. Ferrous oxide X-ray diagram

X-ray phase analysis and chemical analysis shown that ferrous oxide content is 83.8%.

Conclusion

1. Chemical and mineral analysis of polymetallic ore is carried out basis of responses to the support process at various degrees of water whereas transition metal sulphates solubility differ.
2. As a result of sulphate and resulphurization process we have extracted anatase with 62.5 percent titanium dioxide and brookite mineral with 89.6 percent of titanium dioxide as well as mineral with 83.8 percent of ferrous oxide hematite and rare earth oxides with 57.6 percent of cerium oxide. These oxides are identified under various conditions in the thermal processing.
3. We are currently working in research on utilization of high-purity anatase nanopowder (TiO_2) with photocatalytic activitiy for treating waste water in order to eliminate its total bacteria, especially *E. Coli* under UV-A. High-purity anatase nanopowder has been extracted from anatase with 62.5% of titanium oxide by recrystalization process.

References

1. O. Gerald, A.C. Onyedika, Achusim-udenko, Nwoko, Ogwuegbu, Int. J. Chem. Sci. **10**,1 112-130 (2012)
2. J. Zoumei, L. Wang, H. Zhou, Zh. Duan, Titanium Extract and Process, The Mineral, Metals & Materials Society. 122-128, (1997)
3. O. Enkh-Uyanga, B. Munkhtsetseg, J. University of Science and Technology Bulletin. **12**, 83-90 (2015)
4. P.M. Kobylin, H. Sippola, P. Taskinen, CALPHAD: Computer Coupling of Phase Diagrams and Thermochemistry. **35**, 499-511 (2011).
5. B. Munkhtsetseg, G. Burmaa, AIP Conf. Proc. 1542, 1278 (2013)
6. X. Xiong, Z. Wang, F. Wu, J.Advanced Powder Technology. **24**, 60–67 (2013)
7. H. Kominami, Y. Ishii, M. Kohno, S. Konishi, J. Catalysis Letter. **91**, 41-47 (2003).

DOI: 10.1051/epjconf/201714007003

Towards quality by design in pharmaceutical manufacturing: modelling and control of air jet mills

Satyajeet Bhonsale [1,2], *Dries* Telen [1], *Bard* Stokbroekx [2], and *Jan* Van Impe[1*]

[1]BioTeC+ & OPTEC, Department of Chemical Engineering, KU Leuven, Ghent, Belgium
[2]Crystallization Technology Unit, Janssen Pharmaceutica, Beerse, Belgium

Abstract. Milling is an important step in pharmaceutical manufacturing as it not only determines the final formulation of the drug product, but also influences the bioavailability and dissolution rate of the active pharmaceutical ingredient (API). In this respect, the air jet mill (AJM) is most commonly used in the pharmaceutical industry as it is a non-contaminating and non-degrading self-classifying process capable of delivering narrow particle size distributions (PSD). Keeping the principles of Quality by Design in mind, the Critical Process Parameters (CPPs) of the AJM have been identified to be the pressures at the grinding nozzles, and the feed rate which affect the PSD, surface charge and the morphology of the product (i.e. the Critical Material Attributes (CMAs)). For the purpose of this research, the PSD is considered to be the only relevant CMA. A population balance based model is proposed to simulate the dynamics milling operation by utilizing the concept of breakage functions. This model agrees qualitatively with experimental observations of the air jet mill unit present at Janssen Pharmaceutica but further steps for model validation need to be carried out.

1 Introduction

The air jet mill (AJM) is one of the most common equipments used to micronize active pharmaceutical ingredients (API) in the pharmaceutical industry. The huge popularity of the jet mill arises from the fact that it does not involve any moving parts, thus leading to a non-contaminating size reduction process. The primary cause of breakage in the air jet mill is the particle-particle impact along with the particle-wall impact. Pressurized carrier gas is injected tangentially through the nozzles placed around the chamber generating a high speed fluid vortex in the chamber. This vortex has been visualized with help of computer simulations [1] and also experimentally [2]. The material, introduced into the mill by air pushers exploiting the venturi effect, falls into this fluid vortex and is accelerated to (super)sonic velocities (depending on the type of nozzle used). The particles then break through multiple impacts with the wall and other particles. As the dependency of centrifugal force on the particle size ($\sim x^3$) is much stronger than that of the drag force ($\sim x^2$), the breaking particles experience much more reduction in centrifugal force than in drag force. Thus, as the particles break, they move more and more towards the centre until, at one specific particle size range (which depends on the operating conditions used) the drag force exceeds the centrifugal force causing the particles to exit the grinding chamber. Thus, the AJM is considered to be a self-classifying mill.

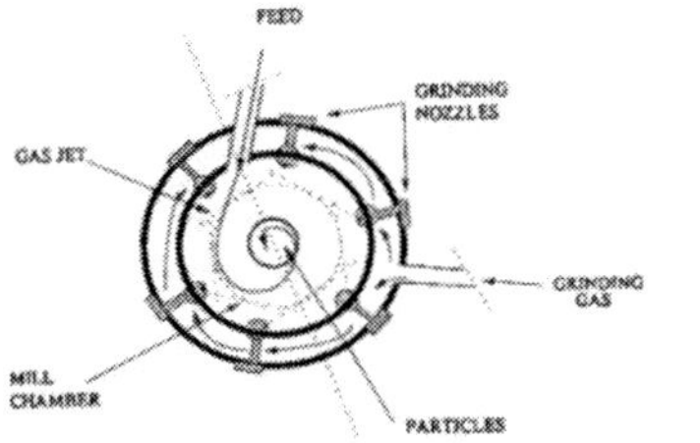

Figure 1. Schematics of the Air Jet Mill [1]

Despite its extensive use, relatively few studies can be found in the literature which deal with modelling or control of the air jet mills, especially for organic compounds. A comprehensive modelling exercise on air jet mills was carried out by Gommeren et al. [3], however, a specially designed jet mill was used which is not realistic in an industrial environment. The parameters in this model are difficult to identify based on experimental data. Moreover, not enough simulation studies were presented to get a comprehensive idea of the model. Some other studies have been aimed at modelling the air jet mill using Markovian theory [4], or steady state population balance model [5], however these are not directed towards control. The output of the mill depends not only on the operating conditions but also on the material characteristics. Considering that every material behaves differently, modelling of milling processes becomes challenging. The present research aims at developing a robust and efficient model for the AJM which could in the

* Corresponding author: jan.vanimpe@kuleuven.be

future be utilized to implement a control strategy to achieve the required particle size distribution (PSD).

2 Mathematical Description

Attempts to model milling processes have been made using various techniques like discrete element modelling (DEM) [6], fracture mechanics based modelling [7], population balance modelling (PBM) [3,5], etc. In this work, the PBM framework is utilized to obtain a model for the air jet mill. PBMs have been widely and successfully used in modelling particulate processes like crystallization, milling, mixing [8-9], etc. For the breakage process in consideration, a size-discrete population balance model has been developed. The bases of using a PBM for comminution are the concepts of breakage rate and breakage distribution function which were first proposed by Epstein [10]. The breakage rate (k_i) defines the rate at which the particles in the size class i break under loading conditions. The breakage distribution function (b_{ij}) defines the probability that a particle breaking from size class j will fall in size class i. The breakage rate is commonly described as [11]

$$k_i = \alpha \left(\frac{x_i}{x_1}\right)^{\gamma} \quad (1)$$

where α and γ are fitting parameters, x_i is the volumetric mean size of the class i, and x_1 is the largest size class. α and γ are the fitting parameters. For milling processes with small time scales, the breakage distribution function can be approximated as [12]

$$b_{ij} = \frac{k_{i-1} - k_i}{k_j} \quad (2)$$

As a jet mill is a self-classifying mill, the output of the mill has a different particle size distribution than that of the material in hold-up. To simulate the internal classification, the concept of Tromp curve is used. It is assumed that the Tromp curve follows a log-normal distribution [3, 13] with respect to the particle size and in its simplest form can be defined as

$$P_i = \frac{1}{1+e^{K(x_i - x_{50})}} \quad (3)$$

where K and x_{50} are the fitting parameters. It is obvious that all the parameters in the model will be influenced by the operating parameters of the mill, namely, solid feed rate and grinding nozzle pressure. Moreover, under dynamic conditions the hold-up inside the mill will increase with time until it reaches a steady state value which is when the mill is said to have reached a stable zone. This hold up will also depend on the operating conditions and will affect the parameters in the model. However, for the purposes of this preliminary simulation study, the parameters are assumed to be constant. Unlike the Gommeren's model, the proposed model does not compartmentalize the mill and an overarching PBM is written over the entire mill. This helps reducing the number of parameters which need to be estimated from experimental studies.

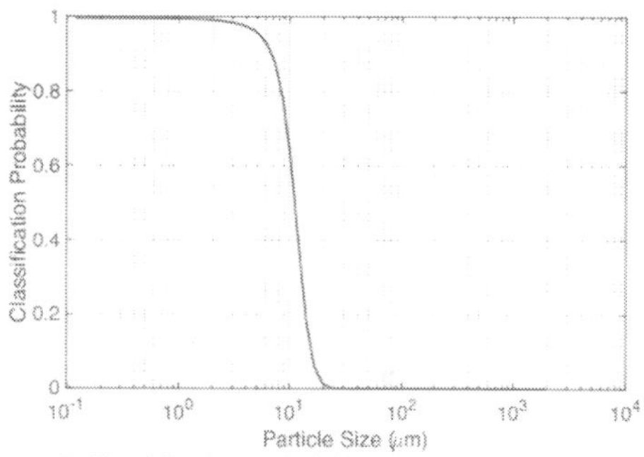

Figure 2. Classification probability as a function of particle size

The population balance model is now described as follows

$$\frac{dm_i}{dt} = f_{in,i}\Phi_{in} - P_i m_i - k_i m_i + \sum_{j=1}^{i} b_{ij} k_j m_j \quad (4)$$

$$\Phi_{out} = \sum P_i m_i \quad (5)$$

$$f_{out,i} = \frac{P_i m_i}{\Phi_{out}} \quad (6)$$

where, m_i is the mass of the particles in size class i within the milling chamber. The total mass holdup inside the chamber is then given by $\sum m_i$. $f_{in,i}$ is the particle size distribution of the feed (% in size class i), Φ_{in} is the solid feed rate to the mill, Φ_{out} is the total production rate of the mill, and $f_{out,i}$ is the particle size distribution of the output. In typical experimental conditions, the solid feed rate and the particle size distribution of the feed are known, while the output particle size distribution is measured. The hold-up inside the mill and the product output rate, however, cannot be measured directly leading to identifiability issues as will be discussed later.

Table 1. Parameter values used for simulation

Parameter	Value
α	5
γ	0.9596
K	0.5
x_{50}	11.2
Φ_{in}	10 (g/s)

3 Simulation Study

As previously mentioned, the breakage parameters of the jet mill depend on the process conditions and in turn on the mass hold up within the milling chamber. The classification of particles within this mill are also affected by these conditions. In this preliminary analysis, time-constant parameters have been used with values mentioned in Table 1. Figure 2 depicts the classification curve when using the parameters defined in Table 1.

For this simulation, the 61 size classes have been considered. The largest size class ($i = 1$) has the value 2000 μm, and a geometric progression with ratio of 1.1795 is used to define all the other size classes. The motivation for the number and values is that the particle size distribution measured by the Insitec online laser diffraction unit (Malvern Instruments Ltd.) is based these values. This will enable the direct introduction of measured data for parameter estimation and model validation in the future.

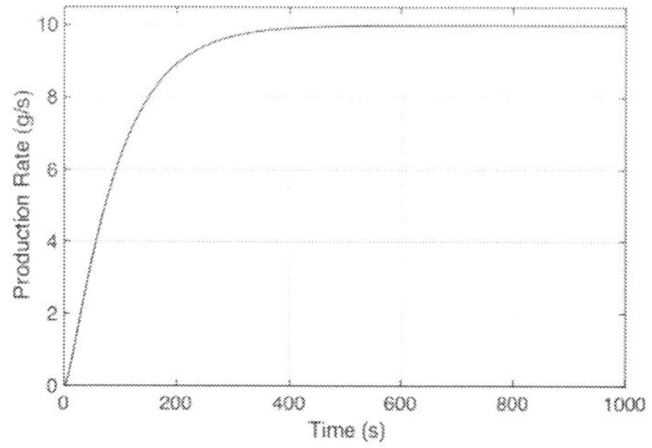

Figure 3. Output production rate of the mill for the defined parameters

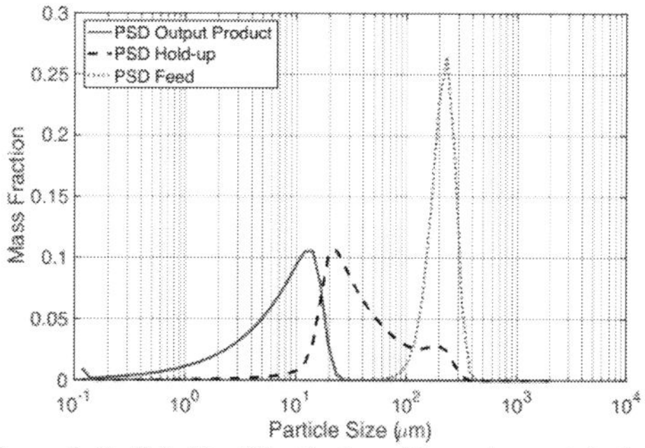

Figure 4. Particle Size Distribution of the output and the hold-up of the mill and the feed used in the simulation

The feed PSD for the simulation is a Gaussian distribution with a mean of 200 μm and a variance of 50 μm and is generated using the MATLAB command 'normpdf'. Figure 3. illustrates the evolution of the product output rate with time for the given parameter values. It can be observed after around 500 s the mill reaches a steady state. This type of dynamic behaviour is also observed on the real mill, albeit on different time scales. Figure 4. depicts the PSD of the output product as well as the PSD of the hold-up at the steady state condition. The shape of the PSD could be explained by the fact that given the structure of P_i, almost all the fine particles escape from the hold-up leading to the long shoulder towards the left in output PSD. Figure 5. shows evolution of the D_{10}, D_{50}, and D_{90} of the output product. The discrete jumps observed in the evolution of these variable is due to their numerical calculations from the cumulative PSD. Since, towards larger size end, the difference between the size classes is relatively larger that at lower end, the jumps are more exaggerated at higher values.

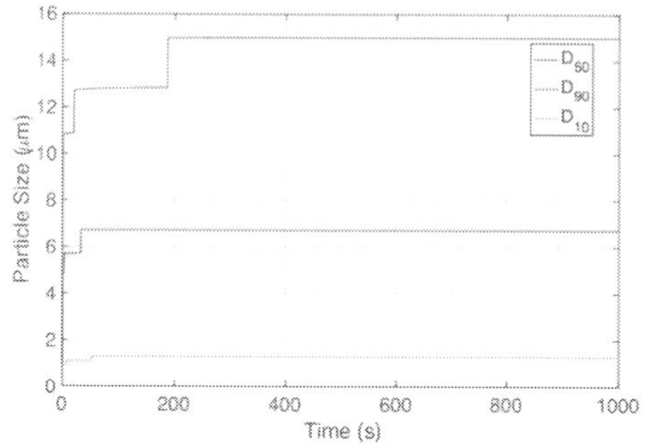

Figure 5. Evolution of particle size indicators D_{10}, D_{50}, D_{90} of the output material with time

4 Discussion

Although the parameters considered in this model are not determined experimentally, it is worth noting that the model structure is able to capture dynamics of the mill which were observed during preliminary experimental analysis. The next step thus would be to estimate these parameters based on experimental data.

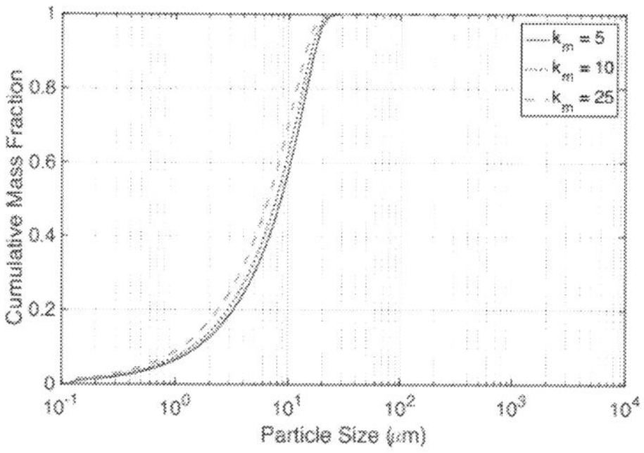

Figure 6. Cumulative particle size distribution of the output material for $k_m = 5$, $k_m = 10$, and $k_m = 25$, with other parameter values constant

The parameter γ in Equation 1 relates to the propensity of a material to break under loading conditions, while the parameter α determines the intensity of breakage which depends on the magnitude of the loading condition. The parameter γ can be assumed to be constant for a particular API [3,11]. The parameter α on the other hand will be influenced by the operating conditions. Similarly, the parameters in the classification curve depend on the operating condition of the mill. This makes the parameter estimation for the model challenging as there will be a huge correlation between the breakage parameters and the classification parameters. It can be observed in Figure 6 that the effect of the parameter α on the output particle size is not as predominant as it is on the production rate and the stability of the mill as can be seen in Figure 7.

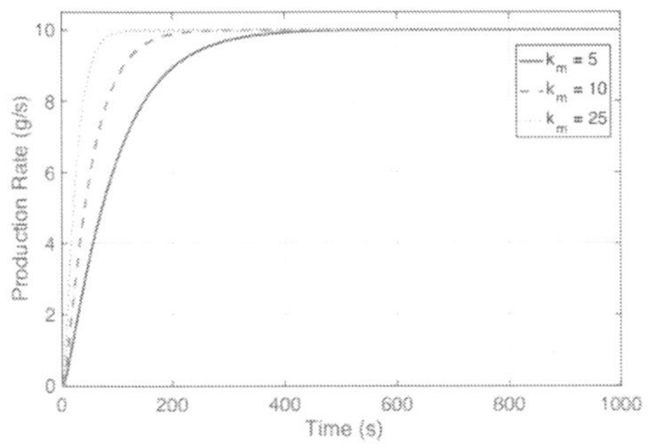

Figure 7. Output Flow Rate of milled material at $k_m = 5$, $k_m = 10$, and $k_m = 25$

Figure 8 shows the output PSD of the milled material at different x_{50} values. It can clearly be observed that this parameter has a more prominent effect on the output.

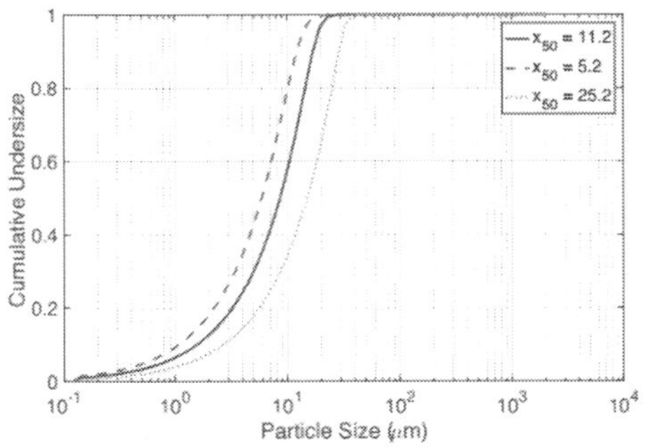

Figure 8. Output PSD of the milled material at $x_{50} = 11.2$, $x_{50} = 5.2$ and $x_{50} = 25.2$ at other constant parameters

5 Conclusions & Future Work

The model structure presented can capture the dynamics of a lab scale air jet mill. This model presents a simplification to a model proposed by Gommeren. The need for simplification arises from the fact that the parameters in the model are not directly identifiable from experimental data and not enough simulation studies were provided in the original study to get a comprehensive understanding of the model. However, the current model also runs into the similar difficulty as the Gommeren's model in that the breakage parameters are not directly identifiable from the measured data. Thus, at this stage a rigorous parameter identifiability analysis need to be carried out which should then be the basis for further model simplifications and experiment designs. The dependence of the breakage and classification parameters on the operating conditions also needs to be investigated. A priori, one can expect the breakage rate parameter, α to increase with increasing pressures and decrease with increasing feed rates. This is also reflected by the simulation shown in Figure 6. The effect on the classification parameters are more complicated. The classification parameters also have an effect on the PSD of the hold-up material inside the mill. This is to be expected as the classification function determines the residence time of a particle within the mill.

In future, other experimental techniques will be investigated to generate data which can be used to identify the breakage and classification parameters easily. Being in an industrial environment, it is important that the experimental time and the amount of materials required needs to be minimized. Hence, advance optimal experiment design techniques will be used to design experiments which can provide informational data which can be used to tune the model to any API that needs to be introduced. It is also envisioned, that the material dependent parameters may be linked to one or more API properties using a black box technique like artificial neural net. Ultimately, the goal of the research is to develop a model that can be easily tuned, and be implemented in a control framework such that given an API, a required PSD can be consistently delivered.

This research was supported by: KU Leuven project PFV/10/002 (OPTEC), project G.0930.13 of the Fund for Scientific Research-Flanders, and BelSPO Interuniversity Attraction Pole IAP VII/19 (DYSCO). SB holds a IWT Baekeland grant IWT150715 for his PhD research.

References

1. S. Teng, P. Wang, L. Zhu, M. W. Young, C. G. Gogos, Powder Technol. **195**, 31 (2009)
2. F. Müller, R. Polke, G. Schädel, Int. J. Miner. Process. **44**, 315 (1996)
3. H. J. C. Gommeren, D. A. Heitzmann, J. A. C. Moolenaar, B. Scarlett, Powder Technol. **108**, 147, (2000)
4. S. Teng, P. Wang, L. Zhu, M. W. Young, C. G. Gogos, Chem. Eng. Sci. **65**, 4323 (2010)
5. D. Starkey, C. Taylor, N. Morgan, K. Winston, S. Svoronos, J. Mecholsky, K. Powers, R. Iacocca, AIChE Journal, **60**, 4086 (2014)
6. T. Brosh, H. Kalman, A. Levy, I. Peyro, F. Richard, Powder Technol. **257**, 104 (2014)
7. Z. Zang, M. Ghadiri, Chem. Eng. Sci. **57**, 3659 (2002)
8. D. Ramakrishna, *Population Balances* (Academic Press, San Diego, 2000)
9. M. Sen, R. Ramachandran, Adv. Powder Technol. **24**, 51 (2013)
10. B. Epstein, Ind. Eng. Chem. **40**, 2289 (1948)
11. L. Austin, Powder Technol. **5**, 1 (1971)
12. H. Berthiaux, C. Varinot, J. A. Dodds, Chem. Eng. Sci. **51**, 4509 (1996)
13. H. Otwinowski, Particul. Sci. Technol. **32**, 399 (2014)

Dissipative lateral walls are sufficient to trigger convection in vibrated granular gases

Giorgio Pontuale[1,*], *Andrea* Gnoli[1,**], *Francisco* Vega Reyes[2,***], and *Andrea* Puglisi[1,****]

[1] *Istituto dei Sistemi Complessi—CNR and Dipartimento di Fisica, Università di Roma Sapienza, Piazzale Aldo Moro 2, 00185 Rome, Italy*
[2] *Departamento de Física and Instituto de Computación Científica Avanzada (ICCAEx), Universidad de Extremadura, 06071 Badajoz, Spain*

Abstract. Buoyancy-driven (thermal) convection in dilute granular media, fluidized by a vibrating base, is known to appear without the need of lateral boundaries in a restricted region of parameters (inelasticity, gravity, intensity of energy injection). We have recently discovered a second buoyancy-driven convection effect which occurs at any value of the parameters, provided that the impact of particles with the lateral walls is inelastic (Pontuale et al., Phys. Rev. Lett. 117, 098006 (2016)). It is understood that this novel convection effect is strictly correlated to the existence of perpendicular energy fluxes: a vertical one, induced by both bulk and wall inelasticity, and a horizontal one, induced only by dissipation at the walls. Here we first review those previous results, and then present new experimental and numerical data concerning the variations of box geometry, intensity of energy injection, number of particles and width of the box.

1 Convection in granular media, from slow dense configurations to dilute gas-like setups

Vibration of granular materials frequently leads to convective patterns [1]. However, the particular mechanism for sustaining and control of a convective dynamics depends upon the particular granular phase under scrutiny [2]. At high packing fraction [3] and low fluidization, "dense convection" has been first identified in the 90's [4–7] and explained through an asymmetric tangential friction at the lateral walls [8–10] or the formation of unstable heaps at the free surface [11]. In granular gases [12–14], on the contrary, the first discovered mechanism is bulk buoyancy-driven thermal convection (BBD-TC), initially observed in simulations [15, 16]. As in molecular liquids [17, 18], it is driven by the buoyancy force associated to temperature/density gradients: in BBD-TC such gradients appear spontaneously because of the bulk inelasticity of grain-grain collisions [19]. This is a bulk effect: indeed, no particular conditions on the lateral or top boundaries are required for its appearance (for instance in [15] and [20] it appears also with elastic lateral walls) and in fact an analytical study of the stability of the granular gas hydrostatic state *without lateral walls* (lateral periodic boundary conditions) has shown the emergence of BBD-TC in a region of system's parameters [21, 22]. When lateral walls are present (elastic or inelastic) a downward flow velocity is always observed near lateral walls, perhaps because of a reduced buoyancy originated from enhanced dissipation [15, 23]. After the first experimental evidence in 3D [24], other experiments in quasi-2D setups have demonstrated the presence of convection in granular gases [25], but with contradicting features: some of them agree quantitatively with analytical theories where lateral walls are absent [26], while others [27] strongly indicate that convection is almost completely killed when wall inelasticity goes to zero, a scenario - incompatible with BBD-TC theories [21, 22] (see [23] for numerical simulations with similar results).

Recently we have demonstrated the existence of a convection mechanism for granular gases which is alternative to BBD-TC and which necessarily appears in all granular gases fluidized by a vibrating base under gravity, with the only necessary condition being the presence of inelastic lateral walls [28]. Such a mechanism has been called dissipative-lateral-wall-induced thermal convection, "DLW-TC". In this recent study we have employed experiments and simulations to isolate DLW-TC from BBD-TC. In Section 2 we review these recent results, while in Section 3 we present new experimental and numerical data. Conclusions are drawn in Section 4.

2 DLW-TC

Let us consider (similarly to the setup sketched in Fig. 1) a 2D low density gas (easily generalized to 3D systems) of identical solid disks of mass m confined by two inelastic

*e-mail: giorgio.pontuale@isc.cnr.it
**e-mail: andrea.gnoli@isc.cnr.it
***e-mail: fvega@unex.es
****e-mail: andrea.puglisi@roma1.infn.it

DOI: 10.1051/epjconf/201714004002

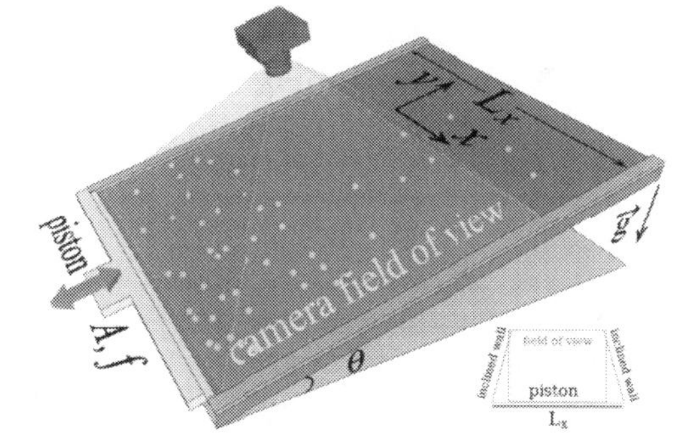

Figure 1. Sketch of the experimental setup. The real length (along y) of the inclined plane, $L_y \gg L_x$, is not marked as it is not essential in the description of the system. In the lower-right corner a sketch of the second setup with inclined lateral walls is shown, with a representation (blue square) of the restricted field of view for the fast camera.

parallel walls. The bottom wall provides energy to the system, for instance through steady vibration or (in numerics) in the form of a thermostat. The presence of a top wall or open top boundary does not change our reasoning. A constant gravity field g acts downwards along the vertical (y) direction. In the dilute limit $p = nT$ (with p the pressure, n the number density, and T the granular temperature) [12]. An outgoing energy flux is always originated at a dissipative wall [29], yielding in our case $\partial T/\partial x \neq 0$ at the lateral walls. Momentum balance in the absence of macroscopic flow (hydrostatic) states that

$$\partial_x p = \partial_x(n(x,y)T(x,y)) = 0 \qquad (1a)$$

$$\partial_y p = \partial_y(n(x,y)T(x,y)) = -mgn(x,y). \qquad (1b)$$

The first equation gives $p(x,y) \equiv p(y)$, which, used in the second equation, sets $n(x,y) \equiv n(y)$ and as a consequence $T(x,y) \equiv T(y)$. This is in contradiction with the horizontal temperature gradient required by dissipation at the lateral walls. Convection must be present.

In [28] we have discussed an experiment (inspired on previous works [30, 31]) under low-gravity conditions (see Fig. 1 and [28], including Supplemental Materials, for details) with N spherical steel beads (diameter d = 1 mm) moving inside a cuboid of dimensions L_x = 175 mm, L_y = 600 mm and L_z = 1.5 mm = 1.5d, thus assuring for a quasi-2D dynamics restricted to the xy plane. A small tilt angle θ with respect to the horizontal guarantees an effective low value for gravity $g_{eff} \approx (5/7)g\sin(\theta)$, where g is Earth's gravity acceleration [32]. The two lateral "walls" are made of polycarbonate (at $x = \pm L_x/2$), characterized by a grain-wall restitution coefficient α_w, the inferior wall (at $y = 0$) is a Plexiglas® piston vibrating with amplitude A and frequency f, giving an average squared velocity $v_0^2 = (A2\pi f)^2/2$. A high speed camera allowed us to reconstruct the average fields $u(x,t)$, $n(x,t)$ and $T(x,t)$. Experiments have been performed in a range of average 2D packing fraction ($\nu_{2D} = N\pi(d/2)^2/(L_xL_y)$) $\nu_{2D} \in [0.1\%, 1\%]$, with a constant amplitude of vibration A = 1.85 mm, a range of piston oscillation frequency $f \in [10,45]$ Hz and a range for the inclination angle $\theta \in [0.011, 0.130]$ radiants, leading to a range of rescaled maximum acceleration $\Gamma = A(2\pi f)^2/g_{eff} \in [70,3800]$. In all our experiments, with the only exception of the lowest number of particles (N = 100, where the mean free path is larger than the box's sides) we observed convection with two convective cells that span the full width of the 2D plane. In all convective cases density and temperature fields display gradients not only in the y but also in the $x$direction (i.e. transverse to gravity): in particular, temperature decreases moving along the horizontal from the lateral walls toward the center of the system, as expected in view of the outgoing energy flux at the boundaries. Most of the qualitative trends appear similar when g or N are increased, suggesting that these two parameters play an analogous role. In particular the centers of the convective cells move toward the lowest corners (left or right) of the system.

All these features have been confirmed, even in a larger range of parameters, by MD numerical simulations: we have employed an event-driven code for inelastic hard disks. Between collisions the disks perform "free" ballistic fligths under the action of effective gravity g_{eff}. When two disks collide, a normal instantaneous collision with constant restitution coefficient $\alpha \in [0,1]$ is performed. When a disk collides with the base wall, the velocity component parallel to the base is left unaltered, while the normal component is set to a new value extracted from a Maxwell-Boltzmann distribution with temperature v_0^2 [33]. When a disk collides with the top wall, an elastic collision is performed. When a disk collides with the lateral walls, an inelastic collision with an infinite mass and with restitution coefficient $\alpha_w \in [0,1]$ is performed. Some simulations with the presence of friction along the ballistic flight (introduced by time-discretized dynamics) - which would be analogous to the rolling/sliding friction between spheres and the setup's floor - have also been performed and discussed in [28], giving very similar results. The simulations clearly show that with the parameters of the experiment and the artificial choice of elastic lateral walls the convection is totally suppressed, as expected for experiments at low gravity [21, 22]. In the simulations we have also shown that the intensity of convection decreases linearly with α_w, in a way similar to the observation of [27] and that the width of the convective cell decreases when g_{eff} increases. When $g_{eff} \sim g$ convection is barely visible: this can explain the results and the phase diagram described in [25]. Interestingly, our simulations allowed us to verify that DLW convection is present even when grain-grain collisions are elastic (α = 1) provided that $\alpha_w < 1$: this suggests that the elastic limit is smooth (a good hint for future theoretical research) and that DLW are sufficient to create not only the horizontal gradient but also the vertical ones, necessary for the developement of convection.

3 New results

In Fig. 2 we report the results of two different experiments with the same number of particles (N = 1000) and same

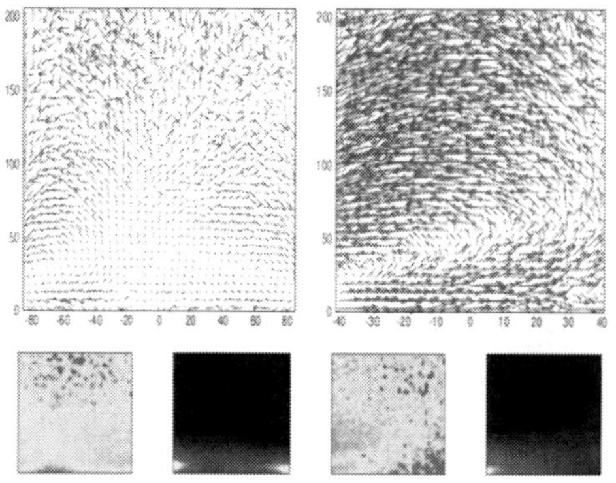

Figure 2. Fields from two experiments with $N = 1000$, $f = 45$ Hz, $A = 1.85$ mm ($v_0 = 370$ mm/s) and (A): perpendicular lateral walls; (B): lateral walls inclined of 12 degrees s (see sketch in the lower-right corner of Fig. 1). Top row: the velocity field with the two convective cells; coordinates are in units of particle's diameters (= 1 mm). Bottom row: each couple of graphs show the (mass-free) temperature T/m (left) and local packing fraction $\nu = n\pi(d/2)^2$ (right) corresponding to the case in the top row. The T scale goes from black (colder) at $T = 0$ to red (hotter) at $T = T_{max}$. The ν scale goes from black (more dilute) at $\nu = \nu_{min}$ to yellow (denser) at $\nu = \nu_{max}$. Values of T_{max}/mv_0^2 are: A) 0.08, B) 0.15. Ranges for (ν_{min}, ν_{max}) are: A) (0.05%, 0.5%), B) (0.05%, 0.5%).

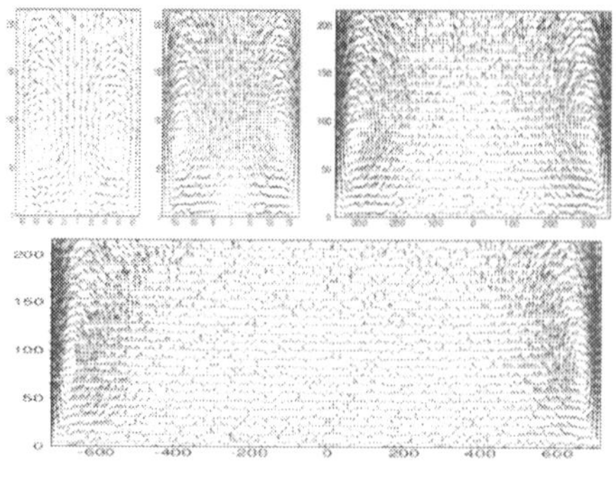

Figure 3. Fields from four simulations with different width and same number of layers at rest ($N\sigma/L_x = 2.78$). From top-left: $L_x = 180$ ($N = 500$), $L_x = 360$ ($N = 1000$), $L_x = 720$ ($N = 2000$), $L_x = 1440$ ($N = 4000$).

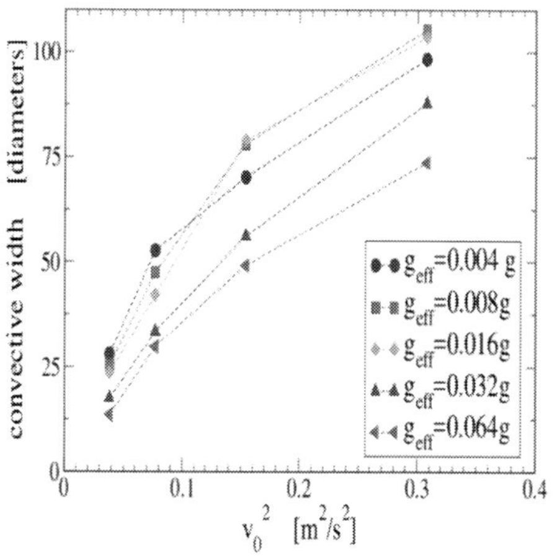

Figure 4. Width of the convection roll ($L_x = 720$ and $N = 2000$ particles) as a function of v_0^2 for different values of g_{eff}.

driving parameters, but different geometry of the box. The experiment on the left is performed with the usual setup whereas on the right we show a different set-up where the lateral walls form an inwards angle of 12 degrees with respect to the bottom wall direction. Apparently, a small change of inclination of the walls leads to quite an important qualitative change in the velocity, density and temperature field, in particular here we observe only a single convective cell. We stress that we have carefully checked the system to be symmetric with respect to inversion of the x direction, but it is possible that small asymmetries escape our scrutiny. We have not a simple interpretation of this result, and in particular of the symmetry breaking. We expect that within long experiments both clockwise and anti-clockwise convections should be observed, separated by "transition times" which could be very long. We underline that the general mechanism for convection discussed after Eq. (1) does not force the presence of two symmetric cells, it only states that a non-convective steady state is forbidden. In particular, it does not prevent an oscillation between single cells with opposite rotations. In Fig. 3 we report results from different MD simulations with the same number of layers at rest $N\sigma/L_x$ (characterising the amount of mass per unit of horizontal length which is the parameters replacing density in an isotropic system such as ours) but different widths of the system, illustrating the remarkable fact that the width of the cell (at fixed layers at rest, gravity and energy driving conditions) is constant and does not depend of the width. This is in perfect agreement with the idea that DLW-TC is a "boundary effect", but nevertheless it spans a macroscopic portion of the system, i.e. a large number of mean free path (see [28] and its Supplemental Materials for a discussion of Knudsen numbers in our system). In Fig. 4 other numerical results allow us to understand, for several values of g_{eff}, the important role of v_0^2 which represents the intensity of vibration of the base piston, i.e. the unique source of energy in the system. For all values of g_{eff} the width of the convective cell increase with v_0^2, even if such a growth slows down at high v_0^2. Consistently with previously reported data [28], the width of the cell decreases monotonically with g_{eff} for low values of v_0: on the contrary, it presents non-monotonicy with an initial growth at the first studied values of g_{eff} followed by a decrease in g_{eff}, for larger values of v_0.

4 Conclusions

In conclusion, we have discussed recent results on a novel thermal (buoyancy-driven) convection effect induced by dissipative lateral walls in vibrated granular gases, which at low gravity can be better isolated from the usual convection induced by bulk (grain-grain) inelasticity [28]. In this manuscript we have presented further evidences about the complexity of this phenomenon. In particular we have shown that the inclination of the lateral walls affects the convection in an unexpected way, and we have shown new numerical results with systems of different widths, supporting the idea that DLW-TC is a "boundary-effect", i.e. it is limited to a (macroscopic) region close to the lateral walls. DLW-TC is replaced by usual thermal convection when the parameters are chosen in the region of instability [21, 22], e.g. for high enough bulk inelasticity, as shown in the Supplemental Materials of [28]. We remark that while many studies in *dense* vibrated granular materials have shown boundary-induced convection, and while other more recent studies in vibrated granular gases have shown the existence of an important role of the lateral wall, the study in [28] is the first to demonstrate the existence of a wall-induced convection in gases which is well distinct from the bulk-inelasticity-induced effect. Such a result is even more fascinating as a theoretical explanation is lacking and cannot be retrieved in the form of a stability analysis (as it was done for bulk thermal convection) because a non-convective simpler base state (either homogeneous or unidimensional [19]) does not exist (i.e., the DLW convective state is the base steady state). This, of course has further implications, for instance, the stability threshold for the BT convection will need to be calculated starting out from the DLWc state and not from a non-convective one.

Acknowledgements

F. V. R. acknowledges support from Grants FIS2016-76359-P (Spanish Gov.) and GRU10158 (Junta de Extremadura, partially financed by ERDF funds).

References

[1] I.S. Aranson, L.S. Tsimring, Rev. Mod. Phys. **78**, 641 (2006)
[2] H.M. Jaeger, S.R. Nagel, R.P. Behringer, Reviews of Modern Physics **68**, 1259 (1996)
[3] B. Andreotti, Y. Forterre, O. Pouliquen, *Granular Media* (Cambridge University Press, 20013)
[4] C. Laroche, S. Douady, S. Fauve, J. Phys. **50**, 699 (1989)
[5] J.B. Knight, H.M. Jaeger, S.R. Nagel, Phys. Rev. Lett. **70**, 3728 (1993)
[6] E.E. Ehrichs, H.M. Jaeger, G.S. Karczmar, J.B. Knight, V. Kuperman, S.R. Nagel, Science **267**, 1632 (1995)
[7] J.B. Knight, E.E. Ehrichs, V. Kuperman, J.K. Flint, H.M. Jaeger, S.R. Nagel, Phys. Rev. E **54**, 5726 (1996)
[8] J.A.C. Gallas, H.J. Herrmann, S. Sokołowski, Phys. Rev. Lett. **69**, 1371 (1992)
[9] J.B. Knight, Phys. Rev. E **55**, 6016 (1997)
[10] E.L. Grossman, Phys. Rev. E **56**, 3290 (1997)
[11] K.M. Aoki, T. Akiyama, Y. Maki, T. Watanabe, Phys. Rev. E **54**, 874 (1996)
[12] J.W. Dufty, Adv. Complex Sys. **4**, 397 (2001)
[13] N.V. Brilliantov, T. Pöschel, *Kinetic Theory of Granular Gases* (Oxford University Press, 2004)
[14] A. Puglisi, *Transport and Fluctuations in Granular Fluids* (Springer, 2015)
[15] R. Ramírez, D. Risso, P. Cordero, Physical Review Letters **85**, 1230 (2000)
[16] P. Sunthar, V. Kumaran, Phys. Rev. E **64**, 041303 (2001)
[17] L.D. Landau, E.M. Lifschitz, *Mécanique des fluides* (Éditions MIR, 1971)
[18] S. Chandrasekhar, *Hydrodynamic and Hydromagnetic Stability* (Dover, New York, 1981)
[19] F. Vega-Reyes, J.S. Urbach, J. Fluid Mech. **636**, 279 (2009)
[20] D. Paolotti, A. Barrat, U.M.B. Marconi, A. Puglisi, Phys. Rev. E **69**, 061304 (2004)
[21] X. He, B. Meerson, G. Doolen, Phys. Rev. E **65**, 030301(R) (2002)
[22] E. Khain, B. Meerson, Phys. Rev. E **67**, 021306 (2003)
[23] J. Talbot, P. Viot, Phys. Rev. Lett. **89**, 064301 (2002)
[24] R.D. Wildman, J.M. Huntley, D.J. Parkar, Phys. Rev. Lett. **86**, 3304 (2001)
[25] P. Eshuis, K. van der Weele, D. van der Meer, R. Bos, D. Lohse, Phys. Fluids **19**, 123301 (2007)
[26] P. Eshuis, K. van der Weele, D. Lohse, D. van der Meer, Phys. Rev. Lett. **104**, 248001 (2010)
[27] C.R.K. Windows-Yule, N. Rivas, D.J. Parker, Phys. Rev. Lett. **111**, 038001 (2013)
[28] G. Pontuale, A. Gnoli, F. Vega-Reyes, A. Puglisi, Phys. Rev. Lett. **117**, 098006 (2016)
[29] P.R. Nott, M. Alam, K. Agrawal, R. Jackson, S. Sundaresan, J. Fluid Mech. **397**, 203 (1999)
[30] A. Kudrolli, M. Wolpert, J.P. Gollub, Phys. Rev. Lett. **78**, 1383 (1997)
[31] A. Kudrolli, J. Henry, Phys. Rev. E **62**, R1489 (2000)
[32] D.L. Blair, A. Kudrolli, Phys. Rev. E **67**, 041301 (2003)
[33] F.J. Alexander, A.L. Garcia, Comput. Phys. **11**, 588 (1997)

Axial segregation in spherical and cylindrical rotating tumblers

Umberto D'Ortona[1,*], *Nathalie* Thomas[2], and *Richard M.* Lueptow[3,4]

[1] *Aix-Marseille Université, CNRS, Centrale Marseille, M2P2 UMR 7340, 13451, Marseille, France*
[2] *Aix-Marseille Université, CNRS, IUSTI UMR 7343, 13451, Marseille, France*
[3] *Department of Mechanical Engineering, Northwestern University, Evanston, Illinois 60208, USA*
[4] *Department of Chemical and Biological Engineering, Northwestern University, Evanston, Illinois 60208, USA*

Abstract. Monodisperse and bidisperse granular flows are studied in rotating tumblers using DEM. In spherical tumblers, flowing particles' trajectories do not follow straight lines but are curved. At the same time particles near the surface drift toward the pole, inducing two global recirculation cells. Combined with radial segregation, drift and curvature compete to impose the axial segregation pattern: Small-Large-Small (SLS) or Large-Small-Large (LSL). Fill level, rotation speed and wall roughness influence drift and curvature, and modify the resulting segregation pattern. In cylindrical tumblers, equivalent recirculation cells occur next to the end walls. A second pair of recirculation cells with a weak drift in the opposite direction appears at the center for long enough tumblers. Unlike the sphere case, curvature and drift in the primary cells combine to push large particles toward the end walls, explaining why large particle bands appear at the end walls for axial segregation in cylinder.

1 Introduction

Mixtures of particles of equal density but different size segregate radially in rotating tumblers in only a few rotations. Subsequently, an axial segregation made of successive bands of small and large particles usually appears. Axial segregation has been studied experimentally and numerically in various geometries (cylinder, sphere, double cone). In cylindrical tumblers, bands of small and large particles emerge with a wavelength of about one tumbler diameter. Large particles always accumulate near the end walls of the tumbler rapidly forming the first bands of segregation. In between them, alternating bands of small and large particles may appear subsequently. In spherical tumblers, the axial segregation is always organised in 3 bands, either in a Small-Large-Small (SLS) or in a Large-Small-Large (LSL) pattern. The choice of the pattern depends on fill level, particles size ratio, sphere-particles size ratio, rotation speed [1], and the wall roughness of the spherical tumbler [2]. The mechanisms for the axial segregation in spherical and cylindrical tumblers will be addressed in this note.

Discrete Element Method (DEM) simulations are perfomed with a standard linear-spring and viscous damper force model for normal forces. In the tangential direction, the classical Cundall and Strack scheme was used [2, 3], as detailed elsewhere.

Figure 1 shows typical situations where the axial segregation patterns are modified due to particle size (a+b), wall roughness (a+c, e+f), or fill level (c+d) in spherical and cylindrical tumblers.

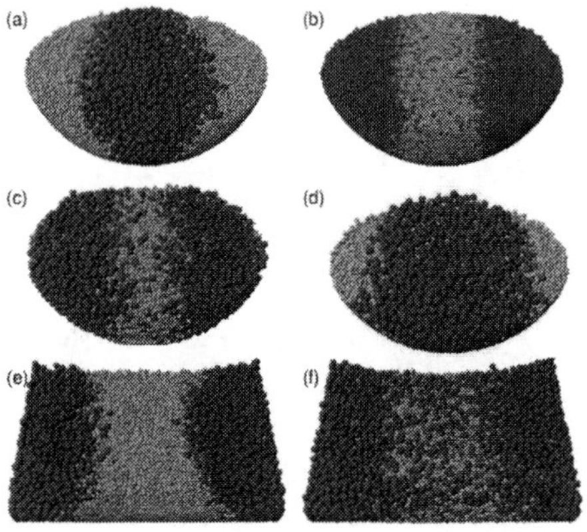

Figure 1. Front view of 14 cm diameter spherical tumblers or 14×14 cm cylindrical tumblers filled at 30% (except 20% for (d)) with an equal volume bidisperse mixture of particles. All systems rotate at 15 rpm and have reached their stationary segregation pattern. (a) 2+4 mm particles with smooth walls, the pattern is SLS, (b) 1+2 mm particles with smooth walls leading to LSL, (c) 2+4 mm particles in a rough tumbler made of 2 mm glued particles, the pattern is reversed to LSL (d) 2+4 mm with 2 mm rough wall tumbler filled at 20%, SLS pattern is obtained, (e) 2+4 mm particles in a smooth cylindrical tumbler, LSL pattern is obtained, (f) 2+4 mm particles in a 2 mm rough cylinder, still LSL but less sharp.

These results have also been obtained experimentally [1]. Fig. 2 shows experiments for conditions equivalent to the simulations (Fig. 1(a, c)) that are obtained in 14 cm

* e-mail: umberto@l3m.univ-mrs.fr

 DOI: 10.1051/epjconf/201714003011

diameter spheres with smooth or rough (interior covered by a monolayer of 2 mm glue particles) walls [2].

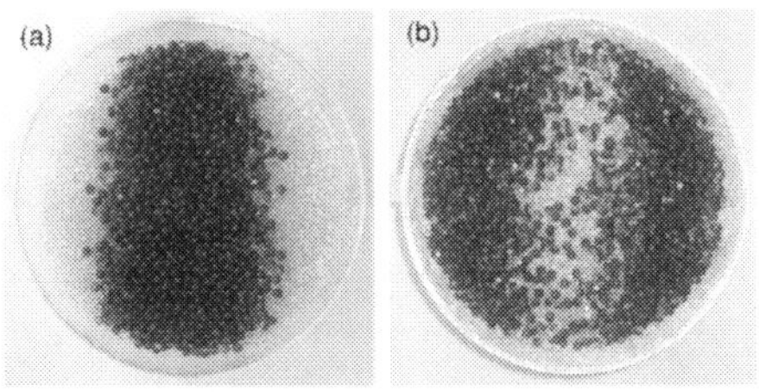

Figure 2. Experimental segregation patterns obtained in a 14 cm spherical tumblers filled at 30% with 2+4 mm particles. Walls are (a) smooth leading to a SLS pattern, or (b) 2 mm rough leading to LSL (pictures from [2]).

2 Spherical tumbler

2.1 Recirculation cells

To gain a better understanding on particle trajectories, simulations for monodisperse flow of 2 mm particles in a 14 cm spherical drum were performed. Figure 3 shows

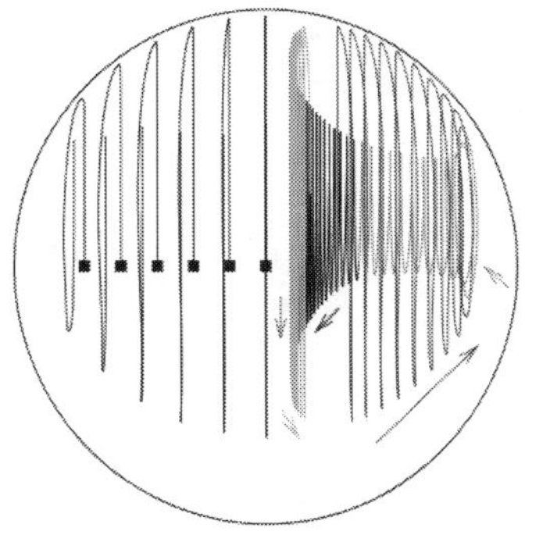

Figure 3. Top view of mean particles trajectories in a 2 mm rough 14 cm spherical tumbler. In the left half, trajectories have been integrated for one revolution. In the right half, one trajectory has been integrated for 50s. Each color corresponds to 10s. The tumbler is filled at 30% and rotates at 15 rpm. Arrows indicate the direction of drift.

mean trajectories of particles in a spherical rotating tumbler with 2 mm rough walls. In the left part, 6 trajectories are integrated for one revolution. They start from the static zone, 3 mm above the bottom wall (squares and vertical straight lines), and enter the flowing layer (curved trajectories), finishing in the static zone again but with a little drift toward the left pole. With this initial position close to the bottom wall, the trajectories flow near the free granular surface. The important facts are: the trajectories in the flowing layer are curved and each trajectory presents a net drift toward the pole.

In the right part of the sphere, one trajectory has been integrated for 50 s. While the particle trajectories are near the surface of the flowing layer (red curve corresponding to 0-10s), the drift is toward the pole. Near the poles the mean trajectory reverses its drift direction flowing near the bottom of the granular layer until it almost reaches the equator (green then dark blue, 10-30s). Then the trajectory reaches the surface of the flow (light blue, 30-50s) and drifts again toward the pole. This trajectory shows that two global recirculation cells are present on a granular flow in a spherical rotating tumbler. Equivalent cells have been obtained in a double cone geometry [4].

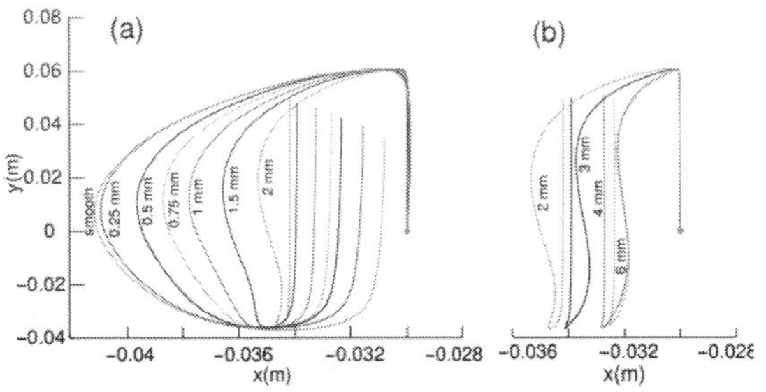

Figure 4. Top view of mean trajectories in 14 cm tumblers of various roughnesses ranging from (a) completely smooth (labeled 0) to a rough wall of 2 mm glued particles and (b) from 2 mm to 6 mm rough walls. The tumbler is filled at 30% with 2 mm particles and rotates at 15 rpm. Note that the horizontal axes are stretched compared to the vertical one.

Figure 4 compares mean trajectories of particles starting from $x = -0.03$m with increasing wall roughnesses (from smooth wall to a rough wall of 6 mm glued particles). The drift, defined as the horizontal distance between the initial position in the static zone and the next pass in that zone, and the curvature, defined as the maximum horizontal distance in one revolution, are modified by the roughness of the wall. With increasing roughnesses, the curvature of the trajectory decreases while the drift increases. For larger roughnesses, the flow trajectories change from simple curves to ones in which the curvature is reversed.

Figure 5 shows particles mean trajectories for two fill levels all along the sphere. For the 20% full tumbler, trajectories are strongly curved for the smooth wall but have almost no drift. For the rough wall of 2 mm glued particles, the curvature is less but the drift is larger. Different roughness of the wall induces very different trajectories. For the 50% full tumbler, the difference in trajectories is smaller.

2.2 Segregation pattern

The curvature and drift of particle trajectories explain the origin of the segregation pattern and its inversion from the SLS to the LSL pattern in terms of the competition between two mechanisms. The first mechanism is a consequence of the drift and the recirculation cells. Particles at the surface of the flow drift toward the poles. In a bidisperse granular flow, due to the classical granular radial segregation, the surface is mainly composed of large

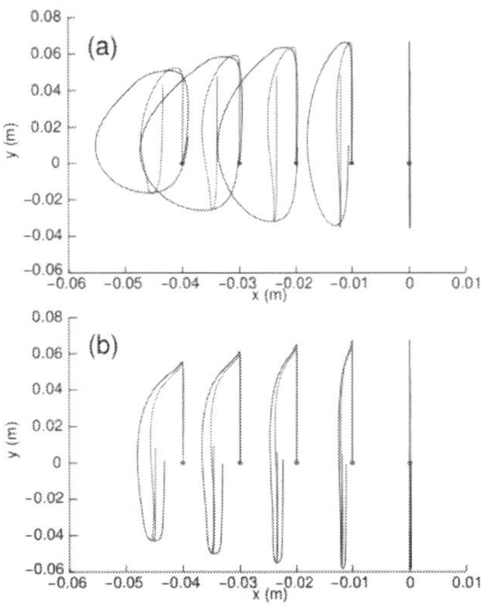

Figure 5. Top view of 2 mm particles mean trajectories at various positions in a 14 cm spherical tumblers filled at (a) 20% and (b) 50%. The spheres have a smooth wall (blue trajectories) or a 2 mm rough wall (red trajectories) and rotate at 15 rpm.

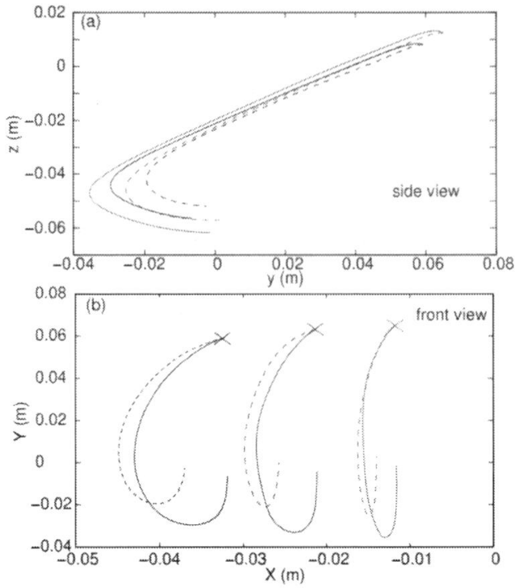

Figure 6. Trajectories of 2 mm (dashed lines) and 4 mm (solid lines) particles in a 14 cm spherical tumblers filled at 30% and rotating at 15 rpm. (a) side view (note that $x - y$ proportions are not the same) and (b) front view. Blue trajectories are omitted in the side view for clarity.

particles. Thus all situations that favor the drift (high fill levels, high roughness) favor the LSL segregation pattern. The second mechanism is linked with the curvature of the trajectories. Fig. 6 shows large (solid lines) and small (dashed lines) particles trajectories starting from the same point (×) in the flowing zone, flowing along a curved path and finishing their trajectories in the static zone. The side view (a) shows that larger particles remain at the surface due to granular segregation and flow further than small particles that reach the static zone sooner. Consequently, due to the curvature and their longer trajectories, larger particles enter in the static zone closer to the equator of the tumbler, while small particles enter closer to the pole (Fig. 6(b)). Thus the curvature combined with granular segregation favors the SLS segregation pattern. All features that increase the curvature (low fill levels, smooth walls, high rotation speeds) favor the SLS pattern.

3 Cylindrical tumbler

The cylindrical tumbler is frequently encountered in industry or in the laboratory. The question that naturally arises is the existence of recirculation cells in cylindrical tumblers.

3.1 Recirculation cells

Figure 7 shows monodisperse particle trajectories in the left half of 3 cylindrical tumblers of various lengths. For all cases, recirculation cells (green) develop close to the end walls. If the tumbler is long enough (b), a second pair of recirculation cells (red) develops at the center of the tumbler. For a longer tumbler (c), the secondary convection cells increase in size while the first cells are almost

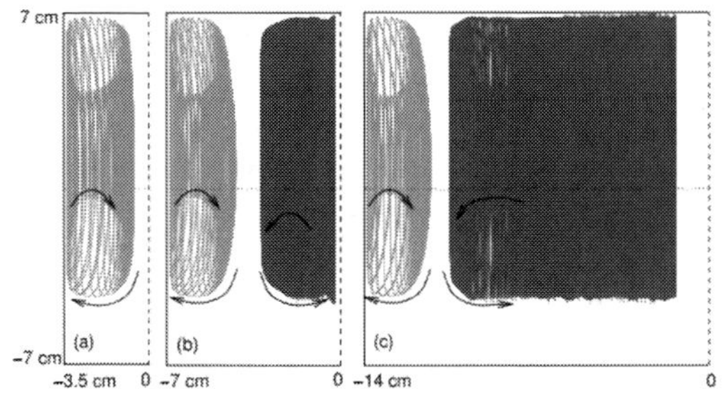

Figure 7. Particle mean trajectories obtained in 14 cm diameter cylindrical tumblers with lengths of (a) 7 cm, (b) 14 cm and (c) 21 cm. Tumblers have smooth walls, rotate at 15 rpm and are filled at 30% with 2 mm particles. The vertical dashed lines show the center of the tumblers; only the left half of the tumbler is shown, since the trajectories are symmetric. Arrows indicate the direction of recirculation.

unaffected. On the other hand, for short tumblers (a), the secondary center cells disappear. Arrows indicate the direction of circulation in the cell. Primary and secondary cells circulate in opposite directions.

To estimate more accurately the size of each recirculation cell, the drift is measured. Since neighbouring cells circulate in opposite directions, the boundary between two cells corresponds to a null drift. Fig. 8(a) shows the drift for tumblers ranging from 7 to 28 cm long. The two regions between the cylinder end walls and the intersections with horizontal axis delimit the first cells. Fig. 8(a) confirms that the extent of the cells next to the end walls

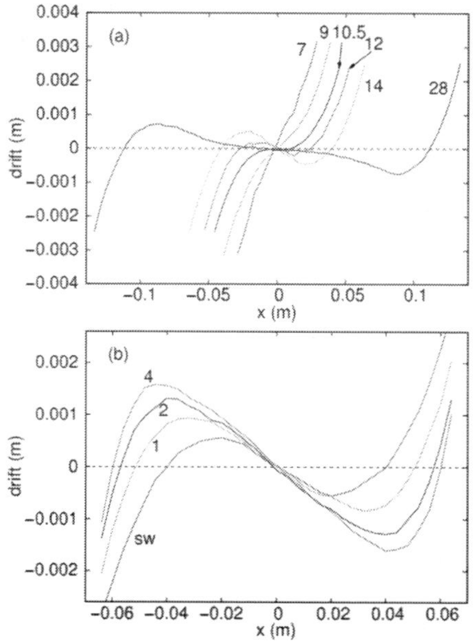

Figure 8. Drift measured along cylindrical tumblers for (a) tumblers of various lengths (7 to 28 cm) and (b) 14 cm cylindrical tumblers made of wall with various roughness (smooth wall (sw) to 4 mm rough). The boundary between two cells occurs at positions of null drift.

(region where drift < 0 in the left half) is almost constant, while the secondary cells (region where drift > 0) fill the empty space between the primary cells, and eventually vanish for tumblers of lengths ≤ 10.5 cm. Fig. 8(b) shows the influence of the wall roughness: with increasing roughness, the size of the cell near the endwall decreases.

3.2 Segregation pattern

The properties of monodisperse particle trajectories in cylindrical tumblers can be used to explain some features of axial segregation in cylinders. As occurs in a spherical tumbler, the drift pushes particles flowing at the surface toward the poles, which are the end walls here. The trajectory curvatures are inverted compared to the sphere. Thus, the two phenomena are not competing, but both lead to a migration of large particles toward the end wall, explaining why experiments and simulations report the fast emergence of bands of large particles next to the end walls.

The size of the large particle band near the end wall is strongly influenced by the size of the primary cells. This is evident in Fig. 9 showing vertical slices through the bed of particles along a 14 cm cylinder filled with a mixture of 2+4 mm particles that undergoes an axial segregation. In the smooth case, the band of large particles next to the end walls captures almost all 4 mm particles leading to a strong segregation, while for the 4 mm rough case, the bands of large particles are narrower, and the excess of large particles fills the surface of the flowing layer. The decrease of size of the band with roughness is directly connected to the decrease of the size of the primary convection cells with roughness (Fig. 8(b)).

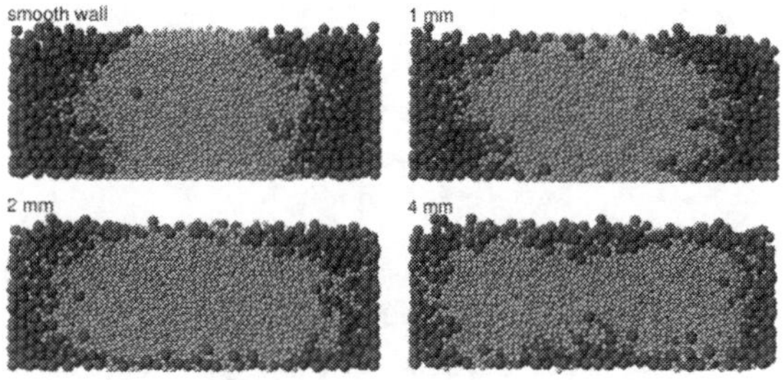

Figure 9. Picture of the segregation pattern (t=200 s) for horizontal 14 cm cylinder tumbler filled at 30% with 2 and 4 mm particles and rotating at 15 rpm. Only particles in a vertical slice including the horizontal axis of rotation are shown. The walls are smooth, or rough made of 1, 2 or 4 mm particles.

4 Conclusions

This study of granular flow in spherical and cylindrical tumblers has demonstrated two competing processes responsible for the axial segregation in rotating tumblers: the drift leading to recirculation cells and the curvature of particle trajectories. Wall roughness, rotation speed, and fill level have an influence in both processes. Drift and curvature combined with radial segregation explain the SLS and LSL segregation patterns and the transition between them in a spherical tumbler. In cylindrical tumbler, primary recirculation cells are present next to the end walls, but secondary cells develop only if the cylinder is long enough. In the primary cells, trajectory drift and curvature combine to send large particles toward end walls, explaining why axial segregation in cylinder always has bands of large particles near the end walls, and why the SLS segregation pattern is never obtained in cylinder.

Acknowledgements

UDO and NT had access to the HPC resource of Aix-Marseille Université financed by the project Equip@Meso (ANR-10-EQPX-29-01).

References

[1] P. Chen, B. J. Lochman, J. M. Ottino and R. M. Lueptow, Phys. Rev. Lett. **102**, 148001 (2009).

[2] U. D'Ortona, N. Thomas, and R. M. Lueptow, Phys. Rev. E **93** 022906 (2016).

[3] U. D'Ortona, N. Thomas, Z. Zaman, and R. M. Lueptow, Phys. Rev. E **92** 062202 (2015).

[4] Z. Zaman, U. D'Ortona, P. B. Umbanhowar, J. M. Ottino, and R. M. Lueptow, Phys. Rev. E **88**, 012208 (2013).

EPJ Web of Conferences **140**, 03012 (2017) DOI: 10.1051/epjconf/201714003012
Powders & Grains 2017

Evolution of the temperature distribution of granular material in a horizontal rotating cylinder

Bereket Yohannes[1], *Heather* Emady[2], *Kellie* Anderson[3], *Maham* Javed[3], *Ingrid J.* Paredes[3], *Fernando J.* Muzzio[3], *William G.* Borghard[3], *Benjamin J.* Glasser[3], and *Alberto M.* Cuitiño[1,*]

[1] *Department of Mechanical and Aerospace, Rutgers University, Piscataway, NJ 08854*
[2] *Department of Chemical Engineering, Arizona State University, Tempe, AZ 85287*
[3] *Department of Chemical and Biochemical Engineering, Rutgers University, Piscataway, NJ 08854*

Abstract. Accurate prediction of the particles' temperature distribution and the time required to heat up the particles is important to maintain good quality products and economical processes for several industrial processes that involve thermal treatment. However, we do not have quantitative models to predict the average temperature or particles' temperature distribution accurately. In this article, we carry out DEM simulations and compute the temporal and spatial evolution of the distribution of the particles' temperature in rotating cylinders. We present typical examples for different particle properties and operating conditions. The temperature distribution follows what is referred to as a uniform distribution with well defined mean and standard deviation values. Our analysis of these statistical parameters can assist in the prediction of the time required to heat up granular materials and the design of efficient processes.

1 Introduction

Several industrial processes involve thermal treatment of granular materials in rotating drums (also referred to as kilns or calciners) [1–5]. The temperature of the particles is raised to a given temperature such that the desired processes take place. For a good product quality and efficient process, it is required to raise the temperature of the particles uniformly with a minimum processing time. However, the temperature of the particles does not increase at the same rate in most cases. The variation of the particles' temperature is caused by 1) heat transfer processes in the radial direction [7, 8] and 2) the dispersion of the particles in the axial direction as the particles flow through the rotating kiln [9]. To achieve economical and high quality production it is essential to understand these two processes in detail.

Several experimental, numerical, and theoretical have been done to study the heat transfer in granular materials in rotating drums [1–8, 10–14]. Based on experiments in rotating calciners and theoretical work based on penetration theory [2, 4, 14], it was shown that the heat transfer coefficient of the bulk granular material can be predicted from the fill level, speed of rotation, and the *bulk* thermal properties of the granular materials and powders. However, there are some difference among the models that were developed in these studies [11]. The heat transfer in granular flows is a complex processes and several factors may have contributed to the variation in these models. One of the factor could be the accuracy in the measurement of the bulk thermal properties. Quantifying the accurate bulk thermal conductivity and heat capacity is very crucial for developing accurate models. Bulk properties are not inherent only to the particles, but also depend on the interstitial fluid (air) and boundary conditions, which might add to the discrepancy in the models. Yet another factor could be the presence of more than one heat transfer mechanisms at the same time[2, 14, 15]. At low temperatures, the heat transfer is dominated by conduction through contact between particles. In addition, conduction through thin gas layer in the vicinity of the contact area between particles and contact area between particles and wall could have a significant contribution to the heat transfer process [14]. At higher temperatures, greater than 700 C, radiation may dominate the heat transfer process. Most of the models in the literature discussed above are limited to low temperature (less than 200 C) experiments and simulations. In the presence of hot air inflow, usually referred to as direct heating processes, the convective heat transfer should be also incorporated in the model [6]. However, developing a complete model that includes all the heat transfer mechanisms is very challenging.

To improve the existing models, it is essential to relate the properties of individual particles to the bulk properties of the granular material. Recently, Emady et al [8] and Yohannes et al [7] introduced dimensionless time parameters that depend on the particle properties and operating conditions. One of the most important time parameters, τ_p, quantifies the time required to heat up a single particle resting on a horizontal wall. τ_p is a function of the thermal, mechanical, physical properties of the particle.

* e-mail: cuitino@jove.rutgers.edu

These two studies showed that the time required to heat up the bulk granular material can be predicted based on τ_p and the time a particle spends in contact with the wall in rotating drums, τ_c.

In this paper, we use 3D discrete element method (DEM) simulations to study the evolution of temperature of individual particles when heat is transferred by conduction through inter-particle and particle-wall contacts. DEM is suited for this purpose, since the method is essentially based on tracking individual particles, and the particles' interaction among themselves and the boundary. In DEM, all the input material properties, including the thermal properties, are properties of the individual particles and no assumption has to be made about the bulk property. This is one of the great advantages of using DEM simulation instead of other methods, such as the finite element method [6], which require bulk material properties as an input. The results from DEM simulations can be used to predict these bulk properties based on particle properties. Using DEM, we show the evolution of the spatial and temporal distribution of particles' temperature and discuss how the results can be used to predict the average temperature and temperature distribution of the bulk granular materials.

2 Numerical Experiments

We run several DEM simulations in rotating drums with periodic boundary condition along the axial direction. The particles are spherical, cohesionless, and elastic. The particle-particle and particle-wall contact force follows Hertz-Mindlin contact theory [16–19]. In this contact theory, the particles are allowed to deform when they are in contact. Due to the deformation of the particles, a contact area is formed through which heat conduction can occur. The contact area is assumed to be circular with a radius of a. In the simulations, the deformation of the particles is represented by the overlap (δ) between a pair of particles (or between a particle and the drum wall). a is computed as $\sqrt{2R\delta}$, where R is the effective radius of a pair of contacting particles. Then, the heat flux (Q_{ij})is computed based on the a as,

$$Q_{ij} = H_c(T_j - T_i) \tag{1}$$

where H_c is the heat conductance, T_i and T_j are the temperatures of particle i and j, respectively. $H_c = 2ka$, where k is the effective thermal conductivity of the particles[20]. Based on the Q_{ij} the rate of change of temperature for any particle is computed as

$$\frac{dT_i}{dt} = \frac{Q_i}{\rho_i C_{pi} V_i} \tag{2}$$

where ρ_i, C_{pi}, and V_i are the density, specific heat capacity, and volume of particle i. Q_i is the summation of all Q_{ij} related to particle i.

The drum wall temperature T_w is kept constant during the entire simulation, while the initial temperature of the particles T_o is set to 298 K. Hence, heat is supplied

Table 1. Material properties

Material property (symbol) [unit]	Value
average particle size (d_p) [mm]	2, 4, 8
particle density (ρ) [kg/m^3]	120 - 15560
shear modulus (G) [Pa]	10^6-10^8
Poisson's ratio (ν)	0.25
thermal conductivity (k) [W/mK]	0.3 - 3000
specific heat capacity (C_p) [J/kg·K]	880
particle initial temperature (T_o) [K]	298
coefficient of restitution (e)	0.2, 0.9
coefficient of friction (μ)	0.4, 0.9

Table 2. operating conditions

Operating condition (symbol) [unit]	Value
drum diameter (D) [cm]	15-120
rotation speed (ω) [rpm]	0.1 - 60
fill level (F_L) [%]	5-45
drum wall temperature (T_w) [K]	578

to the particles only through direct heating from the drum wall, and we did not include any other heat sources such as heat generated due to frictional energy dissipation. The simulations are run until the average temperature of the particles $\bar{T}$ approaches T_w. The thermal, mechanical, and physical properties of the particles were varied in order to understand the effect of each property on the heat transfer process. In addition, the operating condition such as the diameter of the drum, the speed of rotation, and the fill level were also varied. The list of material properties and operating conditions is shown in Tables 1 and 2, respectively. We did not find any noticeable change in the heat transfer process when we changed the coefficients of friction and restitution.

3 Results

Initially, the temperature of the particles that are in contact with the wall increases, and the heat is transferred to the bulk granular materials through inter-particle contacts. This type of heat transfer can be considered as pure heat conduction. In addition, heat is transferred when the heated particles, which were in contact with the wall, are pushed to the surface of the flowing layer of the granular materials. This type of heat transfer can be considered as heat transfer through granular convection [21]. Fig. 1 shows the temperature of the particles at different time steps for one simulation. As can be seen from the figure, the particles in contact with the wall and particles at the surface of the granular materials have higher temperature than the particles in the core of the bulk material. Eventually, the core particles are also heated and the temperature of the particles becomes uniform. The rate of heating and the pattern of the spatial distribution of the particles' temperature depends on the operating condition and the particles properties [7, 8].

DOI: 10.1051/epjconf/201714003012

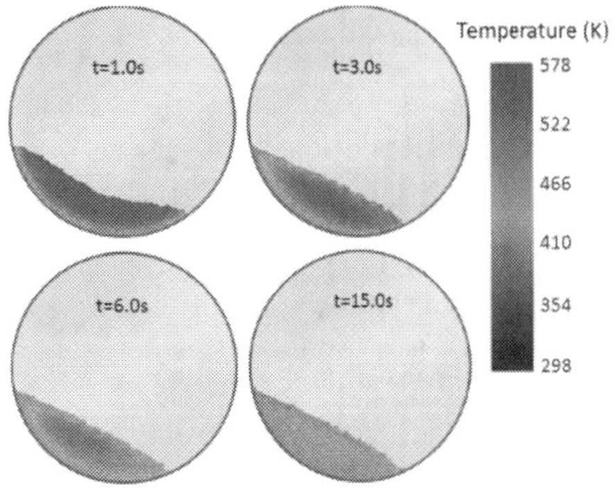

Figure 1. Snapshots from one simulation at different time steps to show the evolution of the particles' temperature. The color bar represents the temperature of the particles. k= 3000W/mk, ω= 3rpm, F_L = 15%, and d_p=2mm.

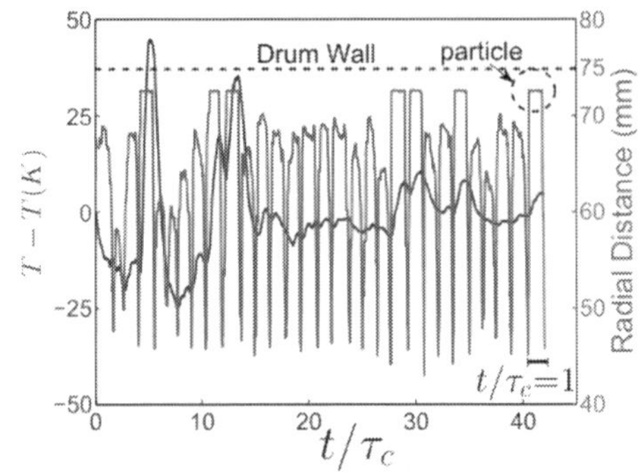

Figure 2. Temperature fluctuation ($T - \bar{T}$) of a particle (on the left, blue) and radial distance of the particle from the center of the drum (on the right, red) vs t/τ_c. The drum has a diameter D = 15 cm (The horizontal dashed line indicates the radius of the drum (75mm)). The radial distance is constant when the particle is in contact with the wall. The contact duration equals to τ_c.

During the process, the temperature of a particle does not increase monotonically during a simulation. Fig. 2 shows the fluctuation of the temperature of one particle with respect to $\bar{T}$. The fluctuations are larger during the early stages of the simulation, when the difference between T_w and $\bar{T}$ is large. The deviations decays continuously as the heating process is continued. Fig. 2 also shows the radial position (radial distance from the center of the drum) of the particle. It can be noted that the temperature of the particle rises significantly when the particle is in contact with the drum wall (when radial distance of the center of the particles is equal to $\frac{D}{2} - \frac{d_p}{2}$). Most of the time, the rise the particle's temperature is followed by a sharp decrease in temperature, which corresponds to the particle being trapped in the bulk or pushed to the surface. If the particle is brought to the surface, the particle tends to circulate back the bottom layer (in contact with the wall) which causes a significant increases of particle's temperature again. On the other hand, if the particle is trapped inside the core, the temperature of the particle increases with $\bar{T}$ of the system ($T - \bar{T} \approx 0$).

Fig. 3 shows a typical plot for the evolution of particles' temperature during the simulation and the volume averaged temperature for the bulk granular material. All particles do not have the same temperature at a given time and they do not follow the same path towards the final temperature. However, the average temperature increases monotonically throughout the simulation. The lower bound of the particles' temperature is also more uniform than the upper bound of the particles' temperature, where deviation from the mean is very large. This indicates that largest changes in the particle's temperature occur when the particles are in contact with the heated wall. When the particles are away from the wall, their temperature drops to the average temperature of the bulk.

Fig. 4 shows the probability distribution function $f(T)$ for the particles' temperature at different time steps. At t=0s, all particles have the same temperature T_o = 298K and $f(T)$ is a delta function. For most of the simulation duration $f(T)$ resembles a uniform distribution (also known as rectangular distribution). However, the width of the $f(T)$ changes during the heating process, and finally reaching a delta function when the temperature of all particles reach T_w = 578 K. Statistically, the width of a uniform distribution is given by minimum temperature T_{min} and the maximum temperature T_{max}. The lower bound of T_{min} is T_o and the upper bound of T_{max} is T_w. However, for the simulation data where $f(T)$ is not perfectly a uniform distribution and where T_{min} and T_{max} are not clearly defined, the average temperature $\bar{T}$ and the standard deviation σ_T of the temperature are the more robust parameters to characterize the temperature distribution. Recently, Yohannes et al [7] showed that $\bar{T}$ and σ_T can be predicted based on ϕ (where $\phi = \frac{\tau_p}{\tau_c}$).

4 Discussion

Accurately predicting the temperature distribution, which includes accurate prediction of $\bar{T}$ and σ_T, is crucial for improving the quality of the final product after processes that involve thermal treatment of particles. Ideally, it is desirable to heat up the particles uniformly, at the same rate. However, it is not always possible to attain the same temperature for all of the particles as shown in Figs. 1, 3, and 4. Yohannes et al [7] have shown that for certain conditions, when the value of ϕ is sufficiently large, σ_T is very low. In general, high ϕ values correspond to slow heating rate. In terms of particle properties, high ϕ values correspond low thermal conductivity, high specific heat capacity, high stiffness, and high density of particles. In terms of operating conditions, high ϕ corresponds to low fill level and high speed of rotation. To improve the uniformity of the final product for a given material, particulary a material with high thermal conductivity and low density, decreasing the fill level or increasing the speed of rotation can be used as a solution.

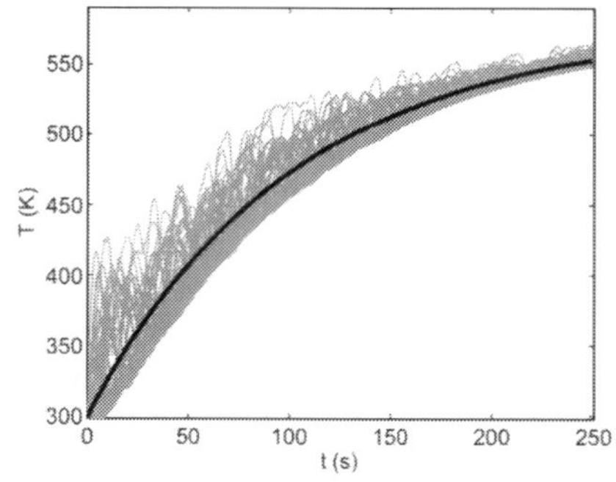

Figure 3. The evolution of individual particle temperature: individual particles temperature (green) and $\bar{T}$ (black). D=15cm, k=30W/mk, ω = 3rpm, F_L = 15%, and d_p=4mm .

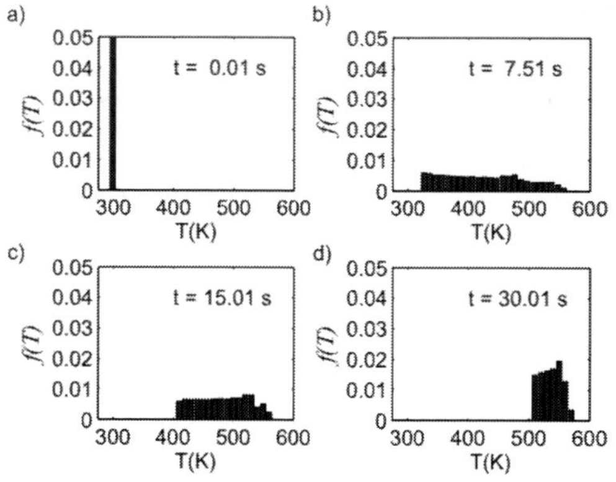

Figure 4. Probability distribution function $f(T)$ of individual particles' temperature at a) t=0.01s, b) t=7.5s, c)t=15.0s, and d) t= 30.0s. D=60cm. k= 3000W/mk, ω= 9rpm, F_L = 15%, and d_p=4mm.

Other mechanisms can also be used to improve the uniformity of the temperature of the particles. One such mechanism is addition of baffles (sometimes referred to as lifters or flights) to the calciners [22, 23]. In these studies, it was shown that the uniformity of the particles' temperature is improved by the addition of baffles, and the uniformity improves significantly when the number of baffles is increased. The baffles enhance the mixing process in the rotating drum and, therefore, reduce the number of particles that remain in the cooler core of the bulk granular materials. Combining the predictive models developed by Yohannes et al [7] and effect of other mechanism, such as baffles and dams, the uniformity of temperature distribution can improved significantly. Most of the studies, including the current one, that attempted to quantify the temperature distribution are limited to conductive heat transfer through contacts between particles. It is also very important to develop such models for radiative and convective heat transfer for granular materials in rotating drums, as these two heat transfer mechanisms are also very common in many processes.

Acknowledgements

We gratefully acknowledge the support of the Rutgers Catalyst Manufacturing Consortium.

References

[1] A. Watkinson and J. Brimacombe, Metallurgical Transactions B **13**, 369 (1982)

[2] G. Wes, A. Drinkenburg, and S. Stemerding, Powder Technology **13**, 177 (1976)

[3] C. Argento and D. Bouvard, International Journal of Heat and Mass Transfer **39**, 1343 (1996)

[4] J. Lehmberg, M. Hehl, and K. Schügerl, Powder Technology textbf 18, 149 (1977)

[5] S. H. Tscheng and A. P. Watkinson, The Canadian Journal of Chemical Engineering **57**, 57 (1979)

[6] A. Boateng and P.Barr, International Journal of Heat and Mass Transfer **39**, 2131 (1996)

[7] Yohannes, B., Emady, H., Anderson, K., Paredes, I., Javed, M., Borghard, W., Muzzio, F. J., Glasser, B. J., and Cuitiño, A. M., Phys. Rev. E **94**, 042902 (2016)

[8] H. N. Emady, K. V. Anderson, W. G. Borghard, F. J. Muzzio, B. J. Glasser, and A. M. Cuitino, Chemical Engineering Science **152**, 45 (2016).

[9] I. J. Paredes, B. Yohannes, H. Emady, B. J. Glasser, W. G. Borghard, F. Muzzio, A. M. Cuitiño, J. Beeckman, S. Ilias, P. Podsiadlo, E. Jezek, J. Baumgartner, Chemical Engineering Science **158**, 50 (2017)

[10] P. Thammavong, M. Debacq, S. Vitu, and M. Dupoizat, Chemical Engineering and Technology **34**, 707 (2011)

[11] A. I. Nafsun, F. Herz, E. Specht, V. Scherer, and S. Wirtz, Experimental Heat Transfer **29**, 520 (2016).

[12] W. L. Vargas and J. J. McCarthy, AIChE Journal **47**, 1052 (2001)

[13] I. Figueroa, W. L. Vargas, and J. J. McCarthy, Chemical Engineering Science **65**, 1045 (2010)

[14] S.-Q. Li, L.-B. Ma, W. Wan, and Q. Yao, Chemical Engineering and Technology **28**, 1480 (2005)

[15] A. Morris, S. Pannala, Z. Ma, and C. Hrenya, International Journal of Heat and Mass Transfer **89**, 1277 (2015).

[16] H. Hertz, J. Reine Angew. Math. **92**, 156 (1882)

[17] R. D. Mindlin and H. Deresiewicz, J. Applied Mech. **20**, 327 (1953)

[18] K. L. Johnson, *Contact mechanics* (Cambridge University Press, Cambridge, 1985), page number

[19] Y. Tsuji, T. Tanaka, and T. Ishida, Powder Tech. **71**, 239 (1992)

[20] G. K. Batchelor and R. W. O'Brien, Proc. R. Soc. London A **355**, 313 (1977).

[21] P. Rognon and I. Einav, Phys. Rev. Lett. **105**, 218301 (2010).

[22] B. Chaudhuri, F. J. Muzzio, and M. S. Tomassone, ibid. **198**, 6 (2010).

[23] M. Li, X. Ling, H. Peng, Z. Cao, and Y. Wang, Can. J. Chem. Eng. doi:10.1002/cjce.22681 (2016)

DOI: 10.1051/epjconf/201714009004

Evolution of Clouds of Migrating Micro-particles with Hydrodynamic and Electrostatic Interactions

Sheng Chen[1] and *Shuiqing* Li[1,*]

[1] *Key Laboratory for Thermal Science and Power Engineering of Ministry of Education, Department of Thermal Engineering, Tsinghua University, Beijing 100084, China*

Abstract. The evolution of a migrating cloud of particles with electrohydrodynamic interactions is numerically investigated. The hydrodynamic interaction is modelled by Oseen dynamics in the limit of small-but-finite particle Reynolds number. The effects of external field and the particle-particle Coulomb repulsion, calculated through pairwise summation, are quantified by a charge parameter, κ_q. With a small or zero κ_q, the cloud is seen to flatten into a toroidal configuration and eventually breaks up into two small clouds, indicating that the external electrostatic field has a similar effect with gravity. Increasing the long-range Coulomb repulsion can delay or even totally prevent the breakup. With sufficiently strong repulsion, the cloud undergoes a self-similar expansion. Finally, we show that the breakup of the cloud is strongly related to the initial inhomogeneity of the particle concentration. This chaotic characteristic, however, is dramatically depressed by the long-range Coulomb repulsion.

1 Introduction

The migration of charged particles in an electrostatic field is ubiquitous in engineering processes, including dust removal in an electrostatic precipitator [1, 2], films fabrication using electrophoretic deposition [3] and self-assembly of colloidal particles [4]. During these process, both hydrodynamic and electrostatic interactions between multiple particles result in a wealth of complex collective behaviours of the particle clouds.

For flows with a sufficiently low Reynolds number, the hydrodynamic interaction among particles is usually simulated by solving the Stokes/Oseen equations. The disturbance of every individual particle on the flow field is linearly summed to obtain the flow field at the position of other particles [5, 6]. Stokes dynamics has been used to investigate the evolution of the settling velocity and the shape of a particle cloud during gravity settling in conditions where inertia is negligible [7]. Cloud settling in the case of small-but-finite Reynolds number was investigated by Pignatel *et al.* It is proposed that the evolution of the cloud deformation is accelerated as the inertia is increased [8]. Yang *et al.* proposed analytical expressions of settling velocities for both spherical and columnar particle clouds, and compared them with the result of Stokes clouds to reflect the effect of the fluid inertia [9].

Most studies are focused on cloud settling under gravity, while the migrating of charged particles in an electrostatic field is rarely involved. The biggest difference between the electrostatic migration process and the gravitational settling process is the need to account for the inter-particle Coulomb repulsion. This long-range repulsion has been found to have significant effects on particle packing structure [10]. However, the migration behavior of particle clouds with a coupling effect of hydrodynamic/electrostatic interactions still remains unclear. In this paper, we try to understand this process in the limit of small-but-finite particle Reynolds number through Oseen simulations with special attention paid to the effect of long-range Coulomb repulsion.

* e-mail: lishuiqing@tsinghua.edu.cn

2 Models and Methods

2.1 Simulation conditions

A sketch of the simulated system is shown in Fig. 1. A spherical cloud with radius R_0 containing $N = 1000$ charged particles is immersed in an unbounded fluid of viscosity μ_f. The fluid is at rest at infinity and the cloud is migrating under an action of a uniform electric field. The radius r_p, density ρ_p, charge q_0 of particles and the strength of the electric field intensity E_0 are also listed in Fig. 1. Besides the hydrodynamic interactions (modelled by Oseen dynamics), we also consider the long-rang electrostatic forces: a uniform external electrostatic force in x direction ($E_0 q_0$) and a pairwise Coulomb repulsion ($q_0^2/4\pi\varepsilon r_{ij}^2$). It should be mentioned that, higher-order multipoles, *e.g.*, dipoles or quadrupoles, decay sufficiently fast with the distance and are ignored in this work due to the large separation between particles. Such interactions, which may have effects on particles dynamic when the concentration of particles in the cloud is high, will be left to future work. During our simulation, the information of

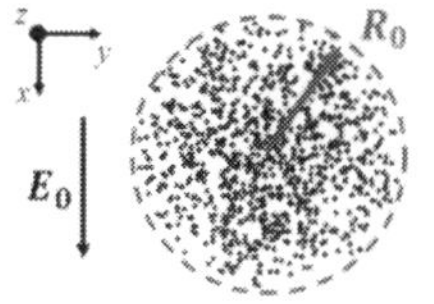

Figure 1. Sketch of the simulated system and main parameters.

all particles (position, velocity and forces) and flow field (velocity and pressure) are simultaneously recorded. Thus the temporal evolution of migrating cloud can be easily captured.

2.2 Oseen Dynamics

For a particle with radius r_p translating with a velocity U_S relative to the surrounding fluid, the generated flow field can be written as:

$$u_r = \frac{U_S r_p^2}{r^2}\left\{-\frac{r_p}{2r}\cos\theta - \frac{3(1-\cos\theta)r}{4r_p}\exp\left(-\frac{Re_S r(1+\cos\theta)}{2r_p}\right) + \frac{3}{2Re_S}\left[1-\exp\left(-\frac{Re_S r(1+\cos\theta)}{2r_p}\right)\right]\right\}$$

$$u_\theta = -\frac{U_S r_p}{r}\sin\theta\left[\frac{r_p^2}{4r^2} + \frac{3}{4}\exp\left(-\frac{Re_S r(1+\cos\theta)}{2r_p}\right)\right]. \quad (1)$$

With the polar axis ($\theta = 0$) coincident with the direction of particle motion. $Re_S = 2r_pU_S\rho_f/\mu_f$ is the instantaneous particle's Reynolds number based on particle slip velocity U_S. Since the particle Stokes number is very small ($\sim 10^{-2}$) in the current simulations, it is reasonable to assume that the particle inertia is negligible. Thus $\mathbf{U}_S$ can be directly calculated from the external forces acting on the particle through $\mathbf{U}_S = \mathbf{F}_{ext}/6\pi\mu_f r_p$. Then the velocity for a particle i inside the cloud can be obtained as a summation of its slip velocity $\mathbf{U}_{S,i}$ and the fluid velocity $\mathbf{u}_i$ at its centre, which is solved by a matrix equation $\mathbf{u}_i = \sum_{j\neq i}\mathbf{W}_{ij}\mathbf{U}_{S,j}$ once the slip velocity of each particle is known.

In this work, the electrostatic forces are the only external forces and the equation of motion for i^{th} particle can be expanded as

$$\mathbf{v}_i = \mathbf{u}_i + \mathbf{U}_{S,i} = \mathbf{u}_i + \frac{1}{6\pi\mu_f r_p}\left(\mathbf{E}_0 q_i + q_i\sum_{j\neq i}\frac{q_j\mathbf{r}_{ij}}{4\pi\varepsilon_0 r_{ij}^3}\right). \quad (2)$$

Here, $\mathbf{E}_0$ is the external field in x direction and q_i is the charge on particles which is identical to q_0 in this work. Normalizing the velocity by the migrating velocity of an isolated particle $U_0 = E_0q_0/6\pi\mu_f r_p$, the length by cloud radius R_0, and the time by R_0/U_0, Eq. 2 becomes

$$\hat{\mathbf{v}}_i = \hat{\mathbf{u}}_i + \left(1\mathbf{e}_x + \kappa_q\frac{1}{N}\sum_{j\neq i}\frac{\hat{\mathbf{r}}_{ij}}{\hat{r}_{ij}^3}\right). \quad (3)$$

Several dimensionless parameters influence migrating process. The most important one is particle Reynolds number $Re_p = E_0q_0\rho_f/6\pi\mu_f^2$, which has been related to the electrostatic properties. Another dimensionless parameter indicated by Eq. 3 is the charge parameter, κ_q, which is defined as

$$\kappa_q = \frac{q_0 N}{4\pi\varepsilon_0 E_0 R_0^2}. \quad (4)$$

κ_q is interpreted as the ratio of the order of magnitude of particle-particle Coulomb repulsion to that of uniform external force.

3 Results and discussions

3.1 Clouds evolutions

The typical evolution of a cloud with different strength of Coulomb repulsion (κ_q) is shown in Fig. 2. The cloud with $\kappa_q = 0$ (Fig. 2(A)), is seen to flatten (at dimensionless time $\hat{t} = 15$) and to transit into a toroidal shape ($\hat{t} = 30$). Then the toroidal cloud expands and finally breaks up into two small clouds ($\hat{t} = 60$) which may undergo repeating beak-up. This is qualitatively in accord with the simulation and the experiments of settling under gravity [7, 8]. The result indicates that the external electrostatic field has a similar effect with gravity.

In the presence of long-range Coulomb repulsion (Fig. 2(B)), the vertical collapse of the cloud still remains. However, instead of an immediate breakup at $\hat{t} = 30$, the cloud remains in an oblate shape. Breakup is delayed to $\hat{t} = 60$. When κ_q further increases, the cloud undergoes a self-similar expansion (Fig. 2(C). No breakup happens during a long-term observation ($\hat{t} \sim 100$). These results suggest that the long-range repulsion makes the cloud more stable during its migration.

3.2 Migrating velocity and aspect ratio

The effect of the long-range repulsion can be further quantified by the temporal evolution of the migrating velocity and the aspect ratio of the clouds (Fig. 3). The migrating velocity has been normalized by the migrating velocity of an isolated particle. The initial migrating velocity at $\hat{t} = 0$ is $U_x/U_0 = 2.13$, which accords well with the analytical prediction $U_x/U_0 = 1.2(r_p/R_0)Nf(l^*) + 1 = 2.12$ of cloud settling under gravity [9]. This result indicates that the uniform electrostatic force plays a similar role as gravity. And the prediction of Yang *et al.*[9] still holds after substituting the migrating velocity of a charged particle ($E_0q_0/6\pi\mu_f r_p$) for the terminal settling velocity of a particle under gravity ($m_pg/6\pi\mu_f r_p$).

The decreasing trends of U_x/U_0 for the three cases are quite similar, however the underlying physical mechanism is different. For the cloud without long-range repulsion, the drop of the U_x/U_0 is mainly due to the evolution towards a torus and the following breakup, which

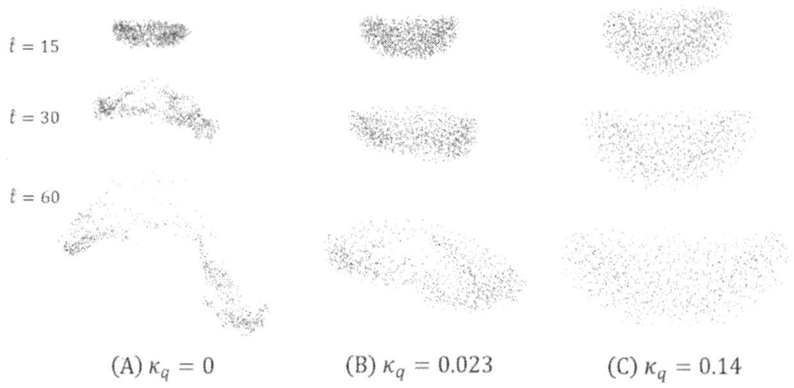

Figure 2. Typical evolution of clouds with $Re_p = 1.27 \times 10^{-3}$ and (A) $\kappa_q = 0$, (B)$\kappa_q = 0.023$ and (C)$\kappa_q = 0.14$.

splits the big cloud into small clouds and weakens the particle-particle hydrodynamic interaction (Fig. 2(A)). As for clouds with long-range repulsion, despite the absence of any breakups, a radial expansion also leads to a decrease of hydrodynamic interaction resulting a smaller migrating velocity. And the decrease rate of U_x/U_0 is larger for cloud with a higher κ_q.

The temporal evolution of the aspect ratio is a more intuitive way to reflect the difference between migrating clouds with and without long-range repulsion. The aspect ratio is defined as $\lambda = R_h/R_v$, where the horizontal radius R_h is defined as the average of the maximum distance from the centre of mass over four quadrants in the horizontal plane and the vertical radius R_v is the distance form the leading particle to the center of mass of the cloud. As shown in Fig. 3(B), the initial increase of λ is caused by the vertical collapse and then remarkably difference can be observed. For cloud without long-range repulsion ($\kappa_q = 0$), λ rapidly decreases from the peak (a) at $\hat{t} = 17$ to the point (b) at $\hat{t} = 33$ (the corresponding snapshots are shown below the figure). It is caused by the breakup of the cloud. Once two small clouds are formed, the inhomogeneity of the distribution of particles inside the two clouds causes a obvious difference between their migrating velocities. Consequently, they separate in the vertical direction, which leads to a increase of R_v i.e. a smaller λ. To some extent, the increasing-decreasing trend of $\lambda \sim \hat{t}$ can be regarded as a sign of cloud's destabilization.

In the presence of long-range repulsion, the decrease of λ is delayed ($\kappa_q = 0.023$) or even completely disappeared ($\kappa_q = 0.14$). This result suggests that the long-range Coulomb repulsion can help the cloud maintain a stable configuration.

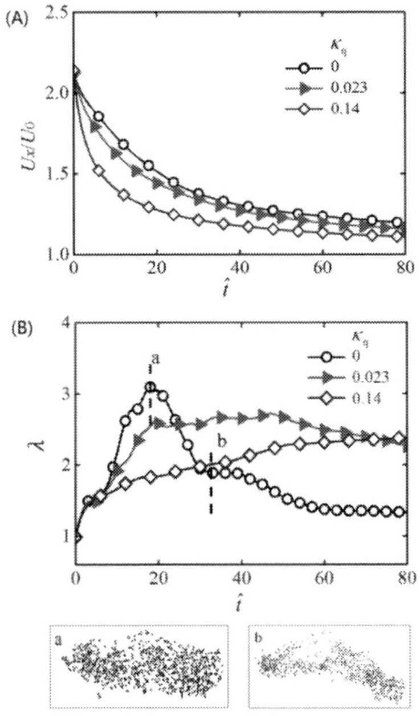

Figure 3. (A) Normalized cloud migrating velocity U_x/U_0 and (B) aspect ratio λ versus time $\hat{t}$. The snapshots below are taken at (a) $\hat{t} = 17$ and (b) $\hat{t} = 33$ in the case of $\kappa_q = 0$.

3.3 Stability of migrating clouds

In this section, we briefly discuss the stability of migrating clouds. It has been reported that long-range hydrodynamic interactions lead to a complex chaotic dynamics and the evolution of a migrating cloud is extreme sensitive to initial configurations[7, 11]. Therefore, we plot the angular probability distribution of particles in the cloud at three typical moment $\hat{t} = 0$, 17 and 33 to show the relationship between initial configurations and clouds' destabilization. As shown in Fig. 4(A), for the case without Coulomb repulsion, the non-uniformity in the angular dis-

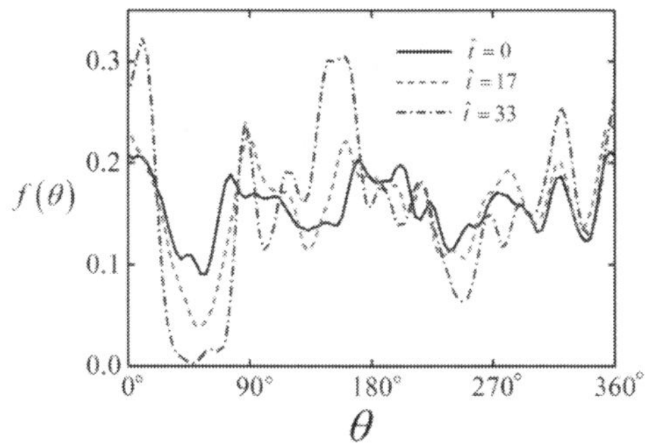

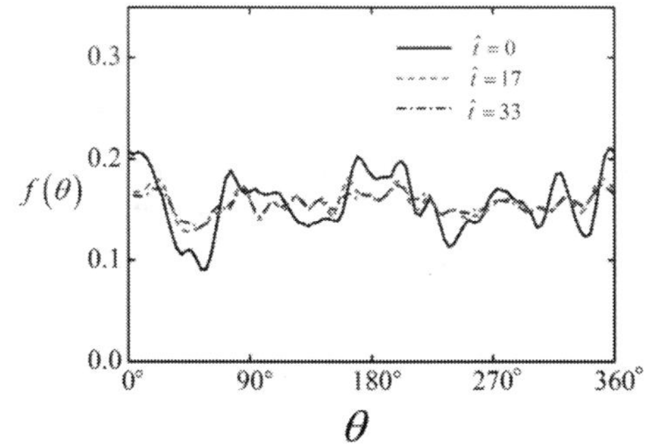

Figure 4. Angular probability distribution of particles in a cloud with (A) $\kappa_q = 0$ and (B) $\kappa_q = 0.14$ at $\hat{t} = 0$, 17 and 33.

tribution is enhanced with time. At $\hat{t} = 33$, the cloud is drastically stretched in the directions with higher particle concentrations(around 0° and 150°). In these particle-rich regions, a higher local concentration leads to a stronger hydrodynamic interaction and a larger local migrating velocity, which eventually induce the breakup. In this regard, we infer that for the migrating cloud, which has a strong chaotic characteristic, the breakup phenomenon is strongly related to the inhomogeneity of initial particle distribution.

On the other hand, for the cloud with a sufficiently strong long-range repulsion ($\kappa_q = 0.14$), the initial inhomogeneity is obviously reduced during the migration. The angular probability distributions at $\hat{t} = 17$ and $\hat{t} = 33$ are essentially identical and are close to the average value ($1/2\pi \approx 0.159$). These results suggest that, an alteration of micro-scale interaction can cause a significant change in the macro behaviour of a particle system. To be specific here, the long-range Coulomb repulsion increases the stability of migrating clouds.

4 Conclusions

In summary, we use Oseen dynamic simulation with electrostatic interactions taken into account to investigate the electrohydrodynamic phenomenon of a migrating particle cloud. It is observed that, the long-range Coulomb repulsion can delay or even totally prevent the breakup of the cloud. Entirely different evolutions of the aspect ratio λ are demonstrated: an increasing-decreasing trend of λ, which can be regarded as a sign of destabilization, is observed for cloud without long-range repulsion, while a cloud with strong Coulomb repulsion undergoes a self-similar expansion with a slowly changing λ. Finally, we show that the breakup of the cloud is strongly related to the initial inhomogeneity of the particle concentration. This chaotic characteristic of a migrating cloud is dramatically depressed by the long-range Coulomb repulsion.

Acknowledgements

This work has been funded by the National Key Research and Development Program (Grant No. 2016YFB0600602) and the National Natural Science Funds of China (No. 51390491)

References

[1] J.S. Marshall, S. Li, *Adhesive Particle Flows* (Cambridge University Press, 2014)

[2] M. Yang, S. Li, G. Liu, Q. Yao, *Electrically-enhanced deposition of fine particles on a fiber: a numerical study using DEM*, in *AIP Conf Proc* (2013), Vol. 1542, pp. 943–946

[3] J. Cordelair, P. Greil, Journal of Materials Science **39**, 1017 (2004)

[4] A. Kumar, B. Khusid, Z. Qiu, A. Acrivos, Physical review letters **95**, 258301 (2005)

[5] I. Proudman, J. Pearson, Journal of Fluid Mechanics **2**, 237 (1957)

[6] E. Guazzelli, J.F. Morris, *A physical introduction to suspension dynamics*, Vol. 45 (Cambridge University Press, 2011)

[7] B. Metzger, M. Nicolas, E. Guazzelli, Journal of Fluid Mechanics **580**, 283 (2007)

[8] F. Pignatel, M. Nicolas, E. Guazzelli, Journal of Fluid Mechanics **671**, 34 (2011)

[9] M. Yang, S. Li, J.S. Marshall, Journal of Aerosol Science **90**, 154 (2015)

[10] S. Chen, S. Li, W. Liu, H.A. Makse, Soft matter **12**, 1836 (2016)

[11] I.M. Jánosi, T. Tél, D.E. Wolf, J.A. Gallas, Physical Review E **56**, 2858 (1997)

Percolation study for the capillary ascent of a liquid through a granular soil

Manuel Antonio Cárdenas-Barrantes[1,*], *José Daniel* Muñoz[1,**], and *Nuno Machado* Araujo[2,***]

[1]*Simulation of Physical Systems Group, Department of Physics, Universidad Nacional de Colombia, Carrera 30 No. 45-03, Ed. 404, Of. 348, Bogota D.C., Colombia.*

[2]*Centro de Física Teórica e Computacional, Departamento de Física, Faculdade de Ciências, Universidade de Lisboa, Campo Grande, P-1749-016 Lisboa, Portugal.*

Abstract. Capillary rise plays a crucial role in the construction of road embankments in flood zones, where hydrophobic compounds are added to the soil to suppress the rising of water and avoid possible damage of the pavement. Water rises through liquid bridges, menisci and trimers, whose width and connectivity depends on the maximal half-length λ of the capillary bridges among grains. Low λs generate a disconnect structure, with small clusters everywhere. On the contrary, for high λ, create a percolating cluster of trimers and enclosed volumes that form a natural path for capillary rise. Hereby, we study the percolation transition of this geometric structure as a function of λ on a granular media of monodisperse spheres in a random close packing. We determine both the percolating threshold $\lambda_c = (0.049 \pm 0.004)R$ (with R the radius of the granular spheres), and the critical exponent of the correlation length $\nu = 0.830 \pm 0.051$, suggesting that the percolation transition falls into the universality class of ordinary percolation.

1 Introduction

Consider a sandpile with its base covered with water. If grains are close enough, water will rise through the interstices of the grains. This capillary rise plays a major role in the transport of fluids across porous media, including water and oil wells. Specially in unsaturated soil structures, like embankments, the capillary rise of water is a real concern, because water can damage the integrity of the structure [1]. The broad spectrum of possible solutions include the use of hydrophobic materials [1] or even the addition of active mechanisms to compensate for the deformations produced by capillary forces [2].

The rising of water through a granular medium is strongly determined by the geometry of the interconnected structure of pores among the grains. First models [3] represented that structure by sites (pore bodies) of arbitrary shape and position interconnected by bonds (pore throats), whose sizes and shapes could be obtained from experimental probes [4, 5]; many properties, like relative permeability [6] or drainage and imbibition [7] can be estimated from this simplified model. More recent works focus on representing the liquid structures among grains (bridges, menisci and pore bodies) as real as possible, based on experiments (as in X-Ray microtomography [8]) or in computer simulations [9, 10]. This modeling is able to reproduce water saturation and drying with the water volume as control parameter and to compute forces and pressures [10]. Together with experiments, they have succeeded identifying trimers (that is, the junction of three liquid bridges and a meniscus) as the minimal building block to build a pathway for rising water [8, 10]. The structure itself must strongly change with the critical length of a capillary bridge (a function of contact angle and surface tension [11]), and some studies in two dimensions have been performed to find when a connected structure first appears as either the contact angle [12] or the liquid volume [13, 14] increases.

The present work investigates how the set of trimers and enclosing pore bodies at disposal for capillary rising changes from a fully disconnected structure to a connected pathway as the critical length of the capillary bridges increases. The goal is to find both the critical half-length λ_c for the transition and the critical exponent ν for the correlation length. The study is performed on the interstices on three-dimensional random close packings of monodisperse spheres. The capillary model of trimers and pore bodies on random close packings is introduced in Sec. 2. Next, Sec. 3 analyses the resulting capillary structures by using tools of percolation theory [15–17]. Finally, Sec. 4 summarizes the main conclusions and discussions.

2 Capillary model

From a microscopic point of view, water can rise if the grains are close enough to build capillary bridges among them. The shape of a capillary bridge between two identical spherical grains is not strongly affected by gravity, but determined by the contact angle θ_c, the liquid-gas surface tension γ, the liquid volume V and the distance S between

*e-mail: macardenasb@unal.edu.co
**e-mail: jdmunozc@unal.edu.co
***e-mail: nmaraujo@fc.ul.pt

DOI: 10.1051/epjconf/201714015007

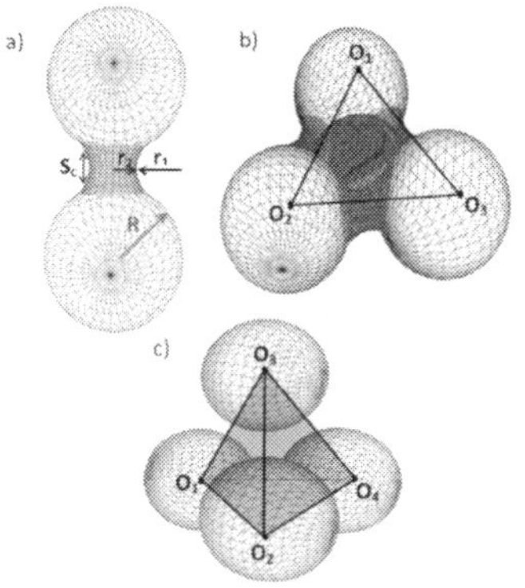

Figure 1. Main capillary structures. (a) Liquid bridge with separation distance S. (b) Trimer, built by three liquid bridges (light blue) and a meniscus (dark blue). (c) A liquid volume enclosed by four trimers.

the grains [18]. There is a maximal distance among grains $S_c = 2\lambda$ a capillary bridge can overcome. If $S < 2\lambda$, a capillary bridge may eventually established for some V; otherwise, there is no possible path between the grains for the water to rise. This limit can be estimated through the toroidal approximation method [19] (Fig. 2) or by numerical simulations.

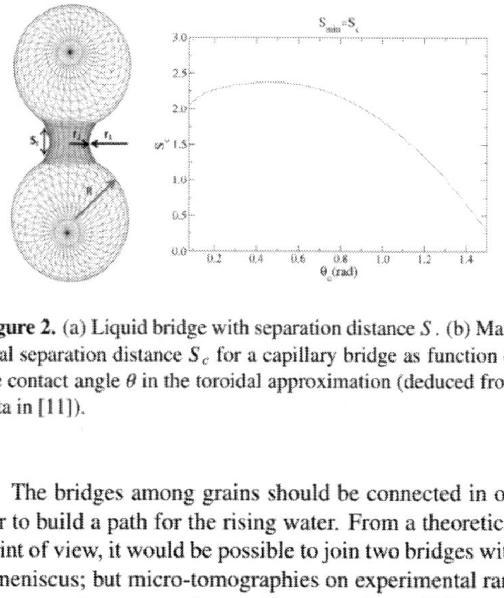

Figure 2. (a) Liquid bridge with separation distance S. (b) Maximal separation distance S_c for a capillary bridge as function of the contact angle θ in the toroidal approximation (deduced from data in [11]).

The bridges among grains should be connected in order to build a path for the rising water. From a theoretical point of view, it would be possible to join two bridges with a meniscus; but micro-tomographies on experimental random granular arrays of monodisperse spheres [8] do not show that kind of structure. On the contrary, the minimal connected structures are trimers, that are the junction of three liquid bridges and a meniscus [10] (Fig. 1). A trimer will eventually form for some water content if three grains are so close together that the distances between any two of them are shorter than S_c and the angles between every two bridges are smaller than $\pi - 2\theta_c$ [11]. Two trimers are assumed connected if they share a bridge, and connected trimers can eventually enclose filling volumes (Fig. 1). Trimers and enclosed volumes form the structure for capillary rising. If S_c is small, the structure is a disconnected set of clusters. On the contrary, if S_c is large enough, there is a connected path across the sample, i.e. a percolating passage for the liquid to rise. The aim of the present work is to characterize the transition across these two regimes - driven by λ, the half maximal length of a capillary bridge - on a monodisperse set of spherical grains by using the standard tools of classical percolation.

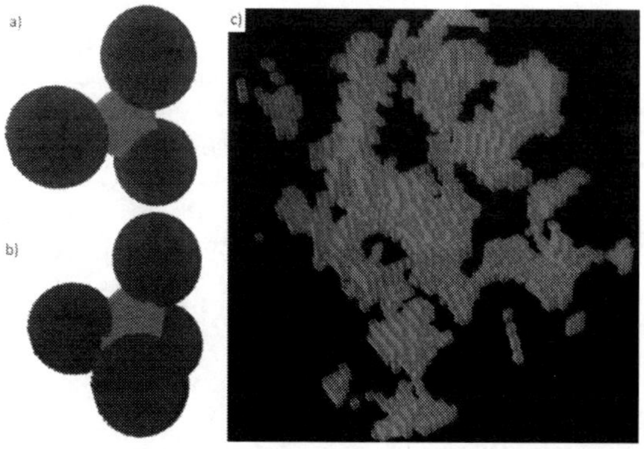

Figure 3. a) Triangle-like liquid structure (trimer). b) Tetrahedron-like liquid structure (enclosed volume). c) Potential capillary rising path (interconnection trimers and enclosed volumes) in a mono-disperse sample of 87 grains with volume fraction $\phi \simeq 0.63$ at a capillary length $\lambda = 0.025R$, R the radius of the grains (The curve contour do not come from the exact shape of trimers and menisci, but from the removed grains).

3 Procedure and Results

The granular medium was modeled as a dense random packing of monodisperse spheres of unitary radius with volume fraction $\phi \approx 0.63$, slightly below the random close packing [20]. The spatial configurations of spheres were generated by using the algorithm of Jodrey and Tory [21] on cubic volumes of sizes $L = 22, 26, 30, 36, 42, 48, 56$ and 64, and configurations were accepted if the overlapping lengths between any two spheres was below 0.015, in units of particle radius. With this procedure, 200 configurations per size for sizes $L \leq 42$ and 500 configurations per size for sizes $L \geq 48$ were generated, corresponding to around 1560 grains each for $L = 22$ and 38200 grains each for $L = 64$.

For each configuration, we identify sets of three and four grains whose interparticle distances are below $S_c = 2\lambda$, and we built triangle-like (trimer) and tetrahedron-like (enclosed volume) structures with those sets of three or four grains, respectively (Fig. 3). The space occupied by the grains themselves is removed. The interconnection of those structures corresponds to possible capillary rising paths. Single bridges are not included, because they are not observed in experiments to contribute to the percolating structure, as discussed in the previous section. The actual shape of the capillary bridges and menisci are not included, because only the connectivity among elements is relevant for the percolation exponents, and not the detailed shape of the percolating elements themselves [15]. Then, the whole structure is discretized with a fine grid of cubic cells of side $a = 0.031$. Once a value for λ was set, each cell was checked to be part of a trimer or an enclosed volume, and trimers become a couple of cells thick. By doing so, the volume that can be occupied by the liquid phase appears as a discrete set of small cubes, as shown in Fig. 3. This allow us to run an efficient site percolation algorithm to study the percolation transition. Two cells are said to be part of the same cluster if they are first neighbors. The Directional Propagation Labelling algorithm (DPL) for GPUs [22] was employed to determine if there was a percolating cluster connecting the lower and upper borders of the cube. It was necessary to implement this algorithm for GPUs because the systems were very large ((800×800×800) cells for $L = 64$) and the procedure becomes very time demanding. For instance, the more traditional algorithm by Hoshen and Kopelman [23] takes around 6 hours on a CPU (Intel(R) Xeon(R) CPU E5-2695 v3 - 2.30GHz) to evaluate percolation and classify the cluster sizes for a single configuration of system size L=36. By contrast, the DPL algorithm takes only around 15 min to analyze the same configuration on an Nvidia(R) Tesla(R) K40, what is 25 times faster. The savings in computational time are even larger for larger system sizes. Finally, a bisection algorithm [24] was implemented to determine the effective critical value λ_c^{eff} where a percolating cluster first appears on that configuration. The cumulative distribution of these effective critical values for each size are the sigmoids (Fig. 4) that will be the incoming data for the finite-size scaling procedure that follows.

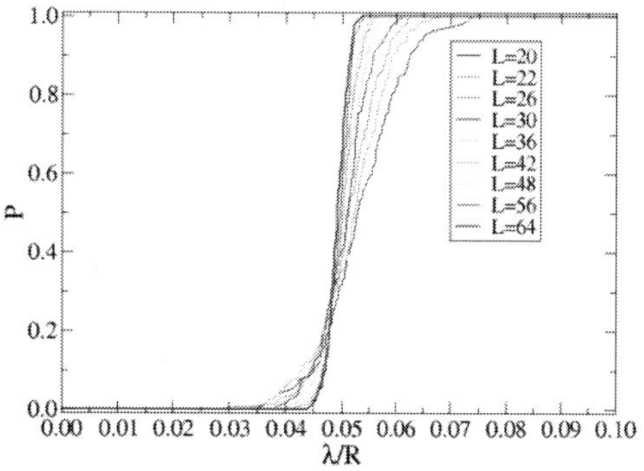

Figure 4. Probability of the capillary structure to percolate as a function of the maximal half-length of a capillary bridge λ.

The critical exponents and the critical half-length λ_c for the transition are obtained by the finite-size scaling procedure proposed by Rintoul and Torquato [25]. The scaling procedure starts by fitting the function

$$f(\lambda_c^{eff}, \Delta(L)) = [1 + \tanh[(\lambda - \lambda_c^{eff}(L))/\Delta(L)]]/2 \quad (1)$$

to the cumulative distribution of each size. The effective critical half-length $\lambda_c^{eff}(L)$ and the width $\Delta(L)$) of the distribution obtained from those fittings are plotted against the system size L to obtain the critical parameters of the transition. Assuming that

$$\Delta(L) \propto L^{\frac{1}{\nu}} \quad , \quad (2)$$

and plotting $\Delta(L)$ against L (Fig 5) gives the critical exponent ν driving the divergence of the correlation length, $\xi \propto [\lambda - \lambda_c]^{-\nu}$. We obtain $\nu = 0.830 \pm 0.051$. Similarly, plotting $\lambda_c^{eff}(L)$ against $L^{-1/\nu}$ (Fig 6) estimates the critical value λ_c for a system of infinite size, because

$$\lambda_c^{\mathrm{eff}}(L) - \lambda_c \propto L^{-\frac{1}{\nu}} \quad . \quad (3)$$

Our result is $\lambda_c = (0.049 \pm 0.004)R$.

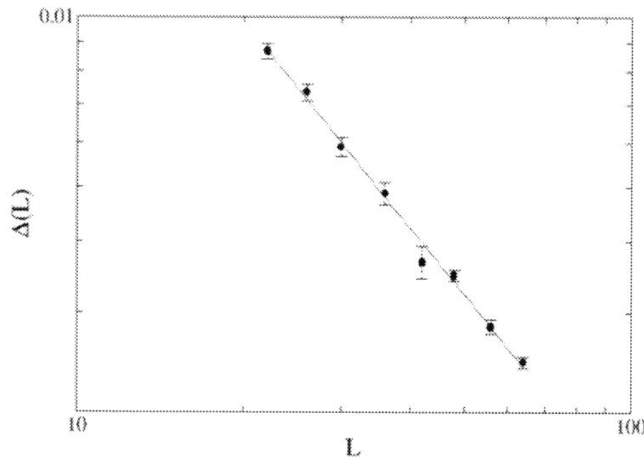

Figure 5. Width of the probability to percolate $\Delta(L)$ as a function of system size L. The line is the best power-law fit, with slope $\frac{1}{\nu} = -1.205 \pm 0.038$.

4 Conclusions and Discussions

This work investigates the capillary ascent of a liquid through a granular soil as a percolation transition driven by the half maximal distance between two grains that can be overpassed by a capillary bridge, λ. Below a critical value λ_c the structure of trimers and enclosed volumes is disconnected, and no water can rise. Above, the water percolates trough the sample. By using standard techniques of percolation analysis, we found $\lambda_c = (0.049 \pm 0.004)R$ and a critical exponent for the correlation length $\nu = 0.830 \pm 0.051$. These results deserve some discussion. First, the finite-size scaling is of very good quality (Fig. 5), suggesting that the technique fits the problem and that there is, indeed, a phase transition. Second, the critical length of a bridge, $2\lambda_c = 0.098(6)R$, is less than one half of the mean distance between neighboring grains in the samples (around $(0.258 \pm 0.007)R$) for a volume fraction $\phi \simeq 0.63$), which can be considered as a naive *a priori* estimation of such parameter. Third, the critical exponent ν is, within error bars,

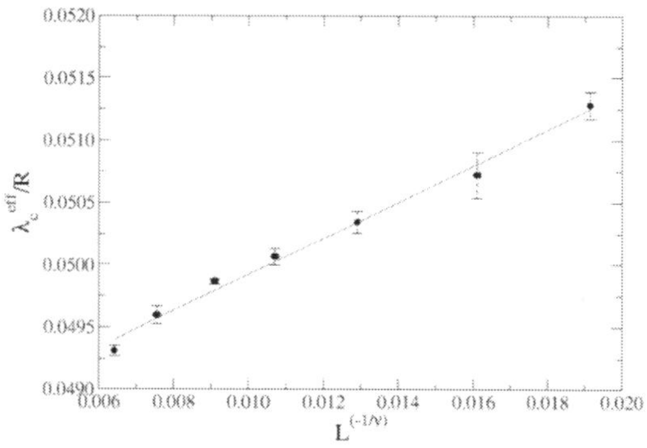

Figure 6. Effective critical half-length of a capillary bridge length $\lambda_c^{eff}(L)$ against $L^{-\frac{1}{\nu}}$. The linear fit (continuous line) estimates a cut with the vertical axis at $\lambda_c = (0.049 \pm 0.004)R$.

the one of ordinary percolation (ν = 0.87619(12) [26]). This result suggests that each facet among three neighboring grains could turn into a trimer at disposal for the capillary rising almost independently from the other facets, a consequence of the random positions for the grains. This similarity should be confirmed by computing other critical exponents, like the critical exponent β for the order parameter, which would be the probability of a cell to belong to the percolating cluster, as usual. Similarly, our study could be reproduced for other void fractions to determine how general are our conclusions. All these are thema for future works.

An usual way to control capillary rising, as we mentioned before, consists in treating some part of the grains with hydrophobic solutions that alter the contact angle, reducing λ (Fig. 2). The key question here is which proportion of the material has to be treated to avoid capillary rising. The simulation would proceed in a similar way as the present one, but with three maximal lenghts λ: one between untreated grains, other between treated grains and a third one between one treated and one untreated grain. Such a future work will be of great interest in geotechnique.

The present study combines capillary structures and percolation theory to investigate capillary rising through a granular medium. It constitutes a novel approach and a new step in the understanding of this rich phenomenon.

Acknowledgments

We thanks COLCIENCIAS Young Researchers Program, Grant 2014-645, the Universidad Nacional de Colombia and the Portuguese Foundation for Sciences and Technology (FCT) under Contracts nos. UID/FIS/00618/2013, EXCL/FIS-NAN/0083/2012, and IF/00255/2013 for finnancial support.

References

[1] N. Khalili, A. Russell, E. Khoshghalb, A, *Unsaturated soils : research and applications*, 1st edn. (Crc Press, 2014)
[2] A. Pozzato, A. Tarantino, Unsaturated Soils: Research and Applications **2**, 1082 (2014)
[3] I. Fatt, Trans AIME **207**, 144 (1956)
[4] H. Vogel, K. Roth, Advances in Water Resources **24**, 233 (2001)
[5] M. Coles, P. Hazlett, W. Soll, E. Muegge, K. Jones, Journal of Petroleum Science and Engineering **19**, 55 (1998)
[6] U. Fischer, M. Celia, Water Resour. Res. **6**, 1089 (1999)
[7] G. Mason, D. Mellor, Journal of Colloid and Interface Science **176**, 214 (1995)
[8] M. Scheel, R. Seemann, M. Brinkmann, M. Di Michiel, A. Sheppard, B. Breidenbach, S. Herminghaus, Nature Materials **7**, 189 (2008)
[9] S. Motealleh, M. Ashouripashaki, D. DiCarlo, S. Bryant, Transport in Porous Media **99**, 581 (2013)
[10] K. Melnikov, R. Mani, F. Wittel, M. Thielmann, H. Herrmann, Cond-mat. Soft **92** (2015)
[11] D. Megias-Alguacil, L. Gauckler, AlChE **55** (2009)
[12] N. Martys, M. Robbins, Phys. Rev. Lett. **66**, 1058 (1991)
[13] M. Cieplak, O. Robbins, Phys. Rev. Lett. **60**, 2042 (1988)
[14] B. Berkowitz, I. Balberg, Water Resources Research **29**, 775 (1993)
[15] D. Staufer, A. Abarony, *Introduction to percolation theory*, 2nd edn. (Taylor and Francis, 1994)
[16] M. Sahimi, *Applications of Percolation Theory*, 1st edn. (Taylor & Francis, UK, 1994)
[17] N. Araujo, P. Grassberger, B. Kahng, K. Schrenk, R. Ziff, The European Physical Journal Special Topics **223**, 2307 (2014)
[18] G. Lian, *Computer simulation of moist agglomerate colisions* (Doctoral Thesis, The University of Aston in Birmingham, 1994)
[19] R. Fisher, J. of Agricultural Science **16**, 492 (1926)
[20] M. Shahinpoor, Powder Technology **25**, 163 (1980)
[21] J. Jodrey, E. Tory, Physical Review A **32** (1985)
[22] K. Hawick, A. Leist, D. Playne, Parallel Computing **36**, 655 (2010)
[23] J. Hoshen, R. Kopelman, Phys. Rev. B **14**, 3438 (1976)
[24] W. Press, S. Teukoisky, W. Vetterlingand, B. Flannery, *Numerical recipes*, 3rd edn. (Cambridge, 2007)
[25] M. Rintoul, S. Torquato, J. Phys. A: Math. Gen. **30**, 585 (1997)
[26] X. Xu, J. Wang, J. Lv, Y. Deng, Frontiers of Physics **9**, 113 (2014)

Structural changes in wet granular matter due to drainage

Prapanch Nair[1,*] and *Thorsten* Pöschel[1]

[1] *Institute for Multiscale Simulation, Friedrich-Alexander-Universität Erlangen-Nürnberg, Germany*

Abstract. Unsaturated wet granular media are usually modelled using force laws based on analytical and empirical results of liquid bridge forces between pairs of grains. These models make ad-hoc assumptions on the liquid volume present in the bridges and its distribution. The force between grains and rupture criterion of the bridge are a function of this assumed volume of liquid, in addition to other parameters like contact angle of the liquid, geometry of the grains and the inter grain distance. To study the initial volume and morphology of liquid bridges, hydrodynamic simulation of dynamic effects leading to formation of liquid bridges at grain scale are indispensable. We use a Smoothed Particle Hydrodynamics algorithm to simulate the hydrodynamics of the evolution of the free surface using a novel freesurface-capillary model, inspired by the molecular basis of surface tension. We present validations for the model and simulations of formation and rupture of liquid bridges.

1 Introduction

Unsaturated wet granular media are encountered in a wide range of engineering applications such as in energy sector, pharmaceutics and food industry. Distribution of liquid as bridges between pairs of grains and in more complex shapes between more than two particles leads to complex constitutive behaviour of the material [1]. The formation of liquid bridges between wet particle also cause grain agglomeration which is either desirable (e.g. wet granulation) or undesirable (wet fluidized beds). A better understanding of the 'formation' of liquid bridges will aid in controlling these processes and in arriving at better input parameters for macroscopic simulations. Studies on liquid bridges in literature usually focus on static bridges, and deformation of bridges during stretching and rupture, quasi-statically [2, 3]. Few experimental and theoretical studies have focused on the initial bridge formation process [4], the dynamics due to liquid transfer from grain to the bridge and the hydrodynamics of the rupture of the bridge [5].

Most research that describes dynamics of liquid bridges do so without considering hydrodynamics owing to the difficulty in resolving such effects at low capillary numbers. Empirical models that resulted from experimental studies were summarized by Herminghaus [6], which are used in a number of industrial scale studies as force laws. Capillary bridge force between grains of different sizes (polydisperse) was studied by Cherblanc et. al. [3], where an analytical rupture criterion for liquid bridges was also proposed. A recent Computational Fluid Dynamics simulation by [5] report the underprediction by analytical models, of the rupture criteria for liquid bridges formed by collision of two wet spheres. Another CFD study [7] presents the physics of formation of liquid bridges for given flow conditions. To our knowledge no theoretical study exists that considers the formation of liquid bridges following drainage. In particular, the volume of liquid 'trapped' after the liquid drains through a pair of particles has not been studied. Current advances in meshless methods for fluid dynamics allow simulations of such scenario, overcoming conventional difficulties in CFD related to large density ratio (between liquid and surrounding gas) and coupling with rigid bodies.

Smoothed Particle Hydrodynamics [8] is extensively used in simulation of complex fluid flow involving interactions with rigid bodies and effects of surface tension. A recent free surface model in SPH [9] allows for accurate simulation of flows involving free surfaces and superimposition of surface tension forces on the particle domain. We present and validate this SPH model in three dimensions. The method is then applied to simulate example cases of rupture of liquid bridge between two solids and the formation of liquid bridge between two grains following liquid draining. Visualizations of dynamic evolution of liquid bridges are presented.

2 Formulation

We outline the governing equations that are being solved, and the numerical method namely Incompressible Smoothed Particle Hydrodynamics (ISPH) with the capillary model, here.

*e-mail: prapanch.nair@fau.de

DOI: 10.1051/epjconf/201714009005

2.1 Governing equations

Incompressible isothermal fluid flow is governed by the momentum conservation equations:

$$\frac{D\mathbf{u}}{Dt} = \frac{1}{\rho}\left(-\nabla p + \nabla \cdot (2\mu\mathbf{D}) + \mathbf{f}^B\right), \qquad (1)$$

where $\mathbf{u}$ is the velocity, p is the pressure, ρ and μ are the density and coefficient of viscosity of the fluid, respectively, $\mathbf{D} = (\nabla\mathbf{u} + \nabla\mathbf{u}^T)/2$ is the deformation rate tensor, $\mathbf{f}^B$ is the body force per unit mass on the fluid element and t is the time. The momentum equation (eq. 1) is the Navier-Stokes equations written in Lagrangian formulation and D/Dt denotes the material derivative. The mass conservation is ensured through the constraint of divergence free velocity field, $\nabla \cdot \mathbf{u} = 0$. In the incompressible version of SPH (as opposed to the Weakly Compressible version [10],) incompressibility is achieved by solving for a pressure field whose gradient ensures a divergence free velocity field.

2.2 SPH approximation

The basics of SPH algorithm and its fundamentals are widely established and are described in a number of publications [8]. For the purpose of brevity, we here present the SPH descritization of the governing equations, together with terms that account for capillary effects directly. The momentum conservation equation in the SPH formulation reads as:

$$\begin{aligned}\frac{D\mathbf{u}}{Dt}_a &= -\sum_b m_b\left(\frac{p_a}{\rho_a^2} + \frac{p_b}{\rho_b^2}\right)\nabla_a W_{ab} \\ &+ \sum_b m_b\left(\frac{2\mu}{(\rho_a\rho_b)}\frac{\mathbf{r}_{ab}\cdot\nabla W_{ab}}{r_{ab}^2 + \eta^2}\right)\mathbf{u}_{ab} + \mathbf{F}_a^{\text{int}} + \mathbf{F}^b. \quad (2)\end{aligned}$$

Here, the first term on the right hand side represents the acceleration due to pressure gradient and the second term represents the viscous dissipation. The equation is written for the conservation of moment for a discretized fluid particle a, and particles in the neighborhood represented by subscript b. The variables p, ρ and μ represent the pressure, density and viscosity of the fluid, respectively. For incompressible flows, the value of ρ remains the same throughout the fluid. The third term on the RHS represents the acceleration due to pairwise-interparticle force superimposed on the discretized particles and is responsible for the effect of surface tension and contact angle dynamics, and will be described in the next subsection. The fourth term on the RHS represents the body force (gravity) acting on the domain.

2.3 Pairwise forces and capillarity

Following the molecular theory of capillarity [11], Tartakovsky and Meakin [12] proposed a surface tension model and methods for applying macroscopic surface tension coefficient to a Weakly Compressible SPH algorithm

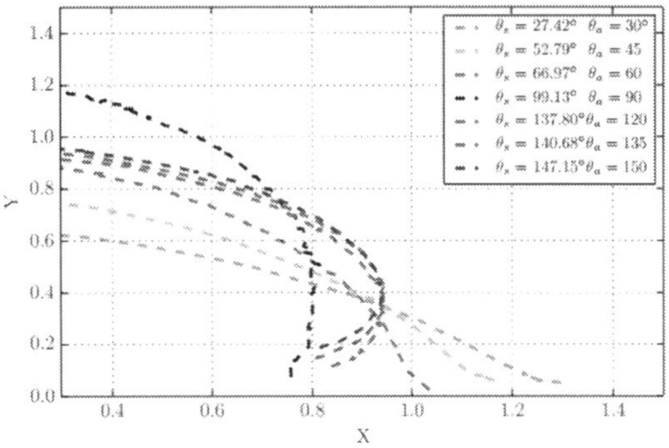

Figure 1: The profile of droplet placed on solid substrate, for different contact angles. θ_s is the contact angle measured by linear regression of six surface particles near the substrate and θ_a is the contact angle given by pairwise force ratio.

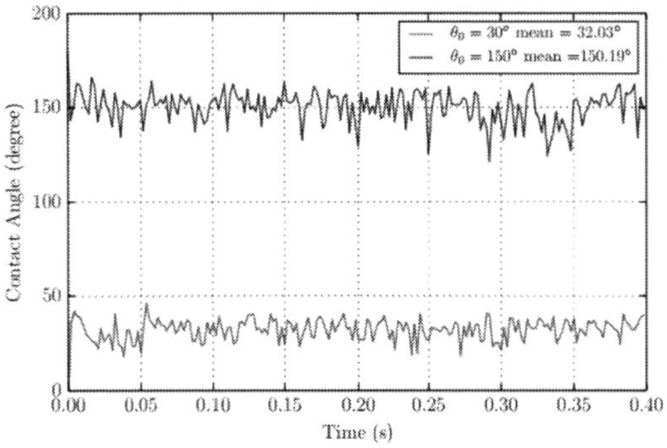

Figure 2: Time variation of the contact angle for the case of $\theta = 30°$ and $\theta = 60°$. The droplets were initiated from the geometry of hemisphere resting on a solid surface.

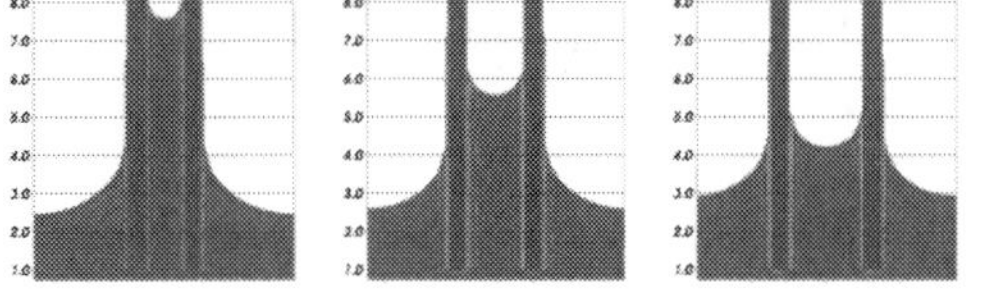

(a) (b) (c)

Figure 3: Capillary rise for different diameters (2D)

with an equation of state based on ideal gas law. In addition to mimicking the effect of surface tension, the superimposition of pairwise forces introduces a 'virial' pressure to the fluid, and this additional pressure can be computed for a given particle system. In the case of incompressible fluids with a free surface (single phase), the virial pressure can be treated as an additive term. Though this needs

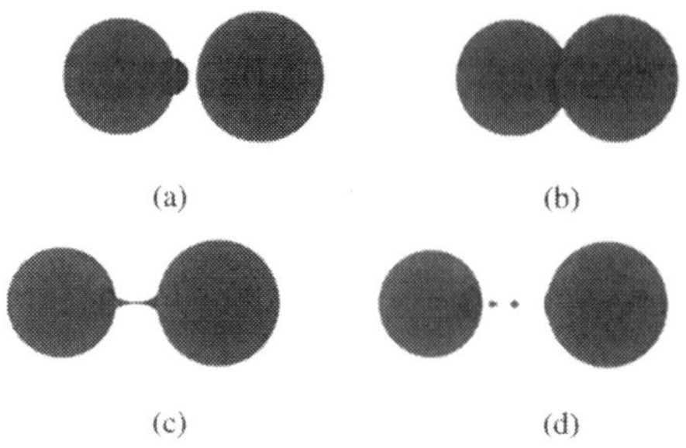

Figure 4: Instances of collision of two rigid spheres with a wet spot on one.

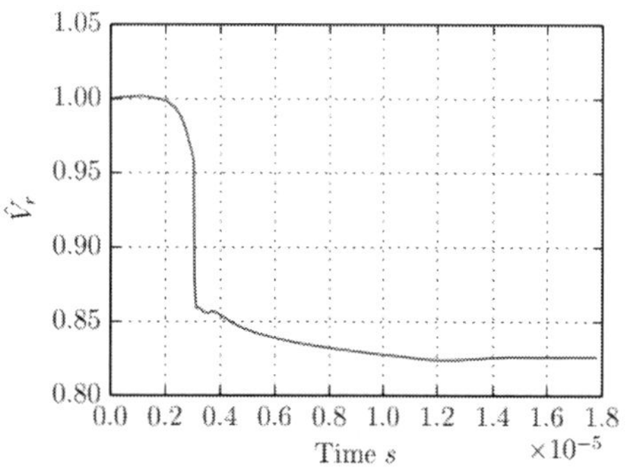

Figure 5: Normalized absolute relative velocity ($\hat{V}_r$) of the colliding grains. The asymptotic value corresponds to the coefficient of restitution

to be corrected for to account for accurate Laplace pressure jump across interfaces, the dynamics of capillary flow would depend only on relative values of pressure, that is, the pressure gradient.

We use the cosine function as given in [12] for the pairwise force and maintain the same cutoff length as the smoothing kernel used for computation of other forces from the continuum model. The pairwise force is given as:

$$\mathbf{F}^{\text{int}}_{ab,\alpha\beta}(r_{ab}) = \begin{cases} -s_{\alpha\beta}\cos\left(\frac{3\pi}{2h}r_{ab}\right) & r_{ab} \leq h \times r_{\text{cutoff}} \\ 0 & r_{ab} > h \times r_{\text{cutoff}} \end{cases}. \quad (3)$$

Using this potential and using the Hardy's [11] formula for integrating stresses at a point in a Lagrangian particle model, the partial surface energy due to pairwise interaction force is given by:

$$\mathbf{T}^{\text{int}} = \frac{1}{8}\pi n_\alpha^2 \int_0^\infty \mathbf{F}^{\text{int}}_{\alpha\alpha}(z)dz. \quad (4)$$

For the 5th order Wendland kernel function [13] used in the current work, this leads to the calibration equation for the surface tension coefficient:

$$\sigma = \frac{0.0134825\pi}{\Delta x} s_{ll} h^5_{\text{ratio}}. \quad (5)$$

Here, h_{ratio} is the ratio of the total cutoff length of the kernel to the initial particle spacing Δx. The strength of the pairwise force is given by s_{ll}. Using different values for strengths of pairwise force, and correspondingly for different ratios for surface energies, different contact angles can be obtained. For the free surface simulations presented here, the contact angle is given by:

$$\cos\theta_0 = \frac{-s_{\alpha\alpha} + 2s_{\alpha\beta}}{s_{\alpha\alpha}}, \quad (6)$$

where θ_0 is the contact angle, $s_{\alpha\alpha}$ is the strength of pairwise potential between liquid and $s_{\alpha\beta}$ is the strength of pairwise potential between particles of different phases, for example, liquid and solid. We integrate the SPH equations using the velocity Verlet integration algorithm [12]

3 Validation and results

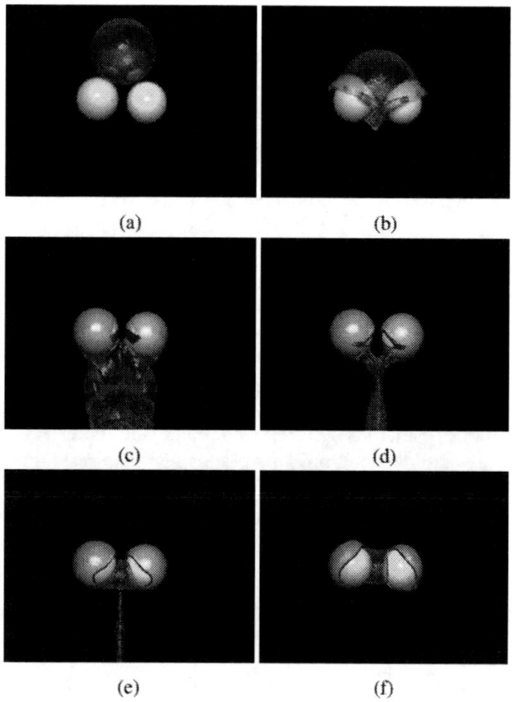

Figure 6: Instance of formation of a liquid bridge following draining of a large liquid drop through a pair of spherical grains

Validation of the surface tension model by an oscillating bubble experiment was performed and will be part of a different detailed publication. Validations of more complex phenomena relevant to the present work, are presented here.

3.1 Contact angle

Equation 6 as a model for wetting phenomena is validated for different contact angles ranging from 30° to 150°. A hemispherical drop is placed on a horizontal solid substrate and is allowed to relax. Figure 1 shows the outer profile of the droplet after a finite time for different contact angles. The time variation of the contact angle after the droplet is allowed to relax from its initial hemispherical position for two extreme contact angles is shown in Fig. 2. The contact angles are measured by linear regression of position of first six surface particles from the substrate.

3.2 Capillary rise in 2D

Table 1: The capillary rise height for different capillary tube diameters (2D)

	Height	
Width (mm)	Analytical	SPH
0.50	5.01	5.12
0.75	2.78	2.95
1.00	1.56	1.31

We perform a numerical experiment with capillary tubes of different diameters immersed in a periodic domain of liquid. The capillary rise height can be analytically derived using balance of pressure due to liquid column and the Laplace pressure jump across curved interfaces. Table 1 shows the comparison of measured capillary rise and the analytical result for different diameters of the capillary tube. The simulations results for different capillary widths are shown in Fig. 3.

3.3 Rupture of capillary bridge

In wet granulation processes, predicting agglomeration of granules is a critical step for successful simulations [14]. In many macro-scale simulations, analytical criteria is used to decide whether capillary bridges formed during collision of two wet particles would sustain for the given approach momenta of the grains. In a recent computational fluid mechanics study [5], it was shown that the analytical cohesion criteria grossly underpredicts the rupture criteria. We perform a similar simulation using the proposed SPH algorithm to corroborate this observation opening up the possibility for deriving accurate agglomeration criteria for complex scenarios involving polydisperse grains and out-of-axis collisions.

Two particles of diameter 50 μm are considered, with a fluid drop of $\approx 1072\mu m^3$ volume on one of the particles. The viscosity and surface tension coefficient of the liquid are 0.001 Nsm^{-2} and 0.071 N/m respectively. The dry particle is imparted a velocity of $5m/s$ towards the wet particle. The moment of rupture is clearly seen in 4, with the formation of two satellite droplets. The coefficient of restitution for the impact can be observed from Fig. 5, as the asymptotic value of normalized relative velocity of the colliding particles. In this case, the restitution coefficient is $e \approx 0.825$.

3.4 Formation of capillary bridge following drainage

The simulation of capillary bridge surfaces in literature usually employ surface energy minimization algorithms [2]. However, the initial liquid volume in the bridge which involves draining of liquid is not given enough attention in literature. We perform simulation of a large liquid droplet impact a pair of grains, and drain through them, resulting in a liquid bridge. Figure 6 shows the different stages of this liquid flow through grains. In Fig. 6b, the liquid drop penetrates the space between grains. In Fig. 6d, a 'Y' shaped configuration of the liquid between the grains pulled down by inertia of liquid is seen. Following this, as see in in Fig. 6e, the bridge begins to be formed when the forked fluid structure straightens and moves upwards due to hydrophilicity of the grain surface. The vertical stem undergoes a rupture at this instant. In Fig. 6f, we see the liquid bridge completely detached and retain a finite fluid volume in it.

4 Conclusion

Using a novel free surface-surface tension model the structural changes of the liquid before formation of liquid bridge following draining of liquid and during its rupture following violent collisions are simulated for example scenarios. These simulations provide confidence for more detailed parametric studies that would result in better force laws, and models to describe liquid bridge formation and rupture.

References

[1] S. Strauch, S. Herminghaus, Soft Matter **8**, 8271 (2012)
[2] C. Semprebon, M. Scheel, S. Herminghaus, R. Seemann, M. Brinkmann, Phys. Rev. E **94**, 012907 (2016)
[3] F. Cherblanc, M.S.E. Youssoufi, C. Saix, F. Soulie, pp. 213–228 (2006)
[4] M. Wu, S. Radl, J.G. Khinast, AIChE Journal **62**, 1877 (2016)
[5] H. Kan, H. Nakamura, S. Watano, Chem Eng Sci **138**, 607 (2015)
[6] S. Herminghaus, Adv Phys **54**, 221 (2005)
[7] Z. Wei, Y. Sun, W. Ding, Z. Wang, Sci China Phys Mech Astron **59**, 694611 (2016)
[8] J.J. Monaghan, Rep Prog Phys **68**, 1703 (2005)
[9] P. Nair, G. Tomar, Comput Fluids **102**, 304 (2014)
[10] E.S. Lee, C. Moulinec, R. Xu, D. Violeau, D. Laurence, P. Stansby, J Comp Phys **227**, 8417 (2008)
[11] J.S. Rowlinson, B. Widom, *Molecular theory of capillarity* (Courier Corporation, 2013)
[12] A.M. Tartakovsky, A. Panchenko, J Comp Phys **305**, 1119 (2016)
[13] K. Szewc, J. Pozorski, J.P. Minier, Int J Numer Meth Eng **92**, 343 (2012)
[14] M. Hussain, J. Kumar, M. Peglow, E. Tsotsas, Chem Eng Sci **101**, 35 (2013)

Dense flow around a sphere moving into a cloud of grains

Philippe Gondret[1], *Sylvain* Faure[2], *Aline* Lefebvre-Lepot[3], and *Antoine* Seguin[1,4,*]

[1]*Laboratoire FAST, Université Paris-Sud, CNRS, Université Paris-Saclay, F-91405, Orsay, France.*
[2]*LMO, CNRS, Université Paris-Sud, Université Paris-Saclay, F-91405, Orsay, France.*
[3]*CMAP, CNRS, Ecole Polytechnique, Université Paris-Saclay, F-91128, Palaiseau, France.*
[4]*SPHYNX/SPEC, CEA, CNRS, Université Paris-Saclay, F-91191, Gif-sur-Yvette, France.*

Abstract. A bidimensional simulation of a sphere moving at constant velocity into a cloud of smaller spherical grains without gravity is presented with a non-smooth contact dynamics method. A dense granular "cluster" zone of about constant solid fraction builds progressively around the moving sphere until a stationary regime appears with a constant upstream cluster size that increases with the initial solid fraction ϕ_0 of the cloud. A detailed analysis of the local strain rate and local stress fields inside the cluster reveals that, despite different spatial variations of strain and stresses, the local friction coefficient μ appears to depend only on the local inertial number I as well as the local solid fraction ϕ, which means that a local rheology does exist in the present non parallel flow. The key point is that the spatial variations of I inside the cluster does not depend on the sphere velocity and explore only a small range between about 10^{-2} and 10^{-1}. The influence of sidewalls is then investigated on the flow and the forces.

1 Introduction

The response of a granular material to a mechanical perturbation arising from the motion of a solid object at its surface or in its bulk is a fundamental issue in many fields such as civil engineering for the resistance of soils to the penetration or extraction of stakes and piles, biophysics for the understanding of animal locomotion in sand which may inspire new robotics, geophysics for the collision phenomena in the impacts of meteorits on planets and asteroids, or in the formation of protoplanetary disks from dust particles. In all these cases, the complex rheology of the granular material plays a key role and needs to be well-understood, which is particularly hard when the packing is dense with a high solid fraction ϕ close to the so-called liquid/solid or jamming transition. Different rheological laws have been proposed for dense granular flows, such as the so-called $\mu(I)$ rheology which relates the local friction coefficient μ and the local packing fraction ϕ to the dimensionless inertial number I [1]. Such a rheological framework gives good results in quasi-steady and quasi-parallel flows but its validity for strongly non parallel flow still needs to be addressed. Even in simple parallel shear flows, the existence of such a local rheology and the influence of solid walls is still under investigation [2]. As a matter of fact, the presence of far boundaries such as fixed or mobile solid walls has shown to have strong and non elucidated actions in different situations [3, 4]. To examine this important question, one must have access simultaneously to the strain and stress fields in the bulk flow which is possible experimentally in some cases [5] but now easier with discrete numerical techniques.

The object of the present paper is to investigate the local rheology of granular matter in a strongly non parallel flow around a moving sphere without any gravity field. We use for this a recent powerful numerical method to investigate such a sphere motion within a bidimensional cloud of spherical particles initially at rest with a solid fraction ϕ_0 far below the jamming point. We first recall the main results reported in [6] in the case where the flow occurs far from any boundaries, before discussing the possible influence of sidewalls.

2 Flow configuration and numerics

The numerical method we use here belongs to the class of "Non-Smooth Contact Dynamic methods" [7] and is described in details in [8]. We simulate the motion of dissipative rigid spheres with a non-elastic impact law (zero restitution coefficient for the collisions) but without any static nor dynamic friction between the spheres. The granular assembly, made of slightly polydisperse spherical grains of mean diameter d_g and density ρ to avoid any possible crystallization, is contained in a rectangular box of size $L_x \times L_y$ with $(x, y) = (0, 0)$ at the center of the box. The aspect ratio of the box has been varied in the range $1 \lesssim L_x/L_y \lesssim 15$. An intruder sphere of larger diameter $d = 10d_g$, which is initially placed at one side of the rectangular box at $x \simeq -L_x/2$ and $y = 0$, is then moved at constant velocity in the x-direction toward the opposite side. The confinement by the sidewalls has been varied

*e-mail: antoine.seguin@u-psud.fr

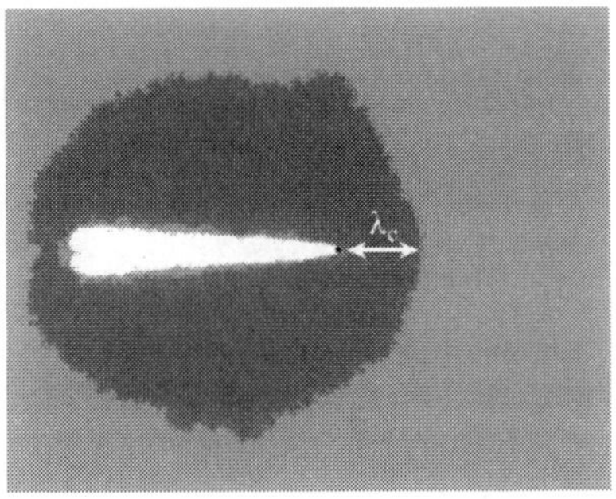

Figure 1. Snapshot of a simulation for the initial solid fraction $\phi_0 = 0.75$. The grains appear in (light) dark blue when there is (no) contact force in between. The larger intruder moving from left to right appear in black. The length λ_c corresponds to the cluster size in front it.

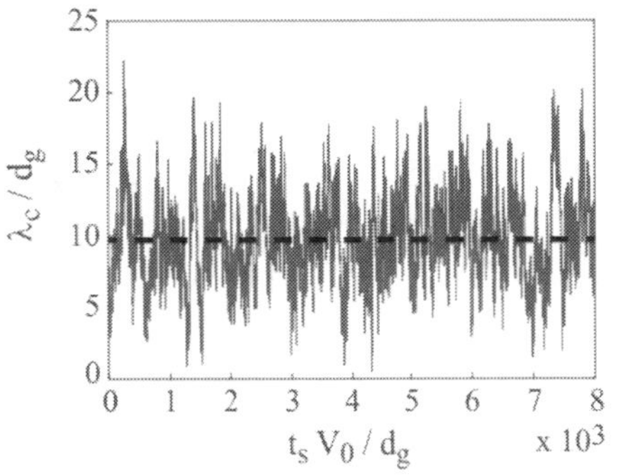

Figure 2. Time evolution of the upstream cluster size λ_c for $\phi_0 = 0.4$ ($L_x = 8000d_g$, $L_y = 100d_g$). The dotted line corresponds to $< \lambda_c/d_g >= 9.7 \pm 3.2$.

in the range $4 \lesssim L_y/d \lesssim 40$. The interaction between the grains and the moving sphere as well as between the grains and the walls is solved using the same law as for the grain-grain interaction. To prepare the granular medium at the initial solid fraction $\phi_0 = N(\pi d_g^2/4)/(L_x L_y - \pi d^2/4)$, we pick randomly N centers of non-intersecting spheres of mean diameter d_g and apply a random force on the grains to ensure an uniform spatial configuration of spheres without any contact. All the explored initial configurations are far below the jamming point $\phi_J \simeq 0.84$.

Starting the intruder sphere motion, the numerical solver gives the velocity v_m of each particle m and the contact force f_{nm} exerted by particle n on particle m, which allows to calculate the corresponding stress tensor $\sigma^m = (4/\pi d_g^2)\sum_n e_{mn} \otimes f_{nm}$, where e_{mn} is the unit vector giving the direction from the center of particle m toward the center of particle n and $\otimes$ is the vector outer product. The local solid fraction ϕ_m around each particle m is computed using a Voronoi tesselation. The local solid fraction ϕ at any point of the domain is then computed by interpolating the ϕ_m values onto a cartesian grid of step $\Delta x = \Delta y = 0.05d_g$, as well as the local velocity field v and stress field σ from the v_m and σ^m values. From the rate of strain tensor $D_{ij} = (\partial_i v_j + \partial_j v_i)/2$, we extract the dilation rate $\dot{\varepsilon} = (1/2)\sum_k D_{kk}$ and the shear rate $\dot{\gamma} = [(1/2)\sum_{i,j}(D_{ij} - \dot{\varepsilon}\delta_{ij})^2]^{1/2}$. From the stress tensor, we extract the pressure $p = -(1/2)\sum_k \sigma_{kk}$ and the shear stress $\tau = [(1/2)\sum_{i,j}(\sigma_{ij} + p\delta_{ij})^2]^{1/2}$. The drag force F exerted by the grains on the intruder can be calculated as $F = -\sum_m f_{m0} \cdot e_x$, where f_{m0} is the contact force exerted by particle m on the intruder and e_x is the unit vector along the x direction of motion.

The spatial variation of the different quantities will be expressed in the polar coordinates (r, θ), where θ is the angle relative to the x direction of motion and r the radial position relative to the moving sphere center.

3 Flow results without sidewalls

As the intruder starts moving, it perturbs progressively the initial grain assembly and a dense cluster zone of touching grains grows around as illustrated in Fig. 1. The position $r_c(\theta, t)$ of the corresponding front delimiting the inner perturbed cluster zone to the outer unperturbed zone is extracted from the simple criterion $p(r > r_c) = 0$ for each θ. A triangular zone with no grain inside, thus appearing in white in Fig. 1, exists in the wake of the intruder as already reported and analyzed in details by [9] in bidimensional experiments. In the present paper, we only focus on the cluster zone upstream the intruder ($-\pi/2 \lesssim \theta \lesssim \pi/2$) which is the key region where the drag force originates, with high stresses and strain rates inside. The upstream extension of the cluster $\lambda = r_c - d/2$ is averaged in the small θ range $-5\,\text{deg} < \theta < 5\,\text{deg}$ around the x-direction of motion. As in one dimensional experiments where a straight rake starts moving [10], we observe that the front position λ moves away linearly in time, with a velocity that is proportional to V_0 and increases with ϕ_0. But in contrast to the one-dimensional configuration of [10], this regime is here only transient until a steady regime is reached with a constant value λ_c despite some fluctuations (Fig. 2). This steady regime appears when the grain flux upstream the intruder is balanced by the grain flux on the intruder sides which can not occur in the one-dimensional configuration of [10] as no grains can circumvent the rake. In the following, we restrict our study to this steady state regime which allows time averaging of the different quantities.

The cluster steady size λ_c is observed to increase with ϕ_0 from only a few grains at low ϕ_0 (e.g. $\lambda_c \simeq 3d_g$ at $\phi_0 = 0.3$) to many grains at high ϕ_0 (e.g, $\lambda_c \simeq 53d_g$ at $\phi_0 = 0.7$), and diverges at the approach of a critical value ϕ_c with the scaling law $\lambda_c \simeq 1.5d_g(\phi_c - \phi_0)^{-2}$ with the value $\phi_c \simeq 0.85 \pm 0.01$ very close to the jamming point ϕ_J. The local solid fraction ϕ in the cluster does not vary significantly with θ in a large azimutal range and displays the about constant value $\phi \simeq 0.83$ in all the cluster for large enough ϕ_0 ($\phi_0 \gtrsim 0.6$), except close to the intruder where ϕ decreases down to about 0.75 and near the front where ϕ decreases down to ϕ_0.

In the present simulations where no gravity acts and no pressure is imposed from any external boundary, no stress scale exists except the kinetic pressure ρV_0^2 arising

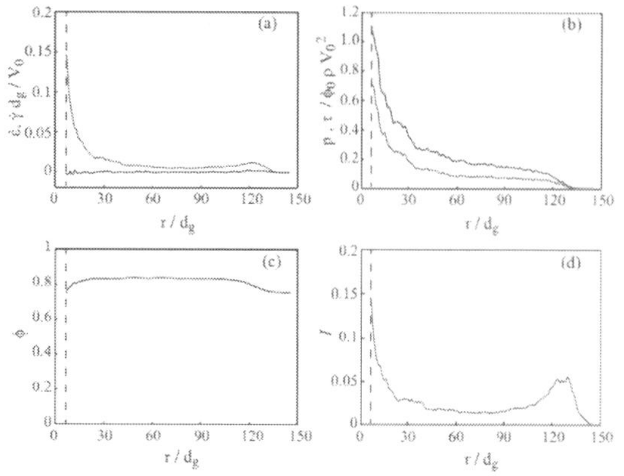

Figure 3. Radial evolution of (a) the dimensionless dilation rate $\dot{\epsilon}$ and shear rate $\dot{\gamma}$, (b) the dimensionless pressure p and shear stress τ, (c) the solid fraction ϕ, and (d) the inertial number I for an initial solid fraction $\phi_0 = 0.75$.

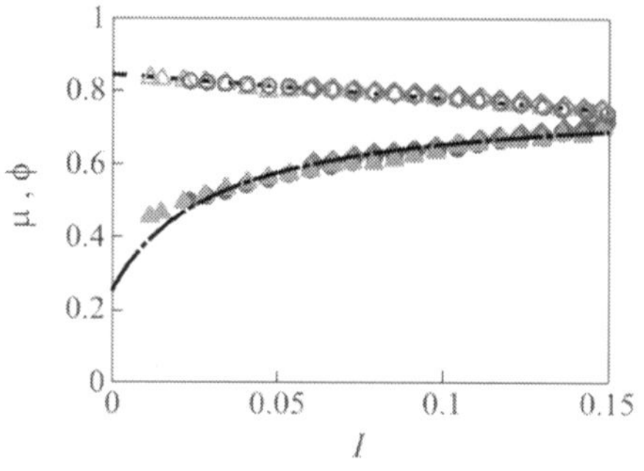

Figure 4. Local friction coefficient $\mu = \tau/p$ (filled symbols) and local solid fraction ϕ (open symbols) as a function of the local inertial number I inside the cluster zone for different initial solid fractions $\phi_0 = 0.5$ (◇), 0.7 (○) and 0.75 (△). Fits of equations $\mu = 0.25 + 0.53/(1 + I/0.03)$ (—) and $\phi = 0.846(1 - 0.8I)$ (- - -).

from collision processes. The present regime corresponds therefore to the inertial high velocity regime found by [11] in their bidimensional experiments of a disk dragged within a monolayer of steel beads and also found by [12] in their numerical simulations of a bidimensional assembly of disks with no friction with the bottom plate. In our simulation, we do not observe any quasi static regime at low velocity where the drag force would be velocity independent or would depend only weakly on the velocity. The existence of such a quasi-static regime which is the most often seen regime [3, 4] arises from the existence of another natural scale of pressure in the system which may come either from gravity and solid friction. We found here that the drag force F increases linearly with the initial solid fraction ϕ_0 with the scaling law $F \simeq 5\phi_0\rho d_g{}^2V_0^2 \simeq \phi_0\rho d d_g V_0^2/2$. Thus, F does not diverge at the approach of ϕ_c in contrast to the cluster size λ_c. This may be explained by the fact that the grains at the outer limit of the growing cluster do not touch any limit boundary in the present simulations.

Let us now detail the flow in the dense cluster around the moving intruder (Fig. 3). The dilation rate $\dot{\epsilon}$ is about zero and thus the flow is incompressible in the overall cluster, except near the cluster rim at the front (Fig. 3a) where there is significant ϕ variation collinear with velocity as mass conservation reads $\phi\ \dot{\epsilon} + \mathbf{v}\cdot\nabla\phi = \mathbf{0}$. The shear rate $\dot{\gamma}(r,\theta)$ is observed to be maximum close to the moving sphere at $r \simeq d/2$ and then strongly decreases away from it with the law $\dot{\gamma}(r,0) \simeq 4(V_0/d_g)[(d_g/r)^2 + \alpha]$ with $\alpha \simeq 0.003$ (Fig. 3a), and also decreases toward the equator. The key point is that its maximal value $\dot{\gamma}(d/2,\theta)$ does not depend significantly on ϕ_0 neither does its radial or azimuthal decreasing rate. This means that an intrinsic flow appears close to the moving sphere within the cluster zone independently of its possible diverging size. This flow has an intrinsic length scale which is thus independent of ϕ_0 and is of the order of $1d$ and thus of about $10d_g$ for the present size ratio $d/d_g = 10$. This flow zone corresponds to the narrow crown of lower solid fraction $\phi < 0.8$ (Fig. 3c).

The local pressure $p(r,\theta)$ is maximal at the sphere surface in front of the moving sphere ($r \simeq d/2, \theta \simeq 0$) and decreases radially away from it with the approximate law $p(r,0) \simeq 5.5\phi_0\rho V_0^2[(d_g/r) + \beta]$ with $\beta \simeq 0.07$ and toward the equator (Fig. 3b). The spatial variations of the shear stress τ are very similar to those of the pressure p with always $\tau < p$ (Fig. 3b).

The strong coupling between τ and p means that the rheological behavior of the grain assembly appears to be of frictional type although no microscopic friction exists in the system. We thus test the possible existence of the local rheology $\mu(I)$ where the local friction coefficient $\mu = \tau/p$ would be linked to the local inertial number $I = \dot{\gamma}d_g/\sqrt{p/\rho\phi}$ [1]. As τ and p have the same scaling in V_0, μ will not depend here on the velocity V_0 neither does I values as both $\dot{\gamma}$ and $\sqrt{p}$ scales as V_0. The only spatial variation of I will come from the weak difference of the spatial scalings of $\dot{\gamma}$ and $\sqrt{p}$, together with the weak spatial variation of ϕ (see $\phi(r)$ shown in Fig. 3d). We see in Fig.3d that I has its highest value of about 0.2 close to the intruder and decreases away from it following the same master curve whatever the cluster size. For the largest cluster size ($\phi_0 = 0.75$), we see that I tends towards a non-zero asymptotic value close to 10^{-2}. We observe only weak azimuthal variations of I around the intruder.

In Fig. 4 where μ and ϕ are plotted as a function of I, all data collapse into master curves which means that a local rheology appears in the present flow. The solid fraction ϕ decreases with the approximate scaling $\phi = \phi_m(1 - aI)$ with $\phi_m = 0.846 \pm 0.002$ and $a = 0.80 \pm 0.03$, and the friction coefficient μ increases as $\mu = \mu_s + (\mu_2 - \mu_s)/(1 + I/I_0)$ with $\mu_s = 0.25 \pm 0.05$, $\mu_2 = 0.78 \pm 0.02$ and $I_0 = 0.03$. The ϕ_m value is very close to the critical value $\phi_c \simeq 0.85$ found for the divergence of λ_c. We believe that ϕ_c and ϕ_m both correspond to the jamming point ϕ_J of the system. The about constant value 0.83 observed for ϕ in the cluster core corresponds to the asymptotic minimum value of $I_m = \alpha/\beta^{1/2} \simeq 0.01$ reached in the flow. The observed $\mu(I)$

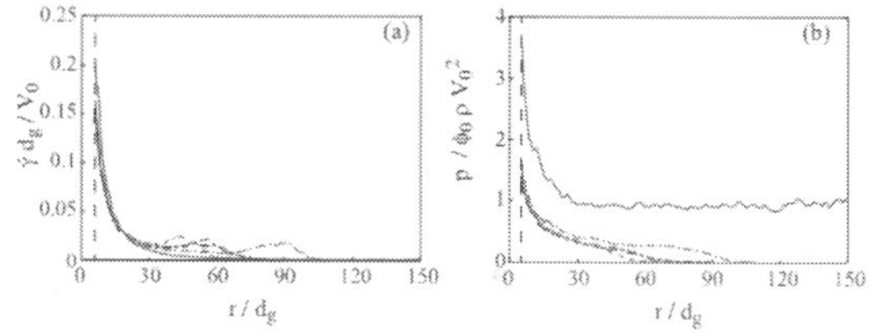

Figure 5. Influence of sidewalls. Radial evolution (a) of the dimensionless shear rate $\dot{\gamma}$ and (b) of the dimensionless pressure p for the same initial solid fraction $\phi_0 = 0.7$ and streamwise extension $L_x/d_g = 550$ but different lateral extensions L_y/d_g corresponding to different confinement ratio $d/L_y = 0.25$ (—) , 0.1 (— - —), 0.05 (— —), 0.025 (– –) and 0.0125 (- - -).

and $\phi(I)$ variations are very similar to what have been already observed in other granular flow configurations [1]. In the present configuration, the inertial number I that naturally arises ranges within only one decade from 0.01 to 0.15. As a consequence μ only varies roughly from 0.45 to 0.70 and ϕ roughly from 0.75 to 0.83. The region of high I (high μ and low ϕ) is close to the moving sphere but it is worth noting that I does not vanish far away. Note that the present natural range of I corresponds to a well-posed behaviour of the $\mu(I)$-rheology [14].

4 Influence of sidewalls

Let us now present some preliminary results on the influence of sidewalls. By decreasing transverse wall gap L_y from $800d_g$ to $40d_g$, we observe that the drag force F increases with the confinement ratio d/L_y as expected from [13], and that the upstream cluster size λ_c also increases. When the sidewalls are close enough($L_y/d \lesssim 10$, the front of the cluster is no more curved but perpendicular to the sidewalls and much far from the intruder. If we look at the local shear rate and pressure (Fig. 5), we observe that $\dot{\gamma}$ is not affected significantly upstream the intruder by the presence of these lateral walls but the pressure is, with a non-zero plateau value that appears far from the intruder in the cluster beyond the rapid pressure decrease close to the intruder. An approximate scaling for the pressure is now $p(r,0) \simeq 5.5\phi_0\rho V_0^2[d_g/r + \beta(d/L_y)]$, where $\beta(d/L_y)$ is a increasing fonction of d/L_y that still needs to be determined.

5 Conclusion

Our simulation results show that a steady state regime is reached in the dense flow around an intruder moving at constant velocity within a cloud of grains. The granular flow in the dense cluster around the intruder is found to obey a local rheology where the local friction coefficient μ and the local solid fraction ϕ are given by the inertial number I even if no microscopic friction between the grains is considered here. With close sidewalls, the spatial variation of stresses are strongly affected with a significative non-zero value that arises for the pressure far upstream the intruder within a long range plateau.

Acknowledgements

This work has been supported by ANR through the project STABINGRAM No. 2010-BLAN-0927-01 and by Investissements d'Avenir LabEx PALM (ANR-10-LABX-0039-PALM) and LabEx LMH (ANR-11-LABX-0056-LMH).

References

[1] A. Andreotti, Y. Forterre, O. Pouliquen, *Granular Media - Between Fluid and Solid* (Cambridge University Press, 2013)
[2] M. Bouzid et al., EPL **109**, 24002 (2015)
[3] A. Seguin, Y. Bertho, P. Gondret, Phys. Rev. Lett. **107**, 048001 (2011)
[4] A. Seguin et al., Phys. Rev. E **87**, 012201 (2013)
[5] A. Seguin et al., Phys. Rev. E **93**, 012904 (2016)
[6] A. Seguin et al., Eur. Phys. J. E **39**, 63 (2016)
[7] F. Radjai, F. Dubois, *Discrete-element modeling of granular materials* (Wiley, 2011)
[8] B. Maury, Numer. Math. **102**, 649 (2006)
[9] E. Kolb et al., Phys. Rev. E **87**, 032207 (2013)
[10] S. R. Waitukaitis et al., EPL **102**, 44001 (2013)
[11] Y. Takehara, K. Okumura, Phys. Rev. Lett. **112**, 148001 (2014)
[12] S. Takada, H. Hayakawa, J. Engin. Mech., **78**, 010301(R) (2008) C4016004 (2016)
[13] A. Seguin et al., Phys. Rev. E **78**, 010301(R) (2008)
[14] T. Barker et al., J. Fluid Mech. **779**, 794 (2015)

Quantifying grain shape with MorpheoLV: A case study using Holocene glacial marine sediments

Isabelle Charpentier[1,⋆], *Alicia B.* Staszyc[2,⋆⋆], *Julia S.* Wellner[2], and *Vanessa* Alejandro[2]

[1] *ICube–Laboratoire des sciences de l'ingénieur, de l'informatique et de l'imagerie, University of Strasbourg and CNRS, 300 Bd Sébastien Brant, CS 10413, F-67412 Illkirch Cedex, France*
[2] *Department of Earth and Atmospheric Sciences, University of Houston, Houston, TX 77204, USA*

Abstract. As demonstrated in earlier works, quantitative grain shape analysis has revealed to be a strong proxy for determining sediment transport history and depositional environments. MorpheoLV, devoted to the calculation of roughness coefficients from pictures of unique clastic sediment grains using Fourier analysis, drives computations for a collection of samples of grain images. This process may be applied to sedimentary deposits assuming core/interval/image archives for the storage of samples collected along depth. This study uses a 25.8 m jumbo piston core, NBP1203 JPC36, taken from a ~100 m thick sedimentary drift deposit from Perseverance Drift on the northern Antarctic Peninsula continental shelf. Changes in ocean and ice conditions throughout the Holocene recorded in this sedimentary archive can be assessed by studying grain shape, grain texture, and other proxies. Ninety six intervals were sampled and a total of 2319 individual particle images were used. Microtextures of individual grains observed by SEM show a very high abundance of authigenically precipitated silica that obscures the original grain shape. Grain roughness, computed along depth with MorpheoLV, only shows small variation confirming the qualitative observation deduced from the SEM. Despite this, trends can be seen confirming the reliability of MorpheoLV as a tool for quantitative grain shape analysis.

1 Introduction

Quantitative grain shape analysis is a strong proxy for determining sediment transport history and depositional environments. As demonstrated in earlier works, grain angularity can be quantified by determining the roughness coefficient (Rc) from Fourier harmonics range [2, 3, 5, 8, 11]. We present MorpheoLV, a well-documented and publicly available code written in Matlab devoted to the computation of roughness coefficients from unique grain images. The new key feature is the opportunity to analyse a collection of samples of grain images all at once to compute roughness variations along a distance or a duration, for instance. Benchmark images proposed in [1] are processed for both validation and interpretation purposes.

A 25.8 m jumbo piston core, NBP1203 JPC36, was taken from a ~100 m thick sedimentary drift deposit from the Perseverance Drift on the northern Antarctic Peninsula continental shelf. Changing ocean and ice conditions throughout the late Holocene are recorded in this sedimentary record. The sediment from this core is composed of laminated, black- to olive-colored diatomaceous mud and ooze. The core's mineralogy is mainly quartz and it contains several horizons of ikaite crystals. Various proxies such as grain shape, grain texture, and grain size can be studied to determine the sediment transport history and assess various proportions of current-transported, iceberg-rafted, and aeolian-transported sediment, thus giving clues on past climate conditions. Microtextures of individual grains, observed in SEM images, exhibit a very high abundance of authigenically precipitated silica that covers the surface of most grains, blurring the broader grain shape. Despite this, MorpheoLV enables the measurement of any small changes in grain shape variation recorded along depth.

2 Roughness calculation with MorpheoLV

MorpheoLV (LV for Light Version) is a small (about 500 lines), well-documented Matlab code derived from MORPHEO [1]. The new features are roughness calculations following [5] and the opportunity to analyse a collection of samples of grain images, all at once. Comparing to MORPHEO, MorpheoLV analyses unique grain images with more speed by a reduction in complexity of the image processing part of the code. Steps of a grain image analysis (outline and centroid determination, polar decomposition and Fourier analysis using `NbHarms` harmonics) are described with detail in [1] and documented in the downloadable archive[1]. Note that images input to MorpheoLV are assumed to come with a jpeg or a tiff extension, while MorpheoLV's output images are png or Matlab figures.

⋆e-mail: icharpentier@unistra.fr
⋆⋆e-mail: alicia.staszyc@gmail.com

[1] http://icube-web.unistra.fr/cs/index.php/Software_download

This convention prevents output images to be considered in any further analysis.

Subsection 2.1 discusses the computation of roughness coefficients from the result of Fourier analysis applied to the pictures of a single clastic sediment grain. The analysis of a sample of grain images, typically grains collected at the same place, the same depth, or observed at the same time, and the analysis of a collection of such samples are then presented in subsections 2.2 and 2.2, respectively. Benchmark images proposed in [1] are discussed in subsection 2.4 to validate the new development and to interpret the case study results.

2.1 Roughness of a single grain

Grain morphology may be quantified from a Fourier expansion, the harmonics coefficients of which contain information [1–3, 5, 8] on the overall shape of the grain and the grain provenance from lower order harmonics (1-10), and on processes such as weathering and sediment transport history from higher order harmonics (11-20).

The roughness coefficient Rc_{a-b} computed from prescribed range (a, b) of harmonics coefficients R_n

$$Rc_{a-b} = \sqrt{0.5} \sum_{n=a}^{b} R_n^2, \quad (1)$$

provides an accurate, dimensionless measure of the grain shape. The harmonics range $a - b$ is chosen in accordance with the grain information sought. As the roughness coefficient increases, the angularity of the grain increases. The harmonics range a, b can be edited in the function `MorpheoLV_Grain('GrainName',NbHarms)`. The parameter `NbHarms` represents the number of computed harmonics.

2.2 Sample analysis

The roughness analysis can be applied to a sample of grains with "common history", collected at the same place, the same depth or observed at the same time for instance, and stored in the same folder. The function, `MorpheoLV_Sample('SampleName',NbHarms)` considers any files in the folder named `'SampleName'` with a jpeg or a tiff extension as part of the grain image sample. MorpheoLV analyses them, individually, as described in subsection 2.1.

The mean roughness coefficient $\overline{Rc_{a-b}}$ and the standard deviation for the sample of grains are output in the Matlab command window. MorpheoLV produces its own histograms to plot the distribution of the roughness coefficient values by using predetermined classes for a fair comparison between the different samples of the same collection. User-defined classes can be edited in the function `MorpheoLV_Histo`.

2.3 Collection analysis

The collection of samples of grain images are assumed to be stored in a "collection/samples/grain images" file system, the main folder being hereafter named `'CollectionName'`. The function `MorpheoLV_Collection('CollectionName',NbHarms)` lists the sample subfolders. For each sample, MorpheoLV evaluates the roughness coefficient of each grain individually using a total of `NbHarms` harmonics in the Fourier analysis.

The provided routine may be adapted to read information, a depth for example, about the location or the time of collection from the folder name or from a complementary file. Means and standard deviations computed for the different samples may be then associated to this key information in a graph to observe their variations, down-core for instance. The naming convention discussed in the case study may serve as an example for such an analysis.

Table 1. Mean roughness coefficients for benchmark images

Samples	$\overline{Rc_{1-10}}$	$\overline{Rc_{11-20}}$	$\overline{Rc_{17-21}}$	$\overline{Rc_{21-40}}$
Angular	7.6e-02	1.0e-02	5.7e-03	6.4e-03
Sub-angular	9.8e-02	1.1e-02	5.2e-03	5.3e-03
Sub-rounded	7.8e-02	6.4e-03	2.7e-03	2.3e-03
Rounded	6.8e-02	1.9e-03	8.4e-04	1.6e-03
Well-rounded	6.0e-02	1.1e-03	5.4e-04	8.9e-04
Round coins	8.9e-04	5.7e-04	1.7e-04	8.4e-04
Dodecagonal	1.4e-03	2.8e-03	9.2e-05	4.2e-04

2.4 Validation tests using MorpheoLV

The Taylor, Russell and Pettijohn chart [7] is split into files of unique grains, taking into account their roundness (well-rounded, rounded, sub-rounded, sub-angular, angular) to sort them into five folders. Similarly, the coin image presented in [1] is split into single images and stored in a folder. The dodecagon coin is taken out from the coin folder to be analyzed alone. These folders and images are part of MorpheoLV's download.

Roughness computations carried out for the seven samples and the dodecagonal coin are reported in Table 1 for different ranges of harmonics. As expected, even if grain samples have very few elements, the mean roughness coefficient allows for a sorting into well-rounded, rounded, sub-rounded, sub-angular classes. As noticed in [1], the distinction between the sub-angular and angular classes of the chart [7] becomes clear only when the higher harmonics ranges are considered. Roughness coefficients for a perfect disc are zero. The uncertainty in the image processing may be thus evaluated through the analysis of the round coin samples. This yields expected small roughness coefficients, less than 1.0e-3 whatever the range of harmonics. The dodecagonal coin results indicate this can be considered as a well-rounded grain for the overall shape and the higher harmonics ranges. Therefore, the main harmonic is the twelfth and this impacts only the roughness coefficient Rc_{11-20}.

3 Case study

The core NBP1203 JPC36 described in Section 1 is chosen as a case study from [11]. Ninety six intervals were

Figure 1. Euhedral ikaite crystal (a), degraded ikaite crystals (b,c,d).

sampled along depth for grain shape analysis. This included only quartz grains that were 63 μm or greater. Interval samples were introduced into the CILAS Laser Particle Size Analyser (LPSA) to capture images of individual quartz grains and to measure the dominant mode of transport from quartz grain roughness. To study grain microtextures, a subset of these 96 intervals were sampled to observe ikaite and quartz grains under the Scanning Electron Microscope (SEM).

3.1 Grain microtexture with SEM

The ikaite crystals found in the core, were studied only under the SEM to determine their elemental composition and to capture an image, see Fig. 1. Grain shape analysis, to provide clues on past climate records, on ikaite crystals was not performed due to the mineral's tendency to degrade to glendonite overtime and its in-situ formation, which does not provide evidence on transport history [12]. SEM images are segmented by adapting the Matlab's tutorial for cell detection, then analysed as described at §2.1. Roughness coefficients reported in Tab. 2. As expected, the overall angularity/roundness property is well captured by the first ten harmonics, while roughness and texture are better described by the higher harmonics. The coefficient Rc_{21-40}, with a very small value for the euhedral crystal, is the most appropriate to reveal the very smooth texture of such a grain. On the other hand, quartz grains, Fig. 2, can reflect whether the dominant mode of transport [4, 5, 14] was glacial, which will result in rougher, angular grains, or nonglacial, which will result in smoother rounder grains, as shown in Tab. 2. Note that the majority of microtextures of individual quartz grains observed in SEM show a very high abundance of authigenically precipitated silica that obscures the original grain shape.

Table 2. Roughness coefficients for the ikaite (I) and quartz (Q) crystals

Grains		Rc_{1-10}	Rc_{11-20}	Rc_{17-21}	Rc_{21-40}
I Euhedral	(a)	1.1e-01	7.9e-03	4.4e-03	2.9e-03
I Degraded	(b)	8.9e-02	9.7e-03	5.5e-03	8.9e-03
I Degraded	(c)	6.7e-02	9.9e-03	4.9e-03	5.5e-03
I Degraded	(d)	3.9e-02	5.2e-03	1.5e-03	3.1e-03
Q Aeolian	(e)	9.6e-02	5.9e-03	2.5e-03	2.5e-03
Q Glacial	(g)	8.0e-02	1.2e-02	4.3e-03	3.6e-03

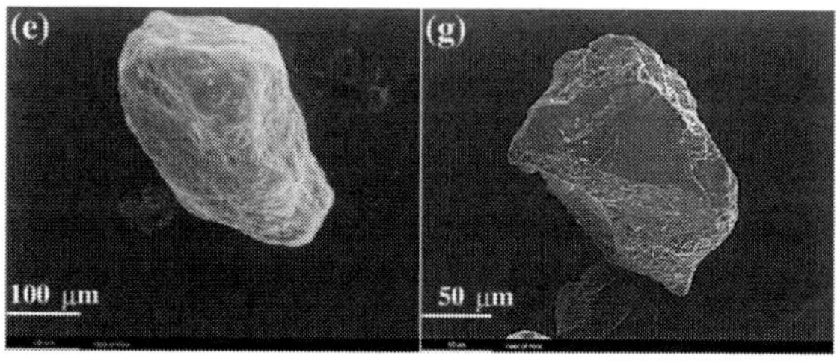

Figure 2. Quartz grains with aeolian (e) or glacial (g) microtextures

Table 3. Mean roughness coefficients for some intervals in NBP1203 JPC36

Depth	$\overline{Rc_{1-10}}$	$\overline{Rc_{11-20}}$	$\overline{Rc_{17-21}}$	$\overline{Rc_{21-40}}$
6.47 m	7.4e-02	8.1e-03	3.6e-03	3.3e-03
18.72 m	6.6e-02	6.1e-03	2.6e-03	4.2e-03
22.96 m	7.1e-02	5.1e-03	2.2e-03	2.9e-03

3.2 Core analysis

By convention in Antarctic geology, samples are named by the ship's name, year, and number of the cruise. In this work, the core folder "NBP1203 JPC36" stores the 96 interval folders for a total of 2319 grains. These secondary repositories are named from the core name and depth information and contain individual images of particles. The depth information may be thus read in the name of the folder under study and is associated to the roughness calculations for the related interval. MorpheoLV, run as described at subsection 2.3, anlyses this core in 15 seconds of an Intel ® Core^{TM} i5-3380M CPU at 2.90GHz.

The mean roughness coefficient values and standard deviations computed for different ranges of harmonics are plotted down-core of NBP1203 JPC36, Fig. 3. In this study, see Tab. 3, the highest mean $\overline{Rc_{17-21}}$ value is 0.0036 and is at 6.46 m, while the lowest mean Rc value is 0.0022 and is at 22.96 m. The median Rc value of 0.0026 at 18.72 m is also shown. Following Tab. 1, the interval at depth 6.46m lies in between the sub-rounded and sub-angular samples, while the grains of interval at depth 22.96m are more rounded and are less rough. Clear trends throughout the core are observed. Using harmonics 17-21, provided reliable quantitative grain shape results.

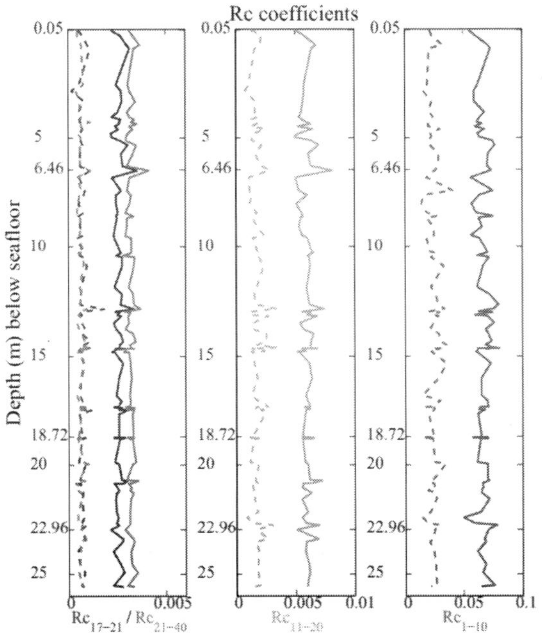

Figure 3. *Rc* coefficients (solid lines) and standard deviations (dashed lines) along depth for different harmonics ranges

3.3 Discussion on transport history

The sediment transport history for the Perseverance Drift was interpreted from entire sediment core's quantitative grain shape results and other proxies, such as grain size and grain texture. Results from the SEM study showed amorphous silica precipitation as the dominant microtexture in the sediment core [6, 9, 10, 13]. The silica precipitation results in smoother grains and covers most of the original microtextures that would give clues about the sediment transport history to the site. However, small changes in grain shape variation are still recorded. These trends are only made clear by being able to observe many grains, as made possible by MorpheoLV. By using other proxies, such as grain size and quantitative pebble analysis, lithostratigraphic facies were assigned and a shift in late Holocene climate from cooler to warmer is observed [11].

4 Conclusion

Using Fast Fourier Transform, MorpheoLV is a reliable method to quantify grain shape by calculating the roughness coefficient of a single grain, a sample of grains with a "common history", or a collection of samples. The validity of MorpheoLV was tested on benchmark images, ikaite crystals, quartz grains, and a sediment core from the Perseverance Drift, which prove the higher harmonic values provide an accurate roughness measurement. In the case study, despite the silica covering the original microtextures, small variations can be seen because of the ability to run many grains as made possible by MorpheoLV. As MorpheoLV is an open source code, the interested user may implement other shape indicators. The full study [11] used multiple proxies, including grain shape and grain microtextures, to help reconstruct late Holocene paleoclimate by studying the sediment transport history to the site.

Acknowledgements

The Perseverance Drift as cored as part of the LARISSA project. We thank all of our LARISSA colleagues.

References

[1] Charpentier, I., Sarocchi, D., Rodriguez Sedano, L.A. Computers and Geosciences **51**, 172–181 (2013)

[2] Ehrlich, R., Weinberg, B., 1970, Journal of Sedimentary Petrology **40**, 205–212 (1970)

[3] Ehrlich, R., Brown, P.J., Yarus, J.M., Przygocki, R.S., Journal of Sedimentary Petrology **50**, 475–484 (1980)

[4] Haines, J., Mazzullo, J., Marine Geology **78**, 227-240 (1988)

[5] Livsey, D.N., Simms, A.R., Clary, W.G., Wellner, J.S., Anderson, J.B., and Chandler, J.P., Journal of Sedimentary Research **83**, 80–90 (2013).

[6] Mahaney, W.C., Atlas of Sand Grain Surface Textures and Applications (Oxford University Press, Oxford, 2002) 237

[7] Müller, G., Methods in Sedimentary Petrology (Hafner Publishing Co, New York, 1967) 183pp.

[8] Murillo-Jimenez, J.M., Full, W., Nava-Sanchez, E.H., Camacho-Valdez, V., and Leon-Manilla, A., Geological Society of America Special Paper **420**, 297–318 (2007)

[9] Narayana, A.C., Mohan, R., Mishra, R., Current Science **99**, 1420–1425 (2010)

[10] Pittman, E.D., Journal of Sedimentary Petrology **42**, 507-519 (1972)

[11] Staszyc, A.B., Charpentier. I., Wellner, J.S., Alejandro, V., A reliable method for quantifying grain shape: A case study in Holocene glacial marine sediments (submitted)

[12] Suess, E., Balzer, Hesse, K.-F., Muller, P.J., Ungerer, C.A., Wefer, G., Science, **216**, 1128–1131 (1982)

[13] Vos, K., Vandenberghe, N., Elsen, J., Earth-Science Reviews **128**, 93–104 (2014)

[14] Wellner, J. S., Anderson, J. B., Ehrmann, W., Weaver, F. M., Kirshner, A., Livsey, D., and Simms, A. R., In: Tectonic, climatic, and cryospheric evolution of the Antarctic peninsula (American Geophysical Union, Wiley Online Library, 2011) 131–151

DOI: 10.1051/epjconf/201714012004

Model of the saltation transport by Discrete Element Method coupled with wind interaction

Luc OGER[1,*] and *Alexandre* VALANCE[1,**]

[1] *Institut de Physique de Rennes, UMR UR1-CNRS 6251, Université de Rennes1, 263, Avenue du Général Leclerc, Campus de Beaulieu,CS 74205, 35042 Rennes Cedex, France*

Abstract. We study the Aeolian saltation transport problem by analysing the collision of incident energetic beads with granular packing. We investigate the collision process for the case where the incident bead and those from the packing have identical mechanical properties. We analyse the features of the consecutive collision process. We used a molecular dynamics method known as DEM (soft Discrete Element Method) with 20000 particles (2D). The grains were displayed randomly in a box (250*X*60). A few incident disks are launched with a constant velocity and angle with high random position to initiate the flow. A wind velocity profile is applied on the flowing zone of the saltation. The velocity profile is obtained by the calculi of the counter-flow due to the local packing fraction induced by the granular flow. We analyse the evolution of the upper surface of the disk packing. In the beginning, the saltation process can be seen as the classical "splash function" in which one bead hits a fully static dense packing. Then, the quasi-fluidized upper layer of the packing creates a completely different behaviour of the "animated splash function". The dilation of the upper surface due to the previous collisions is responsible for a need of less input energy for launching new ejected disks. This phenomenon permits to maintain a constant granular flow with a "small" wind velocity on the surface of the disk bed.

1 Introduction

Drifting sand is a serious threat to the life of many people and may seriously disturb infrastructure and technical installations. The dunes invade houses, agricultural land and perturb the circulation on the roads. It is therefore very important to understand the mechanism of saltation which is the primary way, in which sand grains are propelled by the wind along the surface in short hops [1, 2]. The saltating grains are very energetic and when impact a sand surface, they rebound and consequently eject other particles from the sand bed. The ejected grains, called reptating grains, contribute to the augmentation of the sand flux. Some of them can be promoted to the saltation motion.

The main outcomes of our numerical studies follow. Firstly, we make a short overview of the previous experimental and numerical results concerning the successive collision between incident beads and a granular packing. Then, we describe briefly our mechanical model based on the Discrete Element Method and the special wind conditions used in this study. Finally, we present the main results available from this study.

2 State of the art

Some experimental models of collision between an incident bead and a granular packing have been performed by different authors. They studied the effect of the impact angle [3] and the incident speed [4, 5] on the collision process. The authors have used steel beads (4 mm by Mitha et al [3]), PVC beads (6 mm by Rioual et al. [5]) and sand grains (850 μm by Werner [4]). The experimental results showed that the mean coefficient of restitution (COR) for the impacting bead (defined as the ratio between the rebound speed and the incident speed) depends only on the impact angle. The restitution coefficient decreases with increasing impact angle.The vertical one can be larger than 1 for grazing incident angle. Concerning the ejection process, Mitha et al. [3] showed that the number of ejected beads does not vary significantly when the impact angle increases from 17° to 31° and that the average of vertical speed of ejection is of order of 3 $\sqrt{\mathbf{g}.d}$ (where **g** is the gravitational acceleration and d the bead diameter). However, the number of ejected grains increases with increasing incident speed [5].

In addition to the experiments [6, 7], some simulations of the collision process have been performed [8–10]. The authors made simulation using a discrete element method (DEM). The numerical simulations allowed to extract a law for the incident and ejected beads behaviours The numerical results confirm the experimental result, in particular:

- The dependence of the mean normal restitution coefficient for impacting bead with impact angle. This coefficient is inversely proportional to the sinus of the impact angle.

*e-mail: luc.oger@univ-rennes1.fr
**e-mail: alexandre.valance@univ-rennes1.fr

- The mean number of ejected beads increases linearly with incident speed and impact angle.
-The distribution of the velocities for the ejected beads can be fitted by an exponential law.
These simulations were limited to a single shot and a single grain size but the DEM simulations can be easily extended for multiple grains sizes or for sequential or simultaneous shots of particles on partially immobile packings. This second case is the goal of our present study. We study successive collisions of incident energetic beads with granular packing in the context of Aeolian saltation transport (i.e. with the presence of lateral wind). We have limited our simulation to the 2D case to minimize the sample size and also the global time calculation, but as seen in the 3D splash [10], the lateral dispersion remains symmetric which allows to neglect this direction if we don't take into account the bed packing shape (i.e no counter effect in the flow direction).

3 Description of the numerical simulation

The full description of the different steps of the numerical simulations to mimic the collision process between an incident bead and a packing of identical beads was already made by Oger et al. [10]. So we will briefly present here the two main steps of the numerical simulations. Indeed, in order to save time of computation, we have simplified the technique to create the static packings of disks prior to the active part of the simulation; we perform two consecutive and complementary steps: a pure geometric piling step then a dynamic one.The second step can be separated by an initiation one and a flowing one according how the flow is maintained.

Initial conditions

Firstly, we generated a pure geometrical piling of disks with a Gaussian distribution of diameters to avoid crystallization in which the particles are just touching each other. This small polydispersity is a partial answer of the natural bidisperse natural sand gains in dunes. From a mechanical point of view, this stage corresponds to a dense packing of perfectly rigid, immobile grains (i.e. no overlap yet). The total number of disks used for the packing is between 20.000 and 30.000 (150 − 250 large by 60 − 100 height). Periodic boundary conditions are assigned to the lateral walls during both the piling and the saltating simulations.

DEM model: Soft model approach

Our mechanical model is similar to the two-dimensional formulation of Savage [11]. The *i*th particle is characterized by its radius $\mathbf{r_i}$, the position of its centre (x_i, y_i). A "soft-particle" approach is used, where each particle can have multiple contacts that can persist for extended durations. Both normal and tangential forces develop at the contact between two particles. The normal and tangential contact forces increase as the centres of the particles approach each other.

The normal force F_n at the contact is modelled as viscoelastic. It consists of an elastic contribution (a linear spring K_n) and viscous damping (a linear dashpot coefficient b_n). The force in the tangential direction is modelled

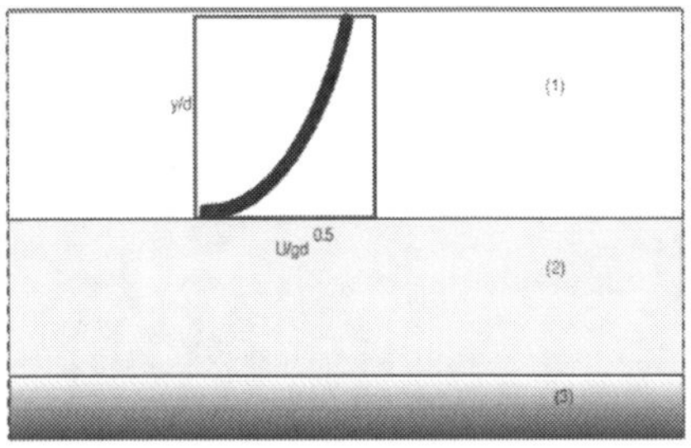

Figure 1. Schematic representation of the disk packing (2). The lower part (3) is the damping zone. The initial lateral wind profile is represented in the upper empty zone (1).

also as a viscoelastic one; a linear spring (with constant K_t) and a linear dashpot (with a constant b_t) are used to generate a tangential contact force. The tangential force is also limited to a maximum value which is chosen according to a Coulomb friction law when slipping can occur according to the inter-grain coefficient of friction. For this mechanical model, the time step calculation remains constant.

It is convenient to cast the governing equations in a non dimensional form and perform the computations based upon these dimensionless equations [11]. Hence, all lengths are nondimensionalized by D, the diameter of the largest sized particles used in the computations. Time is nondimensionalized by dividing by $\sqrt{M/K_n}$, where M is the mass of the largest particle. Velocities are nondimensionalized by dividing by $D/\sqrt{M/K_n}$. This dimensionless approach lets us model either large bead collisions such as in Rioual et al. [5] experiments (6 mm) or sand grains (200 μm) [12–14]. For each particle, the forces acting on it are calculated; then the particle is displaced according to the resultant of all these forces. Local displacements of each particle are recorded. *Wind interaction* For simplicity, in the initiation regime, the wind profile is definedfrom the calculation made by Creyssels et al. (see figure 9 in [6]. Then, but when enough grains are moving, it is possible to calculate the horizontal wind acceleration on the grains and to deduce the wind profile from this profile.

Running conditions

So we assign to each disk, physical parameters corresponding to those of classical sand grains. The disk diameter is taken equal to $200\,\mu m$ and the density is $2500\,kg/m^3$. The coefficient of restitution and the friction coefficient will be selected according to the desired studies (inside a range of 0.75 − 0.95 and 0.2 − 0.4 respectively). Generally the COR for sand can be smaller (0.6-.7) [15] but the choice made here was based on the possibility of comparisons with experiments made in a wind tunnel with glass beads. The spring constants are $K_n = 10^9\,N\,m^{-1}$ and $K_t = 0.3K_n$. The normal dashpot coefficient b_n is calculated according to the restitution coefficient and $b_t = 0.5\,b_n$. The problem and its boundaries are defined as shown in figure 1:
- Perfectly lateral periodic conditions
- The lower part of the packing ((3) in the figure) is defined

DOI: 10.1051/epjconf/201714012004

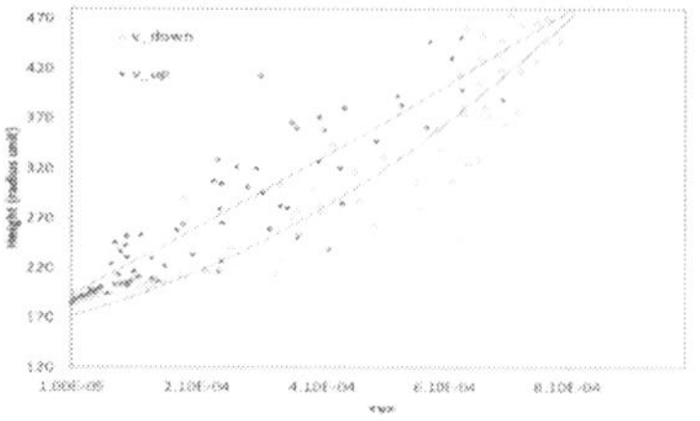

Figure 2. Analysis of the horizontal Velocities for beads going up and down at a given height inside the saltation zone. The difference is linked to the acceleration due to the wind during the parabolic movement of the disks

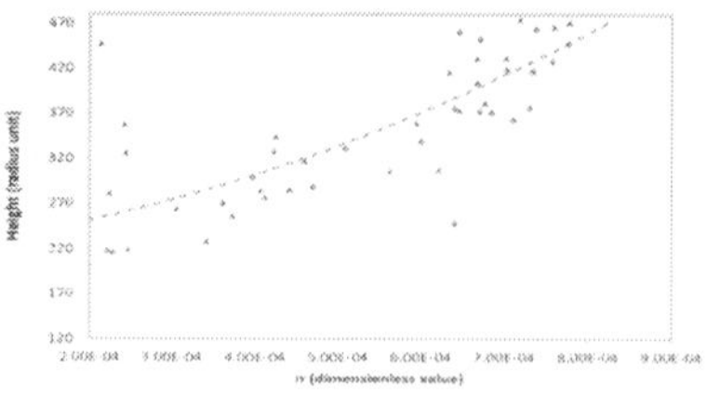

Figure 3. The calculated pressure obtained from the velocities difference according to the equations defined by Creyssels et al. [6].

as a dumping surface in which the two components of the particle velocity are dumped by an amount proportional to the particle altitude: i.e. the particles are less and less mobile as they are deeply located.

- A bead becomes an ejecta if its centre moves, from its initial coordinate, more than one diameter in the vertical upward direction.
- In order to limit the possible flights of moving disks, we have defined a horizontal upper limit like a perfect mirror for the moving disks (same horizontal velocity component and reverse vertical one after touching the mirror). These conditions are biased compared to natural parabolic flights but the number of grains involved in this situation is always very small and avoiding infinite vertical dimension helps a lot in limiting the global time calculation. following some referee comments, incorporation of the results of the cut part of the upper parabolic flight can be made to mimic more precisely the grains acceleration due to the wind.

The initial shooting sequence is activated and the collision process proceeds: Firstly a consecutive series of launched disks at fixed angles and velocities and random horizontal positions is made. This part is to initiate the general movement of the disk knowing that the wind interaction is already present to accelerate the moving disks. Secondly, only wind interactions with the beads are kept. So, after that, a stabilised regime can occurs if the full wind interactions (i.e. wind acceleration and fluidized zone deceleration for the grains) are not too strong. During this stabilisation period, we start the calculi of the horizontal wind acceleration on the disks (Figure 2) and by consequence the pressure applied on them (Figure 3).

4 Results

There are many adjustable parameters in this study that can be tuned such as the restitution and friction coefficients, the stiffness of the grains, the wind velocity at high altitude (i.e. independent of the grain interaction), and so on... In previous researches, we have seen that the amount of moving disks can increase according to the energy flux increase. So the initial amount of input energy has to be limited to control and reach the desired initial stability of the global process.

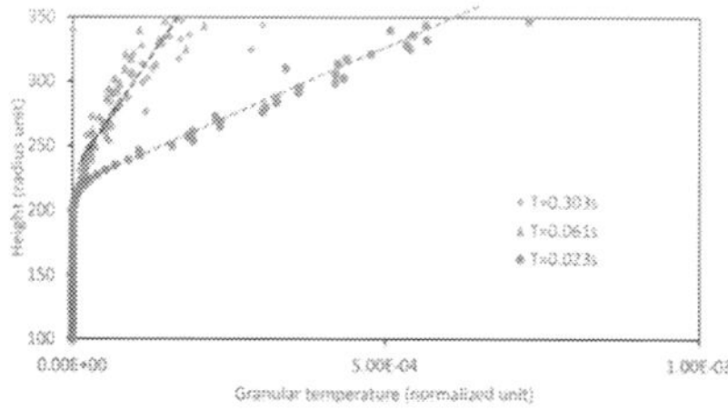

Figure 4. Evolution of the granular temperature versus the height of the saltation zone for three different times of evolution for packing of disks with a restitution coefficient equal to 0.8 and a wind velocity equal to 40 m/s.

As already described, the initial initiation configuration is not defined as a stable one for the grain flow according to the wind velocity but the system will be able to reach this equilibrium if the wind input energy can be exactly counter-balanced by the grain friction and the loss of energy during the splash collision and the flowing grain-grain ones. Of course, this evolution is only possible for given experimental conditions: the wind velocity has to be high enough to compensate the decrease of energy of the moving grains resulting from collisions with other moving grains or with the quasi-static bed of grains, otherwise the system will slow down and stop. Figure 4 confirms this evolution to a stability regime by a decrease of the granular temperature from the regime initiated by the forced energy input (i.e. high granular temperature due to launched grains) and the consecutive regime in which only action and counter-reaction between grains and between grains and wind are taken into consideration.

To perform some series of analysis of the input parameters, we have observed the evolution of the results of the granular temperature by only changing the restitution coefficient of the disks.

Figure 5 confirms that the increase of the restitution coefficient can keep more disks in movement and that this higher amount of moving disks can be maintained even during the stabilisation regime: i.e the granular flux is higher when the restitution coefficient is higher and up to a maximal value in which too many grains are moving inside the dense flowing zone. In this case, a large number of collisions between grains can damp the granular temperature

DOI: 10.1051/epjconf/201714012004

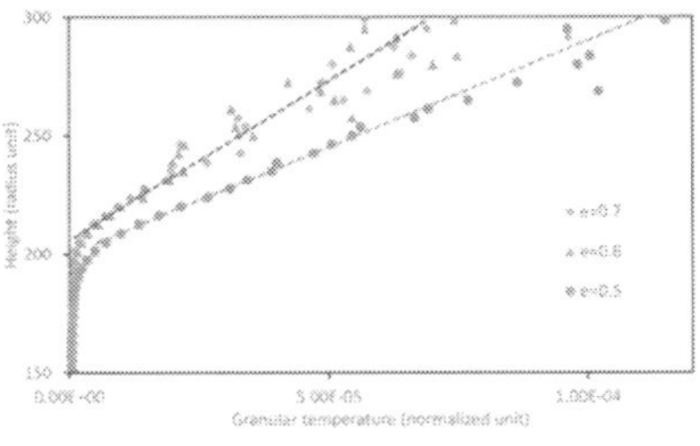

Figure 5. Evolution of the granular temperature versus the height of the saltation zone for three different restitution coefficients at the time t= 0.6s.

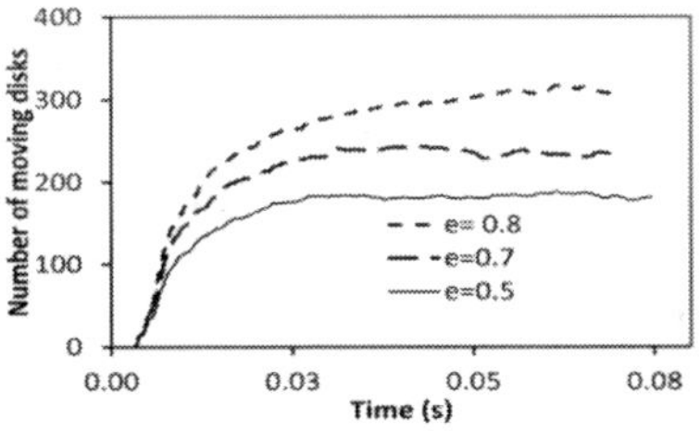

Figure 6. Influence of the restitution coefficient on the number of moving disks versus time with a true wind velocities = 20 m/s.

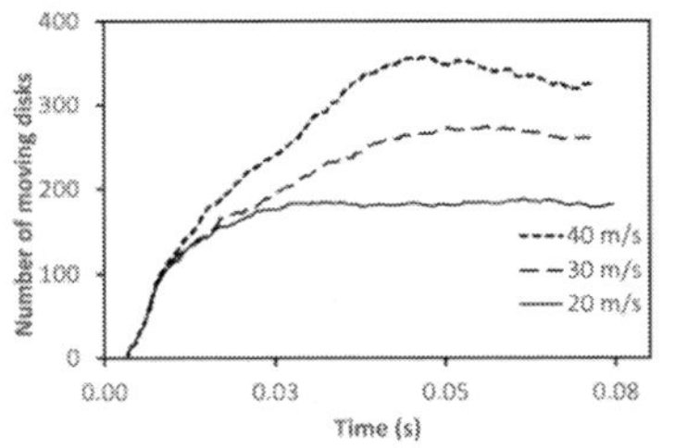

Figure 7. Influence of the velocities effect on the number of moving disks with restitution coefficient = 0.5 .

and generates in consequence a kind of saturation value for the higher possible granular temperature. Figure 6 shows the evolution of the number of moving disks with time up to the apparition of the stabilisation regime obtained with three different values of the restitution coefficients keeping constant all the other parameters of the model. We can see that the range of parameters are correctly selected as they generated a perfect equilibrium regime: i.e. after the initiation and stabilization regime (up to t= 0.15s), the number of moving grains is kept almost constant and much lower than the total number of disks available in the bed. Figure 7 shows the evolution of the number of moving disks versus time with different wind velocities keeping constant all the other mechanical parameters. So we have partially defined the range of possible couples of input parameters in which we can observe the constant stability regime of the saltation. Right now, we have to continue the observation of the transition zone between the dense quasi-static packing and the fully developed saltation one when the system has reached the right saltation regime.

5 Conclusion

Previously, we have demonstrated that using the saltating splash function defined for a unique shoot of a grain on a static grain bed is not enough to handle the full complexity of the saltation process. In the same manner, the energy kept by the bed decreases with higher global energy input. The regime with a constant number of disks inside the flying zone requires an energy balance between grain-wind friction, grain collisions and wind input energy.

Acknowledgements

The authors acknowledge the hospitality of the KITP (UCSB) through the financial support of the National Science Foundation under Grant No. NSF PHY-1125915 and the MPI-PKS in Dresden, where this work was developed. We also thanks J.T. Jenkins for fruitful discussions during the different stages of this work.

References

[1] R.A. Bagnold, *The Physics of blown sand and desert dunes* (Chapmann and Hall, Methuen, London, 1941)

[2] J.T. Jenkins, I. Cantat, A. Valance, Physical Review E **82**, 020301 (2010)

[3] S. Mitha, M. Tran, B. Werner, P. Haff, Acta Mechanica **63**, 267 (1986)

[4] B.T. Werner, Journal of Geology **98**, 1 (1990)

[5] F. Rioual, A. Valance, D. Bideau, Phys. Rev. E **62,2**, 2450 (2000)

[6] M. Creyssels, P. Dupont, A.O.E. Moctar, A. Valance, I. Cantat, J.T. Jenkins, J.M. Pasini, K.R. Rasmussen, J. Fluid Mech. **625**, 47 (2009)

[7] A. Valance, K.R. Rasmussen, A.O.E. Moctar, P. Dupont, Comptes Rendus Physique **16**, 105 (2015)

[8] B. Werner, P. Haff, Sedimentology **35**, 189 (1988)

[9] R.S. Anderson, P.K. Haff, Act. Mech. **[Suppl.]1**, 21 (1991)

[10] L. Oger, M. Ammi, A. Valance, D. Beladjine, The European Physical Journal E **17**, 467 (2005)

[11] S.B. Savage, *Disorder, diffusion and structure formation in granular flows*, in *Disorder and Granular Media*, edited by D. Bideau (North Holland, Amsterdam, 1993), pp. 255–287

[12] B.R. White, J.C. Schulz, J. Fluid Mech. **81, 3**, 497 (1977)

[13] P. Nalpanis, J.C.R. Hunt, C.F. Barrett, J. Fluid Mech. **251**, 661 (1993)

[14] M.A. Rice, B.B. Willets, I.K. McEwan, Sedimentology **43**, 21 (1996)

[15] D. Wang, Y. Wang, B. Yang, W. Zhang, Sedimentology **55**, 461 (2008)

EPJ Web of Conferences **140**, 03014 (2017) DOI: 10.1051/epjconf/201714003014
Powders & Grains 2017

Experiments and numerical modeling for the movement and resuspension of grains

Karina A. Valenzuela Aracena[1], *Jesica G.* Benito[1], *Luc* Oger[2], *Rodolfo O.* Uñac[1], *Irene* Ippolito[3], *Ana M.* Vidales[1,*]

[1]INFAP, Dpto. de Física, CONICET, Universidad Nacional de San Luis, Ejército de los Andes 950, 5700 San Luis, Argentina
[2]IPR, UMR UR1-CNRS 6251, Université de Rennes1, 263 Avenue du Général Leclerc, 35042 Rennes Cedex, France
[3] Universidad de Buenos Aires, Facultad de Ingeniería, Grupo de Medios Porosos, Paseo Colón 850, 1063 C.A.B.A., Argentina

Abstract. We discuss two typical problems involved in particle resuspension phenomena. First, we analyze the onset of movement of a grain deposited on a surface prior to its possible resuspension. In this case, we look for the critical frequency and amplitude needed for starting the movement of the grain when the surface is perturbed by a sinusoidal vertical vibration. We compare results coming from our own experiments and DEM simulations. The agreement is good and allows evaluating the effect of the relative size of the grain on the initiation of movement. The second problem refers to the resuspension of grains, also deposited on a monolayer surface without interactions between them, due to the effect of aerodynamic forces. The detachment of grains as the wind intensity increases is studied through Monte Carlo simulations and the results are compared with experiments belonging to other authors. The role of the particle rolling mechanisms and the reduction factor due to roughness are discussed. Comparison of numerical and experimental data is promising.

1 Introduction

Industrial processes are full of examples where the manipulation of granular matter is implied. Energy is given to grains through different unitary operations such as vibration in conveyor belts, drilling operations, sieving and transport, among many others. Phenomena like bed fluidization, surface cleaning, or the re-entrainment of particles to the atmosphere are based on the existence of a first incipient movement of grains to achieve the detachment of particles and their subsequent dynamic evolution [1-5]. The knowledge of the basic mechanisms causing the destabilization of a deposited particle on a horizontal surface is the key to proceed with its further dynamic description.

The study of the initiation of the movement of a particle when subjected to an external perturbation will of course depend on the scale of sizes involved in the problem. Nevertheless, we ask ourselves if some of the main aspects that are always present in the interaction among particles and surfaces prevail independently of the scale of the problem. One of these parameters will prove to be the roughness.

In the case of the starting of movement of a particle subjected to a continuous perturbation like vertical vibration or to wind bursts, it is mandatory to analyze the adhesion and capillary forces between the particle and the surface, the inertia of the particle when accelerated and its contact interaction with the particles belonging to the surface, especially in cases when rugosity is of the order of the size of the beads deposited on it [6, 7]. This is not straightforward and the inclusion of all the parameters playing a role in the initiation of the particle movement becomes complicated.

In this work, we present part of our last results in studying the initiation of particle movement on a surface in two different scenarios. The first one deals with experiments and simulations (through a Discrete Elements Method, DEM) for spherical particles deposited on a rough surface that is subjected to a mechanical sinusoidal vibration. The idea behind is to find the critical amplitude-frequency pairs needed to initiate the movement of the deposited grains. The second involves simulations for modeling the resuspension of a monolayer of spherical particles deposited on a flat surface subjected to an air flow, and is developed based on a Monte Carlo (MC) method.

2 Incipient movement of vibrated grains

2.1 Experimental procedure

Experiments are performed using two different rough surfaces built by randomly gluing 500 μm and 250 μm glass spheres with epoxy on two smooth cylindrical surfaces, respectively. Each circle has 6 cm in diameter and a thin transparent wall around it to avoid the flow out of deposited particles. The surface is fixed to the arm of a mechanical wave driver whose generator can

* Corresponding author: avidales@unsl.edu.ar

provide a sinusoidal oscillation at well controlled frequency and amplitude. Ten glass beads are randomly deposited over the rough surface and separated to avoid possible contacts or hits among them and/or with the lateral walls. Their diameters go from 1 to 4 millimetres.

The desired amplitude is selected (ranged from 0.7 to 4.2 mm) in the generator and the platform is subjected to an oscillation starting from 0.1 Hz (approximately 0.63 rad/s). The frequency is increased in fixed steps at a rate of 0.1 Hz every second. With the help of a magnifying glass and a camera, we check for the incipient movement of particles and record the critical frequency at which at least half of the deposited spheres start to move from their original positions. Once the critical frequency is reached, the experimental run ends, and the oscillator is turned off. Several runs are performed to average data over 200 different positions.

A new value for the amplitude is chosen and the beads are redistributed again to start the searching of a new critical amplitude-frequency pair. Experiments using spheres with averaged diameters of 1.0, 2.0, 3.0 and 4.0 mm are performed separately.

2.2 Simulation procedure

We use a classical DEM model where the grains are simulated as spheres that can overlap each other, slip and rotate. We follow the two-dimensional formulation of Savage [8] and the three-dimensional extension used by Oger et al. [9]. Both normal and tangential forces develop at the contact between two spheres. In the case of the only presence of compressive forces, simulations represent dry, non-cohesive particle assemblies. The presence of tensile forces at the contacts models wet cohesive particles. In the present work, short range interactions can occur up to the breakage of the adhesive forces (i.e. breaking of a wetting film). No long range interactions are allowed. In order to simulate non-perfectly spherical particles and other local contact effects, rolling resistance is introduced [10] to represent the effects of particle shape on rolling.

Initially, the code creates the rough surface of glued glass spheres, i.e., immobile centres in the referential moving plate. Then, free glass spheres of a given size are randomly deposited on the rough surface for their future motion tracking. Finally, all the system is put under oscillation as in the experiments.

To resemble the experimental conditions, a disordered rough surface is created by a random 3D deposit of spheres with up to three layers. In all cases, all the spheres are sitting on a stable position under gravity with one contact in the case of the first layer, or up to three contacts for the other upper layers. Thus, a free sphere can sit on three underlying contacts distributed at different distances and heights on the rough surface.

2.3 Comparison and analysis of results

To ensure good statistics for data analysis, we check the dependence of the results on the number of free particles used to determine the different thresholds. A number of 200 particles (200 different sitting positions) seems to be a good approximation for both, experiments and simulations.

In Figures 1 and 2 we show the experimental results obtained for the critical frequencies, ω_c, needed to put into incipient movement the glass particles as the amplitude of the excitation is increased. The error bars for the amplitude are indicated and those for the critical frequency are of the order of the symbol size. The results correspond to the four different glass beads diameters. Figure 1 corresponds to a 250 μm rough surface and Figure 2 corresponds to 500 μm. Thus, the size ratios used are, approximately, 4, 8, 12 and 16, in Figure 1, and 2, 4, 6 and 8, in Figure 2.

The points in each figures separate two regions in the plot: the upper part corresponds to the existence of movement of at least 50% of the spheres and the lower one to the absence of movement or to an incipient movement of less than 50% of the particles. Evidently, there is a decrease of the critical frequency as the amplitude increases. The trend can be approximated by an exponential decay, as displayed by the curves in both figures. Independently of their size and the size of the glued spheres, particles subjected to higher amplitudes of oscillation need a smaller critical frequency to initiate their movement. Only a weak dependence of the results on the size ratio is found.

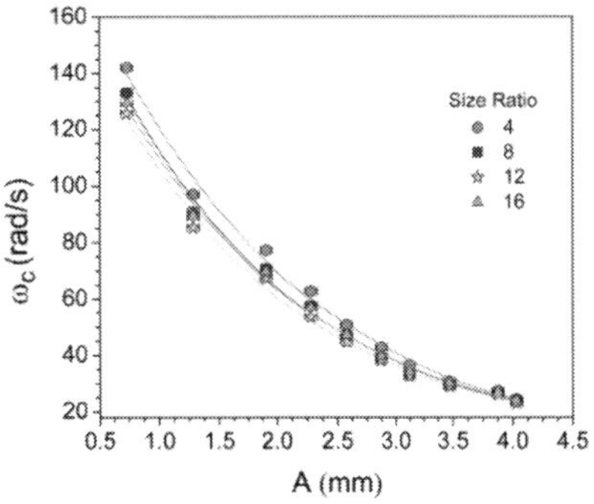

Fig. 1. Critical frequency as a function of the vibration amplitude for glass beads on 250 μm glass surface.

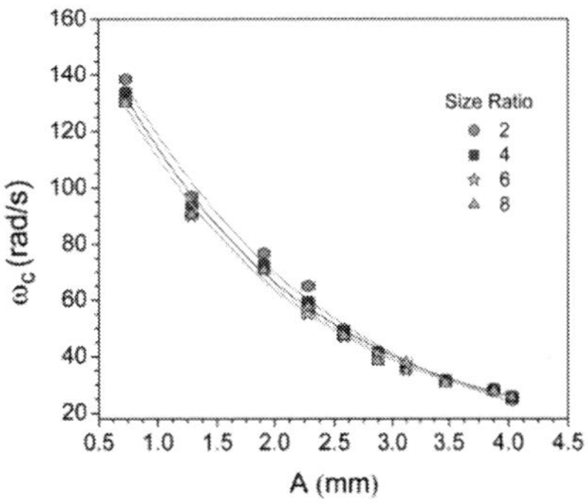

Fig. 2. Critical frequency as a function of the vibration amplitude for glass beads on 500 μm glass surface.

Figure 3 shows the comparison of the frequency-amplitude threshold evolution obtained with our numerical model with two sets of experimental results obtained with 500 μm glued glass beads and similar conditions as in experiments. We observe the same exponential trend and the agreement is quite good for all the range of amplitudes studied. On the other hand, for 250 μm glued glass beads, we can observe a considerable difference between experimental and numerical results although all the other parameters used are the same as in the previous comparison. The results are shown in Figure 4. The difference can be explained taking into account that, in simulations, the number of layers and the mean distance between spheres in these upper layers do not have the topological complexity of the surface generated through real glued spheres, as in the experiments.

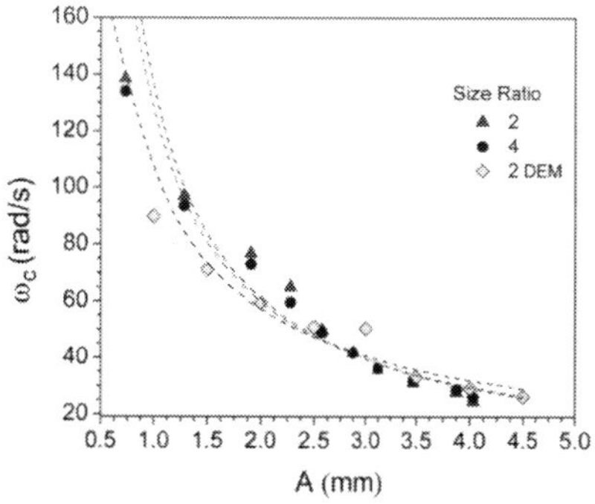

Fig. 3. Comparison between numerical and experimental results for a rough surface of 500 μm glued glass beads.

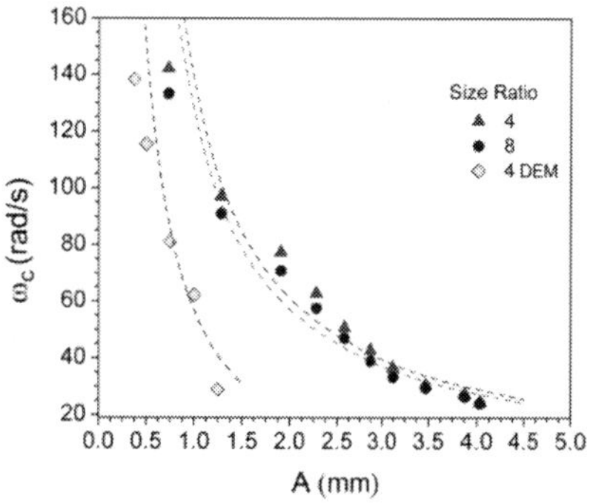

Fig. 4. Comparison between numerical and experimental results for a rough surface of 250 μm glued glass beads.

The roughness of the surface created by the grains in the experiments is such that the characteristic length of this roughness is not dictated by the size of the constituting beads, but rather by the geometrical configuration of those beads in combination with the glue. Thus, the three-contact "tripods" formed by the free particles on top are topologically different from those arranged more horizontally in the simulations. In other words, the scale in the experimental surface structure hides the influence of the size of the constituting spheres. This is not the case in simulations where some level of order is still preserved because no glue effects are considered, thus preventing the formation of larger non horizontal "tripods" for the contacts of the free particles on the top of the surface. A more complex surface could be thought for future simulations for a deeper investigation of this hypothesis.

3 Resuspension of grains

3.1. Simulation model

A model for the resuspension of a monolayer of particles deposited on a flat surface is developed based on a Monte Carlo (MC) simulation. The distance between particles is not relevant in the present model since they do not interact with each other. Spherical particles are attached through an adhesion force that follows a log-normal distribution with a characteristic mean and dispersion, μ_a and σ_a, respectively. A turbulent air flow is assumed to be responsible for the resuspension of particles. A Gaussian distribution of aerodynamic forces is assumed with a mean μ_h and dispersion σ_h. The stochastic process used for resuspension is based on the evaluation of probabilities depending on the ratio between adhesion and aerodynamic forces and using a Metropolis function. Besides, the algorithm does not have to solve any integral equation, which implies quite short simulation times. Although simple, our model accounts for the main features of the resuspension flux observed experimentally and by other models [11, 12].

Typical experiments consist in measuring the resuspended amount of particles as a function of the friction velocity of the wind and especial attention is paid to the role that surface roughness has on the strength of the adhesion forces between particles and surface. Indeed, it has been proved that a strong reduction of adhesion is observed in the presence of a rough surface. In our simulations, this fact is taken into account through a reduction factor, f_r that affect the strength of μ_a as follows

$$\mu_a = F_{JKR} / f_r = 3\pi\gamma R_p/(2f_r) \qquad (1)$$

In this equation, F_{JKR} is the adhesion force estimated by the JKR theory due to Johnson, Kendall, and Roberts [13], i.e., the adhesion force between a particle of radius R_p and the surface is given by $F_{JKR}=3\pi\gamma R_p/2$, where γ is the adhesive surface energy of the particle and substrate, and f_r takes into account the contact geometry in a real surface which is usually characterized by a wide distribution of surface roughness.

We have already proved that results based on a force ratio balance [11] can qualitatively span a wide variety of expected behaviours. Nevertheless, the main mechanism observed in experiments is related to the rolling of particles as an incipient motion before succeeding in the re-entrainment to the fluid phase [14-16]. Thus, the incorporation of the force moments in our MC model is developed, wherein the resuspension probabilities are evaluated from the balance between the adhesion and the aerodynamic moments acting on each particle. In the calculation of moments, the lever arm, a_0, corresponding to the adhesion force moment is crucial to understand that a particle is resuspended even when aerodynamic forces are less strong than adhesion ones. The value of a_0 depends on the deformation of the particle against the surface and on the presence of rugosity.

3.2 Results and discussion

Figure 5 shows the results for the fraction of resuspended particles for two different glass spheres diameters on a glass surface. The solid lines correspond to our simulation results. The open symbols correspond to several repetitions of experiments with 72 μm glass spheres and filled symbols to repetitions with 32 μm glass spheres due to Ibrahim and co-workers [16]. The reduction factor, f_r, is 200 and 43, respectively. Smaller particles present higher adhesion forces, thus the reduction factor is smaller. As said before, the presence of rugosity on the contacting surfaces makes adhesion to decrease and this effect is higher for larger particles. Nevertheless, the main reason why it is feasible destabilizing a particle is related to a_0, i.e., to the distance between the contacting points or "pivots" of the sphere sitting on the rough surface [12]. Thus, it is important to highlight that we have just varied the adhesive forces through the reduction factor, but, the lever arm a_0 in the moment calculation could be varied to obtain similar results. Both parameters allow taking into account roughness effects and, in fact, are connected to each other.

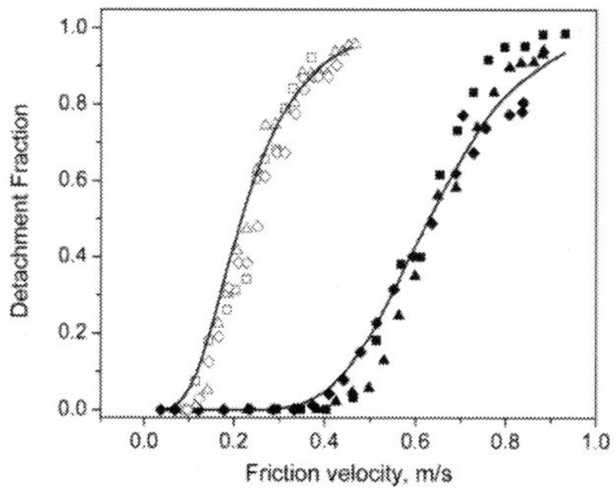

Fig. 5. Comparison between MC results and experimental data in [16]. Open symbols correspond to 72 μm spheres and filled symbols to 32 μm.

4 Conclusions

We have explored two different scenarios to understand the incipient motion of grains subjected to different external perturbations. Although results belong to quite different scales (millimetre in the first case and micrometers in the second), they demonstrate that roughness is the key point to put in motion particles with different sizes. The rugosity intervening in the development of contacts between particles and surface is complex.

In simulations, attention should be paid when generating a rough surface through the arrangement of particles. The way in which free spheres are placed on the surface is crucial to evaluate their chance for movement. In a similar way, care should be taken when roughness factors are included in the description of adhesion forces and the role of the lever arm intervening in the rolling of the particles has to be taken into account in close relation to the rough nature of the surfaces in contact.

Roughness demonstrates to be the key parameter in the incipient movement and resuspension of particles and its role has still to be more deeply investigated.

The authors acknowledge financial support from ECOS-Sud-MINCyT project A15E03, CONICET-PIP N°353 and CyT-UNSL-PROICO N°3-10114.

References

1. G. L. Dybwad, J. Appl. Phys. **58**, 2789 (1985)
2. G. Ziskind, M. Fichman, C. Gutfinger, J. Aerosol Sci. **31**, 703 (2000)
3. W. Theerachaisupakij, S. Matsusaka, M. Kataoka, H. Masuda, Adv. Pow. Technol. **13**, 287 (2002)
4. Y. Mawatari, T. K. Y. Tatemoto, S. Uchida, K. Noda, Pow. Tech. **123**, 69 (2002)
5. N. Tippayawong, I. Preechawuttipong, Aust. J. Basic Appl. Sci. **5** 356 (2011)
6. K. Hein, T. Hucke, M. Stintz, S. Ripperger, Part. Part. Syst. Char. **19**, 269 (2002)
7. F. Soepyan, S. Cremaschi, B. McLaury, C. Sarica, H. Subramani, G. Kouba, H. Gao, Pow. Tech. **292**, 272 (2016)
8. S. B. Savage, *Disorder and granular media* (D. Bideau Ed., North Holland, Amsterdam, 264, 1993).
9. L. Oger, I. Ippolito, A. Vidales, Gran. Matt. **9**, 267 (2007)
10. J. Ai, J.-F. Chen, J. M. Rotter, J. Y. Ooi, Pow. Tech. **206**, 269 (2011)
11. J.G. Benito, K.A. Valenzuela Aracena, R.O. Uñac, A.M. Vidales, I. Ippolito, J. Aerosol Sci. **79**, 126 (2015)
12. J.G. Benito, R.O. Uñac, A.M. Vidales, I. Ippolito, J. Aerosol Sci. **100**, 26 (2016)
13. K. Johnson, K. Kendall, A. Roberts, *Proceedings of the Royal Society of London* (Series A, Mathematical and Physical, **324**, 301, 1971)
14. S. Fu, C. Chao, R. SO, W. Leung, J. Fluids Eng. **135**, 041301 (2013)
15. I. Goldasteh, A. Goodarz, A. Ferro, J. Aerosol Sci. **66**, 62 (2013)
16. A. Ibrahim, P. Dunn, R. Brach, J. Aerosol Sci. **29**, 765 (2003)

An appraisal of the influence of material fabric and stress anisotropy on small-strain stiffness

Hoang Nguyen[1], *Masahide* Otsubo[1], and *Catherine* O'Sullivan[1*]

[1]Department of Civil and Environmental Engineering, Imperial College London, Skempton Building, London, UK

Abstract. This paper examines the effect of a non-isotropic, true triaxial stress state on soil stiffness using discrete element method (DEM) simulations. Samples of uniform spheres with a very stable face centred cubic (FCC) are considered to isolate the effect of stress from stress-induced fabric changes. At the same time the anisotropic nature of the lattice fabric enables the effect of fabric on the observed responses to be explored. The elastic or small strain stiffness was determined by applying small amplitude displacement perturbations to the samples and measuring the resultant shear wave velocity. Two different mean stress levels were considered and at both stress levels the magnitudes of the three principal stresses were varied. The data obtained confirm that the stresses in direction of wave propagation and shear wave oscillation have a measurable influence on shear modulus values. The extent of sensitivity depends on the material fabric. The stress component orthogonal to the plane of wave motion has, however, a less marked effect on shear

Keywords: material fabric, stress anisotropy, small-strain stiffness

1 Introduction

Soil stiffness is stress-dependant. It is important to examine the link between the three dimensional stress state soil experiences in situ and soil stiffness. Roesler and Bellotti *et al.* have proposed expressions linking stiffness and components of stress [1, 2]; however, the challenges of measuring stiffness under general three dimensional stress states have limited examination of their equations. The resultant shortcoming in understanding is addressed here. Lattice packings consisting of uniform spherical particles on a face-centred cubic (FCC) are considered. The mechanical response of uniform spheres on a regular packing differs from that of realistic materials that are polydisperse and randomly packed, however consideration of this simpler type of system has provided useful insight e.g. Rowe [3]. Use of this stable configuration allows the effect of stress anisotropy to be isolated from fabric effects. The FCC packing is cross anisotropic (transversely isotropic) and by examining the stiffness in orthogonal directions the effect of fabric on the observed response can be determined.

2 Background

In situ soil experiences a non-isotropic stress state and the extent of the stress anisotropy also influences the stiffness [1]. Recognizing this dependency Bellotti *et al.* [2] proposed the following relationship

$$G_0 = C_G F(e) p_a^{(1-2n_{prop}-2n_{osc}-2n_{third})} \times \\ \times \sigma_{prop}^{2n_{prop}} \times \sigma_{osc}^{2n_{osc}} \times \sigma_{third}^{2n_{third}} \quad (1)$$

where G_0 is the small-strain shear modulus, $F(e)$ is a normalizing function to eliminate void ratio effects, p_a is the atmospheric pressure, C_G is an experimentally determined material constant. The stresses in the directions of propagation and oscillation of the stress wave are denoted σ_{prop} and σ_{osc}, respectively while the remaining orthogonal stress is denoted σ_{third}. The material exponents n_{prop}, n_{osc} and n_{third} quantify the influence of each principal stress on G_0 .

Table 1. DEM simulation parameters.

Parameter	Value
Particle size, d_p	2.54 (mm)
Particle density, ρ_p	2230 (kg/m^3)
Interparticle friction, μ_p	0.35
Particle shear modulus, G_p	25×10^9 (Pa)
Particle Poisson's ratio, v_p	0.2
Number of particles	3200
Transmitter frequency, f	20 (kHz)
Transmitter amplitude, A	2.5 (nm)

This contribution extends the work of O'Donovan *et al.* [4] who also carried out discrete element modelling (DEM) simulations on FCC packings to assess the effect of anisotropic stress states on G_0 . There was, however,

* Corresponding author: cath.osullivan@imperial.ac.uk

significant scatter in the resulting data, most likely due to the point source excitation used. Only a limited number of stress combinations were considered, consequently, it was difficult to achieve a clear conclusion. This research addresses the gap in understanding by using a planar wall as the source excitation to generate the stress waves with the aim of improving signal quality and considers a broader range of stress combinations.

3 Simulation approach

DEM simulations were carried on FCC packings using a modified version of granular LAMMPS [5]. A simplified Hertz-Mindlin contact model (HM model) was used and simulation parameters are presented in Table 1.

To investigate the shear modulus in different planes, two lattice packings consisting of 3,200 spherical particles were generated. The first sample (4 × 4 × 200 layers) corresponding to samples previously used by Mouraille *et al.* [6] was used to simulate the wave propagation in the z −direction as illustrated in Figure 1.

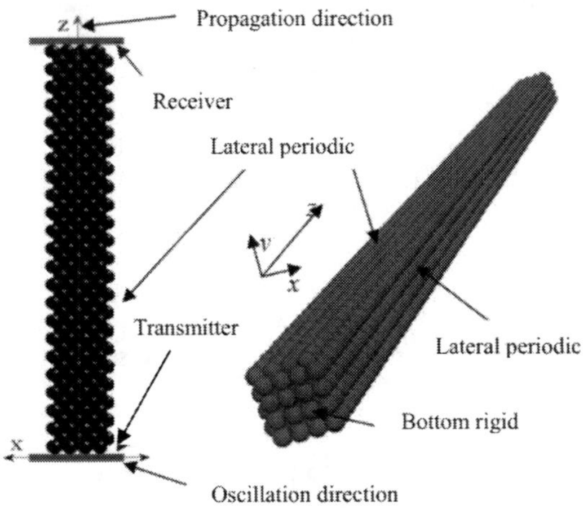

Fig. 1. Simulation configuration to determine the shear modulus G_{zx}.

Maintaining the same lattice structure, the second sample (200 × 4 × 4 layers) with the longest length in the $x-$ direction was created to determining the shear modulus in the horizontal plane. Rigid wall boundaries were applied perpendicular to the direction of wave propagation, while the remaining boundaries were periodic. Various stress states ranging from $p' = 200\ kPa$ to $p' = 1000\ kPa$ were generated by isotropic compression, where p' denotes mean stress. Anisotropic stress states were considered for $p' = 300\ kPa$ and $p' = 1000\ kPa$; the range of stress anisotropies varied within the limits of $\sigma_3 = 0.75p'$ to $\sigma_1 = 1.25p'$, where σ_3 and σ_1 are the minor and major principal stresses.

In experimental geotechnics dynamic, wave propatation tests are commonly used to assess elastic stiffness, DEM simulations of this testing approach can aid the interpretation of such experiments and so this work considered both dynamic and static approaches to determine the stiffness.

Once the desired stress state was achieved, shear waves were generated by applying a single-period sine motion to one rigid wall and the opposite wall was used to analyse received signals. The speed of wave inserted, V_{xy}, was determined by considering changes in the shear stress in the exciting direction of both rigid walls. The corresponding shear modulus, G_{xy} was calculated by assuming elastic behaviour as $G_{xy} = \rho V_{xy}^2$ where ρ is the sample density and is taken to be $\rho = \rho_p/(1+e)$.

The complementary static determination of small strain stiffness firstly used effective medium theory adopting the following expression developed by Chang *et al.* [7] for the HM model using the kinematic assumption.

$$G_0^{EMT-K} = \left[\frac{\sqrt{3}C_N}{\sqrt{2}\pi(1-\nu_P)(1+e)}\right]^{\frac{2}{3}} \times \\ \times \frac{2(5-4\nu_P)}{5(2-\nu_P)} \times G_p^{2/3} p'^{1/3} \tag{2}$$

where C_N is the average contact number per particle.

An expression for $\boldsymbol{G_0}$ for the FCC based on EMT was given by Santamarina and Cascante [8]:

$$G_0^{EMT-FCC} = \frac{(4-3\nu_P)}{2(3-\nu_P)}\left[\frac{3G_p^2 p'}{2(1-\nu_P)^2}\right]^{\frac{1}{3}} \tag{3}$$

Equations 2 and 3 do not account for stress anisotropy and so the following expression proposed by Thornton [9] was also used:

$$\begin{aligned} S_{ijkl}^{Stat} = \frac{N_{cont}}{V}\langle D\rangle\{\langle k^n\, d_p n_i n_j n_k n_l\rangle - \\ -\langle k^t\, d_p n_i n_j n_k n_l\rangle + 0.25(\langle k^t\, d_p n_k n_l\rangle\delta_{ij} \\ +\langle k^t\, d_p n_i n_k\rangle\delta_{jl} + \langle k^t\, d_p n_j n_l\rangle\delta_{ik} + \\ +\langle k^t\, d_p n_i n_l\rangle\delta_{jk})\} \end{aligned} \tag{4}$$

stiffnesses k^n and k^t were directly extracted from the DEM dataset, V is the total volume, n_i is the unit normal in i direction, D is the distance between the two sphere centres, δ_{jl} is Kronecker delta function and N_{cont} is the total number of contacts. Using this approach the shear stiffnesses are given by $G_{zx}^{Stat} = S_{zxzx}^{Stat}$ and $G_{xy}^{Stat} = S_{xyxy}^{Stat}$.

4 Simulation results

To assess whether the simulation approach gives reasonable results, the variation in stiffness with p' for the isotropic case was firstly considered. The data illustrated in Figure 2 show that G_{zx} is proportional to $p'^{0.331}$ and G_{xy} is proportional to $p'^{0.333}$; i.e. in both cases the exponent in the power law relationship linking G and p' is close to the theoretical value of 1/3 which is expected for the Hertz contact model used. Note that the fabric of this sample is transversely isotropic and at each stress level the shear modulus in the vertical plane, G_{zx} is higher than the shear modulus in the horizontal plane,

G_{xy}. By assuming homogeneous and isotropic packing, maximum stiffness, G_0 was also measured by using effective medium theory (EMT) [7, 8]. Data obtained (Figure 2) show that $G_0^{EMT-K} \propto p'^{0.334}$ and $G_0^{EMT-FCC} \propto p'^{0.333}$. The stiffness values obtained using Thornton's tensor [9] are also included and show good agreement with the dynamic measures.

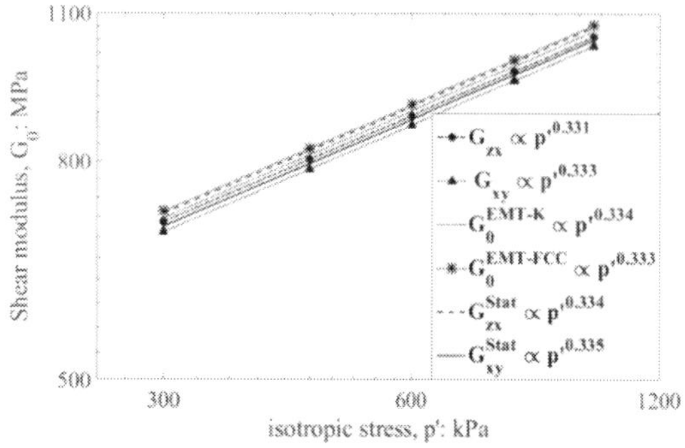

Fig. 2. Elastic stiffness versus isotropic stress, p'.

Sadek *et al.* [10] argued that the stresses in the propagation and oscillation directions, σ_{prop} and σ_{osc}, should have similar effects on the stiffness (G), and so the exponents for σ_{prop} and σ_{osc} in Equation 1 should be equal. Hardin and Blandford [11] also examined the variation in G with the geometric mean of σ_{prop} and σ_{osc}. Consequently, Figure 3 shows the variation in stiffness as a function of the geometric mean of σ_{prop} and σ_{osc}.

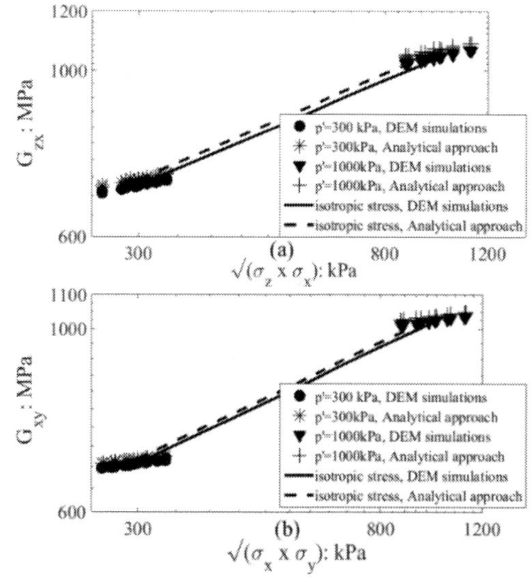

Fig. 3. Stiffness values against the geometric mean of the stresses in propagation and oscillation directions.

Both data from DEM simulations and the analytical approach proposed by Thornton [9] (Equation 4) show that the power law relationship that describes the variation in G with $\sqrt{\sigma_{prop}\ \sigma_{osc}}$ when there is a change in mean stress differs from the power law relationships describing the variation in G with $\sqrt{\sigma_{prop}\ \sigma_{osc}}$ at a constant mean stress. In comparison with the sensitivity to changes in the mean stress, at a given mean stress, the shear modulus increases only slightly with $\sqrt{\sigma_{prop}\ \sigma_{osc}}$. The nature of the relationship varies with orientation, with the G_{zx} values being more sensitive to changes in $\sqrt{\sigma_{prop}\ \sigma_{osc}}$ in both cases.

Hardin and Blandford [11] and Bellotti *et al.* [1] explored whether the variation in G_{zx} with stress level systematically varied with the consolidation stress ratio. For clarity using only the dynamic results the G_{zx} data were grouped according to the value of the stress ratio, $K = \sigma_x/\sigma_z$. When the dataset as a whole is considered (Figure 4a) no dependency on K is evident, the relationships between G_{zx} and $\sqrt{\sigma_z\,\sigma_x}$ are nearly indistinguishable (Figure 4a). This agrees with the observation of Hardin and Blandford [11] who found that the shear modulus is almost independent of the values of K. However, if the data at $p' = 300\ kPa$ are considered independently (Figure 4b), the resultant larger scales reveal that the G_{zx} values are somewhat sensitive to K; the values systematically increase with K for a given value of $\sqrt{\sigma_z\,\sigma_x}$. These observations agree with the findings of Bellotti *et al.* [1] and provide an explanation for the small scatter observed in the G_{zx} : $\sqrt{\sigma_z\,\sigma_x}$ data at a constant p'(Figure 3).

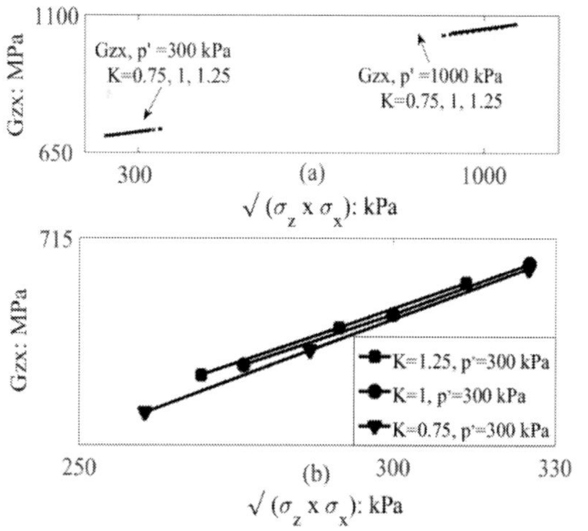

Fig. 4. Shear modulus G_{zx}versus the geometric mean of the stresses in propagation and oscillation directions: (a) for both $p' = 300\ kPa$ and $p' = 1000\ kPa$; (b) $p' = 300\ kPa$.

The influence of the stress in plane orthogonal to wave motion, σ_{third} was also assessed (Figure 5). As expected, σ_{third} has less influence on the elastic stiffness; however, the data do not support the idea that $n_{third} = 0$ as in Sadek *et al.* [10]. As can be seen from Figure 5, the shear modulus decreases as σ_{third} increases while p' remains constant. However, for all these simulations the constraint of p' remaining constant means that an increase in σ_{third} leads to a reduction in either or both σ_{prop} and σ_{osc}; this differs from Sadek *et al.* [10].

When the mean stress is maintained constant it is difficult to understand the influence of each stress component on the response completely as the stress components become interdependent. Therefore, the effect of each stress component on the shear modulus in vertical plane, G_{zx}, was examined from an initial isotropic stress of 300 kPa (Figure 6). The data indicate that σ_{osc} (σ_x) has a slightly larger influence on elastic stiffness than σ_{prop} (σ_z) does. The G_{zx} values increased with increasing σ_{third}; this differs clearly from the case of maintaining p'constant (Figure 5).

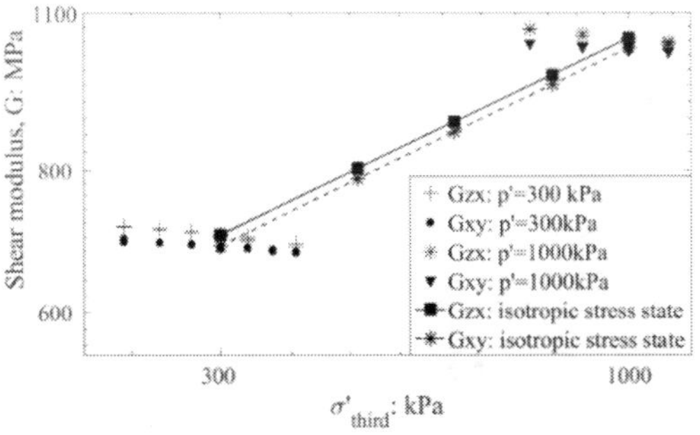

Fig. 5. Stiffness values against the stress orthogonal to wave motion plane.

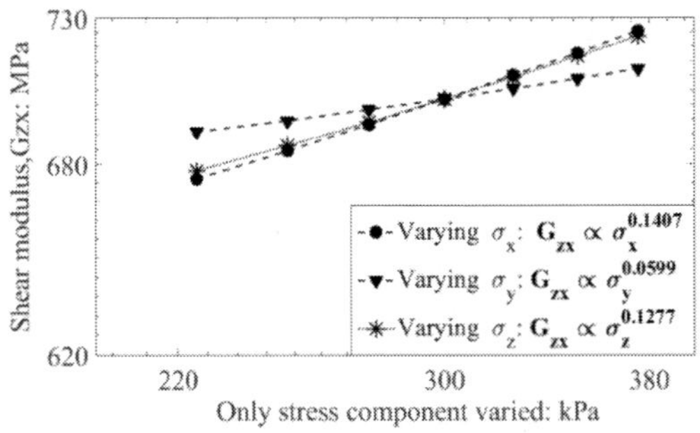

Fig.6. Stiffness values (G_{zx}) from varying one stress and keeping others constant at 300 kPa.

Similarly, the influence of individual stress components on the stiffness in the horizontal plane G_{xy} was also considered. Referring to Figure 7, in this case the exponents n_{prop} and n_{osc} are the same, meaning that σ_{osc} (σ_y) and σ_{prop} (σ_x) have equal influences on G_{xy}. This reflects the isotropic fabric in the horizontal plane. However, the exponent of σ_{third} is only slightly lower than exponents of other principal stresses, which is different from prior studies [1, 2, 10].

5 Conclusion

The effect of stress anisotropy on the shear modulus has been examined. Wave propagation simulations using the discrete element method were carried out considering plane wave propagation which produces clear signals. The simulations were validated using Hertzian theory for the isotropic case. Static, analytical approaches were also used and these confirmed the results observed using the dynamic methods. The data for the more general anisotropic stress states indicate that while G depends on the stresses in both the oscillation and propagation directions the relationship is more complex than hitherto appreciated; different power laws describe the stress dependency at a given mean stress and the stress dependency associated with changes in mean stress. The data indicate a weak dependency of the relationship between G and $\sqrt{\sigma_{prop}\ \sigma_{osc}}$ on the consolidation stress ratio, and so support the earlier conclusions of both [1] and [11]. The dependency of G on σ_{third} may be stronger than hitherto appreciated. These observations were made on lattice, FCC samples. Their applicability to real soil should be explored in future research using random packings with polydisperse particles.

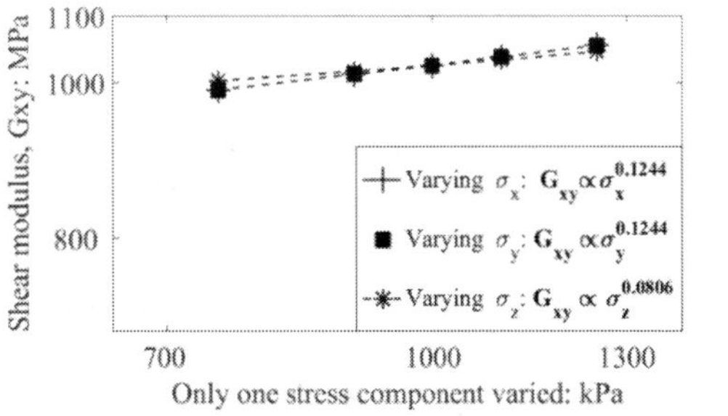

Fig.7. Stiffness values (G_{xy}) from varying only one stress component and keeping others constant at 1000 kPa.

Acknowledgements

The first author has received a Vied-Newton PhD scholarship and a Dixon scholarship.

References

1. S. K. Roesler, *J. Geotech. Eng. Div* **105**, 871 (1979).
2. R. Bellotti, M. Jamiolkowski, D. C. F. Lo Presti, and D. A. O'neill, *Geotechnique* **46,** 115 (1996).
3. P. W. Rowe, *Proc. R. Soc. A* **269,** 500 (1962).
4. J. O. Donovan, C. O. Sullivan, G. Marketos, and D. M. Wood, *Geotech. Lett* **5,** 224 (2015).
5. S. Plimpton, J. Comput. Phys **117,** 1 (1995).
6. O. Mouraille, W. a Mulder, and S. Luding, *J. Stat. Mech. Theory Exp* **2006, 7** (2006).
7. C. S. Chang, A. Misra, and S. S. Sundaram, *Soil Dyn. Earthq. Eng* **10,** 201 (1991).
8. J. C. Santamarina and G. Cascante, *Can. Geotech.* J **33,** 770 (1996).
9. C. Thornton, J. Phys. D. *Appl. Phys* **26,** 1587 (1993)
10. T. Sadek, M. Lings, L. Dihoru, and D. Muir Wood, *Riv. Ital. Geotec* **41,** 69 (2007).
11. B. O. Hardin and G. E. Blandford, *J. Geotech. Eng* **115,** 788 (1989).

Testing the $\mu(I)$ granular rheology against experimental silo data

Luke Fullard[1,*], *Eric* Breard[2], *Clive* Davies[3], *Pierre-Yves* Lagrée[4], *Stéphane* Popinet[4], and *Gert* Lube[2]

[1] *Institute of Fundamental Sciences, Massey University, New Zealand*
[2] *Volcanic Risk Solutions,Massey University, New Zealand*
[3] *School of Engineering and Advanced technology, Massey University, New Zealand*
[4] *Institut Jean le Rond d'Alembert, Université Pierre et Marie Curie, Paris, France*

Abstract. Industrial storage of granular material using silos is common, however, improved understanding of silo flow is needed. Various continuum models attempt to describe the velocity of dense granular flow in silos. Kinematic, and recently, stochastic models, based upon the diffusion of some quantity, perform well when there is a single orifice, and when the yield criterion is satisfied. However, if system stresses are insufficient to satisfy the yield criterion, or if there is a second orifice, these models fail to capture the entire flow behaviour. Advances in granular rheology have allowed a pressure dependent friction law to be defined which can capture the behaviour of granular silo flow including un-yielded zones, flow-rate independence of fill height, the Beverloo flow-rate, and various other phenomena. We performed silo discharge experiments in a flat bottomed planar silo with a single and two adjacent orifices, for two grain types. The velocity was measured using Particle Image Velocimetry. Results were compared to a mathematical model based on the $\mu(I)$ rheology which was shown to qualitatively capture the observed phenomena including plug-like zones where the yield criterion is not satisfied. These preliminary results strongly encourage future investigations into the effect of friction parameters and numerical boundary conditions.

1 Introduction

Although granular material has been stored in silo constructions for thousands of years [1] it is only in the last sixty years that significant progress has been made to understand the physics of silo discharge. The mass flow rate of grains from slot and circular orifices is well described using the Beverloo scaling [2, 3], while the velocity field during silo discharge has been described by kinematic-type models [4–9], stochastic diffusion models [10, 11], discrete modelling [12–14], and more recently, by utilising a pressure dependent viscosity in a Navier-Stokes solver [15, 16]. Experimental studies on silo flow have been conducted using 2D imaging techniques such as particle image velocimetry (PIV) or Lagrangian particle tracking [4, 17–19]. These studies have shown that the velocity field near the orifice of a silo is Gaussian-like (which motivated the use of diffusion-like equations to describe some granular silo flows) and mainly radial, but further from the orifice the flow is more plug-like with neighbouring particles flowing at very similar velocities. Although discrete models have been shown to capture most of the mentioned phenomena, the computational overhead is considered too high for many practical industrial flows, particularly when the particle size is small and the silo is large. As such, continuum models have been pursued to model many industrial granular flows including segregation [20], mixing [6, 21], and blending [22]. Kinematic and stochastic based models describe the Gaussian-like velocity close to the silo orifice well, yet are not predictive of the flow rate, but rather take the flow rate as an input parameter (either explicitly or via a specified velocity far from the orifice). As such, they are unable to describe the Beverloo scaling of flow rate with orifice diameter explicitly, or capture the flow-rate interference of two near-by orifices [23]. It has also been reported that the "diffusion length" parameter in kinematic models is a function of the height in the particle bed [4],which has not been described well by kinematic models. Additionally, since the yield-criterion is assumed satisfied everywhere for these two models, they are unable to describe accurately plug-like flow zones in tall silos. Conversely, models where the so-called $\mu(I)$ rheology [24] is used to define an artificial viscosity in a Navier-Stokes code [15, 16, 25] have proven to capture all these observed phenomena, although comparison to experimental data is still a work in progress. To this end, we present preliminary results of 2D silo drainage experiments with amaranth grains and mustard seeds and compare the observed results with those derived using the $\mu(I)$ model. The experiments (and numerical tests) are performed with a single centrally located slot orifice, and with two adjacent orifices.

*e-mail: L.Fullard@Massey.ac.nz

Figure 1. A formatted image showing the silo test area during an experiment with mustard seed and a centrally located exit orifice.

2 Experimental method and results

Our experimental system is a flat bottomed rectangular silo constructed from perspex on all sides. The silo has width 200*mm*, depth 15*mm* and height 350*mm*, and has a groove near the bottom where we can control the orifice size, geometry, and spacing. The silo is filled with either amaranth grains or mustard seeds and the flow is initiated using a sliding bar mechanism. The flow is recorded using a high-speed camera at 340 frames per second (with the exception of the "Single slot: Amaranthus experiment", which was recorded at 500 FPS), and the resulting images are passed through a particle image velocimetry software, PIVLab [26, 27], to obtain velocity vectors. The experiments are performed once with a single centrally located rectangular slot orifice, and once with adjacent orifices separated by a distance of 80*mm*. The amaranth grains were approximately 1*mm* in diameter, each orifice had a width of 8*mm*. For the mustard seeds, which were approximately 2*mm* in diameter, the slot orifices were 14*mm* in width. One image of the flowing mustard seed is shown in Figure 1. Not visible in the figure is a larger filling silo above the experimental zone which keeps refilling the lower test silo.

Figure 2 displays resultant velocities of the four experiments (two particle types, two orifice geometries). Plotted in the figure is the norm of the velocity vector at five different heights above the exit orifice/s. Apparent in the figures is that close to the orifice the velocity appears to be Gaussian-like, in agreement with other studies. However, further from the orifice, the velocity in the center of the silo is rather in-variant in the horizontal direction and plug-like. This is particularly evident in the double slot amaranth experiment, where the velocity far from the orifice is constant between two wall shear zones. It also appears that one effect of adding two adjacent orifices is to bring the plug flow zone closer to the orifice since, for both particle types, the flow 110*mm* above the orifice is Gaussian-like in the single slot case, but more plug-like in the double slot case. Additionally, there is slip flow at the silo walls for the mustard seeds, but no flow at the walls was visible for the amaranth grains.

3 Numerical model and results

Recent experiments to measure the rheology of dense granular flow [24] allow us to model the bulk friction of granular materials using the following expression

$$\mu(I) = \mu_1 + \frac{\mu_2 - \mu_1}{\frac{I_0}{I} + 1}$$

where μ_1 and μ_2 are the lower and upper limits of friction, I_0 is the inertial number when the friction is half way between μ_1 and μ_2, and I is the granular inertial number, a function of the particle diameter, d, shear rate, $\dot{\gamma}$, particle density, ρ_p, and confining pressure, P;

$$I = \frac{d|\dot{\gamma}|\sqrt{\rho_p}}{\sqrt{P}},$$

By assuming that the volume fraction changes are small, that the deviatoric stress tensor aligns with the strain rate tensor, Jop [28] defined an effective granular viscosity as

$$\eta(|\dot{\gamma}|, P) = \mu(I)P/|\dot{\gamma}|. \quad (1)$$

We note that the yield using Equation 1 is a Drucker-Prager criterion; material will not shear unless the shear stress is greater than the friction multiplied by the pressure. This effective viscosity has been successfully implemented in a Navier-Stokes solver (Gerris Flow Solver [25]) to model dense granular flow in various geometries [15, 16, 29] including silo flow. In these works the viscosity was regularised using the following relation (which we also utilise in this work)

$$\eta = min\left(\frac{\mu(I)P}{D_2}, \eta_{max}\right),$$

where D_2 is the second invariant of the strain rate tensor.It was shown [15, 16] that such a model could account for the Beverloo scaling of the flow rate as a function of orifice width, and the behaviour was well matched with contact-dynamics simulations.

In this investigation, a 2D model was solved using a similar method as in literature [15, 16]. We do not model the flow as a two-fluid flow (that is, we do not investigate the free-surface evolution), but rather, make the assumption that the silo is continuously full and the top of the system is at atmospheric pressure. This procedure allows

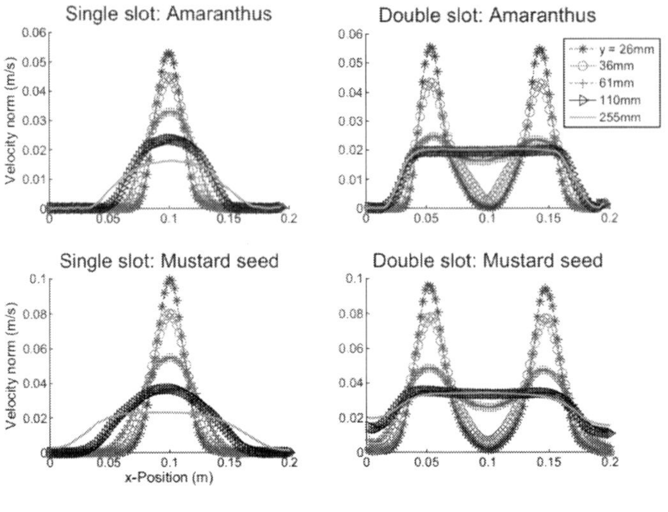

Figure 2. Experimental velocity curves at various heights above the silo orifice for single and two adjacent slot orifices (separated by $80mm$), for two particle types, amaranth grains and mustard seeds.

friction at the side walls to be accounted for, but neglects the friction on the front (viewing side) and back of the silo walls. Although this is expected to have an effect, as a first investigation we neglect this aspect of the flow. The angle of repose of mustard seeds was found to be approximately 29.5°. Therefore, as an initial test of the friction parameters, we chose $\mu_1 = \tan(29.5)$ degrees. The values of I_0 and μ_2 were chosen to be $I_0 = 0.05$ and $\mu_2 - \mu_1 = 0.25$. While these parameters are not measured, we found that they are reasonable approximations. We are pursuing experimental determination of the friction parameters using a set-up similar to Fall et al. [30]. The particle and bulk densities were given as, $\rho_p = 1250kg/m^3$, $\rho_b = 675kg/m^3$ and the particle diameter, $d = 2mm$. On the left and right walls the horizontal velocity was set to zero, and a Navier slip condition was applied for the vertical velocity with slip length $0.5d = 1mm$. On the free surface and at the orifice the pressure was set to zero to simulate atmospheric conditions above and below the silo. The normal derivatives of the velocity components were set to zero above and below the silo. Solids with a no-slip condition were added near the orifice to model the position of the single and double orifice configurations.

Figure 3 displays velocity norm curves for the numerical mustard seed model to be compared with those of Figure 2. Near the orifice the Gaussian-like character of the flow is recovered, and the values of the speed are very close. Far from the orifice the flow is plug-like (although, the regularisation used slightly curves this zone). The addition of a second orifice has the effect of enlarging the plug-flow zone. In the figure, the red curve at $61mm$ above the orifice is Gaussian-like in the single slot case, but plug-like in the double slot case. This was also seen in the experimental results of Figure 2 and indicates that the plug-like zone has extended closer to the bottom of the silo. Such results could be important to predict mixing in silo flows.

Qualitatively the behaviour is similar, yet is not perfect. The numerical model is more diffusive as we move further above the orifice. Far from the orifice the flowing zones are wider (and the shear zones where the velocity is changing rapidly less wide) in the numerical model than experimentally observed. In the numerical model the two orifices interact more strongly than was seen in the experiment. For example, in the center of the silo, at $x = 0.1m$, the interaction between the two orifices appears to be minimal in the experimental case as the velocity is almost zero at this location (for $y = 26mm$, profiles near the bottom of the silo). However in the numerical model this velocity is clearly non-zero, hence the interaction between the two orifices is still influencing the silo discharge. The slip velocity in the experiment was also seen to be larger than was predicted by the numerical model. This was particularly evident far from the orifice. Nevertheless, given that the friction of the front wall is neglected, and the rheological parameters and slip length at this stage remain best guesses, the results are encouraging. Once the friction parameters have been measured for our grains we plan to repeat the numerical tests with and without front wall friction.

4 Discussion

We have presented a first attempt to compare experimentally derived velocity fields using particle image velocimetry analysis, to results of a numerical model with a pressure-dependent viscosity, based upon the $\mu(I)$ granular rheology. We have shown that the numerical model

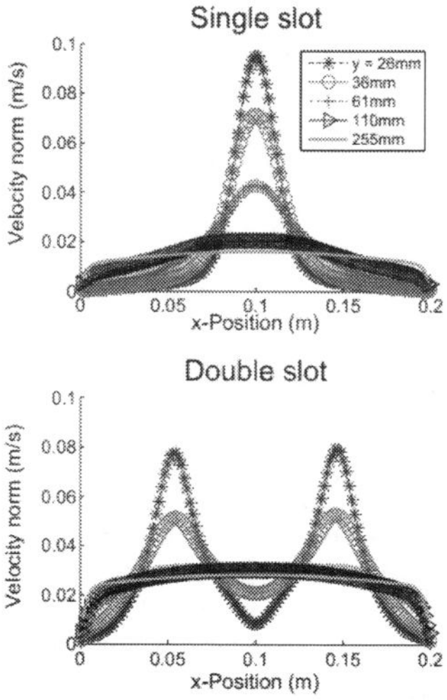

Figure 3. Numerical velocity curves generated using approximate parameters for mustard seed.

based on the $\mu(I)$ model is able to qualitatively capture the observed flow profiles in the silo, but the model is somewhat diffusive. The accurate resolution of plug-flow and Gaussian-like flow zones is critical to predict mixing, segregation, and residence time distributions precisely. Errors in the predicted velocity field will lead to errors in these measures and ultimately lessen the predictive capability of the model. This is particularly important in industries where quality control of granular flow is essential, such as the pharmaceutical and food industries. As such, further work is needed to accurately measure the flow parameters of the grains for the $\mu(I)$ model. The effect of friction on the front wall also needs to be investigated, as does other real-world effects such as cohesion between grain (perhaps generated by a high humidity environment).

However, we are encouraged by the preliminary behaviour of the model. With approximate parameters and several simplifications we have been able to reproduce many of the experimentally observed features of the flow.

In conclusion, the $\mu(I)$ model shows much promise to model industrial-scale granular flow and help in the understanding of granular flow phenomena.

References

[1] Y. Garfinkel, D. Ben-Shlomo, T. Kuperman, Antiquity **83**, 309 (2009)
[2] W. Beverloo, H. Leniger, J. Van de Velde, Chemical engineering science **15**, 260 (1961)
[3] C. Mankoc, A. Janda, R. Arevalo, J. Pastor, I. Zuriguel, A. Garcimartín, D. Maza, Granular Matter **9**, 407 (2007)
[4] J. Choi, A. Kudrolli, M.Z. Bazant, Journal of Physics: Condensed Matter **17**, S2533 (2005)
[5] L. Fullard, C. Davies, Applied Mathematical Modelling **40**, 1494 (2016)
[6] L.A. Fullard, C.E. Davies, G.C. Wake, Advanced Powder Technology **24**, 499 (2013)
[7] R.M. Nedderman, *Statics and kinematics of granular materials* (Cambridge University Press, 2005)
[8] G.J. Weir, Chemical Engineering Science **59**, 149 (2004)
[9] F. Vivanco, F. Melo, C. Fuentes, International Journal of Bifurcation and Chaos **19**, 3533 (2009)
[10] K. Kamrin, M.Z. Bazant, Physical Review E **75**, 041301 (2007)
[11] K. Kamrin, C.H. Rycroft, M.Z. Bazant, Modelling and Simulation in Materials Science and Engineering **15**, S449 (2007)
[12] R. Balevičius, R. Kačianauskas, Z. Mróz, I. Sielamowicz, Advanced Powder Technology **22**, 226 (2011)
[13] R. Kobyłka, M. Molenda, Powder Technology **256**, 210 (2014)
[14] Y. Yu, H. Saxén, Advanced Powder Technology **22**, 324 (2011)
[15] L. Staron, P.Y. Lagrée, S. Popinet, Physics of Fluids (1994-present) **24**, 103301 (2012)
[16] L. Staron, P.Y. Lagrée, S. Popinet, The European Physical Journal E **37**, 1 (2014)
[17] I. Sielamowicz, S. Błoński, T.A. Kowalewski, Chemical Engineering Science **61**, 5307 (2006)
[18] I. Sielamowicz, A. Czech, T.A. Kowalewski, Powder Technology **270**, 393 (2015)
[19] I. Sielamowicz, S. Blonski, T. Kowalewski, Chemical Engineering Science **60**, 589 (2005)
[20] J. Gray, A. Thornton, Proceedings of the Royal Society of London A: Mathematical, Physical and Engineering Sciences **461**, 1447 (2005)
[21] L. Fullard, C. Davies, Particuology **30**, 102 (2017)
[22] K. Johanson, Powder technology **170**, 109 (2006)
[23] X. Zhang, S. Zhang, G. Yang, P. Lin, Y. Tian, J.F. Wan, L. Yang, Physics Letters A **380**, 1301 (2016)
[24] G. MiDi, The European Physical Journal E **14**, 341 (2004)
[25] S. Popinet, Journal of Computational Physics **190**, 572 (2003)
[26] W. Thielicke, E.J. Stamhuis, Journal of Open Research Software **2**, e30 (2014)
[27] W. Thielicke, Ph.D. thesis, Ph. D. Thesis, Rijksuniversiteit Groningen (2014)
[28] P. Jop, Y. Forterre, O. Pouliquen, Nature **441**, 727 (2006)
[29] P.Y. Lagrée, L. Staron, S. Popinet, Journal of Fluid Mechanics **686**, 378 (2011)
[30] A. Fall, G. Ovarlez, D. Hautemayou, C. Mézière, J.N. Roux, F. Chevoir, Journal of Rheology (1978-present) **59**, 1065 (2015)

DOI: 10.1051/epjconf/201714008004

Transport and deposition of cohesive pharmaceutical powders in human airway

Yuan Wang[*], *Kaiwei Chu*, and *Aibing Yu*

Laboratory for Simulation and Modelling of Particulate Systems, Monash University, Clayton, VIC 3800, Australia

Abstract. Pharmaceutical powders used in inhalation therapy are in the size range of 1-5 microns and are usually cohesive. Understanding the cohesive behaviour of pharmaceutical powders during their transportation in human airway is significant in optimising aerosol drug delivery and targeting. In this study, the transport and deposition of cohesive pharmaceutical powders in a human airway model is simulated by a well-established numerical model which combines computational fluid dynamics (CFD) and discrete element method (DEM). The van der Waals force, as the dominant cohesive force, is simulated and its influence on particle transport and deposition behaviour is discussed. It is observed that even for dilute particle flow, the local particle concentration in the oral to trachea region can be high and particle aggregation happens due to the van der Waals force of attraction. It is concluded that the deposition mechanism for cohesive pharmaceutical powders, on one hand, is dominated by particle inertial impaction, as proven by previous studies; on the other hand, is significantly affected by particle aggregation induced by van der Waals force. To maximum respiratory drug delivery efficiency, efforts should be made to avoid pharmaceutical powder aggregation in human oral-to-trachea airway.

1 Introduction

To study the flow of inhaled pharmaceutical powders in human airway is important in estimating the health risk of inhaled toxic pollutant and determining the efficacy of aerosol drugs [1]. Pharmaceutical powders used in inhalation therapy are in the size range of 1-5 microns and are usually cohesive. Understanding the cohesive behaviour of pharmaceutical powders during their transportation in human airway is significant in optimising aerosol drug delivery and targeting.

Numerical simulation is widely used by researchers to study the transport and deposition of particles in human airway [2-4]. Most of the studies are based on the assumption of dilute particle flow and the particle level phenomena are usually neglected. In the real-world drug delivery process, however, a large number of particles are injected within a short time and the local particle concentration can be high; particle aggregation may happen due to the cohesive nature of fine particles. Discrete element method (DEM) allows for a detailed description of the particle aggregation process [5]. CFD-DEM coupling model has gained prominent applications in numbers of industrial systems [6] and drug delivery systems [7]. In coupled CFD-DEM model, the motion of particles is modeled by DEM, under which Newton's second law of motion is applied to each single particle; the fluid flow is described by the local averaged Navier-Stokes equations, which can be solved by CFD. Besides of the capability to account for complex interactions beween particles and particle-fluid, coupling of DEM will also provide new physical insight to the fluid-particle dynamics and particle deposition mechanism in human airways.

In this study, a CFD-DEM model is used to describe the particle transport process in human upper airway. The dominant cohesive force, i.e. van der Waals force, is coupled in and the cohesive behaviour of the particle flow is modelled.

2 Numerical Methods

2.1 Geometry and Mesh

A representative geometry of the adult mouth- throat region is selected for the CFD-DEM model test. The original geometry was developed by Respiratory and Aerosol Dynamics Research Group of Virginia Commonwealth University and described in works by Xi et al [8]. The symmetric plane with cross-sections at several positions and the computational mesh are shown in figure 1. To reduce computation expense, the geometry is down-scaled while the hydrodynamic similarity with real inhalation condition is kept by keeping the same range of the fluid Reynolds number (Re) and particle Stokes number (Stk). The inner space of the model is discretized into a multi-block structured mesh for numerical simulation. The final mesh contains ~112,000 mesh elements, which is determined by a mesh independence study.

[*] Corresponding author: yuan.wang2@monash.edu

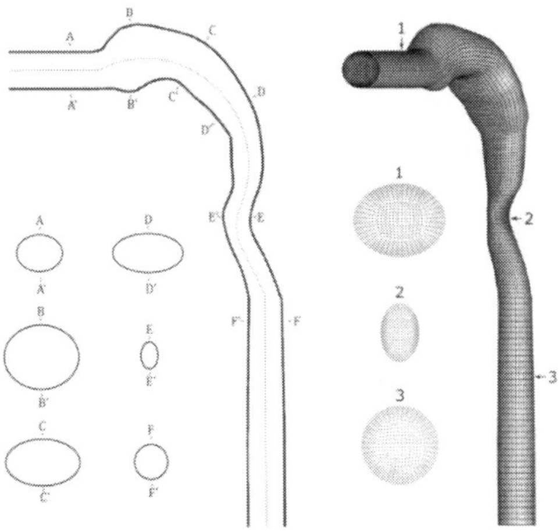

Fig. 1. The symmetric plane with cross-sections at several positions and the computational mesh

2.2 Governing Equations

The governing equations for the CFD-DEM model have been demonstrated in detail by K. Chu et al [9]. For the continuum phase, the mass and momentum over a computational cell are conserved, leading to the governing equation of fluid flow:

$$\frac{\partial \rho_f \varepsilon}{\partial t} + \nabla \cdot (\rho_f \varepsilon u) = 0, \tag{1}$$

$$\frac{\partial (\rho_f \varepsilon u)}{\partial t} + \nabla \cdot (\rho_f \varepsilon u u) = -\nabla P - F_{p-f} + \nabla \cdot (\varepsilon \tau) + \rho_f \varepsilon g + \nabla \cdot (-\rho_f \overline{u'u'}) \tag{2}$$

where ε, ū, u', t, ρf, P, Fp-f, τ and g are, respectively, porosity, mean and fluctuating fluid velocity, time, fluid density, pressure, volumetric fluid-particle interaction force, fluid viscous stress tensor, and acceleration due to gravity.

The particle motion is modelled by discrete element method (DEM) developed by Cundall and Strack [10]. Particle translational and rotational motion is described by Newton's law of motion:

$$m_i \frac{d\mathbf{v}_i}{dt} = f_{p-f,i} + \sum_{j=1}^{k_i} (f_{c,ij} + f_{d,ij} + f_{vdw,ij}) + m_i g, \tag{3}$$

$$I_i \frac{d\omega_i}{dt} = \sum_{j=1}^{k_i} (T_{ij} + M_{ij}) \tag{4}$$

where m_i, I_i, $\mathbf{v}_i$, and $\boldsymbol{\omega}_i$ are the mass, moment of inertia, translational velocity and rotational velocities of particle i, respectively. The inter-particle forces generated by particle collision include the elastic contact force $f_{c,ij}$ and viscous damping force $f_{d,ij}$. The cohesive forces arise from van der Waals force ($f_{vdw,ij}$). Other cohesive forces, e.g., capillary force and electrostatic force are not discussed in current study.

Two-way coupling of CFD-DEM is used in current study, which is achieved by applying Newton's third law in a computational cell. The coupling scheme is well documented in previous literature [11].

2.3 Numerical Setup and Boundary Conditions

The numerical solution of the fluid equations (1) and (2) is carried out by commercial finite-volume-based program Fluent 12.0 (Ansys, Inc.).The particle flow is produced by solving equations (3) and (4) by a time integration method. A DEM code is incorporated with Fluent via user defined functions (UDF).

Particles (5μm in diameter) are injected at mouth inlet with a uniform velocity profile; the air inlet Reynolds number is set to 1000. A uniform pressure boundary condition is applied at the outlets. Other assumptions include incompressible airflow, perfect trapping wall and non-slip wall condition. The physical properties of gas and solid phase are summarized in table 1.

Table 1: Particle and Fluid Properties

Particle Phase	
Particle Diameter (μm)	5
Particle Density (kg/m^3)	1000
Sliding Friction Coefficient	0.3
Damping Coefficient	0.3
Rolling Friction Coefficient	6×10^{-5}
Young's Modulus(N/m)	1×10^{-7}
Poisson's Ratio	0.3
Hamaker Constant (J)	6.5×10^{-20}
Fluid Phase	
Density (kg/m^3)	1.225
Viscosity(kg/m/s)	1.798×10^{-5}

3 Results and Discussion

The particle transport dynamic is well captured by the CFD-DEM model in current study. To stress the influence of dominant cohesive force, i.e. van der Waals force, the flow patterns of non-cohesive and cohesive particle flow are compared. As illustrated in figure 2, under the influence of van der Waals force, particles start to get aggregated in oral cavity and obvious particle clusters are formed(bottom figure) due to the high local concentration; on the other hand, without cohesive force being coupled in, the particles are still well-dispersed in the same region (top figure).

As particles flowing in the airway, more and more particle clusters are formed and the cluster size becomes larger. Figure 3 describes the particle distribution in the lower part of trachea for non-cohesive and cohesive particle flow. It can be seen that for cohesive particle flow (bottom figure), big clusters are formed and the particle distribution is rather non-uniform at the trachea outlet.

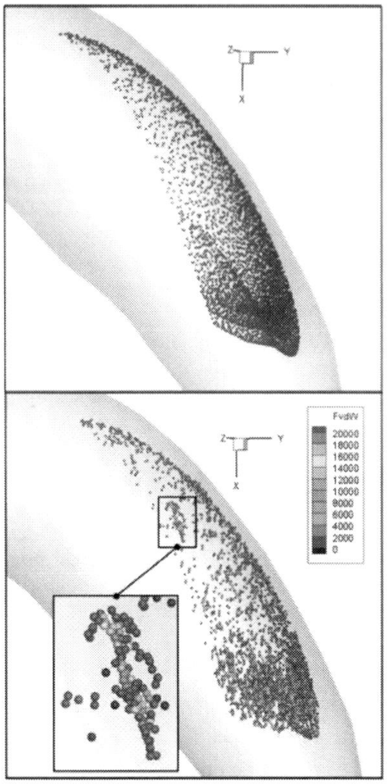

Fig. 2. Comparaison of particle flow pattern in oral cavity between non-cohesive and cohesive particle flow

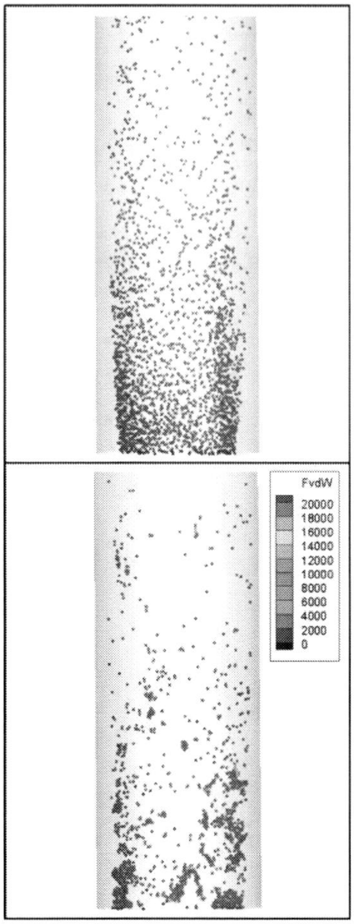

Fig. 3. Comparaison of particle flow pattern in trachea between non-cohesive and cohesive particle flow

In the airway, the wall is assumed to be perfect-trapping wall and thus the clustered particles cannot be deaggregated due to particle-wall impaction. Therefore, the formed clusters can only be deaggregated by fluid drag. In current study, the clusters are assumed to be stable when the particles are bounded by van der Waals force ten times larger than the fluid drag force. Figure 4 illustrates how the mass ratio of the stable clusters to the total particle mass changes with time. It is shown that particle clustered are formed as particles flow in the airway; starting from 0.006s when most of the particles have entered the curved oral cavity, the number of clustered particles are increased more rapidly, which indicates that the curvilinear motion of particles helps the formation of particle aggregation. Later, a steady state is reached where the mass ratio of the clusters no longer changes.

The formation of these particle clusters is predicted to enhance the particle deposition, especially in the subsequent airway bifurcations. As those particle clusters go deeper in the lung, the diameter of each airway is decreasing although their total area is increasing. The ratio of cluster size and airway diameter will get large and the deposition by intersection will become significant.

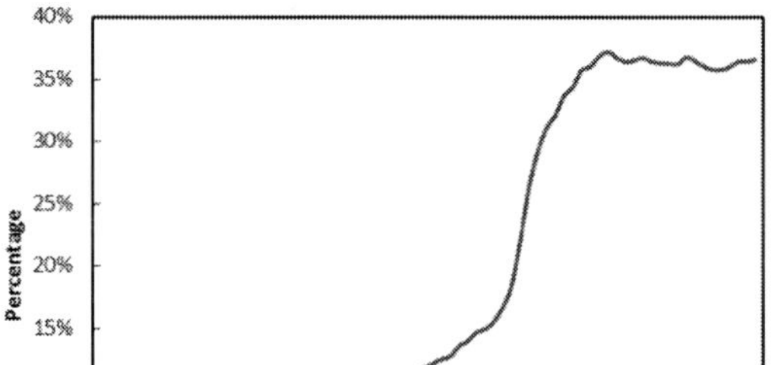

Fig. 4. Mass ratio of the stable clusters to the total particle mass plotted with time

4 Conclusions

In this study, the transport process of cohesive particle flow is modelled and the influence of van der Waals force is discussed. It is concluded that the cohesive force leads to significant aggregation in particle flow and the formed particle clusters will enhance particle deposition in human airway. The deposition mechanism for cohesive pharmaceutical powders, on one hand, is dominated by particle inertial impaction, as proven by previous studies; on the other hand, is significantly affected by particle aggregation induced by van der Waals force. To maximum respiratory drug delivery efficiency, efforts should be made to avoid pharmaceutical powder aggregation in human oral-to-trachea airway.

Acknowledgement

The authors are grateful to the Australia Research Council (ARC) for the financial support.

References

1. C. A. Ruzycki, E. Javaheri, W. H. Finlay, Expert Opin. Drug Deliv. **10**, 3 (2013)
2. B. Soni, S. Aliabadi, Comput. Fluids **88**, 0 (2013)
3. M. C. Piglione, D. Fontana, M. Vanni, Eur. J. Mech. B-Fluid **31**, 0 (2012)
4. Y, Feng, C. Kleinstreuer, J. Aerosol Sci. **71** (2014)
5. B. H. Xu, A. B. Yu, Chem. Eng. Sci. **52**, 16 (1997)
6. S. Limtrakul, A. Boonsrirat, T. Vatanatham, Chem. Eng. Sci. **59**, 22–23 (2004)
7. Z. B. Tong, B. Zheng, R. Y. Yang, A. B. Yu, H. K. Chan, Powder Technol. **240**, 0 (2013)
8. J. Xi, P. W. Longest, Ann. Biomed. Eng. **35**, 4 (2007)
9. K. W. Chu, B. Wang, D. L. Xu, Y. X. Chen, A. B. Yu, Chem. Eng. Sci. **66**, 5 (2011)
10. P.A. Cundall, O.D.L. Strack, Geotechnique **29** (1979)
11. Z. Y. Zhou, S. B. Kuang, K. W. Chu, A. B. Yu, J. Fluid Mech. **661** (2010)

DOI: 10.1051/epjconf/201714006002

DEM simulation of dendritic grain random packing: application to metal alloy solidification

Antonio Olmedilla[1,*], *Miha* Založnik[1,**], and *Hervé* Combeau[1,***]

[1] *Institut Jean Lamour, CNRS – Université de Lorraine, F-54011 Nancy CEDEX, France*

Abstract. The random packing of equiaxed dendritic grains in metal-alloy solidification is numerically simulated and validated via an experimental model. This phenomenon is characterized by a driving force which is induced by the solid-liquid density difference. Thereby, the solid dendritic grains, nucleated in the melt, sediment and pack with a relatively low inertia-to-dissipation ratio, which is the so-called Stokes number. The characteristics of the particle packed porous structure such as solid packing fraction affect the final solidified product. A multi-sphere clumping Discrete Element Method (DEM) approach is employed to predict the solid packing fraction as function of the grain geometry under the solidification conditions. Five different monodisperse noncohesive frictionless particle collections are numerically packed by means of a vertical acceleration: a) three dendritic morphologies; b) spheres and c) one ellipsoidal geometry. In order to validate our numerical results with solidification conditions, the sedimentation and packing of two monodisperse collections (spherical and dendritic) is experimentally carried out in a viscous quiescent medium. The hydrodynamic similarity is respected between the actual phenomenon and the experimental model, that is a low Stokes number, $o(10^{-3})$. In this way, the experimental average solid packing fraction is employed to validate the numerical model. Eventually, the average packing fraction is found to highly depend on the equiaxed dendritic grain sphericity, with looser packings for lower sphericity.

1 Introduction

During metal alloy solidification solid and liquid phases co-exist since solid grains are nucleated in the melt. The solid phase density is slightly higher than that of the melt so the solid particles tend to settle and pack, e.g. aluminum and steel alloys. This forms a porous dendritic network [1] whose properties such as its solid packing fraction are important to better understand in order to improve the solidification macrosegregation modeling of metallic alloys [2, 3].

In this article, we focus on the modeling of the random packing phenomenon of the equiaxed dendritic solid grains during solidification and essentially the study of the geometrical influence though sphericity on the packing fraction. The packing is characterized by particles whose morphology is equiaxed dendrite-like and by a low driving force which is the apparent weight of the solid grains immersed in the melt [4].

For the modeling strategy the DEM is chosen, integrating the solid grain dynamics in sedimentation and packing. Plenty of work has already been carried out on this numerical technique for the case of spherical particles, however, much less work can be found for the case of arbitrary geometry, specially nonconvex particles. Among the different numerical strategies suggested in literature to build a non-spherical morphology, some of them are the spherical-ellipsoidal combination [5], multi-sphere clumping [6] or convex-geometry combinations [7]. In the first strategy the contact detection is based on the contact between two ellipsoids, the second is the simplest contact detection since it is based on the sphere-sphere contact, meanwhile the third strategy is based on the Gilbert–Johnson–Keerthi algorithm to compute the distance between two convex geometries. The second strategy offers the best possibility for the description of arbitrary complex particles. Nevertheless, the computational time is its major weakness.

*e-mail: antonio.olmedilla.aero@gmail.com
**e-mail: miha.zaloznik@univ-lorraine.fr
***e-mail: herve.combeau@univ-lorraine.fr

2 Model description

Our algorithm is designed to be an arbitrary-geometry DEM. We model the equiaxed dendritic grain deposition and packing under solidification conditions. In this case the geometry is characterized by having six principal branches grown along the ⟨100⟩ crystal directions. Subsequently, a secondary set of arms are grown from the principal ones [4]. To simplify an *envelope* wrapping the dendritic branches is considered (see the dendritic envelopes of the experimental model in figure 1a). Eventually the

DOI: 10.1051/epjconf/201714006002

envelope surface is precisely discretized by a multi-sphere clumping approach (see figure 1b).

2.1 Rigid Body Dynamics

The dynamics equations (eqs. from 1 to 4) are solved via the *Verlet* scheme for each *i-th* grain, obtaining position, $\mathbf{r}$; orientation quaternion, q; linear and angular velocities, $\mathbf{v}$ and ω, respectively. Indexes *ir* and *b* correspond to *inertial reference* and *body*, respectively. m and $\mathbf{I}_{G,b}^{-1}$ are referred to the particle mass and inertial tensor about the particle center of gravity, respectively. Finally, $\mathbf{F}_{tot}$ and $\mathbf{M}_{G,tot}$ are the total force and moment applied to the particle center of gravity, respectively.

$$\frac{d\mathbf{r}_{ir}}{dt} = \mathbf{v}_{ir} \tag{1}$$

$$\frac{d\mathbf{v}_{ir}}{dt} = m^{-1}\mathbf{F}_{tot,ir} \tag{2}$$

$$\frac{dq}{dt} = \frac{1}{2}q\omega_b \tag{3}$$

$$\frac{d\omega_b}{dt} = \mathbf{I}_{G,b}^{-1}(\mathbf{M}_{G,tot,b} - \omega_b \times (\mathbf{I}_{G,b}\omega_b)) \tag{4}$$

2.2 Geometry discretization

Spherical elements are set at the particle complex surface in a way that these spheres are internally tangential to the particle surface. This strategy, to face up the particle-particle contact detection, is selected due to their geometrical simplicity and high capacity to describe complex particle shapes.

The dendritic grain morphology is simplified by a geometrical model which consists of 6 quadrilateral pyramids with a central cube. Each of the pyramids is joined to the central cube by its base. We characterize this geometrical configuration by its sphericity, Ψ, which is function of the pyramid apex angle, θ, (see eq. 5). Moreover, the characteristic particle size, l_c, is the distance from a branch tip to opposite branch tip passing along the particle center.

$$\Psi(\theta) = sin\left(\frac{\theta}{2}\right)\left\{\frac{\pi}{6}\left[1 + cotan\left(\frac{\theta}{2}\right)\right]^2\right\}^{1/3} \tag{5}$$

In figure 1b the spherical element approach of complex particle shape description is shown. The spherical elements are uniform and characterized by a diameter d_{sph}. In this specific case, the quadrilateral pyramid apex angle is 45° and $l_c/d_{sph} \approx 20$.

2.3 Contact model

The classical *Kelvin-Voigt* model is used for the normal reaction in case of contact between the *i-th* and *j-th* particles. Thereby, the contribution to the normal reaction of two contacting spherical elements l and m which belong to the *i-th* and *j-th* particles, respectively, is given by eq. 6,

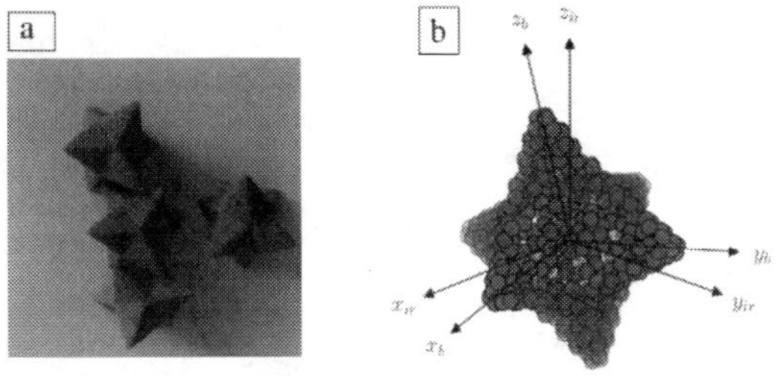

Figure 1. a) Experimental model equiaxed dendritic grain envelope (θ = 45°, d_{eq} = 7 *mm*) and b) numerical equiaxed dendritic grain envelope (θ = 45°, d_{eq} = 7 *mm*). The surface is described with 527 spherical elements ($l_c/d_{sph} \approx 20$). The indexes *ir* and *b* refer to inertial reference and body frames, respectively.

where the stiffness and viscous damping dissipation reactions are summed. In the general case of non-uniform particle contact mechanical properties, the stiffness, K_{lm}, and damping, C_{lm}, coefficients are referred to the $l - m$ pair of contacting spherical elements. The variables δ_{lm} and $\dot{\delta}_{lm}$ refer to the *overlapping* and *overlapping rate* between the pair of spherical elements $l - m$. A critical damping coefficient is used in order to avoid rebounding, i.e. dissipating the overall settling particle kinetic energy at the contact.

$$F_{il,jm}^{n} = K_{lm}\,\delta_{lm} + C_{lm}\,\dot{\delta}_{lm} \tag{6}$$

3 Simulation cases

3.1 Domain and initial conditions

A rectangular prismatic domain is employed whose dimensions are L_x, L_y and L_z, with the direction z considered as vertical and $x - y$ directions forming a horizontal plane. In the simulations presented in this article, the horizontal cross-section of the prism is square, $L_y = L_x$ and the vertical dimension is larger than the horizontal $L_z > L_x$. Additionally, the domain walls are rigid in order to obtain the *wall effect* influence on the packing.

Initially particles are randomly distributed withing the whole domain occupying a low solid fraction ($\approx$ 4%) without initial contacts. In this way, their positions are initialized by a *Random Sequential Addition* algorithm [8]. Meanwhile the particle initial orientations are randomly computed by the *K. Shoemaker* algorithm [9].

Dynamic initial conditions of linear and angular velocities are null.

3.2 Packing protocol

The overall granular system simultaneously sediments and packs by means of a vertical uniform relatively low acceleration ($a = 10^{-5}\ g_o$) over the horizontal domain wall $z = 0$. The objective of this protocol is that of simulating the equiaxed dendritic grain sedimentation and packing with an equivalent inertia-to-dissipation ratio to that in solidification and in the validation experimental model but

DOI: 10.1051/epjconf/201714006002

Figure 2. Random packing of equiaxed dendritic grain envelope (θ = 45°, d_{eq} = 7 mm): a) DEM simulation, b) Experimental validation packing.

decreasing the mathematical stiffness by means of a low driving force and consequently, increasing the integration time step. This protocol considerably accelerates the simulation permitting to achieve higher particle geometry detail.

In the actual solidification phenomenon and in the validation experimental model, the granular system kinetic energy is dissipated through the lubrication forces at the tight fluid layers among the grains. In this numerical model, the kinetic energy dissipation is activated at the grain contact through the critical viscous damping term in eq. 6. The critical contact stiffness is selected in order to reproduce the equivalent inertia-to-dissipation ratio once the driving force is chosen.

3.3 Simulated geometries

- a) Dendrite-like: three different 1000 monodisperse dendrite noncohesive frictionless collections are packed in a $84 \times 84 \times 500$ mm^3 domain. The three of them has an equivalent diameter of d_{eq} = 7 mm. Three different pyramid apex angles are simulated.
  - Dendrites of 45° apex angle and $\Psi = 0.69$ (figure 1b).
  - Dendrites of 60° apex angle and $\Psi = 0.79$.
  - Dendrites of 90° apex angle and $\Psi = 0.90$.
- b) Spherical: 2000 monodisperse noncohesive frictionless particle collection whose diameter is 7 mm in a $84 \times 84 \times 1000$ mm^3 domain.
- c) Ellipsoids of 1.25 : 0.8 : 0.8 axis ratio and $\Psi = 0.97$.

The particle density is uniform and the same for all particle collections (ρ_s = 1400 kg/m^3). The particle density and equivalent diameter between the simulation and the validation experiments are identical.

3.4 Numerical results

Once the granular system achieves the mechanically stable equilibrium, the kinetic system energy is dissipated by means of the viscous damping contact, the average packing fraction at Ω volume, $\langle\phi_s\rangle_\Omega$, is obtained by computing the *Voronoi tessellation* of the granular system by means of the *Voro++ open library* [10] and subsequently computing the average value of a certain volumetric region, Ω, applying eq. 7, where V_{s_i} and V_{cell_i} are the volume of the *i-th* particle and its corresponding *Voronoi cell* volume, respectively. The region Ω_A is defined by the intervals $[0, 84] \times [0, 84] \times [0, h]$ mm^3 where h is the lowest position of the top packing surface whereas Ω_B is defined by $[22, 62] \times [22, 62] \times [10, 20]$ mm^3.

$$\langle\phi_s\rangle_\Omega = \frac{\sum V_{s_i}}{\sum V_{cell_i}}, \ \forall i \,/\, \mathbf{r}_i \in \Omega \tag{7}$$

In the case of spheres ($\Psi = 1$), we obtain a value of $\langle\phi_s\rangle_{\Omega_B} = 0.643$ which corresponds to the result in [11] for sedimentation and pack with relatively low driving force. A slightly looser packing fraction is obtained when the wall effect is accounted $\langle\phi_s\rangle_{\Omega_B} = 0.618$ (see figure 3 for $\Psi = 1$). Besides, the ellipsoidal results are also compared to the same literature reference [11], obtaining a 4% looser packing in our case (see figure 3 for $\Psi = 0.97$).

The three different dendritic geometries are sedimented and packed (see figure 2a) and the same *Voronoi tessellation* procedure is employed to obtain the local volumetric solid packing fraction for later computing the average packing fraction at a desired region (Ω_A and Ω_B).

A clear influence of the equiaxed dendritic grain sphericity on the packing fraction is found: the lower the grain sphericity, the looser the grain packing is. The equiaxed dendritic envelopes whose $\Psi = 0.69$ pack an approximately 38% looser than spheres (fig. 3).

The vessel wall effect (the difference $\langle\phi_s\rangle_{\Omega_B} - \langle\phi_s\rangle_{\Omega_A}$) is shown to slightly increase with decreasing sphericity for equiaxed particles (spheres and dendrites).

A numerical simulation correlation of sequential deposition of spiky nonconvex particles [5] are included in fig. 3 since the particles in this reference are the most similar found to our dendritic envelopes. For larger sphericity $\Psi > 0.85$ our results and this reference results are similar. However, a great behavior difference is found for lower sphericity $\Psi = 0.70$, being the dendritic envelopes denser packed.

Additionally, our numerical packing results of equiaxed dendritic envelopes under low inertia-to-dissipation conditions are validated by means of an experimental model (see sec. 3.5). The numerical and experimental results are compared through $\langle\phi_s\rangle_{\Omega_A}$ since the experimental results include the vessel wall effect. In the case of spheres, our numerical protocol obtains denser packings (7.3% difference) whereas in case of the 45° equiaxed dendritic envelopes a slightly lower numerical fraction is found (6.7%). In this way, arbitrary-geometry DEM and packing protocol is validated to reproduce equiaxed dendritic envelope packings under solidification conditions (low inertia-to-dissipation).

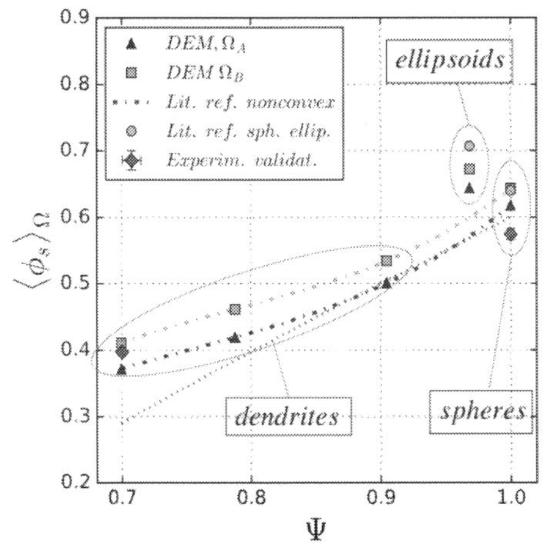

Figure 3. (Color online) Average solid volumetric packing fraction, $\langle \phi_s \rangle_\Omega$, function of particle sphericity. $\langle \phi_s \rangle_{\Omega_A}$ (accounting the vessel wall effect) is shown with *triangular blue markers*) and $\langle \phi_s \rangle_{\Omega_B}$ shown via *squared green markers*). The equiaxed grain results are joined by a dashed green and blue lines, $\langle \phi_s \rangle_{\Omega_B}$ and $\langle \phi_s \rangle_{\Omega_A}$, respectively. The validation experiments (see sec. 3.5) for 45° dendritic envelopes and spheres are shown via red rhomboid markers. The numerical results from literature for spheres and ellipsoids under low gravity conditions [11] (*Lit. ref. ellip.*) and sequential deposition under gravity of spiky nonconvex particles [5] (*Lit. ref. non-convex*) are also included.

3.5 Validation experimental model

In order to validate the DEM solver and packing protocol, a 12200 cornstarch sphere and a 4400 3d-printed 45° apex-angle dendrite monodisperse collections are packed by sedimentation in a cylindrical vertical sedimentation column containing a 99.5% glycerol solution with a vessel diameter to particle equivalent diameter ratio is D/d_{eq} = 17 (see fig. 2b). The density of both particle collections is approximately 1400 kg/m^3. The liquid is chosen in order to obtain a *Stokes* dimensionless number ($St = (\rho_s d_{eq} v_s)/(9\ \mu_l)$) of 10^{-3} which is a characteristic value in solidification. The average solid packing fraction is obtained for both sphere and dendrite collections by carefully measuring the height of the packed system with a 10 repetitions for each geometry.

A 0.575 ± 0.008 and 0.397 ± 0.013 are found for the average solid packing fraction of monodisperse spheres and 45° dendrites, respectively (see fig. 3).

These experimental packings (sedimentation and packing in quiescent viscous fluid) are compared to the literature fluidized packings, which are frequently used to obtain the random loose packing configuration. In this vein, in [12] an average packing fraction of 0.55 is found for monodisperse spheres of diameter of the order of 100 μm and a packing fraction in the interval $0.50 - 0.55$ for tetrahedrons and octahedrons [13].

4 Conclusions

The random packing of equiaxed dendritic grains, phenomenon present in metallic-alloy solidification, is numerically studied by a DEM model. Looser packings are obtained with lower particle sphericity under low inertia-to-dissipation conditions. For the case of $\Psi = 0.70$ equiaxed dendritic envelope, an average solid packing fraction of approximately 0.38 is numerically found, which is much lower than that of the monodisperse spheres.

5 Perspectives

Future improvements of the arbitrary-geometry DEM model are the implementation of the fluid effects on the particles (drag and lubrication forces), particle polydispersity and particle growth (by scaling).

Acknowledgements

This work was supported by the French State through the program "Investment in the future" operated by the National Research Agency (ANR) and referenced by ANR-11 LABX-0008-01 (LabEx DAMAS).

References

[1] I. Steinbach, JOM **65**, 1096 (2013)
[2] H. Combeau, M. Založnik, S. Hans, P.E. Richy, Metallurgical and materials transactions B **40**, 289 (2009)
[3] M. Založnik, A. Kumar, H. Combeau, M. Bedel, P. Jarry, E. Waz, Advanced Engineering Materials **13**, 570 (2011)
[4] J.A. Dantzig, M. Rappaz, *Solidification*, 1st edn. (EPFL Press, 2009)
[5] I. Malinouskaya, V. Mourzenko, J.F. Thovert, P. Adler, Physical Review E **80**, 011304 (2009)
[6] Christophe L. Martin, *Mechanics of Materials with Discrete Element Simulations*, http://simap.grenoble-inp.fr/gpm2/m-christophe-martin--430244.kjsp (2016), online; accessed March 2016
[7] A. Wachs, L. Girolami, G. Vinay, G. Ferrer, Powder Technology **224**, 374 (2012)
[8] A. Donev, Ph.D. thesis, Princeton University (2006)
[9] K. Shoemaker, *Animating rotation with quaternion curves*, in *ACM SIGGRAPH computer graphics* (ACM, 1985), Vol. 19, pp. 245–254
[10] C. Rycroft, Lawrence Berkeley National Laboratory (2009)
[11] G.W. Delaney, J.E. Hilton, P.W. Cleary, Phys. Rev. E **83**, 051305 (2011)
[12] M. Jerkins, M. Schröter, H.L. Swinney, T.J. Senden, M. Saadatfar, T. Aste, Phys. Rev. Lett. **101**, 018301 (2008)
[13] J. Baker, A. Kudrolli, Physical Review E **82**, 061304 (2010)

Comparison between a μFE model and DEM for an assembly of spheres under triaxial compression

Sadegh Nadimi[1,*], *Tom* Shire[2], and *Joana* Fonseca[1]

[1]City, University of London, London, UK
[2]Imperial College London, London, UK

Abstract. This paper presents a simple case of a Face Centred Cubic (FCC) array of 2,000 spheres under triaxial compression to compare the results obtained using the Discrete Element Method (DEM) and a micro finite element model (μFE). This μFE approach was developed so that the internal structure of the soil can be obtained using x-ray computed tomography and converted into a numerical fabric. The individual grains are represented as continuum deformable bodies and the inter-granular interaction based on the defined contact laws. In order to demonstrate the simple contact constitutive behaviour used in this μFE model, the response for two contacting elastic spheres is compared with theoretical equations. The strength at failure of the packing of 2,000 spheres is seen to yield similar values for DEM, μFE and the analytical solution. When comparing the evolving void ratio, a good agreement between the two numerical models was observed for very small strains but as the strain increases, the values start to diverge, which is believed to be related with the rigidity of the grains used in DEM.

1 Introduction

The mechanical behaviour of granular materials is originated from the transmission of forces between the grain to grain contacts [1]. The use of discrete numerical modelling approaches enables the measurement of grain-scale forces and stresses, which it is not a trivial task in experiments. In particular, Discrete Element Method (DEM) has been widely used for the modelling of grain interactions within the assembly, assuming purely elastic and for the most part, spherical grains [2, 3]. The importance of particle shape in force transmission has been highlighted in a number of studies *e.g.* [4, 5] and several attempts have been made to include particle shape in DEM approaches, *e.g.* [6, 7]. Regarding the contact constitutive behaviour, the Hertz [8] and Mindlin & Deresiewicz [9] models, were developed to model two spheres in contact, a case for which the contact evolves from a point to a circle under normal loading. These contact models are therefore of limited application to represent contact interaction of non-spherical grains. Another difficulty in DEM simulations is representing the initial packing (void ratio) of a granular assembly from experiments [10]. An irregular packing of spheres can be prepared using gravity with high computational cost.

A two dimensional microstructure-based finite element (μFE) model, which employ the natural depositional grain scale characteristics of a sand has been proposed by Nadimi *et al.* [11]. The initial packing, grain shape and grain sizes are captured using images obtained from micro Computed Tomography (μCT). The contact behaviour between two irregular grains comes from stress propagation and concentration within the body of the grains and is unique for different contact topology and grain morphology [12]. To consider full contact interaction of irregular grains, the extension of μFE model to three-dimensions (3D) is under development by the authors.

In this paper, the proposed methodology is firstly reviewed. Then, the principles are verified against theoretical solutions. Finally, the response of an assembly of spheres under triaxial loading is compared with DEM simulations.

2 The μFE model

The principle of this μFE model is that the physical phenomena of the deformation of real soil can be investigated using a detailed numerical representation of the constituent grains (obtained from μCT) to simulate grain interaction. Thus, an accurate and tractable representation of the complex geometries of real grains can be acquired and used to enhance our understanding of force and stress transmission between grains. A summary of the methodology for generating a μFE model is shown in Fig. 1. In this Section, each part of methodology is explained in short.

* Corresponding author: sadegh.nadimi-shahraki@city.ac.uk

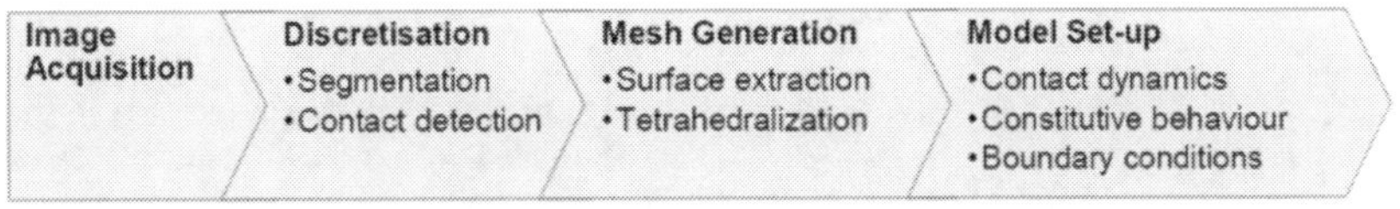

Fig. 1. Flow chart showing the steps involved in the proposed 3D μFE model

2.1 Image acquisition & discretisation

The first step is image acquisition. By using x-ray micro-computed tomography, the internal structure of the soil is obtained including shape, size and fabric [13]. The images are maps of x-ray attenuation based on the density and atomic composition of the material represented by the intensity of each voxel.
In order to obtain the individual grains, the step termed here of discretisation, two steps are involved. Firstly, the solid phase has to be extracted from the void space. A thresholding technique was employed for this part. Successively, the touching grains have to be separated to obtain the individual grains by using a morphological watershed approach [14].

2.2 Mesh generation

The numerical approximation of the problem starts from mesh generation in the μFE framework. A key requirement is to obtain an accurate representation of the object boundary. An advanced surface reconstruction algorithm was employed to extract the grain surface. Triangular iso-surfaces with an adjustable density are extracted from the 3D segmented image. The surface extraction technique employed is a refinement of the constrained Delaunay tetrahedralisation [15]. The second stage is the filling of grain with the tetrahedral elements for the sub-volumes bounded by the iso-surfaces to obtain the volumetric mesh. A MATLAB script has been developed to generate the image-based mesh. Fig. 2 shows an example of two silica sand grains with different features meshed for numerical simulation.

Fig. 2. An example of generated mesh from μCT images of two silica sand grains.

2.3 Model set-up

The generated mesh is imported into the finite-discrete numerical domain [16]. The model is implemented in Abaqus explicit, which uses a dynamic framework [17].
The contacts properties are defined as '*hard contact*' in the normal direction and '*Coulomb friction*' in the tangential direction. Hard contact means that all the force is transmitted when two bodies are in contact. The ability of the grains to deform enables the use of a hard contact in the normal direction, *i.e.* there is no need to use a predefined contact law in this direction. For the case of pure elastic spheres, this assumption is examined in the next section to represent theoretical contact constitutive behaviour *i.e.* Hertz [8] in normal direction, Mindlin & Deresiewicz [9] in tangential direction, Lubkin [18] for torsion, Johnson [19] for rotation.

There is need to define constitutive behaviour of solid part of each grain. This can be as simple as defining elastic modulus and Poisson's ratio or incorporating fracture energy and the breakage potential of each grain. The continuum representation of each individual grain rather than the rigid assumption (used in DEM) enables the measurement of stress propagation and concentration within the grain body.

2.4 Contact interaction

The mechanical interaction between two contacting bodies includes normal loading in combination with tangential, torsional and rotational loading. The three latter loadings are the result of friction and applied normal load. The comparison between theoretical solution and numerical modelling in the framework of the μFE was conducted to verify the contact interaction assumptions. It can be seen from Fig. 3 that the μFE numerical model framework can represent the contact laws with a high precision level by only considering hard contact and Coulomb friction.

3 Triaxial test

An assembly of 2000 spherical grains with a diameter of 2.2 mm was simulated under triaxial conditions (Fig. 4). The face-centred-cubic (FCC) packing of the assembly modelled consists of 10×10 rows of spheres in the XY-plane and 20 rows in Z-direction. Rigid wall boundaries were used. The response was investigated under standard triaxial conditions. Two steps were defined for loading, 1) isotropic compression at 50kPa and 2) shearing under controlled strain. This problem was chosen due to the available theoretical solution, as presented in [20].

3.1 DEM model

The problem was modelled in the commercial DEM code PFC^{3D} [21]. The material properties used are listed in Table 1. Fig. 5 shows the response of the assembly in terms of major principal stress (σ_1) and coordination number (Z) versus axial strain (ε_a).

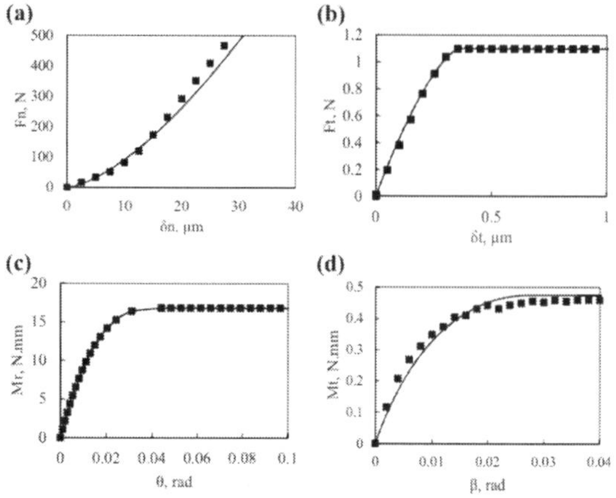

Fig. 3. Comparison between numerical simulation (square marks) and theories (solid lines) for (a) normal load: displacement, (b) tangential load: displacement, (c) rotational moment: angular displacement, and (d) torsional moment: angular displacement.

The strength at so called 'failure' is related to interparticle friction and is σ_1 = 147.6kPa for this arrangement with rigid walls. It can be seen from Fig. 5 that the failure is linked with a drop in coordination number. According to the theoretical solution [20], σ_1 is calculated from the following equation for an infinite number of grains:

$$\frac{2\sigma_1}{\sigma_2+\sigma_3}=\frac{2(1+\mu)}{(1-\mu)} \qquad (1)$$

For the confining stress σ_2=σ_3=50kPa and interparticle friction of μ=0.22, σ_1 is equal 156.4kPa. The slightly lower value obtained from DEM can be related to the packing not being a true FCC close to the rigid boundary and the finite number of grains.

Fig. 4. 3D view of a FCC packed triaxial sample of 2,000 spheres

3.2 μFE model

The FCC model was replicated in the framework of μFE. The so called 'failure' was seen to occur at σ_1 =162.5kPa. As can be seen in Fig 3.a and also noted by [22], the Hertz theory is accurate for small normal deformation. The slight increase in the value of σ_1 can be related to normal force:displacement of deformable spheres which show higher load value than Hertz. Fig. 6a shows a deformed sphere in the μFE framework and this deformation was scaled by a factor of 50 in order to improve visualisation in Fig. 6b.

Regarding void ratio, a good agreement can be seen between DEM and μFE data (Fig. 7). It is expected to see more discrepancy in void ratio for high stress (and strain) level due to rigidity assumption in DEM simulation and DEM void ratio does discount the overlapped volume from the total solid volume.

Table 1. Material properties used for DEM and μFE simulation

Particle density	ρ	2.5 tonne/m^3
Poisson Ratio	υ	0.22 (-)
Shear modulus	G	25.82 GPa
Interparticle friction	μ	0.22 (-)
Young's modulus	E	63 GPa

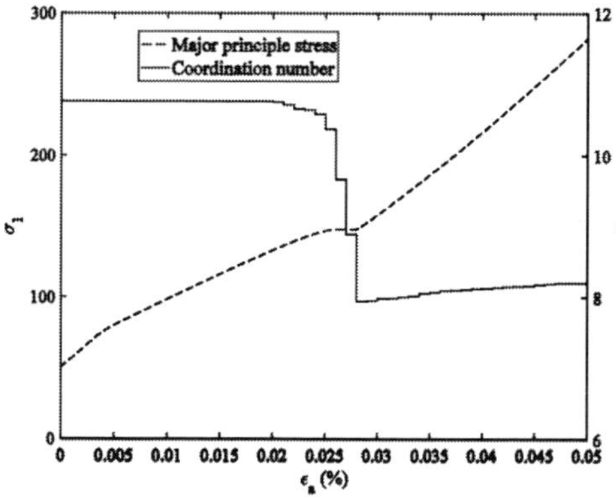

Fig. 5. Major principle stress and coordination number versus axial strain for assembly of spheres under triaxial condition obtained from DEM simulation.

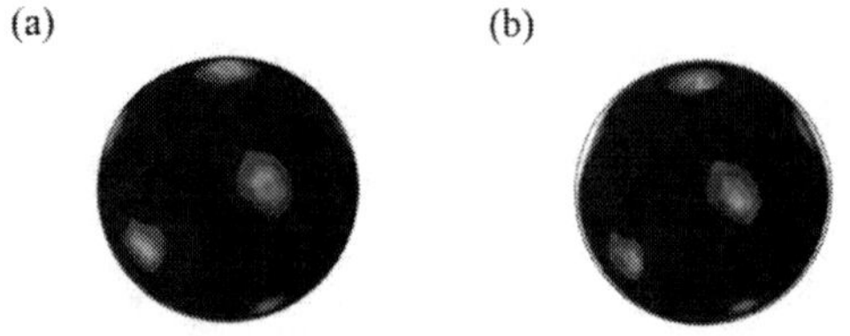

Fig. 6. Stress distribution in a deformed sphere with Z=12; deformation scale factor a) 1 and b) 50.

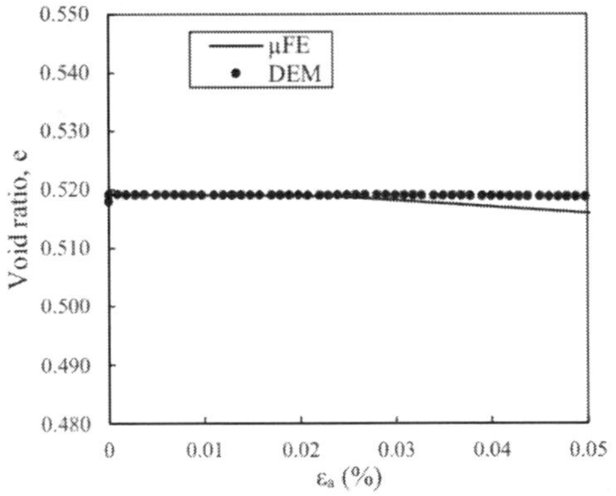

Fig. 7. Comparison of evolving void ratio between DEM and μFE models.

4 Closing remarks

A review on the development and verification of a novel μFE model was presented here. While one of the key features of the μFE model is its ability to include the morphology and fabric of real soil from μCT images, the example presented here uses spheres to simplify the comparison with DEM and analytical solutions.

An important aspect of grain scale modelling is the definition of contact laws. Despite DEM has been used for irregular shapes, it is not possible to predefine contact laws for each irregular contact topology found in natural soil. In the μFE framework, the contact behaviour comes from deformation and stress propagation within the grain, which suggests this as a robust platform to present unique contact response of grains with irregular morphology and contact topology.

It was demonstrated here that the contact constitutive behaviour used for the μFE model for an idealised elastic sphere was precisely matched with theory. The ability of this μFE approach to model large number of grains was shown and the results were seen to compare well with both DEM and the analytical solution. The divergence in the evolving void ratio when compared with the DEM results can be attributed to the deformation of the grains in the μFE model against the rigid grains used in DEM.

The μFE model described here, although computationally expensive, can provide a valuable contribution to unravel fundamental aspects of granular materials behaviour.

The first author would like to express thanks to City, University of London for his doctoral scholarship.

References

1. J. Fonseca, S. Nadimi, C.C. Reyes-Aldasoro, C. O'Sullivan, M.R. Coop, Soils & Foundations **56**, No. 5, pp. (2016).
2. C. O'Sullivan, *Particulate Discrete Element Modelling: A Geomechanics Perspective*. Spon Press (2011).
3. C. Thornton, *Granular Dynamics, Contact Mechanics and Particle System simulations*, Springer (2015).
4. I. Cavarretta, M. Coop, C. O'Sullivan, Géotechnique **60**, pp. 413–423 (2010).
5. F. N. Altuhafi, M. R. Coop, V. N. Georgiannou, J. Geotech. Geoenvironmental Eng., vol. 4016071, (2016).
6. J. Ferellec, G.R. McDowell, Géotechnique **60**, No. 3, pp. 227–232 (2010) [doi: 10.1680/geot.9.T.015].
7. G. T. Houlsby. Computers & Geotechnics, **36**, pp. 953–959, (2009).
8. H. Hertz, Journal für die reine und angewandte Mathematik, **92** (1882).
9. R.D. Mindlin, H. Deresiewiez, Elastic spheres in contact under varying oblique forces, ASME J. Appl. Mech. 20, pp. 327-344 (1953).
10. J. O'Donovan, S. Hamlin, G. Marketos, C. O'Sullivan, E. Ibraim, M. Lings, D. Muir Wood. In Geomechanics from micro to macro (eds K. Soga, K. Kumar, G. Biscontin and M. Kuo). London, UK: Taylor & Francis Group (2015).
11. S. Nadimi, J. Fonseca, R.N. Taylor, In *Deformation characteristics of geomaterials*: Proceedings of the 6th International symposium on deformation characteristics of geomaterials, Buenos Aires, Argentina (2015).
12. J. Fonseca, S. Nadimi, C.C. Reyes-Aldasoro, C. O'Sullivan, M.R. Coop, Soils & Foundations 56 (5): 818-834.
13. J. Fonseca, C. O'Sullivan, M.R. Coop, P.D. Lee, Géotechnique 63, No. 6, (2013).
14. J. Fonseca, C. O'Sullivan, M.R. Coop, AIP Conf. Proc. 1145, 223 (2009).
15. J.R. Shewchuk, Computational Geometry: Theory and Applications, 47, No. 7, pp. 741-778 (2014).
16. A. Munjiza, *The Combined Finite-Discrete Element Method,* Wiley, London (2004).
17. ABAQUS User's Manual 2016, Dassault Systèmes, version 6.14.
18. J. L. Lubkin, ASME J. Appl. Mech., Vol. **73**, pp. 183-187 (1951).
19. K.L. Johnson *Contact mechanics*. Cambridge University Press, Cambridge (1985).
20. C. Thornton, Géotechnique, 29 No. 4, pp. 441-459. (1979).
21. Itasca Consulting Group, Inc. *PFC^{3D}-Particle Flow Code in 3 Dimensions*, ver. 5. Minneapolis: Itasca (2015).
22. Q. J. Zheng, H.P. Zhu, A.B. Yu, Powder Technology 226, pp. 130-142, (2012).

EPJ Web of Conferences 140, 15008 (2017) DOI: 10.1051/epjconf/201714015008
Powders & Grains 2017

Scaling laws for implicit viscosities in smoothed particle hydrodynamics

Claas Bierwisch[1,*] and *Pit* Polfer[1]

[1]*Fraunhofer IWM, Wöhlerstraße 11, 79108 Freiburg, Germany*

Abstract. Smoothed particle hydrodynamics (SPH) is a particle-based method which solves continuum equations such as the Navier-Stokes equations. A periodic fluidic system under homogeneous shear is studied using SPH in the present work. The total pressure of the system and the shear stress contributions from the SPH interaction terms for pressure and viscosity as well as the contribution caused by velocity fluctuations are analyzed. It is found that the pressure and the shear stress contributions obey certain scaling laws depending on physical properties of the system such as compressibility, viscosity and shear rate as well as the spatial resolution. Some of the identified relations resemble scaling laws for the rheology of dense granular flows. These findings render an assessment of the convergence with respect to the spatial resolution of SPH simulations possible. Furthermore, the similarities between numerical SPH particles and physical grains in dense flow provide a deeper understanding of the nature of the SPH method.

1 Introduction

Since its conception in 1977 [1] the SPH method has been used in a wide range of scientific and industrial applications due to its simplicity to implement and its flexibility in modelling (multi-phase) fluids with free surfaces [2, 3]. Despite the successful application of the method, it is still lacking a theoretical proof of its convergence with respect to the spatial discretization, i.e., the SPH particle size.

Several studies focused on the accuracy of SPH shear flow results. Posch et al. [4] described implicit SPH viscosities, i.e., artificial shear stress contributions which have their origin in the particle-based nature of the method. Ellero et al. [5] investigated this phenomenon further and quantified a kinetic contribution which acts as an average Reynolds stress and a potential contribution without continuum analog. Both contributions can become larger than the specified shear viscosity of the system for low viscous flows. Previous work of the present authors [6] showed that the magnitudes of the implicit viscosities depend on the used SPH formulation of the Navier-Stokes viscosity term. Imaeda et al. [7] reported large density errors which were attributed to the local shear deformation of the neighborhood of each SPH particle.

In the present work we analyze the resulting density, pressure and shear stress of SPH shear flow simulations from a different perspective, i.e., by interpreting SPH particles as grains in dense granular flow. This approach enables the mapping of several relations which are valid for granular rheology to SPH simulations and, thus, a detailed quantification of the implicit viscosities.

*e-mail: claas.bierwisch@iwm.fraunhofer.de

2 SPH method

The motion of an incompressible Newtonian fluid in the absence of body forces is described in the moving frame of reference by the Navier-Stokes momentum equation

$$\frac{\mathrm{D}\mathbf{v}}{\mathrm{D}t} = \frac{1}{\rho}\left(-\nabla p + \eta_0 \nabla^2 \mathbf{v}\right), \tag{1}$$

with $\mathrm{D}/\mathrm{D}t$ as the substantial derivative, $\mathbf{v}$ as the velocity, ρ as the density, p as the pressure and η_0 as the dynamic viscosity. Using the SPH summation formalism [8],

$$A(\mathbf{r}) = m \sum_j \frac{A_j}{\rho_j} W(|\mathbf{r} - \mathbf{r}_j|, h), \tag{2}$$

the value of the quantity A at the position $\mathbf{r}$ is given by the summation over the contribution A_j from each particle j at position $\mathbf{r}_j$ with fixed mass m and variable density ρ_j within range of the kernel function in three dimensions, $W(r, h) = f(r/h)/h^3$, where h is the smoothing length. We are using the common cubic spline kernel function,

$$f(q) = \frac{1}{4\pi}\begin{cases} \left[(2-q)^3 - 4(1-q)^3\right], & 0 \le q \le 1, \\ (2-q)^3, & 1 < q < 2, \\ 0, & q \ge 2. \end{cases} \tag{3}$$

Let $\mathbf{r}_{ij} = \mathbf{r}_i - \mathbf{r}_j$, $r_{ij} = |\mathbf{r}_{ij}|$, $W_{ij} = W(r_{ij}, h)$ and $\nabla W_{ij} = \mathbf{r}_{ij}\,(\mathrm{d}f(q)/\mathrm{d}q)/(r_{ij}\,h^4)$.

By applying the SPH summation formalism, Eq. (2), the accelaration of particle i due to the pressure term of the momentum equation can be written as

$$\left(\frac{-1}{\rho}\nabla p\right)_i = -m\sum_j \left(\frac{P_i}{\rho_i^2} + \frac{P_j}{\rho_j^2}\right)\nabla W_{ij} = \frac{1}{m}\sum_j \mathbf{f}_{ij}^{\mathrm{pot}} \tag{4}$$

yielding a conservative (potential) force $\mathbf{f}_{ij}^{\text{pot}}$ between the particles. A weakly-compressible approach is used for closure by calculating the particle pressure from the equation of state,

$$P_i = B\left[\left(\frac{\rho_i}{\rho_0}\right)^{\gamma} - 1\right], \tag{5}$$

with ρ_0 as the reference density, γ as the isentropic exponent and B as a compressibility which limits density fluctuations. The particle mass and the reference density are related by the initial particle spacing s via $m = \rho_0 s^3$. The SPH summation formalism yields the density of an SPH particle i via $\rho_i = m\sum_j W_{ij}$.

For the viscous term of the momentum equation the form proposed by Cleary et al. [9] is used which is based on dissipation due to normal SPH particle collisions,

$$\left(\frac{\eta_0}{\rho}\nabla^2\mathbf{v}\right)_i = m\sum_j \frac{\eta_0\xi}{2\rho_i\rho_j}\left(\frac{\mathbf{v}_{ij}\cdot\mathbf{r}_{ij}}{r_{ij}^2+\epsilon h^2}\right)\nabla W_{ij} = \frac{1}{m}\sum_j \mathbf{f}_{ij}^{\text{diss}}, \tag{6}$$

with $\mathbf{v}_{ij} = \mathbf{v}_i - \mathbf{v}_j$ as the velocity difference. ξ is a dimensionless coefficient in the order of 20. $\epsilon = 0.01$ is used to prevent zero division. Equation (6) yields a dissipative force $\mathbf{f}_{ij}^{\text{diss}}$ between the particles.

The explicit time integration with time step Δt of the total force on particle i, $\mathbf{f}_i = \sum_j(\mathbf{f}_{ij}^{\text{pot}} + \mathbf{f}_{ij}^{\text{diss}})$, is based on the Velocity Verlet scheme. To ensure numerical stability a time step condition is enforced, $\Delta t \le \min(0.25\,h/c_s, 0.05\,h^2\rho_0/\eta_0)$, with the speed of sound $c_s = \sqrt{(}\gamma B/\rho_0)$.

3 Simulation setup and analysis

The simulation setup consists of a cubic three-dimensional cell of volume $V = L^3$. The SPH particles are initially positioned on a simple cubic lattice with lattice constant s filling the volume. Then, an equilibration scheme is used to obtain vanishing kernel derivatives before running the actual simulations [10]. Lees-Edwards boundary conditions are used to impose a simple shear flow with rate $\dot{\gamma}$ on the system [11]. The average stress tensor is evaluated using the Irving-Kirkwood relation [12],

$$\begin{aligned}\sigma^{\text{tot}} &= \sigma^{\text{kin}} + \sigma^{\text{pot}} + \sigma^{\text{diss}}\\ &= \frac{1}{V}\left(m\sum_i \mathbf{v}_i^{\text{rel}}\otimes\mathbf{v}_i^{\text{rel}} + \sum_{i,j>i}\mathbf{r}_{ij}\otimes\mathbf{f}_{ij}^{\text{pot}} + \sum_{i,j>i}\mathbf{r}_{ij}\otimes\mathbf{f}_{ij}^{\text{diss}}\right).\end{aligned} \tag{7}$$

The first term in Eq. (7) is the kinetic contribution with $\mathbf{v}_i^{\text{rel}} = (v_{i,x}, v_{i,y}, v_{i,z} - \dot{\gamma}\, r_{i,x})$ being the deviation of the velocity of particle i from the homogeneous shear flow velocity field. The second term in Eq. (7) is the potential contribution caused by the Navier-Stokes pressure term and the third term in Eq. (7) is the dissipative contribution caused by the Navier-Stokes viscosity term. The shear stress contributions in the simulation are denoted as

$$S_{\text{kind}} = \dot{\gamma}\,\eta_{\text{kind}} = \sigma_{xz}^{\text{kind}}, \tag{8}$$

Table 1. SPH simulation parameters

Parameter	Unit	Values
s	m	$10^{-5}, 10^{-4}, 10^{-3}$
$\dot{\gamma}$	1/s	$10^0, 10^1, 10^2, 10^3$
η_0	Pa s	$10^{-3}, 10^{-2}, 10^{-1}$
B	Pa	$4\cdot10^2, 4\cdot10^4, 4\cdot10^6$

with kind $\in$ {kin, pot, diss}. The total pressure in the system is given by

$$P_{\text{tot}} = \frac{1}{3}\left(\sigma_{xx}^{\text{tot}} + \sigma_{yy}^{\text{tot}} + \sigma_{zz}^{\text{tot}}\right). \tag{9}$$

The density of the system, ρ, is evaluated as the ensemble average of the SPH particle densities ρ_i.

An overview of the ranges of simulation parameters is given in Table 1. The ratio of the simulation volume edge length and the initial particle spacing is $L/s = 10$. The ratio of the smoothing length and the particle spacing is $h/s = 1$. The reference density is $\rho_0 = 10^3\,\text{kg/m}^3$ and the isentropic exponent is $\gamma = 7$. The viscosity model parameter ξ is set to 23.9 according to numerical calibration results. It should be noted that ξ was found to be a function of h/s but not of η_0. A total number of 45 simulations were carried out using only parameter combinations which yield a system Reynolds number $\rho_0\dot{\gamma}L^2/\eta_0 \le 10$ in order to limit the study to the non-turbulent regime. All simulations are run until a shear strain, i.e., the product of shear rate and simulation time, of 50 is reached. For the following analysis all dynamic quantities are time averaged over the stationary second half of each simulation. The software SimPARTIX is used for the SPH simulations.

4 Results

Let us define three stress or energy density scales, respectively, based on the simulation input parameters,

$$\Sigma_{\text{inert}} = \rho_0\dot{\gamma}^2 s^2, \tag{10}$$

$$\Sigma_{\text{shear}} = \dot{\gamma}\eta_0, \tag{11}$$

$$\Sigma_{\text{stiff}} = \gamma B = \rho_0 c_s^2. \tag{12}$$

Equation (10) is the kinetic energy density of the relative shear movement of the SPH particles, Eq. (11) is the viscous shear stress and Eq. (12) is a measure of the inverse compressibility of the model fluid. Based on these stress scales and the total pressure in the system we are able to define a set of dimensionless numbers,

$$\text{Re} = \frac{\Sigma_{\text{inert}}}{\Sigma_{\text{shear}}} = \frac{\rho_0\dot{\gamma}s^2}{\eta_0}, \tag{13}$$

$$\text{Ma} = \sqrt{\frac{\Sigma_{\text{inert}}}{\Sigma_{\text{stiff}}}} = \dot{\gamma}s\sqrt{\frac{\rho_0}{\gamma B}} = \frac{\dot{\gamma}s}{c_s}, \tag{14}$$

$$\text{I} = \sqrt{\frac{\Sigma_{\text{inert}}}{P_{\text{tot}}}} = \dot{\gamma}s\sqrt{\frac{\rho_0}{P_{\text{tot}}}}. \tag{15}$$

The dimensionless numbers are inspired by the particle-based nature of the SPH simulation method. Equation (13)

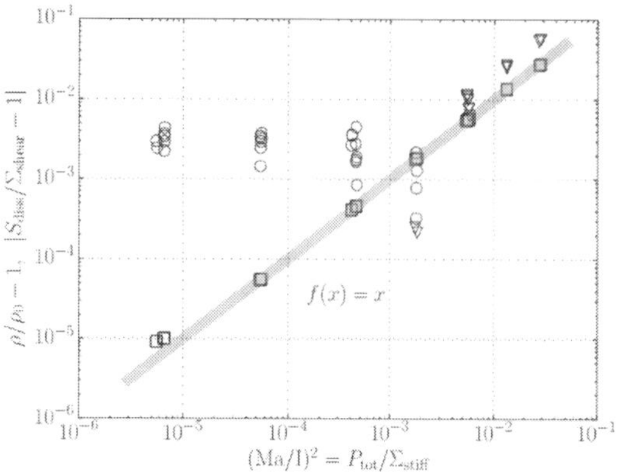

Figure 1. Scaling of density deviation (squares) and dissipative shear stress deviation (circles and triangles). The circles denote positive and the triangles denote negative ordinate values if not taking the absolute value.

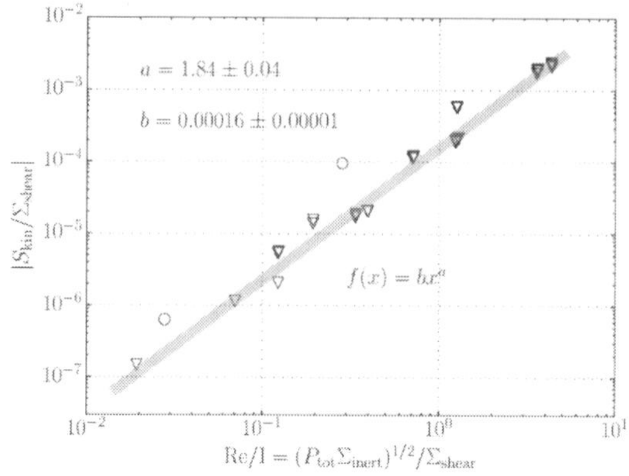

Figure 2. Scaling of kinetic shear stress. The symbols have the same meaning as in Fig. 1.

defines a particle Reynolds number in the shear flow while Eq. (14) defines a particle Mach number by relating the relative particle velocity and the speed of sound in the system. Finally, Eq. (15) defines a particle inertial number which resembles the inertial number of dense granular flows [13, 14]. Note that Re and Ma are only based on input parameters while I is also based on the simulation result via P_{tot}.

Figure 1 displays the linear correlation of the density deviation and the normalized total pressure in the system. This result is not surprising as a Taylor expansion of Eq. (5) around ρ_0 yields $P_i \approx \gamma B(\rho_i/\rho_0 - 1)$. Thus, the following relation apparently holds,

$$P_{\mathrm{tot}} \approx \langle P \rangle \approx \Sigma_{\mathrm{stiff}}(\rho/\rho_0 - 1), \qquad (16)$$

with $\langle P \rangle$ as the time and ensemble averaged pressure form the equation of state.

Furthermore, Fig. 1 shows that the deviation of the dissipative viscosity contribution from the input viscosity is below 0.5 % for a normalized total pressure below $\sim 10^{-3}$ while the dissipative viscosity deviation increases with increasing normalized pressure above $\sim 10^{-3}$. Ideally, $\rho/\rho_0 = S_{\mathrm{diss}}/\Sigma_{\mathrm{shear}} = 1$.

A power-law relation between the relative kinetic shear stress contribution and the ratio of particle Reynolds and particle inertial number was found as presented in Fig. 2. This result is a first indicator for the connection of SPH simulations and dense granular flow. GDR Midi [13] showed that the mean velocity fluctuations in granular plane shear flow scale as $\langle \delta v^2 \rangle \propto \dot{\gamma}^2 d^2/\mathrm{I_g}$, with d as the grain diameter and $\mathrm{I_g}$ as the granular inertial number, $\mathrm{I_g} = \dot{\gamma} d\sqrt{(\rho_0/P)}$. Recalling Eqs. (7), (8), (13) and substituting $\mathrm{I_g} \to \mathrm{I}$ and $d \to s$, we can write

$$\left|\frac{S_{\mathrm{kin}}}{\Sigma_{\mathrm{shear}}}\right| = \frac{\rho_0\langle \delta v^2 \rangle}{\eta_0 \dot{\gamma}} \propto \frac{\rho_0 \dot{\gamma} s^2}{\eta_0 \mathrm{I}} = \frac{\mathrm{Re}}{\mathrm{I}}. \qquad (17)$$

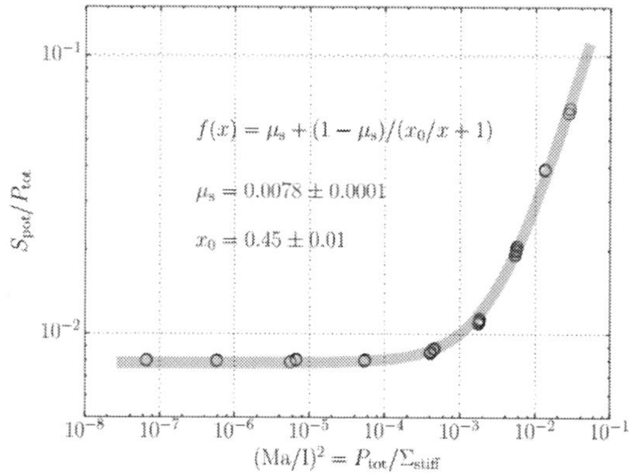

Figure 3. Scaling of potential shear stress. See text for details of the analytic expression fitting the data.

A power-law scaling of the relative kinetic viscosity with Re/I is discernible in Fig. 2, however, with an exponent of ~ 2 and not 1 as predicted by Eq. (17).

Figure 3 displays a scaling relation for the ratio of the potential shear stress and the total pressure as a function of the normalized total pressure. Similar regimes as found for the scaling bahavior of the dissipative shear stress can be identified (compare Fig. 1). Below a normalized total pressure of about 10^{-3} the potential shear stress ratio is constant. Above this value the shear stress ratio increases with the normalized total pressure. Jop et al. [14] used the following expression to model the dependency of the ratio of shear stress and pressure on the granular inertial number $\mathrm{I_g}$ for dense granular flow,

$$S/P = \mu_s + (\mu_2 - \mu_s)/(\mathrm{I_0}/\mathrm{I_g} + 1), \qquad (18)$$

with μ_s and μ_2 as the limiting values at low and high $\mathrm{I_g}$, respectively, and $\mathrm{I_0}$ as a model parameter. A substitution using $S \to S_{\mathrm{pot}}$, $P \to P_{\mathrm{tot}}$, $\mathrm{I_g} \to (\mathrm{Ma/I})^2$ and setting $\mu_2 = 1$ yields an expression which describes the data in

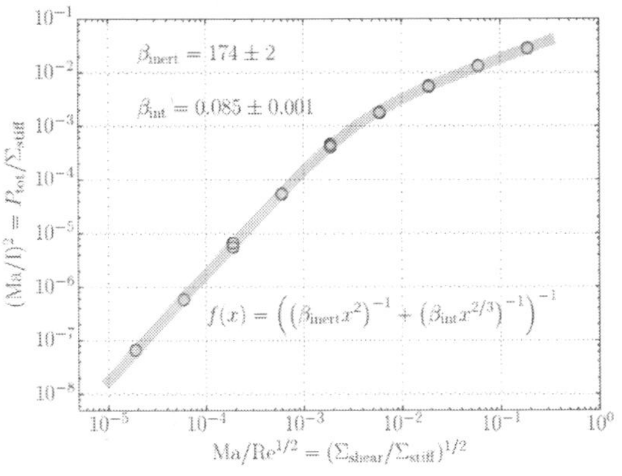

Figure 4. Scaling of total pressure. See text for details of the analytic expression fitting the data.

Fig. 3 well. The substitution of I_g means that the ratio of pressure and stiffness is relevant for the transition in the present system while the ratio of kinetic energy density and pressure is relevent for the granular system.

So far, all of the discovered scaling relations include I or P_{tot}, respectively, as a factor on the abscissa. Thus, it would be desirable to obtain a dependency of I only on input parameters of the simulation. Such a relation exists as shown in Fig. 4. The analytic expression which accurately describes the data collapse was used in a very similar form for modelling the rheology of soft, frictional, noncohesive spheres [15, 16]. Omitting factors related to the solid volume fraction which has no correspondence in the present system, Gu et al. [16] related the normalized total pressure P to the shear rate $\dot{\gamma}$ via

$$\frac{P}{\rho_0 c_g^2} = \left(\left(\beta_{\text{inert}}\hat{\dot{\gamma}}^2\right)^{-1} + \left(\beta_{\text{int}}\hat{\dot{\gamma}}^{m_{\text{int}}}\right)^{-1}\right)^{-1}, \quad (19)$$

with the grain diameter d, the speed of sound c_g, $\hat{\dot{\gamma}} = \dot{\gamma}d/c_g$ and $m_{\text{int}} \approx 2/3$. By interpreting the SPH particles again as grains of diameter s and substituting $P \to P_{\text{tot}}$, $c_g \to c_s$ and $\hat{\dot{\gamma}} \to \text{Ma}/\text{Re}^{1/2}$, an analytic expression is found which describes the simulation data in Fig. 4 accurately. Equation (19) is valid for sheared granular systems below a critical solid volume fraction and describes a transition from an inertial flow regime with Bagnold scaling, i.e., $P \propto \dot{\gamma}^2$, to an intermediate flow regime where the finite grain stiffness plays a role. Such a transition in the stress scaling behavior with respect to the shear rate was previously observed by Campbell [17] in simulations of soft granular media. The substitution of $\hat{\dot{\gamma}}$ indicates that the role of inertial shear motion in the granular system is replaced by viscous shear motion in the SPH simulations with a scaling of $P_{\text{tot}} \propto \dot{\gamma}$.

Interestingly, the transitions in the dissipative shear stress scaling (Fig. 1) and the potential shear stress scaling (Fig. 3) at $(\text{Ma}/\text{I})^2 \approx 10^{-3}$ correspond to the transition between flow regimes in Fig. 4 at $\text{Ma}/\text{Re}^{1/2} \approx 3 \cdot 10^{-3}$.

5 Conclusion

It is demonstrated that deviations of pressure and shear stress in SPH shear flow simulations from their theoretical values can be expressed as functions of dimensionless system parameters, namely a Reynolds number (Re), a Mach number (Ma) and an inertial number (I).

Interestingly, the SPH simulation results contain signatures observed in dense granular flow. The SPH kinetic shear stress scales as power-law of Re/I which corresponds to experimental measurements of the scaling of granular velocity fluctuations albeit with a different power-law exponent [13]. The ratio of SPH potential shear stress and total pressure shows a transition from a constant ratio to a dependency on the total pressure which is also a fundamental feature in the rheology of dense granular flows [14]. The functional dependency of the density or, equivalently, the total pressure on SPH simulation parameters resembles a similar relation found in discrete element simulations of granular flow [16].

The identified scaling relations render quantitative predictions of the accuracy of the stress in SPH simulations possible. Moreover, the influence of the spatial SPH discretization, i.e., the particle size, on the accuracy can be quantified.

References

[1] R.A. Gingold, J.J. Monaghan, Mon. Not. R. Astron. Soc. **181**, 375 (1977)
[2] J.J. Monaghan, Annu. Rev. Fluid Mech. **44**, 323 (2012)
[3] M.S. Shadloo, G. Oger, D. Le Touze, Comput. Fluids **136**, 11 (2016)
[4] H.A. Posch, W.G. Hoover, O. Kum, Phys. Rev. E **52**, 1711 (1995)
[5] M. Ellero, P. Español, N.A. Adams, Phys. Rev. E **82**, 046702 (2010)
[6] P. Polfer, T. Kraft, C. Bierwisch, Appl. Math. Modell. **40**, 2606 (2016)
[7] Y. Imaeda, S.I. Inutsuka, Astrophys. J. **569**, 501 (2002)
[8] J.J. Monaghan, Rep. Prog. Phys. **68**, 1703 (2005)
[9] P.W. Cleary, Appl. Math. Modell. **22**, 981 (1998)
[10] A. Colagrossi, B. Bouscasse, M. Antuono, S. Marrone, Comput. Phys. Commun. **183**, 1641 (2012)
[11] A.W. Lees, S.F. Edwards, J. Phys. C: Solid State Phys. **5**, 1921 (1972)
[12] J.H. Irving, J.G. Kirkwood, J. Chem. Phys. **18**, 817 (1950)
[13] GDR Midi, Eur. Phys. J. E **14**, 341 (2004)
[14] P. Jop, Y. Forterre, O. Pouliquen, Nature **441**, 727 (2006)
[15] S. Chialvo, J. Sun, S. Sundaresan, Phys. Rev. E **85**, 021305 (2012)
[16] Y. Gu, S. Chialvo, S. Sundaresan, Phys. Rev. E **90**, 032206 (2014)
[17] C.S. Campbell, J. Fluid Mech. **465**, 261 (2002)

Simulations of the effect of particle size on texture and force transmission in bidisperse granular composites

Itthichai Preechawuttipong[1] and *Pawarut* Jongchansitto[1,*]

[1]Department of Mechanical Engineering, Faculty of Engineering, Chiang Mai University, 239 Huay Kaew Rd., Muang District, Chiang Mai 50200, Thailand

Abstract. The objective of this study is to investigate the influence of the particle size ratio on texture and force transmission in two-dimensional cohesionless binary granular composties by using molecular dynamics (MD) simulations. Four numerical composite samples, which differ in terms of the particle size ratios, are used in this study. The samples are composed of two constitutive materials with a stiffness ratio of four between the higher one termed as stiff particle and another termed as soft particle. The samples are subjected to an uniaxial confined vertical compression on the upper wall. The results under static conditions show that the particle size ratio mainly affects the contact sub-networks. The coordination number decreases when the particle size ratios (D_{stiff}/D_{soft} = 1.2 - 3.0) increase, contrary to stiff-any case. Considering the spatial arrangement of contact directions, contacts between stiff particles exhibit an anisotropic distribution. On the contrary, the other contacts, i.e. soft-soft and stiff-soft contacts play a role to support the granular system in equilibrium. It is interesting to note that for all the particle size ratios, an exponential distribution and power-law are observed for the strong and weak network, respectively. Furthermore, almost 60% of the entire contacts transmit the weak forces.

1 Introduction

Granular materials are plentiful around us, which can be encountered in daily life and in many industries. They are constituted from grains with a variety of size, density, rigidity, and other physical properties. Due to a great diversity in grains, the behaviours of granular media are generally complex which cannot distinguish from ordinary matter [1, 2]. No studies really deal with "composite" granular media, i.e. previous studies mainly focused on granular media made of only one type of constitutive material. That is the reason why a few studies on this topic can be found in the literature [3]. This causes a lack of knowledge on mechanics of "composite" granular materials. As a consequence, this study aimed at systematically investigating the influence of the particle size on textural properties and force transmission in two-dimensional bidisperse composite granular materials prepared from two types of disk with different rigidities by means of molecular dynamics simulations.

2 Molecular dynamics simulations

The molecular dynamics (MD) method, which belongs to the discrete element method (DEM) proposed in 1971 by Cundall [4], relies on an explicit algorithm. The particles in this method are taken into account as rigid bodies with non-conforming surfaces [5]. Newton's equations of motion, which govern the motions of each particle with respect to time, are numerically integrated by means of a predictor-corrector scheme with Gear's set of corrector coefficients [6].

2.1 Contact force model

The interaction force between two dry particles is divided into a normal force and a tangential force. In the MD method, the local deformation occurred when two solid bodies are in contact can be simply modelled by a virtual overlap δ at the contact (no change in the particle shape). The normal force model used in this study is a simple model so-called the "linear spring-dashpot" [7], which is given by

$$F_n = k_{eff}\delta + \alpha_n v_n \qquad (1)$$

where k_{eff} is the effective contact stiffness, α_n is a normal damping coefficient, and v_n is the normal velocity (time derivative of the virtual overlap).

In the case of the tangential force, a "regularized" form of the exact Coulomb's law of friction [7] is adopted in this study to calculate the tangential force between two contacting particles. This model can be written by

$$F_t = \min\{|\gamma_s v_s|, \mu F_n\} \cdot \text{sign}(v_s) \qquad (2)$$

where γ_s is the tangential viscosity coefficient and μ the coefficient of friction. Note that the rotational motion

* Corresponding author: pawarut.j@cmu.ac.th

due to the tangential force is permitted to be free in this study.

2.2 Performing the simulations

The 2D numerical composite granular sample under study was prepared from two different base materials with differ in rigidities: one is referred to as a "stiff" particle and another to a "soft" particle. The contact stiffness ratio between them is four. The bidisperse sample, which was comprised of 4000 disks, was established by randomly depositing both stiff and soft particles inside a square box consisting of four rigid plane walls. Four different the diameter size ratio (D_{stiff}/D_{soft}) was used in the present work, i.e. D_{stiff}/D_{soft} = 1.2, 1.6, 2.2, and 3.0. The ratio between the number of stiff and soft particles (N_{stiff}/N_{soft}) was fixed at 1.0 for all the diameter size ratios. A confined vertical compressive force of 100 N was then applied on the upper wall, while the others were fixed during this stage. Eventually, an initial dense granular packing under static equilibrium condition was obtained.

Next, a vertical constant force of 1000 N was applied to the initial dense packing on its upper wall. During the compression, the other walls were still stationary. It can be seen that in this study the contacts between the particles can be separated into three types: contacts between stiff-stiff, soft-soft, and stiff-soft particles. The effective contact stiffness k_{eff} in Eq. (1) is dependent of the contact types, which was set underlying on the following relations:

$$k_{eff} = \begin{cases} k_{stiff}, & \text{stiff-stiff and wall-any contacts} \\ k_{soft}, & \text{soft-soft contacts} \\ \dfrac{k_{stiff} \times k_{soft}}{k_{stiff} + k_{soft}}, & \text{stiff-soft contacts} \end{cases} \quad (3)$$

When granular systems reached the static equilibrium conditions, it implies that the simulations were finished. The simulation data at the final of this stage were then statistically analysed through granular textures and force transmissions.

3 Geometrical textures

The geometrical textures were essentially used to define the morphology of the granular media by describing in terms of geometrical orientation of the particles and their contacts in space.

3.1 Coordination number

In this context, the coordination number z is defined as the average number of contact neighbours per particle. It can be calculated from $z = 2N_c/N_p$, where N_c is the total number of contacts and N_p is the total number of particles. The results for all the samples regarding the influence of D_{stiff}/D_{soft} ratios on the coordination number are presented in Fig. 1.

Considering the coordination number z_{all} for the whole of contacts, it slightly decreases with the D_{stiff}/D_{soft} ratios (see the black line). In the similar manner, it is observed that the coordination number $z_{soft\text{-}any}$ for only the contacts between soft and any particles decreases quite greatly with the D_{stiff}/D_{soft} ratios (red line). This reduction of z can be explained rationally by the fact that an increasing of the net area of the stiff particles when the D_{stiff}/D_{soft} ratio increased causes a chance of contact occurrences in the system to be decreased. Note that the net area of the soft particles is still the same for all the D_{stiff}/D_{soft} ratios. On the contrary, when only the contacts between stiff and any particles are considered, the coordination number $z_{stiff\text{-}any}$ seems to be steady for all the ratios (blue line).

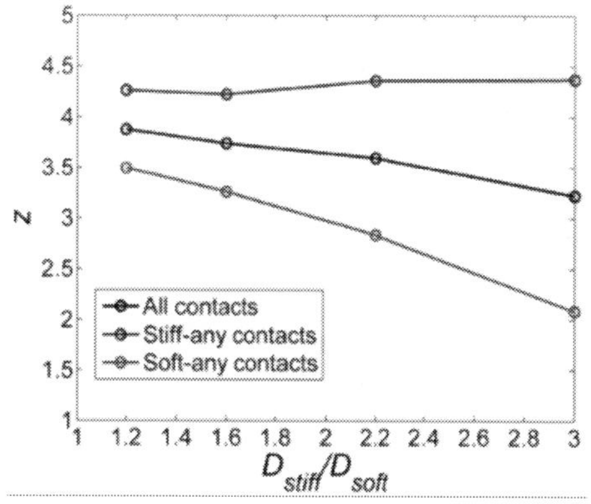

Fig. 1. Coordination number z as a function of the ratio between stiff and soft particle size D_{stiff}/D_{soft}.

3.2 Contact direction

The contact direction is given by the line between the centroids of two particles in contact, measured with respect to the horizontal direction. The angular distribution of the contact directions is used as a statistical quantity to describe and characterize the probability of contacts occurring in each direction of the space. For two-dimensional data, this quantity can be represented by the polar diagrams of the probability distribution $P(\theta)$ of the normal contact directions as shown in Fig. 2. It must be noted that we analysed four different types of contacts: all the contacts, only the stiff-stiff contacts, only the soft-soft contacts, and only the stiff-soft contacts.

First, the spatial arrangement of the particles in the case of D_{stiff}/D_{soft} = 1.2 is taken into account, as presented in Fig. 2 by the green line. In this case, the effect of monodisperse media still exists. That is a reason why we found clearly specific contact orientations, i.e. θ = 0°, 60°, and 120°, for all considerations of the contact types. This phenomenon so-called the "triangular" contact network [8] is logically discovered in the monodisperse media.

Considering next the angular distribution of the contact directions for D_{stiff}/D_{soft} = 1.6, 2.2, and 3.0 represented in Fig. 2 by black, red, and blue line, respectively, it must be noted that applying a confined vertical compression on our granular samples prepared inside a square box leads these samples become a quasi-homogeneous system. Several remarkable points can be shown as follows:

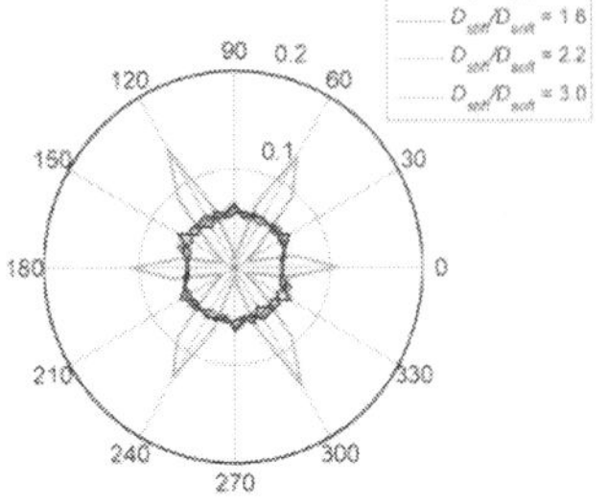

(a) all contacts

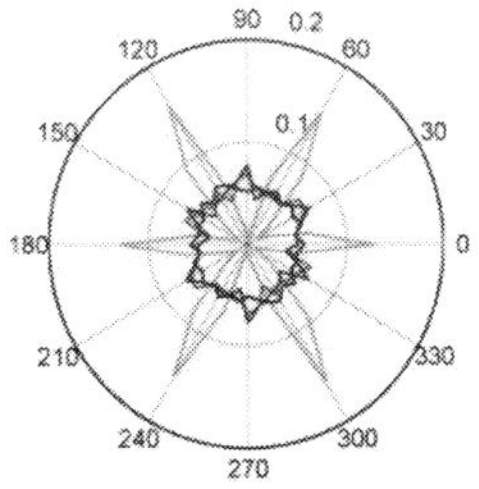

(b) stiff-stiff contacts

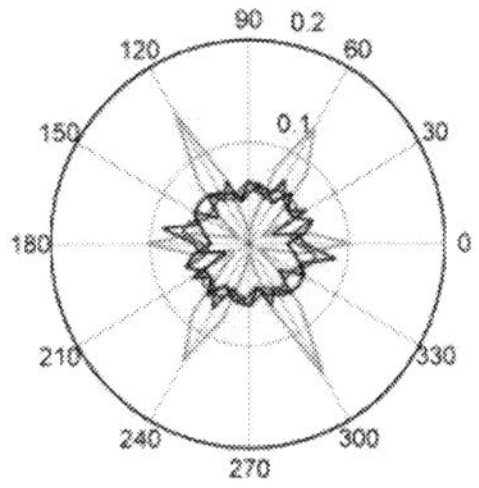

(c) soft-soft contacts

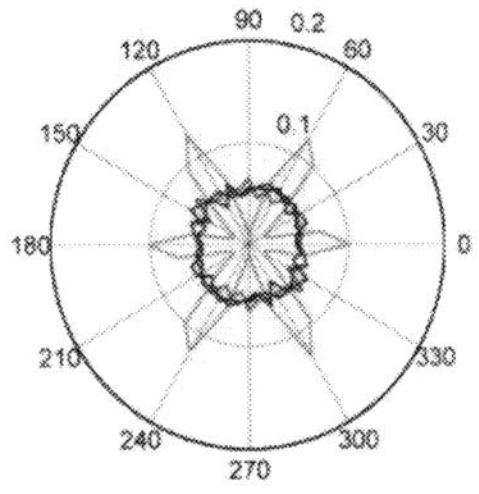

(d) stiff-soft contacts

Fig. 2. Polar diagrams of the probability distribution of the normal contact directions for all the D_{stiff}/D_{soft} ratios. The contact networks in each sample are taken into account separately for: (a) all the contacts, (b) only stiff-stiff contacts, (c) only soft-soft contacts, and (d) only stiff-soft contacts.

- In Fig. 2b, the contact orientation between stiff and stiff particles exhibits an anisotropic distribution for most cases, i.e. specific directions are evidently observed in the directions of $\theta = 30°$, $90°$, and $150°$, excluding the case D_{stiff}/D_{soft} = 1.6. In such case, the contacts between stiff and stiff particles are quite distributed uniformly as observed clearly in both all the contacts (see Fig. 2a) and only stiff-soft contacts (Fig. 2d) for all D_{stiff}/D_{soft} the ratios. We can point out that the specific directions occurred in the contact network between stiff and stiff particles is also caused from the increasing of the net area of the stiff particles, as discussed earlier in the previous section.
- For the contact network between soft and soft particles, the anisotropic distribution is also observed as illustrated in Fig. 2c. It tends to be aligned perpendicular to the compression axis (i.e. the vertical axis), in order to make the granular system in equilibrium by supporting the stiff-stiff contact network.

4 Force transmissions

A strong inhomogeneity of contact forces is a well-known characteristic in granular materials, which is arisen from a force transmission through inter-particle contacts. For all samples, this inhomogeneous contact force is also found in composite granular media, as shown in Fig. 3. It can be observed distinctly that the biggest particles carry the strong forces, similar to what is discovered in granular media made of only one type of base material [9].

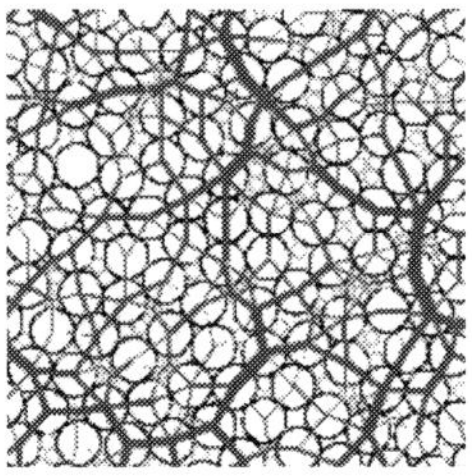

Fig. 3. A snapshot of normal force chains in the sample with D_{stiff}/D_{soft} = 3.0. The stiff particles are represented in black and soft ones in light grey. Note that the magnitude of the normal force is proportional to the thickness of the grey line.

Due to the inhomogeneity of the contact forces, in order to characterize explicitly the distribution of the normal contact forces in composite granular media, the probability distribution function P of the normalized normal contact forces F_{norm} is plotted as shown in Fig. 4. In this figure, the contact force networks can be divided into two complement networks [10]: the strong network presented in Fig. 4a and the weak network presented in Fig. 4b. The former consists of the contacts transmitting the forces greater than the average force $<F>$, contrary to the latter that the contacts carry the forces lower than the average force.

It is obvious that the probability distribution function of the normalized normal forces for all the samples is distributed as an exponential decay for the strong network and a power-law for the weak network. These distribution shapes found in this study for composite granular media are also discovered in non-composite granular media made of only one type of base material [11]. Moreover, both strong and weak networks in the cases of stiff-stiff, soft-soft, and stiff-soft contacts are characterized by the exponential fall-off and the power-law, respectively. It can be that the law of the force distributions for both strong and weak networks is a common feature of granular media, which is not only dependent on the effect of particle size but also the types of contact.

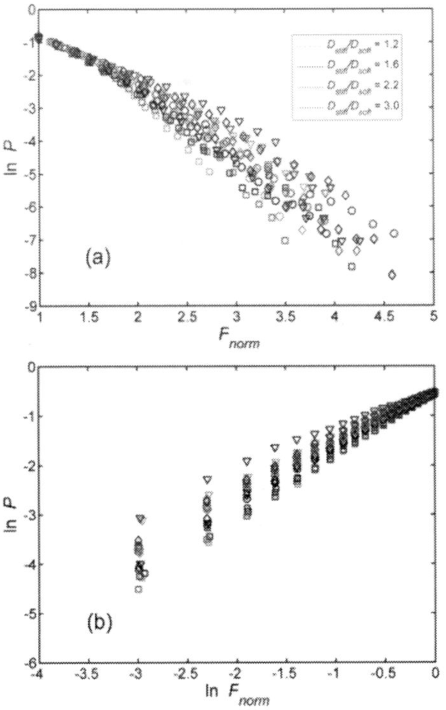

Fig. 4. Probability distribution function P of the normalized normal contact forces F_{norm} for all the samples: (a) the strong network is presented in a semi-logarithmic scale and (b) the weak network is presented in a logarithmic scale. Note that considering the whole set of contacts is represented by the "circle", only stiff-stiff contacts represented by the "square", only soft-soft contacts represented by the "triangle", and only stiff-soft contacts represented by the "diamond".

In addition, it must be noted that the contacts carrying the forces lower than the average force is almost 60% of the total number of contacts. This property is also observed for whatever the type of contact is taken into account.

5 Conclusions

The influence of the particle size on behaviours of 2D composite granular media under a confined vertical compression is methodically investigated in terms of the texture and force transmission. The molecular dynamics (MD) simulations were employed for this purpose. The composite samples were established from two different types of disks with rigidity ratio of four: the larger stiffness is referred to as stiff particles and the lower one as soft particles. Four different ratios between stiff and soft particles (D_{stiff}/D_{soft} = 1.2 - 3.0) were varied in this study. Under static conditions, the numerical results can be concluded that the influence of the particle size ratio on the texture and force transmission is only observed for the contact sub-networks. These can be summarized as follows:

- The coordination number for the whole set of contact decreases when the D_{stiff}/D_{soft} ratios increase, similar to what observed in the soft-any case. On the other hand, the coordination number remains constant in stiff-any case.
- Due to a confined vertical compression, the contact distributions between stiff particles is anisotropic with specific directions in almost cases, while the other types of contact, i.e. soft-soft and stiff-soft contacts, play a role to sustain the granular system in equilibrium. It must be noted that the effect of monodisperse media is observed for the case of D_{stiff}/D_{soft} = 1.2.
- The particle size ratio does not affect force transmissions in composite granular media in terms of the distribution laws and the percentage of the number of contacts characterized for both strong and weak networks. These features are similar to what observed in non-composite granular media.

In a future work, it would be interesting to study the texture and force transmission in composite granular media subjected to a complex mechanical loading such as biaxial compression. Furthermore, the net area of each particle type should be the same for the further study, in order to eliminate its effect.

References

1. H.M. Jaeger, S.R. Nagel, Rev. Mod. Phys. **68**, 1259 (1996)
2. B. Andreotti, Y. Forterre, O. Pouliquen, *Granular media: between fluid and solid*, (Cambridge University Press, 2013)
3. Y. Yang, D. Wang, Q. Qin, Powder Technol. **261**, 272 (2014)
4. P.A. Cundall, *the International Symposium on Rock Fracture*, 47 (1971)
5. K.L. Johnson, *Contact mechanics*, (Cambridge University Press, 1985)
6. M.P. Allen, D.J. Tildesley, *Computer simulation of liquids*, (Oxford Science Publications, 1987)
7. J. Schäfer, S. Dippel, D.E. Wolf, J. Phys. I **6**, 5 (1996)
8. P. Jongchansitto, X. Balandraud, M. Grédiac, C. Beitone, I. Preechawuttipong, Soft Matter **10**, 8603 (2014)
9. C. Voivret, F. Radjai, J.Y. Delenne, M.S. El Youssoufi, Phys. Rev. Lett. **102**, 178001 (2009)
10. F. Radjai, D.E. Wolf, M. Jean, J.J. Moreau, Phys. Rev. Lett. **80**, 61 (1998)
11. F. Radjai, M. Jean, J.J. Moreau, S. Roux, Phys. Rev. Lett. **77**, 274 (1996)

Experimental study of particle size segregation in 2D granular materials under vertical vibrations

Pawarut Jongchansitto[1*], *Thuwachit* Kanyalert[1], and *Itthichai* Preechawuttipong[1]

[1]Department of Mechanical Engineering, Faculty of Engineering, Chiang Mai University, 239 Huay Kaew Rd., Muang District, Chiang Mai 50200, Thailand

Abstract. The influence of particle size on segregation phenomena in two-dimensional binary granular mixtures with a similar solid fraction under vertical vibrations is experimentally investigated in this study. Four different particle size ratios (D_l/D_s) are varied. The granular mixtures are subjected to vertical vibrations under six different dimensionless accelerations (Γ = 1-6), with constant vibrational amplitude of 5 mm. The vertical segregation coefficient (H) is here introduced to analyse the degree of segregation of the mixtures. We found that the Brazil nut effect, where the larger particle rises to the top, is observed for all the tests whenever the segregation is occurred. The degree of segregation of the mixtures is a function of the coefficient H, which can be divided into 3 zones: a compaction zone with particle mixing ($0.0 < H \leq 0.2$), a partially segregated zone ($0.2 < H \leq 0.4$), and an apparently segregated zone ($0.4 < H \leq 1.0$). The segregation begins when $\Gamma \geq 2$ and then tends to increase when Γ is increased. In addition, the segregation is quite clearly found when $D_l/D_s \geq 3$ and $\Gamma \geq 3$.

1 Introduction

Since granular media are composed of grains with various physical properties such as size, density, friction and rigidity, complex behaviours of granular media are generally observed. These behaviours are cannot be simply classified into common substances [1, 2]. In addition, such complex behaviours produce an extensive variety of phenomena that remains only partly understood.

"Segregation" is one of the interesting phenomena observed when a homogeneous bulk solid constituted from a variety of species or components evolves from an initially mixed state to a presumed non-uniform state owing to the relative movement within the materials [3, 4]. This is an unwanted phenomenon often occurring in many industrial areas associated with the handling and processing of granular mixtures such as pharmaceuticals, food processing, geophysics, material sciences, and almost all engineering fields. In fact, it must be realized that a complete particle mixing is the major purpose required in many industrial processes, rather than the unwanted segregation of the mixtures [3-6]. Although we have accepted for a long time that the segregation phenomena in granular media occurs from a difference of their physical properties (e.g. size, density, shape, etc.), a significant factor to control segregation in many industrial processes is the particle size [3, 5-7].

To better understand this problem, the present work aims at experimentally investigating the influence of particle size on the segregation phenomena in 2D granular binary mixtures, which are composed of two species of cylinder differing in diameter size but similar solid fraction, under vertical vibrations. Note that the solid fraction in 2D representation is defined by the ratio of the total particle area to the total container area. The effect of a dimensionless acceleration (Γ), which governs the level of vibrations, is also studied.

2 Experimental setup

A commercial wooden cylinder with a density of about 404 kg/m^3 was used as a sample particle. The length of the cylinder is approximately 60 mm [8]. Four different ratios between large particle size and small particle size (D_l/D_s) were changed for each granular sample as presented in Table 1. The sample was prepared by mixing two species of particle size inside an acrylic container. Note that the particles were sprayed in several colours, in order to distinguish the particle species when experiments were carried out. From the table, it can be noted that the number of particle in each species is not the same, in order to achieve a similar solid fraction between them.

Each granular mixture was then subjected to vertical vibrations provided by a cam-mechanism vibrational testing machine. Figure 1 presents an experimental setup in the present work. The level of the vibration was controlled by the dimensionless acceleration (Γ), which is given by

$$\Gamma = \frac{A\omega^2}{g} = \frac{A(2\pi f)^2}{g} \tag{1}$$

* Corresponding author: pawarut.j@cmu.ac.th

where A is the vibrational amplitude, f is the frequency, and g is the gravitational force. In this study, the vibrational amplitude was fixed at 5.0 mm, while the frequency was progressively changed for obtaining the values of $\Gamma = 1 - 6$.

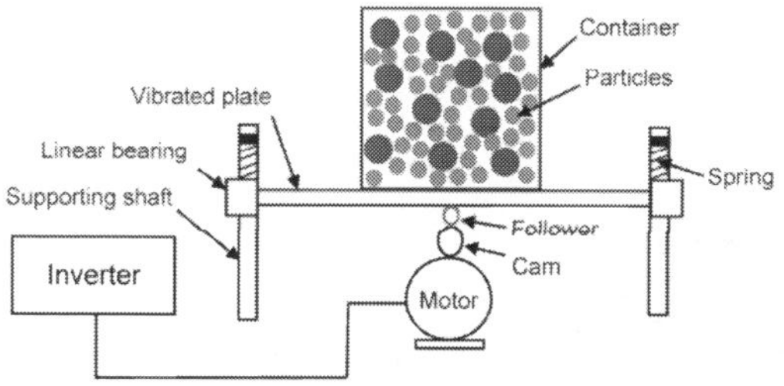

Fig. 1. Schematic drawing of an experimental setup

During the experiments, a photo was taken at each 30s until reach final time at 420s. Each photo was used for finding vertical positions of the centroid of each particle by using open-source software named "ImageJ" [9]. Specifying the vector y_i as the height of particle i with respect to the container base and the total number of particles of species inside the container is N_α, the average height of particle can be thus determined by

$$h_\alpha = \frac{1}{N_\alpha}\sum_{i=1}^{N_\alpha} y_i \qquad (2)$$

Table 1. Tested configurations

D_l/D_s	Diameter size (mm)	Number of particles	Colours
1.2	12	246	●
	10	350	●
2.0	12	126	●
	6	500	●
3.0	12	56	●
	4	500	●
4.0	12	32	●
	3	500	●

These vertical positions were then employed to statistically compute the degree of segregation of the binary mixture in a static state, which is quantified by the vertical segregation coefficient [10, 11]:

$$H = 2\frac{h_l - h_s}{h_l + h_s} \qquad (3)$$

where h_α is the average height of particles of species $\alpha = l, s$. The factor 2 is a multiplication factor arising from the equal solid fraction of each particle species which limits the average height of particles in such species to $0.25 < h_\alpha < 0.75$ and thus $-1 < H < 1$. Note that the Brazil nut effect is initially observed when $H > 0$, while the reverse Brazil nut effect found when $H < 0$. The Brazil nut effect is segregation phenomena providing the upward movement of the larger particles under vibrated granular bed, contrary to the reverse Brazil nut effect. Figure 2 shows an example of a preparation of the granular mixture at an initial stage for all cases. In each D_l/D_s ratio at the initial stage, the vertical segregation coefficient (H) approaches to zero, which means that the system is homogeneous. In addition, it can be seen that although an initial configuration shown in Fig. 2a seems to be inhomogeneous, the value of the vertical segregation coefficient (H) is nearly zero indicating the system is homogeneous.

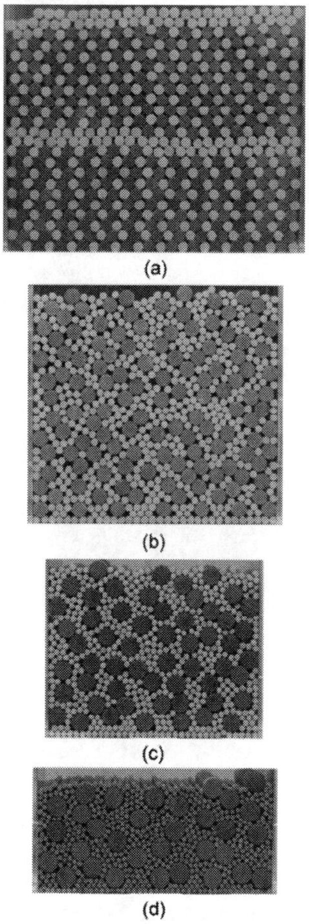

Fig. 2. Binary granular mixture at an initial stage in the case of: (a) $D_l/D_s = 1.2$, (b) $D_l/D_s = 2.0$, (c) $D_l/D_s = 3.0$, and (d) $D_l/D_s = 4.0$ under the level of vibration $\Gamma = 5$

3 Experimental results and discussions

The influence of the particle size ratio (D_l/D_s) and the dimensionless acceleration (Γ) on segregation of the binary granular mixture is then described in this section by means of such vertical segregation coefficient.

The experimental results showed evidently that the Brazil nut effect is observed for all the samples which segregation happens. The vertical segregation coefficient (H) can be employed to characterize the degree of segregation of the mixture. In this case, this segregation degree can be split into 3 zones as follows. In the first zone with $0.0 < H \leq 0.2$, the particles inside the binary mixture are still mixed, as shown in Fig. 3a. The applied vibration also makes the system more compacted. In the second zone, the binary granular mixture starts to segregate where $0.2 < H \leq 0.4$ (see Fig. 3b), so that a partial segregation is observed here. In the third zone, segregation phenomena of the mixture are clearly apparently observed up to reach a complete segregation where $0.4 < H \leq 1.0$, as displayed in Fig. 3c.

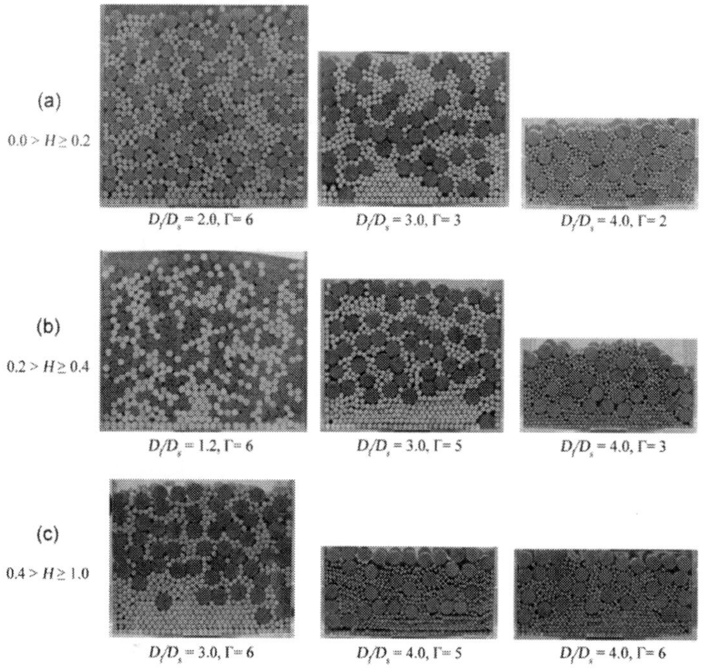

Fig. 3. Examples of segregation zones in different cases of D_l/D_s ratio at 420 s under different levels of vibration: (a) $0.0 < H \leq 0.2$, (b) $0.2 < H \leq 0.4$, and (c) $0.4 < H \leq 1.0$

3.1 The effect of the particle size ratio

Figure 4 presents the influence of particle size ratio (D_l/D_s) on the segregation phenomena in the binary granular mixture. It is interesting to note that most values of the vertical segregation coefficient in the case of D_l/D_s = 1.2 are found in the first zone of segregation. In contrast, when the D_l/D_s ratio is increased to be 3.0 and 4.0, the segregation coefficient values are observed almost in the second zone and the third zone, respectively. It can be also observed that chance of segregation occurrence is higher when the D_l/D_s ratio is increased. The maximum vertical segregation coefficient is equal to 0.56, which occurs at $\Gamma = 6$ in the case of $D_l/D_s = 4.0$.

3.2 The effect of the dimensionless acceleration

The degree of segregation is plotted as a function of the dimensionless acceleration as shown in Fig. 5. No segregation occurs at $\Gamma = 1$ even though we increase the D_l/D_s ratio. In this case, the binary granular mixture remains mixed. Furthermore, the mixture begins to segregate when $\Gamma \geq 2$ for all the particle size ratios. It is clearly seen that the vertical segregation coefficient increases for all the particle size ratios, when the dimensionless acceleration is increased.

From Figs. 4 and 5, it seems that there is something wrong observed in the experimental results. First, the vertical segregation coefficient in the case of $D_l/D_s = 1.2$ at $\Gamma = 2$ is higher than that of the other particle size ratios, despite the fact that its value observed when $\Gamma > 2$ is the lowest, except for $D_l/D_s = 2.0$. Second, the lowest vertical segregation coefficient is found in the case of $D_l/D_s = 2.0$ when $\Gamma \geq 2$. These errors can be explained by the effect of container size that we did not consider carefully in the experimental setup. In other words, an aspect ratio between width and height (w/h) of the container used for each case: the w/h ratio for $D_l/D_s = 2.0$ > the w/h ratio for $D_l/D_s = 3.0$ and 4.0 > the w/h ratio for $D_l/D_s = 1.2$. A difference of the container size may cause a different intensity of bulk convection motion, which is one of main mechanisms behind segregation phenomena [6, 12, 13].

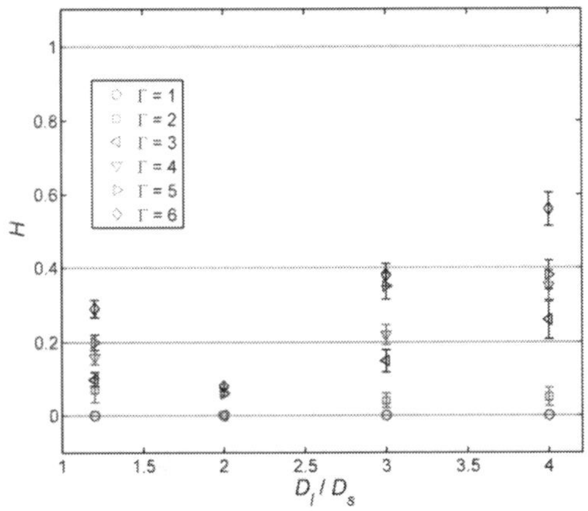

Fig. 4. The vertical segregation coefficient (H) as a function of the ratio between large particle size and small particle size (D_l/D_s) at 420 s

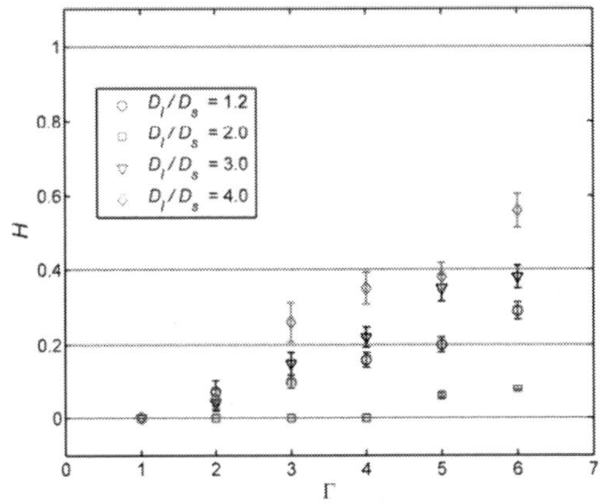

Fig. 5. The vertical segregation coefficient (H) as a function of the dimensionless acceleration (Γ) at 420 s

4 Conclusions

The objective of this study is to experimentally investigate the effects of the particle size ratio (D_l/D_s) and the dimensionless acceleration (Γ) on granular segregation in the binary mixture subjected to vertical vibrations. Note that the solid fraction of each particle species is the same. It was found that for all the samples, the larger particle size migrates to the top of the container. This phenomenon is well-known as the Brazil nut effect. Based on the vertical segregation coefficient (H), we can separate the degree of segregation of the mixture into 3 zones as follows. The particles are still in the mixed state at the first zone lying between $0.0 < H \leq 0.2$. The particles then begin to segregate at the second zone where $0.2 < H \leq 0.4$. In this zone, a partial segregation can be seen. The segregation of the mixture is more clearly observed up to obtain a complete segregation in the third zone where $0.4 < H \leq 1.0$. Interestingly, they start to segregate when $\Gamma \geq 2$ for all the particle size ratios. The vertical segregation coefficient increases with the dimensionless acceleration. The segregation of the mixture is quite clearly observed when $D_l/D_s \geq 3.0$. This segregation phenomenon is clearer when the particle size ratio is increased.

In a future work, it would be interesting to perform the numerical simulations based on the experimental configurations, in order to analyse between experimental and numerical results. Moreover, the simulations would be employed as a supplementary tool for investigating other cases that the experiments cannot be conducted due to their difficulties. Importantly, the effect of container size must be identified thoroughly for the next study.

References

1. H.M. Jaeger, S.R. Nagel, Science **255**, 1523 (1992)
2. H.M. Jaeger, S.R. Nagel, Rev. Mod. Phys. **68**, 1259 (1996)
3. A.D. Rosato, D.L. Blackmore, N. Zhang, Y. Lan, Chem. Eng. Sci. **57**, 265 (2002)
4. D. Pihler-Puzović, T. Mullin, P. Roy. Soc. A **469**, 20130203 (2003)
5. A.D. Rosato, Y. Lan, D.T. Wang, Powder Technol. **66**, 149 (1991)
6. J.B. Knight, H.M. Jaeger, S.R. Nagel, Phys. Rev. Lett. **70**, 3728 (1993)
7. J.C. Williams, Powder Technol. **15**, 245 (1976)
8. F. Calvetti, G. Combe, J. Lanier, Mech. Cohes.-Frict. Mat. **2**, 121 (1997)
9. W.S. Rasband, *ImageJ*, (U.S. National Institutes of Health, Bethesda, Maryland, USA, http://imagej.nih.gov/ij/, 1997-2016)
10. C.R.A. Abreu, F.W. Tavares, M. Castier, Powder Technol. **134**, 167 (2003)
11. M.P. Ciamarra, M.D. De Vizia, A. Fierro, M. Tarzia, A. Coniglio, M. Nicodemi, Phys. Rev. Lett. **96**, 058001 (2006)
12. T. Pöschel, H.J. Herrmann, Europhys. Lett. **29**, 123 (1995)
13. W. Cooke, S. Warr, J.M. Huntley, R.C. Ball, Phys. Rev. E **53**, 2812 (1996)

An Experimental Investigation on the Generation of a Stable Arch in Granular Materials Using a Developed Trapdoor Apparatus

Ali Ahmadi[1] and *Ehsan* Seyedi Hosseininia*[1]

[1]Civil Engineering Department, Faculty of Engineering, Ferdowsi University of Mashhad, Mashhad, Iran

Abstract. This paper discusses the formation of stable arches in granular materials by using a series of laboratory tests. To this aim, a developed trapdoor apparatus is designed to find dimensions of arches formed over the door in cohesionless aggregates. This setup has two new important applications. In order to investigate the maximum width of the opening generated exactly on the verge of failure, the door can be open to an arbitrary size. In addition, the box containing granular materials (or base angle) is able to be set on optional angles from zero to 90 degrees with respect to the horizontal. Therefore, it is possible to understand the effect of different levels of gravity accelerations on the formed arches. It is observed that for all tested granular materials, increasing the door size and decreasing the base angle, both cause to increase the width and height of the arch. Moreover, the shape of all arches is governed by a parabola. Furthermore, the maximum door width is approximately five to 8.6 times the particle size, depending on the internal friction angle of materials and the base angle.

1 Introduction

Arching is one of the most common phenomena encountered in granular materials both in the field and laboratory. Terzaghi [1] defined arching effect as the transfer of pressure from a yielding mass of soil onto adjoining stationary parts. In other words, arching is closely related to the mobilization of shear strength induced by the relative movement in materials. He used trapdoor tests to explain how stress is transferred from yielding parts of a soil mass to adjacent stationary, non-yielding parts that led to the formation of an arching zone.

By using this idea, several works have been carried out to investigate the arching effect as load transferring in different engineering problems, including earth pressure on retaining structures [2], stability of tunnels [3-7], bearing resistance of piles [8], settlement of pile supported embankments [9], load on buried structures [10], and granular flow in hoppers and silos [11-14]. A literature study on the arching effect can be found in Tien [15]. More recently, the numerical Discrete Element Method has been used to explore the application of arching in soil [16, 17]. Approximate theories have been developed for the analysis of granular flow and the design of hoppers [18-20]. Guo and Zhou [19] have studied the formation of critical stable arch over an opening by using trapdoor test. In their tests, the trapdoor width was fixed. Motivated by the aforementioned literature review, this study tries to study the conditions of the formation of stable arches over an opening before the critical (last stable) arch is formed. In this study, the maximum size of opening door related to critical width and height of the arch over the door is explored by considering various gravity acceleration (from zero to 90° base angle of materials box). In the former study (Gou and Zhou's setup), the base angle was fixed at 30°. This developed setup can help designers to select a size of hopper or silo outlet with various base angle of materials.

2 Experimental tests

2.1 Developed trapdoor test

The trapdoor apparatus can be used in order to investigate the formation of arch in granular materials. In the apparatus introduced by Guo and Zhou [19], the base has an angle of 30 degrees to the horizontal. The apparatus is developed in this study as explained as follows. Fig. 1 shows this setup. The apparatus has a box, which can set on arbitrary angle from zero to 90 degrees with respect to the horizontal, so as to apply different gravity accelerations (g) over materials. The box is considered as a space between MDF base plate, lateral frames (called shoulders) and a plexiglass inserted in the fissure existing on the lateral frames. The box is fixed on the base with dimension of 30×40cm. A plexiglass layer can be placed over the box with a distance of average grains diameter plus 1-2 mm such

* Corresponding author: eseyedi@um.ac.ir

that only one layer of particles in the box can be held. Thus, particles move down without any friction with plexiglass and negligible friction with base plate. The particles are stacked only in single layer such that it simulates the plane strain condition. Two transitive segments (called trapdoor blocks), are contrived at the lowest level of materials to access different door width and also use as a stopper for granular materials.

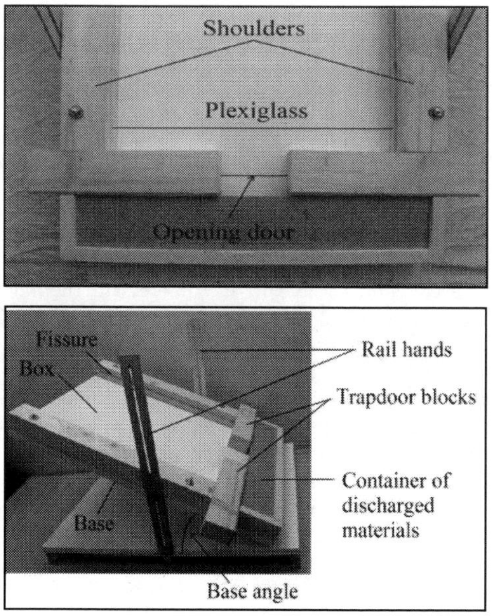

Fig. 1. Developed trapdoor apparatus

2.2 Selected granular materials

Two kinds of granular materials were used in this study, including gravel packing with the average size of 9mm and plastic spherical beads with the diameter of 12mm in two density conditions (i.e. loose and dense). The gradation curves of these materials are shown in Fig. 2.

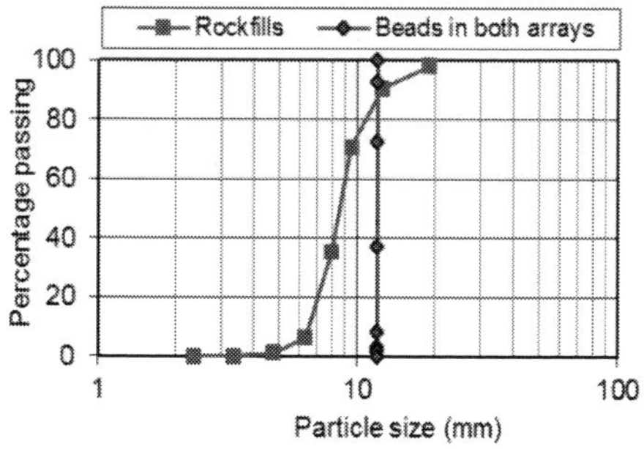

Fig. 2. Gradation curve of granular materials (Gravel: d50≈ 9mm, Beads: d50≈12mm)

The granular materials used in the tests are presented in Fig. 3. In order to access loose and dense packing of beads, two arrays of bead particles are considered: square arrangement as loose state (Fig. 3a) and hexagonal arrangement as dense state (Fig. 3b). The array for gravel is shown in Fig. 3(c). The density of tested materials that were used in the box tests were calculated as 1.58 g/cm^3 for gravel packing and for glass beads as 1.05 g/cm^3 for dense and 0.96 g/cm^3 for loose arrays. The internal friction angles of tested materials were measured by finding the angle of repose (φ). Values of φ are 23°, 37° and 45° for loose beads, dense beads and gravel, respectively.

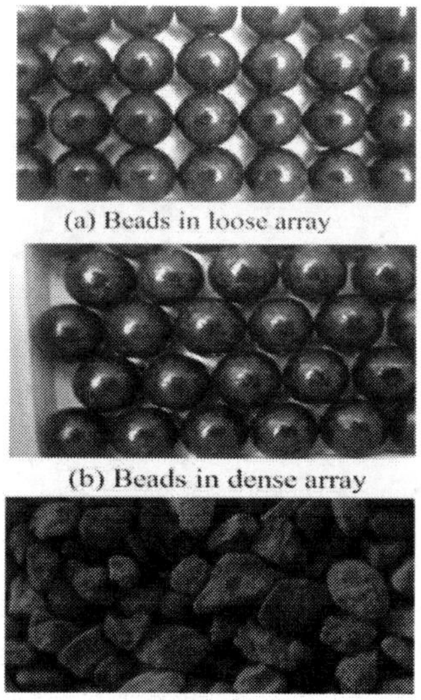

(a) Beads in loose array

(b) Beads in dense array

(c) Rockfills

Fig. 3. Tested granular materials

2.3 Test performance

First, the box is fixed on desired angle, then two trapdoor blocks move away and a distance is created between them. A stiff rectangular rubber as the door fills this space. In second step, granular materials come in the box. It must be considered that entering the materials in the box can be achieved by various manners. For example, gravel is inserted in the box after fixing plexiglass, but beads are first put in loose or dense array in the existing space between segments, shoulders and base plate, then plexiglass is placed in fissures. This method facilitates the process of arranging bead grains as loose or dense array. Finally, rubber door moves down and granular materials discharge and fall into the container. The whole materials do not discharge, but a stable arch forms over the door after a short time. This arch can carry static pressure of upper materials and transfers it to its two bottom supports by arching application. In every run of the test, the distance between two blocks, i.e., opening door width (W) increases 2 mm as compared with previous stage, then the width and height size of the arch corresponding to door width are recorded in sheets relating to specific angle. This process is repeated so that no stable arch forms anymore. On the verge of arch collapse, i.e., no arch can be generated, maximum door width (W_{max}) and the critical arch width and height were measured in this study. It is here noted that the arch width, which is defined as the internal distance between two piers of the arch, might be different from the trapdoor width. This subject is mentioned in the next section. Each test has been repeated at least three times to examine the repeatability of the experiment. The experimental results presented in this study correspond to the assured and repeatable observations.

3 Experimental observations

By performing experimental tests on four base angles (θ) consisting 10°, 40°, 70°, and 90°, considerable results were observed. Fig. 4 shows photographs of typical formed arches.

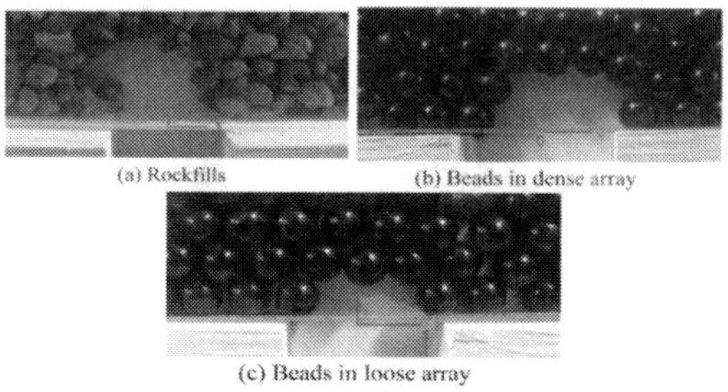

(a) Rockfills (b) Beads in dense array

(c) Beads in loose array

Fig. 4. Typical photographs of arching over door width W=5.6 cm

Arching effect makes a stable arch in ach granular medium. Fig. 5 presents the loci of each particle in stable arch formed over door widths (W) of 2.8 cm and 5.6 cm and base angles of 40° and 90°. Note that x and y values represent the loci of the particles that forms an arch. As shown for all the materials, the arch width is smaller than door width. The arch height is greater for dense array (φ= 37°) compared to loose array (φ= 23°), although the diameter of bead grains is equal in both arrays (with d=12mm). Meanwhile, the arch dimensions of gravel (d50=9mm, φ= 45°) is approximately greater than dense beads. This indicates that the internal friction angle of materials has mobilized the shear strength between particles and it is the most important factor in arch formation for granular materials. Increasing the opening door width causes to discharge more materials and to decrease the surcharge.

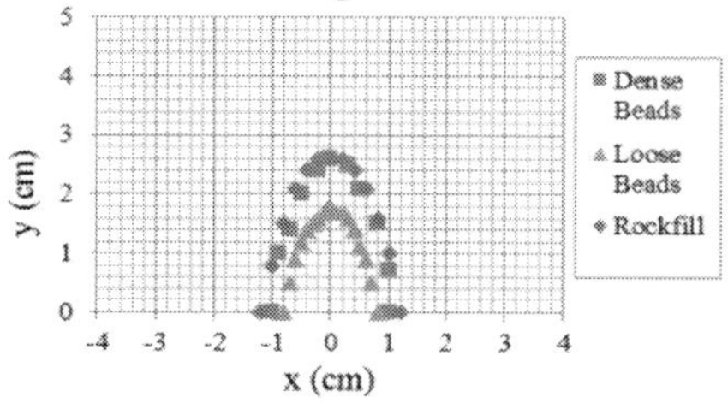

(a) θ=40°, W=2.8 cm

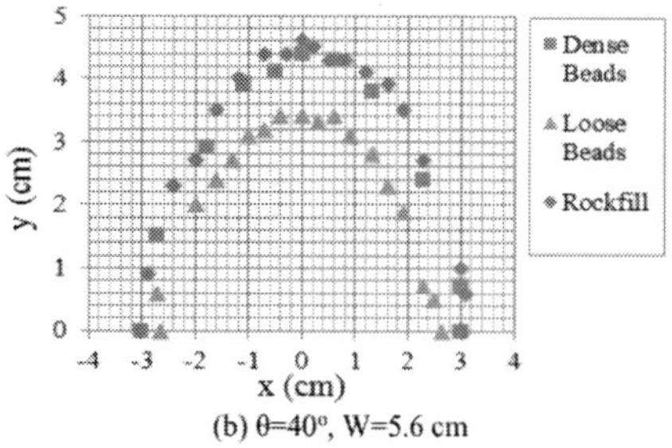

(b) θ=40°, W=5.6 cm

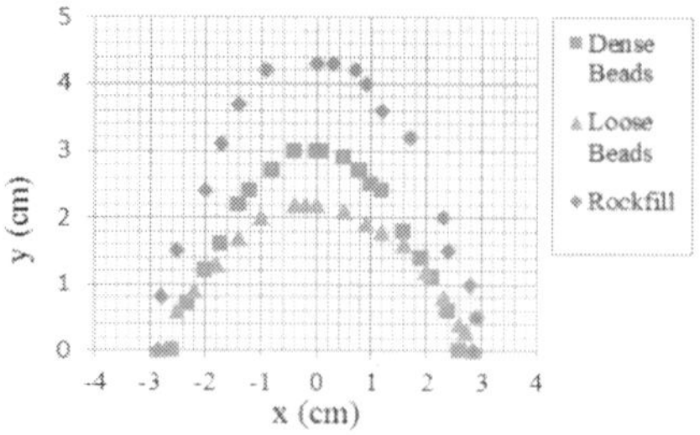

(c) θ=90°, W=5.6 cm

Fig. 5. Comparison between arch shapes formed on determined base angle and door width

In Fig. 6, the variation of the arch height (H) versus the door width (W) is presented. As seen, the arch height increases as the door width increases. This might be as a result of the reduction in surcharge on the arch that causes the arch to endure less pressure and consequently to have greater height. However, the arch height (H) is reduced as the base angle (θ) is increased. The more the base angle inclines, the more effective gravity acceleration (g×sinθ) exerts on granular materials. This increases the effective surcharge on the arch and shortens its height since the arch supports more loads in this condition. This observation has been also shown by numerical simulations [21, 22]. It is also observed that as the door width increases, more materials are discharged and consequently the arch width is greater, until no stable arch can form. The arch forming on the verge of collapse is called as "critical arch". For this arch, the width of door and the arch possess maximum possible size. On the contrary, the height of the arch is not the maximum, but it decreases slightly. The reason might be because the arch cannot endure the materials that have tendency to discharge, and for adoption of this condition, the arch should decrease its height.

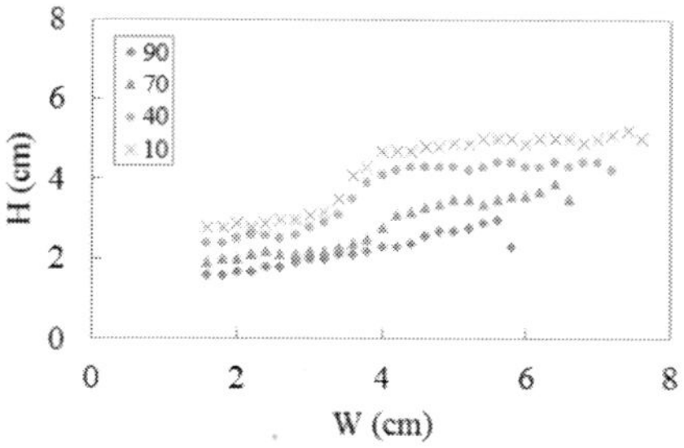

(a) Beads in dense array

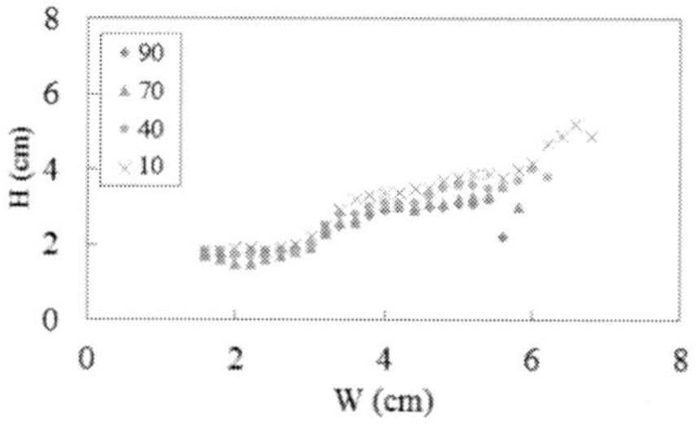

(b) Beads in loose array

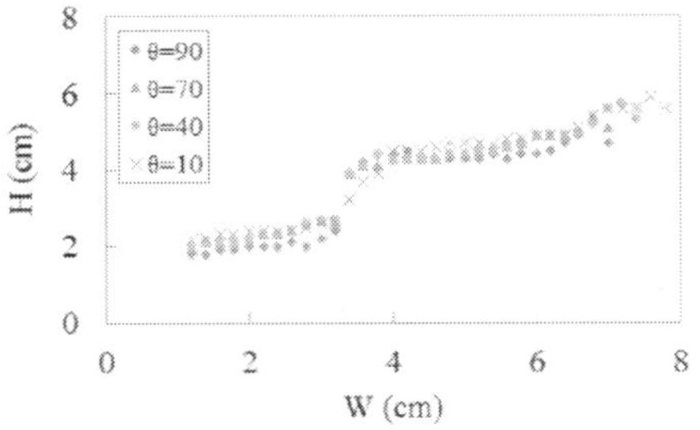

(c) Gravel

Fig. 6. Relation between door width (W) and arch height (H) for different base angles

The maximum door width (W_{max}), which is the width of the door that no stable arch can form over the opening door with greater width, was measured by visual inspection. The W_{max}/d_{50} values versus different base angles for tested granular materials are presented in Fig. 7. As shown, the W_{max} varies from five times the average particle size (d_{50}) for loose beads on the base angle of 90° to 8.6 times d_{50} for gravel on the base angle of 10°. Although there is a considerable debate on the maximum possible width of the outlet [23-27] in the literature, these observed data is comparable with experimental results obtained by Guo and Zhou [17] for granular materials e.g., glass beads (with d_{50}=5.5 mm) and dense sand on the base angle of 30°.

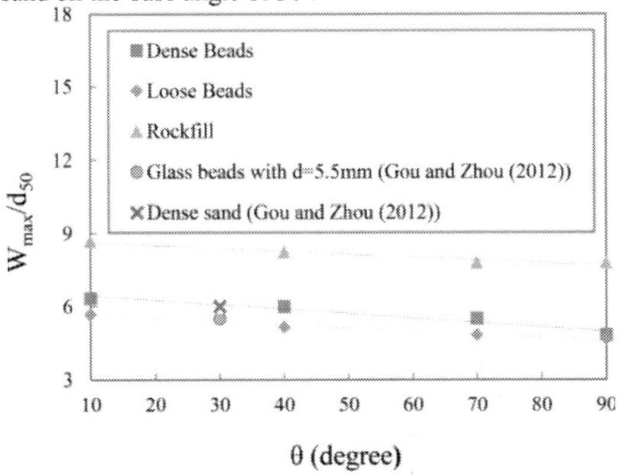

Fig. 7. Maximum opening door size versus different base angles

4 Conclusion

In this study, a developed trapdoor apparatus was implemented. An experimental analysis for ensembles of almost mono-sized particles of granular materials shows that stable arches may form with various widths depending on particle shape and the friction angle. For tested granular materials in this study, the critical width of opening door is approximately five to 8.6 times particle sizes. It is also observed that for all tested granular materials, an increase in door size and a decrease in base angle augment the width and height of the arch. However, on the verge of arch collapse, the height of critical arch decreases. This finding can be useful for granular flow in silos or hoppers.

References

1. K. Terzaghi, Theoretical soil mechanics: J. Wiley and Sons, Inc. (1943).
2. Y.-J. Jiang and I. Towhata, Rock Mechanics and Rock Engineering **1** (2012).
3. H. Akbarzadeh and A. Bayat, Tunn. Undergr. Sp. Tech. 39, **50** (2012).
4. C.N. Chen, W.-Y. Huang, and C.-T. Tseng, Tunn. Underg. Sp. Tech. **26**, 228 (2011).
5. R.P. Chen, L.J. Tang, D.S. Ling, and Y.M. Chen, Comp. Geotech. **38**, 187 (2011).
6. R.-P. Chen, J. Li, L.-g. Kong, and L.-j. Tang, Tunn. Underg. Sp. Tech. **33**, 12 (2013).
7. C.J. Lee, B.R. Wu, H.T. Chen, and K.H. Chiang, Tunn. Underg. Sp. Tech. **21**, 119 (2006).
8. M. Kahyaoglu, G. Imançli, O. Önal, and A. Kayalar, KSCE J. Civ. Eng. **16**, 562 (2012).
9. S.J.M. Eekelen, A. Bezuijen, H.J. Lodder, and A.F. Van Tol, Geotext. Geomembr. **32**, 69 (2012).
10. Y.S. Karinski, A.N. Dancygier, and I. Leviathan, Eng. Struc. **25**, 91 (2003).
11. Miccio, Barletta, and Poletto, Powder Tech. **235**, 312 (2013).
12. A.W. Roberts, Particuology **8**, 591 (2010).
13. A. Garcimartín, I. Zuriguel, L. A. Pugnaloni, and A. Janda, Phys. Rev. E **82**, 031306 (2010).
14. K. To and P.-Y. Lai, Phys. Rev. E **66**, 011308 (2002).
15. H.-J. Tien "A Literature Study of the Arching Effect," SM, MIT (1996).
16. B. Chevalier, G. Combe, and P. Villard, 18ème Congrès Français de Mécanique, Grenoble, France (2007).
17. B. Chevalier, G. Combe, and P. Villard, Acta Geotechnica **7**, 15 (2012).
18. A. Drescher, A.J. Waters, and C.A. Rhoades, Powder Tech. **84**, 165 (1995).
19. Guo and Zhou, Int. J. Num. Analytical Methods In Geomechanics **37**, 1048 (2012).
20. K.K. Rao and P.R. Nott, Cambridge University Press, 7 (2008).
21. K. To, P.-Y. Lai, and H. K. Pak, phys. Rev. Lett., 86, 71 (2001)
22. I. Zuriguel, A. Garcimartín, D. Maza, L. A. Pugnaloni, and J. M. Pastor, Phys. Rev. E 71, 051303 (2005).
23. K. To, Phys. Rev. E 71, 060301-R (2005).
24. A. Janda, I. Zuriguel, A. Garcimartín, L. A. Pugnaloni, and D. Maza, Europhys. Lett. 84, 44002 (2008).
25. C. C. Thomas and D. J. Durian, Phys. Rev. Lett. 114, 178001 (2015).
26. R. Arévalo, I. Zuriguel, D. Maza, and A. Garcimartín, Phys. Rev. E 89, 042205 (2014).
27. R. Arévalo and I. Zuriguel, Soft Matter **12**, 123 (2016)

Thermal properties of an impurity immersed in a granular gas of rough hard spheres

Francisco Vega Reyes[1,2,⋆], *Antonio* Lasanta[1,3], *Andrés* Santos[1,2], and *Vicente* Garzó[1,2]

[1] *Departamento de Física, Universidad de Extremadura, 06071 Badajoz, Spain*
[2] *Instituto de Computación Científica Avanzada (ICCAEx), Universidad de Extremadura, 06071 Badajoz, Spain*
[3] *G. Millán Institute, Fluid Dynamics, Nanoscience and Industrial Mathematics Department of Materials Science and Engineering and Chemical Engineering Universidad Carlos III de Madrid Leganés, Spain*

Abstract. We study in this work the dynamics of a granular impurity immersed in a low-density granular gas of identical particles. For description of the kinetics of the granular gas and the impurity particles we use the rough hard sphere collisional model. We take into account the effects of non-conservation of energy upon particle collision. We find an (approximate) analytical solution of the pertinent kinetic equations for the single-particle velocity distribution functions that reproduces reasonably well the properties of translational/rotational energy non-equipartition. We assess the accuracy of the theoretical solution by comparing with computer simulations. For this, we use two independent computer data sets, from molecular dynamics (MD) and from Direct Simulation Monte Carlo method (DSMC). Both approach well, with different degrees, the kinetic theory within a reasonable range of parameter values.

1 Introduction

The dynamics of matter composed by large sets of macroscopic particles (sizes larger than $\sim 1\mu$m [1]) has always been of interest [2, 3] since particulate matter is present, under a huge variety of forms and composites, in many human-activity and natural processes. Moreover, the activity of this research field has grown enormously over the last decades [4–7], becoming an important area of study in the field of soft matter physics [8, 9]. In particular, the control of granular dynamics has a large impact on a variety of industries, including some high-technology related interesting studies [10–12]. It is worth to mention also that one of the most ubiquitous problems in granular dynamics related industry is the selection of mechanically peculiar particles in a set of otherwise identical grains; i.e., *granular segregation* [13].

The handling of grains usually requires mechanical input. A variety of mechanisms may be used, of which vibration [14] and air-fluidization [15] are the most common procedures in both fundamental research and granular-matter-related industry. Furthermore, the study of granular matter represents an important challenge for fundamental physics since it has been known from quite some time that soft matter [8] theories in which the basic constituents are of molecular/atomic nature (and thus particle collisions are elastic) can eventually apply in a generalized form to granular dynamics. Examples range from symmetry breaking processes like the one described in the Kosterlitz-Thouless-Halperin-Nelson-Young (KTHNY) theory [16–18] (the so-called hexatic phase [19] has also been found in granular systems [20]) to kinetic theory of gases or hydrodynamics [21] at the level of both the associated transport coefficients [22–24] and the hydrodynamic phenomena [25].

We focus here on the kinetic theory aspects of the granular gas dynamics (a low density and thermalized granular system). It is nowadays widely accepted that a low density granular system can be accurately described by either the Boltzmann equation (low density) or the Enskog equation (moderate density), in its inelastic version [26]. It was P. K. Haff who conceptualized the so-called *granular gas*, defining mathematically its reference state as well, back in 1983 [27]. Haff defined this state as a system of macroscopic particles at low density that undergo inelastic binary collisions. The system is isolated and its initial state is spatially uniform, this uniformity being preserved in time. This produces a continuously decaying uniform temperature. The cooling rate of the system, as Haff showed [27], depends on the degree of inelasticity in the collisions. This is nowadays known as *Haff's law* [28], and his reference state for the granular gas is known as the *homogeneous cooling state* (HCS). The HCS has become the theoretical basis for the development of the kinetic theory of granular gases. There is actually a rigorous experimental proof of the existence of the HCS and the accuracy of Haff's law [29].

As we know, the kinetic theory of non-uniform gases accounts for the transport phenomena of non-equilibrium states [30]. Starting out of the work by J. C. Maxwell for a gas under equilibrium state, this theory was laid by D.

⋆e-mail: fvega@unex.es

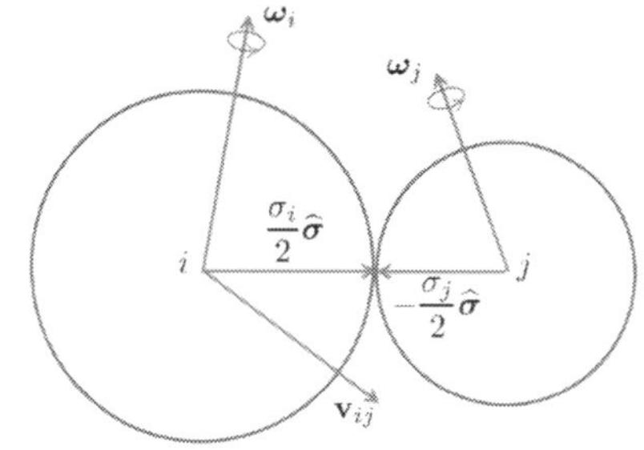

Figure 1. The rotational and translational velocities of two grains at collision. Here, $\mathbf{v}_{ij} = \mathbf{v}_i - \mathbf{v}_j$.

Hilbert [31], who set several key mathematical concepts (notably, that of the *normal solution*, to which the non-uniform gas tends after a quick transient, regardless of the initial state). Afterwards, the theory was simultaneously developed and extended by the works of S. Chapman, D. Enskog and D. Burnett [31–33], and finally Ernst and coworkers developed the Enskog theory for denser systems [34]. Years later, these theories were extended also to the granular gas [22–24, 28], finding consistently the corresponding set of transport coefficients. This, in turn allowed for the deployment of a generalized fluid mechanics, able to additionally describe steady granular flows far from equilibrium.

All of these studies have referred to the simple smooth hard sphere collisional model. However, several important experimental features of granular collisions can only be captured with a more elaborate collisional model [35]. In particular, particle spin (i.e., the grains angular velocities) can have an important role in the global properties of the granular gas dynamics [7], see Fig. 1. A rough hard-sphere collisional model with two velocity-independent coefficients (tangential, β, and normal, α, restitution coefficients) seems to describe more accurately a generic collision between two dry grains [36].

For this reason, and now taking into account the particle rotational degrees of freedom, more recent theoretical studies have revisited homogeneous states like the HCS [37–40], or the steady state produced by a uniform stochastic thermostat [41]. In particular, a recent work [38] develops results for binary mixtures. This includes the case where one of the species is present in a very small proportion, as compared to the other species (the so-called tracer limit). This situation (the tracer limit for a binary mixture of inelastic rough hard spheres) is the one we analyze in the present work. The objective of this work is to present a comparison between the existing theory for this system and new simulation data, obtained from Monte Carlo simulation (direct simulation Monte Carlo method, DSMC) of the Boltzmann-Lorentz equation for the granular impurity and also from molecular dynamics (MD) simulations.

2 Theory and simulation outline

Let us consider two different sets of particles, denoted here with subscript $i = 1, 2$, each of them composed by identical rough hard spheres that collide inelastically. Each set is characterized by the following set of mechanical properties: m_i, σ_i, κ_i, where m_i, σ_i are the particle mass and diameter respectively and κ_i is a dimensionless parameter characterizing the moments of inertia I_i of the particles, i.e., $\kappa_i \equiv 4I_i/(m_i\sigma_i^2)$. The velocities of the points of the spheres which are in contact during the collision are

$$\mathbf{w}_i = \mathbf{v}_i - \frac{\sigma_i}{2}\hat{\sigma}\times\omega_i,$$
$$\mathbf{w}_j = \mathbf{v}_j + \frac{\sigma_j}{2}\hat{\sigma}\times\omega_j, \tag{1}$$

where $i, j = 1, 2$, according to the particle species [36, 38]. Therefore, the relative velocity between colliding particles would be $\mathbf{w}_{ij} \equiv \mathbf{w}_i - \mathbf{w}_j$. In addition to taking into account that linear and angular momentum must be conserved upon collisions, we consider the following relations between pre-collisional and (primed) post-collisional relative velocities

$$\hat{\sigma}\cdot\mathbf{w}'_{ij} = -\alpha_{ij}\hat{\sigma}\cdot\mathbf{w}_{ij},$$
$$\hat{\sigma}\times\mathbf{w}'_{ij} = -\beta_{ij}\hat{\sigma}\times\mathbf{w}_{ij}, \tag{2}$$

where the former relation stands for the normal component and the latter for the tangential components. Coefficients α_{ij}, β_{ij} are the so-called normal and tangential coefficients of restitution. In general, their values depend on the type of binary collision, that in our system can be either gas-gas (α_{22}, β_{22}) or impurity-gas (α_{12}, β_{12}), since in the tracer limit, impurity-impurity collisions are statistically rare events and thus they can be neglected [42]. In our collisional model, the coefficients of restitution are independent of the relative velocities of the collision pair.

The kinetic equations for the granular gas and the impurity respectively

$$(\partial_t + \mathbf{v}\cdot\nabla) f_2(\mathbf{r}_2, \mathbf{v}_2, \omega_2; t) = J_{22}[\mathbf{r}_2, \mathbf{v}_2, \omega_2; t|f_2, f_2], \tag{3}$$
$$(\partial_t + \mathbf{v}\cdot\nabla) f_1(\mathbf{r}_1, \mathbf{v}_1, \omega_1; t) = J_{12}[\mathbf{r}_1, \mathbf{v}_1, \omega_1; t|f_1, f_2]. \tag{4}$$

Here, J_{ij} are the binary collision operators, defined as follows (double primes indicate pre-collisional velocities) [38]

$$J_{ij}[\mathbf{r}_i, \mathbf{v}_i, \omega_i; t|f_i, f_j] \equiv \sigma_{ij}^2 \int d\omega_j \int d\mathbf{v}_j \int d\hat{\sigma}_i \Theta(\hat{\sigma}_i\cdot\mathbf{v}_{ij})(\hat{\sigma}\cdot\mathbf{v}_{ij})$$
$$\times\left[(\alpha_{ij}\beta_{ij})^{-2} f_i(\mathbf{r}_i, \mathbf{v}''_i, \omega''_i; t) f_j(\mathbf{r}_j, \mathbf{v}''_j, \omega''_j; t) - f_i(\mathbf{r}_i, \mathbf{v}_i, \omega_i; t) f_j(\mathbf{r}_j, \mathbf{v}_j, \omega_j; t)\right]. \tag{5}$$

For solution of the kinetic equations (3)–(4), we may use a multi-temperature Maxwellian form of the velocity distribution function that incorporates non-equipartition of the translational and rotational mode [38, 43]:

$$f_i(\mathbf{v},\omega) = n_i\left(\frac{m_i I_i}{4\pi^2 T_i^{\rm tr} T_i^{\rm rot}}\right)^{3/2}\exp\left[m_i\frac{(\mathbf{v}_i-\mathbf{u})^2}{2T_i^{\rm tr}} - \frac{I_i\omega^2}{2T^{\rm rot}}\right]. \tag{6}$$

It is also interesting to recall the definition of the energy production rates

$$\xi_i^{\rm tr} \equiv -\frac{1}{T_i^{\rm tr}}\left(\frac{\partial T_i^{\rm tr}}{\partial t}\right)_{\rm coll}, \xi_i^{\rm rot} \equiv -\frac{1}{T_i^{\rm rot}}\left(\frac{\partial T_i^{\rm rot}}{\partial t}\right)_{\rm coll} \tag{7}$$

These rates are related to the total cooling rate ζ as follows

$$\zeta = \sum_i \frac{n_i}{2nT}\left(T_i^{\rm tr}\xi_i^{\rm tr} + T_i^{\rm rot}\xi_i^{\rm rot}\right). \tag{8}$$

By introducing this form of the distribution function in (3)–(4) and multiplying by v^2, ω^2, we obtain the evolution equations for the translational and rotational temperatures of both species (see [38] for more reference)

$$\partial_t T = -\zeta T, \tag{9}$$

$$\partial_t \frac{T_i^{\rm tr}}{T} = -\left(\xi_i^{\rm tr} - \zeta\right)\frac{T_i^{\rm tr}}{T}, \quad \partial_t \frac{T_i^{\rm rot}}{T} = -\left(\xi_i^{\rm rot} - \zeta\right)\frac{T_i^{\rm rot}}{T}.$$

In the tracer limit, for the impurity we can calculate the functions $\xi_1^{\rm tr}(\frac{T_1^{\rm tr}}{T_2^{\rm tr}}, \frac{T_1^{\rm rot}}{T_2^{\rm rot}}, \kappa_1, m_1/m_2, \alpha_{12}, \beta_{12}, \alpha_{22}, \beta_{22})$ and $\xi_1^{\rm rot}(\frac{T_1^{\rm tr}}{T_2^{\rm rot}}, \frac{T_1^{\rm rot}}{T_2^{\rm rot}}, \kappa_1, m_1/m_2, \alpha_{12}, \beta_{12}, \alpha_{22}, \beta_{22})$ (for the other species can be found in [38]).

A normal state with all energy production rates equal ($\xi_1^{\rm tr} = \xi_2^{\rm tr} = \xi_1^{\rm rot} = \xi_2^{\rm rot}$) is reached after a short transient. In this state the complete time dependence scales with T (because this state is actually the HCS [27, 38]). We have solved equations (9) for this state, obtaining the rotational and translational temperatures of both species. In the tracer limit the abundant species is decoupled from the impurity and the impurity production rates contributions come only from gas-impurity collisions, which in our case can be done analytically. The temperature ratios $T_1^{\rm tr}/T_2^{\rm tr}, T_1^{\rm rot}/T_2^{\rm rot}$ are obtained from the physical solution of a quartic equation. For more details on the theoretical procedures refer to a previous work [38].

We compare our analytical results with an exact numerical solution of (3)-(4), obtained from the Direct Simulation Monte Carlo (DSMC) method [44]. DSMC was implemented as usual (details can be found elsewhere [42]), taking into account roughness in the collisions. Additionally, and since in the kinetic equation and hence in the DSMC method the molecular chaos assumption is inherent, comparison with molecular dynamics (MD) simulation results are insightful and for this reason we also implemented this computational method, by using an event driven algorithm for inelastic rough hard spheres [45].

We discuss the results in the next section.

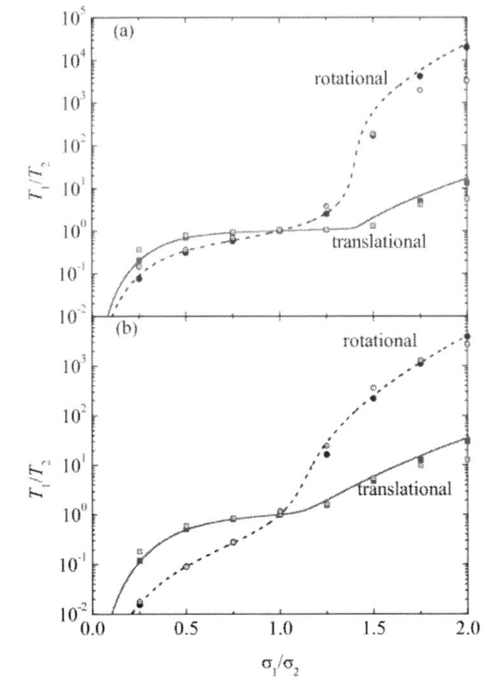

Figure 2. Plot of the temperature ratios $T_1^{\rm tr}/T_2^{\rm tr}$ and $T_1^{\rm rot}/T_2^{\rm rot}$ vs. the relative size σ_1/σ_2 for relative mass $m_1/m_2 = (\sigma_1/\sigma_2)^3$. (a) $\kappa_1 = \kappa_2 = 2/5, \alpha_{22} = \alpha_{12} = 0.9$ and $\beta_{22} = \beta_{12} = -0.5$. (b) $\kappa_1 = \kappa_2 = 2/5, \alpha_{22} = \alpha_{12} = 0.7$ and $\beta_{22} = \beta_{12} = -0.5$. Lines indicate theory results, while solid and open symbols stand for DSMC and MD data, respectively. All MD simulations have a volume fraction $\phi \simeq 0.02$

Results and discussion

We present in Fig. 2 the results of the theory (lines) compared to DSMC (solid symbols) and MD simulations (open symbols) for the case of constant particle mass density ($m_1/m_2 = (\sigma_1/\sigma_2)^3$), uniformly distributed inside all particles ($\kappa_1 = \kappa_2 = 2/5$). For simplicity, we chose equal tangential and normal coefficients of restitution for all types of collisions; i.e., $\beta_{22} = \beta_{12}$ and $\alpha_{22} = \alpha_{12}$. As we see in this figure, a good agreement is obtained between theory and both DSMC and MD simulations, for all the range of relative sizes of gas-impurity particles represented here. Even in the context of the multi-temperature Maxwellian approximation used in (6), the good agreement could be expected since it has been already been observed for a mono-component granular gas of rough particles that temperature is not very sensitive to the details of the high-energy tails of the velocity distribution function. In a forthcoming work, we will present more cases, extending the results presented here.

Acknowledgements

We acknowledge support Grants Nos. FIS2016-76359-P (Spanish Gov.) and GRU10158 (Junta de Extremadura, partially financed by ERDF funds). Computing facilities from Extremadura Research Centre for Advanced Technologies (CETA-CIEMAT), funded by the ERDF, are also acknowledged.

References

[1] R.A. Bagnold, *The Physics of Blown Sand and Desert Dunes* (Dover Publications Inc., New York, USA, 1954)
[2] M. Faraday, Phil. Trans. R. Soc. London **121**, 299 (1831)
[3] O. Reynolds, Phil. Mag. Series 5 **127**, 469 (1885)
[4] H.M. Jaeger, S.R. Nagel, R. Behringer, Rev. Mod. Phys **68**, 1259 (1996)
[5] P.G. de Gennes, Rev. Mod. Phys. **78**, S734 (1999)
[6] L.P. Kadanoff, Rev. Mod. Phys. **71**, 435 (1999)
[7] I.S. Aranson, L.S. Tsimring, Rev. Mod. Phys. **78**, 641 (2006)
[8] P.G. de Gennes, Rev. Mod. Phys. **64**, 645 (1992)
[9] P.M. Chaikin, T.C. Lubensky, *Principles of Condensed Matter Physics* (Cambridge University Press, Cambridge, UK, 2000)
[10] F. Qian, T. Zhang, W. Korff, P.B. Umbanhowar, R.J. Full, D.I. Goldman, Bioinspir. Biomim. **144**, 341 (1990)
[11] C. Li, P.B. Umbanhowar, H. Komsuoglu, D.E. Koditschek, D.I. Goldman, Proc. Natl. Acad. Sci. USA **10**, 056014 (2015)
[12] W.E. Dietrich, J.T. Perron, Nature **439**, 411 (2006)
[13] A. Rosato, K.J. Strandburg, F. Prinz, R.H. Swendsen, Phys. Rev. Lett. **58**, 1038 (1987)
[14] J.S. Olafsen, J.S. Urbach, Phys. Rev. Lett. **81**, 4369 (1998)
[15] W. Losert, L. Bocquet, T.C. Lubensky, J.P. Gollub, Phys. Rev. Lett. **85**, 1428 (2000)
[16] J.M. Kosterlitz, D.J. Thouless, J. Phys. C: Solid State Physics **6**, 1181 (1973)
[17] B.I. Halperin, D.R. Nelson, Phys. Rev. Lett. **41**, 121 (1978)
[18] A.P. Young, Phys. Rev. B **19**, 1855 (1979)
[19] B.J. Alder, T.E. Wainwright, Phys. Rev. **127**, 359 (1962)
[20] J.S. Olafsen, J.S. Urbach, Phys. Rev. Lett. **95**, 098002 (2005)
[21] J.W. Dufty, Adv. Complex Syst. **4**, 397 (2001)
[22] J.J. Brey, J.W. Dufty, C.S. Kim, A. Santos, Phys. Rev. E **87**, 397 (1997)
[23] V. Garzó, J.W. Dufty, Phys. Rev. E **60**, 061306 (1999)
[24] J.F. Lutsko, Pys. Rev. E **72**, 021306 (2005)
[25] F. Vega Reyes, J.S. Urbach, J. Fluid Mech. **636**, 279 (2009)
[26] J.J. Brey, J.W. Dufty, A. Santos, J. Stat. Phys. **87**, 1051 (1997)
[27] P.K. Haff, J. Fluid Mech. **134**, 401 (1983)
[28] N. Sela, I. Goldhirsch, J. Fluid Mech. **361**, 41 (1998)
[29] C.C. Maaß, N. Isert, G. Maret, C.M. Aergerter, Phys. Rev. Lett. **100**, 24800 (2008)
[30] C. Chapman, T.G. Cowling, *The Mathematical Theory of Non-Uniform Gases*, 3rd edn. (Cambridge University Press, Cambridge, 1970)
[31] D. Hilbert, *Foundations of the Kinetic theory of gases* (Pergamon Press, 1972), Vol. 3 of *Kinetic Theory*, chap. Pt. 2, (English translation from the original in Mathematische Annalen, 1912)
[32] S. Chapman, Proc. Roy. Soc. Lond. A **93**, 1 (1916)
[33] D. Burnett, Proc. Lond. Math. **39**, 385 (1935)
[34] H. van Beijeren, M.H. Ernst, J. Stat. Phys **167**, 125 (1979)
[35] S.F. Foerster, M.Y. Louge, H. Chang, K. Allis, Phys. Fluids **6**, 1070 (1993)
[36] G.M. Kremer, *An Introduction to the Boltzmann Equation and Transport Processes in Gases* (Springer-Verlag, Berlin, 2010)
[37] N.V. Brilliantov, T. Pöschel, W.T. Kranz, A. Zippelius, Phys. Rev. Lett. **98**, 128001 (2007)
[38] A. Santos, G.M. Kremer, V. Garzó, Prog. Theor. Phys. **184**, 31 (2010)
[39] F. Vega Reyes, A. Santos, G.M. Kremer, Phys. Rev. E **89**, 020202(R) (2014)
[40] R. Rongali, M. Alam, Phys. Rev. E **89**, 062201 (2014)
[41] F. Vega Reyes, A. Santos, Phys. Fluids **27**, 113301 (2015)
[42] F. Vega Reyes, A. Santos, V. Garzó, J. Stat. Mech. **P07005** (2011)
[43] A. Santos, G.M. Kremer, M. dos Santos, Phys. Fluids **23**, 030604 (2011)
[44] G.I. Bird, *Molecular Gas Dynamics and the Direct Simulation of Gas Flows* (Clarendon, Oxford, 1994)
[45] B.D. Lubachevsky, J. Comp. Phys. **94**, 255 (1991)

Micro-mechanical investigation of the effect of fine content on mechanical behavior of gap graded granular materials using DEM

Habib Taha[1,2,⋆], *Ngoc-Son* Nguyen[1,⋆⋆], *Didier* Marot[1], *Abbas* Hijazi[2], and *Khalil* Abou-Saleh[2]

[1] *GeM Institute, University of Nantes, 58 rue Michel Ange, BP 420, 44606, Saint-Nazaire Cedex, France*
[2] *MPLAB-Multisciplinary Physics Laboratory, Faculty of Sciences, Lebanese University, Hadat-Baabda, Lebanon*

Abstract. In this paper, we present a micro-mechanical study of the effect of fine content on the behavior of gap graded granular samples by using numerical simulations performed with the Discrete Element Method. Different samples with fine content varied from 0% to 30% are simulated. The role of fine content in reinforcing the granular skeleton and in supporting the external deviatoric stress is then brought into the light.

1 Introduction

Granular materials are often used for construction of hydraulic structures such as dikes, levees, dams, etc. Under the action of the fluid flow, gap-graded or widely graded granular materials with an upwardly concave distribution are susceptible to internal erosion, during which fine particles can be detached and transported with the fluid flow through the pore space between coarse particles [1]. The loss of fine particles can impact the mechanical behavior of eroded materials, and then the stability of hydraulic structures. To predict this impact, it is important to deeply understand the effect of the fine content on the mechanical behavior of granular materials. The effect of fine content on the behavior of silty sands has been experimentally investigated by performing triaxial tests under drained conditions [2, 3] or undrained conditions [3, 4]. It was observed that fine content greatly influences the shear strength and the dilatancy of silty sands. Under undrained conditions, Thevanayagam et al. [4] found a threshold value $f_{c,th}$ for fine content f_c. When $f_c < f_{c,th}$, the collapse potential increases with an increase in fine content. The trend is opposite when $f_c > f_{c,th}$. From a micro-mechanical point of view, the dependence of macroscopic properties of widely graded materials on fine content can be attributed to a modification of their micro-structure. According to Thevanayagam et al. [4], for fine content $f_c < f_{c,th}$, contacts between coarse grains play the primary role in the micro-structure, while for $f_c > f_{c,th}$, contacts between fine grains are dominant. However, this is just a conjecture without any experimental verification. It is not clear yet how fine content reinforces granular skeleton and participates in supporting the external loading. It is worth mentioning that it is difficult to investigate experimentally the granular micro-structure. To overcome this difficulty, numerical simulation of granular samples using the Discrete Element Method (DEM) can be performed to investigate the granular micro-structure.

In this paper, we use the DEM to simulate triaxial loading tests on gap graded granular samples with different fine contents. Firstly, simulated granular samples are presented and their macroscopic behavior is analyzed. The effect of fine content on the micro-structure is then investigated.

2 Simulated samples

YADE, an open-source framework for the DEM [5], is used for numerical simulations. Particles are spherical and the particle size distribution follows the idealized gap-graded curve shown in Figure 1. Each sample is composed of a coarse content and a fine content f_c which varies from 0% to 30% with a step of 5%. The gap ratio $G_r = D_{min}/d_{max}$ is defined as the ratio of the minimum diameter for the coarse content to the maximum diameter for the fine content. According to Chang and Zhang [6], internal instability due to seepage flow for a gap graded material might occur for a gap ratio $G_r \geq 3.0$. In our simulations, the gap ratio $G_r = 3$ is chosen. The number of coarse particles is chosen depending on fine content f_c such that the computation time is reasonable and the representativity of simulated samples is guaranteed.

The interaction between two particles i and j is described by two linear springs with stiffnesses k_n and k_s in the normal and tangential directions, respectively. The normal stiffness k_n is defined as $k_n = (k_i k_j)/(k_i + k_j)$ with k_i is the stiffness of particle i. In our simulations, we chose $k_i/d_i = 250$ MPa (d_i is the diameter of particle i) and $k_s = 0.5k_n$. The tangential interaction respects Coulomb's friction law with friction angle $\varphi = 35^o$.

Particles are first generated into a cube composed of six rigid walls. At this stage, each particle diameter is reduced by a factor of 2.0. Particles are then progressively

⋆e-mail: habib.taha@etu.univ-nantes.fr
⋆⋆e-mail: ngocson.nguyen@univ-nantes.fr

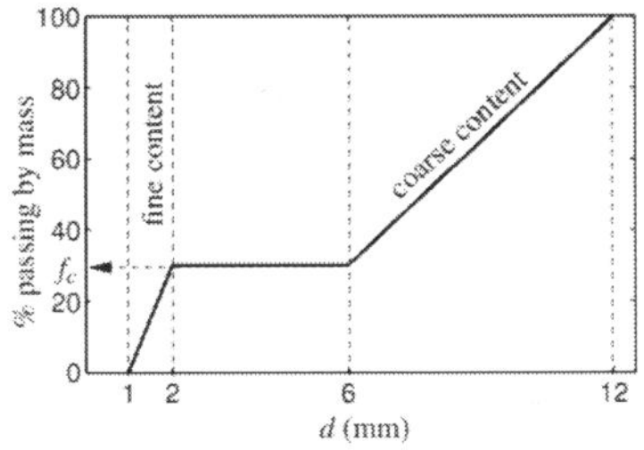

Figure 1. Gap-graded size distribution curve for the simulated samples.

expanded to reach the target size distribution. After that, the box dimensions are slowly reduced until the stresses σ_i, $i = 1, 2, 3$ reach the confinement stress of 100 kPa. During the compaction process, the friction between particles is set to 0 to obtain dense samples. At the end of this process, the friction is reset to its original value.

The density of a fines-coarses mixture can be described by the classical void ratio, e, and the *apparent* intergranular and interfine void ratios, e_c and e_f defined in Equation (1). The intergranular void ratio e_c is defined by considering that the fine grains are inactive so they are a part of voids between coarse grains. The interfine void ratio e_f is defined as the ratio of the effective volume of interfine voids to the solid volume of fine grains.

$$e_c = \frac{e + f_c/100}{1 - f_c/100}, \quad e_f = \frac{e}{f_c/100} \tag{1}$$

Figure 2 shows the void ratios e, e_c and e_f versus fine content f_c: the values of e and e_c are represented in the left axis, while the values of e_f is represented in the right axis. Fine grains fill voids between coarse grains so the classical void ratio e decreases with f_c. Lade et al. [7] experimentally found that the void ratio of sands decreases with fine content when $f_c < 40\%$. We can also see that the intergranular void ratio e_c increases with f_c since fine grains tend to intercalate between coarse grains. On the other hand, the fine fraction becomes denser with increasing fine content.

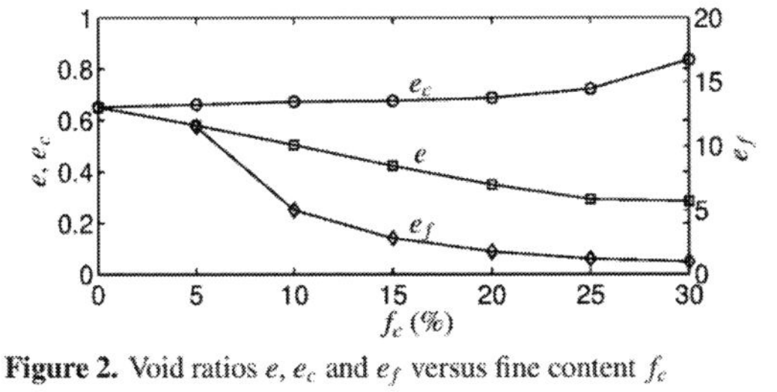

Figure 2. Void ratios e, e_c and e_f versus fine content f_c

3 Triaxial compression tests

Triaxial compression tests are performed on numerical samples. Figure 3 shows the stress ratio q/p (deviator stress $q = \sigma_1 - \sigma_3$ and mean stress $p = (\sigma_1 + \sigma_2 + \sigma_3)/3$ and volumetric strain ε_v versus axial strain ε_{11} for different fine contents. Stress ratio q/p at the peak and critical states are shown in Table 1. It is shown that the effect of fine content on the macroscopic behavior is not significant for $f_c < 20\%$. However, for $f_c \geq 20\%$, the shear strength and the dilatancy increases with f_c and we can see a marked softening behavior .

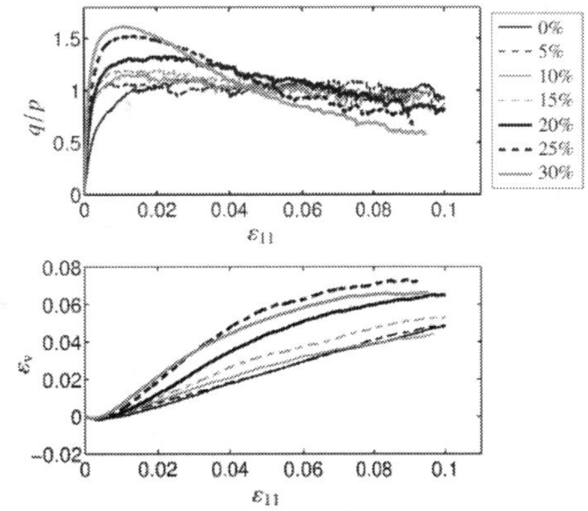

Figure 3. Stress ratio q/p and volumetric strain ε_v versus axial strain ε_{11} for different fine contents f_c.

Table 1. Stress ratios $(q/p)_{\text{peak}}$ and $(q/p)_{\text{critical}}$ at the peak and critical states for different fine contents f_c.

f_c (%)	0	5	10	15	20	25	30
$(q/p)_{\text{peak}}$	1.1	1.1	1.2	1.2	1.3	1.5	1.6
$(q/p)_{\text{critical}}$	0.9	0.9	1.0	0.9	0.8	0.7	0.6

4 Microscopic investigation

Contacts between particles in a gap graded sample can be split into three categories: coarse-coarse ($C-C$), coarse-fine ($C-F$) and fine-fine ($F-F$) contacts. In the following, we analyze the effect of fine content on coordination numbers defined for each category of contacts, on the anisotropy and on the stress transmission.

4.1 Coordination numbers

The coordination number, which is defined as the average number of contacts per particle, is often used to describe the density of a granular sample at the micro-scale. This definition seems to be not appropriate for a gap-graded sample for which the number of contacts on each coarse grain might be much higher than on each fine grain. For such a system, a coordination number for coarse fraction should be separated from the one defined for fine fraction as suggested in [8]. In this study, we define three coordination numbers $\mathcal{N}_C^{C-C}$, $\mathcal{N}_C^{C-F}$ and $\mathcal{N}_F$ which correspond to the respective average numbers of $C-C$ contacts per coarse particle, of $C-F$ contacts per coarse particle and of contacts (both $C-F$ and $F-F$) per fine particle.

Figure 4 shows $\mathcal{N}_C^{C-C}$, $\mathcal{N}_C^{C-F}$ and $\mathcal{N}_F$ versus fine content f_c at the initial, peak and critical states. It can be seen that for $f_c < 20\%$, coarse particles constitute the primary skeleton with N_C^{C-C} dominant with respect to N_C^{C-F} and N_F. This is one of the main reasons which explain why the macroscopic behavior is not significantly affected by fine content $f_c < 20\%$ as shown in Figure 3. For $f_c \geq 20\%$, fine particles come into contact with coarse particles, leading to a quick increase of N_C^{C-F} with f_c. For the sample with $f_c = 30\%$ at the initial state, there is on average 45 fine particles around a coarse particle. Shire et al. [9] obtained a similar result for their simulations with higher gap ratio G_r. Note that N_C^{C-C} at the initial state is not affected by $f_c \leq 30\%$, meaning that the initial primary skeleton formed by coarse particles is not altered by the fine fraction: fine particles actually surround and reinforce this primary skeleton.

The reinforcement of the primary skeleton by fine particles gives the ability to granular samples to dilate better so they can resist better the shear loading. On the other hand, the dilatancy leads to a loss of inter-particle contacts, and hence a reduction of the coordination numbers. For the sample with $f_c = 30$ %, N_C^{C-C} at the peak state becomes lowest among various values of f_c. However, N_C^{C-F} remains still highest (about 35 fine particles around a coarse particle), which compensates for the reduction of N_C^{C-C} and allows this sample to have the best shear strength. After the peak state, N_C^{C-F} reduces drastically. The destruction of the primary skeleton constituted by coarse particles and the drastic reduction in the reinforcement of the primary skeleton by fine particles explain why the sample with $f_c = 30$ % shows a marked softening behavior (Figure 3).

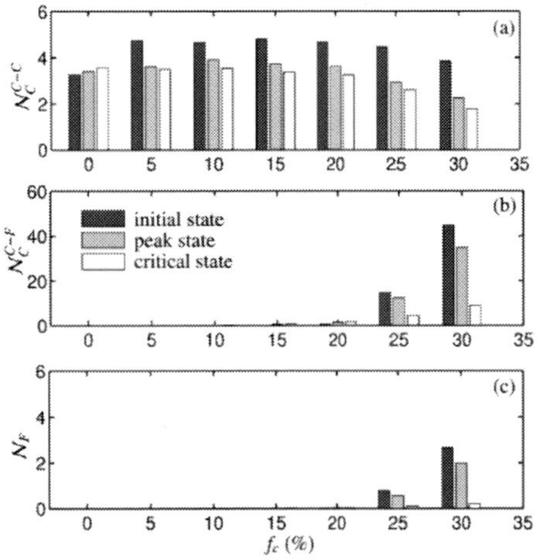

Figure 4. Coordination numbers (a) N_C^{C-C}, (b) N_C^{C-F} and (c) N_F versus fine content f_c at the initial, peak and critical states.

4.2 Anisotropy

One of the particular features of a granular medium is its capability of orienting preferentially its contacts between particles in the major principal stress direction to resist the external deviatoric stress. This feature is called *the induced anisotropy*, which can be described in terms of the distribution of contact orientation by the so-called *fabric tensor* $\boldsymbol{H}$. For a gap graded sample for which the particle size is very different from one to another, the particle size should be taken into account in the fabric tensor $\boldsymbol{H}$:

$$H_{ij} = \sum_k l^k n_i^k n_j^k \quad (2)$$

where superscript k runs over all contacts of the sample; l^k is the length of the branch vector joining the centers of two particles in contact k and $\boldsymbol{n}^k$ is the unit normal vector at contact k. For a triaxial loading, $\boldsymbol{H}$ is approximately diagonal and the two transversal components H_{22} and H_{33} are almost equal. In this case, the anisotropy can be quantified by the *anisotropy index* $H_d = (H_{11} - H_{33})/\text{Trace}(\boldsymbol{H})$.

Figure 5 shows the evolution of the anisotropy index H_d during the triaxial loading for different fine contents f_c. The effect of f_c on the evolution of H_d is not significant for $f_c < 20\%$ but it appears clearly for $f_c \geq 20\%$. At the peak state, the anisotropy index H_d is much higher for $f_c = 20\%$ than for $f_c < 20\%$. Beyond $f_c = 20\%$, the anisotropy at the peak state decreases with f_c although the shear strength increases with f_c as shown in Figure 3. This is quite surprising. However it can be explained by the fact that at a high fine content, the primary skeleton formed by coarse particles is so well reinforced by fine particles that the granular structure can resist the prescribed loading without resort to induce greatly the anisotropy. It is interesting to note that the softening behavior observed for high fine content (Figure 3) is accompanied by a clear decrease of the anisotropy index H_d after the peak state.

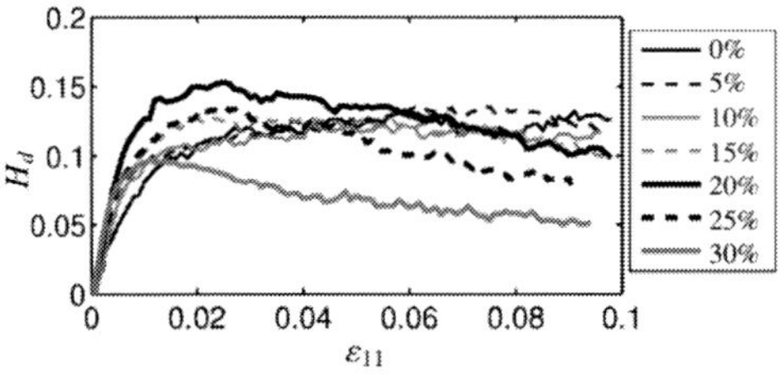

Figure 5. Anisotropy index H_d versus axial strain ε_{11} with different fine contents f_c.

4.3 Stress

Stress applied on the boundary of a granular sample is shared by inter-granular contacts. For a gap-graded sample, three categories of contacts are clearly distinguished. A question that arises here is the participation of each category of contacts in supporting the applied external stress. To answer to this question, we first link the macro-stress tensor $\boldsymbol{\sigma}$ to local information at contacts using the well-known averaging static operator [10]:

$$\sigma_{ij} = \frac{1}{V} \sum_k f_i^k l_j^k, \quad (3)$$

where V is the total volume of the sample; $\boldsymbol{f}^k$ et $\boldsymbol{l}^k$ are contact force vector and branch vector joining two centers of particles in contact. We then decompose stress tensor σ into three parts σ^{C-C}, σ^{C-F} and σ^{F-F} which are the contribution of the respective categories of $C-C$, $C-F$ and $F-F$ contacts. Each part is defined by Equation (3) with superscript k running over the corresponding set of contacts. By doing so, the macroscopic deviator stress q can be decomposed as $q = q^{C-C} + q^{C-F} + q^{F-F}$. The contribution of each category of contacts to q can be calculated from the responding stress tensor, for example the contribution of $C-C$ contacts $q^{C-C} = \sigma_1^{C-C} - \sigma_3^{C-C}$.

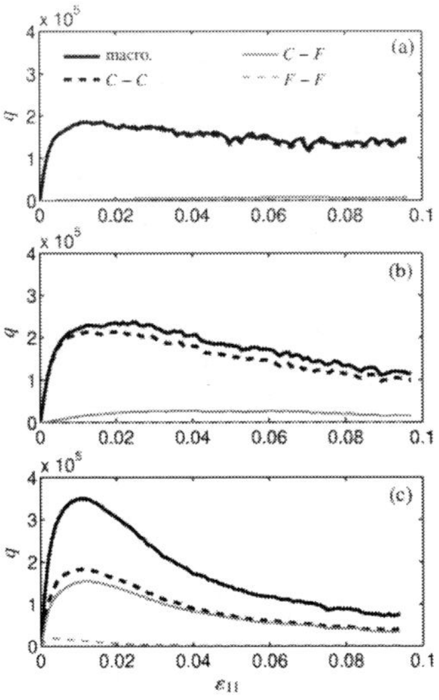

Figure 6. Contribution of each category of contacts $C-C$, $C-F$ and $F-F$ to the deviator stress q for different fine contents: a) $f_c = 10\%$, b) $f_c = 20\%$ and c) $f_c = 30\%$.

Figure 6 shows the macroscopic deviator stress q and its decomposition q^{C-C}, q^{C-F} and q^{F-F} versus axial strain ε_{11} for three fine contents $f_c = 10\%$, 20% and 30%. It can be seen that, for $f_c = 10\%$ and 20%, the applied deviatoric stress is essentially supported by contacts between coarse particles ($C-C$). However, for $f_c = 30\%$, contacts between fine and coarse particles ($C-F$) participate actively in supporting the applied deviatoric stress. For this high value of fine content, the contribution of $C-F$ contacts is almost equal to that of $C-C$ contacts. The contribution of contacts between fine particles ($F-F$) is small for the studied range of f_c. It is interesting to note that q^{C-C} is almost the same whatever the fine content. Combining with the results presented in Section 4.1, we can deduce that the skeleton formed by coarse particles remains almost unchanged and its capability of supporting the external deviatoric stress remains unchanged with increasing fine content. Fine particles which surround coarse particles do not make coarse skeleton stronger. However, $C-F$ contacts take part in supporting the applied deviatoric stress. As shown in Figure 4, as fine content increases, there are more and more fine particles in contact with a coarse particle. As a result, the part of the deviatoric stress supported by $C-F$ contacts increase. This is the reason why the shear strength at high fine content is significantly higher than at low fine content as shown in Figure 3.

5 Conclusions

Despite an idealization of granular samples considered in the numerical simulation, this study has brought several insights into the micro-structure of gap-graded samples. For fine content $f_c < 20\%$, the skeleton formed by coarse particles is dominant and carries essentially the applied deviatoric stress. As a consequence, fine content does not have a significant effect on the macroscopic behavior. However, for $f_c \geq 20\%$, fine particles come into contact with coarse particles, reinforce the granular skeleton and take part in supporting the external stress, which leads to a higher shear strength. This influence of fine content on soil mechanical behavior highlights the necessity to consider the impact of suffusion process which mobilizes the fine particles and may induce instabilities on hydraulic earth structures.

References

[1] Y.S. Sail, D. Marot, L. Sibille, A. Alexis, Eur. J. Environ. Civ. Eng. **15**, 799 (2011)
[2] R. Salgado, P. Bandini, A. Karim, J. Geotech. Geoenviron. Eng., ASCE **126**, 451 (2000)
[3] T.K. Nguyen, N. Benahmed, P.Y. Hicher, M. Nicolas, in *Bifurcation and Degradation of Geomaterials in the New Millennium*, edited by K.T. Chau, J. Zhao (2015), pp. 113–120
[4] S. Thevanayagam, T. Shenthan, S. Mohan, J. Liang, J. Geotech. Geoenviron. Eng., ASCE **128**, 849 (2002)
[5] Šmilauer et al., The Yade Project. DOI 10.5281/zenodo.34073 (http://yade-dem.org/doc/) (2015)
[6] D. Chang, L. Zhang, Soils Found. **53**, 569 (2013)
[7] P. Lade, C. Liggio, J. Yamamuro, Geotech. Test. J. **21**, 336 (1998)
[8] N. Minh, Y. Cheng, Géotechnique **63**, 44 (2013)
[9] T. Shire, C. O'Sullivan, K. Hanley, in *Geomechanics from Micro to Macro*, edited by K. Soga, K. Kumar, G. Biscontin, M. Kuo (2014), pp. 231–236
[10] J. Christoffersen, M. Mehrabadi, S. Nemat-Nasser, J. Appl. Mech. **48**, 339 (1981)

Granular compaction by fluidization

Alexis Tariot, *Georges* Gauthier[1,*], and *Philippe* Gondret[1,**]

[1]*Laboratoire FAST, Univ Paris-Sud, CNRS, Université Paris-Saclay, Orsay, France*

Abstract. How to arrange a packing of spheres is a scientific question that aroused many fundamental works since a long time from Kepler's conjecture to Edward's theory (S. F. Edwards and R.B.S Oakeshott. *Theory of powders*. Physica A, **157**: 1080-1090, 1989), where the role traditionally played by the energy in statistical problems is replaced by the volume for athermal grains. We present experimental results on the compaction of a granular pile immersed in a viscous fluid when submited to a continuous or bursting upward flow. An initial fluidized bed leads to a well reproduced initial loose packing by the settling of grains when the high enough continuous upward flow is turned off. When the upward flow is then turned on again, we record the dynamical evolution of the bed packing. For a low enough continuous upward flow, below the critical velocity of fluidization, a slow compaction dynamics is observed. Strikingly, a slow compaction can be also observed in the case of "fluidization taps" with bursts of fluid velocity higher than the critical fluidization velocity. The different compaction dynamics is discussed when varying the different control parameters of these "fluidization taps".

1 Introduction

Since Edwards' theory on piles of athermal spheres [1], several ways to compact a dry granular pile have been tried and studied using vibrations [2], taps [3–5], shear [4, 6] or heating cycles [7, 8]. Shearing appears the more efficient way to compact [6] with a global solid volume fraction $\phi \simeq 0.66$ reached after about 10^4 cycles which is significantly higher than the random close packing $\phi_{RCP} \simeq 0.64$. To obtain a similar high volume fraction by tapping, one needs a two-step process [3], with first hard taps allowing to reach irreversibly $\phi \simeq 0.63$ and then by decreasing tap intensity to achieve a pile that reach reversibly $\phi \simeq 0.655$. The common feature of all these solicitations is that they originate at wall boundaries and propagate into the bulk of the packing through the grain contacts. A more efficient way of compaction should be a global sollicitation. It is well known that applying an upward viscous flow to a granular pile will generate a global decompaction of the granular bed that ultimately leads to a fluidized bed. Schröter *et al.*[9] showed for instance that cycles of fluidization/sedimentation of a collection of non-Brownian spheres lead to a higher solid fraction of the granular bed at rest for lower fluidization velocities, extending the works of Onada and Liniger[10].

In the present study we extend the work of Schröter *et al.* [9] to short enough upward flow pulses to avoid the final expansion (fluidization) of the granular bed but also to low enough upward continuous flow, below the critical velocity for fluidization. In both cases, an efficient compaction is observed.

*e-mail: georges.gauthier@u-psud.fr
**e-mail: philippe.gondret@u-psud.fr

2 Experimental set-up and procedure

2.1 Apparatus

The experimental set-up is sketched in Fig. 1. The suspen-

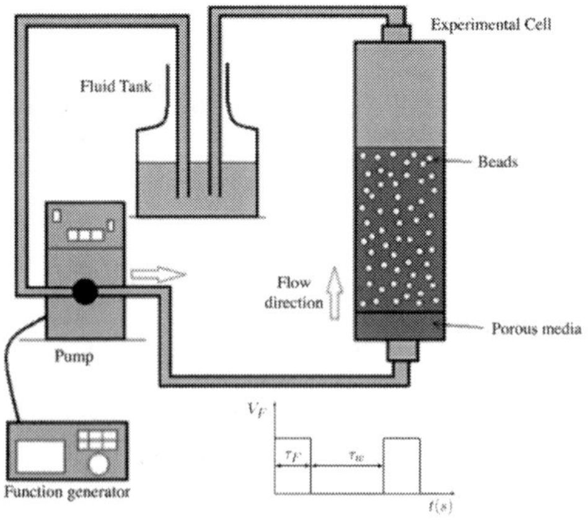

Figure 1. Sketch of the experimental set-up.

sion is contained in a vertical tank of 50 cm high and of rectangular cross-section $S = 8 \times 2\,\text{cm}^2$. The suspension consists in glass spheres of radius $a = 1$ mm and density $\rho_p = 2500\,\text{kg/m}^3$ dispersed in a water-glycerol mixture

(90% in mass of glycerol) of density $\rho = 1235\,\mathrm{kg/m^3}$ and viscosity $\eta = 156 \cdot 10^{-3}$ Pa · s at 24 °C. In these conditions, the typical sedimentation velocity for an isolated sphere is the particle Stokes velocity $U_0 = 2(\rho_p - \rho)ga^2/9\eta = 16.4 \cdot 10^{-3}$ m/s corresponding to a particle Reynolds number $Re_p = \rho U_0 a/\eta \simeq 0.13$ smaller than 1. For the whole set of experiments, the mass M of particles has been kept constant with the value $M_b = 360$ g corresponding to $N = 69590$ particles. An upward volume flow rate Q injected at the bottom of the tank is fixed using a gear pump without any significant noise and without any significant dependance to pressure loss. However to prevent flow rate variation and achieve an uniform flow in the experimental cell, the fluid is injected through a 2 cm high porous media located at the bottom of the cell and which is fitted with the rectangular section of the cell. The porous media opening is $50\,\mu$m with a permeability that is one order of magnitude lower than the permeability of the granular bed and which thus controls the flow. The pump is controlled with a wave form generator which allows the generation of any flow sequence from continuous flow of upward velocity $U_F = Q/S$ to periodic flow bursts with a given upward velocity U_F during the time τ_F separated by the waiting time τ_w with thus the period $T = \tau_F + \tau_w$. In this bursting regime, the mean upward velocity $\overline{U}_F = U_F\tau_F/T$ is thus smaller than the burst velocity U_F and the fluid displacement $U_F\tau_F$ have to be compared to the grain size a. The experimental tank is lightened from behind and images of the cell are taken at a rate of 4 images per second, thanks to a camera located 1 meter from the cell and with its axis perpendicular to the largest cell plates. The solid volume fraction ϕ of the granular bed is measured from the bed height H as $\phi = M_p/\rho_p S H$ through image analysis : From a sequence of individual images, a spatio-temporal diagram of the time evolution of the bed is built which leads to $H(t)$ and $\phi(t)$. The relative precision in the ϕ measurement corresponds to the relative precision in the H measurement which is about 10^{-4} for $H \simeq 30$ cm here.

In a fluidized bed the solid fraction ϕ is imposed by the upward flow of average velocity U_F according to the Richardson-Zaki (RZ) law[11] which reads here $\phi = 1 - (U_F/U_s)^{1/n}$[12]. In our experiments, the evolution of ϕ as a function of U_F/U_s is shown in Figure 2. Our experimental data follows fairly well the RZ law with the exponent value $n = 4.3$ and an experimental Stokes velocity $U_s = 16 \cdot 10^{-3}$ m/s very close to the value expected for a perfectly monodisperse suspension of spheres.

2.2 Procedure

Each granular pile is prepared following the same procedure. A fluidized suspension of solid fraction $\phi = 0.22$ is first obtained by imposing a constant upward velocity $U_F/U_S = 0.359$. This initial upward flow is then switched off so that the suspension settles and achieve a very reproducible granular bed of solid fraction $\phi_0 = 0.605 \pm 0.005$. Note that this ϕ_0 value may depend on some parameters such as the Stokes and particles Reynolds numbers as well as the friction coefficient of the spheres[10, 13]. We define the critical velocity for fluidization U_c as the one that

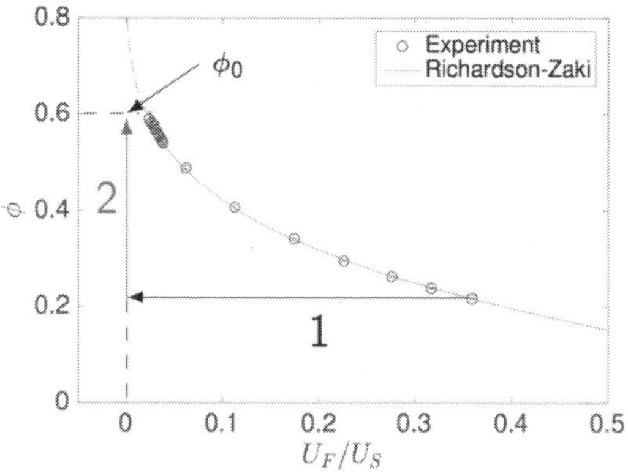

Figure 2. Solid volume fraction ϕ of fluidized bed as a function of the normalized upward flow velocity U_F/U_s. (∘) Experimental data and (—) fit of equation $\phi = 1-(U_F/U_s)^{1/n}$ with $n = 4.3$ from RZ law.

would lead from the RZ law to the initial bed solid fraction ϕ_0 so that $U_c/U_s = (1 - 0.605)^{4.3} \simeq 0.018$. For $U_F > U_c$, the granular bed packing expands to form a fluidized suspension. It has worth noting that the RZ law is for a suspension only granular pile, and thus cannot be extrapolated below the minimum of fluidization. The corresponding curve part of the RZ law are thus shown not in continuous line but in dashed line in Fig. 2. The experimental results present in the following have been obtained with continuous flows of upward velocity $U_F < U_c$ and with burst flows of velocity $U_F < U_c$ and $U_F > U_c$ but $\overline{U}_F < U_c$.

3 Results and Discussion

3.1 Continuous flow

Starting from the granular bed packing of initial volume fraction $\phi_0 = 0.605$, an upward flow of velocity $U_F < U_c$ is now applied and the image processing allows to follow the time evolution of the solid fraction of the packing $\phi(t)$ as reported in Figure 3a for $U_F/U_s = 0.014$. We observe a significant increase of $\phi(t)$ that arises from successive local rearrangements. This increase is first fast at short time from $\phi_0 = 0.605$ to typically $\phi \simeq 0.62$ within about the first 10 s, and then slower as 3 days are required for a subsequent increase up to $\phi \simeq 0.625$. This behavior has to be compared with the one observed by [9] in which the pile reaches a stationary volume fraction after the first flow pulse for $U_F/U_S \approx 0.014$ and after 80 minutes for pulses of lower flow velocity $U_F/U_S \approx 0.0125$. As ϕ_0 depends on the detailed experimental conditions and considering the time evolution observed in Fig. 3a, $\phi - \phi_0$ is now reported in the log-lin plot of Fig. 3b as a function the time t/τ_s normalized with the Stokes time $\tau_S = a/U_s = 0.06$ s which corresponds to the time that would be necessary for an isolated grain to settle along its own radius a. For $\ln(t/\tau_S) \gtrsim 5$, which corresponds to $t \gtrsim 10$ s the data are rather well fitted by the logarithmic

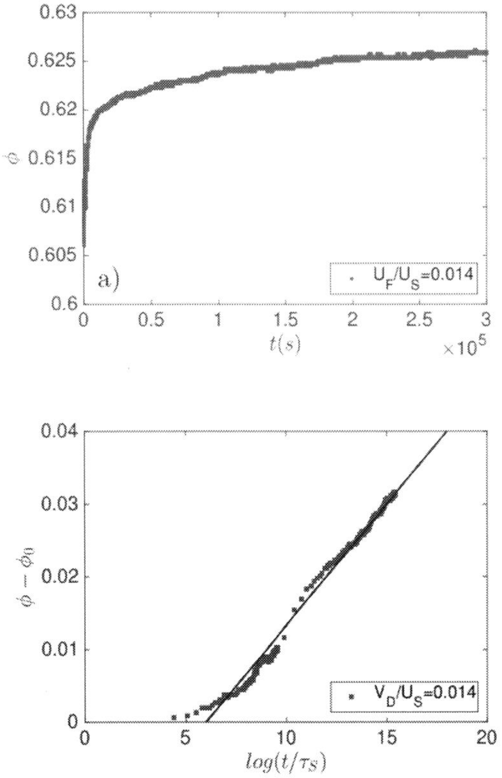

Figure 3. (a) Solid fraction ϕ of the granular bed as a function of time t for $U_F/U_S = 0.014$. (b) Solid fraction deviation from the initial value $\phi - \phi_0$ as a function of the normalized time t/τ_s. ( ) Experimental data and (—) fit of equation $\phi - \phi_0 = \alpha \log(1 + t/\tau)$ with $\tau/\tau_s \simeq 240$ and $\alpha \simeq 0.0022$.

evolution $\phi - \phi_0 = \alpha \ln(1 + t/\tau)$ with the characteristic time $\tau \simeq 15\ s$ and $\alpha \simeq 0.0022$.

By performing experiments at other U_F values, we observed that the characteristic time τ decreases for increasing flow velocity U_F as reported in Fig. 4. Note that it was not possible to perform experiments for $U_F/U_S \lesssim 0.008$ as the pump does not deliver a continuous flow anymore at such low flow rates. The data of Fig. 4 may be fitted by a linear decrease of equity that lead to $\tau = 0$ for $U_F/U_s = 0.0175$. As expected, this value corresponds very well to the critical fluidization velocity $U_c/U_s \simeq 0.018$ defined previously. Note that very close to U_c, we observe an intermittent regime, with successive phases of compaction and decompaction, as already reported in other situations such as in segregation/mixing of bidisperse fluidized suspension [14] or in the avalanche granular flows [15].

3.2 Flow taps

The effect of flow bursts on the solid fraction ϕ of the packing has been studied by varying the burst velocity U_F and

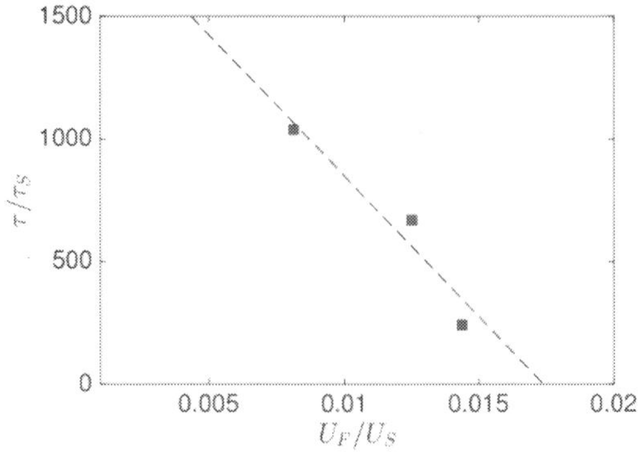

Figure 4. Normalized haracteristic time of compaction τ/τ_s as a function of the normalized upward flow velocity U_F/U_S in the continuous case.

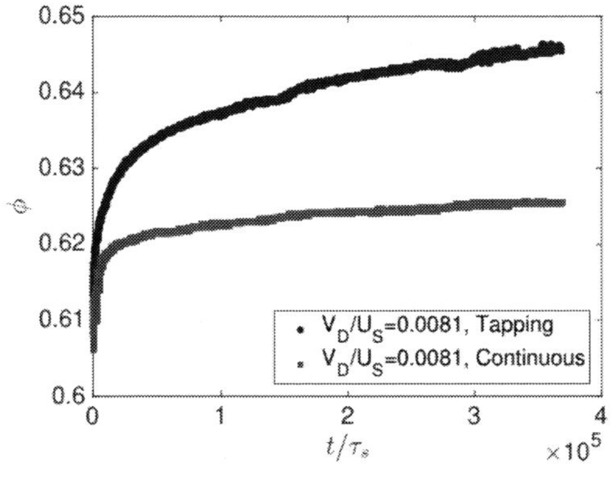

Figure 5. Time evolution of the solid fraction ϕ of the packing for the upward flow velocity $U_F/U_s = 0.0081$ in the (—) burst case and (—) continuous case.

the burst time τ_F but keeping constant the fluid displacement $U_F\tau_F = 0.54$ mm of the order of half the bead radius. The waiting time τ_w between two successive bursts has been kept constant at the value $\tau_w = 12\,s \gg \tau_s$ high enough to allow the subsequent completed sedimentation of the grains that may have been displaced during the burst flow.

Figure 5 displays the evolution of the solid fraction ϕ of the packing as a function of the normalized time t/τ_S for burst flow together with the continuous flow at the same flow velocity $U_F/U_s = 0.0081$ smaller than the critical fluidization U_c. One observe that such a burst flow lead also to a compaction and that the process is surprisingly more efficient than the continuous flow. For instance, at $t/\tau_S \simeq 1.3\,10^5$, $\phi \simeq 0.64$ corresponding to ϕ_{RCP} for the burst flow whereas $\phi \simeq 0.62$ in the continuous flow. Note that the solid fraction of the packing may be larger than the

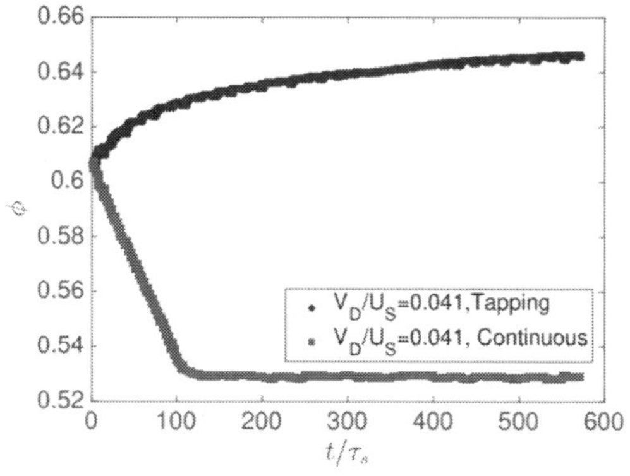

Figure 6. Time evolution of the solid fraction ϕ of the packing for the upward flow U_F/U_s = 0.041 in the (—) burst case and (—) continuous case.

RCP value flow and we observe due some crystallization initiates at the lateral walls before going further in the bulk. Figure 6 displays now $\phi(t)$ for burst flow together with the continuous flow at the same flow velocity $U_F/U_s = 0.041$ now larger smaller than the critical fluidization U_c. In such a case, $\phi(t)$ decreases linearly in the continuous flow before reaching the lower constant value $\phi \simeq 0.53$ as expected from the fluidization process and the corresponding RZ law. But the burst flow still leads to a surprising compaction, which we observed faster for increasing burst flow velocity U_f. Indeed ϕ_{RCP} is reached at $t/\tau_s \simeq 1.3\times10^5$ for $U_F/U_s = 0.008$ but only $t/\tau_s \simeq 2.5\times10^2$ are necessary for $U_F/U_s = 0.041$.

4 Conclusion

We have studied the compaction dynamics of a granular pile when submitted to weak upward viscous flow of velocity U_F. In the case of a continuous flow, the solid fraction ϕ is observed to increase when the flow velocity is below the critical velocity for fluidization U_c, with a logarithmic time evolution controlled by a characteristic time τ decreasing for increasing U_F. In the case of flow bursts, compaction is surprisingly more efficient than in the continuous flow and is also strikingly observed even when the burst flow velocity is larger than the critical velocity ($U_F > U_c$).

Acknowledgements

The authors want to thank A. Aubertin, L. Auffray and R. Pidoux for their help in the design and realization of the experimental set-up, and M. Schröter for his stimulating discussion.

References

[1] S. Edwards, R. Oakeshott, *The theory of powders* (1988)
[2] E.R. Nowak, J.B. Knight, E. Ben-Naim, H.M. Jaeger, S.R. Nagel, Phys. Rev. E **57**, 1971 (1998)
[3] P. Philippe, D. Bideau, EPL (Europhysics Letters) **60**, 677 (2002)
[4] P. Richard, M. Nicodemi, R. Delannay, P. Ribiere, D. Bideau, Nat Mater **4**, 121 (2005)
[5] J.A. Dijksman, M. van Hecke, EPL (Europhysics Letters) **88**, 44001 (2009)
[6] O. Pouliquen, M. Belzons, M. Nicolas, Phys. Rev. Lett. **91**, 014301 (2003)
[7] K. Chen, J. Cole, C. Conger, J. Draskovic, M. Lohr, K. Kelin, T. Scheidemantel, P. Schiffer, Nature **442**, 257 (2006)
[8] T. Divoux, H. Gayvallet, J.C. Géminard, Phys. Rev. Lett. **101**, 148303 (2008)
[9] M. Schröter, D.I. Goldman, H.L. Swinney, Phys. Rev. E **71**, 030301 (2005)
[10] G. Onada, E. Liniger, Phys. Rev. Lett. **64**, 2727 (1990)
[11] J. Richardson, W. Zaki, Trans. Instn. Chem. Engrs. **32** (1954)
[12] N.R. J. Martin, D. Salin, Phys. Fluids **7**, 2510 (1995)
[13] G.R. Farrell, K.M. Martini, N. Menon, Soft Matter **6**, 2925 (2010)
[14] A. Deboeuf, G. Gauthier, J. Martin, D. Salin, New Journal of Physics **13**, 075005 (2011)
[15] R. Fischer, P. Gondret, M. Rabaud, Phys. Rev. Lett. **103**, 128002 (2009)

The sound of avalanches: from a global to a local perspective.

Florine Dubourg[1,†], *Sébastien* Lherminier[1,†], *Ramon* Planet[2], *Kenny* Rapina[1], *Félix* Bunel[1], *Loïc* Vanel[1], and *Osvanny* Ramos[1,⋆]

[1] *Univ Lyon, Université Claude Bernard Lyon 1, CNRS, Institut Lumière Matière, F-69622, Villeurbanne, France*
[2] *Departament de Física de la Matèria Condensada, Universitat de Barcelona, Martí i Franquès 1, E-08028 Barcelona, Spain*

Abstract. We present an experimental study aiming to understand the local origin of the acoustic bursts recorded globally during a two-dimensional shear experiment of a granular medium. In order to explain the power-law distribution of energy of these events (also called avalanches), the analysis focuses on their acoustic frequencies. Experiments with single grains have also been performed, trying to separate the contributions related to collision between neighbours from shear movements between grains. Preliminary results indicate that both frictional sliding and collision-like mechanisms are involved in the origin of the acoustic bursts. The evolution of the structure of the medium around a large acoustic burst is analysed both with high resolution and high-speed imaging. The energy of the acoustic bursts seems related to single and well-localised events with very large variations in energy. However, these local reorganisation processes may eventually bring the global relaxation of the system, which is also captured by a force drop in the resistance to shear.

1 Introduction

Shearing compressed granular materials has often been used as a simplified model of a tectonic fault. In most real [1–3] or numerical experiments [4–7], two parallel plates (mimicking tectonic ones) compress the granular system. As the plates move laterally in relation to each other and at very low speeds, shear stresses build up on the packed grains, and eventually they are liberated through sudden events associated to the reorganisation of the pack. These events –often called avalanches– may be detected from sudden jumps in the global shear force experienced by the plates [2, 8], from acoustics bursts [1, 9] or even from direct visualisation in the case of photoelastic grains [2, 3]. One weakness of most shear experiments, that limits the comparison with real earthquakes, is their lack of statistics. However, we have recently built a shear experiment that, thanks to the implementation of *periodic boundary conditions*, is able to reach significant statistics of more than 10^6 events [9]. The energies of our "avalanches", measured as acoustic bursts, follow a five-decades power law with an exponent of 1.72, which is close to the one of the Gutenberg-Richter law, equal to $5/3 = 1.66$ [10]. Other earthquake laws have also been reproduced. However, many new questions have arisen related to the local origin of the recorded acoustic events.

Are those events the result of "snow-ball" or "domino" effects where many different parts of the system are reached by the event and become part of it? In this case the term avalanche is appropriate. Or instead are the huge variations in energy only associated to single and well-localised events? Like, for example, the breaking of single structures with a large distribution of sizes [11]. Another question focuses on the local mechanism of the globally recorded sound: is it a signature of a collision between grains, or does the frictional sliding between the grains' rough surfaces provoke it? At intermediate scales between the grains and the whole system, the force network plays an essential role in the stability of the granular packing [12–14]. What is the relation between their dynamics and the recorded acoustic bursts? In order to tackle these questions, we have built a simplified version of our experiment consisting of a *classical shear cell*. The experiments will focus on the details of one single avalanche.

2 Main Setup

The experimental setup consists of a monolayer of disks confined into a rectangular shear cell (figure 1) which has an initial size of 172×105 mm^2 ($W \times H$). In order to avoid 3D effects, the cell is confined between two acrylic plates separated by 4.2 mm. The surrounding walls are built in the same material as the grains. The bottom wall is fixed, while the upper and lateral ones are mobile, allowing the shearing of the granular medium. The upper plate is free to move vertically, compressing the medium. A dead load of 120 N is applied during the whole experiment. This plate is also connected to a force sensor (*Interface SML-50*), which is pulled at a constant speed of 100 μm/s. We use cylindrical grains of 4 mm thickness and two different diameters 6.4 and 7 mm (in equal proportion). These discs (the same as the ones used in [9])

[†] These authors contributed equally to this work
[⋆] e-mail: osvanny.ramos@univ-lyon1.fr

DOI: 10.1051/epjconf/201714003015

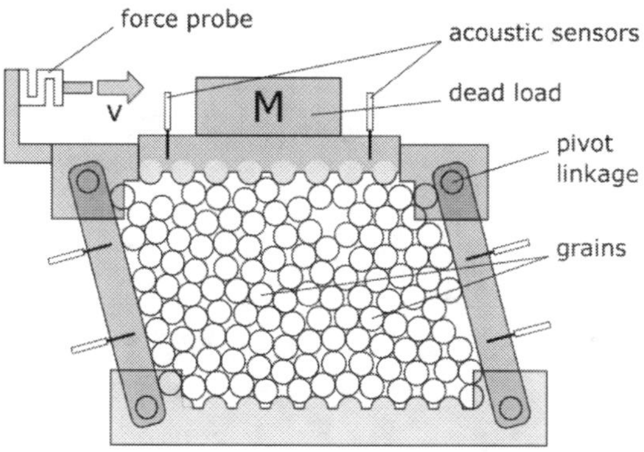

Figure 1. (colour online) Schema of the main experimental setup.

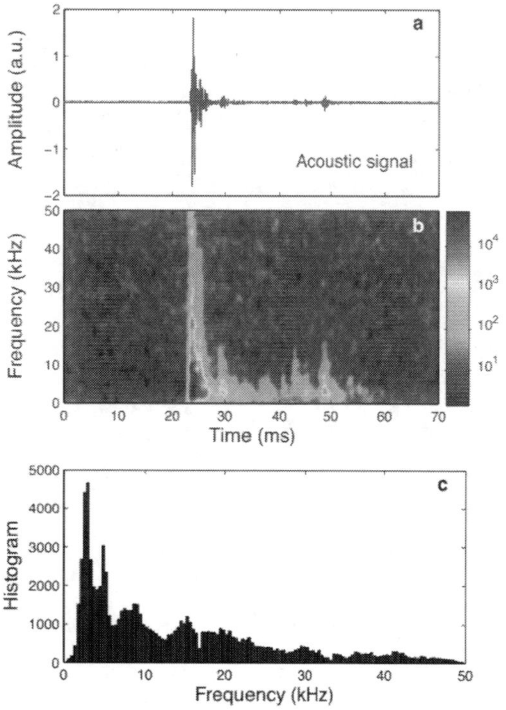

Figure 2. (colour online) **a.** Typical acoustic signal recorded during a shearing experiment, on a 70 ms time window. **b.** The associated spectrogram on 256 frequencies. **c.** Histogram of detected energy peaks on the spectrograms for a set of five different shearing experiments.

are made of *Durus White 430* and have been generated in a *Objet30* 3D–printer. Profile measurements of the surface roughness of the printed grains give a peak-to-peak maximum value of 25 μm on the curved side of the cylinders. The zero-frequency Young modulus of *Durus* material is $E \simeq 100$ MPa. The cell contains 384 grains placed in a disordered configuration, reached thanks to half grains of diameter $d = 6.4$ mm printed in the top and bottom plates with a specific gap of $\sqrt{2}d$ [15]. The translucent and photoelastic character of the grains allows the visualisation of the stress inside the discs when placing the experimental setup between two circular polarizers. We take images of the granular packing with a *Nikon D750* high resolution camera (6016 × 4016 pixels) at 1 fps during the whole experiment and with a *Phantom V711* fast camera (512 × 256 pixels), at 49,000 fps, triggered around one large acoustic event.

Acoustic emissions (AE) are recorded using 6 piezoelectric pinducers (*VP-1.5* from *CTS Valpey Corp.*) inserted in adjusted holes in the upper and lateral walls. Both the force and the acoustics are recorded with a *NI-USB-6366* card at 100 kHz during the whole shear movement.

3 Results

3.1 Acoustic analysis

A typical acoustic signal is shown in the figure 2a. The AE are analysed through a time-frequency method: we computed the spectrogram of the acoustic signal of each sensor. A spectrogram (figure 2b) is obtained by a sliding Fourier transform in a time window of 1.28 ms. It represents the energy of the signal for a set of frequencies linearly distributed between 0 and the Shannon frequency (50 kHz). By thresholding the logarithm of the intensity of the spectrogram (colour code), and finding the local maxima, it is possible to obtain the main frequencies emitted during the avalanches. The histogram of the frequencies emitted during five different shear experiments is presented in the figure 2c. We observe that a few frequencies are more represented, respectively around 2900 Hz, 4800 Hz et 8600 Hz.

3.2 Image analysis

The analysis of high-resolution images (figure 3a) allows the precise detection of the position of the grains and, after a calibration of the photoelasticity, the force experienced by each contact point of every grain inside the packing. The precision of these force measurements is low. However, good enough to detect the grains involved in force chains [12–14] (figure 3). A simple analysis of the intensity variations along a ring excluding the central part of the grain is able to detect the contacts holding a high force value. It is known that acoustic waves propagate preferentially through force chains [14, 15]. Here we show also that the acoustic source of large acoustic bursts is localised along force chains.

By analysing the series of high-speed images we are able to determine the changes in the internal structure. Any change in the force applied at a grain contact will manifest itself by a change in the light intensity recorded by the camera. By examining the differences of intensity pixel by pixel during an avalanche, we are able to resolve the propagation of the reorganisation. On figure 3b, we represent all the detected changes of intensity for respectively a time of 0.25, 1 and 2 milliseconds since the beginning of the avalanche. We observe a good similarity between the pixels that experienced a change of value and the ones corresponding to force chains as detected on the

DOI: 10.1051/epjconf/201714003015

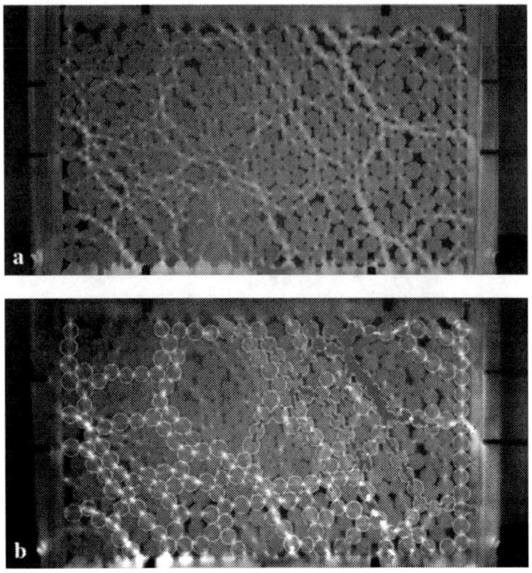

Figure 3. (colour online) **a**. High-resolution image used for force chains detection. A green filter has been used. **b**. High-speed image of the cell with highlighting (white colour) of grains involved in force chains. Colors represent pixels whose intensity changed since the beginning of the avalanche, depending on the time elapsed: red is for 0.25 ms since the beginning of the avalanche, blue for 1 ms, and green for 2 ms.

high-resolution images. We also see that the avalanche begins locally, involving only a few grains belonging to one short force chain, and then the structure relaxes on a large scale during the next few milliseconds. This relaxation is captured also by a force drop in the resistance to shear. The same behaviour have been noticed in all the other analysed events.

3.3 Collisions vs. Shearing

In order to compare the AE produced by either collisions between grains or frictional sliding of their contacts, we designed two experimental setups that each reproduce one of the mechanisms. The first one (figure 4a) consists of one slider with controlled mass freely moving vertically in a guide and with a *Durus* cylinder glued at its bottom. When the slider is released from a certain height, it will accelerate and hit another grain placed on the floor. The AE is recorded thanks to a pinducer stuck to the side of the immobile grain.

The second setup (figure 4b) aims to reproduce the frictional sliding of the grains contacts. A free grain is confined between two pairs of grains, one pair glued to a fixed stand and the other freely moving vertically and supporting a dead load. Thanks to a hole drilled in the free grain, we use a wire to extract it quickly (by hand so without a control of the speed) from its nest. The upper pair of grains is stopped before touching the inferior ones. Again the AE is recorded by a pinducer stuck to one of the still grains.

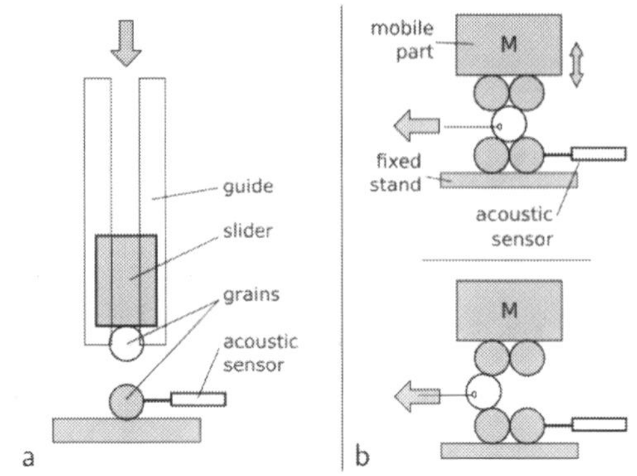

Figure 4. (colour online). Experimental setups used to reproduce the two acoustic emissions mechanisms, respectively the impacts (**a**) and the frictional sliding (**b**).

Acoustic bursts emitted by both mechanisms are analysed in the same way we did for the AE of the shearing experiments, with a time-frequency analysis. These spectrograms are represented on figure 5a–d. We note that the main frequencies of the AE are much better defined in this case, since the wave is much more confined in one or two grains. Again we are able to extract the frequencies emitted in all the experiments performed and plot their histograms on figure 5e–f. We observe different frequencies emitted by the two mechanisms, with a main peak at 2400 Hz for the impacts and one at 4800 Hz for the frictional sliding.

4 Discussion & Conclusion

The first and second peaks appearing in the histogram of figure 2c seem to correspond to the main peaks obtained in the collision of grains (figure 5e) and the frictional sliding (figure 5f) respectively. This result indicates that *both mechanisms, collisions and frictional sliding, play important roles at the origin of the acoustics events*. As frictional forces are responsible for the stability of the system, a local sliding at the contact interface between two grains may be the starting point of the process. Large force values may be required for provoking this failure, which would be consistent with the localisation of the origin of large events along force chains. The sliding of the grain will suddenly change the local distribution of forces, which is equivalent to a collision.

This instability propagates in the system, particularly along force chains (figure 3b) like a "domino" effect, provoking a reorganisation of both the structure and stress in the granular medium. However it concerns just a few dozens of grains in the rare case of a very large event; thus it cannot explain the existence of a power-law distribution of acoustic energies as "snow-ball" effects [16, 17]. Instead, one (or more) of the local events described in the previous paragraph may dominate in terms of energy release, and it is the signature of this local release of energy that accounts for the energy distribution.

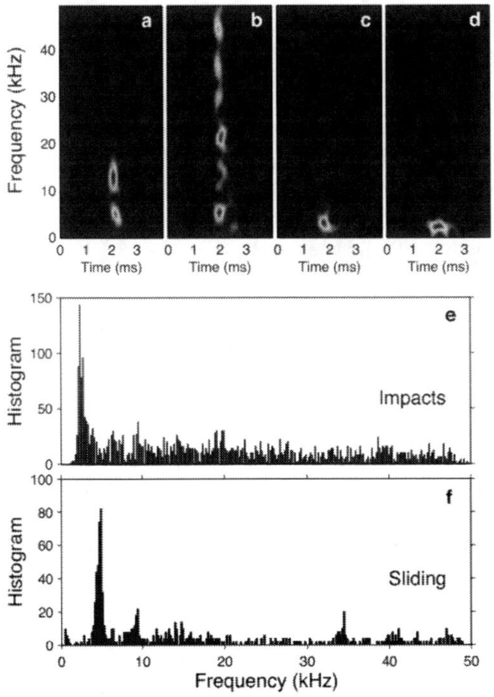

Figure 5. (colour online) **a**. Spectrogram of the AE produce by frictional sliding. **b, c, d**. Spectrograms of the AE produced by impacts of decreasing speeds (resp. 2.8, 0.3 and less than 0.1 m/s). **e, f**. Histograms of frequencies of spectrograms peaks recorded for a group of experiments of impacts and frictional sliding respectively.

In the case in which the main energy contribution comes from a collision-like effect, we could consider the phenomenon as a single and well-localised event. However, the frictional sliding is a much more complicated process [18, 19]. Although the energy captured by the acoustic also comes from single and well-localised events, it is difficult to judge if there is (or not) a microscopical avalanche-like process of breaking of asperities at the origin of the local frictional failure.

In conclusion, our experimental results indicate that both frictional sliding and collision-like mechanisms are involved as the origin of the acoustic bursts recorded globally during the shear of our granular system. The wide distribution (power-law) found for the acoustic energies has to be understood in term of of single and well-localised events with very large variations in energy happening preferentially along force chains. Possibly, these local reorganisations may propagate, bringing a global relaxation of the system.

Acknowledgements

We thank G. Simon for technical support. We acknowledge financial support from the AXA Research Fund and the Chair-CNRS program.

References

[1] P.A. Johnson, H. Savage, M. Knuth, J. Gomberg, C. Marone, Nature **451**, 57 (2008)

[2] K.E. Daniels, N.W. Hayman, Journal of Geophysical Research: Solid Earth **113** (2008)

[3] D.M. Walker, A. Tordesillas, M. Small, R.P. Behringer, K.T. Chi, Chaos: An Interdisciplinary Journal of Nonlinear Science **24**, 013132 (2014)

[4] K. Mair, C. Marone, Journal of Geophysical Research: Solid Earth **104**, 28899 (1999)

[5] O. Sotolongo-Costa, A. Posadas, Physical Review Letters **92**, 048501 (2004)

[6] I. Bratberg, F. Radjai, A. Hansen, Phys. Rev. E **71**, 011301 (2005)

[7] B. Ferdowsi, M. Griffa, R.A. Guyer, P.A. Johnson, C. Marone, J. Carmeliet, Geophysical Research Letters **42**, 9750 (2015), 2015GL066096

[8] R. Planet, S. Lherminier, G. Simon, K. Måløy, L. Vanel, O. Ramos, *Mimicking earthquakes with granular media*, in *22ème Congrès Français de Mécanique* (2015)

[9] S. Lherminier, R. Planet, G. Simon, L. Vanel, O. Ramos, Revista Cubana de Física **33**, 55 (2016)

[10] T.C. Hanks, H. Kanamori, Journal of Geophysical Research B **84**, 2348 (1979)

[11] M. Stojanova, S. Santucci, L. Vanel, O. Ramos, Physical Review Letters **112**, 115502 (2014)

[12] B. Miller, C. O'Hern, R. Behringer, Physical Review Letters **77**, 3110 (1996)

[13] T.S. Majmudar, R.P. Behringer, Nature **435**, 1079 (2005)

[14] E.T. Owens, K.E. Daniels, Europhysics Letters **94**, 54005 (2011)

[15] S. Lherminier, R. Planet, G. Simon, L. Vanel, O. Ramos, Physical Review Letters **113**, 098001 (2014)

[16] P. Bak, C. Tang, K. Wiesenfeld, Physical Review Letters **59**, 381 (1987)

[17] O. Ramos, E. Altshuler, K. Måløy, Physical Review Letters **102**, 078701 (2009)

[18] S.M. Rubinstein, G. Cohen, J. Fineberg, Physical Review Letters **98**, 226103 (2007)

[19] O. Ben-David, G. Cohen, J. Fineberg, Science **330**, 211 (2010)

DOI: 10.1051/epjconf/201714009006

The effect of interfacial properties and liquid flow on the stability of powder islands

Xin Yi Ong[1,*], *Spencer E.* Taylor[2], and *Marco* Ramaioli[1]

[1]Department of Chemical and Process Engineering, University of Surrey, United Kingdom GU2 7XH
[2]Department of Chemistry, University of Surrey, United Kingdom GU2 7XH

Abstract. This study aims at understanding the interplay between the interfacial properties of the powder grains and the characteristics of the liquid flow used to disperse them, in order to obtain an effective dispersion of a powder in a liquid, avoiding air entrainment. The dispersion of grain "rafts" and powder islands "stacks" was investigated both on a static and on a moving air-liquid interface. Powder wicking prevents the formation of a powder island when the grain contact angle is below a critical contact angle. Above the critical contact angle, a powder island forms and grows to a critical depth that depends on grain radius and contact angle. Imposing a flow on the air-liquid interface can either promote water impregnation, reducing the depth of the powder island or destabilise the whole island. In the latter case, the island sinks, forming a heterogeneous powder structure that is wet outside and dry inside.

1 Introduction

A typical example of dispersion of powders in liquids is the reconstitution of beverages from dehydrated powders. Minor changes in the physicochemical properties of the powder can affect dramatically the outcome of this process [1, 2]. Understanding the interplay of grain surface properties (especially contact angle) [3-5] density, particle size [1], addition rate of beads to the interface [6, 7] and liquid flow fields is necessary in order to enhance wettability. When a cylinder or a sphere is at air-liquid interface, stability is governed by forces arising from their mass, capillarity and buoyancy [8].

The question of how an object sinks has drawn much scientific interest. As a result of increased object density or wettability, or decreased surface tension, the weight of an object exceeds the vertical restoring balancing force created by surface tension and the hydrostatic force [9]. Vella & Li [10] proposed the influence of contact line effects may be an important factor during the transition process from floating to sinking. There is a critical impulse speed below which the object is trapped by the interface and floats, or above which it pierces the interface and sinks [6, 7, 9].

During wettability studies, Jurin's theory [11], Washburn's law [12] have been widely employed to explain the process of liquid wicking in a porous medium e.g. layered beads [12, 13]. However, there is a significant difference when wicking occurs in cylindrical capillaries and in a powder beds consisting of spherical particles. The powder bed has indeed varying pore cross section and solid-liquid orientation, which affect the flow behaviours [14, 15]. Several groups [2, 16-18] showed that the critical contact angle, θ_0^*, below which wicking takes place in monodisperse layers is with the range between 51° and 85°, considerably lower than 90° limit for cylindrical capillaries. Jones et al. [19] discussed how the deposition geometry of powders affects the size limits of the assemblies before they sink through the interface by comparing the critical size that "rafts" [20] and "stacks" [2] can achieve. A raft is formed by depositing new spheres away from the spheres that are already at the air-liquid interface, causing the spheres to self-assemble through interfacial deformation and gravity, this being termed the "Cheerios effect" [20]. A stack is formed by the deposition of new spherical grains on top of grains that are already floating at the interface [19]. Raux et al. [2] studied the creation of a powder island (equivalent to a stack) using grains with the contact angle, $\theta > \theta_0^*$ and showed that wicking occurs when the stack depth, h, exceeds a critical stack depth, h^*. Several authors [19, 21, 22] have used the ratio of the density of the deposited particles to those of the surrounding fluids to compare the depth of interfacial deformation triggered by the presence of the particles to the capillary length, i.e.

$$\Delta = \frac{(\rho_S - \rho_B)r_0}{(\rho_B - \rho_A)a} \quad (1)$$

where r_o is the radius of the particle, a is the capillary length $(\gamma/\rho g)^{1/2}$, γ and ρ are the surface tension and density of the liquid, and g is the gravitational acceleration. Through equation (1), these authors found that the maximal density ratio, $\Delta_{max} \approx 1$, allows both raft and stack geometry to remain afloat.

Consistency with most studies cited above, in this study insoluble grains are used to reduce the complexity

* Corresponding author: x.ong@surrey.ac.uk

of the system. The contact angle below which wicking is observed on a static interface is measured for different grain sizes. The effect of liquid flow in the vicinity of the point of impact of the grains is then discussed.

2 Experiments

Different glass bead diameters, d_p were used in this study: <0.106, 0.212-0.300, <0.50 and 0.75-1.00 mm. Bead density, ρ_s was 2500 kg/m^3 and the beads were silanised by treatment with a 5% dichlorodimethysilane solution in *n*-hexane to change their surface properties to hydrophobic, following the literature protocol proposed by Hamlett et al. [17]. The bead contact angle, θ was then tuned by changing the surface tension of the wetting liquid, using mixtures of deionised water (DI) containing different ethanol volume fractions, V_f. The contact angles of the silanised glass beads were measured by depositing them gently onto the air-liquid interface of different solutions, in a 1 cm cuvette (Fig. 1). Each measurement was repeated on ten different beads.

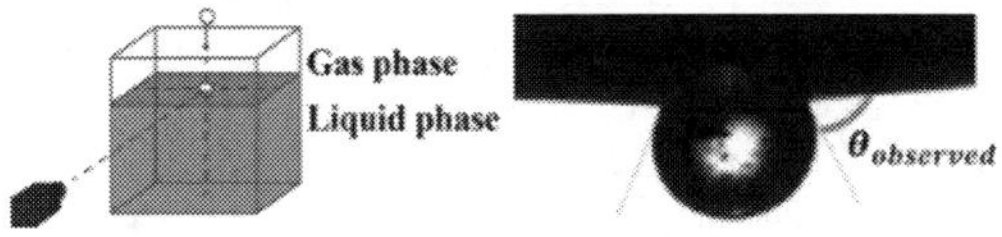

Fig. 1. Contact angle analysis in a cuvette.

The powder islands were created by pouring the beads gently onto a static air-liquid interface from a paper funnel located 30 mm above the initial liquid surface. A raft was initially formed, followed by a central stack (Fig. 2). The evolution of the beads at the liquid-grain interface was recorded using a Basler camera (acA1929-155µm). The average mass flow rate for each size of bead is 0.12 ± 0.01 g/s. The dimensions of this "static cell" were: 15.5, 11.0, 8.0 cm.

A purpose-built "flow cell" was also used, to investigate how the liquid flow affected the stability of rafts and stacks formed at the air-liquid interface. The dimensions of this cell are indicated in Fig. 3. Once a stack had been created, the impellers were rotated at 3.3 rev/s in centre-down direction, depicted in Fig. 3.

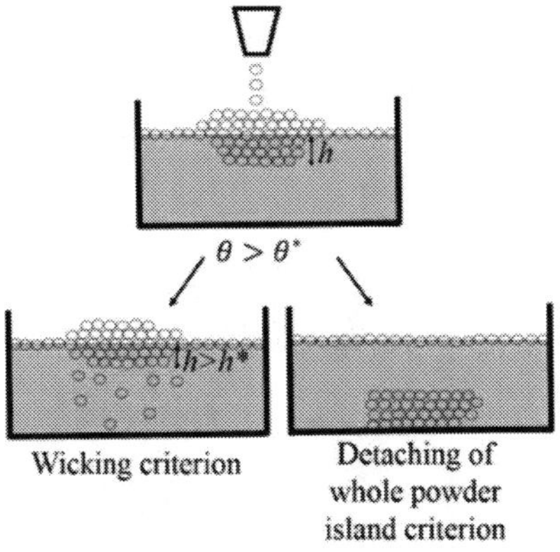

Fig. 2. Experimental setup to investigate the powder islands formed at static air-liquid interface. Wicking (bottom left) and island detachment (bottom right) were alternatively observed.

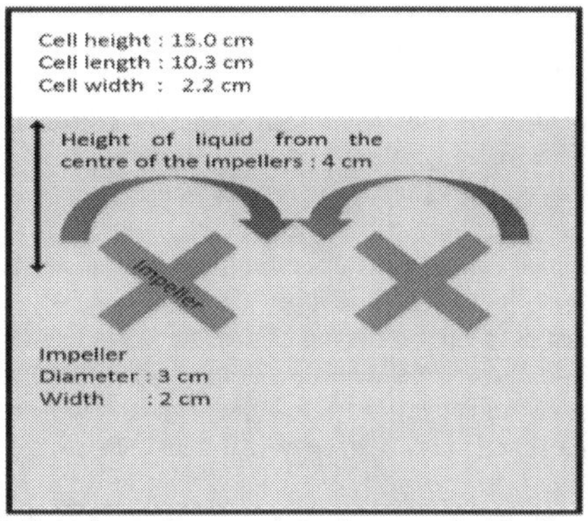

Fig. 3. Experimental setup to investigate powder islands formed at air-liquid interface with a moving fluid.

3 Results

3.1 Contact angle

As shown in Fig. 4, water-ethanol mixtures allow the grains to show contact angles varying from 98 ± 1.5° to 57 ± 3.9° as V_f increases from 0 to 0.5. As V_f increases, surface tension decreases, which leads to decreasing contact angle.

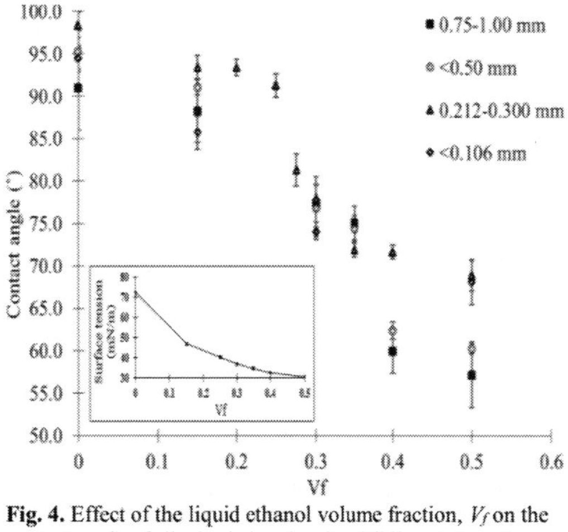

Fig. 4. Effect of the liquid ethanol volume fraction, V_f on the contact angle of silanised glass beads of different diameter, d_p (indicated in the legend). The inset shows corresponding values of surface tension of different V_f.

3.2 Static cell

The top row of Fig. 5 shows pictures of single beads floating at the static interface of different water-ethanol mixtures. The bottom row of Fig. 5 shows pictures of the powder island (stacks) built using the same grains and V_f. These pictures were taken just before the onset of wicking and were used to measure the critical stack depth, h^* above which the liquid wicks in the pores of these stacks of grains. The grain size was kept constant (d_p = 0.75-1.00 mm) and the contact angle decreased with increasing V_f. This in turns resulted in a strong decrease of the critical stack depth. When $V_f \geq 0.40$, which corresponds to θ < 72°, the system can only support rafts instead of stacks for bead size ranging from <0.106 to 1.00 mm.

The critical stack depths are summarised in Fig. 6, separating the non-wicking from wicking or detachment of the whole stacks. As described in Raux et al. [2], when the stack depth is increased, the liquid-air interface in the pores between the beads at the bottom of the stack becomes curved, due to the liquid hydrostatic pressure, $p = \rho g h^*$, inducing a curvature $C = \rho g h^*/\gamma$ and this favours wicking. In this study, the critical stack depth is found to be somehow dependent on the bead size. Smaller beads show a higher critical stack depth, as shown in Fig. 6. The critical stack depths were fitted with straight lines, having slopes of respectively 30, 54, 28, 68 mm/rad for d_p (mm) = 0.75-1.00, <0.50, 0.212-0.300 and <0.106.

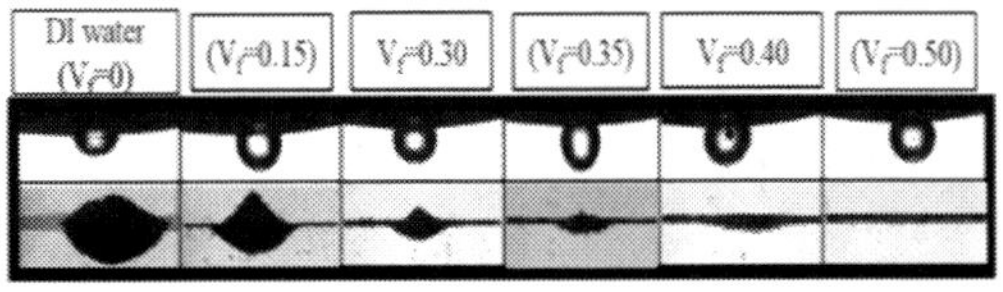

Fig. 5. (Top) Silanised glass beads of diameter d_p = 0.75-1.00 mm placed at the air-liquid interface with different ethanol volume fractions, V_f. (Bottom) Critical stack depth, h^*, formed at different V_f.

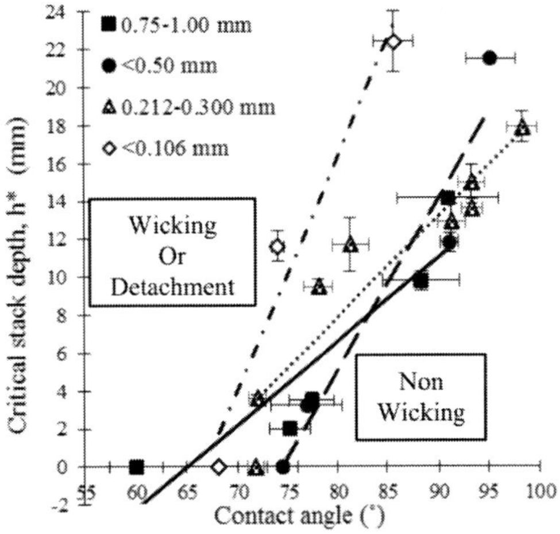

Fig. 6. Effect of the bead contact angle on the critical stack depth. Beads of different diameter are represented using different symbols, as described in the legend.

The theory developed by Raux et al. [2] describes the critical stack depth, h^* formed with bead having contact angle $\theta > \theta_0^*$:

$$h^* = \frac{4a^2}{d_p}\frac{\cos\theta_0^* - \cos\theta}{\cos^2\theta_0^*} \quad (2)$$

where the critical contact angle to obtain a stack composed by two layers is $\theta_0^* \approx 51°$. Fig. 7 shows that the normalised critical stack depth, h^* decreases fairly linearly with cos θ, in qualitative agreement with the theory of equation (2). However, the slope predicted by the model is much larger than the value that can be obtained by the fitted straight line on the experimental slope, as illustrated in Fig. 7 and Table 1. Raux et al. [2] also found a similar inconsistency. The causes of this inconsistency are currently being investigated. Wicking is not always observed when adding grains to an existing stack, as illustrated in Fig. 8, the detachment of the whole stack can also occur. Wicking is observed at the bottom of the stack constructed from larger beads, d_p (mm) = 0.75-1.00 and <0.50 when θ is in the range between 57° and 95° or smaller beads, d_p (mm) = 0.212-0.300 and <0.106 when θ is in the range between 68° and 72°, as shown in Fig. 8 (a) and (b). The whole stack detached from the interface without experiencing wicking for smaller beads, d_p (mm) = 0.212-0.300 and <0.106 when θ is in the range between 74° and 93°.

Table 1. Wicking transition in terms of the critical ethanol volume fraction, V_f^* and corresponding contact angle θ_0^* for different sizes of beads. The slopes obtained from equation 2 and from the experiments are reported.

d_p (mm)	V_f^*	θ_0^*	Model equation (2)	Fitting experimental slope
0.75 -1.00	0.35-0.40	60.0-75.1	-8	-2.2
<0.50	0.30-0.35	74.4-76.8	-28	-1.7
0.212 -0.300	0.35-0.40	71.7-71.9	-20	-0.7
<0.106	0.30-0.35	68.1-74.0	-14	-0.8
0.104	0.73	51.0	-5	-0.6

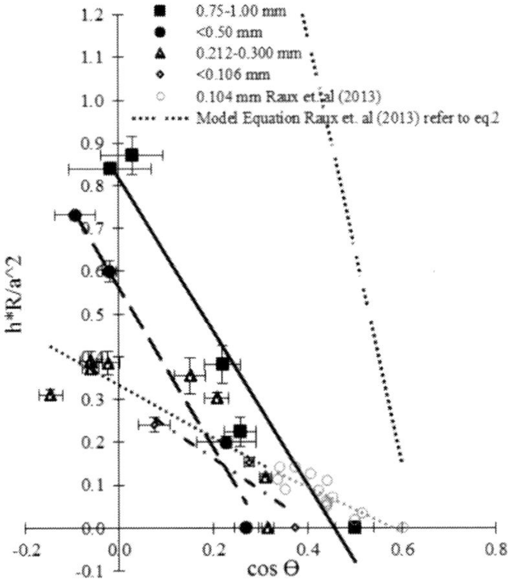

Fig. 7. Normalised critical stack depth vs. the cosine of the bead contact angle. Beads of different diameter are represented using different symbols, as described in the legend.

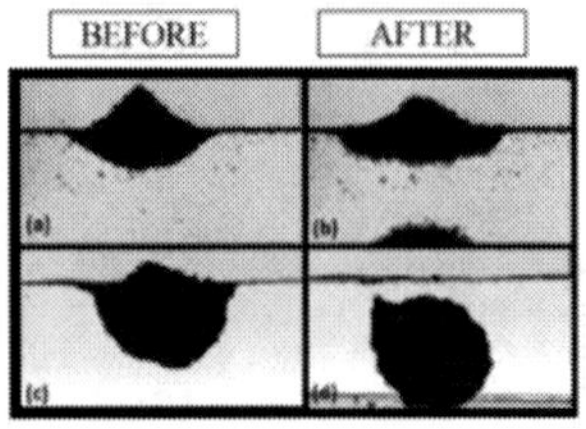

Fig. 8. Images of stacks at V_f = 0.15, (a) - (b) experiencing wicking (for d_p = 0.75-1.00 mm) and (c) – (d) detaching from the interface (for d_p = 0.212-0.300 mm).

The threshold density ratio, Δ, that can be calculated by using equation (1) is in the range between 0.03 and 0.40, thus always smaller than 1. Rafts can indeed be built for all the grains and solutions considered, in agreement with Abkarian et al. [21], who proposed that particle rafts would not sink when $\Delta \leq 1$. However, when $\Delta < 0.1$, the whole stack was observed to detach from the interface, instead of experiencing wicking.

3.3 Flow Cell

The experiments in the flow cell were carried out using two liquid mixture $V_f = 0.15$ and 0.30.

The larger bead fractions d_p (mm) = 0.75-1.00 and <0.50 at $\theta = 88 \pm 3.8°$ and $91 \pm 0.9°$ were always found to experience wicking when the agitation is started. As a result of the wicking induced by the liquid flow, the shape of the stack changed from spherical to flat as it can be seen in Fig. 9 (a) to (b). Then, the whole stack detaches from the interface as a cluster, as illustrate in Fig. 9 (c) to (d).

When $V_f = 0.30$, for all beads, the stack is relatively small and experiences wicking throughout the whole agitation process until the stack disperses, no detachment of stack is observed.

As mentioned above, when $V_f < 0.30$, which corresponds to θ is in the range between 74° and 93°, smaller beads detached from the interface in the static cell, without any agitation. Hence, it was decided to study how shallow stacks with $h < h^*$ behave during agitation. For these shallow stack composed of small beads, adding beads onto the stack pulls the layer of grains at the interface (the raft) toward the stack. Fig. 9 (e) to (h) show that when the liquid is flowing, both sides of the raft progressively shrink toward the stack. The flow destabilises the raft and stack, which fall both into the liquid. On the contrary, with larger beads, the liquid agitation was not sufficient to entrain the raft into the liquid.

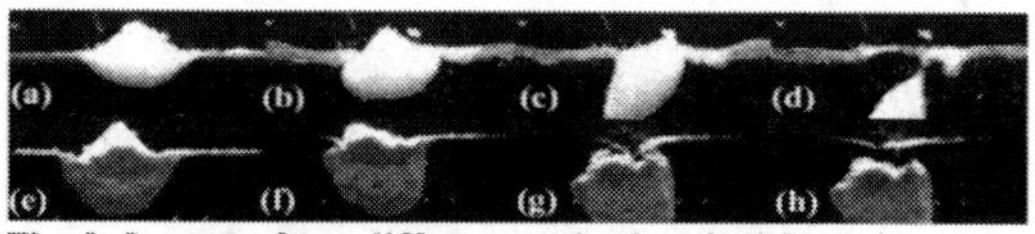

Fig. 9. Images of two different stacks detached from the interface during agitation at $V_f = 0.15$. $d_p = 0.75 - 1.00$ mm, from (a) to (d) and $d_p < 0.106$ mm, from (e) to (h).

4 Conclusions

Two different phenomena have been observed when pouring beads on a static liquid, namely wicking and the detachment of the whole stacks. The critical stack depth on a static liquid decreases with the grain contact angle and when increasing the grain diameter. It was concluded that the difference in stack thickness by forced impregnation is purely due to their particle size and wettability. In agreement with Abkarian et al. [21], wide layers of grains (rafts) can be formed when $\Delta \leq 1$. However, when $\Delta < 0.1$, the stacks were detached from the interface, without any wicking. Introducing flow in the liquid can either promote wicking or destabilise the whole stack. In the second case, the whole stack sinks, forming a heterogeneous powder structure that is wet outside and dry inside. This structure is reminiscent of powder lumps: an undesired outcome, when dispersing most consumer good powders, including beverages and detergents. It is suggested that plotting a state diagram to separate wicking and detachment of stack states would provide some highly interesting relationship. As the correlations between the trendline and points in both Fig. 7 are not obvious, this suggests that further exploration of the critical stack depth by using more different contact angles would be valuable. Lastly, more complex grains such as real powders will be considered for use in this study.

References

1. J. Dupas, L. Forny, and M. Ramaioli, J. Coll. Interface Sci., **448**, 51–56 (2015)
2. P. S. Raux, H. Cockenpot, M. Ramaioli, D. Quéré, and C. Clanet, Langmuir, **29**, 3636–3644 (2013)
3. J. Dupas, E. Verneuil, M. Ramaioli, L. Forny, L. Talini, and F. Lequeux, Langmuir, **29**, 12572–12578 (2013)
4. X. Gao and L. Jiang, Nature, **432**, 36 (2004)
5. D. Vella, D. G. Lee, and H. Y. Kim, Langmuir, **22**, 5979–5981 (2006)
6. D. Vella and P. D. Metcalfe, Phys. Fluids, **19**, 72108 (2007)
7. D. G. G. Lee and H.-Y. Y. Kim, Langmuir, **24**, 142–145 (2008)
8. J. B. Keller, Phys. Fluids, **10**, 3009 (1998)
9. D. G. G. Lee and H.-Y. Y. Kim, Phys. Fluids, **23**, 72104 (2011)
10. D. Vella and J. Li, Phys. Fluids, **22**, 52104 (2010)
11. P. G. de Gennes, F. Brochard-Wyart, and D. Quere, *Capillarity and Wetting Phenomena: Drops, Bubbles, Pearls, Waves* (Springer, 2004)
12. E. W. Washburn, Phys. Rev., **17**, 273–283 (1921)
13. N. Fries and M. Dreyer, J. Coll. Interface Sci., **320**, 259–263 (2008)
14. M. Reyssat, L. Y. Sangne, E. A. van Nierop, and H. A. Stone, EPL Europhys. Lett., **86**, 56002 (2009)
15. W. Bin Young, Colloids Surfaces A Physicochem. Eng. Asp., **234**, 123–128 (2004)
16. S. Ban, E. Wolfram, and S. Rohrsetzer, Colloids Surfaces, **22**, 301–309 (1987)
17. C. A. E. Hamlett et al., Environ. Sci. Technol., **45**, 9666–9670 (2011)
18. N. J. Shirtcliffe, G. McHale, M. I. Newton, F. B. Pyatt, and S. H. Doerr, Appl. Phys. Lett., **89**, 94101 (2006)
19. S. G. Jones, N. Abbasi, A. Ahuja, V. Truong, and S. S. H. Tsai, Phys. Fluids, **27**, 72102 (2015)
20. D. Vella and L. Mahadevan, Am. J. Phys., **73**, 817 (2005)
21. M. Abkarian, S. Protière, J. M. Aristoff, and H. A. Stone, Nature. Commun., **4**, 1895 (2013)
22. D. Vella, P. Metcalfe, and R. Whittaker, J. Fluid Mech., **549**, 215-224 (2005)

Dune formation in dilute phase pneumatic transport system: PIV & PTV based analysis

Sumit Jhalani, *Atharva* Patankar, *Ajay* Makawana, *Manaswita* Bose*

Department of Energy Science and Engineering, Indian Institute of Technology Bombay, Powai, Mumbai, India – 400076

Abstract: Flow of gas-solid mixture through horizontal conveying section show a large variety of phenomena and is broadly classified into dilute and dense unstable regimes. Different types of instabilities are observed in the dense phase flow and are widely studied in literature; however, clustering instabilities are observed in the very dilute regime of flow with volume fraction ~0.001. A recent study has shown that regular, stable dune shaped clusters are formed in a small regime of the dilute phase of conveying. The dunes become unstable as the superficial gas velocity is decreased before it finally leads to the dense mode of conveying. The motivation of the current work is to investigate the velocity distribution on the surface of the stable and unstable dunes and thereby understand the cause behind the formation of the dunes in the conveying section. To that end, particle image and tracking velocimetry techniques are employed with the specific objective to determine the volume fraction and the velocity profile of the solid phase on the stable dune surface. A drastic change in the solid fraction within a few particle diameters from the dune surface suggests that PTV is more appropriate in the bulk whereas PIV is suitable for near surface investigation.

1 Introduction

The flow of suspended solid particles in a gaseous medium through horizontal channel and pipe is widely investigated [1-3], and a large variety of phenomena is reported in the literature. The primary application of the flow of the gas-solid mixture through pipes is pneumatic conveying systems. The design of such systems is in general guided by the classical Zenz plot [4] in which the line pressure drop is plotted as a function of the superficial velocity. The pressure drop shows minima which are known as the saltation point and is typically considered to be the transition point between the dense and the dilute regimes of the transport. Though both dilute [5-8] and dense [9] regimes of the transport systems are well studied in literature; the transition between them is under-examined. It is briefly reported in the literature that the flow becomes unstable near the transition regime and large fluctuation in the line pressure drop is recorded [1, 4, 10]. It was earlier postulated that small unstable clusters form in this regime causing fluctuations in line pressure [1]. In a recent investigation, it is observed that the small regular dune shaped clusters as shown in Figure 1 are formed in the conveying line in the transition regime [11]. The dunes are often stable; however, there is a small region in the flow regime mapping [1] where these dunes are unstable and either their movement in the line or their periodic growth and decay leads to pressure fluctuation [12]. Though the dunes are not common in the context of pneumatic transport systems, they are widely discussed for the atmospheric flows [13]. Quite like the barchans dunes, these clusters are also found to be self-similar in shape [12]. Figure 1b shows that the surface of these clusters is fairly uniform.

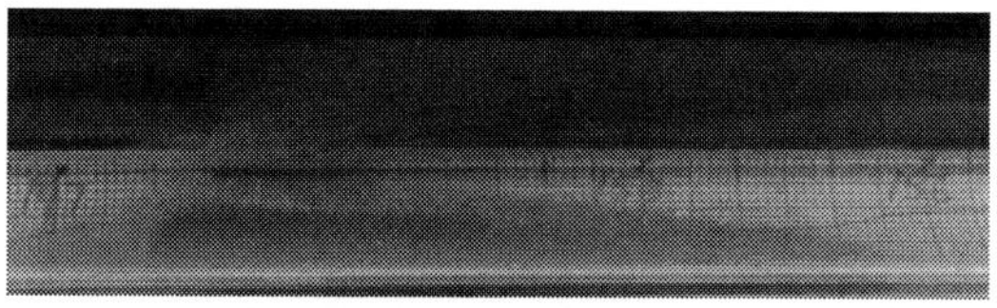

(a)

(b)

Fig.1: A typical stable dune in the conveying line captured with Nikon D70 & AF Zoom-Nikkor 24–85 mm f/2.8–4 D IF lens; (a) The front view; (b) The top view

The formation of either the stable or the unstable dunes in the conveying line is believed to be the result of the particle-particle and fluid-particle interactions, as it is observed for the subaqueous barchan dunes [14-16]. The aim of the current work, therefore, is to develop an

* Corresponding author: manaswita.bose@iitb.ac.in

understanding of the mechanism of the formation of the dunes in the conveying line.

A careful inspection of the dune shaped structures reveals that immediately above the surface the flow is very dilute, but the solid volume fraction of the dune itself is close to the packed bed limit. In the dilute range the particle tracking velocimetry (PTV) is more appropriate; however in the region very near to the surface, especially close the dune tip where the solid volume fraction is high, the particle image velocimetry is expected to predict the velocities more accurately than the PTV. The specific objective of the current work is to determine the velocity profile of the solid phase on the dune surface using PIV and the PTV techniques as appropriate.

2 Experimental Method

2.1 Experimental set-up

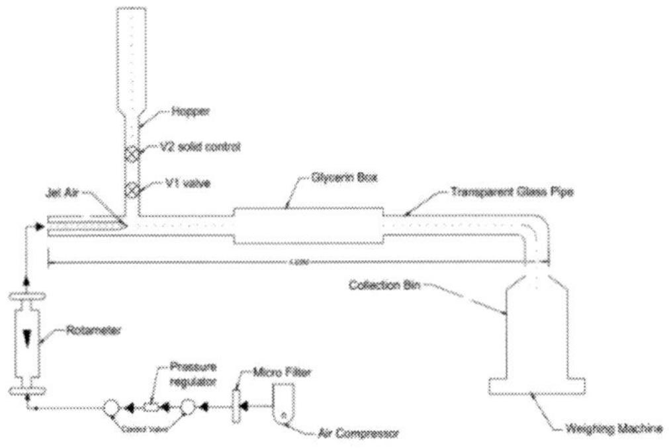

(a)

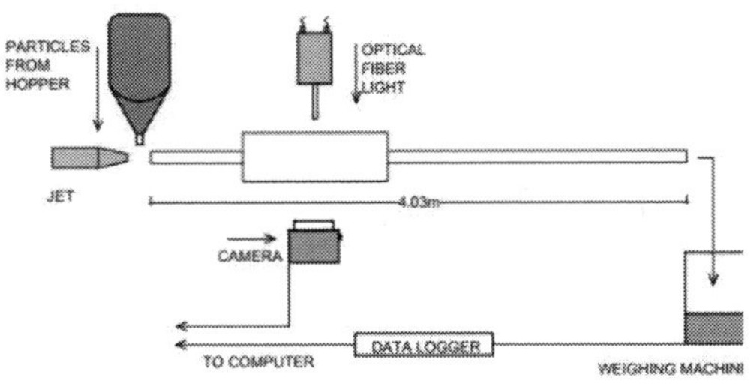

(b)

Fig.2. (a) Schematic of the Experimental set-up; (b) Arrangement of the camera and the illuminator

Figure 2 shows the schematic of the experimental set-up. Particles are fed to the conveying line from a vertically mounted hopper and are conveyed through with the help of a jet ejector pump. The test section is made of borosilicate glass for clear visualisation. The inner diameter of the pipe is 21.5 mm, and the length of the test section is 500 mm. The particles are collected in a bin placed at the end of the line. The collection bin is placed on a load cell.

The solid mass flow is controlled with the help of two valves and the air flow rate is regulated with the help of a ball valve and pressure regulator. The air flow rate is measured with the help of rotameter and the particle mass flow rate is measured using a load cell. The detail of the experimental set-up and the method are described in [11].

2.2 High-Speed Imaging

The motion of the particles is captured using high-speed CMOS camera, Phantom V9.1. The test section is blanketed in a rectangular channel filled with 99% anhydrous glycerine having the refractive index matched with that of borosilicate glass. The test section is illuminated with fibre-optics light source (Fiber-Lite MH100). The camera is placed at the front of the channel. The rear side of the box is covered with a white translucent paper. The light source is placed at the rear end. Undesired reflections are cut down by blocking the light rays with black cloth. The arrangement of the light source and the camera is shown in Figure 2b. The shadows of the particles are captured using the high-speed camera at frame rate ranging between 2000 and 3600 fps. Micro-Nikkor 60 mm f/2.8 lens is used for capturing the images. The depth of the focus of the lens for all the recording is measured twice the particle diameter. The spatial resolution of the images is 0.06 mm. (1mm = 17 pixels). The image quality is enhanced with the help of the CSR of the camera. One sample snap shot of the conveying section is shown in Figure 3a. Experiments are also performed with LRS-0532 DPSS laser sheet for brighter illumination and more precise planar measurement. Figure 3b shows a snapshot captured using the green laser sheet.

2.3 Processing of the Images

Reference images are captured for the calibration as well as for conversion from pixel to real scale. The images are processed with the standard image processing toolbox of MatLab R2012b for filtering of noise. The images are then processed using the algorithms implemented in PIVLab 1.41. Direct cross-correlation is used for determining the local velocity of the particle phase. Dependence of the velocity on the selection of the window size is checked before the final sampling. The final velocity of the solid phase is obtained by averaging the instantaneous velocities for thousands of frames. Analysis using the particle tracking velocimetry are being carried out with the help of a code developed in-house [17]. PTV is performed especially in the region far from the dune surface. The results obtained from both methods are being compared. It is worth to note that the imaging, in this case, needs attention as there is a sudden change in the particle volume fraction in the test section and one also has to address the issue of large light scatter

 DOI: 10.1051/epjconf/201714009007

from the dune surface which forms during the flow. So, a large number of experiments are required even for a single set of flow parameters.

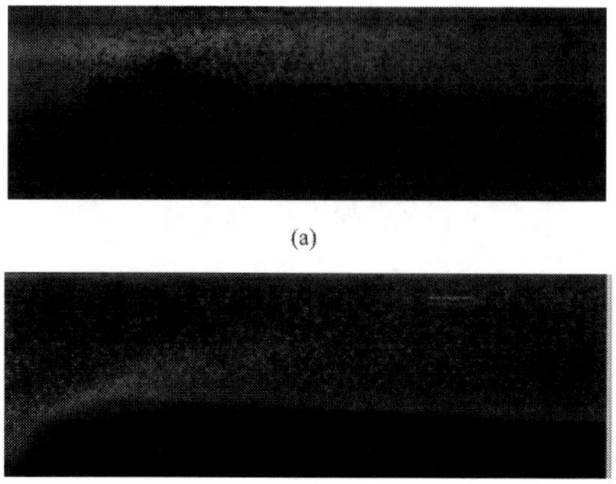

(b)

Fig.3. Snapshot of the Dune using high-speed imaging (a) shadow graphy (b) green laser

2.4 Methodology

Table 1 presents the list of parameters for which the experiments are performed. Once the reference image is captured after the camera and the illuminations are set, the air and the solid flow controlling valves are opened and the experiments are initiated. The recordings of the particle motion are started once the dune is completely developed.

Table 1: List of parameters for which the experiments are conducted

Parameters	Available Range
Air flow rate	6.5 m^3/ hr
Box dimensions (LxBxH)	720mm x 72mm x 90mm
Pipe dimensions (L, D)	3800mm, 21.5mm
Air Velocity	4.91m/sec
Particle density	2500 kg/m^3
Particle diameter (SMD)	166μm
Bulk Volume Fraction	~0.001
Mass loading ratio	3.24

3 Results

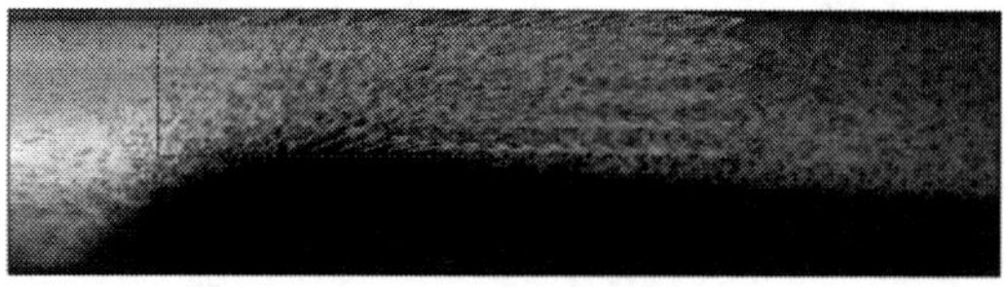

Fig.4. Velocity vectors on the dune surface

The height of the dune changes along its length along the axial direction of the pipe. So the entire frame is divided

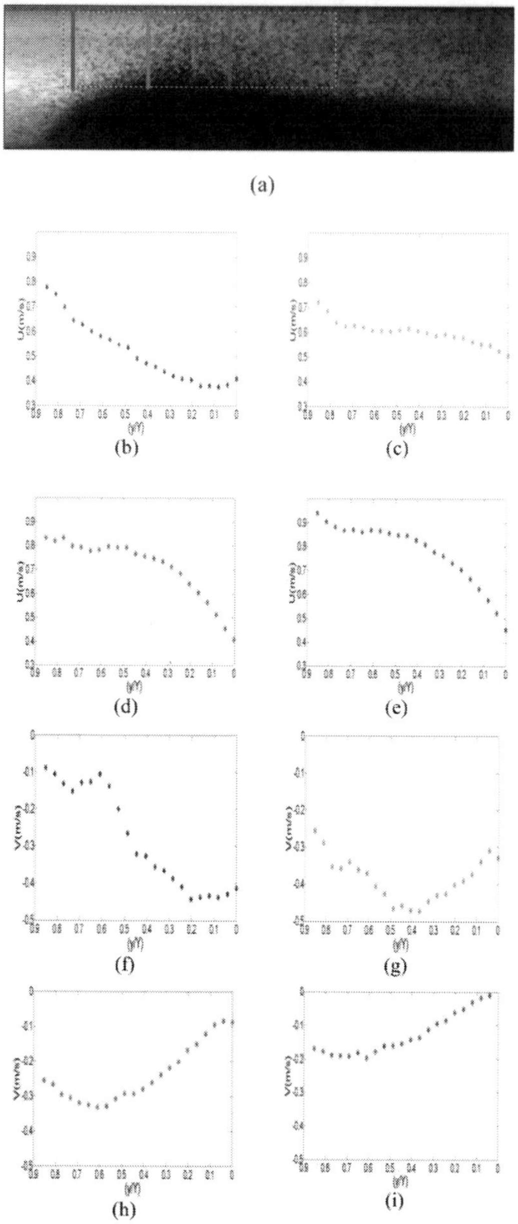

Fig. 5.(a)The grey scale image of the dune showing the positions on the dune surface where the velocity profiles are obtained; Profiles of the (b) – (e) stream-wise (axial) and (f) – (i) cross-stream (gradient direction) components of the particle velocities

in to different regions of interest as shown in Figure 4. The interrogation window and the step size are selected next. The velocity of the particles in each interrogation window at each time instant is then determined. Figure 4 shows the velocity vector plot at one time instant and indicates that there are fluctuations in the particle phase velocity. The size of the interrogation window for this case is selected to be 16 X 16 pixels with a step size of 8 pixels. The mean profiles of the velocity along the radial direction for different axial positions are finally obtained by averaging the particle velocities over a large number of frames ranging between 20,000 and 25,000. Figure 5a shows a typical grey scale image of a dune with the four different locations at which the velocity profiles are obtained.

The colour code is used for clear representation, e.g., blue is used in Figure 5b and 5f which show the profiles for the stream-wise and the cross-stream components of the particle velocities respectively at the location at the dune tip (marked blue in Figure 5a).

Figures 5b to 5e show the stream-wise components and Figures 5f to 5i plot the same for the cross-stream components of the solid phase velocities in the pipe on the dune surface. The stream-wise components have significantly higher magnitude than that of the cross-stream components. The particles have smaller velocity at the surface of the dune and larger as they move away from the surface, which is expected. The large slip of the particles at the solid wall is consistent with the observations reported in [5, 6, 17] for the dilute phase transport. The average solid phase stream-wise velocity (axial) is flatter near the top wall (Figures 5d & 5e) indicating that the flow is dilute; however, this kind of demarcation is not possible near the dune crest (Figures 5b and 5c). Near the surface of the dunes, the particles collide with those on the surface resulting in either rebound or rolling over the surface. Sometimes the particles in flight settle on the dune surface losing the momentum to the clusters; whereas at some other times the collisions result in the ejection of particles from the surface of the dunes; however, for the stable dunes, a steady state is maintained. The detailed dynamics of the particles near the dune surface are not reflected in the profiles of the average velocities of the particles. The distribution of the fluctuating velocities near the dune surface needs to be determined for a better insight.

4 Discussion

Clustering instability in shape of dunes is observed in the dilute regime of the pneumatic transport system. The formation of the dunes is believed to be the result of interaction between the fluctuating motions of the particles and that of the fluid phase turbulence. To understand the reasons behind the formation of dune shaped clusters in the conveying line, the dynamics of the particles are investigated through PIV and PTV algorithms. The profiles of the solid phase volume fraction and the average velocity are determined at the centre of the pipe. The detailed simulation based on the coupled CFD-DEM algorithms [18] along with appropriate contact models [19, 20] need to be carried out for developing more insight into the mechanism of cluster formation. The results of the present experiments can be used as the benchmark for comparing the numerical simulations.

Acknowledgements

The work is supported by Department of Science and Technology through SR/S3/CE/028/2011 and IITB funding through 15IRINTRN026.

References

1. Y. Tsuji and Y. Morikawa, *Int. J. Multiph. Flow*, **8**, 329-341, (1982)
2. M. Hubert and H. Kalman, *Powder Technol.*, **134**, 156-166, (2003)
3. Y. Zheng and A. Rinoshika, *Int. J. Multiph. Flow*, **53**, 114-123, (2013)
4. C. Crowe, Multiphase Flow Handbook, 4–12. Boca Raton, FL: CRC Press, Taylor and Francis Group, (2006)
5. Y. Tsuji, Y. Morikawa, T. Tanaka, N. Nakatsukasa and N. Nakatani, *J. Fluid Mech.*, **13**, 671-684, (1987)
6. S. Lain and M. Sommerfeld, *Int. J. Multiph. Flow*, **39**, 105-120, (2012)
7. D. Kullick, J. Fessler, and J. Eaton, *J. Fluid Mech.*, **277**, 109-134, (1994)
8. T. Tanaka and J. Eaton, *Phys. Rev Lett.*, **101**, 114502-1 (2008)
9. Y. Tomita and H. Asou, *Powder Technol.*, **196**, 14–21, (2009)
10. M. Ochi and K. Ikemori, *JSMEB*, **21**, 1008–1014, (1978)
11. A. Makwana, A. Patankar, and M. Bose, *Part. Sci. Technol.*, **33**, 59-66, (2015)
12. A. Makwana, A. Patankar, and M. Bose, (*Manuscript under preparation*)
13. B. Andreotti, P. Claudin, and S. Douady, *Eur. Phys. J. B*, **28**, 321-339, (2002)
14. E. Franklin and F. Charru, *J. Fluid Mech.*, **675**, 199-222, (2011)
15. F. Charru and M. Franklin, *J. Fluid Mech.*, **694**, 131-154, (2012)
16. R. Hout, *J. Fluid Mech.*, **715**, 389-423, (2013)
17. A. Deshmukh, V. Vasava, A. Patankar and M. Bose, *Powder Technol.*, **298**, 119-129, (2016)
18. Y. Tsuji, T. Kawaguchi and T. Tanaka, *Powder Technol.*, **77**, 79-87, (1993)
19. M. J. Jiang, H.-S. Yu and D Harris, *Comput. Geotec.* **32**, 340–357, (2005)
20. M. Jiang, Z. Shen, J. Wang, *Comput. Geotec.* **65**, 147–163, (2015)

Erosion patterns on a granular bed around a vertical cylinder

Florent Lachaussée[1], *Yann* Bertho[1], *Cyprien* Morize[1], *Alban* Sauret[1,2], and *Philippe* Gondret[1,*]

[1]*Laboratoire FAST, Univ. Paris-Sud, CNRS, Université Paris-Saclay, F-91405, Orsay, France*
[2]*Laboratoire SVI, CNRS, Saint-Gobain, 39 quai Lucien Lefranc, F-93303 Aubervilliers, France*

Abstract. We report on two different patterns that can be observed at the bed surface close to a vertical cylinder when submitted to a strong enough steady water flow. The classical scour pattern observed at the cylinder foot and due to the "horseshoe" vortex around occurs at a critical velocity U_{c_1} below the critical velocity U_{c_0} for erosion without any cylinder, thus under clear-water conditions. But we observe also another pattern downstream the cylinder which consists of two symmetrical ovoid holes that look like "bunny ears". This new scour pattern referred as BES can be observed at lower velocities that the horseshoe scour (HSS), above a critical velocity $U_{c_2} < U_{c_1}$, with a timescale formation much higher that the one of HSS.

1 Introduction

Erosion is encountered in numerous natural or industrial situations. Understanding erosion process is essential for instance to interpret geomorphological patterns on Earth or other planets and to predict the evolution of estuaries and river beds [1]. But quantifying the erosion processes generated at the vicinity of structures referred as scouring is also relevant to a wide range of engineering applications, including the design and risk assessment of hydraulic structures such as bridge piers, off-shore platforms and wind turbines. There has been a huge amount of research carried out to address this question with large scale facilities [2], but the physics of local scour phenomena occurring around structures is far from being understood [3]. This complexity arises both from the complex flow around a cylinder [4] and the complex rheology of granular material [1]. Numerical approaches are far from describing carefully these two complexity levels and also the complex grain/fluid interaction even if key advances have been made with the recent development of coupled methods with Large Eddy Simulation (LES) for the fluid and Discrete Element Method (DEM) for the grain phase [5].

In the present paper, we focus on the simplified case of scouring at the base of a cylindrical structure of circular cross-section. This geometry represents the traditional template for studying scour processes around structures and finds important applications in civil and off-shore engineering. We present experimental results from a small-scale setup that allows easy monitoring and quick change of the different physical parameters.

2 Experimental setup

The setup shown in Fig. 1 consists of a closed channel, with a racetrack shape, of about 3.6 m long and W = 100 mm wide, composed of two linear parts of 900 mm long each separated by two opposite curved parts of radius 300 mm. Water flow (density ρ = 1000 kg m^{-3} and height h = 160 mm) is driven by means of a paddle wheel located in one linear part of the channel and is considered after a honeycomb in the other linear test section where a granular bed composed of glass beads of density ρ_g = 2500 kg m^{-3} and mean diameter d = (270 ± 30)μm is put in a drawer of length 600 mm and 40 mm depth at the bottom of the channel. A vertical cylinder of diameter D = 20 mm exceeds the granular bed from a height H = 90 mm, lower that the water height. This cylinder is fixed at the bottom of the granular bed at the 2/3 of the test section that we will take at the origin of the horizontal frame of reference (x, y), with x the streamwise direction and y the

Figure 1. Picture of the experimental setup.

*e-mail: philippe.gondret@u-psud.fr

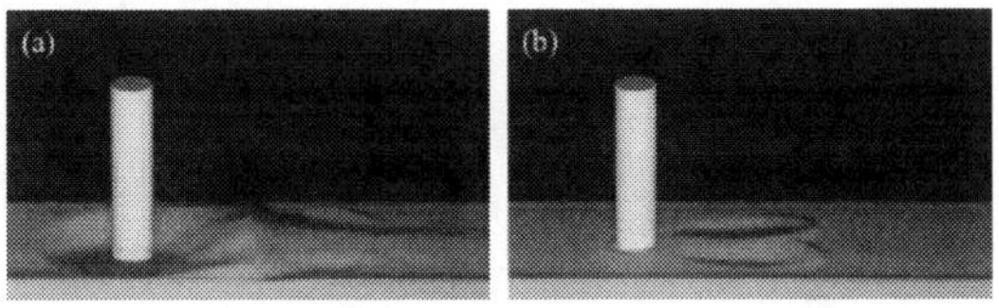

Figure 2. Snapshots of two different erosion patterns close to a cylinder: (a) "Horseshoe" pattern at the cylinder foot for $Re_D \simeq 3.6\,10^3$, $Sh \simeq 8$ and $t \simeq 6\,10^2$ s; (b) "Bunny ears" pattern downstream the cylinder for $Re_D \simeq 2.3\,10^3$, $Sh \simeq 3.3$ and $t \simeq 4.76\,10^4$ s.

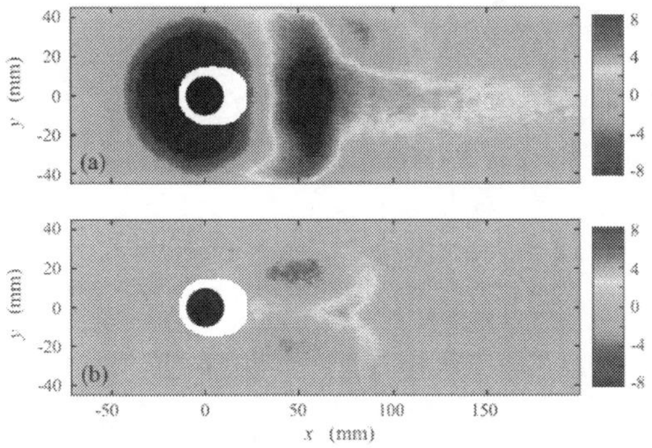

Figure 3. Topography $\xi(x, y)$ of the two different erosion patterns: (a) "Horseshoe" pattern at the cylinder foot for $Re_D \simeq 3.6\,10^3$, $Sh \simeq 8$ and $t \simeq 6\,10^2$ s; (b) "Bunny ears" pattern downstream the cylinder for $Re_D \simeq 2.3\,10^3$, $Sh \simeq 3.3$ and $t \simeq 4.76\,10^4$ s. Dark circles are the top view of the cylinder. White regions are zones that are not detected by the laser profilometer.

transverse direction. The side walls of the test section are made of glass to allow direct visualization of the scouring close to the cylinder. A laser profilometer is mounted on a horizontal translation guide in order to scan the granular bed topography $\xi(x, y)$ from above. To limit signal noise in the ξ measurements that may arise from the fluctuations of the water surface, a horizontal glass wall is put at the water surface above the granular bed.

The flow velocity U is controlled by the paddle wheel rotation in the typical range $0.05 \lesssim U \lesssim 0.25$ m s^{-1} leading to the following typical Reynolds number $Re_l = Ul/\nu$ for the water flow, depending on a length scale l and the kinetic viscosity of water $\nu = 10^{-6}$ m^2 s^{-1}. The Reynolds number based on the channel width W is $Re_W \simeq 10^4$, corresponding thus to a turbulent regime. The Reynolds number of the flow around the cylinder of diameter D is $Re_D \simeq 10^3$, far beyond the critical value 69 for the Bénard-von Kármán unsteadiness for an infinite cylinder between two parallel sidewalls with the present blockage ratio $D/W = 0.2$ [6]. The flow around the cylinder is thus also turbulent with strong unsteady wake vortices. The Reynolds number of the flow around the grains is $Re_d \simeq 50$, corresponding rather to a laminar regime. The Froude number $Fr = U/(gh)^{1/2}$, defined as the ratio of the flow velocity to the velocity of surface waves, is smaller than 0.2 which thus corresponds to subcritical flows. Before any experiment, the granular bed is stirred and then leveled with a ruler for an initial flat horizontal surface. The cylinder is then fixed on its support without disrupting the granular bed, the water flow is turned on at a given velocity U and the laser profilometer is turned on to measure the instantaneous granular bed topography $\xi(x, y, t)$ at regular time intervals Δt adapted to the erosion dynamics without any flow stop. The instantaneous topography $\xi(x, y, t)$ is the difference from the initial topography $\xi(x, y, 0)$ before any flow.

3 Experimental results

Without any cylinder, erosion takes place on the granular bed above the critical value $Sh_0 \simeq 9$ for the global Shields number $Sh = \rho U^2/\Delta\rho g d$ based here on the flow velocity U far above the granular bed. The present Shields number is an inertial Shields number chosen as $Re_d > 1$, which is different from the local Shields number that could be taken by knowing the exact fluid stress at the bed surface [7, 8].

We work here in clear-water conditions ($Sh < Sh_0$) for which no erosion occurs far from the cylinder. The presence of the cylinder disrupts the flow and creates several vortices [2]: a horseshoe vortex at the foot of the cylinder and wake vortices downstream. The erosion threshold of the granular bed is thus lowered due to the presence of these vortices which lead to scouring in the vicinity of the obstacle.

The different patterns observed are reported by imaging from the side and by laser profilometry in figures 2 and 3 respectively. At large enough flow velocity, but small enough to remain in clear-water conditions ($4 \lesssim Sh \lesssim 10$), one observes the classical erosion pattern with a deep scour hole at the foot of the cylinder due to the horseshoe vortex [Figs. 2(a) and 3(a)]. This erosion pattern arises in a short time scale of typically a few minutes in the present setup. At lower flow velocities ($1 \lesssim Sh \lesssim 4$), one observes an other erosion pattern with two elongated holes located a few diameters downstream the cylinder in an amazing "bunny ears" shape [Figs. 2(b) and 3(b)]. This erosion pattern arises typically in a few hours thus with a much larger time scale than the horseshoe pattern.

From the precise topography evolution $\xi(x, y, t)$ of the granular bed, one can extract the time evolution of the maximum scour depth $\delta(t)$ as shown in figures 4(a) and 4(b) for each erosion pattern respectively. In the case of the classical Horseshoe Scour (HSS), $\delta(t)$ increases quickly and tends to reach an almost constant asymptotic value $\delta_\infty \simeq D$ of the order of the cylinder diameter, which is in agreement with the usually reported values [3]. Note that the curves in Fig. 4(a) are truncated as soon as the pit reaches the side walls of the channel. By contrast, $\delta(t)$ increases much more slowly for the Bunny Ears Scour (BES) and reaches a much smaller asymptotic value $\delta_\infty \simeq 0.1D$ than in the HSS case. Note that the position (x_m, y_m) of the maximum scour depth does not evolves with time during the time depth evolution: HSS remains close to the upstream cylinder foot

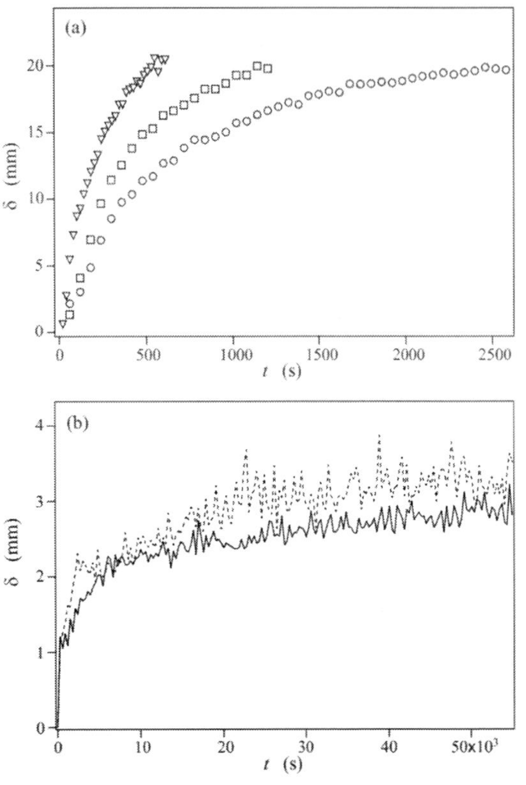

Figure 4. Time evolution of the depth δ for the two erosion patterns at different Reynolds numbers and different Shields numbers. (a) Horseshoe scour at (○) $Re_D \simeq 2.8\,10^3$, $Sh \simeq 5.1$, (□) $Re_D \simeq 3.2\,10^3$, $Sh \simeq 6.4$, and (▽) $Re_D \simeq 3.6\,10^3$, $Sh \simeq 8.0$. (b) Bunny ears scour at (—) $Re_D \simeq 2.1\,10^3$, $Sh \simeq 2.8$ and (– –) $Re_D \simeq 2.3\,10^3$, $Sh \simeq 3.3$.

with $(x_m, y_m) \simeq (-D/2, 0)$ whereas BES is a few diameter downstream with $(x_m, y_m) \simeq (2D, \pm D)$.

The evolution of the depth and position of these patterns as a function of the Reynolds number Re_D or the Shields number Sh will be discussed. The critical Shields number above which HSS appears is larger than the critical Shields number for BES to appear. Note that BES is still observed for large Shields numbers but disturbed by the quicker and stronger HSS development. To investigate the far development of BES, a rigid sabot at the foot of the cylinder will be used. In that case, the HSS development is inhibited as no more granular bed can be eroded by the horseshoe vortex close to the cylinder.

4 Conclusion

With a small scale laboratory setup, the erosion in the vicinity of a vertical cylinder posed on a granular bed is investigated when submitted to a steady flow in clear-water conditions. Besides the well-known scour pattern that may arise quickly at the cylinder foot due to the horseshoe vortex, one observes an other erosion pattern downstream the cylinder with an amazing "bunny ears" shape, that appears much more slowly for weaker velocities. The conditions for this new pattern to appear must be refined together with its relation to the flow structure with the wakes vortices. The effects of the various vortices will be decoupled in our channel by replacing part of the granular bed by a non-erodible rigid floor. Additional experiments with Particle Image Velocimetry means are ongoing to investigate the complex velocity field and its detailed interaction with the granular bed.

Acknowledgements

This work is supported by a public grant of the French National Research Agency (ANR) within the project SS-HEAR (grant No. ANR-14-CE03-001). We are grateful to S. Chateau for preliminary experiments and to E. Wesfreid for fruitful discussions. We thank J. Amarni, A. Aubertin, L. Auffray, and R. Pidoux for their contribution to the development of the experimental setup.

References

[1] A. Andreotti, Y. Forterre, O. Pouliquen, *Granular Media - Between Fluid and Solid* (Cambridge University Press, 2013)

[2] R. Ettema, G. Constantinescu, B. Melville, Evaluation of bridge scour research: Pier scour processes and predictions. Washington, DC: Transportation Research Board of the National Academies (2011)

[3] C. Manes, M. Brocchini, J. Fluid Mech **779**, 309-324 (2015).

[4] C.H.K. Williamson, Ann. Rev. Fluid Mech. **28**, 477 (1996)

[5] S. Kuang, C. LaMarche, J. Curtis, A. Yu, Powder Technol. **239**, 319 (2013).

[6] J.-H. Chen, W.G. Pritchard, S.J. Tavener, J. Fluid Mech. **284**, 23 (1995)

[7] S. Badr, G. Gauthier, P. Gondret, Phys. Fluids. **28**, 033305 (2016)

[8] T. Loiseleux, P. Gondret, M. Rabaud, D. Doppler, Phys. Fluids **17**, 103304 (2005)

Microgravity spreading of water spheres on hydrophobic capillary plates

Laura Steub[1], *Jonathan* Kollmer[1], *Derek* Paxson[2], *Achim* Sack[1], *Thorsten* Pöschel[1], *John* Bartlett[2], *Douglas* Berman[2], *Yaateh* Richardson[2], and *Michel Y.* Louge[2,*]

[1] *Institute for Multiscale Simulation, Friedrich-Alexander-Universität Erlangen-Nürnberg, Erlangen 91052, Germany*
[2] *Sibley School of Mechanical and Aerospace Engineering, Cornell University, Ithaca, NY 14853, USA*

Abstract. We create nearly perfect centimetric spheres of water by splitting a cavity consisting of two metal hemispheres coated with a hydrophobic paint and under-filled with liquid, while releasing the apparatus in free-fall. A high-speed camera captures how water spread on hydrophobic aluminum and polycarbonate plates perforated with cylindrical capillaries. We compare observations at the ZARM drop tower in Bremen with Lattice-Boltzmann numerical simulations of Frank, Perré and Li for the inertial phase of imbibition.

1 Background

The wetting of porous surfaces is crucial to any application where the three states of matter coexist [1]. When a liquid drop touches a dry porous medium, it spreads as if laid on a composite surface. Specifically, the porous surface first behaves as a hydrophobic material, since the liquid must penetrate pores filled with air. As soon as contact is established, part of the liquid is absorbed by capillary forces that are resisted by viscous losses growing with the length of the imbibed region. This process is complicated by the motion of triple gas-liquid-solid contact lines progressively overcoming each capillary. For simplicity, Cassie and Baxter [2] treated the porous surface as an effective material made up of a solid trapping air cavities. Because this approach did not entirely explain the dynamics of liquid imbibition, later studies also considered the balance of viscous forces and capillarity [3–9].

Imbibition of a liquid sphere of density ρ, viscosity μ, surface tension σ and radius R_i always begins with an inertial regime on a time scale $\tau = (\rho R_i^3/\sigma)^{1/2}$ [10]. The relative importance of bulk liquid inertia and viscous forces on the porous surface is measured by a Reynolds number based on drop size and characteristic capillary invasion speed $u_c \sim \sigma/\mu$, known as the Laplace number $\mathrm{La} \equiv \rho R_i \sigma/\mu^2$. Because $\mathrm{La} \propto 1/\mu^2$, it is large even for tiny water drops, and therefore inertia cannot be ignored.

In that context, Frank, Perré and Li [11, 12] observed the role of inertia in lattice-Boltzmann (LB) numerical simulations without gravity. For simplicity, they neglected contact angle hysteresis [13] and modeled the porous medium as a periodic array of cylindrical capillaries perpendicular to the free surface. By following the evolution of the effective contact angle while resolving flow details in and around capillaries, they found that drop spreading and penetration take place simultaneously.

* e-mail: Michel.Louge@cornell.edu

However, to stage a specific viscosity while remaining numerically stable, their LB simulations had to fix the ratio of the temporal increment δ_t and the square of the spatial grid size δ_x [14]. For imbibition, this required $(\delta_t/\tau) \propto (\delta_x/R_i)^2 \mathrm{La}^{1/2}$. Meanwhile, to avoid unphysical compressibility, the LB technique kept the pseudo Mach number $\mathrm{Ma} = u_c\delta_t/\delta_x \propto \mathrm{La}^{3/4}(\delta_t/\tau)^{1/2}$ small by limiting the Laplace number. Despite these challenges, Frank, et al [11, 12] achieved sufficiently high La = 41 or 135 to show that inertia governs imbibition initially by slowing down the rate of decrease in drop height. Once inertia was defeated, they found that drop height decreased at a more rapid rate mitigated by viscous capillary invasion.

Unfortunately, the La values of Frank, et al [11, 12] were smaller than expected. For example, a water drop ($\sigma \simeq 0.072\,\mathrm{J/m^2}$, $\rho = 1000\,\mathrm{kg/m^3}$, $\mu \simeq 8.9.10^{-4}\,\mathrm{kg/m.s}$) of 1 mm diameter has $\mathrm{La} \simeq 4.5.10^4$). Therefore, an objective was to verify that this mismatch was inconsequential. In addition, because pores delay the progression of the air-water interface, we set out to establish whether the contact line pins, as happens at smaller scales [15], particularly with substantial hysteresis between advancing and receding contact angles [16, 17]. To those ends, we designed a microgravity experiment with a drop large enough to resolve details around individual pores. However, we staged solids that turned out to possess hydrophobic advancing angles. Therefore, our results are mainly relevant to the spreading of water on hydrophobic textured surfaces.

We deployed the apparatus at the 4.74 s ZARM microgravity drop tower in Bremen (Germany). After a brief description, we present results and discuss their significance.

2 Experiments

Figure 1 shows components of one of four identical rigs deployed in the payload capsule of the evacuated 120-m

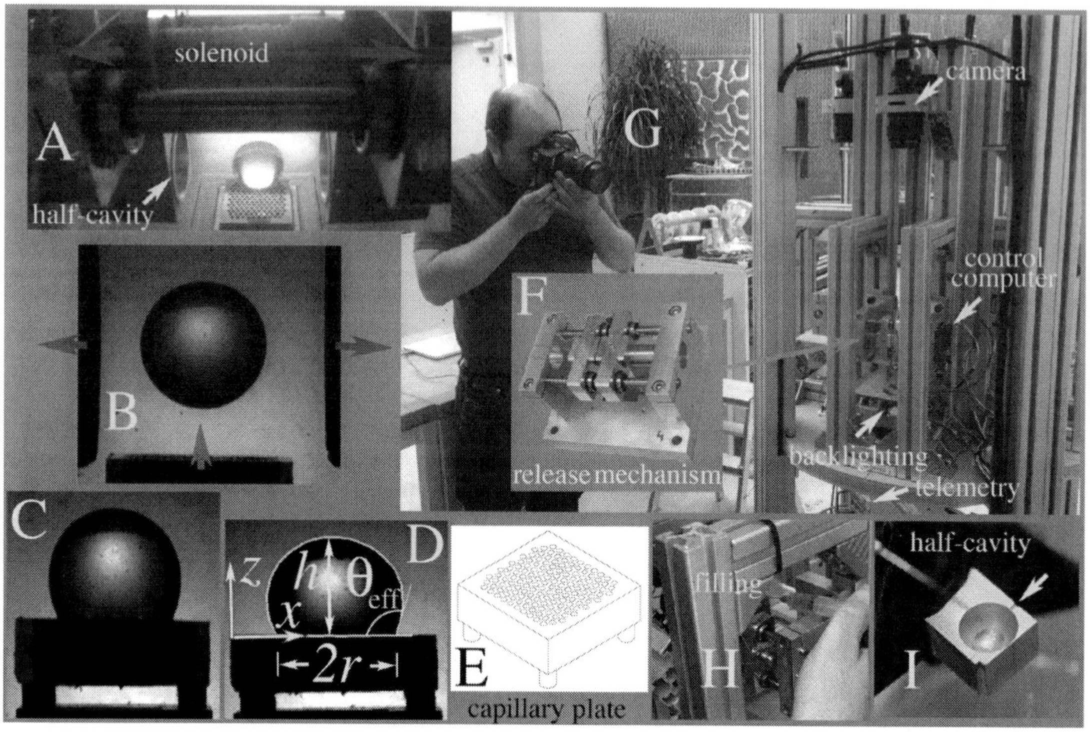

Figure 1. Free-fall apparatus. (A) Viewed by a side camera, the backlit sphere of water just created by the splitting of two hydrophobic hemispherical cavities hovers above the rising capillary plate of aluminum with ϵ = 44%, while the solenoid coil is extended. (B – D) Typical high-speed frames. Red and green arrows represent direction of motion of cavities and plate, respectively. (C) The drop spreads on the capillary plate. (D) Superimposed computer-generated outlines and notation. (E) Perspective drawing of a $20 \times 20\,\mathrm{mm}^2$ aluminum or polycarbonate plate with 146 cylindrical open capillaries of 1 mm on a triangular lattice at ϵ = 44% to 6.4 mm depth. (A plate of similar geometry with ϵ = 11% had 46 holes). (F) Close-up of one of the release mechanisms. (G) Overview of the microgravity payload before insertion into ZARM's hermetic capsule, consisting of four identical apparatuses, each with a different capillary plate (Table 1). (H) Filling the cavity of 19 mm diameter with demineralized water using a curved syringe needle on the closed cavity. (I) Prior test of cavity hydrophobicity with a similar water volume in one of the hemispheres; the arrow points to the filling channel.

free-fall tower. Before sealing the hermetic capsule, demineralized water was introduced to fill roughly 40% of each spherical cavity treated with a hydrophobic coating (e.g. Rust-Oleum® NeverWet®, contact angle ≃ 154°). Underfilling the cavity was essential to let air insinuate itself under the nascent drop and produce a nearly perfect water sphere (radial distortion $\delta R_i/R_i < 2\%$, see Table 1).

In a sequence optimized by trial and error, an Arduino UNO microcontroller delayed cavity opening by 2 s after onboard telemetry detected free-fall to let water relax before splitting the cavity. It then triggered the video camera (2000 fps, 512 × 512 pixels), and turned on LED backlighting last to avoid heating. To split the cavity rapidly without vibration, we designed a mechanism consisting of a solenoid tighly-wound around a permanent neodymium bar magnet, linear bearings, springs and shock absorbers. Once a nearly perfect water sphere hovered motionless, an actuator raised a capillary plate to meet it. We tested plates made up of polycarbonate or aluminum, pierced with open-ended cylinders of 1 mm diameter on periodic triangular arrays with opening surface fraction ϵ = 44% or 11%. Separate measurements on a Rame-Hart 500 Goniometer established that the aluminum had advancing and receding contact angles $\theta_a = 93.5 \pm 1.6\,°$ and $\theta_r = 82.6 \pm 2.2\,°$, respectively, while the polycarbonate had $\theta_a = 103 \pm 5\,°$ and $\theta_r = 90 \pm 2\,°$.

As Fig. 1D shows, the Image Processing Toolbox of MATLAB recorded left and right drop outlines at distances $x_{\min} \equiv \min(x)$ and $x_{\max} \equiv \max(x)$ parallel to the plate and altitude z above it. We also recorded time-histories of drop height $h \equiv \max(z)$ and contact patch radius $r \equiv (x_{\max} - x_{\min})/2$ at $z = 0$. Because the actuator was operated by the microcomputer in open-loop, the plate reached the water sphere with a finite speed u_0 that produced waves. Unfortunately, the absence of images from different camera orientations prevented us from calculating the water volume V above the plate with precision. Instead, we exploited the instantaneous drop outline to find the estimate

$$V \simeq \int_{z=0}^{h} \frac{\pi}{4} \left(x_{\max} - x_{\min}\right)^2 \mathrm{d}z, \qquad (1)$$

which was subject to artificial oscillations as the drop surface lost instantaneous axisymmetry.

 DOI: 10.1051/epjconf/201714016001

3 Results

In a typical test, nearly perfect water spheres were met by the capillary plate, thus initiating a rapid spread illustrated as time-histories of contact patch radius in Fig. 2. The contact line pinned at times ~ τ, marking the end of the spreading phase. Slowly-decaying oscillations, which remained after plate and water touched, later reflected downward, allowing us to trace the effective advancing and receding contact angles $\theta_{a,\text{eff}}$ and $\theta_{r,\text{eff}}$ on the textured surface.

As Table 1 and Fig. 2 show, hardly any water penetrated capillaries, as expected for the hydrophobic advancing angles that our solids turned out to possess. For a pinned spherical cap with $\theta_{a,\text{eff}}$, a reduction δV in initial volume V_i (Table 1) would imply a mean imbibition depth

$$\bar{d} = \frac{3R_i}{8 \times 2^{1/3}\epsilon}\left(\frac{\delta V}{V_i}\right)\frac{(2 - 3\cos\theta_{a,\text{eff}} + \cos^3\theta_{a,\text{eff}})^{2/3}}{\sin^2\theta_{a,\text{eff}}}, \quad (2)$$

or $\bar{d} \simeq 0.5$ mm for Al at $\epsilon = 44\%$, 1.7 mm for Al at $\epsilon = 11\%$, and 0.3 mm for PC at $\epsilon = 44\%$. Therefore, our experiments chiefly tested spreading of water spheres on hydrophobic textured surfaces in the absence of gravity.

Frank, et al [11, 12] fitted such inertial spreading of the contact patch of radius r to the power law

$$\frac{r}{R_i} \simeq C\left(\frac{t}{\tau}\right)^{\alpha}. \quad (3)$$

Figure 2 illustrates similar fits of our data, with constants summarized in Table 1. Figures 3 and 4 compare these results to simulations that Frank, et al [12] conducted without capillaries ($\epsilon = 0$) at different values of $\theta_a = \theta_r$. The agreement implies that the effective advancing contact angle conformed to the model that they recommended,

$$\cos\theta^{\text{mod}}_{\text{a,eff}} = (1 - \epsilon)\cos\theta_a - \epsilon, \quad (4)$$

so long as capillaries were hardly invaded. As Table 1 shows, measurements of $\theta_{a,\text{eff}}$ also support this Cassie-Baxter expression. The data suggest a similar relation for

$$\cos\theta^{\text{mod}}_{\text{r,eff}} = (1 - \epsilon)\cos\theta_r - \epsilon. \quad (5)$$

In short, this behavior of textured surfaces on hydrophobic solids is consistent with the Cassie-Baxter description [2], which Louge [18] identified as the last of six equilibrium regimes of the hysteretic contact line. In regime VI, the internal surface of a cavity on a hydrophobic solid is large compared with its opening, while adjacent cavities remain unconnected [18]. Further microgravity experiments with hydrophilic solids should be conducted to explore adjacent regimes and test whether drop spreading would behave differently.

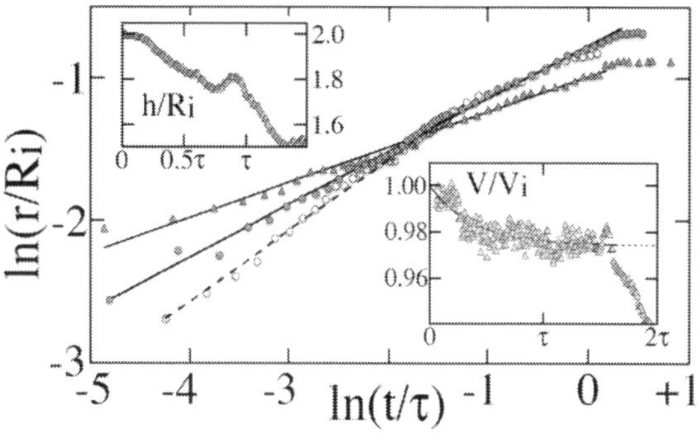

Figure 2. Time histories of $\ln r/R_i$ vs $\ln t/\tau$. Symbols and conditions, see Table 1. Right and left insets show, vs increments in τ, relative volume V/V_i inferred from Eq. (1) and, for the polycarbonate marked by an asterisk in Table 1, relative height h/R_i, before oscillations engulf the entire drop.

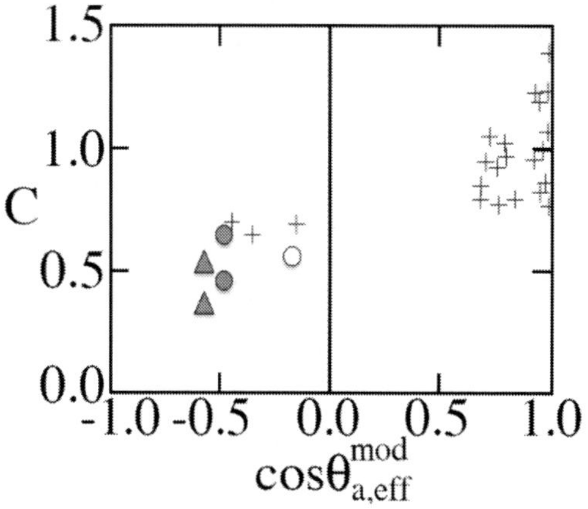

Figure 3. Coefficient C in the power law fit of Eq. (3) vs cosine of the effective advancing contact angle calculated from Eq. (4) and independent measurements of θ_a on the goniometer. Plus signs are from the numerical simulations of Frank, et al. [12] with $\epsilon = 0$ (flat plate). Symbols and conditions, see Table 1.

Acknowledgements

DP and MYL were supported by NPRP grant 6-059-2-023 of the Qatar National Research Fund; LS, JK and TP by the German Science Foundation via the Cluster of Excellence Engineering of Advanced Materials; and LS through the ESA Education 'Drop Your Thesis!' program. The authors are grateful to the ZARM staff, to Nigel Savage, Lily Ha and Lucas Pfeiffer of ESA Education, and to Florian Fuchs for help with experiments, to Patrick Perré, Xavier Frank, Hervé Duval and Floran Pierre for illuminating discussions, to Jin Xu, Carlos Mejia, Nicole Panega, Gilbert Hegermiller, Rukang Huang, Seetha Kolli and Richard Goodwyn for help designing the microgravity payload, and to Anshuman Das and Keya Gangadharan for measuring contact angles with NSF CBET 1637531 support.

References

[1] J. H. Snoeijer, and B. Andreotti, Annu. Rev. Fluid Mech. **45**, 269–292 (2013)

[2] A. B. D. Cassie, and S. Baxter, Trans. Faraday Soc. **40**, 546–551 (1944)

Table 1. Experimental conditions and results for plates made up of solids shown (Al = aluminum, PC = polycarbonate) with capillary surface fraction ϵ (%) at the surface $z = 0$. V_i (ml) and $\delta R_i/R_i$ are, respectively, initial volume of the suspended water sphere and fractional discrepancy in its radii along x and z. u_0 (cm/s) is relative velocity of sphere and plate at impact, persisting until the plate stops at time t_s relative to the inertial time τ. $\delta V/V_i$ is the ultimate visible fraction of water imbibed into capillaries, inferred through Eq. (1), and fitted using a decaying exponential as shown in the right inset of Fig. 2. The constants C and α are fitted to Eq. (3), as shown in Fig. 2. $\theta_{a,eff}$ and $\theta_{a,eff}^{mod}$ are, respectively, the effective advancing contact angle measured once the drop is pinned, and its prediction of Eq. (4) with θ_a independently measured on the goniometer. $\theta_{r,eff}$ and $\theta_{r,eff}^{mod}$ are their receding counterparts, see Eq. (5).

solid		ϵ	V_i	$\delta R_i/R_i$	u_0	t_s/τ	$\delta V/V_i$	C	α	$\theta_{a,eff}$	$\theta_{a,eff}^{mod}$	$\theta_{r,eff}$	$\theta_{r,eff}^{mod}$
Al	●	44	1.26	2.0%	3.1	1.2	3.6 ± 2%	0.65	0.52	118°	118 ± 1°	110°	112 ± 1°
Al	●	44	1.13	0.2%	3.2	1.7		0.46	0.37	116°	118 ± 1°	100°	112 ± 1°
Al	○	11	1.45	0.2%	2.7	0.74	4.7 ± 3%	0.56	0.50	93°	99 ± 2°	93°	90 ± 2°
PC*	▲	44	1.25	1.4%	3.0	1.4	2.6 ± 2%	0.37	0.25	124 ± 1°	125 ± 3°	109 ± 3°	116 ± 2°
PC	▲	44	1.28	0.2%	4.5	0.99		0.54	0.34	110 ± 2°	125 ± 3°	91 ± 1°	116 ± 2°

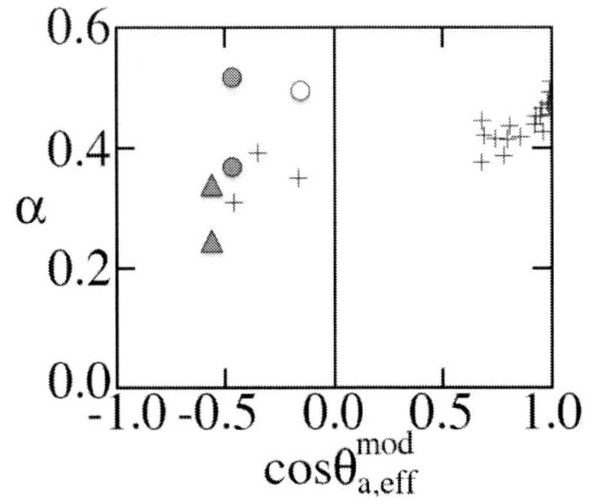

Figure 4. Exponent α in Eq. (3) vs $\cos\theta_{a,eff}$ in Eq. (4). Symbols and conditions, see Fig. 3.

[3] M. Denesuk, G.L. Smith, B.J.J. Zelinski, N.J. Kreidl, and D.R. Uhlmann, Coll. Interface Sci. **158**, 114–120 (1993)

[4] M. Denesuk, B.J.J. Zelinski, N.J. Kreidl, and D.R. Uhlmann, Coll. Interface Sci. **168**, 142–151 (1994)

[5] V. M. Starov, S.A. Zhdanov, and M.G. Velarde, Langmuir **158**, 9744–9750 (2002)

[6] V. M. Starov, S.A. Zhdanov, S.R. Kosvintsev, V.D. Sobolev, and M.G. Velarde, Adv. Colloid Interface Sci. **104**, 123–158 (2003)

[7] S. H. Davis, and L.M. Hocking, Phys. Fluids **11**, 48–57 (1999)

[8] S. H. Davis, and L.M. Hocking, Phys. Fluids **12**, 1646–1655 (2000)

[9] A. Clarke, T.D. Blake, K. Carruthers, and A. Woodward, Langmuir **18**, 2980–2984 (2002)

[10] J. C. Bird, S. Mandre, and H. A. Stone, Phys. Rev. Lett. **100**, 234501 (2008)

[11] X. Frank, and P. Perré, Phys. Fluids **24**, 042101 (2012)

[12] X. Frank, P. Perré, and H.-Z. Li, Phys Rev. E **91**, 052405 (2015)

[13] D. Quéré, Annu. Rev. Mater. Res., **38**, 71–99 (2008)

[14] R. R. Nourgaliev, T. N. Dinh, T. G. Theofanous, and D. Joseph, Int. J. of Multiphase Flow, **29**, 117–169 (2003)

[15] C. Priest, T.W.J. Albrecht, R. Sedev, and J. Ralston, Langmuir **25**, 5655–5660 (2009)

[16] R. E. Johnson, and R. H. Dettre, J. Phys. Chem. **68**, 1744–1750 (1964)

[17] E. Schäffer, and P.-Z. Wong, Phys Rev. E **61**, 5257–5277 (2000)

[18] M. Y. Louge, Phys Rev. E (2017), in press

Dynamics of breaking arches under a constant vibration

Bruno Guerrero[1,*], *Celia* Lozano[1,2,**], *Iker* Zuriguel[1,***], and *Angel* Garcimartín[1,****]

[1]*Dpto. de Física y Mat. Apl., Fac. Ciencias, Univ. Navarra, 31080 Pamplona, Spain*
[2]*Max-Planck-Institut für Intelligente Systeme, 70569 Stuttgart, Germany*

Abstract. Granular flow through an orifice can be suddenly halted by the formation of arches in the vicinity of the outlet, which are stable under the action of gravity. They may be broken when an external driving (for instance, vibration) is applied. With the aim of shedding light on the dynamics of arch destruction, we built an experiment consisting of a vertical two-dimensional silo filled with monodisperse beads, to which a constant vibration is applied. It was previously found that an important parameter to predict the robustness of the arch is the angle between consecutive beads. We focus on long-enduring arches and study the angles among the beads along time. We have found that in many cases the dynamics of the largest angle determines the breaking of the arch; it does not only determine where the "weakest link" is, but also the process that leads to the final destabilization. This is interesting because it can provide information about whether the flow will resume in a well-defined time or not, which is especially useful for industrial processes that have to constantly deal with the possible emergence of clogs.

1 Introduction

Clogging arches that stop the flow of grains at the exit orifice of a silo are a nuisance in industry, and they are at the same time a nice example of clogging in many other systems [1]. It is known that under a sustained vibration, arches can be shattered alleviating the problem [2–4]. In that scenario flow becomes intermittent, alternating between clogs and flow intervals. Then, in order to characterize the flow, it becomes necessary to understand both the clogging and unclogging processes. The first one has been widely studied since 2001 [5] observing that the clogging probability is constant over time yielding an exponential distribution of the flowing times [6–11]. The unclogging process, however, has shown temporal dependence and the distribution of breaking times is compatible with a power law decay [1, 2, 4]. The origin of this is still not fully understood but it could be related with aging, creeping motion or other complex slow dynamics [12]. Incidentally, let us mention that this power law decay of the unclogging times has been also observed in other systems such as colloids, sheep or pedestrians evacuating a room [1, 13, 14].

In previous works [4, 15], it was reported that the angle ϕ between adjacent beads of an arch (see Fig. 1) is an important variable to understand the unclogging process as it affects the robustness of the mechanical structure. In fact if the maximum angle in an arch ϕ_{max} was larger than 180° (that was called a defect) it was found a linear relationship between ϕ_{max} and the vibration intensity at which arches break down [15].

Here we explore the dynamics of arches that collapse when they are submitted to an external vibration. For the vibration parameters used here, most of the arches collapse in a very short time. We select those that last longer than a certain threshold and break down before a temporal upper bound set beforehand (mainly imposed by the experimental constraints). In Fig. 2, we show an example of how much the beads have moved after 20 minutes of sustained vibration; it can also be appreciated how the arch balls move unevenly in space. This issue is important in determining how do arches evolve to resist the external forcing.

The paper is organized as follows, We initially summarize the experimental methods. In the following section we introduce the first results obtained. Finally, we make a few concluding remarks about the future research.

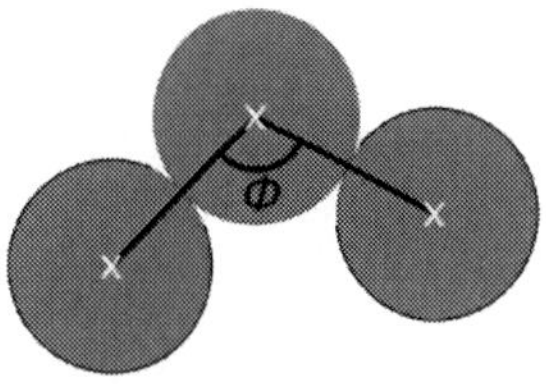

Figure 1. Sketch which illustrates the definition of the angle between adjacent beads ϕ. It is the inner angle formed between the straight lines connecting the center of a bead with the centers of the two adjacent beads, all of them belonging to the arch. The maximum angle ϕ in the arch is called ϕ_{max}.

*e-mail: bguerrero.3@alumni.unav.es
**e-mail: c.lozano@physik.uni-stuttgart.de
***e-mail: iker@unav.es
****e-mail: angel@unav.es

 DOI: 10.1051/epjconf/201714003016

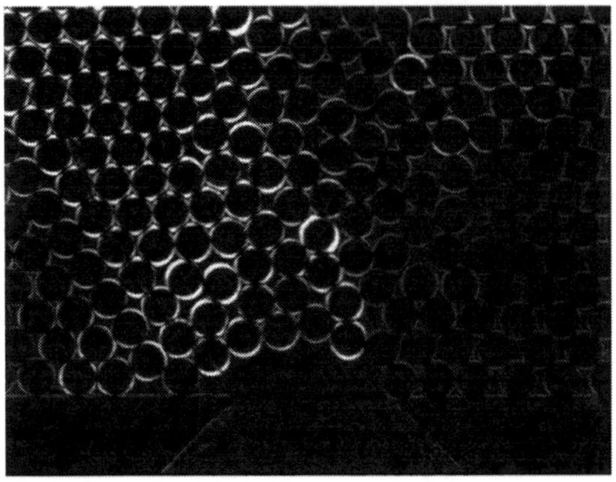

Figure 2. The subtraction of two images of an arch, one before and the other after enduring a sustained vibration for 20 minutes (bright zones correspond to displacements of the beads).

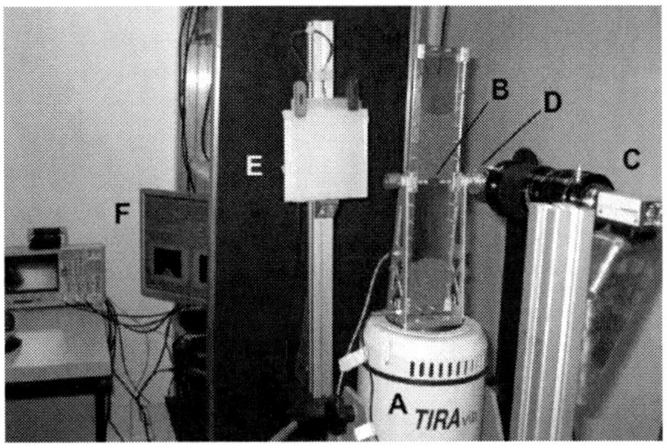

Figure 3. Photograph of the experimental setup: A, shaker; B, silo *(the line points to the orifice)*; C: camera and lens; D: motor to rotate the silo; E: led lighting; F: devices and computer (a photograph of the arch is visible on the computer screen).

2 Experimental set-up and methods

The experimental set-up is the one used by Lozano *et al.* [4]. It consists of a two dimensional silo mounted on top of an electromagnetic shaker (see Fig. 3). The silo is filled with monodisperse steel beads that have a diameter of 1 *mm*. It has an orifice in the middle, and so an electric motor can turn the container to start anew with another run. A camera records the zone near the orifice, detecting the arches. The automated procedure starts with grains flowing through the orifice; when an arch is formed it is detected by a real time image analysis, the vibration is switched on and the image acquisition begins. The run stops when the arch breaks (or after 20 minutes of sustained vibration if the arch has not been shattered by that time).

The improvement and novelty needed for the present measurements is both in the image acquisition stage and in the image processing. In fact, we need as many images as possible for long periods. We have written an *ad-hoc* software to perform this task. With it, we are able to collect snapshots with a standard video camera at a rate of nearly 25 frames per second (the camera rate) for long periods of time. It should be noted that once in a while the operating system flushes the buffer and a comparatively long time lapse (up to a few seconds) between two consecutive images is found; but this only happens occasionally after the experiment has run for many seconds, so this is not really important because it only affects the description at long times –when fast processes are not important. For short times, our temporal resolution is of about 0.05 *s*. The longest time span during which we register images is 20 minutes, so the dynamic range covers four orders of magnitude.

Images are time-stamped and stored in the hard disk, where they are processed at another stage. We have implemented a program for detecting the centres of the beads with subpixel resolution. It was inspired in a code written by M. Shattuck [16], in which the maxima of the convolution of a disk with the image provides the particles' centres (the disk size must be close to the size of the beads). With our code a resolution of 0.1 pixels is attained; a better resolution is feasible, but we did not go further as the stability of the silo during one run is about 1 micron (and 0.1 pixel corresponds to 2 microns). The position changes of the beads we track are typically two orders of magnitude bigger than this (see Fig. 2), so this resolution is enough four our purposes. Note that the height of the base can slightly shift up and down –this is barely visible in Fig. 2– due to the vibration; this item has been dealt with in the image processing stage.

3 Results

It was previously reported [4] that the time needed to break an arch (t_b) follows a probability distribution displaying a power law decay $P(t_b) \propto t_b^{-\alpha}$. Incidentally, this means that if the exponent α is smaller than two, the first moment of the distribution does not converge and the average breaking time $\langle t_b \rangle$ will depend on the size of the sample. It is usually more convenient to work with the complementary cumulative distribution function $P(T \geq t_b)$, because it is less noisy; note that the power law tail for the cumulative distribution has an exponent $\alpha + 1$.

Inspired by the fact that the force needed to break the arches is related to the maximum angle found within the arch at the beginning (when the vibration is switched on), we have sought a dependence between t_b and this ϕ_{max}. We have checked that the subpopulations of arches grouped by ϕ_{max} also have such distributions [4]. For each one of these classes, a power law tail can be fitted with an exponent that is different for each group (see Fig. 4). Groups of arches with a bigger ϕ_{max} display both, a shorter breaking time and a larger α. The fact that the distribution of t_b is to a great extent determined by ϕ_{max} at the moment of the arch formation underscores the notion that the bead associated to this angle will largely dominate the dynamics of the arch breaking.

Assuming that this is so, and taking into account that a larger ϕ_{max} leads to short-lived arches, one can further

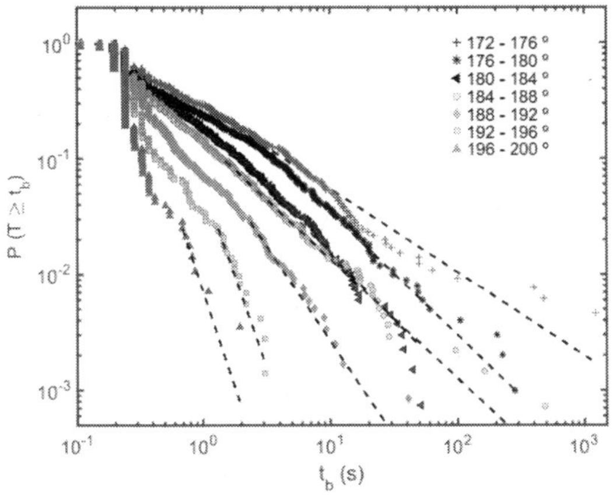

Figure 4. Complementary cumulative distribution function of the breaking times (sometimes also called survival function): it is the probability that an arch endures the vibration for a time equal or longer than t_b. Note the logarithmic scale. These data have been obtained for $R = 4.5\ mm$, $f = 100\ Hz$ and an acceleration of 2.6 g. The power law tails have been fitted (*dashed lines*) using the method described in [17], giving the following exponents (from left, bigger angles, to right, small angles): $\alpha = 4.1, 4.1, 2.6, 2.05, 2.1, 2.1, 1.6$.

hypothetise that

$$\frac{d\phi}{dt} \propto \phi \qquad (1)$$

(or alternatively, $d\phi/dt$ could be proportional to a growing function of ϕ, in which case Eq. 1 would be just a first approximation). If this equation is valid, the dynamics of ϕ_{max} will be the fastest. As changing the angle of one bead would affect the angle of other beads, it can be concluded that if the above statements hold, then the dynamics of the arch is *slaved* to that of the bead associated to ϕ_{max}.

Some previous unpublished observations that we carried out with a high speed camera suggest that indeed arches break in a way that could be described as explained above. In these inspections, we have seen that the bead starting with ϕ_{max} is usually the one that moves faster, and the last stages of the breaking process give something close to an exponential growth for ϕ vs. time. But only weak arches (with a big ϕ_{max}) were accessible to this survey, because the high speed camera is able to store just a short time span –a few seconds. These are usually the arches that have a big initial defect (a bead with a large ϕ_{max}), and consequently break after a short time span.

In the following we will consider the dynamics of long-lived arches. We have fixed the size of the exit orifice at 4.5 mm and recorded the formation of more than three thousand arches, as explained in Sect. 2. Once the arch was formed, the shaker was switched on, imparting the whole silo a constant sinusoidal vibration with a peak acceleration of 0.6 g (where g is the gravity acceleration) and a frequency of 105 Hz. Of all these arches, 174 (just over 5% of the total) withstood the vibration for more than one minute and broke before 20 minutes; we selected these for the following analysis.

After recording the images, we measured the angles associated to each bead in the arch along time. As an example, in Fig. 5, we plot the evolution of the angles of the beads in two arches along time. The accuracy of ϕ attained with our measurements is ± 0.5°. Fig. 5 (a) depicts the case of an angle with 5 beads that breaks slightly after two minutes (just after the photograph shown in Fig. 5 (b)). The bead associated to ϕ_{max} is a defect ($\phi > 180°$), and this angle increases monotonically during all the lifespan of the arch, as shown in Fig. 5 (c). Moreover, the arch breaks precisely there. Note also how $d\phi/dt$ grows bigger as ϕ increases. This is a good example of an arch amenable to be described with Eq. 1.

On the other hand, for the case displayed in Fig. 5 (d-f), the bead starting with ϕ_{max} is not the one where the arch finally breaks down; another bead with smaller ϕ overtakes it and governs the dynamics. Remark, however, that this bead had already begun as a defect, and it is not the only one; thus in this arch there is not a clearly dominant defect governing the dynamics of the whole structure.

It is noteworthy that among the selected 174 arches, 138 (representing 79%) broke at the bead with ϕ_{max}, and only 36 broke at a bead that was *not* the one sporting the maximum angle when the vibration was switched on. Of course, in these cases Eq. 1 does not hold (at least during the whole lifetime of the arch). Note that this is in agreement with the figures given in [15] which correspond to another kind of independent experiments.

The dynamics of the beads in the zone close to the arch (see Fig. 2) are also interesting. For example, although there is an ordered arrangement due to the 2D geometry (see Figs. 2 and 5), just over the arch this order seems to be smaller. We cannot use bidisperse beads to avoid this, but it would be interesting to study compaction in that zone. We know from numerical simulations that flow, stresses and local compaction does not change a lot if there is a small amount of polydispersity. Particle shape should not change a lot these results provided that the particles do not have flat faces.

This opens a new landscape to be explored, namely, how and why do some arches endure the vibration for a long time. For instance, how many of the beads where the arch breaks display the fastest dynamics? Do they all begin as a defect? Are the motions of *all* the beads correlated and effectively slaved to the fastest one? Many more data are needed to answer these questions, but the results shown here demonstrate that the analysis developed can be helpful to tackle them.

4 Conclusions

We have reported the first results of an experiment in which we can record and analyze the dynamics of long-lasting arches when submitted to an external vibration. We have shown how in many cases the initial "weakest link" (the bead with the maximum ϕ) may determine the dynamics of the whole arch. Moreover, it seems that at least in

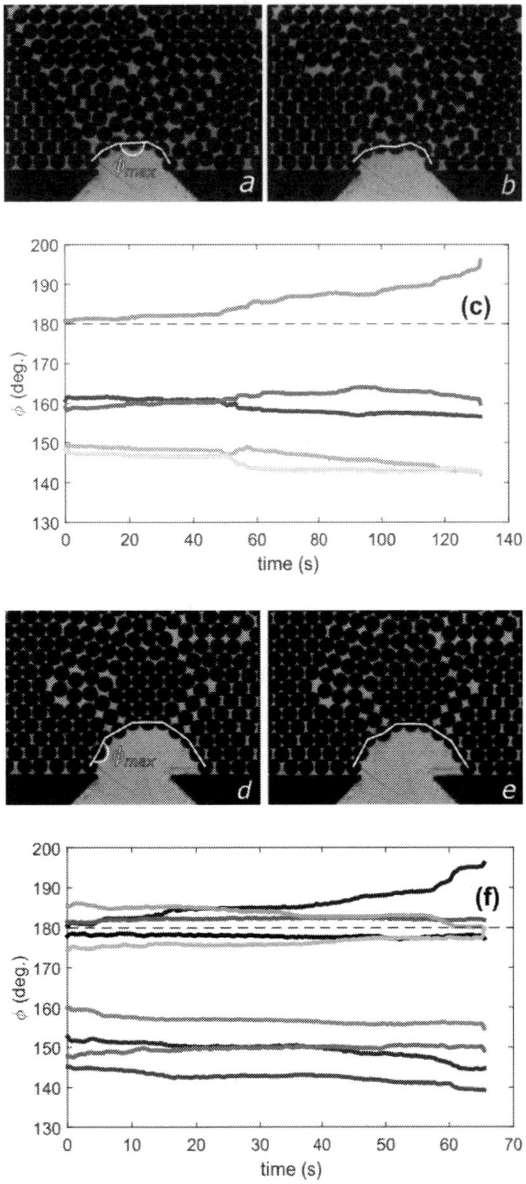

Figure 5. Evolution of the angles ϕ corresponding to the beads in an arch along of time. **(a)** An arch just formed, before the vibration begins. The maximum angle is indicated. Note that the base of the arch (the beads at the left and right ends) are not considered to belong to the arch. The white line joins the centers of the arch beads. **(b)** The last photograph of the arch before breaking. **(c)** The evolution of all the angles for the arch shown in (a) and (b). Here the dynamics is "slaved" to ϕ_{max}. **(d)** Another arch just formed, before the vibration begins. The maximum angle is indicated. **(e)** The last photograph of the arch in (d) before breaking. **(f)** The evolution of all the angles for the arch shown in (d) and (e). In this case, ϕ_{max} is not governing the dynamics. Remark that cases like the latter are much scarcer than the former. In plots (c) and (f), each shade corresponds to one bead, and the thickness of the lines are approximately equal to the experimental accuracy. The dashed horizontal line at $\phi = 180°$ allows to identify defects.

some cases (see Fig. 5 (a-c)) the motion of the other beads in the arch may be correlated to the motion of the fastest one. In some other cases this is not so, and the dynamics of these arches is not trivial and is not governed by the biggest initial defect.

The analysis presented here may allow to identify whether long-lived arches are born with a special configuration, or whether they evolve to become more robust. And it can also be helpful to determine if certain kind of mechanisms (for instance, stick-slip) are present in the process.

Acknowledgements

We thank L.F. Urrea for technical help, and D. Maza and R. Cruz Hidalgo for discussions and comments. This work was funded by Ministerio de Economía y Competitividad (Spanish Government) through Project No. FIS2014-57325. BG also thanks Asociación de Amigos de la Universidad de Navarra for a grant.

References

[1] I. Zuriguel *et al.*, Sci. Rep. **4**, 7324 (2014).
[2] A. Janda *et al.*, EPL **87**, 24002 (2009).
[3] C. Mankoc *et al.*, Phys. Rev. E **80**, 011309 (2009).
[4] C. Lozano, I. Zuriguel, and A. Garcimartín, Phys. Rev. E **91**, 062203 (2015).
[5] K. To *et al.*, Phys. Rev. Lett. **86**, 71 (2001).
[6] I. Zuriguel *et al.*, Phys. Rev. E **68**, 030301 (R) (2003).
[7] K. To, Phys. Rev. E **71**, 060301 (R) (2005).
[8] J. Tang and R. P. Behringer, Chaos **21** 041107 (2011).
[9] S. Tewari, M. Dichter, and B. Chakraborty, Soft Matter **9**, 5016 (2013).
[10] L. Kondic, Granular Matter **16**, 235 (2014).
[11] C. C. Thomas and D. J. Durian, Phys. Rev. Lett. **114**, 178001 (2015).
[12] B. Blanc, J.-C. Géminard, and L. A. Pugnaloni, Eur. Phys. J. E **37**, 112 (2014).
[13] J. M. Pastor *et al.* Phys. Rev. E. **92**, 062817 (2015).
[14] A. Nicolas, S. Bouzat, and M. Kuperman, Phys. Rev. E **94**, 022313 (2016).
[15] C. Lozano, G. Lumay, I. Zuriguel, R. C. Hidalgo, and A. Garcimartín, Phys. Rev. Lett. **109**, 068001 (2012).
[16] M. D. Shattuck, "*Particle tracking*", http://gibbs.engr.ccny.cuny.edu/technical/Tracking/ChiTrack.php
[17] A. Clauset, C. R. Shalizi, and M. E. J. Newman, SIAM Rev. **51**, 661 (2009).

Transient peristaltic transport of grains in a liquid

Marco Marconati[1,*], *Sharvari* Raut[1], *Farshad* Charkhi[1], *Adam* Burbidge[2], *Jan* Engmann[2], and *Marco* Ramaioli[1]

[1] *Department of Chemical and Process Engineering Faculty of Engineering and Physical Sciences, University of Surrey, Guildford GU2 7XH, UK*
[2] *Nestlé Research Centre, Route du Jorat 57, 1000 Lausanne 26, Switzerland*

Abstract. Pumping suspensions and pastes has always been a significant technological challenge in a number of industrial applications ranging from food processing to mining. Peristaltic pumps have become popular to pump and/or dose complex fluids, due to their robustness. During the transport of suspensions with peristaltic pumps, clogging issues may arise, particularly during transient operations. That is a matter of particular concern whenever the pumping device is used intermittently to generate flow only on demand. Further understanding of the transient dynamics of such systems and of the conditions that can lead to jamming would result in more robust peristaltic pump design. To achieve these goals, an experimental setup that simplifies the stator-rotor assembly of a peristaltic hose pump was used. In this setup, a roller transfers momentum to a liquid suspension, upon application of a constant load. The evolution of the velocity of the roller was recorded for different concentrations of mono-dispersed spheres of different diameters. The flow is found not to be strongly dependent on the dispersed particle volume fraction, if the size of the suspended phase is comparable with the hose diameter. Conversely, the flow is strongly slowed down when their size is small and the particle concentration is increased. These findings could help improving the design of peristaltic pumps by a more appropriate sizing, given the diameter of the hose and that of the particles to be transported.

1 Introduction

Peristalsis is defined as the process of progressive contraction and relaxation of membrane that results in the forward net motion of a liquid contained in the membrane. Peristalsis is effective in transporting complex fluids and peristaltic transport is used by nature e.g. the human gastro intestinal tract and reproductive system. In peristaltic transport also solid particles can be moved by the progressive squeezing action at the walls. Physiological examples include transport of kidney stones through the ureter, embryo movement in the uterus, and swallowing of hard objects such as nuts, grains and pills. A number of engineering applications of peristaltic flow are commonly found in the medical field, wastewater treatment, food and pharmaceutical industries, cement, and mining. One basic design comprises a drive-coupled rotor with mounted shoes or rollers, which continually compress and occlude some portion of an elastic tube. This action moves the fluid through the hose providing a precise volume of fluid pumped. That makes these pumps extremely attractive for metering purposes, especially in applications where the highest levels of hygiene have to be ensured. With the right choice of elastomer, hose pumps can cope with corrosive and hazardous fluids without the need for using valuable materials and special alloys in their construction. The increased popularity of peristaltic pumps is also due to other noticeable advantages in respect of classical positive displacement pumping systems: hose pumps are self-priming, reversible, and can cope with concentrated slurries (up to 70% w/W solids [1]). Moreover, peristaltic pumps are also suitable to transport shear sensitive fluids, due to the low shear applied on the fluid. Over the past years, a number of studies have considered numerical simulations of two-dimensional peristaltic flows both in finite and infinite domains solving the Navier-Stokes equation of motion in term of stream functions mainly retaining the approximations of low-Reynolds number and long wavelengths (lubrication theory). Characteristic phenomena of trapping and retrograde flow have been widely discussed [2] [8]. Additionally, several studies were conducted to understand the flow pattern in presence of dispersed solids [6][7][8][10]. The works of Fung [5], Hung [6], and Srivastava [7] constitute fundamental steps in the experimental investigation of peristaltic transport of solids. Numerical studies followed either using a one way coupling in which the fluid is treated as Eulerian and the particle motion as Lagrangian, or considering the full fluid-solid interaction between the suspended and the suspending phases [8] [11]. Despite the relevancy of those studies and the match with the corresponding experiments, there is still a significant lack of understanding the transient flow at start-up. Furthermore, the common assumption of geometrical symmetry is questionable, as significant variations in the flow pattern arise when considering different ampli-

*e-mail: m.marconati@surrey.ac.uk

tudes for the top and bottom walls [9] [2]. Depending on the number of rollers and the hose radius of curvature, an asymmetrical laminar wake, that generates an internal flow recirculation, can occur during the pumping cycle. In presence of suspended solids these recirculation could reduce the homogeneity of the product. Even worse, the presence of rigid objects and particles could clog the hose during start-up. The present study proposes a model, using a simplified hose pump design, to quantify the effect of both particle concentration and size on the measured angular velocity of a semisolid liquid.

2 Materials and Methods

The experimental setup consists of a circular bearing housing containing a thin plastic membrane that can freely deform under the action of a roller driven by a calibrated weight. The plastic membrane contains the liquid and the grains. The center of curvature of the bearing is coaxial with the shaft that supports the roller (Fig. 1). The supporting arm is balanced by a counterweight and the rotational inertia of the system was previously evaluated via free spinning of the arm. In order to investigate the most stringent conditions for peristaltic transport, a total occlusion configuration is used. The liquid volume used in the experiments was 6 mL. The solid phase consisted of rigid mono-dispersed spherical inclusions, whose diameter spanned between 10/12 and 5/12 of the tube maximum cross sectional size. In the particular geometry considered by the study, the ratio of radius of the plastic tube to liquid length ($\alpha = \pi R/\lambda$) is below 0.35, while the maximum value of the Reynolds number, computed by $Re = 0.5 R c \alpha / \nu$ is lower than 1. The density of the liquid phase is 1250 ± 10 kg m^{-3}, while the suspended phase density, measured with gas pycnometry, is 880 ± 40 kg m^{-3}. The liquid carrier consisted of 99.8% pure glycerol (Sigma-Aldrich, CAS Number 56-81-5) whose viscosity at 22°C was measured in compliance to ISO standard 3219 using a Paar Physica rotary rheometer (Model UDS 200, Paar Physica, Germany). Ambient temperature of 22 ± 1°C was constant throughout the duration of the experimental campaign. The experiments were first performed without grains, using three different applied loads: 200 g, 270 g and 400 g (roughly correspondent to 2, 2.7 and 4 N) to drive the roller. Then, solids were incorporated into the thick Newtonian liquid carrier at a volume fraction ϕ_v of either 0.06 or 0.12. The experiments were performed with different sizes of beads, adjusting the number of beads to match the desired volume fraction. The experiment is triggered by the instantaneous release of a pin. The motion of the apparatus and the liquid is recorded with a fast camera at 150 frames per second. The roller is free to move and accelerate till it reaches a final angle of 165° from the horizontal axis, where it is halted by a stopper. The suspension is instead free to exit from the tube. The movies recorded are used to extract the evolution of the position of the roller and the characteristic shape of the liquid through image analysis.

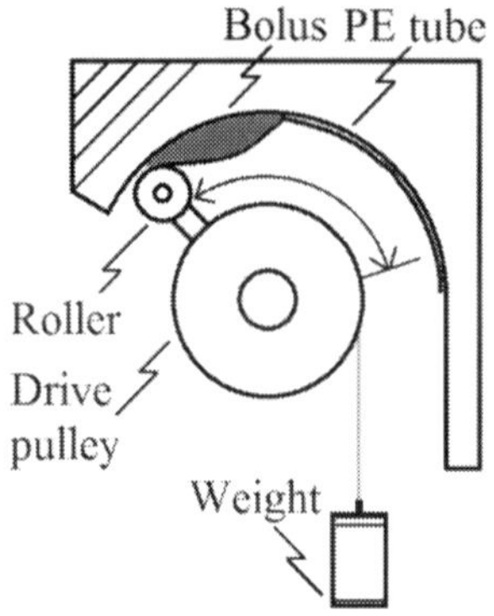

Figure 1. The experimental setup consist of a roller, driven by a set of calibrated masses, imposing a peristaltic deformation onto a thin plastic tube containing the liquid and the beads.

2.1 Theory

A simple mathematic model of the system, based on force balance was previously developed by Hayoun *et al.* The description of the drag force due to viscous dissipation was there derived using a Hagen-Poiseuille approach [12]. That solution well-matched the experimental results in the case of pure Newtonian liquids. Later, Mowlavi *et al.* extended the investigation to shear thinning liquids [13]. In their model, the drag force can be described by the mathematical expression presented in Eq. 1, in which the liquid bolus length and linear velocity are respectively L and ωR_A, μ_0 is the dynamic viscosity of the liquid, and β is a correction factor whose value is independent on liquid viscosity and driving force applied to the system and was experimentally determined by Mowlavi *et al.* [13].

$$F_d \quad = \beta \, 8 \, \mu_0 L \, (\omega R_A) \tag{1}$$

The above drag force was included in the differential equation of motion of the roller. The rotational parameters of the mechanical setup, such as the inertia of the rotating assembly and the friction of the bearing are characterized once for all with empty runs of the device. An additional correction is introduced to take into account the friction of the roller with the plastic membrane thereby the effective driving force momentum is consistently reduced by that fixed amount [13]. As a first approximation the dynamic effect due to the presence of suspended particles, of volume fraction ϕ, was modeled considering a correction term to the viscosity of the liquid carrier (μ_0) obtained from the Krieger-Dougherty law for spherical particles, where the packing ratio ϕ_m was considered that of a simple face-centered cube arrangement of particles (Eq. 2).

$$\mu_{\phi \neq 0} = \mu_0 \, (1 - \frac{\phi}{\phi_m})^{-2.5 \, \phi_m} \tag{2}$$

This approach, although simplistic in term of the actual particle-fluid and particle-particle interactions occurring for the geometries of field of motion considered, can

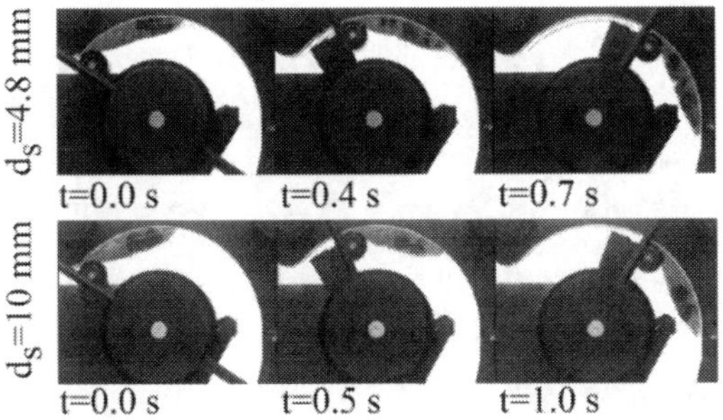

Figure 2. Screenshot taken at identical roller angular position in experiments at low applied load (200 g), with polyethylene beads of d_S =4.8 mm (at ϕ = 0.06 v/V), and d_S=10 mm (ϕ = 0.12 v/V).

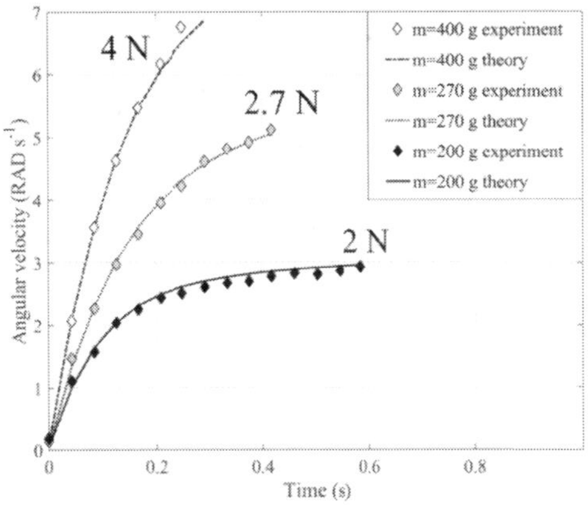

Figure 3. Experimental and theoretical roller angular velocity profiles for the three applied loads (200 g, 270 g, and 400 g) to the drive the roller, in absence of suspended solids (ϕ = 0.00 v/V).

still give useful insights to understand the experimental results.

3 Results

A visual comparison of the effect of inclusion size and volume fraction is reported in Fig. 2. Two extreme conditions of particle size and volume fraction are illustrated as an example, in experiments at low applied load (approximately 2 N). The screenshots show the same roller angle positions, in order to appreciate the different dynamics from the time. The roller movement evolves in a timescale on the order of 1 s in the case of the biggest inclusion at the highest volume fraction of solids (d_S =10 mm and ϕ_v = 0.12). The same force applied to a liquid with the lowest solid concentration (0.06 v/V) resulted in a 30% decrease in the transit time required for the solid-liquid dispersion to reach the end of the plastic membrane in which it is initially contained. The information retrieved from a direct comparison of the screenshots of the videos is made more quantitative using image analysis tools. The roller position was therefore tracked so that the instantaneous velocity profiles can be plotted and more efficiently compared.

A set of experiments were conducted in absence of solid inclusions (ϕ = 0) in order to validate the drag force model and obtain some reference lines for the subsequent trials with dispersed solids. Mean roller angular velocity profiles, together with the theoretical lines obtained with a viscous dissipation expressed in the form of Eq. 1, are illustrated in Fig. 3. The good agreement with the theoretical profiles is there evident comparing the small distance between the data points and this proved to be consistent with the results of Hayoun *et al.* [12]. Later, experiments were run in presence of suspended solid particles. In this case, velocity profiles highlighted an expected reduction in roller angular velocity when the volumetric fraction of particles was increased, consistently to the study by Lonzano [14], who considered a total occluded linear peristaltic pumping system for which Re = 50, α = 0.16 and the ratio of solid inclusion diameter to nominal tube diameter was 1/3. The decrease in angular velocity with increasing the volume fraction of suspended particles is however not linear. Our experimental results suggest that the effect depends both on the applied force to the system and on the size of the suspended particles. A comparison of the average velocity profiles obtained over two repetitions, for different concentrations, applied loads, and grain size, is reported in Fig. 4. Driving the system with a low applied force results in velocity profiles reaching quickly a steady state where the viscous dissipation equilibrates the imposed driving force. This kinematic condition is slightly affected by size of the suspended particles. A higher volume fraction accentuates the viscous dissipation and leads to a lower steady state velocity and a faster onset of the steady state.

Increasing the applied force results always in an increase of angular velocity, but the system takes longer to approach a steady state flow, as the steadily increasing velocity profiles obtained at the highest load configuration demonstrate. It can also be noted that the experimental profiles obtained at the highest load configuration are significantly closer to the theoretical lines drawn considering the increase in viscosity predicted with the Krieger-Dougherty law. Accordingly, the increase in viscosity, compared to that of the suspending carrier, is of the order of 15% in the case of the lowest concentration of suspended particles, and 35% for the highest volume fraction of grains. From Fig. 4 it is also clear that increasing the suspended particle volume fraction affects more the flow of smaller inclusions. Indeed, the velocity profiles for the 10 mm particles are not significantly different when the solid volume is doubled in the same suspending liquid carrier. Finally, the transport of biggest inclusion led generally to lower angular velocity. Although this effect is not strong in the range of inclusion size here considered, it appears consistently in the whole range of applied force considered. The simple model used still captures the ex-

 DOI: 10.1051/epjconf/201714009009

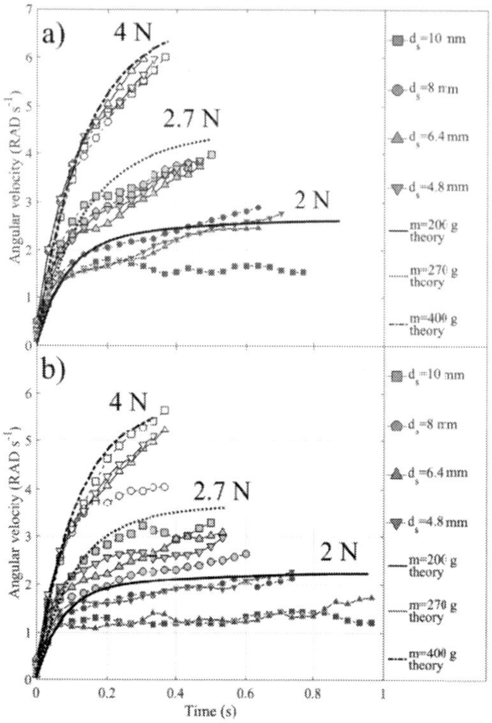

Figure 4. Experimental and theoretical roller angular velocity profiles at different applied loads to the drive pulley, in presence of suspended particles of diameters ranging from 4.8 to 10 mm. The volume fraction of solids is a) $\phi = 0.06$. and b) $\phi = 0.12$.

perimental data points in the first phase of inertial regime, proving the correct choice of the rotational dynamics parameters chosen, and correctly individuates the transition to the viscous regime. The viscous dissipation force is however always under-predicted as the experimental angular velocity data point fall below the corresponding theoretical profiles, suggesting the relevant role of particle-wall and particle-fluid interaction in this type of flow, even in the case of the smallest size of grains considered by the study.

4 Conclusions

This study aimed at investigating the transient start-up of a peristaltic pump transporting a thick Newtonian liquid with suspended hard grains. The experimental results showed that the roller velocity varies significantly as the concentration of suspended solids is increased, in particular in the case of smaller inclusions. Moreover, a significant reduction in pumping speed is always observed comparing the experimental results to the theoretical profiles plotted considering the apparent increase in viscosity predicted using a simple Krieger-Dougherty model. The effect of varying the inclusion size is stronger at lower applied loads, a condition in which the system moves quickly into a steady state flow. These results suggest the existence of a significant interaction of the suspended phase with the continuous liquid phase and highlight the need for developing a theory to describe the transient flow occurring during the start-up of a peristaltic pump in presence of suspended solids. Such theory would help designing robust pumps capable of performing precision metering of suspensions at different concentrations, regardless of the particle size being transported.

References

[1] B. Nesbitt, *Handbook of Pumps and Pumping* (Elsevier, Cambridge, UK, 2006) 38-39.

[2] M. Y. Jaffrin and A. H. Shapiro, Ann. Rev. Fluid Mech. **3**, 13-36 (1971).

[3] J. C. Burns, T. Parkes, J. Fluid Mech. **29**, 731-748 (1967).

[4] S. L. Weinberg, E. C. Eckstein, A. H. Shapiro, J. Fluid Mech. **49**, 461-479 (1971).

[5] Y. C. Fung, C. S. Yih, J. Fluid Mech. **47**, 93-112 (1971).

[6] T. K. Hung, T. D. Brown, J. Fluid Mech. **73**, 77-96 (1976).

[7] L. M. Srivastava and V. P. Srivastava, J. Biomech. Eng. **111**, 157-165 (1989).

[8] J. Jiménez-Lozano, M. Sen, P. F. Dunn, Phys. Rev. **79**, (2009) 041901-9.

[9] L. Fauci, Comput. Fluids **21**, 583-598 (1992).

[10] D. Takagi, N. J. Balmforth, J. Fluid Mech. **672**, 219-244 (2011).

[11] J. Jiménez-Lozano and M. Sen, Phys. Fluids **22**, 043303-10 (2010).

[12] P. Hayoun, J. Engmann, S. Mowlavi, B. Le Reverend, A. Burbidge, M. Ramaioli, J Biomech. **48**, 3922-3928 (2015).

[13] S. Mowlavi, J. Engmann, A. Burbidge, R. Lloyd, P. Hayoun, B. Le Reverend, M. Ramaioli, J Biomech. **49**, 3788-3795 (2016).

[14] J. Jiménez-Lozano, *Peristaltic flow with application to uretral biomechanics* (University of Notre Dame, Notre Dame, IN, 2009) 158-164.

Jamming transition evinced by Voronoi Tesselation

Eduardo Boaventura[1,2,⋆], *Fernando* Ducha[1,⋆⋆], and *A.P.F* Atman[2,3,⋆⋆⋆]

[1]*Graduate program in Mathematical and Computational Modeling - Centro Federal de Educação Tecnológica de Minas Gerais (CEFET-MG). Av. Amazonas, 7675 - Nova Gameleira, Belo Horizonte, MG, Brazil. Cep 30510-000.*
[2]*Physics and Mathematics Department - Centro Federal de Educação Tecnológica de Minas Gerais (CEFET-MG). Av. Amazonas, 7675 - Nova Gameleira, Belo Horizonte, MG, Brazil. Cep 30510-000.*
[3]*Instituto Nacional de Ciência e Tecnologia de Sistemas Complexos (INCT SC).*

Abstract.
We calculated the Voronoi tessellation (VT) of simulations of two-dimensional granular system disturbed by an intruder. The intruder is kept fixed within a rectangular box containing certain amount of grains. Making the box to move against the intruder we calculated the VT for several instants of this movement, analyzing geometrical properties of the Voronoi polygons in function of time. The dependence of stationary values of the polygon areas and number of sides in function of the packing fraction (ϕ) of the granular media has shown that the system displays a possible jamming transition at critical point $\phi_c \approx 80.5\%$.

1 Introduction

The study of granular materials is very important due to its ubiquity and wide range of applications as, for example, pharmaceutical, food or semiconductor industries[1]. They are the second more manipulated material in industry (after water)[2].

In the industry we often have processes involving flow and transport of granular material. An undesired behavior in such processes is the clogging of equipments[3]. Such clogging is related to a possible phase transition of the granular system, from fluid to solid-like state (jammed). That condition is characterized by the presence of mechanical stability, a finite resistance to shear and isotropic deformation. In crystalline solids such stability comes from long-range order but in disordered systems as granular packings there still have not a consensus concerning the mechanism leading to the mechanical stability[4]. We know, however, there must be an external application of stress in the granular system so that it reaches the jammed condition[3]. When granular systems are in jammed phase they are said to belong to a new class of materials, called "fragile matter"[5].

Granular materials may behave as crystalline solids and can also be found acting in similar way as fluids. Considering the possibility of a jamming phase transition for granular materials, it is intuitive the jam can occur only above certain packing fraction. We can also say that increasing the stress applied to the jammed system it may results in unjamming. Thus, it has been proposed a phase

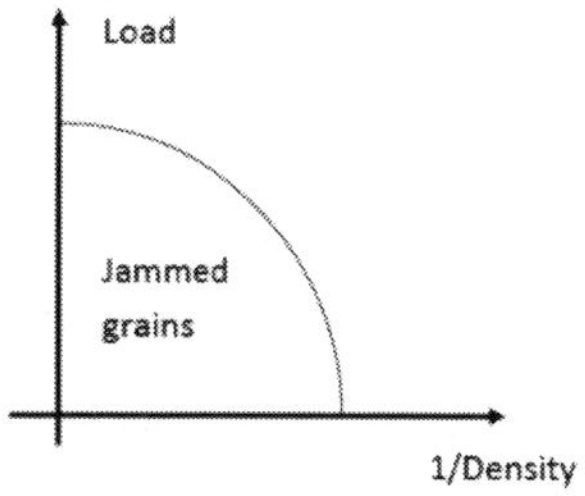

Figure 1. Possible phase diagram for granular material. The system is jammed within the region limited by the curve and the two axes. We can see if one increases the load or decreases the density, the jammed system undergoes a transition to an unjammed phase.

diagram[3] for jamming transition in granular materials which is partially sketched in Figure 1.

A key problem related to the transport and flow of granular materials, their rheology, can be studied by means of immersed objects in their flows. Historically, that was how the scientists studied, in fluid mechanics, behaviors such as drag force, boundary effects, turbulence etc [7] .

The "packing fraction", represented by ϕ, is related to the density of the system, defined, in two dimensions, as the sum of the areas of the grains divided by the area of the box.

In this paper we analyze the behavior of specific Voronoi polygon area which circumscribes the intruder. We also analyze the behavior of polygon areas of the cav-

⋆e-mail: edboaventura@deii.cefetmg.br
⋆⋆e-mail: duchaf@dppg.cefetmg.br
⋆⋆⋆e-mail: atman@dppg.cefetmg.br

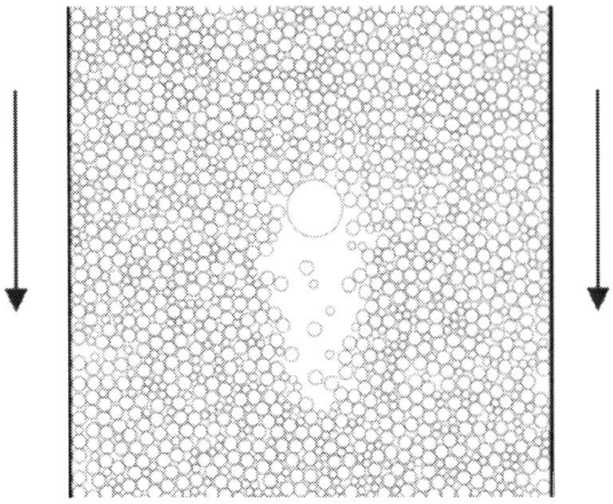

Figure 2. Top view of a little region from the virtual apparatus in some simulation instant. The two arrows outside the box indicate the direction of the box movement. The small circles indicate the grains and the big one indicates the intruder. The empty area behind the intruder is the cavity. ϕ = 0.76.

ity formed behind the intruder during the displacement of the box, Figure 2.

Next, in Section 2, we briefly describe the VT and the features and procedures used in system simulation. In Section 3 the results are presented and in Section 4 our conclusions and perspectives.

2 Methods

The aforementioned Voronoi tessellation, Figure 3, also known as Voronoi diagram, is an usual tool from the computational geometry[8] used to solve numerous problems involving the concept of proximity in the plane or in the space. We will here shortly comment about this technique applied to the Euclidean plane but keeping in mind that its generalization to the 3-dimensional space is easily achieved by extending the following concept: given a set of special points located in a plane - sites - we can partition this plan in regions called Voronoi regions or Voronoi polygons. Each polygon includes one and only one site. The area of the Voronoi polygon associated to the site s corresponds to the area formed by the set of all points of the plane for which s is the closest site among all the sites. The set of all Voronoi polygons of a plan form what we call Voronoi tessellation[9, 10].

To explore the flow behavior of 2D dense and disordered granular material near the jamming transition, using the information given by VT, we simulated a system comprised of a rectangular box containing bi-disperse grains distributed on the box bottom, with random initial velocities. The grains radii have size ratio $r_{max} \approx 1.7r_{min}$, where r_{max} and r_{min} are, respectively, the maximum and minimum radii. The grains with minimum radius in the system represent 4/7 of the total of grains. The intruder radius is 4 times larger than the r_{max}. The longitudinal position of the

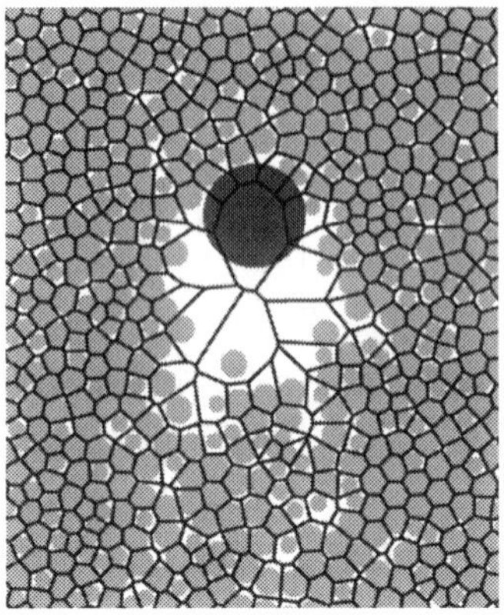

Figure 3. Part of a Voronoi tessellation applied to one of the simulated granular systems. The orange disks represent the grains and the darker (blue) and larger disk is the intruder. (Color online)

intruder is set to 1/4 of the length of the box. After the system reach an stationary state, below a small kinetic energy threshold (1% initial energy), a horizontal displacement is imposed to the box, at the speed of $0,01r_{max}$ by second. This motion is kept until the intruder reaches 3/4 of the box length. While the box moves the intruder offers resistance to its displacement due the interaction between it (intruder) and the grains in the box, producing a drag force. An empty area is formed behind the intruder, Figure 2. As said before this area receives the name of "cavity"[7].

The grains surrounding the intruder impose a drag force on it and it is intuitive that the greater the number of grains around the intruder the greater the value of the drag force and the greater the system density.

The system is simulated by molecular dynamics (MD) which is a type of Discrete Element Method (DEM)[11]. This technique is based on the integration of Newton's equations of the motion. An important observation in this type of system is that the grain collisions are inelastic, so it is introduced a critical damping term for the energy dissipation[12]. The repulsion force between grains during a collision is simulated by Hooke's law. That force is proportional to a little interpenetration (Hertzian contact law)[13] between the interacting grains by normal and tangential stiffness and Coulomb friction condition.

Simulations results were averaged over 5 samples at 12 different packing fractions (from $\phi = 0.75$ to $\phi = 0.84$). We collected data for a time interval from $t = 0$ to, approximately, $t = 100,000$ time steps. We calculated the Voronoi Tessellation for 100 different times (every 1000 steps) for each ϕ. We calculated the Voronoi tessellations considering each grain center as a Voronoi site - Figure 3.

3 Results and discussions

The statistical distribution of polygon areas and polygon number of sides for the ensemble of all the grains in the box did not indicate significant differences on ϕ variation

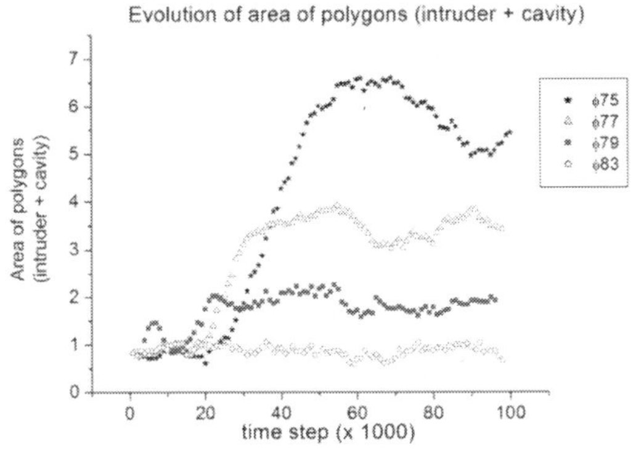

Figure 4. Time evolution of the combined area from the polygons of intruder and cavity.

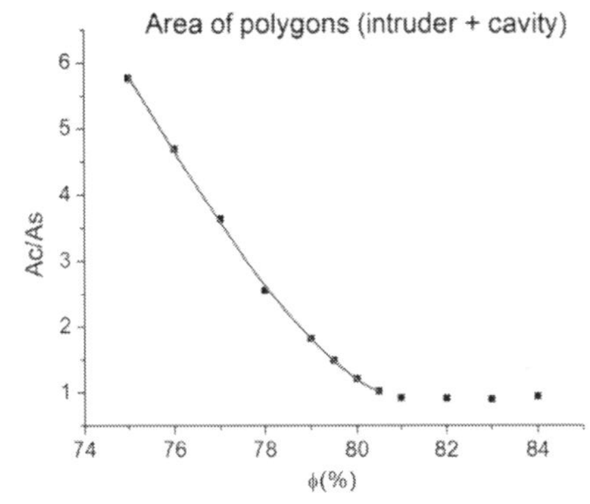

Figure 5. Average of the combined area versus ϕ. Note that an abruptly change on behavior of the areas occurs in the range $0.80 < \phi < 0.81$. The points were fitted by $y = C + A(\phi_c - \phi)^{\beta}$, where $y = Ac/As$, C and A are constants and β is the critical exponent. Ac is the combined area and As is the standard combined area, that is, combined area of $\phi = 84\%$. The value of the parameters are $C = 1.0 \pm 0.2$, $A = 0.5 \pm 0.1$, $\phi_c = 80.5 \pm 0.5$ and $\beta = 1.4 \pm 0.1$.

, that is, by this analysis one can not distinguish between the different $\phi's$ or if could there is evidence of a phase transition from the fluid state to the jammed one.

We believe that the fact of there is no distinction between different $\phi's$ in function of the area distribution occurs because in systems with smaller $\phi's$ the polygons of the majority of the grains are almost as small as the polygons of the grains in systems with largest $\phi's$. Thus, the former (small $\phi's$) has a lower overall density but its local density (out of the cavity) differs little from the local density of the latter (large $\phi's$). We infer that the main difference in the size of the polygons of these two systems (with small $\phi's$ and with large $\phi's$) is more sharply in the region of the cavity, that is, out of the cavity the polygons have almost the same size of area. However, as the amount of the large polygons in the cavity region is too small compared to the whole system, its importance is negligible into the statistics.

Nevertheless, analyzing only cavity polygons together with the polygon of the intruder and comparing the results of small and large $\phi's$ systems we can draw some very interesting conclusions.

In Figure 4 we can see the time evolution of the combined area (intruder polygon + cavity polygons). Note that the smaller the ϕ the greater the values reached by combined areas. Note also the greater the ϕ the sooner the value of combined areas begins their abrupt growth.

In Figure 5, we can see the average of combined area versus ϕ. Note that the linear decrease stops between $\phi = 0.80$ and $\phi = 0.81$. After $\phi = 0.81$ the behavior is approximately constant. A very similar behavior, but inverted in relation to Figure 5, is the evolution of the average drag force on the intruder, Figure 6. That figure shows that the average drag force on the intruder remains approximately constant until $\phi = 0.80$. In the range $0.80 < \phi < 0.81$ the behavior changes abruptly, showing, after $\phi = 0.81$, a linear growth. Those arguments suggest that in this interval, $0.80 < \phi < 0.81$, we can find the critical packing fraction or ϕ_c.

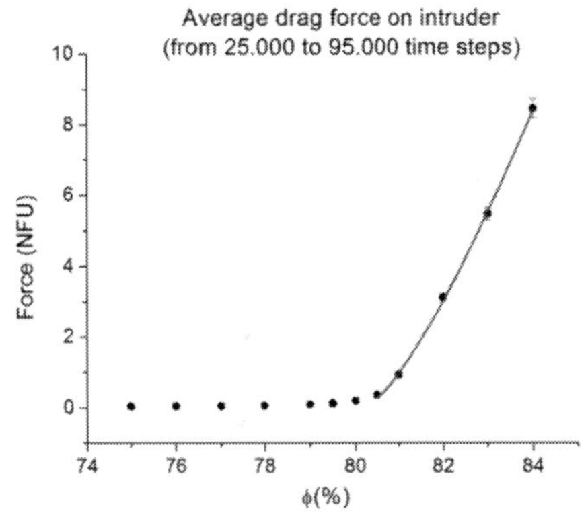

Figure 6. Average of the drag force on the intruder. Here also an abruptly change on behavior of the force occurs in the range $0.80 < \phi < 0.81$. NFU = normalized force unit. The points were fitted by $f = C + A(\phi - \phi_c)^{\beta}$, where f is the force, C and A are constants and β is the critical exponent. The value of the parameters are $C = 0.25$, $A = 1.6 \pm 0.4$, $\phi_c = 80.4 \pm 0.2$ and $\beta = 1.3 \pm 0.1$.

Figure 7 shows the behavior of the number of sides from the intruder polygon in function of ϕ. The data fit gave us the value $\phi = 80.2 \pm 0.1$ for the inflection point. The inflection point can be understood as a critical value, i.e., the point at which one can estimates a phase transition occurs.

We emphasize that the value of the inflection point in Figure 7 is in perfect agreement with the region of behavior change of the system, according to Figure 5 and Figure 6.

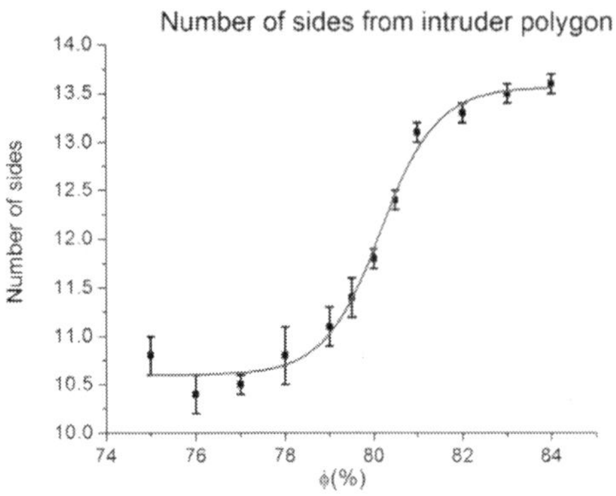

Figure 7. Average of the number of sides from the intruder polygon. The sigmoidal fit provided the following parameters $A_1 = 10.6 \pm 0.1$, $A_2 = 13.6 \pm 0.1$, $\phi_c = 80.2 \pm 0.1$ and $d\phi = 0.7 \pm 0.1$.

4 Conclusions and perspectives

The combined polygon areas (intruder + cavity) decrease as we increase ϕ, Figure 4. Another interesting feature is the moment in which the system begins the abrupt growth of the combined area. This growth occurs at an earlier instant as we increase the value of ϕ, with decreasing area stationary values.

Due the arguments arising from the analysis of the Figure 5, Figure 6 and Figure 7, our calculations show strong evidence that this kind of system goes through a phase transition (from a fluid or unjammed phase for a jammed one). From the results of the three cited figures we calculated $\phi_c = (80.4 \pm 0.2)\%$. We also calculated the average critical exponent as $\beta = 1.4 \pm 0.1$.

We may find practical situations (in industries, scientific studies, etc.) in which the estimation of the jamming transition is difficult by means of force sensor installation. The VT is a functional alternative, requiring the filming of the granular system and a software that calculates the VT's, from this filming.

Calculating the areas or the number of sides of Voronoi polygons is a much simpler technique than directly calculating the area of the cavity. In addition, the calculations involving the polygons are quite accurate. This makes the use of Voronoi tessellation more advantageous than the direct calculation of the cavity area.

The jamming of the system was not observed because our simulation increases (infinitely if necessary) the force (or load) on the box whenever it is needed for continuing moving it but the next step of this research is the analysis of this system with limited forces, that is, we expect to see how the system jams (maybe only for $\phi \geq \phi_c$), according to the maximum possible force as we changed the value of ϕ. We also intend to change the stiffness of grains and dispersity of the system to get new results.

Acknowledgements

We would like to thank the CEFET-MG. APFA thanks CNPq research grant 312312/2015-6.

References

[1] Duran, Jacques. *Sands, powders, and grains: an introduction to the physics of granular materials* (Springer Science & Business Media, New york , 2012), p. 214.

[2] Richard, Patrick and Nicodemi, Mario and Delannay, Renaud and Ribiere, Philippe and Bideau, Daniel, Nature materials **vol. 4**, n.2, p. 121–128 (2005).

[3] Liu, Andrea J. and Nagel, Sidney R., Nature **Vol. 396**, n. 6706, p. 21–22 (1998)

[4] Majmudar, T. S. and Sperl, M. and Luding, S. and Behringer, Robert P., Physical Review Letters **Vol. 98**, n. 5, p. 058001 (2007)

[5] Cates, M. E. and Wittmer, J. P. and Bouchaud, J. P. and Claudin, P., Physical Review Letters **Vol. 81**, n.9, p.1841 (1998)

[6] Atman, APF and Kolb, E and Combe, G and Paiva, HA and Martins, GHB. *POWDERS AND GRAINS 2013: proceedings of the 7th international conference on micromechanics of granular media, AIP Conference Proceedings* (AIP Publishing LLC, Sydney, 2013) vol. 1542, p. 381–384.

[7] Kolb, Evelyne and Cixous, Pierre and Gaudouen, Niels and Darnige, Thierry. Physical Review E **Vol. 87**, n. 3, p. 032207 (2013)

[8] Boaventura, E. C. and Ducha, F. A. and Atman, A.P.F., Proceeding Series of the Brazilian Society of Computational and Applied Mathematics **Vol 4**, n. 1, (2016)

[9] Fortune, Steven. Algorithmica **Vol. 2**, n. 1–4, p. 153–174 (1987)

[10] Okabe, Atsuyuki and Boots, Barry and Sugihara, Kokichi and Chiu, Sung Nok. *Spatial tessellations: concepts and applications of Voronoi diagrams* (John Wiley & Sons, West Sussex, 2009) p. 696

[11] Luding, S., European Journal of Environmental and Civil Engineering **Vol 12**, n. 7-8, p. 785–826 (2008)

[12] Shäfer, J. and Dippel, S. and Wolf, D. E. , Journal de physique I **Vol. 6**, n. 1, p. 5–20 (1996)

[13] Adams, G. G. and Nosonovsky, M. , Tribology International **Vol. 33**, n. 5, p. 431–442 (2000)

 DOI: 10.1051/epjconf/201714003017

Collisional model of the drag force of granular impact

Cacey Stevens Bester[1,*] and *Robert P.* Behringer[1,**]

[1] *Department of Physics, Duke University, Durham, NC 27708, USA*

Abstract. A dense, dry granular target can cause a free-falling intruding object to come to an abrupt stop as its momentum is lost to the grains. An empirical force law describes this process, characterizing the stopping force as the sum of depth-dependent friction and velocity-dependent inertial drag. However, a complete interpretation of the stopping force, incorporating grain-scale interactions during impact, remains unresolved. Here, the momentum transfer is proposed to occur through sporadic, normal collisions with clusters of high force-carrying grains at the intruder's surface. To test this model in impact experiments, we determine the forces acting on an intruder decelerating through a dense granular medium using high-speed imaging of its trajectory. We vary the geometry of the impacting object to infer intruder-grain interactions. As a result, we connect the inertial drag to the effect of intruder shape based on the proposed collisional model. These impact studies serve as an approach to understand dynamic force transmission in granular media.

1 Introduction

Mechanical probing by way of impact serves as a central method to explore momentum transfer in granular media. For instance, when a solid sphere is dropped into dry granular material, the intruder penetrates into the medium and suddenly decelerates to rest. The drag force from granular media is then studied through impact. There is long, rich history of the study of granular impact [1, 2] as well as a natural connection to planetary-scale impact phenomena [3–7]. However, a complete understanding of the physics of granular impact is still lacking.

An empirical force law has been proposed to describe the deceleration $\ddot{z}$ due to the granular target [8]. It compares the force of gravity mg with the velocity-dependent inertial drag $h(z)\dot{z}^2$ and depth-dependent friction force $f(z)$ as

$$m\ddot{z} = mg - h(z)\dot{z}^2 - f(z) \qquad (1)$$

The inertial drag governs much of intruder dynamics during deceleration, and its $\dot{z}^2$ dependence suggests a collisional process. Yet the grain-scale origin of the inertial drag force is not fully determined.

Force transmission of granular impact was recently explored via experiments of intruders impacting a two-dimensional arrangement of photoelastic grains [9], allowing for imaging of the granular response. This reveals a dynamic force network propagating through the granular target. Deceleration of the intruding object was found to be dominated by sporadic collisions with high force carrying clusters. These force chains propagate perpendicular from the intruder surface, indicating that intruder-grain friction is not important for initial deceleration, when $h(z)\dot{z}^2$ dominates. This observation of normally propagating force chains was also observed using intruders with rough and textured surfaces. The inertial drag force of granular media can therefore be described by a collisional model [10].

Here we extend to three-dimensional impact experiments and connect the inertial drag of granular impact to its grain-scale origin. We use high-speed imaging to determine the forces acting on an intruder during impact. As we vary the conical shape of the intruder, its resultant trajectory changes and can be connected to force propagation occurring perpendicular to the surface. As a result, we develop a 3D collisional-based description which is consistent with inertial drag values.

2 Tracking intruder dynamics

A schematic of the experiment setup is displayed in fig. 1(a). In each experimental run, an intruder impacts a granular target of dry, noncohesive grains of either sand or couscous, which are held within a 49 cm length X 38 cm width X 20 cm depth container. The granular bed is large enough to minimize interactions with the container boundary [11], and the packing fraction, the fraction of space that is filled by grains, is held about constant. Before each run, the container is tapped then the granular bed is leveled using a straight edge. The intruder is dropped from a height of up to 2 m, falls under gravity, and impacts the granular target. A thin rod is attached to the intruder for tracking during penetration in the grains.

We methodically vary the intruder-grain interactions during penetration by using cone-shaped intruders. An example is depicted in fig. 1(b), which shows an illustra-

*e-mail: cacey.stevens@phy.duke.edu
**e-mail: bob@phy.duke.edu

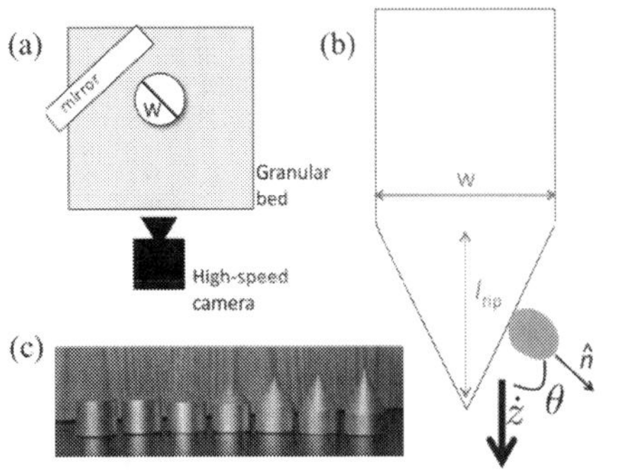

Figure 1. (a) Top view of experiment setup. A mirror is placed at 45 degrees to the impact to measure any tilting during penetration. (b) Diagram of the impact of a conical intruder with a chain of particles within the granular target (orange ellipse). (c) Image of intruders of w=3.8cm, varying s.

tion of a conical intruder colliding with a single cluster of grains (represented by an orange ellipse). At impact, force is expected to propagate perpendicular to the surface with θ being the angle between the intruder's velocity $\dot{z}$ and surface normal n. Cone shapes are defined by its slope s, the ratio of the length of the cone l_{tip} to its width w, and ranges from 0 (cylinder) to 2.1 (a sharp cone) (see fig. 1(c)). With increasing s, we thereby increase θ of expected force propagation. The effect of the cone width w is also explored for w=1.9 cm to w=3.8 cm of s=1 cones. The mass of all intruders is held constant at 101±2 g.

We place a Photron Fastcam SA5 high-speed camera at a side view of the granular bed to record the results. Videos are captured at 30,000 frames per second as the intruder comes to rest within 50 ms. From each frame, we track the position z of the intruder with an accuracy of about 0.04 cm (1 pixel). We then differentiate $z(t)$ to measure velocity $\dot{z}$ and perform a convolution with a gaussian filter of cutoff frequency of 960 Hz to reduce the amplified experimental error from $z(t)$.

The dynamics of $z(t)$ and $\dot{z}(t)$ are shown in fig. 2 for cones of s from 0 to 2.1 as they impact and come to rest in the granular bed of sand. From fig. 2(a), we learn that z approaches and saturates at a maximum penetration depth at about 0.04 s, with z=0 and t=0 at impact. This trend is consistent with different s. Intruders with higher s penetrate deeper, given the same initial energy. We take the derivative of $z(t)$ to find $\dot{z}(t)$ for different s as given in fig. 2(b); $\dot{z}(t)$ declines less rapidly with t as we increase s. However $\dot{z}$ reaches 0 at approximately the same time. We differentiate $\dot{z}(t)$ to find intruder acceleration $\ddot{z}(t)$, which reflects forces exerted by the granular medium. Forces are accordingly determined by high-speed imaging of the paths $z(t)$.

3 3D Collisional model

3.1 Description

We use the change of impact motion from different intruder shapes to verify that normal collisions capture in-

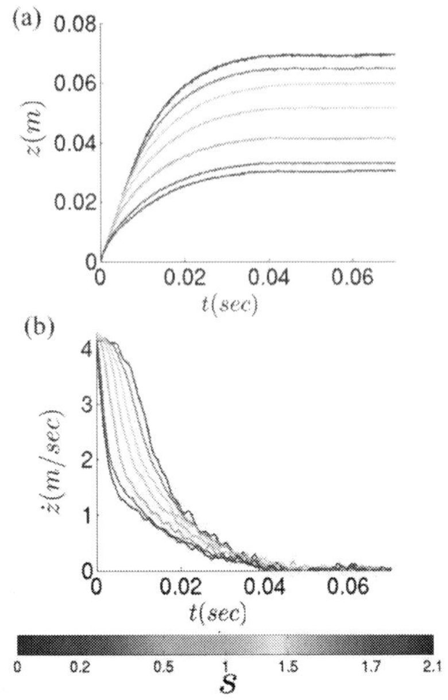

Figure 2. (a) z vs. t, varying s, all impacting sand from the same release height. The maximum penetration depth increases with increasing s. (b) $\dot{z}$ vs. t. The decline of $\dot{z}(t)$ changes with increasing s, but all curves reach $\dot{z}$=0 at approximately the same time. Data sets correspond to s values given by the colorbar.

ertial drag. Since momentum is lost by collisions perpendicular to the surface (fig. 1(b)), different inertial drag values are expected to arise with changing s. Accordingly, we propose an explanation of $h(z)\dot{z}^2$ as a collisional-based model in terms of the intruder shape.

The force due to an intruder-grain collision can be expressed as the ratio of momentum $\Delta p = m_g \dot{z} \cos\theta$ to time $\Delta t = d/\dot{z}\cos\theta$, where m_g and d are the grain mass and diameter, respectively. By assuming these collisions are equally likely across the surface dS, the number of discrete collisions is then given by $dS\, cos\, /d^2$, and total force is calculated by integrating over the surface area. Forces due to collisions along the intruder surface can consequently be expressed as

$$h(z)\dot{z}^2 = \int \frac{\Delta p}{\Delta t} dS \propto \rho_g w^2 \dot{z}^2 cos^2\theta \tag{2}$$

where ρ_g is the grain density. The shape effect due to inertial drag is indicated through w and $cos\ \theta$. With geometric arguments, we can express eq. 2 in terms of s as

$$h(z)\dot{z}^2 \propto \frac{\rho_g w^2 \dot{z}^2}{1+s^2} \tag{3}$$

We thereby connect the inertial drag of the macroscopic force law to a simple scaling relation that depends on intruder shape (s and w) and ρ_g, and we show consistency of $h(z)\dot{z}^2$ measurements with changes to s, w, and ρ_g.

We experimentally determine $h(z)\dot{z}^2$ to demonstrate that eq. 3 connects to the grain-scale origin of inertial drag of 3D granular impact. Using the force law to find

$h(z)$ would require finding z, $\dot{z}$ and $\ddot{z}$ for many trajectories of each intruder. Additionally, acceleration data $\ddot{z}$ can present significant error due to filtering [12]. Accordingly, previous authors have shown that the differential equation can be solved by reframing in the kinetic energy K versus z form where $K = 1/2m\dot{z}^2$ and $m\ddot{z} = dK/dz$ [13, 14]. The approach leads to the following:

$$\frac{dK}{dz} = mg - f(z) - \frac{2h(z)}{m}K \tag{4}$$

Therefore, we acquire $h(z)$ from z and $\dot{z}$ and achieve

$$K(z) = e^{-\int \frac{2h(z)}{m}dz}[K_i + \int (mg - f(z))e^{\int \frac{2h(z)}{m}dz}dz]$$

when eq. 4 is solved with the integrating factor $e^{\int \frac{2h(z)}{m}dz}$.

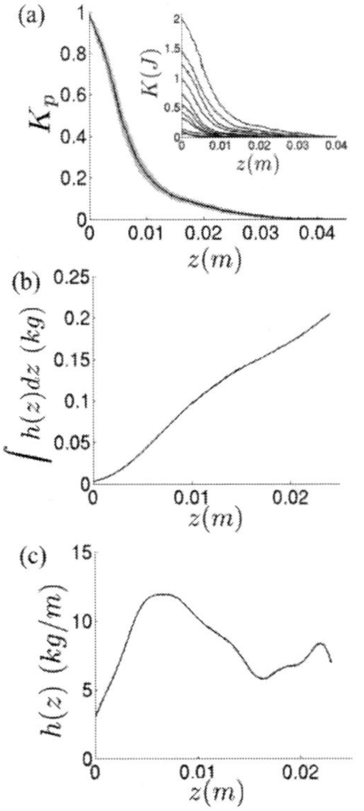

Figure 3. (a) Inset: K versus z of s=0 intruder impacting sand. Each curve represents a trajectory of a different $\dot{z}_i$. Main panel: $K_{p,avg}$ versus z, as determined from all pairs of $K(z)$ trajectories of the inset plot. The gray region surrounding the curve indicates the standard deviation from the mean of K_p. (b) $\int h(z)dz$ versus z (c) $h(z)$ versus z, showing oscillations about a constant value.

We take the difference of two $K(z)$ curves from experimental runs with distinct impact velocity $\dot{z}_i$ to isolate $h(z)$ and obtain the relation

$$K_p = \frac{K_b - K_a}{K_{b,i} - K_{a,i}} = e^{-\int \frac{2h(z)}{m}dz} \tag{5}$$

where a and b indicate indices of different trajectories. By averaging all pairs of $K(z)$ trajectories for a range of $\dot{z}_i$, $h(z)$ is found for each intruder.

Figure 3 shows how we determine $h(z)$ for an s=0 intruder penetrating sand. We start by expressing data as $K(z)$, as shown in the inset of fig. 3(a). There is a rapid decline of K during penetration. The average of differences of all pairs of $K(z)$ trajectories is then determined (see fig. 3(a)), and we use eq. 5 to solve for $\int h(z)dz$, given in fig. 3(b). The solution in this case is a $\int h(z)dz$ curve that is a linear function of z. The derivative of this curve gives $h(z)$ and requires noise reduction by performing a convolution to the data (see fig. 3(c) for result). There are oscillations in $h(z)$ which may connect to the force fluctuations and discretization of collisions [9]; a precise determination of these features would require direct force measurement. To avoid handling noise amplification and possible data distortion introduced by taking derivative of $\int h(z)dz$, we utilize $\int h(z)dz$ for the s, w, and ρ_g comparisons.

3.2 Intruder shape and grain density

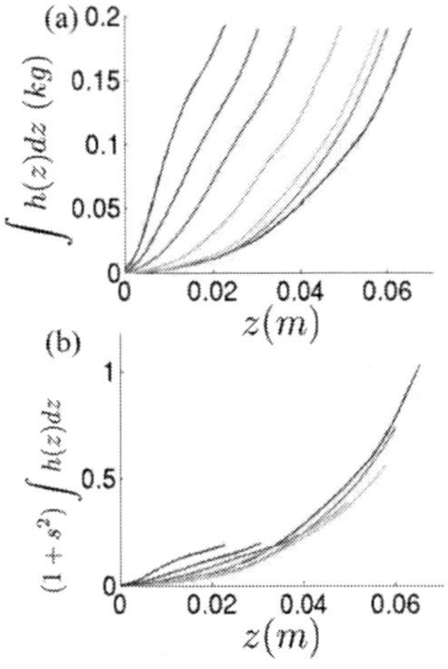

Figure 4. (a) $\int h(z)dz$ versus z for all intruders, ranging from s=0 to s=2.1. Colors correspond with s values of fig. 2. Higher s lowers the curve to smaller inertial drag values. (b) The data collapse to one curve when scaled as a function of s.

Starting with the shape effect, we show that s and w affect $h(z)$ as described by eq.3. In fig. 4(a), $\int h(z)dz$ versus z is plotted as s ranges from 0 to 2.1. Curves shift to lower inertial drag values with increasing s. We also find that $\int h(z)dz$ vs. z plots become increasingly curved as s increases. We propose that the increased curvature results from the changing surface area at low z as a sharper cone is penetrated deeper into the granular target. Since w is increasing at z=0 up to z=l_{tip} and $\int h(z)dz$ is proportional to w^2, $\int h(z)dz$ rises quadratically in this range. After $z > l_{tip}$, $\int h(z)dz$ becomes a linear function of z.

All curves fall onto a single curve when the $\int h(z)dz$ axis is rescaled by a function of s, as shown in fig. 4(b). The expression $1 + s^2$ leads to a collapse of all data as indeed shown in eq. 3. The scaling therefore shows that the collisional model does agree with our 3D experiments.

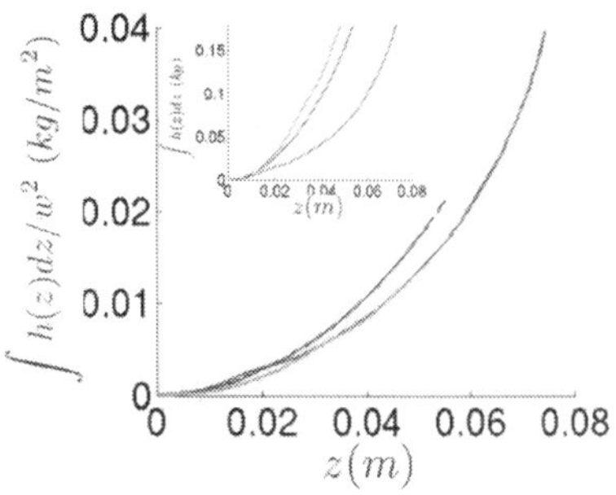

Figure 5. Inset: $\int h(z)dz$ versus z of s=1 intruder, varying width w: 1.9cm, 3.0cm, and 3.8cm. Lower w decreases the curve, but, in this case, the trend in data is not affected. Main panel: When scaled as a function of w, the data collapse to a single curve.

When extending the study to the effect of w, we similarly observe that the collisional model applies. Figure 5 shows $\int h(z)dz$ as a function of z for s=1 cones of three widths: w=1.9cm, w=3.0cm, and w=3.8 cm. We see that $\int h(z)dz$ decreases with decreasing w as displayed in the inset plot. Additionally, the curved-linear transition of each curve connects with l_{tip}. The dependence of inertial drag on w is then displayed by a collapse of data as $\int h(z)dz \propto w^2$. These data rescale to a single curve when plotted as $\int h(z)dz/w^2$.

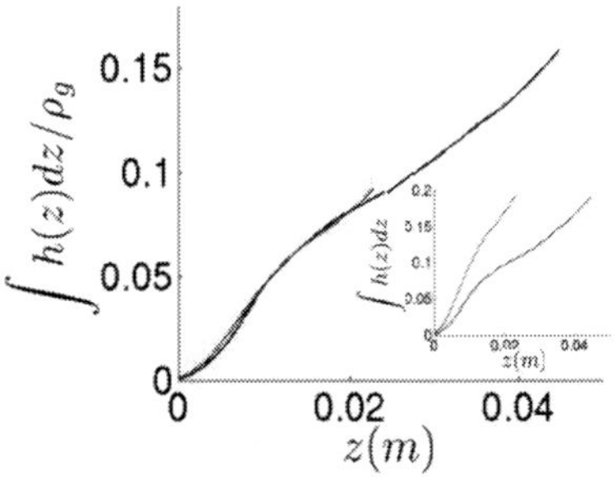

Figure 6. Inset: $\int h(z)dz$ versus z for s=0 intruder, varying ρ_g: sand (red line) or couscous (black line). Higher ρ_g(sand) increases inertial drag coefficient $h(z)$. Main panel: Scaled $h(z)$, $\int h(z)dz/\rho_g$ versus z for the two grains, collapsing data.

Lastly, we vary the granular target (sand or couscous) to study the effect of grain density ρ_g. Both targets are prepared by pouring grains into the container and tapping to settle the surface. The inertial drag coefficient $h(z)$ is determined for an $s = 0$ intruder impacting the sand or couscous target. For s=0, $\int h(z)dz$ is a linear function of z. We show that our collisional model also captures the effect of ρ_g (see Fig. 6). There is a shift of $h(z)$ with ρ_g; we find that $\int h(z)dz$ is larger for higher ρ_g, as expected from the $h(z)$ expression of the collisional model (eq. 3). Accordingly, there is a higher stopping force due to sand, and this affects the maximum penetration depth of the intruder. The two curves collapse to a single curve, when scaled by ρ_g as ascribed in the collisional model.

4 Conclusions

We study the dynamics of a cone-shaped intruder impacting with dry granular media. Granular impact is described, in part, by a velocity-dependent inertial drag force. Based on its dependence on intruder shape, we connect the inertial drag from the granular target to sporadic, normal collisions with force-carrying chains of grains. A collisional model is presented to support our description of inertial drag for 3D impact and is explored through our experimental data. The collapse of the inertial drag coefficient when accounting for intruder shape and granular target demonstrates the validity of the model and shows that intruder-grain friction can be ignored for this term. These scaling results are consistent with our observation of dynamic force transmission in granular media.

Acknowledgements

This work has been supported by Duke University Provost's Postdoctoral Program, NSF-DMR1206351, and NASA grant NNX15AD38G. We appreciate the assistance from Jonathan Bares, Hu Zheng, and Yuchen Zhao.

References

[1] J. Poncelet, *Cours de Mecanique Industrielle* (1829)
[2] H. Katsuragi, *Physics of Soft Impact and Cratering* (Springer, Tokyo, 2016)
[3] C. Li, T. Zhang, D. Goldman, Science **339**, 1408 (2013)
[4] K. Daniels, J. Coppock, R. Behringer, Chaos **14**, S4 (2004)
[5] C. Guttler, N. Hirata, A. Nakamura, Icarus **220**, 1040 (2012)
[6] D. Dowling, T. Dowling, Am. J. Phys. **81**, 875 (2013)
[7] J. Ruiz-Suarez, Rep. Prog. Phys. **76**, 066601 (2013)
[8] H. Katsuragi, D. Durian, Nat. Phys. **3**, 420 (2007)
[9] A. Clark, L. Kondic, R. Behringer, Phys. Rev. Lett. **109**, 238302 (2012)
[10] A. Clark, A. Petersen, R. Behringer, Phys. Rev. E **89**, 012201 (2014)
[11] A. Seguin, Y. Bertho, P. Gondret, Phys. Rev. E **78**, 010301 (2008)
[12] D.M. Meko, *Applied Time Series Analysis* (UA Press, 2015)
[13] M.A. Ambroso, R.D. Kamien, D.J. Durian, Phys. Rev. E **72**, 041305 (2005)
[14] A.H. Clark, R.P. Behringer, EPL **101**, 64001 (2013)

Contact forces between a particle and a wet wall at both quasi-static and dynamic state

Huang Zhang[1], *Sheng* Chen[1], and *Shuiqing* Li[1,*]

[1]Key Laboratory of the Thermal Science and Power Engineering of Ministry of Education, Department of Thermal Engineering, Tsinghua University, Beijing 100084, China

Abstract. The contact regime of particle-wall is investigated by the atomic force microscope (AFM) and theoretical models. First, AFM is used to measure the cohesive force between a micron-sized grain and a glass plate at quasi-static state under various humidity. It is found out that the cohesive force starts to grow slowly and suddenly increase rapidly beyond a critical Relative Humidity (RH). Second, mathematical models of contacting forces are presented to depict the dynamic process that a particle impacts on a wet wall. Then the energy loss of a falling grain is calculated in comparison with the models and the experimental data from the previous references. The simulation results show that the force models presented here are adaptive for both low and high viscosity fluid films with different thickness.

1 Introduction

Wet granular materials are most common in our daily environment as well as in many engineering processes [1,2]. The packing or flow behaviour of wet granular media is closely relevant with the interactions at the grain level [1]. Interstitial liquid between the wet grains induces the granular matter cohesive arising from the capillary forces [3]. The experimental measurements and theoretical derivation to determine the magnitude of capillary force between Geldart-D coarse particles have been well established at quasi-static state [3-5]. For micron-sized grains, other adhesion effects caused by van der Waals force, electrostatic force, etc., become significant and complex [6]. So it is interesting to investigate how capillary force behaves in such a small length-scale. It seems that AFM offers a '*prototypical*' approach to measure the contact forces in these conditions [7].

When the wet granular media is agitated, partially wetted grains are impacting and separating with each other [8]. The thin liquid layer covering a wet grain helps to push or pull the other grain [8]. Due to the hydrodynamic and viscous fluid flows in this liquid film, the first-contact energy loss [9] of wet particles is higher than that of dry ones [10]. Experiments can determine the particle velocities before and after collisions, so that the first-contact energy loss is obtained according to the velocity differences [10]. However, the details of contact mechanism of wet particles are relied on the theoretical analysis [11].

The structure of this paper is organized as following. First, the capillary force between a micron-sized grain and a plate (the plate can be considered as a sphere with zero curvature) is measured by AFM under different humidity. Second, the force models of a grain colliding with a wet wall are established in order to analyze the contacting mechanism, especially the first-contact energy loss between the grain and the wet wall.

2 Experiment setup

Fig. 1 shows the humidity control system and AFM (CSPM5500) setup. The circuit of nitrogen flows going through the water tank is to entrain liquid into the AFM cell, while another circuit of dry nitrogen is to control the magnitude of RH. The RH in AFM cell is adjusted to range from 2% to 85%. RH is measured by VAISALA sensor. A SiO_2 particle with diameter 15.52 μm is glued to the original AFM tip to form a spherical tip.

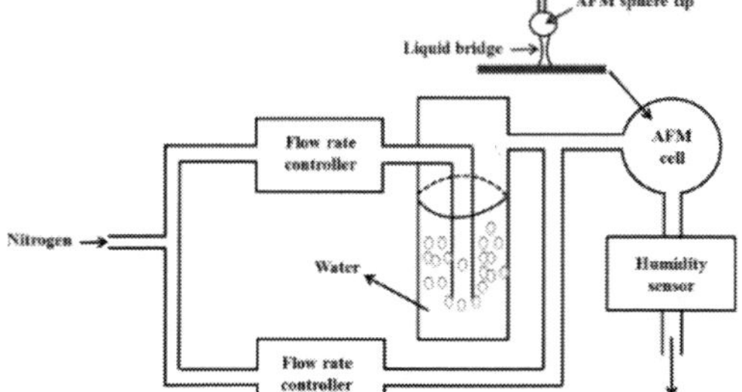

Fig. 1. Humidity control system and AFM setup.

3 Theoretical models

This part focuses on the establishing of the dynamic model that a grain impacts on a wet wall. Fig.2 performs

* Corresponding author: lishuiqing@tsinghua.edu.cn

the penetration and rebound schematic diagram. In this figure, β is the angle between the three-phase contact line and the y axis, and θ is the contact angle of solid and liquid. When the falling particle contacts the liquid layer (stage I), the fluids are draining out in the thin gap. During this process, the particle undergoes the resistance from the surface integration of the fluid stress at the interface between the particle and the liquid layer [12]. After penetrating the liquid film, the grain is colliding with the wall (stage II), and then, the grain rebounds through the fluid layer (stage III). After that, the liquid bridge pulls the particle until it is ruptured.

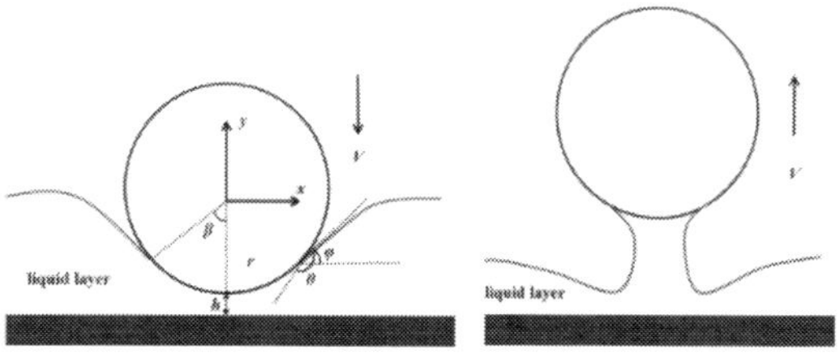

Fig. 2. Schematic diagram of a particle penetrating and separating with a liquid layer.

In stage I, the particle undertakes gravity force, surface tension force and the fluid force resulting from the fluid flows in the thin layer. This fluid force consists of three parts: unsteady, buoyance and resistance term [11,12]. When the particle Reynolds number Re_p is sufficiently small, the buoyance and resistance terms are converted into the lubrication force [3, 11]. The motion equation of the sphere at this stage is shown as below

$$m_p \frac{d\boldsymbol{v}}{dt} = \boldsymbol{F}_g + \boldsymbol{F}_s + \boldsymbol{F}_a + \boldsymbol{F}_b + \boldsymbol{F}_d, \tag{1}$$

where m_p, $\boldsymbol{v}$, $\boldsymbol{F}_g$, $\boldsymbol{F}_s$, $\boldsymbol{F}_a$, $\boldsymbol{F}_b$ and $\boldsymbol{F}_d$ are particle mass, particle velocity, gravity force, surface tension force, unsteady force, buoyance force and drag force, respectively. These forces are expressed as: $\boldsymbol{F}_g = (4/3)\pi r^3\boldsymbol{g}$, $\boldsymbol{F}_s = -2\pi\gamma r\sin\beta\sin\varphi\boldsymbol{v}/|\boldsymbol{v}|$, $\boldsymbol{F}_a = -(1/6)\pi r^3\rho_f(\cos^3\beta - 3\cos\beta+2)d\boldsymbol{v}/dt$, $\boldsymbol{F}_b = -(1/3)\pi r^3\rho_f(2+\cos\beta)(1-\cos\beta)^2\boldsymbol{g}$, $\boldsymbol{F}_d = -(1/2)\pi C_d\rho_f(r\sin\beta)^2|\boldsymbol{v}|\boldsymbol{v}$, where r, $\boldsymbol{g}$, γ, ρ_f, and C_d are particle radius, gravity acceleration, surface tension coefficient, liquid density, and drag coefficient, respectively [11,12]. C_d is related with Re_p, which are defined as below [11]

$$C_d = \frac{24}{\mathrm{Re}_p} + \frac{4}{\sqrt{\mathrm{Re}_p}} + 0.4, \mathrm{Re}_p = \frac{\rho_f |\boldsymbol{v}| d}{\mu}, \tag{2}$$

where d, μ are particle diameter and liquid viscosity. Particularly, if the flow Reynolds number Re_f in the liquid film is small, buoyance force and drag force are eventually lubrication force [3,13]. Re_f, which is estimated by considering mass conservation in the liquid layer, is performed as $\mathrm{Re}_f = \rho_f|\boldsymbol{v}|(rh)^{0.5}/(2\mu)$, where h is the gap distance between the bottom of sphere and the wall [3]. Here, we assume that if Re_f is less than one, $\boldsymbol{F}_b$ and $\boldsymbol{F}_d$ become to be the lubrication force $\boldsymbol{F}_l$, which is shown as $\boldsymbol{F}_l = -6\pi\mu r^2|\boldsymbol{v}|/(h+h_r)$ [11]. Here h_r is the surface roughness, which is used to avoid the physical singularity when h is approaching zero [14].

In stage II, the grain collides with the wall with no interstitial fluid evolves any more [10]. We use $\boldsymbol{v}_1$ and $\boldsymbol{v}_2$ to represent the particle velocities that are just contacting and leaving the wall surface. Therefore, $\boldsymbol{v}_1$ and $\boldsymbol{v}_2$ are correlated with the restitution coefficient between the particle and the dry wall [6,10], which is given as $\boldsymbol{v}_2 = -e_{\mathrm{dry}}\boldsymbol{v}_1$, where e_{dry} is the restitution coefficient for dry condition.

In stage III, the grain starts to leave the wet wall. The forces that the particle undergoes in this period are nearly the same as those in stage I. While, $\boldsymbol{F}_s$ is replaced by the liquid bridge force $\boldsymbol{F}_c$ [4], which is given as below

$$\boldsymbol{F}_c = -2\pi\gamma r\cos\theta\left(1 - \frac{1}{\sqrt{1+\frac{2V_b}{\pi r(h+h_r)^2}}}\right)\frac{\boldsymbol{v}}{|\boldsymbol{v}|}. \tag{3}$$

V_b is the liquid bridge volume, which is estimated as $d^3/16$ [10]. The rupture distance h_{rup} is crucial to determine the breaking point of the liquid bridge. The evolution of h_{rup} is expressed as [5]

$$h_{\mathrm{rup}} = (1+\theta)(1+\sqrt{\mathrm{Ca}})V_b, \tag{4}$$

where Ca is the capillary number that equals to $\mu|\boldsymbol{v}|/\gamma$. Otherwise, β and θ keep constant where $\beta = 120^\circ$ [16] and $\theta = 111^0$ [15] in stage I, and, $\beta = 80^\circ$ [16] and $\theta = 10^0$ [17] in stage III. Additionally, φ equals $\beta+\theta-\pi$ [12].

4 Results and discussions

Fig. 3 gives the AFM measurement of the critical cohesive force F_{crit} under different RH. In general, the magnitude of F_{crit} grows with increasing RH. F_{crit} rises slowly when RH is below the critical RH that is around 65%. While, F_{crit} rises up rapidly as RH is beyond 65%. Actually, the surfaces of both SiO_2 particle and glass plate are not smooth (the average roughness is about 2.98 nm by AFM measurement). As RH ranges from 3% ~ 65%, liquid could only exist within the surface concave due to capillary condensation [7]. Hence, the liquid connections can hardly contribute to the increment of the total cohesive force. When RH is high (beyond 65% in this case), still due to the capillary condensation, liquid could fill up the gap between the micron-sized particle and the plate. Then, the liquid volume V_b in the gap rises with the growing RH. Because the distance between the particle and the wall is constant, F_{crit}, which is now estimated by Eq. (3) [4,7], would become larger and larger with increasing V_b.

The occurrence of critical RH needs further analysis. In equilibrium state, Kelvin equation describes the relationship between the shape of liquid concave and RH, which is shown as below,

$$r_c = \frac{2\gamma V_m}{R_g T\log(\mathrm{RH})}, \tag{5}$$

where r_c, γ, V_m, R_g, and T are liquid concave radius, surface tension, liquid molar volume, ideal gas constant, and temperature. When the condensate liquid just fills up the surface roughness, r_c is estimated as -2.98 nm, whose magnitude is supposed to be the average roughness. We take T as 20°C, so RH is calculated as 69.6% by Eq. (5), which is very close to the critical RH (65%) from Fig. 3. Further, suppose the gap between the particle and the plate is filled with liquid if RH is approaching to 90% or higher. The contact force is regarded as $F_{\text{crit}} \sim \pi\gamma d$ [3]. According to the material properties, $F_{\text{crit}} \sim \pi\times73\times10^{-3}\times15.52\times10^{-6}$ N ~ 3600 nN, which is more or less close to the maximum value of F_{crit} (about 2000 nN) in Fig. 3.

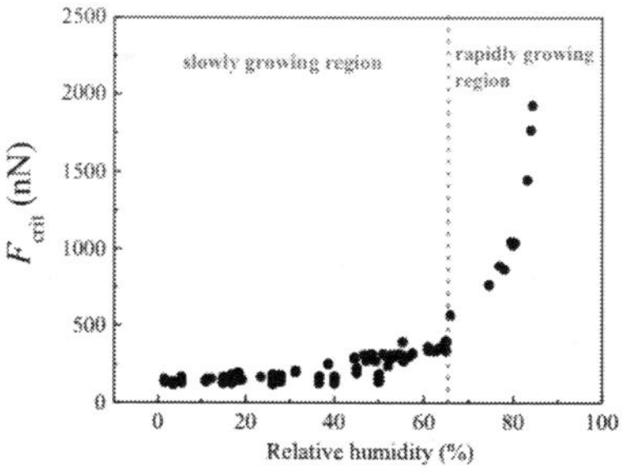

Fig. 3. Critical cohesive force F_{crit} increases with relative humidity by AFM measurement.

Eq. (1) quantitatively capture the essential features of the impacting process in Fig. 2. In stage I, the total distance that the grain moves is the initial thickness of liquid layer δ. Substituting the impacting velocity of particle $\boldsymbol{v}_0$ into Eq. (1), we can get $\boldsymbol{v}_1$ after numerically solving this Ordinary Differential Equation (ODE, and the numerical toolkit used here is ODE45 in MATLAB R2010a). Then, $\boldsymbol{v}_3$ is also obtained from Eq. (1), in which $\boldsymbol{F}_s$ is instead of $\boldsymbol{F}_c$. Therefore, the first-contact energy loss is expressed as $0.5m_p(|\boldsymbol{v}_0|^2- |\boldsymbol{v}_3|^2)$. In order to verify these theoretical models, ΔE_{wet} that indicates the energy loss during the process of the grain interacting with the liquid film, is compared with the experimental data in reference [10]. Obviously, ΔE_{wet} equals to $0.5m_p(|\boldsymbol{v}_0|^2-|\boldsymbol{v}_3|^2-|\boldsymbol{v}_2|^2+|\boldsymbol{v}_1|^2)$. Meanwhile, according to an energy budget analysis referred in ref. [10], ΔE_{wet} mainly consists of viscous dissipation and fluid material acceleration, which are given by [10]

$$\Delta E_{\text{wet}} = \Delta E_{\text{vis}} + \Delta E_{\text{acc}}, \tag{6}$$

$$\Delta E_{\text{vis}} = \frac{3}{2}\pi\mu d^2 \,|\, \boldsymbol{v}_0 \,|\, (\ln\frac{\delta}{h_r} + \ln\frac{h_{rup}}{h_r}), \tag{7}$$

$$\Delta E_{\text{acc}} = \frac{3\rho_f}{2\rho_s}(\frac{\delta}{d} - \frac{3}{5}(\frac{\delta}{d})^2 + \frac{2}{3}(\frac{\delta}{d})^3) m_p v_0^2. \tag{8}$$

Here d_{rup} is considered as $2d$ according to the observation in ref. [10]. The material properties of two kinds of fluid M5 and M50 are given in Table 1. The restitution coefficient e_{dry}, particle roughness h_r, particle density ρ_s and diammeter d are 0.985, 5 μm, 2580 kg/m^3 and 5.5 mm, respectively [10].

Table 1. The material properties of M5 and M50 [10].

Liquid	ρ_f (kg/m^3)	γ (mN/m)	μ (mPa·s)
M5	925	19.2	4.6
M50	965	20.8	48

As seen in Fig. 4(a), ΔE_{wet} predicted by our models is in good agreement with the experimental data. Generally, ΔE_{wet} increases with $|\boldsymbol{v}_0|$. Also, the more thick δ is, the larger ΔE_{wet} is dissipated. Remarkably, the energy loss of the particle impacting into the liquid M50 is much higher than that of the liquid M5. For instance, ΔE_{wet} is around 20 μJ as $|\boldsymbol{v}_0|$ is 1 m/s in the liquid M5 case, which is nearly one third of that in the same case of liquid M50.

Although the results obtained from Eqs. (6-8) qualitatively follow the tendency of ΔE_{wet} versus $|\boldsymbol{v}_0|$ and δ, it deviate a lot in case of liquid M5. However, the accuracy of Eqs. (6-8) is highly promoted when calculating ΔE_{wet} in liquid M50. In contrast, ΔE_{wet} obtained by our model can still agree well with the experimental data of liquid M50, but they are not as well as the results by Eqs. (6-8) in this case. The viscosity of liquid M50 is nearly one thousand times larger than M5. Eq. (7) can be considered as the intergration of $\boldsymbol{F}_l$, which is given as below

$$\Delta E_{\text{vis}} = \int_{h_r}^{d_{rup}} + \int_{\delta}^{h_r} (\boldsymbol{F}_l \cdot d\boldsymbol{x}) = \int_{h_r}^{d_{rup}} + \int_{\delta}^{h_r} (\frac{3\pi\mu d^2 \boldsymbol{v}}{2x} \cdot d\boldsymbol{x}). \tag{9}$$

Eq. (7) is obtained from integrating Eq. (9) by taking $\boldsymbol{v}$ as $\boldsymbol{v}_0$, which should be larger than the particle velocity in actual situation. This is the reason that Eqs. (6-8) could get good results in the viscosity dominated case (M50), but fail in the less viscosity case (M5). However, as seen from Fig. 4, our model is adaptive for both low and high viscosity fluid. Furthermore, our model could predict the particle velocity at any given time, which is convenient to analysis the contacting mechanism of the particle with the liquid layer.

Fig. 4 indicates that ΔE_{wet} depends on $\boldsymbol{v}_0$, μ, and δ, etc. A usually used dimensionless numer to clarify the energy loss process is Stokes numer St, which is defined as $\rho_s d|\boldsymbol{v}_0|/(9\mu)$ [3,10-11,13]. In the main diagram of Fig. 5, the experimental points of $\Delta E_{\text{wet}}/E_0$ (E_0 is the initial kinetic energy of an impact particle) are almost collapsed into a fitting curve, whose form is as $a\text{St}^{-2}+b\text{St}^{-1}+c$. Similarly, the scatter results by our model are also fitted by a curve with the same form as in the main diagram.

As seen in Fig.5, the ratio of ΔE_{wet} to E_0 does not exceed 0.4 when St^{-1} ranges from 0 to 0.005. With increasing St^{-1}, this ratio grows up to around 0.8. The law of the fitting curve is not occasional, which is intrinsically shown as following: $\Delta E_{\text{wet}}/E_0$ is direct propotion to e_n^2 [10], where e_n is the coefficient of restitution. Further, $e_n \sim (1\text{-St}^{-1})$ [3] to lead that the curve of $\Delta E_{wet}/E_0$ is in the form of $a\text{St}^{-2}+b\text{St}^{-1}+c$.

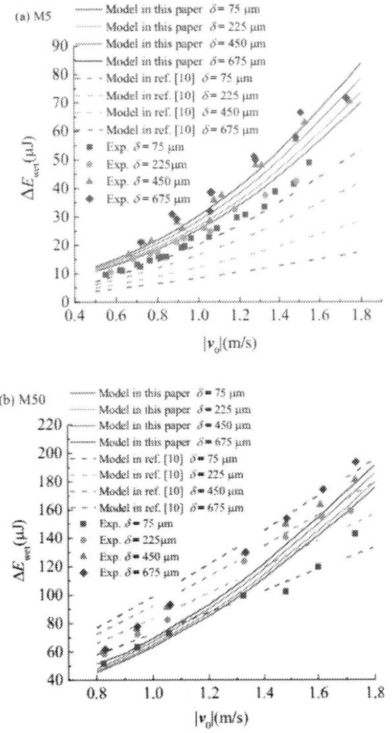

Fig. 4. ΔE_{wet} caclulated by the model in this paper is compared with both the model as shown in Eqs. (6-8) and experimental data in reference [10].

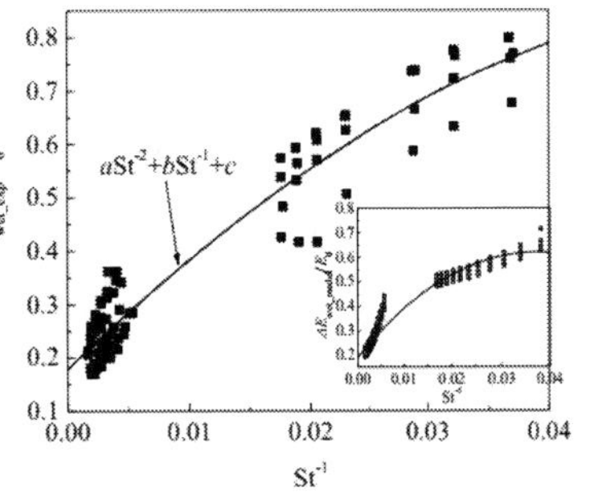

Fig. 5. The ratio of ΔE_{wet} to E_0 varies with St^{-1} by experiment in reference [10] (inset is the result predicted by our theoretical model).

5 Conclusions

In this paper, the forces that the particle contacts with a wet wall in both quasi-static and dynamic state are performed. First, capillary force is measured by AFM between a micron-sized particle and a glass plate. It is found out that the capillary force begins to grow slowly with RH, and then increases rapidly. This result is explained by capillary condensation that liquid only exists in the concave of the surface roughness below a critical RH, but it fills up the gap beyond a critical value.

Second, the colliding process that a falling grain impacting onto a wet wall is analyzed. Compared with the methodology of energy budget referred in the previous literature, the force models presented here could not only calculate the first-contact energy loss, but also study the contact forces and particle velocities at any given time. The simulation results show that our model agrees well with the experimental data and are adaptive in both low and high viscosity fluid cases.

Acknowledgements

This work was funded by the China Postdoctoral Science Foundation and the Major State Basic Research Development Program of China (Grant No. 2016YFC0203705).

References

1. S. Herminghaus, Adv. Phys. **54**, 221 (2005).
2. J. Delenne, V. Richefeu, F. Radjai, J. Fluid Mech. **762**, R5 (2015).
3. B. Andreotti, Y. Forterre, O. Pouliquen, *Granular media between fluid and solid* (Cambridge University Press, New York, 2013).
4. O. Pitois, P. Moucheront, X. Chateau, J. Colloid Interface Sci. **231**, 26 (2000).
5. O. Pitois, P. Moucheront, X. Chateau, Eur. Phys. J. B **23**, 79 (2001).
6. J. Marshall, S. Li, *Adhesive particle flow a discrete-element approach* (Cambridge University Press, New York, 2014).
7. R. Jones, H. Pollock, J. Cleaver, C. Hodges, Langmuir **18**, 8045 (2002).
8. P. Liu, R. Yang, A. Yu, Chem. Eng. Sci. **86**, 99 (2013).
9. S. Chen, S. Li, M. Yang, Powder Technol. **274**, 431 (2015).
10. F. Gollwitzer, I. Rehberg, C. Kruelle, K. Huang, Phys. Rev. E **86**, 011303 (2012).
11. S. Antonyuk, S. Heinrich, N. Deen, H. Kuipers, Particuology **7**, 245 (2009).
12. D. Lee, H. Kim, Langmuir **24**, 142 (2008).
13. J. Simeonov, Acta. Mech. **227**, 565 (2016).
14. E. Izard, T. Bonometti, L. Lacaze, J. Fluid Mech. **747**, 422 (2014).
15. D. Vella, Langmuir **22**, 2972 (2006).
16. A. Wang, Q. Song, Q. Yao, Atmos. Environ. **115**, 1 (2015).
17. R. Zhu, S. Li, Q. Yao, Phys. Rev. E **87**, 022206 (2013).

Looking into the evolution of granular asteroids in the Solar System

Paul Sánchez[1,⋆], *Daniel* Scheeres[2,⋆⋆], *Masatoshi* Hirabayashi[3,⋆⋆⋆], and *Simon* Tardivel[2,⋆⋆⋆⋆]

[1] *Colorado Center for Astrodynamics Research, University of Colorado Boulder, 431 UCB, Boulder, CO 80309-0431, USA*
[2] *Aerospace Engineering Sciences, University of Colorado Boulder, 429 UCB, Boulder, CO 80309-0429, USA*
[3] *Department of Earth and Atmospheric Sciences, Purdue University, 550, West Lafayette, IN 47907, USA*

Abstract. By now it has been accepted that most of the small asteroids in the Solar System are granular aggregates kept together by gravitational and possibly, cohesive forces. These aggregates can form, deform and disrupt over millennia subjected to different internal and external factors that would ultimately determine how they evolve over time. Parameters such as porosity, cohesive and tensile strength, angles of friction, particle size distributions, stress states, heterogeneity and yield criteria among others, determine how these granular systems will react when subjected to different, changing, external factors. These external factors include solar photon momentum, gravitational tides, micro- and macro-impacts and are believed to have produced and shaped the current asteroid population. In our research we use a combination of Soil Mechanics theory, Soft-Sphere Discrete Element Method (SSDEM) Simulations and Orbital Mechanics in order to understand how simulated, homogeneous and heterogeneous, ellipsoidal and spherical gravitational aggregates, a crude but useful representation of an asteroid, evolve when rotated to the point of disruption. Then, we compare our results to the shapes of observed asteroids as well as to the disruption patterns of a few active asteroids. Our results lead us to believe that the different shapes of observed asteroids as well as their unique disruption patterns could give us clues about their internal structure, strength and geophysical properties in general.

1 Granular Asteroids and YORP

In the last decade, the study of asteroids as granular aggregates has gained importance. Even though it was originally assumed that the smallest of asteroids were monolithic rocks with a bare surface, the pictures obtained by NASA Galileo, Near-Showmaker (or just NEAR) and JAXA Hayabusa space missions revealed a substantial *layer of unconsolidated rocky material and dust (regolith)* covering the surface of (951) Gaspra, (243) Ida, (433) Eros [1, 2] and (25143) Itokawa. Additionally, observations made by the Hubble Space Telescope of the disruption events of active asteroids P2013/R3 [3] and P2013/P5 [4] imply that not only asteroid surfaces are covered by regolith, but that their internal structure is not monolithic. This is what we define as a granular asteroid, a naturally occurring self-gravitating granular system.

Of the many events that can disrupt a granular, non-monolithic asteroid, here we will focus only on the disruption that could be caused by rotation. The angular velocity of this rotation, in the context of NEOs, can be accelerated by the Yarkovsky-O'Keefe- Radzievskii-Paddack (YORP) effect which results from an interplay between solar radiation pressure and the shape of an asteroid [5, 6]. Very succinctly, when the photons that come from the Sun impact on the surface of an asteroid, they will be reflected, absorbed and re-emitted from it anisotropically due to its irregular shape. This will produce a net torque that affects the angular velocity of the asteroid. Over millennia, this change in the angular velocity will produce the deformation or disruption of the asteroid [7]. A clear signature of the YORP effect was first observed in asteroid 54509 YORP [5, 8] and (1862) Apollo [9].

If the strength of granular asteroids were governed by gravitational forces alone, a plot of their size (or absolute magnitude H, more appropriately) versus spin rate would show a very clear limit for the latter, which could give us a way to estimate the bulk density of these bodies if an albedo (proportion of reflected sunlight) of 0.2 is assumed. This is exactly what happens for most asteroids, but not for all. The observed spin barrier shows that larger asteroids have a limiting rotation period of ≈2.2 h [10]. Small asteroids (H<21.5mag) with sizes <150 m can rotate much faster as was first described by [10] and reconfirmed by later surveys [11, 12]. This is what has led scientists to believe that cohesive forces are important for these granular systems [13–16].

2 Cohesive Forces

However, if we say that cohesive forces are important, we need to understand where they come from and what their effects are, if any at all, for the evolution of asteroids. If we

⋆e-mail: diego.sanchez-lana@colorado.edu
⋆⋆e-mail: dscheeres@colorado.edu
⋆⋆⋆e-mail: thirabayashi@purdue.edu
⋆⋆⋆⋆e-mail: simon.tardivel@colorado.edu

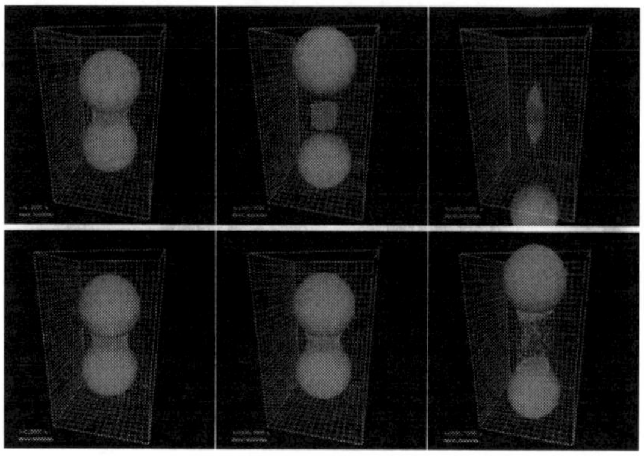

Figure 1. SSDEM simulations. Time evolution of two self-gravitating granular bridge: (top) cohesionless and (bottom) cohesive. The latter needed a pulling force about 40 times greater than the one on top to fail catastrophically [17].

suppose that asteroids are free of volatiles that could bond their constituent particles and provide the tensile strength needed, we are left with another option and this is van der Waals forces. In a terrestrial environment, these forces are mostly negligible, except for powders as the weight of each particle is comparable to their cohesive attractive force. In the asteroid environment however, where ambient gravity is in the order of milli- to micro-g, cohesion and weight become comparable for larger, centimeter size, particles [15]. An asteroid, of course, is not made up only of pebbles, but it has particles that range from microns to tens of meters. A system as such is impossible to simulate with the available computational resources. However, simpler systems, that show the effect of cohesive forces on a self-gravitating granular system, can be constructed and studied. Fig. 1 shows one such systems.

Fig. 1 shows a granular equivalent to a liquid bridge. In these self-gravitating systems, we have two 1 m boulders, joined together by a bridge formed by thousands of centimeter size particles. The system on top is cohesionless, whereas the one at the bottom is cohesive. The boulders are pulled in opposite directions by a force increasingly stronger than their mutual gravitational attraction to measure the tensile strength of the bridge. This study [17] showed that the tensile strength of a cohesive aggregate is inversely proportional to the average particle size of the particles forming it (porosity, Hamaker constant, coordination number and others are considered constants). Basically, the smaller dust and regolith in an asteroid could be seen as a weak cement, a matrix in which the larger boulders are embedded. This effect is as important as gravity for small asteroids, but being independent of mass, it looses its preponderance for larger asteroids where gravitational effects dominate. This study also showed that it should be possible to simulate real size asteroids with much less particles if we simply add a particle-particle cohesive force that depends on a predetermined average particle size. This being so, we could now be in a position to carry out realistic simulations of asteroids with levels of tensile strength that are related to their physical properties (particle size, porosity, etc.). This would help us understand the effects of different levels of strength on self-gravitating cohesive aggregates that are spun up to deformation/disruption levels.

3 Disruption Patterns

Exact solutions for the deformation of self-gravitating aggregates have been obtained for spherical and ellipsoidal shapes, but these solutions could not predict how a disruption process would occur. To solve this problem Sánchez and Scheeres[18] carried out a survey on angle of friction and tensile strength for ellipsoidal and spherical self-gravitating aggregates which were quasi-statically spun up to disruption. Fig. 2 shows some of the produced disruption patterns.

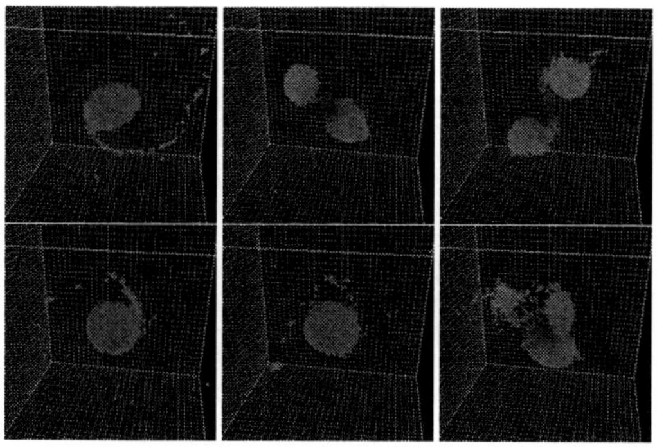

Figure 2. Different disruption patterns for ellipsoidal (top) and spherical (bottom) aggregates (angle of friction ≈35°) with different tensile strengths (σ^a_{yy}). From left to right: 0 Pa, 1.63 Pa and 16.05 Pa. Size ≈142 m [18].

This study revealed that if the angle of friction of an aggregate defined how much it would deform, tensile strength would determine how it would lose mass. This is, low or no tensile strength would result in a particle-by-particle detachment and high tensile strength, in big chunks being ejected. For an angle of friction of ≈35° and at high enough values, the aggregate would be catastrophically disrupted. Incidentally, this could also provide a possible mechanism for the formation of binary asteroids and asteroid pairs, though this is still to be corroborated. In spite of this, the characteristic shapes of asteroids (1999 KW4, 1950 DA and Geographos) could be observed in the simulated aggregates, though the latter were much more deformed. Deformation could be diminished by strengthening the aggregates, but this would result in the ejected piece being to large to form a satellite. The solution to this dilemma could come in the form of aggregates with a core that is stronger than the outer shell; this is discussed next.

4 The Core of a Gravitational Aggregate

Most studies about small asteroids have made the common assumption that their structure was homogeneous. However, there is no reason to believe that this is so, especially if we observe the heterogeneity of their surfaces [19, 20]. Additionally, previous studies carried out by Hirabayahi and Scheeres [21, 22] had shown that aggregates with denser, stronger cores would strengthen the whole structure and reproduce surface slopes similar to those found in 1999 KW4. It was also shown that for such a bicone shape, a homogeneous structure would first fail at its centre and not at the surface which, in the case of asteroid 1950 DA, would result in a weak interior due to shearing [23]. On the other hand, the results of having a core and shell with different cohesive strengths were yet to be studied.

For this study [24], the same simulation setup shown in fig. 2 was used, with the difference that the tensile strength of the shell was 5 times weaker than its core. Only spherical shapes were used and the aggregates were spun up to disruption rates. The size of the core was changed in order to observed the variations in shape. From this study it was concluded that the failure conditions derived by the analytical model were consistent with those by the SSDEM [24]. The results showed that an aggregate with homogeneous cohesion would always fail internally before it fails superficially. To have surface shedding, therefore, the progenitor body needed a strong core so that only the surface region failed structurally. The simulations also exposed the dynamics of the initial failure phase. The primary result was that for all the cases, a granular spheroid would fail locally through a severe material flow on its surface. This came from the fact that since the initial body configuration was not perfectly uniform, which we believe is more realistic, some surface regions were more susceptible to failure than other regions. Interestingly enough, an aggregate with a core with a radius of $0.7R_b$ (R_b is the radius of the aggregate) would produce slopes that match asteroid 1999 KW4 almost perfectly [25]. For an aggregate with a core of radius $0.9R_b$, deformation would be minimal and the ejected matter would leave a hole close to the equator.

Aggregates with weak cores are also interesting to study as this should also be possible. We again start from the same configuration as before, weaken the core with respect to the shell and spin up the aggregate to induce its disruption. As expected, the aggregates fail structurally earlier than their homogeneous counterparts, but for a core of $0.5R_b$ something interesting happens. The shape into which the aggregate evolves in strikingly similar to that of asteroid Itokawa; this can be observed in fig. 3. This simulation shows that the deformation of the body causes the subsurface material of the shell to be exposed (magenta particles). This exposed material is located on the larger lobe of the body and towards the neck and also in the back of the body. These are the exact places where asteroid Itokawa shows areas with smaller regolith.

If we apply all these results to the recent observations of active asteroids P/2013 R3 and P/2013 P5, we could conclude that the former needed a strong core to form the observed tails and that the latter had at best a homogeneous interior. Asteroid 1999 KW4 could have a strong core, and asteroid Itokawa could possibly have a weak interior [6]. As encouraging as all these results and comparisons are, we still have not observed evidence that the formation mechanism here proposed has actually happened. However, there is possibly a piece of evidence that might have been mistaken for a crater when it could be in fact evidence of fission via rotation.

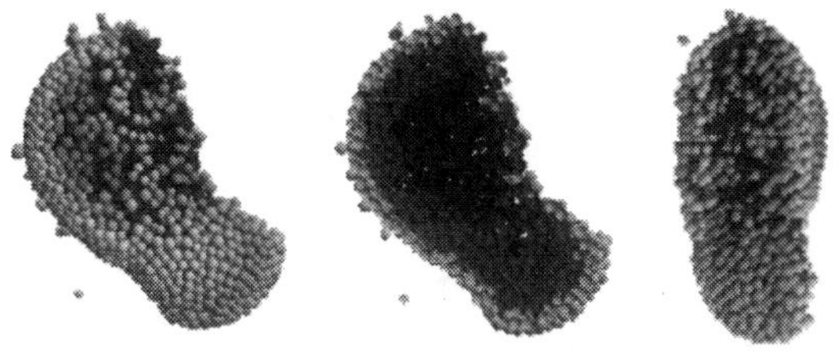

Figure 3. Final stage of an initially spherical aggregate with a $0.5R_b$ weak core. The image at the left and right show the view from the side and bottom respectively. The centre image is a cross-sectional cut. The blue particles form the core, the magenta particles form the inner part of the shell.

5 The curious case of 2008 EV5

Asteroid 2008 EV5 is 450 m in diametre and has a symmetrical, wide bicone shape with one peculiarity, it has a 20 m deep cavity on its equatorial ridge. It has been previously explained that this feature could have been formed by an impact [26]. However, in the best case scenario, the probability of having an impact at that precise location and in the required time frame, taking into account the YORP effect, is extremely low (<0.01%). Though not impossible, it is highly improbable and so we propose a different explanation.

As the shape model of this asteroid is known to great resolution, it is possible to compute its gravity field with fine accuracy. As angular velocity is increased, the effective force felt by a particle at its surface, sum of the gravity acceleration and centrifugal acceleration, becomes increasingly tenuous. Eventually, for an asymmetric asteroid, some regions see this balance reverse, and they feel an outward pull. The key observation we made here is that these pulled-out regions were limited to a very specific and identified portions of the asteroid.

We then make the assumption that the cavity in 2008 EV5 was initially filled with material (see fig. 4). In that case, the filled cavity is *precisely* the portion of the asteroid that feels this outward pull at high spin rate. We calculated that such a piece would require under 1 Pa of tensile strength to be fissioned, leaving the observed cavity behind. This begs another question, how could the equatorial region have such a low tensile strength when in other asteroids this has been measured at values between 50-250 Pa?

We have already explained that at high enough spin rates, aggregates with a strong core would produce material flow towards the equatorial region. This flow should be conducive to particle size segregation [27], leaving the larger ones to form the equatorial ridge. This flow would produce a region formed by larger particles in which

 DOI: 10.1051/epjconf/201714014004

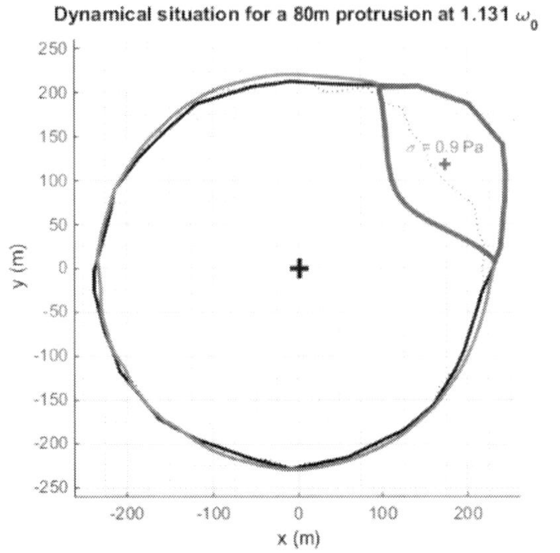

Figure 4. Planar cut of asteroid 2008 EV5. The ridge line is in green. The red line represents the region feeling an outward pull. The black line represents the altered asteroid contour, and the dotted line is the actual asteroid shape.

porosity as well as the average particle size have increased, allowing for a lower tensile strength. A similar argument could be done for asteroid 2000 DP107 [28].

6 Conclusions

In this paper we have presented a study of the effects of cohesive forces and heterogeneities in self-gravitating aggregates and their application to small asteroids in the Solar System. We have shown how van der Waals forces between the grains that form an asteroid, and other bulk properties, are very important to determine the tensile strength of these aggregates. In turn, its specific value in comparison to gravitational forces could determine specific disruption patterns at elevated spin rates. If we now take into account that asteroids can be heterogeneous, our study suggests that the current asteroid shapes and disruption patterns should reflect their internal structure.

References

[1] R.J. Sullivan, P.C. Thomas, S.L. Murchie, M.S. Robinson, Asteroids III pp. 331–350 (2002)

[2] M.S. Robinson, P.C. Thomas, J. Veverka, S.L. Murchie, B.B. Wilcox, Meteoritics and Planetary Science **37**, 1651 (2002)

[3] D. Jewitt, J. Agarwal, J. Li, H. Weaver, M. Mutchler, S. Larson, The Astrophysical Journal Letters **784**, L8 (2014)

[4] D. Jewitt, J. Agarwal, H. Weaver, M. Mutchler, S. Larson, The Astrophysical Journal Letters **778**, L21 (2013)

[5] P.A. Taylor, J.L. Margot, D. Vokrouhlický, D.J. Scheeres, P. Pravec, S.C. Lowry, A. Fitzsimmons, M.C. Nolan, S.J. Ostro, L.A.M. Benner et al., Science **316**, 274 (2007)

[6] S. Lowry, P. Weissman, S. Duddy, B. Rozitis, A. Fitzsimmons, S. Green, M. Hicks, C. Snodgrass, S. Wolters, S. Chesley et al., Astronomy and Astrophysics **562**, 48 (2014)

[7] D.P. Rubincam, Icarus **148**, 2 (2000)

[8] S.C. Lowry, A. Fitzsimmons, P. Pravec, D. Vokrouhlický, H. Boehnhardt, P.A. Taylor, J.L. Margot, A. Galád, M. Irwin, J. Irwin et al., Science **316**, 272 (2007)

[9] M. Kaasalainen, J. Ďurech, B. Warner, Y. Krugly, N. Gaftonyuk, Nature **446**, 420 (2007)

[10] P. Pravec, A.W. Harris, Icarus **148**, 12 (2000)

[11] B.D. Warner, A.W. Harris, P. Pravec, Icarus **202**, 134 (2009)

[12] P.A. Taylor, E.S. Howell, M.C. Nolan, A.A. Thane, *The Shape and Spin Distributions of Near-Earth Asteroids Observed with the Arecibo Radar System*, in *American Astronomical Society Meeting Abstracts #220* (2012), Vol. 220 of *American Astronomical Society Meeting Abstracts*, p. 128.02

[13] K.A. Holsapple, Icarus **172**, 272 (2004)

[14] K.A. Holsapple, Icarus **205**, 430 (2010)

[15] D. Scheeres, C. Hartzell, P. Sánchez, M. Swift, Icarus **210**, 968 (2010)

[16] D.P. Sánchez, D.J. Scheeres, Icarus **218**, 876 (2012)

[17] P. Sánchez, D.J. Scheeres, Meteoritics & Planetary Science **49**, 788 (2014)

[18] P. Sánchez, D.J. Scheeres, Icarus **271**, 453 (2016)

[19] A. Fujiwara, J. Kawaguchi, D.K. Yeomans, M. Abe, T. Mukai, T. Okada, J. Saito, H. Yano, M. Yoshikawa, D.J. Scheeres et al., Science **312**, 1330 (2006)

[20] H. Yano, T. Kubota, H. Miyamoto, T. Okada, D. Scheeres, Y. Takagi, K. Yoshida, M. Abe, S. Abe, O. Barnouin-Jha et al., Science **312**, 1350 (2006)

[21] M. Hirabayashi, Icarus **236**, 178 (2014)

[22] D. Scheeres, Icarus **247**, 1 (2015)

[23] M. Hirabayashi, D.J. Scheeres, The Astrophysical Journal Letters **798**, L8 (2015)

[24] M. Hirabayashi, D.P. Sánchez, D.J. Scheeres, The Astrophysical Journal **808**, 63 (2015)

[25] P. Sánchez, Proceedings of the International Astronomical Union **10**, 111 (2015)

[26] M.W. Busch, S.J. Ostro, L.A.M. Benner, M. Brozovic, J.D. Giorgini, J.S. Jao, D.J. Scheeres, C. Magri, M.C. Nolan, E.S. Howell et al., Icarus **212**, 649 (2011), 1101.3794

[27] N. Thomas, Phys. Rev. E **62**, 961 (2000)

[28] S.P. Naidu, J.L. Margot, P.A. Taylor, M.C. Nolan, M.W. Busch, L.A.M. Benner, M. Brozovic, J.D. Giorgini, J.S. Jao, C. Magri, The Astronomical Journal **150**, 54 (2015)

Modeling Segregation in Modulated Granular Flow

Richard M. Lueptow[1,2,*], *Zhekai* Deng[2], *Hongyi* Xiao[1], and *Paul B.* Umbanhowar[1]

[1]Department of Mechanical Engineering, Northwestern University, Evanston, IL 60208, USA
[2]Department of Chemical and Biological Engineering, Northwestern University, Evanston, IL 60208, USA

Abstract. We consider unsteady segregating granular flows, focusing on stratification in bidisperse bounded heap flow. Experiments indicate that periodically changing the feed rate of particles falling onto the upstream portion of the heap results in a stratified segregation much like that which occurs at low feed rates, but with more regular stratified layers of large and small particles and a higher overall feed flow rate. Experiments clarify how a front of large particle at a high feed rate deposits a layer of large particles that is subsequently covered over by a layer of small particles during a low feed rate, thus generating layers of large and small particles. The stratification can be modelled using a version of an advection-segregation-diffusion equation with a segregation velocity. Simply repeatedly switching the model from a low volume flow rate to a high volume flow rate instantaneously along the entire length of the flowing layer results in stratification patterns that are similar to those observed in experiments. Using a modulated feed rate offers the potential to intentionally create extended layers of bidisperse granular materials to enhance the effective degree of mixing of the deposited materials at heap length scales.

1 Introduction

Flowing granular materials can segregate due to particle size or density differences [1-3] to form many different pattern including stratified layers of particles [4-7]. The degree of segregation and nature of the segregation pattern depends on the flow conditions and the differences between particle species.

Previous research on segregating granular materials has focused largely on steady feed rates [5, 7]. However, here we consider the impact of modulating the feed rate in order to intentionally manipulate the segregation to generate a stratified segregation pattern, both experimentally and based on the application of an advection-segregation-diffusion model that has been successfully implemented in steady segregating flows in quasi-2D bounded heaps. We will show that by controlling the feed rate of mixed size bidisperse particles, the segregation can be controlled so that well-ordered stratification of the small and large particles is achieved, as shown in Fig. 1. Such stratification may be useful in providing a better overall homogeneity at scales larger than the stratification than can be achieved without stratification for application in industrial situations such as filling a hopper.

2 Experiments

We consider flow in a quasi-2D, one-sided bounded heap, shown in Fig. 1. Well-mixed size bidisperse spherical glass particles (confirmed by direct sampling) are fed from a screw feeder into the system at the upper left corner and flow down the slope in a thin layer, segregating as they flow until they reach the bounding wall at the right side. The surface of the heap rises as particles fill the container.

Fig. 1. A stratified pattern (enhanced contrast) generated by feeding a size bidisperse mixture (0.05 cm red particles and 0.20 cm blue particles) onto a quasi-2D bounded heap at alternating feed rates of q_f =37.0 cm^2/s for t_f =3 s and q_s=0.4 cm^2/s for t_s=62 s.

For constant feed rates, the particles form irregular stratified layers at low feed rates, but segregate completely into two distinct regions with small particles upstream and large particles downstream for moderate flow rates [7]. At high enough flow rates, the particles remain mixed [7], but such high flow rates are sometimes difficult to accomplish in practical situations.

---
* Corresponding author: r-lueptow@northwestern.edu

On the other hand, if a stratified pattern can be forced to occur, the particles are essentially mixed at the length scale of the container. It is this situation, an example of which is shown in Fig. 1, that we investigate here.

2.1 Apparatus

The experimental apparatus consists of a pair of parallel vertical rectangular plates – an aluminum back wall and a glass front wall for visualization. Spacers between the plates fix the gap at either T=1.27 cm or T=1.5 cm and the width at W=69 cm, comparable in length to the flowing layer length in small industrial scale silos, as described previously [7]. Mixed glass particles are dropped vertically into the gap between the front and back walls near the left bounding wall. An auger feeder is used to feed the particles at alternating 2D flow rates: a fast flow rate q_f for duration t_f and a slow flow rate q_s for duration t_s. We considered mixtures of 1) 0.050±0.004 cm particles and 0.200±0.007 cm particles; and 2) 0.118±0.007 cm and 0.295±0.004 cm particles.

Videos of the entire bed of particles are recorded at 60 frames/s with a pixel corresponding to slightly less than 1 particle diameter (~0.1 cm) during each experiment. The free surface of the flowing layer of particles is identified by locating the transition point of the image intensity in each pixel column of the image. A time average over 0.1 s (6 video frames) and a space average over 0.5 cm are applied to reduce noise in the data.

2.2 Results

A typical experimental result in which the feed rate is modulated so it alternates between fast and slow values is shown in Fig. 1 for an equal volume mixture of small red particles and large blue particles. As a result of the feed flow modulation, layers of small and large particles appear. If the flow is not modulated, but operates at the same average flow rate, irregular stratification occurs [7]. Under conditions of higher flow rates, feed flow modulation results in stratification, while an equivalent constant feed rate results in a completely segregated pattern in which the small red particles deposit on the fixed bed upstream of the large blue particles resulting in a distinct vertical boundary between the red and blue particles [7]. Though completely mixed particles is typically optimal, in many applications layers of particles like that shown in Fig. 1 is preferable to completely segregated particles.

To clarify the nature of the segregation, we consider first the single transition that occurs when the flow rate changes from low to high or from high to low. In particular, we track the time evolution (time dependence) of the surface of the flowing layer at different locations along the length of the flowing layer as the change in flow rate propagates along the length of the flowing layer.

Consider the results shown in Fig. 2(a) for the transition from a low flow rate to a high flow rate at time t=5 s. In this figure, each curve represents the height z_h of the surface of the flowing layer at each of 16 locations along the length of the flowing layer as a function of time. Thus, each solid curve represents the time evolution of the surface height: the top curve corresponds to the position furthest upstream and the lowest curve corresponds to the position furthest downstream. The slope of the curve represents the velocity at which the surface is rising at that specific streamwise location. Note that if the local flow is not changing, the curves would be straight lines indicating a steady rise velocity. A change in the flow rate and associated rise velocity of the heap surface is evident as a change in the slope of the curve.

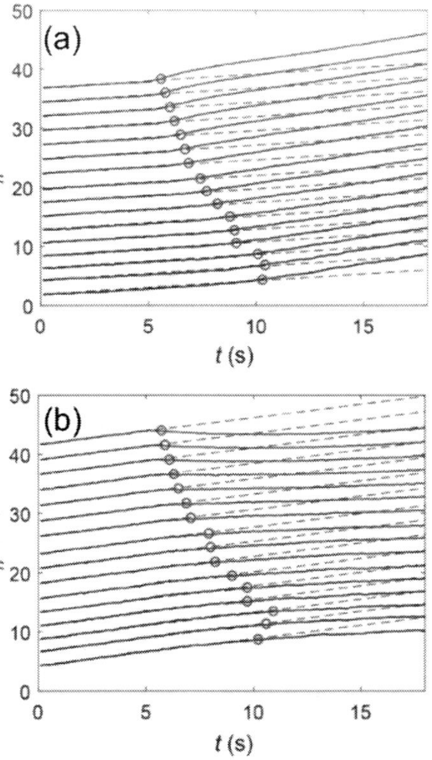

Fig. 2: The time evolution of surface height (solid lines) at various horizontal positions (from x=8 cm [red] to x=68 cm [blue] with interval Δx=0.04 m) using the bidisperse mixture (0.118 cm and 0.295 cm particles) during (a) a single flow rate transition from q_s =15.9 cm²/s to q_f =37.2 cm²/s at t=5 s and (b) a single flow rate transition from q_f =31.1 cm²/s to q_s =12.4 cm²/s at t=5 s. Dashed lines depict the surface heights that would have occurred based on the initial rise velocity. The red circles mark where z_h deviates from its expected value (based on the initial rise velocity) by more than one large particle diameter.

At the most upstream position (uppermost curve) the vertical position of the flowing layer responds almost immediately to an increase in the flow rate at t=5 s with a change in slope at the point marked with a circle. The deviation between the uppermost solid curve and the dashed line demonstrates how much faster the surface at this point is rising after the flow rate increases. The change in flow rate does not propagate to the most downstream measurement location until about 5 seconds later, as indicated by the circle marking the change in slope of the lowest curve. Based on the circles marking the flow rate transition at all the streamwise locations, it is clear that the change in flow rate propagates downstream and that it decelerates as it travels downstream. This is not unexpected, given that a steady feed flow rate results in a flow rate that decreases to zero at the end wall of the bounded heap.

The surface height at various streamwise locations as the feed flow rate transitions from a high rate to a low rate at time t=5 s is shown in Fig. 2(b). In this case the slope of the curves decreases when the transition in the flow rate reaches any particular streamwise location. Again the transition is almost immediate at the most upstream position (uppermost curve), but delayed by about 5 s at the most downstream location (lowermost curve).

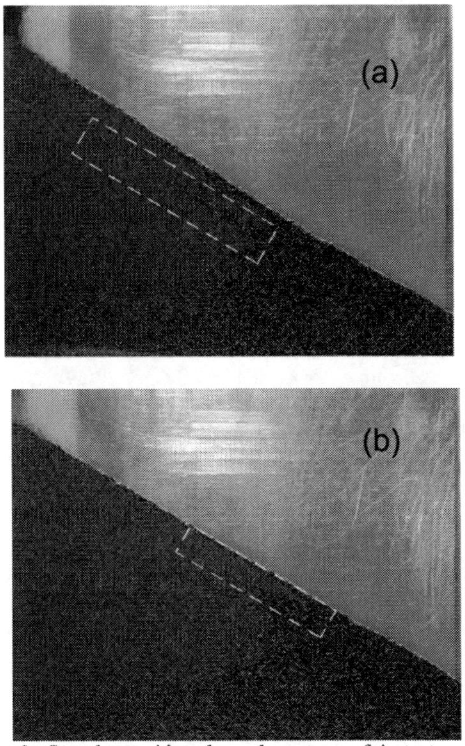

Fig. 3: Snapshots with enhanced contrast of instantaneous segregation patterns of a bidisperse mixture (0.118 cm red and 0.295 cm blue particles) taken at ~10 s after the transitions for (a) a single flow rate transition from q_s =15.9 cm^2/s to q_f =37.2 cm^2/s and (b) a single flow rate transition from q_f =31.1 cm^2/s to q_s =12.4 cm^2/s.

The segregation patterns that occur during a single flow rate transition are also instructive. Fig. 3(a) shows a large particle layer (blue) extending upstream into the small particle region ~10 s after a transition from a low to a high feed flow rate. This is a result of a wedge of material deposited after the transition from the low flow rate to the high flow rate propagating downstream. The large particles accumulate at the front of the faster flowing material much like described previously for single avalanches [8, 9], and some are buried as the high speed flow surges downstream, thus forming the blue layer.

In Fig. 3(b), a small particle layer (red) extends into the large particle region (blue) ~10 s after the transition from a low flow rate to a high flow rate. From Fig. 2(b), it is clear that the surface height in the upstream positions actually decreases during the first few seconds of the transition, which leads to the previously deposited small particles being carried downstream with the slower flowing material. This also means that the newly added particles are not deposited upstream, but are carried downstream. Together this extends the small particle layer further into the large particle region.

3 Model

A model based on the advection-diffusion equation with an additional term for segregation has been successful for modeling steady, developing, and transient granular flows of size segregating and density segregating materials [10-14]:

$$\frac{\partial c_i}{\partial t} + \underbrace{\frac{\partial (u c_i)}{\partial x} + \frac{\partial (w c_i)}{\partial z}}_{\text{advection}} + \underbrace{\frac{\partial (v_{p,i} c_i)}{\partial z}}_{\text{segregation}} - \underbrace{\left[\frac{\partial}{\partial x}\left(D\frac{\partial c_i}{\partial x}\right) + \frac{\partial}{\partial z}\left(D\frac{\partial c_i}{\partial z}\right)\right]}_{\text{diffusion}} = 0 ,$$

where c_i is the concentration of species i, u and w are the mean granular velocity in the streamwise (x) and surface-normal (z) directions, respectively (note that this coordinate system is different than that for the experiments). The first term in the equation is the rate of change of species concentration c_i, the second and third terms account for advection due to mean flow, the fourth term accounts for segregation due to the species specific percolation velocity $v_{p,i}$, which is the component normal to the free surface, and the bracketed terms account for collisional diffusion. A key aspect of this bidisperse model is the dependence of the percolation velocity of species i on the local shear rate and the local concentration of the other particle species [12, 15].

Here we apply this model using conditions for fully-developed steady flow on a quasi-2D bounded heap [16], but instantaneously change the velocity field (and thereby the shear rate) and flowing layer thickness from that for a low flow rate to that for a high flow rate and vice versa. In this way, the deposition of particles corresponding to the modulated feed flow rate (alternating between low to high) can be modeled, though this approach does not account for the propagation of the flow rate change along the length of the flowing layer evident in Fig. 2. Nevertheless, it provides a first order approximation of the effect of a modulated feed flow rate on the particle segregation and subsequent deposition on the heap, which is expected to be valid when the switching time between different flow rates is long compared to the relaxation time.

Fig. 4 compares experiments (extracted from an image like that in Fig. 1) and model results calculated using an operator splitting approach [12, 17]. In the operator spliting approach at each time step the advection step is solved with a matrix mapping method [18], and the segregation/diffusion step is solved with an implicit Crank-Nicolson method. It is clear that even the simple application of the model (alternating between high and low flow rates) represents the stratification in the experiments quite well, particularly in terms of the wavelength of the stratified layers. Work to refine this model is currently under way, primarily focused on the non-uniform local deposition rate.

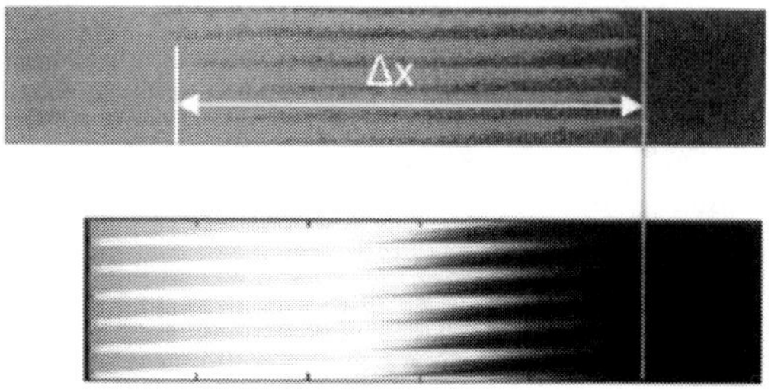

Fig. 4: Comparison of concentration fields from experiment (upper) and continuum model solved with steady state parameters for bidisperse modulated bounded heap flow (lower). Experiments used 0.05 cm (light) and 0.2 cm (dark) particles. Modulated feed flow conditions: q_f =23.6 cm^2/s for t_f=5 s and q_s=2.0 cm^2/s for t_s=20 s. The extent of the model domain is reduced on the left to account for feed zone effects in the experiment.

4 Summary

Modulation of the feed flow rate can be used to force a stratified segregation pattern for a size bidisperse material in bounded heap flow. This may have significant advantages over the usual fully segregated pattern that otherwise occurs. For instance, if the bounded heap is formed in a hopper that is to be emptied later, a stratified segregation pattern will be mixed as it exits the hopper, whereas a fully segregated pattern will result in smaller particles exiting before larger particles exit.

Of course, further work is necessary to fully characterize the impact of the feed flow parameters (high and low flow rate magnitudes and durations) so they can be optimized for stratification. A better understanding of the mechanism described briefly with respect to Fig. 3 would aid in further optimizing the parameters for stratification. And the advection-diffusion-segregation model can be adapted to account for the propagation of the change in the flow rate down the heap, rather than the simple first order approach demonstrated in Fig. 4. Nevertheless, these results demonstrate the potential for modelling feed flow modulation and using it to intentionally generate stratified patterns in segregating granular flows.

The authors gratefully acknowledge the support of the US National Science Foundation, Grant No. CBET-1511450.

References

1. S.W. Meier, R.M. Lueptow, J.M. Ottino, Adv. Phys. **56**, 757 (2007).
2. I.S. Aranson, L.S. Tsimring, Rev. Modern Phys., **78**, 641 (2006).
3. S.B. Savage, C.K.K. Lun, J. Fluid Mech., **189**, 311 (1988).
4. J.C. Williams, Powder Tech., **2**, 13-20 (1968).
5. H.A. Makse, S. Havlin, P.R. King, H.E. Stanley, Nature **386**, 379 (1997).
6. J.M.N.T. Gray and K. Hutter, Cont. Mech. Thermodyn. **9**, 341 (1997).
7. Y. Fan, Y. Boukerkour, T. Blanc, P.B. Umbanhowar, J.M. Ottino, R.M. Lueptow, Phys. Rev. E **86**, 051305 (2012).
8. O. Pouliquen, J.W. Vallance, Chaos **9**, 621 (1999).
9. J.M.N.T. Gray and C. Ancey, J. Fluid Mech. **629**, 387 (2009).
10. J.M.N.T. Gray and V.A. Chuganov, J. Fluid Mech. **569**, 365 (2006).
11. A.R. Thornton, J.M.N.T. Gray, A.J. Hogg, J. Fluid Mech. **550**, 1 (2006).
12. Y. Fan, C.P. Schlick, P.B. Umbanhowar, J.M. Ottino, R.M. Lueptow, J. Fluid Mech. **741**, 252 (2014).
13. D.R. Tunuguntla, O. Bokhove, A.R. Thornton, J. Fluid Mech., 749, 99 (2014).
14. H. Xiao, P.B. Umbanhowar, J.M. Ottino, R.M. Lueptow, Proc. Royal Soc. A **472**, 20150856 (2016).
15. C.P. Schlick, Y. Fan, A.B. Isner, P.B. Umbanhowar, J.M. Ottino, R.M. Lueptow, AIChE J. **61**, 1524 (2015).
16. Y. Fan, P.B. Umbanhowar, J.M. Ottino, R.M. Lueptow, Proc. Royal Soc. A **469**, 20130235 (2013).
17. C.P. Schlick, I.C. Christov, P.B. Umbanhowar, J.M. Ottino, R.M. Lueptow, Phys. Fluids **25**, 10 (2013).
18. M.K. Singh, O.S. Galaktionov, H.E.H. Meijer, P.D. Anderson, Comput. Chem. Eng. **33**, 8 (2009).

Changes to Grain Properties due to Breakage in a Sand Assembly using Synchrotron Tomography

Tabassom Afshar[1,*], *Mahdi Disfani*[2], *Guillermo Narsilio*[2], and *Arul Arulrajah*[1]

[1] Department of Civil & Construction Engineering, Faculty of Science, Engineering & Technology, Swinburne University of Technology, Hawthorn, Victoria, Australia
[2] Department of Infrastructure Engineering, Melbourne School of Engineering, The University of Melbourne, Parkville, Victoria, Australia

Abstract. Grain breakage is of paramount importance for understanding the behaviour of granular materials used in various engineering applications, such as pavements, roads, rail tracks, the oil and gas industry, and mineral processing. Changes to grain properties of a uniformly graded sand specimen stemming from breakage during compression were studied with the aid of three-dimensional Synchrotron Radiation-based Micro-Computed Tomography. The fast scanning and high-resolution 4D imaging were utilised to capture images from the interior body of the granular assembly during loading. The fractal distribution of the sand assembly showed that breakage becomes dominant in smaller grains rather than larger ones, where an increase in the amount of newly generated fine fragments leads to high coordination number surrounding the larger grains. More importantly, the results of morphological changes in the particulate assembly revealed that there is a reversal trend in the grain morphology evolution with increasing stress. The sand grains tended to create more spherical fragments with higher aspect ratio whereas by increasing the stress this trend completely shifted.

1 Introduction

Grain breakage is common in granular materials, such as powders, soils, and manufactured aggregates, when subjected to high loading or large deformations. While breakage is desirable in some industries such as comminution in mineral processing [1], it may cause accelerated degradation of civil infrastructure, such as roads and pavement layers or ballast rail tracks [2], or decreases in the recovery rate in the oil and gas industry due to proppant crushing [3]. Grain breakage brings about changes in characteristics of granular materials, especially in grain size distribution and grain morphology. These shifts affect the mechanical behaviour of granular media and most prominently the material's crushing strength against further breakage [4].

While there are several studies on single-particle fracture mechanism and crack propagation (e.g. [5] and [6]), due to testing difficulties and the time-consuming steps of X-ray imaging, research on the impact of breakage on grain properties (particularly changes in grain morphology) in confined assemblies of individual grains is limited.

In this study, in order to obtain a better understanding of sand behaviours in relation to grain-scale damage, a novel loading apparatus capable of conducting compression tests at the high stress on assemblies of grains was designed and developed. Synchrotron Radiation-based X-ray Micro-Computed Tomography (SR-μCT) was used to capture high-resolution 4D images (i.e. 3D monitoring over time) from a sand assembly subjected to compression at different loads.

2 Experimental Method

A loading apparatus was developed in order to apply one-dimensional compression to the confined granular sand sample. The apparatus consists of two major parts: a sample chamber and a loading and data acquisition system (Figure 1). The sample chamber was made of a high-strength radiolucent tube of aluminium with a thickness of 5 mm, 25 mm internal diameter and 25 mm height. The load was applied manually using a reaction frame and screw. A Linear Variable Differential Transducer (LVDT) was used to measure deformation while a load cell continuously recorded the force.

The synchrotron facilities used in this study, in contrast with laboratory scanners, provide a monochromatic and highly culminated source of X-ray allowing high resolution imaging and also much more rapid scanning of samples loaded in high pressure metallic chambers. The vertical beam size of Hutch 3B of the Australian Synchrotron used in this study is approximately 25 mm, and the scans were conducted using a Ruby detector. Ruby is a custom designed detector of the Australian Synchrotron [7]. Resolutions of 11.6 μm were obtained in this study for the 25 mm samples tested.

* Corresponding author: tafshar@swin.edu.au

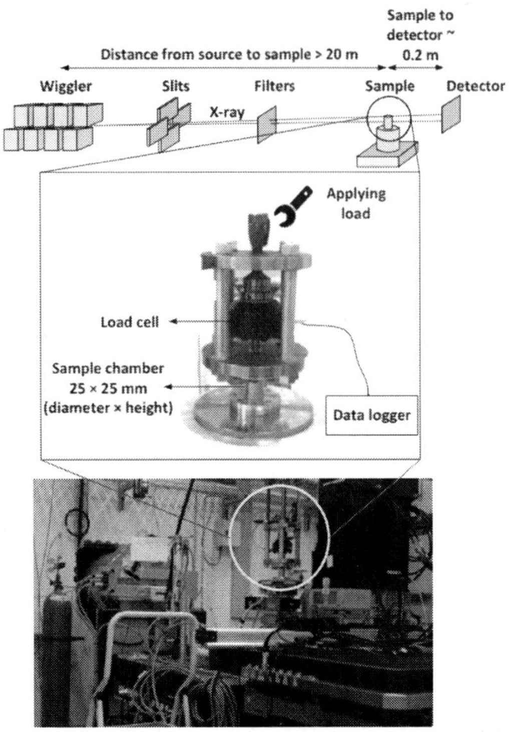

Fig. 1. Schematic diagram of the experimental layout and the loading setup used for the experiments.

3 Grain property evolution due to breakage

In order to gain a better understanding of soil behaviours in relation to grain-scale damage, the changes to the grading and grain morphology of a sand specimen were examined. A uniformly graded sand specimen, 1.18 mm < grain diameter d < 2.36 mm, was scanned under 0, 2.5, 5, and 10 kN compressive loading (i.e. 0, 5.1, 10.19, and 20.38 MPa). The grain size and shape distribution after each loading sequence were obtained from precise 3D reconstruction of 2159 CT horizontal slices using Avizo 9 (Figure 2a-insert). Image processing techniques including image thresholding, filtering, and segmentation, were used to separate and label each fragment in the assembly.

3.1 Soil grading and fractal distribution

While initial grain size distribution is one of the main factors governing the mechanical behaviour of a granular assembly, the way it changes under loading can predict the material behaviour, especially the potential for further breakage. The grain size distribution of the confined sand assembly under different compressive loading is illustrated in Figure 2a. Only fragments with a diameter larger than 0.075 mm were analysed due to inaccuracy in segmenting and evaluating characteristics of extremely small fragments. After scanning under 10 kN loading, the sample was recovered and put through sieve analyses. As shown in Figure 2a, there is a slight discrepancy between grain size distribution measured by 3D reconstruction of the sample and the results from the sieve analysis, particularly between the fine contents. This discrepancy is related to the loss of fine grains during extruding the sample from the chamber at the end of the scanning.

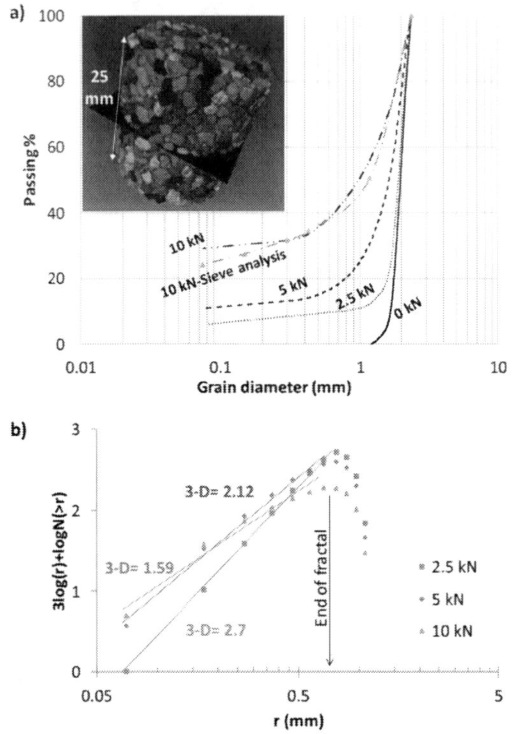

Fig. 2. Sand specimen under different loading levels (0, 2.5 kN, 5 kN, and 10 kN): a) Changes in grain size distribution, b) The fractal distribution.

The sample grading clearly changed from uniform to a well graded distribution as stress increased and breakage progressed further (Figure 2a). Altuhafi and Coop [4] also reported that toward the terminal state, the soil gradation eventually tends to shift to well-graded, which explains the fact that very well-graded and fractal soil experiences no dramatic breakage. Fractal theory, first used by Hartmann [8] to study crushing processes in meteorites, is widely used to quantify the amount of breakage occurring under loading or after large displacement. It is widely accepted that fragmentation is a fractal phenomenon suggesting that it is also a scale-invariant mechanism, and the fractal dimension (D) describes the changes in grain size distribution. Referring to Eq. 1, the original fractal dimension can be calculated based on number of grains, i.e. $N(>r)$ which is the cumulative number of fragments with radius larger than r [9]. However, due to the difficulty in counting grains, Einav [2] proposed a method based on the mass of grains smaller than a certain size.

$$3\log(r) + \log N(>r) \propto (3-D)\log(r) \quad (1)$$

With the aid of CT scanning, numbers of grains and newly generated fragments were directly and precisely measured. The radius was calculated by measuring the volume of each grain and defined as the radius of a sphere having the same volume as the grain. The fractal condition is clear in Figure 2b which shows (*3-D*) from Eq. 1, and the fractal dimension rises from 0.3 to 1.41 with the increase in compressive stress, confirming that as further breakage happens in the sample the gradation tends to be more fractal. This is the same trend observed by Turcotte [10], Einav [2], and Zhao et al. [9]. Nevertheless, it seems some large grains did not experience remarkable fragmentation, and the fractal region has ended for grains with diameter above 1.5 mm; at all stress levels the upper bound of fractality is the same. This observation contrasts with the notion of size effects in strength of material, where breakage survival probability of a grain decreases as its size increases [11]. McDowell, Bolton and Robertson [12] also stated that while crushing strength of a grain depends on its size, high coordination number (number of neighbouring grains) reduces the probability of grain fractures, and attempts to model grain breakage based on these two effects (i.e. grain size and coordination number) are very limited. Therefore, the termination of fractality in Figure 2b is related to the fact that when stress increases, more fine fragments are generated by breakage, resulting in a denser packing and higher coordination number around the larger grains. Consequently, breakage becomes more dominant in smaller grains and less likely for larger ones.

3.2 Morphology

Statistical and quantitative analyses of 3D labelled images are presented in Figures 3 and 4 in terms of grain shape/morphology evolution. Figure 3a shows the changes in Aspect Ratio (AR) distribution due to breakage, which is the ratio of the short axis of the best fitted ellipsoid to its long axis. Figure 3b demonstrates the true sphericity distribution, defined as the ratio of the surface area of a sphere having the same volume as the fragment to the actual surface area of the fragment [13]. Figure 3 indicates that after the first loading sequence (2.5 kN or 5.1 MPa) both AR and true sphericity increased. This observation suggests that grains show a tendency to move away from morphological extremes during crushing. Abbireddy and Clayton [14] also reported similar results on changes in grain forms during a triaxial shear test. Nonetheless, under higher levels of stress a reverse trend was observed in both true sphericity and aspect ratio distributions as both of these shape factors tend to decrease (Figure 3). Figure 4 also illustrates the evolution of mean values of AR and true sphericity versus Hardin's relative breakage, which is calculated as the ratio of total breakage to breakage potential (more information can be found in Hardin [15]). A slight decrease in the mean value of true sphericity and a more notable drop in average AR were measured at higher stress levels (i.e. 5 kN or 10.19 MPa and 10 kN or 20.38 MPa). It is interesting that newly created fragments under higher stress levels are less spherical and have lower aspect ratios than the initial grains. Takei et al. [16] and Zhao et al. [9] also noted slight reduction in sphericity and AR of different individual grains in a single particle crushing test. Referring to Figure 3a, despite the severe fragmentation under 10 kN loading, no noticeable change was measured in the fragments' AR compared to the AR distribution of fragments generated under 5 kN loading, supporting the notion of a terminal state [17]. Furthermore, to quantify the uniformity of shape distribution, the Relative Distribution Factor (*RDF*) was calculated for each shape factor distribution (Eq. 2):

$$RDF = AR_{90} / AR_{10}$$

$$RDF = True\ Sphericity_{90} / True\ Sphericity_{10} \qquad (2)$$

where AR_{90} and AR_{10} or $True\ Sphericity_{90}$ and $True\ Sphericity_{10}$ are the value of shape factor at which 90% and 10% of fragments have a smaller value, respectively. Figure 4 reveals that at first there has been a slight fall in *RDF* of AR, and a larger fall in *RDF* of true sphericity. This means that the grains during the first phases of splitting and breakage tend to create fragments with approximately uniform shapes. Afterwards, *RDF* of both shape factors tend to gradually increase while the sample was experiencing more severe fragmentation under higher loading level. This implies that due to the complex mechanisms of breakage, during severe fragmentation newly generated fragments have more diverse shapes. It should also be noted that true sphericity at every loading stage distributed normally with equal mean, median, and mode values. On the contrary, considering the skewness of AR distribution shown in Figure 3a, the distribution tends to skew to the right, toward lower values of AR, with increased crushing in the sample.

Overall, morphological changes during breakage showed a reversal trend as stress increased, with a shift from more spherical and bulky to less spherical and flaky shapes.

3.3 Particle shape versus size

The literature shows that particle size and shape are interrelated. Domokos et al. [18] showed that the shape of fragments generated under (significant) dynamic loading on 'large' 15-150 mm particles obeys universal scaling laws and that the larger crushed particles tend to be elongated. However, Altuhafi and Coop [4] and Sun et al. [19] obtained contrasting results (increase in sphericity and aspect ratios of sand and ballast particles as particle size increased), for smaller diameter particle assemblies under different crushing loads than that of [18].

Figure 5 shows AR of newly generated fragments in relation to their size, i.e. fragment diameter. Particles with d>1.5 mm were removed from the analysis since they did not experience fragmentation (Section 3.1). The figure shows a slight decrease in AR as fragment size decreases, in line with [4] and [19].

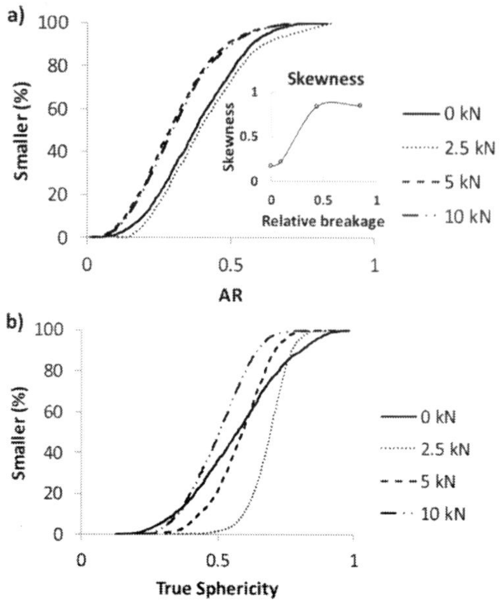

Fig. 3. Morphology evolution of the sand grains under different loading levels: a) Aspect Ratio, b) True sphericity distribution.

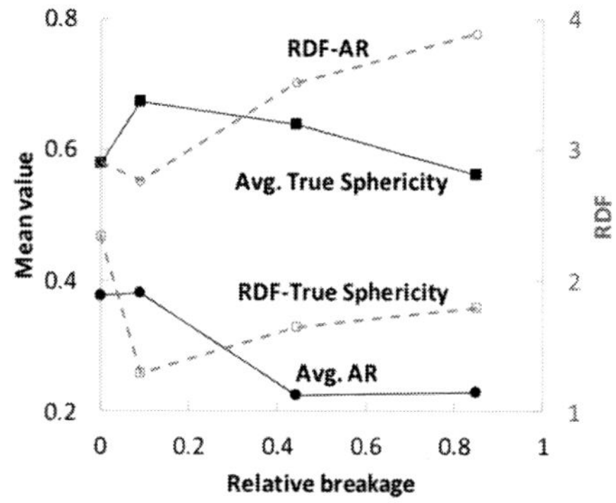

Fig. 4. Changes in statistical properties of shape factor distributions of sand fragments due to breakage: Mean values of AR and true sphericity, and Relative Distribution Factor.

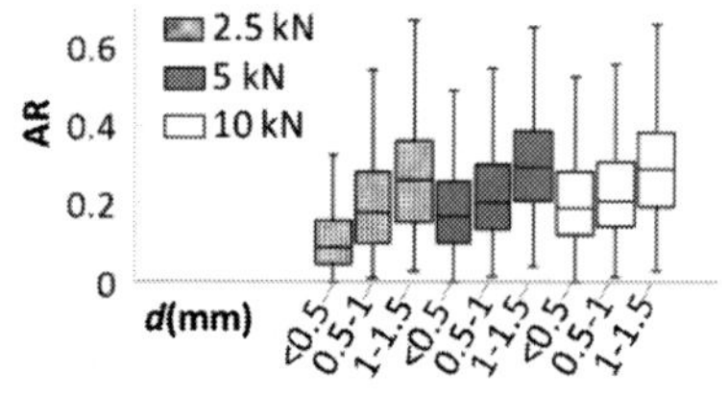

Fig. 5. Relationships between Aspect Ratio and particle size.

Conclusion

A novel loading apparatus was developed to perform constraint compression tests on a sand specimen while being scanned using cutting-edge Synchrotron tomography. Synchrotron tomography was used to study alterations in grain characteristics of a sand specimen. The fractal distribution of the sand assembly due to breakage demonstrated that breakage becomes dominant in smaller grains rather than larger ones where an increase in the amount of newly generated fine fragments causes high coordination number surrounding the larger grains.

The results of morphological changes also reveal that there is a reversal trend in the grain shape evolution with increasing stress. The sand grains tended to generate more spherical fragments with higher aspect ratio whereas by increasing the stress this trend reversed. Due to severe breakage and splitting under higher stress levels, less spherical fragments with lower aspect ratio compared to the original grains were created.

This research was undertaken on the Imaging and Medical Beam Line at the Australian Synchrotron, Victoria, Australia.

References

1 Y. Ghorbani, M. Becker, J. Petersen, S. H. Morar, A. Mainza, and J.-P. Franzidis, Minerals Engineering **24** (12), 1249 (2011).
2 I. Einav, Journal of the Mechanics and Physics of Solids **55** (6), 1274 (2007).
3 W. Zheng and D. D. Tannant, Canadian Geotechnical Journal (ja) (2016).
4 F. Altuhafi and M. R. Coop, Géotechnique **61** (6), 459 (2011).
5 T. Afshar, M. M. Disfani, A. Arulrajah, G. A. Narsilio, and S. Emam, Powder Technology **308**, 1 (2017).
6 M. B. Cil and K. A. Alshibli, Géotechnique **64** (5), 351 (2014).
7 C. Hall, D. Hausermann, A. Maksimenko, A. Astolfo, K. Siu, J. Pearson, and A. Stevenson, Journal of Instrumentation **8** (06), C06011 (2013).
8 W. K. Hartmann, Icarus **10** (2), 201 (1969).
9 B. Zhao, J. Wang, M. Coop, G. Viggiani, and M. Jiang, Géotechnique **65** (8), 625 (2015).
10 D. Turcotte, Journal of Geography Research **91** (12), 1921 (1986).
11 D. R. Jones and M. F. Ashby, *Engineering materials 2: an introduction to microstructures, processing and design.* (Butterworth-Heinemann, 2005).
12 G. McDowell, M. Bolton, and D. Robertson, Journal of the Mechanics and Physics of Solids **44** (12), 2079 (1996).
13 H. Wadell, The Journal of Geology, 250 (1935).
14 C. Abbireddy and C. Clayton, Granular Matter **17** (4), 427 (2015).
15 B. O. Hardin, Journal of Geotechnical Engineering **111** (10), 1177 (1985).
16 M. Takei, O. Kusakabe, and T. Hayashi, Soils and Foundations **41** (1), 97 (2001).
17 A. Narsilio and J. Santamarina, Geotechnique **58** (8), 669 (2008).
18 G. Domokos, F. Kun, A. Á. Sipos, and T. Szabó, Scientific reports **5**, 9147 (2015).
19 Y. Sun, B. Indraratna, and S. Nimbalkar, Géotechnique Letters **4** (July-September), 197 (2014).

DOI: 10.1051/epjconf/201714008005

Warren-Spring based model for the shear yield locus of cohesive biomass powders

Clément Vanneste-Ibarcq[1,*], *Thierry* Melkior[1], *Alain* de Ryck[2]

[1]CEA-LITEN, laboratoire de préparation des bioressources, 17 rue des martyrs 38054 Grenoble Cedex 9, France
[2]Centre RAPSODEE, UMR CNRS 2392, Ecole des Mines d'Albi, 81013 Albi CT Cedex 09, France

Abstract. The objectives of this work are to determine accurately the cohesion of biomass powders from simple measures and to propose a new method for the description of the yield locus of powders with easy to measure parameters. The cohesion of 32 powders (wood, other biomasses and inorganic powders) have been analysed with two methods. The first method is the determination of the yield locus from shear tests at 3 kPa, performed with a powder rheometer, which gives an access to parameters such as cohesion (Y-intercept) and traction (X-intercept). The second method is the measurement of avalanche angles in a rotating drum. A linear relation is found between this angle and the cohesion length, ratio of the cohesion derived from the yield locus and the aerated density. Finally, a model is proposed for the prediction of the cohesion and the yield locus at 3 kPa, using only 2 parameters easy to measure: the avalanche angle and the aerated density.

1 Introduction

Energy from biomass should replace in the future a part of the energy we consume. Some biomass conversion technologies, such as biomass gasification, require the biomass to be finely ground. However, biomass powders are cohesive powders. They may present flow issues (consolidation and arching in hoppers for example) [1].

Thus, the design of the process units has to be made cautiously in order to avoid flowing issues. A way to characterize the flowing properties of powders under stress is the yield locus, deduced from the shear test. Important parameters can be derived from this test, such as the unconfined yield stress (representative of the arching tendency of a powder).

In the method of silo design proposed by Jenike [2], the yield locus is fitted with a straight line according to the Mohr-Coulomb failure criterion:

$$\tau = \mu\sigma + C \qquad (1)$$

τ: Shear stress (Pa)
μ: coefficient of static friction (-)
σ: Normal stress (Pa)
C: Cohesion (Pa)

However, cohesive powders exhibit curved yield loci towards lower shear stresses [3] that cannot be described with this model.

Several models for the description of curved yield loci have been proposed [4, 5, 6]. The Warren-Spring equation [3] is commonly used with cohesive powders:

$$(\tau / C)^n = \frac{\sigma + T}{T} \qquad (2)$$

τ: Shear stress (Pa)
C: Cohesion (Pa)
n: Warren-Spring Index (-)
σ: Normal stress (Pa)
T: Tensile stress (Pa)

The Warren Spring equation presents three unknowns: C, T and n (with $1 < n < 2$). Then it cannot be used directly, as different sets of parameters can fit the data. To reduce the number of degrees of freedom, a parameter has to be set. In the work reported by Ashton et al. [3], where the Warren-Spring model has been proposed, the authors measure separately the tensile stress with a tensile test. However, this implies the use of an external experiment, and the hypothesis that the measured tensile stress corresponds to the yield locus tensile stress.

Measuring the avalanche angle in a rotating drum is another way to characterize the flowing properties of powders, under low stresses. The angle of avalanche is defined as the angle to the horizontal of the free surface formed in a slowly rotating drum [1]. A powder showing a high avalanche angle will have a worse flowability than a powder with a low angle. This measure is easy to set up. It is a constant for given setup and method. It is mostly used to compare the flowability of powders. It is not linked yet to powder properties.
The objectives of this study are:

* Corresponding author: clement.vanneste-ibarcq@cea.fr

• To be able to determine the cohesion of a biomass powder from simple measures (density, avalanche angle…).
• To propose a new method for the determination of the yield locus of powders with easy to measure parameters.

2 Materials and methods

2.1 Materials

In this work, thirty-two powders have been characterised. Twenty-two are wood powders (deciduous and resinous trees). These powders are obtained by drying and grinding (with different mills) wood chips. Six are other biomass powders: industrial sugar, wheat semolina (two grain sizes), grape grounds and two microalgae (dried and ground). Four are inorganic powders: talcum from Fisher Chemicals, beach sand, olivine and alumina.

2.2 Methods

2.2.1 Shear testing

The yield locus of the powders is measured using a FT4 powder rheometer (Freeman tech), an automated torsional shear tester. The main advantage of this tester is that the powder is aerated with a propeller before the test. Thus, the powder is conditioned, and the results do not depend on the operator manipulations.

The powder is filled in an 85 mL vessel, aerated with the propeller and consolidated with a piston to 3 kPa. Then the vessel is split (the top layer of the powder is removed, the surface is flat). Finally, the shear cell is placed on the top of the powder sample, which is consolidated and then sheared until it yields for different normal stresses.

The yield loci are measured at preconsolidation stress of 3 kPa. Twelve points are measured, the normal stresses applied go from 2.5 kPa down to 0.05 kPa (2.5, 2, 1.75, 1.25, 1, 0.8, 0.5, 0.3, 0.1, 0.05, 0.01. The last one is never reached). The shear stress corresponding to the yield of the powder at each normal stress is the yield locus.

2.2.2 Aerated density

The aerated density is measured with the FT4. The powder sample is placed in the 42.5 mL vessel and aerated with the propeller. The vessel is then split (the exact volume of the sample is known) and weighed.

2.2.3 Avalanche angle

The avalanche angle is measured with the Revolution Powder analyser from Mercury Scientific Inc. It consists in a see-through cylindrical drum (10cm diameter, 3.3cm length). 79 mL of powder are introduced into the drum, then placed in the analyser. The drum rotates at 0.6 rpm and a camera records the movements of the powder in the drum. The avalanche angle is measured by a software. It is the angle of the top part of the L-shaped surface of the powder before it collapses due to the rotation of the drum. 150 avalanches are recorded. The chosen avalanche angle is the median of the angle distribution.

3 Results and discussions

3.1 Shear test

An example of yield locus of a wood powder (oak, d_{50} = 22 μm) at 3 kPa is given in the figure 1. The curvature of the locus is obvious.

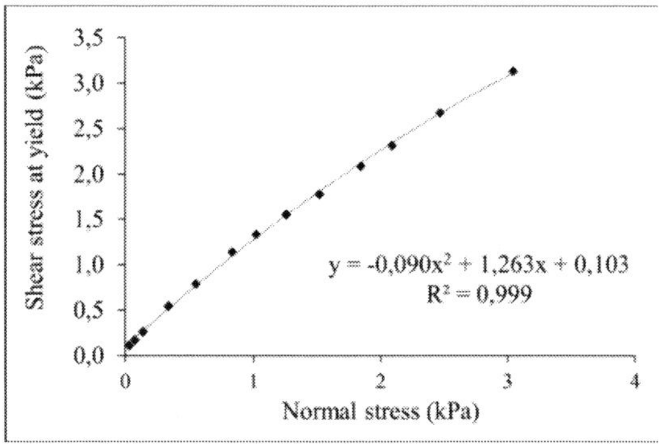

Fig. 1. Yield locus of a wood powder.

The cohesion C is the Y-intercept of the yield locus. The traction T is the X-intercept.

To avoid the use of an additional experiment to measure C or T (as proposed in the work of Ashton et al.), the data obtained from the shear test are fitted with a second order polynomial. Indeed, this equation fits correctly the data. The Y-intercept of the polynomial is considered as the cohesion C.

The parameters T and n are determined by fitting the Warren-Spring equation to experimental data, with the method of least squares (with the condition $1<n<2$).

A relation has been found between C and T using this method as shown in the figure 2. This relation enables to get rid of the unknown T, but requires the aerated density of the powders to be measured. However the density is easier to determine than the tensile stress.

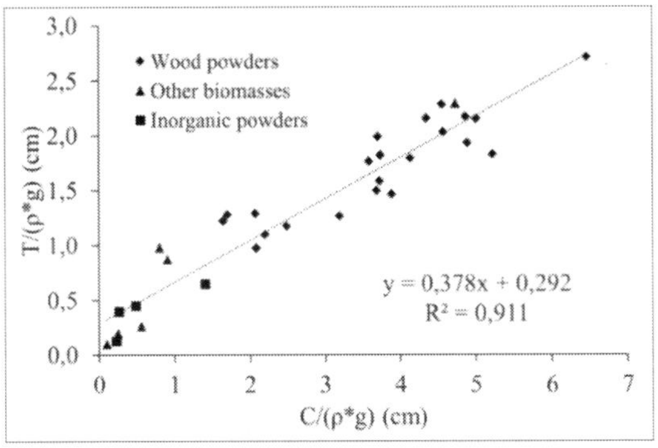

Fig. 2. Relationship between cohesion length and traction length for 32 different powders.

The relation between T and C can be written as follows:

$$T = aC + b\rho g \tag{3}$$

C: Cohesion (Pa)
T: Tensile stress (Pa)
ρ: Powder aerated density (kg/m^3)
g: Acceleration of gravity (m/s^2)
a: Constant (-)
b: Constant (m)

With a = 0.378, b = 2.92 mm, $R^2 = 0.911$ at 3kPa

Thus, T is a function of C and the density of the powder. This relation is then combined with (1) and the following equation is obtained:

$$(\tau / C)^n = \frac{\sigma}{aC + b\rho g} + 1 \tag{4}$$

This equation has two unknowns, C and n. The loci are correctly described by equation (4), as presented in the figure 3 for a wood powder (oak, d_{50} = 22 μm).

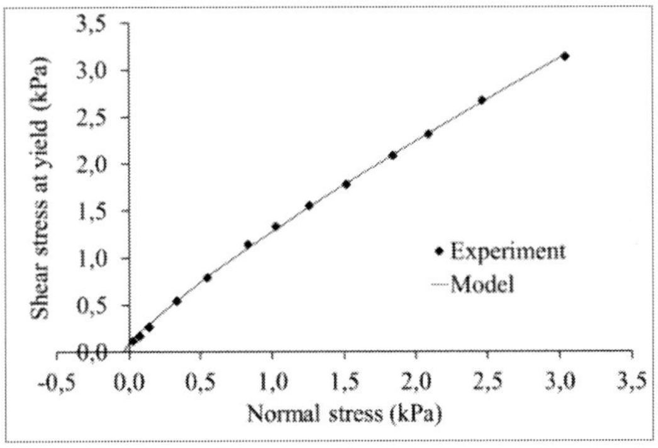

Fig. 3. Yield locus of a wood powder preconsolidated at 3kPa, experimental data and model.

3.2 Avalanche angle

3.2.1 Relationship between the avalanche angle and the cohesion length

The avalanche angle, cohesion at 3kPa consolidation and aerated density are measured for each of the 32 powders. The avalanche angle is then plotted versus l (see figure 4):

$$l = \frac{C}{\rho g} \tag{5}$$

l: Cohesion length (m)
C: Cohesion at 3kPa (Pa)
ρ: aerated density (kg/m^3)
g: Acceleration of gravity (m/s^2)

This length estimates the influence of the cohesive forces acting within the powder over the weight. The larger it is, the greater are the cohesive forces compared to the weight. Figure 4 shows a linear relationship between the avalanche angle and the cohesion length.

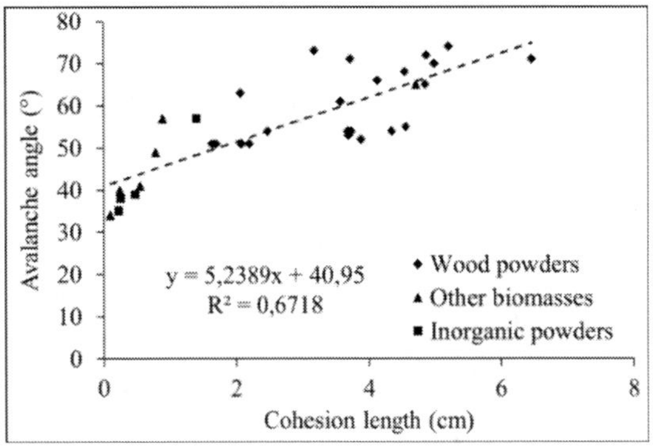

Fig. 4. Relationship between the angle of avalanche and the cohesion length for the 32 powders.

Thus, it is possible to approximate the cohesion of a powder by measuring its avalanche angle and aerated density using the following relation:

$$Angle = \alpha l + \beta \tag{6}$$

Angle: Avalanche angle (°)
α: Constant (°/cm)
l : Cohesion length (cm)
β: constant (°)

With α = 5.24, β = 40.95, $R^2 = 0.672$

The relative gap between the measured and calculated cohesions are presented in the figure 5.

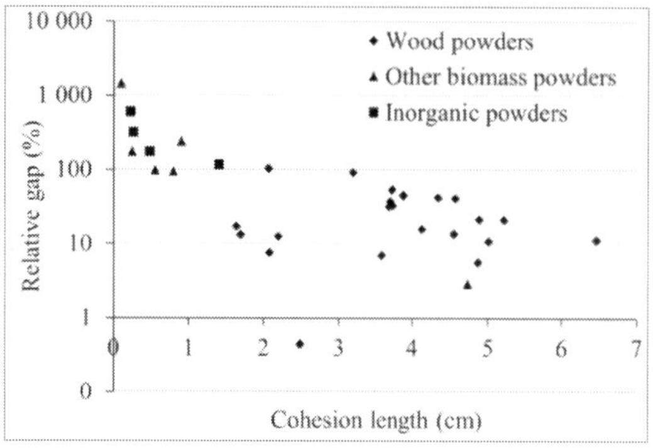

Fig. 5. Relative gap between measured and calculated cohesions as a function of the cohesion length, for the 32 considered powders

For small cohesion lengths (<1.5cm), the cohesions are small and the error on the calculated cohesion is therefore large. However, for larger cohesion length (>1.5cm), the relation gives an acceptable approximation of the cohesion. Then, to use this relation, the cohesion length must be larger than 1.5 cm, which corresponds to an avalanche angle higher than 50° (using equation 7).

3.2.2 Link with the shear test

The equations found in the previous parts can be used to develop a model for the description of the yield locus.

The Warren-Spring equation has three parameters. C and T have been expressed with easy to measure parameters (aerated density and avalanche angle). The last parameters n has to be expressed as well.

For all the powders with l > 1.5 cm, the preconsolidation normal stress is in average $\sigma_p = 3050$ Pa (0.02 kPa standard deviation) and the shear stress is approximately constant with τ_p = 3050 Pa (0.13 kPa standard deviation). Thus it is admissible to consider:

$$n = \frac{\ln(\sigma_p / T + 1)}{\ln(\tau_p / C)} \qquad (7)$$

By combining the Warren-Spring equation (2) with equations (3), (6) and (7), measuring the avalanche angle and the density of a powder is enough to model its yield locus. An example is presented in the figure 6 for a wood powder. This model is forced to go through the point (3.05; 3.05). Then it does not reach the highest points if the preconsolidation point is not exactly (3.05; 3.05). However, the lowest stresses are well described.

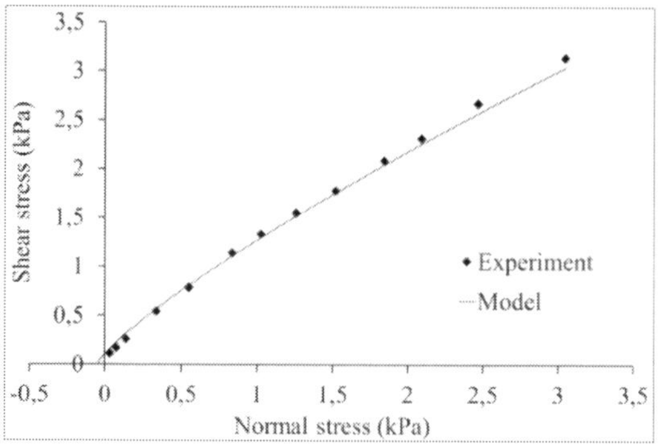

Fig. 6. Yield locus of a wood powder: experiment and model.

4 Conclusions

Based on the study of 32 powders, mainly made of wood:

- An approximation equation (4) of the yield locus has been proposed, based on the Warren-Spring model. The proposed model has two unknowns, C and n, with a condition on n. These parameters can be determined by fitting.
- A linear relation (6) between the avalanche angle, the cohesion and the aerated density of powders is proposed.
- For cohesive biomass powders with an avalanche angle higher than 50°, the yield locus at 3kPa preconsolidation can be modelled by measuring their aerated density and avalanche angle. These parameters are easier to obtain than a yield locus.

This work has highlighted a way to determine the cohesion and the yield locus of cohesive biomass powders from parameters which are easy to measure (aerated density and avalanche angle). It enables to compare the cohesions and flowability of powders in the situation of a confined flow, typically the discharge of a silo.

References

1. H. G. Merkus, G. M. H. Meesters, *Production, Handling and Characterization of Particulate Materials* (Springer, 2015)
2. A. W. Jenike, *Storage and Flow of Solids* (University of Utah, 1964)
3. M. D. Ashton, D. C.-H. Cheng, R. Farley, F. H. H. Valentin, *Rheol. Acta* **4**, 3 (1965)
4. M. Hirota, K. Takenaka, K. Iimura, M. Suzuki, *Adv. Powder Technol.* **18**, 3 (2007)
5. J. L. Amorós, G. Mallol, C. Feliu, M. J. Orts, *Chem. Eng. Sci.* **66**, 18 (2011)
6. J.-M. Geoffroy, J. T. Carstensen, *Powder Technol.* **76**, 2 (1993)

Shear-banding predicted by a constitutive model with a structural parameter in cylindrical Couette flows

Mathieu Jenny[1,2,⋆], *Sébastien* Kiesgen de Richter[1,2], *Nicolas* Louvet[1,2], *Naima* Gaudel[1,2], and *Salaheddine* Skali-Lami[1,2]

[1] *Université de Lorraine, LEMTA, UMR 7563, Vandoeuvre-lès-Nancy, F-54500, France*
[2] *CNRS, LEMTA, UMR 7563, Vandoeuvre-lès-Nancy, F-54500, France*

Abstract. Dense suspensions of non-Brownian particles may partially behave as thixotropic yield stress fluids. We study the flow of such fluids between two concentric cylinders using a phenomenological structural kinetics model. The structural kinetics approach balances the simplicity of phenomenological continuum approaches with a simplified model for structure against the complexity a more fundamental model based on particle micromechanics. A modified version of Houska's model, which includes a diffusive term for the structural parameter, is considered. Depending on the breakdown rate of the structural parameter, shear-banding may be observed. For shear-banding in steady flows, the stress selection depends on the diffusion of the structural parameter. If there is no structural diffusion, the displacement of the interface between the flowing and the static regions fixes the stress at the interface during the transient flow. In the cases of very small diffusive coefficients, the stress at the fluid / solid interface converges to a limit value which is different from the yield stress of the structured material as expected without any diffusion. Nevertheless, the inner torque and the flow profile are quite similar in both cases and the differences are localized near the fluid / solid interface. For shear-banding, the gradients of the structural parameter and the strain rate are very abrupt but the continuity is preserved by the diffusion.

1 Introduction

Dense suspensions of non-Brownian particles exhibit a very rich behavior such as yield stress and shear-banding [1, 2] that remain incompletely understood. Shear-banding has been studied in visco-elastic context [3, 4] but recent work [5] exhibits a different mechanism for shear-banding predicted by a visco-plastic thixotropic model. In that context, no-stress jump is associated with shear-banding. The selection of the stress at the flowing-plug interface does not demand to add stress diffusion in the model as it would be expected in visco-elastic context [6]. Fardin *et al.* [7] show that the length based on this stress diffusion is at the order of magnitude of the molecular size in worm-like micelles. Thus, the diffusion coefficient is expected to be very small ($\sim 10^{-10}$). For non-Brownian suspensions, the velocity profiles obtained by Fall *et al.* [2] suggests that the diffusion length is at the order of magnitude of the particle size, if such diffusive mechanism plays a role in the flow. In our framework, one question arise: does solutions without diffusion term differ from those with a weak diffusion?

It has been shown that the coexistence of static and flowing regions can be associated with the existence of a yield stress which is often related to the existence of a rigid network between the elements of the fluid which has to be broken for the system starts to flow. Shear banding occurs, in many cases, in systems which exhibit a competition between at least two mechanisms: a breakdown process due to an external applied shear and a spontaneous restructuring of the fluid due to the non-linear interaction between its elements [2, 8]. Ovarlez *et al.* 2009 [9] and Coussot & Ovarlez 2010 [10] recently discuss the physical origin of shear localization and shear banding in complex fluids. Authors highlight the existence of a discontinuity of the shear rate profile during shear banding contrary to shear localization where the shear rate goes to zero continuously as one approaches the static region. In the last case, the shear rate is zero on both sides of the static and flowing regions.

The existence of a critical shear rate $\dot{\gamma}_c$ in thixotropic yield stress fluids is often explained in terms of an underlying decreasing branch of the flow curve at low shear rates [8]. It is generally admitted that the shear banding may appear in case of a non-monotonic composite curve [3]. Several scenarii can be distinguished: either the shear banding occurs in a flowing material, either it occurs in localized flows between the solid-like and the flowing regions. We focus on the shear banding in localized flows. In such a scenario, the constitutive relation of the material is a decreasing function of the strain rate between 0 and $\dot{\gamma}_0$ [see 4, for instance].

⋆e-mail: mathieu.jenny@univ-lorraine.fr

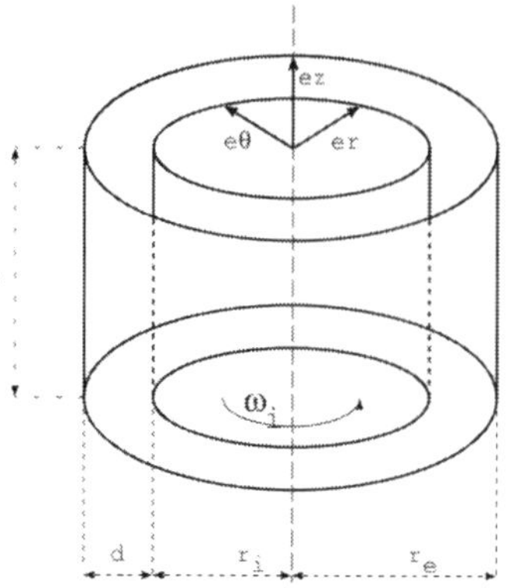

Figure 1. Taylor-Couette geometry. $\mathbf{e}_r$, $\mathbf{e}_\theta$ and $\mathbf{e}_z$ are the unit vectors of the cylindrical coordinates system (r, θ, z).

Here, we study shear-banding in cylindrical Couette flows (see geometry on fig. 1) according to the Houska's model [11] modified in order to include a diffusion term. Results obtained with and without structural diffusion are compared, especially when the diffusion coefficient is evanescent.

2 CONSTITUTIVE EQUATIONS AND NUMERICAL APPROACH

2.1 Modified Houska's model

The Houska's model [11] is built from the Herschel-Bulkley model, commonly used for non-elastic yield stress fluids, considering that the consistency and the yield stress depend linearly on the structural parameter λ. To non-dimensionalize the constitutive equations of the flow in a cylindrical Couette geometry, we choose the following references: the density of the fluid ρ, the velocity of the inner cylinder v_i and the gap width d. Thus, the structural parameter λ is determined by the kinetic equation:

$$\frac{\partial \lambda}{\partial t} + \mathbf{v}.\nabla\lambda = a^\star(1-\lambda) - b^\star\lambda\dot{\gamma}^m + \mathcal{D}^\star\Delta\lambda. \quad (1)$$

where the non-dimensional building and breakdown parameters are

$$a^\star = \frac{ad}{v_i} \quad \text{and} \quad b^\star = b\left(\frac{v_i}{d}\right)^{m-1}, \quad (2)$$

respectively. The non-dimensional structural diffusion coefficient reads:

$$\mathcal{D}^\star = \frac{\mathcal{D}}{v_i d}. \quad (3)$$

The thixotropic breakdown index m is taken equal to 1 in the following. The vector $\mathbf{v}$ denotes the non-dimensional velocity of the fluid.

For the non-Newtonian fluids, several choices may be done for the reference viscosity. In this paper, the reference viscosity is the plastic viscosity of the fluid at the given reference strain rate v_i/d. The corresponding structural parameter λ_{ref} is given by eq. (1) at equilibrium:

$$\lambda_{ref} = \frac{1}{1 + b^\star/a^\star} \quad (4)$$

Thus, the reference viscosity built using the reference strain rate v_i/d reads:

$$\mu_{ref} = \mu_0(1 + \Delta K^\star \lambda_{ref}) \quad (5)$$

where $\mu_0 = K(v_i/d)^{n_c-1}$ can be recognized as the standard reference viscosity of a power law fluid with a consistency K and a shear-thinning index n_c. $\Delta K^\star = \Delta K/K$ is the reduced thixotropic consistency factor.

Using the previous reference dimensions, the Navier-Stokes and mass conservation equations for incompressible fluids are:

$$\frac{\partial \mathbf{v}}{\partial t} + (\mathbf{v}.\nabla)\mathbf{v} = -\nabla p + \frac{1}{Re}\nabla.\bar{\bar{\tau}} \quad (6)$$

$$\nabla.\mathbf{v} = 0, \quad (7)$$

where p stands for the reduced pressure. The Reynolds number is defined using the reference viscosity (5):

$$Re = \frac{Re_0}{1 + \Delta K^\star \lambda_{ref}} \quad \text{where} \quad Re_0 = \frac{\rho v_i d}{\mu_0}. \quad (8)$$

Thus, the reduced stress tensor reads:

$$\bar{\bar{\tau}} = \left[\left(\frac{1 + \Delta K^\star \lambda}{1 + \Delta K^\star \lambda_{ref}}\right)\dot{\gamma}^{n_c} + Bn\left(\frac{1 + \tau_1^\star \lambda}{1 + \tau_1^\star \lambda_{ref}}\right)\right]\frac{\bar{\bar{\dot{\gamma}}}}{\dot{\gamma}}, \quad (9)$$

where $\dot{\gamma}$ and $\bar{\bar{\dot{\gamma}}}$ are the non-dimensional strain rate and strain tensor. The equation (9) involves the Bingham number which is the ratio between the yield stress and the plastic viscous stress:

$$Bn = Bn_0\frac{1 + \tau_1^\star \lambda_{ref}}{1 + \Delta K^\star \lambda_{ref}} \quad \text{where} \quad Bn_0 = \frac{\tau_0}{K(v_i/d)^{n_c}}. \quad (10)$$

One can recognize the standard Bingham number Bn_0 of a Herschel-Bulkley fluid with a yield stress $\tau_y = \tau_0$. The last parameter, $\tau_1^\star = \tau_1/\tau_0$, is the reduced thixotropic yield stress.

2.2 Boundary conditions for the flow

The inner and outer radii of the Couette setup are defined by $r_i = \frac{\eta}{1-\eta}$ and $r_e = \frac{1}{1-\eta}$ with $\eta = r_i/r_e$ the radii ratio. We only consider the case where the inner cylinder rotates and the outer cylinder is fixed, *i. e.* $\mathbf{v} = v(r)\mathbf{e}_\theta$ in the cylindrical basis. Thus, the boundary conditions are:

- At the inner radius, $v(r_i) = 1$.
- At the outer radius of the flowing zone, $v(r_o) = 0$.
- In our case, there is a material limit at $r = r_e$. Thus, the outer radius r_o is given by the following criterion: If $\tau(r_e) \geq \tau_y$, $r_o = r_e$, else, $\tau(r_o) = \tau_y$.
- If $\mathcal{D}^\star \neq 0$, we have to add the condition of no-diffusion through the surface of the cylinders, *i. e.* $\partial\lambda/\partial r = 0$ at $r = r_i$ and $r = r_e$.

 DOI: 10.1051/epjconf/201714011003

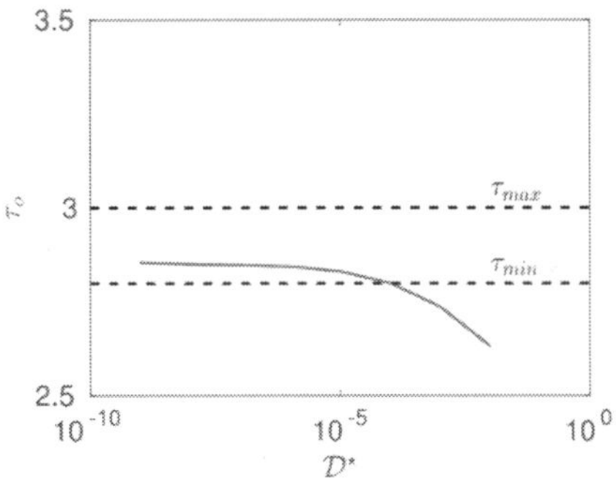

Figure 2. Stress τ_o at the interface for $b^\star = 2$ (shear-banding) compared to the minimum value of the composite curve τ_{min} and the yield stress of the fully structured material τ_{max} *vs* the diffusion coefficient $\mathcal{D}^\star$. $Bn = 2$, $n_c = 1$, $\Delta K^\star = 1$, $\tau_1^\star = 1$, $a^\star = 1$.

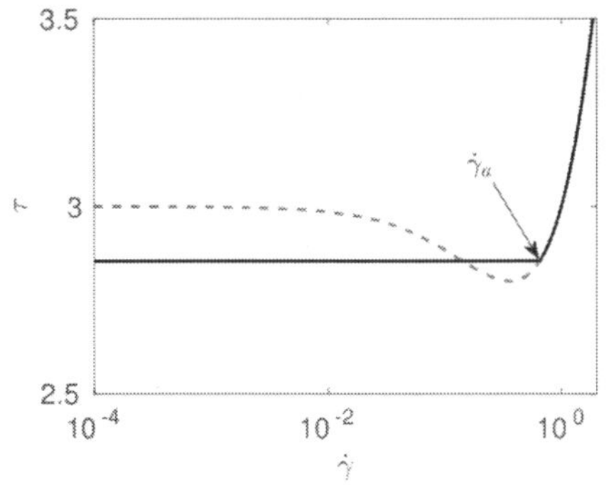

Figure 3. Composite curve (blue dashed line) and local stress in the fluid region (dark full line) *vs* the strain rate $\dot\gamma$ with $\mathcal{D}^\star = 10^{-9}$ (fig. b). $Bn = 2$, $n_c = 1$, $\Delta K^\star = 1$, $\tau_1^\star = 1$, $a^\star = 1$, $b^\star = 2$.

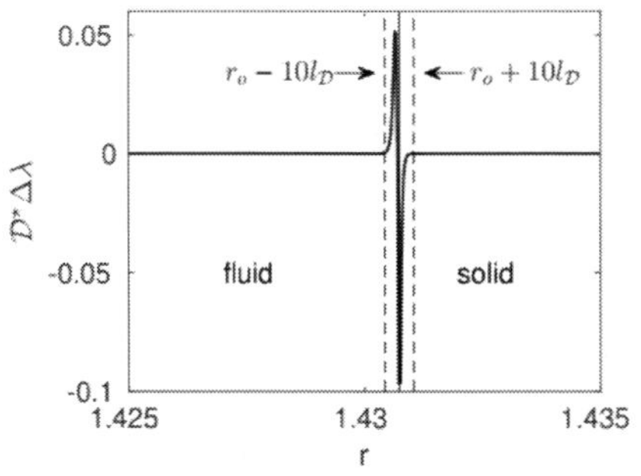

Figure 4. Values of the diffusion term in the gap with $Bn = 2$, $n_c = 1$, $\Delta K^\star = 1$, $\tau_1^\star = 1$, $a^\star = 1$, $b^\star = 2$ and $\mathcal{D}^\star = 10^{-9}$. $l_\mathcal{D}$ denotes the diffusion length.

2.2.1 Numerical method

To perform the numerical solution, we use a finite difference method for the spatial discretization. For the derivative operations, the standard second order centered scheme is used. The numerical method used for the spatial discretization is quite well established and is similar to the ones used, for instance, in [5, 12–14]. In order to solve the transient flow, we use an Arbitrary Lagrangian-Eulerian (ALE) method [15, 16] for the interface tracking in the partially flowing cases.

3 Stress selection in shear-banding

3.1 Diffusive stress selection

In the following, we choose the parameters in order to have a localized and shear-banded flow in the gap, *i. e.* $Bn = 2$ and $b^\star = 2$. Considering the stress value at the fluid/solid interface (figure 2), we remark that the stress tends toward a limit value for very small coefficients $\mathcal{D}^\star < 10^{-4}$ or 10^{-5} depending on the breakdown parameter $b^\star$. In our case, beyond $b^\star = 5$, the weakness of the structure dominates and the stress ends to decrease at $r = r_o$ when $b^\star$ increases.

Nevertheless, for very small diffusive coefficients $\mathcal{D}^\star$, the limit value of the selected stress τ_o at $r = r_o$ is between the yield stress of the structured material and the minimum value reached by the composite curve (figs. 2 and 3). The local stress (full line in fig. 3) is a monotonic growing function of $\dot\gamma$ but the variation of the stress in the range of strain rate between 0 and the apparent strain rate $\dot\gamma_a$ at the fluid-solid interface is almost flat. Thus, this value would be seen as an apparent yield stress, different from the one observed without shear-banding. In fact, the strain rate falls from the apparent strain rate $\dot\gamma_a$ down to zero over a distance which is in the order of magnitude of $10l_\mathcal{D}$, $l_\mathcal{D} = \sqrt{\mathcal{D}^\star/a^\star}$ denoting the diffusion length. This length caracterizes the thickness of the interfacial region where the diffusion term is not negligible (fig. 4).

The diffusion ends to smooth the solution for moderate values of $\mathcal{D}^\star$ but, for very small values of $\mathcal{D}^\star$, the sharpness of the field is indistinguishable from the case $\mathcal{D}^\star = 0$. Moreover, the velocity profile obtained, with only 50 nodes, without any diffusion by setting the value of the selected stress at the interface is indistinguishable from the one provided by the diffusive model (fig. 5). In comparison, up to 16000 nodes are needed both in the flowing and the static regions to catch the diffusion length scale around the interface for $\mathcal{D}^\star = 10^{-9}$. The non-diffusive model still provides a solution close to the diffusive model for small diffusion coefficients (fig. 5): the interface position r_o is only 2.1% closer to the inner cylinder.

3.2 Transient flows

During the transient flow, the inertial term depending on the velocity of the interface insures the continuity of the solution across the flowing / solid-like interface. Thus the stress is well defined at $r = r_o$. A brutal cracking of the solid phase of the material at the start of the rotation would allow other stress values than the yield stress but such scenario is beyond the frame of the model considered here. Thus, when $\mathcal{D}^\star = 0$, the steady equations admit a physical solution where the stress at the interface is the yield stress of the fully structured material even in shear-banded

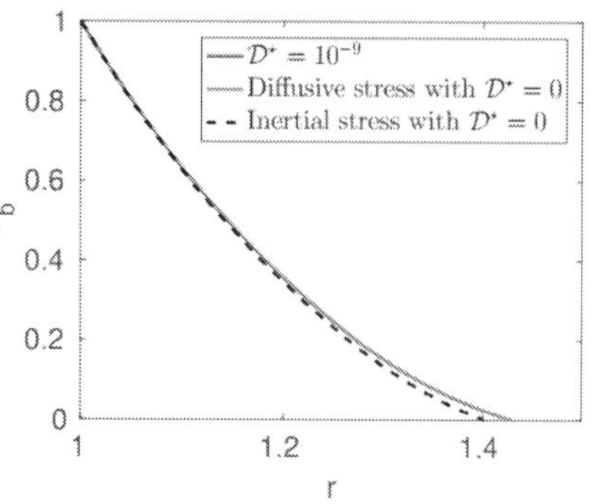

Figure 5. Velocity profile of the steady flow with $Bn = 2$, $n_c = 1$, $\Delta K^\star = 1$, $\tau_1^\star = 1$, $a^\star = 1$, $b^\star = 2$ for a large gap $\eta = 0.5$. The computations use 16000 nodes for $\mathcal{D}^\star = 10^{-9}$ and only 50 nodes for the calculation without diffusion term where the stress is arbitrary set at the interface.

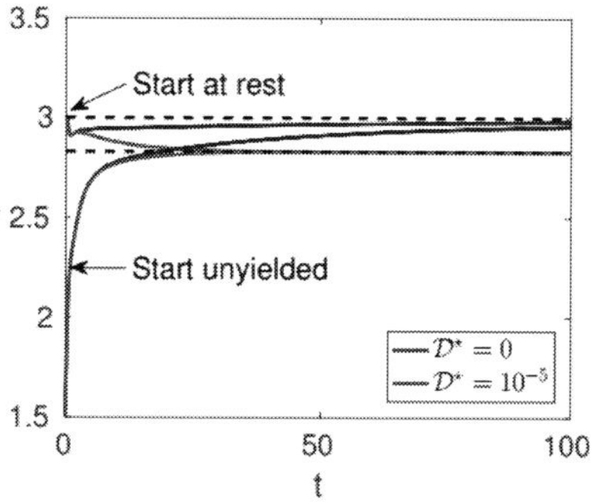

Figure 6. Stress τ_o at the interface ($r = r_o$). The dashed lines represent the steady limit for $\mathcal{D}^\star = 0$ and $\mathcal{D}^\star = 10^{-5}$ respectivelly. Transient flows with $Re = 10$, $Bn = 2$, $n_c = 1$, $\Delta K^\star = 1$, $\tau_1^\star = 1$, $a^\star = 1$ and $b^\star = 2$ for a large gap $\eta = 0.5$.

flows [5]. Other solutions would imply that the stress at the interface is different from the yield stress.

If $\mathcal{D}^\star \neq 0$, $\partial\tau_o/\partial t < 0$ before reaching the steady state in the 'start at rest' protocol and the characteristic time of the transient flow is smaller than for $\mathcal{D}^\star = 0$, even for very small $\mathcal{D}^\star$ (fig. 6).

4 Conclusion

The diffusive term of the structural parameter selects the stress in case of shear-banded flows. When the diffusion coefficient $\mathcal{D}^\star$ is small, typically less than 10^{-5}, the selected value of the stress does not vary effectively with $\mathcal{D}^\star$ and a limit value is reached. In case $\mathcal{D}^\star = 0$, there is another mechanism to select the stress at the interface in shear-banded flows, assuming that it is always the yield stress of the material. This is imposed by the inertial part of the transient flow where the motion of the interface smooths the transition from the fluidized region to the static region. One can notice that other assumptions for the stress value are beyond the model we consider here. The case $\mathcal{D}^\star = 0$ may have a meaning for non-Brownian and non-colloidal suspensions although that it is commonly admitted that for colloidal suspensions $\mathcal{D}^\star \neq 0$. Prediction of shear-banding and the influence of the structural diffusion are of practical interest to determine the yield stress.

In case of shear-banding, we show that the flows with and without the structural diffusion are different but the solutions remain close to each other. The best way to determine if $\mathcal{D}^\star \neq 0$ is to track the shear stress at the interface in shear-banding cases using a planar geometry, used by S. Fielding for instance, where the stress is homogeneous in the gap.

References

[1] G. Ovarlez, F. Bertrand, S. Rodts, Journal of Rheology **50**, 259 (2006)
[2] A. Fall, F. Bertrand, G. Ovarlez, D. Bonn, Phys. Rev. Lett. **103**, 178301 (2009)
[3] S.M. Fielding, arXiv preprint arXiv:1512.04715 (2015)
[4] T. Divoux, M.A. Fardin, S. Manneville, S. Lerouge, arXiv preprint arXiv:1503.04130 (2015)
[5] M. Jenny, S. Kiesgen de Richter, N. Louvet, S. Skali-Lami, Y. Dossmann, Phys. Rev. Fluids **2**, 023302 (2017)
[6] P. Olmsted, O. Radulescu, C.Y. Lu, Journal of Rheology (1978-present) **44**, 257 (2000)
[7] M.A. Fardin, O. Radulescu, A. Morozov, O. Cardoso, J. Browaeys, S. Lerouge, Journal of Rheology (1978-present) **59**, 1335 (2015)
[8] P.D. Olmsted, Rheologica Acta **47**, 283 (2008)
[9] G. Ovarlez, S. Rodts, X. Chateau, P. Coussot, Rheologica Acta **48**, 831 (2009)
[10] P. Coussot, G. Ovarlez, The European Physical Journal E **33**, 183 (2010)
[11] J. Šesták, R. Žitný, M. Houška, Journal of Food Engineering **2**, 35 (1983)
[12] M. Jenny, B. Nsom, Phys. Fluids **19**, 108104 (2007)
[13] A.M. Philippe, C. Baravian, M. Jenny, F. Meneau, L.J. Michot, Phys. Rev. Lett. **108**, 254501 (2012)
[14] M. Pourjafar, E. Chaparian, K. Sadeghy, Meccanica **50**, 1451 (2015)
[15] C. Hirt, A. Amsden, J. Cook, Journal of Computational Physics **14**, 227 (1974)
[16] J. Donea, S. Giuliani, J. Halleux, Computer Methods in Applied Mechanics and Engineering **33**, 689 (1982)

Blending liquid into a flowing dry granular material

Guillaume Saingier[1,*], *Alban* Sauret[1], and *Pierre* Jop[1]

[1] *Surface du Verre et Interfaces, UMR 125 CNRS/Saint-Gobain, 93303 Aubervilliers, France*

Abstract. We study experimentally the growth dynamics of a horizontal wet granular aggregate produced by accretion when a dry granular jet impacts a wet substrate. The tomographic imaging demonstrates that the wet aggregate is fully saturated and its cohesion is related to the capillary suction due to the pressure drop at the liquid/air interface. We highlight that the accretion process is characterized by two different growth dynamics depending on the hydrostatic depression in the material. At low depression, the growth dynamics exibits a "diffusive" regime whereas the dynamics becomes linear for higher depressions. A competition between the viscous displacement of the fluid into the granular material and the sticking dynamics is proposed to understand the transition in the growth velocity.

1 Introduction

Granular materials are commonly used as raw materials in the building industry [1]. In particular, the preparation of many building materials such as concrete, mortar or plaster requires mixing granular matter with liquid and additives in order to obtain new chemical or physical properties. For a large added volume of water, a mixing stage easily produces a homogeneous granular suspension in which the grains freely move in the interstitial fluid. By contrast, the addition of smaller amounts of liquid provides a strong cohesion in the material through the formation of liquid bridges between grains [2–4]. For intermediate liquid amounts, spatial heterogeneities can appear during the blending process and generate bubbles or saturated aggregates, which will locally impact the mechanical properties of the final material. Most studies on wet granular matter are focused on the fluid flow dynamics and the repartition of the liquid into a static granular packing [5, 6]. In this case, the material can be assimilated to a porous medium. Therefore, the liquid transport is essentially led by capillarity in the porosity of the material and exhibits well-known imbibition dynamics described in 1D by the Lucas-Washburn law [7].

Recently, dynamical situations were studied through experiments of granular towers built by dripping a granular suspension on a dry substrate [8] or by pouring dry sand on a wet substrate [9]. A cohesive granular structure is erected in each situation but the physcial origin and the growth dynamics is different. On one side, the tower growth is related to a competition between liquid drainage and dripping frequency. On the other, the dynamics is limited by the accretion of dry grains on the wet structure. However, this accretion process is still not understood and the origin of the growth dynamics and the capture mechanism were not explained. Here, we investigate the accretion process between liquid and grains, using a model experiment of liquid incorporation into a flowing granular material. We focus on an experiment of accretion at a constant height to remove the effect of gravity and keep the hydrostatic pressure constant in the material. We present an experimental set-up that allows modifying the hydrostatic pressure in the wet granular material and highlighting the growth regimes.

2 Experimental set-up

We performed experiments of granular accretion using spherical glass beads of diameters d_g = 300-350 μm poured at constant flow rate and constant velocity over a wet substrate composed of the same beads (see Fig. 1). The wet granular layer of height h_g = 10 mm is prepared in a polypropylene tube of 10 mm diameter connected to a water tank to soak the grains. The wet substrate is separated from the liquid by a porous membrane and the water level can be tuned to modify the hydrostatic pressure into the granular substrate. We note Δh the distance between the water level and the top of the wet substrate. Dry beads are poured using an inox funnel at a constant flow rate (Q_g = 1.1 g/s). The funnel is connected to a inox tube of length L = 200 mm with a flexible polyethylene end to inject the grains with an angle of about 45°. The granular aggregate grows horizontally on a thin plexiglas plate of width w = 3 mm to hold it while avoiding the accumulation of dry grains in the front of the aggregate during the experiments. The extremity of the tube is held at 10 mm above the plexiglas plate such that the grains kept a constant velocity (v_g $\simeq$ 1.6 m/s) to impact the tip of the aggregate. The grain container is placed on a translating support allowing us to shift it while the aggregate is growing along the support such that the distance between the growing aggregate and

*e-mail: guillaume.saingier@saint-gobain.com

 DOI: 10.1051/epjconf/201714009011

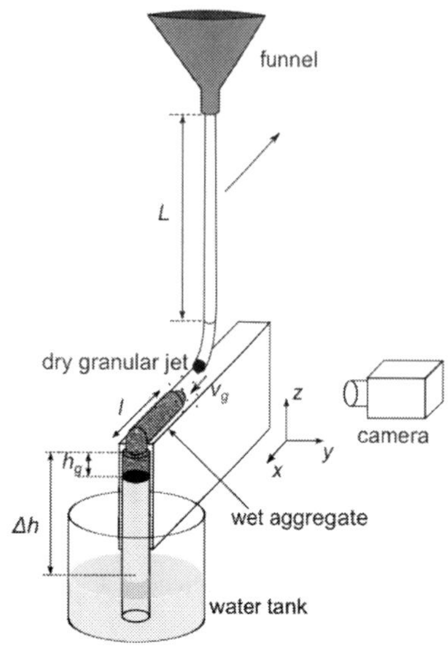

Figure 1. Schematic of the experimental set-up.

the extremity of the tube remains constant. The growth of aggregate is filmed at low acquisition frequency (f_{acq} = 0.5 Hz) with a CCD camera and analyzed by image processing with ImageJ and Python scripts [10, 11]. By using this experimental set-up, homogeneous aggregates are built horizontally.

3 Experimental results

In a typical experiment, the granular jet is oriented toward the wet substrate and an aggregate grows horizontally on the support. The width of structure is about the width of the support and the section remains nearly constant in each experiment. By increasing the height Δh, the growth dynamics can be changed drastically. To understand this phenomenon, we proposed to study the internal structure of the wet aggregate and its growth dynamics.

3.1 Microstructure of the aggregate

To investigate the microstructure of the aggregate, X-ray tomography experiments were carried out with a laboratory tomograph available at Saint-Gobain Research. The X-ray radiations were generated by an electron gun bombing a tungsten plate which produces photons at a peak energy of 40 kev. The spatial resolution was 7.5 μm and the detector fields of view was 13 mm x 10 mm. To increase the contrast between the different phases (grains, liquid and air), an aqueous zinc iodide solution, of 0.5 g/mL, was used as liquid to build the aggregate [6]. A tomographic acquisition consisted of 1395 projections acquired in about 40 min and equally separated to cover a total angle of 180°.

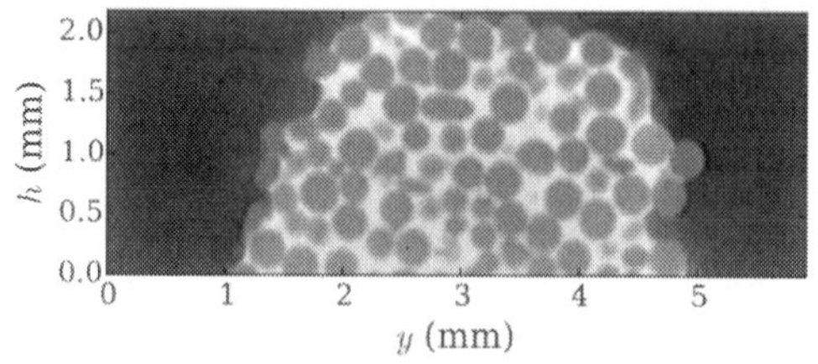

Figure 2. Cross-section through a 3D tomogram of a wet aggregate. The liquid is doped by adding zinc iodide to improve the contrast between the different phases. The liquid is colored in white whereas the glass beads appear in gray and the air in black.

The 3D reconstruction was performed from the set of 2D projection using the filtered backprojection algorithm.

The analysis of the tomographic reconstructions shows that the wet aggregate is fully saturated without almost no bubble in the granular assembly (see Fig. 2). The aggregate is in the capillary state and the cohesive interactions between the particles result from the capillary depression at the interface with the air. From this observation, we conclude that the hydrostatic pressure is the same at each point of the horinzontal aggregate growing at constant Δh. At the liquid/air interface, the capillary drop associated to the local curvature of the menisci balances the hydrostatic pressure:

$$p_0 - \rho g \Delta h = p_0 - \frac{2\gamma \cos\theta}{r_c}, \tag{1}$$

where γ = 72 mN/m is the surface tension of the water, θ is the contact angle of water on a glass bead and r_c is the radius of curvature of the meniscus at the interface. Consequently, the local curvature of a meniscus increases with Δh in our experiment.

3.2 Growth dynamics

We observed that the aggregate grows slower when Δh increases, so that the accretion process is less efficient in this case. An example of snapshots from the experiments is presented in Fig 3. By performing several tests at different altitude and by measuring the length ℓ of the wet granular structure in function of the time t, we can determine the growth dynamics of the aggregate. These curves are plotted in Fig. 4 for different Δh at Q_g and v_g constant. A change of growth dynamics is clearly observed when Δh increases with a transition from a "diffusive" regime where ℓ is proportional to $t^{1/2}$ to a linear regime where ℓ is proportional to t.

To understand the existence and the difference between the two regimes, we consider the local mechanism of the aggregate growth. The aggregate is built by trapping dry grains at the liquid interface of the wet material.The liquid needs to penetrate into the granular packing until being accessible to the granular jet. When Δh is small, the radius of curvature of the menisci is large and we can assume that

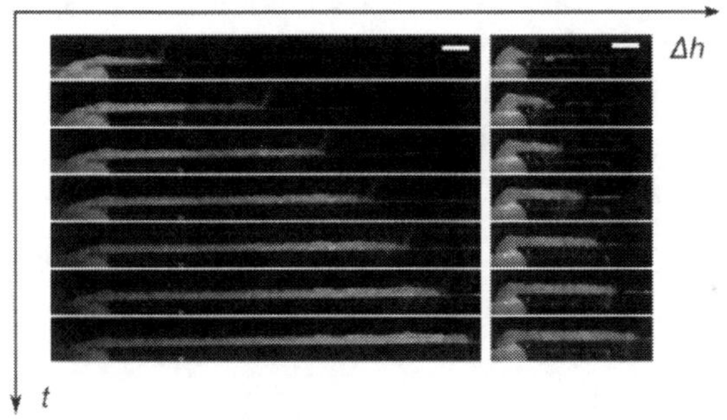

Figure 3. Snapshots of the growing process at different times and for two altitudes Δh. (left) $\Delta h = 0.5$ cm - (right) $\Delta h = 4.9$ cm. Snapshots are seperated by 40 s. Scale bar is 1 cm.

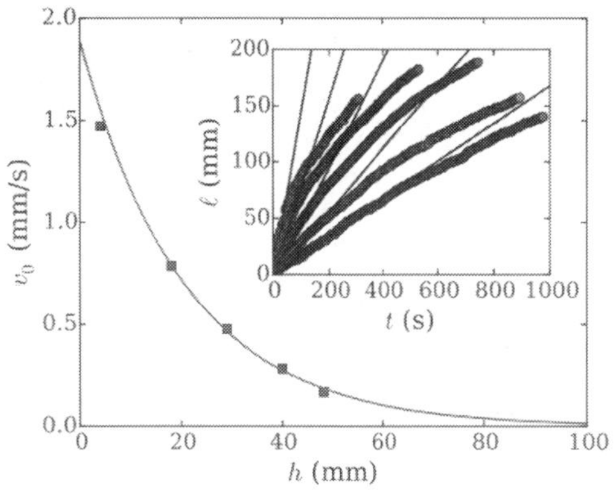

Figure 5. Evolution of the initial velocity of growth as a function of the distance Δh between the water level and the top of the aggregate. The solid line is obtained by fitting the experimental data with an exponential expression. Inset: Length of the aggregate ℓ as a function of the time t for different Δh.

the liquid is available to catch a large fraction of grains. The growth rate of the aggregate is limited by the viscous displacement of the fluid into the granular packing until it reaches its equilibrium position. We can model this liquid-limited regime with the Darcy's law without gravity. The Lucas-Washburn law is obtained:

$$\ell_{visc}(t) = \sqrt{\frac{2k\Delta pt}{\eta}}, \tag{2}$$

where k is the permeability of the packing (in m^{-2}), η is the dynamic viscosity (in Pa.s) and $\Delta p = p_c - \rho g \Delta h$ is the capillary pressure reduced by the hydrostatic depression. The pressurep_c is defined as $2\gamma \cos\theta/(\rho_l r_p)$ where r_p is the pore radius. This "diffusive law" is qualitatively in agreement with the first regime identified in our results (see Fig. 4). For large Δh, the radius of curvature of the menisci is reduced and the liquid is less accessible to the grains impacting the aggregate. In this case, the efficiency of the capture process limits the growth of the aggregate. Considering the growth rate as the ratio of grains stuck over the total amount of grains impacting the aggregate,

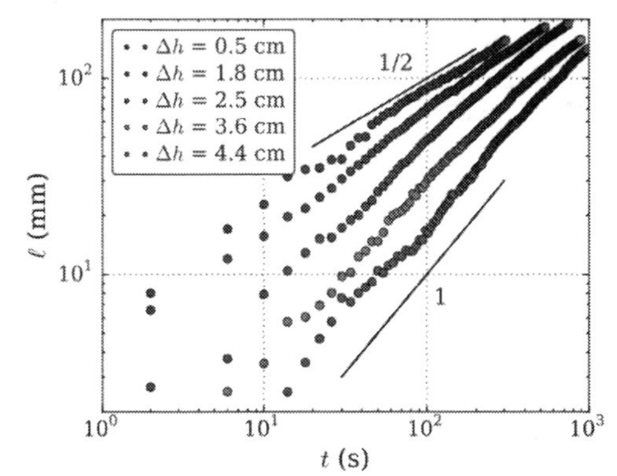

Figure 4. Time evolution of the length of the aggregate for several Δh in log-log representation. The power laws lines curves are indicative.

we can write:

$$\frac{d\ell}{dt} = \frac{Q_g}{\rho_s \phi S} P_{capt}, \tag{3}$$

where ϕ is the compacity of the packing, S is the section of the aggregate and P_{capt} is the ratio of stuck grains over the total injected grains. Assuming that P_{capt} is constant at fixed Δh, the length of the aggregate grows in the capture-limited regime as:

$$\ell_{capt}(t) = \frac{Q_g}{\rho_s \phi S} P_{capt} t. \tag{4}$$

In summery, the "diffusive" behavior corresponding the liquid-limited regime and the linear behavior corresponding to the capture-limited regime are qualitatively plotted on the Fig. 4.

To estimate the efficiency of the accretion process at different altitude Δh, we derived the initial growth velocity v_0 of the aggregate. Indeed, the beginning of the aggregate growth is limited by the capture process on the wet substrate since the liquid is already available. The evolution of v_0 is plotted in Fig. 5 as a function of Δh. We observe that v_0 decreases with the altitude and is well-fitted by an exponential expression:

$$v_0(h) = v^\star \exp\left(-\frac{\Delta h}{h^\star}\right), \tag{5}$$

where $v^\star$ is the capture-limited growth velocity at $\Delta h = 0$ and $h^\star$ is the length scale of variation of the sticking probability. Similar variations were observed for the growth velocity of a vertical granular towers built upon the accretion from a dry diluted granular jet [9]. The value of $h^\star$, computed from pressure-controlled experiments of horizontal aggregate, is compatible with measurements performed on vertical granular towers. This result indicates that the logarithmic law observed in the growth of vertical aggregate is linked to the variation of the hydrostatic pressure in the wet material.

4 Discussion

The transition between the two growth regimes is expected when the growth rates associated to each limiting mechanism are comparable. For a constant Δh, at the beginning of the growth, the capture of dry grains is slow enough, compared to the liquid transfer, and limits the growth rate. As the length of the aggregate increases, the viscous transfer of liquid slows down and becomes limiting for the growth rate. Equating the capture and viscous growth rates leads to a critical length L_c determining the linear-diffusive transition and defined as:

$$L_c = \frac{k}{2\eta} \frac{\rho_s \phi S}{Q_g P_{capt}} \Delta p. \quad (6)$$

This transition drastically depends on the efficiency of the accretion process, which appears to decrease with the hydrostatic depression, controlled by Δh. Consequently, the transition length L_c increases when Δh increases because of the decrease of the sticking probability. For transitional values ($\Delta h = 1.8$ cm, $\Delta h = 2.5$ cm), an intermediate exponent between 1/2 and 1 is identified, corresponding to the smooth linear-diffusive transition. Furthermore, for low values of Δh ($\Delta h < 1$ cm), the critical length is close to the height h_g of the substrate and the transition is not observed. In this case, nearly all the growth dynamics is dominated by the viscous displacement of liquid. Further investigations are now required to identify the impact of the fluid properties (surface tension, viscosity) and the grains velocity on the sticking probability.

5 Conclusion

The blending of liquid into a dry granular flow is considered in a model situation to create a wet aggregate growing by granular accretion at the interface between dry and wet grains. The tomographic imaging confirmed that the aggregate is in the capillary regime and its cohesion is related to the capillary suction associated to the pressure drop at the liquid/air interfaces. Our experimental set-up allowed us to control the hydrostatic pressure in the wet substrate to isolate the accretion process. We highlighted two regimes of accretion are identified associated to the fluid flow into the porous matrix and the efficiency of the grains capture process. In particular, the sticking probability is found to be constant at a fixed height. It confirms the importance of the hydrostatic depression in the dynamics and the locality of the capture mechanism. Our work may be useful for industrial applications to improve or control agglomeration processes such as the wet granulation.

Acknowledgements

We are grateful to William Woelffel for his help and his advice for the tomographic acquisitions.

References

[1] J. Duran, *Sands, Powders, and Grains: An Introduction to the Physics of Granular Materials* (Springer-Verlag, New-York, 2012).

[2] S. Herminghaus, Adv. Phys. **54**, 221-244 (2005).

[3] N. Mitarai & F. Nori, Adv. Phys. **55**, 1-45 (2006).

[4] A. Kudrolli, Nat. Mater. **7**, 175-175 (2008).

[5] M. Reyssat, L. Y. Sangue, E. A. Van Nierop & H. A. Stone, Europhys. Lett. **86**, 56002 (2009).

[6] M. Scheel, R. Seeman, M. Brinkman, M. di Michel, A. Sheppard, B. Breidenbach & S. Herminghaus, Nat. Mater. **7**, 189 (2008).

[7] W. Washburn, Phys. Rev. **17**, 273 (1921).

[8] J. Chopin & A. Kudrolli, Phys. Rev. Lett. **107**, 208304 (2011).

[9] F. Pacheco-Vázquez, F. Moreau, N. Vandewalle & S. Dorbolo, Phys. Rev. E **86**, 051303 (2012).

[10] https://imagej.nih.gov/ij/

[11] https://www.python.org/

Nonaffinity in amorphous solids close to the jamming transition

Roberto Arévalo[1] and *Massimo* Pica Ciamarra[1,*]

[1] *Division of Physics and Applied Physics, School of Physics and Mathematical Sciences, Nanyang Technological University, 21 Nanyang Link, Singapore, 637371.*

Abstract. Nonaffinity is known to be an integral part of the response of amorphous solids. Its role is particularly relevant in particulate systems close to their jamming transition, where it dominates the elastic response. Thus, to determine the elastic properties of amorphous solids it is essential to rationalize the features of their non-affine response. Via numerical simulations we investigate the relation between the non affine response and the vibrational properties of model amorphous materials. We show that, contrary to previous speculations, modes below the Boson peak are those mostly responsible for the nonaffine response.

1 Introduction

Crystalline solids are currently very well understood in terms of an ordered array of atoms that repeats itself in space. Linear elasticity and dislocations account for the deformations of the solid when submitted to external stresses. The situation is rather different in amorphous solids [1]. Lacking an ordered structure, the constituents of an amorphous solid perform nonaffine movements when the boundaries are deformed. These nonaffine displacements are not contemplated in linear elasticity theory and lead to the softening of the system.

In a system of particles interacting through repulsive forces, such as granular materials [2, 3], nonaffinity might dominate the elastic response. Specifically, when particles are packed at high density they form a solid able to resist shear and compression. As the density or packing fraction is decreased, the nonaffinity increases, which, in turn, affects the elastic moduli. At a critical packing fraction ϕ_c the system ceases to be rigid, as the elastic moduli go to zero. Measuring the distance to the transition in terms of $\Delta\phi = \phi - \phi_c$, in harmonic solids the bulk modulus is constant but drops to zero discontinuously at the critical point. While the shear modulus vanishes continuously as $\Delta\phi^{1/2}$. The ratio between both moduli defines a length scale that diverges as $\Delta\phi$ vanishes.

Here we investigate the relation between the nonaffine response and the normal modes of the dynamical matrix [4], *i.e.*, the Hessian or matrix of second derivatives of the potential respect to displacements of the particles. In first place, we analyze the features of the density of vibrational states $D(\omega)$ which is known to differ from that of crystals. The latter presents a decay $D(\omega) \propto \omega^{d-1}$ where d is the dimensionality of the system. In amorphous solids, instead, $D(\omega)$ shows an increased number of low frequency modes forming a plateau known as the boson peak. The power-law decay, that we will refer to as Debye scaling, is recovered only below a frequency $\omega^* \propto \Delta\phi^{1/2}$ [2, 3]. Modes below this frequency are supposed to be plane waves, while above, they are heterogeneous. It has been suggested [5] that the modes controlling the nonaffine response of granular systems are those lying in the plateau and with frequency close to ω^*.

Here we investigate this speculation in light of recent results showing that the density of states does not follow Debye scaling up to zero frequency. Indeed, it has been recently [6] observed that a very low frequencies the density of states crossovers towards a $D(\omega) \propto \omega^4$ regime below a frequency ω_{MGM}. The corresponding modes are described as quasilocalized in 3D, with localization being independent of frequency for the lower modes. This regime had been predicted for general glasses at high density [7–10]. After showing that this crossover is also present in a 2D dimensional model of frictionless granular system, we show that the heterogeneity of the elastic regime is due to the modes in the frequency range $[\omega_{MGM}, \omega^*]$, and not to the modes with $\omega \geq \omega^*$ as previously speculated.

2 Methods

We simulate [11] a 50 : 50 mixture of bidisperse soft disks in *2D* with size ratio $R_1/R_2 = 1.4$. We begin by placing the disks at random positions and minimizing the energy using conjugate gradient method to a target packing fraction. We use this configuration to compute and subsequently analyze the dynamic matrix. Next, we apply a small deformation to the box such that we remain in the linear regime. For simple shear, we apply a strain of magnitude $\delta\gamma = 10^{-5}$. For compression, we increase the packing fraction of the system by an amount $\delta\phi = 2 \cdot 10^{-6}$. After the deformation we record the intermediate state of

*e-mail: massimo@ntu.edu.sg

the system, including the unbalanced forces on each particle. Finally, we minimize the energy again using the conjugate gradient method, thus obtaining the final configuration. The particle displacements occurred during this last deformation are their nonaffine displacements.

Our system consists of $N = 1000$ particles interacting by means a harmonic potential $V(r_{ij}) = \frac{\epsilon}{2}\left(R_i + R_j - r_{ij}\right)^2$, where r_{ij} is the distance between contacting particles and R_i is the radius. We use periodic boundary conditions and, in the case of shear, the Lees-Edwards boundary condition is applied in the transversal direction. The dynamical matrix is diagonalized using the SLEPc library. We run around 4000 simulations for each density, and eigenvectors are computed for around 2000.

The nonaffine displacements can be computed from the normal modes using [5]:

$$\langle \| \delta\vec{R}_{ij} \|^2 \rangle = \sum_{\omega} \omega^{-4} \langle \delta R(\omega) | \delta F \rangle^2 \langle \| \delta\vec{R}_{ij}(\omega) \|^2 \rangle. \quad (1)$$

Here $\vec{R}_{ij}$ is the nonaffine part of the relative displacement of neighbouring particles. $|\delta F\rangle$ are the unbalanced forces, $|\delta R(\omega)\rangle$ is the normal mode with frequency ω. This expression can be simplified further since modes are very heterogeneous and uncorrelated, then $\langle \| \delta\vec{R}_{ij}(\omega) \|^2 \rangle \sim \langle \| \delta\vec{R}_i(\omega) \|^2 \rangle = 1/N$, where the last equality is due to normalization. Using again low spatial correlation of the modes, one derives $\langle \delta R(\omega) | \delta F \rangle^2 \sim \omega^2 \gamma^2$ (see [5] for more details). With these approximations, one finally has

$$\frac{\langle \| \delta\vec{R}_{ij} \|^2 \rangle}{\delta\gamma^2} \sim \sum_{\omega} \frac{\omega^{-2}}{N} \sim \int_{\omega > \omega^*} \frac{D(\omega)}{\omega^{-2}} d\omega \sim \frac{1}{\omega^*} \sim \delta\phi^{-1/2}.$$

The key assumption of this estimation is that the dominant contribution comes from modes around ω^* where the density of states is a constant, while modes at lower frequencies give subdominant contributions [5]. Thus, the above integral is only evaluated for $\omega > \omega^*$. While this result is in agreement with measurements of the displacement field under shear, and allows to rationalise the scaling of the shear modulus, its underlying assumption has never been explicitly checked. In the remainder of the paper we perform this check, and we will use as a measure of the non-affine response normalized by the deformation applied, *i.e.*, $\chi = \langle \| \delta\vec{R}_{ij} \|^2 \rangle / \delta\phi^2$ and $\chi = \langle \| \delta\vec{R}_{ij} \|^2 \rangle / \delta\gamma^2$, for compression and shear respectively.

3 Results

We compute the theoretical values of the nonaffine displacements using Eq.1 after diagonalizing the dynamical matrix. We compare these values with those obtained the from the actual relative displacement of the particles occurring when minimizing the energy after imposing the shear or the compression deformation. Results are shown in Fig.1. The data show the expected the scaling $\Delta\phi^{-1/2}$ for shear and we obtain a scaling $\Delta\phi^{-1}$ for compression.

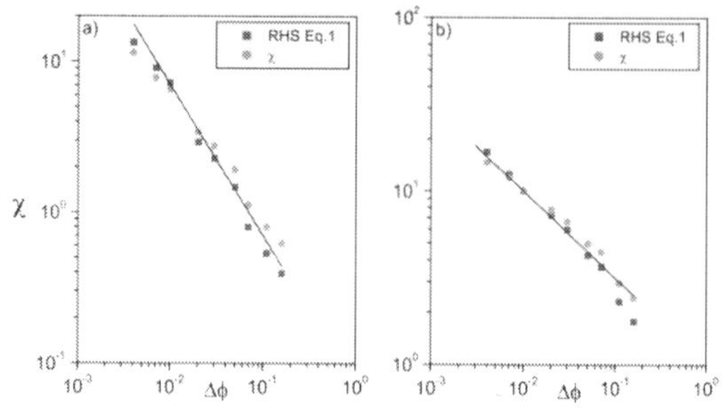

Figure 1. Full sum in the RHS of Eq.1 compared with the LHS computed from the actual displacements of the particles. Left: compression. Right: shear. Solid lines are added as guide to the eye, and have slopes 0.5 (right plot) and 1 (left plot).

The assumptions used to obtain Eq.1 work very well and turn systematically better at lower densities.

Similar critical-like scalings have been found for related systems. Granular packings under quasi-static shear have been shown to display critical behavior in the dynamical quantities, see *e.g.* [12–16]. In [17] additional exponents and hyper-scaling relations between them were reported. Related studies of the incremental response of amorphous solids can be found, *e.g.*, in [18].

Length scales have also been discussed recently. The length $l \propto \Delta\phi^{-1/2}$ relating bulk and shear moduli was first rationalized in [19] using a variational argument. Subsequent simulations [20] found that the same length scale described the response of granular systems to point forces. However, in [21] a length scale $l \propto \Delta\phi^{-1/4}$ was identified studying the response to dipolar forces. A relation between length scales and normal modes was established in [22], where two lengths scaling as $\Delta\phi^{-1/2}$ and $\Delta\phi^{-1/4}$ were related to longitudinal and transversal vibrations, respectively. Finally, similar scalings have been related to stability under compression [23] and shear [24].

Before considering how modes with different frequency contribute to the nonaffine response, we analyse the density of states. In Fig.2 we show the low frequency regime $D(\omega)$ for all the densities studied. A plateau is apparent for densities smaller than $\sim$ 0.87. This is the well-known departure of amorphous solids from the Debye scaling. At some lower frequency ω^* the linear regime reappears. Finally, below a frequency $\omega_{MGM} = 2\pi\sqrt{\mu/\rho}/L$ [6], the density of states decreases as ω^4 for all densities. The latter limiting frequency is the minimum Goldstone mode, *i.e.*, the frequency of the largest plane wave that can be fit in the system. Hence, it depends on the system size N, the density ρ and the shear modulus μ. For harmonic solids, given the dependence of the shear modulus on the density, one has $\omega_{MGM} \propto \Delta\phi^{1/4}$. In Fig.2b we rescale the horizontal axis by the minimum Goldstone mode, and the vertical axis to collapse the data below ω_{MGM}. At higher frequencies the data do not collapse because the frequency ω^* above which the density of states reaches a plateau scales as $\Delta\phi^{1/2}$.

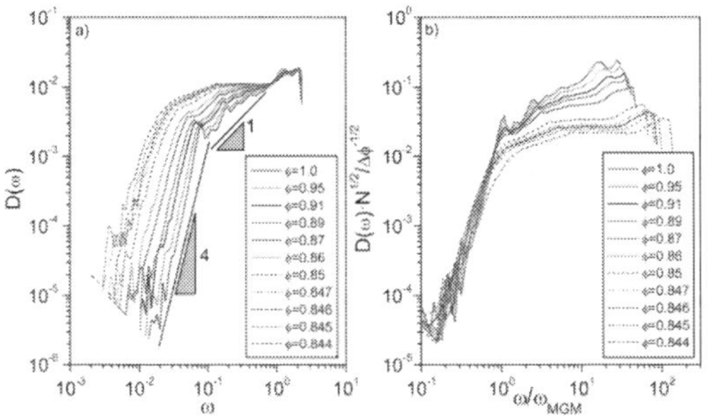

Figure 2. Density of states at different densities. The solid lines, added as guides to the eye, have slopes 4 and 1. b) Rescaling to achieve the collapse of the data close to ω_{MGM} . Close to ω^* the collapse does not hold because the scaling of this frequency with $\omega^* \propto \Delta\phi^{1/2}$, and not $\omega_{MGM} \propto \Delta\phi^{1/4}$

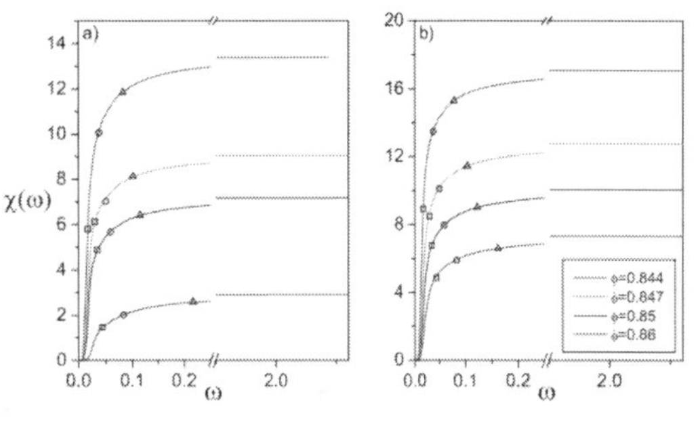

Figure 4. Contribution chi(omega) to the non affine parameter of the modes with frequency smaller than omega. The nonaffine displacement is computed using the RHS of Eq.1. Squares mark the position of ω_{MGM}, circles mark ω^*, and triangles mark the 90% of the plateau value. Left: compression. Right: shear.

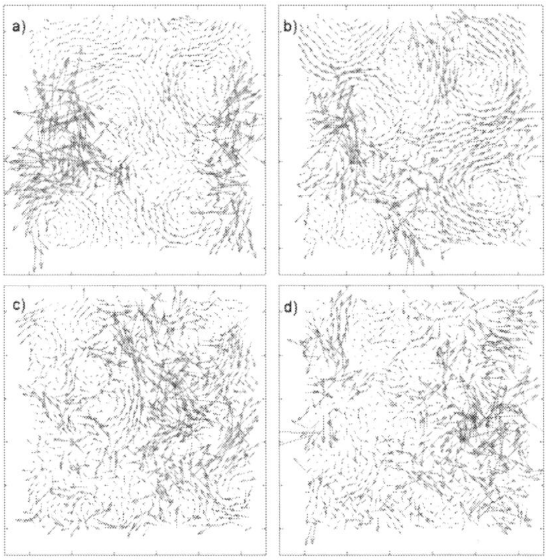

Figure 3. Examples of eigenmodes for $\phi = 0.847$. a) $\omega < \omega_{MGM}$. b) $\omega_{MGM} < \omega < \omega^*$ close to the lower. c) $\omega_{MGM} < \omega < \omega^*$ close to ω^*. d) $\omega > \omega^*$.

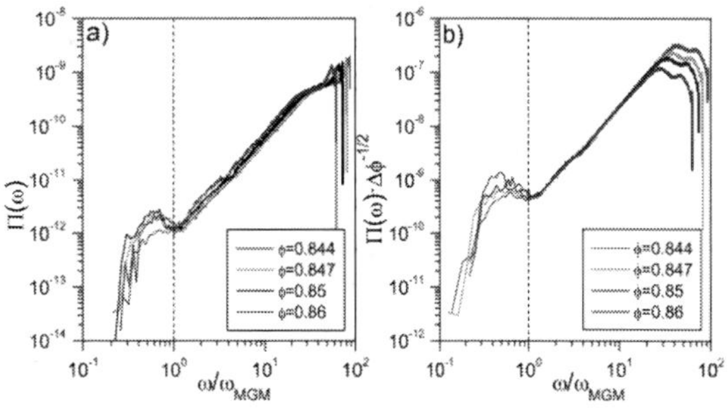

Figure 5. Product $\Pi(\omega)$ for the analyzed densities normalized by the minimum Goldstone mode. Left: compression. Right: shear.

Let us take a look at the structure of the eigenmodes. In Fig.3 we represent four different eigenmodes for $\phi = 0.847$. In Fig.3a the mode has frequency $\omega < \omega_{MGM}$. In Fig.3b,c modes have a frequency in the range $\omega_{MGM} < \omega < \omega^*$, such that Fig.3b is close to the lower frequency, while Fig.3c is close to ω^*. Finally, Fig.3d shows a mode in the plateau. The structures in Fig.3a,b look rather similar, with vortices and streams of relatively large displacements mixed with areas of very small displacements. Neither of them seems a neat plane wave and can be more aptly described as *quasi-localized* or *hybrid* [25–27]. In Fig.3c,d we see that at higher frequencies vortices smear out and large displacements shrink, making the field look more homogeneous and more similar to a plane wave.

In Fig.4 we show the cumulative contribution to the nonaffine parameter, $\chi = \langle\| \delta\vec{R}_{ij} \|^2\rangle/\delta\phi^2$ for shear, and $\chi = \langle\| \delta\vec{R}_{ij} \|^2\rangle/\delta\gamma^2$, where $\langle\| \delta\vec{R}_{ij} \|^2\rangle$ is computed using Eq.1. The modes up to ω_{MGM} give a small contribution that, in addition, is expected to vanish in the thermodynamic limit. Modes in the Debye regime (up to ω^*) give the greatest contribution. This is clearly the steepest part of the sum function, and is followed shortly after by a gentle transition to a plateau. This result is different from the conclusion of the estimation given in [5], where the main contribution to nonaffinity is attributed to modes above ω^*.

Converting the sum in Eq.1 into an integral, as it is done in [5], we can gain insight into which modes are contributing more to the nonaffinity. Simplifying the notation, let us write $R^2 = \int_0^\infty \omega^{-4}\Pi(\omega) D(\omega) d\omega$ where $\Pi(\omega)$ stands for the product of the two averages inside summation. We plot this function in Fig.5 scaled in the horizontal axis by ω_{MGM} (marked by the vertical lines) and in the vertical axis to achieve a collapse of the data. The complicated shape below the minimum Goldstone mode should not concern us since it will not contribute in the thermodynamic limit. Above ω_{MGM} there is a small plateau and afterwards the function grows quadratically with the frequency.

In order to perform the integral we consider the intervals $[\omega_{MGM}, \omega^*]$ and $[\omega^*, \infty]$. Note that to take into account the transition from a plateau to a power law of $\Pi(\omega)$ we need to introduce a frequency ω_2 in the first interval. Since the data appear collapsed in Fig.5 we know that $\omega_2 \sim \omega_{MGM} \propto \Delta\phi^{1/4}/L$. For $D(\omega)$, as discussed in reference to Fig. 2, we use the scaling shown in Fig. 2a close to ω_{MGM}. While close to ω^* we use the scaling imposed by this frequency, *i.e.*, $D(\omega) \propto \omega/\Delta\phi^{1/2}$. Finally, note that due to the normalization of the modes $\Pi(\omega)$ scales like $1/N$. Summarising, we split the integral in three parts:

i) $[\omega_{MGM}, \omega_2]$ with $\Pi(\omega) \propto \Delta\phi^{1/2}/L^2$ and $D(\omega) \propto N^{-1/2}\Delta\phi^{1/2}(\omega/\omega_{MGM})$; *ii)* $[\omega_2, \omega^*]$ with $\Pi(\omega) \propto \left(\omega/\Delta\phi^{1/4}\right)^2 \Delta\phi^{1/2}/L^2$ and $D(\omega) \propto \omega/\Delta\phi^{1/2}$; *iii)* $[\omega^*, \infty]$ with $\Pi(\omega) \propto \left(\omega/\Delta\phi^{1/4}\right)^2 \Delta\phi/L^2$ and $D(\omega)$ a constant.

Plugging all the scalings in the integral, we have (for shear):

$$\chi \sim \frac{1}{L^2}\left(-\Delta\phi^{-1/2}L^2 + \Delta\phi^{-1/2}L^2\right) + \frac{1}{L^2}ln\left(\Delta\phi^{1/4}L\right) - \frac{\Delta\phi^{1/2}}{L^2}\left(0 - \Delta\phi^{-1/2}\right) \sim \Delta\phi^{-1/2}.$$

This is effectively the correct scaling for shear. For compression $\Pi(\omega)$ is scaled by $\Delta\phi^0$ instead of $\Delta\phi^{1/2}$, so we obtain immediately $R^2 \sim \Delta\phi^{-1}$.

Summarising, in this work we have carried out a detailed investigation of the contribution of normal modes to the nonaffine displacement. We have shown that modes below ω^* cannot be disregarded since they give an important contribution, and set the correct scaling of nonaffinty with density. Modes above ω^* increase the value of the nonaffinity but the scaling is unaffected. There remains to investigate the thermodynamic limit. We expect our conclusions to hold since in that case the Debye regime is larger, because $\omega_2 \sim \omega_{MGM}$ decreases faster that ω^*.

We have seen that the structure of the modes changes slowly a one moves from low to high frequencies. As a consequence, modes below the minimum Goldstone mode and even those in the Debye regime cannot be neatly described as plane waves. It would be interesting to investigate the relationship, if any, between the structure of the modes and their contribution to nonaffinity.

References

[1] G. Biroli and P. Urbani, Nature physics (2016).

[2] M. van Hecke, J. Phys.: Condens. Matter **22**, 033101 (2010).

[3] A.J. Liu and S.R. Nagel, Annual Reviews of Condensed Matter Physics **1**, 14.1 (2010).

[4] A. Lemaître and C. Maloney, Journal of Statistical Physics **123**, 415 (2006).

[5] M. Wyart, H. Liang, A. Kabla, and L. Mahadevan, Phys. Rev. Lett. **101**, 215501 (2008).

[6] E. Lerner, G. Düring, and E. Bouchbinder, Phys. Rev. Lett. **117**, 035501 (2016).

[7] U. Buchenau, Y. M. Galperin, V. L. Gurevich, and H. R. Schober, Phys. Rev. B **43**, 5039 (1991).

[8] V. Gurarie and J. T. Chalker, Phys. Rev. B **68**, 134207 (2003).

[9] V. L. Gurevich, D. A. Parshin, and H. R. Schober, Phys. Rev. B **67**, 094203 (2003).

[10] D. A. Parshin, H. R. Schober, and V. L. Gurevich, Phys. Rev. B **76**, 064206 (2007).

[11] S. Plimpton, J. Comp. Phys., **117**, 1-19 (1995).

[12] C.E. Maloney and A. Lemaitre, Phys. Rev. E **74**, 016118 (2006).

[13] P. Olsson and S. Teitel, Phys. Rev. Lett. **99**, 178001 (2007);

[14] C. Heussinger, P. Chaudhuri and J.L. Barrat, Soft Matter **6** 2050 (2010).

[15] A. Tanguy, F. Léonforte and J.L. Barrat, Eur. Phys. J. E **20**, 355 (2006).

[16] E. Lerner and I. Procaccia, Phys. Rev. E **79**, 066109 (2009).

[17] R. Arévalo and M. Pica ciamarra, Soft Matter **8** 2728 (2014).

[18] I. Agnolin and J-N. Roux, Phys. Rev. E **76**, 061304 (2007); P-E.Peyneau and J-N. Roux, Phys. Rev. E **78**, 041307 (2008)

[19] M. Wyart, Ann. Phys. Fr. **30** 1 (2005); M. Wyart, L.E. Silbert, S.R. Nagel, T. A. Witten, Phys. Rev. E **72**, 051306 (2005).

[20] W.G. Ellenbroek, E. Somfai, M. van Hecke and W. van Saarloos, Phys.Rev.Lett. **97**, 258001 (2006); W.G. Ellenbroek, M. van Hecke and W. van Saarloos, Phys. Rev. E **80**, 061307 (2009).

[21] E. Lerner, E. DeGiuli, G. Dürig and M. Wyart, arxiv:1312.2146v1.

[22] L.E. Silbert, A.J. Liu and S.R. Nagel, Phys. Rev. Lett. **95**, 098301 (2005).

[23] C.P. Goodrich, W.G. Ellenbroek and A.J. Liu, Soft Matter **9** 10993 (2013).

[24] S.S. Schoenholz, C.P. Goodrich, O. Kogan, A.J. Liu and S.R. Nagel, Soft Matter **9** 11000 (2013).

[25] A. Tanguy, B. Mantisi and M. Tsamados, EPL **90**, 16004 (2010).

[26] A. Tanguy, JOM **67**, 1832 (2015).

[27] N. Xu, V. Vitelli, A. J. Liu and S. R. Nagel, EPL **90**, 560001 (2010).

Fragmentation modeling of a resin bonded sand

William Hilth[1,*] and *David* Ryckelynck[1,**]

[1] *Mines ParisTech, PSL Research University, MAT - Centre des Matériaux, CNRS UMR 7633, BP 87 91003 Evry, France*

Abstract. Cemented sands exhibit a complex mechanical behavior that can lead to sophisticated models, with numerous parameters without real physical meaning. However, using a rather simple generalized critical state bonded soil model has proven to be a relevant compromise between an easy calibration and good results. The constitutive model formulation considers a non-associated elasto-plastic formulation within the critical state framework.

The calibration procedure, using standard laboratory tests, is complemented by the study of an uniaxial compression test observed by tomography. Using finite elements simulations, this test is simulated considering a non-homogeneous 3D media. The tomography of compression sample gives access to 3D displacement fields by using image correlation techniques. Unfortunately these fields have missing experimental data because of the low resolution of correlations for low displacement magnitudes. We propose a recovery method that reconstructs 3D full displacement fields and 2D boundary displacement fields. These fields are mandatory for the calibration of the constitutive parameters by using 3D finite element simulations. The proposed recovery technique is based on a singular value decomposition of available experimental data. This calibration protocol enables an accurate prediction of the fragmentation of the specimen.

1 Introduction

In casting foundry, resin bonded sand is used in the casting mold to create the internal cavities of cylinder heads for instance. After the solidification, the casting sand has to be extracted. This decoring operation is most commonly carried out by mechanical means, knocking on the part with pneumatic hammers to fracture the cores first, and then vibrating to eliminate the fragmented pieces. Our main aim is to develop a mechanical model able to anticipate the fragmentation of the cemented sand. Nevertheless, this model has to be simple enough to be easily calibrated and used in finite elements simulations.

The literature concerning resin bonded sand is limited and mostly consists in industrial studies varying the resin or its amount, or the sand itself. The model highlighted here is adapted from the Clay And Sand Model (CASM) developped by Yu [1] and the cementention model developped by Gens and Nova [2]. This approach, called Cemented Clay And Sand Model (CCASM), has succesfully been applied to cemented sands (see Rios *et al.* [3]).

The model is calibrated and validated with the use of standard tests realized in collaboration with INSA of Lyon. It is completed with the study of an uniaxial compression test observed by tomography. 3D FEM calculations should be able to predict the fracture surfaces initialization of the cemented sand, therefore the comparison with a non-homogeneous 3D strain field is particularly crucial.

*e-mail: william.hilth@mines-paristech.fr
**e-mail: david.ryckelynck@mines-paristech.fr

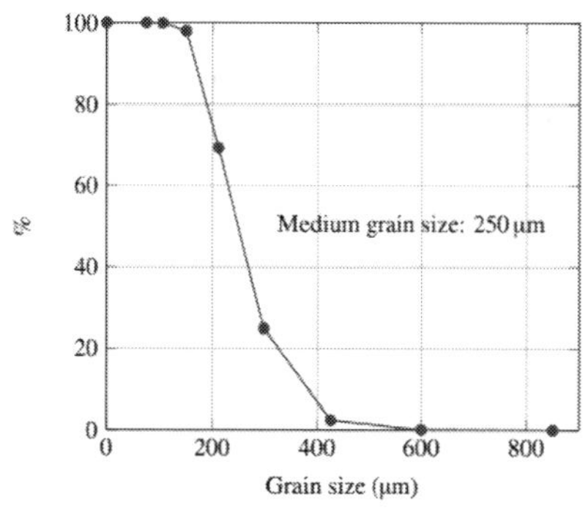

Figure 1. Grain size distribution

This study is focused on a polyurethane bonded silica sand (1%wt. of resin). The grain size distribution is plotted in Figure 1.

2 Model

2.1 Yield surface and plastic flow

The constitutive elasto-plastic model, called Cemented Clay And Sand Model (CCASM), is the extension for cemented soils of the Clay And Sand Model developped by Yu. The cementation is taken into account by the introduction of a scalar variable b named bonding parameter. This

variable and its evolution enable to model the intergranular bonding and its decay.
The bonding parameter amplifies the yield surface, assuming that its shape is the same as the yield surface of the unbonded sand. The amplification of the yield surface is controled by the isotropic compression and tensile yields:

$$p_o = p_c(1+b)$$
$$p_t = \alpha p_c b$$

where p_c is the isotropic compression yield for the uncemented sand and α a constant scalar parameter for the traction resistance. We assume here, for simplicity, a linear elasticity. The yield surface of the CCASM is:

$$f(\underline{\sigma}; p_c, b) = \left(\frac{q}{M(p+p_t)}\right)^n + \frac{1}{\ln r}\ln\left(\frac{p+p_t}{p_c(1+b)+p_t}\right) \quad (1)$$

where p is the hydrostatic pressure, q the equivalent von Mises stress and M is the stress ratio $\eta = q/(p+p_t)$ at critical state. For $p \to -p_t$ we impose $f = 0$.
In the critical state framework, the soil behavior is assumed to rely on the stress ratio:

- $\eta > M$: contraction of the soil.
- $\eta = M$: shearing can occur without any changes in stress or volume. In this case, the incremental volumetric plastic strain is zero.
- $\eta < M$: dilatation of the soil.

The yields surfaces for the bonded and unbonded soil are plotted in Figure 2. For each configuration, the Critical State Line ($\eta = M$) is plotted in black.
The expression of f is the sum of two terms:

- $\left(\frac{q}{M(p+p_t)}\right)^n$ that controls the behavior of the core when sheared
- $\frac{1}{\ln r}\ln\left(\frac{p+p_t}{p_c(1+b)+p_t}\right)$ that controls the hydrostatic closure of the yield locus

n and r are constant coefficients that control the shape of the yield. Increasing n tends to bring closer the yield locus to the critical state line and closes more abruptly the yield locus. The spacing ratio r enables to modulate the intersection of the yield locus and the Critical State Line, hence controlling the softening/hardening transition of the model.
Rowe flow rule for bonded soil has been developped for this model, based on the ratio between the incremental plastic volumetric strain and the incremental plastic shear strain:

$$\frac{\dot{\varepsilon}_v^p}{\dot{\varepsilon}_s^p} = \frac{9(M-\eta)}{9+3M-2M\eta} \quad (2)$$

We note here that at the critical state, we have $\dot{\varepsilon}_v^p = 0$ as stated before. Moreover, for $\eta > M$, $\dot{\varepsilon}_v^p < 0$ inducing a dilatation of the soil (alternatively, for $\eta < M$, the soil contracts).

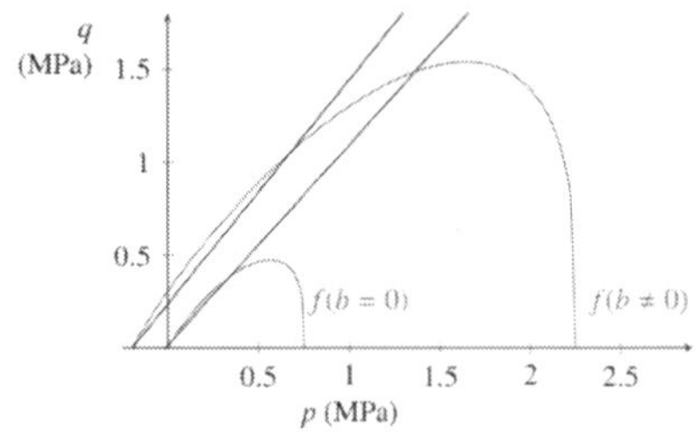

Figure 2. Yield surfaces in the (p,q) plane

2.2 Evolution of the parameters

The model has a hardening variable p_c and a plastic strain damage mesure, h, which ensures the irrecoverable decrease of b.
The evolution of the preconsolidation pressure is controled by the classical critical state rule:

$$\frac{\dot{p}_c}{p_c} = \frac{v_0}{\lambda - \kappa}\dot{\varepsilon}_v^p \quad (3)$$

where κ and λ are compressibility parameters of the uncemented sand and v_0 the initial specific volume. Combining this hardening rule and equation 2, we can see that for a contraction of the soil p_c decreases, leading to a softening of the bonded soil.
The bonding parameter b decrease exponentially relatively to the plastic strain damage mesure evolution as:

$$b = b_0 \exp(-h)$$
$$\dot{h} = h_0|\dot{\varepsilon}_v^p| + h_1|\dot{\varepsilon}_s^p|$$

where h_0 and h_1 are two material parameters that control the sensibility of the material to plastic shear or volumetric strain.
Because of the decay of the intergranular bonding, we modeled the decrease of the effective stiffness modulus with a damage parameter D that only depends on the evolution of b.

$$E = E_{ini}(1-D)$$
$$D(b;k_0,k_1) = 1 - \frac{\sinh(k_0(b-k_1)) + \sinh(k_1k_0)}{\sinh(k_0(b_0-k_1)) + \sinh(k_1k_0)} \quad (4)$$

where k_0 and k_1 are constant parameters and b_0 the initial value of b. The evolution of D with respect to b/b_0 is plotted in Figure 3. We considered here that the damage of the media is only caused by the intergranular debonding. No damage mechanisms due to the sand itself (involving p_c for instance) are considered. The damage law was calibrated using cycled uniaxial compression tests and assessing the loss of stiffness (E/E_{ini}) during plastic loading.

3 Validation procedure

3.1 Idealized uniaxial compression test

The first step in the calibration and validation of the CCASM is to model an idealized test and compare the

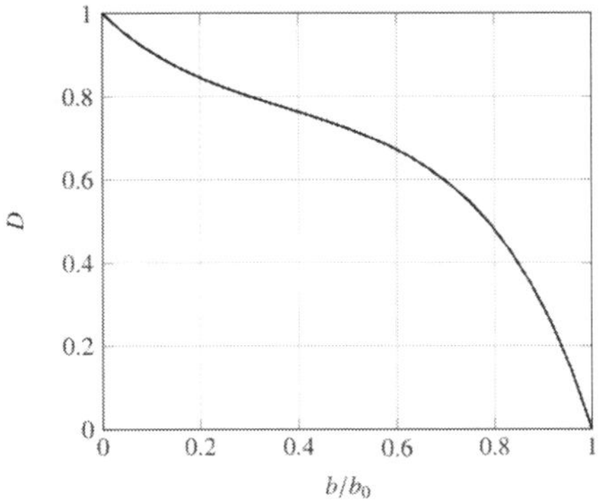

Figure 3. Evolution of D with respect to b/b_0

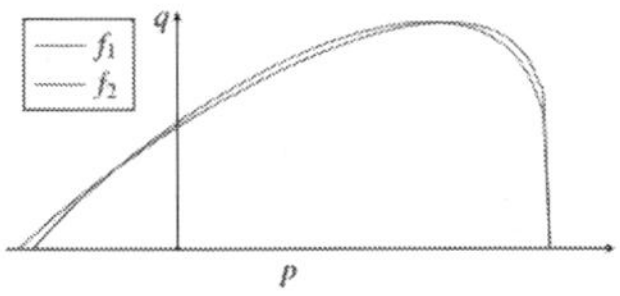

	f_1	f_2
M	1.4	1.1
n	3.8	4.75
r	6000	2500
α	0.42	0.46

Figure 4. Redoundancy forecast in the parameters space

results with laboratory tests. The literature generally focuses on strain-stress curves or the evolution of plastic shear/volumetric strain during triaxial compression tests. This approach usually gives intervals of admissible parameters especially for the ones involved in the formulation of the yield surface. Given the uncertainties due to the tests, whether in the mesures themselves or in test repeatability (variability of the samples, initial inhomogeneity, ...), it is almost impossible to produce a single set of parameter values that can calibrate the model with only the data of standard tests. See for instance Figure 4 that shows two sets of parameters for the yield surface that lead to nearly the same yield locus.

In Figure 5, strain-stress curve for an uniaxial compression test are plotted. The model doesn't take into account for the pre-peak hardening as we considered the material elastic until the peak. This hypothesis tends to overevaluate the plastic strain. A modification of the CCASM with a multisurface approach can be imagined with one surface modeling the pre-peak hardening and the CCASM modeling the post-peak behavior (see for instance the Revised Modified Cam-clay model by Ami Saada *et al.* [6]). The damage law captures precisely the loss of stiffness through the loading. In Figure 6 the axial plastic strain and the damage parameter value taken from a cycled uniaxial compression test is compared to the FEM simulation.

Moreover, focusing only on these macroscopic curves doesn't allow to compare the fracture surfaces between tests and FEM simulations. For instance, for uniaxial compression test, the strain and stress fields during the post-peak domain are widely inhomogeneous as the localisation appears. This phenomenon makes troublesome to

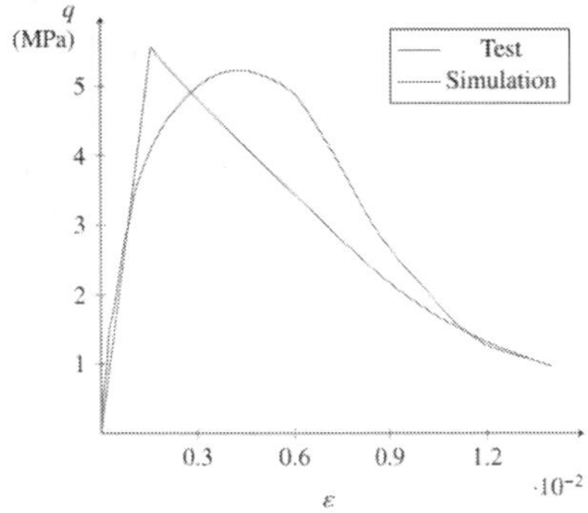

Figure 5. Comparison of strain-stress curves for uniaxial compression test

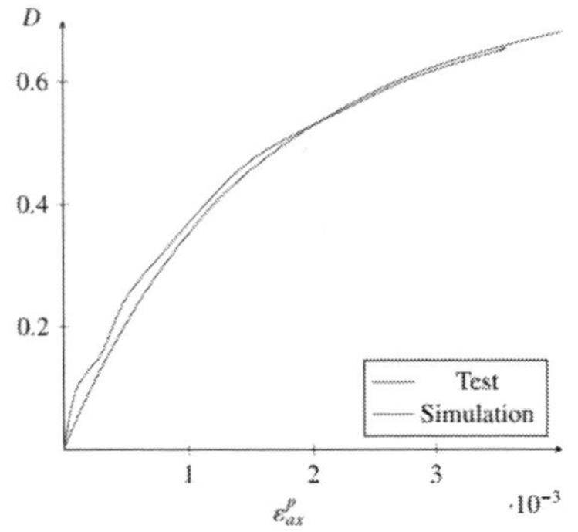

Figure 6. Evaluation of the damage law

evaluate the stress or strain with only the data of macroscopic curves. The parameters involved in the plastic flow or hardening rules are not obvious to calibrate. Thus, the constitutive model must be evaluated not only on the basis of standard laboratory tests but with the comparison of 3D inhomogeneous displacement fields.

3.2 Uniaxial compression test studied in tomography

As stated before, the model must predict correctly the fracture surfaces. Realistic displacement fields, taken from an uniaxial compression test on a cuboid observed in tomography, are considered as Dirichlet conditions for a FEM calculation. Unfortunately, these fields have missing experimental data because of the low correlations resolution for low displacement magnitudes. Moreover, the displacement fields are noisy. A method for reconstructing and filtering the data has been developped. This algorithm is based on the Gappy POD method [4, 5]. This method is:

1. Build a data base of m points, or snapshots, where all the information is known at every time step

2. Build a proper order decomposition base V of these snapshots

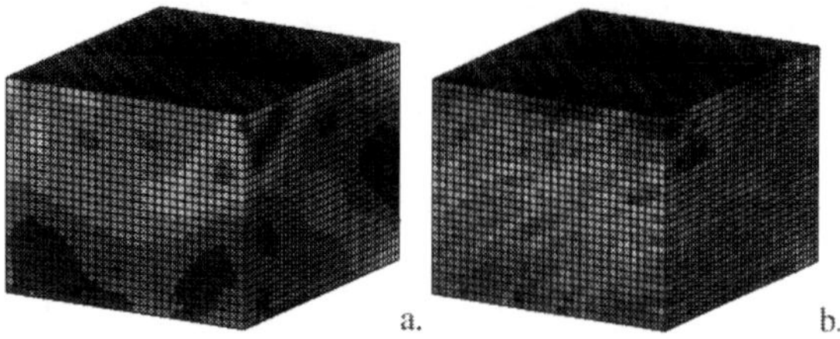

a. b.

Figure 7. a. ε_s of the model, b. ε_s Gappy POD

3. At each points where the information is missing, apply the base V in order to reconstruct the gappy field

4. Enrich the snapshots matrix with the new reconstructed points and continue the process until all the points are reconstructed

We used these reconstructed displacement fields as Dirichlet conditions at the top and the bottom of the cuboid sample. The simulation is performed until the post-peak domain. We compare the strain field calculated to the one determinated with the total displacement field applied to all the cuboid sample. Figure 7 shows the total shear strain, ε_s, for both approaches, as the fragmentation of the media is mostly due to shear strain. Figure 7 a. is displayed at a time step before Figure 7 b..

Qualitatively, the simulation using the CCASM model forecasts the first slip in the top corner. A second conjugate shear band at 45° is also predicted. This band is actually supposed to appear after. Figure 7 b. shows the first time step of initialization of this band at a later stage of the loading. As pointed out before, the constitutive model tends to overevaluate the plastic strain. This explains the anticipation of this shear band.

The study of this FEM calculation has allowed a deeper comprehension of the parameters, especially those involved in the plastic flow. The resin-bonded sand possesses at the same time the caracteristics of a coesive material (via the intergranular bonding) and a frictional material (as a sand). Those different behaviors are modeled respectively by the evolution of b (or h) and p_c. The study of the displacement field taken from the uniaxial compression test enabled to distinguish more accurately the relative importance of the decohesion or the frictional contraction of the soil. Both phenomena lead to a softening of the material that are fused in the macroscopic strain-stress curve.

4 Conclusion

The model was adapted and calibrated using a 3 steps method:

1. Adaptation of the Cemented Clay And Sand Model as described by Rios *et al.* (2016). The model was modified to the resin-bonded sand by considering a linear elasticity, changing the damage law and the plastic flow rule
2. Determination of intervals of correct parameters using macroscopic strain-stress curves taken from standard laboratory tests. These tests only gave qualitative informations on the fracture surfaces of the cemented sand subjected to various loading

3. Deeper study of realistic 3D displacement fields that enabled a better comprehension of the various parameters of the model, notably the impact of the decay of the intergranular bonding in the macroscopic behavior of the material.

The use of 3D FEM calculations is a relevant approach for calibrating and validating a model, especially for a softening material presenting strong inhomogeneities as studied here. This method is particularly suitable when the final aim is to anticipate the fracture modes of the media.

Acknowledgements

The authors would like to acknowledge the Agence Nationale pour la Recherche for their financial support for the FIMALIPO project.

References

[1] Yu H.-S., International Journal for Numerical and Analytical Methods in Geomechanics, **22**, 621-653 (1998)
[2] Gens A. and Nova R., Geotechnical engineering of hard soils-soft rocks, **1**, 485-494 (1993)
[3] Rios, S. and Ciantia, M. and Gonzalez, N. and Arroyo, M. and Viana da Fonseca, A., Computers and Geotechnics, **73**, 100-108 (2016)
[4] Everson, R. and Sirovich, L., J. Opt. Soc. Am. A, **12**,1657-1664 (1995)
[5] Carlberg, K. and Bou-Mosleh, C. and Farhat, C., International Journal for Numerical Methods in Engineering, **86**, 155-181 (2011)
[6] Ami Saada, R. and Bonnet, G. and Bouvard, D., International Journal of Plasticity, **12**, 273-294 (1996)

DOI: 10.1051/epjconf/201714005005

DEM investigation on characteristics of rolling resistance for modelling particle shape

Lunlun Zhou[1], *Xihua* Chu[1,*], and *Yuanjie* Xu[1]

[1]School of Civil Engineering, Wuhan University, Wuhan, 430072, China

Abstract. To examine the capability of rolling resistance to model the effects of particle shape, two sets of samples, composed of binary clumped particles and circular particles with rolling resistance, are tested in DEM simulation. The coefficient of rolling friction is estimated based on the energy dissipation. The effects of rolling resistance and particle shape on the shear strength, deformation behavior and non-coaxiality are compared. The numerical results show that rolling resistance reproduces well the effect of particle shape on the peak strength. However, other macro-properties, such as residual strength, elasticity modulus, poisson's ratio, dilatancy and non-coaxiality, introduced by rolling resistance both exist certain differences compared with the effect of particle shape. The discrepancies is thought to be due to the increasing compressibility of samples as the particle shape becomes more elongated, which cannot be reproduced by increasing rolling friction.

1 Introduction

The discrete element method (DEM) has been a valuable tool in the study of mechanical behavior of granular materials. Most DEM studies simulate granular systems as assemblies of 2D disks or 3D spheres. However, real granular materials are almost composed of particles of complicated shapes. And it has been shown that the non-spherical shape of particles has significant effects on shear strength, deformation behavior and micro-structural properties of granular systems [1].

One of the major challenges for DEM is to simulate non-spherical particles in an efficient manner. When modeling complicated particle shapes in DEM, it brings more complex geometric algorithm, contact judgment and contact model. All these cause sharply rising in computational cost. Shape is thus an expensive thing to model in DEM. Consequently, rolling resistance is typically used in DEM as a simple way of modeling the effects of irregular particle shape.

In the conventional discrete element formulation, the interactions between two particles consist only of contact forces in the normal and tangential directions. Particles are assumed to rotate freely and offer no resistance to rotation. The technique (using rolling friction to simulate shape) incorporates a rotational frictional torque to inhibit particle rotation. The torques acting on the two particles in contact are usually applied as a couple of action and reaction. The magnitude and direction of this torque depend on the relative rotation between the particles in contact. In reality, the torque produced by particle shape is an eccentric torque caused by the normal contact force, as shown in Figure 1 [2]. The magnitude and direction of this eccentric torque are not relative to the torque acting on the adjacent particle in the contact.

As above, the rolling frictional torque and the real eccentric torque have different action mechanisms on the microscopic level. The question then is what effects of particle shape on macroscopic mechanical behaviors can be captured by using rolling resistance. This question is the primary concern of this paper. In this study, two sets of samples, composed of binary clumped particles and composed of circular particles with rolling resistance, are tested in DEM simulation. The capability of rolling resistance to model the effects of particle shape is examined by comparing the effects of rolling resistance and particle shape on the shear strength, deformation behavior and non-coaxiality of granular materials.

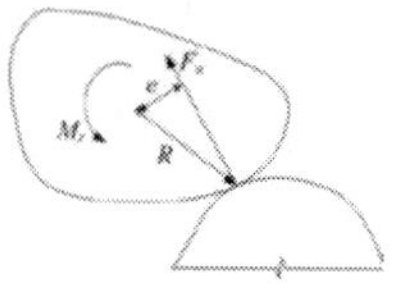

Fig. 1. Contact between particles with irregular shapes [2].

2 Rolling resistance model

Many researchers have suggested various contact models incorporating rolling resistance [3-5]. The model used in this study was initially developed by Iwashita and Oda [3] and was identified to be superior in terms of its behavior for static materials [4]. The rolling resistance torque applied to contact is defined in an incremental way by the following

* Corresponding author: Chuxh@whu.edu.cn

$$M_{r,t+\Delta t} = \begin{cases} M_{r,t} - k_r \omega_{rel} \Delta t \text{ provide that} \\ \left| M_{r,t} - k_r \omega_{rel} \Delta t \right| < \mu_r R_r \left| F_n \right|, \\ \mu_r R_r \left| F_n \right| \frac{M_{r,t} - k_r \omega_{rel} \Delta t}{\left| M_{r,t} - k_r \omega_{rel} \Delta \right|} \text{ otherwise.} \end{cases} \quad (1)$$

Where $M_{r,t}$ is the rolling resistance torque at time t, Δt is the time increment, ω_{rel} is the relative angular velocity of the particles in contact, F_n is the normal contact force at the contact, $R_r=R_iR_j/(R_i+R_j)$ is the common radius, R_i, R_j is the radius of the two particles in contact respectively, k_r is the rolling stiffness and $k_r \approx k_s R_r^2$ according to the suggestion of Iwashita and Oda [3], μ_r is the coefficient of rolling friction. This rolling resistance model defines that the rolling resistance torque is not exceeding the maximum value $M_{r,max}=\mu_r R_r |F_n|$.

3 DEM simulation method

Two sets of DEM simulations are carried out using the DEM software PFC2D. In the first set of simulations, the samples compose of binary clumped particles. The clumped particles consist of two overlapping circular particles of the same radius. And varying relative spacing, δ, is used to characterize different particle shapes, where δ is the value of the spacing between the centers of the constitutive particles normalized by the radius of the constitutive particles. In the second set, circular particles with rolling resistance are used to simulate the effect of particle shape. Different equivalent rolling frictions, μ_r, are estimated corresponding to the different clump shapes.

3.1 Rolling friction and shape

The coefficient of rolling friction, μ_r, can be estimated in two different ways [1-2]. In the first approach [2], the selection of the coefficient of rolling friction is based on the normalized average contact eccentricity of non-spherical particles, but the DEM result showed that half the rolling friction estimated from the normalized eccentricity is appropriate to capture the similar behavior of the case of non-spherical particles. The second approach[1], used in this study, defined the coefficient of rolling friction based on the energy dissipation associated with relative rotation between particles in contact. And this calculated coefficient is exactly one half of the value calculated by the first approach.

Considering a binary clumped particle and a circular particle with rolling friction roll on a flat surface with a vertical force F_n exerted at the center of mass, as shown in Figure 2. The horizontal force T is applied at the center of mass to roll the particles. The work required to roll the binary clumped particle one round is

$$W_c = \delta R_1 F_n. \quad (2)$$

Where R_1 is the radius of the constitutive particles of the clumped particle. In the case of circular particle with rolling friction, the work required is

$$W_r = 2\pi M_{r,max} = 2\pi \mu_r R_2 F_n. \quad (3)$$

Where R_2 is the radius of the clumped particle. Assuming the work required is equal, the coefficient of rolling friction can be given by

$$\mu_r = \delta R_1 / 2\pi R_2. \quad (4)$$

Assuming that clumped particle and circular particle have the same volume, the relationship between R_1 and R_2 is

$$R_1/R_2 = \sqrt{\pi/(2\pi + \delta/2\sqrt{1-\delta^2/4} - 2COS^{-1}(\delta/2))}. \quad (5)$$

For each different clump shape, the equivalent rolling friction is calculated and outlined in Table 1. Other parameters used in DEM simulations are outlined in Table 2.

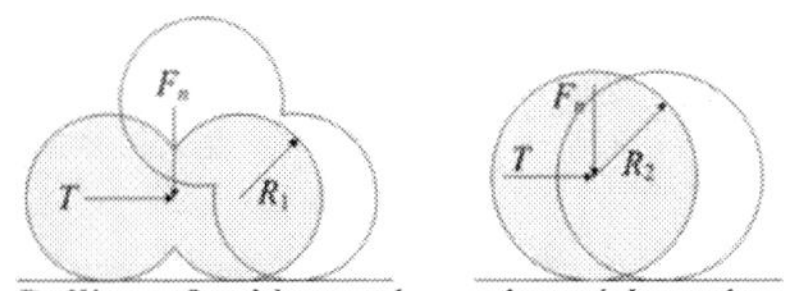

Fig. 2. Rolling of a binary clumped particle and a circular particle with rolling friction on a plane [1].

Table 1. Clump spacing and corresponding rolling frictions.

Relative spacing δ	Rolling friction μ_r
0.49	0.07
0.79	0.11
1.10	0.14
1.45	0.18
1.76	0.20

Table 2. Parameters used in DEM simulations.

Particle size: mm	0.75-1.0
Size distribution	Uniform
Porosity	0.14
Particle density: kg/m^3	2650
Particle sliding friction	0.5
Particle normal stiffness: N/m	5.0×10^8
Particle tangential stiffness: N/m	2.5×10^8
Particle rolling stiffness: N·m	500

3.2 Testing procedure

Two types of tests, biaxial tests and simple shear test, are conducted to investigate the stress-strain behaviors and non-coaxiality of the two sets of samples.

In the biaxial tests, the samples consist of 5792 particles or clumps, and its initial dimension is 180×90 mm. Four frictionless walls are created as the rigid boundaries. The stiffness of the horizontal walls is set to be the same as the particles, and the stiffness of vertical walls is one tenth of the stiffness of particles. The samples are first consolidated isotropically with the confining pressure of 1.0 Mpa and then sheared under the strain rate of 0.02%/min until the axial strain reached 12%. The lateral walls are controlled by a servo-control algorithm to keep the confining pressure constant during shear stage.

In the simple shear tests, the samples consist of 5148 particles or clumps, and its initial dimension is 120×120 mm. The stiffness and friction coefficient of the wall boundaries are the same as the particles. The samples are first subjected to anisotropic consolidation with the vertical normal stress of 1.2 Mpa and horizontal normal stress of 0.8 Mpa. Then the two vertical walls are rotated at a constant angular velocity, ω, of 0.02 rad/min about the mid-points of the two walls. The two horizontal walls are translated with a compatible horizontal velocity in opposite directions, and the vertical velocities of the horizontal walls are adjusted using a servo-control algorithm to keep the vertical normal stress constant. The test continues until the shear strain, γ=sin(ωt), reached 25%.

4 DEM results and analysis

4.1 Stress-strain characteristics

The curves of shear strength, $q=\sigma_1-\sigma_2$, and volumetric strain, ε_v, versus axial strain, ε_1, obtained from the biaxial tests of the two sets of the samples are plotted in the Figure 3 and 4, respectively. In the case of clumped particles, the volumetric strain curves decline gradually as the relative spacing increases. The corresponding shear strength curves transforms gradually from strain softening to strain hardening. However, in the case of circular particles with rolling resistance, there are no significant differences between the volumetric strain curves for varying rolling frictions, and the corresponding shear strength curves show distinct strain hardening characteristic for all rolling frictions.

In order to compare the effects of rolling resistance and particle shape in more detail, the macro-parameters for the two sets of smaples obtained from Figure 3 and 4 are outlined in Table 3. The symbol "CP" and "RL" mean the case of clumped particles and circular particles with rolling resistance, respectively. It can be seen that the peak shear strengths, q_{max}, are both increasing with the increasing relative spacing and rolling friction in the two cases, and the values of q_{max} in the case of clumped particles are coincidence with the data in case of rolling resistance basically. The residual shear strength, q_{res}, increases gradually as the shape becomes less circular, whereas it almost keeps constant as rolling friction increases. The elasticity modulus, E, in the case of circular particles with rolling resistance increases with rolling friction and is up to 550 Mpa, which is much higher than that of the clumped particles. The poisson's ratio, v, of the samples of clumped particles decrease with relative spacing, while the evolution of v in the rolling resistance case is not clear. The values $v$of former case is generally higher than the latter case. The peak volumetric contraction, ε_{vmax}, increases in the first case, but keeps constant in the latter case, and the dilatancy of the samples of clumped particles is significantly less than that of the circular particles with rolling resistance.

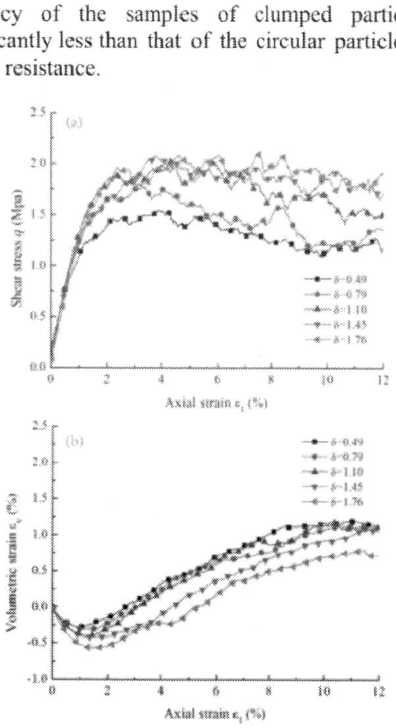

Fig. 3. Stress-strain relations for binary clumped particles of varying relative spacing in biaxial tests: (a) q versus ε_1, and (b) ε_v versus ε_1.

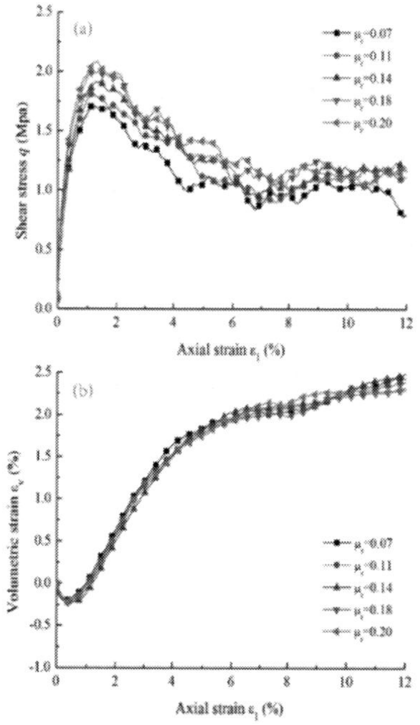

Fig. 4. Stress-strain relations for circular particles with varying rolling frictions in biaxial tests: (a) q versus ε_1, and (b) ε_v versus ε_1.

The macro-properties introduced by rolling resistance are somewhat different from the effects of particls shape, as above. The reason of these discrepancies is realatied to the increasing compressibility of samples as the particle shape becomes more elongated, which cannot be reproduced by increasing rolling friction.

Table 3. Comparison of the macro-parameters for the two sets of smaples.

$\delta(\mu_r)$		0.49 (0.07)	0.79 (0.11)	1.10 (0.14)	1.45 (0.18)	1.76 (0.20)
q_{max}(Mpa)	CP	1.54	1.79	2.02	2.07	2.09
	RL	1.71	1.81	1.94	2.02	2.08
q_{res}(Mpa)	CP	1.16	1.21	1.69	1.87	1.84
	RL	1.03	1.1	1.18	1.13	1.13
E(Mpa)	CP	140	150	148	137	138
	RL	470	497	470	529	550
v	CP	0.3	0.31	0.27	0.24	0.13
	RL	0.18	0.14	0.21	0.13	0.16
ε_{vmax}(%)	CP	0.28	0.31	0.41	0.42	0.57
	RL	0.19	0.21	0.23	0.22	0.21

4.2 Non-coaxiality

In the simple shear test, the direction of the major principal stress and strain rate will rotate during shearing process. Their orientation angle can be obtained easily by the following [6]

$$\tan\theta_\sigma = 2\sigma_{xy}/(\sigma_y\text{-}\sigma_x),\ \tan\theta_{\Delta\varepsilon} = 2\Delta\varepsilon_{xy}/(\Delta\varepsilon_y\text{-}\Delta\varepsilon_x). \quad (6)$$

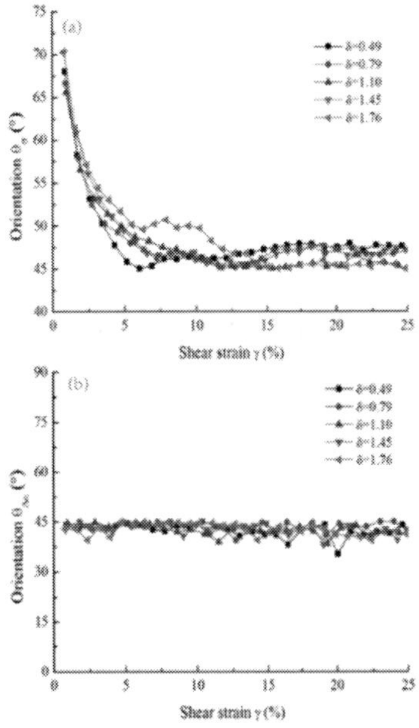

Fig. 5. Evolutions of (a) major principal stress directions, and (b) major principal strain rate directions for binary clumped particles of varying relative spacing in simple shear tests.

For the two sets of the samples, the evolutions of the major principal stress and strain rate directions (inclination to horizontal) are shown in the Figure 5 and 6, respectively. It is noted that the a slower rate in approaching coaxiality is observed in the sample of clumped particles with a large relative spacing. However, the rolling friction has no obvious effect on the non-coaxiality before shear strain reached about 5%. After γ=5%, the evolutions of the principal stress directions start to emerge differences, but the regularity is not obvious.

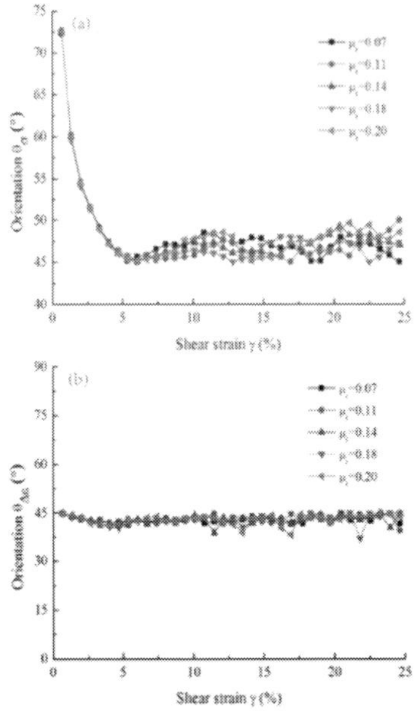

Fig. 6. Evolutions of (a) major principal stress directions, and (b) major principal strain rate directions for circular particles with varying rolling frictions in simple shear tests.

5 Conclusion

Two sets of samples, composed of binary clumped particles and circular particles with rolling resistance, are tested in DEM simulation. Rolling resistance reproduces well the effect of particle shape on the peak shear strength. However, different macro-properties, such as residual strength, elasticity modulus, poisson's ratio, dilatancy and non-coaxiality, introduced by rolling resistance are observed compared with particle shape. The discrepancies are assumed to be due to the fact that the increasing compressibility of samples, as the particle shape becomes more elongated, cannot be reproduced by increasing rolling friction.

References

1. N. Estrada, E. Azema, F. Radjai, A. Taboada. Rev. E **84**, 011306 (2011)
2. C.M. Wensrich, A. Katterfeld, D. Sugo. Powder. Technol. **217**, (2012)
3. K. Iwashita, M. Oda. J. ENG. MECH. ASCE. **124**, (1998)
4. J. Ai, J.F. Chen, J.M. Rotter, J.Y. Ooi. Powder. Technol. **206**, (2011)
5. Y.C. Zhou, B.D. Wright, R.Y. Yang, B.H. Xu, A.B. Yu. Physica. A. **269**, (1999)
6. J. Ai, P.A. Langston, H.S. Yu. Int. J. Numer. Anal. Met. **38**, (2014)

Grain breakage under uniaxial compression, through 3D DEM modelling

François Nader[1,*], *Claire* Silvani[1,**], and *Irini* Djeran-Maigre[1,***]

[1] *Univ Lyon, INSA-Lyon, SMS-ID, F-69621, France*

Abstract. A breakable grain model is presented, using the concept of particles assembly. Grains of polyhedral shapes are generated, formed by joining together tetrahedral subgrains using cohesive bonds. Single grain crushing simulations are performed for multiple values of the intra-granular cohesion to study the effect on the grain's strength. The same effect of intra-granular cohesion is studied under oedometric compression on samples of around 800 grains, which allows the evaluation of grain breakage model on the macroscopic behaviour. Grain size distribution curves and grain breakage ratios are monitored throughout the simulations.

1 Introduction

Rockfill structures are becoming more common in civil engineering, mainly for the benefits they present on a practical and economical scales. These structures are subjected to loads that can lead to grain breakage, which in turn modifies the grain size distribution of the material, and therefore its behaviour.

To simulate the behaviour of this type of material, the discrete element method is usually chosen for its ability to capture the discrete nature characteristic of grains. Many breakage models are suggested in the literature that can be grouped into two principal concepts. The first concept is based on replacing a grain with a number of smaller grains when an induced stress threshold is reached. This concept was used by Äström and Hermann 1998 [1], Tsoungui 1999 [2], and Ben-Nun and Einav 2010 [3] who modelled grains as discs. McDowell and De Bono 2013 [4] used the same concept with three-dimensional spherical grains. Cantor et al. 2014 [5] modelled a grain as a rigid polygon that split in the direction of the principal stress.

The second concept consists of assembling subgrains using cohesive bonds to form a grain. The breakage of a grain is the consequence of the breakage of the cohesive bonds. Different types of subgrains have been studied in the literature. Silvani et al. 2009 [6] applied a Mohr-Coulomb cohesive law between rigid discs to form circular grains and simulated oedometric tests. Cheng et al. 2003 [7] applied cohesive bonds between small spheres to form spherical grains, removing 20% of the subgrains to induce defects inside the grain and reduce its strength. Nguyen et al. 2015 [8] used a two-dimensional Voronoï tessellation to divide a circular grain into smaller polygonal particles and joined them together using cohesive bonds. Cantor et al. [9] 2016 used a three-dimensional Voronoï tessellation to divide a spherical grain into polyhedral particles joined together by a cohesive Tresca law.

In the work presented hereafter, the second concept is adopted, which consists of assembling rigid tetrahedral subgrains using a Mohr-Coulomb cohesive law to generate three-dimensional polyhedral grains. The grain generation algorithm is described, single crushing and oedometric compression simulations are presented for different values of the intra-granular cohesion. The goal of the study is to create a model in which the angularity of grains is taken into account, an aspect that cannot be studied when circular or spherical grains are used, and to show the effect of grain breakage and intra-granular cohesion on the macroscopic behaviour.

2 Grain generation

Grain shape plays a major role in the distribution of contact forces inside granular media, and in the overall macroscopic behaviour. To take this into account, polyhedral grains are generated. The generation algorithm takes in input the maximum grain dimension $2r_{max}$, the elongation $e = \frac{r_{max}}{r_{min}}$ where $2r_{min}$ is the minimum grain dimension, the number of half-planes N_{hp} and the number of vertices N_v in each half-plane. The center point of the polyhedron is placed along with the top and bottom vertices. Then in a vertical half-plane, N_v vertices are placed at a random distance r from the center, such as $r_{min} < r < r_{max}$ (figure 1.a). The second half plane is then selected by rotating around the vertical axis, and the same process is repeated until the N_{hp} half-planes are filled. The vertices are joined to form the polyhedron. Figure 1.b shows a sample of 30 randomly generated polyhedra. The meshing software *Gmsh* is used to divide the polyhedron into elementary tetrahedra, that are then joined together using a Mohr-Coulomb law to form the breakable polyhedral grain.

*e-mail: francois.nader@insa-lyon.fr
**e-mail: claire.silvani@insa-lyon.fr
***e-mail: irini.djeran-maigre@insa-lyon.fr

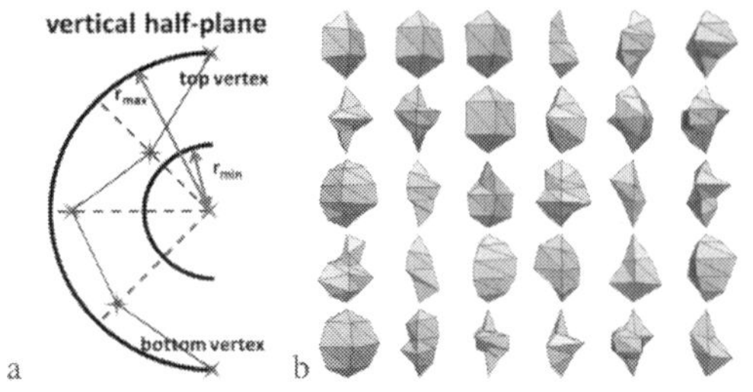

Figure 1. a. Grain generation – b. Sample of randomly generated polyhedra

3 Single grain crushing

Single grain crushing simulations are carried out to evaluate the grain's strength for different values of the intra-granular cohesion.

The open-source software $LMGC^{90}$ is chosen to run the simulations. This software implements the discrete element method called Non-Smooth Contact Dynamics (NSCD) [10, 11]. This method takes into consideration the dynamics of each element, as well as the interactions between the bodies, in order to describe the overall behaviour of the medium.

A breakable polyhedral grain is placed between two rigid plates. An increasing downward vertical load is applied on the upper plate until the grain breaks (figure 2). The simulated grain has a maximum size of $4cm$ divided into 12 subgrains. Multiple values of the intra-granular cohesion are tested, varying from $1kPa$ to 10^4kPa. The grain's reaction on the upper plate increases and reaches a peak value that is considered equal to the grain's breakage force. Figure 3 shows the evolution of the grain's breakage force as a function of the cohesion. Each point on the curve represents the average breakage force of a set of 15 single crushing simulations, run on grains having a similar size, the same order of magnitude of subgrain size, but different number of vertices. Error bars represent standard deviation.

The breakage force increases from $13N$ to $1500N$. This increase can be explained by the fact that when the value of the intra-granular cohesion increases, the force needed to break the bond increases, which in turn leads to an increase in the breakage force.

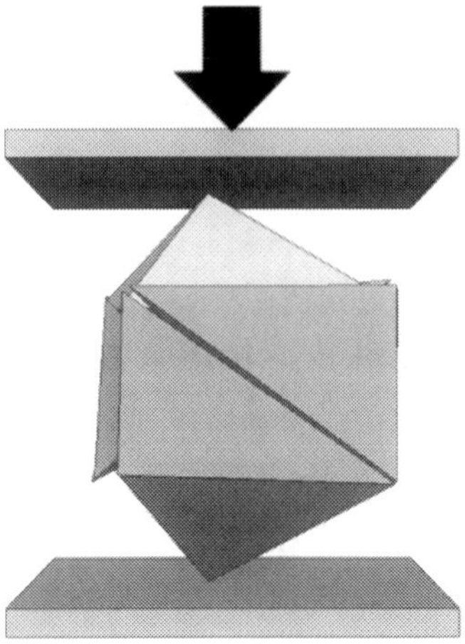

Figure 2. Single grain crushing under vertical load

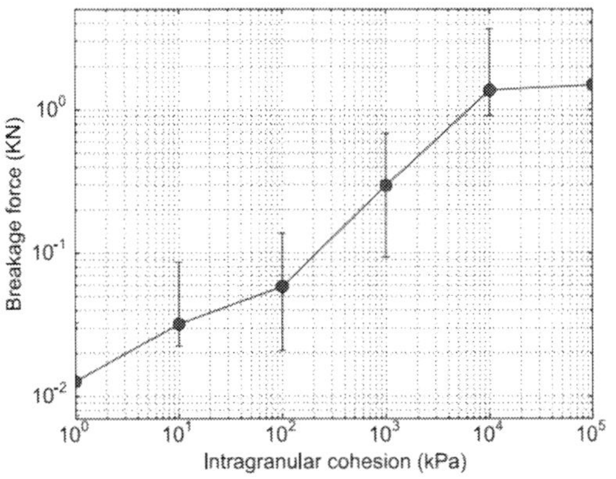

Figure 3. Breakage force as a function of the intra-granular cohesion

4 Oedometric simulations on a multigrain sample

A series of oedometric compression simulations are conducted on a sample of 855 grains, deposited by gravity in a rectangular box of a horizontal section of $30cm \times 30cm$, and a height of $26cm$. All the grains have a maximal grain size of $4cm$, with a number of vertices varying randomly between 8 and 14. All grains have an elongation equal to 1.0 (regular polyhedra), and each grain is divided into 8 to 12 tetrahedral subgrains (figure 4). Gravity is not taken into account during compression.

Figure 5 shows the evolution of the void ratio as a function of the vertical stress, on a semi-logarithmic scale, for three values of the intra-granular cohesion: $1kPa$, 10^2kPa and 10^4kPa. Many grain/plate contacts appear and disappear during compression, generating noise in the curves involving the vertical stress, the void ratio and the vertical strain. To cancel the noise and have clearer curves, the median value of these parameters is calculated over an interval of numerical steps. A slight decrease in the void ratio is observed at the beginning until the applied stress reaches $10kPa$, where the slope becomes steeper indicating a more important decrease in the void ratio with the increasing vertical stress. The difference between the three simulations is not clearly noticeable on these curves, therefore additional parameters are explored. Figure 6 shows the evolution of the vertical stress as a function of the vertical strain. The median values are shown in this figure. The effect of the cohesion is more visible: for a stronger cohesion, a bigger vertical stress is needed to obtain the same vertical strain, which means that a higher intra-granular cohesion reduces the compressibility of the sample. The effect of the intra-granular cohesion on the macroscopic behaviour of the studied material could be clearer. This can be a result of the boundary conditions imposed by the oedometric compression. this can also explain why the vertical strain strain reached in the oedometric compression does not show the decrease in the compressibility usually seen in this type of material.

To further analyse the behaviour, the percentage of the remaining cohesive bonds is monitored during compression (figure 7). It is clear that grains that have a lower intra-granular cohesion break for low values of vertical strain, therefore under lower loadings, which is in agreement with the observations made in section 3.

The evolution of the grain size distribution is monitored throughout the simulations. Figure 8 shows a typical evolution of the grain size distribution (in this case for a cohesion of $10^4 kPa$) for multiple vertical strains. Due to the uniform initial granular size distribution, as well as the rigid nature of the model, the maximal grain size decreases during compression as can be seen when comparing the successive grain size distribution curves. The final state curve is defined as the state where all cohesive bonds are broken, so the elementary tetrahedral subgrains become grains.

To quantify this evolution of the grain size distribution, the breakage ratio is computed at multiple vertical stresses, using the formula proposed by Einav 2007 [12] :

$$B_r = \frac{B_t}{B_p} \quad (1)$$

where B_r is the breakage ratio, B_t is the area between the initial and the current distribution, and B_p is the area between the initial and the final distribution. Einav 2007 [12] suggests the use of a fractal distribution $F(d)$ for the final state $F(d) = \left(\frac{d}{d_M}\right)^{3-D}$ where d is the grain size, d_M the maximum grain size and D the fractal dimension. Instead, the final distribution where all the cohesive bonds are broken is chosen because it is indicative of the limitation of the numerical model, since it represents the last state the simulated sample of grains can reach. Figure 9 shows the evolution of this breakage ratio as a function of the intra-granular cohesion, for two values of the vertical stress. Lower values of the cohesion show higher breakage ratios, which means that more grains are broken for the same value of the vertical stress. This result is also in agreement with the previous conclusions: stronger cohesions lead to less grain breakage under the same applied stress.

5 Conclusion

The results of this study can be summarised by the following conclusions :

A grain generation algorithm is created. It allows the generation of complex polyhedral shapes of grains. A meshing of the generated polyhedron is used to turn it into a breakable grain.

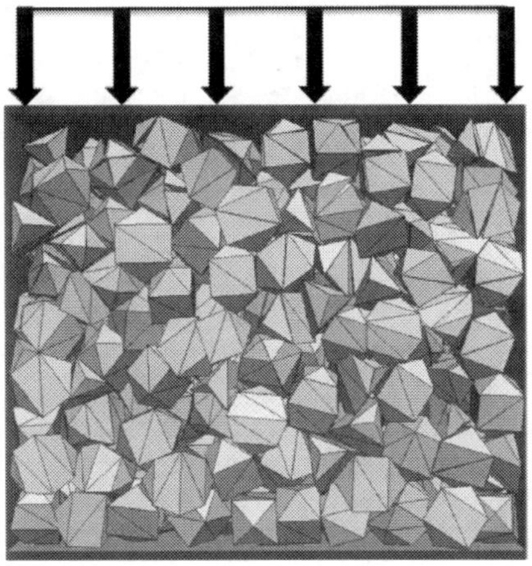

Figure 4. Sample of breakable grains during the oedometric compression

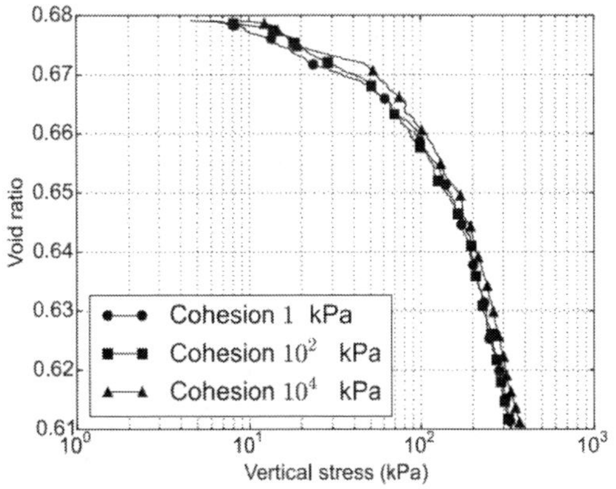

Figure 5. Oedometric curves for for three different values of the intra-granular cohesion (median values)

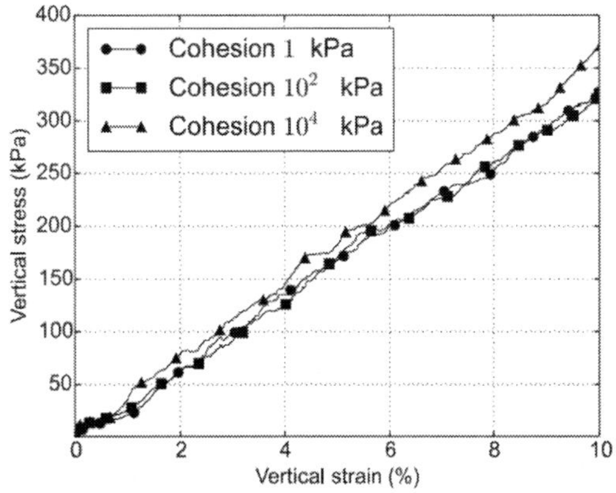

Figure 6. Vertical stress as a function of vertical strain for for three different values of the intra-granular cohesion (median values)

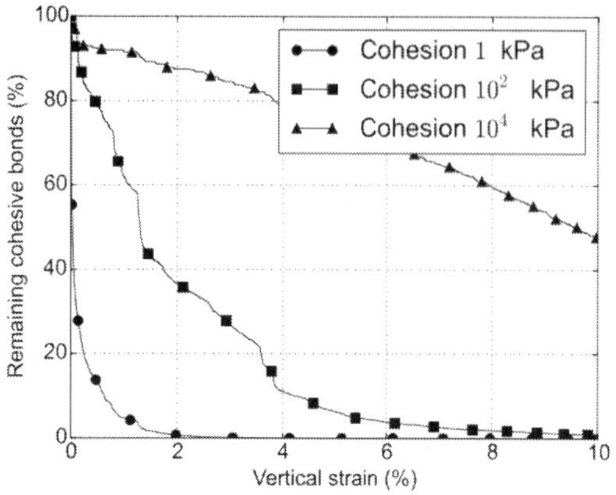

Figure 7. Remaining cohesive bonds as a function of vertical strain for three different values of the intra-granular cohesion

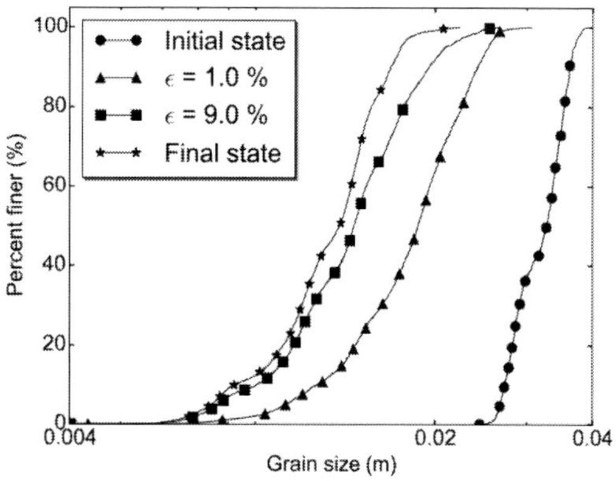

Figure 8. Grain size distribution throughout an oedometric compression

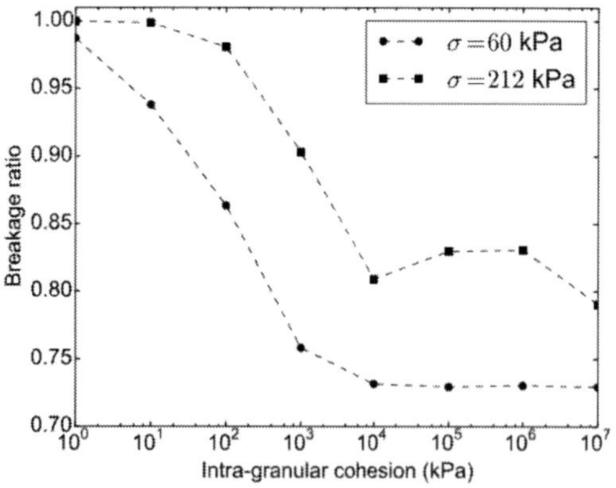

Figure 9. Breakage ratio as a function of intra-granular cohesion

Regular polyhedra are selected to conduct the single grain crushing and the oedometric compression simulations.
A single grain's strength depends on the intra-granular cohesion joining its subgrains: the higher the cohesion, the stronger the grain.
The oedometric simulations also show the effect of intra-granular cohesion: the higher the cohesion, the lower the compressibility of the sample. Nevertheless, this effect is limited as a result of the boundary conditions imposed by the oedometric conditions, the ratio between the oedometric cell size and the grain size, as well as the meshing of the grain (its division into subgrains). Triaxial simulations might allow additional rearrangement of grains, thus leading to different macroscopic behaviour. Other potential solutions are increasing the size of the oedometric cell or refining the grain meshing. On the microscopic scale, the effect of the intra-granular cohesion is clearer. A higher applied stress is needed to break grains that have a higher intra-granular cohesion. These observations, along with the evolution of the grain size distribution throughout the compression, show that this numerical model, that joins subgrains using cohesive bonds, manages to simulate grain breakage.
Further work is planned to study the effect of the meshing and a broader initial granular size distribution on the macroscopic behaviour.

References

[1] J. Åström, H. Herrmann, Eur. Phys. J. B **5**, 551 (1998)
[2] O. Tsoungui, D. Vallet, J. Charmet, Powder Technol. **105**, 190 (1999)
[3] O. Ben-Nun, I. Einav, Philos. Trans. R. Soc. London, Ser. A **368**, 231 (2010)
[4] G. McDowell, J. De Bono, Géotechnique **63** (2013)
[5] D. Cantor, N. Estrada, E. Azéma, Comput. Geotech. **67**, 150 (2015)
[6] C. Silvani, T. Désoyer, S. Bonelli, Int. J. Numer. Anal. Methods Geomech. **33**, 665 (2009)
[7] Y. Cheng, Y. Nakata, M. Bolton, Géotechnique **53**, 633 (2003)
[8] D.H. Nguyen, P. Sornay, E. Azéma, F. Radjai, Geomechanics from Micro to Macro **1**, 275 (2014)
[9] D. Cantor, E. Azéma, P. Sornay, F. Radjaï, Comput. Part. Mech. pp. 1–10 (2016)
[10] J.J. Moreau, in *Nonsmooth Mechanics and Applications* (Springer, Vienna, 1988), pp. 1–82
[11] M. Jean, Comput. Methods Appl. Mech. Eng. **177**, 235 (1999)
[12] I. Einav, J. Mech. Phys. Solids **55**, 1274 (2007)

DOI: 10.1051/epjconf/201714009012

An experimental study on particle effects in liquid sheets

Alban Sauret[1,*], *Anthony* Troger[1], and *Pierre* Jop[1]

[1] *Surface du Verre et Interfaces, UMR 125 CNRS/Saint-Gobain, 93303 Aubervilliers, France*

Abstract. Many industrial processes, such as surface coating or liquid transport in tubes, involve liquid sheets or thin films of suspensions. In these situations, the thickness of the liquid film becomes comparable to the particle size, which leads to unexpected dynamics. In addition, the classical constitutive rheological law for suspensions cannot be applied as the continuum approximation is no longer valid. Here, we consider experimentally a transient particle-laden liquid sheet that expands radially. We characterize the influence of the particles on the shape of the liquid film and the atomization process. We highlight that the presence of particles modifies the thickness and stability of the liquid sheet. Our study suggests that the influence of particles through capillary effects can modify significantly the dynamics of processes that involve suspensions and particles confined in liquid films.

1 Introduction

Suspensions of solid non-brownian particles in a liquid are encountered in various geophysical, industrial or biological systems. Whereas most of the past studies have characterized the transport of particles in bulk flow or the rheology of suspensions [1, 2], situations in which the thickness of the liquid layer becomes comparable to the particle size are less understood. This is particularly relevant in industrial processes, such as coating by liquid curtains. When the coating is performed with a suspension, the thickness and stability of the liquid sheet can be largely modified by the presence of particles [3]. Indeed, the particles are able to deform the liquid interface, which leads to complex interactions with the interfaces and between the particles.

The presence of particles in liquid film of comparable size can be especially important for its stability as it has been shown to modify the pinch-off of viscous fluid, when extruded from a nozzle to eventually generate a droplet. In a pure liquid, *i.e.*, without particles, the liquid filament becomes thinner and thinner until the pinch off occurs owing to interfacial tension effects. Various studies have then considered the detachment of a droplet of granular suspension and have shown that the initial pinch-off can be described by its effective viscosity [4, 5]. However, at later stages, the dynamics of the particles become important for the detachment of the drop when the liquid filament becomes of the same size as a particle. Typically, the detachment becomes accelerated and depends on the particle volume fraction and size. This uncommon situation cannot be explained with a classical Newtonian fluid approach or classical rheology of diluted suspensions. Indeed, the presence of particles perturbs locally the thinning of the filament. It then leads to the acceleration of the droplet formation compared to the detachment observed for a pure fluid.

However, the pinch-off is due to a 1D flow and more complex situations remain poorly understood. In particular, the influence of particles in self-suspended liquid films, which constitutes a 2D situation, could lead to a modification of its stability [3]. In this study, we consider experimentally a classical model system in which a drop impacts at the center of a flat target of size comparable to the drop diameter [6]. Upon impact the liquid flattens into a self-suspended liquid sheet that expands radially. Such experiments have been performed with Newtonian and non-Newtonian fluids to characterize the radial expansion of the liquid sheet and its fragmentation into droplets [7, 8] and with emulsions to investigate the influence of the liquid dispersed phase on the dynamics of the liquid sheet [9].

2 Experimental methods and phenomenology

The generation of the transient liquid sheet is based on a method adapted from the original experiments of Rozhkov *et al.* [6]. In our experimental set-up, shown in Fig. 1, a drop impacts the flat cross-section of a solid cylindrical target of diameter $d_t = 6\,\mathrm{mm}$. The drop is formed by pushing the suspension through a needle positioned vertically above the target. The impact velocity u_0 on the target is directly set by the height of fall of the drop. The suspension is manually extruded from a syringe using a microcontroller linear stage to obtain monodisperse drops of diameter $d_0 = 3.9\,\mathrm{mm}$. The drop then falls on the target and spreads into a liquid sheet. A thin aluminium coaxial cylinder around the target keeps the angle of ejection of

*e-mail: alban.sauret@gmail.com

 DOI: 10.1051/epjconf/201714009012

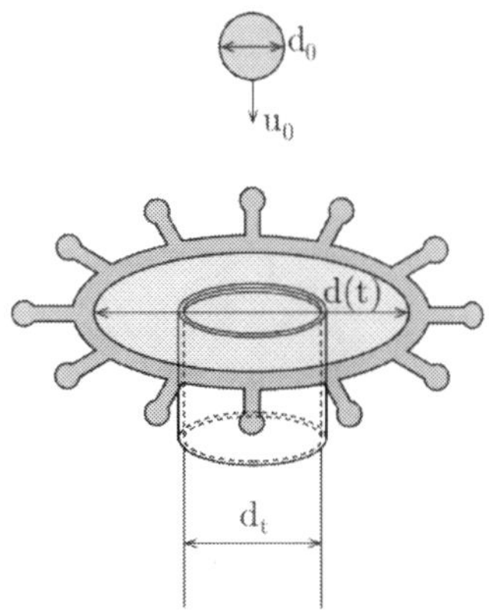

Figure 1. Schematic of the experimental set-up: a drop of diameter d_0 impacts the cylindrical target (diameter d_t) at a velocity u_0, leading to the formation of a transient liquid sheet of diameter $d(t) = 2R(t)$.

the liquid at 90° and the liquid sheet horizontal. The target is mounted on a hollow plexiglas plate and illuminated from below using a LED Panel. We record the spreading of the sheet from the top using a high speed camera (Phantom v611) with a macro lens (Nikon 105 mm) typically at 9000 fps.

The suspensions are prepared with spherical polystyrene PS particles (Dynoseeds) of density $\rho \simeq 1058\,\mathrm{kg\,m^3}$ dispersed in a mixture of 25%wt glycerol/water solution such that the particles are neutrally buoyant. We use particles of diameter $d = 140\,\mu$m that are larger than the mean thickness of the liquid sheet. The viscosity of the glycerol/water mixture at $T = 24$°C is $\eta_0 = 1.85$ mPa.s and its surface tension is $\gamma = 70\,\mathrm{mN.m^{-1}}$. The liquid viscosity is chosen small enough for the liquid sheet to reach a diameter of few centimeters. It requires that both viscous and capillary effects remain small compared to the inertial effects. In term of dimensionless number, this condition is described trough the Weber number

$$We = \frac{\rho\, u_0^2\, d_0}{\gamma}, \qquad (1)$$

which compares the magnitude of inertial and capillary effects and the Capillary number

$$Ca = \frac{\eta\, u_0}{\gamma}, \qquad (2)$$

which describes the ratio of viscous and capillary effects. Here, we have used a liquid phase with a low viscosity and work with diluted suspensions ($\phi = [0\,\%;\,30\,\%]$). Therefore, all of our experiments are performed in the regime where $We \gg 1$ and $Ca \ll 1$. The experiments presented in this paper are performed at a Weber number $We \simeq 490$.

Two typical experiments, without and with particles, are shown in Fig. 2 and illustrate the dynamics of the liquid sheet from the impact of the drop on the target at $t = 0$ ms to the retraction and the atomization into droplets after about 10 ms. We observe that the geometry of the liquid sheet during the spreading satisfies to the cylindrical symmetry of the geometry: a thin liquid sheet surrounded by a rim extends horizontally, suspended in air until it reaches the maximum diameter that results from a balance between the inertial and capillary effects. As shown on this figure, the value of the maximum diameter depends on the presence or absence of particles but also on the size of the drop and the impactor as well as the impact speed. Once the liquid sheet reaches its maximum diameter, the film retracts under the effect of capillary forces and destabilizes, losing its cylindrical symmetry. We also note that during most of the process, especially during the recessing phase, droplets are generated.

3 Spreading of particle-laden liquid sheets

We first characterized the evolution of the thickness of the liquid sheet in the absence of particles. We have implemented an optical method based on the absorption of light to measure the thickness of the sheet. This method relies on grayscale images of the sheet recorded with a high-speed camera. Using the Beer-Lambert law and a preliminary calibration, we can obtain the time and spatial evolution of the liquid thickness [10]. The results obtained via this method are shown in Fig. 3 and indicate that the thickness of the sheet varies between $20\,\mu$m and $160\,\mu$m. The particles used in this study, of diameter $140\,\mu$m, are larger than the liquid thickness and will therefore be subject to capillary forces. We also observe that the thickness of the liquid sheet decreases with the distance r from the center of the impactor at both short time and long time.

We consider the expansion dynamics of liquid sheet in the absence of particles to compare our results with previous predictions who considered the limit of long times and were able to show that the time evolution of the diameter of the liquid sheet is described by the expression [7]:

$$\frac{R(t) - r_0}{r_0} = \sqrt{\frac{2}{3}\, We}\, \frac{t}{\tau} \left(1 - \frac{t}{\tau}\right)^2 \qquad (3)$$

where $\tau = \sqrt{\rho\, d_0^{\,3} / (6\,\gamma)}$. The spreading dynamics is partially in agreement with the theoretical modeling as shown in Fig. 4. In this model, we expect a prefactor $a = \sqrt{2\,We/3} \simeq 18$, whereas our experiments suggest that $a = 15$. Note that this discrepancy can be explained by the characteristics of the impactor, and in particular its wettability and the presence of the outer cylinder [10].

We have systematically studied the effect of the $140\,\mu$m diameter particles on the spreading dynamic of the liquid sheet as shown in Fig. 4. The particles are larger than the liquid sheet thickness and are therefore subject to capillary force. We observe that, although the evolution are similar in all cases, there is a strong decrease of the maximum spreading diameter, even at low concentrations. We can also notice that the lifetime of the liquid sheet is greatly reduced by the presence of particles. More specifically, the expansion phase is slightly modified, but the total life is greatly reduced. A first assumption can be made to

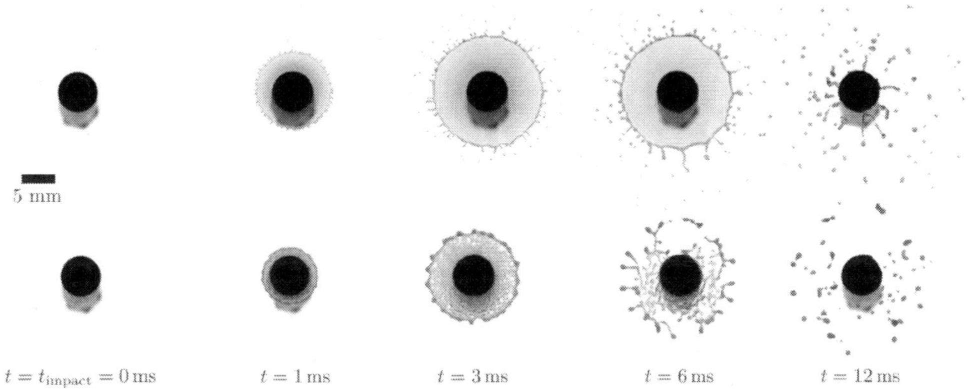

Figure 2. Successive pictures showing the time-evolution of a liquid sheet for the liquid (top pictures) and a suspension with $\phi = 0.35$ (bottom).

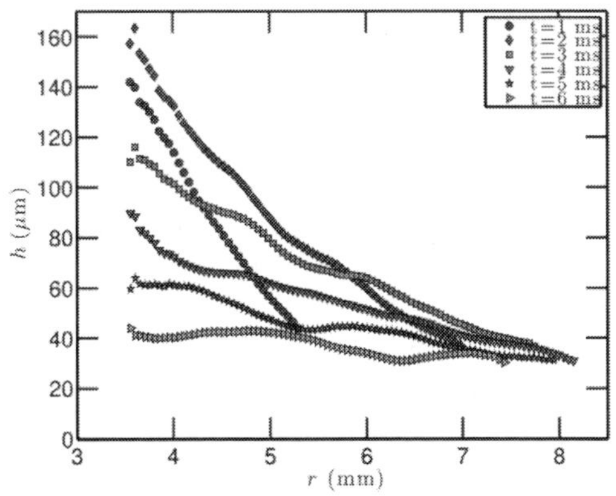

Figure 3. Evolution of the liquid sheet thickness formed by the drop impact of a pure liquid on the target at $t = 0\,$ms taken at various time.

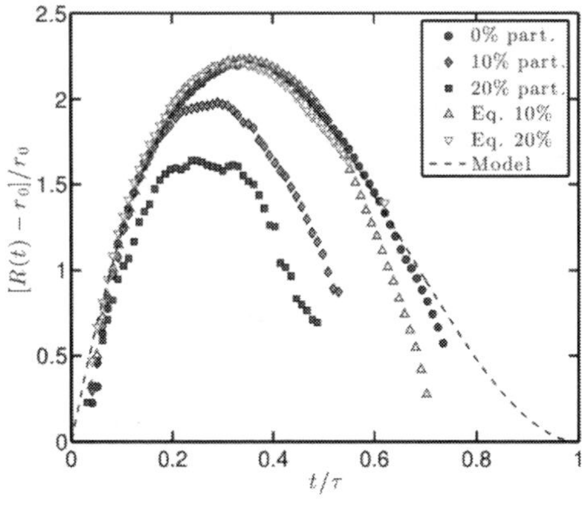

Figure 4. Evolution of the rescaled radius of the liquid sheet as a function of the dimensionless time t/τ for the pure liquid, suspensions with 10%wt and 20%wt of particles. The measurements performed with Newtonian liquids of same viscosities as the 10% and 20% suspensions are also reported. The dashed-line is the prediction given by eq. (3) with a prefactor $a = 15$

explain these observations: the increase in viscosity due to the addition of particles leads to the change of dynamics. To quantify this influence, we performed similar experiments using water/glycerol mixture that have an effective dynamic viscosity identical to the suspensions. The time-evolution of this Newtonian liquid sheet is shown in Fig. 4 and illustrates that the increase in viscosity induced by the particles cannot account for the differences observed. Indeed, for water/glycerol concentrations corresponding to suspensions with $\phi \leq 20\%$, no difference is observed. Therefore, the modification of the spreading dynamics is an intrinsic local effect of the particles. Qualitatively, the presence of particles larger than the thickness of the sheet creates a deformation of the interface that limits its expansion and dynamics [11].

4 Fragmentation of the particle-laden liquid sheet

During the spreading and the fragmentation of the particle-laden liquid sheet, droplets are generated from two different ways. The first mechanism takes place upon impact, during the expansion and the retraction of the liquid film. The droplets are created by the outer rim and have relatively substantial size. Then, droplets are also generated when the liquid film ends its retraction phase and smaller droplets are then created.

We reported in Fig. 5 the resulting droplet size distributions for increasing particles concentration. It shows that the distributions share a similar trend even if they tend to be larger and shift to larger radii as the concentration in particles increases. We also notice that the larger the con-

DOI: 10.1051/epjconf/201714009012

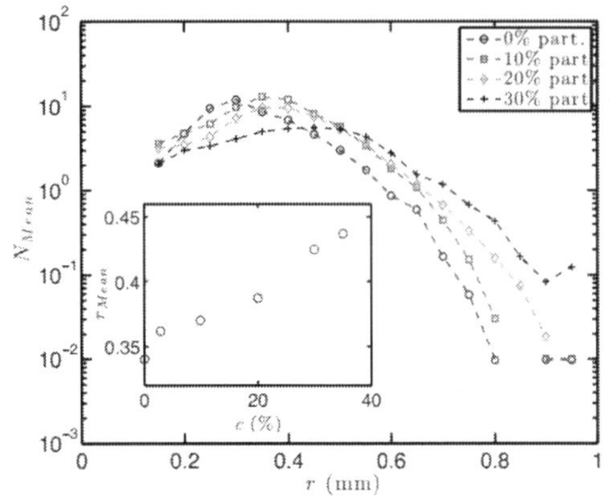

Figure 5. Mean number of droplets per impact N_{Mean} of radius r for different concentrations of particles. Each curve contains over a hundred drop impacts. Inset: mean radius r_{Mean} of the droplets resulting from the atomization of the liquid sheet for varying particle concentration c.

centration is, the less numerous are the droplets emitted at each impact and these droplets tend to have a large diameter. The evolution of the mean radius r_{Mean} of the droplets generated during the fragmentation process is plotted in the inset for Fig. 5 as a function of the particle concentration. The mean droplet size increases with the particle concentration.

We have performed the same experiments with Newtonian liquid of increasing viscosity. The viscosities of the solution without particles are taken equal to the viscosity of the different suspensions ($\phi \leq 30\%$). We did not notice a significant change in droplet size for Newtonian fluid with similar viscosities. Therefore, the influence of the particles goes beyond the increase in viscosity of the suspension, modifying the generation process of the droplets compared to a classical Newtonian liquid [11].

5 Conclusion

In study, we have experimentally studied the formation and atomization of a particle-laden liquid sheet formed upon impact of a drop on a small target. The presence of spherical, neutrally buoyant, non-colloidal particles introduces various effects that make the dynamics of the liquid sheet different from that of a pure liquid.

Our experimental results have first illustrated the influence of the particles on the spreading of the liquid film. We have characterized the spreading dynamics of the sheet such as its radius and thickness in the presence and absence of particles. We have been able to compare the theoretical prediction for a Newtonian fluids and quantify the deviation to this simple law for a particle-laden liquid sheets.

We have also investigated the destabilization of the liquid sheet and its atomization in droplets with and without particles. Our results suggest that the particles do not alter the physical mechanism of droplet generation, but they modify the mean radius of the drops generated. We have also observed a widening of the distribution and a shift towards larger drop radii when increasing the particles concentration. These effects can not be explained using the effective viscosity and are directly related to the discrete nature of the particles.

Acknowledgements: the authors acknowledge the support from the French ANR (project ProLIFIC ANR-16-CE30-0009) and partial support from a CNRS PICS grant n°07242.

References

[1] E. Guazzelli & J. F. Morris, *A physical introduction to suspension dynamics* (Cambridge University Press, 2011) 256 pp

[2] M. M. Denn & J. F. Morris, Annu. Rev. Chem. Biomol. Eng. **5**, 203-228 (2014)

[3] F. O. Addo-Yobo, M. J. Pitt & H. A. Obiri, AIChE J. **57**(8), 2007-2024 (2011)

[4] R. J. Furbank & J. F. Morris, Phys. Fluids **16**(5), 1777-1790 (2004)

[5] C. Bonnoit, T. Bertrand, E. Clément & A. Lindner, Phys. Fluids **24**(4), 043304 (2012)

[6] A. Rozhkov, B. Prunet-Foch & M. Vignes-Adler, Proc. Roy. Soc. A **460**(2049), 2681-2704 (2004)

[7] E. Villermaux & B. Bossa, J. Fluid Mech. **668**, 412-435 (2011)

[8] B. Keshavarz, E. C. Houze, J. R. Moore, M. R. Koerner, G. H. McKinley, Phys. Rev. Lett. **117**(15), 154502 (2016)

[9] C. Vernay, L. Ramos & C. Ligoure, Phys. Rev. Lett. **115**(19), 198302 (2015)

[10] C. Vernay, L. Ramos & C. Ligoure, J. Fluid Mech. **764**, 428-444 (2015)

[11] L. A. Lubbers, Q. Xu, S. Wilken, W. W. Zhang & H. M. Jaeger, Phys. Rev. Lett. **113**(4), 044502 (2014)

Fabrication of composites via spouted bed granulation process and simulation of their micromechanical properties

Eduard Eichner[1], *Maksym* Dosta[1], *Stefan* Heinrich[1] and *Gerold A.* Schneider[2]

[1]Hamburg University of Technology, Institute of Solids Process Engineering and Particle Technology, 21073 Hamburg, Germany
[2]Hamburg University of Technology, Institute of Advanced Ceramics, 21073 Hamburg, Germany

Abstract. In this contribution numerical simulation of Young's modulus of copper-polymer composites is presented. For the simulation of the composites the Bonded-Particle-Model was applied. The model allows representing of the structure of composite materials realistically. The polymer matrix, which surrounds the particles, was represented as network of solid bonds connecting copper particles. Simulation results were validated based on mechanical determination of modulus of elasticity. The modulus of elasticity was approximated in experiments as well as in simulation by four-point-bending tests. It was observed, that obtained simulation results are in good agreement with experimental results.

1 Introduction

Beside of one-component materials like pure metals there exist also materials, which consist of two or more components, which are arranged in a particular way. These materials are called composites. Naturally occurring composites consist mostly of ceramics and polymers, which are structured in several hierarchical levels [1-2]. The structure of the composites shows that ceramic particles are uniformly distributed in a polymer matrix and the content of ceramics is significantly higher than that of polymer. The high content of ceramic in polymer matrix is significant for reinforcement of the polymer. Polymer in the composites ensures the slight ductility of the materials. One of the best investigated natural materials is nacre. Nacre consists of small ceramic platelets. The platelets are surrounded by a polymer matrix (see fig. 1). The ceramic content in nacre is about 95 vol.%. Due to such a high amount of ceramic and further aspects like high aspect ratio of ceramic particles, nacre has very good mechanical properties. Strength of nacre reaches values up to 120 MPa and modulus of elasticity is up to 70 GPa.

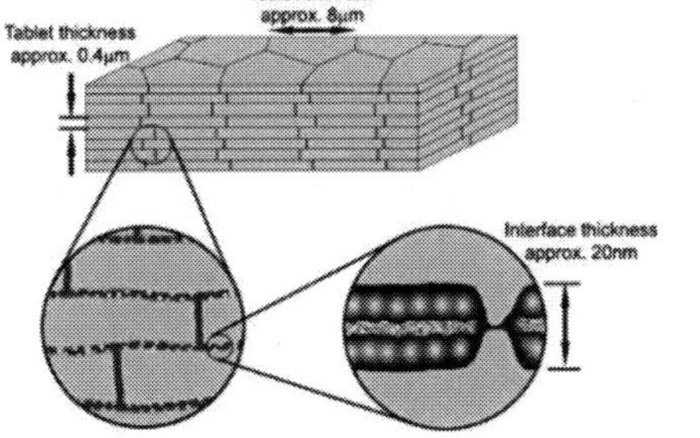

Fig. 1. Structure of nacre [2].

The good mechanical properties of natural materials gave an impulse for the reproducing of the natural structures. Therefore, in recently years, many attempts have been made to reconstruct the structure of the natural material and to reach a high ceramic content. One possible route for the reproducing of natural composite materials is the spouted bed spray granulation process. The application of the process has allowed reaching of very high ceramic contents in composite materials [3] and also reproducing of the hierarchical structure [4].

Spouted bed is a special form of fluidized bed with many applications in particle technology, such as drying, granulation, coating and sometimes also agglomeration. For processing of fine and cohesive particles a special spouted bed was designed. It could be successfully applied also for fabrication of ceramic-polymer composites [3, 4]. During processing, solid particles are covered with a polymer layer. Afterwards, coated agglomerates are pressed to get a bulk material. During warm pressing the melting of polymer occurs and it can fill the gaps between particles.

For the characterization of final material, its composition and its mechanical properties are evaluated. The determination of mechanical properties is carried out by four-point bending tests. Bending tests deliver strength and modulus of elasticity for investigated composites.

Numerical modelling of composite material plays important role for optimization of material structure and allow minimizing experimental effort. The mechanical properties of composites can be simulated by discrete element method (DEM). Initially, DEM was developed as a method for numerical simulation of granular materials consisting of ideally spherical particles. In

* Corresponding author: eduard.eichner@tuhh.de

DEM each particle is considered as a separate element and Newtonian equations of motion are solved for each particle [5]. As a consequence, motion and deformation behaviour, and stress state of single particle and of particle collectives can be calculated [6]. Many different extensions of DEM and additional contact models were developed in recent years. One of these extensions is the Bonded-Particle-Model (BPM). In the BPM materials are represented as particles, which are connected to each other by solid or liquid bonds. For the simulation of composite materials the BPM is probably one of the best suitable models, because composites consist of discrete particles, which are surrounded by a polymer matrix. The BPM represent the particles surrounded polymer by solid polymer bonds. This allows reflecting the real material structure with high precision. Nevertheless, the numerical model does not directly correspond to the real material. Therefore, the mechanical properties of polymer cannot be just taken from experimental tests to simulations and additional adjustment is needed.

In this contribution the BPM was used to simulate mechanical properties of copper-polymer composites. Solid bonds among particles were described with elastic solid beam model [7].

2 Methods

DEM simulations are carried out with an in-house developed software MUSEN. In the first stage, a bending beams consisting of primary particles and solids bonds were generated. The dimensions of simulated bending beams were approximately 10 times smaller than in the experiments. Dimensions of the beam are summarized in table 1.

Table 1: Dimensions of the bending beam.

Beam length (x)	Beam height (z)	Beam width (y)
4270 μm	235 μm	256 μm

In the experiments and simulations investigated composite material was consisting of copper primary particles and polyvinyl butyral polymer. The particles in experiments were not monodisperse therefore in simulation a particle size distribution was generated. Particles had a size between 6 and 40 μm and they were distributed in 4 fractions, so the mean diameter of particles was 30 μm as in experiments.

After generation of copper particle in the bending beam, they were connected to each other by polymer bonds. So the continuous polymer matrix surrounding the particles in experiments is represented by polymer bonds, which connect particles. During bending tests the relative motion of aggregated primary particles causes forces and moments in the solid bonds [8]. In each simulation step the forces and moments in the bonds are calculated as an increment to the corresponding values from previous time step [9]. The forces in tangential and normal direction are determined as:

$$\bar{F}_{b,t}^{i} = T \cdot \bar{F}_{b,t}^{i-1} - \bar{v}_{rel,t} \cdot \Delta t \cdot k_t \cdot A_b \qquad (1)$$

$$\bar{F}_{b,t}^{i} = k_n \cdot A_b (L_{init} - L_{cur}) \cdot \bar{r}_n \qquad (2)$$

where L_{init} and L_{cur} are the initial and current length of the bond, A_b is the bond cross-cut surface area, $\bar{r}_n$ is the unit vector between bonded particles, k_n and k_t are the normal and tangential stiffness of the bond, and Δt is the time step. For taking into account motion of the particles in the space, the values calculated during the previous iteration are multiplied by a transformation matrix T. The length of the bond L is calculated from distance between centres of particles L_{pp}, radii of the particles R_1, R_2 and radius of the bond R_b:

$$L = L_{pp} - \sqrt{R_1^2 - R_b^2} - \sqrt{R_2^2 - R_b^2} \qquad (3)$$

The moments in bonds in tangential and normal direction are calculated as:

$$\bar{M}_{b,t}^{i} = T \cdot \bar{M}_{b,t}^{i-1} - \bar{\omega}_{rel,t} \cdot \Delta t \cdot k_n \cdot I \qquad (4)$$

$$\bar{M}_{b,n}^{i} = T \cdot \bar{M}_{b,n}^{i-1} - \bar{\omega}_{rel,n} \cdot \Delta t \cdot k_t \cdot J \qquad (5)$$

where $\bar{\omega}_{rel,t}$ and $\bar{\omega}_{rel,n}$ are relative rotational velocities between two bonded particles in tangential and normal direction, respectively. I is the moment of inertia and J is the polar moment of inertia of the bond cross section [9].

The breakage criteria for bonds has following definition: if shear or tensile stress exceed the corresponding bond strength, then the bond breaks and is removed from the calculation procedure [9].

For the simulation of four-point-bending tests two upper and two lower supports were added to the simulation. The upper supports were moved down with a constant loading speed of 10 mm/s. Lower supports were stationary. Thereby beam was bended. A sketch of the bending beam is presented in fig. 2.

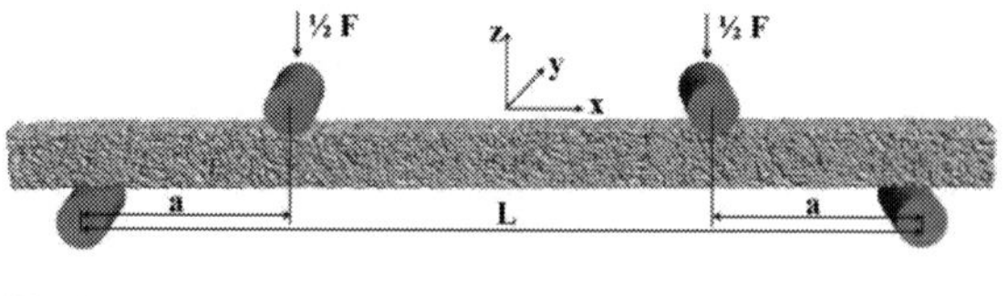

Fig. 2: Bending beam for DEM simulation. L=3.6 mm, a=0.25*L.

The modulus of elasticity for experiments and for simulations was calculated according to:

$$E = \frac{F * L^3}{8bh^3d} \qquad (6)$$

Thereby F is bending force, L distance between lower supports, b width of sample, h height of sample and d deflection.

Properties of copper particles and steel supports were identical to experiments and are listed in table 2.

Table 2: Properties of copper particles and steel supports for DEM simulations.

Property	Unit	Copper	Steel
Density	kg/m^3	8930	7860
Modulus of elasticity	GPa	120	200
Poisson's ratio	-	0.34	0.3

In contrast to the fixed parameters of particles, properties for polymer bonds had to be adjusted to reproduce experimental results of four-point-bending tests [10]. This had to be done, because the presentation of continuous polymer matrix surrounding the copper particles by cylindrical polymer bonds, which connect two particles, deviates from reality slightly.

3 Results

For the adjusting of polymer properties, a composition with a copper content of 65.5 vol.% was used. The rest material in the sample was the polymer, whereby also the porosity of about 5 vol.% was taken into account. The modulus of elasticity at the composition was calculated by linear interpolation between two in experiments determined modulus of elasticity at compositions with copper contents of 60.6 and 69 vol.% respectively. Beside of modulus of elasticity also normal and tangential strength in polymer, and coordination number were adjusted. The coordination number describes an average number of solid bonds connected to single particle. By variation of the parameters in simulations best suitable parameters were determined. These parameters are presented in table 3.

Table 3: Coordination number of particles, modulus of elasticity, and normal and tangential strength of polymer.

Coordination number [-]	6
E_{PVB} [GPa]	6
$\sigma_{n,t}$ [MPa]	150

In further simulations these parameters were used. Figure 3 shows the force distribution in the bending beam at different loadings and deflections. The upper beam is unloaded, the beam in the middle is slightly loaded and the lower beam is fully loaded.

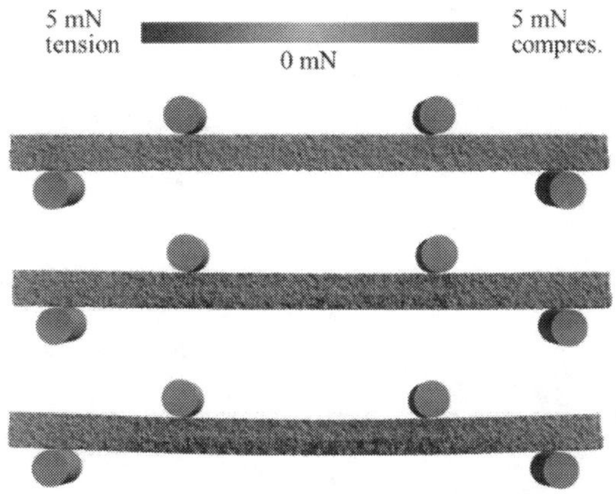

Fig. 3: Force distribution in polymer bridges of bending beam: unloaded (above), slightly loaded (middle) and loaded (below).

Especially at high loading three areas can be clearly differentiated. The upper part of the beam is compressed. This can be seen on the positive forces in the upper part. Negative forces in the lower part of the beam mean that there act tensile forces. The part in the middle of the beam stays unloaded also at high loadings. This loading distribution in a bending beam correlates to experimental and theoretical loading distribution.

After the adjustment of the parameters for the polymer, simulations at other compositions were carried out with the parameters. For further simulations four different compositions were chosen. For the simulations compositions with a copper content of 62, 64, 67 and 69 vol.% were used. Higher copper content led to a higher number of particles in the beam. For each simulation the porosity was set to 5 %. This corresponds to mean porosity in experiments. The rest in the beam was polymer. The parameters for simulations are summarized in table 4.

Table 4: Parameters for simulations of modulus of elastic in copper-polymer composites.

Parameters	**sample 1**	**sample 2**	**sample 3**	**sample 4**
Copper content [vol.%]	62	64	67	69
PVB content [vol.-%]	34	31	28	26
Porosity [%]	5	5	5	5
Number of particles	14570	15650	16190	17269

Modulus of elasticity determined by simulation and comparison between simulation and experimental results are presented in figure 4.

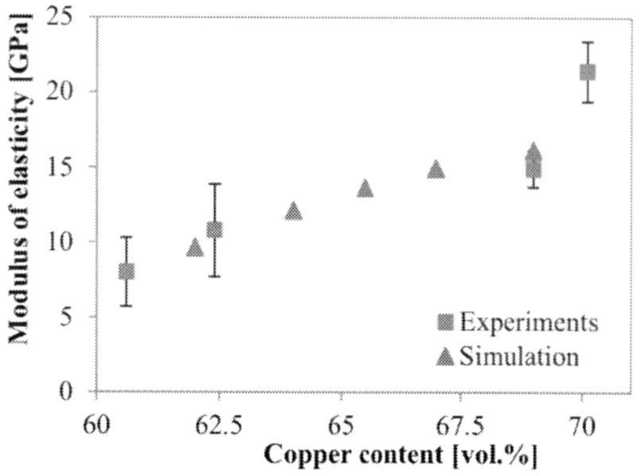

Fig. 4: Comparison between experimental [10] (blue) and DEM simulations (red) results. Here the dependency modulus of elasticity on copper content in copper-polymer composites is illustrated.

Figure 4 shows that the simulation could reproduce experimental values of modulus of elasticity for copper-polymer composites very well. The adjustment of simulation parameters was done only for one copper content. The moduli of elasticity at lower and higher copper contents could be reproduced automatically with high accuracy. This proves that the proposed modelling approach could be effectively applied for simulations of mechanical behaviour of composite materials.

4 Conclusions

In this contribution the applicability of the Bonded-Particle-Model for numerical modelling of mechanical properties of composite materials was investigated. The microscale model was validated on experimental results obtained from four-point-bending tests. Afterwards, it was successfully applied for composites with different higher and lower copper contents. The obtained results show that the model can be effectively applied for investigation of modulus of elasticity in composite materials.

We gratefully acknowledge financial support from the German Research Foundation (DFG) via the collaborative research center SFB986 "M3".

References

1. H. Gao, Int. J. Fract. **138**, 101-137, (2006)
2. F. Barthelat, H.D. Espinosa, Exp. Mech. **47**, 3, 311-324 (2007)
3. M.F.H. Wolff, V. Salikov, S. Antonyuk, S. Heinrich, G.A. Schneider, Compos. Sci. Technol. **90**, 154-159 (2014)
4. K. Brandt, M.F.H. Wolff, V. Salikov, S. Heinrich, G.A. Schneider, Sci. Rep. **3**, 2322 (2013)
5. A. Spettl, M. Dosta, S. Antonyuk, S. Heinrich, V. Schmidt, Adv. Powder Technol. **26**, 1021-1030 (2015)
6. P.A. Cundall, O. Strack, Geotechnique **29**, 1, 47-65 (1979)
7. M. Dosta, S. Dale, C.R. Wassgren, S. Heinrich, J.D. Litster. Powd. Techn. **299**, 87-97 (2016)
8. D.O. Potyondy, P.A. Cundall, Int. J. Rock Mech. & Min. **41**, 1329–1364 (2004)
9. M. Dosta, S. Antonyuk, S. Heinrich, Ind. Eng. Chem. Res. **52** (33), 11275–11281 (2013)
10. E. Eichner, V. Salikov, P. Bassen, S. Heinrich, G.A. Schneider, Powder Technol. DOI: 10.1016/j.powtec.2016.12.028

Influence of temperature and velocity on the emptying of silo hoppers

Marcus Ripp*, *Robert* Hesse and *Sergiy* Antonyuk

Institute of Particle Process Engineering, University of Kaiserslautern, Germany

Abstract. For the design of bulk material silos, a detailed knowledge of the flow properties of bulk materials is necessary. If the dynamic processes in a silo and, above all, the transition from a stationary bulk material to a flowing bulk material, the designer faces difficult challenges. In the following work a new measuring method is presented, with which it is possible to measure important information about the behavior of bulk goods in silos.

1 Introduction

The flow behaviour of bulk solids depends on a number of parameters. The flow parameters required for the design of silos are determined with shear cells at room temperature [1]. The results of the measurements with shear cells could use for the comparison of the flow properties of the bulk solids and for optimization of the silo discharge process. To predict the flow behavior in a silo it is necessary to know the flow property at the transition from passive to active dynamic stressing state or the transition of adhesion to sliding friction. Also the conditions of the slip-stick effect are important to know.
Investigations show, that there is an influence of the velocity [1] and the temperature [2, 3] on the flow inside the bulk silo. These effects cannot be completely estimated in standard shear cells. Therefore, it cannot be assumed that there is always coulomb friction.
In the work a novel experimental setup will be presented which provides further details on the flow behavior of bulk materials.

2 Measurement setup

The newly developed experimental setup is shown in Fig. 1. A tube (diameter 100 mm) was clamped in a tensile testing machine (universal Zwick/Roell testing machine, Fig. 2), and the tube was closed from below by means of a piston which can be displaced in the tube. Various bulk materials were charged into the tube (240mm) through a recess in the upper part, which also functions as prevention for a potential vacuum during the shearing process. The bulk material was subsequently loaded with weights from above to simulate a higher filling level to obtain a uniform normal stress in the tube walls (Fig. 3). The tube was withdrawn from the bulk material column and the wall friction forces between the tube and filled bulk material were measured by force transducers.

During the removal of the tube with a defined velocity the particle-wall friction force and displacement are measured. To perform the measurements at different temperatures, the tube is covered with a heating jacket. The construction allows a variation of the diameter and the material of the tube.

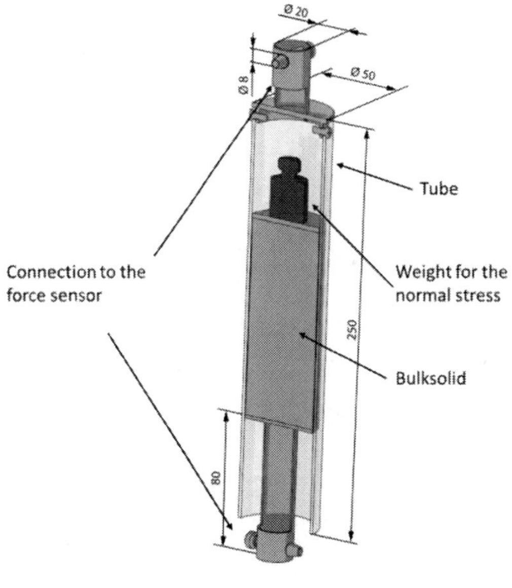

Fig. 1: Experimental setup for dynamic shear test with bulk solid (all dimensions in mm)

* Corresponding author: ripp@mv.uni-kl.de

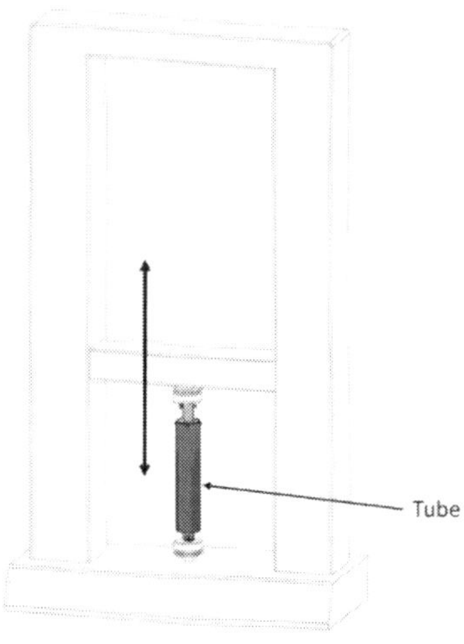

Fig. 2: Tension testing machine with the tube measuring cell

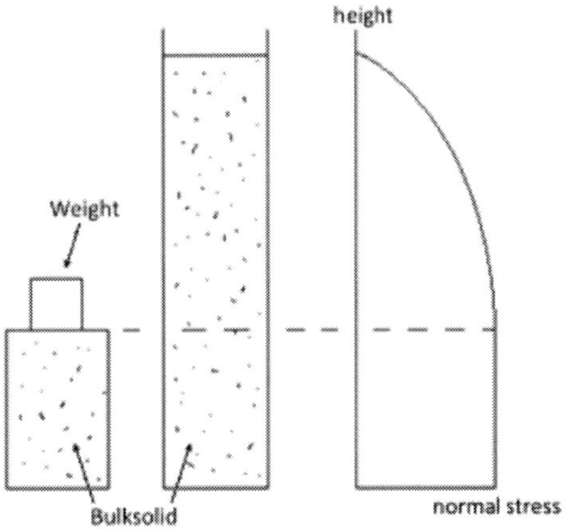

Fig. 3: Normal stress in the walls for varying bulk solid heights

Various plastic granules were used in the tests (Fig. 4 and 5).

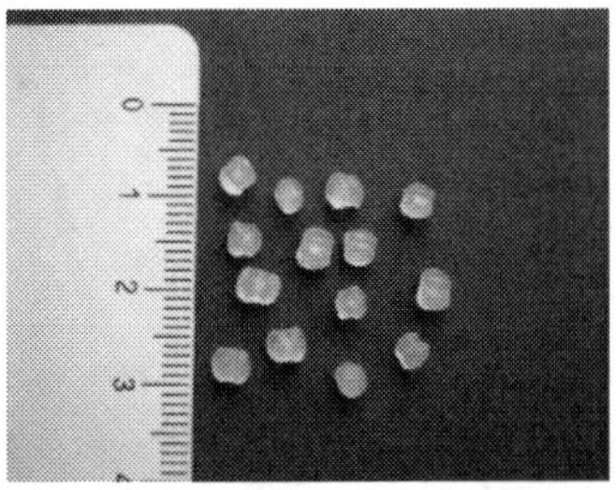

Fig. 4: Polyethylene Pellets (PE)

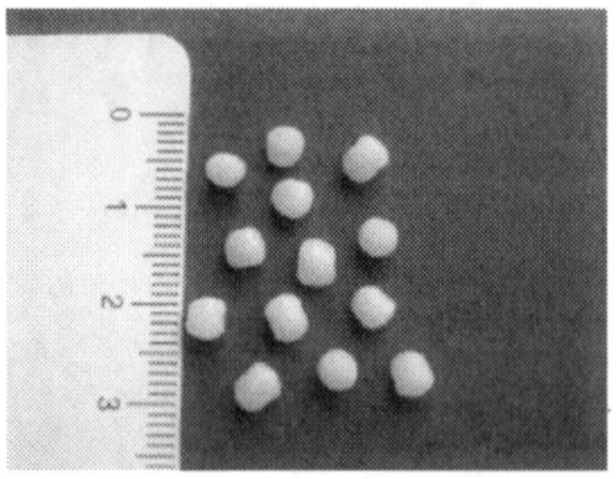

Fig. 5: Polypropylene Pellets (PP)

3 Experimental results

The results of the measurements with Polyethylene Pellets show very clearly that relatively large stickings have to be overcome during the start-up of the tube. A constant value for the sliding friction is established only after a travel distance of approximately 10 mm (Fig. 6). This sticking friction at the beginning of the discharge of a bulk silo could lead to problems during the operation of a bulk silo. In the further course of the measurement, a slip-stick effect can be recognized which is to be analysed in further experiments.

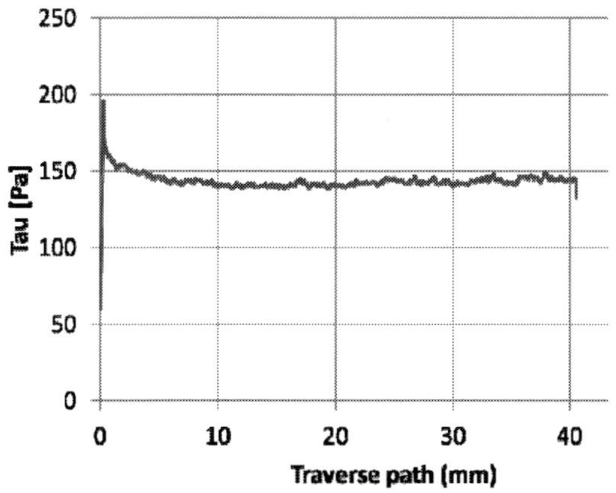

Fig. 6: measured wall shear stress (Tau) versus shear displacement for polyethylene granules at a constant velocity of 40 mm/min. (PE)

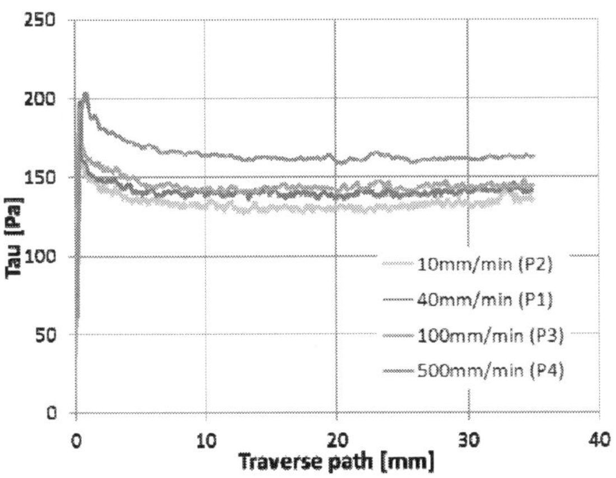

Fig. 7: Different drawing speeds of the tube with PE granules

Fig. 7 shows that different values for the sliding friction are set at different drawing speeds. This is a clear indication that there is no Coulomb friction in this case. If these results are transferred to tests in shear cells, the shear cell tests have to be adapted to the flow velocities in the silo-application.

In Fig. 8, results are shown in which the temperature of the measuring device was varied with a defined pull speed. Polypropylene (PP) was tested as granules.

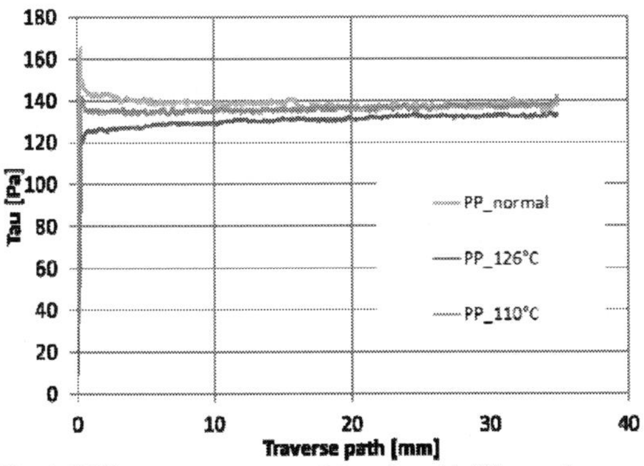

Fig. 8: Different temperatures of the tube with PP granules

The results show that the temperature has a significant influence on the wall shear stress in the pipe. Especially at the beginning of the measurement, a large difference of the measured results is to be determined depending on the temperature. The highest peak at the beginning of the movement of the tube was obtained for the material at the room temperature. With increasing the temperature, the effect of the sticking friction at the beginning of the movement is decreased. These results show very nicely that the wall friction depends on the surface properties of the particles. The surface of the particles becomes softer with increasing temperature, which leads to a reduction in the adhesion friction at the start of the measurement. Conversely, the stiffness of the plastic particles is responsible for the force at the start of the measurement. It is astonishing that, in the further course of the measurement, there is almost no difference in the sliding friction.

4 Conclusion

In bulk material technology, it is important to know the flow behavior of bulk solids very precisely. For many years, shear cell testers have been used, but they do not always give all the necessary information. In addition, it is often assumed that the friction between particles and particles and silo walls is a Coulomb friction. This is not always correct.

A new measuring system was developed at the Institute of Particle Process Engineering, University of Kaiserslautern, Germany, to provide more detailed information on the behavior of bulk solids in silos.
The measurements showed that a sharp increase in the shear stress can be observed with plastic granules at the beginning of the flow process. A slip-stick effect is also shown during the further course of the measurement. These measurements are a basis for further research, which will be supported by simulation methods in the future. It is planned that conclusions can be drawn about the actual processes in bulk material silos and thus the construction of bulk material silos can be made safer. This applies in particular to the dynamic behavior of bulk goods in silos and the transfer of vibrations to the silo construction.

In further experiments measurements were carried out at different temperatures with the new experimental apparatus. This showed a clear influence of the temperature on the flow behavior. These attempts to interpret the results of measurements at room temperature also clearly show that the influence of the temperature on the flow behavior of bulk materials is not always negligible.

References

1. D. Schulze, Powder and Bulk Solids, Behavior, Characterization, Storage and Flow (Springer Berlin Heidelberg New York, 2008)
2. M. Ripp, S. Ripperger, Chemical Engineering Science, **65**, 13 (2010)
3. I. Tomasetta, D. Barletta, M. Poletto, Advanced Powder Technology, **24**, 3 (2013)

Trends in the evolution of particle morphology with size in colluvial deposits overlying channel iron deposits

Sandra Linero[1,2*], *Stephen* Fityus[1], *John* Simmons[3], *Arcesio* Lizcano[4] and *Jessica* Cassidy[1]

[1] University of Newcastle, Faculty of Engineering and Build Environment, University Dr Callaghan NSW2308, Australia
[2] SRK Consulting (Australasia) Pty Ltd, 10 Richardson St WA6005, Australia
[3] Sherwood Geotechnical and Research Services, 5/8 Grebe St Peregian Beach QLD4573, Australia
[4] SRK Consulting (Canada) Inc, 1066 West Hastings St, BC V6E 3X2, Canada

Abstract. Size limitations of testing equipment often impliy that samples of coarse granular materials must be scalped or scaled, to reduce the size of the constitutive particles, before they can be tested either by triaxial or direct shear in the laboratory. The objective of the investigation is to evaluate the particle shapes in a natural sample of colluvial sediments, to identify potential correlation(s) between shape and size, that could impact shear strength of scaled samples. The material investigated is derived from eroded ancient sedimentary rocks from the Pilbara region of Australia. The fragments have a particle shape ranging from slabs to sub-equant blocks. The observation indicates that there is an increase in the tendency for slab-shapes in larger particles. Therefore, scaling inevitably alters the characteristic shapes of the material particles as it implies substituting larger (slabs) particles by smaller (sub-equant) particles. Changes in particle shape distribution may induce changes in material fabric and shear strength and therefore may need to be considered when scaling samples.

1 Introduction

The mechanical behaviour of granular materials is affected by the shape of the constituent particles. Experimental work suggests that the shear strength increases with particle angularity [1][2]. Micro-mechanical investigations using DEM suggest that the shear strength decreases with increasing shape polydispersity [3].

The objective of this investigation is to evaluate the particle shapes in a natural sample of colluvial sediments, to identify potential correlations between shape and size.

When a sample of granular material has very large particles, compared with the dimensions of the available shear testing device, manipulation of the particle size distribution (PSD) is required to obtain a maximum particle size (d_{max}) that suits the available equipment. Parallel scaling is a common technique to reduce the maximum particle size of a granular sample (Figure 1).

Any of the common techniques for material scaling (parallel scaling or scalping) change the original particle size distribution and, potentially, the particle shape distribution. This is significant, however, because changes in particle shape distribution may conduce to a change in material fabric and shear strength. Therefore, potential correlation between particle- shape and size in natural granular materials may need to be considered when scaling samples for laboratory testing.

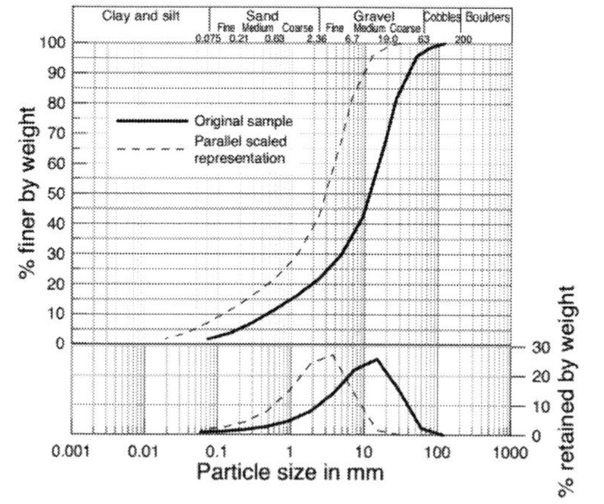

Fig. 1. PSD for the investigated sample (d_{max} 120 mm) and its scaled representation (d_{max} 30mm): Cumulative function (above), and variation of weight percentage as a continuum function of grain size (below)

2 The sample

The sample is an overburden material composed of eroded and transported particles (colluvium) derived from a sedimentary Precambrian-aged Banded Iron Formation (BIF), with province-wide distribution, located at the Pilbara region of Australia [4]. The BIF formations consist essentially of alternating beds and

* Corresponding author: sandra.lineromolina@uon.edu.au

laminae of hematite or magnetite and chert. The sample material occurs in what are known as channel iron deposits, one of the three major styles of iron ore mineralisation that are developed and commercially exploited in the Hamersley Province [5]. Locally, the colluvium derived from the BIF outcrops forms density-separated deposits in incised channels, with denser iron-rich fragments (the orebody) in the bottom of the channel, overlain by chert-rich fragments (overburden).

Thin layering of banding of less than 1 mm to few centimetres can be observed in saw cut and polished fragments (Figure 2). Parting occurs preferentially along laminations. Parting along fractures sub-perpendicular to the lamination can occasionally be recognized. The evaluation of the relationship between particle size and lamination composition was not part of this research.

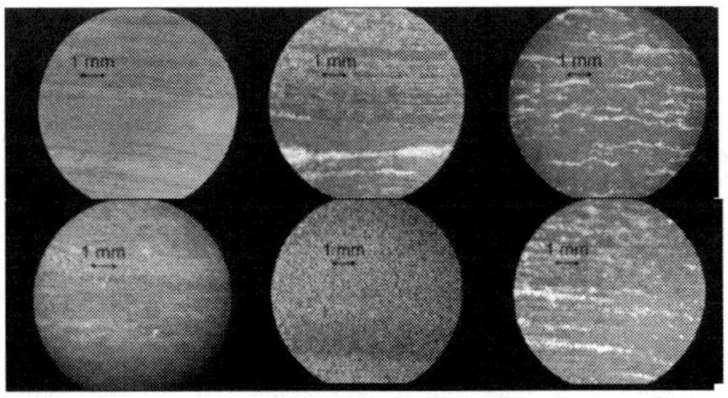

Fig. 2. 30X magnification of polished fragments (grey grains are hematite and magnetite; red and orange material is chert.)

A sample of 16.8 kg was obtained by dividing 1.4 tonnes of material using a rotary divider. The particle size distribution (PSD) was determined by dry sieving, using a standard set of meshes with opening sizes between 75 mm and 0.075 mm.

Twelve different size fractions (Table 1) covering 95% of the mass of the material were defined for analysis.

Table 1. Weight and number of particles retained by sieve and number of particles selected for shape determination.

No	Sieve Size (mm)	Retained mass (gr)	Retained mass (%)	No of particles Retained	No of particles Analysed
	75	231	1.37	1 (1)	
1	50	199	1.18	1 (1)	24(*)
2	26.5	2346	13.89	44 (1)	40
3	19.0	2388	14.14	128 (1)	117
4	13.2	2556	15.14	372 (1)	43
5	9.5	1763	10.44	671 (1)	161
6	6.5	1230	7.28	1,230 (2)	51
7	4.75	980	5.80	2,571(2)	70
8	2.36	1344	7.96	16,850 (3)	75
9	1.18	1020	6.04	51,298 (3)	20
10	0.6	789	4.67	531,388 (3)	20
11	0.3	731	4.33	2,334,654 (3)	20
12	0.15	617	3.66	16,621,585(3)	20
	0.075	387	2.29	-	-
	Resid.	305	1.81	-	-

(*) Population augmented to improve observations

The numbers in parentheses in the Table 1 indicate how the number of particles was determined: (1) counting, (2) extrapolating the number of particles counted in 200 g of material, (3) dividing retained weight by average particle weight. The average particle weight was calculated from averaged volume and density of particles. The average volume was determined based on the average dimensions of the particles using a correction factor. The latter was back-calculated from a sub-sample of counted particles assuming similarity of large and small sizes.

Roundness determination was conducted on ten particles of each size fraction.

3 Measurement protocols

The shape of the particles was evaluated based on "elongation" and "flatness", considering the standard measurement protocol proposed by Krumbein [6]. Those descriptors are defined as the elongation (breadth to length) ratio (I/L) and flatness (thickness to breadth) ratio (S/I), respectively, where L is the greatest distance apart of two parallel planes tangential to the particle surface, I is the maximum dimension perpendicular to L, and S is the maximum dimension perpendicular to L and I. L, I and S define an imaginary box that contains the particle, as shown in Figure 3. This information allows classification of the shape of the particles in 8 different categories (illustrated in Figure 5) [7].

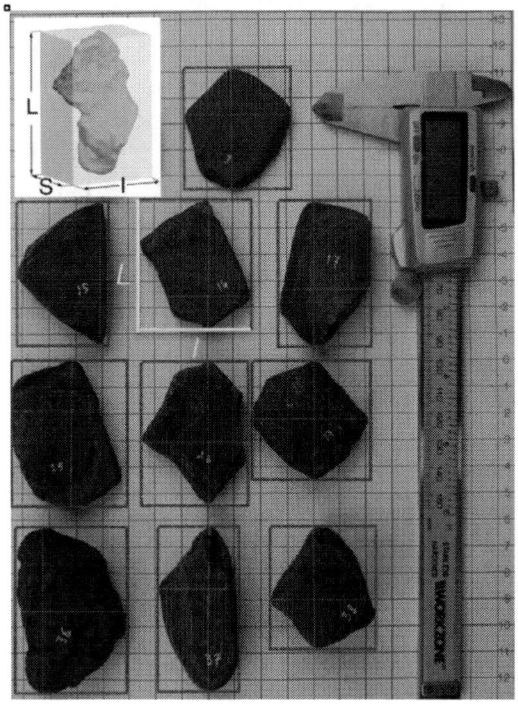

Fig. 3. Characteristic dimensions of the particles, L=Length, I=Breath and S=Thickness. Insert imaged modified from [8].

The Krumbein protocol is well suited for rather ellipsoidal particles. If the distances L, I, and S, are instead defined as those of the smallest rectangular box that contains the particle, then elongation and flatness ratio may vary especially for rather square parallelepiped pieces.

A calliper was used to determine dimensions of particles larger than 2.36 mm. Each was placed on graph paper exposing the greatest area projection. By orienting the length of the particle with one of the lines of the grid, length (L) and breadth (l) orthogonality was ensured.

Smaller particles were placed under a microscope and photographed. The length and breadth of those small particles were determined by analysing the microscope images. Their thickness was defined using the microscope's depth measuring device.

The roundness was determined using the roundness coefficient Rr by Wadell [9] which evaluates the relative sharpness of particle corners. A corner is that part of the outline of the projection area which has a radius of curvature equal to or less than the radius of curvature of the maximum inscribed circle of the same projection area (Figure 4). The roundness coefficient is defined as the average radii of the corners divided by the radii of the inscribed circle, on the largest projected area of the particle. The coefficient allows classification into four categories: angular $0<Rr<0.13$; sub-angular $0.13<Rr<0.25$; sub-rounded $0.25<Rr<0.5$ and rounded $0.5<Rr<1$.

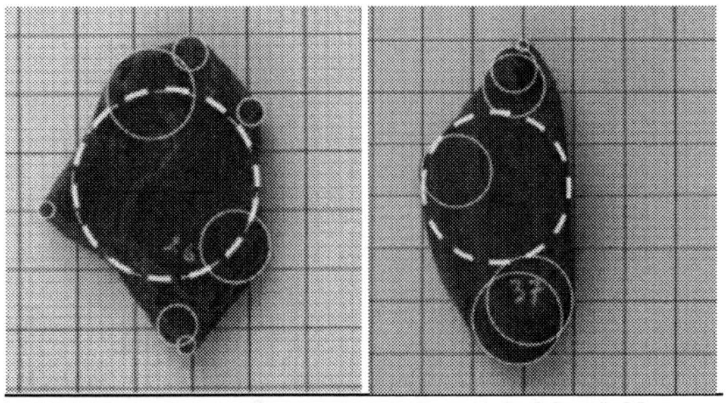

Sub-angul. Rr=0.17 Sub-rounded Rr=0.33

Fig. 4. Roundness determination showing maximum inscribed circle (thick dashed line) and corner circles (thin solid line).

4 Results

4.1 Particle shape

From the flatness vs. elongation data in Figure 5, most particles classify as slabs, and flat to sub equant blocks. The breadth to length ratios (elongation) are mostly greater than 0.5, with a mean of around 0.75, reflecting a tendency for the two largest dimensions to be relatively close in value. There is a much greater spread in flatness ratio compared with elongation ratio, with values relatively evenly distributed between 0.2 and 1.0, reflecting greater variability between the smallest dimension and the two larger dimensions. This outcome can be attributed to the sedimentary nature of the material, and the preferential weakness imparted by the laminations, resulting in a tendency for the minimum dimension being variable, and perpendicular to the laminations.

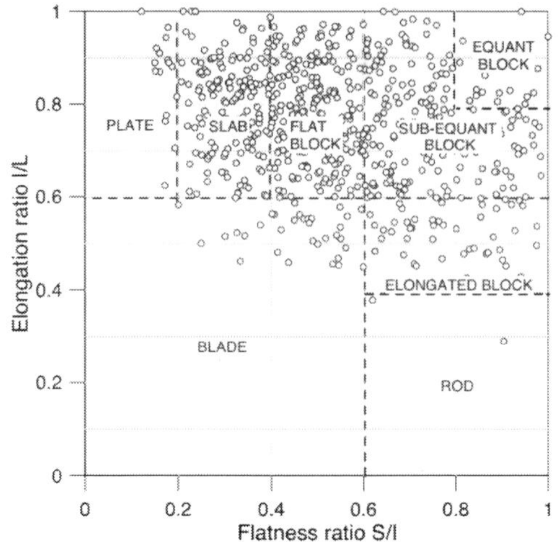

Fig. 5. Elongation vs Flatness for all the analysed particles.

The distribution of data by size fraction is illustrated in Figures 6, 7 and 8. The notched boxes are drawn to represent the interquartile range of the data. The horizontal line inside indicates the median value. The lower and upper quartiles are shown as horizontal lines either side of the boxes. "Whiskers" above and below each box show the locations of the 5 and 95 percentile. The proportion of material mass by size fraction is also illustrated in the figures for the analysed sample and for its scaled representation, for comparison.

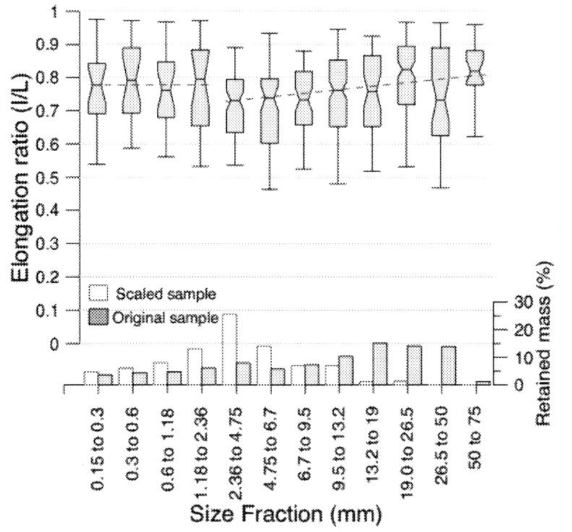

Fig. 6. Box plot showing the variability of elongation ratio, calculated from the form dimensions L and I for the different size fractions compared to the represented proportion of mass.

Trends of increasing elongation ratio and decreasing flatness ratio with size are observed for the particles between 2.36mm and 75mm (gravels and cobbles). In this range, larger particles tend to be less elongated and flatter than smaller particles (Figure 6 and 7). However, the elongation trend is not very significant, with median values varying less than 10% in the range. The variation in flatness ratio is more pronounced with up to 25% difference between larger and smaller pieces in the rage.

The elongation and flatness characteristics of the sand-sized particles (between 0.15mm and 2.3 mm) seems to be persistent across the investigated sand sizes, or, at least, a trend is not observed.

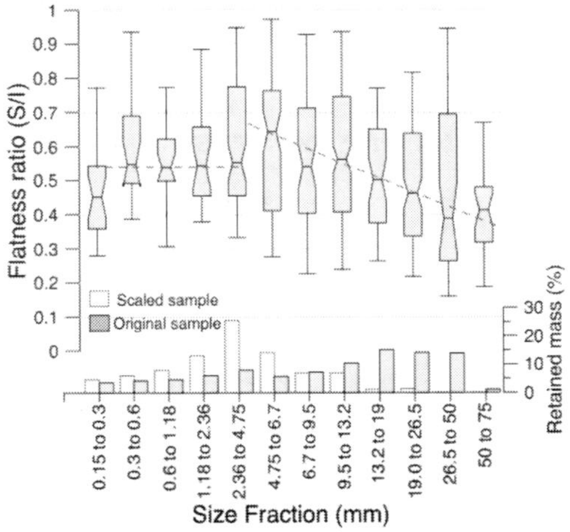

Fig. 7. Box plot showing the variability of flatness ratio, calculated from the form dimensions *S* and *I* for the different size fractions compared to the represented proportion of mass.

4.2 Particle roundness

Most of the particles classify as sub-rounded. A correlation with particle size or a particular trend cannot be identified from results.

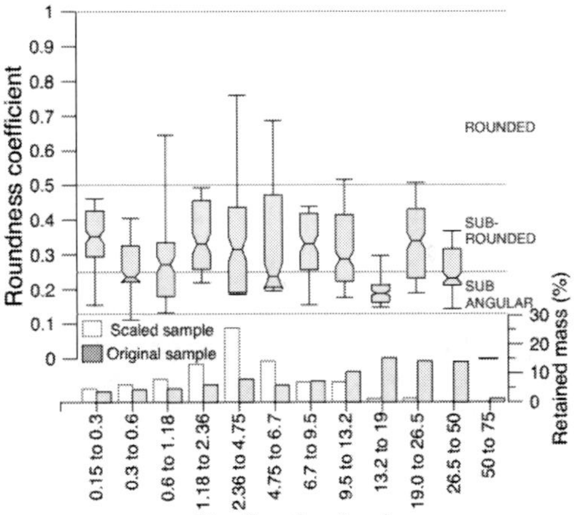

Fig. 8. Roundness data for the different size fractions compared to the represented proportion of mass.

5 Discussion

The elongation ratio of the particles composing the original and the scaled sample is not very different. Most of the material (by weight) both in the original and the scaled samples (sizes 9.5mm to 50 mm and 1.18 mm to 6.7 mm respectively) contains particles that have relatively similar elongation ratios. However, the differences in flatness ratio are more significant with up to 20% difference in median values when comparing the dominant size fractions, by weight, for the original samples and the scaled sample.

In any case, the scaled sample is less polydisperse in shape and its particles tend to be more regular compared with the original sample.

6 Conclusions

For the colluvial material studied, the trends in shape across particle size scales are small but potentially significant. Larger particles of the analysed material tend to be more platy or discoidal (less elongated and flatter) than smaller particles. Scaling inevitably alters the characteristic shapes of the material particles because it involves removing platy particles and replacing them with blockier pieces. The magnitude of the alteration depends on the scaling factor. The larger the degree of scaling, the larger the changes to particle shape induced by scaling.

The questions remain: how much impact do the changes in shape due to scaling have on shear strength, or alternatively, how much change in shape can be tolerated before it has a significant impact on measured strength. Resolving these questions may require both physical testing and DEM numerical modelling.

The fabric of the particles characterised by lamination may explain the variability in flatness ratio, as there is a preferential tendency for the particles to break along laminations, which are characteristically oriented perpendicular to their thickness (minimum dimension). Therefore, the origin of the material is important to the appreciation of shape effects related to scaling.

No trend is observed with size in particle roundness across size ranges, so, roundness is unlikely to add a shape effect when scaling, in the particular case of the studied colluvial material.

Thanks to Fortescue Metals Group Ltd for providing the material and funding this research along with SRK Consulting and the University of Newcastle.

References

1. T. Leps, J Soil Mech Found Div, **96** 1159-1170 (1970)
2. W. Holtz, H. Gibbs, J Soil Mech Found Div, **82** 1-22 (1956)
3. D. Nguyen, E. Azema, P, Sornay, F. Radjai, Phys. Rev. E, **91** 032203 (2015)
4. Z. Lewy, IJG, **3** 226-236 (2012)
5. C. Simpson, D. Storey, D. Kepert, R. Boyd, N. Nitschke, *Iron Ore Conference* (2011)
6. Krumbein, W.C. J. Sed. Petrol., **11**, 64–72 (1941)
7. S. Blott, and K. Pye, Sedimentology, **55**, 31–63 (2008).
8. Bagheri 2015, Powder Technology, **270** 141-153 (2015)
9. H. Wadell, The Journal of Geology, 40, 443–451 (1932)

EPJ Web of Conferences **140**, 13006 (2017) DOI: 10.1051/epjconf/201714013006
Powders & Grains 2017

Coupling in-situ X-ray micro- and nano-tomography and discrete element method for investigating high temperature sintering of metal and ceramic powders

Zilin Yan[1], *Christophe L.* Martin[1,*], *Didier* Bouvard[1,**] *David* Jauffrès[1], *Pierre* Lhuissier[1], *Luc* Salvo[1], *Luis* Olmos[2], *Julie* Villanova[3], and *Olivier* Guillon[4]

[1] *Univ. Grenoble Alpes, CNRS, SIMAP, 38000 Grenoble, France*
[2] *Universidad Michoacana de San Nicolás de Hidalgo, IIMM and INICIT, Fco. J. Mujica S/N, Ed. C-2 C.U., Morelia, Michoacán, C.P. 58060, Mexico*
[3] *ESRF The European Synchrotron, CS 40220, 38043 Grenoble Cedex 9, France*
[4] *Forschungszentrum Jülich Institute of Energy and Climate Research, D-52425 Jülich, Germany*

Abstract. The behaviour of various powder systems during high temperature sintering has been investigated by coupling X-ray microtomography and discrete element method (DEM). Both methods are particularly relevant to analyse particle interactions and porosity changes occurring during sintering. Two examples are presented. The first one deals with a copper powder including artificially created pores which sintering has been observed in situ at the European synchrotron and simulated by DEM. 3D images with a resolution of 1.5 μm have been taken at various times of the sintering cycle. The comparison of the real displacement of particle centers with the displacement derived from the mean field assumption demonstrates significant particle rearrangement in some regions of the sample. Although DEM simulation showed less rearrangement, it has been able to accurately predict the densification kinetics. The second example concerns multilayer ceramic capacitors (MLCCs) composed of hundreds of alternated metal electrode and ceramic dielectric layers. The observation of Ni-based MLCCs by synchrotron nanotomography at Argon National Laboratory with a spatial resolution between 10 and 50 nm allowed understanding the origin of heterogeneities formed in Ni layers during sintering. DEM simulations confirmed this analysis and provided clues for reducing these defects.

1 Introduction

X-ray microtomography is a non-destructive method used to observe the microstructure of materials in 3D [1]. This technique is particularly powerful for particulate materials where the pore phase is transparent to X-rays. Although technologically challenging, in situ observation of a material during a high temperature thermal treatment, as for example powder sintering, is particularly attractive. With third generation synchrotrons, typical resolution is 1 μm and typical scanning time is a few tens of seconds, which is convenient for analysing the dynamics of the sintering of metal powders. More recently, novel nano-tomography configurations allow the antagonism between high resolution (sub-micronic) and rapid scan times (tens of seconds) to be resolved. This opens new avenues for the observation of micronic ceramic particles during sintering. Discrete Element Methods (DEM) are relevant for modelling powder sintering since they explicitly take into account the particulate nature of the material. DEM models for sintering should include the specific contact forces that mimic interparticle bonding at high temperature. The writing of these contact force includes surface energy and diffusion coefficients. The overall problem must be treated as the minimization of the total surface energy of the packing. A difficulty of DEM modelling is the availability of appropriate data for both feeding and validating simulations. In situ X-ray tomography is clearly appropriate for this purpose as it can provide quantitative data on particle assembly dynamics during sintering. Here, we report on two examples of coupling synchrotron tomography experiments and discrete element simulations.

*e-mail: christophe.martin@grenoble-inp.fr
**e-mail: didier.bouvard@grenoble-inp.fr

2 DEM model

The details of the DEM model may be found in [2]. The main ingredients, specific to the high temperature sintering problem are recalled here. Particles are modeled as spheres that may indent each other due to surface energy minimization. Parhami and McMeeking' model, which assumes that grain boundary and surface diffusions are the major mechanisms of mass transport, is used [3]. The normal contact force between two sintering particles writes:

$$N_s = \frac{\pi a_s^4}{8\Delta_b}\frac{\mathrm{d}\delta}{\mathrm{d}t} - \pi\gamma_s\left[8R^*\left(1-\cos\frac{\psi}{2}\right) + a_s\sin\frac{\psi}{2}\right] \quad (1)$$

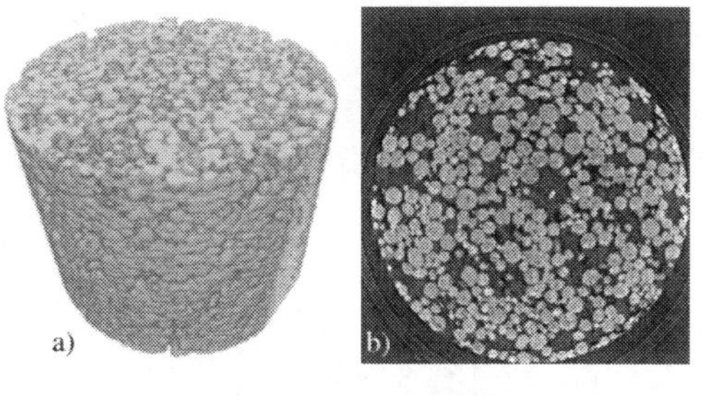

Figure 1. 3D rendering (a) and virtual slice (b) of the image of a 1 mm diameter copper sample with artificially created pores.

where a_s is the contact radius, δ is the indentation, Δ_b is a temperature dependent diffusion coefficient, γ_s is the surface energy, R^* is the effective radius of the two particles and ψ is the dihedral angle that defines the equilibrium shape of the contact. The first term on the rhs of eqn (1) is the normal viscous term while the second term, which is tensile, is the sintering term. A tangential force, opposes the tangential relative velocity at the contact $\left(\frac{du}{du}\right)$:

$$T_s = -\eta \frac{\pi a_s^2 R^{*2}}{2\Delta_b} \frac{du}{dt} \quad (2)$$

where η is a is a viscous parameter with no dimension. Because of material transport, the contact radius a_s writes:

$$a_s = 4R^*\delta \quad (3)$$

which differs by a factor 4 from its equivalent in elasticity.

No external load is applied on the particle packing and particles motion is only dictated by surface minimization (γ_s term in eqn (1)).

3 Copper based powders

Experiments were carried out at the European Synchrotron in Grenoble with copper based powder systems sintered up to 1060°C in a furnace located between the X-ray source and the detector. 3D images with a resolution of 1.5 μm and a scanning time of 1 min were recorded at various stages of the thermal cycle [4]. Among the various investigated systems, we focus on the one with artificially created pores (so-called porous), which mimics materials with large, controlled porosity as filters, SOFC electrodes, bone implants [5]. Copper powder with 0-63 μm size was first mixed with 10 vol.% of NaCl salt particles with a diameter between 40 and 63 μm as pore formers. This mixture was then poured in a 1 mm diameter quartz capillary and pre-sintered at 500°C during 30 minutes. The capillary was removed after cooling and the salt was dissolved by keeping the sample in distilled water for 5 hours. Fig. 1 shows a 3D rendering and a virtual slice of an image of the sample. Sintering was achieved at 1050°C under reducing atmosphere (He - 5 at.% H2) to prevent particle oxidation.

The raw 3D images were converted to binary images by thresholding and particle segmentation was achieved by a watershed procedure. Fig. 2a compares the displacement of each particle center projected in a horizontal plane with the one predicted by the mean field hypothesis which assumes uniform isotropic shrinkage. The real displacement roughly follows the mean field assumption with significant deviations in a few regions that evidence particle rearrangement.

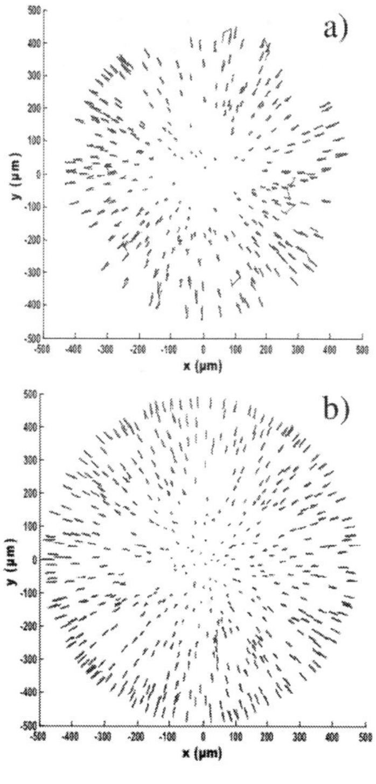

Figure 2. Particle centre trajectories during sintering (in blue) compared with estimated mean field trajectories (in red). a) from microtomography images, b) from DEM simulation.

In parallel, DEM simulations of the sintering have been performed. Building a realistic DEM packing directly from a 3D image was not satisfactory because of the non-regular particle shape. A customary fabrication method, involving the real particle size distribution of particles and pore formers as well as the initial density of the material to be modeled, which have been deduced from microtomography data, was thus worked out [6]. As shown in Fig 2, the results of porous sample simulation are more consistent with the mean field assumption than the microtomography results. Fig. 3 shows experimental and numerical densification kinetics of both a regular copper sample and the porous sample. The diffusion coefficient has been fitted to get the best agreement between experimental and simulation curves for the regular material. Then, with the same coefficient the sintering of the porous sample was also nicely predicted, which suggests that local rearrangement does not play a significant role in the densification process.

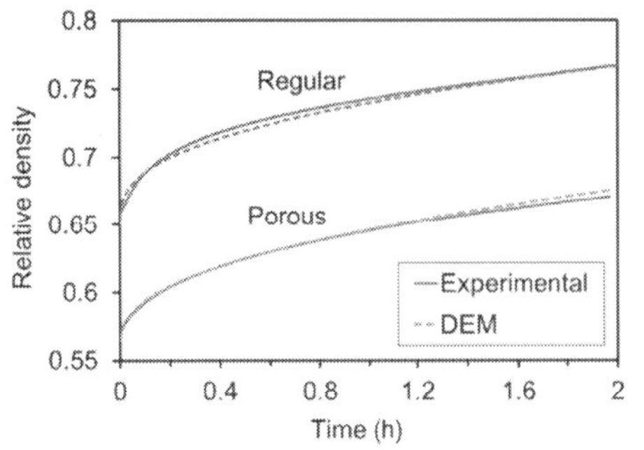

Figure 3. Comparison of experimental and numerical densification kinetics of porous copper sample during sintering.

4 Multilayer ceramic capacitors

Multilayer ceramic capacitors (MLCCs) are composed of hundreds of alternated metal electrode and ceramic dielectric layers. Several trillions of them are fabricated each year for electronic parts. Up-to-date MLLCs comprise typically 1 μm thick layers made of nano-sized powders. Using the synchrotron radiation provided at Argon National Laboratory, non-sintered (green) and sintered MLCCs composed of Ni electrode and BaTiO3 (BT) dielectric layers have been observed with a spatial resolution between 10 and 50 nm [8]. The green parts were first machined into 60 μm cone shape with a special drilling machine (Fig. 4a). The tip was then milled by dual-beam FIB-SEM into a 20 μm-diameter cylinder of 20 μm height (Fig. 4b). Fig. 4c-f show ex-situ 2D projections and 3D images before and after sintering. The bright areas correspond to the Ni phase while dark ones correspond to the BT phase. Reverse colour code is used in Fig. 5, which shows the in-situ evolution of an MLCC during sintering. Temperature calibrating was carried out by using nano-gold particles, which melting was monitored in-situ.

Both post-mortem and in-situ observations indicate that the particulate nature of the layers dictates the microstructural evolution [7]. This is particularly true of the Ni layer, which only contains a few submicronic particles along its thickness. The discontinuities observed in the Ni layers were found to originate from initial heterogeneities of nickel powders (Fig. 6). The heterogeneity grows into a full defect because of the constraint imposed by the two layers that sandwich the Ni layer. These ceramic layers do not sinter since their sintering temperature is higher. DEM simulations were carried out to refine this scenario.

Particle size and size distribution measured by tomography were used as inputs for simulations. We have observed that the free sintering of a Ni monolayer leads to a final homogeneous microstructure. When the metal Ni layer is sandwiched between two non-sintering BT layers (Fig. 7), discontinuities appear, thus demonstrating that a constraint is necessary for their nucleation.

Fig. 7 indicates that particles that form clusters in the final microstructure were initially closely packed. As sintering proceeds, local sintering shrinkage mismatch in-

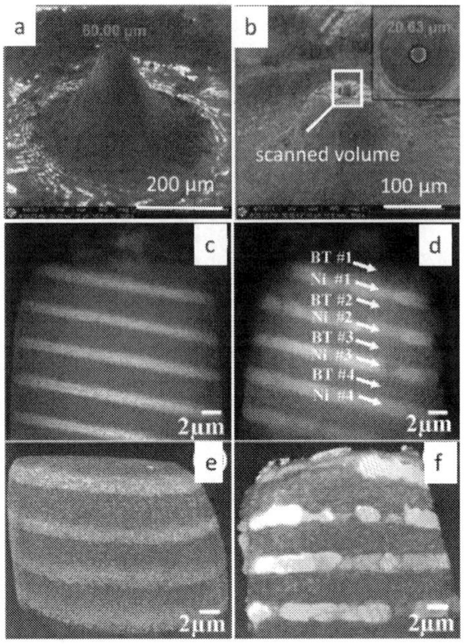

Figure 4. (a-b) MLCC sample preparation for x-ray tomography; 2D projection images (c) before and (d) after sintering, (d), reconstructed 3D microstructure (e) before and (f) after sintering.

Figure 5. In-situ observation of MLCC sintering at increasing temperature. Au particles are used both for image alignment and for calibrating the furnace temperature.

duces some initially contacting particles to detach from each other. In particular, inspection of eqn (1) shows that small particles sinter faster than large ones. Indeed, we have observed that contacts between large particles were more prone to 'de-sintering' than contacts between small ones. Regions with initially low coordination number and with some large particles at their boundaries detach, re-

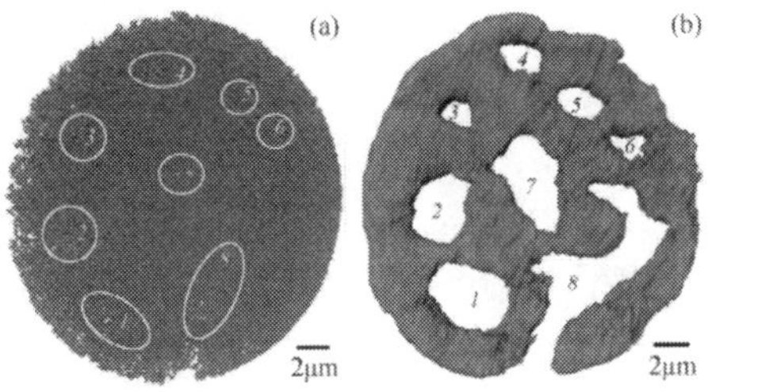

Figure 6. Nickel layer evolution with sintering. Low packing density regions lead to large discontinuities after sintering.

sulting in pore growth. Conversely, regions with high coordination number sinter rapidly leading locally to pore shrinkage. Although particle coarsening is not included in the model (the mechanism that makes large particles grow by eating away smaller ones), it provides a sound scenario for explaining the evolution of the real microstructure pictured in Fig. 6a to the one in Fig. 6b.

The DEM simulations presented here may also be used to propose potential improvements by varying some of the process parameters [8]. We observed that increasing the Ni layer thickness had a beneficial effect on the discontinuity level because it decreases the constraint that non-sintering layers. However, this is not a option for the MLCC industry, which wants to pack more dielectric layers per unit volume, not less. A more viable route is to decrease the sintering driving force for Ni particles until both the ceramic (BT) and the metal (Ni) layers are at a sufficiently high temperature to sinter at similar rates. In that case, the Ni layer has formed only a small amount of discontinuities when the constraint imposed by the non-sintering BT layer is relieved. In that regard, increasing the heating rate to reach the sintering temperature (1150°C) should be effective since it reduces the constrained sintering time of the Ni layer. DEM simulations indeed show that increasing the heating rate from 5 K/min to 50 K/min decreased discontinuity from 6 to 1 %. Another option is to use BT as additives in the Ni layer. This has the desired effect of retarding the sintering of the metal layer in the temperature range where the BT layer does not sinter. The addition of 10 vol. % of BTO nano-particles in the Ni layer decreases the amount of discontinuity from 6 to 4 % at 5 K/min.

5 Conclusion

Both study cases show the interest of coupling microtomography and discrete element modelling to better understand and describe the sintering behaviour of metal and ceramic powders. The occurrence of particle arrangement during sintering copper material with artificially created pores has been assessed. DEM simulation proved that this rearrangement does not play a key role in densification. Concerning multilayer ceramic capacitors, nanotomography observation allowed understanding the origin of defects created during sintering. DEM simulation confirmed this analysis and provided clues for reducing this damage. The versatility of the experimental configuration described here is of great interest for investigations of powders where visualization of thermally activated phenomena is necessary at a fine scale. Current progresses to improve further the acquisition parameters toward even smaller resolution (< 100 nm) together with smaller scan times (< 1 s) will certainly qualify in-situ tomography as a method of choice for small powders at room and high temperature.

References

[1] L. Salvo, P. Cloetens, E. Maire, S. Zabler, J.J. Blandin, J.Y. Buffière, W. Ludwig, E. Boller, D. Bellet, C. Josserond, Nuclear Instruments and Methods in Physics Research, Section B: Beam Interactions with Materials and Atoms **200**, 273 (2003)

[2] Z. Yan, C.L. Martin, O. Guillon, D. Bouvard, Scripta Mater. **69**, 327 (2013)

[3] F. Parhami, R.M. McMeeking, Mech. Mater. **27**, 111 (1998)

[4] A. Vagnon, J.P. Riviere, J.M. Missiaen, D. Bellet, M.D. Michiel, C. Josserond, D. Bouvard, Acta Mater. **56**, 1084 (2008)

[5] L. Olmos, T. Takahashi, D. Bouvard, C.L. Martin, L. Salvo, D. Bellet, M.D. Michiel, Philosophical Magazine **89**, 2949 (2009)

[6] L. Olmos, C.L. Martin, D. Bouvard, D. Bellet, M.D. Michielz, J. Am. Ceram. Soc. **92**, 1492 (2009)

[7] Z. Yan, O. Guillon, C.L. Martin, S. Wang, C.S. Lee, D. Bouvard, Appl. Phys. Lett. **102**, 223107 (2013)

[8] Z. Yan, C.L. Martin, O. Guillon, D. Bouvard, C. Lee, Journal of the European Ceramic Society **34**, 3167 (2014)

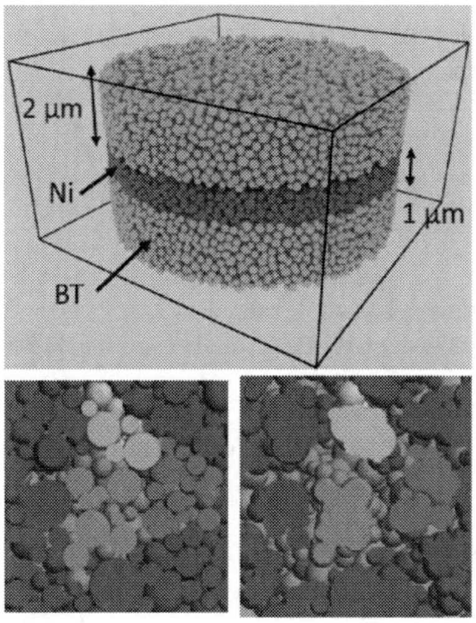

Figure 7. Nickel layer evolution with sintering. Low packing density regions lead to large discontinuities after sintering.

Influence of particle deformation on the plastic flow of ductile granular materials

Nouha Abdelmoula[1,2], *Barthélémy* Harthong[1,2,⋆], *Didier* Imbault[1,2], and *Pierre* Dorémus[1,2]

[1] *Univ. Grenoble Alpes, 3SR, F-38000 Grenoble, France*
[2] *CNRS, 3SR, F-38000 Grenoble, France*

Abstract. A multi-particle finite-element method was proposed to study the elastic-plastic behaviour of ductile powders composed of highly deformable elastic-plastic particles. The focus was put on the study of the uniqueness of the direction of plastic strain increment vectors for a given stress state on the plastic limit, which was assessed using a spherical stress-probing method. Results revealed a non-uniqueness of the direction of plastic flow in a small region of the stress space located in the vicinity of the loading point. The direction of plastic flow was almost unique elsewhere on the plastic limit. The non-uniqueness was explained using a combination of two distinct mechanisms for plastic deformation involving two very different plastic limits.

1 Introduction

This paper is concerned with the mechanical behaviour of granular assemblies involving highly deformable elastic-plastic particles. Such assemblies will be referred to as 'ductile powders'. The main targeted application is powder compaction. There is a need for continuum-mechanics-based constitutive models which would be able to predict the evolution of stress and strain with a reasonable accuracy in localised high-shear zones where cracking is susceptible to happen.

To this end, a numerical model involving an assembly of meshed spheres interacting through contact conditions is studied in the framework of the finite-element method; this approach will be named Multi-Particle Finite-Element Method (MPFEM) in the following. This method allows for the simulation of many loading paths that are not attainable to experimental studies due to technical difficulties. By studying the response of the numerical sample (which is considered as a model material for ductile powders) to such loading paths, insight is given on the mechanical behaviour of such materials.

Previous studies [1–3] showed the ability of the MPFEM in deriving yield surfaces and describing strain-hardening mechanisms and their relation to the evolution of the microstructure. The focus of the present work is the study of plastic flow.

2 Method

2.1 Numerical model

In this study, a polydisperse assembly of 50 randomly-distributed spherical particles was built into the commercial finite-element software ABAQUS using an explicit time integration scheme. Let the dimensions of the box be taken as 1x1x1 mm, then the stresses presented in the following figures are expressed in MPa. Particles were meshed using approximately 3300 quadratic tetrahedral elements per particle, which allowed to keep a reasonable accuracy in the description of contact surfaces. The effect of the number of particles was studied and results showed that yield surfaces were decreasing in size when increasing the number of particles but were keeping the same shape and orientation. The influence of polydispersity was not studied. It is important to understand that the use of only 50 spheres in the numerical model implies that the following results do not mean to be quantitative, general results for assemblies of ductile particles. The purpose of the present paper is solely to observe the behaviour of the 50-spheres sample and provide explanations of this behaviour. Only the understanding of the observed behaviour can be transferred to real materials.

Individual particles were assigned a constitutive model associating linear, isotropic elasticity (with elastic modulus E = 10 GPa and Poisson coefficient ν = 0.435) and Mises plasticity with an isotropic power-law strain-hardening:

$$\sigma_Y = k\left(\varepsilon^{pl}\right)^n + \sigma_0 \tag{1}$$

where $k = 15.5$ MPa and $n = 0.35$ are material parameters, and σ_0, σ_Y and ε^{pl} are, respectively, the initial and current (Mises) yield stress and the equivalent plastic strain. These parameters were calibrated on a lead alloy corresponding to the experimental tests performed by Chen [4]. Particle/particle and particle/plane contact interactions were modelled by means of a penalty contact algorithm and a classical Coulomb friction law with a friction coefficient $f = 0.1$. Chen [4] observed that the forces on the planes

⋆e-mail: barthelemy.harthong@3sr-grenoble.fr

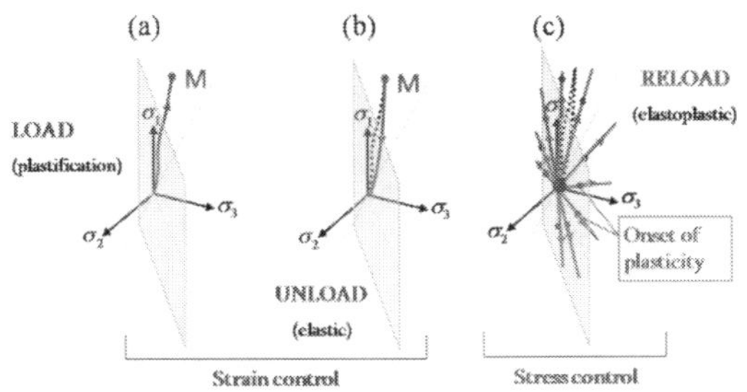

Figure 1. Method for probing yield surfaces: (a) First loading, (b) unloading, (c) reloading in various directions.

increased by approximately 10% when raising the particle/plane friction coefficient from 0 to 0.1 at a relative density $D = 0.97$ (much less than 10% for lower values of D); and another 10% when raising it from 0.1 to 0.25. No contact cohesion was introduced.

The particles were enclosed in a cubic box made of six rigid planes which were allowed to move along their normal, so that initially parallel planes remained parallel and initially perpendicular planes remained perpendicular throughout the loading. As a result, distortional strains were all kept to zero. Using such boundary conditions, it was checked that tangent force components on the rigid planes were negligible compared to normal components, meaning that no rotation of either stress or strain tensors was applied. As a result, the numerical sample being initially isotropic, stress and strain principal directions were assumed to remain coaxial throughout the simulation, so that the present analysis is restricted to the 3D space of principal stresses or strains, in the particular case of coaxiality of the principal axes of stress and strain.

The numerical assembly was compacted using either isotropic straining (referred to as isotropic compaction) and uniaxial straining along x direction (referred to as closed-die compaction). Boundary conditions were either stress-driven (the total normal force on each plane was controlled to reach desired stress values) or strain-driven (normal displacements were imposed to reach desired strains).

2.2 Probing of yield surfaces

The numerical sample was loaded following the isotropic or closed-die loading path and unloaded. From the unloaded state, a series of reloading steps were applied in every direction of the stress space to probe yield surfaces (Fig. 1).

Following [2], the detection of yield points during the reloading steps was based on the dissipated energy W, equal to the sum of plastic and frictional dissipation within the numerical model. The onset of plasticity was defined as the point where the total dissipation in the assembly reached a threshold value equal to $W_0 + 0.3\% W_0$, W_0 being the value of the total dissipation at the end of the unloading step.

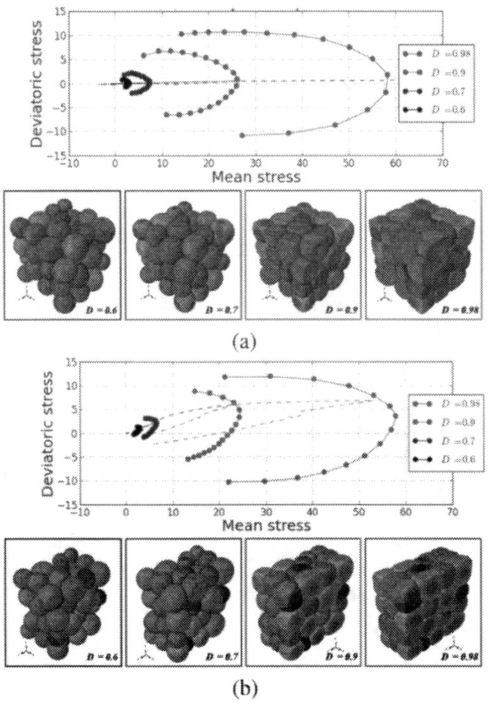

Figure 2. Evolution of yield surfaces from initial to various final relative density values, $D = 0.6, 0.7, 0.9$ and 0.98, with a monotonic loading: (a) isotropic compaction and (b) closed die compaction. The dotted line is the initial loading/unloading path. The snapshots represent the deformed sample corresponding to each yield surface (the colour plot is the displacement magnitude).

Fig. 2 shows the yield surfaces in a deviatoric stress/mean stress diagram with various values of final relative density D for the 50-particles numerical sample which was submitted to isotropic and closed die compaction.

2.3 Spherical stress probing method

In classical elasto-plasticity, the incremental plastic strain obeys the flow rule equation:

$$d\boldsymbol{\varepsilon}^{pl} = \lambda \frac{\partial g(\boldsymbol{\sigma}, \chi)}{\partial \boldsymbol{\sigma}} \qquad (2)$$

where $d\boldsymbol{\varepsilon}^{pl}$ is the increment of plastic strain, $\boldsymbol{\sigma}$ the stress tensor, χ the strain-hardening variables, g the plastic potential and λ the plastic multiplier. Assuming g is regular, equation (2) states that for a given state of stress and strain-hardening, the direction of the incremental plastic strain $d\boldsymbol{\varepsilon}^{pl}$ is unique.

In the present work, the existence of a plastic potential, and more generally, the incremental plastic flow could be studied on the numerical sample by using a spherical stress probing method [5]. This method consisted firstly in loading the numerical sample to a given state of stress $\boldsymbol{\sigma}_0$ on the yield surface (i.e. at the elastic limit). Secondly, small stress increments $d\sigma$ are applied in every direction of the stress space (Fig. 3a). Thirdly, the sample is unloaded back to σ_0.

Assuming that no plastic dissipation occurs during unloading, the incremental plastic strain $d\varepsilon^{pl}$ was assimilated to the difference in total strain after loading and unloading.

The plastic strain response envelope is the curve formed by the tips of the incremental plastic strain vectors corresponding to the response of the sample to the probing stress increments $d\sigma$ (Fig. 3b). In the present paper, only stress increments in the Rendulic plane were applied.

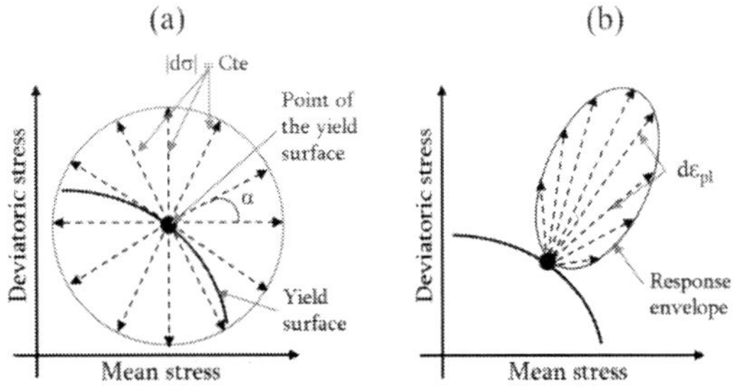

Figure 3. Spherical stress probing method: (a) stress probes and (b) plastic strain response envelopes.

Referring back to equation (2), if the flow rule applies and if a regular potential function exists, the direction of the plastic strain increment $d\varepsilon^{pl}$ does not depend on the direction of the stress increment $d\sigma$. In such a case, the plastic strain response envelope is a line segment, since the direction of $d\varepsilon^{pl}$ is unique and entirely defined by σ and χ. If the response envelope of the plastic strain is not a straight line, then $d\varepsilon^{pl}$ depends on the direction of $d\sigma$, which means that the flow rule equation (2) is not valid.

3 Results and discussion

To improve readability of the results, plastic strain response envelopes were plotted in deviatoric strain/volumetric strain graphs, then normalised and superimposed to deviatoric stress/mean stress graphs, following the common representation.

Fig. 4 shows normalised plastic strain response envelopes for isotropic and closed die compaction of the sample for a constant stress increment of magnitude $d\sigma$ in the case of a relative density $D = 0.7$. The value of $d\sigma$ was chosen as small as possible, but large enough so that the incremental plastic strain response was significantly larger than numerical noise. Fig. 4 show that most plastic strain increment vectors are relatively close to line segments, and are approximately normal to the yield surface, suggesting the validity of an associated flow rule for such materials.

However, in the vicinity of the tip of the yield surfaces, the direction of plastic strain increments was far from being unique. Close to the tips of closed-die-compaction yield surfaces, the plastic strain response envelopes were oriented toward a direction which significantly differed

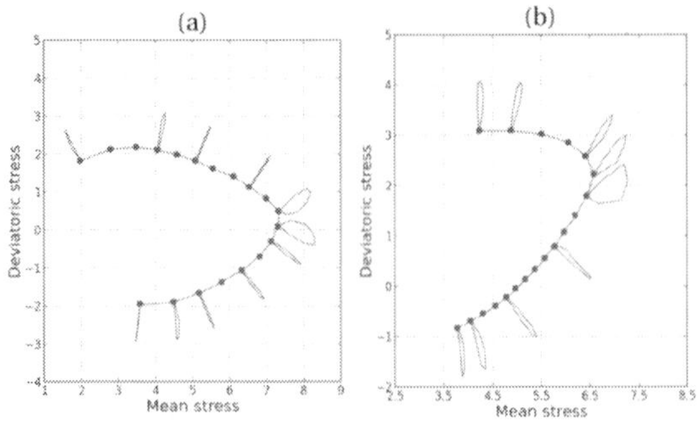

Figure 4. Normalised plastic strain response envelopes for some points of the yield surface with final relative density $D = 0.7$ for (a) isotropic and (b) closed die compaction.

from the normal to the yield surface. The non-uniqueness of the direction of plastic flow was confirmed by results obtained for other relative densities, ($D = 0.6$ and $D = 0.9$) which are not shown here to keep the paper brief.

The results indicated a correlation between the non-uniqueness of the direction of plastic strain increments and the particular stress state corresponding to the loading point. As pointed out in [6] and [3], the stress state corresponding to the loading point is actually the stress state for which all contact surfaces between particles are approximately normal to contact forces, because it is the stress state which created the contact surfaces (at least for monotonous loadings such as isotropic or closed-die compaction). In other words, the loading point is the state of stress for which shear stresses at contacts are minimum in average.

As a consequence, sliding between particles was less likely to happen for the stress states close to the loading point than for other stress states which mobilise more shear stresses at contact surfaces. This resulted in the fact that the ratio of frictional dissipation over plastic dissipation was minimum at the loading point, as can be seen in Fig. 5, and indicating that for such a stress state, particles were blocked. As a result, the macroscopic plastic strain was mainly, if not solely due to plastic deformation of particles.

Furthermore, the contact surfaces which increased in size during densification opposed an increasing resistance to normal stresses. This phenomenon results in the fact that the sample develops an increased resistance to the stress state of the loading point compared to other stress states. In the present case, this phenomenon resulted in plastic strain increment vectors being very small in magnitude (which was observed even though not shown here).

A plausible explanation for the non-uniqueness of the direction of plastic flow can be proposed on the basis of a simplified analogy. The analogy consists in a block sliding on a plane under the effect of a force F (Fig. 6). Let F increase in magnitude until the block reaches the limit of sliding (the analogous of the yield point). At this stage, let a small force dF be added in an arbitrary direction (the analogous of the probing stress increment). For some di-

DOI: 10.1051/epjconf/201714002004

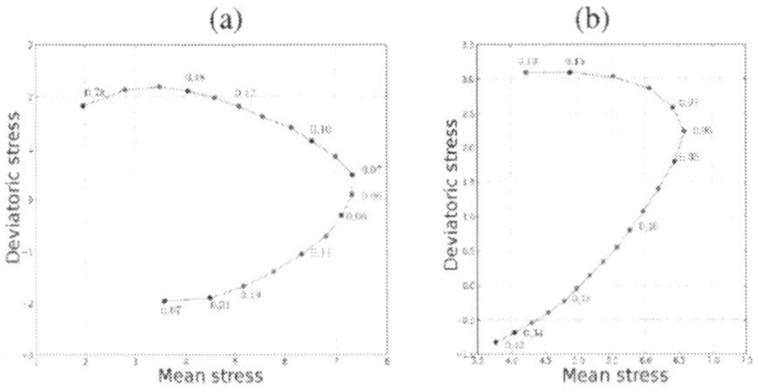

Figure 5. Evolution of the frictional dissipation / plastic dissipation ratio with the stress state, in the case of a relative density $D = 0.7$: (a) isotropic compaction, (b) closed die compaction. Green dotted curves are the yield surfaces.

rections of dF, the block will pass the sliding limit and move on the plane, resulting in a small displacement du (the analogous of the plastic strain increment vector).

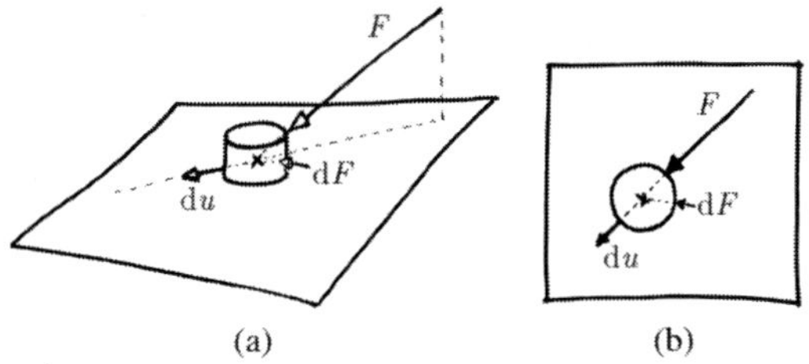

Figure 6. Sliding block analogy: (a) general view, (b) top view.

In such an analogy, the block represents an average particle, and the plane is the contact structure which defines the possible directions in space along which the particle is allowed to move. The force F represents the stress state. The motion of the block will be the result of 1– the direction of the total force F + dF and 2– the orientation of the plane. dF being very small, the approximation that $F + \mathrm{d}F = F$ is valid and it follows that du is uniquely defined by F. In other words, the block moves in a direction defined by the projection of F on the plane (Fig. 6a).

However, the previous reasoning is no more valid when the force F is normal to the plane. In such a case, if dF remains extremely small, it is not possible to reach the sliding limit. However it is possible to imagine that if the friction coefficient is low enough, dF can be large enough to make the block pass the sliding limit while still very small compared to F. In such a case, the direction of slip, du, is purely determined by the direction of dF since the projection of F on the plane is zero. This result implies that the direction of slip is non-unique: it depends on dF and not solely on F.

To improve the analogy, it is possible to introduce a deformable plane such that a very small component of displacement is allowed in the direction of the normal to the plane, but this component is so small that in the case where F is normal to the plane, it remains comparable to the component due to dF alongside the plane.

The sliding block analogy suggests that the requirement to observe the non-uniqueness of the direction of plastic flow is mainly to have two deformation mechanisms with very different plastic limits (in the analogy, the movement against the plane and alongside the plane). In the case of the 50-particle numerical sample studied here, the two mechanisms at stake were thought to be the shearing and normal loading at contact surfaces. As mentioned above, the direction of the stress space corresponding to the loading path is the one for which the sample shows the best resistance. Let this direction be called a "strong" direction, meaning a direction in which contact surfaces are the largest in average, while other stress directions are "weak". When an average particle is loaded in a "strong" direction, the corresponding deformation or displacement of the particle is very small, whereas it can be significantly deformed by a relatively small stress when loaded in a "weak" direction. The "strong" and "weak" directions are equivalent to the directions normal and parallel to the supporting plane in the analogy. In such a case, the direction of plastic strain is influenced by the direction of the stress increment, hence the non-unique behaviour.

4 Conclusion

A numerical assembly of 50 deformable particles with elastic-plastic behaviour was studied with the aim of characterising incremental plastic flow. Results show that the direction of plastic flow exhibited a well-marked non-uniqueness for some stress states. The non-uniqueness was correlated with the fact that particles displacements were restrained and stresses at contacts were mainly normal. These results suggested the existence of to distinct mechanisms for macroscopic plastic strains in such materials, so that a very high stress in a particular direction could lead to a very small plastic strain while very small stresses in other directions could lead to plastic strains of similar magnitude, leading to the non-uniqueness of plastic flow direction.

References

[1] A.T. Procopio, A. Zavaliangos, Journal of the Mechanics and Physics of Solids **53**, 1523 (2005)

[2] I. Schmidt, A. Trondl, T. Kraft, A. Wonisch, *Simulation of the material behavior of metal powder during compaction*, in *Proceedings of the Institution of Mechanical Engineers, Part E: Journal Process Mechanical Engineering* (Institution of Mechanical Engineers, 2010), Vol. 224, pp. 187–194

[3] B. Harthong, D. Imbault, P. Dorémus, Journal of the Mechanics and physics of Solids **60**, 784 (2012)

[4] Y. Chen, Ph.D. thesis, Institut Polytechnique de Grenoble (2008)

[5] F. Calvetti, G. Viggiani, C. Tamagnini, Rivista Italiana di Geotecnica **34**, 11 (2003)

[6] N.A. Fleck, L.T. Kuhn, R.M. McMeeking, Journal of the Mechanics and Physics of Solids **40**, 1139 (1992)

What is fluidity? Designing an experimental system to probe stress and velocity fluctuations in flowing suspensions

Marcel Workamp[1], *Sepideh* Alaie[1], and *Joshua A.* Dijksman[1]*

[1] *Physical Chemistry and Soft Matter, Wageningen University & Research, The Netherlands*

Abstract. We develop a method to investigate the microscopic origin of granular fluidity. We design a Couette cell in which we can probe the flow of soft hydrogel suspensions. As we drive the suspension with a rheometer, we have access to global flow characteristics. In addition, the Couette cell has been modified to have a transparent bottom and lid, allowing for imaging of suspension characteristics in transmission, for example flow fields. We can also use transmission imaging to probe local stresses in the suspension: we use hydrogel particles composed of gelatin, which through its photoelastic properties gives access to local stress fluctuations. We thus have access to all local microscopic variables that are relevant in the understanding of granular suspensions. We show here that our setup can indeed visualize stress fields inside the suspensions and perform flow field measurements in transmission mode. We compare the profiles of the gelatin suspension to that of a polyacrylamide hydrogel suspension, to assess robustness of observed phenomenology. We find that the flow profiles for both types of hydrogels are different; gelatin suspensions feature narrower and less rate-dependent flow profiles. We speculate on the origin of the observed difference by considering the frictional properties of the suspended particles.

1 Introduction

The flow behavior of dry granular materials and dense suspensions can generally be viewed as consisting of two regimes. In the slow flow limit or quasi-static regime, grain-to-grain interactions are frictional and the shear stress is rate-independent. At higher driving rates, the material becomes more fluid-like: inertia or the viscosity of the interstitial fluid starts to play a role and the stress increases are well described by a power law originating from collisional or viscous energy loss in this flow regime. Such behaviour can be captured by the Herschel-Bulkley (HB) model that describes yield-stress fluids:

$$\tau = \tau_0 + k\dot{\gamma}^n, \quad (1)$$

where τ denotes the shear stress, τ_0 the yield stress, $\dot{\gamma}$ the shear rate, k a proportionality constant and n a power law index. The HB model is very effective in capturing the macroscopic flow response of granular materials and suspensions. However, to understand and describe the local flow behavior of these materials it has been shown to be ineffective. In particular, it is well-known that HB fluids can form shear bands (for a review see Ref. [1]), especially when a spatially inhomogeneous stress field is applied. One way to apply such a stress field is in a Couette geometry, where a cylinder rotates coaxially in a second cylinder. The shear stress τ in such a system decays as $\sim 1/R^2$, with R the radius. Although one may expect the flow to stop abruptly at $R(\tau < \tau_0)$, the flow profile for a yield stress fluid is smooth [1–4].

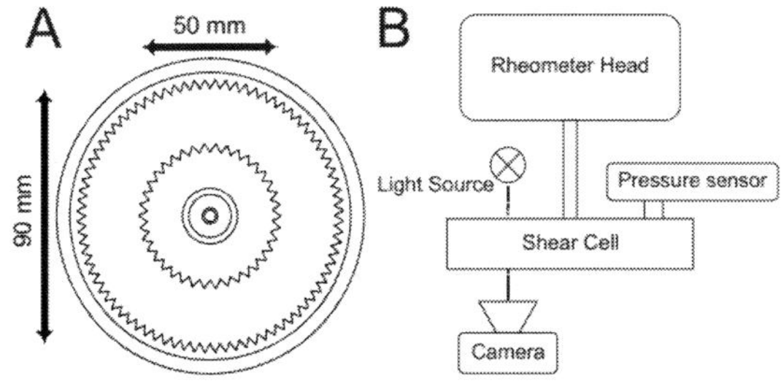

Figure 1. (A) Schematic of the Couette cell. The cell has a height of 20 mm. (B) Schematic overview of the experimental setup. A pressure sensor is mounted to the lid of the cell. We image the suspension in transmission mode using an LED lamp and diffuser, and an 8-bit camera.

We can view flows of granular materials in terms of two dimensionless quantities. A ratio of two time scales provides us with $I = \dot{\gamma}d\sqrt{\rho_p/P}$, with d the particles' diameter, ρ_p its density and P the local pressure. The rheology is then described by the friction coefficient $\mu = \tau/P$ to be a simple function of I [5]. This theory can be extended to explain the rheology of suspensions [6]. However, such local descriptions fail to accurately describe flows in the quasistatic regime, such as the smoothly decaying velocity

*e-mail: joshua.dijksman@wur.nl

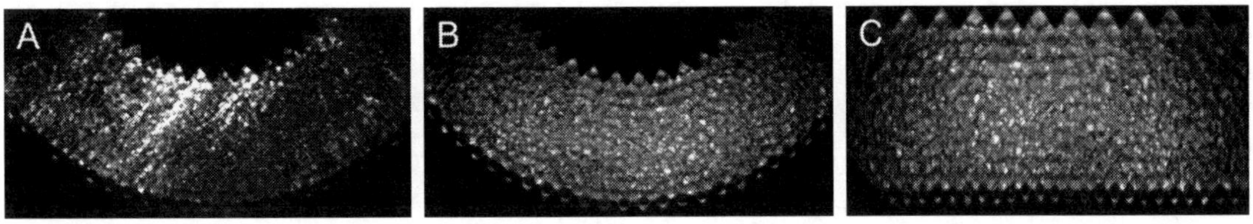

Figure 2. (A) Example of an image of the photoelastic response of the gelatin particles. Stresses are anisotropic and point against the direction of shear (inner cylinder rotating clockwise). (B) Image of a suspension without polarizers. (C) Same as (B) after transformation of the coordinate system.

field discussed before. To solve these issues, a fluidity theory first proposed for emulsions [7], was adapted to granular materials [8]. An additional so-called granular fluidity field g may be used to predict nonlocal effects in multiple geometries [8, 9]. Fluidity modeling requires remarkably few fitting parameters, yet the microscopic meaning of the fluidity field is as yet subject of debate. The fluidity field somehow models the fluctuations present in even a slowly flowing granular material, yet are these fluctuations of velocities, stresses, or something else?

Here we propose an experimental environment in which we can probe the role of local stress and velocity fluctuations in the flow behavior of a three-dimensional suspension. We perform experiments in Couette flows of gelatin particles. The gap of the Couette cell is approximately 10 particle diameters wide (see Section 2), resulting in a highly heterogeneous stress field, making nonlocal effects probable. Gelatin is photoelastic [10], giving us access to local stress fluctuations, by imaging the particles in transmission mode. As the particles are transparent but not refractive index matched, we also have access to the velocity profiles of the particles. In this paper we will demonstrate that we can indeed optically resolve an average local stress field inside a flowing suspension. Since particle image velocimetry (PIV) of granular materials is typically done by imaging only a surface, it is not evident that transmission mode imaging can also resolve flow fields. In the second part of this paper we show that we can indeed obtain velocity measurements from transmission imaging. To see how universal the flow profiles of the gelatin suspension are, we then compare them to those of another typical hydrogel, polyacrylamide.

2 Materials & Methods

2.1 Experimental setup

We use a custom Couette geometry with fixed volume to drive the flow of a hydrogel particle suspension. The inner cylinder diameter d_i = 50 mm, while the outer cylinder diameter d_o = 90 mm (Figure 1A). To avoid slip, both inner and outer walls are made rough with teeth of approximately 2.5 mm. The height of the shear cell h = 20 mm. The bottom and top plate of the cell are made of smooth acrylic glass. We measure the vertical confining pressure using a load cell on part of the top plate. We use a stress-controlled rheometer (Anton-Paar MCR 300) to drive the inner cylinder at a fixed rotation rate. We image the suspension from below with an 8-bit camera, while illuminating from the top using an array of LEDs and a diffuser (Figure 1B). We can add polarizers to image the photoelastic response, of which we show an example in Figure 2A.

2.2 Hydrogel particles

We synthesize spherical hydrogel beads of roughly 2 mm in diameter to perform the experiments. We use a method described in detail elsewhere [11]. We use two different materials; gelatin and polyacrylamide (PAAm). In short, the gelatin particles are made by dispersing a hot solution of gelatin (10 wt%) in a column of cold oil (decane), thereby solidifying the droplets. The resulting particles are cross-linked using glutaraldehyde [12] to avoid dissolution of the gelatin when they are suspended in water for the flow experiments. The PAAm particles are made in a similar way: we disperse a solution of acrylamide (20 wt%), bis-acrylamide (1 wt%) and ammonium persulfate (1 wt%) in a column of oil (kerosene) with a catalyst (TEMED, 2 wt%). As we have control over the droplet size, we can make the suspension bidisperse to avoid crystallization. We have measured the Young's modulus of

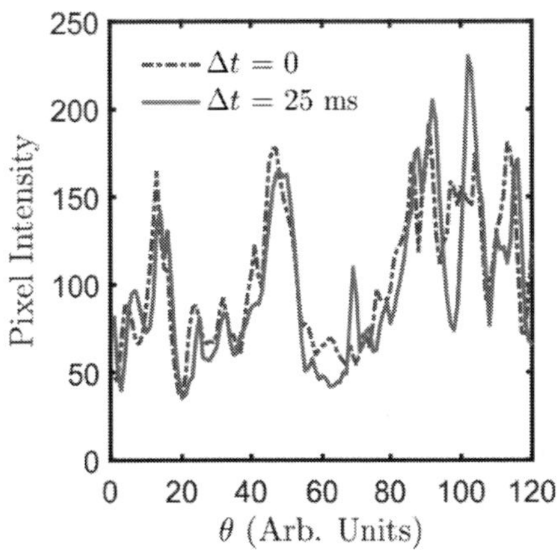

Figure 3. Intensity profiles of a row of pixels at two distinct times, showing that the signal is correlated. The profile at later time is shifted to the right.

 DOI: 10.1051/epjconf/201714003020

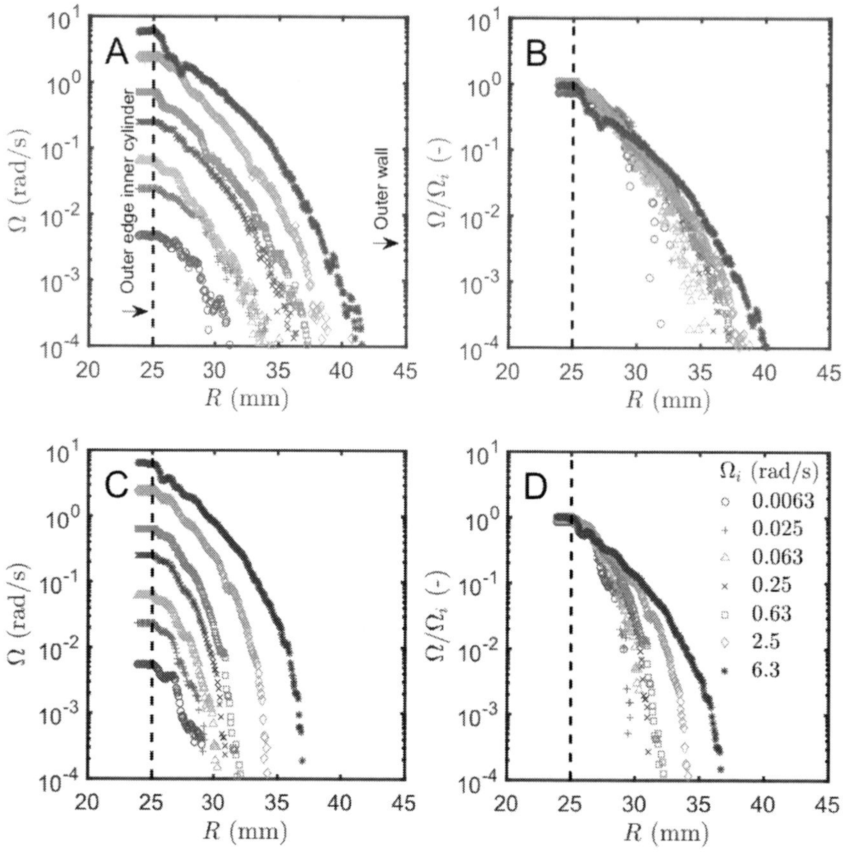

Figure 4. (A,B) Flow profiles for the PAAm particles. (C,D) Flow profiles for the gelatin particles. Legend applies to all panels.

both particle types to be approx. 3×10^5 Pa. The particles thus differ only in their chemical composition, which may result in differences in their surface structure or other physico-chemical properties.

2.3 Experimental procedure

We perform experiments in suspensions at fixed volume fraction, characterized by the pressure exerted on the top plate. We measure this confining pressure at a low rotation rate of the inner cylinder ($\Omega_i = 6.3\times10^{-3}$ rad/s) and make sure all experiments are carried out at a pressure $P^p \approx 200$ Pa. As the particles are porous and their swelling depends on the solvent, density matching of these particles is challenging. We thus use water as the interstitial fluid. The maximum hydrostatic pressure resulting from the density difference $\Delta\rho$ between the particles and water can be estimated $P_g = \Delta\rho g h \approx 20$ Pa. The pressure due to the density mismatch is always much smaller than any applied confining pressure, so we can neglect the role of P_g. Before our measurements, we pre-shear the suspension at $\Omega_i = 62$ rad/s for 10 seconds. After waiting for 30 seconds, we then start driving the suspension at a fixed Ω_i ranging from 6.3×10^{-3} - 6.3 rad/s, thus spanning three decades in driving rate. For each rate, we image the suspension for 10 seconds at 119 frames/second. To ensure that we measure in the steady-state, we start recording only after the inner cylinder has made about one full rotation.

2.4 Particle Image Velocimetry

We first convert our recorded images such that the coordinates are transformed from Cartesian to polar (Figure 2B+C). This allows for finding the intensity as a function of the angle θ at given radius R, corresponding to a row of pixels. Typical for PIV, we find the correlation between intensity profiles to get an estimate of the velocity at given R. Although we image in transmission mode, we can still find significant correlation between intensity profiles; an example is shown in Figure 3. By finding the shift of the signal for different time steps Δt, we find the angular velocity $\Omega(R)$ — this is the standard method to do PIV.

3 Results & Discussion

We plot the results of our analysis in Figure 4. In Figure 4A, we plot the velocity profiles of the PAAm suspension, showing that the shear band widens as we increase the driving rate of the inner cylinder Ω_i, as expected and similar to the findings in Ref. [13, 14]. For the gelatin suspension (Figure 4C), the shear is much more localized; particles in the outer half of the gap show very little movement, even at the highest rates applied. This finding shows that hydrogel material properties affect flow properties in this dense limit. We can find another difference in Figure 4B+D, where we have normalized the angular velocity Ω with the applied driving rate Ω_i. For the PAAm suspension (Figure 4B), the width of the shear band increases as Ω_i increases. However for the gelatin suspension (Figure 4D), the shear band has very similar width, except for the two highest rates. We believe the differences in flow profiles between these two similar materials to be related to their frictional behavior. It is known that hydrogels may have very complex frictional behavior [15], where the friction *coefficient* may depend on the load as well as the sliding velocity. Gel friction depends on the chemical structure of the gel [16]. Thus it is likely that the two types of gel examined here have different frictional behavior, resulting in different flow profiles.

4 Conclusion

We demonstrate that we can successfully measure stress fields and perform PIV on suspensions of hydrogel particles in transmission mode, using a custom Couette cell. We find that, although composed of very similar particles, gelatin and polyacrylamide suspensions have different flow profiles. We speculate on the origin of the observed difference by considering the frictional properties of the suspended particles. The gelatin suspension is a potent new experimental system for suspension rheology, as we have access to both local stress and velocity fluctuations. As we also drive the suspension flows with a rheometer, we can measure both global and local flow characteristics, which provides an excellent test bed for the fluidity model.

References

[1] P. Schall, M. van Hecke, Annual Review of Fluid Mechanics **42**, 67 (2009)
[2] P. Coussot, J. Raynaud, F. Bertrand, P. Moucheront, J. Guilbaud, H. Huynh, S. Jarny, D. Lesueur, Physical review letters **88**, 218301 (2002)
[3] G. Ovarlez, S. Rodts, A. Ragouilliaux, P. Coussot, J. Goyon, A. Colin, Physical Review E **78**, 036307 (2008)
[4] H. De Cagny, A. Fall, M.M. Denn, D. Bonn, Journal of Rheology (1978-present) **59**, 957 (2015)
[5] F. da Cruz, S. Emam, M. Prochnow, J.N. Roux, F. Chevoir, Physical Review E **72**, 021309 (2005)
[6] F. Boyer, É. Guazzelli, O. Pouliquen, Physical Review Letters **107**, 188301 (2011)
[7] J. Goyon, A. Colin, G. Ovarlez, A. Ajdari, L. Bocquet, Nature **454**, 84 (2008)
[8] K. Kamrin, G. Koval, Physical Review Letters **108**, 178301 (2012)
[9] D.L. Henann, K. Kamrin, Proceedings of the National Academy of Sciences **110**, 6730 (2013)
[10] A. Kuske, G.S. Robertson, *Photoelastic stress analysis* (Wiley-Interscience, 1974)
[11] M. Workamp, S. Alaie, J.A. Dijksman, Review of Scientific Instruments **87**, 125113 (2016)
[12] A. Bigi, G. Cojazzi, S. Panzavolta, K. Rubini, N. Roveri, Biomaterials **22**, 763 (2001)
[13] N. Huang, D. Bonn, Journal of fluid mechanics **590**, 497 (2007)
[14] G. Koval, J.N. Roux, A. Corfdir, F. Chevoir, Physical Review E **79**, 021306 (2009)
[15] J.P. Gong, Soft matter **2**, 544 (2006)
[16] J.P. Gong, Y. Osada, in *High Solid Dispersions* (Springer, 2010), pp. 203–246

A Granular System of Ellipses under Linear Shear

Dong Wang[1,*], *Hu* Zheng[1,2,**], and *Robert P.* Behringer[1,***]

[1] *Department of Physics & Center for Nonlinear and Complex System, Duke University, Durham, NC, USA*
[2] *School of Earth Science and Engineering, Hohai University, Nanjing 210098, China*

Abstract. Shear of granular systems of disks (in $2D$) and spheres ($3D$) has been studied extensively. However, less is known about systems of non-spherical particles, i.e., ellipses and polygons, etc. Here we studied a quasi $2D$ granular system composed of ellipses in a novel apparatus, which provided homogeneous simple shear by utilizing the weak friction between ellipses and the deformable base of the apparatus. Shear jamming, which was first reported for a disk system [1], was also observed for our ellipse system. We contrast shear jamming for systems of disks and systems of ellipses. By tracking the positions and orientations of the ellipses, we observed that the ellipses gradually rotated to align along a preferred direction as the shear strain increased. We also applied Voronoï tessellation to the packing of ellipses, showing that the distribution of the local density changed during shear. By contrast, such a change in the distribution was not observed in a disk system [2].

1 Introduction

Granular systems have shown to be of great interest and benefit to theoretical study and industrial applications. Of special interest is the response of a granular system to shear. Systems composed of disks (in $2D$) or spheres (in $3D$) under shear have been studied extensively in theory and experiments, including simulations [1, 3–10]. Though simple and able to capture important features of granular systems, disk/sphere models are often different from real situations, i.e., particles have an anisotropic or even irregular shape. Much is less known for this type of systems.

One of the most simple anisotropic shapes is an ellipse. There have been studies on ellipses, with comparisons to disks, i.e., jamming transitions and dynamical properties, both in experiments and simulations [11–14]. Most of these studies either focus on systems above jamming or considers dynamics. On the other hand, Bi *et al.* have shown that a granular system of frictional disks below the isotropic jamming point can be jammed by applying a certain amount of shear strain quasi-statically, a process called shear jamming [1]. Therefore, it is necessary to test shear jamming in a system of ellipses and compare the results with disks.

In this work, we use a system of quasi-$2D$ elliptical particles made of photoelastic materials and apply quasi-static simple shear to it homogeneously. Note that simple shear can be decomposed in to a rotation followed by pure shear, which consists of compression in one direction and dilation in the perpendicular one. The photoelastic property of particles enable us to visualize and determine pressure on each individual particle. We find that shear jamming still occurs in a granular system of ellipses. Using UV bars on the surfaces of particles, we determine their centers and orientations. The evolution of the probability distribution of particle orientations shows that particles tend to align during shear to a direction that is perpendicular to the compression of shear. Furthermore, by applying Voronoï tessellation to the packing, we see a dramatic change and asymmetry in the distribution of the Voronoï cell areas, indicating a spacial inhomogeneity.

2 Experimental Methods

The granular system is composed of more than 3000 monodisperse ellipses made of photoelastic material, which reveal force information on individual particles if put between a pair of crossed polarizers. An example of a polarized image is shown in Fig. 1(c), with particles that bear forces lighting up and ones that are force free remaining dark. Each particle has dimensions of: major axis 1.04cm and minor axis 0.57cm. We also obtain an image taken with unpolarized white light, as shown in Fig. 1(b). An invisible UV bar, only illuminated by UV light, was drawn at the center of the particle surface, as shown in Fig. 1(d), to accurately find the major axis orientation (indicated as orientation in the following text) and center of a particle.

To apply a homogeneous simple shear to the granular system, we used a bottom-assisted shear apparatus utilizing friction between the base and particles. The bottom is split into small slats of width 1.27cm, moving with the wall, to ensure that every particle feels the uniform shear profile. That is shear is applied homogeneously throughout the system, and not only from the boundaries. Such a

*e-mail: dw148@duke.edu
**e-mail: hz64@phy.duke.edu
***e-mail: bob@phy.duke.edu

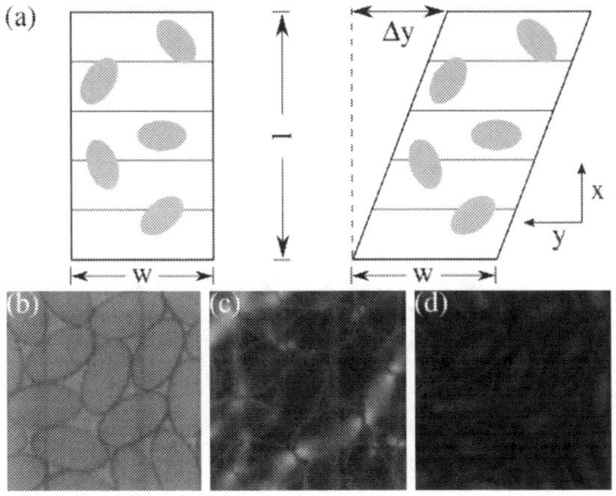

Figure 1. (a) Sketch of the shear cell setup with the Cartesian coordinate system illustrated. Starting as a rectangular box, which has a length l = 62.1cm and a width w = 29.7cm, the cell is deformed to a parallelogram, with the deformation characterized by shear strain: $\gamma = \Delta y/l$. White light illuminates the shear cell from below while UV light illuminates the shear cell from above. Bottom: examples of images of the same part from a shear experiment taken (b) with white light on and without polarizer, (c) with white light on and with polarizer and (d) with white light off and UV light on.

design is sketched in Fig. 1(a). For a typical experiment, starting from a rectangular box of width w = 29.7cm and length l = 62.1cm, the system was sheared along the width side by a small strain step, 0.0027, followed by enough relaxation time for the particles to come to rest. Then we obtained three pictures in sequence: a polarized image to record forces, a normal and a UV image to determine positions and orientations of particles. Such a process of small amplitude shear, relaxation and image acquisition is repeated 100 times to achieve a total strain (γ) of 0.27.

3 Results and Discussion

3.1 Shear Jamming

Shear jamming is still observed in the case of ellipses, and that is qualitatively similar to what is observed for disks. According to the simulations [15], the isotropic jamming point, ϕ_J, for a frictionless bidisperse ellipse system with an aspect ratio 1.82(= 1.04/0.57) is between 0.885 and 0.89. The densities we tested are all well below this value. Typical results from an example of ϕ = 0.847 (the same experiment for the results shown in Figs. 3 and 4) are shown in Fig. 2. Starting from a completely stress-free state, as shown in Fig. 2(a), where no particle is bearing any forces, the system gradually comes to a fragile state, where particles form a force network, force chains, mainly in the compression direction, as shown in Fig. 2(b) [1, 6]. Such a special state is mechanically stable; however, it cannot sustain more than a small stress if the shear is reversal. This feature distinguishes itself from a jammed

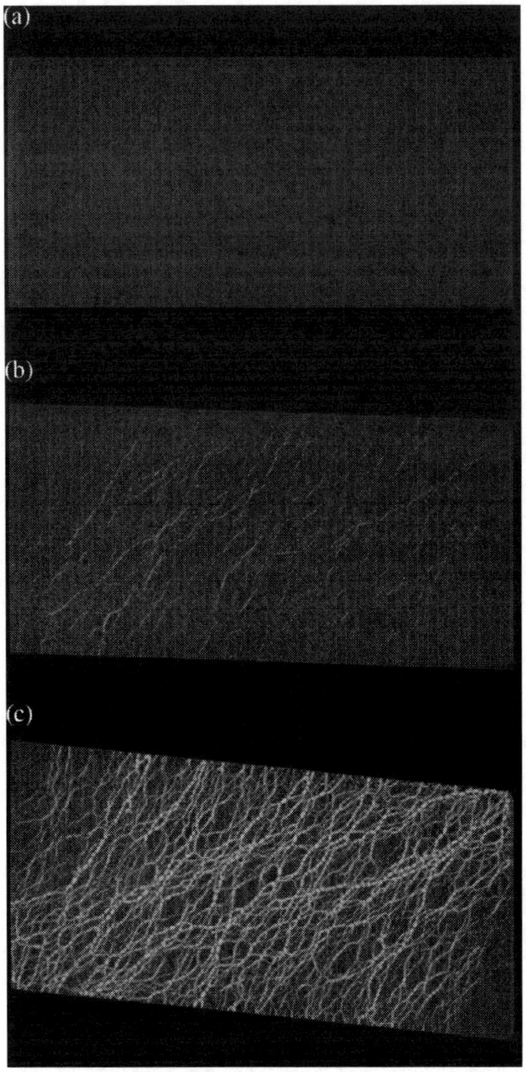

Figure 2. Three snapshots of polarized images from one shear experiment (ϕ = 0.847) showing that the system undergoes transitions from (a) unjammed to (b) fragile and eventually to (c) shear jammed states. The corresponding strains from (a) to (c) are γ = 0, 0.027, 0.135.

state, where the system can sustain stress from any direction. As γ increases more, force chains in the compression direction bend and make extra contacts, and eventually percolate all over the system, bringing the system to a jammed state, as shown in Fig. 2(c).

In order to investigate the difference for shear jamming between disk and ellipse systems, we probe structural changes in the ellipse system and contrast it with disks. Specifically, we look at the probability distribution function (PDF) of orientations of ellipses, denoted as θ in the following, and apply Voronoï tessellation to the packing, with results shown and discussed below.

3.2 Orientation Distribution

When computing the probability distribution function of the orientations of ellipses, only the ellipses whose distances to the nearest boundary are twice or more the minor axis are taken into consideration to minimize the boundary effect. After such an elimination of boundary particles, approximately 2500 ~ 3000 particles are used in the calculation, which still gives reliable statistical results.

For the ease of plotting, θ is confined in the range $(0°, 180°)$ and then extended to $(0°, 360°)$ by adding $180°$ to the original values, where $\theta = 0°$ corresponds to the positive x axis in the inset of Fig. 1(a). Results from a few selected γ are shown in Fig. 3. Note at $\gamma = 0$ in Fig. 3(a), there are more ellipses oriented around $80°$, while supposedly there will be no preferred orientations if particles are put randomly. The main reason is: in order to achieve a stress free state at a relatively high ϕ, particles tend to align with their neighbors, which hence creates a local order. However, such a built-in order does not affect the final conclusion, as seen in the following.

As γ increases, we can clearly see that: first, the distribution remains non-uniform; second, the most probable θ changes from $\sim 80°$ at $\gamma = 0$ to $\sim 110°$ with a slightly higher probability at $\gamma = 0.15$, and then stays around the same value for the rest of shear. The first feature indicates that shear induces or keeps a preferred direction in an ellipse system. The second feature indicates that the system reaches a steady state. Such a feature is reminiscent of the work by Börzsönyi *et al.*, where they showed a granular system of elongated particles formed a certain angle with the streamline of shear, unaffected by the shear rate [14]. Furthermore, both features show up for all the other densities tested.

As mentioned in Sec. 1, there exists a rotation associated with simple shear. The amount of rotation is given by $tan^{-1}(\gamma/2)$, which in our case is no larger than $tan^{-1}(0.27/2) = 7.7°$. From Fig. 3, we clearly see that the change of the preferred θ is more than $15°$. Therefore, such a change cannot be only induced by the rotation associated with shear. Note that an ellipse is more stable if the forces it bears are mainly perpendicular to its major axis. Combining this with the fact that shear induces force anisotropy with larger forces mainly in the compression direction, particles are naturally expected to align their major axis perpendicular to the compression direction. This expectation coincides with our observation that the most probable θ is around $110°$, which is very close to the perpendicular direction of compression direction $(45 - tan^{-1}(\gamma/2) + 90° = 135° - tan^{-1}(\gamma)$, which is $120°$ for $\gamma_{max} = 0.27)$.

3.3 Voronoï Tessellation

To further verify the structural change of the system under shear, a Voronoï tessellation is applied to the packing. In this approach, we use the distance transformation tool in Matlab to determine a Voronoï tessellation associated with a particle, in which the distance of any point to that particle is less than that to any other particle [16]. We follow the

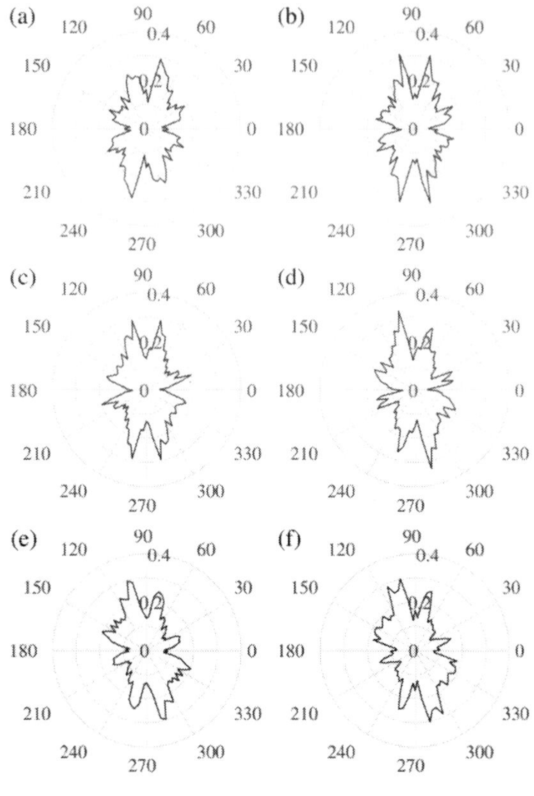

Figure 3. The probability distribution function of the orientation of ellipses from the same experiment in Fig. 2, excluding particles that are close to the boundary at different strain γ: (a) $\gamma = 0$, (b) $\gamma = 0.0513$, (c) $\gamma = 0.1026$, (d) $\gamma = 0.1539$, (e) $\gamma = 0.2052$, and (f) $\gamma = 0.2565$. In the figures, the orientations are first fixed to the range $(0°, 180°)$ and then duplicated by adding $180°$ to have a range in $(0°, 360°)$.

same procedure in Sec. 3.2 to minimize boundary effects, and the results are shown in Fig. 4.

Fig. 4(a) shows the evolution of the PDF of the Voronoï area, $Avoro$, with shear. The peak value, P_{max}, and corresponding $Avoro$, of the PDF, shift to a lower value as γ increases. In comparison, Ren *et al.* reported that this PDF did not change during shear for a disk system [2]. In addition, Aste *et al.* showed an invariant Gamma distribution of the Voronoï volume in static granular spheres of different packing fraction by rescaling the Voronoï volume [10]. Here we applied the same rescaling: $A_{rescaled} = (Avoro - A_{min})/(\langle Avoro\rangle - A_{min})$, where A_{min} is the minimum and chosen to be 1.048 times the actual area of an ellipse. Assuming the data to be the Gamma distribution, given by

$$P(A_{rescaled}) = \frac{1}{\Gamma(k)\theta^k} A_{rescaled}^{k-1} e^{-A_{rescaled}/\theta}, \qquad (1)$$

k and θ can be determined by the mean $(k\theta)$ and variance $(k\theta^2)$. One thing to note is that due to the rescaling, the mean is always 1, namely, $k\theta = 1$.

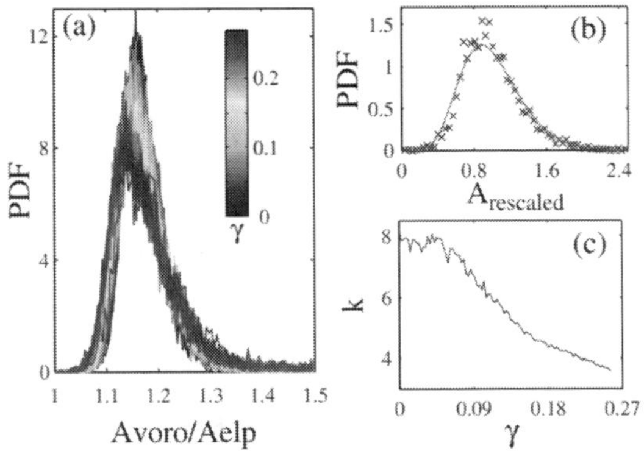

Figure 4. (a) Probability distribution function (PDF) of the Voronoï area of ellipses in the unit of the area of an ellipse, *Avoro*/*Aelp*, from the same experiment in Fig. 2, as a function of shear strain γ, indicated by different colors. The inset shows the color bar for γ. (b) The Gamma distribution fit (red line) to the original data (blue crosses) at $\gamma = 0$ for the rescaled Voronoï area: $A_{rescaled} = (A - A_{min})/(\langle A \rangle - A_{min})$. (c) The evolution of k from the Gamma distribution fit as γ increases.

An example of the Gamma distribution fit is shown in Fig. 4(b) for $\gamma = 0$. Since $k\theta$ is fixed regardless of γ, we only look at k here, plotted against γ as shown in Fig. 4(c). Starting from a value around 8, k stays the same for small γ and then gradually decreases as γ increases. The skewness of the Gamma distribution is proportional to $1/\sqrt{k}$, hence the PDF of *Avoro* becomes more and more asymmetric as γ increases.

4 Conclusion and Outlook

To summarize, applying homogeneous shear to a quasi $2D$ granular system of ellipses causes shear jamming for an ellipse system well below isotropic jamming point. Under applied strain, particles tend to align with their major axes approximately perpendicular to the compression direction of shear. Such an alignment is seen in the probability distribution of the orientations. A structural inhomogeneity, indicated by this alignment, is further validated by the probability distribution of the Voronoï cell area, where the most probable value shifts to a smaller value. By contrast, such a inhomogeneity induced by shear is not observed with disk granular systems. Therefore, by introducing anisotropy into the shape of elements that make up the packing, the response of the packing shows a substantially different behavior, especially structurally.

We note that Schaller *et al.* showed an application of Voronoï tessellation in frictional ellipsoids packing to explain the local origin for global contact number [12], which can serve as a criterion to distinguish jammed states from unjammed ones. For the future work, we would like to apply similar analysis to see if there exists a connection between Voronoï area and contact number, and hence shear jamming. In addition, a more detailed analysis of force networks is required to better distinguish ellipses from disks during the shear jamming transition.

Acknowledgments

This work was supported by NSF grant DMR1206351, NASA grant NNX15AD38G and the William M. Keck Foundation. DW and RB would like to thank a Triangle MRSEC fellowship. HZ thanks NSFC Grant No. 41672256 and the NSF (Jiangsu, China) Grant No. BK20140845 for financial support.

References

[1] D. Bi, J. Zhang, B. Chakraborty, R.P. Behringer, Nature **480**, 355 (2011)
[2] J. Ren, J.A. Dijksman, R.P. Behringer, AIP Conference Proceedings: Powders & Grains **1542**, 527 (2013)
[3] S.F. Edwards, R.B.S. Oakeshott, Physica A **157**, 1080 (1989)
[4] A.J. Liu, S.R. Nagel, Nature **396**, 21 (1998)
[5] C.S. O'Hern, L.E. Silbert, A.J. Liu, S.R. Nagel, Phys. Rev. E **68**, 011306 (2003)
[6] M.E. Cates, J.P. Wittmer, J.P. Bouchaud, P. Claudin, Phys. Rev. Lett. **81**, 1841 (1998)
[7] T.S. Majmudar, R.P. Behringer, Nature **435**, 1079 (2005)
[8] T.S. Majmudar, M. Sperl, S. Luding, R.P. Behringer, Phys. Rev. Lett. **98**, 058001 (2007)
[9] L.E. Silbert, Soft Matter **6**, 2918 (2010)
[10] T. Aste, T. Di Matteo, M. Saadatfar, T.J. Senden, M. Schröter, H.L. Swinney, EPL **79**, 24003 (2007)
[11] M. Mailman, C.F. Schreck, C.S. O'Hern, B. Chakraborty, Phys. Rev. Lett. **102**, 255501 (2009)
[12] F.M. Schaller, M. Neudecker, M. Saadatfar, G.W. Delaney, G.E. Schröder-Turk, M. Schröter, Phys. Rev. Lett. **114**, 158001 (2015)
[13] J. Tang, R.P. Behringer, EPL **114**, 34002 (2016)
[14] T. Börzsönyi, B. Szábo, G. Törös, S. Wegner, J. Török, E. Somfai, T. Bien, R. Stannarius, Phys. Rev. Lett. **108**, 228302 (2012)
[15] C.F. Schreck, N. Xu, C. O'Hern, Soft Matter **6**, 061305 (2012)
[16] N.N. Medvedev, Dokl. Akad. Nauk **337**, 767 (1994)

Numerical models for fluid-grains interactions: opportunities and limitations

Amir Esteghamatian[1,*], *Mona* Rahmani[1,**], and *Anthony* Wachs[1,***]

[1] *Department of Mathematics, University of British Columbia, 1984 Mathematics Road, Vancouver, BC, Canada V6T 1Z4*

Abstract. In the framework of a multi-scale approach, we develop numerical models for suspension flows. At the micro scale level, we perform particle-resolved numerical simulations using a Distributed Lagrange Multiplier/Fictitious Domain approach. At the meso scale level, we use a two-way Euler/Lagrange approach with a Gaussian filtering kernel to model fluid-solid momentum transfer. At both the micro and meso scale levels, particles are individually tracked in a Lagrangian way and all inter-particle collisions are computed by a Discrete Element/Soft-sphere method. The previous numerical models have been extended to handle particles of arbitrary shape (non-spherical, angular and even non-convex) as well as to treat heat and mass transfer. All simulation tools are fully-MPI parallel with standard domain decomposition and run on supercomputers with a satisfactory scalability on up to a few thousands of cores. The main asset of multi scale analysis is the ability to extend our comprehension of the dynamics of suspension flows based on the knowledge acquired from the high-fidelity micro scale simulations and to use that knowledge to improve the meso scale model. We illustrate how we can benefit from this strategy for a fluidized bed, where we introduce a stochastic drag force model derived from micro-scale simulations to recover the proper level of particle fluctuations. Conversely, we discuss the limitations of such modelling tools such as their limited ability to capture lubrication forces and boundary layers in highly inertial flows. We suggest ways to overcome these limitations in order to enhance further the capabilities of the numerical models.

1 Introduction

The understanding of the dynamics of suspension flows is still incomplete, although major advances have been achieved over the past 50 years. Complementary to experiments and theory, numerical simulation has become an indispensable tool to get some more insight into the intricate momentum transfer between the continuous fluid and dispersed solid particles. Suspension flows exhibit a large range of scales: from the scale of an individual particle to the scale of the process. It is hence natural to adopt a multi-scale description of these flows. A nowadays worthwhile research path to follow is to adopt a micro/meso/macro separation of scales with corresponding numerical models at these different scales combined to a bottom-up strategy. The bottom-up strategy, also called upwards cascade of knowledge, involves transferring the knowledge acquired at the smallest scale with high-fidelity simulations to the larger scales as a tool to improve numerical models at these large scales.

At the micro scale, Particle Resolved Simulation (PRS) has been extensively used to study fundamental aspects of suspension flow dynamics. PRS models contain very few assumptions and are hence expected to supply high-fidelity simulation results on momentum, heat and mass transfer in suspension flows. Body-fitted techniques as, e.g., Arbitrary Lagrangian-Eulerian, are nowadays not favored as the remeshing step is computationally too expensive in 3D. Among fixed mesh techniques, the most popular PRS methods include Immersed Boundary method (IB), lattice-Boltzmann method (LB) and Distributed Lagrange Multiplier/Fictitious Domain (DLM/FD) method (see [2] for an extended discussion). Initially developed to model momentum transfer, they can easily be extended to model heat and mass transfer too.

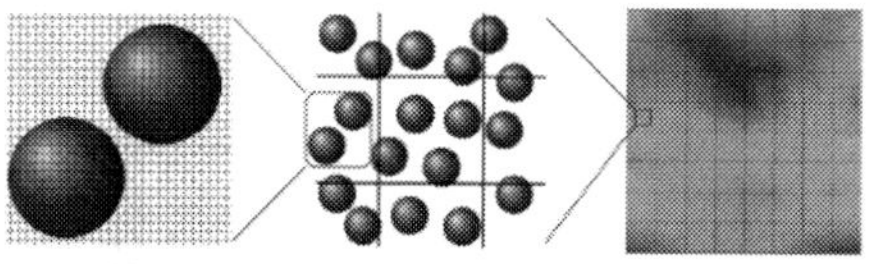

Figure 1: A discrete view on a grid of a micro/meso/macro multi-scale approach

At the meso scale, mass and momentum conservation equations are averaged in control volumes encompassing a few particles and momentum transfer between the two phases is modeled using appropriate drag laws. Meso-scale methods provide a good basis for intermediate size simulations of suspension flows with a relatively wide range of Reynolds number and porosity. They represent an attractive alternative to PRS but the accurate treatment of

*e-mail: amir.est@gmail.com
**e-mail: mrahmani@math.ubc.ca
***e-mail: wachs@math.ubc.ca

fluid/particle interactions is the key to reliable numerical predictions.

Finally, macro-scale models correspond to different variants of the Euler/Euler model as, e.g., the Two Fluid Model (TFM). In this family of models, the solid phase is somehow assumed to behave as a fluid and both phases are solved in an average way based on closure laws for momentum transfer. Although quite advanced, the main drawback of this type of approach is its deficiency to model particle/particle interactions very well due to the use of (non necessarily extremely reliable) correlations for solid phase pressure and viscosity.

In this paper we introduced our numerical models at the micro scale and the meso scale and discuss how to transfer knowledge from micro to meso in the case of a bubbling fluidized bed as an illustration of what can be achieved with such a multi-scale approach. Finally, we discuss the current limitations of the different models and the next stages in the development of more advanced models for suspension flows.

2 Numerical models

In the following section, the mathematical aspects of both micro- and meso-scale models are described. We assume that the fluid is Newtonian and that the flow is incompressible. In both models the solid phase is treated in a direct fashion by solving Newton-Euler equations for each individual particle. All parameters in this section are dimensionless, while dimensional variables are distinguished by a "star" symbol. To non-dimensionalize the set of equations, the following scales are introduced: L_c^* for length, V_c^* for velocity, L_c^*/V_c^* for time, $\rho_f^* V_c^{*2}$ for pressure, V_c^{*2}/L_c^* for gravity acceleration, $\rho_f^* V_c^{*2}/L_c^*$ for distributed Lagrange multiplier (for the micro-scale model only) and $\rho_f^* V_c^{*2} L_c^{*2}$ for contact force.

2.1 A DLM/FD formulation as a micro scale model

Our micro-scale approach is based on the DLM/FD formulation of [1] and is implemented in a Finite Volume fashion (see [2] for more details). The dimensionless and non-variational form of combined momentum continuity and momentum equations are as follows:

$$\frac{\partial \boldsymbol{u}}{\partial t} + \boldsymbol{u}\cdot\nabla\boldsymbol{u} = -\nabla p + \frac{1}{\mathcal{R}e}\nabla^2\boldsymbol{u} - \boldsymbol{\lambda}, \text{ over } \Omega \qquad (1)$$

$$(\rho_r - 1)V_p\left(\frac{d\boldsymbol{U}}{dt}\right) - \int_{\mathcal{P}} \boldsymbol{\lambda}\, dx = \sum_j \boldsymbol{F}_{cj} + \mathcal{F}r(\rho_r - 1)V_p\frac{\boldsymbol{g}^*}{|\boldsymbol{g}^*|}, \text{ over } \mathcal{P} \qquad (2)$$

$$(\rho_r - 1)\left(\boldsymbol{I}_p\frac{d\boldsymbol{\omega}}{dt} + \boldsymbol{\omega}\times\boldsymbol{I}_p\cdot\boldsymbol{\omega}\right) - \int_{\mathcal{P}} \boldsymbol{r}\times\boldsymbol{\lambda}\, dx = \sum_j \boldsymbol{R}_j\times\boldsymbol{F}_{cj}, \text{ over } \mathcal{P} \qquad (3)$$

$$\boldsymbol{u} - (\boldsymbol{U} + \boldsymbol{\omega}\times\boldsymbol{r}) = 0, \text{ over } \mathcal{P} \qquad (4)$$

$$\nabla\cdot\boldsymbol{u} = 0, \text{ over } \Omega \qquad (5)$$

where $\boldsymbol{u}$, $\boldsymbol{U}$, $\boldsymbol{\omega}$ stand for the fluid velocity vector, the particle translational velocity vector and the particle angular velocity vector, respectively. The solid domain and whole fluid/particle domain are denoted by $\mathcal{P}$ and Ω, respectively. $\boldsymbol{\lambda}$ represents the distributed Lagrange multiplier vector, $\boldsymbol{F}_c$ the contact force, $V_p = M_p^*/(\rho_s^* L_c^{*d})$ the particle volume, M_p^* the particle mass, d the dimension of the system, $\boldsymbol{I}_p = \boldsymbol{I}_p^*/\rho_s^* L_c^{*d+2}$ the particle inertia tensor, $\boldsymbol{r}$ the position vector and $\boldsymbol{R}$ the vector connecting the gravity center and the contact point. The following dimensionless numbers are also introduced:

$$\text{Reynolds number } \mathcal{R}e = \frac{\rho_f^* V_c^* L_c^*}{\mu^*} \qquad (6)$$

$$\text{Froude number } \mathcal{F}r = \frac{g^* L_c^*}{V_c^{*2}} \qquad (7)$$

$$\text{Density ratio } \rho_r = \frac{\rho_s^*}{\rho_f^*} \qquad (8)$$

2.2 A two-way Euler/Lagrange (TWEL) meso scale model

Based on the principle of locally averaged Navier-Stokes equations proposed by [3], fluid phase variables are solved on grid cells the volume of which is an order of magnitude larger than that of a particle. The dimensionless fluid conservation equations read:

$$\frac{\partial \varepsilon}{\partial t} + \nabla\cdot\varepsilon\boldsymbol{u} = 0 \qquad (9)$$

$$\frac{\partial \varepsilon\boldsymbol{u}}{\partial t} + \nabla\cdot(\varepsilon\boldsymbol{u}\boldsymbol{u}) = -\nabla p - \boldsymbol{F}_{fp} + \frac{1}{\mathcal{R}e}\nabla\cdot(2\varepsilon\boldsymbol{D}) \qquad (10)$$

where $\boldsymbol{D}$ and ε denote the strain rate tensor and the fluid volume fraction, respectively. The pressure gradient term contains the hydrodynamic pressure only and $\boldsymbol{F}_{fp}$ represents the fluid/particle interaction force. The most general expression of $\boldsymbol{F}_{fp}$ is:

$$\boldsymbol{F}_{fp} = \int_{\partial\mathcal{P}} \boldsymbol{\tau}\cdot\boldsymbol{n}\, dS \qquad (11)$$

where $\boldsymbol{\tau}$ is the point-wise fluid stress tensor and $\boldsymbol{n}$ is the normal vector to the solid boundary. Since point-wise variables (pressure and viscous stress) on the solid boundary are not resolved in meso-scale models, the surface integration over $\partial\mathcal{P}$ cannot be carried out and a closure law is used instead to evaluate $\boldsymbol{F}_{fp}$.

Among the different contributions to the total hydrodynamic force (drag, Basset, added mass, Magnus, Saffman), the drag force is generally the order 1 contribution. Many drag correlations are available for spheres in the literature. They all predict roughly the same values. For instance, the correlation of [4] is widely used. Introducing a particle Reynolds number computed with the superficial slip velocity $|\boldsymbol{u}^* - \boldsymbol{U}^*|\varepsilon = V_c^*$ as a characteristic velocity and the particle diameter $d^* = L_c^*$ as a characteristic length:

$$\mathcal{R}e_p = \frac{\rho_f^*|\boldsymbol{u}^* - \boldsymbol{U}^*|\varepsilon d^*}{\mu^*} \qquad (12)$$

the drag correlation of [4] reads as follows:

$$F^*_{drag} = 0.5 C_{d,su} \rho^*_f \frac{\pi d^{*2}}{4} |u^* - U^*|(u^* - U^*)\varepsilon^{-\chi} \quad (13)$$

$$\chi = 3.7 - 0.65 \exp(-\frac{(1.5 - \log_{10} \mathcal{R}e_p)^2}{2}) \quad (14)$$

$$C_{d,su} = \frac{24}{\mathcal{R}e_p}(1 + 0.15 \mathcal{R}e_p^{0.687}) \quad (15)$$

where C_d is the drag coefficient for a single unhindered particle for $\mathcal{R}e_p$ less than 1000 ([5]) and the pre-factor $\varepsilon^{-\chi}$ accounts for the presence of neighboring particles.

2.3 Lagrangian tracking of particles with collisions using a DEM/soft-sphere approach

The Newton-Euler equations for each individual particle are as follows:

$$\rho_r V_p \frac{dU}{dt} = F_b + F_c + F_{fp} \quad (16)$$

$$\rho_r (I_p \frac{d\omega}{dt} + \omega \times I_p \omega) = T_c + T_{fp} \quad (17)$$

where $F_b = (\rho_r - 1)\mathcal{F}r V_p \frac{g}{|g^*|}$ denotes the buoyancy force, F_c the inter-particle or particle-wall contact force and F_{fp} the fluid/particle interaction force. Likewise, T_c and T_{fp} represent the contact-induced and the fluid-induced torque acting on the particle. The contact force and torque are computed by a spring-dashpot-frictional slider DEM/soft-sphere model (see [6]). For the micro-scale model, since the flow field is resolved at the particle level, no closure law is used for the fluid/particle interaction force F_{fp} and torque T_{fp}, and these quantities can be computed a posteriori using the Lagrange multipliers as:

$$F_{fp} = \int_{\mathcal{P}} \lambda dx + V_p \frac{dU}{dt} \quad (18)$$

$$T_{fp} = \int_{\mathcal{P}} (r \times \lambda) dx + I_p \frac{d\omega}{dt} + \omega \times I_p \omega \quad (19)$$

For the meso-scale model, however, the fluid/particle interaction force contains several terms:

$$F_{fp} = F_{drag} + F_{lift} + F_{ba} + F_{am} \quad (20)$$

For high density ratios ($\rho_r > 8$), we can safely neglect the added mass force (F_{am}) and the Basset force (F_{ba}). The lift force (F_{lift}) acting on a particle is due to the rotation induced pressure distribution around it. This rotation can be the result of either the fluid local shear rate (i.e. Saffman force) or the particle own rotation (i.e. Magnus force). Unfortunately, correlations for these two forces available in the literature are limited to dilute regimes. Thus, to our knowledge, all meso-scale models in the literature (including ours) neglect the lift force. Finally, the fluid-induced torque T_{fp} is also assumed to be zero in the meso-scale model, in the absence of both an appropriate correlation and reliable fundamental understanding of the importance of particle rotation in suspension flows.

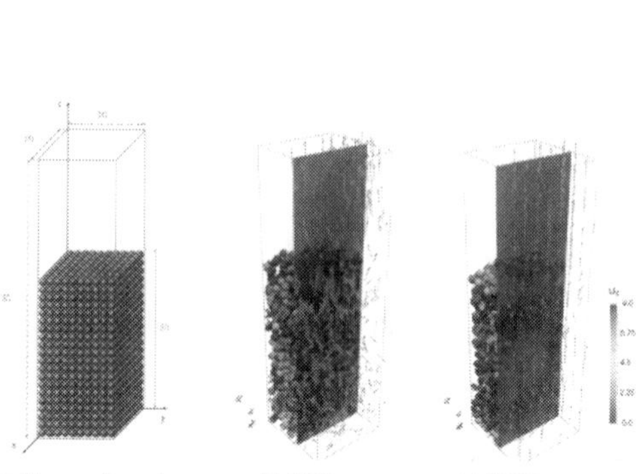

(a) Flow configuration (b) PRS (c) TWEL

Figure 2: Flow configuration and flow field with the two models.

3 Results

We consider the fluidized bed shown in FIG. 2(a). It contains 2000 monodisperse spheres and has a dimensionless size of $10 \times 10 \times 35$. We impose periodic boundary conditions on the lateral walls and a homogeneous and constant in time upwards inlet velocity at the bottom wall. Dimensionless parameters are set to $\rho_r = 10$, $u^*_{in}/u^*_{mf} = 3$ (where u^*_{mf} is the minimum fluidization velocity) and $Re_{in} = 6$. We simulate this flow configuration with the micro scale model and the meso scale model. Assuming that micro-scale simulation results are highly reliable, we directly compare the two sets of data produced by the two models and identify what is flawed in the meso-scale model.

Integral measures as average expansion of the bed (FIG. 2(b) and FIG. 2(c)) and pressure drop across the bed are well predicted by both models. Based on an extensive analysis of both Eulerian and Lagrangian particles statistics, we observe that particles motion is underpredicted by the meso-scale TWEL model in particular in the transverse direction. We identify two causes for this underprediction. First, our simple Bounding Cube (BC) projection kernel (FIG. 3(a)) is not accurate enough and using a Gaussian kernel (GK, FIG. 3(b)) enables us to recover part of the proper particles motion. The GK also decouples particle size and grid size, giving rise to the opportunity to use grid size smaller than particle diameter. Second, filtering PRS data reveals that for a pair of (Re, ε) the corresponding drag coefficient is not a deterministic (unique) value but a distribution. This led to the definition of a new stochastic drag law C'_d that models the sub-grid fluctuations. The total drag coefficient C^t_d is hence defined as $C^t_d = \bar{C}_d + C'_d$ where $\bar{C}_d$ is the deterministic value given by, e.g., [4]. The parameters of the stochastic drag model C'_d are extracted from the filtered PRS data, such that no parameter is empirical. In FIG. 4, we plot the granular temperature in the vertical z and transverse x directions as a measure of particles fluctuations, parametrized by the level of sophistication of the TWEL model. It is very visible that we obtain a hierarchy of solutions: (i) BC, (ii) GK, and (iii) GK + stochastic drag, where the most sophisticated version of the model, i.e., GK + stochastic drag, is almost able to re-

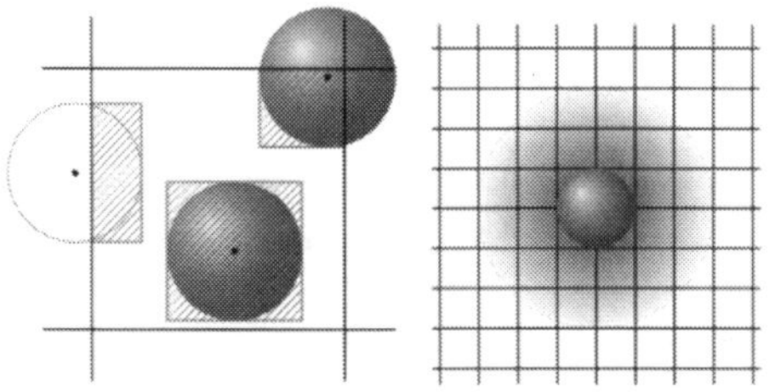

(a) Bounding Cube (b) Gaussian kernel

Figure 3: Projection operators: bounding cube and Gaussian kernel

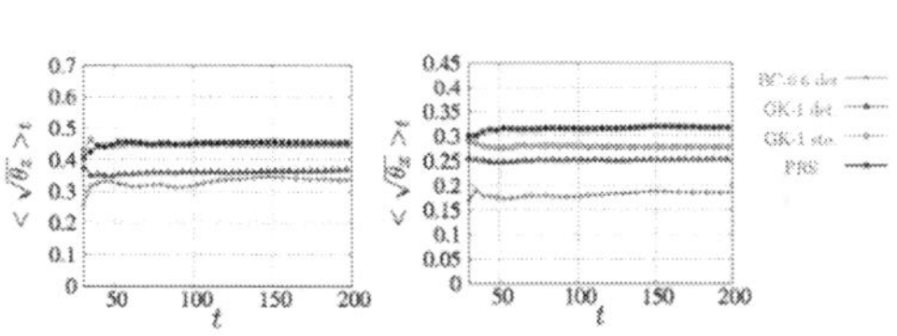

(a) (b)

Figure 4: Time-evolution of average granular temperature θ in (a) vertical direction and (b) transverse direction with different variants of the TWEL model

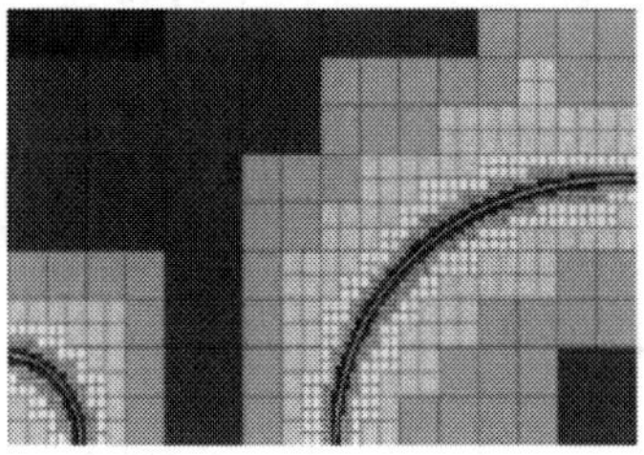

Figure 5: Quadtree Adaptive Mesh Refinement

cover the PRS solution as plotted in Fig. 4(a) and Fig. 4(b). This is very promising.

4 Discussion and Perspectives

Advanced numerical models embedded in a multi-scale approach create many opportunities to gain novel physical insight into fluid-grains interactions that govern suspension flows. We have shown an example in which using high-fidelity PRS data is extremely helpful in identifying and partly curing some defects of meso-scale TWEL models. This analysis can be conducted for heat (and mass) transfer as our DLM/FD method (like IB or LB) and TWEL model extend easily to heat (and mass) transfer too. This offers the opportunity to examine a broder range of flow problems. Although very promising and efficient, this type of multi-scale approach also exhibits limitations.

The first limitation is computational. In fact, PRS are extremely expensive. The PRS of the fluidized bed presented above, that contains 2000 spheres only, requires a mesh comprises at least 50 million of cells (i.e. at least 24 cells over the particle diameter) and runs over a long physical time to acquire converged statistics. This translates into weeks of computing on, e.g., 256 cores of a supercomputer. Highly scalable codes with fast algorithms are hence sheer game-shifters. Very soon, computations with a billion of cells will not be uncommon anymore to gain even more understanding into the dynamics of suspension flows. The main limitation of micro-scale models is the local resolution. In other words, the smallest length scale resolved by the model is controlled by the grid size. This creates intrinsic limitations in terms of what can be accurately capture as, e.g., boundary layers or lubrication force. For dense suspensions, most PRS are not able to fully captured lubrication forces and hence rely on analytical corrections, that exist for spheres only. In dilute and semi-dilute flows, the upper bound in terms of particle Reynolds number is generally around 500, as above 500 the boundary layer is too thin to be properly captured by a constant grid size mesh, unless at least 100 cells are used over the particle diameter. Similarly, the extension to non-spherical and presumably complex shape puts even more stress on the fineness of the mesh. One remedy to this issue of local mesh size is to implement these micro-scale models with an Adaptive Mesh Refinement (AMR) technique as illustrated in Fig. 5. This would offer the opportunity to refine the mesh close to the particle surface without increasing the total number of cells. One interesting problem that one could attack with such a tool is a dense suspension of non-spherical particles, for which there is no lubrication force correction available in the literature and hence lubrication force would be fully computed/captured by the solution of the coupled Navier-Stokes/Newton's equations.

At the meso scale level, we have shown how to improve TWEL models based on PRS data. Our impression is that the accuracy of TWEL models relies mainly on the projection operator (BC, GK, others ?) and on drag laws conceptually novel (with respect to the deterministic laws introduced over the past 50 years). This is crucial to recover the right level of particles motion. This is a rewarding research path to follow.

References

[1] R. Glowinski, T.W. Pan, T.I. Hesla, D.D. Joseph, International Journal of Multiphase Flow **25**, 755 (1999)

[2] A. Wachs, A. Hammouti, G. Vinay, M. Rahmani, Computers & Fluids **115**, 154 (2015)

[3] T.B. Anderson, R. Jackson, Industrial & Engineering Chemistry Fundamentals **6**, 527 (1967)

[4] R. Di Felice, International Journal of Multiphase Flow **20**, 153 (1994)

[5] L. Schiller, Z. Naumann, Vdi Zeitung **77**, 51 (1935)

[6] A. Wachs, L. Girolami, G. Vinay, G. Ferrer, Powder Technology **224**, 374 (2012)

[7] C.T. Crowe, M. Sommerfeld, Y. Tsuji, *Multiphase flows with droplets and particles* (CRC press, 1997)

EPJ Web of Conferences **140**, 06004 (2017) DOI: 10.1051/epjconf/201714006004
Powders & Grains 2017

A parallel Discrete Element Method to model collisions between non-convex particles

Andriarimina Daniel Rakotonirina[1,*], *Jean-Yves* Delenne[3,**], and *Anthony* Wachs[1,2,***]

[1] *Department of Mathematics, University of British Columbia, Vancouver, BC, Canada V6T 1Z4*
[2] *Department of Chemical & Biological Engineering, University of British Columbia, Vancouver, BC, Canada V6T 1Z2*
[3] *IATE, UMR 1208 INRA – CIRAD – Montpellier Supagro – Université Montpellier 2, Montpellier, France*

Abstract. In many dry granular and suspension flow configurations, particles can be highly non-spherical. It is now well established in the literature that particle shape affects the flow dynamics or the microstructure of the particles assembly in assorted ways as e.g. compacity of packed bed or heap, dilation under shear, resistance to shear, momentum transfer between translational and angular motions, ability to form arches and block the flow. In this talk, we suggest an accurate and efficient way to model collisions between particles of (almost) arbitrary shape. For that purpose, we develop a Discrete Element Method (DEM) combined with a soft particle contact model. The collision detection algorithm handles contacts between bodies of various shape and size. For non-convex bodies, our strategy is based on decomposing a non-convex body into a set of convex ones. Therefore, our novel method can be called "glued-convex method" (in the sense clumping convex bodies together), as an extension of the popular "glued-spheres" method, and is implemented in our own granular dynamics code Grains3D. Since the whole problem is solved explicitly, our fully-MPI parallelized code Grains3D exhibits a very high scalability when dynamic load balancing is not required. In particular, simulations on up to a few thousands cores in configurations involving up to a few tens of millions of particles can readily be performed. We apply our enhanced numerical model to (i) the collapse of a granular column made of convex particles and (i) the microstructure of a heap of non-convex particles in a cylindrical reactor.

1 Introduction

The Discrete Element Method has been extensively used over the past 40 years for the numerical modelling of granular material dynamics. When combined with a soft-sphere approach, it is conceptually simple, easy to implement and has shown to supply computed results of satisfactory accuracy. DEM simulations have overall helped a lot to gain novel physical insight into granular flows and generally contributed significantly to advance the understanding of the intricate mechanisms involved (force chains, dilatance, blockage, compaction, etc). Although conceptually simple, the implementation of DEM is however more challenging for large systems and systems containing non-spherical, angular and potentially non-convex particles. In the following, we suggest to treat the former by distributed computing and the latter by an extension of our numerical strategy for convex bodies to non-convex bodies.

The rest of the paper is organized as follows. In Section 2, we shortly recall the basic governing equations. Then we elaborate in Section 3 on our glued-convex method to treat (almost) any body shape. Section 4 provides the main ingredients for a simple though extremely efficient domain decomposition based parallel implementation. We conclude by illustrating the computing capabilities of our code Grains3D in Section 5 and discuss what can be achieved in terms of novel physical insight with such a numerical tool in Section 6.

2 Numerical model

The motion of the granular material is determined by applying Newton's second law to each particle $i \in< 0, N-1 >$, where N is the total number of particles. The rigid body motion assumption leads to the decomposition of the velocity vector $\boldsymbol{v}$ as $\boldsymbol{v} = \boldsymbol{U} + \boldsymbol{\omega} \wedge \boldsymbol{R}$, where $\boldsymbol{U}$, $\boldsymbol{\omega}$ and $\boldsymbol{R}$ denote the translational velocity vector of the center of mass, the angular velocity vector of the center of mass and the position vector with respect to the center of mass, respectively. The complete set of equations to be considered is the following one:

$$M_i \frac{d\boldsymbol{U}_i}{dt} = \boldsymbol{F}_i \quad , \quad \boldsymbol{J}_i \frac{d\boldsymbol{\omega}_i}{dt} + \boldsymbol{\omega}_i \wedge \boldsymbol{J}_i \boldsymbol{\omega}_i = \boldsymbol{M}_i \tag{1}$$

$$\frac{d\boldsymbol{x}_i}{dt} = \boldsymbol{U}_i \quad , \quad \frac{d\boldsymbol{\theta}_i}{dt} = \boldsymbol{\omega}_i \tag{2}$$

where M_i, $\boldsymbol{J}_i$, $\boldsymbol{x}_i$ and $\boldsymbol{\theta}_i$ stand for the mass, inertia tensor, center of mass position and angular position of particle i. $\boldsymbol{F}_i$ and $\boldsymbol{M}_i$ are the sum of all forces and torques applied

*e-mail: adr.rkt@gmail.com
**e-mail: jean-yves.delenne@supagro.inra.fr
***e-mail: wachs@math.ubc.ca

on particle i, respectively, and can be further decomposed in purely granular dynamics (i.e., without accounting for any external forcing as e.g. hydrodynamic or electrostatic) into a torque-free gravity contribution and a contact force contribution as:

$$\boldsymbol{F}_i = M_i\boldsymbol{g} + \sum_{j=0,j\neq i}^{N-1} \boldsymbol{F}_{ij} \tag{3}$$

$$\boldsymbol{M}_i = \sum_{j=0,j\neq i}^{N-1} \boldsymbol{R}_j \wedge \boldsymbol{F}_{ij} \tag{4}$$

where $\boldsymbol{F}_{ij}$ is the force due to collision with particle j and $\boldsymbol{R}_j$ is a vector pointing from the center of mass of particle i to the contact point with particle j. In our model, $\boldsymbol{F}_{ij}$ comprises a normal Hookean elastic restoring force, a normal dissipative force and a tangential friction force.

The set of equations Eq. (1) is integrated in time using a second order leap-frog Verlet scheme. Rotations of particles are computed using quaternions for computational efficiency as well as to avoid any gimbal lock configurations. The collision detection algorithm is a classical two-step process. Potential collisions are first detected via a linked-cell list and then actual collisions are determined using a GJK algorithm. Our GJK-based collision detection strategy enables us to consider any convex shape and size. For more detail, see [1] and the references therein.

3 Extension to non-convex bodies

In [1], we suggested an original and efficient way to detect collisions between particles of arbitrary convex shape. Our strategy utilizes the Gilbert-Johnson-Keerthi algorithm to compute the minimal distance between two convex bodies and a sequence of shrinking-swelling of bodies. This 3-step collision detection strategy has proven to be well adapted and to preserve shape angularity (unlike, e.g., a superquadric representation). Once combined with a soft-sphere approach in our granular solver Grains3D, the method constitutes a powerful tool to compute granular flows. In [1], we applied it to cubes, cylinders and tetrahedra in (i) the filling of a small container and (ii) rotating drum dynamics.

We achieve the extension to more complex shapes and in particular to non-convex bodies by recognizing that any non-convex body can be decomposed into a set of convex bodies as shown in Fig. 1. We call the original convex body the composite particle and the set of convex bodies the elementary components. Our method is thus an extension of the popular glued-sphere method [2]. We call it the glued-convex method, in the sense that non-convex bodies are considered by clumping convex bodies. There is however a major improvement compared to the glued-sphere approach: because elementary components are convex bodies the original shape of the non-convex particle is perfectly accounted for (no rounded corner, no artificial rugosity). The main asset is the ability to use the same collision detection strategy as for convex bodies (with a linked-cell spatial sorting and circumscribed spheres). In fact, if

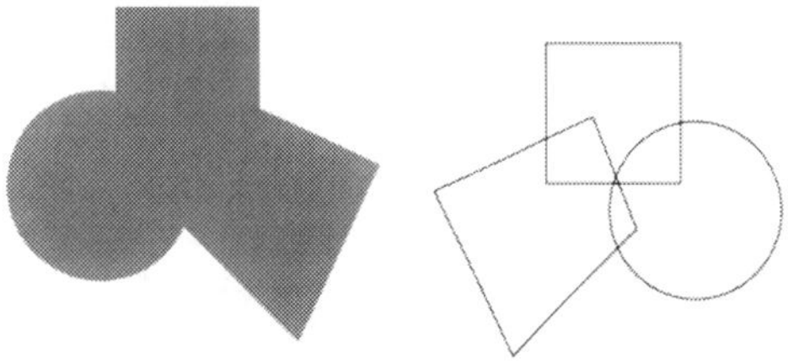

Figure 1: Decomposition strategy illustrated in two dimension

two non-convex bodies touch (or overlap), then there exists at least one pair of convex elementary components that touch. In practice, to detect collisions between two non-convex bodies, we test potential collisions over all pairs (E_i, E_j), $i = 0, ..., N-1$, $j = 0, ..., M-1$ of elementary (convex) components, where N is the total number of elementary components of the first composite particle and M is the total number of elementary components of the second composite particle. The only real drawback of the glued-convex method is that the computing cost theoretically scales as $N \times M$.

Finally, since mass properties (volume, center of mass, tensor of inertia) of a complex non-convex particle cannot be calculated analytically, we employ a parallel numerical integration method to evaluate these quantities. In practice, we embed a non-convex body in a regular cartesian fine mesh and compute the integrals numerically based on a simple cell center in/out criterion. From a 500^3 mesh, mass properties are computed with an error less than 0.1% and this takes less than a few minutes in parallel on a supercomputer. These mass properties are computed once for a non-convex shape and are stored in a database.

4 A MPI/Domain Decomposition parallel implementation

A key to simulate systems with a large number of particles is distributed computing [3, 4]. Our parallel strategy is classical and is based on a constant in time domain decomposition technique. We employ a cartesian domain decomposition. Each process hosts a single subdomain and we hence define a cartesian MPI communicator using the `MPI_Cart_create` command. On each subdomain, we construct a cartesian linked-cell list with an additional layer of cells at the boundary with neighbouring subdomain to serve as an overlapping zone. This overlapping zone hosts clone particles used to compute collisions with particles located on a neighbouring subdomain (process). As a consequence, cells in a linked-cell list are tagged based on their location on the subdomain: 0 = interior, 1 = buffer and 2 = clone, as illustrated on Fig. 2(a). At each time step, clone particles are either created, deleted or updated. All particles are tagged based on the cell they belong to. Hence they consistently change status as they

DOI: 10.1051/epjconf/201714006004

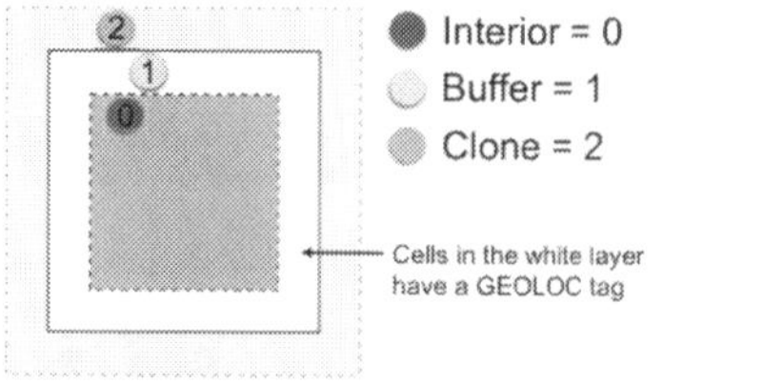

(a) 2D illustration of the status of a particle depending on their location on the domain, i.e., depending on the tag of the cell it belongs to

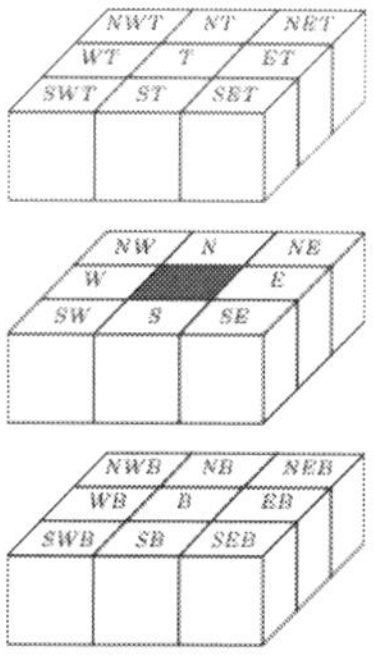

(b) GEOLOC tag of cells in the buffer zone. N, S, W, E, T and B denote respectively the North, South, West, East, Top and Bottom directions.

Figure 2: Tags (status and geolocalisation) of particles in the linked-cell grid.

move in the subdomain. Corresponding operations are performed on neighbouring subdomains when a particle change status, e.g., if a particle moves from an interior cell (tag = 0) to a buffer cell (tag = 1), a clone particle (tag = 2) is automatically created on the neighbouring subdomain.

Not only cells (and hence particles belonging to these cells) are tagged in terms of their status but cells in the buffer zone are also tagged in terms of their location with respect to the neighbouring subdomains using a second tag, named GEOLOC for geographic location, that takes the 26 following values (whose meaning is rather obvious on a 3D cartesian grid as can be seen in Fig. 2(b)): EAST & WEST in the x direction, NORTH & SOUTH in the y direction, TOP & BOTTOM in the z direction are the main neighbours, NORTH_EAST, NORTH_WEST, SOUTH_EAST, SOUTH_WEST, NORTH_BOTTOM, NORTH_TOP, SOUTH_BOTTOM, SOUTH_TOP are the edge neighbours, and NORTH_WEST_TOP, NORTH_WEST_BOTTOM and so on are the corner neighbours. The aftermath is an exact tailoring of sent MPI messages with the appropriate information only, thus reducing the size of each sent message to the minimum, as illustrated in Fig. 3.

5 Results

5.1 Parallel performance

We illustrate the parallel performance of Grains3D on a granular column collapse (also called granular dam break

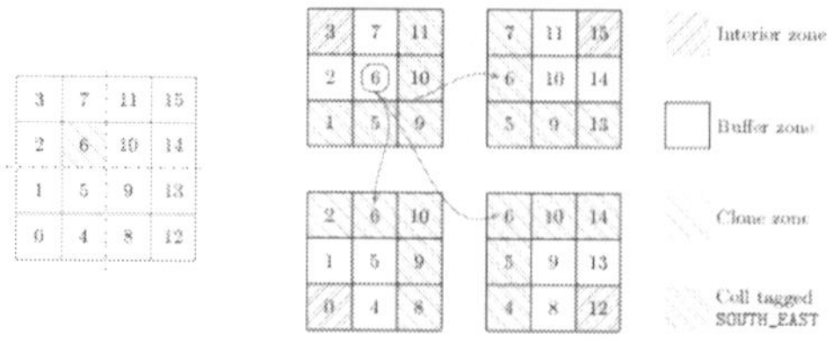

Figure 3: 2D illustration of inter-process communication for a particle tagged SOUTH_EAST.

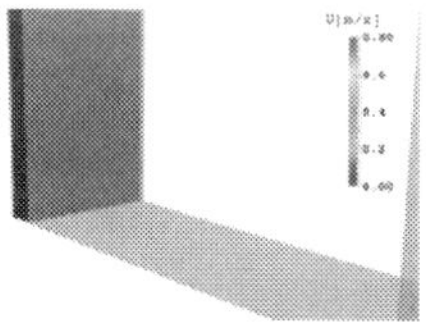

(a) $t = 0.0s$

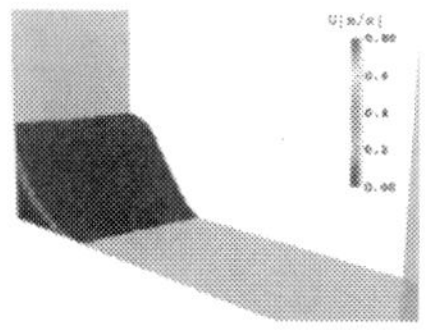

(b) $t = 0.3s$

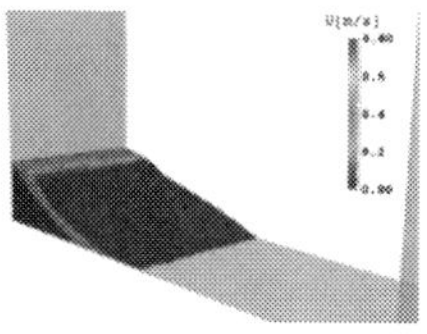

(c) $t = 0.5s$

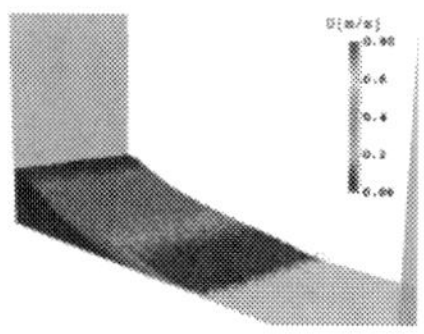

(d) $t = 0.7s$

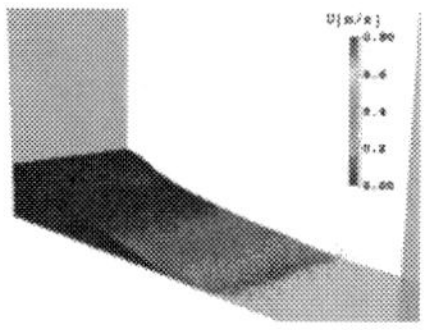

(e) $t = 0.8s$

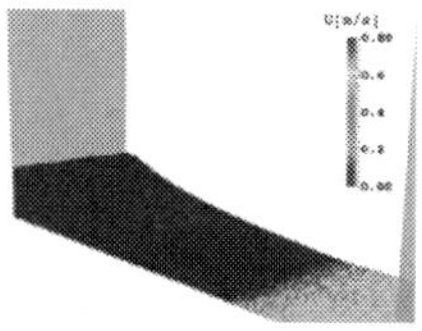

(f) $t = 1.5s$

Figure 4: 3D view of the granular dam break flow with $1,627,500$ icosahedra.

in the literature). The column is made of icosahedral particles that mimic sand grains. We perform weak scaling tests that involve increasing the number of cores and the size of the system simultaneously. In other words, we set the load of particles per core constant. Here, the load of particles per core is set to $102,850$ and we perform computations with up to $52,080,000$ icosahedra on 512 cores. The domain is periodic in the transverse direction and the initial aspect ratio of the column is set to 7.3. Fig. 4 shows snapshots of the collapse. The parallel performance of Grains3D is presented in Fig. 5. Grains3D exhibits a very satisfactory scaling factor of around 0.9 up to 512 cores (that corresponds to the largest system containing 52 million of icosahedron).

5.2 Filling a container with non-convex shapes

We expand here the results obtained in [1] on the filling of a cylindrical reactor with particles of various shape but

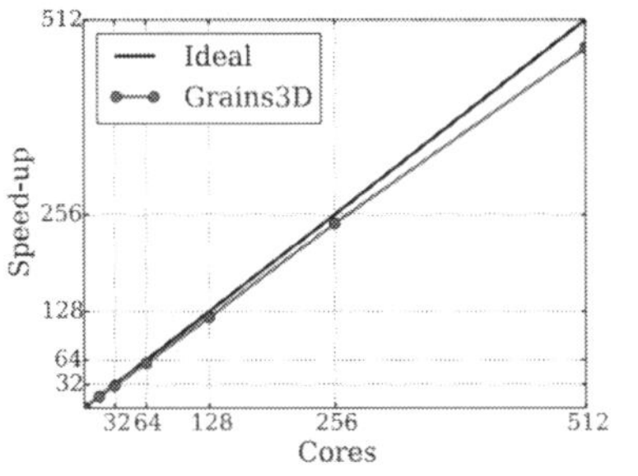

Figure 5: Weak scaling parallel performance of Grains3D in granular dam break computations up to 512 cores.

(a) Sphere (top) and Cylinder (bottom) (b) Cube (top) and Tetrahedron (bottom) (c) "2D-cross" (d) "3D-cross"

Figure 6: Packing of 250 particles of six different shapes in a cylindrical container. Shapes in (a,b) are from [1]

same volume (spheres, cylinders, cubes and tetrahedra). Here we create two new shapes: a 2D cross and a 3D cross of same volume. These two new shapes are indeed non-convex and are modelled by clumping beams of square cross section together (2 in 2D and 3 in 3D). We run the same simulations for these two new cross-like shapes as in [1] for the 4 aforementioned convex shapes. All physical and numerical parameters are the same, only the shape changes. Please also note that the reactor diameter to particle equivalent diameter is pretty small, which promotes high porosity close to the wall. The height of the pack of particles is a measure of the level of compaction, or conversely of porosity, in the system. It is quite spectacular in Fig. 6 that 2D and even more 3D crosses lead to a level of porosity 2 to 4 times higher than for any of the isometric convex shapes. This nicely illustrates the effect of non-convexity on the microstructure of a pack of particles.

6 Discussion and Perspectives

The GJK-based collision detection strategy suggested in [1] already enabled us to examine granular flows with particles of multiple size and multiple convex shapes. We have now extended the modelling capabilities of our code Grains3D to non-convex shapes and massively parallel computing. Our glued-convex method for the extension to non-convex bodies is very robust and accurate as it uses the same tools as for convex bodies. Its computational cost is however quite high as it scales with the product of the number of elementary components in two colliding particles. But it also opens up unprecedented perspectives in the numerical modelling of granular flows. In particular, highly non-convex particles as 3D crosses provide a sort of cohesion to the granular media that significantly modifies the way it flows.

In systems in which the load of particles per core does not change too much, our parallel strategy shows very satisfactory weak scaling properties. Systems of up to a few hundreds of millions of convex particles can now be computed on a few hundreds to a few thousands of cores in a reasonable time (i.e., of the order of a few days). Systems with a few tens of millions of non-convex particles are also attainable. High Performance Computing and supercomputers combined to highly scalable codes are really becoming game-shifters in the field of granular flow simulation. The next stage in terms of improving even further our numerical model is to develop a dynamic load balancing algorithm in order to offer a good scalability even in systems with a high level of particle volume fraction heterogeneity in space and in time. This is an on-going work in our group.

Acknowledgements

This work was granted access to the HPC resources of CINES under the allocations 2013-c20132b6699 and 2014-c20142b6699 made by GENCI.

References

[1] A. Wachs, L. Girolami, G. Vinay, G. Ferrer, Powder Technology **224**, 374 (2012)

[2] M. Abbaspour-Fard, Biosystems Engineering **88**, 153 (2004)

[3] J.H. Walther, I.F. Sbalzarini, Engineering Computations **26**, 688 (2009)

[4] K. Iglberger, U. Rüde, Multibody System Dynamics **25**, 81 (2011)

Effects of vertical vibration on surface intruder loading in a multiple-size granular system

Wenqing Hou[1], *Shiping Zhang*[1], *Ping Wu*[1*], *Li Li*[1] *and Li Wang*[2*]

[1]School of Mathematics and Physics, University of Science and Technology Beijing, 100083 Beijing, China
[2]School of Energy and Environmental Engineering, University of Science and Technology Beijing, 100083 Beijing, China

Abstract. We investigate the behaviors of the large porous-alumina particles on the free surface of the small glass-grain system under vertical vibration. The experiments are performed using cylindrical container with diameter Φ = 240 mm, loaded with small glass beads to a static depth h =100 mm. We control the shaker to vibrate at the various frequencies f, from 40 to 80 Hz, and dimensionless acceleration $\Gamma = (4\pi^2 Af^2)/g$ from 2 to 6. When the glass granular system is at rest, porous alumina particles are placed on its free surface. A parameter critical frequency f_u is defined to distinguish two kinds of diffusion in particles. When the vibration frequency is less than f_u, quasi-two-dimensional surface diffusion can occur in porous alumina particles, surface granules clusters formed under certain condition. The frequency and dimensionless acceleration of the vibration are varied to view their effect on the clustering and surface particle-distribution. When the vibration frequency is larger than f_u, the surface diffusion disappears and a three-dimensional diffusion appears.

1 Introduction

Surface movement of a vertically vibrating granular system has many unique characteristics. When subjected to vertical vibration, granular systems show many interesting behaviors including pattern formation[1-3], surface fluidization[4-6], convection rolls[7-11], particle separation[12-15], spontaneous heap formation[16-20] and energy dissipation[21]. Detail experiments have classified that surface particles in vibrating system have obvious spontaneous collective effect performance for regularly pattern which is far away from chaotic distribution[16-18]. The investigation of microscopic mechanisms about collective behavior have been done by many authors[22]. In the present study we focused attention on the multiple-size granular system, and discovered a special cluster phenomenon in two-dimensional (2D) multiple-size particle system of a vertically vibration and carried on research on collective cells in different situation.

2 Experiment

2.1 Experiment Setting

We paid attention on the behavior of the large porous-alumina particles on the free surface of the small glass-grain system under vertical vibration. We used porous activated alumina particles as upper granules, which is a kind of spherical multihole desiccating agent comprised of activated alumina powder. The sine vertical vibration is usually represented as:

$$z = A\sin(2\pi f t) \tag{1}$$

where A stands for vibration amplitude, f is the frequency of vibration.

Surface effects were concerned during the experiment. Walking behavior of these porous alumina particles under vertical vibration was observed through a camera. The important parameters in this experiment are: (i) the diameter of substratum glass particles Φ_1 and the diameter of porous alumina particles Φ_2. Two kinds of glass grains were used, Φ_1=1.25mm (1.0-1.5mm) named as g_1 and Φ_1 =0.9mm (0.8-1.0mm) called g_2. Besides, two distinct size alumina granular was a_1 with diameter Φ_2=4mm and a_2 with diameter Φ_2=8mm; (ii) the vibrating parameters: the frequency of vibration f and dimensionless acceleration Γ:

$$\Gamma = \frac{4\pi^2 Af^2}{g} \tag{2}$$

where g is the acceleration due to gravity. Moreover, there were other conditions which could impact on the results during the experiment. For example, the shape of the container, the height of glass particle layers h at rest and so on. To curtail the extra effect, we used a cylindrical container which the internal diameter was Φ=240mm and the static height of glass layer h was always 100 mm. Besides, the same glass under layer particle were used during the experiment. We poured 570±25 (15g) alumina particles each time.

* Ping Wu: pingwu@sas.ustb.edu.cn; Li Wang: liwang@me.ustb.edu.cn

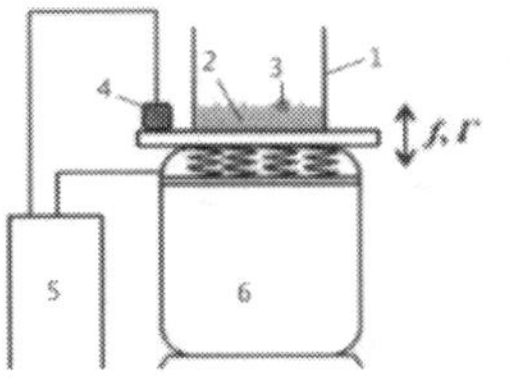

Fig. 1. Schematic of the experimental settings: 1) plexiglass container, 2)glass granular, 3) large porous alumina particles, 4) horizontal balance detector, 5) vibration controlling systems through PC, 6) vibrator apparatus

2.2 Experiment Results

2.2.1 Spontaneous cluster

In the experiment upper single-layer alumina particles a_1 were poured into the surface of statically vibrated system which was consisted of small glass particles g_1 (Φ_1 =1.25mm, Φ_2=4mm, f=30Hz, Γ=4). The alumina particles a_1 moved radially under the influence of surface wave. We encountered a phenomena in which there were some clusters of particles accompanied by the radial movement in surface behaviors, as shown in Fig.2.

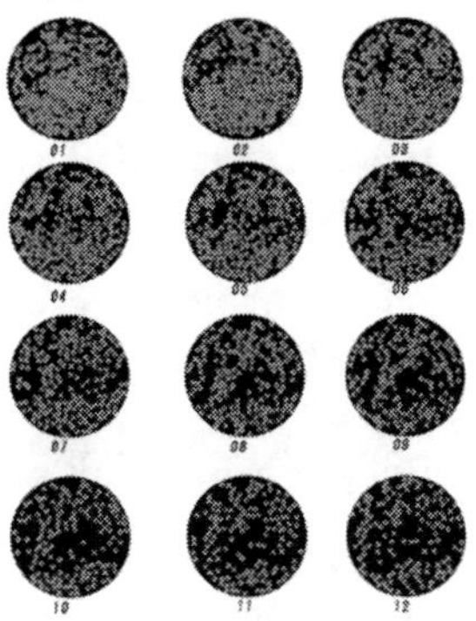

Fig.2. Time-series of the upper alumina particles a_1(red) surface behavior for under layers glass beads g_1(black) at Γ=4, f =30.0 Hz(Φ_1 =1.25mm, Φ_2=4mm). The time interval was 0.5s.

Particles collided and merged with one another, forming a particle cluster. The smallest clusters were consisted of two particles. When smallest cluster collided with another particle, the cluster was able to merge with it and "grew up". The life span of cluster cell could be several seconds. The evenly distribution of surface particles had been shattered and alumina particles layer disintegrate into several parts swiftly. However, the general trend of these clusters and particles was radial motion, particles touched the wall of the container and "sank" in glass grains in the end. When cluster phenomenon occurred in alumina particles, they divided into several cells. The dimension of those cells could be different in various conditions (f, Γ, Φ_1, Φ_2), we needed to define a parameter to describe those clusters under same condition. The mean number of particles in one cluster cell under same condition (f, Γ, Φ_1, Φ_2) was defined as n. We also defined a parameter p which represented the probability of particles involved in a cluster. Each (n, p) parameters combination show a surface conformation of grains system.

To study the relationship between (n, p) and vibration parameters (f, Γ), a camera was used to record the surface conformation change and we counted the grains to keep statistic, as shown in Fig.3 and Fig.4. The unified parameter ε, which contains Γ and t, marks the activation energy that work on the alumina particle:

$$\varepsilon = \Gamma \cdot t \tag{3}$$

where Γ is the dimensionless acceleration and t is the time that recorded. We studied the relationship between (n, p) and vibration parameters (f, Γ) in cluster phenomenon clearly situation, for example, when f=40-80Hz, alumina particles a_1 tiled on glass grains g_1. Captured the pictures when ε=12 and plotted the statics results as shown in Fig. 3 and Fig. 4.

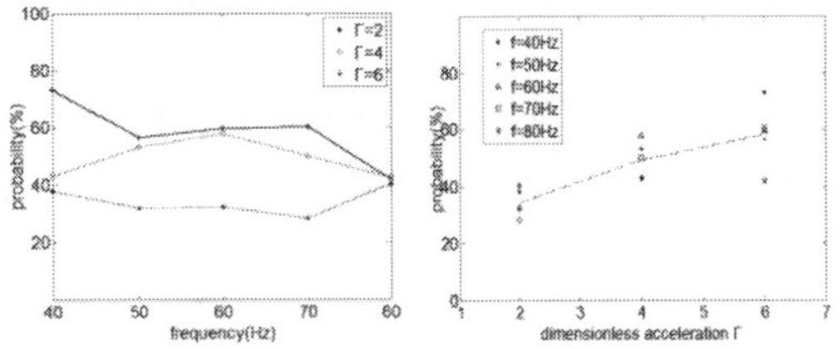

(a) (b)

Fig.3. Diagram for surface alumina grains a_1 cluster probability (p) and vibration parameter (f, Γ) on glass grain g_1, ε=12, Φ_1 =1.25mm, Φ_2=4mm. (a) p-f relationship; (b) p-Γ relationship.

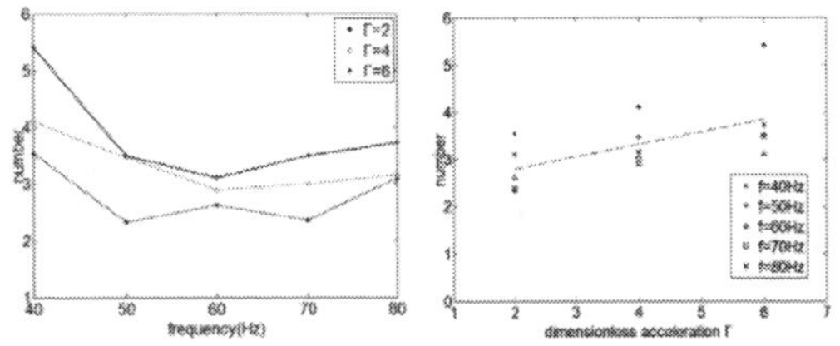

(a) (b)

Fig. 4. Diagram for surface alumina grains a_1 cluster cell mean particle number (n) and vibration parameter (f, Γ) on glass grain g_1, ε=12 ,Φ_1 =0.9mm, Φ_2=4mm. (a) n-f relationship; (b) n-Γ relationship.

According to Fig.3 and Fig.4, both two vibration parameters have effects on the scale and the forming probability of cluster cells. However, we could not find brief descriptions of these influences. As cluster phenomenon occurred, surface particles trended to form big cells when Γ was smaller, nonetheless, the probability of cluster is lower. By contrast, small cells (cells consisted by 2~3 particles) are more likely to form when Γ is quiet big, meanwhile, the particles were more likely to move individually. What could be said is gamma effected the change more apparent, perhaps, because gamma is directly proportional to short-range repulsive force. Higher frequency probably could only offer more opportunity for particles to recombine clusters, which maybe not play the main part in the time-averaged situation. Dynamics model building is still in progress.

2.2.2 Multiple-size particles spontaneous cluster

When the alumina particles were multiple-size, we could observe the spontaneous cluster effect more obviously, see Fig.5. The insertion of larger alumina particles accelerated the formation of cluster cells. Large particles were rapidly packed around by large grains which are close-packed.

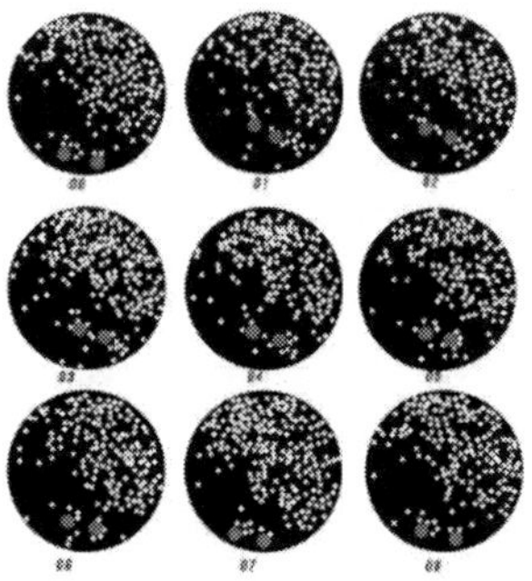

Fig. 5. Time-series of the multiple-size alumina particles (a_1 red and a_2 blue) surface behavior for under layers glass bead s(g_1 black) at dimensionless acceleration Γ=4, f=30.0 Hz (Φ_1 =1.25mm, Φ_2=4mm(red), 8mm(blue)). The time interval was 1.0s.

2.2.3 Three-dimensional diffusion

A parameter critical frequency f_u is defined to distinguish two kinds of diffusion in particles. When the vibration frequency was bigger than f_u, the surface diffusion disappeared and a three-dimensional diffusion appeared, as shown in Fig.6. Upper alumina particles sank in the center of glass grains surface, compared with intense convection roll[7] in glass-layer. However, when 2D diffusion appeared, alumina particles only descended when they touch the wall of the container. Besides, the experimental results shown that the critical frequency f_u is related to the size and density of particles, but irrelevant with the vibrating intensity parameter Γ. Granular segregation was first reported in 1939 by Brown[26]. Under certain vertical vibration conditions, large particles in bi-disperse granular mixtures will move to the top of the bed, which is the so-called Brazil nut effect (BNE). Conversely, under certain different vibration conditions, large particles can sink into the granular bed, which is the so-called reverse Brazil nut effect (RBNE)[23]. Kong had indicated that frequency is the decisive factor of such kind of convert between these two effects[22]. An explanation of this phenomenon was proposed in previous studies: as vibration frequency increases, the granular bed is less compact than before, so that big particles sink into the assembly. Similar phenomenon had been observed in our experimental, however, f_u is too large for our instrument to work steadily. We didn't careful researches about "sink" phenomenon.

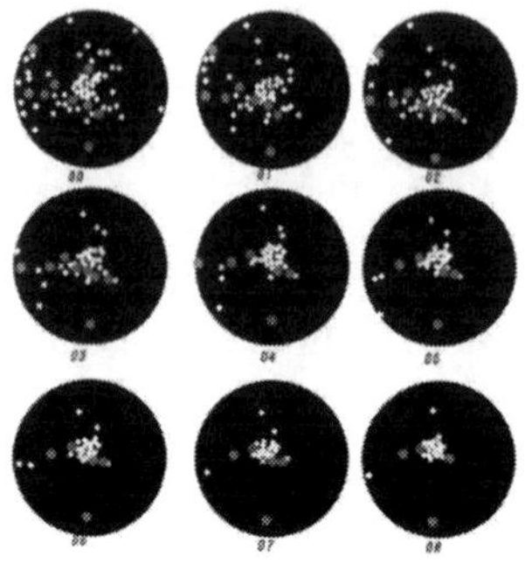

Fig. 6 Time-series of the alumina particles (a_1 red and a_2 blue) 3D diffusion on glass grains a_1, at dimensionless acceleration Γ=4, f=100.0 Hz(Φ_1=1.25mm, Φ_2=4mm). The time interval was 1.0s. The colored grains "disappeared" in the center because they descended to the glass layer.

3 Discussion

Based on Fig.3 and Fig.4, both two vibration parameters have effects on the scale and the forming probability of cluster cells. According to the statistic results, we found that although identical particles random walked on the glass beads under the condition of $f < f_u$, there was a great chance of surface grains cluster phenomenon instead of even distribution.

When comes to pursuing particle dynamics, the collective motion of granular materials can be described by the excluded volume effect and the dissipation of kinetic energy between granules [19]. A simulation model is described below. A short-range repulsive force was defined as interacting between two particles: the excluded volume is expressed by a short-range repulsive force, which is directly proportional to Γ in our experiments, interacting between particles[24]:

$$f_{ij} = \frac{-\partial \Phi_{ij}}{r_{ij}} \tag{4}$$

where Φ_{ij} takes the form of a discrete short-range repulsive pair potential:

$$\Phi_{ij} = \begin{cases} \varepsilon[(\frac{d}{r_{ij}})^{12} - (\frac{d}{r_{ij}})^{6} + \frac{1}{4}], if |rij| < r_0 \\ 0, orherwise \end{cases} \tag{5}$$

where $r_{ij} = r_j - r_i$ and $r_0 = 2^{\frac{1}{6}} d$, d is the characteristic length representing the particle size. Identical grains have same pair potential function in the initial state, considered that all alumina particles could acquire same energy ε during impacting with the glass substrate once. The energy was stored in this distribution is:

$$E = \int_{i,j=1}^{n} \Phi_{ij} \tag{6}$$

When $r_{ij}=d$, two particles tangent to each other, the potential of these particles will reach a maximum value. Our discussion is motivated by the "friendship paradox" that "*your friends have more friends than you do*"[25]. During the vibration, the original distribution has N alumina particles which impact each other x_i times. The mean collision times of the system can be determined as $\bar{x} = \sum x_i / n$ and the mean transferred energy is $E_1 = \sum x_i e / n$, where e is the transform-energy unit per collision. Thus, the mean number of particles which impact with others' impact number is $x_w = \sum x_i^2 / \sum x_i$, which is the weighted average of the impact time of each particle. $E_2 = \sum x_i^2 e / \sum x_i$.

$$x_w = \bar{x} + \frac{\sigma^2}{\bar{x}} \tag{7}$$

where σ^2 is variance of the system. Weighted average is bigger than the mean number of the whole system, that means those alumina particles which has ever impact with other particles has transfer more energy. In the other words, there is some probability that some of grains have less probability to connect with others. Some of grains form cluster cells and reduce the chance of collision. When the short-range repulsive force is larger, the energy transfer will be more efficient. Therefore the difference of connecting probability between particles could be more obvious. From what has been discussed above, we may safely explain some of our experiment results.

4 Conclusions

We investigated the behavior of the large porous alumina particles on the free surface of the small glass-grain system under vertical vibration. When frequency was lower than critical frequency f_u, the large grains walked radially and quasi-two-dimensional surface diffusion occurred. For uniform size alumina grains, spontaneous cluster phenomenon had been found and the statistic properties of cluster cells had been researched. When frequency was higher than critical frequency f_u, 2D diffusion disappear and 3D convection appear. The alumina grains sank in the center of glass beads surface. Lastly, although not presented here, for experiments in high frequency, we found that the parameter f_u was not related with vibration parameters. Besides, it was related to the size and density of particles.

This article was sponsored by National Natural Science Foundation of China (No. 51476009).

References

1. Istafaul H A, Meheboob A, AIP Conference Proceedings **1542**, 775 (2013)
2. Peter E, Ko van der W, Devaraj van der M, Robert B, and Detlef L, Physics of Fluids **19**, 123301 (2007);
3. Eric C, Loic V, Jean R, and Jacques D, Phys. Rev. E **53**, 2972 (1996)
4. Stephen W, Jonathan M. H, and George T. H. J, Phys. Rev. E **52**, 5583 (1995)
5. Thorsten P, Thomas S and Clara S, Phys. Rev. E **62**,1361(2000)
6. Vincent M, Lydéric B, and Annie C, Soft Matter **10**, 6984-6989 (2014)
7. Frank R, AIP Conference Proceedings **1542**, 771 (2013)
8. Frank R and Ralf S, AIP Conference Proceedings **1145**, 741 (2009);
9. Sakon K, Charles S. Campbell, Physics of Fluids **20**, 103303 (2008);
10. Zhang F, Wang L, Liu C, Wu P, Zhan S, Physics Letters A **378**, 1303-1308 (2014)
11. Hu K W, Xie Z A, Wu P, Sun J, Li L, Jia C, Zhang S P, Liu C P, Wang L, Soft Matter **10**, 4348-4359 (2014)
12. Xie Z-A, Wu P, Zhang S P, Chen S, Jia C, Liu C P, Wang L, Physical Review E **85**, 061302 (2012)
13. Xie Z-A, Wu P, Yang W, Zhao J, Zhang S, Li L, Chen S, Jia C, Liu C, Wang L, Powder Technol. **260**, 1-6 (2014)
14. Sun J, Liu C, Wu P, Xie Z-A, Hu K, Wang L, Phy. Rev. E **94**, 032906 (2016)
15. Liu C P, Wang L, Wu P, Jia M, Physical Review Letters **104**, 188001 (2010)
16. Gauri R. P, Sagar A. Pa, Anil D. G, V. Sitaramamb, Physica A **270**, 288-294 (1999)
17. Krishna K, Eric F , Kapil M S, Physica A **270** , 97-104 (1999)
18. Nicholas C, Tom De W, Kristof V, Ilse C, Applied and Environmental microbiology **73**, 1809-1824 (2007)
19. Keiko M Ai, Tetsuo A, Phys. Rev. Lett. **77**, 4166 (1996)
20. Xie Z-A, Wu P, Wang S, Huang Y M, Zhang S P, Chen S, Jia C, Liu C P, Wang L, Soft Matter **25**, 35-57 (2013)
21. Li L, Wu P, Rehman A, Wang L, Zhang S, Xie Z-A, Powder Technol. **307**, 84-89 (2017)
22. Xiang-Z K, Mao-B H, Qing-S W, Phys. Lett. A **356**, 267-271 (2006)
23. Shinbrot T , Muzzio F J, Phys. Rev. Lett. **81**, 4365-4368 (1998)
24. Keiko M. Aoki ,Tetsuo Akiyama, Phys. Rev. E 52, 3288 (1995)
25. Scott L. Feld, The American Journal of Sociology, 96,1464-1477(1991)
26. Brown R.L, Inst. Fuel 13, 15(1939)

On the progressive nature of grain crushing

Matteo O. Ciantia[1*], *Gema* Piñero[2], *Jian* Zhu[1] and *Tom* Shire[1]

[1] Dept. Civil and Environmental Engineering, Geotechnics Section, Imperial College London, UK
[2] Institute of Telecommunications and Multimedia Applications, Universitat Politecnica de Valencia, Spain

Abstract. In this work acoustic emission (AE) is used as experimental evidence of the progressive nature of grain crushing. Stress controlled high pressure oedometric compression test are carried out on 1.2 mm monodisperse samples of glass beads. It was observed that the granular assembly starts to experience particle breakage at a vertical stress of about 25MPa. When this yield pressure is exceeded the glass beads start to break emitting loud impulsive sound and the vertical displacement increases rapidly. The load was increased beyond the yield stress and at each increment while the vertical stress remained constant the sample continued to emit sound. The emission of sound at a constant vertical stress indicates that crushing is a progressive failure mechanism; once the first crushing event occurs, the structure starts to rearrange causing other crushing events to occur and additional settlement. In particular, two signal processing algorithms are used on the samples of the acoustic signal to obtain two additional metrics of the crushing evolution. The first is the cumulative energy versus time. The second is the number of crushing events versus time, which is based on the automatic detection of the peaks of the sound signal envelope. There is a clear correlation between the cumulative acoustic energy emitted and the observed sample displacement. Using laser scanning, the evolution of the particle size distribution and particle shape are measured in detail so that a link between the acoustic data and the crushing intensity is established. The crushing intensity was controlled using materials with different strengths.

1 Introduction

It is well acknowledged that most solids emit low-level seismic signals when they are deformed or stressed [1]. Such phenomenon of generating signals is termed as acoustic emission (AE) activity. AE are the mechanically induced elastic waves typically generated by grain rearrangement, grain inter-friction and crack formation for granular geomaterials [2]. The measurement and quantification of AE was initially developed for the attempt to predict the rock bursts for mining industry, and recently increasingly utilized for laboratory-scale studies for granular materials given their discrete nature [3,4]. Since the measurements of macroscopic stress and total deformation do not allow the grain-scale inspections of the microscopic responses of the particles, the application of acoustic emission techniques provides a new approach to looking inside the aggregates when they are stressed. As emphasized by Hardy Jr [1], although AE does not directly determine the basic mechanical parameters like stress and strain, it does provide a complement for the mechanical measurements of stress and displacement so that the interactions among particles can be revealed. Hidalgo et al. [5] used acoustic emission to study the force chain evolutions of granular materials with a 140mm diameter cylinder sample while smaller sized specimens were also used to study the particle size distribution evolutions of powders [6]. Frequency and attenuation of measured acoustic events can be related to certain physical and mechanical processes of granular materials under shearing or compression. Measurement of AE are realized by detecting and monitoring the generation and the propagation of elastic waves due to the release of stored elastic strain energy during material deformation and abrupt grain interaction. The total energy of the AE is dominated by its sources, and usually described in terms of attenuation and signal pulse frequency [2].

Grain crushing is an elastic strain energy release mechanism that produces AE with a loudness and frequency within the range of audible sound. It is hence possible to detect and record crushing events within a stressed geomaterial using normal microphones. Whereas to capture other source mechanisms of AE such as grain friction and rearrangement more advanced sensors are required.

Experimental [7] and numerical [8] studies have shown how grain crushing in granular media is a progressive type of mechanism. Upon a small load increment, if a grain crushes, its load must be redistributed to the remaining intact grains. This load redistribution may lead to secondary failures of elements and triggers avalanches of failures before the load is distributed to a sufficient number of intact elements to return to an equilibrated state [2].

* Corresponding author: m.ciantia@imperial.ac.uk

Due to the discrete nature of acoustic emissions associated with failure of individual grains, the progressive material degradation should also cause avalanches of AE. The focus of this paper is to hence experimentally demonstrate how the progressive nature of crushing in granular materials can be indirectly observed through the recording and analysis of AE. High pressure one-dimensional compression tests were performed. In addition to stresses and strains, size and shape evolution was also monitored. To control the level of crushability, the tests were run on mixtures of crushable and non-crushable spherical particles. This was complimented by AE analysis of the sound emitted during compression. A simple methodology to estimate the number of crushing events occurring within the sample is also proposed.

2 Testing methodology

2.1 Materials tested

Spherical glass and steel beads were tested. The physical properties of the materials are given in Table 1. Sphericity (perimeter of equivalent circle ÷ real perimeter) and aspect ratio (ratio of Ferer min and max diameters) were determined from the QicPic laser image analysis instrument [9]. Both materials are very spherical and have uniform particle size distributions (low R_D values, where $R_D = D_{90}/D_{10}$). The glass beads have a slightly larger median diameter. The steel beads are uncrushable under the stresses applied here.

Table 1 Properties of granular materials tested

Property	Glass Beads	Steel Beads
Density (-)	2.60	7.81
Shear Modulus (GPa)	70	200
Nominal Strength (MPa)	70	1000
Median diameter, D_{50} (mm)	1.29	1.0
Relative distribution, R_D (-)	1.13	1.02
Sphericity, S^{PQ} (-)	0.93 – 0.95	0.93 – 0.95
Aspect ratio, A_r (-)	0.96 – 0.99	0.98 – 0.99

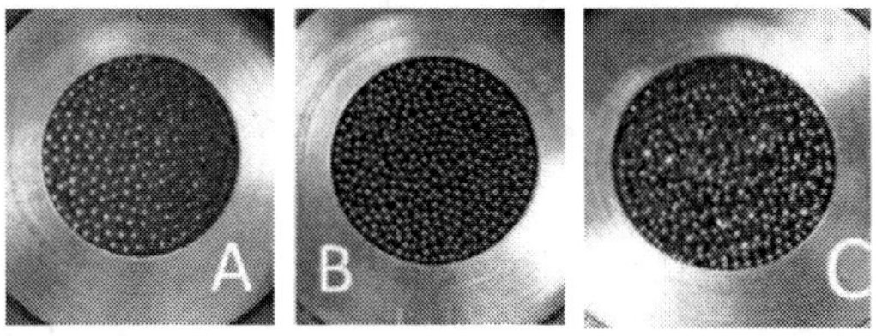

Fig. 1. Picture of the glass (G), steel (S), and mixture (M) samples in the oedometric cell.

In the following, results from tests on three samples are presented: (i) glass beads only (G), (ii) steel beads only (S), (iii) 50% glass and 50% steel by volume (M) and in Fig. 1 photographs of the three samples before the high pressure oedometric compression test are reported.

2.2 One-dimensional compression

A load-controlled oedometer apparatus was used to carry out the one-dimensional compression tests. A 20mm diameter confining ring was used to enclose the bead mixtures. Vertical loads were applied in stages using dead-weights, up to a maximum vertical pressure of 49.7 MPa. Samples were prepared using a combination of pouring and light tapping. The mixed sample was thoroughly mixed prior to placement to minimise particle segregation. The initial void ratios are given in Table 2. Displacement was monitored throughout the loading stage. Each loading stage was maintained until the vertical deformation was less than 0.003mm/s for at least 20 minutes. Following unloading the particle size and shape characteristics of the glass fraction was determined using the QicPic instrument [9].

Table 2. Properties of granular mixtures.

	Glass	Steel	Mixture
Mass (gr)	9.13	27.46	18.19
Glass fraction (% mass)	100	0	33.35
Glass fraction (% vol)	100	0	50
Initial void ratio, e_0 (-)	0.593	0.59	0.603
Final void ratio, e_f(-)	0.225	0.531	0.353
Yield stress, σ_y (MPa)	27	-	30

2.3 Acoustic emission analysis

AE were recorded and analysed for each test using the embedded microphone of a MacBook Air and Audacity®, a free open source digital audio recording and editor software (http://www.audacityteam.org/). The laptop was placed about 30cm from the oedometer, the sampling frequency was $f_s = 44100$ Hz, resulting in a sampling time of $T_s = 1/f_s = 22.7$ µs and a signal bandwidth between 0 and 22050 Hz for the recorded sound. The sound signal of the G test is shown in Fig. 2a, the units are arbitrary. The maximum range permitted is between -1 and +1. It can be appreciated that the signal has a lot of peaks, but also bursts of energy between the peaks. The detailed shape of peaks and bursts can be appreciated in Fig. 2b where three bursts are captured. Their peak amplitudes are different, but their burst durations are quite similar. Every burst has a duration around 60 ms. Two measures of acoustic emission were attained, namely sound cumulative energy (E_{cu}) and the number of acoustic events (N_c) during a loading increment. If $s(t)$ is the sound signal, the recorded signal can be described by its discrete-time signal $s(n)$:

$$s(n) = s(nT_s) = s(t)\big|_{t=nT_s}, n = 0,1... \quad (1)$$

Therefore, the discrete-time cumulative energy is expressed as:

$$E_{cum}(n) = \sum_{k=0}^{n} |s(k)|^2 \quad (2)$$

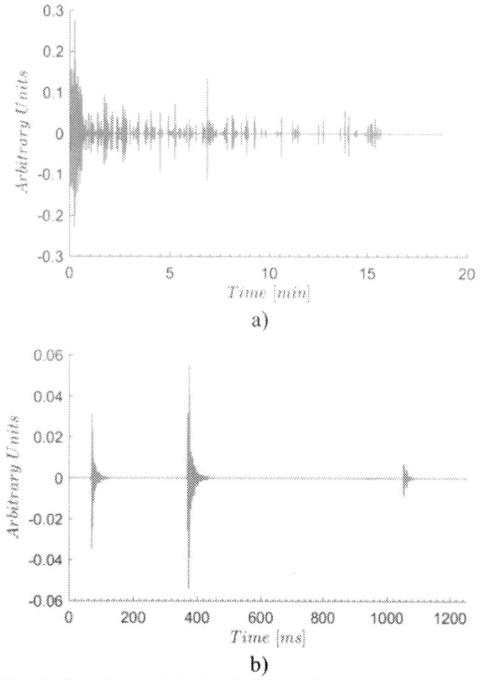

Fig. 2. Sound signal for the G test; a) 12 minutes recording and b) 1200ms sound signal segment detail showing three crush events.

To avoid unwanted low-frequency noise in lab influencing the calculation of E_{cu}, a high-pass-filtered (HPF) signal of cut-off frequency f_c = 8000 Hz and an attenuation of 60 dB in the stop band The resulting signal is called $s_{HP}(n)$. In Fig. 2b a sound segment of the original AE recording is shown where three bursts (each one associated with a crushing event) can be identified. It can be appreciated that the three follow a similar pattern but with different amplitudes. The high-pass filtered signal presents similar behaviour, and hence the crush events can be identified by a certain pattern (i.e. signal shape) although the amplitude may vary. Therefore, signal $s_{HP}(t)$ can be expressed as

$$s_{HP}(t)=\sum_{k=1}^{N_c} A_k p(t-\tau_k) \quad (3)$$

where A_k is the amplitude of the kth repetition of the pattern $p(t)$, τ_k is the time instant when $p(t)$ starts and N_c is the number of crush events. We can rewrite the above equation in discrete-time

$$s_{HP}(n)=\sum_{k=1}^{N_c} A_k p(n-n_k) \quad (4)$$

Where $p(n)=p(t)\big|_{t=nT_s}=p(nT_s)$, index $n_k=[\tau_k f_s]$ and [.] means the nearest integer to the argument. If we call $R_p(n)=E[p(k)p(k+n)]$ the autocorrelation function of the pattern, then $R_p(n)$ can be estimated as:

$$R_p(n)=\frac{1}{N_p-|n|}\sum_{k=0}^{N_p-1-|n|} p(k)p(k+|n|), -N_p<n<N_p \quad (5)$$

where N_p is the length of $p(n)$ in number of samples. Therefore, the cross-correlation function of the discrete-time signal $s_{HP}(n)$ and the pattern $p(n)$ is computed as

$$x(n)=\sum_{k=1}^{N_c} A_k p(n-n_k)p(-n)=\sum_{k=1}^{N_c} A_k R_p(n-n_k) \quad (6)$$

where $x(n)$ is a new signal that enhances the peaks of the original $s_{HP}(n)$ signal. At this point the amplitude envelope of $x(n)$ is computed by means of a Hilbert transform [11], and a crushing event is summed up to the count when a peak of the envelope is above a certain threshold. A parametric study assessing the influence of threshold on the Ncr-time curve was performed and for this type of AE signals, the method proposed resulted to be independent from the threshold value chosen.

3 Results

The compression curves for the three samples are shown in Fig. 3. It is clear how particle breakage induces large irreversible deformations which are proportional to the amount of glass beads in the sample. In Fig. 5 it is clear how the crushing induced deformation evolves with time, with rapid crushing on yield (point A, σ_z=33 MPa), followed by significant further displacement over time. Upon a single stress increment of 5 MPa a displacement of about 2 mm is recorded (Point B).

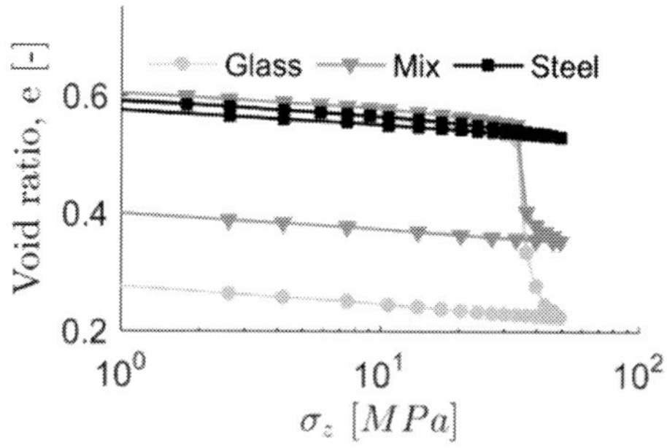

Fig. 3. Load unload oedometric compression; comparison of the void ratio evolution for the G, S and M tests.

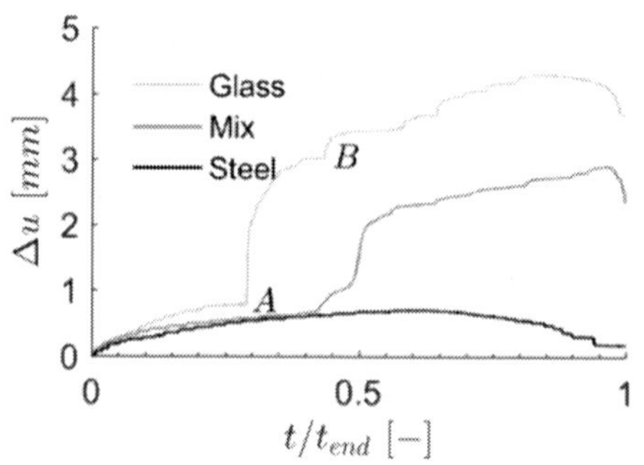

Fig. 4. Temporal evolution of displacements for the three tests.

The progressive nature of the crush induced deformation is well reflected in the AE analyses in terms of both E_{cu} and N_{cr}, as shown in Fig. 5 for the first yield of the glass sample G. This clearly demonstrates that AE analysis can capture the energy release during particle crushing. Following [9] QicPic laser image analyses can be performed on the sample at the end of the experiment and the particle size distribution (PSD) and sphericity evolution can be used as another indicator of crushing evolution (Fig. 6).

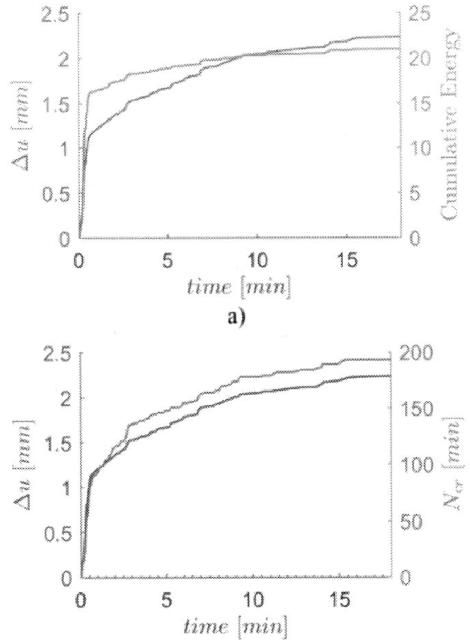

Fig. 5. Comparison of temporal evolution of a) cumulative energy and b) number of crushing events for the of Glass sample test from point A to B in **Fig. 4**.

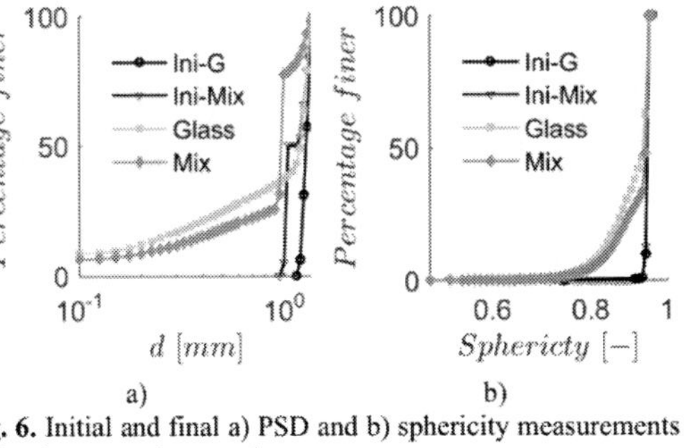

Fig. 6. Initial and final a) PSD and b) sphericity measurements of the three samples tested.

4 Conclusions

In this paper two signal processing algorithms are used to obtain two additional metrics of crushing evolution. The first is the cumulative energy versus time, and the second is the number of crushing events versus time. This last measurement is based on the automatic detection of the peaks of the sound signal envelope. The strong correlation between the time displacement curves and the time E_{cu} and N_{cr} indirectly confirm the progressive nature of grain crushing. Upon a small load increment, if a grain crushes a sharp burst is emitted, the load carried by the element is redistributed to the remaining intact grains leading to secondary failures of elements triggering avalanches of crushing episodes. The determination of the number of crushing events is a novel measure that may be used as an extra variable for DEM crushing model validation.

Acknowledgments

This study was undertaken as part of an Imperial College London Junior Research Fellowship Research Grant.

References

1. Hardy JR, H.R. (2005). Acoustic emission microseismic activity: volume 1: principles, techniques and geotechnical applications. CRC Press
2. Michelmayr, G.; Cohen, D. and Or, D. (2012). Sources and characteristics of acoustic emissions from mechanically stressed geologic granular media—A review. Earth-Science Reviews, 112(**3**), 97-114.
3. Michlmayr, G., Cohen, D., & Or, D. (2013). Shear-induced force fluctuations and acoustic emissions in granular material. Journal of Geophysical Research: Solid Earth, 118(**12**), 6086-6098.
4. Sha Luo S., Diambra A., Ibraim E. (2016) Application of acoustic emission on crushing monitoring of individual soil particles in uniaxial compression test. 32nd European Conference on Acoustic Emission Testing Prague, Czech Republic, September 07-09, 2016
5. Hidalgo, R.C., Grosse, C.U., Kun, F., Reinhardt, H.W. & Herrmann, H.J. (2002). Evolution of percolating force chains in compressed granular media. Physical Review Letters, 89(**20**), 205501-4.
6. Carson, G., Mulholland, A.J.; Nordon, A.; Tramontana, M.; Gachagan, A. and Hayward, G. (2008). Particle sizing using passive ultrasonic measurement of particle–wall impact vibrations. Journal of Sound and Vibration, 317(**1**), 142-157.
7. Tapias, M., Alonso, E. E., & Gili, J. (2016). Compressibility, grain breakage and time-dependent behavior of gap-graded aggregates of sugar cubes. Soils and Foundations.
8. Ciantia, M. O. Arroyo, M., Calvetti, F. & Gens, A. (2016) A numerical investigation of the incremental behaviour of crushable granular soils. Int. J. Numer. Anal. Meth. Geomech. 40(**13**) 1773–1798.
9. Cavarretta, I., Coop, M., & O'Sullivan, C. (2010). The influence of particle characteristics on the behaviour of coarse grained soils. Géotechnique, 60(**6**), 413-423.
10. Rabiner, L.R., Gold, B. (1975) Theory and application of digital signal processing, ed. Prentice-Hall, Englewood Cliffs, N.J.

DOI: 10.1051/epjconf/201714004004

Cluster splitting in granular segregation driven by horizontal shaking

Angel Garcimartin[1,*], *Ignacio* Larrea[1], *Celia* Lozano[2], and *Iker* Zuriguel[1]

[1]*Departamento de Física, Facultad de Ciencias, Universidad de Navarra, E-31080 Pamplona, Spain.*
[2]*Max-Planck-Institut für Intelligente Systeme, Heisenbergstraße 3, 70569 Stuttgart, Germany.*

Abstract. In a recent work [C. Lozano et al. Phys. Rev. Lett **114**, 178002 (2015)] segregation in an horizontally shaken granular layer was studied by analysing the particle-particle interactions in the simplest case possible of a two particles cluster. There, it was found that all clusters are transient (they eventually split if one waits long enough) and the probability distribution function of the separation times displays a power law tail, indicating that the splitting probability is not constant over time. Here, we extend this study to clusters of 3, 5, 10 and 20 particles where we also observe the power law decay of the distribution of cluster splitting time. In addition, we observe a weak increase of the average cluster splitting time with the cluster size, suggesting that interaction forces are non-additive. Finally, we show interesting statistics on the way in which clusters break suggesting that escaping of individual particles in the cluster borders is more likely than cluster breakage in subclusters of similar size.

1 Introduction

If a mixture of grains is fluidized, granular segregation may occur driven by differences in particle size, density, shape, surface properties, restitution coefficient, rolling friction, and so on [1–3]. Sometimes segregation occurs when similar particles tend to move towards a particular location. The canonical example of this is the Brazil nut effect in vertical shaking [4]. A different case is when there is not such distinct spatial affinity of any of the mixture constituents for a given place. Then, segregation necessarily occurs due to the existence of collective aggregation interactions which are stronger among similar particles than for different ones. The consequence of this is the formation of segregated patterns at locations that can be different from one experiment to another. Examples of this are the band segregation in a long drum [5] or stripe formation in an horizontally shaken granular mixture [6].

The origin of these particle-particle interactions can be of different nature, ranging from excluded volume effect [7, 8] to long range fluctuation induced forces [9, 10]. These may appear due to differences of either pressure [11], density [12], or velocity fluctuations [13] in the region between clustering particles and the surroundings. In a previous work [14], we took a well-known system that leads to segregation as it is an horizontally shaken shallow layer of a binary mixture of dry particles [15–20] and we analysed the interaction among pairs of spheres within the bulk of other, different kind of particles. In such conditions, stripe formation has been systematically found in the direction perpendicular to the forcing, a behavior also observed in other systems [21–23]. The particle-particle interaction study was done by extensive measurements of the dynamics of the pairs of particles when they come together but also by the cluster properties, and the particle separation [14]. The results suggest that the dynamics is controlled by a competition between attraction forces and random agitation, both depending on the packing fraction within the system determined by the amount of bulk particles. As the packing fraction increases, the random motion of the particles reduces and the attraction becomes stronger. Interestingly, even for the densest case (which

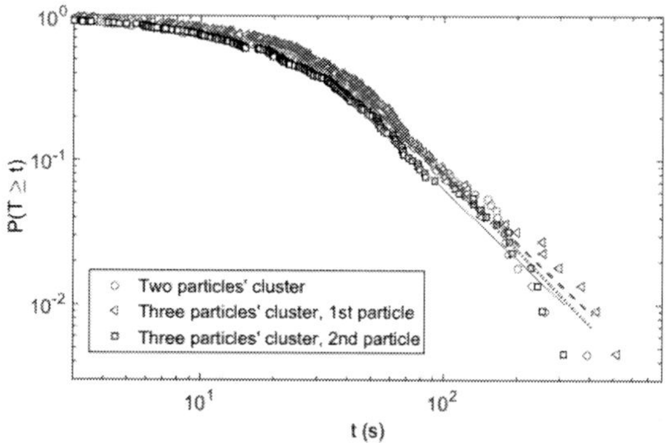

Figure 1. Complementary CDFs of the time that a cluster takes to split. As indicated in the legend the three different symbols indicate 1) the splitting time of a two particles' cluster, 2) the splitting time of the triplet which is given by the time that it takes to the first particle of the cluster to separate from the other two; and 3) the splitting time of the two particles' cluster which remains after the splitting of the triplet. Lines are power law fits of the distribution tails.

*e-mail: angel@unav.es

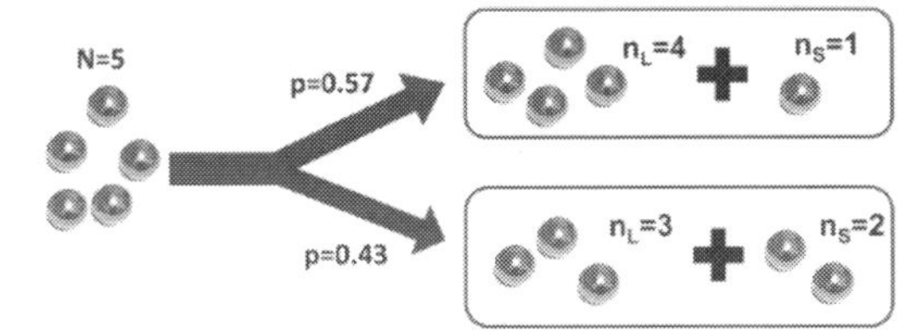

Figure 2. Sketch illustrating the two possible ways in which a 5-particles cluster splits into two smaller ones.

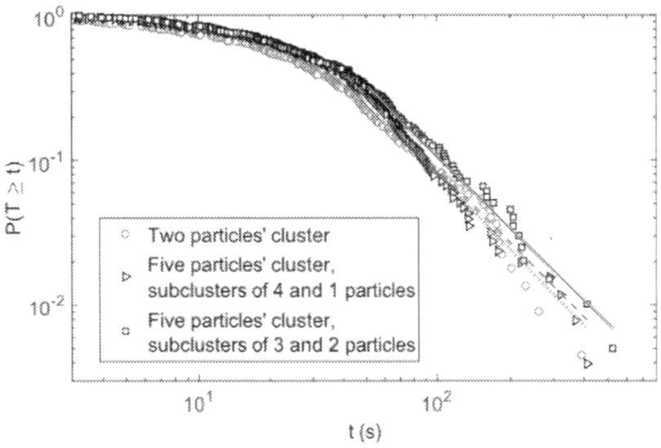

Figure 3. Complementary CDFs of the time that a 5 particles cluster takes to split compared with the case of 2 particles in the same conditions. Two different distributions are shown depending on the way in which the cluster splits: either in two subclusters of 4 and 1 particles or in 3 and 2 as shown in Fig. 2 and indicated in the legend.Lines are power law fits of the distribution tails.

leads to a very segregated pattern) the two particles' clusters eventually split in a finite amount of time. Moreover, it was reported that the distribution of these splitting times is a power law decay, a result that is robust as it does not depend on particle's properties.

2 The experiment

Despite all the knowledge gained about particle-particle interactions in our previous work and other recent ones using a similar approach [24–26], its extrapolation to a system with many particles has not been implemented. Obviously, the two particles case represents the foundation upon which to build any theory or model aiming to explain the global segregation. The gap, however, is perhaps still too large and therefore, we believe that the study of the properties of a few particles' clusters could be useful as an intermediate step. This is precisely the goal of this preliminary work where we have studied the splitting process of clusters analysing both the time that it takes to the cluster destruction and the way in which the clusters split in terms of the size of the subclusters produced.

The experimental setup is the same as the one described in [14] in which we are going to work with poppy seeds (kidney shaped, with an average diameter of $d = 1 \pm 0.2$ mm) as bulk particles with a concentration $C = 0.91$. Then, we prepare isolated clusters of phosphor-bronze spheres ($d = 1.5$ mm) of size N, ranging from $N = 2$ to $N = 20$ particles. From the initial cluster, we measure the time that it takes until it breaks apart. This condition is fulfilled when any of the particles conforming the cluster separates more than $\Delta - d > 12$ mm from the closest one, with Δ being the distance among the particles centers.

3 Results

In Fig. 1 we show the survival time distribution (also called Complementary Cumulative Distribution Function CDF) for a triplet together with the reference case of a duplet in the same experimental conditions. Also, we show the survival distribution of the duplet that remains formed after the separation of the first particle of the triplet. Interestingly, all the distributions look very similar without remarkable differences in the power law exponent $\alpha = 2.5$. Indeed, the average splitting time of the duplet and the triplet are almost identical: $\langle t \rangle = 44$ s for the duplet and $\langle t \rangle = 45.8$ s for the triplet. Remark that the notation $\langle t \rangle$ is used in this paper to define an average time. The stability of the duplet remaining after the destruction of the triplet seems to be slightly lower than the standard (isolated) duplet as the distribution shown in Fig. 1 goes systematically below the other ones. In effect, the average splitting time of this remaining duplet is $\langle t \rangle = 35.8$ s, which is considerably lower than the standard case. This behavior is attributed to the effect that the already escaped particle might have on the other two, causing an effective attraction that helps to the separation of the two particles' remaining cluster.

Independently on what happens with the stability of the remaining cluster, the results obtained for the splitting times of the triplet and the standard duplet prompt us to think that the forces among particles are non-additive. To confirm this, we have performed the same sort of experiments with clusters of $N = 5$ particles. In this case, we are not going to look to the dynamics of the remaining clusters as the diversity of situations is rather large. Indeed, there is not a single way of splitting. As sketched in Fig. 2 the remaining subclusters might be in a configuration of $n_L = 4$ and $n_S = 1$ or, otherwise, $n_L = 3$ and $n_S = 2$, where n_L is the number of particles in the larger remaining cluster and n_S the number of particles in the smaller one. Note that due to the criteria we use to decide the time at which the cluster has been broken, the remaining clusters at that precise time is necessarily two. Remarkably, the probability of giving rise to a 4 + 1 configuration, $p(n_S = 1) = 0.57$, is larger than the 2 + 3 one, $p(n_S = 2) = 0.43$. Again, for these 5-particles clusters, the survival distribution of the splitting times (Fig. 3) evidences a power law with similar exponents that for the 2 and 3 particles scenario, yet it should be mentioned that the average splitting time increases a bit.

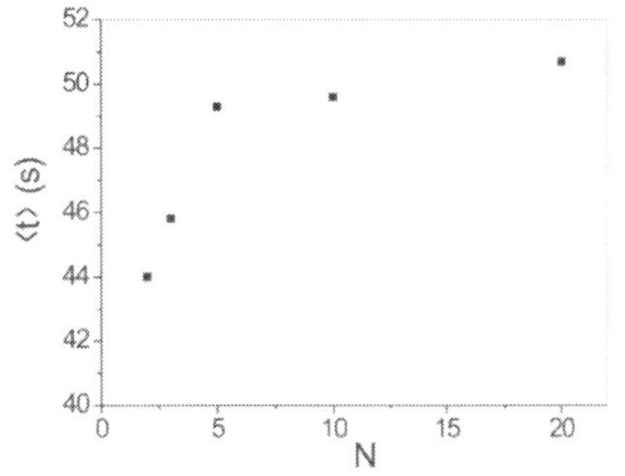

Figure 4. Average splitting time $\langle t \rangle$ of clusters formed by different number of particles N.

Having realized the slight tendency to augment the average splitting time with the cluster size, we have performed more experiments with clusters of $N = 10$ and $N = 20$ particles. The data shown in Fig. 4 suggest that, if this trend exists, it becomes negligible for sufficiently large clusters where the splitting time is independent on the cluster size. This result, which is still not fully understood, definitively shows the non-additive character of the interactions among particles in these experimental conditions.

Finally, we have analysed the way in which the $N = 10$ and $N = 20$ clusters split in terms of the size of the subclusters remaining after the initial cluster destruction. The outcomes are shown in Fig. 5a and 5b for $N = 10$ and $N = 20$ respectively. As in the case of 5 particles, the clusters necessarily split in two and the size of the smallest one (n_S) is taken as descriptor. Clearly, clusters break rather asymmetrically, in the sense that the most probable situation is to find a very large and a very small subclusters. This phenomenology was already glimpsed from the results of the 5-particles cluster but it becomes robustly confirmed with the results of these larger clusters. Also, the graphs seem to suggest that the difference between the $19 + 1$ configuration and $18 + 2$ is more important than the one between the successive configurations. These results evidence that the way in which clusters break is mostly due to escaping of individual particles from the borders instead of cluster breakage in subclusters of similar size.

Another subtle feature observed in Fig. 5 is that cluster splitting in subclusters of even number of particles is slightly less favoured than splitting in subclusters of odd number of particles. In effect, the probabilities obtained for $n_S = 2, 4$ (in the case of $N = 10$) and $n_S = 4, 6, 8, 10$ (in the case of $N = 20$) are underpopulated when compared with the values that the global trend suggest. We speculate that this phenomenology could be related with a preferred particle arrangement.

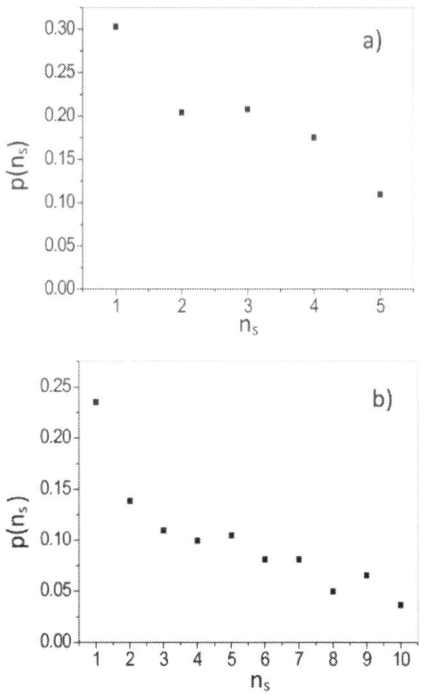

Figure 5. Probability that a cluster of a) $N = 10$ and b) $N = 20$ particles splits in a configuration where the smaller subcluster is composed of n_s particles.

4 Conclusions

In this work we have performed a preliminary experimental analysis of the splitting process of clusters of phosphorbronze particles within a sea of poppy seeds. The way in which this study is performed (looking at isolated clusters) supposes a clear differentiation from previous works where the dynamics of patterns with many clusters was investigated, hence observing interactions among clusters as the key factor determining the arising collective behavior. Based on previous knowledge of particle-particle interaction in duplets, we have extended the analysis to many particles' clusters. The first remarkable result is that the time that it takes until any of the particles of the cluster (or a group of them) leaves it does not depend very much on the cluster size. Also it is interesting to see that the clusters tend to split in subclusters of very different size, with a strong tendency to observe a single particle escaping from the other ones. This trend seems to be accentuated when the original cluster size enlarges as the probability of getting $n_s = 1$ with respect to $n_s = 2$ increases: it is 32% higher for $N = 5$, for $N = 10$ it is 48% higher, and 70% higher for $N = 20$. From these results it is obvious that further investigation of the geometrical properties of the clusters is necessary. In fact, it is observed that the clusters adopt elongated shapes, mostly perpendicular to the forcing direction. This information, seems to be a crucial ingredient to include in any model that aims to reproduce the complexity in the cluster dynamics reported here.

Finally, we would like to mention a collateral result of this work in relation with the stability of the remaining clusters, a measurement that has been only performed for the clusters of three particles. We show that the time that the remaining duplet takes to split is considerably smaller than the one obtained for an isolated duplet. This is attributed to the attraction that the already escaped particle induces on the duplet, hence speeding up the splitting process.

Acknowledgements

We thank T. Mullin for his help and discussions. This work was funded by Ministerio de Economía y Competitividad (Spanish Government) through Project No. FIS2014-57325.

References

[1] I. S. Aranson and L. S. Tsimring, Rev. Mod. Phys. **78**, 641 (2006).
[2] A. Kudrolli, Rep. Prog. Phys. **67**, 209 (2004).
[3] J. M. Ottino and D. V. Khakhar, Annu. Rev. Fluid Mech. **32**, 55 (2000).
[4] M. E. Möbius, B.E. Lauderdale, S.R. Nagel, H.M. Jaeger, Nature **414**, 270 (2001).
[5] K. M. Hill, A. Caprihan, and J. Kakalios, Phys. Rev. Lett. **78**, 50 (1997).
[6] T. Mullin, Science **295**, 1851 (2002).
[7] T. Shinbrot and F. Muzzio, Nature, 410, 251 (2001).
[8] M. Bose, P. R. Nott and V. Kumaran, Europhys. Lett. **68** 508 (2004).
[9] C. Cattuto, R. Brito, U. M. Marconi, F. Nori, and R. Soto, Phys. Rev. Lett. **96**, 178001 (2006).
[10] M. R. Shaebani, J. Sarabadani, and D. E. Wolf, Phys. Rev. Lett. **108**, 198001 (2012).
[11] S. Aumaitre, C. A. Kruelle, and I. Rehberg, Phys. Rev. E **64**, 041305 (2001).
[12] D. A. Sanders, M. R. Swift, R. M. Bowley, and P. J. King, Phys. Rev. Lett. **93**, 208002 (2004).
[13] I. Zuriguel, J. F. Boudet, Y. Amarouchene, and H. Kellay, Phys. Rev. Lett. **95**, 258002 (2005).
[14] C. Lozano, I. Zuriguel, A. Garcimartín, and T. Mullin, Phys. Rev. Lett. **114**, 178002 (2015).
[15] T. Mullin, Phys. Rev. Lett. **84**, 4741 (2000).
[16] P. M. Reis and T. Mullin, Phys. Rev. Lett. **89**, 244301 (2002).
[17] P. M. Reis, G. Ehrhardt, A. Stephenson and T. Mullin, Europhys. Lett. **66**, 357 (2004).
[18] P. M. Reis, T. Sykes, and T. Mullin, Phys. Rev. E **74**, 051306 (2006).
[19] D. Pihler-Puzović and T. Mullin, Proc. Roy. Soc. A **469**, 20130203 (2013).
[20] M. P. Ciamarra, A. Coniglio, and M. Nicodemi, Phys. Rev. Lett. **94**, 188001 (2005).
[21] H. Löwen, Soft Matter **6**, 3133 (2010).
[22] R. Moosavi, M. Maleki, M. Reza Shaebani, J. C. Ruiz-Suárez and E. Clément, Europhys. Lett. **107**, 34006 (2014).
[23] R. Möbius and C. Heussinger, Soft Matter **10**, 4806 (2014).
[24] G. M. Rodríguez-Liñán, Y. Nahmad-Molinari, G. Pérez-Ángel, PLoS ONE **11(5)**, e0156153 (2016).
[25] R. A. López de la Cruz and G. A. Caballero-Robledo, Journal of Fluid Mechanics, Volume **800** 248-263 (2016).
[26] J. C. Burton, J-J. Liétor-Santos, arXiv:1604.05360v2

Influence of particle size in silo discharge

Diego Gella[1,*], *Diego* Maza[1], and *Iker* Zuriguel[1]

[1]*Departamento de Fisica y Matematica Aplicada, Facultad de Ciencias, Universidad de Navarra, E-31080 Pamplona, Spain*

Abstract. Recently Janda et al. [Phys. Rev. Lett. **108**, 248001 (2012)] reported an experimental study where it was measured the velocity and volume fraction fields of 1 mm diameter stainless steel beads in the exit of a two-dimensional silo. In that work, they proposed a new expression to predict the flow of granular media in silos which does not explicitly include the particle size as a parameter. Here, we study if effectively, there is not such influence of the particle size in the flux equations as well as investigate any possible effect in the velocity and volume fraction fields. To this end, we have performed high speed motion measurements of these magnitudes in a two-dimensional silo filled with 4 mm diameter beads of stainless steel, the same material than the previous works. A developed tracking program has been implemented to obtain at the same time both, the velocity and volume fraction. The final objective of this work has been to extend and generalize the theoretical framework of Janda et al. for all sizes of particles. We have found that the obtained functionalities are the same than in the 1 mm case, but the exponents and other fitting parameters are different.

1 Introduction

Despite it has been studied since years ago, the discharge of silos through holes is still an open problem in granular matter physics. It is well known that the flow rate of particles W is a power law of the outlet size $D = 2R$, so that the exponent depends on the dimensionality of the system. The most widely expression used to fit that magnitude was proposed by Beverloo et al. [1] in 1961 from a dimensional analysis,

$$W = C\rho_b \sqrt{g}(D - kd_p)^{n-1/2} \tag{1}$$

where C is a fitting constant, n is the dimensionality and $\rho_b = \rho\phi_b$ is the bulk density of the material, being ϕ_b the bulk packing fraction and ρ the density of the pure material. The factor $\sqrt{g}$ arises from the concept of *free fall arch* developed by Brown and Richards [2]. This idea assumes the existence of a parabolic [3] or hemispheric [4] region over the outlet above which the particles are subject to contact stress and under which they fall freely under the action of gravity [5]. Instead of D, Eq.(1) includes a reduced outlet size $D - kd$ involving a term proportional to the particle size. This term, also supported by Brown and Richards [2] and known as *empty annulus*, is associated with the existence of a space of indefinite size near the outlet borders where particles are forbidden to pass through. Although the ideas above are difficult to justify from a physical point of view, they have been used in diverse systems[6] and Beverloo's expression has been proved to be useful to fit the flow rate in different circumstances[7].

However, over the years the community has found some inaccuracies in the expression of Beverloo. In 1971, Hunginton and Roney [8] reported that materials dilated differently when were near the exit and defined a new *flowing density* ρ_f to be inserted in Eq.(1) instead ρ_b. This ρ_f was obtained as the ratio of the mass flow rate to the volumetric flow rate calculated at the top of the silo. In fact, packing fractions ϕ vary considerably from point to point within the silo [9], so there is a potential error up to 20% in the flow rate if the choice of ρ_f is not the appropiate.

In 2007, Mankoc et al. [10] showed that the flow rate is not possible to be fitted to the Beverloo's expression if a wide span of outlet sizes is regarded. So, they proposed a new empirical expression,

$$W = C' \sqrt{g}[1 - \tfrac{1}{2}e^{-b(D-d)}](D - d_p)^{n-1/2} \tag{2}$$

where b and C' are fitting parameters. Apart from setting k to 1, the main novelty of Eq.(2) is the inclusion of an exponential term, which was suggested to arise from a dependence of ρ_f (at the exit) on the outlet size.

In 2012, Janda et al. [11] reported an experimental work where they measured vertical velocity v_z and packing fraction ϕ profiles for several outlet lengths at the exit of a two-dimensional silo by means of high speed image adquisition. The authors found that these profiles were self-similar and could be collapsed by means of the expressions:

$$v_z(x) = v_c(1 - (x/R)^2)^{1/\nu} \tag{3}$$

$$\phi(x) = \phi_c(1 - (x/R)^2)^{1/\mu} \tag{4}$$

*e-mail: dgella@alumni.unav.es

where ν and μ are fitting exponents, and x is the horizontal position measured from the center of the outlet. v_c and ϕ_c represent, respectively, the vertical velocity and packing fraction measured at the center of the outlet and are fitted to the following expressions in function of R:

$$v_c(R) = \sqrt{2g\gamma R} \quad (5)$$

$$\phi_c(R) = \phi_\infty(1 - \alpha_1 e^{-R/\alpha_2}) \quad (6)$$

where γ, α_1, α_2 and ϕ_∞ are fitting parameters. Despite it has been recently proved [12] that the free fall arch idea is quite different from what it was initially thought, the square root dependence of the velocity on gravity is still recovered by Eq. (5). Furthermore, the fact that data are well fitted by Eq.(6) is the experimental proof of the flowing density dependence on R with the form suggested in [10]. By using mass conservation arguments, the mass flow rate could be calculated assuming density and velocity profiles as continuum functions by integrating $\int \rho\phi(x)v_z(x)dx$. Thus, the proposed expression of the mass flow rate for a two-dimensional system is

$$W = C''\phi_\infty[1 - \alpha_1 e^{-R/\alpha_2}]R^{3/2} \quad (7)$$

where C'' depends on the curvature of the profiles, that is, on the exponents μ and ν. From the original Beverloo expression (Eq. (1)), the reduced aperture $D-kd_p$ was substituted in Eq. (7) by the actual dependence of ϕ and v_z on x and R. It should be noted that the particle size is not explicitly included in Eq. (7). This implies that the mass flow rate would be independent on the particle size unless there was some dependence through the fitting parameters of Eqs. (3-6). Janda et al. found the values of them for 1 mm diameter stainless steel beads in their experimental work. In a more recent work, Zhou et al. [13] performed numerical simulations of a two-dimensional flat-bottomed silo with 2 and 6 mm diameter disks. They used the equations proposed by Janda et al. to adjust the data, encountering self-similarity in the profiles. Also, their packing fractions fitted successfully to Eq. (6), obtaining parameters that described some dependence on the particle size. However, they could not fit the velocity data to Eq.(5) and proposed an alternative expression for this magnitude. The aim of our work is therefore trying to generalize Eq. (7), checking the validity of the equations (3-6) and analysing whether the fitting parameters change with the particle size.

2 Experimental Setup

The experimental setup has been designed to be a rescaled version of that used by Janda et al. in [11]. Instead of the 1 mm diameter beads, we have used 4 mm diameter spheres of the same type of stainless steel. Therefore, the only variable that has been changed is the beads size, implying a projected area 16 times larger than in the previous case, and a weight that increases 64 times.

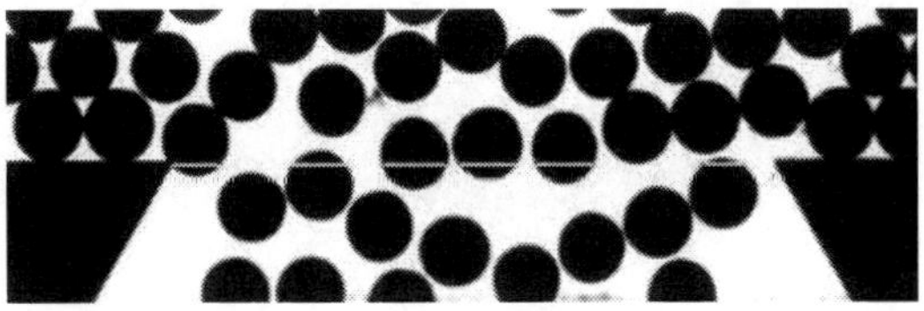

Figure 1. Example of a frame of a video taken for an outlet size $R = 1.72$ cm. The parts of the spheres occupying the exit line are highlighted in yellow.

We have built a two-dimensional flat-bottomed silo from two safety glass sheets of 1600×700 mm^2. Between them, we have stuck two 4 mm-width aluminium flat bars at its lateral edges. They have been supplemented with a small piece of paper in such a way that just a single layer of 4 mm diameter beads fits inside. At the bottom of the silo two 4 mm width movable blade-shaped pieces of stainless steel have been placed to adjust the size of the outlet conveniently. Finally, there is a hopper at the top of the silo that facilitates manual loading. The experiment is protected by means of an aluminium framework to make it stable and to avoid glass bending, ensuring that the grains arrange in a monolayer.

The magnitudes related to the particles flow rate have been measured by analysing a set of films taken by means of a '*Photron FastCam-1024 PCI 100K*' high speed camera, which is able to record up to 10^6 frames per second at low resolution. The camera has been placed in front of the silo, and a 30 W led panel light has been placed behind it to maximize the contrast between the beads and the voids. As an example, in Fig. 1 we show a frame of one of the several videos we have taken. These videos have been processed by means of a developed program in matlab, whose code consists in the following: first, the images are binarized and the centroid positions of every particle have been determined at each frame. In this work we are interested only in measuring magnitudes at the position of the exit, so for calculating the velocity of the beads we have focused on the particles that cross the outlet. In practice we find particles whose centroid is over the line of pixels shown in Fig. 1 at one frame and below it at the following. Thus, the velocity of each ball is the result of computing the displacement of the particles divided by the time interval between two frames. Furthermore, the flow rate has been calculated simply by counting the centroids that have passed through that pixel line and dividing it by the total acquisition time.

In order to obtain the packing fraction we also take account of every particle that overlaps the pixel line at each frame. Since we know the size of the beads and the pixel length, the program calculates, from the centroid positions, how much each ball contributes to the outlet pixel line by a simple trigonometric relationship; that is, it obtains the length of the chord of each ball that is onto the outlet line. Then, the outlet is divided in as many cells as is required, and a value of 1 is assigned to each cell if it is occupied and a value of 0 if not. In Fig. 1 it is possi-

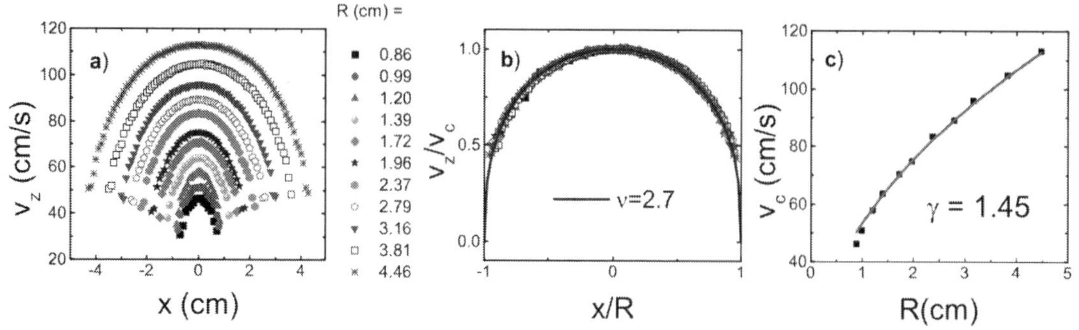

Figure 2. a) Vertical velocity profiles for several outlet sizes. b) Normalized vertical velocities. The solid line is the best fit according to Eq. (3) with $\nu = 2.7$. c) Dependence of the velocity at the center of the exit with the outlet size. The solid line corresponds to the best fit of Eq. (5) with $\gamma = 1.45$.

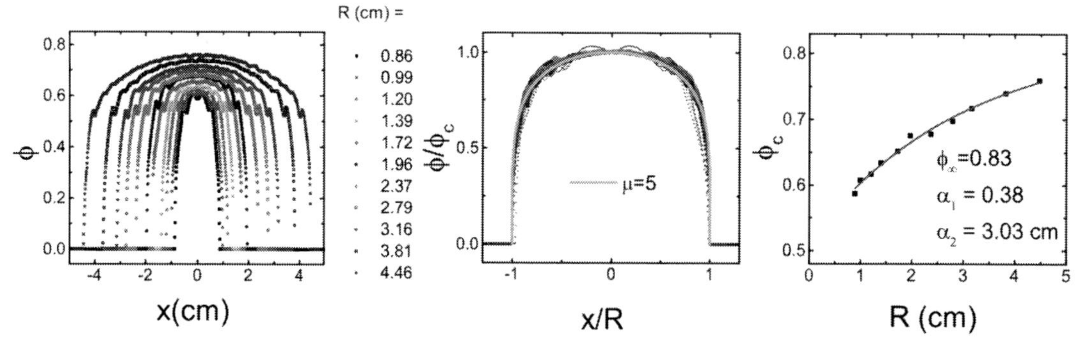

Figure 3. a) Packing fraction profiles for different outlet sizes. b) Normalized packing fraction profiles. The solid line is the best fit according to Eq. (4) with $\mu = 5$. c) Dependence of the packing fraction at the exit center with the outlet size. The solid line corresponds to Eq.(6) with the fitting parameters showed inside the graph.

ble to observe in yellow the segments which overlap the exit line. The packing fraction profile is finally calculated by time-averaging these amounts over all frames in each position.

In comparison to the work of Janda et al.[11], this way of processing the films has supposed an important enhancement in spatial resolution, as it will be appreciated in the results.

3 Results

The experimental data of vertical velocity profiles at the outlet line for a range of apertures from $R = 0.86$ cm to 4.46 cm are shown in Fig. 2.a. It is clearly noted that the curves shape is the same for each hole size. In fact, they have resulted to be self-similar and have been successfully collapsed in the curve governed by Eq. (3), just as in Ref. [11] as it is depicted in Fig. 2.b. However, our curve is slightly wider, since we have found an exponent $\nu = 2.7$, a little greater of that of Janda et al.[11] and similar to that obtained in [13] (see Table 1). The velocities at the center as a function of R are represented in Fig. 2.c. The correspondence between the data and the solid line confirms the validity of Eq. (5), although in our case the parameter γ has resulted to be 1.45, a bit greater than those reported in [11] or [12].

Table 1. Comparison among the parameters obtained in this work, in [11] and in [13].

	4 mm (this work)	1 mm [11]	2 and 6 mm [13]
ν	2.7	2	2.63
μ	5	4.55	5.26
γ	1.45	1.07	*
ϕ_∞	0.83	0.83	0.81
α_1	0.38	0.50	0.45
α_2	3.03 cm	0.33 cm	$3.84d_p$

On the other hand, the experimental packing fraction profiles for the same range of apertures are plotted in Fig 3.a. The increase in resolution enables us to appreciate a small ripple in each curve at a position that coincides with a distance to the border of approximately one particle radius. In the curves corresponding to small outlets, some secondary ripples are observed with a wavelength in the

* In their work, Zhou et al.[13] proposed another expression for the velocity as a function of R.

order of the particles size, evidencing that this property is consequence of the discrete nature of our system. Moreover, all the curves exhibit the same shape and they have been collapsed by Eq. (4), as it is shown in Fig 3.b. The exponent obtained in this case has been $\mu = 5$, not much different from the exponents of the Refs. [11] and [13], as it is shown in Table 1. The dependence of the packing fraction at the exit center is represented in Fig. 3.c. Although the experimental data seem to be compatible with the form of Eq.(6) there are some features that should be considered. First, the asymptotic limit (ϕ_∞) is similar to the ones previously reported, hence appearing to be universal. Second, the parameters obtained ($\alpha_1 = 0.38$ and $\alpha_2 = 3.03$ cm) are quite different than in [11], in agreement with a slower rise of the value of ϕ with R for larger particles. Thus, our value of α_2 does not reproduce the linear behavior with the particle size that is suggested in [13] (see Table 1). In general terms, regarding the comparison among the parameters obtained in different works, it is not easy to establish a direct relationship of the parameters to the particle size.

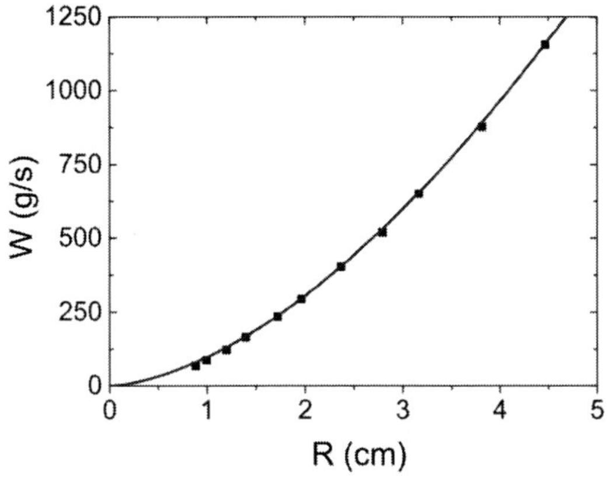

Figure 4. Experimental results of the flow rate. The solid line is the function of Eq.(7) with the parameters calculated in the analysis of the velocity and packing fraction profiles as shown in Figs. 2 and 3.

Finally, the experimental results of mass flow rate in function of R are represented by the dots in Fig 4. The solid line corresponds to Eq.(7) using the parameters included in the first column of Table 1 and fits quite well to the experimental data. That proves the validity of Eq.(7) for 4 mm stainless steel spheres. Curiously, although the parameters obtained here and in [11] are rather different, the flow rate values are similar, at least in the range of measurement.

4 Conclusions

The study of the flow rate in a two dimensional silo with 4 mm stainless steel beads has served to extend the validity of the expressions proposed by Janda et al. However, the parameters found seem to show up a dependence on the particle size which origin is not clear yet. Although for big apertures the density tends to the same asymptotic value in both cases, the velocity scaling seems to depend on the beads size. This fact suggests that for sufficiently large orifices the dependence of the of the mass flow rate on the particle size will become evident. Thus, more work is needed to study this question and to unveil the origin of the issue.

Acknowledgements

This work was funded by Ministerio de Economía y Competitividad (Spanish Government) through Project No. FIS2014-57325. DG acknowledges Ministerio de Economía y Competitividad (Spanish Government) for the scholarship BES-2015-074436 within the same project.

References

[1] W.A. Beverloo, H.A. Leninger and J.J. van de Velde, Chem. Eng. Sci. **15**, 260 (1961)
[2] R.L. Brown and J.C. Richards, *Principles of Powders Mechanics* (Pergamon Press, Oxford, 1970)
[3] B.P. Tighe and M. Sperl, Gran. Matt. **9**, 141 (2007)
[4] J.E. Hilton, P.Y. Lai and H.K. Pak, Phys. Rev. E **84**, 011307 (2011)
[5] R.M. Nedderman, U. Tüzün, U. Savage and S.B. Houlsby, Chem. Eng. Sci. **37**, 1597-1609 (1982)
[6] M.A. Aguirre, J.G. Grande, A. Calvo, L.A. Pugnaloni and J.-C. Géminard, Phys. Rev. Lett. **104**, 238002 (2010)
[7] H. Pacheco-Martinez, H.J. van Gerner and J.C. Ruiz-Suárez, Phys. Rev. E **77**, 021303 (2008)
[8] A.P. Hunginton and N.M. Roney *Chemical Engineering Tripos. Part 2* (Research Project Report, University of Cambridge, 1971)
[9] U. Tüzün, *PhD Thesis* (Univ. of Cambridge, 1970)
[10] C.Mankoc, A. Janda, R.Arévalo, J.M. Pastor, I. Zuriguel, A. Garcimartín and D.Maza, Gran. Matt. **9**, 407 (2007)
[11] A. Janda, I. Zuriguel and D. Maza, Phys. Rev. Lett. **108**, 248001 (2012)
[12] S.M Rubio-Largo, A. Janda, D. Maza, I. Zuriguel and R.C. Hidalgo, Phys. Rev. Lett. **114**, 238002 (2015)
[13] Y. Zhou, P. Ruyer and P. Aussillous, Phys. Rev. E **92**, 062204 (2015)

Modelling polymeric deformable granular materials - from experimental data to numerical models at the grain scale

Maxime Teil[1,2,⋆], *Barthélémy* Harthong[1,2], *Didier* Imbault[1,2], and *Robert* Peyroux[1,2,⋆⋆]

[1] *Univ. Grenoble Alpes, 3SR, F-38000 Grenoble, France*
[2] *CNRS, 3SR, F-38000 Grenoble, France*

Abstract. Polymeric deformable granular materials are widely used in industry and the understanding and the modelling of their shaping process is a point of interest. This kind of materials often presents a viscoelastic-plastic behaviour and the present study promotes a joint approach between numerical simulations and experiments in order to derive the behaviour law of such granular material. The experiment is conducted on a polystyrene powder on which a confining pressure of 7MPa and an axial pressure reaching 30MPa are applied. Between different steps of the *in-situ* test, the sample is scanned in an X-rays microtomograph in order to know the structure of the material depending on the density. From the tomographic images and by using specific algorithms to improve the images quality, grains are automatically identified, separated and a finite element mesh is generated. The long-term objective of this study is to derive a representative sample directly from the experiments in order to run numerical simulations using a viscoelactic or viscoelastic-plastic constitutive law and compare numerical and experimental results at the particle scale.

1 Introduction

The use of polymeric granular materials having a deformable behaviour is of common use in the industry. The behaviour of such materials, when mechanically loaded under relatively large stresses, and at a given temperature and moisture content, is ruled by grain kinematics but also, in a very significant manner, by the deformability of the grains.

In the present study, an analysis of the mechanical behaviour under mechanical loading is developed, using in complementary ways, X-rays tomography and in situ triaxial compression test. The aim is to be able to observe each grain of the experimental sample at each step of the experiment.

The approach developed here consists in generating digitized grains from the images obtained through tomography, meshing these grains and integrating these meshed geometries into a finite element model. After applying consistent boundary conditions and mechanical loading, a grain-by-grain comparison of indicators such as contact areas and their orientations should allow an accurate analysis of mechanical behaviour of an average set of grains by using a 3D multi-particle finite element method (MP-FEM). That method has been already used on deformable materials [1–3]. Some studies also took into account the same kind of material, but they used only 2D models [4, 5].

⋆e-mail: maxime.teil@3sr-grenoble.fr
⋆⋆e-mail: robert.peyroux@3sr-grenoble.fr

2 Material and experiment

2.1 Material

The deformable granular material used in that work is a polymeric powder composed of polystyrene grains supplied by Goodfellow (UK). The size of the grains is between 150μm and 500μm and their polyhedral shape is very random due to the freeze-fracturing process (see paragraph 2.2).

2.2 Experiment

Compression

The experiment performed in this study was a triaxial compression test. Three tests were performed at different confining pressures: 1, 2 and 7MPa. The results presented here were obtained by applying a constant confining pressure of 7MPa while the axial stress reached 30MPa. The tested powder (approximately 2.5g) was poured into a flexible membrane made of latex, having a thickness of 400μm in average and a diameter of approximately 11mm. The length of the sample before loading was approximately 22mm. The confining pressure was applied using a hydraulic pump and held to its set point with an accuracy of ±0.05MPa. The axial pressure was applied by a displacement-controlled piston and was measured by a 5kN force sensor. The compressive load was applied in subsequent 700μm displacement steps at a velocity of 2μm/s. Figure 1 shows the typical curves of the experiment. X-ray tomography scans are performed at each step,

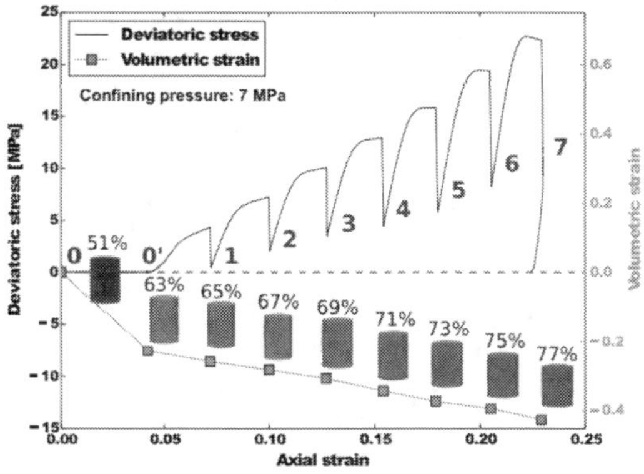

Figure 1. Deviatoric stress, volumetric strain and mean density (with 3D density maps) measured during the triaxial compression test as functions of the axial strain

after a time of at least 2 hours allowing the time-dependent response of the viscoelastic material to vanish. The duration of the scans was approximately 2 hours. Hence, the total duration of one compression test was approximately 40 hours.

Microtomography

The scans were performed on the 3SR-lab microtomograph (manufactured by RX Solutions, with a Hammatsu L12161-07 source, and a Varian flat panel detector). The pixel size was set to 9μm and the beam was obtained by applying a voltage of 80kV and a current of 125μA. 3D images were captured using 1440 projections per scan and averaged with 8 images per projection. Figure 2 shows the quality of a non-processed image.

Density and mechanical analysis

The deviatoric stress is plotted on figure 1 as a function of the axial strain. The figure also shows the almost constant increase of the mean relative density of the sample under compression and, thus, the decrease of the volumetric strain. The measurement of the relative density was made by thresholding the 3D images and calculating the ratio of the volume of voxels corresponding to the solid phase and the total volume. This process was done on subvolumes, the size of which was increased until the measured density stabilised so as to define the size of a representative element. Then, the determination of the relative density was calculated on all the subvolumes of this size within the sample. The volumetric strain was calculated as:

$$\varepsilon_v = \ln\left(\frac{V_t}{V_t^0}\right) = \ln\left(\frac{D^0}{D}\cdot\frac{V_g}{V_g^0}\right) = \ln\left(\frac{D^0}{D}\right) \quad (1)$$

Where V_t^0, V_g^0 and D^0 are, respectively, the total volume, the volume of grains and the density before loading; V_t, V_g

Figure 2. X-rays radiography without load (step 0)

and D the same quantities at a given compression step. It is considered, here, that V_g equals V_g^0.
It is observable that, when using a confining pressure of 7MPa, the deformable behaviour of the grains allows the sample to reach a relative density of 77% with an axial strain of 25%. No dilatancy phenomenon is observable, which is explained by the ability of the particles to adapt to the load by changing their shape.

3 Treatment of the experimental data

3.1 Image processing

As shown in figure 2, the quality of the reconstructed images was not suitable for direct use. Indeed, the contrast between the grains and the porosity was poor and a significant noise was visible.
Figure 3 shows the histogram (number of voxels as a function of the grey scale value) of the original volume. The presence of two peaks was due to the bimodal character of the sample (solid material and voids). The width of the peaks was mainly due to the experimental noise (visible on the corresponding picture). To improve the contrast between the two phases, the noise was reduced by applying a blurring filter (3D total variance). The effect of the filter was to achieve a better separation between the peaks and to improve their magnitude as seen on the lower histogram. As a result, one clearly observes the decrease of noise inside the grains (dark dots) and outside (bright dots) on the lower image. The signal being much closer to a binary image, the determination of the threshold in the processing algorithm was made easier and more accurate.
Note that if the blurring filter efficiently reduced the noise, it also blurred the edges of the grains.

3.2 Segmentation

The aim of the segmentation was to identify and separate the grains. Grains with realistic shapes and an accurate contact detection can be achieved by this means. For this paper, the segmentation was first tested on the two grains presented on figures 3 and 4.
The segmentation algorithm was based on the diffusion of markers. The script uses the random walker function from

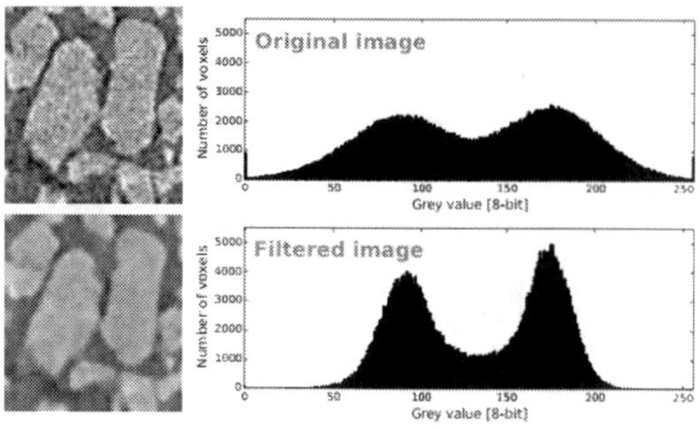

Figure 3. 3D blurring filter effect on images and histograms

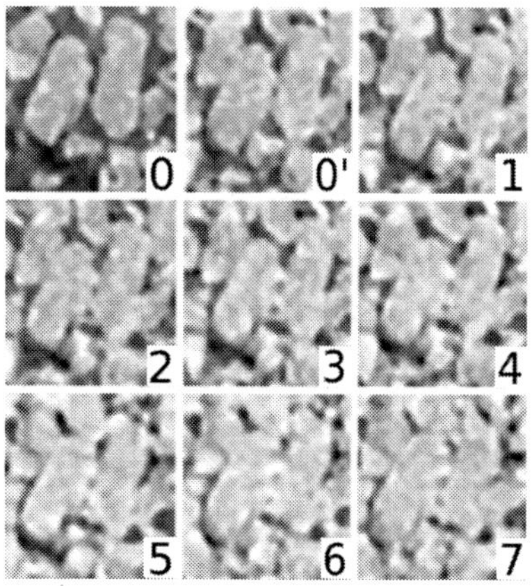

Figure 4. Evolution of two grains during compression

the python module of scikit-image [6].
The first task consisted in selecting what is solid material and what is void. An automatic threshold was chosen for that purpose. Two methods were tested: Kittler [7] and Otsu [8] methods. Even if both of them gave realistic and adequate binary images, they led to slightly different values of threshold. To take advantages of both Otsu and Kittler methods, an averaged threshold was defined (equation 2):

$$T = T_{\text{Otsu}} \times \alpha + T_{\text{Kittler}} \times (1 - \alpha) \qquad (2)$$

where T, T_{Otsu}, T_{Kittler} are respectively the obtained threshold value, the one from Otsu and the one from Kittler. The variable α is a parameter chosen between 0 and 1 in order to get images as realistic as possible on a visual aspect. In most cases α was approximately 0.4.
Once the images were binarised, the residual noise had to be removed. To do that, opening (erosion and dilation) and closing (dilation and erosion) processes were performed to remove all the black or white dots that were isolated. It was then necessary to divide each image into three categories:

- the "sure background", obtained by dilation of the solid material;

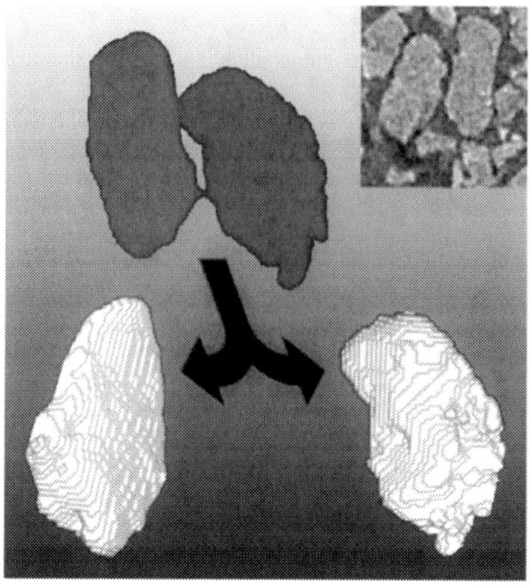

Figure 5. Identification and segmentation of two grains

- the "sure foreground", which give markers obtained by generating a 3D euclidian distance transform (EDT) and selecting the local brightest parts;
- the "unknown region", which is not the "sure foreground" or the "sure background".

Every marker in the sure foreground was labelled and the random walker was applied in a specific order: the labels were growing in the unknown region depending on the neighbourhood and, then, the background propagated toward the grains. The propagation rate depends on the blurred original images and a given "penalization coefficient".
Some inconsistencies appeared near the contact areas depending on the order of the labelled markers and the distance between the markers and the contact areas: the boundary of one grain could be taken as the boundary of another grain. To avoid that, an opening process was made on the isolated grains, so that the very small defects around the grain were removed. The surfaces of the new grains were then smoother and the volume could be considered unchanged.
Contacts were characterized as the borders between labels after propagation of the markers.

4 Toward a mechanical discrete analysis

As mentioned in paragraph 3.2, it was possible to obtain the volume of the grains from the tomography images. The aim is to introduce a certain number of grains (approximately 5×5×5) into a finite element code. For the purpose of using the numerical grains, it was first needed to mesh them.
The mesh generation was performed thanks to the free matlab/octave-based toolbox *iso2Mesh* [9], partially based on the software project CGAL 3.5 3D mesher [10]. The grains were meshed by using the so-called "cgalmesh" method, which allowed to mesh the surface, to rearrange

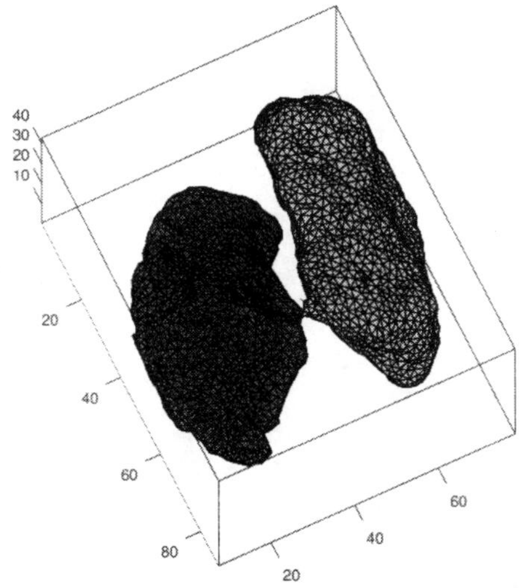

Figure 6. Mesh generation of the two grains with *iso2mesh*

the elements of the surface and to mesh the volume. This method also gives the opportunity to manage the size of the triangular surface faces by choosing the radius of the Delaunay spheres. The maximum volume of the tetrahedral elements can also be chosen in order to manage the volumetric density of elements.

Figure 6 shows the mesh generation of the two grains studied in the previous paragraphs when using a radius of the Delaunay spheres of 13.5μm and a maximum volume of 729μm^3. Considering these parameters, slightly more than 30500 nodes and 13500 tetrahedral quadratic elements were sufficient to accurately mesh the two grains. The digitized and meshed neighbouring grains will be introduced into a MP-FEM code where a viscoelastic or viscoelastic-plastic constitutive law will be considered for the material behaviour and a Coulomb friction model will be used. Periodic boundary conditions or flexible membranes will be chosen as boundary conditions. Contact areas and their orientations should be efficient indicators to compare the real and the numerical granular material at a same density. The analysis of the displacement or strain fields will also be taken into account.

5 Conclusion

This study presented a methodology designed for modelling the mechanical behaviour of a granular material using X-ray tomography as an experimental characterization tool at the grain scale. Indeed, from the 3D images of the grain assembly structure, it is possible to detect and isolate the grains and to generate a 3D mesh for each grain. A region of meshed particles can then be considered and introduced into a finite element code to simulate its mechanical behaviour.

To this end, we used already existing tools or methods, in order to remove as much as possible the experimental noise without damaging the borders of the grains and to perform an accurate detection of the contact zones between the grains.

The next step will consist in studying the mechanical behaviour of the granular material by using, in a MP-FEM code, a certain number of meshed grains and applying a load comparable to the one obtained with the triaxial compression test. The constitutive law will be based on viscoelastic or viscoelastic-plastic models. The comparison between the real region (observed in microtomography) and the numerical one (resulting from simulation) should allow to understand the mechanical behaviour of the granular material.

Acknowledgements

The authors acknowledge the financial supports of PolyNat Carnot Institute and the labex Tec21 (Investissements d'Avenir - grant agreement n°ANR-11-LABX-0030). They also acknowledge the technical support of Edward Andò, Pascal Charrier and Jean-Benoît Toni for microtomography. Moreover, all that work would not be as consistent as it was without the use of the very useful tool Kalisphera [11] which enabled to validate all the algorithms by working on spheres in contact.

References

[1] G. Gustafsson, H.A. Haggblad, P. Jonsen, Powder Technology **239**, 231 (2013)

[2] B. Harthong, D. Imbault, P. Dorémus, Journal of the Mechanics and Physics of Solids **60**, 784 (2012)

[3] Y. Chen, D. Imbault, P. Dorémus, *Numerical Simulation of Cold Compaction of 3D Granular Packings*, in *Progress in Powder Metallurgy* (Trans Tech Publications, 2007), Vol. 534 of *Materials Science Forum*, pp. 301–304

[4] D.T. Gethin, X.S. Yang, R.W. Lewis, Computer Methods in Applied Mechanics and Engineering **195**, 5552 (2006)

[5] R.W. Lewis, D.T. Gethin, X.S. Yang, R.C. Rowe, International Journal for Numerical Methods in Engineering **62**, 853 (2005)

[6] L. Grady, IEEE Transactions on Pattern Analysis and Machine Intelligence **28**, 1768 (2006)

[7] J. Kittler, J. Illingworth, Pattern Recognition **19**, 41 (1986)

[8] N. Otsu, IEEE Transactions on Systems, Man, and Cybernetics **9**, 62 (1979)

[9] Q. Fang, D.A. Boas, *Tetrahedral mesh generation from volumetric binary and grayscale images*, in *2009 IEEE International Symposium on Biomedical Imaging: From Nano to Macro* (2009), pp. 1142–1145

[10] P. Alliez, C. Jamin, L. Rineau, S. Tayeb, J. Tournois, M. Yvinec, in *CGAL User and Reference Manual* (CGAL Editorial Board, 2016), 4.9 edn., `http://doc.cgal.org/4.9/Manual/`

[11] A. Tengattini, E. Andò, Measurement Science and Technology **26**, 095606 (2015)

Processing and properties of Titanium alloy based materials with tailored porosity and composition

Jose Luis Cabezas-Villa[1,2], Luis Olmos[1], Jose Lemus-Ruiz[1], Didier Bouvard[2,*], Jorge Chavez[3], Omar Jimenez[3], and Victor Manuel Solorio[4]

[1]Universidad Michoacana de San Nicolás de Hidalgo, IIMM and INICIT, Fco. J. Mujica S/N, Morelia, Michoacán, C.P. 58060, México
[2]Univ. Grenoble Alpes, CNRS, SIMAP, 38000 Grenoble, France
[3]Universidad de Guadalajara, Departamento de Ingeniería de Proyectos, José Guadalupe Zuno # 48, Jalisco, C.P. 45100, México
[4]Tecnológico Nacional de México, Posgrado de Ciencias en Metalurgia ITM, Av. Tecnológico # 1500, Morelia, C.P. 58120, México

Abstract. This paper deals with powder processing of Ti6Al4V titanium alloy based materials with tailored porosity and composition. Ti6Al4V powder was mixed either with salt particles acting as space holder, so as to provide two-scale porosity, or with hard TiN particles that significantly modified the microstructure of the material and increased its hardness. Finally an original three-layer component was produced. Sample microstructure was observed by SEM and micro-tomography with special interest in pore size and shape, inclusion distribution and connectivity. Compression tests provided elastic modulus and yield stress as functions of density. These materials are representative of bone implants subjected to complex biological and mechanical conditions. These results thus open avenues for processing personalized implants by powder metallurgy.

1 Introduction

Powder metallurgy is the ideal route to produce parts with complex composition, microstructure and geometry. The critical step is sintering, during which the final properties and dimensions are obtained. Nevertheless, the possibility of fabricating components including several regions with various characteristics (dense, porous, composite) has hardly been considered. This is because of the difficulty of cosintering materials with different densification rates without generating cracks at the interfaces. Yet, the interest of such "architectured" materials with multifunctional properties is increasing in various fields, in particular in biomedical engineering for bone replacement. Titanium alloys have been proposed as an alternative to enhance the lifespan of orthopedic implants because of their excellent combination of mechanical and corrosive properties. In the past two decades, many works have been devoted to defining the most appropriate method to fabricate dense and porous Ti6Al4V parts by powder metallurgy [1,2]. In this work, we investigated the sintering of various systems based on biocompatible Ti6Al4V titanium alloy with the objective of producing architectured layered materials with suitable features allowing for their use as advanced bone implants. First, materials with two scale porosity have been fabricated from the mixture of the powder with salt particles acting as space holders. Next, the sintering of Ti6Al4V powder mixed with TiN particles has been investigated. Finally, we have been able to produce a three-layer component comprising a close-to-dense layer, a highly porous layer and a reinforced layer. The microstructure of every component has carefully been characterized by SEM and microtomography and its mechanical properties have been evaluated by compression and microhardness tests.

2 Materials and methods

A prealloyed powder of Ti-6Al-4V with spherical particles smaller than 20 µm, produced by Raymor, Quebec Canada, was used as the matrix material. Those powders were mixed (1) with particles of ammonium bicarbonate ((NH4)HCO3) with irregular shape and size distribution between 100 and 500 µm, provided by Alfa Aesar (USA), for acting as space holder in porous samples, (2) with angular TiN particles with particle size below 45 µm, supplied by AEE Inc., to produce composites. Powders were mixed in a Turbula during 30 minutes. Then, 10 wt% of polyvinyl alcohol (PVA) was added as a binder. Next, the mixture was poured into an 8 mm diameter stainless steel die and pressed at 500 MPa using an Instron 1150 universal machine to obtain cylindrical compacts of 10 mm height. The fabrication of the three-layer sample includes the following steps: a layer of regular powders was poured into the die; then, the mixture to produce the porous layer was poured above; finally, the powder mixture to form the composite layer was poured onto the porous layer. This resulting sample was compacted at the same pressure.

In every case, space holder particles were removed at 150 °C during 6h in air and the PVA was thermally extracted by a slow heating treatment at 5 °C/min up to a temperature of 500°C with a dwell time of 40 min in argon. Afterwards, each compact was sintered in a vertical dilatometer Linseis L75V between 1150 and 1260 °C during 1 h in argon. The weight density of every sintered compact was calculated from mass and dimension measurements. The sintered compacts were cut and the surface was metallographically prepared with SiC papers for grinding and alumina powders (up to 50 nm particle size) for final polishing. The microstructure of sintered samples was observed with a Mira3 TESCAN Field-Emission Scanning Electron Microscope (FE-SEM). 3D imaging was performed by computed micro tomography (CMT) with a ZEISS Xradia 510 Versa microCT. The beam intensity was 120 kV, which was enough to pass through Ti6Al4V samples of 8 mm diameter. 1600 projections were taken around 360° with a CCD camera of 2048x2048 pixels providing a pixel resolution of around 10 μm. Due this resolution matrix pores cannot be seen. The treatment of 3D images included filters to eliminate the noise and thresholding to distinguish between the solid and porous phases. Different morphological operations like opening, closing and erosion were achieved to improve the quality of the binary images. Finally, the pore volume fraction was calculated by dividing the number of voxels corresponding to the pore phase by the total number of voxels in the image. The pore size distribution was estimated by measuring the pore volume accessible to an octahedral structural element with increasing size, as proposed by various authors for interconnected porosity [3].

3 Results and discussion

3.1 Porous compacts

Fig. 1 shows the relative density of all sintered samples as a function of the volume fraction of salt. It also includes the density of the green compacts. It can be noticed that the initial relative density decreases almost linearly with increasing volume fraction of salt. It goes from 0.26 for the material with 50 vol.% of salt to 0.7 for the material without salt. The final density also decreases with increasing fraction of salt and it increases with increasing sintering temperature. The minimum value, 0.28, was obtained for the sample with 50 vol.% of salt sintered at 1150°C. Note that the samples obtained without salt have a density around 0.95.

The pore size distribution estimated from 3D images (Fig. 3c and d) is shown in Fig .2 for three porous samples with 10, 20 and 30 vol.% of salt. The particle size distribution of the initial salt particles has been added as a reference value to analyze the pore size obtained after sintering. It can be seen that the pore size distribution is quite similar for all of porous samples, confirming the uniform distribution of salt particles inside of samples. However, it is notable that the d_{50} diameter of pores after sintering is about 35% smaller than the one of the salt particles. This difference should result from pore changes occurring during compression and sintering steps.

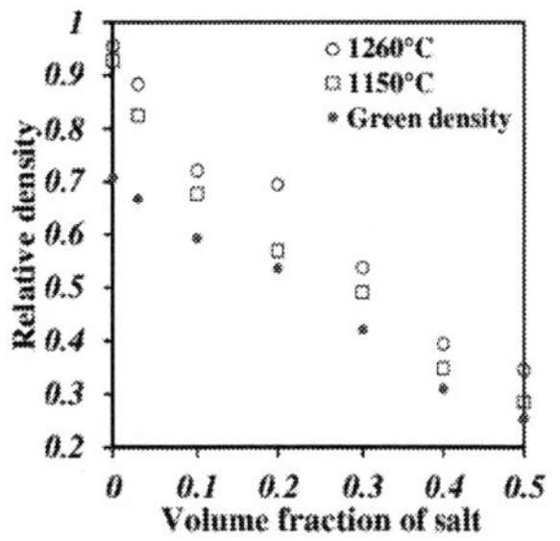

Fig. 1. Relative density as a function of the volume fraction of salt in the green state and after sintering

Fig.3 shows the microstructure of sintered samples observed by SEM and CMT. The matrix composed of Ti6Al4V particles is well sintered around the large pores, although, interparticle pores are visible. It appears that the matrix is less dense when the volume fraction of salt is higher (Fig. 3b) which suggests that sintering is perturbed by the presence of large pores. Fig.3b suggests that large pores are connected with each other through interparticle pores. These microstructures are different from those obtained by EBM because this technique melts the particles [4]. Figs 3c and 3d show 3D renderings of the porosity inside the samples. Pseudo-colors are used to identify interconnected pores. It can be noticed that large pores are isolated from each other in the specimen with 10 vol.% of salt. Salt particles appear to be randomly distributed inside the sample because very

 DOI: 10.1051/epjconf/201714013007

few particle agglomerates are observed. On the other hand, with 30 vol.% of salt, the large pores are fully interconnected (Fig. 3d), which proves that these particles formed a continuous network inside the sample before being eliminated. Additional images showed that the transition between large pore separation and interconnection occurs at 20 vol.%. Indeed, non-detected interparticle pores may connect large pores with each other even when 10 vol.% of salt are used, as suggested by Fig. 3b.

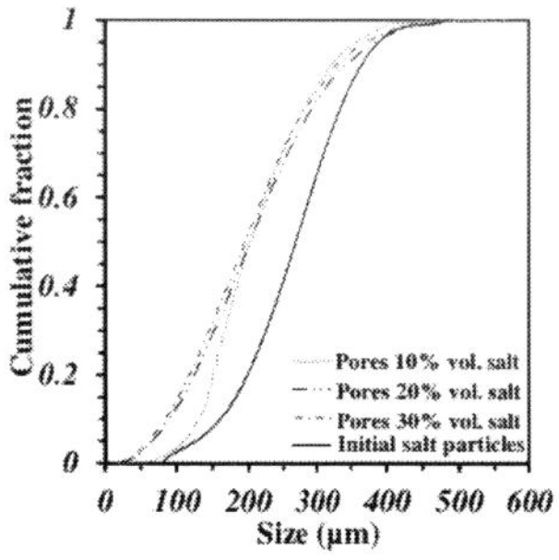

Fig. 2. Pore size distribution in 4 samples with 10, 20 and 30 vol.% of space holders and in three-layer sample.

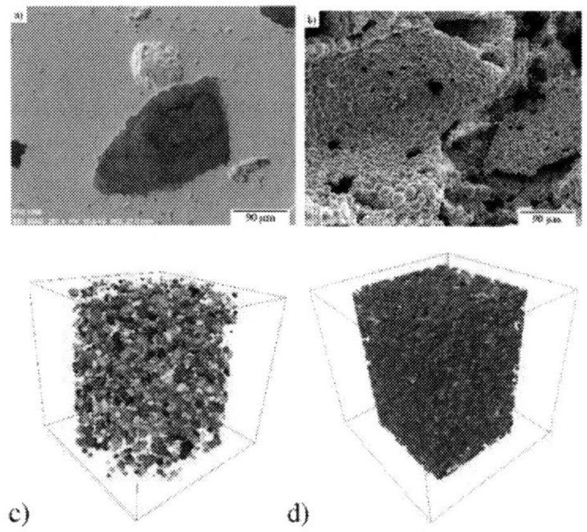

Fig. 3. Microstructure of porous samples, SEM micrographs a) 10 vol.% of salt and b) 50 vol.%, 3D rendering of porosity, c) 10 vol.%, and d) 30 vol.%.

3.2 Composite compacts

Fig. 4 shows SEM micrographies of fractured composite samples. TiN particles were found embedded in the Ti64 matrix as isolated particles (Fig. 4a) or they formed small agglomerates when their amount increased (Fig. 4b). From Fig. 4a it can be observed that pores are nearly spherical and the shape of initial particles is indiscernible when the relative density is close to 90% for the highest sintering temperature. On the other hand, larger irregular pores are observed when the relative density is lower than 80% and the spherical shape of Ti-64 particles remains discernible, which indicates a lower densification, for the lowest sintering temperature.

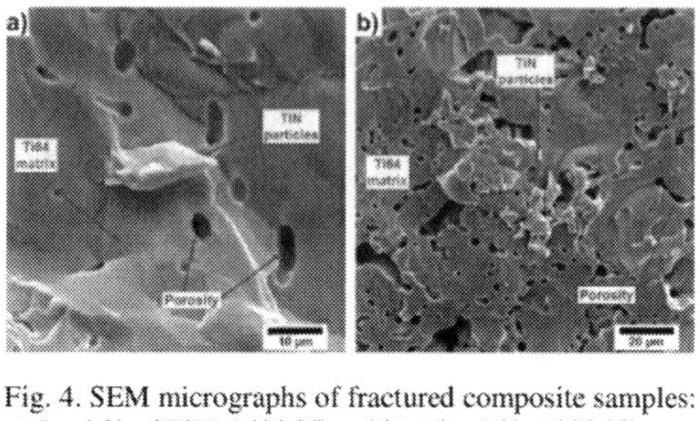

Fig. 4. SEM micrographs of fractured composite samples: a) 5 vol.% of TiN, 1400 °C and b) 15 vol.%, 1200 °C.

Vickers microhardness tests were performed on polished cross-sections of the sintered samples by means of a Future Tech FM-800 microindenter with a load of 500 gf and a dwell time of 20 s. The microhardness of the composite samples is shown in Fig. 5. For each fraction of reinforcement, the microhardness increases with increasing relative density. For a given density it increases with increasing volume fraction of reinforcement particles. With regard to the samples with 89% relative density, adding only 10 vol.% of TiN particles multiplied the microhardness by a factor of 1.8. All reinforced materials exhibit higher hardness than unreinforced ones, although the latter are denser. SEM analysis showed that the addition of TiN particles caused a change in the microstructure of Ti6Al4V matrix from lamellar to equiaxed grains, which is also characterized by a decrement in the amount of β phase with increasing temperature. Thus, the rise in microhardness is associated to both the presence of hard inclusions and the refinement of the microstructure within the matrix.

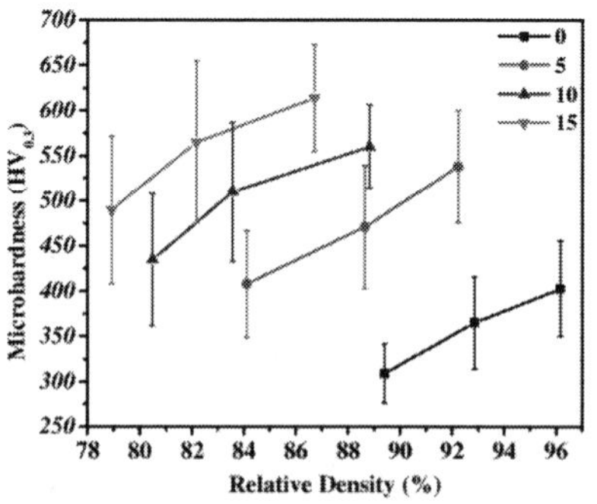

Fig. 5. Microhardness as a function of the relative density for 0, 5, 10 and 15 vol.% of inclusions.

3.3 Three-layer sample

Fig. 6a shows a cross section of the three-layer sample. The three layers, named dense, porous and composite, respectively, are clearly observed from the top to the bottom in the image. Larger pores and reinforcing particles are randomly distributed inside the matrix, which seems to be denser at the interface between the dense and porous layers. Larger interparticle pores are seen at the edges, which is mainly caused by the constrained sintering due to the different densification rates. Fig. 6b shows a 3D rendering of the porosity. It is observed that large pores remain connected after sintering. Cracks formed at the interface between the porous and composite layers as well as larger porosities at the interphase between porous and dense layers. Far from the interphases, sintering was normal. Therefore, the porosity was reduced and a disconnected network of pores was obtained.

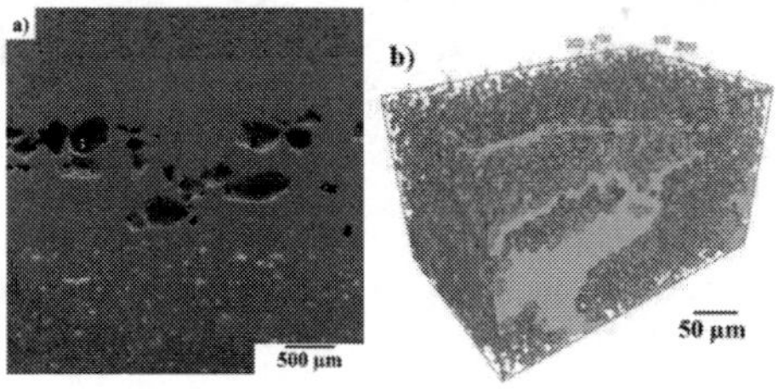

Fig. 6. Microstructure of the three-layer sample, a) SEM cross-section, b) 3D rendering of porosity.

3.4 Mechanical properties

After sintering, the bottom and top sample surfaces were polished and compression tests were performed under ASTM D695-02 using an Instron 1150 universal mechanical testing machine at a displacement rate of 0.5 mm.min^{-1} [5]. The Young's modulus (E) and the yield strength (σ_y) were estimated from the stress-strain curve. It was found that E and $\sigma_{0.2\%}$ values increased with increasing relative density in reasonable accordance with different model proposed. Fig. 7 depicts E as a function of $\sigma_{0.2\%}$ for all samples fabricated in this study. The points show a linear trend, which is surprising since the samples cover a wide range of microstructure. However, it should be noted that the points corresponding to the composite samples lie below the fitted line, indicating that TiN particles increase the yield stress at equivalent Young's modulus.

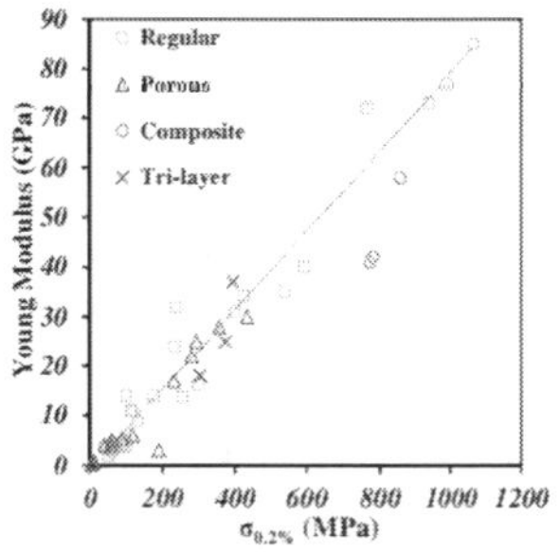

Fig. 7. Young's modulus as a function of yield strength.

4 Conclusions

This paper presented a unique set of data on powder processing and characterization of titanium alloy based compacts with complex microstructures. It demonstrated that physical characteristics and mechanical properties of Ti6Al4V alloy can be controlled by the addition of space holders and/or hard inclusions. The fabrication of an original three-layer sample also showed the feasibility of producing components with specific and localized features by traditional powder metallurgy. The mechanical properties obtained for the different materials span across wide ranges of Young's modulus and yield stress, which are consistent with cortical bone properties. This kind of materials could thus be used for fabricating tailored bone implants with microstructure and properties close to those found in real human bones.

Acknowledgments

The authors wish to thank SEP-CONACYT-ANUIES, ECOS Nord (Grant 262907) and Universidad Michoacana de San Nicolás de Hidalgo for offering financial support and facilities.

References

[1] Y. Yan, G.L. Nash, P. Nash, Int. J. Fatigue **55**, 81 (2013).

[2] M.M. Dewidar, J.K. Lim, J. Alloy. Compd. **454**,442 (2008).

[3] L. Olmos, C.L. Martin, D. Bouvard, D. Bellet, M. Di Michiel, J. Amer. Ceram. Soc. **92**, 492 (2009).

[4] H. Galarraga, D.A.Lados, R.R. Dehoff, M.M. Kirka, P. Nandwana, Addit. Manuf. **10**, 47 (2016).

[5] M. Guden, E. Celik, E. Akar, S. Cetiner, Mater. Charact. **54**(4), 399 (2005).

EPJ Web of Conferences 140, 09014 (2017) DOI: 10.1051/epjconf/201714009014
Powders & Grains 2017

Modelling the normal bouncing dynamics of spheres in a viscous fluid

Edouard Izard[1,*], *Laurent* Lacaze[2] and *Thomas* Bonometti[2]

[1] ArcelorMittal R&D Maizières, Voie Romaine, F-57283, Maizières-Lès-Metz, France
[2] Institut de Mécanique des Fluides de Toulouse (IMFT) - Université de Toulouse, CNRS-INPT-UPS, Toulouse FRANCE

Abstract. Bouncing motions of spheres in a viscous fluid are numerically investigated by an immersed boundary method to resolve the fluid flow around solids which is combined to a discrete element method for the particles motion and contact resolution. Two well-known configurations of bouncing are considered: the normal bouncing of a sphere on a wall in a viscous fluid and a normal particle-particle bouncing in a fluid. Previous experiments have shown the effective restitution coefficient to be a function of a single parameter, namely the Stokes number which compares the inertia of the solid particle with the fluid viscous dissipation. The present simulations show a good agreement with experimental observations for the whole range of investigated parameters. However, a new definition of the coefficient of restitution presented here shows a dependence on the Stokes number as in previous works but, in addition, on the fluid to particle density ratio. It allows to identify the viscous, inertial and dry regimes as found in experiments of immersed granular avalanches of Courrech du Pont et al. *Phys. Rev. Lett.* **90**, 044301 (2003), e.g. in a multi-particle configuration.

1 Introduction

Experiments on the bouncing motion of particles in a fluid have been carried out during the past decade. [1-3] and [4] studied the normal rebound in a fluid of a particle on a wall and of two particles respectively. In both configurations, the effective restitution coefficient $\varepsilon = -V_R/V_T$, defined as the ratio between the relative velocity of rebound V_R and the relative terminal velocity before contact V_T, is lower than the restitution coefficient in the dry case, noted ε_{max}. The presence of the fluid squeezed between the objects during the bouncing motion tends to dissipate the initial kinetic energy of the particle system. The ratio $\varepsilon/\varepsilon_{max}$ is shown to be a function of a single parameter, namely the Stokes number St which compares the inertia embedded in the bouncing motion, which is composed of the particle inertia and added mass of the fluid (see [5]), with the fluid viscous dissipation and reads $St = (\rho_p + 0.5\rho)V_T D/(9\mu)$ where ρ_p and ρ are the particle and fluid density, respectively, D is the sphere diameter and μ is the dynamic viscosity of the fluid. Experiments of a dense avalanche in a fluid by [6] revealed that beyond the Stokes number, the particle to fluid density is also a pertinent dimensionless number to distinguish the observed granular regimes, namely viscous, inertial and dry. Interestingly, one may note that the dependence on the density ratio has not been explicitly reported in the bouncing configuration, for which the reported coefficient of restitution ε was shown to vary only with St.

In the present work, we simulate the bouncing motion of one and two spheres using a coupled fluid-particle method (see [7]). An immersed boundary method (IBM) resolves the fluid flow in the presence of moving solid grains while the discrete element method (DEM) models the contact and lubrication interactions between particles. Simulations allow to investigate fluid to particle density ratio in the range $O(10)$ to $O(10^3)$ difficult to explore in experiments due to the nature of the fluid and particulate elements. This article is structured as follows. First, the numerical technique is described. Then, simulations of normal bouncing of a sphere on a wall and of two spheres in a fluid are presented. We finally discuss the definition of the effective restitution coefficient of particles bouncing in a fluid.

2 Numerical methods

In this section, the coupled IBM/DEM is briefly presented. The reader may refer to [7] for more details and validation about the numerical approach.

2.1 Fluid flow calculation

The IBM (see [8]) solves the following modified Navier-Stokes equations (1)-(2) for a Newtonian fluid in the entire computational domain, e.g. in the fluid domain as well as inside the particles, using a Cartesian system of coordinates,

* Corresponding author: izard.edouard@gmail.com

$$\nabla \cdot \boldsymbol{V} = \boldsymbol{0}, \quad (1)$$

$$\rho \frac{\partial \boldsymbol{V}}{\partial t} + \rho \boldsymbol{V} \cdot \nabla \boldsymbol{V} = \rho \boldsymbol{g} + \nabla \cdot [\mu(\nabla \boldsymbol{V} + {}^t\nabla \boldsymbol{V})] - \nabla P + \rho \boldsymbol{f} \quad (2)$$

$\boldsymbol{V}$ and P are the local fluid velocity and pressure, $\boldsymbol{g}$ is the gravity vector and $\boldsymbol{f} = \alpha(\boldsymbol{U} - \boldsymbol{V})/\Delta t_f$ is a body-force source term formulated such as it accounts for the presence of the solid particle in the fluid. $\boldsymbol{U}$ and α denote the local velocity of the solid object and the solid volume fraction, respectively, and Δt_f is the fluid time step. Note that in the fluid domain, $\boldsymbol{f} = \boldsymbol{0}$ and the regular Navier-Stokes equations are solved. The hydrodynamic force applied by the fluid upon the particle p of volume ϑ_p is calculated as

$$\boldsymbol{F}_h = -\frac{\rho \rho_p}{\rho_p - \rho} \int_{\vartheta_p} \boldsymbol{f} dV. \quad (3)$$

2.2 Particle motion calculation

In the case of normal bouncing, the particles motion is calculated through the resolution of the Newton's equations (4) for the particle linear momentum which reads

$$m_p \frac{d\boldsymbol{u_p}}{dt} = m_p \boldsymbol{g} + \boldsymbol{F}_h + \boldsymbol{F}_c + \boldsymbol{F}_l, \quad (4)$$

where u_p and m_p are the translational velocity and particle mass of the particle p, respectively. The total force applied on the solid particle can be divided in four contributions, namely the particle weight; the hydrodynamic force $\boldsymbol{F}_h$, defined in (3); the normal contact force $\boldsymbol{F}_c$ and a lubrication force $\boldsymbol{F}_l$ which are defined in the following.

Solid-solid interactions are modelled via a DEM method, namely the soft-sphere method [9]. Let us consider the normal unit vector $\boldsymbol{n}$ which links the centre of mass of the two solids embedded in a contact interaction. An overlap δ_n between the solid objects during the contact is allowed and permits to compute the normal contact force $\boldsymbol{F}_c = F_c \boldsymbol{n}$ with a damped mass-spring model as follows

$$F_c = \begin{cases} 0 & \text{if } \delta_n > 0 \\ max\left(0, -k_n \delta_n - \gamma_n \frac{d\delta_n}{dt}\right) & \text{otherwise.} \end{cases} \quad (5)$$

For a contact interaction between two objects i and j of mass m_i and m_j, k_n and γ_n are the normal stiffness and the damping coefficient, respectively, and are defined as a function of the dry restitution coefficient $\varepsilon_{\max}$ and the contact time t_c by $\gamma_n = -2m_* ln(\varepsilon_{max})/t_c$ and $k_n = m_* \pi^2/t_c^2 + \gamma_n^2/4m_*$ where $m_* = m_i m_j/(m_i + m_j)$ is the effective mass. If one considers a contact interaction of the particle i with a wall, m_* is reduced to m_i.

The flow structure may not be entirely captured by the IBM when two solid objects approach or separate from each other. As shown by several authors (see [10-11] for instance), this issue is overcome using a lubrication force $\boldsymbol{F}_l = F_l \boldsymbol{n}$ which is defined as

$$F_l = \begin{cases} -\frac{6\pi\mu(u_i.n - u_j.n)}{\delta_n + \eta_e} R_*^{\ 2} & \text{if } 0 \le \delta_n \le \frac{R_*}{2} \\ 0 & \text{otherwise,} \end{cases} \quad (6)$$

where $R_* = R_i R_j/(R_i + R_j)$ is the effective radius of the particles i and j and η_e is a so-called effective roughness length which accounts for particle surface asperities and can range in $[10^{-6}R_*; 10^{-3}R_*]$ [11]. Note that R_* is reduced to R_i, if one considers the lubrication interaction of the particle i with a wall. A time step $\Delta t_c = t_c/100$ is set to resolve the time integration of (4).

3 Spheres bouncing in a viscous fluid

IBM/DEM simulations of bouncing of one spherical particle on a wall and between two spherical particles in a fluid are presented for a range of Stokes and Reynolds number ranging from 1 to 10^4 and 1 to 10^3 respectively. The density ratio $r_d = \rho_p/\rho$ ranges from 3.6 to 900. Throughout this work, contact parameters are set to $\varepsilon_{max} = 0.97$, $t_c\sqrt{g/D} = 8.10^{-3}$, and the effective roughness length is fixed to $\eta_e/R_* = 4.10^{-4}$. Note that the dry coefficient of restitution ε_{max} is kept constant in the present study as its dependence on the bouncing dynamics was found to be trivial in the experimental results reported in the literature (see [2] for instance).

Simulations of a sphere p bouncing on a flat wall are considered here. The fluid grid resolution is $\Delta x = D/20$. A no-slip boundary condition is imposed on the wall where bouncing occurs while a free-slip boundary condition are applied on the other walls. The time step of the fluid phase resolution is set to $\Delta t_f\sqrt{g/D} = 2.10^{-2}$.

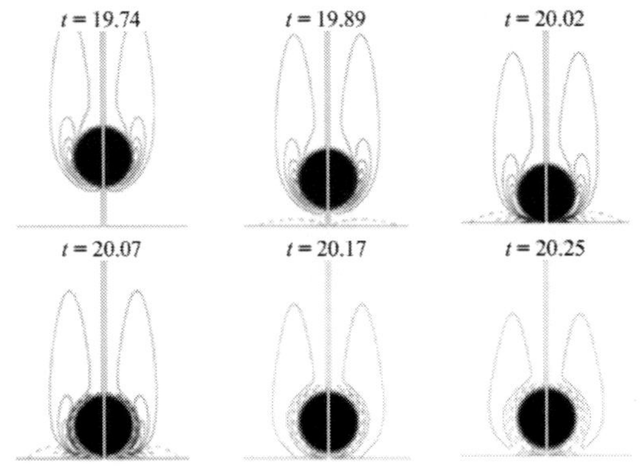

Fig. 1. Iso-vorticity contours from −17.8 to 17.8 of a sphere bouncing on a wall with with $\rho_p/\rho = 8$, $St = 53$ and $Re \approx 60$. Positive (resp. negative) values are continuous (resp. dashed) lines. Images are inspired from [11] and have been mirrored.

In the following, t denotes the dimensionless time for which the characteristic time $(D/g)^{1/2}$ has been used. A typical time evolution of the flow structure near bouncing is depicted in Fig. 1 while the particle velocity versus time is plotted in Fig. 2. At $t \sim 19$, the particle reaches the near wall area at a constant velocity V_T. Then at $t = 19.89$, the particle starts to decelerate due to the fluid film drainage characterized by some vorticity appearing near the wall surface. The particle inertia is large enough for the particle-wall rebound to occur with the considered particle roughness. The bouncing occurs at $t = 20.02$. Afterwards, the particle takes off from the wall but is decelerated by (i) its own weight, (ii) the drag induced by the fluid entrained in the particle wake and (iii) the lubrication force stemming from the fluid film recovering from the particle and the wall. On a short time scale just after the bouncing the deceleration is mostly controlled by (ii) and (iii).

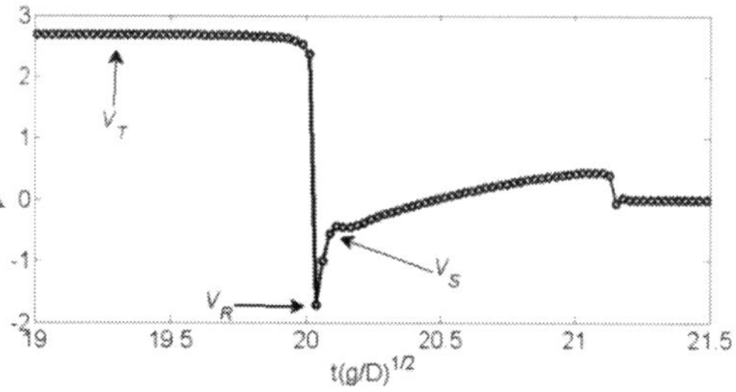

Fig. 2. Time evolution of the sphere velocity (same case as in Fig. 1). The terminal velocity V_T and the rebound velocities V_R and V_S are represented.

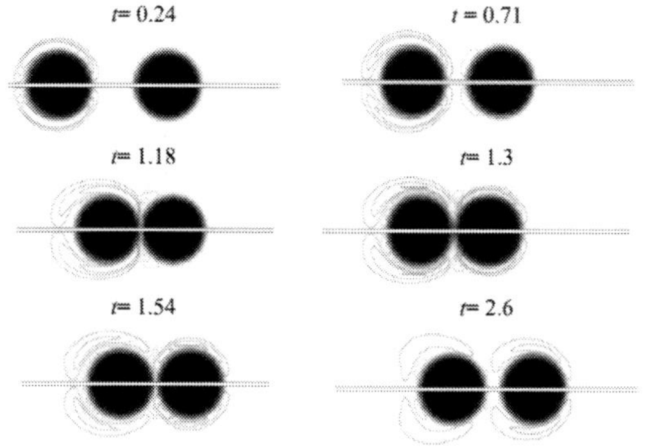

Fig. 3. Iso-vorticity contours from −4.25 to 4.25 over time of two spheres bouncing with $\rho_p/\rho = 8$, $St = 140$ and $Re \approx 175$. Positive (resp. negative) values are continuous (resp. dashed) lines. Images are mirrored. The particle on the left (right) is labelled particle 1 (2).

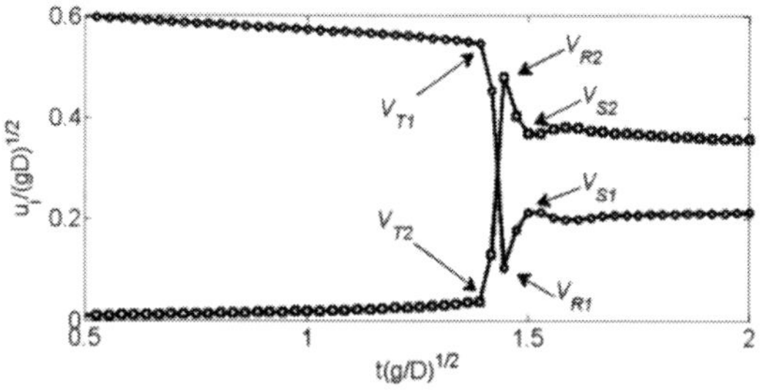

Fig. 4. Time evolution of the particle velocities (same case as in Fig. 3) during the bouncing of two spheres. The terminal velocity V_{Ti} and rebound velocities V_{Ri} and V_{Si} of particle i are represented.

The case of a sphere labelled 1 approaching another one labelled 2 initially at rest in a fluid is now investigated with the same numerical tool. The grid resolution is $\Delta x = D/20$. Here, gravity is set to zero. An example with $\rho_p/\rho = 8$, $St = 140$ and $Re \cong 175$ is shown in Fig. 3 and Fig. 4 where the structure of the flow and the time evolution of the velocity of the two particles are presented near the bouncing event, respectively. At $t = 0.24$, particle 2 is not influenced by particle 1. Vorticity is generated on particle 2's surface at $t = 0.71$ as particle 1 is approaching at a distance which is less than a radius. The bouncing starts at $t = 1.18$ after which part of the kinetic energy of particle 1 is transfered to particle 2. Afterwards, the separation of particle 2 from particle 1 entrains fluid in-between the two particles. At later times, particles move with slowly decreasing velocities during the time of the simulation, due to viscous dissipation at the particle surface.

4 Discussion about the definition of the effective coefficient of restitution

During particle rebound, several characteristic velocities may be defined. First, as for the bouncing of one particle on a wall, we classically define the terminal velocity V_T as the velocity when the particle is not influenced by the wall presence and the rebound velocity V_R as the particle velocity just after contact (see Fig. 2). A coefficient of normal restitution of bouncing can be defined as $\varepsilon = -V_R/V_T$ and only depends on the Stokes number [1, 2]. Additionally, one may remark a drastic change in the particle acceleration after contact which arises from the various forces acting on the particle during taking off. A velocity V_S may then be defined as a second characteristic velocity of rebound (see Figs. 2). The corresponding effective restitution coefficient can then be defined as $e = -V_S/V_T$. In the case of the bouncing motion of two spheres, [4] show that the normalised coefficient of restitution also depends on the Stokes number defined as $St_B = \rho_p D(V_{T1} - V_{T2})/(9\mu)$ with V_{Ti} the velocity of the i–th particle before bouncing. Again, no influence of the density ratio was reported in this study. As in the case of the normal bouncing of a single particle on a wall, one can define in this case a restitution coefficient of bouncing of two particles as $\varepsilon = -(V_{R1} - V_{R2})/(V_{T1} - V_{T2})$ where V_{Ri} is the rebound velocity after contact (see Fig. 4). Again, we remark a change in the particle acceleration when particle 2 takes off after contact. At that time, we define the velocities V_{S1} and V_{S2} which stand for a second characteristic velocity of rebound of particle 1 and 2 (see Fig. 4). We here propose an alternative definition of the effective restitution coefficient of bouncing for two particles, namely $e = -(V_{S1} - V_{S2})/(V_{T1} - V_{T2})$. Note that same notations for ϵ and e are used between the different bouncing configurations for clarity.

As shown in Fig. 5a, the present IBM/DEM simulations are in good agreement with available experimental data in both configurations confirming the exclusive dependence of $\varepsilon/\varepsilon_{max}$ with the Stokes number. Nevertheless, this exclusive Stokes dependency is not observed with the new effective coefficient of restitution e, proposed here, scaled by ε_{max} (see Fig.

5b). One may observe some plateaus for Stokes numbers higher than approximately 300 which depend on the particle to fluid density ratio $r_d = \rho_p/\rho$. A new scaling of e proposed here is $\varepsilon_{eff} = \varepsilon_{max}/(1 + r_d^c/r_d)$ with $r_d^c = 13$. This scaling makes the data collapse on a master curve for e/ε_{eff} as a function of the Stokes number (see Fig. 5c). With this new definition of e, the energy dissipation during bouncing is shown to be dependent not only on the dry restitution coefficient ε_{max} and the Stokes number St, as is the case for ε, but also on the particle to fluid density ratio r_d. Note that the quite simple influence of ε_{max} reported here is simply generalized from the results obtained from ε similar to the ones obtained in the experiments. However, this dependency would probably deserve a deeper investigation (both experimental and numerical) regarding the new definition of the effective restitution.

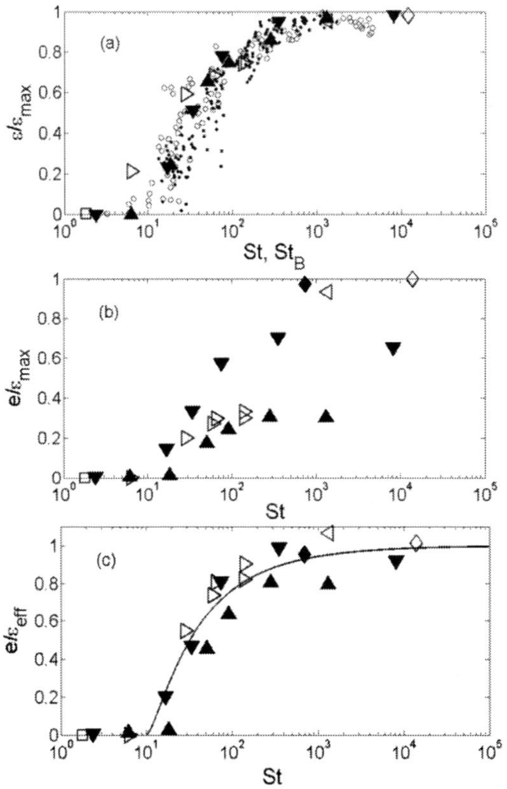

Fig. 5. (a) Effective coefficient of normal restitution $\boldsymbol{\varepsilon/\varepsilon_{max}}$ and (b-c) new effective coefficient of normal restitution $\boldsymbol{e}$ scaled by $\boldsymbol{\varepsilon_{max}}$ and $\boldsymbol{\varepsilon_{eff}}$, respectively, as a function of Stokes number. Present simulations of the bouncing motion of a sphere on a wall with $\rho_p/\rho = [8,\ 32,\ 900]$ = [▲, ▼, ◆] and two spheres $\rho_p/\rho = [3.6, 8, 90, 900]$ = [□, ▷, ◁, ◇]. Experiments of [1] and [4] are represented by • and ∘, respectively. The solid line is the analytical model proposed in [11] for $\boldsymbol{\varepsilon/\varepsilon_{max}}$ which was straightforwardly used for $\boldsymbol{e/\varepsilon_{eff}}$.

This new definition of the coefficient of restitution and in particular its new scaling allow to make an analogy with the regimes observed in immersed granular avalanches [6]. Firstly, when $St \ll St_c$ where St_c is some critical Stokes number of the order of 10, there is no rebound and $e = 0$ regardless of ρ_p/ρ. This may be interpreted as the viscous regime. Alternatively, when $\rho_p/\rho \gg r_d^c$, there is no influence of the density ratio on e which depends only on the Stokes number. In this case, when $St \gg St_c$ we have $e = \varepsilon_{max}$ and the regime is the so-called dry or free-fall regime. Finally, when $\rho_p/\rho \sim r_d^c$ and $St \gg St_c$, the regime would presumably be inertial as in the immersed granular avalanches of Courrech du Pont *et al.* [6]. The present estimation of the density ratio $r_d^c = 13$ corresponding to the transition between the inertial and dry regimes is close to the experimental value of 16 predicted in [6].

5 Conclusion

The bouncing of a sphere on a wall or another sphere in a viscous fluid has been simulated with an IBM/DEM method. The method is shown to adequately reproduce experimental results for the whole range of investigated parameters. A novel effective restitution coefficient is proposed which accounts for both the solid contact and the whole hydrodynamics effects just before and after the bouncing. With this coefficient both ρ_p/ρ and the Stokes numbers are observed to play a role in the dynamics of bouncing in line with the viscous, inertial and dry regimes found in avalanche experiments of Courrech du Pont *et al.* [6]. The analytical model proposed in [11] for $\varepsilon/\varepsilon_{max}$ has been straightforwardly extended to account for this influence of the density ratio by defining a new effective scaling factor such as: e/ε_{eff} with $\varepsilon_{eff} = \varepsilon_{max}/(1 + r_d^c/r_d)$, $r_d = \rho_p/\rho$ and $r_d^c = 13$.

References

1. G. G. Joseph, R. Zenit, M. L. Hunt, A. Rosenwinkel, J. Fluid Mech. **433**, 329-346 (2001)
2. P. Gondret, M. Lance, L. Petit, Phys. Fluids **14**, 643-652 (2002)
3. A. Ten Cate, C. H. Nieuwstad, J. J. Derksen, H. E. Van Der Akker, Phys. Fluids **14**, 4012-4025 (2002)
4. F-L. Yang, M. L. Hunt, Phys. Fluids **18**, 121506 (2006)
5. D. Legendre, R. Zenit, C.Daniel, P. Guiraud, Chem. Eng. Sci. **61**, 3543–3549 (2006)
6. S. Courrech du Pont, P. Gondret, B. Perrin, M. Rabaud, Phys. Rev. Lett. **90**, 044301 (2003)
7. E. Izard, T. Bonometti, L. Lacaze, The Journal of Computational Multiphase Flows **6(4)**, 391-405 (2014)
8. Y. Yuki, S. Takeuchi, T. Kajishima, J. Fluid Sci. Technol. **2**, 1–11 (2007)
9. P. A. Cundall, O. D. L. Strack, Geotechnique **29**, 47–65 (1979)
10. J. C.Brandle De Motta, W.-P. Breugem, B. Gazanion, J.-L. Estivalezes, S. Vincent, E. Climent, Phys. Fluids **25**, 083302 (2013)
11. E. Izard, and T. Bonometti, L. Lacaze, J. Fluid Mech. **747**, 422-446 (2014)

Rheology of dense granular chute flow: simulations to experiments

S Bharathraj[1,*], *T G* Murthy[2,**], and *V* Kumaran[3,***]

[1]*Chemical Engineering, Indian Institute of Science, Bangalore, India*
[2]*Civil Engineering, Indian Institute of Science, Bangalore, India*
[3]*Chemical Engineering, Indian Institute of Science, Bangalore, India*

Abstract. Granular chute flow simulations reveal an interesting transition from a random disordered structure to an ordered one with hexagonally ordered sheets of spherical particles, when the base roughness is modulated. Two types of base roughness are considered. The first is a fixed base, where glued spherical particles form the base, and the base roughness is varied by changing the ratio of diameters of the base and flowing particles. In the second sinusoidal base, a smooth wall with sinusoidal height variation is used; the amplitude and wavelength of the base modulation determine the base roughness. The transition is studied as a function of these roughness parameters. For the fixed base, there is a critical base particle diameter below which ordered states are observed. For the sinusoidal base, the critical amplitude increases linearly with the wavelength at lower wavelengths, reaches a maximum depending on the height of the flowing layer, and then decreases as the wavelength is further increased. There is flow for angles of inclination from $15° \leq \theta \leq 25°$ for the ordered state and $20° \leq \theta \leq 25°$ for the disordered state. Flow confinement by sidewalls also influences the rheology of the system and we see that the ordering is induced by the sidewalls as well. Experiments on chute flow at low angles indicate the presence of two types of rheology depending on the system height. A transition is observed from an erodible base configuration, where a dead zone at the bottom supports a free surface reposing at the top, to a Bagnold rheology with considerable slip at the bottom.

1 Introduction

Granular chute flow has been extensively used as a phenomenological model for understanding the mechanics of landslides, and industrial processes such as flow over a conveyor belt. Discrete element simulations have provided extraordinary insight into the mechanics of chute flow from the inter particle interactions through the overall flow profile. Although the results are quite sensitive to the input i.e. the model and parameters, several investigations over the years have provided a firm theoretical underpinning for this problem.

These simulations on chute flow have shown that the base configuration plays a pivotal role in the rheology of the granular ensemble and have also been verified through experiments [1][2]. The arrangement of the base in the simulations can broadly be divided into three categories, a bumpy base made of glued particles, an erodible base with movable or unconstrained particles and a smooth base with no particles but a smooth wall interacting with the flowing particles. In case of a bumpy base, the glued particles at the base are arranged with a random or an ordered structure, and the size of the glued base particles are also varied relative to the flowing ones. All these modulations effectively change the base roughness in the system. The methodology of constructing the bumpy base has been explained in detail by Silbert [3], Jenkins and Richman [4]. Silbert reported the effect of the base configuration on the rheology, where hexagonally ordered glued particle bases resulted in ordering of the flowing particles, which organized into non-interacting layers [5]. Transition was also observed between states as a function of the base particle diameter.

The effect of randomly arranged fixed bases on the flow dynamics was reported by Kumaran and Maheshwari [6] for a linear contact model. A sharp discontinuous transition from a random disordered structure to one with hexagonally ordered layers of flowing sheets of particles was observed, when the fixed base particle diameter decreased from 0.62 to 0.61, relative to the flowing particles (of diameter 1). These observations were independent of the angle of inclination and system depth. The ordered states exhibit steady flow for a chute angle, θ, between $15 < \theta < 25$ and the disordered states flow between $20 < \theta < 25$. Both these states conform to the Bagnold rheology where the stress was proportional to the strain rate squared, (it should be noted here, that this is a dimensional necessity for a system with instantaneous contacts, as the inverse of strainrate presents the only relevant time scale in the system). The ordered state was found to have significantly higher mean and fluctuating velocities. The Bagnold coefficients are an order of magnitude higher for

*e-mail: sagarbharathraj@gmail.com
**e-mail: tejas@civil.iisc.ernet.in
***e-mail: kumaran@chemeng.iisc.ernet.in

the ordered state due to lower strainrates. The volume density was also slightly higher for the ordered state due to better confinement. The number density and volume density profiles also show nice crests and troughs spaced by a particle diameter, indicative of the layering in the system. The ordering is quantified through the 2D order parameter q_6, which is defined as follows:

$$q_m = \frac{1}{N}\sum_{i=1}^{N}\sum_{j\neq i}\exp\left(im\theta_{ij}\right). \quad (1)$$

where N is the total number of particles in the system and m is 6 to check for a hexagonally ordered system. Thus, q_6 is 1 for a hexagonally ordered system and close to 0 for a disordered system. The evolution of these two states was studied by Kumaran and Bharathraj [7], who proposed a boundary layer formulation for the development of the ordered state. It was observed that the ordered state, during its evolution, has two distinct stages and has two distinct regimes during the first stage in the system. In the first stage, there is propagation of shear from the bottom and the shear front also carries with it the order front, which supports a disordered plug flow above it with no internal shear. The velocity in the shear region was found to obey the Bagnold law and the maximum velocity increases linearly with time in this first stage. When the shear and order simultaneously encompass the entire system, there is much slower increase in the maximum velocity, which is the second stage of development. The rate of propagation of this front was found to be dependent only on the angle of inclination of the chute. In contrast, the evolution of the disordered state showed the presence of shear stresses throughout the system [7]. It was also shown that the transition was observed for the Hertzian contact model [8]. The energy balance in the system reveals that the energy due to shear was balanced by the dissipation due to inelastic collisions except at the bottom and top layers, where there is a zone of conduction approximately up to 5 particle diameters. One consequence of the same is the scaling of the granular temperature as the square of the strainrate.

Extensive experiments on granular rheology have revealed very interesting physics of various flow instabilities, and regimes. We present a series of experiments that were carried out on a glass chute, using glass beads for small chute angles. We further present in this paper a series of simulations examining the effect of another type of base modulation, wherein a smooth wall at the bottom is provided a periodic structure of the form, $w_0 + a \cdot sin(2\pi p/\lambda)$, where w_0 is the reference position of the wall, a is the amplitude of the position perturbation, λ is the wavelength, p is either the flow direction x, or the span direction, y. We report the transition as a function of the wavelength, amplitude and direction of the perturbation. We specifically seek to highlight the variation in the rheology for such a smooth wall problem from that of a fixed base configuration. The evolution of the ordered and disordered states induced by these different base modulations in addition, to preliminary observations on flow confinement in the presence of side walls are also presented here [9].

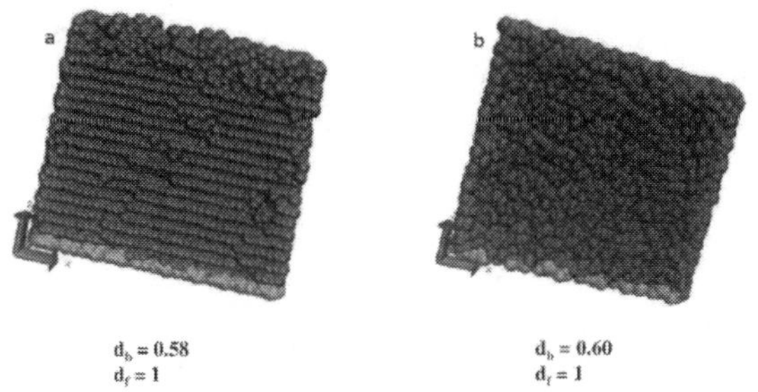

Figure 1. Snapshot of (a) an ordered and (b) a disordered system with flowing particles of diameter, d_f and a bumpy base with glued particles of diameter, d_b

2 Experimental and Simulation methodology

The simulation cell consisted of monodisperse spherical particles stacked up in all the three dimensions, with periodic boundary conditions in the flow(x) and span(y) directions. The flow is confined by a base at the bottom and a free surface at the top as discussed in the previous section. LAMMPS, an open source molecular dynamics package, implementing Newtonian mechanics for individual particles with inter-particle contact laws, is used for simulations with the Hertzian inter-particle contact model. All the relevant quantitites in the system are non dimensionalised with suitable system parameters. All masses and lengths are scaled by the mass m and diameter d of the flowing particles, and all times by $\sqrt{d/g}$. Details of the contact model and its implementation are available in previous studies [3][6][7][8]. The important quantities monitored are the average velocity in the flow direction, V_x, the average coordination number, Z, the number and volume densities, n,ϕ, the order parameter, q_6 were all computed from these simulations and are compared to experiments for certain conditions. The averaging techniques used to calculate these parameters and the input simulation parameters is available [6][7].

A schematic of the ordered and disordered system is given in figure 1. Experiments were carried out on a glass chute with dimensions $150cm \times 10cm \times 10cm$. Glass beads of diameter $2 - 2.5mm$ were used as a model system for these experiments. Imaging was carried using a high frame rate camera(PCO 1200HS). The images obtained from the camera were analysed using PTVlab [14] and velocity fields are obtained after calibration.

3 Results and discussions

3.1 Transition due to a periodic smooth base

Transition from a random disordered structure to a hexagonally ordered one, occurs when different types of forcing (or modulations) are provided at the base. For the set of simulations performed on a rough bumpy base, this transition occurs at critical base particle diameter of 0.59 for the

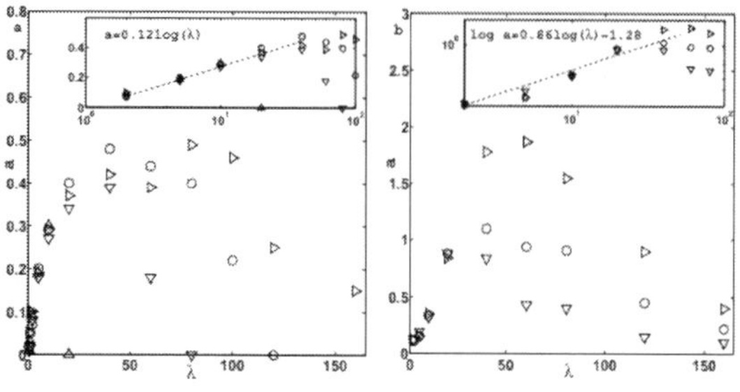

Figure 2. Transition amplitude, a, as a function of the wavelength, λ, of the bottom smooth wall for modulations in the : (a) flow direction, x, and (b) span direction, y, with the corresponding log curves in the inset with the model fit (- -), for different system heights, z = 18(△), 25(▽), 37(○), 75(▷) at an angle of inclination, 22°

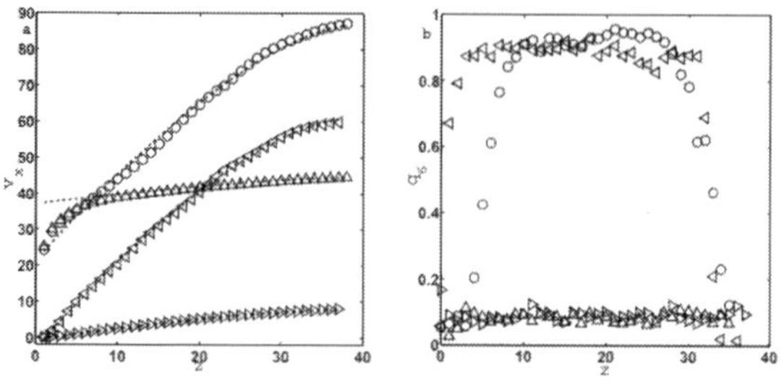

Figure 3. (a) Average flow velocity, v_x with the Bagnold fit (- -), (b) averaged 2D order parameter, q_6 variation with system depth, z, for an ordered state generated by a sinusoidal smooth base (○), a rough base (◁), and a disordered state generated by a sinusoidal smooth base (△), a rough base (▷), at an angle of inclination, 22°

Hertzian model. As the bumpy base has a random (irregular) structure, the base roughness is also aperiodic. This transition is studied as a function of the amplitude, wavelength and direction of the modulation, as shown in figure 2. As seen in the figure 2, the transition amplitude, a, increases linearly with wavelength, λ at lower wavelengths irrespective of the system depth. The amplitude then reaches a critical value and then decreases with wavelength, depending on the system depth. As is evident from the figure, at a given wavelength, the transition amplitude is much larger for the span wise (y) modulation than the flow wise (x).

We further compare the velocity profiles of the ordered-disordered states for the sinusoidal base with the profiles obtained with the fixed base. For this, we plot the flow velocity, v_x, and the order parameter, q_6, profiles as a function of the depth, z, for the ordered and disordered states obtained with the two types of base modulations as shown in figure 3. The average velocity profile in figure 3(a) reveals that all the configurations follow the Bagnold rheology, except for the bottom few layers of the disordered state generated by the sinusoidal base. A significant difference being the slip generated by the smooth wall in both these states, it is interesting to note that when this profile is subtracted from the bulk, we approximately recover the profiles of the fixed base states. The averaged order parameter, q_6, is approximately 1 in the bulk for the ordered states and 0 for the disordered states, as expected. However, in case of the sinusoidal bases, the profile reveals the presence of a few layers of disorder at the bottom, up to around 5 particle diameters, where the hexagonal order is imperfect. This feature in the ordered states of the smooth wall case is universal, and is found to be independent of the wavelength, amplitude and direction of the modulation.

Another feature, which is seen only in the smooth wall modulations, is the tendency of ordering in the disordered states, just above the critical transition amplitude. The order seems to propagate from the bottom to the bulk, but seems to be disrupted before it reaches the top surface. In the evolution of the ordered state, the propogation of the order front can be treated in a boundary layer formulation, wherein we define the height up to which the q_6 is 0.5 or higher, as the boundary layer thickness. It was reported that the order front has a universal slope of 1 particle diameter in 13.5 dimensionless timesteps, dependent only on the angle of inclination [7]. Here, we find the same observation even for sinusoidal bases, apart from the disordered bottom layers.

3.2 Effect of confinement with sidewalls

The effect of boundaries on the flow rheology has been well documented, as outlined earlier. We present a series of simulations on a system confined between sidewalls [10]. In figure 4, the effect of variations in the span on the flow velocity is presented. As seen in figure 4, a significant difference exists in the magnitude of the velocity at the wall and away from the wall. In case of the system with a total span of 20 particle diameters, a dead zone measuring up to 5 particle diameters exists at the bottom, and the average flow velocity is lower compared to a system with periodic boundaries. The overall flow profile is convex, indicating an erodible base configuration. For the remaining systems, we see that there is a significant slip at the bottom and a concave profile with height, indicative of the Bagnold rheology. The magnitudes of the velocity in the span and height directions is less than 1% of the flow direction magnitudes.

The profile variations with the spanwise co-ordinate, y, also show interesting features. The percolation effect of the sidewalls on the flow velocity does not seem to affect a small system (i.e. with total span of 20 particle diameters) and is reasonably evident at lower depths for the remaining systems. The presence of side walls in the simulations also increases the minimum angle of inclination required for steady flow, from 15° to 30° for a system with a span of 20 particle diameters. For the system confined by smooth sidewalls and a bottom wall, a 3D ordering was also observed. The effect of the system height, aspect ratio, different types of sidewalls, ordering effects are discussed elsewhere.

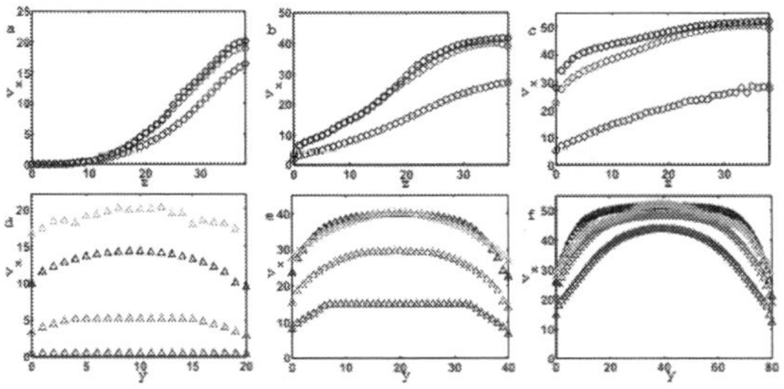

Figure 4. Average flow velocity, v_x variation with height, z, and span, y, for a system with total span = 20 (a)(d), 40(b)(e), 80(c)(f) at different locations; y=wall(○), centre(○), in between wall and centre(○); z=10(△), z=20(△), z=30(△), z=40(△) for an angle of inclination, 30°

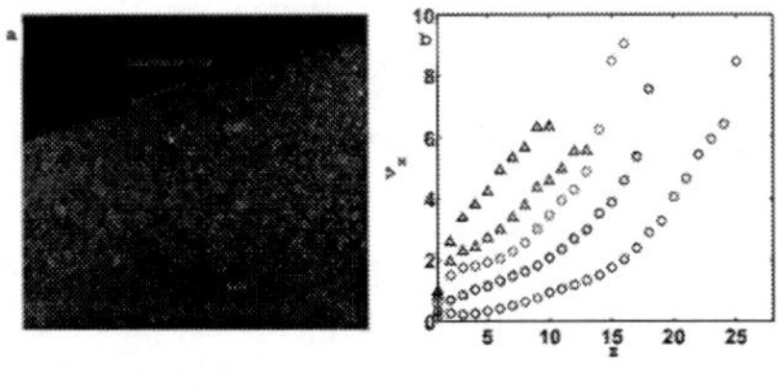

Figure 5. (a) Image obtained from the high frame rate camera of the glass beads flowing through the glass chute and (b) the variation of the steady flow velocity, v_x with system height, z for different system heights, z>25(○), 15 < z < 20(○), z=15(○),10 < z < 15(△), z <10(△) for an angle of inclination, 11°

3.3 Experiments at low angles

Experiments on granular chute flow have been performed mostly at intermediate and higher angles of inclinations [11] [12][13]. The rheology at lower angles, 9° – 16° show some intriguing features. One such feature is the existence of regimes in the flow which depend on the total system height. Figure 5(b) clearly shows the change in the curvature of the velocity profile with the total system height. For a system height greater than 20 particle diameters, there is a prominent dead zone in the system, which extends to a height of 5 particle diameters and the velocity profile in the bulk is convex, which resembles the one obtained in figure 4(a). Below this height, the flow undergoes a transition as is evident in the figure. For systems with less than 15 particle diameters, the profile is concave adhering to the Bagnold rheology. It was observed that the free surface of the flow always occurs at the repose angle, irrespective of the chute angle, and is supported by a dead zone at the bottom. Once the flow height reduces below a critical value, the profile changes from an erodible convex one to a Bagnold concave one, which is also dependent on the slip velocity at the bottom.

4 Conclusions

The effect of different types of base modulations on the rheology is studied, which shows a transition from a random disordered state to one with hexagonally ordered 2D sheets, depending on the critical values of the base roughnesses. This transition is universal, irrespective of the contact model. The ordered and disordered states generated by the base modulations have many similarities and unique features. Flow confinement using sidewalls influences the rheology. The percolation effect of the sidewalls was observed as a function of different span depths. Experiments at very low angles indicate the existence of different flow regimes in the system, which seems to be influenced by the total height of flow and the slip at the bottom.

References

[1] Pierre Jop, Yoel Forterre, Olivier Pouliquen, Nature Letters **441**, 727-30(2006)

[2] Olivier Pouliquen, Physics Of Fluids **11**, 542-48(1999)

[3] Leo Silbert, Deniz Ertaz, Gary S. Grest, Thomas C. Halsey, Dov Levine, Steven J. Plimpton, Physical Review E **64**, 051302 1-14(2001)

[4] J.T Jenkins, M.W Richman, J. Fluid Mech **171**, 53-69(1986)

[5] Leornardo E Silbert, Gary S. Grest, Steven J. Plimpton, Physics Of Fluids **14**, 2637-45(2002)

[6] V. Kumaran, S. Maheshwari, Physics Of Fluids **24**, 053302 1-23(2012)

[7] V. Kumaran, S. Bharathraj, Physics Of Fluids **25**, 070604 1-26(2013)

[8] P. A. Cundall, O. D. L. Strack, Geotechnique **29**, 4765 (1979)

[9] J. Baker, T. Barker, J.M.N.T. Gray, J. Fluid. Mech **787**, 367-395(2016)

[10] S. Bharathraj, V. Kumaran, (unpublished)

[11] Leornardo E Silbert, James W. Landry, Gary S. Grest, Physics Of Fluids **15**, 1(2003)

[12] Thomas. G. Drake, J. Geophys. Res. **95**, 8681-8696(1990)

[13] Emmanuel Azanza, Francois Chevoir, Pascal Moucheront, J. Fluid. Mech. **400**, 199-227(1999)

[14] W. Brevis, Y. Nino, G.H Jirka, Experiments in Fluids, **50(1)**, 135-147(2011)

DOI: 10.1051/epjconf/201714011005

Quantitative predictions of the μ(I) rheology in 2D and 3D granular column collapse: scaling laws and quasi-static vs. inertial regimes

Rudy Valette[1,*], *Stéphanie* Riber[1], and *Elie* Hachem[1]

[1]MINES ParisTech, PSL Research University, CEMEF, CNRS UMR 7635, CS 10207 rue Claude Daunesse, 06904 Sophia-Antipolis Cedex, France

Abstract. We introduce an accurate numerical method for the computation of 2D and 3D granular collapse under gravity flows using the $\mu(I)$ rheology. We show that this rheology can capture the two experimentally observed types of spreading and the corresponding scaling laws, and we provide an accurate sensitivity analysis to rheological constants. Finally, we underline the role of time-dependency on the spreading dynamics.

1 Introduction

Granular column collapse under gravity (Fig. 1) is a benchmark problem for several applications from geophysics to industry.

It has been shown experimentally, by Lube et al. [1] for 2D and Lube et al. [2] and Lajeunesse et al. [3] for 3D granular columns of initial aspect ratio $a=h_i/r_i$ (initial height/initial width or radius), that for large enough values of a, the final width or radius r_∞ reached after spreading, obeys the scaling law $(r_\infty - r_i) / d_i \approx a^{2/3}$ in 2D and $a^{1/2}$ in 3D flows. Moreover, it was shown [1] that the normalized distance-time data (t, x) plot (see Fig. 1) exhibits a universal shape, independently of the grain type (fine or coarse sand, rice, sugar, ...).

For shear flows in the dense granular regime, da Cruz et al. [4] shown, by dimensional arguments on discrete numerical models, that the inertial number $I = |\dot{\gamma}| d \sqrt{\rho_s / p}$ (where $|\dot{\gamma}|$, d, ρ and p are the norm of the strain rate tensor, the grain diameter, the grain density and the pressure) governs the transition between a low friction *quasi-static* regime and a high friction *inertial* regime.

2 Flow model and numerical method

Then, Jop et al. [5] proposed a (computationally lighter) pressure p dependent viscoplastic continuum model called $\mu(I)$, where the extra-stress $\tau_{ij} = (\mu(I) p / |\dot{\gamma}|) \dot{\gamma}_{ij}$ is defined using an effective friction coefficient $\mu(I) = \mu_s + (\mu_F - \mu_s)/(I_0 / I + 1)$ that varies from μ_s (*quasi-static* regime) to μ_F (*inertial* regime) when the inertial number reaches the (material dependent) parameter I_0. So far, the $\mu(I)$ model has been successfully applied for the prediction of benchmark steady experimental flows [6]. Concerning complex transient flows of granular column collapse, even though recent papers have shown that discrete numerical simulations can match the experimental scalings [7, 8], few studies have addressed the pertinence of the $\mu(I)$ continuum model in 2D [8] and, to our knowledge, 3D studies are still missing.

In this paper, we study the pertinence of the $\mu(I)$ model to predict the case of 2D and 3D granular column collapse flows. We use a time-dependent regularization algorithm to solve this model using a finite element method, combined with anisotropic mesh adaptation to capture accurately the quasi-static vs. inertial flow zones, and using a variational multiscale method [9].

A Level-Set method, based on self-reinitialization of the signed distance function, aims to capture and follow efficiently the interface between the fluid/air domains.

3 Results

3.1. Column collapse simulations

Our simulations show that the $\mu(I)$ model can reproduce all the qualitative features of the experimentally observed flows (see Fig. 1), such as spreading, front flow acceleration, steady and deceleration phases, as well as a remaining flat top for low aspect ratios and the appearance of surface waves for high aspect ratios. Our computations have been validated using a comparison to 2D continuum and discrete simulations from [8], in particular an accurate deceleration phase is predicted (see Fig. 2).

Moreover, 2D and 3D results show that most of the flow is in a *quasi-static* regime, the *inertial* regime being only reached in a small volume at the front flow. However, this thin and low pressure front flow contributes

* Corresponding author: rudy.valette@mines-paristech.fr

significantly to the spreading distance, in the deceleration phase. 2D and 3D initial aspect ratio scaling laws for spreading distance are then quite accurately retrieved: exponents 0.97 and 0.69 in 2D, and exponents 1.00 and 0.54 in 3D (see Fig. 3 and Fig. 4).

3.2. Shear bands

For large friction coefficients μ_s, the $\mu(I)$ model exhibits non-unique solutions at early times of the flow (acceleration phase): mesh dependent short length scale instabilities (shear bands) appear, which is consistent with the theoretical predictions of [10]. However, these instabilities do not subsist during the steady front velocity and deceleration phases.

3.3 Sensitivity to rheological constants

Sensitivity analysis on rheological constants was performed for quasi-static μ_s and inertial μ_F friction parameters, in order to verify the universality (no dependency on grain type) of flow features such as relative spreading vs. aspect ratio, and normalized distance-time.

As shown on Fig. 5, sensitivity to quasi-static friction is of order ~ $1/\mu_s$ for low and high aspect ratios. The whole final surface shape is significantly affected, as shown on Fig. 6, where the final global slope is strongly dependent on μ_S.

Sensitivity to the inertial friction parameter is subtler (of order ~ $1/\mu_F$), but noticeable (see Fig. 7). As expected, the final surface shape is changed only at larger radii (see Fig. 8), as it was pointed out that the inertial regime is reached at the front flow.

An additional sensitivity, not discussed in experimental papers, is due to the time-dependency. Indeed, the time scale of the flow being $\sim\sqrt{h_i/g}$ (where g is the gravity acceleration), a collapse from a larger initial height h_i, for fixed materials properties and aspect ratio, should benefit to the inertial regime, as the strain rate would scale as $\sim 1/\sqrt{h_i}$ and the pressure as $\sim h_i$. Consequently, the inertial number would scale as $\sim 1/h_i$, which would mean a lower friction for larger columns, keeping the aspect ratio constant. A similar effect on inertial number can be deduced by considering the grain diameter (which is an internal length scale not taken into account in the experimental results) rescaled by the column height. Indeed, our computations show that when increasing the initial height, one still get a power law $\sim a^{1/2}$ with aspect ratio, but also a larger relative spreading (see Fig. 4, red dots).

4 Conclusion

A quantitative dependency of granular collapse flow features to material (rheological) constants was computed in the light of the $\mu(I)$ rheology. This gives complementary information to experimental literature, which reported these features in log plots, where a subtle but real sensitivity exists. Moreover, we have pointed out that the relative spreading should be time-dependent. This property will be investigated in a future experimental study.

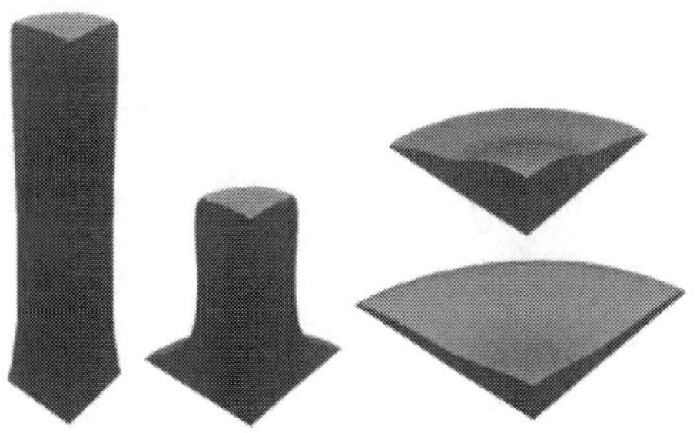

Fig. 1. $\mu(I)$ computation of the shape of a granular column at time t = 20%, 40%, 60% and 80% of total collapse, for an initial ratio a=7; rheological parameters were taken from [5]

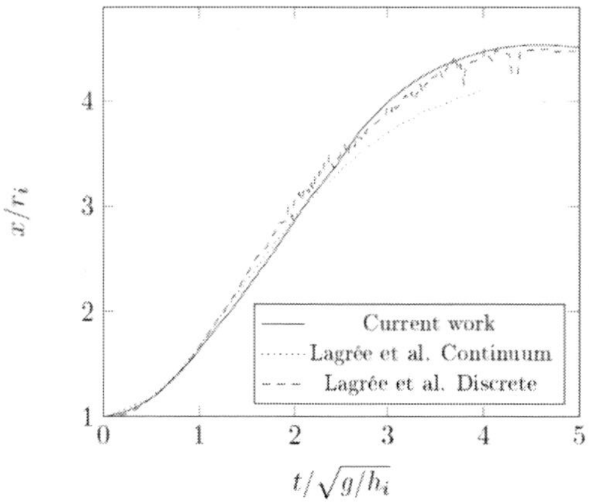

Fig. 2. Normalized distance-time computations: current work $\mu(I)$ (−) ; Lagrée et al.[8] $\mu(I)$ (..) ; Lagrée et al. [8]

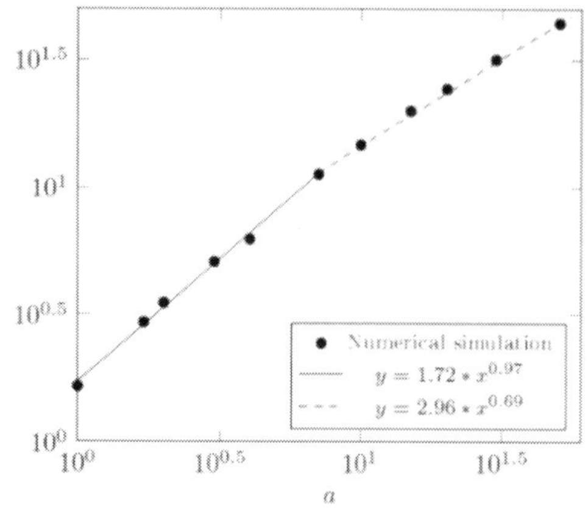

Fig. 3. Relative spreading distance vs. initial aspect ratio for 2D computations

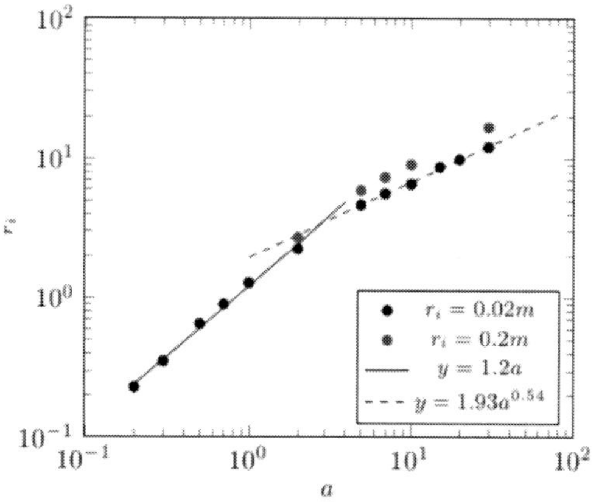

Fig. 4. Relative spreading distance vs. initial aspect ratio for 3D computations – Influence on the initial column height (for each value of aspect ratio a, two different radii give two different column heights)

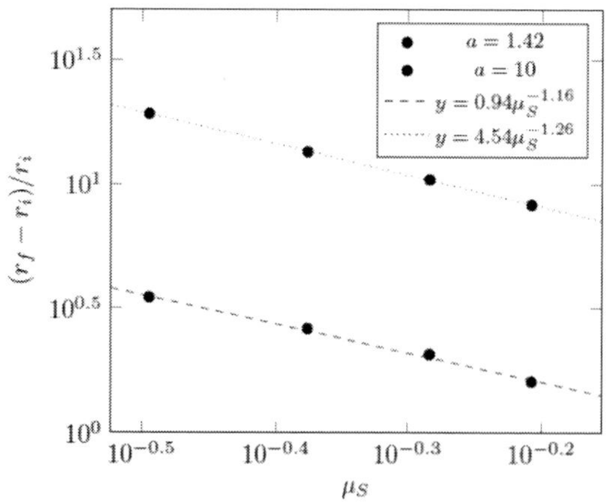

Fig. 5. Variation of relative spreading with quasi-static friction coefficient for different aspect ratios

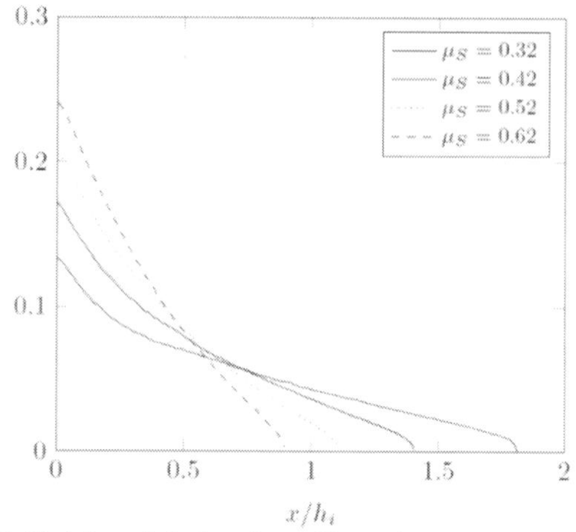

Fig. 6. Variation of final surface with quasi-static friction coefficient for an aspect ratio a=10

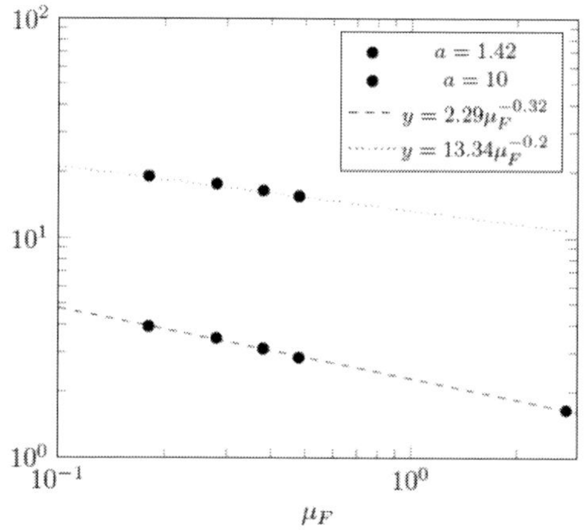

Fig. 7. Variation of relative spreading with inertial friction coefficient for different aspect ratios

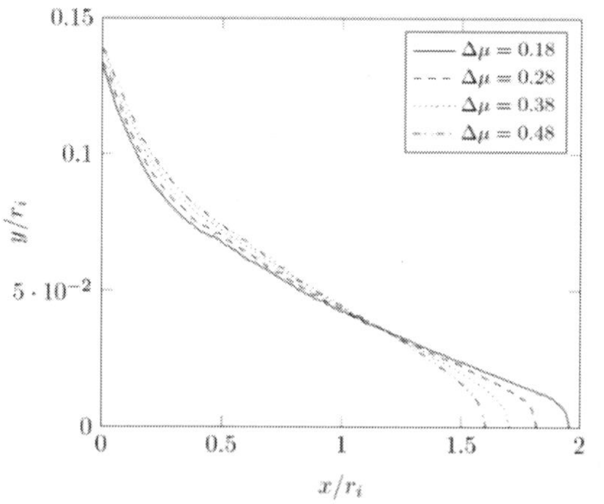

Fig. 8. Variation of final surface with inertial friction coefficient for an aspect ratio a=10

References

1. G. Lube, H. E. Huppert, R. S. J. Sparks, A. Freundt, Phys. Rev.E, **72**, 041301 (2005)
2. E. Lajeunesse, A.Mangeney-Castelnau, J. P. Vilotte, Phys. Fluids, **16**, 2371 (2004)
3. G. Lube, H. E. Huppert, R. S. J. Sparks, M. A. Hallworth, J. Fluid. Mech., **508**, 175 (2004)
4. F. da Cruz, S. Emam, M. Prochnow, J.-N. Roux, F. Chevoir ,Phys. Rev. E, **72**, 021309 (2005)
5. P. Jop, O. Pouliquen, Y. Forterre, Nature, **441**, 727 (2006)
6. G.D.R. Midi, Eur. Phys. J. E, **14**, 341 (2004)
7. L. Lacaze, R. R. Kerswell, Phys. Rev. Let., **102**, 108305 (2009)
8. P.-Y. Lagrée, L. Staron, S. Popinet, J. Fluid Mech., **686**, 378 (2011)
9. S. Riber, R. Valette, Y. Mesri, E. Hachem, Computers and Fluids, **138**, 51 (2016)
10. T. Barker, D.G. Schaeffer, P. Bohorquez, J. M. N. T. Gray, J. Fluid Mech., **779**, 794 (2015)

EPJ Web of Conferences **140**, 03023 (2017)
Powders & Grains 2017
DOI: 10.1051/epjconf/201714003023

Influence of flight shape on discharging profiles of granular material in rotary dryer

Ilham Benhsine[1,*], *Mustapha* Hellou[1,**], *Franck* Lominé[1,***], and *Yves* Roques[2,****]

[1] *INSA, LGCGM, EA3913, F-35708, Rennes, France*
[2] *AII / CERA, BP10180, F-35803, Dinard, France*

Abstract. In this work, Discrete Element Method (DEM) is used in order to calculate the motion of granular material in rotating dryers. We are particularly interested in analysing the effect of flight shape on the behaviour of spherical particles in the cross section of the dryer. We will be using two segments flights and three different profiles : a straight flight (180° between both segments), an angled flight (with an angle of 120°) and a right-angled flight (90°). The results show that the profile of the flight affects significantly the motion of the particles in the cross section of the dryer. Changing the angle between the segment's flight, changes the flight loading and thus the material hold-up which leads to different discharging profiles of the flight. For a right angled flight, the range of the discharge angle increases leading to a more uniformized cascade pattern in time and an enlarging of the area occupied by the curtains of particles. The specific durations (discharging time, falling time) are also determined and studied as a function of the flight shape.

1 Introduction

Drying granular materials can be defined as the removal of moisture from a solid product. The moisture commonly consists of water. Materials are dried to ease handling, to preserve materials for storage and shipment and mostly they must be dried to bring their moisture content to a specific value before being sold [1].

Rotating dryers are widly used in industry to dry variety of solids [1]. They consist of long drum driven at low speed. The rotation speed, ω, is selected according to the desired mode of bed motion which is characterized by the Froude number $\omega^2 R/g$ [2], where R is the inside radius of the drum and g is the acceleration due to gravity. Industrial rotary dryer is a tube of length L of circular cross section of a diameter D, with aspect ratio L/D about 10. The drum can be horizontal or slightly inclined [3], [4], [5].

Wet granular material is introduced into the upper inlet of the drum. Then the material accumulates and progress through the drum due to the combined action of the rotation of the drum, the slope of the drum -if present- and the hot gas flow passing through the drum. The dried product leaves the drum from its other extremity.

The dryers are usually fit with flights, which consist of slab made of steel distributed on the periphery of the cross section and periodically along the tube. Flights carry the solid particles and disperse them into the hot gas flow. Flights used in industrial dryers can have complex shape. However, for theoretical and experimental purposes, simple flight shapes are frequently considered composed by one , two or three segments ([6], [7], [8], [5], [9]).

Several topics that interest scientists and engineers in connection with heat and mass transfer arise in the problem of rotary dryers. The most important of them are the behaviour of the material bed, the holdup of the material by the flights, the material curtains and the mean residence time. While the behaviour of the granular bed material is greatly studied by the DEM approach [10], [11], the other topics are treated essentially with theory or experiments. Thus, most of the studies on the subject of the holdup of the material use the approach of bulk material [12], [6], [8], [5], [9]. Experiments also showed that the mean residence time is affected by the flight's shape. A drum fitted with right-angled flights has a higher residence time than a drum fitted with straight flights [13].

In order to study the effect of flight shape on the behaviour of the granular material, we use EDEM, a high-performance software for bulk material flow simulation based on the Discrete Element Method. This study is conducted to investigate the influence of the flight shape on the holdup. And it is also extended to the analysis of the falling length and falling time of particles in the cross section of the dryer. The flight profiles chosen are: straight flight, large angled flight and right angled flight.

2 Description of the problem and simulation parameters

We are interested to study the granular motion in a transverse cross section of the drum. Due to the periodicity

*e-mail: ilham.benhsine@insa-rennes.fr
**e-mail: mustapha.hellou@insa-rennes.fr
***e-mail: franck.lomine@insa-rennes.fr
****e-mail: yves.roques@roullier.com

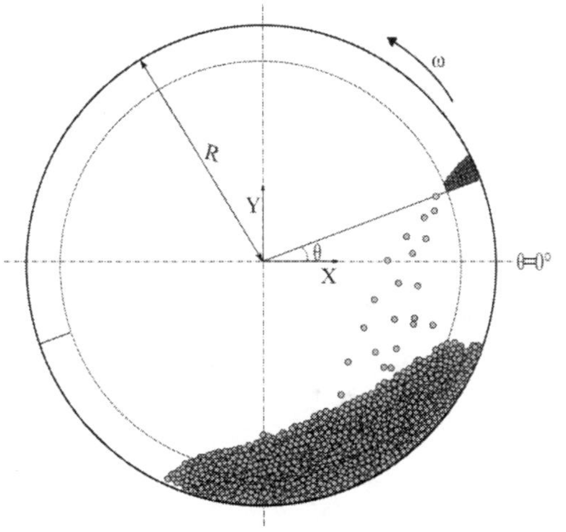

Figure 1. Drum's section used during the simulations

of the granular motion along the tube [7], we have used a portion of the rotating drum with thickness of 0.01 m. The diameter is also reduced in order to minimize the particles number. The diameter, D, is equal to 0.16 m (*Fig.* 1). This portion is driven with rotation speed ω = 16 rpm. The Froude number is equal to 0.02. It corresponds to the cascade regime for a filling degree $\tau > 0.1$ according to [14]. Where τ express the volumetric filling percentage of drum's load ($\tau = \frac{V_{Material}}{V_{Drum}}$).

In industry, the number of the flights is calculated to insure a maximum holdup. In the present study, in order to avoid interaction between the flights, we have used only two opposite flights. Thus the portion of the tube is fitted with flights composed of two segments which have the same length l of 0.0075 m. Three profiles of flights are considered for which the angle α_1 between the segments is 180°, 120° and 90° (Fig.2). The angle α_0 of fixation of the flights on the drum wall is 90°. Each flight is drawn in Fig.2 at $\theta = 0°$, where the flight is supposed to be filled at it's maximum hold-up [7], [6]. In the approach of bulk material, the free surface is an inclined line with an angle equal to the dynamic angle of repose, Φ, which is equal to the static angle of repose (Φ_0) at $\theta = 0°$. This dynamic angle depends on the rotation speed, drum's radius and friction coefficient as it was demonstrated by Schofield and Glikin [15] by studying the equilibrium of a particle about to fall from the flight. The volume of material in the flight can be calculated by the formula given in [5] for angled flight with two segments or in [9] for right angled flight. For three segments flight, a similar formula can be found in [8].

In our simulations, the particle-particle and particle-wall collision are modelled through the soft-sphere approach. This was used to simulate particle-to-particle and particle-to-wall contact interaction. The model is used due to its accurate and efficient force calculation. It is based on Hertzian contact theory when calculating the normal force

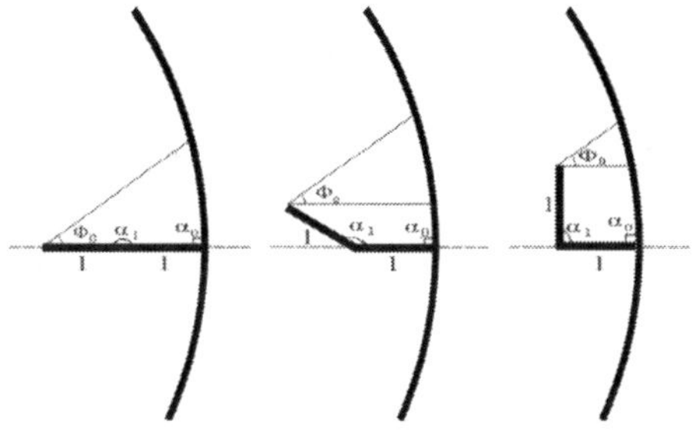

(a) Straight flight (α_0 = 90°, α_1 = 180°) (b) Large angled flight (α_0 = 90°, α_1 = 120°) (c) Right angled flight ($\alpha_0 = \alpha_1$ = 90°)

Figure 2. Three differents flight's shapes

Table 1. Physical parameters of the particle and of the tube used in the numerical simulations

	Particle	Tube
Poisson's ratio	0.3	0.3
Shear modulus (Pa)	2.00E+07	7.00E+07
Density (kg/m^3)	2600	7800
Coefficient of restitution	0.1	0.24
Coefficient of static friction	5.45E-01	5.45E-01
coefficient of rolling friction	0.01	0.01

component and the work of Mindlin-Deresiewicz when determining the tangential force [16], [17].

Particles are considered spherical and the dryer is made of steel. Particles have a radius of $r = 0,00075$ m resulting in the use of 9558 particles to acheive $\tau = 15\%$ as a filling degree, with a porosity of 0.44, corresponding to a random loose packing [18]. This value has been verified later on with an image analysis software on some images of our configuration obtained by EDEM. Characteristics of the materials used are shown in table 1.

3 Results

In order to increase the drying efficiency, it is recommended to increase the density of material in the airborne phase (particles flying through the cross-section of the drum) particularly in the upper half of the drum [5], [4]. Thus most of the studies consider the quantity of particles in the upper half of the drum as a design criterion [15], [7], [6]. We have applied this criterion and have calculated the variation of the particles number in this upper half of the drum.

In Fig. 3, we see the change of volume of material on the flight from $\theta = 0°$ up to complete discharge. To compare all flight's shapes we have drawn the relative volume of particles in the flight. This volume is calculated according to a constant porosity, equal to 0.44.

In order to compare the behaviour of the flights, we have divided this volume by the capacity of each flight for $\theta = 0°$. In this figure, we have also drawn the holdup of right angled flight with the approach of bulk material developed by Sunkara and al. [9]. The angle α_1 between

DOI: 10.1051/epjconf/201714003023

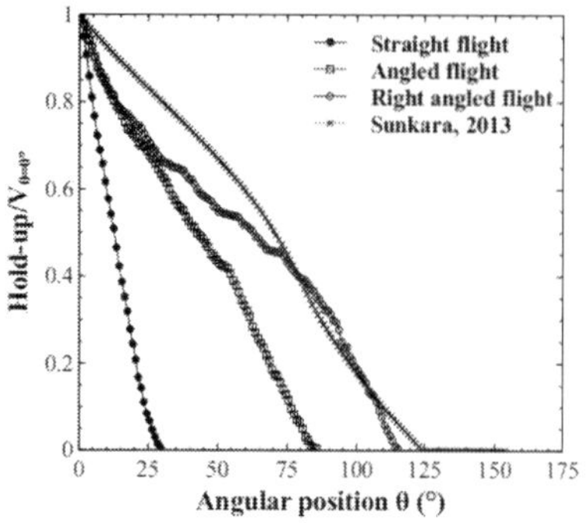

Figure 3. Relative hold-up in the flight, where $V_{\theta=0^\circ}$ is the hold-up at $\theta = 0^\circ$: for a straight flight $V_{\theta=0^\circ} = 111\ \mathrm{cm^3/m}$, for an large angled flight $V_{\theta=0^\circ} = 119\ \mathrm{cm^3/m}$ and for a right angled flight $V_{\theta=0^\circ} = 72.8\ \mathrm{cm^3/m}$

the two segments influences the discharging profile of the flight. The time the flight takes to be completely empty is more important for an angle $\alpha_1 = 90^\circ$. This favours the presence of particles in the cross section of the tube for a long period. Even though straight flight carries more particles ($V_{\theta=0^\circ} = 111\ \mathrm{cm^3/m}$). Its angular discharging angle, for which the flight is completely empty, is small.

In figure 4, we present the flight's cascading rate which varies from one flight to another. The number of particles leaving the flight is more or less constant over time for a right angled and a large angled flight.

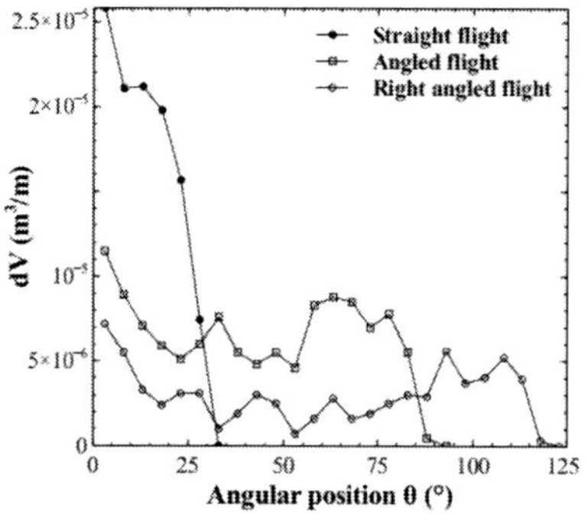

Figure 4. Flight cascading rate as a function of the angular position for different flight's shape

Snapshots of the behaviour of the granular material in the drum are presented in Fig. 5. These snapshots show that the flight profile affects the behaviour in the cross-section of the drum. The particles are concentrated in a small zone for the straight flight. However for the angled flight, a more extended zone is observed. Fig. 5 also illustrate both the hold-up in the flight and the particle curtain, which appear to have a constant width for angled flights and change shape for a straight flight. A straight flight has the maximum initial volume, but a Right-angled flight carries the particles higher. So for further characterization of the particles movement, we calculate the falling length and falling time. The falling length is the distance traveled by the particle from the flight's tip up to the contact point with the bed of material or the drum's wall. And the falling time is the time the particle takes to travel the falling length.

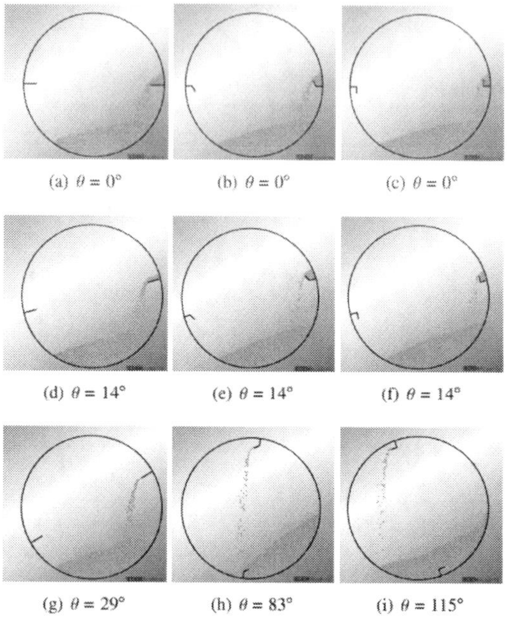

(a) $\theta = 0^\circ$ (b) $\theta = 0^\circ$ (c) $\theta = 0^\circ$

(d) $\theta = 14^\circ$ (e) $\theta = 14^\circ$ (f) $\theta = 14^\circ$

(g) $\theta = 29^\circ$ (h) $\theta = 83^\circ$ (i) $\theta = 115^\circ$

Figure 5. Snapshots of the granular behaviour in the rotating drum for three shapes of the flight

The dispersion of the falling length and the falling time is not the same from one flight shape to another, leading to different discharging patterns. Since the number of particles leaving the flight is important and depends on the angular position of the flight, we calculate the average falling length and falling time over an angular range of 1° to present the results and to ease the comparison between the flight's shapes (Fig. 6, 7). The more we reduce the angle α_1 between the segments of the flight, the greater the range of the discharging flight. Moreover, when calculating the average falling length and falling time of all particles leaving each the flight, the results show an increase in the falling length and the falling time up to 50% and 70%, respectively, when using a right-angled flight instead of a straight one.

4 Conclusion

The shape of the flight influences the discharging pattern of the granular material retained in the flight. Compared to the other flights an angled flight with 120° offers a large angular discharging range while carrying an important quantity of material compared to a right-angled flight. The falling length and the falling time of the particles also increases up to a maximum value then they start decreasing for a large-angled flight and a Right-angled flight.

DOI: 10.1051/epjconf/201714003023

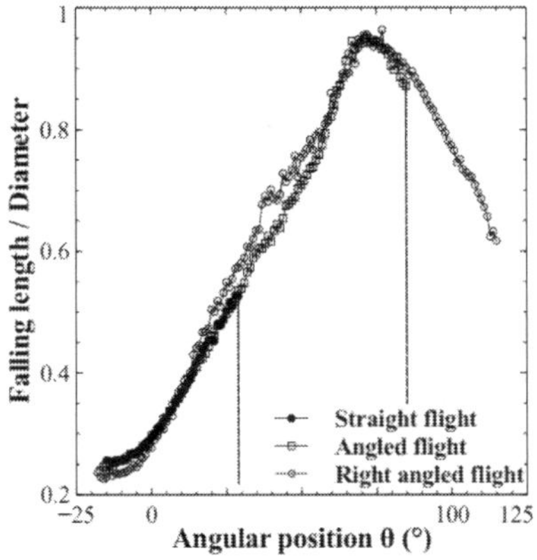

Figure 6. Falling length of the particles, for three different flight profiles. Vertical lines show the angular discharging angle

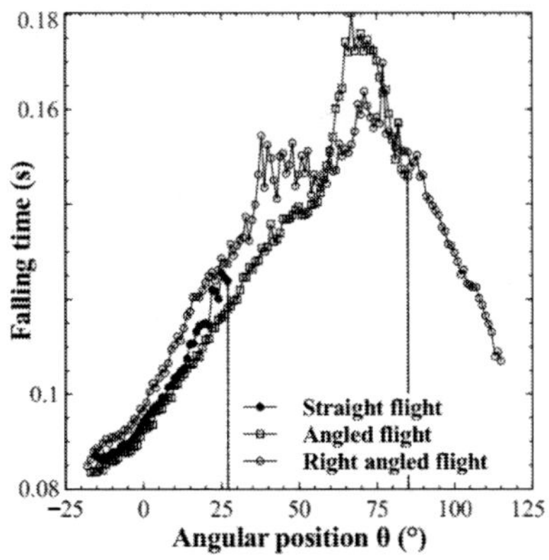

Figure 7. Falling time of the particles, for three different flight profiles. Vertical lines show the angular discharging angle

Compared to the other flight's shape, the angled flight offers a compromise between the angular discharging range and the falling length and falling time. It allows to get greatest falling length and time without decreasing afterwards as with a right-angled flight. The angular discharging rate is optimum compared to a straight flight.

Although the comparaison between the hold-up given by Sunkara's equations [9] and DEM's simulations shows the same tendancy. However, DEM's simulations offer a more accurate calculations of the flight's hold-up since the DEM takes into account the interactions between particles and the particle's size unlike the bulk material's approach.

Notations

D : dryer diameter (m)
L : length of the dryer (m)
r : particle radius (m)
l : length of the flight segment (m)
ω : dryer rotational seed (rad/s)
τ : drum's filling degree (%)
θ : angular position of the tip of the flight (°)
α_0 : angle between the fist segment and the wall (°).
α_1 : angle between the segments of the flight (°)
Φ : the dynamic angle of repose (°)
Φ_0 : the static angle of repose (°).

References

[1] R. B. Keey, *Drying : Principles and practice* (Pergamon Press, 1972)
[2] H. Henein and JK. Brimacombe and AP. Watkinson, Metallurgical transactions B **14**, 191-205 (1983)
[3] A. S. Mujumdar, *Handbook of industrial drying* (Crc Press, 2014)
[4] R. G. Sherritt, and R. Caple, The Canadian Journal of Chemical Engineering **71**, 337-346 (1993)
[5] M. H. Lisboa and D. S. Vitorino and W. B. Delaiba and J. R. D. Finzer, and M. A. S. Barrozo. Brazilian Journal of Chemical Engineering, **24**, 365-374 (2007)
[6] C.G.J. Baker, Drying Technology **6**, 631-653 (1988)
[7] P. G. Glikin, Chemical Engineering Research and Design **56a**, 120-126 (1978)
[8] D. Revol and C.L. Briens and J.M. Chabagno, Chemical Powder Technology **121**, 230-238 (2001)
[9] K. R. Sunkara and F. Herz, and E. Specht and J. Mellmann and R. Erpelding. Powder technology, **234**, 107-116 (2013)
[10] M, Riccardo and De R. S. Kiesgen and M. Guillain and W. Gabriel. Powder technology, **286**, 732-739 (2015)
[11] P. Van Liedekerke and E. Tijskens and E. Dintwa and F. Rioual and J. Vangeyte and H. Ramon. Powder technology, **190**, 348-360 (2009)
[12] S.J. Friedman and W.R. Marshall, Chemical engineering progress **6**, 482-493 (1949)
[13] N. A. S. Bongo and S. Vitu and M. Clausse and J-L. Dirion and M. Debacq, Powder Technology **269**, 554-565 (2015)
[14] J. Mellmann, Powder Technology **6**, 251-270 (2001)
[15] F. R. Schofield and P. G. Glikin, Trans I Chem E **40**, (1962)
[16] R. D. Mindlin, Journal of Applied Mechanics **16**, 259–268 (1949)
[17] R.D. Mindlin and H. Deresiewicz, Journal of Applied Mechanics **20**, 327–344 (1953)
[18] F. Lominé, *Ecoulement de Particules dans un Milieu Poreux* (PHD dissertation, Université de Rennes I, France, 2007)

2D ultra-fast MRI of granular dispersion by a liquid jet

Yunan Peng[1], *Andrew J.* Sederman[1], *Marco* Ramaioli[2], *Eric* Hughes[3], *Lynn F.* Gladden[1], and *Mick D.* Mantle[1,*]

[1]Department of Chemical Engineering and Biotechnology, University of Cambridge, Pembroke Street, Cambridge CB2 3RA, UK
[2]Department of Chemical and Process Engineering, University of Surrey, Guildford GU2 7XH, UK
[3]Department of Chemistry, Durham University, Lower Mountjoy, South Road, Durham DH1 3LE, UK

Abstract. This paper illustrates the application of ultra-fast magnetic resonance imaging (MRI) as a non-invasive tool to study the dispersion of a dry, static granular bed by the injection of a liquid. Spatial distributions of undispersed grains (poppy seeds) and injected water were independently imaged at sub-millimetre resolution in 2D with ultra-fast MRI techniques. A liquid jet was observed above the bottom injection orifice, complementing optical imaging. Co-registration of the grains and water images enables the interaction of the static grains and of the liquid jet to be observed for the first time *in situ*. This visualization of the dispersion process can be used to identify optimal process parameters for a fast and uniform dispersion and to validate quantitatively numerical granular-fluid simulations [1].

1 Introduction

Whereas the steady fluidization of grains by a fluid has been studied extensively, the transient dispersion of grains initially packed in a heap is not well understood. Dispersion is of interest across many applications: pigments are dispersed into paints, pharmaceuticals powders are dispersed by inhalers, whilst the problem of how to disperse, hydrate [2] and dissolve dehydrated powders is key for many food products [3-5].

Recent advances in magnetic resonance imaging (MRI) hardware and software means it is now possible to study these granular-fluid systems non-invasively and both the solid (granular bed) and liquid may be visualized. This new tool could be vital in investigating optically opaque systems [6,7]. This paper describes how 2D MRI is used to achieve a better understanding of liquid jetting in granular dispersion. This paper reports the first use of MRI to study a three-phase dispersion system. Specifically, we show the interaction of the grains and the liquid during granular dispersion. When a liquid jet is injected upwards into a dry static bed of poppy seeds its decompaction and progressive destruction can be monitored in time using two-dimensional magnetic resonance imaging techniques. The results are compared with optical methods and the capabilities and complementarity of these techniques are discussed. In addition, the quantitative measurement of the jetting angle from the MRI data are discussed. Finally, co-registration of 2D liquid and 2D static grain MR images reflects the good reproducibility of the dispersion experiments which are carried out.

2 Materials and Methods

In this study, a Perspex dispersion cell with PEEK end-caps, shown in Figure 1, was filled with 4 g of blue poppy seeds (diameter ~ 1 mm) and then vertically tapped manually 100 times to create a reproducible dry bed packed under gravity. Two distributors made from Scotch-Brite™ were placed below the injection and exit holes to keep the mass of seeds within the system constant. Water was injected at 200 ml min^{-1} into the dispersion cell from below, using a constant pressure liquid injection rig relying on compressed air to supply a steady injection pressure and hence liquid flow into the dispersion cell. Excess water leaves the system through the exit hole at the top of the cell at atmospheric pressure, before being evacuated by a peristaltic pump. Both optical imaging and MRI were used to record the dispersion experiments. A high-speed, high resolution digital camera (Photron FASTCAM MC1) was used to obtain optical side-on 512 pixels × 512 pixels images of the dispersion cell at 125 fps. The MRI of dispersion experiments were carried out in a wide bore 9.4 T (400 MHz ^{1}H resonance frequency) superconducting magnet equipped with a Bruker AV spectrometer. A 30 mm birdcage radio frequency (r.f.) coil was used to excite and detect the ^{1}H nuclei present in the water and the oil of the poppy seeds. Spatial resolution was achieved using a 3-orthogonal axis (x, y, z) gradient system capable of delivering a maximum gradient strength of 1.48 T m^{-1}.

Two MRI protocols were used to observe the water and grains within the dispersion cell: (i) 2D velocity compensated Fast Low Angle Shot (vc-FLASH) [8,9], which highlights the liquid only; (ii) 2D standard Rapid Acquisition with Relaxation Enhancement (RARE) [10,11], used to image changes to static poppy seeds.

Details of NMR parameters used can be found in Table 1. The 2D vc-FLASH imaging sequence was used to image the liquid component. In this gradient echo sequence, the NMR signal from poppy seeds is two orders of magnitude lower than it is for water due to the

* Corresponding author: mdm@cam.ac.uk

short T_2^* of the poppy seeds [12], and thus the liquid phase signal dominates. The central vertical plane of the dispersion cell was studied, allowing the injection hole and its immediate surroundings to be visualised. Velocity compensation was introduced by designing specific gradient waveforms to remove artifacts due to the flow-dependent phase shift of the signal of the ^1H nuclei in motion [13].

2D RARE is a spin-echo imaging protocol which utilizes total excitation of the system; however, it suffers heavily from motion artifacts [14]. The inherent nature of the liquid in the system is fast flowing, whilst the granular bed is initially static. For this reason, static poppy seeds in the system can be imaged with the RARE sequence. Additionally, manganese salt (Mn^{2+}) doping is used to reduce the T_2 of the water and hence suppress the water signal [15]. These images enable visualization of the static poppy seed regions during the liquid injection.

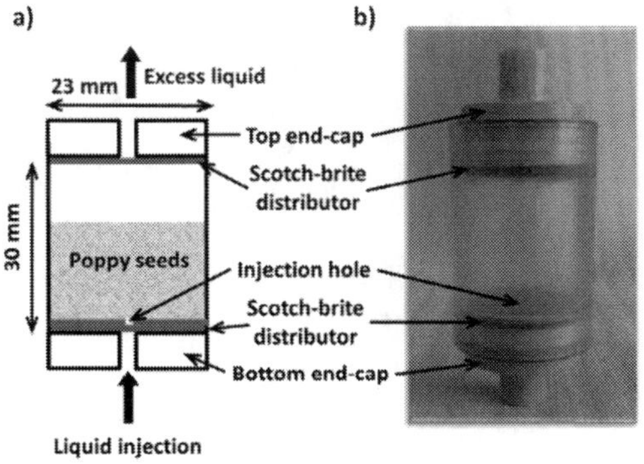

Fig. 1. a) Schematic, and b) photo of dispersion cell used in this study.

Table 1. Parameters for MRI protocols.

	Liquid imaging	Static grain imaging
	2D vc-FLASH	2D RARE
Field-of-view (fov) / mm	50 × 50	50 × 50
Matrix size / pixel	64 × 32	64 × 64
Spatial resolution / µm	781 × 1560	781 × 781
Slice thickness / mm	1	1
Echo time / ms	1.3	110
Repetition time / ms	3.0	500
Image acquisition time / ms	100	180
Temporal resolution / (fps)	10	2
Injected liquid	2.5 mM Gd^{3+} solution (T_1 = 29 ms, T_2 = 23 ms)	15 mM Mn^{2+} solution (T_1 = 7 ms, T_2 < 1 ms)
Granular material	Poppy seeds (T_1 = 450 ms, T_2 = 78 ms)	

3 Results and Discussion

3.1 Optical imaging

The capture of optical images using a high-speed, high-resolution camera allows the dispersion of poppy seeds in the geometry shown in Figure 1 to be appreciated.

Figure 2 shows a time series of images obtained where water is injected at 200 ml min^{-1} into a 4 g bed of poppy seeds in the dispersion cell. Initially a dry static bed of poppy seeds can be seen, before decompaction of the poppy seed bed between t = 0 s and t = 0.2 s. Water is then flowing through the centre of the packed bed between t = 0.2 s and t = 1.0 s. After t = 0.6 s, a water layer forms above the bed. This layer of water is turbulent with air bubbles and poppy seeds present and increases in height until it reaches the top of the dispersion cell at around t = 2.0 s. The system reaches a steady state after t = 2.0 s for this dispersion experiment with a turbulent region of water, poppy seeds and air bubbles formed above the remains of the poppy seed bed. The liquid injection was continued for 5 minutes, but the system did not evolve significantly after 2.5 s.

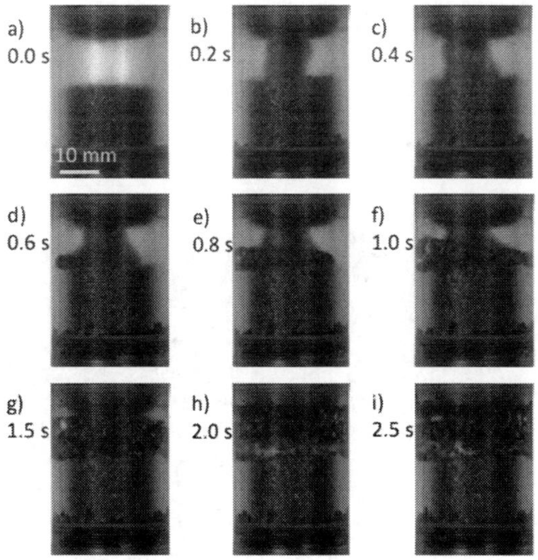

Fig. 2. A selection of optical images recorded at 512 pixels × 512 pixels during the dispersion of poppy seeds by water. At 0.0 s water enters the bottom of the cell. Note the dispersion cell is of constant radius, however the difference in refractive index between water and air gives the illusion that wetted areas of the dispersion cell appears wider.

3.2 Liquid imaging: 2D vc-FLASH

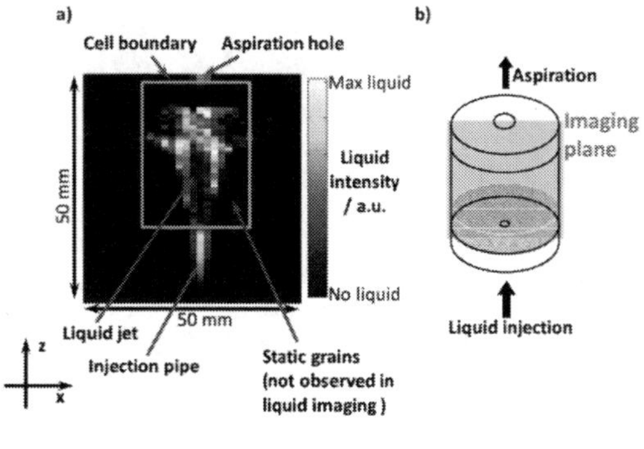

Fig. 3. a) A typical snapshot of 2D selective liquid imaging of the central vertical slice of a dispersion experiment. The jet can be seen rising from the bottom of the cell (green boundary) in a plume. b) Schematic of the vertical imaging plane.

Figure 3 a) shows a steady state 2D vc-FLASH image 4.0 s after the start of liquid injection into a bed of poppy seeds in the central vertical slice through the dispersion cell depicted in Figure 3 b). A well-developed steady state liquid jet is shown. The dark areas in the 2D vc-FLASH map correspond to areas without liquid signal; the two large areas around the bottom circumference of the cell are the regions of stagnant poppy seeds, which are not seen in liquid imaging due to its much lower signal than water. It is important to note that although a scale bar with two limits is given, reduced pixel intensity is also likely due to signal loss as a result of the high shear present within a voxel, especially with the high liquid flow rates involved. For example, at the central core of the liquid jet shows low signal, which is likely due to the water moving too fast to be imaged rather than lack of water. The superficial velocity for water flowing through the 1 mm injection hole at 200 ml min^{-1} is 1.1 m s^{-1}. Therefore, during the 1.3 ms echo time, a voxel of water will move on average 1.43 mm. This does not compare favourably with the spatial resolution of 781 μm in the vertical direction and is a reason for signal loss.

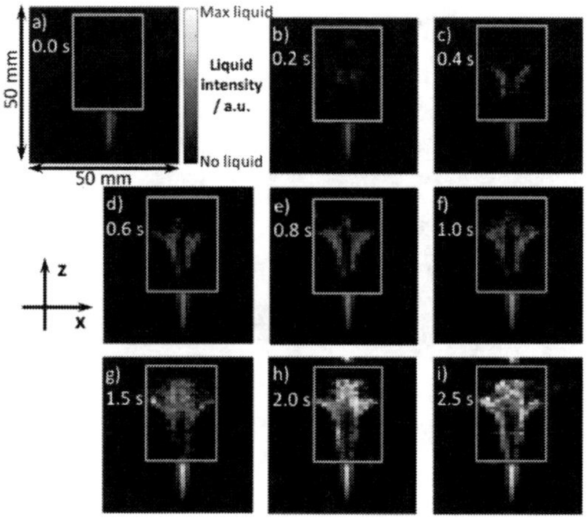

Fig. 4. A selection of liquid-imaging snapshots of the central vertical slice acquired using 2D vc-FLASH for a dispersion experiment with water jetted into a granular bed of poppy seeds. Temporal resolution: 10 fps. Spatial resolution: 0.781 mm × 1.560 mm. The time is normalized such that t = 0 s corresponds to water reaching the bottom of the dispersion cell.

Figure 4 shows a time series of 2D image vc-FLASH snapshots during the approach to steady state, where water is injected from the bottom into the poppy seeds bed. The water injected into the dispersion cell was observed to penetrate through the middle of the granular bed. The water can already be seen to have penetrated completely through the poppy seeds by t = 0.6 s as signal is seen from the top of the dispersion cell. The initial liquid jet then forms from the top-down between t = 0.4 s and t = 0.8 s. At this point, a steady jet has formed and this stabilizes over the next second as the intensity of the jet increases; the system is then defined to have reached steady state at t = 2.0 s as very little change is seen in the jet thereafter. The increase in signal intensity is because of lower velocity attenuation at steady state compared with in the initial dry dispersion cell. As more water is injected into the dispersion cell, the distribution area of liquid jet widens, therefore the vertical velocity profile of the jet reaches a lower maximum.

The results show the symmetry of the jet is as expected at steady state, although it was unable to provide further insight into determining how the initial changes in the dispersion cell are taking place, due to the much shorter timescale of this phenomenon, i.e. of the order of hundreds of microseconds.

The 2D spatial maps have allowed the internal water flow to be directly visualized non-invasively and *in situ*. When injecting a liquid in dry grains, after the jet has broken through the granular bed and reached the top of the cell, the liquid is not directly aspirated through the exit hole, but rather cascades down the side of the dispersion cell. It then builds up above the regions where grains are static and it wicks in the grains downwards. The liquid is therefore partly above the dry granular regions, as shown in Figure 4 at t = 2.0 s and 2.5 s.

3.3 Grain imaging: 2D RARE

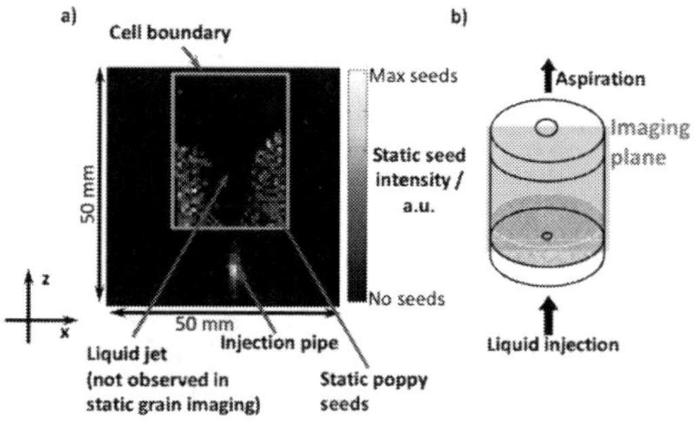

Fig. 5. a) A typical 2D RARE snapshot of the central vertical slice during a dispersion experiment. The static grains are pushed towards the outer edge of the bottom of the cell. Liquid is not visible, except for an artefact below the dispersion cell, at the injection pipe. b) Schematic of the vertical imaging plane.

Two dimensional RARE was used to produce 2D spatial maps of the static poppy seeds in the same central vertical plane as shown in Figure 3 b) for liquid imaging with 2D vc-FLASH. Figure 5 a) shows the corresponding 2D RARE image to the liquid spatial map in Figure 3 a). The image is taken t = 4.0 s after the start of liquid injection into the bed of poppy seeds in the central vertical slice through the dispersion cell depicted in Figure 5 b). With the liquid signal suppressed, the signal observed arises from the poppy seed oil in static, or stagnant, poppy seeds. When taken together with Figure 3 a), it can be concluded a liquid jet at the centre of the dispersion cell is observed with a ring of stagnant seeds around the circumference at the bottom of the cell. The fluid intensity is lower at the bottom of the cell, where the stagnant seeds are imbibed with some fluid. Water can be observed on top of the regions of static seeds. Figure 6 shows a time series of snapshots in an equivalent experiment to that shown in Figures 2 and 4. Here, a growing vacant region can be seen in the centre of the granular bed, while static grains progressively disappear. The co-registration presented in the following section helps interpret this pattern more clearly.

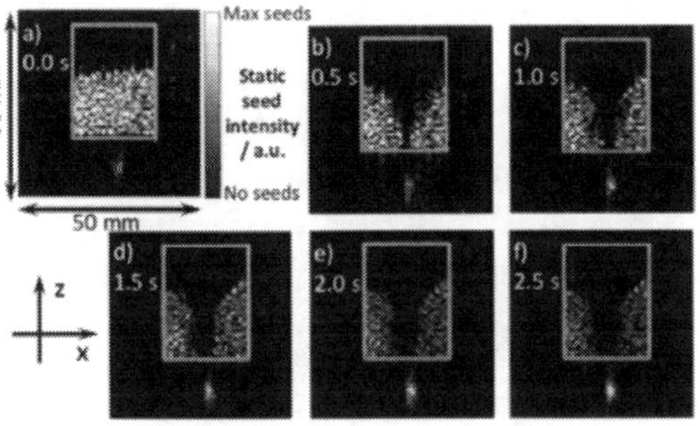

Fig. 6. A selection of grain-imaging snapshots of the central vertical slice acquired using 2D RARE. Dispersion of a granular bed of poppy seeds with injected water. Temporal / Spatial resolution: 0.781 mm × 1.560 mm; 10 fps.

3.4 Co-registration: 2D liquid and grain imaging

With the two dimensional spatial maps of liquid and powder imaging, it is possible to study the liquid-grain interactions within an optically opaque system. The co-registration of these spatial maps has proven the reproducibility of the dispersion experiments carried out, with the static bed-liquid jet interface being directly visualised.

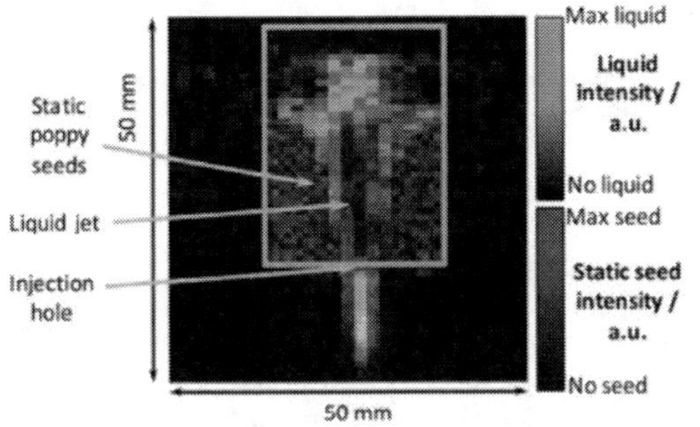

Fig. 7. Co-registration of 2D liquid and grain spatial maps, obtained after 2.5 s of water injection at 200 ml min^{-1} into 4 g of poppy seeds. The water jet is shown in cyan, the stationary poppy seeds in red and the dispersion cell is outlined in green.

A co-registration image of 2D liquid and static grain spatial maps is shown in Figure 7. The cyan channel represents the water jet and the red channel represents the static grains in the system. A ring of stagnant poppy seeds is seen around the outer edge at the bottom of the dispersion cell. Although the two spatial maps are obtained using two different methods and acquired during separate experiments, the liquid jet–static poppy seed boundary overlaps well. This image shows both the injection of water into the system and the static granular bed dispersed and eroded away. It directly reflects simulations of the central vertical plane of the dispersion cell [1], where a low porosity jet is immediately surrounded by high porosity regions towards the circumference of the bottom of the dispersion cell. This data complement optical methods to provide a fundamental understanding of the process taking place during powder dispersion.

This powerful MRI technique could also be used to extend the physical understanding of the formation of air jets in a liquid saturated granular medium [16,17] or of wicking of a dry powder [18].

4 Conclusions

The dispersion of a dry static granular bed by liquid injection was investigated using ultra-fast MRI. Whilst optical methods are limited by the opaqueness of these 3D dispersion set-ups, sub-millimetre resolution images of two phases during dispersion were achieved using ultra-fast MRI protocols. MRI can characterise quantitatively the dispersion due to liquid jetting and how the surrounding regions of static poppy seeds evolve over time. Good reproducibility for dispersion of poppy seeds was achieved. This enables the co-registration of the liquid and static grain spatial maps, enabling in turn granular dispersion to be studied *in situ*.

The authors are grateful to Nestlé Research for financial support and to EPSRC Loan Pool for contributing towards the optical imaging of dispersion.

References

1. M. Robinson, M. Ramaioli, S. Luding, AIP Conf. Proc. *Powder and Grains 2013*, **1542**, 1122 (2013)
2. J. Dupas, E. Verneuil, M. Ramaioli, L. Forny, L. Talini, F. Lequeux, Langmuir **29**, 40 (2013)
3. P. Schuck, A. Davenel, F. Mariette, V. Briard, S. Méjean, M. Piot, Int. Dairy J. **12**, 51 (2002)
4. L. Forny, A. Marabi, S. Palzer, Powder Technol. **206**, 72 (2011)
5. J. Dupas, L. Forny, M. Ramaioli, Journal of Colloid and Interface Science **448**, 51 (2015)
6. M.D. Mantle, A.J. Sederman, Prog. Nucl. Magn. Reson. Spectrosc. **43**, 3 (2003)
7. C.R. Müller, D.J. Holland, A.J. Sederman, M.D. Mantle, L.F. Gladden, J.F. Davidson, Powder Technol. **183**, 53 (2008)
8. A. Haase, J. Frahm, D. Matthaei, W. Hanicke, K.D. Merboldt, J. Magn. Reson. **67**, 258 (1986)
9. J. Frahm, A. Haase, D. Matthaei, J. Comput. Assist. Tomogr. **10**, 363 (1986)
10. J. Hennig, A. Nauerth, H. Friedburg, Magn. Reson. Med. **3**, 823 (1986)
11. J. Hennig, H. Friedburg, Magn. Reson. Imaging **6**, 391 (1988)
12. T.M. Nguyễn, A.J. Sederman, M.D. Mantle, L.F Gladden, Phys. Rev. E **84** (2011)
13. J.M. Pope, S. Yao, Concepts Magn. Reson. **5**, 281 (1993)
14. B. Madore, R.M. Henkelman, J. Magn. Reson. Imaging **4**, 577 (1994)
15. A.P. Koretsky, A.C. Silva, NMR Biomed. **17**, 527 (2004)
16. M. Jankov, G. Løvoll, H.A. Knudsen, K.J. Måløy, R. Planet, R. Toussaint, E.G. Flekkøy, Transport Porous Med. **84**, 569 (2010)
17. B. Sandnes, E.G. Flekkøy, H.A. Knudsen, K.J. Måløy, H. See, Nat. Commun. **2**, 288 (2011)
18. P. S. Raux, H. Cockenpot, M. Ramaioli, D. Quéré, C. Clanet, Langmuir, **29** 11, (2013)

 DOI: 10.1051/epjconf/201714012006

Testing Occam's razor to characterize high-order connectivity in pore networks of granular media: Feature selection in machine learning

Joost van der Linden[1], *Antoinette* Tordesillas[2,*], and *Guillermo* Narsilio[1]

[1] *Department of Infrastructure Engineering, The University of Melbourne, Australia*
[2] *School of Mathematics and Statistics, School of Earth Sciences, The University of Melbourne, Australia*

Abstract. A perennial challenge for the characterization and modelling of phenomena involving granular media is that the internal connectivity of, and interactions between, the pores and the particles exhibit hallmarks of complexity: multi-scale and nonlinear interactions that lead to a plethora of patterns at the mesoscale, including fluid flow patterns that ultimately render a permeability of the granular media at the macroscale. A multitude of physical parameters exist to characterize geometry and structure, including pore/particle shape, volume and surface area, while a rich class of complex network parameters quantifies internal connectivity of the pore and particles in the material. A large collection of such variables is likely to exhibit a high degree of redundancy. Here we demonstrate how to use feature selection in machine learning theory to identify the most informative and non-redundant, yet parsimonious set of features that optimally *characterizes* the interstitial flow properties of porous, granular media, e.g., permeability, from high resolution data.

1 Introduction

Porous, granular materials are complex systems that embody rich patterns and dynamics. Applications involving physical flow, such as hydrocarbon recovery and geothermal energy, rely on estimations of the (coupled) hydraulic, thermal and mechanical material properties. For instance, significant experimental evidence has shown that transport properties of materials, such as permeability, are strongly influenced by the presence of concentrated zones of inelastic deformation such as shear bands, compaction bands, fractures and joints (e.g., [1]). Localized deformation can strongly influence flow pathways, potentially becoming either barriers or conduits for flow depending on the attendant evolution of local pores (e.g., see [3] and references therein). Many of these patterns occur naturally in frictional soft materials, especially in rocks and soil, due to material discontinuities and heterogeneity [2]. Microstructural grain rearrangements can alter permeability not only through changes in the geometry and size distribution of individual pores but also their connectivity [3].

Almost invariably, these complex and iterrelated processes can only be captured in a high-dimensional multivariate parameter space. Such a dataset is generated in this work to characterize permeability using the data-driven framework introduced in [4]. The framework fuses proven finite-element and discrete-element methodology with modern advances in statistics (machine learning) and complex systems (complex networks), towards a data-driven 3D analysis of multiscale and nonlinear phenomena in granular, porous media. Furthermore, the framework can be used to address the coupled evolution of the solid grain and interstitial pore phases through a study of two classes of interdependent networks in a single platform, *viz.* one that represents the grain contacts while the other represents the pores, thus advancing the approach in [3] for planar systems.

As shown in [4], the high-dimensional parameter space of variables related to permeability is likely to exhibit a high degree of redundancy. Occam's razor dictates all but the *most relevant* and *least redundant* of these variables should be retained to explain a given phenomenon of interest. To this end, a ranking of the variables in order of relevance and redundancy is crucial for predictive modelling and ultimately control. We apply two such algorithms in this paper, quantify the redundancy, and highlight several highly relevant parameters for the permeability.

2 Methods

The methods and dataset used here are based on the methodology outlined in [4]. In this paper, we summarize the corresponding framework. For a comprehensive discussion of all model equations, parameters and assumptions involved, we refer to the aformentioned study. The employed data generation process is shown in Figure 1.

2.1 Discrete element modeling

Our discrete-element model is a simple proxy for sands and sedimentary rocks (i.e., Ottawa sand and sandstone) [2, 5, 6]. Batches of 400 soft spheres are dropped in a rectangular container with periodic boundary conditions for

*e-mail: atordesi@unimelb.edu.au

DOI: 10.1051/epjconf/201714012006

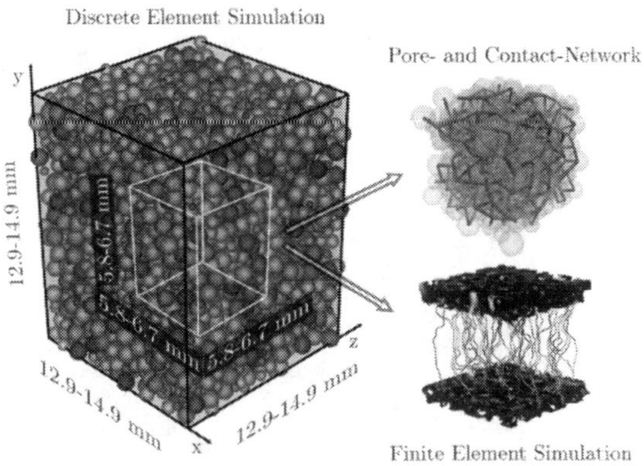

Figure 1. A discrete element simulation is performed first, to generate a compressed packing of spheres. From this packing, a subsample is selected to generate a weighted pore network and weighted contact network. Fluid flow is simulated in this subsample in a finite element simulation, to compute the permeability.

the vertical walls. Each batch is released at equal height from the top of the existing packing, and allowed to settle under gravity, simulating the air pluviation method of sample preparation. After dropping 4000 particles, the resulting packing is subjected to an isotropic confining pressure, leading to changes in porosity. To generate a large (synthetic) dataset, this process of gravity deposition followed by triaxial compression is repeated 1536 times, each time randomly selecting the particle radius distribution ($U(0.5 - \alpha, 0.5 + \alpha)$ mm, $\alpha \in U(0.0, 0.3)$), friction angle ($\theta \in U(5.7°, 31°)$) and confining pressure (10^n Pa, $n \in U(5, 7)$). The Young's modulus is set at the relatively low value of 10^8 Pa. Although this choice is lower than the usual quartz value, it does allow for a simple approximation of a wide range of porosities. At high compression rates, the particle overlap models sintered materials such as sandstones [2].

2.2 Finite element modeling

A representative element volume (REV) is extracted from the center of each packing. Next, we run a finite element simulation for each REV, imposing a pressure difference of 1 Pa over the top and bottom boundaries and solving the governing Navier-Stokes equations to simulate incrompressible fluid flow. The permeability of each assembly is computed by averaging the vertical velocity component over the top and bottom boundaries, and subsequently averaging as in [7].

2.3 Complex networks

In van der Linden et al. [4] we constructed both a contact network and a pore network. For details of the pore network construction approach, we refer to the aformentioned paper and our upcoming work [8]. In this study, for simplicity, we omit the contact network results. Every edge in the pore network is weighted with the throat conductance, derived from the Hagen-Poiseuille equation in a simple tube model of the corresponding (adjacent) pores (see [4], Appendix 1).

We compute a range of complex network parameters for the pore network. These parameters, in various ways, capture the connectivity and topology in the network. For a detailed overview, we refer to the seminal work by Newman [9], and for a brief overview of the equations employed in this work, see [4]. For each node, we compute the *degree* as the number of adjacent edges (also known as the coordination number), and the *weighted degree* as the sum of the weights of adjacent edges. The *network density* is the fraction of the actual number of edges over the number of edges if the network were to be fully connected. Furthermore, define $d(i, j)$ as the length of the (*shortest*) path from node i to node j that minimizes the summed weights of the traversed edges. The *network diameter* is the length of the 'longest' shortest path between all pairs of nodes. We also compute the *node betweenness centrality* for a particular node as the fraction of all shortest paths in the network passing through this node. Similarly, we compute *edge betweenness centrality* as the fraction of all shortest paths passing through a particular edge. As such, betweenness centrality captures the 'importance' of a node/edge for the flow through the network. Lastly, we compute the *closeness centrality*, which can be understood as the 'centrality' of a particular node, meaning that the node is well-connected to the rest of the network. Averaged over all nodes, high closeness centrality indicates a well-connected network (i.e. large pores/throats) for which we hypothesize a higher value of the permeability.

2.4 Features

Having generated the required dataset, the final step in our framework is to collate all physical (1-13) and network (14-20) parameters in a 'feature set', as shown in Table 1. Table 1 can be interpreted as our 'initial guess' of 20 features that characterize the permeability to some degree. Connectivity is highly relevant for the permeability [10], yet none of the traditionally used physical features (1-13) capture this aspect. This is why we include the network features (14-20). For any of the distributions (bracket notation in Table 1), we compute the average (denoted with μ) to obtain a single parameter. In the analysis, we consider the natural logarithm of $p, [T]_K, [B]_{A_c}, \mu[G^p]_{\kappa_w}$ and $\mu[G^p]_{C^c}$, because these features span multiple orders of magnitude.

2.5 Feature selection

For each packing, we obtain the values of the parameters listed in Table 1 (the features) and the value of the permeability obtained from the finite element simulations. Given such a high-dimensional dataset of features and a 'target variable' (permeability), we can apply machine learning techniques and, in particular, feature selection, to (1) uncover the most 'important' features for the permeability,

Table 1. Feature set. The notation $[X]_a$ is used to denote a distribution of parameter a on entity X (T for throat, P for pore, B for particle, G for graph).

#	Notation	Attribute	Units
1	e	global void ratio	
2	p	confining pressure	[Pa]
3	ssa	specific surface area	$[m^{-1}]$
4	$[T]_K$	throat conductance	$[m^3Pa^{-1}s^{-1}]$
5	$[T]_e$	throat void ratio	
6	$[P]_e$	pore void ratio	
7	$[B]_{A_c}$	particle contact area	$[m^2]$
8	$[T]_V/[P]_V$	throat/pore volume ratio	
9	$[P]_{A_s}$	pore surface area	$[m^2]$
10	α	grain size range	[m]
11	c_u	coefficient of uniformity	
12	c_c	coefficient of curvature	
13	θ	friction angle	[deg]
14	G^p_ρ	network density	
15	G^p_D	network diameter	$[m^{-3}Pa\ s]$
16	$[G^p]_\kappa$	degree	
17	$[G^p]_{\kappa_w}$	weighted degree	$[m^{-3}Pa\ s]$
18	$[G^p]_{C^B_{edge}}$	edge betwnness centrality	
19	$[G^p]_{C^B_{node}}$	node betwnness centrality	
20	$[G^p]_{C^C}$	closeness centrality	$[m^3Pa^{-1}s^{-1}]$

(2) eliminate redundant features, and (3) predict the target variable for new, previously unseen packings. In this paper, we focus on the first and second task. For a discussion of the third task, refer to [4] for preliminary results.

Two feature selection algorithms are considered in this paper. The *minimum redundancy, maximum relevance* (mRMR) [11] feature selection algorithm minimizes redundancy in the set of selected features, while simultaneously maximizing dependency between selected features and the target variable. Relevance and redundancy is calculated in terms of mutual information, a measure of mutual dependence between two variables. The mRMR algorithm uses an incremental search procedure, in which the most relevant feature is selected first. In subsequent iterations, the most relevant feature that also minimizes redundancy with the previously selected feature(s) is added to the feature set. The mRMR method is myopic, meaning that conditional dependencies between features are ignored.

In contrast, the *RReliefF* [12] feature selection method estimates a feature's importance while taking conditional dependencies into account (i.e. non-myopic). RReliefF does so by rewarding features that separate dissimilar values of the target variable, and penalizing features that separate similar values of the target variable. A score is assigned based on the degree to which a feature uniquely explains the target variable. Consequently, RReliefF implicitely penalizes redundancy by dividing up the score awarded for explaining a corresponding portion of the target variable among the redundant features.

3 Results

Before considering the correlations between features and the permeability, it is instructive to analyze the degree of redundancy *within* the feature set. Figure 2 shows the linear correlation matrix, providing an overview of the inter-feature correlations. Broadly, three groups can be distinguished.

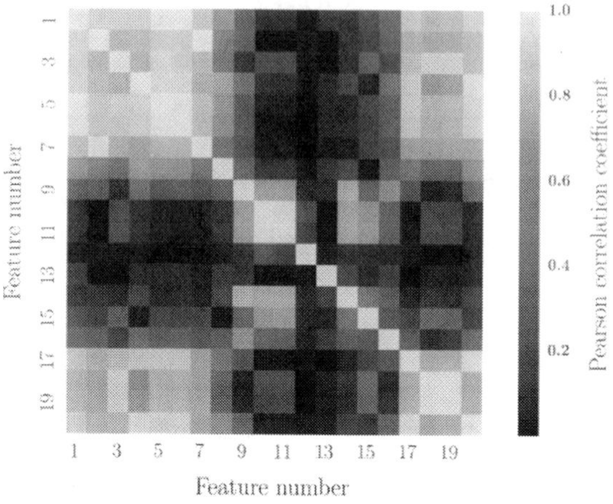

Figure 2. Inter-feature correlations.

The first group of features (1-9) are strongly inter-correlated, with an average inter-correlation coefficient of 0.77. These features all relate to the degree of compression and the void ratio. Higher compression results in lower global and local void ratio, as well as higher average inter-particle contact area. Features 1-9 are weakly correlated with the second group: features 10-15. The latter group of features is less correlated with the sample compression and the resulting effect on the void ratio. The third group of features, 16-20, is both inter-correlated (0.66 on average) and correlated with the compression-related features in the first group (1-9). This demonstrates that the connectivity in the samples is broadly interlinked with the degree of compression.

A limitation of using the Pearson correlation coefficient to measure inter-feature correlations is that it does not pick up on non-linear relationships. The coefficients of curvature and uniformity, for example, have a clearly defined non-linear (parabolic) relationship, yet the Pearson correlation coefficient is only 0.20. We do find noteworthy linear correlations, though, between the average pore surface area and the network density (0.90), and between the coefficient of uniformity and the network diameter (0.79). These correlations warrant further investigations.

Having observed clear redundancy in the feature set, we apply the mRMR and RReliefF feature selection methods to apply 'Occam's razor' and analyze the full feature set from Table 1 to identify the most relevant (for the permeability) and non-redundant feature set. For ReliefF, the scores converge as the fraction of data approaches 1.0. Both the local (pore and throat) void ratio and global void ratio are represented in the top 5, consistent with well-known importance of the void ratio for the permeability. A high RReliefF score indicates high relevance of the feature for the permeability and/or strong conditional dependence on the permeability, given other features. The confining

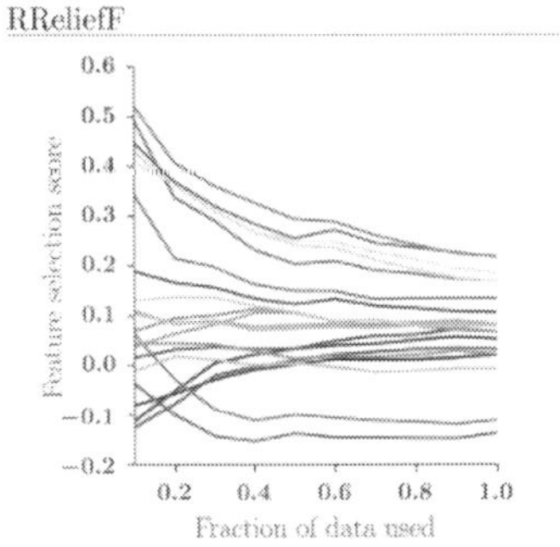

Final ranking				Score
1	log	μ	$[T]_K$	.2187
2	log	μ	$[P]_e$	.2172
3			e	.1856
4	log	μ	$[G^p]_{C^c}$	.1691
5		μ	$[T]_e$	.1681
6	log	μ	$[G^p]_{k_w}$	.1347
7			ssa	.1064
8		μ	$[G^p]_{C^*_{idw}}$	.0826
9			G^p_ρ	.0790
10		μ	$[T]_V/[P]_V$	.0788
11		μ	$[G^p]_{C^*_{idw}}$	.0780
12		μ	$[P]_{A_t}$	.0671
13			θ	.0665
14			c_u	.0531
15			α	.0323
16		μ	$[G^p]_\kappa$	.0226
17			c_c	.0205
18			G^p_Ω	−.0073
19	log		p	−.1116
20	log	μ	$[B]_{A_t}$	−.1373

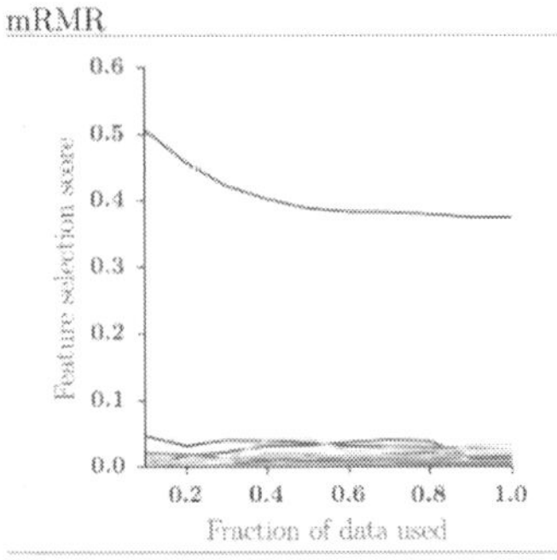

Final ranking				Score
1	log	μ	$[G^p]_{C^c}$	.3744
2		μ	$[T]_e$	.0329
3	log	μ	$[P]_e$	.0248
4			e	.0247
5		μ	$[T]_V/[P]_V$	.0185
6	log		p	.0172
7	log	μ	$[T]_K$	.0154
8	log	μ	$[G^p]_{k_w}$	.0108
9		μ	$[G^p]_{C^*_{idw}}$	.0088
10		μ	$[G^p]_\kappa$	.0066
11		μ	$[P]_{A_t}$	.0048
12	log	μ	$[B]_{A_t}$	.0040
13			θ	.0025
14			ssa	.0019
15			G^p_ρ	.0000
16			α	.0000
17		μ	$[G^p]_{C^*_{idw}}$	.0000
18			G^p_Ω	.0000
19			c_u	.0000
20			c_c	.0000

Figure 3. (Figures) Convergence of the selection scores, for RReliefF (left) and mRMR (right). (Tables) Final ranking and scores when 100% of the data is used, for RReliefF (left) and mRMR (right).

pressure is strongly correlated with the various void ratios (see Fig. 2), yet p appears at the bottom of the RReliefF ranking. We conclude that, according to RReliefF, p is neither able to explain variance in the permeability by itself, nor does it explain the permeability well when conditioned on other features. A similar conclusion applies to the coefficients of uniformity c_u and curvature c_c, friction angle θ and grain size range α, although we do show in [4] that manually picked combinations of certain features (such as ssa and c_u) can explain part of the variance in permeability.

Recall that the mRMR algorithm follows an incremental search procedure. The convergence of the feature selection scores in Figure 3 for mRMR shows that after picking the first feature ($[G^p]_{C^c}$) subsequent choices are highly penalized for redundancy. In other words, the pore network closeness centrality is highly relevant (in terms of mutual information) for the permeability, and none of the other features are able to explain the permeability much further.

4 Conclusion

The data-driven framework is applied to generate and analyze a set of physical and connectivity descriptors of the permeability. Using feature selection algorightms, we (objectively) show that the initially proposed feature set is highly redundant and can be reduced to a few void ratio descriptors and the pore network closeness centrality. The latter, in particular, captures both the connectivity and (through the conductance weighting) the pore- and throat-geometry of the packings. Upcoming research focuses on characterizing the thermal conductivity and pore space evolution in shear zones [13] using this general framework.

Acknowledgements

The authors acknowledge the support of the Australian Research Council (FT140100227, DP120104759), the U.S. Air Force (AFOSR 15IOA059), and the U.S. Army Research Office (W911NF-11-1-0175, W911NF-15-1-0527).

References

[1] M.A. Etheridge, V.J. Wall, S.F. Cox, R.H. Vernon, Journal of Geophysical Research: Solid Earth **89**, 4344 (1984)

[2] S. Torquato, *Random Heterogeneous Materials: Microstructure and Macroscopic Properties*, Vol. 16 (Springer-Verlag, New York, 2002)

[3] S. Russell, D. Walker, A. Tordesillas, Journal of the Mechanics and Physics of Solids **88**, 227 (2015)

[4] J.H. van der Linden, G.A. Narsilio, A. Tordesillas, Physical Review E **94**, 022904 (2016)

[5] H. Zhu, Z. Zhou, R. Yang, A. Yu, Chemical Engineering Science **63**, 5728 (2008)

[6] G. Narsilio, J. Kress, T. Yun, Computers and Geotechnics **37**, 828 (2010)

[7] G. Narsilio, O. Buzzi, S. Fityus, T. Yun, D. Smith, Computers and Geotechnics **36**, 1200 (2009)

[8] J. van der Linden, A. Sufian, G. Narsilio, A. Russell, A. Tordesillas (2016), unpublished manuscript

[9] M.E.J. Newman, *Networks: an introduction* (Oxford University Press, New York, NY, 2010)

[10] J.T. Fredrich, W.B. Lindquist, International Journal of Rock Mechanics and Mining Sciences **34**, 368 (1997)

[11] H.C. Peng, F.H. Long, C. Ding, IEEE Transactions on Pattern Analysis and Machine Intelligence **27**, 1226 (2005)

[12] M. Robnik-Šikonja, I. Kononenko, Machine Learning **53**, 23 (2003)

[13] A. Tordesillas, J. van der Linden, G. Narsilio (2016), unpublished manuscript

Self-Structuring of Granular material under Capillary Bulldozing

Guillaume Dumazer[1,*], *Bjørnar* Sandnes[2], *Monem* Ayaz[1,3], *Knut Jørgen* Måløy[1], and *Eirik* Flekkøy[1]

[1] *Physics Department, University of Oslo, P. O. Box 1048 Blindern Oslo, Norway*
[2] *College of Engineering, Swansea University, Bay Campus, Swansea SA1 8EN, UK*
[3] *Institut de Physique du Globe de Strasbourg, Université de Strasbourg/EOST, CNRS, 5 rue Descartes, F67084 Strasbourg Cedex, France*

Abstract. An experimental observation of the structuring of a granular suspension under the progress of a gas/liquid meniscus in a narrow tube is reported here. The granular material is moved and compactifies as a growing accumulation front. The frictional interaction with the confining walls increases until the pore capillary entry pressure is reached. The gas then penetrates the clogged granular packing and a further accumulation front is formed at the far side of the plug. This cyclic process continues until the gas/liquid interface reaches the tube's outlet, leaving a trail of plugs in the tube. Such 1D pattern formation belongs to a larger family of patterning dynamics observed in 2D Hele-Shaw geometry. The cylindrical geometry considered here provides an ideal case for a theoretical modelling for forced granular matter oscillating between a long frictional phase and a sudden viscous fluidization.

1 Introduction

Transport of oil, water and sand in sub-sea pipelines [1], of blood cells suspended in plasma [2, 3] give two extreme examples of the broad range of situations featuring the multiphase transport of granular material in confined geometries. Pneumatic transport offers a well established field for the transport of granular particles immersed in a single phase [4]. Such a two phase system shows a broad range of flow regimes depending on the filling fraction and flow rate. Plug flow, fluidized flow, bubbly flow, turbulent fluidization, cluster flow, annular flow and suspended flow gives the variety of flow regimes in vertical pneumatic conveying [5]. Such observed flow regimes involve the competing effects of viscosity and friction [6–8], respectively destabilizing and stabilizing the granular phase.

The case of dense granular suspension has recently been actively studied as it features complex rheological behaviour such as shear thinning, shear thickening, yield stress and shear banding [9–12]. The frictional interaction and contact time between granular particle and the viscous interaction with the carrying fluid are the two factors determining the behaviour of dense suspensions. The geometrical confinement is then another key factor as it produces extra frictional forces on the granular flow. The frictional interaction with boundaries induces contacts network spanning over a large part of the system.

In multiphase flows, the existence of one or several meniscii introduce capillary forces as a third factor. Consequently such flows provide a rich source of instabilities and pattern formation processes [13–15].

To simplify the geometrical complexity, we observed the progress of a gas/liquid meniscus along a tube filled with sedimented granular mixture of liquid and glass beads [16]. The confinement chosen is approaching a one-dimensional geometry. We are reporting here the response of a confined granular material under capillary bulldozing.

2 Materials and Methods

A long horizontal tube of size $L = 1.0$ m with constant diameter $D = 2$ mm is filled with a sedimented mixture of liquid and glass beads, with reduced filling fraction $\varphi = \phi_0/\phi_c$, defined as the ratio between the volume fraction ϕ_0 of grains in the grain-water mixture, and the random close packing ϕ_c. Glass beads are sieved to control the polydispersity. The tube is pumped at constant withdrawing rate I_0 and the pressure at the tube's outlet is recorded, see Fig. 1 (a). The tube's diameter is chosen to not exceed the capillary length of the liquid used, ensuring that the meniscus fills the tube's aperture, without spreading along the tube, see Fig. 1 (b). The inner surface of the tube were treated with a silanization solution to achieve an hydrophobic coating ensuring a contact angle $\geq 90^o$ between the meniscus and the confining walls. The experiment is imaged against an LED screen. At the early stages of the withdrawing, the meniscus progresses along the tube, perturbs the sedimented granular material, see Fig. 1 (b), and forms a growing accumulation front, see Fig. 1 (c). At slow withdrawing rates the capillary

*e-mail: guillaume.dumazer@fys.uio.no

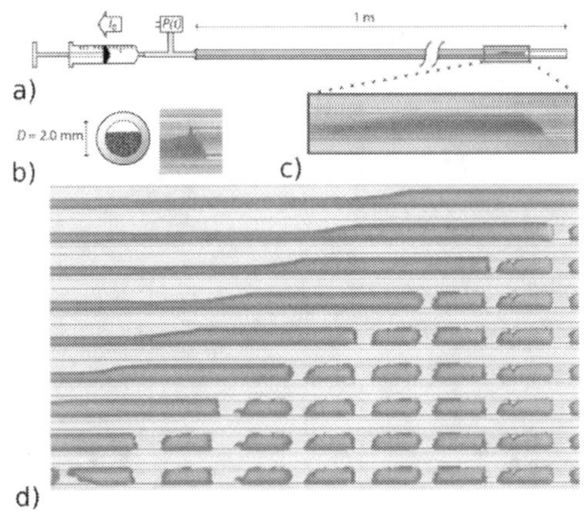

Figure 1. (a) Schematic view of the experimental setup. The sedimented glass beads and liquid mixture is withdrawn by a syringe pump at constant withdrawing rate I_0, withdrawing pressure is recorded at the tube's outlet. (b) (left) initial cross-section schematics, and (right) zoom over the meniscus just after the withdrawing starts. (c) accumulation front after start of withdrawal. (d) series of pictures showing the accumulation front progress, and the trail of plugs left behind, time develops downwards.

bulldozing leads to the intermittent splitting of the growing accumulation front, and a series of granular plugs are left in the tube, see Fig. 1 (d).

3 Plus Formation

The formation of a single plug is imaged after correcting the variations of illumination by subtracting a blank image of the screen, see Fig. 2 (a). Images are then thresholded to clearly distinguish the glass beads from the background. The number of glass beads' pixels are counted as a function of tube's axis x. Repeating this operation for every image recorded during the experiment we get a dimensionless signal $s(x,t)$ with 0 value for no glass beads, 1 value for a fully filled tube's cross-section, s_0 value for unperturbed sedimented glass beads. By mapping $s(x,t)$ with a colorscale on a spatio-temporal diagram with time developing downwards, and tube's axis horizontally, we get a useful visualization of a single plug formation, see Fig. 2 (b). The formation of a single plug can be decomposed into 4 steps: (i) driven by the syringe pump the meniscus progresses against the sedimented grains which slowly compactifies into a growing accumulation front (t_1, t_2). The frictional interaction between the granular material and the confining wall increases. The Janssen's description of stress in confined granular material [14, 17, 18] suggests here an exponential increase of the pressure with the accumulation front length $P \approx \sigma_0(\varphi, \mu, D)\exp(L/\lambda)$. In this description a slice of compacted granular material is considered. The radial stress σ_{rr} normal to the cylindrical confining walls is assumed to be proportional to the longitudinal stress σ_{xx}, $\sigma_{rr} = K\sigma_{xx}$ with the Janssen's coefficient K. The frictional length $\lambda = D/(4\mu K)$ is derived from the force balance over the slice of granular material. (ii) the pressure imposed by the syringe pump reaches the pore pressure of the compacted granular material, $P_{pore} = 4\gamma/d_{pore}$ with the surface tension γ, allowing the meniscus to bend into interstitial pores of diameter d_{pore}. Air quickly percolates through the compacted glass beads and the liquid is pushed away towards the tube's outlet (t_3, t_4). The rapid liquid flow can be considered as a Darcy flow. The viscous forces acting on the compacted grains counteract the frictional interactions, which eventually leads to destabilize the still-immersed packing, and to split the drained glass beads which forms a plug (t_4). (iii) the elastic energy stored in the system during the slow compaction phase is released and the accumulation front rapidly grows (t_4, t_5) before settling down as the next slow compaction phase starts (t_5, t_6). This cycling process is repeated until the meniscus reaches the tube's outlet. The intermittent nature of the plugs formation can be seen from the spatio-temporal diagram obtained over a larger fraction of the tube, see Fig. 3 left for the smallest withdrawing rates.

4 Flow Regimes

The formation of plugs corresponds to an instability resulting from the competition between frictional and viscous forces. At low withdrawing rates the granular material settles easily and frictional forces can be established among granular particles as well as with the confining walls. Increasing the withdrawing rate corresponds to increasing the viscous forces in the system. Viscous drag on particles prevent frictional interactions to be established and the granular packing gets fluidized more easily. We observed the various behaviours of the confined glass beads for larger values of withdrawing rates $I_0 \in [0.01, 30]$ mL.min^{-1}, and reported in Fig. 3 the corresponding spatio-temporal diagrams obtained. As the withdrawing rate is lower than ≈ 0.3 mL.min^{-1} the frictional forces are large enough to form the pattern of granular plugs separated by gaps. Larger withdrawing rate, $I_0 \in [0.3, 3]$ mL.min^{-1} weakens frictional interactions. The accumulation front is getting larger as stronger fluidizations occur, and fewer plugs are formed. Another flow regime is obtained for large flow rate $I_0 \leq 3$ mL.min^{-1}. The viscous forces dominates and prevent any frictional interactions between grains. Glass beads get suspended and no plugs are left in the tube. These three distinct regimes are denoted respectively the frictional, transitional and viscous flow regimes.

5 Phase Diagram

The size of the glass beads influences the balance between friction and visosity. Small glass beads get easily suspended whereas large glass beads are more stable under viscous flow. The viscous drag of a flow through a porous media is indeed proportional to its permeability. A bed

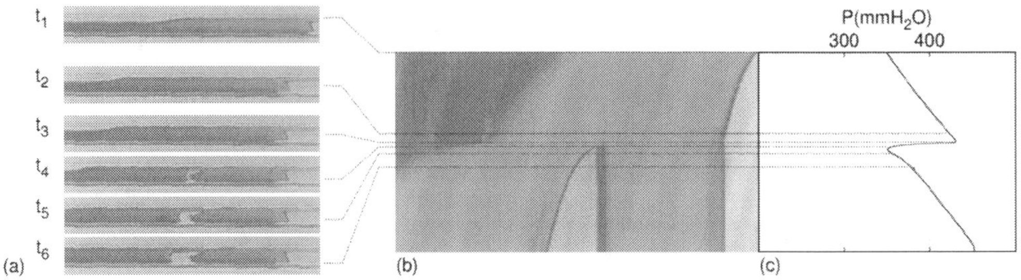

Figure 2. Focus over one plug formation obtained for I_0 = 0.01 mL.min^{-1}, filling fraction φ = 0.5, beads size d = [75 – 100] μm illustrated by (a) images and (b) spatio-temporal diagram. (c) pressure measured at the tube's outlet in mmH$_2$O during the plug formation.

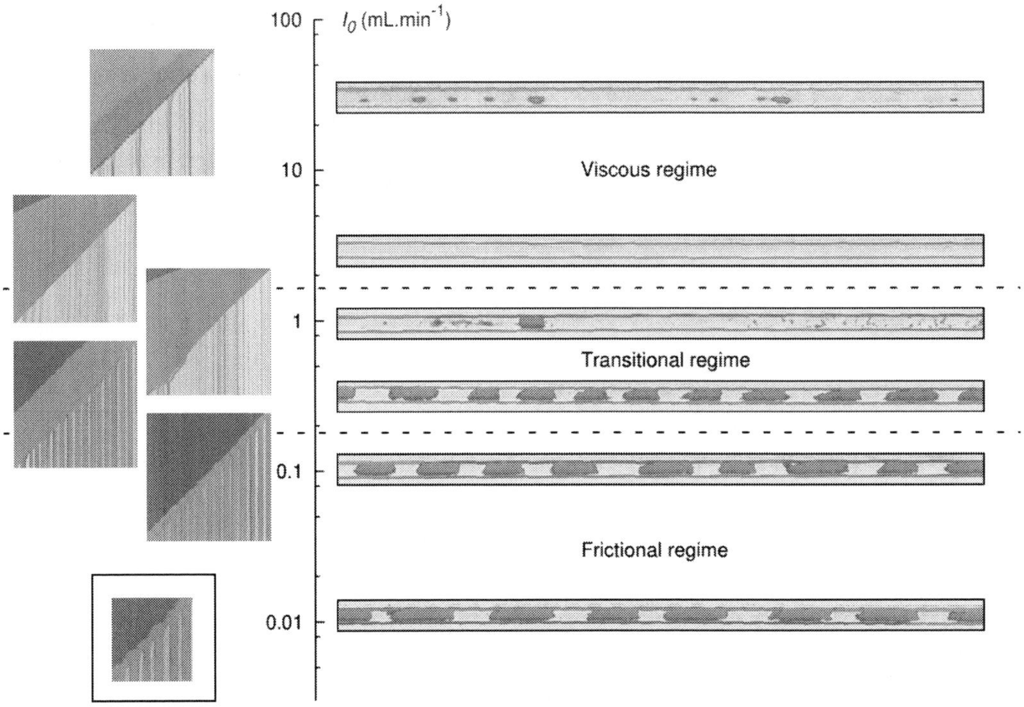

Figure 3. (left column) spatio-temoral diagrams obtained with withdrawing rate increase in the range [0.01 – 30] mL.min−1. The reduced units correspond here to a fraction of the total withdrawing time τ required to empty the tube of size L at withdrawing rate I_0. The same filling fraction φ = 0.5 and beads size d = [75, 100] μm is used for all observations. Note that at the lowest withdrawing rate I_0 = 0.01 mL.min^{-1}, the total withdrawing time τ exeeds the recording time so a smaller fraction of the total spatiotemporal diagram is shown. (right column) corresponding images of the final states obtained. Frictional, transitional and viscous flow regimes are distinguished by the dashed lines according to the qualitative differences in the observations.

of granular particles has a permeability increasing as the particles size decreases, as suggested by the the theoretical Carman-Kozeny relationship. So glass beads were sieved into 5 size ranges [53 – 75], [75 – 100], [100 – 150], [150 – 200], and [200 – 300] μm. A systematic exploration of the three flow regimes described above has been conducted for every size ranges. Figure 4 gives a phase diagram summarizing the observations with squared beads size horizontally, and imposed flow rate vertically. Observations corresponding to frictional, transitional, or viscous flow regime are pointed respectively with disks, open circles, and solid squares.

6 Conclusion

We observed the self structuring of a confined granular material and liquid mixture under capillary stress. A gas/liquid meniscus driven along a narrow tube by a

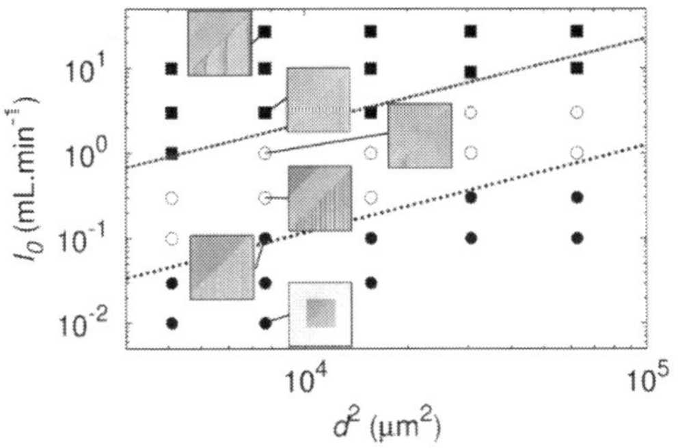

Figure 4. Phase diagram featuring the various flow regimes obtained when varying the withdrawing rate $I_0 \in [0.01, 30]$ mL.min^{-1}, and the bead sizes $d \in [53-75]$, $[75-100]$, $[100-150]$, $[150-200]$, and $[200-300]$ μm obtained after sieving. Disks, circles and squares represent respectively frictional, transitional and viscous flow regime. Spatio-temporal diagrams observed in fig 3 are indicated. Dashed lines are used for flow regime demarcations.

syringe pump bulldozes sedimented glass beads and forms regular granular plugs spaced with empty gaps, or suspends and evacuates the granular material from the tube. Three flow regimes are put in evidence as determined by the imposed flow rate as well as the granular particles sizes. The simplified quasi one-dimensional setup presented here provides a useful tool to interprete the complex behaviour of confined granular material in 2D Hele-Shaw cells under capillary stress [13–15].
The complexity of the patterns observed in confined frictional fluid derives from a mixture of capillary, frictional and viscous interactions. Our setup is constraining the frictional fluid by reducing one degree of freedom in comparison to regular Hele-Shaw cells. A subsequent structuring is still occurring as the confinement preserves the internal fluid self-structuring.

Acknowledgements

This work has been supported by the Norwegian Research Council through the FRINAT Grant No. 213462/F20. B. S. ackowledges support from EPSRC Grant No. EPL/013177/1 and Sêr Cymru National Research Network in Advanced Engineering and Materials Grant No. NRN141.

References

[1] J. Trondvoll and E. Fjær, Int. J. Rock Mech. Min. Sci. Geomech. Abstr. **31**, 393 (1994)
[2] D. Gidaspow and J. Huang, Ann. Biomed. Eng. **37**, 1534 (2009)
[3] J. B. Freund and M. M. Orescanin, J. Fluid Mech. **671**, 466 (2011)
[4] O. Molerus, Powder Technol. **88**, 309 (1996)
[5] E. Rabinovich and H. Kalman, Powder Technol. **207**, 119 (2011)
[6] M. Strauss, S. McNamara, and H. J. Herrmann, Granular Matter **9**, 35 (2007)
[7] S. Shaul and H. Kalman, Powder Technol. **256**, 310 (2014)
[8] F. B. Soeypan, S. Cremaschi, C. Sarica, H. J. Subramani, and G. E. Kouba, AIChE J. **60**, 76 (2014)
[9] F. Boyer, E. Guazzelli, and O. Pouliquen, Phys. Rev. Lett. **107**, 188301 (2011)
[10] A. Fall, F. Bertrand, G. Ovarlez, and D. Bonn, Phys. Rev. Lett. **103**, 178301 (2009)
[11] R. Seto, R. Mari, J. F. Morris, and M. M. Denn, Phys. Rev. Lett. **111**, 218301 (2011)
[12] E. Brown and H. M. Jaeger, Rep. Prog. Phys. **77**, 046602 (2014)
[13] B. Sandnes, E. G. Flekkøy, H. A. Knudsen, K. J. Måløy, and H. See, Nat. Comm. **2**, 288 (2011)
[14] H. A. Knudsen, B. Sandnes, E. G. Flekkøy, and K. J. Måløy, Phys. Rev. E **77**, 023301 (2008)
[15] C. Chevalier, A. Lindner, M. Leroux, and E. Clément, J. NonNewton. Fluid **158**, 63 (2009)
[16] G. Dumazer, B. Sandnes, M. Ayaz, K. J. Måløy, and E. G. Flekkøy, Phys. Rev. Lett. **117**, 028002 (2016)
[17] H. A. Janssen, Z. Ver. Dtsch. Ing. **39**, 1045 (1895)
[18] B. Marks, B. Sandnes, G. Dumazer, J. A. Eriksen, and K. J. Måløy, Front. Phys. **3**, 41, (2015)

A spheropolygonal-based DEM study into breakage under repetitive compression

Guien Miao[1,*], *Fernando* Alonso-Marroquin[1], and *David* Airey[1]

[1]School of Civil Engineering, The University of Sydney, NSW Australia 2006

Abstract. Experimental breakage studies have often focused on comparing grading and particle shape data from the beginning and end of a test, but one major advantage of DEM simulations is that, although the data are still discrete, more information on intermediate stages is available. This paper describes a repetitive compression test using a 2D aggregate-based DEM model comprised of spheropolygonal particles (formed by the Minkowski sum of a circle and a polygon, viz. sweeping a circle around the edges of the polygon) that are connected by beams and compares the behaviour with experimental data on the breakage of Barrys Beach carbonate sand. The one-dimensional repetitive compression test was performed on 20 particles—each consisting of over 100 sub-particles—which were generated from the outlines of particles of Barrys Beach carbonate sand. Particle breakage was described through the breakage of beams (particle bonds), allowing the evaluation of changes in the compressibility and grading. It was noted that the simulation compared well with the experimental behaviour of Barrys Beach carbonate sand.

1 Introduction

Breakage causes particle shapes to tend away from extremes and particle size distributions to tend towards increasing well-gradedness [1]. Consequently, substantial changes occur to the macroscopic properties—such as density, compressibility and strength—of a soil. Thus, an understanding of how both particle shape and grading evolve as a result of breakage under compression will provide further insight into how the behaviour of a soil will change as a result of very large applied loads.

The Discrete Element Method, DEM, is a discontinuum model that was proposed by Cundall & Strack [2]. The method allows the exploration of the interactions between particles within a granular assembly as the modelling of individual particles within DEM reflects the discrete nature of particles in real granular media, where particles only interact at points of contact. Many DEM studies use circular particles; however, a significant advantage of spheropolygonal particles [3] is that they eliminate the complications associated with the formulation of rolling resistance to account for particle shape when using circular particles. The use of spheropolygons in DEM also offers the opportunity to model a variety of shapes; most significantly, it allows for variation in the angularity of particles away from perfectly round (circular) or perfectly angular (polygonal) particles through the variation of the radius of the circle that is swept around the polygon (spheroradius).

Breakage can be modelled in DEM via using solid-bridge bonds/beams [4]. The bond's properties determine its reaction against imposed forces, giving continuity to the bond behaviour under all relevant deformations: axial, tangential and rotational.

2 Methodology

Repetitive compression involves a number of cycles of monotonic compression with sample reconstitution between cycles. It is a methodology which induces more particle movement than monotonic compression and therefore results in greater amounts of breakage [5]. In this study, experimental and 2D DEM data in the case of repetitive compression is compared to demonstrate the utility of the model in providing data on the breakage behaviour of soils under one-dimensional compression.

2.1 Experimental Repetitive Compression

A uniformly graded sample of Barrys Beach sand weighing 30 g was placed in a one-dimensional compression cylinder with an internal diameter of 35 mm and repetitively compressed to stresses of 95 MPa. A carbonate sand was used in this study as the brittleness and the irregular shape of the particles of carbonate sands [7] allow for a higher amount of breakage for a given load and thus the final state can be reached more quickly. The uniformity of the soil also induces maximal breakage [1].

2.2 DEM Repetitive Compression

A one-dimensional repetitive compression test was performed on 20 particles—each consisting of over 100 sub-particles—which were generated from the outlines of

* Corresponding author: guien.miao@sydney.edu.au

20 particles of Barrys Beach carbonate sand. The DEM model used in this study is a two-dimensional spheropolygonal-based model, shown in Fig 1. The discretisation of each grain is the only source of disorder in this model. The size effects of particle strength are accounted for via the larger particles consisting of more sub-particles and therefore containing a greater number of sites where failure can occur. 2D DEM studies are advantageous in providing qualitative information on the behaviour of granular assemblies without the computational complexity of 3D models [3].

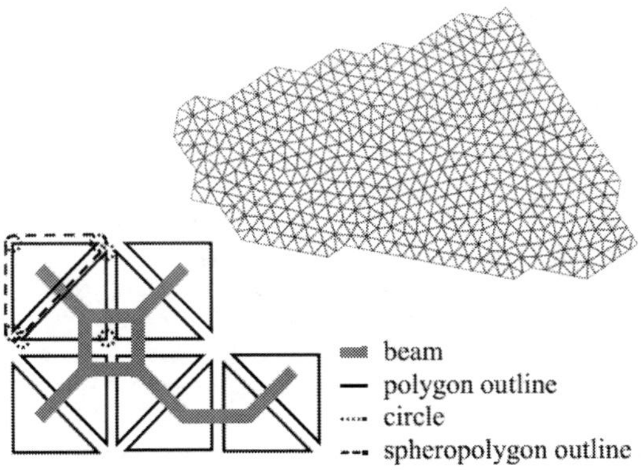

Fig. 1. Example of a triangulated particle with a schematic diagram of a solid-bridge bond (beam) model

Two models were used in this study, a non-convex particle model (as detailed in [3]) and a convex particle model. The non-convex model was used to place the particles without breakage, whilst the convex model was used to model the subsequent breakage of the assembly under compression. The contact forces between two particles, i and j, are described by (1) for the non-convex model and (2) for the convex model.

$$F_c^{ij} = \sum_{k,l} F_{V_{i,k}E_{j,l}} + \sum_{m,n} F_{V_{j,m}E_{i,n}} \quad (1)$$

$$F_c^{ij} = \sum_k F_{V_{i,k}P_j} + \sum_l F_{V_{j,l}P_i} \quad (2)$$

where $V_{i,k}$ refers to the vertex of particle i with index k, which is in contact $E_{j,l}$, that is, the edge of particle j with index l or in contact with a convex polygonal particle P_j.

The breakage model consists of a contact model and a beam/bond model and the interaction between contact forces and beam forces is described by (3), (4) and (5).

$$\boldsymbol{F_n} = \sum_c (\boldsymbol{F_c^{n,e}} + \boldsymbol{F_c^{n,v}}) + \boldsymbol{F_b} \quad (3)$$

$$\boldsymbol{F_t} = \sum_c (\boldsymbol{F_c^{t,e}} + \boldsymbol{F_c^{t,v}}) + \boldsymbol{V_b} \quad (4)$$

$$M = \sum_c \boldsymbol{F_n} \times \boldsymbol{r_a} + \sum_c \boldsymbol{F_t} \times \boldsymbol{r_b} + M_b \quad (5)$$

where F_n is the overall normal force, consisting of elastic and viscous components of the contact forces ($F_c^{n,e}$ and $F_c^{n,v}$ respectively) and a beam component (F_b); F_t is the overall tangential force, also consisting of elastic and viscous components of the contact forces ($F_c^{t,e}$ and $F_c^{t,v}$ respectively) and a beam component (V_b); and M is the torque caused by the normal and tangential forces with a beam component (M_b).

Renzo & Di Maio explored different contact models and noted that, on a macroscopic level, the agreement with experimental results was better with a simpler model [7]. For this reason, a relatively simple normal contact model, as described by (6), was used:

$$F_c^{n,e} = \frac{k_n \delta^3}{(\delta - \delta_m) l_c} \quad (6)$$

where k_n is the normal contact stiffness, δ is the overlap between two particles, δ_m is the maximum overlap (which is defined as the sum of the two spheroradii of the particles) and l_c is a characteristic length (which is required for dimensional compatibility and given a nominal value of 1 cm). The contact force is proportional to the area of the contact, which is a function of δ^2, and tends to infinity at $\delta = \delta_m$. It creates adhesive forces at small overlaps with repulsive forces at large overlaps when combined with a simple first-order (Euler-Bernoulli) beam as shown in Fig. 2. The tangential contact model is a standard frictional contact model.

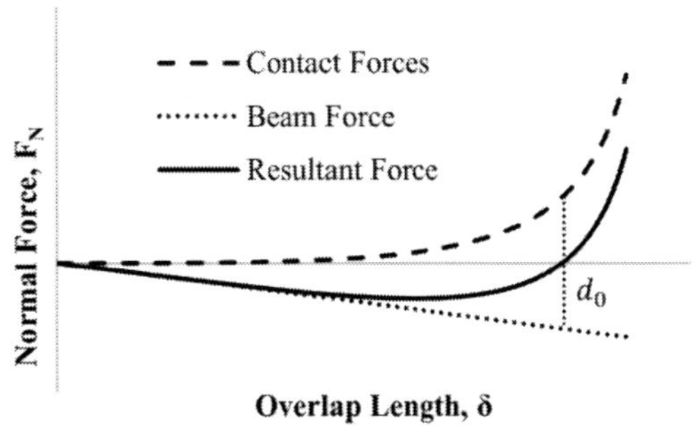

Fig. 2. The normal forces applied to a given sub-particle in the range of $0 \leq \delta_0 \leq 0.9\delta_m$

The failure criterion that governed the breakage of the beams is shown in (7). The criterion was non-linear and allowed breakage only under tension. The criterion reflects the equivalent effect of axial stress on the plastic moment capacity and combined moment and shear failure criterion [9] for the failure of beams.

$$\sqrt{\frac{|M|}{M_f}} + \frac{\sigma_p}{\sigma_f} + \left(\frac{\tau_p}{\tau_f}\right)^2 \geq 1 \quad (7)$$

where σ_p and τ_p are the axial (tensile) and shear stresses respectively when projected onto vectors normal and tangential to the failure plane.

3 Compression Curves

Fig. 3 shows a strong similarity between the experimental and DEM compression curves. As the simulation was two-dimensional, a modified definition of void ratio, $e = A_v/A_s$, was used. In both cases, with increasing cycles of compression, the void ratio at high stresses decreases to an asymptotic value and the slope of the limiting compression curve, LCC, (the slope of the compression curve after yield) becomes less steep with increasing cycles (breakage). In the compression curves from the DEM simulation, it should be noted that the first two cycles are similar as the first cycle was terminated early. While the change in the LCC at early cycles is associated with the occurrence of breakage, the increasing similarity of the final LCCs is an indication of the cessation of breakage.

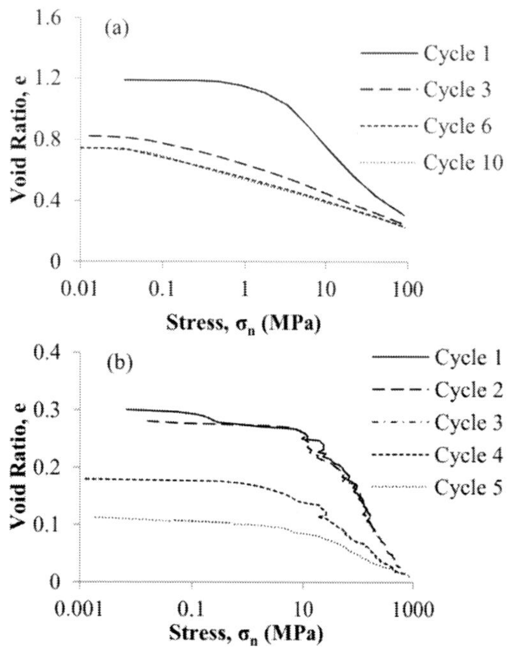

Fig. 3. Compression curves for a repetitively compressed assembly obtained from (a) an experiment and (b) DEM

The stiffness of the material increases with each cycle of repetitive compression. This is due to breakage causing a reduction in particle irregularity and an increasingly well-graded particle size distribution, as can be seen in Fig. 4. Decreasing particle irregularity has been shown by [9] to increase the soil stiffness; it has also been shown that a well-graded sand will have a lower LCC gradient—or equivalently, compression index, C_c—than a uniform grading of the same material [10]. These changes produce a particle assembly which has a greater packing potential so particles are less free to move translationally or rotationally and the assembly therefore becomes stiffer.

Of interest is that the limiting compression curve decreases in slope and begins to reduce in slope at very high stresses (or alternatively, at very low void ratios) in the case of the DEM simulation. This behaviour was not noted in the experimental results due to limitations in the apparatus which did not allow higher stresses to be achieved; however, this behaviour is in line with the expected behaviour of the LCC as it cannot continue infinitely downwards as a void ratio of zero is a limit by definition. Ueda et al. noted similar behaviour in their monotonic compression tests; however, their comminution limit was reached at a significantly higher void ratio [11]. The mean sub-particle diameter used in their study was 10 mm, which is higher than the maximum mesh size of 5 mm size used in this study. It would be expected that the use of a larger minimum particle size would result in the comminution limit being reached at a higher void ratio because the minimum particle size constrains the potential amount of breakage as well as the reduction in void space.

It should also be noted that the void ratios for the simulation lie significantly below the range of void ratios for Barrys Beach sand (e_{min} = 1.25; e_{max} = 1.61). The minimum void ratio for circular (two-dimensional) disks can be shown to be 0.103 by geometry whilst the densest packing of (three-dimensional) spheres is achieved by close-packing, yielding a void ratio of approximately 0.35. Thus, even in ordered assemblies, the transition from 2D to 3D results in a substantial increase in the void ratio. The initial void ratio for DEM simulation is furthermore significantly higher than the minimum void ratio for circular disks due to the irregular particle shapes and particle arrangement. For disordered assemblies of irregular particles, it would be expected that the difference in packing between 2D and 3D void ratios would be greater due to the greater variation in shape allowing a significantly more open structure in three dimensions.

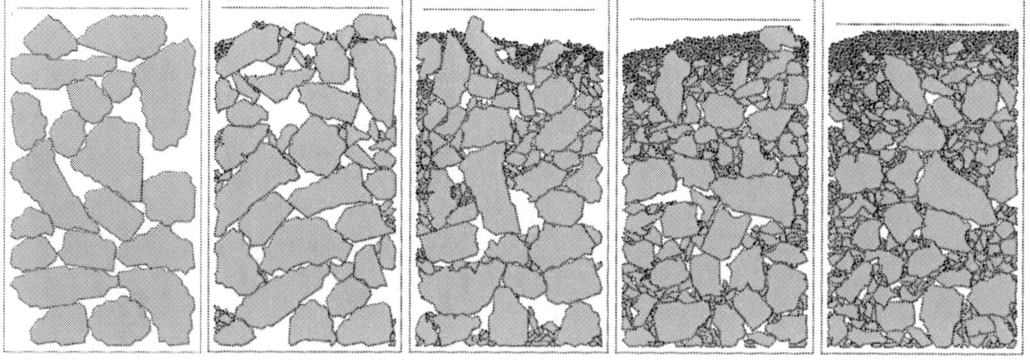

Fig. 4. Visualisation of particle assemblies at the start of each cycle of repetitive compression

4 Grading

Fig. 5 shows the evolving gradings with increasing cycles of breakage for both the experimental and DEM cases. Although the particle sizes produced are not similar (due to limitations in the minimum particle size for the simulation), there are still key similarities between the two sets of data. It can be seen that the greatest amount of breakage occurs within the first cycle and that the difference in grading for later cycles reduces. This occurs in spite of the grading change in first cycle being less than expected due to early termination of the test. This reflects the behaviour of the compression curves shown in Fig. 3, where the greatest changes in void ratio occur in early cycles and differences in later cycles are smaller.

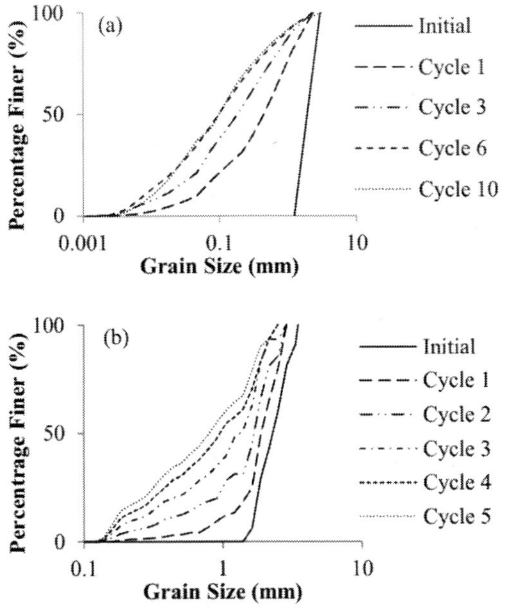

Fig. 5. Evolution of grading for a repetitively compressed assembly obtained from (a) an experiment and (b) DEM

Furthermore, in both cases, the largest particles undergo complete breakage, which confirms the findings of Bridgwater et al. that 100% attrition of particles is possible when using analyses which capture relatively small changes [12]. Nevertheless, there is a difference in the curvature of the experimental and numerical gradings. Although the experimental grading is single-curvature (with no inflection point) after a single cycle, it becomes double-curvature after more cycles. Conversely, the DEM grading remains single-curvature throughout the test, in part due to the complete breakage of the coarsest particles.

Now, Palmer & Sanderson's model suggests that when each grain has an equal probability of fracture, a (single-curvature) mono-fractal particle size distribution evolves [13]. Therefore, the transition from a single-curvature grading to a double-curvature grading suggests that there is an unequal probability of breakage which is skewed towards the breakage of larger particles in the case of the experiment, resulting in preferential breakage of the larger particles. The complete breakage of the coarsest particles in the DEM simulation suggests that there is indeed a preferential breakage of the coarsest particles; however, better particle cushioning in the experimental gradings may have contributed to the protection of the coarsest particles from complete breakage. In fact, it can be seen from Fig. 4 that the particle placement did not result in an even distribution of fines, which may have caused a bias towards greater breakage. The simulation produces gradings that are linear in semi-log scale, which is similar to those from the experiments of Lőrincz et al., who noted a significant loss of fines during their tests [5].

5 Conclusion

As in the experimental data, the compression curves showed an increase in the stiffness of the soil and downward shift with increasing cycles. Furthermore, the grading was also noted to become increasingly well-graded and finer with increasing stress and increasing cycles as in the experiment. Interestingly, in contrast to the experimental data on the Barrys Beach sand, which showed the formation of double-curvature grading, the grading in the simulation tended towards a single-curvature grading that is linear in semi-log space. This may be attributed to insufficient sample-mixing in the DEM simulation and may better reflect situations where fines are lost from the sample during experiments. Despite limitations in the model (2D with only a few particles), the similarities between the simulation and the experimental data demonstrate the utility of the model for a qualitative analysis of breakage behaviour.

References

1. K. Lee, I. Farhoomand, Can. Geotech. J **4**, 1 (1967)
2. P.A. Cundall, O.D.L. Strack, Géotechnique **29**, 1 (1979)
3. F. Alonso-Marroquin, Y. Wang, Granul. Matter **11**, 5 (2009)
4. F. Kun, H.J. Herrmann, Comput. Methods Appl. Mech. Eng. **138**, 1 (1996)
5. J. Lőrincz, E. Imre, M. Gálos, Q.P. Trang, K. Rajkai, S. Fityus, G. Telekes, Int. J. Geomech. **5**, 4 (2005)
6. J.P. Carter, D.W. Airey, M. Fahey, *A Review of Laboratory Testing of Calcareous Soils* (1999)
7. A. Di Renzo, F.P. Di Maio, Chem. Eng. Sci., **59**, 3 (2004)
8. D. Drucker, R. Providence, *The Effect of Shear on the Plastic Bending of Beams* (1955)
9. G.C. Cho, J. Dodds, J.C. Santamarina, J. Geotech. Geoenviron. **132**, 5 (2006)
10. M.R. Coop, Géotechnique **40**, 4 (1990)
11. T. Ueda, T. Matsushima, Y. Yamada, Granul. Matter **15**, 5 (2013)
12. J. Bridgwater, R. Utsumi, Z. Zhang, T. Tuladhar, Chem. Eng. Sci. **58**, 20 (2003)
13. A. Palmer, T. Sanderson, Proc. R. Soc., Lond., Ser. A **433**, 1889 (1991)

The Experiment of the Clog Reduction in a Plane Silo

Ai-Le Sun[1] and *Jie* Zhang[1,2,*]

[1] *Department of Physics and Astronomy, Shanghai Jiao Tong University, Shanghai, China 200240.*
[2] *Institute of Natural Sciences, Shanghai Jiao Tong University, Shanghai, China 200240.*

Abstract. The flow of particles may be clogged when they pass through a narrow orifice. Many factors can change the probability of clogging, such as the outlet size, the presence of obstacles and external perturbation, but the detailed mechanisms are still unclear. In this paper, we present an experimental study of reduction of the clogging probability in a horizontal plane silo, which consists of a layer of elastic particles transported on an annular flat plate rotating with a constant angular velocity passing through a hopper structure. We found the exponential distributions of the avalanche size for different sizes of orifice and the power law tails of the passing time between two particles. We did not confirm whether there was a critical size of orifice above which the clogging became impossible. We explored the effect of the obstacle on the probability of clogging: and if we chose a proper obstacle placed at a proper position, the probability of clogging could be reduced by a factor of about seven.

1 Introduction

Granular materials are very common in our daily life, such as sand, powder, cement, where the motions of particles due to thermal fluctuations are negligible[1, 2]. When lots of granular particles pass through a small orifice, the flow of particles could be abruptly arrested, i.e. clogged. This phenomenon is ubiquitous and has be observed in many systems, such as granular flows in a hopper [3–7], dense suspension flows in a micro-channel [8], emulsions [9, 10], dense colloidal systems [11], and living animals [12]. In addition, when in an emergency, people rush to escape through a narrow exit, the clog may appear [13].

Many factors can affect clogging, such as the size of orifice, presence of an obstacle, and external vibrations. Previous studies have revealed the exponential distribution of avalanche size. To and colleagues [3] proposed a restricted random walk model to explain the exponential distribution by analyzing the arches of particles. This model provides a good fitting to the data. However, the motion of particle is not necessary like a restricted random walk and the friction between the particles is large in many systems. Another problem is whether there is a critical outlet size that the probability of clog decreases to zero. CC.Thomas[14] proposed that the critical outlet size does not exist. To further study this problem, we designed a novel apparatus where the particles can be automatically refilled so that the experiment can run continuously and particles in the hopper will never be exhausted. In addition, one interesting thing is that if an obstacle is placed at a proper position, the silo will have much less chance to clog[4, 15]. In this paper, we will also present some preliminary results related to the reduction of clogging by an obstacle.

* e-mail: jiezhang2012@sjtu.edu.cn

2 Experimental Setup

In order to probe the key factors that influence the probability of clogging, we designed a 2D annular silo that allowed the particles to be recycled, so that the experiment could run continuously. The silo was filled with elastic particles, which were bidisperse. The Young's modulus of particles is 6 MPa. The total number of particles was ~ 1400, including ~ 600 large particles of a diameter of 14 mm and ~ 800 small particles of a diameter of 12 mm in order to avoid crystallization. [16]

As shown in Fig. 1, the radius of the large circle was $R_2 = 0.5m$ and the radius of the small circle was $R_1 = 0.1m$. The particles were driven by the frictional force between the particles and the bottom plate that was rotating at a constant angular velocity of $\omega = \frac{\pi}{45}$ rad/sec. There were two Basler cameras mounted above the apparatus to record the motion of particles, and the internal stress information of particles. When the silo was clogged for a sufficiently long time, we would apply a manual perturbation to break the arch formed near the orifice by turning on a small vibrator. One important control parameter of this system is the size of the orifice in mean particle diameters. We systematically changed the size of orifice and the position of obstacle. Through the imaging analysis, we extracted the location of particles in the system, and used the result to track the motion of the particles.

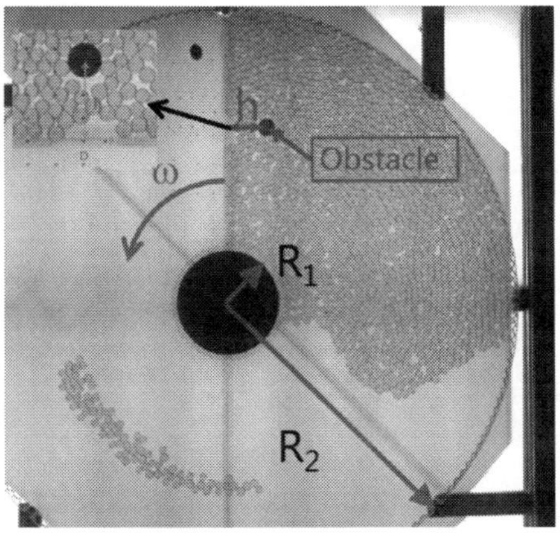

Figure 1. The experimental setup. The apparatus is essentially a horizontal silo, which consists of an annular plate of a large radius of 50 cm and a small radius of 10 cm. Particles were driven by the frictional forces between the particles and the plate, which rotated counter-clockwise with an angular velocity of $\frac{\pi}{45}$ rad/s. A hopper structure is visible on the plot along with an obstacle (the black blob on the figure).Particles clog the orifice intermittently.

3 Results

3.1 The time lapse of particles

To define a clog, we need to analyze the interval time between two consecutive particles passing through the orifice. What will be the proper interval time for us to affirm that the silo is clogged? The data were obtained by analyzing images of particles passing through the orifice. The flow could be intermittent. Figure 2(a) plots the survival or reliability function of the interval time between two consecutive particles passing through the orifice. We found that the survival had a power law tail, which is consistent with the results of Zuriguel et al.[17]. The difference is that in their experiment the sheep could adjust themselves to get out of the exit if a clog happened, but in particulate system the particles can not adjust themselves unless a perturbation is applied.

Based on the above results, we can choose a proper critical interval of time Δt_c to define the clog. As Fig. 2(b) shows, we can choose different Δt_c and the distribution of avalanche size remains almost the same. Here the x axis is normalized by the average avalanche size, and the y axis is the probability of the corresponding avalanche size (the average of frequency). When Δt_c is large, the distribution is exponential, and only when Δt_c is sufficiently small, i.e. $\Delta t_c = 0.25s$, the distribution becomes different. As we mentioned before we would apply perturbations to break permanent clog in order to perform a new measurement. The frequency of $\Delta t > 10s$ is rather small and we can classify them in one bin. That means the choice of Δt_c does not strongly influence the results of statistics. So we simply chose the $\Delta t_c = 10s$ as the critical time in our later experiments.

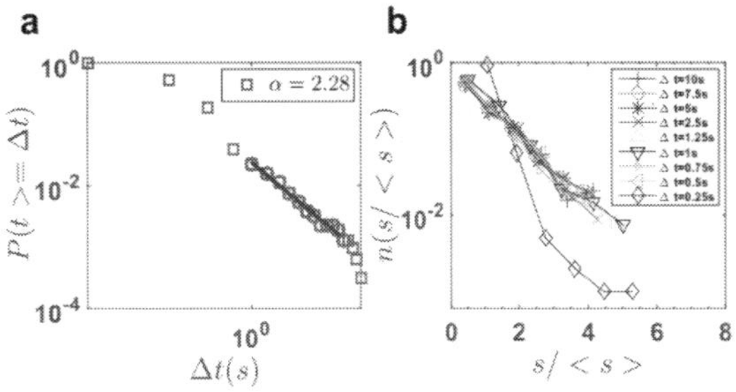

Figure 2. (a)The cumulative probability distribution function versus time interval Δt. (b) The probability distributions of avalanche size s normalized by its mean $\langle s \rangle$ for a set of different time intervals Δt.

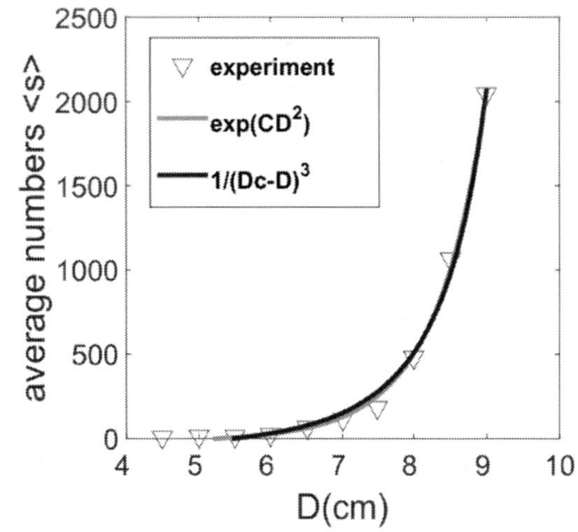

Figure 3. The mean avalanche size $\langle s \rangle$ versus the size of outlet. An exponential fitting(green) and a power law fitting(blue) can fit the data equally well.

3.2 The effect of outlet size

We found the avalanche size was an exponential distribution, consistent with the early finding of To et al. [3]. The model assumes that the probability of one particle passes through the outlet as a constant, denoted as p, so the probability of passing through n particles and then the flow being stopped can be written as:

$$p(n) = p^n(1 - p) \tag{1}$$

So the average size of an avalanche is

$$< n >= \frac{\sum n * p(n)}{\sum p(n)} \tag{2}$$

We measured the avalanche sizes for 10 different outlet sizes by fixing the angular velocity. we selected the critical time interval $\Delta t_c = 10s$ to define the clog. Every point is an average of nearly 150 avalanches. As Fig. 3(b) shows, we can see the influence of the outlet size. When the diameter of outlet enlarged, the average avalanche size increased as expected. When the outlet size exceeded 8cm, the clog happened rarely. In our experiment, the flow continued for a long time, but the clog still happened after a sufficiently long time. That means if the flow was clogged, and sometimes the arch of particles was very stable unless we applied a perturbation to the system. This is much more difficult to be observed in systems of finite size because it needs to refill particles into the system once the particles are empty.

3.2.1 *Critical outlet*

The problem is whether there is a critical outlet size that the flow can never be stopped. As we have mentioned early that in our system the probability of clog decreased as the outlet size increased. we fit our data by two means: (1)an exponential fitting and (2)a power law fitting.

First, we applied exponential function to fit the data. In 2D systems, the average size of an avalanche is proportional to $exp(CD^2)$, [14](C is a constant related to the system.)(in 3 dimension is $exp(CD^3)$.) In addition, we used a power law to fit the data. The hypothesis is that there is a parameter, named D_c, which is the critical outlet size and the average avalanche size is divergent. From the figure, we can see that both descriptions are comparable in fitting the experimental results.

If the power law were better, there would exist a critical outlet size and if the exponent fitting were better, there would not exist a critical outlet size. Based on the fitting, we calculated the critical outlet size to be around 10.8cm. Now the difficulty is that if D is larger than 9.5cm, we could hardly observe the clogging, although our system can be automatically refilled, the interval between twice clog is increasing sharply and exceed our observation time .Thus we did not confirm whether there was a critical size of orifice above which the clogging became impossible.

3.3 The effect of the height of obstacle

The average avalanche size will increase if a proper obstacle is put at an appropriate position. [4, 15] The obstacle can influence the clogging process significantly. The shape and size of obstacle play significant roles here.

In order to probe the influence of obstacle, we measured 11 different positions for a fixed obstacle size and the outlet size. We found that the exponential distribution is still valid with the presence of an obstacle. It is reasonable to assume that the probability of clog without an obstacle equals the one when the obstacle is far from the outlet. As Fig. 4 shows(when $D = 5.5cm$), when the distance of obstacle relative to the outlet was larger than 9 cm, the influence of obstacle could be neglected. When the distance decreased to about 5cm, the average avalanche size had a maximum value, about three times of that without an obstacle. When the distance decreased below 5 cm,

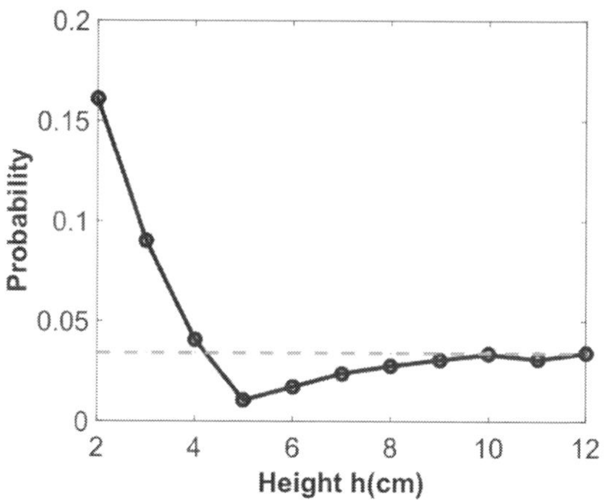

Figure 4. The probability of clogging versus the height h of obstacle that is defined in Fig. 1.The dash line is the probability of no obstacle.

the probability of clog would increase and even exceed the probability without an obstacle. The main reason is when the obstacle was close to the orifice, the obstacle stopped flow of the particles, so that the particles could only form arches at the two sides of the obstacle rather than the front of obstacle. In conclusion, if we choose proper obstacle and put it on an appropriate position, the probability can be diminished by a factor of 3.

The reason of reducing clog by an obstacle is still controversial but we tend to agree with the idea that the obstacle relieves the pressure of particles. The obvious parameter that influences the clog is the position of obstacle. The main zone is the front of the obstacle, so if the zone is sufficiently larger the pressure can be supported by the obstacle, the probability of clog will decrease. In our experiment, when h>5cm, particle could form the arches at the orifice, so the pressure would decrease and the force chain was relatively weak. In addition, the regime was relatively large, so that the arches were usually unstable and could be destroyed by small perturbations of particles coming from the upstream flow. Therefore the probability of clog was small. When h was too large, the shield of the obstacle would become weaken gradually until the height reached infinite, the effect of obstacle would vanish completely.

Note that the factor of the clog reduction probability was about 3, smaller than the previous research. We changed the diameter of the obstacle, and measured the average avalanche size at h=5cm. As a result, we found the ratio became larger, about 7. Limited by our system, we could not use a bigger obstacle. In addition, we think the variety of particle is the main difference from others' work. So we will use other types of particles in future.

DOI: 10.1051/epjconf/201714003024

4 Conclusion

In this paper, we presented an experimental study of reduction of the probability in a plane silo by changing the ratio of outlet vs particles by the presence of an obstacle. The setup consisted of a layer of elastic particles transported on an annular flat plate of a constant angular velocity passing through a hopper structure. By tracking the motion of particles we get an analysis of the reduction of clogging in a two dimensional plane silo.

In this experiment, we placed an obstacle at different locations near the exit of the silo in order to investigate the influence of the obstacle to the clogging process. We found that a system driven by frictional force is quite different from a system driven by gravity. Since the particles were recycled in the system, we could explore the evolution of the clogging process in an infinitely long time scale, in principle. We measured the time between two consecutive particles which passed through the outlet and found the time distribution had a power law tail.

We studied the influence of the outlet size and the position of obstacle. By fitting the data, we found an exponential fitting and a power law fitting are comparable in describing the statistics of clogging. Now there is a debate about whether there is a critical outlet size above which the clogging is impossible. Based on our experiment data we couldn't confirm whether there is a critical outlet size.We think when the outlet size becomes bigger, the probability of clogging will become smaller until we can not obverse the clog, therefore it is hard to prove in experiment.

Then we put an obstacle on the front of the outlet and found the probability of clogging had been changed significantly. In our experiment, the probability of clogging could be reduced by a factor of 7 or more. The reduction of the probability of clogging is possibly caused by the reduction of the pressure at the down stream of the obstacle, and we will explore it in our future work. The size of the obstacle has a significant influence on the clog. If we choose proper obstacle, the factor may become bigger.

Acknowledgements

J.Z. acknowledges support from the award of the Chinese 1000-Plan (C) fellowship and National Natural Science Foundation of China (11474196).

References

[1] H.M. Jaeger, S.R. Nagel, R.P. Behringer, Rev. Mod. Phys. **68**, 1259 (1996)
[2] P.G. de Gennes, Reviews of modern physics **71**, S374 (1999)
[3] K. To, P.Y. Lai, H. Pak, Physical review letters **86**, 71 (2001)
[4] I. Zuriguel, A. Janda, A. Garcimartín, C. Lozano, R. Arévalo, D. Maza, Physical review letters **107**, 278001 (2011)
[5] D.J. Durian, Phys. Rev. Lett. **75**, 4780 (1995)
[6] I. Zuriguel, T. Mullin, J.M. Rotter, Phys. Rev. Lett. **98**, 028001 (2007)
[7] J. Tang, R.P. Behringer, EPL (Europhysics Letters) **114**, 34002 (2016)
[8] K.L. Rock, C. Gramm, L. Rothstein, K. Clark, R. Stein, L. Dick, D. Hwang, A.L. Goldberg, Cell **78**, 761 (1994)
[9] X. Hong, K.W. Desmond, D. Chen, E.R. Weeks, arXiv preprint arXiv:1503.07569 (2015)
[10] X. Hong, M. Kohne, E.R. Weeks, arXiv preprint arXiv:1512.02500 (2015)
[11] C.P. Ortiz, R. Riehn, K.E. Daniels, Soft Matter **9**, 543 (2013)
[12] I. Zuriguel, D.R. Parisi, R.C. Hidalgo, C. Lozano, A. Janda, P.A. Gago, J.P. Peralta, L.M. Ferrer, L.A. Pugnaloni, E. Clément et al., Scientific reports **4** (2014)
[13] I. Zuriguel, Papers in physics **6**, 0 (2014)
[14] C. Thomas, D. Durian, Physical review letters **114**, 178001 (2015)
[15] C. Lozano, A. Janda, A. Garcimartín, D. Maza, I. Zuriguel, Physical Review E **86**, 031306 (2012)
[16] D. Helbing, A. Johansson, J. Mathiesen, M.H. Jensen, A. Hansen, Physical review letters **97**, 168001 (2006)
[17] A. Garcimartín, J. Pastor, L. Ferrer, J. Ramos, C. Martín-Gómez, I. Zuriguel, Physical Review E **91**, 022808 (2015)

The relationship between particle strength and particle breakage in loaded gravels

Stephen Fityus [1*] and *John* Simmons [1,2]

[1]The School of Engineering, University of Newcastle, University Drive, Callaghan, Australia, 2308.
[2] Sherwood Geotechnical and Research Services, Peregian Beach QLD, Australia 4573

Abstract. The particle breakage behaviour of coarse grains during shearing of a gravelly granular assemblage changes in accordance with the magnitude of the confining stress acting on the assemblage. The transition from particle rearrangement and sliding to particle breakage is studied for two gravelly materials subjected to a wide range of compressive and shear stresses (10kPa to 3.5MPa). The tested materials are a gravel fraction derived from crushed sedimentary sandstones and shales, and a gravel derived from a crushed rhyodacitic volcanic. At low confining stresses, breakage due to compression and shear are both small, with only some fine fragments produced. Confining stress increase results in a stage being reached where particle damage starts to increase greatly, under both compressive and shearing loads. The results indicate that, as stresses become very large, the particle breakage due to compression continues to increase, but the breakage due to shearing under the same vertical stress actually decreases. This is attributed to the attainment of a fabric which, through extreme damage during compression, allows subsequent shearing with reduced additional particle damage.

1 Introduction

The shear strength of granular materials with crushable particles is determined by the complex contribution of several phenomena including particle rubbing, sliding and rolling, as well as particle breakage, which may involve comminution of thin corners/edges or complete particle failure [1]. Generally, rubbing, sliding and rolling dominate under lower stresses, whereas breakage becomes more prevalent as applied stresses increase and dilation is suppressed [2]. As a result, it is believed that the shear strength envelope is non-linear [3,4].

However, the relationship between the particle strength and the stress conditions which would cause the particles to become damaged (as opposed to simply rearranged) during shearing are poorly understood.

In this study, two crushed rock materials, of different strength, were subjected to direct shear testing over an extreme range of stresses, and the resulting particle breakage was inferred from the amount of fines produced. From this, insights are gained for the relationship between particle strength, confining stress and particle damage.

2 Methods

The samples tested in this research were a relatively weak sedimentary rock, dominated by siltstone, derived from the waste rock from a coal mine in the Permian Sydney Basin of Australia, and a rhyodacite ignimbrite taken from a quarry in the Carboniferous volcanics of the New England Fold Belt. Unconfined compressive strength (UCS) measurements of samples of the sedimentary rocks in the sampled geological sequence ranged from 25 to 40 MPa with a mean value of around 35 MPa. UCS tests on samples from the rhyodacite quarry ranged from 150 to 177 MPa with a mean value of around 170 MPa.

For the purposes of this study, damage was characterised by the production of fines, where fines were defined as any particles passing the 4.75mm sieve. So as to avoid issues with possible size effects [5][6], each sample was prepared to have the same range of initial particle sizes, ranging between 4.75 and 19.6mm.

Tests were performed using a 300 x 300mm Prolab Direct Shear Machine (DSM) with a 1MPa capacity (Fig 1a). Tests were carried out with normal stresses ranging from 9 kPa to 12 MPa. So as to achieve stresses greater than 1 MPa, concrete and steel inserts with a 167mm circular sample space were manufactured, that could be inserted into the 300mm shear boxes, reducing the area of the sample by about 75% (Fig 1b). For stresses less than 100kPa, the mass of the shear box and insert became significant, and hydraulic control of the DSM was unable to apply and maintain the stress to sufficient accuracy. So, in the case of the 9 and 48 kPa tests, a special lightweight top box was fabricated and the normal load was applied using deadweights (Fig 1c). For the stress of 12 MPa only, an additional 104mm diameter insert was used to further reduce the sample area.

Samples were carefully prepared by sieving the required fractions from bulk samples. The basic fractions

* Corresponding author: stephen.fityus@newcastle.edu.au

of each material are shown in Fig 2. A fraction mixture with roughly ¼ of each size was selected. After several bench trials, a target mass was determined to allow samples to be prepared to the same consistent, loose initial density for each test.

a) Prolab direct shear machine, with inserts installed

b) sample insert under hydraulic load

c) lightweight top box and loading weights

Fig. 1. Features of the testing arrangement.

Siltstone

21.3%	23.4%	23.4%	31.9%
4.75-6.7mm	6.7-9.5mm	9.5-13.2mm	13.2-19.6mm

Rhyodacite

21.6%	23.5%	23.5%	31.4%
4.75-6.7mm	6.7-9.5mm	9.5-13.2mm	13.2-19.6mm

Fig. 2. Basic fractions used to create the initial samples.

Acknowledging that significant particle breakage may occur in the application of the normal compressive stress, even before the sample is sheared, particle breakage during shear was estimated as follows:

1. Prepare a new sample, apply compression, determine fines.
2. Prepare a new sample, apply compression, shear, determine fines.
3. Estimate fines due to shear as the difference between samples tested by compression and shear (2.), and compression only (1.).

Determination of fines after compression and/or shearing involved careful recovery of all the tested material and a sieve analysis. After testing under 3.5 and 12 MPa, the compression of the siltstone in particular, was so substantial that the recovered sample was effectively a monolith. Before it could be sieved, it had to be broken up. This was achieved very carefully by gently teasing out crushed and recompacted material that had clearly been reconstituted by the test process, whilst not breaking down primary particles that had not been damaged.

3 Results

Figure 3 shows examples of samples of both tested materials, after testing by compression and shearing, under low and very high stress. The intense compression of the siltstone under 12 MPa is clearly evident.

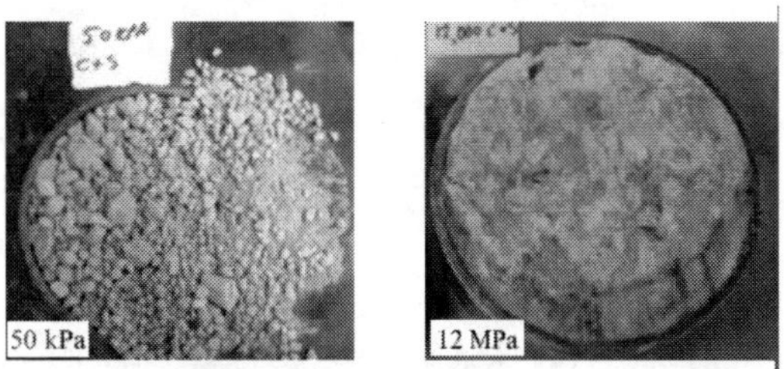

Siltstone

Rhyodacite

Fig. 3. Characteristics of samples after testing.

Fig 4 shows the particle crushing results for both material types, with a number of outcomes evident:

- The weaker siltstone material produces more fines than the rhyodacite for a given stress, as expected.
- A higher proportion of fines was produced during confined compression than during shear.
- The percentage fines produced by confined compression is small under very low stress (9 kPa) and increases slowly with increasing stress, until reaching some "threshold" stress, whereafter the percentage fines produced increases rapidly, but at a steady rate with increasing log stress.
- The percentage fines produced during shear is small under very low stress (9 kPa) and increases steadily with increasing log stress, until reaching some "critical" stress. Beyond the critical stress, the percentage fines produced decreases with increasing stress, despite the fines produced during compression increasing logarithmically.

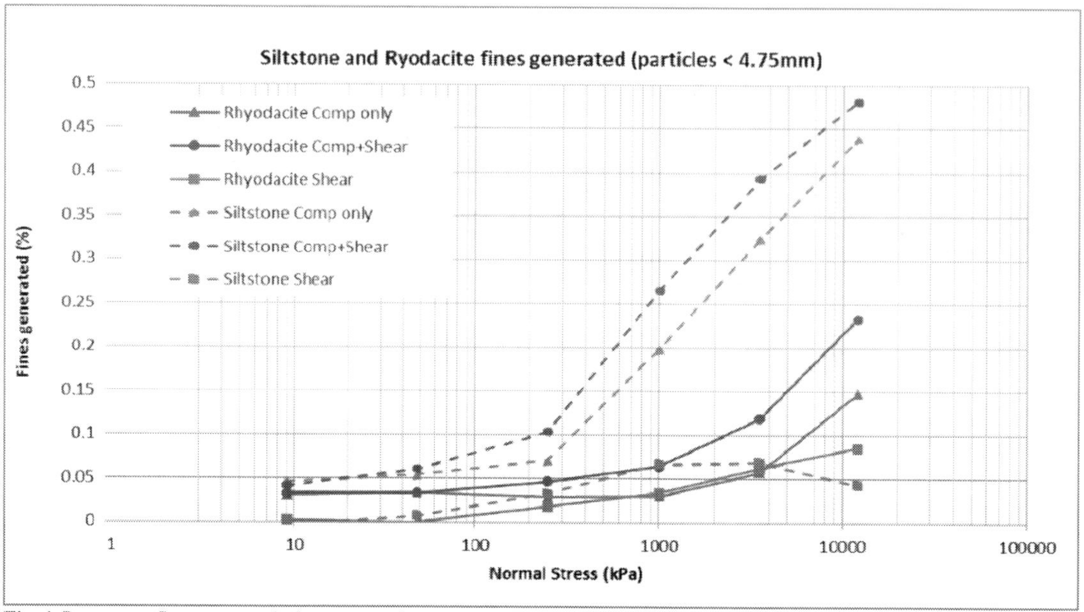

Fig. 4. Percentage fines generated after compression, compression and shear and shearing for the siltstone and rhyodacite samples.

The normal displacement vs shear displacement responses for each material are shown in Fig 5. It is evident that dilation ceases in the weaker siltstone material between normal stresses of 48 and 250 kPa, but it persists in the stronger rhyodacite material up to stresses of between 1 and 3.5MPa.

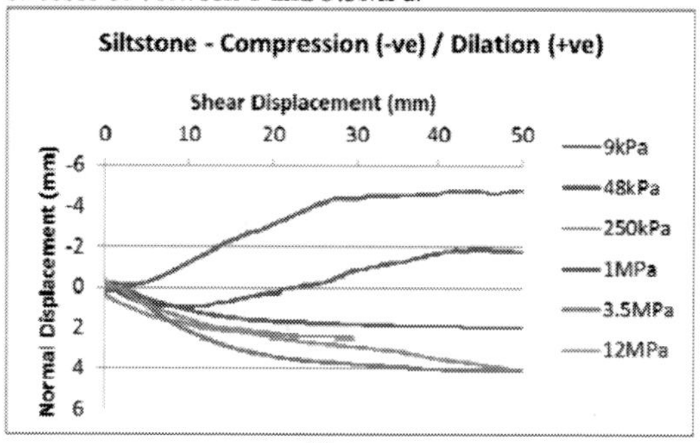

Siltstone

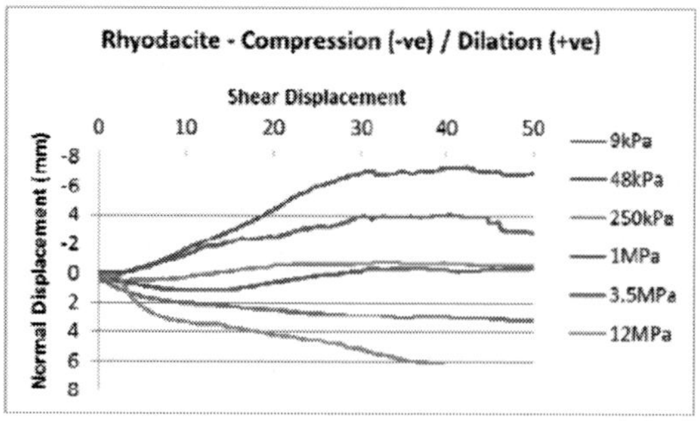

Rhyodacite

Fig. 5. Normal displacement vs shear displacement curves.

The peak shear stress obtained during each tests is plotted against normal stress in Fig 6. It is apparent that the shear stresses are proportionally larger at lower stresses, and that this is more so the case for the stronger rhyodacite material.

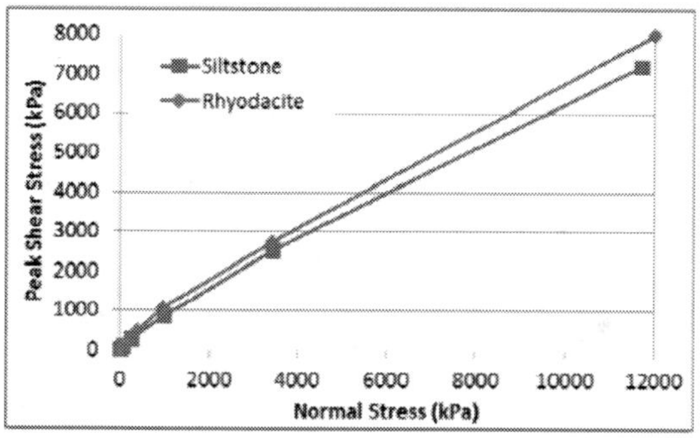

Fig. 6. Peak shear stress vs normal stress curves.

4 Discussion and Analysis

The "threshold" stresses for each material can be estimated by fitting two tangent lines to the obvious linear segments of the compression and compression + shear curves in Figure 4, taking their point of intersection as the threshold stress. For the siltstone material, the threshold stress is close to 250kPa for both compression and compression + shear, whilst for the rhyodacite it is around 2.3MPa for compression only and 1.3MPa for compression + shear. These values represent 0.7, 0.8 and 1.4% of the UCS of the corresponding parent rock materials, respectively. The values of 0.7 and 0.8 for the threshold stress ratio under compression are consistent, and although they fall far short of being definitive values, they give some insight of the relationship between particle material strength and the role of particle breakage in highly stressed rockfill materials.

The threshold stress is a prominent feature of the results involving sample compression, but it is not reflected in the derived results for the fines produced during shearing only. On this basis it is interpreted as representing that stress under which the response to compression transitions from particle rearrangement with chipping and crushing of particle points and edges, to full particle breakage, with crushing and splitting. It is also evident that the transition between particle breakage modes is gradual, with the threshold stress being an idealisation of a more transitional phenomenon.

The phenomenon of a "critical stress", where the percentage of fines produced during shearing reaches a peak value, was not anticipated, but is readily explained. In order to achieve a shearing mechanism under high stresses, particles must be damaged, as they cannot be rearranged after being densely compacted during heavy compression. However, with increasing vertical stress, the amount of particle damage during compression (application of the normal stress) becomes so great that it reaches an extent where a shearing mechanism can be achieved without having to break an increasing number of particles.

The critical stress was only clearly reached for the siltstone material for the stresses applied, and its value is somewhere between 1 MPa and 3.5 MPa; probably around 2 MPa. This corresponds to around 6% of the siltstone UCS. The data in Figure 4 suggests that the critical stress for the rhyodacite is in excess of 12 MPa and that the percentage fines produced during shear at the critical stress will be greater for the stronger rhyodacite than for the weaker siltstone.

The curvature in the relationship between peak shear strength and normal stress (Fig 3) is consistent with that observed by many previous authors [3],[4], and its interpretation is consistent with the generally accepted view that the shear strength under low stress is higher due to the phenomenon of dilation. The greater curvature in the rhyodacite results is entirely consistent with the persistence of dilation in the rhyodacite to significantly higher stresses (Fig 5). The results are consistent with [7] who found for crushed rocks, the shear strength parameters decreased consistently up to normal stresses of 1MPa for strong materials.

In evaluating these results, it should be appreciated that the measure of particle breakage adopted here is a relatively simple one: creation of fragments smaller than 4.75mm. The full results (not presented here) indicate that there was corresponding degradation in the larger particles of all sizes, with the proportion of the largest particles decreasing steadily with increasing log normal stress. Additional insights can be gained from a consideration of the full set of results, though space does not permit this here.

5 Conclusions

As expected, the particle breakage of crushed rock materials increases with increasing applied stress, in a manner which reflects the strength of the parent rock material. Under compressive loading, a threshold normal stress is recognised where the rate of particle breakage suddenly increases rapidly, corresponding to a transition from particle sliding and chipping, to full particle crushing and splitting. The threshold stress appears to correspond to around 0.7% of the material UCS.

With increasing normal stress, the particle breakage which occurs due to application of the normal stress continues to increase steadily, whereas the breakage due to subsequent shearing seems to reach a peak amount (at some critical stress) before actually decreasing for even higher stresses. This behaviour is attributed to the role of the compressive stress in achieving so much breakage under high stresses, that a shearing mechanism can then be achieved which requires less additional breakage than would be required under lower stresses. For the weaker siltstone material, the critical stress was found to be around 6% of the UCS.

The authors would like to acknowledge the excellent work of Mr Robert Dun in obtaining the experimental results, and the funding support of Australian Coal Research Limited.

References

1. Ovalle, C., Dano, C., Hicher, P.-Y., Cisternas, M. (2015). Experimental framework for evaluating the mechanical behavior of dry and wet crushable granular materials based on the particle breakage ratio. Canadian Geotechnical Journal 52, No. 5, 587-598.
2. Marsal, R. (1973). Mechanical properties of rockfill dams in Embankment dam engineering: Casagrande Volume., eds. Hirschfeld, R.C., Poulos, S.J. Wiley, New York, pp 109-200.
3. Leps, T.M. (1970). Review of shearing strength of Rockfill. J. Soil Mech. Found. Div. ASCE 96, No. 4, 246 1159-1170
4. Indraratna, B., Wijewardena, L., Balasubramaniam, A. (1993). Large-scale triaxial testing of greywacke rockfill. Géotechnique 43, No. 1, 539-543
5. Ovalle, C., Dano, C., Hicher, P.-Y. (2013). Experimental data highlighting the role of surface fracture energy in quasi-static confined comminution. International Journal of Fracture, Vol. 182, No. 1, 123-130.
6. Weibull, W. (1939). A statistical theory of the strength of materials. Proceedings of Royal Swedish Inst. Eng. Res. 151
7. Anagnosti, P. and Popovic, M. (1982) Evaluation of shear strength for coarse-grained granular materials. *Fourteenth Congress on Large Dams*, Rio de Janeiro, ICOLD. Vol. Q55 pp. 753-767.

Stress-Strain diagrams for non-convex particles

Hans-Georg Matuttis[1,*], *Masaki* Nawa[1], and *Dominik* Krengel[1]

[1] *Department of Mechanical Engineering and Intelligent Systems, Graduate School of the University of Electro-Communications, 1-5-1 Chofugaoka, Chofu, 182-8585 Tokyo,Japan*

Abstract. While most granular materials in nature and technology consist of non-convex particles, the majority of discrete element (DEM) codes are still only able to cope with convex particles, due to the complexity of the computational geometry and the occurrence of multiple contacts. We have reengineered a code for convex polygonal particles to model non-convex particles as rigidly connected clusters. Constricting non-convex particles along the symmetry axes by 30% leads to an increase of the materials strength of up to 50%.

1 Introduction

Most discrete element methods use convex particles: Tangential forces are taken up only by (dry, Coulomb-) friction forces. Technically, in many applications like railway ballast, non-convex grains are preferred, as the higher strength is well known. For computer simulations, the controlled modeling of non-convex shapes is tedious. For clusters of round particles [1, 2] or ellipsoids[3], the continuous variation of the non-convexity and the influence of interlocking is difficult to quantify. Rigid body simulations [4] correspond to the limit of vanishing strain, which for strength tests (like the bi-axial compression we want to investigate here) are as unsuitable as softly connected particles[5]. We have reengineered a simulation of "soft" convex polygons in rigid non-convex clusters, overlap algorithms and force laws for non-convex are unfeasible.

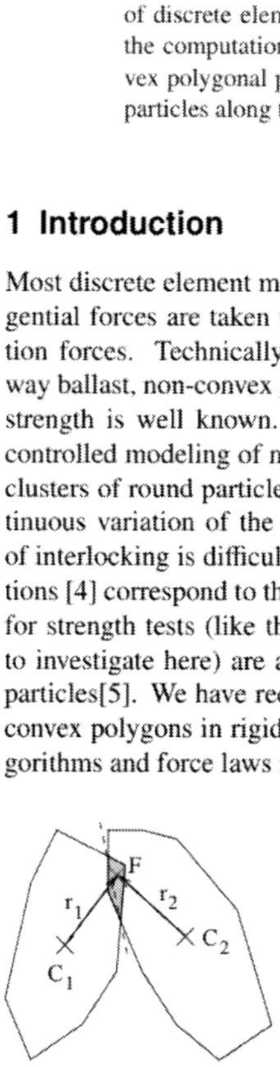

Figure 1. Overlap computation with centers of mass C_1, C_2 (black ×), force point F (center of the overlap polygon, gray ×) and contact vectors r_1, r_2. The dashed line indicates the tangential contact direction, given by the intersection points of the outlines.

2 Force computation

(Justifications and more details about for the interaction law beyond the short summary here can be found in Ref. [6].) For convex polygons, the elastic force is proportional to the overlap area A and the Young's modulus (in units of [N/m] for two dimensions). The "characteristic length" is

$$l = 4\frac{r_1 r_2}{r_1 + r_2}, \tag{1}$$

for the contact vectors r_1, r_2 (see Fig. 1) between the centers of mass and the contact point. This definition fixes the propagation speed of (linear) deformations perpendicular to the sides of rectangles in space-filling packings to the sound velocity $c = \sqrt{Y/\rho}$. Then the elastic force is

$$F_{el} = \frac{YA}{l}, \tag{2}$$

the normal dissipation in analogy with the linear oscillator

$$F_{diss} = \gamma\sqrt{mY}\frac{\dot{A}}{l}, \tag{3}$$

with the reduced mass m of the contacting particles with a damping constant γ. In contrast to the linear oscillator, in the DEM the dissipative force may not overcompensate and reverse the elastic force, which introduces unphysical attraction and noise. Therefore, a cutoff is needed so that the normal force cannot become attractive. Finally, a static (Coulomb-) friction force with friction constant μ is necessary to maintain the basic properties of physical grains (particles sticking on a slope $\tan\alpha < \mu$, forming stable heaps also on flat surfaces). As friction law, we use the Cundall-Strack-model [7] for Coulomb friction where the tangential friction force between two particles is incremented proportional to their relative sliding distance, and a cutoff μN is introduced proportional to the normal force $N = F_{el} + F_{diss}$ (including the truncation mentioned

Table 1. Parameters for the computation of the interaction for 1m-long rod like particles.

Young's modulus	10^7[N/m]
Density of particle	2640[kg/m^2]
Density of wall particle	5000[kg/m^2]
μ (particle-particle, particle-wall)	0.6
Damping constant γ	0.3

*e-mail: hg@mce.uec.ac.jp

 DOI: 10.1051/epjconf/201714006005

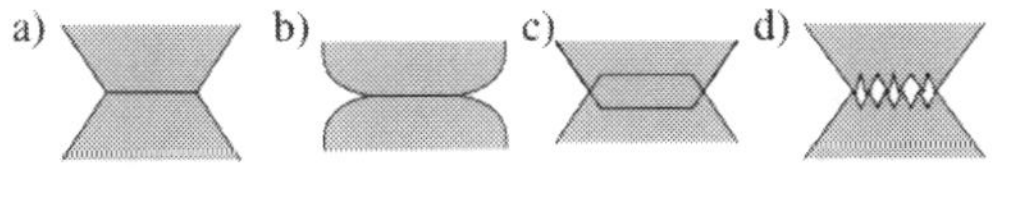

Figure 2. Rigid contact in a), idealized deformed contact in b), DEM overlap in c) and actual contact with micro asperities in d).

above). For polygons, Coulomb friction is much more effective than for round particles which can escape sliding friction by rolling. The parameters are given in Tab. 1. For the theoretical contact time of a square particle

$$T = \pi \sqrt{m_{\min}/Y}$$

with the smallest mass $m_{\min}$ and Young's modulus Y, our simulation is stable with integration time-steps below $1/10T$. Y in DEM-simulations should not be understood as that of the bulk Y_{bulk}, but as the effective elasticity of the actual contact in the presence of micro-roughness as in Fig. 2 d): The actual contacts are much "softer" due the thinned out material density at the interstices, so the use of $Y < Y_{\text{bulk}}$ is appropriate.

For elastically connected monomers, one is faced with a choice between two evils: Either one has to use very soft connections to preserve the time-step for the monomers [5]. or uses more rigid connections with much smaller integration time-steps. Instead, we implement rigid clusters where the the equations of motion (including rotation) act on the center of mass of the (rigid) cluster, while the interaction laws (forces, torques) between monomers of different clusters are the ones above for convex polygons, as layed out in Ref. [6]. Additionally, only a data structure is needed which assigns monomers to clusters and vice versa. The definition of the characteristic length (see Fig.3) is dealt with in Sec. 4. Different from clusters of round particles [1], polygonal monomers of the same particle (see Fig. 4) don't overlap, so there is no local variation of the Young's modulus due to multiple overlaps, with unforseeable consequences for the stability.

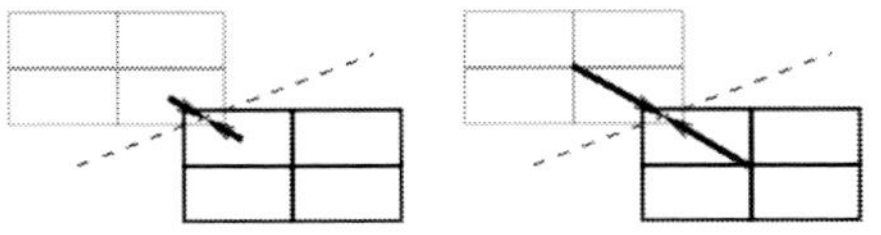

Figure 3. Possible definitions of contact vectors and characteristic lengths, from the center of mass of the monomers (left) and the centers of mass from the clusters (right).

3 Simulation Geometry

For biaxial compression (see Fig. 5) we have implemented an approximately quadratic cell with 1400 particles. In experimental three-dimensional triaxial compression, a rubber membrane under water-pressure forms the side-walls and the stress is measured on the floor. Instead, we use straight walls with a force equilibration and measure the

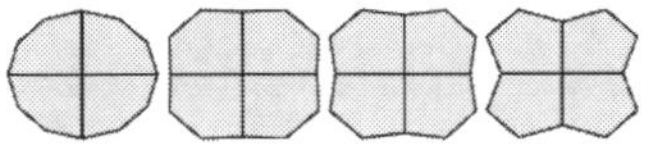

Figure 4. Composite particles with various degrees of non-convexity, obtained by constricting the corresponding convex particle with the same elongation by 0%, 15 %, 20 % and 30 %, abbreviated in the following as c-0%, c-15% ....

stress σ_1 on the ceiling. (This may lead so deviations from Ref. [8] where the measurement is on the bottom. The pressure on the floor is not equal to the lid pressure plus the weight of all particles combined because arching with the frictional contacts on the left and right wall deflects a part of the weight.) The bottom is fixed, while the lid moves down at constant velocity (0.01[m/s]) and the side walls are held at constant pressure σ_3 = 10000[N/m]. (For artifacts resulting from more rectangular geometries or one side wall fixed, see [8]). The granular particles move according to Newton's equation of motion (2nd order Gear Predictor corrector as time integrator) while the positions x of the side-walls are determined from the external pressure p and the sum of the forces on the side-walls $\sum F$ as

$$dx = \frac{dt^2}{\mathrm{M}}\left(p \cdot l_y + \sum F\right), \quad (4)$$

$$x = x + dx, \quad (5)$$

with $M = 0.002 \times m_{\text{total}}$ to mimic the adaption of a rubber membrane to the whole system. (In Ref. [8], we had used $\mathrm{M} = 0.02 \times \mathrm{m}_{\text{total}}$, the differences are insignificant.) Trying position the side-walls at the equilibrium position with a zero-order approximation of Newton's equations of motion [9] fails as for a constantly deforming cell, there is no equilibrium position, but an (on average) constant drift.

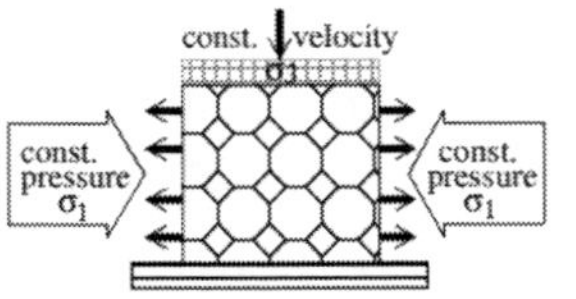

Figure 5. Compression cell for biaxial compression.

4 Convex particles

For a base radius of r = 0.0055 [m], we set up regular dodecagons with elongations ϵ = 1.0, 1.4, 1.8. and a size distribution for the half axes

$$a = r \cdot (1 + 0.5(\mathtt{rand}_a - 0.5)) \cdot \sqrt{\epsilon} \quad (6)$$

$$b = r \cdot (1 + 0.5(\mathtt{rand}_b - 0.5)) / \sqrt{\epsilon} \quad (7)$$

with different, equally distributed random numbers $\mathtt{rand}_{...}$. The resulting size distributions (see Fig. 6 for non-elongated particles, the ones for elongated ones were similar) were sufficiently wide to prevent ordering in lattice-like structures, which could decrease the material strength considerably due to slipping along the "crystal axes" of the particle packing. Throughout the paper, the same sequence of random numbers is used, so initially, the particle areas are the same, only the elongations and the degree

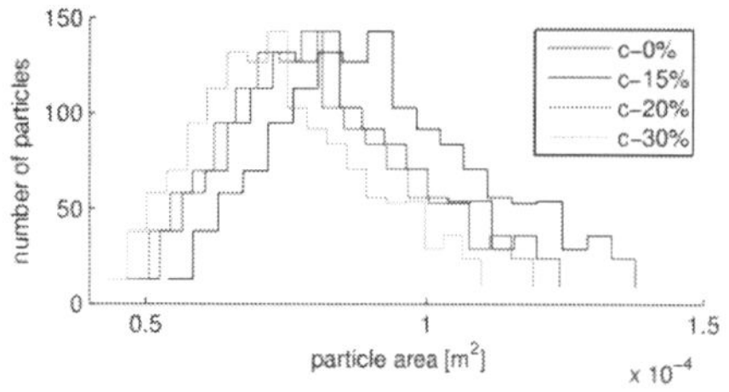

Figure 6. Size (area) distribution of non-elongated convex particles obtained from eq. (6,7).

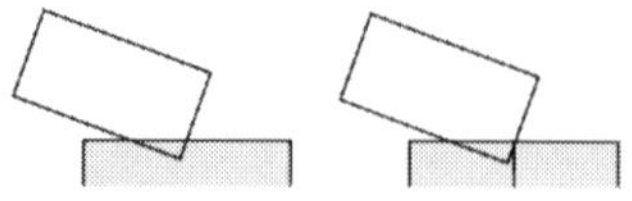

Figure 8. Non-composite convex particles experience tangential resistance to sliding (left) only via Coulomb friction, while composite particles (right) may interlock with a normal component.

of convexity are different. For increasing elongation, typical stress-strain curves and the behavior of the volume (inverse proportional to the packing density) are shown in Fig. 7. For all systems, the initial positions and relative particle sizes were generated with the same random number seeds. In this and the following graphics, the value for every thousand time-step is shown (the difference to a plot with every time-step was insignificant). The curves are aligned with respect to the linear regime. The peak of the stress-strain diagram for the dodecagons will usually not appear for circles. In the upper part of Fig. 7, one clearly sees the typical patterns of stress-strain curves for dense granular materials, the linear regime with constant gradient, the peak strength and the failure regime. In the lower part, we have shown the volume (area) occupied by the cell: The peak strength in the upper part corresponds to the minimal volume.

One issue for the use of composite particles is the definition of the characteristic length: It is either possible to define it with respect to the center of mass of the contacting monomers or with respect to the center of mass of the whole particle, see Fig. 3. We will discuss both possibilities and the relation of the results in this paper. The modellization is stable if penetrating contacts are treated as outlined in Ref. [6]. When the approach it generalized for polyhedra in three dimensions, accordingly the most demanding task is the appropriate treatment of penetrating contacts. Our simulation advantages over that for clusters of round particles: Clusters for the particles and simple "monomer" polyhedra for smooth walls allows an implementation with the minimum amount of particles, without the remaining doubt whether the walls are smooth enough. When we go from simple to composite convex particles, with the same outline, there is the question whether the properties of the aggregate will change due to the internal "seams" in the particles as shown in Fig. 8. As one can see in Fig. 9, for composite particles, the corresponding stress increases slightly, and the stability (less fluctuations, delayed onset of the "breakdown") is slightly improved. (On a side-note: This means that a surface uneveness in spherical beads, e.g. scratches on glass beads or flashes on plastic beads, may increase the mechanical stability in comparison to simulations of spheres with ideal surfaces.) The definition of the inverse characteristic length with respect to the monomers of half of the size of the original particles (blue line) corresponds to an effective doubling of the Young's modulus (green line).

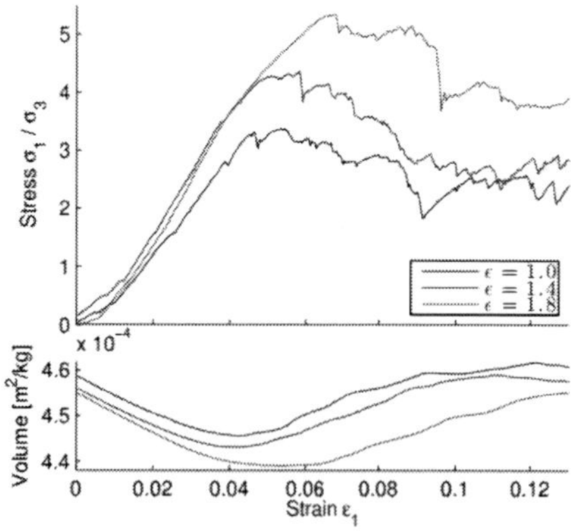

Figure 7. Stress-strain curves (measured at the lid) (above) and variation of the volume (below) for increasing particle elongation of a system with smooth size dispersion.

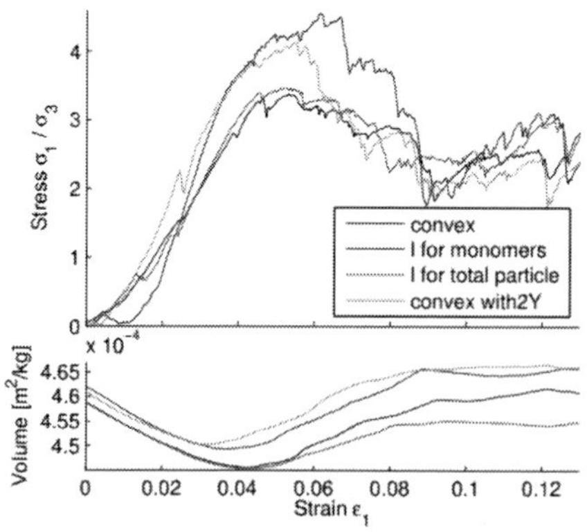

Figure 9. Stress strain diagram for non-composite dodecagons with the base Young's modulus Y (red) as well as $2Y$ (green) and for composite polygons of the same shape with two different definitions of the inverse characteristic length.

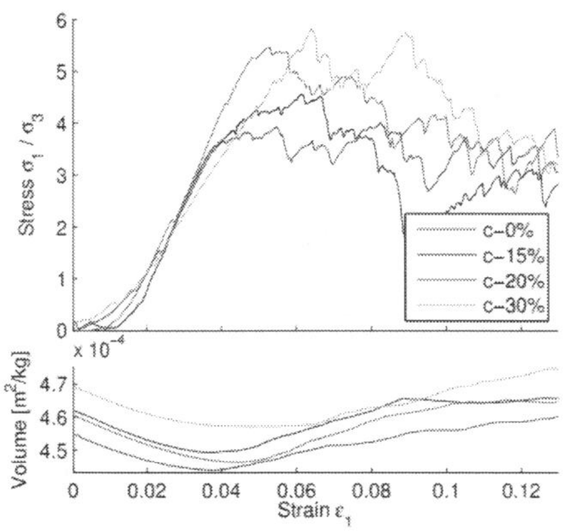

Figure 10. Stress strain diagram (above) and volume behavior (below) for non-convex non-elongated (elongation 1.0) particles.

5 Non-convex particles

With the non-convex shapes from Fig. 4 and respective elongations of 1, 1.4 and 1.8, we obtain the stress-strain diagrams in Figs. 10, 11 and 12. The material strength (gradient of the linear regime, position of the maximal obtainable stress) increase with particle elongation and non-convexity, as do the fluctuations in the data. As in the case of convex particles, the same seeds are used for the initialization, so that the relative volume (area) of the particles is the same throughout all simulations. For an increase of the non-convexity for the same elongation, the material strength increases up to 50The effect comes from the tangential forces which for non-convex particles are picked up and deflected also by the interlocking surface geometries via the elastic normal forces. This mechanism is more stable against load, slip and vibrations than the Coulomb friction force between smooth surfaces alone. Nevertheless, the Young's modulus must be sufficiently large so that interlocking asperities are not "pulled through" each other. For the external pressure uses, a Young's modulus of 1/4th would have annihilated effect. So for non-convex particles, a prudent choice of the Young's modulus is more important than for the convex case.

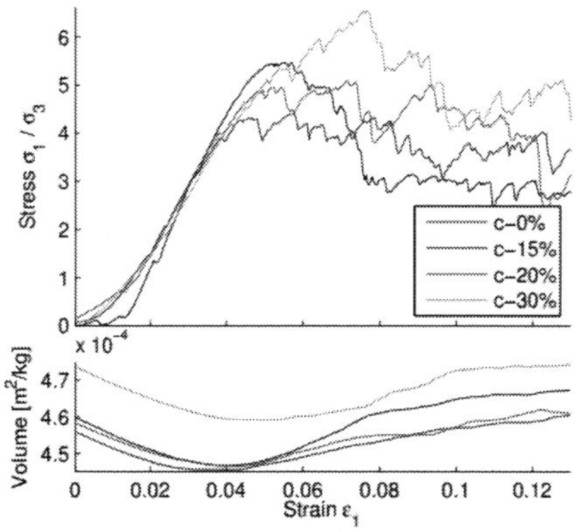

Figure 11. Stress strain diagram (above) and volume behavior (below) for non-convex particles with elongation 1.4.

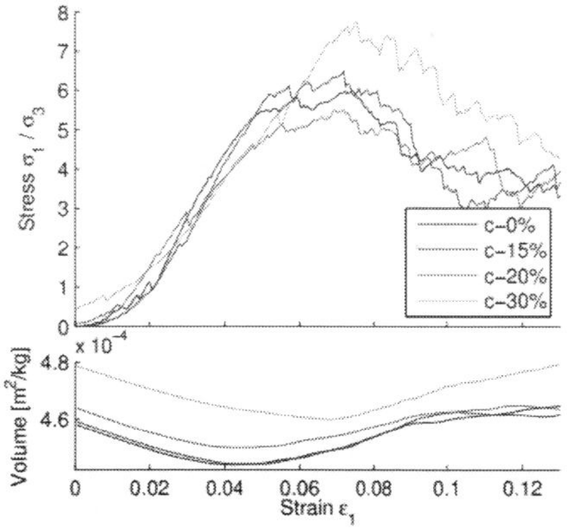

Figure 12. Stress strain diagram (above) and volume behavior (below) for non-convex particles with elongation 1.8.

6 Summary and conclusions

We have successfully adapted a simulation for convex polygonal particles to non-convex clusters of such particles. For polydisperse dodekagons inscribed in ellipses, a variation of the elongation from 1 to 1.8 increases the material strength by a factor of ≈ 1.7. Constricting the particle shape along the symmetry axes by up to 30% increases the material strength by a factor of ≈ 1.5, so a total variation of the strength of up to a factor of ≈ 2.5 is possible.

References

[1] T. Matsushima, C.S. Chang, Granular Matter **13**, 269 (2011)

[2] D. Negrut, A. Tasora, H. Mazhar, T. Heyn, P. Hahn, Multibody System Dynamics, **27**, 95 (2012)

[3] J. F. Peters, M. A. Hopkins, R. Kala, R. E. Wahl, Engineering Computations, **26**, 645 (2009)

[4] K.-W. Lim, K. Krabbenhoft, J. E. Andrade, José E., Computational Particle Mechanics **1**, 257 (2014)

[5] T. Pöschel, V. Buchholtz, Physics Review Letters, **71**, 3963 (1993)

[6] H.G. Matuttis, J. Chen, *Understanding the discrete Element Method*, 1st edn. (Wiley, 2014)

[7] P.A. Cundall, O.D.L. Strack, Géotechnique **29**, 47 (1979)

[8] S.A.M.E. Shourbagy, S. Morita, H.G. Matuttis, Journal of the Physical Society of Japan **75**, 104602 (2006)

[9] S.H. Ng, H.G. Matuttis, Theoretical and Applied Mechanics Japan **59**, 323 (2011)

Phosphonated superplasticizers synthesis: adsorption, dispersion and fluidification ability on calcium carbonate suspensions.

Tramaux A. [1], Azéma N.[1,*], David G.[2], Negrell C.[2], Poulesquen A.[3], Haas J.[3], Remond S.[4]

[1] *Centre des Matériaux de l'Ecole des Mines d'Ales, France.*
[2] *Laboratoire Ingénierie et Architecture Macromoléculaire, Institut Charles Gerhardt de Montpellier.*
[3] *Laboratoire de Physicochimie des Matériaux Cimentaires, Commissariat à l'Energie Atomique de Marcoule, Bagnols-sur-Cèze.*
[4] *Mines Douai, LGCgE – GCE, F-59508 Douai, France.*

Abstract. Superplasticizers are chemical admixtures used to fluidize cement pastes. Among them, polycarboxylate are comb-like copolymers with carboxylic acid and poly(ethylene glycol) side chains. We present here the synthesis of various polycarboxylate-like copolymers including phosphonic acid functions. Their adsorption, dispersion and fluidification efficiencies are evaluated on calcium carbonate suspensions, a mineral material commonly used to simulate early-age cement behaviour.

1. Introduction

Improving concrete fluidity is a matter of importance for concrete placing. More fluid the material is, easier the filling of the formwork. Although this high fluidity can be reached by increasing the water to cement ratio (W/C), an excess of water will lead to the weakening of long term mechanical properties. Moreover, due to drying shrinkage, cracks in material can occur. In order to avoid such inconvenient, concrete industry has developed admixtures named "superplasticizers"[1]. These admixtures can lower the viscosity of concrete without any change of the W/C ratio[2]. Since their invention in 1930, superplasticizers have become ones of the most used admixtures in the concrete industry and their use is almost systematic in today's concrete formulation. Using specific mixture proportions, it is even possible to prepare "self compacting concrete", that can be placed without any vibrations, under their own weight.

A major innovation in the superplasticizer field was in 1980 the development of polycarboxylate[3]. These admixtures were comb-like copolymers, petro-based, composed by a succession of methacrylic acid units in the main chain, and a substitution of a part of these methacrylic acid units by poly(ethylene glycol) methyl ether methacrylate, forming side chains from the main chain. Much more efficient than previous kinds of superplasticizers, polycarboxylate are still today the most used and also the most studied. Indeed, by changing structural parameters (DPn, molecular weight, side-chains ratio or length) it is possible to design various polycarboxylates, more or less efficient. Because of the wide variety of cement Portland compositions, the concrete industry has reported many issues of incompatibility between cement and admixtures, leading to a loss of efficiency of superplasticizers or other undesired effects[4]. One of the strongest interests of polycarboxylate is that their important structure diversity is useful to assure a good compatibility with cement compositions, even in presence of mineral additions (fly ash, slag, silica fume).

Origin of the fluidization action of polycarboxylate is decomposed in two steps: firstly, adsorption of the comb-like copolymer onto surface of the cement grain. Carboxylic acids are anionic functions fully deprotonated in alkali environment which have high affinity for positively charged surfaces, such as ettringite phases for example. Second step is the formation of a polymeric crown around particles, which generates steric hindrance and acts against the van der Walls attractions. Particles are thus dispersed, water trapped in agglomerates is released and very fine particles are generated. From a macroscopic point of view, the cement paste becomes more fluid and easier to place.

Even if polycarboxylate are widely used in building industry, various studies try to enhance their performances with structural modifications of macromolecular architecture, including the nature of anionic function. Some works tried to replace the carboxylic acid by phosphonic acid, with good results[5]. One of the interests for improving the affinity of the "anchor group" for mineral surfaces is the ionic competition resistance. A high ion concentration can drastically decrease superplasticizer adsorption efficiency and thus its fluidification effect. This occurs often in high sulfate content cement paste.

In this study we present the synthesis of various comb-like copolymers, whose macromolecular structures are similar to standard polycarboxylate, but with phosphonic acid functions instead of carboxylic acid moieties. Adsorption, dispersion and fluidification efficiencies of these superplasticizers are evaluated on calcite suspensions, a model inert material widely used to simulate early-age cement behaviour. Finally, effect of

*Corresponding author: nathalie.azema@mines-ales.fr

sulfate concentration is also investigated through addition of sodium sulfate.

2. Materials and methods

2.1. Calcite

Model mineral material chosen, calcite, is the most stable allotropes of calcium carbonate. It has been purchased from Omya company (ref: OMYA BL). This low soluble white powder is composed of monomorphic angular particles from 0.3 to 40 μm diameter.

2.2. Calcite suspensions preparation

Initial suspensions were prepared mixing calcite, superplasticizer aqueous solution and the solution of interest (deionized water, limewater or limewater with sulfate ions) keeping constant the water/solid ratio (W/S=0.5). The superplasticizer dosage corresponds to the mass of solid superplasticizer inserted over the mass of calcite. After at least 10 minutes of magnetic stirring, suspensions are diluted according to measurement requirements.

2.3. Superplasticizers preparation

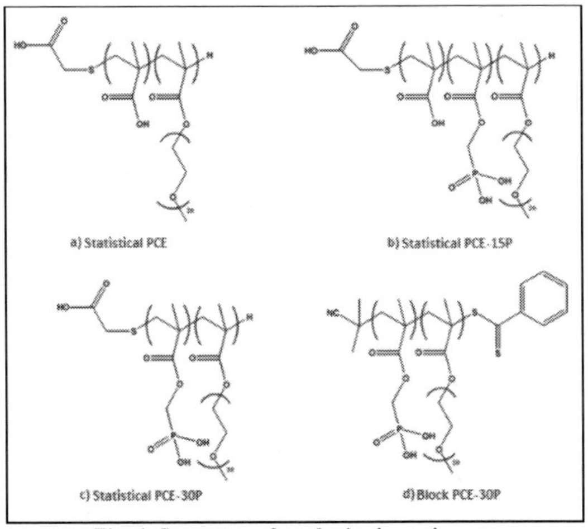

Fig. 1. Structures of synthetized copolymers

Four comb-like copolymers have been prepared according to two synthesis routes: three of them (a, b and c) are random copolymers and have been synthetized through conventional radical copolymerisation, using thioglycolic acid as chain transfer agent. The last one (d) is a block copolymer prepared by RAFT (Reversible Addition-Fragmentation Chain Transfer) copolymerisation, with CIDB as RAFT agent. These copolymers have been synthetized with the idea to keep constant macromolecular parameters such as DPn, side chains length and PEG side chains/anionic functions ratio. Once synthetized, these copolymers are conditioned in 100 g/L aqueous solutions.

2.4. Adsorption analysis

Adsorption analyses were performed thanks to a Vario TOC cube analyser, from Elementar Company. After centrifugation of initial suspension, supernatant is filtered, diluted and then analysed. The polymer concentration is determined from integration of the area under the curve. Finally, thanks to the difference between amount of copolymer added and measured, the amount of copolymer adsorbed is expressed in mg per gram of calcite.

2.5. Rheological measurements

From a concentrated calcite suspension (W/S=0.36) admixtured with 0.05 % superplasticizer, rheological measurements were performed with an AR2000 Ex rheometer from TA instruments Company. The geometry employed was rotating vane. After a short pre-shear, suspension is sheared at $100s^{-1}$ during 10 minutes. η_{10} is the average viscosity of suspensions during these 10 minutes of analysis.

2.6. Suspension stability evaluation

Suspensions stability is investigated with a concentrated suspension light analyser, Turbiscan MA2000 from Formulaction Company. A 3 factor dilution with the liquid of interest (W/S=3.5) is performed from initial calcite suspension before being inserted in a glass tube. Turbiscan analyses light transmitted and backscattered as a function of column height and time. Analyses last 30 minutes, with a scan every minute.

3. Results and discussion

3.1. Macromolecular architectures description

Statistical PCE is a comb-like copolymer whose structure corresponds to a standard polycarboxylate. Statistical PCE-30P has almost the same structure, except that all carboxylic acids are replaced by phosphonic acids. Statistical PCE-15P is intermediate, with half of anionic functions which are carboxylic acids and half are phosphonic acids. Block PCE-30P has a structure similar to statistical PCE-30P, but it is a block copolymer, expected to behave differently from the others. To our knowledge, it is the first time that a block copolymer is synthetized for superplasticizer applications[5].

Table 1. Copolymers structural parameters. *Mn and DPn are theoretical, according to reactant quantities employed.

Superplasticizer	Methacrylic acid ratio	Phosphonic acid ratio	PEG side chain ratio	PEG side chain length	Mn*	DPn*
Statistical PCE	30 %	-	70 %	20	≈ 38500	57
Statistical PCE-15P	15 %	15 %	70 %	20	≈ 35000	50
Statistical PCE-30P	-	30 %	70 %	20	≈ 37000	52
Block PCE-30P	-	30 %	70 %	20	≈ 38500	52

3.2. Superplasticizers adsorption capacity

Suspensions were prepared with 0.5 % superplasticizer dosage. Fig.2 presents adsorption capacity of the four superplasticizers, in mg of copolymer adsorbed per grams of calcite.

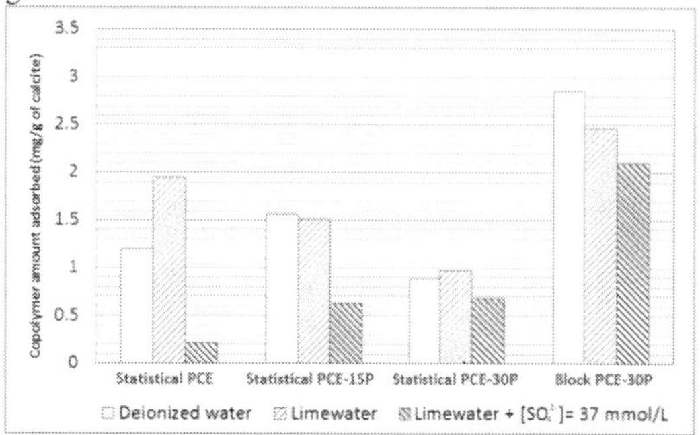

Fig. 2. Amount of copolymer adsorbed onto calcite in three different environments.

Using limewater is useful to get a pH of 12.8, close to cement pore solution pH. We can notice that few differences are visible between water and limewater. Only statistical PCE seems to adsorb in higher amount in limewater. This could be explained by the full deprotonation of COOH moieties in alkali environment, increasing affinity of the macromolecule for calcite. In all liquids studied, block copolymer provides much greater adsorption ability than statistical ones probably due to its different macromolecular structure close to surfactant ones. On the other hand, in water and in limewater adsorption ability of statistical copolymers seems to be less effective when carboxylic acid groups are substituted by phosphonic acid but not in presence of sulfate ions where the trend is reversed, due to the less sensitivity of phosphonic acid to ionic competition[6]. The more the macromolecule includes phosphonic acid, the less its adsorption is affected by sulfate ions. Moreover, even if the block PCE-30P seems to adsorb in higher proportions than others, we propose that not necessarily reveal higher affinity for mineral surfaces but may be due to another mode of adsorption of the "anionic block".

3.3. Rheology measurements

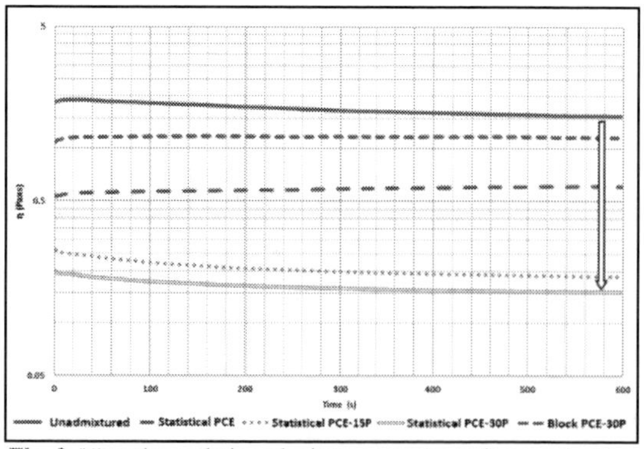

Fig. 3. Viscosity evolution of calcite suspensions admixtured with 0.05% of superplasticizer. Acquisitions in deionized water.

Fig.3 displays the viscosity of calcite suspension in deionized water when admixtured with 0.05 % of superplasticizer. As a general trend, addition of superplasticizer decreases viscosity as substitution of carboxylic acid by phosphonic acid group, dividing the viscosity by a factor 10 for statistical PCE-30P. Despite the high adsorption ability of block PCE-30P, its adsorption in $CaCO_3$ suspensions does not lead to a significant decrease of viscosity, as statistical copolymers.

Table 2 summarizes viscosities of suspensions in different solutions.

Table 2. Viscosities averaged over 600 seconds of calcite suspensions in three different environments.

Copolymer	η_{10} (Pa×s) Deionized water	Limewater	Limewater + [SO_4^{2-}] =37mmol/L
-	1.68	1.52	2.54
Stat. PCE	0.58	0.35	2.51
Stat. PCE-15P	0.20	0.16	1.45
Stat. PCE-30P	0.16	0.13	0.35
Block PCE-30P	1.17	0.85	1.12

Same observations are made in all solutions studied, deionized water, limewater and limewater with sulfate ions. For water and limewater solutions, results seem to indicate that non adsorbed copolymer could have a role in particles dispersion[7] because viscosity decreases as well adsorption ability decreases. In presence of sulfate, all superplasticizers fluidizing efficiencies are lowered in agreement with lowest values of adsorption amounts. In particular statistical PCE, which doesn't contain phosphonic acid, undergoes drastic decrease of its efficiency. Its viscosity is almost the same than unadmixtured suspension.

3.4. Calcite suspension stability

Turbiscan MA 2000 allows the monitoring of transmitted and backscattered light intensity of calcite suspensions as a function of time. From the backscattered profile, we have developed a suspension stability index, named "A", taking into account all destabilizations phenomena. A is the "Area difference" of the two backscattered signal intensity curves, between t=6 and t=0 minutes. These two times have been chosen arbitrarily, to exhibit maximum differences between formulations, especially at intermediate dosages. When suspension is not stabilized, the signal evolves rapidly and the value of A is high. When suspension is stabilized, at high superplasticizer dosage, sedimentation is greatly slowed down, and the value of A is minimal. We consider that for a value below 50 area units, the suspension is stabilized in analysis time.

Fig.4. displays evolution of A according to amount of superplasticizer used. At high dosages, (>0.2%) all copolymers lead to a stable suspension, except block copolymer until 0.5 % for the same effect. Overall, block PCE-30P presents a lesser dispersion efficiency than others copolymers, in good agreement with rheological measurements. At low dosages (between 0.01 and 0.1%),

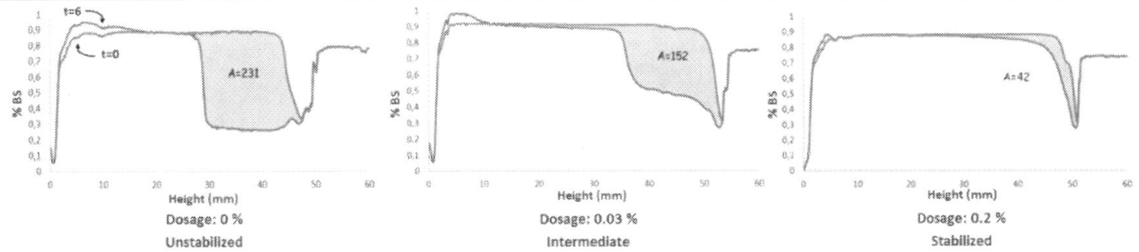

Fig. 5. Three examples of calculation of A with several dosages of statistical PCE in limewater. A is in area unit.

we can notice that the value of A is higher than in the case of unadmixtured suspension. This observation, may be probably explained by bridging phenomena[8]. Bridging is the multiple adsorption of one single macromolecule onto two or more particles at the same time when surfaces are not saturated. This phenomenon, well-known to occur at low polymer concentration, leads to formation of agglomerates, thus particles sedimentation.

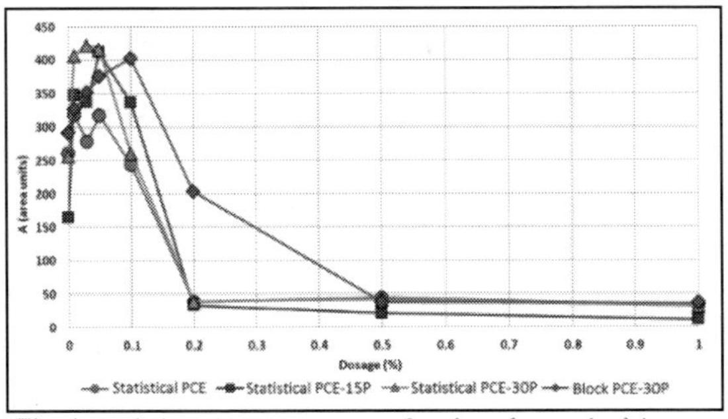

Fig. 4. Evolution of A parameter as a function of superplasticizer dosage. Acquisitions in deionized water.

Table 3 summarizes the amount of copolymer to use in order to get a value of A below 50 area units namely stabilized systems. As it was observed in adsorption analysis, few differences are noticeable between limewater and deionized water. Block copolymer have lowest dispersion ability in agreement with rheological results but an improvement seems to be observed in limewater. In presence of sulfate ions, statistical PCE-30P keeps its efficiency whereas statistical PCE is no more able to lower A, which remains higher than 50 area units even with 1% dosage. Statistical PCE-15P, whose structure is intermediate between PCE and PCE-30P, display intermediate dispersion efficiency. These last results confirm that in presence of sulfate ions, superplasticizers have lowest fluidizing efficiencies.

Table 3. Superplasticizer dosage required (%) to reach a value of A below 50 area units.

Copolymer	Deionized water	Limewater	Limewater + $[SO_4^{2-}]$ =37mmol/L
Stat. PCE	0.2	0.2	-
Stat. PCE-15P	0.2	0.2	0.5
Stat. PCE-30P	0.2	0.2	0.2
Block PCE-30P	0.5	0.1	0.2

4. Conclusion

Four comb-like copolymers have been synthetized with different macromolecular architectures but with same molar mass, side chain length and anionic functions ratio. Their adsorption, dispersion and fluidification efficiencies have been investigated on calcite suspensions prepared in three different solutions: deionized water, limewater and limewater with sulfate ions. In water and limewater, few differences were observed between statistical copolymers. Only block copolymer leads to better adsorption ability but has in return the lowest effect on fluidization and dispersion of calcite suspensions. Although statistical copolymers present a lower adsorption ability, their dispersion and fluidization efficiency are more important probably due to non-adsorbed copolymer role.
Ionic competition leads to a lowest adsorption, dispersion and fluidification efficiencies of superplasticizers in presence of sulfate ions, however an increase of phosphonic acid group rate in statistical macromolecular structures make superplasticizers less sensitive to ionic competition.

Acknowledgement

This work was financed by LabEx CheMISyst of Balard Chemistry Center of Montpellier and ARMINES association (Paris), with a financial support from CEA of Marcoule (France).

References

[1] F. Winnefeld, S. Becker, J. Pakusch, and T. Götz, "Effects of the molecular architecture of comb-shaped superplasticizers on their performance in cementitious systems," *Cem. Concr. Compos.*, vol. 29, pp. 251–262, 2007.
[2] D. Nagrockiene, I. Pundienė, and A. Kicaite, "The effect of cement type and plasticizer addition on concrete properties," *Constr. Build. Mater.*, vol. 45, pp. 324–331, 2013.
[3] T. Hirata, A cement dispersant, in: JP Patent 84,2022, 1981, S59-018338.
[4] S. Hanehara and K. Yamada, "Interaction between cement and chemical admixture from the point of cement hydration, absorption behaviour of admixture, and paste rheology," *Cem. Concr. Res.*, vol. 29, pp. 1159–1165, 1999.
[5] J. Plank, E. Sakai, C. W. Miao, C. Yu, and J. X. Hong, "Chemical admixtures — Chemistry, applications and their impact on concrete microstructure and durability," *Cem. Concr. Res.*, vol. 78, pp. 81–99, 2015.
[6] F. Dalas, S. Pourchet, A. Nonat, D. Rinaldi, S. Sabio, and M. Mosquet, "Fluidizing efficiency of comb-like superplasticizers: The effect of the anionic function, the side chain length and the grafting degree," *Cem. Concr. Res.*, vol. 71, pp. 115–123, 2015.
[7] A. Lange and J. Plank, "Contribution of non-adsorbing polymers to cement dispersion," *Cem. Concr. Res.*, vol. 79, pp. 131–136, Jan. 2016.
[8] F. Lafuma, K. Wong, and B. Cabane, "Bridging of colloidal particles through adsorbed polymers," *J. Colloid Interface Sci.*, vol. 143, no. 1, pp. 9–21, 1991.

Modeling of light dynamic cone penetration test – Panda 3® in granular material by using 3D Discrete element method

Quoc Anh Tran[1,2,⋆], *Bastien* Chevalier[1,⋆⋆], *Miguel* Benz[2], *Pierre* Breul[1], and *Roland* Gourvès[2].

[1] *Université Clermont Auvergne, CNRS, Institut Pascal, F-63000, Clermont-Ferrand, France.*
[2] *Sol-Solution, Géotechnique Réseaux, Riom, France.*

Abstract. The recent technological developments made on the light dynamic penetration test Panda 3® provide a dynamic load–penetration curve $\sigma_p - s_p$ for each impact. This curve is influenced by the mechanical and physical properties of the investigated granular media. In order to analyze and exploit the load-penetration curve, a numerical model of penetration test using 3D Discrete Element Method is proposed for reproducing tests in dynamic conditions in granular media. All parameters of impact used in this model have at first been calibrated by respecting mechanical and geometrical properties of the hammer and the rod. There is a good agreement between experimental results and the ones obtained from simulations in 2D or 3D. After creating a sample, we will simulate the Panda 3®. It is possible to measure directly the dynamic load–penetration curve occurring at the tip for each impact. Using the force and acceleration measured in the top part of the rod, it is possible to separate the incident and reflected waves and then calculate the tip's load-penetration curve. The load–penetration curve obtained is qualitatively similar with that obtained by experimental tests. In addition, the frequency analysis of the measured signals present also a good compliance with that measured in reality when the tip resistance is qualitatively similar.

1 Introduction

The light dynamic penetrometer with variable energy Panda is widely used to determine the mechanical properties of soils. In the beginning, this device was able to measure only the tip resistance (q_d) for each impact by using the Dutch formula with the kinetic energy measured by the strain gauges placed in the head of the apparatus [1]. In recent years, the latest version - Panda 3® has been developed by Benz [2]. For each impact, we will measure the force $F_A(t)$ and acceleration $A_A(t)$ at the measuring point - (A). Then we are able to rebuild the dynamic load–penetration curve $\sigma_p - s_p$ from these measurements by using the theory of decoupling the waves propagating in the rod [3]. The exploitation of this curve gives us not only the tip resistance (q_d) but also other mechanical parameters of the soil such as the wave velocity, the damping coefficient, the elastic modulus [2].

In order to validate the method applied to the Panda 3® and better understand the mechanism of penetration involved in this kind of test, there are many numerical studies based on Discrete Element Method (DEM) in two dimensions (2D) [2, 4–7].

In this paper, we will present a numerical model in three dimensions (3D) for reproducing Panda 3® test with PFC^{3D} from Itasca [8].

At first, the theory of decoupling waves propagating in the rod will be briefly presented in the section 2. We will talk in detail about the numerical model in the section 3. Then, the section 4 will show some results obtained with the numerical model in order to validate the wave analysis used for Panda 3®. Finally, the values of accelerance ($A/F(\omega)$) obtained by the numerical model and by experimental tests of Panda 3® with similar amplitude of tip resistance ($q_d \approx 2MPa$) are compared.

2 Theory of decoupling of waves propagating in the rod

In this section, we will presents briefly the method of wave separation applied on Panda 3®. For more details about wave theories applied on Panda 3®, the reader can refer to the works of Benz [2].

In order of separate ascending and descending waves in the rod, three assumptions are made: (*i*) the rod is linear elastic; (*ii*) the stress is uniform throughout all of cross-sections; (*ii*) the propagation velocity of wave in the rod does not depend on the frequency [9].

When the hammer of mass M hits the head of the penetrometer with the velocity v_m, a compression wave $u(x,t)$ will be generated and will propagate at constant speed c_{rod} towards the cone by the following equation:

$$\frac{\partial^2 u(x,t)}{\partial t^2} = \frac{1}{c_{rod}^2} \cdot \frac{\partial^2 u(x,t)}{\partial x^2} \quad (1)$$

⋆e-mail: qatran@sol-solution.com
⋆⋆e-mail: bastien.chevalier@uca.fr

Table 1. Micro-parameters used for the contact

Contact Type	$E^*[MPa]$	$\kappa^*[-]$	$\mu[-]$	$\beta_n[-]$
Hammer vs Anvil	305.04	2.7058	0.0	0.0
Rod balls vs Tip	206e3	2.7058	0.0	0.0

The general solution of the Eq.1 is represented by the the superposition of two elementary waves, u_{down} and u_{up}, respectively going down and up.

$$u(x,t)=u_{down}(\xi)+u_{up}(\eta) \quad (2)$$

where $\xi = t - x/c_{rod}^2$ and $\eta = t + x/c_{rod}^2$.

During the wave propagation through the rod, the wave $u(x,t)$ causes in each point x along the rod a deformation $\epsilon(x,t)$ and particle velocity $v(x,t)$ also represented by the superposition of two elementary waves:

$$\begin{aligned}\epsilon(x,t)&=\epsilon_{down}(\xi)+\epsilon_{up}(\eta)\\ v(x,t)&=v_{down}(\xi)+v_{up}(\eta)\end{aligned} \quad (3)$$

By measuring at the point of measure A, located close to the anvil, the deformation $\epsilon(x_A,t)$ and/or acceleration $a(x_A,t)$ for each impact, we are able to describe entirely the dynamic phenomenon in each point x along the rod [2]. We will present in the subsection 4.1 the validation of the theory of wave separation.

3 Numerical Model

We present here first the calibration of the impact in the numerical model, by respecting mechanical and geometrical properties of the hammer and the rod (subsec. 3.1). After creating a sample (subsec. 3.2), we will simulate the penetration test Panda 3® and present the results in the subsection 3.3.

3.1 Calibration of the impact

In order to respect mechanical and geometrical properties of the hammer and the rod [2], the anvil is modeled by a sphere of 3 *cm* in diameter with a mass equal to 0.75 *kg*. The rod is made by a set of contacting spheres of 1.4 *cm* diameter aligned vertically. The length of the rod L_{rod} is first set to 8 *m* in order to cancel the superposition of waves at the measurement point A. The standard hammer is modeled by a sphere of 6 *cm* in diameter with a mass of 1.69 *kg*.

A linear model was used for the contact between the hammer and the anvil. Contact bond was used to model the contact inbetween rod spheres and between rod spheres and the tip. The table 1 shows microparameters used for each type of contact where: E^* is the effective modulus, κ^* the stiffness ratio ($\kappa^* = k_n/k_s$) at the contact, μ the friction coefficient and β_n the normal critical damping ratio.

Figure 1 shows the evolution of impact force measured at the measurement point A versus time for different impact velocities V_I. The impact duration is order of about 2.0 *ms*. Note that impact sphere is deleted just after contact loss with the anvil.

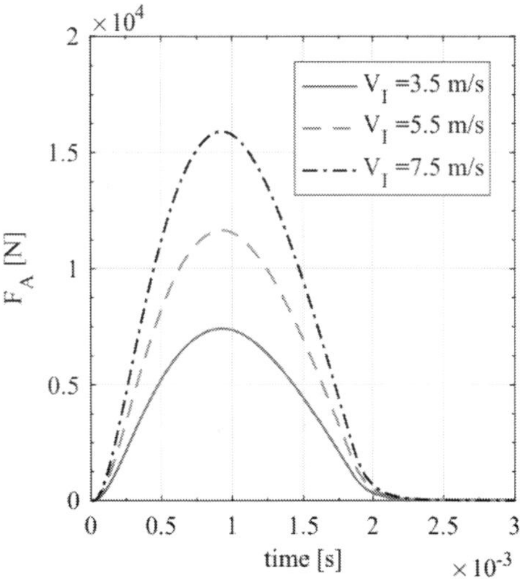

Figure 1. Force F_A at the measuring point A versus time for different impact velocities (V_I) with rod length $L_t = 8m$.

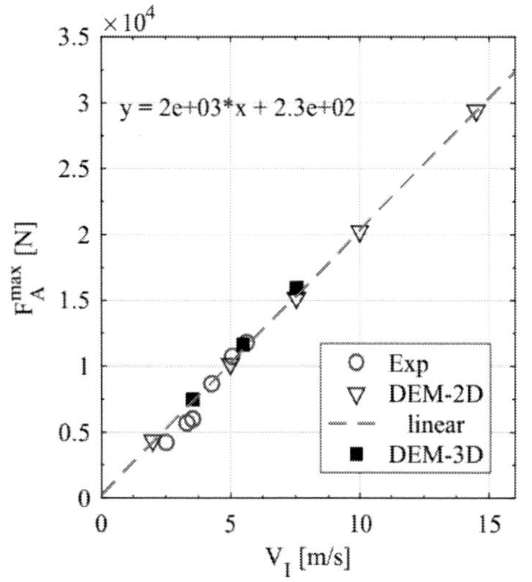

Figure 2. Maximal measured force F_A^{max} obtained by experimental test (○), simulation tests in 2D (▽) [2] and simulation tests in 3D (□).

Figure 2 presents the comparison of the maximal force F_A^{max} at the measurement point A obtained either by the model, the DEM-2D model proposed by Benz, and the experimental test with Panda 3® [2]. The similarity between the three curves is noticeable.

3.2 Sample creation

Once the mechanical and geometrical properties of the hammer are set, the rod and tip can be calibrated. We will

DOI: 10.1051/epjconf/201714016003

Figure 3. Example of dense assembly of $N_p = 389167$ particles ($D_p^{max} = 0.7cm$, $R_p^{dia} = 2$) in a cylindrical sample with $H_s = D_s = 40cm$.

Table 2. Micro-pararameters used for the contact involving particles

Contact Type	$E^*[MPa]$	$\kappa^*[-]$	$\mu[-]$	$\beta_n[-]$
Particle vs particle	200	3.3	0.7	0.10
Particle vs rod	200	3.3	0.0	0.00
Particle vs tip	200	3.3	0.3	0.00
Particle vs walls	2	3.3	0.0	0.00

create a sample of particle for simulating the penetration test Panda 3®.

Figure 3 shows an example of dense particle assembly with $N_p = 389167$ particles in a cylindrical sample with the height H_s and the diameter D_s equal to 40 cm. Particle diameters are distributed uniform in terms of diameters between maximal and minimal diameter presenting a ratio of 2. The maximal particle diameter D_p^{max} is equal to 0.7 cm. Linear contact model is used between particles. The minimal porosity measured at the centre of the sample is 0.36.

The rigidity of sample walls is smaller than the one of particles (Tab. 2) in order to reduce the influence of boundary condition. A vertical confining stress of 40 kPa is maintained constant during the simulation.

The table 2 shows the micro-pararameters used for the contact involving particles.

3.3 Simulation procedure

After creating a particle assembly, the rod and tip will be driven in the sample at first with constant velocity ($V = 1m/s$) until a depth of 15 cm. This penetration rate is much higher than that one used in the standard CPT (V = 2 ± 0.5 cm/s) in order to reduce the calculation time. In addition, we created a cylindrical walls around rod particles in order to avoid contacts between the rod spheres and the particles of the soil which would disturb the propagation of the wave during penetration [2, 4].

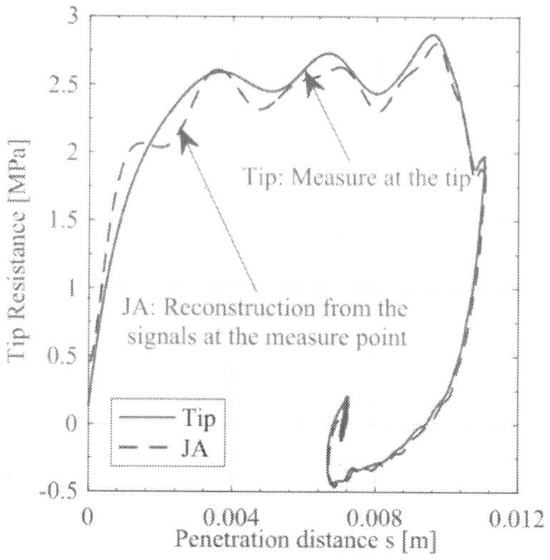

Figure 4. Validation of applied theories on Panda 3®: *Tip*: tip resistance measured directly at the tip; *JA*: tip resistance calculated from the force F_A and acceleration Acc_A at the measurement point *A*

Once the whole system is stabilized, we will create a sphere representing the hammer that will hit the anvil with a velocity of 3.5 m/s.

Some simulation results will be presented in next section for validating the signal processing proposed for the Panda 3® and then exploit the signals obtained in the frequency domain.

4 Simulated results

4.1 Validation of the theory of wave decoupling

Figure 4 shows the load–penetration curve $\sigma_p - s_p$ measured directly at the tip and calculated with the theory of wave decoupling presented before. The good conformity between the two curves demonstrates the relevance of the decoupling theory applied on Panda 3® test results.

4.2 Exploitation of measured signals

In order to determine the mechanic properties of granular materials, Escobar proposed to analyse in the frequency domain the measured signals obtaining by Panda 3® [5]. Despite it is difficult to compare between numerical and experimental results, we present here a qualitative comparison of the properties measured when the tip resistance is comparable ($q_d \approx 2MPa$).

In the frequency domain, the accelerance curve $A/F(\omega)$ is the ratio between the Fourier transform amplitude of accelaration - $|A(\omega)|$ and the one of force $|F(\omega)|$, at the measurement point A [5],

$$A/F(\omega)=\frac{|A(\omega)|}{|F(\omega)|} \quad (4)$$

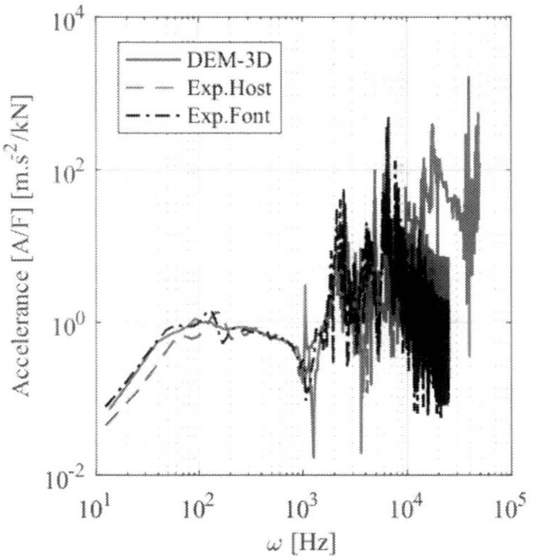

Figure 5. Comparaison of accelerance curve $A/F(\omega)$ obtainted by the numerical model (DEM-3D) and by experimental tests of Panda 3® for Hostun (Exp.Host) and Fontainebleau sands (Exp.Font) for tip resistance values around 2 MPa

Figure 5 presents the accelerance versus frequency obtainted with numerical model and with experimental tests of Panda 3® for Hostun and Fontainebleau sands. There is a good agreement between these curves when frequency is smaller than 1000 Hz. It shows that the numerical model is able to properly simulate the dynamic penetration test on granular media.

5 Conclusion

We have presented in this paper the simulation of the light dynamic penetration test Panda 3® by using the DEM in 3D in order to validate the methods of analysis proposed for the Panda 3®. A qualitative comparaison of the values of accelerance in the frequency domain showed a good correspondence between numerical and experimental results. It means that the numerical model presented is able to properly simulate dynamic penetration tests in granular media.

It would be interesting to study the mechanisms of penetration and to establish a relationship between the micromechanical parameters of the modeled soil and the mechanical parameters obtained by the signals measured by Panda 3®.

References

[1] R. Gourvès, *Pénétromètre dynamique léger à énergie variable*, (LERMES CUST, University Blaise Pascal, Clermont-Ferrand, France, 1991)

[2] M. A. Benz-Navarrete, *Mesures dynamiques lors du battage du pénétromètre PANDA 2* (PhD thesis, Université Blaise Pascal-Clermont-Ferrand II, 2009)

[3] Casem, D, W.Fourney, P.Chang, Polymer testing , **22**, 155-164 (2003)

[4] E. Escobar, M. Benz, R.Gouvès, P. Breul, Proc. 7th Int. Conf. Micromechanics of Granular Media, **1542(1)**, 389-392 (2013)

[5] E. Escobar, *Mise au point et exploitation d'une nouvelle technique pour la reconnaisance des sols: le PANDA 3* (PhD thesis, Université Blaise Pascal-Clermont-Ferrand II, 2015)

[6] Q. A.Tran, B. Chevalier, P. Breul, Proc. 7th Int. Conf. Micromechanics of Granular Media, **1542(1)**, 225-228 (2013)

[7] Q. A. Tran, *Modélisation numérique du comportement des milieux granulaires à partir de signaux pénétrométriques: approche micromécanique par la méthode des éléments discrets* (PhD thesis, Université Blaise Pascal-Clermont-Ferrand II, 2015)

[8] Itasca Consulting Group, Inc. *PFC-Particle Flow Code, Ver. 5.0.* Minneapolis: Itasca (2014)

[9] C. Bacon, Int. Journal of impact engineering, **13(4)**, 527-539 (1993).

Thermomechanical properties and fracture of resin-bonded-sand cores – Experimental study and application in aluminium foundry

Claire Menet[1,*], *Pascal* Reynaud[1], *Gilbert* Fantozzi[1], *Delphine* Thibault[2] and *Adrien* Laforêt[3]

[1] MATEIS, INSA Lyon, 69621 Villeurbanne, France
[2] MONTUPET, 60290 Laigneville, France
[3] MONTUPET, 36130 Diors, France

Abstract. Sand cores are used to produce internal cavities of metallic cast parts with complex shapes like automotive cylinder heads. Foundry cores are granular materials made of sand grains aggregated with binder bridges. In the cold box coring process, the binder is a polyurethane resin. It is noteworthy that during the casting of the liquid metal, the polymer binder is seriously damaged. This kind of materials has been poorly investigated so far. This study aims for a better understanding of the mechanical behaviour and fracture of cores subjected to various loads and thermal ageing. Particularly, the focus is on the decoring step, which consists in removing the sand by hammering and vibration of the metallic part after casting. This major project, generated from the collaboration of the aluminum casting company Montupet, and two laboratories Centre des Matériaux (CdM) and MATEIS, includes both experimental and numerical activities in order to model the decoring step of cylinder heads based on empiric data. Here, the experimental part of the work is presented.

1 Introduction

Cold box coring process is used in foundry industry for decades. It consists of mixing silica sand with a both phenolic resin component and an isocyanate hardener in organic solvents in low amount (around 1 wt.%). Then the mixture is injected into a permanent mould, and after catalysis, the binder is reticulated and the sand core has sufficient mechanical properties to be handled and used for the metal casting. The liquid metal (near 700°C in the case of aluminum alloys) is cast and then cooled at a controlled rate. The sand is finally removed during the decoring step by hammering and vibration of the whole part. Nowadays, the different parameters of the process are based on the long experience in foundry history but have not been scientifically optimized. However, the decoring step becomes ever more critical with the decreasing size of the automotive motors, which induces ever more complex shapes of the metallic parts. Indeed, sand core fragments should be evacuated through the pipes of the metallic part. A numerical modelling of this step is essential to determine parameters such as the position, the frequency, the intensity of the hammer hits, and increase development time. Before implementing such a numerical model, it is necessary to have a complete characterization of the thermo-mechanical behaviour and fracture mechanisms of sand cores, depending on the thermal ageing induced in particular by the casting and cooling temperature, the binder nature and content for example.

Granular materials with a solid binder in low proportion are not common materials and only few studies deal with foundry sand cores, and even less with their mechanical properties [1]. Some peculiar properties of granular matter have already been pointed out like the stress transfer through a network of contact chains [2,3]. Several papers propose experimental and/or numerical models to improve the understanding of the effects of contacts and cementation of the contacts in granular materials [4,5,6]. Another non-trivial phenomenon is the strain localisation developed under compression in the form of shear bands. These bands are associated with volumetric effects: dilation or contraction depending on the confining pressure (respectively low or high). Localisation has been detected in the case of both uncemented [7] and cemented [8] granular materials.

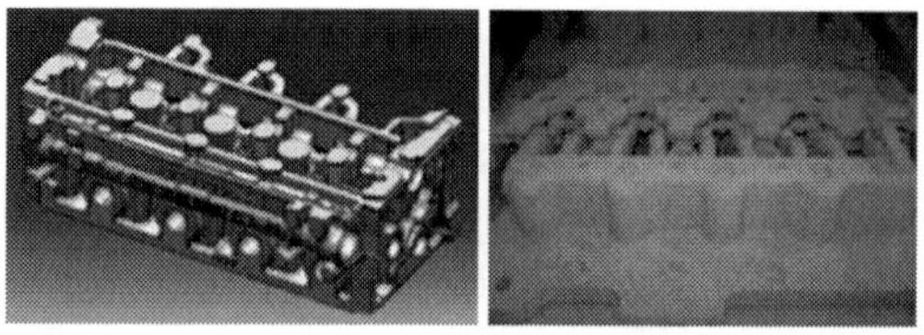

Fig. 1. An aluminum cylinder head casted with a sand casting process – Montupet (left) and the associated sand core of the bottom part of the cylinder head (right)

* Corresponding author: claire.menet@insa-lyon.fr

This experimental study requires to consider the specific characteristics of sand cores: a plastic organic resin binder in low proportion, a difference in stiffness between grains and binder (that preventing grains from breaking), and particularly the thermal degradation of the binder.

2 Materials and Methods

Cores samples have been made in Montupet Laigneville following the industrial coring process. Silica sand grains have a rounded shape and a diameter between 50 and 250µm. The commercial binder is a polyurethane resulting of the polyaddition of a formo-phenolic resin and the MDI isocyanate. Its content is lower to 1,5 wt.%.

The mechanical response of the material is examined under uniaxial compression, 4-point bending and shear tests. Compression tests have been performed on cylindrical samples with a diameter of 30 mm and a height of 45 mm. The displacement is measured with an extensometer (25 mm). Samples for the bending tests are 220 mm long bars with a section of 22.5x22.5 mm². The displacement is measured with a LVDT sensor. For the shear test, double-notched parallelepipeds are compressed. The shape of the sample is described in Fig.2. All the tests have been performed with an electromechanical testing machine (Zwick), with a controlled displacement of 0.5 mm/min. Both monotonous and cyclic tests have been performed. For the cyclic loading, a new cycle is started every 50 µm of crossbar displacement.

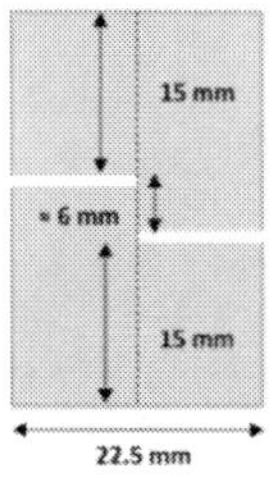

Fig. 2. Double-notched sample used for the shear test

To simulate the thermal degradation induced by the casting, some samples are aged 30 min in a furnace at a temperature noted θ under atmosphere conditions. The tests are performed at room temperature.

Some compressive tests have been followed by X-ray tomography. The resolution of the images is 17µm. Strain fields are calculated via the image correlation software Ufreckles (developed by J. Réthoré at LaMCoS) with a finite element method.

3 Influence of the thermal ageing

Sand cores require specific properties because they need to be tough enough to withstand handling and the weight of the metal without deforming or breaking. However, after the casting, the cores have to lose their strength to be removed easily from the cavities. The thermal degradation of the resin binder is the key of these contradictory properties. In this following section, the effects of this degradation on the mechanical properties of the core will be investigated.

The evolution of polyurethane resin as a function of the temperature has been studied with DSC and TGA tests [9]. Two domains of weight loss appear: the first one is located between 100°C and 200°C and corresponds to an evaporation of the solvent. The second one, starts near 250°C and is much broader. It is linked to several oxidation reactions and finally to a pyrolysis of the resin.

3.1. SEM observations

The fracture surface of a core broken under bending is observed under SEM. Grains are intact, the fracture is inter-granular and occurs via the rupture of the binder bridges. Fig.3. reveals the impact of the degradation on the aspect and fracture surface of the bonds. First of all, bubbles appear from 200°C, become ever more numerous and at 350°C some bridges are actually hollow. Indeed, at this temperature some torn off bridges have been noticed. These bubbles create defects into the bonds, which become weaker and moreover decrease the contact area between resin and grain. Another significant point is the differences on the fracture surfaces: the signs of plasticity (torn aspect) noticeable for the as-received sample are not anymore present for aged ones.

There is an agreement between the SEM observation and the TGA domains of degradation: damages occur from 150°C and get worse after 250°C.

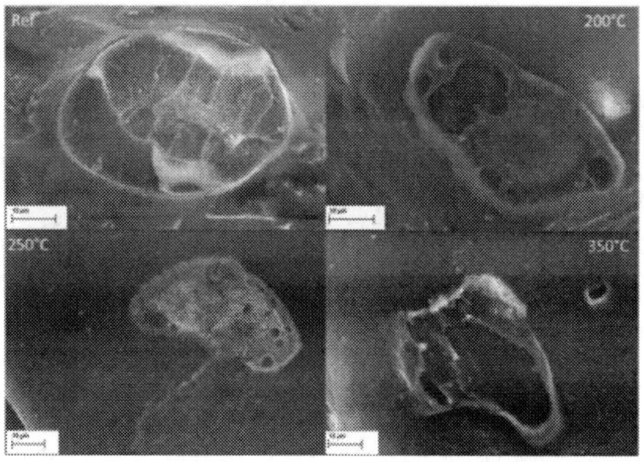

Fig. 3. Fracture surfaces of broken binder bridges depending on various ageing temperatures: unaged, 200°C, 250°C and 350°C.

3.2 Mechanical resistance

As-received and thermally aged samples are tested under compression, bending and shear. The rupture resistances R under this various loads are indexed below in fig.4. R is defined as the peak compressive stress for

DOI: 10.1051/epjconf/201714008006

compression tests, as the peak tensile stress in the middle point of the lower sample face for 4-point bending, and as the peak shear stress for double-notched shear test.

$$\begin{cases} R\,(compression) = F/D \\ R\,(bending) = \dfrac{3F(L_{inf} - L_{sup})}{2HB^2} \\ R\,(shear) = \dfrac{F}{hB} \end{cases} \qquad \textbf{(Eq. 1)}$$

With F the applied load, D the diameter of the sample, H the height of the sample, B the sample width, L_{inf} and L_{sup} respectively the spans between the lower and top bending supports, h the distance between the two notches of the shear sample (fig.2.).

Unsurprisingly, the resistance to compression is much higher than bending or shear ones because the granular skeleton supports the stresses. The low resistance under shear stress is related to the formation of shear bands in compression tests.
The evolution of the mechanical resistance follows the scheme of the degradation of the binder with a drop of the resistance from around 200°C. A slight increase in the properties could be detected at 150°C, it could be explained by a stiffening of the material due to the solvent evaporation and a possible continuation of the reticulation process.

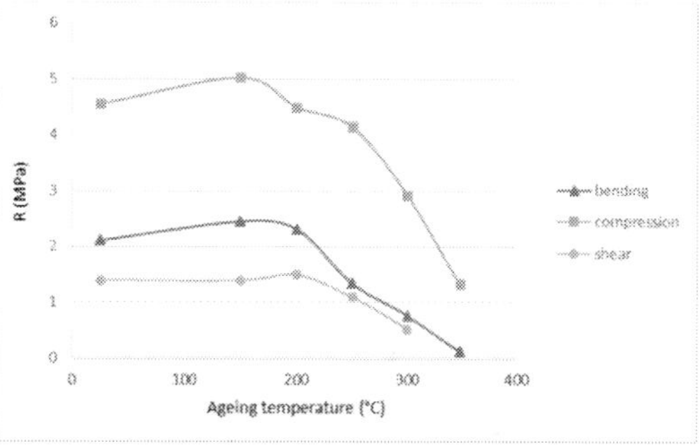

Fig. 4. Breaking resistance under bending, compression and shear as a function of θ.

3.3 Modification of the properties

As noticed in the 3.1. section, thermal degradation not only creates bubbles which weaken the material but there is also a loss of plasticity. An example is given in Fig.5. with stress-strain curves in compression for an unaged sample and a 250°C-aged one. If the breaking resistances are similar for the both cases, the overall behaviour is totally different.

For θ = 250°C, the fracture is more brittle and the plasticity is limited. Plasticity is evaluated during the load with the residual strain at the end of an unload cycle. For aged samples, the residual strain before the peak stress is halved. Secant modulus of the cycles decrease sharply after the peak stress for the aged sample, whereas it decreases linearly for unaged ones. Likewise, the ratio of the area of the loops per the elastic energy increases a lot after the peak only in the case of aged samples. This ratio represents the dissipated energy during a cycle (friction of the grains mostly). Evolutions of secant modulus are generated by grain reorganisation, which modifies the stress-supporting network and also by eventual damages (bond ruptures). The brutal modification of the cycle characteristics after the maximal stress for aged samples may be the cause of a serious damage leading to the rupture of the sample. For unaged sample, the plasticity of the bonds permits a progressive grain reorganisation leading more slightly to the rupture.

For bending tests, the fracture is brittle for both unaged and aged samples. A really slight plasticity is detected for unaged samples, and disappears totally at 250°C of ageing. After an ageing at 350°C, the material loses its resistance against tension.

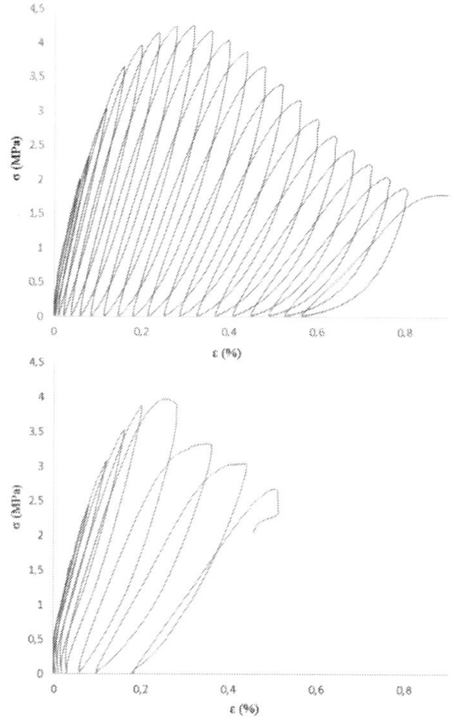

Fig. 5. Stress-strain curves for cycled compressive tests for an unaged sample (top) and a 250°C-aged one (bottom)

4 Fracture

The initiation and development of cracks are really concerning issues to study the decoring process. For bending tests, fracture is generated in the tension zone. A unique crack crosses the bar: which is thin and skirts clustered grains. For information, fracture propagation was examined through a toughness assessment test on notched bars submitted to bending. This test indicates that the critical size of defects is in the order of the millimetre, corresponding to the size of a skirted cluster.

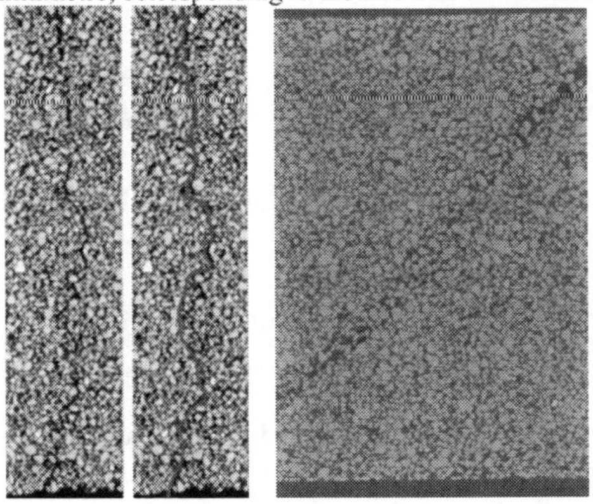

Fig. 6. Tomography stacks: (left) fracture in tension with a horizontal tensile direction; (right) shear zone in compression with a vertical compressive direction

It is common for granular materials with no/or low cementation to present localisation bands under compressive loads, and sand cores are not an exception to the rule [7, 8]. A shear band with a thickness of around 4 grains could be observed in Fig.6.

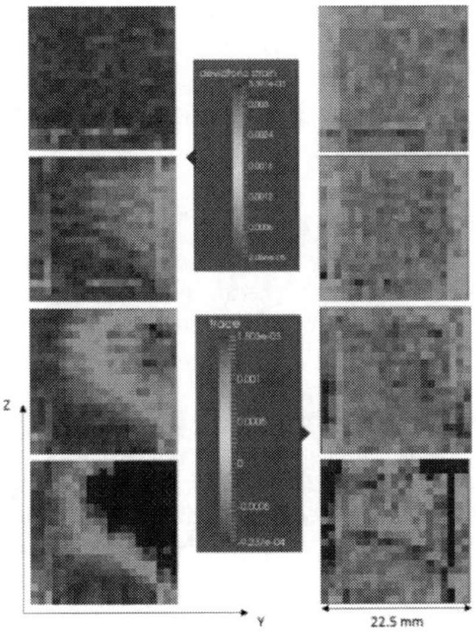

Fig.7. Deviatoric strain (left) and 3*volumetric strain (right) of an unaged sample during a compressive test. The pictures correspond to an applied load of 28%, 56%, 73% and 81% of the maximal load (from top to bottom).

Some compressive in-situ tests have been performed in a X-ray tomograph. It appears that it is difficult to distinguish cracks from pores. In fact, no crack is noticeable before the final fracture of the sample. According to literature [7-10], shear bands appear before the peak stress. The Figure 7 illustrates the deviatoric and the volumetric strain evolution during a compressive test. The sample tends to dilate on its edges (the number of binder bridges per grains is lower there). After the elastic limit, the deviatoric strain concentrates in the whole top corner, it is also possible to distinguish in this area a slight contraction followed by a slight dilation. The erosion of the structure plays an essential role for uniaxial compression.

5 Conclusion and perspectives

In this paper, the mechanical behaviour of sand cores has been examined through different methods. Macroscopically, monotonous and cyclic tests inform about the damages and the dissipated energy occurring during bending or compression tests. Microscopically, SEM pictures permit an analysis of the fracture surface, when in-situ tomography underlines the strain localisation. Three domains appear: an elastic one, then a plastic one, where localisation, damage and erosion start. A crack appears at the peak stress and propagates immediately for bending and monotonous compression. The fracture could be more controlled in the case of cyclic compression (stress relaxation). The originality of this study stands in the consideration of the thermal degradation of the organic binder. Correlated tests link this degradation to a loss of mechanical resistance and an embrittling.To continue the study of this material and establish a behaviour law, different results of cycled tests and creep under various loads must be gathered for samples with varying parameters: proportion of the binder, temperature of the test, and nature of the binder.

Acknowledgment

This research project has been funded and supported by the ANR agency within the frame of FIMALIPO project. Thanks are particularly due to W. Hilth and Prof. D. Ryckelynck from the CdM for fruitful discussions.

References

1. B.J. Stauder, H. Kerber, P. Schumacher, Journal of Materials Processing Technology **237** (2016)
2. A. Drescher, G. De Josselin de Jong, J. Mech. Phys. Solids **20** (1972)
3. V. Topin, J.Y. Delenne, F. Radjai, L. Brendel, F. Mabille, Eur. Phys. J. E **23** (2007)
4. J.Y. Delenne, V.Topin, F. Radjai, Acta Mechanica **205** (2009)
5. M. Jiang, W. Zhang, Y. Sun, S. Utili, Granular Matter **84** (2013)
6. N.S. Nguyen, H. Magoariec, B. Cambou, C.R. Mecanique **342** (2014)
7. J. Desrues, E. Andò, C.R. Physique **16** (2015)
8. Y.H. Wang, S.C. Leung, J. Geotech. Geoenviron. Eng. **134**, 7 (2008)
9. G. Jomaa, P. Goblet, C. Coquelet., V. Morlot, Thermochimica Acta **612**, (2015)

Capturing of the internal mechanics of liquid-granular flows comprised of polydisperse spherical particles

Devis Gollin[1,*] and *Elisabeth* Bowman[1,**]

[1] *The University of Sheffield, Department of Civil & Structural Engineering, Western Bank, Sheffield, S10 2TN, UK*

Abstract. This paper presents a series of flume experiments designed to examine the motion and arrest of concentrated granular-fluid flows, with a view to understanding the role of polydispersity in debris flows. A non-intrusive technique is used to investigate the internal behaviour of small scale experimental flows. Three different particle size distributions comprised of polydisperse spherical particles and one with the finer component made of angular particles were analysed. The choice of using spherical shaped particles was made to improve the visualization of the internal mechanics without reducing overmuch the complexity involved in the study of these flows. We examined and compared the internal velocities of the flows and their depositional spreads. While the optical performance of the non-intrusive technique was improved, some of the characteristics commonly seen in these types of granular flows were not observed. Velocity profiles obtained in the body of the flows were similar in shape but with differences in velocity magnitude depending on the amount of fines and the angularity of the particle in one case. Depositional runouts between flows were similar at low inclinations when little internal energy was supplemented to the system or when the viscous effects dominated the mechanics at steeper angles.

1 Introduction

Debris flows constitute a significant natural hazard in regions of mountainous terrain. They are often described as a succession of non-uniform surges with particle sizes ranging from clay to boulders. Flow segregation can occur where larger particles are preferentially transported to the snout which is characterized by a reduced water content and greater friction resistance to motion than the dispersion behind it. The rear interior, where large amount of suspended fines tend to accumulate, provides a mobile core which advects some of its downstream momentum into the resistive, coarse-grained snout, thereby providing the force to drive the snout forward. This mixture of sediment stops when approaching gently sloping surfaces. These areas of the deposit spread or runout are generally of most interest in terms of human risk. How far such flows will travel is critical for delineating areas at risk and to design protective measures.

The inherent complexity and the multitude of heterogeneities within granular flows make them very hard to study in the field. However, fundamental characteristics of mobile granular flows can be reproduced in laboratory experiments, albeit with some simplifications. This has shed some light on their overall behaviour, but the mechanics of their movement is still far from understood. Hence debris flows and more in general granular flows are subject of intense research from an experimental perspective which can also be supported by numerical and theoretical analyses

The combined effects of mixture composition and problem geometry on internal behaviour and the effects on the runout can be reproduced in laboratory flumes. Such an approach was previously adopted by Sanvitale and Bowman [1] . Several flows were studied using an innovative non-intrusive technique that allowed the visualization of a transparent medium, thus the internal dynamics of a flow, while matching closely the properties of real debris flows. Here a similar apparatus was used, but with modifications for this particular work. Based on the previous results, more simplified solid particles are employed in order to improve the visualisation technique while trying to keep an approximation to debris flow behaviour. The choice of the particles leads to a better visualization of the flow interior. However, we show that there are fundamental limitations on the choice of the shape and size of particles.

2 Background

Iverson [2] has argued that the dynamic similarity between natural and small-scale debris flows is probably unattainable since small models cannot reproduce the same stresses as those of real large debris flows. However, Bowman and Sanvitale [3] showed that comparison between the two scales does appear feasible and careful laboratory testing performed on small scale flumes can still inform on the mechanics behind debris flow behaviour. It was shown that the well-graded nature of debris flow is not

*e-mail: devis.gollin@sheffield.ac.uk
**e-mail: e.bowman@sheffield.ac.uk

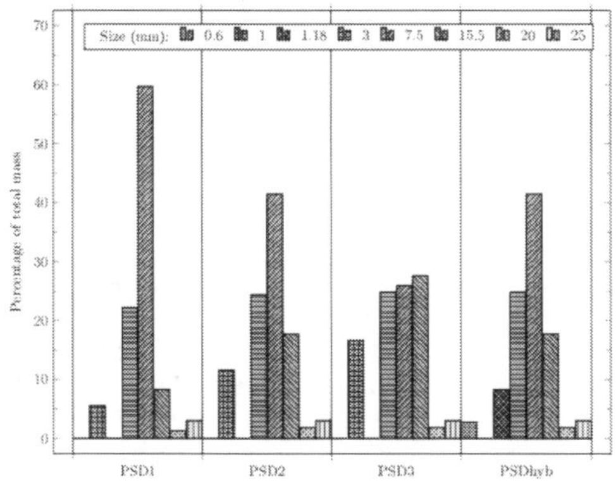

Figure 1: Particle size distributions

incidental, but an important factor affecting their mobility and the capacity to travel far before the final deposition phase. The correct choice of the particle size distribution (PSD) enabled important aspects of debris flow behaviour to be replicated even at small scale, such as the effects of flow segregation and its influence on final deposit shape and grading profile.

An experimental debris flow consisting of well-graded materials may generate low frictional resistance within the flow interior as the flow segregates with agitated finer particles reducing the propensity for fluid to exit from the mass. Hence pore pressures may be greater than hydrostatic and effective stress greatly reduced. This is different from typical experiments characterized by a relatively uniformly-graded particle size distribution. In this case segregation cannot occur and the mixtures maintain a steady frictional resistance. The fluid is evenly distributed throughout the flow so that momentum is lost quickly due to the uniformly distributed friction resistance and the reduced effect of pore pressure. Comparatively, for systems with the same masses of solids and fluid, well-graded mixtures show higher velocity, smaller flow thickness and a larger runout extent than uniform flows [4].

These distinct behaviours may explain the difference in behaviour at large and small scale that is often observed. If well-graded materials are used for small scale work, high overall mobility can be achieved in this configuration despite the lower velocity attained.

Here small scale experiments considering the flow of transparent materials down an inclined chute are described. These are characterized by particle size distributions made of few, but distinctly different, particle diameters in the attempt to step away from monodisperse realizations but at the same time, only gradually increase the complexity of the investigations.

3 Experimental set-up

The flume used in this study is similar to that used by Sanvitale and Bowman [1]. The apparatus is a sloping rectangular 15 cm wide and 200 cm long channel that can tilt from horizontal up to 45°. At the upstream end, the material is held in a rectangular tank. A double-slider gate mechanism enables control of the releasing flow height that develops downstream. Attached at the bottom of the inner gate is a flexible flap with holes that holds a small pocket of liquid. At the opening of the gate this liquid exits first removing any entrainment of air bubbles which would be otherwise trapped in the solid phase and carried down the channel, thus creating unwanted visible micro lenses in the images and interfering with the laser plane. The rough base has been made via 3D-printing using a plastic material (Nylon SLS). Semi-spherical particles of diameter 3 mm at the circular edge are organized in a hexagonal pattern and positioned at the bottom. Near the end of the flume a planar laser light is shone 50 mm away from the sidewall in order to illuminate the transparent material. At the same fixed location measurements were taken for all flows. The lit surface of the granular flow is captured by a high-speed camera (Phantom Miro 310) at a frame rate of 500 fps. A typical example of images for these particular flows is shown in Fig 2. Once collected, the images are processed via an open-source particle image velocimetry (PIV) software [5] and velocity profiles extracted for each flow. A flat deposition area is situated at the end of the channel where the material is collected and measurements of final runout length and area recorded.

The visualization of the flow interior is done via planer laser induced fluorescence (PLIF), which requires the matching of the solid and liquid refractive indices (RI). A fully saturated flow is required for this method to work properly. The materials used are crushed angular particles borosilicate glass with size 0.6 and 1.18 mm, respectively, and spherical particles of the same material with diameters ranging from 1 to 25 mm. Given the focus was on image quality here, manufactured spherical particle which are nearly flawless, enabled an improvement to that achieved previously using angular glass particles. However, the cost of very fine spherical particles was prohibitive, hence fines were made of angular materials. The fluid (a hydrocarbon oil produced by Cargille laboratories) was refractively matched to the glass and a small amount of fluorescent dye (Nile Red) added so that to distinguish the particles from the brighter background. All tests were conducted at a controlled room temperature of 23°. This was found optimal for matching the two refractive indices as the RI for the liquid varies with the temperature.

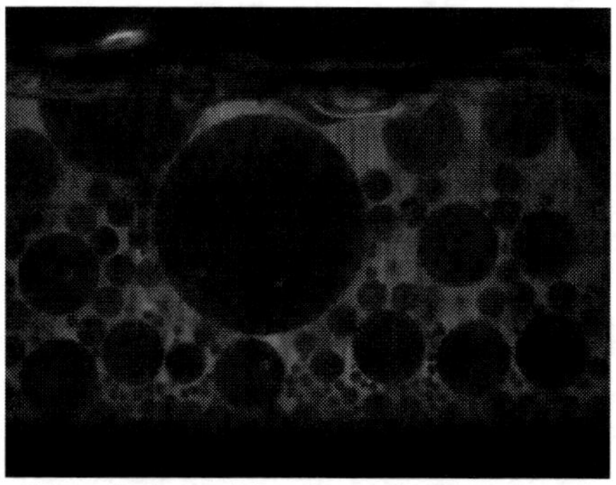

Figure 2: Internal flow image for PSD1 at 15°

3.1 Methodology

All tests were carried out with the same experimental conditions. The opening height of the gate was kept constant. Three chute inclinations were used, respectively, 15°, 20° and 25°. A total mass of 18 Kg for each mixture was loaded inside the hopper and then saturated so that the initial sediment volumetric concentration (Cs) was 0.57 (i.e. porosity = 0.43). The four particle size distributions (PSD) used in this study are shown in Fig 1. They are presented as histograms showing the percentage of mass for each particle size over the total mass. The smallest and largest sizes were limited due to practical reasons, however we are able to approximate those used by Sanvitale and Bowman [1] (in Fig. 8). PSD1, PSD2 and PSD3 are composed of spherical particles only. For each of these three we progressively increased the amount of finer materials (1 mm) from 5.53% for PSD1, 11.2% for PSD2 and 16.6% for PSD3, reduced the number of particle in the middle of the distribution (7.5 mm) and increased the number of larger particles (15.5 mm). The two largest sizes, 20 and 25 mm were kept constant throughout due to the limited number of particles available. PSD_{hyb} is the same as PSD2 but all 1 mm particles (i.e. 11.2% of the total mass) were substituted by angular particles of 0.6 and 1.18 mm in order enhance expected pore pressure effects caused by addition of fines.

4 Results

For the flows released at an inclination of 15°, velocity profiles obtained by averaging over the body of the flows

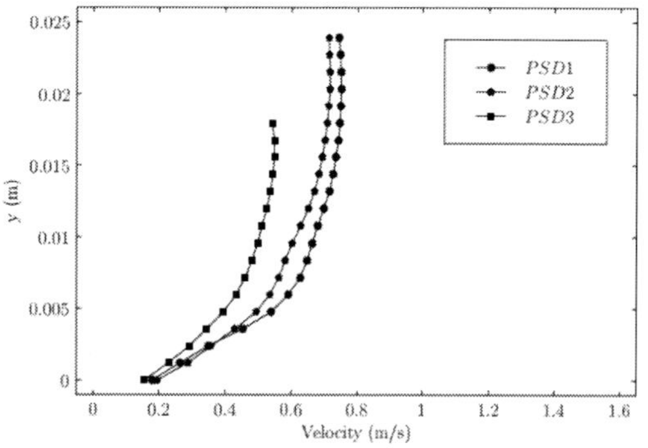

Figure 3: Velocity profiles for flows released at 15°

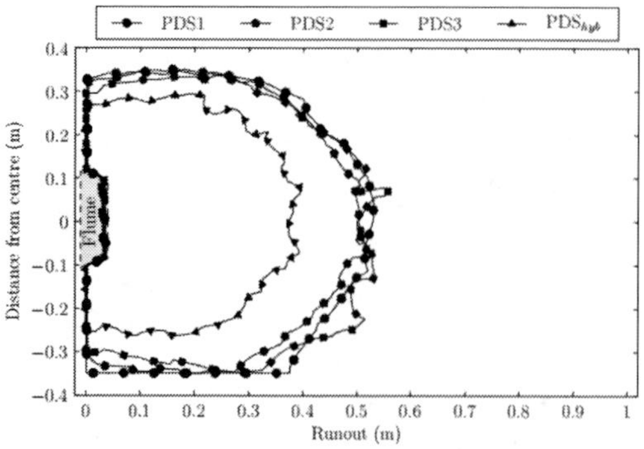

Figure 4: Depositional runouts for flows released at 15°

(i.e. avoiding the noisy front and the tail of the surge) are shown in Fig. 3. In all cases a slip velocity developed at the bottom boundary. PSD1 and PSD2 have similar flow heights and velocity magnitudes compared to PSD3. PDS_{hyb} was not able to develop sufficient momentum which resulted in a thick layer of material being deposited over the channel. Only a short recording was possible before the material stopped, which did not allow a velocity profile to be determined.

From the analysis of flow images it was revealed that after the coarser but still saturated front smaller particles segregated to the bottom of the flow leaving larger ones to travel at higher velocities (e.g. Fig. 2). This occurred for all inclinations. In addition, fine materials (smaller than 1 and 1.18 mm) were partially trapped and slowed by the rough base made of 3 mm semi-spheres.

The profiles of the final deposit are shown in Fig 4. These were similar to each other for PSD1, PSD2 and PSD3. If the velocities are compared to the runout lengths, some changes would be expected for the slower PDS3. However, the length and shape of the spreads were very similar indicating that not enough inertial energy was induced into the system and that the material was progressively accumulating without showing characteristic depositional behaviour. This was different for PSD_{hyb}, possibly due to the increased frictional resistance provided by the angular fine particles leading to an even shorter runout. Note that the choice to use 15° was led by previously tested flows (not shown here) conducted with monodisperse 3 mm dispersions which developed much higher velocities.

To increase the energy in the system and produce different behaviours, the inclination needed to be increased. While 20° showed intermediate behaviours between 15° and 25°, here we focus on flows released at 25° only. Velocity profiles obtained from the body of the flows are depicted in Fig. 5. Similarly to the flows at 15°, a distinct slip velocity was generated at the bottom boundary. For PSD1, PSD2 and PSD3 the magnitudes of this quantity decreased progressively with an increasing amount of fines. PDS_{hyb} travels slower than PSD3 due to the increased friction. Note that the all profiles tend to reverse back at the free surface (i.e. with slower velocities than the middle part), which is unexpected for these flows, in which we would expect a maximum at the free surface. This was caused by the changing of the saturation line (also the limit for the PLIF technique). For these faster flows, occasional and local reduction of the saturation line is caused by the crossing of big particles transported at the free surface. These are interpreted by PIV as areas with lower velocity. Hence averaging results in an apparent reversing of the velocity profile at the top of the flow. This was also noticed when the average was done for shorter periods of time (e.g. for 30 or 50 frames only).

The flows analysed at an inclination of 25° produced the depositional areas shown in Fig. 6. PSD1 had the longest runout and fines segregated close to the bottom boundary. This resulted in larger particles dominating the mechanics during runout producing a longer spread. Conversely, from the analysis of flow images for PSD2 and

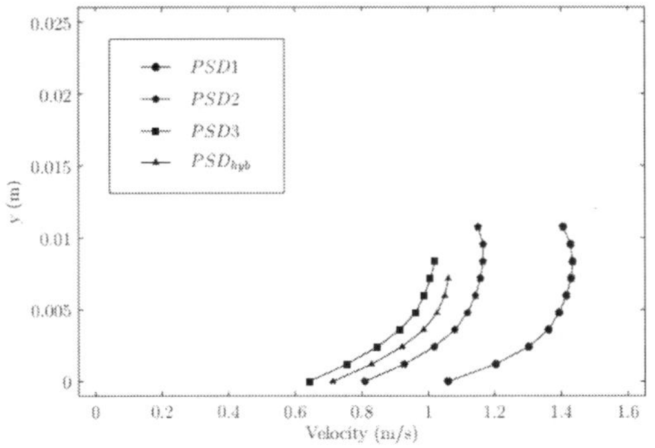

Figure 5: Velocity profiles for flows released at 25°

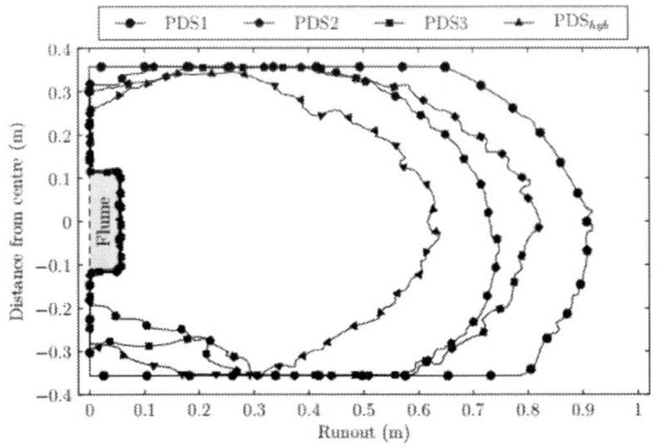

Figure 6: Depositional runouts for flows released at 25°

PSD3 it was clear that fine particles were more evenly distributed throughout the flow height. They have shorter and comparable runouts, with PSD3 being somewhat longer. That is, increasing the amount of fine particles (< 1 mm for these three PSDs) considerably slowed the flows. Therefore, the expected changes of runout caused by adding fines to increase pore pressure (i.e. via accumulation of fines in the body and rear) did not occur, instead viscous and vertical segregation effects appeared to dominate. Again, PDS_{hyb} produced shortest runout due to the increased frictional resistance of fine angular particles.

5 Conclusion

This paper presented the results obtained for polydisperse granular flows characterized by different particle size distributions and particle shapes. Velocity profiles were obtained from internal cross sections of the flows. The use of spherical particles allowed us to improve the optical performance of the PLIF technique. However, some of the typical depositional characteristics for these types of granular flows were not observed. The extents of the runout areas were similar for low energy flows at an inclination of 15°. In contrast, for 25° it was clear that flows with the highest amount of fines were affected by strong viscous effects and vertical segregation. For both inclinations, PSD_{hyb} (a modified PSD2 with angular particles) had the shortest runout due to increased fictional resistance.

Perhaps the display of different depositional behaviours could be achieved by a more careful choice of the PSD compositions. Moving from uniform to well-graded distribution could be done by varying the amount of fine and large particles around a specific point (e.g. a mean diameter (D_{50})), as done previously [1]. For these tests we were limited in the amount of large particles available while considerably increasing the amount of fines. That is, by just increasing the fines content we only induced additional viscous effects. In addition, large step changes in the specific particle sizes – e.g. beyond Terzaghi's filter limit [6] - may have enabled particle size segregation to spontaneously occur at the commencement of the flows. Finally, while we considered velocity profiles obtained in the body of the flow, it would be interesting to compare runout versus front velocity as in [3] but this velocity was difficult to obtain due to the controlled flow release adopted to improve the flow visualization.

Ultimately, other substantial effects remain to be investigated. In case of particle vertical segregation, frictional and viscous gradients develop across the flow. Local effects may be generated, for example, the variations of global bulk motion of the interstitial fluid relative to the granular material as well as different distribution of the stresses connected to the particle rearrangement. The degree of segregation and the related effects may be quantified by estimating solid concentration profiles. The bottom also plays an important role. Very bumpy base may increase the frictional resistance due to small particles trapped in the gaps at shallow angles but also generate strong agitation and remixing at stepper ones. To clarify the boundary effect, the same flows should be tested at different bottom boundary. Undertaking further investigations could lead to new insights into internal and depositional mechanics of these granular flows.

Acknowledgements

The first author is supported by an Engineering and Physical Sciences Research Council (EPSRC) DTA Scholarship.

References

[1] Sanvitale, N and Bowman, E.T., International Journal of Physical Modelling in Geotechnics **12**(4), 129-142 (2012)

[2] Iverson, R. M., Reviews of Geophysics **35**(3), 245-296 (1997)

[3] Bowman, E.T. and Sanvitale, N. 17th International Conference on Soil Mechanics and Geotechnical Engineering, 470-473 (2009)

[4] Sanvitale N., Bowman E.T., Genevois R. 5th International Conference on Debris-Flow Hazards Mitigation: Mechanics, Prediction and Assessment, 375-384 (2009)

[5] Thielicke, W. and Stamhuis, E.J. Journal of Open Research Software. **2**(1), e30 (2014).

[6] Skempton, A. W. and Brogan J.M, Géotechnique. **44**(3), 449-460(1994).

EPJ Web of Conferences 140, 12007 (2017) DOI: 10.1051/epjconf/201714012007
Powders & Grains 2017

Erasure of memory in paste by irradiation of ultrasonic waves

Akio Nakahara[1,*], *Ryota* Yoneyama[1], *Maruto* Ito[1], *Yousuke* Matsuo[1] and *So* Kitsunezaki[2]

[1]Laboratory of Physics, College of Science and Technology, Nihon University, Funabashi 274-8501, Japan
[2]Research Group of Physics, Division of Natural Sciences, Faculty of Nara Women's University, Nara 630-8506, Japan

Abstract. Densely packed colloidal suspension, called paste, remembers the direction of applied forces, such as vibration and flow, and these memories kept in paste can be visualized as morphology of desiccation crack patterns. For example, when the paste remembers the direction of vibration, all primary cracks propagate in the direction perpendicular to the direction of initial vibration. On the other hand, when the paste remembers the direction of flow, all primary cracks propagate along the direction of initial flow. These results indicate that external forces imprint easy-breakable direction into paste as memories. Therefore, by controlling memories in paste, we can tune to produce various types of crack patterns, such as cellular, radial, lamellar, ring, spiral and lattice structures. Recently we have found that memories in paste can be erased by the irradiation of ultrasonic waves to paste as we obtain only isotropic and cellular crack patterns without any anisotropy related to memory effect. This method can be applied to increase the breaking strength of dried paste by homogenizing microstructure in paste.

1 Introduction

The morphology of crack patterns contains variety of information, such as a position of crack nucleation and the direction of crack propagation, but it is a difficult work to derive more useful information from them. Here we deal with crack patterns of dried paste. We study what kinds of information we can get from these crack patterns and try to control the morphology of crack patterns to prevent fatal crack formation.

When a densely packed colloidal suspension, called paste, is poured into the container and is dried at room temperature and at low humidity for few days, the paste shrinks as the water evaporates, and if the paste is stick to the bottom of the container, desiccation crack patterns emerge so as to release the stress induced by the shrinkage [1]. Figure 1 shows the typical example of desiccation crack pattern, such as the one of calcium carbonate paste dried in laboratory.

Usually the morphology of desiccation cracks is isotropic and cellular as is shown in Fig. 1. The final spacing between cracks is proportional to the thickness of the paste as long as the paste is not so thick [2].

On the other hand, when the thickness of the paste is large, a crack front propagates from the surface toward the inside along the drying direction, and, as for the case of starch paste, the columnar joint of hexagonal structure emerges, which looks similar to the columnar joint of cooled lava [1, 3-6]. In each case, materials shrink under drying or cooling, and the directions of drying or cooling gradient produce anisotropic crack structure along the gradient, such as columnar joints. The diameters of these columns are determined by the value of drying or cooling gradient. Therefore, the directions of these columns indicate in which direction the paste was dried and the lava was cooled.

Can we control the direction of crack propagation without any drying gradient? If we can, we can protect products from serious damages even if we cannot stop crack formation.

In this manuscript, we report that, even without any drying gradient, we can control and get regular and anisotropic desiccation crack patterns, such as radial, lamellar, ring and spirals. We have found that paste remembers the direction of external forces and the morphology of desiccation crack patterns is determined by the memory stored in paste. These results mean that, if we imprint or rewrite the memory in paste, we can order cracks to propagate along any directions we want [7]. Then we report our latest findings that memories in paste can be erased by the irradiation of ultrasonic waves. We will conclude with its further application to control the breaking strength of materials.

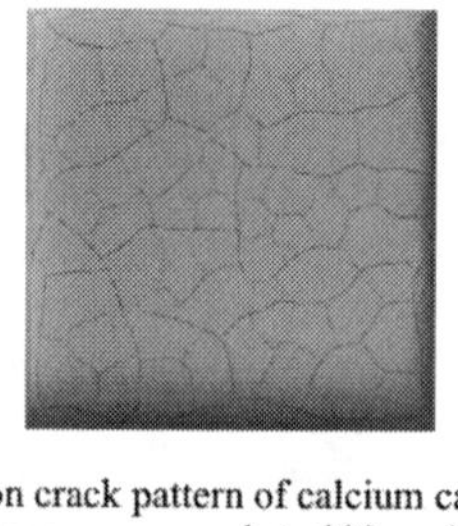

Fig. 1. Desiccation crack pattern of calcium carbonate paste. The size of the figure corresponds to 200mm*200mm.

* Corresponding author: nakahara.akio@nihon-u.ac.jp

DOI: 10.1051/epjconf/201714012007

2 Memory effect of paste

One day we performed drying experiments of calcium carbonate paste. The median diameter of colloidal particles is 2μ m. 3kg of calcium carbonate powder is mixed with 1.5kg of distilled water and the mixture is poured into the circular acrylic container with 500mm in diameter. Since the paste is water-poor, we had to vibrate the circular container in the angular direction to spread the paste homogeneously throughout the container and to get a flat surface of the paste. The thickness of the paste is 13mm after the vibration. Then, we stopped vibration, dried the paste at room temperature and at low humidity of 30%. Two days later desiccation cracks appear, but surprisingly the pattern was not the isotropic and cellular one, the pattern was radial, as is shown in Fig. 2(a).

At first, we thought that we get radial crack pattern because we used a circular container. However, we noticed that we had vibrated the paste to spread it before drying, so we did similar drying experiment but this time we vibrated the paste horizontally in one direction. Then we get lamellar cracks as is shown in Fig. 2(b) and all primary cracks run in the direction perpendicular to the direction of the initial vibration.

Therefore, we realized that the water-poor paste with plasticity can remember the direction of the vibration applied to it, and the memory of the vibration can be visualized as morphology of desiccation crack patterns. When the paste remembers the direction of the vibration, all primary cracks run in the direction perpendicular to the direction of the initial vibration [7].

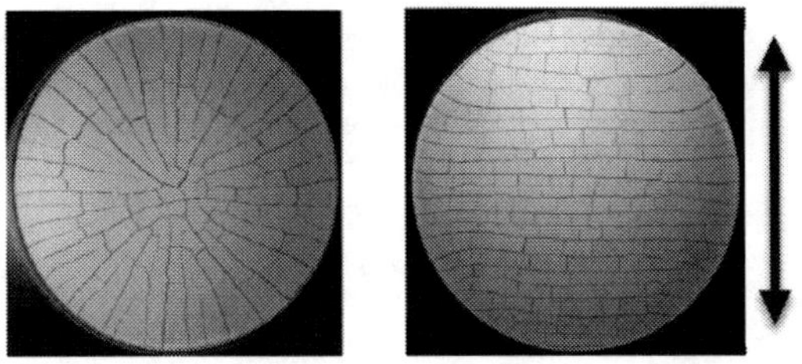

(a) Radial crack pattern **(b) Lamellar crack pattern**

Fig. 2. Desiccation crack patterns of calcium carbonate paste. The diameter of each container is 500 mm. (a) Radial crack pattern. (b) Lamellar crack pattern. The arrow indicates the direction of the initial vibration.

2.1. Memory of vibration

To reveal the condition when paste remembers the direction of the vibration, we performed systematic experiments by controlling two parameters, solid volume fraction of particles in paste and the strength of the initial vibration, and the result is shown in Fig. 3.

To prepare the paste, we mix calcium carbonate powder of 360g with water and set the paste into a square acrylic container with each side 200mm. After the container is horizontally vibrated in one direction for 1 min., the paste is dried at room temperature and desiccation cracks appear in two days.

In Fig. 3, open circles in regions A and C represent isotropic and cellular cracks, while solid squares in region B represent lamellar cracks with the direction perpendicular to the direction of the initial vibration. The solid line denotes Yield stress line on which the shear stress induced by the initial vibration equals the yield stress of the paste. Here the yield stress of paste is an increasing function of the solid volume fraction in paste and is estimated by rheological measurements using Rheometer Physica MCR-301 (Anton Paar, Graz, Austria). We find that, in region B when the water-poor plastic paste is vibrated at strength just above the yield stress of the paste, the paste experiences and remembers the motion of vibration, and the memory of vibration stored in paste can be visualized as morphology of desiccation cracks [7-9]. This experimental result means that the plasticity of paste plays important role in memory effect. Theoretical approaches based on residual tension models are proposed to explain how the memory of vibration remains as a form of plastic deformation in paste [1, 10-12].

The boundary between regions B and C, denoted by broken line, corresponds to hydrodynamic instability from the motion of vibration to flow. We find that, once calcium carbonate paste flows, it cannot remember any motions.

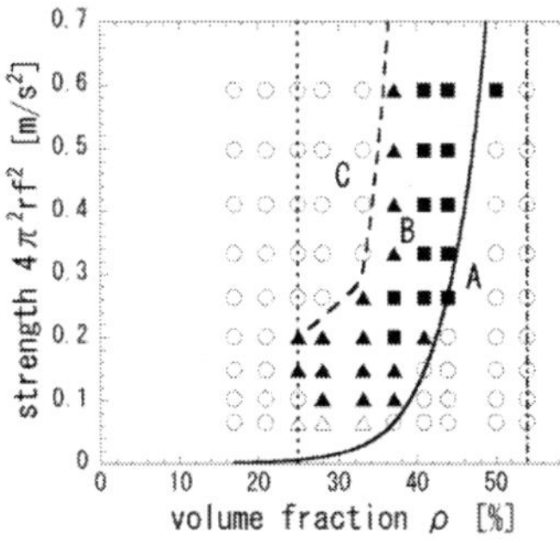

Fig. 3. Morphological phase diagram of desiccation crack patterns of calcium carbonate paste, as function of the solid volume fraction in paste and the strength of the initial vibration. Open circles represent isotropic and cellular cracks, solid squares represent lamellar cracks, the direction of which is perpendicular to the direction of the initial vibration (memory of vibration), and solid and open triangles represent combinations of cellular and lamellar cracks. The vertical dotted line denotes Liquid-Limit, the vertical dash-and-dotted line denotes Plastic-Limit, the solid line denotes Yield-stress line, and the broken line denotes the boundary between regions B and C. Calcium carbonate paste remembers the direction of initial vibration when it was vibrated in region B [7].

2.2 Memory of flow

After the discovery of the memory effect of vibration, we had performed many drying experiments by using various type of paste, and found that some pastes remember not only the direction of vibration but also the direction of flow. When the paste remembers the flow direction, the primary desiccation cracks run in the direction parallel to the flow direction. Figure 4 show the spiral crack pattern produced by using the memory of spiral flow motion. To imprint the spiral motion into the paste, we had translated the container along a circuit with a radius of 15mm clockwise.

To understand the mechanism how such pastes remember the flow direction, we performed systematic experiment to obtain a morphological phase diagram of desiccation cracks as functions of solid volume fraction and the strength of the initial vibration applied to the container. The experimental result using the paste of magnesium carbonate hydroxide is shown as Fig. 5.

Figure 5 shows that this paste has two type of memories. In region B when the water-poor paste is vibrated at strength just above the yield stress of the paste, the paste shows the same type of memory effect of vibration that the calcium carbonate paste has. On the other hand, in region C when the water-medium paste is fluidized by the vibration of the container, the paste remember the direction of the flow motion, and the primary desiccation cracks run in the direction parallel to the initial flow [13].

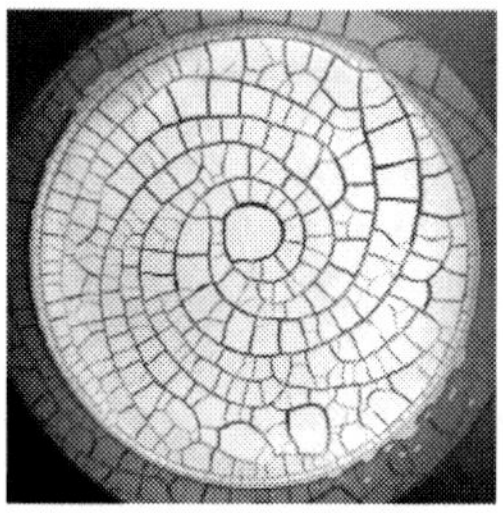

Fig. 4. Spiral crack pattern of paste of magnesium carbonate hydroxide. The diameter of the container is 500mm [13].

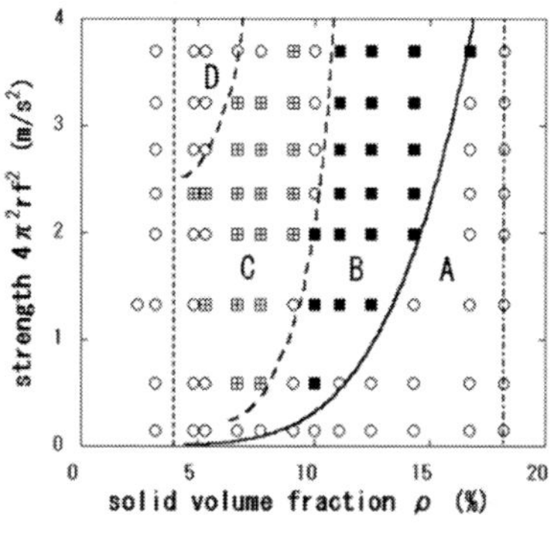

Fig. 5. Morphological phase diagram of the desiccation crack patterns of paste of magnesium carbonate hydroxide, as function of the solid volume fraction in paste and the strength of the initial vibration. Open circles represent isotropic and cellular cracks, solid squares represent lamellar cracks, the direction of which is perpendicular to the direction of the initial vibration (memory of vibration), and open squares with plus inside represent lamellar cracks, the direction of which is parallel to the direction of the initial flow. The vertical dotted line denotes Liquid-Limit, the vertical dash-and-dotted line denotes Plastic-Limit, the solid line denotes Yield-stress line, and broken lines denote the boundary between regions B and C and that between regions C and D. Paste of magnesium carbonate hydroxide can remember not only the direction of vibration but also the direction of the initial flow [13].

Note that calcium carbonate paste remembers the direction of vibration but cannot remember flow direction. We find that colloidal particles in calcium carbonate paste are charged positively in water. On the other hand, as for the paste of magnesium carbonate hydroxide which can remember directions of vibration and flow, colloidal particles are not charged in water. To screen the Coulombic repulsion we added sodium chloride to calcium carbonate paste and confirmed that the paste gets the ability to remember flow direction. Therefore, the ability to remember flow direction strongly depends on the type of the interaction among colloidal particles. Without Coulombic repulsion colloidal particles form clusters via short-ranged inter-particle attraction, and the elongation of these clusters along flow direction plays a key role in keeping memory of flow in paste [14].

Recently it has been shown that a memory in paste can be measured, without any information of crack patterns, by checking the anisotropic stress before cracks are formed [15]. This result is expected to be applicable to detect the precursor of how cracks will be formed.

3 Erasure of memory by ultrasonic waves

In this section, we report our recent challenge to erase the memory of vibration in paste. First we heated the paste to destroy microstructure related to memory by thermal fluctuation, but it was difficult to control and homogenize the drying gradient under heating, so we just obtained some irregular crack patterns induced by irregular drying. It had been reported that the jamming of granular particles is suppressed by the irradiation of focused ultrasonic waves [16], Therefore, in the next challenge, we irradiated ultrasonic waves to the paste to destroy microstructure by strong acoustic oscillation and erased the memory in paste.

3.1. Experimental method

We prepared calcium carbonate paste with a solid volume fraction of 35%, which remembers the direction of initial vibration. We poured the paste into a circular stainless container of 100 mm diameter and the thickness of the paste is 10 mm in depth. First we vibrated the paste horizontally for 1 min. at a frequency of 40rpm and the amplitude of 15mm by using Triple Shaker NR-80 (Taitec Inc., Tokyo, Japan). Then we lifted up a cubic water tank of 500mm in each side until the surface of the water touch the bottom of the circular container from below. At the bottom of the cubic water tank we had placed an ultrasonic wave generator Advantage Multi Sonic Cleaner (Otari Inc., Tokyo, Japan) with a frequency of 37kHz, 71kHz or 102kHz and with a continuous control of the sound pressure. We irradiated ultrasonic waves to the paste for 1 min., removed the water tank, and dried the paste at room temperature and low humidity. The sound pressure of ultrasonic waves, which is irradiated to the paste, was measured by Hydrophone AQH-100 (Aquasound Inc., Kyoto, Japan).

3.2 Effect of ultrasonic waves on crack patterns

When we just vibrated a paste in one direction for 1 min. and then dried it, we obtain a lamellar crack pattern shown in Fig. 6(a), the direction of which is perpendicular to the direction of the initial vibration. On the other hand, if the paste is vibrated for 1 min., irradiated by ultrasonic waves of 71kHz and 25kPa for 1 min, and then dried, an isotropic and cellular crack pattern appears, as is shown in Fig. 6(b). This result indicates that the irradiation of ultrasonic waves erases the memory of vibration in paste and therefore the isotropic and cellular desiccation crack pattern appears.

We systematically studied the effect of applying ultrasonic waves by controlling two parameters, a frequency and a sound pressure of ultrasonic waves. Figure 7 shows that, if the sound pressure exceeds a threshold value indicated as a solid line, the irradiation of ultrasonic waves weakens or erases the memory in paste. It also shows that the value of the threshold is an increasing function of the frequency. The pressure difference between both sides of each particle at the threshold sound pressure is estimated of the order of the yield stress of the paste. The mechanism to erase a memory in paste is now under investigation.

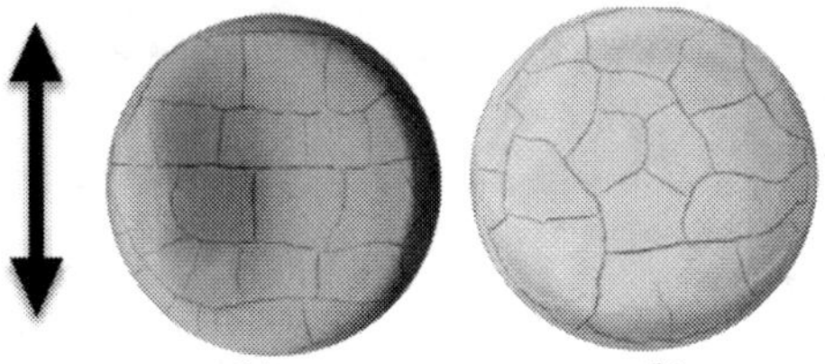

(a) (b)

Fig. 6. Desiccation crack pattern of calcium carbonate paste. The diameter of the circular container is 100mm in (a) and (b). (a) The paste is vibrated in one direction for 1 min., then dried. (b) The paste is vibrated for 1 min., irradiated by ultrasonic waves of 71kHz and 25kPa for 1 min, and then dried.

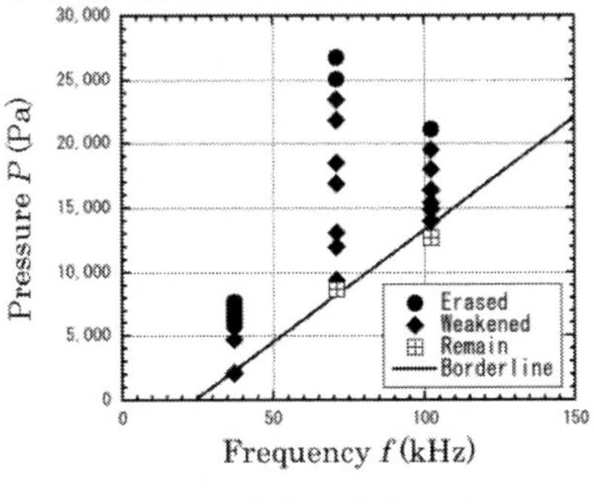

Fig. 7. Effect of the irradiation of ultrasonic waves to paste, as a function of frequency and pressure of applied ultrasonic waves. The paste had been vibrated and had a memory of vibration before it was irradiated by ultrasonic waves. The solid circle denotes isotropic and cellular desiccation crack pattern, the result of which indicates that the memory in paste is erased by the irradiation of ultrasonic waves. The solid rhomboid corresponds to the case where the ultrasonic waves weaken the memory in paste and the open square with plus inside means there is no effect of applying ultrasonic waves.

4 Conclusions

We have found that the memory in paste can be visualized as morphology of desiccation crack patterns. This result has two meanings. First, if you observe the crack pattern, you can estimate what happened to the paste before it dries. Second, by imprinting and rewriting the memories in paste, we can control the direction of crack propagation and protect products from serious damages. We have also fond that we can erase the memory in paste by irradiation of ultrasonic waves. Since the memory effect produces easy-breaking directions in paste, the erasure of memory in paste will increase the breaking strength of paste without any easy-breaking directions. We consider that this method can be applicable to produce durable construction material by erasing the memory of pouring into molds.

We acknowledge H. Onozaki, Ooshida Takeshi, Sebastien Manneville and Ferenc Kun for valuable discussions. This work was supported by Grant-in-Aid for Scientific Research (KAKENHI) (B)22340112, (C)16K05485 and (C)26400395 of Japan Society for the Promotion of Science (JSPS), by JSPS and HAS under Japan-Hungary Research Cooperative Program and by CST Research Fund of Nihon Univ.

References

1. L. Goehring, A. Nakahara, T. Dutta, S. Kitsunezaki, S. Tarafdar, *Desiccation Cracks and their Patterns: Formation and Modelling in Science and Nature*, (Wiley-VCH, Weimheim, Germany, 2015), and references therein.
2. A. Groisman, E. Kaplan, Europhys. Lett., **25**, 415 (1994).
3. G. Mueller, J. Volcanol. Geother. Res., **86**, 93 (1998).
4. A. Toramaru, T. Matsumoto, J. Geophys. Res., **109**, B02205 (2004).
5. L. Goehring, S. W. Morris, Europhys. Lett., **69**, 739 (2005).
6. T. Mizuguchi, A. Nishimoto, S. Kitsunezaki, Y. Yamazaki, I. Aoki, Phys. Rev. E, **71**, 057122 (2005).
7. A. Nakahara, Y. Matsuo, J. Phys. Soc. J., **74**, 1362 (2005) ; J. Stat. Mech.: Theory Exp., P07016 (2005).
8. S. Kitsunezaki, J. Phys. Soc. Jpn. 78, 064801 (2009).
9. H. Nakayama, Y. Matsuo, O. Takeshi, A. Nakahara, Euro Phys. J. E, **36**, 1 (2013).
10. M. Otsuki, Phys. Rev. E, **72**, 046115 (2005).
11. O. Takeshi, Phys. Rev. E, **77**, 061501 (2008) ; J. Phys. Soc. Jpn., **78**, 104801 (2009).
12. S. Kitsunezaki, J. Phys. Soc. Jpn. 79, 124802 (2010) ; Adv. Powd. Tech., **22**, 311 (2011).
13. A. Nakahara, Y. Matsuo, Phys. Rev. E, **74**, 045102(R) (2006).
14. Y. Matsuo, A. Nakahara, J. Phys. Soc. Jpn., **81**, 024801 (2012).
15. S. Kitsunezaki, A. Nakahara, Y. Matsuo, Europhys. Lett., **114**, 64001 (2016).
16. P. Lidon, N. Taberlet, S. Manneville, Softmatter, **12**, 2315 (2016).

Effect of particle shape on the efficiency of granular dampers

Hamzeh Pourtavakoli[1],*, *Eric J. R.* Parteli[2],**, and *Thorsten* Pöschel[1],***

[1] *University of Erlangen-Nuremberg, Erlangen, Germany*
[2] *University of Cologne, Cologne, Germany*

Abstract. Granular dampers, containers partially filled with granular material, are often applied for attenuating mechanical vibrations in a broad range of systems. However, the role of the particle shape on the performance of the damper has remained largely uncertain and is investigated here by means of particle-based simulations. It is found that, for large excitation amplitudes (collect-and-collide regime), particle shape nearly does not affect the damper's performance. For low excitation amplitudes (gas-like regime), a dependence on the average dissipated energy per cycle on the particle shape is found. In this regime, the spherical particle geometry leads to the highest damper's efficiency.

1 Introduction

Many authors have studied the influence of the parameters of the driving [1, 2], material properties and particle size [3–6] on the performance of granular dampers. However, the role of particle shape has remained largely out of the focus of previous investigations. This role is examined in the present work by means of numerical simulations using the Discrete Element Method (DEM), which consists of simultaneously solving Newton's equations of translational and rotational motion for all particles as described in detail in Ref. [7]. We used DEM to perform a systematic study by investigating a diversity of particle shapes covering a broad range of shape parameter values (specified below).

Since granular dampers perform particularly well in applications where the acceleration due to gravity can be neglected (as gravity tends to demobilize the granulate [1]), we focus here mainly on the study of granular dampers in microgravity conditions.

2 Numerical experiments

Although most DEM models are designed for particulate systems composed of spherical particles, geometrically complex particles can be modeled using the multisphere method, where spherical particles are arranged to form composite particles [8–11]. In Table 1 we present different particle shapes chosen to cover a broad range of values of the following shape parameters [12, 13]:

- *aspect ratio*, that is, the ratio of the Feret's minimum diameter to the Feret's maximum diameter

$$\eta \equiv D^{\mathrm{F}}_{\mathrm{min}}/D^{\mathrm{F}}_{\mathrm{max}}\,, \tag{1}$$

*e-mail: hamzeh.pourtavakoli@fau.de
**e-mail: eric.parteli@uni-koeln.de
***e-mail: thorsten.poeschel@fau.de

Table 1. Multisphere particles used in the numerical experiments (first row) along with associated symbols (second row) used in figs. 2-4. The corresponding shape parameters are listed in the subsequent rows. For particles #2 – #11 these parameters are calculated with respect to the plane *i*-*j* (see coordinate system on the bottom right hand corner). Bottom: particle #8 with the spherocylinders encompassing the particle's constituent beads (first sketch from left to right) as well as the definition of *P*, *A*, P_c and A_c for this particle.

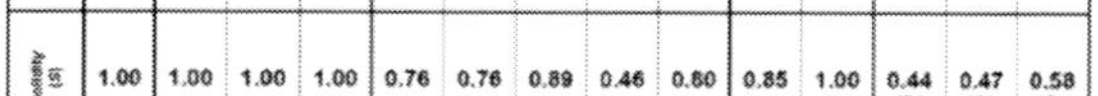

particle shapes	1	2	3	4	5	6	7	8	9	10	11	12	13	14
convexity (C_x)	1.00	1.00	1.00	1.00	0.91	0.76	0.74	0.77	0.69	0.94	1.00	0.89 (C_{x1})	0.95 (C_{x2})	0.97 (C_{x3})
solidity (S)	1.00	1.00	1.00	1.00	0.76	0.76	0.89	0.46	0.80	0.85	1.00	0.44 (S_1)	0.47 (S_2)	0.58 (S_3)
aspect ratio (η)	1.00	0.50	0.25	0.13	1.00	1.00	1.00	1.00	1.00	0.67	0.57	ij plane 0.67 jk plane 1.00 ik plane 1.00	ij plane 1.00 jk plane 1.00 ik plane 1.00	ij plane 1.00 jk plane 1.00 ik plane 1.00
circularity (C)	1.00	0.92	0.75	0.58	0.71	0.62	0.65	0.48	0.62	0.77	0.88			

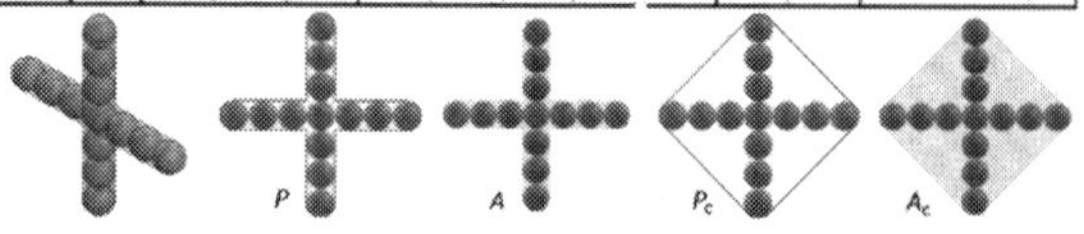

- *circularity*,

$$C \equiv \sqrt{4\pi A/P^2}, \tag{2}$$

where *A* and *P* denote the area and perimeter of the particle, respectively.

- *convexity*,

$$C_x \equiv P_c/P\,, \tag{3}$$

where P_c is the hull perimeter.

- *solidity*,

$$S \equiv A/A_c\,, \quad (4)$$

where A_c is the convex hull area. That is, A_c denotes the area inside the convex hull around the two-dimensional particle shape.

The sketches at the bottom of Table 1 show how to define P, P_c, A and A_c taking as example particle #8.

Moreover, for particle shapes #12 – #14 the center-of-mass positions of the constitutent beads do not lie on the same plane, that is, these complex particles are three-dimensional objects. Therefore, for these particles, the convexity and solidity are defined as follows:

- *convexity*,

$$C_{x3} = A_{c3}/A_3, \quad (5)$$

- *solidity*,

$$S_3 = V_3/V_{c3}, \quad (6)$$

where A_3 (V_3) is the sum of the total surface area (volume) of all spherocylinders enclosing the constituent spheres of the composite particle, while A_{c3} (V_{c3}) is the area (volume) of the three-dimensional convex hull or the smallest three-dimensional envelope enclosing the particle shape. Moreover, while the circularity is a measure inherent to two-dimensional objects [13], the aspect ratio of particle shapes #12 – #14 can be computed for each one of the planes defined in Table 1 and the results are displayed in this table.

In the DEM model, the multispheres particles are treated as rigid bodies, the translational and rotational motion of which results from the forces acting on the constituent beads. These forces are computed using non-cohesive inter-particle force model based on Cundall and Strack [14] and described in detail elsewhere [15]. In the model, the beads are treated as viscoelastic and friction is neglected [15]. The model equations have been implemented in the DEM library of Ref. [16], and the values of the model parameters are listed in table 2 (see Ref. [15] for more details about the model).

Table 2. Numerical values of the parameters used in the simulations. For particle-wall collisions the Poisson's ratio used was $\nu = 0.37$. The dimensions of the damper (L, W and H, defined with respect to the axes x, y and z, respectively) are as in Ref. [17].

parameter	symbol	value
particle material density	ρ_p	7800 kg/m^3
particle diameter	d	4 mm
Young's modulus	Y	10^8 Pa
Poisson's ratio	ν	0.3
box length	L	100 mm
box width	W	50 mm
box height	H	50 mm
timestep	Δt	10^{-6} s

3 Results and discussion

Figure 1 shows snapshots of the simulations of the damper oscillation. As we can see from this figure, for an amplitude of 10 mm, the granular system is in a gas-like regime, in which only a few particles interact with the oscillating walls during one oscillation period [17]. By contrast, for the amplitude of 80 mm, the system is in the collect-and-collide regime [17]. These regimes are separated by an optimal operation mode — the critical amplitude of the damper's oscillation — for which the energy dissipation is maximal.

The damping efficiency associated with a given particle shape can be characterized by computing the relative energy dissipated per cycle [17], $E_{\rm diss}/E_{\rm max}$, and is shown in Fig. 2 as a function of the oscillation amplitude for all particle shapes investigated. As we can see in this figure, for all particle shapes, the value of damping efficiency ($E_{\rm diss}/E_{\rm max}$) increases linearly with the amplitude $A_{\rm damp}$ in the gas-like regime, for $0 < A_{\rm damp} \lesssim 20$ mm. Moreover, for $A_{\rm damp} = A_{\rm crit} \approx 30$ mm a maximal value is reached, whereas the system enters the collect-and-collide regime as the amplitude becomes larger than 30 mm. As can be seen from Fig. 2, the critical amplitude $A_{\rm crit}$ is nearly independent of the particle shape. We also see in Fig. 2 that the results for the different particle shapes collapse well for the collect-and-collide regime. In this regime, the behavior of the system can be described by by means of a one-particle model [18, 19].

For amplitudes in the gas-like regime (in particular when $A_{\rm damp} \ll A_{\rm crit}$), we observe an interesting behavior in Fig. 2, that is the curves do not collapse as well as they do in collect-and-collide regime ($A_{\rm damp} > A_{\rm crit}$). In the gas-like regime, we expect the dissipative behavior of the system to be related with the granular temperature [7, 20], or equivalently the spread of the granular system, since in this regime the bulk of particles is located around the center of the damper (thus far from the driving walls). This means that the dissipated energy is due to collisions of only a few particles with driving walls, which are far away from the bulk.

The simplest quantity which we can compute to investigate this behavior is the standard deviation of the positions of the particles along the oscillation axis relative to the particles' center-of-mass position. We thus investigate the time-averaged standard deviation of the particles' positions $x_i(t)$ relative to the center-of-mass position $x_{\rm cm}(t)$ of the granular material, $\sigma_x = \left\langle \sum_{i=1}^{N} (x_i(t) - x_{\rm cm}(t))^2/N \right\rangle_t^{1/2}$ (with $N = 473$ particles), for an amplitude $A_{\rm damp} = 10$ mm, which is well within the range of amplitudes characterizing the gas-like regime. Figure 3 shows the damping efficiency as a function of σ_x for all particle shapes investigated. As we can see, by considering all complex particles shapes investigated, a linear increase of the damping efficiency with σ_x is observed, whereas this behavior is found for all amplitudes below 15 mm, that is when the system is well within the gas-like regime. The best fit to the simulation data associated with the complex particle shapes using the equation

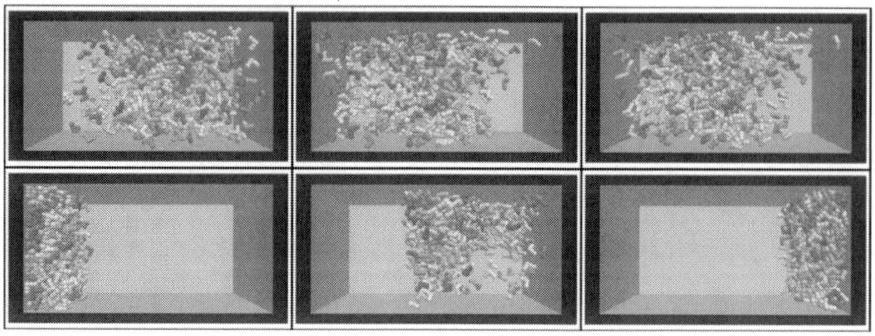

Figure 1. Snapshots of the damper simulations using particle shape #5 in Fig. 1 at oscillation amplitudes $A_{\text{damp}} = 10\,\text{mm}$ (top: gas-like regime) and $A_{\text{damp}} = 80\,\text{mm}$ (bottom: collect-and-collide regime). Shown are snapshots corresponding to the oscillation phases $\pi/2$, π and $3\pi/2$ (from left to right). In the simulations, the damper's horizontal position evolves in time according to the equation $x(t) = A_{\text{damp}}\sin(2\pi\nu t)$ (where $\nu = 4\,\text{Hz}$ is the oscillation frequency [17]).

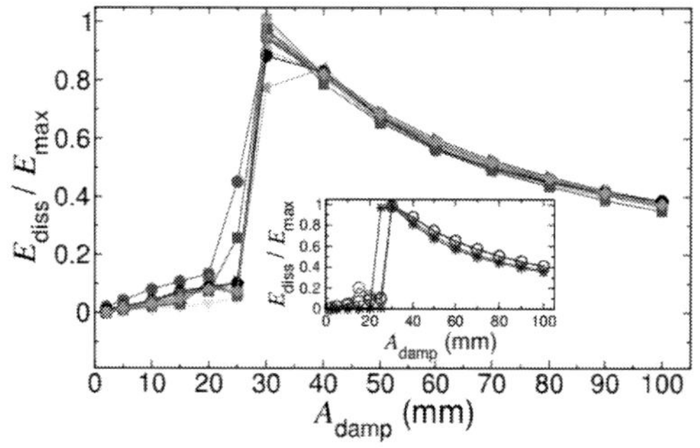

Figure 2. The main plot shows the damping efficiency, $E_{\text{diss}}/E_{\text{max}}$, as a function of the oscillation amplitude. The symbols correspond to different particle shapes as defined in Table 1. The inset shows the corresponding results for particles #1 (circles) and #4 (stars) of Table 1 obtained from simulations with frictional interactions, with $\mu = 0.2$ (red) and $\mu = 0.5$ (blue).

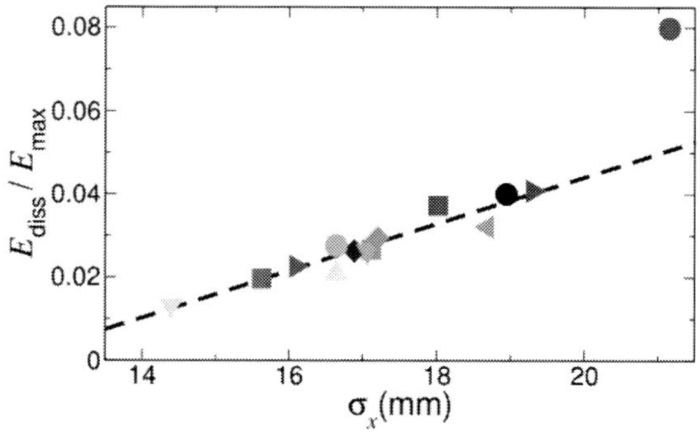

Figure 3. Dependence of the damping efficiency on the time-averaged standard deviation of the particles' positions along the oscillation axis (x) relative to the center-of-mass position of the particulate system. An amplitude of oscillation $A_{\text{damp}} = 10\,\text{mm}$ (gas-like regime) has been considered. Particle shapes are denoted to different symbols as in Table 1. The dashed line represents the fit to the simulation data (with exception of the spherical particle) using Eq. (7).

$$\frac{E_{\text{diss}}}{E_{\text{max}}} = a + b\sigma_x \qquad (7)$$

gives $a \approx -0.0688$ and $b \approx 0.00565\,\text{mm}^{-1}$, with correlation coefficient 0.95. This best fit is denoted by the continuous line in Fig. 3.

By considering one type of particle shape (the rods), we can see a dependence of the damping efficiency on some of the shape parameters. We see in Fig. 4a that the damping efficiency increases linearly with the aspect ratio η of the rods, whereas the best fit to the data using the equation $E_{\text{diss}}/E_{\text{max}} = a_r + b_r\eta$ gives $a_r \approx 0.0097$ and $b_r \approx 0.069$ with correlation coefficient 0.997. Moreover, in Fig. 4b we see that the damping efficiency increases non-linearly with circularity (C) of the rods. The best fit using the equation $E_{\text{diss}}/E_{\text{max}} = a_c + b_c C^{\alpha}$ gives $a_c \approx 0.023$, $b_c \approx 0.057$ and $\alpha \approx 13.2$ with correlation coefficient 0.994.

While the present manuscript focuses on the damping efficiency based on the conditions of the experiments of Refs. [17, 18, 21], the physics of the granular system in the gas-like regime should be investigated more systematically considering different particle numbers and system sizes. Moreover, it would be interesting to investigate the dissipative behavior of the different particle shapes in more detail, especially close to the critical amplitude where dissipation is maximal (see Fig. 2).

4 Conclusions

By means of DEM simulations under consideration of complex particle shapes, we have shown that particle shape affects the performance of granular dampers in the regime where the amplitude of oscillation is low (gas-like regime). In this regime, we could find an increase of damper's efficiency with circularity and aspect ratio when the particle shape is the rod. Moreover, independently of

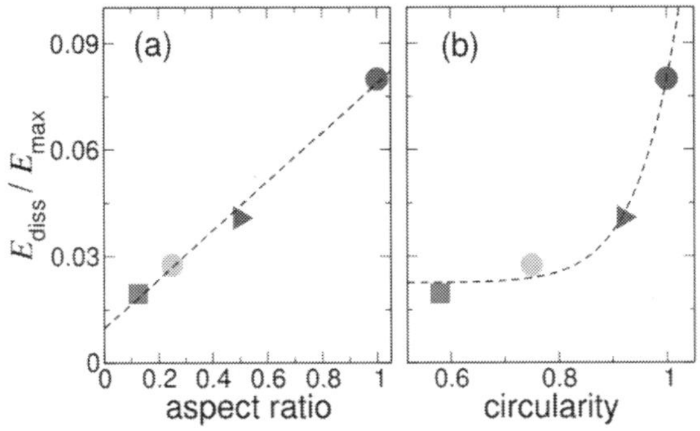

Figure 4. Damping efficiency E_{diss}/E_{max} as a function of aspect ratio η (a) and circularity C (b) for the rods. An amplitude of oscillation A_{damp} = 10 mm (gas-like regime) has been considered. The symbols denote different particle shapes according to Table 1. The dashed lines correspond to the fits $E_{diss}/E_{max} = 0.0097 + 0.069\eta$ (a) and $E_{diss}/E_{max} = 0.023 + 0.057C^{13.2}$ (b).

the particle shape, we found that the average energy dissipated by cycle of the damper in the gaseous regime increases with the standard deviation of the particle's position. This can be understood by noting that this standard deviation is a measure of the granular temperature, which plays an important role for the system's dynamics. Interestingly, the spherical particle shape leads to the highest efficiency in the gas-like regime.

Furthermore, we found that the critical amplitude above which the system abandons the gas-like regime and enters the collect-and-collide regime is independent of the particle shape. Moreover, in the collect-and-collide regime, the dependence of the damper's efficieny on the amplitude of oscillation is nearly independent of the particle shape.

Based on our results, the following question is raised: Does particle shape matter for the damper's efficiency? As a matter of fact, most dampers operate with nearly spherical particles, and our simulations suggest no improvement in the damper's efficiency is obtained if complex particles are used. This work should be now continued by including effect of friction. However, we have obtained results for spheres and rods using Coulomb friction coefficients μ = 0.2 and μ = 0.5, from which we found little dependence of E_{diss}/E_{max} on μ (see Fig. 2). Moreover, the effect of cohesion following the model of Ref. [22] as well as particle material properties (stiffness and normal dissipation behavior) on the damping efficiency should be investigated in future work.

Acknowledgements

We thank Mohammad Farahani for discussions. We thank the German Research Foundation (DFG) for funding through the Cluster of Excellence "Engineering of Advanced Materials", ZISC, FPS, the Collaborative Research Center SFB814, and grant PO472/20-2. We grate-fully acknowledge the computing time granted by the John von Neumann Institute for Computing (NIC) and provided on the supercomputer JUROPA at Jülich Supercomputing Centre (JSC).

References

[1] C. Salueña, T. Pöschel and S. E. Esipov, Phys. Rev. E **59**, 4422-4425 (1995)

[2] C. Salueña, S. E. Esipov, T. Pöschel and S. Simonian, Proc. SPIE **3327**, 19-26 (1998)

[3] W. Chen, M. Hou, K. Lu, Z. Jiang and L. Lam, Phys. Rev. E **64**, 061305 (2001)

[4] K. S. Marhadi and V. K. Kinra, J. Sound Vib. **283**, 433-448 (2005)

[5] X. M. Bai, L. M. Keer, Q. L. Wang and R. Q.Snurr, Granular Matter **11**, 417-429 (2009)

[6] M. Sánchez, G. Rosenthal and L. A. Pugnaloni, J. Sound Vib. **331**, 4389-4394 (2012)

[7] T. Pöschel and T. Schwager, *Computational Granular Dynamics* (Springer, Heidelberg, 2005)

[8] T. Pöschel and V. Buchholtz, Phys. Rev. Lett. **71**, 3964-3966 (1993)

[9] J. A. C. Gallas and S. Sokołowski, Int. J. Mod. Phys. **7**, 2037-2046 (1993)

[10] E. J. R. Parteli, AIP Conf. Proc. **1542**, 185-188 (2013)

[11] E. J. R. Parteli and T. Pöschel, Powder Technology **288**, 96-102 (2016)

[12] ISO (2008), ISO 9276-6:2008, *Representation of results of particle size analysis — Part 6: Descriptive and quantitative representation of particle shape and morphology* (ISO, Geneva, 2008)

[13] E. Olson, Journal of GXP Compliance **15(3)**, 85-96 (2011)

[14] P. A. Cundall and O. D. L. Strack, Geotechnique **29**, 47-65 (1979)

[15] H. Pourtavakoli, E. J. R. Parteli and T. Pöschel, New J. Phys. **18**, 073048 (2016)

[16] C. Kloss, C. Goniva, A. Hager, S. Amberger and S. Pirker, Prog. Comput. Fluid Dy. **12**, 140-152 (20112)

[17] A. Sack, M. Heckel, J. Kollmer, F. Zimber and T. Pöschel, Phys. Rev. Lett. **111**, 018001 (2013)

[18] M. N. Bannerman, J. Kollmer, A. Sack, M. Heckel, P. Müller and T. Pöschel, Phys. Rev. E **84**, 011301 (2011)

[19] A. Sack, M. Heckel, J. Kollmer and T. Pöschel, Granular Matter **17**, 73-82 (2015)

[20] K. Harth, U. Kornek, T. Trittel, U. Strachauer, S. Hörne, K. Will and R. Stannarius, Phys. Rev. Lett. **110**, 144102 (2013)

[21] J. Kollmer, M. Tupy, M. Heckel, A. Sack and T. Pöschel, Phys. Rev. Applied **3**, 024007 (2015)

[22] E. J. R. Parteli, J. Schmidt, C. Blümel, K.-E. Wirth, W. Peukert and T. Pöschel, Scientific Reports **4**, 6227 (2014)

EPJ Web of Conferences 140, 03025 (2017) DOI: 10.1051/epjconf/201714003025
Powders & Grains 2017

Heaping and secondary flows in sheared granular materials

Ralf Stannarius[1,*], *David* Fischer[1,**], and *Tamás* Börzsönyi[2,***]

[1] *Institute of Experimental Physics, Otto von Guericke University, Universitätsplatz 2, 39106 Magdeburg, Germany*
[2] *Institute for Solid State Physics and Optics, Wigner Research Center for Physics, Hungarian Academy of Sciences, P.O. Box 49, H-1525 Budapest, Hungary*

Abstract. Split-bottom cylindrical containers are well established devices in experiments where granular materials are continuously sheared. They are characterized by localized broad shear bands. In such shear experiments, shape-anisotropic grains develop a secondary flow profile in the radial direction. A macroscopically observable consequence is the formation of a heap in the center of the container. This effect is found for all investigated types of prolate and oblate grains, but it is completely absent for spherical particles. There are qualitative differences in the behavior of short (moderate aspect ratio < 8) and long grains. The fill height of the granulate in the container considerably affects the time scale for heap formation, while the height of the heap is less dramatically influenced. Under reversal of the shear direction, the heap collapses within a few rotations, before the secondary flow in the reverse direction establishes a new heap.

1 Introduction

Granular flows are present in countless processes encountered in geology, transport of materials like sand, coal or ores, and agricultural products. It also affects our everyday life in the kitchen or offices whenever we stir or shake grainy matter or pour it out of a container. However, while flow in Newtonian fluids has been well understood for centuries, and even many non-Newtonian flow phenomena are well characterized in liquids, granular flows have still preserved their mysteries to a great deal. Different aspects of granular dynamics were discussed in books and review articles (e.g. [1–5]). Peculiarities of shape-anisotropic granular matter have been reviewed recently [6], with focus on specific packing problems, shear flow, alignment and orientational order.

Recently, the phenomenon of secondary flow has attracted the interest of researchers in different fields of granular matter physics (e.g. [7–20]). One prominent example where secondary flow effects manifest themselves is the heaping of elongated grains under shear in a cylindrical geometry [19]. The formation of these structures as well as the dynamics of the secondary flow and its dependence on particle shapes and geometrical parameters of the setup have been studied in detail by Fischer et al. [20]. Wortel et al. [19] gave the following description of the phenomenon: Elongated grains form a heap in the center, which reaches a saturation height after a certain number of rotations. When the sense of rotation is reversed, the heap first collapses but thereafter reappears at a comparable time scale as from the disordered initial state. The fill level of the container has considerable influence on the phenomenon. The optimal fill height for heap formation is when the granulate at the surface above the revolving disk rotates with approximately half the disk velocity. A convective flow was detected by means of X-ray CT. Fischer et al. [20] showed that heaping is also found in oblate material. They demonstrated the influence of the fill height on the time scales of heap formation and presented the results with numerical simulations. We compare the experimental data for different rotationally symmetric grain shapes (cylindrical and ellipsoidal) and discuss the dependence of the secondary flow on these parameters and the geometry of the experiment.

*e-mail: ralf.stannarius@ovgu.de
**e-mail: david.fischer@ovgu.de
***e-mail: borzsonyi.tamas@wigner.mta.hu

2 Experimental method

Cylindrical containers with split-bottom have been introduced [4, 21, 22] to study granular materials under continuous shear. The central part of the bottom can be rotated with respect to the outer container, the granulate develops a broad shear zone [21]. Observation and analysis of the particle arrangements and dynamics can be performed optically at the granular surface [19–23], or in an index-matching fluid with a Laser sheet [24], destructively by excavation [25–27], or non-invasively by X-ray CT [26, 27, 30, 31] and Magnetic Resonance Imaging [28, 29].

Two geometries are explored in our investigation, the first one shown in Figure 1a is the conventional split-bottom experiment. All quantitative results presented here are obtained in this geometry. The second one shown in Figure 1b includes a vertical cylinder above the center of the rotating disk, much smaller than the disk size. Some qualitative results with this geometry are described below.

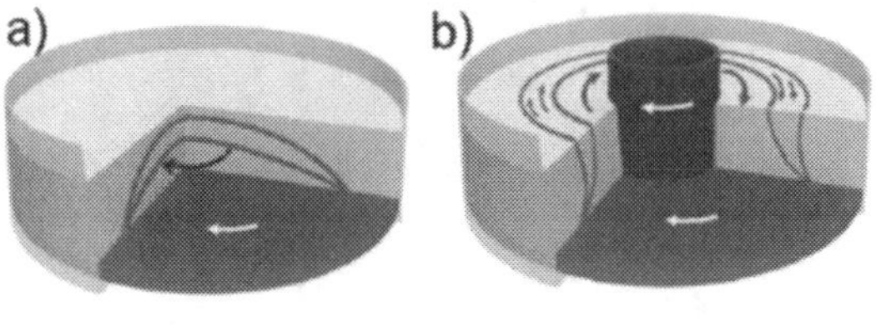

Figure 1. Sketch of the experimental setup and the location of the shear zones. a) In the first geometry of a simple rotating bottom plate, the shear zone can form a closed dome when the fill level of the container is low, or it forms an open ring as in b) when the fill level is sufficiently high. In the geometry b) with the inner cylinder, the shear zone has always an open ring shape.

The central part can be rotated by a motor at constant rates up to about 27 rpm, the outer part of the container remains fixed. The investigated materials and their properties are listed in Table 1.

Table 1. Lengths ℓ, diameters d and aspect ratios Q of the investigated grain types. The images reflect realistic size ratios.

	material	ℓ [mm]	d [mm]	Q
	plastic cyl.	24	2	12
	wooden peg	25	5	5
	glass cyl.	6.6	1.9	3.5
	airsoft ball	-	6	1
	pea	-	7	≈ 1
	lentil	2.5	7	0.36

Figure 2. Top view of the experiment with elongated cylinders (Q = 12) after the heap is fully developed. The distortion of the Laser reflection lines indicates the height profile in different regions. Here, multiple Laser lines visualize the asymmetry of the heap.

The top of the granular surface is illuminated with a He-Ne Laser sheet at an incident angle of 45°. A camera mounted vertically above the container allows us to record the horizontal displacement of the Laser reflection. From this shift, the radial profile of the emerging heap is calculated [20]. With an array of Laser lines (Figure 2), accomplished by a combination of a line Laser and a diffraction lattice, multiple scans of the granular surface can be recorded simultaneously during the shear experiment. This allows us to obtain more details when the heap profile is not axially symmetric.

3 Results

First, we discuss experiments in the geometry of Figure 1a: The formation of the heap extracted from the optical analysis of the surface profile is shown in Figure 3 for particles of different aspect ratios. We define the heap height h_{max} as the highest elevation of the granular surface in the container, relative to the initial, uniform fill level. The parameters for the graph of the $Q = 3.5$ glass rods are fill height $h_0 = 75$ mm, rotation rate of the bottom disk $\omega = 4.7$ rpm, for the $Q = 5$ pegs $h_0 = 75$ mm, $\omega = 3$ rpm, for the lentils $h_0 = 70$ mm, $\omega = 10$ rpm, and for the long $Q = 12$ plastic rods $h_0 = 90$ mm, $\omega = 5$ rpm. Heaping is observed for all investigated anisometric grain types, there is no clear trend in the final heap height vs. particle aspect ratio. For the perfectly spherical airsoft balls, $h_0 = 75$ mm, for the peas $h_0 = 80$ mm. Rotation rates of the bottom disk are 5 rpm in both cases. The fluctuations of the surface are less than one particle diameter in both cases.

In all experiments with anisometric grains except the long ($Q = 12$) rods, the peak appears directly above the central axis of the bottom disk, and the heap is axially symmetric (peculiarities of the $Q = 12$ material will be discussed in the next paragraph). For all other rods, the characteristic times of heap formation are comparable. The peak shape after the initial transient is stationary (see also Figure 5 below). This confirms earlier observations with different grain types (see Figures 6, 7, 9, 11 in Ref. [20]).

The long rods with aspect ratio $Q = 12$ behave qualitatively different. The heap forms above the rotating bottom disk (i. e. within the outer diameter of the disk), but not necessarily axially symmetric with respect to the rotation center. This asymmetry is evident in the profiles of the multiple Laser lines in Figure 2. A typical temporal evolution of the surface profile along a central cross section, illuminated by the single Laser sheet, is shown in the space-time plot of Figure 4: The off-axis heap rotates with the precession of the granulate, and moreover it performs slow structure changes (slow in the sense that it requires several dozen rotations of the disk for relevant shape changes of the heap). The heap shape is not stationary neither in the laboratory frame nor in the local co-rotating frame of the granular surface. Not only the position of the heap changes, but also the heap height fluctuates considerably. The short-period shifts of the peak in Figure 4 reflect the precession of the off-axis heap around the center. It is accompanied by a modulation of the absolute height with a period of about 15 revolutions of the bottom disk. This is also reflected in the fluctuations in Figure 3.

The contrast to the behavior of the shorter grains is evident in the comparison with Figure 5, top. There, the

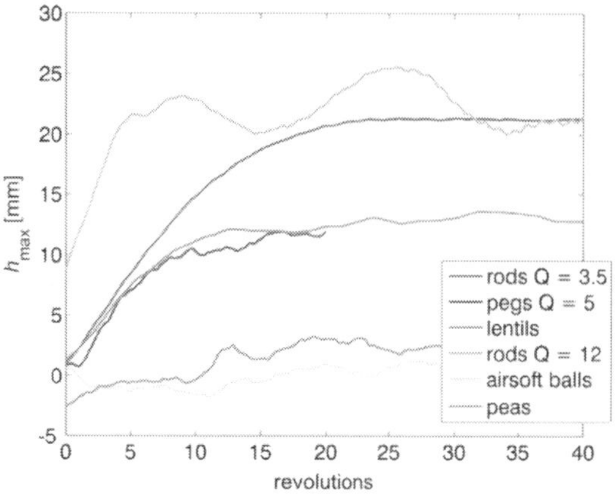

Figure 3. Evolution of the heap from the initial flat surface. The heap height h_{max} is defined as the highest point of the granular bed relative to the initial fill level. All measurements are performed with the optimum fill heights of the container (upper surface above the rotating disk rotates with approximately half the rate of the disk). The Q = 12 cylinders show large long-term fluctuations not only of the position of the peak, but also of its magnitude.

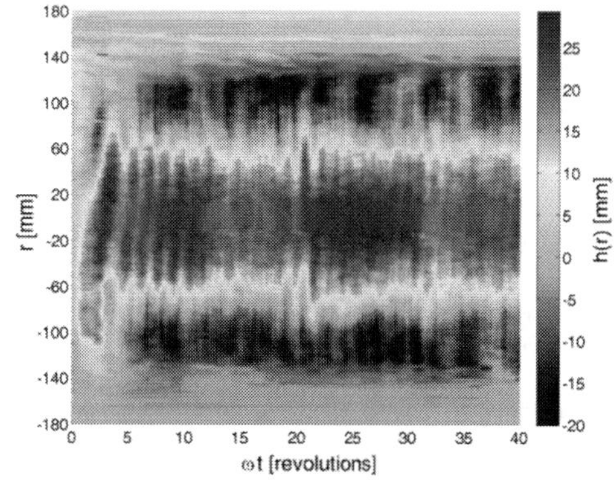

Figure 4. Temporal evolution of the cross section profile for plastic rods of Q = 12. It is evident that the heap is not located directly above the rotation axis. It rotates with the precession frequency of the granular surface but also undergoes slow structure reorganizations on the time scale of dozens of rotations. h_0 = 90 mm, ω = 5 rpm

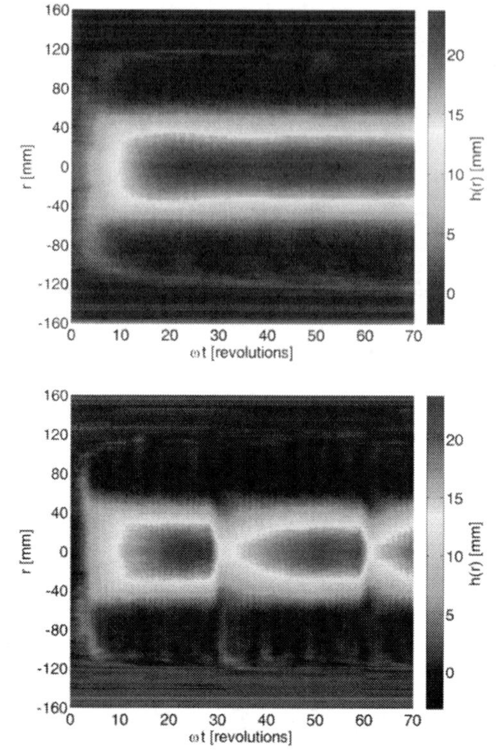

Figure 5. Evolution of the heap profile of Q = 3.5 glass rods under constant shear (top) and during successive reversal of the shear direction every 30 revolutions (bottom). The heap collapses almost instantly after shear reversal, but thereafter recovers with a rate comparable to the initial growth rate. After the heap has formed, it remains stationary. This is in sharp contrast to the dynamics of the long rods shown in Figure 4.

heap is central and symmetric and it does not noticeably fluctuate in shape. The bottom image of Figure 5 shows collapse and recovery of the peak when the rotation direction of the bottom disk is reversed. Peak formation is reproducible and independent of the shear direction. After each reversal, the peak re-appears with the same profile and same height, in sharp contrast to the Q = 12 rods where the positions and heights of the heap are not exactly reproducible.

Finally, we describe qualitatively the experiments with a central vertical cylinder (Figure 1b). In this modified geometry, the heap formation is suppressed. If the cylinder has a large diameter (e.g. 16 cm cylinder above a 39 cm bottom disk), then one finds only a shallow wall around the cylinder, no heap, and the surface profile slightly decreases from that wall to the cylinder surface (Figs. 4, 11 in Ref. [31]). But even with a narrow cylinder, with a diameter of the order of a few grains, we find only a circular wall around it, with a slight depletion towards the cylinder surface. Obviously, the vertical flow in the convection vortex is inhibited at the cylinder surface, and thus the heaping process is much less effective there. The convection still takes place, but the roll ends inwards at some distance from the vertical cylinder wall. This distinguishes the observed heaping in anisometric granular matter from the at first glance very similar phenomenon of rod climbing (*Weissenberg* effect [33]) in some complex liquids. There, the rotation of a rod inside the liquid leads to the climbing of the liquid upwards at the rod surface.

4 Discussion and Summary

It has been demonstrated [20] that heaping in cylindrical split-bottom containers is characteristic for both prolate and oblate particles, while for all spherical or near-spherical grains the surface remains flat, without heaping. These experiments have been extended to high aspect ratios. The results give evidence that the heaping phenomenon changes qualitatively for the material with aspect ratio $Q = 12$, in contrast to the materials with lower aspect rations, no stationary central heap is formed. Instead, we find fluctuating off-axis heaps. We note that this cannot be the consequence of the large ratio of particle length and bottom disk size, since this ratio is even larger for the $Q = 5$ pegs, which nevertheless form a stationary central heap. It is rather an indication that the dynamic behavior of such materials cannot simply be extrapolated from the dynamics of less anisometric grains. At large aspect ratios, qualitatively new features have to be taken into account. This agrees with studies of hopper discharge of the same grain types [32].

Comparison of the heap formation process in the geometries of Figure 1a,b emphasizes the importance of details of the shear geometry. On first glance, the observed heaping reminds one of the Weissenberg (rod-climbing) effect in complex fluids. However, the latter effect is even qualitatively different from the heaping observed here. A completely different physical origin has to be sought, irrespective of some striking phenomenological similarities. The effect described here should be studied in detail in other geometries, in order to reveal whether the heaping phenomenon is restricted to the cylindrical split-bottom container, or whether it is also present in a linear shear cell (two L-shaped sliders). Such experiments will help to clarify the nature of the convection, which is still not understood satisfactorily.

Acknowledgements

Financial support from the DAAD/MÖB researcher exchange program (Grants No. 57141022 and 64975), from DFG (Grant STA 425/38), and from OTKA (Grant No. OTKA NN 107737) is acknowledged. The authors thank Torsten Trittel for technical assistance and helpful advice.

References

[1] H. M. Jaeger, S. R. Nagel, and R. P. Behringer, Rev. Mod. Phys. **68** 1259-1273 (1996)

[2] J. Duran, *Sands, Powders, and Grains,* (Springer New York, 2000)

[3] I. S. Aranson and L. S. Tsimring, Rev. Mod. Phys. **78** 641-692 (2006)

[4] J. A. Dijksman and M. van Hecke, Soft Matter **6** 2901 - 2907 (2010)

[5] M.C. Marchetti et al., Rev. Mod. Phys. **85** 1143-1189 (2013)

[6] T. Börzsönyi R. and Stannarius, Soft Matter **9** 7401 (2013)

[7] J. B. Knight, H. M. Jaeger, and S. R. Nagel, Phys. Rev. Lett. **70** 3728 (1993)

[8] R. Khosropour, J. Zirinsky, H. K. Pak, and R. P. Behringer, Phys. Rev. E **56** 4467 (1997)

[9] S. Laín, M. Sommerfeld, and B. Quintero, Brazil. J. Chem. Eng. **26** 583 (2009)

[10] V. Vidyapati et al., Powder Techn. **220** 7-14 (2012)

[11] M. Alletto and M. Breuer, Int. J. Multiphase Flow **55** 80-98 (2013)

[12] M. Harrington, J. H. Weijs, and W. Losert, Phys. Rev. Lett. **111** 078001 (2013)

[13] N. Murdoch et al., Phys. Rev. Lett. **110** 018307 (2013)

[14] J. Havlica, K. Jirounkova, T. Travnickova, M. Kohout, Powder Techn. **280** 180-190 (2015)

[15] K. P. Krishnaraj and P. R. Nott, Nature Comm. **7** 10630 (2016)

[16] T. M. Yamada and H. Katsuragi, Planetary and Space Sci. **100** 79-86 (2014)

[17] F. Zhang, Li Wang, C. Liu, P. Wu, and S. Zhan, Phys. Letters A **378** 1303-1308 (2014)

[18] T. M. Yamada, K. Ando, T. Morota, and H. Katsuragi, ICARUS **272** 165-177 (2016)

[19] G. Wortel, T. Börzsönyi, E. Somfai, S. Wegner, B. Szabó, R. Stannarius, and M. van Hecke, Soft Matter **11** 2570-2576 (2015)

[20] D. Fischer, T. Börzsönyi, D. Nasato, T. Pöschel, and R. Stannarius, New J. Phys., **18**, 113006 (2016)

[21] D. Fenistein and M. van Hecke, Nature **425** 256 (2003)

[22] D. Fenistein, J. W. van de Meent, and M. van Hecke, Phys. Rev. Lett. **92** 094301 (2004)

[23] R. Moosavi et al., Phys. Rev. Lett. **111** 148301 (2013)

[24] S. Slotterback, M. Mailman, K. Ronaszegi, M. van Hecke, M. Girvan, and W. Losert, Phys. Rev. E **85** 021309 (2012)

[25] D. Fenistein, J.-W. van de Meent, and M. van Hecke, Phys. Rev. Lett. **96** 118001 (2006)

[26] T. Börzsönyi et al., Phys. Rev. Lett. **108** 228302 (2012)

[27] T. Börzsönyi et al., Phys. Rev. E **86** 051304 (2012)

[28] X. Cheng et al., Phys. Rev. Lett. **96** 038001 (2006)

[29] K. Sakaie, D. Fenistein, T. J. Carroll, M. van Hecke, and P. Umbanhowar, Europhys. Lett. **84** 38001 (2008), *Europhys. Lett.* **84** 49902 (2008)

[30] S. Wegner, T. Börzsönyi, T. Bien, G. Rose, and R. Stannarius, Soft Matter **8** 10950-10958 (2012)

[31] S. Wegner et al., Soft Matter **10** 5157-5167 (2014)

[32] A. Ashour, S. Wegner, T. Trittel, T. Börzsönyi, and R. Stannarius, Soft Matter **13** 402 (2017)

[33] K. Weissenberg, Nature **159** 310-311 (1947).

Dense, inhomogeneous shearing flows of spheres

Diego Berzi[1,*], and *James T.* Jenkins[2]

[1]Dept. of Civil and Environmental Engineering, Politecnico di Milano, 20133 Milano, Italy
[2]School of Civil and Environmental Engineering, Cornell University, 14850 Ithaca, NY, USA

Abstract. We make use of recent extensions of kinetic theory of granular gases to include the role of particle stiffness in collisions to deal with pressure-imposed shearing flows between bumpy planes in relative motion, in which the solid volume fraction and the intensity of the velocity fluctuations are not uniformly distributed in the domain. As in previous numerical simulations on the flow of disks in an annular shear cell, we obtain an exponential velocity profile in the region where the volume fraction exceeds the critical value at which a rate-independent contribution to the stresses arises. We also show that the thickness of the inertial region, where the solid volume fraction is less than the critical value, and the shear stress at the moving boundary are determined functions of the relative velocity of the boundaries.

1 Introduction

Recently [1], it has been quantified how the particle stiffness influences the frequency of collisions responsible for the rate-dependent components of particle stresses, and induces the development of rate-independent components to the stresses associated with permanent deformations. That theory has been successfully tested against discrete numerical simulations of simple shearing, in which the velocity fluctuations and the solid volume fraction were uniformly distributed. It has been also briefly shown [1] how that theory can be applied to nonhomogeneous shearing flows, to explain the presence of an exponentially decaying velocity profile in very dense regions (creeping flow [2, 3]). Here, we use a simplified version of the theory of ref. [1] to solve for the flow of identical spheres sheared between two parallel, bumpy planes, one moving at constant velocity and one at rest. To mimic the results of previous discrete numerical simulations of disks in an annular shear cell [4], we take the pressure constant in the flow and the shear stress inversely proportional to the square of the distance from the moving boundary. Suitable boundary conditions allow us to obtain solutions to the flow that are in qualitative agreement with the numerical simulations.

2 Theory

An assembly of identical spheres of mass density ρ_p and diameter d is sheared between two parallel planes, with the upper plane moving at constant velocity, in the absence of gravity. The two planes are made bumpy by gluing a layer of spheres (identical to those of the flow) in a regular, hexagonal array. The bumpiness ψ of the planes is, then, determined by the mean interparticle distance between the glued grains [5]. We take x and y to be the directions parallel and perpendicular to the planes, respectively. The x-component of the particle velocity, the only one present, is u, while ν is the solid volume fraction. The distance, R, between the planes and the velocity, V, of the moving plane are the control variables of the system. The flow configuration is depicted in Fig. 1.

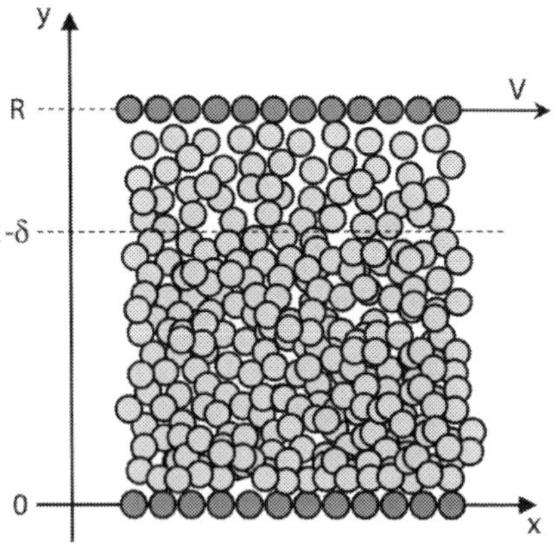

Fig. 1. Sketch of the flow configuration with the frame of reference.

To mimic the annular shear cell of Koval et al [4] with the above described configuration, we take the distributions of particle stresses to be those determined from the numerical simulations: a constant particle pressure p; and for the particle shear stress,

* Corresponding author: diego.berzi@polimi.it

$s = S\left[R/(R-y)\right]^2$, where S is the particle shear stress at $y = R$. As observed in the simulations, the flow domain can be in general divided into two sub-regions: an inertial region adjacent to the moving boundary of thickness δ and a quasi-static region of thickness R-δ (Fig. 1). Koval et al. [4] distinguished the two regions on the basis of the local value of the stress ratio s/p being larger or lower than the minimum (yield) value in simple shearing. Here, based on the analysis reported in ref. [1], we identify the inertial (quasi-static) region with the region where the solid volume fraction is less (larger) than the critical value, ν_c, necessary to develop elastic stresses associated with permanent deformations of the aggregate [6].

2.1 Inertial region

In the inertial region, we use the constitutive relations for the pressure and the shear stress of kinetic theory [7] in the dense limit [8],

$$p = \rho_p 2(1+e)\nu^2 g_0 T; \tag{1}$$

and

$$s = \rho_p d \frac{8J\nu^2 g_0}{5\pi^{1/2}} T^{1/2} u', \tag{2}$$

where: e is the coefficient of normal restitution (the negative of the ratio of pre- to post-collisional relative velocity along the line of centres of two colliding grains); g_0 is the radial distribution function at contact, here taken to be the dense limit of that suggested in ref. [9], $g_0 = 2/(\nu_c - \nu)$, to fit the results of numerical simulations of simple shearing when e is less than 0.95; T is the granular temperature (one third of the mean square of the velocity fluctuations); and $J = (1+e)/2+\pi(1+e)^2(3e\text{-}1)/[96\text{-}24(1\text{-}e)^2\text{-}20(1\text{-}e^2)]$. Here, and in what follows, a prime indicates a derivative with respect to y.
The fluctuation energy balance is

$$Q' = su' - \Gamma, \tag{3}$$

where Q is the flux of fluctuation energy:

$$Q = -\frac{4M\nu^2 g_0}{\pi^{1/2}} \rho_p d T^{1/2} T', \tag{4}$$

with $M = (1+e)/2+9\pi(1+e)^2(2e\text{-}1)/[128\text{-}56(1\text{-}e)]$; and Γ is the collisional rate of dissipation:

$$\Gamma = \frac{12(1-\varepsilon^2)\nu^2 g_0}{\pi^{1/2} L} \rho_p T^{3/2}, \tag{5}$$

where $\varepsilon = e\text{-}3/2\mu \exp(\text{-}3\mu)$ is an effective coefficient of restitution that takes into account the role of particle surface friction, μ [10]. The quantity L in eq. (5) is the correlation length of extended kinetic theory [11, 12] to take into account the breaking of molecular chaos at volume fractions larger than 0.49. Here, as in ref. [13], we use:

$$L = \left[\frac{4J}{15(1-\varepsilon^2)}\right]^{1/2} \left[1 + \frac{26(1-\varepsilon)}{15} \frac{\nu - 0.49}{0.64 - \nu}\right]^{3/2} \frac{d^2 u'}{T^{1/2}}. \tag{6}$$

Without loss of generality, we now take the particle density, diameter and pressure to be unity. As in ref. [8], we derive eq. (1) with respect to y, and use the expression of the radial distribution function at contact and eq. (4) to obtain a differential equation for the solid volume fraction,

$$\nu' = \frac{\pi^{1/2} Q}{4M\left[\nu^2 g_0\right]_\nu T^{3/2}}; \tag{7}$$

where the subscript indicates the derivative with respect to ν and T is, from eq. (1), $T = \left[2(1+e)\nu^2 g_0\right]^{-1}$. A differential equation for the particle velocity is obtained by inverting eq. (2),

$$u' = \frac{5\pi^{1/2} s}{8J\nu^2 g_0 T^{1/2}}. \tag{8}$$

Finally, the differential equation for the energy flux is, from eq. (3) with eqs. (5) and (8),

$$Q' = \frac{5\pi^{1/2} s^2}{8J\nu^2 g_0 T^{1/2}} - \frac{12(1-\varepsilon^2)\nu^2 g_0 T^{3/2}}{\pi^{1/2} L}, \tag{9}$$

where L is calculated from eqs. (6) and (8). The system of the 3 differential equations (7)-(9) allows the determination of the distribution of ν, u and Q in the inertial region once appropriate boundary conditions are provided.

At $y = R$, we use the boundary conditions derived by Richman [5] for the flows of spheres over bumpy planes: a slip velocity given by

$$u_B = \left(\frac{\pi}{2}\right)^{1/2} S \left\{ \frac{1 - \left[5(1+e)/2^{5/2} J\right](1+B)\sin^2\psi}{(2/3)\left[2\csc^2\psi(1-\cos\psi) - \cos\psi\right]} + \frac{5(1+e)}{2^{3/2} J} \right\} T_R^{1/2}, \tag{10}$$

where $B = 12\pi/2^{1/2}[1+5/(8\nu g_0)]$; and an energy flux given as

$$Q_B = S u_B - \frac{2^{3/2}(1-\cos\psi)\csc^2\psi}{\pi^{1/2}} T^{1/2} \frac{(1-\varepsilon)}{L_R}. \tag{11}$$

In eqs. (10) and (11), T_R and L_R are the granular temperature and correlation length evaluated at $y = R$.

At $y = \delta$ (the interface between the inertial and the quasi-static region), we use the boundary condition for the energy flux at an erodible boundary [14], modified with the introduction of the correlation length,

$$Q_E = -2T_{R-\delta}^{1/2}\left[\frac{3M(1-\varepsilon)}{\pi L_{R-\delta}}\right]^{1/2}, \quad (12)$$

where $T_{R\text{-}\delta}$ and $L_{R\text{-}\delta}$ are the granular temperature and correlation length evaluated at $y = \delta$. We also fix the solid volume fraction there to be equal to a given value ν_E, less than ν_c. We cannot use ν_c as a boundary condition for the solid volume fraction, because the radial distribution function at contact is singular there, and so would be the stresses [15]. Including the role of the particle stiffness in collisions would allow for the stresses to be non-singular at $\nu = \nu_c$ [1]. Taking ν_E less than ν_c is a simpler, but equivalent, alternative to that approach. Finally, we set the value u_E of the particle velocity at $y = \delta$, based on the analytical solution to the flow in the quasi-static region (see next Section). The five boundary conditions $u(y = R) = V\text{-}u_B$, $u(y = \delta) = u_E$, $\nu(y = \delta) = \nu_E$, $Q(y = R) = -Q_B$ and $Q(y = \delta) = Q_E$ permit the solution (using the Matlab® routine 'bvp4c') of the system of differential equations (7)-(9) and the determination of the thickness δ and the value of the shear stress S at $y = R$ as part of the solution.

2.2 Quasi-static region

In the quasi-static region, i.e., $y \le \delta$, we assume, as in ref. [1], that the stresses have an elastic component, associated with permanent deformations of the particles, and an inertial component, associated with the momentum exchange due to the fluctuating motion of the particle centres of mass. We assume that the ratio of the shear stress to the pressure is equal to the ratio of eq. (2) to eq. (1) even in the quasi-static region (this allows the reproduction of the results of numerical simulations of simple shearing [1]). With $p = d = 1$, this gives

$$s = \frac{4J}{5\pi^{1/2}(1+e)}\frac{du'}{T^{1/2}}, \quad (13)$$

We also assume that the fluctuation energy balance in this quasi-static region reduces to a balance between the diffusion and the collisional dissipation of energy, i.e., that the energy production due to the work of the inertial component of the shear stress is negligible. The constitutive relations for the energy flux and the dissipation rate are [1]

$$Q = -\frac{5M\nu_c}{6}E^{1/2}T', \quad (14)$$

and

$$\Gamma = \frac{5(1-\varepsilon^2)\nu_c}{2L}E^{1/2}T, \quad (15)$$

respectively, where E is the Young's modulus of the particles. For simplicity, we take L in eq. (15) to be equal to the its value in simple shearing [1],

$$L = 1 + \frac{26(1-\varepsilon)}{15}\frac{\nu_c - 0.49}{0.64-\nu_c}. \quad (16)$$

Using eqs. (14)-(16) in the energy balance, and neglecting the work of the shear stress, gives

$$T'' = \frac{1}{\lambda^2}T, \quad (17)$$

where $\lambda^2 = LM/\left[3(1-\varepsilon^2)\right]$. The analytical solution of eq. (17) is

$$T = T_{R-\delta}\exp\left(-\frac{R-\delta-y}{\lambda}\right), \quad (18)$$

where, as already mentioned, $T_{R\text{-}\delta}$ is the granular temperature at $y = R\text{-}\delta$. Using eq. (18) in eq. (13), with the distribution of the shear stress with the distance from the planes measured in the numerical simulations, and integrating, with the boundary condition $u = 0$ when $y = 0$, gives

$$u = \frac{5\pi^{1/2}(1+e)}{4J}ST_{R-\delta}^{1/2}2\lambda \times\left[\frac{R^2}{(2R-y)^2}\exp\left(-\frac{R-\delta-y}{2\lambda}\right)-\frac{1}{4}\exp\left(-\frac{R-\delta}{2\lambda}\right)\right], \quad (19)$$

When $y = R\text{-}\delta$, eq. (19) provides the value of the velocity u_E to be used as boundary condition in the numerical solution of the inertial region (see previous Section):

$$u_E = \frac{5\pi^{1/2}(1+e)}{4J}ST_{R-\delta}^{1/2}2\lambda\left[\frac{R^2}{(R+\delta)^2}-\frac{1}{4}\exp\left(-\frac{R-\delta}{2\lambda}\right)\right]. \quad (20)$$

3 Results and conclusions

We now obtain the results of the present theory for a representative set of parameters. We take: $e = 0.7$ and $\mu = 0.5$, so that $\nu_c = 0.587$ [6]; $\nu_E = 0.586$ and $\psi = \pi/6$. We use different values of total cell thickness R and plane velocity V and make qualitative comparisons with the numerical simulations on disks in an annular shear cell [4].

Fig. 2 shows the particle velocity, scaled with the velocity of the moving plane, as a function of the distance from the moving plane. As in ref. [4], it is possible to appreciate the slip velocity at $y = R$ and the exponential behaviour (eq. (19)) in the quasi-static region; when $R = 50$ and $V = 2.5$, it is also possible to notice the change in the concavity of the velocity profile in the inertial region. The thickness of the inertial region, marked by the knee in the velocity profiles of Fig. 2, apparently increases with increasing R and V.

The behaviour of δ with V is actually non-monotonic, as shown in Fig. 3 for the case $R = 50$. The thickness δ is zero when V is less than 0.1, and has a maximum when V is near 2.5; for large values of V, it decreases slightly and seems to saturate for even larger values. This is qualitatively similar to the relation between the shear stress at $y = R$ and the velocity of the moving plane depicted in Fig. 4, although in that case the decrease after the peak is much more pronounced. Koval et al. [4] obtained monotone increase of δ and S with V. However, they did not perform simulations for V larger than 2.5, and their figures 9a and 3b seem to confirm that the measurements are approaching a maximum at that value.

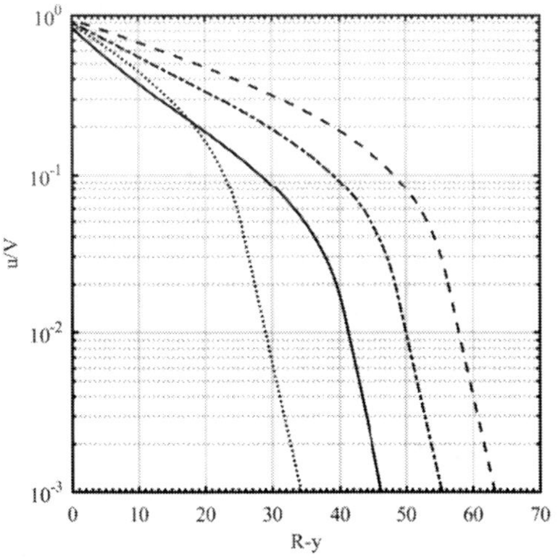

Fig. 2. Velocity profiles for: $R = 50$ and $V = 2.5$ (solid line); $R = 100$ and $V = 2.5$ (dashed line); $R = 200$ and $V = 2.5$ (dot-dashed line); $R = 50$ and $V = 1$ (dotted line).

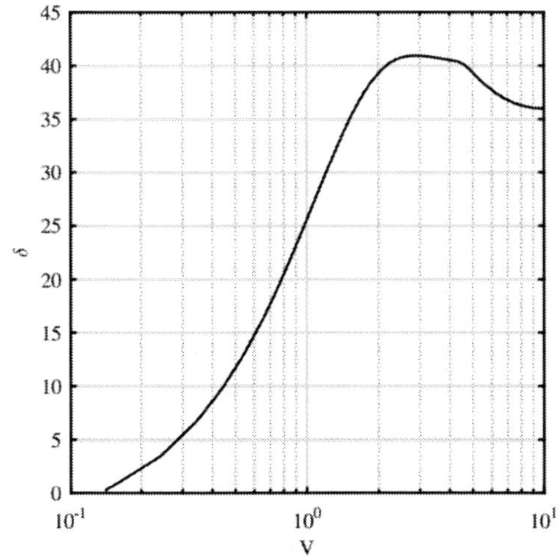

Fig. 3. Thickness of the inertial region as a function of the velocity of the moving plane for $R = 50$.

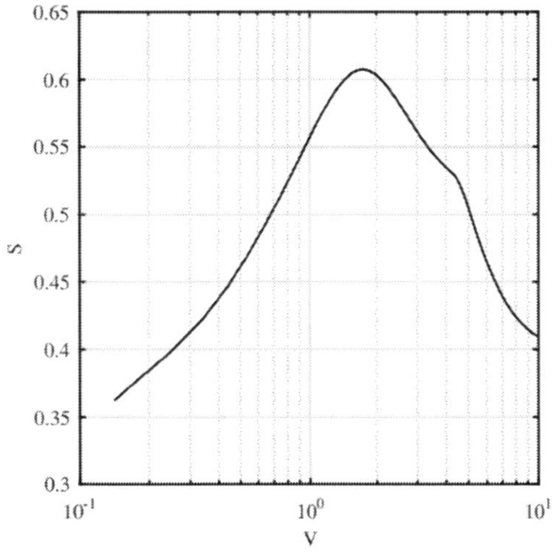

Fig. 4. Shear stress at $y = R$ as a function of the velocity of the moving plane for $R = 50$.

Although the results of the present theory are in good qualitative agreement with 2D discrete numerical simulations, 3D discrete numerical simulations are required to fully test the capability of the theory developed in ref. [1] to reproduce dense granular flows in nonhomogeneous configurations. We postpone these comparisons to future work.

References

1. D. Berzi, J.T. Jenkins, Soft Matter **11**, 4799 (2015)
2. T.S. Komatsu, S. Inagaki, N. Nakagawa, S. Nasuno, Phys. Rev. Lett. **86**, 1757 (2001)
3. P. Richard, A. Valance, J.-F. Métayer, P. Sanchez, J. Crassous, M. Louge, R. Delannay, Phys. Rev. Lett. **101**, 248002 (2008)
4. G. Koval, J.-N. Roux, A. Corfdir, F. Chevoir, Phys. Rev. E **79**, 021306 (2009)
5. M.W. Richman, Acta Mech. **75**, 227 (1988)
6. S. Chialvo, J. Sun, S. Sundaresan, Phys. Rev. E **85**, 021305 (2012)
7. V. Garzó, J.W. Dufty, Phys. Rev. E **59**, 5895 (1999)
8. D. Berzi, J.T. Jenkins, Phys. Fluids **27**, 033303 (2015)
9. D. Vescovi, D. Berzi, P. Richard, N. Brodu, Phys. Fluids **26**, 053305 (2014)
10. S. Chialvo, S. Sundaresan, Phys. Fluids **25**, 070603 (2013)
11. J.T. Jenkins, Granul. Matter **10**, 47 (2007)
12. J.T. Jenkins, D. Berzi, Granul. Matter **12**, 151 (2010)
13. D. Berzi, Acta Mech. **225**, 2191 (2014)
14. J.T. Jenkins, E. Askari, J. Fluid Mech. **223**, 497 (1991)
15. D. Berzi, D. Vescovi, Phys. Fluids **27**, 013302 (2015)

Coupled flow and deformations in granular systems beyond the pendular regime

Chao Yuan[1,*], *Bruno* Chareyre[1,**], and *Felix* Darve[1,***]

[1] *Université Grenoble Alpes, Laboratoire 3SR, UMR CNRS 5521 BP 53, 38041, Grenoble, France*

Abstract. A pore-scale numerical model is proposed for simulating the quasi-static primary drainage and the hydro-mechanical couplings in multiphase granular systems. The solid skeleton is idealized to a dense random packing of polydisperse spheres by DEM. The fluids (nonwetting and wetting phases) space is decomposed to a network of tetrahedral pores based on the Regular Triangulation method. The local drainage rules and invasion logic are defined. The fluid forces acting on solid grains are formulated. The model can simulate the hydraulic evolution from a fully saturated state to a low level of saturation but beyond the pendular regime. The features of wetting phase entrapments and capillary fingering can also be reproduced. Finally, a primary drainage test is performed on a 40,000 spheres of sample. The water retention curve is obtained. The solid skeleton first shrinks then swells.

1 Introduction

The situation of two immiscible fluids - say, water and air - going through a deformable granular material is widely encountered in nature and in many areas of engineering and science. Modeling such multi-phase systems has long been restricted to the so called pendular regime, in which the wetting phase (i.e., water in air/water systems) is only present in the form of bridges connecting grain pairs [1]. The volume content of wetting phase is constrained to a very low level of saturation (usually less than 5%). The hydro-mechanical couplings for highly saturated states, capillary regime and funicular regime, for instance, have been rarely addressed.

In this work, we present a pore-scale numerical model to simulate the quasi-static primary drainage and the hydro-mechanical couplings, where the wetting phase is drained from a fully saturated state to a residual state but beyond the pendular regime. The basic idea is to combine the discrete element method (DEM) and a pore-network model for, respectively, the granular solid phase and two fluids. The drainage criterion and procedure can be established based on the decomposition of the pore space. The capillary effects, i.e., the fluids forces on grains, can be formulated as well. Then motion of the grains can be predicated by the DEM solver and consequently a macro-scale deformation of the skeleton is obtained. In the end of this paper, we will apply the model to simulate the primary of a random packing in order to obtain the water retention curve and the mechanical response.

*e-mail: chao.yuan@3sr-grenoble.fr
**e-mail: bruno.chareyre@3sr-grenoble.fr
***e-mail: felix.darve@3sr-grenoble.fr

2 Numerical model

2.1 Network

We consider that the solid skeleton is represented by a dense random cuboid packing of poly-disperse spheres, which is created by DEM [2]. The void space is decomposed into large pores connected by narrow throats using Regular Triangulation (RT) method. RT method generalizes the Classical Delaunay Triangulation (CDT) to weight points, where the weight accounts for the size of each solid sphere[3]. The difference between RT method and CDT method is shown in Fig.1. The merit of RT method is that its dual Voronoi graph is entirely contained in the void space, and therefore is appropriately used to describe the flow path. For the boundaries, we represent the rigid confining walls of the cuboid sample by fictitious spheres with infinite radii (*i.e.*, assuming $R \to +\infty$, P'_1, P_1, P_2 and P'_2 become aligned) in order to keep the consistency of the geometrical algorithm.

A typical network generated by RT method is shown in Fig.2 (a). Based on the decomposition, we define the void volume in a tetrahedron as a pore body and its four related facets as four pore throats (see Fig.2 (b)). The throat does not enclose any volume, but it will play a key role when defining the criterion of the drainage.

2.2 Governing equations

2.2.1 Assumptions

We propose the model to simulate the quasi-static hydro-mechanical behaviors of multiphase granular materials. The system contains of a pair of immiscible wetting and non-wetting fluid phases and a solid phase, which are

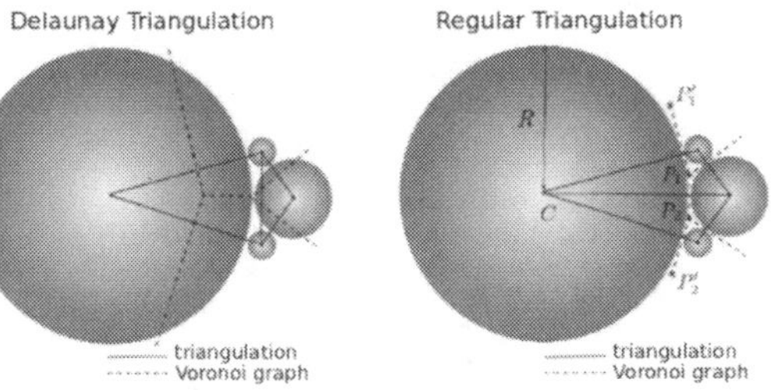

Figure 1. Comparison of 2D triangulation and dual Voronoi graph obtained from classical Delaunay and regular Delaunay: classical Voronoi graph has branches inside discs, while regular Voronoi gives all branches in the void space.

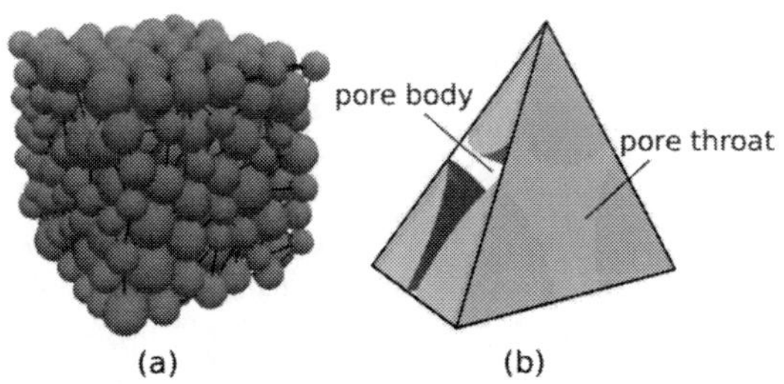

Figure 2. (a) Definition of pore network for packing of polydisperse spheres, generated by regular triangulation. (b) A pore geometry defined by tetrahedral element of the finite volume decomposition.

noted W-phase, NW-phase and S-phase, respectively. The S-phase is perfectly wetting by the W-phase. The "quasi-static" is restricted by defining the capillary number C_a approaching 0. ($C_a = \frac{\mu v}{\gamma}$, where μ is the viscosity of receding phase, i.e., W-phase; v is the average velocity of the receding phase; and γ is the interfacial tension between the two fluids.) In such case, the viscous effects are negligible.

We consider the state of a single pore is completely either drained or undrained. The saturation is simply binary (0 or 1 depending on which phase is present). Although practically a pendular bridge is formed between two neighboring solid grains, we neglect its volume and capillary effects currently.

For the solid phase, we assume that the motion (mainly sliding) of the rigid solid grains induced by the capillary forces is reversible. Consequently, the macro-scale deformation of the skeleton is of small amplitude and elastic. We assume such deformation will not cause large changes on the RT network connectivity and the fluid phases pressure. To conclude, we present a one-way coupling.

2.2.2 Fluid phases

Locally, the displacement of W-NW interface is controlled by the relation between local capillary pressure p^c (the pressure difference between the NW-phase and the W-phase: $p^c = p^n - p^w$) and a local threshold value of pressure p_e^c (termed "entry capillary pressure"). If $p^c > p_e^c$, the local pore will be drained instantaneously. For drainage, p_e^c is determined by the geometry of the pore throat.

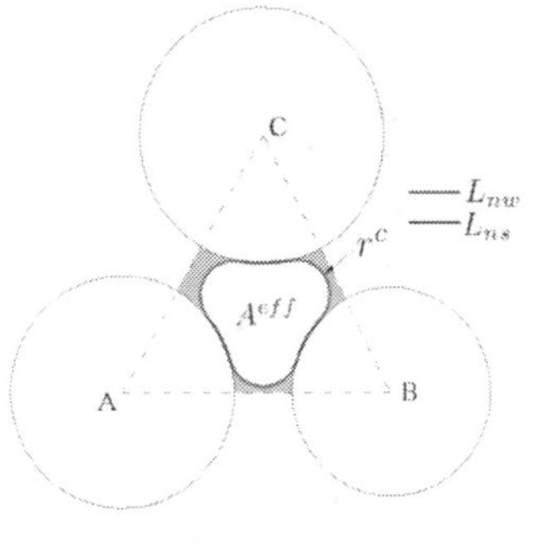

Figure 3. Geometry of a meniscus. r^c is the curvature of meniscus; L_{nw} is the length of contact line between NW and W phases; L_{ns} is the length of contact line between NW and S phases. $L^{eff} = \sum(L_{nw} + L_{ns})$, in which L^{eff} is termed effective entry pore throat perimeter. A^{eff} is the planar project area of NW-W interface, and is termed effective entry pore throat area.

Young-Laplace equation defines a precise relation between capillary pressure, p^c, interfacial tension, γ^{nw}, and mean curvature of the NW-W interface, C,

$$p^c = 2C\gamma^{nw}. \quad (1)$$

But C is difficult to define for an interface near a pore throat of complex geometry when p^c approaching p_e^c. Approximations are necessary.

We propose to calculate p_e^c by following MS-P (Mayer-Stowe-Princen) method, which employs the balance of forces on the NW-W interface ([4, 5]). The balance can be written as (see Fig.3),

$$\sum \vec{F}(p^c) = \vec{F}^c(p^c) + \vec{F}^t(p^c) = 0 \quad (2)$$

where, $\vec{F}^c$ is the capillary force acting on pore throat section domain (i.e., A^{eff}); and $\vec{F}^t$ is the total tension force along contact lines (i.e., L^{eff}). p_e^c is the value of p^c such that $\sum \vec{F}(p^c) = 0$. It can be numerically solved. More details for determining p_e^c can be found in [6].

Overall, the drainage follows the invasion percolation model by [7] and [8]. Initially, the pores are fully saturated by W-phase. The sample connects to NW and W reservoirs by two opposite boundaries. Drainage starts by increasing p^c. The NW-phase invades through the pore throats which separate the phases connected to the two reservoirs.

The first displacement of the W-NW interface occurs through the throat with lowest p_e^c, leading to the evacuation of the first pore. As soon as this pore is drained the NW-phase reaches new throats, possibly triggering a recursive cascade of displacements and invading more than one pore for one single value of p^c, until no more throats satisfy $p_e^c < p^c$. Then a state of equilibrium is obtained and ready for the invasion of next p^c. As the NW-phase is invading, the W-phase may form clusters of pores which are disconnected from the W reservoir. We assume that there

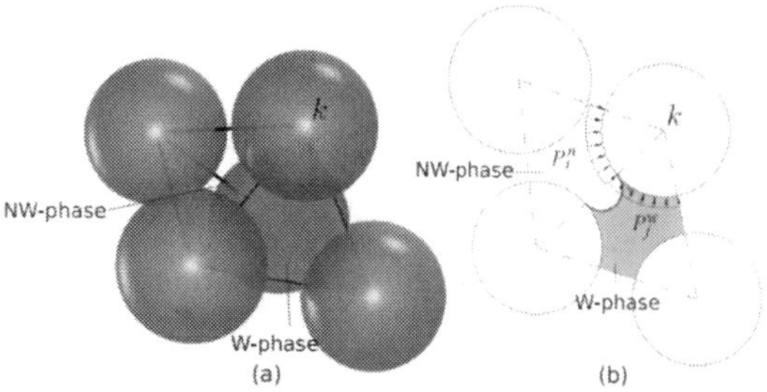

Figure 4. (a) Decomposition of the fluid space based on regular triangulation method. Particle k is incident with a NW pore and a W pore; (b) Fluids pressure and interfacial tension acting on particle k. (in 2D for clarity).

are no flux exchanges between individual separated clusters, thus the disconnected regions will remain saturated by a fixed amount of the W-phase throughout subsequent drainage.

2.2.3 Solid phase

During the drainage, the motion of the fluids and interface will cause fluid forces acting on the solid grains, leading to their movement (mainly sliding motion between contact grains). Consequently, a macro-scale deformation is formed. Considering a single solid grain k, the fluids forces can be written by,

$$\vec{F}_k = \vec{F}^f_k + \vec{F}^t_k = \int_{\partial\Phi_k} p^n \vec{n}\,\mathrm{d}s + \int_{\partial\Theta_k} p^w \vec{n}\,\mathrm{d}s + \int_{\partial\Gamma_k} \gamma \vec{t}\,\mathrm{d}l, \tag{3}$$

where $\partial\Phi_k$ denotes the contact surface between NW-phase and particle k, $\partial\Theta_k$ denotes the contact surface between W-phase and particle k, and $\partial\Gamma_k$ denotes the W-NW-S contact lines.

Based the RT topology, the domains of $\partial\Phi_k$, $\partial\Theta_k$ and $\partial\Gamma_k$ can be the decomposed and expressed by the geometrical terms of the incident tetrahedral (see Fig.4). Therefore, Eq.3 can be written in terms of individual fluid pore attributions,

$$\vec{F}_k = \sum_{(kj)_{incident}} \{\vec{F}^f_{kj} + \vec{F}^t_{kj}\}, \tag{4}$$

where kj represents the tetrahedron j incidental with particle k. $\vec{F}^f_{kj}$ and $\vec{F}^t_{kj}$ represent the corresponding forces induced by fluid pressure and surface tension incidental with tetrahedron j. The algorithm follows [9]. After solving the total force on each particle, the DEM solver is employed to predicate its motion. Currently, we are using the traditional Cundall's linear elastic-plastic law [10] to define the contact relations.

2.2.4 Implementation

The model has been implemented in Yade platform, where it is called 2PFV (two-phase pore-scale finite volumes) package. The RT network generation has been implemented in C++ [11] using the CGAL library [12]. A set of functions has been implemented for solving the hydraulic-mechanical couplings (eq.2 and eq.4). As a whole, the model is freely available as part of the open-source discrete element code Yade-DEM.

Table 1. Particle size distribution of porous medium

Weight(%)	Diameter(mm)
30	1-1.4
35	0.850
35	0.600

3 Simulation and results

3.1 Numerical setup

We apply the model to simulate the primary drainage of a cuboid packing in order to obtain the water retention curve (WRC) and to observe the mechanical evolution of the solid skeleton. The sample contains 40,000 particles with random positioning, which is generated by using the DEM software Yade [2]. The particle size distribution (PSD) is shown in Table.1 and the porosity is 0.34 (replicating the parameters of [6]).

The simulation is performed the oedometer test conditons, where only the one-dimensional deformation induced by the drainage can be obtained. The rigid confining walls are set to prevent the lateral displacement of the skeleton. The initial pressure state of the fluid phases is: $p^c = p^n - p^w = 0$. The sample is fully saturated. Drainage is controlled by decreasing p^w and keeping p^n constant. The results are put in dimensionless forms. p^c is represented by: $\overline{p^c} = \frac{p^c \overline{D}}{\gamma^{nw}}$, where $\overline{p^c}$ is termed normalized capillary pressure, and $\overline{D}$ is the average sphere size.

3.2 Results and discussion

We observe the drainage process by cutting a slice of the sample, as shown in Fig.5. The top boundary connects to NW-reservoir and the bottom boundary connects to W-reservoir. The invasion starts from the pores with larger throat, where the entry capillary pressure is smaller. In the beginning, the fluid phase distribution show strong gradients of saturation. The capillary fingering can be observed. After the NW-phase penetrates the sample, the distribution tends to homogeneous. At the end, the remaining W-phase is in the form of disconnected clusters in short-term entrapped by the NW-phase.

The results of the normalized capillary pressure - W-phase saturation - one dimensional strain ($\overline{p^c} - s^w - \epsilon_{11}$) relationship are reported in Fig.6. From the WRC ($\overline{p^c} - s^w$), we can find that the sample is fully saturated when $\overline{p^c} < 7.0$; the NW-phase invades the pores gradually during $7.0 < \overline{p^c} < 11.0$, leading to a small amplitude change of

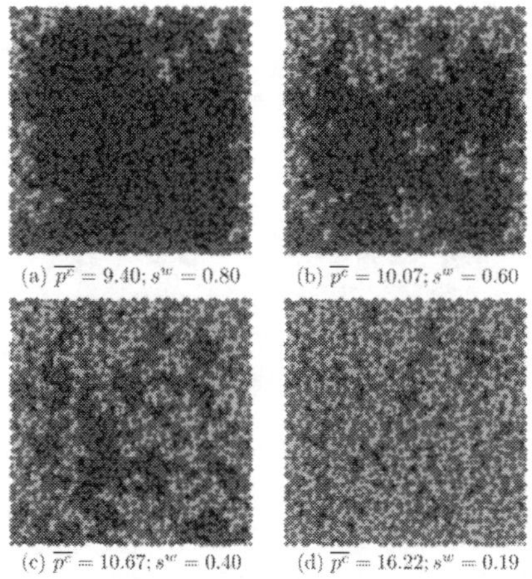

Figure 5. The process of drainage (40,000 particles), NW-phase invade from top. Brown (gray) is solid phase, blue (black) is W-phase, and light cyan (white) is NW-phase, see colour version of this figure in the HTML.

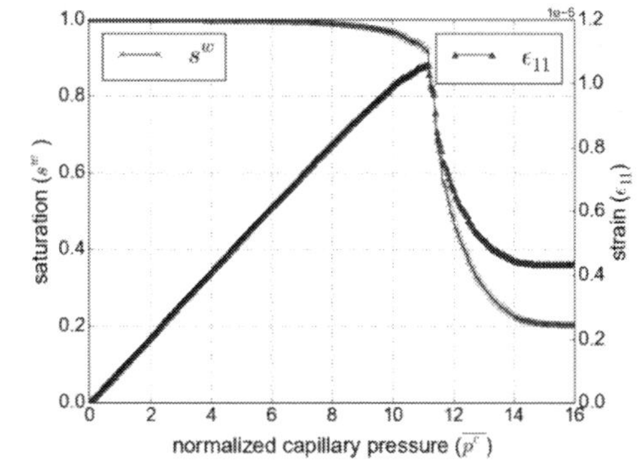

Figure 6. The relationship between normalized capillary pressure-saturation-strain ($\overline{p^c} - s^w - \epsilon_{11}$) during drainage simulation under the oedometer condition.

W-phase content; then s^w drops quickly from 0.9 to 0.4 with narrow range of increase of $\overline{p^c}$, which can be explained by Haines jumps going through large cluster of pores; finally a residual saturation of 0.2 is achieved due to the entrapments.

From the $\epsilon_{11} - \overline{p^c}$ curve, it can be found that the relation is strictly linear when $\overline{p^c} < 7.0$ (i.e., the sample is fully saturated). It can be explained by Terzaghi's effective stress principle for saturated conditions [13]: $\sigma' = \sigma - p^w\mathbf{I}$, where $\mathbf{I}$ the identity tensor. The decrease of p^w leads to the increase of effective stress σ'. Then the sample keeps shrinking until reaching the peak during $7.0 < \overline{p^c} < 11.0$, but the relation loses linearity due to the invasion of NW-phase.

With the receding of W-phase, the sample becomes dry. The effective stress for dry scenario is: $\sigma' = \sigma - p^n\mathbf{I}$, where $p^n = 0$ and keeps constant. It is the same as in the initial state (*i.e.*, $p^n = p^w$). Thus, the sample starts swelling. Since the residual W-phase is entrapped in the pore, the deformation of skeleton induced by the capillary effects will not be completely eliminated even the $\overline{p^c}$ keeps increasing.

4 Conclusions

A pore-scale numerical model for the simulation of primary drainage and the hydro-mechanical couplings has been developed. The underlying idea is to build an explicit and efficient link between fluid space and solid skeleton by a network topology. For the fluids, the local invasion criterion and global logic of drainage has been defined. For the solid phase, the capillary forces acting on grains has been formulated. The model has been implemented in open-source code Yade-DEM. A drainage tests has been performed under the oedometer conditions. The water retention curve has been obtained. The $\overline{p^c} - s^w - \epsilon_{11}$ curve illustrated the evolution of the deformation during drainage, in which the solid skeleton showed first shrinking and then swelling.

References

[1] L. Scholtès, P.Y. Hicher, F. Nicot, B. Chareyre, F. Darve, International Journal for Numerical and Analytical Methods in Geomechanics **33**, 1289 (2009)

[2] V. Šmilauer et al., *Yade Documentation 2nd ed* (The Yade Project, 2015), http://yade-dem.org/doc/

[3] H. Edelsbrunner, N. Shah, Algorithmica **15**, 223 (1996)

[4] R.P. Mayer, R.A. Stowe, Journal of Colloid Science **20**, 893 (1965)

[5] H. Princen, Journal of Colloid and Interface Science **30**, 359 (1969)

[6] C. Yuan, B. Chareyre, F. Darve, Advances in Water Resources **95**, 109 (2016), pore scale modeling and experiments

[7] R. Chandler, J. Koplik, K. Lerman, J.F. Willemsen, Journal of Fluid Mechanics **119**, 249 (1982)

[8] D. Wilkinson, J.F. Willemsen, Journal of Physics A: Mathematical and General **16**, 3365 (1983)

[9] C. Yuan, B. Chareyre, Computer Methods in Applied Mechanics and Engineering (Accepted) (2016)

[10] P.A. Cundall, O.D.L. Strack, Géotechnique **29**, 47 (1979)

[11] B. Chareyre, A. Cortis, E. Catalano, E. Barthélemy, Transport in Porous Media **92**, 473 (2012)

[12] J.D. Boissonnat, O. Devillers, M. Teillaud, M. Yvinec, *Triangulations in CGAL (Extended Abstract)*, in *Proceedings of the Sixteenth Annual Symposium on Computational Geometry* (ACM, New York, NY, USA, 2000), pp. 11–18

[13] E. Catalano, B. Chareyre, E. Barthélémy, International Journal for Numerical and Analytical Methods in Geomechanics **38**, 51 (2014)

DOI: 10.1051/epjconf/201714012008

Investigation of micro-structural phenomena at aggregate level in concretes using DEM

Michal Nitka[1], and *Jacek* Tejchman[1,*]

[1]Faculty of Civil and Environmental Engineering, Gdańsk University of Technology, Gdansk, Poland

Abstract. This paper presents numerical analyses of concrete beams under three-point bending. The discrete element methods (DEM) was used to calculate fracture at the aggregate level. Concrete was described as a four-phase material, which was composed of aggregate, cement matrix, interfacial transitional zones (ITZs) and macro-voids. The beam micro-structure was directly taken from our experiments using x-ray micro-tomography. 3D simulations were carried out with real aggregate modelled as sphere clusters. Numerical results were compared with laboratory outcomes. The special attention was laid on the fracture propagation and some micro-structural phenomena at the aggregate level.

1 Introduction

The fracture process is a fundamental phenomenon in quasi- brittle materials like concrete. It is very complex since it consists of main cracks with various branches, secondary cracks and micro-cracks. During fracture, micro-cracks first arise in a hardening region on the stress–strain curve which change gradually during material softening into dominant distinct macroscopic cracks up to damage. The fracture process strongly depends upon a heterogeneous structure of materials over many different length scales, changing e.g. in concrete from the few nanometres (hydrated cement) to the millimetres (aggregate particles). In order to properly describe fracture, material micro-structure has to be taken into account since its effect on the global results is pronounced. At the meso-scale, concrete may be considered as a composite material wherein four important phases may be separated: cement matrix, aggregate, interfacial transition zones ITZs between the aggregate and cement matrix and macro-voids.

The main objective of this study is to investigate a complex fracture process in concrete beams under bending at the aggregate level under 3D conditions using the discrete element method (DEM) with angularly-shaped aggregate particles. Discrete models (if they are enough consistent) might progressively replace experimental tests to study the influence of concrete meso-structure (aggregates size, aggregate shape, aggregate roughness, aggregate/mortar volume, macro porosity, etc.) on the concrete behaviour. The disadvantages of DEM are: enormous computational cost and a difficult calibration procedure with respect to geometric and mechanical properties of ITZs. In the calculations, the concrete micro-structure was assumed based on 3D images by means of x-ray micro-tomography using the micro-tomograph Skyscan 117, which represents a new generation in high-resolution desktop x-ray micro-tomography systems [1]. The paper is a continuation of our research outcomes presented in [1], [2] which concerned the crack propagation in concrete beams under 2D conditions only.

2 Experimental results

Laboratory tests were carried out with notched concrete beams (height 80 mm, depth 40 mm and length 320 mm with the notch 8 mm high and 3 mm wide) [1], [2]. Concrete was prepared as a mix of the cement matrix and aggregate (river sand and gravel). The mean aggregate diameter was 2 mm and maximum aggregate diameter was equal to 16 mm. The aggregate volume was 75%. Quasi-static three-point bending tests were performed with the controlled notch opening displacement rate (CMOD). The CMOD gauge with the length of 5 mm was located in the notch at the beam bottom. The maximum vertical force was equal to 2.18 kN for CMOD equal to 0.017 mm.

Figure 1 shows the 3D μCT image of the cracked cuboidal specimen (80×50×40 mm^3), cut out from the beam mid-part after one test [1]. The main crack was strongly curved mainly due to presence of aggregate particles. Its shape changed along the specimen depth in spite of the fact that 2D boundary value problem (plane stress) was considered. The discrete macro-crack mainly propagated through ITZs (which were the weakest phase in concrete) and sometimes through macro-voids. It might very rarely propagate through a single weak aggregate particle. The effect of macro-voids on the crack shape was small. The discrete crack was created by

* Corresponding author: tejchmk@pg.gda.pl

bridging the interfacial micro-cracks. It possessed many small branches. ITZs around aggregate particles were characterized by a very non-uniform porous structure and presence of separated small sand particles. They appeared mainly aggregate particles but sometimes they were also visible around larger cement matrix particles. The width of ITZs changed between 30 and 50 lm. The width was independent of the aggregate diameter.

a) b)

Fig. 1. Images of cracked cuboidal specimens cut out from beam obtained by means of 3D μCT: a) general view and b) image with separated phase (blue colour – aggregate, green colour – cement matrix, red colour – crack and air void) [1].

3 Numerical Results

3.1 Discrete element method

DEM simulations shown in the paper were carried out using the open-source code YADE [3]. A linear normal contact model under compression was used. The interaction force vector representing the action between two elements in contact was decomposed into normal and tangential vectors. The normal and tangential forces were linked to the displacement through the normal K_n and tangential K_s stiffness, respectively:

$$\mathbf{F_n}=K_n U\mathbf{N} \tag{1}$$

and

$$\mathbf{F_s}=\mathbf{F_{s,prev}}+K_s\Delta\mathbf{X_s}, \tag{2}$$

where U denote the overlap between elements, $\mathbf{N}$ denotes the normal vector at the contact point, $\Delta\mathbf{X_s}$ is the increment of the relative tangential displacement and $\mathbf{F_{s,prev}}$ is the tangential force from the previous iteration. The contact forces satisfied the cohesive frictional Mohr-Coulomb equation:

$$||\mathbf{F_s}|| - F_{s,max} - ||\mathbf{F_n}||\tan\mu \leq 0 \text{ (before contact breakage)}, \tag{3}$$

and

$$||\mathbf{F_s}|| - ||\mathbf{F_n}||\tan\mu \leq 0 \text{ (after contact breakage)}, \tag{4}$$

where μ is the inter-particle friction angle and $F_{s,max}$ denotes the cohesive force between elements. The normal force might be negative down to the minimum tensile value of $F_{n,min}$. The cohesive force and tensile force were assumed as a function of the cohesive stress C, tensile normal stress T and element radius R:

$$F_{s,max}=C\times R^2 \quad \text{and} \quad F_{n,min}=T\times R^2. \tag{5}$$

Concrete was described as four-phase material which consisted of aggregate, cement matrix, interfacial transitional zones (ITZs) and macro-voids [1]. All aggregate particles (from 2 mm to 16 mm) were modelled as grain clusters composed of pure spheres connected to each other as rigid bodies. All aggregate particles (larger than 2 mm) included ITZs. ITZs were solely modelled as the contacts between aggregate and cement matrix (thus they had no a physical width). Cement matrix was simulated with spheres of the diameter 0.5-2 mm (without ITZ). Macro-voids were assumed as empty spaces. The remaining beam region (outside the meso-part close to the notch) was modelled with spheres with the diameter of 2-8 mm in order to reduce calculation time (Fig.2).

Fig. 2. Front face of concrete beam with meso-region of 50×80×40 mm^2 close to notch (dark grey corresponds to aggregate and light grey is related to cement matrix).

The following parameters of the elastic modulus of grain contact E, cohesion C and tensile strength T were assumed in DEM calculations: E_{cm}=29.2 GPa, C_{cm}=140 MPa and T_{cm}=25 MPa for the cement matrix, E_{itz}=20.4 GPa, C_{itz}=100 MPa and T_{itz}=17.5 MPa for ITZs and E_{macro}=36.1 GPa, C_{macro}=140 MPa, and T_{macro}=25 MPa for the region outside the meso-region. Thus, ITZs were the weakest phase. The calibration procedure was based on uniaxial compressive and tensile tests [4]. The remaining DEM parameters were constant for all phases and regions: Poisson's ratio of grain contact ν=0.2, inter-particle friction angle μ=18°, damping parameter α=0.08 and mass density ρ=2600 kg/m^3. The calculations were performed under quasi-static conditions (inertial number was I<10^{-3}). The entire beam included about 155'000 spherical elements with 150'000 in the meso-region (20'000 for aggregate and 130'000 for cement matrix).

3.2 Numerical 3D results

The evolution of the resultant vertical force F against CMOD from DEM is shown in Fig.2 as compared to the experimental curve. The computed maximum vertical force was F_{max}=2.19 kN for CMOD=0.017 mm and was in agreement with the experimental value. The calculated residual force was however too high.

 DOI: 10.1051/epjconf/201714012008

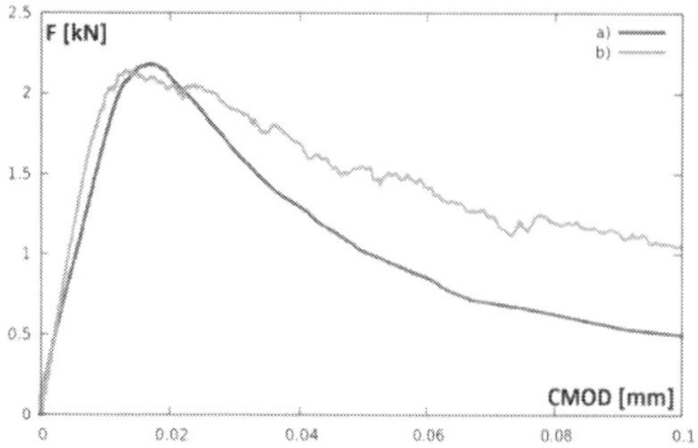

Fig. 3. Evolution of vertical force F against displacement CMOD in beam under 3-point bending: a) experimental curve and b) 3D DEM.

In Fig. 4, the numerical final crack path was shown for CMOD=0.1 mm in 2 different vertical cross-sections.

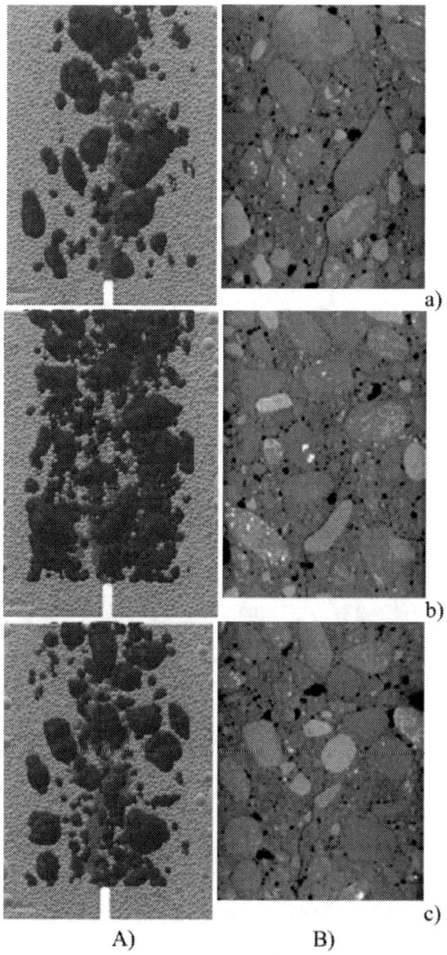

Fig. 4. Final crack path in concrete beam above notch in 2 different cross-sections from: A) 3D DEM and B) experiment (10 mm from beam front side (a), beam mid-depth (b) an 10 mm from beam rear side (c), red colour denotes cement spheres with broken contacts, dark grey denotes aggregate and light grey denotes cement matrix).

The crack propagation was similar as in experiments. The crack was strongly curved along height and depth due to a stochastic distribution of aggregate (Fig. 5). The crack high varied along the beam depth from 52 mm up to 58 mm (50-55 mm in the experiment). The width of the crack varied between 0.05-0.1 mm above the notch up to 0.01 mm at the crack tip.

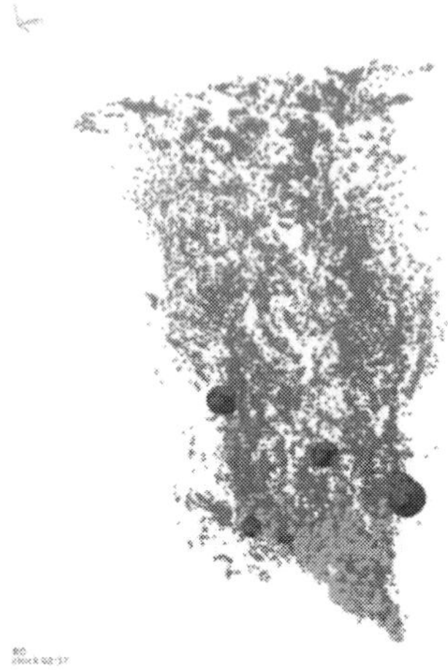

Fig. 5. Calculated cracked region (empty space) above notch from 3D DEM for CMOD=0.1 mm (blue colour corresponds to crack width larger than 0.2 mm, green colour to crack width 0.05-0.2 mm and red colour to crack width smaller than 0.05 mm)

The evolution of the broken contact number is shown in Fig. 6a. The initial contact number in the beam micro-region was about 640'000 with the coordination number of about 10. The breakage started early for about CMOD=0.007 mm (the peak force was for CMOD=0.017 mm). Finally, more than 7'000 contacts were broken that corresponded to 1% of all existing contacts. The global void ratio of the meso-region initially showed contractancy and later dilatancy for CMOD>0.01 mm (Fig. 6b).

The evolution of the normal contact force network between spheres is demonstrated in Fig. 7 (the red colour corresponds to the high compressive normal contact forces (higher than the average ones) and blue colour to the high tensile normal contact forces (lower than the average ones). The remaining contact forces were marked in green. The line thickness denotes the magnitude of the normal contact force. The external vertical force F was transmitted via a network of normal contact forces which formed pronounced force chains. Their distribution was strongly non-uniform due to the grain rearrangement after the crack development. The maximum compressive contact force was about 30 N and the maximum tensile contact forces was about -20 N. Some pronounced compressive forces also existed in the damage zone parallel to the crack due to shear along a rough crack surface (aggregate interlocking).

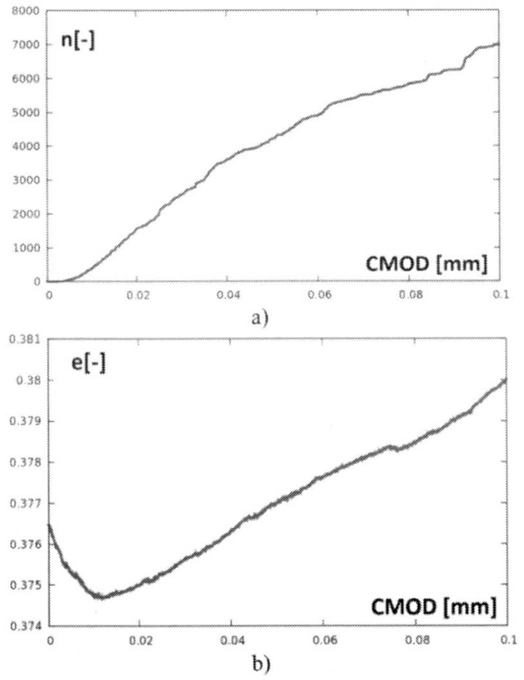

Fig. 6. DEM analyses: a) evolution of broken contacts and b) evolution of global void ratio in 3D meso-region.

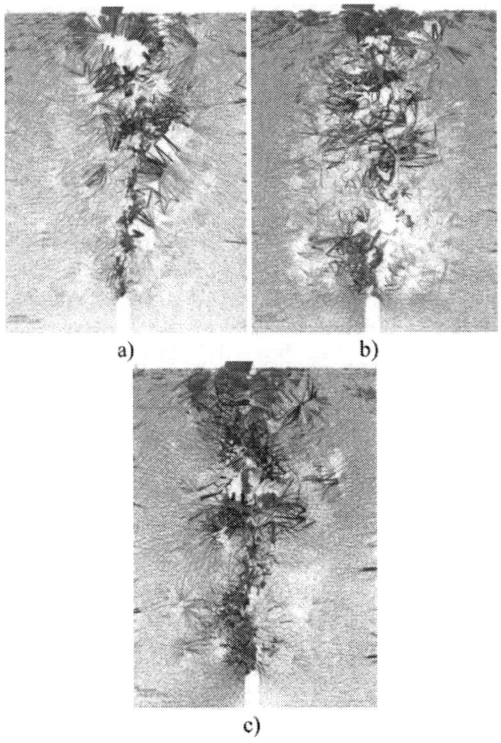

Fig. 7. Distribution of normal contact forces between elements in 3 different cross-sections: a) 10 mm from beam front side, b) beam mid-depth and c) 10 mm from beam rear side (red colour denotes high compressive forces, blue colour – high tensile forces and green colour denotes remaining forces).

4 Conclusions

The 3D calculation results for four-phase concrete at the aggregate level using the discrete element model showed satisfactory agreement with experimental observations. They evidently showed that DEM was able to realistically capture the beam strength and fracture in concrete beams at the macro-level under condition that the real aggregate shape and location were mapped. DEM realistically followed a fracture process including the occurrence of micro- and macro-cracks during their onset, formation and propagation (including phenomena of crack bridging, crossing, branching and closing). In addition, it allowed for investigating different interesting micro-structural events occurring at the aggregate level (e.g. force chains, contact breakage, particles rotations).

The experimental discrete macro-crack above the notch was strongly curved and depended on the concrete micro-structure. It was created by bridging the interfacial micro-cracks. It mainly propagated through the weakest contact zones (ITZs) between the cement matrix and aggregate. The width of ITZs varied between 30 and 50 μm. Their porosity was strongly uniform. The crack rarely propagated through weak aggregate particles. Some small crack branches were also visible. The crack shape was different with the beam depth. Single micro-cracks also occurred far beyond the macro-crack.

It was found that the mechanical properties of ITZs which were cracks' attractors had a pronounced influence on the crack strength and macro-cracking. The weaker strength of ITZs and their higher number increased the beam ductility (due to a longer crack propagation way).

The external load was transmitted via a network of normal contact forces which formed force chains of a different intensity. The compressive normal contact forces (connected to the tangential contact forces) also developed along the macro-crack due to aggregate inter-locking.

The research work has been carried out within the project: "Innovative ways and effective methods of safety improvement and durability of buildings and transport infrastructure in the sustainable development" financed by the European Union (POIG.01.01.02-10-106/09-01) and the project "Experimental and numerical analysis of coupled deterministic-statistical size effect in brittle materials" financed by the National Research Centre NCN (UMO-2013/09/B/ST8/03598).

References

1. Ł. Skarżyński, J. Tejchman, Strain **52**, 26-45 (2015)
2. Ł. Skarzyński, M. Nitka, J. Tejchman, Engng. Fract. Mech. **147**, 13-35 (2015)
3. J. Kozicki, F. Donze, Computer Methods in Applied Mechanics and Engineering **197**, 4429-4443 (2008)
4. M. Nitka, J. Tejchman, Granular Matter **17**, 145-164 (2015)

A model experiment to study swallowing of spherical and elongated particles

Marco Marconati[1,*], *Sharvari* Raut[1], *Farshad* Charkhi[1], *Adam* Burbidge[2], *Jan* Engmann[2], and *Marco* Ramaioli[1]

[1] *Department of Chemical and Process Engineering, University of Surrey, Guildford GU2 7XH, UK.*
[2] *Nestlé Research Center, Route du Jorat 57, 1000 Lausanne 26, Switzerland.*

Abstract. Swallowing disorders are not uncommon among elderly and people affected by neurological diseases. For these patients the ingestion of solid grains, such as pharmaceutical oral solid formulations, could result in choking. This generally results in a low compliance in taking solid medications. The effect of the solid medication size on the real or perceived ease of swallowing is still to be understood from the mechanistic viewpoint. The interplay of the inclusion shape and the rheology of the liquid being swallowed together with the medication is also not fully understood. In this study, a model experiment was developed to study the oropharyngeal phase of swallowing, replicating the dynamics of the bolus flow induced by the tongue (by means of a roller driven by an applied force). Experiments were performed using a wide set of solid inclusions, dispersed in a thick Newtonian liquid. Predictions for a simple theory are compared with experiments. Results show that an increase in the grain size results in a slower dynamics of the swallowing. Furthermore, the experiments demonstrated the paramount role of shape, as flatter and more streamlined inclusions flow faster than spherical. This approach can support the design of new oral solid formulations that can be ingested more easily and effectively also by people with mild swallowing disorders.

1 Introduction

Ingestion of grains and solid oral medications can be particularly challenging for people affected by swallowing disorders. While aspiration of liquids can expose the patients to the risk of aspiration pneumonia, swallowed solids could potentially become trapped in the larynx folds, hence leading to a potential risk of choking. Moreover, retained solid oral formulations may even start dissolving in the pharyngeal and upper esophageal regions and cause local injuries resulting in pain, esophagitis and ultimately ulceration, stricture, and perforation if the dissolved medicine irritates the mucosa of the esophagus [1]. Size and shape are deemed the most important reason that limit acceptability of prescriptions among the individuals who find issues in taking tablets and capsules [2]. In particular the insurgence of compliances, such as aspiration and choking during the oropharyngeal phase of deglutition, becomes increasingly common as the size of the oral prescription is increased [3]. Indeed studies reveal that the level of acceptability of an oral medication, albeit dependent upon the age subset, is greatly reduced when its diameter is above 8 mm [4]. It has also been shown that size is a major player in determining the esophageal transit time, smaller tablets flowing faster. however, there is a lack of quantitative studies on the duration of the oropharyngeal phase of swallowing and the effect of the size of the oral prescription. Other than size, shape is also an important factor to be taken into account in pill design. Previous studies suggest that oval tablets may be easier to swallow and have faster esophageal transit times than round tablets of the same weight [5]. However, it was shown that flat tablets have greater tendency to adhere and stick to the esophageal epithelium than capsule-shaped tablets [6], hence potentially leading to esophagitis. Finally, factors such as density and pill surface characteristics may all affect the swallowability and esophageal transit of tablets and capsules. This study considered the oral transit time of tablets and pills to quantify the importance of size and shape. In order to lay the basis for a mechanistic understanding of these phenomena, before moving to panel tests, the study has been carried out using an *in vitro* model of the oral cavity. The same model was previously successfully used to evaluate the effect of bolus rheology on its flow and those results were successfully compared with *in vivo* ultrasound observations [7].

*e-mail: m.marconati@surrey.ac.uk

2 Materials and Methods

Different sizes and shapes of solid inclusions were used in the experiments. The effect of size is investigated using calibrated plastic beads, of diameters ranging from 4.8 to 10 mm. Commercially available capsules and a pharmaceutical tablet were also later employed to study the effect of shape on the kinematics of the bolus. Empty capsules *size* 00 and *size* 3 were purchased and filled with sucrose so that their density would match that of the polyethylene beads ($0.875 \pm 0.050\,\mathrm{kg\,m^{-3}}$). The *size* 00 capsule was used as a limiting case, as it is currently the maximum size

Table 1. List of the solid inclusions dispersed in glycerol and used during the experiments with the *in vitro* model

Type of inclusion	Size	Weight
sphere	⌀ = 4.8 mm	0.049 g
sphere	⌀ = 6.35 mm	0.116 g
sphere	⌀ = 8 mm	0.225 g
sphere	⌀ = 10 mm	0.462 g
capsule *size* 00	L = 23.5 mm ⌀ = 8.5 mm	0.912 g
capsule *size* 3	L = 15.5 mm ⌀ = 5.8 mm	0.246 g
ellipsoidal tablet	H=5 mm L=W=11 mm	0.436 g

for a generic drug suggested by the Food and Drug Administration. The *size* 3 capsule was instead specifically chosen because it has the same volume and mass of a 8 mm bead, hence allows for direct comparison of shape effect. Finally, an ellipsoidal uncoated tablet was used to provide a preliminary assessment of the relative deviation of the other objects from a reference over the counter pain relief. The inclusion dimensions and volume are reported in Tab. 1.In all cases the solids were dispersed in 99.98% w/w glycerol (SG 1.25, μ = 1.05) that was used throughout the experimental campaign as suspending liquid carrier.

2.1 The *in vitro* model experiment

In this study an *in vitro* model experiment is used to replicate the essential features of the peristaltic flow, observed *in vivo* with both ultrasound and video fluoroscopy measurements. The oral cavity is substituted with a thin plastic membrane that sticks to the upper palate. The liquid bolus is driven to flow by a rigid roller deforming the lower boundary of the membrane. The roller movement mimics the wave of muscular contraction traveling from the front to the back of the tongue. A set of calibrated weights is used to modulate the applied driving force (see Fig. 1). The occlusion is complete. Moreover, the associated wavenumber, defined by the ratio of bolus radius to its length is on the order of 0.15. During the study, three different applied forces were investigated, by attaching the masses m = 200, 270, and 400 g to the driving pulley (approximately 2, 2.7, and 4 N). Using these three forces resulted in maximum linear velocities of the bolus below 1 m/s thereby the Reynolds number, defined as $Re = R v/\nu$, resulted to be less than 0.5. Whilst previous experimental and numerical studies already considered similar Reynolds numbers and wavenumber [11] [10] and the propagation of a single peristaltic wave over a linear trajectory, the peristaltic transport of macroscopic solid inclusions at amplitude ratios close to unity had not been studied, as far as the authors are aware. The total inertia of the rotating arm was chosen to match the *in vivo* the ultrasound measurements obtained in a previous study [7]. In the experimental setup the roller movement is triggered under the release of a pin and the experiment is then recorded using a fast camera (Model ac1920-155 μm, Blaser, Germany) at acquisition rate of 150 frame per second (fps). Image processing tools are subsequently employed to extract from each frame the position of the roller with respect of the center of rotation, and the solid inclusion position. Velocity profiles are obtained by means of raw data filtering, followed by three-point numerical differentiation of the instantaneous roller angular positions. The velocity profiles were compared with the theoretical values obtained solving the differential equation describing the motion in absence of solid inclusions. The relative displacement between the solid inclusion and the bolus was quantified by the difference of the instantaneous roller and bead angular positions.

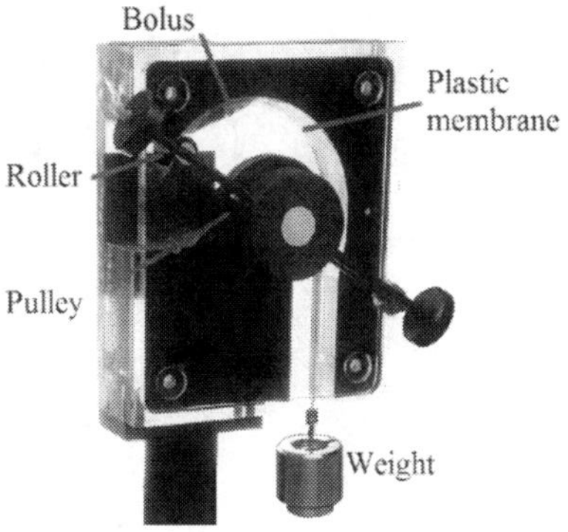

Figure 1. Schematics of the *in vitro* setup.

2.2 Theory

Previous studies considered *in vitro* swallowing of homogeneous liquids, either Newtonian or shear thinning, to address swallowing disorders [8] [7] . The acceleration and motion of the bolus was also successfully described by a simple mathematical model, considering the torque generated by the applied force, the inertia of the system and the drag force generated by the liquid bolus flow. The viscous dissipation was computed based on a modified Hagen-Poiseuille flow, corrected by a factor β whose value is independent on the applied force to the system and the viscosity of the liquid. This theory has then been modified to account for suspended solids, considering the apparent restriction in the effective longitudinal section of the bolus experienced in proximity of the hard inclusion. The flow field was thereby simplified, considering an annular flow of length $L = L_{bolus}$, where the inner shell has a diameter equivalent to the inclusion considered (R_i), while the outer shell diameter is fixed and correspondent to the diameter of the bolus (R_e). Boundary conditions on velocity apply at the walls: the inner cylinder (i.e. the inclusion) was supposed to move longitudinally, in the direction of swallowing, with velocity equal to to that of the roller v_{roller}, while a no-slip condition applies to the external shell (i.e. the plastic membrane in which the bolus is contained). A similar problem was previously considered by Charles in the modeling of pipeline flow of cylindrical capsules [12]. The theory there presented, validated on experiments at low Reynolds numbers [13], indicates that the pressure drop experienced in transporting a infinitely long cylindrical neutrally buoyant object without axial offset, can be

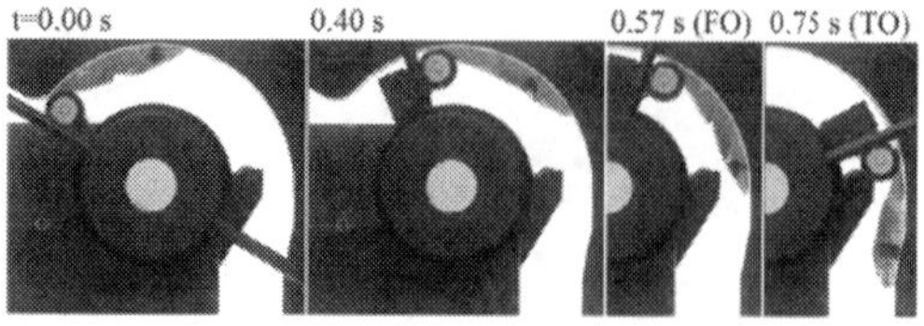

Figure 2. Screenshots of the *in vitro* swallowing of a single plastic sphere of 6.35 mm diameter suspended in glycerol at external applied force of 2 N.

related to the pressure drop in a Poiseuille flow, with identical mean velocity, corrected by a factor dependent upon of the ratio of radii $k = R_i/R_e$ of the inclusion and the pipe. On this basis, the same correction was applied to relate the viscous dissipation of the bolus flow in presence of solid inclusions to that of the homogeneous fluid in a Hagen-Poiseuille flow.

$$F_d^* = \left(1 - k^4\right)^{-1} \beta\, 8\pi \mu L v_{roller} \tag{1}$$

Eq. 1 correctly reduces to the theory presented by Hayoun *et al.* [7] when the cross sectional radius of the inclusion in the direction perpendicular to the flow is reduced ($R_i \to 0$).

3 Results

The objective of the experiments was to understand the effect of inclusion size and shape on the flow, when varying the applied force. Some characteristic events during the bolus flow are illustrated in Fig. 2, in the case of a 6.35 mm spherical inclusion swallowed at the lowest applied force (2 N). The event FO and TO are of particular importance, as they represent respectively the exit of the front of the bolus from the plastic tube (Front Out - FO), and the moment in which the roller reaches the final position that corresponds to the tail of the bolus leaving the tube (Tail Out - TO). The time when these events occur depends on the size and shape of inclusions and on the applied load. The evolution of the roller angular velocity in time, till bolus front out obtained for different inclusions and applied loads is illustrated in Fig. 3. For a given applied force, lines of different colors represent the experimental results different diameters of spherical inclusions. These can be also compared with the theoretical curves (thin solid lines), obtained solving the rotational trajectory of the system, using the viscous dissipation described in Eq. 1. The theory predicts a monotonic increase of roller angular velocity with time. This had been compared favorably against *in vitro* experiments carried out in absence of any solid inclusion [7]. In these conditions, an initial inertial regime was observed where the angular velocity is increasing linearly with time, followed by a viscosity-dominated regime, in which the drag force becomes predominant and the bolus velocity becomes constant. Conversely, experiments either at higher applied loads, or lower viscosities, only showed the inertial regime. The velocity profiles of the experiments with inclusions, as shown in Fig. 3, generally exhibit similar features. In particular, the inertial regime, below $t = 0.1s$, is still well predicted by the theory and the theoretical curves are close to the initial slope of the experimental data. Experiments at reduced external forces (2 N) have been previously demonstrated to best match *in vivo* ultrasound observations [8] and are therefore the conditions closest to a human swallow. The *in vitro* results show that angular velocity decreases, when increasing the size of the spherical inclusion. This is well reproduced by the theory, in particular in the case of the biggest inclusion (spherical bead d =10 mm), when the velocity approaches a constant asymptote in the viscous regime. The experimental velocity profiles obtained using the highest load configuration (4 N) show no significant effect of the inclusion size. The theory captures well the experimental curves, with the exception for the biggest inclusion, where the theory over predicts the viscous dissipation. The intermediate load (2.7 N) shows a trend similar to the lower load and the 10 mm inclusion dynamics is significantly slower than the others. Inclusion shape has also a paramount impact on bolus kinematics, as can be appreciated comparing the profiles for spherical and elongated objects (Fig. 4). The capsule *size* 3 has almost identical volumes and mass (Table 1) than that of the d =8 mm sphere, yet the capsule flows significantly slower than the bead. Conversely, the ellipsoidal tablet is heavier and longer than the d =10 mm bead, but it flows as fast as a d =8 mm bead. In all cases the theoretical values obtained using the inclusion cross section as R_i in Eq. 1 over predicts the velocity profile. This suggests that more complex models are required to account realistically for the inclusion shape. The relative motion the solid inclusion within the bolus was measured using image analysis. This can clarify the deviation from considering that the velocity of the inclusion v_i equals the linear velocity of the roller, as assumed in the computation of the viscous dissipation term of Eq. 1. The profiles of $\Delta\alpha$ as a function of time highlight a noticeable reduction in the angular distance between the inclusion and the roller in both the extremes cases of 4.8 mm and 10 mm bead diameters. In those cases angular deviations of up to -30° from the initial position are experienced. Conversely, no significant trends are observed in the case of the intermediate size beads (6.35 and 8 mm), the *size* 3 capsule, and the ellipsoidal tablet. The backward motion of the smallest solid inclusion is of particular interest, as it confirms the fact that small objects are less efficiently transported by peristalsis [10]. On the other hand, the fact that the 10 mm bead is also moving backwards in the bolus frame of reference, is mainly deemed to the non negligible wall friction experienced when running experiments with the biggest inclusions. Increasing the external applied load up to 4 N does not alter the bead relative position: the 4.8 mm and the 10 mm beads consistently move backwards, while no noticeable variations in bead position are observed for both the 6.35 and the 8 mm beads. The measured velocity of the inclusion is therefore always lower or equal to that of the roller. Furthermore, the less effective peristaltic transportation of smaller in-

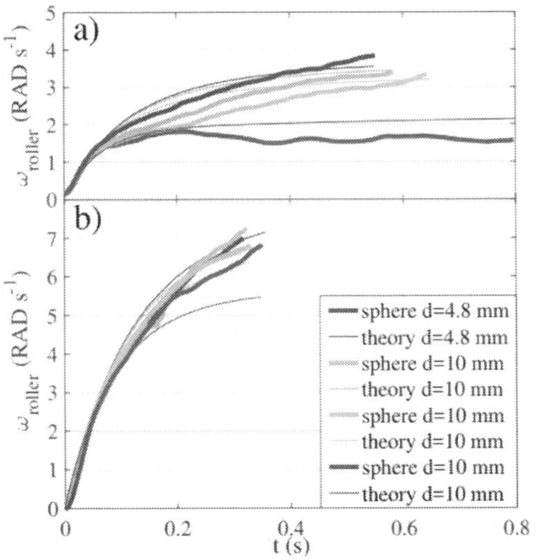

Figure 3. Evolution with time of the angular velocity of the tail of the bolus for different diameters of spherical inclusions and two applied forces: a) 2 N, b) 4 N. Theoretical values are illustrated with thin solid lines.

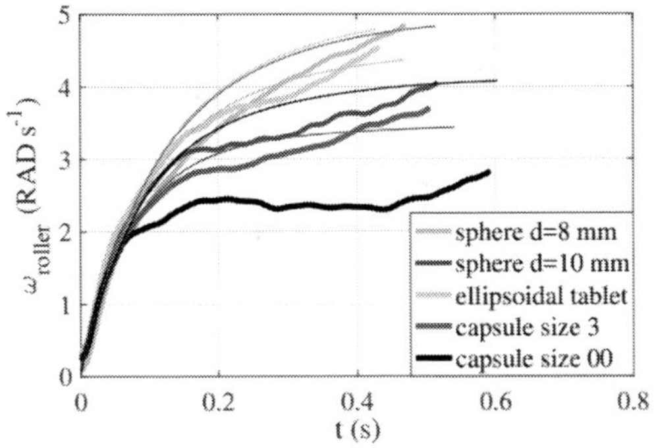

Figure 4. Evolution of the roller velocity with spherical and non-spherical inclusions at 2.7 N load. Thin lines represent the theoretical values obtained assuming R_i equal to the cross sectional radius of the inclusion in Eq. 1.

clusions is of particular interest from the clinical prospective, as it highlights the need for further liquid swallows to transport the hard object into the esophagus.

4 Conclusions

Building on the results obtained in previous studies [7] and [8], this article proposes a novel approach to study the swallowing dynamics of hard particles in liquids with high viscosity during the oropharyngeal phase of human swallowing. Spherical and non-spherical inclusions in size ranging 4.8 to 10 mm were considered. The experimental results demonstrated that inclusions affect significantly the peristaltic flow, in particular at low applied forces. Inclusion size influences bolus velocity: the bigger the diameter of the inclusion, the longer the oral transit time. A simple mathematical model was used to successfully evaluate the viscous dissipation at very low Reynolds numbers for big enough inclusion, although the drag force is over predicted at higher applied loads. The shape of the inclusion influences the dynamics of the system too. Inclusions of same volumes and mass, but significantly different shape show a different oral transit times. Finally, the efficiency of peristaltic transport is higher for inclusion of intermediate size while both small beads and very large inclusions flow backwards, toward the trailing edge of the bolus. This phenomenon could potentially lead to residues of inclusions in the pharynx. These results on the oral phase of swallowing of hard solid inclusions, can support the development of novel pill designs, before costly and invasive clinical trials.

References

[1] F. J. Collins, H. R. Matthews, S. E. Baker, J. M. Strakova, Brit. Med. J. **2**, 132-133 (1979).

[2] FDA guidance for industry, *Size, shape, and other physical attributes of generic tablets and capsules* (US FDA, Rockville, MD, 2015).

[3] J. Kelly, G. D'Cruz, D. Wright, J. Adv. Nurs. **66**, 82-91 (2010).

[4] D. P. Brotherman, T. O. Bayraktaroglu, R. J. Garofalo, J. Am. Pharm. Ass. **44**, 587-593 (2004).

[5] H. Hey, F. Jorgensen, K. Sorensen, H. Hasselbelch, T. Wamberg, Brit. Med. J. **285**, 1717-1719 (1982).

[6] M. Marvola, M. Rajaniem, E. Marttila, K. Vahervuo, A. Sothmann, J. Pharm. Sci. **72**, 1034-1036 (1983).

[7] P. Hayoun, J. Engmann, S. Mowlavi, B. Le Reverend, A. Burbidge, M. Ramaioli, J. Biomech. **48**, 3922-3928 (2015).

[8] S. Mowlavi, J. Engmann, A. Burbidge, R. Lloyd, P. Hayoun, B. Le Reverend, M. Ramaioli, J. Biomech. DOI 10.1016/j.jbiomech.2016.10.011 (2016).

[9] H. Chisaka, Y. Matsushima, F. Wada, S. Saeki, Dysphagia **21**, 275-279 (2006).

[10] T. K. Hung and T. D. Brown, J. Fluid Mech. **73**, 77–96 (1976).

[11] L. J. Fauci, Comput. Fluids **21**, 583-598 (1992).

[12] M. E. Charles, Can. J. Chem. Eng. **41**, 46-51 (1963).

[13] H. S. Ellis, Can. J. Chem. Eng. **42**, 1-8 (1964).

EPJ Web of Conferences **140**, 03026 (2017) DOI: 10.1051/epjconf/201714003026
Powders & Grains 2017

Discrete Element Method simulations of standing jumps in granular flows down inclines

Ségolène Méjean[1,2,*], *Thierry* Faug[1], and *Itai* Einav[2,3]

[1] *Université Grenoble Alpes, Irstea, UR ETGR, 2 rue de la Papeterie BP76, F-38402 St-Martin-d'Hères, France*
[2] *School of Civil Engineering, The University of Sydney, Sydney NSW 2006, Australia*
[3] *Departement of Civil Environmental and Geomatic Engineering, Faculty of Engineering Science, University College London, Gower Street, London WC1E 6BT, United Kingdom*

Abstract. This paper describes a numerical set-up which uses Discrete Element Method to produce standing jumps in flows of dry granular materials down a slope in two dimensions. The grain-scale force interactions are modeled by a visco-elastic normal force and an elastic tangential force with a Coulomb threshold. We will show how it is possible to reproduce all the shapes of the jumps observed in a previous laboratory study: diffuse versus steep jumps and compressible versus incompressible jumps. Moreover, we will discuss the additional measurements that can be done thanks to discrete element modelling.

1 Introduction

A standing jump is a rapid change in height and velocity in a free-surface flow, demarcating supercritical from subcritical flows. The phenomenon is well-known for hydraulic jumps on a smooth horizontal bottom, where the jumps obey Bélanger equation: $h_*/h = (\sqrt{1+8Fr^2}-1)/2$, where h and h_* hold respectively for the height before and after the jump, and $Fr = \overline{u}/\sqrt{gh}$ is the Froude number of the flow before the jump, with $\overline{u}$ the depth-averaged velocity and g the acceleration of gravity. However, this equation is not suitable for a granular flow which can occur only down an incline because of the frictional nature of granular materials, and which may be compressible [1]. A recent study [2] confirmed the deviation from Bélanger equation for grains, and even for unfrictional incompressible flows like water when the bottom is rough or inclined. Some applications like the design of avalanche protection dams need an accurate prediction of the geometry of the jumps formed in a flow of dry granular materials whatever the incoming regime [1]. As highlighted in [2], the jump height ratio does not depend on Fr only but is rather a function of a number of parameters:

$$\frac{h_*}{h} = f\left(Fr, \frac{L}{h}, \frac{\rho_*}{\rho}, \tan\zeta - \mu_e, \beta, \beta_*, k, k_*\right), \qquad (1)$$

where L is the jump length, ζ the incline slope, ρ and ρ_* hold for the density before and after the jump, respectively, and μ_e is the effective friction (see [2]). The coefficients β, β_*, k and k_* will be defined in Sec.2.3. In particular, Eq.1 suggests that h_*/h can be predicted for any incoming flow and boundary conditions provided L/h is known.

*e-mail: segolene.mejean@irstea.fr

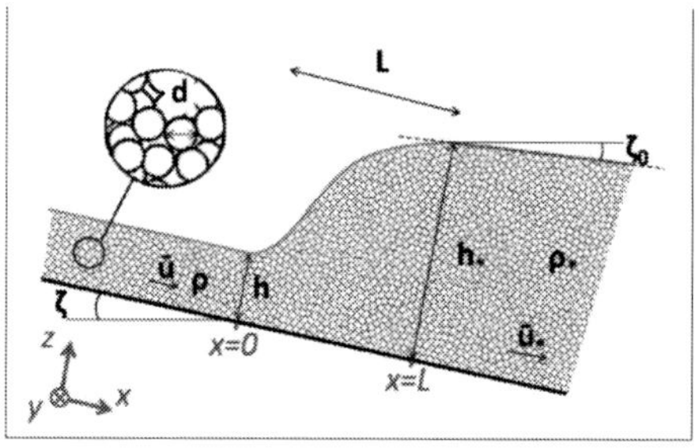

Figure 1: Sketch of a granular jump and notations

Our study aims to reproduce numerically a standing jump formed in a flow of granular materials thanks to the Discrete Element Method (DEM), in order to measure precisely the macroscopic parameters needed for Eq.1, and to decipher the internal structure of the jump. In granular media, the jump is defined as the part of the flow between the incoming flow, where the free-surface is parallel to the incline, and the outgoing flow, where the slope of the free-surface is constant and equal to the critical slope ζ_0 below which no flow can occur, as proposed in [1]. For the largest ζ, the flow is slightly accelerating (non-uniform, but still at steady state). In those cases, the slope angle of the free-surface is however very small compared to the changes in the jump, and is not considered here (like in [1]). The main variables defining a jump are summarized in Fig.1. The depth-averaged velocity after the jump is noted $\overline{u}_*$, and the mean diameter of the grains is d.

Section 2 describes the DEM contact-laws, the numerical set-up used to create a jump, and the techniques to measure all the variables of interest. Section 3 shows that this method allows to create a wide variety of jumps, and presents some of the first results. Finally, a short conclusion is given on the main challenging issues this granular jump numerical set-up will allow to investigate.

2 DEM simulations of standing jumps

This section describes how granular flows are simulated thanks to DEM, using YADE open-source software [3].

2.1 Microscopic contact laws

In order to model the grains interacting each other, we chose the classical following viscoelastic law:

$$\begin{cases} \mathbf{F_n} = k_n\delta_n\mathbf{n} + c_n\dot{\delta_n}\mathbf{n} \\ \Delta\mathbf{F_s} = k_s\dot{\delta_s}\mathrm{dt}\ \mathbf{s} \\ F_s \leq \mu F_n \end{cases} \tag{2}$$

In this equation, $\mathbf{F_n}$ and $\mathbf{F_s}$ are the normal and tangential forces. $\mathbf{F_n}$ is the sum of a linear spring (of stiffness k_n, and proportional to the normal overlap δ_n) and a dashpot of damping coefficient c_n that depends on the restitution coefficient e. $\mathbf{F_s}$ is incremented at each time step dt as a linear spring (of stiffness k_s proportional to the derivative of the tangential overlap $\dot{\delta_s}$) restricted to a Coulomb threshold force defined by the friction coefficients μ and μ_b for grain-grain and grain-wall interactions, respectively. In this study, the tests are carried out in two dimensions (2D), with only one particle across the width. This is made by blocking one degree of freedom in translation and two degrees of freedom in rotation for each particle. k_n and k_s are calculated from the values of Young's modulus E—which was taken equal to 1×10^6 to respect the limit of rigid grains condition, and Poisson coefficient ν taken equal to 0.3: $k_n = Ed/2$ and $k_s = \nu k_n$, considering one diameter d for all particles. A polydispersity of 15% around the chosen diameter $d = 4$ cm is considered to avoid crystallization effects. The bed-friction coefficient was $\mu_b = 0.25$. The grain density $\rho_P = 2500$ kg m^3 to mimick the density of glass beads usually used in laboratory models. The time-step was $dt = 3 \times 10^{-4}$s.

2.2 DEM set-up to produce the jumps

The numerical set-up is shown in Fig.2. A reservoir permanentely filled up with grains feeds an incline. At the exit, a gate initially retains the grains. When there are enough grains in the incline, the gate moves up. Every 150 time steps, the outflow discharge is calculated and compared with the inflow. The gate then slightly moves, up or down to adjust the discharge and guarantee a steady state. This is a slight difference from the laboratory model where the gate was adjusted by the operator and fixed [1]. The strong advantage of DEM is that many control variables can be changed to see their effect on the jump: the slope angle ζ, the height H of the reservoir exit (that controls the inflow), but also the grain diameter, the interparticle friction and the restitution coefficient. All parameters have a default value, and each parameter was varied—the other ones being fixed at the default value (see Tab.1).

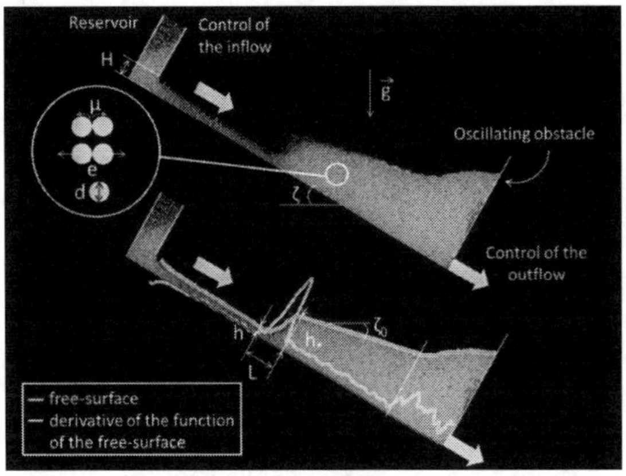

Figure 2: DEM numerical set-up to produce standing granular jumps. ($\zeta = 30°, H = 0.5, d = 2cm, \mu = 0.5, e = 0.5$)

Table 1: Default values (first line) and ranges (second line) of the varying parameters for the 53 preliminary tests.

ζ [°]	H [m]	d [cm]	μ [−]	e [−]
20	0.35	4	0.5	0.5
[15 − 40]	[0.35 − 1.1]	[1 − 6]	[0.05 − 1]	[0.2 − 1]

2.3 Measurement techniques

2.3.1 Geometry of the jump

When a simulation runs, the first step is to reach a steady state. Then, the free-surface is calculated by discretising the incline, and identifying the highest grain in each cell. This is made a hundred times at several time steps to obtain a smooth time-averaged free-surface at continuum scale. The latter allows to identify the beginning of the jump, which is the location where the free-surface is not parallel to the bottom anymore—when the derivative of the function describing it becomes non-zero (Fig.2). We also identify the end of the jump, where the slope of the free-surface becomes constant—meaning that the derivative is constant. Once a steady state is reached and both the beginning and the end of the jump are well identified, a new simulation is done to record the variables of interest.

2.3.2 Velocity profiles

An example of velocity profiles recorded before and after the jump is shown in Fig.3. The raw DEM data sets exhibit a noticeable scattering caused by the (discrete) fluctuating nature of the flowing grains, but the mean velocities are well-described at continuum scale and they can be fitted by a Bagnold profile with basal slip, as already suggested in [1, 4], which is given by the following equation:

$$u(z) = u_b + A\sqrt{gh}\left[\left(\frac{h - z_0}{d}\right)^{3/2} - \left(\frac{h - z}{d}\right)^{3/2}\right] \tag{3}$$

where u_b denotes the sliding velocity of the basal layer, which is taken at $z = z_0 = 1.5d$ (following [1, 4]) and A

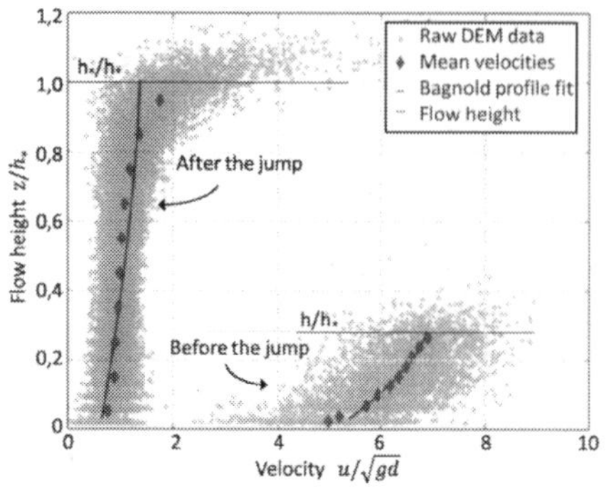

Figure 3: Velocity profiles across flow-depth before and after the jump for one test: $\zeta = 20°$, $H = 0.5$m, $d = 4$cm, $\mu = 0.5$, $e = 0.5$. Blue diamonds are velocities averaged over height increments; green lines are fits with Eq.3

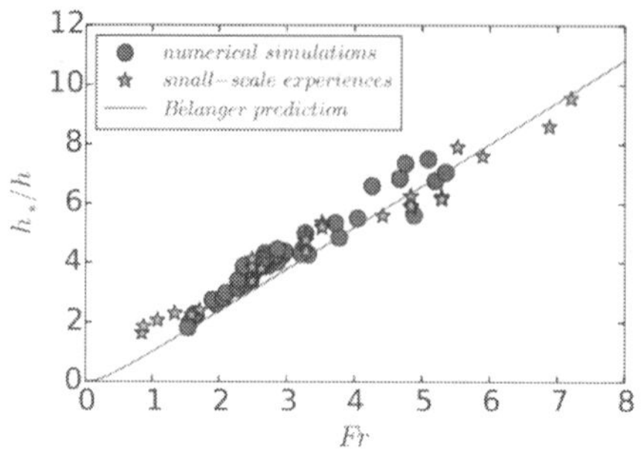

Figure 5: h_*/h versus Fr obtained with the present numerical experiments (red circles) compared to the laboratory experiments conducted by [1] (green stars).

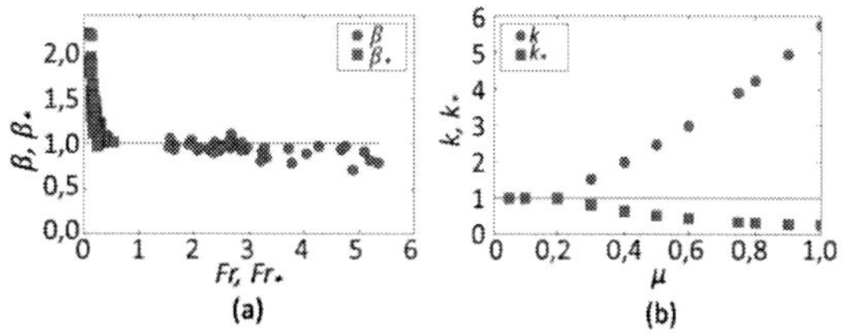

Figure 4: (a) Boussinesq coefficients β or β_* versus the Froude number Fr or Fr_*. (b) Earth pressure coefficients calculated from Eq.4 versus the grain-grain friction μ.

is the Bagnold pre-factor, which was let free in the fits of Fig.3. The hypothesis of the Bagnold profile seems very appropriate for the velocity of the incoming flow. But Fig.3 shows that it is less satisfying for the outgoing flow, because of a change in concavity next to the free-surface that will need further investigation in the future.

Measuring accurately the velocities thanks to DEM makes it possible to calculate precisely the Boussinesq coefficient, β before the jump (or β_* after the jump), which is defined by the relation $\beta = \overline{u^2}/\bar{u}^2$. The variations of β and β_* with Fr and Fr_* respectively are given in Fig.4(a). It clearly shows that the usual assumption that $\beta = 1$ is very accurate for median values of the Froude number that we find most of the time in the incoming flows, while it may become an underestimate for very low values of the Froude number as found in the downstream flow.

2.3.3 Earth pressure coefficients k,k_*

The earth pressure coefficients, noted k and k_* before and after the jump respectively, relate the normal stresses through the relation $\sigma_{xx} = k\sigma_{zz}$. In this study, k and k_* were not directly measured but derived from soil mechanics concepts, as proposed by Savage and Hutter [5]:

$$k_{pass/act} = \frac{2\left(1 \pm \sqrt{1-(1+\tan^2\varphi_s)\cos^2\varphi_e}\right)}{\cos^2\varphi_e} - 1, \quad (4)$$

where $\varphi_s = \tan^{-1}(\mu_b)$ is the bed friction angle and $\varphi_e = \tan^{-1}(\mu)$ the internal friction angle. When μ becomes lower than μ_b, we can consider $k = k_* = 1$. The variations of k and k_* with μ according to Eq.4 are plotted in Fig.4(b). Because the inter-particle friction coefficient used in DEM is known to be different from the macroscopic (internal) friction coefficient ([6]), our assumption for Equation 4 remains questionable. Future work will consider shear-dependent friction coefficients, such as developed by [7].

3 A rich variety of granular jumps

The method presented in Sec.2.2 allows to create a wide range of jumps, and confirms the results obtained in the laboratory [1] as shown in Fig.5. We could observe steep or diffuse jumps, compressible or incompressible jumps, and identify the presence or absence of a recirculation zone. Figure 6 displays pictures of jumps—with the internal streamlines drawn, obtained for different ζ and H/d.

3.1 Jump steepness

The two characteristic length-scales of a jump are its height ratio h_*/h and its relative length L/h (if not neglected like in Bélanger equation). Combining the two leads to the steepness coefficient $S = (h_* - h)/L$ with respect to the slope of the bottom. S can grow by increasing ζ, and weakly depends on H/d (Fig.7). It is also correlated with the presence of recirculation as a very steep jump will be instable, thus promoting the recirculation. A larger opening of the reservoir leads to a higher discharge and a denser jump, more stable. As such, the recirculation appears for higher ζ and S. It is worthwhile to note that jumps with recirculation were also observed for very low μ, typically smaller than μ_b. This result–likely to unravel the key role played by friction in the birth of jump recirculation, will need further investigation in the future.

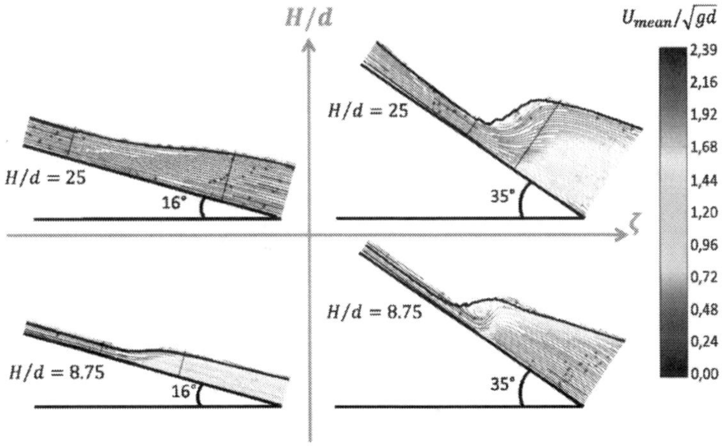

Figure 6: $(H/d, \zeta)$ phase diagram for granular jumps. In each picture of the jumps, the streamlines are drawn, thus allowing us to identify the presence or absence of a recirculation zone

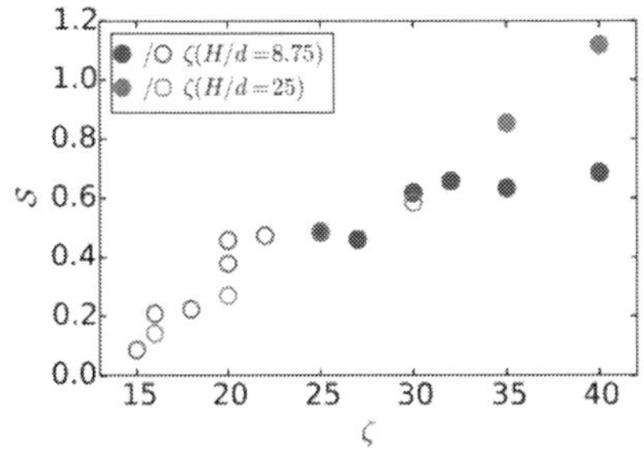

Figure 7: Steepness S versus the slope ζ for two values of H/d (d = 4 cm). Empty—full—circles represent the jumps without—with—a recirculation zone.

3.2 Jump compressibility

The granular jump compressibility is defined as the ratio ϕ_*/ϕ (where $\rho = \rho_P\phi$) between the volume fractions before and after the jump. ϕ_* was systematically found to be close to 0.81 (whatever the incoming flow and boundary conditions), which is nearly equal to the 2D granular random close packing. Then, once the incoming flow is dilute ($\phi < \phi_*$), a compressible jump is obtained. Our simulations showed that it is possible to tend toward an incompressible jump ($\phi_*/\phi \mapsto 1$) by different ways: increasing H/d, decreasing ζ, decreasing d or decreasing μ.

4 Conclusion

The current paper presented preliminary DEM simulations of standing jumps formed in flows of granular materials down an incline. The first results are consistent with the recent findings in laboratory tests [1]. This novel numerical set-up now offers the possibility to study in detail how the jump geometry and its internal structure are influenced by a number of parameters, and will allow to investigate the energy dissipation inside the jump, as proposed in [2].

Acknowledgements

The authors are thankful to the labex Tec21, which is part of the "Investissement d'avenir" ANR program, and to the "Région Rhône-Alpes-Auvergne" through their "ExploRa'Doc" program for funding. TF is grateful to the financial support by the People Programme (Marie Curie Actions) of the EU FP7 under REA grant Agreement No. 622899 (FP7-PEOPLE-2013-IOF, GRAINPACT).

References

[1] T. Faug, P. Childs, E. Wyburn, I. Einav, Physics of Fluids **27**, 073304 (2015)

[2] S. Méjean, T. Faug, I. Einav, Journal of Fluid Mechanics (2017)

[3] V. Šmilauer, E. Catalano, B. Chareyre, S. Dorofeenko, D. J., A. Gladky, J. Kozicki, C. Modenese, L. Scholtès, L. Sibille et al., in *Yade Documentation* (The Yade Project, 2010)

[4] N. Brodu, P. Richard, R. Delannay, Physical Review E **87**, 022202 (2013)

[5] S.B. Savage, K. Hutter, Journal of Fluid Mechanics **199**, 177 (1989)

[6] A. Taboada, N. Estrada, F. Radjai, Physical review letters **97**, 098302 (2006)

[7] P. Jop, Y. Forterre, O. Pouliquen, Nature **441**, 727 (2006)

Pomelo, a tool for computing Generic Set Voronoi Diagrams of Aspherical Particles of Arbitrary Shape

Simon Weis[1,*], *Philipp W. A.* Schönhöfer[1,2,**], *Fabian M.* Schaller[1], *Matthias* Schröter[3], and *Gerd E.* Schröder-Turk[1,2]

[1] *Institut für Theoretische Physik I, Universität Erlangen-Nürnberg, Staudtstraße 7, 91058 Erlangen, Germany*
[2] *School of Engineering and Information Technology, Murdoch University, 90 South Street, Murdoch, WA 6150, Australia*
[3] *Institute for Multiscale Simulation, Nägelsbachstrasse 49b, 91052 Erlangen, Germany*

Abstract. We describe the development of a new software tool, called "Pomelo", for the calculation of Set Voronoi diagrams. Voronoi diagrams are a spatial partition of the space around the particles into separate Voronoi cells, e.g. applicable to granular materials. A generalization of the conventional Voronoi diagram for points or monodisperse spheres is the Set Voronoi diagram, also known as navigational map or tessellation by zone of influence. In this construction, a Set Voronoi cell contains the volume that is closer to the surface of one particle than to the surface of any other particle. This is required for aspherical or polydisperse systems. Pomelo is designed to be easy to use and as generic as possible. It directly supports common particle shapes and offers a generic mode, which allows to deal with any type of particles that can be described mathematically. Pomelo can create output in different standard formats, which allows direct visualization and further processing. Finally, we describe three applications of the Set Voronoi code in granular and soft matter physics, namely the problem of packings of ellipsoidal particles with varying degrees of particle-particle friction, mechanical stable packings of tetrahedra and a model for liquid crystal systems of particles with shapes reminiscent of pears.

The analysis of geometries and structures on a micro scale level is an important aspect of granular and soft matter physics to attain knowledge about many interesting properties of particle packings, including contact numbers, anisotropy, local volume fraction, etc. [1–3]. A well-established concept is the so called *Voronoi Diagram*. Here, the system is investigated by dividing the space into separate cells in respect to the positions of the center of the particles. A cell assigned to a certain particle is defined as the space (or region of space) that contains all the volume closer to the center of this specific particle than to any other one (see figure 1 left). This partition of space, however, only yields precise results for monodisperse spheres as the construction fails otherwise due to morphological properties of the objects. For nonspherical or polydisperse particles the classical Voronoi diagram is of limited usefullnes, as shown in figure 1 (center) for a system of bidisperse spheres. A generalized version of the Voronoi Diagram, the *Set Voronoi Diagram* [4], also known as navigational map [5] or tessellation by zone of influence [6], has to be applied. In this case the cells contain all space around the particle which is closer to the particle's surface than to the surface of any other particle. Figure 1 (right) shows the Set Voronoi Diagram of a mixture of differently shaped particles.

Here we introduce a software tool called *Pomelo*, which calculates Set Voronoi Diagrams based on the algorithm described in [4]. Pomelo is particularly versatile due to its ability to generically handle any arbitrary shape which can be described mathematically. For instance objects neither have to be convex nor simply connected. In the first step of the algorithm (figure 2 left) a triangulation of the particle's bounding surface is generated to sample its shape. *Pomelo* offers functionality for some common particle shapes, but the user is can also specify generic particle shapes. After the discretisation of the surface the system is tesselated by calculating the classical Voronoi diagram of all surface points of this triangulation. This is shown in figure 2 (center). To get the Set Voronoi tessellation of the system, cells belonging to points on the same particle surface are merged to a single cell in the last step (figure 2 right). The resulting partition represents the Set Voronoi Diagram of the system. With this algorithm, systems and mixtures of particles of any arbitrary shape can be treated.

1 Pomelo

Pomelo is a generic Set Voronoi tool written in `c++11` and licensed under GPL3. *Pomelo* can be downloaded, see ref. [7], as well as all instructions regarding setting up, building and using *Pomelo*. The system requirements are `g++ 4.9.2` or `clang++ 3.5.0-10` or any higher version.

*e-mail: simon.weis@fau.de
**e-mail: P.Schoenhoefer@murdoch.edu.au
SW and PS have contributed equally to the work described in this project

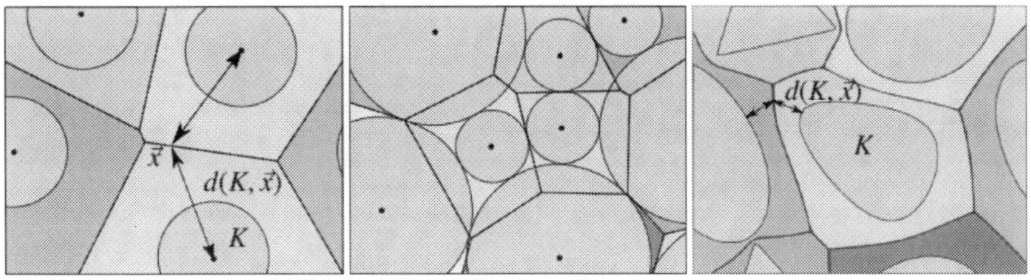

Figure 1. Left: The Voronoi Diagram of a system of monodisperse spheres. Center: The Voronoi Diagram is not suitable for a system of bidisperse spheres, as Voronoi cells are overlapping with particles. Right: The Set Voronoi Diagram of a mixture of differently shaped objects.

Figure 2. Sketch of the Set Voronoi algorithm from left to right. Generating points on particle surface. Calculating Voronoi Diagram of surface points. Merging cells belonging to the same particle.

While *Pomelo* can directly handle common particle shapes (mono- and polydisperse spheres, tetrahedra, ellipsoids and spherocylinders), it also provides a `generic` mode. The latter works for any shape which surface can be described mathematically. The following two sections will describe how to use *Pomelo* in both cases. To use *Pomelo* in `generic` mode (see section 1.2), `lib-lua 5.2` or higher is required.

1.1 Common Particle Shapes

Particle shapes that are intrinsically supported by *Pomelo* are mono- and polydisperse spheres, tetrahedra, spherocylinders and ellipsoids. *Pomelo* comes with a set of demos and tests.

One test case is a set of polydisperse spheres in a cubic cell. The input is a `xyzr` file. Its first line is the number of particles. The second line is a comment which contains some information (separated by comma) required by *Pomelo*. This includes boundary conditions (periodic/non periodic in each axis), box size, shrink and the number steps in ϕ and θ for discretizing the sphere's surface. Every following line describes one sphere in the packing with its parameters (coordinates and radius). To run the demo, call *Pomelo* with the command line argument `-SPHEREPOLY`, the path to the `xyzr` file and the desired output folder.

Pomelo's output are the vertices, faces and cells of the Set Voronoi Diagram. The output can be set to different formats, like the POLY file format, off (for geomview) or a gnuplot readable format for easy visualisation.

1.2 `Generic` Mode

Using *Pomelo* in generic mode allows to calculate the Set Voronoi Diagram for generic particles. The input is a `position` file, which is a list of all particles which are described by a set of parameters each. The parameters are a complete description of the particles surface within the packing. The `read` file is key for *Pomelos* versatility. It allows the user to create a surface triangulation based on the parameters given in the `position` file with a script written in the lua language [8]. The script tells *Pomelo* how to create a surface triangulation with the particles parameters as given in the `position` file. It can be fully customized by the user to match any specific particle shape. This allows the user to create even systems composed of a mixture of diferent particles. The surface triangulation of all particles will then be used to calculate the Set Voronoi Diagram as described above.

Pomelo's demos include a variety of examples on how to use the `read` script to handle different types of particles.

2 Applications of Set Voronoi Diagrams

2.1 Ellipsoid Packings

To demonstrate the use of Set Voronoi Diagrams the local packing fraction which is defined as $\Phi_l = \frac{v_e}{v_i}$ with v_i being the volume of the Set Voronoi Cell and $v_e = \frac{4}{3}\pi abc$ the volume of the particle is calculated. It has been shown that the probability distribution of local packing fractions is universal for sphere packings [1] and jammed pro- and oblate ellipsoid packings [9].

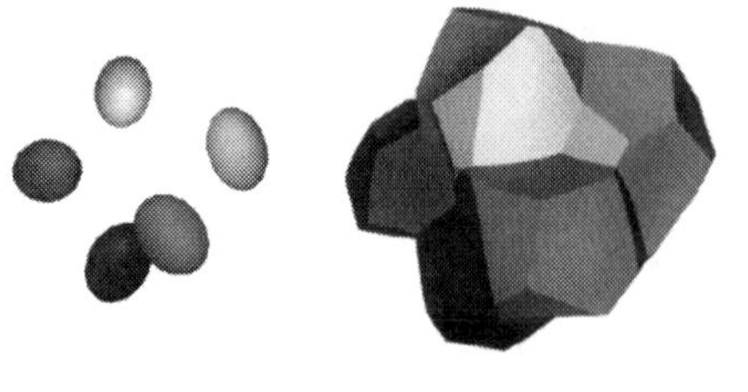

Figure 3. Ellipsoids (α = 1.4) in an isotopic system (left) and ther Set Voronoi Cells (right).

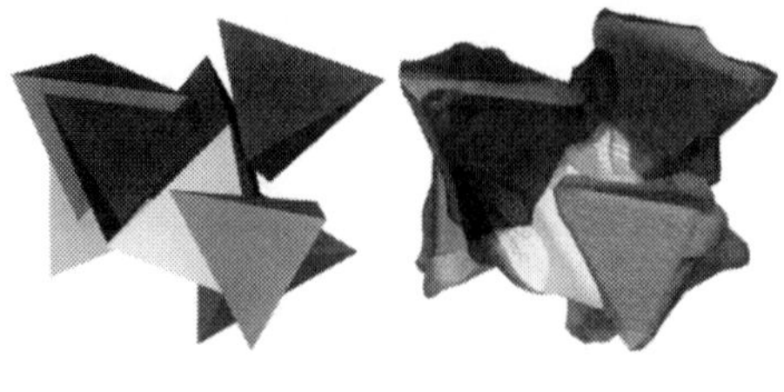

Figure 5. Tetrahedra of one of the experiments (left) and their Set Voronoi cels (right).

Experimental granular packings of triaxial ellipsoids (three distinct axis lengths a, b, c) have been measured at various global packing fractions using X-ray tomography, see figure 3. Particle positions, orientations and size have been determined and the Set Voronoi Diagrams have been calculated for each packing. Here the particles are shrinked by *Pomelo* to improve the quality of the cells. The statistical distribution of local packing fractions Φ_l for spheres ($\alpha = 1.0$) and triaxial ellipsoids ($\alpha = 1.1$, $\alpha = 1.4$) is shown in figure 4. To first order, the distribution can be collapsed on a master curve, as shown in Ref. [9], by subtracting the global volume fraction $\Phi_g = \frac{<v_p>}{<v_l>}$ and scaling the distribution by $\sigma(\Phi_l)$, see figure 4. Thus the functional form of the distribution is invariant to global packing fraction Φ_g and aspect ratio α, to first order.

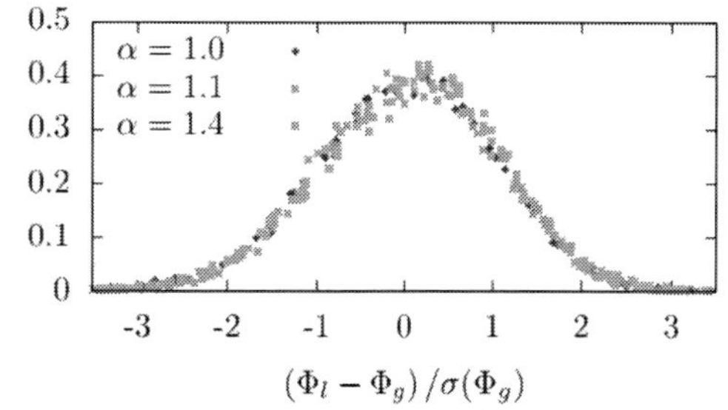

Figure 4. Probability distribution of rescaled local packing fractions Φ_l for packings of spheres (particle aspect ratio α = 1.0) and triaxial ellipsoids (α = 1.1, α = 1.4). Global packing fractions range from 0.62 to 0.65.

2.2 Tetrahedra

Packings of tetrahedra can also be treated with *Pomelo* without using the generic mode. The data was obtained by X-ray tomography [10]. Each tetrahedra is described by the positions of its 4 vertices. *Pomelo* is able to process the data directly. This gives the advantage to treat not only tetrahedra but pyramids in general.

The Set Voronoi tessellations of tetrahedra packings (figure 5) with different packing fraction and contact numbers are calculated with *Pomelo*. The local packing fraction is shown for those systems in figure 6.

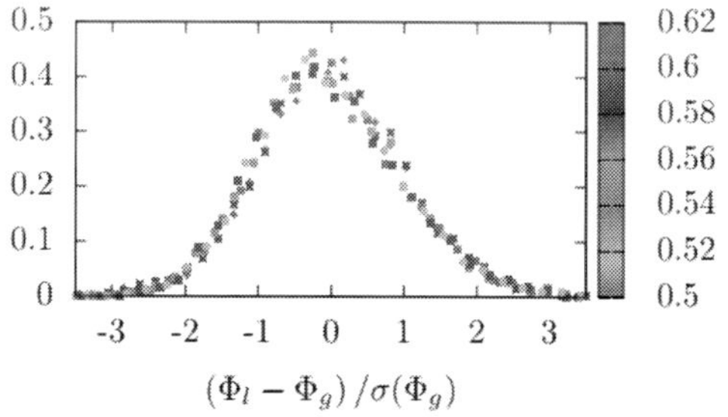

Figure 6. Probability distribution of rescaled local packing fractions Φ_l for different systems of tetrahedra. The colorbar and the point's color show the global packing fraction of the system.

2.3 Pear Shaped Particles

The third example is the calculation of the Set Voronoi Diagrams of pear shaped or tapered particles. The shape of these particles is described by the aspect ratio α and the degree of tapering α_θ. By using two Bézier curves forming the bottom and top part of the pear and rotating them around their symmetry axis the surface of the particle is generated (figure 7) [11]. A triangulation algorithm of this surface is implemented within the `read` file, which allows *Pomelo* to process pear shaped particles in the generic mode. The `position` file provides the posi-

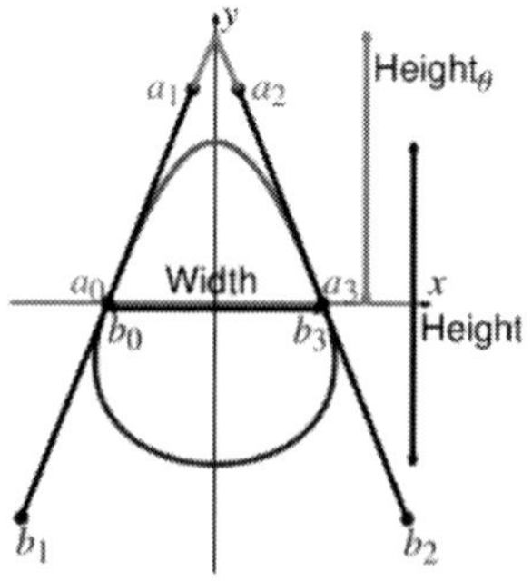

Figure 7. Pear shaped by two Bézier curves. Both are determined by the aspect ratio $\alpha = \frac{\text{Height}}{\text{Width}}$ and the degree of tapering $\alpha_\theta = \frac{\text{Height}_\theta}{\text{Width}}$.

tion, orientation, size, aspect ratio and degree of tapering for every individual particle.

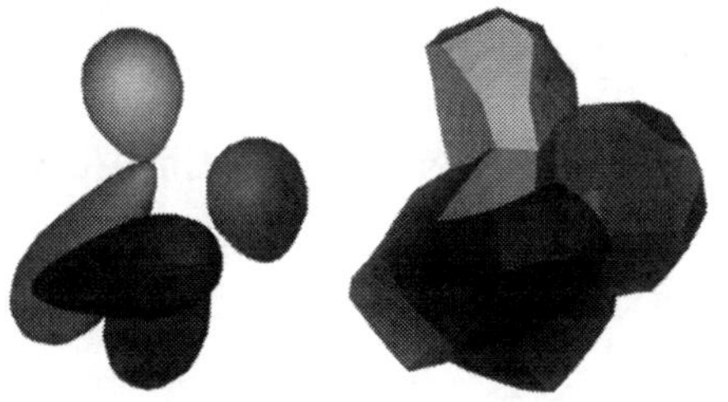

Figure 8. Pears (α = 3.0, α_θ = 3.8) in an isotropic system (left) and their Set Voronoi cells (right).

Systems of hard pear shaped particles (α = 3.0) with different degrees of tapering are generated using Molecular Dynamics (see figure 8) [11]. The statistical distribution of the local packing fractions Φ_l for different pear systems (α_θ = 3.0, Φ_g = 0.48 ; α_θ = 3.8, Φ_g = 0.50 ; α_θ = 6.0, Φ_g = 0.54) is shown in figure 9. Similarly to the sphere and ellipsoid packings the distributions of the different pear systems collapse on a master curve. Accordingly the distribution is not only invariant to global packing fraction Φ_g but also to degree of tappering α_θ for pears. However, the data of ellipsoids/spheres and pears do not collapse on the same curve.

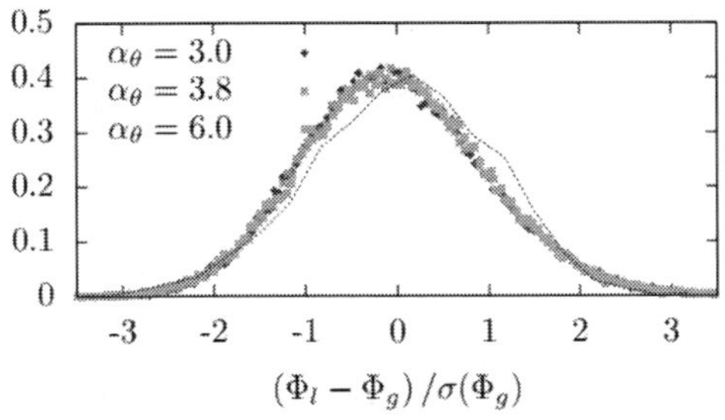

Figure 9. Probability distribution of rescaled local packing fractions Φ_l for pears (α = 3.0) with different degree of tapering and packing fraction $(\alpha_\theta, \Phi_g) \in \{(3.0, 0.48); (3.8, 0.50); (6.0, 0.54)\}$. The dashed line shows the master curve of the sphere distributions.

3 Outlook

We have shown that *Pomelo* is applicable for a variety of systems, including spheres, ellipsoids, tetrahedra and pear shaped particles. We have illustrated its use to calculate Set Voronoi volume distributions. However, it is conceivable to analyse other interesting measures like Minkowski tensors, by piping the Set Voronoi Cells into corresponding analysis tools [9, 12, 13]. It is yet also unclear how similar volume distributions of different systems are.

One way to improve the calculation of the Set Voronoi Diagram is to implement an adaptive sampling of the particle's surface. By comparing the distance between two neighboring surface points to the distance to the face of the Set Voronoi Cell it is possible to obtain an estimate on where the surface triangulation needs a better resolution and where a coarse resolution is good enough. This estimate will be used to change the density of the surface triangulation of the particles.

References

[1] T. Aste, T. Di Matteo, M. Saadatfar, T. J. Senden, M. Schröter, H. L. Swinney, EPL **79**, 24003 (2007)

[2] M. van Hecke, J. Phys: Condens. Matter, **22**, 0033101 (2010)

[3] P. Wang, C. Song, Y. Jin, K. Wang, H. Makse, J. Stat Mech **2010**, P12005 (2010)

[4] F. M. Schaller, S. C. Kapfer, M. E. Evans *et al.*, Philosophical Magazine **93**, 3993-4017 (2013)

[5] V. Luchnikov, N. Medvedev, L. Oger, J. Troadec, Phys. Rev. E **59**, 7205 (1999)

[6] E. Preteux, J. Math. Imaging Vision **1**, 239 (1992)

[7] `http://theorie1.physik.uni-erlangen.de/research/pomelo/index.html`

[8] `https://www.lua.org/`

[9] F. M. Schaller, S. C. Kapfer, J. E. Hilton *et al.*, EPL **111**, 24002 (2015)

[10] M. Neudecker, S. Ulrich, S. Herminghaus, M. Schröter PRL **111**, 028001 (2013)

[11] F. Barmes, M. Ricci, C. Zannoni, D. J. Cleaver, Phys. Rev. E. **68**, 021708 (2003)

[12] G. E. Schröder-Turk, W. Mickel, S. C. Kapfer *et al.*, Adv. M. **23**(22-23), 2535–2553 (2011)

[13] G. E. Schröder-Turk, W. Mickel, S. C. Kapfer *et al.*, NJP **15**, 083028 (2013)

Combined effect of moisture and electrostatic charges on powder flow

Antonella Rescaglio[1], *Julien* Schockmel[1], *Nicolas* Vandewalle[1], and *Geoffroy* Lumay[1,*]

[1]*CESAM-GRASP, University of Liége, Belgium*

Abstract. It is well known in industrial applications involving powders and granular materials that the relative air humidity and the presence of electrostatic charges influence drastically the material flowing properties. The relative air humidity induces the formation of capillary bridges and modify the grain surface conductivity. The presence of capillary bridges produces cohesive forces. On the other hand, the apparition of electrostatic charges due to the triboelectric effect at the contacts between the grains and at the contacts between the grains and the container produces electrostatic forces. Therefore, in many cases, the powder cohesiveness is the result of the interplay between capillary and electrostatic forces. Unfortunately, the triboelectric effect is still poorly understood, in particular inside a granular material. Moreover, reproducible electrostatic measurements are difficult to perform. We developed an experimental device to measures the ability of a powder to charge electrostatically during a flow in contact with a selected material. Both electrostatic and flow measurements have been performed in different hygrometric conditions. The correlation between the powder electrostatic properties, the hygrometry and the flowing behavior are analyzed.

1 Introduction

When two materials are rubbed, electric charges are exchanged at the surfaces [1, 2]. This contact electrification is one of the oldest scientific subject. Thales of Miletus (624 - 546 BC) discovered that amber and wool are electrified when rubbing together. It is interesting to note that the word 'electron' comes from the greek word for 'amber' and the term 'triboelectric' means 'rubbing amber' in greek. Despite the numerous studies dedicated to this subject, the fundamental mechanisms behind this triboelectric effect is not fully understood. The charging of objects composed by the same material [3] and charging of powders [2, 4, 5] are two examples of poorly understood subject. Even the basic question related to the nature of the transferred charges (electrons, ions or material) is still debated. The difficulties are related to the non-equilibrium character of the tribolectectric dynamic and to the variety of mechanisms behind this effect. Different mechanisms could induce triboelectric charges inside a granular material: impact, rolling and sliding (or rubbing) [6].

On the one hand, the triboelectric effect is useful for many applications (electrophotography, powder coating processes, triboelectric separators, ...), but on the other hand, the triboelectrification causes complications. The apparition of electric charges inside a granular material can produce sparks and cause explosion hazards in dispersed fine particles. In pharmaceutical industries, the contact charging during particles flows induces agglomeration and segregation witch cause dosage problems.

*e-mail: geoffroy.lumay@ulg.ac.be

In combination with electrostatic charges, moisture is known to affect both static and dynamic behaviors of a granular materials [7–9]. However, the effect of moisture is far to be obvious. Indeed, moisture influences both surface grains conductivity and capillary bridges formation. For low relative air humidity, the electrical conductivity necessary for charge dissipation is reduced and the relative permittivity comes close to vacuum permittivity. Therefore, the electric charges created by triboelectric effects lead to uncontrolled electric field, electrostatic forces between the grains and/or between the grains and the container. For high relative air humidity, the electrical conductivity increases, liquid bridges may be formed at the contacts between the grains and the relative air permittivity increases. Therefore, the electrical charges are dissipated more easily. However, the apparition of liquid bridges induces cohesive forces inside the packing [10]. At intermediate relative humidity the cohesion is expected to decrease. In addition to these effects, the humidity could also modify the friction due to lubrication effects [11].

In the present paper, we describe an experimental method to measure the quantity of electric charges created inside a powder during a bulk flow in contact with a selected material. This device is called a powder triboelectrometer. In addition to the electrostatic measurement, a flow measurement based on the classical rotating drum method is used to study the influence of the relative air humidity on powder flow. In particular, we show how the association of capillary and electrostatic forces modifies the macroscopic properties of powders. Finally, we discuss how the results can be used to optimize industrial applications involving powders like pneumatic conveying.

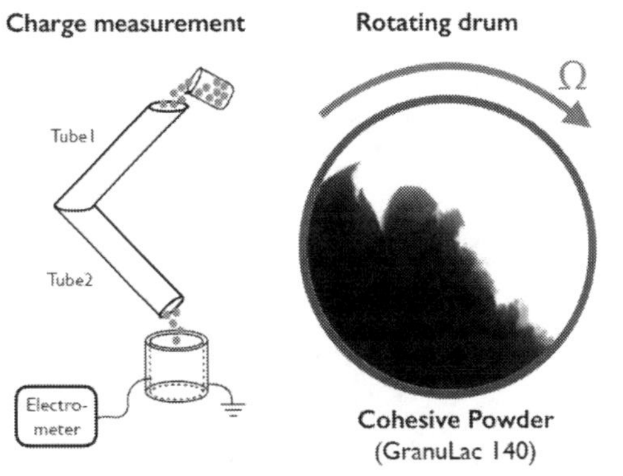

Figure 1. Sketch of the experimental set-ups used to perform the measurements. (left) The powder triboelectrometer measures the electrostatic charge inside the powder after a flow in contact with a selected material. (right) The powder flowing properties are measured in a rotating drum.

2 Set-up

2.1 Powder triboelectrometer

Figure 1 (left) shows the experimental set-up used to measure the total electrostatic charge created inside a granular material during a flow in contact with a selected material. The sample is poured manually (the feeding could be automated) in a V-tube and flows to a Faraday cup. The V-tube is an assembly of two tubes (Tube1 and Tube2) of length L=350mm and internal diameter D=47mm to form a V shape with an angle of 90°. The tube material can be selected for each tube. In the present study, we used 316L stainless steel (S.S.) and ABS. The faraday cup is connected to a customized electrometer able to measure electrostatic charges ranging from 0.1 nC to 1 μC. At the end of the flow, the total value of the electric charge Q present in the powder is measured and the charge density Q/m is computed, where m is the sample mass, is measured.

The V-tube geometry has been selected to combine the different mechanisms leading to tribo-electrification: (i) friction between the grains, (ii) friction between the grains and the wall and (iii) impacts of the grains on the wall at the connection between the two tubes.

The device used to perform the present study is the prototype of the GranuCharge presently commercialized by GranuTools.

2.2 Flow measurement

Experimentally, the most practical geometry to study a free surface flow of granular materials is the rotating drum. This flow geometry has been extensively studied with non cohesive granular materials and with cohesive powders [12–14]. In the case of cohesive grains, the flow is intermittent in the whole range of rotating speeds.

An aluminum cylinder of internal diameter D = 84mm and length L = 20 mm with glass side walls is half-filled with the powder. The cylinder rotates around its axis at an angular velocity Ω producing the flow of the grains (see Figure 1 (right)). The rotating speed $\Omega = 8RPM$ has been selected to obtain a roughly flat interface without fluctuation with a non-cohesive granular material. Therefore the fluctuations of the interface obtained with a cohesive powder is related to the existence of cohesive forces. From 50 images of the rotating powder bed separated by 0.5s, the average position of the powder/air interface and the fluctuations (standard deviation) around this average interface are calculated. The standard deviation is integrated over the whole interface to obtain the total interface fluctuation σ [15, 16].

The device used to perform the present study is the prototype of the GranuDrum presently commercialized by GranuTools.

2.3 Conditioning

Before the flowability measurement in the rotating drum, the powder is conditioned during a time T in a slowly rotating reactor. The reactor is a 316L stainless steel tube of internal diameter D = 48mm and length L = 320mm. The tube axis is placed horizontally. The tube is rotating around its axis at 0.66 RPM. A flux F = 0.3 l/min of air with controlled relative humidity flows through the reactor. The slow motion of the reactor allows the humid air to penetrate inside the bulk. The accessible values of the relative humidity range from RH=0% to RH=95%. Directly after this preparation, the drum cell is half filled by the powder and is closed hermetically.

2.4 Powders

Two lactose powders produced by the company Meggle and widely used in pharmaceutical industries have been selected: GranuLac 140 (Lactose1) and InhaLac 230 (Lactose2). Granulac 140 powder comes from milling process with an average grains diameter of 65 μm and a standard deviation of 55 μm (see insert of Figure 2). Inhalac 230 powder is a milled and sieved lactose powders with an average grains diameter of 98 μm and lower standard deviation of 39μm. We received the samples in small sealed bags in order to perform the measurements without powder aging.

3 Results and discussion

A typical flow of lactose grains at the V-tube output is shown by Figure 2. The grains are electrostatically charged and are repealing each other.

Figure 3 shows the charge density Q/m obtained with the lactose powders for different relative air humidities and for different materials in contact (Stainless steel and ABS). The different relative air humidity values have been obtained by switching off the humidity regulator of the laboratory. Therefore, we took advantage of the Belgian meteorological fluctuations. The sign of the charge depends

Figure 2. Flow of a lactose powder between the V-tube output and the Faraday cup. The insert shows a micrograph of a few lactose (GranuLac140 from Meggle) grains.

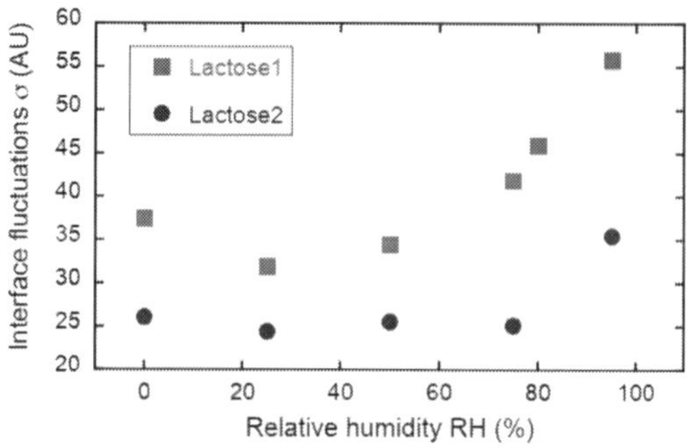

Figure 4. Fluctuations σ of powder/air interface measured in the rotating drum at a rotating speed Ω =8 RPM as a function of the relative humidity RH for a residence time in the conditioner fixed to T = 150 minutes.

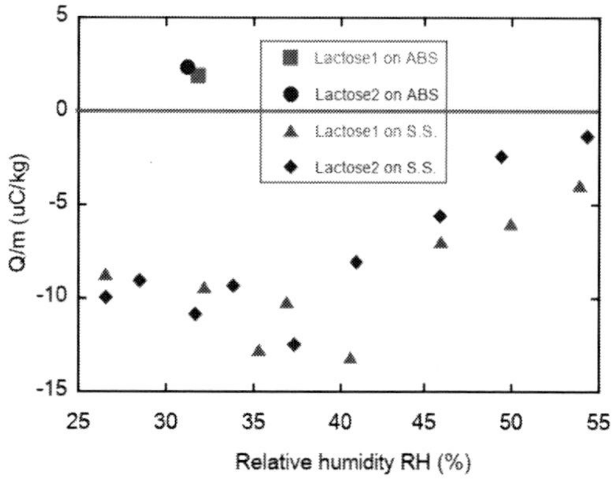

Figure 3. Charge density Q/m obtained with the powder tribo-electrometer for different relative air humidities and for different materials in contact (Stainless steel and ABS)

on the material in contact: positive for lactose on ABS and negative for lactose en stainless steel. Therefore, in a tribo-electric series, lactose powders should be placed between stainless steel and ABS. Moreover, the absolute value of the charge density decreases drastically when the relative air humidity goes from 40% to 50%. This observation is in agreement with the results obtained very recently by Schella *et al* [17] in a shaken cell.

In order to study the influence of relative air humidity on powder cohesiveness, the samples of powder has been conditioned at fixed relative air humidity during T = 150 minutes before filling the rotating drum. The cohesiveness is found to decrease slightly in the range of relative humidity from 0% to 40% (see Figure 4). Therefore, the cohesive forces inside the bulk decrease and the flowability is enhanced. Afterward, the cohesion increase importantly from RH=50% to RH=100%. Then, the relative humidity that optimizes the flowability of lactose grains is situated between 30% and 50%. For low humidity conditions, the cohesion is induced by the apparition of electric charges on the surface of powder grain *(see Figure 3)*. For the extremal case RH=0% the powder was sticking on the drum surfaces due to these electrostatic charges. For high humidity conditions, the condensation of water at the surface of the grains leads presumably to the formation of capillary bridges at the contact between the grains, increasing powder cohesiveness. For the relatively short residence times in high humidity conditions considered in the present study, we do not expect water absorption inside the grains.

Globally, the cohesiveness of Lactose1 is higher than the cohesiveness of Lactose2 (see Figure 4) in the whole range of RH. Indeed, the average grain size is lower in Lactose1. Therefore, the ratio between the cohesive forces acting on a grain and the grain weight is higher in Lactose1.

During powder pneumatic transport, it is well known in industries that the pipe material influences the apparition of electrostatic charges. Stainless steel is usually used. Some parts (bends, flexible connections, ...) are sometimes made in plastics. The problems due to the apparition of electrostatic charges are difficult to predict and the approach is mainly empirical. The measurement device considered in the present paper can be used to predict the apparition of electrostatic charge in these devices. Moreover, the association of different materials can be considered. The measurement with lactose powders has been made in four configurations: two S.S. tubes, two ABS tubes, an ABS tube followed by a S.S. tube and a S.S. tube followed by an ABS tube. Figure 5 shows that the tube order plays an important role on the final charge. In one configuration (S.S. tube followed by an ABS) the final charge is reduced. Moreover, the final charge is not a simple combination of the charge obtained separately with S.S. and ABS. The non-linearity of the charging mechanism and the difference of flowing speed in the two tubes could explain this difference.

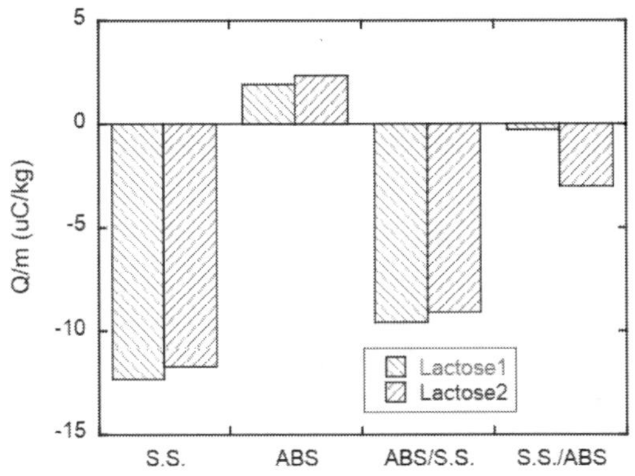

Figure 5. Charge density Q/m obtained with the powder tribo-electrometer in four tubes configurations: two S.S. tubes, two ABS tubes, an ABS tube followed by a S.S. tube and a S.S. tube followed by an ABS tube.

4 Conclusion

We have shown that the relative air humidity affects strongly the flow of powders. The powder cohesiveness increases for both dry and wet conditions. When the air relative humidity is low, grain-grain friction inside the flow induces the appearance of electrical charges in the packing, inducing cohesive forces between the grains. The condensation in high humidity conditions implies the formation of liquid bridges between contacting grains which induce also cohesive forces. Taking both effects (triboelectricity and capillary) into account, the cohesion is minimized for intermediate values of the relative humidity, i.e. between RH=30% and RH=50%.

The powders triboelectic properties have been measured with a device formed by a V-tube and a Faraday cup. The measurement has been repeated for different hygrometric conditions and for different materials (stainless steel and ABS) in contact with the powder. The electrostatic charge created inside a powder during a flow decrease drastically when the relative hair humidity becomes higher than 40%. This observation is coherent with the flow behavior.

Finally, we have shown that the association of different materials in a device can reduce the charge acquired by the powder after a flow inside the device.

Acknowledgements

We Thanks Walloon region (Fonds de maturation - convention 1318086 and BEWARE - convention 1410265) for the financial support.

References

[1] D. J. Lacks, Nature Physics **6**, 324 (2010)
[2] S. Matsusaka, H. Maruyama, T. Matsuyama, and M. Ghadiri, Chemical Engineering Science **65**, 5781 (2010)
[3] S. R. Waitukaitis, V. Lee, J. M. Pierson, S. L. Forman, and H.M. Jaeger, Phys. Rev. Lett. **112** 218001 (2014)
[4] P. M. Ireland, and K. Nicholson, Minerals Eng. **24**, 914 (2011)
[5] E. Mersch, G. Lumay, F. Boschini, and N. Vandewalle, Phys. Rev. E **81**, 041309 (2010)
[6] P. M. Ireland, Powder Technol. **198**, 189 (2010)
[7] J. E. Fiscina, G. Lumay, F. Ludewig, and N. Vandewalle, Phys. Rev. Lett. **105**, 048001 (2010)
[8] E. Emery, J. Oliver, T. Pugsley, J. Sharma, and J. Zhou, Powder Technol. **189**, 409 (2009)
[9] N. Vandewalle, G. Lumay, F. Ludewig, and J. E. Fiscina, Phys. Rev. E **85**, 031309 (2012)
[10] L. Bocquet, E. Charlaix, S. Ciliberto, and J. Crassous, Nature **396** 735 (1998)
[11] Q. Xu, A. V. Orpe, and A. Kudrolli, Phys. Rev. E **76**, 031302 (2007)
[12] J. Rajchenbach, Phys. Rev. Lett. **65**, 2221 (1990)
[13] R. Fischer, P. Gondret, and M. Rabaud, Phys. Rev. Lett. **103**, 128002 (2009)
[14] A. Castellanos, J.M. V., A.T. Perez, A. Ramos and P.K. Watson, Phys. Rev. Lett. **82**, 1156 (1999)
[15] G. Lumay, F. Boschini, K. Traina, S. Bontempi, J.-C. Remy, R. Cloots, and N. Vandewalle, Powder Technology **224**, 19 (2012)
[16] F. Boschini, V. Delaval, K. Traina, N. Vandewalle, and G. Lumay, Int. J. of Pharmaceutics **494**, 312 (2015)
[17] A. Schella, S. Herminghaus, and M. Schroter, arXiv:1609.04639 (2016)

From continuum analytical description to discrete numerical modelling of localized fluidization in granular media

Eduard Puig i Montellà[1,*], *Marcella* Toraldo[1], *Bruno* Chareyre[1,**], and *Luc* Sibille[1,***]

[1] *Grenoble Institut of Technology (G-INP), University Grenoble Alpes (UGA), 3SR, F-38000 Grenoble, France.*

Abstract. We present analytical and numerical results on localized fluidization within a granular layer subjected to a local injection of fluid. As the injection rate increases the three different regimes previously reported in the literature are recovered: homogeneous expansion of the bed, fluidized cavity in which fluidization starts developing above the injection area, and finally the chimney of fluidized grains when the fluidization zone reaches the free surface. The analytical approach is at the continuum scale, based on Darcy's law and Therzaghi's effective stress principle. It provides a good description of the phenomenon as long as the porosity of the granular assembly remains relatively homogeneous. The numerical approach is at the particle scale based on the coupled DEM-PFV method. It tackles the more heterogeneous situations which occur at larger injection rates. A direct link is evidenced between the occurrence of the different regimes of fluidization and the injection aperture. Finally, the merging of chimneys in case of two injection points is investigated.

1 Introduction

Fluidization refers to the fluid-solids systems in which the solid phase in a granular material is subjected to behave like a fluid by an upward seepage flow [1]. Fluidization is employed in a wide variety of industrial processes such as heat transfer, petroleum refining, coal conversion and water treatment [2, 3].

A specific case concerns very localized influx of fluid, inducing a spatial heterogeneity of the phenomenon inside the grains packing, this situation is known as localized fluidization.

A typical configuration for laboratory experiments on localized fluidization is shown in figure 1. The granular layer is initially at static equilibrium then subjected to one or several local injections of fluid through the bottom face of the box (figure 1). The previous works on such configurations evidenced three successive regimes during a gradual increase of the injection rate [4–7]. At very low rates, the bed is stable. Larger rates cause bed expansion even before any fluidization zone can be observed (expansion regime). For a yet larger rate, the hydrodynamic forces exerted on some particles are sufficient to counterbalance their weight, triggering movements above the injection point in the so called "fluidized" zone (cavity regime). Eventually, the height of the fluidized zone increases with the injection rate, until it reaches the top of the granular layer (chimney regime). Numerical simulations were found to match qualitatively with available experimental data [8].

*e-mail: eduard.puigmontella@3sr-grenoble.fr
**e-mail: bruno.chareyre@3sr-grenoble.fr
***e-mail: luc.sibille@ujf-grenoble.fr

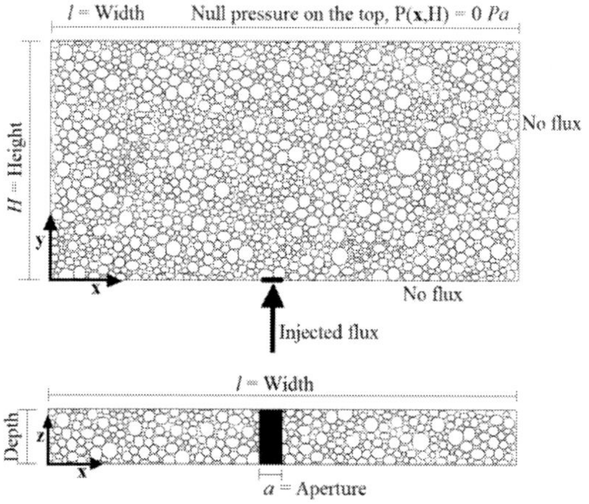

Figure 1. Sample geometry and boundary conditions.

Crude simplifications are required to find an analytical solution due to the nonlinear process. Nevertheless, it will be seen that a continuum scale model can give some insight into some governing mechanisms as the analytical model predicts the transitions between the different regimes and it provides a simple framework to explain the main trends. In this work, we provide closed form solutions for the field of initial pore pressure in the problem of figure 1.

In order to overcome this drawback and obtain a more realistic simulation of the process, a numerical model of grain-fluid systems is proposed. In this work we use a

Table 1. Physical and geometrical variables of the problem. The dimensions are defined in the [FLT] (force, length, time) system.

Variable	Dimension	SI units
Discharge per unit depth (q)	$L^2 \cdot T^{-1}$	$[m^2/s]$
Pressure (P)	FL^{-2}	$[Pa]$
Height (H)	L	$[m]$
Length (l)	L	$[m]$
Grain diameter (D)	L	$[m]$
Dynamic viscosity (μ)	$FL^{-2}T$	$[Pa \cdot s]$
Aperture(a)	L	$[m]$
Weight density, solid (γ_s)	FL^{-3}	$[N \cdot m^{-3}]$
Weight density, fluid (γ_w)	FL^{-3}	$[N \cdot m^{-3}]$
Porosity (n)	–	$[-]$

coupling between the discrete element method and a pore-scale finite volume method (DEM-PFV) [9, 10] to simulate the complete process from the expansion regime to the chimney regime.

2 Methodology

2.1 Problem statement

As a model system, we consider a layer of quasi mono-disperse spheres immersed in a viscous fluid inside a rigid box. The system is subjected to gravity and the density of the solid particles is larger than that of the fluid. The physical variables of the problem are summarized in table 1.

2.2 Dimensionless variables

Based on the above variables we introduce the so-called submerged density $\gamma' = (1-n)(\gamma_s - \gamma_w)$, and the reference vertical effective stress:

$$\sigma'_0 = (1-n)(\gamma_s - \gamma_w)H = \gamma' H. \qquad (1)$$

Normalization by this reference pressure leads to the following dimensionless group, where the normalized form of each variable is denoted by the "$*$" symbol. Note that the dimensionless fluid pressure is a normalized *excess* pore pressure.

- Normalized fluid pressure

$$p^* = \frac{P + \gamma_w(y - H)}{\sigma'_0} \qquad (2)$$

- Normalized flux

$$q^* = \frac{q\mu}{D^2 \sigma'_0} \qquad (3)$$

- Normalized coordinates

$$x^* = \frac{x}{l} \qquad y^* = \frac{y}{H} \qquad (4)$$

- Normalized aperture

$$a^* = \frac{a}{l} \qquad (5)$$

2.3 Theoretical model

Fluidization refers to situations in which the total stress tensor $\boldsymbol{\sigma}$ in a saturated material and the pressure of the pore fluid P are such that the *effective stress* tensor vanishes or has at least one vanishing eigen value.

Then, following [8], the key part of the theoretical modeling is to determine the spatial distribution of pore pressure within the specimen to identify the locations in which it reaches (or exceeds) the total stress such that:

$$\sigma'^* := \frac{y - H}{H} + p^*(x, y) \geq 0, \qquad (6)$$

Assuming Darcy's flow in a rigid porous layer (figure 1), the pressure at any point of the specimen can be obtained as [8]:

$$P + \gamma_w y = \frac{q}{2\pi K} \int_{-a/2}^{a/2} \sum_{j=-\infty}^{\infty} \sum_{i=-\infty}^{\infty} -1^{|j|} \left[ln\left(\sqrt{(x - i\,l - s)^2 + (y - j2H)^2} \right) \right] ds \qquad (7)$$

where K is the hydraulic conductivity and q is the influx rate at the injection point.

2.4 Numerical model

A quantitative approach of the problem requires a full coupling where the solid phase can be deformed under the action of the fluid, and fluid flow depends on this deformation. To this aim, a micro-hydromechanical model, combining the discrete element method (DEM) and a pore scale finite volume (PFV) formulation of the viscous flow of an incompressible pore fluid [9, 10] is used. This method accounts the effect of drag forces but lubrication terms are not included.

3 Results and discussion

3.1 Analytical solution for a single injection

Figure 2 shows the typical the pressure field after Eq. 7 for an aperture of a^*=0.1 (Eq. 5). Near the injection area, the pressure contours tend to concentric half-circular shapes due to the quasi-radial flow distribution. On the other hand, the isolines are horizontal near the side walls, consistently with the no-flux condition (figure 1).

Following section 2.3, the fluidized zone can be identified with respect to the sign of the effective stress when the injected flux is increasing. Therein the extent of the fluidized zone can be defined through the shape of the null-pressure isoline plotted in figure 3. The different regimes described in [5] can be distinguished from this series of plots. Low flux values correspond to the expansion regime in which no fluidization zone is detected (figure 3*a*). As the flux increases, pore pressure keeps building up until it balances the total stress. At this point a fluidized zone starts developing above the injection area, corresponding to the cavity regime (figure 3*b*). Eventually, the fluidized zone reaches the top of the specimen, leading to a chimney of fluidized grains (figure 3*c*). The solution with a wider

DOI: 10.1051/epjconf/201714009019

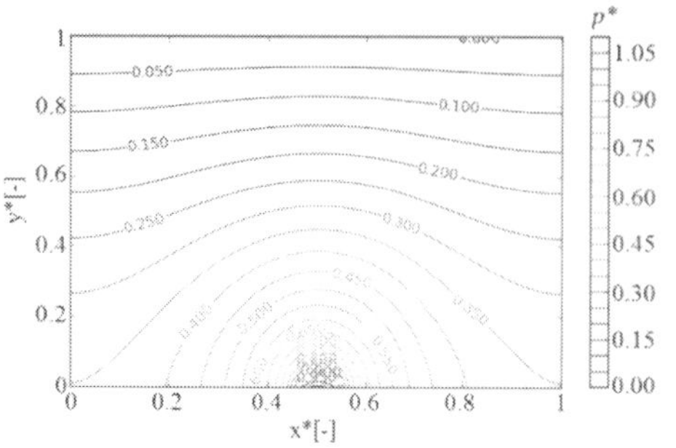

Figure 2. Dimensionless pressure field for an injection rate of q^* = 0.00067 and aperture of a^* = 0.1.

injection area (not shown in the present work) evidences that a slightly larger flow rate is required to initiate the fluidization above a wide aperture. Besides, distinguishing the cavity and the chimney regime in this situation is made uneasy by the fact that large aperture tends to fluidize the granular layer simultaneously at every point in space.

3.2 Numerical simulations for a single injection

Figure 4 shows the evolution of porosity within the granular layer when the injected flux is increasing. The large porosity values at the top of the layer are artifacts and should be disregarded: the particles of the free surface have large porosity values by definition as their Voronoi cells enclose some void space above the free surface. Figure 4 highlights the heterogeneous changes of porosity as localized fluidization starts developing: a cavity appears for q^*=1.16 × 10^{-3} , followed by a chimney for q^*≥1.45 × 10^{-3} . In both cases the particles located in the regions of low porosity are moving and have only transient contacts with each other, while the particles of the dense regions are static and contribute to a permanent contact network.

A "bubble" of high porosity is shown in figure 4*c*. Actually this bubble is not fixed in time and space (the figure is only a snapshot at a given time) but tends to move up until it reaches the free surface. Then, another bubble appears at the bottom and move up in a cyclic manner. This is in clear contrast with the cavity regime in plot (*b*) in which the porosity field is stationary.

After setting the injection rate back to zero and reaching a final equilibrium state, an irreversible increment of porosity ($n \approx$ 0.42 locally) is observed above the injection point and throughout the layer (figure 4*d*).

3.3 Aperture size dependency

As mentioned before (see figure 3) and confirmed in figure 5, large apertures (a^* close to 1) lead to the entire liquefaction of the granular assembly rather than forming a cavity and a chimney of fluidization (cavity regime gets narrower near a^* = 1.). The aperture dependency is summarized in the diagram of figure 5 where the different fluidization

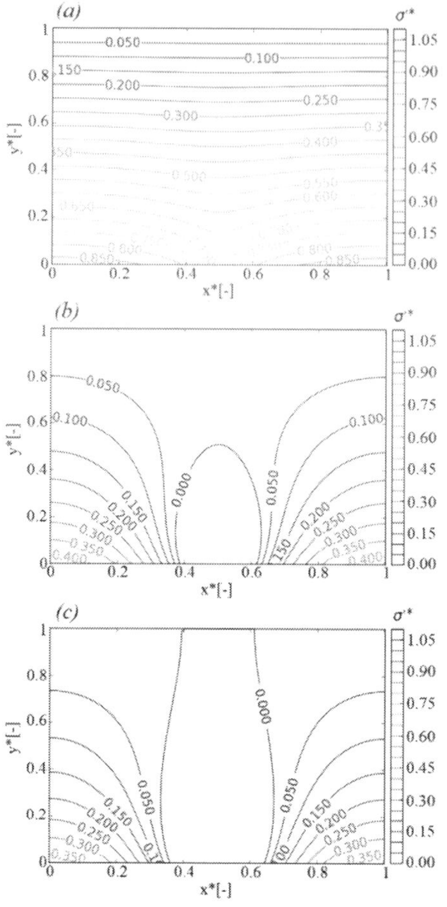

Figure 3. Evolution of the dimensionless effective stress σ^* field when the injected flux increases. Injection aperture a^*=0.1.The normalized fluxes are q^* = 0.00019 (*a*), q^* = 0.00105 (*b*), q^* = 0.00114 (*c*).

regimes are identified from the numerical approach for different combinations of flow rate and aperture

3.4 Two injection points case

Fluidized cavities developing above the injection points are strongly disturbed when two orifices of injection are close to each other. Based on the analytical approach, figure 6 evidences effective stress field at the time when the cavities resulting from two punctual sources are about two merge forming a single chimney. On the contrary, chimneys have individual behavior if the two injection points are sufficiently far from each other.

Figure 7 shows the threshold separating the two behaviors that have been described. Small distances between the injection points lead to a single chimney generated by the two cavities (left part of the diagram in figure 7). On the contrary, cavities are not affected or attracted to each other when the distance between the orifices increases (right part of the diagram in figure 7).

 DOI: 10.1051/epjconf/201714009019

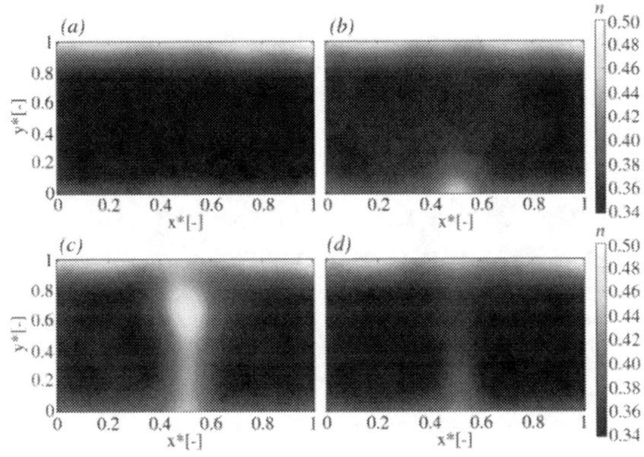

Figure 4. Evolution of the porosity for narrow aperture ($a^* = 0.1$). (a) Static regime, $q^* = 0$. (b) Cavity regime, $q^* = 0.00116$. (c) Chimney regime, $q^* = 0.00145$. (d) $q^* = 0$.

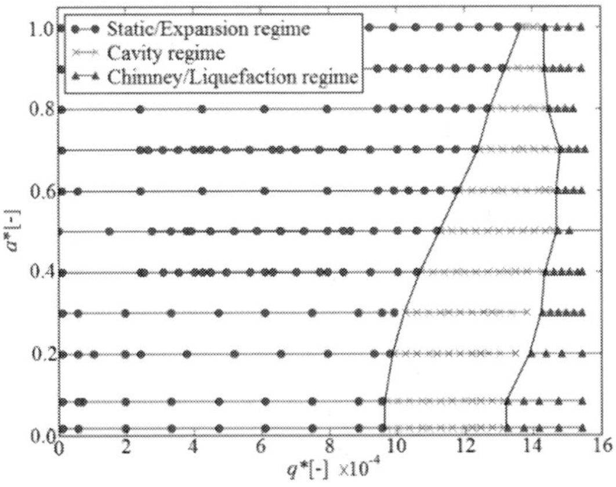

Figure 5. Occurence of the different fluidization regimes (static/expansion, cavity or chimney) depending on aperture a^*. (Determined from the numerical coupled simulations).

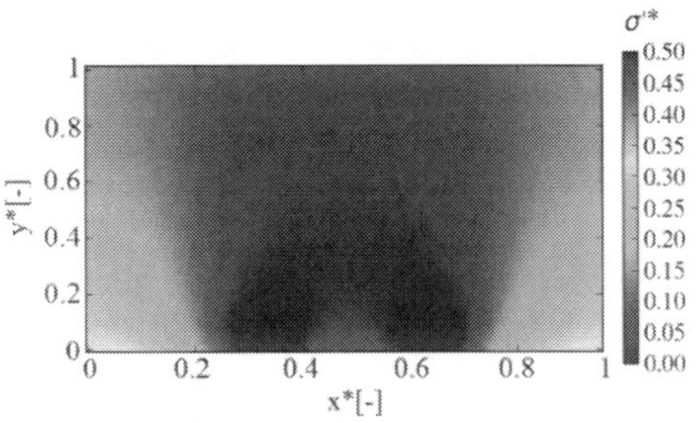

Figure 6. Dimensionless effective stress field for an injection rate of $q^* = 0.00057$ and two punctual sources spaced $\frac{1}{6}$.

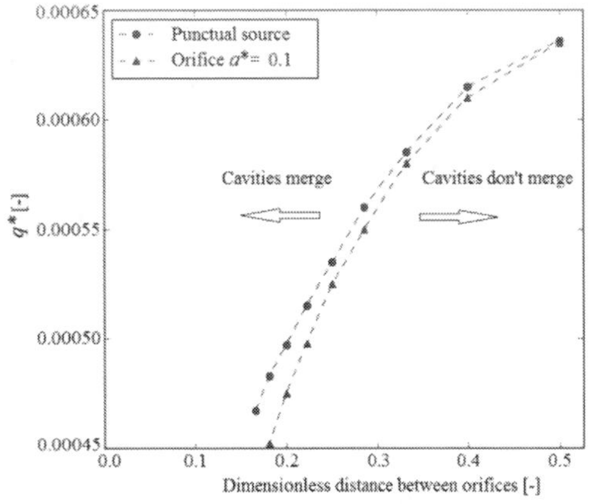

Figure 7. Occurence of cavities merging for two punctual sources and two sources with an aperture of $a^* = 0.1$.

4 Conclusions

Both analytical and numerical approaches were presented to characterize the localized fluidization of a granular layer. They are qualitatively in good agreement, with respect to the description of the different fluidization regimes. However, these approaches deviate for large fluxes when a homogeneous permeability field can no longer be assumed. From these approaches, dependency of the fluidization regimes (mainly cavity and chimney) on injection aperture (for a single aperture) and on distance between injection point (for a set of two injections) is given by tracking the vanishing of the effective stress within the granular layer.

References

[1] Payne, Fred C and Quinnan, Joseph A and Potter, Scott T, *Remediation hydraulics* (CRC Press, 2008)

[2] Peng, Yimin and Fan, LT, Chemical Engineering Science, **52**, 2277–2290, (1997)

[3] Weisman, Richard N and Lennon, Gerard P, Journal of waterway, port, coastal, and ocean engineering **120**, 468–487, (1994)

[4] Ngoma, J and Philippe, P and Bonelli, S and Delenne, JY and Radjai, F, Geomechanics from Micro to Macro: Proc. of the Int. Symposium on Geomechanics from Micro to Macro, **4**, 1571–1576, (2014)

[5] P. Philippe and M. Badiane, Physical Review E **87**, 042206, (2013)

[6] Zoueshtiagh, Farzam and Merlen, Alain, Physical Review E **75**, 056313, (2007)

[7] Xilin Cui, Jun Li, Andrew Chan, and David Chapman, Powder Technology **254**, 299-306, (2014)

[8] E. P. Montellà, M. Toraldo, B. Chareyre and L. Sibille, Physical Review E, **94**, 052905, (2016)

[9] Catalano, E and Chareyre, B and Barthélemy, E, International Journal for Numerical and Analytical Methods in Geomechanics **38**, 51–71, (2014)

[10] Chareyre, B and Cortis, A and Catalano, E and Barthélemy, E, Transport in porous media **94**, 595–615, (2012)

Mesostructural investigation of micron-sized glass particles during shear deformation – An experimental approach vs. DEM simulation

Lutz Torbahn[1,*], *Alexander* Weuster[2], *Lisa* Handl[3], *Volker* Schmidt[3], *Arno* Kwade[1], and *Dietrich E.* Wolf[2,**]

[1] *Institute for Particle Technology (iPAT), Braunschweig University of Technology, Braunschweig, Germany*
[2] *University of Duisburg-Essen and Center for Nanointegration Duisburg-Essen (CeNIDE), Duisburg, Germany*
[3] *Institute of Stochastics, Ulm University, Ulm, Germany*

Abstract. The interdependency of structure and mechanical features of a cohesive powder packing is on current scientific focus and far from being well understood. Although the Discrete Element Method provides a well applicable and widely used tool to model powder behavior, non-trivial contact mechanics of micron-sized particles demand a sophisticated contact model. Here, a direct comparison between experiment and simulation on a particle level offers a proper approach for model validation. However, the simulation of a full scale shear-tester experiment with micron-sized particles, and hence, validating this simulation remains a challenge. We address this task by down scaling the experimental setup: A fully functional micro shear-tester was developed and implemented into an X-ray tomography device in order to visualize the sample on a bulk and particle level within small bulk volumes of the order of a few micro liter under well-defined consolidation. Using spherical micron-sized particles (30 μm), shear tests with a particle number accessible for simulations can be performed. Moreover, particle level analysis allows for a direct comparison of experimental and numerical results, e.g., regarding structural evolution. In this talk, we focus on density inhomogeneity and shear induced heterogeneity during compaction and shear deformation.

1 Introduction

Shear flow of granular media is ubiquitous in nature and of industrial importance when it comes to handling and processing of bulk solids. The interesting shear rate dependence of granular flow is accompanied by a rate-independent regime at slow, quasi-static deformation. Strain localization within the bulk, often referred to as failure zone or shear band, represents one unique feature of this regime [1]. A crucial task is the prediction of the critical shear stress a resting bulk solid can withstand. This shear stress depends on the preparation-procedure and is also known as yield locus. Utilizing non-associate flow rules, plasticity theory provides a framework to model quasi-static flow. However, in most cases only the volume fraction is taken into account to characterize the structure.

Modern technology offers a variety of powerful tools to further investigate the interplay between structure and mechanical behavior of granular media. On one hand, the discrete element method (DEM) provides detailed information on the behavior of small samples presuming a certain particle interaction. On the other hand, sophisticated experimental setups are capable of determining the properties of micron-sized particles [2] and of picturing the inner structure of a bulk solid non-destructively via computer tomography [3]. Using this methodology, we critically explore the potential of DEM simulations and micro-sized experiments by performing a direct comparison considering the bulk structure as well. For this task, a fully functional micro-sized shear tester was developed [3], which can be fitted into an X-ray tomography device. As will be shown, this setup allows for an in-situ shear band detection during an experiment at constant normal load with only a few μl of sample volume.

*e-mail: l.torbahn@tu-braunschweig.de
**e-mail: dietrich.wolf@uni-duisburg-essen.de

2 Experiments

For this study we used micron-sized particles made of borosilicate glass microspheres (CoSpheric LLC, USA). The glass spheres are characterized by a narrow monodisperse particle size distribution with median value $x_{50,3} \approx 30$ μm. Coequally an almost spherical shape of particles is observed. Both features benefit the subsequent computational image-based analysis of tomographic data and ease the comparison to the DEM simulation using ideal spheres. Besides, a non-negligible surface roughness and a slightly cohesive behavior is identified.

2.1 X-ray microtomography and micro shear-tester

In order to measure the spatial powder structure and behavior, a nondestructive image-based analysis is realized by using a high-resolution X-ray microtomography device

(MicroXCT-400, Zeiss (Xradia), Germany). The operating parameter of the X-ray source amounts to an acceleration voltage of 50 kV with a constant current intensity of 200 μA. This choice of parameters results in high contrast images, and hence, supports the accurate computational image-based analysis afterwards. Considering a shear cell diameter of 2 mm, a ten-fold optical magnification was adequate to image the entire cell with a resolution of 2.2 μm (2x binning). In addition, an intensity gray-scale projection is formed by detecting X-rays passing through the sample. A record of 2000 projections with varying sample orientation provide the basis for a precise three-dimensional reconstruction and smooth the way for a following detailed computational sample analysis.

The experimental setup contains a torsional micro shear-tester with a cylindrical shaped sample chamber [4]. In comparison to common shear devices it allows for investigating very small volumes on the order of 6 μl and features both powder densification up to 20 kPa and shear deformation of small angle increments from millidegree up to theoretically unlimited shear displacement. Vertically, the sample chamber is confined by a pair of structured pistons whereas a filigree glass capillary with wall thickness of 50 μm acts as a sidewall. Another characteristic feature of this modern shear device is the decoupled and most precise determination of the acting forces and torque behavior whereas the whole system is theoretically frictionless using advanced technology of magnetic spring and air bearing mechanisms.

In order to prevent agglomeration during the preparation process, powder is sieved into the capillary. After the filling process, the sample is compressed to the desired normal load of 0.5 kPa. During shear deformation the sidewall and the upper piston rotates stepwise in angle increments of 0.5° up to 9.5° shear deformation. Afterwards, the angle increment is raised up to 5° until the final shear deformation of 39.5° is achieved. The shear deformation is performed in a quasi-static regime using an angle velocity of $\omega = 0.1°/\text{s}$. Between the shear steps tomographic data is recorded.

2.2 Computational analysis - Image segmentation and particle tracking

The advanced computational analysis of tomographic data bridges the gap between experiments and DEM simulations regarding particle positions, and hence particle movement over time. First, particle positions (and radii) are extracted from the image data by using a marker-based watershed transformation [5]. In this context, a Gaussian filter and the IsoData algorithm from ImageJ [6] are applied to smooth and binarize the image data, respectively. Besides, due to few hollow particles, small disconnected pores are filled with solid artificially.

On the basis of particle centroids calculated from these segmented binary images, a particle tracking is realized. Several time steps with a small angle increment of 0.5° are used to apply a tracking algorithm that approximately minimizes the sum of squared displacements for each time step [7]. To speed up calculations, particle displacements which exceed a predefined threshold are not considered to be possible during optimization. In this study the tracking threshold is set to s = 19.8 μm. This simplifying assumption is valid for small shear angle increments of 0.5°, where the largest particle displacements between two time steps are expected to be 9 μm (at the capillary sidewall). However, it does not hold for later time steps of the experiment, when shearing was performed in steps of 5°. Here tracking is still possible, but it requires prior information. For this purpose, we estimate a local angle of rotation directly from the image data as a function of the radius and height in the sample. The basic idea for this approach, which has been proposed in [4], is to rotate each slice of an image stack by a number of different angles and to determine for which angle it agrees best with the corresponding slice of the subsequent image stack. We refined this approach by subdividing each image slice into disjoint rings of equal area which are rotated independently, and captures variations in radial direction. The quality criterion for comparing two image slices is their correlation. Using this image-based local shear deformation and an adaption of [7], we can track a large number of particles even in steps of 5°. We validated the method based on the time step from 5° to 10° of shearing, where information in small steps is available, and found that 98.9 % of the tracks identified by our method agreed with the tracking based on 0.5° steps.

3 Simulations

3.1 Contact model

A soft particle DEM approach is utilized to simulate the system of interest, i.e. trajectories are calculated by numerically solving Newton's equations of motion. Particles interact via pairwise forces ($\vec{F}_{ij}$) and torques, which, in general, depend on their relative position ($\vec{r}_{ij} = \vec{r}_i - \vec{r}_j$), velocity, angular velocity, diameter (d_i, d_j) and the contact history to account for friction. In consideration of the external load, we use a damped Hertz contact model [8] and add adhesion as well as rolling and torsional friction. The normal part of the contact model in dependence of the particle overlap ξ is given by

$$F_n(\xi, \dot{\xi}) = \frac{4\,E_{\text{eff}}\,\sqrt{r_{\text{eff}}}}{3}\,\xi^{\frac{3}{2}} + \gamma_n\,\sqrt{\xi}\dot{\xi} - F_{\text{adh}}\frac{r_{\text{eff}}}{\langle r_{\text{eff}}\rangle} \quad (1)$$

with F_{adh} being a constant, $E_{\text{eff}} = E/(2(1-\nu^2))$ the effective elastic modulus, γ_n a viscous constant and $r_{\text{eff}} = \frac{1}{2}d_i d_j/(d_i + d_j)$ the reduced radius of the particles i and j. Although a linear dependence of adhesion force on r_{eff} is employed, we expect minor influence due to the narrow particle size distribution.

In order to implement Coulomb friction, together with rolling and torsional friction, we use the approach presented in [9], which suppresses each contact mobilization mode (sliding, rolling and twisting) individually by using a linear spring-dashpot, until a threshold is exceeded. This threshold, as well as the resisting force (or torque) in case of mobilization, is proportional to $|F_n + F_{adh}|$. For simplicity, the same friction coefficients are used in the static and dynamic case.

3.2 Parameters and calibration

In principle, all microscopic parameters could be determined by sophisticated experiments [2]. However, the cumbersome work of measuring every parameter experimentally can be circumvented by utilizing microscopically undetermined parameters to calibrate the contact model. This enables a correct description of the bulk's macroscale behavior [10], e.g. stress response to applied strain. Is it possible to predict the structural behavior, e.g. Poisson's ratio ν as well? We pursue this task by intentionally keeping the contact model and the parameter choice as simple as possible, exploring its predictive power as well as its limitations.

Particle stiffness and adhesion force are set according to the experimentally determined values. As we study the bulk's response to a quasi-static deformation, viscous parameters should have a negligible influence. For computational convenience, we use the high damping coefficient $\gamma_n = \sqrt{4/3\, E_{\text{eff}}\, m_{\text{eff}}}$, where m_{eff} is the reduced mass of the colliding particles. A typical choice for the tangential stiffness is $k_t = 2/7k_n$, where $k_n(\xi) = \partial F_n^{(el)}/\partial\xi$ with $F_n^{(el)}$ being the elastic part of equation (1). Here we use $k_n = 4/3E_{\text{eff}}x_{50,3}/2$, approximating the contact radius with the median particle radius and therefore overestimating the stiffness. All springs utilized for friction are damped, using a viscoelastic constant of $2\sqrt{k_t m_{\text{eff}}}$. As twisting and sliding are closely related frictional modes, the torsion friction coefficient, $\mu_{\text{tor}}/r_{\text{eff}}$, is set equal to the Coulomb friction coefficient μ. Furthermore, a load dependent rolling friction coefficient $\mu_{\text{rol}} = \sqrt{2r_{\text{eff}}\xi}$ is employed. This choice of parameters leaves the Coulomb friction coefficient, μ, as a free parameter to calibrate the contact model.

We model the experimental particle size distribution with a log-normal distribution and draw particle diameters according to the corresponding probability function. Only diameters between 25 μm and 50 μm are accepted, resulting in a slightly sharper cut-off at small particle sizes and an interquartile distance of $x_{75,3} - x_{25,3} \approx 5$ µm, in comparison to experiment $x_{75,3} - x_{25,3} \approx 10$ µm.

Calibration of the contact model is done by executing plane shear simulations under variation of μ at a fixed normal stress $\sigma = 15$ kPa and comparing the steady state macroscopic friction coefficient μ_{macro} to the experimental findings. These calibration simulations are done with $N = 10.000$ particles, confining walls perpendicular to the shear direction and periodic boundary conditions in all other directions (see [3, 10] for details). Fitting an exponential saturating function for $\mu_{\text{macro}}(\mu)$ as suggested in [11] results in $\mu \approx 0.58$.

3.3 Setup and preparation

The simulation setup mimics the experimental micro shear-tester: A cylinder with diameter of 1.78 mm is used to confine the sample in horizontal direction, while two pistons (one movable, one fixed) close the simulation cell in vertical direction. Both pistons are structured similar to the experiment, i.e., with six vanes arranged in a star shape. As there is a small gap between vanes and outer cylinder in the experiment ($\approx$ 0.025 mm), we introduce a finite distance between the outer edge of the piston and the cylinder wall of $\approx$ 0.1 mm. This slight gap enlargement is introduced to avoid jamming of single particles between the piston and capillary. Particle-wall interaction is described by the visco-elastic part of the contact model Equation (1), using the same material parameters.

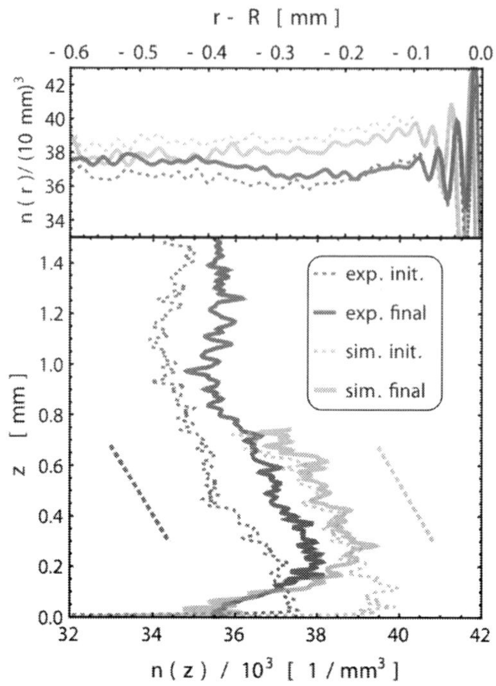

Figure 1. Axial and radial density profiles of the initial and final configuration. The density gradient is represented by the dotted lines. $z = 0$ marks the vanes top edge

The preparation procedure includes the following steps: A random, porous particle configuration with volume fraction $\approx$ 0.2 is created. Potential overlaps are removed in a relaxation run that uses a viscous background friction. After relaxation, this omnipresent viscous term is removed and a normal load of 0.5 kPa is applied to the lower piston, compressing the material. Following compression, a constant angular velocity ω is added in order to shear the sample. To achieve a quasi-static deformation, we use the criterion $\omega\sqrt{\rho_s\, x_{50,3}^2/\sigma} \approx 10^{-4}$.

4 Results and discussion

We compare density profiles of the initial (compacted) and final (sheared) configuration as well as the temporal evolu-

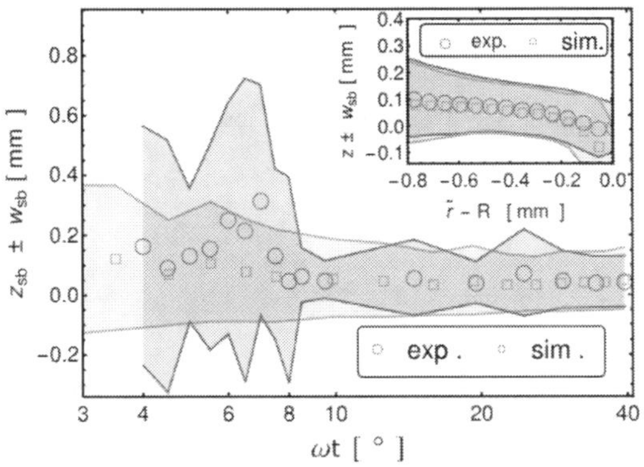

Figure 2. Shearband position and width over time. Inset: Shearband localization for steady state ($\omega t = 10°$).

tion of shearband position and width. Assuming rotational invariance, all quantities are given as a function of height (z) relative to the vanes top edge and distance to the central axis of the chamber (r).

The number density $n(r, z)$ of particle centers in axial and radial direction is displayed in Fig. 1. Due to the different filling height, only the range 0 mm $\leq z \leq$ 0.7 mm was considered for the radial distribution.

Densities in experiment and simulation are of the same order of magnitude. The initial states differ to much higher extent due to different filling procedures and configurations for experiment and simulation. In contrast, after steering, the differences between experiment and simulation are quite small, showing that the steering can be well described by the simulation. However, the number density increases with identical gradient according to height in experiment and simulation.

This gradient reflects the outer wall's influence, which seems to be equal at first sight. However, the radial profile reveals a denser region in the chamber's center in the experiment, which is even more compacted during shear. Simulation results show the opposite: Slightly increasing density in the vicinity of the sidewall and dilation during shear. Of course, in both experiment and simulations, a dilation zone ($z \leq$ 0.2 mm) as well as a compaction zone ($z >$ 0.2 mm) can be identified. Layering is observed in both experiment and simulations. Besides wall influences, the systematically higher density suggests that the contact model is not able to reproduce the structure, in terms of volume fraction, fully correct. A model refinement addressing this issue is work in progress.

The shear-induced dilation zone ($z \leq$ 0.2 mm), which can be identified in the axial density profile, corresponds to the shear band position. This can be shown by analyzing particle displacements. We determine the shear-band position (z_{sb}) and width (w_{sb}) by fitting [1]

$$v_{\tilde{\varphi}}(z) = \frac{1}{2} + \frac{1}{2}\,\mathrm{Erf}\left(\frac{z - z_{sb}}{w_{sb}}\right) \qquad (2)$$

to the normalized angular velocity at fixed radial distances $\tilde{r}$ to center of rotation, which deviates slightly from the central axis in the experiment. The temporal evolution of both quantities, averaged over $\tilde{r}$, is depicted in Fig. 2. The Inset shows the shearband position as a function r in the steady state. Deviations in shear band width at small deformation might be correlated to the experiment control, while the deviation in the radial dependence of the shearband position is caused by the larger gap used in simulations. Despite these differences, the shear band width ($\approx 8x_{50,3}$) and position is identical and seem to be robust with respect to details of the contact model parameters. Note that the number density inside the shear band is also equal in experiment and simulation.

In this study a correlation between local density reduction and a shearband formation above the lower piston is verified. Moreover, we could image the temporal sequence of shearband formation on a particle level using an efficient tracking algorithm that allows for a direct comparison of micron-sized glass particles in experiment and DEM simulations. Besides comparing mechanical properties of the bulk, this structural analysis increases the validation depth for the chosen contact model.

Acknowledgements

This work was funded and prepared within the priority program SPP 1486 by the Deutsche Forschungsgemeinschaft (DFG).

References

[1] D. Fenistein, J.W. van de Meent, M. van Hecke, Phys. Rev. Lett. **92**, 094301 (2004)

[2] R. Fuchs, T. Weinhart, J. Meyer, H. Zhuang, T. Staedler, X. Jiang, S. Luding, Granul. Matter **16**, 281 (2014)

[3] S. Strege, A. Weuster, H. Zetzener, L. Brendel, A. Kwade, D.E. Wolf, Granul. Matter **16**, 401 (2014)

[4] S. Strege, Ph.D. thesis, TU Braunschweig (2014)

[5] E.R. Dougherty, *Mathematical Morphology in Image Processing* (Marcel Dekker Inc., New York, 1993)

[6] T.W. Ridler, C. S, IEEE T. Syst. Man Cyb. **8**, 630 (1978)

[7] J.C. Crocker, D.G. Grier, J. Colloid Interf. Sci. **179**, 298–310 (1996)

[8] N.V. Brilliantov, F. Spahn, J.M. Hertzsch, T. Poeschel, Phys. Rev. E **53**, 5382 (1996)

[9] S. Luding, Granul. Matter **10**, 235 (2008)

[10] A. Weuster, S. Strege, L. Brendel, H. Zetzener, D. Wolf, A. Kwade, Granul. Matter **17**, 271 (2015)

[11] A. Taboada, N. Estrada, F. Radjaï, Phys. Rev. Lett. **97**, 098302 (2006)

Secondary flows in slow granular flows

Peter Varun Dsouza[1,*], *Krishnaraj* K.P.[1,**], and *Prabhu R* Nott[1,***]

[1] *Indian Institute of Science, India*

Abstract. Recent findings by Krishnaraj and Nott [1] show that a granular material sheared in a cylindrical Couette cell at low shear rates forms a single secondary vortex. The vortex spans the entire width of the Couette cell and has a sense opposite to the centrifugally driven Taylor-Couette vortex in a Newtonian fluid – it is in fact shown to be driven by shear-induced dilation. Krishnaraj and Nott [1] show that the vortex also explains a rheological anomaly observed earlier [2], wherein all components of the stress on the outer cylinder increase nearly exponentially with depth from the free surface. In this study, we test the robustness of this vortex by varying the parameters of the grain contact model. We show that the presence of a free surface is not essential for the formation of the secondary vortex. The vortex forms even when a rigid plate of finite weight confines the granular column at the top. We find that as the shear rate is increased, an additional centrifugally-driven vortex appears. This new vortex keeps growing until, at Savage number close to one, the dilation-driven vortex disappears. We also present the variation of the wall stresses at the inner cylinder with depth. Finally, we argue that the secondary flow can also help to understand the rheological behaviour observed in geometries such as the split-bottom Couette device [3].

1 Introduction

The cylindrical Couette device (Figure 1) is a standard setup used for measurement of the rheological behaviour of fluids. However, adapting the Couette apparatus to study particle flows is not straightforward. Gutam et. al. [2] have shown that on shearing the material, all components of the stress along the outer wall increase exponentially from the free surface. They also show that the component of the shear stress in the vertical direction of gravity changes sign upon commencement of shear. Krishnaraj and Nott [1] showed the presence of a system-spanning secondary vortex in this geometry and showed that it explains the stress anomaly (Figure 2). Understanding such a flow is crucial in developing continuum models for granular flows. In this paper we test for the robustness of this secondary flow. We do this by varying the simulation parameters. We also show how the flow behaves under load or high shear rates.

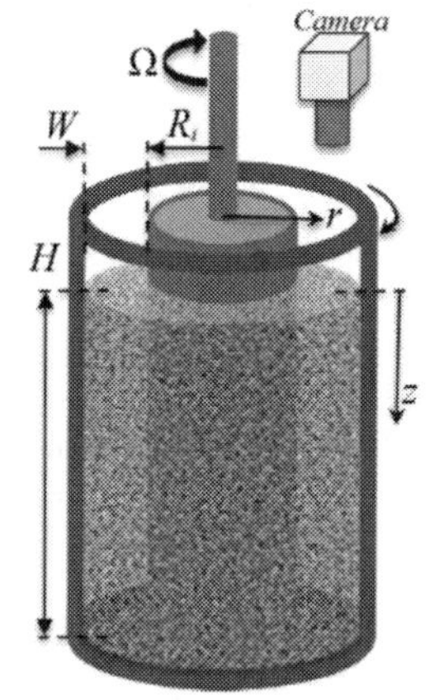

Figure 1: A schematic of the cylindrical Couette.

2 Methods

Granular flows can be classified into slow, intermediate and rapid flows. This is given by the Savage number Sa, the ratio of inertial stress to the system stress scale. The Savage number is related to the Inertial Number, I, which are given by

$$Sa = I^2 = \frac{\rho g W}{N} \qquad (1)$$

where ρ is the particle density, W is the system length scale and N is the stress scale. Rapid flows are given by $Sa \sim 1$. In most of our studies, the inner cylinder is rotated with a shear rate $\dot{\gamma}$, which corresponds to a Savage number of order 10^{-6}. For higher shear rates, refer to Section 4.

The Couette device used in the experiments is a custom built setup. It is filled with particles to a height, H, of 27 cm. The annular gap, W, used in this paper is 4 cm. We measured the stress along the outer wall for glass beads (bulk density, ρ = 1600 kg/m^3, average diameter d_p = 0.833 mm) and for mustard seeds (ρ = 720 kg/m^3, d_p = 1.2 mm). We measured the stress profile along the outer wall

*e-mail: varun@chemeng.iisc.ernet.in
**e-mail: krishnaraj@chemeng..iisc.ernet.in
***e-mail: prnott@chemeng.iisc.ernet.in

DOI: 10.1051/epjconf/201714003028

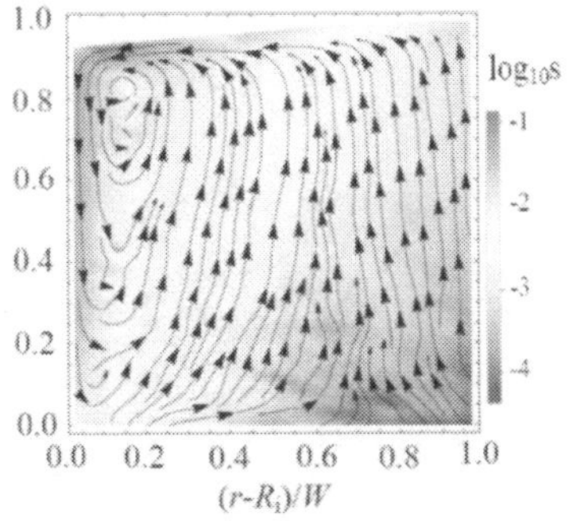

Figure 2: Streamlines of the secondary flow at steady state for a fill height $30d_p$. The background colour is $\log_{10} s$ where $s \equiv (v_r^2 + v_z^2)/V_W$, is the magnitude of the secondary flow. V_W is the velocity of the inner cylinder surface. [1]

and the velocity profile of the particles at the top surface. The velocity profile was obtained by video imaging the top surface and carrying out Particle Image Velocimetry (PIV) analysis using the PIVlab MATLAB toolbox [4]. The ring used to confine the top surface had a window made of optical glass for imaging. The simulations method was the same as in [1].

3 Effect of confinement with a load

3.1 Stress profile

We tested to see if the presence of a free surface is important for the formation of the secondary flows. To test this, we placed a freely floating ring in the annular gap of the Couette. This ring is allowed to move freely along the z direction but restricted along r and θ. This is done such that the weight of the ring falls only on the granular material. This also allows the material to dilate freely. Figure 3 shows the normal stress, σ_{rr}, and shear stress, σ_{rz}. We still see the exponential rise of σ_{rr}, and that σ_{rz} has a sign opposite to a static column.

3.2 Kinematics

The exponential decay of the tangential velocity, V_θ, is a characteristic of slow particle flows in a Couette. The inward movement of particles with velocity, V_r, along the top surface is necessary to close the vortex. With a confinement, the top surface is no longer traction free making it more difficult to close the vortex. Figures 4(**a**) and 4(**b**) shows how the velocity profile changes under confinement. The material forms a slope on the surface to some distance from the inner cylinder up on reaching steady state. Without confinement, the material slopes all the way till the outer wall. The particles close to the outer wall no longer move inward because of the applied traction. How far the material slopes is dependent on the confining weight. The particles that lie on this slope are not restricted by the confinement's traction and can flow inward.

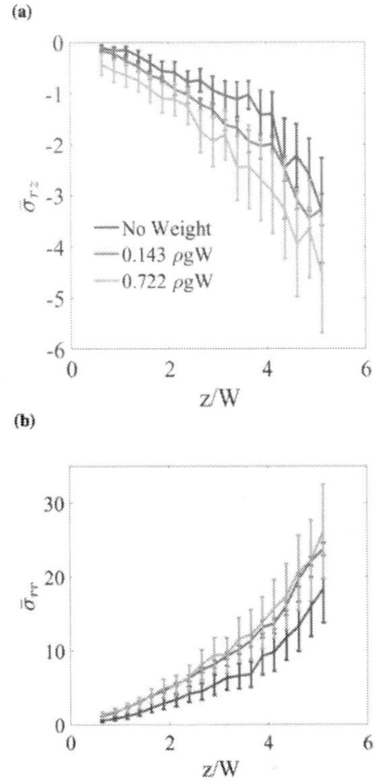

Figure 3: (**a**) Shear and (**b**) normal stress on the outer wall of a sheared, cylindrical Couette. The stresses, $\bar{\sigma}$, are scaled by $\rho g W$ and the distances are scaled by the Couette gap, W. Each data point is represented as mean ± standard deviation.

4 Varying the simulation model parameters

The model used for the DEM simulations was proposed by Silbert et. al. [5]. This model assumes that the normal contact between particles can be represented by a linear spring with elastic constant k and a dash-pot with a damping constant γ in parallel. The tangential contact is similar but has a Columbic slider with coefficient μ to represent friction in series with the parallel, spring and dash-pot system. Consider two particles i and j. The normal force, F_n, and tangential force, F_t, between them are

$$\mathbf{F}_n^{(i)} = k_n \delta \mathbf{n} - m_{eff} \gamma_n \mathbf{v}_n \quad (2a)$$

$$\mathbf{F}_t^{(i)} = \begin{cases} -k_t \Delta \mathbf{s} - m_{eff} \gamma_t \mathbf{v}_t & \text{if } |F_n^{(i)}/F_t^{(i)}| < \mu \\ -\mu |F_n^{(i)}| \frac{\mathbf{v_t}}{|\mathbf{v_t}|} & \text{otherwise} \end{cases} \quad (2b)$$

Here, δ is the overlap between the particles, $\mathbf{n}$ is the unit normal joining the two centers in the direction of i, $m_{eff} \equiv (1/m_i + 1/m_j)^{-1}$ is the effective mass, $\mathbf{v_n}$ and $\mathbf{v_t}$ are the normal and tangential relative velocities at contact and $\mathbf{\Delta s}$ is the tangential displacement.

In the following subsections we show that the secondary flow is not dependent on the model parameters

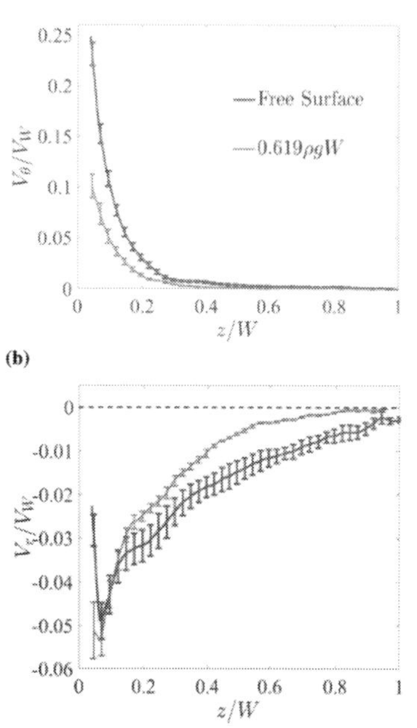

Figure 4: **(a)** Tangential and **(b)** radial velocity profiles along the top surface with and without confinement. The velocities are scaled by the inner cylinder velocity, V_W, and the distances are scaled by the Couette gap, W. The error bars are one standard deviation in length.

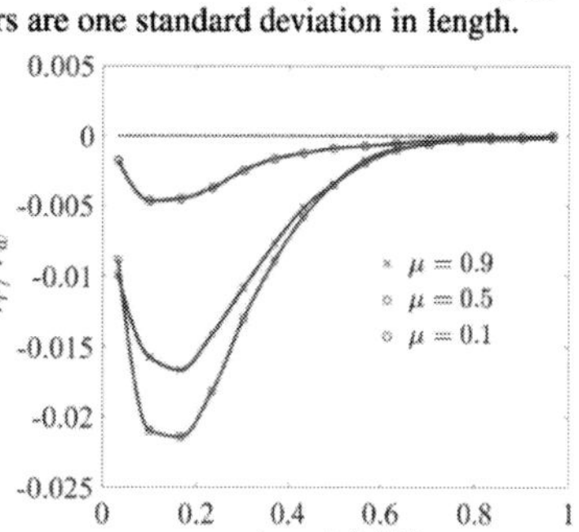

Figure 5: Radial velocity along the free surface for different values of μ

used. We vary μ and $\dot{\gamma}$ in this paper. The damping coefficient ϵ is related to the restitution coefficient by Equation 3. We found varying ϵ from 0.4 to 0.9 in steps of 0.1 had no change in our results. Krishnaraj and Nott have shown how k affects the system in [1].

$$\epsilon_{n,t} = \exp(-\gamma_{n,t} t_{col}/2) \tag{3}$$

where t_{col} is the collision time given by

$$t_{col} = \pi(2k_n/m - \gamma_n^2/4)^{-1/2} \tag{4}$$

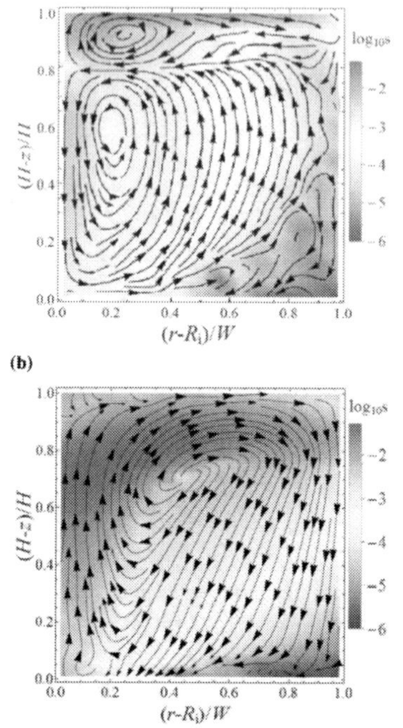

Figure 6: Streamlines for **(a)** $Sa = 2 \times 10^{-2}$, **(b)** $Sa = 1$.

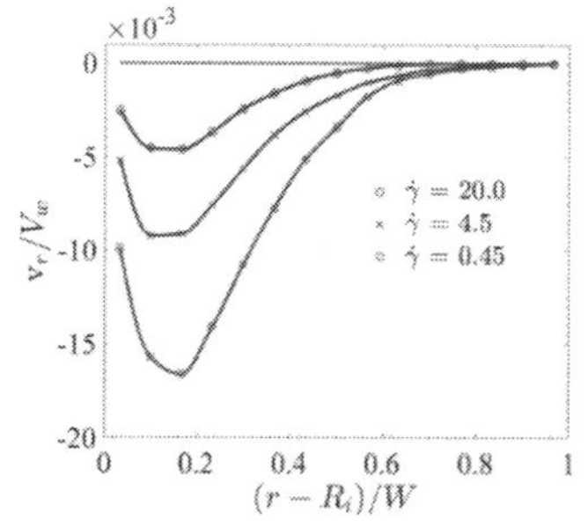

Figure 7: Radial Velocity at the top surface for different shear rates.

Figure 5 shows how the radial velocity profile changes along the top surface of the material. We can see how changing the value of μ does not change the direction of v_r, the radial velocity. The secondary flow is a characteristic of slow and intermediate flows. Increasing the shear rate, $\dot{\gamma}$ from 0.45 s^{-1} in the slow flow regime ($Sa = 2 * 10^{-6}$) (Figure 2) to the inertial regime ($Sa \equiv 1$) replaces the dilation-driven vortex with a centrifugal vortex (Figure 6).

5 Stress on the inner cylinder

Gutam et. al. [2] have reported the stress exerted on the outer wall of the Couette. Figure 8 shows the normal stress on the inner cylinder obtained from the DEM simulations. We can see that even here, there is an exponential increase in the normal stress with respect to depth from the free

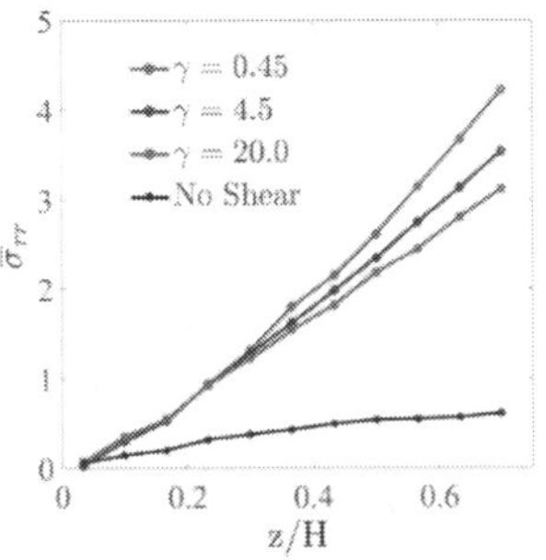

Figure 8: Normal Stress along the inner cylinder for different shear rate and for a static column.

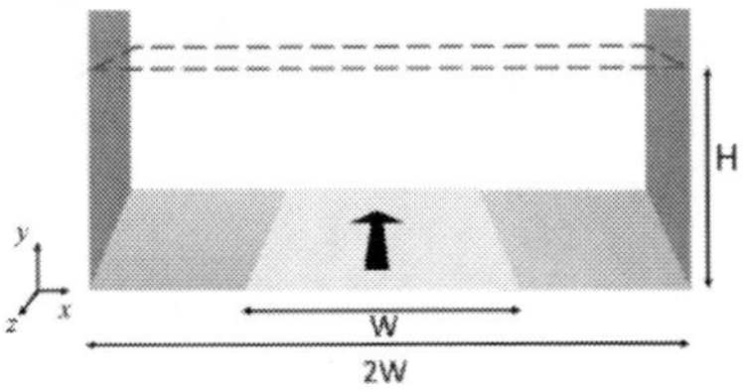

Figure 9: Schematic of a Cartesian equivalent of the split-bottom Couette.

surface. The stress along the vertical axis of inner cylinder was computed using DEM simulations. The radial wall stress profiles show similar exponential rise like the outer wall radial stress. Interestingly the sign of inner wall axial stress also changes after initiation of shearing the material. The direction of the radial vector pointing from the material to wall is opposite to the case of the outer wall. Hence the traction imposed by the material on the wall in both cases push the walls upwards against the direction of the gravity vector. Our results here further extend the that the anomalous wall stress profiles on the outer wall reported in previous study is true for the inner wall of the Couette as well.

6 Other systems with secondary flows

Secondary vortex formation has been seen in other studies [6, 7] though none of them could discern their cause. Further work could to characterize the vortex build-up using techniques similar to those suggested by [8]

The secondary vortex is dilation-driven, so we expect to see similar flows in other geometries. To test this, we carried out DEM simulations of the geometry shown in Figure 9. This is a box with periodic walls in the x and z directions. The bottom wall consists of two parts. The yellow section moves with a velocity $10\sqrt{gd_p}$ in the $-z$ direction. The grey sections are stationary. Gravity acts along $-y$. This is a Cartesian equivalent of the split-bottom Couette used in [3].

We pour particles up to a height of H. At steady state, we get the velocity profile shown in Figure 10. These are the streamlines seen in the $x-y$ plane, averaged over the depth of the system. Here as well, we can see flows similar to those seen in the Couette. This shows that the secondary flow we obtain is not an artefact of the Cylindrical Couette. Further work is needed to understand these results.

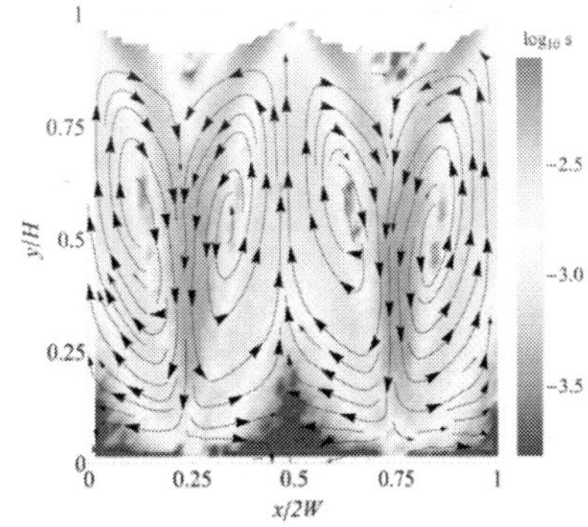

Figure 10: Streamlines obtained in the split-bottom Couette

7 Conclusion

We have tested and shown that the secondary vortex in a Couette still persists despite changes in the model parameters. We also tested the robustness of the flow under a confining weight and in the absence of a traction free surface. Finally, we also showed that such flows persist in other geometries too.

References

[1] Krishnaraj K.P., P.R. Nott, Nat. Commun. **7**, 10630 (2016)

[2] K.J. Gutam, V. Mehandia, P.R. Nott, Phys. Fluids **25.7**, 070602 (2013)

[3] K. Nichol, M. van Hecke, Phys. Rev. E **85.6**, 061309 (2012)

[4] W. Thielicke, E.J. Stamhuis, J. Open Res. Software, **2.1**, (2014)

[5] L.E. Silbert, et al., Phys. Rev. E, **64**, 051302 (2001)

[6] R. Khosropour, J. Zirinsky, H.K. Pak, R.P. Behringer. Phys. Rev. E, **56.4** 4467 (1997)

[7] N. Murdoch, et al. Phys. Rev. Lett. **110.1**, 018307 (2013)

[8] P.G. Rognon, T. Miller, B. Metzger, I. Einav, Granular Matter **17.2** 177-188 (2015)

DOI: 10.1051/epjconf/201714012009

Are polygonal faults the keystone for better understanding the timing of fluid migration in sedimentary basins?

Aurélien Gay [1,*]

[1] Laboratoire Géosciences Montpellier (GM), CNRS UMR 5243, Place Eugène Bataillon, 34000 Montpellier, France

Abstract. The initial sediment lithification starts with complex interactions involving minerals, surface water, decomposing organic matter and living organisms. This is the eogenesis domain (0 to 2 km below the seafloor) in which the sediments are subject to physical, chemical and mechanical transformations defining the early fabric of rocks. This interval is intensively prospected for its energy/mining resources (hydrocarbons, metal deposits, geothermal energy). In most basins worldwide it is composed of very fine-grained sediments and it is supposed to play the role of a seal for fluids migration. However, it is affected by polygonal faulting due to a volume loss during burial by contraction of clay sediments with a high smectite content. This process is of high interest for fractured reservoirs and/or cover integrity but it is not well constrained giving an uncertainty as this interval can either promote the migration of deeper fluids and the mineralized fluids intensifies diagenesis in the fracture planes, rendering this interval all the more impermeable. The next challenge will be to define where, when and how does this polygonal fault interval occur and this can only be done by understanding the behavior of clay grains and fluids during early burial.

1 Introduction

On the Gjallar Ridge, the 800 m thick Middle Oligo-Pliocene sequence is affected by small, closely-spaced intraformational faults, showing small throw values with a maximum of 30 msTWT and spacing ranging from 200 to 1000 m (Fig. 1). Some of these faults almost reach the modern seabed, thus representing the top of a recently active polygonal fault system [1-5] (Fig. 2). Closely spaced polygonal faults form in fine-grained clay-rich sediments and they display a polygonal pattern in plane view [6-8]. Such fault systems have been recognised in many basins all over the world: in the North Sea (6-7, 9], in the China Sea [10-11], in the Australian Eromanga Basin [12], in the Lower Congo Basin [13], in the northern Danish Central Through [14] and in the Canadian Atlantic margin [15]. They are usually interpreted as layer-confined because they occur within sub-horizontal intervals generally associated with lithological variations of the host sediments. Distinct fault tiers are defined by their own fault frequency and spacing due to different mechanical properties of the intervals [6-7, 16-18]. However, as already shown by studies in the North Sea, polygonal faults do not abruptly end at a specific stratigraphical horizon [6, 16]. Other mechanisms different from lithology variation have been referenced in the literature as responsible for polygonal fault initiation and propagation, such as diagenetic transformations [4-5] or reactivation by sediment loading [2, 13] for example. Four hypotheses are actually proposed to explain the formation of these polygonal faults: i) syneresis related to colloidal properties of such fine-grained sediments [8], ii) density inversions and associated hydrofracturing [19], iii) smectite-rich clays causing residual friction at low burial depth [20-21] and iv) grain dissolution in uncemented media inducing a decrease in horizontal stress that leads to shear failure and shear strain localization [22-23]. Polygonal faults are linked to fluid flow as they form pathways for upward-fluid migration from deeper levels [13, 21, 24]. However, the mineralized fluids intensifies diagenesis in the fracture planes, rendering this interval all the more impermeable [25]. Thus the kinematic model of polygonal fault growth in which the propagation of faults is discontinuous during basin infilling leads to a 4D interpretation of the whole fault system which will be one of the next challenge in basin modelling.

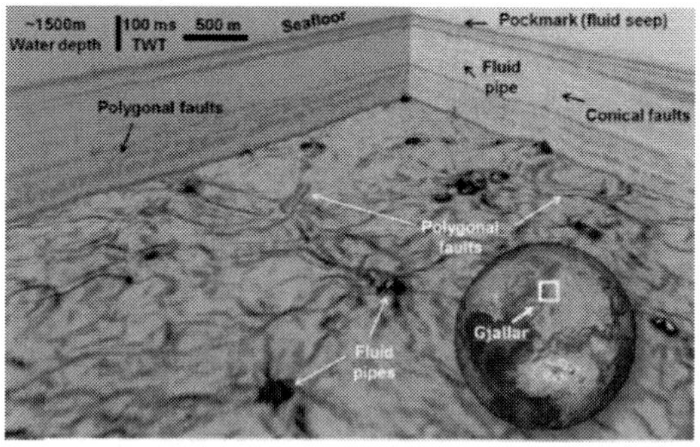

Fig. 1. The polygonal network in the Gjallar Basin (Norway) from 3D seismic data, showing the relation with conical faults and fluid pipes (Variance attribute).

2 Architecture and initiation

The first hypothesis for polygonal faults formation considers a finite bed length extension and the development of conjugate shear planes with extensional offsets due to the loss of pore fluids. This process leads to a thinned sequence, which is partly compacted [26].

* Corresponding author: author@e-mail.org

The second hypothesis considers that there is no change in the horizontal length and shear planes represent an apparent extension [26]. The consequences are that the complex polygonal fault systems in mudstone dominated sequences formed due to volumetric contraction during compactional dewatering [2, 13, 26]. This process is believed to occur in response to fluid expulsion during early compaction, leading to a bulk volume loss. The low permeability of fine-grained argillaceous sediments would impede the pore fluid expulsion during burial by an increase in pore fluid pressure in the sealed layer initiating failure.

Increased dewatering during burial implied more displacement along faults, which would continue to grow as long as the dewatering cell contracted volumetrically. This mechanism induces a component of tensile stress within the sediment mass resulting in the reduction of the horizontal effective stress and the Mohr circle can intersect the failure envelope in the shear domain leading to the propagation of non-vertical faults [8, 26]. The progressive reduction in bed length through contraction is balanced by an incremental increase in the amount of displacement distributed along the faults.

On the seismic data, major polygonal faults exhibit breached relays or overlapped fault-segments. Their fault planes show a succession of throw maxima separated by horizontal throw minima located in the relay zones [25]. This pattern is interpreted in the brittle domain as a reactivation by dip-linkage of two individual sub-parallel fault segments [27-28] (Fig. 2).

However, recent studies have shown that the linkage zone is consistent with boundaries of mechanical units and the throw maxima correspond to the centre of units [25]. It suggests that the nucleation of polygonal faults occurs at the centre of a mechanical unit, when the threshold of compaction is reached allowing shear localisation on fault. Then faults grow in any direction, both upward and downward. They can connect to underlying pre-existing intervals of polygonal faults or reaching the seabed forming pockmarks and allowing fluid to escape into the water column.

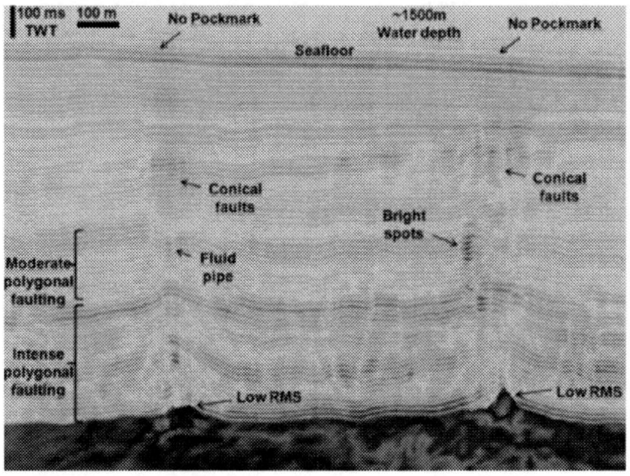

Fig. 2. 2D seismic profile showing anomalies, such as conical faults, fluid pipes, bright spots and pockmarks, due to the upward migration of fluids through the polygonal fault interval.

3 Processes

In most basin models, compaction and related phenomena are assumed to be due to mechanical effects in the first 0-2000m in response to the progressive loading of sediments [29]. As sediment porosity decreases with compaction, so does the permeability, due to a set of processes involving particle re-orientation and fluid expulsion leading to the decrease of void spaces between particles [29-30]. Overpressure development in sedimentary basins is thus directly related to lithology, sedimentation rate, thermal expansion of fuids, transformation of clay minerals and hydrocarbon generation or bacterial methanogenesis [31]. Among these factors, grain size and sedimentation rates are of primary importance in controlling fuid pressure development in a basin.

Normal faulting related to burial and sediment loading is not a common process that has been proposed for rocks deformation, which is probably because soft sediments like clays have a specific behaviour. The main factor controlling the development of polygonal fault systems is the grain size [8]. However, the dominant orientation of faults as well as their bending in the slope direction suggests the interaction of gravity-driven shearing of the compacting interval. The appearance of polygonal fault system reaching the seafloor could thus be indicative of the present day strain state within the sedimentary column [32].

A radial stress tension related to the volumetric contraction of clay could allow to reduce the lithostatic mean stress, also increasing the differential stress, and then to reach shear failure along a Coulomb-type conventional envelope. However this process assumes (i) that the faulted material has a common frictional envelope, (ii) that the lithostatic stress state prior to faulting due to compaction was sub-critical, i.e. close to shear failure, and/or (iii) that the stress tension related to the volumetric change is large enough. In such a non-lithified and uncompacted clay sediment, the coefficient of Earth pressure at rest (Ko) is generally large (0.45–0.8) implying a little differential stress due to the vertical loading [33-36]. This therefore suggests that if a radial horizontal tension is added due to volume contraction, it needs to be very large to increase the differential stress and get the failure envelope. However, as mentioned above, the behaviour of clays is actually quite far from the simplistic Coulomb type or Druker–Prager Model, and faulting comes frequently with non-isochoric shear, i.e. with compaction, as for example what happens in clay smears [37-38]. Yielding of clay materials have been described as following Cam–clay behaviour in q–p diagram (q is differential stress and p is mean stress) in which the yielding envelope has a cap, i.e. it is closed for relatively high mean stresses (Fig. 3a). This part of the material envelope has been described in plasticity theory as a behaviour allowing strain localisation [39-41] especially in soft materials as for example porous sandstones [42]. This model allows distributed pure compaction for high mean stress and localised pure shear at the critical state point (intersection of Coulomb type envelope with the cap). The interesting window of this cap envelope is the intermediate behaviour between these two end member points, along which localisation can start as shear compactional banding (i.e. clay smearing and subsequent compaction). This part of the envelope has a large probability to be crossed by a burial curve since it is below the Coulomb-type part of the envelope (see the location of the white star in Fig. 3b) and that strain hardening due to compaction, even moderate, must favour a non-linear stress path on the q–p diagram. Therefore, the conjunction of (i) possible localisation along a cap side of the clay envelope, (ii) a linear Coulomb-like part of the envelope with low coefficient of friction and (iii) burial with horizontal contraction and work hardening (bold dashed line) give mechanical support for the initiation of shear localisation related to burial. This could be also favoured by perturbed local stresses at the tip or the vicinity of another fault.

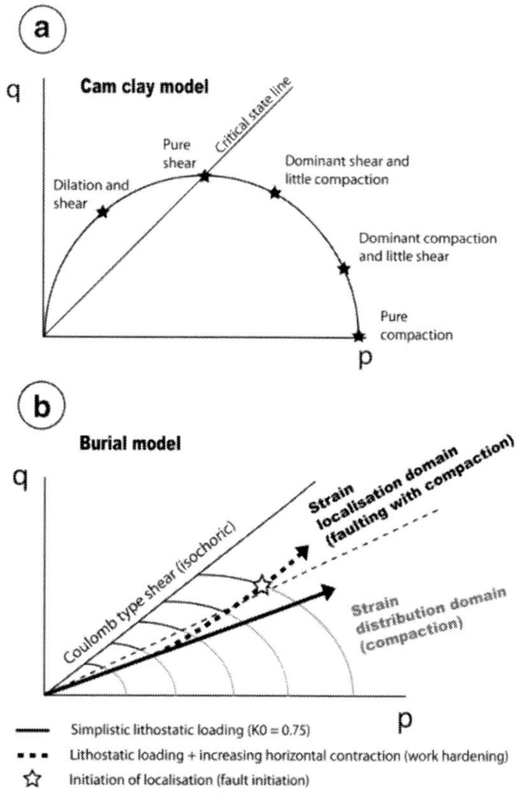

Fig. 3. (a) Diagram of a Cam–clay model (modified after Rudnicki, 2004) in which differential stress (q) versus mean stress (p) diagram shows a yielding envelope for clay materials with a cap, allowing behaviour of strain localisation. (b) Diagram of burial model in which yielding envelope is crossed by a burial curve corresponding to the addition of compaction and horizontal contraction of clay sediments leading to shear-compactional strain (white star).

The process of shear localisation on a cap part of the clay yielding envelope is consistent with the low dip angle of the polygonal faults observed (50° to 60°), which is attempted for shear banding having little compaction and/or material having low friction coefficient. This does not preclude these shear compaction strains to evolve into "real" faults (isochoric shear) by strain hardening of the material as the result of the progressive compaction (bold dashed line reaching the Coulomb-type line, Fig. 3b). This mechanical concept suggests that progressive sediment loading acts as a centrepiece in the initiation and propagation of polygonal faults [2, 13, 43]. It also implies that, at great burial depths, the compaction through the process of volumetric contraction ends and the dissolution–precipitation mechanisms become dominant [13, 44]

4 Conclusion

3-D seismic data provide new insights on the margin history and improve the understanding of post-depositional processes that affect the sedimentary column. During eogenesis in mud-dominated sediments numerous polygonal faults are initiated. This process is of high interest for fractured reservoirs and/or cover integrity, even in the extraterrestrial exploration [45-46]. However, at present day, polygonal faults have never been sampled nor modelled giving an uncertainty in their ability for driving fluids from deeper levels [47]. The next challenge will be to define where, when and how does this polygonal fault interval occur and this can only be done by understanding the biphasic behaviour of clay grains and fluids during early burial.

References

1. J.P.V. Hansen, J.A. Cartwright, M. Huuse, O.R. Clausen, Bas. Res. **17** (2005)
2. A. Gay, C. Berndt, JGS of London **164** (2007)
3. R.J. Davies, M.T. Ireland, J.A. Cartwright, Bas. Res. **21** (2009)
4. R.C. Neagu, J.A. Cartwright, R.J. Davies, J. of Str. Geology **32** (2010)
5. M.T. Ireland, R.J. Davies, N.R. Goulty, D. Carruthers, Sedimentology **58** (2011)
6. J.A. Cartwright, Geology **22** (1994)
7. L. Lonergan, J.A. Cartwright, R. Jolly, J. of Str. Geology **20** (1998)
8. D.N. Dewhurst, J.A. Cartwright, L. Lonergan, Mar. and Petr. Geology **16** (1999)
9. J.A. Clausen, R.H. Gabrielsen, P.A. Reksnes, E. Nysaether, J. of Str. Geology **21**(10) (1999)
10. C. He, C. Tang, D. Huang, S. Shi, Min. Sc. and Tech. **20** (2010)
11. Q. Sun, S. Wu, F. Lü, S. Yuan, J. of As. Earth Sciences **39** (2010)
12. J.A. Cartwright, L. Lonergan, Expl. Geophysics **28** (1997)
13. A. Gay, M. Lopez, P. Cochonat, G. Sermondadaz, Bas. Res. **16** (2004)
14. J.A. Clausen, J.A. Korstgård, J. of Str. Geology **15** (1993)
15. D.M. Hansen, J.W. Shimeld, M.A. Williamson, H. Lykke-Andersen, Mar. and Petr. Geology **21** (2004)
16. J.A. Cartwright, Mar. and Petr. Geology **11**(5) (1994)

17. R.V. Ackermann, R.W. Schlische, M.O. Withjack, J. of Str. Geology **23** (2001)
18. R. Soliva, A. Benedicto, J. of Str. Geology **27** (2005)
19. J. Watterson, J. Walsh, A. Nicol, P.A.R. Neil, P.G. Bretan, JGS of London **157** (2000)
20. N.R. Goulty, JGS of London **159** (2002)
21. N.R. Goulty, Petr. Geoscience **14** (2008)
22. H. Shin, J.C. Santamarina, J.A. Cartwright, Geology **36** (2008)
23. H. Shin, J.C. Santamarina, J.A. Cartwright, JGR **115**(B7) (2010)
24. C. Berndt, S. Bunz, J. Mienert, Geol. Society of London Sp. Pub. **216** (2003)
25. D. Laurent, A. Gay, C. Baudon, C. Berndt, R. Soliva, S. Planke, R. Mourgues, S. Lacaze, F. Pauget, M. Mangue, M. Lopez, Mar. Geol. **332-334** (2012)
26. J.A. Cartwright, L. Lonergan, Bas. Res. **8** (1996)
27. C. Childs, A. Nicol, J. Walsh, J.J. Watterson, J. of Str. Geology **18** (1996)
28. C. Baudon, J.J. Cartwright, J. of Str. Geology **30** (2008)
29. G. Vasseur, I. Djeran-Maigre, D. Grundberger, G. Rousset, D. Tessier, B. Velde, Mar. and Petr. Geology **12** (1995)
30. A. Maltman, *The Geological Deformation of Sediments* (1994)
31. Z. Yu, I. Lerche, Mar. and Petr. Geology **13**(2) (1996)
32. G.W. Tuckwell, L. Lonergan, R. Jolly, J. of Str. Geology **25** (2003).
33. E. Earl, J. of Agr. Eng. Res. **68** (1997)
34. E. Voznesensky, S. Nordal, Soil Dyn. and Earthquake Eng. **18** (1999)
35. J.K. Mitchell, K. Soga, *Fund. of Soil Behavior* (2005)
36. D.M.D. James, JGS of London **162** (2006)
37. K.J. Weber, G. Mandl, W.F. Pilaar, F. Lehner, R.G. Precious, OTC **10** (1978)
38. G. Yielding, B. Freeman, T. Needham, AAPG Bull. **81** (1997)
39. T.F. Wong, P. Baud, E. Klein, Geophys. Res. Letters **13** (2001)
40. J.W. Rudnicki, JGR **109**(3) (2004).
41. B. Dehandschutter, S. Vandycke, M. Sintubin, N. Vandenberghe, L. Wouters, J. of Str. Geology **27** (2005)
42. R.A. Schultz, R. Siddharthan, Tectonophy. **411** (2005)
43. S. Reiche, B.O. Hjelstuen, H. Haflidason, Mar. Geol. **284**, 1 (2011)
44. M.J. Jiang, F.G. Zhang, Y.G. Sun, Computers and Geotechnics 57 (2014)
45. L. Petracchini, M. Antonellini, A. Billi, D. Scrocca, Frontiers in Earth Science 3, 67 (2015)
46. D.Z. Oehler, N. Mangold, B. Hallet, A.G. Fairén, L. Le Deit, A.J. Williams, R.S. Sletten, J. Martínez-Frías, Icarus 277 (2016)
47. D.Z. Oehler, The Plan. Rep. **8-12** (2013)

Tracer experiments in periodical heterogeneous model porous medium

Samer Majdalani[1,*], *Carole* Delenne[1,2,**], and *Vincent* Guinot[1,2,***]

[1]*Univ. Montpellier, HSM (UMR 5569, CNRS, IRD, UM), CC0057, 163 rue Auguste Broussonnet 34090 Montpellier.*
[2]*Inria Sophia, Team Lemon*

Abstract. It is established that solute transport in homogenous porous media follows a classical 'S' shape breakthrough curve that can easily be modelled by a convection dispersion equation. In this study, we designed a Model Heterogeneous Porous Medium (MHPM) with a high degree of heterogeneity, in which the breakthrough curve does not follow the classical 'S' shape. The contrast in porosity is obtained by placing a cylindrical cavity (100% porosity) inside a 40% porosity medium composed with 1mm glass beads. Step tracing experiments are done by injecting salty water in the study column initially containing deionised water, until the outlet concentration stabilises to the input one. Several replicates of the experiment were conducted for $n = 1$ to 6 MHPM placed in series. The total of 116 experiments gives a high-quality database allowing the assessment of experimental uncertainty. The experimental results show that the breakthrough curve is very different from the 'S' shape for small values of n, but the more n increases, the more the classical shape is recovered.

1 Introduction

To predict the fate of contaminants in natural soils, it is important to understand solute transport in porous media. What characterizes natural soils is their high degree of heterogeneity, which impacts solute transport mechanisms. The classical Advection-Dispersion (AD) model is generally used to describe solute transport in porous media, and it is characterized by a typical 'S' shape breakthrough curve. The AD model is capable of reproducing tracer experiments in homogeneous porous media, but it fails to reproduce tracer experiments in heterogeneous porous media because breakthrough curves do not have any more the typical 'S' shape. In the literature, many laboratory tracer experiments in porous media can be found [1] but they are generally related to homogeneous porous media. Therefore, laboratory experiments on heterogeneous porous media are very few in comparison to those done on homogeneous porous media (see e.g. [2–6]).

In order to understand solute transport mechanisms, we use laboratory tracer experiments on Model Heterogeneous Porous Media (MHPM), built with a high degree of heterogeneity to guarantee that the effects of material heterogeneity are clearly visible. Moreover, we study the effect of periodicity of MHPM on breakthrough curves (and thus on solute transport behaviour) by assembling several MHPM in series.

Our experimental curves are characterized by a very fine sampling (one data every 5 seconds), as well as a high number of replicates in order to eliminate statistical errors in breakthrough curves. We can resume the features of our study by the following characteristics: *i)* high quality tracer experiments, *ii)* high degree of heterogeneity of the porous media, and *iii)* effect of periodicity of the heterogeneous porous media on solute transport.

In the following, experimental setup is detailed in section 2 and results are given in section 3 in terms of experiment and comparison with the classical advection-dispersion model. Section 4 provides conclusions about this experimental work.

2 Experimental setup

Each Model Heterogeneous Porous Medium (MHPM) is built in a PVC column of 10 cm in diameter, 15 cm in length. As shown in Figure 1, it consists of a cylindrical cavity with 100% porosity placed in the centre of the column and surrounded by a medium composed of 1 mm glass beads (40% porosity). Four inlets and outlets occupy the centre of gravity and the three vertices of an equilateral triangle. The centre of gravity is on the central axis of the column and the three vertices are 1 cm away from the column periphery. The internal cavity is a tube that is punched on top and bottom and covered with a 500 μm sieve. The sieve impedes the glass spheres from entering inside the cavity and guaranties that only water (and solute) can transfer through it. A similar sieve is also used to prevent glass spheres from clogging the inlets and outlets, possibly blocking the flow.

The MHPM initially contains deionised water. The tracer used in the experiments is composed of deionised water supplemented with NaCl at $C_0 = 0.1$ Mol/l. A constant discharge (7.5 l/h) of this salty water is prescribed at

*e-mail: samer.majdalani@umontpellier.fr
**e-mail: carole.delenne@umontpellier.fr
***e-mail: vincent.guinot@umontpellier.fr

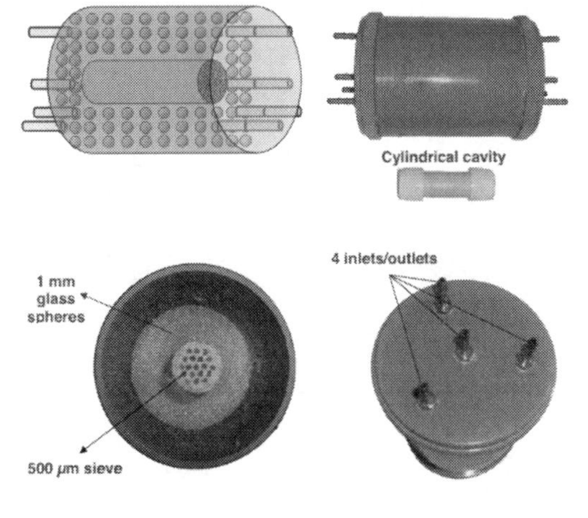

Figure 1. A Model Heterogeneous Porous Medium (MHPM).

the upstream boundary by a peristaltic pump (Gilson MP3) and is equally distributed between the four inlets.

The tracer exits the column through the four outlets that converge in a unique one after which the concentration is measured using a conductimeter (WTW TetraCon 325). Measurements are saved on a data logger (Campbell CR1000) every 5 s until the outlet concentration $C(t)$ stabilizes to $C_0(t)$. The duration $T_{\max}$ for one MHPM of 15 cm length is about 15 minutes.

To study the periodical effect on solute transport in heterogeneous media, 12 of the MHPM presented in [7] were used and experiments were conducted with $n = 1$ to 6 MHPM placed in series. An important difference with [7] stems in the in-between MHPM connection. Indeed, after each MHPM a convergence flow outlet is followed by a divergence inlet for the next MHPM. The period of the periodic heterogeneous porous media is thus defined by one MHPM. Figure 2 represents the general sketch of the experimental setup with an example of a series of $n = 4$ MHPM. At the end of the setup, the cumulative outlet volume V is measured by weighting the effluent every 5 s and is also saved on the data logger.

To avoid possible biases arising from asymmetry in the column, four replicates of each experiment were conducted after a 90° rotation of the column. Using different combinations of the 12 MHPM enabled from 8 to 48 replicates depending on the value of n, and thus a total of 116 experiments. The characteristics of the column combinations used for each value of n are summarized in Table 1, as well as the number of replicates and the amount of time required by each experiment.

3 Results

3.1 Experimental results

For each value of n, the minimum, maximum and average breakthrough concentrations obtained from the R replicates are computed as:

Table 1. Experiment replicates.

n	V	L	$T_{\max}$	R
1	V_0	L_0	15 min	48
2	$2V_0$	$2L_0$	30 min	24
3	$3V_0$	$3L_0$	40 min	16
4	$4V_0$	$4L_0$	50 min	12
5	$5V_0$	$5L_0$	60 min	8
6	$6V_0$	$6L_0$	70 min	8

n is the number of MHPM in the column, V is the pore volume (V_0 = 461 ml), L is the total length of the porous medium (L_0 = 15 cm), $T_{\max}$ is the total time for one experiment and R the number of replicates.

$$C_{\min}(t) = \min_{i=1...R} C_i(t), \quad C_{\max}(t) = \max_{i=1...R} C_i(t) \tag{1}$$

$$\bar{C}(t) = \frac{1}{R}\sum_{i=1}^{R} C_i(t) \tag{2}$$

where $C_i(t)$ is the output concentration for replicate i at time t.

Figure 3 shows the experimental breakthrough curves for $n = 1$ to 6 MHPM in series. For the sake of clarity, the normalized concentration is given as: $C^*(t) = C(t)/C_0$. The minimum and maximum statistical variation for the different replicates is represented by the grey envelop and the black line is the mean breakthrough curve deduced from all the replicates. It is worth mentioning the very high number of data points in comparison with other works found in the literature. This enables to draw conclusions that are not biased by a too sparse database.

Results show that for $n = 1$ (only one MHPM), the breakthrough curve is far from the classical 'S' shape of a homogeneous porous medium. As expected however, the more the periodicity increases, the more we recover this shape. The behaviour for $n = 1$ stems from the high contrast in porosity between the cylindrical cavity (100%) and its surrounding medium (40%). The salty water thus flow preferentially through this cavity, bypassing the granular medium. A consequent part of the volume flows with higher velocity and thus faster reaches the outlet. As seen in Figure 1, the fact that one of the outlets is aligned with the cylindrical cavity may enhance this behaviour.

Another important result obtained thanks to the repetition of experiments is that the breakthrough curves strongly differ from one replicate to another. The amplitude of the min- max interval seems to decrease for larger n, but this might be due only to the smaller number of experiments as explained in [7].

3.2 Modelling attempt

It is well known that the Advection-Dispersion model, based on the classical Fickian operator, may be used to recover 'S'-shaped breakthrough curves obtained in homogeneous porous media. However, attempting to calibrate this model against field or laboratory data involving heterogeneous porous media has been seen to lead to contradictory conclusions (see e.g. [2, 8, 9], that may however

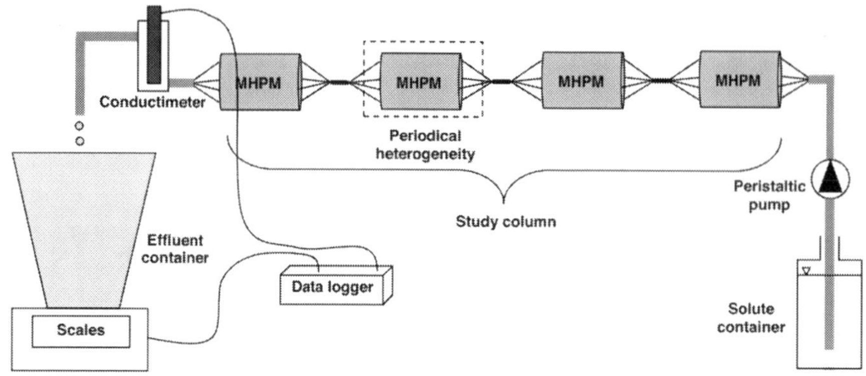

Figure 2. Experimental setup with an example of a series of $n = 4$ columns.

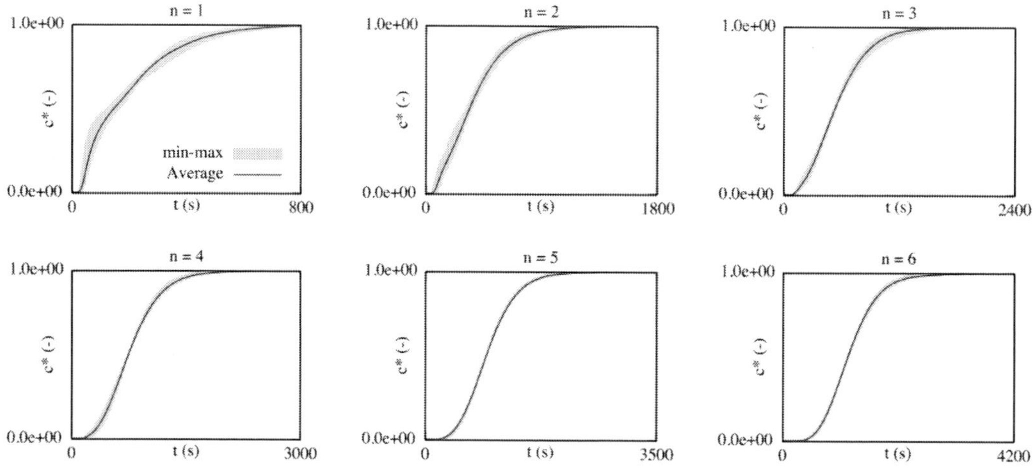

Figure 3. Experimental breakthrough curves for $n = 1$ to 6 MHPM.

be tempered considering the bias induced by experimental noise.

The Advection-Dispersion model is defined as follows:

$$\frac{\partial C(x,t)}{\partial t} + u\frac{\partial C(x,t)}{\partial x} - D\frac{\partial^2 C(x,t)}{\partial t^2} = 0 \quad (3a)$$

$$C(x,0) = 0 \ \forall x \in [0, L] \quad (3b)$$

where D is the dispersion coefficient and u the flow velocity.

Using a substantial amount of experiments enabled us to show in a previous work [7] that no significant trend can be detected for the dispersion coefficient as a function of distance, given the experimental uncertainty.

In the present work, the model was calibrated in D and u against the six experiments ($n = 1$ to 6) simultaneously. The aim being to reproduce every breakthrough curves with the same set of parameters. Result that best fits the six curves is obtained with $D = 4.25 \cdot 10^{-5}$ m^{2}s^{-1} and $u = 7.6 \cdot 10^{-4}$ ms$^-$1 (see Figure 4). As expected, this model fails to reproduce the curves for small values of n.

The high quality database presented in this study, will be used in forthcoming work to perform reliable calibration of several other models, that may be more prone to reproduce atypical behaviours of breakthrough curves in strongly heterogeneous porous media.

4 Conclusions

Our experimental results show that periodicity impacts the behaviour of solute transport in heterogeneous porous media. This highlights the importance of observation scale of solute transport. If the observation scale is larger than the heterogeneity scale, we will recover the classical 'S' shape breakthrough curve. Conversely, if the observation scale is comparable to the heterogeneity scale, the classical behaviour is lost and the breakthrough curve is no more 'S'-shaped.

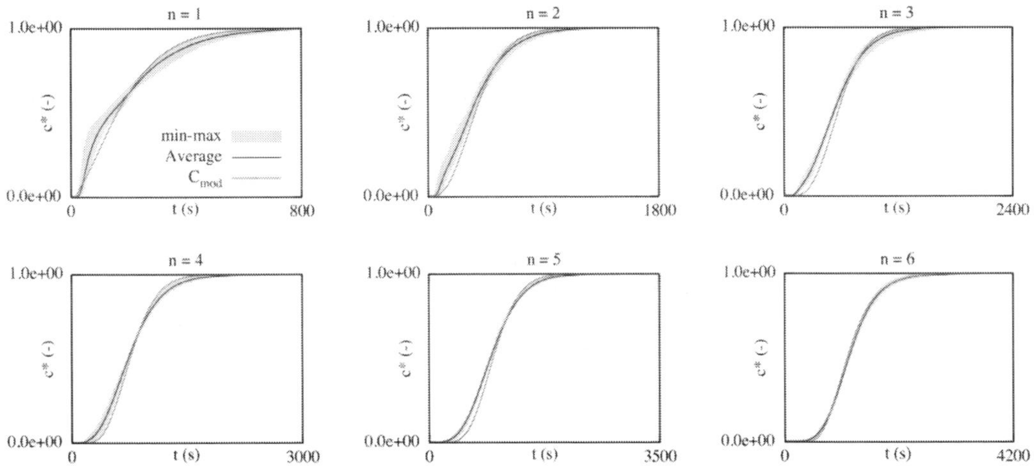

Figure 4. Modelling Results obtained with the classical Advection-Dispersion model.

An important characteristic of this work is the high number of replicates for each experiment as well as the great amount of measurements. This highlights the strong variability of the experimental results, even in the case of artificial and well controlled MHPM. Consequently, extreme caution is required when dealing with natural media.

References

[1] Silliman, S.E., Zheng, L., Conwell, P., 1998. The use of laboratory experiments for the study of conservative solute transport in heterogeneous porous media. Hydrogeol. J. 6, 166-177.

[2] Silliman, S.E., Simpson, E.S., 1987. Laboratory evidence of the scale effect in dispersion of solutes in porous media. Water Resour. Res. 23 (8), 1667?1673.

[3] Saiers, J.E., Hornberger, G.M., Hervey, C., 1994. Colloidal silica transport through structured, heterogeneous porous media. J. Hydrol. 163, 271?288.

[4] Li, L., Barry, D.A., Culligan-Hensley, P.J., Bajracharya, K., 1994. Mass transfer in soils with local stratification of hydraulic conductivity. Water Resour. Res. 30 (11), 2891-2900.

[5] Huang, K., Toride, N., Van Genuchten, M.T., 1995. Experimental investigation of solute transport in large, homogeneous and heterogeneous, saturated soil columns. Transp. Porous Media 18, 283-302.

[6] Golfier, F., Quintard, M., Woodd, B.D., 2011. Comparison of theory and experiment for solute transport in weakly heterogeneous bimodal porous media. Adv. Water Resour. 34, 899-914.

[7] Majdalani, S., Delenne, C., Chazarin J.P., Guinot, V. Solute transport in periodical heterogeneous porous media: Importance of observation scale and experimental sampling. Journal of Hydrology, 520, 52-60, 2015.

[8] Irwin, N.C., Botz, M.M., Greenkorn, R.A., 1996. Experimental investigation of characteristic length scale in periodic heterogeneous porous media. Transp. Porous Media 25, 235-246.

[9] Sternberg, S.P.K., Cushman, J., Greenkorn, R.A., 1996. Laboratory observation of nonlocal dispersion. Transp. Porous Media 23, 135-151.

An hybrid multiscale model for immersed granular flows

Matthieu Constant[1,*], *Frédéric* Dubois[2,3,**], *Jonathan* Lambrechts[1,***], and *Vincent* Legat[1,****]

[1] *Université catholique de Louvain, Avenue Georges Lemaître 4-6 bte L4.05.02, 1348, Louvain-la-Neuve, Belgique*
[2] *Laboratoire de Mécanique et Génie Civil - CNRS/UM, Montpellier, France*
[3] *Laboratoire de Micromécanique et Intégrité des Structures - CNRS/UM/IRSN, France.*

Abstract. The intensive use of immersed granular flows has motivated a large amount of work in this field of research. We present here a new numerical model to represent accurately and efficiently mixtures of fluid with grains. On one hand, the motion of the grains is solved at the grain scale by a contact dynamics method that consider grains as discrete elements. On the other hand, the fluid flow is computed from a continuous model of the mixture at a greater scale. The link between the two scales is provided by an interaction force based on the Darcy's law for porous media and used as a closure relation in the equations of the model. Some results of simulations in a two dimensional space are provided to prove the efficiency of the implementation.

1 Introduction

A lot of numerical methods have been developed to understand and predict immersed granular flows i.e. mixture of grains in suspension within a fluid [1]. We can separate the different types of physical models for immersed granular flows with respect to the scale at which the flow is modeled [2]. At large scale, the grains are introduced in the fluid in an implicit way. On one hand, some models consider the medium as a mixture described as a non newtonian fluid [3]. On the other hand, the grains can be introduced via a dissipative force of friction [4]. This force is used as a closure relation in the momentum equations and depends largely on the considered problem [5]. At a smaller scale, direct methods consider the two phases distinctly. The dynamics of the fluid is solved in an Eulerian way while the motion of the grains is solved in a Lagrangian way. This can be numerically done by remeshing the fluid phase at each time step [6]. Other approaches use methods of penalty or Lagrange multipliers to avoid the numerical cost imposed by the successive creations of a mesh [7]. Finally, a Lattice-Boltzmann approach can be used to simulate immersed granular flows without a mesh [8].

Discrete elements methods (DEM) are commonly used to describe trajectories of grains in a Lagrangian way for dry granular flows [9]. Specific models are able to solve grain-grain and grain-boundary collisions. We can distinguish several types of DEM. Collision methods solve contacts sequentially. They are not applicable to dense granular media where the time between two collisions is very short. Other methods solve all the contacts that happened during a given time step. They are based on two models of grains that change the representation of the contacts. The *smooth grain* model allows a slight deformation or an interpenetration of the grains. The elastic, plastic and friction forces between the grains are deduced from these deformations/interpenetrations. On the contrary, the *nonsmooth grain* model totally bans such deformations. An implicit resolution of the movement equations of the grains and the contact forces is needed for this method.

In this paper, we present an hybrid method [10] that can be viewed as a compromise between direct methods and methods based on continuum mechanics. The motion and contacts of grains are solved using a DEM at the grain scale while the fluid flow is obtained from a continuous description of the mixture at a greater scale using finite elements method. It is then possible to solve problem faster while describing accurately the physics of grains.

2 Model

Our model is based on a well-known hypothesis of the continuum mechanics: the separation of scales. The fluid flow characteristics are deduced from a continuous model of the mixture at a larger scale than the grains. The incompressible Navier-Stokes equations are averaged on elementary subsets of the domain. These elementary subsets are the sum of a fluid volume Ω_f and a solid volume Ω_p. The size of the grains have to be negligible in comparison with the size of the subsets.

Let us define the porosity that corresponds to the volume fraction of the fluid in an elementary subset:

$$\phi = \frac{\Omega_f}{\Omega}$$

*e-mail: matthieu.constant@uclouvain.be
**e-mail: frederic.dubois@umontpellier.fr
***e-mail: jonathan.lambrechts@uclouvain.be
****e-mail: vincent.legat@uclouvain.be

The averaged Navier-Stokes equations on an elementary subset are:

$$\frac{1}{\Omega}\int_{\Omega_f}\left(\frac{\partial\rho}{\partial t}+\nabla\cdot\rho\boldsymbol{v}\right)\mathrm{d}\Omega=0$$

$$\frac{1}{\Omega}\int_{\Omega_f}\left(\frac{\partial\rho\boldsymbol{v}}{\partial t}+\nabla\cdot(\rho\boldsymbol{v}\boldsymbol{v})-\nabla\cdot\boldsymbol{\sigma}+\rho\boldsymbol{g}\right)\mathrm{d}\Omega=0$$

where ρ is the density of the fluid, $\boldsymbol{\sigma}$ is the Cauchy tensor, $\boldsymbol{v}$ is the velocity and $\boldsymbol{g}$ is the gravity. The intrinsic fluid volume average of the velocity on an elementary subset is defined by:

$$\langle\boldsymbol{v}\rangle_f=\frac{1}{\Omega_f}\int_{\Omega_f}\boldsymbol{v}\,\mathrm{d}\Omega$$

The velocity is then splitted in this average and a perturbation with zero mean value:

$$\boldsymbol{v}=\langle\boldsymbol{v}\rangle_f+\boldsymbol{v}'$$

The intrinsic solid volume average operator can defined in a similar way. Then, using some average theorems [11, 12], we derive the average form of the Navier-Stokes equations:

$$0=\nabla\cdot\left[\phi\langle\boldsymbol{v}\rangle_f+(1-\phi)\langle\boldsymbol{v}\rangle_p\right]$$

$$\begin{aligned}0=\;&\rho\frac{\partial\left(\phi\langle\boldsymbol{v}\rangle_f\right)}{\partial t}+\rho\nabla\cdot\left(\phi\langle\boldsymbol{v}\rangle_f\langle\boldsymbol{v}\rangle_f\right)+\phi\nabla\langle p\rangle_f-\phi\rho\boldsymbol{g}\\&-\mu\nabla^2\left(\phi\langle\boldsymbol{v}\rangle_f\right)+\mu\frac{\nabla\left(\phi\langle\boldsymbol{v}\rangle_f\right)\cdot\nabla\phi}{\phi}-\mu\langle\boldsymbol{v}\rangle_f\frac{\nabla\phi\cdot\nabla\phi}{\phi}\\&\underbrace{-\nabla\cdot\left(\frac{\mu}{\Omega}\int_s\boldsymbol{n}\boldsymbol{v}\,\mathrm{d}s-\rho\phi\langle\boldsymbol{v}'\boldsymbol{v}'\rangle_f\right)-\frac{1}{\Omega}\int_s(\mu\boldsymbol{n}\cdot\nabla\boldsymbol{v}'-\boldsymbol{n}p')\,\mathrm{d}s}_{\boldsymbol{F}}\end{aligned}$$

where μ is the dynamic viscosity, $\boldsymbol{n}$ is the outward normal to the interface s of the two phases.

Finally, this mathematical model is closed as we replace those last terms by a force $\boldsymbol{F}$ that integrates the interactions of the fluid with all the grains in the representative volume. The interaction force between a grain i and the fluid can be written at the grain scale:

$$\boldsymbol{F}_i=-V_{p_i}\nabla\langle p\rangle_f-\boldsymbol{D}_i+\boldsymbol{g}V_{p_i}\frac{\rho_{p_i}-\rho}{\rho_{p_i}}$$

where $\boldsymbol{D}_i$ is the drag force, V_{p_i} is the volume of the grain i and ρ_{p_i} is its density. In our implementation, if the mixture is assumed to be composed of identical grains, we can use the formula given by Di Felice and Rotondi [13]

$$\boldsymbol{D}_i=f(\phi)\,C_d\left(\mathrm{Re}_{p_i}\right)\frac{A_{p_i}\rho}{2}\left|\boldsymbol{v}_{p_i}-\langle\boldsymbol{v}\rangle_f\right|\left(\boldsymbol{v}_{p_i}-\langle\boldsymbol{v}\rangle_f\right)$$

where A_{p_i} is the projected area of the grain on a plane perpendicular to the grain motion and Re_{p_i} is the grain Reynolds number:

$$\mathrm{Re}_{p_i}=\frac{2r\rho\phi}{\mu}\left|\boldsymbol{v}_{p_i}-\langle\boldsymbol{v}\rangle_f\right|$$

To take into account the effect of the neighbor grains, we use a function $f(\phi)$ given by Wen and Yu [14]:

$$f(\phi)=\phi^{-1.8}$$

while the drag coefficient $C_d\left(\mathrm{Re}_{p_i}\right)$ is provided by Dallavalle et al. [15]

$$C_d\left(\mathrm{Re}_{p_i}\right)=\left(0.63+\frac{4.8}{\mathrm{Re}_{p_i}^{0.5}}\right)^2$$

3 Results

Since the theoretical researches conducted by Brinkman [17] in 1947 on the fluid flow in an immersed swarm of grains, a lot of experiments have been realized to understand the observed micro- and macroscopic behaviors. An overview of these results can be found in [16, 18].

Just after being introduced in the fluid, the swarm falls and some grains escape from the closed volume of the swarm at the rear of the motion. They form a sort of tail while the lower part of the swarm forms something like a helmet giving to the whole grains the form of a mushroom. Twenty years later, Nitsche and Batchelor [19] performed numerical simulations to better understand this leakage of grains. They characterized the trajectories of the grains inside the swarm by decomposing the dynamics in two types of kinematics: a circular motion due to the toroidal velocity field inside the swarm and a random motion due to the hydrodynamical interactions. The leakage is then produced by the random motions that lead the grains outside the swarm where the streamlines are not closed.

The grains in the tail slowly fall because they are no longer dragged by the swarm [16]. The tail breaks away from the hemispherical part. Then, the center of the swarm is almost empty but the fluid can't cross it because of the overpressure in front of the motion. This overpressure point change into a ring below the swarm that becomes a torus and the fluid will then be able to cross it [18]. The torus becomes unstable and divides into two new smaller swarms that can repeat the same process if they are made with enough grains.

Simulations in two dimensions are performed with our model to compare the results with the simulations and experiments of Machu et al. [16]. Figure 1 shows the results for glass beads falling in glycerin. We clearly see the different steps corresponding to the experimental observations. Initially, we release a spherical swarm (circular in 2D) in the fluid. In figure 1a, the leakage of grains at the rear of the motion leads to the formation of a reverse mushroom. The tail we obtain in simulations is smaller than the one shown by Machu et al. [16] but it could be explained by comparing the number of grains in our swarm. In two dimensions, the number of grains is much less than in three dimensions. The computed velocity field looks like a cross-section of the toroidal field described by Metzger et al. [18]. The situation in Figure 1g shows the tail is no longer attached to the swarm. The overpressure point also divides in two part that can be again considered as a cross section of the ring described above (see Figure 1i).

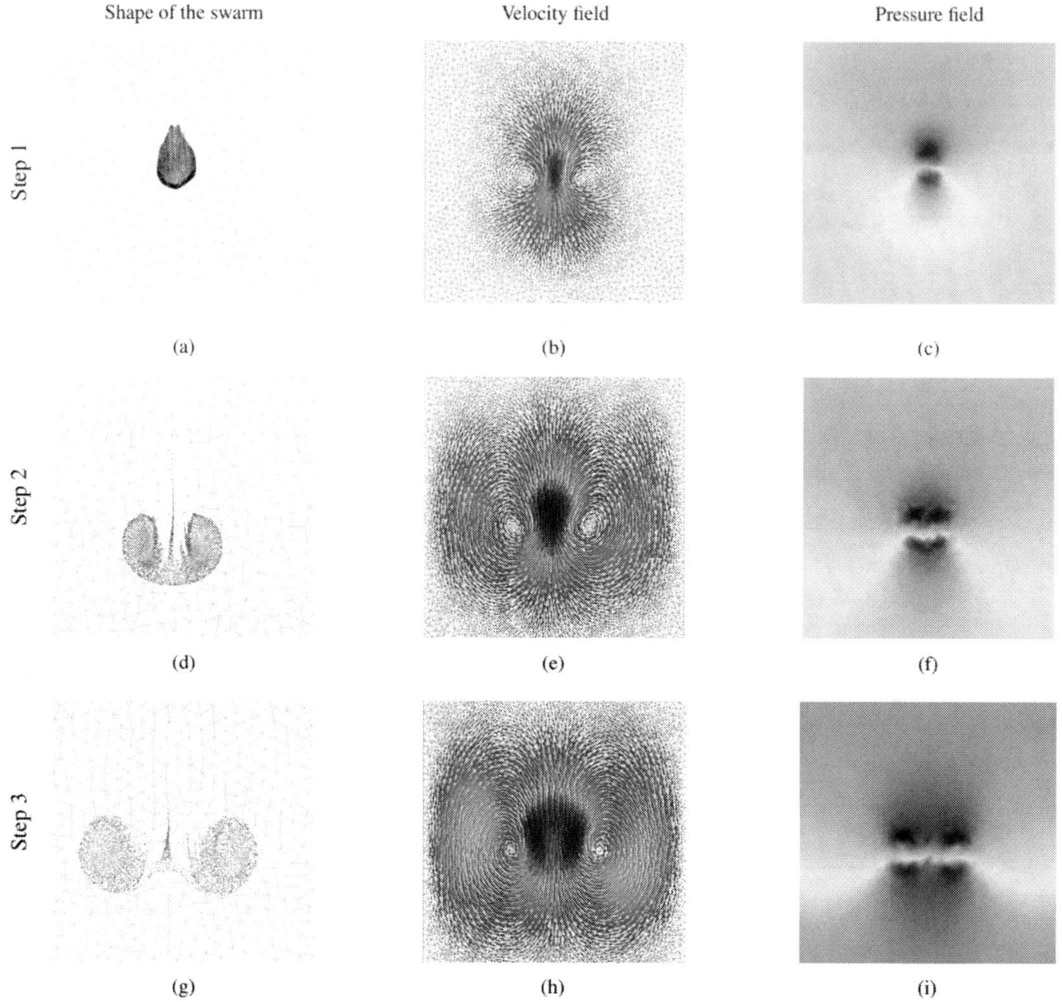

Figure 1: Evolution of a swarm made up of glass beads falling in glycerin

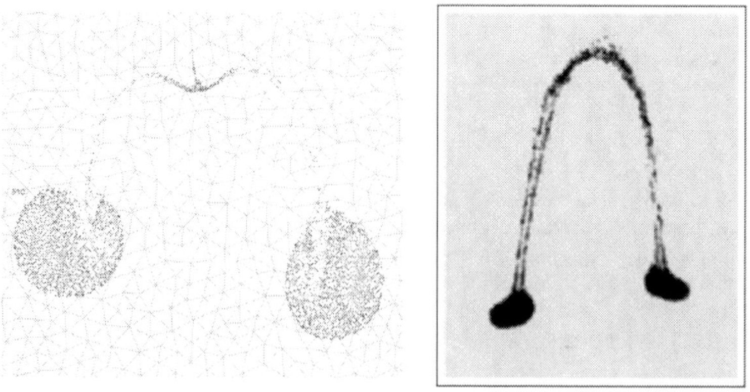

Figure 2: Division of the swarm (comparison with the experiment of Machu et al. [16])

 DOI: 10.1051/epjconf/201714009021

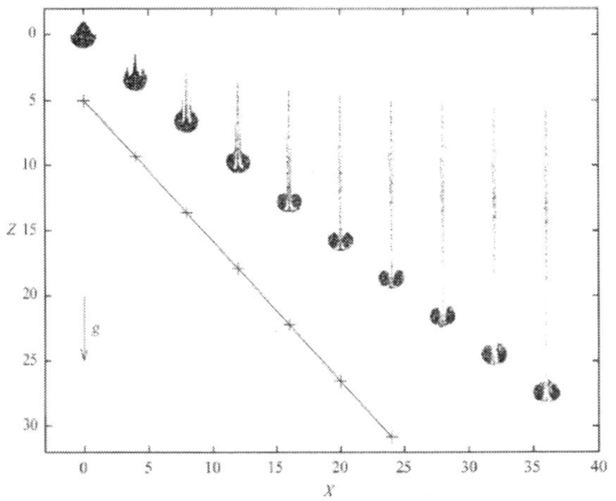

Figure 3: Simulation of Machu et al. [16]

The division of the swarm in two new smaller swarms is shown in Figure 2. Finally, it appears that the numerical simulation provides a quite similar behaviour than the results of Figure 3.

4 Conclusion

We develop an hybrid multiscale model to compute immersed granular flows. This model solves the contacts between grains with a DEM. We consider a non-smooth contact dynamics method and inelastic collisions. The fluid is solved at a greater scale by averaging the Navier-Stokes equations on an elementary subset containing the mixture of grains within the fluid. These averaged equations are solved by a finite elements method. With this model, we can compute a wide range of immersed granular flows from pure fluids to porous media with a reasonable computational time while the physics of the grains is described accurately.

An interaction force between grains and the fluid link the two scales of the model. This force is introduced in the momentum equations to replace some perturbation and interaction terms appeared during the averaging. This interaction force depends on the drag force that is determined empirically. The Rotondi and Di Felice's law neglects some important characteristics of the mixture such as the diversity of the grains encountered in real flows. More complex formula can be used to incorporate those effects.

Those preliminary results are very promising and demonstrate the qualities of the model. Firstly, we are able to compute the forces and constraints applied on each grain and their macroscopic motions can be compared with some simulations and experiments found in literature. Secondly, we can represent the flow induced by the displacements of the grains that can't be obtained in real experiments. And finally, we have shown that we could obtain accurate results using a representation of the fluid at a greater scale than the grains. Of course this is just a preliminary test of our model and other simulations should be achieved to show the real benefit of the contacts representation. The next step consists of generalize our approach in three dimensions in order to obtain better quantitative comparisons with experiments.

References

[1] H.P. Zhu, Z.Y. Zhou, R.Y. Yang, A.B. Yu, Chemical Engineering Science **62**, 3378 (2007)
[2] E. Izard, Ph.D. thesis, Université de Toulouse (2014)
[3] S.A. Rzadkiewicz, C. Mariotti, P. Heinrich, Physics and Chemistry of the Earth **21**, 1 (1996)
[4] D. Gidaspow, *Multiphase Flow and Fluidization: Continuum and Kinetic Theory Descriptions* (Academic Press, 2012)
[5] M. Pailha, O. Pouliquen, Journal of Fluid Mechanics **633**, 115 (2009)
[6] H.H. Hu, International Journal of Multiphase Flow **22**, 335 (1996)
[7] S.L. Dance, M.R. Maxey, Journal of Computational Physics **189**, 212 (2003)
[8] N.Q. Nguyen, A. Ladd, Physical Review E **66**, 046708 (2002)
[9] F. adjaï, F. Dubois, *Discrete-element modeling of granular materials* (Wiley-Istel, 2011)
[10] B.H. Xu, A.B. Yu, Chemical Engineering Science **52**, 2785 (1997)
[11] S. Whitaker, Chemical Engineering Education (1985)
[12] S. Whitaker, Transport in porous media **1**, 3 (1986)
[13] R. Di Felice, M. Rotondi, International Journal of Chemical Reactor Engineering **10** (2012)
[14] C.Y. Wen, H.Y. Yu, Mechanics of fluidization **62**, 100 (1966)
[15] J.M. DallaValle, A. Klemin, *Micromeritics: the technology of the particles* (Pitman Publishing Corporation, 1943)
[16] G. Machu, W. Meile, L.C. Nitsche, U. Schaflinger, Journal of Fluid Mechanics **447**, 299 (2001)
[17] H.C. Brinkman, Applied Science Research **A1**, 27 (1947)
[18] B. Metzger, M. Nicolas, E. Guazzelli, Journal of Fluid Mechanics **580**, 283 (2007)
[19] J.M. Nitsche, G.K. Batchelor, Journal of Fluid Mechanics **340**, 161 (1997)

Experimental added modal damping induced by confined granular media on a single degree of freedom system

Antoine Sternberger[1,2,*], *Adrien* Pelat[2,**], and *Jean-Michel* Génevaux[2,***]

[1] *IRT Jules Verne, Chemin du Chaffault, 44340 Bouguenais, FRANCE*
[2] *LAUM UMR 6613, Université du Maine, Avenue Olivier Messiaen, 72085 Le Mans Cedex 9, FRANCE*

Abstract. The use of granular media to induce vibration energy's dissipation in lighter huge industrial structures permits to decrease the mass of the structure and consequently to spare the construction's cost and to satisfy oil consumption. In fact, when the structure in which the granular media is in contact overtakes an acceleration threshold, relative movements of the grains appears which lead to a dissipation of energy. When the grains are confined inside a cavity, the dissipation's level depends on several parameters (the acceleration's amplitude, the frequency, the grain's characteristics, the cavity's dimensions, the cavity's filling ratio, the fluid between the particles, etc.). This study quantifies the influence of several parameters by exciting uniformly a given volume of grains. A modal damping coefficient of a single degree of freedom system (SDOF) can be thus calculated as a function of the preceding parameters.

1 Motivations

Weight reduction of structure is an important issue for the transport industry in a way of doing some savings. Consequently to this mass reduction, an increase of the vibration is observed. This can lead to a shorter lifetime of the industrial structure by the weakening done trough this evolution.

An original way of structures' damping is the use of granular material. The dissipation is due to the contact between the particles and beween the particles and the surface of the structure. The transmission (and the loss) of the energy has been studied in several cases [1, 2]. Numerous advantages of this treatment (little influence of the temperature, easy installation...) make it very attractive for the industry. For example the granular media can be placed inside a cavity which is fixed on the primary system. [3].

Firstly, the mass of the added beads lead to a vibration reduction. This mass effect is completed by damping if relative movement of grains appears inside the granular media. However, this loose media adds a strong non-linear effect which has to be taken into account [4]. The effect of the granular treatment will therefore be influenced by the acceleration's amplitude.

To deduce the damping from the micro-scale of the beads to the macroscopic scale of the structure need a realistic non-linear model using a huge numer of parameters. Consequently, most of the works are based on an equivalent viscous coefficient's of a single degree of freedom model: it is adjusted to the experiment in one particular case [5].

The aim of this paper is to investigate experimentally the influence of several parameters on the modal damping coefficient above an acceleration's thresholds to induce relative movement between the beads [6]. The finality would be to create a adjustable equivalent viscous damping coefficient which could depend on the principle parameters of the system.

After having presented the experimental setup, this document will present the result of several parametric studies. First, the comparison between a not deformable mass and a grains is performed. Then, the sensitivity to a variation of the mass of grain and of the acceleration amplitude is studied. Finally, an observation of the modal damping's coefficient shows the efficiency of using grains to damp vibrations on the tested single degree of freedom system.

2 Experimental setup

In this study the behaviour of a single degree of freedom system is investigated (see Fig.1). It is composed of a rigid cavity whose mass is $m_{\text{cavity}} = 3.140$ kg, guided in translation in the gravity direction and suspended on a stiffness. The granular material used is glass beads with a diameter of $\varnothing = 2$ mm. This cavity set to vibration by the use of a shaker which excites the system with a sinusoidal

*e-mail: antoine.sternberger@irt-jules-verne.fr
**e-mail: adrien.pelat@univ-lemans.fr
***e-mail: jean-michel.genevaux@univ-lemans.fr

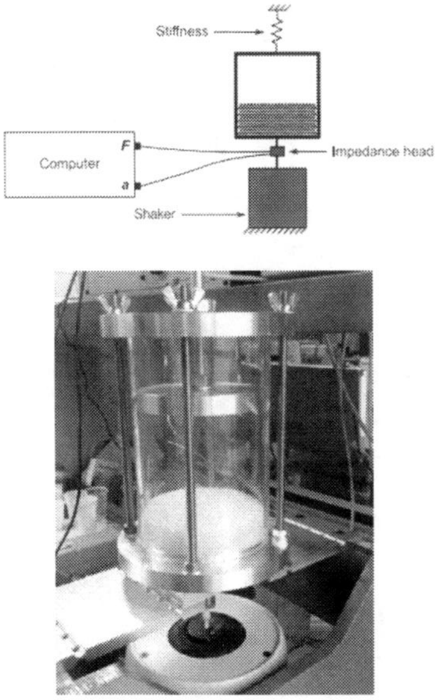

Figure 1. UP : Drawing of the setup - DOWN : Experimental set-up with granular media inside the cavity.

signal at an increasing step by step frequency. During each frequency increase, the acceleration amplitude is kept constant. An impedance head (B&K 8001) is used to get the acceleration and force signal at the same point. The Frequency Response Function (FRF), in kg^{-1}, is the ratio between acceleration and force.

Generally, the damping delivered by a plastic contact between two grains is considered as a friction damping. However the huge amount of grains can lead to an other interpretation (see Fig. 2).

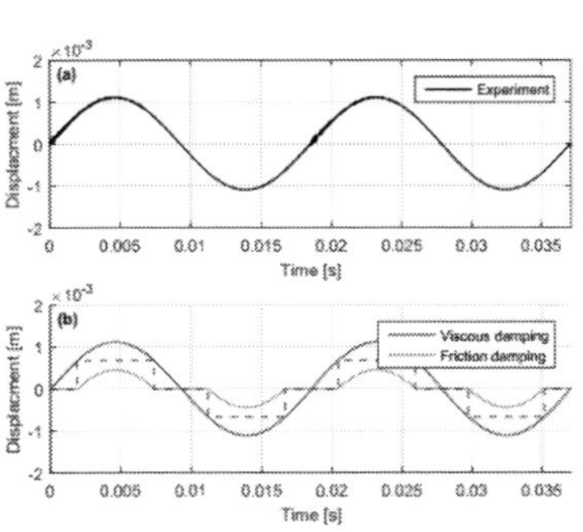

Figure 2. Influence of the several damping properties, comparison between viscous and frictional damping.

The figure 2-a shows a time capture of the container filled with grains. The figure 2-b shows a comparison in the time domain of the threshold effect where the friction force breaks and let the movement appears. The times without movement, can be called stick-times. This observation shows that no stick-times are present and allows us to consider an equivalent viscous damping induced by the amount of granular media. Indeed, if at the microscopic scale stick-times are present, they do not appear simultaneously in the whole volume. Therefore they are not detected by the sensors. The equivalent *modal loss factor* is also deduced by the *circle fit method* (see Fig.3). This method represents the complex FRF in a Nyquist graph.

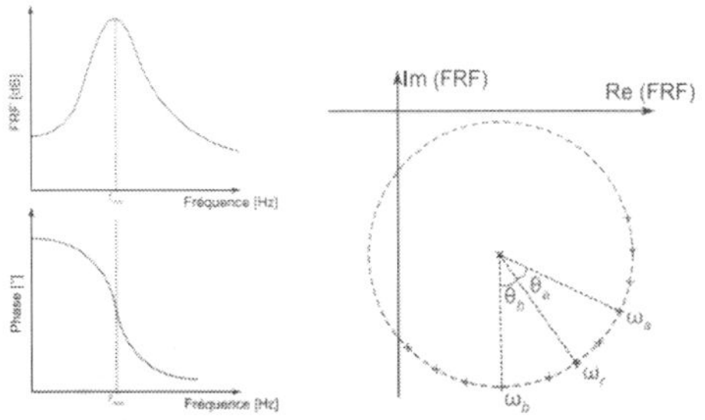

Figure 3. Frequency Response Function in the complex form.

The modal loss factor is deduced by the following equation [7] p.313 :

$$\zeta = \frac{\omega_a^2 - \omega_b^2}{2\omega_r[\omega_a \tan(\theta_a/2) + \omega_b \tan(\theta_b/2)]}. \quad (1)$$

3 Experimental analysis

3.1 The use of granular media and its damping representation

The figure 4 shows for a mass ratio of added material $r_2 = \frac{m_{\text{material}}}{m_{\text{cavity}}} = 15\ \%$, only a small decrease of 1 dB of the maximal amplitude of the FRF if a rigid mass is used: this solution needs a huge ratio of added mass to achieve large decrease of amplitude.

By opposite, the use of a granular material with the same mass ratio is examined. As a matter of fact, even if the granular material is compacted, a 2 dB decrease of FRF amplitude is observed compared to the undeformable mass. Interaction between the grains is also acting in compacted volume of grain. Note that the quality factor appears to be constant between these two cases. This effect is increased when the granular material is loosed inside the cavity. In that case, a reduction of more than 20 dB of the amplitude of the FRF is achieved and the quality factor strongly decrease, indicating that damping occurs.

3.2 Variation of the added mass on the primary system

In a second part, the variation of added mass on the system is investigate. It has been shown previously that the not deformable mass was not the most efficient way to

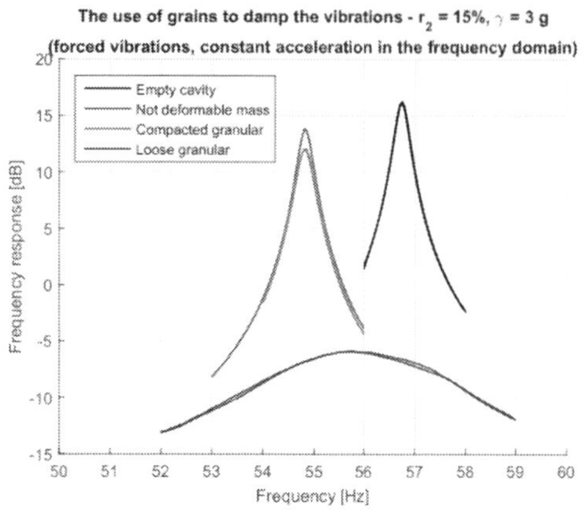

Figure 4. Comparison between the use of an inderformable mass and granular material (ref$_{FRF}$ = 1 kg^{-1}).

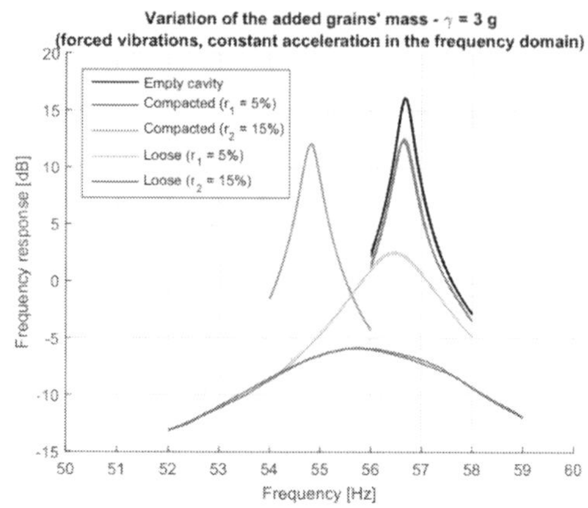

Figure 5. Variation of the grains' mass ratio on the resonanced system (ref$_{FRF}$ = 1 kg^{-1}).

damp the vibrations. In this part, two mass ratio are added separately with values of $r_1 = 5$ % and $r_2 = 15$ % and each are tested in two configurations: with compacted grains and with loose grains (see Fig.5).

Firstly, the behaviour of the compacted granular material is comparable with the behaviour of the added not deformable mass: when the mass increase, the resonance frequency ω_0 declines, as shown in Eq. 2, where ω_0 is the resonance frequency, k the stiffness of the SDOF system and m the mass of the SDOF.

$$\omega_0 = \sqrt{\frac{k}{m}}. \quad (2)$$

However, with the same amount of granular material the non-compaction of the grain induces a huge damping. For example, with the mass ratio r_2 where the amplitude of the peak of the FRF is reduced of almost 15 dB for an acceleration $\gamma = 3g$, with g the gravity acceleration. The loose granular material add grain interactions to the mass effect.

3.3 Variation of the acceleration

The use of a granular material leads to deal with non-linearities according to the driving amplitude of acceleration γ. The higher amplitude of acceleration is, the higher dissipation becomes. A stiffening of the system can also be seen (see Fig. 6). Indeed, neither the contact forces between the grains are proportional to γ, nor the the mean stream of the grains in the bulk. With the same experimental setup by only increasing the amplitude of acceleration γ, the resonance peak decrease by nearly 15 dB.

3.4 Evolution of the modal damping coefficient

The amount of damping is calculated by expression (Eq.1). This calculated value has to be compared to the empty system, which has no damping's treatment. The

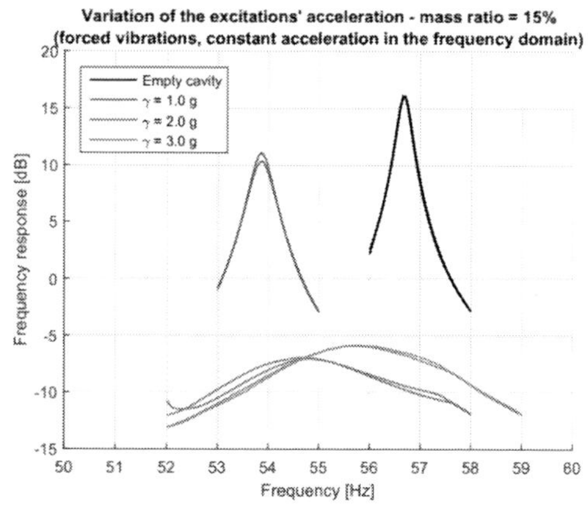

Figure 6. Variation of the acceleration's amplitude for a set-up with loose granular material (ref$_{FRF}$ = 1 kg^{-1}).

results are presented with a grain's mass ratio $r_1 = 5$ % and $r_2 = 15$ % (see Fig.7).

The modal damping coefficient of the bare structure is rather equal from 0.3 % to 0.4 % independently of the acceleration. In the case of a mass ratio r_1, the different configurations don't show an important variation of the damping coefficient except with the Loose configuration where an variation is shown.

Moreover, the use of the mass ratio r_2 gives larger variation of the modal damping coefficient. Firstly, an increase of the measured damping appears above a threshold value of acceleration $\gamma_t = 1g$.

For loose granular media, the damping coefficient reach a high value (4 %), which reach up to 10 times the value obtained by the not deformable mass (0.4 %). By increasing the acceleration's amplitude above 2.5 g, a

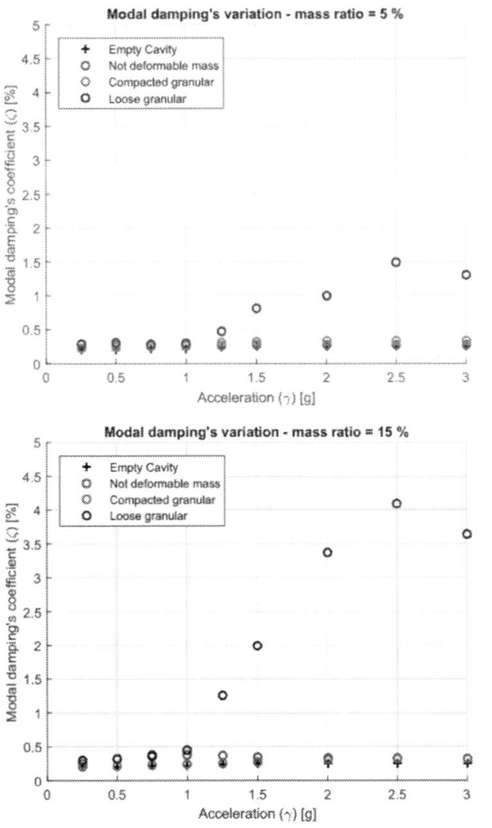

Figure 7. UP : System's modal damping, mass ratio = 5% / DOWN : System's modal damping, mass ratio = 15%.

reduction of this coefficient is observed. At this particular value of acceleration, measured signals are no longueur sinus-like curves due to the hight number of particles interactions. This lead to a bias in the modal loss factor calculation. However the Energyloss (calculate with the help of the Lissajous method) continues to increase with the acceleration's amplitude.

Finally, the most efficient way to damp the vibrations is to have a cavity's volume much larger than the volume of the granular material to let the grains interact under the vibration and so dissipate the vibration's energy.

4 Summary and conclusion

This paper describes experimentally the effect of a granular media on the damping of a single degree of freedom system. Glass grains with a diameter of $\varnothing = 2$ mm are used. The studied system consist of a SDOF on which a cavity is fixed, inside which a granular media or a not deformable mass are placed.

To be efficient the granular media have to be not confined and the acceleration must be greater than $\gamma_t = 1g$. The more is the mass of granular media, the more efficient is the damping. For a mass ratio of $r_2 = 15$ % the added damping can be 9 time the damping of the bare system. Moreover, because the damping depends strongly on the acceleration's amplitude, the amount of granular material has to be optimised to obtain a chosen damping target.

Acknowledgements

This study is part of the VIBROLEG project managed by IRT Jules Verne (French Institute in Research and Technology in Advanced Manufacturing Technologies for Composite, Metallic and Hybrid Structures). The authors wish to associate the industrial and academic partners of this project; respectively Airbus, CETIM, Daher, DCNS Research, General Electrics, STX France, Bureau Veritas, le Laboratoire d'Acoustique de l'Université du Maine and l'Université du Maine.

References

[1] C.N. Bapat, S. Sankar, Journal of Sound and Vibration **99**, 85 (1985)

[2] M. Saeki, Journal of Sound and Vibration **251**, 153 (2002)

[3] M. Heckel, A. Sack, J.E. Kollmer, T. Pöschel, Physica A: Statistical Mechanics and its Applications **391**, 4442 (2012)

[4] W. Liu, G.R. Tomlinson, J.A. Rongong, Journal of Sound and Vibration **280**, 849 (2005)

[5] G. Michon, A. Almajid, G. Aridon, Journal of Sound and Vibration **332**, 536 (2013)

[6] J.M. Génevaux, N. Dauchez, O. Doutres, Journal of Sound and Vibration **326**, 150 (2009)

[7] D. Ewins, *Modal testing: theory, practice and application - Second Edition*, Mechanical engineering research studies: Engineering dynamics series (Research Studies Press Ltd., 2000), ISBN 9780863800177

Vibration Induced Phenomena in Granular Media in Microgravity

Eric Opsomer[1,2,⋆], *Martial* Noirhomme[1], and *Nicolas* Vandewalle[1]

[1] *University of Liege, Research Unit CESAM, GRASP, B-4000 Liege, Belgium*
[2] *Université Paris Diderot, Sorbonne Paris Cité, MSC, F-75013 Paris, France*

Abstract. In order to study the dynamical behavior and the handling properties of granular materials under microgravity conditions, an ESA Topical Team is developing the VIP-Gran instrument whose multiple functionalities allow for the study of Vibration Induced Phenomena in Granular media in low gravity. Here, we present an overview of VIP-Gran's evolution, from the original idea to the latest encouraging and fascinating results. At first, we give a description of the instrument and the different investigated topics. Then, we present numerical simulations that we performed in order to prepare our experiments and tackle fundamental questions concerning granular gases. Finally, we give an insight on the first experimental results from parabolic flight campaigns and confront them with preliminary works and theoretical models.

1 Introduction

A granular gas is a paradigm for dissipative systems exhibiting spectacular behaviors such as inelastic collapse or clustering [1, 2]. The latter structures raise fundamental questions regarding their possible relations to thermodynamics principles [3]. Indeed, the classical kinetic theory of gases is unable to predict the behavior of such systems. The interest of the scientific community for granular gases has been strongly increased when sounding rocket experiments were performed in the late nineties by Falcon et al. [4] and confirmed for the first time the formation of clusters in microgravity. Since then, the European Space Agency formed a Topical Team (TT) of about 20 people and started the development of VIP-Gran (Vibration Induced Phenomena in Granular media) [5], an instrument designed for the study of the dynamics and the statistical mechanics of an ensemble of particles that interact through dissipative collisions. For dilute systems, the physical mechanisms and experimental conditions for the formation of clusters are deeply studied. For dense systems, phenomena such as convection [6], jamming [7] and segregation [8] are investigated. The prototype of the instrument is available and has already flown on two parabolic flight campaigns (PFC) of the European Space Agency (ESA).

In this article, we present an overview of VIP-Gran's evolution, from the original idea to the latest encouraging and fascinating results [5]. After a brief description of the instrument, we present the numerical simulations that we performed in order to prepare the experiments and give an insight on the first experimental results.

⋆e-mail: eric.opsomer@ulg.ac.be

2 Setup and experimental parameters

The VIP-Gran instrument has a simple but polyvalent design. Granular media is enclosed in an exchangeable container where two opposite walls act as pistons and assure a continuous energy injection into the system. The container is exchangeable since it is connected to the kinematic chain of the instrument via magnetic joints. A bead feeding mechanism allows the injection of additional grains into the cell in order to increase the density of the granular media. Data is recorded using pressure sensors and accelerometers mounted on the pistons. Moreover, the experiments are recorded by two high speed cameras.

Currently, two container geometries have been designed for VIP-Gran. Both present a length of 60 mm and a depth of 30 mm but differ in their height h. The value of h is 30 mm in the case of the 3D container and 5 mm in the case of the quasi-2D container. Parameters such as the amplitudes A, the frequencies f and the phase shifts φ of the pistons can be tuned in order to expose the granular media to several types of driving and different acceleration levels. The average distance L between the oscillating walls can also be modified for controlling the volume of the cell.

3 Numerical simulations

Given the need for microgravity, the efficient preparation of the experiments for VIP-Gran relies strongly on predictive numerical simulations. Discrete Element Method (DEM) algorithms are a good candidate for such task. Indeed, this numerical model is broadly used for the simulation of granular materials [9, 10] and has been validated through the two last decades. All following simulations were realized thanks to a homemade soft sphere DEM software in which normal forces F_n are evaluated via a linear spring-dashpot model using a restitution coefficient $\varepsilon = 0.9$ and a viscosity constant η. Tangential forces F_t are proportional to the relative tangent velocities at contact while being bounded as follows: $||F_t|| \leq \mu F_n$, where $\mu > 0$ is the friction coefficient. For further information on our numerical model see previous works [11].

3.1 Defining the gas cluster transition

For a better understanding of the different dynamics of granular gases, it is of great interest to be able to anticipate the occurence of density inhomogeneities [12]. Indeed, being a consequence of the dissipative collisions within a granular gas, clustering is expected once the density of the system exceeds some threshold value. In order to link this threshold to the geometry of the container we realized a first series of simulations modifying the filling number of the system for different values of L, h and A. The granular media is composed of N beads of radius R that are excited with a frequency $f = 10$ Hz.

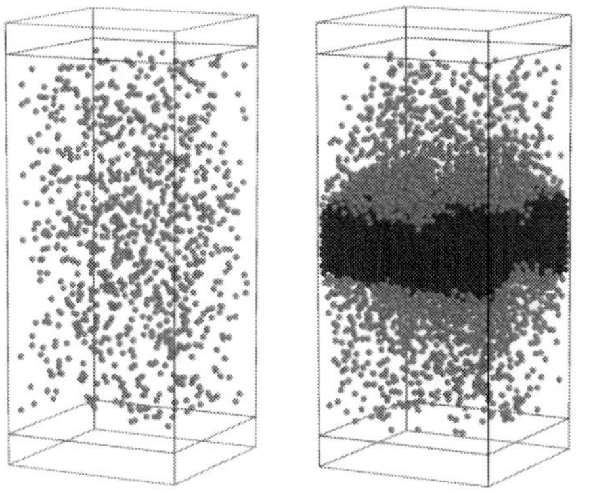

Figure 1. Simulation of clustering within in the 3D container with $R = 0.5$ mm and $\delta/R = 80$. (Left) Granular gas obtained for a packing fraction of $\phi = 1\%$. (Right) Clustered system obtained for a packing fraction of $\phi = 10\%$. The bulk is colored using a local density criterion.

For most simulated systems, clustering could be observed and a *phase diagram*, describing the encountered regimes, could be established. As shown in Figure 2, the occurence of clustering can be linked to two dimensionless parameters. One is the packing fraction $\phi = Nv_g/V$, with v_g the volume of a grain and V the maximum volume of the container. The other is the reduced system size δ/R, where δ is the average distance between two points chosen randomly on each piston. Transitions points are obtained by testing the uniformity of the density distribution along the driving axis. Symbols correspond to different sets of parameters all described in the table. Systems situated above (below) the curve present a cluster (gaseous) regime. The solid curve is obtained by balancing energy propagation and dissipation due to successive collisions. A detailed description can be found in [13].

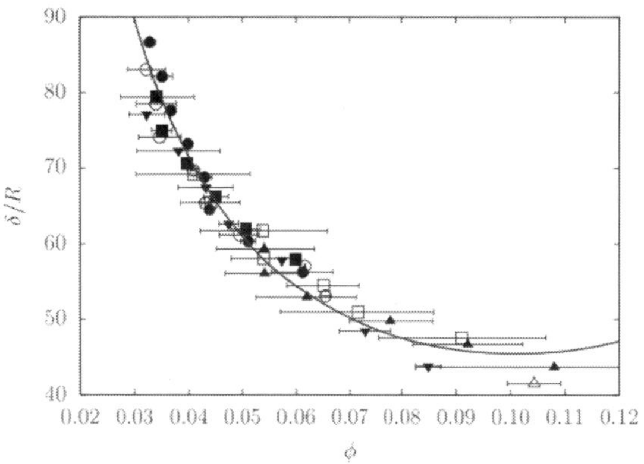

	A (mm)	L (mm)	h (mm)	R (mm)
●	4.0	30.0 - 47.5	30	0.5
○	5.0	30.0 - 47.5	30	0.5
■	6.0	35.0 - 47.5	30	0.5
□	5.0	32.5 - 47.5	30	0.6
▲	5.0	35.0 - 47.5	30	0.7
△	5.0	47.5	30	1.0
▼	5.0	30.0 - 47.5	15	0.5

Figure 2. Phase diagram describing the regimes for dilute driven granular media. Transitions points are obtained by testing the uniformity of the density distribution along the driving axis. Symbols correspond to different sets of parameters described in the bottom table.

These first simulations allowed us to describe the frontier between the gaseous and the cluster regime. It is now possible to predict, for both cell geometries, the apparition of a cluster. This new ability is important for the VIP-Gran instrument since the large field of parameters that had to be explored can be reduced dramatically regarding this results.

3.2 Clustering in a thermal gradient

The previous simulations were realized using a symmetric driving (same amplitudes and frequencies were used for both pistons). By using an asymmetric driving, it is possible to model a *thermal gradient* in the cell. In order to avoid a continuous phase shift, we decided to keep identical frequencies of oscillation on both pistons and to modify only their respective driving amplitudes noted A_c and A_h. We realized simulations for different container geometries and various amplitudes. As one can see in Figure 3, the cluster always forms near the cold piston. Moreover, its average position is determined by the amplitude ratio. Indeed, let us imagine the cluster as one solid particle, floating in the the cell about its equilibrium position $\mathcal{P}$. In order to stay there, the energetic impulses that impact the bulk from both sides must not only be of equal intensity but must also arrive at the same moment. With regards to the axis on Figure 3, the average position of the cluster is then given by

$$\mathcal{P} = \frac{L - \sigma}{2} \frac{1 - a}{1 + a}, \quad (1)$$

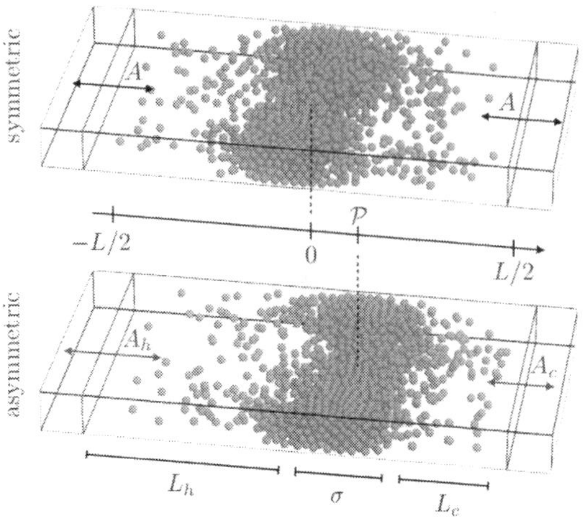

Figure 3. Simulation of a cluster under asymmetric driving in the quasi-2D container. The bulk forms always near the cold piston. Its average position $\mathcal{P}$ can be tuned using the amplitude ratio $a = A_c/A_h$.

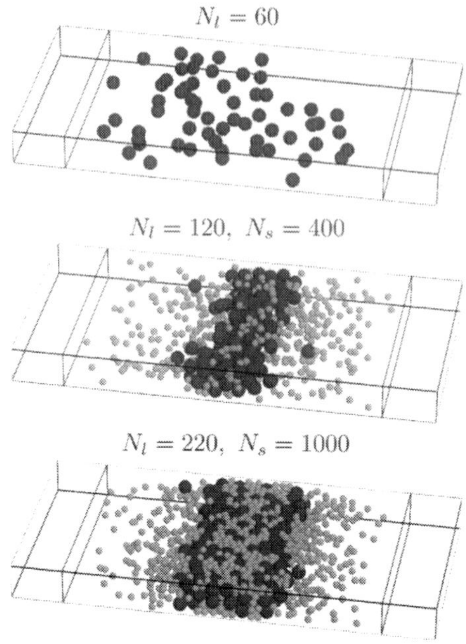

Figure 4. Simulation of driven bi-disperse granular media in the quasi-2D container. From top to bottom: Granular gas, Cluster made of large particles surrounded by a gas of small ones and Stripy structure.

where σ is the width of the cluster and $a = A_c/A_h$ is the amplitude ratio. Since σ remains constant for a given cluster, a can be tuned in order to move the bulk precisely along the driving axis.

Controlling the amplitude ratio of the piston's movements in the VIP-Gran instrument allows us to modify the clustering position and thus to achieve *granular transport* in this kind of geometry. Moreover, the cluster is found to oscillate about its position of equilibrium with a characteristic frequency that can be linked to the mass of the bulk. For more information see previous works [14].

3.3 Segregation in bidisperse granular systems

Using different grains sizes impacts strongly on the dynamics of the system. For instance, the energy transmission from the pistons towards the center of the cell will be altered by the different encountered pairs of collisions. In order to study the possibility of segregation in microgravity, we realized simulations of bi-disperse granular media composed of N_s small beads of radius $R_s = 0.5$ mm and N_l large beads of radius $R_l = 1.0$ mm. Both species have the same density so that mass and size vary together. Depending on the filling conditions of the system, different structures appear as shown in Figure 4. For very low packing fractions, a granular gas is observed. Particles of any kind are distributed uniformly over the entire container. For higher packing fractions, *segregation* occurs. Large grains are forced into the center of the cluster while most of small grains are found in the surrounding gas phase. Finally, changing the mixing proportion of small and large particles can modify the morphology of the cluster so that stripy structures with alternatively high concentrations of each granular species arise.

We showed that shaking bi-disperse granular media in microgravity can lead to segregation even though mechanisms such as convection and percolation rely all on the presence of gravity. Moreover, size and mass effects are found to be in competition in our system which leads to the formation of complex patterns in the bulk. A detailed description of all our simulations as well as a complete phase diagram describing all regimes can be found in [15].

4 Parabolic flight experiments

Thanks to the previous simulations, the TT was able to prepare a series of promising experiments for PFC 63 and PFC 64 of the ESA. These campagnes are essential for a future implementation of VIP-Gran in the International Space Station (ISS). Indeed, in addition to the validation of our numerical models, parabolic flights help us to prepare for parasitic effects such as g-jitter. For instance, finding a efficient initialization protocol in order to homogenize the granular media before each shaking is still to be found.

Let us now have a look at the first experimental results concerning clustering. Given the shape of transition curve in our phase diagram, we investigated the occurence of clustering by tuning either ϕ or δ all else equal. For both scenarios, clustering is reported as one can see in Figure 5. Snapshots correspond to 3D systems with respective packing fractions $\phi_a = 1.5\%$, $\phi_b = 8.5\%$, $\phi_c = 7.5\%$, $\phi_d = 7.5\%$ and respective system sizes $\delta_a = 35$ mm, $\delta_b = 35$ mm, $\delta_c = 22$ mm, $\delta_d = 31$ mm. The studied granular media is composed by bronze spheres of radius $R = 0.5$ mm that are excited with an amplitude $A = 2$ mm

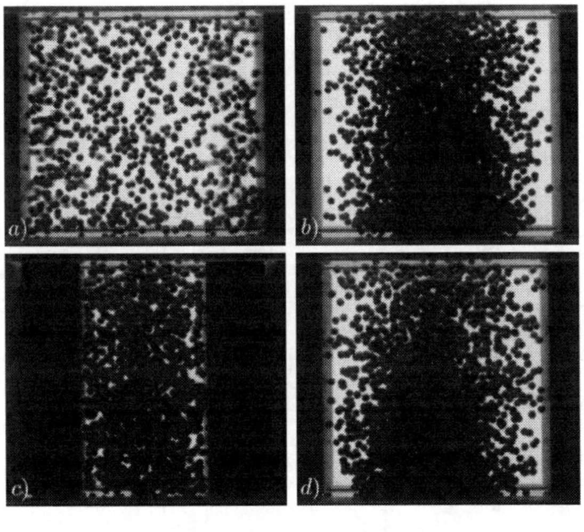

Figure 5. From snapshot *a*) to *b*) clustering is induced by an increase of the packing fraction ϕ for a fixed cell geometry. From snapshots *c*) to *d*) clustering is triggered by an increase of the system size δ while keeping a constant packing fraction.

and a frequency $f = 20$ Hz. In the top row, clustering is triggered by adding more particles into a container of fixed geometry. In the bottom row, clustering is induced by an increase of the system size while the packing fraction is kept constant constant by injecting additional grains. The transition has been tested for several packing fraction and system sizes. So far, the majority of the experimental results corroborate our model.

During PFC 64, the TT also explored experimentally the possibility of segregation in microgravity. As in our simulations, we used different concentrations of small and large bronze beads. In order to assure a good visibility of the phenomenon, experiments were realized within the quasi-2D container. The three expected regimes could be observed and are presented in Figure 6. Snapshots *e*),

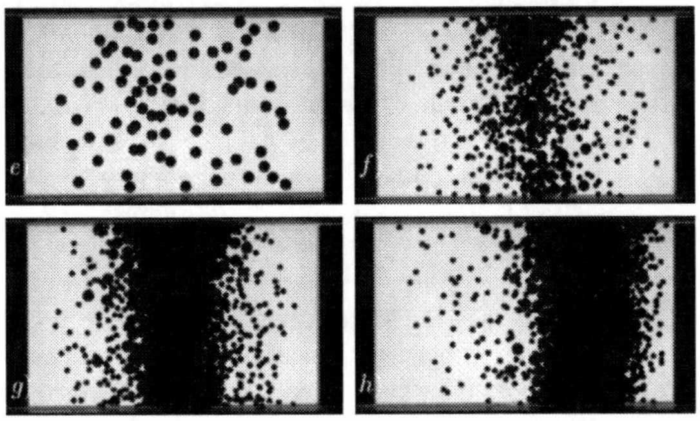

Figure 6. Segregation in bi-disperse granular media. *e*) Granular gas of large particles. *f*) Cluster of large particles surrounded by a gas of small particles. *g*) Stripy structure. *h*) Bi-disperse cluster exposed to asymmetric driving.

f), and *g*) correspond to a system with constant values of $\delta = 41$ mm, $A = 3$ mm and $f = 20$ Hz. The experiment starts with an initial loading $N_s = 0$ and $N_l = 80$, leading to a granular gas. After the injection of 500 small grains, a cluster of large beads surrounded by a gas of small ones appears. Finally, 500 additional small beads and 150 large ones are injected and a stripy structure arises. Snapshot *h*, corresponds to a system with asymmetric driving conditions. A shift of the cluster's position towards the cold piston is noted, however, a precise measurement of $\mathcal{P}$ was not yet realized.

5 Conclusion

Thanks to numerical simulations, the Topical Team prepared successfully the first microgravity experiments of the VIP-Gran instrument. The collected data has been used to tackle fundamental questions concerning clustering, segregation and the handling of granular matter in low gravity. The developed theoretical models, coupled to the experimental data from parabolic flight campagnes will be crucial for the implementation of VIP-Gran within the ISS.

Acknowledgements

We acknowledge the European Space Agency and Prodex-Belspo for financial support. Specials thanks to the entire Topical Team of VIP-Gran (see involved people in [5]).

References

[1] I. Goldhirsch and G. Zanetti, Phys. Rev. Lett. **70**, 1619 (1993).
[2] T. Pöschel and N. V. Brilliantov, *Granular gas dynamics*, (Springer, Berlin, 2003).
[3] F. Rouyer and N. Menon, Phys. Rev. Lett. **85**, 3676 (2000).
[4] E. Falcon, R. Wunenburger, P. Evesque, S. Fauve, C. Chabot, Y. Garrabos and D. Beysens, Phys. Rev. Lett. **83**, 440 (1999).
[5] More information on http://spacegrains.org.
[6] N. Murdoch, B. Rozitis, K. Nordstrom, S. F. Green, P. Michel, T.-L. de Lophem and W. Losert, Phys. Rev. Lett. **110**, 029902 (2013).
[7] M. Neudecker, S. Ulrich, S. Herminghaus and M. Schröter, Phys Rev. Lett. **111**, 028001 (2013).
[8] A. Kudrolli, Rep. Prog. Phys. **67**, 209 (2004).
[9] T. Pöschel and T. Schwager, *Computational Granular Dynamics*, (Springer,Berlin, 2005).
[10] S. Luding, Gran. Matt. **10**, 235 (2008).
[11] E. Opsomer, F. Ludewig and N. Vandewalle, Phys. Rev. E. **84**, 051306 (2011).
[12] J. S. Olafsen and J. S. Urbach, Phys. Rev. E. **60**, R2468 (1999).
[13] E. Opsomer, F. Ludewig and N. Vandewalle, Europhys. Lett. **99**, 40001 (2012).
[14] M. Noirhomme, E. Opsomer, N. Vandewalle and F. Ludewig, Eur. Phys. J. E. **38**, 9 (2015).
[15] E. Opsomer, N. Vandewalle, M. Noirhomme and F. Ludewig, Eur. Phys. J. E. **37**, 115 (2014).

EPJ Web of Conferences 140, 03029 (2017) DOI: 10.1051/epjconf/201714003029
Powders & Grains 2017

Force fluctuations experienced by a boundary wall subjected to a granular flow in two distinct systems

François Kneib[1,⋆], *Thierry* Faug[1,⋆⋆], *Frédéric* Dufour[2], and *Mohamed* Naaim[1]

[1] *Univ. Grenoble-Alpes, Irstea, ETGR, 38402 St-Martin d'Hères, France*
[2] *Univ. Grenoble-Alpes, CNRS, 3SR, F-38000 Grenoble, France*

Abstract. The present paper describes a numerical study of force fluctuations experienced by a boundary wall subjected to a granular flow on two distinct systems, namely a lid-driven cavity and an immersed wall system. Though the two systems exhibit different time-averaged dynamics, the force fluctuations experienced by the boundary wall show robust features in terms of the shapes of the probability density distributions and auto-correlation functions, under a wide range of boundary confinement pressure and shearing velocity imparted to the granular flow at the top of the system. This study identifies the key link between the grain-wall force fluctuations and the $\mu(I)$–rheology while moving from quasistatic to inertial regime.

1 Introduction

The problem of an assembly of grains in interaction with a boundary wall is relevant for a number of real-world applications involving granular materials, such as the transport of particle in food processing or mining industry, the impact of large-scale avalanche-flows with civil engineering structures in geophysics, the design of technological earthmoving equipments used to work with the soils in agriculture, etc. Understanding the force experienced by a boundary wall in interaction with a granular flow is crucial but little is known about the fluctuating part of the granular force under slow to fast flow regimes. The distributions of grain-grain or grain-wall contact forces (noted f) in static or quasistatic systems received a great attention, as reported in [1, 2] and a number of references therein. A salient feature in static granular media is a relatively robust exponential decrease at large forces, while the distributions at small forces may show very different behaviours (absence or presence of a plateau—even a peak, non-zero probability at $f = 0$), depending on many parameters associated with the shear history of the system. By contrast, only a few studies were devoted to force fluctuations at stake in flows of granular materials (see for instance [3]).

The present study investigates the distributions of grain-wall forces, as well as force time-series autocorrelation, when a granular flow interacts with a rigid boundary wall, under a wide range of macroscopic boundary shearing velocity U and confinement pressure P imparted to the granular bulk. The approach is based on the cross-comparison between two distinct granular flow systems simulated by discrete element method (DEM). The two systems and their time-averaged dynamics are briefly described in section 2. Section 3 is devoted to the grain-wall force distributions on the entire height of the wall and on portions of wall. The analysis of force time-series autocorrelation is addressed in section 4. Finally, a conclusion is given in section 5.

2 Two distinct granular systems studied

2.1 Flow-wall geometries and methods

This section gives a brief summary of the two systems studied, both involving a boundary wall that is subjected to a planar and gravity-free granular flow. The first system, namely the granular *lid-driven cavity*, is a closed system for which the grains are forced to stay inside the cavity, thus producing a perpetual recirculation of the grains in the whole volume. The second system, namely the *immersed wall* granular system, is an opened system which enables the grains to escape by passing over an immersed wall, while other grains are trapped behind the wall-like obstacle. Both systems are simulated thanks to DEM, using the open-source YADE software [4].

The contact forces between the spherical particles of diameter d were modelled using a damped linear spring law for the normal direction and a linear spring with a static Coulomb threshold law for the tangential direction. Four microscopic parameters were required for the simulations presented here: the normal stiffness k_n, the tangential stiffness k_t, the interparticle friction coefficient μ and the normal restitution coefficient e. All simulations were conducted in the limit of rigid grains, varying k_n and k_t accordingly, with respect to the macroscopic pressure imparted by the top wall to the system. We considered

⋆e-mail: francois.kneib@irstea.fr
⋆⋆e-mail: thierry.faug@irstea.fr

$\mu = 0.5$ and $e = 0.5$. Full information on how the values of those parameters were implemented is given in our initial study of the lid-driven cavity system [5].

For both systems, we defined the two dimensionless following parameters: $N_U = U/\sqrt{gd}$ and $N_P = P/(\Phi\rho_P gH)$ where H is the height of the system (equal to $30d$ with $d = 1$ mm in the simulations presented here), and ρ_P was the particle density taken equal to 2500 kg m^{-3} (corresponding to the density of glass beads generally used in small-scale laboratory experiments). The systems lengths L were set to five times the height of grains: $L/H = 5$. A polydispersity of $\pm15\%$ was given to the grains in order to avoid crystallization. Note that we considered typical pressures $\Phi\rho_P gH$ (where Φ was a typical volume fraction taken equal to 0.6 for a dense three-dimensional packing) and typical velocities $\sqrt{gd}$ in order to define N_P and N_U, though the systems simulated by DEM were gravity-free in the present study. This choice was made for convenience if laboratory experiments under gravity are developed in the future and compared to the DEM simulations presented in the paper. For both systems, N_U and N_P were typically varied in the ranges [1; 20] and [0.01; 100], respectively.

Based on the boundary conditions imposed at the top in terms of shearing velocity U and confinement pressure P, a macroscopic inertial number can be defined as the ratio between the typical time $t_P = d\sqrt{\rho/P}$, associated with the top confinement pressure P, and the typical time $t_U = H/U$, equal to the inverse of the macroscopic shear rate over the height of the sample: $I_M = \frac{U}{H} d\sqrt{\rho/P}$.

2.1.1 The granular lid-driven cavity

The two-dimensional granular lid-driven cavity simulated by DEM consists of a planar assembly of spheres trapped into a box made of four walls, as sketched in Fig. 1. The perpetual recirculation of the grains within the cavity (closed system) produced streamlines showing the formation of a vortex in the whole volume of the cavity. The mean force on the right sidewall (see Fig. 1) was found to fit the empirical following scaling:

$$\overline{F} = \left(r + \frac{\Delta r}{1 + I_M^0/I_M}\right) PLd, \qquad (1)$$

where $r = 0.53$, $\Delta r = 0.33$ and $I_M^0 = 0.2$ in this study (see detail in [5, 6]). The values of r and Δr depend on the mechanical properties of the grains, for instance the interparticle friction, as reported in [6] where results from DEM simulations with $\mu = 0.27$ are presented too.

2.1.2 The immersed wall granular system

The immersed wall granular system consists of a horizontal flow of spheres passing over a wall of height $h = 20d$, as sketched in Fig. 2. This system was simulated by DEM using periodic conditions. The immersed wall-like obstacle led to the formation of a dead zone behind the wall, this dead zone co-existing with an overlying sheared granular

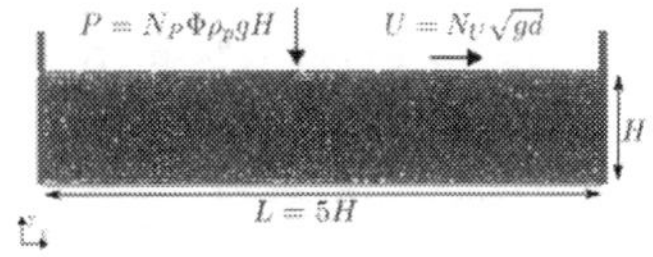

Figure 1. Geometry and boundary conditions for the granular lid-driven cavity simulated by DEM

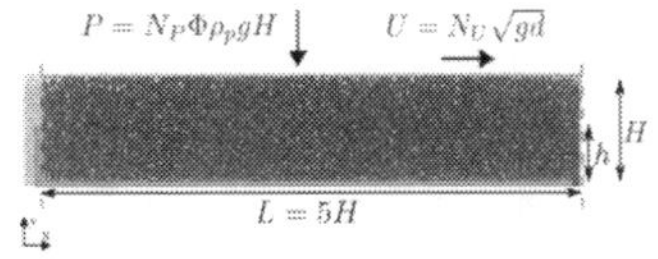

Figure 2. Geometry and boundary conditions for the immersed wall granular system simulated by DEM. Periodic conditions were used, thus explaining the presence of a wall on the left side (lighter-colored area), which is a duplicate of the wall (of interest in our study) on the right side

flow. As a result, the mean force on the wall was found to obey the empirical following relation:

$$\overline{F} = \left(r^* + \frac{\Delta r^*}{1 + I_M^{*0}/I_M^*}\right) Phd, \qquad (2)$$

where $r^* = 1.4$, $\Delta r^* = 1.9$ and $I_M^{*0} = 0.3$ in the present study (see detail in [7]). The values of r^* and Δr^* depend on the mechanical properties of grains. I_M^* is the macroscopic inertial number defined with the difference between the height of the system (the height of grains) and the immersed wall height: $I_M^* = \frac{U}{H-h} d\sqrt{\rho/P}$.

2.2 Distinct time-averaged dynamics

Because the two systems presented above are by construction different, they show two different time-averaged dynamics. First, a standing vortex-like structure is formed inside the cavity volume [5, 6], while longitudinal streamlines with a curvature in the vicinity of the wall take place in the immersed wall system [7]. Second, the mean force scalings allowed us to identify different relevant spatial scales specific to each system. The relevant pressure force is PLd for the cavity (see Eq. (1)), while it is Phd for the immersed wall system (see Eq. (2)). The relevant spatial scale for the macroscopic inertial number is the system height H for the cavity (see Eq. (1)), while it is the height difference $H - h$, namely the height of the granular layer overflowing the wall, for the second system (see Eq. (2)). Much more details about the specific features of the lid-driven cavity and the immersed wall system can be found in [5, 6] and [7], respectively.

However, we show in the following how these two distinct systems exhibit very similar—not to say, for some aspects, identical—key features in terms of the force distributions and the force time-series autocorrelations on the boundary wall, and we will identify a robust signature of the transition from quasistatic to dense inertial granular flow regime.

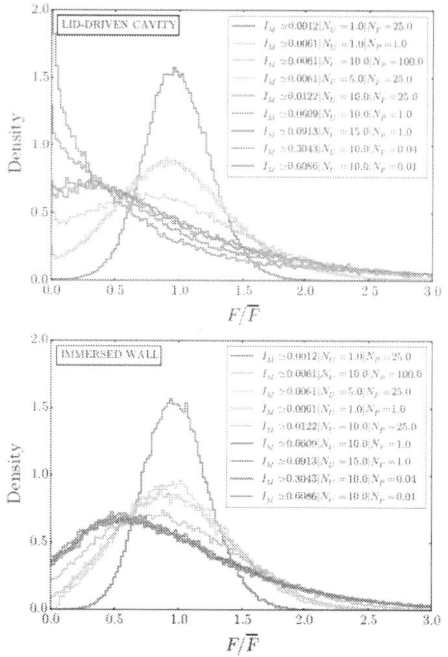

Figure 3. Distributions of total force experienced by the boundary wall for the lid-driven cavity (top panel) and the immersed wall system (bottom panel), for different values of I_M

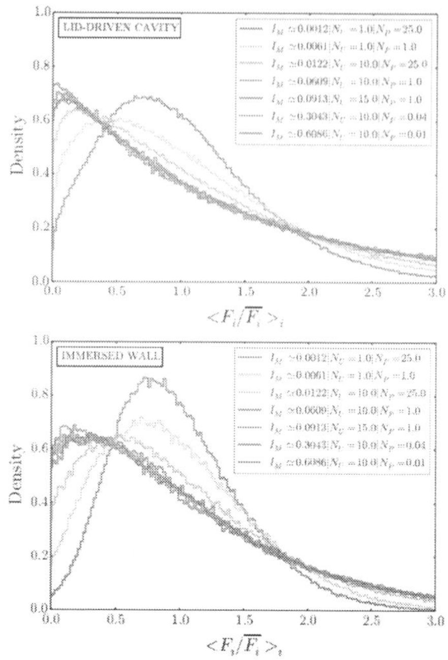

Figure 4. Distributions of the force experienced by a portion of the boundary wall for the lid-driven cavity (top panel, H split into 10 portions) and the immersed wall system (bottom panel, h split into 5 portions). The curves drawn in both graphs correspond to the mean value obtained from all the vertical locations i along the wall, excluding the extreme locations at the top and the foot of the boundary wall

3 Grain-wall force distributions

3.1 At the scale of the entire wall

Figure 3 shows the distributions of total force on the boundary wall for the lid-driven cavity (top panel) and the immersed wall system (bottom panel) for different values of I_M. First, it is worthwhile to note that the distributions are entirely controlled by the value of I_M, whatever the pair (U, P) (see the curves for $I_M = 6.1\ 10^{-3}$ obtained with three different pairs (U, P) for the two systems). Second, it is found that they generally evolve from an exponential-like form at high forces (much larger than the mean) with the presence of a plateau/peak at small forces for high I_M, toward a Gaussian-like form at the lowest I_M. We observe a difference between the cavity and the immersed wall at the highest I_M: the cavity exhibit purely exponential shapes without any peak, while the peak at small forces is still present in the immersed wall system.

3.2 At the scale of a portion of wall

For both systems and a given I_M, our simulations showed that the distribution of the force F_i on each portion i of the boundary wall was independent of the vertical location along the boundary wall, excluding the extreme locations at the top and the foot of the wall (not shown here: details can be found in [6] for the cavity and [7] for the immersed wall). As a result, we could average the distributions and extract robust mean values of distributions as functions of I_M, as depicted in Fig. 4. The top panel shows the results for the lid-driven cavity, while the bottom panel holds for the immersed wall system.

The force distributions shown in Fig. 4 collapse nicely into one single master curve, independent of I_M, once I_M becomes greater than 10^{-2}, whatever the granular system considered. This remarkable result, showing that there exists a threshold value of I_M which is independent of the system geometry, suggests that the distributions on portions of wall are robust to identify the transition between quasistatic and dense inertial granular flow regimes.

4 Force time-series autocorrelation

The autocorrelation function $C(t, t + \Delta t)$ of force time-series on the wall was analysed for the cavity and the immersed wall system. It was generally observed that $C(t, t + \Delta t)$ dropped quickly (exponentially) to zero over a time scale Δt_c and could fluctuate around zero (see details in [6, 7]). Note that very similar $C(t, t + \Delta t)$ functions were extracted from an experimental study on the force experienced by an intruder dragged into a granular medium in an annular geometry [8].

Well-defined fluctuations, associated with the transmission of the frequency U/d linked to the roughness of the wall at the top shearing the granular sample, could

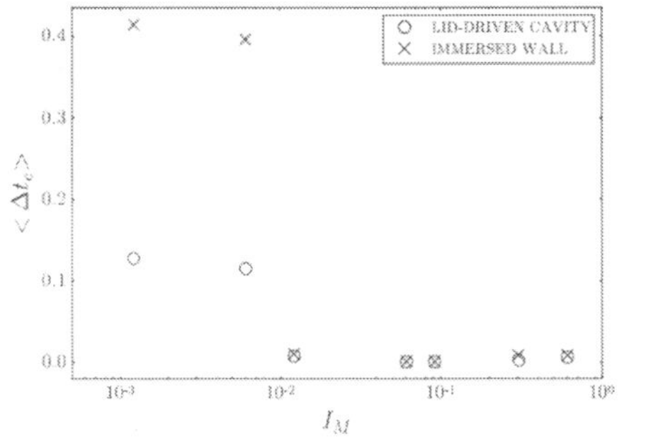

Figure 5. Critical time $\langle \Delta t_c \rangle$ beyond which the force time-series become weakly correlated, as a function of I_M. We report the mean over the $\Delta t_c{}^i$ values for $i \in [1; 10]$ in the case of the lid-driven cavity (cross symbols) and for $i \in [1; 5]$ in the case of the immersed wall system (circle symbols)

sometimes be identified in the case of the cavity, as detailed in [6]. By contrast, this effect was never detected for the immersed wall. This observation illustrates the fact that the cavity can produce high confinement pressure on the right top corner (see [5] for much more discussion).

We could not find any clear relation between the critical time Δt_c beyond which the force time-series became weakly correlated, and the inertial number I_M, considering the entire height of the boundary wall subjected to the granular flow. However, by considering the autocorrelation function for portions of wall (as done for the distributions in section 3), we found that each $\Delta t_c{}^i$ (the critical time corresponding to a vertical location i along the boundary wall) systematically had a non-zero value for I_M below 10^{-2}, while it was nil beyond $I_M = 10^{-2}$. This result being robust whatever the system considered, we defined the critical time $\langle \Delta t_c \rangle$ averaged over all the locations i considered along the boundary wall. The latter is plotted in Fig. 5 as a function of I_M for the two granular systems. It is remarkable that both curves display very similar behaviours.

The autocorrelation function $C(t, t + \Delta t)$ can be seen as a measure of the memory of the system through time Δt. High values of C indicate that a future state of the system (in terms of force applied to the wall) depends on the past state of this system, while it does not when C tends toward zero. As such, we can conclude from Fig. 5 that a future state of the system may strongly depend on the past state below $I_M \simeq 10^{-2}$ while it does not beyond this critical I_M. The latter result being robust whatever the granular system considered, this reveals another clear signature of the transition from the quasi-static regime (system with a memory) to the dense inertial regime (full loss of memory).

Note that the non-zero value of $\langle \Delta t_c \rangle$ below $I_M \simeq 10^{-2}$ is higher for the immersed wall than for the cavity. This difference may be explained by the fact that the grains against the boundary wall are moving along the cavity wall (perpetual recirculation), while they are not in the immersed wall system because they are trapped in the stagnant zone formed against the immersed wall. Therefore, memory effects may be more pronounced in the immersed wall system.

5 Conclusion

The present paper paid a particular attention to the force fluctuations experienced by the boundary wall, considering a wide range of macroscopic boundary conditions in terms of shearing velocity and confinement pressure imposed to each system. By comparing the results between the two distinct systems, we could extract some key features which appeared to be independent of the system at stake. In particular, our study showed that the distributions measured on portions of wall (of size around $3d$) moved from an exponential decrease at large forces with a shoulder at small forces, which were independent of I_M for I_M greater than 10^{-2}, toward I_M-dependent Gaussian-like distributions while I_M was decreased. Moreover, we could detect a system memory for the lowest I_M, which disappeared while increasing I_M beyond 10^{-2}. Those features were obtained from two distinct granular flow systems, characterized by a different time-averaged dynamics in terms of streamlines and mean force scalings on the boundary wall. Those results then appear as a robust signature of the transition between quasistatic and dense inertial granular flow regimes, revealing and quantifying the link between the $\mu(I)$–rheology proposed by [9] and the force fluctuations experienced by a boundary wall.

Acknowledgements

This work has been partially supported by the LabEx Tec21 (Investissements d'Avenir: grant agreement No. ANR11-LABX-0030). Thierry Faug and Mohamed Naaim acknowledge the financial support by the People Programme (Marie Curie Actions) of the EU 7th FP under REA Grant Agreement No. 622899 (FP7-PEOPLE-2013-IOF, GRAINPACT).

References

[1] S. J. Antony, Physical Review E **63**, 011302 (2000)
[2] F. Radjai, Comptes Rendus Physique **16**, 3 (2015)
[3] E. Gardel, E. Seitaridou, K. Facto, E. Keene, K. Hattam, N. Easwar, and N. Menon, Philosophical Transactoions of the Royal Society A **367**, 5109–5121 (2009)
[4] https://yade-dem.org/doc/
[5] F. Kneib, T. Faug, F. Dufour, and M. Naaim, Computational Particle Methods **3**, 293 (2016)
[6] F. Kneib, T. Faug, F. Dufour, and M. Naaim (submitted manuscript)
[7] F. Kneib, T. Faug, F. Dufour, and M. Naaim (manuscript in preparation)
[8] J. Geng and R. Behringer, Physical Review E **71**, 011302 (2005)
[9] P. Jop, O. Pouliquen, Y. Forterre, Nature **441** 727 (2006)

Mechanical behaviour of weak snow layers: modelling a porous structure of sintered grains

Tijan Mede[1,*], *Guillaume* Chambon[1], *Pascal* Hagenmuller[2], and *François* Nicot[1]

[1] *University Grenoble Alpes, Irstea, 2 Rue de la Papeterie, 38402, Saint-Martin-d'Hères, France*
[2] *Météo France – CNRS, CNRM, CEN,1441 Rue de la Piscine, 38400 Saint-Martin-d'Hères, France*

Abstract. Weak layers of snow are thin layers of low density and cohesion that consist of a complex network of sintered ice grains. These layers are very fragile and their collapse is considered to be the primary cause of dry slab snow avalanche release. This study reports on an attempt at modelling the mechanical response of these weak snow layers with discrete element (DEM) simulations, using x-ray tomographical images of real snow samples as input data on the microstructure of the material. An original method for representing irregular grain shapes in the DEM is first introduced. The method is based on utilizing the medial axis concept to represent an arbitrary grain shape with an optimal set of overlapping spheres. A thorough study of the effect of the grain approximating technique parameters on the key geometrical features of the grains is then carried out. Finally, the functionality of the model is demonstrated by performing an oedometric test on an image of a real snow sample.

1 Snow as a granular material

Weak snow layers are thin layers of low cohesion and density, naturally occurring in the snowpack. The understanding of their mechanical behaviour is considered to be of great importance as they are believed to have the decisive effect on a dry slab avalanche release. These layers can be viewed as porous structures of sintered ice grains, that collapse under loading and disintegrate into an cohesionless assembly of grains, withdrawing support to the overlaying slab layer of snow.

Field observations show that weak layers are usually formed by very distinct snow grain types [1]. While it is clear that the mechanical character of the weak layer is defined by its microstructure i.e. shape and geometrical arrangement of the snow grains, the precise relation between these properties has not yet been quantified. The extremely fragile nature of these layers makes systematic laboratory experiments notoriously difficult [2, 3].

We therefore propose studying the effect of the microstructure on the mechanical behaviour of snow by conducting numerical experiments on a numerical model of a weak layer. The model, naturally has to account for the actual microstructure of the weak layer and has to able to follow the material through different stages - from a porous network of sintered grains to the final stage of a decomposed granular material.

The discrete element method (DEM) is an efficient tool for exploring the mechanics of granular materials. Johnson and Hopkins [4] were the first to demonstrate the usefulness of DEM in studying the mechanics of snow. Their simulation was however conducted on a granular deposition of idealized grain shapes and did not account for the actual microstructure of snow.

Recently Hagenmuller [5–7] managed to combine DEM approach with x-ray tomography images of snow in order to come up with a method which could account for the microstructure of the material, while having the potential to follow the large deformations and rearrangements within the snow matrix. To this end the tomographical image of snow is divided into individual grains by detecting weak mechanical points i.e. neck regions in the snow matrix. These grains are modelled as the basic elements of the simulation, as separate unbreakable entities, which are sintered into the original matrix of intact snow. When applying mechanical loading to the snow structure, the sintered bonds break and the ice matrix gradually transforms into a cohesionless granular medium. The present study is based on this particular approach to modelling the weak layer of snow, where a novel type of grain geometry representation has been added to improve the computing efficiency of the model.

2 Approximation of grain geometry

Since the objective of the study is the quantification of the relation between microstructure and the overall mechanical response of the medium, the particular representation of the grains in the DEM simulation is of paramount importance. A grain approximating method is needed that

*e-mail: tijan.mede@irstea.fr

 DOI: 10.1051/epjconf/201714006008

allows on one hand an accurate representation of grain shapes, but on the other hand also allows a gradual simplification of grain shapes in order to study the importance of the accuracy of the grain representations in the simulation. All this needs to be accomplished at the minimum possible amount of utilized discrete elements in order save on the processing time, which is substantial in a simulation of this type.

Usually, the grains in DEM simulations are represented by simple elements such as spheres or ellipsoids and the irregularity of grain shapes is taken into account through parameters such as rolling friction, that simulate grain interlocking [8]. However, estimating the rolling friction solely from grain shapes remains very problematic [9]. To directly model the shape of grains, discrete elements can be modelled as polyhedrons - this however proves to be computationally very expensive due to complex contact detection and calculation of contact forces. A very efficient manner of representing arbitrary grain shapes is by clumping together a subset of simpler discrete elements – usually spheres – to approximate the overall grain shape. Hagenmuller et al. [7] used a very straight-forward approach and meshed the grains in the latter manner by simply arranging the spheres on an orthogonal grid. This approach resulted in a large number of discrete elements and very long computational times. In the present article, we present a sphere overlapping grain approximating method that allows us to substantially diminish the number elements needed for grain approximation, while keeping the key geometrical parameters of grains (Figure 1).
Although the overlapping sphere technique had already previously been used for grain approximation in several different cases [10, 11], our demands on diminishing the number of discrete elements as much a possible while still keeping control over morphological features of grains led us to devise a novel optimized approach. It is based on the concept of medial axis which has been utilized in order to compress the information, needed to reconstruct a binary image [12].

The grain approximating technique is applied by first calculating the medial axis of each grain in order to produce the skeleton of the sample, upon which the approximating spheres will be placed. Placing the center a sphere to every point of the medial axis at this point would result in a perfect reconstruction of the grain image, but would on the other hand contain redundant information due to the discrete nature of the image [12]. The medial axis is therefore filtered by removing the spheres which are completely covered by the union of all the other spheres. By starting this filtering process with the smallest spheres and continuing towards the largest sphere, one promotes the use of large spheres for grain reconstruction.

Approximation is carried out via two controlling parameters - the smallest sphere radius R_{min} limits the size of the smallest sphere used for grain approximation and disregards the smaller reconstructing spheres. The delete ratio D deletes all the medial axis points in the distance DR (where R is the radius of the sphere) around the center of an approximated sphere. The spheres are created in the order from the largest to the smallest.

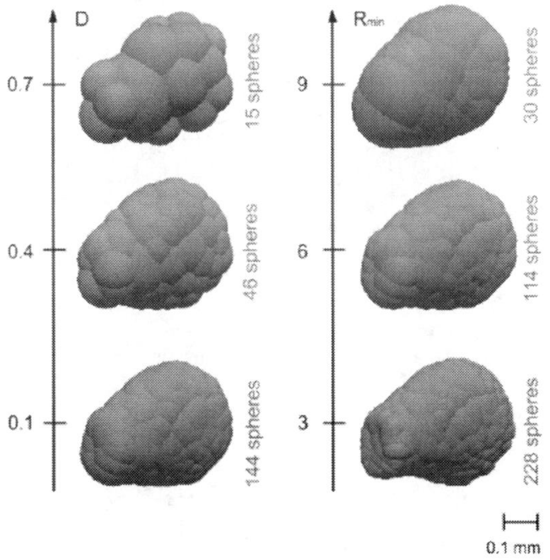

Figure 1. The effect of the two grain approximating parameters. The left column demonstrates the effect of increasing the parameter D, while the right column demonstrates the effect of increasing the parameter R_{min} in terms of approximating grain geometry as well as the number of spheres used for approximation.

Effectively, the parameter D sets the density of the spheres that form the grain surface - the larger the value D, the larger the artificial grain surface roughness, that emerges as a consequence of approximating an originally smooth surface with a set of spheres as shown in the left column of Figure 1. On the other hand, the parameter R_{min} rounds off the sharp edges of the grain and disregards the details on grain surface with a characteristic size smaller that R_{min} as shown in the right column of Figure 1. Hence, one can view the D parameter as the parameter that controls the artificial grain roughness and R_{min} as the parameter of grain detail resolution.

3 Sensitivity analysis of geometrical features to approximation algorithm parameters

An extensive sensitivity analysis was performed in order to assess the influence of the two grain approximating parameters (D and R_{min}) on the key geometrical properties of the grains: the volume, the anisotropy and the surface roughness of the grains.

In the present paper, only the study of the volumetric parameter will be presented in details. One is interested in the volumetric error, defined as the volumetric difference between the original and the approximated grains, divided

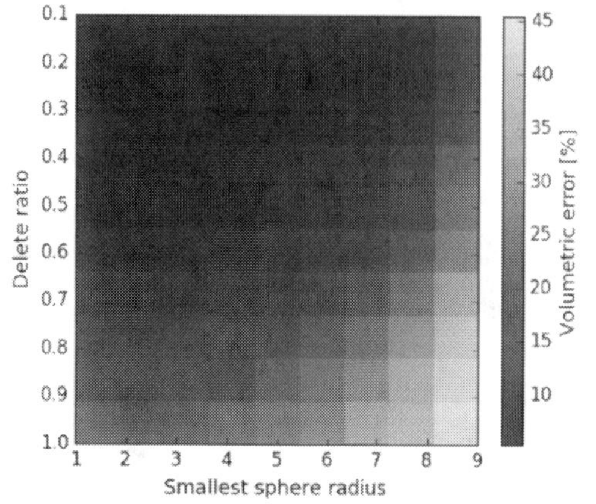

Figure 2. The effect of the two grain approximating parameters on the volumetric error of approximating the given set of grains.

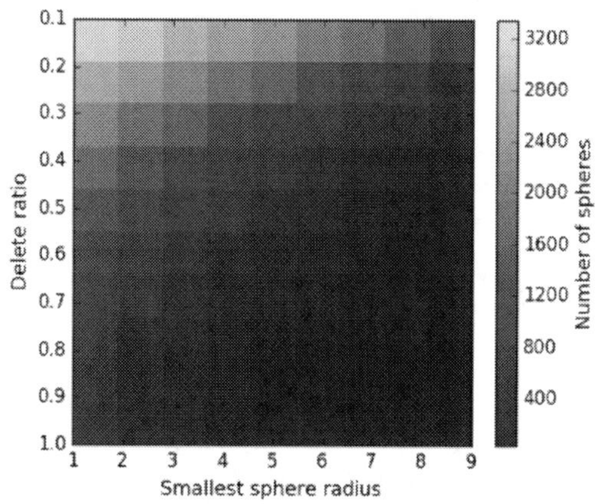

Figure 3. The effect of the two grain approximating parameters on the number of spheres, needed to approximate the given set of grains.

by the total volume of the original grains. The effect of the two grain approximating parameters on the volumetric error of a subsample of 10 grains is presented in Figure 2, whereas the resulting number of spheres is presented in Figure 3.

As one might expect, decreasing the parameters D and R_{min} will increase the number of spheres, needed for grain approximation (Figure 3), due to a rising density of spheres along the grain surface as well as adding increasingly smaller spheres to represent smaller details on the grain surface. On the other hand, this will also diminish the volumetric error of the approximation (Figure 2). Increasing the parameters D and R_{min} will inevitably produce the opposite effect - diminishing the number of spheres and increasing the volumetric error.

4 Mechanical behaviour

YADE DEM tool [13] is used to perform the numerical simulation of a classical oedometric compression test on a cubic sample of roundgrained snow with a sidelength of 4,9 mm and initial void ratio of 2.79 (left side of the Figure 4) in order to demonstrate the functioning of the devised micromechanical model.
The specimen is approximated with the choice of parameters $D = 0.7$ and $R_{min} = 6$, resulting in 19238 discrete elements needed to represent the specimen consisting of 1393 grains. It is confined by a set of rigid walls and compressed with a constant velocity of $10^{-2}ms^{-1}$. A cohesive elastic contact law with the brittle regime of cohesion breaking has been used.

Cohesion along the contact surface between the grains is imposed by fist estimating, from the original image, the area of the bond between each pair of grains in contact. Each pixel on the contact surface of a grain is assigned the nearest approximating sphere. Afterwards for each pair of pixels in contact, a cohesive bond is created between the two pertaining approximating spheres. If more than one pair of pixels in contact points to the same pair of approximating spheres, the magnitude of the cohesive bond is multiplied accordingly.

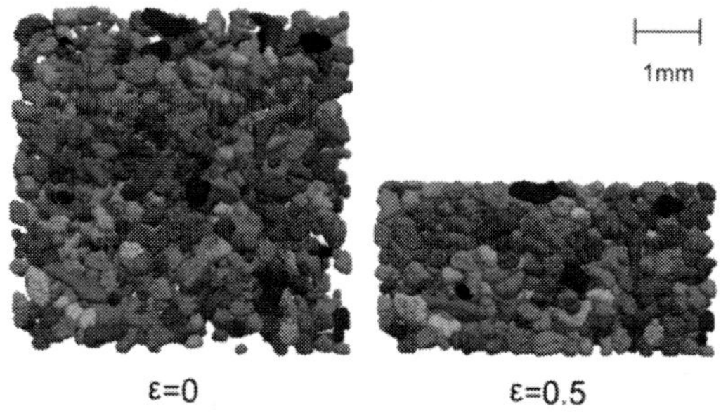

Figure 4. Images of the snow sample. The left image shows the sample prior to mechanical loading ($\varepsilon = 0$) while the one on the right shows the sample at the strain of $\varepsilon = 0.5$.

The stress-strain curve of the oedometric compression simulation on the snow sample is shown in the Figure 5, together with the corresponding evolution of cohesive grain bond braking (plotted as the ratio between the number of broken bonds and the number of cohesive bonds in the intact sample). A zoom into the primary phase of the curves is provided in the upper left corner of the plot. One can note that the specimen undergoes a short elastic phase in the beginning of the simulation, followed by a peak ($\varepsilon \approx 0.001$) and subsequent softening $0.001 < \varepsilon < 0.01$. The amount of broken cohesive bonds remains very low in the elastic phase, rises rapidly at the peak stress, and continues to grow during the softening phase. Further advancement of the mechanical loading results in a multitude of smaller peaks marking the gradual breakage of cohesion, allowing for grain rearrangement and compaction.

The diminishing void ratio results in a gradually increasing resistance to compaction for strains $\varepsilon > 0.3$. The specimen in the final stages of compaction is shown in the right side of Figure 4.

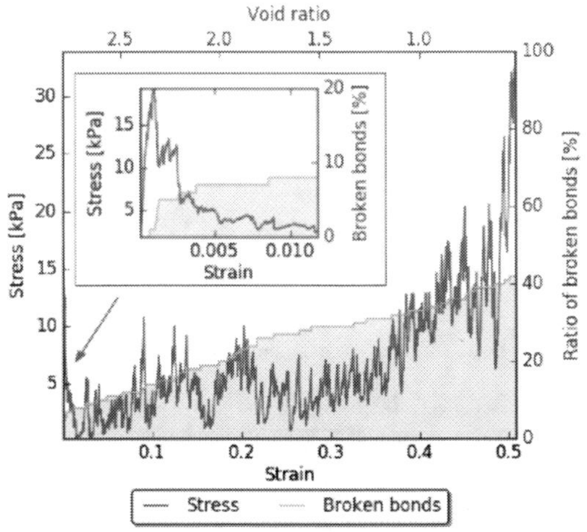

Figure 5. The stress-strain curve of the oedometric simulation on the given snow sample (blue curve) and the evolution of the ratio of broken cohesive bonds between the grains (yellow curve). The corresponding void ratio is marked at the top of the plot. An additional zoom into the primary phase of the simulation is added in the top left corner of the plot.

5 Conclusion

In the present paper we show development of a micromechanical model, capable of of predicting the mechanical response of a sintered porous material under large-strain loading. The development is based on a pre-existing model, where a novel method of irregular grain representation is added, which results in a substantial diminishment of the number of discrete elements needed for the simulation. An original approach to modelling the cohesion between the grains is also utilized that insures a homogenity of the cohesion along the contact surface of the grains. The parameters of the grain approximation are chosen by performing a thorough sensitivity analysis and studying the effect on the key geometrical features of the approximated grains - the volumetric error of the approximation, the average ratio of the anisotropy of the approximated grains with respect to the anisotropy of the grain images and the amount of the artificial roughness of the approximated grains due to sphere representation of grain surfaces.

The preliminary mechanical evaluation of the model has been conducted in order to demonstrate the functioning of the model on a case of a simple oedometric compression of a snow sample. The results, although preliminary, exhibit all the crucial phases that are to be expected of performing compression on a sintered porous structure: a short elastic phase, followed by a multitude of peaks, marking the gradual breakage of cohesion between the grains. The dimininishig void ratio eventually leads to a rise in mean stresses as the sample gets increasingly compacted.

In our future work, we intend to use this model, in order to study the effect of the microstructure of a weak layer in snow on its mechanical behaviour. Systematic numerical experiments will be conducted on a number of different weak layer sample images in order to determine the dependency between the morphological properties and the peak force and subsequent softening of the weak layer.

References

[1] J. Schweizer, J. Jamieson, M. Schneebeli, Reviews of Geophysics **41**, 2 (2003)

[2] I. Reiweger, J. Schweizer, Geophysical Research Letters **37** (2010)

[3] I. Reiweger, J. Schweizer, Cryosphere **7**, 1447 (2013)

[4] J. J.B., H. M.A., Journal of Glaciology **51**, 432 (2005)

[5] P. Hagenmuller, G. Chambon, F. Flin, S. Morin, M. Naaim, Granular Matter **16**, 421 (2014)

[6] P. Hagenmuller, N. Calonne, G. Chambon, F. Flin, C. Geindreau, M. Naaim, Cold Regions Science and Technology **108**, 72 (2014)

[7] P. Hagenmuller, G. Chambon, M. Naaim, Cryosphere **9**, 1969 (2015)

[8] J. Ai, J.F. Chen, J.M. Rotter, J.Y. Ooi, Powder Technology **206**, 269 (2011)

[9] C. Wensrich, A. Katterfeld, Powder Technology **217**, 409 (2012)

[10] E. Parteli, *DEM simulation of particles of complex shapes using the multisphere method: Application for additive manufacturing* (2013), Vol. 1542, pp. 185–188

[11] J.F. Ferellec, G. McDowell, Granular Matter **12**, 459 (2010)

[12] D. Coeurjolly, J. Hulin, I. Sivignon, Theoretical Computer Science **406**, 72 (2008)

[13] V. Šmilauer et al., *Yade Documentation 2nd ed* (The Yade Project, 2015), http://yade-dem.org/doc/

A micromechanical approach of suffusion based on a length scale analysis of the grain detachment and grain transport processes

Antoine Wautier[1,2,3,*], *Stéphane* Bonelli[2], and *François* Nicot[3]

[1] *AgroParisTech-ENGREF, 19 avenue du Maine, 75732 Paris, France*
[2] *Irstea UR RECOVER, 3275 Rte Cézanne, CS 40061, 13182 Aix-en-Provence Cedex 5, France*
[3] *Université Grenoble Alpes, Irstea, UR ETGR, 2 rue de la Papeterie-BP 76, F-38402 St-Martin-d'Hères, France*

Abstract. Suffusion is the selective erosion of the finest particles of a soil subjected to an internal flow. Among the four types of internal erosion and piping identified today, suffusion is the least understood. Indeed, there is a lack of micromechanical approaches for identifying the critical microstructural parameters responsible for this process. Based on a discrete element modeling of non cohesive granular assemblies, specific micromechanical tools are developed in a unified framework to account for the two first steps of suffusion, namely the grain detachment and the grain transport processes. Thanks to the use of an enhanced force chain definition and autocorrelation functions the typical lengths scales associated with grain detachment are characterized. From the definition of transport paths based on a graph description of the pore space the typical lengths scales associated with grain transport are recovered. For a uniform grain size distribution, a separation of scales between these two processes exists for the finest particles of a soil

1 Introduction

Under various circumstances, fully saturated granular materials are subjected to internal flows which may modify their microstructure and consequently their overall hydraulic and mechanical properties. At the microscale, this process consists in a rearrangement of particles driven by three elementary mechanisms, namely the detachment of grains from the granular skeleton, their transport through the pore network and possibly their reattachment to the granular skeleton farther away. At a larger scale, this process results in the selective erosion of the smallest particles of a soil and is referred to as suffusion within the geomechanics community [1]. Despite many attempts to provide criteria to assess the suffusion susceptibility of a soil, the onset of the three cited elementary mechanism is still poorly understood.

With the recent development of fluid/grain coupling schemes and thanks to ever increasing computation power, it is now possible to address this issue from a micromechanical point of view. However, fully coupled three dimensional simulations of the suffusion of granular materials are still very demanding in terms of computation time and the number of particles which can be simulated in practice is limited.

In order to assess the ability of numerical simulations to be representative of the phenomenon at the microscale, this paper investigates the possibility of providing a quantitative definition of the length scales associated with the grain detachment and the grain transport mechanisms. First, the grain detachment mechanism is investigated thanks to the distinction between chained and non-chained particles [2] as the detachable particles should not be heavily stressed. Thanks to the computation of autocorrelation functions [3] a length scale associated with this mechanism is defined. Then, the grain transport mechanism is addressed with the description of the pore space in terms of a graph composed of pores and constrictions [4]. A propagation criterion is then introduced in order to define a length scale associated with this mechanism.

For the sake of illustration, these micromechanical tools are tested on a numerical sample composed of 10 000 spheres with radii uniformly distributed between $r_{\min} = 3.6 \times 10^{-5}$ m and $r_{\max} = 10\, r_{\min}$. A Discrete Element Method (DEM) is used in order to simulate the mechanical behavior of the individual grains and the simple elasto-frictional contact law introduced in [5]. This sample is first prepared in a dense configuration with a void index of 0.6 thanks to the radius expansion technique and is then subjected to a drained triaxial test up to 20 % of axial strain (denoted as ε_{zz}) and under a confining pressure of 100 kPa. This procedure enable the definition of a collection of microstructure configurations presenting the same global grading curve but exhibiting different grain arrangements subjected to various loading states.

2 Mesoscale analysis of the grain detachment mechanism

The microscale analysis of the grain detachment mechanism is underpinned by two governing ideas. First, the

*e-mail: antoine.wautier@irstea.fr

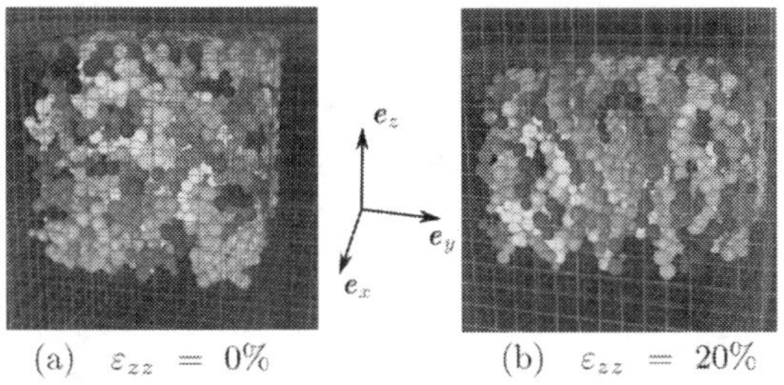

Figure 1. Force chains visualization under an isotropic confining pressure of 100 kPa (a) and at the end of the triaxial test (b).

most detachable grains should also be the least stressed in comparison with the local additional forces applied by the fluid. Then, the mesoscale associated with the grain detachment process should be linked to the inter-distance between the heavily stressed particles of the grain assembly. As a result, the solid fraction of a granular media can be divided into two groups of particles based on the force chain definition introduced in [2] according to the following three characteristics:

- Chained particles have a higher principal stress than the mean particle principal stress;
- Their principal stress direction is aligned with the geometrical contact direction (less than 45° deviation);
- Force chains are composed of at least three contacting particles.

A typical visualization of the force chains is visible in Figure 1 for the dense sample introduced in the Sect. 1 under an isotropic confining pressure of 100 kPa (Figure 1.a) and in the critical state at the end of the drained triaxial test (Figure 1.b). At the beginning of the triaxial test, the force chains are distributed in an isotropic way which is consistent with the fact that no principal direction of loading exists. Once the deviatoric loading is applied, the force chains tend to align in the vertical direction of the macroscopic principal stress. It should be also notice here that few particles belongs to the force chains as only 25 to 30 % of the total number of the particles composing the whole sample are visible in Figure 1. As the force chains align in the vertical direction to resist the deviatoric loading the number of chained particles also decreases by a sixth. The dual comment is that in the critical state, groups of loose particles bounded by force chains are larger than at the beginning of the triaxial test and that grain detachment is more likely to be observed in the critical state than in the initial state.

From the partition of the solid fraction of our sample between chained and non-chained grains a mesovolume is conceptually introduced to account for the grain detachment as a small box bounded by force chains and filled with non-chained grains. An estimate of the three dimensions of this mesovolume is then defined thanks to two-point spatial autocorrelation functions [3].

If Ω denotes the domain occupied by the chained particles of a given sample of volume V, the spatial autocorrelation

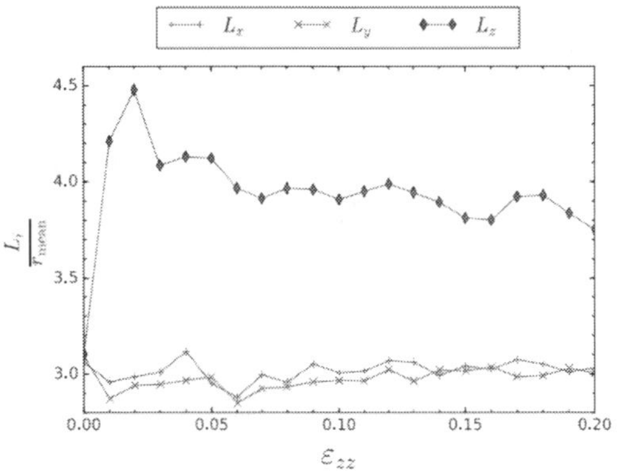

Figure 2. Strain evolution of the three autocorrelation lengths L_i ($i \in \{x, y, z\}$) during a drained triaxial test. The values are normalized by the grain mean radius r_{mean}.

function C is defined for any vector $\boldsymbol{h} = (h_x, h_y, h_z)$ as the joint probability that a point $\boldsymbol{x}$ and the translated point $\boldsymbol{x} + \boldsymbol{h}$ simultaneously belong to chained particles:

$$C : \begin{cases} \mathbb{R}^3 & \mapsto \mathbb{R} \\ \boldsymbol{h} & \mapsto \mathbb{P}\{\boldsymbol{x} \in \Omega \cap \boldsymbol{x} + \boldsymbol{h} \in \Omega, \forall \boldsymbol{x} \in V\} \end{cases} . \quad (1)$$

A dimensionless form of this function can be defined as $\tilde{C}(\boldsymbol{h}) = \dfrac{C(\boldsymbol{h}) - C(0)^2}{C(0) - C(0)^2}$. In this form, $\tilde{C}$ decreases from 1 to 0 as $\|\boldsymbol{h}\|$ goes from 0 towards $+\infty$, and the rate of decrease provide a typical distance over which the microstructure considered is correlated. Based on the integral range theory [3], the volume of the elementary box linked to the grain detachment mechanism is then defined by integrating $\tilde{C}$ over $\mathbb{R}^3$. With the use of an exponential fit as

$$\tilde{C}_{\text{fit}}(\boldsymbol{h}) = \exp\left(-2\pi^{1/3}\sqrt{\left(\frac{h_x}{L_x}\right)^2 + \left(\frac{h_y}{L_y}\right)^2 + \left(\frac{h_z}{L_z}\right)^2}\right) \quad (2)$$

the dimensions of the mesovolume associated with grain detachment are simply characterized by the three introduced length scales L_x, L_y and L_z.

In Figure 2, the strain evolution of these three autocorrelation lengths is plotted with respect to the axial strain ε_{zz} applied during the triaxial test introduced in Section 1.

As the horizontal autocorrelation lengths remains more or less constant around three times the grain mean radius r_{mean}, the vertical autocorrelation length increases up to a peak value of $4.5\, r_{\text{mean}}$ before decreasing and stabilizing around $4\, r_{\text{mean}}$ for $\varepsilon_{zz} > 15\%$. It is worth underlining here that these observations are closely related to the macroscopic mechanical response of the considered sample (not shown here) and is typical of a dense granular assembly. Indeed, the increase in the vertical autocorrelation corresponds to a lengthening of the force chains along this direction which accounts for the initial hardening of our dense sample. The following decrease in the vertical autocorrelation accounts for the destruction of force chains and thus for the stress softening regime observed classically for dense granular materials at large strain levels during triaxial tests.

 DOI: 10.1051/epjconf/201714009022

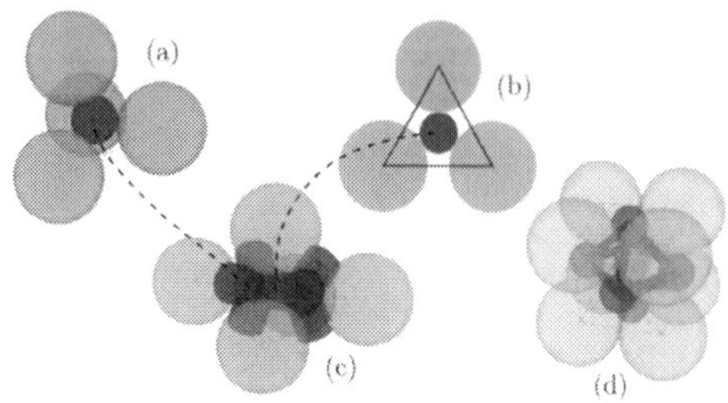

Figure 3. Visualization of different pore networks for an assembly of 5 grains (b) and 9 grains (d). Pores are represented as inscribed spheres (a) and constrictions as inscribed cylinders between pores (c).

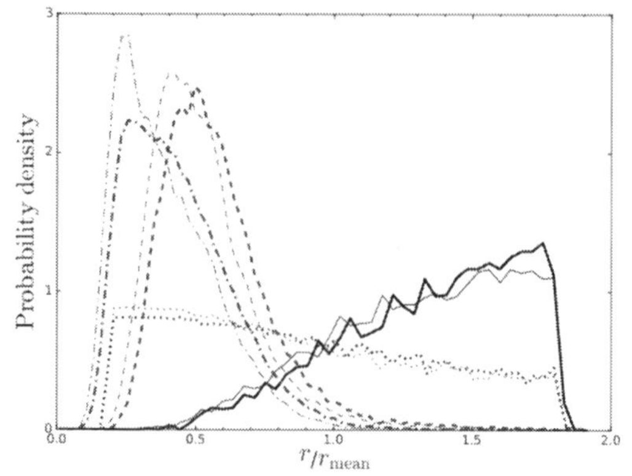

Figure 4. Probability densities for a constriction (dash-dotted line), a pore (dashed line), a chained grain (solid line) or a non-chained grain (dotted line) to be of a given radius. The thick line corresponds to $\varepsilon_{zz} = 20\%$, while the thin line to $\varepsilon_{zz} = 0\%$.

3 Mesoscale analysis of the grain transport mechanism

If the grain detachment mechanism is closely linked to the stress transmission through the solid phase of granular materials, the grain transport mechanism is fundamentally linked to the properties of the dual phase, namely the pore space. In the rest of this study, potential electrostatic interactions between particles are disregarded and it is assumed that the ability of a grain to be transported is purely based on geometrical considerations. As a result, the complex description of the pore space can be reduced to the definition of a graph composed of pores (the nodes of the graph) and constrictions (the edges of the graph). The definition of this graph is derived from a tessellation of the granular assembly as in [4] and [6], and Figure 3 provides an illustration of this type of graph for simple grains assemblies.

The pores are defined following the level 0 analysis introduced in [4]. Their positions are defined at the center of each tetrahedron of a regular Delaunay triangulation and their radii as the radius of the largest interior sphere in the associated tetrahedra (Figure 3.a). The constrictions are modeled as cylinders joining two adjacent pores and their radii are defined as the radius of the largest interior circle on the common face of the two tetrahedra defining the constriction (Figure 3.b).

In order to assess the possibility for the grains composing a granular material to be transported, it is interesting to compare the grain radius distribution to the pore and constriction radius distributions characterizing the void space. In Figure 4 the probability densities for a fictive particle of a given radius to find a pore or a constriction large enough to fit in are plotted together with the probability densities corresponding to the radius distribution of the chained and non-chained grains identified in the previous section. These probability densities are given at the beginning and at the end of the triaxial compression introduced in Section 1.

The comparison between the constriction probability density and the non-chained grain probability density in Figure 4 highlights that a large number of non-chained grains are small enough to be transported through the pore network as 33 to 34 % of these grains have a radius smaller than the mean constriction radius plus one standard deviation. On the contrary, very few chained grains are small enough to fit in the pores and constrictions.

While observing the strain evolution of the constriction probability density, the contractive/dilative behavior typical of dense granular materials is recovered. Indeed, the probability densities are shifted toward larger radius values, which is related to an overall increase of the pore space volume.

Even though many particles are identified in Figure 4 as potentially transportable by analyzing the statistical properties of the pore network, the definition of a mesoscale associated with the grain transport mechanism requires knowing the spatial distribution of pores and constrictions together with the fluid flow direction. If the definition of the pore network proposed in this section provide the spatial distribution of the pores and constrictions under the form of a non-oriented mathematical graph, there is a need to define a propagation criterion on this graph to incorporate the forcing imposed by the internal flow with respect to grain transport. For a given flow direction (taken along the direction $\boldsymbol{e}_x$ in Figure 1 for the rest of this study), the definition of such propagation criterion is achieved thanks to the combined used of the DEM-PFV model implemented in YADE software [7]:

- First, the DEM-PFV scheme is used to generate a pressure map in the pore network under a fluid flow imposed by a small pressure drop between the sample boundaries in the x direction;

- Then, based on the Hagen-Poiseuille flow velocity profile for a circular tube and the expression of the drag force of a uniform flow acting on a sphere for low Reynolds numbers, a flow intensity indicator is defined in each constriction as $\frac{\Delta p}{\ell} \times R^2$ with ℓ being the length of the constriction, Δp the pressure drop between the two adjoining pores and R the constriction radius;

 DOI: 10.1051/epjconf/201714009022

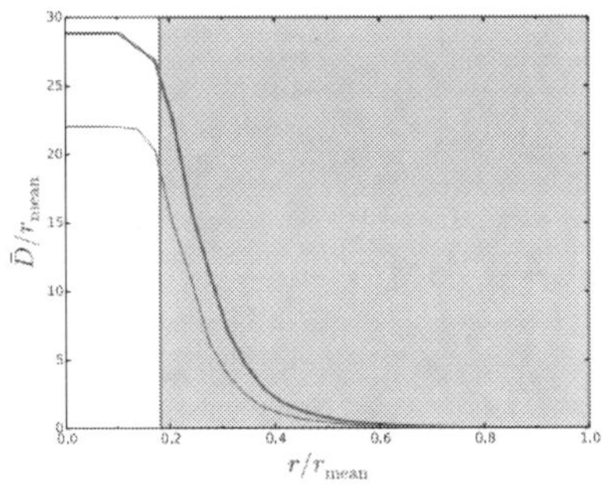

Figure 5. Mean travel distance for different radius thresholds. The thick line corresponds to $\varepsilon_{zz} = 20\%$, while the thin line to $\varepsilon_{zz} = 0\%$. The grey domain in the background corresponds to the radius values for the grains composing the sample.

- Finally, a transport path is defined for a given particle of radius r and for a given starting pore as the path of maximum flow intensity on the restriction of the pore network to pores and constrictions of radii larger than r.

For a given radius r, a mean travel distance $\bar{D}(r)$ is then computed from statistical averaging of the length of all the existing transport path. This quantity stands for a measure of the typical length associated with grain transport mechanism. In Figure 5, $\bar{D}(r)$ is plotted for the two same sample configurations used in Figure 4.

For these two configurations, the mean travel distance decreases from $22 - 29\ r_{mean}$ to zero. If no transport is possible for particles larger than roughly $0.5\ r_{mean}$ the smallest particles of the considered sample have good chances to be transported over large distances going up to $25\ r_{mean}$ representing roughly half of the total length of the sample. The comparison between these typical length scales associated with grain transport with the ones associated with grain detachment obtained in the previous section highlights a separation of scale between the two mechanisms provided that the detached grains are found among the smallest particles of the sample considered.

4 Conclusion and outlooks

In this paper, specific micromechanical tools are developed to investigate the susceptibility of a polydisperse assembly of spherical particles to grain detachment and grain transport. By simultaneously considering the size distributions of the non-chained particles and the constrictions of the pore network, the ability of a fluid flow to modify the microstructure of the samples is analyzed with respect to the fraction of both potentially detachable and transportable particles which represents approximately 25% of the non-chained ones. During a drained triaxial test, it is found that the microstructure of the dense sample considered evolves towards a configuration more favorable to observe the detachment and the transport of grains within the sample volume. Indeed, as the axial stress increases during the triaxial test, the number of chained particles tends to decrease which results in larger groups of potentially detachable particles. In the meantime, as our sample exhibit a dilative behavior, the pores and constrictions get larger which results in an increase of the travel distances accessible to detached particles. As a result, for a same grading curve, our micromechanical analysis predicts a higher suffusion susceptibility of our sample at the critical state than in the initial dense configuration at the beginning of the triaxial testing.

By carefully considering the spatial distribution of particles participating in stress transmission and the spatial distribution of constrictions enabling particle transport within the pore space, two mesoscales are introduced to study the grain detachment and the grain transport processes from a micromechanical point of view. It was shown that the typical length scale associated with grain transport can be up to ten times larger than that associated with grain detachment. As a result, a scale separation should exist between these two processes provided that the detached grains are found among the smallest particles of the sample considered.

Eventually, as the order of magnitude of the introduced length scales is still relatively small compare to the total length of the considered sample ($48r_{mean}$), full 3D modeling of the suffusion process is accessible and can be used to confirm this scale separation as well as to explore the local conditions leading to the occurrence of grain detachment and transport. This would pave the way for the definition of a homogenized law for suffusion thanks to the use of representative elementary volumes sufficiently big with respect to these two elementary mechanisms.

References

[1] S. Bonelli, *Erosion of geomaterials* (John Wiley & Sons, 2012)

[2] J.F. Peters, M. Muthuswamy, J. Wibowo, A. Tordesillas, Physical review E **72**, 041307 (2005)

[3] T. Kanit, S. Forest, I. Galliet, V. Mounoury, D. Jeulin, International Journal of solids and structures **40**, 3647 (2003)

[4] N. Reboul, E. Vincens, B. Cambou, Granular Matter **10**, 457 (2008)

[5] P.A. Cundall, O.D. Strack, Geotechnique **29**, 47 (1979)

[6] E. Vincens, K.J. Witt, U. Homberg, Acta Geotechnica **10**, 291 (2015)

[7] B. Chareyre, A. Cortis, E. Catalano, E. Barthélemy, Transport in porous media **94**, 595 (2012)

DOI: 10.1051/epjconf/201714002006

Force measurements in stiff, 3D, opaque granular materials

Ryan C. Hurley[1,★], *Stephen A.* Hall[2], *José E.* Andrade[3], and *Jonathan* Wright[4]

[1] *Lawrence Livermore National Laboratory, Livermore, CA 94550, USA*
[2] *Division of Solid Mechanics, Lund University, Lund 22100, Sweden*
[3] *California Institute of Technology, Pasadena, CA 91125, USA*
[4] *European Synchrotron Radiation Facility, Grenoble 38000, France*

Abstract. We present results from two experiments that provide the first quantification of inter-particle force networks in stiff, 3D, opaque granular materials. Force vectors between all grains were determined using a mathematical optimization technique that seeks to satisfy grain equilibrium and strain measurements. Quantities needed in the optimization – the spatial location of the inter-particle contact network and tensor grain strains – were found using 3D X-ray diffraction and X-ray computed tomography. The statistics of the force networks are consistent with those found in past simulations and 2D experiments. In particular, we observe an exponential decay of normal forces above the mean and a partition of forces into strong and weak networks. In the first experiment, involving 77 single-crystal quartz grains, we also report on the temporal correlation of the force network across two sequential load cycles. In the second experiment, involving 1099 single-crystal ruby grains, we characterize force network statistics at low levels of compression.

1 Introduction

The stress, stability, and transport properties of granular materials are controlled by the network of contact forces between grains (e.g., [1, 2]). Researchers have studied these networks for decades in an effort to build multiscale models for granular media. However, experimental capabilities for measuring forces in stiff, 3D, frictional materials remain limited. Such data are required to validate models and provide access to broader ranges of materials and environments than are currently available in computations.

Photoelastic measurements on 2D disk assemblies provided the earliest visualizations of force networks in granular solids and continue to furnish new insight into the relationship between these networks and mechanical properties [3]. Measurements in 3D systems include those using microscopy with emulsion droplets [4], X-ray computed tomography (XRCT) with rubber spheres [5], and laser scanning microtomography with hydrogels [6]. To move beyond the spherical, frictionless, soft 3D grains explored in past experiments, XRCT and 3D X-ray diffraction (3DXRD) have recently been combined to explore the contact network and grain strains in stiff, 3D, frictional grains [7, 8]. Recent work has also enabled the calculation of the full force network in these materials from the XRCT and 3DXRD data [9].

In this paper, we discuss the force measurement capabilities originally described in [9] in the context of two experiments. Both experiments used XRCT and 3DXRD to furnish the contact network and tensor grain strains needed to infer inter-particle forces in a compressed granular material. In the first experiment, 77 single-crystal quartz grains were subjected to two compression cycles. We discuss force network statistics and correlation across load cycles. In the second experiment, 1099 ruby grains were subjected to three compression cycles. We discuss preliminary force network inference for this experiment.

2 Experiments

Both experiments described in this paper were performed at beam line ID11 of the European Synchrotron Radiation Facility (ESRF). The experiments used a custom-built loading device [9] to compress granular samples in the beam line, as shown in Fig. 1a. In the first experiment, described in more detailed in [9], 77 nearly spherical single-crystal quartz grains with diameters ranging from 220 to 310 μm were compressed through two load cycles. In the second experiment, 1099 nearly spherical single-crystal ruby grains with diameters ranging from 100 to 120 μm were compressed through four load cycles. In each experiment, at load increments of 5 to 30 N, strain was held constant while the granular sample was fully illuminated by a 1.5 mm wide by 1.5 mm tall incident X-ray beam and rotated, in small increments, over 180° for 3DXRD and transmission radiography scans. 3DXRD data was analyzed using techniques from [10] to determine average tensor grain strains, $\bar{\epsilon}_\alpha$, for each grain, α, at each load increment. Average grain stresses, $\bar{\sigma}_\alpha$, were computed using the elastic stiffness constants, C_{ij}, of single-crystal quartz from [11] and single-crystal ruby from [12].

★e-mail: hurley10@llnl.gov

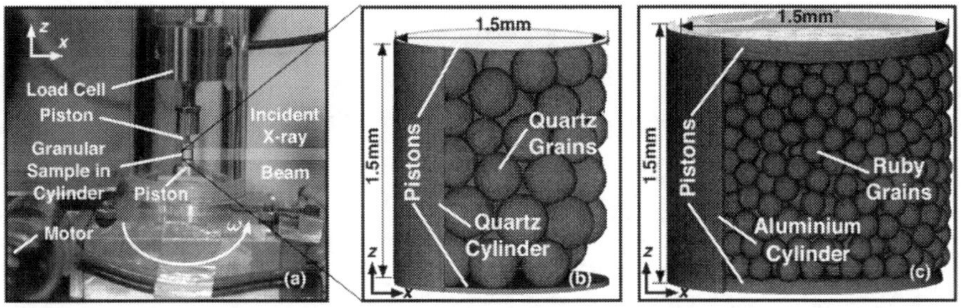

Figure 1. (a) Custom-built loading device positioned in beam line ID11 at ESRF, showing orientation of incident X-ray beam, rotation direction, and laboratory coordinate frame. (b) XRCT image of 77 single-crystal quartz grains in an amorphous quartz cylinder (also reported in [9]). (c) XRCT image of 1099 single-crystal ruby grains in an aluminium cylinder.

Transmission radiography images were used to construct 3D XRCT images of the first and second experiments using the ASTRA [13] and PyHST2 [14] toolboxes, respectively, with a resolution of approximately 1.5 μm. These images were processed using algorithms such as binarization and topological watershed in Matlab to obtain Figs. 1b and 1c, showing segmented grains. From these images, grain volumes and centroids were determined. Contact locations were determined by first finding all pixels whose 26-pixel neighborhood contained a pixel belonging to distinct grains (or a grain and a boundary), before thresholding to retain only pixel clusters with volume greater than 0.003% of the smallest grain in the assembly. Results were not sensitive to this threshold as long as it was small.

3 Force inference

Force vectors were inferred at every inter-particle contact point, at each load step, in each experiment using the measurements from 3DXRD and XRCT as inputs into the multi-objective optimization procedure from [9]. This optimization procedure seeks to satisfy grain equilibrium and stress measurements by solving

$$\begin{aligned} \boldsymbol{f} &= \arg\min(|\boldsymbol{K}_s\boldsymbol{f} - \boldsymbol{b}_s|_2 + \lambda|\boldsymbol{K}_e\boldsymbol{f}|_2) \\ &\quad \text{subject to: } \boldsymbol{f}_n \geq 0;\ \mu\boldsymbol{f}_n \geq \boldsymbol{f}_t \end{aligned} \tag{1}$$

where $\boldsymbol{f}$ contains all force vectors, $\boldsymbol{K}_s$ and $\boldsymbol{K}_e$ contain contact point locations, μ is the friction coefficient (0.7 for quartz [15], 0.1 for ruby-ruby and 0.4 for ruby-aluminum [12]), $\boldsymbol{b}_s$ contains volume-averaged grain stress tensor components multiplied by grain volumes, $\boldsymbol{f}_n$ contains all normal force magnitudes, $\boldsymbol{f}_t$ contains all shear force magnitudes, λ is a tradeoff parameter, and $|\cdot|$ represents the 2-norm. The first 2-norm contains all equations relating volume-averaged stress to boundary forces on each grain, or $V_\alpha\bar{\sigma}_{ij}^{(\alpha)} = \sum_{\beta=1}^{N_c^\alpha} x_i^{(\beta)} \otimes f_j^{(\beta)}$, where N_c^α is the number of contacts on grain α. The second 2-norm contains equations of linear and angular force balance for each grain. Forces that minimize these terms are those that most closely satisfy balance of forces while also producing grain stresses that most closely agree with 3DXRD measurements. The constraint requires all inter-particle forces to be positive and obey a Coulomb friction constraint. Solutions to Eq. (1) are unique (determinate) when $3N_c^T \leq 12N_p^T$, where N_c^T is the total number of contacts and N_p^T is the total number of grains in the sample. This condition is fulfilled for both samples examined here and will generally be fulfilled for systems with a coordination number close to 8 if the number of grain-grain contacts significantly exceeds the number of grain-boundary contacts. Full solution details are given in [9].

We note that the high moduli of the materials studied ($E_{33} \approx 103$ GPa for quartz; $E_{33} \approx 495$ GPa for ruby) limit relative grain deformation to less than 5 pixels for quartz and less than 1 pixel for ruby at maximum sample compression and thus prohibit the use of local deformation as a means for inferring accurate normal contact forces from a known contact law.

4 Experiment 1: 77 quartz grains

The load cell data recorded during the first experiment are shown in Fig. 2a and illustrates that the sample of 77 quartz grains was subjected to two compression cycles. Figure 2b shows the evolution of volume-averaged grain stress for the sample, $\bar{\boldsymbol{\sigma}}$, obtained by computing $\bar{\boldsymbol{\sigma}} = \frac{1}{V}\sum_{\alpha=1}^{77}\bar{\boldsymbol{\sigma}}_\alpha V_\alpha$, where V_α is the volume of grain α and V is the total volume (including pore space) of the granular sample between loading pistons. We observe a consistent trend between $\bar{\boldsymbol{\sigma}}$ components and the load cell data. We also observe that $\bar{\sigma}_{zz}$ is greater than both $\bar{\sigma}_{xx}$ and $\bar{\sigma}_{yy}$ throughout most of the experiment; the lateral pressure coefficients, $k_0^{(1)} = \bar{\sigma}_{zz}/\bar{\sigma}_{xx}$ and $k_0^{(2)} = \bar{\sigma}_{zz}/\bar{\sigma}_{yy}$, are less than unity. However, $k^{(1)} > 1$ and $k^{(2)} > 1$ when the sample is fully unloaded, implying that a residual stress is presented in the system, a feature sometimes found in overconsolidated soils.

Results from force inference, using Eq. (1), for load steps 15 and 24 are illustrated in Fig. 3a and 3b, in which force chain structures can clearly be observed. Figure 2c shows the sum of vertical forces on the bottom, top and side boundaries. The sums exhibit the same trend as the load cell data in Fig. 2a, but are lower in magnitude due to friction between the load piston and quartz cylinder [9]. Figure 2d illustrates the mean, maximum, and maximum

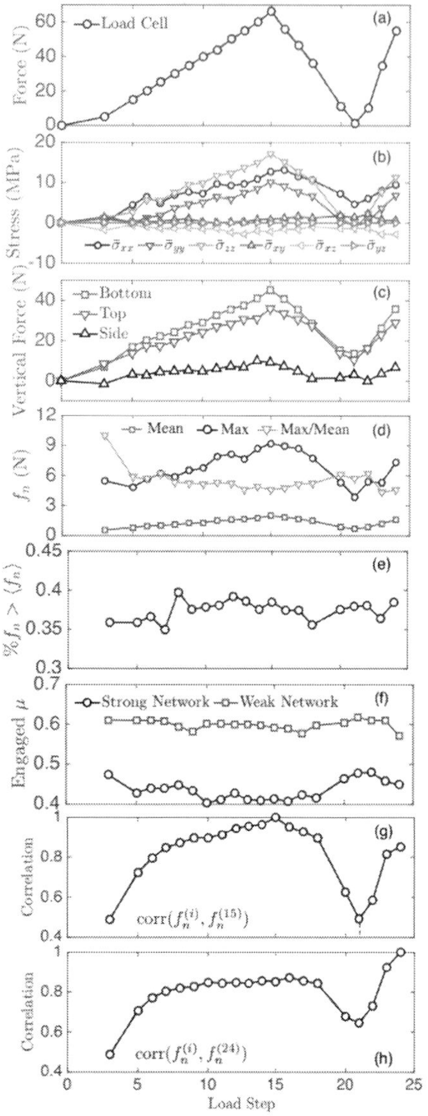

Figure 2. Results of first experiment. (a) Load cell data. (b) Volume-averaged stress evolution. (c) Sum of inferred vertical forces on bottom, top and side boundaries. (d) Statistics of inferred normal forces. (e) Percentage of normal forces above the mean. (f) Engaged friction in strong and weak force networks. (g) Correlation of entire force network with force network at load step 15. (h) Same as (g) but for load step 24.

divided by mean normal force magnitude, f_n, in the entire force network. The mean and maximum values of f_n exhibit the same trend as the applied load while the maximum divided by mean exhibits the inverse trend. This observation suggests that the largest values of f_n in the system are closer to the mean at high loads, implying a more homogeneous response. This finding was corroborated by a decreasing value (with increasing load) of the force distribution exponent and the Gini coefficient in [9].

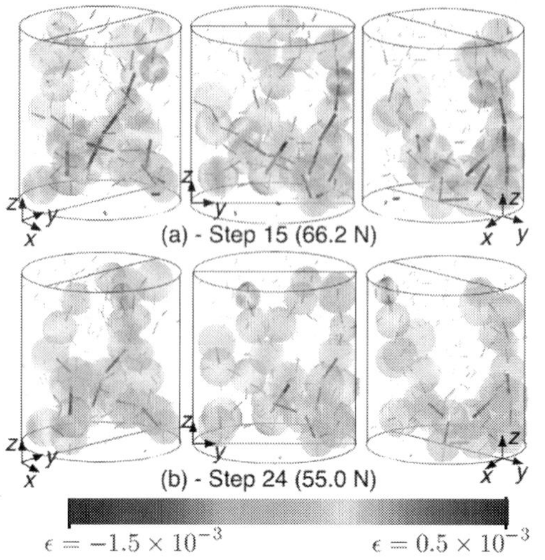

Figure 3. (a) Three views of the force network at load step 15. All forces are plotted as lines but only grains containing a normal force satisfying $f_n \geq 2\langle f_n \rangle$ are shown for clarity. Grains are 30% opaque and plotted with strain ellipsoids [8, 9]. (b) Same as (a) but for load step 24.

Figure 2e illustrates that the force network can be partitioned at each load step into a strong network ($f_n > \langle f_n \rangle$) containing about 38% of contacts and a weak network ($f_n \leq \langle f_n \rangle$) containing about 62% of contacts, which is quantitatively consistent with simulations [1]. Fig. 2f illustrates that the weak network features a higher value of engaged friction, $\langle f_t/(\mu f_n)\rangle$, than the strong network, also consistent with simulations [16] and suggesting that the weak network dissipates more energy in the system.

To explore force network reversibility and the influence of structural packing features on forces, we compute the correlation coefficient of forces across load steps by

$$\mathrm{corr}\left(f_n^{(i)}, f_n^{(j)}\right) = \frac{\sum_{k=1}^{N_c}\left(f_k^{(i)} - \langle f^{(i)}\rangle\right)\left(f_k^{(j)} - \langle f^{(j)}\rangle\right)}{\sum_{k=1}^{N_c}\left(f_k^{(i)} - \langle f^{(i)}\rangle\right)^2\left(f_k^{(j)} - \langle f^{(j)}\rangle\right)^2} \quad (2)$$

where $f^{(i)}$ is a an ordered vector containing all values of f_n for step i, N_c is the number of contacts in all load steps, and forces are 0 if their corresponding contact does not exist in a given step. We observe in Fig. 2g that the force network at peak load, step 15, is highly correlated (corr > 0.75) to previous steps after step 5. This suggests that only minor relative changes between forces occur as the applied load increases from 1/3 of its peak to its peak. Furthermore, we observe in Fig. 2h that the force networks in steps near the first peak load (steps 7 through 18) are highly correlated with the network at the second peak load (step 24), indicating a significant amount of reversibility in the force network across load cycles. We suspect that these trends occur because the force network is determined by the structural features of the packing, which change minimally throughout the experiment in this confined sample.

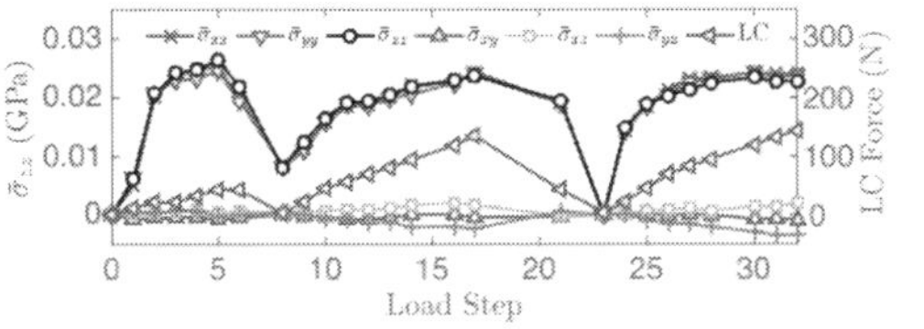

Figure 4. Load cell (LC) and grain stress data from second experiment. Steps with missing data points

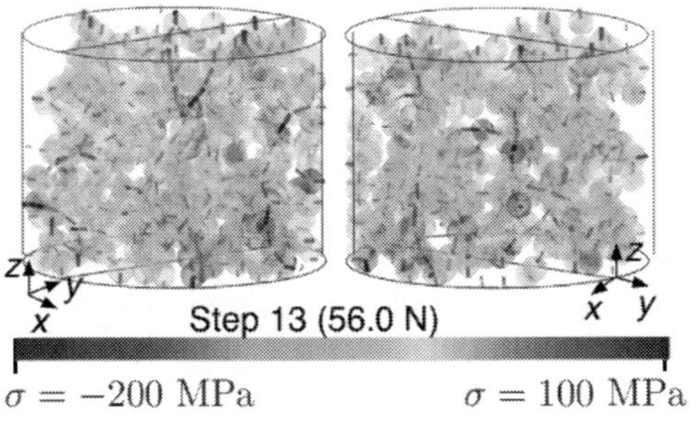

Figure 5. Two views of the force network at load step 13. Only forces for which $f_n \geq 2\langle f_n \rangle$ and corresponding grains are shown for clarity. Grains are 30% opaque and plotted with stress ellipsoids [8, 9].

5 Experiment 2: 1099 ruby grains

The load cell and grain stress data from the second experiment are shown in Fig. 4 and illustrate that the sample of 1099 ruby grains was subjected to three compression cycles. During the final compression cycle, the grains were compacted until widespread fracture occurred. We observe that the stress $\bar{\sigma}_{zz}$ along the compression axis falls below $\bar{\sigma}_{xx}$ and $\bar{\sigma}_{yy}$ after the first load cycle. This behavior was also seen during unload in the first experiment and suggests a residual lateral stress is presented in the system.

Results from force inference using Eq. (1) for load step 13 is illustrated in Fig. 5. Force chain structures are visually observed in these figures but we leave their full quantification for future publication. Here, we report that the normal forces very closely exhibit an exponential decay above the mean and partition into strong and weak networks, with the strong network ($f_n > \langle f_n \rangle$) containing approximately 42% of all contact points.

6 Discussion and conclusions

We have presented results from two experiments that provide the first measurements of inter-particle force networks in stiff, 3D, frictional and opaque granular materials. The results demonstrate the potential for the approach to validate multiscale models for granular media and explore a broader range of granular material responses (e.g., fracture and communition) than is currently available computationally. Ongoing analysis of data generated from larger assemblies will shed light on local stability and rearrangements around grain fractures or strain localizations.

Acknowledgements

The authors acknowledge the ESRF for synchrotron beam time for proposals ma1913 and ma2665. R.C.H. and J.E.A acknowledge U.S. AFOSR Grant No. FA9550-12-1-0091 and U.S. DTRA Grant No. HDTRA1-12-1-0041. S.A.H. acknowledges a Marie Curie FP7 integration grant within the 7th EU Framework Programme. R.C.H. acknowledges support from Lawrence Livermore National Laboratory's Lab (LLNL) Directed Research and Development program under grant 17-LW-009. Part of this work was performed under the auspices of the U.S. Department of Energy by LLNL under Award No. DE-AC52-07NA27344.

References

[1] F. Radjai, D.E. Wolf, M. Jean, J.J. Moreau, Physical review letters **80**, 61 (1998)
[2] B. Gilles, C. Coste, Physical review letters **90**, 174302 (2003)
[3] T.S. Majmudar, R.P. Behringer, Nature **435**, 1079 (2005)
[4] J. Brujić, S.F. Edwards, I. Hopkinson, H.A. Makse, Physica A: Statistical Mechanics and its Applications **327**, 201 (2003)
[5] M. Saadatfar, A.P. Sheppard, T.J. Senden, A.J. Kabla, Journal of the Mechanics and Physics of Solids **60**, 55 (2012)
[6] N. Brodu, J.A. Dijksman, R.P. Behringer, Nature communications **6** (2015)
[7] K. Alshibli, M.B. Cil, P. Kenesei, U. Lienert, Granular matter **15**, 517 (2013)
[8] S. Hall, J. Wright, Géotechnique Letters **5**, 236 (2015)
[9] R. Hurley, S. Hall, J. Andrade, J. Wright, Physical Review Letters **117**, 098005 (2016)
[10] J. Oddershede, S. Schmidt, H.F. Poulsen, H.O. Sørensen, J. Wright, W. Reimers, Journal of Applied Crystallography **43**, 539 (2010)
[11] P. Heyliger, H. Ledbetter, S. Kim, The Journal of the Acoustical Society of America **114**, 644 (2003)
[12] E.R. Dobrovinskaya, L.A. Lytvynov, V. Pishchik, *Sapphire: material, manufacturing, applications* (Springer Science & Business Media, 2009)
[13] W.J. Palenstijn, K.J. Batenburg, J. Sijbers, *The ASTRA tomography toolbox*, in *13th International Conference on Computational and Mathematical Methods in Science and Engineering, CMMSE* (2013), Vol. 2013
[14] A. Mirone, E. Brun, E. Gouillart, P. Tafforeau, J. Kieffer, Nuclear Instruments and Methods in Physics Research Section B: Beam Interactions with Materials and Atoms **324**, 41 (2014)
[15] J. Byerlee, Pure and applied geophysics **116**, 615 (1978)
[16] N. Guo, J. Zhao, Computers and Geotechnics **47**, 1 (2013)

Sticking properties of ice grains

M. Jongmanns[1,⋆], *M. Kumm*[1], *G. Wurm*[1], *D. E. Wolf*[1], and *J. Teiser*[1]

[1] *University of Duisburg-Essen, D-47048 Duisburg, Germany*

Abstract. We study the size dependence of pull-off forces of water ice in laboratory experiments and numerical simulations. To determine the pull-off force in our laboratory experiments, we use a liquid nitrogen cooled centrifuge. Depending on its rotation frequency, spherical ice grains detach due to the centrifugal force which is related to the adhesive properties. Numerical simulations are conducted by means of molecular dynamics simulations of hexagonal ice using a standard coarse-grained water potential. The pull-off force of a single contact between two spherical ice grains is measured due to strain controlled simulations. Both, the experimental study and the simulations reveal a dependence between the pull-off force and the (reduced) particle radii, which differ significantly from the linear dependence of common contact theories.

1 Introduction

In a standard scenario of planet formation, collisions between particles and aggregates thereof are important [1, 2]. Beyond the snowline in protoplanetary disks, water condenses into ice [3–5]. Therefore, collisions between ice particles occur. The interaction between these particles is fundamental for e.g. the formation of comets. Whether ice aggregates grow in mutual collisions depends on their sticking properties. It is commonly assumed [2], that ice aggregates can grow, if the kinetic collision $E_{\mathrm{C}}^{\mathrm{coll}}$ energy does not exceed

$$E_{\mathrm{C}}^{\mathrm{coll}} = 0.4 \cdot F_{\mathrm{C}}\delta_{\mathrm{C}}. \tag{1}$$

Here, δ_{c} is a critical displacement and F_{C} is the pull-off force needed to separate two particles. To understand the early process of planet formation, it is crucial to know the pull-off force F_{C}. Common contact theories (such as the JKR theory [6]) propose a linear dependence between F_{C} and the reduced radius $R_{\mathrm{red}} = \left(\frac{1}{R_1} + \frac{1}{R_2}\right)^{-1}$ (R_1 and R_2 being the radii of two contacting particles). However, it is not known if this dependence applies for values of R_{red} at the micro- or nanometer-scale for water ice. For example, in [7] Aumatell and Wurm investigated ice contacts at the nanometer-scale and found a big discrepancy between measured torques and those resulting from standard contact theories; in their work, the theoretical and experimental values differ up to a factor of 1000. To gain a better understanding of the critical pull-off force for small ice contacts, we have performed laboratory experiments (with micrometer ice particles, see section 2) and numerical simulations (with nanometer ice particles, see section 3). We compare the experimental and numerical results in section 4.

⋆e-mail: Malte.Jongmanns@uni-due.de

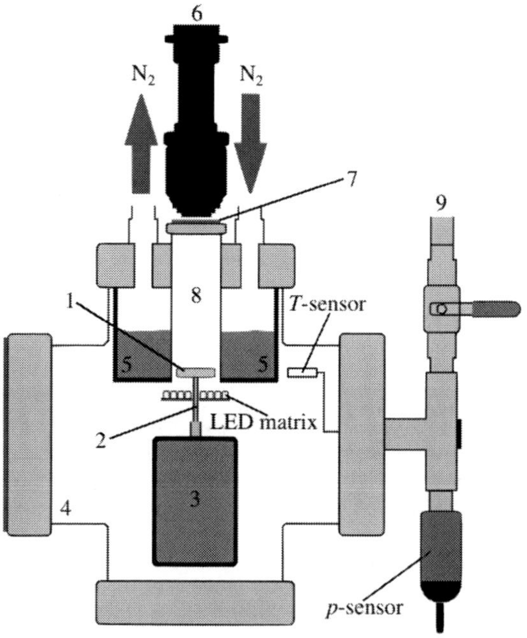

Figure 1. Our experimental setup as described in the text. (1): ice flywheel, (2): electric motor axis, (3): electric motor, (4): vacuum chamber, (5): liquid nitrogen-filled reservoir, (6): high-speed camera (not fully shown), (7): window, (8): cooling reservoir channel, (9): vacuum pump (not fully shown)

2 Experiment

Our experimental setup is shown in figure 1. A water ice flywheel (1) is mounted onto the axis (2) of an electric motor (3) and placed in the middle of a vacuum cham-

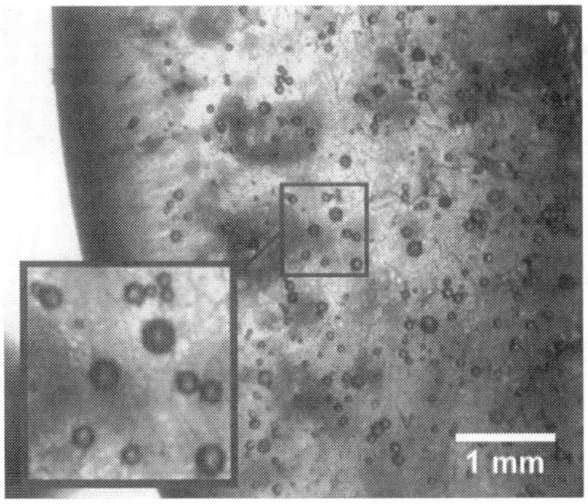

Figure 2. A snapshot of the flywheel taken by the highspeed camera. The inset shows ice grains on the surface of the ice flywheel in the static case (non-rotating) at higher resolution.

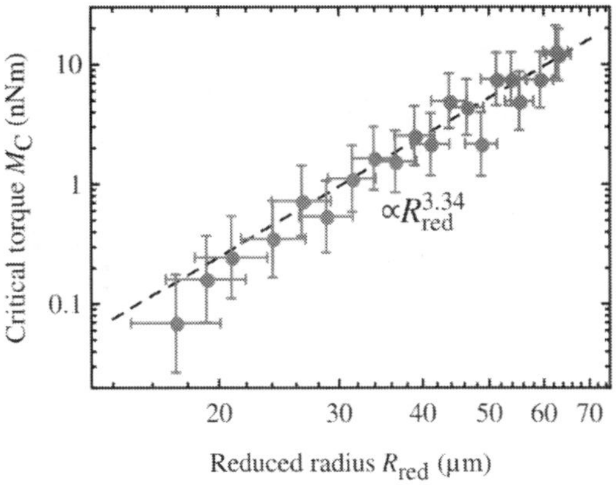

Figure 3. Measured M_C values on a log-log scale for averaged ice grain radii.

ber (4). A low temperature environment (of about −60 °C) is provided by means of a liquid nitrogen-filled reservoir (5). Spherical water ice grains of an average size of (36.4 ± 12.1) μm are brought onto the surface of the flywheel by spraying liquid water into the chamber (see figure 2). With the aid of a highspeed camera (6), it is possible to view the (spinning) ice flywheel through a window above the wheel (7). The view is along the channel of the cooling reservoir (8). The chamber is evacuated using a vacuum pump (9). Pressures of about $p = 5 \times 10^{-2}$ mbar are reached within this process. Therefore, we examine experimentally hexagonal ice Ih.

The motor spins with a frequency ω. We observed that, eventually (depending on ω), the ice grains are removed sideways from the flywheel because of a centrifugal force $F_{cf} = m\omega^2 r_{tr}$. Here, m is the mass of the ice grains and r_{tr} the distance from the rotation axis of a certain spherical ice particle. We calculate the mass by $m = \frac{4}{3}\pi R^3 \cdot \rho_{ice}$ where $\rho_{ice} = 925\ \frac{kg}{m^3}$ [8] denotes the density of the ice and R the radius of the particle.

Due to the fact that the sticky spherical water ice particles were removed in the *tangential* direction from the surface of the ice flywheel, this process is linked to a critical torque $M_C = R \cdot F_{cf}$, where $R = R_{red}$ is the radius of a certain spherical particle. We measure R by analyzing the highspeed camera images. Our measured data is shown in figure 3 in a $M_C\,(R_{red})$-plot. In this plot, our data points (∼ 100) have been averaged by using a bin width $\Delta R = 5$ μm. The corresponding fit function of our raw data (without averaging) can well be described by a power law $R_{red}^{3.34\pm0.40}$. According to [2], there is a linear dependence between the critical torque M_C and the critical pull-off force F_C. Therefore, we obtain experimentally:

$$F_C^{exp} \propto R_{red}^{3.34\pm0.40} \qquad (2)$$

3 Simulations

In our second approach to study the sticking properties of ice contacts, we have performed molecular dynamics (MD) simulations. The simulations were carried out using a coarse-grained *model-Water*(mW)-potential [9] which proved successful in modeling ice [10]. All simulations were conducted in the canonical NVT-ensemble. The temperature T was controlled by means of the Nosé-Hoover chain thermostat [11]. An integration time step of Δt = 10 fs was used and each system was relaxed for 0.3 ns.

The goal of our MD simulations is to calculate the pull-off force F_C for different ice contacts. This is achieved by conducting strain controlled simulations of specific configurations as will be described in the following. In our study, the ice contacts are modeled as two touching spherical ice grains. This is realized by placing mW-particles into a perfect hexagonal diamond lattice which is shaped as two touching half-spheres of the same radii R. By setting the lattice parameters to $a = b = 4.51$Å and $c = 7.34$Å [12], our diamond lattice corresponds to the structure of ice Ih at $T = 223.15$K. The crystal's $\vec{a}$-, $\vec{b}$- and $\vec{c}$-axis point in $[2\bar{1}\bar{1}0]$-, $[\bar{1}2\bar{1}0]$- and [0001]-direction, respectively. The configuration is enclosed by two walls which consist of the same lattice as the half-spheres. The walls are slightly bigger than the diameter of the spheres and as thick as two cutoff radii of the used potential. Such a start configuration is shown in figure 4 (a) for spheres of radii $R/a = 14$. The walls are needed for two reasons: first, to stabilize the configuration and second, to calculate the pull-off force F_C as it will be described in the next paragraphs. During the relaxation, the walls have the constraint, that they can only move (independently) "as a whole" in lateral direction. To implement this constraint, we neglect interactions of wall-particles among each other. After the relaxation, the positions of the walls are fixed to control the strain ε of the configuration. Before the actual strain controlled simulations and the calculation of F_C are explained, we describe how the system evolves during its relaxation.

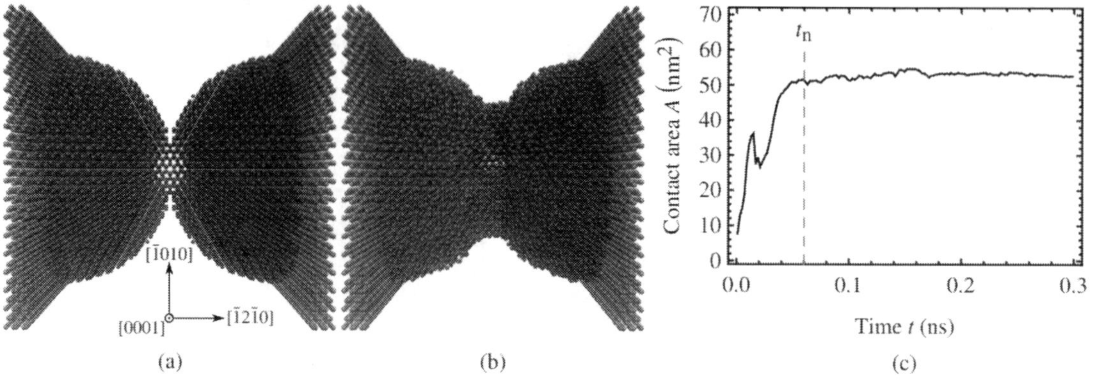

Figure 4. The geometry of the contact is modeled as two touching spheres (dark red and dark blue). (a) and (b) show the configuration for the specific case $R/a = 14$ with the view on the (0001)-plane in each case. The walls are shown in lighter colors. (a) shows the initial system. (b) shows the system after its relaxation without any applied strain. Here it can be seen, that after the relaxation the contact is shaped as a neck. (c) shows the temporal evolution of the contact area A during the relaxation. The time t_n indicates the complete formation of the neck.

During the relaxation of the system, we investigated an attraction of both spheres due to van-der-Waals-interactions. This attraction ends after a time t_n, which is far smaller than the total relaxation time. At t_n, the contact between the two ice grains is shaped as a cylindrical neck and does not change its form for the remaining relaxation. In figure 4 (b), such a neck is shown for two spheres of radii $R/a = 14$. We examined the formation of the neck independently of R. To understand the formation of the neck in more detail, it is instructive to investigate a $A(t)$-plot. Here, A is the contact area between the two ice grains and is calculated by $A = N_{cp}A_p$. N_{cp} is the number of contacting mW-particle pairs and A_p their two-dimensional projected surface area [13]. To obtain N_{cp}, we count the mW-particle pairs at the grain contact whose distance is smaller than $d = 3.3\text{Å}$. This value was calculated by means of a radial distribution function as it is described in detail in [13]. To obtain A_p, we assume the mW-particles to be perfect spheres with the volume of a water molecule $V = 32.53\text{Å}^3$ [8] and set $A_p = \left(\frac{3}{4}\sqrt{\pi}V\right)^{2/3}$. The temporal evolution of the calculated contact area during the relaxation is shown in figure 4 (c). As can be seen, the contact area at $t = 0\,\text{ns}$ is not zero. This is due to the discrete nature of the particle system; it is not possible to initialize a lattice in such a way, that the two spheres only touch at one point. After $t = t_n \approx 0.06\,\text{ns}$ (indicated by the dashed red line), the contact area stays constant for the remaining relaxation. At this point, the neck has been formed completely and does not change its size during the remaining relaxation; thus $\frac{dA}{dt} \approx 0$ is measured. The contact area of a configuration is therefore the mean value of A measured between $A(t = t_n)$ and $A(t = 0.3\,\text{ns})$.

After the relaxation, strain is imposed in $\vec{x}$-direction to measure F_C as follows: The right end of the left wall is $x = 0$. Then, after a time $\tau = 0.1\,\text{ns}$, we multiply the x-position of each mW-particle within the spheres by $\lambda = 1.001$. By doing this, the configuration is strained by 0.1% every $\tau = 0.1\,\text{ns}$, which defines a strain rate of $\dot{\varepsilon} = 1 \times 10^{10}\,\text{s}^{-1}$. Every τ, the position of the right wall is matched to the new length of the configuration. During the strain controlled simulations, the walls are fixed in their position. This allows us to measure the strain-force F_s which is needed to keep the configuration in its strained state: F_s is calculated by summing up all x components of the forces on all right wall-particles i: $F_s = \sum_{i\in r.\,\text{wall}} f_{i,x}$. The forces $f_{i,x}$ on the right wall-particles arise due to interactions with the mW-particles of the right sphere. The value of F_s is calculated at each time step. This allows us to measure force-strain diagrams $F_s(\varepsilon)$. If we perform a full strain controlled simulation, the $F_s(\varepsilon)$-plot will have a maximum value at a critical strain ε_c. The pull-off force F_C is defined as $F_C = F_s(\varepsilon_c)$.

By changing the radii R of the spheres, the size of the contact neck changes. This allows us to measure a dependence between F_C and A. We have measured this dependence for various system parameters. For example, we have changed the strain rate $\dot{\varepsilon}$, the temperature T, the direction in which strain is imposed with respect to the crystal axes, the lattice orientation of each ice grain (which leads to a grain boundary at the contacting area) and the geometry of the contact. The change in the geometry of the contact is motivated by the cylindrical shape of the neck (see figure 4 (b)); thus, we have also modeled the contact as a cylinder with radius R_{cyl} and length $L = 1.5R_{cyl}$. A detailed analysis of all these system parameter dependencies will be given elsewhere. Here, we only give the main results. For all these different systems, we measured: $F_C \propto A$. Representative for all different systems, figure 5 shows a $F_C(A)$-plot for the following specific system: $\dot{\varepsilon} = 1 \cdot 10^{10}\text{s}^{-1}$, $T = 223.15\text{K}$, strain direction is along the crystal's $\left[\bar{1}2\bar{1}0\right]$-axis, no grain boundary and both contact geometries (with sphere radii of $R/a = \{7, 8, 12, 13, 14, 15, 16, 18, 20\}$ and cylinder radii of $R_{cyl}/a = \{4, 6, 8, 10, 12, 14\}$). The number of mW-particles vary in this specific study between $N = 3698$ (smallest cylinder) and $N = 121944$ (biggest spheres). Figure 5

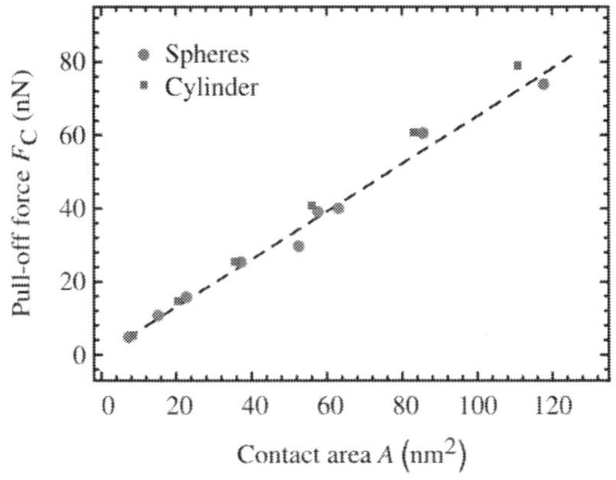

Figure 5. Measured F_C values as a function of the contact area A for both contact geometries. The black dashed line shows a linear fit for the sphere geometry.

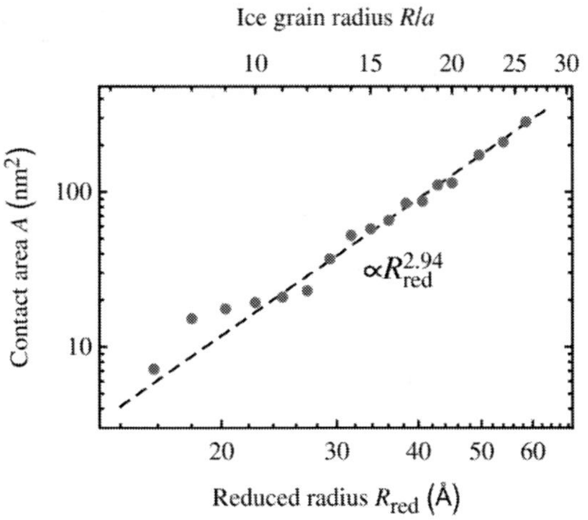

Figure 6. Measured A values on a log-log scale for different ice grain radii.

shows the already mentioned relation $F_C \propto A$ and the fact, that the contact neck can be modeled as a cylinder (showing that the contact neck is indeed cylindrical).

To compare the simulated result with the experimental result, we have measured a dependence between the reduced ice grain radii R_{red} and the area A of the contact neck. Since the ice grains are both of the same size, the reduced radii are $R_{\text{red}} = 0.5R$. Figure 6 shows the $A\,(R_{\text{red}})$-plot on a log-log scale for the same system as mentioned above (with additional sphere radii up to $R/a = 26$, which corresponds to $N = 256912$ mW-particles). The interesting result of the $A\,(R_{\text{red}})$-plot is: $A \propto R_{\text{red}}^{2.94\pm0.073}$. Since we have always measured $F_C \propto A$, the result of the simulations is:

$$F_C^{\text{simul}} \propto R_{\text{red}}^{2.94\pm0.073} \tag{3}$$

4 Conclusions

The experimental and simulated results (equations 2 and 3) reveal a non-linear dependence between the pull-off force F_C and the reduced radii of the ice grains R_{red}. Both power laws are similar and match within the standard error. Our results differ so clearly from the common linear dependence [6], that we have to conclude, that ice grains at the nano- and micrometer-scale might not be described approximately by common contact theories. This has an interesting impact on the understanding of the early formation of e.g. icy planets, asteroids or comets: The pull-off force defines, whether particles are kept together after a collision. Therefore, our results reveal, that the growth of dust aggregates at the nano- and micrometer-scale might proceed differently than previously thought.

References

[1] J. Blum, G. Wurm, Annual Review of Astronomy and Astrophysics **46**, 21 (2008)

[2] C. Dominik, A. Tielens, The Astrophysical Journal **480**, 647 (1997)

[3] J.P. Williams, L.A. Cieza, Annual Review of Astronomy and Astrophysics **49**, 67 (2011)

[4] H. Terada, A.T. Tokunaga, N. Kobayashi, N. Takato, Y. Hayano, H. Takami, The Astrophysical Journal **667**, 303 (2007)

[5] M. Honda, A. Inoue, M. Fukagawa, A. Oka, T. Nakamoto, M. Ishii, H. Terada, N. Takato, H. Kawakita, Y. Okamoto et al., The Astrophysical Journal Letters **690**, L110 (2008)

[6] K. Johnson, K. Kendall, A. Roberts, Proceedings of the Royal Society of London A: Mathematical, Physical and Engineering Sciences **324**, 301 (1971)

[7] G. Aumatell, G. Wurm, Monthly Notices of the Royal Astronomical Society **437**, 690 (2014)

[8] V.F. Petrenko, R.W. Whitworth, *Physics of ice* (OUP Oxford, 1999)

[9] V. Molinero, E.B. Moore, The Journal of Physical Chemistry B **113**, 4008 (2008)

[10] E.B. Moore, V. Molinero, Nature **479**, 506 (2011)

[11] G.J. Martyna, M.L. Klein, M. Tuckerman, The Journal of chemical physics **97**, 2635 (1992)

[12] K. Röttger, A. Endriss, J. Ihringer, S. Doyle, W. Kuhs, Acta Crystallographica Section B: Structural Science **50**, 644 (1994)

[13] S. Solhjoo, A.I. Vakis, Computational Materials Science **109**, 172 (2015)

Granular flow fluctuations in a conical hopper

Guilhem Mollon[1,*], and *Jidong* Zhao[2]

[1]LaMCoS, INSA Lyon, UMR 5259, Université de Lyon, CNRS, 69621 Villeurbanne Cedex, France
[2]Department of Civil and Environmental Engineering, Hong Kong University of Science and Technology, Clearwater Bay, Hong Kong

Abstract. In this article, we present simulations of 3D hopper flows through conical hoppers, with a focus on time fluctuations of flow-related quantities (velocity, coordination number and solid fraction). Dedicated post-processing techniques are used to obtain local and yet statistically meaningful measurements of these quantities. Simulations show that the fluctuations that were already observed in 2D hopper flows are still present in 3D. A parametric study also demonstrates that the frequency of the fluctuations is strongly related to the stiffness parameter of the contact law used in the simulations.

1 Introduction

Fluctuations of granular flows in conical hoppers have been reported both experimentally [1-3] and numerically [4-6] in the granular science literature. They are often attributed to intermittent arch networks above the outlet of hoppers which trigger high-frequency instationarities of the flow [6], or stress-related static avalanches [1]. In a previous work in 2D [3], we demonstrated that these instationarities take the form of organized waves which propagate upstream in the hopper. However, conclusions drawn in 2D do not necessarily apply in 3D, since the stability, strength and dynamics of force chains are rather different in both topologies. The 3D case is investigated here in a similar DEM framework, in order to determine (i) if these organized instationarities do exist in a 3D topology and (ii) if they do behave as acceleration waves propagating upwards. Besides, the results of the first stage of a comprehensive parametric study are presented and their implications on discrete modelling of granular flows are detailed.

2 Simulations

In order to study flow fluctuations, simulations are driven in the standard DEM framework. Spherical particles are used, with a unit density, a unit average diameter (this diameter is used all along this paper as a unit for distances) and a moderate polydispersity (i.e. a coefficient of variation of the grains volumes equal to 0.4). Contacts between grains and with walls are penalized in the normal and tangential direction by a contact stiffness K_n. Normal kinetic energy is damped by a viscous dashpot with a dissipation parameter of 0.2 (with the classical definition, see e.g. [5]). In the tangential direction, Coulomb friction is considered, with a coefficient of 0.4. The equations of motion are solved with a velocity-Verlet explicit scheme. Since linear contact laws (in displacement and velocity) are used, classical scaling rules make it possible to generalize these results to other geometries (in contrast with the remarks of [7], into which nonlinear stiffness and normal damping are used). The hopper has a conical shape, with an opening angle equal to 70°. Its outlet is a circle of diameter 10. It is filled with approximately 50,000 grains which are deposited under gravity. The outlet is then opened a t=0, triggering the flow through the hopper.

Fig. 1 shows the proportion of material passed through the outlet along time, for various values of the contact stiffness ranging from $K_n=10^4$ to 10^8. It appears from this figure that the flow rate is constant (i.e. independent from the height of material in the hopper) and does not depend on the contact stiffness.

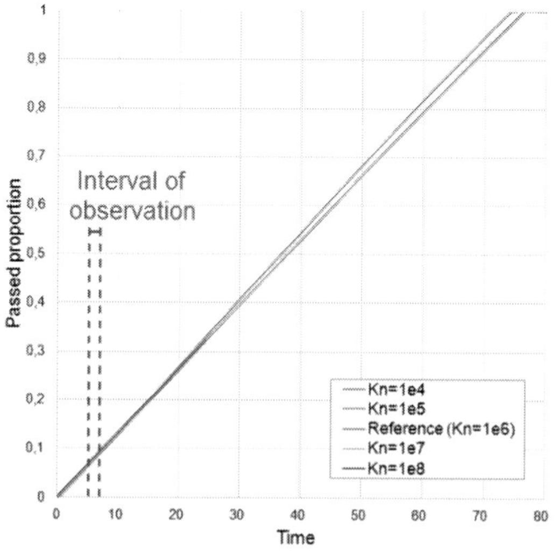

Fig. 1. Proportion of material passed through the hopper in time, for different values of the contact stiffness.

---
[*] Corresponding author: guilhem.mollon@gmail.com

3 Fluctuations

In the remainder of this article, attention is paid to a narrow time period (between instants t=5 and t=7), illustrated in Fig. 1 and called hereafter the "interval of observation". Fig. 2 shows two snapshots of the velocity fields (i.e. velocity norms of each grain), at two close instants in this interval, for the simulation corresponding to $K_n=10^6$. From these snapshots, it clearly appears that the velocity field is not stationary.

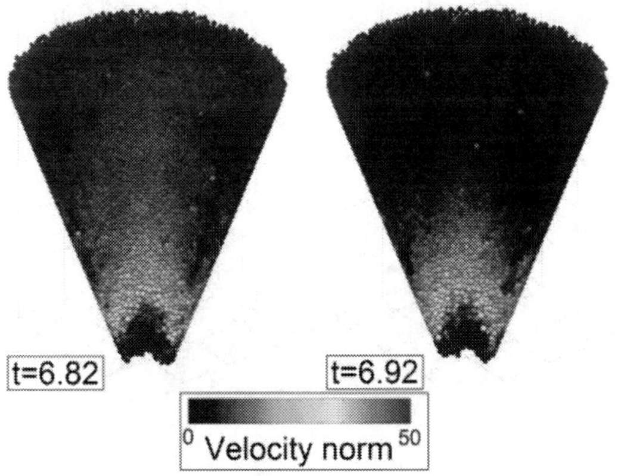

Fig. 2. Time fluctuations of the velocity field.

In order to characterize in a more quantitative way these fluctuations, appropriate data processing is needed. Indeed, some relevant data are attached to grains (e.g. velocity and coordination number, i.e. number of contacts per grain), while some others are attached to collections of grains (e.g. solid fraction). In any case, all these quantities do follow the grains in their motion, as always in a Lagrangian framework. Conversely, a proper statistical analysis requires an Eulerian description, where quantities are attached to fixed spatial locations. For this purpose, two different approaches are used:

-grain-related quantities (velocity and coordination number) are interpolated at fixed locations between grains using moving least squares, in a spherical domain with a radius equal to 5 distance units around the desired location.

-A local solid fraction at a given location *(x,y,z)* of space is computed by a weighted averaging technique which complies with the principle of mass conservation:

$$F_s(x,y,z)=(\pi d^2)^{3/2}\ \Sigma V_i e^{-ri(x,y,z)/d} \qquad (1)$$

In this expression V_i corresponds to the volume of the grain *i*, $r_i(x,y,z)$ corresponds to the distance from the centre of this grain to the location *(x,y,z)*, and the summation is executed on all the grains located in a sphere of radius *d=8* around this location.

These post-processing techniques are used to analyse the evolutions of velocity, coordination number and solid fraction along a vertical profile located on the axis of revolution of the hopper. This is performed every 0.002 time units during the interval of observation, in order to draw an accurate picture of the fluctuations of the hopper flow. Acquired data are summarized in Fig. 3.

In this figure, it clearly appears that the fields of velocity, coordination number and solid fractions are submitted to organized variations, albeit to various extents. Velocity and coordination number seem to obey to unpredictable fluctuations in the lower part of the hopper (i.e. below a height of 5 average diameters for coordination number, and 15 for velocity) and to evolve towards organized and periodic patterns above this height. Regarding velocity, the intensity of these fluctuations is rather limited, while the coordination number fluctuates more strongly, especially in the upper part of the flow.

Solid fraction, in contrast, shows extremely faint fluctuations (Fig. 3c.-left), with an interplay between two patterns (Fig. 3c.-right): some slow variations related to the downward motion of organized clusters of grains, and some rapid fluctuations with the same period as those of velocity and coordination number.

4 Influence of contact stiffness

In many mechanical systems, especially in continuum and structural mechanics, natural oscillations are related to the dynamic interplay between density and stiffness (like for example in the spring-mass classical system). In a granular system composed of rigid grains, the only stiffness (i.e. ratio between a force and a displacement) in the system is the contact stiffness. This parameter is well-known for controlling the local oscillations, because the period of the most frequent eigenmode of the system drives the time step used for resolution, and hence the simulation cost.

To investigate the influence of the contact stiffness on the fluctuations, this parameter is varied from 10^4 to 10^8, i.e. over 4 decades. Fig. 4 shows two space-time diagrams corresponding to the evolution of the coordination number on the vertical axis of revolution of the hopper, for values of the stiffness equal to 10^5 and 10^7.

It appears that the contact stiffness has a large influence on the flow fluctuations. When the stiffness is small (i.e. soft contacts), the fluctuations occur at a lower frequency, and the velocity of propagation of the upstream "waves" (i.e. the slope of the patterns observed in the space-time diagram) is quite low. On the contrary, when the stiffness is high (i.e. hard contacts), fluctuations are very frequent, and there is no noticeable upwards propagation of the pattern (or this propagation occurs too fast to be observed at this sampling rate and the fluctuations appear synchronized in space). In both cases, the amplitudes of the variations seem to be rather similar.

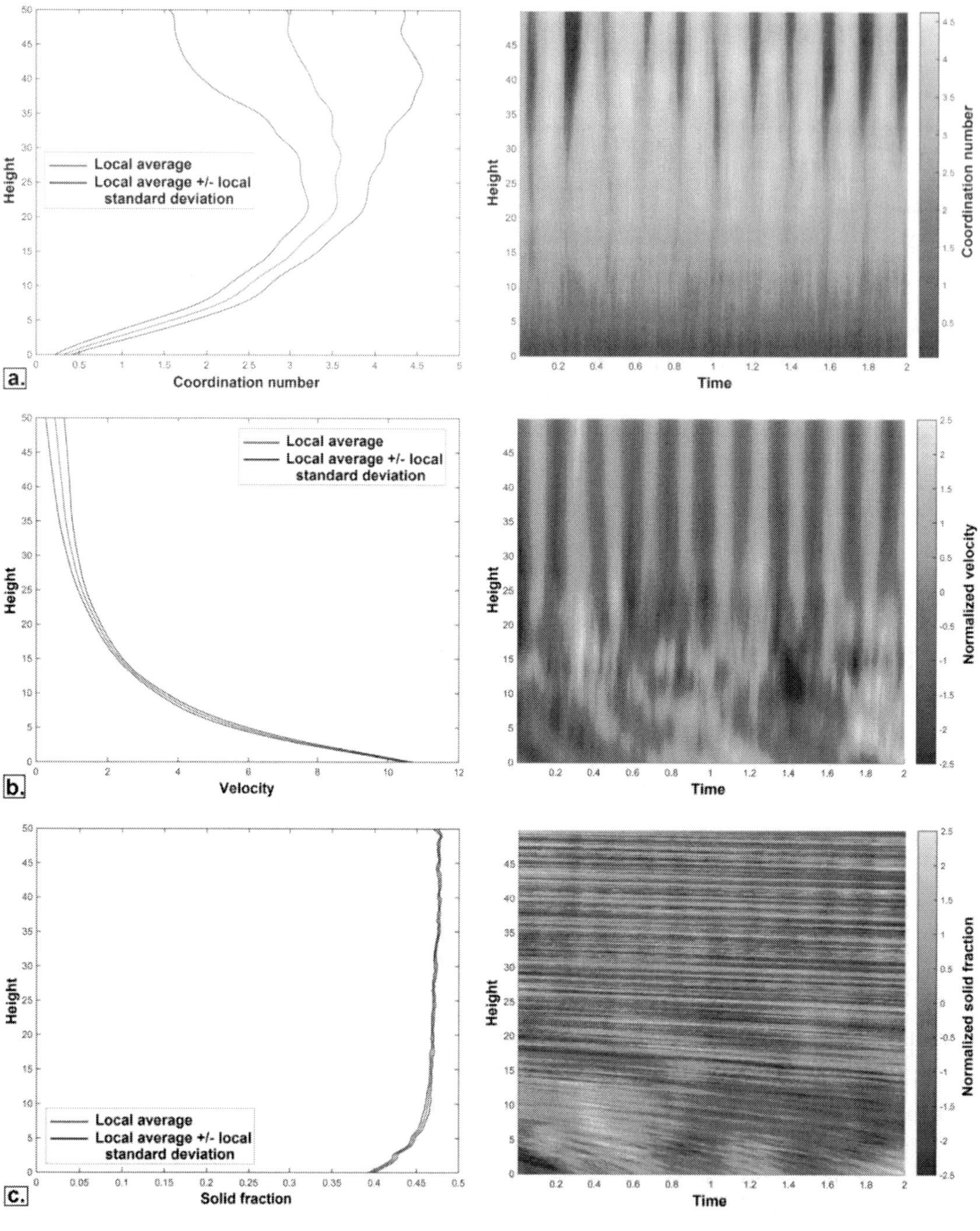

Fig. 3. Statistics of several quantities during the interval of observation. Left: Time-averaged values (plus and minus one time-related standard deviation) as a function of the vertical position ; Right: Space-time diagrams ; a. Coordination number ; b. Velocity ; c. Solid fraction. For b. and c., quantities were normalized in the space-time diagrams to enhance the visibility of fluctuations: to each local instantaneous value, the local time-average was subtracted, and the result was divided by the local time-related standard deviation.

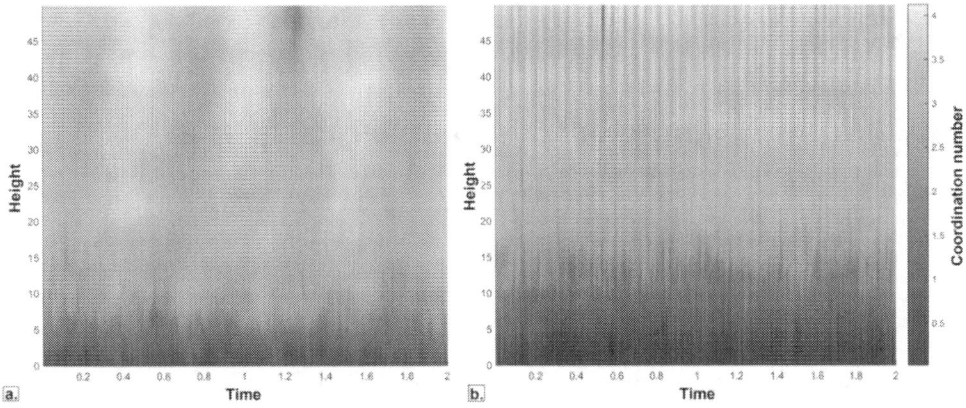

Fig. 4. Space-time diagrams of the coordination number for two different values of the contact stiffness. a. Kn=10^5 ; b. Kn=10^7.

Fig. 5 provides the evolution of the main frequency of the fluctuations (obtained by computing the autocorrelation function of the time evolution of the coordination number at a height of 40 average diameters and by running a peak search) as a function of the contact stiffness. This figure clearly shows that the period is proportional to the inverse square root of the contact stiffness.

This behaviour is typical of eigenmodes in structural mechanics, but is rather surprising in the framework of granular mechanics because the fluctuations observed here are not local, but are at the scale of the system. Besides, this result seems to indicate that the same eigenmode is activated in all the five simulations, whatever the value of the contact stiffness. The dynamic mechanisms leading this system to evolve towards established excitations of a given mode of vibration are yet to be understood.

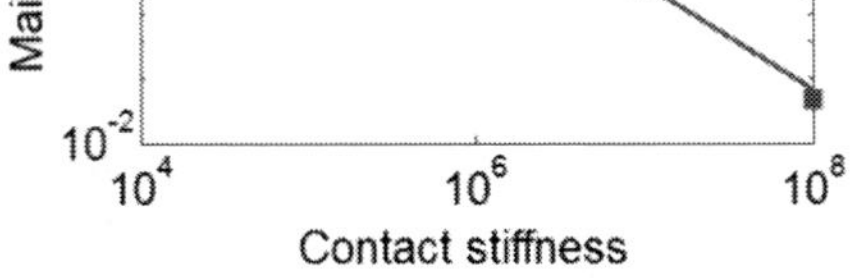

Fig. 5. Influence of the contact stiffness on the main period of the fluctuations.

5 Conclusion and prospects

The preliminary results reported in this communication indicate that granular flows through 3D conical hoppers are spontaneously submitted to organized and periodic fluctuations, in a way similar to the 2D case. It appears that the main period of these fluctuations scales with the inverse square root of the contact stiffness.

This finding gives rise to several questions, among which an important one is related to the way contact stiffness is considered in a modelling framework. Indeed, this parameter may be either considered as a physical parameter (accounting for the deformability of the grains, in which case it is related to their Young modulus and Poisson ratio) or as a numerical parameter (as a penalization of the contact interpenetration, like in FEM contacts). The latter situation is more common, because (i) accounting for a realistic stiffness of the grains usually leads to very stiff contacts and to prohibitive computational costs, and (ii) it is often found that this parameter has no influence on the physical phenomena of interest for the modeller (see e.g. Fig. 1).

Based on the results of the present paper, it appears that, if fluctuations are the actual phenomenon that the modeller wishes to investigate, then the contact stiffness should be considered as a physical parameter and chosen appropriately to reflect the actual stiffness of the grains.

References

1. P. Claudin, J.P. Bouchaud, PRL 78(2), 231-234 (1997)
2. E. Gardel, E. Seitaridou, K. Facto, E. Keene, K. Hattam, N. Easwar, N. Menon, Phil. Trans.. Roy. Soc. A **367**, 5109-5121 (2009)
3. A. Garcimartin, I. Zuriguel, A. Janda, D. Maza, PRE 84, 031309 (2011)
4. P.W. Cleary, M.L Sawley, Appl. Math. Model. **26**, 89-111 (2002)
5. R. Arévalo, A. Garcimatin, D. Maza, Eur. Phys J. E 23, 191-198 (2007)
6. G. Mollon, J. Zhao, Gran. Mat. **15**(6), 827-840 (2013)
7. T. Pöschel, C. Salueña, T. Schwager, PRE 64, 011308 (2001)

A multibody strategy for deformable grains

Guilhem Mollon[1,*]

[1]LaMCoS, INSA Lyon, UMR 5259, Université de Lyon, CNRS, 69621 Villeurbanne Cedex, France

Abstract. This article presents a numerical framework dedicated to the simulation of granular materials with highly deformable grains. This framework is based on a multibody meshfree strategy, which makes it possible to account for the constitutive model of the material composing each grain and for possibly complex contact laws (e.g. adhesion, friction, etc.). The main principles of the approach are first presented, and two illustrative cases are then detailed in order to emphasize its potential in several domains of the granular science.

1 Introduction

Discrete-Element modelling (DEM) has become a widely-used numerical method for fine and predictive modelling of granular materials in several industrial and scientific fields (e.g. geomechanics, food industry, chemistry, etc.). In this framework, each grain is considered as a rigid body which follows the laws of motion and interacts with other bodies by the means of contacts. In some situations, however, the assumption of rigid bodies enters into contradiction with the observed physical behaviours. This is the case when the material composing the grains is relatively soft (e.g. ductile powders), or when this material is submitted to very high loadings (e.g. tribology). To go beyond this limitation, this article presents a novel numerical framework which makes it possible to reproduce the compliance of the material at the grain scale. Two illustrative cases are then proposed in order to emphasize the potential of the approach.

2 Numerical framework

The numerical framework proposed hereafter is based on the following choices, some of which being detailed in [1]:

- The granular material is composed of a set of individual grains, and each grain is delimited by a closed boundary. The present approach is restricted to two dimensions.

- Each grain is a deformable body, and the displacement field is interpolated on its domain using a meshfree description based on a certain number of field nodes, with two degrees of freedom in displacement each (Fig. 1). The shape functions are based on Moving Least Square (MLS) interpolation, as first proposed in the Element-Free Galerkin framework [2] (Fig. 2).

- Contact forces are computed on the basis of a two-pass node-to-segment penalized formulation, hence considering the interaction between each field node located on the boundary of a given grain and the boundaries of the surrounding grains.

- At a given time step, a classical weak-form of the continuum mechanics equations is considered for the whole problem (like in Finite Element Method). Numerical integration is performed on each grain using a Gauss quadrature on a triangular mesh. An adaptive explicit scheme is used for time integration.

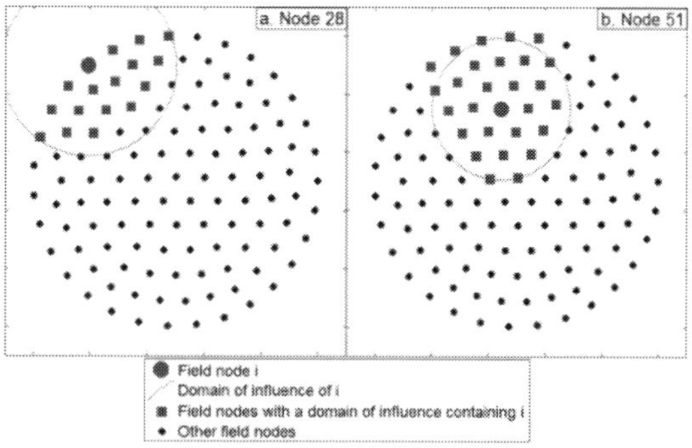

Fig. 1. Meshfree discretization of a circular grain, focusing on two particular nodes. a. node 28 (boundary node) ; b. node 51 (inner node).

Fig. 2. MLS shape functions of node 51.

* Corresponding author: guilhem.mollon@gmail.com

3 First illustrative case - Tribology

The third body concept is often used to describe the thin layer of degraded material which separates contacting bodies during sliding contact (Fig. 3). This third body may be either natural (i.e. produced by the degradation of the contacting surfaces and by matter entering into the contact) or artificial (i.e. placed there for a given purpose, e.g. grease). In any case, it is submitted to high loading (up to several GPa) and strong shearing. If such phenomena as friction and wear are to be fully understood, an accurate modelling of the third body is hence crucial. In recent years, simulation methods based on DEM have been successful in understanding several properties of the third-body behavior [3]. The next stage, however, is to render each third body particle compliant in order to be able to deal with important phenomena such as complex constitutive laws, plasticity, Poisson effect, surface adhesion, etc.

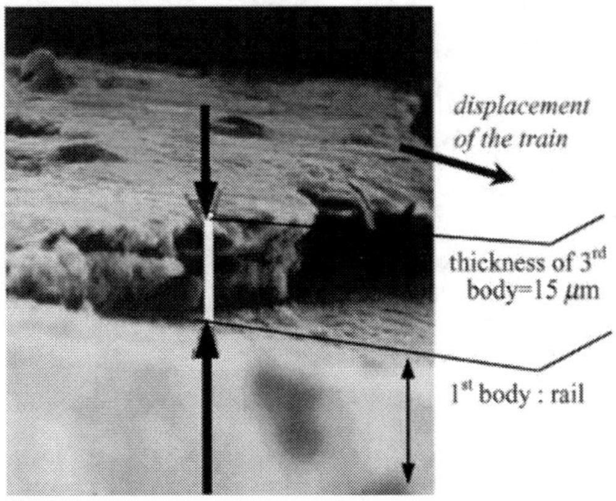

Fig. 3. Third body layer in a wheel-rail contact (from [4]).

To illustrate the potential use of multibody meshfree modelling in this framework, a simple study case is described in Fig. 4. In this typical tribological situation, two deformable bodies (so-called "first bodies") are located below and above a collection of 500 deformable discs which represent the third body. The lower boundary is fixed, the upper boundary is submitted to an imposed vertical displacement (in order to compress the third body) and to a horizontal velocity (in order to shear it continuously in the same way as in a sliding contact). Periodic boundaries are applied on each lateral side. An arbitrary roughness is introduced at the interfaces between first and third bodies.

The system is hence composed of 502 bodies discretized by roughly 75,000 field nodes, reaching a moderate size by modern computational standards (i.e. about 150,000 degrees of freedom). First bodies and third body particles follow an hyperelastic constitutive model, and the contact law is adhesive in order to reflect the typical behaviour of solid interfaces at sub-micrometric scale. In the case presented here, low third body stiffness and moderate contact adhesion are considered.

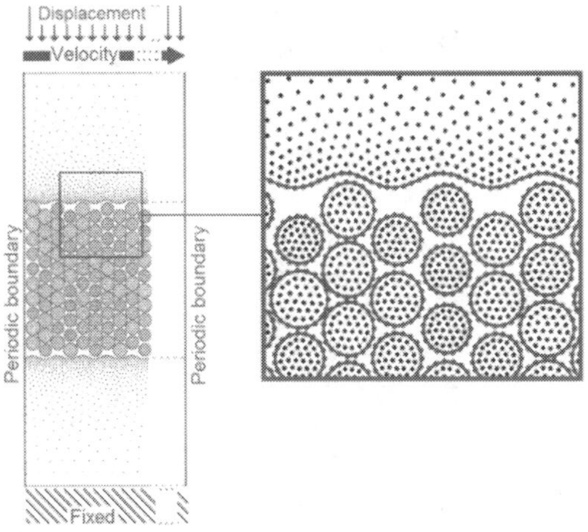

Fig. 4. Layout of the first illustrative case related to tribology.

Some numerical results are presented in Fig. 5, which shows four successive stages of the simulation. Situation (a) corresponds to the initial state of the system, (b) corresponds to the first stages of the loading, (c) to the installation of a flow regime in the third body, and (d) to the final stage. For these four instants, both stress and velocity fields are provided

These results show that, in the first stages of the loading, the stress field looks similar to the force network which is usually observed in hard granular materials (i.e. force chains mostly oriented in the direction of the main effort). However, when the loading is high enough to strongly deform the grains, this network evolves towards a continuous and complex stress pattern. The use of the same discretization method for the first and third bodies also makes it possible to observe the stress patterns in the first bodies, and to study their degradation on a quantitative basis.

The velocity fields show that, in this specific case, the velocity jump between the two first bodies is firstly accommodated by a rather smooth velocity gradient (i.e. bulk shearing in the third body), although Fig. 5c. shows that this gradient is actually controlled by localized slipping events. However, after a period of sliding, the accommodation regime evolves naturally towards a single slip line, with a moderate macroscopic roughness. Simultaneously to this change of regime, quantitative data (not provided here) show a reduction of the rate of energy dissipation in the system, i..e. a reduction of the apparent coefficient of friction. This kind of behaviour is consistent with some tribological observations, and could not be observed with rigid grains.

Other simulations with different parameters (stiffness of the third body, adhesion, loading, velocity, etc.) led to a large variety of local phenomena (bulk-shearing, wall-slip, rollers, stick-slip, etc.), which reflect the varied phenomenology which is observed in experiments.

 DOI: 10.1051/epjconf/201714015010

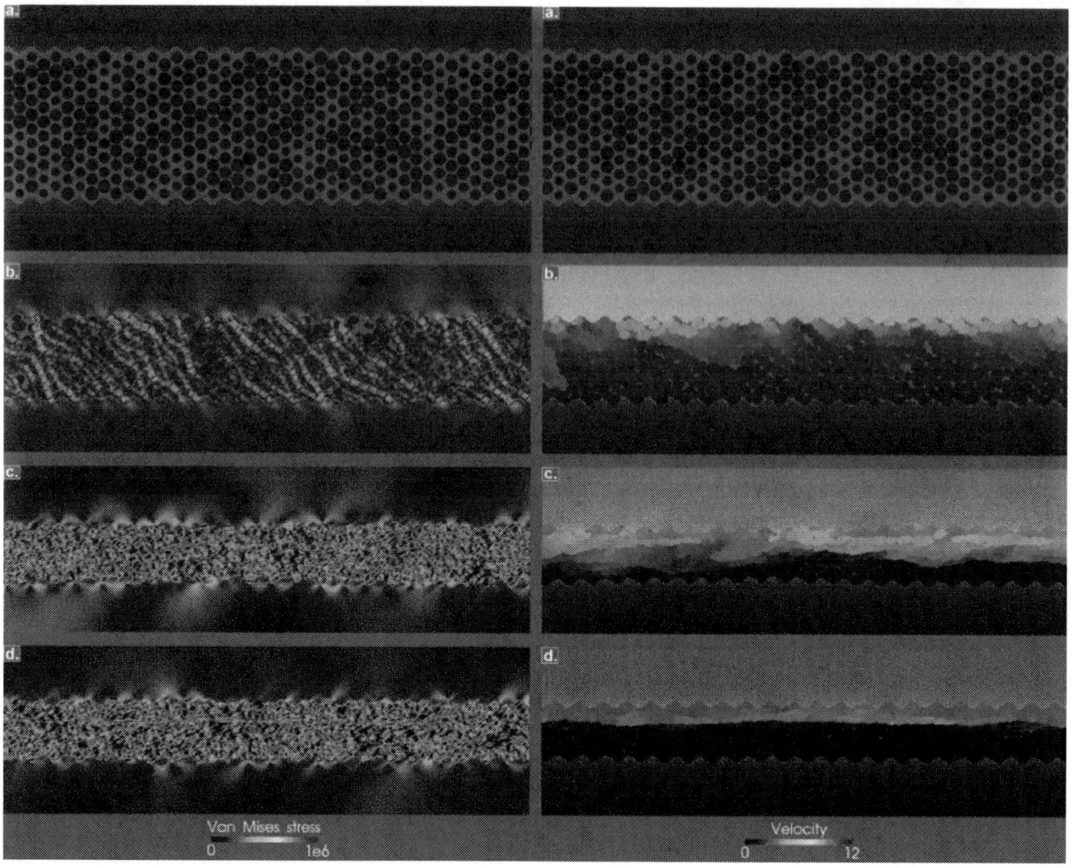

Fig. 5. Four snapshots of a tribological simulation. Left: Von-Mises stress field ; Right: Velocity norm field (arbitrary units).

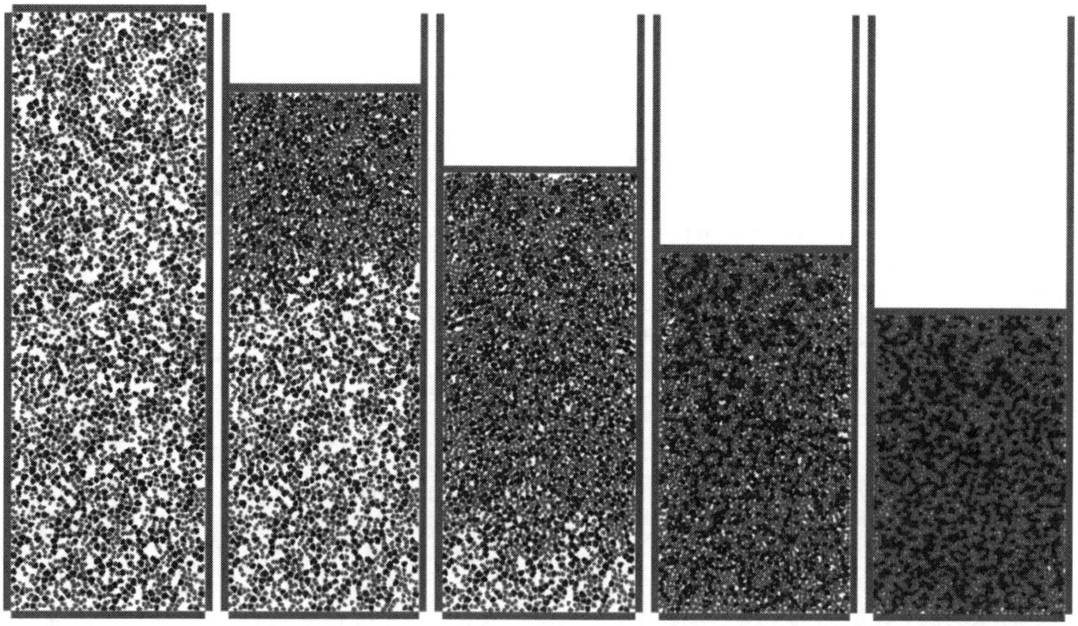

Fig. 6. Simulation of the compaction of a mixed powders with 3000 grains. In red: rigid bodies ; In blue: deformable bodies.

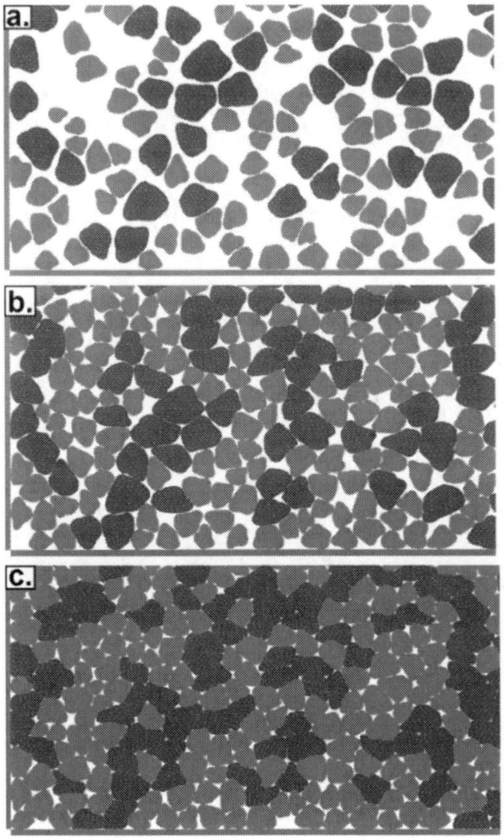

Fig. 7. Zoom on the lower-left corner of the simulation shown in Fig. 6. ; a. Initial state ; b. Beginning of powder compaction ; c. Final state.

4 Second illustrative case - Mixed powders compaction

The second illustrative case aims at showing the ability of the numerical framework to deal with rigid and deformable bodies within the same simulation. For this purpose, a powder compaction setup is simulated. The granular sample is generated with realistic shapes on a Fourier-Voronoi basis (see [5]). It is composed of 2000 rigid grains (as in a DEM simulation) and 1000 deformable grains (discretized using the meshfree technique presented above). The granular sample is positioned between four rigid walls, and the upper wall is submitted to an imposed downwards velocity (Fig. 6). This situation is typical of the compaction of a powder with two types of grains composed of materials with very different Young moduli.

The interactions between rigid and deformable grains is handled in a natural way in the code, since the contour of the rigid grains is discretized by a set of nodes and of segments: the same contact algorithm may then be applied for any couple of grains, either rigid or deformable. Coulomb friction with a coefficient of 0.5 is considered for all the contacts of this simulation.

Fig. 7 shows a zoomed view on the lower-left corner of the compacting box, at three different stages of the simulation. Stage (a) corresponds to the initial state, stage (b) corresponds to a compaction stage where the compliant grains are not yet strongly deformed, and stage (c) corresponds to the final state of the simulation. From this figure, it is evident that the numerical framework is able to capture the difference of behaviour between the two populations of grains: the final distribution of the voids is concentrated around the contacts between rigid grains, while deformable grains are able to "fill" the voids and to strongly reduce the local porosity.

5 Conclusion and prospects

A novel numerical framework dedicated to highly deformable granular materials was presented in this article. Two illustrative cases were provided in order to demonstrate the applicability of the code to a wide variety of situations where the compliance of the grains should be considered. The next stages of the work will be dedicated to a convergence study (in terms of time and space discretizations) and to experimental validations.

Future versions of the code will include several features which will enable to represent in a more accurate manner the local physics of highly loaded contacts with a third body, for tribological applications:

-Enriched behaviour of the grains (plasticity, viscosity, breakage, etc.) and of the first bodies (degradation, fatigue, etc.)

-Enriched contact laws based on physical models (e.g. JKR, Van der Waals, etc.)

-Heat creation and diffusion

-Fluid/grain coupling

A particular attention will also be paid to the computational efficiency of the code, in order to be able to use it in a High Performance Computing (HPC) framework.

References

1. G. Mollon, Int. J. Num. Meth. Eng., *accepted* (2016)
2. T. Belitchko, Y.Y. Lu, L. Gu, Int. J. Num. Meth. Eng. **37**, 229-256 (1994)
3. M. Champagne, M. Renouf, Y. Berthier, ASME J. Tribol. **136**(2), 021603 (2014)
4. S. Descartes, C. Desrayaud, E. Niccolibi, Y. Berthier, Wear **258**, 1081-1090 (2005)
5. G. Mollon, J. Zhao, Gran. Mat. **14**, 621-638 (2012)

System size effects on the mechanical response of cohesive-frictional granular ensembles

Saurabh Singh[1,*], *Ramesh Kannan* Kandasami[1,**], *Rupesh Kumar* Mahendran[2,***], and *Tejas* Murthy[1,****]

[1] *Indian Institute of Science, Bangalore, India - 560 012*
[2] *Indian Institute of Technology Madras, Chennai, India - 600 036*

Abstract. Shear resistance in granular ensembles is a result of interparticle interaction and friction. However, even the presence of small amounts of cohesion between the particles changes the landscape of the mechanical response considerably. Very often such cohesive frictional (c-ϕ) granular ensembles are encountered in nature as well as while handling and storage of granular materials in the pharmaceutical, construction and mining industries. Modeling of these c-ϕ materials, especially in engineering applications have relied on the oft-made assumption of a "continua" and have utilized the popular tenets of continuum plasticity theory. We present an experimental investigation on the fundamental mechanics of c-ϕ materials specifically; we investigate if there exists a system size effect and any additional length scales beyond the continuum length scale on their mechanical response. For this purpose, we conduct a series of 1-D compression (UC) tests on cylindrical specimens reconstituted in the laboratory with a range of model particle–binder combinations such as sand-cement, sand-epoxy, and glass ballotini-epoxy mixtures. Specimens are reconstituted to various diameters ranging from 10 mm to 150 mm (with an aspect ratio of 2) to a predefined packing fraction. In addition to the effect of the type of binder (cement, epoxy) and system size, the mean particle size is also varied from 0.5 to 2.5 mm. The peak strength of these materials is significant as it signals the initiation of the cohesive-bond breaking and onset of mobilization of the inter particle frictional resistance. For these model systems, the peak strength is a strong function of the system size of the ensemble as well as the mean particle size. This intriguing observation is counter to the traditional notion of a continuum plastic typical granular ensemble. Microstructure studies in a computed-tomograph have revealed the existence of a web patterned 'entangled-chain' like structure, we argue that this ushers an additional length scale as well as presents a system size effect.

1 Introduction

In nature, granular materials when stored or under the action of natural forces, develop inter particle cohesion. They develop organic or inorganic bonds between the particles, which changes the mechanical behaviour at the ensemble significantly. These materials acquire their strength from cohesive bonds, in addition to the frictional resistance due to interaction at contacts. For engineering of these materials, i.e. to understand the mechanical behaviour, weak c-ϕ granular materials are treated as a "continua" with no consideration of dependency on length scales [1–4]. In case of brittle and quasi-brittle materials such as glass, rock, concrete etc. scaling is well understood and is considered in any numerical modeling which is inspired from Weibull's statistical model [5]. It is well known that the fabric or structure of c-ϕ granular materials change with changes in amount of cohesion/binder present [10]. This proportion of binder and the arrangement of particles due to the presence of binders, brings forth fabrics in c-ϕ materials such as a matrix bound structure (asphalt - wherein individual particles are suspended in a matrix of the binder), void bound structure (concrete, mortar, rocks - wherein the voids in-between particles are filled with the binder) and contact bound structure(weak c-ϕ materials, geological materials such as sandstones). It has been widely accepted that materials with void bound structure follow Weibull's statistical scaling model, in that with increase in the size of the system (or specimen) the strength reduces and has been well adapted into the mechanics and design of such systems. In this study we seek to understand if scaling other than the continuum scaling exists in weak c-ϕ materials (i.e. materials with a contact bound structure) with varying system size and particle size.

2 Background

In order to model weakly cohered frictional particulate ensembles (in short c-ϕ materials), two broad approaches have been practiced. In the first approach, models for frictional granular materials were used and modified to incorporate additional features exhibited due to the presence of

*e-mail: saurabh@civil.iisc.ernet.in
**e-mail: ramesh.k.kannan@gmail.com
***e-mail: dmrupeshkumar@gmail.com
****e-mail: tejas@civil.iisc.ernet.in

a weak cohesion between the particles. In order to accomplish this, utilising the framework of the mathematical theory of continuum plasticity, the features of a plastic model are modified, i.e. the plastic potential function, yield function are modified to take care of cohesion by considering this cohesion as an additional confinement on the ensemble. In other words, the yield surface and plastic potential surfaces were translated along hydrostatic axis to include this cohesion or cementation in the material [3, 4]. While another approach has been less phenomenological, by observing the governing mechanism and the evolving mechanism in the inter particulate cohesion to reach failure [1, 2]. In a typical study of this kind, [1, 2] assumed that net load taken by the ensemble can be thought as a combination of resistance of particles at the contact and bonds which is true at both micro (inter particle) and macro (ensemble or continuum) levels. Separate elastic-plastic evolution laws were considered for two phases and then combined using the micro-mechanical equilibrium. The model parameters for such constitutive models were obtained by performing "continuum" elemental tests on cylindrical specimens of standard dimension (38x76mm). These elemental tests can be relied upon to provide an appropriate response of the material only if the response is that of a true or a typical "continuum material". In that, the overall mechanical response along with failure strength is not affected by changing the size of the system under consideration. This has been clearly documented through simulations and experiments for the case of a purely frictional granular material [6]. However, thus far, no study has thrown light if such continuum assumptions and scale independence is valid for these weak c-ϕ materials (or materials with a contact bound structure). As stated earlier, in case of brittle and quasi brittle materials such as (c-ϕ material with void bound structure) concrete, rock show significant scaling in the strength of the material. With increase in specimen size, peak strength (failure strength) decreases. This appears akin to weibull's distribution which was initially coined for failure of a chain like structure with links. Where failure of a link is considered as failure of the entire chain. With increasing number of links probability of failure increases which resembles the behaviour of these quasi-brittle materials. That is, with increase in size of a specimen, the structural strength decreases (or the probability of failure increases). Bazant et al.[7] has provided an extension of this in a framework of fracture mechanics.

In recent times, discrete element simulations for weak c-ϕ materials have been carried out [8]. The results show a systematic transition from elastic behaviour followed by bond breakage which in turn leads to dilation, as a result of the frictional rearrangement of the particulates. A change in the stress distribution from web patterned force chains to columnar structure of force chains was also shown from these simulations. A computed tomographic scan image of an initial contact bound structure is shown in the figure 1.

3 Experimental

In these experiments, we used two model granular materials - sand and glass ballotini and two materials for imparting inter grain cohesion - cement and epoxy, in effect, we used three c-ϕ systems sand+cement (cemented sand), sand+epoxy (SE) and glass beads+epoxy (GBE).

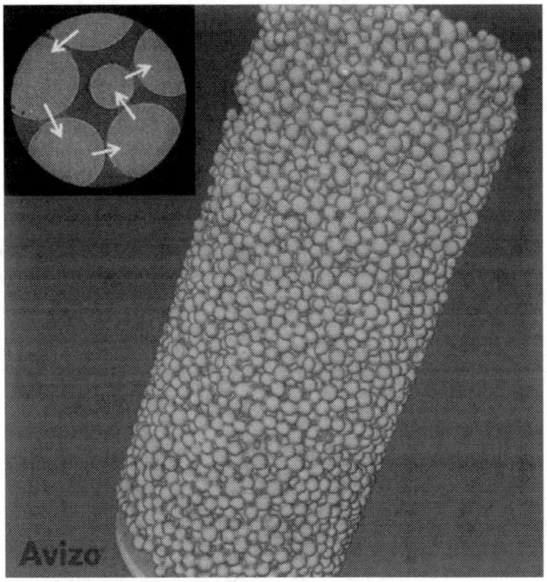

Figure 1. Computed tomography image of a weakly cemented specimen (glass ballotini + 1% epoxy) showing a contact bound structure. The inset shows the chain like cemented structure

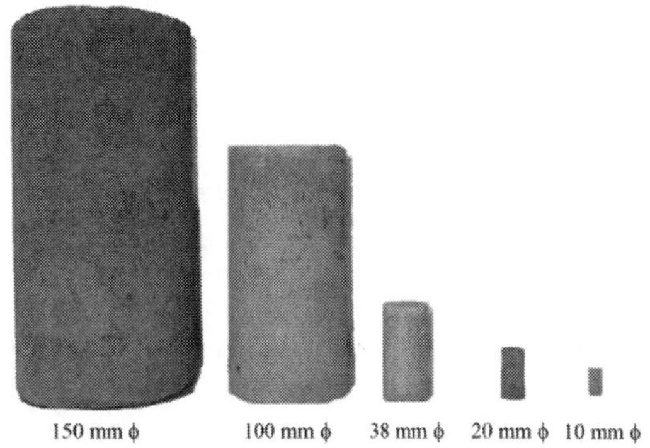

Figure 2. Artificially reconstituted weakly cemented (sand + 4% cement) with specimen/ system size varying from 150 to 10 mm dia keeping the particle size, density and aspect ratio constant

For the case of cemented sands, specimens were reconstituted by mixing quartzitic sand (specific gravity 2.65 and mean grain size 0.45 *mm*) with ordinary Portland cement (OPC) at a density of 1.4 g/cc. Three types of cemented sand specimens were prepared using 2%, 4%, and 8% OPC by weight of sand. For each specimen the ratio of height to diameter (aspect ratio) was kept constant at two. To understand the system size effect specimens with diameter of 10 *mm*, 20 *mm*, 38 *mm*, 100 *mm*, 150 *mm* were prepared. All the specimens were prepared to a desired

DOI: 10.1051/epjconf/201714008007

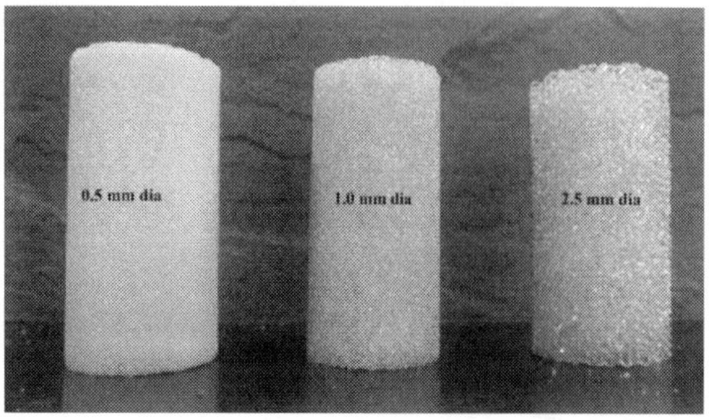

Figure 3. Artificially reconstituted weakly cemented (glass ballotini + 1% epoxy) specimens with particle size varying from 0.5 to 2.5 mm dia while keeping the system size and density constant

packing fraction and were cured under moist conditions to achieve a contact bound structure and a characteristic compressive strength. Figure 2 shows the prepared specimen at different diameters.

For GBE, specimens were prepared by thoroughly mixing glass beads (specific gravity 2.5) with 1% epoxy uniformly to create ensembles of density of 1.5 g/cc. Three types of specimens with particle diameter 0.5 mm, 1.0 mm, 2.5 mm were chosen to understand the particle size effect on the failure strength of weak c-ϕ granular materials. Specimens were cured by heating in oven at 50°C for 48 hours. Figure 3 shows the specimens prepared to study particle size effect. Similar procedure was followed to prepare SE specimen.

All the uniaxial compression tests were performed at a constant strain rate of 0.5% per min.

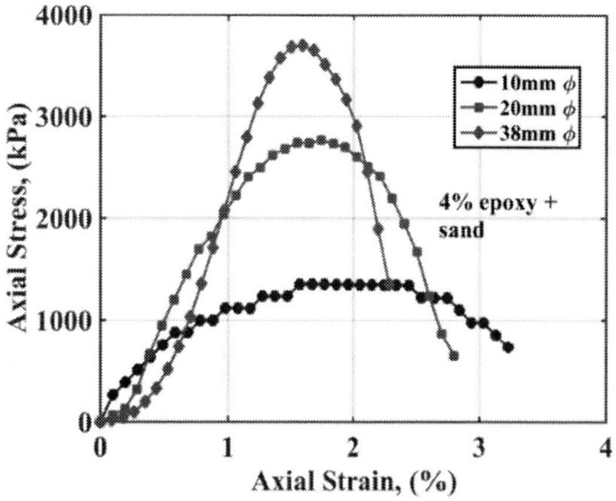

Figure 4. Stress strain response for cemented sand sample with specimen diameters of 10 mm, 20 mm, 38 mm

4 Results

A typical stress-strain plot is shown in figure 4 with varying specimen dimension (or with increasing number of particles) for SE specimen. The increase in specimen dimension not only reflects in the change of peak strength but also other features of stress behaviour such as elastic stiffness, failure stress, hardening, and softening are significantly affected. Other parameters such as mean particle size (0.45 mm), density (1.4 g/cc), binder content (4%) were kept constant for studying system size effect.

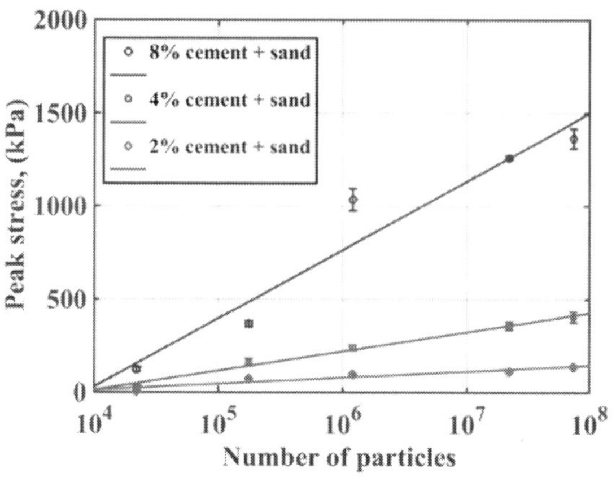

Figure 5. The peak strength increases with increase in the percentage of cementation as well as increase in the number of particles across the cross section or as the system size increases

This effect of system size on cemented sand specimens was studied for various amounts of cohesion of 2%, 4%, 8%, it should be noted that even with increase in the binder content, a contact bound structure was ensured. A plot of peak strength with number of particles (or specimen dimension) is presented in figure 5 along with a linear fit for each binder content. With increase in binder content normalized peak strength with number of particles (slope of fit lines) increases. This linear fit can be interpreted as the number of particles directly controlling the failure strength.

Figure 6 shows a typical stress-strain curve for GBE with 2% binder content. For understanding the effect of particle size; the density of the ensemble, specimen dimension and binder content were kept constant. With increasing particle diameter the failure strength correspondingly decreases. We refer to this as the particle size effect. We further studied this by varying binder content, however, ensuring a contact based structure. A plot of peak strength is presented for 1% and 2% epoxy in figure 7. The peak strength is shown to increase with increasing number of particles and binder content. The normalized peak strength with number of particles appears to be constant. Similar results were obtained for sand+epoxy specimen with changing mean grain size.

5 Discussion

Brown et. al [9] present a set of experiments on granular chains, especially highlighting the shear stiffening of

DOI: 10.1051/epjconf/201714008007

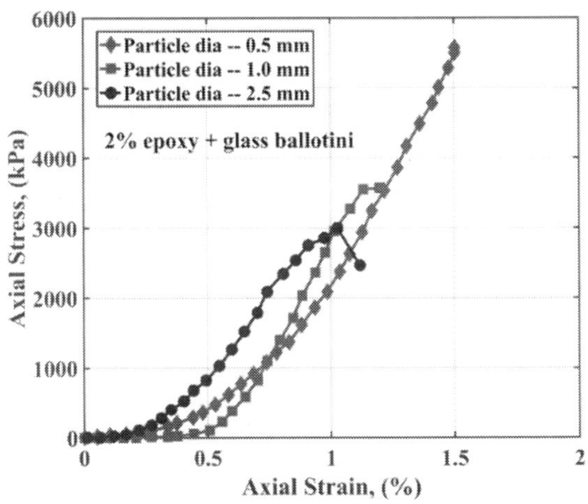

Figure 6. Stress strain response for GBE with changing particle dia for specimen diameter of 38 *mm*

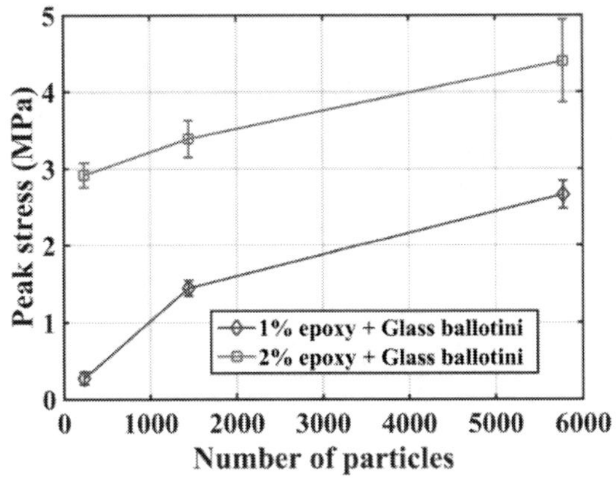

Figure 7. The peak strength increases with increase in the number of particles across the cross section or as the particle size decreases

random packed granular chains. In order to explain the interesting phenomenology of their experiments, the authors suggest that the increase in the length of a granular chain contributes to increased entanglement, in turn contributing to higher shear stiffening (or hardening). A similar increase in failure strength with specimen size (or number of particles) is observed in our experiments. We conjecture that the structure of weak c-ϕ materials can be construed as an ensemble comprising of multiple granular chains. Given that the structure of the ensemble is complex, we further suggest that this c-ϕ ensemble can be modelled as a series of knotted (entangled) granular chains. Brown et al. [9] further defined a minimum loop circumference (mlc) which was a function of particle diameter. With increase in the particle diameter the minimum loop circumference decreased for a given chain length whereas this minimum loop diameter also increases as the particle size decreased. The entanglement, or the minimum loop circumference governed the overall mechanical behaviour of the ensemble. If the chain length was smaller than mlc then material behaviour resembled a typical granular ensemble. If chain length was sufficiently larger than the mlc then significant strain stiffening was observed within a test along with an increase in peak strength as the chain length increases.

For weak c-ϕ materials considered in our experiments, system size or the particle size both affect the length of a "hypothetical granular chain". As the system size increases, we further conjecture that the lengths of these granular chain structures also increases, in other words, the possibility of obtaining more entangled structure increases. Similarly with increase in particle size although the chain length would remain the same, however, the equivalent mlc decreases which further causes reduction in entanglement measure.

Figure 5 and figure 7 represent the system size and particle size effect, respectively. The results are plotted between the peak strength and number of particles to support our argument of minimum loop circumference.

We present exploratory set of experimental results, and the first evidence of the existence of a second length scale beyond the traditional continuum length scale in weakly cohered-granular ensembles. Ongoing work is implementing these systems through extensive tomography for understanding the microstructure, and the manifestation of this microstructure at the ensemble level.

References

[1] Abdulla, Ali A and Kiousis, Panos D, International Journal for Numerical and Analytical Methods in Geomechanics **21**, 549–568 (1997)

[2] Vatsala, A and Nova, R and Murthy, BR Srinivasa, Journal of Geotechnical and Geoenvironmental Engineering **127**, 679–687 (2001)

[3] Gao, Zhiwei and Zhao, Jidong, Computers and Geotechnics **41**, 57–69 (2012)

[4] Lade, Poul V and Kim, Moon K, International Journal of Solids and Structures **32**, 1963–1978 (1995)

[5] Weibull, Waloddi, Journal of applied mechanic **103**, 293–297 (1951)

[6] Jefferies, M.G., Been, K. and Hachey, J.E., Canadian Geotechnical Conference **43**, (1990)

[7] Bazant, Zdenek P and Xi, Yunping and Reid, Stuart G, Journal of engineering Mechanics **117**, 2609–2622 (1991)

[8] Wang, YH and Leung, SC, Canadian Geotechnical Journal **45**, 29–44 (2008)

[9] Brown, Eric and Nasto, Alice and Athanassiadis, Athanasios G and Jaeger, Heinrich M, Physical review letters **108**, (2012)

[10] Sowers, GB and Sowers, GF, *Introductory soil mechanics and foundations* (The Macmillan Co., New York, 1961) 26–27

EPJ Web of Conferences **140**, 02007 (2017) DOI: 10.1051/epjconf/201714002007
Powders & Grains 2017

A 2D Fourier tool for the analysis of photo-elastic effect in large granular assemblies

Danuta Leśniewska*

Koszalin University of Technology, Department of Civil, Geodetic and Environmental Science, 74-453 Koszalin, Poland

Abstract. Fourier transforms are the basic tool in constructing different types of image filters, mainly those reducing optical noise. Some DIC or PIV software also uses frequency space to obtain displacement fields from a series of digital images of a deforming body. The paper presents series of 2D Fourier transforms of photo-elastic transmission images, representing large pseudo 2D granular assembly, deforming under varying boundary conditions. The images related to different scales were acquired using the same image resolution, but taken at different distance from the sample. Fourier transforms of images, representing different stages of deformation, reveal characteristic features at the three ('macro-', 'meso-' and 'micro-') scales, which can serve as a data to study internal order-disorder transition within granular materials.

1 Introduction

Fourier transforms are the basic tool in constructing different types of image filters, mainly those reducing optical noise. Some DIC or PIV software also uses frequency space to obtain displacement fields from a series of digital images of a deforming body. There exists one more possible way of utilizing 2D Fourier transforms for granular materials – they can be regarded as a direct subject of analysis themselves [1], similarly to Laue patterns in crystallography (physical Fourier transforms). The Lauegrams are used to measure internal structure of crystals from distances between some ordered intensity peaks. The idea of the approach proposed in this paper comes from the certain similarities between poly-crystals and granular materials - the former also show some kind of structural order, especially if so called 'force chain' network is studied. Investigation of 'force chains' structure in granular materials started about 1920 and was continued by several researchers up to now – the lack of space does not allow to discuss their achievements here.

The important problem still left to be solved is the quantitative interpretation of the structure of 'force chain' network, observed in photo-elastic experiments. Some objective tool is necessary to measure characteristic distances and angles od these networks and to connect them with the way granular material deforms. The additional problem is related to scale dependence – the force chain network seem to be the macro-scale feature, but it is necessary to determine some physical limits of the phenomenon. To illustrate the problem, the paper presents selected series of 2D Fourier transforms of photo-elastic transmission images, representing large (consisting of ~1e+6 grains) granular assembly, acquired during small scale geotechnical model tests.

2 Granular assembly and test details

The assembly investigated in this work consisted of small, uniformly graded glass granules (mass median diameter D_{50}~1mm, [2,3]) and can be classified as pseudo 2D, because it is placed in a narrow rectangular box (30cm x 27cm x 2cm), enforcing plane strain condition on macro (sample) scale.

It has to be remembered that individual grains are spherical and form several layers, none of them perfectly plane. The detailed description of the test layout was published elsewhere [2,3].

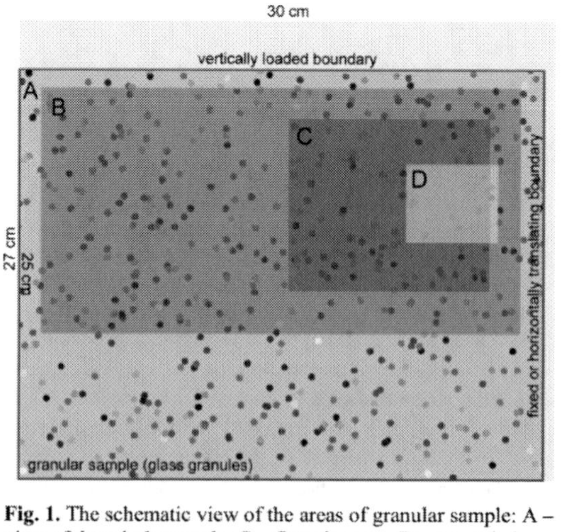

Fig. 1. The schematic view of the areas of granular sample: A – view of the whole sample, C – first close-up, D – second close-up. B – part of the area A, used in the analysis.

* Corresponding author: danuta.lesniewska@tu.koszalin.pl

The granular assembly deforms under varying boundary conditions (top boundary is loaded vertically and right boundary translates horizontally, Fig.1). Principle of integrated photo-elasticity is used (transparent grains immerged in immersion liquid – clove oil here). Specific type of circular polariscope was employed, equipped with rectangular polariser and analyser and thus giving much better quality of photo-elastic images of granular sample under load, than standard polariscopes.

The photographs presented in Fig.2, column A, were taken using RGB scientific grade camera and later transformed to grayscale images to calculate Fourier transforms. Fourier transforms of images, representing different stages of real sample deformation (Fig.2, column B) and reverse Fourier transforms of artificial images (Fig.3, column B) were calculated using commercial DaVis8 software by LaVision.

Separate images of three different areas of the sample (B, C and D, Fig.1) were analysed and are presented in this paper. They represent:

(1) the whole width of the sample ('macro' scale, the area of an image equal to ~250x200 D_{50}, B, ~28cm x 15cm, Fig.1),

(2) medium size part of the sample ('meso-scale', the area of an image equal to ~100x80 D_{50}, C, ~10cm x 8cm, Fig.1),

(3) (3) small part of the sample, representing the interior of the expected shear band ('micro-scale', the area of an image equal to ~40x30 D_{50}, ~5cm x 4cm, Fig.1).

The area B was selected to be analysed instead of the whole sample view A, to exclude any possible influence of some optical imperfectness found in the lower part of the images, on the shape of Fourier transforms.

The imperfection was created by slightly different colouring of the immersion liquid used in the tests. The images related to different scales were acquired using the same image resolution, but were taken at different distances from the granular sample. As a result the average grain is represented by ~10x10 pixels in the areas A and B, ~25x25pixels in the area C and ~50x50 pixels in the area D (Fig.1). The last resolution gives the opportunity to discern the edges of individual grains and to study the sample at micro scale.

3 Discussion of Fourier transforms

Fig.2, column A, shows different areas of the granular sample, photographed in circularly polarised light, using full camera resolution (~2400x2000 pixels). All images show photo-elastic response of the granular sample averaged over the sample thickness (about 20 layers of grains).

3.1 Real images

Fig.2 (column A) represents some typical experimental data – photo-elastic response of 2cm thick granular sample, consisted of uniformly graded glass granules. Due to dimensional uniformity of grains some crystallization effect can be expected.

The granular sample undergoes growing external loading on the top boundary (Fig.2, column A, rows 1-2, 6-7), some unloading (Fig.2, column A, rows 3 and 6) and some uniform displacement of the right boundary (Fig.2, column A, rows 3, 5, 7). Development of complex 'force chain' system, demonstrating clear geometrical structure can be seen in all cases, depending mainly on the value of external load. Dark bands appearing at the right side of the images coincide with shear band location, what was checked by PIV (DIC) method, [2,3]. Such dark, arched band is most clearly visible on Fig.2 column A, row 4 and 5. Darker colour on the image means weaker 'force chains', suggesting that some force chains must have been broken within the shear band.

It is worthy of noting that even after total unloading some strong 'force chains' remain stable at the bottom of the sample (Fig.2, column A, row 6), thus the sample does not fully return to the initial isotropic state.

The order of the 'force chain' system, clearly visible in Fig.2, column A, row 2 and others, by a naked eye (two privileged directions suggesting induced stress anisotropy) failed to be captured by any standard image analysis tools. Possible reason for that is the fact, that the 'force chains' are numerous and very thin structures (usually much thinner than the average grain diameter) and cannot be turned into continuous structures by any edge finding filter or image segmentation algorithm.

In effort to extract some quantitative information concerning organization of force chain network the photo-elastic images were transformed into frequency spectrum – Fig.2, column B shows Fourier transforms of images shown in Fig.2, column A.

It can be observed that Fourier transforms change with the value of external load at 'macro scale' (Fig.2, column A, rows 1-3), but any significant difference exist only between the first phase of relatively low value of external load (p = 0.2 MPa, Fig.2, column B, row 1) and the higher load levels

There exists the ring-shaped distribution in the power spectrum of Fourier transforms throughout the whole loading range at 'macro scale' – it is well known from crystallography that such a ring demonstrates periodicity at intermediate spatial frequencies, that is equivalent in all directions in the image (in case of this paper this is most probably the manifestation of the sample 'crystallization' due to significant mono dispersity of glass granules). In course of the loading process other objects appear in Fourier power spectrum, most likely related to emerging and stabilizing force chain network.

Fig.2, rows 4-6 represent the 'meso-scale' of the sample. If the Fourier transforms are compared to ones, obtained for 'macro-scale' (Fig.2, rows 1-3), it seems obvious, that the effect of 'crystallization' (ring-shaped distribution) is observable only at bigger scale. Fig.2B (row 6) shows that the sample unloading is equivalent to the transition from more anisotropic to more isotropic state. Fig.2B (row 7) suggests that at the scale close to a shear band width the stress anisotropy is less distinct, than at larger scales.

Area and the test conditions (Fig.1)	**A.** System of 'force chains' photographed in circularly polarized light (dark field)	**B**. Fourier transforms corresponding to images from column A
1. Area B in Fig.1 Vertical surcharge of the top boundary **p=0.2MPa**, right boundary fixed		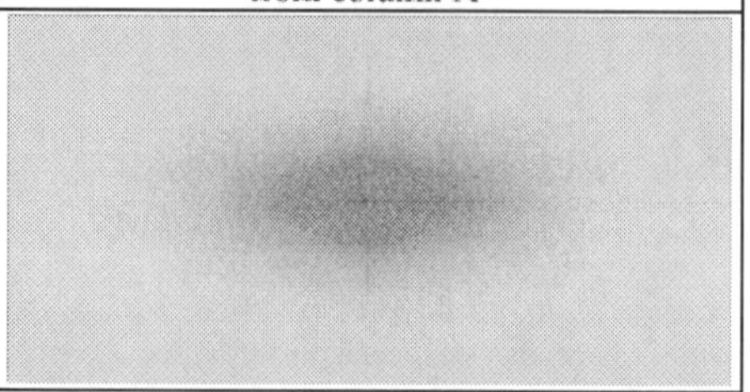
2. Area B in Fig.1 Vertical surcharge of the top boundary **p=1.6MPa**, right boundary fixed		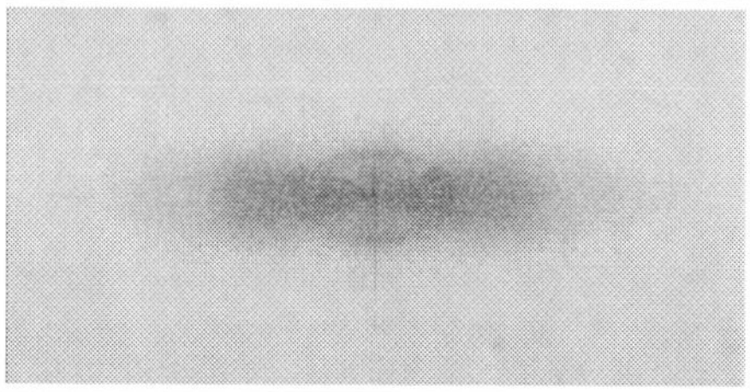
3. Area B in Fig.1 Vertical surcharge of the top boundary **p=0.8MPa** (unloading), right boundary displacement of **d=1.25mm**		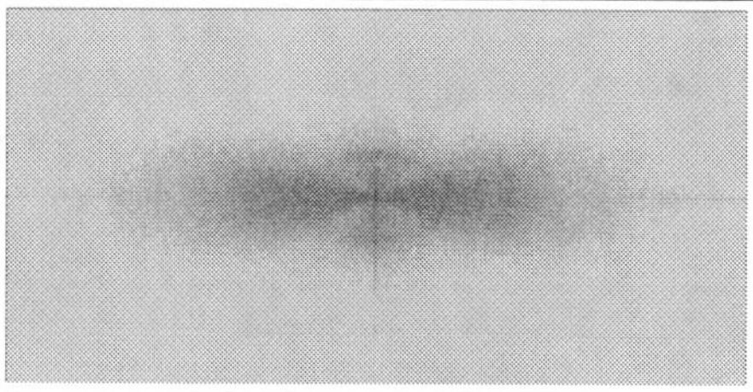
4. Area C in Fig.1 Vertical surcharge of the top boundary **p=1.6MPa**, right boundary initial displacement **d=1.25mm** - no displacement increment		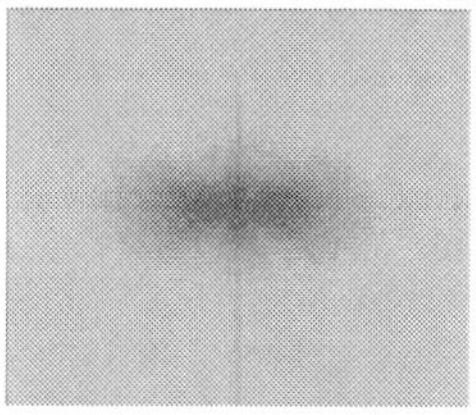
5. Area C in Fig.1 Vertical surcharge of the top boundary **p=0.8MPa**, right boundary initial displacement d_0=1.25mm, added **d=3.75mm**.		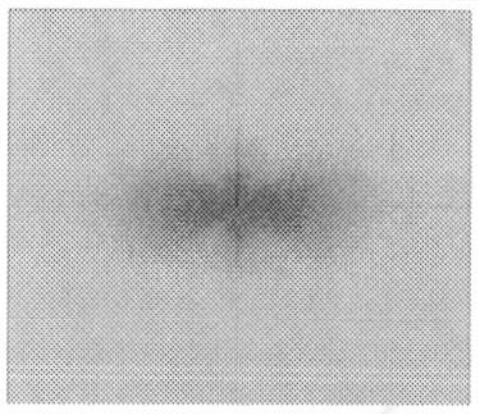
6. Area C in Fig.1 Vertical surcharge of the top model's boundary **p=0MPa** (unloading), right boundary initial displacement d_0=1.25mm, total **d=3.75mm**.		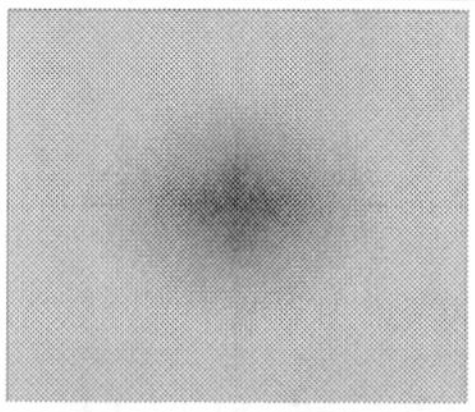
7. Area D in Fig.1 Vertical surcharge of the top model's boundary **p=1.6MPa** - unloading, right boundary initial displacement d_0=5mm, added **d=0.0625mm**.		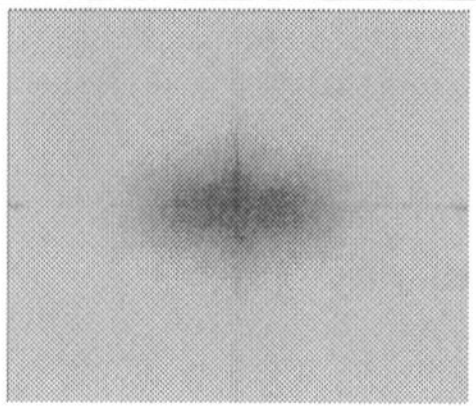

Fig. 2. Examples of photo-elastic images taken at different scales (column A) and their Fourier transforms (column B).

DOI: 10.1051/epjconf/201714002007

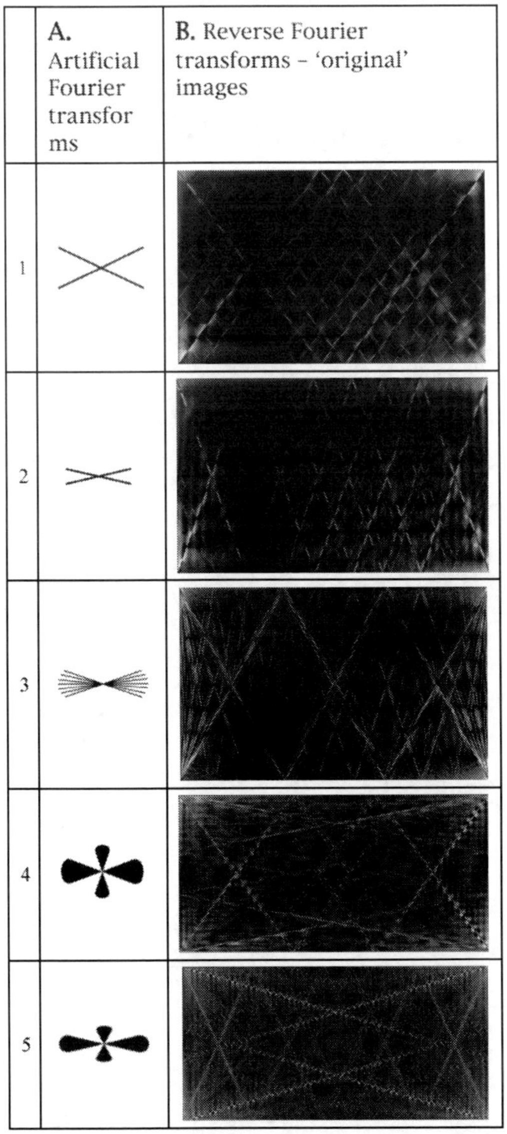

Fig. 3. Idealized components of Fourier transforms (A), sketched on the base of Fig.2, and their reverse (B), showing structures corresponding to different arrangement of the real force chain networks.

3.2 Artificial images

As the geometric features visible in Fig.2A (rows 2 and 3) are quite regular, it is possible to recognize the ones directly related to force chains and their arrangement. It can be agreed that Fourier transforms of Fig.2B (rows 2 and 3) consists of a ring-shaped area and a double fan, which can have complicated internal structure, but is most probably related to anisotropic force chain networks, visible in Fig.2, column A. At 'meso scale' Fourier transforms show lack of ring-shaped object and instead the second, smaller, vertically oriented double fun is present (Fig2B, rows 4-6). Column A of Fig.3 contains artificial images of features which could be recognized in Fig.2B (artificial Fourier transforms). Idealised elements of the fun present at macro-scale are postulated in column A, rows 1-3. The reverse of Fourier transforms calculated in Fig.3 (column B, rows 1-3) show that the angle of the fun is responsible for the range of directions of 'force chains' allowed in given experimental conditions – this range is visibly limited. The external directions (the edges of the fun) define the main 'force chain' network (Fig.3B, rows 1 and 2), the inside of the fun is responsible for the set of weaker 'force chains' directions (Fig.3B, row 3). Also some information on changing length of 'force chains' could be deduced from the dimensions of double fan. Two double funs, characteristic for 'meso scale' are approximated as continuous features in Fig.3A, rows 4 and 5. Corresponding reverse Fourier transforms are shown in Fig.3B, rows 4 and 5. Their structure suggests the presence of second set of two privileged directions in the granular sample, more horizontally oriented. Such family of lines exists in real images – they are dark, in contrast to bright 'force chains' and best visible in reality in Fig.2A, row 7 ('micro-scale'), but also present at 'meso scale'. Their exact nature has to be find yet.

4 Summary and conclusions

Fourier transforms of 'force chain' systems in granular materials reveal characteristic features at the three scales, which can serve as a data to study internal order-disorder transition within granular materials. Simplified artificial 'Fourier transforms', after reversing them to artificial 'original images', give insight into different components of 'force chains' network, produced by collective photo-elastic response of the real assembly. Comparison between artificial and real images at three scales are presented in the paper to the limited extent. Proposal of determining some characteristic material lengths, like average dimensions of 'force chains' mesh, average characteristic distance between neighbouring 'force chains', average length of 'force chains' at different stages of deformation, is under preparation.

Acknowledgement

This work was done within the frame of the project nr.2011/03/B/ST8/05865, Polish National Science Centre.

References

1. T. Prum, R.H. Torres, Integr. Comp. Biol. 43, 591-602 (2003)
2. D. Lesniewska, D. Muir Wood, J Eng Mech 135:1038–1054 (2009)
3. D. Muir Wood, D. Lesniewska, Granul Matt 13:395–415 (2011)

Automated discrete element method calibration using genetic and optimization algorithms

Huy Q. Do, *Alejandro M.* Aragón, and *Dingena L.* Schott*

Faculty of Mechanical, Maritime and Materials Engineering, Delft University of Technology, Mekelweg 2, 2628CD Delft, The Netherlands

Abstract. This research aims at developing a universal methodology for automated calibration of microscopic properties of modelled granular materials. The proposed calibrator can be applied for different experimental set-ups. Two optimization approaches: (1) a genetic algorithm and (2) DIRECT optimization, are used to identify discrete element method input model parameters, *e.g.*, coefficients of sliding and rolling friction. The algorithms are used to minimize the objective function characterized by the discrepancy between the experimental macroscopic properties and the associated numerical results. Two test cases highlight the robustness, stability, and reliability of the two algorithms used for automated discrete element method calibration with different set-ups.

1 Introduction

Use of the discrete element method (DEM) has been rising within the powder, raw material and bulk material handling industry for analyzing, and designing material handling systems. However, a major barrier to the effective use of DEM for industrial applications is selecting appropriate input parameters so that simulations can accurately reproduce the behavior of real systems. A calibration approach is commonly chosen for the determination of these parameters that are not easily measured in experiment. Calibration using DEM simulations is actually an iterative process of adjusting input parameters such that the macroscopic results of simulations and experiments are equivalent. "Trial and error" empirical procedures are very time-consuming and unpractical, particularly for DEM models involving many parameters being calibrated. Some attempts have been recently made to improve the calibration procedure, *e.g.*, experimental design [1-4], artificial neutral network training [5], and Latin hypercube sampling and Kriging [6]. The readers are referred to [7] for a complete review on calibration. This study aims to improve the DEM calibration procedure using two different optimization methods: genetic algorithms (GAs) and the DIRECT.

2 Calibration methods

The flowchart of the calibration method is illustrated in Fig. 1. This iterative process couples the optimization algorithms with DEM modelling. In each iteration, the optimization algorithm uses the objective function values to update the design parameters, *e.g.*, the coefficients of rolling (μ_r) and sliding (μ_s) friction. The DEM model inputs these parameters to provide the simulated macroscopic result, *e.g.*, the angle of repose (*AoR*) that is required to compute the objective function. If the predefined stopping criterion has been reached the process terminates, otherwise the next iteration will be performed.

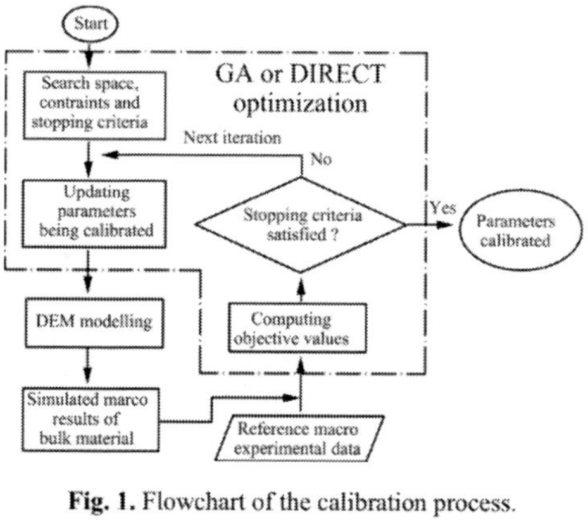

Fig. 1. Flowchart of the calibration process.

2.1 Genetic algorithms

Genetic algorithms are search and optimization methods inspired by Darwin's theory of evolution of natural selection and genetics. In GAs, an initial population of candidate solutions evolves through generations (iterations) towards individuals with a better fitness by applying genetic operators such as selection, crossover, and mutation. The fitness function is measured by an objective function of the optimization problem being solved. In each iteration, the better fit individuals are stochastically selected from the current population for breeding to create a new generation. The population is

* Corresponding author: d.l.schott@tudelft.nl

then updated with the new generation and used for the next iteration of the process. The algorithm terminates when either fitness values of the individuals satisfying a predefined criterion have been found or a predefined maximum number of generations has been obtained or the best fitness value has reached a plateau such that successive iterations no longer produce any improvement.

For further details of GA's theory and applications, the readers are referred to the work of Goldberg [8]. The core components of GA's implementation for our calibration problems are described as the follows:

Initial population and encoding: each calibrated parameter, *i.e.*, μ_r and μ_s, is binary {0,1} encoded and represents a gene. The concatenation of these two genes creates an individual. Initially, a group of individuals is randomly generated and should be seeded on the entire range possible solutions.

Fitness function: defined as the discrepancy between DEM model results and experimental measurement data.

Selection: depending on their fitness values, more fit individuals are chosen for crossover.

Crossover: takes chromosomes from selected parents and produces offsprings for the next generation.

Mutation: promotes genetic diversity of the solution set and aims at preventing the GA from converging to a local optimal solution.

2.2 DIRECT optimization

Another calibration approach used in this study is DIRECT [9], that is an algorithm for finding the global minimum of a multivariate function subjected to simple bounds without using gradient information. The algorithm was implemented as a subroutine within an open-source library for nonlinear optimization, providing a common interface for several optimization algorithms [10].

DIRECT is a modification of the standard Lipchizian approach, that eliminates the need to specify a Lipschitz constant. The concept of the algorithm is to carry out simultaneous searches on both global and local levels using all possible constants. The DIRECT means "DIviding RECTangles"; this is an iterative process partitioning the search space into a set of hyper-rectangles containing a single sampling point at the centre of each rectangle. Initially, the algorithm normalizes the search space by transforming it into the unit hypercube. The objective function is then evaluated at the centre point of this cube. In the first iteration, the unit hypercube is then divided into small hyper-rectangles whose centre points are also sampled. In next iterations, all potentially optimal hyper-rectangles are further divided into smaller rectangles whose centre points are sampled. During each iteration, the Graham's scan algorithm is used to find a convex hull that identifies a set of potentially optimal hyper-rectangles. The iterative process continues until either the predefined number of iterations has been reached or a predefined criterion for the objective function or for the variables have been achieved.

3 DEM simulations and reference results

To illustrate the calibration approach, LIGGGHTS [11], an open source software, is chosen to conduct the DEM simulations using Hertz-Mindlin contact law with a rolling resistance model [12]. To numerically reproduce the experiments involving hundreds of thousands of sand grains with their actual size, spherical shape is used to reduce the computational complexity. Hourglass and rectangular container simulations on quartz sand grains are carried out to calibrate two main microscopic contact model parameters, *i.e.*, the coefficients of rolling and sliding friction. The rolling friction coefficient of spheres is here introduced to include the shape effect [13]. The experimental measurement data and DEM simulation set-ups of previous work [14] in our research group have been used. Some properties of quartz sand used as the DEM modelling input parameters are listed in Tables 1 and 2. Three macroscopic experimental results are being used here as the references: *AoR* of 41.57° and the discharging time of 6.56 s for the hourglass test, and *AoR* of 36° for the rectangular container test. The readers are referred to [14] for further details of these two set-ups.

Two sample test cases for calibrating μ_r and μ_s are proposed. In the first sample test case, only the hourglass modelling for 25 g of sand with the hourglass neck diameter of 5 mm is used to provide simulated values of *AoR* and discharging time with varying μ_r and μ_s. The DEM simulations were conducted at 36 sample points in the μ_r-μ_s plane using an evenly spaced regular grid. The contour plots in Fig. 2 and 3 illustrate the *AoR* and discharging time as functions of the two variables being calibrated, respectively. Compared against the measured values of *AoR* at 41.57° and discharging time at 6.56 s, the reference result of the friction coefficients is visualized in Fig. 4, which corresponds to the intersection of the two contour lines.

The second sample test case uses the previous hourglass investigation and another set of 6x6 DEM simulations on the rectangular container to provide simulated values of *AoR* with varying μ_r and μ_s. The contour plot in Fig. 5 illustrates the *AoR* as functions of these two variables. Compared against the measured values of discharging time at 6.56 s for the hourglass and *AoR* at 36° for the rectangular container, the intersection of the two contour lines as shown in Fig. 6 is considered as the reference result.

Table 1. Input particle size distribution of sand for DEM simulations [14].

Particle diameter (μm)	300	425	500	600
Mass fraction (%)	6.21	18.29	26.05	49.45

Table 2. Input properties of sand for DEM simulations [14].

Particle density (kg/m^3)	2653
Young's modulus (N/m^2)	5e7
Poisson ratio	0.3
Coefficient of restitution	0.9
Coefficient of rolling friction	0.0-0.5
Coefficient of sliding friction	0.2-07

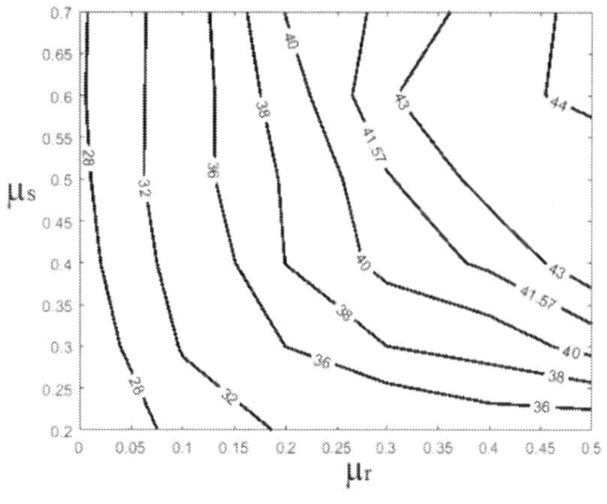

Fig. 2. *AoR* in degree as a function of the rolling and sliding coefficients from the hourglass simulations [14].

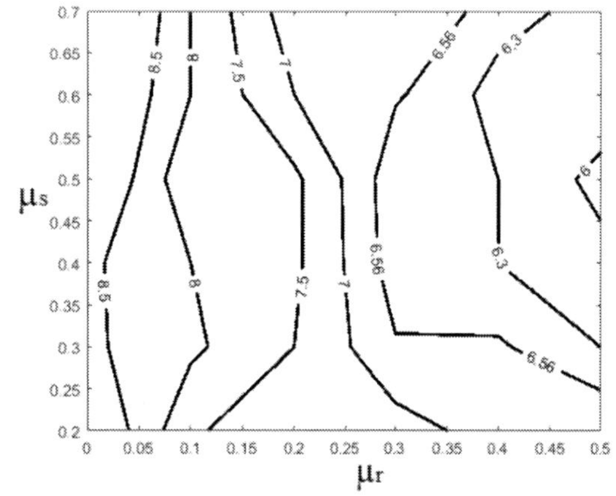

Fig. 3. Discharging time in second as a function of the rolling and sliding coefficients from the hourglass simulations [14].

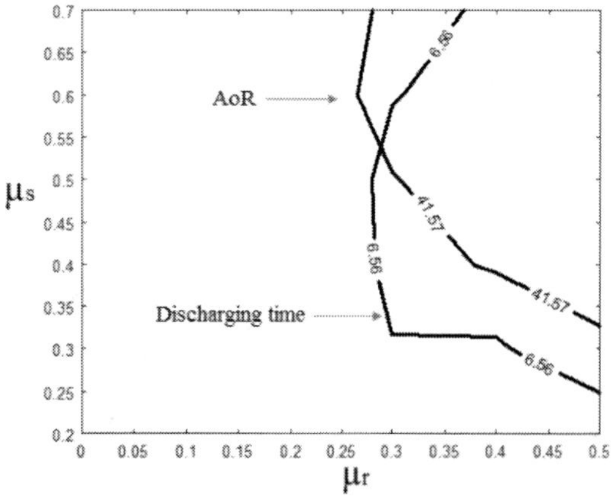

Fig. 4. Reference result defined by the intersection of simulated two contour-lines of *AoR* and discharging time referred from the experimental results for hourglass [14].

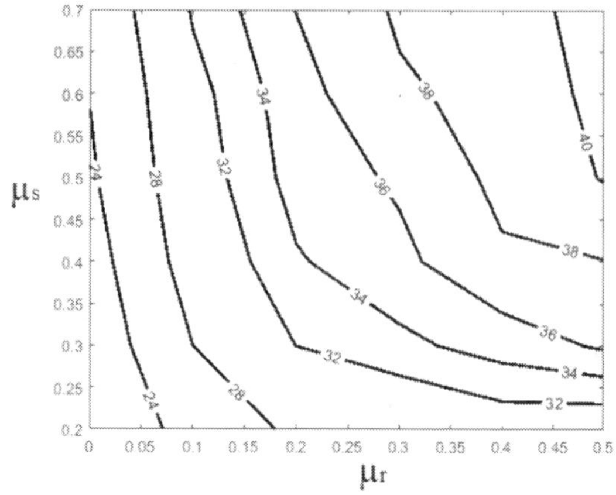

Fig. 5. *AoR* in degree as a function of the rolling and sliding coefficients from the rectangular container simulations.

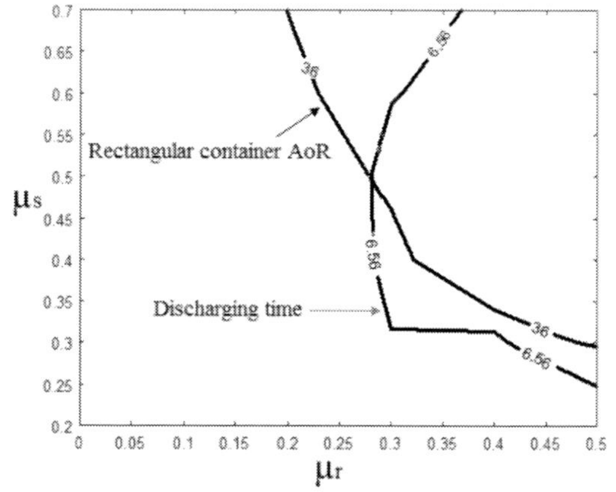

Fig. 6. Reference result defined by the intersection of simulated two contour-lines of hourglass discharging time and rectangular container *AoR* referred from the experimental results.

4 Results and Discussion

DEM simulations are time-consuming. To reduce computational time during the calibration process, the grids of 6x6 sample points in Fig. 2 and 3 are reused for bilinear interpolating values of *AoR* and discharging time for arbitrary μ_r and μ_s.

4.1 Calibration using hourglass - first sample test case

4.1.1 Result from GA

Table 3. Input parameters for the GA.

Population size	500
Maximum of generation	30
Chromosome length	20
Crossover probability	1.0
Mutation probability	0.005

Input parameters for the GA are listed in Table 3. A population size of 500 individuals is found to be large enough to be distributed stochastically on the search space. Using its length of 20 bits, a chromosome allocates 10 bits for encoding μ_r in the range from 0.0 to 0.5 and 10 other bits for encoding μ_s in the range from 0.2 to 0.7.

The fitness function ($\mathbf{F}_{fit}$) has been chosen as the total discrepancies on *AoR* and discharging time (*Tdis*) between the numerical results and experimental measurement data from the hourglass as:

$$\mathbf{F}_{fit} = |AoR_{sim} - 41.57| + |Tdis_{sim} - 6.56| \quad (1)$$

100 GA runs are repeatedly performed to search for the possible solutions considering stochastic oscillation. After 30 iterations of all runs, the optimal fitness value has reached a plateau of 0.020± 0.015, such that successive iterations no longer produce any improvement. The computed rolling and sliding friction coefficients associated with the optimal fitness are 0.284±0.002 and 0.543±0.003, respectively. The result is in good agreement with the reference result as shown in Fig. 4.

4.1.2 Result from DIRECT

The objective function is defined the same way as the fitness function in equation (1). The relative tolerance of the two optimization variables are set to 10^{-3} to define a stopping criterion for the algorithm.

Terminating after 259 evaluations, DIRECT outputs the optimal objective value of 0.01, corresponding to the rolling and sliding friction coefficients of 0.286 and 0.541, respectively. These results are in good agreement with GA's outcomes in 4.1.1.

4.2 Calibration using hourglass and rectangular container - second sample test case

4.2.1 Result from GA

The fitness function has now been redefined as the total discrepancies between the numerical results and experimental measurement data on discharging time from the hourglass and *AoR* from the rectangular container as:

$$\mathbf{F}_{fit} = |Tdis_{sim1} - 6.56| + |AoR_{sim2} - 36| \quad (2)$$

100 GA runs are carried out targeting the new fitness function. After 30 iterations of all runs, the computed rolling and sliding friction coefficients are 0.277±0.004 and 0.502±0.003, respectively, and their associated optimal fitness value is 0.007±0.005. As expected, the result is in good agreement with the reference result as shown in Fig. 6.

4.2.2 Result from DIRECT

The objective function is defined the same way as the fitness function in equation (2). DIRECT has terminated after 275 evolutions. The calibrated values of the coefficients of rolling and sliding friction and their associated optimal value are 0.281, 0.496 and 0.015, respectively; this result is identical to the GA's outcome in 4.1.2.

5. Conclusions

This research has demonstrated that the micro properties of quartz sand in DEM model can be automatically calibrated using two methods, *i.e.*, GA and DIRECT optimizations. Both of them are capable of identifying accurately the optimal values within a fairly small number of iterations. The algorithms are robust and stable since they need to evaluate only the fitness or objective function without computing its derivatives. Interpolation grids of the bulk properties resulted from DEM simulations can be incorporated within GA and DIRECT optimizations to reduce the computational cost. The findings of this paper show that the calibration methods can be deployed for a large set of unknown input parameters and other experimental set-ups.

References

1. J. Yoon, Int. J. Rock. Mech. Min. Sci., **44**(6), 871-889 (2007)
2. J. Favier, D. Curry, R. LaRoche, *Proc. 6th World Congr. Particle. Technol.*, Nuremberg, Germany (2010)
3. M. Johnstone, *Calibration of DEM models for granular materials using bulk physical tests*, PhD thesis, University of Edinburgh, UK (2010)
4. K.J. Hanley, V. O'Sullivan, J.C. Oliveira, K. Cronin, E.P. Byrne, Powder. Technol., **210**(3), 230-240 (2011)
5. L. Benvenuti, C. Kloss, S. Pirker, Powder. Technol., **291**, 456-465 (2016)
6. M. Rackl, K.J. Hanley, Powder. Technol., **307**, 73-83 (2017)
7. C.J. Coetzee, Powder. Technol., **310**, 104-142 (2017)
8. D.E. Goldberg, *Genetic algorithm in search, optimization and machine learning*, Adisson-Wesley publishing, Reading Massachusetts, USA (1989)
9. D.R. Jones, C.D. Perttunen, B.E. Stuckman, J. Optim. Theory. Appl., **79**(1), 157-181 (1993)
10. http://ab-initio.mit.edu/NLopt
11. C. Kloss, C. Goniva, A. Hager, S. Amberger, S. Pirker, Prog. Comput. Fluid. Dy., **12**(2/3), 140-152 (2012)
12. J. Ai, J. F. Chen, J. M. Rotter, J. Y. Ooi, Powder. Technol., **206**(3), 269-282 (2011)
13. N. Estrada, E. Azéma, F. Radjai, A. Taboada, Phys. Rev. E, **84**(1), 011306 (2011)
14. S. M. Derakhshani, D.L. Schott, G. Lodewijks, Powder. Technol., **269**, 127-138 (2015)

EPJ Web of Conferences **140**, 15012 (2017) DOI: 10.1051/epjconf/201714015012
Powders & Grains 2017

Periodic cells for large-scale problem initialization

Matteo O. Ciantia[1*], *Marcos* Arroyo[2], *Ningning* Zhang[2], and *Sacha* Emam[3]

[1] Dept. Civil and Environmental Engineering, Geotechnics Section, Imperial College London, UK
[2] Dept. Civil and Environmental Engineering, Geosciences Division, UPC, Spain
[3] Itasca Consultants, S.A.S., 64 Chemin des Mouilles, 69130 Écully, France

Abstract. In geotechnical applications the success of the discrete element method (DEM) in simulating fundamental aspects of soil behaviour has increased the interest in applications for direct simulation of engineering scale boundary value problems (BVP's). The main problem is that the method remains relatively expensive in terms of computational cost. A non-negligible part of that cost is related to specimen creation and initialization. As the response of soil is strongly dependant on its initial state (stress and porosity), attaining a specified initial state is a crucial part of a DEM model. Different procedures for controlled sample generation are available. However, applying the existing REV-oriented initialization procedures to such models is inefficient in terms of computational cost and challenging in terms of sample homogeneity. In this work a simple but efficient procedure to initialize large-scale DEM models is presented. Periodic cells are first generated with a sufficient number of particles matching a desired particle size distribution (PSD). The cells are then equilibrated at low-level isotropic stress at target porosity. Once the cell is in equilibrium, it is replicated in space in order to fill the model domain. After the domain is thus filled a few mechanical cycles are needed to re-equilibrate the large domain. The result is a large, homogeneous sample, equilibrated under prescribed stress at the desired porosity. The method is applicable to both isotropic and anisotropic initial stress states, with stress magnitude varying in space.

1 Introduction

The discrete element method (DEM) was first proposed to study fundamental aspects of soil response [1] and that is still the main field of application of this numerical method in soil mechanics [2]. Typically, small cubic or cylindrical samples are generated to obtain a representative elementary volume (REV), which is later tested to explore different links between micromechanical features and mesomechanical response. Element tests may be performed using boundaries that mimic conditions typical of physical soil tests (e.g. rigid walls, membranes) or else using periodic boundaries (e.g. [5]). Periodic boundaries have the advantage of eliminating boundary-effects thus offering direct access to material responses.

Soil mechanical responses are strongly dependent on state. At the mesosocopic level this is typically described by stress and porosity. Ensemble properties such as particle size distribution (PSD) are also significant. Hence attaining a specified initial state of stress and porosity for given PSD, is a crucial step of DEM modelling in soil mechanics. Different procedures are available for initialization. Those used more often include the Radius Expansion Method [1], the Fixed Point Method [3], the Isotropic Compression Method [4], the Modified Isotropic-compression Method [5] and the multi-layer Under Compaction Method [6]. These procedures have been tried and tested with some success to create REV's suitable for element testing, but pose important problems when the method is used to address engineering scale geotechnical problems. Indeed, when large models -large in terms of model dimension to particle size- are assembled the computational cost of system initialization quickly becomes prohibitive. Moreover, specimen homogeneity is difficult to attain [7, 8]. Because periodic boundaries do not appear naturally in the definition of most engineering-scale BVP the possibilities they offer for model building have not been explored before.

In the following the cell repetition method (CRM), a simple but efficient procedure to initialize large-scale DEM models, is presented. In the CRM periodic cells are first generated matching a desired PSD. The cells are then equilibrated at low-level isotropic stress at the target porosity. Once the cell is in equilibrium, it is replicated in space in order to fill the model domain. For uniform stress states only a few mechanical cycles are needed to re-equilibrate the large domain after the domain is thus filled. By means of contact force scaling, the procedure is then extended for cases when the stress state varies in magnitude bot not in orientation (i.e. the principal stress directions are uniform within the domain). This allows to initialize a DEM model under prescribed k_0 conditions, applying a gravitational field and simultaneously controlling the horizontal stress (hence k_0 where $k_0 = \sigma'_h/\sigma'_v$). As far as we know this is a novel procedure, with important implications for the simulation of BVP. All the simulations described in this paper were performed using the PFC3D software [9].

* Corresponding author: m.ciantia@imperial.ac.uk

2 Periodic cells and REV

We use the problem of REV generation as an example of the potential of the method for DEM initialization under homogenous stress fields.

2.1 REV Identification: traditional approach

The REV size is selected to be as small as possible to ease computation, but big enough so that the observed mechanical response is stable with further size increases. This criterion is usually applied by trial and error comparing performance of different-sized cells. Small (with N = 1118 particles), medium (N = 8884) and large periodic cells (N = 29885), were prepared from a PSD with mean radius of 0.5 mm and standard deviation of 0.2. Using the Isotropic Compression Method [4], by setting the contact friction to zero, a dense state (porosity of 0.365) at 100 *kPa* of isotropic pressure was obtained (Fig. 1). Triaxial compression tests were then performed on the three cells using a linear contact model with normal contact stiffness of 1000 *MPa*, *kn/ks* ratio of 2.5 and contact friction coefficient (μ) of 0.25. The three samples are first loaded isotropically to 1 *MPa* and then sheared until the axial strain reaches a value of 30%. The results are shown in Fig. 2.

Fig. 1. Periodic cell sample of 1118 (left), 8884 (centre) and 29885 (right) spherical particles (figures not in scale).

All samples tested showed a similar response and the large fluctuations clearly visible for the small sample reduce as the cell size (particle number) is increased. Depending on the application, either the medium or the large models would be selected as REV. What is here most relevant is to note that the equilibrated initial state for this test was obtained after 35, 85 and 240 minutes for the small medium and large samples respectively.

2.2 REV from periodic cell repetition

Now that the REV minimum dimension has been identified, to exploit the idea of the CRM, the equilibrated small sample (which was discarded as a REV) is replicated 8 and 27 times in space in order to fill a cubic cell with twice and three times its initial size respectively. These two new samples are now tested under the same triaxial conditions as the precedent ones. The results are shown in Fig. 3. It appears that the new samples, created from replicas of the precedent small cell are both behaving as REV's. The time needed to generate these samples in an equilibrated condition from replicas of the small cell resulted to be 5 and 10 minutes respectively. If we add the 35 minutes needed to generate the small cell, it appears that we have obtained medium and large samples in 40 and 45 minutes, or, respectively, 47% and 19% of the time spent with the standard procedure.

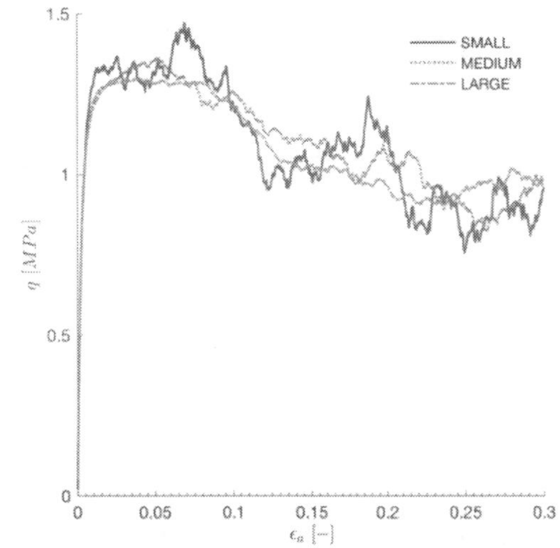

Fig. 2. Deviatoric stress (q) axial strain (ε_a) response of the small, medium and large DEM models under triaxial compression at cell pressure of 1 *MPa*.

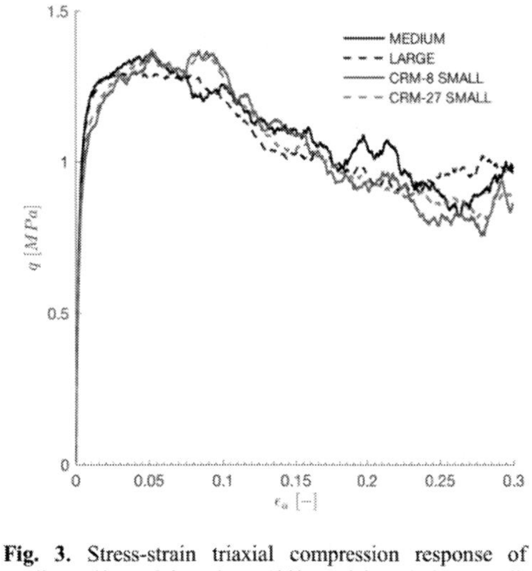

Fig. 3. Stress-strain triaxial compression response of the medium (8k particles), large (30k particles), 8 times replicated small and 27 times replicated small periodic cell samples.

3 Periodic cells and BVP

To illustrate the computational efficiency of the CRM and a possible application of this approach, the CRM is here used to initialize a large scale 2D domain, of base B=αL and height H=βL, under a 10g gravitational field. The intent is to use the DEM as a virtual centrifuge to simulate spudcan penetration test. Spudcan penetration is a challenging problem in offshore geotechnical engineering, for which centrifuge model testing has been regularly employed [12] and that is currently simulated through a variety of continuum-based methods [13].

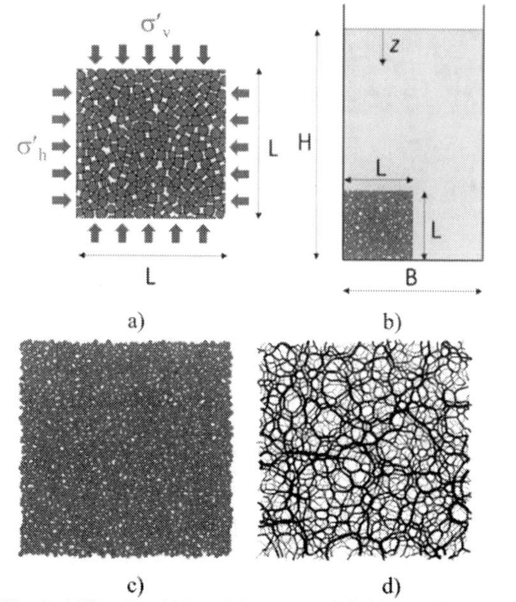

Fig. 4. a) 2D cell and b) model geometry definition, c) 2511 disk periodic cell and d) corresponding CF for the equilibrated sample.

The approach followed to initialize this BVP for DEM based simulation is schematized in Fig. 4 and can be summarized as follows:

- Generate a LxL 2D periodic cell and equilibrate a sample under a stress state such that i) the effective vertical stress corresponds to the vertical stress the ground should have at a depth of H and ii) the horizontal stress corresponds to the target k_0.
- Replicate the cell α times in the horizontal direction and β times in the vertical direction.
- For each replica added, loop through all the contacts and scale the contact force, CF by a ratio z/H where z is the distance from the ground level to the position of the contact.
- Install gravity and fix a thin layer of particles in contact with the horizontal boundaries.
- Cycle to equilibrate.

Fig. 5 illustrates some aspects of the equilibrated models, namely the distribution of vertical particle stresses and the contact force distribution. The time required to prepare the model was of roughly 30 minutes. The geometrical dimension of the problem and the number of particles used to generate the model are summarized in **Error! Reference source not found.**. To compare the computational advantage of using the CRM, a new sample, having the same number of disks was generated at a higher porosity and equilibrated under gravitational filed. The time required to equilibrate the sample under 1g resulted to be around 11000 minutes (just more than 1 week). It is clear that for such application the CRM is really efficient compared to the pluviation method used to recreate k_0 conditions. The response of the model to spudcan penetration is illustrated in figures 6 and 7. The response features illustrated appear qualitatively correct.

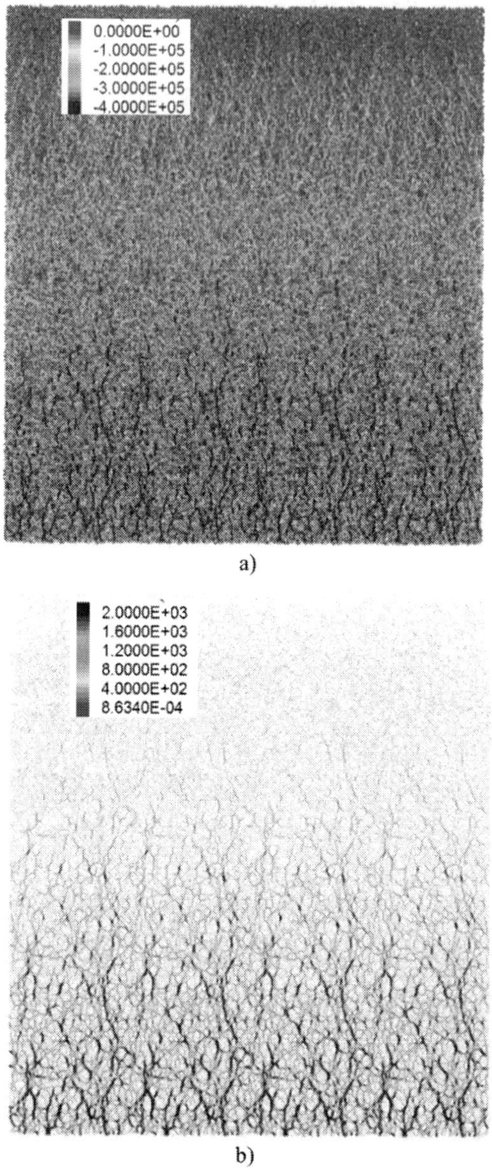

Fig. 5. Model showing (a) the disks coloured by particle σ'_{yy} stress and (b) contact forces (CF) scaled by CF magnitude.

Table 1

H (cm)	B (cm)	L (cm)	Number of disks per cell	Number of disks
100	100	25	2511	45910

4 Conclusions

In this paper a novel approach for fast initialization of DEM models has been presented. The cell repetition method (CRM) builds upon a simple idea: using equilibrated periodic DEM models to fill larger spaces. This idea is complemented by force scaling to initialize anisotropic stress fields of magnitude variable in space. Although the anisotropic case has been illustrated with a 2D example, extension to 3D appears straightforward. The computational efficiency, control of initial conditions and homogeneity of the generated specimen give the CRM characteristics that make it very attractive for simulations requiring large scale DEM models.

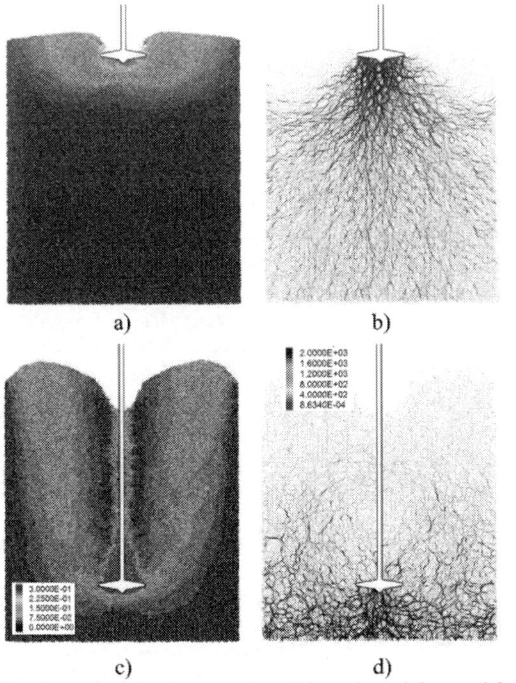

Fig. 6. model showing a,c) the disks coloured by particle displacement and b,d) contact forces scaled by CF magnitude after 0.1m and 0.9m of spudcan penetration in 10g gravitational field respectively.

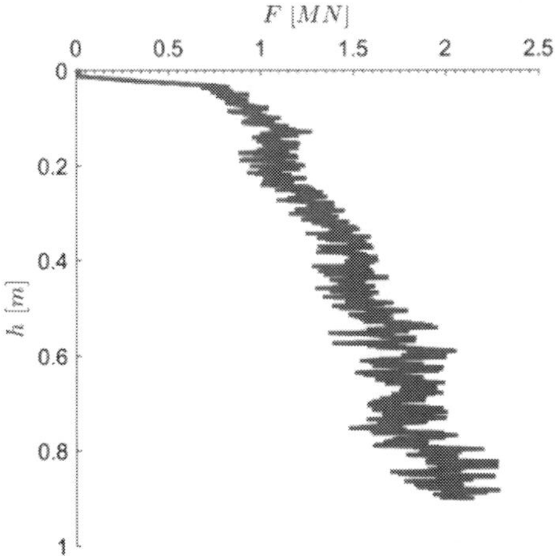

Fig. 7. Force displacement curve during spudcan penetration.

Acknowledgments

The work presented here has been supported by the EU PROJECT RISE GEO-RAMP (H2020 MSCA-RISE-2014 GA 645665), by the Spanish ministry of Economy (SIMGESUB BIA2014-59467-R) and as part of an Imperial College London Junior Research Fellowship Research Grant.

References

1. P. A., Cundall & O. D. L. Strack (1979). A discrete numerical model for granular assemblies. *Géotechnique*, 29(**1**), 47–65.
2. O'Sullivan, C. (2011). Particulate discrete element modelling. Taylor & Francis.
3. Katsuki S, Ishikawa N, Ohira Y, Suzuki H. (1989) Shear strength of rod material. *Journal of Civil Engineering*, 410, 1–12 [in Japanese].
4. Rothenburg L, Bathurst RJ. (1992). Micromechanical features of granular assemblies with planar elliptical particles. *Géotechnique,* 42(**1**), 79–95.
5. Thornton C. (2000). Numerical simulation of deviatoric shear deformation of granular media. *Géotechnique*, 50(**1**), 43–53.
6. Jiang, M. J., Konrad, J. M., & Leroueil, S. (2003). An efficient technique for generating homogeneous specimens for DEM studies. *Computers and geotechnics*, 30(**7**), 579-597.
7. Butlanska, J., Arroyo, M. & Gens, A. (2009) Homogeneity and symmetry in DEM models of cone penetration, Powders & Grains 2009, Nakagawa & Luding (eds), AIP Conference Proceedings 1145, 425-429
8. Ciantia, M. O., Arroyo, M., Butlanska, J., & Gens, A. (2016). DEM modelling of cone penetration tests in a double-porosity crushable granular material. *Computers and Geotechnics*, 73, 109-127.
9. Itasca Consulting Group. (2016). PFC3D: Itasca Consulting Group, Inc. (2014) PFC — Particle Flow Code, V.5.0 Minneapolis: Itasca Consulting Group.
10. Ciantia, M., Arroyo, M., Calvetti, F. & Gens, A. (2015) An approach to enhance efficiency of DEM modelling of soils with crushable grains *Géotechnique,* 65(**2**) 91-110.
11. Ciantia, M. O., Arroyo, M., Calvetti, F., and Gens, A. (2016) A numerical investigation of the incremental behavior of crushable granular soils. *Int. J. Numer. Anal. Meth. Geomech*., 40, 1773–1798.
12. Gaudin, C., Cassidy, M. J., Bienen, B., & Hossain, M. S. (2011). Recent contributions of geotechnical centrifuge modelling to the understanding of jack-up spudcan behaviour. *Ocean Engineering*, 38(**7**), 900-914.
13. Tho, K. K., Leung, C. F., Chow, Y. K., & Swaddiwudhipong, S. (2010). Eulerian finite-element technique for analysis of jack-up spudcan penetration. *International journal of geomechanics*, 12(**1**), 64-73.

Patterns of granular convection and separation in narrow vibration bed

Chuanping Liu[1,*], *Ping* Wu[2], *Li* Wang[1], *Lige* Tong[2], Shaowu Yin[1]

[1]School of Energy and Environment Engineering, University of Science and Technology Beijing, 100083, Beijing, China
[2]School of Mathematics and Physics, University of Science and Technology Beijing, 100083, Beijing, China
[3]Beijing Engineering Research Centre of Energy Saving and Environmental Protection, 100083, Beijing, China

Abstract. Granular convection/separation of single and binary component particles are studied in a narrow vibration bed, respectively. With filling the single light particles (molecular sieve beads), the bed exhibits five different states successively by increasing the vibration frequency f from 15Hz to 70 Hz (vibration strength $\Gamma>3$), as the global convection, symmetrical heap, unsymmetrical heap, local convection and pseudo solid. Comparatively, the granular bed of the single heavy particles (steel beads) is only in pseudo solid state at the above vibration condition. By filling binary component particles (molecular sieve and same size steel beads) instead of the single component, the bed shows similar convection state with that of the single molecular sieve beads, and the heavy steel beads are aggregated in the centre of convention roll as a core. Varying the initial distribution of binary component particles, the final convection and separation are not influenced, although the aggregation process of steel beads changes.

1 Introduction

Granular matter wildly exist in industry and agriculture. Due to the complex force chain structure among particles and the dissipation of kinetic energy during inelastic collision, granular matter usually exhibits interesting behaviors different with traditional liquid and solid, such as convection [1,2] and separation [3,4].

Under externally vibration, binary mixture would show diffrent types of separation. Size segregation is known as the Brazil nut (BN) effect when large particles rise to the top; by contrast, in the reverse effect, called the reverse BN (RBN) effect, large particles sink to the bottom. Separation may appear in particle species with different density besides different size. King et al. [5,6] investigated equal-sized glass–copper system and found that heavy copper powder rises to the bed top, whereas light glass powder sinks to the bed bottom at low vibration frequencies. However, at high vibration frequencies, the mixture separates into the "Sandwich" configuration, with copper powders distributed at the middle of the bed. Shi et al. [7] found that light particles tend to rise and form a pure layer on the top of the granular bed, with a mixed part of heavy particles and some light particles at the bottom of the bed; these particles form a partially separation state. Besides "BN", "RBN" and "Sandwich" separation, several studies [8-11] reported that clusters can be formed in the granular bed.

Convection is a representative mechanism used to explain granular separation [4]. With the upward stream of granular convection, the large particles rise to bed top; however, they cannot sink back with the downward stream, as the large particles size is larger than the convection zone width of the downward stream, resulting in "BN" separation. Only mild convection rates result in size separation because strong convections can enhance mixing [12]. Yang et al. [9] indicated that two convection cells exist in a vibrated system, and heavy particles concentrate near the cell centers. Hsiau [11] investigated the effect of container geometry on density separation. At the few filling layers, heavier particles migrate to the bottom of the bed. For the high filling layers, the bulk convection motion is the main mechanism that affects separation; in this case, heavier particles accumulate in the center of convection.

Ehrichs et al. [13] measured the velocity distribution of the particles in cylinder vibration bed by NMR. The particles move upward in the middle of the bed and downward along the wall, resulting a pair of symmetric convection. In the experiments of Wong et al. [14], for the frequency range from 14Hz to 18Hz, an unsymmetrical convection roll with the formation of heap against one of the sidewall was observed. And Akiyama's research [15] indicated that the interstitial gas plays a non-negligible role on granular heap. According to the simulation performed by Majid [8], vibration parameters and container geometries also are the important factors effecting the granular convection. Eshuis [16] and Zhang [17] observed multiple patterns of granular motion and presented their phase diagram by varying the vibration amplitude and frequency.

Serval literatures [4,9,11] indicated that the separation, especilly density separation, is induced by

* Corresponding author: cpliu@ustb.edu.cn

granular convection. This paper aimed to further study the granular convection and separation under vertical vibration, and discuss the mutual relations between convection and separation. Firstly, the single component particles [13X molecular sieve (13XMS, sintered aluminosilicate adsorbing material) / steel beads] are filled in vibration bed, and different types of convection patterns are observed at our experimental condition, which are influenced by both the vibration parameters, especially vibration frequency. Secondly, the 13XMS and the same size steel beads are initially layered in bed. The convection patterns and separation process of binary component particles are analyzed by comparing with that of single component.

2 Experimental setting

The following experiments are performed in a narrow rectangle glass bed of size 200 mm (length) × 5 mm (width) × 220 mm (height). As shown in Figure 1, the bed is fixed at a vibration platform of a Brüel & Kjær standard vibrating system (LDS V555), which can generate sinusoidal vibration in the form of $z = A\sin(2\pi ft)$, where A and f are the vibration amplitude and frequency, respectively. The experimental particles include spherical 13XMS and steel beads. The physical properties of the two kinds of particles are listed in Table 1. Bulk density (ρ_b) is the ratio of the total mass of single particle piles to the volume that these particles occupy, including the voids. A charge-coupled device (CCD) camera was fixed at the front of the bed to record particle motions in vibration bed.

The single component or the binary component particles are initially filled into glass bed with a certain granular layer height (h). In our experiments, vibration frequency (f) is adjusted from 20 Hz to 70 Hz, and dimensionless vibration acceleration (Γ) from 0 to 8, where, $\Gamma = A(2\pi f)^2/g$ and g is gravitational acceleration. We recorded the particle motion at the bed front by using a CCD camera. Basing on the continuous images from the 200 fps CCD recordings, we got the movement of the particles in vibration bed.

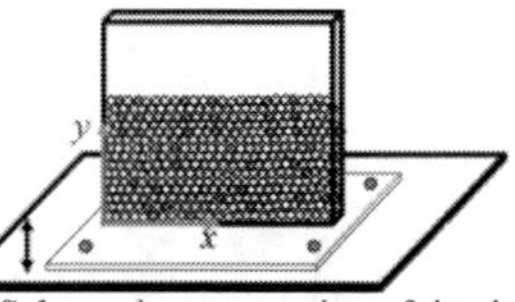

Figure 1. Schematic presentation of the device.

Table 1. Physical properties of the two kinds of particles selected in our experiment.

Material	Diameter d (mm)	Material density ρ_m (g/cm^3)	Bulk density ρ_b (g/cm^3)	Sphericity Φ
13XMS	2.00 ± 0.10	1.47 ± 0.08	0.82 ± 0.08	0.92
Steel beads	2.00 ± 0.05	8.60 ± 0.05	4.82 ± 0.26	0.96

3 Experimental results

3.1 Convection of single component

We filled the 13XMS beads into the bed with 160 mm height (h), and then vibrated it at Γ = 4.0. The granular bed exhibits five different states by increasing the vibration frequency gradually from 15 Hz to 70 Hz [Figure 2(I)~(VI)].

(1) Global convection. When the vibration frequency f < 25 Hz, the whole granular bed is completely fluidized under large amplitude vibration. As shown in Figure 2(I), The particles at the middle of the bed rise to the bed top quickly. At the bed surface, particles bound from bed center to sidewall and then sink along the bed wall. The whole bed shows one pair of symmetrical strong global convection roll.

(2) Symmetrical heap. By increasing frequency (25 Hz < f < 34 Hz), the vibration amplitude decreases and the particles in bed are cooled. Comparing with the global convection, although the granular convection always exists, its strength weakens obviously. The bottom center of the granular bed starts to be solidified, that is, the particles in the bottom center vibrate in place and cannot migrate to another position. Meanwhile, a symmetrical granular heap occurs at the bed top. The convective particles rise from the bed middle to the heap peak, and then flow downward at the heap surface, as shown in Figure 2(II).

(3) Unsymmetrical heap. When the frequency increases further (34 Hz < f < 47 Hz), the solidified zone extends to the whole bed bottom and only the particles at the upper bed take part in convection. The original unsymmetrical form of the heap is broken, resulting in one pair of unsymmetrical convection roll with the rising point deviating from the center of the bed, as shown in Figure 2(III). One roll extends over a larger volume of the bed than the other, and the convection strength of the larger rolls is higher than that of the smaller one.

(4) Local convection. When the frequency 47 Hz < f < 66 Hz, most of the bed is solidified except the bed top near the wall. The granular convection only occur at the top corner of the bed and the bed surface re-becomes flat, as shown in Figure 2(IV).

(5) Pseudo solid. When the frequency f > 66 Hz, the whole bed is almost solidified. All the particles vibrate in place and the bed surface about keeps flat, as shown in Figure 2(V).

By varying both the vibration frequency f and acceleration Γ, we can got the phase diagram of the above five patterns of granular motion, as shown Figure 3(a). When Γ<1.2, the granular bed is in pseudo solid state as the vibration strength is too weak. When Γ>3, the patterns are mainly influenced by the vibration frequency, and the vibration acceleration is a minor

DOI: 10.1051/epjconf/201714003031

factor. The Γ only less affects the f range of each granular motion pattern.

By filling the steel beads into the bed with h=160 mm, and vibrating the bed at Γ = 4.0, the granular bed only exhibits Pseudo solid state during the frequency 15 Hz < f < 70 Hz. The bed surface keeps flat, as shown in Figure2(VI), being same with Figure 2(V). As the steel beads are much heavier than 13XMS, the bed filled heavy steel beads cannot be fluidized at same vibration condition. The steel beads vibration together and keep relative still each other. Changing the vibration parameter, the steel beads hardly generate any convection during 15 Hz < f < 70 Hz and 0<Γ<8. The phase diagram of steel beads only has the pseudo solid state, as shown in Figure 3(b).

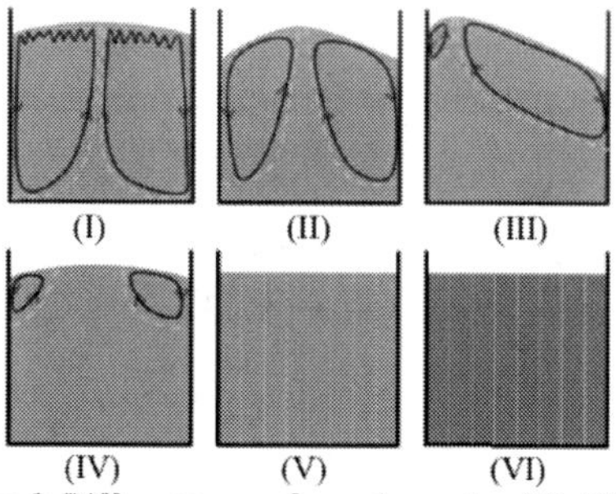

Figure 2. Different types of granular motion [(I)~(V) for 13XMS and (VI) for steel beads]

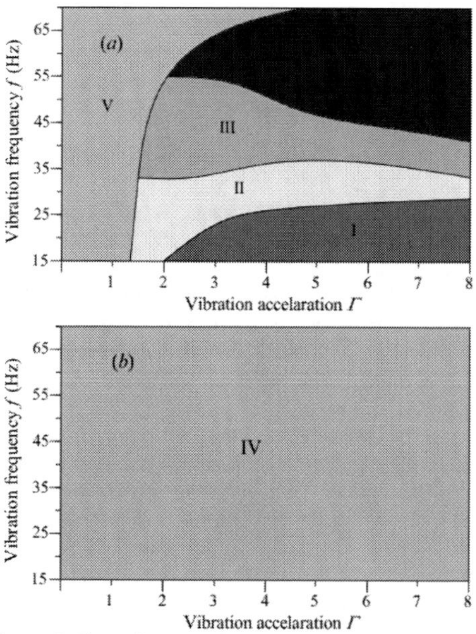

Figure 3. Phase diagram of granular motion [(a) for 13XMS and (b) for steel beads]

3.2 Convection and separation of binary component

We filled the 13XMS beads into the bed with 144 mm height (h) and then covered 16 mm height steel beads layer on the 13XMS [Figure 4(a1)]. Once vibrating the bed at Γ = 4.0 and f = 60Hz, the granular convection occurs, being similar with Figure 2(IV). The steel beads on the bed top gradually move toward sidewall and aggregate at the top corner [Figure 4(a2)]. The 13XMS beads move upward from the bed center, then over the steel beads layer [Figure 4(a3)]. As the 13XMS beads cover on steel beads and do convective motion, the steel beads are wrapped in the 13XMS. After the vibration starts 5 min, the bed exhibits a stable distribution, as shown in Figure 4(a4), Most of the heavy steel beads cluster in the center of the convection, as a core of convection roll.

The 16 mm steel beads layer was initially placed below the 13XMS beads [Figure 4(b1)]. By vibrating the bed at Γ = 4.0 and f = 60Hz, we recorded the steel beads movement. The steel beads at the bottom firstly rise slowly and move through the 13XMS bed [Figure4 (b1)~(b4)]. After about 10min, they reach the bed surface, as Figure 4(b5). Then the steel beads move toward sidewall and finally format a convection core [Figure 4(b6)].

By varying the vibration frequency, the granular convection and separation of the 13XMS beads also change. Figure 5(a)~(d) show the stable convection and separation at f = 20 Hz, 30 Hz, 40 Hz and 60 Hz. As the volume ratio of steel beads is much less than that of 13XMS, the effect of steel beads have hardly effect on granular convection, which is almost same with that of single 13XMS bead [Figure 2(I)~(IV)]. In the solidified zone, there are few steel beads. The steel beads are wrapped in the granular convection roll as a granular core. The much strong the convection is, the looser the core (steel beads) is.

We filled the steel beads into the bed with 144 mm height (h) and then covered 16 mm height 13XMS layer on the steel beads [Figure 6(a1)]. The bed was vibrated at Γ = 4.0 and f = 60Hz. At the beginning, the steel beads are in solidified state, and the 13XMS beads at the bed surface move toward sidewall, then downward along the sidewall [Figure6(a2)]. Afterward, the 13XMS beads reach the bottom corner, as Figure 6(a3). After about ten minutes, the bed reaches stable state. It is observed that the 13XMS beads move downward along the sidewall and migrate upward slowly thought the steel beads based on the mass conservation laws. According to Figure 2(VI), the bed of single steel beads cannot generate convection at this vibration condition. However, by adding a small number of light 13XMS beads, the bed is activated. The 13XMS beads near the sidewall play the role of lubrication.

When the 16 mm 13XMS beads layer was initially placed below the steel beads [Figure 6(b1)], the 13XMS beads first rise through the steal beads bed and reach the bed surface [Figure6(b2)~(b4)]. Then these 13XMS beads move along the sidewall, and finally format a stable distribution same with Figure 6(a4).

Figure 6(a)~(d) show the stable convection and separation at f = 20 Hz, 30 Hz, 40 Hz and 60 Hz. The convection patterns of the binary component also are similar with as that of single 13XMS bead [Figure 2(I)~(IV)]. By comparison, the convection strength is weak as a large number of steel beads are difficult to fluidize.

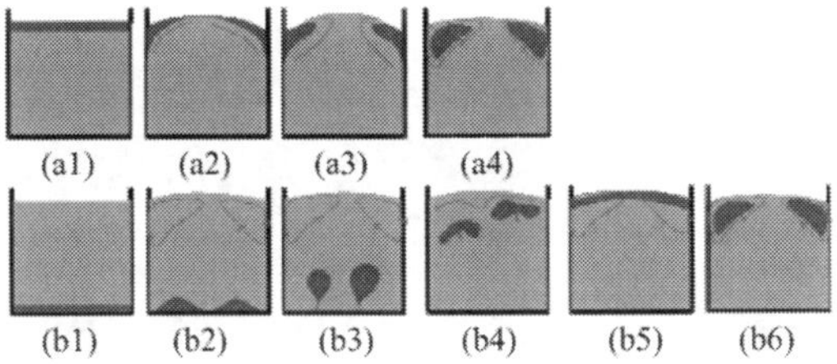

Figure 4. Formation process of steel beads core. The dark presents steel beads (10% volume ratio) and the peak 13XMS (90% volume ratio).

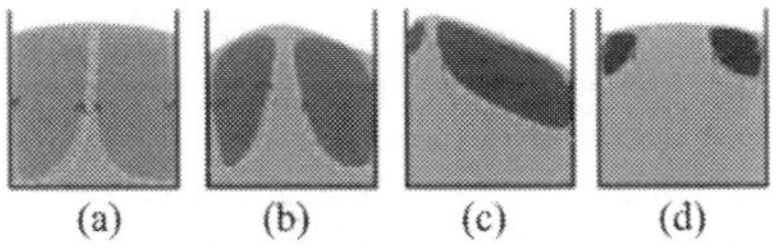

Figure 5. Separation of 13XMS and steel beads. The volume ratio of 13XMS and steel beads are 90% and 10%, respectively.

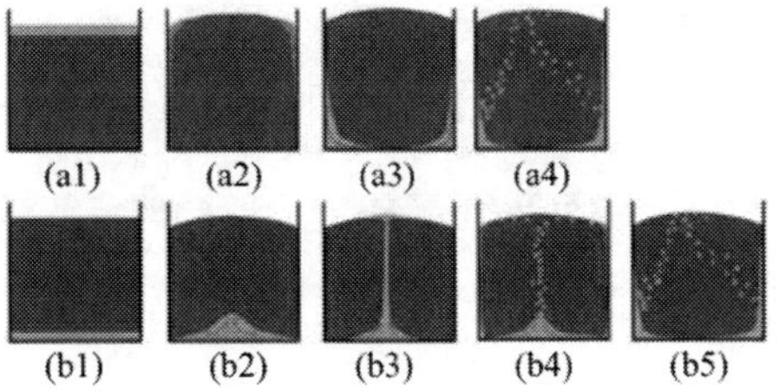

Figure 6. Formation process of steel beads core. The dark presents steel beads (90% volume ratio) and the peak 13XMS (10% volume ratio).

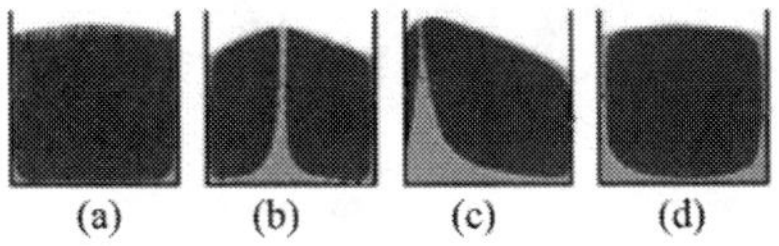

Figure 7. Separation of 13XMS and steel beads. The volume ratio of 13XMS and steel beads are 10% and 90%, respectively.

4 Conclusion

By vertically vibrating single light particles (13XMS beads) in a narrow bed, the bed shows five different types patterns of granular motion, as global convection, symmetrical heap, unsymmetrical heap, local convection and pseudo solid state, by increasing the vibration frequency from 15 Hz to 70 Hz ($\Gamma > 3.0$). The alteration between these patterns is mainly influenced by vibration frequency, and the acceleration is a minor factor. However, the heavy particles (steel beads) cannot be fluidized at these vibration condition and the bed only shows pseudo solid state.

When a small quantity of heavy particles (10% volume ratio steel beads) are added into the 13XMS beads bed, the granular convection is hardly effected. And the steel beads cluster in the center of the convection, as a core of convection roll. The initial distribution of binary particles will influence the aggregation process of steel beads, however, the final convection and separation are not influenced.

When 10% volume ratio 13XMS beads are added into the steel beads bed, the 13XMS will make the bed flowing. The bed changes from the pseudo solid to global convection, symmetrical heap, unsymmetrical heap or local convection, which is decided the convection of 13XMS. Comparing with the single 13XMS, the binary granular bed has a weak convection strength.

Acknowledgement

The authors are grateful for support from the National Natural Science Foundation of China (Grants No. 51506006).

References

1. Y. H. Taguchi, Phys. Rev. Lett., 69, 1367 (1992)
2. M. A. Keiko, A. Tetsuo, Phys. Rev. Lett., 77, 4166 (1996)
3. A. D. Rosato, K. J. Strandburg, F. Prinz, R. H. Swendsen. Phys. Rev. Lett., 58, 1038(1987)
4. J. B. Knight, H. M. Jaeger, Phys. Rev. Lett., 70, 3728 (1993)
5. N. Burtally, P. J. King, M. R. Swift, Science, 295, 1877, (2002)
6. P. Biswas, P. Sánchez, M. R. Swift, P. J. King, Phys. Rev. E, 68, 050301 (2003)
7. Q. Shi, G. Sun, M. Hou, K. Lu, Phys. Rev. E, 75, 061302 (2007)
8. T. Akiyama, K. M. Aoki, K. Yamamoto, T. Yoshikawa, Granular Matter, 1, 15 (1998)
9. S. C. Yang, Powder Technol., 164, 65 (2006)
10. S. S. Hsiau, W. C. Chen, Adv. Powder Technol., 13, 301(2002)
11. C. H. Tai, S. S. Hsiau, C. A. Kruelle, Powder Technology, 204, 255 (2010)
12. D. Brone, F. J. Muzzio, Phys. Rev. E, 56, 1059 (1997).
13. E. E. Ehrichs, H. M. Jaeger, G. S. Karczmar, J. B. Knight, V. Y. Kuperman, S. R. Nagel, Nature, 267, 1632 (1995)
14. Y. S. Wong, C. H. Gan, C. H. Wang, X. Fan, D. J. Parker, A. Ingram, J. P. K. Seville, Phys. Fluids, 18, 043302 (2006)
15. M. Majid, P. Walzel, Powder Technol., 192, 311 (2009)
16. P. Eshuis, K. van der Weele, D. van der Meer, R. Bos, D. Lohse, Phys. Fluids, 19, 123301 (2007)
17. F. Zhang, L. Wang, C. Liu, P. Wu, S. Zhan, Phys. Lett. A, 378, 1303 (2014)

Micromechanics of soil responses in cyclic simple shear tests

Liang Cui[1,*], *Subhamoy* Bhattacharya[1] and *George* Nikitas[1]

[1]Department of Civil and Environmental Engineering, University of Surrey, Guildford, GU2 7XH, United Kingdom

Abstract. Offshore wind turbine (OWT) foundations are subjected to a combination of cyclic and dynamic loading arising from wind, wave, rotor and blade shadowing. Under cyclic loading, most soils change their characteristics including stiffness, which may cause the system natural frequency to approach the loading frequency and lead to unplanned resonance and system damage or even collapse. To investigate such changes and the underlying micromechanics, a series of cyclic simple shear tests were performed on the RedHill 110 sand with different shear strain amplitudes, vertical stresses and initial relative densities of soil. The test results showed that: (a) Vertical accumulated strain is proportional to the shear strain amplitude but inversely proportional to relative density of soil; (b) Shear modulus increases rapidly in the initial loading cycles and then the rate of increase diminishes and the shear modulus remains below an asymptote; (c) Shear modulus increases with increasing vertical stress and relative density, but decreasing with increasing strain amplitude. Coupled DEM simulations were performed using PFC2D to analyse the micromechanics underlying the cyclic behaviour of soils. Micromechanical parameters (e.g. fabric tensor, coordination number) were examined to explore the reasons for the various cyclic responses to different shear strain amplitudes or vertical stresses. Both coordination number and magnitude of fabric anisotropy contribute to the increasing shear modulus.

1 Introduction

Designing foundations for Offshore wind turbine (OWTs) are challenging as these are dynamically sensitive structures in the sense that natural frequencies of these structures are very close to the forcing frequencies [1]. A designer apart from predicting the global natural frequency of the structure, must also ensure that the overall natural frequency due to dynamic-soil-structure-interaction does not shift towards the forcing frequencies [2, 3]. The global natural frequency is dependent on the soil-structure stiffness, thus an investigation of soil stiffness under cyclic loading is desired. Majority of current OWTs are supported by monopile foundation, which is a large steel tube typically 30-40m in length and 3.5-6m in diameter. Unlike the slender piles for offshore structure, monopile tends to rotate rather than bend under lateral load or overturning moment. Therefore, the interactions between the monopile and the surrounding soil could be represented by cyclic simple shear tests. Cyclic responses of soil stiffness have been studied intensively [4, 5], but few of them were focused on the micromechanism underlying these responses.

The aim of this paper is to study the soil stiffness responses under cyclic simple shear loading with a focus on the micromechanics. Experimental cyclic simple shear tests on typical silica sand (RedHill 100) is first presented. DEM simulations performed using PFC2D are then described, and the macroscopic stiffness and volumetric responses are validated against experiments. The micromechanical parameters are then analysed to find the relationships between the micromechanical parameters and macroscopic responses.

2 Experimental tests

Cyclic simple shear apparatus was used for testing cylindrical samples of 50mm in diameter and 20mm in height. The sample height-to-diameter ratio is 0.4, agreeing with the ASTM D6528 [6] requirements. The tested soil is RedHill 110 Sand with d_{50}=0.18mm (PSD curve shown in Fig. 1), a specific gravity G_s=2.65 and minimum and maximum void ratio of 0.608 and 1.035 respectively. Cyclic simple shear tests with different vertical stress (σ), relative density (D_r), and shear strain amplitude (γ_{max}) were performed.

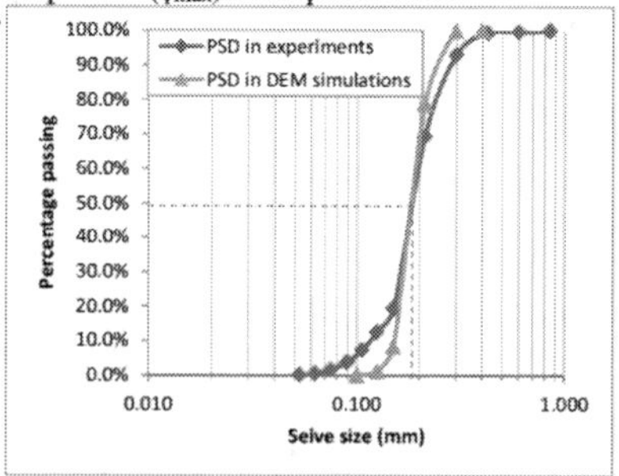

Fig. 1. Particle size distribution (PSD).

The variation of shear modulus and accumulated vertical strain are illustrated in Figs. 2 and 3,

* Corresponding author: l.cui@surrey.ac.uk

respectively. The shear modulus increases rapidly in the initial loading cycles and then the rate of increase diminishes and the shear modulus remains below an asymptote. The increase of shear modulus is underlain by the consistent contractive responses of all samples. The shear modulus increases with increasing vertical stress and relative density, but decreases with increasing strain amplitude as expected. It can also be observed that vertical accumulated strain is proportional to the shear strain amplitude but inversely proportional to the relative density of soil. The results correlated quite well with the observations from scaled model tests with different types of offshore wind turbine foundations [2, 3, 7]. The only abnormal trend observed is that the accumulated vertical strain decreases with increasing vertical stress. This is mainly due to the fact that more soil consolidation was resulted by the application of higher vertical stress and the initial relative density has increased to a value much higher than 50%. It could be deduced that the initial relative density has more significant impact on the vertical strain than the vertical stress does. It also shows the difficulties in controlling the relative density for various vertical stresses in sample preparation.

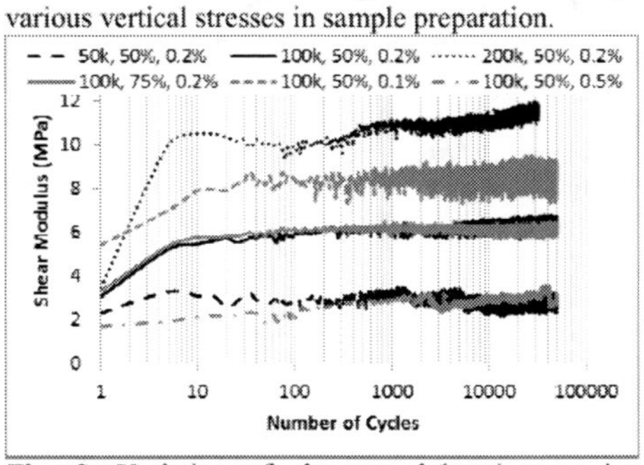

Fig. 2. Variation of shear modulus in experiments (Legend: σ, D_r, γ_{max}).

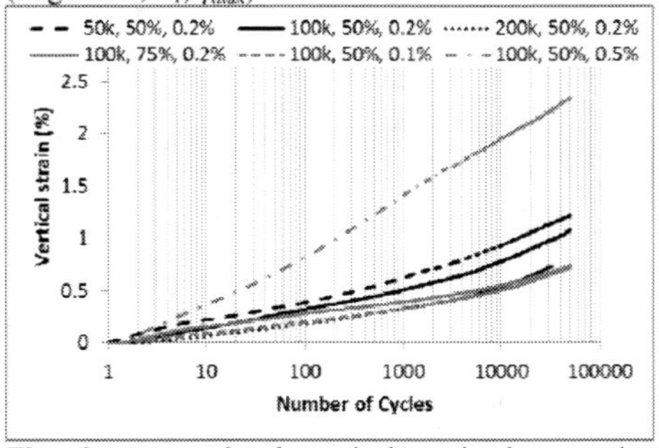

Fig. 3. Accumulated vertical strain in experiments (Legend: σ, D_r, γ_{max}).

3 DEM simulations

3.1. Description of DEM setup

DEM simulations of cyclic simple shear tests were performed using PFC2D [8]. The sample generated for testing is about 20mm in height and 50mm in width, containing 8000 disks with size ranging from 0.1mm to 0.3mm and d_{50}=0.18mm, matching the value in experiments. The PSD curve is also given in Fig. 1. Particle density is 2650 kg/m^3. Inter-particle and particle-boundary frictional coefficient, μ, is 0.5. Normal stiffness of particle is 8.0×10^7 N/m and shear stiffness of particle is 4.0×10^7 N/m. Normal and shear stiffness of boundary are 4.0×10^9 N/m. Two sets of samples were generated with radius expansion approach followed by K_0 consolidation with μ =0 and 1.0 to generate dense and loose samples, respectively. The void ratios of dense samples at consolidation pressure of 50kPa and 100kPa are 0.185 and 0.181, respectively. The void ratios of loose samples at consolidation pressure of 50kPa and 100kPa are 0.215 and 0.227, respectively. Note that due to different boundary velocities in the consolidation stage, the loose sample reached slightly lower void ratio at 50kPa than that at 100kPa. 6000 cycles of shear strain in the range of (0, 0.5%) were applied to each sample. 12 measurement circles were defined within the sample to measure the stress, void ratio and coordination number.

3.2 Macro-scale responses

The shear modulus (G) was calculated as the ratio of the difference between shear stress at maximum shear strain (τ_{max}) and shear stress at zero strain (τ_0) to the maximum shear strain (γ_{max}). The variations of shear modulus and vertical strain of the four samples under cyclic loading are illustrated in Fig. 4 and Fig. 5, respectively. The magnitudes of shear moduli in the DEM simulations are in the same range as the values in the experimental tests. For both loose samples, there is a clear increase in the shear modulus under cyclic load. For the dense sample with vertical stress σ=100kPa, shear modulus also increases slightly; however, for the dense sample with vertical stress σ=50kPa, shear modulus decreases. After 1000 cycles, the shear moduli of the two loose samples exceed the modulus of one dense sample at low confining pressure.

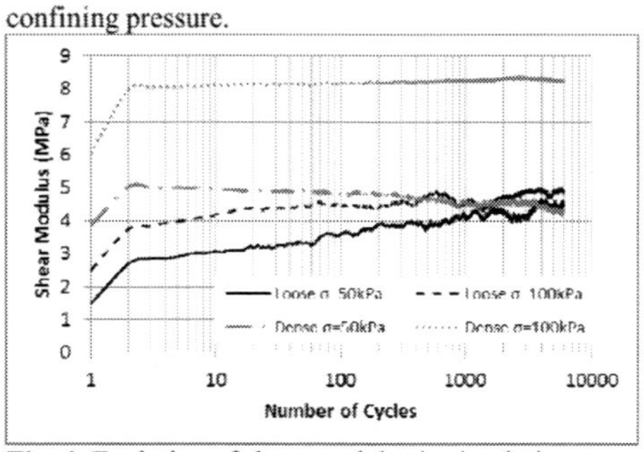

Fig. 4. Evolution of shear modulus in simulations.

The increasing shear modulus could be explained by the densification of loose samples as observed in Fig. 6, which illustrates the evolutions of void ratios in these four samples. However, the different variation trends of stiffness for the two dense samples could not relate directly to the slight dilations of both samples. Moreover, the void ratios of the two loose samples were still much higher than the two dense samples at the end of 6000 loading cycles, but their shear moduli have exceeded the value for the dense sample with σ=50kPa. These observations should be explored in more details by considering their particle-scale parameters.

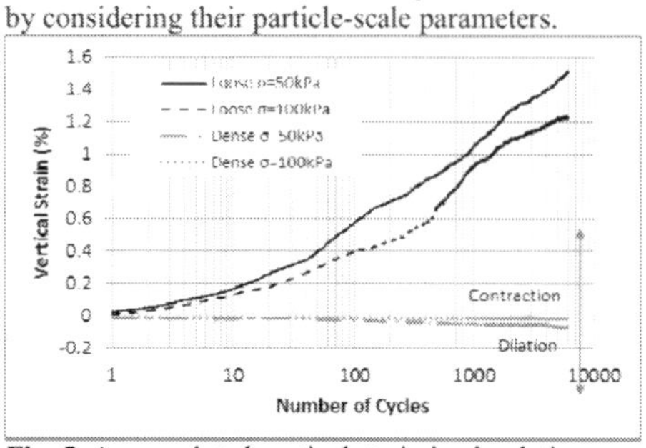

Fig. 5. Accumulated vertical strain in simulations.

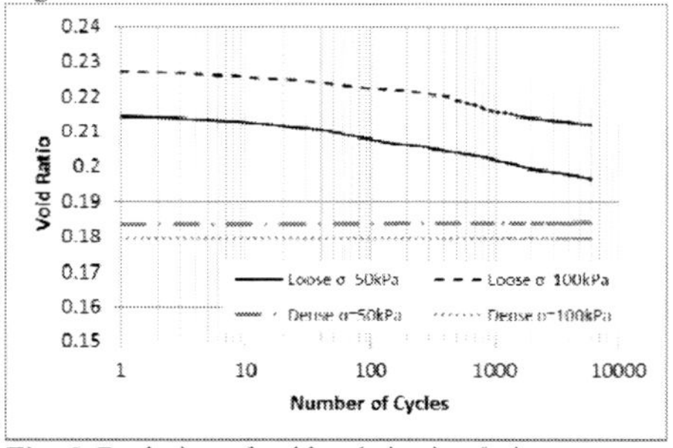

Fig. 6. Evolution of void ratio in simulations.

3.3 Particle-scale responses

3.3.1 Coordination number (N_c)

The coordination number (N_c) is the average number of contacts surrounding each particle. It has a strong relation with the stress level within the sample [9]. The evolution of N_c under cyclic loading for the four samples are shown in Fig. 7. It is clear that the initial low N_c corresponds to initial low shear stress (thus low shear modulus). The increase in shear modulus for the two loose samples is related to the increase in N_c and the decrease in shear modulus for the dense sample at σ=50kPa agrees with the reduce in N_c. However, the evolution N_c for the dense sample at σ=100kPa does not follow this trend. There must be other factor dominating the increase of shear modulus for the fourth sample.

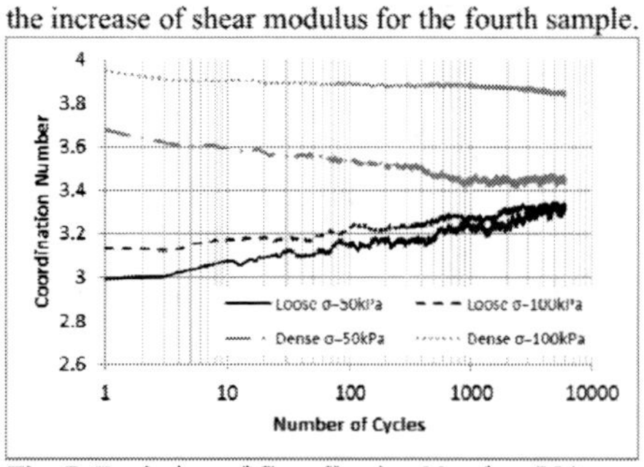

Fig. 7. Evolution of Coordination Number (N_c).

3.3.2 Fabric

There are many evidences of the impact of fabric anisotropy on characteristics of granular materials, e.g. [10, 11]. It is worth to analyse the evolution of soil fabric in the current cyclic loading conditions. The spatial distribution of directions of particle contact normals can be quantified using the Fourier approximation [12] as:

$$E(\theta)=\frac{1}{2\pi}[1+a\cos 2(\theta-\theta_a)] \tag{1}$$

where a is a parameter defining the magnitude of anisotropy and θ_a defines the direction of the principal fabric.

The histogram of spatial distribution of directions of particle contact normals for the dense sample at σ=100kPa is illustrated in Fig. 8. The Fourier approximation function is indicated by the red ellipse. The major principle fabric of the initial sample is oriented around the horizontal (θ_a=15.0°) as a result of K_0 consolidation. When sheared to the maximum strain, the major principle fabric direction rotated to the diagonal direction (θ_a≈130.0°); when sheared back to zero strain, the major principle fabric direction rotated to the opposite diagonal direction (θ_a≈40.0°). These large rotations of principle fabric in both dense samples only changed in the first loading cycle and remained the same in the remaining cycles. However, the rotation of principle fabric direction occurred slowly in both loose samples. The major principle fabric direction at zero strain rotated slowly from its original direction to θ_a≈40.0°.

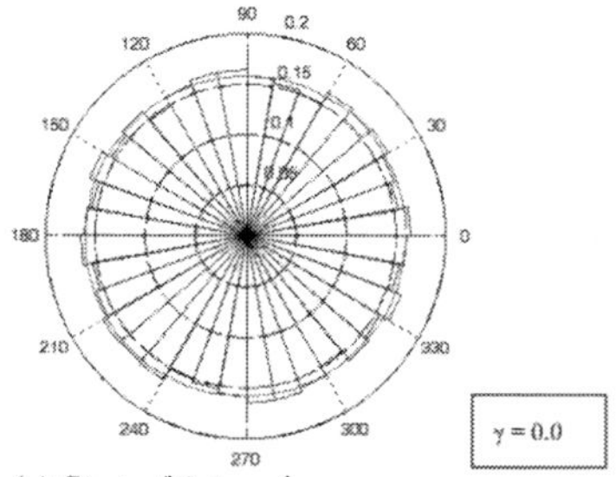

(a) Start of 1st cycle

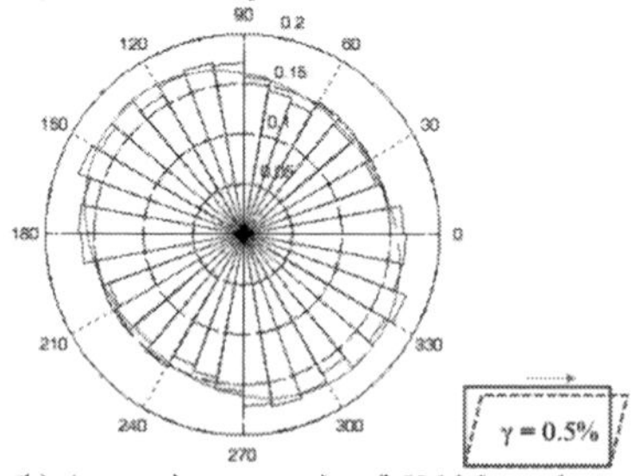

(b) At maximum strain of 6000th cycle

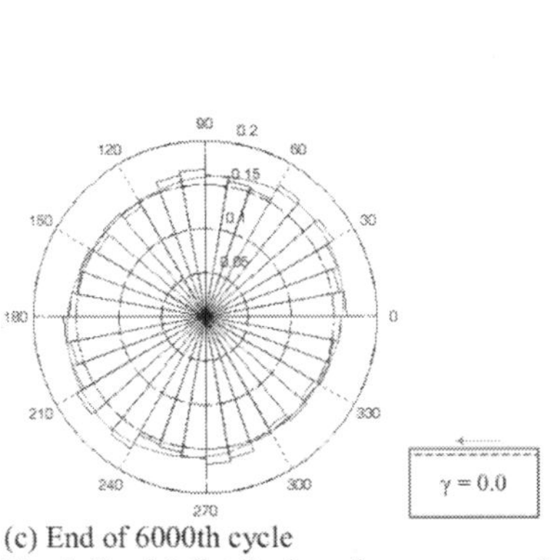

(c) End of 6000th cycle

Fig. 8. Spatial distribution of contact normal.

The difference in the principle fabric directions between the maximum strain and zero strain are plotted in Fig. 9. It is clearly shown that the rotation of principle fabric directions evolved slowly in loose samples but only changed in the first cycle in dense samples. It may be the key reason for the significant volumetric contraction and shear modulus increase in loose samples.

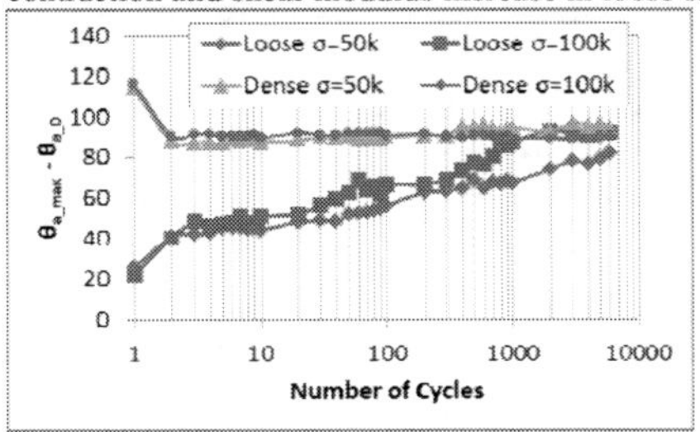

Fig. 9. Difference of major fabric direction between maximum strain and zero strain.

The magnitude of fabric anisotropy is quantified by a in Eq (1). It has been demonstrated previously [10, 11] that larger magnitude of anisotropy can result in higher shear stress. The current four simulations also confirmed this trend. Therefore, the shear modulus, which is determined by τ_{max} - τ_0, is underlain by magnitudes of anisotropy at corresponding strain values, i.e. a_{max} - a_0. The evolutions of a_{max} - a_0 for the four simulations are shown in Fig. 10. It is clear that a_{max} - a_0 for the dense sample with σ=50kPa drop significantly to negative values, which agrees with the fact that the τ_{max} - τ_0 for this sample also drop moderately.

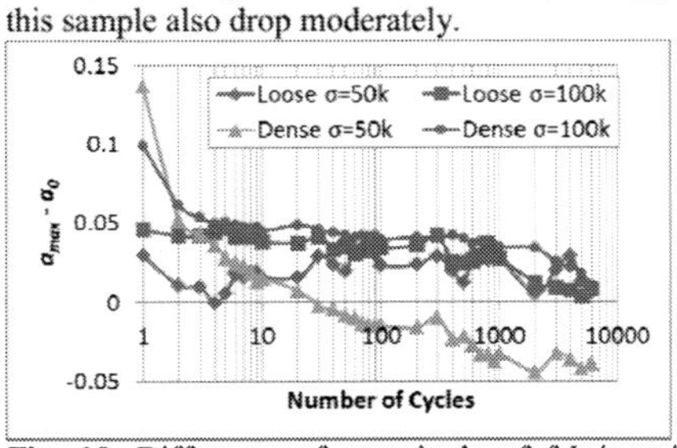

Fig. 10. Difference of magnitude of fabric anisotropy between maximum strain and zero strain.

4 Discussions and Conclusions

Both experimental tests and DEM simulations of cyclic simple shear tests were performed to explore the variations of soil characteristics under cyclic loading and its underlying micromechanism. It has been found that:

- Shear modulus for loose soil increases rapidly in the initial loading cycles as a result of soil densification, and then the rate of increase diminishes when void ratio approaches constant;
- The shear modulus increases with increasing vertical stress and relative density, but decreases with increasing strain amplitude as expected.
- Vertical accumulated strain is proportional to the shear strain amplitude but inversely proportional to the relative density of soil;
- Initial relative density has more significant impact on the volumetric response than vertical stress does;
- Higher shear stress level and shear modulus is correlated to higher coordination number;
- The shear stress level and shear modulus is directly correlated to magnitude of fabric anisotropy.

More parametric studies will be carried out in the future work, including two-way cyclic loadings with various ratio of maximum strain to minimum strain. Non-circular disks will be considered as well. An empirical model for evolution of shear modulus will be established to include the impact of all macroscopic and microscopic parameters.

References

1. S. Bhattacharya, S. Adhikari, Soil Dyn. Earthq. Eng. **31**, 805 (2011).
2. S. Bhattacharya, J.A. Cox, D. Lombardi, D. Muir Wood, P. I. Civil Eng.-Geotec., **166**, 159 (2012).
3. D. Lombardi, S. Bhattacharya, D. Muir Wood, Soil Dyn. Earthq. Eng. **49**, 805 (2013).
4. M. Vucetica, A. Mortezaie, Soil Dyn. Earthq. Eng. **70**, 60 (2015)
5. H. Liu, D. Zou, J. Liu, Int. J. Numer. Anal. Met, **38**, 1503 (2014).
6. ASTM D6528 – 07, (2007).
7. P. Cuéllar, S. Georgi, M. Baeßler, W. Rücker., Granul. Matter, **14**, 11 (2012).
8. Itasca, PFC2D manual 4.0 (2008).
9. L. Cui, C. O'Sullivan, S. O'Neill, Geotechnique, **57**, 831 (2007).
10. X. Li, H.S. Yu, Acta Mech., **225**, 2345 (2014).
11. L. Cui, C. O'Sullivan, Geotechnique, **56**, 455 (2006).
12. L. Rothenburg, R.J. Bathurst, Geotechnique, **39**, 601, (1989).

The influence of the fractal particle size distribution on the mobility of dry granular materials

Luis E. Vallejo[1,*], *Jairo M.* Espitia[2], and *Bernardo* Caicedo[2]

[1]Department of Civil Engineering, University of Pittsburgh, Pittsburgh, USA
[2]Departamento de Ingenieria Civil, Universidad de los Andes, Bogota, Colombia.

Abstract. This study presents an experimental analysis on the influence of the particle size distribution (*psd*) on the mobility of dry granular materials. The *psd* obeys a power law of the form: $N(L>d)=kd^{-D_f}$, where N is the number of particles with diameter L greater than a given diameter d, k is a proportionality constant, and D_f is the fractal dimension of the *psd*. No laboratory or numerical study has been conducted to date analysing how a fractal *psd* influences the mobility of granular flows as in the case of rock avalanches. In this study, the flow characteristics of poly-dispersed granular materials that have a fractal *psd* were investigated in the laboratory. Granular mixtures having different fractal *psd* values were placed in a hollow cylinder. The cylinder was lifted and the distance of flow of the mixture was measured with respect to the original position of the cylinder. It was determined that the distance of flow of the mixtures was directly related to their fractal *psd* values. That is, the larger the distance of flow of the mixture, the larger is the fractal *psd* of the granular mixture tested. Thus, the fractal *psd* in dry granular mixtures seems to have a large influence on the easiness by which dry granular mixtures move in the field.

1 Introduction

Granular flows in the form of dry rock avalanches are widespread in nature and are among the most impressive, powerful and destructive phenomena with respect to their capacity to modify the natural landscape and devastate engineering structures. An important feature of these flows is their extremely high mobility (capable of moving up to tens of kilometers) on slopes of very small inclinations. Among the many explanations for the "hypermobility" of rock avalanches are the following [1]:

(1) Air cushion theory. It is suggested that a sheet of rock slide debris slides on a cushion of air, trapped when the slide is catapulted into an air trajectory by a ramp.

(2) Fluidization by trapped air. A similar air entrapped process is thought to cause a full or partial fluidization of the debris by means of upward flow of air.

(3) Fluidization by vapor. It has been demonstrated that rock avalanche movement expends sufficient energy to vaporize pore water. The pressure induced in the granular material by this vaporized water aids its mobility.

(4) Rock melting. It has been found that specimens of molten rock near the sliding surface of a rockslide in Switzerland. Frictional heat can be produced in exceptionally thick slide masses to melt igneous rock which in turn causes a corresponding reduction of the friction angle in experiments.

(5) Fluidization by dust dispersion. It has been proposed that during rock avalanche dense dispersion of rock dust acts as pore fluid among the larger clasts.

(6) Acoustic fluidization. Vibrations produced on the sliding surface by rapid movement over uneven ground could reduce the friction angle of the granular debris. Direct shear tests of sand conducted on a vibrating table have shown that such a phenomenon does exist.

(7) Lubrication by liquefied saturated soil. The high mobility of the Elm Slide was explained by the effects of mud, entrained by the rockslide from loose valley deposits, liquefied under the weight of the debris.

(8) Fragmentation and spreading. Studies of rock avalanches have indicated that the component material is pervasively fragmented with rock fragments of different sizes immersed in a matrix constituted of finely comminuted rock granules.

From field measurements, Crosta [2] have determined that the particle size distribution in various rockslide deposits is fractal in nature (Table 1).

An analysis of Table 1 indicates that the particle size distribution in the rock avalanches analyzed by Crosta et al. [2] is fractal in nature. The fractal dimension, D_f, of the particle size distribution (*psd*) of the materials forming part of rock avalanches varied in value between 1.90 and 3.50. According to Tyler [3] a particle size distribution with $D_f = 0$, reflects a distribution composed solely of particles of equal diameter; a D_f between 0 and 3 reflects a particle size distribution with a greater number of larger grains; while a $D_f > 3$ reflects a particle size distribution dominated by smaller grains. Thus, the different values of D_f reported in Table 1 seems to indicate different levels of fragmentation in the

* Corresponding author: vallejo@pitt.edu

material forming part of rock avalanches.

Table 1. Fractal dimension of size distribution, D_f, for rock avalanche deposits [2]

Location	Lithology	Fractal Dimension, D_f
Flims, Switzerland	limestone	1.90 – 2.95
Campo di Giove, Italy	limestone	2.65 – 2.72
Coal dumps, Canada	coal	2.62 – 2.89
Mount St. Helens, United States	volcanic	3.00 – 3.50
Thurwieser, Italy	dolostone	2.65 – 2.86
Mt. Cook, New Zealand	sandstone	2.73
Val Pola gabbro, Italy	diorite	2.20 - 310

The fractal dimension, D_f , shown in Table 1 was obtained by Crosta [2] using the following relationship developed by Turcotte [3]:

$$N(L>d)=kd^{-Df} \quad (1)$$

where, N is the number of particles with diameter L greater than a given diameter d, k is a proportionality constant, and D_f is the fractal dimension of the *psd*

No laboratory or numerical study has been conducted to date analysing how a fractal particle size distribution influences the mobility of granular flows as in the case of rock avalanches. In this study, the flow characteristics of poly-dispersed granular materials that have a fractal particle size distribution were investigated in the laboratory.

2 Laboratory program

2.1 Equipment and materials used

For the laboratory program a hollow Plexiglass cylinder measuring 51.5 mm in internal diameter, 61.5 mm in external diameter, and 136 mm in height was used for the angle of repose experiments for granular mixtures having a particle size distribution that was fractal in nature. The cylinder was placed on top of a sheet of sand paper with a coefficient of friction equal to 0.8. The interior volume of the cylinder was equal to 283,248.5 mm^3. This interior volume was filled with a mixture of rough granular materials. The average fractal dimension of the *profiles* of the rough granular materials was about 1.15. Fig. 1 shows a grain size distribution for the 4 different types of granular materials used in the experiments. Table 2 indicates the average diameter D_{50} of the materials used in the experiments. First, the granular materials made of uniform particles with D_{50}=5mm were placed in the interior volume of the hollow cylinder until they completely filled it. The cylinder filled with the uniform particles was then

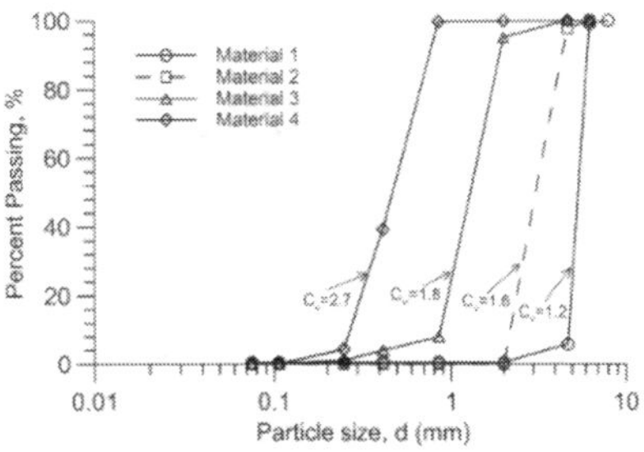

Fig.1. Particle size distribution of the granular materials used in the experiments

Table 2. Granular materials used in experiments

Granular Material	Average Diameter, D_{50} (mm)
Material 1	5.5
Material 2	3.4
Material 3	1.4
Material 4	0.5

lifted on top of the horizontally inclined sand paper. The lifting velocity of the cylinder was equal to 2 cm/sec. After release, the granular material formed a cone with sides inclined at the angle of repose of the granular material (for the uniform granular material with D_f = 0 this average angle, θ was equal to 46.5°). Besides the angle of repose, the lengths of spreading (L_1 and L_2) of the sand into the sand paper was also measured (Fig. 3). The angle of repose was calculated as the average of the two sides of the cone; the length of the spreading was also obtained as the average of the two spreading lengths measured on the two sides from the tip of the resulting cone (the average value of L for the sample with D_f = 0 was equal to 6.2 cm).

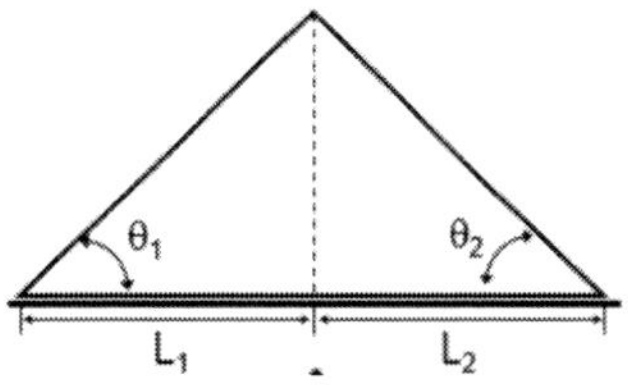

Fig.2. Angles of repose and spreading lengths in the angle of repose experiments.

In order to evaluate the influence of the fractal particle size distribution, D_f, on the values of the angle of repose and the length of spreading (L_1 and

DOI: 10.1051/epjconf/201714003032

L_2, Fig. 2), mixtures of the granular materials with the average grain diameters shown in Table 2 were prepared in the laboratory.

Six mixtures having different fractal particle size distribution were prepared in the laboratory. Mixture number 1 was the one made of uniform particles with D_{50}= 5.5 mm and D_f = 0. The particle size distributions with their respective fractal dimensions of the remaining five mixtures are shown in Fig. 3.

The fractal particle size distribution of the five mixtures was obtained by plotting on a log-log paper the cumulative number of the particles in the mixtures and the size of the particles (Fig. 4).

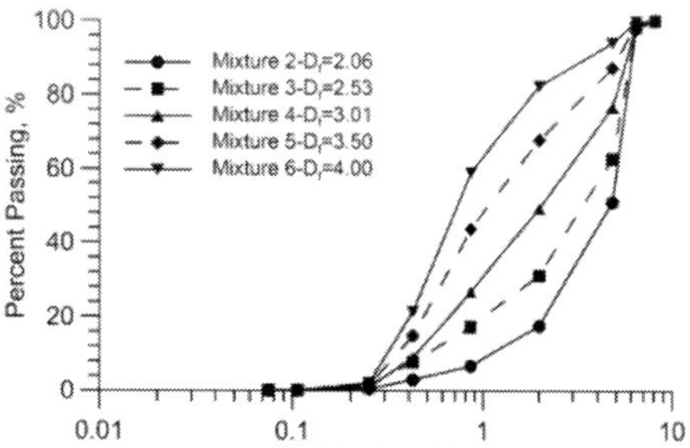

Fig.3. Particle size distribution of the five mixtures with different fractal dimensions of the five used to measure the angle of repose and advancement length (Fig. 2) in the laboratory experiments.

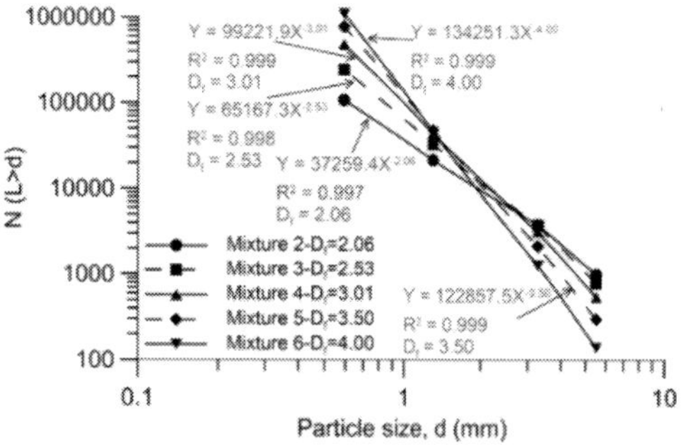

Fig.4. The number of particles, N, versus the particle size, d, method used to obtain the fractal dimension, D_f, of the mixtures with five different fractal particle size distribution shown in Fig. 3. In the resulting plots, $N = y$, and $X = d$, and the power is D_f as depicted by Eq. (1).

2.2 Angle of repose and length of advancement for the granular mixtures with varying fractal particle size distributions

Following the angle of repose experiments on the uniform material with D_{50} = 5.5 mm and D_f = 0, the same type of experiments were performed on the mixtures with different fractal particle size distributions (Figs. 3 and 4). The mixtures after filling the hollow cylinder were released on top of the sand paper sheet and the angles of repose (θ) of the two sides of the resulting cone were measured (Fig. 2). The average value of the two lengths of spreading (L_1 and L_2, Fig. 2) were also measured.

Figure 5 shows the angles of repose that the mixtures with D_f = 2.06 and D_f = 3.50 acquired when the mixtures were released on the horizontally inclined sand paper. The lifting velocity of the cylinder was equal to 2 cm/sec. Also, the average value of the two lengths of spreading (L_1 and L_2, Fig. 2) was measured and shown in Fig. 5.

Fig.5. (a) The angles of repose of the mixture with D_f = 2.06. The average length of spreading, L = 6.02 cm. (b) The angle of repose of the mixture with D_f = 3.50. The average length of spreading L =7.1 cm.

Similar results as those shown in Fig. 5 were obtained for the mixtures with the other four fractal particle size distributions (Figs 3 and 4).

A summary of the laboratory results with respect to the average measured angles of repose, θ, the average length of advancement, L (Fig. 2) as a function of the fractal particle size distribution, D_f, are shown in Figures 6 and 7.

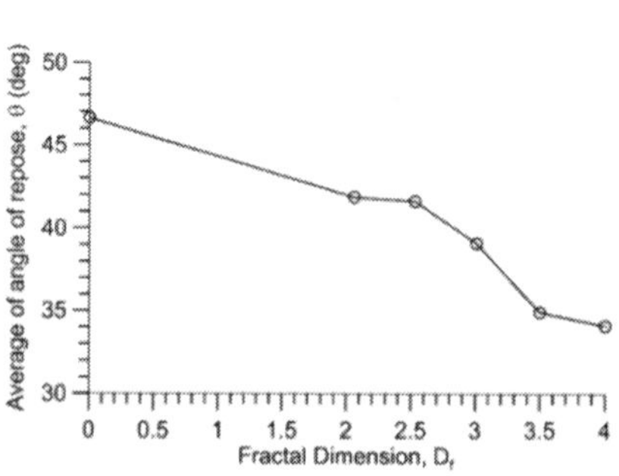

Fig.6. Relationship between the average angle of repose and the fractal dimension of the particle size distribution of the mixtures.

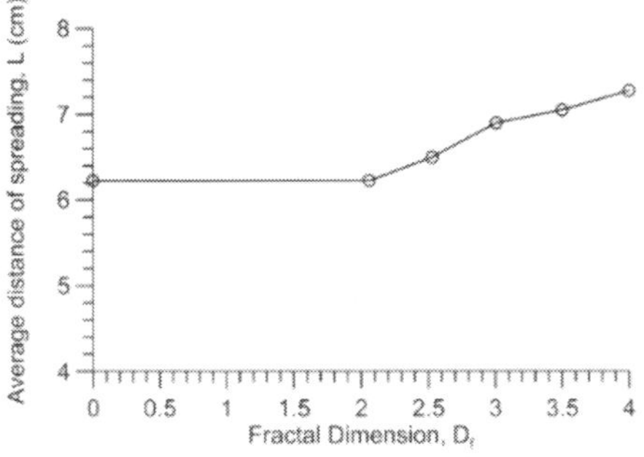

Fig.7. Relationship between the average distance of spreading, L, and the fractal dimension of the particle size distribution of the mixtures.

3 Analysis of the results

An analysis of the results shown in Figures 6 and 7 indicates that the fractal particle size distribution values (D_f) of the mixtures have a large influence on the angle of repose values (θ), and the average length of spreading (L) of the mixtures. It was determined that the higher were the fractal dimension values of the size distribution of the mixtures, the lower were the values of the angle of repose. Also, it was determined that the higher were the fractal dimension values of the size distribution of the mixtures, the higher were the values of the spreading distances of the mixtures.

According to Tyler [4] a granular material with particle size distribution with D_f = 0, reflects a distribution composed solely of particles of equal diameter (D_{50}= 5.5 mm, Table 2); a D_f between 0 and 3 reflects a particle size distribution with a greater number of larger grains (made mostly of particles with D_{50} equal to 5.5, and 3.4mm, Table 2); while a $D_f > 3$ reflects a particle size distribution dominated by smaller grains (mixtures made mostly of particles with D_{50} equal to 1.4 and 0.5 mm, Table 2). The laboratory results shown in Fig. 7 indicates that granular mixtures with high fractal particle size distribution values can spread to a large distance on ground slopes with zero inclination (behaving as if the frictional values between the particles is very low) [1].

A DEM study of flows of granular materials made of two different size particles indicated that the smaller size particles improved the fluidity of the mixture [5]. The same seems to occur in the flow of granular mixtures with a fractal particle size distribution. The presence of a great number of the small grain size particles in the mixtures ($D_f > 3$, Table 2) seems to contribute to the rotation of the particles and not to the shearing friction between during the flow of the particles (Fig. 8). The increase in number of the small grain size particles in the granular mixtures seems to cause not only a decrease in their angle of repose (θ), but also to increase the spreading distance (L) of the mixtures.

The analysis presented in this study focused on the effect of the fractal particle size distribution on the angle of repose and the length of spreading of dry granular mixtures. The effects of important parameters such as the average diameter of the grains, their shapes and degree of roughness on the angle of repose and spreading lengths were not considered in this study. The effect of these parameters could be important not only on the angle of repose but on the spreading of the granular mixtures [6,7].

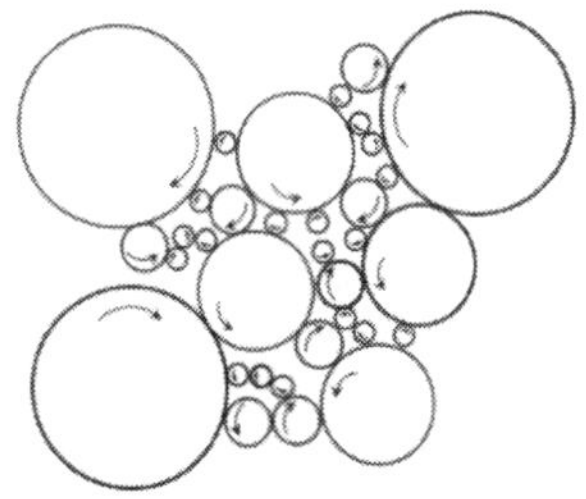

Fig.8. Relationship between the average distance of spreading, L, and the fractal dimension of the particle size distribution of the mixtures.

4 Conclusions

From the laboratory experiments on the angle of repose of granular mixtures with a fractal particle size distribution the following conclusions can be reached:

(1) The angle of repose of the mixtures decreased in value as the fractal dimension of the particle size distribution increased in value.

(2) The spreading of the mixtures on a horizontal plane increased in value as the fractal dimension of the particle size distribution increased in value.

(3) The mobility of the granular mixtures was improved by the presence of large number of smaller size particles in the mixture. The presence of these small size particles promoted the rotation of the particles in the mixtures.

References

1. F. Legros. Eng. Geology, **63** (2002)
2. G. B. Crosta, P. Frattini, and N. Fusi. J., Geophys. Res., **112** (2007)
3. D.L. Turcotte, J. of Geophys. Res., **91**(1986).
4. S.W. Tyler, *Fractal Applications to Soil Hydraulic Properties* (Thesis, University of Nevada, Reno, 1990)
5. E. Linares-Guerrero, C. Goujon, and R. Zenit, J. Fluid Mech., **593** (2007).
6. J.P. Hyslip, and L.E. Vallejo, Eng. Geology, **48** (1997).
7. L.E. Vallejo, Geotechnique, **45** (1995).

Clogging and unclogging of many-particle systems passing through a bottle-neck

Iker Zuriguel[1,⋆], *Álvaro* Janda[2], *Roberto* Arévalo[3], *Diego* Maza[1], and *Ángel* Garcimartín[1]

[1]*Departamento de Física, Facultad de Ciencias, Universidad de Navarra, E-31080 Pamplona, Spain*
[2]*Particle Analytics Ltd, Alrick Building, Max Born Crescent, The King's Buildings, Edinburgh, EH9 3BF, UK*
[3]*Division of Physics and Applied Physics, School of Physical and Mathematical Sciences, Nanyang Technological University, 25 Nanyang Link, 637371, Singapore*

Abstract. When a group of discrete particles pass through a narrowing, the flow may become arrested due to the development of structures that span over the size of the aperture. Then, it is said that the system is clogged. Here, we will discuss about the existence of a phase diagram for the clogged state that has been recently proposed, arguing on its usefulness to describe different systems of discrete bodies ranging from granular materials, to colloidal suspensions and live beings. This diagram is built based on the value of a flowing parameter which characterizes the intermittent flow observed in all these discrete systems provided that there is an external or internal energy supply. Such requirement, which is necessary to destabilize the clogging arches, is absent in a standard static silo, which is therefore examined as a particular case. This view will help to understand some *a priori* inconsistencies concerning the role of driving force in the clogging process that have been found in the last years.

Consider a container full of grains with an orifice at the bottom which is continuously shaken up and down. If the relative size of the outlet with respect to the particles is not too large, an arch can be spontaneously developed arresting the flow. If the shaking intensity is strong enough, this clog could be eventually destroyed resuming the flow [1]. As a result, we have an intermittent flow of grains where avalanches of grains alternate with lapses of time during which the orifice is blocked. In a series of papers, it has been reported that the avalanche duration distribution decays exponentially suggesting a Poisson process governed by a constant probability of clogging over the whole duration of the avalanche [2, 3]. On the other hand, the lapses of time during which the orifice is blocked have been proved to display a power-law decay [3, 4] indicating a temporal dependence of the probability that the arches become destabilized. The origin of the power-law decay for the unclogging process is still not fully understood [5] although it could be related with creeping motion [6] due to friction.

1 Flowing parameter

Considering the different distributions observed for the clogging and the unclogging processes, the intermittent flow of grains through apertures (Fig. 1) can be characterized by the the flowing parameter Φ which is nothing but the average fraction of time that the grains are flowing:

$$\Phi = \frac{\langle t_f \rangle}{\langle t_c \rangle + \langle t_f \rangle} \tag{1}$$

where $\langle t_f \rangle$ is the average duration of the avalanches and $\langle t_c \rangle$ the average duration of clogs [4]. Note that $\langle t_f \rangle$ is always well defined as it corresponds to the first moment of an exponential distribution. Oppositely, $\langle t_c \rangle$ may not be defined: as stated above the clogging duration distribution always exhibits a power law decay $t_c^{-\alpha}$, the first moment of which only converges if $\alpha > 2$.

Having this in mind we can distinguish among three different scenarios. When the flow is continuous, and the silo never clogs, the flowing time is equal to the total measuring time, hence $\Phi = 1$ (Fig. 1a). When intermittencies appear $\Phi < 1$ and in this case, we can distinguish two situations. If the distribution of clogging times has an exponent $\alpha > 2$, then the average clogging time $\langle t_c \rangle$ is well defined and $0 < \Phi < 1$ (Fig. 1b). Otherwise, when $\alpha \leq 2$, the average of this distribution is not defined. In practical terms the measured $\langle t_c \rangle$ will increase with the temporal window in which the measurements are performed. In an hypothetical infinitely long measurement, one would obtain $\Phi = 0$ (sketched in Fig. 1c). At this point it should be stressed that differentiating among the scenarios represented in Fig. 1b and Fig. 1c is not straightforward unless the distribution of clogging times is obtained with sufficient statistical significance. To this end, the method introduced by Clauset et al. [7] has been employed in all the cases reported here. The flowing parameter Φ has an additional advantage from a practical point of view: if we

⋆e-mail: iker@unav.es

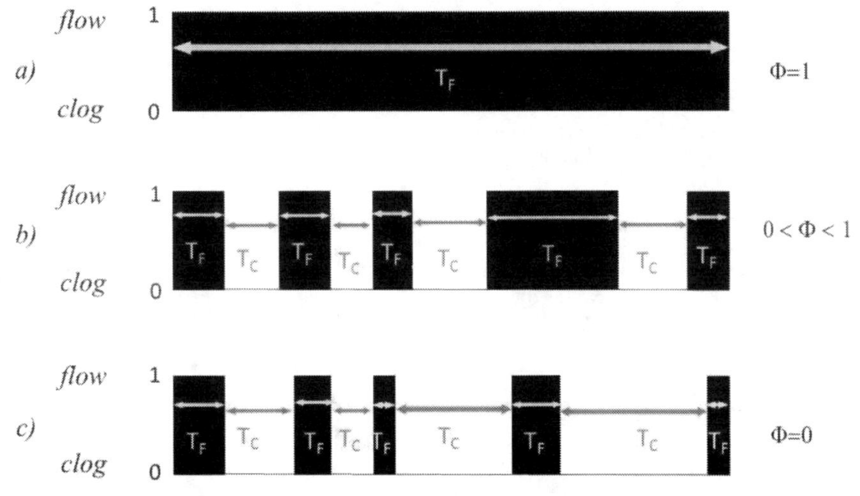

Figure 1. Scheme of the different flow situations that can be observed in the discharge of a vibrated silo. At each time the system is either flowing (*signal* = 1) or arrested (*signal* = 0). On the right, the value of the flowing parameter Φ that characterizes the type of flow. $\Phi = 1$, continuous flow; $0 < \Phi < 1$, intermittent unclogged flow; and $\Phi = 0$, clogged system. Note that, as explained in the text, the differences among the two last scenarios cannot be recognized unless a statistical analysis of the clogging time distribution is performed.

consider the flow rate during the flowing interval (W_i) we can straightforwardly obtain the average flow rate (considering intermittencies) as $W = \Phi W_i$. This is so because the flow rate during the flowing interval seems to be practically independent on the vibration amplitude (assuming that this is not too large) [2]. In addition, this W_i corresponds to the flow rate it will be obtained in an static silo, a topic that has been largely studied from Beverloo [8] until very recently (see [9] and references therein).

In Fig. 2 we show results of the flowing parameter versus the outlet size for a hopper vibrated continuously at the base and discharged eccentrically as explained in [3]. Different symbols are used for different vibration intensities Γ. The excitation can be approximated to a sinusoidal oscillation which intensity is:

$$\Gamma = \frac{A 4\pi^2 f^2}{g} \tag{2}$$

where A is the amplitude, f the frequency, and g the gravity acceleration. In our works, we typically set a constant frequency and change A. In Fig. 2 the flowing parameter was calculated by measuring the average of both, the flowing and the arrest intervals. In all the cases, special attention was paid to the distribution of clogging times. For the case of $\alpha \leq 2$ the value for Φ was set to zero even though, in the finite temporal window in which the measurement was performed, some flowing intervals were observed, as in Fig 1c.

2 Unclogging diagram

From the results displayed in Fig. 2 it becomes evident that for all the intensities of vibration studied, a transition from $\Phi = 0$ to $\Phi > 0$ occurs for a given outlet size. As it could be anticipated, the smaller Γ is, the larger the outlet at which the transition takes place. From these data, we can sketch a 2D phase diagram such as in Fig. 3 where we use stars to represent points where $\Phi = 0$ and open circles for $\Phi > 0$. The dashed curve suggests the possible transition between two phases: a clogged one when $\Phi = 0$, and an unclogged one when $\Phi > 0$. Note that the latter can be intermittent or continuous depending on wether $\Phi < 1$ or $\Phi = 1$. The study of such intermittent to continuous flow transition is not the aim of this manuscript; instead we will focus in the transition from unclogged to clogged based on the properties of the intermittent flow.

Indeed, the same features of the intermittent flow described above have been observed in other bidimensional silos, and also in disparate systems such as colloids flowing through an orifice, sheep passing through a narrow gate and pedestrians evacuating a room through a narrow door. In all cases, the distribution of flowing times is exponential and the one of the clogging times displays a power law decay. Of course, depending on the system, the variables affecting the value of the exponent α are different (table 1).

In a 2D flat bottomed silo vibrated as a whole we found that the transition from clogged to unclogged state can be achieved by increasing the outlet size and increasing the external vibration [4]. This confirmed the results presented above for the 3D case and enlarged the scope of this phenomenon, which seems to occur independently of the dimensionality of the problem, the geometry of the outlet (symmetric orifice in a flat bottom or asymmetric hopper) and the nature of the vibration (local or global). In addition, in this 2D flat bottomed silo we observed that the transition from clogged to unclogged can be also attained by reducing the amount of grains above the orifice. This suggests that the weight of the material above the arch prevents the clog destruction (for a given vibration intensity and outlet size). To confirm this idea, we performed independent experiments in a tilted 2D hopper which inclination (measured by the angle that the silo plane makes with the gravity direction) could be varied at will. Interestingly, for a given value of shaking intensity and outlet size, the system was clogged when the silo was vertical and the transition to unclogged appeared when the inclination became more horizontal. This confirms that arches with smaller load above them are broken easier than the ones with more load.

As stated above, another system in which the intermittent flow was observed was a suspension of colloidal particles passing through a small orifice. Simulations were performed with the lattice-Boltzmann method, incorporating 128 discrete solid spherical particles (more information about the numerical method can be found in [4]). The fluid was submitted to a uniform force perpendicular to a wall with a circular orifice at the center, mimicking a pressure driven flow that pushes the colloidal particles through an opening. As a consequence, an accumulation of the colloidal particles was observed in the neighbourhood of the orifice which eventually caused the formation of a clog. Long time simulations were required to obtain sufficient statistics of the clogging events. To this end, in each simulation the colloidal particles were allowed to cross several times the system size, being reinjected at the backwards of the system each time they crossed the outlet. For each simulation conditions, the passage time of around 10^5 colloidal particles was registered. Two different scenarios were simulated with different temperatures (which controls the interplay between colloidal particle fluctuations and the intensity of the driving force). The flow was observed to be intermittent, displaying (as in the case of silos) power law decays for the clogging times and exponential distributions for the flowing intervals. As expected, low temperatures lead to clogged states ($\phi = 0, \alpha \leq 2$), and high temperatures to unclogged states ($\phi > 0, \alpha > 2$).

With the idea of further extending this description to other systems, we performed experiments of sheep entering a barn through narrow gates. Details of these tests can be found in [4, 10–12] but they basically consisted on registering the entrance (one per day) of 100 sheep to a barn where food is placed. The recording was performed with two videocameras (one inside the barn and the other outside). From the movies, spatiotemporal diagrams were made in which the passage time of each sheep was measured. Again, the intermittent flow was studied evidencing power law decays for the clogging times and exponential distributions for the flowing intervals. For safety reasons, we never observed a transition to a clogged phase, but several parameters were shown to affect the exponent of the power law. The first one was the gate size: the smaller the gate, the lower the exponent, hence suggesting that for a sufficiently small door, it would be possible to reach $\alpha \leq 2$.

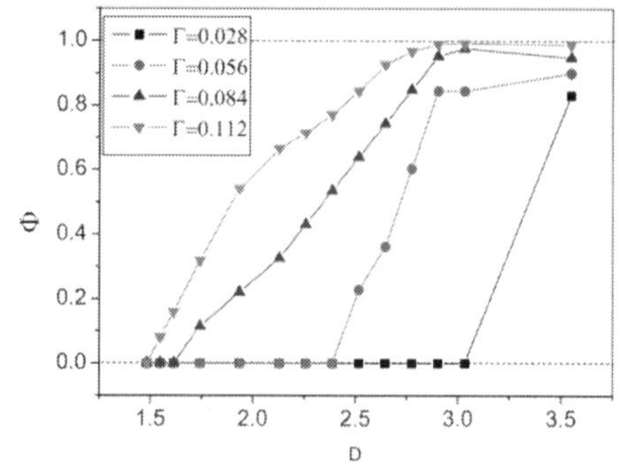

Figure 2. Flowing parameter Φ versus the rescaled outlet size D for different values of vibration intensity Γ in a locally vibrated eccentrically discharged hopper as the one described in [3].

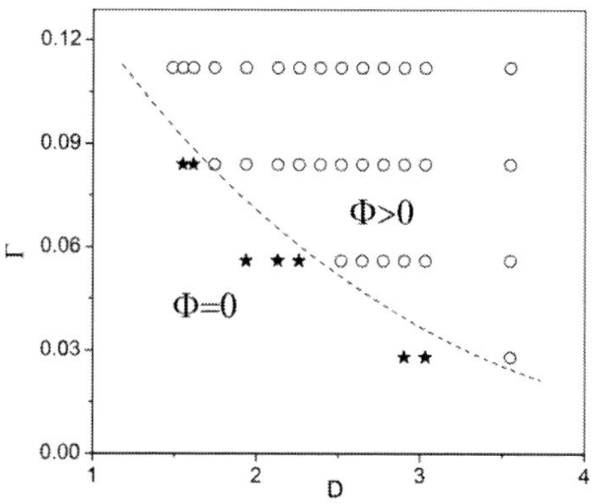

Figure 3. Phase diagram for the plane $\Gamma - D$ obtained for a locally vibrated eccentrically discharged hopper [3]. Stars indicate points where $\Phi = 0$ (clogged phase) and circles show positions where $\Phi > 0$ (unclogged phase). The dashed line is a guide to the eye suggesting a possible boundary between the two phases.

Another one was the presence of an obstacle in front of the door. Interestingly, if the obstacle is placed at an appropriate position the exponent increases and the flow improves (there are fewer clogs and they last less time). Of course, if the obstacle is too far from the door, its effect is negligible and if it is too close it could be even detrimental as clogs appear between the obstacle and the door (instead of at the very door) [11]. The effect of the obstacle in the flow rate through narrow doors has been traditionally attributed to the waiting-room effect: the obstacle controls the inflow of animals preventing overcrowding the constriction region. Interestingly, we show that it also has an effect on facilitating resuming the flow, surely due to a pressure reduction at the door which may also be linked to an increase of sheep mobility that can lead to a raising of sheep "temperature". The final variable that was shown to affect the exponent is the competitiveness of the sheep [12].

Table 1. Parameters that have been observed to affect the exponent α of the power law decay of clogging times in the different systems studied. Arrows indicate the direction in which the variables have to be modified in order to increase α and improve the flow. The variables marked in bold are the ones in which the transition from clogged ($\alpha \leq 2$, $\phi = 0$) to unclogged ($\alpha > 2$, $\phi > 0$) has been observed. In the other ones, we expect such transition but it cannot be attained experimentally for safety reasons. All the variables in each column have been grouped with a single name following the spirit of [15]. Italics are used to denote that the obstacle effect is not definitively proved to be linked with a reduction of compatible load or an increase of the temperature.

	Typical length (Λ)	Incompatible Load (I_L)	Compatible Load (C_L)
2D granular silo	↑ **outlet size**	↑ **external global vibration**	↓ **height of the layer of grains**
2D granular hopper			↓ **inclination**
3D granular eccentric hopper	↑ **outlet size**	↑ **external local vibration**	
colloids		↑ **temperature**	
sheep	↑ gate size	*placing an obstacle*	*placing an obstacle* & ↓ motivation
pedestrians (simulations)	↑ **gate size**	↑ **internal noise**	↓ **motivation**
pedestrians (experiments)	↑ gate size		↓ motivation

Indeed, tests performed during warm conditions (summer time at 3 pm Spanish time) lead to a reduction of clogging events compared with standard measurements carried out in cooler circumstances (in the morning during the rest of the year). This was attributed to a noticeable reduction of the eagerness of sheep in warmer days which implies that the higher the competitiveness of the animals the less efficient is the entrance. Indeed, this behavior was already predicted for human evacuation through narrow doors by Helbing et al. [13] and was given the name of "Faster is Slower" suggesting that the faster the pedestrians aim to escape, the slower is the process.

In other series of experiments performed in the University of Navarra, this "Faster is Slower" effect was proved experimentally with real people [12]. A group of around 80-90 volunteers were instructed to evacuate a room with different competitiveness. This was quantified by measuring the velocity of the people at the beginning of the evacuation (before the density in front of the door built up and clogs started to appear) and a good correlation was observed: the higher the competitiveness, the higher the initial velocity. In addition, pioneer pressure measurements were performed at the doorjamb, also showing an increase of the registered values as the competitiveness grew higher. As in previous cases, intermittencies in the flow rate were analysed and the outcomes evidenced an exponential tail of bursts sizes and a power law decay for the clogging times. Moreover, pedestrian evacuations also evidenced that the larger the door size the higher the exponent of the power law tail [14] as it happens for sheep and granular media.

Finally, numerical simulations of pedestrian evacuations were used to confirm that both, reducing competitiveness and enlarging the door size, improve the flowability (as they lead to an increase of the exponent of the power law decay). In addition, these simulations revealed that increasing the internal noise of the simulated bodies did also produce an increase of the exponent of the power law decay. Contrary to the experiments, these simulations could be extended to situations where clogs were much more stable, and the transition to $\alpha \leq 2$ ($\phi = 0$) attained.

All the findings concerning the ways in which different variables affect the exponent value in such disparate systems are summarized in table 1. The arrows indicate the direction in which the variables should be tuned to increase α and bold letters are used to stress the cases where the transition from clogged to unclogged has been obtained. Considering these outcomes, we have grouped the different specific variables in three general groups. In the first column of table 1, we place those variables that are related to the size of the orifice (or the relation between the size of the orifice and the size of the particles). In the second column we find parameters that are related to the excitation that particles intrinsically have or at which they are submitted. This is a kind of temperature or, more generally speaking, an *Incompatible Load* (I_L) in the sense that it is a stress made in a direction which is different from that of the force that caused the blockage. Finally, in the third column we have grouped variables related to a *Compatible Load* (C_L) indicating that there are forces acting in the same direction that the one driving the system towards the clogged situation. Note that the terminology of Compatible Load and Incompatible Load has been taken from [15] which was indeed the work that inspired A. Liu and S. R. Nagel to propose the groundbreaking idea of the jamming phase diagram in [16].

Considering the grouped variables in table 1 and the shape of the phase diagram obtained for the plane $\Gamma - D$ in Fig. 3, we can speculate about the existence of a clogging phase diagram as the one sketched in Fig. 4. The variables at play will be the ones mentioned above: a lengthscale, Λ; the Incompatible Load, I_L; and the Compatible Load, C_L. Increasing Λ and I_L, and reducing I_L leads to a transition from the clogged to the unclogged phase. Therefore, in analogy with the packing fraction in the jamming diagram [16], the compatible load is written as $1/C_L$ in the clogging phase diagram. It is important to remark that this clogging phase diagram is just a concept which is proposed based on the findings obtained in many different systems. Now, it becomes necessary to characterize quantitatively each of the diagram planes with the aim of assessing the kind of transition that clogging is. It would not be surprising that the specific nature of the transitions depend on the system. However, the unified proposed scenario allows to point out some generic issues that are, in our opinion, relevant. The presence of a characteristic length scale

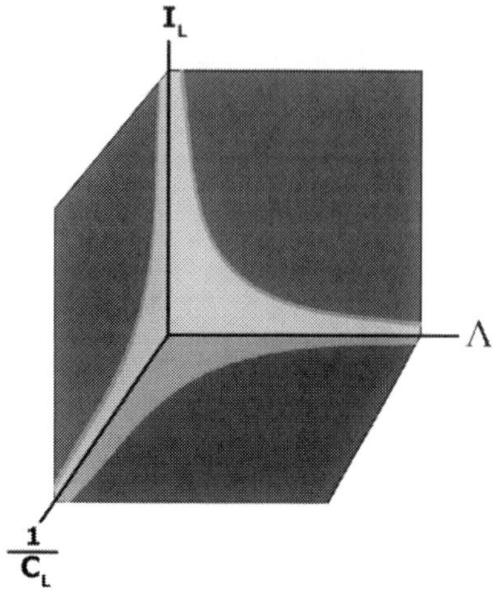

Figure 4. Clogging phase diagram proposed in [4]. Within the light surface the system would be in a clogged phase and outside it, in an unclogged phase. The generic variables are the Length Scale, Λ; the Incompatible Load, I_L; and the Compatible Load, C_L.

stresses the finite-size nature of clogging, where the outlet size competes with the scale of the structures formed near the outlet. Surely, such a scale will strongly depend on the specific properties of the many-body system we are dealing with: cohesive or faceted grains [18–22], for example, will imply a larger characteristic size. The insight gained in this work also allows us to explain the increased stability of clogs exhibited by suspensions when are subjected to higher fluid velocities [17] or helps to understand the origin of Faster is Slower effect found in pedestrian dynamics.

3 The static silo as a special case

Up to now, we have described the intermittent flow in a vibrated silo and defined a phase transition depending on whether the first moment of the distribution of clogging times converges or not. Then, we have extended the analysis of this property of the intermittent flow to other different systems leading to the proposal of a generic clogging phase diagram. Now, we can go back to the case of the silo and study what it seems to be the simplest case of a standard non-vibrated one. The first issue that becomes evident is that, once there is an arch that blocks the outlet, the flow becomes forever arrested. In practical terms, this constitutes a drawback as the distribution of clogging times is not accessible any more: all clogs last an infinite time. Therefore, it could naively thought that the problem is simpler and, in order to decide if we are in a clogged phase or not, it suffices to build a silo with an orifice at the bottom and see if, eventually, there is an arch that arrests the flow. Paradoxically, the situation is much more complex than expected and, despite the numerous efforts devoted to unveil if there is a transition from clogged to unclogged regimes, there is not a robust and definitive answer yet.

At this time, it has been shown that both the avalanche duration and the avalanche size distributions display an exponential tail analogous to the flowing times interval distribution observed in the vibrated silo [23–35]. This exponential tail, which extends to other systems such as fluid driven particle flow [36, 37] has been explained in terms of the existence of a constant probability of clogging [28, 38, 39]. Only for the case of very elongated orifices (slits) a power law decay was observed which was justified as a convolution of many exponential distributions with different exponents [40].

Concerning the existence of a transition from a clogged to an unclogged phase, the most studied variable is the mean avalanche size $\langle s \rangle$ which has been shown to abruptly increase with the outlet size [24]. Remarkably, in a three-dimensional silo an augment of the outlet diameter from $D = 2$ to $D = 5$ (measured in number of particles' diameters) leads to an increase of the avalanche size in around 7 orders of magnitude. The first fitting proposed to explain this sudden growth was a critical power law that diverges at a given critical value D_C. This implies that, above this value, clogging events are not possible. Subsequently, it was proposed that other fittings were equally valid. In particular it was suggested that $\langle s \rangle$ could grow exponentially: e^{CD^2} in a two-dimensional system and e^{CD^3} in a three-dimensional one (where C is a constant) [30]. For the two-dimensional case there seems to be a consensus about the appropriateness of both fittings, and indeed there are models that justify only the nondivergent approach [31, 32]. Otherwise, the three-dimensional silo is still controversial as some results plea in favor of the nondivergent scenario [32] whereas others [31] evidence that the e^{CD^3} fitting is not valid.

Having this controversy in mind and the methodological difficulty we find to define the transition, we can still see how the static silo could fit into the generic framework presented in Fig. 4. Clearly, if such transition exists, it must be abrupt as the flowing parameter could only take values of $\Phi = 0$ (if there is clogging) or $\Phi = 1$ if there is not clogging. The intermittent unclogged regime ($0 < \Phi < 1$) does not exist as, once the silo clogs, nothing can resume the flow. This is so because the Incompatible Load (vibration) is 0 due to the absence of external vibration. At this point we would like to note that previous to the arrest of the flow, the energy of the particles within the silo can be considerably large. Therefore, if we want to strictly explore the plane $I_L = 0$, an experiment where the grains are extracted without velocity (in a quasistatic way) is necessary.

A possibility that has been explored in the recent years is to increase the driving force (i.e. the gravity) and see the effect on the clogging probability. This has been done by means of numerical simulations [33, 34] in a 2D non-vibrated silo. In both works, we used soft-particle molecular dynamics simulations of equally sized disks. The restoring force in the normal direction of collision depends linearly on the particles overlap $\xi = d - r_{ij}$, with stiffness

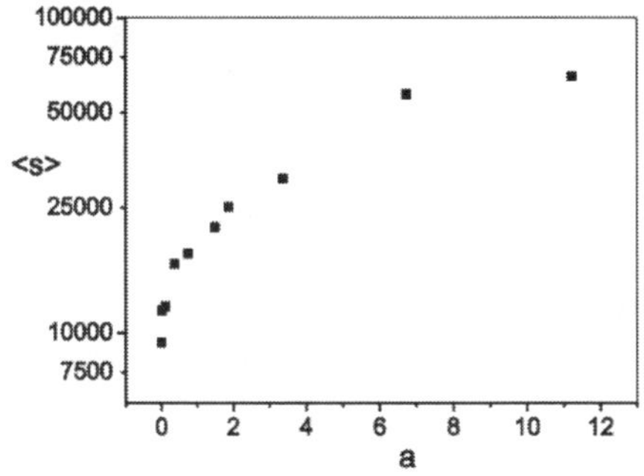

Figure 5. Average avalanche size (in logarithmic scale) versus the external field a measured in units of the gravity acceleration g. Numerical results, obtained with the model explained in [34] are displayed for for a 2D silo with $D = 5$.

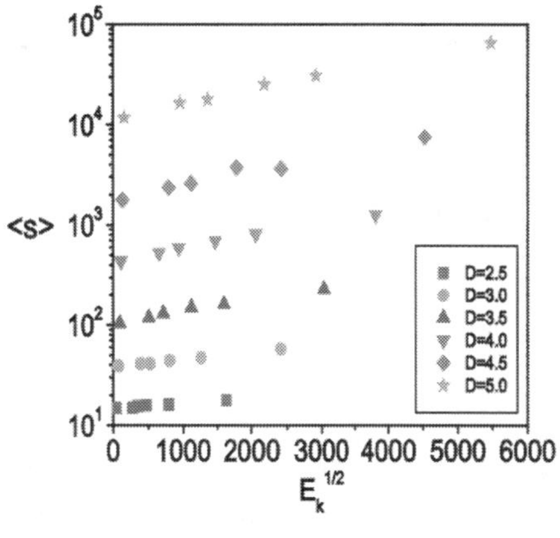

Figure 6. Numerical results of the average avalanche size (in logarithmic scale) versus the kinetic energy E_k within the whole system for different outlet sizes as indicated in the legend.

$k_n = 10^5(mg/d)$, where $d = 1\ mm$ is the particle diameter, $m = 1$ is the mass and r_{ij} the distance between the centers of the particles. Additionally, there is a dissipative force proportional to the relative normal velocity of the colliding grains, with damping parameter $\gamma_n = 300(m\sqrt{g/d})$. We implement static friction placing a spring in the direction tangential to the normal joining the centers of the particles. The elongation of this spring is obtained integrating the relative velocity of the surfaces in contact. The parameters are $k_t = (2/7)k_n$ and $\gamma_t = 200(m\sqrt{g/d})$. The friction coefficient is set to $\mu = 0.5$, and the gravity to a times the value of g. In order to optimize the computing time, we use a integration step $\delta = 10^{-4}\sqrt{d/g}$ for simulations with $a \simeq 1$. For $a \simeq 3$ the integration step is $\delta/3$, and for simulations at $a \simeq 6$ and $a \simeq 10$ we use a step of $\delta/6$. For simulations at $a \simeq 10^{-3}$ we increase the time step to 10δ to assure that collisions are simulated with sufficient accuracy. Importantly, note that in the different simulated scenarios the values of k_n, k_t, γ_n, and γ_t were constant (independent on the value of a). Also, it should be stressed that the selected value of k_n assures that in all the simulations the elastic deflections are very small compared to particle radii.

Despite that within this framework one may expect that results obtained with different driving forces might not depend on gravity (dimensionless results should only depend on dimensionless combinations of the input parameters) it is observed that increasing gravity leads to an increase of the mean avalanche size, i.e. a reduction of clogging probability [33, 34]. Indeed, this increase of the avalanche size with gravity depends on the outlet size (the larger the outlet the stronger the effect) and is not very pronounced: for the largest outlet size studied (Fig. 5) an increase of $\langle s\rangle$ in one order of magnitude is obtained by increasing the driving force a in four orders of magnitude (from $a = 10^{-3}$ to $a = 10$ times the gravity value). The origin of this increase was discarded to be caused by the fact that the higher the external field, the stronger the weight that the arches have to resist [33]. Instead, it was argued that the origin was related to the increase of the average kinetic energy of the flowing grains caused by the augment of the gravity which is, perhaps, not compensated by a corresponding augment of the dissipation (recall that both dissipative terms γ_n and γ_t were chosen to be constant). In this sense, it was proposed that two conditions must be fulfilled in order to observe the clogging phenomenon: 1) an arch should be formed at the orifice; 2) the arch should resist until all the kinetic energy in the system dissipates. Increasing the driving force leads to a linear increase of the kinetic energy of the grains that has to be dissipated and therefore to an increase of the probability that an arch gets destabilized before the whole system comes to a rest. Based on this idea, in Fig. 6 we represent the mean avalanche size (in logarithmic scale) versus the square root of the total kinetic energy of the grains within the silo. Clearly, for all the outlet sizes the higher the kinetic energy, the larger the avalanche size. In addition, it is observed that there seems to be a limit avalanche size when the kinetic energy vanishes (for very low driving forces). This limit avalanche sizes clearly depends on the outlet size. Nevertheless, the growth rate of the avalanche with the kinetic energy, seems to be rather independent on the outlet size, yet a small dependence has been observed: the higher the outlet size is, the faster is the growth rate.

From all these numerical outcomes, it was concluded that in a static silo the higher the driving force the larger the avalanche size. This somehow contradicts the behaviour expected from the clogging transition diagram proposed in Fig. 4. Indeed, it can be definitively stated that in a static silo, faster is faster, and also the faster the grains flow, the less likely that a clog develops. Having said that, it is also true that gravity does not play a crucial role in clogging development if it is compared to the effect that it has on the flow rate [41]. Indeed, the most important parameter determining clogging is the outlet size: increasing the outlet size from 4.5 to 5 has a stronger effect in the avalanche size than increasing the driving force from $a = 10^{-3}$ to $a = 10$. In any case, the fact that the driving force effect is related to the kinetic energy of the grains suggests that care should be taken when looking at the results of $\langle s \rangle$ vs the outlet size in the traditional way. And this is so because when we modify the outlet size we are increasing the required size of the clogging structures, but we are also increasing the kinetic energy within the system which is related with the above-mentioned probability of destabilization of an arch.

4 Conclusions

In this manuscript we have explained the origin and features of a recently proposed clogging phase diagram. This phase diagram differentiates among clogged and unclogged phases based on a statistical property of the flow intermittencies: in some situations the mean time of the lapses during which the flow is arrested cannot be defined, as the first moment of the distribution does not converge. The origin of this non-convergence hinges on the distribution of these clogging times which has been shown to be compatible with a power law decay. Independently of whether it is or not a power law, the distribution of clogging times is undoubtedly not an exponential as it would be expected if the probability that the arch becomes destabilized were constant over time. Otherwise, the unclogging probability decreases with time suggesting some kind of aging of the blocking structure which is probably associated to the dynamics of the particles behind it. Let us note that in a recent paper where clogging was observed in more diluted situations, the unclogging time distributions were observed to decay exponentially [42]. This reflects that the unclogging process in such a dilute situation is different to the dense systems analysed here.

The clogging phase diagram is useful to understand several features of the flow through constrictions of many different systems, ranging from colloids to pedestrians. In such systems, the common ingredients that are necessary to observe the clogging transition seem to be: 1) a bottleneck small enough in order to allow the formation of a geometrical structure spanning over its size; 2) a force that drives the system towards the outlet which is also necessary to stabilize the clogging structures; and 3) an internal or external agitation that is able to break the clogging structure and resume the flow. Paradoxically, if the latter condition is not fulfilled the problem becomes more complex as the arches cannot be destroyed and the strategy followed to study the transition (which is basically focused in the unclogging process) is not useful anymore. Otherwise, in a static silo the only process that can be studied is the clogging one.

At this point let us stress the importance of separating both processes, clogging and unclogging, to fully understand the dynamics of the flow of particles through bottlenecks. Clogging seems to occur randomly, with a probability that does not depend on time, whereas unclogging reveals temporal dependence. Indeed, distinct variables may affect both processes in different ways. For example, increasing the orifice size always improves the flow, preventing clogging and facilitating unclogging. However, the effect of increasing the driving force in the flow is not unique: it is beneficial because it prevents the development of clogging (as in the static silo), but it is also prejudicial because it precludes unclogging as the formed arches are stronger.

This different effect of the driving force in the clogging and unclogging processes causes the Faster is Slower effect. For low driving forces, the flow rate augments with it as the velocity of the particles increases (and clogging decreases). But eventually, for a given driving force, the arches that form become so strong that the dynamics becomes controlled by the unclogging process. Therefore, increasing the driving force leads to a reduction of the flow rate. From this reasoning, it becomes obvious that in a static silo, Faster is Slower cannot be observed as this effect is linked to the unclogging process.

Acknowledgements

We thank Ignacio Pagonabarraga, Daniel Ricardo Parisi, Luis Ariel Pugnaloni, Raúl Cruz Hidalgo, Celia Lozano, and Eric Clément for helpful discussions and LuisFer Urrea for technical help. This work has been funded by Ministerio de Economìa y Competitividad (Spanish Government) through Project No. FIS2014-57325.

References

[1] C. Lozano, G. Lumay, I. Zuriguel, R. C. Hidalgo, and A. Garcimartín, Phys. Rev. Lett. **109**, 068001 (2012).
[2] C. Mankoc, A. Garcimartín, I. Zuriguel, D. Maza and L. A. Pugnaloni, Phys. Rev. E **80**, 011309 (2009).
[3] A. Janda, D. Maza, A. Garcimartín, E. Kolb, J. Lanuza and E. Clément, Europhys. Lett. **87**, 24002 (2009).
[4] I. Zuriguel, D. R. Parisi, R. C. Hidalgo, C. Lozano, A. Janda, P. A. Gago, J. P. Peralta, L. M. Ferrer, L. A. Pugnaloni, E. Clément, D. Maza, I. Pagonabarraga and A. Garcimartín, Scientific Reports **4**, 7324 (2014).
[5] A. Nicolas, S. Bouzat, M. Kuperman, Phys. Rev. E **94**, 022313 (2016).
[6] B. Blanc, J. -C. Géminard, L. A. Pugnaloni, Eur. Phys. J. E **37**, 112 (2014).
[7] A. Clauset, C. R. Shalizi, and M. E. Newman, SIAM review **51**, 661-703 (2009).
[8] W.A. Beverloo, H.A. Leninger and J.J. van de Velde, Chem. Eng. Sci. **15**, 260 (1961)
[9] S.M. Rubio-Largo, A. Janda, D. Maza, I. Zuriguel and R.C. Hidalgo, Phys. Rev. Lett. **114**, 238002 (2015)
[10] A. Garcimartín, J. M. Pastor, L. M. Ferrer, J. J. Ramos, C. Martín-Gómez, and I. Zuriguel, Phys. Rev. E **91**, 022808 (2015).
[11] I. Zuriguel, J. Olivares, J. M. Pastor, C. Martín-Gómez, L. M. Ferrer, J. J. Ramos and A. Garcimartín, Phys. Rev. E **94**, 032302 (2016).

[12] J. M. Pastor , A. Garcimartín, P. A. Gago, J. P. Peralta, C. Martín-Gómez, L. M. Ferrer, D. Maza, D. R. Parisi, L. A. Pugnaloni and I Zuriguel, Phys. Rev. E. **92**, 062817 (2015).

[13] D. Helbing, I. J. Farkas , and T. Vicsek. Nature **407**, 487 (2000).

[14] A. Garcimartín, D.R. Parisi, J.M. Pastor, C. Martín-Gómez, and I. Zuriguel, J. Stat. Mech. **2016**, 043402 (2016).

[15] M. E. Cates, J. P. Wittmer, J. P. Bouchaud, and P. Claudin, Phys. Rev. Lett. **81**, 1841 (1998).

[16] A. J. Liu, and S. R. Nagel, Nature **396**, 21-22 (1998).

[17] J. R. Valdes, and J. C. Santamarina, Can. Geotech. J. **45**, 177-184 (2008).

[18] T. Kanzaki, M. Acevedo, I. Zuriguel, I. Pagonabarraga, D. Maza, and R.C. Hidalgo. The European Physical Journal E **34**, 133 (2011).

[19] D. Höhner, S. Wirtz, and V. Scherer. Powder Technology **226**, 16-28 (2012).

[20] F. Alonso-Marroquín, Á. Ramírez-Gómez, C. González-Montellano, N. Balaam, D. A. H. Hanaor, E. A. Flores-Johnson, Y. Gan, S. Chen, and L. Shen. Granular Matter **15**, 811-826 (2013).

[21] J. Tang, and R. P. Behringer. Europhysics Letters **114**, 34002 (2016).

[22] T. Borzsonyi, E. Somfai, B. Szabó, S. Wegner, P. Mier, G. Rose and R. Stannarius. New J. Phys. **18**, 093017 (2016).

[23] K. To *et al.*, Phys. Rev. Lett. **86**, 71 (2001).

[24] I. Zuriguel, A. Garcimartín, D. Maza, L. A. Pugnaloni, and J. M. Pastor, Phys. Rev. E **71**, 051303 (2005).

[25] J. Tang and R.P. Behringer, Chaos **21** 041107 (2011).

[26] L. Kondic, Granular Matter **16**, 235 (2014).

[27] S. Tewari, M. Dichter, and B. Chakraborty, Soft Matter **9**, 5016 (2013).

[28] I. Zuriguel *et al.*, Phys. Rev. E **68**, 030301 (R) (2003).

[29] C. C. Thomas and D. J. Durian, Phys. Rev. E 87, 052201 (2013).

[30] K. To, Phys. Rev. E **71**, 060301 (R) (2005).

[31] A. Janda, I. Zuriguel, A. Garcimartín, L. A. Pugnaloni, and D. Maza, Europhys. Lett. **84**, 44002 (2008).

[32] C. C. Thomas, D. J. Durian, Phys. Rev. Lett. **114**, 178001 (2015).

[33] R. Arévalo, I. Zuriguel, D. Maza, and A. Garcimartín, Phys. Rev. E **89**, 042205 (2014).

[34] R. Arévalo, and I. Zuriguel. Soft Matter **12**, 123-130 (2016).

[35] S. Mondal, M. M. Sharma, *Role of flying buttresses in the jamming of granular matter through multiple rectangular outlets*, Granular Matter **16**, 125-132 (2014).

[36] A. Guariguata, M. A. Pascall, M. W. Gilmer, A. K. Sum., E. D. Sloan, C. A. Koh, D. T. Wu, *Jamming of particles in a two-dimensional fluid-driven flow*, Phys. Rev. E **86**, 061311 (2012).

[37] P. G. Lafond, M. W. Gilmer, C. A. Koh, E. D. Sloan, D. T. Wu, A. K. Sum., *Orifice jamming of fluid-driven granular flow*, Phys. Rev. E **87**, 042204 (2013).

[38] D. Helbing, A. Johansson, J. Mathiesen, M. H. Jensen, A. Hansen, *Analytical approach to continuous and intermittent bottleneck flows*, Phys. Rev. Lett. **97**, 168001 (2006).

[39] T. Masuda, K. Nishinari, A. Schadschneider, *Critical Bottleneck Size for Jamless Particle Flows in Two Dimensions*. Phys. Rev. Lett. 112, 138701 (2014).

[40] S. Saraf, S V Franklin *Power-law flow statistics in anisometric (wedge) hoppers*, Phys. Rev. E **83**, 030301(R) (2011).

[41] S. Dorbolo, L. Maquet, M. Brandenbourger, F. Ludewig, G. Lumay, H. Caps, N. Vandewalle, S. Rondia, M. Mélard, J. van Loon, A. Dowson, S. Vincent-Bonnieu, Granular Matter **15**, 263 (2013).

[42] I. G. Tejada, L. Sibille and B. Chareyre, EPL **115**, 54005 (2016).

Clogging of granular material in vertical pipes discharged at constant velocity

Diego López-Rodríguez[1,*], *Iker* Zuriguel[1,**] and *Diego* Maza[1,***]

[1] *Universidad de Navarra (Spain)*

Abstract. We report an experimental study on the flow of spherical particles through a vertical pipe discharged at constant velocity by means of a conveyor belt placed at the bottom. For a pipe diameter 3.67 times the diameter of the particles, we observe the development of hanging arches that stop the flow as they are able to support the weight of the particles above them. We find that the distribution of times that it takes until a stable clog develops, decays exponentially. This is compatible with a clogging probability that remains constant during the discharge. We also observe that the probability of clogging along the pipe decreases with the height, i.e. most of the clogs are developed near the bottom. This spatial dependence may be attributed to different pressure values within the pipe which might also be related to a spontaneous development of an helical structure of the grains inside the pipe.

1 Introduction

In mining, it is quite frequent that the transport of rocks and ore between different levels is done through underground excavated tubes or ore passes [1]. This method has been traditionally employed in underground mining, but nowadays it is becoming extended to open-cast mining. The reasons are of different nature and transcendence, ranging from economical to environmental. Indeed, using gravity driving force to deliver material trough a pipe supposes important savings in the economical costs related to the transportation with trucks through the typically long and winding roads that go from the top of the mine to the bottom. In addition, dust production is considerably reduced when comparing pipe transport with the traditionally implemented strategy of throwing the material from one bench to the other. Other advantages of this pipe transport system are lessening the visual impact of traditional open-cast mines and reducing the noise levels in nearby populated areas. Apart from transport, ore passes also serve for low cost storage which is essential to efficient mine operation.

A typical ore pass has three basic parts: (1) the pipe itself connecting two or more levels in a mine; (2) the top-end facilities to incorporate material, for example diggers, trucks and (sometimes) crushers; (3) bottom-end structures such as crushers and belts which are necessary to control material extraction. The bad operation of any of the parts is critical to the economic performance of mining operations.

Despite its great potential, the design of ore passes do not seem to receive the same engineering attention as

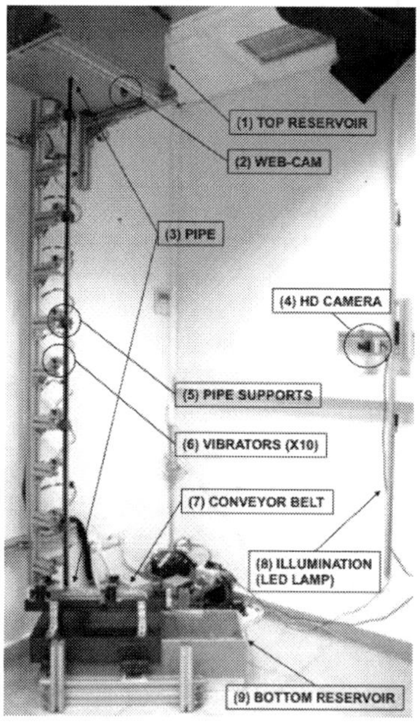

Figure 1. Experimental setup. In the image: (1) top reservoir from which the pipe is filled; (2) web-cam used to determine the time at which a clog occurs; (3) two meters long transparent pipe; (4) camera that takes pictures of the entire pipe; (5) pipe holders; (6) 10 vibrators controlled independently used to break the clogs; (7) conveyor belt that extracts the granular material; (8) lamp led; (9) bottom reservoir.

*e-mail: dlopez.14@alumni.unav.es
**e-mail: iker@unav.es
***e-mail: dmaza@unav.es

other mining infrastructures such as silos. Indeed, from the scarce works available [2–4] it becomes evident that many different aspects should be considered. For example, choosing the inclination of ore passes -which typically ranges from vertical to 30°- is a practice based in a few empirical observations. The cross sections may be square, rectangular or circular and the diameter (or the side length) should be large enough to guarantee the absence of hang-ups. These hang-ups are typically caused by the formation of a stable structure of coarse material and are, by far, the most important problem associated to this, otherwise, convenient method to transport raw material in mines [4].

The development of clogs that cause arrest of the flow has been widely studied for the case of silos after the original work of K. To et al. [5]. Typically, clogging in silos is characterized by means of the avalanche size, defined as the number of beads that flow out of the silo from the destruction of a clogging arch until the development of a new one. We know that, for a given outlet and particle size, the avalanche size distribution is an exponential tail [6–9] suggesting that the probability of clogging remains constant over the whole avalanche duration. Also, it is known that the avalanche size grows very quickly with the outlet size. In this sense, although the existence of a critical outlet size above which clogging may never occur was originally proposed [7], further works have suggested that this transition does not necessarily exist [8, 9].

Going back to the pipe flow case, Janda et al. [10] experimentally evidenced the development of clogs in an inclined pipe full of a non-spherical ballast of limestone that was discharged at constant speed by means of a belt at the bottom. Their findings are compatible with a clogging probability that remains constant during the discharge. In addition, they show that the amount of material discharged before the development of a clog increases as the tube diameter is enlarged, the velocity of the conveyor belt grows, and the tube tilt deviates from the vertical. In another work, Verbücheln et al. [11] reported a numerical work where clogging was observed in a system of spheres falling freely under the action of gravity within a confined cylinder with periodic boundary conditions in the direction of gravity. At the beginning, grains accelerated but eventually, a limit velocity was achieved due to collisions. Then, depending on the ratio between the pipe diameter and the beads, the flow became intermittent giving rise to clogs. Interestingly, they show that inducing very well ordered patterns by means of the introduction of an inner helical texture in the pipe leads to a notable reduction of clogging.

In this work we design a experimental setup to analyze in detail the phenomenon of clogging in a vertical tube. First, we will show that clogging of spheres in perfectly vertical tubes is possible, an issue that has been only observed in numerical simulations [11]. Then, we will study both the temporal and spatial dependence of the clogging development. We are not aware of any work that characterizes the position at which clogs occur along the pipe and, as an improvement from previous work [10], we have implemented a system that allows this kind of measurement.

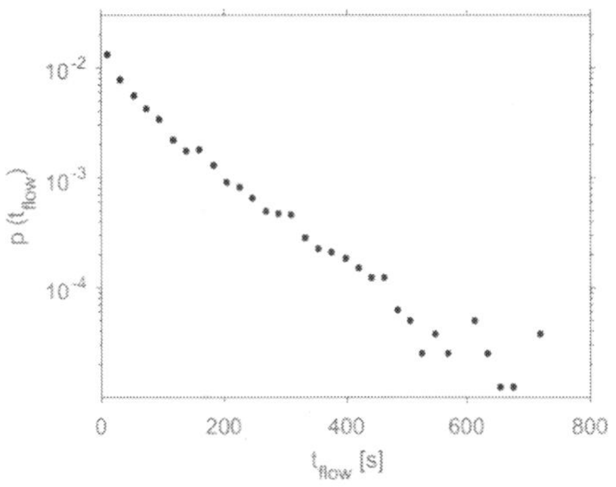

Figure 2. Probability distribution of avalanche duration. Note the semi-logarithmic scale.

2 Experimental setup

The experimental setup (Fig. 1) consisted of a long narrow pipe full of grains which were extracted at a constant rate by means of a conveyor belt placed at the bottom of the pipe. The granular material was formed by mono-disperse spherical glass beads of 0.6 *cm* diameter. The pipe was made of transparent methacrylate to allow the observation of clogs. The inner diameter was 2.2 *cm* which supposes a ratio between pipe and particle diameters of 3.67. The pipe had a length of 200 *cm* and the inclination with respect the horizontal was 90°, i.e. it was vertically oriented. The supporting structure was made of aluminum profiles fixed to the wall of the laboratory in order to prevent any movement or vibration.

The material was extracted with a conveyor belt perpendicular to the pipe. The gap between the pipe exit and the belt was 1 *cm*. This configuration avoids the formation of clogs between the end of the pipe and the surface of the belt. The width of the belt, 6 *cm*, and the distance from the end of the pipe to the rear part of the belt, 6 *cm*, were kept constant.

The velocity of the belt, which determines the flow-rate, was controlled by digital hardware allowing a straightforward modification of this parameter. The velocity was set at $v = 0.215\ cm/s$ in all the experiments shown here. At the end of the conveyor, the granular material is accumulated in a box with capacity for the 10^5 particles of glass beads employed (25 kilograms). Once all the beads have flown out of the pipe, they are poured in another reservoir of similar size placed at the top of the pipe. From there, the material is refilled into the pipe through an orifice of the same diameter than the pipe. As the ratio between the pipe diameter and the particle size is 3.67, clogs appear rather frequently at that top bottleneck [6]. To prevent such clogs, we build an autonomous system controlled by an Arduino® consisting on an inverted cone that moves up and down with an amplitude of 5 *cm* in a period of 3 seconds. This device was placed inside the top reservoir, just above the orifice, hence leading to the breakage of all the arches formed at that upper orifice and allowing a constant refilling of the pipe.

The measuring protocol was: first, the pipe was filled through the reservoir at the top. Then, the conveyor was switched on and the material started to flow through the pipe by gravity. This instant was registered and defined the beginning of the avalanche, t_{start}. When eventually, a hanging arch clogged the pipe at any position, all the material above it stopped. This precise moment was detected by means of a web-cam, placed at the top of the pipe which performed a real-time analysis of a 3.5 cm height region. We compared the pixel values of two consecutive frames (every second) and we decided that a clog exist whenever these two images were identical. To ensure that the clog was stable we continued extracting material during the time that the conveyor needs to transport the distance of two pipe diameters. Therefore, we defined the duration of an avalanche: $t_{flow} = t_{end} - t_{start}$, where t_{end} is the time defined by the two equal consecutive images.

A second camera (1600x1200 pixels resolution and 15 frames per second) recorded the entire pipe with the objective of finding the position of the clog, h_{clog} with respect to the bottom of the pipe. To this end, we analyzed the value of the pixels in a rectangular region of 8 pixels width and almost 1600 high (spanning the whole length of the pipe). The regions of the pipe with and without granular material were easily identified because, under proper illumination, beads were shown as white pixels, whereas empty regions appeared as black. We define the positions of the clogs at these places where in a slice of 1.2 cm height, all the pixels were black.

After detecting the clog position (sometimes more than one could be identified), the hanging arches were destroyed. To this end we employ a set of 10 electric shakers placed at the back of the pipe. From them, only the nearest to the (already identified) clog was activated for a lapse of 0.5 s. The shaker hit several times the tube with a tuned intensity that was strong enough to break the clog. After this, the conveyor was switched on, resuming the extraction of material and the measurement loop started again. All these operations were controlled with a computer, allowing the complete automation of the experiment.

3 Experimental results

We start presenting the avalanche duration distribution obtained for the given set of parameters specified above: a 200 cm length and 2.2 cm inner diameter tube full of 0.6 cm diameter glass beads discharged at a constant velocity of 0.215 cm/s. As displayed in Fig. 2 the distribution looks like an exponential tail, in agreement with previous results obtained in silos [6–9] and for faceted particles discharged through an inclined pipe [10]. The explanation given to justify this tail in all these different works relies on the existence of a constant probability of clogging over the whole duration of the avalanche. Here, we can argue the same despite that, contrary to the silo (where clogging always happens above the outlet), in the pipe, clogging can appear at different positions.

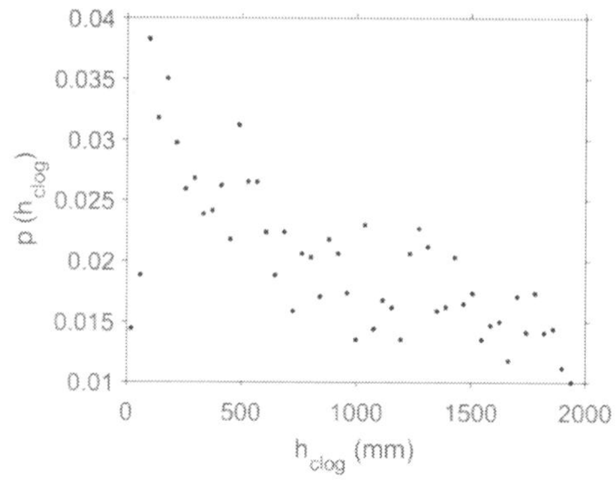

Figure 3. Probability of clogging at different positions within the pipe, measured as the height h from the bottom. The probability has been obtained in 50 bins of 4 cm each.

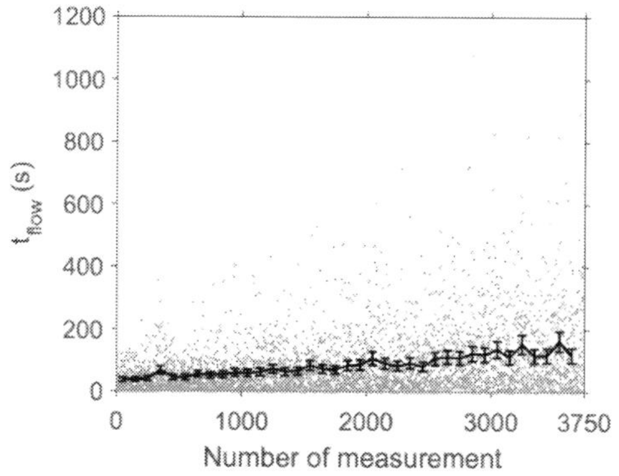

Figure 4. t_{flow} versus number of measurement. Gray dots represent the experimental results of t_{flow}. The black line links the averages of 100 consecutive measurements of t_{flow}. The error bars are the 95% confidence intervals of these averages assuming an exponential distribution.

In the next stage we analyze, precisely, where the hanging clogs appear. In Fig. 3 we show the probability of finding a clog at different heights of the pipe, h, measured from the bottom. Clearly, if we exclude the lower 4 cm we can state that clogging is more frequent near the bottom and reduces monotonously with height. The decrease looks nonlinear but more statistics than the 3738 clogs obtained are needed to venture any deeper conclusion.

Up to now, we have shown that in a very long pipe full of grains, clogs are possible even when the granular material is formed by spheres. This was not trivial as in previous experimental works, faceted anisotropic particles were used, and those are known to clog more as they align each other and with the walls [12–14]. On the contrary, hanging clogs of spherical particles are necessarily stabilized by frictional forces: at some position of the arch, a bead should be hanging below the equator of their neighbours (or below the equator of a neighbour and the wall).

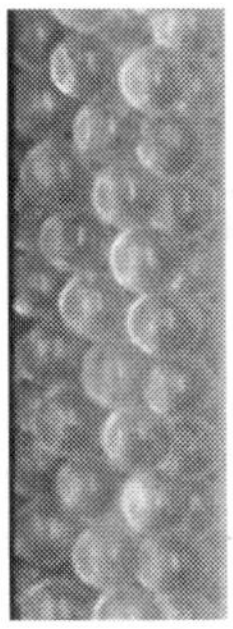

Figure 5. Helical arrangement of glass beads formed spontaneously inside the pipe.

These kind of arches have been also found in static silos and have been demonstrated to be very weak, in the sense that they are easily destabilized [15]. We believe that this arch weakness and strong dependence on frictional forces is also behind an undesirable effect that we have found in our experiments: the avalanche duration is not constant over time (at least under these experimental conditions). This is readily appreciated in Fig. 4 where we plot each avalanche duration in chronological order. The black line represents the averages for 100 consecutive avalanche durations. At this moment, we do not fully understand this effect but we believe that it could be related with the fact that, over time, the glass beads or the methacrylate tube wear out. In such case, a little amount of dust could lubricate the material and delay the formation of clogs.

4 Conclusions

In this work we have shown that spherical beads flowing within a vertical narrow tube are able to display clogging events provided that the tube diameter is small enough in comparison with the particles size. As previously reported, we observe that the duration of clogs displays an exponential distribution. Moreover, the experimental improvements performed to observe the position of the hanging arches allows to find a net spatial dependence of them: clogging is more frequent in the lower parts of the tube. The origin of this is not clear but it could be speculated that clogging could increase with pressure and, in such narrow tube, the pressure does not really saturates as predicted by Janssen [16]. An alternative explanation that might also affect the way in which pressure propagates within the pipe could be based in the spontaneous formation of helical patterns observed inside the tube (Fig. 5). Indeed, recent findings have shown that these helical patterns lead to clogging reduction [11]. Undoubtedly, a careful investigation of such structures and the implications that they have for pressure transmission is necessary. Interestingly, these helices seem to be generic, being the focus of interest in disparate scenarios such as in granular media aiming for packing optimization [17]; in botany concerning the ordering of the leaves of a plant (Phyllotaxis) [18]; or in the self assembly of nanospheres inside carbon nanotubes [19].

Acknowledgments

Diego López thanks Asociación de Amigos de la Universidad de Navarra for a scholarship. This work was funded by Ministerio de Economía y Competitividad (Spanish Government) through Project No. FIS2014-57325.

References

[1] W. A. Hustrulid and W. A. Hustrulid and R. C. Bullock, *Underground Mining Methods: Engineering Fundamentals and International Case Studies* (Society for Mining and exploration Inc., 2001) 627-634.
[2] V. Aytaman, Canadian Mining Journal **81**, 77-81 (1960).
[3] J. Hadjigeorgiou, J. F. Lessard and F. Mercier-Langevin, Afr. Inst. Min. Metall **105**, 809-816 (2005).
[4] J. Hadjigeorgiou and J. F. Lessard, Rock Mechanics and Mining Science **44**, 820-834 (2007).
[5] K. To *et al.*, Phys. Rev. Lett. **86**, 71 (2001).
[6] I. Zuriguel, L. A. Pugnaloni, A. Garcimartín and D. Maza, Phys. Rev. E **68**, 030301 (R) (2003).
[7] I. Zuriguel, A. Garcimartín, D. Maza, L. A. Pugnaloni and J. M. Pastor, Phys. Rev. E **71**, 051303 (2005).
[8] K. To, Phys. Rev. E **71**, 060301 (R) (2005).
[9] C. C. Thomas and D. J. Durian, Phys. Rev. Lett. **114**, 178001 (2015).
[10] A. Janda, I. Zuriguel, A. Garcimartín and D. Maza, Granular Matter **17**, 545 (2015).
[11] F. Verbücheln, E. Parteli and T. Pöschel, Soft Matter **11**, 4295-4305 (2015).
[12] T. Kanzaki, M. Acevedo, I. Zuriguel, I. Pagonabarraga, D. Maza and R.C. Hidalgo. The European Physical Journal E **34**, 133 (2011).
[13] J. Tang and R. P. Behringer. Europhysics Letters **114**, 34002 (2016).
[14] A. Ashour, S. Wegner, T. Trittel, T. Börzsönyi, and R. Stannarius, Soft Matter **13**, 402-414 (2017).
[15] C. Lozano, G. Lumay, I. Zuriguel, R. C. Hidalgo and A. Garcimartín, Phys. Rev. Lett. **109**, 068001 (2012).
[16] H. A. Janssen Versuche über Getreidedruck in Silozellen. Zeitschr. d. Vereines deutscher Ingenieure **39**, (1895) 1045.
[17] A. Mughal, H. K. Chan, D. Weaire and S. Hutzler. Phys. Rev. E **85**, 051305 (2012).
[18] R. Erickson, Science **181**, 705-716 (1973).
[19] A Khlobystov, D. Britz, A. Ardavan and A. Briggs, Phys. Rev. Lett. **92**, 245507 (2004).

Stress and strain rate analysis of the FT4 Powder Rheometer

Colin Hare[1*], and *Mojtaba* Ghadiri[2]

[1]Department of Chemical and Process Engineering, The University of Surrey, Guildford, UK
[2]School of Chemical and Process Engineering, The University of Leeds, Leeds, UK

Abstract. The Freeman FT4 Powder Rheometer has been reported to describe well the powder flow behaviour in instances where other techniques fail. We use DEM to simulate the FT4 operation for slightly cohesive large glass beads at a range of strain rates. The curved impeller is shown to be beneficial in comparison to a flat blade as the variation of shear stress across the blade is reduced. The shear stress in front of the blade correlates well with flow energy (which the device measures) for a range of tip speeds and is shown to increase approximately linearly with tip speed when operating beyond the quasi-static regime.

1 Introduction

Poor understanding and control of powder flowability is a major cause of process downtime and reduction in process performance. The onset of flow in a quasi-static state is largely understood, with methods available for design of storage bins and hoppers [1]. However many processes operate at higher strain rates and are concerned with dynamic flow, such as mixing and die filling. The only commercially available devices which characterise powder flow in the dynamic regime are the FT4 Powder Rheometer of Freeman Technology and that of Anton-Paar. Here we address the analysis of the dynamics of the FT4, whilst that of Anton-Paar has been reported by Salehi-Kahrizsangi *et al.* [2]. In the FT4 a twisted-blade impeller is rotated and driven downwards into a column of powder. The device directly measures the torque on the blade, T, and the force on the base of the vessel, F_{base}, in order to determine 'flow energy',

$$E_{flow} = \int_0^h \left(\frac{T}{R\tan\alpha} + F_{base} \right) dh \qquad (1)$$

where h is the penetration depth, R the radius of the impeller and α the helix angle to the horizontal of the impeller movement. The FT4 measurement has often been cited to correlate with powder flow performance in processes, however the meaning of flow energy and how it can be used to better design or control powder processes is not well understood. Here we use DEM to simulate the FT4 device operation in order to determine the stress and strain distribution in the bed under a range of operational speeds.

2 Stress analysis

The FT4 operation is simulated using EDEM (DEM Solutions), using spheres of 1.7 – 2.1 mm diameter. The linear elastic-plastic and adhesive contact model of Pasha *et al.* [3] is used to represent cohesive spheres with a surface energy of 29 mJ/m^2, in order to correlate with the experimental work of Hare *et al.* [4]. The particle-particle sliding and rolling friction coefficients are set at 0.3 and 0.01, respectively, to best represent the experimental glass beads in their work. The spheres are generated in a 50 mm diameter vessel and allowed to settle under gravity, after which the particles whose centre is greater than 80 mm above the base are removed, thus leaving approximately 25,000 particles. The bed diameter and height are identical to the experimental setup.

The 48 mm diameter impeller is rotated in an anti-clockwise direction and driven downwards into the bed using a tip speed of 100 mm/s and helix angle of 5° to the horizontal, until a penetration depth of 70 mm is reached, 10 mm above the base of the bed.

The normal and shear stresses in measurement cells in front of the blade (Figure 1) are estimated using Eq 2

$$\sigma_{ij} = \frac{1}{V}\sum_1^N F_{ij} \cdot r \qquad (2)$$

where V is the cell volume, N is the number of particles in the cell, F is the force acting in direction i on face j [4], using Cartesian coordinates. The principal stresses are then determined from the eigenvalues of these nine stress tensors, and the deviatoric (shear) stress,

* Corresponding author: c.hare@surrey.ac.uk

$$\tau_D = \frac{\sqrt{(\sigma_1-\sigma_2)^2+(\sigma_1-\sigma_3)^2+(\sigma_2-\sigma_3)^2}}{\sqrt{6}} \quad (3)$$

where σ_1, σ_2 and σ_3 are the major, intermediate and minor principal stresses, respectively. The hydrostatic (confining) stress, σ_H, is taken as the average of the three principal stresses.

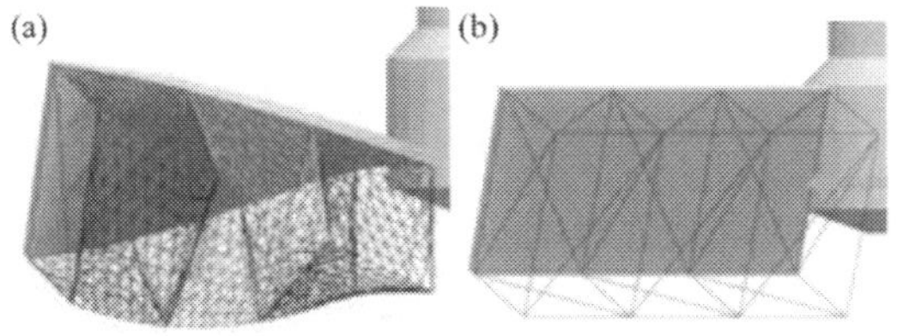

Fig. 1. Measurement cells (a) curved blade, (b) flat blade

Hare *et al.* [4] showed that the shear stress in front of the curved blade is approximately constant across its radial length. Here we analyse the dynamics of a flat blade set at the same inclination angle with respect to the horizontal plane as the tip of the real, curved blade (Fig. 1), by DEM simulation. Figure 2 shows the 10-point moving average of the deviatoric stress against penetration depth for these two blades. It can be seen that with the flat blade the stress increases with radial position across the blade, from approximately 90 to 160 Pa at the final penetration depth of 70 mm. In contrast, the stresses near the shaft and tip of the curved blade at this penetration depth show much less deviation, with values of 100 and 120 Pa, respectively.

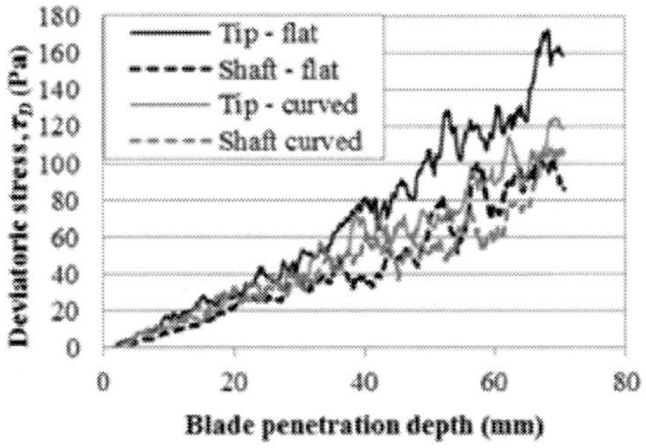

Fig. 2. Shear stress against penetration depth for curved and flat blades

For some powders the flow energy is observed to change with tip speed. The FT4 is capable of operating up to a tip speed of 300 mm/s. Using the curved blade we simulate the FT4 operation using tip speeds of 40 - 1200 mm/s, with a fixed helix angle of 5°. In all cases the shear stress is found to be approximately constant across the blade length, and increasing linearly with penetration depth (though notable scatter is observed as in Fig. 2). A linear regression is applied to the shear stress against penetration depth, to estimate the maximum shear stress occurring in front of the blade, τ_{max}. Figure 3 shows the maximum shear stress and flow energy for the range of impeller speeds tested.

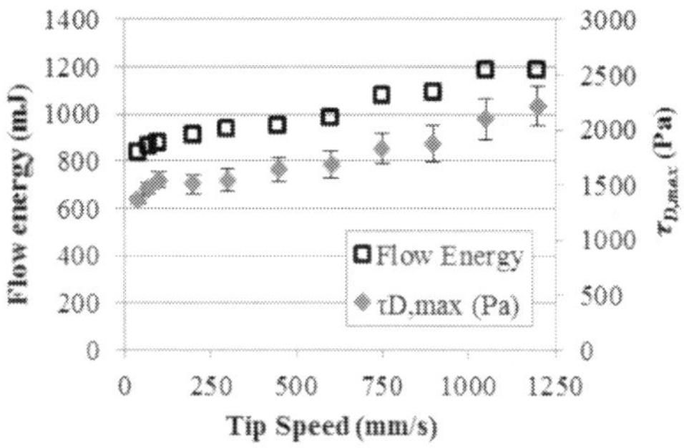

Fig. 3. Maximum shear stress and flow energy for various impeller tip speeds

The maximum shear stress and the flow energy increase with tip speed through the entire range tested. The shear stress error bars represent one standard deviation from the linear regression of shear stress against blade penetration depth (e.g. Figure 2), at maximum penetration depth. The error bars indicate that the increase in shear stress with impeller speed is statistically significant. There is a strong correlation between flow energy and shear stress for the range of impeller speeds tested.

3 Strain rate analysis

In addition to the stress profile in the bed, there is a need to understand the strain rate regime of operation in the FT4 tests. Figure 4 shows the particle velocity distribution for impeller tip speeds, v_{tip}, of 70, 600 and 1050 mm/s, at a penetration depth of 10 mm. The darkest colour represents the tip speed and the lightest is 10% of the tip speed. At an impeller speed of 70 mm/s the bed is densely packed and a small fraction of the bed is mobilised. As the impeller speed is increased there is a clear dilation of the bed, with particles being lifted high above the initial bed height at impeller speeds of 600 and 1050 mm/s.

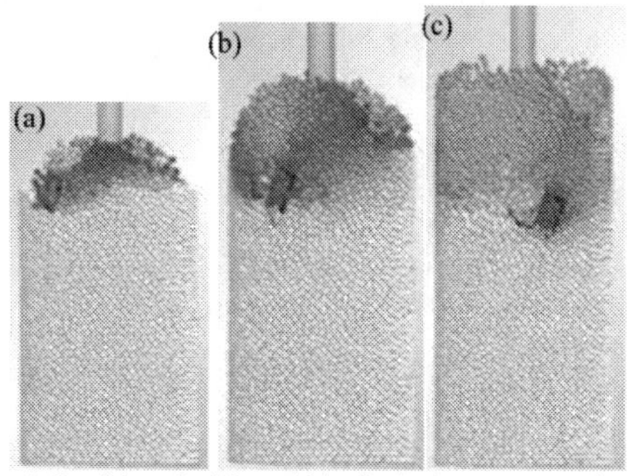

Fig. 4. Particle velocities within the bed at a penetration depth of 10 mm with impeller tip speeds (a) 70 mm/s (b) 600 mm/s and (c) 1050 mm/s

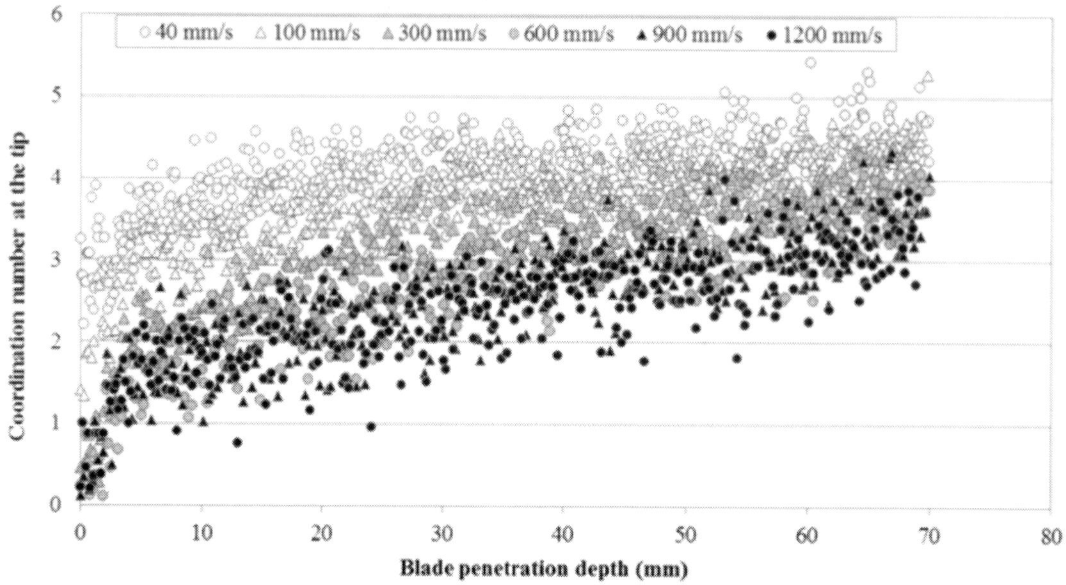

Fig. 5. Coordination number in the measurement cell near the tip for impeller speeds of 40, 100, 300, 600, 900 and 1200 mm/s

The evolution of coordination number, Z, within the measurement cell near the tip during the FT4 test with a range of impeller speeds is shown in Figure 5. For all impeller speeds the coordination number increases with penetration depth. At penetration depths of a few mm the increase in Z with depth is sharp, particularly at greater impeller speeds. This may be largely due to an increasing number of particles being lifted and dispersed high above the bed. At penetration depths of 10 mm (as shown in Figure 4) and beyond, there is a more gradual increase in Z with penetration depth. This is expected to be partly due to the previously lifted powder settling on top of the bed; resisting the upward movement of particles out of the measurement cell, in addition to the fact that as the active region of the bed descends, it is under an increasing hydrostatic stress due to the weight of the bed above. At an impeller speed of 40 mm/s Z is approximately 4-4.5 at a penetration depth of 70 mm. An increase of impeller speed to 100 mm/s causes negligible change in Z, whereas at an impeller speed of 300 mm/s there is notable change as Z drops below a value of 4. As the impeller speed is increased to 1200 mm/s there is a continual drop in Z.

The length of a shear band is estimated from Figure 4 by considering the number of particles running parallel to the blade from the last particle with velocity v_{tip} to the first particle with velocity $0.1v_{tip}$, above and below the blade. The strain rate, γ, is calculated using Eq 4.

$$\gamma = dv/\left(n_{band} \cdot d_p\right) \tag{4}$$

where dv is the velocity differential across the shear band ($0.9v_{tip}$), n_{band} is the number of particles in the shear band and d_p is the average particle diameter. The strain rate is calculated using this approach at penetration depths of 25, 38, 51 and 64 mm, and the average taken. For all impeller speeds the particles immediately in front of the blade have maximum velocity, equal to the tip speed. The dimensionless strain rate, γ^0, introduced by Tardos *et al.* [6] is calculated using Eq 5.

$$\gamma^0 = \gamma\sqrt{d_p/g} = \frac{dv}{n_{band}}\sqrt{\frac{1}{d_p \cdot g}} \tag{5}$$

where g is gravitational acceleration.

The strain rates at the given tip speeds are shown in Table 1. It is suggested that at low dimensionless strain rates the bed is in the quasi-static regime and beyond values of 0.15-0.25 the system is in the intermediate regime [6], thus implying that at the maximum tip speed of the FT4 (300 mm/s), the bed is in this regime. It is worth noting from Eqs 4 & 5 that if the number of particles in a shear band remains constant, a reduction in particle size leads to an increase in dimensionless strain rate, thus suggesting that for fine particles the FT4 may be capable of operating in the dynamic regime. However, for cohesive powders the number of particles in the shear band may increase due to clustering, so further work is needed to investigate this.

The flow regime is commonly assessed by the inertial number,

$$I = \gamma \cdot d_p\sqrt{\rho/\sigma_H} \tag{6}$$

where ρ is the particle density of the bed. Figure 6 shows the inertial number in the measurement cell near the tip of the blade against penetration depth for a range of impeller speeds. In all cases the inertial number decreases as the impeller penetrates further into the bed, with an initially rapid decline followed by a more gradual reduction beyond a depth of 5-10 mm. The bed is in the dense flow regime up to an impeller speed of 900 mm/s ($1 \times 10^{-3} < I \le 1 \times 10^{-1}$), whilst at greater impeller speeds the bed is in the collisional flow regime ($I > 1 \times 10^{-1}$). The regime boundaries differ significantly between the inertial number approach and that of Tardos *et al.* [6], with the inertial number approach suggesting the system enters the dense and collisional flow regimes at much lower impeller speeds.

Table 1. Strain rates at various tip speeds

Impeller speed (mm/s)	Average n_{band}	Strain rate, $\dot{\gamma}$ (s^{-1})	Dimensionless strain rate, $\dot{\gamma}^{0}$
40	4.8	4.6	0.06
70	4.1	8.6	0.12
100	4.4	12	0.17
200	4.8	23	0.32
300	4.3	37	0.51
450	4.8	57	0.79
600	4.9	75	1.04
750	4.6	99	1.38
900	4.8	100	1.42
1050	5.4	110	1.58
1200	5.4	130	1.75

4 Conclusions

The FT4 Powder Rheometer operation was simulated using DEM with slightly cohesive spheres of approximately 2 mm diameter. The curved blade of the FT4 is shown to minimise the variation of shear stress across the blade length, as compared to a flat blade of the same inclination angle. The shear stress at maximum penetration depth is shown to correlate well with the flow energy for a wide range of impeller tip speeds. The shear stress and the flow energy both increase slightly with tip speed throughout the entire range of impeller speeds tested. Beyond a tip speed of 300 mm/s the bed is found to be in the intermediate/dense flow regime using the approach of Tardos *et al.* [6]. It is noteworthy that if the number of particles in a shear band is independent of particle size, the impeller tip speed corresponding to this flow regime boundary is expected to reduce as particle size is reduced. In contrast, the inertial number approach suggests that even at a tip speed of 40 mm/s the bed is in the dense flow regime, and enters the collisional regime at a tip speed beyond 900 mm/s.

Acknowledgements

The authors would like to express their thanks to Messrs Tim Freeman and Jamie Clayton of Freeman Technology, UK, for their support of this work, and Dr Marty Murtagh of Corning inc. for useful discussions.

References

1. A.W. Jenike, Powder Technol., **1**, 237-244 (1967).
2. H. Salehi-Kahrizsangi, D. Shütz, D. Barletta, M. Poletto, AIChE Annual Meeting, Nov 13-18, San Francisco, USA (2016).
3. M. Pasha, S. Dogbe, C. Hare, A. Hassanpour, M. Ghadiri, Granul. Matter, **16**, 151-162 (2014).
4. C. Hare, U. Zafar, M. Ghadiri, T. Freeman, J. Clayton, M.J. Murtagh, Powder Technol., **285**, 123-127 (2015).
5. K. Bagi, Mech. Mater., **22**, 165-177 (1996).
6. G.I. Tardos, S. McNamara, I. Talu, Powder Technol., **131**, 92-98 (2003).

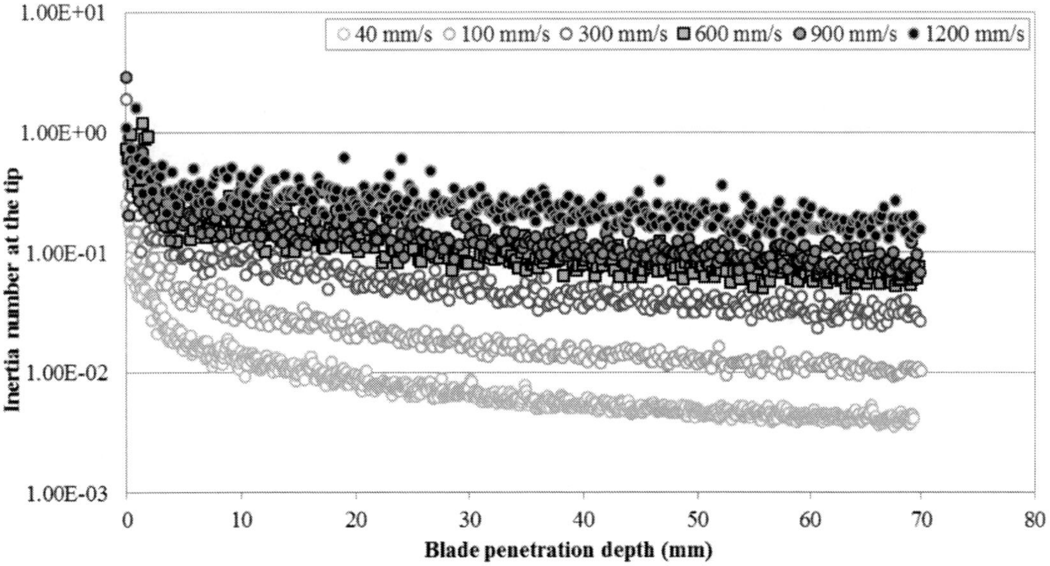

Fig. 6. Inertia number in the measurement cell near the tip for impeller speeds of 40, 100, 300, 600, 900 and 1200 mm/s

Effect of grain shape on the jamming of two-dimensional silos

Ezequiel Goldberg[1,⋆], *C. Manuel* Carlevaro[1,2,⋆⋆], and *Luis A.* Pugnaloni[3,⋆⋆⋆]

[1] *Universidad Tecnológica Nacional – FRBA, UDB Física, Mozart 2300, C1407IVT Buenos Aires, Argentina.*
[2] *Instituto de Física de Líquidos y Sistemas Biológicos (CONICET La Plata, UNLP), Calle 59 Nro 789, 1900 La Plata, Argentina.*
[3] *Dpto. Ingeniería Mecánica, Facultad Regional La Plata, Universidad Tecnológica Nacional, CONICET, Av. 60 Esq. 124, 1900 La Plata, Argentina.*

Abstract. We present discrete element method simulations of the discharge of silos in two dimensions. We study the effect of the grain shape on the clogging of small apertures, considering regular polygons and disks of equal mass. In particular, we analyze the avalanche size distribution and the jamming probability for disks, triangles, squares, pentagons, hexagons and heptagons as a function of the aperture size. We show that the jamming probability presents a non-linear response as a function of the number of vertexes of the polygons.

1 Introduction

The clogging (or jamming) of the flow of grains through small apertures is usually an undesired phenomenon in different industrial processes. The search for a prediction of a minimum suitable opening size that warrants that clogging will not occur in the flow has been driven by practical reasons. It is generally accepted that, for spherical grains, a circular opening larger than five particle diameters will rarely clog. Whether this is a sharp transition is still under debate [1, 2]. The friction coefficient of the grain–grain interaction seems to play little role for spheres [3]. However, one should expect that the shape of the grains is a key factor for the ability of grains to clog. Indeed, Zuriguel et al. [3] have shown that rice (elongated grains) and pasta pellets (short cylindrical grains) do present a larger tendency to jamming. Despite these initial investigations, there has not been systematic studies on the effect of particle shape on clogging.

In this work, we present results of two-dimensional simulations of grains with shape of regular polygons (triangles, squares, pentagons, hexagons and heptagons) in addition to circular grains. We focus on the clogging events during the discharge of the grains trough small openings in the base of a flat bottom silo. To compare results from different shapes we use as a criterion that the grains must have the same mass. We will show that the intuitive prediction that polygonal grains are more prone to clogging than disks is indeed confirmed. However, triangular particles show a marked reduction in clogging if compared to squares, pentagons and hexagons.

⋆e-mail: ezequiel.goldberg@outlook.com.ar
⋆⋆e-mail: manuel@iflysib.unlp.edu.ar
⋆⋆⋆e-mail: luis.pugnaloni@frlp.utn.edu.ar

2 Simulation

We simulate a 2D silo by means of a discrete element method (DEM) using the Box2D library [4]. This package uses a constraint solver. At each time step of the dynamics, a series of iterations (typically 20) are used to resolve penetrations between bodies through a Lagrange multiplier scheme [5]. After detecting overlaps, the inelastic collision at each contact is solved and new linear and angular velocities are assigned. The equations of motion are integrated through a symplectic Euler algorithm. The time step δt used to integrate the equations of motion is 0.025 d/g; with d the diameter of our circular grains and g the acceleration of gravity. Solid friction is also handled by means of a Lagrange multiplier scheme that implements the Coulomb criterion. Box2D has been previously used to simulate continuous silo discharges [6], and tapping of granular systems [7].

The silo consist of a box $20d$ wide and $180d$ high. Gravity acts in the negative vertical direction. 2000 grains are used in each discharge, and these are not reinjected after they leave the silo. After grains initially placed at ran-

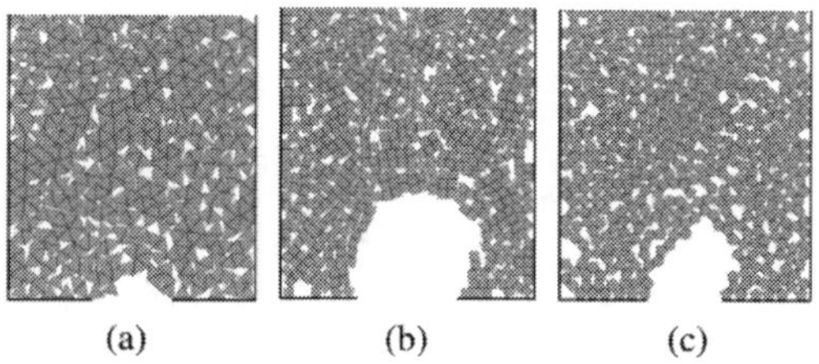

Figure 1. Snapshots of the simulations after a clog is detected: (a) triangles, (b) squares and (c) pentagons. Only the region close to the aperture is shown.

 DOI: 10.1051/epjconf/201714006009

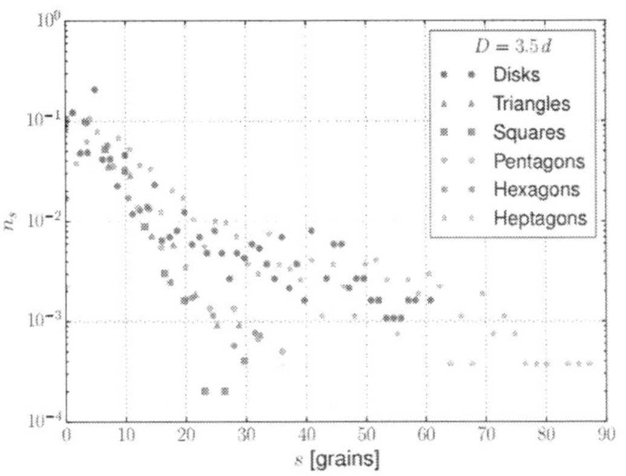

Figure 2. Normalized distribution of avalanche sizes $n_s(D)$ for an aperture of width $D = 3.5d$ for different grain shapes (see legend).

dom without overlaps settle, an aperture opened at the center of the flat base of the silo allows the grains to discharge. For small apertures, clogging is observed. We consider a clog is permanent if more than 5 seconds have passed without any grain falling through the aperture. For each clogging we record the number of particles that passed through the aperture. We do not resume the discharge by perturbing the clogged system, instead, we carry out a new simulation by filling a new silo using different random initial positions for the grains. We record, for each aperture size, and for each grains shape, 1500 clogs. For large aperture sizes the entire system discharges without clogging. In these cases we record that an avalanche larger that 2000 grains has occurred.

Grains of different shapes corresponding to regular polygons (triangles, squares, pentagons, hexagons and heptagons) are used in the simulations apart from the circular grains. All grains have the same material density and the same area, hence the same weight. Each simulation contains only one type of grain and these are monosized. Figure 1 shows some snapshots of systems after a clog. The restitution coefficient is set in all cases to $\epsilon = 0.05$ and the friction coefficient μ is set to 0.5. The significantly low restitution coefficient allows for a rapid dissipation of the kinetic energy during the initial filling of the silo. The material density ρ of the grains is set to 0.01 Kg/m^2

3 Results

We consider the distribution of avalanche sizes $n_s(D)$ which is the number of avalanches of s grains. Figure 2 shows the normalized n_s for an aperture $D = 3.5d$, for each grain shape. As we can see, most polygonal shapes show a similar size distribution of the avalanches. There is however a marked difference with circular grains, which tend to deliver much larger avalanches for a given orifice. Interestingly, heptagons seem to have avalanches similar to disks. This may be due to the polygon having enough edges to behave close to a disk. We have not explored

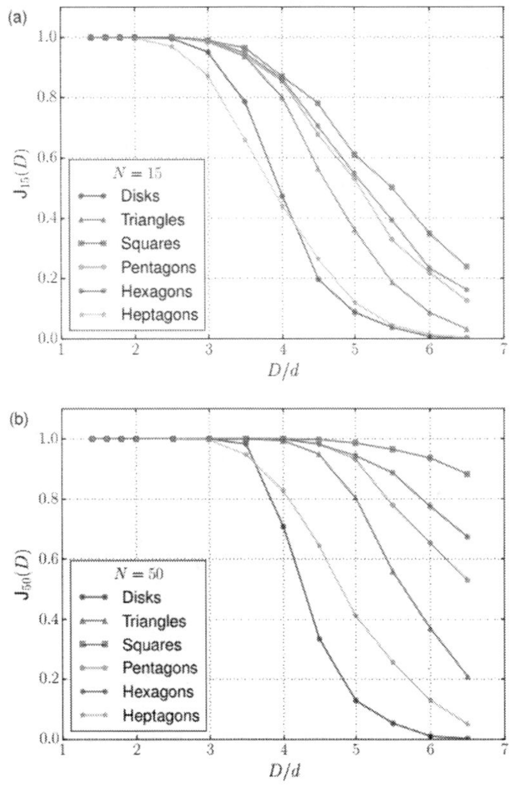

Figure 3. Jamming probability for the different grain shapes for (a) $N = 15$, and (b) $N = 50$.

polygons with larger number of vertexes to confirm if this is the case. The distributions have an exponential tail, consistent with all previous results on disks and spheres (see for example Refs. [1, 3, 8]

We also calculate the jamming probability $J_N(D)$ defined as [3]

$$J_N(D) = \frac{\sum_{s=N}^{\infty} s n_s(D)}{\sum_{s=1}^{\infty} n_s(D)}. \tag{1}$$

$J_N(D)$ corresponds to the probability that an avalanche smaller that N grains discharges thought an aperture of size D before the system clogs. Figure 3 shows $J_{15}(D)$ and $J_{50}(D)$ for different grain shapes. As observed in previous studies, $J_N(D)$ is one for small D and falls to zero when D increases. An increase in N also leads to a shift of the curves to the right, indicating a larger probability of jamming before N grains flow freely. For disk, there have been a number of studies (both experimental [1, 8] and numerical [9–11]). The values of jamming probability we find for disks are sensibly larger than those reported in most previous studies. However, these previous works have followed a protocol where avalanches are triggered after a clog by perturbing the blocking arch. In a few cases, the system is emptied and made to flow again into the container keeping the aperture open [8]. In our case the silo is emptied and refilled with the aperture closed, which is opened only after the grains are static. Uñac et al. [12] have conducted a similar type of protocol as the one used here, but the system is tapped before opening the aperture to achieve a desired initial packing fraction. They found that clogging maw vary dramatically depending on the initial packing fraction before starting the avalanche.

Figure 3(a) shows a clear non-linear response when the number of vertexes of the grains are increased. From the most prone to jamming, the grain shapes can be ordered as: squares, hexagons, pentagons, triangles, disks and heptagons. It is unexpected that heptagons present a lower jamming probability than disks. However, this is observed only for small D. For values of D where $J_{15}(D)$ is small, heptagons have a larger jamming probability. If one is interested in clogging developing after a larger number of grains have passed, the peculiar behavior of heptagons is partially lost. This can be seen in Fig. 3(b). For $J_{50}(D)$, for example, heptagons present a larger tendency to jamming than disk for most values of D. It is important to bear in mind that, when the orifice is opened, a few grains initially supported by the base fall through right away, affecting the statistics of jamming for small N. However, if one focuses on large avalanches, a steady flow develops before an arch is formed. Hence, results for large N are less affected by the statistics of the flow of the initial group of grains.

A non-linear behavior on the number of vertexes of the grains has also been found in the packing fraction of tapped columns of regular polygons [7] and tapped packings of aggregates [**?**]. However, the packing fraction is significantly determined by the ability of a polygonal shape to tessellate the space, and one can naturally expect that tessellating shapes will yield higher packing fractions. In the case of clogging, the ability of interlocking thanks to the vertexes seems more significant. It is worth mentioning that hexagons and pentagons display very close values for $J_N(D)$ which make difficult to decide which clogs more. If their order is swapped in the list, the only singular shape becomes the triangle, which is much less prone to clogging than squares, pentagons and hexagons. We have carried out some preliminary simulations using a different "unclogging" protocol (where arches are broken every time they form to trigger a new avalanche) and we have found that in this case triangles present the larger jamming probability. A detailed discussion on the protocol dependence of clogging will be presented elsewhere.

4 Conclusions

We have shown that grains having polygonal shapes present significant differences in their clogging probability when they flow through small apertures. In general, the lower the number of vertexes of the regular polygon grains the larger is their tendency to jamming. However, triangles display a peculiar behavior, showing lower jamming probabilities than squares, pentagons and hexagons.

We have observed that the avalanche size distributions present exponential tails as it has been observed for circular and spherical grains in the past.

It is important to bear in mind that we have chosen to compare grains of different shapes keeping the mass (area) constant. However, other criteria may be used such as keeping the same diameter or perimeter for all shapes. As we have shown for the flow rate through large apertures this choice alters the results to a large extent [6].

Acknowledgements

This work has been supported by ANPCyT (Argentina) through grant PICT 2012-2155 and UTN (Argentina) through grants PID MAUTNLP0002184 and PID IFI1871.

References

[1] A. Janda, I. Zuriguel, A. Garcimartín, L. A. Pugnaloni and D. Maza, Eur. Phys. Lett. **84**, 44002 (2008).

[2] C. C. Thomas and D. J. Durian, Phys. Rev. Lett. **114**, 178001 (2015).

[3] I. Zuriguel, A. Garcimartín, D. Maza, L. A. Pugnaloni and J. M. Pastor, Phys. Rev. E **71**, 051303 (2005).

[4] Box2D Physics Engine, www.box2d.org

[5] E Catto, *Iterative dynamics with temporal coherence*, available at http://box2d.googlecode.com/files/GDC2005 Erin-Catto.zip (retrived on October 2010).

[6] E. Goldberg, C. M. Carlevaro and L. A. Pugnaloni, Papers in Physics **7**, 070016 (2015).

[7] C. M. Carlevaro and L. A. Pugnaloni, J. Stat. Mech. P01007 (2011).

[8] K. To, Phys. Rev. E **71**, 060301 (2005).

[9] C. F. M. Magalhães, J. G. Moreira and A. P. F. Atman, Phys. Rev. E **82**, 051303 (2010).

[10] G. Pérez, Pramana **70**, 989 (2008).

[11] R. Arévalo, I. Zuriguel, Soft Matter **12**, 123 (2016).

[12] R. O. Uñac, A. M. Vidales and L. A. Pugnaloni, J. Stat. Mech. P04008 (2012).

[13] L. K. Roth and H. M. Jaeger, Soft Matter **12**, 1107 (2016).

Granular rheology: measuring boundary forces with laser-cut leaf springs

Zhu Tang[1,*], *Theodore A.* Brzinski[1,2,**], and *Karen E.* Daniels[1,***]

[1] *Department of Physics, North Carolina State University, Raleigh, NC 27695 USA;*
[2] *Department of Physics, Haverford College, Haverford, PA 19041 USA*

Abstract. In granular physics experiments, it is a persistent challenge to obtain the boundary stress measurements necessary to provide full a rheological characterization of the dynamics. Here, we describe a new technique by which the outer boundary of a 2D Couette cell both confines the granular material and provides spatially- and temporally- resolved stress measurements. This key advance is enabled by desktop laser-cutting technology, which allows us to design and cut linearly-deformable walls with a specified spring constant. By tracking the position of each segment of the wall, we measure both the normal and tangential stress throughout the experiment. This permits us to calculate the amount of shear stress provided by basal friction, and thereby determine accurate values of $\mu(I)$.

1 Introduction

It is an open question what constitutive equations best describe flows of dense cohesionless granular materials [1–3]. There has come to be a consensus that two dimensionless parameters play a key role: interial number I and the friction μ. Each of these can be defined at the particle scale. The inertial number is given by

$$I \equiv \frac{|\dot{\gamma}|d}{\sqrt{P/\rho}} \tag{1}$$

where ρ is density of the solid granular material, d is their diameter, $\dot{\gamma}$ is the local shear rate, and P is the local pressure. Higher values of I correspond to rapid flows, and lower values to slower flows. The ratio

$$\mu \equiv \frac{\tau}{P} \tag{2}$$

is the local ratio of tangential stress τ to the normal stress (pressure P), and $\mu(I)$ is observed to be a good empirical descriptor of the rheology of the system [1] for well-developed flows. However, both $\dot{\gamma}$ and μ can exhibit strong spatial and temporal gradients. Recently, non-local extensions to the $\mu(I)$ rheology [2, 3] aim to provide a theoretical framework for capturing such features as the transition from inertial to creeping flow [4], boundary-driven shear-banding [5–7], and fluidization due to non-local perturbations [8–10].

However, direct comparisons between experiments and theory have been hampered by the difficulty measuring the tensorial stress within a granular material. Here, we report a new design for a 2D annular Couette cell which can measure the shear and normal stresses at the boundaries (see Fig. 1). This design has several advantages. First, the spring walls can be cut from standard acrylic sheets, making them cheaper and more convenient than using photoelastic particles. Second, the shape of the wall can be easily customized to have a particular spring constant by changing the thickness or length of the springs. Third, because photoelastic materials are no longer required, experiments on ordinary granular materials are made possible.

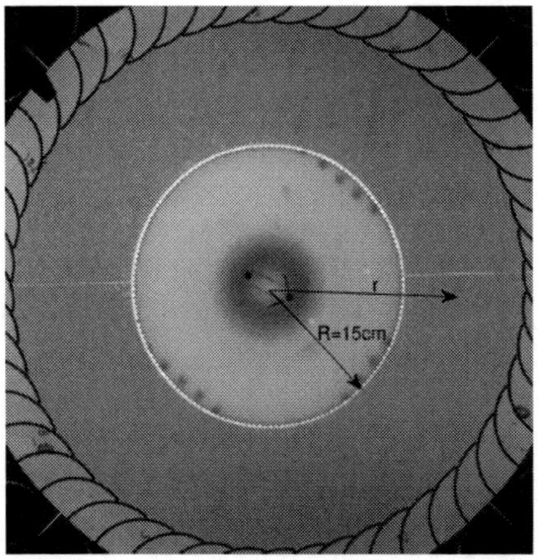

Figure 1. Top view of annular Couette experiment with ≈ 5000 disk-shaped particles confined between a rough inner ring of radius R_i = 15 cm and a segmented outer ring of radius R_o = 28 cm. Each of the leaf springs which make up the outer ring can allow dilation under stresses imposed by the dilating granular material.

*e-mail: ztang2@ncsu.edu
**e-mail: tbrzinski@haverford.edu
***e-mail: kdaniel@ncsu.edu

Below, we provide a description of the method for making quantitative boundary stress measurements using walls of this type. This involves (1) calibrating a single leaf spring, (2) using image cross-correlation to measure the displacement of each spring tip, and (3) calculating the stress as a function of time and azimuthal position by combining these two measurements at multiple positions around the outer wall. We close by presenting sample measurements of the $\mu(I)$ rheology made using this method.

2 Method

2.1 Apparatus

We develop our technique using a standard annular Couette geometry, which has the advantage of allowing continuous shear of a granular material to arbitrary total strain. The apparatus consists of a rotating inner disk (R_i = 15 cm) and a fixed outer wall (R_o = 28 cm) made up of 52 leaf springs which can dilate slightly (a few mm) and thereby provide a stress measurements at the outer wall. The granular material is about 5000 circular and elliptical disks, of diameter $d \approx 5$ mm and thickness 3 mm. A photograph of the apparatus from above is shown in Fig. 1.

2.2 Calibrating the spring wall

To provide a calibration for our experiments, we cut a single spring of the same acrylic, but with a short "handle" attached, and performed force-displacement measurements using an Instron materials tester. This process is shown in Fig. 2. One set of measurements was taken with the single spring oriented for normal compression, and a second set with tangential shear, while simultaneously recording a video of the dynamics. In both cases, we measured the x- and y-displacements of the spring tip as a function of applied force and observed a linear response. A least-squares fit to the data provides values for the calibration constants. These are $C_{n,x}$ = 1.43 ± 0.02 mm/N (x-deformation under normal force), $C_{n,y}$ = 5.06 ± 0.05 mm/N (y-deformation under normal force), $C_{t,x}$ = 3.49 ± 0.05 mm/N (x-deformation under tangential force), and $C_{t,y}$ = 6.86±0.08 mm/N (y-deformation under tangential force). All errors are reported from the fit.

2.3 Measuring wall deformation

To determine the normal pressure and the shear stress on the rim, we can make measurements of tip displacements (dx, dy, rotated into the appropriate coordinate system) for each of the leaf springs around the rim. Here, we illustrate the principle using a single leaf spring. First, we extract a subregion of the overhead image (see Fig. 1) in the vicinity of the tip. To determine its displacement, we also extract the same subregion from an image of the experiment taken without particles. Sample images are shown in Fig. 3ab. However, these raw tip-images also contain a piece of the neighboring spring. Using image-segmentation and masking, we remove the neighboring spring (cd). Using these two images, a simple image cross-correlation (e) determines the tip displacements (dx, dy) by fitting with sub-pixel resolution.

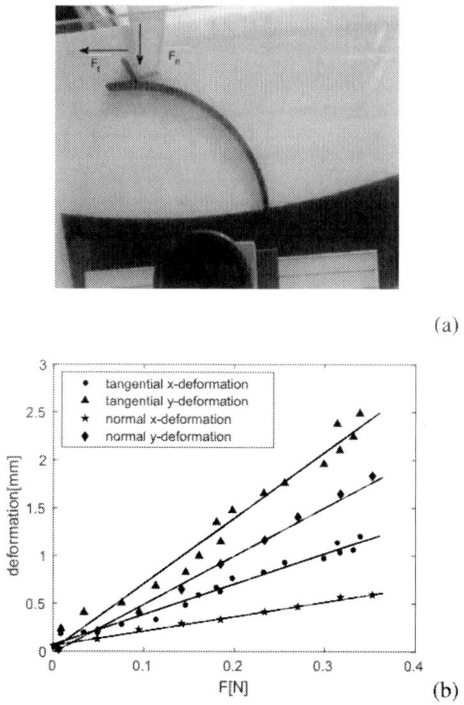

Figure 2. (a) Photograph of a single leaf spring mounted in the Instron for calibration, oriented to measure normal forces. (b) Summary of calibration data for both x- and y-displacements of the tip as a function either pure-normal or pure-tangential applied force.

2.4 Measuring wall stresses

For each measured pair of tip deformations (dx, dy), we can use the calibration from §2.2 to calculate the vector force on the tip:

$$\begin{aligned} C_{t,x}F_t + C_{n,x}F_n &= dx \\ C_{t,y}F_t + C_{n,y}F_n &= dy. \end{aligned} \quad (3)$$

For the example show in Fig. 3, this provides F_n = 0.899 N and tangential force F_t = 0.251 N.

Since the thickness of the spring wall is w = 3 mm, and the segment length of each leaf spring is L = 33 mm, we can convert this to the shear stress $\tau(R_o)$ measured at the outer wall, and pressure P:

$$\begin{aligned} \tau(R_o) &= \frac{F_t}{wL} \\ P &= \frac{F_n}{wL}. \end{aligned} \quad (4)$$

This provides a segment-averaged estimate for the normal and shear stresses along that particular leaf spring.

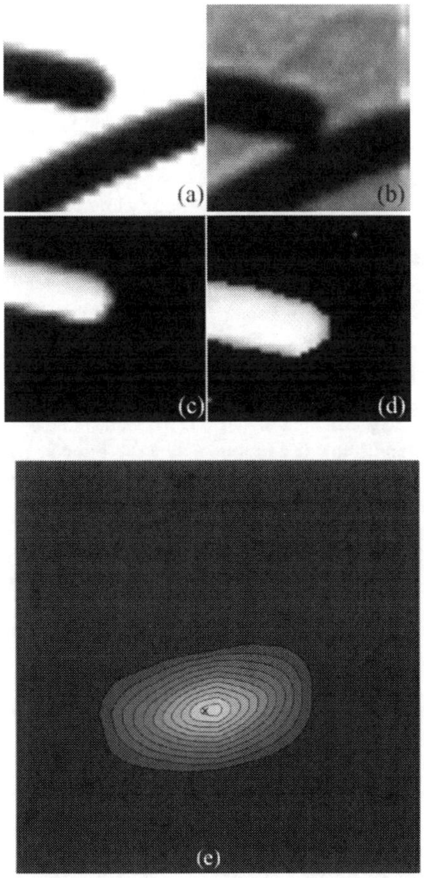

Figure 3. Closeup images of the spring tip with (a) and without (b) particles, and these same tips isolated using image-processing (c,d). (e) The contour plot for of cross-correlation values between images (c,d), with the maximum value corresponding to the (dx, dy) displacement.

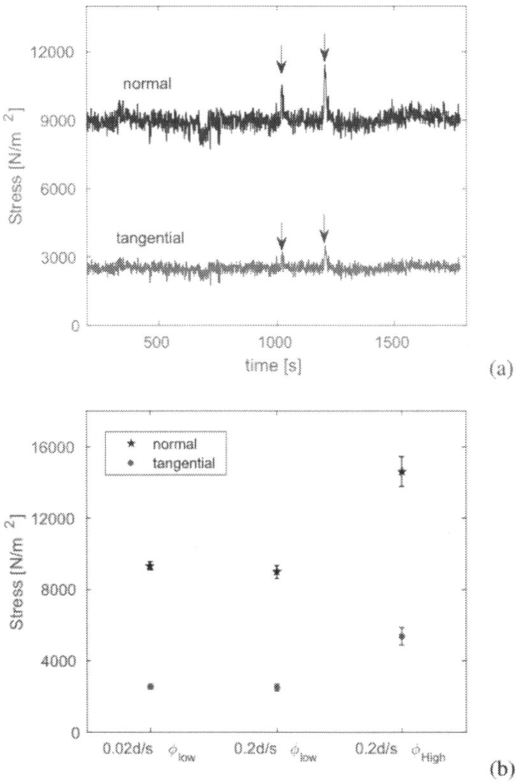

Figure 4. (a) Time series of normal and tangential stresses for a single leaf spring, Case 1 experiment. (b) Time-average normal and tangential stress and standard error for all three cases by one sensor.

3 Results

To illustrate how this method can be used to provide rheological measurements, we perform 3 sample runs for two packing fractions (ϕ) and two rotation rates (specified by the speed v of the inner wall at its rim). For Case 1 and 2, $\phi = 0.816 \pm 0.003$, and for Case 3, $\phi = 0.840 \pm 0.003$. The error is propaged from errors in the particle and apparatus size measurements. Case 1 and 3 are taken at $v = 0.2$ d/s, and Case 2 at $v = 0.02$ d/s, to provide a set of controlled comparisons among the three experiments.

3.1 Stress measurements

Using the methods of §2.4, we measure the pressure and the shear stress for a single spring as a function of time (see Fig. 4a). We observe that both values fluctuate around a well-defined mean value, punctuated by brief spikes in both. These can be seen near 1000 s and 1200 s, as indicated by the arrows. These are likely due to the transient loading of force chains, which have strong spatial variations on length scales similar to that of a single spring. Measurements at other springs have different average values that this sample, further indicating the presence of spatial heterogeneities. Future work combining boundary measurements with photoelastic measurements are planned.

For the three sample cases, we can determine how the time-averaged normal and tangential stresses vary according to ϕ and v. This data is shown in Fig. 4b. We observe that Case 1 and 2 (same ϕ) have similar values for both normal and tangential stress. We observe that Case 1 and 3 (same v), illustrate that both stress values increase with packing fraction, as would be expected.

3.2 Rheological measurements

By combining the wall stress measurements with particle-tracking, we can determine the $\mu(I)$ rheology throughout the granular material. We demonstrate this using values for the two wall stresses measured at four equally-spaced leaf springs around the outer wall. Improved statistics would be obtained for using all 52 available springs (for

 DOI: 10.1051/epjconf/201714003035

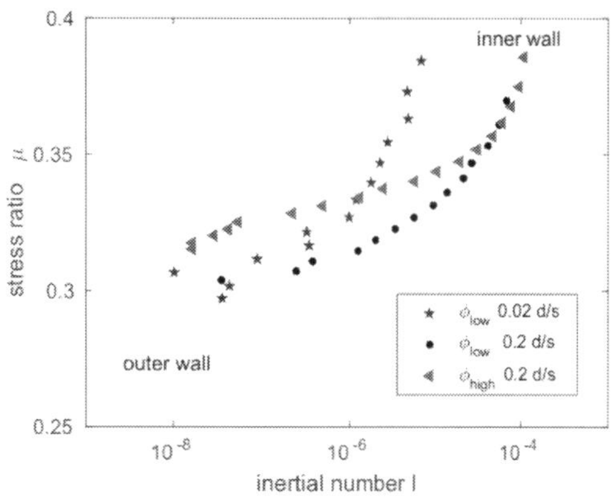

Figure 5. Sample $\mu(I)$ curves for all 3 cases.

which code is under development). In addition, we record the shear stress S measured at the inner wall via a Cooper Instruments torque sensor placed in line with the drive shaft.

If there were no friction with the supporting plate, the tangential component of the stress would from a maximum at the inner wall according to $\tau(r) = S\left(\frac{R_i}{r}\right)^2$, where R_i is the radius of the inner wall. Empirically, we model the effect of this friction with

$$\tau(r) = S\left(\frac{R_i}{r}\right)^2 + \tau_f \qquad (5)$$

where τ_f is a constant chosen so that $\tau(R_o)$ matches the measured value for the tangential stress at the outer wall. From force-balance, we approximate that the average pressure P is independent of radial position r. Thus, the dimensionless stress ratio is $\mu(r) = \tau(r)/P$.

For the three cases, we find the following values for the frictional stress τ_f: Case 1 has 1870 N/m^2; Case 2 has 1630 N/m^2, and Case 3 has 3580 N/m^2. The errors on all are ± 10 N/m^2, calculated from the standard error in Fig. 4a, averaged over the 4 sensors.

To measure the inertial number I (Eq. 1), we track the particles using video taken at 1 Hz over the whole system [11, 12]. We azimuthally-average the velocity profile $v(r)$ and then use Fourier deriviates to calculate $\dot\gamma(r) = \frac{1}{2}\left(\frac{\partial v}{\partial r} - \frac{v}{r}\right)$. We use the value of P determined from the wall stress measurements, and plot $\mu(r) = \tau(r)/P$ parametrically for all values of r.

As shown in Fig. 5, the $\mu(I)$ rheology depends on both the packing fraction and the rotation rate of the inner disk. We observe that for the same inner wall speed (Case 1 and 3), the $\mu(I)$ curves agree for large I (close to the driving wall). When the inertial number is low, all curves (but particularly Case 1 and 2) approach a constant value $\mu \approx 0.3$ which is dominated by the basal friction.

4 Conclusion

We find that laser-cut leaf springs provide a convenient method to both confine a granular material, and measure the boundary wall stresses. In future work, we are expanding the image-processing to provide measurements over the full outer wall, and performing a comparison with photoelastic force measurements. Finally, the $\mu(I)$ measurements provided by Fig. 5 will allow for quantitative investigation of the utility of nonlocal rheology models [2, 3] to describe granular rheology of real materials.

Acknowledgements

We thank Michael Shearer, Dave Henann, and Ken Kamrin for useful discussions about the project, and Austin Reid for help creating the boundary wall designs. We are grateful to the National Science Foundation (NFS DMR-1206808) and International Fine Particle Research Institute (IFPRI) for financial support.

References

[1] Y. Forterre, O. Pouliquen, Annu. Rev. Fluid Mech. **40**, 1 (2008)

[2] K. Kamrin, G. Koval, Physical Review Letters **108**, 178301 (2012)

[3] M. Bouzid, M. Trulsson, P. Claudin, E. Clément, B. Andreotti, Physical Review Letters **111**, 238301 (2013)

[4] G. Koval, J.N. Roux, A. Corfdir, F. Chevoir, Physical Review E **79**, 021306 (2009)

[5] GDR Midi, European Physical Journal E **14**, 341 (2004)

[6] D. Fenistein, M. van Hecke, Nature **425**, 256 (2003)

[7] X. Cheng, J.B. Lechman, A. Fernandez-barbero, G.S. Grest, H.M. Jaeger, G.S. Karczmar, M.E. Mobius, S.R. Nagel, Physical Review Letters **96**, 38001 (2006)

[8] K. Nichol, A. Zanin, R. Bastien, E. Wandersman, M. van Hecke, Physical Review Letters **104**, 078302 (2010)

[9] K. Reddy, Y. Forterre, O. Pouliquen, Physical Review Letters **106**, 108301 (2011)

[10] E. Wandersman, M. Van Hecke, EPL (Europhysics Letters) **105**, 24002 (2014)

[11] *Hough transform*, https://www.mathworks.com/help/images/ref/imfindcircles.html

[12] D. Blair, E. Dufresne, *The matlab particle tracking code repository*, http://site.physics.georgetown.edu/matlab/

Numerical analysis of air effect on the powder flow dynamics in the FT4 Powder Rheometer

Wenguang Nan[1,2], *Vincenzino* Vivacqua[1], *Mojtaba* Ghadiri[1*], and *Yueshe* Wang[2]

[1]Institute of Particle Science and Engineering, University of Leeds, Leeds LS2 9JT, UK
[2]State Key Laboratory of Multiphase Flow in Power Engineering, Xi'an Jiaotong University, Xi'an 710049, China

Abstract. The FT4 powder rheometer of Freeman Technology is widely used nowadays in industry for characterisation of particle flow under dynamic conditions of shear strain rate. It measures the work (termed flow energy) required to penetrate a rotating impeller into a powder bed. However, little is known about its underlying powder mechanics, i.e. the relationship between the flow energy and the prevailing local shear stress. This has recently been studied, but only for very simple and ideal systems amenable to analysis by DEM. We analyse the effect of gas flow through the powder bed on the flow behaviour of cohesionless particles in FT4 by DEM-CFD simulation. The results show that the relative particle velocities induced by the mean shear speed, is of the same order as that produced by the root of granular temperature. The shear stress in both cases with and without gas flow could be quantified by the inertial number. The flow energy correlates well with the shear stress in front of the blade, and both increase with the inertial number and could be significantly reduced by the upward gas flow.

1 Introduction

The study of powder flow as a function of strain rate is of great interest in fast feeding, conveying, mixing and packaging processes. Conventional devices for assessing the flowability of powders, such as shear cell [1, 2], usually operate at low levels of shear strain rate, i.e. in the quasi-static flow regime. However, in many processes the powder is subjected to higher values of strain rate, and the characterisation of the flow behaviour needs to be carried out in dynamic conditions as in a Couette device [3, 4] in order to obtain meaningful measurements.

Recently, the FT4 powder rheometer of Freeman Technology (Tewkesbury, UK) has received attention for its capability to operate in the dynamic conditions, where a twisted blade rotates at a set tip speed whilst translating in or out of the powder bed [5]. The input work required to drive the rotating blade into the powder bed, termed 'flow energy', is measured and advocated to represent the ease with which powder flows. This instrument is nowadays widely used for powder rheometry in industry and academia [6-8]. Several studies have focused on the analysis of the powder mechanics of FT4 by numerical simulations. Bharadwaj et al. [9] used the Discrete Element Method (DEM) to simulate the effect of friction coefficient and particle shape on the particle flow in FT4, and showed that the flow energy was sensitive to the particle shape and friction coefficients. Hare et al. [10] simulated the dynamics of FT4 powder rheometer for a bed of glass beads made cohesive by silanisation and showed that the shear stress increased with the penetration depth. To evaluate the effect of aeration on the flow properties, a controlled gas flow (usually air) is optionally introduced into the particle bed through a porous stainless steel disc at the base of the vessel. As a result, the blade rotating and moving down into the bed will encounter less resistance and the measured flow energy will decrease. However, the effect of gas flow on the particle flow at high strain rates in FT4 rheometer has not so far been analysed.

We report on our numerical analysis of the dynamic behaviour of the particle bed under aeration conditions by DEM-CFD simulation. By varying the permeating air velocity and impeller tip speed, the effect of air drag on the particle flow under different strain rates is analysed. This is of particular relevance to fine and low density particles, as their motion is affected by air drag. Besides the flow energy, the stresses within the powder bed immediately in front of the blade are also analysed. It provides a step towards understanding the dynamics of the FT4 powder rheometer in the presence of air.

2 Simulation methods and conditions

The FT4 powder rheometer used in this work comprises a cylindrical glass vessel with 25 mm diameter and a corresponding 23.5 mm diameter stainless steel blade, as shown in Fig. 1(a) and (b). As the initial packed bed is almost uniform, only the downward test is considered to reduce the computational time, where the blade rotates anti-clockwise and moves into the powder bed while pressing down. In this way, the blade action is more

* Corresponding author: M.Ghadiri@leeds.ac.uk

compacting than slicing, resulting in shearing the bed, whilst applying normal stresses to the particle bed by the blade. The variation of the vertical translational velocity u and rotational velocity ω with the tip speed u_{tip} used in the simulation is shown in Table 1, where a helix angle of -5° is used:

Table 1. Values of the translation and rotational velocity.

u_{tip}, m/s	0.025	0.05	0.10	0.25	0.50	1.00
u, mm/s	2.18	4.36	8.72	21.8	43.6	87.2
ω, rad/s	2.12	4.24	8.48	21.2	42.4	84.8

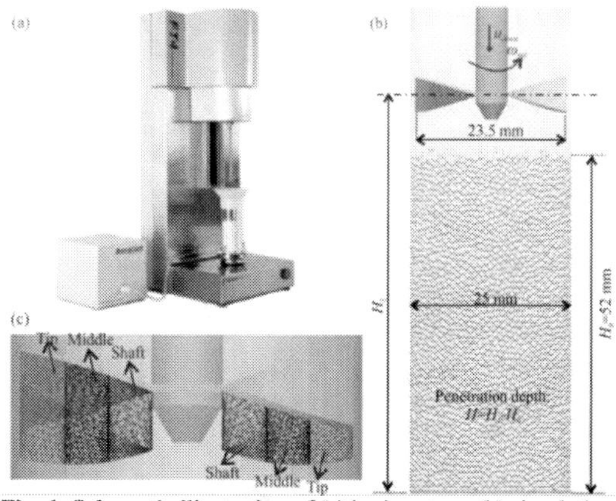

Fig. 1. Schematic illustration of (a) basic set up, (b) simulation system and (c) measured cells.

This process is simulated by DEM-CFD, where particles are modelled as discrete elements and their motions are tracked individually by solving Newton's laws of motion, whilst the fluid phase is continuum. Using the poured packing method, 36,000 spheres with a flat distribution of diameters in the size range of 0.85-1.0 mm are generated to form a packed bed with height of 52 mm and porosity of 0.41. The material properties and interaction parameters are listed in Table 2 and Table 3, respectively. For DEM-CFD simulations, air with density of 1.2 kg/m^3 and viscosity of 1.8×10^{-5} Pa·s is used. The gas flow is introduced from the bottom of the vessel and three superficial gas velocities are simulated, 0.1, 0.2 and 0.3 m/s. The effect of blade motion on the air flow is considered by using the method of dynamic mesh. For the case with no gas flow, only DEM simulation is used, as the particles are large and the interstitial fluid drag is negligible.

Table 2. Material properties in simulations.

Material property	Particles	Impeller blade	Cylinder vessel
Density, ρ (kg/m^3)	1000	7800	2500
Shear modulus, G (Pa)	1×10^8	7.3×10^{10}	2.4×10^{10}
Poisson's ratio, v	0.35	0.3	0.3

Table 3. Contact interaction parameters in simulations.

Interaction property	Particle-particle	Particle-blade/vessel
Friction coefficient, μ	0.5	0.25
Restitution coefficient, e	0.6	0.6

3 Results and discussions

Snapshots of the simulated flow pattern are shown in Fig. 2. For small penetration depth as shown in Fig. 2(a) and (b), the free surface changes with time as the blade moves. As the blade is driven down further, the free surface of the particle bed becomes flat when the penetration depth is larger than 20 mm, as shown Fig. 2(c) and (d). The impeller motion mainly affects the particles locally around the blade, whilst particles remote from the impeller are essentially stationary. These results suggest that the particle flow in front of the blade can be representative of particle behaviour. The measurement cells in front of the blade spans the width of the blade as also adopted by Hare et al. [10] are used, as shown in Fig. 1(c). The cell-averaged granular temperature [11] of particles θ is normalised by the strain rate γ, which is calculated by the particle velocity gradient, given as:

$$\theta=\sqrt{\frac{1}{3N}\sum_{p=1}^{N}\left[(u_p-u_{avg})^2+(v_p-v_{avg})^2+(w_p-w_{avg})^2\right]} \quad (1)$$

$$S=\frac{\gamma d}{\theta} \quad (2)$$

where N is the particle number in the considered measurement cell; u_p, v_p and w_p are the velocity components of particle p in x, y and z directions, respectively; u_{avg}, v_{avg} and w_{avg} are the corresponding mean value of u_p, v_p and w_p, respetively; d is the particle diameter. The strain rate is only determined by the impeller tip speed and is not sensitive to the gas flow rate. The variation of the non-dimensional granular temperature with the tip speed and gas flow rate is shown in Fig. 3. The gas flow rate does not have much influence on the granular temperature of particles. S is almost constant (around 2.0) and shows a little decrease when the strain rate is low. It suggests that the relative particle velocities induced by the mean shear speed $d\gamma$, is of the same order as that produced by the granular temperature θ.

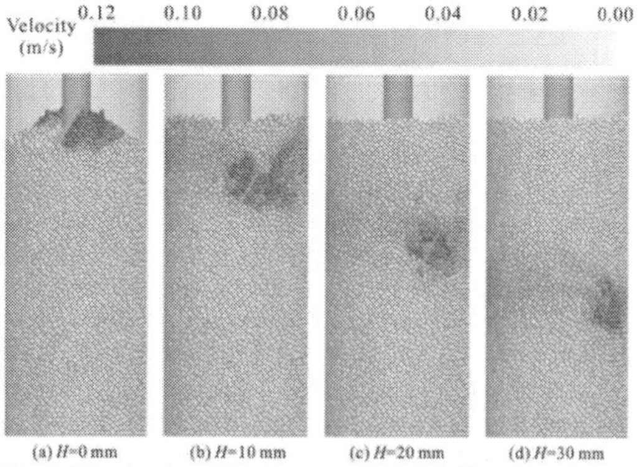

Fig. 2. The simulated flow pattern of particle flow (u_{tip}=0.1 m/s and u_f=0.0 m/s).

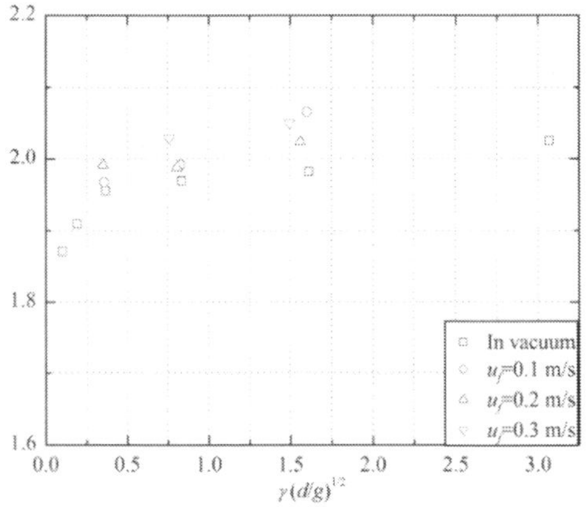

Fig. 3. The variation of non-dimensional granular temperature.

3.1 Stress analysis

The stress of particle flow is also analysed immediately in front of the blade, and the stress tensor is given by:

$$\sigma_{ij} = \frac{1}{V}(\sum_{p\in V} m_p \delta v_i \delta v_j + \sum_{c\in V} f_{ij} \cdot r_{ij}) \quad (3)$$

where V is the cell volume; m_p is the mass of particle p; δv_i and δv_j are the fluctuation velocities of particle p; f_{ij} is the contact force at contact c and r_{ij} is the corresponding branch vector. Based on the stress tensor, three principal stresses are calculated: major one σ_1, intermediate one σ_2 and minor one σ_3. The normal stress p and shear stress τ are then given as:

$$p = \frac{\sigma_1 + \sigma_2 + \sigma_3}{3} \quad (4)$$

$$\tau = \frac{\sqrt{(\sigma_1 - \sigma_2)^2 + (\sigma_1 - \sigma_3)^2 + (\sigma_2 - \sigma_3)^2}}{\sqrt{6}} \quad (5)$$

Similar to the characterisation of the flow regime, the inertial number I [11] is also used, given as:

$$I = \gamma d \sqrt{\rho_p / p} \quad (6)$$

The inertial number is interpreted as the ratio between the inertial timescale $d/(p/\rho_p)^{0.5}$ and the macroscopic deformation timescale $(1/\gamma)$. It follows that the inertial number decreases with the blade penetration depth due to the increased normal stress.

The variation of the shear stress with the inertial number is shown in Fig. 4. The shear stress decreases with the inertial number, in the form of negative power law. The two cases with air flow and without accounting for its effect (i.e. vacuum) both show the same trend and collapses to one curve. It suggests that the evolution of the shear stress could be described by the inertial number. To further depict this mechanism, the "viscosity" τ/γ (the resistance to deformation rate by the shear stress) under different tip speeds is shown in Fig. 5, where the cases with air flow are not included for simplicity as they have the same trend with the cases without air flow. It is interesting to note that $\ln(\tau/\gamma)$ shows a linear decrease with $\ln(I)$ in all cases. The "viscosity" could be described by the following equation:

$$\tau / \gamma = a + bI^{-n} \quad (7)$$

where a, b and c are parameters which vary with the tip speed. Their value calculated by the least square fitting is shown in Table 4. n is almost 2.0, indicating that $\tau \sim I^{2}$. With the increase of the tip speed, b increases whilst a decreases, suggesting that the dependence of τ on I is stronger for larger tip speed.

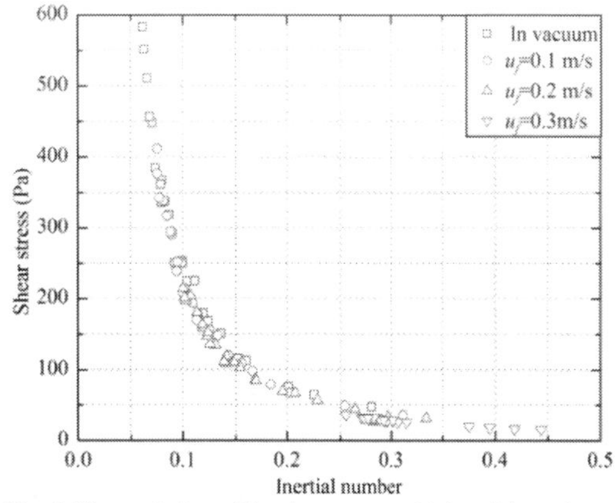

Fig. 4. The evolution of the shear stress with inertial number.

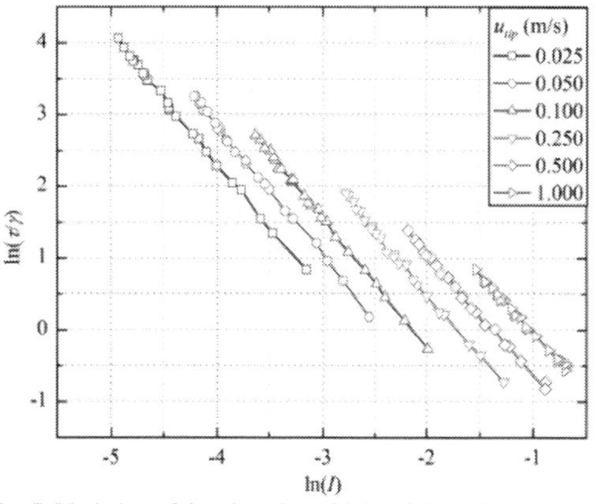

Fig. 5. Variation of the viscosity with inertial number.

Table 4. The values of the fitting parameters.

u_{tip}	0.025	0.05	0.1	0.25	0.5	1.0
a	1.269	0.664	0.194	0.171	0.158	0.079
b	0.0026	0.0054	0.0135	0.0301	0.0575	0.1558
n	2.024	2.008	1.933	1.933	1.929	1.727

3.2 Flow energy

The total input work E, which is also referred to as 'flow energy' of the powder is given as:

$$E = \int_0^H \left(\frac{T}{R\tan(\alpha)} + F\right) dH \quad (8)$$

where T and F are the torque and axial force required for the blade motion, respectively; R is the blade radius; α is the helix angle; $H = H_1 - H_0$ is the penetration depth as shown in Fig. 1(b), where H_1 is the average vertical position of the blade and H_0 is the initial bed height. The flow energy is normalised by the potential energy of particles ($m_b gH$, where m_b is the total particle mas above the blade) and then averaged for the last 20 mm depth.

The variation of non-dimensional flow energy with the inertial number is shown in Fig. 6. At very low inertial numbers, the flow energy is almost constant as the particle flow is in the quasi-static regime. At large inertial numbers, the flow energy increases with the inertial number in the form of power law. The shear stress (normalised by dividing it by $m_b g/A$, where A is the cross-section area of the vessel) shows a similar trend and is not shown here for simplicity. The results clearly show that there exists a direct relationship between the flow energy and shear stress, as shown in Fig. 7. The flow energy increases with the shear stress in a linear trend for both cases of vacuum and gas flow. In summary, the flow energy as determined by FT4 is well correlated with the shear stress. The latter is of course a function of the strain rate.

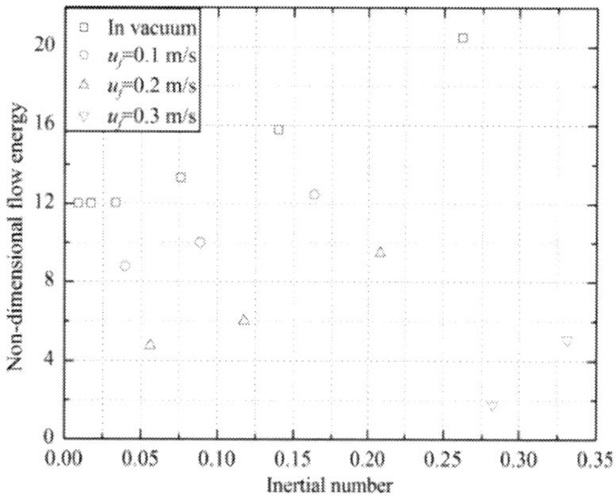

Fig. 6. Variation of the flow energy with ineritial number.

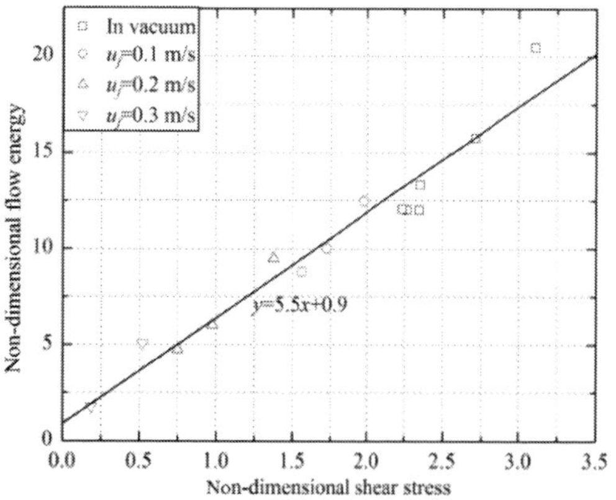

Fig. 7. The variation of flow energy with shear stress [12].

4 Conclusions

The rheological behaviour of spherical cohesionless particles subjected to shearing flows by an impeller in the presence of an upward gas flow has been analysed under high shear strain rates using the FT4 instrument by DEM-CFD simulations. The bulk particles are subjected to the standard downward test procedure of FT4 rheometer, where a rotating blade is driven into the particle bed. The main results from the present study are summarised as follows:

1) The relative particle velocities induced by the mean shear speed $d\dot{\gamma}$, is of the same order as that produced by the granular temperature θ.

2) The evolution of shear stress could be quantified by the inertial number, and an equation is proposed to quantify the effect of tip speed on the pseudo-viscosity of particle flow.

3) The flow energy and shear stress both increase with the inertial number and decrease significantly with the gas flow. The flow energy correlates well with the shear stress in front of the blade for all conditions including permeating air, implying that the flow energy could also be used to map out the flow regimes.

Acknowledgments

The authors are grateful to the scholarship from China Scholarship Council (CSC) for the first author. We are thankful to DEM Solutions, Edinburgh, UK, for providing a special license for the EDEM software for use in this work, and also to Drs Colin Hare and Mehrdad Pasha for their generous help in setting up the computational facility. We also thank Freeman Technology for making available an FT4 instrument for the experiment, as well as technical support and comments from Dr Doug Millington-Smith and Mr Tim Freeman.

References

1. J. Schwedes, Granular Matter **5**, 1-43 (2003).
2. A.W. Jenike, Bulletin No. 108, Utah State University, (1961).
3. G.I. Tardos, S. McNamara, I. Talu, Powder Technology **131**, 23-39 (2003).
4. M.K. Langroudi, S. Turek, A. Ouazzi, G.I. Tardos, Powder Technology **197**, 91-101 (2010).
5. R. Freeman, Powder Technology **174**, 25-33 (2007).
6. H. Kinnunen, G. Hebbink, H. Peters, J. Shur, R. Price, AAPS PharmSciTech **15**, 898-909 (2014).
7. A. Vasilenko, B.J. Glasser, F.J. Muzzio, Powder Technology **208**, 628-636 (2011).
8. J.G. Osorio, F.J. Muzzio, Drug Dev Ind Pharm **39**, 1464-1475 (2013).
9. R. Bharadwaj, W.R. Ketterhagen, B.C. Hancock, Chemical Engineering Science **65**, 5747-5756 (2010).
10. C. Hare, U. Zafar, M. Ghadiri, T. Freeman, J. Clayton, M.J. Murtagh, Powder Technology **285**, 123-127 (2015).
11. G.D.R. MiDi, Eur Phys J E Soft Matter **14**, 341-365 (2004).
12. W.G. Nan, M. Ghadiri, Y.S. Wang, Cheimcal Engineering Science **162**, 141-151(2017).

Jamming Transition: Heptagons, Pentagons, and Discs

Yuanyuan Xu[1,*], *Jonathan* Barés[2,**] *Yiqiu* Zhao[2,***] and *Robert P.* Behringer[2,****]

[1] *Taishan College, Shandong University, Jinan, Shandong 250100, China*
[2] *Department of Physics and Center for Nonlinear and Complex Systems, Duke University, Durham, North Carolina 27708, USA*

Abstract.
The jamming behavior of a system composed of discs has been well documented. However, it remains unclear how a granular system consisting of non-spherical particles transitions between unjammed and jammed states. Here, we present compression experiments to study the jamming transition of 2D granular materials composed of photoelastic heptagonal particles and compare these results to data for discs and pentagons. We determine the critical packing fraction of heptagons and make a comparison to discs and pentagons. In the experiment, we subject 618 heptagonal particles to cyclic compression. We track the motion (inlcuding rotations) of the particles, and we measure forces on particles by photoelasticity. We observe a power law relationship between the average contact number (Z) and the pressure (P). Furthermore, we classify the type of contacts by the relative orientation of pairs of contacting particles (creating point-to-face and face-to-face contacts), and we explore the evolution of the contacts during jamming.

1 Introduction

Jamming occurs when materials like foams, glasses and granular materials become mechanically rigid and stable. The jamming transition has become an important research topic in physics and chemistry, and it has a wide range of applications in industrial productions and environmental events. In recent years, many interesting features have been found in systems composed of discs or ellipses [1] [2] [3] [4] [5][6]. Simulations for polygons and other complex shapes have also been carried out [7] [8] [9] [10][11]. We provide an experimental study of more complex particles, pentagons and heptagons, to bridge the gap between simulation and physical experiment. We focus particularly on heptagons, but we make comparision to discs and pentatons.

In this work we focus on a system of heptagons that we compress and decompress via many small steps. The system consists of 618 heptagonal particles in two dimensions. We use photoelastic techniques [1] to measure the system pressure and to study the formation of contacts and force chains. Image processing methods and mathematical criteria are applied to both count the number of contacts and classify the type of the contact. We find the relation between the number of point-to-face contacts and face-to-face contacts, which is crucial to understanding force transmission near jamming.

*e-mail: xuyy95@outlook.com
**e-mail: jb427@phy.duke.edu
***e-mail: yiqiu.zhao@duke.edu
****e-mail: bob@phy.duke.edu

2 Experimental setup

The experiments consists of biaxial compression of quasi-2D systems of photoelastic particles. We have three samples: 618 heptagons with side length 6.93 mm and thickness 6.35 mm; 784 pentagons with side length 9.40 mm and thickness 6.35 mm; 895 discs with radius 6.35 mm and thickness 5.56 mm. The inter-particle friction coefficient is 0.7 for discs and 0.9 for pentagons and heptagons. We draw a dot and a bar on the top of each particle in UV-fluorescent ink and these markers give us the position and the orientation of the particle when the system is illuminated with UV light. Figure 1 shows a schematic of the experiment. All particles are randomly placed on a horizontal Plexiglas sheet (38.60 cm × 38.32 cm) lubricated by a thin layer of talcum powder and prepared in a stress-free state initially. Two movable walls (blue arrows) driven by linear motors compress the system quasi-statically via small steps (0.4 mm per step) until the system is in a strongly jammed state. Then, the system is decompressed step by step to the initial packing fraction. After each step, the camera above the container records three types of images (see Fig. 2). From UV light images, we extract the center and the orientation of each particle. We then use the white light images to fine tune the particle positions. The photo-elastic image gives the global pressure using a gradient squared measure[2]. Contacts between particles are detected using the *Shadow Overlap* method devised by Moreau [12] and then the true force-transmitting contacts can be screened out by using the photoelastic image and examining the region around each possible contact [2].

 DOI: 10.1051/epjconf/201714006010

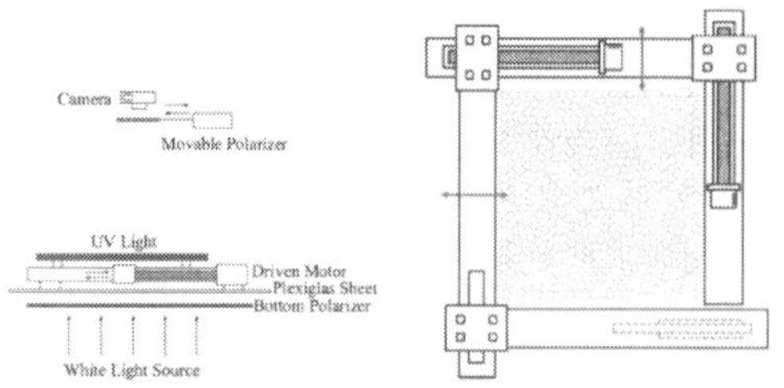

(a) (b)

Figure 1: Schematics of the biaxial compression experiments. (a) Side view of the setup. (b) Top view of the setup.

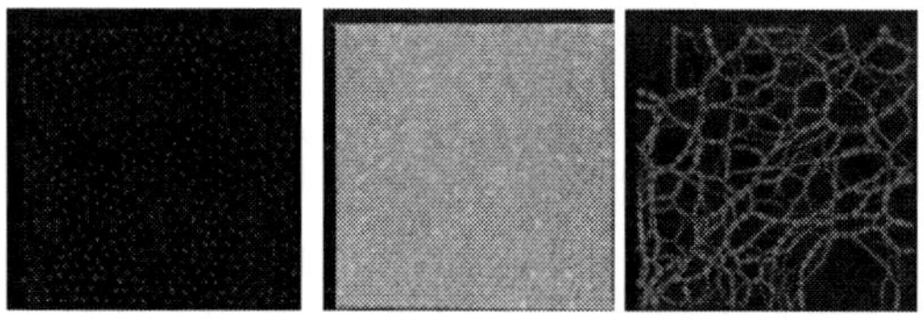

(a) (b) (c)

Figure 2: (a) UV light image: white light is off and UV light is on. (b) White light image: only white light is on. (c) Photo-elastic image: white light is on and the crossed polarizer is moved under the camera.

3 Results

3.1 Critical packing fraction

Figure 3 shows the pressure versus packing fraction for heptagonal particles during one cycle of compression/decompression. There exists a well defined point in the decompression process where there is a sharp transition and we define the packing fraction ϕ_c at this point as the jamming or critical packing fraction. The global pressure, determined by G^2, in decompression is higher than that in the compression for larger ϕ's ($\phi > \phi_c$). We made 20 runs with the same initial packing fraction (see Fig. 4a) and the average value of ϕ_c for heptagonal particles is 0.767. We also carried out similar experiments for discs and pentagons. Figure 4b shows a comparison of the average ϕ_c for heptagons, pentagons, and discs. We find that $\bar{\phi}_c^{Hept} < \bar{\phi}_c^{Pent} < \bar{\phi}_c^{disc}$. As a comparison, Figure 5 presents G^2 vs. packing fraction in pentagon decompression experiments. The inserted panel of Figure 5 shows the scatter of the critical packing fractions. Except perhaps for the first few cycles, there is no systematic variation of ϕ_c with cycle number.

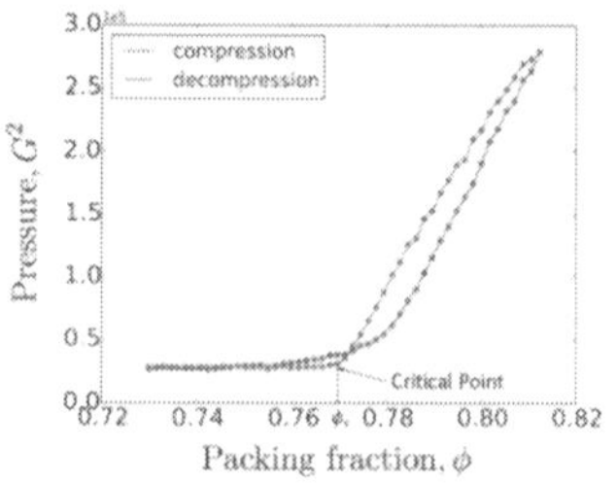

Figure 3: Evolution of the global pressure, measured by G^2, during one cycle. Blue dots are compression process and red dots are decompression process.

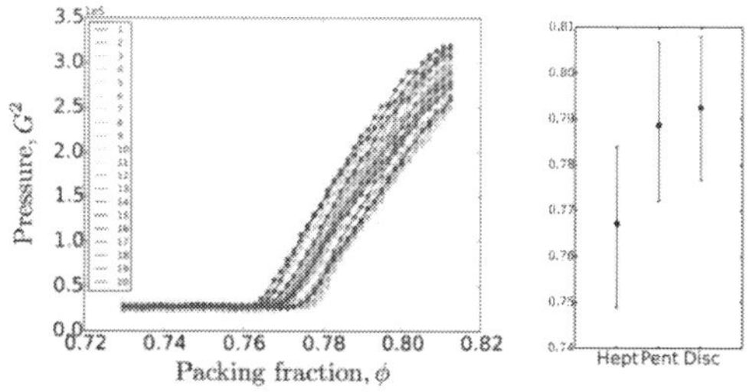

(a) (b)

Figure 4: (a) Results from 20 runs for heptagons starting from the same boundary configuration. Different colors denote different runs. (b) A comparison of the average ϕ_c for the three particle types: 1) 618 heptagons with side length 6.93 mm, 2) 784 pentagons with side length 9.40 mm, and 3) 895 discs with radius 6.35 mm. Note that the blue dot represents the average value and the green bar represents the range of values.

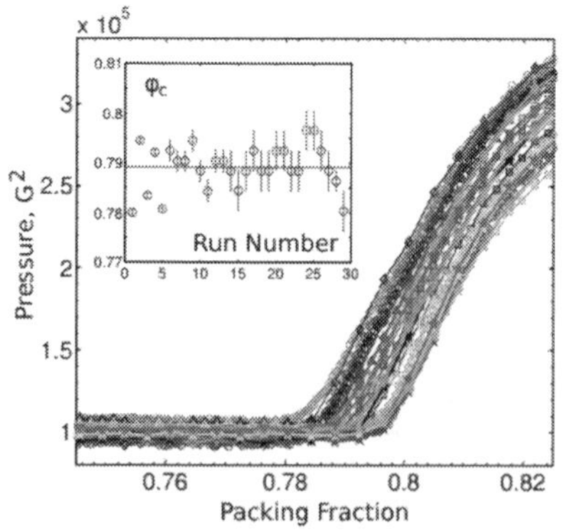

Figure 5: Main panel: results from runs of pentagon decompression starting from same packing fraction. Inserted figure: the value of critical packing fraction detected from each run.

3.2 Contact number

The average contact number Z is computed by counting only the force bearing heptagons, i.e., each particle has at least 2 contacts [3]. Figure 6a shows the evolution of the average contact number during one cycle of compression. We find that the ϕ dependence of Z is qualitatively similar to the ϕ dependence of G^2 (Fig. 3). From a log-log plot of Z vs. G^2 (see Fig. 6b), we observe a power law relationship between the average contact number and the global pressure, i.e., $Z - Z_c \propto (G^2 - G_c^2)^\alpha$. We made 10 runs starting from the same packing fraction and found $\alpha = 0.253$ (standard deviation $\sigma = 0.017$).

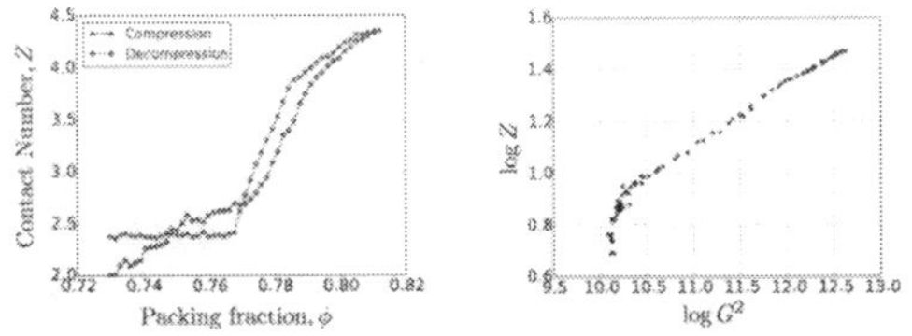

(a) (b)

Figure 6: (a) Average contact number Z vs. packing fraction ϕ. (b) Log-log plot of Z vs. G^2. The data follow a straight line for larger enough values.

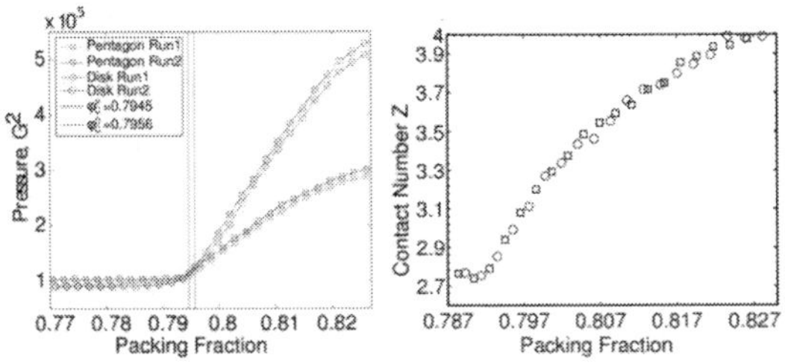

(a) (b)

Figure 7: (a) Runs from discs and pentagons possessing similar critical packing fraction. (b) The average contact number for pentagons (squares) and discs (circles) show qualitatively similar variation with ϕ as for same shape with heptagons. The color indicates same runs as in (a)

For comparison, we also studied the variation of the average contact number during the decompression process for discs and pentagons. To control variables, we choose two runs with similar critical density, one from discs and the other from pentagons. As shown from Figure 7, we see qualitatively similar behaviors for all three kinds of particles: discs, pentagons and heptagons.

3.3 Classification of contact

Previous simulations indicate that strong forces are mainly carried by face-to-face (flat) contacts while weak forces are mostly carried by point-to-face (point) contacts [7]. In order to understand the structure of the contact network, we studied the evolution of the flat and point contacts (see Fig. 8a). Figure 8b shows the number of the point contacts versus the number of the flat contacts and the data points lie close to a straight line.

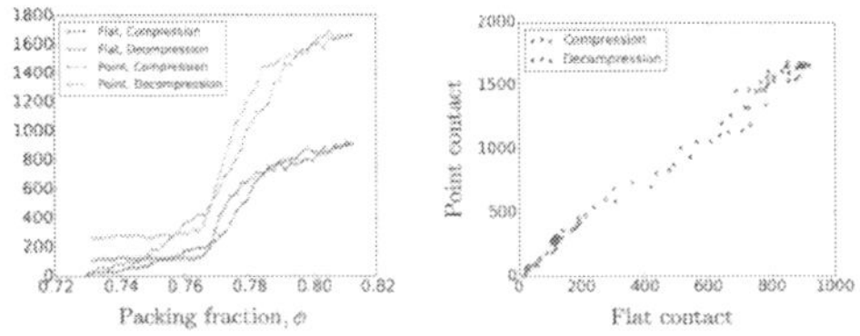

(a) (b)

Figure 8: Data for contact types. (a) Blue curve shows the number of the face-to-face contacts and green curve shows the number of the point-to-face contacts. (b) Number of the point-to-face contacts vs. number of the face-to-face contacts.

We made 10 runs starting from the same packing fraction and the data is fitted by a straight line. From Figure 9, we found that the ratio of the point-to-face contacts to face-to-face contacts is about 1.834 in the compression process and 1.857 in the decompression process.

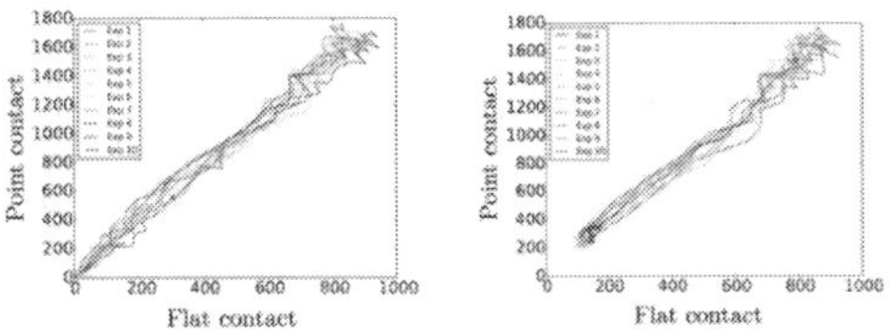

(a) (b)

Figure 9: (a) Compression process: mean slope $k = 1.834$, standard deviation $\delta = 0.090$. (b) Decompression process: mean slope $k = 1.857$, standard deviation $\delta = 0.081$.

The inverse problem of non-spherical photoealstic particles has not been developed. To estimate the role of flat and point contacts in the force network, we calculate the ratio of flat and point contacts for the particles that bear a pressure larger than the system average. The result for a typical heptagon run is presented in Figure 10. All runs of the experiments possess qualitatively similar properties. In this run, the system was compressed during 1 to 50 experimental steps, and decompressed during 51 to 100 experimental steps. The ratio calculated from particles possessing pressure larger than average (G^2 > *average*) remains close to the ratio calculated from the whole system during all the experimental steps, indicating no special role for large force vs small force contacts during heptagon jamming transition.

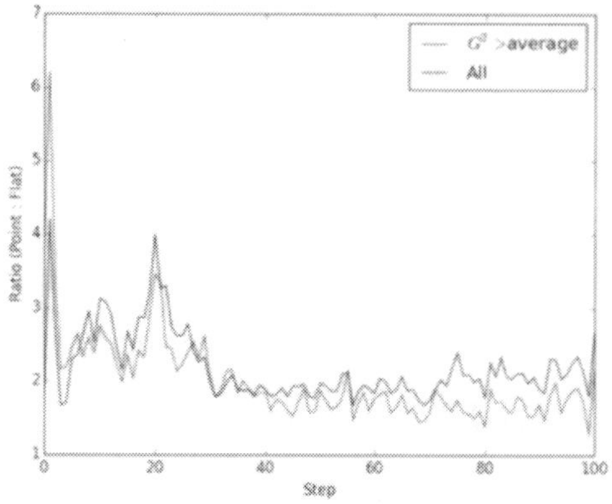

Figure 10: The point (point-to-face) to flat (face-to-face) contact ratio for heptagon particles in a particular run. Step 1 to 50 corresponds to compression process and step 51 to 100 corresponds to decompression process.

4 Conclusions

We present data for the critical packing fraction in the decompression process for pentagons, heptagons, and discs. We observe a power law relationship between the average contact number Z and the global pressure G^2 which follows $Z - Z_c \propto (G^2 - G_c^2)^{0.253}$. In addition, the ratio of the point-to-face contacts to face-to-face contacts is around 1.85. In particular, we find the ratio calculated from particles with pressure larger than average remains the same as the ratio calculated from all particles. However, the inverse problem of non-spherical photoelastic particles should be developed to give a more precise conclusion.

Acknowledgements

This work was supported by NSF grants DMR1206351, and DMS1248071, NASA grant NNX15AD38G, and a grant from the William M. Keck Foundation.

References

[1] T.S. Majmudar, R.P. Behringer, Nature **435**, 1079 (2005)

[2] T. Majmudar, D. University, *Experimental Studies of Two-dimensional Granular Systems Using Grain-scale Contact Force Measurements* (Duke University, 2006), ISBN 9780549109396

[3] T.S. Majmudar, M. Sperl, S. Luding, R.P. Behringer, Phys. Rev. Lett. **98**, 058001 (2007)

[4] S. Farhadi, R.P. Behringer, Physical review letters **112**, 148301 (2014)

[5] S. Farhadi, R.P. Behringer, Phys. Rev. Lett. **112**, 148301 (2014)

[6] C.S. O'Hern, L.E. Silbert, A.J. Liu, S.R. Nagel, Phys. Rev. E **68**, 011306 (2003)

[7] E. Azéma, F. Radjaï, R. Peyroux, G. Saussine, Phys. Rev. E **76**, 011301 (2007)

[8] T. Schilling, S. Pronk, B. Mulder, D. Frenkel, Physical Review E **71**, 036138 (2005)

[9] E. Azéma, F. Radjaï, Physical Review E **81**, 051304 (2010)

[10] Y.L. Duparcmeur, A. Gervois, J.P. Troadec, Journal of Physics: Condensed Matter **7**, 3421 (1995)

[11] F. Alonso-Marroquin, H. Herrmann, S. Luding, Proceedings of IMECE, New Orleans, Loisiana, Nov. 2002, CD: ASME2002-32498 (2002)

[12] G. Saussine, C. Cholet, P. Gautier, F. Dubois, C. Bohatier, J. Moreau, Computer Methods in Applied Mechanics and Engineering **195**, 2841 (2006)

 DOI: 10.1051/epjconf/201714003037

Influence of polydispersity on the microstructure of frictionless disc packings under simple shear

Nicolas Estrada[1,*] and *William-Fernando* Oquendo[2,**]

[1] *Departamento de Ingeniería Civil y Ambiental, Universidad de los Andes, Bogotá, Colombia*
[2] *Departamento de Ingeniería Industrial, Universidad de la Sabana, Chía, Colombia*

Abstract. By means of 2D Contact Dynamics simulations, we explored the effects of the shape and size span of the grain size distribution (GSD) on the microstructure of a sheared granular packing, in terms of its packing fraction and the connectivity of its contact network. We focused on GSDs that can be represented by a power law, which are widely used in several engineering and industrial contexts. The interest in power law GSDs originates in the works of Fuller and Thompson in 1907 and 1919, in which it was found that a maximum density is obtained for a power law GSD with an exponent of 0.5. This disagrees with recent DEM results where the densest packing is obtained for linear cumulative volume distributions. In order to explore this discrepancy, we performed systematic simulations in which we varied both the size span (the ratio between the largest and the smallest diameter) and the exponent (shape) of the distribution. We find that the exponent equal to 0.5 produces the highest density for all size spans. Furthermore, the proportion of rattlers and coordination were also analyzed, showing that the system's connectivity is strongly affected by both the size span and the shape of the distribution.

1 Introduction

It is well known that the grain size distribution (or polydispersity) plays an important role on the dynamics of granular materials. For instance, polydispersity strongly affects the material's packing fraction, which in turn affects its mechanical response. Therefore, it is of paramount importance to choose the appropriate grain size distribution (GSD) when the material's design targets some specific property, as it is done for the granular phase of composite materials such as Portland and asphalt concrete, or in the granular bases in roadways pavements.

The relationship between polydispersity and packing fraction has been studied for more than a century. The first experimental works are probably those of Fuller and Thompson in 1907 [1] and Taylor in 1919 [2]. Fuller and Thompson found that the optimal GSD (i.e., that with the highest density) is the one for which the cumulative volume distribution is described by

$$\rho = (d/d_{\max})^{0.5}, \tag{1}$$

where ρ is the mass percentage of particles with diameter smaller than d, and $d_{\max}$ is the maximum diameter of the sample. In the years that followed, several researchers explored different materials, used in different industrial contexts, finding smaller values for the optimal exponent. In particular, the optimal exponent decreased as the amount of fines in the samples increased [3–6]. A useful extension, which includes a minimum diameter $d_{\min}$ was proposed in [7] and reads

$$\rho = \left(\frac{d - d_{\min}}{d_{\max} - d_{\min}}\right)^{\eta}, \tag{2}$$

in which $d_{\min}$ and $d_{\max}$ control the size span of the GSD, while η controls its shape. For example, $\eta = 0$ corresponds to a mono-sized material.

Numerous theoretical studies have also been conducted in order to develop analytical expressions that allow for predicting the packing fraction, given a certain GSD. However, these expressions only give good predictions for discrete GSDs (binary or ternary mixtures) [8–17].

Another possible approach is to build and analyze arrangements of particles according to a geometrical rule, as those presented in [13, 18–25]. These models represent a tool of great value since they allow for exploring a wide range of systems, configurations and size distributions, which is something that cannot be easily done experimentally or theoretically. In particular, in [23] a large set of GSDs were built and compared by using the cumulative β-distribution, finding that the GSD that produces the densest packing is the uniform distribution by volume fractions (i.e., $\eta = 1$ in eq. (2)). This disagrees with the optimal exponent of $\eta = 0.5$ proposed in [1, 2], although these experiments were based on mixtures of fine and coarse grains, and the model in [23] is two dimensional.

During the past decades, several works have used the Discrete Element Method (DEM) to study the effects of

*e-mail: n.estrada22@uniandes.edu.co
**e-mail: william.oquendo@unisabana.edu.co

DOI: 10.1051/epjconf/201714003037

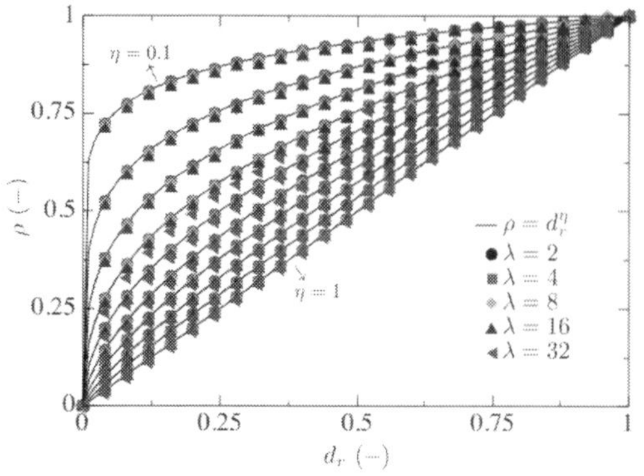

Figure 1. (Color online) Grain size distributions for all samples. The black line corresponds to eq. (2)

the GSD on different properties of granular packings [26–31]. In this paper, we use DEM to explore the influence of both the size span and the shape of the GSD on the packing fraction and connectivity, for two-dimensional samples of frictionless discs under simple shear.

2 Numerical procedure

The numerical samples were composed of approximately 10000 discs, with a GSD that followed a power law such as that presented in Eq. 2. The size span of the GSD was described by the size ratio $\lambda = d_{max}/d_{min}$, and the shape of the GSD was controlled by the exponent η. Systems were built for the following set of parameters: $\lambda \in (2, 4, 8, 16, 32)$ and $\eta \in (0.1, 0.2, ..., 1)$. For this purpose, the range of diameters (d_{min}, d_{max}) was divided into ten subranges, and, inside these subranges, a uniform distribution by number of grains was used.

Figure 1 shows the GSDs (i.e., the cumulate area distributions) of all samples; ρ is the area percentage of particles with a reduced diameter smaller than $d_r = (d - d_{min})/(d_{max} - d_{min})$. It can be seen that the built GSDs closely follow the theoretical curves. For $\lambda = 32$, only samples with $\eta \geqslant 0.4$ were built, since it was seen that smaller values of η would require a larger number of grains in order to obtain a good fit (i.e., a maximal vertical offset of 0.03) with the theoretical curve.

All samples were periodic on the horizontal direction. These samples were first compressed under a constant stress σ_c, and then the systems were sheared horizontally with shear speed v_s under the same confining stress. The inertial number I was defined as

$$I = \frac{\dot{\gamma}\langle d \rangle}{\sqrt{\sigma_c/\varrho}}, \tag{3}$$

where $\dot{\gamma}$ is the shear rate and ϱ is the grains' density. In all shear tests $I \simeq 10^{-4}$, which means that these tests can reasonably be considered as quasi-static [32].

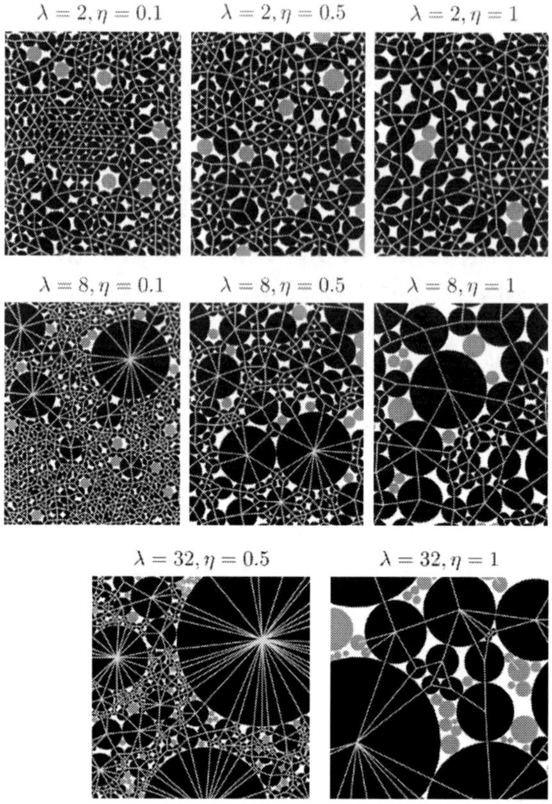

Figure 2. (Color online) Contact network for $\eta = 0.1, 0.5, 1.0$ and $\lambda = 2, 8, 32$. Yellow (white) lines join the center of particles in contact. The length scale is the same on all snapshots.

3 Results

3.1 Contact network

Figure 2 shows the contact network for eight systems with different values of the size span λ and exponent η of the GSD. For $\eta = 0.1$ a big part of the system is almost monodisperse and spontaneous crystallization occurs to some degree. Large size spans are characterized by a disordered structure but also by a higher proportion of rattlers (i.e., grains that do not participate in the force-carrying backbone). For $\eta = 0.5$ it seems that the amount of smaller particles is appropriate to fill better the voids between the larger particles. Simultaneously, the contact network's connectivity is also enhanced.

3.2 Packing density

The packing density was quantified by means of the packing fraction, defined as

$$\nu = \frac{A_g}{A}, \tag{4}$$

where A_g is the area covered by the grains, and A is the total area of the sample. Figure 3 shows ν as a function of

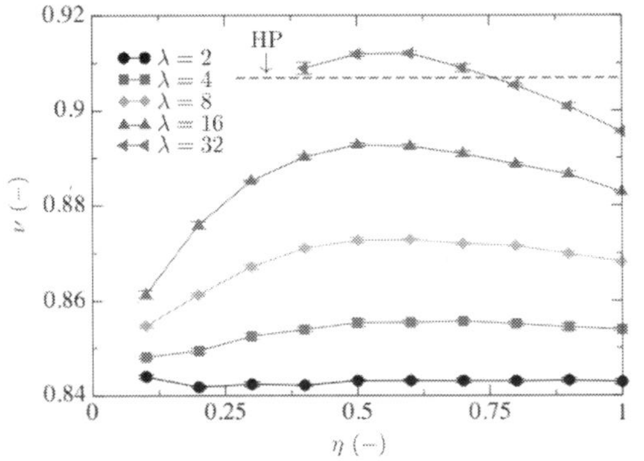

Figure 3. (Color online) Packing fraction ν as a function of η for all values of λ. Error bars indicate the standard deviation from several measures at the steady state. The dashed line indicates the packing fraction of a set of mono-disperse discs packed on a hexagonal arrangement - HP ($\nu_{HP} = \pi/2\sqrt{3} \simeq 0.9069$)

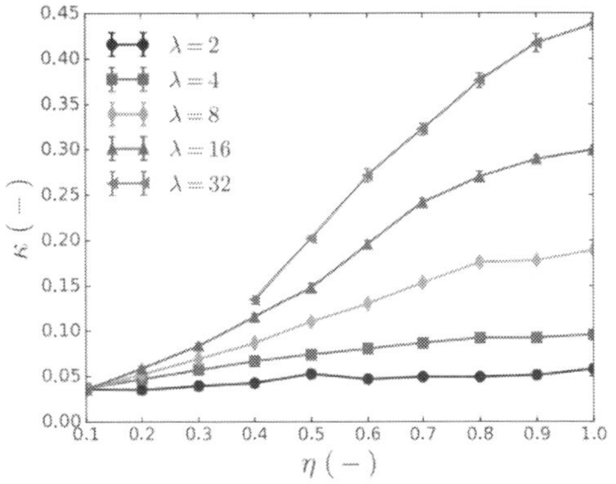

Figure 4. Proportion κ of rattlers as a function of η for all values of λ. Error bars indicate the standard deviation from several measures at the steady state.

η for all values of λ. First, it can be seen that ν increases with λ, as shown also in [28, 31], as small particles fill the pores between the larger particles. Second, it can be seen that ν varies non-monotonically with η, showing a maximum around $\eta \simeq 0.5$. Therefore, the maximum density is obtained when the GSD follows a power law with an exponent close to 0.5, as suggested by Fuller and Thompson [1, 2], and not for the uniform distribution by volume fractions, as suggested by Voivret [23]. This discrepancy could be due to the fact that the set of GSDs explored in [23] did not include a power law of exponent 0.5. In addition, the samples analyzed in [23] were built geometrically, while the systems analyzed in this paper were built mechanically. Third, it is also shown that ν can exceed the packing fraction of system of mono-disperse discs in hexagonal packing (i.e., $\nu = \pi/2\sqrt{3} \simeq 0.9069$).

3.3 Connectivity

Figure 4 shows the proportion κ of rattlers as a function of η for all values of λ. It can be seen that κ increases with both λ and η. For $\lambda = 32$ and $\eta = 1$, almost 45% the particles are rattlers, which means that the force-carrying backbone is only composed by approximately half of the particles. Figure 5 shows the average coordination number $\langle n_c \rangle$ as a function of η for all values of λ. It can be seen that $\langle n_c \rangle$ decreases with both λ and η, as a result of the increasing number of rattlers. It is also shown $\langle n_c \rangle$ for the force-carrying backbone, which is very close to 4, as is expected for a two-dimensional system of frictionless discs. It is interesting to note that systems with $\eta = 0.5$ are notably better connected than those with $\eta = 1$, which should have an important effect in macroscopic mechanical properties such as Coulomb cohesion and Stiffness.

4 Conclusions and perspectives

In this paper we explored the effects of the shape and size span of the grain size distribution (GSD) on the mi-

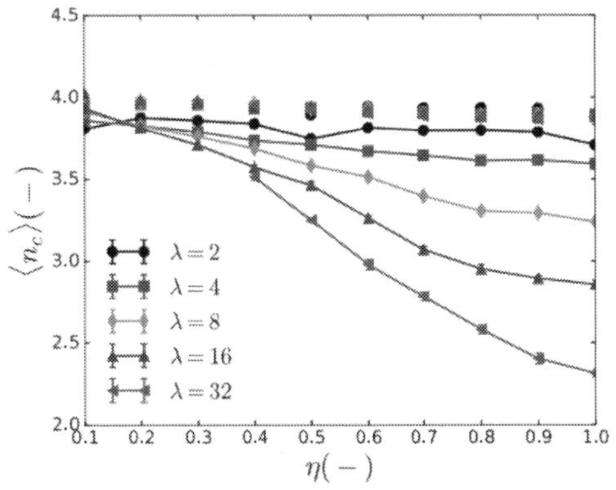

Figure 5. Coordination number $\langle n_c \rangle$ including all particles (lines and points) and for the force-carrying backbone only (only points). Error bars indicate the standard deviation from several measures at the steady state.

crostructure of a sheared granular packing. This microstructure was described in terms of the packing fraction and the connectivity of the contact network.

We showed that the GSD that produces the densest packing corresponds to a power law with an exponent close to $\eta = 0.5$ (see equation (2)), in agreement with the results of Fuller and Thompson [1, 2] for 3D experiments. Our results enhance and extend the validity of the Fuller and Thompson optimal GSD [1, 2] to 2D systems of frictionless discs. Qualitatively it was suggested that the high packing fraction obtained with this distribution is due to a higher proportion of small particles that allows for efficiently filling the voids between the larger particles, an effect which increases with increasing size span.

Or results disagree with those presented in [23], where the densest packing was found to be the the uniform distribution by volume fractions (i.e., a power law GSD with $\eta = 1$). It is important to note that the set of GSDs explored in [23] did not include a power law of exponent 0.5. Moreover, the samples analyzed in [23] were built geometrically, while the systems analyzed in this paper were built mechanically.

Finally, we showed that both the size span and shape of the GSD increase the proportion of rattlers and decrease the coordination of the force-carrying backbone. It is also shown that the distributions with $\eta = 0.5$ are not only denser than those with $\eta = 1$, but also better connected. This is important because it could have a major effect on macroscopic mechanical properties such as Coulomb cohesion and stiffness.

This work can be extended by exploring the role of the size span and shape of the GSD in a three dimensional system. Furthermore, we are currently exploring the links between the GSD and other macroscopic properties beyond shear strength. This should be useful to better understand the effects of polydispersity and use this knowledge for industrial applications in which the GSD can, or even needs to be, controlled.

Acknowledgements

The authors would like to thank J. P. Carmona and J. M. Torres for their help with numerical simulations.

References

[1] W.B. Fuller, S.E. Thompson, Transactions of the American Society of Civil Engineers **LIX**, 1 (1907)

[2] F.W. Taylor, S.E. Thompson, *A treatise on concrete, plain and reinforced* (John Wiley & Sons, 1919)

[3] A.N. Talbot, F.E. Richart, Illinois Univ Eng Exp Sta Bulletin (1923)

[4] J.M.S. Shilstone, Concrete International: Design and Construction **12**, 33 (1990)

[5] P.N. quiroga, D.W. Fowler, Tech. Rep. 104-2, International Center for Aggregates Research, Austin, TX (1994)

[6] H.J.H. Brouwers, H.J. Radix, Cement and Concrete Research **35**, 2116 (2005)

[7] J.E. Funk, D.R. Dinger, *Predictive process control of crowded particulate suspension* (Springer US, 1994)

[8] C.C. Furnas, Tech. Rep. 2894, Department of Commerce, Bureau of Mines (1929)

[9] R.F. Fedors, R.F. Landel, Powder Technology **23**, 225 (1979)

[10] N. Ouchiyama, T. Tanaka, Industrial & Engineering Chemistry **23**, 490 (1984)

[11] T. Stovall, F. de Larrard, M. Buil, Powder Technology **48**, 1 (1986)

[12] A.B. Yu, N. Standish, Powder Technology **55**, 171 (1988)

[13] T. Aste, Physical Review E **53**, 2571 (1996)

[14] H. Brouwers, Physical Review E **74**, 031309 (2006)

[15] H. Brouwers, Physical Review E **84**, 042301 (2011)

[16] H. Brouwers, Physical Review E **87**, 032202 (2013)

[17] H.J.H. Brouwers, Physical Review E **89**, 052211 (2014)

[18] A. Bezrukov, D. Stoyan, M. Bargiel, Image Anal. Stereol **20**, 203 (2001)

[19] M. Wackenhut, H. Herrmann, Physical Review E **68**, 041303 (2003)

[20] H.J. Herrmann, R. Mahmoodi Baram, M. Wackenhut, Physica A: Statistical Mechanics and its Applications **330**, 77 (2003)

[21] K. Sobolev, A. Amirjanov, Advanced Powder Technology **15**, 365 (2004)

[22] K. Sobolev, A. Amirjanov, Powder Technology **141**, 155 (2004)

[23] C. Voivret, F. Radjaï, J.Y. Delenne, M. El Youssoufi, Physical Review E **76**, 021301 (2007)

[24] J.F. Jerier, V. Richefeu, D. Imbault, F.V. Donzé, Computer Methods in Applied Mechanics and Engineering **199**, 1668 (2010)

[25] S.D.S. Reis, N.A.M. Araújo, J.S. Andrade, jr., H.J. Herrmann, EPL (Europhysics Letters) **97**, 18004 (2012)

[26] P. Stroeven, M. Stroeven, Cement and Concrete Research **29**, 1201 (1999)

[27] M. Wackenhut, S. McNamara, H. Herrmann, The European Physical Journal E **17**, 237 (2005)

[28] C. Voivret, F. Radjaï, J.Y. Delenne, M. El Youssoufi, Physical Review Letters **102**, 178001 (2009)

[29] B. Yohannes, K.M. Hill, Chaos: An Interdisciplinary Journal of Nonlinear Science **82**, 061301 (2010)

[30] D.H. Nguyen, E. Azéma, F. Radjaï, P. Sornay, Physical Review E **90**, 012202 (2014)

[31] D.H. Nguyen, E. Azéma, P. Sornay, F. Radjaï, Physical Review E **91**, 032203 (2015)

[32] GDR-MiDi, Eur. Phys. J. E **14**, 341 (2004)

Contour fractal analysis of grains

Giulia Guida[1,*], *Francesca* Casini[2], and *Giulia MB* Viggiani[2]

[1]Università degli Studi Niccolò Cusano, Dipartimento di Ingegneria, 00166 Rome, Italy
[2]Università degli Studi di Roma "Tor Vergata", Dipartimento di Ingegneria Civile e Ingegneria Informatica, 00133 Rome, Italy

Abstract. Fractal analysis has been shown to be useful in image processing to characterise the shape and the grey-scale complexity in different applications spanning from electronic to medical engineering (*e.g.* [1]). Fractal analysis consists of several methods to assign a dimension and other fractal characteristics to a dataset describing geometric objects. Limited studies have been conducted on the application of fractal analysis to the classification of the shape characteristics of soil grains. The main objective of the work described in this paper is to obtain, from the results of systematic fractal analysis of artificial simple shapes, the characterization of the particle morphology at different scales. The long term objective of the research is to link the microscopic features of granular media with the mechanical behaviour observed in the laboratory and *in situ*.

1 Introduction

The shape of a particle can be expressed in term of three independent properties: overall shape, angularity, and roughness (surface texture) [2]. These properties correspond to a three-tiered hierarchy of observational scales and their characterization may lead to a better understanding of the behaviour of granular materials. Data analysis performed by [3] on a wide range of experimental data for granular materials, shows that an increased shape irregularity leads to more open particle configurations, therefore to an increase of maximum and minimum void ratios, respectively e_{max} and e_{min}. Irregularity causes also a decrease of stiffness, an increase of one-dimensional compressibility, an increase of its sensitivity to the state of stress, and an increase of critical state friction angle, φ_{cs}.
Overall shape, angularity and roughness are essentially independent properties, because one can vary widely without necessarily affecting the other two.
While many well established definitions of overall shape have been proposed in the literature, [2,4,5,6,7] the definition of angularity and roughness are still vague. Due to its fractal nature, roughness implies that there is no characteristic scale on the surface itself [3].
To be quantified, both angularity and roughness require more sophisticated techniques than overall shape, such as the acquisition of images by microscopy and computer based image analysis [8,9].
The challenge of this work is to find a quantitative approach to the characterisation of particle shape at different scales using the fractal analysis method.

2 Fractal analysis method

The fractal analysis contour method consists in an automatic computation of the particle perimeter on a 2D image, using sets of successive sticks of the same length. As the length of the sticks decreases, the computed perimeter tends to its "true" value. The length of the boundary of the particle is not strictly "true" because it depends on the image resolution adopted. This method is called fractal due to self-similar length of the sticks across the different scales.
Starting by the particle 2D image, the method consists in the following steps:

- extraction of the coordinates of the particle boundary (pixels);
- computation of the particle equivalent diameter, *D*, defined as the diameter of the circle having the same area as the particle;
- definition of the relative stick lengths as a fractions of the particle equivalent diameter *D*;
- measurement of the particle perimeter starting from the first coordinate of the particle contour, counting the number of sticks to run all the boundary, for each stick length;
- normalisation of the computed perimeter by the equivalent diameter;
- plotting the measured relative perimeters as a function of the stick lengths.

The hypothesis is that the relatively large stick lengths give information about the overall shape of the particle, the intermediate stick lengths about its angularity and the small sticks about its roughness.
In the following, the method outlined above is applied first to artificial simple shapes, in order to study the effect of overall shape, angularity, and roughness on the pattern of the relative perimeter – relative stick length

* Corresponding author: giulia.guida@unicusano.it

plot. Smooth shapes are adopted to study the influence of the overall shape (from isometric to elongated) and the angularity (from circles to squares). These are rendered increasingly rough by adopting regular zig-zagged profiles of different magnitude and frequency, to investigate the role of roughness. Finally, the method is used to describe the contour of a real particle of a Light Expanded Clay Aggregate (LECA).

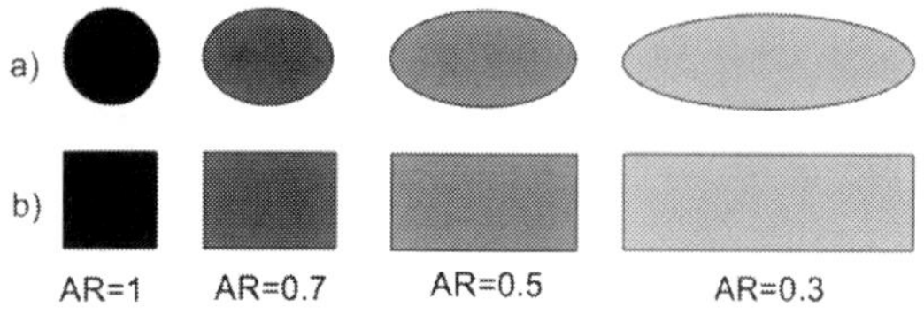

Fig. 1. a) family of circles and b) family of rectangles with AR equal to 1, 0.7, 0.5, and 0.3.

3 Simple shapes

3.1 Smooth particles

In order to study the effect of overall shape and angularity, four ellipses (Figure 1.a) and four rectangles (Figure 1.b) with different elongations are considered first. The elongation of each shape is quantified by the Bounding Box Ratio (*BBR*) defined as the ratio between the sizes of the rectangle circumscribed to the shape. *BBR* is a dimensionless parameter ranging between 0 (extremely elongated shapes, such as *e.g.*, lines) and 1 (isometric shapes, such as *e.g.*, circles or squares). The stick lengths adopted in the calculations are fractions of the shape equivalent diameter $D=2(\pi A_p)^{0.5}$, where A_p is the area of the shape. Each stick length is used as the unit to compute the perimeter of the shape. Figure 2 shows the computed values of the relative perimeter, p/D, as a function of the logarithm of the relative stick length, b/D, for: (a) the family of ellipses, and (b) the family of rectangles. As expected, the computed perimeter increases with decreasing stick length.
The stick value b in which the asymptotic perimeter is reached depends by the particle curvature. The black stars in Figure 2 show the point at which the asymptotic value of the perimeter is reached within a tolerance of 1%. For the ellipses (Figure 2.a) these points correspond to a relative stick length ranging between 0.13-0.24 and increasing with *BBR*. For the rectangles (Figure 2.b) these points correspond to a relative stick length of 0.03, essentially independent from *BBR*. This indicates that the value of the relative stick length at which the asymptotic perimeter is reached depends on the shape angularity: the rectangles are more angular than the ellipses and their angularity is independent of *BBR*, while the curvature of the ellipses, somehow related to their angularity, increases with decreasing *BBR*.
The asymptotic value of the relative perimeter contains information about the shape isometry: the circle, that is the most isometric shape, is characterized by the minimum attainable asymptotic value of the relative perimeter, that is $p/D = \pi = 3.14$, and this increases with increasing elongation of the shape.
The oscillations in Figure 2, at large values of the relative stick length, are related to the changes of curvature of the shape along its boundary and/or to its angularity, because for angular shapes or shapes where there are strong variations of curvature along the boundary, the computed relative perimeter depends significantly on the position of the first point along the boundary. In fact, no oscillations occur in circular shapes, as every start point is equivalent to another. Information about macro and medium-scale shape descriptor can be obtained from relative stick lengths between 0.1 and 1.

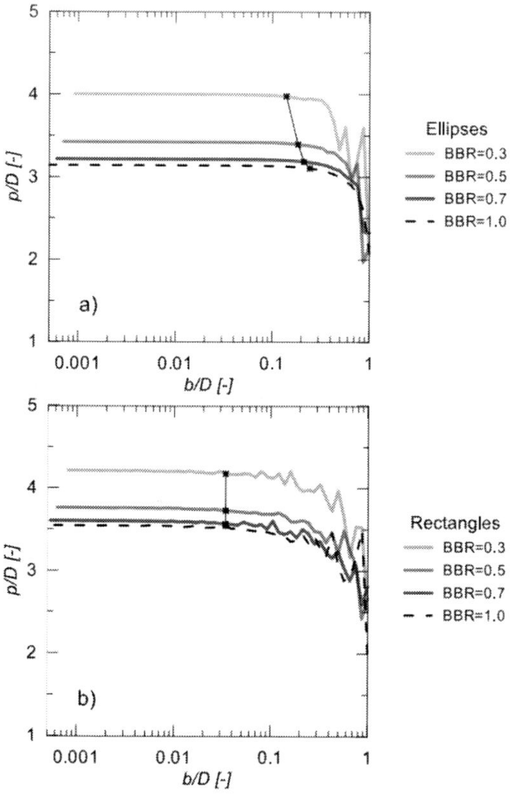

Fig. 2. Fractal analysis results for a) ellipses and b) rectangles

3.2 Rough particles

In order to study the effects of roughness, a family of circles with artificial roughness are examined in this section. The artificial roughness was generated by translating each even point externally and each odd point internally, to obtain a saw tooth boundary. Three different frequencies of roughness with three different amplitudes were considered. The frequency is related to the number of points included in the shape (100, 500 and 1000): the larger the number of points the smaller the distance between two following asperities. The amplitudes of the saw teeth considered in this study were of 0.2%, 0.6% and 1.0% of the semi major side of the

bounding box. Figure 3 shows an example of the saw tooth boundaries. In Figure 3.a, the amplitude is fixed to 0.6% and the number of points varies from 100 to 1000 points. In Figure 3.b the number of points is fixed to 500 and the amplitude varies from 0.2% to 1.0%.
Both the frequency and the amplitude of roughness affect the trend of the results of the fractal analysis. Figure 4.a shows the effect of the amplitude of the artificial roughness on the resulting curve; in this case the frequency was fixed at 500 points. The slope of the curves, for $b/D < 0.04$, increases with roughness amplitude. The effect of frequency (number of points), is shown in Figure 4.b where the amplitude is fixed at 1.0% and the frequency takes values between 100, and 1000 points. The final value of the relative perimeter is affected by the frequency while the slope to attain it is constant. The asymptotic value of the perimeter is reached when the stick length is comparable with the image resolution, and, therefore, it depends on the number of points (or pixels) defining the particle shape. It is interesting to note in Figures 4 that, as b/D increases (>0.04), all the curves overlap. This means that roughness does not affect the morphology of the shape at the meso-macro scale.

4 Real particle analysis

The procedure described above is applied to describe the fractal characteristics of the contour of a LECA particle extracted from a 2D image of the grain. The procedure followed is:
- elaboration of the Scanning Electron Microscopy (SEM) micrograph of the particle with MATLAB;
- segmentation of the image used to identify the contour of the particle;
- fractal analysis of the contour: definition of the stick lengths used to measure the grain perimeter;
- representation of the results in the dimensionless plane *p/D vs b/D*.

4.1. LECA material

LECA is an artificial granular material used in many civil engineering applications due to its very low unit weight. The reason of lightness is the high level of particle intra-porosity, that confers to its fragments a very irregular shape, sharp angularity and high roughness (*e.g.* [10]). Figure 5.a shows the SEM image of a LECA particle used for the fractal analysis. The particle dimensions are between the sieve range of 0.25-0.50 mm, and its Bounding Box Ratio is BBR=0.45.

4.2 Results and comparisons

Figure 5. b-d show the perimeter of the particle defined with decreasing stick lengths, corresponding to the stars in Figure 6. In step (b) the relative stick length is $b/D = 0.569$ and the evaluation of the particle perimeter is not accurate, but can still provide an indication about the overall shape of the particle (macro-scale). Step (c), corresponds to a relative stick length $b/D = 0.160$, and provides information about the angularity or change of curvature of the boundary. Step (d), where $b/D = 0.006$, identifies the asymptotic value of the perimeter and provides information on the amplitude of roughness.

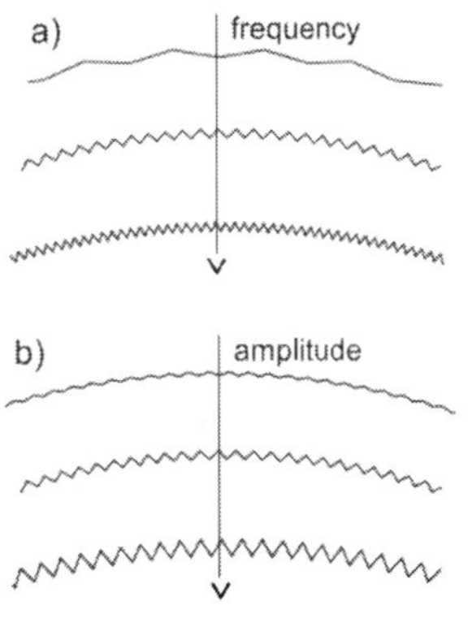

Fig. 3. Artificial roughness: a) varying frequency (100 to 1000 points) at constant amplitude (0.6%), and b) varying amplitude (0.2% to 1.0%) at constant frequency (500 points).

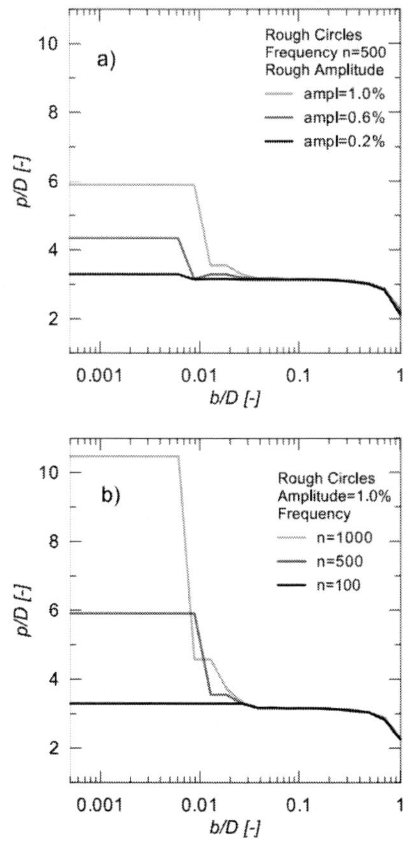

Fig. 4. Fractal analysis results for artificial rough circles. a) effects of amplitude. b) effects of frequency.

The fractal analysis performed on the particle of LECA (Figure 6) results in a more irregular trend than those obtained for the simple shapes in Figure 4. Starting from values of *b/D* close to 1, the oscillations are more extended due to the angularities of the shape. A change in the slope of the curve is detectable for a value of *b/D* ~ 0.1 where there is a point of inflection in the curve. This may represent the attainment of the horizontal asymptote for the smooth shapes. From the results in Figure 6, the inflection point occurs close to point (c), and the corresponding *b/D* and *p/D* values provide information on the angularity and the elongation of the particle, as highlighted in section 3.1.

The final part of the curve is characterized by a longer path, with less constant slope compared with the artificial rough particles. In fact, the latter only include one artificial roughness, with a specific amplitude and frequency. In the case of the LECA particle, different families of roughness characterize the morphology with different amplitudes and frequencies, proper of the fractal nature of the particle roughness.

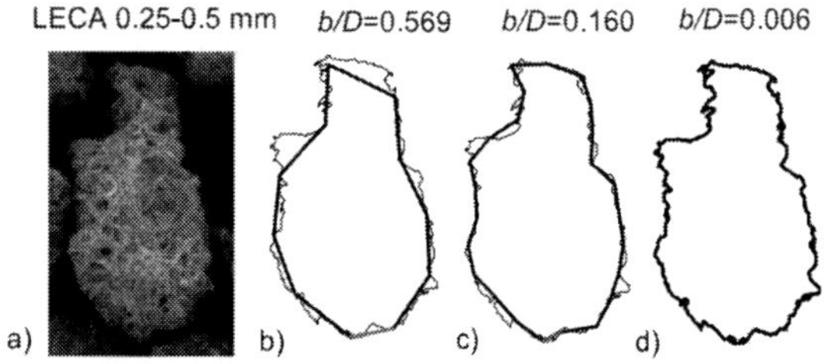

Fig. 5. LECA particle. a) original SEM image. b), c), d) different definitions of the particle perimeter using decreasing stick lengths.

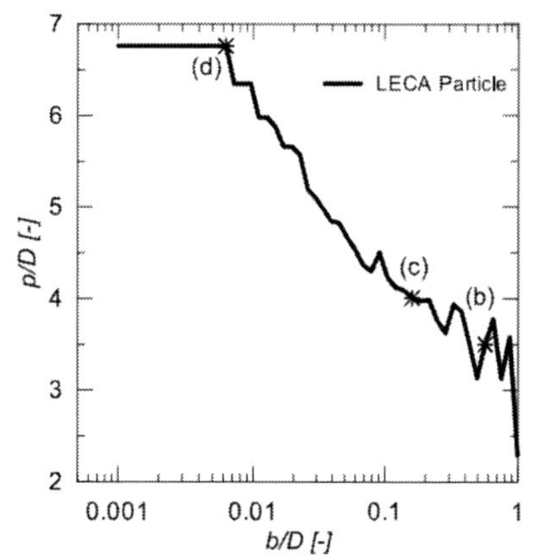

Fig. 6. Fractal analysis results on a LECA particle.

5 Conclusions

The fractal analysis of particle contour is a promising tool to study particles morphology at different scales. It consists of measuring the length of the perimeter of the particle using stick of decreasing length, and plotting the computed values against the logarithm of the stick lengths. Different shapes were examined to understand the effects of morphology at different scales, and the results obtained from this simple shapes were used to guide the interpretation for a particle of LECA.

Each stick length conveys information on the associated morphological shape. Particular attention was paid to the characterisation of the particle's roughness, connected to the slope corresponding to the micro stick lengths ($b/D < 0.1$). The idea is that the slope of the curve relative to small stick length, can quantify the particle micro-roughness and the slope of the curve relative to big stick lengths, more than $b = 0,1D$ can quantify the macroscopic roughness. Further, the difference between the values of *p/D*, relative to sticks length maximum, of $b/D = 1.0$, and the one relative to the inflection point, *b/D* ~ 0.1, can quantify the macro isometry of the shape. Good analogies are between macro roughness and the literature parameter of regularity [5] and between the macro isometry value and the circularity [11]. The approach must be validated over a wider range of real particles.

References

1. R. M. Rangayyan, T. M. Nguyen, *Fractal analysis of contours of breast masses in mammograms* (Journal of Digital Imaging), **20**(3), 223-237 (2007).
2. P. J. Barrett, *The shape of rock particles, a critical review* (Sedimentology), **27**, 291-303 (1980).
3. G. C. Cho, J. Dodds, J. C. Santamarina, *Particle shape effects on packing density, stiffness, and strength: natural and crushed sands*, (Journal of geotechnical and geoenvironmental engineering), **132**(5), 591-602 (2006).
4. W.C. Krumbein, *Measurement and geological significance of shape and roundness of sedimentary particles* (Journal of Sedimentary Research) **11**(2) (1941).
5. W.C. Krumbein, L. L. Sloss, *Stratigraphy and sedimentation*, 2nd Ed. Freeman and Company, San Francisco (1963).
6. M.C. Powers, *A new roudness scale for sedimentary particles*, (J. Sediment. Petrol.), **23**(2), 117-119 (1953).
7. H. Wadell, *Volume, shape and roudness of rock particles*, (J. Geol.), **40**, 443-451 (1932).
8. E. T. Bowman, K. Soga, W. Drummnond, *Particle shape characterization using Fourier descriptor analysis* (Geotechnique), **51**(6), 545-554 (2001).
9. T.P. Meloy, *Fast Fourier transforms applied to shape analysis of particle silhuettes to obtain morphological data*, (Powder Technol.), **17**, 27-35 (1977).
10. G. Guida, M. Bartoli, F. Casini, G.M.B. Viggiani. *Weibull Distribution to Describe Grading Evolution of Materials with Crushable Grains.* (Procedia Engineering), **158**, 75-80 (2016).
11. E. P. Cox, *A method of assigning numerical and percentage values to the degree of roundness of sand grains.* (Journal of Paleontology), **1**(3), 179-183 (1927).

Twisting, an alternative strategy to compact granular materials

Karol Asencio[1], *Iker* Zuriguel[1], and *Diego* Maza[1,*]

[1] *Granular Media Lab. Departamento de Física y Matemática Aplicada. Universidad de Navarra. 31080, Pamplona, Navarra, Spain.*

Abstract. Nowadays, the common method to pack granular materials is to tap the ensemble against the gravity. Despite the apparent simplicity of that method, the asymptotic states reached by the tapped systems have strongly dependences on parameters like the shape of the tapping pulse, the container geometry or the ratio between lateral and axial dimensions. Beyond the restrictions imposed by the system boundaries, the particle shape (like rods or tetrahedrons) plays a central role in the evolution and the final state of the ensemble. In this work, we introduce an unconventional method for compacting granular ensembles by applying a sequence of alternating counterrotating pulses or "twists". By using spherical particles we analyze the efficiency of this method to achieve highly packed configurations.

1 Introduction

In almost the totality of granular storage processes, it is desirable to optimize the packing fraction in order to minimize the volume of the container. This, in principle, simple task implies to implement sophisticated technological solutions to drive the material to its optimal packing. The reason lies in the inherent complexity associated to the process of arrangement (or rearrangement) of any group of particles. From a fundamental point of view, the packing problem involves several challenges concerning: a) the existence of an well defined mean value of the volume fraction compatible with the history and boundary conditions of the packing; b) the mechanism that conduits any group of grains to this asymptotic value; and c) the dependence of this asymptotic value on the excitation parameter. Indeed, even in the case of monodisperse beads where a well defined maximum packing fraction has been recently predicted [1, 2], the geometry of the boundaries imposes important restrictions to this limit [3] implying that its calculation becomes a complex optimization problem [4]. Apart from the inherent difficulties of calculating this limit, the dynamical evolution to this magnitude and the possibility of approaching it asymptotically, are also relevant issues. Moreover, bead properties like friction and boundary conditions impose extra restrictions on the possibility of reaching such theoretical limit. Indeed, the states exhibited by an ensemble of particles that determine the typical average for the packing fraction, have relevant statistical features that have merited the introduction of novel theoretical frameworks to describe them [5].

*e-mail: dmaza@unav.es

2 Experimental setup

Nowadays, tapping is considered the canonical protocol employed to drive granular systems into statistically stationary equilibrium states, compatible with the boundary conditions of the system. In spite of the fact that this method is profusely used, its validity to bring the system to states statistically homogeneous is not completely clear in many circumstances and the results extracted from the experimental observations could appear contradictory. The most popular example of this is the differences between the "Chicago" and the "Rennes" experiences: both deal with the problem of packing grains by tapping but they are contained in cells with very different lateral dimension [6]. In the former case, a tall and narrow tube is filled with beads of a diameter one tenth smaller than the tube diameter; in the second, the particles' diameter was two orders of magnitude smaller than the diameter of the shallow cylindrical cell employed. In both setups, a large number of tapping pulses had to be applied to reach a stationary packing fraction: Moreover, the asymptotic volume fraction seems to have a monotonous dependence with the amplitude of the applied pulse. Nevertheless, the temporal evolution of the packing fraction with the number of taps is notably different: in the first case an inverse logarithmic growth seem to fit the results, whereas in the second, an stretched exponential was shown to be better to fit the temporal evolution. Moreover, significative differences exist in the figures obtained for the asymptotic packing fraction attained with the same tapping amplitude. Indeed, for tapping amplitude slightly larger than $\Gamma = 3$ (where Γ is the maximum acceleration applied to the setup normalized by the gravity), $\phi_{max} \approx 0.635$ with the "Chicago" setup and $\phi_{max} \approx 0.614$ with the "Rennes" one. The explanation given was that the strong confinement of the Chicago setup caused crystallization against the disorder induced by the convective regime developed when a large cell is used.

In this work we explore the possibility of obtaining highly packed arrangements when an large aspect ratio cell is used. In order to do that, we implement an alternative strategy to the tapping protocol which consists on twisting alternately a cylindrical container. Under this excitation, the granular sample self-organizes increasing its volume fraction by arranging the beads in superimposed layers that tends to reach its maximal packing fraction (see inset of Fig. 1). The temporal evolution of the volume fraction is analyzed for different twist amplitudes. Note that a similar strategy has been used previously but combined with tapping pulses [7]. The experimental setup consists of a cylindrical container of radius $R = 87.0 \pm 0.1$ mm made in transparent plexiglass with a base of aluminium (see Figure 1). In order to induce compaction the whole cylinder is rotated in alternating angular directions by a gearbox connected to a $1/2$ HP electrical motor which was controlled by an electronic variator. This allowed the variation of the angular velocity at will. The range of angular velocities goes from 2.3 s^{-1} to 17.8 s^{-1}, which correspond to tangential velocities at the border of the cylinder v, between 200 mm/s and 1550 mm/s. This twisting produces an effective "shaking"; a perturbation that appears when the cell inverts its rotational direction. Just for comparison with other tapping experiments, we normalize the cell tangential acceleration, $\alpha = 2\mathrm{v}/\Delta t$ (where Δt is the time necessary to invert the rotation) by the gravity acceleration: $\Gamma = \alpha/g$. The values of Γ explored are in the range $0.3 < \Gamma < 1.01$, notably smaller than the used in other compaction experiments. Two type of beads are used, delrin spheres (radius $r = 1.5$ mm) and PVC beads ($r = 3$ mm). Delrin beads have a mass $m_D = 0.018$ g while PVC beads are one order of magnitude heavier, $m_P = 0.200$ g: the purpose of this was to compare the influence of inertial force on the final result. After the first twist, the free surface of the material becomes almost flat independently of the configuration with which the grains were introduced in the cell. Therefore, the mean volumetric packing fraction , ϕ can be determined from the apparent volume of the material. That is, the packing fraction is calculated as the ratio between the volume of the beads poured in the cell and the volume of a cylinder of height equal to the mean value of the free surface. To determine this profile, a linear camera registered a lateral line of pixels of the cell rotating at small velocity. Then, the height of the packing is extracted from the profile of the free surface is obtained by image processing technics as is displayed in Fig. 2. Therefore, the relevant height fluctuations of the sample are due to some of the isolated beads resting on the top layer. Nevertheless, the maximum relative standard deviation observed in any typical experimental situation is below 5%. Moreover, the free surface is only lightly deformed for the range of accelerations applied. Hence, although a systematic error exist in the estimation of the packing fraction due to the height overestimation, this error equally affects all the experimental results independently of the Γ value.

Figure 1. Experimental setup. An electrical motor rotates a cylindrical cell filled with spherical beds in alternating directions. The arrangements of the particles near the lateral surface could be obtained by using a line scan camera, and the free surface is registered by a digital camera. Inset: Upper view of a single layer of particles manually compacted near its maximum packing fraction.

Using a high speed camera we have also tested that, when the angular velocity is constant, in the frame of reference of the cylinder, particles do not move. Therefore, only when the angular velocity inverts its sign there is a stop-start motion that induces a shear stress within the material in the azimuthal direction. It is at this precise moment when particles move with respect to each other and therefore rearrangements occur. Moreover, the vertical load of the material also induces a torsional stress in the vertical direction; due to this, the free surface of the material suffers a net displacement respect to the base of the cell between successive twists.

3 Experimental results

Fig.3 shows the packing fraction evolution of delrin particles as a function of the number of twists applied to the cell. The two extreme values of Γ studied in this work have been presented. Like in similar experiments, we performed five experimental repetitions for each Γ in order to check the repeatability of the results. Let us now discuss carefully each of these curves comparing the twisting results with previous, tapping ones. Clearly, the twisting protocol provides a very valuable tool to induce the compaction of the grain ensemble, primarily by using small amplitudes. Indeed, after 10^4 twists the volume fraction obtained for $\Gamma = 0.31$ is $\phi = 0.66$ which is notably larger than the densest packing ($\phi < 0.59$) obtained for the smallest amplitude ($\Gamma = 1.4$) explored in the Chicago experiment [9]. The same argument could be extended to the results introduced by the "Rennes" group [8]. In this latter case, the final volume fraction is even lower for the smallest reported tapping amplitude. Note that these values of ϕ obtained for the smallest tapping intensities in that pioneering works [8, 9] were notably improved by performing an annealing procedure; that is, first tapping at high intensities and then reducing the intensity. In our

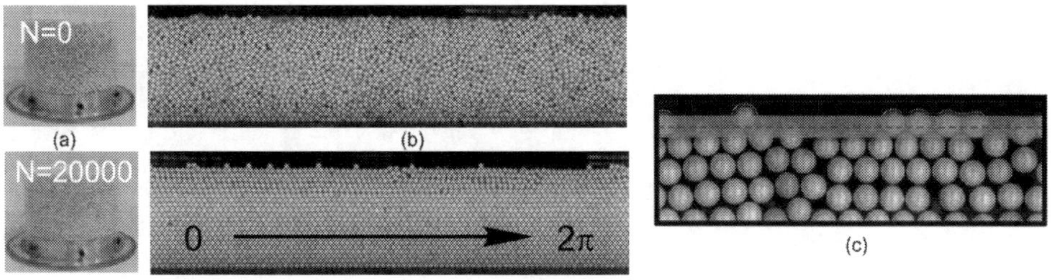

Figure 2. a) Pictures showing the initial (top) and final (down) states for a layer of 6 *mm* PVC beads submitted to 20000 twists. b) The image reconstructions obtained with a line scan camera showing all the particles in the external layer for these states. c) Detail of the top layer. Dashed line indicates the calculated mean height and the shadow region represent the interval defined by the standard deviation of the profile.

case, this annealing process seems to be not necessary to reach really well packed arrangements, although this should be deeply investigated in future works. Furthermore, preliminary results shown in Fig. 3 suggest that the packing fraction growth is logarithmic, until it reaches a cutoff value beyond which the packing fraction remains almost constant. For the case of $\Gamma = 0.31$ only the logarithmic growth is readily visible as it seems that, after 30000 twists, the system has not reach the cutoff value. For the data corresponding to $\Gamma = 1.01$, however, both regimes are clearly evidenced. When this large amplitude is applied, the volume fraction decreases during the first twists applied because some beads jump to the highest positions in the perimetrical layer of particles. Nevertheless, after a few twists, the packing fraction starts to increase as the sample arranges in a more compacted structure. At this stage, the volume fraction evolves between $\phi = 0.6$ to $\phi = 0.64$ in less than 500 twists. Clearly, this growth is markedly larger than the analogous in the tapping experiences using accelerations around the unity. After reaching the cutoff value, the packing slightly increases revealing important fluctuations which are related with the competition between domains as the ones shown in the right panel of Fig. 2. This observation suggests that the geometrical frustration induced by the ratio between the cylinder and particle sizes could be behind the fluctuations of the packing fraction around a well defined asymptotic mean value. As this ratio also determines the maximum volume fraction possible, in order to generalize the observed features, in Fig. 4 we compare the temporal evolution of packings of delrin ($r = 1.5$ *mm*) and PVC beads ($r = 3$ *mm*). Clearly, PVC beads present systematically lower volume fractions than delrin beads, although the global evolution of both type of particles is almost the same: a rapid growth until a cutoff value is reached. Beyond this point, the asymptotic volume fraction fluctuates around a mean value. Despite this similarity, their dynamics are not completely equal in both cases. Indeed, for larger spheres (PVC), the volume fraction fluctuations around the stationary state are more important than for the smaller ones (Delrin). Visual inspection of the external layer evidences the development of grain boundaries between ordered domains.Therefore, the fluctuations observed when the largest explored twist

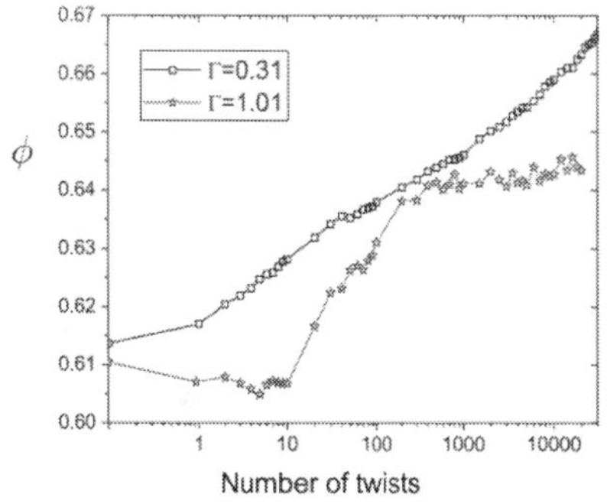

Figure 3. Volume fraction evolution of delrin particles as a function of the number of twists for the two limit accelerations studied: $\Gamma = 0.31$ (square symbols) and $\Gamma = 1.01$ (star symbols) as shown in the legend. Note the logarithmic scale in the x axis. Clearly, for low tapping amplitudes the volume fraction evolves quasi-monotonically, but for larger values of Γ a different trend is observed: after a short interval of strong particle rearrangements (where the volume fraction even decreases) a rapid growth of the packing fraction is produced, followed by a (quasi) stationary state. Dotted lines are guides for the eyes.

amplitude is used seems to be related with the development of defect lines inside the ordered layers. This observation is compatible with the ideas introduced in [14, 15] where the microstructure of the arrangements is numerically analyzed in order to justify the evolution and fluctuations of the volume fraction. Importantly, we have also observed the sudden jumps reported in [15] and at a preliminary stage it is possible to associate these jumps to the competition between different orientational domains inside the system. This fact will be reported elsewhere.

Finally, let us note that the asymptotic limit of the volume fraction seems to reduce monotonically as the twist amplitude is increased in line with "Chicago" results. At this point, it should be stressed again that in [8, 10] this behavior was obtained only if an a annealing process was applied for the smallest intensities. In our case, we do not need such an annealing process as very high volume

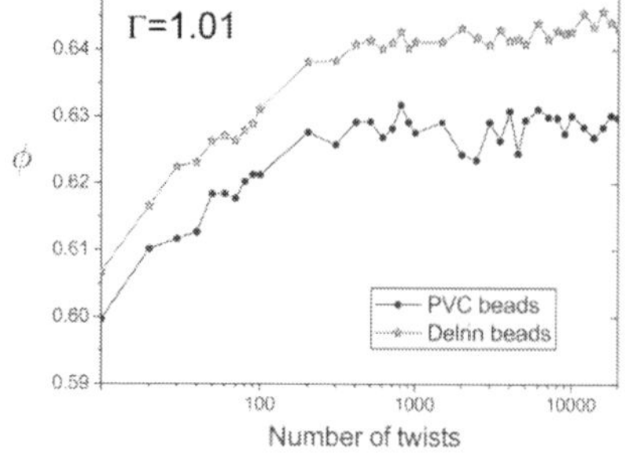

Figure 4. Volume fraction evolution for two different beads size: 5.95 *mm* diameter PVC beads (circle symbols) and 3 *mm* diameter delrin beads (star symbols). Note the logarithmic scale in the x axis. Despite the difference in the magnitude of the volume fraction, related with the difference in the diameter of the beads, the temporal evolution shows the same trends. Nevertheless, the strong fluctuations observed for large beads seem to be associated with the competition of ordered domains.

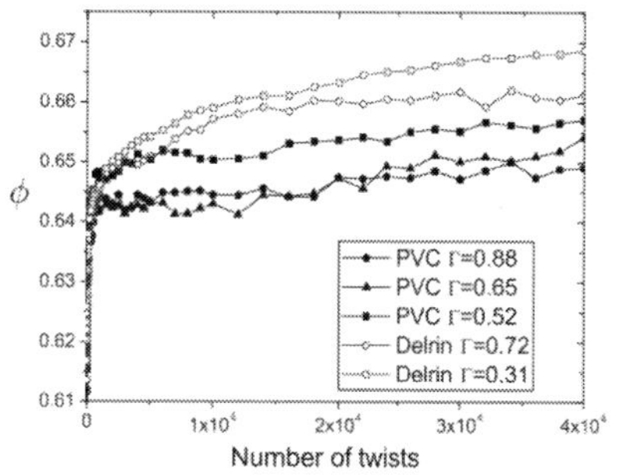

Figure 5. Volume fraction evolution for different twist amplitudes and the two types of beads used in this work as indicated in the legend. Clearly, independently of the mass of the beads, the asymptotic packing fraction follows a monotonous dependence with the magnitude of the perturbation applied: the stronger the intensity, the lower the final packing fraction. The highest volume fractions (obtained for the smaller intensities) have been achieved without applying any annealing protocol.

fractions could be reached if a sufficiently large number of twist cycles is applied as evidenced in Fig. 5.

4 Conclusions

In this work we have provided experimental evidence of the usefulness of applying alternating cycles of rotation (or *"twists"*) to compact granular materials. Until now, the compaction of a granular ensembles has been mainly achieved by shaking the system which implies the application of sudden and large amplitude oscillations to small samples of material. Nevertheless, our protocol provides a simple mechanism to induce well ordered arrangements by shearing the material without need to provide external perturbation against the gravity force field (a condition needed in tapping). Indeed, the compaction of the system is only induced by the grains inertia during the reversal of the rotational motion of the container. Importantly, the observed asymptotic values of the packing fractions are larger than the ones previously reported in the literature. In addition, for low twist amplitudes the volume fraction seems to approach a limiting value compatible with the geometry of the boundary conditions, independently on the initial conditions and without applying any kind of annealing protocol [16]. Finally, let us mention that the comparison with the tapping results has been performed assuming the equivalence between twist and tap acceleration but this extreme will be confirmed using other parameters in line with the ideas introduced in [11, 12].

Acknowledgments

This work was funded by Ministerio de Economía y Competitividad (Spanish Government) through Projects No. FIS2011-26675 and FIS2014-57325. K. Asencio thanks Asociación de Amigos de la Universidad de Navarra for a scholarship.

References

[1] T. C. Hales, *An overview of the Kepler conjecture.* arXiv:math/9811071v2, (1998).
[2] Denis Weaire, Tomaso Aste, *The Pursuit of Perfect Packing, 2nd Ed.* (Taylor& Francis, NY, London, 2008).
[3] See http://mathworld.wolfram.com/SpherePacking.html and references therein.
[4] http://hydra.nat.uni-magdeburg.de/packing/
[5] D. Bi,S. Henkes, K. Daniels, and B. Chakraborty *Annu. Rev. Condens. Matter Phys.* **6**, 63 (2015).
[6] P. Richard, M. Nicodemi, R. Delannay, P. Ribière, and D. Bideau. *Nature Materials* **4**, 121 (2005).
[7] K. Daniels, and R. Beheringer. *Phys. Rev. Lett.* **94**, 168001 (2005).
[8] P. Philippe, and D. Bideau. *Europhys. Lett.* **60**, 677 (2002).
[9] J.B. Knight, C. G. Fandrich, C.Ning Lau, H.M. Jaeger, and S. R. Nagel. *Phys. Rev. E* **51**, 3957 (1995).
[10] E. Nowak, J. Knight, E. Ben-Naim, H. Jaeger, S. Nagel. *Phys. Rev. E* **57**, 1971 (1998).
[11] G. Lumay, and N. Wandewalle. *Phys. Rev. E* **74**, 021301 (2006).
[12] J.A. Dijksman, and M. van Hecke. *Europhys. Lett.* **88**, 44001 (2009).
[13] E. Ben-Naim, J. Knight, E. Nowak, H. Jaeger, and S. Nagel. *Physica D* **123**, 380 (1998).
[14] A. D. Rosato, and D. Tacoub. *Powder Technology* **109**, 255 (2000).
[15] A. Rosato, O. Dybenko, D. Horntrop, V. Ratnaswamy, and L. Kondic. *Phys. Rev. E* **81**, 061301 (2010).
[16] O. Caravente, and J.C. Ruiz-Suárez *Phys. Rev. Lett.* **95**, 018001 (2005).

Effect of the characteristics of granular media on dynamic penetration test - Numerical study using 2D - DEM.

Quoc Anh Tran[1,*], *Bastien* Chevalier[1,**], and *Pierre* Breul[1]

[1] *Université Clermont Auvergne, CNRS, Institut Pascal, F-63000 Clermont-Ferrand, France*

Abstract. The paper presents a numerical study on the influence of the properties of granular media on light dynamic penetration test results. A numerical model based on the Discrete Element Method (DEM) in 2 dimensions has been carried out. The model is able to reproduce the penetration tests in dynamic conditions. The model provide for each impact, the load–penetration curve that presents the evolution of reaction forces of granular media acting on the tip of the penetration test. The load – penetration curves are investigated in terms of tip resistance and analyzed by using the Discrete Fourier Transform in order to study the oscillations of the curves. The influence of the properties of granular media on the load-penetration curve has been investigated. We concentrated on the influence of particle friction and sample compacity. Besides the known results obtained in the literature by the dynamic tip resistance analysis for macroscopic response, we found that for a particle friction greater than 0.5, tip resistance and oscillations of signal are stable. Concerning the sample compacity, although variation of volume fraction is tight (ϕ = 0.817 – 0.844), a clear variation in terms of tip resistance and amplitude spectrum is highlighted.

1 Introduction

In the practice of geotechnical engineering, light dynamic penetration testing device such as Panda penetrometer can be used to mechanically characterize soils [1, 2]. Based on recent technological developments made on this device, it is possible to record for each impact the curve of tip stress or tip force versus the tip penetration distance [3]. This curve, referred to as load - penetration curve, provides information on dynamic tip resistance but also on additional mechanical parameters involved during the driving of the tip [3, 4].

In order to validate the methods of analysis and to better analyze the load - penetration curve obtained, many works focused on penetration tests in dynamic conditions with impacts by using DEM in 2D and in 3D [3–7].

In this paper, we will present at first the numerical model and parameters used to reproduce the dynamic penetration test in granular media presenting different properties, using DEM, as well as the method used to analyze the load - penetration curve [6, 7]. The following section explains how the load - penetration curve obtained from dynamic cone penetration tests was analyzed. At the scale of the contact or of the particle, the shapes of the load - penetration curves are analyzed in terms of both frequency and amplitude of signal oscillation by using Discrete Fourier Transform (DFT). Finally, the influence of material parameters is discussed: particle friction coefficient, volume fraction.

*e-mail: qatran@sol-solution.com
**e-mail: bastien.chevalier@uca.fr

2 Numerical Model

The numerical model in two dimensions is used with Itasca software PFC2D [8] to reproduce the dynamic penetration test. The description of the model and the effect of the impact velocity are addressed in the papers of Tran [6, 7].

In a rectangular with size of 0.60 x 0.45 *cm*, we generated 10 000 cylindrical particles of 0.4 *cm*, where the ratio between maximal and minimal particle diameters D_{max}/D_{min} is equal to 2; D_{max} = 0.7 *cm*. The impacting velocity is equal to 1.25 $m.s^{-1}$ in order to obtain an average penetration distance that is representative of experimental tests (Fig. 1). The Table 1 summarizes the main parameters of the model.

The effect of friction coefficient was studied, with a volume fraction corresponding to random close packing: friction equal to 0.00 was used during sample generation. The friction coefficient - μ varies from 0.05 to 1.00.

The effect of volume fraction was also studied for different friction coefficient were used in the radius expansion phase, in order to reach different values of volume fraction. In the phase of penetration test, particle friction was equal to 1.0. The volume fraction of particle assembly before penetration test - ϕ varies from 0.817 to 0.844.

DOI: 10.1051/epjconf/201714016004

Table 1. Parameters of the model

Parameter	*Symbol*	*Value*	*Unit*
Width box	L	0.60	m
Height box	H	0.45	m
Particle density	ρ	2700	$kg.m^{-3}$
Normal contact stiffness	k_n	1.250e8	$N.m^{-1}$
Tangential contact stiffness	k_s	9.375e7	$N.m^{-1}$
Rod friction coefficient	μ_{rod}	0.0	–
Tip friction coefficient	μ_{tip}	0.3	–

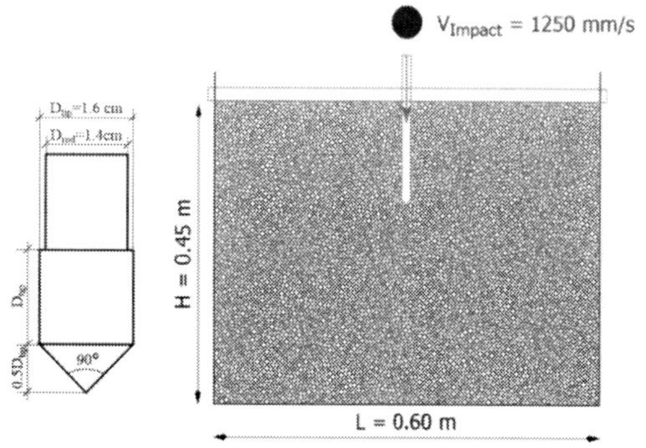

Figure 1. Dynamic penetration test presented by Tran [6, 7] with tip details.

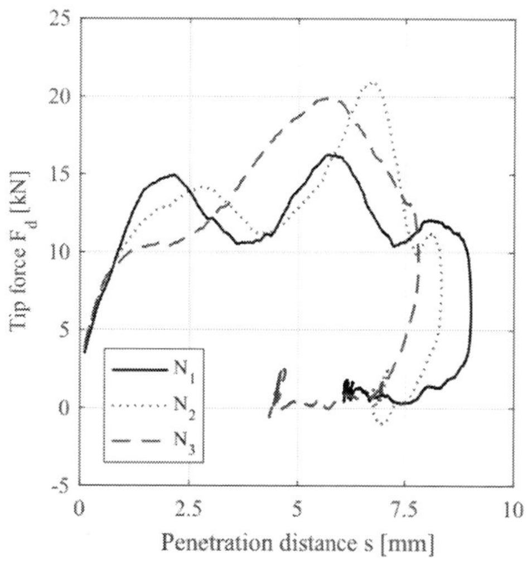

Figure 2. Load penetration curves examples obtained for different impacts by numerical model [6, 7].

3 Investigation of load - penetration curve

3.1 Macroscopic exploitation - Tip resistance

Tip force F_d is defined as the vertical component of the force applied by the granular material on the tip as it drives in the granular with a penetration distance s. Figure 2 shows examples of load - penetration curves $F_d = f(s)$ obtained for 3 impacts with the numerical model. The response obtained with the model is similar to the one classically obtained experimentally [4].

The dynamic tip resistance R_d of the granular material for one impact was calculated as the average tip force F_d for penetration distance between 0 and s_{max}:

$$R_d = \frac{1}{s_{max}} \cdot \int_{t=0}^{t_{s_{max}}} F_d(t)\, \mathrm{d}s(t) \tag{1}$$

Then, $< R_d >$ is the average value of dynamic tip resistances obtained for the five impacts and for the three samples.

3.2 Frequency analysis by using Discrete Fourier Transform

As the properties of the granular material change, the shape and size of these variations change too. Consequently, it seems that the load - penetration curve can provide not only the tip resistance of granular media but also information on the granular material properties.

Fourier transformation provides a powerful way for study discrete data acquisition in the frequency domain. We define the band - pass of frequency from f_{trans} to $f_{limited}$, for obtaining the full oscillation information of load - penetration curve (Fig.3). The frequency f_{trans} is the first local minimum value of the amplitude spectrum that corresponds to a transition zone where amplitude spectrum decreases rapidly, $f_k < 100\,m^{-1}$. The frequency $f_{limited}$ provides the most significant information to rebuild the signal with inverse DFT and to filter the signal noise. $f_{limited}$ is the frequency having an relative error (RE) smaller than 10 % (Eq.2).

The relative error RE between the reconstructed signal of the k first components - $F_d^k(s)$ and the original signal - $F_d(s)$, with N - points DFT, descripted by the following equation has been computed:

$$RE = \frac{\sum_{i=1}^{N} |F_d^i - F_d^i|_{0:k} \times (s^i - s^{i-1})}{\sum_{i=1}^{N} F_d^i \times (s^i - s^{i-1})} \tag{2}$$

In order to quantify the oscillations of signal for each material and in order to compare them to each other, we defined the coefficient Af as the average value of the product of the frequency f_k and the associated amplitude A_k detected in the band - pass range $[f_{trans}, f_{limited}]$:

$$Af = \frac{\sum_{i=1}^{n} A_k^i \times f_k}{n} \tag{3}$$

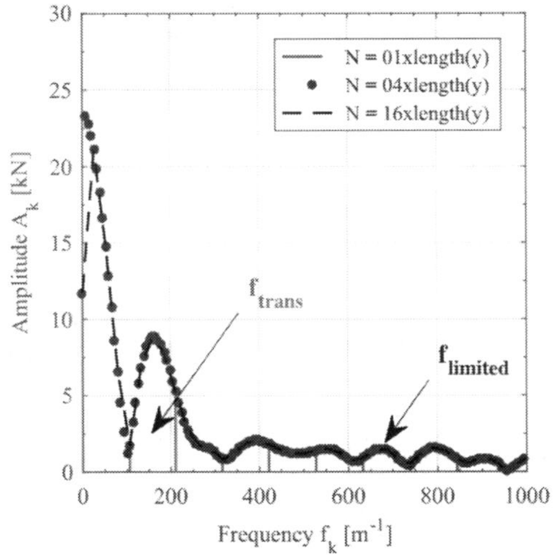

Figure 3. Example of the amplitude spectrum of one impact with difference N - points DFT. The amplitude spectrum becomes stable when N - points DFT increases.

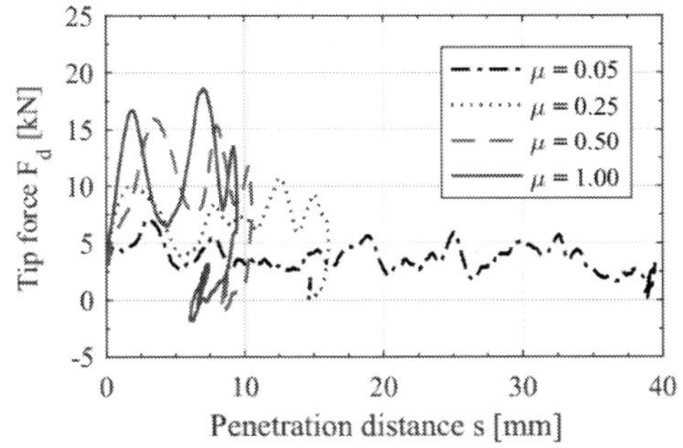

Figure 4. Load - penetration curve examples obtained in a dynamic penetration test for different particle friction coefficients μ = 0.05 - 1.00.

with n the number of frequencies detected in band-pass range $[f_{trans}, f_{limited}]$.

4 Effect of particle friction

Figure 4 shows, for each value of particle fiction coefficient - μ, the load - penetration curve obtained for one impact. We observe that the magnitude of tip force F_d is similar for μ equal to 0.50 and to 1.00. In addition, the amplitude of oscillations of the curve increases as μ increases. Figure 5 presents average value of $< R_d >$ and of $< Af >$ when μ varies from 0.05 to 1.00. For greater values of μ (≥ 0.5), average tip resistance $< R_d >$ presents a plateau; average value of $< Af >$ seems similar when $\mu \geq 0.5$. Thus, this tends to show that for materials having a friction angle greater than 45 degrees regardless the shape of the particles, the influence of the variation of friction

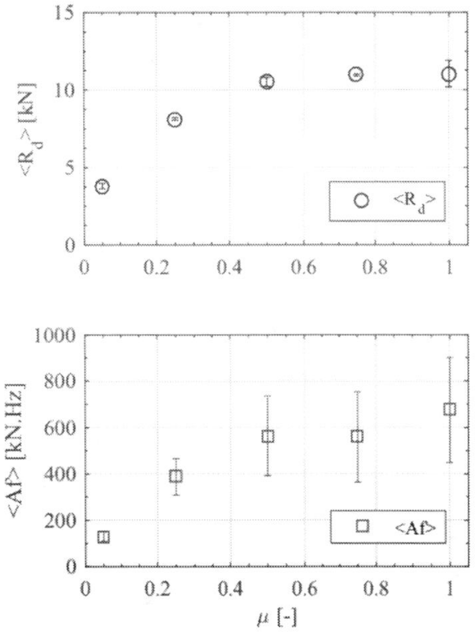

Figure 5. $< R_d >$ and $< Af >$ as function of particle friction μ.

becomes negligible on the tip resistance and shape of load – penetration curve.

5 Effect of particle assembly density

Figure 6 shows the load - penetration curves of five successive impacts for different values of volume fraction. Higher volume fraction corresponds to higher penetration resistance of the material. As the impact energy is constant, the maximal penetration distance decreases as volume fraction increases. Also, the curve oscillate with greater amplitude for higher volume fraction ϕ. Figure 7 shows that the tip resistance $< R_d >$ increases with volume fraction ϕ, even though the range of ϕ is quite narrow from 0.817 to 0.844. Tip resistance $< R_d >$ is very sensitive to volume fraction as already observed by other authors [1, 2]. Thus, the tip resistance $< R_d >$ is pertinent parameter for the evaluation of the density. By the DFT analysis, we found that it seems that when ϕ increases, the average value of $< Af >$ increases. It means that the volume fraction have an effect on the oscillation of load - penetration curve (Fig. 7).

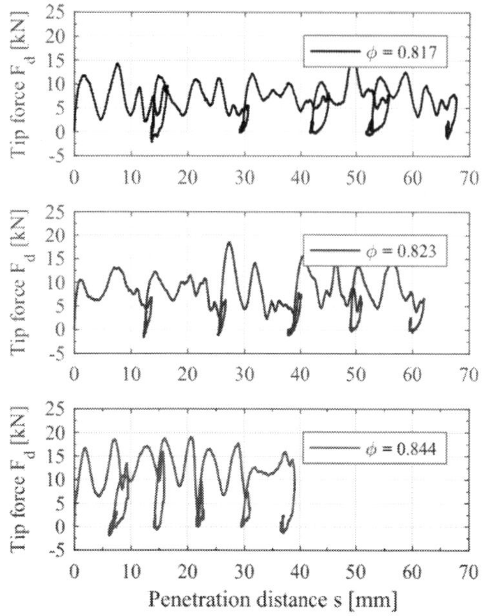

Figure 6. Load - penetration curve examples obtained in a dynamic penetration test for different granular volume fractions ϕ = 0.817 - 0.844.

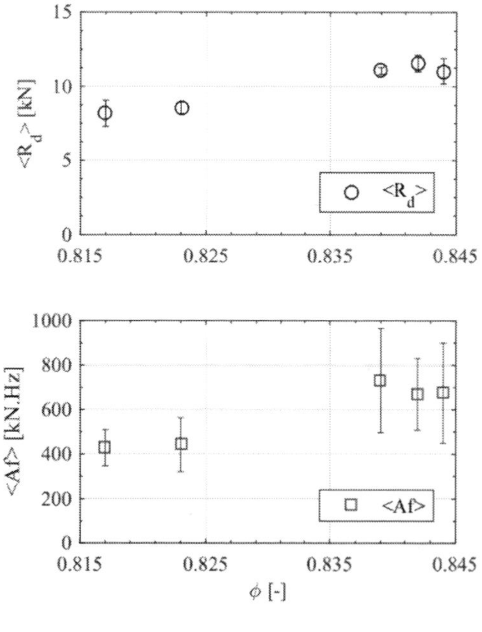

Figure 7. $< R_d >$ and $< Af >$ as function of volume fraction ϕ.

6 Conclusion

In this paper, we presented the influence of characteristics of granular media (μ, ϕ) on a numerical dynamic penetration test of Panda 3. For each characteristic material, the penetration tests are analyzed in terms of the average tip resistance $< R_d >$ in a classical analysis on the one hand, and on the other from the DFT analysis that allows us to go further on investigation of the load - penetration curves in terms of signal oscillation analysis.

The influence of the variation of friction becomes negligible on the tip resistance and shape of load – penetration curve when friction angle is greater than 45 degrees. Concerning the volume fraction, a clear variation in terms of tip resistance and amplitude spectrum was observed when ϕ varies from 0.817 to 0.844.

The analysis of the signal presented in this paper could be useful to extract more information from the penetrometer test results, about the physical properties of the material.

References

[1] L. Chaigneau, *Caractérisation des milieux granulaires de surface à l'aide d'un pénétromètre* (PhD thesis, Université Blaise Pascal-Clermont-Ferrand II, 2001)

[2] P. Breul, M. Benz, R. Gourvès, G. Saussine, Proc. 6th Int. Conf. Micromechanics of Granular Media, **1145(1)**,173-176 (2009)

[3] M. A. Benz-Navarrete, *Mesures dynamiques lors du battage du pénétromètre PANDA 2* (PhD thesis, Université Blaise Pascal-Clermont-Ferrand II, 2009)

[4] E. Escobar, *Mise au point et exploitation d'une nouvelle technique pour la reconnaisance des sols: le PANDA 3* (PhD thesis, Université Blaise Pascal-Clermont-Ferrand II, 2015)

[5] J. C. Quezada, P. Breul, G. Saussine, F. Radjai, Comput Geotech, **55**, 248-253 (2014)

[6] Q. A. Tran, *Modélisation numérique du comportement des milieux granulaires à partir de signaux pénétrométriques: approche micromécanique par la méthode des éléments discrets* (PhD thesis, Université Blaise Pascal-Clermont-Ferrand II, 2015)

[7] Q. A. Tran, B. Chevalier, P. Breul, Comput Geotech, **71**, 12-18 (2016)

[8] P. A. Cundall, O. D. L. Strack, Géotechnique, **29(1)**, 47-65 (1979).

Microstructural description of shear-thickening suspensions

Abhinendra Singh[1,*], *Jeffrey F.* Morris[1,2,**], and *Morton M.* Denn[1,2,***]

[1] *Benjamin Levich Institute, City College of New York, New York, NY 10031, USA.*
[2] *Department of Chemical Engineering, The City College of New York, New York, NY 10031.*

Abstract. Dynamic particle-scale numerical simulations are used to study the variation of microstructure with shear stress during shear thickening in dense non-Brownian suspensions. The microscale information is used to characterize the differences between the shear thickened (frictional) and non-thickened (lubricated, frictionless) states. Here, we focus on the force and contact networks and study the evolution of associated anisotropies with increase in shear stress. The force and contact networks are both more isotropic in the shear-thickened state than in non-thickened state. We also find that both force and structural anisotropies are rate independent for both low and high stress, while they are rate (or stress) dependent for the intermediate stress range where the shear thickening occurs. This behavior is similar to the evolution of viscosity with increasing stress, showing a clear correlation between the microstructure and the macroscopic rheology.

1 Introduction

Suspensions are dispersions of rigid particles in fluids. The fluid allows numerous possible interactions between the particles, but here we consider only hydrodynamic interactions and short-range or contact forces. Even with this simple system, a rich rheology is found, including shear thinning or shear thickening, normal stress differences, particle migration, shear jamming and yield stress behavior [1–3]. For moderate volume fractions, the viscosity increases with increase in shear rate (or shear stress) and is termed *continuous shear thickening* (CST). However, at high volume fraction the viscosity can increase by orders of magnitude at a critical shear rate in what is known as *discontinuous shear thickening* (DST) [1, 3].

The mechanism for DST has been of interest for a number of years, with an order-disorder transition [4] and hydrodynamic clustering [5] postulated. Recent numerical and experimental studies have linked DST to a transition between 'lubricated' low-viscosity and 'frictional' high viscosity states as the shear stress increases [1, 2, 6, 7]. The friction requires contact and this scenario thus requires an added repulsive force F_0 to stabilize the material in the lubricated regime. F_0 sets the stress scale F_0/a^2 for particles of size a for the transition from the lubricated to frictional behavior. We explore features of this lubricated-frictional transition through numerical simulations.

Suspension rheology is often discussed in the framework of shear-induced microstructure [2, 8]. For smoothly shearing suspensions, representative of the lubricated state, the pair distribution is strongly anisotropic, with accumulation of pairs near contact in the compressional quadrant [2, 8]. Other particulate systems, such as dry granular materials, show similar behavior; isotropically generated assemblies develop anisotropic structure under shear, which results in the finite shear strength of the material [9, 10]. Recent studies [1, 2] have shown that, when contact with friction is simulated, the contact network in the shear-thickened state is more isotropic than the one in the non-thickened state; this is seen especially in the normal stress response, as the normal stress differences grow through the DST transition, but less so than the isotropic particle normal stress (particle pressure) [1]. Here, we report the evolution of both force and contact network anisotropies, toward the goal of bridging between the microscopic and macroscopic scales in dense suspension flows.

2 Model system

We simulate a two-dimensional or monolayer suspension of non-Brownian spherical particles. We consider the motion to be inertialess, so that the equation of motion reduces to a force balance between hydrodynamic ($\mathbf{F}_H$) and contact ($\mathbf{F}_C$) forces:

$$0 = \mathbf{F}_H(\mathbf{r}, \mathbf{u}) + \mathbf{F}_C(\mathbf{r}), \tag{1}$$

where $\mathbf{u}$ and $\mathbf{r}$ are many-body position and velocity vectors ($\mathbf{u} \equiv \dot{\mathbf{r}}$). Rather than employing a repulsive force to set the stress scale for the transition to the frictional interactions, we embed this scale in the contact force as noted below.The hydrodynamic forces generally consist of two terms: Stokes drag on the particle due to its motion relative to the surrounding fluid, $-\boldsymbol{R}_{FU}\cdot(\mathbf{u} - \mathbf{u}^\infty)$, and resistance associated with the bulk straining motion, $\dot{\gamma}\boldsymbol{R}_{FE} : \hat{\boldsymbol{E}}^\infty$, where

* e-mail: asingh@ccny.cuny.edu
** e-mail: morris@ccny.cuny.edu
*** e-mail: denn@ccny.cuny.edu

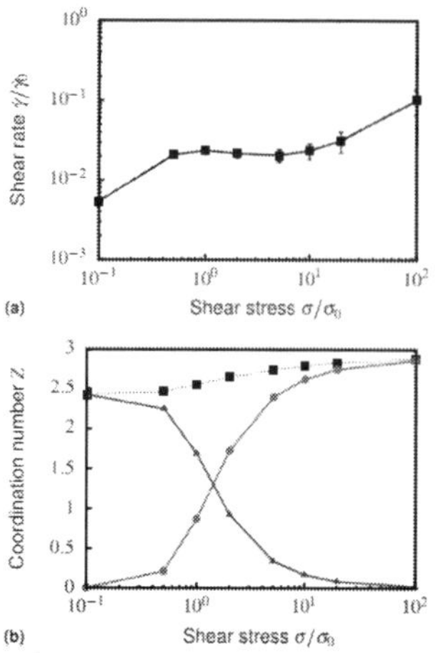

Figure 1. Macroscopic and microscopic characterization. (a) The macroscopic shear rate as a function of stress for $\phi_A = 0.78$ with $N = 2000$ particles. (b) Number of contacts as functions of stress. Squares show all contacts, triangles represent frictionless contacts , while circles represent frictional contacts.

$\dot{\gamma}\hat{\boldsymbol{E}}^{\infty}$ is the rate of strain (details are elsewhere [1, 2]). The hydrodynamic forces scale linearly with shear rate $\dot{\gamma}$, and thus their contribution to the relative viscosity, given by $\eta_0\eta_r = \sigma/\dot{\gamma}$, with σ the shear stress and η_0 the fluid viscosity, has no rate dependence for purely hydrodynamic interactions. Similarly, frictional interactions do not introduce a rate dependence. To incorporate rate dependence we employ a Critical Load Model (CLM) with an inter-particle threshold normal force F_{CL} used to activate friction between particles [1, 2]. This force scale sets the stress scale F_{CL}/a^2 for the transition between 'frictionless' interactions that take place via a liquid film and those making unlubricated 'frictional' contacts at high stress. CLM can be thought of as a small Debye length limit of the electrostatic repulsion model [2]. The interparticle contacts are defined by a linear spring model in both normal and tangential directions, as commonly used in granular physics [1, 2, 11]. The tangential force between two particles is subjected to Coulomb's friction law $|\mathbf{F}_{\mathrm{C}}^{t}| \leq \mu_p|\mathbf{F}_{\mathrm{C}}^{n}|$. In this study we use $\mu_p = 1.0$. The strain rate and stress are non-dimensionalized by $\dot{\gamma}_0 \equiv F_{\mathrm{CL}}/6\pi\eta_0 a^2$ and $\sigma_0 \equiv \eta_0\dot{\gamma}_0$.

3 Results and discussion

In Fig. 1a, the average steady state shear rate is shown as a function of the applied shear stress for an area fraction $\phi_{\mathrm{A}} = 0.78$. The simulations are performed at controlled shear stress σ, and the shear rate is thus variable [12]. The error bars represent the standard deviation taken over many samples in a simulation of typically $\dot{\gamma}t = 20$ strain units. At small and large shear stresses ($\sigma/\sigma_0 < 1$ and $\sigma/\sigma_0 \geq 20$), the shear rate is linear with σ, i.e. the suspension viscosity is rate independent, while it shows a non-monotonic S-shaped flow curve for the intermediate values of shear stress. A rate-controlled simulation at this volume fraction would display discontinuous shear thickening, because such S-shaped curves are not accessible under a rate-controlled simulation [6, 12]. Figure 1b shows the evolution of the direct particle-particle coordination number or number of contacts per particle $Z = N_{\mathrm{C}}/N$ as a function of stress. The number of frictional contacts increases with an increase in the shear stress, whereas the number of frictionless contacts decreases. For low stress all contacts are frictionless, but at high stress contacts are predominantly frictional, which is consistent with the evolution of frictional contacts previously reported [2]. The number of contacts in the shear-thickened state exceeds the frictionless interactions that were present in the non-thickened state.

We can gain more insight by examining the contact network anisotropy. There are several methods that have been used to quantify the structural anisotropy of a granular packing [9, 10]. One common method is to consider the probability distribution $P(\mathbf{n}_{ij})$ of the normal vectors $\mathbf{n}_{ij}$ oriented from particle i to particle j. In a two dimensional setting, the unit vector $\mathbf{n}$ can be described by its orientation angle θ, measured here relative to the flow direction. Now the anisotropy can be characterized by a probability density function $P(\theta)$ of finding a vector aligned along the direction θ. This probability density function is π-periodic, as the contact normal could be defined in either direction. A Fourier expansion with only even terms provides an exact representation of the geometrical anisotropy

$$P(\theta) = 1/2\pi\left\{1 + \sum_{n=1}^{\infty} a_{2n}\cos 2n\,(\theta - \theta_c)\right\}, \qquad (2)$$

where a_i (with $i = 2, 4, 6..$) are the anisotropy descriptors along principal direction θ_c. When $a_i = 0$, the distribution is isotropic and $P(\theta)$ is a constant equal to $1/2\pi$ (as indicated by the dashed line in Fig. 2a). Figure 2 displays a polar representation of $P(\theta)$ for various values of stress (only a few are shown here for the sake of clarity). Each of these probabilities is normalized by the respective number of interactions in each class. Before we discuss our results, we introduce the nomenclature used. We use 'contact' when particles touch each other (distance between their centers is less than the sum of their radii), and 'interaction' is used otherwise (hydrodynamic forces would fall here). The generalized network including both contacts and interactions (non-contact hydrodynamic lubrication and direct particle-particle contacts) is shown in Fig. 2a. The network is highly non-uniform, with no preferred orientation at low stress and tending to become isotropic with an increase in stress (for high stress the data almost collapse on the dashed circle). We observe at low stress a large accumulation of pairs aligned along the flow direction, consistent with examination of structures seen in sheared Brownian suspensions [13].

To gain better insight the general network is split into various classes, i.e. lubrication and direct particle-particle contacts (which are further divided into frictional and frictionless contacts). For low stress values, both hydrodynamic and particle-particle contacts have peaks in $P(\theta)$ along the flow direction (Figs. 2b and 2c), and as the stress

 DOI: 10.1051/epjconf/201714009023

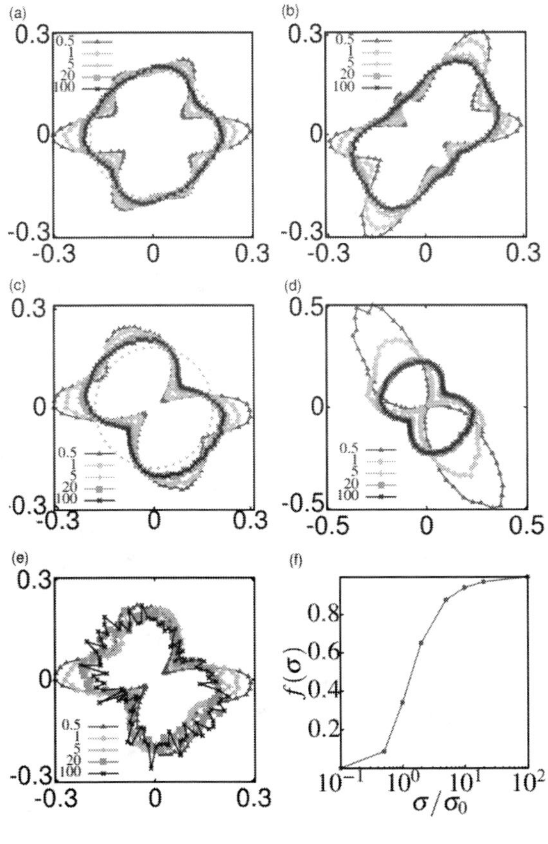

Figure 2. Polar representation of the probability density function $P(\theta)$ of the contact normal vectors for (a) generalized network, (b) only lubrication interaction, (c) all direct particle-particle contacts, (d) frictional contacts, and (e) frictionless contacts. Different symbols represent simulations with σ/σ_0 as shown in the legend. All the data shown are sampled in the steady state. (f) Fraction of frictional contacts $f(\sigma)$ as a function of shear stress. As can be seen from Fig. 1, $\sigma = 0.5$ corresponds to non-thickened state, while $\sigma \geq 20$ correspond to the thickened state, and $\sigma = 1, 5$ belong to the transitional state. In Figs. (a-e) the direction of shear is from right to left at the *top* and left to right at the *bottom*.

increases the magnitude of these peaks decreases. With increasing stress, the lubrication interactions develop the elongation axis as the dominant direction of anisotropy, while the direct particle-particle contacts are found to align along the compressional axis; i.e., the directions of anisotropy in the two types of interactions are orthogonal to each other. Eventually, at high enough values of the imposed stress, the contact network tends to become closer to isotropic.

We further categorize the direct particle-particle contacts as frictional and frictionless, as shown in Figs. 2d and 2e, respectively. As these polar plots are normalized by the total number of interactions in each class, Figs. 2d and 2e should be analyzed together with Fig. 2f, where we show the fraction of frictional contacts $f = N_C^f/N_C$ (this is the

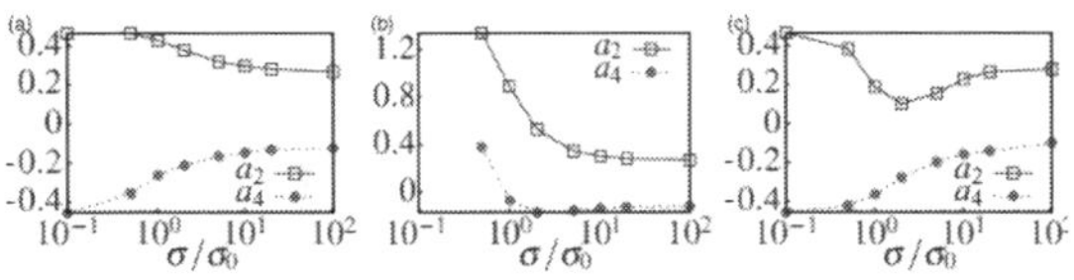

Figure 3. Contact anisotropy descriptors a_2 and a_4 as functions of stress for (a) all direct particle-particle contacts, (b) frictional contacts, and (c) frictionless contacts. All the data shown are sampled in the steady state.

ratio between contacts for which the normal force exceeds the critical load force F_{CL} and total number of particle-particle contacts). The frictional contacts are rare at low stress and are concentrated in small force chains along the compression direction, whereas the frictionless contacts behave similarly to the lubrication interactions (being aligned along the flow direction). As the stress increases, the number of frictional contacts increases (Fig. 2f), and at the same time these contacts tend to become more isotropically distributed. Figure 3 shows the contact anisotropy descriptors a_2 and a_4 as functions of shear stress for all, frictional and frictionless contacts.

Opposite signs of a_2 and a_4 signify the orthogonal anisotropic directions (a_2 captures $P(\theta)$ along the compression direction, a_4 along the elongational axis). Interestingly, a_4 is large for low stress and diminishes with increasing stress, signifying it is required to capture the non-uniform and highly anisotropic distributions at low stress as seen in Figs. 2c, 2e, and 2f. We observe that the contact anisotropy (for all and frictionless contacts) is almost a constant for low stress $\sigma < 1$ (as the number of frictional contacts for $\sigma = 0.1$ is almost zero, we do not have $P(\theta)$ for that case). For all the cases shown, the contact anisotropy decreases with increasing stress and once again saturates for high stress $\sigma > 20$ the magnitudes of both a_2 and a_4 decrease and tend to saturate to constant values with increase in stress). θ_c for all types of contacts nearly coincide and show a slight decrease $\theta_c \in [135, 130]$ (in degrees) with increasing stress.

Next we examine the anisotropy of the force network in the non-thickened (frictionless) and thickened (frictional) states. The angular distribution of normal contact forces between particles can be represented by average force $\langle f_n \rangle(\mathbf{n})$ as a function of contact normal direction $\mathbf{n}$. This has been previously defined by [9] as $\langle f_n \rangle(\theta) = \dfrac{1}{N_c(\theta)} \sum_{c \in S(\theta)} f_n^c$, where f_n^c represents the normal contact force acting at the contact c and $N_c(\theta)$ represents the number of contacts in $S(\theta)$ (set of contacts with direction θ). Similar to the contact normal vector, the normal force could also be defined in either direction, hence the probability density is π-periodic. In the steady state, $\langle f_n \rangle(\theta)$ can be described by a similar function as used for contact orientation distribution in Eq. (2):

$$\langle f_n \rangle(\theta) = f_0 \left\{ 1 + \sum_{n=1}^{\infty} b_{2n} \cos 2n \left(\theta - \theta_f \right) \right\}, \quad (3)$$

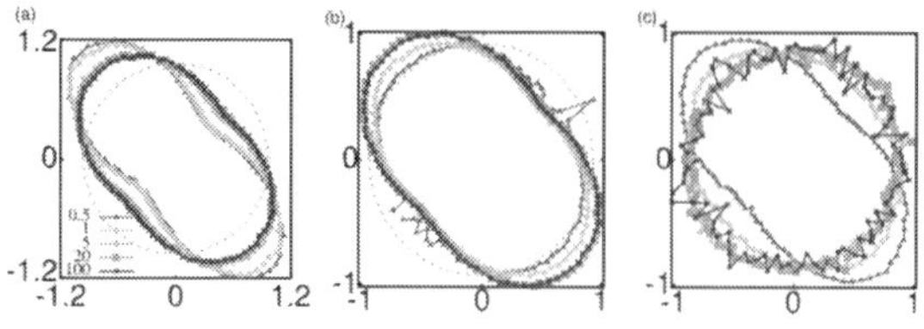

Figure 4. Polar representation of the angle-averaged normal forces $\langle f_n(\theta)\rangle$ for (a) all, (b) frictional, and (c) frictionless direct particle-particle contacts for various stress values. All the data shown are sampled in the steady state.

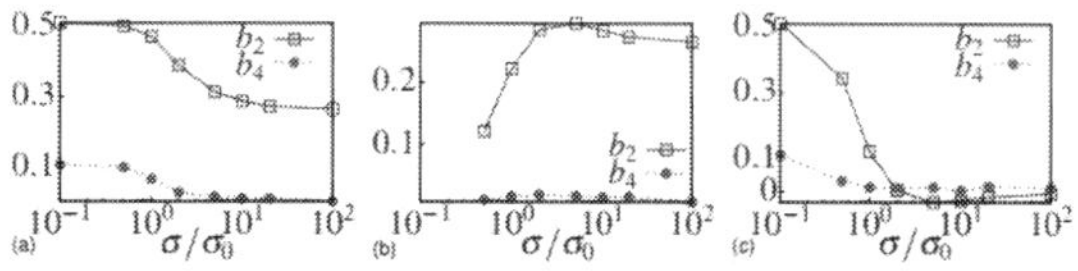

Figure 5. Normal force anisotropy descriptors b_2 and b_4 as functions of stress for (a) all direct particle-particle contacts, (b) frictional contacts, and (c) frictionless contacts. All the data shown is sampled in the steady state.

where f_0 is the mean of the magnitude of normal contact forces, b_i (with $i = 2, 4, 6..$) are the anisotropy descriptors of the normal force network, and θ_f denotes the direction of the maximum average normal force. In Fig. 4, $\langle f_n\rangle(\theta)$ (for only direct particle-particle contact forces) is displayed in polar coordinates. Once again, these distributions are normalized by the mean force for each class of contacts. We observe that all particle-particle contacts display pronounced force anisotropy, unlike $P(\theta)$ (as shown in Figs 2c). The force polar plots do not show any alignment of force chains along the flow direction as previously seen in the case of contact normal polar representation. We also observe that the overall force network tends to become less anisotropic with increasing stress. The polar representations of frictional and frictionless normal forces exhibit highly contrasting features. The anisotropy in the frictional force network increases with increase in the applied shear stress. On the other hand, the force network of frictionless contacts is initially anisotropic and the anisotropy decreases with increase in stress. Figure 5 shows the evolution of force anisotropy parameters b_2 and b_4 with the applied shear stress σ. The overall normal force network is initially anisotropic and its anisotropy decreases and tends toward constant values of b_2 and b_4 with increasing stress. The force anisotropy of frictionless normal forces decreases to zero with σ. At the same time, the anisotropy of frictional normal forces increases sharply initially and then saturates as a function of σ. θ_f for the three types of contacts are nearly identical ($\theta_f \simeq 3\pi/4$) and are independent of stress.

4 Conclusions

We have studied shear thickening in dense suspensions by considering the microstructure and its correlation to the bulk rheology. Estimation of anisotropy parameters revealed that the anisotropy of both contact and force networks plateau for low and high stress values and show a rate dependence for intermediate stress values, much as the viscosity does. This shows that the rate dependence in the viscosity can be attributed to both increase in the number of frictional contacts as well as the local rearrangements in contact and force networks.

Acknowledgements

This work was supported, in part, under National Science Foundation Grants CNS-0958379, CNS- 0855217, ACI-1126113 and the City University of New York High Performance Computing Center at the College of Staten Island. JFM was supported by NSF 1605283.

References

[1] R. Seto, R. Mari, J.F. Morris, M.M. Denn, Phys. Rev. Lett. **111**, 218301 (2013)
[2] R. Mari, R. Seto, J.F. Morris, M.M. Denn, J. Rheol. **58**, 1693 (2014)
[3] E. Brown, H.M. Jaeger, Rep. Prog. Phys. **77**, 046602 (2014)
[4] R.L. Hoffman, J. Rheol. **42**, 111 (1998)
[5] N.J. Wagner, J.F. Brady, Phys. Today **62**, 27 (2009)
[6] M. Wyart, M.E. Cates, Phys. Rev. Lett. **112**, 098302 (2014)
[7] B.M. Guy, M. Hermes, W.C.K. Poon, Phys. Rev. Lett. **115**, 088304 (2015)
[8] J.F. Morris, Rheol. Acta **48**, 909 (2009)
[9] E. Azéma, F. Radjai, R. Peyroux, G. Saussine, Phys. Rev. E **76**, 011301 (2007)
[10] A. Singh, V. Magnanimo, K. Saitoh, S. Luding, New J. Phys. **17**, 043028 (2015)
[11] S. Luding, Gran. Matt. **10**, 235 (2008)
[12] R. Mari, R. Seto, J.F. Morris, M.M. Denn, Phys. Rev. E **91**, 052302 (2015)
[13] J.F. Morris, B. Katyal, Phys. Fluids **14**, 1920 (2002)

The 4-parameter Compressible Packing Model (CPM) including a critical cavity size ratio

Gerard Roquier[1,*]

[1]Laboratoire Navier, 2 allée Kepler, Cité Descartes, 77420 Champs-sur-Marne (France).
[*]gerard.roquier@laposte.net

Abstract. The 4-parameter Compressible Packing Model (CPM) has been developed to predict the packing density of mixtures constituted by bidisperse spherical particles. The four parameters are: the wall effect and the loosening effect coefficients, the compaction index and a critical cavity size ratio. The two geometrical interactions have been studied theoretically on the basis of a spherical cell centered on a secondary class bead. For the loosening effect, a critical cavity size ratio, below which a fine particle can be inserted into a small cavity created by touching coarser particles, is introduced. This is the only parameter which requires adaptation to extend the model to other types of particles. The 4-parameter CPM demonstrates its efficiency on frictionless glass beads (300 values), spherical particles numerically simulated (20 values), round natural particles (125 values) and crushed particles (335 values) with correlation coefficients equal to respectively 99.0%, 98.7%, 97.8%, 96.4% and mean deviations equal to respectively 0.007, 0.006, 0.007, 0.010.

1 Introduction

The packing density is often an optimized physical value for high performing and sustainable materials. Packing density models call upon to geometrical interaction coefficient functions of the size ratio: the *wall effect b* and the *loosening effect a* [1-7] to which can be added [2-3, 5] a *wedging effect c*. Because the packing density depends on the packing process efficiency and on the particle shape, two strategies can be adopted.

The first strategy [2-3, 5] consists in taking that into account in the mathematical expressions used for the interaction coefficients. It means that an adjustment by regression analysis is each time necessary to find the best functions of these parameters. Three sets have been proposed by the authors: the first one for the spherical particles, the second and the third ones for respectively the compacted and uncompacted angular particles. Recently, the interaction function for the *wedging effect c* has been expressed in two separated parts [5].

The second strategy [4, 7] consists in taking into account of the packing process efficiency by a *compaction index K* and of the particle shape and of the surface texture through a *critical cavity size ratio* x_0. Below this value, a small particle can be inserted into a small cavity created by touching coarser particles without disturbed their arrangement. Beyond this value, the local volume of the coarse class decreases: a *loosening effect* occurs. Thus, the *4-parameter Compressible Packing Model* (*CPM*) [4, 7] has been developped by incorporating an original theory about the *wall effect b* and the *loosening effect a*. The 4 parameters are: *a*, *b*, *K* and x_0. Even if the expressions of *a* and *b* are more sophisticated, the adjustment of *K* and x_0 has proven effective for spherical particles for which x_0 = 0.2, in harmony with the tetrahedral cavern theory ($x_0 \approx 0.224$). It remains to be able to adjust easily x_0 for other types of particles, especially angular ones.

2 Wall effect, loosening effect, interference effect

Fig. 1 represents the "void ratio" (e=1/ϕ*-1) of a binary mixture *vs* the "volume fraction of fine particles". In the *"coarse dominant"* field, the *insertion mechanism* is represented by AM and, in the *"fine dominant"* field, the *substitution mechanism* by FM. The *wall effect* is highlighted by FH which is located above FO. In the AK section, the *loosening effect* is only localized around a particle insufficiently fine to insert into a cavity created by some touching coarser particles. The more global *interference effect* is materialized by the AKJI section.

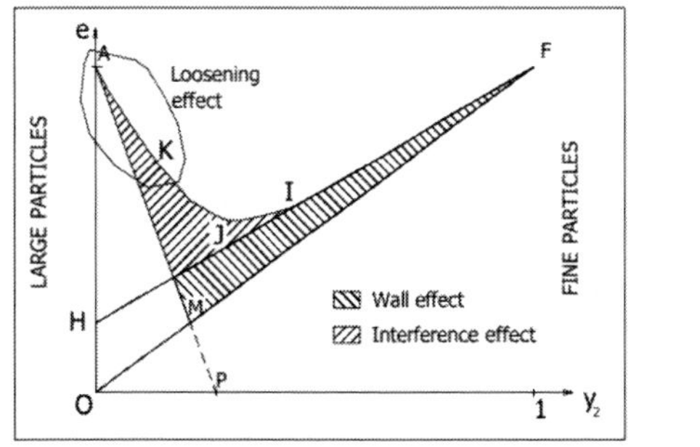

Fig. 1. Illustration of wall effect, loosening effect, interference effect.

3 The 4-parameter CPM

The reference frame is constituted by the CPM [1]. The real packing density corresponds to that obtained for a randomly dense packing. It depends on the process used to fill and to compact the material inside the mould. If we consider a perfect placing process where each particle is placed one by one in its ideal location, the *compaction index K* tends to infinity and the packing density reaches the virtual packing density. For a binary mixture "with geometrical interactions", the virtual packing density γ obtained by the CPM can be written:

$$\gamma = Min\left(\gamma_1 = \frac{\beta_1}{1-\left(1-\frac{\beta_1}{\beta_2}a_{12}\right)y_2}\,;\gamma_2 = \frac{\beta_2}{1-\left(1-\beta_2+b_{21}\beta_2\left(1-\frac{1}{\beta_1}\right)\right)y_1}\right) \quad (1)$$

where γ_1 and γ_2 are respectively the virtual packing densities of the binary mixture in the *"coarse dominant"* field and in the *"fine dominant"* field, β_1 and β_2 the virtual packing densities of the coarse size class 1 and of the fine size class 2, y_1 and y_2 their volume fractions by reference of the total solid volume, a_{12} the *loosening effect* coefficient, b_{21} the *wall effect* coefficient. When the size ratio $x=d_2/d_1$ (fine/coarse) is equal to 1, a total interaction occurs:

$$b_{21}(1)=1 \text{ and } a_{12}(1)=1 \quad (2)$$

The real packing density ϕ^* is calculated by introducing the *compaction index K*. The expression (3) is an implicit equation of ϕ^*, with one and only one positive root.

$$K = \frac{\frac{y_1}{\beta_1}}{\frac{1}{\phi^*}-\frac{1}{\gamma_1}} + \frac{\frac{y_2}{\beta_2}}{\frac{1}{\phi^*}-\frac{1}{\gamma_2}} \quad (3)$$

3.1. Wall effect theory for spheres

The *wall effect* is the further amount of voids in the packing of small particles against the wall of a coarser one. The disturbance is studied in a portion delimited by two concentric spheres. The first one has a diameter d_1. The second hypothetical one is chosen in such a way that the boundary condition $b_{21}(1)=1$ is respected. That's why we call this theory: ROAD (ROund ADjustable cell) theory.

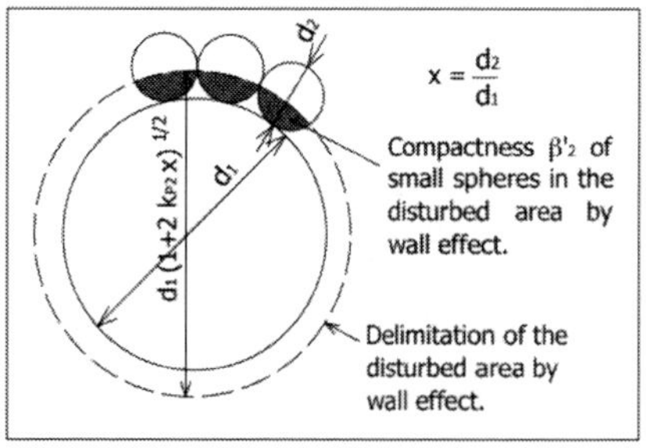

Fig. 2. Definition of the spherical reference cell for studying the wall effect.

The diameter of the sphere delimiting the disturbed volume by the *wall effect* around a coarse sphere is:

$$d_{hyp} = d_1\sqrt{1+2k_{P2}\,x} \quad (4)$$

Let α_2 be the real packing density of the fine class 2. We can deduce its virtual packing density β_2 by involving *K*:

$$\beta_2 = \alpha_2\frac{(1+K)}{K} \quad (5)$$

The number of fine spheres against a coarse one is calculated from the Spherical Square Model (SSM), not presented here, as part of the dense virtual packings:

$$N_{12,SSM}^{dense}(x) = \frac{\pi(1+x)}{x\arcsin\left(\frac{x}{1+x}\right)} \quad (6)$$

The packing density β'_2 of the fine particles in the disturbed volume by the *wall effect* is then deduced:

$$\beta'_2(x) = \frac{\pi(1+x)}{4x\left((1+2k_{P2}x)^{\frac{3}{2}}-1\right)\arcsin\left(\frac{x}{1+x}\right)}\times$$
$$\times\left(2(1+2k_{P2}x)^{\frac{3}{2}}-3(1+2k_{P2}x)\left(1+\frac{k_{P2}x}{1+x}\right)+\left(1+\frac{k_{P2}x}{1+x}\right)^3-x^3+\frac{3k_{P2}x^3}{1+x}+x^3\left(1-\frac{k_{P2}}{1+x}\right)^3\right) \quad (7)$$

In the case where $\beta_1=\beta_2$, the *wall effect* coefficient is:

$$b_{21}(x) = \frac{(\beta_2-\beta'_2(x))}{(1-\beta_2)}\left[(1+2k_{P2}x)^{\frac{3}{2}}-1\right] \quad (8)$$

It remains to be determined k_{P2} by respecting the boundary condition $b_{21}(1)=1$, leading to the numerical solution, of the following equation:

$$(\beta_2-6)(1+2k_{P2})^{\frac{3}{2}}+\frac{9}{2}k_{P2}^2+18k_{P2}+5=0 \quad (9)$$

The equation to be solved being of the 3rd degree, the value of k_{P2} which is coherent with those presented in the following table should be kept:

Table 1. Examples of values of k_P as a function of the virtual packing density β.

β	0.65	0.68	0.70	0.72	0.734
k_P	0.4466	0.7543	0.9253	1.1476	1.4729

An example of the *wall effect* coefficient b_{21} as a function of the size ratio x is given Fig. 3. The results from the ROAD theory have been achieved for a *compaction index K*=9 corresponding to a packing process *"vibration + compression"* (Table 2) and a real packing density of the fine class 2 α_2=0.641 corresponding to the *Random Close Packing RCP* state.

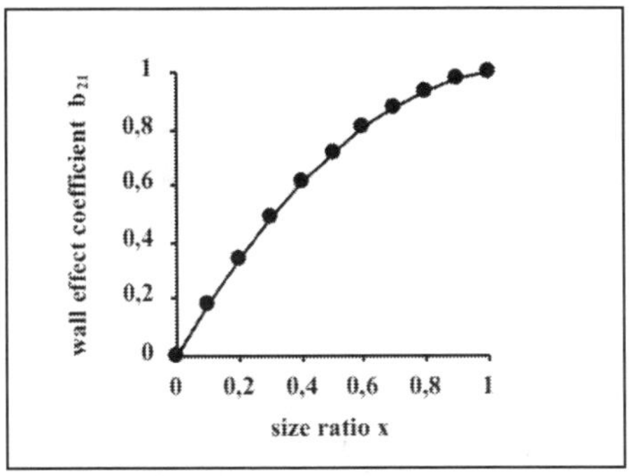

Fig. 3. ROAD theory: α_2=0.641 – Vibration + compression.

3.2 Loosening effect and interference effect theory for spheres

The concept of *critical cavity size ratio* x_0 is now introduced: below this value, the intrusion of small quantities of finer spheres does not disturb the bed of coarser spheres; beyond this value, the fine spheres cannot be placed without disturbing this one. The concept of *local isotropic expansion* of the coarse particle skeleton around a small one is adopted when $x>x_0$. To be consistent with the *wall effect* theory, a spherical reference cell is used.

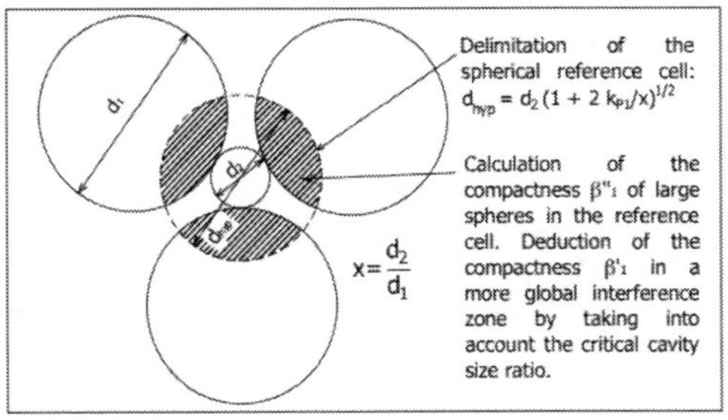

Fig. 4. Definition of the spherical reference cell for studying the loosening effect.

The diameter of the sphere delimiting the disturbed volume by the *loosening effect* is:

$$d_{hyp} = d_2 \sqrt{1 + \frac{2\,k_{P1}}{x}} \tag{10}$$

Let α_1 be the real packing density of the large class 1. We can deduce its virtual packing density β_1:

$$\beta_1 = \alpha_1 \frac{(1+K)}{K} \tag{11}$$

The number of coarse spheres against a fine one is calculated from the Spherical Square Model (SSM):

$$N^{dense}_{21,SSM}(x) = \frac{\pi(1+x)}{\arcsin\left(\frac{1}{1+x}\right)} \tag{12}$$

The packing density β''_1 of the coarse spheres in the reference cell is:

$$\beta''_1(x) = \frac{\pi(1+x)}{4\left(1+\frac{2k_{P1}}{x}\right)^{\frac{3}{2}} \arcsin\left(\frac{1}{1+x}\right)} \times$$
$$\times\left(2\left(1+\frac{2k_{P1}}{x}\right)^{\frac{3}{2}} - 3\left(1+\frac{2k_{P1}}{x}\right)\left(1+\frac{k_{P1}}{1+x}\right) + \left(1+\frac{k_{P1}}{1+x}\right)^3 - \frac{1}{x^3} + \frac{3k_{P1}}{x^2(1+x)} + \left(\frac{1}{x} - \frac{k_{P1}}{1+x}\right)^3\right) \tag{13}$$

The packing density of the large spheres in the more global interference zone is:

$$\beta'_1(x) = \frac{\beta''_1(x)}{\beta''_1(x_0)\left(1+\frac{(x-x_0)}{(1-x_0)}\left(\sqrt[3]{\frac{2\,\beta''_1(1)}{\beta''_1(x_0)}} - 1\right)\right)^3}\,\beta_1 \tag{14}$$

When $x=x_0$, $\beta'_1(x_0)/\beta_1=1$: *the loosening effect* and the *interference effect* do not occur. When $x=1$, $\beta'_1(1)/\beta_1=0.5$: the continuity between the *"coarse dominant"* field and the *"fine dominant"* field is provided when the volume fractions of each class are equal and when all particles become identical. The volume fraction of the fine particles at the *"eutectic"* point, corresponding to the crossing point into the *"fine dominant"* field, is:

$$\phi^*_2(x) = \beta_2 + \left((1-\beta_2)\left(1-b_{21}(x)\right) - 1\right)\beta'_1(x) \tag{15}$$

The *loosening effect* coefficient is deduced from this expression:

$$a_{12}(x) = \frac{\beta_1 - \beta'_1(x)}{\phi^*_2(x)} \; if \; x \geq x_0 \; and \; a_{12}(x) = 0 \; if \; x \leq x_0 \tag{16}$$

k_{P1} remains to be determined by respecting the boundary condition $a_{12}(1)=1$ leading to solve the equation (9) by replacing k_{P2} by k_{P1} and β_2 by β_1.

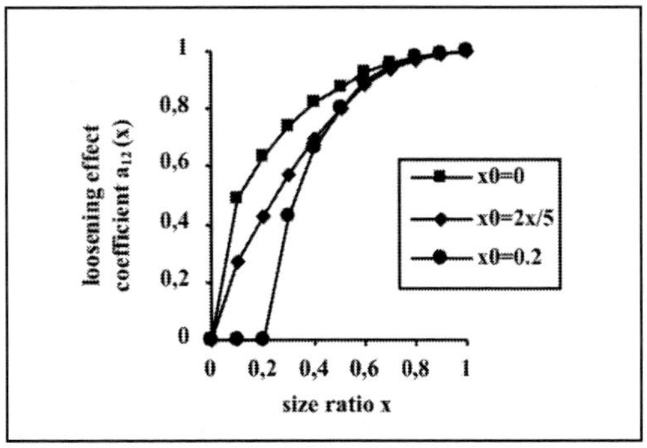

Fig. 5. ROAD theory: α_1=0.641 – Vibration + compression.

3.3 Compaction index K and critical cavity size ratio x_0

K (Table 2) and x_0 (Table 3) need now to be calibrated by an analysis on 780 values on binary mixtures.

x_0 depends on the shape and on the finish surface of the materials. For equivalent sizes, the *loosening effect* created by a small particle in a matrix of coarse ones is more important when this particle is irregularly shaped.

Table 2. Compaction index K.

Packing process	Pouring	Vibration or compaction	Vibration + compression	Optimized filling + vibration + compression
K	4.7	5.6	9	15

Table 3. Critical cavity size ratios x_0 (x: size ratio).

Type of particles	Crushed particles			Round natural particles	Frictionless glass beads
	Very Angular / very rough	Angular / rough	Angular		
x_0	**0**	**0.02**	**0.1**	$x_0=2x/5$	**0.2**

The *4-parameter CPM* has been evaluated [7] by comparison with 300 values of packing densities on frictionless glass beads, 20 on spherical particles numerically simulated, 125 on round natural particles and 335 on crushed particles (Table 4).

Table 4. Mean deviation ξ and correlation coefficient r. Comparison between different models.

Models / Materials (780 values)	**Original CPM [1]**		**3PPM [2, 3]**		**4-parameter CPM [4, 7]**	
	ξ	r	ξ	r	ξ	r
Glass beads	0.012	0.9754	0.009	0.9863	***0.007***	***0.9904***
Simulation	0.012	0.8783	0.008	0.9598	***0.006***	***0.9877***
Round natural	0.009	0.9619	0.012	0.9534	***0.007***	***0.9788***
Crushed	0.013	0.9408	0.013	0.9455	***0.010***	***0.9642***

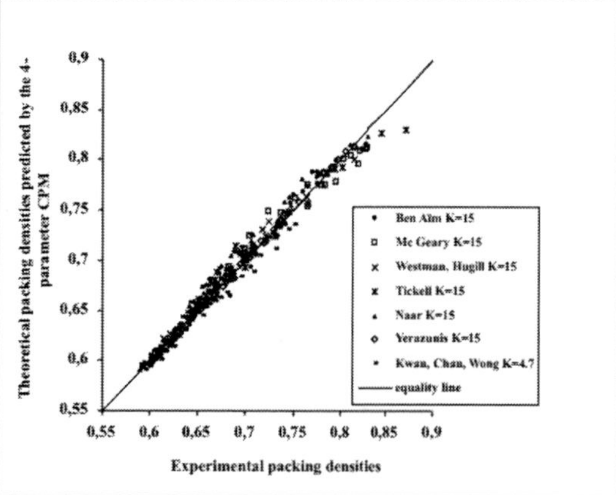

Fig. 6. Theoretical predictions by the 4-parameter CPM for frictionless glass beads with x_0=0.2.

The *4-parameter CPM* has been compared with the *original CPM* [1] and with the *3-parameter particle packing model* (3PPM) [2, 3] (Table 4). It provides reliable predictions, as well as for low and high packing densities, especially at the optimum.

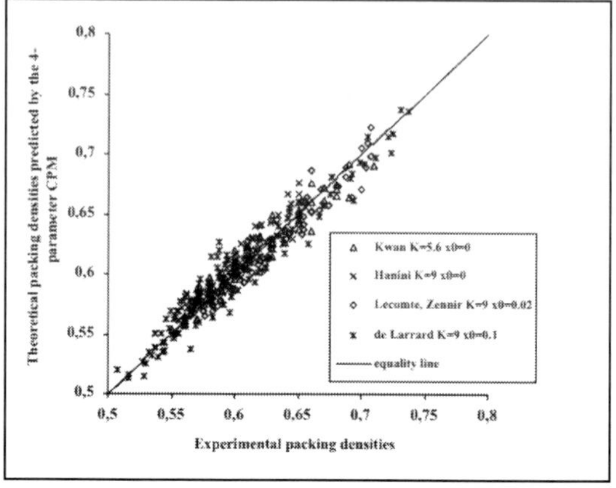

Fig. 7. Theoretical predictions by the 4-parameter CPM for crushed particles.

4 Conclusion

The *4-parameter CPM* offers a new theory on the *wall effect* and on the *loosening effect* studied by considering elementary juxtaposed cells. The effect of a coarse or a fine particle is then examined on the basis of a foreign sphere surrounded by dominant class neighbours. It includes the introduction of a *critical cavity size ratio* depending on the type of materials. This reveals that there would not be a single curve for the geometrical interactions. The limitations of the proposed model are of two types. Firstly, the interaction coefficients are determined by considering the same virtual packing density for each granular class. Secondly, it remains to characterize the shape, the angularity and the surface texture of the particles by reference to test methods. However, the results obtained suggest that efforts will not have been in vain.

References

1. F. de Larrard, *Concrete mixture proportioning: a scientific approach*, E & FN Spon, London (1999)
2. A.K.H. Kwan, K.W. Chan, V. Wong, Powder Technol. **237** (2013) 172-179
3. A.K.H. Kwan, V. Wong, W.W.S. Fung, Powder Technol. **274** (2015) 154-162
4. G. Roquier, Powder Technol. **302** (2016) 247-253
5. O. Koutný, J. Kratochvíl, J. Švec, J. Bednárek , Procedia Eng. **151** (2016) 198-205
6. Y. Knop, A. Peled, Constr. Build. Mater. **102** (2016) 44-50
7. G. Roquier, *Etude de la compacité optimale des mélanges granulaires binaires: classe granulaire dominante, effet de paroi, effet de desserrement*, Thèse soutenue à l'Université Paris-Est le 15 Fevrier 2016

Vibrational Collapse of Hexapod Packings

Yuchen Zhao[1,*], *Jingqiu* Ding[2], *Jonathan* Barés[3], *Hu* Zheng[1], *Karola* Dierichs[4], *Achim* Menges[4], and *Robert* Behringer[1]

[1] *Department of Physics and Center for Nonlinear and Complex Systems, Duke University, Durham, NC, USA.*
[2] *Department of Physics, Nanjing University, Nanjing, China.*
[3] *Université de Montpellier, Montpellier, France.*
[4] *Institute for Computational Design, University of Stuttgart, Stuttgart, Germany*

Abstract. Columns made of convex noncohesive grains like sand collapse after being released from a confining container. However, structures built from non-convex grains can be stable without external support. In the current experiments, we investigate the effect of vibration on destroying such columns. The change of column height during vertical vibration, can be well characterized by stretched exponential relaxation when the column is short, which is in agreement with previous work, while a faster collapse happens when the column is tall. We investigate the collapse after the fast process including its dependence on column geometry, and on interparticle and basal friction.

1 Introduction

Jamming of granular matter is a wide-spread phenomenon [1, 2]. One can see piles of sand and rocks, packing of apples and oranges in grocery stores. The jamming of soft, frictionless spheres has been studied in a detail [2], and yet new ways of jamming emerge constantly [3, 4]. One of the recent explorations is the jamming of non-convex particles [5–11]. In a typical experiment, if one prepares a packing of the particles in a vertical tube and then removes the tube confinement, the cylindrical packing can retain its shape without confinement, and possibly sustain compression and shear [10]. This new phenomenon is not only interesting from the packing and jamming theory point of view [12–15], but also has significant implications, such as material and architectural design [16–18]. Open questions include: what are the physics that causes this jamming and how can one quantitatively describe the effect of particle shapes on the ability to jam and design new materials for applications based on particle level properties [19]. In particular, the work by Gravish *et. al.* [8] on staple packings is relevant here. Their experiment involves collapse of a column under external vibration. During collapse, the centroid height of the column follows a stretched exponential relation: $h(t) = h(0)\exp[-(t/\tau)^{\beta}]$ [20]. Further, the entanglement between particles was quantified. It is interesting to ask whether the packing of other extreme shaped particles has a similar response to vibration.

Here we studied the collapse of columns made with hexapod (star-shaped) particles under vibration of the base supporting the column (Fig.1). In the first part of the paper, we present our experimental setup, preparation and data characterizing the collapse of hexapod columns. In the second part, we confirm the applicability of a stretched exponential relation for collapse of hexapod columns under vibration. Although a stretched exponential describes the collapse dynamics where the column height is low enough, we observe deviation from this relation when the initial centroid height of the column is large enough. We also studied the effect of interparticle and basal friction on the stretched exponential relation. The final section provides a discussion and conclusions.

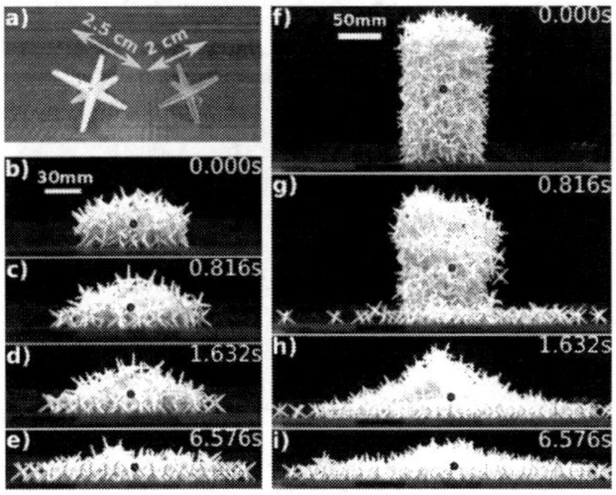

Figure 1. (a) Nylon (left) and acrylic (right) particles and snapshots from two trials ((b)-(e) and (f)-(i)) in $\mathcal{N}1$. The centroid height of the packing $h(t)$ are measured and marked as a red dot in each image. (b)-(e): collapse of a column with $h(0) = 11.6$ mm. (f)-(i): collapse of a column with $h(0) = 71.4$ mm. Both columns were prepared with $D = 102$ mm and sat on a Teflon sheet base with low friction coefficient. A sinusoidal oscillation with frequency $f = 58.6 \pm 0.3$ Hz and $\Gamma = 2.0 \pm 0.2$ was applied.

* e-mail: yz172@phy.duke.edu

DOI: 10.1051/epjconf/201714006011

2 Experimental Set-up

We used two kinds of hexapods (Fig.1 (a)), which are the same as those used in [10]. Each has the same hexapod shape: six equal-length arms with square cross-section lying symmetrically in the xyz direction. One kind is made of cast acrylic, which has a low friction coefficient, 0.4 ± 0.1, between any pair of particle arms. The other is made of laser-sintered white nylon PA2200, and has friction coefficient 1.0 ± 0.3. The arm tip-to-tip length is $s = 20$ mm and $s = 25$ mm for the acrylic and nylon hexapods, respectively. In the following we refer to the acrylic particles as low friction (LF) particles, and the nylon high friction (HF). Columns were prepared by first randomly pouring particles into a straight PVC tube (inner diameter D) from 40 cm above the base. The process was the same as [10]. Each tube was 30 cm tall. Then the tube was lifted vertically without disturbing the packing. The tube rested on either a Teflon surface, which had a friction coefficient 0.3 ± 0.1 between particles and the surface, or a sand paper surface, which had a higher friction coefficient such that particles only rolled when we tilted the surface. We refer to the Teflon and sand paper as low friction (LF) and high friction (HF) bases.

We prepared columns with different diameters D and initial heights, which were chosen to be around the critical height h_c in our previous results [10]. Columns were placed on a vibration plate and we applied sinusoidal vibrations to induce collapse (See Fig.2). The sinusoidal vibration had frequency $f = 58.6 \pm 0.3$ Hz and dimensionless vibration strength $\Gamma = A\omega^2/g = 2.0 \pm 0.2$, which were kept unchanged in all runs. On top of the plate, we glued either a Teflon sheet or sand paper as described earlier.

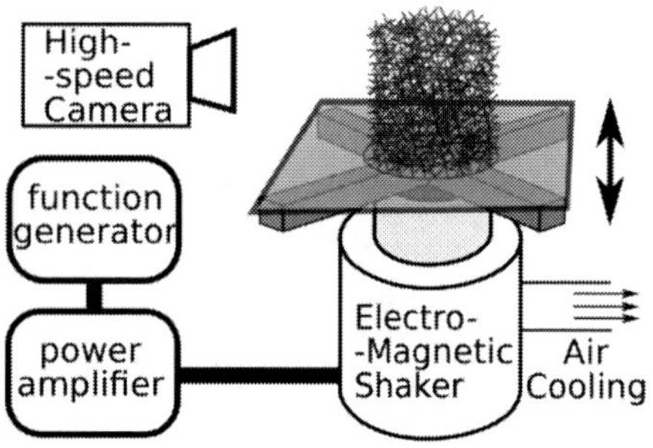

Figure 2. Schematic of the setup: The vibration generator consisted of an electromagnetic shaker that drove a 45 cm-by-45 cm square plate, that was screwed to a pair of square, aluminum tubes that strengthen the base. The driving sinusoidal signal was generated by a function generator, which was then amplified by a power amplifier. The shaker was cooled by air. We measured the vibration amplitude by imaging the vibrating plate via a high-speed camera (Photron FASTCAM SA5). The first peak of the vibration was large relative to later vibration, but contributed insignificantly to the packing compared to the total amount of energy from the vibration.

All the collapse dynamics were captured by high speed video imaging at 250 frame/s on the LF base and 60 frame/s on the HF base. Homemade Python codes were used to post-process the images and the centroid height $h(t)$ of the packing in the images was found (Fig.1). Because the packings were almost axially symmetric, the centroid height was close to the center of mass of the packing. We summarized the experiments in Table.1.

3 Results

3.1 Relaxation dynamics

In Fig.1, we present several representative snapshots during column collapses on the LF base. Fig.1 (b) to (e) describes the collapse process of an initially short column relative to its width. Particles on the edge of the column fell off within the first second, and formed a cone shape (Fig.1(c)). Then, the cone shortened and widened slowly until there was almost a single layer of particles on the vibrating plate. Fig.1(f)-(i) shows that a column with large $h(0)$ collapsed differently. It took about 1.6 second for the column to gradually shorten and widen as in Fig.1(f) and (g). After a cone was formed, the packing collapsed in a similar way as the initially short column. We also observed this difference with LF hexapod, although the first of the two-step process happened faster than HF, tall columns (data not shown).

In Fig.3, we present the collapse dynamics of short columns and test the assumption of a stretched exponential relation for both HF and LF hexapods, and two different column diameters. By plotting the dimensionless centroid height $h(t)/h(0)$ as $-\ln[h(t)/h(0)]$ vs. t. on a log-log scale, the assumption $h(t) = h(0)\exp[-(t/\tau)^\beta]$ would be shown as a straight line. We found these collapses are well approximated by stretched exponentials. The exponent β is between 0.5 and 1, which is similar to previous work by Gravish *et. al.* [8]. However, the dependence of β on initial diameter or particle friction properties are not clear here. Pictorially, these collapse processes resemble Fig.1(b)-(e). For HF hexapods, the collapse dynamics from two different diameters overlap within errorbars. The situation is not clear for LF hexapods. The characteristic relaxation times are $\tau = 3.0$ s and $\tau = 1.3$ s for HF and LF hexapods, respectively. These results agree with our previous study [10] showing that columns made of HF hexapod are more stable than LF ones.

3.2 Deviation from relaxation dynamics

As we increased $h(0)$, however, we observed that the dynamics of $h(t)/h(0)$ significantly deviated from a stretched exponential relation. As in Fig.4(a), the data are well fitted to a straight line for low $h(0)$, but gradually bend near time T_b as the initial centroid height increases, and the collapse dynamics depends on $h(0)$. Before T_b, collapse is relatively fast, corresponding to a massive falling of those particles which constitute the top or middle of the column, while the bottom of the column retains its original shape (Fig.1(g)). After T_b, particles formed a cone-shaped pile and then slowly relaxed under vibration (Fig.1(h,i)) as for

Table 1. Parameters used for vibration-collapse experiments: particle and base friction properties, column geometry.

$\mathcal{N}$	Hexapod	Bases	D (mm)	$h(t=0)$ (mm)	D/s	$h(t=0)/s$
1	HF	LF	102, 130	$12 \to 72$	4.1, 5.2	$0.48 \to 2.9$
2	LF	LF	102, 130	$7 \to 62$	5.1, 6.5	$0.35 \to 3.1$
3	HF	HF	102	$31 \to 78$	4.1	$1.2 \to 2.9$
4	LF	HF	102	$14 \to 66$	5.1	$0.7 \to 3.3$

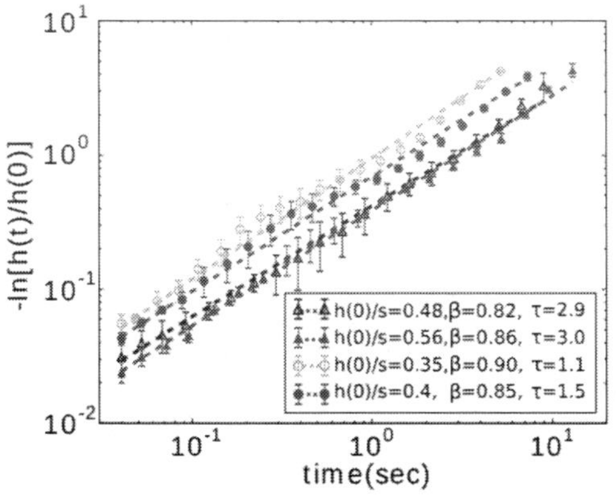

Figure 3. The negative ln[] of dimensionless centroid height $h(t)/h(0)$ vs. time t on a log-log scale. Data from $\mathcal{N}1$ and $\mathcal{N}2$. Markers: triangle for frictional particles and circle for frictionless particles; filled symbols for $D = 130mm$ and empty symbols for $D = 102mm$. Each data curve is an average of 2 or 3 repeated independent trials, with the errorbars indicate variation between trials. For fitting, we take 20 data points evenly spaced on the logarithmic time scale, and fit them into a straight line with the form: $y(x) = \beta x - \beta \ln(\tau)$ which minimizes the least square error. The fitting parameters are in the legend of the plot, and the unit of τ is second. The change of number of points used in the fitting does not change only slightly change the fitting parameters but not our conclusions.

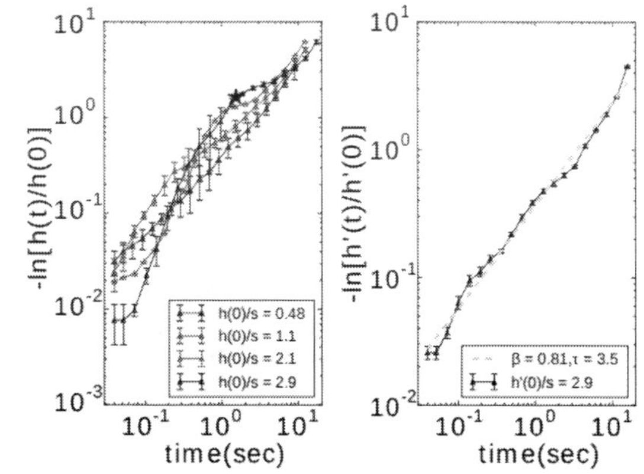

Figure 4. (a): Relaxation of $h(t)$ for columns with various $h(0)$ and diameter $D = 102$ mm. Each data is an average of 2-3 independent, repeated trial. The data for $h(0)/s = 0.48$ is the same one used in Fig.3. The black star on the data $h(0)/s = 2.9$ indicates the T_b, whose snapshot is shown in Fig.1(g). (b): Dynamics after the T_b for the $h(0)$ = data. Formally, We define $h'(t') = h(t' + T_b)$, (t' >= 0) and fit a straight line into $-\ln[h'(t')/h'(0)]$ vs t' on a log-log scale, as in Fig.3.

short columns. We identify these bending points T_b to be roughly the point where the cone-shape pile was formed and particles stopped rolling down from the top. One such snapshot at T_b is Fig.1(h). Such a two-step process happens for LF hexapods, as well as both hexapods on the HF base (not shown).

We do not have a specific form for the collapse before reaching T_b. For data after the bending point, we took T_b and $h(T_b)$ at the bending point to be the new initial condition for the collapse, and fitted to a stretched exponential relation. The assumption behind shifting the starting point of the relaxation is that, there are two distinct collapse processes before and after T_b. We observed experimentally that the collapse of a cone-shape pile (Fig.1 (b-e,h and i)) could be stopped immediately after the vibration was turned off, while the collapse of taller columns could not (Fig.1f and g, and it is also true for LF hexapod experiments). After T_b, the pile configurations were more stable than the column. Our assumption is supported by the data shown in Fig.4(b) for experiments $\mathcal{N}1$. In Fig.4(b) we fit the dynamics after T_b for curve $h(0)/s = 2.9$. The choice of T_b here corresponds to the moment when the cone was formed. We obtain similar fitting parameters as for the short columns ($\beta = 0.8$ and $\tau = 3.0$ s) (Fig.3), which indicates the collapse of the column after T_b likely shares similar mechanisms as the collapse of short columns.

3.3 Effect of friction

Our assumption is further supported by experiments $\mathcal{N}3$, e.g. HF particles on HF base, shown in Fig.5 and insert. In the insert, the whole collapse dynamics are presented. The separation between fast collapse and slow relaxation is clear, and more apparent with increasing $h(0)$, which agrees with the experiments on LF base. We define new time zeros at T_b and fit the data to stretched exponentials, which are shown in the main figure. Overall, the good fits suggest that the collapse processes for cone-shape piles can again be approximated by stretched exponentials on the HF base. One difference is, that the characteristic relaxation times τ are 10 times larger than the ones for LF base, and the exponent β is systematically smaller, around 0.5 to 0.6. We attribute the increase of relaxation time to drag with the frictional base. A packing relaxes by expanding in the horizontal directions. On the frictional base, the horizontal expansion is retarded. Thus, under the same external vibration, the relaxation is slower.

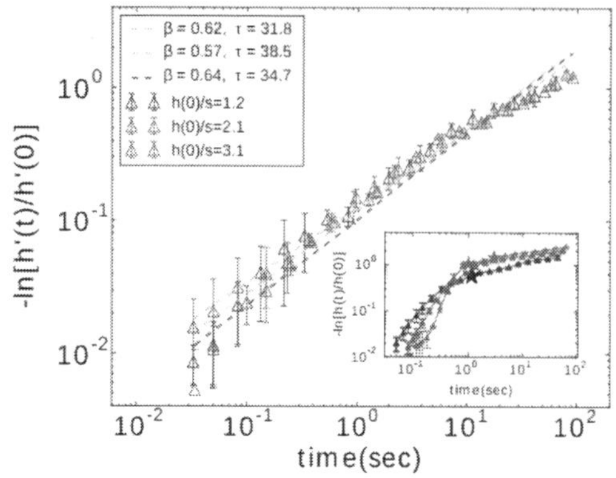

Figure 5. Main: Collapse dynamics after T_b for HF hexapod on HF base. Each color of triangles represents one $h(0)$, as shown in the legend. Dashed lines indicate a stretched exponential relation. Insert: Collapse dynamics of these runs from the start of the sinusoidal vibration. The star markers indicate the choice of T_b, which is around time of the cone-shape piles form.

4 Conclusions

In this paper, we quantitatively investigated the collapse dynamics of columns of hexapods (star-shaped particles) under constant sinusoidal vibration. For short and flat columns made of high or low friction particles, the dynamics of the centroid heights of the columns can be well fitted to stretched exponential relation for different column diameters. We also find that the inter-particle friction increases the characteristic relaxation time. The collapse dynamics departs from a stretched exponential when the column is tall initially. A tall column undergoes a two-step process, where initially it collapses fast into a cone-shape pile, and afterwards its relaxation is described using a stretched exponential fit with similar parameters as the short and flat columns. We also investigate the effect of friction on the base that supports the column. High friction surfaces increase the characteristic relaxation time significantly.

Acknowledgments

We thank Dong Wang for useful discussions. Y.Z., and R.P.B. are funded by the U.S. National Science Foundation (grants NSF-DMR1206351 and DMS1248071), NASA (grant NNX15AD38G) and the W. M. Keck Foundation. K.D. would like to thank the ITASCA Education Partnership Program (IEP).

References

[1] Liu, Andrea J., Nagel, Sidney R., Nature **396**, 21 (1998)
[2] M. van Hecke, Journal of Physics: Condensed Matter **22**, 033101 (2010)
[3] Bi, Dapeng, Jie, Zhang, Chakraborty, Bulbul, Behringer, Robert P., Nature **480**, 355–358 (2011)
[4] C. Zhao, K. Tian, N. Xu, Phys. Rev. Lett. **106**, 125503 (2011)
[5] A. G. Athanassiadis, M. Z. Miskin, P. Kaplan, N. Rodenberg, S. H. Lee, J. Merritt, E. Brown, J. Amend, H. Lipson, H. M. Jaeger, Soft Matter **10**, 48 (2014)
[6] B. Saint-Cyr, J.-Y. Delenne, C. Voivret, F. Radjai, P. Sornay, Physical Review E **84**, 041302 (2011)
[7] E. Brown, A. Nasto, A. G. Athanassiadis, H. M. Jaeger, Physical Review Letters **108**, 108302 (2012)
[8] N. Gravish, S. V. Franklin, D. L. Hu, D. I. Goldman, Physical Review Letters **108**, 208001 (2012)
[9] M. Trepanier, S. Franklin, Physical Review E **82**, 011308 (2010)
[10] Y. Zhao, K. Liu, M. Zheng, J. Barés, K. Dierichs, A. Menges, R.P. Behringer, Granular Matter **18**, 24 (2016)
[11] K.A. Murphy, N. Reiser, D. Choksy, C.E. Singer, H.M. Jaeger, Granular Matter **18**, 26 (2016)
[12] Y. Jiao, F. H. Stillinger, S. Torquato, Physical Review E **79**, 041309 (2009)
[13] J. de Graaf, R. van Roij, M. Dijkstra, Physical Review Letters **107**, 155501 (2011)
[14] S. V. Franklin, Europhysics Letters **106**, 58004 (2014)
[15] K. Desmond, S. Franklin, Physical Review E **73**, 031306 (2006)
[16] S. Keller, H.M. Jaeger, Granular Matter **18**, 29 (2016)
[17] K. Dierichs, A. Menges, Architectural Design **85**, 86 (2015)
[18] M.Z. Miskin, H.M. Jaeger, Soft Matter **10**, 3708 (2014)
[19] H.M. Jaeger, Soft Matter **11**, 12 (2015)
[20] J.C. Phillips, Reports on Progress in Physics **59**, 1133 (1996)

DOI: 10.1051/epjconf/201714003038

Avalanches in a granular stick-slip experiment: detection using wavelets

Aghil Abed Zadeh[1,⋆], *Jonathan* Barés[2,⋆⋆], and *Robert P.* Behringer[1,⋆⋆⋆]

[1] *Duke University, Durham, NC, 27708, USA*
[2] *Laboratoire de Mécanique et Génie Civil, Université de Montpellier, CNRS, Montpellier, France*

Abstract. Avalanches have been experimentally investigated in a wide range of physical systems from granular physics to friction. Here, we measure and detect avalanches in a 2D granular stick-slip experiment. We discuss the conventional way of signal processing for avalanche extraction and how statistics depend on several parameters that are chosen in the analysis process. Then, we introduce another way of detecting avalanches using wavelet transformations that can be applied in many other systems. We show that by using this method and measuring Lipschitz exponents, we can intelligently detect noise in a signal, which leads to a better avalanche extraction and more reliable avalanche statistics.

1 Introduction

Many systems, under slow loading, exhibit intermittent and discrete events of broad sizes, known as crackling noise [1, 2]. This behavior has been experimentally investigated in a wide range of physical systems [3–5] from earthquakes in seismic activities to Barkhausen noise in ferromagnets [6, 7]. Slowly sheared granular media are examples of these systems. They respond to shear stress with discrete slip avalanches, also called stick-slip events [8–11]. The avalanches statistics of these systems and their universal and non-universal quantities are key properties of these systems. In this paper, we introduce a method for avalanche detection based on wavelet transformation to extract more reliable statistics. This method can be used for many other systems for a better avalanche detection.

As shown in figure 1, to observe the physics of stick-slip for slowly sheared granular materials, we have developed a novel apparatus, which is capable of detecting avalanches both at the global and local scales. In our experiments, a constant speed stage pulls a frictional slider of length 25 cm which rests on a vertical bed of bi-disperse cylindrical photo-elastic particles with diameteres of 0.4 cm and 0.5 cm (number ratio of 2.7) in a 2D granular system. The verticle walls are low frictional to reduce the efect of boundary friction. The slider is designed to have very high friction with the top layer of granular media. The stage is connected to the slider by a spring, the stiffness of which can be tuned. We measure the force on the spring by a force sensor attached to it, which provides a global measure of shear stress. As we want to fix the system stiffness and pressure at a certain chosen value; it is not possible to modify the device to force the slider to remain horizontal. However, to make sure the slider always stay in the same inclination; we have calibrated its wait and initial angle so that it may tilt only few degrees. That may add some noise to the system; however, we think is not significant. The photo-elastic response also provides us a measure of local stress; however, in this paper, we focus on the global response of the system.

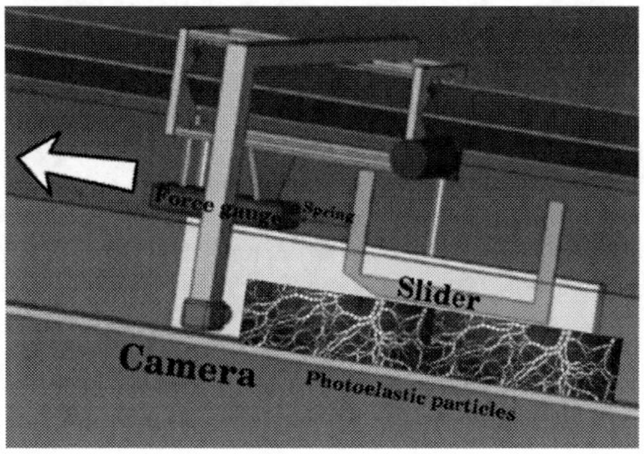

Figure 1. Side view of apparatus. A bi-disperse photo-elastic granular media is sheared by a slider that is moved by the force gauge connected to a spring. The slider exhibits a stick and slip dynamics. The force on the slider is recorded via a force gauge and the internal structure and stress is imaged via high speed cameras (not treated in this paper).

For a given pulling stage speed v, spring stiffness k, and granular bed height l, the evolution of the force F, applied to the slider is measured with a frequency of $100Hz$ in our experiment. To quantify the global avalanche statistics based on $F(t)$, we use specific signal processing. In experimental data, noise is unavoidable, but the way noise is detected or reduced can be crucial to quantitative re-

⋆e-mail: aghil.abed.zadeh@duke.edu
⋆⋆e-mail: jonathan.bares@umontpellier.fr
⋆⋆⋆e-mail: bob@phy.duke.edu

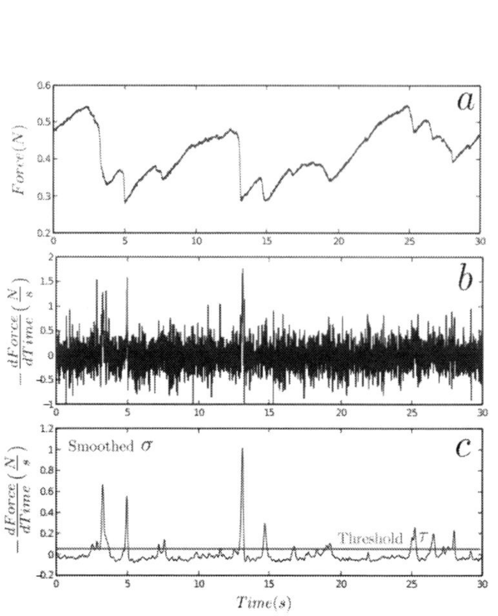

Figure 2. a) A typical force gauge signal $F(t)$ from the experiment with spring stiffness of $k = 70N/m$ and stage speed of $v = 1mm/s$ acquired at $100Hz$. b) Point to point derivative of force signal. The signal-to-noise ratio decreases dramatically when differentiated. c) The smoothed derivative of force signal. Smoothing is applied by a convolution of part b with a Gaussian function of standard deviation $\sigma = 0.05s$. The threshold τ (red line) determines the beginnings and endings of avalanches.

sults. Conventionally, these noise spikes are removed using Fourier transform domain filters, including averaging over many data points. The latter is a low-pass filter in Fourier transform domain and increases the signal-to-noise ratio. However, the trade-off is loosing the edge information. Moreover, if the duration of an event in a signal is of the order of the duration of a noise spike, the event may break into several events and make them hardly distinguishable; or some noise spikes may be detected as events. These flaws can be reduced by applying a wavelet transform domain filter that selectively distinguishes noise spikes from events.

In the first part of this paper, we summarize a conventional way of avalanche detection in a typical signal of our experiment. We discuss how several chosen parameters in the signal analysis process can affect the statistics. Then, we introduce a wavelet transform domain filter and how to selectively detect noise spikes locally using Lipschitz exponents [12], without loosing other signal information. This method may provide a tool for a better avalanche extraction in many other experimental systems.

2 Conventional method

The slow shearing of a granular materials leads to intermittent slow loading and fast unloading of the medium that is called an avalanche. Figure 2a shows a typical signal $F(t)$ from the force gauge. The fast drops in F (like at time $t = 13s$) correspond to slip avalanches of the slider and similar behavior in observables have been detected in many physical systems [5, 8, 13, 14]. Hence, to detect an avalanche, we need to find where the slope of the force is negative. However, if we use point to point derivatives, as shown in figure 2b, the presence of noise may make it impossible to find the avalanches. To overcome this problem conventionally, as depicted in figure 2c, the signal is smoothed and then a threshold τ is chosen for finding where it crosses the smoothed data. This defines the beginnings and endings of avalanches. To smooth the signal from our data, we applied a convolution as presented in equation 1, in which h is the derivative of the force signal F and g is a normalized discrete Gaussian function. This method requires two variables, the standard deviation of the smoothing function, σ, and the threshold, τ. The statistics of avalanches vary by changing these two variables.

$$(h * g)[m] = \sum_{t=1}^{n} f[t]g[m-t] \quad (1)$$

$$g[t] = \frac{1}{\sigma\sqrt{2\pi}} \exp(-\frac{t^2}{2\sigma^2}) \quad (2)$$

Figure 3 shows the probability density function of slip sizes for different values of the threshold τ and standard deviation σ (as in fig.2c). The slip size is the displacement of the slider during an avalanche, which is calculated using the force signal $\Delta x = \Delta F/k$, in which k is the stiffness of the spring. As shown in the figure, the probability density function (PDF) is affected by τ and σ. Indeed, both the exponent and the non-universal quantities such as lower cut-offs and prefactors depend on these two parameters. As large avalanches are easier to detect and less affected by low-pass filters, the statistics of them do not change dramatically by changing these two variables. However, small avalanches statistics are more affected by these parameters as they contain more high frequency information. This variation can also be seen in other avalanche properties such as the average shape of avalanches (not shown here).

3 Wavelet transformation method

Wavelet transforms have been widely used in signal processing in computer science for detection of different types of singularities in a signal [12]. A wavelet transformation of a signal is simply its convolution with a function g as in equation 1. The convoluting wavelet that we use here is proportional to the first derivative of a Gaussian function, as in equation 3, which is a widely used wavelet for edge detection problems [12].

$$g[t] = \frac{t}{\sigma^2\sqrt{2\pi}} \exp(-\frac{t^2}{2\sigma^2}) \quad (3)$$

Each local maximum in the signal wavelet transformation cooresponds to a "singularity". The dependence of the local maxima on σ determines the type of the singularity [12]. Typical noise is associated with a back and forth

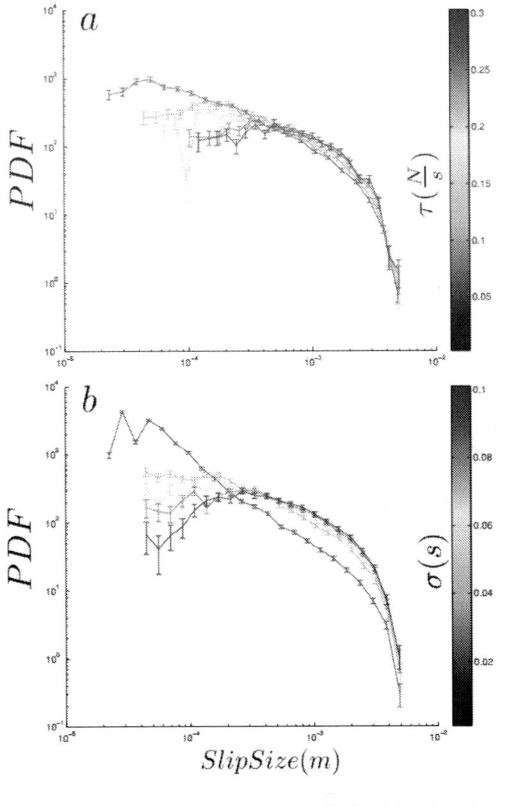

Figure 3. Probability density functions (PDF) of slip sizes during avalanches for: a) Several different values of threshold τ (for fixed $\sigma = 0.05$) and b) Several different values of standard deviation σ (for fixed $\tau = 0.1$) . These two parameters affect the statistical measurements of the force signal.

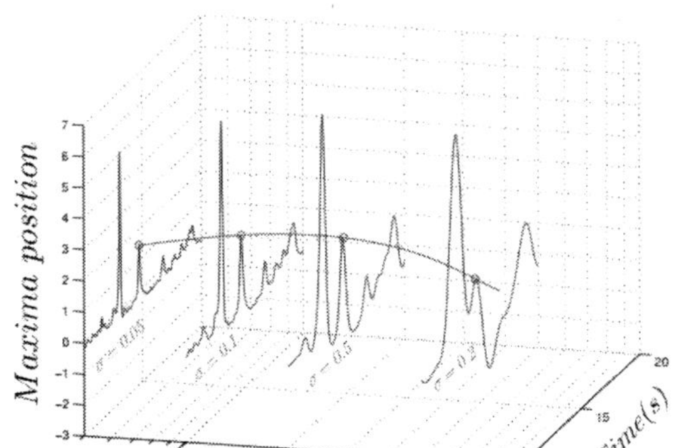

Figure 4. Wavelet transformation of the force signal F of figure 2a with the wavelet given in equation 3 as a function of σ. One of local maxima is marked by red circles. We track the position of such local maxima to detect whether it corresponds to an avalanche or not. The avalanche detection is based on the evolution of the maxima position vs σ.

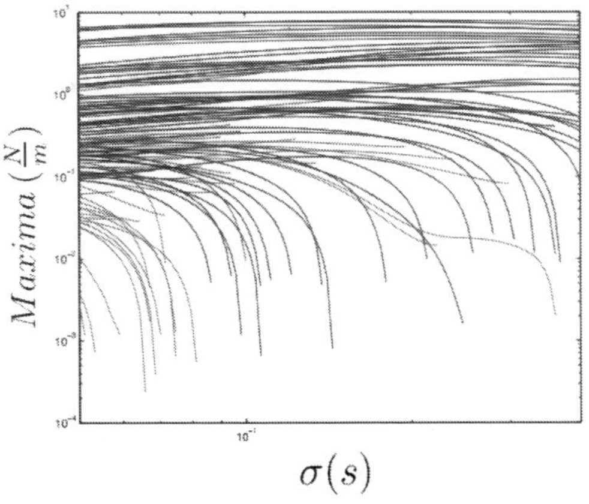

Figure 5. Position of the local maxima (as in fig.4) values as a function of σ for a typical force signal F. Maxima that are detected to belong to avalanches are marked blue. See text for more details.

peak which resembles a Dirac delta function, whereas, avalanches are sharp unique variations (slow loading and rapid unloading), which resemble step function. This difference can be captured by well chosen wavelets. Figure 4 depicts the wavelet transformation of the force signal F (of fig 2a) as a function of σ. We track each local maximum, one of which is marked red in figure, vs. σ to detect the singularity type.

The evolution of local maxima vs. σ for the figure 2a is shown in figure 5. Each line in figure 5 belongs to a local maximum. The slope of each line is called a Lipschitz exponent of the singularity cooresponding with the maximum of the signal derivative. This exponent is non-negative for an edge function and negative for a Dirac delta function [12]. Hence, it is also non-negative for a singularity corresponding to avalanche events in our case, and negative for those of noise. Moreover, to detect the avalanches more precisely,we need to have non-negative Lipschitz exponent and a clear exponential decaying tail. By using the Lipschitz exponent and detecting the decaying tail, we can extract the maxima that correspond to the avalanches, which are marked blue in figure 5.

This wavelet transformation method can improve on avalanche detection over the conventional methods, without losing high frequency information. It also provides a better detection of small avalanches. Figure 6 shows the detected avalanches in a force signal F. The avalanches are marked as blue and each avalanche is a connected set of blue points in the figure. Using this method, we can then calculate the probability density function of avalanche sizes and other avalanche statistics. Figure 7 shows the PDF of avalanche slip size for instance. This figure is calculated using several thousand detected avalanches for a spring stiffness of $k = 70N/m$ and stage speed of $v = 1mm/s$. The distribution obeys a power-law for about two decades with a clear lower cut-off and an exponen-

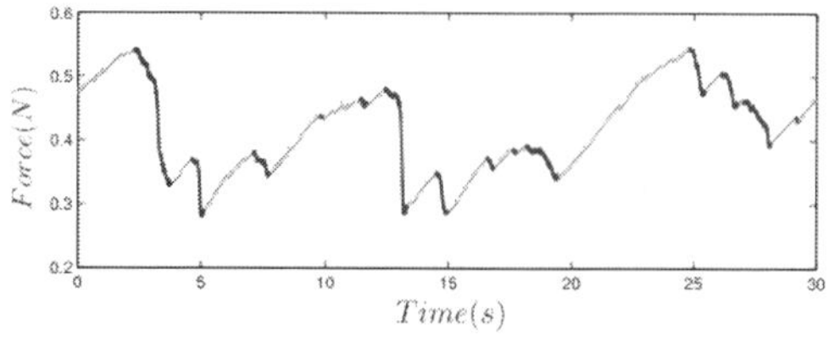

Figure 6. The detected avalanches in a force signal sample using the wavelet transformation method. Avalanches are marked blue. This method has the advantage of better detection of small avalanches and ignoring the noise spikes occurring during avalanches. These noise spikes can break the avalanches into several ones in the conventional method.

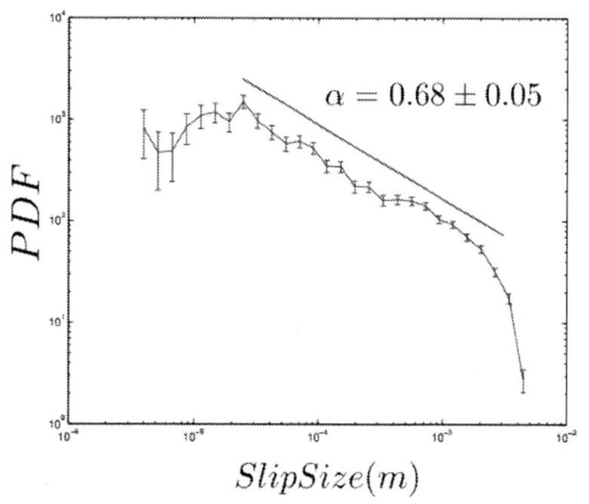

Figure 7. Probability density functions (PDF) of slip sizes during avalanches, detected by the wavelet transformation method. The PDF obeys a power-law for more number of decades compared to conventional method (fig 3), as a result of avalanche detection using wavelets.

tial upper cut-off at slip size $2 \times 10^{-5}m$ and $2 \times 10^{-3}m$ respectively. Because of better detection of avalanches, the wavelet transformations method provides a smaller lower cut-off and larger number of power-law decades, compared to the conventional method (fig 3). This local noise detection and removal provides a more reliable exponent calculation in the PDF of avalanche sizes over the conventional method, which usually uses global smoothing of the data.

4 Conclusion

Our experimental apparatus enables us to study the avalanches in a granular stick-slip experiment. The slow shear of granular system is followed by intermittent fast unloading of shear stress that is measured by a force signal. This force signal is conventionally smoothed (low-pass filter) to increase the signal-to-noise ratio for avalanche detection. However, this conventional method changes the statistics of avalanches as it removes all low frequency information. We introduced a smarter way of avalanche extraction from measured quantities using wavelet transformation. Similar methods have been tested in many other problems and here it provides a better avalanche detection and consequently more reliable statistics, by detecting the noise locally in the signal. The method introduced here can be applied in many other systems for a more precise avalanche extraction.

Acknowledgements

We acknowledge support from NSF-DMS-1248071, NASA Grant NNX15AD38G, and William M. Keck Foundation.

References

[1] J.P. Sethna, K.A. Dahmen, C.R. Myers, Nature **410**, 242 (2001)
[2] R. Dickman, M.A. Munoz, A. Vespignani, S. Zapperi, Brazilian Journal of Physics **30**, 27 (2000)
[3] A. Kabla, J. Scheibert, G. Debregeas, Journal of Fluid Mechanics **587**, 45 (2007)
[4] R. Planet, S. Santucci, J. Ortín, Journal of Contaminant Hydrology **120-121**, 157 (2011), reactive Transport in the Subsurface: Mixing, Spreading and Reaction in Heterogeneous Media
[5] J. Barés, M.L. Hattali, D. Dalmas, D. Bonamy, Physical Review Letters **113**, 264301 (2015)
[6] O. Perkovic, K.A. Dahmen, S.P. Sethna, Physical Review Letters **75**, 4528 (1995)
[7] R.A. White, K.A. Dahmen, Physical Review Letters **91**, 085702 (2003)
[8] D. Howell, R.P. Behringer, C. Veje, Physical Review Letter **82**, 5241 (1999)
[9] F. Dalton, D. Corcoran, Physical Review E **63**, 061312 (2001)
[10] K.E. Daniels, N.W. Hayman, Journal of Geophysical Research: Solid Earth **113** (2008)
[11] M. Otsuki, H. Hayakawa, Physical Review E **90**, 042202 (2014)
[12] S. Mallat, W.L. Hwang, IEEE transactions on information theory **38**, 617 (1992)
[13] E. Aharonov, D. Sparks, Journal of Geophysical Research: Solid Earth **109**, n/a (2004)
[14] A. Petri, A. Baldassarri, F. Dalton, G. Pontuale, L. Pietronero, S. Zapperi, The European Physical Journal B **64**, 531 (2008)

Dynamics of granular segregation in quasi two-dimensional system

Sandip H. Gharat[1,⋆]

[1] *Department of Chemical Engineering, Gharda Institute of Technology Lavel, Khed, Ratnagiri 415 708 India.*

Abstract. Segregation during flow of granular materials is important from an industrial point of view. Considerable work has been done on granular segregation in heap flows by continuous pouring. We studied the flow and segregation of granular mixtures during heap formation in a quasi two-dimensional rectangular bin by intermittent pouring. The heap formed by repeatedly pouring a fixed mass of the mixture. Each feeding results in the formation of a layer of the mixture on the surface of the heap. The system is a simplified model for the feeding of raw materials to a blast furnace, which is widely used for the manufacture of iron and steel. Experiments were carried out to study the dynamics of granular materials during heap formation. The number density, area fraction and average velocity of small and big particles are plotted across the flowing depth with time. Results shows that larger particles are always on top flowing over small particles. During flow small particles easily percolate through the gaps between the large particles. A thin layer of small particles is also observed at the free surface. Here the system never reached a steady state as we are pouring the mixture intermittently and system is closed. The velocity increases initially and then decreases towards the end. The number density (i.e. area fraction) profile changes for small and big particles during flow. Image analysis is done to detect the position of each particle on the side wall. Each experiment is repeated six times to get average data.

1 Introduction

Granular materials segregate due to difference in their sizes [1-6]. Segregation has a significant impact in industry as well as in various natural processes [7-9]. Although granular segregation on a heap has been studied earlier, in most of the cases feeding is continuous [10-15], flows on a heap by intermittent feeding are not available. In this paper, we consider the segregation during heap formation in a quasi two dimensional rectangular bin for a mixture of different sized spherical particles having equal density by means of high speed video imaging. The objective is to study the dynamics of the segregation process.

2 Materials used

Stainless steel (*SS316*) balls of different sizes (1 and 2 mm) but equal in density are used as a model of granular materials.

3 Experiments

Experiments were carried out in a quasi two dimensional rectangular bin as shown in Fig.1. It consists of two vertical glass plates (52 cm x 29 cm), aluminum divider and an auxiliary hopper. The side walls (glass plates) are transparent to facilitate imaging. The gap between two vertical glass plates is 10 mm and the aluminum spacer forms a bin in which the heap is formed. An aluminum divider forms an inclined chute for feeding the mixture. The auxiliary hopper is placed on the top of the bin for supply of the feed. The position of the aluminum divider can be adjusted to control the flow rate, gap (k) and the height of fall to the top of the heap (h). A halogen lamp (1000 watt) is used as a light source. The setup is leveled properly to ensure that side walls are vertical and the base is horizontal.

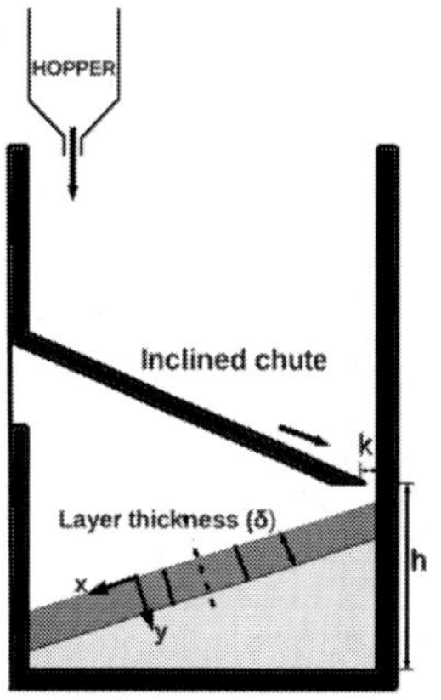

Figure 1. Schematic view of heap formation, showing feeding arrangement, feeding point, layer formed after pouring, height of feeding point, thickness of layer. The coordinate system employed in the analysis is shown.

⋆e-mail: sandipgharat78@gmail.com

 DOI: 10.1051/epjconf/201714003039

3.1 Experimental Method

An equal % of fixed mass of a binary mixture of different sizes is poured intermittently to form a heap. The process is repeated until the bottom part of a heap is filled. The particles were randomly mixed at the start of the experiment. A measurement was then taken for each flowing layer poured above the bottom part of a heap. A small region of flowing layer at the center of the heap (shown in dotted rectangle in figure 2a) was captured with high speed camera (*Phantom 663*) at 1000 frames per second. The exposure time was kept very low. In a typical capture image the particles are seen as bright white spots (a cluster of white pixels) due to reflection of light. The particles were detected using computer codes. The position of the particle was taken as mean position of the pixels present in that cluster. Small and big particles were identified based on the radius of gyration (R_g). The radius of gyration (R_g) is used to characterize the size of the cluster calculated as

$$\bar{R_g} = \sqrt{\frac{\sum_{i=1}^{N}[(X_i - X_0)^2 + (Y_i - Y_0)^2]}{N}} \quad (1)$$

where, X_i and Y_i are the coordinates of each pixel in the cluster (see Figure 1 for definition of coordinate system) X_0 and Y_0 are center of mass of each cluster given by

$$X_0 = \sum_{i=1}^{N} \frac{X_i}{N}$$

$$Y_0 = \sum_{i=1}^{N} \frac{Y_i}{N}$$

where X_0, Y_0 is taken to be the mean position of the particle.Figure (2c) shows the detected small and big particles for the original captured image. Position of each particle in all images were detected to get trajectories of each particle (small and big). From the trajectories, the velocity of each particle is determined. The co-ordinate axis is shown in figure (2a). For analysis we have taken a region of length 20 mm (10 mm downstream and upstream from the center) and depth of 15 mm (flowing layer thickness) to understand the flow of binary granular materials. The region of interest is divided into bin of 1 or 2 mm to the flow direction. The total number of images for each flowing layer are 7500 (±300). The total images counted is based on the material seen in the first image till it finishes. For analysis we have taken a set of 200 images starting from image no. 500 to 699 (where velocity of particle is sufficient) till the last set of 7000-7199 with the gap of 500 from staring image. (e.g., 500-699, 1000-1199 and so on). During flow, void space between the particles increases due to which small particles can easily percolate through a gap of larger particles. Deposition of small particle in turn increases the depth as flow progresses. Variation in the y-position affects the velocity profile therefore it is necessary to shift the y-position with the reference to get correct analysis of flow of particles. Hence, for each set of images the y-position is kept constant by shifting the co-ordinate (y) with the reference surface after each pouring i.e. passage of each flow.

4 Image Analysis

Image analysis is used to detect the position and size of the particles from captured images. The particles appear as tiny bright spots on a black background. A typical captured image is shown in Figure 2(b).

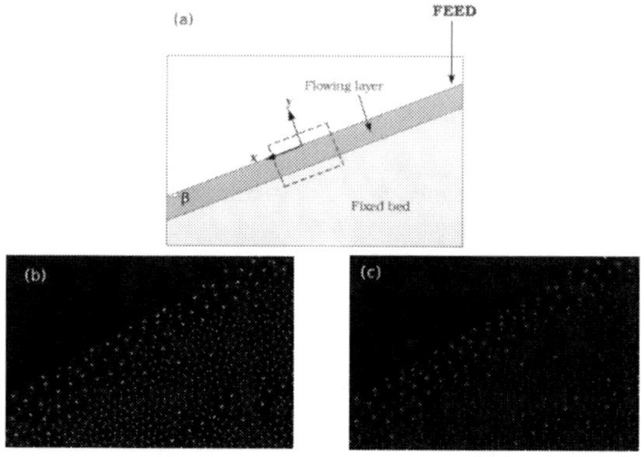

Figure 2. (a) Schematic view of the system: a flowing layer with a mixture of 50% by volume of small (1 mm) and 50% by volume of big (2 mm) spherical steel particles. The free surface makes an angle β with the horizontal. The coordinate system used has its origin at the center as shown. The rectangular region denoted by the dashed line is captured with a high-speed camera at 1000 fps and analyzed. (b) Typical captured image: The particles are seen as bright spots for large particles and (c) Identified large (red on-line) and small (blue on-line) particles for the image shown in (b) using image analysis.

The captured gray scale images are converted into black and white images by thresholding. The threshold value is chosen such that the less illuminated particles from the inner layers are eliminated and particles from the front layer appear as a clusters of white pixels on a black background. During the thresholding process, individual pixels in an image are marked as object pixels if their value is greater than some threshold value (assuming an object to be brighter than the background) and as background pixels otherwise. The captured image after thresholding shows individual white clusters of pixels (figure 2(b)). We next scan the threshold image for a white pixel, starting from the top left corner, in both the right and the downward directions. Pixels adjacent to this pixel are scanned to identify white pixels and these are assigned to the cluster. All neighbouring pixels of pixels in the cluster are then scanned to again identify white pixels which are assigned to the cluster. The process is repeated iteratively until all pixels belonging to the cluster are identified and stored. Similarly all the white clusters in that image are identified.

The particles detected in the threshold images as white clusters are identified as large or small based on the radius of gyration of the cluster. Clusters with values of R_g greater than or equal to 3.5 pixels are taken to be large particles and small particles otherwise. Detected particles are

 DOI: 10.1051/epjconf/201714003039

shown in figure 2(c) in red (big particles - 2 mm) and blue (small particles - 1 mm).

For analysis we have chosen a co-ordinate system with its origin (x, y) at the top left corner of the free surface layer, such that the $+ve$ x-axis is opposite to the flow direction and $+ve$ y-axis is in the downward direction. The default and transformed co-ordinate axis is shown in figure 1. The new position of the particles are calculated using the following transformation.

$$x = X\cos\theta + Y\sin\theta - X_c \quad (2)$$
$$y = Y\cos\theta - X\sin\theta - Y_c \quad (3)$$

Here x, y and X, Y gives the position according to the new and old coordinate system respectively. θ is the angle the free surface makes with the horizontal. X_c, Y_c are coordinates of the origin.

5 Results and Discussion

We performed experiments by intermittently pouring mixtures of granular materials on a heap. The measurement were taken at the center of flowing layer.

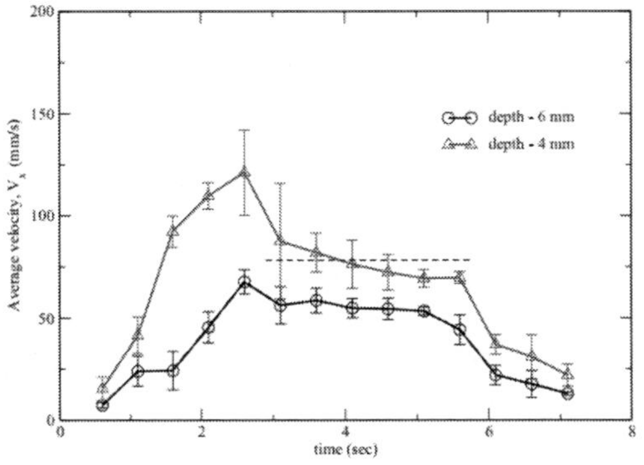

Figure 3. The mean velocity (v_x) for small and big particles across the depth within the flowing layer versus time.

The average velocity (v_x) are plotted for two different y-position with time as shown in figure 3. The velocity (v_x) increases initially and is then constant for a certain time and decreases in the end. The velocity decreases as the flowing depth increases, as expected. The nature of velocity profile for different y-position is the same. The average velocity (v_x) and area fraction along the depth (y-position) are plotted for the steady state region as shown (dotted line) in figure 3 and discuss below. The average velocity profile and area fraction for an equal % of small and big particle in the mixture within flowing layer at the center is shown in figure 4. The position having an total area upto 0.02 is shown in the graphs and total area fraction 0.5 (actual value is 0.47) is taken as the free surface (dotted line shown in fig. 4). Gray & Kokelaar (2010) and Weiderseiner et. al. (2011) showed that the flowing layer may have a different apparent concentration that the stationary mix. The average velocity (v_x) for large particles is less

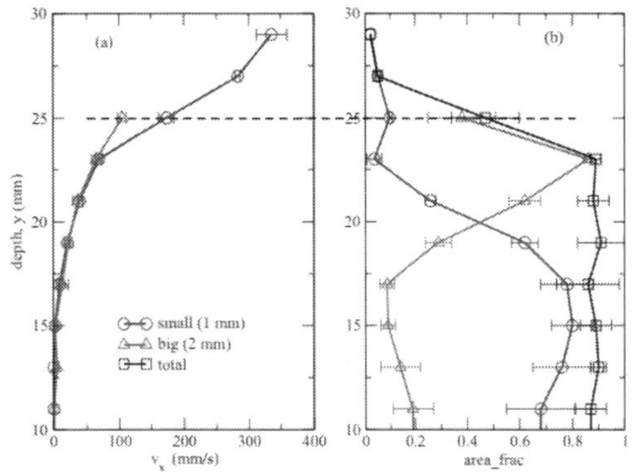

Figure 4. (a) The average velocity (v_x) for small and big particles within flowing layer at the center. (b) The area fraction for small, big particles and total area fraction within flowing layer for mixture of 50% small and 50% big particles.

than for smaller particles near the free surface and almost the same deeper in the bed (Figure 4(a)). The velocity (v_x) is zero near the interface and the area fraction is half of the bulk value at the free surface. The layer between the two is considered as the flowing layer.

The area fraction profile (figure 4b) shows there is partial segregation within the flowing layer as the fraction of large particles is higher near the free surface than the small particles. There is a slight deviation in total area fraction value (0.8 to 0.9) into the bed. The area fraction for large particles is higher than for small particle within the flowing and this value decreases suddenly to zero near the free surface.

Figure 5a shows the variation in number density profile for big particles within the flowing layer across depth (y) with time for a set of 50 images. We used equal % of small and big particles as the initial mixture. The volume of pouring and the height of the divider from the bottom of the bin are 200 g and 22 cm respectively. There is a variation in number density with time. We also checked the number density profile (n) for two additional sets, one is for 100 images (figure 5b) and another is for 200 images (figure 5c).

Figure 5(d) shows the profiles of the number density at the three sets of images. The profiles are identical within experimental error, verifying that the all sets of images gives almost same result.

6 Conclusions

The segregation of equal density, different sized particles is studied to understand dynamics of segregation. The area fraction profile shows segregation within the flowing layer. These results shows that larger particles accumulate near the free surface flowing over small particles. During flow small particles easily percolate through the gaps between large particles. The total area fraction in the deep bed is between 0.8 to 0.9 which is close to hexagonal close packed value of 0.9068. For mixed region the value

DOI: 10.1051/epjconf/201714003039

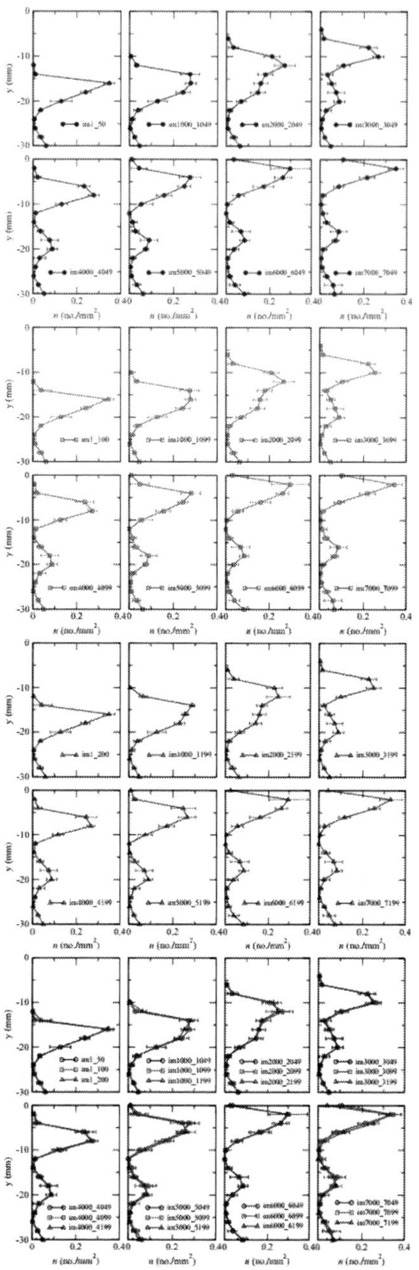

Figure 5. Variation of number density (n) profile within the flowing layer for big particles with time for a set of (a) 50 images (b) 100 images and (c) 200 images (d) 50, 100 & 200 images. Equal % of small (1 mm) and big (2 mm) particles. Volume of pouring 200 g.

of the area fraction is higher than the hexagonal closed packing. In Hexagonal closed packing layers of particles (sphere) are closely packed so that particles (sphere) in alternating layers overlie one another. Packing fraction (efficiency) is defined as how much space being occupied by the particles present in a unit cell. The area fraction decreases near the free surface for both types of particles. The velocity profiles are linear over most of the layer depth with an exponential decay near the base of the flowing layer [16]. Velocity increases with time and decreases towards the end.

Acknowledgements

This work was done at IIT Bombay and Project was sponsored by TataSteel Jamshedpur, India.

References

[1] Makse, H. A., R. C. Ball, H. E. Stanley, and S. Warr, Phys. Rev. E **58**, 3 (1988)

[2] Changxing, L., Z. Zhou, R. Zou, D. Pinson, and A. Yu, AIP Conf. Proc. **1542**, 767-770 (2013).

[3] Sandip H. Gharat and D. V. Khakhar, AIP Conf. Proc. **1542**, 755-758 (2013).

[4] Cizeau, P., H. A. Makse, and H. E. Stanely, Phys. Rev. E, **59**, 4408 (1999).

[5] Dolgunin, A. N. Kudy, and A. A. Ukolov, Powder Tech., **96**, 211-218 (1998).

[6] Gray, J. M. N. T. and K. Hutter, Contin. Mech. Thermodyn., **9**, 341–345 (1997)

[7] Drahun, J. A. and J. Bridgwater, Powder Technol., **36**, 39 (1983).

[8] Fan, Y., Y. Boukerkour, T. Blanc, P. B. Umbanhowar, J. M. Ottino, and R. M. Lueptow, Phys. Rev. E, **86**, 051305 2012).

[9] Gray, J. M. N. T. and A. R. Thornton, Proc. R. Soc. A, **461**, 1447-1473 (2005).

[10] Hajra, S. K. and D. V. Khakhar, Granul. Matter, **13**, 475-486 (2011).

[11] Jenkins, J. T. and D. K. Yoon, Phys. Rev. L, **88**, 19 (2002).

[12] Khakhar, D. V., J. J. McCarthy, and J. M. Ottino, Phys. Fluids, **9**, 3600-3614 (1997).

[13] Khakhar, D. V., J. J. McCarthy, and J. M. Ottino, CHAOS, **9**, 3 (1999).

[14] Williams, S. C., Powder Technol., **2**, 1320 (1968)

[15] GRAY, J. M. N. T. and C. ANCEY, J. Fluid Mech. **629**, 387–423 (2009)

[16] GDR MiDi, Eur. Phys. J. E **14**, 341–365 (2004)

DOI: 10.1051/epjconf/201714016005

An innovative device for powders classification based on combined aerodynamic and electrostatic separation of particles

Bruno Piriou[1], *Claire* Mayer-Laigle[2], *Firas* Maalel[2], *Jorys* Plissot[2], and *Xavier* Rouau[2,*]

[1]CIRAD, unité de recherche Biomasse-Energie, TA-B 114/16, 34060 Montpellier cedex, France
[2]INRA-CIRAD-University of Montpellier-Supagro Montpellier, unité mixte de recherches IATE, 34060 Montpellier cedex, France

Abstract. An innovative separator has been developed which aims at sorting out powders fractions according to aerodynamic and electrostatic properties of particles. The prototype is composed of a projection unit, a separation module and a classification/collection zone. The projection part is ensured by a pressurized gun. The separation module is conceived to disperse particles according to front and transverse directions. At the output of the gun, electrodes can be disposed to deviate the trajectory of charged particles. The classification part is made of a series of collecting bins. Starch, fine ash powders, and mixtures of these two materials were used to test the working of the separator. The efficiency of the separation was assessed by yields and particle sizes distribution of the collected fractions in the different bins. An aerodynamic classification was achieved along the main axis, with the finest particles crossing larger distances. With a blend of starch and ash, particles from both materials were unevenly distributed in the classifyier, opening the way to a possible separation based simply on aerodynamic properties. When an electric deviation was applied in addition to the projection of powders, the distribution of particles in the collecting part was modified.

1 Introduction

In the concept of biorefinery, fractionation of lignocellulosic biomass is a prerequesite for further valorization into energy, chemicals or synthons, and materials [1]. Dry processing, that yields intermediate powdered products, is particularly attractive as it has low environmental impact with no water and solvent consumption nor generation of other effluents [2]. Dry fractionation includes a step of dissociation, generally obtained by grinding and milling, and a step of separation or classification which allows the sorting out of distinct fractions. Several techniques of classification can be used, taking advantage of different properties of the powder components: sieving based on particle sizes, aerodynamic sorting based on particle densities and shapes, and electrostatic sorting based on chemical compositions of particles. These techniques are usually employed separately but may be considered jointly to increase the resolution of the separation process [3]. In the present work we have developed an innovative separator which aims at sorting out mixtures of lignocellulosic powders according to the aerodynamic and electrostatic properties of individual particles. The device has been sized following the potential performances of a pressurized machine designed for plasma spraying of ceramic fine powders, in using corn starch and rice husk ash as references. The separative efficiency of the device was assessed with a mix of these two powdered materials.

2 Materials and methods

2.1 Description of the equipment

The prototype is composed of a projection unit, a separation module and a classification/collection zone (Figure 1). The projection part is ensured by a pressurized gun controlled to expel powders at fixed flow rates (Sulzer Metco 9MP-CL20, Oerlikon). The powder is fluidized in a vibrating feed hopper then directed to a pressurized channel and projected throuh a diffuser designed to deliver a ribbon-like shaped blast (dimensions of the diffuser: 0.051 m height, 0.003 m width, 0.00015 m^2 section). The air flow was set at 15-18 l/mn to maintain the powder flow rate at 20 g/mn. In these conditions, the maximum front distance of projection was approximately 1 m for all tested powders. The lateral extent of the deposit was approximately 0.4 m. The separation module is

*Corresponding author: xavier.rouau@inra.fr

composed of a plexiglass housing allowing powders to disperse through the diffuser according to front and transverse directions. This module has 1.2 m length and 0.5 m width. Preliminary experiments have shown that part of the particles stuck onto the walls of the housing. Therefore, the walls were covered inside with an earthed wire mesh to avoid unwanted electrostatic phenomenon. At the output of the diffuser, two electrodes of 25 cm length and 5 cm height connected to a high voltage generator (0-20 kV) can be disposed to create an electrical field able to deviate the trajectory of particles which have been triboelectrically charged by rubbing in the projecting part of the gun. The classification part is made of a series of 18 collecting bins (25 x 10 cm) covering the whole separation area and allowing to distinguish fractions deposited according to the main projection axis and also right and left sides (9 bins on the right, 9 bins on the left). Powders fractions collected in the different bins were weighted separately in order to determine their distribution and their median particle sizes (d_{50}) were measured with a laser granulometer (Mastersizer 2000, Malvern, France).

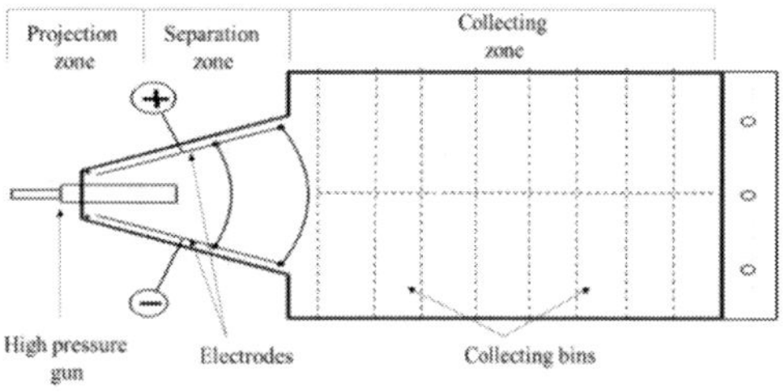

Fig. 1. Top view scheme of the separator

2.2 Powder samples

The maize starch powder (Maizena, Unilever, France) had an average particle size (d50) of 20 µm and a true density of 1.2. The ash was obtained from full calcination (900°C, 3h) of rice husk (silo de Tourtoulen, Arles, France). It exhibited a d_{50} of 27 µm and a true density of ~2. The ratio of starch and ash masses recovered in the bins after separation was assessed by the measurement of mineral contents of the powder blends in weighting residual ash after calcination in a furnace at 900°C during 2 h, assuming a mineral content of 0% for starch and 100% for ash.

2.3 Regime of the projection flow

Calculations were made to determine the Reynolds number in different air velocity conditions, using the following relations:

$$v\_p = v\_air = D\_air/S*60 \quad (1)$$

$$v = \mu\_air/\rho\_air \quad (2)$$

$$R = (v\_air*l)/v \quad (3)$$

with v_p et v_air as particles and air velocities, respectively (m/s), D as air flow rate (m^3/min), S as diffuser cross section (m^2), v as kinematic viscosity (1.46 E-05 m^2/s), µ as air dynamic viscosity (1.79 E-05 kg/m/s), rho as air volumic mass (1.22 kg/m^3), R as the Reynolds number

3 Results and discussion

3.1 Laminarity of the powder flows

The powders used in the experiments had mean particle sizes around 20 µm. It was necessary to maintain a laminar flow to allow constant speed of the particles and a consequent good control of the forces applied to the powder flow. The flow can be considered as laminar if the Reynolds number is below 2000. Then, their speed at the diffuser outlet was considered identical to the air flow Results obtained with different air flow rates are reported in Table 1.

Table 1. Air velocity, Reynolds numbers and flow regimes obtained for increasing air flow rates at the exit of the diffuser (temperature: 288 K).

Air flow rate (m^3/min)	Air velocity (m/s)	Reynolds number	Régime
0.010	1.09	223	Laminar
0.011	1.20	246	Laminar
0.012	1.31	268	Laminar
0.013	1.42	290	Laminar
0.014	1.53	312	Laminar
0.015	1.63	335	Laminar
0.016	1.74	357	Laminar

Therefore, the regime of particle flows was laminar whatever the air flow rate in the range used in these experiments.

3.2 Aerodymanic experiments

200 g starch and ash powders were separately applied to the separator to test the behaviour of the fine particles. A typical separation curve is shown in Figure 2, with ash powder as an example. Fractions of the powder were collected in the bins according to the distance from the diffuser, and their granulometry was measured. It was observed that the main part of the material was recovered at distance comprised between 10 and 30 cm from their starting point. These particles had a median particle size of ~20 µm corresponding to the powder nominal d_{50}. For longer distances, decreasing amounts of powder fractions

with regularly decreasing particle sizes were collected. This distribution is in accordance with the particle size distribution curve of this powder obtained from laser granulometry. Interestingly, a small-sized particle fraction was recovered at a very short distance from the diffuser. This results probably from turbulent flowing at the exit of the diffuser and the formation of eddied which attracts small particles at first place. A similar curve was obtained with starch powder, with a slight shift towards larger distances of recovery.

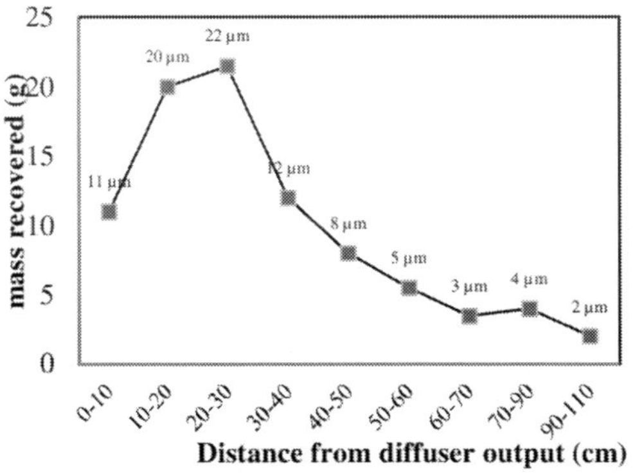

Fig. 2. Mass distribution and median particle sizes (d50) of ash powder particles as a function of the distance from diffuser exit (powder flow rate: 20 g/min, air flow rate: 18 l/min).

A 50/50 blend of starch and ash was then separated in the device with the same operating conditions. The result is shown in Figure 3. The 10-40 cm zone from the diffuser was clearly enriched in ash particles, whereas starch particles were more concentrated in the more distant zone (50-110 cm). In the shortest zone (0-10 cm), approximately equal amounts of starch and ash were present. This distribution evidences that the behaviour of particles is different in blends compared to mono-component powders. Here it is seen that the denser ash-rich fractions repel the lighter starch-rich fractions to longer distances of recovery. This opens the way to a partial separation of the two components in a simple manner.

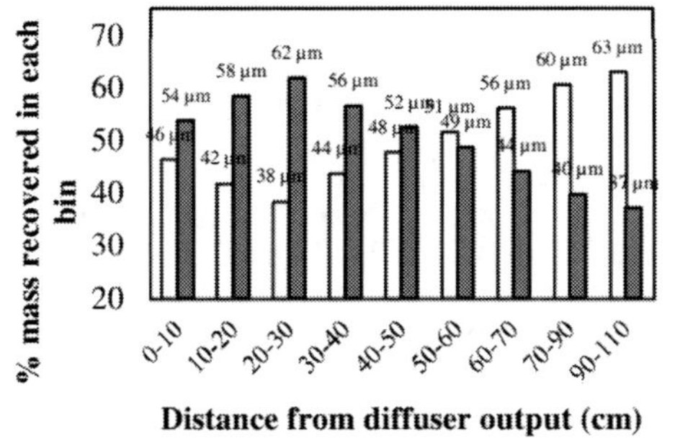

Fig. 3. Mass ratio distribution of starch (white) and ash (grey) from a 50/50 blend separation according to the projection distance from the exit of diffuser (powder flow rate: 20 g/min, air flow rate: 18 l/min)

3.3 Electrostatic experiments

Preliminary experiments combining pressurized projection and electric deviation of powders have been carried out. Both starch and ash powders were used separately for these experiments. When a 10 kV tension was applied between the electrodes, it was observed that the powder clouds were deviated as compared to purely aerodynamic tests and that part of the particles were going to stick onto the charged electrodes. These first results already demonstrate the interest of coupling dynamic aero- and electro-separations, but must be completed by further experiments that are currently under realization. These experiments use mixtures of starch and ash powders and will be interpreted by comparing to the separation obtained by the aerodynamic set up alone.

4 Conclusion

A prototype innovative device for the classification of complex fine powders of lignocellulosic biomass was designed. The instrument included elements for the projection, the classification and the collection of fine particles. It was dimensioned following experiments carried out with pure starch and ash powders. Mixtures of these two powders were used to assess the separation performances of the equipment. With simple projection of powders, a partial classification of the mixture was obtained so that fractions enriched in starch or in ash could be collected. The coupling with an electric deviation of the tribo-electrically charged particles was also set up, additionnal separative phenomena were observed, in preliminary experiments. These results point out a possible new mode of recovering partially purified fractions from complex and heterogeneous vegetal powders. When the different geometries and conditions of use will be finalized, the prototype equipment would be ready for scale-up development for a potential usage in plant biomass bio-refinery.

References

1. V. Menon, M. Rao, Trends in bioconversion of lignocellulose: Biofuels, platform chemicals & biorefinery concept, Prog. Energy Combust. Sci., **38**,522-550 (2012)

2. A. Barakat, H. de Vries, X. Rouau, Dry fractionation process as an important step in current and future lignocellulose biorefineries: A review, Bioresour. Technol., **134**, 362-373 (2013)

3. M.A.I. Schutyser and A.J. van der Goot, The potential of dry fractionation processes for sustainable plant protein production. Trends Food Sci. Technol., **22**, 154-164 (2011)

 DOI: 10.1051/epjconf/201714008008

The theory of granular packings for fine soils

Calixtro Yanqui

San Agustin National University, Arequipa, Peru

Abstract. Fine soils are substances made of tiny particles, for which, besides their geometrical assemblage, the electrochemical forces are important. But the broad experience in Soil Mechanics has shown that, excluding the time, they behave like granular materials. By using the scanning electron microscopy, it has been shown by several authors that clay assemblages are structures of the second order, made of solid particles grouped in clusters. Applying the theory of packings of spheres in contact to the intra-cluster structure as well as to the entire mass of soil, a fundamental equation is obtained. The plastic limit is found to be a constant for a geological formation, because it is obtained by eliminating the capillary water and transforming the clay mass into a first order structure. The liquid limit is a inherently property of the second order, due to the development of clusters. Consequently, there is a linear relationship between the plasticity index and the liquid limit, with a slope of unity, as can be deduced from the abundant data reported by several authors around the world.

1 Introduction

In spite of its importance, a structural model to describe quantitatively the fine soils behavior has not been developed yet, mainly due to platelet shape of their particles, as well as their complicated assemblage, and the effect of bonding. However, the consolidated and drained shear tests have shown that bonding is a time dependent variable, whose effect becomes null after a long time. In this paper, the theory of granular packings, that has proved to be useful to explain several features of sands, is adapted to take into account the other two features.

2 Assemblage of fine grains

Fine soils are made of very small particles. Qualitatively, it is known that, if two clay particles are brought them toward each other, they exert forces on each other; basically, repulsion due to their electrical charge, and attraction, due to the van der Waals or secondary bonding force. According to the predominance of some force, fine soils flocculate or disperse; causing the soil to have very open assemblage. From recent studies on real clay soils using the scanning electron microscopy [1], the individual particles appear to be aggregated or flocculated together in an assembly of small units called domains, which in turn group together to form clusters, that are large enough to be seen by a simple microscope (Fig. 1). Other authors classify the fabric of clays in two classes: elementary particle arrangements, and groups of clay platelets [2]. Powers [3] arrives to the same conclusion, when studying the structure of the fresh cement paste, defining two kinds of evaporable water: the adsorbed water, held by the individual particles by short-range forces, and the capillary water, filling the much larger pores. As early as 1962, Olsen [4] showed that, for explaining the hydraulic conductivity, fine-grained soils should be modeled as clusters of spherical shape.

Fig. 1. Clay structure, modified from Collins and McGown [1]

3 Theory of granular packings

A grain may be defined by two diameters: the equivalent diameter, D, and the diameter of contact between two grains, D_c. So that, a set of grains can be assembled in packings of equivalent spherical grains, whose centers are at the points of the crystallographic Bravais lattices whose edges are of the same length, and equal to the diameter of contact between the spheres, D_c (Fig. 2): the trigonal bipyramid and the rhombohedron. According to the principle of multiplicity, the spatial orientation adopted by this rhombohedron depends on the cinematic conditions imposed by the applied stress.

* Corresponding author: cyanquimurillo@hotmail.com

3.1 Phase relationships

In general, it is always possible to choose a parallelepiped as a primitive cell whose vertices are the centers of eight neighboring spheres, not necessarily all in contact. Now, the total volume of this cell is $V=abc\sin\alpha\cos\zeta$, and the volume of the grain, $V_s=(\pi/6)D^3$; (Fig. 2.a). Thus, the voluminosity of the parallelepiped can be found from the expression [5]:

$$\upsilon = \frac{6abc}{\pi D^3}\sin\alpha\cos\varsigma \qquad (1)$$

where υ stands for the voluminosity of the packing of spheres; *a*, *b*, and *c*, are the lengths of the edges of the parallelepiped; α, the acute angle forming by the two edges of the horizontal face; and ς, the azimuth angle of the non horizontal edge. It is worth to mention at this point that there are sophisticated models to describe some additional features of granular packings, such as the grain shape and the polidispersivity [6] [7] [8].

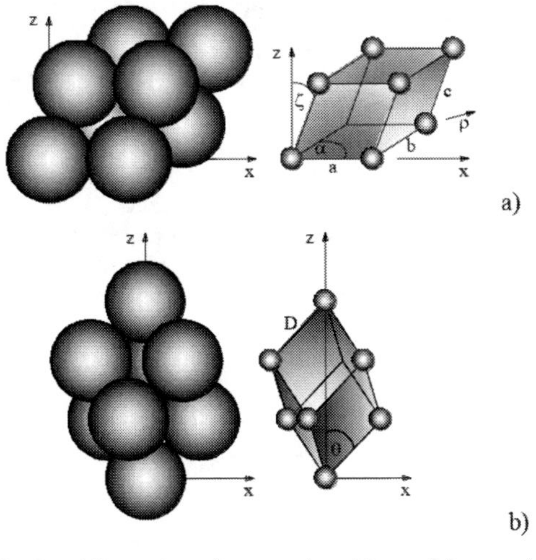

Fig. 2. a) For $a\neq b\neq c$, the general packing and its associated crystallographic cell, and specially, for $a=b=c=D_c$, the distortional rhombohedron, b) the elongational rhombohedron

3.2 Mechanically allowable packings

In engineering, the final object of the packings is the study of the stress-strain relationship of the granular matter. For it, packings are to be cinematic and statically admissible. The first condition ensures the continuous deformation of the mass, and the second, the stress transfer form one grain to another. Consequently, from the fourteen Bravais cells, only the monoclinic, the rhombohedric and the bipyramidal cells are acceptable. But, only the rhombohedron allows the configuration of chains of contact forces, and the development of large deformations, regarding the experimental behavior of soil. However, the cell of this packing must not necessarily match the basic parallelepiped used to calculate the phase relationship.

Distortional kinematics of a packing requires that one of the faces of the rhombohedron be horizontal and the principal diagonal be on the plane of mirror symmetry (Fig. 2.a). Under these conditions, the azimuth angle becomes the shear angle. So that $\varsigma = \beta$. Besides, the maximum deformation capacity is achieved when $\alpha=90°$, and the wedges of the parallelepiped becomes $a=b=c=D_c$. Then, defining the textural parameter $\chi=(D_c/D)^3$, equation (1) may be specialized to:

$$\upsilon = \frac{6}{\pi}\chi\cos\beta \qquad (2)$$

For elongational kinematics of a packing, the principal diagonal of the rhombohedron is vertical, and the lines of contact are axisymmetric with respect to this axis, according to the angle θ (Fig.2.b). Then, $\varsigma = \theta$. These conditions are only possible if the base of the basic parallelepiped is made of two equilateral triangles, resulting in $\alpha = 60°$; $a = b = D_c\sqrt{3}\text{sen}\theta$ and $c = D_c$. Then,

$$\upsilon = \frac{9\sqrt{3}}{\pi}\chi\sin^2\theta\cos\theta \qquad (3)$$

3.3 Packings duality

Analysis of equations (2) and (3) shows that, for two physical values of the angle of distribution, there is a single value of the voluminosity. This property has been called duality of packings and it is one of the fundamental concepts in interpreting the behavior of soils. Indeed, these functions show two branches, separated by a peak value: the first associated with the increasing of the voluminosity, describing a dilative state, and the second, related to the decreasing of the voluminosity, revealing a contractive state.

3.4 Extreme states

In addition, due to its discrete nature, the angles involved in these functions have physical meaning in a limited range, whose boundaries are associated with the densest states, which are achieved when the number of contacts between spheres is increased. For a distortional packing, the loosest state is attained when $\beta=0°$, and the voluminosity is 1.91; and the two densest states are obtained when $\beta = 45°$ and $\beta = -45°$, for which the voluminosities coincide with each other and are 1.35, In terms of the volume fraction, the extreme values are 0.523 and 0.74, respectively. For an elongational packing, the loosest state is found when $\cos\theta=1/\sqrt{3}$, being the voluminosity 1.91; the densest state for a dilative packing is attained when $\sin\theta=1/\sqrt{3}$; being the voluminosity 1.35; and the densest state for the contractive packing is obtained when $\cos\theta= 1/3$, being the voluminosity 1.47. So that, the extreme values of the volume fraction are 0.523, 0.74 and 0.68, respectively.

3.5 Submerged ordered packings

The special behavior of the submerged fine solid particles can be described by the Gouy-Chapman theory of the diffuse double layer. Accordingly, a water layer of very small thickness, a_p, is adsorbed onto the solid particle due to chemical interactions and Coulomb electrical attraction, increasing the diameter of contact to: $D_p = D+2a_p$, (Fig. 3). This layer displays colloidal properties, and, consequently, originates the plasticity of fine soils. The voluminosity of the elongational packing of these bigger spheres may be found as:

$$\upsilon_w = \frac{9\sqrt{3}\chi}{\pi}\left(1+\frac{2a_p}{D}\right)^3 \sin^2\theta\cos\theta \qquad (4)$$

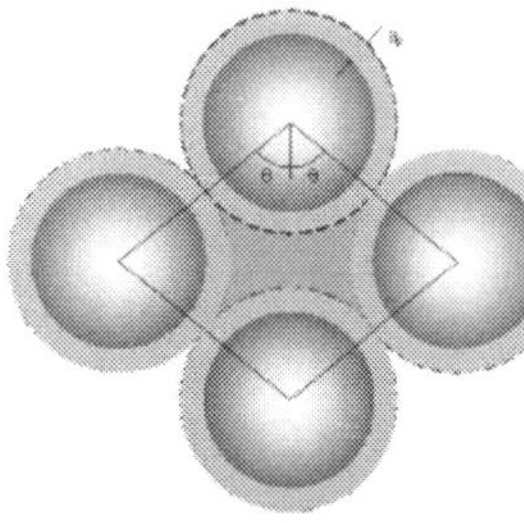

Fig. 3. Two dimensional packing of submerged tiny spheres

4 Structure of fine soils

Since the beginning of the Soil Mechanics, the authors recognize that structure of clays is related to the assemblage of clusters. This means that fine soils may modeled as a packing of packings, as a second order structure (Fig. 4). Within the cluster solid particles are modeled as spheres of contact assembled in a first order packing, whereas, all the soil mass is modeled as a second order packing of spherical clusters of contact. Hence, by definition, the voluminosity of the fine soil is found as $\upsilon = V/V_s$. If Vc is the volume of the cluster, this relationship can be written as: $\upsilon = (V_c/V_s)(V/V_c)$, or:

$$\upsilon = \upsilon_1\upsilon_2 \qquad (5)$$

Fig. 4. Two dimensional packing of spherical clusters, which at the same time, are packings of spherical particles

where υ_1 is the first order voluminosity, and υ_2, is the second order voluminosity. In general, υ is related to the saturated water content, w, by the equation $\upsilon = Gw+1$, being G the specific gravity of the solid particles. In terms of this quantity, υ_1 is associated to the adsorbed water, whereas υ_2, to the capillary water. Equation (5) is the fundamental equation to be used in analyzing the plasticity parameters standardized by the Soil Mechanics, and the Granular Mechanics of fine soils. The plastic limit is obtained by distorting a clay ball exerting some kind of compression to transform it into a cylinder by rolling it several times until fissures. This phenomenon occurs when the capillary water is totally eliminated, and the soil assemblage becomes a first order alone. Consequently, the plastic limit stands for the densest state of plastic soils, and depends only on the properties of the particles, for what it remains constant for a geological formation. The liquid limit, which is obtained by hitting a loose sample on a flat surface, is supposed to stands for the densest state of the second order structure. Numerically, if there is not capillary water, the soil is at the plastic limit, $\upsilon_2=1$, and $\upsilon = \upsilon_1 = 1.35\chi$ = constant; if the soil is at the liquid limit, $\upsilon_1=1$, and $\upsilon = \upsilon_2$, that depends on the sample of the geological formation.

5 Validation of results

The difference between these two limits is called plasticity index and is denoted by I_p.

$$I_p = w_l - w_p \qquad (6)$$

As the plastic limit is a constant, this quantity varies linearly with the liquid limit. It is worth to mention that this observation was already known by the pioneers of the soil mechanics. The contribution of the theory of granular packings is the establishment of the slope of the line as the unity, if the soil comes from the same geological formation.

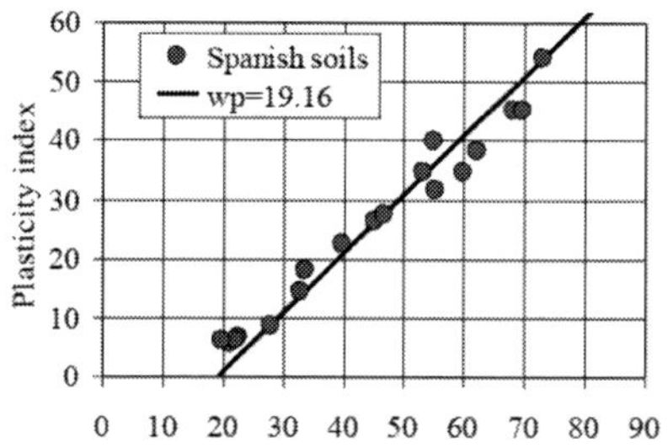

a)

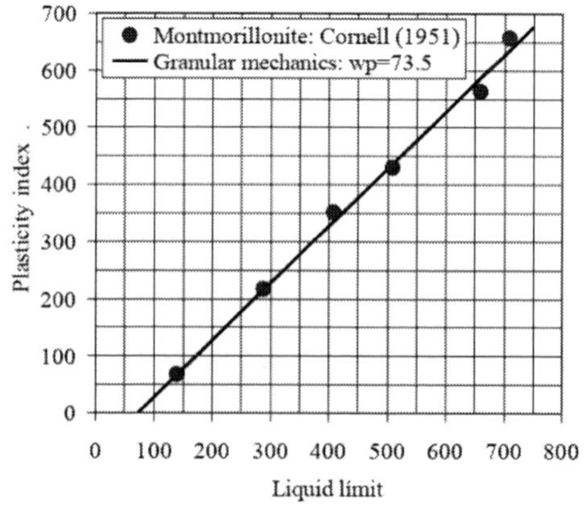

b)

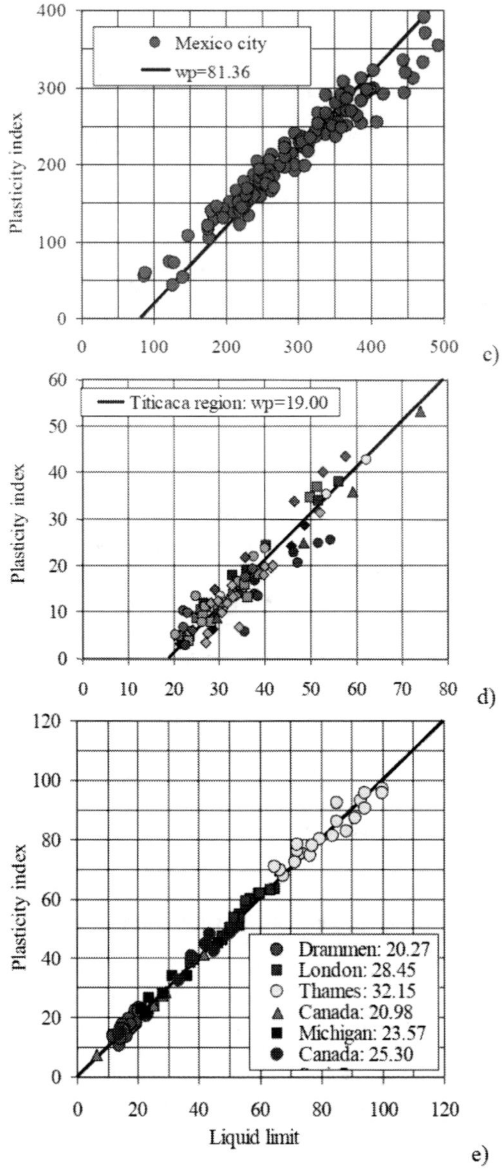

Fig. 5. Relationship between the liquid limit and the plasticity index for: a) Spanish soils [9] b) montmorillonite [10], c) Mexico city clay [11], d) Southern Peru clays, e) several clays around the world, compiled by Lambe and Whitman [12]

Geologically, these findings may be interpreted as follows; the plastic limit defines the nature of the fine particles, and the liquid limit stands for the effect of some other geological feature. As a matter of fact, the experimental data of some clay formations from different places of the world, usually obtained during normal subsoil prospecting work, and reported by various authors, fit very well with equation (6), where w_p is the average of all site plastic limits. Besides, it is interesting to notice that the experimental plasticity index never is zero, revealing the jump from the plastic limit to the liquid limit, associated to the development of clusters or, on the opposite way, capillary water voids.

6 Conclusions

Fine soils are made of assemblages of clusters of tiny particles, displaying structures of the second order. In spite of this complication, fine soils can be described using the theory of granular packings. Within the cluster, the packing is the same as the one of coarse grains. The entire soil mass may be considered as a packing of clusters, modeling the structure of the second order. The plastic limit is obtained in such a way that the capillary water disappears, becoming a property of the clay composition and, therefore, a constant for the geological formation. The liquid limit is obtained in loose samples. So, it stands for the second-order structure of fine soils. As long as the plastic limit is a constant, the plasticity index is linearly related to the liquid limit, with a slope of unity. The data reported by several authors confirm this finding.

References

1. R.N. Yong, D.E. Sheeran, Fabric unit interaction and soil behaviour. *Proc. Symp. On Soil Structure*, Sweden, 176-183 (1973)
2. K. Collins, A. McGown, The form and function of microfabric features in a variety of natural soils. *Geotechnique XXIV*, 223-254, (1974)
3. T.C. Powers, *The properties of fresh concrete*. John Wiley. New York. 1968.
4. H.W. Olsen, Hydraulic flow throw saturated clay. *Proceedings Ninth Nat. Conf. on Clays and clay minerals*, 131-161 (1962)
5. C. Yanqui, The theory of packings for coarse soils. *AIP Conf. Proc.* 1542, 341-344 (2013)
6. T. Aste, Circle, sphere, and drop packings. *Physical Review E*, Vol. 3, No. 3, 2571-2579. (1996)
7. F. de Larrard, T. Sedran. Optimization of ultra high performance concrete by the use of packing model. Laboratorie Central des Ponts and Chaussées (1996)
8. G. Mollon, J.D. Shao, Fourier-Voronoi based generation of realistic samples for discrete modelling of granular materials. *Granular Matter*. 14(5), 621-638 (2012).
9. J.A. Jimenez, J.L de Justo, *Geotecnia y cimientos*, Tomo I. Editorial Rueda. Madrid (1975)
10. Cornell University, *Final report on soil solidification research*. New York (1951)
11. R.J. Marsal, Unconfined compression and vane shear tests in volcanic lacustrine clays. *Proc. ASTM Conf. on Soils for Eng. Purposes*, Mexico (1959)
12. T.W. Lambe, R.V. Whitman, *Soil Mechanics*. John Wiley and Sons. New York, (1969).

Granular mechanics of normally consolidated fine soils

Calixtro Yanqui

San Agustin National University, Arequipa, Peru

Abstract. In this paper, duality is demonstrated to be one of the inherent properties of granular packings, by mapping the stress-strain curve into the diagram that relates the pore ratio and the localization of the contact point. In this way, it is demonstrated that critical state is not related to the maximum void ratio, but to a unique value related to two different angles of packing, one limiting the domain of the dense state, and other limiting the domain of the loose state. As a consequence, packings can be dilative or contractive, as mutually exclusive states, except by the critical state point, where equations for both granular packings are equally valid. Further analysis shows that stresses, in a dilative packing, are transmitted by chains of contact forces, and, in a contractive packing, by shear forces. So that, stresses, for the first case, depend on the initial void ratio, and, for the second case, are independent. As it is known, normally consolidated and lightly overconsolidated fine soils are in loose state, and, hence, their strength is constant, because it does not depend on their initial void ratio; except at the critical state, for which, the consolidated-drained angle of friction is related to the plasticity index or the liquid limit. In this fashion, experimental results reported by several authors around the world are confronted with the theory, showing a good agreement.

1 Introduction

Clays behaves in different way when submitted to a shearing stress, according to the degree of overconsolidation. Normally consolidated or lightly overconsolidated saturated clays are observed to show gradual increase in shear stress as the shear strain increases until an approximately constant shear stress, $\tau=\tau_{cs}$, is reached, and to compress until an approximate voluminosity, $\upsilon=\upsilon_{cs}$, is attained. Heavily overconsolidated clays are observed to show a rapid increase in shear stress reaching a peak value, τ_p, at low shear strains, and then show a decrease in shear stress with increasing shear strain, ultimately attaining the value $\tau=\tau_{cs}$. These clays are observed to compress initially and then to expand until a voluminosity $\upsilon=\upsilon_{cs}$ is attained. The quantities τ_{cs} and υ_{cs} define the critical condition, at which the shear stress and the voluminosity do not change any more. The behavior of the first soil is called "contractancy", and that of the second was named by Reynolds "dilatancy" [1], [2]. In this paper, only normally consolidated and lightly overconsolidated clays are studied.

As occurs with sands, this peculiar behavior of clays is a consequence of their particulate nature. So that, normally consolidated and lightly overconsolidated clays may be analyzed by means of the theory of granular packings, having account of the second order of their structure, for what υ_1 is the packing of clay particles and υ_2, the packing of clusters [3]:

$$\upsilon = \upsilon_1 \upsilon_2 \tag{1}$$

As it has been shown previously, the minimum voluminosity of the packing of particles remains constant for a geological formation, and, hence, the packing of the second order governs the mechanical behavior of clays.

2 Mechanics of granular packings

The strain of the basic cell whose vertexes are the center of the contact spheres, considered for the moment to be rigid, may be described in terms of the azimuth angles of assemblage, regarding the principle of multiplicity of packings. In this path, it has been established that the rhombohedron is the appropriate lattice for studying the clay behavior [3].

2.1 Distortional packing

Because its kinematics is ease to understand, the distortional or shearing packing of contact spheres is to be described in some detail (Fig.1). Once the relationship between the voluminosity, υ, and the azimuth angle β of the line of contact is established [3], the change of volume with the shearing strain is obtained as its derivative:

$$\frac{d\upsilon}{d\beta} = \frac{6\chi}{\pi}\sin\beta \tag{2}$$

According to this equation, there is not change of volume during pure shearing when β is zero, that is, when the packing is at the loosest state. If β is positive,

* Corresponding author: cyanquimurillo@hotmail.com

the packing dilates, and, if β is negative, it compresses. This model, proposed by several authors, seems to be appropriate to describe the behavior of a soil assuming the friction between grains as the friction angle at the critical state, φ_{cs}. The static balance on the two vertical spheres, in contact gives the friction angle of the soil as: $\varphi = \varphi_{cs} + \beta$ [1]. But this equation does not match the experimental data. This is due to the following mistakes: first, the static equilibrium is established along the inclined chain of forces only; and second, the critical state is supposed to be related to the loosest state.

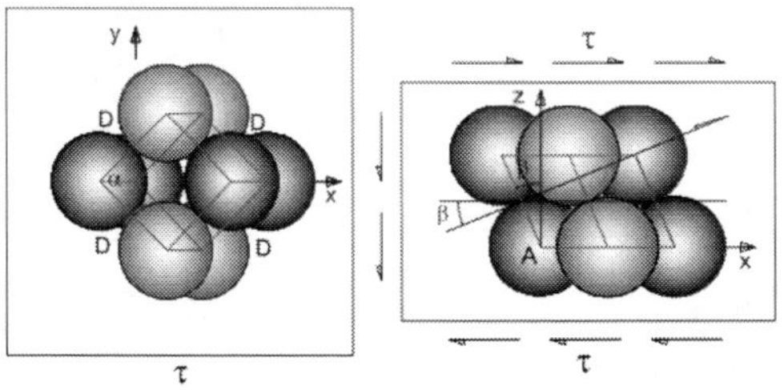

Fig. 1. Dilative shearing packing: a) plan view, and b) section showing the mechanical elements

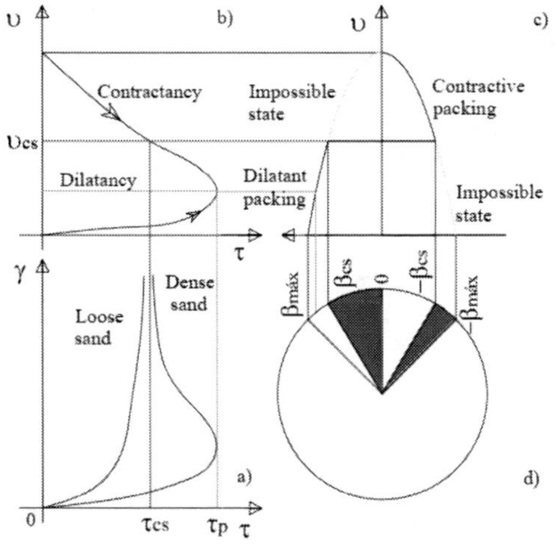

Fig.2. Mapping of the soil shear strength and the parameters of contact packing

The first mistake is overcome by including the horizontal forces of contact, regarding the law of reciprocity of the shear stress. To overcome the second mistake, it is necessary to map the stress-strain plot onto the lower contact sphere of the basic cell, through the Casagrande´s diagram, and the volumetric strain plot (Fig. 2). From this mapping, it is seen that the voluminosity at the critical state is not related to the maximum voluminosity, but with two points of contact, defined by the angles $|\beta_{cs}|$, one for the dilative branch and other for the contractive branch, both related to the same voluminosity, υ_{cs}.

Further analysis of the υ-sinβ diagram shows that, on the subtended elliptical arc, there are four sections, separated by five points: two of them associated to the densest states, two defined by the critical state, and one related to the loosest state. Of these four sections, only two are physically possible: one for the dilative packing and one for the contractive packing, both of them connected by the voluminosity at the critical state, υ_{cs}. In synthesis: for a dilative packing, contact is possible if $\beta_{cs} \leq \beta \leq 45°$ and is impossible if $0 \leq \beta \leq \beta_{cs}$; and, for a contractive packing, contact is possible if $-\beta_{cs} \leq \beta \leq 0$ and is impossible if $-45° \leq \beta \leq -\beta_{cs}$.

2.2 Elongational packing

The above analysis also applies to triaxial compression test. In this case, the basic cell must meet the conditions of axial symmetry and large elongation. This happens when the principal diagonal of the rhombohedron is vertical, and its eight spheres occupy four horizontal layers [3]. The change of volume may be related to the axial shearing strain through the equation:

$$\frac{d\upsilon_\phi}{d\theta} = \frac{9\sqrt{3}}{\pi}\sin\theta\left(3\cos^2\theta - 1\right) \qquad (3)$$

where θ is the azimuth angle of the line of contact. Equation (3) highlights the dual character of the elongational packing. For a dense or dilative packing, contact is possible if $35.26° \leq \theta \leq \theta_{cs1}$ and is impossible if $\theta_{cs1} \leq \theta \leq 54.74°$. For a loose or contractive packing, contact is possible if $54.74° \leq \theta \leq \theta_{cs2}$ and is impossible if $\theta_{cs2} \leq \theta \leq 70.53°$; where θ_{cs1} and θ_{cs2} are the angles corresponding to the critical state voluminosity, υ_{cs}.

3 Shear strength of a clay sample

In the classic Casagrande´s diagram, curves refer to the initial loosest and densest states of a soil sample. But the complete description of all the initial states of a sample submitted to the shear test requires an infinite family of curves. In figure 3.a, it has been plotted the curves corresponding to ten initial states of a sample; five for the dense state and five for the loose state, separated each other by the line $\upsilon = \upsilon_{cs}$. It can be seen that the shear stress of each dense state is associated with a peak value, after which the response curve decreases asymptotically to the critical stress; whereas for the loose state, the maximum shear stress is the critical strength, τ_{cs}. In figure 3.b, the graph of the initial voluminosities against the respective maximum values of the shear strength is shown. There is observed that the voluminosity at the critical state separates two regions: one where the strength of the soil is a curve and is always greater than the critical strength; and other where the strength is constant and equal to the critical strength. That is the reason by which the shear strength of clays has been successfully correlated equally with the liquid limit, the plasticity index, the saturated natural water content, among others. Particularly, in soil mechanics, it has extensively used the plasticity index and the liquid limit. It can be concluded that shear strength of a clay sample, such as the consolidated-drained friction angle, is a constant.

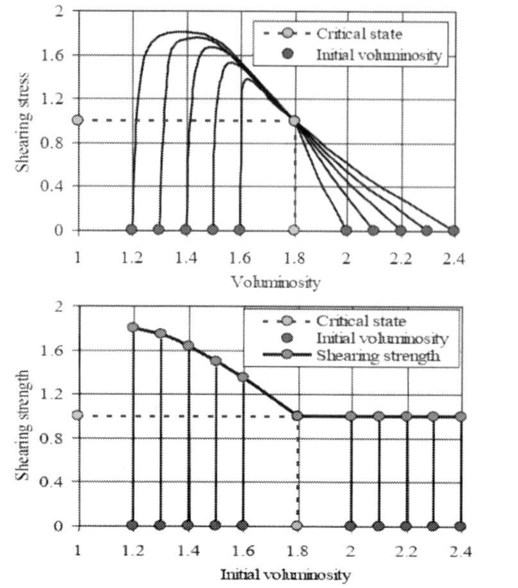

Fig. 3. Illustrative results on clay sample: a) Generalized Casagrande's diagram, b) The shear strength as a function of the initial voluminosity of a soil sample, showing the critical initial voluminosity that separates the shearing strengths for a dilative and contractive packing respectively.

4 Shear strength of a clay formation

For clay soil samples coming from the same geological deposit, soil engineers have noticed that the shear strength exhibit a correlation with the voids contents of the constituent layers. This means that the shear strength is constant for any sample of the same layer, but is different for the samples coming from different layers of the same geological formation. As demonstrated above, it is not possible to establish a relationship between the saturated water content and the friction angle because, this is independent of the voluminosity for normally consolidated clays, except by one point: the critical state, where equations of dilative and contractive states are completely valid. Therefore, for the critical state, the contractive soil can be described by using the hypothesis and methods of dilative soils. So this, the entire geological formation is described by the equation [3]:

$$w = \frac{1}{G}\left[\frac{9\sqrt{3}}{\pi}\chi_1 \sin^2\theta_{cs}\cos\theta_{cs} + \chi_2\right] \qquad (4)$$

5 Elementary granular statics

Structural cinematic parameters of soil are usually not known. For this reason, it is necessary to establish a relationship between them and any known parameter. In this path, simplest stress states can be derived, such as the geostatic and the infinite mass slopes. For dense soils, the transmission of the stresses happens as a network of chains of contact forces. These properties can be evidenced, for example, in photographs taken on the two - dimensional specimens of birefringent packing of disks regularly or randomly distributed [4]. They can also be obtained through simulations based on molecular dynamics. It has been shown [5] that stresses obtained from chains of contact forces between grains, assumed as linear in an infinite soil slope, leads to the following relationships:

$$\frac{\tan^2\theta_{cs}}{n-1} = K_0 = \frac{1}{1+n\tan^2\varphi_{cs}} \qquad (5)$$

where n represents the dimension of the packing, φ_{cs}, the consolidated and drained angle of internal friction, and K_0, the coefficient of lateral pressure at rest. Substituting (5) in (4), a general equation for the clay formation is attained. Particularly, the saturated water content can be replaced by the liquid limit, w_l, or by the plasticity index, $I_p = w_l - w_p$, where w_p is the plastic limit. Besides, it must be noticed that the general equation is bounded as follows: $\varphi_\mu \leq \varphi_{cs} \leq \varphi$, where φ_μ is the particle-to-particle shear strength, that plays the same roll respect to φ_{cs} as done by φ_{cs} respect to φ. For instance, recognizing that the plastic limit remains constant for a geological formation, the following equation holds:

$$I_p = \frac{9\sqrt{3}\chi_{1p}}{\pi G}\left[\cos^2\varphi_{cs}\sqrt{3-2\cos^2\varphi_{cs}} - \chi_{2p}\right] \qquad (6)$$

6 Comparison with experimental data

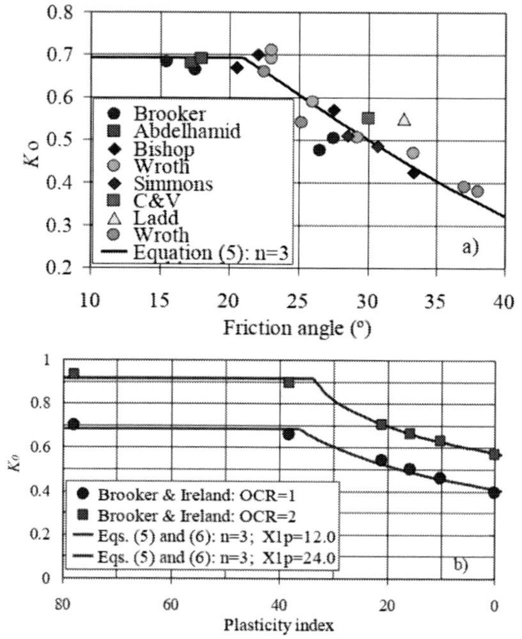

Fig. 4. Relationship between the coefficient of lateral pressure, K_0, and: a) the consolidated-drained friction angle at the critical state for different clay samples, as compiled by Holtz and Kovacs [6], b) the plasticity index for two clay samples [7].

Equations (5) and (6) may be confronted with the experimental data reported by several authors around the world. In figure 4.a, the K_0 values are compared with the consolidated-drained friction angle, φ_{cs} for clays of diverse origin. In this case, the curve has two sections, one that is independent of φ_{cs} and comes from the particle-to-particle friction angle, φ_{μ}, and another that corresponds to the critical state. In figure 4.b, K_0 values are related to the plasticity index, I_p, for two clay samples whose overconsolidation ratios (OCR) are 1.0 and 2.0. It is seen that the curve shows the same duality, with $I_p = 35$ as common point, approximately. In figures 5 and 6, the data reported by several authors, linking the friction angle at the critical state with the plasticity index and the liquid limit are presented to show the good correspondence with equations (6) and (5), respectively.

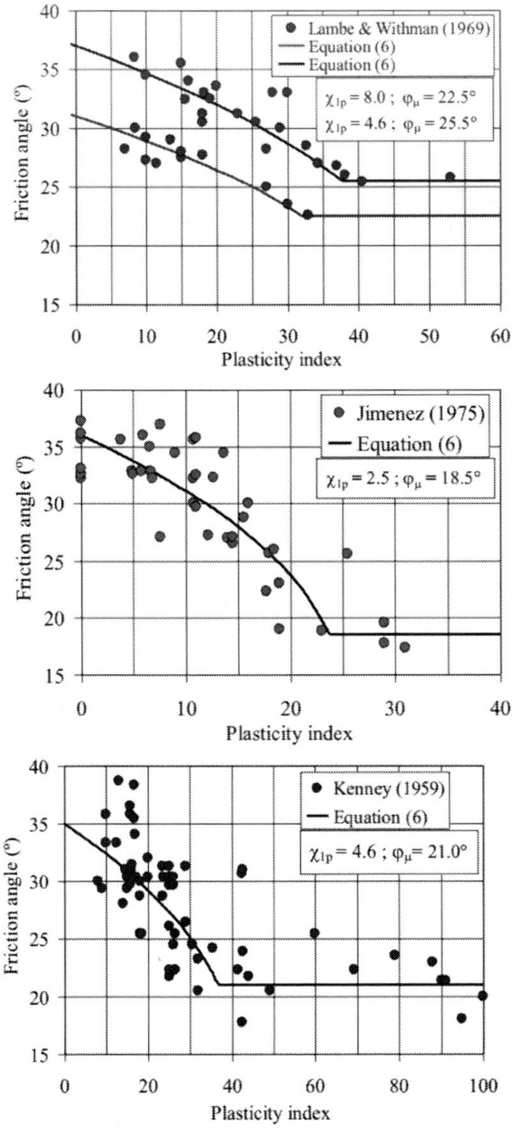

Fig. 5. Comparison of equation (6) with the experimental data reported by some authors: a) Marine clays [2], b) Spanish compacted clays [8], c) Several clays samples, as compiled by Kenney [9]

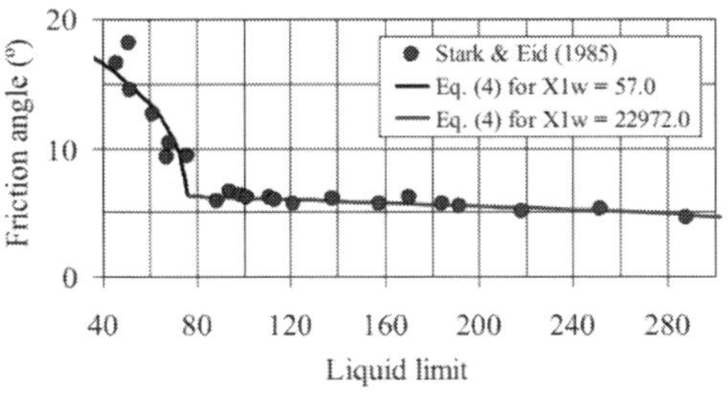

Fig. 6. Comparison of equation (4) with the experimental data reported by Stark and Eid [10] for a clay sample, showing the critical state and the residual shear strengths

7 Conclusions

Normally consolidated and lightly overconsolidated clay can be modeled as contractive packings. Mapping of the Casagrande´s diagram into the circumference of contact shows that the voluminosity at the critical state does not correspond to the loosest state of a packing. The shear strength of contractive packings is constant and independent of the initial water content. Loose and dense states are exclusive one from the other, except at the critical state. For this reason, clays belonging to the same geological formation can be studied by using the equation of the dilative packing at the critical state. Within this context, the theory of linear chains of contact forces can be used to obtain a relationship between the consolidated-drained friction angle and the packing angle, as well as the coefficient of lateral stress at rest. The experimental data for specific geological formations reported by several authors fits very well the theoretical curve. Even for broad regions, the correlation is good.

References

1. M. Budhu, *Soil Mechanics and Foundations*. Third edition. John Wiley and Sons, New York, (2011)
2. T.W. Lambe, R.V. Whitman, *Soil Mechanics*. John Wiley and Sons. New York, (1969)
3. C. Yanqui, The theory of granular packings for fine soils. Powders and grains (2017)
4. T.S. Majmudar and R.P. Behringer, Contact force measurements and stress- induced anisotropy in granular materials. *Nature* **435**: 1079, (2005)
5. C. Yanqui, Granular mechanics of the critical state of coarse soils. *Powders and grains* **1542**, pp 197-200 (2013)
6. R.D. Holtz, W.D. Kovacs, *An introduction to geotechnical engineering*. Prentice Hall, (1981)
7. E.W. Brooker, H.O. Ireland, Earth pressures at rest related to stress history.*Canadian G. J.* **11** (1965)
8. J.A. Jimenez, J.L de Justo, *Geotecnia y cimientos*, Tomo I. Editorial Rueda. Madrid (1975)
9. T.C. Kenney, Discusión, *Proc. ASCE* **85**, 67-89, (1959)
10. T.D. Stark, H.T. Eid, Drained residual strength of cohesive soils. J. Geotech. Eng. ASCE, **120**, 856-871, (1994)

Inelastic wedge billiards

Martin Martinez and *Jeffrey S.* Olafsen*

Department of Physics, Baylor University, Waco, TX 76798 USA

Abstract. Billiards are simple systems used to investigate Hamiltonian dynamics in physics. When real billiards are examined experimentally, the energy dissipated in each collision must be replaced by an external stimulus to maintain the dynamics. We focus on a specific system of a driven billiard using a wedge shaped boundary to examine nonlinear and chaotic behavior. Mathematical models such as the logistic map are simple low dimensional systems that exhibit nonlinear and chaotic behavior as a single parameter is varied. This logistic map can then be used to identify a very specific mathematical parameter known as the Lyapunov exponent, which helps in identifying chaos more clearly. In the current experiment, the dynamics of a particle free to move near a horizontally shaken vertical boundary will be examined for the presence of chaos. The goal of the research is to extract a Lyapunov exponent between any two trajectories in the system. In addition, the manner in which the dynamics evolve freely through dissipative collisions provides a testbed for measurements of the velocity dependent coefficients of restitution for the billiard. A better description of hard sphere coefficients of restitution would be beneficial to a host of experiments and numerical simulations in granular physics.

1 Introduction

Gravitational billiards, where the shape of the particle boundary governs the observed dynamical motion of the particle [1–3], continue to be a rich system for studying instability and chaos in Hamiltonian dynamics. Not surprisingly, parabolic boundaries demonstrate stable orbits whose amplitude is determined by the energy of the billiard [1], while a wedge boundary demonstrates both unstable and chaotic behavior dependent upon the value of the vertex angle of the wedge [2]. An even more interesting case is that of a hyperbolic boundary that demonstrates parabolic behavior for orbits near its origin and wedge behavior for orbits near its asymptotes [3]. The motion of a particle in a wedge has been experimentally demonstrated for an optical billiard with ultra-cold particles [4].

Previously, we have demonstrated an experimental manifestation for a billiard in a driven experiment [5] for two-dimensional parabolic, wedge, and hyperbolic geometries where the particle undergoes inelastic collisions with the confining boundaries in the presence of a gravitational potential. This prior study has led to several simulations of dissipative gravitational billiards [6-8]. In addition, the study of dissipative billiards can contribute to a deeper understanding of the dissipation in granular systems [9]. Inelastic collisions have been shown to lead to clustering [10] and inelastic collapse [11], and a prior investigation analytically and numerically investigated the behavior of a particle of variable elasticity within an effective wedge boundary to search for effects of clustering and inelastic collapse [12]. For a wedge boundary, evidence is obtained of stable, unstable, period doubling, and chaotic dynamics for the billiard. While the energy injection to the system is maintained at a fixed value, the inelasticity allows the particle to move between nearby trajectories, demonstrating a variety of orbits at fixed energy input.

2 Experiment

In both the prior investigation and this study, the experiment consists of a steel frame with an interior square of 16.5 cm × 16.5 cm into which a 3.2 mm thick wedge, parabolic, or hyperbolic aluminum insert can be placed vertically with respect to gravity to examine the motion of a 3.1 mm diameter chrome steel ball within the selected boundary. The coefficient of restitution between the chrome steel ball and the aluminum boundary is approximately 0.9 but is known to be velocity dependent [13]. One of the primary motivations of this study is to reconcile results from simulations that are at odds with this assumption [6-8].

Sheets of Plexiglas constrain the motion of the particle to a thin 2D layer and at the same time allow for observation of the particle's motion. The vertical cell is seated on a linear bearing that constrains the motion of the cell to a horizontal line along the plane of the boundary (see Figure 1). A wheel and linking armature converts the rotational motion of the Dayton 1/8 hp dc motor into a sinusoidal horizontal motion of the cell on the linear slide. The frequency of oscillation, ν, was varied between 4–7 Hz, and the peak-to-peak amplitude of oscillation was held constant at A = 2.54 cm. The system is started from rest and the particle's motion is soon ballistic across the cell. The energy lost in the

* Corresponding author: Jeffrey_Olafsen@baylor.edu

collisions is compensated by the horizontal driving of the cell whose peak acceleration is given by $\Gamma = A\omega^2$, where $\omega = 2\pi\upsilon$. Thus, the energy input to the system is increased or decreased by increasing and decreasing the driving frequency. The images are obtained via a 256×256 pixel 1000 frame per second (fps) Dalsa CA-D6 8-bit camera. For imaging purposes, a second identical chrome steel particle is imbedded in each of the boundary inserts beneath the vertex of each surface, so that high-speed CCD imaging can track both the motion of the cell and the free particle moving ballistically within the cell. The tracking of the particle imbedded in the boundary beneath the vertex allows for an independent measurement of the shaking frequency over the entire data set. Programs written in IDL are used to track the free particle's trajectory and extract the particle's velocity before and after these collisions. A mapping is then made to relate the velocity and position of one collision point to the velocity and position at the following collision point.

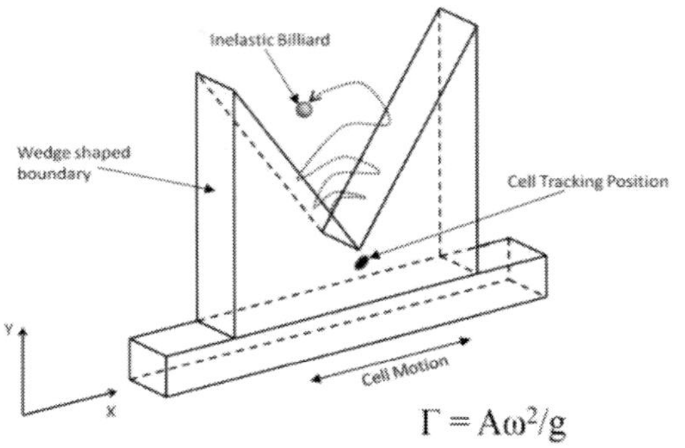

Fig. 1. A schematic of the experimental cell.

To examine the long term behavior of the dynamics, the height at one collision can be plotted versus the height at the following collision, producing a map of the motion. To understand the goals of this study, we reproduce the results of our prior study [5] in the left column of Figure 2. The dashed lines in the figure are the $y_n = y_{n+1}$ return map that would denote stable period-one orbits. The right column of Fig. 2 shows the time of flight between collisions versus the time of flight from the next collision. Figure 2(a) shows the y positions for the case of the parabola driven at 5.4 Hz. The trajectory of the particle is not simply a single stable orbit, but rather accesses several nearby stable orbits, resulting in an elongation of the stable fixed point along the direction of the slope 1 line. Similarly, this leads to a slight smearing of the collision times as can be seen in Fig. 2(b). Figure 2(c) shows the map of y positions of the particle collisions within the wedge driven at 6.6 Hz. The wedge is this study had a half angle of 28.5 degrees. The motion is clearly unstable and the particle is continuously driven to the top of the wedge. In Fig. 3(d), the unstable behavior is represented in terms of the collision times as well as an apparent period-two behavior, a coexisting stable dynamic that one would expect for the selected vertex angle. In Fig. 3(e) – 3(h) we examined the y positions of the particle collisions within the hyperbolic boundary when driven at 4.5 Hz, which was the focus of the prior study.

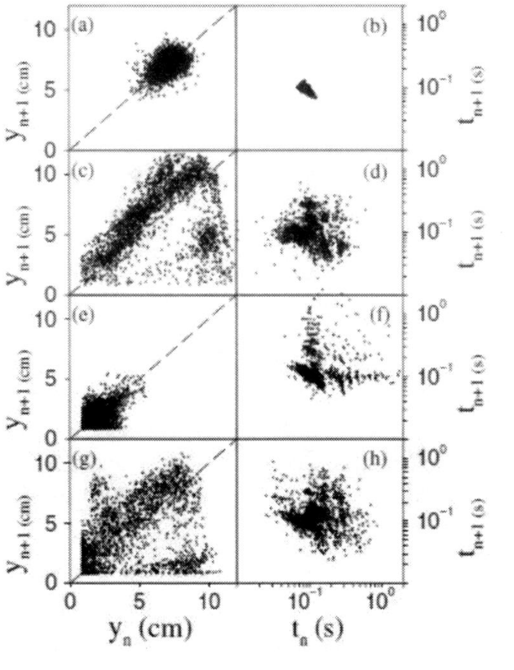

Fig. 2. Previous results from parabolic (a) & (b), wedge (c) & (d), and hyperbolic (e) – (h) boundaries. Reproduced from [5]. The mappings of the collision heights (left column) and the times of flight (right column). In (a) and (b) we see the parabola at 5.4 Hz, (c) and (d) show the wedge at 6.6 Hz, (e) and (f) show the hyperbola at 4.5 Hz, and (g) and (h) show the hyperbola at 5.8 Hz.

For this study, the decision to focus on the wedge geometry was two-fold: varying the half angle of the wedge geometry allows for both the exploration for chaotic behavior as well as to further investigate the inelasticity of the particle within the experiment. In this study, we used three different wedges (shown in Figure 3) with half angles of (a) 28.5, (b) 45 and (c) 60 degrees, respectively. The use of the same 28.5 degree wedge was seen as a manner in which to check the consistency of the results between this study and the prior investigation. While the experimental details of the driving along the horizontal linear bearing and the orientation of the cell vertically to the direction of gravity naturally lends itself to a cartesian description, the collision dynamics lend themselves more naturally to a frame rotated by the value of the half angle of the wedges as shown in Figure 3(d). Within the frame of camera, data is taken that allows the accelerated motion of the free particle in the vertical direction and the constant linear momentum in the horizontal direction to be observed as shown in Figure 4. Through a simple rotation, the dynamics may easily be resolved in the natural directions normal (n) and tangential (t) to each of the three wedge surfaces. In Figure 3(d) we show such a rotation for the right hand side of the cell for the 60 degree wedge. For the left hand side of the wedge boundary (not shown) the normal vector is of course perpendicular to the wedge boundary, but the tangential direction points away from the vertex to the left. This choice results in a tangential direction on both the left and right hand sides of the vertex that is in the direction of increasing potential energy.

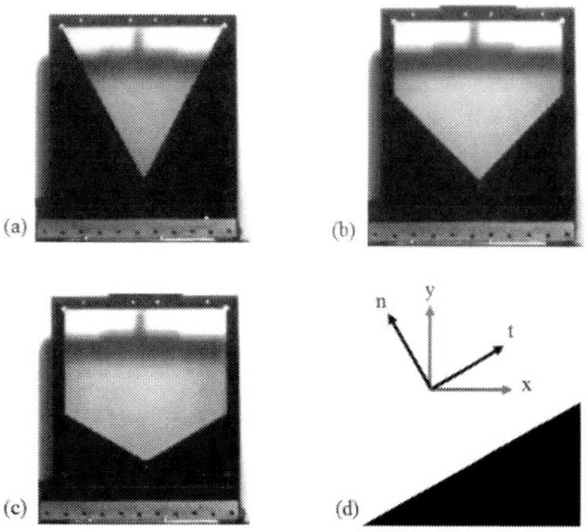

Fig. 3. The wedges used in this study with half angles of (a) 28.5 (b) 45 and (c) 60 degrees respectively. In (d) the rotation of the camera frame (x,y) into the boundary frame of normal (n) and tangential (t) directions is demonstrated for the right side of the 60 degree half angle wedge. Here the cells are photographed with a white background to show the wedges. In the experiment, a black background is used to more easily track the free billiard and the particle imbedded in the boundary beneath the vertex to allow the motion of the cell to be tracked.

3 Results

In Figure 4, we demonstrate the results of tracking both the free particle and the identical chrome steel ball imbedded in the wedge boundary for the cell with a half angle of 60 degrees. In the figure, we show a subset of 4000 frames of data from more than a quarter million images taken for the cell at a shaking frequency of 5.78 Hz. In arbitrary units, we show the results for the y-position (y_{ball}), and the x-position (x_{ball}) of the billiard and the oscillating x-position of the cell (c_{cell}) determined from tracking the imbedded particle.

The free particle demonstrates the expected accelerated motion in the y direction, constant momentum in the x direction and horizontal oscillation of the cell. In addition, the dynamics demonstrates repeating trajectories of the particle climbing along the wedge. Two very similar trajectories in the evolution of the y-position are in Figure 4 between frames 400 and 1900 and again between frames 2200 and 2800.

One method of probing the dynamics is to examine the time of flight between collisions of the billiard with the boundary. The results for one such analysis is shown in Figure 5. Here, we plot the histogram of the time between collisions for the entire data set at 5.78 Hz. In the inset of Figure 5 we demonstrate the results of the tracking program that identifies the ballistic trajectories by locating the ends of each parabolic track. While not shown in this paper, analysis demonstrates that the collisions of the billiard occur nearly uniformly throughout the shaking cycle, showing a lack of coherence between the billiard dynamics and the driving.

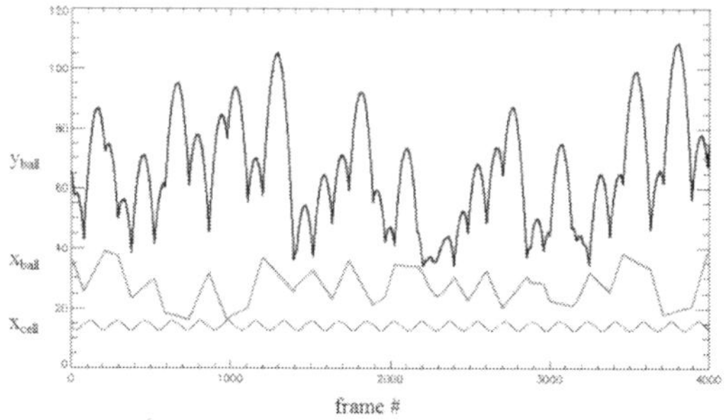

Fig. 4. From top to bottom, the y position of the billiard (y_{ball}), the x position of the billiard (x_{ball}) and the horizontal motion of the driven cell (x_{cell}). To show all three tracks on a single plot, the results are plotted in arbitrary units.

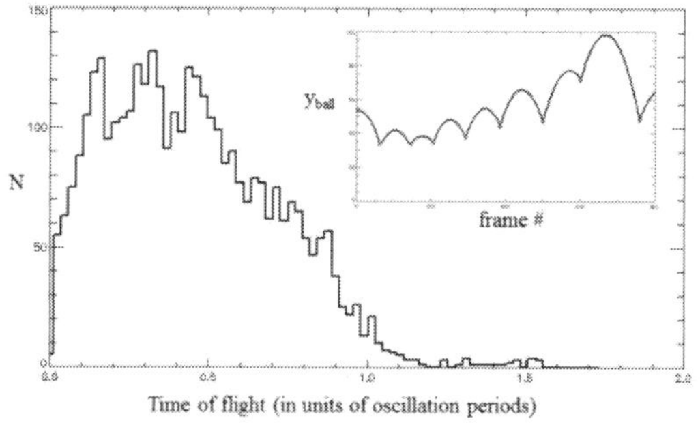

Fig. 5. Histogram for the time of flight of the billiard in the 60 degree wedge driven at 5.78 Hz. The time of flight (determined by identifying collisions from the ballistic motion of the billard as shown in the inset) is plotted in units of the period of oscillation.

In Figure 5, three distinct peaks are apparent in the distribution of the time of flight, potentially indicating chaotic behaviour in the dynamics of the billiard. The three peaks are approximately at 0.15,0.30 and 0.45 of the value of the shaking period (173.5 ms). The three peaks are distinct in what would otherwise be a nearly uniform distribution of the time of flight over half the period of oscillation of the cell. However, collisions occur throughout the entire period of oscillation of the cell and even sometimes over more than one cycle of the shaking (time of flight > the oscillation period).

The conversion of the dynamics into the normal and tangential directions lends well to the analysis of the coefficients of restitution [14-16], one of which is in the normal direction to the collision and the other in the direction tangent to the collision direction (the motion in the third direction is constrained by the 2D nature of the cell). In addition to the individual coefficients of restitution (CORs) based upon the change in relative speed in normal and tangential speeds, it is also possible to quantify the dissipation in the system by examining the total change in energy of the particle before and after the collision. Hence, we define three coefficients as follows:

$$r(v_n) = -1*(v_n^+/v_n^-) \quad (1)$$

$$r(v_t) = (v_t^+/v_t^-) \quad (2)$$

$$r(E) = (\Delta E/E^-)^{1/2} \quad (3)$$

where the +/- velocities are the relative speeds of the billiard after and before the collision, respectively. The normal COR has a -1 prefactor to address the change in direction of the particle in the collision, however the tangential COR has no such constraint. So, the normal COR should be a value between 0 and 1but the tangential COR can have values between ±1. As the energy is positive, it should also have a value between 0 and 1. The energy COR is the square root of the ratio of the change in energy to the energy before the collision.

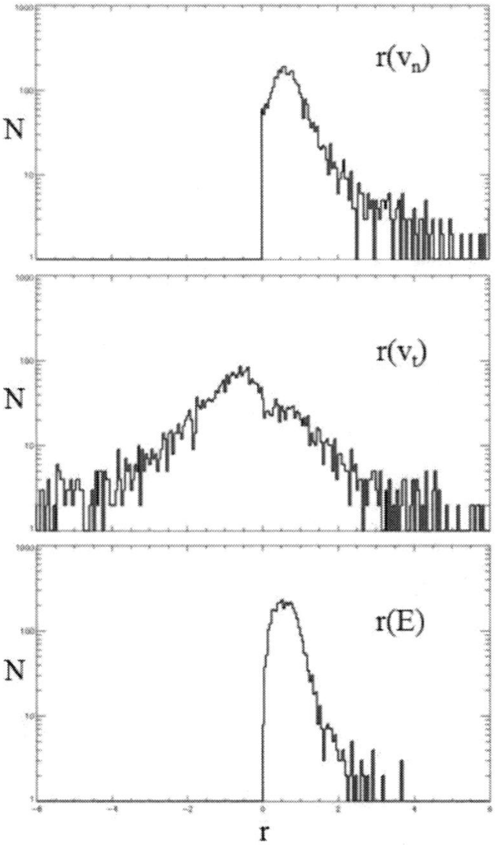

Fig. 6. From top to bottom, the COR calculated for the normal component of the collision velocities after to before the collision, calculated from the tangential velocity component, and from the energy change before and after the collision.

The results in Figures 6 and Figure 7 demonstrate that while the COR in this experiment behaves approximately as expected, there are values obtained that are larger than unity. This indicates that the rotational degree of freedom of the billiard is an important quantity. While it is not measured in this experiment, the rotational degree of freedom acts as an additional source of energy that "leaks" momentum into the normal and tangential velocities during the collisions with the boundary [16].

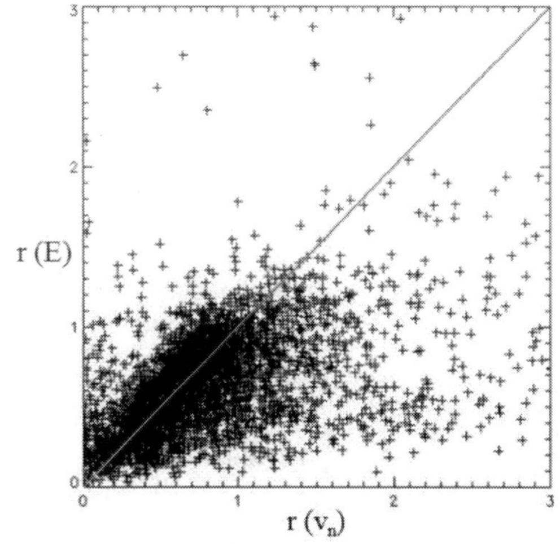

Fig. 7. The COR based on the total translational kinetic energy of the particle versus the COR based on the normal velocities.

References

1. H. Korsch and J. Lang, J. Phys. A **24**, 45(1991).
2. H. Lehtihet and B. Miller, Physica **21D**, 93 (1986).
3. M. Ferguson, B. Miller and M. Thompson, Chaos **9**, 841 (1999).
4. V. Vilner, J. Hansen, W. Campbell, and M. Raizen, Phys. Rev. Lett. **86**, 1514 (2001).
5. S. Feldt and J. S. Olafsen, Phys. Rev. Lett. **94**, 224102 (2005).
6. A. Z. Gorski and T. Srokowski, Acta Phys. Pol. B **37**, 2561 (2006).
7. A. E. Hartl, B. N. Miller, and A. P. Mazzoleni, Phys. Lett. A **37**, 3682 (2011).
8. A. E. Hartl, B. N. Miller, and A. P. Mazzoleni, Phys. Rev. E **87**, 03290 (2013).
9. N. V. Brilliantov, F. Spahn and J-M. Hertzsch, and T. Pöschel, Phys. Rev. E **53**, 5382 (1996).
10. I. Goldhirsch and G. Zanetti, Phys. Rev. Lett. **70**, 1619 (1993).
11. S. McNamara and W. R. Young, Phys. Rev E **53**, 5089 (1996).
12. E. Grossman and M Mungan, Phys. Rev. E **53**, 6435 (1996).
13. A. Kudrolli, M. Wolpert, and J. P. Gollub, Phys. Rev. Lett. **78**, 1383 (1997).
14. P. Müller, M. Heckel, A. Sack, and T. Pöschel, Phys. Rev. Lett. **110**, 254301 (2013).
15. P. Müller, D. Krengel, and T. Pöschel, Phys. Rev E **85**, 041306 (2012).
16. T. Schwager, V. Becker and T. Pöschel, Eur. Phys. J. E **27**, 107 (2008).

EPJ Web of Conferences 140, 03040 (2017) DOI: 10.1051/epjconf/201714003040
Powders & Grains 2017

Pulling an intruder from a granular material: a novel depinning experiment

Yue Zhang[1,*] and *Robert* Behringer[1,**]

[1] *Department of Physics and Center for Nonlinear and Complex Systems, Duke University, Durham, North Carolina 27708, USA*

Abstract.
Two-dimensional impact experiments by Clark et al. [2] identified the source of inertial drag to be caused by 'collisions' with a latent force network, leading to large fluctuations of the force experienced by the impactor. These collisions provided the major drag on an impacting intruder until the intruder was nearly at rest. As a complement, we consider controlled pull-out experiments where a buried intruder is pulled out of a material, starting from rest. This provides a means to better understand the non-inertial part of the drag force, and to explore the mechanisms associated with the force fluctuations. To some extent, the pull out process is a time reversed version of the impact process. In order to visualize this pulling process, we use 2D photoelastic disks from which circular intruders of different radii are pulled out. We present results about the dynamics of the intruder and the structures of the force chains inside the granular system as captured by slow and high speed imaging.

1 Introduction

Impact and the corresponding energy loss in granular material have attracted attention for some time[1]. Recently, there has been considerable work[2][3][4][5][6]about impact in granular material. When an intruder hits a granular material, such as sand, the velocity of the intruder decreases due to the force exerted by the granular particles. Impact in granular materials is particularly common in many applications, including ballistic applications, meteorite strikes and spaceship landings.

The goal of past research has been to understand the dynamics of the impact in this complex system of granular material. There are many methods used to extract this force law empirically. A common version of the force law is

$$m\ddot{z} = mg - f(z) - h(z)\dot{z}^2 \qquad (1)$$

To observe the intruder motion and the impactor dynamics, Clark et al. used the apparatus sketched in figure 1. This consists of filling photoelastic grains between two glass plates to form a bed on which different shapes of intruders impinge. As a result, stressed regions appear brighter while stress-free areas appear darker. In the visualization of the internal stresses, the photoelastic response of the system, calibrated through intensity, provides a quantitative measure of the local pressure.

With this photoelastic method, Clark et al. established a collisional model describing the microscopic origins of the impact in granular material. In the collisional model, Clark et al. assume that the particles of granular material have formed force chains that randomly collide with the intruder. They derived equations for the force exerted on the intruder, and the results match quite well with the experimental data. In this model, both the multiplicative factor in the inertial force($\dot{z}^2$ term of equation) depends on the intruder and grains properties. In addition, multiplicative fluctuations in both static and inertial forces exerted on the intruder were discovered in these experiments [2] [7] [8].

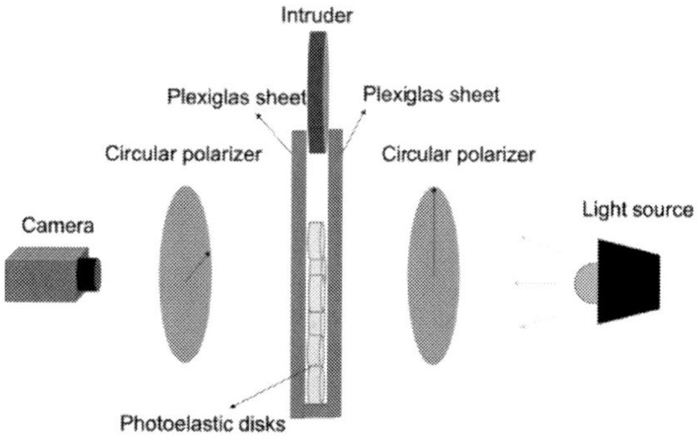

Figure 1. Schematic of apparatus of an impact experiment

However, the characteristics of the force fluctuation still remains unknown. As a comparison and supplement, we carried out controlled pull-out experiments to explore the mechanisms associated with these fluctuations and the nature of the static force. The pull out process is, to some extent, the time reverse of the impact process. The pull out process of an object buried in a granular material also has many applications for pilings and buildings.

*e-mail: yue.zhang@phy.duke.edu
**e-mail: bob@phy.duke.edu

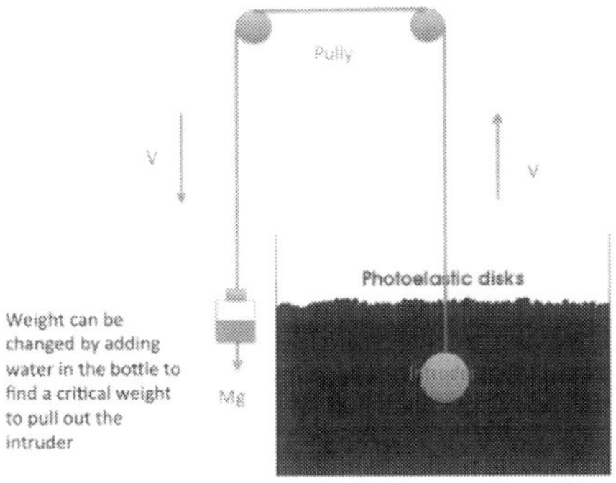

Figure 2. Images of main apparatus for pull experiment

2 Pull Out Experiment

For the pull out problem, we immediately come across several questions. First, what is the smallest force needed to trigger the pull out process? Second, how does the granular material respond? Third, what are the dynamics of the intruder? In order to answer these questions, we have carried out pull-out experiments using photoelastic techniques and image processing.

2.1 Apparatus and Techniques

The sketch above (fig.2) shows the main apparatus used in this experiment. A bidisperse mixture of about 3000 photoelastic particles (with thickness about 0.32cm, diameters about 0.89cm and 0.56cm) are sandwiched between two sheets of transparent Plexiglas. The reason for using bidisperse particles instead of monodisperse particles is to avoid crystallization in this granular system. The distance between the two Plexiglas sheets is 0.41cm so that the particles can move smoothly while loosely constrained by Plexiglas sheets. The size of each Plexiglas sheet is 122cm x 92cm. We prepare the buried intruder in each run by tipping the sheets, placing the intruder in a given position and letting particles rain back in. As a result, the particles are randomly packed. The intruder is pulled by a thread connected to a bottle whose weight can be changed by adding water to provide different pulling forces.

The recording techniques used in a pull out experiment is almost the same as in an impact experiment (fig.1), which also can visualize the force structures inside the granular system. Light from several bulbs passes through the first circular polarizer, one Plexiglas sheet, the layer of photoelastic particles, the other Plexiglas sheet and the second circular polarizer before arriving at a camera.

Then we acquire high speed video and single images. From these images we first obtain the position of the intruder. From this we track the trajectory of the intruder and derive the velocity and acceleration. The second type of information is the photoelastic visualization of force chains from the polarized image. From the force chains, we can estimate the granular structure and force acting on the intruder.

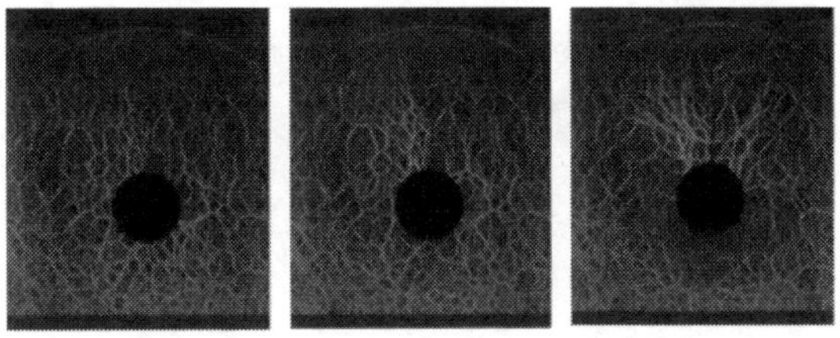

Figure 3. Force chains build up when adding load

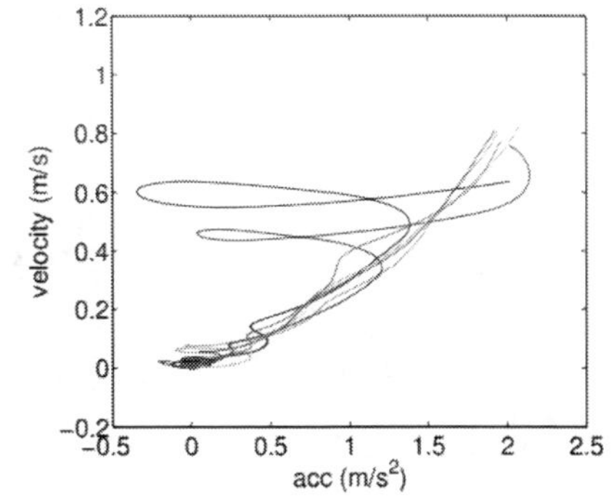

Figure 4. The graph of velocity vs. acceleration. Different colors stand for different runs with the same circular intruder R=6.4cm.

2.2 Procedures and Results

In a typical experiment, the intruder is buried at a given position in a 2D Plexiglas bed, which is filled with randomly packed photoelastic disks. Then we add load to the puller by adding small amount of water to the bottle. The process is repeated until the intruder is pulled out. As is shown in figure 3, the force chains build up step by step.We can determine the smallest pulling force needed to pull out the intruder.

Then we conduct dynamic pull experiments with the smallest pulling force. High speed video visualizes the fast process once the intruder begins to move. At first, the intruder is stuck, but then it starts to accelerate and escapes quickly from the bed. We track the intruder so that we obtain velocity and acceleration curves, as shown in figure 4. For the dynamic pull out experiment, a fast camera taking images with a speed of 1000fps is used to capture the dynamic pull process which lasts about 1 second.

As is shown in figure 5, the main part of the v vs. a curves can be fitted with exponential functions $v = a * \exp(bt) - c$, indicating a linear relationship between velocity and acceleration as in figure 4. The slope in figure 4 is the reciprocal of the b factor in the exponential fitting function of velocity. Some loops can be found in figure 4, which is because the grains reorganize during the process creating renewed resistance to the intruder. During that time, the acceleration drops significantly.

Additionally, to visualize what is happening during the fluctuations of the acceleration, we plot space-time graphs

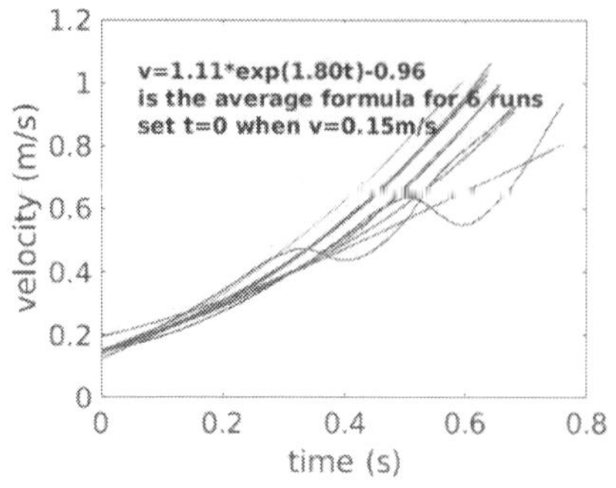

Figure 5. The graph of velocity vs. time. Different colors stand for different runs with the same circular intruder R=6.4cm.

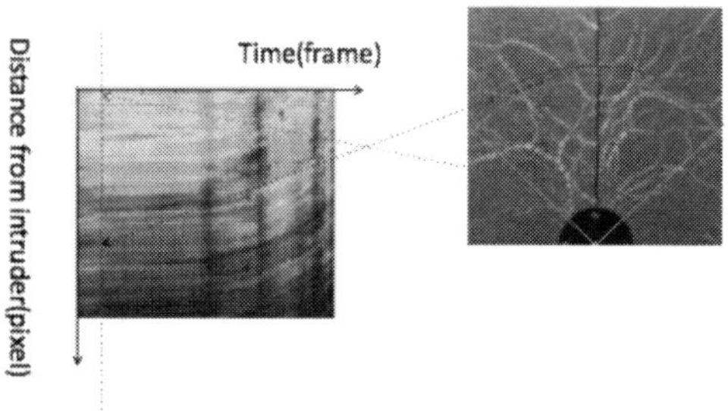

Figure 6. Space-time graph and its physical meaning.

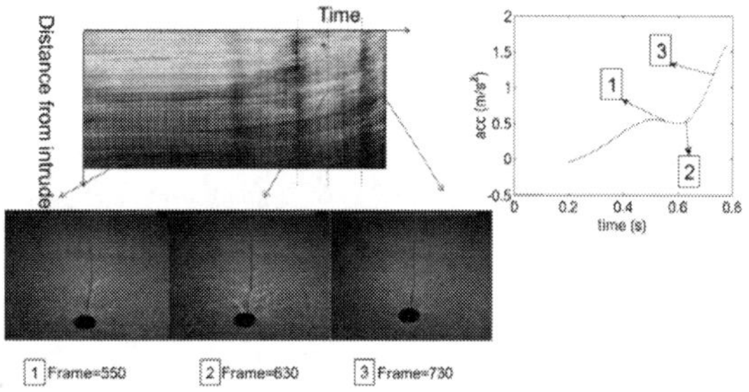

Figure 7. Discontinuities in space-time graph corresponding to fluctuations in acceleration

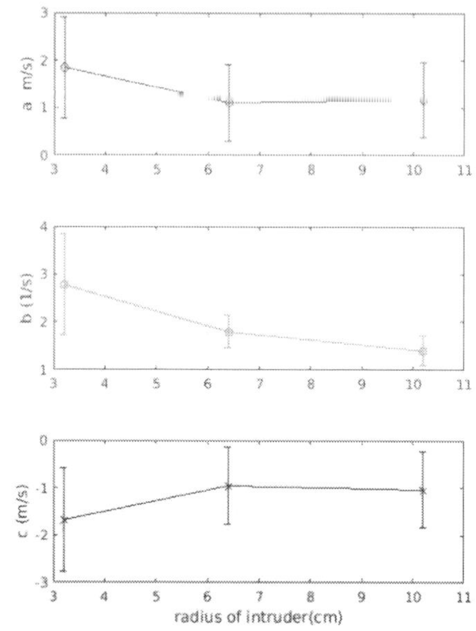

Figure 8. Fitting parameters(a, b, and c) vs. radius of intruder

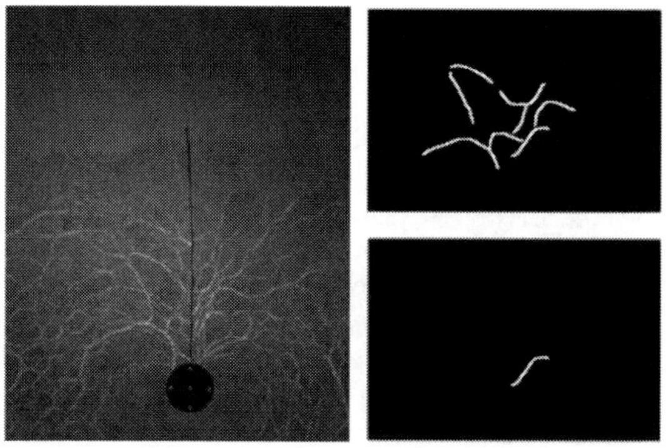

Figure 9. Extraction of single force chain: the left image is an original image from the dynamic pull-out experiments, the upper right image shows the force chains in a chosen box area above the intruder, and the lower right image shows one of those force chains in that box area.

of the image intensity. As shown in the left graph of figure 6, the x-axis is time with unit of frames, and the y-axis is the distance from above the intruder in units of pixels. The right graph of figure 6 is part of a typical experimental image. We average the intensity on an arc, which gives us one point value in the space time graph.

Discontinuities in the space time graph are shown in figure 7. At those moments the force chains break and during those moments, there is a local fluctuation in acceleration.

After repeating the experiment with different circular intruders, we also find b, the reciprocal of the slope in the velocity vs. acceleration graph, which decreases as the radius increases, as shown in figure 8. However, parameters a and c in the fitting function of velocity are independent of the intruder radius. In order to understand what is happening inside the granular system, we also extract the force chains (bright parts) from high speed photoelastic images, as shown in figure 9.

Then we calculate each force chain's curvature, and obtain the distribution of curvatures. The force chain curvatures in the first and last half of the runs obey the same distribution for different intruder sizes, as shown

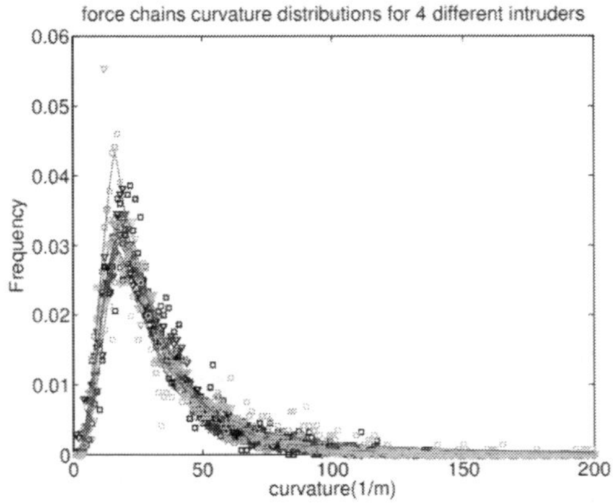

Figure 10. Force chains curvature distribution for 4 different intruders

in figure 10. Different colors stand for different intruder sizes, while triangle "▽" and square "□" represent first and last half run, respectively. After normalization, the distributions can be collapsed into a single fitting function $y = a * \exp(-b * |log(x) - c|^d)$, where a =0.03, b=1.3, c=2.7 and d = 1.9. This curvature distribution and its effect on the stick-slips during the pull out process remains a subject for future investigation.

3 Conclusion

We have observed the force chain build-up in a 2D granular system during pre-pull experiments, and also investigated the dynamics of the intruder during the pull-out process. The velocity and the acceleration of the intruder follow a linear relationship, as the velocity increases exponentially with time. The b factor in the exponential function decreases with the radius of the intruder. Additionally, fluctuations in the acceleration have been observed and explained through simultaneous breaking of force chains in the granular system. The curvature of each force chain can be calculated, and those curvatures are found to follow the same distribution function for different intruder sizes in both first and last half experimental run.

Acknowledgements

We would like to acknowledge NASA grant NNX15AD38G, NSF grants DMR1206351 and DMS124807, and the William M. Keck Foundation for the support of this research.

References

[1] Poncelet, J.V. Cours de M'ecanique Industrielle. Paris, 1829.

[2] Clark, A. H., et al. , "Particle scale dynamics in granular impact." Phys Rev Lett, 109, 2012.

[3] Iskander, Bless, Omidvar. Rapid Penetration into Granular Media, Visualizing the Fundamental Physics of Rapid Earth Penetration. Elsevier, 2015.

[4] Tsimring, Volfson. Powders and Grains, 2005.

[5] Katsuragi, Durian. Nature Physics, 2007.

[6] Nordstrom, K. N., et al. , "Granular dynamics during impact." Phys Rev Lett, 112, 2014.

[7] Clark, A. H., et al. , "Granular impact dynamics: Fluctuations at short time-scales." 445-448, 2013.

[8] Clark, A. H., et al. , "Collisional model for granular impact dynamics." Physical Review E 89(1), 2014.

DOI: 10.1051/epjconf/201714006012

Mechanical analysis of the dry stone walls built by the Incas

Jaime Castro[1,*], *Luis E.* Vallejo[2,**], and *Nicolas* Estrada[1,***]

[1] *Departamento de Ingeniería Civil y Ambiental, Universidad de los Andes, Bogotá, Colombia*
[2] *Department of Civil and Environmental Engineering, University of Pittsburgh, Pittsburgh PA, USA*

Abstract. In this paper, the retaining walls in the agricultural terraces built by the Incas are analyzed from a mechanical point of view. In order to do so, ten different walls from the Lower Agricultural Sector of Machu Picchu, Perú, were selected using images from Google Street View and Google Earth Pro. Then, these walls were digitalized and their mechanical stability was evaluated. Firstly, it was found that these retaining walls are characterized by two distinctive features: disorder and a block size distribution with a large size span, i.e., the particle size varies from blocks that can be carried by one person to large blocks weighing several tons. Secondly, it was found that, thanks to the large span of the block size distribution, the factor of safety of the Inca retaining walls is remarkably close to those that are recommended in modern geotechnical design standards. This suggests that these structures were not only functional but also highly optimized, probably as a result of a careful trial and error procedure.

1 Introduction

The Inca Empire was the largest pre-Columbian civilization of the American continent, with a surface close to one million square kilometres and a population close to twelve million people during the sixteenth century. Their development relied on a remarkable infrastructure network of bridges, roads, agricultural terraces, buildings, and water conduction systems. Most of these structures were built with stone blocks, more than 500 years ago. Yet, many of them remain in good conditions without any significant maintenance, demonstrating a noteworthy engineering expertise.

Amongst the different types of structures built by the Incas, one of the most important was the system of agricultural terraces. These terraces were an essential part of the Inca food supply system [1], allowing the development of a variety of crops throughout the Andes, especially maize, potatoes, and quinoa.

Despite their importance and engineering splendor, Inca agricultural terraces have been seldom studied from an engineering perspective. The works of Wright and Valencia-Zergara must be noted [2–5], in which water management and drainage in terraces of Machu Picchu, Tipon, and Moray were analyzed. Soil composition and stratification were then studied by Wright and Dr. James Self [6]. However, investigations in which these systems are analyzed from a mechanical point of view remain scarce. Probably, the first researcher to carry out this kind of analyses was Vallejo in 2014 [7].

*e-mail: jj.castro429@uniandes.edu.co
**e-mail: vallejo@pitt.edu
***e-mail: n.estrada22@uniandes.edu.co

The purpose of this work was to analyze the mechanical stability of the retaining walls used in the agricultural terraces built by the Incas. This was carried out by means of a careful image analysis coupled with a simple mechanical model, which allowed for highlighting the functionality and optimality of these retaining walls, even while built by means of a trial and error procedure.

In the following, the location and description of the analyzed walls are presented in Secs. 2 and 3. The procedures followed for analyzing the mechanical stability of these systems is presented in Sec. 4, and the paper ends in Sec. 5 with a brief conclusion.

2 Location of the analyzed walls

In order to carry out this investigation, ten retaining walls were selected from the Lower Agricultural Sector of Machu Picchu, Perú. These walls were selected by analyzing simultaneously images from Google Street View and Google Earth Pro. The images from Google Street View were necessary to identify the blocks that compose each of the walls, and the images from Google Earth Pro were used to define the length scale. Figure 1 shows the location of nine of the analyzed walls.

3 Description of the analyzed walls

As a first step, the retaining walls were qualitatively analyzed and compared to masonry walls used in other type of Inca structures. Figures 2 (a) and (c) show pictures of two walls. The first one corresponds to wall 6 in Fig. 1 (i.e., a wall used in an agricultural terrace), and the second one is a typical masonry wall used in a house. Figures

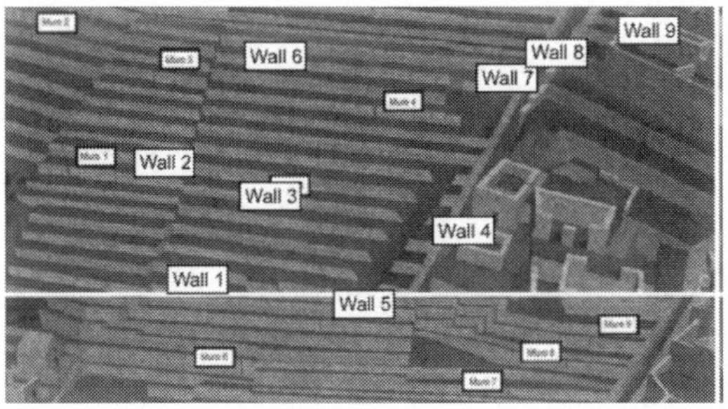

Figure 1. Location of nine of the analyzed walls in a Google Earth Pro image. The walls height, indicated by the red lines, varies from 1.58 to 3.40 m, with a average value of 2.02 m.

2 (b) and (d) show the result of a DEM simulation of the same systems [1]. The red lines represent the normal forces between particles, the thickness being proportional to the force intensity.

The microstructure observed in both types of walls (i.e., retaining wall and masonry wall) is notably different. The retaining wall is characterized by a disordered arrangement, made up of blocks of different shapes and sizes. The size of these blocks varies from 4 cm up to 1.5 m, and the mass varies approximately from 100 g up to 7 tons, which must have implied serious construction difficulties. On the other hand, the masonry wall is characterized by an ordered arrangement, formed by blocks with similar shape and size, comparable to a modern-day masonry wall. The blocks in the masonry wall have a manageable size, so that they can be easily carried and placed by one person, which is very practical from a construction point of view.

In order to compare the block size distributions of the different walls, it is convenient to analyze the cumulate mass distributions, also known as the material's grading curves. Figure 3 shows the grading curves for the ten analyzed retaining walls, as well as for two masonry walls (i.e., walls 11 and 12). It can be seen that the grading curves of the retaining walls cover a much broader range of sizes, while the grading curves of masonry walls are basically uniform (i.e., all blocks have approximately the same size). The degree of uniformity of the block size distributions can be quantified by means of the uniformity coefficient $C_u = D_{60}/D_{10}$, where D_{60} is the diameter of the block for which 60% of the blocks are smaller, and D_{10} is the diameter of the block for which 10% of the blocks are smaller. For the analyzed retaining walls, C_u varied from 1.80 to 4.41, with an average value of 2.76. For the masonry walls, the average C_u was 1.58. This confirms that the retaining walls are characterized by a block size distribution with a large size span, whereas the masonry walls used in buildings are characterized by a much narrower range of sizes.

[1]Simulations were carried out using the LMGC90 software, developed in Montpelllier, which is well suited for simulating large assemblies of rigid grains of arbitrary shape.

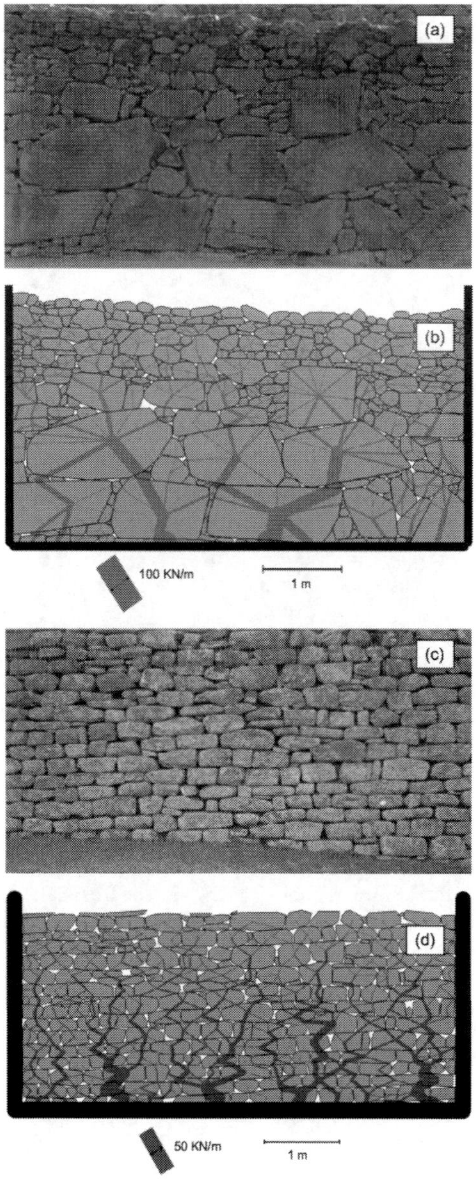

Figure 2. Picture (a) and DEM simulation (b) of a retaining wall in an agricultural terrace (i.e., wall 6 in Fig. 1). Picture (c) and DEM simulation (d) of a masonry wall in a house.

4 Stability of the retaining walls

The mechanical stability of the analyzed retaining walls was assessed by means of the safety factor against sliding and overturning. The initial necessary step was to identify the mechanical parameters of the soil and the rock blocks. Most of these values must be assumed, since it is not possible to extract soil samples in order to conduct laboratory tests, because of the archeological importance of these systems. The analyses presented in this paper were based on the values suggested in [8]. The unit weight of the rock

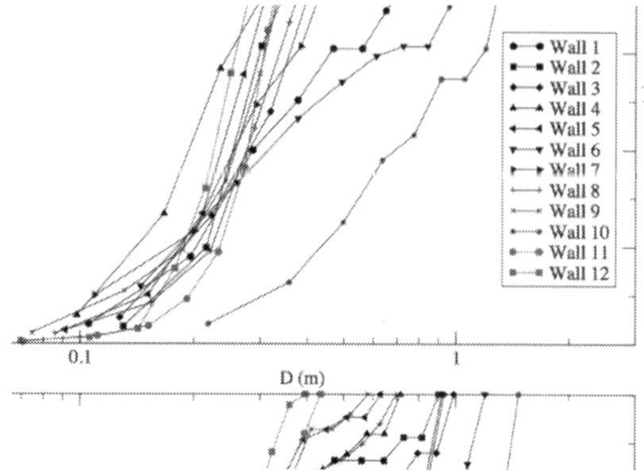

Figure 3. Grading curves for ten analyzed retaining walls used in agricultural terraces (i.e., walls 1 to 10) and for two masonry walls (i.e., walls 11 and 12).

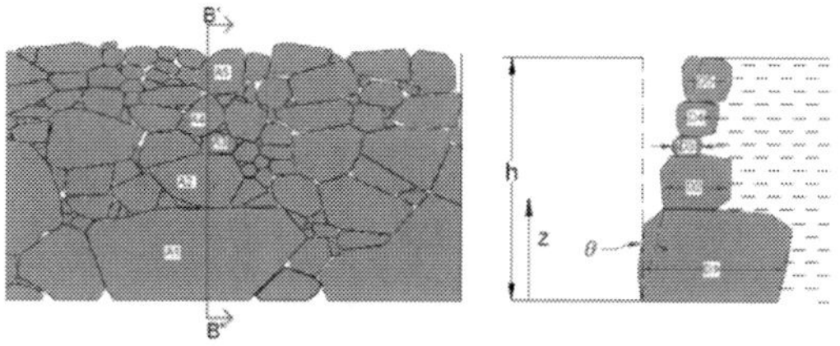

Figure 4. Left: Schematic representation of a retaining wall. Right: Internal structure assumed for section B'B'.

blocks was $\gamma_b = 25.36$ KN/m^3, and the dry unit weight of the soil was $\gamma_s = 15.71$ KN/m^3. The friction angle between blocks was $\phi_b \in (33°, 38°)$, and the soil's internal friction angle was $\phi_s \in (20°, 30°)$.

The next step was to assume the structure of the wall inside the terrace. In order to to this, the block's area A in the face of the wall (i.e., the area of the particles shown in Figs. 2(b) and (d)) was measured, and the diameter $D = \sqrt{4A/\pi}$ of a circle having the same area was calculated. Then, it was supposed that the depth of the particles in the direction perpendicular to the face of the wall was D. This structure is shown schematically in Fig. 4. Figure 4 also shows the inclination θ of the wall towards the terrace, which was assumed in the range (7°, 15°) as reported in [9], as well as the total height h and the height z.

The final step was to calculate the factor of safety SF for two possible failure modes: sliding $_s$ and overturning $_o$.

For the sliding failure mode, SF_s was calculated as f_r/f_d, where f_r is the resisting force and f_d is the driving force [2]. When evaluating a possible failure plane at any height z, f_r is the friction force between the portion of the wall above z and that below z. f_d is the force applied by the soil on the portion of the wall above z, which was calculated with the Rankine theory, by considering a

[2] $SF > 1$ corresponds to a stable system and $SF < 1$ to an unstable system.

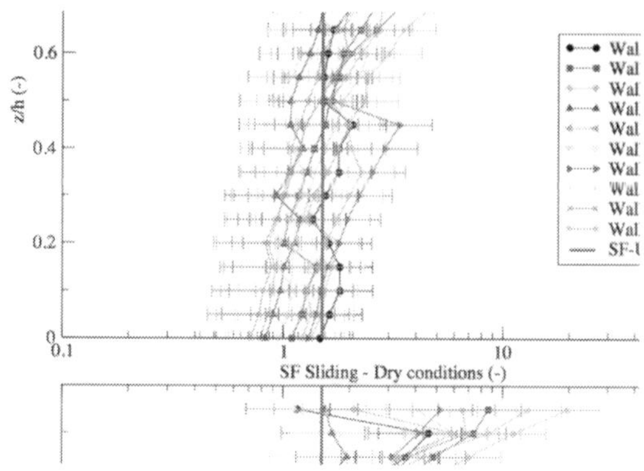

Figure 5. Factor of safety SF_s for the sliding failure mode as a function of the normalized height z/h. The green line at $SF_s = 1.5$ indicates the factor of safety recommended in modern geotechnical design standards in USA. Solid symbols indicate the average value and error bars indicate the maximal and minimal values.

dry soil. This condition is appropriate since the walls were built without any mortar to bind the blocks together, ensuring that the walls behave as drained systems. SF_s is then given by

$$SF_s = \frac{\sum(\gamma_b A_i D_i)\tan\phi_b}{K\gamma_s(h-z)^2/2}, \qquad (1)$$

where the sum runs over all blocks with center of gravity above z. A_i and D_i are the area and the diameter of block $_i$, respectively, and K is the coefficient of lateral earth pressure, which was varied between the active earth pressure $(1-\sin\phi_s)/(1+\sin\phi_s)$ and the at-rest earth pressure $1-\sin\phi_s$.

Figure 5 shows SF_s as a function of the normalized height z/h for all analyzed walls. Firstly, it can be seen that SF_s changes with z/h, being larger for the top portion of the walls and smaller at the base of the walls. This is to be expected, since f_r increases linearly with depth (i.e., with $h-z$) while f_d increases as $(h-z)^2$. Thus, the walls must necessarily become unstable at a certain depth $h-z$. Secondly, it can be seen that the evolution of SF_s with z/h is not gradual, but exhibits strong fluctuations. These fluctuations are due to particularly large blocks, which drastically increase W_b and, thus, f_r. This effect is important, because it allows for building higher walls than those that could be built using only medium-sized blocks, such as those used in masonry walls. Thirdly, it is remarkable that SF_s in most of the walls' height is close to 1.5 (indicated by the vertical line in the figure). This is the factor of safety recommended in modern geotechnical design standards in North America [10–12]. Finally, it can be observed that SF_s at the base of the walls is close to unity, evidencing a highly optimized design, even while reached by means of a trial and error procedure.

For the overturning failure mode, SF_o was calculated as M_r/M_d, where M_r is the resisting moment and M_d is the driving moment, with respect to a point o' located at the face of the wall. When evaluating a possible failure at

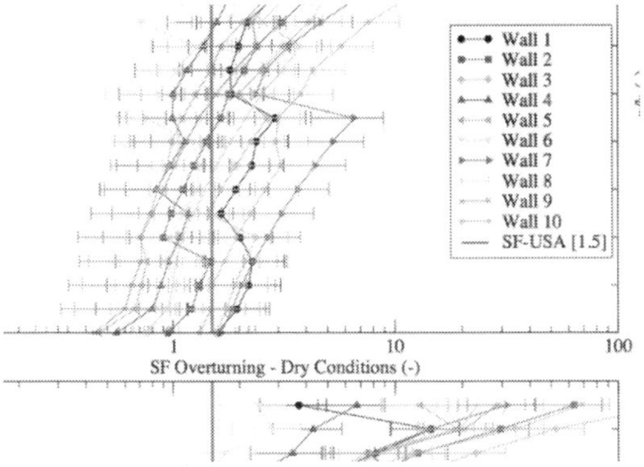

Figure 6. Factor of safety SF_o for the overturning failure mode as a function of the normalized height z/h. The green line at $SF_o = 1.5$ indicates the factor of safety recommended in modern geotechnical design standards in USA. Solid symbols indicate the average value and error bars indicate the maximal and minimal values.

any height z, M_r is the moment exerted by the weight of the blocks above z with respect to o'. M_d is the moment of the force applied by the soil on the portion of the wall above z with respect to o'. SF_o is then given by

$$SF_o = \frac{\sum[(\gamma_b A_i D_i)(D_i/2 + (z_i - z)\tan\theta]}{K\gamma_s(h-z)^3/6}, \quad (2)$$

where the sum runs over all blocks with center of gravity above z and θ is the inclination of the face of the wall.

Figure 6 shows SF_o as a function of the normalized height z/h for all analyzed walls. As for SF_s, it can be seen that large blocks play an important role in keeping SF_o above unity, SF_o for most of the walls' height is close to 1.5, and $SF_o \simeq 1$ at the base of the walls, evidencing the optimal design of these retaining structures.

5 Conclusion

The purpose of this work was to analyze the mechanical stability of the retaining walls used in the agricultural terraces built by the Incas. For so doing, ten retaining walls from the Lower Agricultural Sector of Machu Picchu were selected and their mechanical stability was analyzed. In order to illustrate these analyses, the structure of the retaining walls was also analyzed and compared with that of masonry walls used in houses and temples.

Firstly, our results show that retaining walls used in terraces have a characteristic microstructure that is significantly different from those observed in masonry walls. Specifically, when building retaining walls, the Incas avoided ordered patterns and used a block size distribution with a large size span (i.e., the blocks size varies from blocks that can be carried by one person to large blocks weighing several tons).

Secondly, the mechanical stability of the retaining walls was analyzed by estimating the factor of safety, considering two possible failure mechanisms: sliding and overturning, as it is usually done for geotechnical retaining structures. It was shown that, thanks to the large size span of the block size distribution, the factor of safety fluctuates inside a range of values that are remarkably close to those that are recommended in modern geotechnical design standards. It is also shown that the stability condition at the base of the walls is close to limit equilibrium, as long as the system is considered drained. This could explain why the Incas did not use any mortar to bind the blocks together, allowing the phreatic water to drain freely.

Finally, our conclusions suggest that Inca retaining walls are not only functional but also highly optimized, probably as a result of a careful trial and error design procedure.

Acknowledgements

We thank Jaime Wills and Luisa Orozco for fruitful discussions, and Emilien Azéma for his assistance with the LMGC90 platform.

References

[1] C. Graber, Smithsonian Magazine (2011)
[2] K. Wright, Journal of Hydraulic Engineering (1997)
[3] K. Wright, A. Valencia Zegarra, C.M. Crowley, Tech. rep., Instituto Nacional de Cultura del Perú, Wright Paleohydrological Institue, Wright Water Engineers (2000)
[4] K. Wright, A. Valencia Zegarra, R.M. Wright, G. McEwan, Tech. rep., Instituto Nacional de Cultura del Perú, Wright Paleohydrological Institue, Wright Water Engineers (2001)
[5] K. Wright, Water environment and technology pp. 78–87 (2008)
[6] K. Wright, A. Valencia Zegarra, W.L. Lorah, Journal of Irrigation and Drainage Engineering **125** (1999)
[7] L.E. Vallejo, M. Fontanese, Geo-Congress 2014 Technical Papers (2014)
[8] T.R. West, *Geology Applied to Engineering* (Waveland Press, Long Grove, Illinois, 2010)
[9] T.G.S. of London, J. Dixon, Geoscientist (December 2014)
[10] E.E.C. Ltd., Tech. rep., City of Nanaimo (2014)
[11] C.D. of Transportation, Tech. rep., State of California (2004)
[12] D. of Public Works-Building, S. Division, Tech. rep., County of Los Angeles (2011)

DOI: 10.1051/epjconf/201714003041

A differential equation for the flow rate during silo discharge: Beyond the Beverloo rule

Marcos A. Madrid[1,*], *José R. Darias*[2,**], and *Luis A. Pugnaloni*[1,***]

[1] *Dpto. Ingeniería Mecánica, Facultad Regional La Plata, Universidad Tecnológica Nacional, CONICET, Av. 60 Esq. 124, 1900 La Plata, Argentina.*
[2] *Laboratorio de Óptica y Fluidos, Universidad Simón Bolívar, Apartado Postal 89000, Caracas 1080-A, Venezuela.*

Abstract. We present a differential equation for the flow rate of granular materials during the discharge of a silo. This is based in the energy balance of the variable mass system in contrast with the traditional derivations based on heuristic postulates such as the free fall arch. We show that this new equation is consistent with the well known Beverloo rule, providing an independent estimate for the universal Beverloo prefactor. We also find an analytic expression for the pressure under discharging conditions.

1 Introduction

The discharge of grains though an opening at the base of a silo has been considered in a number of studied since the 19th century (see for example [1–9] and references therein). The most salient feature is the fact that the flow rate does not depend on the height h of the material in the container, in clear contrast with the behavior of viscous fluids. This has been sometimes attributed to the pressure saturation observed in static silos, however, discharging silos have a continuously evolving pressure [10].

If the discharge orifice is circular and large enough to avoid clogging [11], the mass flow rate Q is described by the so called Beverloo rule [1, 12]

$$Q = C\rho_b \sqrt{g}(D_o - k\,d)^{5/2}, \tag{1}$$

where D_o is the diameter of the opening, ρ_b the bulk density of the granular sample, g the acceleration of gravity and d the diameter of the grains. Here, k and C are two fitting dimensionless constants. Interestingly, while k may vary up to a factor of 2, depending on the grains used, $C \approx 0.58$ for virtually any material tested [12]. However, some deviations are observable for very low friction materials [13].

The Beverloo rule is generally explained based on heuristic models such as the "free fall arch" model and the concept of "empty annulus" [12]. However, these two models have been recently challenged [14, 15]. Staron *et al* have also shown that the Beverloo rule can be obtained if the Navier–Stokes equations are solved for a plastic fluid including a constitutive equation for the effective friction based on the $\mu(I)$-rheology [16].

*e-mail: marcosamadrid@googlemail.com
**e-mail: jrdarias@usb.ve
***e-mail: luis.pugnaloni@frlp.utn.edu.ar

In this work, we use the global energy balance for the granular material inside a discharging silo and a constitutive relation to derive a simple differential equation for the mass $M(t)$ in the silo. We show that for a simple discharge the equation is consistent with the Beverloo rule, providing an independent estimate for the prefactor without fitting data. In the process of validating the differential equation, we also find a functional form for the pressure in the silo in the dynamic regime which is rather different to the well known Janssen equation.

2 Energy balance

Consider a cylindrical silo of radius R_s, discharging through an opening of radius R_o (see Fig. 1). The energy balance requires that, at any time,

$$W_g = \dot{K}_{in} + W_{out} + \dot{E} + W_D, \tag{2}$$

where W_g is the power injected by the force of gravity acting on the grains, $\dot{K}_{in}$ is the rate of change of the kinetic energy of the grains inside the silo, W_{out} is the power loss due to the grains that leave the silo at a velocity v_{out}, $\dot{E}$ is the rate of change of the elastic energy of the grains, and W_D is the dissipated power due to the non-conservative interactions between grains and between grains and walls.

During the discharge, the mass $M(t)$ in the silo can be written as

$$M(t) = mN(t) = \rho_b A_s z(t) = 2\rho_b A_s z_{cm}(t), \tag{3}$$

where m is the mass of one grain, $N(t)$ is the number of grains in the silo, ρ_b is the apparent density in the bulk, $A_s = \pi R_s^2$ is the cross section of the silo, $z(t)$ is the head of material, and $z_{cm}(t) = z(t)/2$ is the center of mass of the granular column. We have assumed that the density is

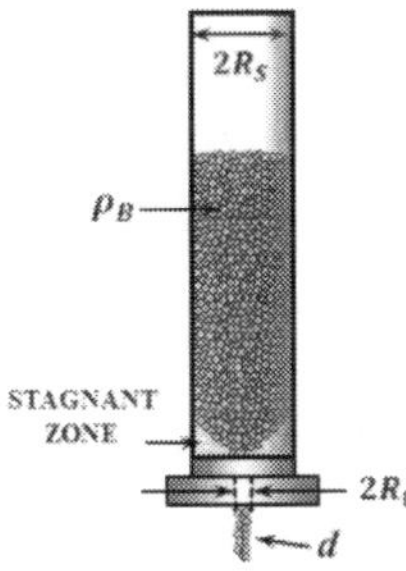

Figure 1. Sketch of the discharge of a silo.

homogeneous throughout the column. Therefore, the flow rate $q(t)$ in particles per unit time is

$$q(t) = \frac{Q(t)}{m} = -\frac{\dot{M}(t)}{m} = -\frac{2\rho_b A_s}{m} v_{cm}(t), \tag{4}$$

where $v_{cm}(t)$ is the velocity of the center of mass of the granular column.

2.1 Internal kinetic energy ($\dot{K}_{in}$)

The kinetic energy of the grains inside the silo is

$$K_{in}(t) = \frac{1}{2}\sum_{i=1}^{N(t)} m v_i^2(t), \tag{5}$$

where the sum includes all particles in the silo at time t and $v_i(t)$ is the velocity of particle i. In terms of the center of mass $K_{in}(t)$ can be expressed as

$$K_{in}(t) = \frac{1}{2}M(t)v_{cm}^2(t) + \frac{1}{2}\sum_{i=1}^{N(t)} m_i[\boldsymbol{v}_i(t) - \boldsymbol{v}_{cm}(t)]^2, \tag{6}$$

where the term $[\boldsymbol{v}_i(t) - \boldsymbol{v}_{cm}(t)]^2$ can be neglected according to results obtained from DEM simulations (data not shown), hence

$$K_{in}(t) \approx \frac{1}{2}M(t)v_{cm}^2(t). \tag{7}$$

Then, the rate of change of K_{in} is

$$\dot{K}_{in}(t) = M(t)v_{cm}(t)\dot{v}_{cm}(t) + \frac{1}{2}v_{cm}^2(t)\dot{M}(t). \tag{8}$$

which can be written, using Eqs. (3) and (4), as

$$\dot{K}_{in}(t) = \frac{m^2}{4\rho_b^2 A_s^2}\left[M(t)q(t)\dot{q}(t) - \frac{m}{2}q^3(t)\right]. \tag{9}$$

2.2 Gravitational energy (W_g)

The gravitational potential energy of the particles inside the silo using Eq. (3) is

$$U_g(t) = M(t)gz_{cm}(t) = \frac{gM^2(t)}{2\rho_b A_s}. \tag{10}$$

The power injected is therefore

$$W_g(t) = -\dot{U}_g(t) = \frac{mg}{\rho_b A_s}M(t)q(t). \tag{11}$$

2.3 Discharge energy (W_{out})

The cumulative kinetic energy that was removed from the system by time t due to the particles that exit through the opening is

$$K_{out}(t) = \frac{1}{2}mv_{out}^2[N_0 - N(t)], \tag{12}$$

where v_{out} is the mean velocity of the grains that exit the system and N_0 is the initial number of grains in the silo [$N_0 - N(t)$ corresponds to the number of grains that have left the silo]. Hence, the power removed from the system is

$$W_{out} = \dot{K}_{out}(t) = \frac{1}{2}mv_{out}^2 q(t). \tag{13}$$

If we consider that the mass flow rate is $mq(t) = \rho_o A_o v_{out}$, with ρ_o the apparent density at the opening and A_o the cross section area of the opening, the velocity v_{out} can be put in terms of $q(t)$ and A_o, then

$$W_{out} = \frac{m^3 q^3(t)}{2\rho_o^2 A_o^2} \tag{14}$$

2.4 Dissipated energy (W_D)

It has been shown that for a shear cell of thickness L the tangential stress τ necessary to develop a flow at velocity v of a granular sample can be put in terms of the inertial number I as [17]

$$\tau = \mu(I)P, \tag{15}$$

where P is the confining pressure, $\mu(I)$ is the effective friction coefficient, $I = \frac{v}{L}\frac{d}{\sqrt{P/\rho}}$ is the inertial number that characterizes the flow if the gains are stiff and $L \gg d$, and ρ is the density of the material the grains are made of.

Here, we assume that the flow in the silo can be represented by a shear flow, where v is the velocity of the free surface, L is the silo radius and P is the mean hydrostatic pressure in the column (i.e., 1/3 of the trace of the stress tensor). In a silo, the shear rate is different at different heights. To be more rigorous, one should measure the shear rate layer-wise. This can be done by measuring the difference in mean velocity of the grains at the center of the layer and the grains close to the walls and dividing by the silo radius. For a flat bottomed silo discharging with a mass flow regime, close to the free surface the shear rate is negligible. However, close to the base, the shear rate is maximum since the velocity at the walls is zero (due to the stagnant zone) and the velocity at the outlet is the highest. Taking the velocity at the free surface, as we do here, is representative of the shear rate at a height equal to the height of the stagnant zone. At this height, the velocity in the center of the layer is close to the velocity of the free surface, whereas the velocity at the walls fall to very low values as the grains hit the stagnant zone. With this approximation, the power dissipated by friction is

$$W_D(t) = \tau(t)A(t)v(t) = \mu(I)P(t)A(t)v(t), \tag{16}$$

where $A(t)$ is the area of frictional contact between the grains and the silo. This area includes the lateral walls, which is connected to the height of the column at time t, plus a fixed area that should be proportional to the base of the silo, i.e., $A(t) = 2\pi R_s z(t) + \alpha\pi R_s^2 = 2\pi R_s \frac{M(t)}{\rho_b A_s} + \alpha A_s$.

Then, using Eq. (4) and $v(t) = 2v_{cm}(t)$, we obtain

$$W_D(t) = \mu(I)P(t)\left[2\pi R_s \frac{M(t)}{\rho_b A_s} + \alpha A_s\right] q(t)\frac{m}{A_s \rho_b}. \quad (17)$$

3 Differential equation

If the grains are stiff, the variation in the elastic energy are expected to be small and we can neglect $\dot{E}$. Notice also that $\dot{K}_{in}$ [Eq. (9)] falls quadratically with A_s in comparison with the remaining terms of Eq. (2) that fall linearly with A_s or are independent of A_s. Therefore, for a wide silo $\dot{K}_{in}$ vanishes. Plugging in Eqs. (11), (14), and (17) into Eq. (2) and considering Eqs. (3) and (4), we obtain

$$\begin{aligned} gM(t) &= \frac{\rho_b A_s}{2\rho_o^2 A_o^2}\dot{M}^2(t) \\ &+ \mu(I)P(t)\left[\frac{2\pi R_s}{\rho_b A_s}M(t) + \alpha A_s\right] \end{aligned} \quad (18)$$

Solving for $\dot{M}$ and using $A_o = \pi D_o^2/4$, being $D_o = 2R_o$, the diameter of the opening

$$\begin{aligned} \dot{M}(t) &= \frac{\pi\sqrt{2}}{4}\rho_o\sqrt{g}D_o^2\left[\frac{M(t)}{\rho_b A_s}\right. \\ &\left.-\mu(I)\frac{P(t)}{g}\left(\frac{2\pi R_s}{\rho_b^2 A_s^2}M(t) + \frac{\alpha}{\rho_b}\right)\right]^{1/2}. \end{aligned} \quad (19)$$

We note that Eq. (19) is similar to the Beverloo rule Eq. (1). The exponent 5/2 is not apparent since D_o is squared in Eq. (19). We will come back to this point below. Before, it is important to mention that Eq. (19) includes in the prefactor ρ_o instead of ρ_b. However, if we approximate roughly $\rho_o \approx \rho_b/2$, then the proportionality constant becomes $C = \frac{\pi\sqrt{2}}{8} \approx 0.56$. This has to be compared with the value obtained by fitting the Beverloo experiments that led to $C = 0.58$, which has been shown to be remarkably insensitive to the material properties [12].

Equation (19) is a first order differential equation for $M(t)$ that can be closed with an initial condition such as $M(t = 0) = M_0$. To solve this equation it is necessary to know $\mu(I)$, $P(t) = P(M(t))$ and α. In the rest of this paper, instead of directly solving Eq. (19), we estimate $\mu(I)$ and then extract an expression for $P(M(t))$ based on the condition that the flow rate must be constant throughout the discharge, as observed experimentally. This allows to fit the value of α using pressure data from DEM simulations of the discharge.

To estimate $\mu(I)$, consider that if $R_s \to \infty$, then $I \to 0$. Da Cruz *et al.* have shown that in the quasistatic limit (i.e., $I \to 0$) $\mu(I = 0) \approx 0.26$ for all material properties if the grain–grain friction coefficient is above 0.4 [17]. We have carried out DEM simulations using YADE [18], for a silo where R_s = 15 mm (i.e., 3 cm in diameter) and D_o = 15 mm, and a granular sample with d = 1 mm, ρ = 2000 kg/m^3, ρ_b = 1160 kg/m^3 and the friction coefficient of the grain–grain and grain–wall interaction is 0.55. The inertial number estimated as $I = \frac{v}{R_s}\frac{d}{\sqrt{P/\rho}}$ varies during the discharge in the range $9.8\times10^{-3} < I < 1.6\times10^{-2}$, which is in the quasistatic limit consistent with an effective friction coefficient 0.26 [17].

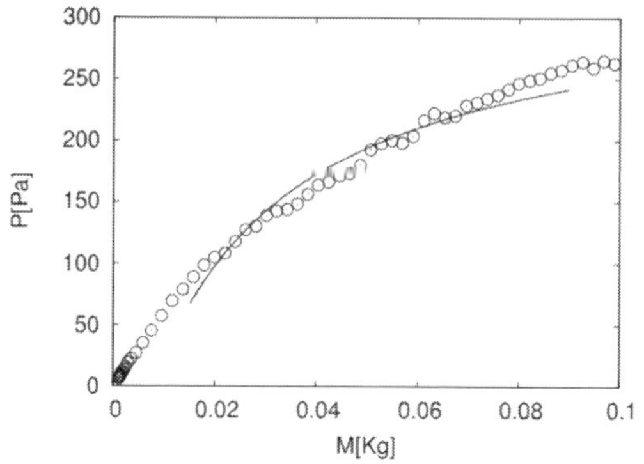

Figure 2. Pressure (isotropic component P = Tr(σ)/3 of the stress tensor) as a function of the mass M inside the silo during the discharge. Results obtained via a DEM simulation for a silo (symbols) and fit using Eq. (20) (line).

As we mentioned, Eq. (19) does not show the classical 5/2 exponent for D_o, nor the correction $-kd$ for the "empty annulus". However, the expression under the square root in Eq. (19) must compensate the difference between D_o^2 and the traditional Beverloo factor $(D_o - kd)^{5/2}$. During the discharge it is known that Q is constant. Hence, the radicand in Eq. (19) must be also a constant γ. Therefore, the pressure can be written as

$$P(t) = \left[\frac{M(t)}{\rho_b A_s} - \gamma\right]\frac{g\rho_b^2 A_s^2}{\mu(I)[2\pi R_s M(t) + \alpha\rho_b A_s^2]}. \quad (20)$$

If the radicand γ has to compensate the difference with the Beverloo rule, then $\gamma = (D_o - kd)^5/D_o^4$. For our simulations with D_o = 15 mm and d = 1 mm, and considering a typical value of k = 1.4, we obtain $\gamma \approx 9.2 \times 10^{-3}$.

Figure 2 shows a fitting of Eq. (20) to the aforementioned DEM data of the silo discharge. The pressure is calculated as the average of Tr(σ)/3 over the entire volume of the granular column a any given time during the discharge. Here, σ is the stress tensor. We took, following the discussion above, $\mu(I) = 0.26$ and $\gamma = 9.2 \times 10^{-3}$. As we can see, the only fitting parameter is α whose best fit value is $\alpha = 3.6 \pm 0.2$. The fit is only applied to a section of the process since the assumption of a constant flow rate is not justified in the final stage of the discharge. The fit is qualitatively fair. In particular, Eq. (19) has a saturation of the pressure that is much slower than the traditional Janssen law for static silos. This is consistent with the results found here for the simulations and also results reported in experiments of discharging silos [19].

We speculate that the value of α found may be "universal" in the sense that it does not depend significantly on D_o, R_s, and material properties. However, this remains a subject of investigation.

4 Conclusions

We have shown that an energy balance, coupled with a constitutive equation for the rheology of a granular sample leads to a simple differential equation for the flow rate of a discharging silo. The equation resembles the Beverloo rule and yields a proportionality constant $C = \frac{\pi\sqrt{2}}{8} \approx 0.56$ which is very close to the experimental fitted value of 0.58.

The assumption of a constant flow rate leads to an expression for the pressure in the silo that is more suitable for these dynamic conditions than the Janssen law derived for static silos. We have to bear in mind, however, that the pressure in the silo is not homogeneous and a more elaborated analysis should include the local pressure in the description rather than the global pressure as we have done in this work.

The derived differential equation allows to explain the known phenomenology of silo discharge without the need of heuristic postulates such as the "free fall" arch and the "empty annulus". Moreover, this equation also opens the way to consider more complex situations in silo discharge, such as forced flow or vibrated discharges, that have not been explored from a theoretical perspective.

Acknowledgements

This work has been sup-ported by ANPCyT (Argentina) through grant PICT 2012-2155 and UTN (Argentina) through grants PID MAUTNLP0002184 and PID IFI1871.

References

[1] W Beverloo, H Leniger, J Van de Velde, Chem. Eng. Sci. **15**, 260 (1961).

[2] R L Brown, J C Richards, Trans. Inst. Chem. Eng. **37**, 108 (1959).

[3] J. Duran, *Sands, Powders and Grains*, Springer, New York, (2000)

[4] R. M. Nedderman, U. Tüzün, S.B. Savage and G.T. Houlsby, *J. Chem. Eng. Sci.* **37**, 1597-1609 (1982); U. Tüzün, G. T. Houlsby, R. M. Nedderman, S. B. Savage, *J. Chem. Eng. Sci.***37**, 1691-1709 (1982); S. B. Savage, R. M. Nedderman, U. Tüzün and G.T. Houlsby, *J. Chem. Eng. Sci.* **38**, 189-195 (1983).

[5] B.P. Tighe and M. Sperl, *Gran. Matt.* **9**, 141 (2007)

[6] G. H. Ristow, *Physica A* **235**, 319 (1997).

[7] E. Goldberg, C. M. Carlevaro, L. A. Pugnaloni, Papers in Physics **7**, 070016 (2015).

[8] T J Wilson, et al., *Granular discharge rate for submerged hoppers*, Papers in Physics **6**, 060009 (2014).

[9] C Mankoc, A Janda, R Arévalo, J Pastor, I Zuriguel, A Garcimartín and D Maza, Granular Matter **9**, 407 (2007).

[10] M A Aguirre, J G Grande, A Calvo, L A Pugnaloni, J-C Géminard, Phys. Rev. Lett. **104**, 238002 (2010).

[11] I. Zuriguel, L. A. Pugnaloni, A. Garcimartín and D. Maza, Phys. Rev. E **68**, 030301(R) (2003).

[12] R M Nedderman, *Statics and kinematics of granular materials* (Cambridge University Press, Cambridge, 2005).

[13] L Kondic, *Simulations of two dimensional hopper flow*, Granular Matter **16**, 235 (2014).

[14] A. Janda, I. Zuriguel and Diego Maza, Phys. Rev. Lett. **108**, 248001 (2012).

[15] S. M. Rubio-Largo, A. Janda, D. Maza, I. Zuriguel and R. C. Hidalgo, Phys. Rev. Lett. **114**, 238002 (2015).

[16] L. Staron, P.-Y. Lagrée, and S. Popinet, *Phys. Fluids* **24**, 103301 (2012).

[17] F. da Cruz, S. Emam, M. Prochnow, J. N. Roux, and F. Chevoir, Phys. Rev. E **72**, 021309 (2005).

[18] V. Šmilauer, E. Catalano, B. Chareyre, S. Dorofeenko, J. Duriez, A. Gladky, J. Kozicki, C. Modenese, L. Scholtès, L. Sibille, J. Stránský, K. Thoeni , Yade Documentation. The Yade Project. (http://yade-dem.org/doc/) (2010).

[19] C. Perge, M. A. Aguirre, P. A. Gago, L. A. Pugnaloni, D. Le Tourneau, and J. C. Géminard, Phys. Rev. E **85**, 021303 (2012).

Shear localization and effective wall friction in a wall bounded granular flow

Riccardo Artoni[1,*] and *Patrick* Richard[1,**]

[1]*LUNAM Université, IFSTTAR, MAST, GPEM, F-44340 Bouguenais, France*

Abstract. In this work, granular flow rheology is investigated by means of discrete numerical simulations of a torsional, cylindrical shear cell. Firstly, we focus on azimuthal velocity profiles and study the effect of (i) the confining pressure, (ii) the particle-wall friction coefficient, (iii) the rotating velocity of the bottom wall and (iv) the cell diameter. For small cell diameters, azimuthal velocity profiles are nearly auto-similar, i.e. they are almost linear with the radial coordinate. Different strain localization regimes are observed : shear can be localized at the bottom, at the top of the shear cell, or it can be even quite distributed. This behavior originates from the competition between dissipation at the sidewalls and dissipation in the bulk of the system. Then we study the effective friction at the cylindrical wall, and point out the strong link between wall friction, slip and fluctuations of forces and velocities. Even if the system is globally below the sliding threshold, force fluctuations trigger slip events, leading to a nonzero wall slip velocity and an effective wall friction coefficient different from the particle-wall one. A scaling law was found linking slip velocity, granular temperature in the main flow direction and effective friction. Our results suggest that fluctuations are an important ingredient for theories aiming to capture the interface rheology of granular materials.

1 Introduction

The rheology of granular materials is relevant to many industrials applications (grain transport and storage) and to natural events (avalanches, mud-slides...). Several geometries have been used to probe the rheology of granular systems (inclined plane, shear cell, confined gravity-driven flows...); however, a satisfying modeling of 3D flows is still lacking. This is particularly true for wall-bounded configurations, in which the nonlocality of the rheology near a boundary and the non trivial boundary conditions increase the complexity of the physics. In this work we present discrete element simulations of a wall-bounded three dimensional dense granular flow. The flow configuration studied in this paper, which can be referred as *torsional shear flow*, consists in a cylindrical geometry filled by the studied system, where the bottom wall rotates and the upper and cylindrical wall are fixed. In this work the flow is at fixed normal stress, i.e. the upper wall is free to move vertically under the action of the imposed normal force and the reaction of the particles contained in the cylinder. This configuration is interesting because simple velocity profiles were obtained for viscous and viscoelastic fluids, and therefore it is tempting to consider it as a granular rheometer. On the other hand, secondary flows were already observed for newtonian fluids in this geometry. The main objective of this work is therefore to characterize the granular flow in such a configuration. A particular focus will be given on the kinematics of the flow and on the flow regimes as a function of the main system parameters, and on the effective wall friction, and related scalings, at the cylindrical wall.

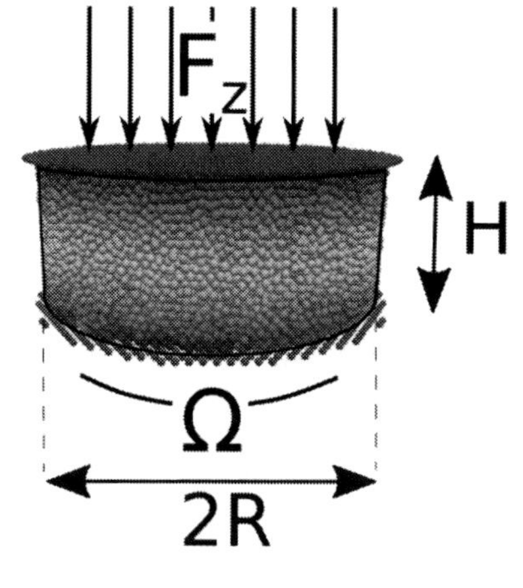

Figure 1. Sketch of the torsional flow configuration. F_z is the imposed normal force, and Ω the angular velocity of the bottom bumpy wall.

2 Simulation method

Numerical simulations are performed using the non smooth contact dynamics method [1], as implemented in the LMGC90 open source framework [2]. The flow configuration studied in this paper, which can be referred as

*e-mail: riccardo.artoni@ifsttar.fr
**e-mail: patrick.richard@ifsttar.fr

torsional shear flow, consists in a cylindrical geometry where the bottom wall rotates and the upper and cylindrical wall are fixed. In this work the flow is at fixed normal stress, i.e. the upper wall is free to move vertically under the action of the imposed normal force and the reaction of the particles contained in the cylinder. The top and bottom walls are bumpy, while the cylindrical wall is smooth but frictional. Gravity acts on the system along z. The top wall cannot move on the x- and y- directions but is free to move in the z-direction, simply according to the balance between its weight, the externally applied force and the force exerted by the grains. Simulations were performed with N slightly polydisperse spheres (uniform number distribution in the range $0.9d - 1.1d$) interacting through perfectly inelastic collisions and Coulomb friction ($\mu_p = 0.5$). Each bumpy wall was composed of spheres with the same properties of the particles (positioned randomly with a surface density of approximately 0.2 spheres / d^2). We chose perfectly inelastic grains to maximize dissipation and thus save computation time. Interactions of particles with the flat walls were also perfectly inelastic and frictional (with a coefficient of friction μ_{pw}).

We performed several simulations varying the angular velocity of the bottom bumpy wall Ω, the force applied to the upper bumpy wall F_z, the particle wall friction coefficient μ_{pw}, and the radius of the cylinder R (while maintaining approximately the same system height $H \sim 20d$). The first two parameters can be made dimensionless for example by considering a particle Froude number $\tilde{\Omega} = \Omega R / \sqrt{gd}$ and the ratio between the total force mass exerted by the top wall and the weight of the grains, $\tilde{F} = \frac{Mg+F_z}{Nmg}$ where m is the average particle mass, and $M = 100m$ is the mass of the top wall. In particular, the investigated ranges correspond to $\mu_{pw} = 0 - 0.3$, $\tilde{\Omega} = 0.12 - 2.4$, $\tilde{F} = 0.2 - 100$, $R = 12 - 36d$ (corresponding to $N = 10000 - 90000$). Our simulations lie in a range of inertial number [3, 4] from 0 (for $r = 0$) to 10^{-1}, which corresponds to the quasistatic and dense regimes of flow. It is important to stress that, due to the cylindrical geometry, the inertial number is an increasing function of the radial coordinate.

3 Kinematic profiles

In the following, we report results on azimuthal velocity profiles. In order to compute such profiles, several snapshots of particle positions and velocities were extracted from the simulations at different times. Due to the axial symmetry of the flow geometry, averages were computed with respect to a grid in the (r,z)-plane; for each grid point the average velocity was computed by a space-time weighted average [5]. The time average was performed as a simple average of the space-weighted averages; for space averaging, on the other hand, a Heaviside step function around the grid point with diameter equal to one particle diameter was chosen. This corresponds to performing averages on toroidal volumes.

Autosimilarity of the profiles end effect of aspect ratio H/R. Figure 2 displays azimuthal velocity profiles

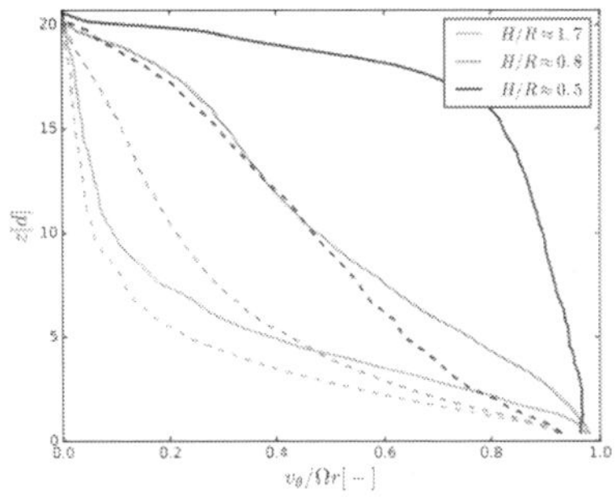

Figure 2. Azimuthal velocity profiles along z, rescaled by the velocity of the bottom wall, for $\tilde{\Omega} = 1.2$, $\tilde{F} = 0.2$, $\mu = 0.3$, three different values of the cell radius, and two radial positions, $1d$ from the cylindrical wall (dashed lines), and $10d$ far from the wall (solid lines).

rescaled by the velocity of the bottom plate, for $\tilde{\Omega} = 1.2$, $\tilde{F} = 0.2$, $\mu = 0.3$, three values of cell aspect ratio (at constant system height H). At first, the figure highlights the dependence of the velocity profile on the radial position: for large aspect ratios ($H/R = 1.7$), the velocity profile near the cylindrical wall is similar to the inner velocity profile. In this case, the velocity profiles are nearly autosimilar (see [6] for the dependance of the velocity with r). For smaller aspect ratios, the cylindrical wall slows down particle motion (in a well defined zone, not shown here, less than $10d$ thick, close to the wall), and this effect is stronger the smaller aspect ratio. Bulk profiles display a bottom localization for the large aspect ratio, a top localization for the small aspect ratio, and a more linear profile for the intermediate value. In the following, the effect of wall friction, pressure, bottom plate velocity will be evaluated for $H/R = 1.7$, and therefore, due to the near autosimilarity the profiles, rescaled azimuthal velocity profiles will be averaged in the radial direction.

Effect of wall friction coefficient and confining pressure. Figure 3 shows the joint effect of the particle-wall friction coefficient μ_{pw} and of the normal force applied to the top wall on the flow profiles, for $\tilde{\Omega} = 1.2$ and $H/R = 1.7$. For the sake of simplicity, due to the nearly autosimilarity of the profiles for this value of R, the profiles displayed in the Figure correspond to the average of the rescaled velocity profiles, $v_\theta/\Omega r$. First of all, profiles display shear localization with an exponential decay. However, the localization pattern depends strongly on wall friction. If we look at the behavior for low values of confining pressure (solid lines), for low wall friction, shear is localized near the top wall, while for strong friction shear is localized near the bottom wall. For intermediate values of the wall friction coefficient, a more uniform velocity profile prevails. As the wall friction coefficient, confinement pressure has a strong effect on velocity profiles: for negligible wall friction, increasing the normal force widens the top localized shear band. We can imagine that for very large confinement pressures and no wall friction the ve-

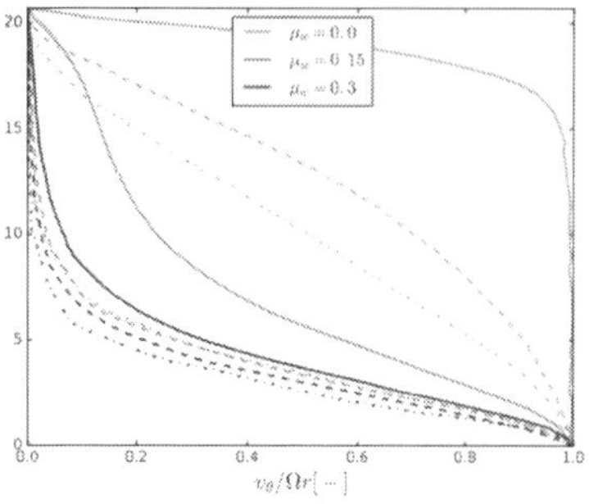

Figure 3. Average normalized azimuthal velocity profiles as a function of applied pressure, for $\tilde{\Omega} = 1.2$, $H/R = 1.7$, and three different values of the wall friction coefficient, and three different values of the force applied to the top wall : $\tilde{F} = 0.2$ (solid lines), $\tilde{F} = 8$ (dashed lines), $\tilde{F} = 100$ (dot-dashed lines).

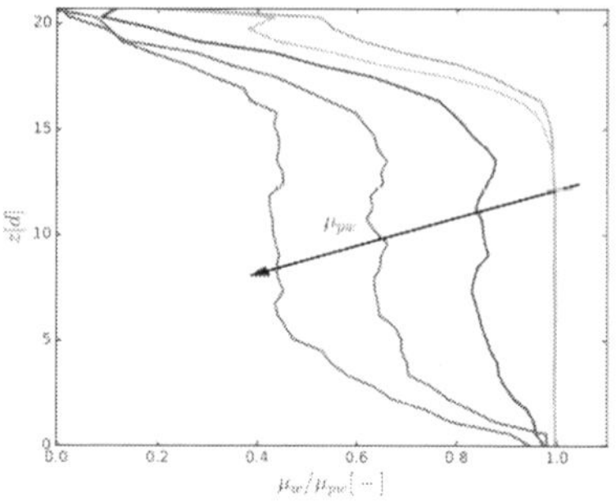

Figure 4. Profiles along z of the effective wall friction coefficient rescaled by the particle-wall friction coefficient μ_{pw} $(0.05, 0.1, 0.15, 0.2, 0.3)$, for $\tilde{\Omega} = 1.2$, $H/R = 1.7$, and different values of the wall friction coefficient μ_{pw}.

locity profiles will become linear. On the other hand, for larger values of the wall friction coefficient, increasing the normal force induces a strengthening of the bottom localization pattern with a decrease in the shear band width. The difference of the strain localization originates from the competition between dissipation at the sidewalls and dissipation in the bulk of the system [6, 14].

Effect of angular velocity and relevance of a local rheology. In the range of velocities considered here, the driving speed has little effect on the shape of velocity profiles (not shown). In [6], a model based on the $\mu(I)$ scaling and a minimum energy principle was developed. It was shown that, due to the presence of gravity, stresses are expected to increase with depth, and therefore the bulk energy dissipation is higher when shear is localized at the bottom. On the other hand, wall energy dissipation is higher if shear is localized near the top wall, because the plug flow zone contributes largely to the power needed to let particles slide at the wall, given that nearly all the particles at the wall slip with a large velocity. Changing the system parameters may change the relative balance of the two terms and therefore the shear localization pattern. The model was shown to predict the effect of pressure and wall friction, but wasn't able to capture the non-influence of the angular velocity of the bottom wall on the velocity profiles, at least without an ad-hoc correction.

This was explained by thinking at the combined effect of the rotating velocity and pressure. If the behavior of the system was described only by the characteristic inertial number and the wall friction coefficient, as a simple dimensional analysis could suggest, then we would expect that decreasing the rotational velocity by one order of magnitude should be the same as increasing the pressure by two orders of magnitude. But, as discussed above, the rotating velocity has no effect, while results are very sensitive to pressure, particularly for low wall friction. This means that the global inertial number and the wall friction coefficient are not sufficient for predicting the behavior of the system. It will be interesting to see if recent modelings of non-locality in granular flows [7–13] are able to capture the full behavior of our system.

4 Wall forces

We adapt here an analysis which we recently developed for a similar geometry [14]. We performed averages on slices with a thickness $2d$ in the vertical direction. The streamwise effective wall friction coefficient at the lateral walls is estimated for each slice as the ratio of the average force in the flow direction θ and the average force in the direction normal to the wall r: $\mu_w = \langle F_\theta \rangle / \langle F_r \rangle$ [15]. This corresponds to the stress ratio $\sigma_{r\theta}/\sigma_{rr}$ at the wall. Profiles of velocity fluctuations at the wall in the θ direction (which are related to granular temperature) are also calculated as $T_\theta = \langle (v_\theta - \langle v_\theta \rangle)^2 \rangle$, where $\langle v_\theta \rangle$ is the average azimuthal velocity of particle touching the wall for each slice. The fluctuations are correctly computed with respect to the average velocity value extrapolated at particle center [5].

Wall friction profiles. Figure 4 displays profiles of the effective wall friction coefficient, for $\tilde{\Omega} = 1.2$, $H/R = 1.7$, and different values of the wall friction coefficient μ_{pw}. It is clear that, as it was seen for 2D flows [16] or in 3D confined flows [14], μ_w/μ_{pw} decreases with μ_{pw}. These profiles are strongly related to velocity profiles: for low μ_{pw} (when shear is localized at the top), far from the shear band, grains rotate as a solid block with angular velocity Ω. In this case there are no stick events, so the effective coefficient of friction in the plug zone is equal to μ_{pw}. In the shear zone, stick-slip events may emerge, and therefore $\mu_w < \mu_{pw}$ [15]. For higher μ_{pw} values we find a decreasing profile of μ_w vs z in the shear band, a plateau in the center of the creep zone (intermediate z) and a decrease again when approaching the top wall. This can be explained by the fact that, reasonably, stick-slip events become more and more probable when we approach the creep zone [15].

Wall friction scaling. In [14], we found that μ_w/μ_{pw} scaled with a dimensionless parameter defined as the ratio between the slip velocty and the square root of the stream-

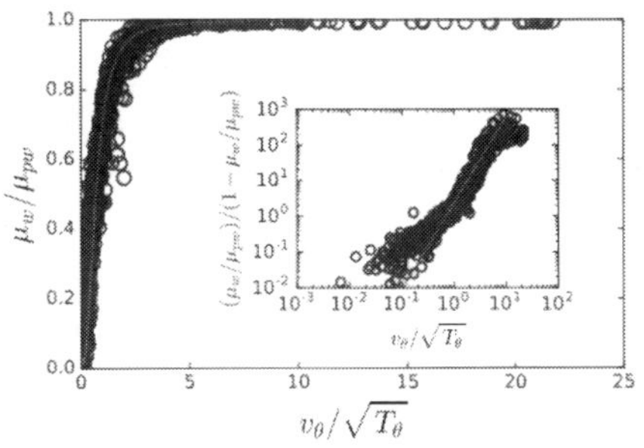

Figure 5. Rescaled effective wall friction coefficient versus the dimensionless slip parameter $v_\theta/\sqrt{T_\theta}$. The inset shows that a power law relationship holds between $\frac{\mu_w/\mu_{pw}}{1-\mu_w/\mu_{pw}}$ and $v_\theta/\sqrt{T_\theta}$. The solid line is a fit to Eq. 1, with $A = 0.7$ and $B = 2.4$.

wise velocity correlation. The origin of this scaling was postulated to be the presence of force and velocity fluctuations, yielding stick-slip events. Here we test the scaling for the present flow configuration. This test is shown in Fig. 5, where we display the rescaled effective friction coefficient μ_w/μ_{pw} as a function of $v_\theta/\sqrt{T_\theta}$. The scaling performs very well on several orders of magnitude. The functional form of the scaling appears to be the same as compared to the empirical one used in [14]:

$$\frac{\mu_w}{\mu_{pw}} = \frac{\left(v_\theta/\sqrt{T_\theta}\right)^B}{A + \left(v_\theta/\sqrt{T_\theta}\right)^B}, \tag{1}$$

where $A = 0.7$ and $B = 2.4$ are best-fit parameters.

5 Conclusions

In this work, the rheological properties of granular matter submitted to torsional shear were investigated by means of discrete numerical simulations. The torsional shear flow configurations consists in a a shear cell made of a cylinder filled by grains which are sheared by a bumpy bottom and submitted to a vertical pressure which is applied at the top. An analysis of the kinematics of the system was presented. We focused on azimuthal velocity profiles and studied the effect of the confining pressure, the particle-wall friction coefficient, the rotating velocity of the bottom wall and the aspect ratio of the cell. For small cell diameters (large aspect ratio) profiles are nearly auto-similar, i.e. they are almost linear with the radial coordinate. Regimes differing by their strain localization features were observed : we find that shear can be localized (i) at the bottom or (ii) at the top of the shear cell, or that (iii) it can be even quite distributed. This interesting behavior originates from the competition between dissipation at the sidewalls and dissipation in the bulk of the system. A purely local rheology based on the inertial number I cannot predict the effect of the rotating velocity of the bottom wall: our results point out the need for an alternative rheology.

We also characterized the effective friction at the cylindrical wall. The strong link between effective wall friction, wall slip and fluctuations of forces and velocities was pointed out. At flat frictional walls, even if the system is globally below the sliding threshold, force fluctuations trigger slip events, leading to a nonzero slip velocity and an effective wall friction coefficient scaling with a sliding parameter. This confirms our previous findings in another flow configuration [14]: a scaling law was found linking slip velocity, granular temperature in the main flow direction and the effective friction. Our results suggest that velocity and force fluctuations can be an important ingredient for theories aiming to capture the interface rheology of granular materials.

Acknowledgements

The numerical simulations were carried out at the CCIPL (Centre de Calcul Intensif des Pays de la Loire) under the project MTEEGD.

References

[1] M. Jean, Computer Methods in Applied Mechanics and Engineering **177**, 235 (1999)
[2] M. Renouf, F. Dubois, P. Alart, Journal of Computational and Applied Mathematics **168**, 375 (2004)
[3] GDR-MiDi, Eur. Phys. J. E **14**, 341 (2004)
[4] F. da Cruz, S. Emam, M. Prochnow, J.N. Roux, F. Chevoir, Phys. Rev. E **72**, 021309 (2005)
[5] R. Artoni, P. Richard, Phys. Rev. E **91**, 032202 (2015)
[6] R. Artoni, P. Richard, Computational Particle Mechanics pp. 1–10 (2016)
[7] K. Kamrin, G. Koval, Phys. Rev. Lett. **108**, 178301 (2012)
[8] D.L. Henann, K. Kamrin, Proceedings of the National Academy of Sciences **110**, 6730 (2013)
[9] K. Kamrin, D.L. Henann, Soft Matter **11**, 179 (2015)
[10] D.L. Henann, K. Kamrin, Phys. Rev. Lett. **113**, 178001 (2014)
[11] J. Jenkins, D. Berzi, Granular Matter **12**, 151 (2010)
[12] J. Jenkins, D. Berzi, Granular Matter **14**, 79 (2012), 10.1007/s10035-011-0308-x
[13] R. Artoni, A. Santomaso, P. Canu, Phys. Rev. E **83**, 051304 (2011)
[14] R. Artoni, P. Richard, Phys. Rev. Lett. **115**, 158001 (2015)
[15] P. Richard, A. Valance, J.F. Métayer, P. Sanchez, J. Crassous, M. Louge, R. Delannay, Physical Review Letters **101**, 248002 (4) (2008)
[16] R. Artoni, A. Santomaso, M. Go', P. Canu, Phys. Rev. Lett. **108**, 238002 (2012)

Wear of sharp aggregates in a rotating drum

Ivan Deiros Quintanilla[1,2,3*], *Gaël* Combe[1,2], *Fabrice* Emeriault[1,2], *Jean-Benoît* Toni[1,2], *Charles* Voivret[3], and *Jean François* Ferellec[3]

[1]Université Grenoble Alpes, 3SR, F-38000 Grenoble, France
[2]CNRS, 3SR, F-38000 Grenoble, France
[3]SNCF Réseau, Direction Ingénierie & Projets, 93574 La Plaine Saint Denis, France

Abstract. Aggregates constituting ballast layer wear due to the continuous passage of trains and during the necessary maintenance operations of the track. In order to develop efficient solutions for ballasted tracks design and maintenance, a proper knowledge of the degradation laws of ballast grains is needed. In tribology, the amount of wear due to friction when two surfaces are in contact is classically predicted by Archard's equation. However, due to the continuous evolution of grain angularity and roughness, at the macro-scale wear coefficient cannot be assumed to remain constant, but will depend on the state of degradation of the grain surface. In order to adjust the model to this particular case, the Micro-Deval Attrition test is used. The rotating drum is stopped at intermediate stages and the amount of generated fine particles is measured. Thus the curve of mass loss along time is built. These results are then linked to Archard's model using the values of contact forces and relative displacements extracted from discrete element simulations. Finally, a morphology analysis is performed tracking shape and roughness parameters at different stages of degradation using X-ray tomography and a laser profilometer.

1 Introduction

In ballasted railway tracks, ballast layer plays a crucial role in the transmission of dynamic stresses imposed by the circulation of trains to the geotechnical structure supporting the track. The performance of ballast strongly depends on the shape, size and mineralogical nature of the grains composing it. Throughout the life of the track ballast particles wear, due to the constant friction between grains, steadily reducing the angularity and surface roughness of the grains and producing very fine particles (d < 0.5 mm) that will eventually induce a loss of track performance. The interlock between particles is reduced limiting the anchorage of sleepers and the transfer of loads to the platform, and the presence in excess of fine particles renders maintenance operations ineffective and reduces the permeability of the track.

After some years of service of the first high-speed service in France, ballast has proven not to be resistant enough, showing a faster degradation than expected. This has been partially compensated over the years by the strict selection of good quality granite for the aggregates. This selection is performed in the quarry, where some mineral samples are subjected to an attrition test called Micro-Deval (MDA) [1, 2]. During this test, 10 kg of grains are turned at 100 rpm inside a drum during 140 minutes. Particles retained at the 1.6 mm sieve are then weighted and compared to the initial weight through a dimensionless coefficient C_{MDA} determining the resistance to abrasion of the material.

2 Archard model

Wear produced by particles in contact is a very complex process depending on the loading conditions at the contact scale. In tribology, Archard model [3] has been classically used to predict the flux of eroded mass when two surfaces in contact move relatively one to the other. Archard assumes that the volume W of fine particles produced is proportional to the normal force (f_n) and the relative displacement between surfaces (s).

$$W = (k / H) \cdot f_n \cdot s \quad (1)$$

The product $f_n{\cdot}s$ is directly related to the work produced by friction forces. On the other side, the proportionality is often expressed as k/H, where H is the hardness of the softest material and k is a dimensionless constant called wear coefficient. In the case of study, H can be considered as constant since the material is granite for both surfaces and does not change its properties when eroded. However, surfaces in contact suffer changes in shape, losing their roughness and angularity, and consequently reducing the severity of wear at the contact interface. At the macro-scale, wear coefficient cannot be assumed to be constant, but will depend on the state of degradation of the surface.

In order to adjust Archard model to the particular case of irregular sharp aggregates, the MDA test is used, linking the variables extracted from discrete element simulations of the test with the results of the real experimental test.

* Corresponding author: ivan.deiros@3sr-grenoble.fr

3 Micro-Deval

3.1 DEM simulation

In order to better understand the wear process occurring within the Micro-Deval device, a discrete element simulation of the test [4, 5], using a Non-Smooth Contact Dynamics approach (NSCD) [6], has been performed and compared with the experimental results.

The drum has been numerically built following the European code [1, 2]: 40 cm long with a 20 cm inner diameter. The grains have been randomly chosen from a database of 1000 different modelled ballast grains. This database has been built following the correct ballast grain-size distribution using polyhedral shapes containing from 12 to 70 faces and from 8 to 37 vertices. In order to cover the required 10 kg of ballast, 109 grains are required with a density of 2700 kg/m^3.

Collisions are considered perfectly inelastic and grains are modelled as rigid bodies. Wear is not modelled during the simulation. Thus, only 35 seconds of test have been simulated, since system conditions remain the same along time.

Friction coefficient between ballast grains has been stablished as 0.8, following the results of experimental tests (Figure 1), and between ballast and the metallic drum arbitrarily set at 0.4. It is noteworthy that friction coefficient does not change significantly between new and eroded grain surfaces, especially when dealing with small relative displacements.

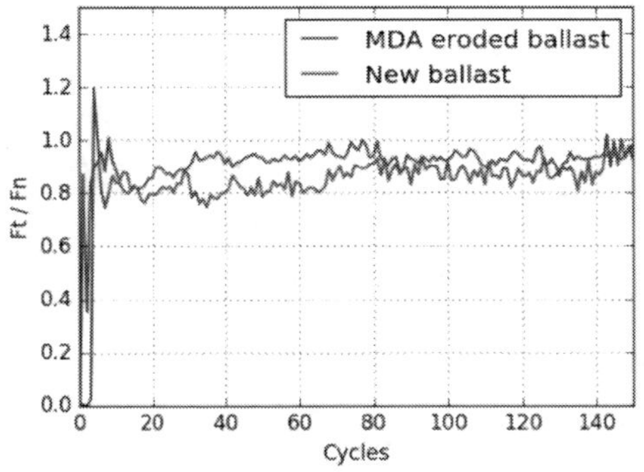

Fig. 1. Evolution of the friction coefficient between two grains along shearing cycles in BCR3D shearing device [7].

Simulations show that the MDA test is a very dynamic process in which grains turn very fast at almost the same speed as the cylinder. The grains kinematics analysis highlights two plausible wear mechanisms: gently degradation due to persistent intergranular contact with tangential displacements accumulation when the grains go upwards inside the drum, and an aggressive but short degradation due to impact after a phase of free fall.

In order to estimate the amount of wear produced in a contact, it is necessary to establish which fraction of the contact energy is absorbed by each body. Since the main goal is to obtain a global approximation of the wear, in the case of ballast-ballast contacts it is acceptable to consider that each grain absorbs half of the energy. When the ballast is in contact with the drum, the relative displacements are high due to the lower friction coefficient, leading to a higher friction work since it is calculated, following Archard equation, over f_n and normal forces do not change. However, this friction energy is mostly absorbed by the drum in the form of heat, leading to a lower degradation of the ballast grains than that due to contacts between grains. Following empirical observations in other MDA tests with only one grain and metal balls, the amount of energy in ballast-metal contacts causing grain degradation can be estimated between 10% and 15% of the friction energy of the contact. For this paper, 15% is considered.

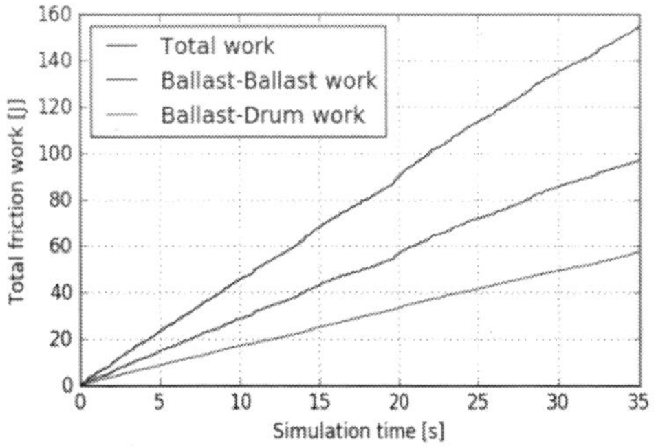

Fig. 2. Evolution of the friction work absorbed by 109 ballast grains along the NSCD simulation of a Micro-Deval test.

Figure 2 shows the amount of friction work absorbed by grains along the simulation time in each type of contact. Due to the homogeneity of the test throughout time (conditions not changing and contacts happening at the same rate), the evolution of the friction work is linear, as expected. Taking into account all types of contacts, the friction work rate is constant and equal to 4.4 J/s absorbed by all ballast grains.

3.2 Experimental campaign

The data extracted from the numerical simulation needs to be compared to the actual experimental test and the amount of generated fines. This test is performed without water to be as close as possible to the simulation conditions. In order to extract the curve of mass loss during the test, the drum is stopped at intermediate stages. Grains are then cleaned and the sample is weighted. Particles bigger than 1.6 mm are extracted and weighted apart, so the amount of generated fines can be measured as the difference between the weight of the sample before and after the test minus the amount of small particles bigger than 1.6 mm.

Figure 3 shows the rate of generation of fine particles during the test, stopping the drum every 5 and 20 minutes with three different ballast samples. This curve is plotted for top-quality granite used in high-speed lines and for granite for conventional lines. A curve fit is added to each set of experimental points. This function has not been chosen for any particular physical reason but as a best-fit choice, which has turned out to be a power law for all three Micro-Deval tests.

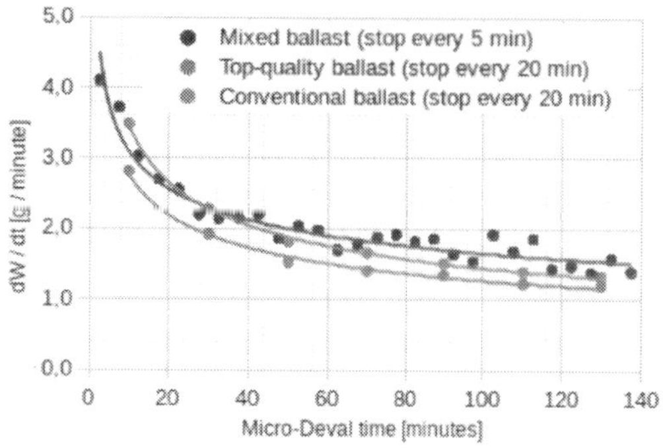

Fig. 3. Rate of fines production during a Micro-Deval test.

Two well-distinguished phases are identified: at the beginning of the test, the wear rate is higher, but it decreases rapidly following approximately the fitted power law. After the first hour of test, the production of fines tends to stabilise to a fixed wear rate of 1.2 ~ 1.3 g / minute (Table 1).

Table 1. Production of fines during a Micro-Deval test.

	Top-quality ballast	Conventional ballast
First 20 min	2.82 g / minute	3.49 g / minute
Last 20 min	1.20 g / minute	1.33 g / minute

This quasi-linear production of fines observed after the initial stages of the test is coherent with the constant rate of friction work extracted from the NSCD simulations.

From a macro-scale point of view, the wear coefficient k/H of the Archard model is not constant at the beginning of the test but it can actually be considered as constant after the first hour initial phase.

3.3 Linking results

Taking into account the friction work extracted from NSCD simulations and the production of fines observed in the experimental tests, it is then possible to link both results using Archard's model to get a first approximation of the wear coefficient.

After the first phase of the MDA test, an averaged constant wear rate of 1.3 g/minute can be assumed. Using the work rate extracted from the NSCD simulations, *k*/*H* will have an approximate value of $1.8{\cdot}10^{-9}$ Pa^{-1}. Wear production will be given by:

$$W = (k / H) \cdot f_n \cdot s = 1.8{\cdot}10^{-9} \cdot E_{fr} \qquad (2)$$

where W is the volume of fine particles generated in m^3 and E_{fr} is the friction energy absorbed by the grain.

This expression, though, is only valid after the initial phase of degradation of the grains. In order to better understand the initial degradation and to be able to assess when the constant rate is reached, it is necessary to study how the morphology of the grain evolves during the process.

4 Evolution of morphology

Grain morphology plays a key role in the behaviour of the ballast layer as well as in the production of fine particles. Some authors [8] use three different scales to characterize shape and texture. At large scales, sphericity, elongation and flatness are usually used to provide a general idea of the grain shape. The intermediate scale is focused on describing the roundness of the grain and sharpness of angles and borders, and it is usually described using the parameter called angularity. Finally, roughness is the parameter used to describe, at the micro-scale, the surface texture.

Large scale parameters are classically calculated using the three characteristic lengths (shortest, intermediate and longest) of a rectangular prism in which the grain is inscribed [9, 10]: sphericity (ψ), shape factor (*SF*), flatness ratio (*FR*), elongation ratio (*ER*) and flat-elongation ratio (*FE*). For the intermediate parameters, angularity and form indexes are defined using harmonic analysis [11]. Finally, surface roughness is assessed by scalar parameters quantifying the height of the surface points relative to a reference line (average roughness R_a, RMS roughness R_q, total roughness R_t) or plane (S_a, S_q, S_t) [12]. In addition, a fractal approach to roughness is considered (D) [13].

In order to study the evolution of the different morphology parameters, five ballast grains have been identified and submitted to a more aggressive MDA test: 3h of test with only 10 grains so the higher intensity of the impacts produce accelerated erosion. Before and after the MDA test, the samples have been scanned using X-ray tomography to obtain a 3D image of the whole grain, and with a laser profilometer over one flat face of the grain to assess the roughness.

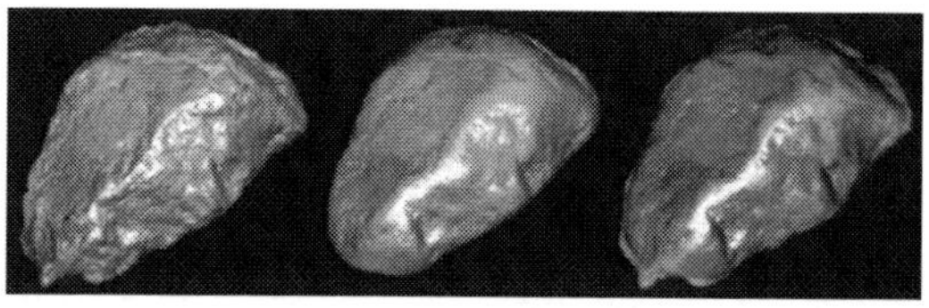

Fig. 4. X-ray images of a grain (sample S12) before and after a Micro-Deval Attrition test, and a comparison of both images highlighting the eroded areas.

In figure 4, it can be observed how sharp corners disappear first, due to either wear or breakage of weaker parts, reducing considerably the angularity of the grain. It is also shown how roughness is dramatically reduced leaving a smooth surface. However, the general shape (elongation, sphericity, etc.) of the grain remains without remarkable changes.

Figure 5 shows the evolution in percentage of all shape parameters in comparison with their initial values. It is clearly seen how the angularity of all grains decreases in a significant amount. Large scale parameters, however, remain with approximately the same values. These parameters should tend to values close to 1 if the erosion is more aggressive, since the grains tend to reach a spherical-ellipsoidal shape. However, the MDA test is not able to reproduce such a big amount of erosion and ballast in railway tracks is actually substituted much before these erosion levels are reached.

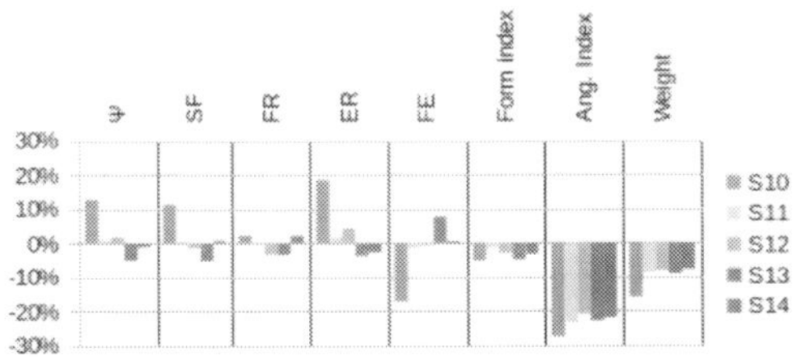

Fig. 5. Evolution of the different shape parameters. S1X is the name of the different grain samples.

Figure 6 shows the evolution in percentage of the different roughness parameters and fractal dimension. As expected, these parameters are clearly reduced during the process (except for sample S13, which presented an already relatively smooth surface before the erosion). Empirical observations show evidence that roughness is lost during the first minutes of the MDA test, corresponding to the higher production of fine particles at the beginning of the test.

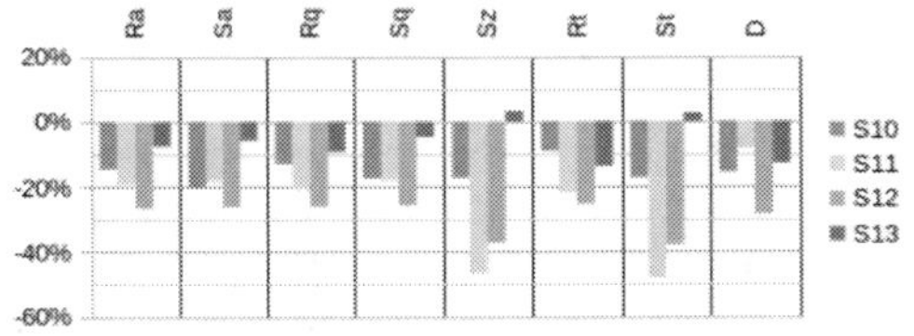

Fig. 6. Evolution of the different roughness parameters.

Some studies [14] confirm the existence of at least two phases in the wear process: first angularity is progressively lost without changing the main axis dimensions and it is afterwards that the general shape is slowly rounded. The results shown in this paper suggest the existence, for sharp and rough grains, of an initial phase corresponding to a fast degradation at early stages, which leads to a three-phase wear process corresponding to the three morphology scales.

Friction wear of ballast grains is a local phenomenon concerning only the first two phases: the micro-texture of the grain and its angularity. This behaviour is in accordance with the two phases observed in the rate of generation of fine particles during the MDA test.

5 Conclusions

Discrete Element simulations (NSCD) of the Micro-Deval Attrition test have shown that the amount of energy absorbed by the grains follow a linear tendency over time. However, experimental tests have demonstrated that the production of fine particles is not linear during the whole test, but it is more aggressive at the beginning and becomes almost linear after the first half of the test producing about 1.3 g/minute.

Following Archard, it is possible to relate the linear phase of the wear process to the wear coefficient k. In order to understand this wear process, grain morphology has been tracked and analysed.

The morphology study has shown that the smaller the scale is the faster the parameters are smoothed. Hence large scale parameters have a very slow evolution, depending to a large extent on the grains crushing. The intermediate scale, mainly represented by angularity, plays a very important role in the evolution of shape since sharp angles and edges are rounded and weak pointed tips break easily. Finally, micro roughness is lost during the very first minutes of the test.

This evolution is coherent with the experimental curve of mass loss. The higher production of fine particles at the beginning of the Micro-Deval Attrition test can be related to the polishing of the surface texture and weaker edges and vertices, losing very quickly the roughness and reducing faster the angularity. Afterwards, the process becomes more linear since faces are already polished and edges and vertices present larger surfaces to resist the contact forces. From this point, angularity is reduced in a slower but constant process leading to an almost linear production of fines.

As a perspective for further studies, the morphology description of a grain, especially concerning roughness and angularity, should be associated to a specific point in the wear rate curve.

The authors gratefully acknowledge Denis Tunkin for his contribution to the study of shape evolution during his Master thesis. The Laboratoire 3SR is part of the LabEx Tec 21 (Investissements d'Avenir – grant agreement n° ANR-11-LABX-0030).

References

1. AFNOR, *NF EN 1097-1* (Saint Denis, 1996)
2. AFNOR, *NF EN 13450* (Saint Denis, 2003)
3. J.F. Archard, J. Appl. Phys., **24**, 981-988 (1953)
4. I. Deiros, C. Voivret, G. Combe, F. Emeriault, *Proc. of the Third International Conference on Railway Technology* (Civil-Comp Press, UK, 2016)
5. I. Deiros, C. Voivret, G. Combe, F. Emeriault, Procedia Engineering, **143**, 1016-1023 (2016)
6. M. Jean, Computer Methods in Applied Mechanics and Engineering, **177(3-4)**, 235-257 (1999)
7. J. Hans, M. Boulon, Int. J. Numer. Anal. Meth. Geomech., **27(6)**, 513-548 (2003)
8. J.K. Mitchell, K. Soga, *Fundamentals in Soil Behavior* (John Wiley & Sons, New Jersey, 2005)
9. W.C. Krumbein, J. Sedimentary Petrology, **11**, 64-72 (1941)
10. R.D. Barksdale, M.A. Kemp, W.J. Sheffield, J.L. Hubbard, Transp. Research Record, **1301**, 107-116 (1991)
11. E.J. Garboczi, Cement and Concrete Research, **32(10)**, 1621-1638 (2002)
12. B. Bhushan, *Modern Tribology Handbook* (CRC Press, USA, 2000)
13. A. Majumdar, B. Bhushan, J. Tribology, **112(2)**, 205-216 (1990)
14. G. Domokos, D.J. Jerolmack, A.A. Sipos, Á. Török, PLOS one, **9(2)**, e88657 (2014)

Shear test on viscoelastic granular material using Contact Dynamics simulations

Juan Carlos Quezada[1,*], *Loba* Sagnol[1,2], and *Cyrille* Chazallon[1]

[1] *ICUBE, UMR 7347, CNRS, INSA de Strasbourg, Strasbourg, France*
[2] *Institut Grund- und Straßenbau, Hochschule Karlsruhe Technik und Wirtschaft, Karlsruhe, Germany*

Abstract. By means of 3D contact dynamic simulations, the behavior of a viscoelastic granular material under shear loading is investigated. A viscoelastic fluid phase surrounding the solid particles is simulated by a contact model acting between them. This contact law was implemented in the LMGC90 software, based on the Burgers model. This model is able to simulate also the effect of creep relaxation. To validate the proposed contact model, several direct shear tests were performed, experimentally and numerically using the Leutner device. The numerical samples were created using spheres with two particle size distribution, each one identified for two layers from a road structure. Our results show a reasonable agreement between experimental and numerical data regarding the strain-stress evolution curves and the stress levels measured at failure. The proposed model can be used to simulate the mechanical behavior of multi-layer road structure and to study the influence of traffic on road deformation, cracking and particles pull-out induced by traffic loading.

1 Introduction

Granular materials are widely used as filling materials since of their mechanical properties such their shear strength and porosity. Examples include construction foundations, earthworks and transport infrastructures such as railway ballast and asphalt concrete.

Asphalt is a complex multi-phase mixture comprises bitumen, graded mineral aggregate and air. This mixture relies on an interlocking the solid phase, conformed by crushed aggregates, for its strength with bitumen to provide a cohesion to the mixture. Herein, the solid phase distributes the forces induced by traffic loading. This material has a typical viscoelastic behavior, where the mechanical performance of the mixture is a function of strain rate and temperature. This macroscopic behavior is mainly dependent of the micro-scale behavior of the asphalt mixture.

Laboratory trials allow to identify the overall response of asphalt mixture, but struggle to obtain a micromechanical insight of these materials. Numerical simulation using continuum mechanics approaches shows the same difficulties regarding the identification of micro-scale properties. A way around this barrier is to use a discrete approach, which has been widely used to model the behavior of granular materials. This method allows to simulate the interaction of a collection of rigid or deformable bodies in contact. Over the past two decades, the discrete element method (DEM) has been used by many researchers to simulate the microstructure of asphalt mixtures. These studies have modeled the asphalt using simple spherical particles, irregular particles created with clumps of spheres or by generating samples using imaging techniques [1–3].

In this work, a viscoelastic contact law is developed and implemented in a DEM code to model asphalt mixtures. To validate the method and to calibrate the contact model parameters, laboratory shear tests were performed. The initial section (Sect. 2) of this paper presents the experimental setup and laboratory tests. Then, the numerical procedures are outlined (Sect. 3) together with the simulation method, particle properties, and preparation protocol. Finally, in Sect. 4 the results from numerical shear test are analyzed and compared to experimental data to validate the proposed contact model.

2 Experimental procedures

2.1 Experimental setup

Experimental procedures were performed using the Leutner shear test on asphalt samples. The specimen cores were extracted from an actual double-layer asphalt pavement. This pavement structure test comprises a 60mm thick layer for the binder course and a 60mm thick layer for the surface course. The particle size distribution of these layers were 0/10 and 0/16 respectively. This pavement structure was constructed on an unbound granular base course. From this asphalt pavement 5 cores were extracted with the following dimensions: 120mm height and 150mm diameter.

*e-mail: juan-carlos.quezada@insa-strasbourg.fr

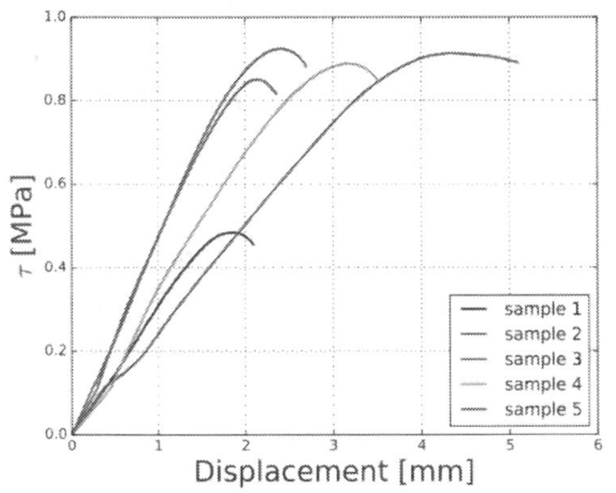

Figure 1. Shear stress-displacement curves obtained for experimental trials.

2.2 Monotonic shear test

The shear test were carried-out with the Leutner device [4]. This trial is a destructive monotonic shear test with pure interlayer shearing. During the test the binder course layer of the core was fixed whereas the surface course layer was loaded at a controlled rate of 50 mm/min at 25 °C until failure was reached. The equipment records the shear force applied at the interface between the two layers and the displacement of the surface layer.

Figure 1 displays the shear stress τ as a function of the imposed displacement. The experimental results show a variability in terms of the maximum applied shear stress and measured displacement at failure. From the shear stress-displacement curves, an average value of 0.898MPa is obtained for the shear-stress and 2.37mm for the displacement at failure.

3 Numerical procedures

3.1 Contact dynamics method

The numerical simulations were performed using the Contact Dynamics (CD) method with spherical particles [5–7]. The CD method is a discrete element approach for the simulation of non-smooth granular dynamics with contact laws expressing basically the mutual exclusions and dry friction between particles without introducing the elastic or viscous regularization often used in explicit methods such as molecular dynamics [8–11] or the distinct element method [12]. The non-smoothness refers to various degrees of discontinuity in the velocities and contact forces arising in a system composed of rigid particles. In this method, the equations of motion for each particle are formulated as differential inclusions in which velocity jumps replace accelerations [13, 14]. The unilateral contact interactions and Coulomb friction law are treated as complementarity relations or set-valued contact laws. At each time step, all the velocities and contact forces in the system are determinate simultaneously from the equations of dynamics. This kinematic problem is solved by an iterative procedure based on the non-linear Gauss-Seidel method. For our simulations, we used the LMGC90 software, which is capable of modeling a collection of rigid or deformable particles of various shapes by different algorithms [15].

3.2 Description of the contact model

To model an asphalt mixture under a shear loading, a viscoelastic contact law is implemented during the shear test. The chosen contact law was the Burgers model which is a more accurate model of viscoelastic behavior for asphaltic materials [16–18]. It describes finely the creep, relaxation and dynamic properties of asphalt mixtures and has become a commonly used contact model in the DEM simulation of asphalt mixtures. This model consists of the Maxwell model combined in series with the Kelvin-Voigt model. The time dependent normal contact stiffness K_n is given by:

$$K_n = \left(\frac{1}{K_m} + \frac{t}{\eta_m} + \frac{1}{K_k}\left(1 - e^{-t/t_r}\right)\right) \quad (1)$$

where K_m is Maxwell normal stiffness, η_m is Maxwell normal viscosity, K_k is Kelvin-Voigt normal stiffness, η_k is the Kelvin-Voigt normal viscosity, t is the loading time and $t_r = \eta_k/K_k$ is the relaxation time. The parameter values used in our simulation were 2×10^6 N.m^{-1} for K_m, 2×10^6 N.m^{-1} for K_k, 2×10^7 N.s.m^{-1} and 2×10^7 N.s.m^{-1} for η_m and η_k respectively. This set of values generates the best fit regarding the shear stress-displacement curve obtained for the numerical sample in comparison with experimental data. Assuming a perfectly isotropic behavior, tangential stiffness K_t and K_s are estimated as function of normal stiffness as: $K_t = K_s = K_n/(2(1+\nu))$ where ν is the Poisson's ratio. As a sake of simplicity, this ratio was set to 0.3.

This contact law was adapted to the CD formulation to be implemented in the LMGC90 software. The equations of dynamics can be written in a compact form by means a matrix representation in the contact frame:

$$\mathbf{U}^+ = \mathbf{U}^- + \mathbf{W}\mathbf{R} \quad (2)$$

where $\mathbf{U}^+$ and $\mathbf{U}^-$ the left-limit and right-limit contact velocities, respectively, $\mathbf{W}$ is the Delassus local matrix and $\mathbf{R}$ the local impulsion at contact. In this form, the accelerations are replaced by velocity jumps defined by $\mathbf{U}^+ - \mathbf{U}^-$ and the equations of motion take the form of an equality between the change of momentum and the average impulse during the time-step H. The coefficients W_{ij} from the Delassus local matrix are the inverse reduced inertia. To incorporate our viscoelastic model in the local frame formulation, we added complementary components in this local matrix W^*_{ii} for normal and tangential components (i stand for n, t or s):

$$W^*_{ii} = W_{ii} + 1/(H^2 K_i) \quad (3)$$

Table 1. Particle size distribution for each layer composing the sample.

diameter(mm)	2	5.6	8	11.2	16
binder course (%)	22.9	20.1	24.0	29.8	3.2
surface course (%)	28.4	32.4	36.6	2.7	0.0

To model the brittle behavior of this material, a yield stress criterion is imposed to identify when cohesive contacts are lost. This criterion is verified at each time step, by calculating the generated stress at contact as $K_i \times (\Delta_i)/A$, where K_i is the stiffness at contact, Δ_i the contact displacement during a time-step and A is the smaller projected surface of particles at contact. When shear or tensile stress reaches the yield value of 2.7×10^6 Pa, the bond between particles is broken and a frictional contact law is applied instead.

3.3 Sample preparation

The numerical samples are composed of spherical rigid particles with two different size distributions. Each particle has a diameter ranging from 2mm to 10mm for the surface course layer and from 2mm to 16mm for the binder course layer (see Tab. 1). Here, the graded curve was cut at 2mm to avoid modeling all the fines, to reduce simulation time. For each particle size, a specific number of particles are created according to the particle size distribution. Then these particles are disposed randomly within a cubic lattice with 150mm diameter and 1.6m height. The bulk density for all particles is 2600 kg.m^{-3}. The coefficient of friction between the particles is set to $\mu = 0.7$ which is a typical value used for rock crushed aggregates.

The preparation protocol consists in first pouring the particles of the bottom layer by gravity into a cylindrical box with 150mm diameter with zero particle-wall friction. The binder course layer obtained by this procedure is subjected to vibrations of small amplitude (6mm) by applying a vertical sinusoidal displacement on the top wall during 1s with a frequency of 30Hz, obtaining a final layer of 60mm height. This procedure is repeated for the surface course layer. Finally the resulting sample is confined between two plates inside a cylinder box.

Following this entire process, five samples were prepared. The cylindrical samples have the same geometrical properties as experimental samples: 150mm diameter and 120mm of total height. The number of particles is about 5200 for all samples, with a packing fraction ranging from 0.584 to 0.59. Figure 2 shows a snapshot of the resulting numerical sample.

3.4 Direct shear test modeling

To model the Leutner shear test, the cylinder box is replaced by two cylinders, each one with 60mm height and separated each other of 1mm. The shear test is performed by imposing a velocity of 50mm/min to the upper cylinder until reach a displacement of 2.5mm. The shear stress τ

Figure 2. Snapshot of a numerical sample after the preparation stage.

is obtained by measuring the total reaction of contacts between particles and the upper cylinder normalized by the cross section of the sample. The time step was 2×10^{-4}s in all simulations, and at most 15000 time steps were needed to obtain a displacement of 2.5mm. The CPU time was 2×10^{-2}s per particle and per time step on a Dell computer of speed 3.3GHz.

4 Validation of the numerical modeling

4.1 Shear test results

Figure 3 depicts different stages during the numerical test. At the beginning, particles from each layer are bonded and stay in equilibrium. When velocity is imposed at the upper cylinder, the upper layer of the sample start to moves as a monolithic assembly, cutting the sample in two halves. The individual velocity of each particle is mapping such as the total reaction between particles and the upper cylinder.

Figure 4 gives the results of the evolution of shear stress τ as a function of the measured displacement for each numerical sample. For all samples, it is possible to identify a slope at the beginning of the measured displacement until reach failure produced by the shear loading. Once the failure is reached, the system evolves showing a post-peak relaxation. This residual stress value is generated by frictional forces between unbounded particles. The maximum value of shear stress is obtained for a displacement comprised between 2mm and 2.3mm, where this range of values is quite similar to the measured displacement values obtained for the experimental tests.

4.2 Comparison with experimental results

In the aim to validate the numerical model, table 2 provides the shear stress and displacement values at failure, for numerical and experimental tests. The average numerical values show a good agreement with experimental data in both shear stress and displacement measured at failure, where the relative error between them is found smaller than 7%. It can be considered as an encouraging result for a validation of the proposed model for the study of asphalt mixtures.

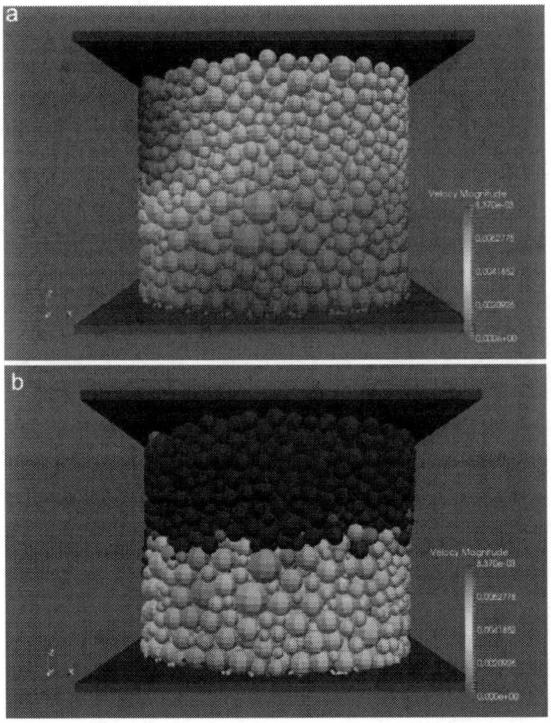

Figure 3. Sketch of average velocity of grains for different stages during the shear test: a) At the beginning of the shear test; b) At failure.

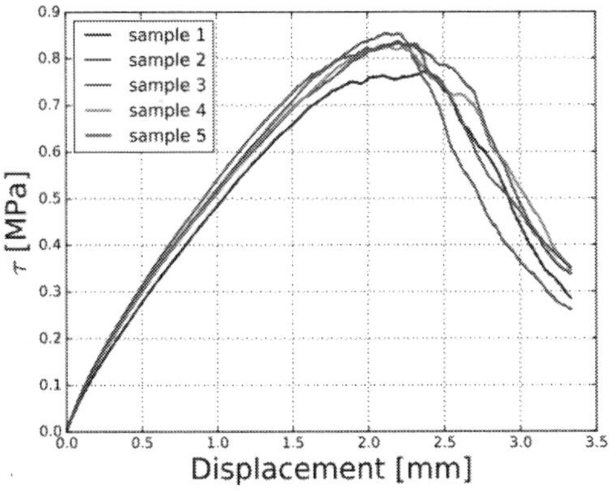

Figure 4. Shear stress τ evolution as a function of imposed displacement

Table 2. Average values of shear stress τ and peak displacement d at failure through the experiments and the simulations.

	τ [MPa]	d [mm]
experiments	0.898	2.37
simulations	0.837	2.2

5 Concluding remarks

Our results show that implemented viscoelastic contact model is able to reproduce experimental data from shear test, regarding shear force and maximum displacement at failure. The general results of these simulations for each configuration are in a good agreement with the average values through experiments despite the intrinsic variability within experimental trials. The results of this preliminary work can be considered as encouraging for the validation of the proposed contact model to study the mechanical behavior of multi-layer road structures. The next step in this work will be the inclusion of irregular particles to model actual aggregates to obtain more realistic asphalt mixtures.

References

[1] A.C. Collop, G.R. McDowell, Y.W. Lee, Granular Matter **8**, 175 (2006)
[2] S. Adhikari, Z.P. You, International Journal of Pavement Research and Technology **1**, 94 (2008)
[3] W. Cai, G. McDowell, G. Airey, Soils and Foundations **54**, 12 (2014)
[4] R. Leutner, Bitumen **41** (1979)
[5] J. Moreau, European Journal of Mechanics A/Solids **supp.**, 93 (1994)
[6] M. Jean, Computer Methods in Applied Mechanic and Engineering **177**, 235 (1999)
[7] F. Radjai, V. Richefeu, Mechanics of Materials **41**, 715 (2009)
[8] T. Pöschel, V. Buchholtz, J. Phys. I France **5**, 1431 (1995)
[9] N.V. Brilliantov, F. Spahn, J.M. Hertzsch, T. Pöschel, Physical review E **53**, 5382 (1996)
[10] H. Herrmann, S. Luding, Continuum Mechanics and Thermodynamics **10**, 189 (1998)
[11] F. Radjaï, F. Dubois, *Discrete-element modeling of granular materials* (Wiley-Iste, 2011)
[12] P.A. Cundall, O.D. Strack, Geotechnique **29**, 47 (1979)
[13] J.J. Moreau, *Unilateral contact and dry friction in finite freedom dynamics*, in *CISM* (1988), pp. 1–82
[14] B. Brogliato, *Nonsmooth mechanics* (Springer, London, 1999)
[15] F. Dubois, M. Jean, *LMGC90*, in *Actes du sixieme colloque national en calcul des structures* (2003), Vol. 1, pp. 111–118
[16] J. Betten, *Creep mechanics* (Springer Science & Business Media, 2008)
[17] Y. Liu, Z. You, in *Pavements and materials: Modeling, testing, and performance* (2009), pp. 26–36
[18] A. Zbiciak, Bulletin of the Polish Academy of Sciences: Technical Sciences **61**, 65 (2013)

Coupled DEM-CFD Analysis of the Initiation of Internal Instability in a Gap-Graded Granular Embankment Filter

Kenichi Kawano[1], *Tom* Shire[2], and *Catherine* O'Sullivan[2,*]

[1]Kajima Corporation, Tokyo, JAPAN
[2]Department of Civil and Environmental Engineering, Imperial College London, UK

Abstract. Internal instability is a form of internal erosion that can occur in embankment dams or flood embankments where the finer fraction of the material is washed out under the action of seepage flow; if undetected this process can progress to cause embankment collapse. Gap-graded materials are particularly susceptible. Skempton and Brogan [1] proposed that a key contributor to instability is the reduced stress transmitted by the finer fraction and that the magnitude of this reduced stress could be inferred from the hydraulic gradients observed at the initiation of particle migration in experiments. Here Skempton and Brogan's hypothesis is assessed at the particle scale using a discrete element method (DEM) model coupled with computational fluid dynamics (CFD). This contribution discusses validation of the coupled DEM-CFD software prior to describing the simulation of a permeameter experiment. The simulation generated particle-scale data at the initiation of instability by considering a gap-graded sample subject to at a hydraulic gradient of 1.0 (upward flow). The results provide insight into the instability mechanism, most notably showing that while the particles that move under seepage flow do indeed transmit relatively small effective stress, a finite proportion of the particles that move transfer relatively large stresses.

1 Introduction

Internal instability is a form of internal erosion which can occur in water-retaining earth structures such as embankment dams and flood levees. Internally unstable soils typically have fairly distinct coarse and fine fractions (i.e. gap or broadly-graded soils), and it is this fine fraction which can be preferentially eroded by seepage flow [2]. While initially this phenomenon might not be associated with any noticeable deformation of the embankment, the migration of particles can cause changes in local permeability and the overall flow field and can progress to a scenario involving significant deformations, possibly causing embankment breach. In recent years there have been a number of experimental research studies considering this mechanism (e.g. [1,3–5]) which have hypothesized the fundamental particle-scale interactions that trigger this instability. A particular hazard of internal instability is the erosion of the fine particles which can initiate at much lower hydraulic gradients than would be predicted by the continuum analyses used in engineering practice which assume a homogenous soil. Skempton and Brogan [1] attributed this phenomenon to the lower proportion of the effective stress transmitted by the finer particles in comparison with the coarser particles. They proposed that this could be quantified using the stress reduction factor, α:

$$\sigma'_{fine} = \alpha \, \sigma' \qquad (1)$$

where σ'_{fine} is the effective stress transferred by the fines and σ' is effective overburden stress. Skempton and Brogan [1] inferred α by comparing the hydraulic gradient at which erosion was observed in a permeameter test with no surface loading (i_c) to the the hydraulic gradient required for zero effective stress from Terzaghi's classical theory ($i_{c,heave}$):

$$\alpha = \frac{i_c}{i_{c,heave}} \qquad (2)$$

In order for the fines to carry reduced effective stress they must be insufficient to fill the voids between the coarser particles. Depending on sample porosity the fines should just fill the voids between the coarse particle at a fines content of between 24% and 29% by volume [1,6].

Shire et al. [6] carried out a micro-scale study of the stress distribution within gap-graded soils using the discrete element method (DEM). By directly measuring the stress within coarse and fine fractions of the soil they demonstrated that α is a function of fines content, packing density and the size ratio between coarse and fine particles. However, as no water was applied they were unable to directly link α to the hydraulic gradient at which erosion initiates, i.e. they could not confirm that Equation 2 is valid.

In this paper the initiation of internal instability was modelled using a coupled discrete element method-computational fluid dynamics (DEM-CFD) approach, where DEM was used to model the soil particles and CFD the seepage of water. The simulation approach and

* Corresponding author: cath.osullivan@imperial.ac.uk

validation are first introduced, then results for a representative sample are analysed in detail.

2 Simulation approach

2.1 Methodology

The simulations were carried out using PFC3D with the CCFD add on as detailed in [7]. Using the DEM data the stress tensor for each particle, $\bar{\sigma}_{ij}^{p}$ was calculated using the method described in [8]. The average stress tensor for the entire volume, $\bar{\sigma}_{ij}$ is then:

$$\bar{\sigma}_{ij} = \frac{1}{V}\sum_{p=1}^{N_p} \bar{\sigma}_{ij}^{p} V^{p} \tag{3}$$

where V = total volume (particles + voids); N_p = number of particles; $\bar{\sigma}_{ij}^{p}$ = average stress in particle p; and V^p = volume of particle p. The stress tensor for the finer fraction $\bar{\sigma}_{ij}^{f}$ is given as follows:

$$\bar{\sigma}_{ij}^{fine} = \frac{(1-n)}{\sum_{N_p,fine} V^p} \sum_{p=1}^{N_{p,fine}} \bar{\sigma}_{ij}^{p} V^{p} \tag{4}$$

where n =sample porosity; $N_{p,fine}$ = number of fine particles. As an isotropic stress state is applied here, the stress reduction factor, α, is given in terms of mean normal stress p' = $1/3(\sigma_{ii}')$:

$$\alpha = p'_{fine} / p' \tag{5}$$

The coarse-grid approach to DEM-CFD coupling proposed by [9,10] was adopted here. The CCFD software discretises the fluid domain into a regular fluid grid, which is overlain on the DEM domain. The averaged Navier-Stokes equations are solved to obtain average values of fluid pressure and velocity for each cell within the grid and the empirical Di Felice [11] expression is used to calculate the fluid-particle interaction force, which is added as an additional force term in the DEM simulation. Full details of the coupling scheme are available in [7].

2.2 Validation using a single column of particles

The coupled software system was validated using the case proposed by Suzuki et al. [12] in which 100 monodisperse spherical particles were placed at diameter intervals along the centre line of a vertical square column, which was filled with water. Free slip, prescribed velocity and prescribed pressure conditions were imposed at the lateral, base and top boundaries respectively. Suzuki et al. [12] derived an expression for the distribution of pressure in the system using the Ergun expression for drag. Following their approach a similar expression was derived for the Di Felice drag expression as detailed in [13]. Representative data comparing the values obtained using this analytical expression with DEM simulation results are presented in Figure 1. For the case illustrated, the model parameters are identical to those used by Suzuki et al. [12], with the exception of the fluid cell length, which was twice the length that used by [12]. After the particles were generated, a gravitational body force and buoyancy were applied and the system was cycled until the particles came to rest. Then, an upward fluid velocity of v_f = 0.005 m/s was applied at the bottom boundary and a pressure of P = 0 Pa was applied at the top boundary. The distributions of water pressure in excess of hydrostatic pressure are compared in Figure 1. An exact match between the analytical and experimental data was not expected as the analytical expression is based on application of the heat equation to a continuum material. Consequently, the agreement is considered to be reasonable as the DEM-CFD generated data were within 5% of the analytical values. The excess pore water pressure values obtained from the analytical expression derived with the Ergun drag expression are also included to show the sensitivity of the predicted system response to the drag expression used.

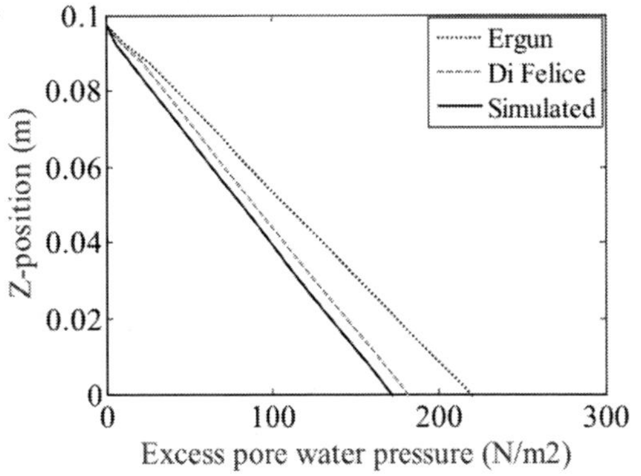

Figure 1. Distribution of excess fluid pressure from the validation case

2.3 Permeameter test simulation

For the main simulations a Hertz-Mindlin contact model was used. The shear modulus was 27.0 GPa, Poisson's ratio was 0.3, particle density was 2,670 kg/m^3 and coefficient of interparticle friction was 0.3. The particle timestep was 2.0×10^{-8} s and the fluid timestep was 1.0×10^{-3} s. A sample of 20,000 spherical particles was generated by creating a non-contacting cloud of particles to a specified particle size distribution (PSD) within a rigid-walled cubical cell. The particle size distribution is shown in Figure 2. The particles were isotropically compressed to a mean normal stress of p' = 50 kPa. Gravity was then applied to the sample. The PSD is classified as borderline internally unstable according to the empirical guidelines proposed by Kézdi [14], with a D'_{15}/d'_{85} = 4.6 where D'_{15} is the diameter of the coarse fraction for which 15% by mass is smaller and d'_{85} is the diameter of the fine fraction for which 85% by mass is smaller.

Once the sample was in equilibrium under gravity the coupled DEM-CFD simulations were initiated, details of which can be found in [13]. Free slip and prescribed pressure conditions were used on the lateral and top/bottom boundaries respectively. The sample was allowed to come into equilibrium under gravity and buoyancy. A hydraulic gradient of i = 1.0 was then applied across the top and bottom boundaries to induce an upward flow of water.

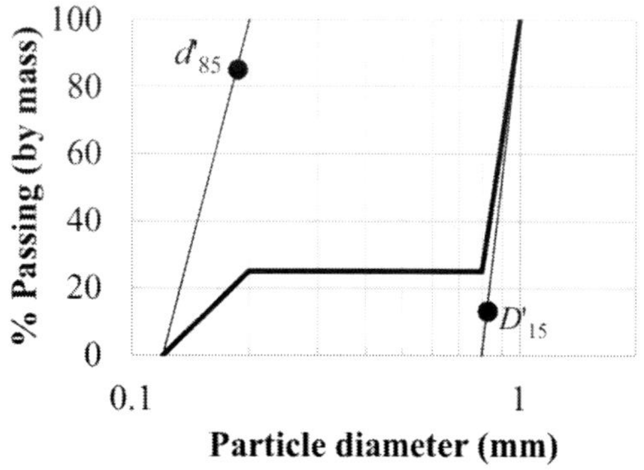

Figure 2. Particle size distribution of sample

3 Results

3.1 Soil fabric

Referring to Table 1 the soil fabric was quantified by considering porosity, average coordination number and α. Data from Shire et al. [6] are also included relate to a second DEM sample isotropically compressed using the open-source DEM code Granular LAMMPS [15] with periodic boundaries (no gravity) and having with the same PSD and input variables. Despite the differences in sample preparation and boundary conditions the porosity and stress-reduction factor values agree well. However the coordination number for the current study is much higher than that from [6], in which gravity was neglected so that non-stress transmitting particles remained suspended within the voids.

Figure 3 shows the cumulative distribution of normalised particle mean normal stresses $\frac{\overline{p^p}}{p'}$. It can be seen that the fine particles (~25% by volume) carry minimal stress for both datasets. These data along with the α values indicate that there are insufficient fine particles to fill the voids between the coarse particles, leaving the fine particles understressed within these voids. The coarser particles (75% by volume) have a heterogeneous stress distribution, which is also independent of sample preparation method.

Table 1. Fabric variables for this study and Shire et al. [6].

Study	Porosity	Coordination number	Stress reduction factor (α)
Current study	0.239	4.66	0.11
Shire et al. [6]	0.258	0.15	0.18

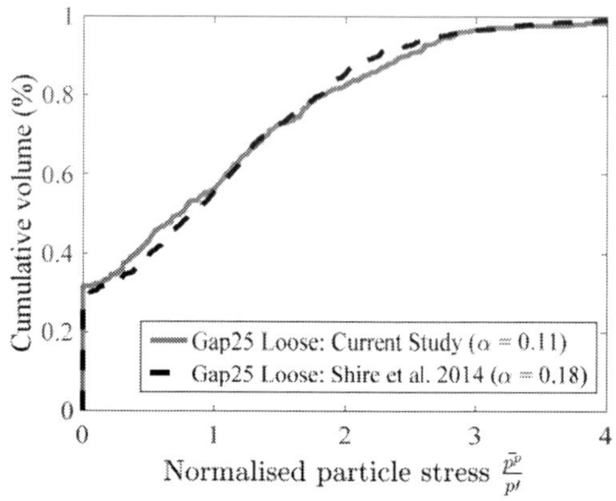

Figure 3. Cumulative distribution of particle stresses

3.2 Fluid-particle interaction

Upon application of a hydraulic gradient of i = 1.0, the permeability was determined to be 2.3×10^{-4} m/s, similar to the value which would be obtained from the empirical Hazen formula [16]. Figure 4 shows the change in coordination number with time following the application of the hydraulic gradient. There was a large initial drop as understressed fine particles began to move due to the upward flow of water. The coordination number then increased as these mobile fine particles became trapped at the void boundaries between coarse particles. However, after around 0.45 s the coordination number became stable at a value lower than the initial value, indicating that some particles were not retained within the voids and are therefore potentially erodible. Figure 5 presents the relationship between the net change in connectivity and displacement. The histogram represents how many particles had a given connectivity change. Generally, particles which had a net reduction in connectivity moved further than those which gained contacts, demonstrating that fine particles which lose contacts are less likely to be retained by the coarser particles. The relationship between particle stress and displacement is shown in Figure 6. It can be clearly seen that particles with lower initial stress undergo much more displacement than those with higher stress. However, some particles with higher initial stress undergo significant displacement. These highly stressed but erodible particles could be significant as their erosion could lead to buckling of force chains and volumetric strain of the soil.

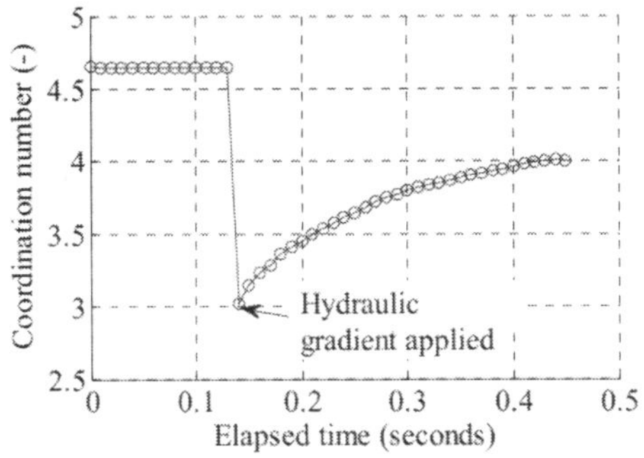

Figure 4. Change in coordination number following application of fluid flow

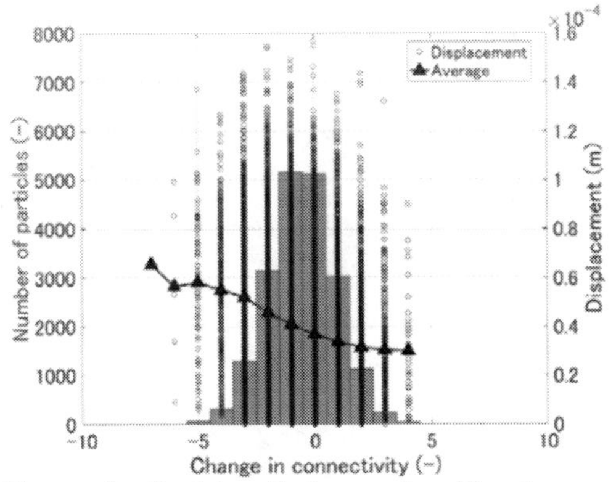

Figure 5. Particle displacement with change in connectivity. Histogram shows how many particles had a given connectivity change.

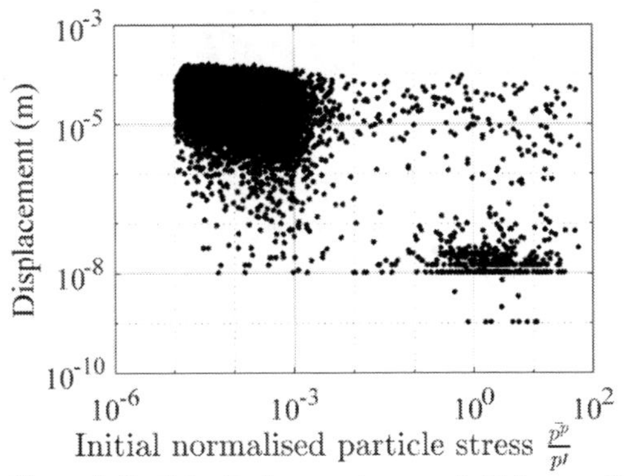

Figure 6. Particle displacement versus initial normalised particle stress.

4 Conclusions

This paper presents the results of a coupled DEM-CFD analysis of internal instability. Following software validation, a permeameter test on a borderline internally unstable gap-graded soil was simulated. The conclusions are:

(a) The coupled DEM-CFD simulations were successfully validated using a case proposed by [12]. Permeability values were also close to those suggested by the empirical Hazen rule [16].

(b) In coupled DEM-CFD simulations the values of parameters such as fluid pressure are quite sensitive to the choice of empirical drag expression, e.g. [11,17].

(c) The stress transfer characteristics of the DEM sample were insensitive to preparation method.

(d) On application of fluid flow many fine particles were mobilised, resulting in a large drop in coordination number.

(e) More mobile fine particles tended to have a larger net drop in connectivity, suggesting they were less likely to be retained by the coarse particles.

(f) Most particles which underwent significant displacement had low initial stress. However, some highly stressed particles were displaced, and this could cause volumetric strain in a real soil.

Acknowledgement

The first author was funded by Kajima Corp., Japan.

References

[1] A. W. Skempton and J. M. Brogan, Géotechnique **44**, 449 (1994).

[2] ICOLD, *Bulletin on Internal Erosion of Dams, Dikes and Their Foundations: Volume 1.* (Paris, 2015).

[3] R. J. Fannin and R. Moffat, Géotechnique **56**, 497 (2006).

[4] L. Ke and A. Takahashi, Soils Found. **54**, 713 (2014).

[5] D. S. Chang and L. M. Zhang, J Geotech Geoenviron. **139**, 1454 (2013).

[6] T. Shire, ; C O 'sullivan, K. J. Hanley, and R. J. Fannin, J Geotech Geoenviron. **140** (2014)

[7] Itasca Consulting Group, (2014).

[8] D. O. Potyondy and P. A. Cundall, Int. J. Rock Mech. Min. Sci. **41**, 1329 (2004).

[9] Y. Tsuji, T. Kawaguchi, and T. Tanaka, Powder Technol. **77**, 79 (1993).

[10] B. H. Xu and A. B. Yu, Chem. Eng. Sci. **52**, 2785 (1997).

[11] R. D. Di Felice, Int. J. Multiph. Flow **20**, 153 (1994).

[12] K. Suzuki, J. Bardet, M. Oda, and K. Iwashita, J Geotech Geoenviron. **133**, 104(2007).

[13] K. Kawano, Numerical Evaluation of Internal Erosion due to Seepage Flow, Imperial College London, 2016.

[14] Á. Kézdi, *Soil Physics: Selected Topics* (Elsevier, Amsterdam, Netherlands, 1979).

[15] S. Plimpton, J Comput Phys **117**, 1 (1995).

[16] A. Hazen, in *Am. Civ. Eng. Handb.* (John Wiley & Sons, Inc., New York, 1930).

[17] S. Ergun, Chem. Eng. Prog. **48**, 89 (1952).

 DOI: 10.1051/epjconf/201714003043

Discharge flow of a granular media from a silo: effect of the packing fraction and of the hopper angle

Mebirika Benyamine[1], *Pascale* Aussillous[2,*], and *Blanche* Dalloz-Dubrujeaud[2]

[1] *Laboratoire FIMAS, Université de Béchar, B.P 417 Béchar 08000, Algérie*
[2] *Aix-Marseille Univ., CNRS, IUSTI UMR 7343, 13013 Marseille, France*

Abstract. Silos are widely used in the industry. While empirical predictions of the flow rate, based on scaling laws, have existed for more than a century (Hagen 1852, translated in [1] - Beverloo et al. [2]), recent advances have be made on the understanding of the control parameters of the flow. In particular, using continuous modeling together with a mu(I) granular rheology seem to be successful in predicting the flow rate for large numbers of beads at the aperture (Staron et al.[3], [4]). Moreover Janda et al.[5] have shown that the packing fraction at the outlet plays an important role when the number of beads at the apeture decreases.
Based on these considerations, we have studied experimentally the discharge flow of a granular media from a rectangular silo. We have varied two main parameters: the angle of the hopper, and the bulk packing fraction of the granular material by using bidisperse mixtures. We propose a simple physical model to describe the effect of these parameters, considering a continuous granular media with a dilatancy law at the outlet. This model predicts well the dependance of the flow rate on the hopper angle as well as the dependance of the flow rate on the fine mass fraction of a bidisperse mixture.

1 Introduction

The prediction of the mass flow rate of a granular media discharged from a silo has been widely studied because of its practical interest in several industries, for example to ensure reproducible and efficient handling in plant operations (food, mining, ceramic, paint, pharmaceutical... industries). One of the simplest and widely used expression is the one proposed by Berverloo [2] for flat-bottomed silos. Following the experimental observation that the flow rate is mainly controlled by the outlet diameter D, it is based on dimensional analysis with a concept of a free fall arch at the outlet [1] and of an "empty annulus" [6] considering a reduced outlet length due to the particle size. More recent works proposed a continuous modeling with a mu(I) granular rheology which gives good results for a large amount of beads at the aperture [3], [4]. However, the control parameters of the flow are not well understood. In this paper we investigate the influence of the discharge angle of the hopper, and the bulk packing fraction of the granular material (using bidisperse mixtures). Our results are interpreted through a physical and analytical model (based on the work of Janda et al.[5]) considering a modification of the density close to the outlet due to the flow.

2 Experimental methods

In order to measure the flow rate and to observe the granular flow, the experimental setup is made of a rectangular silo (height H = 1m, width L = 60 mm, and thickness W which can be varied from 3.5 to 10 mm) made with two parallel plates (metal/PMMA for the flat bottom experiments and glass/glass for the study with several discharge angles) (Fig. 1). The rectangular outlet located at the centre of the bottom spans the thickness W and its length D can be varied from 1.8 to 26 mm. The discharge angle θ of the outlet to the horizontal can also be varied from θ=0° to θ=80°. When opening manually the outlet, the grains fall out of the silo and are collected in a vessel. The temporal evolution of the mass $m(t)$ is recorded using

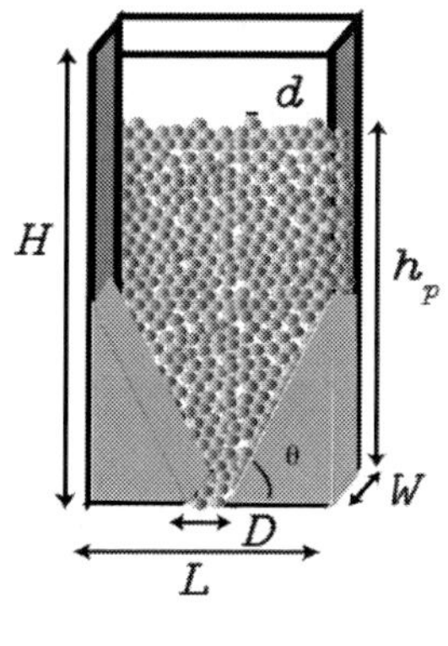

Figure 1. Experimental setup: rectangular silo with adjustable outlet D and discharge angle θ.

*e-mail: pascale.aussillous@univ-amu.fr

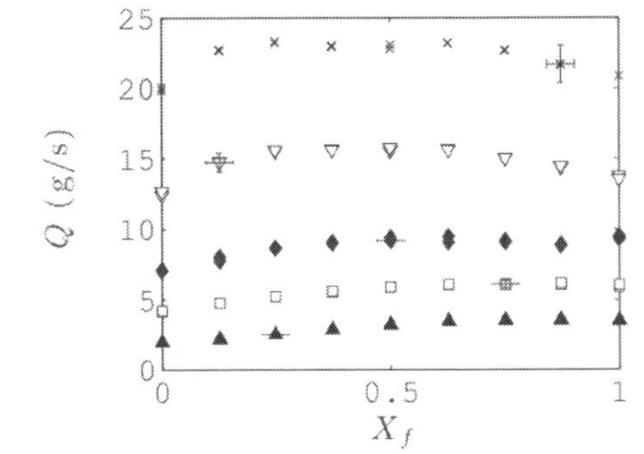

Figure 2. Flow rate versus X_f for binary mixtures (r = 6.3, W = 3.5mm) with D(mm): 3.5(▲), 5.4(□), 7.4(♦), 10(▽), 3.3(×).

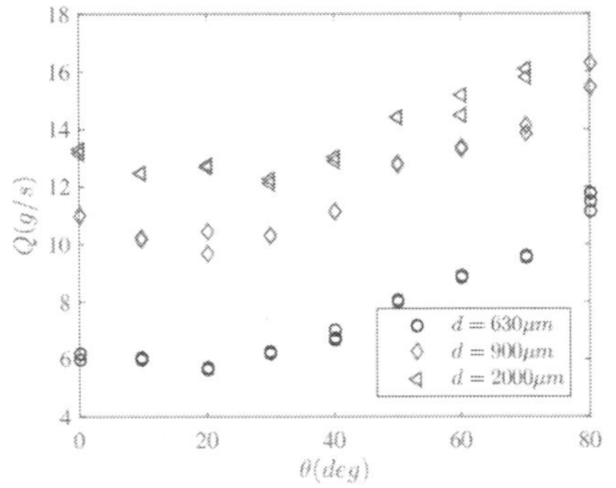

Figure 3. Flow rate versus the discharge angle θ for $D = 8mm$

an electronic scale. A high speed camera allowed us to visualise the flow at the wall. All the experiments have been carried out in the stationary flow regime.

The granular material consists of smooth spherical glass beads (density $\rho = 2500\ kg.m^{-3}$) with different diameters d from 80 to 4500 μm (with a dispersion of ±10%). With the particles and aperture available, the number of beads across the aperture (D/d) ranges between 1 and 325 (clogging occurs for D/d < 4.1 [7]). In order to control the bulk volume fraction, different binary mixtures were tested (size ratio of coarse particles diameter on fine particules diameter $r = d_c/d_f$ between 2.2 to 6.3). Controlling the mass fraction X_f of the fine particles, the bulk volume fraction varies from 0.58 to 0.73. To ensure the homogeneity of the bi-disperse mixture in the silo, ten layers of the desired fine mass fraction (X_f), previously mixed by hand, were poured successively and carefully into the silo. The packing fraction ϕ_b for each experiment has been calculated measuring the height and the weight of the grain assembly in the silo. Mono-disperse and bi-disperse experiments have been carried out, for the bi-disperse mixtures, all the experiments have been performed with an angle $\theta = 0^o$.

3 Results and discussion

In our configuration the Beverloo law becomes:

$$Q = C\rho\phi_b g^{1/2} W(D - kd)^{3/2} \quad (1)$$

Where ϕ_b is the initial bulk particle volume fraction and C and k are fitting parameters. This expression fits well all our experimental results for mono-disperse experiments, but it is difficult to modify this expression in the case of a binary mixture even if ϕ_b is known. Indeed, for a bidisperse mixture, we observe that the flow rate always exceeds the rate predicted by a simple mixture law between the flow rate of the purely coarse particles and the purely fine particles (Fig. 2). A mixture law to express the mean diameter of the particules is not suitable to predict the flow rate.

We also measure the flow rate when the discharge angle is increased from θ=0° to θ=80°. In agreement with the literature [8],[9], the flow rate increases significantly for a discharge angle $\theta > 20^o$ as shown in Fig 3.

Thus, to interpret our results, we have adapted the model proposed by Janda *et al.* [5]. This model is based on the experimental measurement of the velocity and the density profile at the outlet. The author found that both these profiles are self similar. Thus the flow rate is given by the velocity v_o and the density ϕ_0 at the outlet, $Q = \rho \iint_{S_o} \phi(x)v(x)dx = C\rho WD\phi_0 v_0$. Measuring this quantity these authors have shown that for a given outlet diameter, the particle velocities are the same whatever the particle diameter ($v_0 \approx \sqrt{gD}$). Moreover, they have found that, instead of considering a reduce outlet, a dilatancy at the outlet should be taken into account ($\phi_0 \approx \phi_b[1 - \alpha_1 e^{-\alpha_2 D}]$). A recent work [10] has shown that close to the outlet, the particles are accelerated leading to the scaling of the velocity, but not in an idealised free fall. We adapted this model to propose a new expression for the flow rate :

$$Q = C\rho\phi_0 \sqrt{gD^3}W = C\rho\phi_b G(D/d) \sqrt{gD^3}W \quad (2)$$

with the dilatancy function (ϕ_0/ϕ_b),

$$G(D/d) = [1 - \alpha e^{-\beta \frac{D}{d}}] \quad (3)$$

Equation 2 fits well our experimental results as shown on Fig 4 and allowed us to calculate the parameters C = 0.9, α = 0.66 and β = 0.11.

This is then possible to calculate the experimental dilatancy function given by :

$$G(D/d) = \frac{\phi_o}{\phi_b} = \frac{Q}{C\rho\phi_b \sqrt{gD^3}W} \quad (4)$$

The evolution of the dilatancy function $G(D/d)$ with D/d (which represents the number of bead at the aperture) is given in fig. 5 for a mono-disperse flow in a flat bottom silo and is well fitted by equation 3. This result was also obtained numerically using discrete simulations by Zhou et al. [11].

DOI: 10.1051/epjconf/201714003043

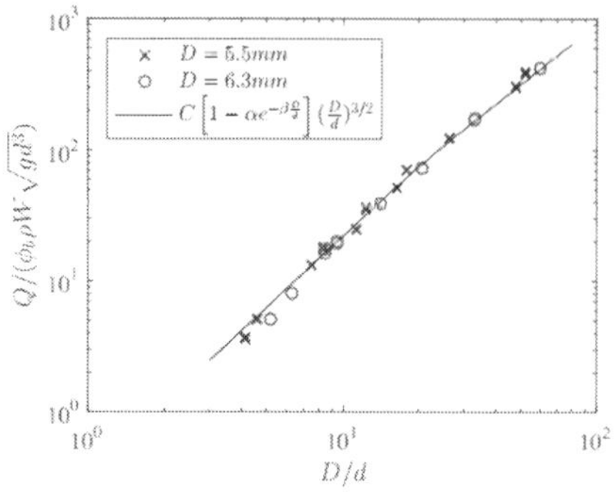

Figure 4. Dimensionless flow rate versus D/d for two outlets size D. The black line represents equation 2 with $C = 0.9$, $\alpha = 0.66$ and $\beta = 0.11$

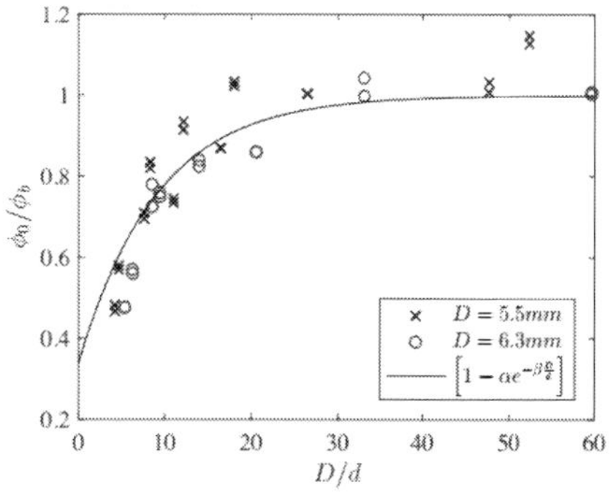

Figure 5. Dilatancy at the outlet for $W = 3.5$mm. Theoretical curve (equation 3) obtained with $\alpha = 0.66$ and $\beta = 0.11$

3.1 Bidisperse mixtures

We assume that for a bidipserse mixture, the same hypothesis are still and independently valid for each size of particles d_i. Each population of particles has his own dilatancy function $G(di/D)$. The dilatancy at the outlet is then simply given by $\phi_0 = \phi_{0f} + \phi_{0c}$, and the flow rate for a bi-disperse mixture becomes :

$$Q = C[X_f G(\frac{D}{d_f}) + (1 - X_f)G(\frac{D}{d_c})]\rho\phi_b W \sqrt{gD^3} \quad (5)$$

Equation 5 suggests that the dilatancy at the outlet is given by a simple mixture law between the dilatancy factor of the fine and the coarse particles. Note that ϕ_b depends also on the fine mass fraction, which implies a complex variation of the flow rate with X_f as seen in Fig. 2. This model fits well our experimental results represented in Fig. 6, where the dilatancy at the outlet is plotted versus the fine mass fraction for three mixtures. More details can be found in [7].

Our model suggests that a mixture diameter cannot be simply defined to adapt the monodisperse expressions to predict the bidisperseflow rate as commonly seen in the literature [12], [13].

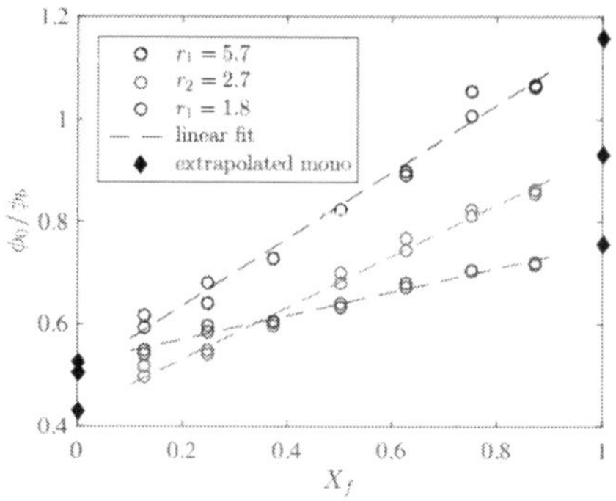

Figure 6. Linear evolution of the dilatancy parameter with X_f for bidisperse Mixture. Monodisperse results have been excluded from the linear fit, and ♦ represent the extrapolated point for monodisperse flow.

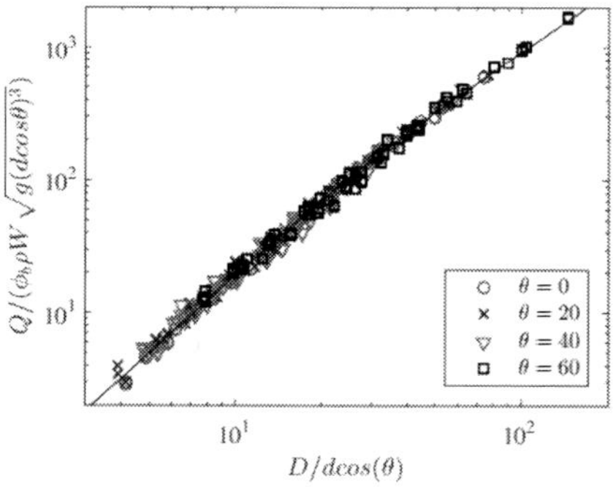

Figure 7. Dimensionless flow rate versus $D/(dcos\theta)$ for several discharge angles. The black line represents equation 7 with $C = 0.92$, $\alpha = 0.80$ and $\beta = 0.085$.

3.2 Discharge angle

In this part, all the presented results have been obtained with monodisperse material, varying the discharge angle and presented on Fig. 3. We made the assumption that the self similarity hypotheses are still valid for the velocity and the density profile, and that the modification of the flow rate is only due to the dilatancy factor which must depend on the discharge angle. Indeed, to allow the granular media flowing through the outlet, a singular region exist above the outlet, where the particle are accelerated. We suppose that this region is delimitated by an arch of radius of curvature $\mathcal{R}$, corresponding to the convergence of the velocity stream through the outlet. For a discharge angle of $\theta = 90^o$ (i.e. a vertical tube), the velocity stream are parallel and this radius of curvature is infinite. We suppose that the the evolution of $\mathcal{R}$ is continuous with the discharge angle and the simplest function we can suggest is $\mathcal{R} = (D/2)cos(\theta)$, which recovers the typical length scale $D/2$ for the flat bottom. Thus, a new expression is proposed for the dilatancy function:

$$G(D/d, \theta) = [1 - \alpha e^{-\beta \frac{D}{dcos\theta}}] \quad (6)$$

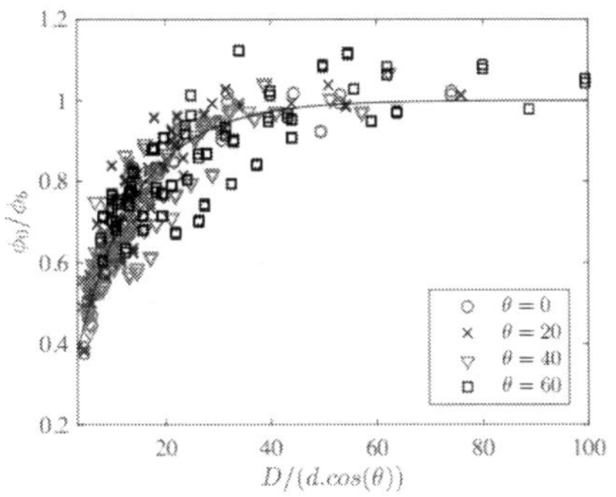

Figure 8. dilatancy at the outlet versus $D/(dcos\theta)$ for several discharge angles. The black line represents equation 6 with $\alpha = 0.80$ and $\beta = 0.085$.

Using this function, we obtain for the flow rate

$$Q = CG(\frac{D}{dcos\theta})\rho\phi_b W\sqrt{gD^3} \qquad (7)$$

In figures 7 and 8 we have plotted the dimensionless flow rate and the dilatancy function, for several angles of discharge, as a function of $D/(dcos\theta)$. In this representation the data collapse fairly well and are well fitted by equations 6 and 7 with the parameter $C = 0.92$, $\alpha = 0.80$ and $\beta = 0.085$ obtained in our apparatus using the flat bottom data.

4 Conclusion

Using experiments we have proposed a model which predicts the flow rate of a granular media discharging from silos taking into account the discharge angle of the hopper and the bulk packing fraction of the granular material through a dilatation factor. The perspective of this work is to validate the hypothesis proposed thanks to velocity and density profiles measurement close to the outlet.

Acknowledgements

This work has been carried out in the framework of the Labex MEC (ANR-10-LABX-0092) and of the A*MIDEX project (ANR-11-IDEX-0001-02), funded by the "Investissements d'Avenir" French Government program managed by the French National Research Agency (ANR).

References

[1] B. Tighe and M. Sperl, Granul. Matter **9**, 141 (2007)
[2] W. A. Beverloo, H. A. Leniger, and J. V. de Velde, Chem. Eng. Sci. **15**, 260 (1961)
[3] L. Staron, P.-Y. Lagrée, and S. Popinet, Phys. of Fluids, **24**,103301 (2012)
[4] L. Staron, P.-Y. Lagrée, and S. Popinet, Eur. Phys. J. E **37**,5 (2014)
[5] A. Janda, I. Zuriguel, and D. Maza, PRL **108**, 248001 (2012)
[6] R. L. Brown and J. C. Richards, Trans. Inst. Chem. Engrs. **38**, 243 (1960)
[7] M. Benyamine and M. Djermane, B. Dalloz-Dubrujeaud and P. Aussillous, PRE **90**, 032201 (2014)
[8] R. M. Nedderman, Cambridge University Press, (1992).
[9] H. F. Rose and T. Tanaka, The Engineer (London) **208** Oct. 23 (1959)
[10] S.M. Rubio-Largo, A. Janda, D. Maza, I. Zuriguel, and R. C. Hidalgo, PRL **114**, 238002 (2015)
[11] Y. Zhou, P. Ruyer, and P. Aussillous, PRE **92**, 062204 (2015)
[12] S. Humby, U. Tüzün, and A. B. Yu, Chem. Eng. Sci. **53**, 483 (1998).
[13] P. Arteaga and U. Tüzün, Chem. Eng. Sci. **45**,205(1990)

Unifying Suspension and Granular flows near Jamming

Eric DeGiuli[1,3,*] and *Matthieu* Wyart[2,**]

[1]*Institute of Physics, École Polytechnique Fédérale de Lausanne (EPFL), CH-1015 Lausanne, Switzerland*

Abstract. Rheological properties of dense flows of hard particles are singular as one approaches the jamming threshold where flow ceases, both for granular flows dominated by inertia, and for over-damped suspensions. Concomitantly, the lengthscale characterizing velocity correlations appears to diverge at jamming. Here we review a theoretical framework that gives a scaling description of stationary flows of frictionless particles. Our analysis applies *both* to suspensions and inertial flows of hard particles. We report numerical results in support of the theory, and show the phase diagram that results when friction is added, delineating the regime of validity of the frictionless theory.

1 Introduction

Microscopic description of particulate materials such as grains, emulsions or suspensions is complicated by the presence of disorder, and by the fact that these systems are often out-of-equilibrium. One of the most vexing problems is how these materials transition between a flowing and a solid phase. When this transition is driven by temperature, it corresponds to the glass transition where a liquid becomes a glass, an amorphous structure that cannot flow on experimental time scales. Here we focus instead on athermal systems driven by an imposed stress, such as granular flows, and consider both the case where inertia is important (such as in aerial granular flows) or not (such as over-damped suspensions). We focus primarily on the case of hard particles.

Empirical constitutive relations have been proposed to describe such dense flows in the limit of hard particles [1–5]. Two important dimensionless quantities are the packing fraction ϕ and the stress ratio $\mu \equiv \sigma/p$ (also called the effective friction), where σ is the applied shear stress and p the pressure carried by the particles. For inertial flow, dimensional analysis implies that both quantities can only depend on the strain rate $\dot{\epsilon}$, p, the particle diameter D and the mass density of the hard particles ρ via the inertial number $I \equiv \dot{\epsilon} D\sqrt{\rho/p}$. One finds empirically that the constitutive relations $\mu(I)$ and $\phi(I)$ converge to a constant as $I \to 0$, corresponding to the jamming transition where flow stops. We define $\mu(0) \equiv \mu_c$ and $\phi(0) \equiv \phi_c$, which are system-specific and depend on particle shape, poly-dispersity, friction coefficient, etc. Near jamming, the constitutive relations are observed to be singular with:

$$\delta\mu \equiv \mu(I) - \mu_c \propto I^{\alpha_\mu} \quad (1)$$

$$\delta\phi \equiv \phi_c - \phi(I) \propto I^{\alpha_\phi} \quad (2)$$

As jamming is approached the dynamics becomes increasingly correlated in space [6, 7]. By considering the dominant decay [8] of the velocity correlation function, one can define a length scale ℓ_c:

$$\ell_c \sim I^{-\alpha_\ell} \quad (3)$$

Similar dimensional arguments have been made for dense suspensions of non-Brownian particles [5, 9]. In that case the relevant dimensionless number is the viscous number $J = \eta_0\dot{\epsilon}/p$, where η_0 is the viscosity of the solvent. Empirically one finds similar relations:

$$\delta\mu \equiv \mu(J) - \mu_c \propto J^{\gamma_\mu} \quad (4)$$

$$\delta\phi \equiv \phi_c - \phi(J) \propto J^{\gamma_\phi} \quad (5)$$

$$\ell_c \sim J^{-\gamma_\ell} \quad (6)$$

These relations imply that the viscosity $\eta = \sigma/\dot{\epsilon}$ of the suspension diverges as jamming is approached. Indeed Eq.(4) implies that $\sigma \sim p$ near jamming (in our scaling arguments below we may thus exchange freely σ and p), so that $J \propto \eta_0/\eta$. Eq.(5) then implies that:

$$\frac{\eta}{\eta_0} \propto (\phi_c - \phi)^{-1/\gamma_\phi}, \quad (7)$$

as often written in the suspension literature [3, 5, 10]. When both viscosity and inertia are present, a transition from viscous to inertial flow occurs as strain rate $\dot{\epsilon}$ is increased at fixed volume fraction [11–13]. This defines a cross-over strain rate

$$\dot{\epsilon}_{v\to i} \propto \frac{\eta_0}{\rho D^2}(\phi_c - \phi)^{\gamma_\epsilon}, \quad (8)$$

where the prefactors follow from a dimensional analysis.

Empirical values found for the exponents in Eqs.(1-7) are reported in Table 1. They seem not to depend on dimension, which we thus did not report in our table. In the case of inertial flow they appear to depend on the presence

*e-mail: eric.degiuli@epfl.ch
**e-mail: matthieu.wyart@epfl.ch

Table 1. Predicted critical exponents *vs.* values from experiments and numerical simulations, with and without frictional interactions. The values extracted in Ref.[16] correspond to simulations closest to hard spheres (the "roughness parameter" of that reference is 10^{-4}). When available, error bars are indicated by the notation 0.38(4) = 0.38 ± 0.04, 2.77(20) = 2.77 ± 0.20, etc.

Regime	Relation	Prediction	Experiment	Frictionless Sim'n	Frictional Sim'n
Inertial	$\delta\mu \sim I^{\alpha_\mu}$	$\alpha_\mu = 0.35$	1 [14]	0.38(4) [15]	0.81(3) [16], 1 [17], 1 [18], 1 [12]
	$\delta\phi \sim I^{\alpha_\phi}$	$\alpha_\phi = 0.35$	1 [14]	0.39(1) [15]	0.87(2) [16], 1 [18], 1 [12]
	$L \sim I^{-1/2}$ $\epsilon_v \sim I$	1/2 1	0.33 [19], 0.3 [6] 1 [19]	0.48 [15], 0.5 [20] 1.1 [20]	0.5 [1], 0.2 [20] 0.95 [20]
Viscous	$\eta \sim \lvert\delta\phi\rvert^{-1/\gamma_\phi}$	$\gamma_\phi^{-1} = 2.83$	2 [5], 2 [21]	2.6(1) [22], 2.77(20) [23], 2.56[24], 2.77 [25]	1.5 [26]
	$\delta\mu \sim J^{\gamma_\mu}$	$\gamma_\mu = 0.35$	0.38 [27], 0.5 [5] 0.42 [27, 28]	0.37 [16], 0.25 [23], 0.32 [25]	0.5 [12]
	$\delta z \sim J^{\gamma_z}$ $\ell_c \sim \lvert\delta\phi\rvert^{-\gamma_\ell/\gamma_\phi}$ $L \sim J^{-1/2}$ $\epsilon_v \sim L^{-2} \sim J$ $dL/d\epsilon \sim -L^3$ $\dot{\epsilon}_{v\to i} \sim \delta\phi^{\gamma_{\dot{\epsilon}}}$	$\gamma_z = 0.30$ $\gamma_\ell/\gamma_\phi = 0.43$ 1/2 (-2,1) 3 $\gamma_{\dot{\epsilon}} = 2.83$	 0.30 [29] 1 [11]	0.30[25] 0.6(1) [7] 0.5[25], 0.5 [26] $\epsilon_v \sim L^{-2}$[25], $\epsilon_v \sim J$ [30] 3 [25]	0.33 [26]

of friction, whereas for suspended particles exponents appear to be similar with and without friction. In this work we focus initially on theory for frictionless particles, and in a second section, discuss the effect of friction.

Currently there is no accepted microscopic theory describing quantitatively these singular behaviors, in particular Eqs.(1-7). Observations support that as jamming is approached, particles form an extended network of contacts, and that the stress is dominated by contact forces [2, 5, 15, 31]. In this work we review a framework to describe flow in such situations. The detailed arguments are presented in [25]; here we discuss the essential ingredients of the theory, and focus on the effect of friction, discussed in more detail in [20, 26].

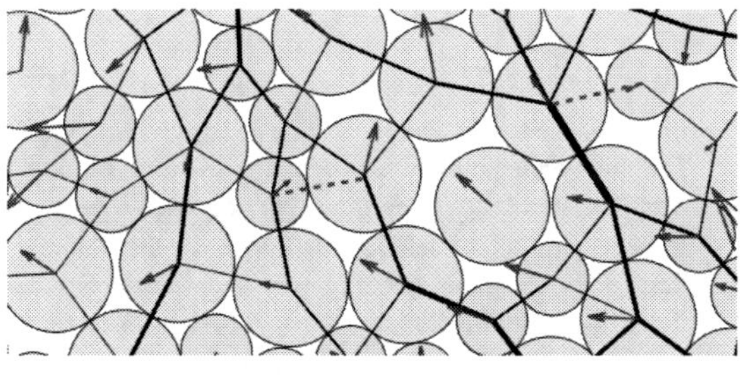

Figure 1. Illustration of solid destabilization: several weak contacts, indicated by red dashed lines, are opened. This induces a space of extended, disordered floppy modes, one of which is shown (arrows). Line thickness indicates force magnitude in the original, stable solid.

We attack the problem in two steps. First, we isolate the microscopic quantities that control flow. Then, we compute the scaling properties of these quantities by performing a perturbation around the solid phase. The idea is to consider the solid in the critical state, i.e. carrying the maximal stress ratio possible $\mu = \mu_c$, corresponding to a packing fraction $\phi = \phi_c$. Next, one adds an additional kick to the system, corresponding to a small additional stress ratio $\delta\mu$. As a result, some contacts between particles will open, forces will be unbalanced, and the system will start to flow (see Fig. 1). Our key assumption is that flowing configurations are similar to a solid that is thus destabilized. This approach enables us to propose a full scaling description of the problem, and to predict the exponents entering Eqs.(1-7) in good agreement with observations in the absence of static friction. Moreover, our approach predicts several other properties singular near jamming: the speed of the particles, the strain scale of velocity decorrelation, and the coordination of the contact network. The first two quantities are accessible experimentally, and provide an additional experimental test of our views.

When an additional stress ratio is imposed to a marginal solid, naively one expects the contacts carrying the smallest forces to open. However, only a vanishingly small fraction of weak contacts are significantly coupled to external stresses [32]. We call them *extended* contacts, because perturbing such contacts mechanically lead to a spatially extended response in the system, as shown in Figure 2 [32, 33]. In a packing only those contacts lead to plasticity when stress is increased, or when a shock (say a collision) occurs in the bulk of the material [32, 34]. The density of extended contacts as a function of the force f in the contact follows

$$P(f) \propto \frac{1}{p^{1+\theta_e}} f^{\theta_e}. \tag{9}$$

Numerically it is found that $\theta_e \approx 0.44$ both in two and three dimensions [32, 33], suggesting that this quantity may be independent of dimension.

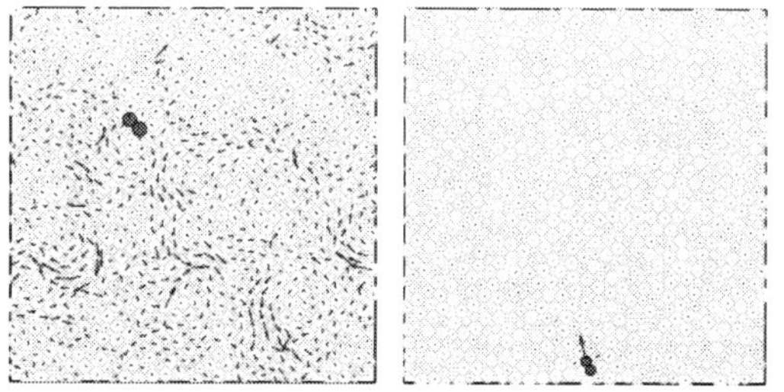

Figure 2. Extended *vs* localized contacts. When a contact is opened from an isostatic packing, the resulting deformation (arrows) can either be *extended*, as shown at left, or *localized*, as shown at right. Localized contacts are more numerous, but only extended contacts couple strongly to an imposed shear stress. Reproduced from [32] by permission of The Royal Society of Chemistry (RSC).

Moreover, its value does not depend on the preparation protocol of the isostatic state: up to error bars, equal values are found from compression of hard spheres [32], shear-jammed hard disks [35], and decompression of soft spheres [33, 36] [1]. The exponent θ_e can be shown to control the stability of the solid phase [32, 37, 38]. Recently replica calculations in infinite dimension on the force distribution [36, 39, 40] led to the prediction [33]:

$$\theta_e = 0.423..., \tag{10}$$

within the error bar of our measurements. In our proposed scaling description all exponents can be expressed in terms of θ_e, in particular:

$$\alpha_\mu = \alpha_\phi = \gamma_\mu = \gamma_\phi = \frac{3+\theta_e}{8+4\theta_e} \approx 0.35 \tag{11}$$

$$\gamma_{\dot\epsilon} = \frac{8+4\theta_e}{3+\theta_e} \approx 2.83 \tag{12}$$

$$\alpha_\ell = \gamma_\ell = \frac{1+\theta_e}{8+4\theta_e} \approx 0.15 \tag{13}$$

Empirically it was noticed that $\gamma_\mu = \gamma_\phi$ and that $\alpha_\mu = \alpha_\phi$, which our arguments rationalize.

[1] We include here works where θ_e was not directly measured, but inferred from the force distribution exponent θ_ℓ by the marginal stability relation $\theta_e = 2\theta_\ell/(1-\theta_\ell)$. See [33].

2 Theory for Frictionless Particles

2.1 General Approach

We argue that several dimensionless quantities that characterize the microscopic dynamics under flow critically affect rheological properties. As jamming is approached, the assembly of particles acts as a lever: due to steric hindrance, the typical relative velocity between adjacent particles V_r becomes much larger than the characteristic velocity $\dot\epsilon D$ where $\dot\epsilon$ is the strain rate and D the mean radius of the particles [41, 42]. We thus define the amplitude of this lever effect L as:

$$L = \frac{V_r}{\dot\epsilon D}. \tag{14}$$

Another fundamental quantity, particularly relevant for inertial flow, is the strain scale ϵ_v beyond which a particle loses memory of its direction relative to its neighbors. ϵ_v can be extracted from the decay of the autocorrelation function $\langle V_r^\alpha(0)V_r^\alpha(\epsilon)\rangle$, where the average is made over all pairs of adjacent particles α. As the packing fraction ϕ increases toward jamming, collisions are more frequent per unit strain (due to the increase of relative particle motion L), and each collision affects the motion of the particles on a growing length scale. These two effects imply that ϵ_v vanishes rapidly near jamming.

We now argue that dissipation is entirely governed by L in overdamped suspensions, and by both L and ϵ_v in inertial flows. In both cases the power injected into the system at the boundaries, which is simply $\mathcal{P} = \Omega\sigma\dot\epsilon$ at constant volume, must be dissipated in the bulk.

In a dense suspension we expect dissipation to be governed by local mechanisms such as lubrication [43]. Lubrication forces are singular for the ideal case of perfectly smooth spheres, but not for rough particles where they must be cut off. Thus the viscous force exchanged by two neighboring particles must dimensionally follow $F \sim \eta_0 V_r D^{d-2}$, leading to a power dissipated $\mathcal{P}/N = C\eta_0 V_r^2 D^{d-2}$ where d is the spatial dimension, C is a dimensionless constant that depends on the particle shape and roughness, and N is the number of particles. Equating the power dissipated to the power injected, one gets that for a given choice of particles:

$$\frac{\eta}{\eta_0} \propto 1/J \propto L^2 \tag{15}$$

implying that the divergence of viscosity is governed by L. This result holds by construction in simple models of dissipation in suspension flows [41, 42, 44].

Concerning inertial flows, we suppose that the restitution coefficient characterizing a collision between two particles is smaller than one, and that collisions dominate dissipation. Then each time two neighboring particles change relative direction, a finite fraction of their relative kinetic energy $E_c \sim MV_r^2$ must be dissipated, where M is the particle mass. Then the total power dissipated must follow $\mathcal{P} \propto N\dot\epsilon E_c/\epsilon_v$. Using Eq.(14) and balancing power injected and dissipated, one gets $\sigma/(\dot\epsilon^2 D^2 \rho) \sim L^2/\epsilon_v$ where ρ is the mass density of the particles, so that the inertial number I follows:

$$I \sim \frac{\sqrt{\epsilon_v}}{L} \tag{16}$$

To our knowledge Eq.(16) has not been proposed before, and could be tested empirically.

2.2 Perturbation around the solid

To obtain a complete description of flow, one must therefore express L and ϵ_v in terms of control parameters such as $\delta\mu$ or $\delta\phi$. To achieve this goal, we make the assumption that the contact network of configurations in flow is similar to that of jammed configurations at μ_c immediately after increasing the stress ratio by $\delta\mu > 0$. The detailed arguments in this section are presented in [25]; here we discuss the physical ingredients of each argument and summarize the results.

The coordination z of the network of contacts is a key microscopic quantity that distinguishes flowing from jammed configurations. At jamming the coordination is just sufficient to forbid motion, corresponding to $z_c = 2d$ for frictionless spheres [45–47]. As illustrated in Fig.(1), the kick of amplitude $\delta\mu$ opens a fraction $\delta z \equiv z_c - z$ of the contacts, allowing collective motions of the particles for which particles do not overlap, but simply stay in contact, the so-called called *floppy modes*. Using a virtual work theorem to compute the work done after a contact is opened, we find [25] that the lever amplitude is directly related to the density of floppy modes δz, in particular

$$L \sim \delta z^{-(2+\theta_e)/(1+\theta_e)} \tag{17}$$

To see how many contacts are opened when an increment of stress ratio $\delta\mu$ is added to an isostatic packing, we can use the known behavior of the shear modulus near jamming. This results in

$$\delta z \sim \delta\mu^{(2+2\theta_e)/(3+\theta_e)} \tag{18}$$

Jointly Eqs.(17,18) predict a relationship between lever amplitude and stress ratio:

$$L \sim \delta\mu^{-(4+2\theta_e)/(3+\theta_e)} \sim \delta\mu^{-1.41} \tag{19}$$

Eqs.(15,19) lead to a prediction for the exponent γ_μ entering in the constitutive relation $\mu(J)$. Together with previous results showing that $\ell_c \sim 1/\sqrt{\delta z}$ [8, 48], we obtain expressions for γ_ℓ and α_ℓ, corresponding to:

$$\ell_c \sim \delta\mu^{-(1+\theta_e)/(3+\theta_e)} \sim \delta\mu^{-0.41} \tag{20}$$

both for inertial and viscous flows.

We now consider the characteristic strain scale ϵ_v at which velocities decorrelate. This is fixed by a geometrical argument: we use the fact that in dense flows, when a grain has an unbalanced net contact force, $\vec{F}$, the ensuing motion will tend to make the remaining contacts of the grain align along $\vec{F}$ (see Fig.3). Since forces are repulsive, this further increases the unbalanced contact force. The increase in force is proportional to the typical contact force, pD^{d-1}, as well as to the rotation of the contacts, of magnitude $\sim Ld\epsilon$, thus

$$\frac{dF}{d\epsilon} \sim pD^{d-1}L, \tag{21}$$

where L is the dimensionless magnitude of the velocity fluctuation. This equation can also be derived formally, see [25]. In inertial flow, unbalanced forces are proportional to accelerations, $F = pI^2 dL/d\epsilon$, which leads to

$$\frac{d^2L}{d\epsilon^2} \propto \frac{L}{I^2}. \tag{22}$$

Eq.22 indicates that there is a characteristic strain scale $\epsilon_v \approx I$ in which a velocity fluctuation grows by an amount proportional to its initial magnitude. In steady flow, such growth must be destroyed by collisions on the same strain scale, since the latter reorganize the direction of particle motion. Hence ϵ_v is indeed the scale of decorrelation of particle velocities. Together with Eqs.(16,19) this result leads to a prediction for the exponent α_μ characterizing the constitutive relation $\mu(I)$.

For viscous flows, we still have Eq.(21), but now unbalanced contact forces are equal to the viscous drag forces, since the total net force vanishes. In this case $\vec{F}_i \approx \eta_0 \vec{V}_i^{na}$, where ^{na} indicates the non-affine part. Assuming $\vec{V}_i^{na} \approx V_r$, we get $F/p \approx (\eta_0\dot{\epsilon}/p)L = LJ$, leading to

$$\frac{dL}{d\epsilon} \propto \frac{L}{J}. \tag{23}$$

By a similar argument as above, this implies a characteristic strain scale $\epsilon_v \approx J$ in which velocity fluctuations are created and destroyed.

One missing link to obtain a full scaling description of the problem is how the packing fraction depends on other control parameters. If we make the additional assumption that isotropic packings of frictionless particles in the thermodynamic limit have a finite (although presumably small) dilatancy, then this implies the scaling relation $\delta\phi \sim \delta\mu$, known to agree well with observations [25]. This result enables us to predict the exponents α_ϕ and γ_ϕ entering the constitutive relation for $\phi(I)$ and $\phi(J)$, leading to a complete scaling description of rheological properties near

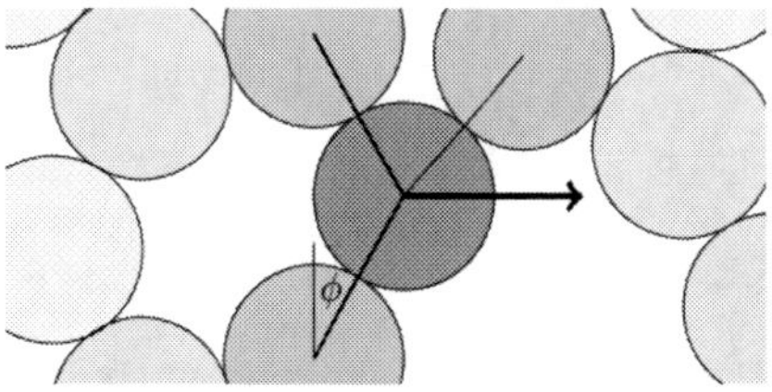

Figure 3. Illustration of geometrical nonlinearity. If the central grain has an unbalanced contact force as indicated by the arrow, then the ensuing flow will tend to align the contact normals of the dominant contact forces (thick lines), i.e. the angle ϕ will increase. Geometrically, $d\phi/dt \propto V$, the velocity of the particle.

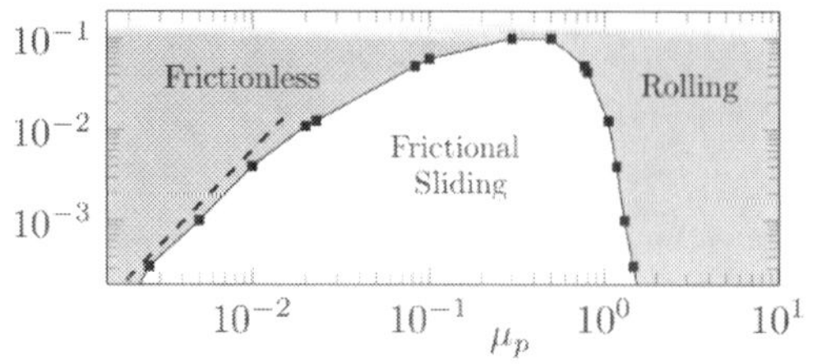

Figure 4. Phase diagram of dense homogeneous inertial frictional flow. In the frictionless and rolling regimes, most energy is dissipated by inelastic collisions, while in the frictional sliding regime energy dissipation is dominated by sliding. Along the phase boundary, grains dissipate equal amounts of energy in collisions and in sliding. For $I \gtrsim 0.1$, one enters the dilute regime [18]. The dashed line has slope 2.

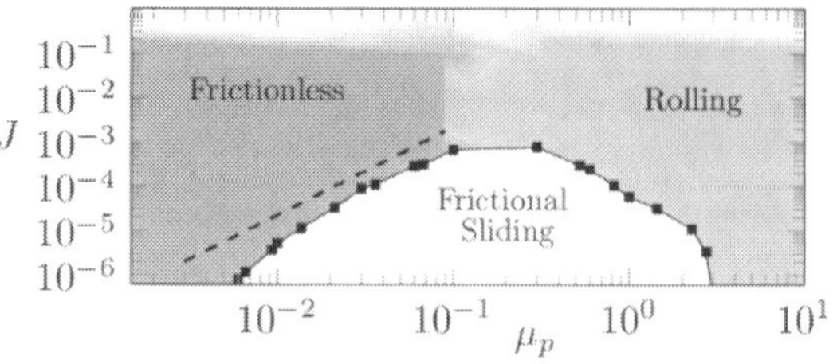

Figure 5. Phase diagram of dense non-Brownian suspension flow. In the Frictionless and Rolling regimes, the dominant source of dissipation is viscous drag, whereas in the Frictional Sliding regime, dissipation is dominated by sliding friction. The dashed line has slope 2.

jamming for frictionless particles. In particular the divergence of viscosity with packing fraction in suspensions is expected to follow:

$$\frac{\eta}{\eta_0} \sim (\phi_c - \phi)^{-(8+4\theta_e)/(3+\theta_e)} \sim (\phi_c - \phi)^{-2.83} \quad (24)$$

Finally, by a straightforward analysis one can extract the strain rate when one expects a transition from viscous to inertial flow, leading to Eq.(8) [25].

Our results are compared to previous empirical and numerical observations in Table 1. Overall we find a very good agreement between observations and predictions for frictionless particles. However, friction appears to change exponents for inertial flows, and may also affect viscous flows.

3 Effect of Friction

3.1 Phase Diagram

In the theory for frictionless particles, the fact that the power injected and the power dissipated exactly compensate on average plays a crucial role in relating the strain rate to geometrical quantities, as expressed in Eqs.15,16. Central to the arguments are the dominant dissipation mechanisms, assumed above to be viscous dissipation in over-damped dynamics, and grain inelasticity in inertial dynamics. However, when particles are frictional, energy will also be dissipated by sliding at frictional contacts, in addition to the other dissipation mechanisms above. As Coulomb friction coefficient μ_p and shear rate are varied, it is not obvious *a priori* in which regimes the sliding dissipation rate $\mathcal{D}_{slid}$ will dominate over inelastic collisions in inertial flows, and viscous dissipation in viscous flows; we call these latter sources of dissipation $\mathcal{D}_0$.

This question can be straightforwardly investigated with numerical simulations. To map out the phase diagram as μ_p and shear rate are varied, we considered, separately, the parameter space (μ_p, I) for inertial dynamics [20], and (μ_p, J) for viscous dynamics [26]. The main results are the phase diagrams presented in Figs.4,5. For both viscous and inertial dynamics, we find 3 phases: (i) a 'Frictionless' regime in which $\mathcal{D}_0$ dominates, most contacts are sliding, and investigated quantities are consistent with the theory for frictionless particles, in particular L, ϵ_v, $\delta\mu$, and $\delta\phi$; (ii) a 'Rolling' regime in which again $\mathcal{D}_0$ dominates, but most contacts are rolling, not sliding. Here L, $\delta\mu$, and $\delta\phi$ appear to be consistent with the frictionless theory, particularly in the limit $\mu_p \gg 1$. Finally, in between these regimes we find (iii) a 'Frictional Sliding' regime in which $\mathcal{D}_{slid} > \mathcal{D}_0$. Here $\epsilon_v \sim I(J)$ in inertial (viscous) dynamics, as predicted by Eqs.21,22,23, which hold in the presence of friction, but other quantities have new scalings: for example, L has a weaker divergence, and $\delta\mu$ has a much less dramatic behavior at small strain rate.

The crossover from 'Frictionless' to 'Frictional Sliding' regimes can be predicted from the theory above. Indeed, $\mathcal{D}_{slid}$ is simply the sum over sliding contacts of $\vec{f}^T_\alpha \cdot \vec{U}^T_\alpha$, where T indicates the tangential component. Denoting by χ the fraction of contacts that are sliding, we have $\mathcal{D}_{slid} \approx N_C \chi \mu_p \langle f_N \rangle L\dot{\epsilon}$, where N_C is the total number of contacts and $\langle f_N \rangle \sim p$ is the typical normal force. Inside the frictionless regime, nearly all contacts are sliding, $\chi \approx 1$. Therefore we find that sliding friction constitutes a fraction

$$\frac{\mathcal{D}_{slid}}{\mathcal{P}} \sim \frac{N\mu_p p\dot{\epsilon}L}{\Omega\sigma\dot{\epsilon}} \sim \frac{\mu_p}{\mu_c}L \quad (25)$$

of the total energy dissipation rate (which must equal $\mathcal{P}$ in steady flow). Since L is diverging as jamming is approached (Eq.19) we thus expect a transition to a regime dominated by frictional sliding, both for inertial and viscous flows, as observed. From Eqs.(15,16) and $\epsilon_v \sim I$, it follows that $L \sim I^{-1/2}$ and $L \sim J^{-1/2}$ in frictionless inertial and viscous dynamics, respectively, thus leading to a crossover 'Frictionless' to 'Frictional Sliding' at $I \sim \mu_p^2$ and $J \sim \mu_p^2$. These scalings are plotted as dashed lines in Figs.4,5, and are consistent with the data.

3.2 Frictional Theory

Still missing is a complete theory of the Frictional Sliding regime. We expect Eqs.21,22,23 to be always valid, and indeed the associated prediction $\epsilon_v \sim I$ appears to hold in the Frictional Sliding regime as well. However, the velocity fluctuations and rheological properties present new scalings. In particular, in all cases the velocity fluctuations are large and their spatial correlations are long-range as jamming is approached. For frictionless particles, a single characteristic scale characterizes these velocity fluctuations. By contrast, in [26] we provided evidence that several scales are required to describe the kinetics of frictional particles. For example, the relative velocity between contacting particles differs from the amplitude of non-affine velocities. This additional complexity is plausibly caused by the intermittent and non-extensive localization of strain observed in several studies [49–52]. A similar localization is observed in the plasticity of soft particles [53], whose connection to hard frictional particles is, however, unclear. A possible factor causing the difference between frictional and frictionless particles relates to the criticality of the marginal solid: for hard frictionless spheres, Maxwell counting leads to a precise identification of the marginal state, which is exactly isostatic. By contrast, frictional solids at jamming are generally hyperstatic [54–56]. How these issues are related are outstanding problems.

4 Conclusion

We reviewed a theory for flow of hard, frictionless particles, both in viscous and inertial dynamics. In a first step, we related the power dissipated in flow of frictionless particles to certain microscopic kinetic quantities. The latter control singularities in the rheological properties near jamming. In a second step, we have computed these quantities, using a perturbation around the solid phase. Our main hypothesis is that configurations in flow are similar to jammed configurations at maximum stress ratio μ_c, destabilized by an additional stress increment $\delta\mu$. In this approach, the properties of the solid phase are central, in particular the fact that the density of contacts which can couple to external forces is singular at small forces, and characterized by a non-trivial exponent θ_e. Our description of flow can thus be thought as that of a jammed solid, populated by elementary excitations corresponding to the opening of weak contacts, of density δz.

When friction is added, the theory holds in a finite region of parameter space, whose boundary is well predicted by the theory. Theory for the regime in which sliding friction dominates is, however, lacking. In our view, a central question for the future is what controls the stability of isostatic frictional systems, how these respond to an additional stress ratio $\delta\mu$, and how the combination of finite softness, inertia and friction qualitatively affects the flow curves [57].

Acknowledgements

We are pleased to thank our collaborators on this project: Gustavo Düring, Edan Lerner, Jim McElwaine, and Martin Trulsson.

References

[1] G. MiDi, The European Physical Journal E: Soft Matter and Biological Physics **14**, 341 (2004-08-01)
[2] F. da Cruz, S. Emam, M. Prochnow, J.N. Roux, F.m.c. Chevoir, Phys. Rev. E **72**, 021309 (2005)
[3] J.J. Stickel, R.L. Powell, Annu. Rev. Fluid Mech. **37**, 129 (2005)
[4] P. Jop, Y. Forterre, O. Pouliquen, Nature **441**, 727 (2006)
[5] F. Boyer, E. Guazzelli, O. Pouliquen, Phys. Rev. Lett. **107**, 188301 (2011)
[6] O. Pouliquen, Physical review letters **93**, 248001 (2004)
[7] P. Olsson, S. Teitel, Phys. Rev. Lett. **99**, 178001 (2007)
[8] G. Düring, E. Lerner, M. Wyart, Physical Review E **89**, 022305 (2014)
[9] A. Lemaître, J.N. Roux, F. Chevoir, Rheologica acta **48**, 925 (2009)
[10] M. Wang, J.F. Brady, Physical review letters **115**, 158301 (2015)
[11] A. Fall, A. Lemaitre, F. Bertrand, D. Bonn, G. Ovarlez, Phys. Rev. Lett. **105**, 268303 (2010)
[12] M. Trulsson, B. Andreotti, P. Claudin, Physical review letters **109**, 118305 (2012)
[13] D. Vågberg, P. Olsson, S. Teitel, Physical Review Letters **112**, 208303 (2014)
[14] Y. Forterre, O. Pouliquen, Annual Review of Fluid Mechanics **40**, 1 (2008)
[15] P.E. Peyneau, J.N. Roux, Physical review E **78**, 011307 (2008)
[16] P.E. Peyneau, Ph.D. thesis, Ecole des Ponts ParisTech (2009)
[17] M. Bouzid, M. Trulsson, P. Claudin, E. Clément, B. Andreotti, Phys. Rev. Lett. **111**, 238301 (2013)
[18] E. Azéma, F. Radjai, Physical review letters **112**, 078001 (2014)
[19] N. Menon, D.J. Durian, Science **275**, 1920 (1997)
[20] E. DeGiuli, J. McElwaine, M. Wyart, Phys. Rev. E **94**, 012904 (2016)
[21] G. Ovarlez, F. Bertrand, S. Rodts, Journal of rheology **50**, 259 (2006)
[22] P. Olsson, S. Teitel, Physical review letters **109**, 108001 (2012)
[23] P. Olsson, S. Teitel, Physical Review E **83**, 030302 (2011)
[24] T. Kawasaki, D. Coslovich, A. Ikeda, L. Berthier, Physical Review E **91**, 012203 (2015)
[25] E. DeGiuli, G. Düring, E. Lerner, M. Wyart, Physical Review E **91**, 062206 (2015)
[26] M. Trulsson, E. DeGiuli, M. Wyart, Physical Review E **95**, 012605 (2017)
[27] R. Lespiat, S. Cohen-Addad, R. Höhler, Phys. Rev. Lett. **106**, 148302 (2011)
[28] C. Cassar, M. Nicolas, O. Pouliquen, Physics of Fluids **17**, 103301 (2005)

[29] M. Souzy, H. Lhuissier, E. Villermaux, B. Metzger, Journal of Fluid Mechanics **812**, 611 (2017)
[30] P. Olsson, Phys. Rev. E **81**, 040301 (2010)
[31] S. Gallier, E. Lemaire, F. Peters, L. Lobry, Journal of Fluid Mechanics **757**, 514 (2014)
[32] E. Lerner, G. During, M. Wyart, Soft Matter **9**, 8252 (2013)
[33] E. DeGiuli, E. Lerner, C. Brito, M. Wyart, Proceedings of the National Academy of Sciences **111**, 17054 (2014)
[34] G. Düring, E. Lerner, M. Wyart, Physical Review E **94**, 022601 (2016)
[35] E. Lerner, G. Düring, M. Wyart, EPL (Europhysics Letters) **99**, 58003 (2012)
[36] P. Charbonneau, E.I. Corwin, G. Parisi, F. Zamponi, Physical Review Letters **114**, 125504 (2015)
[37] M. Wyart, Phys. Rev. Lett. **109**, 125502 (2012)
[38] M. Müller, M. Wyart, Annual Review of Condensed Matter Physics **6**, 177 (2015)
[39] P. Charbonneau, J. Kurchan, G. Parisi, P. Urbani, F. Zamponi, Nature communications **5** (2014)
[40] P. Charbonneau, J. Kurchan, G. Parisi, P. Urbani, F. Zamponi, Journal of Statistical Mechanics: Theory and Experiment **2014**, 10009 (2014)
[41] E. Lerner, G. Düring, M. Wyart, Proceedings of the National Academy of Sciences **109**, 4798 (2012)
[42] B. Andreotti, J.L. Barrat, C. Heussinger, Phys. Rev. Lett. **109**, 105901 (2012)
[43] R.C. Ball, J.R. Melrose, Physica A: Statistical and Theoretical Physics **247**, 444 (1997)
[44] D. Vågberg, P. Olsson, S. Teitel, Physical Review Letters **113**, 148002 (2014)
[45] A.V. Tkachenko, T.A. Witten, Phys. Rev. E **60**, 687 (1999)
[46] C.F. Moukarzel, Phys. Rev. Lett. **81**, 1634 (1998)
[47] J.N. Roux, Phys. Rev. E **61**, 6802 (2000)
[48] G. Düring, E. Lerner, M. Wyart, Soft Matter **9**, 146 (2013)
[49] C. Heussinger, Physical review E **88**, 050201 (2013)
[50] S. Henkes, D.A. Quint, Y. Fily, J.M. Schwarz, Physical Review Letters **116**, 028301 (2016)
[51] M. Maiti, A. Zippelius, C. Heussinger, EPL (Europhysics Letters) **115**, 54006 (2016)
[52] P. Kharel, P. Rognon, arXiv preprint arXiv:1605.00337 (2016)
[53] C.E. Maloney, Phys. Rev. Lett. **97**, 035503 (2006)
[54] M. Wyart, S. Nagel, T. Witten, EPL (Europhysics Letters) **72**, 486 (2005)
[55] A.J. Liu, S.R. Nagel, W. van Saarloos, M. Wyart, *The jamming scenario: an introduction and outlook* (Oxford University Press, Oxford, 2010)
[56] M. van Hecke, Journal of Physics: Condensed Matter **22**, 033101 (2010)
[57] M. Otsuki, H. Hayakawa, Physical Review E **83**, 051301 (2011)

Experimental study of erosion by suffusion at the micro-macro scale

Cong Doan Nguyen[1], *Nadia* Benahmed[1,*], *Pierre* Philippe[1], *Elizabeth Victoria* Diaz Gonzalez[2]

[1]Irstea, RECOVER Research Unit, Aix-en-Provence, France, France

[2] Autonomous University of the State of Mexico, Mexico

Abstract. Internal erosion is a complex phenomenon which represents one of the main sources of risk to the safety of earth hydraulic structures such as embankment dams, dikes and levees. Its occurrence may cause instability and failure of these structures with consequences that can be dramatic. Erosion by suffusion corresponds to the process of detachment and transport, under the action of hydraulic flow, of the finest soil particles within the porous media formed mainly of large grains. Its occurrence usually causes change of the initial microstructure and hence a change in the physical, hydraulic and mechanical characteristics of the soil. In this study, we present first an experimental characterization of the erosion mechanism during its occurrence within a granular soil. Particular emphasis was put on the role of hydraulic conditions in triggering of fines migration. Thereafter, we present a preliminary microstructural characterization of the erosion process through direct visualization by optical techniques of particles migration using crushed glass samples as model materials.

1 Introduction

The suffusion phenomenon corresponds to the process of detachment and transport under the action of hydraulic flow of the finest soil particles within the porous media formed mainly of large grains. This phenomenon, often termed as internal instability, was first observed in studies related to base soil-filter compatibility against seepage flow. Thereafter, many empirical methods to evaluate the potential of internal instability of soils based on analysis of the grain size distribution curve were proposed [1, 2, and 3]. The hydraulic gradient governing the onset of internal erosion is named "critical hydraulic gradient", and has been widely studied to better understand the hydraulic conditions leading to the initiation of internal erosion. According to Terzaghi [4], the critical hydraulic gradient is the value at which there is a balance between grain weight and the forces induced by the flow. Later, Skempton and Brogan [5] argued that the "segregation piping" occurs at the hydraulic gradient one third to one fifth of the Terzaghi's critical gradient for a homogeneous granular material of the same porosity. However, these critical hydraulic gradients are measured just before the onset of erosion for which the hydraulic forces generated by seepage flow are large enough for detached particles to pass through. That is to say, detachment of particles inside the sample without washout, clogging and grain redeposition are not taken into consideration. Consequently, the real hydraulic gradient causing the initiation of erosion (first movement of particles) could be underestimated.

The aim of this study is to establish, on one hand, the hydraulic conditions under which the erosion is triggered; and on the other hand, the real mechanism of erosion by suffusion, including the detachment of fine particles, their transport or redeposition, and the clogging phenomenon. First, this is done by following the evolution of the hydraulic gradient at macroscopic scale during suffusion tests carried out on real soil samples. Secondly, a novel direct visualization approach by optical techniques of particles detachment and migration was used. In this case, crushed glass particles and optically matched oil were used instead of soil particles and water, respectively.

2 Experimental Investigations

2.1 Materials used

The mixture of coarse and fine silica Hostun sand (HN 1/2.5 and HN 34, respectively, ρ_s=2650 kg.m^{-3}) was selected to perform the suffusion tests on widely graded cohessionless real soil.

For the optical techniques, a mixture of crushed Borosilicate glass was used (ρ_s=2330 kg.m^{-3}). Borosilicate glass rods with diameters of 22 mm, purchased from Verresatine manufacturer, were crushed manually and the resulting powder was sieved through different sizes of mesh stainless steel to obtain particles ranging from 0.5 to 20 mm as required by the target particle size distribution (*cf. 2.3*).

* Corresponding author: nadia.benahmed@irstea.fr

To ensure that internal erosion will occur during the suffusion tests, the vulnerability of the two mixtures, prepared with fines content FC of 20% for sand and 15% for crushed glass, was evaluated and assessed as internally unstable following the methods proposed in the literature [1-3]. The grain size distributions of the used mixtures, both of soil and crushed glass particles, are presented in Fig.1.

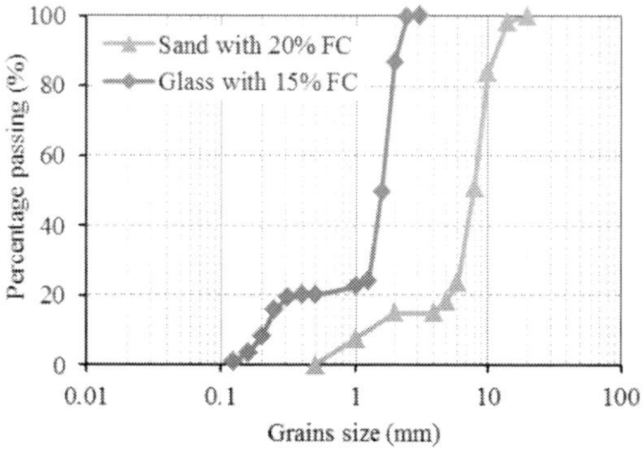

Fig. 1. Grain size distribution curves of the used mixtures.

2.2 Suffusion test on granular soil

To perform suffusion erosion tests, a newly developed suffusion permeameter is used. It is constituted by a cylindrical seepage cell made in Plexiglass to allow visual observation of particles migration, water tank, fines collector and pump. The cell has an inner diameter of 70 mm and height of 140 mm to fit triaxial sample size for further mechanical behavior investigations which are not presented in this paper (Fig. 2).

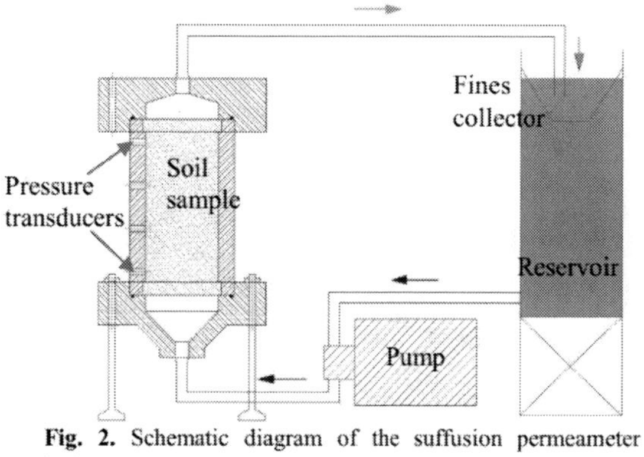

Fig. 2. Schematic diagram of the suffusion permeameter device.

The soil samples were reconstituted at an initial relative density of 40% by moist tamping method, in seven layers at fixed height and mass for each with respect to the target density. They were first flushed with CO_2, followed by de-aired water in the upward direction to achieve fully saturated state. Thereafter, the erosion test was performed by applying a flow at controlled rate thanks to pump. The initially imposed flow velocity is in the range of 0.01-0.03 $cm.s^{-1}$, and then increased by increments until the movement of fine particles is triggered. The corresponding hydraulic gradient is determined from the pressure gradient measured by two pressure transducers installed at upper and lower part of the seepage cell. The increasing flow rate procedure is illustrated in Fig. 3.

2.3 Suffusion test on crushed glass material

The optical techniques used in this study is called Refractive Index Matching (RIM) and consists in immersing the solid particles in a liquid having the same optical index ($n \approx 1.473$) to create a transparent granular material [6]. In order to visualize specifically the movement of the finest particles under the action of the flow, 1% of fine particles of the crushed glass, whose behavior is expected to be representative, were colored in blue in order to be opaque, hence visible,.

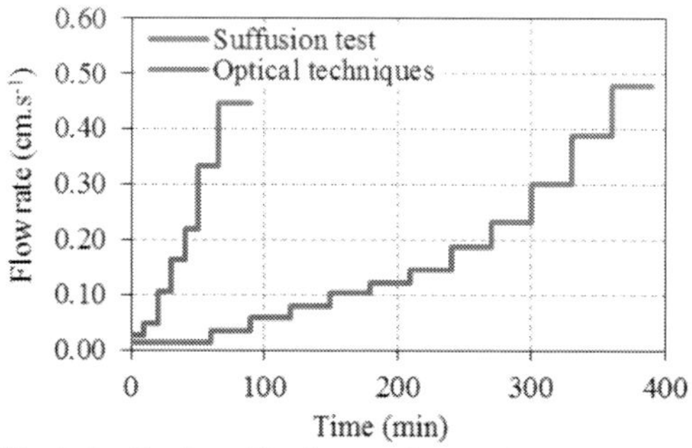

Fig. 3. Application of the flow rate by step during internal erosion test.

The RIM liquid is here a mixture of two mineral oils whose dynamic viscosity and density are equal to $\mu=28\times10^{-3}$ $kg.m^{-1}.s^{-1}$ and $\rho=850$ $kg.m^{-3}$, respectively. To avoid any refraction issue at the walls, a square box is used here (Fig. 4) contrary to other previous permeameters, usually cylindrical. An upward fluid flow is applied through the solid matrix by means of a gear pump, with the flow rate being increased by stages until internal erosion is initiated and subsequently progresses. The time evolution of the erosion phenomenon was recorded by a camera. The increasing flow rate procedure is illustrated in Fig. 3. As can be noted, the imposed flow rates are approximately of the same order of magnitude as those needed for the previous suffusion test with sand and water. In this test, the next stage was proceeded only once suffusion was no more observed.

The glass and oil materials used in this experiment have different physical properties to those of sand and water mixtures. To realistically mimic internal erosion behavior in those model materials by creating similar hydraulic conditions than for sand and water mixtures, the sizes of the crushed glass particles need to be up-scaled compared to the sand particle size distribution.

For this purpose, one can refer to the pioneer studies of fluvial sediment transport by Shields in 1936 which showed, analytically and experimentally, that the initiation of grain movement is governed by the ratio of the force exerted by the fluid flow on a grain and the buoyant weight of the grain $\Delta\rho g d^3$ ($\Delta\rho = \rho_s - \rho$; where ρ_s is density of a grain, ρ *is* density of fluid and d is grain diameter):

$$Sh = \frac{\tau_f d^2}{\Delta \rho g d^3}$$

Here, the hydrodynamical force is expressed as the shear-stress τ_f multiplied by the cross section of the grain. The experimental observations show that there is a critical value of the Shields number Sh_c above which some grains start to move.

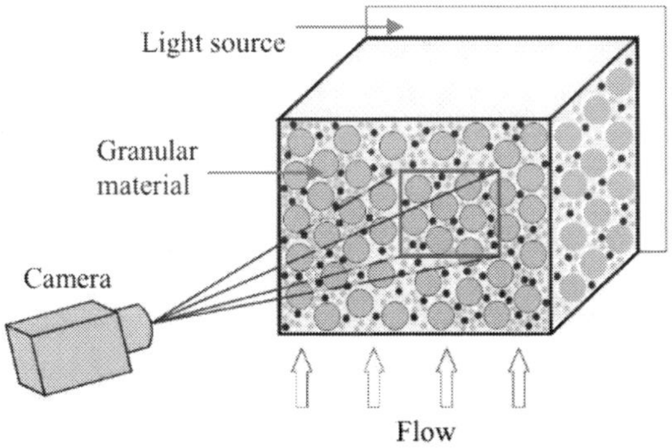

Fig. 4. Schematic diagram of the seepage test assembly used in optical techniques.

In both of our granular systems, the fine particles that will be first eroded are not, or very weakly, contributing to force chains. The only resistant forces against fluid flow action are consequently their buoyant weight and friction. Consequently, it is still relevant here to use the Shields number. Considering then a same critical value for the Shields number, the crushed glass particles were required to be scaled up to about 9 times to account for the higher viscosity of the RIM oil compared to water. Due to practical requirements for the size of the apparatus, the glass particles were scaled up only to a factor about 6 in this study. Scaled particle size distribution for crushed glass samples is shown in Fig.1.

3 Results and Discussion

3.1 Erosion of granular soil

Fig. 5 presents the variation of hydraulic gradient versus time. At first stage, when the flow velocity is very low (v = 0.03 to 0.06 cm.s^{-1}), no loss of fines was observed and the hydraulic gradient remains nearly constant around 0.2. When the flow rate is increased (v = 0.08 cm.s^{-1}), the hydraulic gradient in the beginning remains fairly unchanged (0.2) owing probably to the increase of the permeability, and then shows a sudden increase, around 0.3. During this increase, no eroded mass was observed. Hence, this unexpected increase might be induced by the occurrence of clogging at constrictions by particles transport inside the sample leading to a decrease of the permeability.

During further steps (v = 0.10 and 0.12 cm.s^{-1}), small movements of fine grains occurred and the erosion became dominant. A very thin layer of fine particles covers the top surface of the sample. Besides, a few sand spots appeared after increasing the flow rate and a slight movement of the fine grains was observed around those spots. The later could be a mark of preferential flow path appearance inside the sample.

In general, the critical hydraulic gradient for internal erosion as defined in the literature is the hydraulic gradient for which the erosion phenomenon can be observed by naked eyes. It means the hydraulic gradient i = 0.3 (v = 0.1 cm.s^{-1}) in the present case. However, on the basis of Figure 4, we suppose that the detachment of fine particles inside the sample occurred before this value (for v = 0.06 and 0.08 cm.s^{-1}), even if no washout was observed.

3.2 Particles migration on crushed glass test

The same phenomenon as for the sand was observed in the test with the model particles of crushed glass. Fine particles start to move approximately at the same flow velocity ($v \approx$ 0.1 cm.s^{-1}) than with soil-water, which confirms that the up-scale shows good coherence between model material and natural material.

However, looking at the critical hydraulic gradient, the first motion of fine particles is observed at very low hydraulic gradient: i = 0.07 << i_{cr} defined in literature. For instance, Skempton & Brogan [4] have proposed $i_c \sim$ 0.2 to 0.35.

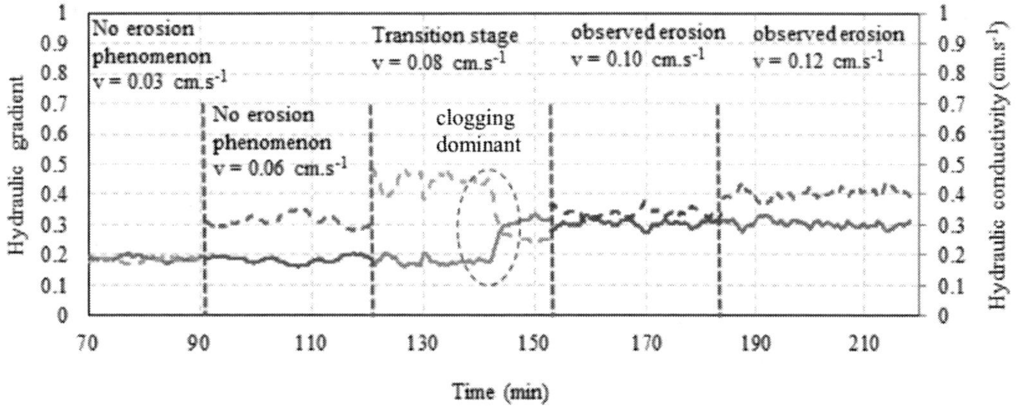

Fig. 5. Evolution of hydraulic gradient (continuous line) and hydraulic conductivity (dashed line) during the internal erosion test.

By observing longer the erosion process, detachment and transport cease after a short time (1–2 minutes), corresponding to the case with the decrease of instantaneous hydraulic gradient observed previously. As predicted above, it can be seen from Figure 6 and 7 that detachment and transport occurs when the flow velocity is high enough ($v \approx 0.1$ cm.s^{-1}) to create a force able to move the particles. Then, depending on tortuosity, grain shape and constriction size ratio, the particles will be either transported away or rapidly clogged.

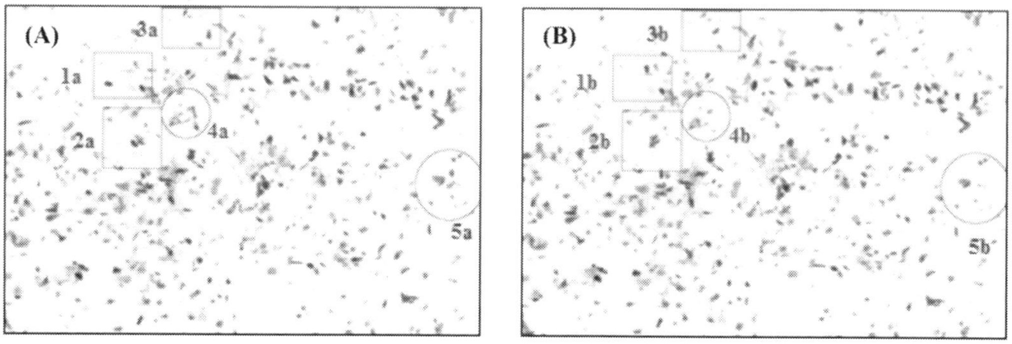

Fig. 6. Movement of fine particles during erosion under a hydraulic gradient i=0.07; (A) at t = 30 min, and (B) at t = 40 min. Displacement and clogging (rectangles), wash-out (circles).

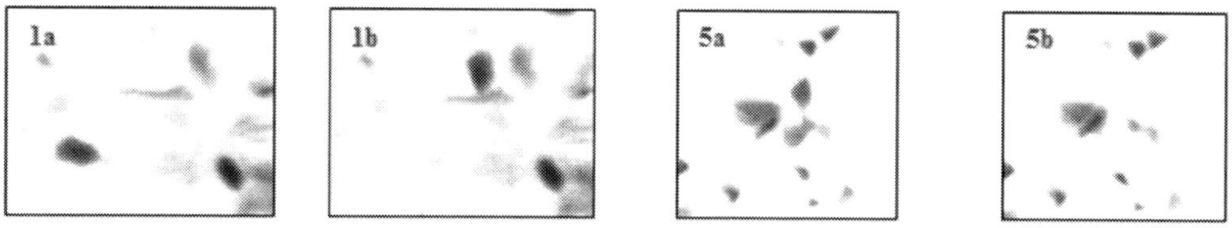

Fig. 7. Zoom of particles movement; (1a) and (5a) initial position, (1b) and (5b) final position.

4 Conclusions

This work was tackled by suffusion tests on real soil samples which showed that the erosion phenomenon could occur before reaching the hydraulic gradient corresponding to the observed wash-out of fine particles. It was also suspected the existence of clogging phenomenon which is reflected by an increase of hydraulic gradient.

The microstructural characterization through direct visualization by optical techniques of suffusion process on crushed glass particles confirms that, while no washout of eroded particles are observed with naked eyes, the particles movement, hence the erosion onset, may take place inside the soil mass under a very small hydraulic gradient about 0.07 (much smaller than that defined in literature which is around 0.2). The optical technique permits also the visualization of the microstructure changes by the mobilization and the redeposition of eroded fine particles at the pore scale. Consequently, clogging and grains redeposition, which representing a problem of great importance for several disciplines, need to be taken into consideration in the analysis of internal erosion.

From this preliminary work, and beside of the onset of internal erosion evaluation based on critical hydraulic gradient, it appears a real need for the description of critical flow velocities and critical hydraulic shear stresses at the pore scale. This could be addressed by coupling the RIM technique described above with the Planar Laser Induced Fluorescence technique (PLIF) [6].

This research was carried out with a financial support from the PACA region. The authors are also grateful to their socio-economic partnership with the engineering company SAFEGE (Suez Group).

References

1. Burenkova, VV.: Assessment of suffusion in non-cohesive and graded soils. Paper presented at the Filter in geotechnical and hydraulic engineering. Balkema, Rotterdam. (1993).
2. Kenney, TC. and Lau, D.: Internal stability of granular filters. Canadian Geotechnical Journal, 22(2), 215-225. (1985).
3. Kezdi, A.: Soil physics: selected topic. (1979).
4. Skempton, AW, & Brogan, JM. (1994). Experiments on piping in sandy gravels. Geotechnique, 44(3), 449-460.
5. Terzaghi, Karl. (1939). 45th James Forrest Lecture, 1939. Soil Mechanics-A New Chapter In Engineering Science. Journal of the ICE, 12(7), 106-142.
6. Béguin, R., Philippe, P. and Faure Y.-H. (2013). Pore-scale flow measurements at the interface between a sandy layer and a model porous medium: Application to statistical modeling of contact erosion. J. Hydr. Eng., 139(1), 1-11.

Investigation of vertical size segregation in bedload sediment transport with a coupled fluid-discrete element model

Philippe Frey[1,*], Raphaël Maurin[1,2], Lamia Morchid-Alaoui[1], Shashank Gupta[1], Julien Chauchat[3]

[1]Université Grenoble Alpes, Irstea, UR ETNA, 2 rue de la Papeterie-BP 76, F-38402 St-Martin-d'Hères, France
[2]Université de Toulouse, INPT, UPS, Institut de Mécanique des Fluides de Toulouse, Allée Camille Soula, 31400 Toulouse, France
[3]Univ. Grenoble Alpes, LEGI, G-INP, CNRS, F-38000 Grenoble, France

Abstract. In order to gain understanding of kinetic sieving-type segregation in bedload sediment transport, numerical experiments of two-size particle mixtures were carried out, using a validated coupled fluid-discrete element model developed at Irstea. A 3D 10% steep domain consisting at initial time of a given number of layers of 4 mm particles deposited on top of a coarser 6 mm particle bed, was submitted to a turbulent and supercritical fluid shear flow (Shields numbers of 0.1 and 0.3). The elevation of the centre of mass of the infiltrated fine particles is observed to follow the same logarithmic decrease with time, whatever the initial number of fine layers. This decrease is steeper for a higher Shields number. The main result is that this typical behaviour is related at first order to the shear rate depth profile.

1 Introduction

Bedload transport by opposition to suspension is defined as the part of sediment transport remaining close to the bed in stream channels. It concerns the coarser grain material shear-driven by the turbulent water flow. It has major consequences for public safety, often aggravating flooding issues, for the management of water resources, and for environmental sustainability. Bedload transport investigations have mainly been focused on mass flux and its correlation with the water flow. Yet after more than a century of work we have no satisfactory theory for sediment transport. Empirical sediment transport formulas often poorly compare with field measurements, by sometimes order of magnitudes.

An important reason for our limited ability to predict sediment flux is the absence of general understanding of grain-grain interactions in stream channels [1, 2].

Most particularly, in mountains, steep slopes drive an intense transport of a wide range of grain sizes leading to size segregation, also named grain size sorting [3]. This poorly understood phenomenon largely modifies fluxes and results in patterns that can be seen ubiquitously in nature.

Focusing on vertical size segregation, the infiltration of fine sediment into an immobile coarse bed has been extensively studied in fluvial geomorphology because of ecological considerations especially related to the reproduction of salmonids [4]. However when the coarse bed is moving, another type of infiltration named kinetic sieving can take place even with a coarse to fine diameter size ratio close to unity. Smaller grains may then move downward in the bed through voids that are opened by the general motion. It appears that this phenomenon which is important during floods has rarely been investigated in stream channels [5, 6].

In industrial processes, usually quite energetic, kinetic sieving-type segregation takes place, and has been more largely studied. Indeed, to be able to perfectly mix input materials and thus to counteract the natural segregation remains a challenge.

At Irstea, a coupled fluid-discrete element model has been developed and validated to study bedload transport [7]. In this paper we present a study of vertical size segregation and kinetic sieving using the model which has been extended to simulate two grain sizes. Following classical theories such as [8], segregation rates are analysed in term of shear rates.

2 Coupled fluid-discrete element model

A two-phase Eulerian-Lagrangian flow model developed at Irstea is used in this research. It is composed of a three dimensional discrete element model (DEM), describing each individual particle coupled with a one dimensional Reynolds average Navier Stokes (RANS) model [9].

The 3D DEM uses the open source code Yade [10] with a classical spring-dashpot contact law. The friction coefficient is set to 0.4, and a restitution coefficient of 0.5 is taken consistently with the experimental validation [7].

In the RANS model, the water flow is considered to be stationary and unidirectional in the streamwise direction so that the fluid profile depends only on the depth (1D).

The momentum equation for the fluid is composed of the gravity term, the viscous stress, the turbulent

* Corresponding author: philippe.frey@irstea.fr

Reynolds stress and the fluid-particle interaction forces (restricted to drag and buoyancy). In our turbulent flow, the viscous stress is negligible. The turbulent stress is based on the eddy viscosity concept using an integral mixing length formulation, which allows recovering the law of the wall in clear fluid, while damping the turbulence down to zero at maximum packing fraction in the granular bed.

The drag coefficient depends on the local particle Reynolds number taking into account hindrance effects [9]. The fluid and particle velocity at the bottom of the domain and the fluid velocity gradient at the imposed free surface elevation are all set to zero.

The effect of the particles on the fluid phase is taken into account through the drag force and the solid volume fraction. Both the effect of the particle on the fluid and of the fluid on the particle are considered (two-way coupling). The model has been validated with turbulent bedload transport experiments, in terms of both the sediment transport rate and granular depth profiles. More details can be found in [7, 11]. This model has also been used to analyse the dense granular rheology in turbulent bedload transport [12].

Our 3D numerical domain is made of a 10% (5.71°) inclined channel filled with water, in which glass spherical particles (density of 2,500 kg/m^3) are deposited by gravity. At initial time, a given number N of layers of 4 mm diameter particles (N = 0.05, 0.5, 1 and 2) are deposited on top of a bed made of 7 layers of 6 mm diameter particles (Fig 1a). A layer is here defined as the number of particles necessary to occupy a volume corresponding to a depth of one diameter times the cell width and length, taking into account the maximum packing fraction.

The elevation of the water free-surface is imposed and creates a fluid flow driven by gravity on the top of the granular bed. The momentum input from gravity is balanced by the Reynolds shear stress in the dilute suspension layer and by the grain grain interactions in the sediment bed. The flow is turbulent (bulk Reynolds number of about 10^4), hydraulically rough (particle Reynolds of order 10^3) and supercritical (Froude greater than 1).

Both a low and a high free surface elevation have been set respectively at 10.3d and 14d where d is the coarse diameter (d=6mm), which yields respective Shields numbers Sh of 0.1 and 0.3. The Shields number, classically used in sediment transport studies, is the dimensionless bottom fluid shear stress. It is defined as $Sh = \frac{\tau_b}{(s-1).d}$ with s the glass/water density ratio (s=2.5). τ_b is unambiguously defined as the maximum of the fluid turbulent Reynolds shear stress.

Fig. 1. Infiltration of two fine particle layers into the coarse bed a. t=0s b. t=70s (Sh=0.3)

Bi-periodic boundary conditions are imposed for the DEM, and only steady uniform conditions are analyzed. The unidirectional character of the fluid flow allows averaging spatially over the streamwise and spanwise directions, and analyzing the results only over the depth.

Due to kinetic sieving, the fine particles infiltrate through the coarser bed (Fig 1b). The centre of mass z_c of the infiltrated particle cloud was calculated over time. To be able to follow the fine particles infiltrating in the lowermost coarser layers, it was necessary to simulate up to 10^4 seconds (3 hours) for the Sh=0.1 case.

3 Results

Both the fluid and particles starting from rest, the system follows a transient behaviour before reaching a characteristic steady state with a constant sediment transport rate.

Fig. 2 shows the evolution over time of the centre of mass of the fine particles z_c for both Shields number of 0.1 and 0.3 (the origin of z_c is at the bottom of the domain). The initial centre of mass position z_c is dependent of the number N of fine particle layers since the number of coarse layers is kept constant and equal to 7 in the different simulations.

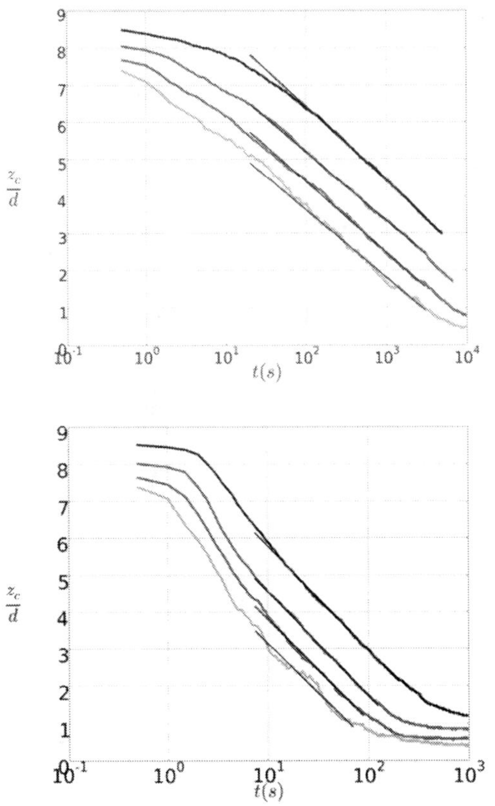

Fig. 2. Time evolution of the fine particles centre of mass for a number of layers N of 0.05, 0.5, 1 and 2, from light yellow (N=0.05 bottom) to dark black (N=2 top). See text for the slope of the fit **a**. Shields of 0.1 **b**. Shields of 0.3.

As one would expect, the infiltration evolution depends on the Shields number. For *Sh*=0.3, the cloud of fine particles reached the bottom of the domain between 10^2 and 10^3s, whereas this duration was above 10^4s for the *Sh*=0.1 case.

Considering the transient behaviour, both the number of fine particle layer deposited N and the Shields number influence the evolution of the fine particles centre of mass (Fig. 2). During this period, the size-segregation phenomenon is coupled with the evolution of the system to the steady state, making a clear analysis complicated.

Therefore, we rather focus on the steady state, which corresponds to the infiltration of fine particles through the coarse granular bed in quasi-static motion. In this region, the fine particles behaviour does not depend any more on the number N (provided the cloud is not yet affected by the bottom) and their centre of mass position z_c decreases linearly with $\log(t)$ (Fig.2).

This linear trend was fitted on all curves (Fig.2). For a given Shields number, the slope was found to be the same. For *Sh*=0.1 (Fig. 2a), the mean slope (out of the four N cases) amounts to –0.83 (min of -0.84 and max of -0.80). For *Sh*=0.3 (Fig. 2b), the mean is -1.19 (min of -1.22 and max of -1.16). Surprisingly the segregation rate does not depend of the concentration of fines, even for N= 0.05 corresponding to a low number of fine particles initially scattered on top of the coarse bed, though the evolution is a bit less smooth than for higher N. As could be expected, the segregation rate (given by the absolute slope) increases with the Shields number.

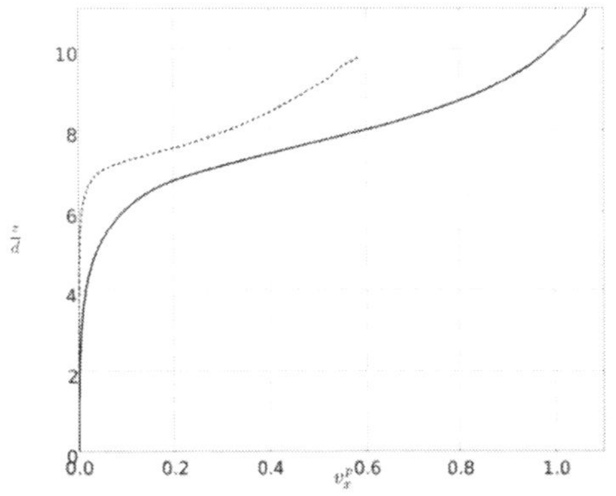

Fig. 3. Depth profiles of the streamwise particle velocity (m/s) for both Shields numbers of 0.1 (dashed) and 0.3 (plain).

Following theories such as Savage and Lun [8] (see also a simplified description in [13]), we will consider herein that the shear rate plays the major role. the segregation rate is taken proportional to the shear rate $\frac{dz_c}{dt} \propto -\dot{\gamma}$. Streamwise coarse-only particle velocity depth profiles appear in Fig. 3 for both Shields numbers. An exponential decrease is observed in the quasi-static part of the bed [11] as evidenced on Fig. 3 and more precisely on Fig. 4 (semi-logarithmic): $v^p{}_x \propto e^{z/a}$. This means that the shear rate $\dot{\gamma} = \frac{dv^p{}_x}{dz}$ also decreases exponentially. As a result, the segregation rate $\frac{dz_c}{dt} \propto -e^{z/a}$. Separating variables and integrating from time t=0 and initial position gives $z_c = -a\log(t/a + b)$ with b an integration constant. When t>>1, $z_c = -a\log(t) + c$.

According to this simple calculation, the fitted slope a appearing on Fig. 2 should be the same as that of the exponential decay of the particle velocity profile. Velocity profiles were fitted over the exponential domain in Fig. 4a (*Sh*=0.1) and Fig. 4b (*Sh*=0.3) yielding mean values of respectively 0.80 (min of 0.76 and max of 0.86) and 1.07 (min of 1.01, max of 1.12). Those values compare very well with the segregation evolution values (Fig. 2). For *Sh*=0.1, the difference of slopes is only 4% while it amounts to 11% for *Sh*=0.3.

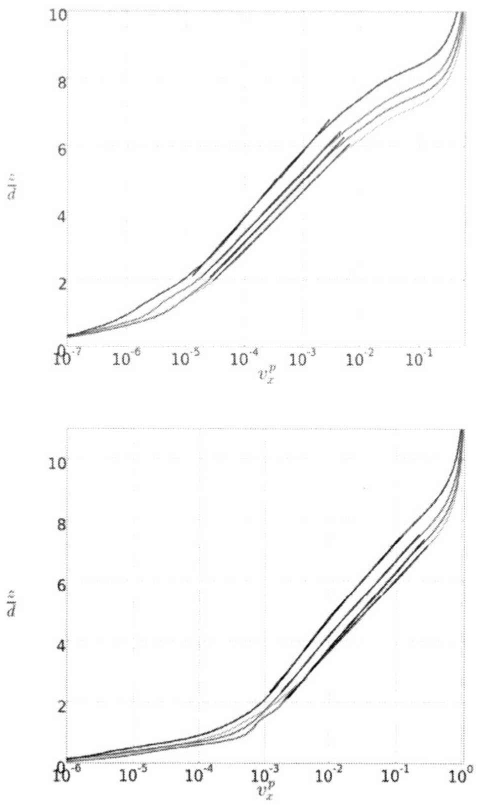

Fig. 4. Depth profiles of the streamwise particle velocity (m/s) with fitting in the exponential domain (see text for fit values) from light yellow (N=0.05 bottom) to dark black (N=2 top). **a**. Shields of 0.1 **b**. Shields of 0.3.

4 Conclusion

After the initial transient period, the elevation of the centre of mass of the infiltrated fine particles is observed to follow the same logarithmic decrease with time, whatever the initial number of fine layers, from a few scattered fine particles up to two full fine layers. This decrease is steeper for a higher Shields number.

The main finding is that this typical behaviour is related at first order to the shear rate exponential depth profile present in the quasi-static region. These results confirm that the shear rate is the main driver of the kinetic sieving-type observed segregation in our bedload experiments.

While this is consistent with established theoretical framework for the largest number of fine particle layers, one would expect a different type of behaviour for isolated particles. Indeed an asymmetry between fine particles surrounded by large particles vs. large particles surrounded by fine particles has been reported [14]. Future work will therefore be devoted to the study of single fine particle behaviour in a similar system.

A better understanding of size segregation poorly known processes will ultimately improve sediment transport and river morphology modelling.

This research was supported by Irstea (formerly Cemagref), the labex OSUG@2020, the French Institut National des Sciences de l'Univers programs EC2CO-BIOHEFECT and EC2CO-LEFE MODSED, and by ANR (the French national research agency) project ANR-16-CE01-0005.

References

1. P. Frey and M. Church, Science **325**, 1509 (2009).
2. P. Frey and M. Church, Earth Surface Processes and Landforms **36**, 58 (2011).
3. V. Hergault, P. Frey, F. Métivier, C. Barat, C. Ducottet, T. Böhm, and C. Ancey, Experiments in Fluids **49**, 1095 (2010).
4. A. R. Dudill, P. Frey, and M. Church, Earth Surface Processes and Landforms, DOI: 10.1002/esp.4080 (in press).
5. V. Bacchi, A. Recking, N. Eckert, P. Frey, G. Piton, and M. Naaim, Earth Surface Processes and Landforms **39**, 1075 (2014).
6. A. Recking, P. Frey, A. Paquier, and P. Belleudy, Journal of Geophysical Research-Earth Surface **114**, F03010 (2009).
7. R. Maurin, J. Chauchat, B. Chareyre, and P. Frey, Phys. Fluids **27**, 113302 (2015).
8. S. B. Savage and C. K. K. Lun, J. Fluid Mech. **189**, 311 (1988).
9. T. Revil-Baudard and J. Chauchat, J. Geophys. Res.-Oceans **118**, 619 (2013).
10. Šmilauer et al., *Yade Documentation 2nd ed. The Yade Project. DOI 10.5281/zenodo.34073 (http://yade-dem.org/doc/)*, 2015).
11. P. Frey, Earth Surface Processes and Landforms **39**, 646 (2014).
12. R. Maurin, J. Chauchat, and P. Frey, J. Fluid Mech. **804**, 490 (2016).
13. B. Andreotti, Y. Forterre, and O. Pouliquen, *Granular Media: Between Fluid and Solid* (Cambridge University Press, 2013).
14. K. van der Vaart, P. Gajjar, G. Epely-Chauvin, N. Andreini, J. Gray, and C. Ancey, Phys. Rev. Lett. **114**, 10.1103 (2015).

Clay particles as binder for earth buildings materials: a fresh look into rheology of dense clay suspensions

Gnanli Landrou[1,*], *Coralie* Brumaud[1], and *Guillaume* Habert[1]

[1]Chair of Sustainable Construction, Civil Engineering Department, ETH Zürich-8093 Zürich, Switzerland

Abstract. In the ceramic industry and in many sectors, clay minerals are widely used. In earthen construction technique, clay plays a crucial role in the processing. The purpose of this research is to understand and modify the clay properties in earth material to propose an innovative strategy to develop a castable earth-based material. To do so, we focused on the modification of clay properties at fresh state with inorganic additives. As the rheological behaviour of clays is controlled by their surface charge, the addition of phosphate anion allows discussing deep the rheology of concentrated clay suspensions. We highlighted the thixotropic and shear thickening behaviour of a dispersed kaolinite clay suspensions. Indeed, by adding sodium hexametaphosphate the workability of clay paste increases and the behaviour is stable during time after a certain shear is applied. Moreover, we stress that the aging and the shift in critical strain in clay system are due to the re-arrangement of clay suspension and a decrease of deformation during time. The understanding of both effect: thixotropy and aging are crucial for better processing of clay-based material and for self-compacting clay concrete. Yet, studies need to pursue to better understand the mechanism.

1 Introduction

In the ceramic industry and in many sectors clay minerals are widely used and of importance to the processing [1]. In earthen construction, clay acts like binder for earth as it displays colloidal interactions and adhesion forces between fine sand and grains [2]. While extensive researches exist on clay dispersion [3, 4]; few researches applied the knowledge on earth building material. However due to the current environmental challenges buildings with earth regains interest as it is fully recyclable, is adaptable in most climates and societies; and it has the lowest environmental impact with respect to cement [5, 6]. Howbeit, its development is hindered by the time and cost required for current earth construction techniques [7].

The objective of this study is therefore to transform earthen architecture by providing a material that is as easy and as cheap to use as current concrete products thanks to novel technologies borrowed to science and technology of cement and concrete. This mean investigates the influence of additive as sodium hexametaphosphate on clay in order to improve the workability of the earth-based material

To do so, a careful control of the rheology of the clays requires a better understanding of colloidal interactions between particles and knowledge transfer from the fundamental physics of grain and colloidal interactions to the civil engineering [8, 9].

As unique binder in earth, clay; negatively charged particle; can have its surface interaction changed with the help of dispersants coming from cement and ceramic industries. Among dispersants commonly used in ceramic processing, sodium phosphate is known to guarantee an efficient dispersion of natural clay materials [10, 11]. However, the rheological behavior access so far for dispersed clay suspensions are difficult to transpose in earth.

Even though several studies have been done on the influence of sodium hexametaphosphate (NaHMP) on concentrated clay suspensions, the stability and control of behaviour of concentrated clays suspensions when additives are introduced [11-15], few studies explain the effect of time and the thixotropic effect due to the percentage of dispersant added with respect to viscoelastic properties of clays.

In this paper, we focus on the modification of clay properties with inorganic additive in order to deflocculate the clay during the casting process. While doing so, we highlight the ability of sodium phosphate to fluidify the earth material by reducing the yield stress. Moreover, we stress the thixotropic behaviour of dispersed clay through the adsorption of phosphate on the positive edge and the stability of the suspension during time. Furthermore, we show that during time the re-organization of clay particles can allow a phase transition from dispersed particles to agglomerate while keep the reversibility of the system.

2 Experimental observations

The clay used is a kaolin FP80 powder supplied from Alberto Luisoni AG, Switzerland. The characteristics of kaolin FP80 is reported in table 1. The XRD measurement reveals that more than 87% of kaolin FP80

* Corresponding author: glandrou@ethz.ch

 DOI: 10.1051/epjconf/201714013010

is kaolinite clay; specifically, non-swelling clay minerals.

Table 1: The physical characteristics of kaolin FP80.

specific surface area SSA_{BET} ($m^2.g^{-1}$)	mean diameter (μm)	density ($g.cm^{-3}$)
9.10	8.69	2.62

All rheological measurements were performed on dispersed clay pastes. The ratio water to clay was kept constant (W/C=0.69) and the amount of dispersant sodium hexametaphosphate (NaHMP) was kept at 0.2wt% with respect to the amount of solid.

2.1 Thixotropy and shear thickening

To study the influence NaHMP and to investigate it stability during time on concentrated kaolinite paste, we conducted at room temperature (23 ± 0.1°C) rheological measurements. The sequence used for yield stress measurement comprise the following steps: first, we applied an increasing shear rate ramp from 0.1 s^{-1} to 300 s^{-1} (with a logarithmic distribution of shear rate) for 300s followed by a decreasing shear ramp from 300 s^{-1} to 0.1 s^{-1} for 300s. The yield stress of the reference mixture (W/C=0.69) is 1000Pa and an addition of 0.2wt% NaHMP, reduce considerably the yield stress.

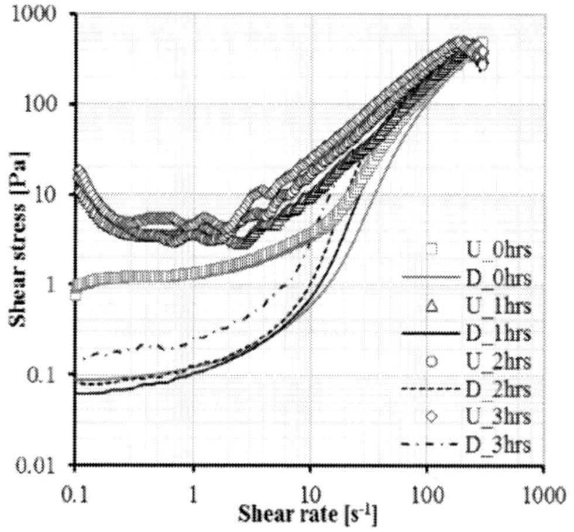

Figure 1: Flow curve of kaolinite (W/C=0.69) + 0.2% NaHMP, evolution of shear stress as a function of shear rate during time. U represents ramp up and D ramp down.

Figure 1 illustrates the evolution of shear stress versus shear rate measured every one hour for a suspension of kaolinite dispersed with 0.2wt% NaHMP. The yield stress determined from ramp down fluctuate in the same range which means that the structural change is reversible when a sufficient shear rate is applied. Besides the static stress determined from the ramp up after one hour resting time is 10Pa. The increase of static stress measured at low shear rate (0.1s^{-1}) from t_0 to t_1 suggest a reorganization between clay particles and formation of new interaction forces during time. Additionally, we observed that for constant rest time, the same static stress is measured. The results confirm the reversible behaviour mentioned above and as the process is repeatable. We suggested that at high volume fraction and when dispersed, clay suspensions display thixotropic behaviour. It means clay particles, at rest, have time to re-organize and re-structure due to Brownian motion. Moreover, zeta potential measurement showed a partial coverage (0.2wt% NaHMP) of clay edge charge. Then the presence of residual positive charge induces weak Van der Walls attractive forces. This behaviour typical of thixotropic fluid is discussed for others colloidal system [16, 17]. The dispersion mentioned is due to the adsorption of phosphate ions at the edge of the clay which result on the increase of negative surface charge of clay particles [14] and the increase of electrostatic repulsion forces.

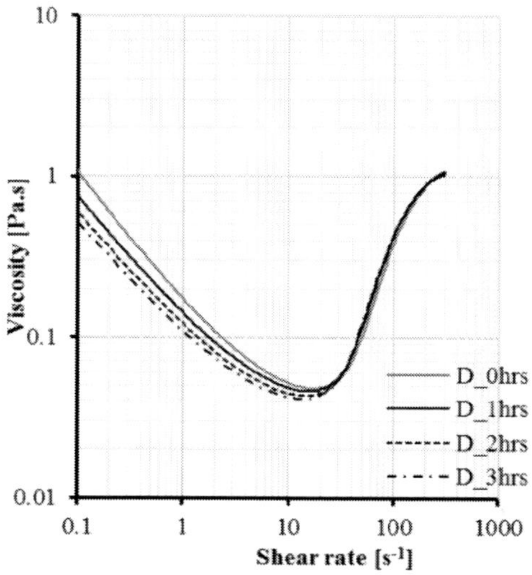

Figure 2: Flow curve of kaolinite (W/C=0.69) + 0.2% NaHMP, evolution of viscosity as a function of shear rate during time. D represents the ramp down.

Furthermore, we observe in figure 2 that dense and dispersed clay suspension displays a shear thickening behaviour as at shear rate above 10s^{-1} the viscosity increases. Without dispersant, shear thinning behaviour is observed. We assume that the surface chemistry of clay, the orientation of clay particles affect the transition from shear thinning to shear thickening. As the shear thickening behaviour is inexistence without dispersant, we concluded that the shear thickening is hidden by a yield stress and it can be recovered with appearance of a threshold yield stress as shown in [18]. Besides, previous studies showed that this phenomenon depends on the shape of particles and the volume fraction [19], so further experiments are ongoing to understand the driven mechanism and the influence of volume fraction and shape.

Yet we suggest in our case that the onset of shear thickening is linked to change in the microstructure of the suspension. At low shear, the clay particles are orientated parallel into layer which facilitate the flow of colloid then at high shear, the suspensions organized into clusters jam due to the percolation of clusters This suggest a hydrodynamic interaction between dispersed clay clusters formed at high shear. Similar observation is discussed in [20] as the authors stress on the effect of surface chemistry on the shear thickening.

2.2 Aging: phase transition

To understand the aging process of dispersed kaolinite clay suspension, we carried out vane test measurement in order to identify the deformation range and transition from liquid dispersed to liquid agglomerated.

Yet, to study the aging behaviour, to determine the interaction range, to classify the microstructure change during time and cover flow onset no matter the critical strain, a compromise was however to be reached between the duration of the test and its accuracy . It was not possible to apply a constant shear rate to measure all the strain and so we developed the following protocol. The sequence starts with an initial pre-shear at $100s^{-1}$ during 30s to ensure that all the samples were in the same state of stress and strain, followed by different waiting time (0, 0.5, 2, 6 and 12hrs). Then a constant shear rate of 10^{-3} s^{-1} was applied during 50s followed by a logarithmic increase of shear rate from 10^{-3} to $0.5*10^{-3}$ s^{-1} during 10s. A constant shear rate of 0.005 s^{-1} was finally applied during 16min to determine late the flow onset and late deformation.

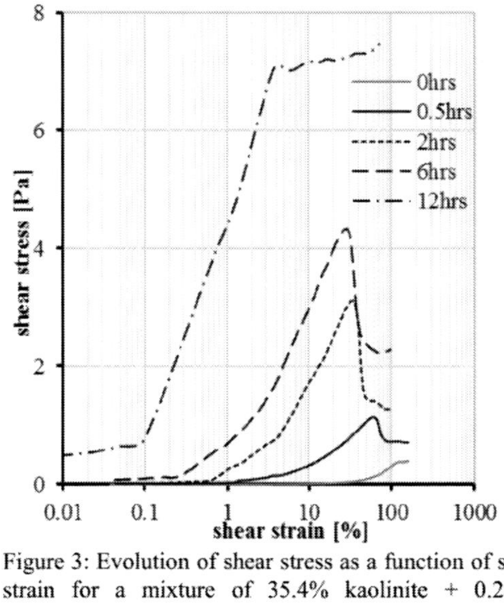

Figure 3: Evolution of shear stress as a function of shear strain for a mixture of 35.4% kaolinite + 0.2wt% NaHMP. Vane test measurements carried out at different waiting time.

In figure 3, we reported the evolution of shear stress as a function of strain for a mixtures of kaolinite paste (W/C=0.69) dispersed with 0.2wt% of NaHMP. The measurements are performed at different resting time. We can observe that at t_0, right after mixing, the strain is above 100% and the static stress is lower than 1Pa meaning that the material is fluid, deformable and the distance of interaction is considered as long range. It is suggested that the clay interaction forces are controlled by face (-) / face (-) and face (-) / edge (-). Moreover, during time the critical strain decreases while the static stress increases so we can propose that the aging allow the agglomeration of clay particles and the rigidification of colloidal network. The shift in strain and the increase of static stress confirm the reorganization of clay particles driving by Brownian motion and the existence of few positive charges. This enables the agglomeration. Similar behaviour are observed in colloidal gel and mentioned in [21, 22]. The authors describe the phenomenon as glass-glass transition during aging of colloidal clay. The transition from long range interaction due to the presence of dispersant to short range colloidal interaction between clay particles but with lower stress in comparison to reference paste is still unclear however, we should keep in mind that the agglomeration is reversible and the kinetics of agglomeration is dependent of the percentage of dispersant introduce.

3 Discussions

In the present studies, the samples are covered during the experiments to avoid effect of evaporation. The results obtained from Fig 1 and 2 show that the NaHMP activity is stable during time which highlight the robustness and efficiency of the dispersant. Moreover, we can observe that when deflocculate, the clay particles are non-organized and are randomly assembled with weak attractive forces which explain the low yield stress and higher critical strain, above 100% deformation at t_0 Fig 3. However, during time the increase of static stress determine at flow onset and the decrease of shear strain suggests a change in the microstructure. Indeed, the samples microstructure become more structured and ordered Fig 3.

These findings point out that there are microstructural change or reaction occurring during time when the reference paste is deflocculated. This observation appears odd as the sample still fluid after 6 hours as shown in Fig 1. The increase of shear stress is then associated to aging and new interaction between deflocculate clay platelets face (-) / (-) face, or face (-) / (-) edge) and face (-) / (+) edge.

Yet, the shift in strain suggests the formation of soft clusters or agglomerate easily breakable at higher shear stress. In fact, as reported in Fig 1, we observed that the waiting time increases the shear stress from 0.1Pa to 10Pa and the process is reversible. The hypothesis of formation of cluster and agglomerate seems to support the shear thickening behaviour but more analysis need to be done to confirm the origin of shear thickening.

4 Conclusion

We have conducted extensive experimental studies in order to get insight on the rheological behaviour of dispersed kaolinite clay and the stability of NaHMP. The results reveals that NaHMP is robust, efficient and stable dispersant during time. More we highlight the thixotropic behaviour and rigidification of the system with time due to the different type of interaction in clay (face-face or face-edge).

Besides, the findings reveal that the aging and the re-arrangement of clays particles are responsible to the net increase of inter-particles forces with a decrease of critical strain during. However, further works are undergoing to understand the shear thickening behaviour.

The authors would like to acknowledge the Swiss National Science Foundation (SNSF) for it financial support.

References

1. F. Bergaya,G. Lagaly, *Handbook of clay science* **5**. (2013)
2. D. Gelard, L. Fontaine, S. Maximilien, C. Olagnon, J. Laurent, H. Houben, H. Van Damme, *Proceedings of the International Symposium on Earthen Structures. IIS Bangalore*, (2007)
3. J. L. Amorós, V. Beltrán, V. Sanz, J. C. Jarque, *Appl. Clay Sci.*, **49**, 33 (2010)
4. P.-I. Au,Y.-K. Leong, *Colloids Surf. A: Physicochem. Eng. Asp.*, **436**, 530 (2013)
5. J.-C. Morel, A. Pkla, P. Walker, *Constr. Build. Mater.*, **21**, 303 (2007)
6. A. Shukla, G. N. Tiwari, M. S. Sodha, *Renew. Energy.*, **34**, 755 (2009)
7. R. F. Anger, Laetitia, *Bâtir en terre: du grain de sable à l'architecture* (2009)
8. K. Kovler,N. Roussel, *Cem. Concr. Res.*, **41**, 775 (2011)
9. J. Yammine, M. Chaouche, M. Guerinet, M. Moranville, N. Roussel, *Cem. Concr. Res.*, **38**, 890 (2008)
10. F. Andreola, E. Castellini, J. M. F. Ferreira, S. Olhero, M. Romagnoli, *Appl. Clay Sci.*, **31**, 56 (2006)
11. F. Andreola, E. Castellini, G. Lusvardi, L. Menabue, M. Romagnoli, *Appl. Clay Sci.*, **36**, 271 (2007)
12. E. Castellini, R. Andreoli, G. Malavasi, A. Pedone, *Colloids Surf. A: Physicochem. Eng. Asp.*, **329**, 31 (2008)
13. E. Castellini, C. Berthold, D. Malferrari, F. Bernini, *Appl. Clay Sci.*, **83–84**, 162 (2013)
14. G. Landrou, C. Brumaud, F. Winnefeld, R. Flatt, G. Habert, *Materials*, **9**, 330 (2016)
15. A. Perrot, D. Rangeard, A. Levigneur, *Mater. Struct.*, 1 (2016)
16. J. Mewis,N. J. Wagner, *Adv. Colloid Interface Sci.*, **147–148**, 214 (2009)
17. N. Roussel, G. Ovarlez, S. Garrault, C. Brumaud, *Cem. Concr. Res.*, **42**, 148 (2012)
18. E. Brown, N. A. Forman, C. S. Orellana, H. Zhang, B. W. Maynor, D. E. Betts, J. M. DeSimone, H. M. Jaeger, *Nat Mater*, **9**, 220 (2010)
19. H. A. Barnes, *J. Rheol.*, **33**, 329 (1989)
20. W. Yang, Y. Wu, X. Pei, F. Zhou, Q. Xue, *Langmuir*, (2017)
21. R. Angelini, E. Zaccarelli, F. A. de Melo Marques, M. Sztucki, A. Fluerasu, G. Ruocco, B. Ruzicka, *Nat. Commun.*, **5**, (2014)
22. E. Zaccarelli, *Journal of Physics: Condensed Matter*, **19**, 323101 (2007)

Electrical transport in granular metals

Chongpu Zhai[1], *Dorian* Hanaor[2], *Gwénaëlle* Proust[1] and *Yixiang* Gan[1,*]

[1]School of Civil Engineering, The University of Sydney, Sydney, Australia
[2]Institute for Material Science and Technologies, Technische Universität Berlin, Germany

Abstract. In this paper, we studied the frequency-dependent AC conductance of randomly packed stainless steel spheres by means of impedance spectroscopy. Two types of power-law behaviour have been observed: (a) at low frequencies, the dependence of the measured impedance on the applied load; (b) at high frequencies, the dependence of the impedance modulus on frequency. Under different loading conditions, the imaginary parts of the measured conductances exhibit respective peaks at critical frequencies, corresponding to the onset of conductance dispersion. Using these critical points as scaling parameters to normalize the measured conductance, results in the spectra from different loading levels collapsing onto a single master curve. Both the electron tunnelling and capacitive paths among particles contribute to the conduction in granular metallic media, resulting in well-characterized universal behaviour.

1 Introduction

The electrical conduction of granular materials is important in many engineering applications, such as powder metallurgy [1], thermoelectric systems [2], tactile sensing devices [3], and soil moisture content measurements [4]. Generally, these systems comprise assemblies of large numbers of conductive and/or insulating particles in contact. In order to evaluate the overall electrical properties of such systems, they can be considered as complex networks of randomly positioned electrical components, which can be regarded as resistor-capacitor (RC) networks [5-7]. Electrical responses of such RC networks subjected to alternating current (AC) have been investigated using experimental [8-10], numerical [11,12], and theoretical approaches [7,13,14]. These studies have shown that electrical characteristics, including conductivity, impedance, and permittivity exhibit power-law type dependences on the AC frequency, ω, known as the 'universal dielectric response' (UDR) [15], within a certain range of frequencies. For example, the conductivity can be expressed as:

$$\sigma'(\omega) = \sigma_{dc} + A\omega^n, \quad (1)$$

where $\sigma_{dc} = \lim_{\omega\to 0} \sigma'(\omega)$ is the direct current (DC) conductivity of the material, A is a constant, and n is the fractional exponent which is typically $0 \le n \le 1$. Both A and n are temperature dependent. Jonscher's UDR successfully describes most disordered dielectric systems when AC conductivity and permittivity show a dispersive region beyond a critical frequency. Moreover, it has been shown in a range of systems, from semiconductors to polymers, that the conductivity spectra collapse into a single master curve, suggesting the validity of the time-temperature superposition principle [15-17], which can be expressed by the following scaling law

$$\sigma'(\omega)/\sigma_{dc} = F(\omega/\omega^*), \quad (2)$$

where F is a temperature-independent scaling function and ω^* is the characteristic frequency corresponding to the onset of the conductivity dispersion, which in turn is a scaling parameter [15-17]. Temperature affects the density of charge carriers in ionic conductors or the electron hopping probabilities in granular metals [18]. This change of conduction properties between grains can actually be achieved also by varying the mechanical loading, which influences the inter-particle distance and contact area between particles, thus affecting the conduction behaviour [19-21]. These systems have been simulated using random RC networks and the electrical responses have been found to follow the universal power law [22-25]. This supports that the emergent scaling phenomenon develops from the network nature of the system [15,22] without referring to element properties.

The aim of this paper is to examine the frequency-dependent collective electrical responses of packed stainless steel grains, considered as conductor-insulator elements. We focus on the evolution of established networks under varying compressive loads. This study demonstrates the importance of understanding, at the element level, the origins of the intrinsic electrical characteristics of multi-phase materials or components which can be represented by a RC network.

2 RC network formed by grain contacts

In this paper, randomly packed conductive spheres are considered. The electrical characteristics of this system can be simplified to an electrical network formed

* Corresponding author: yixiang.gan@sydney.edu.au

by a large number of RC components in random positions. A single electrical contact between two grains can be represented by a resistor and/or a capacitor, depending on the inter-particle force and distance. The electrical conduction through isolated spheres is capacitive as the DC current could not tunnel through the interval. As for the slightly contacting spheres, a thin surface oxide layer acting as a resistive film with typical thickness ranging from 1 to 100 nm [26,27], tends to form and cover the metallic material. Consequently, these dielectric layers cause contacts to be non-conductive, under conditions of low loads, effectively acting as capacitors.

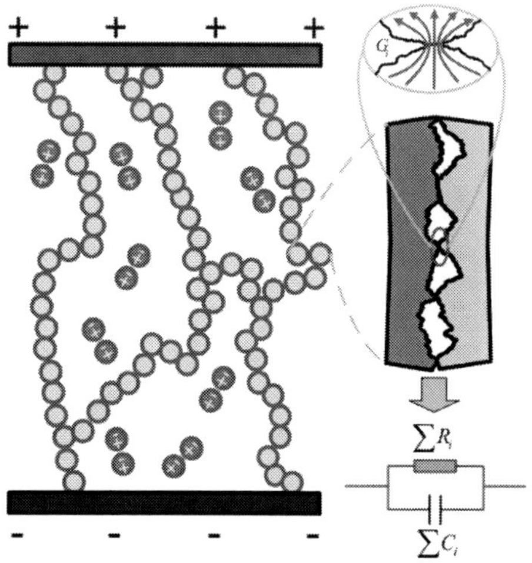

Fig. 1. Schematic of the electrical conduction through packed spheres. Typical percolating chains are indicated by connecting grey particles. Typical capacitors are shown by pairs of close but non-contacting spheres.

For contacting spheres, surface roughness plays an important role in the conduction properties [28]. As shown in Fig. 1, the contacting interfaces between grains tend to exhibit complex geometries and structures governing physical properties and interfacial phenomena [19] and giving rise to constriction resistance [29]. The constriction resistance due to the convergence and divergence of current flow at a single contact, results in a high resistance state. When two balls are compressed together with sufficient pressure, surface asperities deform and even penetrate the oxide layer thus forming small metal-to-metal contact patches. This leads to the low conduction of electrical current with electrons being transported in different ways depending on the size of contacting spots, i.e., Sharvin, Wexler, and Maxwell mechanisms [21,25,29,30]. When the radius of the contacting asperities, is smaller than the average electron mean free path, electrons can only travel ballistically across the micro-contacts [20]. Additionally, tunnelling through the oxide layer contributes to the conduction at this stage [21]. Under increasing compression, the numerous independent contacting spots increase in size, merging into larger contact areas [19]. When the effective radius of a contacting asperity becomes larger than the mean free path of electrons, Maxwell contact [31] will be the dominant transport mechanism, resulting in a relative low resistance. With further increasing inter-particle force, large zones of metal-to-metal contact appear, likely establishing metal chains along several contacting grains, presenting inductive properties [25]. Noticeably, the inductive phase may not appear for metals with thick oxide layers even with high levels of compression [13]. On the other hand, the non-contacting zones at the interface act effectively as capacitors in parallel, which is also true for contacts without oxide layers. The contacting spheres can be then regarded as a resistor and a capacitor in parallel.

3 Experimental setup

In this paper, experiments were performed with commercial stainless steel ball bearings (Material: AISI 304; Precision Grade: G200) with diameters of 2 mm. Spheres of 2400 g were randomly packed in a nonconductive cylinder (Ultra High Purity Al_2O_3, Taylor Ceramic Engineering, Australia) of 85 mm in height, H, and with an inner diameter, D, of 98 mm. The packed spheres were positioned between two metallic plates acting as electrodes. The impedance and phase angle were simultaneously measured using an impedance analyzer (Agilent, 4294A), with a measurement frequency ranging from 40 Hz to 20 MHz. The electrical responses were monitored under different loading levels, with each subsequent loading level being 4/3 times that of the previous one, reaching a maximum load of 250 N (roughly 10 times of the self-weight of sphere assembly). Before each experimental sequence, the container was properly shaken to produce fresh contacts. The amplitude of the voltage applied to the systems was fixed to a low level with the peak-to-peak voltage value being 200 mV, in order to avoid voltage-induced surface modification [29]. For each given load, a one-minute equilibration time was given to minimize the variation of the measured impedance due to the creep deformation. It should be noted that the effects of the necessary cables and connections (which are in series with the sample under testing) have been subtracted through calibration processes.

4 Results and discussion

The typical measured impedance modulus of the studied granular system under different levels of compression is presented as a function of frequency in Fig. 2, with the phase angle shown in the insert. A larger load results in higher conductance and larger phase angle across the entire studied frequency range. The negative values of phase angle obtained indicate the capacitive properties. For relatively higher frequencies, a linear dependence in a log-log plot emerges over two orders of magnitude, which corresponds to power law dependence. For a certain range of frequencies, the absolute value of the exponent of the observed power-law function seemingly decreases with increasing pressure. We will come back to this point in the following discussion.

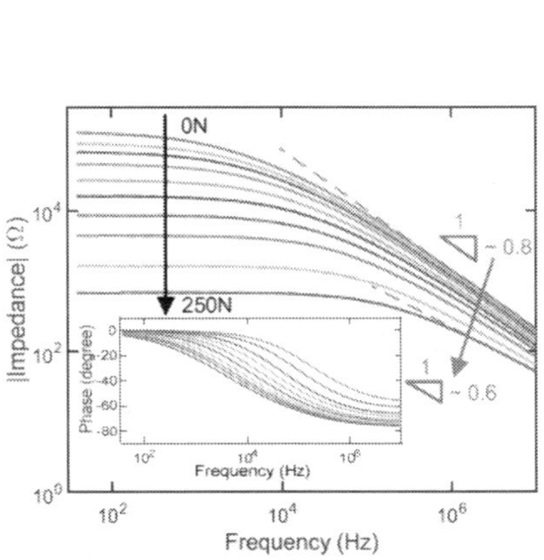

Fig. 2. Typical dependence of the measured impedance with respect to the frequency for the studied spherical grains. The phase angle variation with frequency is shown in the insert. Each individual curve corresponds to a given compression level with the arrow showing the increasing trend.

The relationships shown in Fig. 2 are attributed to the intricate electrical network formed by the elements, regardless of the physical or chemical properties of the material involved [5,7,8]. The frequency dependence of the conductance can be divided into three regions. At low frequencies, the network conductance is dominated by random percolation paths of resistors across the network. At intermediate frequencies, where the conductance of the capacitors and resistors become comparable, both types of components contribute to the network conductivity. In this frequency range, conductance rises as a power function of frequency. Importantly, the power law exponent is equal to the capacitive component proportion [11,22,32]. At high frequencies, the conductance of the capacitors would far exceed those of the resistors and thus these components can be regarded as short circuits. Consequently, the network conductance is determined by the random distribution of resistors, effectively bound together by shorting capacitors, reaching saturated values. However, the frequencies at this stage cannot be reached within the present methodology.

It should be noted that various mechanisms of surface structure evolution have been observed during electrical conduction through rough interfaces, including dielectric breakdown of oxide layers, localised current-induced welding, chemical disorder arising with random composition and oxidation processes in corrosive environments, and surface diffusion [26, 29]. These phenomena are, however, outside the scope of the present work.

The dependence of resistance on applied pressure, measured at low frequencies is presented in Fig. 3. Here the resistance, the real part of measured impedance, was averaged over numerous tests with error bars indicating the measurement range obtained at the selected measurement frequency (40 Hz). The non-dimensional electrical resistance of the sphere assembly decreases by several orders of magnitude under increasing stress. A larger applied load results in larger inter-particle force, thus a better conductance through the contact, therefore, the assembly. The dimensionless resistance and pressure can be used to estimate the contact properties between two grains.

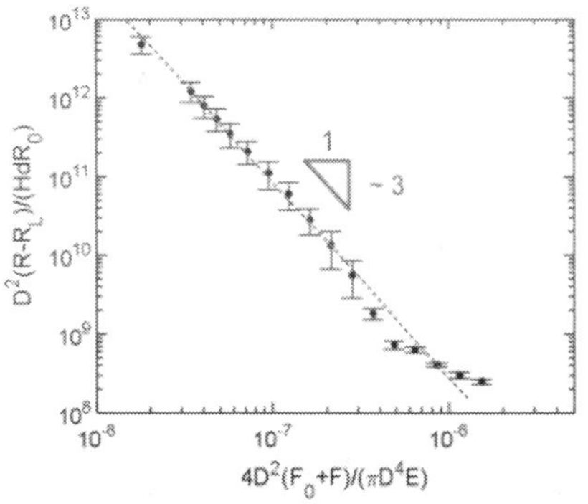

Fig. 3. Non-dimensional stress-dependent electrical resistance of two packed bed types under various loading conditions at a frequency of 40 Hz. Here we assume the spheres are packed in layers. The contact resistance between spheres in each layer are considered to be in parallel and layers in series. E is the Young's Modulus of the tested material (203 GPa), D is the inner diameter of the container, d the sphere diameter, R_L is the total resistance of the cables and connectors, R the measured resistance, R_0 the theoretical resistance of the bulk material for an identical volume as that of the packed bed, F_0 is the sum of half self-weight and the weight of the top cap and F is the applied compressive force.

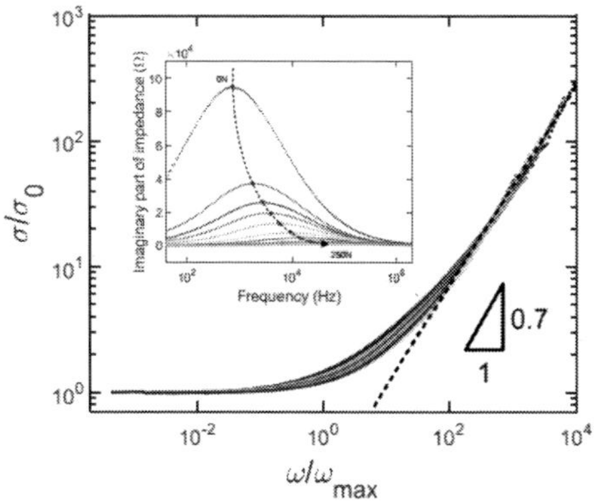

Fig. 4. Scaling plot of σ/σ_0 versus ω/ω_{max}, with σ being the module of conductance, σ_0 the conductance measured at a frequency of 40 Hz, ω_{max} the critical frequency corresponding to the onset of the conductivity dispersion. The insert shows the imaginary part of the measured impedance as a function of frequency with the red dots marking the peaks at critical frequencies. Each individual curve corresponds to a given compression with the arrow showing the increasing trend.

The electrical impedance across the considered frequency range has been analysed by plotting the imaginary part of impedance (Z_i) versus frequency at different loading levels, as is shown in the insert of Fig. 4. This provides information on the dielectric processes taking place in the studied media. The critical frequency at the peak shifts to higher frequency with increasing load, indicating the existence of load-dependent relaxation process in the system. The critical frequencies along with the corresponding values of Z_i are further included, as scaling parameters, in the conductance spectroscopy. The collapse of all the curves at different loading levels onto a single master curve has been observed, exhibiting an evident universal behaviour. Moreover, the exponent extracted from the dispersive region of the master curve has a value of around 0.7. The explanation for this specific value needs to be traced back to the sphere contacts in the studied systems. According to Equation (1), the exponent can be directly related to the fraction of capacitive phases [15,22,32,33]. For our loading range, each pair of contacting spheres can be considered as a resistor and a capacitor in parallel, configuring the basic structure of a RC network, with the capacitor percentage being around 50%. Additionally, the non-contacting and slightly touching sphere pairs, acting effectively as capacitors, add extra capacitive phases (~ 20 %) into the RC network.

5 Conclusion

The scaling behaviour of AC electrical responses of granular stainless steel has been studied under increasing load. Both the resistive and capacitive paths among particles contribute to the conduction in granular metals, resulting in well-characterized universal behaviour. The real part of the measured impedance at low frequency follows a power law relationship with respect to the applied load. For various loading conditions, the frequency-dependent conductance is found to obey the emergent scaling power law, with a universal exponent value of about 0.7, which tends to be unaffected by the applied compressive load.

References

1. R. M. German, Solar-Terrestrial Physics (Solnechno-zemnaya fizika), 568 (1996).
2. L. E. Bell, Science **321**, 1457 (2008).
3. S. Stassi, V. Cauda, G. Canavese, and C. F. Pirri, Sensors **14**, 5296 (2014).
4. S. R. Evett, J. A. Tolk, and T. A. Howell, Vadose Zone Journal **4**, 1020 (2005).
5. K. Murphy, G. Hunt, and D. P. Almond, Philosophical Magazine **86**, 3325 (2006).
6. P. Watts, W. Hsu, V. Kotzeva, and G. Chen, Chemical physics letters **366**, 42 (2002).
7. J. Clerc, G. Giraud, J. Laugier, and J. Luck, Advances in Physics **39**, 191 (1990).
8. M. Creyssels, E. Falcon, and B. Castaing, Physical Review B **77**, 075135 (2008).
9. B. Castaing and C. Laroche, Europhysics Letters **65**, 186 (2004).
10. S. Dorbolo, M. Ausloos, and N. Vandewalle, Physical Review E **67**, 040302 (2003).
11. R. Bouamrane and D. P. Almond, Journal of Physics: Condensed Matter **15**, 4089 (2003).
12. R. K. H. Galvão, S. Hadjiloucas, K. H. Kienitz, H. M. Paiva, and R. J. M. Afonso, Circuits and Systems I: Regular Papers, IEEE Transactions on **60**, 624 (2013).
13. C. Grimaldi, P. Ryser, and S. Strässler, Journal of applied physics **92**, 1981 (2002).
14. R. Kawakami Harrop Galvao, K. H. Kienitz, S. Hadjiloucas, G. C. Walker, J. W. Bowen, S. Figueredo Carreiro Soares, and U. Araujo, Dielectrics and Electrical Insulation, IEEE Transactions on **20**, 995 (2013).
15. A. K. Jonscher, nature **267**, 673 (1977).
16. B. Roling, Solid State Ionics **105**, 185 (1998).
17. J. C. Dyre, P. Maass, B. Roling, and D. L. Sidebottom, Reports on Progress in Physics **72**, 046501 (2009).
18. H. Bakkali, M. Dominguez, X. Batlle, and A. Labarta, Scientific Reports **6** (2016).
19. C. Zhai, Y. Gan, D. Hanaor, G. Proust, and D. Retraint, Experimental Mechanics **56**, 359 (2016).
20. C. Zhai, D. Hanaor, G. Proust, L. Brassart, and Y. Gan, Extreme Mechanics Letters (2016).
21. L. Kogut and K. Komvopoulos, Journal of applied physics **95**, 576 (2004).
22. D. Almond, C. Budd, M. Freitag, G. Hunt, N. McCullen, and N. Smith, Physica A: Statistical Mechanics and its Applications **392**, 1004 (2013).
23. D. P. Almond, C. J. Budd, and N. J. McCullen, in *Approximation Algorithms for Complex Systems* (Springer, 2011), pp. 3.
24. D. P. Almond and C. Bowen, Physical review letters **92**, 157601 (2004).
25. C. Zhai, D. Hanaor, G. Proust, and Y. Gan, in *2016 IEEE 62th Holm Conference on Electrical Contacts* (IEEE, 2016), pp. 9.
26. G. Proust, D. Retraint, M. Chemkhi, A. Roos, and C. Demangel, Microscopy and Microanalysis **21**, 919 (2015).
27. E. Kikuti, R. Conrrado, N. Bocchi, S. R. Biaggio, and R. C. Rocha-Filho, Journal of the Brazilian Chemical Society **15**, 472 (2004).
28. J. Greenwood, British Journal of Applied Physics **17**, 1621 (1966).
29. C. Zhai, D. Hanaor, G. Proust, and Y. Gan, Journal of Engineering Mechanics, B4015001 (2015).
30. A. Mikrajuddin, F. G. Shi, H. Kim, and K. Okuyama, Materials Science in Semiconductor Processing **2**, 321 (1999).
31. R. L. Jackson, E. R. Crandall, and M. Bozack, in *2012 IEEE 58th Holm Conference on Electrical Contacts (Holm),* (IEEE, 2012), pp. 1.
32. D. P. Almond, C. Bowen, and D. Rees, Journal of Physics D: Applied Physics **39**, 1295 (2006).
33. K. Falconer, *Fractal geometry: mathematical foundations and applications* (John Wiley & Sons, 2013).

How granular vortices can help understanding rheological and mixing properties of dense granular flows

Pierre Rognon[1,⋆], *Prashidha* Kharel[1,⋆⋆], *Thomas* Miller[1,⋆⋆⋆], and *Itai* Einav[1,2,⋆⋆⋆⋆]

[1] *Particles and Grains Laboratory, School of Civil Engineering, The University of Sydney, Sydney, New South Wales 2006, Australia.*
[2] *Department of Civil, Environmental & Geomatic Engineering, University College London, London WC1E 6BT, UK.*

Abstract. Dense granular flows exhibit fascinating kinematic patterns characterised by strong fluctuations in grain velocities. In this paper, we analyse these fluctuations and discuss their possible role on macroscopic properties such as effective viscosity, non-locality and shear-induced diffusion. The analysis is based on 2D experimental granular flows performed with the stadium shear device and DEM simulations. We first show that, when subjected to shear, grains self-organised into clusters rotating like rigid bodies. The average size of these so-called *granular vortices* is found to increase and diverge for lower inertial numbers, when flows decelerate and stop. We then discuss how such a microstructural entity and its associated internal length scale, possibly much larger than a grain, may be used to explain two important properties of dense granular flows: (i) the existence of shear-induced diffusion of grains characterised by a shear-rate independent diffusivity and (ii) the development of boundary layers near walls, where the viscosity is seemingly lower than the viscosity far from walls.

1 Introduction

Dense granular flows involved in geophysical processes and industrial applications are typically comprised of a considerable number of grains. Often, a continuum description of their rheological behaviour and mixing properties is required, as predicting the dynamics of every individual grains would exceed the best computational capabilities. However, continuum-based predictions critically rely on effective parameters such as effective viscosity and effective diffusivity. It is therefore necessary to establish reliable and accurate models rationalising how these parameters depend on the grain properties and on the flow conditions.

Effective viscosity ν [m^2/s] and diffusivity D [m^2/s] both involve a length scale and a time scale. Considering grains of size d [m] subjected to a shear strain rate $\dot{\gamma}$ [s^{-1}] naturally leads to expressing the following models, referred to as Bagnold's scaling [1–4]:

$$D \propto d^2\dot{\gamma} \quad (1)$$
$$\nu \propto d^2\dot{\gamma} \quad (2)$$

While these scalings are generally valid, there are multiple evidences pointing out that they are not always sufficient to capture the observed rheological and mixing behaviour. Specifically, both effective diffusivity and viscosity were showed to strongly depend on the microstructure of the grains within the flow. One approach developed to model such dependencies consists in introducing a function of the solid fraction ϕ in each of the scalings (1) and (2). However, it was showed that these functions tend to diverge near a maximum packing fraction, which makes them difficult to measure accurately [2, 3]. Moreover, while the solid fraction is an elementary property of the flow microstructure, its signification in terms of internal kinematics and dynamics is limited. As a result, pinpointing the internal mechanisms underpinning both scaling, and thus both effective diffusivity and viscosity remains a great challenge.

An alternative approach developed to model and explain the dependencies of the viscosity and diffusivity in terms of flow microstructure consists in expressing these quantities as a function of a length scale ℓ as:

$$D \propto \ell^2\dot{\gamma} \quad (3)$$
$$\nu \propto \ell^2\dot{\gamma} \quad (4)$$

where ℓ represents the spatial extent of some microstructural organisation of the grains. However, the precise nature of the microstructural process relevant to diffusivity and viscosity remains somewhat elusive.

In this paper we discuss the relevance of considering granular vortices as a microstructural entity to better understand the effective diffusivity and viscosity in dense granular flows. In section 2, we present some observations and measurements of granular vortices in experimental and simulated granular flows. In section 3, we discuss

⋆e-mail: pierre.rognon@sydney.edu.au
⋆⋆e-mail: prashidha.kharel@sydney.edu.au
⋆⋆⋆e-mail: thomas.Miller@smec.com
⋆⋆⋆⋆e-mail: itai.einav@sydney.edu.au

how granular vortices may explain observations of shear rate independent diffusivities. In section 4, we discuss how granular vortices could help rationalising some observed non-local behaviour of the granular viscosity.

2 Shear induced granular vortices

Within dense granular materials subjected to shear, grain trajectory does not follow a laminar streamline. Instead, grains exhibit strong fluctuations of velocities [5–9]. These fluctuations were reported to form spatially correlated patterns in many different flow configurations. Here we focus on one particular quantity characterising these kinematic fields, the local vorticity w. This section presents experimental and numerical results with the aim of highlighting that, within dense granular flows, the local vorticity is highly heterogenous and present some degree of spatial correlation from which vortices can be identified.

2.1 Stadium shear experiments

The Stadium shear device, showed on Figure 1a, was designed to continuously shear an assembly of bi-dimensional grains in a plane shear geometry [10]. Grains are nylon cylinders 10 mm high, of diameters 12, 15, and 20 mm. They are set vertically on a horizontal glass panel, and enclosed into a belt. The belt is set in motion by a motor/sprocket system, subjecting the central zone to a plane shear flow with a prescribe shear rate $\dot{\gamma} = V/H$; a system of constant force springs prescribes a normal stress σ, thereby prescribing the inertial number I:

$$I = \dot{\gamma} t_i = \dot{\gamma} d \sqrt{\frac{\rho}{\sigma}} \tag{5}$$

where ρ is the grain density. Grains can move under the sprocket and recirculate into the central region, enabling us to perform shear deformations as large as desired. Individual grain trajectories are measured during the flow using a particle tracking algorithm.

Figure 1a shows a snapshot of the field of velocity fluctuations of the grains during one experiment. In some locations, the velocity fluctuations seemingly form nearly close loops streamlines, delineating zones rotating coherently. Using a circulation-based method introduced in [11], it is possible to extract individual vortices from such velocity fields. Such vortices are represented on Figure 1a. These are characterised by a large distribution of sizes and short life time [11].

2.2 DEM simulations

Besides these experiments, a discrete element method was used simulate the plane shear flow between rough walls (see Figure 1b). Grains are subjected to a prescribed shear rate $\dot{\gamma} = V/H$ and normal stress σ. Grains are cohesionless discs that interact through elastic, dissipative and frictional contacts characterised by a Young's modulus $E = 1000\sigma$,

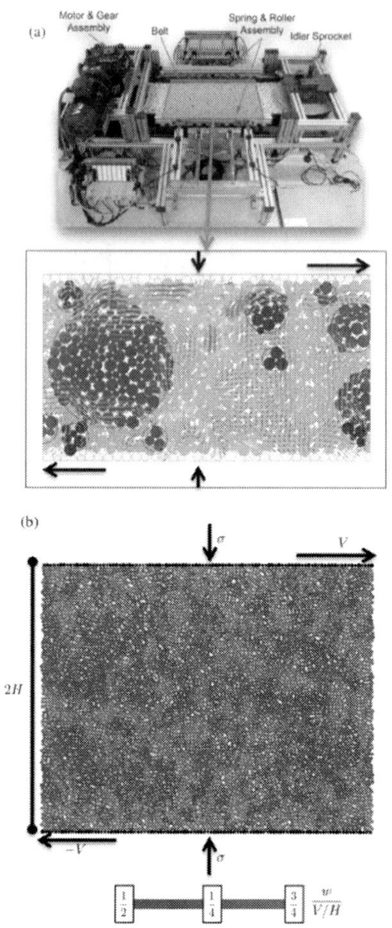

Figure 1. Granular vortices in plane shear of dense granular materials. (a) Granular vortices detected in flows produced by the Stadium Shear Device using a circulation-based method. Arrows denote individual grain velocity fluctuations, defined as the grain instantaneous velocity minus the time-average flow velocity at the grain location. Large blue circles denote detected vortices, including grains coloured in red for vortices rotating clockwise and blue for vortices rotating anti-clockwise. (b) Vorticity field measured in granular flow simulated with a discrete element method.

a coefficient of friction of 0.5 and a coefficient of normal restitution of 0.5.

Figure 1b shows a snapshot of the grains during the flow, coloured according to the level of vorticity in their surrounding. The local vorticity for each grains was measured as the skew-part of the local velocity gradient tensor, estimated using the neighbouring grains located within a distance of $2.5d$ [12]. It appears that the local vorticity is not only highly heterogeneous, but also exhibits some spatial correlation, evidencing large patches of grains with a similar vorticity. This suggests that these patches of grains experience a nearly purely rotational rotation, with little or no shear.

DOI: 10.1051/epjconf/201714003044

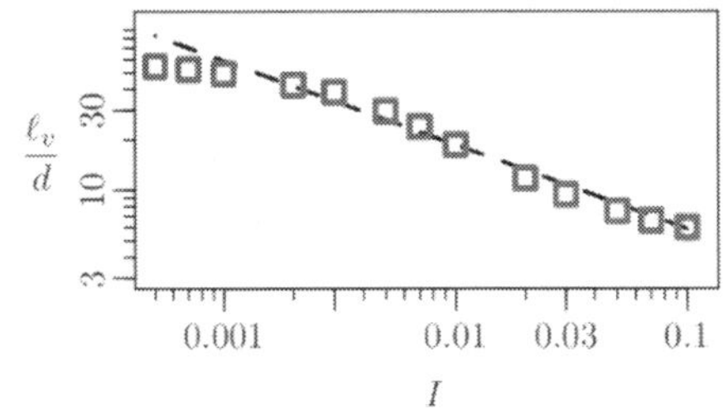

Figure 2. Average vortex size as a function of the inertial number, measured in simulated periodic plane shear flow (H/d = 100). Symbols: simulation data; dashed line: $y \propto x^{-\frac{1}{2}}$

2.3 Vortex size and inertial number

A striking feature of granular vortices is that their average size within a flow depends on the inertial number. Figure 2 shows the average size of vortices measured in simulated plane shear flows at different inertial numbers. As a way to avoid wall effects, these simulation were performed using bi-periodic Lees-Edwards boundary conditions, as in [13]. On most of the explored range of inertial number, the average vortex size is found to scale like:

$$\ell_v \propto \frac{d}{I^{\frac{1}{2}}} \tag{6}$$

For too small inertial numbers, the size of the vortices seems to saturate, which we attribute to be an effect of the finite size of the simulated system: even in a periodic domain, vortices cannot be larger than H.

3 Mixing

An important consequence of the kinematic fluctuations in shear granular flows is that grains may move in the direction transverse to the flow. This shear-induced diffusion may be seen as resulting from a random walk, where grains move in the y direction over a typical distance scaling like their own size d at a frequency $\dot{\gamma}$. Such a random walk would lead to a effective diffusivity scaling like (2).

Figure 3 shows the grain diffusivity measured in a number of simulated flow in a periodic domain for different inertial numbers. While a linear dependency with D and $\dot{\gamma}$ is evidenced at high inertial numbers, a breakdown of this scaling is observed at lower inertial number, with the diffusivity seemingly plateauing and becoming independent of the shear rate [13, 14]. For even lower inertial number, the diffusivity becomes linearly dependent on the shear rate again.

These features may be explained by considering the vortex size as a typical length scale in (3). Accordingly, the diffusivity would scale like $D \propto \ell_v^2 \dot{\gamma}$, or:

$$\frac{D}{d^2/t_i} \propto \left(\frac{\ell_v}{d}\right)^2 I \tag{7}$$

Considering the scaling of the vortex size with the inertial number 6 would then lead to $D \propto d^2/t_i$ [15] and capture the shear rate independent behaviour of the diffusivity. This scenario also predicts the recovery of a shear rate dependent viscosity at lower inertial number, given that the vortex size would then become limited by the system size and thus stop increasing as I decreases.

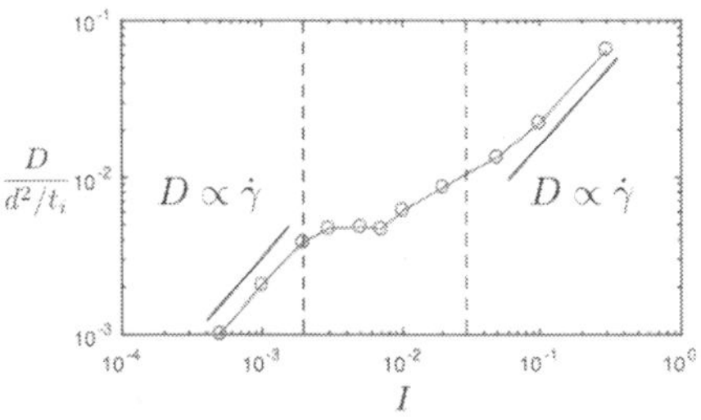

Figure 3. Effective diffusivity as a function of the inertial number measured in simulated plane shear flow in a periodic domain (H/d = 100). Symbols: data from DEM simulations; solid black lines: $y \propto x$; blue dashed lines: approximation of the regime where the diffusivity is nearly shear rate independent.

4 Viscosity and non-locality

The stadium shear experiments and the simulated plane shear flow between rigid walls presented in figure 1 produce homogeneous shear and normal stress across the sheared layer. Accordingly, assuming that the constitutive behaviour of the material is the same across the layer, one would expect that the shear rate $\dot{\gamma}(y)$ should be constant across the layer, $\dot{\gamma}(y) = V/H$. However, experimental and simulated flows exhibit a large variation of shear rate across the layer, as illustrated on figure 4. The shear rate is much larger near the walls than in the central region. This indicates that the effective viscosity of the material is larger in the centre and smaller near the walls [12, 16].

Strikingly, the measured vortex size across the shear layer, $\ell_v(y)$, also exhibits large variations (see figure 4). Vortices are, in average, larger in the central region and smaller near walls. This can be attributed to a finite size effect: a vortex centred at a distance Y from a wall cannot be larger than Y.

The correlation between vortex size variation and shear rate variation near walls suggests that vortices could be a microstructural entity affecting the viscosity of the material. In order to assess this possibility, we considered introducing the local vortex size as a typical length scale in the granular viscosity (4):

$$\nu(y) \propto \ell_v^2(y)\dot{\gamma}(y) \tag{8}$$

Applying the very definition of the viscosity, $\nu \propto \frac{\partial \tau}{\partial \dot{\gamma}}$ where τ is the shear stress, to this model leads to $\partial\tau \propto \ell_v^2(y)\dot{\gamma}(y)\partial\dot{\gamma}$. Integrating along the transverse direction y leads to the following relationship between the local shear rate and the vortex size:

$$\frac{\dot{\gamma}(y)}{V/H} \propto \frac{d}{\ell_v(y)} \tag{9}$$

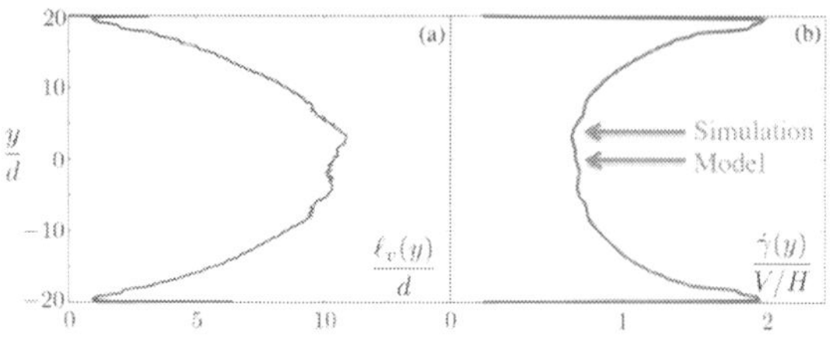

Figure 4. Granular vortices and effective viscosity. (a) Average granular vortex size measured at different location within a simulated plane shear flow between wall ($H = 20d$, $I = 10^{-2}$); (b) shear rate profile measured in the same flow, and predicted by using a viscosity based on the vortex length scale as per Eq. (8) (the two curves are on the top of each other) [12]).

Figure 4 shows that the shear rate profile predicted using (9) perfectly matches the numerical results. This supports the conclusion that vortices and their associated length scale may be an important microstructral parameter that influences the effective viscosity.

Beside long range wall effects, a number of non-local behaviours were observed in dense granular flows, as well as in dense flow of glassy materials, in presence of gradient of stress or shear rate. Different continuum non-local models were developed to capture such effects [12, 17–20], which all rely on a characteristic length ℓ_{nl}, sometimes referred to as cooperativity length, describing the spatial extent of the non-locality. Empirically, it was showed that this length should scale like:

$$\ell_{nl} \propto \frac{d}{I^{\frac{1}{2}}} \tag{10}$$

Strikingly, this scaling with the inertial number matches the measured scaling of the vortex size in (6). This suggests that vortices could potentially be a microstructural origin of non-local behaviour at the continuum scale. In a different paper part of these proceedings [21], we specifically focus on this connection and propose a mechanism that allows us to derive a continuum non-local fluidity model similar to that introduced in [19] simply by considering the existence of vortices.

5 Conclusion

The conclusion of this paper is that (i) there is a spontaneous development of spatially correlated vorticity within dense granular flows, and (ii) the resulting granular vortices may be a useful microstructural indicator to better understand and rationalise effective properties such as effective diffusivity and viscosity. Beyond granular materials, such analysis should be applicable to other glassy materials such as foams and emulsions, in which similar velocity fluctuations patterns have been observed.

While the purely kinematic analysis presented here has led to a number of useful insights, it is expected that it is not the only possible approach to explain the observed macroscopic behaviours. For instance, one could focus on other microstructural kinematic quantities such as granular temperature, or on dynamics properties such as force chains or distribution of local stresses. These approach could potentially lead to complementary insights and further the understanding of glassy material behaviour.

References

[1] Y. Forterre, O. Pouliquen, Annu. Rev. Fluid Mech. **40**, 1 (2008)

[2] B. Andreotti, Y. Forterre, O. Pouliquen, *Granular media: between fluid and solid* (Cambridge University Press, 2013)

[3] C.S. Campbell, Journal of Fluid Mechanics **348**, 85 (1997)

[4] J. Ottino, D. Khakhar, Annual Review of Fluid Mechanics **32**, 55 (2000)

[5] F. Radjai, S. Roux, Physical review letters **89**, 064302 (2002)

[6] O. Pouliquen, Physical Review Letters **93** (2004)

[7] M.R. Kuhn, K. Bagi, International journal of solids and structures **41**, 5793 (2004)

[8] F. da Cruz, S. Emam, M. Prochnow, J.N. Roux, F. Chevoir, Physical Review E **72**, 021309 (2005)

[9] G. Combe, V. Richefeu, M. Stasiak, A.P.F. Atman, Phys. Rev. Lett. **115**, 238301 (2015)

[10] T. Miller, P. Rognon, B. Metzger, I. Einav, Physical Review Letters **111**, 058002 (2013)

[11] P. Rognon, T. Miller, I. Einav, Granular Matter **17**, 177 (2015)

[12] P.G. Rognon, T. Miller, B. Metzger, I. Einav, Journal of Fluid Mechanics **764**, 171 (2015)

[13] P. Rognon, I. Einav, Physical Review Letters **105** (2010)

[14] Y. Fan, P.B. Umbanhowar, J.M. Ottino, R.M. Lueptow, Physical review letters **115**, 088001 (2015)

[15] D. Griffani, P. Rognon, B. Metzger, I. Einav, Physics of Fluids (1994-present) **25**, 093301 (2013)

[16] L. Staron, P.Y. Lagrée, C. Josserand, D. Lhuillier, Physics of Fluids (1994-present) **22**, 113303 (2010)

[17] J. Goyon, A. Colin, G. Ovarlez, A. Ajdari, L. Bocquet, Nature **454**, 84 (2008)

[18] O. Pouliquen, Y. Forterre, Philosophical Transactions of the Royal Society A: Mathematical, Physical and Engineering Sciences **367**, 5091 (2009)

[19] K. Kamrin, G. Koval, Physical Review Letters **108**, 178301 (2012)

[20] M. Bouzid, M. Trulsson, P. Claudin, E. Clément, B. Andreotti, Physical Review Letters **111**, 238301 (2013)

[21] P. Kharel, P. Rognon, *Partial jamming and non-locality in dense granular flows*, in *Powders and Grains 2017* (2017)

EPJ Web of Conferences **140**, 03045 (2017) DOI: 10.1051/epjconf/201714003045
Powders & Grains 2017

Modeling flow arrest using a non-local rheology?

Guilherme Salvador-Vieira*, *Lydie* Staron**, *Stéphane* Popinet, *Stéphanie* Deboeuf, and *Pierre-Yves* Lagrée

[1] *Sorbonne Universités, UPMC Univ Paris 06, CNRS, UMR 7190, Institut Jean Le Rond ∂'Alembert, F-75005 Paris, France*

Abstract. Using an incompressible Navier-Stokes solver with a modified non-Newtonian viscosity, we implement the $\mu(I)$-rheology and study the transient arrest for thick granular layers. In order to study the stopping height phenomenology H_{stop}, we implement the steady-state-only approximate version of the non-local model from [6, 7] and successfully incorporate this feature in the solver. We compare the velocity profiles for granular layer thicknesses close to H_{stop} given by the model with discrete simulations, and discuss the ability of the non-local model to describe the dynamics of the flow close to stopping height. We find that for flows with thickness close to H_{stop}, the velocity profiles given by the model do not reflect the behavior of discrete simulation: discrete velocity profiles remain self-similar when approaching H_{stop}, while the non-local rheology predicts a change of concavity near the bottom.

1 Introduction

The flow of granular matter on a rigid or erodible bed seems the simplest configuration one can think of, yet its intrinsic complexity has long been a challenge to understanding and modeling. As a matter of fact, surface flows and chute flows still form a much studied system. Those flows have a lot of industrial and geophysical applications. The existence of a finite range of slope values in which the flow is stationary and the evolution of the rheological properties of the flow within this range is well captured by the $\mu(I)$-rheology [3]. However, other features require the introduction of non-local effects in the rheological model, which allow for local information on the shear rate to propagate in the system: a distant shear, for instance, or a solid boundary.
In this contribution we are interested in the arrest of granular chute flows, either during the transient evolution towards arrest, or during steady flow close to the critical thicknesses H_{stop} and H_{start} [11]. In particular, we try to address the relevance of non-local modeling to describe the arrest mechanisms.

2 Modeling non-local effects

2.1 The $\mu(I)$-rheology and the non-local fluidity

$\mu(I)$-rheology

We consider the case of a dense granular flow in the xz plan, tilted at an angle θ and oriented in the x direction, which develops a steady velocity profile $u(z)$ as illustrated in Fig. 1. As the flow develops, by analogy with the

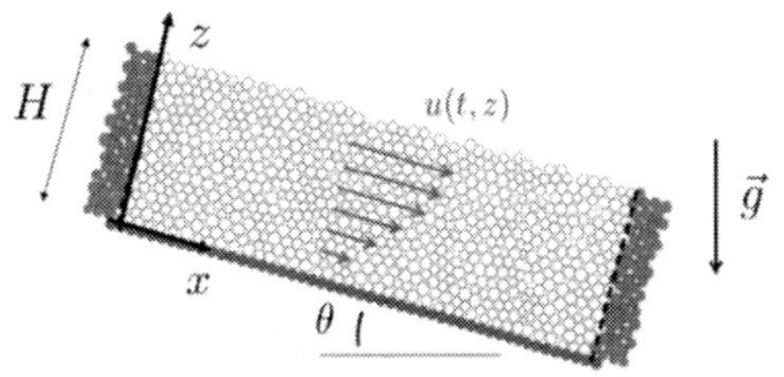

Figure 1. Diagram of the physical problem studied: dense granular flow down an inclined plane. Gray particles correspond to the periodicity of the domain.

Coulombic friction, the local normal stress p and the local tangential stress τ are found in practice and assumed in the model to be proportional:

$$\tau = \mu p, \tag{1}$$

μ being the analogue of a coefficient of friction, that may vary from point to point in a granular flow. The strength of the $\mu(I)$-rheology lays in the fact that it relates the effective coefficient of friction μ to a dimensionless number reflecting the local state of the granular packing, the inertial number I [3], defined as:

$$I = \frac{|\dot{\gamma}|\, d}{\sqrt{p/\rho}}, \tag{2}$$

where $|\dot{\gamma}| = |\mathrm{d}u/\mathrm{d}z|$ is the shear rate, d is the mean diameter of the grains and ρ their density. The normal stress p is set to zero at the surface ($z = H$). The following nonlinear expression on I is adopted to account for the shape of the $\mu(I)$ dependence [5]:

$$\mu(I) = \mu_s + \frac{\mu_2 - \mu_s}{I_0/I + 1}, \tag{3}$$

*e-mail: guilherme.salvador-vieira@polytechnique.edu
**e-mail: lydie.staron@upmc.fr

where the values of the coefficients μ_s, μ_2 and I_0 are material-dependent. By analogy with Newtonian fluids, we define an effective granular viscosity η_{eff} proportional to the shear rate, such that

$$\tau = \mu p = \eta_{\mathrm{eff}} \, |\dot{\gamma}| \, ,$$

from which we deduce the expression for the non-Newtonian viscosity used in the model:

$$\eta_{\mathrm{eff}} = \frac{\mu p}{|\dot{\gamma}|}. \quad (4)$$

Therefore, the granular viscosity depends at the same time on the material characteristics, on the inertial number, on the local normal stress and on the shear rate itself. The previous equations have been written for a pure shear flow, see [5] for a 3D generalized formulation.

Non-local model

The non-local model implemented in this work, from [6, 7], consists in introducing a granular fluidity defined by:

$$g = \frac{|\dot{\gamma}|}{\mu}, \quad (5)$$

and solving for each time step the following additional set of equations with values obtained from the previous local formulation, in order to obtain a global (non-local) value for the granular viscosity (written here for a pure shear flow: z dependence only):

$$g_{\mathrm{loc}} = |\dot{\gamma}| \, / \, \mu_{\mathrm{loc}}(I), \quad (6)$$

$$-\left[\xi\left(\mu_{\mathrm{loc}}(I)\right)\right]^2 \, \mathrm{d}^2 g/\mathrm{d}z^2 + g = g_{\mathrm{loc}}, \quad (7)$$

$$\mu = |\dot{\gamma}| \, / \, g, \quad (8)$$

$$\eta_{\mathrm{eff}} = \mu p \, / \, |\dot{\gamma}|. \quad (9)$$

The field g is a state parameter and ξ is the plastic cooperativity length, proportional to the grain diameter d and calibrated with a constant A:

$$\xi(\mu) = A \sqrt{\frac{\mu_2 - \mu}{\Delta\mu \, (\mu - \mu_s)}} d.$$

In planar shear, $\mathrm{d}^2 g/\mathrm{d}z^2 = 0$ and the above reduces appropriately to the local rheology ($g = g_{\mathrm{loc}}$), but in the presence of gradients, the second-derivative term spreads fluidity based on ξ.
An important characteristic of the model is the imposition of zero-shear on the wall, which corresponds $g(z = 0) = 0$, condition that will be propagated by the model. See [6, 7] for more details.
We implemented the steady-state-only approximate version of the model in Basilisk[1], and solved the momentum balance along the x and z axis, together with 3. Full details on the resolution will be found in [8]. Velocity profiles for steady flows with various values of H/d are then compared.

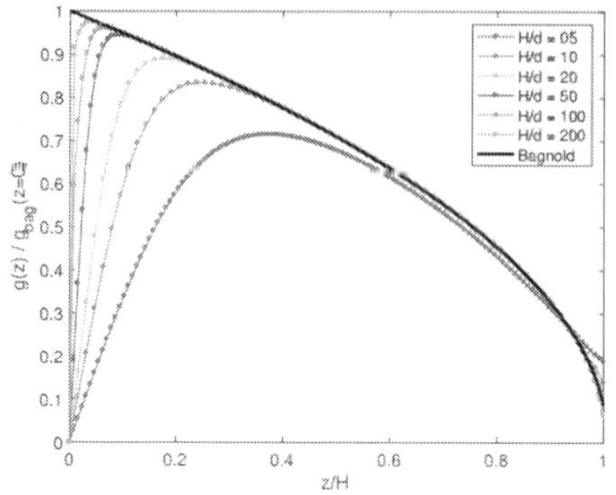

Figure 2. Numerical resolution of the granular fluidity g shape varying the number of grains. In black, the fluidity without the non-local model. $g(z = 0) = 0$ is imposed when solving (7).

2.2 Asymptotic analysis for a chute flow

The previous system can be solved numerically. Fig. 2 shows the variation of the re-scaled g profiles

$$\bar{g}(\bar{z}) = g(\bar{z}) / \max(g_{\mathrm{bag}}), \quad \bar{z} = z/H,$$

with the layer thickness H/d. g_{bag} stands for the fluidity profile associated to the Bagnold solution, corresponding to the local formulation, for which $u(\bar{z}) \propto 1 - (1 - \bar{z})^{3/2}$.
We observe that the profiles superimpose on the characteristic Bagnold profiles far from the wall, but close to the wall the condition $\bar{g}(\bar{z} = 0) = 0$ creates a boundary layer whose relative width $\bar{z}_{\mathrm{bl}} = \mathrm{argmax}_{\bar{z}}(\bar{g})$ decreases with H/d. We want to use the fluidity profile to characterize the velocity profile u in this boundary layer.
$\bar{g}$ depends on two variables: the column height H/d and the reduced ordinate $\bar{z}$, so $\bar{g} = \bar{g}(H/d, \bar{z})$. From Fig. 2, we observe that in the boundary layer region $\bar{g}$ can be well approximated by a straight:

$$\bar{g}(H/d, \bar{z}) \approx \frac{\partial \bar{g}}{\partial \bar{z}}(H/d, 0) \cdot \bar{z} \quad \text{for } \bar{z} < \bar{z}_{\mathrm{bl}}.$$

In addition, one can verify linear dependence between $\partial \bar{g}/\partial \bar{z}$ on the wall and H/d:

$$\frac{\partial \bar{g}}{\partial \bar{z}}(H/d, 0) \propto H/d,$$

so

$$\bar{g}\,(H/d, \bar{z}) \propto (H/d) \cdot \bar{z} \quad \text{for } \bar{z} < \bar{z}_{\mathrm{bl}}. \quad (10)$$

Thus the velocity profile $u(H/d, \bar{z})$ on the boundary layer can finally be obtained from the definition of fluidity $\partial u/\partial z = \mu g$, which gives

$$u(H/d, \bar{z}) = \int_0^{\bar{z}} \mu(H/d, \zeta) \cdot g(H/d, \zeta) \, \mathrm{d}\zeta.$$

Now, assuming μ nearly constant (it should actually equal $\tan(\theta)$) and using (10), we finally get

$$u(H/d, \bar{z}) \propto \mu H/d \cdot \bar{z}^2 \quad \text{for } \bar{z} < \bar{z}_{\mathrm{bl}}. \quad (11)$$

[1] open source, available and documented on the website [2]

The model proposed in [6, 7] to take into account non-local effects introduces a boundary layer close to the wall, whose relative width decreases with H/d and whose growth behaves like $H/d \cdot \bar{z}^2$. Hence, imposing a zero shear rate at the boundary is crucial to recover the H_{stop} phenomenology.

Note that using the steady-state approximate version of the non-local model does not allow for a true arrest of the flow (by contrast with the full model, see [7]). Hence, a cut-off value for the velocity must be introduce to obtain an estimate of H_{stop}.

3 Discrete simulations

A contact dynamics algorithm is applied to simulate the flow of perfectly rigid grains interacting through contact friction and non-elastic collisions [4, 9]. We consider 2D simple chute flow configurations. The grains are perfect disks of diameters distributed around the mean value d within an interval of $\pm 0.4d$. This slight size difference allows for the introduction of generic disorder without inducing significant size segregation at the time scales considered. They form a bed of height H and width L with periodic boundary conditions in the x direction (see illustration in Fig. 1), tilted at an angle θ. H is varied between $4d$ and $35d$ to recover the H_{start} / H_{stop} phenomenologies, as well as to evidence finite size effects in the shape of the velocity profiles. The numerical parameters setting the values of contact friction and restitution were kept constant and set to $\mu = 0.5$ and $e = 0.5$.

The bottom of the flow is made of fixed glued grains of diameter d, forming a bumpy bottom condition. The dimensional values used for the simulations are $d = 5 \times 10^{-4}$ m, gravity $|\vec{g}| = 9.81$ m/s^2 and grains density $\rho = 0.1$ kg/m^2.

In the following, we perform series of discrete simulations to obtain insights on the structure of granular flows close to arrest and systematically compare them to the outcome of continuum modeling.

4 Comparing discrete and continuum flow behavior

4.1 Transient arrest

In a first configuration, we consider a discrete layer of grains with $H = 35d$ flowing at an angle $\theta = 20°$. For this set of height and slope, the flow is steady and the velocity profiles is consistent with Bagnold's prediction [1]. We then tilt instantaneously the layer at a lower angle $\theta = 15°$, for which the flow is no longer sustainable and comes to rest. During the deceleration phase, the velocity profile is plotted at consecutive instants averaged over a time interval $T_S = 0.15T_D$, where T_D is the full duration of the deceleration phase (Fig. 3). We observe how the velocity profile evolves from a Bagnold-like shape to lower velocity nearly linear shape, and finally changes convexity before vanishing.

The same evolution is simulated using the continuum $\mu(I)$-rheology described above and implemented in Basilisk [2] with corresponding parameters, *without introducing non-local effects*. For the same time periods spanning the deceleration phase, the velocity profiles are plotted in Fig. 4. They compare well qualitatively with the discrete evolution shown in Fig. 3, and in particularly we observe that the change in the flow structure is reproduced. This suggests that the transient arrest of a granular flow does not require introducing non-local effects, as shown in [10]; the I dependence from the $\mu(I)$ is a sufficient physical ingredient.

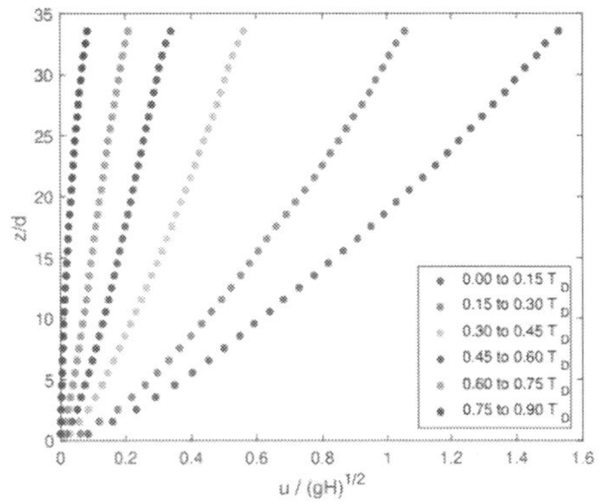

Figure 3. Discrete simulation: evolution of the velocity profile for a thick-layer flow towards arrest.

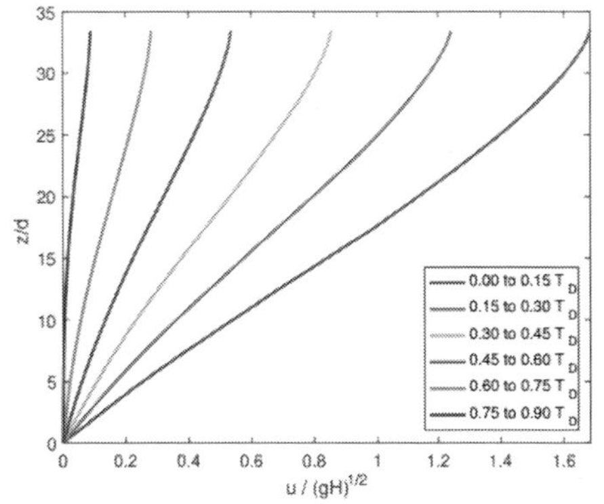

Figure 4. Continuum simulation: evolution of the velocity profile for a thick-layer flow towards arrest.

4.2 Steady flow close to H_{start} / H_{stop}

In the previous configuration, the flow evolves towards arrest in time, and eventually comes to rest. We now investigate the case of a layer whose thickness is close to the critical case for which no flow occurs at all, but which is nevertheless able to flow in steady regime. Thereby we want to evidence the possible emergence of changes in the flow structure that could be seen as for-runners of the arrest condition.

We consider discrete granular layers tilted at an angle $\theta = 20°$, and of different thicknesses $H = 8d, 9d, \ldots, 16d$. For $H = 8d$ and $9d$, no flow is observed, but for all

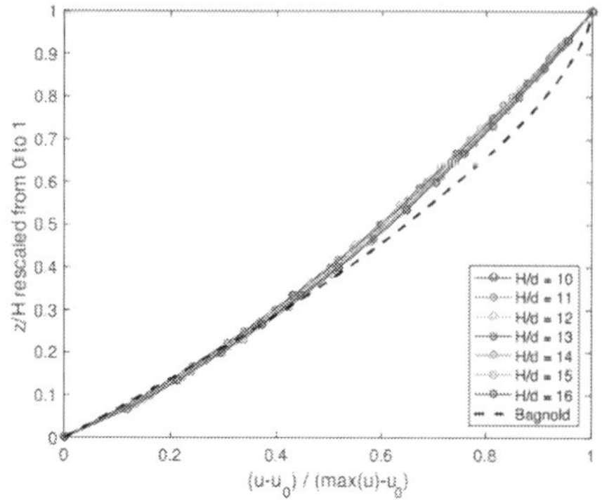

Figure 5. Discrete simulation, normalized velocity profiles for flows with H close to H_{start} at an angle $\theta = 20°$. Profiles observed are self-similar.

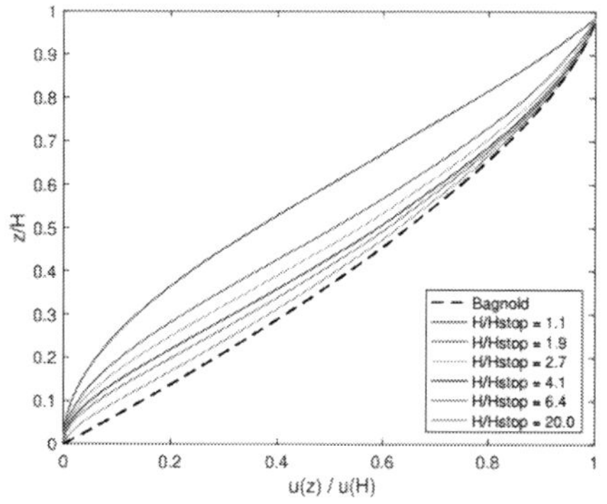

Figure 6. Continuum simulation, normalized velocity profiles for a different material than the one from Fig. 5. At an angle $\theta = 24°$, behavior as $H \to H_{\text{stop}}$ is characterized by a change of concavity near the wall. Here $H_{\text{stop}} \approx 2d$.

other cases a steady flow develops, giving an estimate of $H_{\text{start}} \approx 10d$. Thus, we are studying configurations for which H can reach remarkably close to H_{start}.

For those cases in which the layer flows, namely $H/d = 10$ to 16, we superimpose the velocity profiles after re-scaling them as $\tilde{u}(z) = (u(z) - u_0) / (\max(u) - u_0)$, where u_0 is the bottom slip velocity (Fig. 5). We recover self-similar profiles resembling the Bagnold prediction, with no non-local effect noticeable on the velocity profile shapes.

We now perform a similar experiment applying continuum modeling. Introducing non-local effects in the model is necessary to recover the H_{stop} phenomenology. For the continuum flow of a granular material similar to the one from [7] at $\theta = 24°$, Fig. 6 shows the change on the velocity profile shape for values of H close to H_{stop}. Profiles are re-scaled by the max velocity value at the top of the layer, $u(H)$, and superimposed. As $H \to H_{\text{stop}}$, there is a change of concavity near the wall, as expected from the analysis in Section 2.2. We notice a qualitative difference of behavior between discrete results for H_{start} and continuum results for H_{stop}.

5 Conclusion

Using a continuum model, we were able to reproduce the behavior of granular chute flows down an inclined plane for the steady regime and for the transient arrest using the local $\mu(I)$-rheology model. These results were tested against discrete contact dynamics simulations. The local model, however, is not enough to reproduce the stopping height phenomenology. Thus we implemented a non-local model based on the propagation of a physical quantity called granular fluidity [6, 7], which allowed us to get the desired H_{stop} for granular media spreading a condition of zero shearing on the contact with the wall.

We studied the impact of the steady-state-only approximate version of the model from [6, 7] on the velocity profile shapes as $H \to H_{\text{stop}}$. In order to question whether the observed change of concavity near the bottom is physical, we performed discrete numerical simulations using contact dynamics for H close to H_{start}.

The discrete simulations exhibit consistent self-similar Bagnold profiles, until a sharp bifurcation causing the grains not to flow anymore occurs. No non-local effect was perceptible on the shape of the velocity profiles. These preliminary results suggest that the flow arrest may not be fully understood within the framework proposed in [6, 7]. Additional work need be done, including solving the full time-dependent version of the model. It is worth stressing that close to H_{stop}, only few grain diameters do compose the layer, situation that might be challenging for continuum modeling in general.

References

[1] R. A. Bagnold, Proc. of the Royal Society, London, Series A., Vol. 295, pages 219-232 (1966).

[2] Basilisk solver, http://basilisk.fr/

[3] Gdr Midi, Eur. Phys. J. E 14-4, 341-365 (2004)

[4] M. Jean and J.-J. Moreau, Proc. Cont. Mech. Int. Symp., 31 (1992).

[5] P Jop, Y Forterre, O Pouliquen Nature 441 (7094), 727-730 (2006)

[6] K Kamrin and G Koval, Physical Review Letters, 108(17), p.178301 (2012)

[7] K. Kamrin and D.-L. Henann, Soft matter, 11(1), pp.179-185 (2015)

[8] P.-Y. Lagrée, L. Staron & S. Popinet, *J. Fluid Mech.*, **686**, pp 378-408 (2011).

[9] Moreau, J.-J. In Novel approaches in civil engineering, pp. 1-46. Springer Berlin Heidelberg, (2004)

[10] S. Parez, E. Aharonov and R. Toussaint, *Phys Rev E* 93, 042902 (2016)

[11] O. Pouliquen, Phys. Fluids 11, 542 (1999)

Granular front for flow down a rough incline: about the value of the shape factor in depths averaged models

Pierre-Yves Lagrée[1]*, *Guillaume* Saingier[2], *Stéphanie* Deboeuf[1], *Lydie* Staron[1], and *Stéphane* Popinet[1]

[1]*Sorbonne Universités, UPMC Univ Paris 06, CNRS, UMR 7190, Institut Jean Le Rond d'Alembert, F-75005 Paris, France*
[2]*SVI (Saint-Gobain/CNRS UMR 125), Aubervilliers, France*

Abstract. The thin layer (or Boundary Layer scaling) approximation of the Navier Stokes equation with $\mu(I)$ rheology for dry granular flows over an inclined plate is presented. It is called "Granular RNS/P" (Reduced Navier Stokes/ Prandtl). Integrated over the depth, it gives the classical depth averaged equations (Shallow Water type equations, popularized by Savage Hutter). But the process of depth averaging needs hypothesis on the velocity shape profile trough a "shape factor coefficient". In the case of the displacement of the front of an avalanche down a rough incline, we present comparisons of "Granular RNS/P" simulations versus depth averaged model (which compares well with experimental data). One conclusion is that the shape factor goes from 5/4, the Bagnold value, to a unit value, corresponding to a flat velocity profile, in a region very close to the front of the avalanche.

1 Introduction

Considering dry avalanches, landslides, rockfalls, snow avalanches, debris flow, or general geophysical flow as depth averaged thin layer equations is a very common and powerful point of view. These simplified continuous models contain the dominant physics: gravity driven and Coulomb friction dominated flow. Therefore the system is smaller and faster to solve numerically. Indeed, the case of dry avalanches (the one we will look at here) is up to now too demanding for complete continuous flows such as in [10]. The obvious reason being the small aspect ratio of the avalanches, they are thin and long. This induces numerical difficulties.

As a practical example of such flows reproduced in the laboratory, we have revisited Pouliquen [12, 13] seminal work (small-scale model of a granular layer of glass beads flowing on a rough incline, see figure 1). In [17] the experiment set up is reproduced. In this paper, larger Froude numbers are tested. It is shown that the solution for the global shape of the front of the avalanche proposed by [7, 12, 13] is no more valid for Froude numbers of order one. The analysis is based on depth averaged equations in which several hypothesis are changed. One is about the "shape factor" (defined as α latter, see Eq. 3). The usual approximation is to take a shape factor of value one (supposing a "plug flow"). In [17] it is shown that depth averaged equations may be solved with an exact implicit solution with a any shape factor α value. From comparisons with experiments, the value $\alpha = 5/4$ (corresponding to a Bagnold flow) is found to be the best one. The effect is clear for Froude numbers of order one. For small Froude, the value of α plays no role. This good comparison fails in the vicinity of the font it self. If one uses the value $\alpha = 5/4$, the very front of the avalanche is not sharp. It presents an unphysical exponential decay. If $\alpha = 1$, the very front of the avalanche is sharp. It presents a finite slope, which is more physical. This is one of the reasons for the use of $\alpha = 1$ in the literature.

In this work, we will present a set of equations "Granular Reduced Navier Stokes/ Prandtl" using the frictional rheology $\mu(I)$ ([6, 8]). These equations have been proposed by [4]. This set comes when reducing the number of terms in Navier Stokes thanks to thin layer approximation (Prandtl Boundary layer scaling). We present a numerical method to solve this system. An integration across the depth gives then the classical depth averaged model. We will then numerically solve these "Granular RNS/P" equations and compare to depth averaged solutions. The outcome of the simulation will be the "shape factor", it will be a result rather than an hypothesis. We will see that it changes from $\alpha = 5/4$ to $\alpha = 1$ in the close neighborhood of the front. This explains why the results of [17] are good for $\alpha = 5/4$, and removes the unphysical exponential decay as $\alpha \to 1$ at the front.

2 Models

2.1 Granular RNS/P equations

First we present the equations. We start from the continuous Navier Stokes incompressible model (as in [10]) with the $\mu(I)$ rheology. We write the thin layer system of equations associated as introduced by [4] recently. It means that some derivatives are dropped in the equations leading

*e-mail: pyl@ccr.jussieu.fr

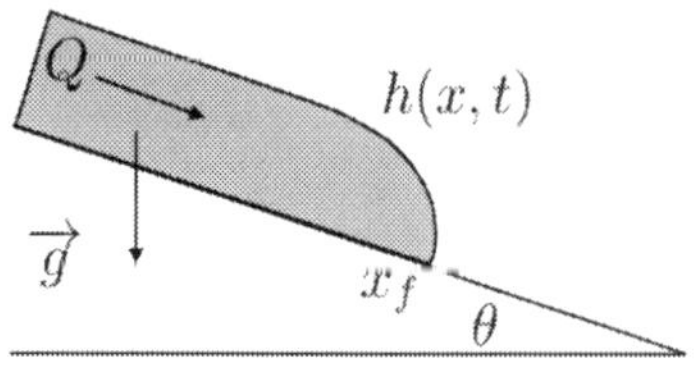

Figure 1. A rough sketch of the experimental setup of ref [17] describing a granular avalanche of hight $h(x,t)$ of discharge $Q(x,t)$ along a plate at angle θ. The position of the front is x_f.

to a kind of boundary layer system of equations. It results in the following system with dimensions, we call it "Granular RNS/P" (Granular Reduced Navier Stokes/ Prandtl):

$$\begin{cases} \dfrac{\partial u}{\partial x} + \dfrac{\partial v}{\partial y} = 0 \\ \rho(\dfrac{\partial u}{\partial t} + \dfrac{\partial u^2}{\partial x} + \dfrac{\partial uv}{\partial y}) = -\rho g Z_b' - \dfrac{\partial p}{\partial x} + \dfrac{\partial \tau_{xy}}{\partial y} \\ 0 = -\rho g - \dfrac{\partial p}{\partial y}. \end{cases} \quad (1)$$

$Z_b(x)$ is the topography on which flows the granular fluid. The constitutive law for the stress is here $\tau_{xy} = \nu_{eq}\dfrac{\partial u}{\partial y}$ with $\nu_{eq} = \frac{\mu(I)p}{\partial u/\partial y}$, it depends on the inertial number $I = \frac{d}{\sqrt{p/\rho}}\dfrac{\partial u}{\partial y}$ (where d is the diameter of the grains) through a given $\mu(I)$ function:

$$\mu(I) = \mu_0 + \frac{\Delta\mu}{1 + I_0/I},$$

see GDR MiDi [6], Jop et al. [8] for details on $\mu(I)$-rheology. The boundary conditions are no slip at the wall $u = v = 0$ in $y = 0$, and no stress at interface $p = 0$ and $\partial_y u = 0$ in $y = h(x,t)$, plus any initial velocity field. It is clearly a set of equations deduced from Navier Stokes equations in which we have incompressibility, the convective term, the hydrostatic pressure term and the dominant part of the derivative of the stress which follows a Druker-Prager law.

This system is reminiscent to the RNS/P system ([3, 9]). This terminology comes from the efforts to reduce the set of Navier Stokes in adding only the transverse derivative of the viscous forces in the equations (see Rubin et al. [15, 16]). In simple words, these equations are no more than the Prandtl Boundary Layer equations (see the fundamental book of Schlichting [19]), with the gravity terms as sources and different boundary conditions.

2.2 Numerical discretisation

System (1) is a kind of Boundary Layer system. Numerical resolution of steady boundary layer equations presents several traps, for instance it is well know that there is a singularity when steady equations are solved with a given pressure (Goldstein singularity, [19]). Furthermore, the unsteady boundary layer equations are either instable or may even lead to a finite time singularity if pressure is given. Fortunately, Audusse et al. [1] proposed recently a resolution of system (1) extended by [4]. The idea of resolution is in fact a clever transverse discretization of (1) in several layers, each layer being a Saint-Venant one (so that in this paper, there is not any reference to boundary layer theory but it is focussed on Shallow Water literature). Those layers are solved by finite volumes. Each layer interacting one with the two others: the upper and the lower, through the viscous force (viscosity ν_{eq}), and by cross exchange of mass and momentum (in reference [1] this exchange is done by the functions G_i which are the difference between the velocity of the moving layer and the real transverse velocity). This implementation has been shown to be very robust.

The Multilayer Saint-Venant has been implemented in the Gerris flow solver suite by Popinet [14] (and by Ghigo et al. in arterial flow [5]). The viscosity is here $\nu_{eq} = \mu(I)p/(\frac{\partial u}{\partial y})$ with $I = d(\frac{\partial u}{\partial y})/\sqrt{p/\rho}$.

2.3 Depth averaged models (or Saint-Venant Savage-Hutter 1D models)

We have defined a full 2D problem (1) and we have proposed a numerical method to solve it. The usual model is the 1D depth averaged model. The 1D depth averaged model is a simplification of system (1). Indeed, integrating equation (1) from $y = 0$ to $y = h(x,t)$, we have without new hypothesis:

$$\begin{cases} \dfrac{\partial h}{\partial t} + \dfrac{\partial}{\partial x}\displaystyle\int_0^h u dy = 0 \\ \rho(\dfrac{\partial}{\partial t}\displaystyle\int_0^h u dy) + \dfrac{\partial}{\partial x}(\displaystyle\int_0^h u^2 dy + g\dfrac{h^2}{2}) = -\rho g Z_b' h - \tau_{xy}|_0. \end{cases} \quad (2)$$

At this point of the analysis comes the choice of the velocity profile to close the system to obtain the 1D model. In aerodynamics, the integral equations arising are the Von Kármán equations, and the closure maybe the Pohlhausen one (see [19]). In hydrodynamics, the integral equations arising are the Saint-Venant (Shallow Water) equations, and the closure involves plug flow and turbulent friction (or half a Poiseuille in viscous films). In granular flows, the integral equations arising are the Savage-Hutter equations. First the basal friction term $\tau_{xy}|_0$ has to be modeled, usually since [18] by a Coulomb friction: $\tau_{xy}|_0 = \mu\rho g h$ (with a constant friction coefficient μ). Second, the shape of the profile has to be defined to link the flux of velocity (the volume flow rate) to the flux of kinetic energy, defining the shape factor (as a constant in those models):

$$\alpha = (h\int_0^h u^2 dy)/(\int_0^h u dy)^2. \quad (3)$$

The usual approximation is to use a plug profile so that if we define the discharge as $Q = \displaystyle\int_0^h u dy$ then $\displaystyle\int_0^h u^2 dy$ is approximated by Q^2/h (i.e. $\alpha = 1$ in the following equation, if the profile is supposed to be a Bagnold one, then

$\alpha = 5/4$, see [17]):

$$\begin{cases} \dfrac{\partial h}{\partial t} + \dfrac{\partial}{\partial x} Q = 0 \\ \rho \dfrac{\partial}{\partial t} Q + \dfrac{\partial}{\partial x}(\alpha \dfrac{Q^2}{h} + g\dfrac{h^2}{2}) = -\rho g Z_b' h - \mu \rho g h. \end{cases} \tag{4}$$

These are the Savage-Hutter Saint-Venant equations for granular flows [18]. This hierarchy of simplifications is justified by the fact that the avalanches are very thin compared to their longitudinal extent. Indeed, numerical simulation of Navier Stokes equations with $\mu(I)$ rheology (as in [10]) is very difficult, and numerical simulations of these thin layer approximations ([4] or [18]) are more tractable.

The approximation $\alpha = 1$ is used in nearly every paper as it helps the numerics and as it gives a sharp front (Poulinquen [12, 13]).

2.4 Analytical solution of the front shape, comparison with experiments

It has been shown in [17] that an analytical solution of Savage-Hutter Saint-Venant system (4) may be obtained in the case of the flow over an inclined flat plate with $Z_b' = -\tan\theta$. It is an implicit solution were the position is function of the reduced hight:

$$\frac{(x - u_0 t)(\tan\theta - \mu_0)}{h_{max}} = \mathcal{F}(\frac{h(x,t)}{h_{max}}, \frac{\tan\theta - \mu_0}{\Delta\mu}, \alpha, Fr), \tag{5}$$

valid for any α, see reference [17] for details and for the exact expression of $\mathcal{F}$ which involves arc tangent and logarithm functions. This implicit analytical solution extends the Gray and Ancey's one [7] valid only for plug flows or for small Froude.

In [17] a flume of 2 meters long and of 40 cm of width, with a variable inclination angle starting by a silo has been build. Several avalanches at various angles, and with various initial height of aperture of the silo have been performed (see reference for details). With optical techniques (laser sheet), profiles of the front span wise and stream wise were measured. The aim was to explore larger Froude number values. The constant front velocity as in [12] was reobtained. Very good comparisons between measured fronts and analytical solution Eq. 5 for values of Froude of order one and $\alpha = 5/4$ have been obtained. The effect of increase of inertia (by increasing the angle) is clearly reproduced by the analytical solution.

One of the drawbacks of the solution Eq. 5 is the existence of a precursor film with an exponential decay, which is not observed in the experiments. For $\alpha = 1$, there is no such precursor film.

3 Discussion

In the depth averaged formulation of system (4), the shape factor is constant by construction. With this formulation an analytical solution may be obtained, this is Eq. 5. It allows quantitative comparisons with experimental data as just explained and presented in [17]. The sole problem being this spurious exponential decay, which is not observed in the experiments. Hence, now we turn to the numerical solution of (1) in the case of the front. We compute the shape factor as a result, it is no more an hypothesis. On figure 2 several computed shape factors are ploted for different values of Froude number. We observe that indeed the computed α from (1) is mostly equal to Bagnold's value 5/4. Nevertheless, close to the front it goes to one (indeed, longitudinal velocity at the very front is more like a plug flow instead of a Bagnold profile).

On each plot of figure 3, three curves are presented. First is the analytical solution with $\alpha = 1$ or $Fr = 0$ (that is the same) corresponding to the Savage-Hutter Saint-Venant solution [7, 13]. This is for reference. Second is the present numerical solution of system (1). Third is the analytical solution with $\alpha = 5/4$ and the corresponding Fr value from the second. As we can notice, the second and the third are very close. It is only very near the front that we see the differences. As the value of α is fixed to 5/4, the solution of equation (5) presents a slight difference when approaching the front. Furthermore, the solution of equation (5) presents the exponential unphysical precursor layer.
The agreement between (system (1) and equation (5) with $\alpha = 5/4$) and Savage-Hutter Saint-Venant (equation (5) with $\alpha = 1$) is only good for small values of Froude number ($Fr \lesssim .5$). Then, the higher the Froude the smaller the slope of the front and the larger the departure from system 1 to $\alpha = 1$ models.

4 Conclusion and perspectives

This paper is devoted to the solution of a reduced system (1) issued from the non-Newtonian $\mu(I)$-rheology of

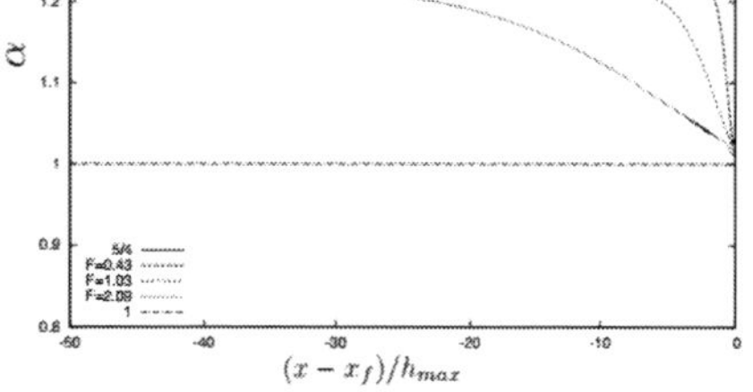

Figure 2. Numerical resolution of "Granular RNS/P". Plot of α the shape factor ratio of flux and kinetic flux (defined by equation 3) at a fixed time as a function of $(x - x_f)/h_{max}$ the position before the front divided by the value of the thickness of the avalanche for three values of the Froude number (upper curve $Fr = 0.43$, middle curve $Fr = 1.03$, lower curve $Fr = 2.09$). Far from the front, the value starts from the Bagnold 5/4 value and goes towards 1 at the front on a scale of several thicknesses. The smaller the Froude, the better is the approximation $\alpha = 5/4$, except close to the front.

DOI: 10.1051/epjconf/201714003046

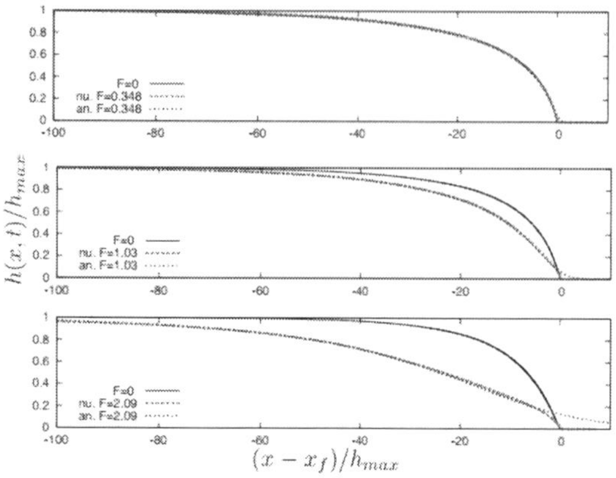

Figure 3. In red $h(x, t = 150)/h_{max}$ as function of $(x - x_f)/h_{max}$ at same time ($t = 150$) for various increasing θ of various increasing initial thickness giving a Froude from 0.348 to 2.09 (from top to bottom). The analytical solution (equ. 5) is plotted as well in blue for the corresponding values of θ and Fr (with $\alpha = 5/4$). For reference the analytical solution (equ. 5 with $\alpha = 1$ or $F = 0$) of the usual Savage-Hutter Saint-Venant resolution is plotted in black and labelled $F = 0$ (colors online).

the Navier-Stokes equations leading to kind of Prandtl equations. This system is showed to reproduce the expected behavior in the case of avalanche: in steady conditions far from entrance and far from the front we obtain the Bagnold solution. Our method allows to focus on the front, we notably compute the local value of the shape factor α. It starts from 1 in the very beginning of the front. This corresponds to a very first plug flow. It goes then to 5/4 the Bagnold expected value at constant height. The numerical resolution uses the robust multi layer approach in finite volumes as proposed by [1] and [4]. The numerical resolution is shown to be very close to the analytical solution established in [17], this analytical solution was itself very close to experimental data. The conclusion is that the usual Savage-Hutter Saint-Venant equations (with $\alpha = 1$) are not a very good description of the avalanche front at Froude of order one. A better description is with $\alpha = 5/4$, but there is a small unphysical exponential tail. A better description is this Granular RNS/P description. Note that reduced system (1) seems not to have the instability, or "ill posed" behavior found by [2] in the full system.

Finally, those equations are in between full resolution (Navier-Stokes) and a too much simplified one (depth averaged Savage-Hutter Saint-Venant). They create the missing link between a too complicated Navier-Stokes resolution (useful to focus on details) and the Savage-Hutter Saint-Venant. This latter depth averaged method is too crude maybe, but it is fast and efficient enough to compute a reasonable and realistic description of an avalanche on an actual topography [11].

References

[1] E Audusse, M.-0. Bristeau, B. Perthame, and J. Sainte-Marie, ESAIM: M2AN 45 (2011) 169-200

[2] T. Barker, D.G. Schaeffer, P. Bohorquez & J. M. N. T. Gray (2015) J. Fluid Mech. 779 , 794-818

[3] F. Chouly, P.-Y. Lagrée (2012): Applied Mathematical Modelling 36 (2012), pp. 6061-6071

[4] E. D. Fernandez-Nieto, J. Garres-Diaz, A. Mangeney & G. Narbona-Reina J. Fluid Mech 798, 643-681

[5] A. Ghigo, Jose-Maria Fullana, and Pierre-Yves Lagrée Multi-ring simulation of blood flow in an artery XXIV ICTAM, 21-26 August 2016, Montreal, Canada

[6] GDR MiDi, Eur. Phys. J. E., **14**, 341-365 (2004).

[7] J. M. N. T. Gray and C. Ancey, J. Fluid Mech. 629, 387-423 (2009).

[8] P. Jop, Y. Forterre & O. Pouliquen, Nature, **44**, 727-730 (2006).

[9] P.-Y. Lagrée & S. Lorthois (2005): Int. J. Eng Sci., Vol 43/3-4 pp 352-378.

[10] P.-Y. Lagrée, L. Staron, S. Popinet (2011): Journal of Fluid Mechanics, pp1-31 , doi:10.1017/jfm.2011.335

[11] Kuo, C. Y., Tai, Y. C., Bouchut, F., Mangeney, A., Pelanti, M., Chen, R. F., and Chang, K. J., (2009) Eng. Geology 104(3-4), 181-189.

[12] O. Pouliquen, Phys. Fluids, 1999a, v11, pp. 542-548

[13] O. Pouliquen, Phys. Fluids, v 11, pp. 1956-1958

[14] S. Popinet http://gerris.dalembert.upmc.fr/gerris/tests/tests/lake.html

[15] S.G. Rubin, P.K. Khosla (1990) Computing Systems in Engineering 1, 549-562.

[16] Annu. Rev. Fluid Mech. 1992. 24: 117-44

[17] G. Saingier, S. Deboeuf & P.-Y. Lagrée, Physics of Fluids (2016).

[18] S. B. Savage & K. Hutter, **199**, 177-215 (1989).

[19] H. Schlichting, ed Mc Graw Hill, New-York, 748p (1987).

DOI: 10.1051/epjconf/201714004006

Clustering impact regime with shocks in freely evolving granular gas

Masaharu Isobe[1,*]

[1] *Graduate School of Engineering, Nagoya Institute of Technology, 466-8555, Japan.*

Abstract. A freely cooling granular gas without any external force evolves from the initial homogeneous state to the inhomogeneous clustering state, at which the energy decay deviates from the Haff's law. The asymptotic behavior of energy in the inelastic hard sphere model have been predicted by several theories, which are based on the mode coupling theory or extension of inelastic hard rods gas. In this study, we revisited the clustering regime of freely evolving granular gas via large-scale molecular dynamics simulation with up to 16.7 million inelastic hard disks. We found novel regime regarding on collisions between "clusters" spontaneously appearing after clustering regime, which can only be identified more than a few million particles system. The volumetric dilatation pattern of semicircular shape originated from density shock propagation are well characterized on the appearing of "cluster impact" during the aggregation process of clusters.

1 Introduction

Clustering in the granular gas is one of the fascinating phenomena in the granular physics [1–3]. In general, a freely cooling granular gas with an inelastic hard sphere (IHS) model is known to evolve from the initial homogeneous equilibrium state to the inhomogeneous clustering state, at which the energy decay deviates from so-called "Haff's law" [4]. Due to simplicity of this model, there have been much investigated by various methodologies with a wide coarse-graining levels from the microscopic (molecular dynamics (MD)) and the mesoscopic (Enskog-Boltzmann equation), to the macroscopic (Navier-Stokes (NS)-like continuum equations and Ginzburg-Landau theory). However, there are few studies based on theoretical approach after clustering regime.

Previously [5–8], we have investigated the macroscopic statistical properties on the freely evolving quasi-elastic IHS model by performing a large-scale event-driven molecular dynamics with mainly 512^2 [5], 4096^2 [7] in two-dimensional system (2D) and 128^3 [8] in three dimensional system (3D). These extensive numerical simulations have revealed a quite similarity with fluid Navier-Stokes turbulence, such as enstrophy cascade (Kraichnan-Leith-Bachelor theory), Kolmogorov scaling and Bose-Einstein condensation, which are well-known concept in the fluid turbulence [9]. Reynolds number with Taylor's microscale before clustering grow up and was estimated as $R_\lambda \sim 327$ (2D) and ~ 1470 (3D) at the onset of clustering regime [7, 8] respectively, which are sufficiently high and reasonable that the ideal granular gas system shows the specific laws of turbulent behaviors.

At the initial stage of the cooling process, density and velocity filed remain homogeneous. However, the kinetic energy (or granular temperature) decreases via inelastic collisions in time. In this homogeneous cooling state (HCS), the granular temperature $T(t)$ decays according to the Haff's law,

$$T^*(t) = \frac{T(t)}{T(0)} = \frac{1}{(1+t/\tau_0)^2} \equiv \exp(-2\tau_0\tau) \qquad (1)$$

where τ is collision number and τ_0 is a fitting parameter, which can be estimated explicitly by the kinetic theory in dilute systems. The reduced temperature $T^*(t)$ for freely cooling IHS model is equivalently the reduced total energy in the hard sphere system since there is no potential energy.

In the clustering regime, due to inhomogeneity in density, $T^*(t)$ deviate from Haff's law [4]. The asymptotic behavior of $T^*(t)$ after clustering regime has been predicted by several theories, which are based on the mode coupling theory [10] and the extension of inelastic (sticky) hard rods gas [11]. Those theoretical predictions show the asymptotic behavior as $T^*(\tau) \sim A\tau^{-d/2}$ $(d \geq 2)$ and $\sim Bt^{-d/2}$ $(2 \geq d \geq 4)$, respectively, where $A = A(r, \nu)$, B and r are independent constant, d is dimension. In 2D case, both theories predict $T^* \sim \tau^{-1}$ and $\sim t^{-1}$, respectively [12]. In this paper, we performed large-scale and long-time event-based molecular dynamics simulation [13, 14] in IHS model with up to 16.7 million particles to clarify the validity of asymptotic behaviors predicted by these theories, and especially focus on inhomogeneous clustering regimes in the dense systems, at which the theoretical approach cannot be resolved.

2 Model and numerical method

We address the simple IHS model without any external force (such as gravity), i.e., so-called "freely cooling granular gas". At microscopic molecular dynamics level, the

*e-mail: isobe@nitech.ac.jp

IHS model can completely be characterized by three independent parameters: the total number of disks N, the packing fraction ν, and the restitution coefficient r. In case of $r = 1$, the system is govern by only solid-fluid phase transition, so-called "Alder transition" (See, recent progress in Refs. [15, 16]). The whole systems with a side length L in the two dimensional (2D) square systems include monodisperse hard disks up to $N = 4096^2$ with a radius of disk σ, in which we adopted the periodic boundary conditions. The relations between L and packing fraction ν are $L/(2\sigma) = \sqrt{\pi N/\nu}/2$. Initially, the system is prepared for the equilibrium states by a long preparation run in an elastic mono-disperse system with $r = 1$. For the dissipation, the normal restitution coefficient $r < 1$ of sequential binary instantaneous inelastic collisions is introduced (i.e., parallel to the relative position of the two colliding disks in contact), where the momentum is conserved and energy is dissipated. The collisons are evolved by an efficient event-based algorithms (See, in Refs. [13, 14]).

Because of monotonically decreasing total energy in the evolving process, the physical time scale is slowing down due to dissipation between inelastic collisions. The time rescaled system is also often considered, in which the particle velocities are rescaled so as to keep the energy of the whole system being constant. This procedure has a direct relation to the thermostat in the the methodology of molecular dynamics. In general, thermostat methods are categorized by deterministic or stochastic ones. This time rescaling procedure resembles "velocity scaling method". This procedure do not affect the particle trajectories in the hard sphere system, which just affect the time scale of relative velocities between future collides particles, i.e., next collision time. In the rescaled time system, the unit of evolving time in the original system t is modified as new scaled time $t_s = \int_0^t \beta(t)^{-1} dt$, $\beta(t) = \sqrt{T(0)/T(t)}$, where $T(t)$ is the granular temperature as a function of t [17, 18].

3 Cluster impact regime & shock wave

3.1 Asymptotic energy decay after Haff's law

It is known that the freely cooling granular gas systems become unstable and several spatiotemporal scales appear. The standard evolution scenario is that spatial instabilities occur from the initial homogeneous cooling state (HCS: 1st stage) to the velocity field (shearing regime: 2nd stage), and then, to the density field (clustering regime: 3rd stage). Finally, global structures in the system (final attractor regime: 4th stage) appear as a non-equilibrium steady state, or local inelastic collapse occurs, which depends on the restitution coefficient. Since we only focus on the quasi-elastic large system, i.e., $r \sim 1$, inelastic collapse do not occur in our present simulation.

In Fig. 1, the typical decaying process of energy (or granular temperature) in the simulation with two packing fractions $\nu = 0.10$ and 0.60 as a functions of $t_s^* = t_s(1 - r^2)/(4t_0)$ are shown, where t_0 is the mean free time in equilibirum for each density. The solid red curve shows the result obtained by the simulations at $(N, \nu, r) =$

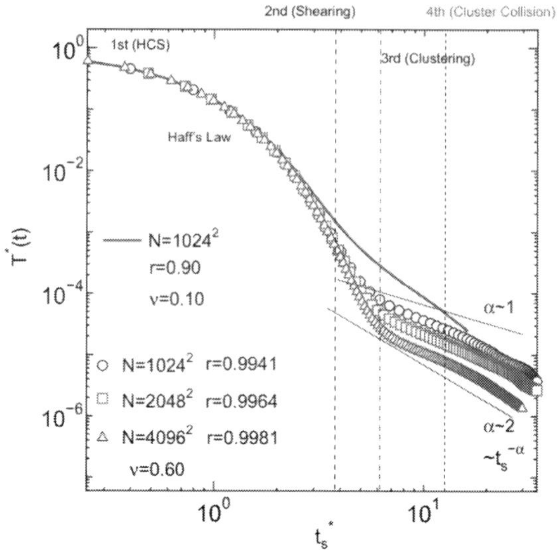

Figure 1. The evolving process of the reduced temperature $T^*(t)$ in terms of rescaled (reduced) time t_s^* with 4 sets of parameter are shown.

$(1024^2, 0.10, 0.90)$. At first, we confirmed that the energy decay is consistent with the Haff's law before clustering regime, in which we confirmed the relationship of $t \sim \exp(t_s)$. After clustering regime, the decay of granular temperature deviates from the Haff's law, at which we found empirical relationship for time t, rescaled time t_s and collision number τ are $t \sim t_s^2$ and $\tau \sim t_s$, respectively. We estimate $\alpha \sim 2$ in the form of $T^*(t_s) \sim t_s^{-\alpha}$ for asymptotic power decay of temperature, which leads to $T^*(t) \sim t^{-1}$. The results are fairly consistent with the prediction of Ref. [10].

In the dense case ($\nu = 0.60$), we performed simulations wtih three sets of parameter by changing system size as $(N, r) = (1024^2, 0.9941)$, $(2048^2, 0.9964)$, and $(4096^2, 0.9981)$, respectively. The largest system consists of up to about $N = 16.7$ million hard disks (2D). Before clustering regime, the relationship between t and t_s are also consistent with the Haff's law as shown in the case of $\nu = 0.10$. After clustering regime, we observe the exponents $\alpha \sim 1$, which are obviously smaller than that of the dilute case. Moreover, during the growing process of the clustering after $t_s^* \sim 10$, the exponent seems to have a crossover to the different larger values, i.e., the energy decaying rate is increasing. These peculiar behaviors have not predicted by present theories and suggest unknown physical mechanisms regarding the dynamics of clusters, which are increasing inelastic collisions (i.e. decreasing energy) per unit time after clustering stages.

3.2 Collision rate

To investigate the reason for the deviation from the theories, we focus on the collision rate after the clustering regime. In Fig. 2, the time evolution of the collision rate

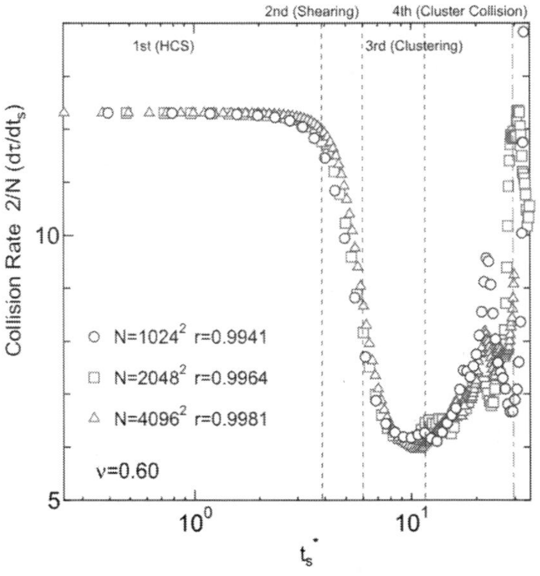

Figure 2. The evolving process of the collision rate for different system sizes in terms of rescaled (reduced) time t_s^* are shown.

$(2/N)(d\tau/dt_s)$ in terms of t_s^* is shown, where $d\tau$ is the amount of binary collision number occurring during small time interval dt_s. We found that the data obtained by the simulation with three different system sizes show almost universal scaling curves and the rapid decreasing collision rate after clustering begins since particles move collectively as a cluster, where velocity correlation near the particle takes positive values along to the flow. However, the collision rate increases again around $t_s^* \sim 10$, which correspond to crossover time that temperature decay shows the different exponents of power decay as shown in Fig. 1.

3.3 Volumetric dilatation and shoke waves

To clarify the origin of the rapid increase of collision rate, we focus on the spatial distribution of the rate of volumetric dilatation $Q(\mathbf{x})$,

$$Q(\mathbf{x}) = \frac{1}{2}\langle(\nabla \cdot \mathbf{u}(\mathbf{x}))^2\rangle = \frac{1}{2}\langle(div\, \mathbf{u}(\mathbf{x}))^2\rangle \quad (2)$$

where $\mathbf{u}(\mathbf{x})$ is the velocity field at $\mathbf{x}$ (i.e., in 2D case, $\mathbf{u}(\mathbf{x}) = (u_x(x, y), u_y(x, y))$). This properties are related to the compressibility of the fluid. In case of the incompressible fluids, the volumetric dilatation rate should be zero, i.e., $Q = 0$. This have higher values if the strong change is occurred in spatial differentiation of velocity field such as shock wave.

In Fig. 3, we visualize $Q(\mathbf{x})$ after well developed clustering regime, where the corresponding density field is also shown as the reference (monotone snapshot drawn as upper right of Fig. 3). We can clearly identify the high Q regions (shown as red) in not only the edge of cluster but also *inside* of the cluster. Four series of snapshots for Q show the growing semicircular shape density shocks, shown inside of the solid open circles in Fig. 3, which are excited from the contact point of giant impact between clusters. (See, higher values of $Q(\mathbf{x})$ drawn as red in Fig. 3.). These shocks propagates inside the clusters which cause the rapid increase of collision rate. These results indicated that the theoretical prediction of the asymptotic behavior of energy decay after the clustering regime seems not to be valid at high density and large-scale system.

4 Discussion

Although we confirmed that the theories seem to be consistent in relatively dilute and small system, we found the new regime regarding on collisions between "clusters" spontaneously after the clustering regime in dense case. In the new regime, the power decay of energy shows clearly slower than that of the dilute case. Furthermore, the power decay itself is also violated in the long time limit. Since the collision rate in the new regime increases rapidly, the novel two-step energy relaxation as a clustering scenario is expected as follows: At first, several string-like clusters are nucleated with an aggregation of inelastic hard spheres, which are distributed separately in space. Those nucleated clusters move actively, and deformed with complex behaviors, and gradually aggregating or coarsening. Next, a certain amount of cluster begins to collide each other in the time regimes of $t_s^* \sim 10$. This giant impact between clusters preceded correctively and made round shape density wave propagation (shocks) on the cluster. This shock propagation results in the increase of collision rate, which causes a rapid growth of dissipation. At that time, the volumetric dilatation pattern of semicircular shape originated from shock density propagation in the clusters after cluster impact can be detected. This regime called as "clustering impact regime" can only be identified in more than a few million particles system. In our analysis of volumetric dilatation pattern of semicircular shape and collision rate clarify that the cause of giant impact between cluster is a completely novel physical mechanism in the field of freely cooling granular gas. This has not known yet, since "clustering impact regime" can be detected only in the systems more than several million particles systems. These findings might be relevant in the field of the late stage of planetary formation. With these results, the asymptotic predictions of energy (granular temperature) in the long time limit should be modified in dense and large systems. Theoretical breakthrough in correspondence with simulation results might be demanded. We will address the results for the other packing fraction and 3D case in the near future.

Acknowledgements

MI was supported by JSPS KAKENHI Grant Number JP26400389. Part of the computations was performed using the facilities of the Supercomputer Center, ISSP, Univ. of Tokyo.

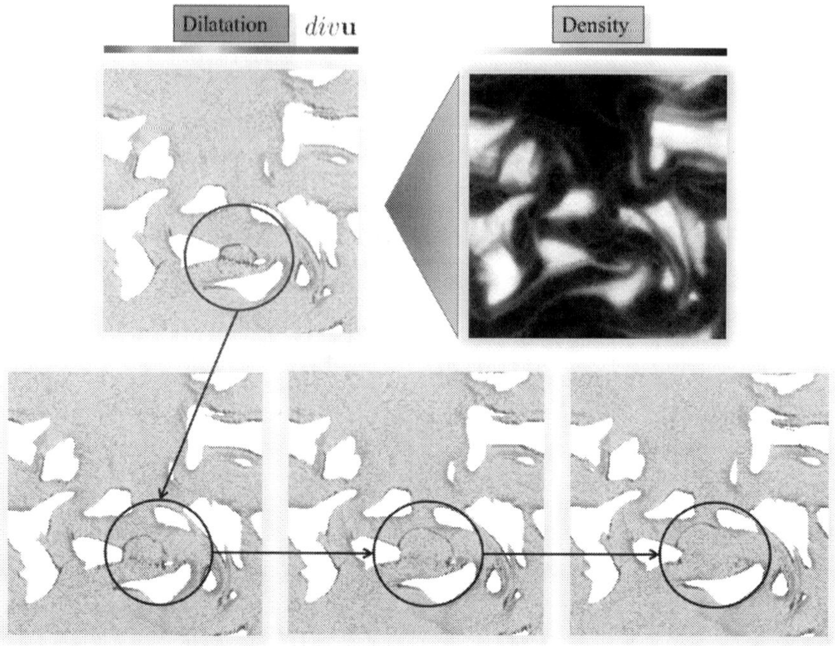

Figure 3. Four time series of snapshots for the evolving process of the spatial distribution of the rate of volumetric dilatation $Q(\mathbf{x})$ after clustering impact are shown, in which the propagation of semicircular shape density shock wave in the clusters appears. The corresponding density field is also shown as the reference at upper right.

References

[1] I. Goldhirsch, Annu. Rev. Fluid Mech. **35**, 267 (2003).

[2] N. Brilliantov and T. Pöschel, *Kinetic Theory of Granular Gases* (Oxford University Press), Oxford, 2004

[3] T. Pöschel, and T. Schwager, *Computational Granular Dynamics*, Springer, 2004.

[4] P. K. Haff, J. Fluid Mech. **134**, 401 (1983).

[5] M. Isobe, Phys. Rev. E **68**, 040301(R) (2003).

[6] M. Isobe, Proceedings in Powders and Grains 2005, Garcia-Rojo, Herrmann and McNamara(eds), 1123 (2005).

[7] M. Isobe, Int. J. Mod. Phys. C **23**, 1250032 (2012).

[8] M. Isobe, J. Phys: Conf. Ser. **402**, 012041 (2012).

[9] P. Tabeling, Phys. Rep. **362**, 1 (2002).

[10] R. Brito and M. H. Ernst, Europhys. Lett. **43**, 497 (1998).

[11] E. Ben-Naim, S. Y. Chen, G. D. Doolen and S. Redner, Phys. Rev. Lett. **83**, 4069 (1999).

[12] J. A. G. Orza, R. Brito, and M. H. Ernst, *cond-mat/0002383* (2000).

[13] M. Isobe, Int. J. Mod. Phys. C **10**, 1281 (1999).

[14] M. Isobe, Mol. Sim. **42**, 1317 (2016).

[15] M. Engel, J. A. Anderson, S. C. Glotzer, M. Isobe, E. P. Bernard, and W. Krauth, Phys. Rev. E **87**, 042134 (2013).

[16] M. Isobe and W. Krauth, J. Chem. Phys. **143**, 084509 (2015).

[17] T. S. Komatsu, J. Phys. Soc. Jpn. **69**, 5 (2000).

[18] R. Soto, M. Mareschal, and M. Malek Monsour, Phys. Rev. E **62**, 3836 (2000).

Segregation in inclined flows of binary mixtures of spheres

Michele Larcher[1,*], and *James T.* Jenkins[2]

[1] Faculty of Science and Technology, Free University of Bozen-Bolzano, Bozen-Bolzano, 39100, Italy
[2] School of Civil and Environmental Engineering, Cornell University, Ithaca, NY, 14853, USA

Abstract. We outline the equations that govern the evolution of segregation of a binary mixture of spheres in flows down inclines. These equations result from the mass and momentum balances of a kinetic theory for dense flows of inelastic spheres that interact through collisions. The theory employed for segregation is appropriate for particles with relatively small differences in size and mass. The flow of the mixture is assumed to reach a fully developed state much more rapidly than does the concentrations of the two species. We illustrate the predictions of the theory for a mixture of spheres of the same diameter but different masses and for spheres of different diameters but nearly the same mass. We show the evolution of the profiles of the concentration fractions of the two types of spheres and the profiles in the final, steady state. The latter compare favourably with those obtained in discrete-element numerical simulations.

1 Introduction

Larcher & Jenkins [1] predict the evolution of concentration profiles in dry flows of a binary mixture of inelastic spheres in dense inclined flows. For the mixture, they employed the extension of kinetic theory for identical, inelastic spheres developed by Garzo & Dufty [2], modifying the expression for energy dissipation to take in to account the formation of particle clusters [3-5]. The theory employed for segregation is appropriate for particles with relatively small differences in size and mass. In doing this, they assumed that the flow of the mixture reaches a fully developed state much more rapidly than does the concentrations of the two species. This is consistent with the hypothesis of dense flow, because the rate at which segregation takes place is proportional to the average distance between the edges of spheres, while the rate at which momentum is transferred in the mixture is inversely proportional to this distance [1].

As do Larcher & Jenkins [1], we consider a mixture of spherical particles with two different radii, r_A and r_B, with $r_{AB} = r_A + r_B$, masses m_A and m_B, with $m_{AB} = m_A + m_B$, number densities, n_A and n_B, with $n = n_A + n_B$, and mass densities, $\rho_A = m_A n_A$ and $\rho_B = m_B n_B$, with $\rho = \rho_A + \rho_B$. We predict the evolution of the profiles of the species concentration fraction, c_A/c, where $c_A = 4\pi n_A r_A^3/3$ and $c = c_A + c_B$, for two combinations of particle radii and particle masses in time-dependent segregation under gravity. The flow is inclined by an angle ϕ, the gravitational acceleration is g, and the strength T of the velocity fluctuations of the mixture is equal to one-third the mean of the square of their magnitude. We compare the steady states obtained with the profiles measured by Tripathi & Khakhar [6] in discrete numerical simulations.

2 Governing equations

2.1 Segregation

Larcher & Jenkins [1] characterize the segregation of two types of spheres using the measure $X = (n_A - n_B)/(2n)$ of segregation and show that when the differences $\delta r = (r_A / r_B) - 1$ and $\delta m = (m_A - m_B)/m_{AB}$, in radii and masses are small, the evolution of the segregation is uniform inclined flow is governed by

$$\rho \frac{\partial X}{\partial t} + \frac{\partial}{\partial y}\left[\frac{m_A n}{4}\left(1-4X^2\right)\left(\tilde{v}_A - \tilde{v}_B\right)\right] = 0, \qquad (1)$$

where y varies across the flow from 0 at the base to h at the upper surface, and

$$\tilde{v}_A - \tilde{v}_B = -D_{AB}\left[\left(6.2\delta m - 4.4\delta r\right)G\frac{T'}{T} - (\delta m - 1.5\delta r)\frac{m_{AB} g\cos\phi}{T} + \frac{X'}{0.25 - X^2}\right], \qquad (2)$$

with the prime denoting a y-derivative, gives the difference of the diffusion velocities for the two species normal to the bed; the diffusion coefficient, D_{AB}, is given by

$$D_{AB} \equiv \frac{\pi^{1/2}}{16}\frac{r_{AB}}{G}\left(\frac{2T}{m_{AB}}\right)^{1/2}, \qquad (3)$$

* Corresponding author: michele.larcher@unibz.it

in which

$$G = 5.69c\frac{(c_M - 0.49)}{(c_M - c)} \quad (4)$$

is the product of the granular concentration, c, and the radial distribution function for dense flows [2]. The value of the singularity, c_M, in the radial distribution function is taken to be 0.58. This provides the best fit to the simulations of Tripathi & Khakhar [3, 6].

We take N_A, V_A, N_B and V_B to be the total numbers and volumes of spheres of species A and B, respectively, and $N = N_A + N_B$ and $V = V_A + V_B$. In the discrete numerical simulations, the total number fraction N_A/N or, equivalently, the total volume fraction V_A/V of spheres is specified. In order to incorporate this information [3], the calculation is effected in terms of the variable $\zeta \equiv Xn/\bar{n}$, where

$$X \doteq \frac{N}{\bar{c}}\left(\frac{\bar{c}_A}{N_A} + \frac{\bar{c}_B}{N_B}\right)\zeta, \quad (5)$$

using

$$\bar{\zeta} = \frac{(\bar{n}_A - \bar{n}_B)}{2\bar{n}} = \frac{(N_A - N_B)}{2N} \quad (6)$$

In these expressions, the overbar denotes depth-averaged quantities.

With equations (2-6), eq. (1) can be solved for the variable ζ; then, the time evolution of the concentration profiles of the two species can be evaluated through eq. (5) and

$$c_A = -\frac{(2X+1)(r_A/r_B)^3 c}{(2X-1) - (2X+1)(r_A/r_B)^3}, \quad (7)$$

$$c_B = -\frac{(2X-1)c}{(2X-1) - (2X+1)(r_A/r_B)^3}. \quad (8)$$

2.2 Mixture fields

The mixture temperature, velocity, and concentration are determined, as outlined by Larcher & Jenkins [3]. The profile of the granular temperature, T, is

$$T = \frac{m_{AB}(h-y)}{4(1+e)G} g\cos\phi(1 + 2\bar{X}\delta m), \quad (9)$$

where e is an effective coefficient of restitution [3], $J = (1+e)/2 + (\pi/4)(3e-1)(1+e^2)/[24 - (1-e)(11-e)]$ [2] and

$$G = \left\{ \frac{4J}{5\pi^{1/2}} \frac{1}{1+e}\left[\frac{30}{J}(1-e^2)\right]^{1/3} \frac{1}{\tan\phi} \right\}^9 \times\left[1 + 3\bar{X}(\delta r + \delta m)\right]. \quad (10)$$

The profile of the mixture flow velocity, u, is

$$u = \frac{5\pi^{1/2}}{6J}\frac{1}{r_{AB}}\left(\frac{1+e}{2G} g\cos\phi\right)^{1/2} \times\left[h^{3/2} - (h-y)^{3/2}\right]\tan\phi\left(1 - \bar{X}\delta r\right), \quad (11)$$

where the slip velocity has been neglected because it does not affect significantly the prediction of granular segregation [3].

The concentration of the mixture, c, can be obtained by inverting eq. (4) and making use of eq. (10):

$$c = \frac{c_M G}{G + 5.69(c_M - 0.49)}. \quad (12)$$

It should be noted that the mixture velocity, temperature and concentration are influenced by the size and mass differences of the two types of spheres through the parameters δm and δr and by their volume fraction through the depth-averaged segregation measure $\bar{X}$:

$$\bar{X} = \frac{1}{4}\left[\frac{V_A}{V} - \left(1 - \frac{V_A}{V}\right)\left(\frac{r_B}{r_A}\right)^3\right]\left[1 + \left(\frac{r_B}{r_A}\right)^3\right]. \quad (13)$$

3 Predictions

For an inclination of $\phi = 25°$, we predict the evolution from a perfectly mixed state of the concentration fractions for two systems: spheres of equal size, but different masses (Fig. 1); and approximately equal masses, but different sizes (Fig. 2).

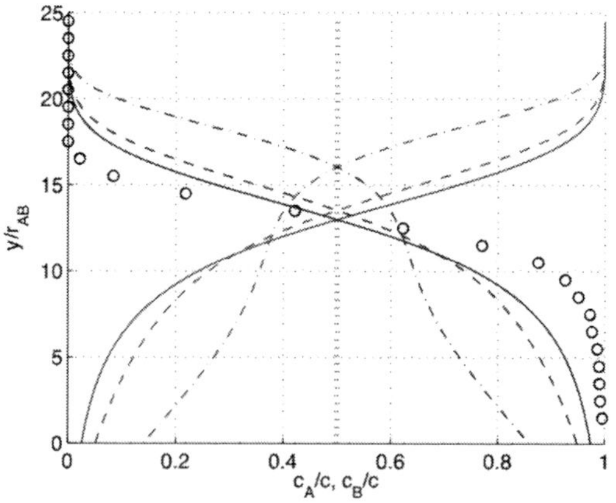

Fig. 1. Predicted profiles, c_A/c, of the more massive spheres (decreasing left to right, blue lines) and of the less massive spheres, c_B/c, (increasing left to right, red lines) at four equally spaced time intervals: $\Delta\tau = 19800$, $r_A/r_B = 1$, $m_A/m_B = 3$, $V_A/V = 0.5$. The black circles indicate the steady profile of the more massive spheres in the discrete numerical simulations of Tripathi & Khakhar [6]. The initial states (dotted lines) evolve through two subsequent intermediate states (dash-dotted and dashed lines) to the steady state (solid lines).

The curves are at equally spaced time intervals $\Delta\tau$, with time made dimensionless by $(r_{AB}/g)^{1/2}$, from a homogeneously mixed state to the final steady state through two subsequent intermediate states. The black circles indicate the steady profile of the more massive (Fig. 1) or the larger (Fig. 2) spheres in the discrete numerical simulations of Tripathi & Khakhar [6].

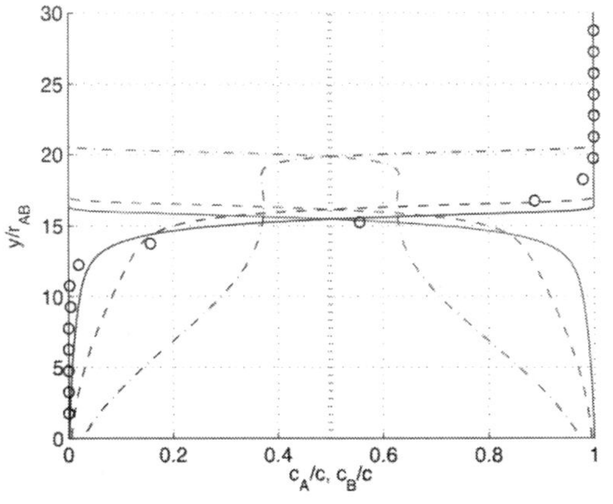

Fig. 2. Predicted profiles of c_A/c for the larger spheres (increasing left to right, blue lines) and for the smaller spheres, c_B/c, (decreasing left to right, red lines): $\Delta\tau$ = 4950, e = 0.7, r_A/r_B = 1.5, V_A/V = 0.5. The black circles indicate the steady profile of the larger spheres in the discrete numerical simulations of Tripathi & Khakhar [6]. The initial states (dotted lines) evolve through two subsequent intermediate states (dash-dotted and dashed lines) to the steady state (solid lines).

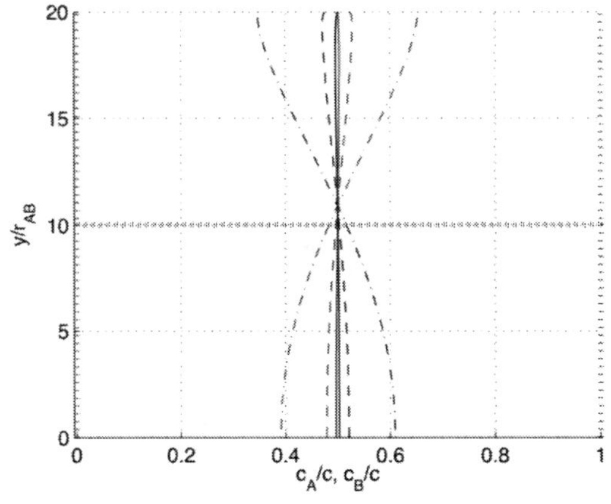

Fig. 3. Predicted profiles of c_A/c for the large, heavy spheres (decreasing left to right, blue lines) and for the smaller spheres, c_B/c, (increasing left to right, red lines) at four equally spaced time steps from the initial state (dotted lines) to the final, mixed state (solid lines): $\Delta\tau$ = 30000, e = 0.7, r_A/r_B = 1.0375, m_A/m_B = $2(r_A/r_B)^3$, V_A/V = 0.5, ϕ = 25°, h/r_{AB} = 20. Initially, all the large, heavy particles are in the lower half of the flow.

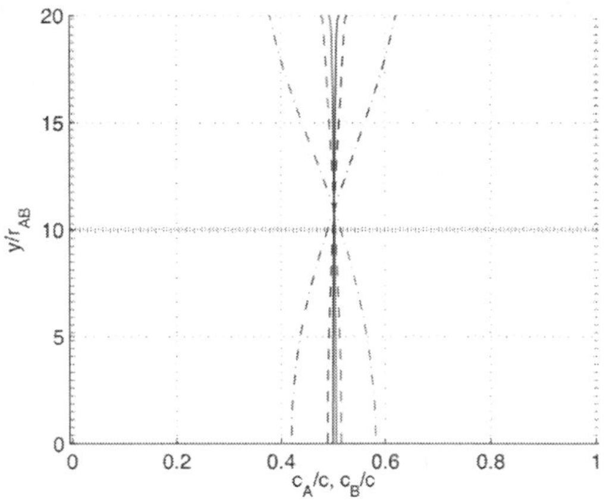

Fig. 4. Predicted profiles of c_A/c for the large, heavy spheres (increasing left to right, blue lines) and for the smaller spheres, c_B/c, (decreasing left to right, red lines) at four equally spaced timesteps from the initial state (dotted lines) to the final, mixed state (solid lines): $\Delta\tau$ = 36000, e = 0.7, r_A/r_B = 1.0375, m_A/m_B = $2(r_A/r_B)^3$, V_A/V = 0.5, ϕ = 25°, h/r_{AB} = 20. Initially, all the small, light spheres are in the lower half of the flow.

For bimodal mixtures of spheres with the same density but different size, the larger particles tend to concentrate at the top of the flow (Fig. 2). In the case of mixtures of species with the same size but different density (Fig. 1), the more massive spheres migrate downwards. Similar findings were shown in free surface flows [7], rotating tumblers [8, 9] and rotating tubes [10].

Due to the competing role of size and mass described above in the evolution of segregation, a particular mixture of large heavy and small light spheres might produce a perfect mixing. This is shown in the next two figures, where two types of spheres, completely separated initially, evolve towards a perfect mixing. The theory predicts a slightly faster evolution if the large, heavy particles are initially concentrated in the lower half of the flow (Fig. 3). In the opposite case (Fig. 4), the steady state condition is exactly the same, although it is reached in a longer time.

4 Discussion

We have outlined a theoretical model for the time evolution of particle segregation of dense, collisional channel flows driven by gravity in the absence of sidewalls. An extension of kinetic theory for dense, dissipative flows was used to derive the profiles of velocity, granular temperature and concentration for the mixture, while a kinetic theory for a binary mixture of nearly elastic spheres was used to describe the time evolution of the concentration profiles for the two types of spheres.

We tested the theory against the numerical simulations of Tripathi and Khakhar [6] and showed good agreement both for the case of particles of the same size and different mass and for particles of different size and nearly the same mass.

The theory predicts a larger influence of size than of mass on the time needed to reach a steady state: the relative concentration of particles with a size ratio of 1.5 and nearly the same mass (Fig. 2) evolves approximately four times faster than particles of the same size having a mass ratio equal to 3 (Fig. 1). The predictions of perfect mixing reinforce the fact that the size dominates mass. In this case, particles with a density that is twice as large as that of the other species, need only be about 4% larger, in order to inhibit the segregation mechanism.

References

1. M. Larcher, J.T. Jenkins, J. Fluid Mech. **782**, 405-429 (2015)
2. V. Garzo, J.W. Dufty, Phys. Rev. E **14**, 341-365 (2004)
3. M. Larcher, J.T. Jenkins, Phys. Fluids **25**, 113301 (2013)
4. J.T. Jenkins, Granul. Matt. **10**, 47–52 (2007)
5. J. T. Jenkins, D. Berzi, Granul. Matt.**12**, 151–158 (2010)
6. A. Tripathi, D.V. Khakhar, Phys. Fluids **23**, 113302 (2011)
7. J. A. Drahun, J. Bridgewater, Powder Technol. **36**, 39–53 (1983)
8. G. Felix, N. Thomas, Phys. Rev. E **70**, 051307 (2004)
9. K. M.Hill, Y. Fan, J. Zhang, C. V. Niekerk, E. Zastrow, S.C. Hagness, J.T. Bernhard, Granul. Matt. **12**, 201–207 (2010)
10. G. Metcalfe, M. Shattuck, Physica A **233**, 709–7171996 (1996)

DOI: 10.1051/epjconf/201714008010

Mean shear resistance at steady-state for wet glass beads impact of liquid content and particle size

Haithem Louati, *Driss* Oulahna, and *Alain* de Ryck*

Rapsodee Research centre, Mines Albi, CNRS University of Toulouse, Campus Jarlard, F-81013 Albi Cedex 09, France

Abstract. The flow behaviour of a granular media is due to their weight, frictional contact forces between them, and external forces exerted by the walls. If their size is lower than 50 microns, the Van-der-Waals forces between them may also influence their flowability. When adding some wetting liquid, we introduce attractive forces between the particles, whose order of magnitude may overcome the particle weight and V-d-W interactions. This leads to a cohesive behaviour. The shear stress to start the flow is greater than in the dry case but the steady-state flow is also perturbed by the presence of liquid bridges. This later phenomenon has been recently quantitatively studied for 70-110 μm glass beads with a non-volatile liquid, with experimental results for different normal stresses (up to 12 kPa) and liquid content (up to 20 % in volume). These results have been compared to a heuristic model, based on the model for capillary bridges and the simplest hypothesis for the granular bed texture depending on the stresses applied. We extend this study with new results concerning smaller glass beads 12-40 μm in diameter and larger liquid fraction for 70-110 μm glass beads using experimental and theoretical approaches.

1 Introduction

The introduction of a small amount of wetting liquid in a granular media leads to enhance its shear resistance if the particle size is typically lower than a few mm (the capillary forces exerted by the liquid menisci are then greater than the particle weight). This behaviour also occurs for a large liquid content till the full filling of the interparticle space. The strengthening is due firstly to the capillary pressure raising from the curvature of liquid-air interface imposed by the solid geometry and fluid incompressibility; and secondly to the liquid-air interface at the solid surface. Depending on the specific proportion of liquid, Newitt and Conway [1] defined the pendular, funicular and capillary regimes for separated liquid bridges, partially and fully filling of the space between particles. This behaviour has been studied by the observation of the limit of static configurations by Hornbaker *et al.* [2], Fraysse *et al.* [3] and Bocquet *et al.* [4]. To quantify more this effect in a dynamic situation, we have performed in a previous study by Louati et al. [5], a series of experiments where the shear resistance of 70-110 μm glass beads, wetted by Polyethylene Glycol of 400 g molar weight, is measured in steady-state conditions for different normal loadings. Depending on both parameters (liquid fraction V_R and normal stress σ), the apparent behaviour is either frictional (shear stress proportional to the normal stress) or presents an apparent cohesion (for normal stresses $\sigma > 4$ kPa). These two behaviours have been semi-quantitatively explained by evaluating the number of capillary bridges renewed during shear at low and high normal stresses. These effects depend on the particle size. We present here how the size reduction affects the shear resistance in steady-state for partially wet granular media, with a liquid content up to the saturation of the interparticles space (60% in our experiments).

* Corresponding author: alain.deryck@mines-albi.fr

2 Materials and methods

The glass beads used as model materials in this study are supplied from Sovitec® and sieved to obtain a narrow size distribution in the range of 70-110 μm and 12-40 μm in diameter. These sizes of glass beads are chosen relatively small to enhance the effect of the capillary forces over the gravitational forces. The surface of the material, analysed using SEM, shows small asperities in the range of 1 to 2 μm regardless the size of glass beads. This information is needed for the calculation of the capillary forces according to Adams and Perchard Approach [6], for the minimal distance separation between the beads.

Polyethylene glycol 400 (PEG 400) from Alfa Aesar was used to prepare the wet glass beads samples. This liquid wets well the glass surface with an angle of contact close to 0°. The density of PEG 400 is 1.128 g/cm^3, and the surface tension is 56.74 mN/m. The fraction of liquid is expressed as the volume of liquid relative to the volume of the glass beads sample, denoted by V_R%. The fractions of liquid studied vary from 0.01% to 20% for the partially state of saturation for both sizes of glass beads. V_R is then increased up to 60% for the 70-110 μm glass beads to explore more saturation regimes.

In order to study the shear behaviour of the wet glass beads, a Schulze shear cell was used. This device consists of an annular shear cell of external and internal diameters of 10 and 20 cm respectively. The shear cell filled with the glass beads rotates at constant velocity of 2.3 mrad/s, while the lid is maintained fixed. The lid has bars to avoid wall slippage and the shear plane is occurred at the level of these bars after rupture of the granular material (see Fig. 1). The shear test consists of measuring the shear stress τ under different normal stresses σ applied to the granular material. We investigate the shear response at the steady-state situation. Further information about this device and the shear test procedure can be found in [5].

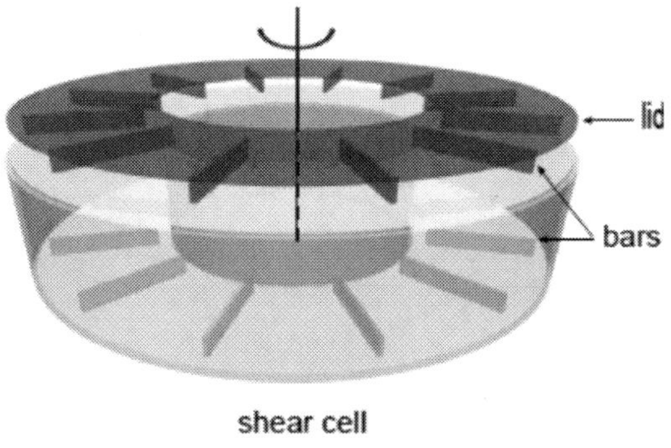

Fig. 1. Schema of the annular shear cell.

3 Results and discussions

3.1. Impact of the particle size on the apparent friction of partially wet glass beads at low normal stress

Figure 2 shows the apparent friction (the so-called bulk friction coefficient) calculated from the slope of the curves given by the shear stress as a function of the normal stress from the shear test experiments. The data are plotted in blue triangles for the 70-110 μm glass beads, and in green circles for the smaller beads (12-40 μm). The data are obtained for low normal stress σ less than 4 kPa where the shear stress τ is proportional to σ for both particle sizes, or linear with a small apparent cohesion (studied in section 3.2.1). The error bars indicate the incertitude of the measure. For the small glass beads, this frictional behaviour extends on the whole range of normal stresses applied in our experiments and the green circles represents therefore the shear behaviour up to σ = 12 kPa. This is not the case for the 70-110 μm glass beads, for which a slope change is observed in the τ- σ curve around σ = 4 kPa. Beyond that value, the 70-110 μm glass beads present a larger apparent cohesion during steady-state shear, studied further in section 3.2. This difference between the 12-40 and 70-110 μm at higher normal stresses is detailed in [7].

From fig. 2, we observe a similar behaviour of the apparent friction for an increase of the liquid fraction for both sizes: a quick increase of the apparent friction from 0.4 to around 0.6 for a small increase of the liquid fraction (less than 5 %) and then a saturation around 0.65. This assumes that the liquid bridges formed between grains contribute to the increase of the apparent friction of the granular material during shearing: in addition to the grain-grain friction, the rupture of the liquid bridge is necessary to provoke the shearing.

At low liquid filling, the liquid bridge strength is mainly independent on the particle diameter but induced by the surface roughness size since the latter fixes the curvature of the liquid bridge. Since the surface asperities are similar in size around one μm for both particle sizes, it is expected a similar apparent friction. This low content stands for quite low liquid content (0.1 % or 0.01 % for respectively the small and larger beads). For larger liquid content, the capillary bridges have a curvature fixed by the particle diameter and the capillary force remains mainly constant, giving this saturation or a liquid content from 5 to 20 %. Finally, the independency of the saturation on the particle size indicates that during shearing at low normal stresses, the weakest capillary bridges are broken and/or the others bonds partially mobilized. This phenomenon needs to be further studied in order to identify the mechanisms. As yet mentioned, this independency of the shear behaviour on the particle size does not occur at high normal stresses, and discussed in the section 3.2.

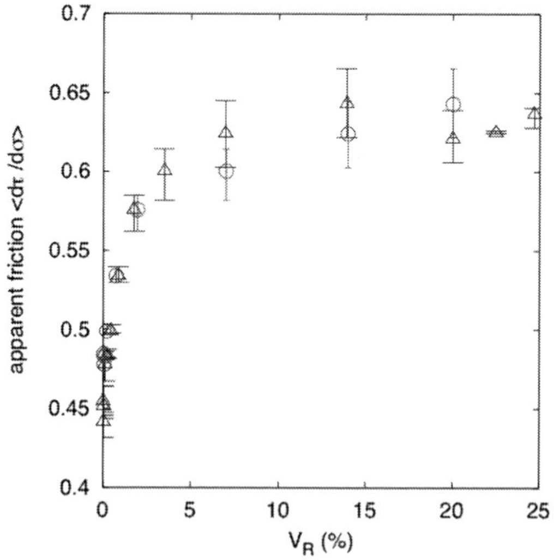

Fig. 2. Variation of the apparent friction at low normal stresses (σ < 4 kPa) with the liquid fraction for two sizes of glass beads (blue triangles: 12-40 μm glass beads; green circles: 70-110 μm).

3.2 Investigation of the tensile strength of wet glass beads for different saturation states

For the 70-110 μm glass beads, the frictional behaviour ends for a normal stress around 4 kPa and the shear stress τ —normal stress relationship at higher σ is best fitted by:

$$\tau = \mu(\sigma + T_S)$$

where μ is an apparent friction close to the one observed for dry beads and T_S is an apparent tensile strength dependent on the liquid content V_R. In order to investigate this tensile strength, experimental and theoretical approaches have been considered. Theoretically, the tensile strength is estimated using Rumpf model [4] for mono-sized spherical particle in the pendular regime (partially saturation of liquid), which states:

$$T_{S,p} = \frac{1-\varepsilon}{\pi} \kappa \frac{F_C}{d^2} \quad (1)$$

where $T_{S,p}$ is the tensile strength in the pendular regime, ε is the void fraction of the wet glass beads, κ is the coordination number, d is the particle diameter and F_C is the capillary force. The latter is calculated using liquid bridge model, as described by Adams and Perchard [1, 2].

In the capillary regime, where the liquid fills all the gaps between grains, the capillary force act through the capillary pressure at the agglomerate surface. Schubert proposed a model to estimate this tensile strength [9, 10]:

$$T_{S,c} = a' \frac{1-\varepsilon}{\pi} \kappa \frac{\gamma}{d} S \quad (2)$$

where $T_{S,c}$ is the tensile strength in the capillary regime, a' is a dimensionless factor ranging from 6 and 8 depending on the particle rearrangement and S is the liquid saturation given as the volume of liquid relative to the volume of void in the glass beads. In the funicular regime, it has been reported by Pierrat and Caram [10] that zones with separated liquid bridges and zones with glass beads totally saturated of liquid coexist. Therefore the capillary force arises from the liquid bridges together with the capillary pressure. The tensile strength is given as:

$$T_{S,f} = T_{S,c} \frac{S_2 - S}{S_2 - S_1} - T_{S,p} \frac{S - S_1}{S_2 - S_1} \quad (3)$$

where S_1 is the limit saturation between the pendular and the funicular regimes and S_2 is the limit saturation between the funicular and the capillary regimes.

The Tensile strength has been investigated in our previous work [1]. We have distinguished between two regimes depending on the applied normal stress to the wet glass beads.

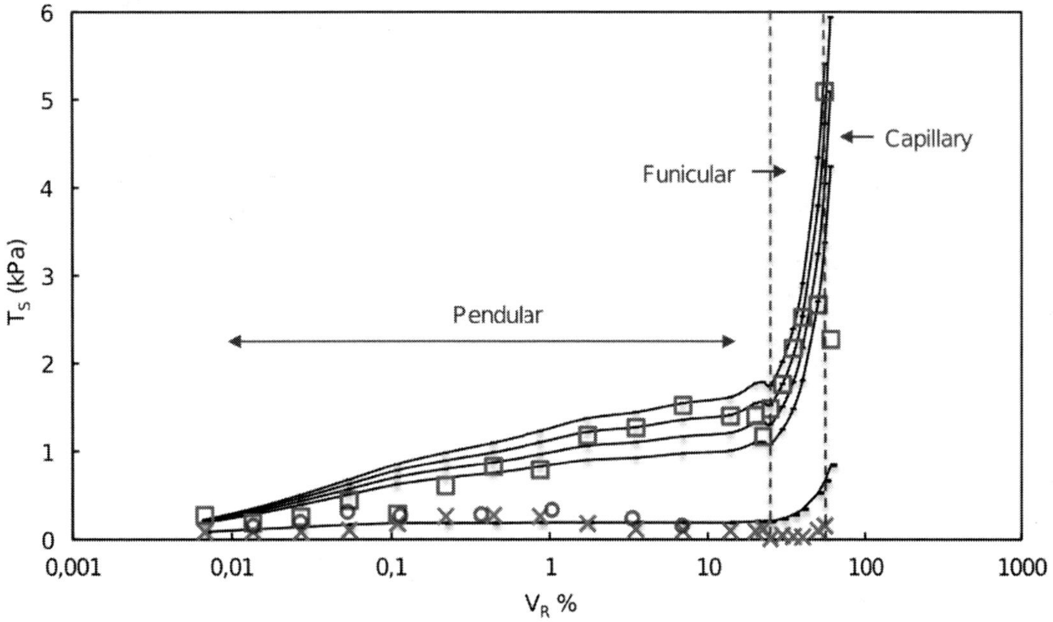

Fig. 3. Experimental and theoretical tensile strength as a function of the liquid fraction on a semi-logarithmic scale in the 70-110 μm case. Green squares and blue crosses: experimental data for the 70-110 μm glass beads at high normal stresses (**σ > 4 kPa**) and low normal stresses (**σ < 4 kPa**) respectively. Blue circles: experimental data for the 12-40 μm glass beads. Continuous lines: theoretical tensile strength for different coordination numbers (1,5,6,7 and 8) shown from bottom to top.

3.2.1 Low normal stress regime for both sizes

At low normal stress (σ < 4 kPa) for the largest beads or for the whole range of normal stresses for the smallest beads, the tensile strength remains low (see blue crosses and circles respectively in Fig. 3) due to a loose configuration of the granular material, inducing a small number of liquid bridges broken explained by rolling rearrangements of glass beads. This small residual tensile strength or apparent cohesion is of order the one we compute with the hypothesis of one capillary bridge per particle broken in the shear plane (lowest continuous line in Fig. 3).

3.2.2 High normal stress regime for largest size

At high normal stress (σ > 4 kPa), the tensile strength increases more significantly with the liquid fraction compared to the one at low normal stress (green squares in Fig. 3). This is explained by the increase of the number of liquid bridges broken in the shear plane when the wet glass beads bed occurred a dense configuration due to the applied normal stress. In the range labelled *Pendular*, the square data are compared with 4 continuous lines obtained from Eq. 1 with a coordination number variation from 5 to 8. The lines are not perfectly smooth because we have used the experimentally measured values for the void fraction ε. (We remind that this regime is not attained for the smallest particles).

In the continuation of these observations, we investigate the tensile strength for other saturation regimes implying eqs. 1, 2 and 3. Figure 3 shows a regime change of the tensile strength variation in the funicular regime. This appears in the graph as a remarkable increase of the tensile strength at high applied normal stress (squares). We assume that the normal stress applied to the wet glass beads affects their packing density. Indeed, agglomerates may be formed in the funicular regime. Thus, at low normal stress, the tensile strength required to separate agglomerates is lower compared to the case of packed granular bed at high normal stress. Moreover, the rolling rearrangement may be enhanced in the presence of agglomerates at low normal stress. This analysis based on the applied normal stress calls attention to the granular bed texture, which affects largely the wet granular materials.

The transition from the funicular to the capillary regime is difficult to identify. The tensile strength supposes to increase further in the capillary regime and vanish after that indicating the slurry regime [11, 12]. Theoretically, the capillary regime is not supposed to be reached for liquid fraction up to 60%. However, we observe a decrease of the tensile strength for the experimental point at high normal stress. This means that under high applied normal stresses, the void fraction in the granular bed is reduced and consequently the saturation of liquid is modified.

4 Conclusions and perspectives

In this study we have been focusing on the effect of the particle size and the liquid content on the shear behaviour of wet glass beads. It was observed that decreasing the size of particles leads to increase the shear resistance of the wet granular material due to the increasing number of broken liquid bridges in the shearing plane. The liquid bridges help increase the apparent friction. Thus, a saturation regime of the apparent friction was observed for the large glass beads after adding a small amount of liquid. While, the apparent friction continues to increase for the small glass beads since the saturation in the number of liquid bridge is not reached yet.
The effect of the liquid content was investigated by looking at the tensile strength variation with the liquid fraction. It was observed a regime change from the pendular to the funicular states of saturation. This is appeared as a remarkable increase of the tensile strength at high normal stress in the funicular regime. However, this tensile strength decreases at low normal stress, which highlights the effect of packing on the behaviour of wet granular materials. The packing of the wet glass beads was investigated using X-ray tomography analysis and has been submitted for publication [7].

References

1. D.Newitt,J.Conway-Jones, Trans. Inst. Chem. Eng. **36**, 422–442 (1958).
2. D. Hornbaker, R. Albert, I. Albert, A. Barabasi, P. Schiffer, Nature **387**, 765 (1997).
3. N. Fraysse, H. Thomé, L. Petit, Eur. Phys. J. B **11**, 615–619 (1999).
4. L. Bocquet, É. Charlaix, F. Restagno, C. R. Phys. **3**, 207–215 (2002).
5. H. Louati, D. Oulahna, A. de Ryck, Powder Technol. **278**, 65-71 (2015)
6. M. Adams, V. Perchard, Inst. Chem. Eng. Symp. **91**, 147-160 (1985)
7. H. Louati, D. Oulahna, A. de Ryck, *submitted to Powder Technol.*
8. H. Rumpf, Intersci. Publ. 379-418 (1962)
9. H. Schubert, Powder Technol. **37**, 105-116 (1984)
10. P. Pierrat, H.S. Caram, Powder Technol. **91**, 83-93 (1997)
11. N. Mitarai, F. Nori, Adv. Phys. **55**, 1-45 (2006)
12. W. Pietsch, *Agglomeration process* (Wiley-VCH, Weinhein, 2002)

Effect of moisture content on the flowability of crushed ores

Francisco Cabrejos[1,*]

[1]Jenike and Johanson Chile S.A., Avda. Libertad 798, Office #501, Viña del Mar, Chile

Abstract. In many mining and industrial processes where large quantities of non-degrading bulk materials such as crushed ores are handled, silos, hoppers, stockpiles and chutes are widely used because they are economical and reliable (if properly designed and operated). However, they are not free of trouble and may experience flow problems such as arching, ratholing, erratic flow, limited storage capacity, limited discharge flow rate, caking, segregation and/or flooding. Moisture content and fine particles significantly affect the flowability of most ores, increasing their cohesive strength and turning them more prone to these problems. The purpose of this article is to highlight a proven, scientific method that can be utilized to ensure reliable storage, flow and discharge of bulk solids in these equipment based on Jenike's flow-of-solids theory and laboratory testing. Knowledge of the flow properties of the material handled provides a design basis to ensure mass flow, avoid arching and prevent the formation of "ratholes". The effect of an increase in water content of the ore is discussed with experimental results.

1 Introduction

Today, bulk solids handling is one of the largest industrial activities in the world. According to the British Material Handling Board more than 16 billion tons of common and industrial materials were transported, handled, and stored in bulk during 2013, usually several times. Millions of tons of ores, concentrates, coal, flyash, fertilizers, agricultural products and salt are mined, handled, transported, stored, crushed, clasified, and/or processed every day of the year in many industries.

In the case of Chile, mining is the most important industrial activity in the country. Chile produced approx. 5.8 million tons of fine copper in 2013, which represents about 50% of our total exports. To produce 5.8 million tons of fine copper in one year, more than 2 million tons of ores must be mined, handled, loaded, transported, stored, crushed, clasified, milled and processed every day, not including the overburden and the low grade ores transported to dumps. Clearly, bulk solids handling is of vital importance to the nation´s development.

Bulk solids are defined as an ensemble of many solid and discrete particles of different size and/or shape, but with the same common nature. For design purposes, it is not enough to simply define a bulk material by its name, origin, source or history. The main characteristics and flow properties will be described in this paper, including testing and standards available to properly characterize a bulk material from a solids handling viewpoint.

As opposed to liquids, bulk solids can hold shear stresses even being at rest, and they may also be cohesive and/or compressible, which explains why they behave so differently. An important characteristic of bulk materials is their ability to form a pile when particles are dropped from some height over a flat surface, forming a repose angle as shown in Figure 1.

Fig. 1. Repose angle in a coarse ore conical stockpile [1].

2 Characteristics of bulk solids

From the viewpoint of flow, bulk materials may or may not gain cohesive strength when handled in stockpiles, silos, bins and hoppers depending on the combination of several factors. If the material does gain cohesive strength, then problems of arching or ratholing may occur - depending on the shape of the hopper, dimensions of the outlet, wall angles, wall liner, and the flow pattern developed by the bulk solid in the storage equipment (mass flow or funnel flow).

The flowability of most bulk solids and ores are affected by the following characteristics and parameters:

- Particle size and size distribution,
- Maximum particle size and content of fines,
- Moisture content,
- Storage time at rest and under pressure,
- Consolidating pressure (i.e. height of the pile or silo),
- Temperature and weather conditions,
- Presence of oils, clays, talcs and/or additives,
- Chemical and mineralurgical nature of the material.

* Corresponding author: fcabrejos@jenike.com

The size of the particles is one of the main characteristics of a bulk material from the viewpoint of handling, storage and transport. The most simple and economic technique to determine the size distribution - widely used in industry for years - is sieving.

Another characteristic needed to describe a bulk material is particle density. This parameter corresponds to the real density of the material, i.e. its weight per unit volume, and it is measured by volumetric displacement in a calibrated tube. In the case of water-soluble materials, other liquids should be used such as acetone or alcohol. This parameter is required to estimate minimum transport velocity and pressure drop in pneumatic conveying, and for flow modeling with DEM.

The **moisture content** of a bulk material (dry basis) is defined as the weight of the surface or free water contained in the material (W_w) over the weight of the sample material minus the weight of the surface or free water (W_s), and is expressed as a percentage, as follows:

$$MC = \frac{W_w}{W_s} \cdot 100\% \qquad (1)$$

A small sample is dried in an oven at 110° ± 5°C until the material does not loose any more weight, based on the standard procedure given by ASTM D 2216. To determine the total moisture content of coal samples, standard ASTM D 3302 should be used.

This parameter strongly affects the flowability of bulk materials, may create arching and/or ratholing flow problems, and reduces the live storage capacity in silos and stockpiles, as will be described next.

3 Flow properties of bulk solids

About five decades ago, Dr. Andrew Jenike proposed a flow-of-solids theory to properly design silos, bins and hoppers [2]. This theory is based on the flow properties of the material handled, which can be determined in a laboratory at the expected operating conditions to be encountered in the plant. The main flow properties required to characterize a bulk solid include:

- Cohesive strength and internal friction,
- Wall friction and chute angle,
- Compressibility (bulk density) and permeability.

Tests should be performed on the minus 1/4" mesh fraction of a representative sample of the bulk solid (100% - 6.3 mm) since typically this fine fraction is what governs the flow behavior of the whole material. Tests are often conducted at different moisture contents, and for continuous flow as well as for different storage times at rest, to determine how these parameters affect the material´s flowability. Flow function tests (today standard ASTM D 6128 using Jenike´s direct shear tester, or ASTM D 6773 using the rotational shear tester proposed by Dr. Dietmar Schulze) are used to determine the cohesive strength of a material and its flowability.

More details about other tests to determine wall friction and chute angles on different wall liners, compressibility and permeability can be found in [3].

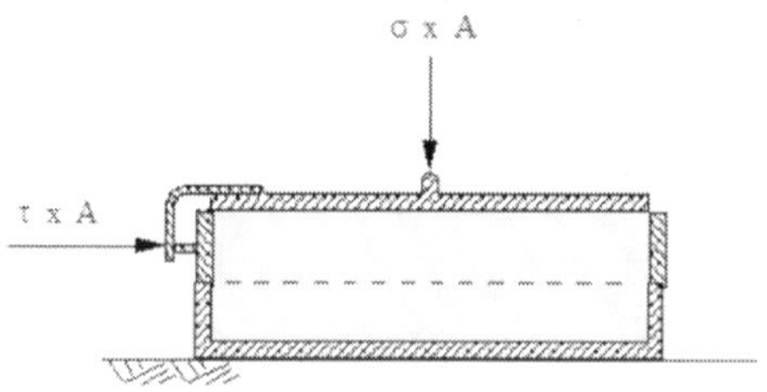

Fig. 2. Direct uniaxial shear test proposed by A. Jenike [2].

4 Effect of moisture content

As mentioned above, the moisture content of a bulk solid strongly affects its flowability, which in turn, increases its tendency to arch and/or rathole, leading to flow stoppages and loss of live storage capacity in storage equipment such as silos, hoppers, stockpiles and chutes.

4.1 Arching

Arching in a silo or hopper can be due to the size of the particles and/or to the cohesive nature of the material, mainly due to its moisture content. To avoid the problem of cohesive arching, the discharge outlet should be large enough and be designed following Jenike´s theory [2] considering the cohesive strength of the material. Based on our experience with crushed ores tested in our laboratory over the years, Figure 3 shows schematically how the minimim outlet dimensions of a conical hopper increases with moisture content.

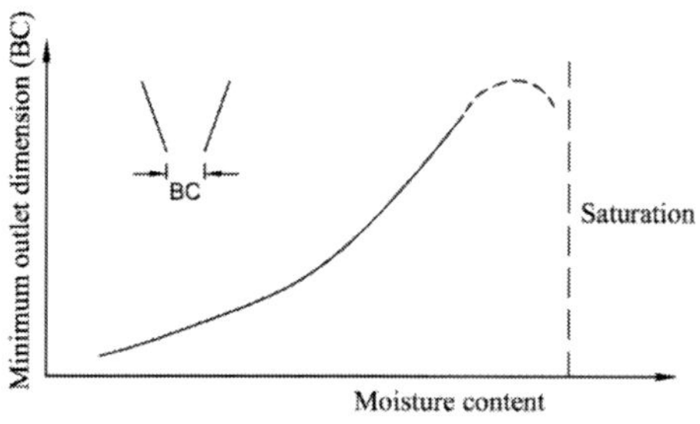

Fig. 3. Minimum outlet dimension of a conical hopper as a function of moisture content.

4.2 Ratholing

The formation of ratholes is typical in stockpiles and funnel flow silos and hoppers with small discharge outlets when handling cohesive materials, as shown in Figure 4. Ratholes may be stable or unstable. In the case of a stable rathole, the material "hangs up" and forms a steep, almost vertical wall, leading to flow stoppage and loss of live storage capacity. In the case of an unstable rathole, flow from the hopper may become erratic. As the rathole collapses, falling material impacts the hopper outlet and may cause a cohesive arch to form, thus stopping flow again.

To avoid the problem of ratholing and to ensure reliable flow, mass flow hoppers should be used to expand the flow channel to a diameter greater than the critical rathole diameter. The critical rathole diameter is a function of the cohesive strength of the material, which in turn increases with moisture content, storage time at rest and consolidation pressure (effective head).

Fig. 4. Typical formation of a "rathole" in a stockpile.

4.3 Cohesive strength

Figure 5 shows the results of flow function instantaneous tests on a sample of the fine fraction of a crushed copper ore (100% minus 1/4"), tested at different levels of moisture content to determine how sensitive this material is to an increase in water content. Time tests (24-hours or more) could be also done to simulate the maximum storage time at rest and under pressure of the material in a silo, hopper or stockpile. This figure also includes the original classification proposed by Jenike [2], and it can be seen that this particular ore is easy-flowing if handled at 1% moisture and for instantaneous flow conditions but becomes cohesive at higher moisture levels.

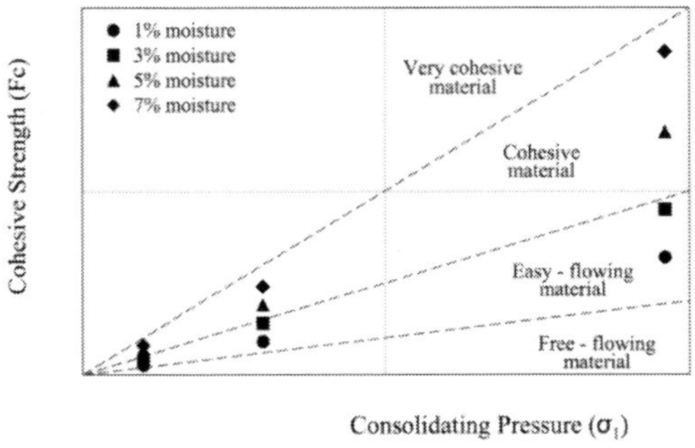

Fig. 5. Instantaneous flow function of a particular crushed copper ore (fines minus 1/4") tested at four moisture levels.

4.4 Drawdown angle

Gravity discharge tests illustrate the range of occurence for the repose angle (α_R) and the drawdown angle (α_{DD}) of a bulk material, as shown schematically in Figure 6. The repose angle is formed when discharging a material from some height and onto a flat surface, while the drawdown angle corresponds to the slope of the remaining walls of material left after gravity discharge ceases. Both angles are measured from the horizontal and are due to the ability of bulk solids to withstand shear stresses (liquids cannot).

As an example, Figure 7 shows the drawdown angle formed by the fine fraction minus 1/4" of a crushed copper ore at 1% moisture (top), i.e. just a few degrees steeper than the repose angle, and about 70° at 5% moisture (bottom). The live storage capacity of stockpiles and flat bottom silos strongly depends on this parameter, which in turn depends on the moisture content and the amount of fine particles in the ore.

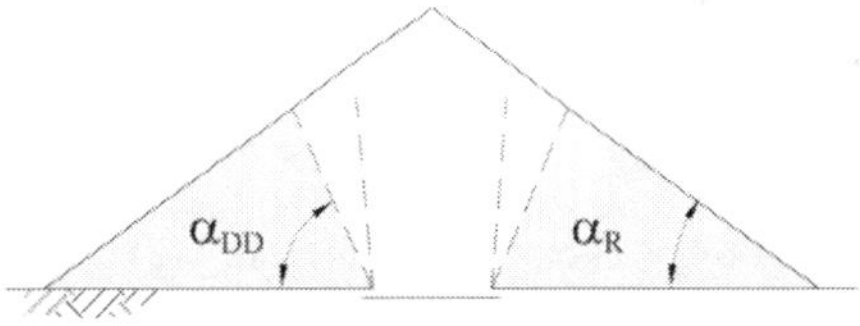

Fig. 6. Repose and drawdown angles of a bulk material.

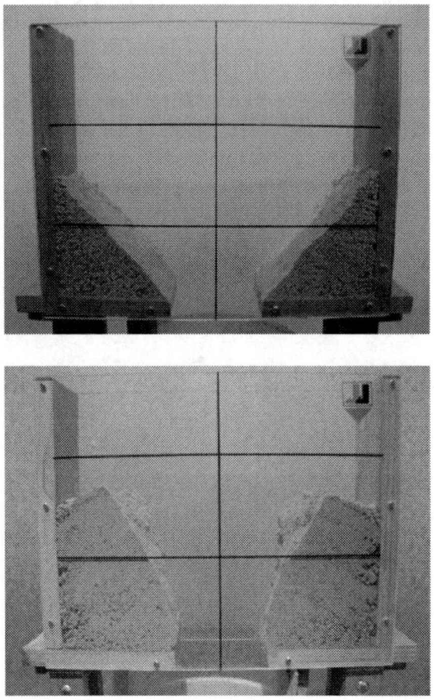

Fig. 7. Drawdown angle of a crushed copper ore tested at 1% and 5% moisture in a slice 2D model of a flat bottom bin.

4.5 Wall friction and chute angle

Wall friction tests should be performed on different types of wall liners to determine the minimum angle for mass flow. The coefficient of friction is also required for bin loads and structural design of silos, hoppers and feeders. Chute tests should be performed on different types of wall liners to determine the minimum slope of a plane surface to maintain flow of material after impact.

Along with the wall friction angle, this is an important parameter in the design of sliding chutes and belts, to prevent adhesion problems between the bulk material and the wall liner selected, as illustrated in Figure 8.

Fig. 8. Adhesion of copper concentrate at 10% moisture onto a plate of carbon steel.

4.6 Moisture content as function of particle size

Some interesting results of a test done with a primary crushed copper ore handled in a local mine are given in the next figures. Figure 9 shows the cumulative sieve analysis of the ore while Figure 10 the moisture content of each fraction of the ore tested. The moisture content of the complete sample is 3.8%, but clearly, the fines contain a higher proportion of water due to their larger surface area, as compared with the coarser particles.

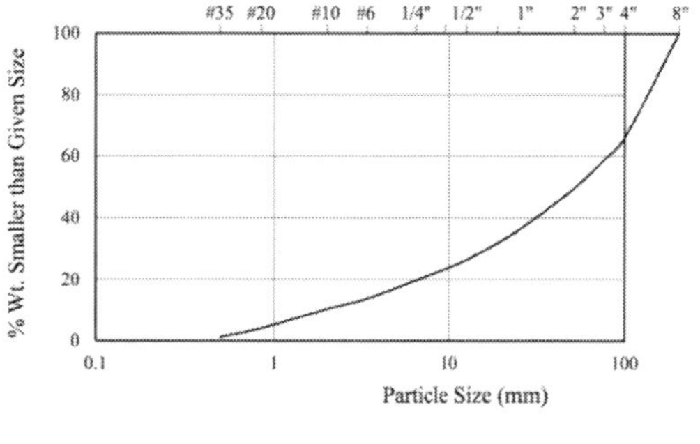

Fig. 9. Particle size distribution of a primary crushed ore.

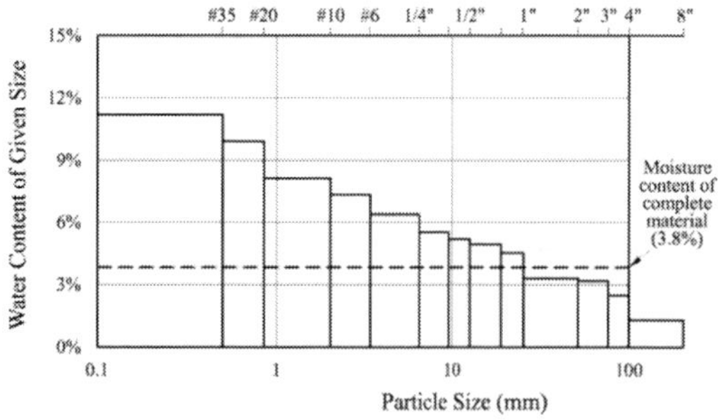

Fig. 10. Moisture content of the primary crushed ore shown in Figure 9 (determined per sieve size).

4.7 Relationship for water content of the fines

The test described next was developed by J&J engineers to better understand how the water gets distributed between the fine and coarse particles in a sample of crushed ore. Fines are defined here as the fraction below mesh 1/4" as per ASTM D 6128. A sample of 50 kgs. of dry ore containing a known percentage of fines and a top size of 4" was placed in a drum. A given amount of water was sprayed on the ore surface and the moisture content of the full size ore was calculated using equation (1). Next, the drum was closed, rotated during 3 days, and the moisture content of the fines was determined.

The test was repeated varying the fines content of the ore, and the complete results are summarized in Figure 11. For example, the moisture content of the fine fraction (minus 1/4") of a full size ore with 25% of fines is about 2 times larger as compared to the full size ore.

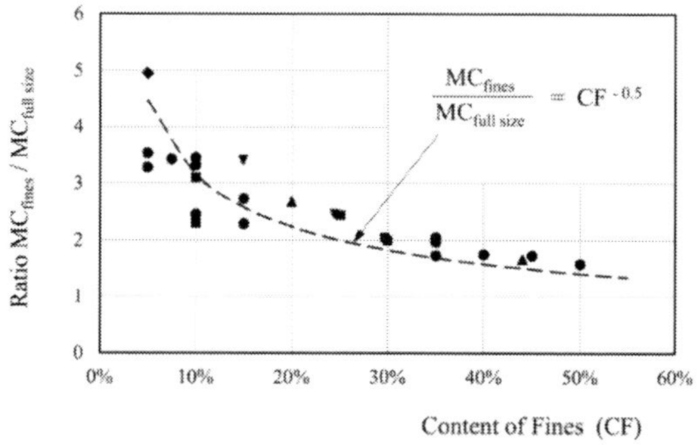

Fig. 11. Effect of moisture content of the minus 1/4" fraction of the ore as a function of moisture and content of fines of the ore.

5 Conclusions and recommendations

The moisture content strongly affects the flowability of most crushed copper ores, increasing their cohesive strength and turning them more prone to create flow problems such as arching and ratholing.

Fine particles of a crushed ore have a higher moisture content - as compared with the coarser particles - due to their larger surface area. This article presented the results of a simple test to determine how much water this fine fraction could contain, based on the percentage of fines (minus 1/4") and the moisture content of the full size ore.

Finally, testing provides the best method to characterize a bulk material and its handling properties for proper equipment and system selection, design and operation of silos, hoppers, stockpiles and chutes [3].

References

1. F. Cabrejos and A. Del Campo, *Ensure Reliable Flow - How to Design Gravity Reclaim Stockpiles*, BSH, Vol. 29, No. 6 (2009), pp. 350-355
2. A. Jenike, *Storage and Flow of Solids*, University of Utah Engineering Experiment Station, Bulletin No. 123, November 1964
3. www.jenike.com

Lift on side by side intruders of various geometries within a granular flow

M. F. Acevedo-Escalante[1,*] and *G. A.* Caballero-Robledo[1]

[1]CINVESTAV-Monterrey, PIIT, Nuevo Leon, 66600, Mexico

Abstract. Obstacles within fluids have been widely used in engineering and in physics to study hydrodynamic interactions. In granular matter, objects within a granular flow have helped to understand fundamental features of drag and lift forces. In our group, we have studied numerically the flow mediated interaction between two static disks within a vertical granular flow in a two-dimensional container where the flow velocity and the distance between obstacles were varied. Attractive and repulsive forces were found depending on flow velocity and separation between intruders. The simulations evidenced a relationship between the average flow velocity in a specific section ahead of the obstacles and the attractive-repulsive lift. On the other hand, it was showed that the lift force on an object dragged within a granular medium depends on the shape of the intruder. Here we present experimental results of the interaction between two side-by-side intruders of different shapes within a vertical granular flow. We built a quasi-two-dimensional container in which we placed the intruders and using load cells we measured lift and drag forces during the discharge process for different flow velocities.

1 Introduction

The interaction of obstacles within a flow is widely studied for theory and applications in fluid mechanics. In granular matter, similar experiments where obstacles are immersed in a granular medium, have been made because they are useful to understand mechanical properties in this area.

Granular matter community has thoroughly studied forces on objects immersed in granular flows [3-6]. Flow velocity and its fluctuations have shown to play an important role in attractive and repulsive regimes in flow mediated interactions between obstacles, and lift forces have been associated with a Bernoulli-like phenomenon: a pressure drop associated to flow velocity.

On the other hand, lift forces have been shown to depend on the shape of the immersed object which is due to the "asymmetry in pressure caused by gravity between the lower and the upper parts of the object" [2].

Recently, it was shown numerically that in a static pair of obstacles within a granular flow exists attractive and repulsive lift forces depending of the flow velocity and the separation between them [1]. Among other things, it is found that the mechanism to explain these interacting forces is not a Bernoulli-like effect.

In the present work we show the experimental results for drag and lift forces on obstacles of different shapes placed into a gravitational granular flow.

2 Experimental Setup

The experimental setup consists of a vessel built with two 125x50x1.3 cm^3 polycarbonate sheets separated by a gap of 1.3 cm (Fig. 1a and Fig. 1b).

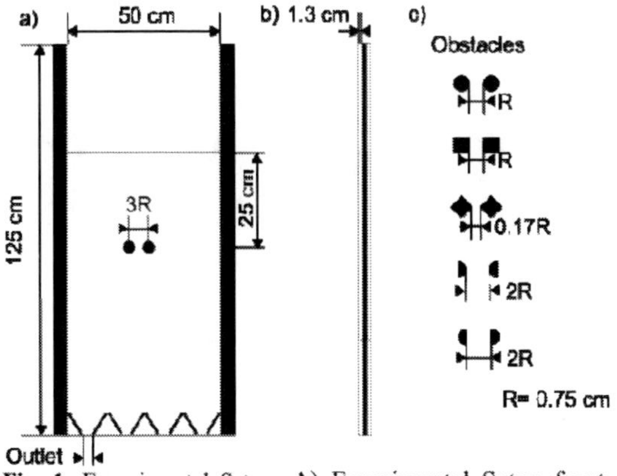

Fig. 1. Experimental Setup. A) Experimental Setup front View. B) Experimental Setup Side View. C) Pair of Obstacles used.

The vessel is filled with 6 kg of tapioca seeds of 3.3 ± 0.2 mm in diameter using a hopper placed at the top of the vessel. The gap between the sheets is approximately four times the size of the seeds.

The mass flow is controlled by a system of outlets of variable area placed at the bottom of the container. The

* Corresponding author: acevedo.escalante@gmail.com

area of the outlets is controlled by varying the angle of the sticks that form the outlets, as can be seen in Figure 1a. The mass flow was measured by collecting the mass flowing through the bottom outlets in a container placed on a scale.

At a height of 55 cm from the bottom of the vessel a pair of obstacles are connected to a pair of uncoupled load cells of 780g capacity which measure drag and lift forces. The pairs of obstacles that were used in the experiments are shown in Fig. 1c: cylinders of radius R; squares of side R placed horizontally and diagonally; and semicircles with the round side facing the centre and towards the walls. The experimental procedure used to measure the forces was as follows: A strip is placed on the bottom of the container to prevent flow. The data logger is initialized and by filling from the top, the container is completely filled. After about 10 seconds the base strip is removed and the system begins to flow. When the system is emptied, data collection is completed. This procedure is repeated for different mass flows and for each pair of obstacles. Finally, particles velocity was measured using a high speed camera to film the system flowing free of obstacles and analysed with a Particle Image Velocimetry (PIV) algorithm using the open source software ImageJ. This software consists in identify the positions of the particles in two sequential images and determine the mean velocity of the particles.

3 Experimental Results

3.1. Mass Flow

Figure 2 shows the results obtained for all configurations of pairs of obstacles. First, this plot shows that the mass flow is similar in all the outlets used in these experiments and, second, the plot is not linear. This last result was expected because it is consistent with Beverloo's Law for the relation between the quantity of particles falling through an outlet and outlet's area [8].

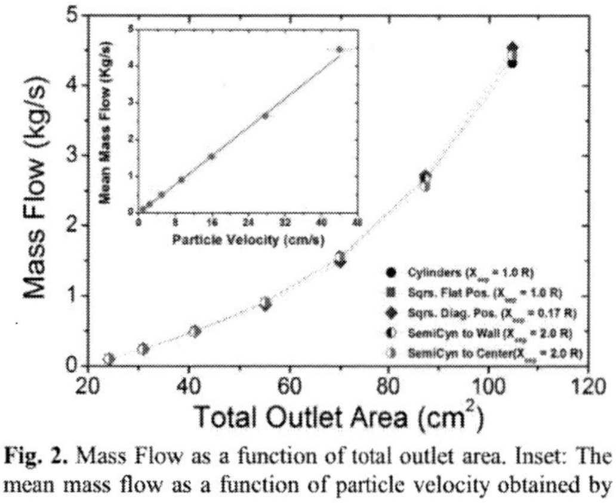

Fig. 2. Mass Flow as a function of total outlet area. Inset: The mean mass flow as a function of particle velocity obtained by PIV.

Our PIV analysis could be tested by plotting the velocity measured by this method and the mass flow directly measured with the scale at the outlet. The relation between these two quantities is linear and is shown in the inset of the Fig. 2.

3.2 Drag Force

In this study we define Drag force as the force in vertical direction perpendicular to the direction of the flow. In Fig. 3 Drag force as a function of flow velocity is plotted.

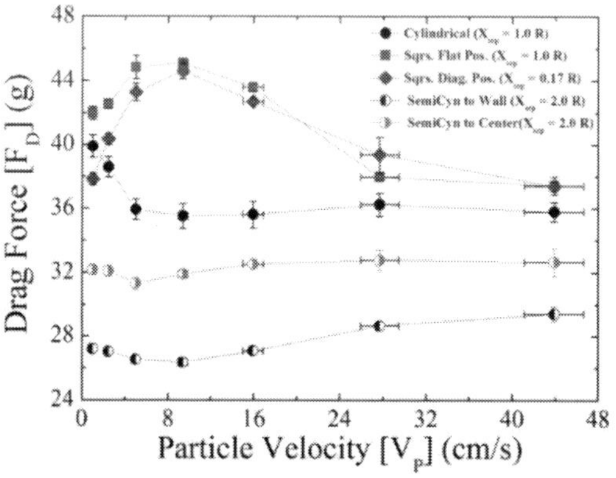

Fig. 3. Drag Force as a Function of Particle Velocity for each pair of obstacles placed in the container.

According to these results, there can be identified the following behaviours: the square obstacles (flat and inclined) have a similar behaviour between them. It can be seen that there is a maximum drag in $V_P = 9.37 \pm 0.47$ cm/s and then decreases to a value $F_D \sim 37$g for the maximum experimental particle velocity reached ($V_P = 43.86 \pm 2.75$ cm/s). When the cylindrical obstacles are placed, an inverse behaviour is observed. In this case, the plot suggests that the minimum F_D is observed in the same value of the maximum drag for the square obstacles. Finally, the drag measured for the pair of semi-cylindrical obstacles shows a similar behaviour that the observed for the cylindrical obstacle, presenting a minimum approximately in 4.98 ± 0.27 cm/s $< V_P < 9.37 \pm 0.47$ cm/s. In these cases F_D (SC to Walls) $<$ F_D (SC to Centre) $<$ F_D (Cylinders). At Low velocities, in the case squares obstacles we thing that in the flat sides the probability of jamming and the arch formation is high than the other cases, because the friction between the particles and the flat face are acting along all the surface unlike in the round obstacles (complete and halves) where the friction acts in a single point. Then, when the velocity is increased; apparently the drag force for all cases trends at the same value, since the probability of jamming is decreased.

3.3 Lift Force

When the lift force $F_L < 0$ the lift is attractive, *i. e.* the force is towards the centre of the vessel, and when $F_L > 0$, the lift is repulsive, *i. e.* the force is towards the walls.

DOI: 10.1051/epjconf/201714003048

The behaviour of F_L as a function of flow velocity for each pair of obstacles can be observed in Fig. 4. For the cylindrical, square, and semi-cylindrical obstacles oriented to the centre, the lift is entirely repulsive.

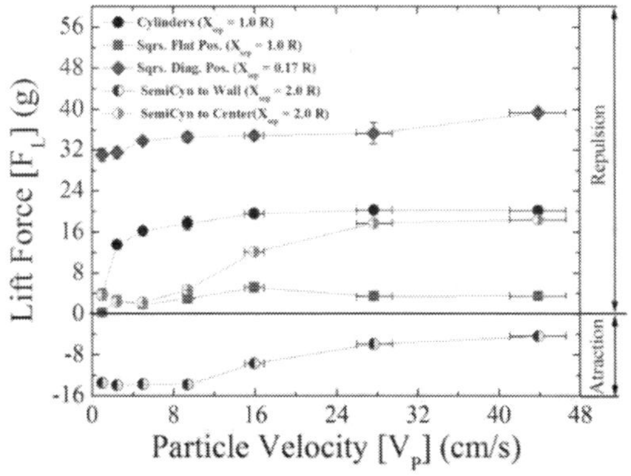

Fig. 4. Lift Force as a function of particle velocity for each pair of obstacles placed in the container.

To understand these results, it is shown in [7] that in these separations the probability of jamming is high. This implies that the quasi-static arches formed between the obstacles tends to separate the obstacles. It can be observed that the maximum lift force corresponds to the diagonal squares, which is in agreement with this hypothesis. In this case, the probability of jamming is very high since the separation is very small (X_{sep} = 0.17 R), and the lift is almost constant, independent of flow velocity. The cylinders have an interesting behaviour: at low velocities the lift is small, but when the flow velocity increases, the lift increases approaching asymptotically to a constant value.

The semicircles oriented towards the centre show that at low velocities the lift force is small. When the flow velocity is increased the lift force seems to reach a minimum, and then it increases and tends to values near those of the lift for cylindrical obstacles.

The squares in flat position show the lowest attractive lift force. This curve seems to confirm that the probability of jamming between the obstacles is high and thus a repulsive lift force is favoured. The position of obstacles also explains the low lift force value since in a flat position it is not easy to form a strong arc, and the few that are formed are probably formed in the edges.

Finally, the semicircles oriented towards the walls present values of lift force entirely attractive. This behaviour is consistent with the results reported by Y. Ding et al. [2]. Interestingly, at small flow velocities the lift force is the most attractive but as the flow velocity increases the lift force tends to zero. We can speculate that at low velocity the packing fraction is high, and the granular medium exerts a larger force toward the centre. When the velocity is increased, the packing fraction is decreased, and the arcs formed between obstacles become important and they can begin to balance the force exerted by the medium. However, according this results, the arcs cannot completely balance the force from the medium and the resulting lift force is still attractive.

4 Conclusions

According with the results presented here we can conclude the following: 1) the drag forces are larger using faceted obstacles than rounded obstacles; and 2) the lift force is also affected by the form of the obstacles. These results suggest that attractive or repulsive forces can be altered by changing the shape of the obstacles while maintaining the global symmetry.

For future work we intend to study of the drag and lift forces varying the separation between obstacles, using the same shapes used in the present work.

Acknowledgments

M. F. Acevedo–Escalante thanks CONACYT for financial support. This work has been supported by Conacyt, Mexico, under grant 180873.

References

1. R.A. Lopez de la Cruz and G. A. Caballero-Robledo. J. Fluid Mech. **800** (2016): 248-263.
2. Y. Ding, N. Gravish, and D. I. Goldman. PRL **106**.2 (2011): 028001.
3. F. Pacheco-Vazquez and J.C. Ruiz-Suarez. Nat. Commun. **1**, 123.
4. J. M. Solano-Altamirano, G. A. Caballero-Robledo, F. Pacheco-Vazquez, V. Kamphorst and .C. Ruiz-Suarez. PRE **88**, 032206
5. D. I. Goldman. Rev. Mod. Phys. **86**, 943-958.
6. I. Zuriguel, J. F. Boudet, Y. Amarouchene and H. Kellay. PRL **95**, 258002
7. I. Zuriguel, L. A. Pugnaloni, A. Garcimartín and D. Maza. PRE **68**, 030301.
8. Beverloo W A., Leniger H. A. and Van de Velde J. J. Chem Eng. Sci. **15**, 260-296.

Particle-based simulations of powder coating in additive manufacturing suggest increase in powder bed roughness with coating speed

Eric J. R. Parteli[1,*] and *Thorsten* Pöschel[2,**]

[1] *University of Cologne, Cologne, Germany*
[2] *University of Erlangen-Nuremberg, Erlangen, Germany*

Abstract. We have developed the first particle-based numerical tool to simulate the coating of powder particles in additive manufacturing devices. Our Discrete Element Method considers realistic particle shapes and incorporates attractive interaction (van-der-Waals) forces between the particles. From simulations of powder coating using a roller as coating device, we find that the surface roughness of the powder bed scales with the square of coating speed. Moreover, we find that using fine, highly polydisperse powders may lead to larger powder bed roughness, compared to process simulations using coarser powders, due to the formation of agglomerates resulting from cohesive forces.

1 Introduction

Additive manufacturing (AM), which consists of building parts by successively melting powder layers by means of a laser or electron beam, has been pointed out as one of the pilars of the third industrial revolution. In particular, parts of nearly arbitrarily complex geometries can be efficiently produced without the waste of material involved in conventional machining processes [1–5].

However, to achieve mass production a quantitative understanding of the effect of powder particle shape, material properties and powder coating speed on the packing characteristics of the deposited powder layer is required. Since it is difficult to achieve this understanding by experiments alone, a particle-based tool for the simulation of the coating process, which allows for the process investigation under different powder systems and process parameters, is required. This numerical tool should account for a realistic description of inter-particle forces, particle size distributions and complex geometric shapes of the powder particles, which play major role for static and dynamic characteristics of granular systems. Here we develop such a numerical tool, based on the Discrete Element Method (DEM), and apply it for the first time to simulate powder coating using a roller as coating device. The results of our simulations show a suprising dependence of the powder bed packing behavior on particle size distributions, and suggest that increasing the coating speed may cause decrease in part quality. As we will show, DEM simulations prove useful to shed light on properties of the powder transport that are challenging to investigate in experiments.

*e-mail: eric.parteli@uni-koeln.de
**e-mail: thorsten.poeschel@fau.de

2 Numerical experiments

The first step to achieve reliable numerical simulations of the powder application in AM is the accurate representation of the complex geometric shapes of the particles. This representation is accomplished here using the multisphere method, which consists of building clumps of spherical particles to model the target complex shape (see e.g. Refs. [6–9]). Fig. 1 shows some examples where the target shapes were obtained from light microscope images of commercially available PA12 powder as described in detail in Ref. [10].

The translational and rotational motion of the multisphere particles are dictated by the forces on the constituent spheres, that is gravity, inter-particle forces and forces between particles and the walls of the device. Given the small particle sizes involved in the process ($1-100\,\mu$m, see Fig. 1), attractive interaction forces must be modeled and incorporated in the DEM. In Ref. [11] we have developed a particle-based model which takes into account contact forces, as well as inter-particle attractive interactions. The contact forces are modeled by applying a modified Cundall and Strack model [12] which considers elastic and dissipative forces in the normal collision direction as well as in the tangential direction (that is, frictional forces) [13–15]. Recently, we have extended this modeling to incorporate attractive particle interaction forces [10, 11]. Our improved DEM model thus accounts for both adhesion at contact and non-bonded van der Waals forces, following the models in Refs. [16, 17]. As shown in Ref. [11], simulation predictions for the solid fraction of different types of powder covering a broad range of particle size distributions relevant for AM process agreed quantitatively with experimental results.

 DOI: 10.1051/epjconf/201714015013

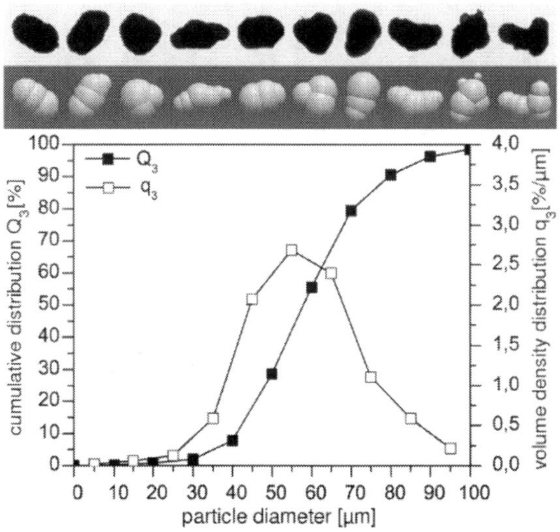

Figure 1. The upper figure shows light microscope images of PA12 powder (top) along with the corresponding model particles (bottom), while the graph shows the particle size distribution.

We have implemented the inter-particle interaction model in the DEM solver of Ref. [18] (LIGGGHTS), as this solver includes an interface to CAD programs from which we can reproduce the complex geometries of the AM devices. In the simulations, these geometries are interpreted as (dynamic) boundary conditions for the particles, whereas the force model described above has been adapted and implemented to compute particle-wall interactions (considering one of the interaction partner of infinite mass and radius). Further extensions of the DEM solver were made to incorporate the model of Ref. [8] for the correct computation of inertia tensor of the multisphere particles.

Simulations are performed with 5761 complex particles, which are modeled with the multisphere method, with the shapes shown in Fig. 1 and material parameters listed in Tab. 1 (see Ref. [10] for an in-depth discussion of the model parameters). In total, 24,771 constituent spheres are involved. Initially, the volume on both sides of the part is filled with powder particles, whereas the surface of the part is free of powder. The powder is then transported by the roller onto the part, represented by the green obstacle Fig. 2, and approximated with the multisphere method. In the simulations, the part has length (in the transport direction) of about 1 mm.

3 Results and discussion

Fig. 2 shows a simulation snapshot indicating the main components of our DEM model for powder application in AM. The roller moves in positive x direction with translational velocity V_R thereby rotating counter-clockwise with a nominal rotational velocity $\omega_R \approx 165\,\text{rpm}$. While we did not find significant dependence of the results presented

Table 1. Numerical values of the parameters used in the simulations [10].

parameter	symbol	value
particle material density		$1000\,\text{kg/m}^3$
Young's modulus	Y	2.3×10^7 Pa
Poisson's ratio	ν	0.40
Coulomb's friction coefficient	μ	0.50
surface energy density	γ	$0.1\,\text{mJ/m}^2$
Hamaker constant	A_H	0.2×10^{-21} J

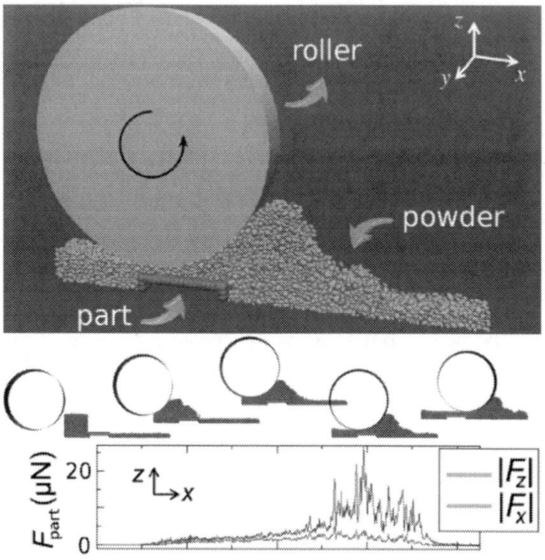

Figure 2. On top we see one snapshot of the DEM simulation while on the bottom we see the evolution of the vertical and longitudinal components of the total force on the part as the roller moves to the right (corresponding position of the roller is indicated by the snapsthos on top of the plot). The part has length 1 mm and 0.4 mm in x and y directions, respectively, while periodic boundary condition in y direction is applied.

here on ω_R (see next Section for further discussion), we find that the packing behavior of the applied powder depends strongly on V_R, as we will discuss below.

We started our investigation by analyzing the evolution of the stress exerted by the granular particles transported by the roller on the part (see Fig. 1. The study of the forces within the granular system and its interactions with the device's walls and part is important as future modeling will focus on breakage and particle deformation. It is well known that stresses in granular systems are characterized by force chains which display complex dynamical behavior as the granular material is subject to external shearing or compaction. We have found (see Ref. [10]) that stress distribution within the granular bulk varies strongly over time. A fingerprint of this variation is the behavior of the total force on the part shown in the plot of Fig. 1. We see strong fluctuations in the total force that reach more than an order of magnitude, both in the vertical and in the horizontal direction, as the roller moves in the transport direction.

Moreover, we see that both components F_x and F_z nearly vanish when the roller has covered 50% of the part's length (note that time in the plot is indicated by the position of the roller relative to the bed in the upper snapshots). This result indicates the time instant from which the lowermost granular layer in the powder bed deposited on the part becomes nearly shielded from the stress due to the roller motion. However, we expect very different behavior for the plot in Fig. 2 if load in the vertical direction is exerted by the roller on the powder bed. Because this load is typically applied in experimental conditions, it constitutes one important extension to our model, whereas its effects on the total force on the part and the characteristics of force chains in the system are matter of ongoing research (first results will appear soon [20]).

Moreover, the transport dynamics largely affects the packing behavior of the powder particles deposited on the powder bed. To investigate this behavior, and since it is desired to enhance production speed while obtaining flat, densely packed powder beds, we calculated the roughness of the powder bed as a function of the roller's velocity V_R. To quantitatively characterize this roughness, we divided the powder bed on the part in 4 longitudinal slices of equal width (in y direction) and computed the mean standard deviation of the maximal powder bed height $h(x)$ for all slices (where x is the transport direction). The result is shown by the blue squares in Fig. 3. We see a clear increase of the bed roughness — which we call δ — with the roller's translational velocity V_R. In particular, we find that the expression

$$\delta = a + b\,|V_R|^2 \tag{1}$$

fits well the simulation results thereby leading to $a \approx 6.05\,\mu$m and $b \approx 3.9 \times 10^{-4}\mu$m. Although this result remains to be confirmed experimentally, there is experimental evidence for increase in the roughness of powder lines deposited (poured) by a nozzle as a result of increase in coating speed.

We thus addressed the question of how particle size distribution affects the powder roughness. To investigate this issue, we modified the size distribution shown in the plot of Fig. 1 by removing all particles with diameter smaller than 60μ (but keeping the same total mass of powder particles in the numerical experiment).

The result for surface roughness δ as a function of V_R is denoted by the circles in Fig. 3. We see that δ increases with V_R as it does for the original size distribution, but overall values of δ obtained using the modified (coarser) powder are smaller. This is suprising result if we consider the well-known effect that deposition of highly polydisperse granular systems leads to higher solid fractions than for nearly monodisperse systems, as small particles fill the pores between larger particles. However, we can understand our results by noting that attractive inter-particle forces (adhesion and non-bonded van der Waals) lead to the formation of large agglomerates in the powder associated with the blue squares in Fig. 3, due to presence of very

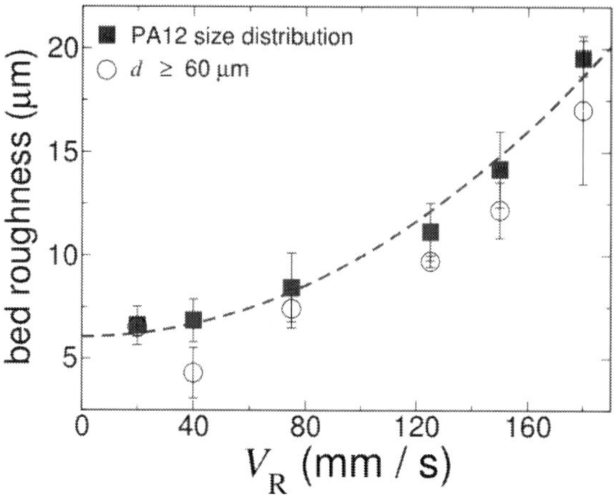

Figure 3. Roughness of the deposited powder layer as a function of the translational velocity of the roller for constant rotational velocity of 165 rpm.

small particles < $30\,\mu$m. For such particles, gravitational forces are largely compensated by cohesive forces, so that such small particles are transported as larger, irregular clumps or coated on the surface of much larger particles. In the modified powder size distribution, the phenomenon of agglomerate formation is less pronounced as van der Waals forces become less efficient in keeping agglomerates stable when particle size increases beyond $60\,\mu$m (see Refs. [11] for a discussion on the effect of polydispersity in the solid fraction of fine powders and Ref. [10] for a more detailed analysis of the deposited powder bed in AM).

We have thus found that a way to decrease the roughness of deposited powder beds in additive manufacturing is to modify the particle size distribution so that smallest particles are filtered out from the system. This surprising result — which could be exploited to improve the quality of powder beds in AM without necessity of changing process parameters — should be now verified in experiments. Moreover, we should remark that the solid fraction may be substantially improved if vertical load is exerted by the roller on the powder bed, or when the device is subject to vibrations, which remains to be investigated.

4 Conclusions and outlook

We have presented a particle-based numerical tool for the simulation of powder application in additive manufacturing (AM) devices. This tool accounts for a realistic description of complex particle shapes, inter-particle forces and particle size distributions of powder systems of different materials used in AM. By applying our tool to investigate the coating process using the roller as coating device, we have found a surprising result, that is the roughness of the applied powder bed increases with the coating speed. This result, which is consequence of the cohesive inter-particle forces leading to large particle agglomerates, suggests that faster production speed may not be a good option in mass production since low solid fraction and large bed roughness may negatively influence part quality. We refer the reader to Ref. [10] for further discussions on our results as well as for a more detailed presentation of our DEM model and numerical experiments. It would be interesting to perform systematic experiments using different roller translational velocities to confirm the predictions from our simulations.

Our numerical tool provides a means to investigate particle-based processes in AM that are difficult to be studied experimentally, such as stress distribution within the powder bulk, on the part and on the coating device. We have seen that such stress distributions are characterized by force chains and large stress fluctuations that are characteristics of granular materials [19]. However, the interparticle force model should be now extended to incorporate electrostatic interactions and tribocharging, which are still poorly understood but may largely affect the transport process depending upon the powder material and process parameters.

Moreover, several questions and challenges constitute matter of ongoing [20] and future work. For instance, how do vibrations applied to the device affect the packing properties of the bed? How does the packing density depend on particle material properties or shape? What is the effect of the load of the coating system on the powder bed on the relation between coating speed and surface roughness? In particular, we have found that the rotational velocity of the roller does not play a significant role for the solid fraction of the deposited powder, but does this result change if load is applied?

We have shown how DEM simulations may help to find the answers to such questions. Following the aforementioned model improvements, our research thus advances with the vision to achieve a numerical tool that can reliably complement and largely replace systematic powder coating experiments, as well as derive strategies to optimize the production process.

Acknowledgements

We thank the Institute for Polymer Technology and the Institute for Particle Technology of the University of Erlangen-Nuremberg for the light microscope images and particle size distribution of Fig. 1, the German Research Foundation (DFG) for funding through the Cluster of Excellence "Engineering of Advanced Materials", ZISC, FPS, the Collaborative Research Center SFB814, and grant PO472/20-2. We gratefully acknowledge the computing time granted by the John von Neumann Institute for Computing (NIC) and provided on the supercomputer JUROPA at Jülich Supercomputing Centre (JSC).

References

[1] I. Campbell, D. Bourell and I. Gibson, Rapid Prototyp. J. **18**, 255 (2012).

[2] R.D.Goodridge, C.J.Tuck and R.J.M.Hague, Prog. Mater. Sci. **57**, 229 (2012).

[3] P. Heinl, A. Rottmair, C. Körner and R.F. Singer, Adv. Eng. Mater. **9**, 360 (2007).

[4] B. Wendel, D. Rietzel, F. Kühlein, R. Feulner, G. Hülder and E. Schmachtenberg, Macromol. Mater. Eng. **293**, 799 (2008).

[5] T. Stichel, T. Laumer, T. Baumüller, P. Amend and M. Schmidt, In: Proceedings of the Polymer Processing Society 29th Annual Meeting - PPS-29, Nuremberg, Germany, 2013.

[6] T. Pöschel and V. Buchholtz, Phys. Rev. Lett. **71**, 3964 (1993).

[7] J. A. C. Gallas and S. Sokołowski, Int. J. Mod. Phys. **7**, 2037-2046 (1993).

[8] E. J. R. Parteli, AIP Conf. Proc. **1542**, 185-188 (2013).

[9] H. Pourtavakoli, E. J. R. Parteli and T. Pöschel, New J. Phys. **18**, 073048 (2016).

[10] E. J. R. Parteli and T. Pöschel, Powder Technology **288**, 96-102 (2016).

[11] E. J. R. Parteli, J. Schmidt, C. Blümel, K.-E. Wirth, W. Peukert and T. Pöschel, Scientific Reports **4**, 6227 (2014).

[12] P.A. Cundall, O.D.L. Strack, Geotechnique **29**, 47 (1979).

[13] H. Kruggel-Emden, E. Simsek, S. Rickelt, S. Wirtz and V. Scherer, Powder Technol. **171**, 157 (2007).

[14] H.P. Zhang and H.A. Makse, Phys. Rev. E **72**, 011301 (2005).

[15] T. Pöschel and T. Schwager, *Computational Granular Dynamics* (Springer, Heidelberg, 2005).

[16] H.C. Hamaker, Physica **4**, 1058 (1937).

[17] K.L. Johnson, K. Kendall and A.D. Roberts, Proc. R. Soc. Lond. A Math. Phys. Sci. **324**, 301 (1971).

[18] C. Kloss, C. Goniva, A. Hager, S. Amberger and S. Pirker, Prog. Comput. Fluid Dy. **12**, 264 (2012). www.liggghts.com

[19] R.C. Hidalgo, H.J. Herrmann, F. Kun, E.J.R. Parteli, in: F. Mallamace, H. Stanley (Eds.), *The Physics of Complex Systems* (IOS Press, Amsterdam 2004), pp. 153-171.

[20] D. Schiochet Nasato, T. Pöschel and E.J.R. Parteli, Proc. ICAT (accepted, 2016).

[21] A.B. Yu, J. Bridgwater and A. Burbidge, Powder Technol. **92**, 185 (1997).

[22] C.H. Liu, S.R. Nagel, D.A. Schecter, S.N. Coppersmith, S. Majumdar, O. Narayan and T.A. Witten, Science **269**, 513 (1995).

Tuning strain of granular matter by basal assisted Couette shear

Yiqiu Zhao[1,*], *Jonathan* Barés[1,2,**], *Hu* Zheng[1,3,***], and *Robert* Behringer[1,****]

[1]*Department of Physics, Duke University, Durham, NC, 27708*
[2]*Laboratoire de Mécanique et Génie Civil, Université de Montpellier, Montpellier, France 34090*
[3]*School of Earth Science and Engineering, Hohai University, Nanjing Jiangsu, China 210098*

Abstract. We present a novel Couette shear apparatus capable of generating programmable azimuthal strain inside 2D granular matter under Couette shear. The apparatus consists of 21 independently movable concentric rings and two boundary wheels with frictional racks. This makes it possible to quasistatically shear the granular matter not only from the boundaries but also from the bottom. We show that, by specifying the collective motion of wheels and rings, the apparatus successfully generates the desired strain profile inside the sample granular system, which is composed of about 2000 photoelastic disks. The motion and stress of each particle is captured by an imaging system utilizing reflective photoelasticimetry. This apparatus provides a novel method to investigate shear jamming properties of granular matter with different interior strain profiles and unlimited strain amplitudes.

1 Introduction

Granular matter can be jammed by shear. A jammed granular system can sustain finite stress without irreversible deformation [1]. If the packing fraction ϕ of a granular system is larger than a particular value ϕ_S, the system can transition from an unjammed state to a shear jammed state after a finite amount of shear strain [1–5]. In simulations, the shear strain is usually defined as the affine deformation driving individual particles and can be easily tuned [2, 5]. However, in the experimental case the shear strain is usually given by the deformation of the confining boundaries [1, 3, 6, 7]. This boundary strain, although a practical tool, is not always useful to describe the deformation inside the granular matter. The shear jamming transition is achieved by percolation of a force network inside the granular system under shear [1]. The evolution of this force network is strongly related to the interior strain of the granular system [1, 6]. Therefore the analysis of a stress-strain relation based on boundary strain may not be able to probe the intrinsic constitutive relation of granular matter undergoing a shear jamming transition.

To understand the interior strain of the granular matter, we present a novel apparatus that can generate boundary Couette shear with programmable interior strain field. The interior strain of the granular system corresponds to the displacements of each particle. This displacement field, or strain field will be called in this paper the resultant profile. And the collective motion of boundaries and bottom rings used to apply shear will be called the driving profile.

*e-mail: yiqiu.zhao@duke.edu
**e-mail: jb@jonathan-bares.eu
***e-mail: hz64@duke.edu
****e-mail: bob@phy.duke.edu

In this work, we compare the resultant and driving profiles for three kinds of shear: (*i*) boundary Couette shear (*ii*) linear shear and (*iii*) power-law shear with exponents $n = 0.5$ and $n = 2$. We show that this novel Couette shear apparatus leads to resultant profiles keeping the features of the driving profiles. In particular, we show that a linear shear results in a homogeneous density field even for very large boundary shear strains (~ 500%). By extracting stress information from the photoelastic patterns, we track the evolution of force network and probe the shear jamming transition. This apparatus provides a novel method to investigate the role played by the interior quasistatic strain of granular matter during shear jamming transition.

2 Experiment

The granular system studied here consists of about 2000 frictional (friction coefficient: $\mu \sim 0.6$) bi-disperse photoelastic disks. The bigger ones have a radius Rb = 7.95mm, while the smaller ones have a radius Rs = 6.35mm. The size ratio is kept constant for all experiments: $N_{big}/N_{small} = 1/3$. The particles have reflective bottoms (see fig. 1(b)), making it possible to view the photoelastic patterns from the top using reflective photoelasticimetry [8]. We use a cooling system to ensure the room temperature remains the same to get rid of any possible thermal deformation of the particles. The system consists of 12 cooling fans to exhaust hot air heated by step motors and one thermal meter to monitor the temperature (fig. 1(d)). The room temperature is kept at 70°F, at which the particles are free from any residual stress.

The particles are placed randomly onto the Couette shear apparatus shown in fig. 1. This Couette shear appara-

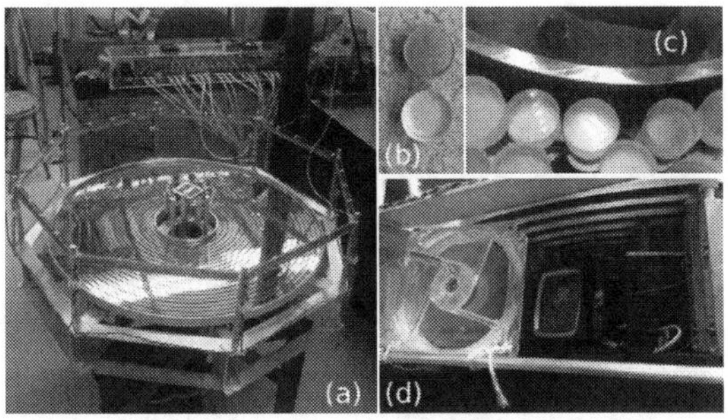

Figure 1: Experimental setup. (a) The Couette shear apparatus with outer boundary diameter 70.9cm and inner boundary diameter 17.5cm. Concentric rings can be rotated independently. (b) The photoelastic particles. The upper particle is up side down, showing its painted reflective bottom. (c) Blue rack attached to to inner and outer boundaries to increase their roughness and apply boundary shear. (d) Cooling system. Left: fans to exhaust hot air. Right: Stepper motors controlling the motion of rings.

tus is made of two boundary wheels and 21 independently movable rings. Each ring has a width that is the same as the diameter of the smaller particles. As the ring rotates, it carries the particles on it. However, the friction between particle and ring ($\sim 4 \times 10^{-3} N$) is smaller than any interparticle forces are visible as photoelastic patterns. The inner wheel and the inner-most ring of the apparatus are joined and the outer wheel moves with the out-most ring. Each ring but the inner-most one is controlled by a stepper motor beneath. Both boundary wheels have racks attached to apply boundary shear (see fig. 1(c)).

To apply shear to the system, both outer boundary and bottom rings move to drive the system. The shear is quasistatic, which means we shear the system step by step and take pictures between steps when the rings are stopped. For each step, the shear rate applied by the rings is in the order of $10^{-3} s^{-1}$. The motions of rings are controlled by 20 stepper motors. We call the relation between the number of steps a motor moves, $N(r)$, and the radial position of the ring controlled by the motor, r, the driving profile. The displacement of the ring is $N(r) \cdot d$, where $d = 0.01431(9)$mm is the same for all rings. The driving strain applied by this ring is defined as $\epsilon(r) = N(r) \cdot d/r$ (see fig. 2 insert). The strain applied by the outer-most ring will be called the boundary strain.

The driving profile determines the collective motion of all the rings and boundaries. If $N(r) = c \cdot r$, we call this driving profile a linear profile, and the shear applied to the system in this way a linear shear. Similarly, if $N(r) = c \cdot r^n$, we call the driving profile a power-law profile with exponent n, and this type of shear a power law shear with exponent n.

Each experimental run is composed of a series of quasi-static shear steps. Fig. 2 presents two driving profiles for one quasi-static steps. The driving profile is the same for all quasi-static steps of a given experimental run.

For every quasi-static steps, the motors move with different speed so that they start and stop at same time. The boundary shear displayed in fig. 2 means that only the outer-most ring and the outer boundary fixed to it are moving. This boundary shear can be regarded as a power-law shear with a very large exponent. As an example, the red curve demonstrates a fit with exponent $n = 100$.

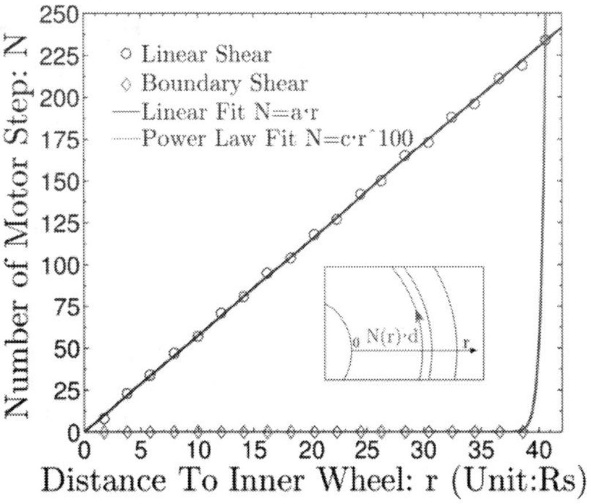

Figure 2: The driving profiles of linear and boundary shears. Main panel: The motor steps needed to generate linear and boundary shear for one quasi-static shear step. Inset: The azimuthal displacement of a ring at radial distance r is $N(r) \cdot d$, where $d = 0.01431(9)$mm. The driving strain caused by this ring is $\epsilon(r) = N(r) \cdot d/r$.

The imaging system of the experiment is shown in fig. 3(a). A 20Mp CCD camera (Canon® 70D) lays 3m above the Couette shear apparatus to image the whole system. Between two consecutive quasi-static steps, the camera is triggered to take two pictures. The first one is taken under UV light (see fig. 1(a)). The particles are labeled by UV sensitive marks on their top, which permits us to find their position during picture post-processing. Such a picture is presented in fig. 3(b). Another picture is taken using reflective cross-polarized light coming from the green LED light surrounding the camera. This green light picture shows the photoelastic response of each particle (see fig. 3(c)).

3 Result

We call the displacements of particles resulting from the driving profile a resultant profile. The resultant profiles of the driving profiles shown in fig. 2 are displayed in fig. 4. In these experiments, the granular system under shear has a packing fraction of $\phi = 0.7304$. The resultant profiles show the displacements of particles after 398 quasi-static shear steps, corresponding to 500% boundary strain. As shown in fig. 4, the azimuthal displacements of particles generated by linear shear are very well fitted by a straight line. However, this linear fit does not exactly goes

 DOI: 10.1051/epjconf/201714003049

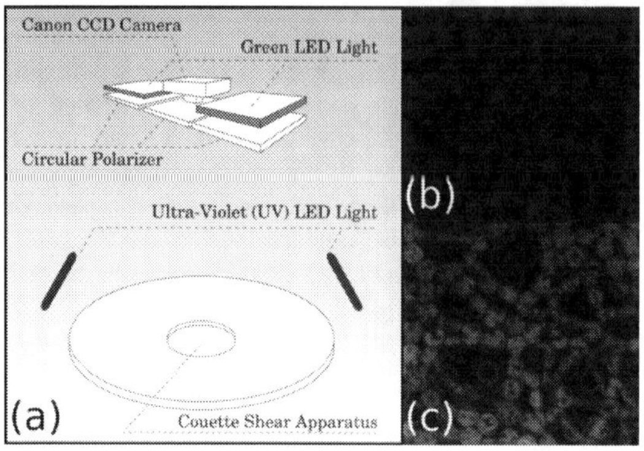

Figure 3: (a) Imaging system of the experiment. Particles placed on the shear apparatus can be luminated by two kinds of lights: (*i*) UV light from the UV LEDs surrounding the shear appratus and (*ii*) Green polarized light from the green LEDs surrounding the camera. (b) UV light image. Labels drawn on the particles using UV sensitive ink are luminated. Those labels are used to find centers and orientations of the particles. (c) Green light image. Reflective photoelasticity permits to observe stress inside particles. Green light is used because it is optimal for the circular polarizers we used.

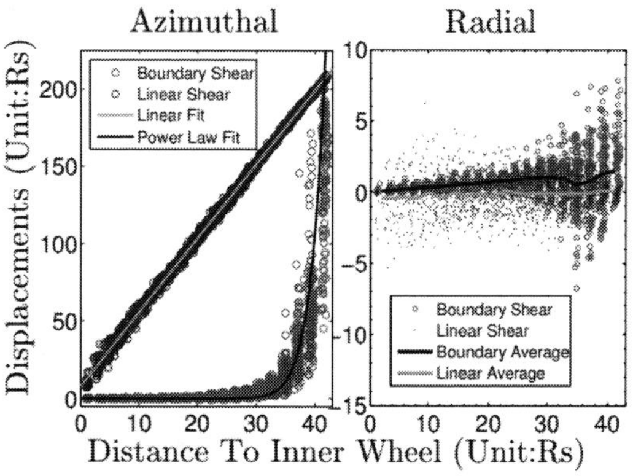

Figure 4: Resultant profiles of linear and boundary shears after 500% boundary strain. This example system has a packing fraction of $\phi = 0.7304$ (see also fig. 2). Left panel shows the azimuthal displacements of particles *v.s.* their initial position. The linear fit slightly deviates from the origin. The power-law fit gives an exponent $n = 16.93(34)$. The right panel shows the radial displacements of particles *v.s.* their initial positions. The positive direction is centripetal direction. Boundary shear results in a collective centripetal motion while linear shear does not.

through origin. This means there is a small global rotation of the whole system. The red circles in fig. 4 show the displacement of particles under boundary shear. For azimuthal displacements, the power law-fit gives an exponent $n = 16.9(7)$, which is significantly less than the driving profile ($n = 100$). But it keeps the feature of driven profile: the deformations are localized close to the outer boundary.

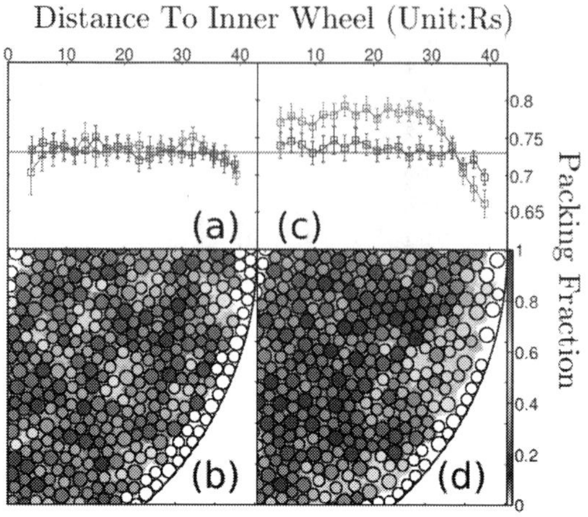

Figure 5: Shear-band study. (a) and (c) present the radial distribution of packing fractions before and after 500% boundary strain for systems under linear shear (a) and boundary shear (c). The blue squares show the density distribution along the radial direction before shearing, while the red ones display the density after shearing. The blue lines indicate the system packing fraction ($\phi = 0.7304$). In (a) no low density area is created. But in (c) a localized low density band (shear band) is created after shear. (b) shows the local packing fraction computed from Voronoi cells after 500% linear shear. It is compared to the case where the system is sheared from boundaries in (d).

The right panel of fig.4 shows the radial displacements of particles. The green curve shows the average displacements from the linear shear resultant profile. It fluctuates around zero which indicates there is no collective flow along the radial direction. However, as shown by the black curve, which is for boundary shear, the average radial displacement is non-zero and positive (towards ring center). This collective radial motion results in a shear band near the outer wheel. Fig. 5 shows the radial density (packing fraction) profile of the sample granular system before and after linear shear (a) and boundary shear (c). Blue squares show the density distribution before shear while red ones show the density distribution after 500% shear strain. The data points are averaged from the local packing fractions defined as particle area divided by the Voronoi area associated with the same particle. Error bars are at a 95% confidence interval. We remark that after 500% shear strain, the boundary shear generates an obvious shear band (lower density area) with width around $10 \cdot Rs$, while the system under linear shear remain homogeneous everywhere except for small fluctuations due to the intrinsic randomness of the granular systems.

We present the green light images for the sample system after 500% boundary strain under linear shear (see fig. 6(a)) and boundary shear (see fig. 6(b)). There are

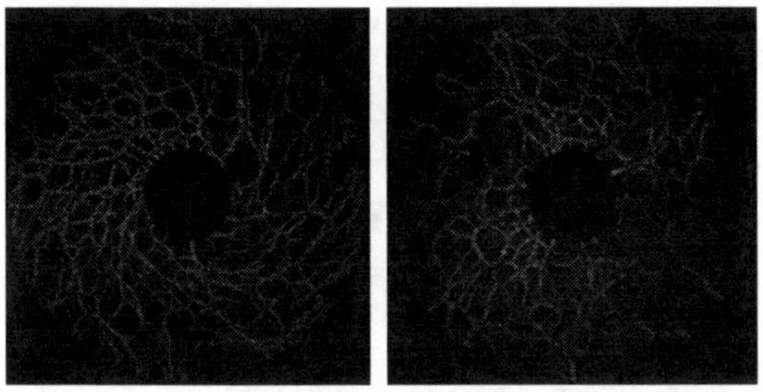

(a) Linear Shear (b) Boundary Shear

Figure 6: The green light cross polarized image for system with ϕ = 0.7304 under linear (a) and boundary (b) driven profile after 500% boundary strain. The particle with brighter intensity bears larger force.

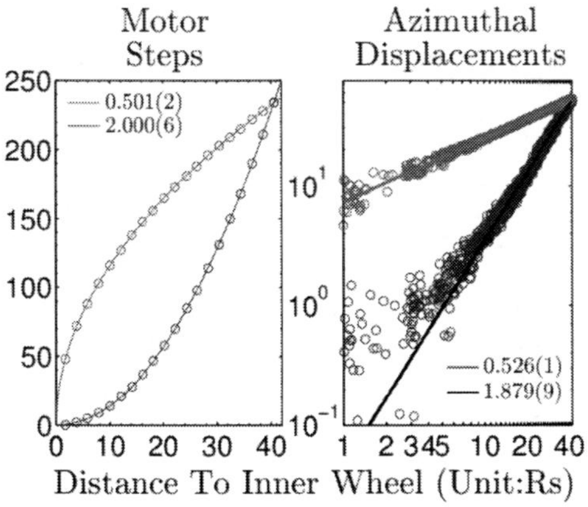

Figure 7: Driving and resultant profiles for a granular system with packing fraction ϕ = 0.7257 after 125% boudnary shear strain. The left panel shows the number of motor steps for each ring needed to generate the power-law driving profiles. The right panel shows the resultant azimuthal displacements for individual particles (Unit: Rs). The power-law fits show that driving profiles with exponents 0.501(2) and 2.000(6) generate resultant profiles with exponents 0.526(1) and 1.879(9) respectively.

two key observations from fig. 6: (*i*) the apparatus is able to generate resultant profiles retaining the features of the driving profiles even if the inter-particle forces are much larger than basal friction forces; (*ii*) by changing the interior strain we generate force networks with significantly different structures under the same boundary strain. Both observations confirm the ability of our apparatus to be used to study the relation between interior strain and shear jamming, and shear flows.

Besides linear and boundary shear, we can generate arbitrary programmable driving profiles. We present two more examples. Fig. 7 presents the driving and resultant profiles for a system with packing fraction ϕ = 0.7257 under power-law shear with exponents n = 0.5 and n = 2. The left panel shows the driving profile for an individual quasi-static step. And the right panel shows the resultant particle azimuthal displacement. Power-law fitting of the resultant profile shows that an n = 2 driving profile results in an n = 1.879(9) resultant profile, and an n = 0.5 driving profile results in an n = 0.526(1) resultant profile. Although the resultant exponents deviate slightly from the driving exponents, they are close enough to retain the main features of the driving profiles.

4 Conclusion and outlook

We have described a novel experimental system that can tune the interior deformation of granular systems under boundary Couette shear with basal assistance. The shear is applied by a novel Couette shear apparatus consisting of 21 independently movable rings and two boundary wheels. We present resultant profiles for three kinds of driving profiles: (*i*) boundary Couette shear, (*ii*) linear shear, and (*iii*) power-law shear with exponents 0.5 and 2. The resultant profiles keep the features of the driving profiles. In particular, we showed that under linear basal assisted shear, the granular system does not form shear bands even under very large boundary strains (~ 500%). This setup makes it feasible to investigate shear jamming properties of granular system by changing interior strains.

Acknowledgement

This work is supported by NSF-DMR1206351, DMS1248071, NASA NNX15AD38G.

References

[1] B.C. Dapeng Bi, Jie Zhang, R.P. Behringer, Nature **480**, 355 (2011)
[2] N. Kumar, S. Luding, Granular Matter **18**, 58 (2016)
[3] H. Zheng, J.A. Dijksman, R.P. Behringer, EPL (Europhysics Letters) **107**, 34005 (2014)
[4] J. Zhang, T.S. Majmudar, M. Sperl, R.P. Behringer, Soft Matter **6**, 2982 (2010)
[5] M. Baity-Jesi, C.P. Goodrich, A.J. Liu, S.R. Nagel, J.P. Sethna, ArXiv e-prints (2016), 1609.00280
[6] J. Ren, J.A. Dijksman, R.P. Behringer, Phys. Rev. Lett. **110**, 018302 (2013)
[7] B. Utter, R.P. Behringer, Phys. Rev. E **69**, 031308 (2004)
[8] J.G. Puckett, K.E. Daniels, Phys. Rev. Lett. **110**, 058001 (2013)

Evolution of force networks in dense granular matter close to jamming

Lou Kondic[1,*], *Miroslav* Kramár[2,**], *Lenka* Kovalčinová[1,***], and *Konstanin* Mischaikow[3,****]

[1] *Department of Mathematical Sciences, New Jersey Institute of Technology, University Heights, Newark, NJ 07102, USA*
[2] *Advanced Institute for Materials Research, Tohoku University, Sendai, Japan*
[3] *Department of Mathematics and BioMaPS Institute, Hill Center-Busch Campus, Rutgers University, 110 Frelinghusen Rd, Piscataway, NJ 08854, USA*

Abstract. When dense granular systems are exposed to external forcing, they evolve on the time scale that is typically related to the externally imposed one (shear or compression rate, for example). This evolution could be characterized by observing temporal evolution of contact networks. However, it is not immediately clear whether the force networks, defined on contact networks by considering force interactions between the particles, evolve on a similar time scale. To analyze the evolution of these networks, we carry out discrete element simulations of a system of soft frictional disks exposed to compression that leads to jamming. By using the tools of computational topology, we show that close to jamming transition, the force networks evolve on the time scale which is much faster than the externally imposed one. The presentation will discuss the factors that determine this fast time scale.

1 Introduction

The interaction between particles that build dense granular matter (DGM) is characterized by the presence of a force network, mesoscopic structure that spontaneously form as a granular system is exposed to shear or compression. These networks are crucial for understanding the static and dynamic properties of DGM, and have been recently studied extensively, using the tools that vary from statistical methods [1–3] to network type of analysis [4–6] and application of topological methods [7–12].

Force networks exhibit complicated time dependent structures. Earlier approaches, used to analyze these structures, usually involved rather arbitrary choices of the force thresholds and often depended on ad hoc descriptions of the structures. To avoid these problems we define the force network [11] as a collection of sets $\{X_\theta\}_{\theta\in\mathbb{R}}$ such that for $\theta \geq 0$ the set X_θ is the part of the contact network on which the force interactions between the particles exceed the value θ. In the series of papers [9–12], we make use of an algebraic topological technique that rigorously describes geometric structures of the sets $\{X_\theta\}_{\theta\in\mathbb{R}}$ over all force levels θ. This approach provides a quantitative description of the force networks. Moreover, it allows us to define a concept of distance between force networks, so we can quantitatively compare them.

In [12] we focused on comparing force networks in simulated granular systems built from soft, possibly inelastic and/or frictional particles exposed to steady slow compression. In that work we found that force networks' evolution was characterized by large distances between consecutively sampled states of the systems that were approaching jamming transition from below, and the evolution was much smoother for jammed systems. In particular close to jamming transition, we found rather dramatic evolution of the force networks in all considered systems (frictional or not, with elastic or inelastic particles).

The following unexpected findings obtained in [12] motivate the present peper. Distances between force networks at consecutive states of the system sampled at the rate used in [12] are comparable to the distances between force networks at the same packing fraction obtained from different realizations, see Figure 1. Sec. 2 provides a brief description of the plotted quantities while more in depth discussion of the figure is presented in Sec. 3. This result suggests that force networks evolve on a time scale that is much faster than the inverse sampling rate and therefore, the main topic of the present paper is the analysis of the system on extremely fast time scales. Furthermore, we discuss how the time scale characterizing evolution of force networks depend on two externally imposed time scales: the slow one imposed by external driving (compression in the case considered here) and the fast one imposed by the temporal discretization used in our simulations. We show that variation of these two scales provides further insight into the mechanisms governing force network evolution.

2 Methods

Here we give a brief overview of the techniques used; the reader is referred to [12] for more detailed description of the protocol and definitions of various quantities.

*e-mail: kondic@njit.edu
**e-mail: miroslav.kramar.1@gmail.com
***e-mail: lk58@njit.edu
****e-mail: mischaik@math.rutgers.edu

 DOI: 10.1051/epjconf/201714015014

We consider a square domain in two spatial dimensions exposed to slow compression by the moving walls. The particles are polydisperse, soft, inelastic disks interacting by normal and tangential forces, with the latter ones modeled by Cundall-Strack type of interaction [13]. All quantities are expressed using the average particle diameter, d, as the length scale, the binary particle collision time $\tau_c = \pi \sqrt{d/(2gk_n)}$ as the time scale, and the average particle mass, m, as the mass scale. Here k_n (in units of mg/d) is the spring constant used for modeling (linear) repulsion force between colliding particles, and g is gravity (used only for the purpose of scaling). For the initial configuration, particles ($\approx$ 2000) are placed on a square lattice and given random initial velocities. Except if specified differently, the walls move at a uniform inward velocity $v_c = 2.5 \cdot 10^{-5}$. We integrate Newton's equations of motion for both the translational and rotational degrees of freedom using a 4th order predictor-corrector method with time step $\delta t = 1/50$.

We use persistent homology [14] to quantitatively describe the structure of force networks. Briefly, each network is represented by two persistence diagrams that quantify the changes in geometry of the sets X_θ for all force levels θ. The β_0 (β_1) persistence diagram encodes the appearance and disappearance of connected components (loops) by recording all values at which a connected component (loop) appears in a set X_θ and all values of the force level at which the corresponding connected component (loop) disappears. For a rather brief discussion of persistence, we refer the reader to [12] and for a more in-depth presentation to [11]. There are different types of well defined distances between the persistence diagrams. Each of these distances measures differences between the force levels at which distinct topological features appear and disappear. In particular, the distance between two β_0 persistence diagrams corresponding to different force networks measures differences between the force values at which distinct connected components/clusters are formed and merged together. In a loose sense, this corresponds to measuring differences between the force chains. The results reported in this paper do not depend on the particular choice of the distance and thus, for brevity, we will focus only on the so called Wasserstain distance, d_{W^1}, between the β_0 persistence diagrams (see [11] for a detailed definition).

3 Results

We start by recalling the results obtained for slow sampling rate of approximately 100 samples covering the packing fraction range [0.63, 0.9]. Since the evolution of force networks varies from realization to realization, and the distances between considered states of the system may be rather noisy, we considered the averages over a large number (20) of realizations [12]. Figure 1 shows that the distances between two consecutive states of a given realization are comparable to the distances between two states resulting from different realizations. This suggests that the memory of the system is lost on the time scale faster than the one defined by the slow sampling rate.

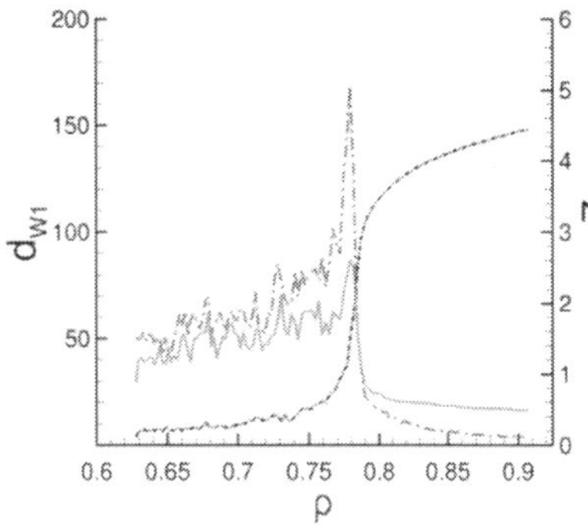

Figure 1. Wasserstein distance, d_{W^1}, measuring differences between force networks. The dash-dotted (green) line shows the distances of the networks averaged over 20 realization and the solid (orange) line shows the distances between different realizations. The dotted (black) line shows the number of contacts, Z; the rapid rise is a sign of jamming transition.

To proceed, we consider much faster sampling rates. We also focus on presenting and analyzing the results from a single realization, since averaging the results over realizations obscures part of the information about force network evolution in a given system. Figure 2 shows d_{W^1} distance for four sampling rates r_1, r_{10}, r_{100} and r_{1000}; here the index of a rate corresponds to the number of computational steps of duration δt between two consecutive sample points. The difference in the packing fraction, ρ, between two consecutive states sampled at the rate r_i is $\Delta\rho = i \cdot \delta\rho$ where $\delta\rho \approx 4 \cdot 10^{-8}$ is the change of ρ during one computational step. Note that r_1 corresponds to the fastest possible sampling rate, at which the state of the system is recorded at every time step. First, we note that, as expected, for faster sampling rates, there is a large number of points such that the distance is very small, suggesting that the force networks do not evolve much during a significant portion of the data shown. However, contrary to expectations, there is an *increase* of the peaks' sizes as the sampling rate is increased from r_{1000} to r_{100} and r_{10}, followed by a gentle decrease as the rate is increased from r_{10} to r_1. This essentially means that one needs to sample at the rate comparable to r_{10} (at which the peaks in d_{W^1} distance are the largest) to capture the corresponding large events present in networks' evolution. In turn, this implies that large changes between the force networks occur on the time scales shorter than τ_c (capturing the system evolution with the sampling rate given by $1/\tau_c$ would correspond to r_{50}).

The number of data points shown in Fig. 2 is large; e.g., Fig. 2a shows about 10^6 data points. Since it is impractical to work with such a large number of data points, we proceed by carrying out a running average of the data shown in Fig. 2, and then proceed with considering a subset of the data extracted from the region very close to the jamming transition (that occurs at $\rho \approx 0.78$ for the present

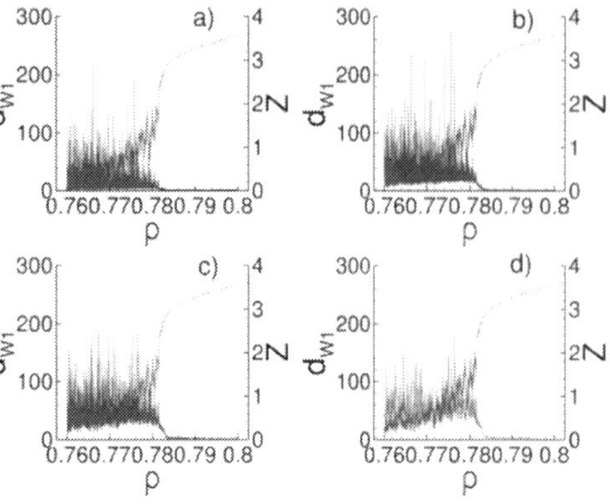

Figure 2. Wasserstein distance d_{W^1} (red) between β_0 persistence diagrams computed for different sampling rates: a) r_1 b) r_{10}, c) r_{100} and d) r_{1000}. The number of contacts between particles, Z is shown in blue.

system). Figure 3 shows the running averages for d_{W^1} distance. The averaging interval is given by $10^4\delta\rho \approx 4 \cdot 10^{-4}$. This averaging interval corresponds to the averaging time of about $200\tau_c$ which for a common granular system may range between a μsec and a msec. The actual number of points used for computing varies from 10^4 for the fastest sampling rate r_1 to 10 for the slowest sampling rate r_{1000}.

Before jamming transition a typical distance between consecutive states grows as the sampling rate decreases, but at a rate that is typically much slower than expected. Note that the consecutive sampling rates differ by one order of magnitude. However, the differences between averaged distances is one order of magnitude only for the sampling rates r_1 and r_{10} and becomes significantly smaller for slower sampling; the results are actually overlapping for r_{100} and r_{1000}. This finding suggests, consistently with the results shown in Fig. 2, that the time scale on which force networks evolve is comparable to τ_c or even shorter.

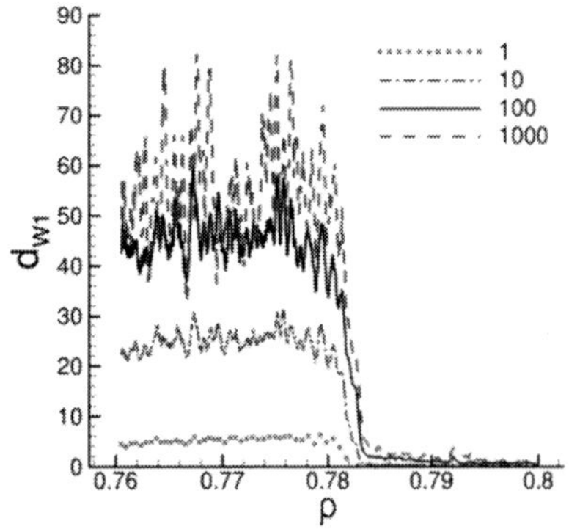

Figure 3. Running average of d_{W^1}-distances over $10^4\delta\rho$. The curves correspond to the sampling rate used to compute the distances shown in Fig. 2.

The fast time scale on which force networks evolve raises new questions. Recall that δt is the time interval needed for information about possible interactions to travel across a particle and hence δt is related to the speed of sound, v_s, in the material. In particular, the approximate speed of sound, determined by the numerical time step, is given (in physical units) by $d/(\delta t\tau_c)$, and therefore, information about the features present in force networks that evolve on the time scale given by δt essentially propagates by the speed of sound. Thus, it is natural to ask whether the results may be influenced by δt.

Figure 4 shows an influence of decreasing δt on d_{W^1} distance in a small interval based at $\rho \approx 0.78$. Consequences of decreasing the δt are two fold. On one hand, as δt decreases, on average d_{W^1} decreases (the number of points at which d_{W^1} is very small increases as δt decreases). On the other hand, we notice that the peaks' heights do not necessarily decrease with δt, and more importantly, the peaks still relax on the scale corresponding to δt: for smaller δt's, the peaks relax faster (see the insets in Fig. 4). This finding is consistent with the general discussion above regarding the relevance of the speed of sound, and is showing clearly that in the considered system the force networks evolve on the time scale determined by the speed of sound in the material.

Another question to ask is related to the influence of the compression speed, v_c, which defines a slow time scale in the problem. As v_c is reduced, one may expect that the distance between the consecutive states of the system becomes smaller. As we show next, such trend is indeed observed in our results.

Figure 5 shows d_{W^1} distance at the same ρ as in Fig. 4, but now for four (decreasing) v_c's. We see that, indeed, on average d_{W^1} decreases as v_c is decreased, meaning that there are many δt intervals when the force networks essentially do not evolve. However, the height of the peaks does not necessarily decrease with v_c; we even find examples of larger peaks for slower compression. This finding can be understood by realizing that as the system is compressed slower, the particles have higher chance to rearrange, leading in some cases to large changes in the force network and increased distances between the consecutive states.

4 Conclusions

The force networks are found to evolve on the time scale that is much faster than the one introduced by externally imposed driving. As a consequence, an analysis of dynamics of force networks based on slow sampling rates may provide incomplete information about the networks' evolution, since slow sampling may miss large events that occur on faster time scales. In particular, we find that the time scale on which evolution occurs is specified by d/v_s, where d is a typical particle diameter and v_s the speed of sound in the material. It will be of interest to see to which degree such behavior can be seen in experimental systems. Furthermore, it should be noted that in the present work we focus on the systems very close to jamming transition; it remains to be seen whether such fast evolution as uncov-

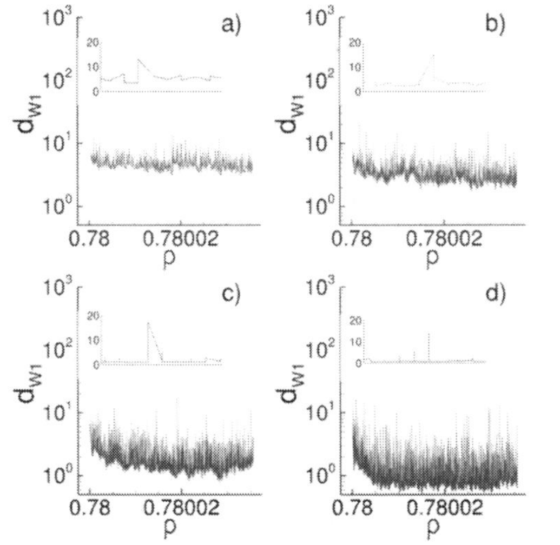

Figure 4. d_{W^1} distance between β_0 persistence diagrams corresponding to the consecutive force networks (one computational time step apart) for $\rho \approx 0.78$ as δt is reduced, with the part a) using the reference value $\delta t = 1/50$, and the parts b), c), d) using $\delta t/2$, $\delta t/4$, and $\delta t/8$, respectively. The insets show the distances for few consecutive time steps, with the shown range of 5×10^{-7}. The compression speed, v_c, is kept fixed.

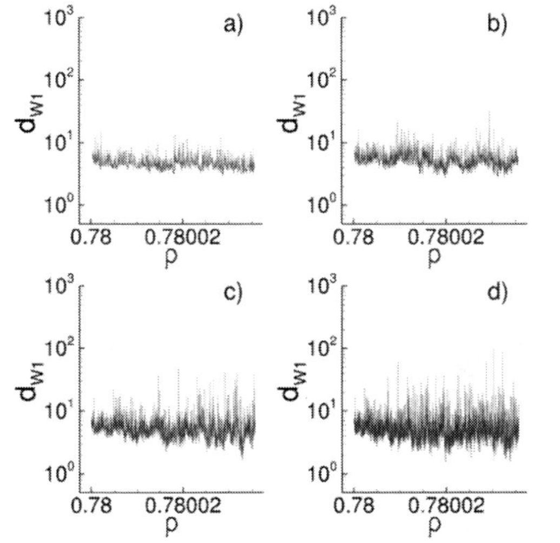

Figure 5. d_{W^1} distance between β_0 persistence diagrams corresponding to the consecutive force networks (one δt apart) for $\rho \approx 0.78$ as the compression speed is reduced, with the part a) using the reference value, v_c, and the parts b), c), and d) using $v_c/2$, $v_c/4$ and $v_c/8$, respectively. δt is kept fixed.

ered here may be seen further away from jamming, and/or for different type of external forcing.

The reported results rely on the use of computational homology that allows us to quantify the differences between force networks as a granular system evolves. One particular strength of the computational homology approach is that it is system-independent and can be applied to any system, independently of the physical properties of the considered system, such as the type of interaction between the particles, or the number of dimensions of the physical space in which they live. Therefore, natural next step will be to consider particulate systems characterized by different physical properties (friction, polydispersity, cohesion, shape) or different geometry (2D versus 3D). Furthermore, the approach can be as well applied to experimental systems, such as those built from photoelastic particles, where detailed information about the force networks is available. Further analyses of such systems will be the subject of our future works.

Acknowledgements

This work was supported by the NSF Grant DMS-1521717 and by DARPA contract No. HR0011-16-2-0033.

References

[1] J. Peters, M. Muthuswamy, J. Wibowo, A. Tordesillas, Phys. Rev. E **72**, 041307 (2005)
[2] A. Tordesillas, D.M. Walker, Q. Lin, Phys. Rev. E **81**, 011302 (2010)
[3] A. Tordesillas, D.M. Walker, G. Froyland, J. Zhang, R. Behringer, Phys. Rev. E **86**, 011306 (2012)
[4] D.S. Bassett, E.T. Owens, K.E. Daniels, M.A. Porter, Phys. Rev. E **86**, 041306 (2012)
[5] M. Herrera, S. McCarthy, S. Sotterbeck, E. Cephas, W. Losert, M. Girvan, Phys. Rev. E **83**, 061303 (2011)
[6] D. Walker, A. Tordesillas, Phys. Rev. E **85**, 011304 (2012)
[7] R. Arévalo, I. Zuriguel, D. Maza, Phys. Rev. E **81**, 041302 (2010)
[8] R. Arévalo, L.A. Pugnaloni, I. Zuriguel, D. Maza, Phys. Rev. E **87**, 022203 (2013)
[9] L. Kondic, A. Goullet, C. O'Hern, M. Kramár, K. Mischaikow, R. Behringer, Europhys. Lett. **97**, 54001 (2012)
[10] M. Kramár, A. Goullet, L. Kondic, K. Mischaikow, Phys. Rev. E **87**, 042207 (2013)
[11] M. Kramár, A. Goullet, L. Kondic, K. Mischaikow, Physica D **283**, 37 (2014)
[12] M. Kramár, A. Goullet, L. Kondic, K. Mischaikow, Phys. Rev. E **90**, 052203 (2014)
[13] P.A. Cundall, O.D.L. Strack, Géotechnique **29**, 47 (1979)
[14] H. Edelsbrunner, J.L. Harer, *Computational topology* (AMS, Providence, RI, 2010)

Of cuts and cracks: data analytics on constrained graphs for early prediction of failure in cementitious materials

Sanath Kahagalage[1], *Antoinette* Tordesillas[1,2,*], *Michal* Nitka[3], and *Jacek* Tejchman[3]

[1] *School of Mathematics and Statistics, The University of Melbourne, Australia*
[2] *School of Earth Sciences, The University of Melbourne, Australia*
[3] *Faculty of Civil and Environmental Engineering, Gdańsk University of Technology, Poland*

Abstract. Using data from discrete element simulations, we develop a data analytics approach using network flow theory to study force transmission and failure in a 'dog-bone' concrete specimen submitted to uniaxial tension. With this approach, we establish the extent to which the bottlenecks, i.e., a subset of contacts that impedes flow and are prone to becoming overloaded, can predict the location of the ultimate macro-crack. At the heart of this analysis is a capacity function that quantifies, in relative terms, the maximum force that can be transmitted through the different contacts or edges in the network. Here we set this function to be solely governed by the size of the contact area between the deformable spherical grains. During all the initial stages of the loading history, when no bonds are broken, we find the bottlenecks coincide consistently with, and therefore predict, the location of the crack that later forms in the failure regime after peak force. When bonds do start to break, they are spread throughout the specimen: in, near, and far from, the bottlenecks. In one stage leading up to peak force, bonds collectively break in the lower portion of the specimen, momentarily shifting the bottlenecks to this location. Just before and around peak force, however, the bottlenecks return to their original location and remain there until the macro-crack emerges right along the bottlenecks.

1 Introduction

There is no one-size-fits-all approach in data analytics. The best approach for analyzing data on current and historical behaviour of a system for the purposes of predicting future behaviour depends crucially on the problem context. When it comes to transmission and flow problems, techniques based on graphs or networks are arguably among the best in uncovering nonlinear and multiscale patterns from microstructural data (i.e., information on the individual units of the system), judging from their broad applications in transportation, telecommunication, supply chain and logistics, disease spread etc [1, 2]. Recently, network-based data analytics has also shown promise in studies of transmission patterns in granular media, but these were confined to unbonded grain assemblies [3–6].

The study reported here concerns the characterization and modelling of force transmission and failure propagation in cementitious granular materials (e.g., concrete, rock, bone, asphalt etc.) For a recent review of the state-of-the-art and challenges in the characterization and prediction of failure for this class of materials, see Lefort et al [7]. A promising new approach in damage mechanics, highlighted in [7], is the use of advanced statistical techniques to extract a length scale from spatial data on mesoscale damage events in acoustic emission experiments. They employed a spatial point pattern analysis that saw its first application to data sets for granular systems in [8]. Here, we similarly build upon a technique that has shown demonstrable promise in describing multiscale failure patterns from microstructural data in unbonded granular systems. Different from [7, 8], however, our aim here is to characterize *and predict the ultimate macro-crack pattern*. To achieve this, we go beyond descriptive analytics, since these use data mining to deliver insights into current and past behaviour only. Instead, we use network flow theory to uncover patterns from data that can help forecast in the pre-failure regime an aspect of a future event, viz., the location of the macro-crack.

In related past work, data from X-ray CT experiments and discrete element simulations, containing information at the level of individual grains [3–6], were analyzed using a network flow theory to identify bottlenecks and major flow pathways in force transmission for bonded and unbonded granular systems. This approach leads to an optimization problem in which the amount of flow through the network is maximised, considering all feasible paths from source to sink, for a given capacity distribution (i.e., the maximum flow units that can be transmitted across each edge of the network). The approach has shown promise in the prediction of: (a) the location and pattern of failure (localised versus diffused) [5, 6], and (b) force chains [6], without information on the contact forces that govern these mechanisms. Beyond prediction, these studies also delivered insights into the underlying structure and dynamics

*e-mail: atordesi@unimelb.edu.au

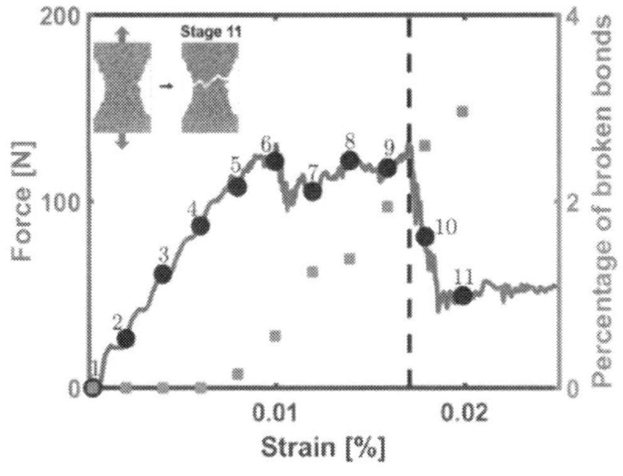

Figure 1. Data from DEM model. Strain evolution of the vertical tensile force in a 'dog-bone' concrete specimen submitted to uniaxial tension. We track behavior across 11 stages of loading: the strain state corresponding to the peak vertical force is shown by the vertical dashed line. A crack that splits the specimen into two pieces develops at stage 11. Right axis shows the evolution of damage as measured by the percentage of broken bonded contacts in the specimen (no damage occurs in stages 1-4).

of shear bands and force chains. In particular, findings in [5] for natural, synthetic and virtual granular assemblies demonstrated that flow bottlenecks emerge early in the loading history, progressively degrade, and persists in the location where the shear band eventually forms.

This study is part of a broader effort to investigate force transmission and failure in bonded granular media [9, 10]. In contrast to unbonded grains [5], force transmission pathways in brittle or quasi-brittle bonded granular materials remain essentially fixed prior to fracture. Accordingly, our aim is to determine the extent to which the distribution of bond strengths and contact topology *per se* govern the ultimate crack pattern. In particular, *can the ultimate crack pattern be predicted even before any damage by bond breakage occurs?* Because there is minimal loss, and ensuing frictional rearrangement, of contacts prior to peak stress in the sample, we neglect friction and any topological constraint that promotes frustration of rotations: the contact network sustains little "rewiring" and alteration of the available routes for transmission. Consequently, in this optimization problem, the flow through the network is regulated and maximised on highly constrained routes where the only change to the available pathways is through blockage of flow at edges where the corresponding bond breaks. Essentially no new routes form in the pre-failure regime where the prediction of failure matters.

2 Data

We discuss here the data set from a discrete element (DEM) model of an unconfined concrete specimen, described as a random heterogeneous one-phase material, subjected to quasi-static, uni-axial tension [9]. A linear normal contact model under compression was used. The contact tangential forces and normal forces satisfy

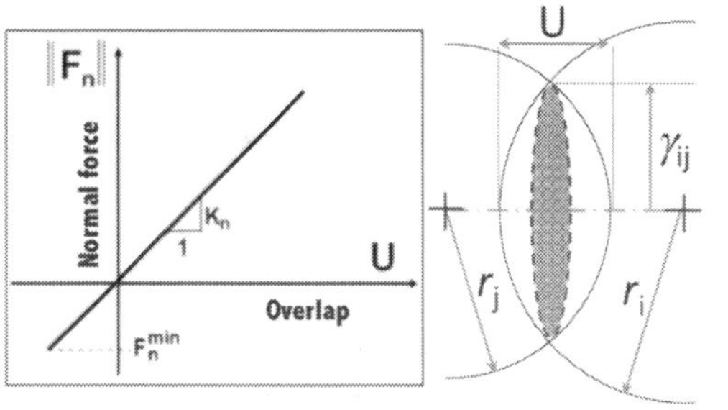

Figure 2. The normal tensile force depends on the overlap U; the bond is broken if this force reaches F_n^{min}. The circular contact area (red) has radius γ_{ij}.

the cohesive-frictional Mohr-Coulomb equation using the inter-particle friction angle, permissible cohesive and tensile force. The bonded specimen has the dog bone shape with 0.15 m height, 0.10 m along the top and bottom, 0.06 m at mid-height and one layer of grains in depth. It contains 4942 grains with a wide distribution of radii: minimum is 2.7895×10^{-4} m, maximum is 5.3064×10^{-3} m.

The evolution with strain of the macroscopic force, along with the number of broken bonds, is shown in Figure 1. We analyse force transmission in 11 equilibrium stages: stages 1 to 9 are in the pre-peak force regime, while stages 10 to 11 are in the post-peak failure regime. The initial number of bonds is 12,350: bonds do not break until stage 5 from which the population of broken bonds steadily increases to around 2% just before peak force. The ultimate crack that forms in stage 11 span the 'neck' of the specimen (inset, Figure 1).

3 Method

We apply the method developed in [3, 5] to the data described in Section 2 as follows. For each equilibrium state, we formulate a network flow model of force transmission based on two input information: the contact network and the contact area over all the contacts. A network $G = \{V, E, C, s, t\}$ is constructed where: G is a weighted-directed graph, V is the set of intermediate vertices i that represents the grains, E is the set of edges (i, j) that connect those vertices i and j in V that correspond to grains in physical contact (bonded or unbonded frictional contact), $C = \{c_{(i,j)},\ (i, j) \in E\}$ is the set of scalar edge weights that represent the flow capacities (maximum units of flow that can be transmitted through the different edges in E), and s and t are two fictitious vertices that direct the flow from the source node s to the sink node t.

Here the edge capacities $c_{(i,j)}$ represent the force bearing capacities of the corresponding contacts; hence $c(i, j) = c(j, i) \geq 0,\ \ \forall (i, j) \in E$. We choose the form of this capacity function to be in accordance with the DEM model from which the data, described in Section 2, was generated. As discussed in [9], and illustrated in Figure 2, the tensile force bearing capacity α_{ij} depends on the overlap $U = U_{ij}$ at the contact between the bonded spherical grains corresponding to nodes i and j:

$$\alpha_{ij} = \begin{cases} k_n U > F_n^{min} & \text{if the contact is bonded} \\ 0, & \text{if the contact is unbonded} \end{cases} \quad (1)$$

where k_n is the normal stiffness of the contact and F_n^{min} is the minimum tensile force as calibrated from uniaxial tension tests. The overlap is $U_{ij} = r_i + r_j - d_{ij}$, where r_i and r_j are the radii of the grains corresponding to the vertices i and j, respectively, and d_{ij} is the distance between the grains' centroids. Following eqn. 1, we propose an edge capacity $c(i, j)$ that depends solely on the size of the contact area, assumed here to be circular of radius γ_{ij}:

$$c_{ij} = \begin{cases} \pi\gamma_{ij}^2 = \pi \times \left[r_i^2 - \left(\frac{r_i - r_j + d_{ij}}{2}\right)^2 \right] & \text{if the contact is bonded} \\ 0. & \text{if the contact is unbonded} \end{cases} \quad (2)$$

The larger the contact area, the stronger the bond, the higher the edge capacity.

Next we decide a direction of flow. Since the specimen is submitted to uniaxial tension in the vertical direction, we direct the flow from the top wall to the bottom wall, assigned as the source node s and sink node t, respectively. Note that G is symmetric, $c_{(i,j)} = c_{(j,i)}$; thus switching the source and sink assignments will not alter the flow. Following [5], we minimize boundary wall effects by connecting the top layer of grains (grains within two mean diameter away from the top wall) to s while the bottom layer of grains (grains within two mean diameter above the bottom wall) to t. These artificial edges are assigned an infinite capacity so that only those contacts between constituent grains form candidates for bottlenecks.

To find the bottlenecks at each equilibrium state, we direct the maximum possible flow in G from s to t by solving an optimisation problem known as the *maximum flow-minimum cut* [2]. For a given feasible flow f of value $|f|$, the maximum flow problem is to find the maximum flow value $|f|_{max}$ as follows. **MAXIMIZE** $|f|$: $|f| = \sum_{i\in V} f(s,i) = \sum_{j\in V} f(j,t)$ subject to the following constraints:

1. *Flow Capacity* – For any edge $(i, j) \in E$, the flow cannot exceed its capacity: $f(i, j) \leq c(i, j)$.
2. *Flow Conservation* – For any vertex $i \in V$, the sum of the flows entering i must be the same as the sum of the flows exiting i: $\sum_{i:(i,j)\in E} f(i, j) = \sum_{j:(j,i)\in E} f(j, i)$.

The above is a dual of the minimum cut problem. Given a cut that partitions G into two disjoint sets of vertices S and T with $s \in S, t \in T, S \cup T = V \cup \{s, t\}$, the minimum cut problem is to find that cut with the minimum capacity c_{min} as follows. **MINIMIZE** c: $c(S,T) = \sum_{u\in S, v\in T} c(u, v)$. By the maximum flow-minimum cut theorem [2], the bottlenecks of force flow are the contacts corresponding to the edges in the minimum cut with capacity $|f|_{max} = c_{min}$.

4 Results

There is no damage in the specimen across stages 1 to 4, since bonds do not break until stage 5 (Figure 1). During these initial stages of the loading when the specimen is damage-free (stages 1-4 of Figures 1 and 3), the bottlenecks coincide consistently with, and therefore predict, the location of the ultimate crack that forms in stage 11 after peak force (Figure 4). Damage is progressive from stage 5 onwards, as seen in the monotonic increase in the percentage of broken bonds (Figure 1). The initial sites of damage are spread throughout the specimen (stages 5 and 6 of Figure 3): here the bottlenecks remain in the same location as before, which is in the location of the crack that later forms in stage 11 (Figure 4). In the subsequent stages 7 to 8, bonds on the lower part of the specimen collectively break, momentarily shifting the bottlenecks to this new location (stages 7 and 8 of Figures 3 and 4). Just before and around peak force, however, the bottlenecks return to their original location, reflecting the spread of damage there until the ultimate crack emerges right along the bottlenecks.

5 Conclusions

We developed a network flow approach for mining patterns from microstructural DEM data in a cementitious granular material, here a 'dog-bone' concrete specimen submitted to uniaxial tension. Force transmission bottlenecks, a subset of contacts that are prone to becoming overloaded beyond their capacity, can predict the location and pattern of the ultimate crack in the nascent stages of loading *before* damage initiates inside the specimen. The extent to which this data analytics technique can predict failure patterns in a range of multiphase concrete specimens and loading conditions (e.g., compression, 3-point bending tests [9, 10]) will be reported in [11].

Acknowledgement

This work was supported by the Australian Research Council (DP120104759), the US Army Research Office (W911NF-15-1-0527) and US Air Force Office of Scientific Research (FA2386-15-1-4059).

References

[1] R.K. Ahuja, T.L. Magnanti, J.B. Orlin, *Network flows : theory, algorithms, and applications.* (Englewood Cliffs, N.J. : Prentice Hall, c1993., 1993), ISBN 013617549X

[2] D.P. Bertsekas, *Network optimization : continuous and discrete models.*, Optimization and neural computation series. 8 (Athena Scientific, 1998), ISBN 1-886529-02-7

[3] A. Tordesillas, A. Cramer, D.M. Walker, *Minimum cut and shear bands*, in *POWDERS AND GRAINS 2013: Proceedings of the 7th International Conference on Micromechanics of Granular Media* (AIP Publishing, 2013), Vol. 1542, pp. 507–510

[4] Q. Lin, A. Tordesillas, Journal of Industrial and Management Optimization **10**, 337 (2014)

[5] A. Tordesillas, S. Pucilowski, S. Tobin, M.R. Kuhn, E. Andò, G. Viggiani, A. Druckrey, K. Alshibli, Europhysics Letters **110**, 1 (2015)

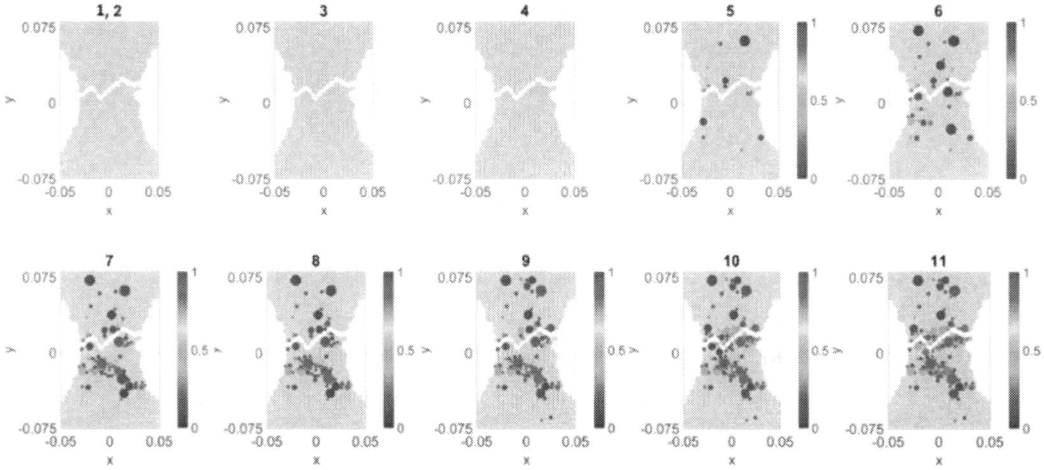

Figure 3. Spatial distribution of normalized damage sustained by individual grains, measured in terms of the percentage of number of broken bonded contacts relative to the initial number of bonded contacts (stages 1-4 are identical since no damage occurs across these stages). Solid gray grains are the grains with no broken bonds. The separation between the grains on either side of the crack that develops during failure (stage 11) is artificially increased to aid a visual comparison with the grains that sustained damage.

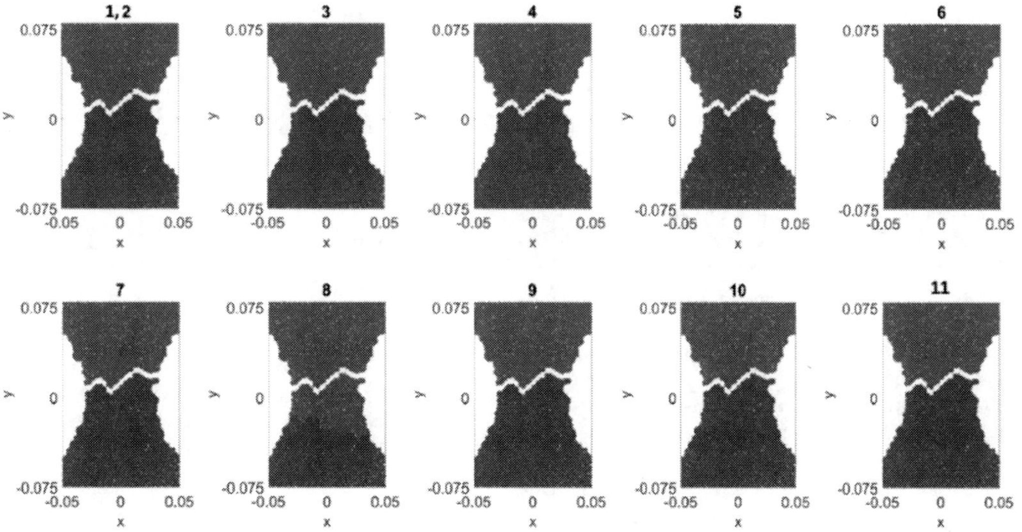

Figure 4. Spatial distribution of the predicted location of the crack, as identified from the bottlenecks in force transmission (red-blue interface). Grains on either side of the actual crack that later develops at stage 11 are artificially separated to aid visual comparison of the predicted crack location versus the actual crack location. Note the bottlenecks in stage 8 capture the spread of damage that momentarily dominates in the lower portion of the specimen just before peak force (recall stages 7-8 in Figure 1).

[6] A. Tordesillas, S.T. Tobin, M. Cil, K. Alshibli, R.P. Behringer, Physical Review E **91**, 062204 (2015)

[7] V. Lefort, G. Pijaudier-Cabot, D. Grégoire, Engineering Fracture Mechanics **147**, 449 (2015)

[8] A. Tordesillas, S. Pucilowski, L. Sibille, F. Nicot, F. Darve, Philosophical Magazine **92**, 4547 (2012)

[9] M. Nitka, J. Tejchman, Granular Matter **17**, 145 (2015)

[10] Ł. Skarżyński, M. Nitka, J. Tejchman, Engineering Fracture Mechanics **147**, 13 (2015)

[11] A. Tordesillas, S. Kahagalage, M. Nitka, C. Ras, J. Tejchman (2017), in preparation

DOI: 10.1051/epjconf/201714009026

Viscosity of particle laden films

Yousra Timounay[1,*] and *Florence* Rouyer[2,**]

[1] Université Paris-Est, Laboratoire Navier (UMR CNRS 8205, École des Ponts ParisTech, IFSTTAR) 2, allée Kepler, 77420 Champs-Sur-Marne, France

[2] Laboratoire Navier, Université Paris-Est, Laboratoire Navier (UMR CNRS 8205, École des Ponts ParisTech, IFSTTAR), 5 bd Descartes, 77 454 Champs-Sur-Marne, France

Abstract. We perform retraction experiments on soap films where large particles bridge the two interfaces. Local velocities are measured by PIV during the unstationnary regime. The velocity variation in time and space can be described by a continuous fluid model from which effective viscosity (shear and dilatational) of particulate films is measured. The 2D effective viscosity of particulate films η_{2D} increases with particle surface fraction ϕ: at low ϕ, it tends to the interfacial dilatational viscosity of the liquid/air interfaces and it diverges at the critical particle surface fraction $\phi_c \simeq 0.84$. Experimental data agree with classical viscosity laws of hard spheres suspensions adapted to the 2D geometry, assuming viscous dissipation resulting from the squeeze of the liquid/air interfaces between the particles. Finally, we show that the observed viscous dissipation in particulate films has to be considered to describe the edge velocity during a retraction experiment at large particle coverage.

1 Introduction

In many applications, as for example encapsulation or foams, hydrophobic particles are placed at liquid/air interfaces as they are expected to create an armor-like protective layer [1, 2]. Single liquid interfaces laden by hydrophobic particles, called particle rafts, are known to exhibit properties of both liquid interfaces and solid membranes [3, 4]. Such hybrid objects made of microscopic grains present size dependent properties as it is the case for granular media. A lot of studies are dedicated to the static regime and the Jansenn effect for example [5] of particle laden interfaces, whereas particulate soap films have been less studied [8]. In a previous study, we showed that the opening and retraction of particulate soap films depend on the particles position at the interfaces as well as on the constraints on the film edges [9]. Particles that bridge the two film's interfaces can inhibit the expansion of a hole because the rim formation and the opening of the capillary bridges between the grains are unfavorable. When the edge is constrained by a mobile stick, we observed a stationary regime during which the edge's velocity can be predicted at low particle surface fraction by a modified Taylor Culick model without any adjustable parameter. This model balances out surface tension that pulls the edge with the inertia of the collected particles. Larger is the particle surface fraction, lower is the retracting velocity. However, at larger particle surface fraction, experimental velocities are lower than expected, meaning that an additional resistive force slows down the particulate film retraction. We thus perform PIV measurements during the unstationary regime when surface tension accelerates the mobile stick to probe momentum diffusion and highlight the viscous character of particulate films.

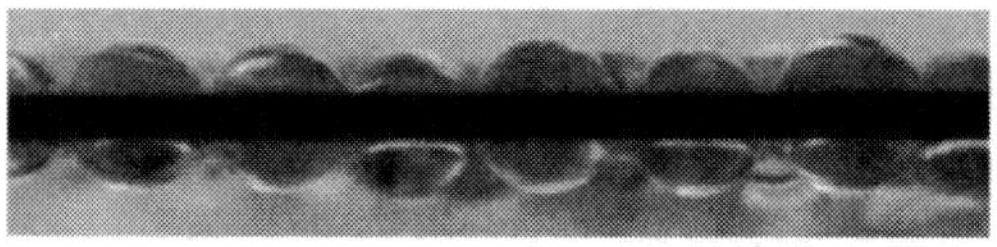

Figure 1. Side view of a particulate film, particle size d = 250 μm

2 Particulate films retraction experiment

Particulate films are created by pulling out a frame from a soap solution laden by hydrophobic particles at its liquid/air interface. The frame is a rectangle of 5 mm width noted a and 20 mm length. The soap solution is a Sodium Dodecyl Sulfate solution in water with a concentration close to the Critical Micellar Concentration 0.2% w/w and a density ρ_l = 1000 kg/m^3. The polystyrene beads are spherical and monodisperse, their diameter d can either be equal to 250 μm or 590 μm and their density ρ_p = 1050 kg/m^3. The contact angle between the particles and the liquid/air interface at equilibrium is equal to 95° ± 10°. Inside the film, the particles bridge the two interfaces, and the angle of contact is slightly deviated from its value at equilibrium (Figure 1).

*e-mail: yousra.timounay@ifsttar.fr
**e-mail: florence.rouyer@u-pem.fr

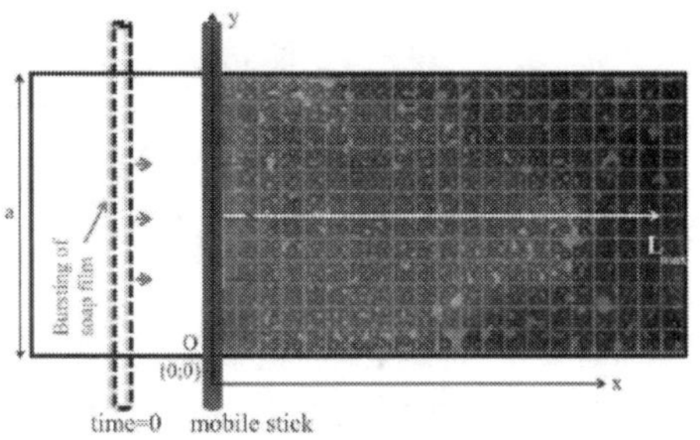

Figure 2. Sketch of the retracting film experiment with PIV image. The local velocities increase from blue to red.

A mobile stick is placed at the surface of the horizontal frame, perpendicularly to the wires, to pull apart a particulate film and a bare film (Figure 2). The particle coverage ϕ of the film is calculated as $N\pi(d/2)^2/S$ where N is the number of particles present over a surface S of the covered film that corresponds to the area of the film after excluding the peripheral areas near the edges (we subtract $2d$ to the real width and length of the particulate film). The particles are counted by image analysis. Right after bursting the bare film on the left of the mobile stick, the stick is pulled by surface tension and accelerates. Motion propagates through the particulate film before the stick velocity reaches a stationary velocity noted V_s. Particle Image Velocimetry (PIV) allows to measure local velocity in the plane of the film (Figure 2) by comparing successive images taken by a fast camera placed above the retracting film. This technique requires to partition the images into small areas called interrogation regions. In our case, these areas are squares of two to three particle diameters side. PIV measurements we performed show that the local velocity varies essentially along the direction perpendicular to the stick (x-axis), whereas no variation is observed in the direction parallel to the stick (y-axis) except at the proximity of the wires. To get rid of these edge effects, we don't consider in the following the local velocities in the two squares along each of the wires perpendicular to the stick. In Figure 3, we plot the velocity averaged over y and normalized by V_s as a function of the distance to the mobile stick (x). The velocities $V(x,t)$ increase with time and decrease over a characteristic length that increases with time. We measured the distance at which the velocity is half the velocity of the stick and show that this distance varies as the square root of time (Figure 3-Insert). This behavior is characteristic of momentum diffusion i.e. viscosity. In the next section, a model for measuring effective viscosity from velocity profiles is presented.

3 Continuous model in 2D for effective viscosity measurements

A continuous model for 2D compressible viscous fluid is considered for modeling local velocities of particle laden films. Note that the particulate films have a constant mass

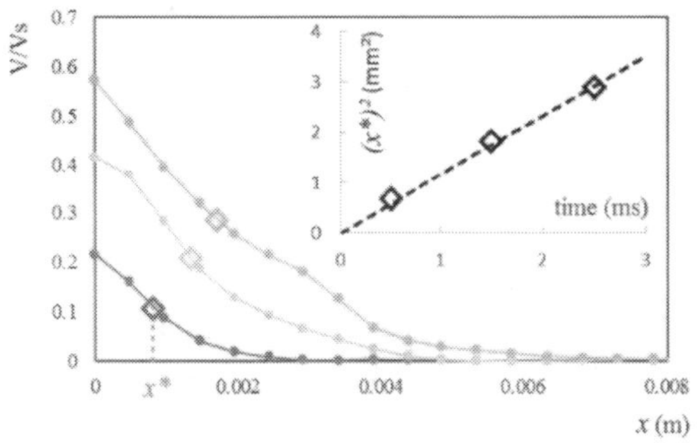

Figure 3. Velocity profiles as a function of the distance to the stick, 5 , 15 and 25 ms after the stick motion starts. Insert: square of distance for which velocity is half the velocity of the stick (diamonds in the main graph) versus time.

while their surface is varying during experiments, so we assume that the corresponding 2D viscous fluid is compressible. The momentum equation is written in a mobile frame with the stick as the origin. As the compression is along the x-axis and since the velocity varies along this axis (we neglect velocity variation along y axis), the momentum equation is written as :

$$\frac{\partial V}{\partial t} + (V - V_{stick}(t))\frac{\partial V}{\partial x} = \nu \frac{\partial^2 V}{\partial x^2} \qquad (1)$$

where ν is the effective kinematic viscosity and $V(x,t)$ is the velocity of the particle laden film along x at a distance x from the stick and at a interval of time t from the beginning of stick motion i.e the film retraction.

The differential equation 1 is numerically solved assuming that the film is at rest initially, and assuming the continuity of velocities at the boundaries. These conditions can be written as : $V(x,0) = 0$, $V(0,t) = V_{stick}(t)$ and $V(L_{max},t) = 0$, where L_{max} is the distance from the stick to the opposite edge of the frame. The experimental velocity profiles are fitted by numerical solutions with only one adjustable parameter that equals the kinematic viscosity ν. Fits are shown in figure 4.

For films covered homogeneously with large particles bridging both interfaces as illustrated in figure 1, PIV measurements and velocity profiles analysis are done with particle coverage ϕ varied from 60% up to 84%. Moreover, few additional experiments are done for films with no large particles but tracers made of small polystyrene particles of diameter equal to 40 μm at small surface concentrations (less than 1%) that are positioned entirely inside the film. For these experiments with particle-free films, PIV measurements and velocity profiles analysis lead to a kinematic viscosity $\nu_0 \simeq 1.8 \cdot 10^{-3}$ m²/s. A dynamic 2D-viscosity η_{02D} is thus deduced considering a "2D density" for the liquid film that writes as $\rho_{02D} = \rho_l e$, where $e = 45$ μm is the liquid thickness of the particle-free film [9]. The calculated value $\eta_{02D} = \nu_0 \rho_{02D} \simeq 7.9 \cdot 10^{-5}$ Pa.s.m is in the order of the interfacial dilatational viscosity of the liquid/air interfaces that delimit the liquid film [10]. We thus highlight that the dominant source of dissipation is the de-

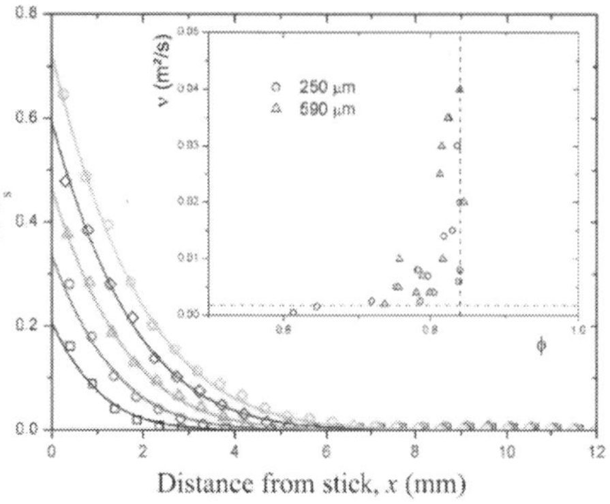

Figure 4. Velocity profiles as a function of the distance to the stick every 5 ms. The points are obtained from PIV measurements and the lines correspond to the numerical solutions of equation (1) for a particulate film where $d = 250\ \mu$ m and $\phi = 0.71$. Insert: the symbols correspond to the kinematic viscosity ν of particle laden films as a function of ϕ, the dotted horizontal and vertical lines correspond to $\nu_0 = 1.8 \, . \, 10^{-3}$ m^2/s and $\phi_c = 0.84$, respectively.

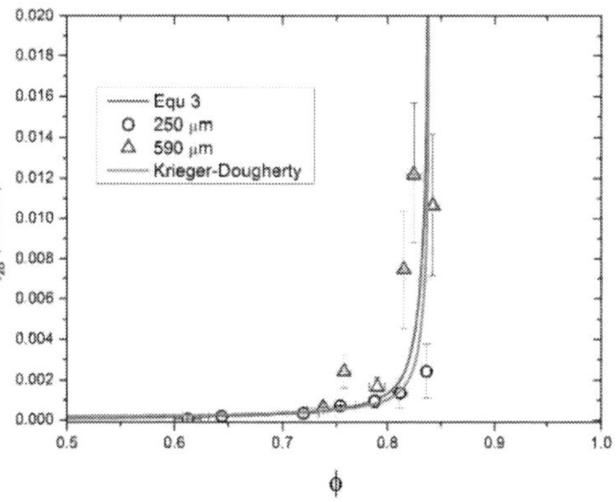

Figure 5. Effective dynamic viscosity versus particle surface fraction.

formation of the interfaces and not the deformation of the liquid inside the film.

4 Effective viscosity of particulate films: Results and Discussion

We now focus on the effective kinematic viscosity of the granular sheet and its variation with particle coverage. The effective kinematic viscosity of the granular sheet increases with particle surface fraction (figure 4-insert) from 10^{-3} to 4.10^{-2} m^2/s. These values are more than a thousand times the kinematic viscosity of water. At low ϕ, ν tends to the value ν_0 measured for particle-free films. The kinematic viscosity diverges at the critical packing fraction $\phi_c \simeq 0.84$ approaching jamming transition [11]. This critical behavior is also observed for drag friction that acts on a disk in a 2D dry granular medium [12]. Moreover, this divergence of viscosity is also characteristic of concentrated suspensions of hard spheres in a viscous liquid and indicates a transition from liquid-like behavior to solid-like behavior [13].

An effective dynamic 2D-viscosity η_{2D} is deduced from the kinematic viscosity as presented above for liquid films considering a "2D density" for particulate films $\rho_{2D} = (1-\phi)\rho_l e + \phi\rho_p \frac{2}{3} d$. The dependence of the effective dynamic 2D-viscosity with particle surface fraction is plotted in figure 5 and compared to "classical" laws for concentrated suspensions.

First, the experimental law of Krieger-Dougherty is tested : $\eta_{2D} = \eta_{02D}(1-\frac{\phi}{\phi_c})^{-k\phi_c}$ with $\eta_{02D} = 7.9 \, . \, 10^{-5}$ Pa.s.m, $\phi_c = 0.84$ and $k = 1$. This law has no adjustable parameter. Indeed, the values of ϕ_c and η_{02D} are justified previously. Moreover the value of $k = 1$ is consistent for $\phi \to 0$ with the model of Khair [14] in 2D that develops Einstein correction for the effective viscosity of dilute suspensions in a compressible fluid. This law is plotted in figure 5. Second, we adapt the 3D theoretical model of Mills for concentrated hard spheres [15] to the 2D geometry of the film. We write that the viscous dissipation of the particulate film $\frac{dE_v}{dt}$ per unit time and unit volume is due to viscous dissipation in the fluid film squeezed between the particles:

$$\frac{dE_v}{dt} \sim \eta_{2D}\dot{\epsilon}^2 \sim \eta_{02D}\left(\frac{\dot{\epsilon}d}{\Delta l}\right)^2 \omega \qquad (2)$$

where $\dot{\epsilon}$ is the average strain rate for the particulate film, $\Delta l = d\left(1-\sqrt{\phi/\phi_c}\right)$ is the average distance between the particles and $\omega = 2\phi\left(1-\sqrt{\phi/\phi_c}\right)$ is the surface fraction of the fluid. Equation 2 leads to an effective viscosity that diverges at ϕ_c and that writes :

$$\eta_{2D} \sim \eta_{02D}\frac{\phi}{1-\sqrt{\phi/\phi_c}} \qquad (3)$$

Equation 3 is plotted in figure 5 with the parameters fixed previously : $\phi_c = 0.84$, $\eta_{02D} = 7.9 \, . \, 10^{-5}$ Pa.s.m, and a prefactor of one-half.

The experimental data agree relatively well with both viscosity laws for concentrated 2D-suspensions presented above, this result suggests that the effective viscosity of particulate films is mainly due to the deformation of the liquid interfaces between the particles during the particulate film retraction.

Finally, we check if the effective viscosity presently measured is the source of dissipation responsible of the low retraction velocities measured at large particle coverage that are overestimated by the model based on the simple balance of inertia and surface tension. For this purpose, we calculate the viscous force per unit length F_η that the particulate film applies on the mobile stick during the film retraction at a time t_s when $V(0,t_s) = V_s$ (t_s corresponds to the end of acceleration of the stick and the beginning of the stationary regime):

$$F_\eta = \eta_{2D}\frac{e}{d}\left(\frac{\partial V}{\partial x}\right)_{(0,t_s)} \qquad (4)$$

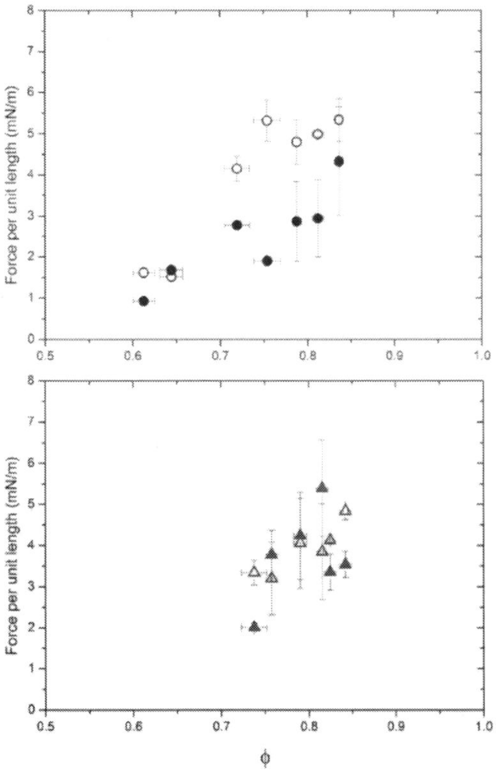

Figure 6. Open symbols and full symbols correspond to the resistive linear force F_r and to the linear viscous force F_η respectively. Particles diameter is equal to 250 μ m (top) and 590 μ m (bottom).

In figure 6, the force F_η is plotted versus ϕ. As expected, this force increases with particle coverage.

We now compare this force to the force F_r per unit length that resists the motion of the stick and that has to be added to the momentum balance on the stick to describe the stick motion during its stationary regime :

$$\beta\gamma - F - F_r = V_s \frac{dm}{dt} \quad (5)$$

Where $1 < \beta < 2$ because of the angle of the menisci that link the liquid film to the stick and $\beta\gamma - F$ is equal to the pulling force resulting from the surface tension of the two interfaces reduced by contact effects of the stick on the wires of the frames. Assuming that $\beta\gamma - F$ is similar in the case of particle-free films and particulate films, we deduce the value of F by measuring the retraction velocity of particle-free films. We find $F \simeq 10$ mN/m.

In figure 6, the linear force F_r is plotted versus ϕ, and compared to F_η. The two forces are quiet comparable. We thus conclude that viscous dissipation at the liquid interfaces that are highly solicited during particulate films retraction is responsible of the slowed down dynamics for particle surface fractions larger than 60%.

5 Conclusion

Retraction of particulate films where large particles bridge both interfaces is studied through a microscopic approach at the particles scale and a macroscopic one at the scale of the film. We show that such particulate films behave like a compressible viscous fluid which viscosity diverges approaching jamming transition. The 2D-effective viscosity of particulate films is comparable to viscosity model of hard sphere suspensions in 2D, suggesting that the dissipations result from the squeeze of the liquid bridges between the particles during a retraction experiment. The dissipations due to the interfacial dilatational viscosity of the liquid-air interfaces seems to be dominant relative to the dissipations in the bulk of the liquid connecting the particles. Moreover, the divergence of the effective viscosity at 0.84 that corresponds to the jamming surface fraction of a 2D granular media can be seen as a manifestation of the granular character of particulate films we present in this article.

Acknowledgments

We thank X. Chateau and O. Pitois for discussions. We acknowledge J.Bico for welcoming us in his laboratory for complementary experiments. We are grateful to IFSTTAR for financial support, and Judith Vatteville from LaVision for technical help.

References

[1] Aussillous, P. and Quere, D., Nature, **411**, 6840, 924-927 (2001)
[2] Stocco, A. et al., Soft Matter, **7**, 2, 631-637, (2011)
[3] Vella, D., Aussillous, P., Mahadevan, L., EPL (Europhysics Letters), **68**, 212 (2004)
[4] Planchette, C., Lorenceau, E., Biance, A.-L., Soft Matter, **8**, 2444 (2012)
[5] Cicuta, P., Vella, D., Physical review letters, **102**, 138302 (2009).
[6] Abkarian, M., et al. Phys. Rev. Lett. **99**, 188301 (2007)
[7] Pitois, O., Chateau, X., Eur. Phys. J. E, **38**:48 (2015)
[8] Horozov, T.-S., Current Opinion in Colloid & Interface Science, **13** 134–140 (2008)
[9] Timounay, Y., Lorenceau, E., Rouyer, F., EPL (Europhysics Letters), **111**(2), 26001 (2015)
[10] Kao,R., Edwards,D., Wasan, D., Chen, E., J. Colloid Interface Sci., **148**(1), 247–256 (1992)
[11] Majmudar, T. S., Sperl, M., Luding, S., Behringer, R. P., Physical Review Letters, **98**(5), 058001 (2007)
[12] Takehara, Y., Okumura, K., Phys. Rev. Lett. **112**,148001 (2014)
[13] Krieger, I. M., Dougherty, T. J., Transactions of The Society of Rheology, **3**, 1, 137-152 (1959)
[14] Khair, A.S., J. Colloid Interface Sci., **302**(2):702-3 (2006)
[15] Mills, P., Snabre, P., Eur. Phys. J. E, **30**, 309-316 (2009)

EPJ Web of Conferences **140**, 03050 (2017) DOI: 10.1051/epjconf/201714003050
Powders & Grains 2017

Free flowing and cohesive powders agitation in a cylindrical convective blender- kinetics experiments and Markov chain modelling

Léonard Legoix[1], *Mathieu* Milhé[1,*], *Cendrine* Gatumel[1], and *Henri* Berthiaux[1]

[1] Université de Toulouse, Centre RAPSODEE, CNRS UMR 5302, Ecole des Mines d'Albi, Campus Jarlard, 81013 Albi Cedex 09, France

Abstract. An original methodology for studying powder flow in a cylindrical convective blender has been developed. A free-flowing and a cohesive powder were studied, at a fixed stirring speed, in rolling regime. For both powders, three apparent flow mechanisms were evidenced: convection in the volume swept by the blades, diffusion/shearing between the agitated zone and the stagnant one, as well as in the stagnant zone itself, and avalanches at the powder bed surface between agitated and stagnant zones. After defining six zones in the blender, tracing experiments were carried out by placing appropriate tracers in different starting zones and sampling the whole bed at different stirring times, which lead to mixing kinetics of the powders into themselves. A Markov chains model of the blender allowed the quantification of the three mechanisms respective magnitude by fitting the experimental data. This simple model has a good agreement with the free-flowing powder data, but is not able to represent well the observations for the cohesive powder. Bed consolidation should probably be taken into account for this kind of powders and thus a linear Markov model is not sufficient.

1 Introduction

In spite of the recurring use of convective blenders in the pharmaceutical industry, their global modelling has not been carried out yet, particularly when focusing on fine or cohesive powders behaviour. A DEM approach is difficult to take on with particle sizes about 10μm leading to overall populations of 10^{14} when batches of 100kg are considered.

Moreover, when dealing with batch planetary blenders, which combine two circular movements (namely rotation of the blades and gyration of their axis), the task is yet more cumbersome. For these reasons, a systemic approach has been undertaken in previous works, based on rheological observations at pilot scale [1]. This lead to the setup of a cylindrical convective blender (Fig. 1): this device allows carrying out rheological studies, flow observations and determination of mixing kinetics.

It should be noted that the term "mixing" is used in the sense of self-mixing of the powder in this study: the tracer employed for the mixing experiments are supposed to have the same flow behaviour and particle size than the powders studied.

The objective of this work is to study the differences in flow behaviour between a free-flowing and a cohesive powder when agitated by straight blades in a cylindrical convective blender, both experimentally and numerically thanks to a systemic modelling based on Markov chains.

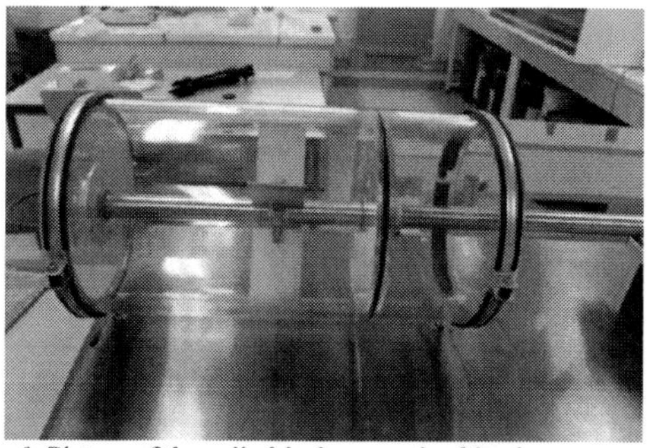

Fig. 1. Picture of the cylindrical convective blender

2 Experimental determination of mixing kinetics

2.1 Powders studied

The free-flowing powder used was semolina, the cohesive powder was lactose. Both have a polygonal shape, their physical characteristics are summed up in Table 1. The mean diameter d_{50} was determined by laser diffraction (Mastersizer3000, Malvern) under an air pressure of 3.5 bar, the Carr Index CI thanks to a volumenometer Erweka, according to eq. 1, where ρ_b and ρ_t are respectively the bulk density and the tapped density.

* Corresponding author: mmilhe@mines-albi.fr

$$CI = (\rho_t - \rho_b)/\rho_t \quad (1)$$

Table 1. Powders characteristics

Powder	Semolina	Lactose
d_{50} (μm)	312	26
ρ_b (kg/m^3)	679	495
ρ_t (kg/m^3)	720	638
CI (%)	5.8	22.5

2.2 Experimental protocol

Each experiment was performed with a stirring speed of 6 rad/s, corresponding to a rolling regime (the powder flows along the rotating blades and is not projected). For both powders, the blender was filled at 83%.

The blender sides were set asymmetrically and the volume was virtually divided into six zones of identical volume as illustrated in **Fig. 2**. Before an experiment, the blender was filled with the powder studied in five zones and with the corresponding tracer in the sixth zone.

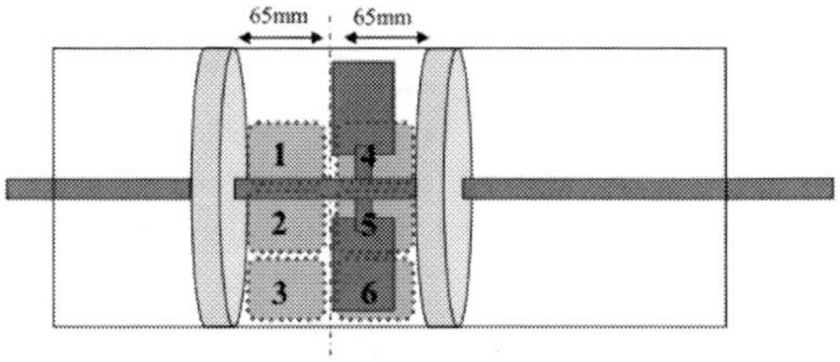

Fig. 2. Illustration of the 6 transversal zones considered in the blender

For each tracer initial position (zone 1, 3 and 4 in this study), three to four experiments were performed with different agitation durations. At the end of each experiment, the whole bed is sampled zone by zone, and the tracer mass fraction in each zone is determined (see next section).

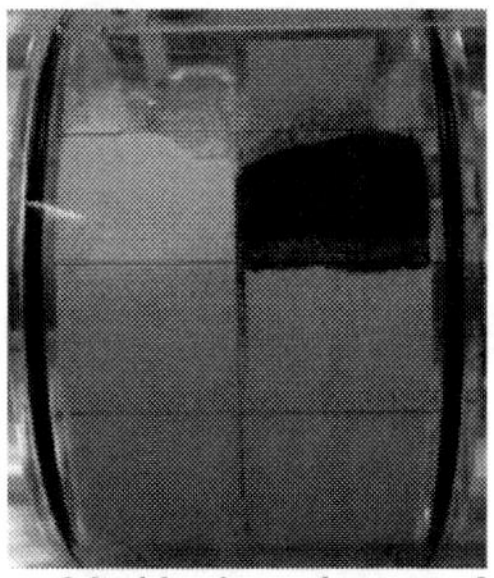

Fig. 3. Front view of the blender at the start of an experiment with the semolina tracer initially in zone 4.

Each series of experiment thus constitutes a mixing kinetics of the powder into itself, assuming the tracer behaves the same as the powder.

2.3 Tracer analysis

The tracer used for free-flowing experiments was semolina coloured thanks to an iodine ethanol solution: this tracer has the same d_{50} than the original semolina and was detected by image analysis. A given zone sampled from the blender was homogenised and poured on a 19 cm diameter plate. 8 pictures were then taken at random locations with an x50 magnifying microscope and the images were analysed thanks to a Matlab© routine. The tracer concentration was obtained thanks to a calibration curve.

The tracer used for cohesive experiments was talc, since no colorimetric techniques proved satisfying for tracing purposes. The talc used had characteristics close to that of lactose. Its mass fraction in the samples was determined by thermogravimetry analysis: each sampled zone was homogenised and 12 samples of 2g each were sampled randomly from it, before being heated at 650°C for 2h. The remaining talc was then weighed again and its initial mass fraction deduced since this compound showed no weight loss at this temperature.

3 Markov chain modelling

Powder mixing operations have been studied and modelled with Markov chains in previous works for continuous mixers [2,3] and batch mixers [4]. In this work, it was chosen to model the evolution of the tracer mass fraction in the six zones by a homogeneous and linear Markov chain: the transition probability from a zone (a state in Markov formalism) to another does not depend on the time or on the amount of tracer in this zone.

3.1. Mechanisms considered

Based on observations, four mass transfer mechanisms have been taken into account in this model:
- convection between zones 1 and 2, and between zones 2 and 3, due to blade rotation.
- avalanching between zone 1 and 4, due to powder bed level differences after each blade pass.
- horizontal diffusion between zones 2 and 5, and between zones 3 and 6.
- vertical diffusion between zones 1 and 2, and between zones 2 and 3.

These mechanisms are respectively represented by the coefficients c, a, d_h and d_v hereafter. They represent the fraction of powder moving from a zone to another one due to the different mechanisms.

3.2 Markov chain

Let denote S_n the vector describing the mass fraction of a tracer (or any extensive quantity related to the powder) in the different zones of the powder bed in the blender at the nth transition. According to the Markov property, the evolution of S_n can be calculated from S_{n-1} with eq. 2, where P is the matrix of transition probabilities.

 DOI: 10.1051/epjconf/201714003050

$$S_n = P \, . \, S_{n-1} \tag{2}$$

The matrix P is constructed with the coefficients defined above (section 3.1): each term $P_{i,j}$ is the proportion of $S_n(j)$ moving to zone i: it follows that it must be comprised between 0 and 1. The diagonal terms $P_{j,j}$ are deduced from the other thanks to a mass balance, i.e. the sum of each column must be equal to 1. The matrix P is symmetric since all mechanisms are supposed to occur in both directions with the same magnitude.

$$P = \begin{pmatrix} 1-d_v-a & d_v & 0 & a & 0 & 0 \\ d_v & 1-2d_v-d_h & d_v & 0 & d_h & 0 \\ 0 & d_v & 1-d_v-d_h & 0 & 0 & d_h \\ a & 0 & 0 & 1-a-c & c & 0 \\ 0 & d_h & 0 & c & 1-d_h-2c & c \\ 0 & 0 & d_h & 0 & c & 1-d_h-c \end{pmatrix} \tag{3}$$

The characteristic time defining a transition is the time needed by the blades to make ¼ of a rotation: every 1/4 of a rotation, the quantity described in S_n is redistributed according to eq. 2.

3.3 Coefficients identification

The four coefficients have been identified by fitting the model to the experimental mixing kinetics for both powders. Knowing the experimental mixing times n_{exp} and the corresponding tracer mass fractions S_n^{exp} for each experiment of a series with initial tracer position in zone i, the best coefficients are obtained by minimizing the objective function in eq. 4.

$$\Delta = \sum_{i=1,3,4} \left(\sum_{n_{exp}} \left(S_n - S_n^{exp}\right)^2 \right) \tag{4}$$

4 Results

The experimental results are first used to obtain model coefficients for the two powders studied. For semolina, four experimental values of S^{exp} were obtained, with agitation durations ranging from 4 to 300 s. For lactose, only three values were obtained in the same range.

4.1 Model coefficients

The two sets of coefficients obtained for free-flowing semolina and cohesive lactose are gathered in Table 2. The coefficients c and d_v must be lower than 0.5 in order to avoid negative terms in P, which would be physically impossible. Both sets of coefficients show the same trend: convection is responsible for most part of mass transfer between the zones, while avalanches only concern about 3% of the powder in zones 1 and 4. These two coefficients reflect the difference between the powders: the lactose is clearly less prone to flow due to its higher cohesion. This confirms the physical basis for this model, in spite of strong assumptions.

Diffusion mechanisms showed to have an influence 3 to 4 order of magnitude lower than convection by the blades. However, for semolina, the fact that d_h is null is linked to the value of c because $P_{5,5} = 1 - d_h - 2c$ and not necessarily because this particular mixing mechanism is not involved. The value of d_h obtained for lactose is actually the accuracy of the optimization procedure used so it might not be reliable.

Table 2. Model coefficients for both powders

Powder	Semolina	Lactose
c	0.4999	0.4511
a	0.0347	0.0279
d_h	0	0.0001
d_v	0.0071	0

4.2 Free-flowing case

Simulations and experimental data for semolina are compared in figures 4 and 5, corresponding respectively to initial tracer position in zones 4 and 1. Transition number is represented in logarithmic scale for visualization purposes. In both cases, the agreement is satisfying.

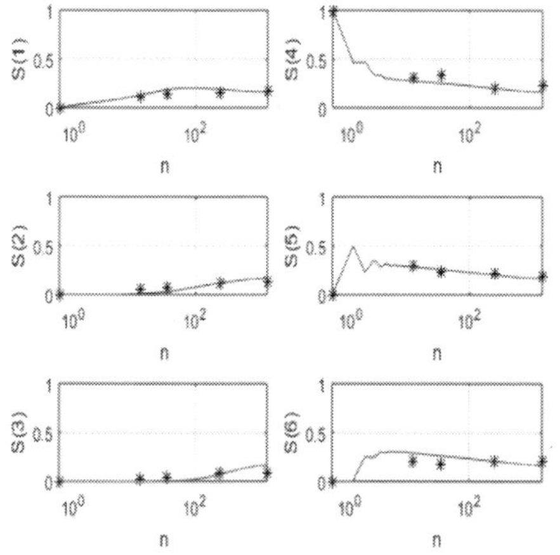

Fig. 4. Comparison of simulations and experimental data for semolina, with initial tracer position in zone 4

An oscillating behaviour can be observed in zones 4 to 6 during the first transitions, mainly because most of the powder in zone 5 is supposed to be exchanged with zones 4 and 6 at each transition.

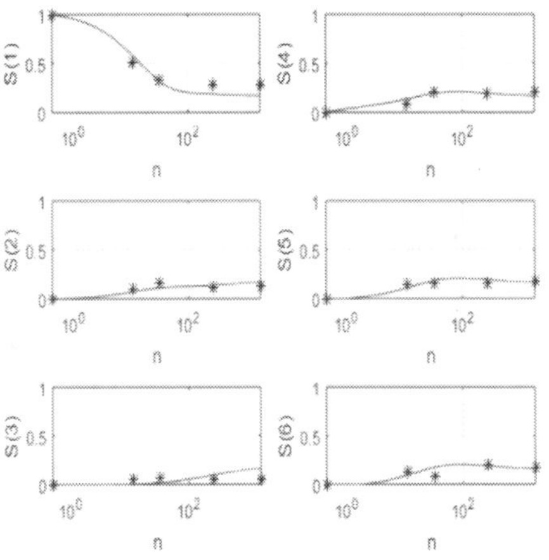

Fig. 5. Comparison of simulations and experimental data for semolina, with initial tracer position in zone 1

4.3 Cohesive case

Simulations and experimental data for lactose are compared in figures 6 and 7, with the same initial tracer positions. In this case, the agreement is less good, particularly when the tracer is initially in zone 1. The set of coefficients identified does not allow describing correctly powder flows in the blender

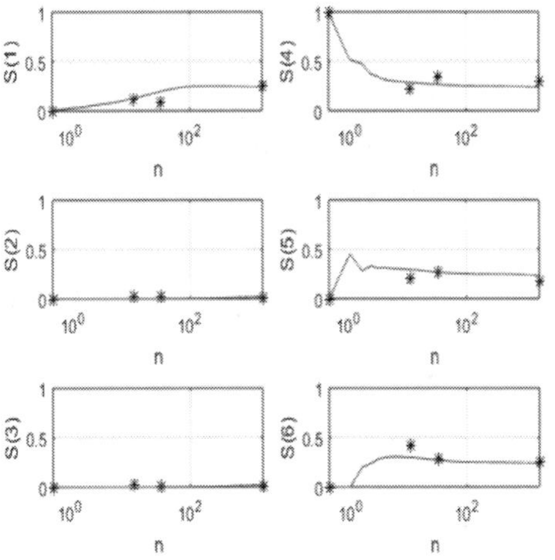

Fig. 6. Comparison of simulations and experimental data for lactose, with initial tracer position in zone 4

While experimental data show a stationary behaviour after the very first transitions, the model suggests a longer evolution and a better homogenization of the tracer between zones 1, 4, 5 and 6. This could be due to a non-linear behaviour of the powder bed, with more intense flows between zones at the beginning of the blending operation and virtually no powder exchanges when the bed is consolidated after a few rotations of the blades. This behaviour cannot be predicted by a linear Markov chain, in which the transition probabilities do not depend on transition number (and thus time).

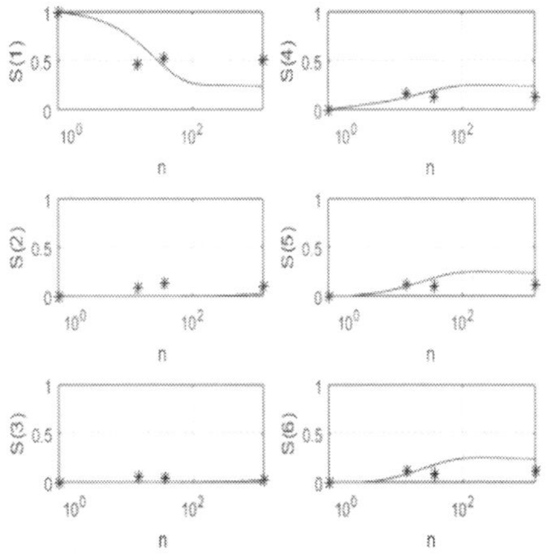

Fig. 7. Comparison of simulations and experimental data for lactose, with initial tracer position in zone 1

5 Conclusion

The influence of powder cohesion in convective blending has been characterized through mixing kinetics experiments. A simple model based on Markov chains is able to reproduce satisfactorily the self-mixing of a free-flowing powder. However, the non-linear behaviour of cohesive powders is more problematic and requires more advanced modelling.

The authors would like to thank Luana Grillet and Asma Toualoua for their valuable contribution to the experimental results.

References

1 L. Legoix, C. Gatumel, M. Milhe, H. Berthiaux, Rheology of cohesive powders in a pilot scale planetary blender, Powder Technol. **305** (2017).
2 K. Marikh, H. Berthiaux, V. Mizonov, E. Barantseva, D. Ponomarev, Flow Analysis and Markov Chain Modelling to Quantify the Agitation Effect in a Continuous Powder Mixer, Chem. Eng. Res. Des. **84** (2006).
3 C. Ammarcha, C. Gatumel, J.L. Dirion, M. Cabassud, V. Mizonov, H. Berthiaux, Transitory powder flow dynamics during emptying of a continuous mixer, Chem. Eng. Process. Process Intensif. **65** (2013).
4 M. Aoun-Habbache, M. Aoun, H. Berthiaux, V. Mizonov, An experimental method and a Markov chain model to describe axial and radial mixing in a hoop mixer, Powder Technol. **128** (2002).

Influence of stress-path on pore size distribution in granular materials

Arghya Das[1,*] and *Abhinav* Kumar[1]

[1]Department of Civil Engineering, Indian Institute of Technology Knapur, Kanpur - 208016, India

Abstract. Pore size distribution is an important feature of granular materials in the context of filtration and erosion in soil hydraulic structures. Present study focuses on the evolution characteristics of pore size distribution for numerically simulated granular assemblies while subjected to various compression boundary constrain, namely, conventional drained triaxial compression, one-dimensional or oedometric compression and isotropic compression. We consider the effects initial packing of the granular assembly, loose or dense state. A simplified algorithm based on Delaunay tessellation is used for the estimation of pore size distribution for the deforming granular assemblies at various stress states. The analyses show that, the evolution of pore size is predominantly governed by the current porosity of the granular assembly while the stress path or loading process has minimal influence. Further it has also been observed that pore volume distribution reaches towards a critical distribution at the critical porosity during shear enhanced loading process irrespective of the deformation mechanism either compaction or dilation.

1 Introduction

Pore size distribution (psd) is one of the key microstructural components of geomaterials that influences various flow related geotechnical problems such as erosion, piping, infiltration as well as filter design [1]. Even the nature of soil moisture interaction, which is crucial for unsaturated soils, depends on the distribution of pore spaces [2].

Estimation of psd in a granular geomaterials requires intense geometrical understanding of the amorphous pores [3,4]. Triangulation or tessellation (either Voronoi or Delaunay) approach is widely used for the determination of psd in granular materials. Chan and Ng [5] used tetrahedral tessellation to estimate the pore size distribution (psd) and the pore network. Extensive image analysis technique was employed by Sweeney & Martin [6] to capture the psd of a simulated particulate assembly involving discrete element method (DEM). Reboul et al. [7] introduced a novel pore merging criteria among the neighbouring pores and improved the psd estimation process. Alternatively, a maximum ball algorithm was introduced by Silin and Patzak [8] and further improved by Dong and Blunt [9] to quantify the pore network. Recently Gao et al. [10] and Vincens et al. [11] performed rigorous analysis such as identifying the specific geometry of pore spaces, and the connectivity using a tube network of pore spaces for the realistic characterization of psd.

Despite a large number of noteworthy contributions in the area of psd analysis, only few have attempted to study the effect of external loading on the evolution of psd(s). Kuhn & Bagi [12] identified the association of pore volume evolution with the deformation mechanism of granular materials using void cell distortion technique. Sánchez et al. [14] argued that the variation in pore size or constriction size distribution is a result of a microstructural alteration under external mechanical loadings. Some studies on the pore distribution by Kang et al. [3] and Shire et al. [15] demonstrated the relationship between pore orientation and principal stress anisotropy during the shearing of granular materials. In the present study, we numerically simulate various loading conditions in DEM and simultaneously track the psd evolution. The aim is to capture the characteristics of psd at various stages of deformation in a granular material, precisely to understand the effects of stress path.

2 Numerical simulations of compressive loading

We employ Particle Flow Code in Three-Dimensions (PFC-3D, ver. 5, 2015) that implements DEM for the present numerical analysis. The code generates cylindrical (76 mm in height and 38 mm in diameter) assembly of rigid spherical particles. Two different packing of particles are constructed, dense specimen with porosity 0.4 consists of 4300 particles and loose specimen with porosity 0.45 consists of 4080 particles. Rigid boundary walls and linear inter-particle contact law is considered for the simulations. Other relevant parameters for DEM simulations are listed in Table 1.

* Corresponding author: arghya@iitk.ac.in

Table 1. DEM parameters.

Parameters	Value
Normal stiffness to ball	1×10^8 N/m
Tangential stiffness to ball	0.667×10^8 N/m
Wall stiffness	1×10^7 N/m
Particle friction coefficient	0.25
Wall friction coefficient	0.0
Density of the balls	2650 kg/m^3
Maximum and minimum ball radius	2 mm and 1 mm

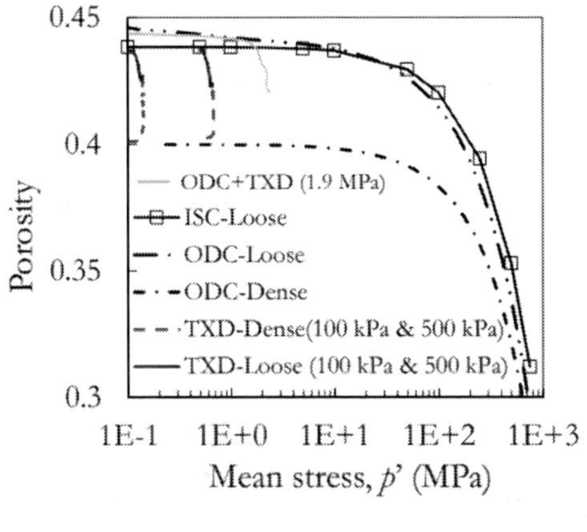

Fig. 1. Volumetric response.

The loading program consists of conventional drained triaxial compression test (TXD), oedometric compression test (ODC), combined oedometric and triaxial compression (ODC+TXD), and isotropic compression test (ISC). TXDs and ODCs are performed on both loose and dense specimens, while ODC+TXD and ISC are performed only on the loose specimen. 500 kPa and 100 kPa initial isotropic confinement is applied prior to the deviatoric compression in TXDs. Fig. 1 shows the volume compression curves for all the three loading programs. The volumetric responses of TXDs follow dilative and contractive nature in the case of dense and loose specimens respectively. In both the TXDs, irrespective of the initial density, a residual porosity is achieved, which is an indication of the critical-state condition. On the other hand, the volume compression curves in ODCs and ISC converge towards the normal compression line for both the loose and the dense samples.

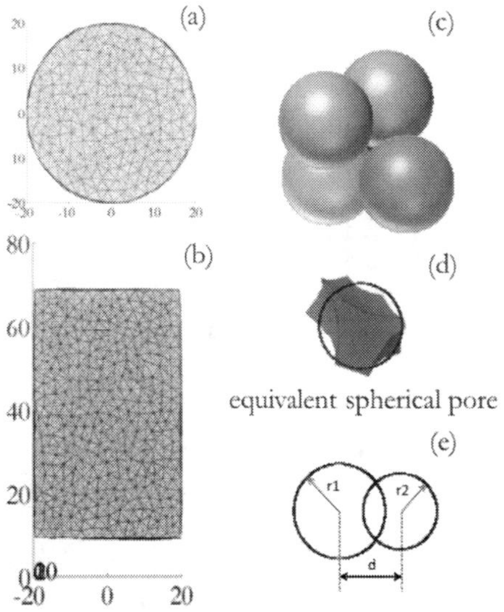

Fig. 2. Pore size estimation: (a-b) cross sectional view of Delaunay tessellation; (c-e) pore merging criteria.

3 Evolution of pore size distribution

3.1 Algorithm for pore size analysis

In order to quantify the pore sizes within the particle assembly generated from DEM, a Delaunay tessellation based triangulation is adopted (see the typical sample cross section after triangulation in Fig. 2a). In this process, every particle center virtually connects with three other neighbouring particle centers forming a tetrahedron (Fig. 2c). The void space is the volume of tetrahedron after subtracting the solid part occupied by those four particles (Fig 2d). Further, we employ a merging technique, which initially proposed by Reboul et al. [7] to identify the overlapped void spaces within the tetrahedrons. The algorithm identifies the overlapped pores by measuring the central distance between the equivalent void spheres. If the sum of the radii of any two such void spheres is greater than the central distance, those two void spaces are merged into a single void space (Fig. 2e). Thereafter, the search for overlapped pores is further carried out to the next neighbouring Delaunay cell and so on. However, unlike Gao et al. [10], we do not distinguish between the various levels of pore overlap to identify the discrete pore size and constriction size.

3.2 Evolution of pore size distribution in TXD

Fig. 3(a-b) represents the initial and final psd(s) for triaxial compression tests carried out for two different initial confinements. The evolving psd(s) indicate irrespective of the initial particle packing density of the specimens, near-identical probability density distribution is attained at the critical state i.e. at porosity 0.426 for

100 kPa confinement and at 0.42 porosity for 500 kPa confinement. In the case of dense specimens, the sample dilates during shearing, eventually the small pores diminish, and more number of larger pores evolve and hence modal pore density reduces. Alternatively, in the case of the loose specimen, disintegration of big pores during contraction generates new small pores and modal pore density slightly increases.

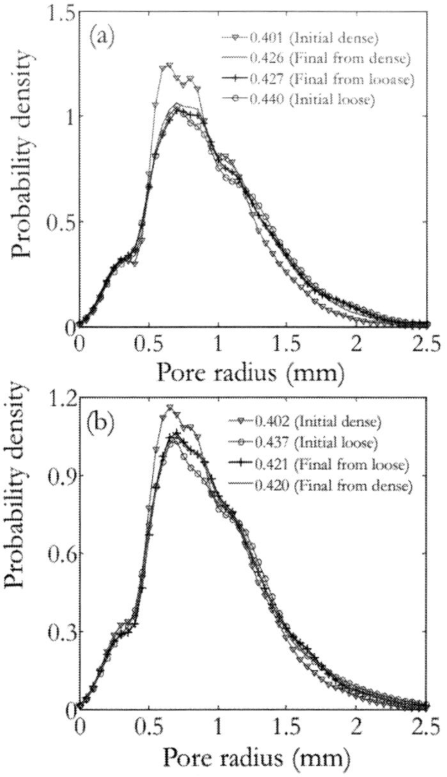

Fig. 3. PSD evolution during triaxial drained compression at the initial confinement of (a) 100 kPa; (b) 500 kPa.

3.3 Evolution of pore size distribution in ODC

Oedometric compression results in a contractive pore volume change for both loose and dense sample (see Fig. 4). However, unlike TXDs, here the evolution of psd(s) demonstrates significant alteration, in terms of the shape of the curve, modal values, and the peak. Such distinct shift in the modal values (~0.25 mm) in ODCs indicates complete alteration in pore distribution and not just a local pore rearrangement as previously observed by Reboul et al. (2008) for randomly packed spheres in the absence of external loading. The climbing peak value in the psd(s) is the evidence of increasing smaller pores because of pore compression and large pore disintegration. The porosity reduction in ODC also decreases the relative dispersion in psd. Additionally, multimodal nature in the psd(s) at the low porosity (0.3 to 0.24) condition is also noticed. ISC and dense ODC type loading show qualitatively similar trend in the psd evolution as the loose ODCs, thus not presented here.

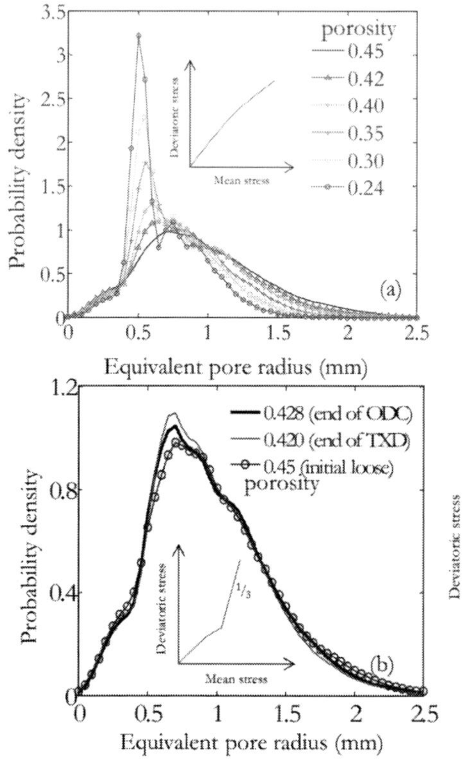

Fig. 4. PSD evolution during (a) oedometric compression for loose specimen; and (b) combined oedometric + drained triaxial compression for the loose specimen.

4 Stress path dependency on psd

Stress path dependency on the characteristics of psd(s) is analysed by comparing the pore distributions evolved through various loading programs at identical porosities. However, such choice may lead to large variation in stress-states. In Fig. 3 it has already been demonstrated that at the critical state condition both the loose and dense samples converge to an unique psd during TXDs. Similar to the TXDs, the pore distributions obtained from ODCs are compared in Fig. 5a for loose and dense samples at porosity 0.35. Despite, completely different stress conditions in these cases, the psd(s) are matching.

Interestingly one can notice the differences in the pds(s) obtained from TXD and ODC given in Fig. 5b at porosity ~0.42. Despite the similarity in the overall shape of the psd(s), they are not as identical as we observed at the critical state in TXD. Similarly, the comparison given in Fig. 5c between the psd(s) from oedometric compression and isotropic compression show dissimilar characteristic at porosity 0.4.

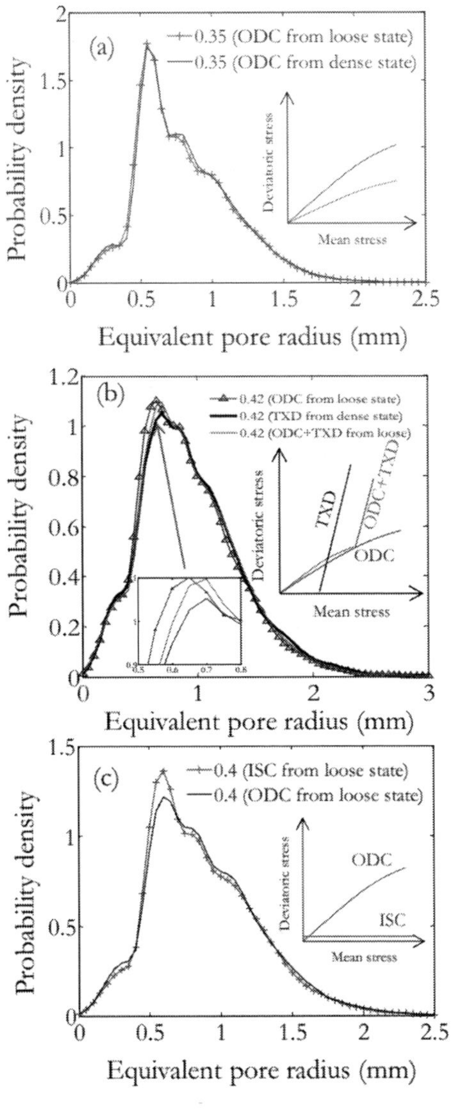

Fig. 5. Comparison of psd(s) obtained from: (a) loose and dense ODC; (b) TXD (500 kPa), ODC and ODC+TXD (1.9 MPa); (b) ISC and ODC.

These observations indicate psd(s) emerged from different loading process are not same even at identical porosity and stress state. Similar stress paths possibly leads to similar microscopic stress orientation within granular assemblies. Earlier studies [3,16,15] already established that principal stress direction governs pore orientation. Hence, it is likely that the relative magnitude of the principal stresses dictates the pore sizes as well. However, the influence of the stress path on the variation of the pore distribution is less as compared to the porosity variation.

5 Conclusions

In the present study, we demonstrate that porosity is the major factor that dictates the characteristics of pore size distribution in deforming granular materials irrespective of its initial packing. In addition, minor influence of stress path on the pore size is also noticed. Numerical simulations using DEM show that, a unique pore distribution is obtained at the critical-state during drained triaxial compression. Similarly, oedometric compression on loose and the dense specimens converses to identical pore distribution at the same porosity level. However, variation in stress path leads different pore distribution even at the same porosity level.

References

1. B. Indraratna, A. K. Raut, and H. Khabbaz, J. Geotech. Geoenvironmental Eng. **133**, 266 (2007).
2. X. Li and L. Zhang, in *Exp. Unsaturated Soil Mech.*, edited by T. Schanz (Springer Berlin Heidelberg, 2007), pp. 401–412.
3. D. H. Kang, J. Choo, and T. S. Yun, Comput. Geotech. **49**, 53 (2013).
4. H. J. Vogel and A. Kretzschmar, Geoderma **73**, 23 (1996).
5. S. K. Chan and K. M. Ng, Powder Technol. **54**, 147 (1988).
6. S. M. Sweeney and C. L. Martin, Acta Mater. **51**, 3635 (2003).
7. N. Reboul, E. Vincens, and B. Cambou, Granul. Matter **10**, 457 (2008).
8. D. Silin and T. Patzek, Phys. A Stat. Mech. Its Appl. **371**, 336 (2006).
9. H. Dong and M. J. Blunt, Phys. Rev. E **80**, 036307 (2009).
10. S. Gao, J. N. Meegoda, and L. Hu, Int. J. Numer. Anal. Methods Geomech. **36**, 1954 (2012).
11. E. Vincens, K. J. Witt, and U. Homberg, Acta Geotech. (2014).
12. M. R. Kuhn and K. Bagi, Int. J. Solids Struct. **41**, 5793 (2004).
13. V. Ogarko, N. Rivas, and S. Luding, J. Chem. Phys. **140**, 211102 (2014).
14. J. Sánchez, G. Auvinet, and B. Cambou, in *Geomech. from Micro to Macro*, edited by K. Soga, K. Kumar, G. Biscontin, and M. Kuo (2014), pp. 73–78.
15. T. Shire, C. O'Sullivan, D. Barreto, and G. Gaudray, Géotechnique **63**, 85 (2013).
16. D. H. Kang, T. S. Yun, Y. M. Lau, and Y. H. Wang, Comput. Geotech. **39**, 98 (2012).

DOI: 10.1051/epjconf/201714011007

FEM × DEM: a new efficient multi-scale approach for geotechnical problems with strain localization

Trung Kien Nguyen[1,*], *Albert* Argilaga Claramunt[1], *Denis* Caillerie[1], *Gaël* Combe[1,**], *Stefano* Dal Pont[1], *Jacques* Desrues[1], and *Vincent* Richefeu[1]

[1]*Grenoble-INP, UGA, CNRS UMR5521, Laboratoire 3SR, 1270 rue de la Piscine, Domaine Universitaire, Saint-Martin-d'Hères, BP53, 38041 Grenoble Cedex 9, France*

Abstract. The paper presents a multi-scale modeling of Boundary Value Problem (BVP) approach involving cohesive-frictional granular materials in the FEM × DEM multi-scale framework. On the DEM side, a 3D model is defined based on the interactions of spherical particles. This DEM model is built through a numerical homogenization process applied to a Volume Element (VE). It is then paired with a Finite Element code. Using this numerical tool that combines two scales within the same framework, we conducted simulations of biaxial and pressuremeter tests on a cohesive-frictional granular medium. In these cases, it is known that strain localization does occur at the macroscopic level, but since FEMs suffer from severe mesh dependency as soon as shear band starts to develop, the second gradient regularization technique has been used. As a consequence, the objectivity of the computation with respect to mesh dependency is restored.

1 Introduction

Since its appearance in the late 70s, the Discrete Element Method (DEM) has become well known as an effective method for modeling granular material at the grain scale. Despite the advantages of the DEM approach, it seems extremely challenging in most cases to model directly with DEM boundary value problems (BVP) involving the actual size of both the grains and the structure, because of the tremendous amount of grains and current abilities of computers.

On the contrary, the Finite Element Method (FEM) is appropriate for the modeling of BVP. However, seeking for a constitutive model that can account for all the relevant properties of granular materials, or a significant part of them, is difficult - and even not possible. Such mathematical (often very sophisticated) models are necessarily phenomenological, and involve a number of parameters for which the physical meaning can be tricky.

To face this issue in the field of granular matters, the FEM × DEM multi-scale approach has been recently proposed by various researchers [1–3]. Its basic concept lies on replacement of the phenomenological model by the constitutive response of the material with the use of DEM computations.

Using this approach, the actual grain scale can be considered even with large size macroscopic problems; without facing the intractable issue of dealing with trillions of grains in a DEM-mapped full field problem. Micro-scale related features, such as inherent and induced anisotropy of the granular material, or softening/hardening, emerge naturally from the micro-scale DEM model.

Localized failure is a major issue to be considered for modeling BVP related to geotechnical engineering. In this study, we present a (plane strain-FEM) × (3D-DEM) model of BVP with cohesive-frictional granular materials and an emphasis on strain localization. To avoid FEM mesh dependencies for shear bands, an internal length is introduced by means of the 2nd gradient regularization technique. Thanks to this technique, associated with a high performance computing ability of the code (MPI parallelization), the tool is now able to address some real-size geotechnical structures by accounting grain-scale features and strain localization.

The paper is laid out as follows: Section 2 presents the principle of FEM × DEM method. Section 3 gives some results related to biaxial tests and hollow cylinder cases modeled by this innovative approach. Some conclusions are drawn in Section 4.

2 Multi-scale modeling

The resolution of a BVP by the FEM requires a constitutive law used at each integration point of the mesh. This law is supposed to express the local stress as a function of the history of the local deformation gradient. In our case, this constitutive law is not formulated mathematically according to a plasticity theory, nor it comes from a mathematically formulated phenomenological law; instead, it results directly from the response of a DEM volume element (VE) as a function of the history of local deformation gradient provided by the FEM (see figure 1). The

*e-mail: trungkien.nguyen@3sr-grenoble.fr
**e-mail: gael.combe@3sr-grenoble.fr

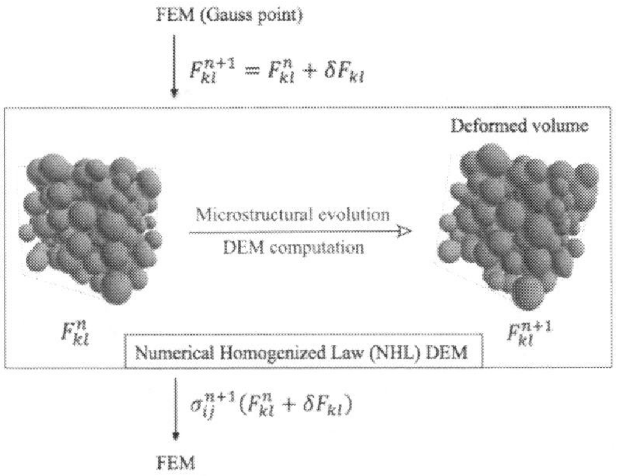

Figure 1. A sketch of the concept of DEM constitutive law

latter response can be considered to be similar to that of a constitutive law with a large (but finite) number of internal variables such as: grain geometries, contact network and contact forces. The procedure to build this numerical homogenized law is detailed in [3, 4].

2.1 3D discrete element micro-model

We describe only some main features of the micro model. The numerical model of granular behavior is herein obtained by a DEM approach using periodic boundary conditions (PBC) (see chapter 7 in [5]). The VE associated with each Gauss point is a packing of spheres. All spheres interact via a linear elastic force-law in the normal direction and the Coulomb friction law in the sliding direction. Accordingly, the normal repulsive contact force f_{el} is related to the overlap distance as $f_{el} = k_n \cdot \delta$, where k_n is a normal stiffness. The intensity of the friction force f_t evolves incrementally and proportionally to the tangential relative displacement - with a tangential stiffness k_t - up to a limit constant value $\pm\mu f_n$ when sliding occurs (μ is the Coulomb friction coefficient). In order to model cohesive-frictional granular materials, a local cohesion is introduced between the grains by adding an attractive (negative) force f_c to f_{el}. As suggested by [6], this local cohesive force can be set in regards to the mean level of reduced pressure $p^* = f_c/(\sigma_0 \cdot a^2)$, where a is the mean diameter of grains and σ_0 is the 3D isotropic pressure of VE.

All these features put together provide a *numerical homogenized law* (NHL) with a Mohr-Coulomb-like response, but it is actually much richer since it is able to handle very subtle responses involving, *e.g.*, volume changes, non-associated flow, hardening/softening, fabric dependency and cyclic loading.

2.2 Coupling with the FE macro-model

The FEM×DEM approach has been implemented in the FEM code Lagamine [7], which is able to perform finite strain analysis. The implementation consists in pairing the DEM code as another "constitutive law", and solving some specific difficulties linked to convergence issues. The numerical integration of the constitutive law requires solving a nonlinear system of equations when the mechanical behavior is non-linear, which is the case for granular materials. In order to solve the non-linear system of equations, an incremental-iterative Newton-Raphson strategy is adopted [8]. This requires computing a consistent tangent operator C, where the word "consistent" relates to the consistency with respect to the stress update algorithm as defined in Eq. 1 [9, 10].)

$$C_{ijkl} = \frac{\partial d\sigma_{ij}}{\partial d\epsilon_{kl}} \tag{1}$$

The consistent tangent operator can be derived analytically for a simple law, but for more complex laws, numerical differentiation needs to be adopted. However, a convergence problem is encountered when the consistent tangent operators are calculated in a phase of pronounced softening. Indeed, the scattered response provided by DEM calculation in the VE (and inherent to the nature of discrete matters) may cause some convergence difficulties. A specific modified Newton-Raphson algorithm has been developed in the code Lagamine to overcome these issues. Instead of the tangent operator defined above, we use an operator extracted from the underlying granular structure of the VE as described in [11, 12]. Since the relation by Kruyt and Rothenburg is only valid for 2D problem, an extended solution for the 3D case is proposed (Eq. 2) . It results in the following form of elastic stiffness operator:

$$C_{ijkl} = \frac{1}{V}\sum_{c\in V}(k_n n_i^c l_j^c n_k^c l_l^c + k_t t_i^c l_j^c t_k^c l_l^c + k_t w_i^c l_j^c w_k^c l_l^c) \tag{2}$$

where V is the volume of granular packing. k_n, k_t are normal and tangential contact stiffness, respectively; l is the branch vector connecting two centers of particles in contact; **n**, **t** and **w** are three vectors that define the contact forces directions: **n** is the normal direction, **t** and **w** are arbitrarily positioned in the sliding plan so that $\mathbf{w} = \mathbf{n} \times \mathbf{t}$.

3 Examples of BVPs

Although the NHL DEM model implemented in the study is a 3D model, it can be used in plane strain, plane stress and 3D FEM models. In this paper, we discuss plane strain BVPs, using 2D finite elements.

3.1 Biaxial Tests

Figure 2a presents the geometry and boundary conditions of biaxial test simulations. The constant confined pressures σ_2 is applied on the lateral sides and a vertical displacement at constant velocity is imposed on the top surface of the sample.

As shown in previous work of the same authors [3] or in ref. [13], it is well known that implementing strain softening constitutive laws in FEM produces mesh dependency: the deformation concentrates in zones as narrow as the mesh permits, independent of any material parameter.

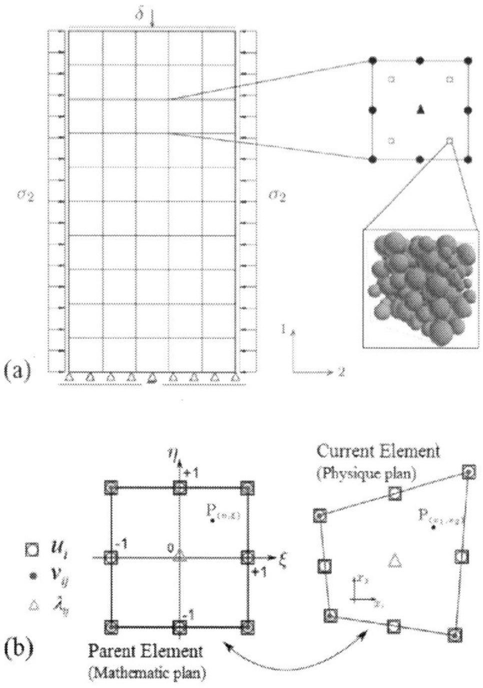

Figure 2. (a) Spatial discretization of the problem; (b) element enriched for the 2nd gradient technique [15]

In order to restore a mesh independent behavior in such computations, an enrichment of the constitutive model following [13, 14] is used. This enrichment involves the addition of a second gradient term to the constitutive model, the classical (first gradient) part being still given by the NHL DEM described above. The domain is discretized by using 128 (1st case) or 512 (2nd case) 2D elements second gradient (see figure 2b). These elements use eight nodes for the macro kinematic field (displacement field u_i), four nodes for the micro-kinematic field (micro deformation gradient v_{ij}) and one node for a Lagrange multiplier λ_{ij} which is used to impose the equality of the microscopic and the macroscopic deformation gradient. Details about the second gradient method can be found in [13] or [14].

At the micro scale level, identical VE of 1000 spheres are used at each Gauss point. The volume is isotropically loaded and all contacts are compressive but with adhesion. This represents a cohesive-frictional granular material. The microscopic parameters have been chosen according to classical *dimensional analysis*: *(i)* the stiffness ratio is $k_n/k_t = 1$, *(ii)* the cohesion force f_c is defined so that $p^* = 1$, *(iii)* the normal stiffness k_n is defined to obtain a stiffness parameter $\kappa = k_n/\sigma_0 = 500$ and *(iv)* the intergranular coefficient of friction is $\mu = 0.5$

The macroscopic responses of the biaxial tests are given in figure 3, with comparison to DEM simulation on the single VE of 1000 spheres. The responses show that the 2nd gradient model regularize the post-peak responses, especially after $\varepsilon_{11} = 5\%$, as the softening response of

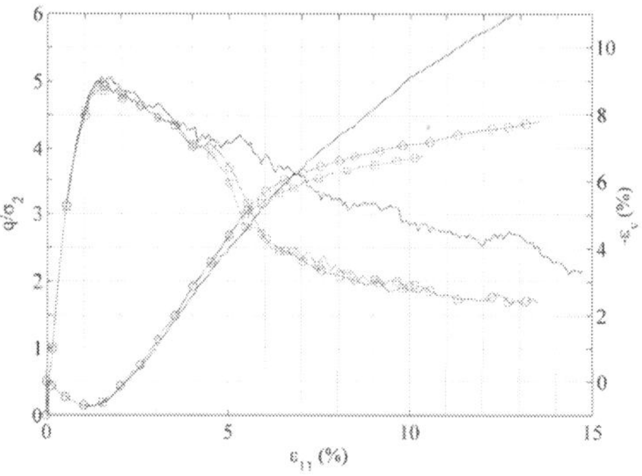

Figure 3. Stress-strain responses of biaxial tests (FD means FEM × DEM computation): DEM (continuous line), FD 128 elements (continuous line with □) and FD 512 elements (continuous line with ○)

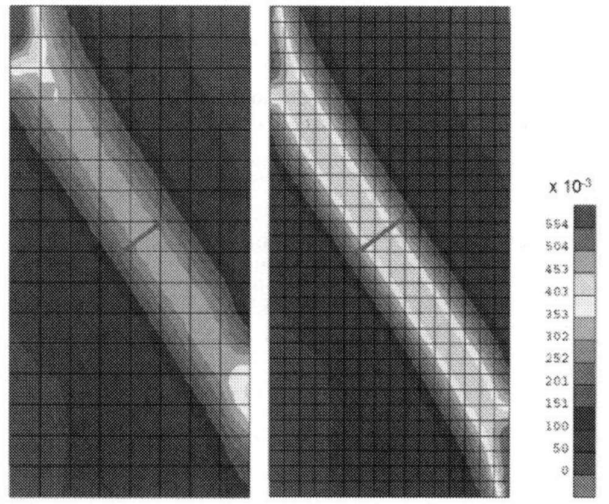

(a) 128 elements (b) 512 elements

Figure 4. Map of the second invariant of strain tensor at $\varepsilon_{11} = 10\%$ for (a) 128 elements and (b) 512 elements

two cases are almost identical. It is confirmed again in figure 4 that uses the 2nd invariant of strain tensor to exhibit the shear band at $\varepsilon_{11} = 10\%$. In first gradient model, shear bands tend to concentrate into a band of one element width [3]. As result of 2nd gradient model effect, shear band consists now of few elements. From this figure, it can be checked that the observed width of the shear band is intrinsic and independent of the mesh size.

3.2 Hollow Cylinder

These simulations predict the response of the material in the mid-place section of a pressuremeter, assuming plane strain condition in the axial direction. Because of the symmetry of the problem, only one fourth of a plane section is modeled. The domain is discretized using either first gradient element (Q8: 8 nodes, 4 Gauss points) or 2nd gradient element (see figure 5d). The computation was performed

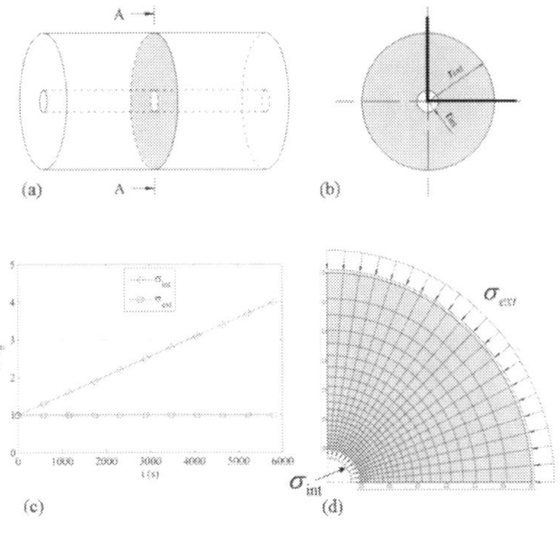

Figure 5. Numerical model setup: (a) hollow cylinder; (b) section A-A; (c) mechanical loading; (d) FE mesh and boundary conditions

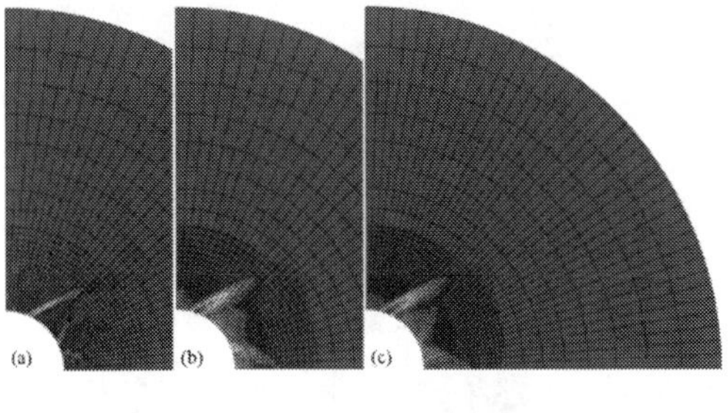

Figure 6. Multiscale modeling of hollow cylinder pressurized

as follows: starting from a homogeneous state of isotropic compression, the internal pressure (σ_{int}) was increased up to 4 times of the initial isotropic stress, while the external pressure (σ_{ext}) was kept constant. Numerical model setup and other related informations are given in figure 5.

Figure 6 shows the deformation mode in the model at the end of the loading in terms of the second invariant of strain tensor: (a) mesh with 952 standard Q8 elements (1st gradient, *i.e.* no regularization); (b,c) enriched elements (2nd gradient), two different meshes using 870 (b) or 1280 (c) elements. In the three cases presented in this figure, strain localization has taken place, organized in shear bands originated at the internal wall and progressing significantly inside the cylinder. This is the result of the inherent strain softening exhibited by the materials. In (b) and (c), the thickness of the shear bands does not depend on the mesh size, while in (a) it concentrates into a single layer of elements. These results demonstrate the capacity of this novel multi-scale approach to account for complex failure patterns. It is shown again that the second gradient method mitigates efficiently the mesh dependency problem observed with classical (first gradient) constitutive models.

4 Conclusion

A numerical law based on Discrete Element Modeling has been presented. This is accomplished in the framework of numerical homogenization applied to granular media. The model has been implemented in a finite element code. Due to scattering in the response of a model by DEM, some difficulties in the determination of consistent tangent operators occur, inducing convergence problems in the Newton-Raphson interactive processes in the FEM code. An alternative solution using an elastic stiffness operator instead of the tangent operator is shown to solve the difficulty. Thanks to the multi-scale FEM × DEM numerical approach, examples of BVPs involving strain localization is proposed, simulating biaxial tests and pressurized hollow cylinder cases. Strain localization is observed at macro level. The second gradient method is shown to solve efficiently the mesh dependency problem.

References

[1] H.A. Meier, P. Steinmann, E. Kuhl, Tech. Mech. **28** (2008)
[2] M. Nitka, G. Combe, C. Dascalu, J. Desrues, Granular Matter **13(3)** (2011)
[3] T.K. Nguyen, G. Combe, D. Caillerie, J. Desrues, Acta Geophys. **62(5)** (2014)
[4] J. Desrues, T. K. Nguyen, G. Combe, D. Caillerie, *Bifurcation and degradation of geomaterials in the new millenium* (Springer, 2015)
[5] F. Radjai, F. Dubois, *Discrete-element Modeling of Granular Material* (Wiley, 2011)
[6] F. A. Gilabert, J.N. Roux, Castellanos, Phys. Rev. **75(1)** (2007)
[7] R. Charlier, PhD Thesis, University of Liège (1987)
[8] T. J. Ypma, SIAM review **37(4)** (1995)
[9] R. de Borst, O. M. Heeres, Int. J. Num. Anal. Meth. Geomech. **26(11)** (2002)
[10] A. Perez Foguet, A. Rodriguez Ferran, A. Huerta, Int. J. Num. Meth. Engng. **48(2)** (2000)
[11] N. P. Kruyt, L. Rothenburg, Int. J. Engr. Sci. **36** (1998)
[12] N. P. Kruyt, L. Rothenburg, Mech. Mat. **36** (2001)
[13] T. Matsushima, R. Chambon, D. Caillerie, Int. J. Num. Meth. Engng. **54(4)** (2002)
[14] R. Chambon, D. Caillerie, N. El Hassan, Eur. J. Mech. Solids. **17(4)** (1998)
[15] F. Marinelli, Y. Sieffert, R. Chambon, Int. J. Solids Struct. **54** (2015)

DOI: 10.1051/epjconf/201714003051

Granular flows down inclined and vibrated planes: influence of basal friction

Naïma Gaudel*, Sébastien Kiesgen de Richter**, Nicolas Louvet, Mathieu Jenny, and Salaheddine Skali-Lami

[1]Laboratoire d'Energétique et de Mécanique Théorique et Appliquée (LEMTA), Université de Lorraine-CNRS, UMR 7563, Vandœuvre-lès-Nancy, 54504, France

Abstract. We present an experimental study about granular avalanches when external mechanical vibrations are applied. The results of the flow properties highlight the existence of two distinct regimes: (i) a gravity-driven regime at large angles where scaling laws are in agreement with those reported in the literature for non-vibrating granular flows and (ii) a vibration-driven regime at small angles where no flow occurs without applied vibrations. The flow in this regime is well described by a vibration-induced activated process. We also propose an empirical law to capture the evolution of the thickness of the deposits as a function of the vibration intensity and the inclination angle.

1 Introduction

Free surface flows of granular matters down inclined planes are often encountered in geophysical situations, engineering applications or pharmaceutical manufactures. Vibratory solutions are often used in these industries to control the transport, the storage, or the mixing properties of the granular materials. Recent works have focused on the influence of vibrations on the rheology of granular flows in confined configurations [1–4]. The present work focuses on free surface flows properties of a beads assembly when both vibrations and gravity act, and on the influence of the basal friction.

2 Experimental methods

Our experimental setup is sketched in Fig. 1. Grains are released from a tank of volume 900 cm^3 and from an adjustable height δ. Experiments are performed for four different beads systems (Table 1). Samples are forced by transverse sinusoidal vibrations that we quantify through the relative acceleration $\Gamma = A(2\pi f)^2/g$, where g is gravity, f is the frequency and A the amplitude of vibrations. A CCD camera recording the flow is placed above the plane, at 25 cm from the aperture. The particle image velocimetry (PIV) provides us surface velocity u_s measurements. The beads fall inside a reservoir resting on a scale: this provides us mean flow velocities $\bar{u}$ measurements. The local flow height h is measured using the deviation of a laser sheet inclined at a small angle of incidence ($\approx 3°$) over the layer. The influence of θ, Γ, and d/λ on the flows are studied.

*e-mail: naima.gaudel@univ-lorraine.fr
**e-mail: sebastien.kiesgen@univ-lorraine.fr

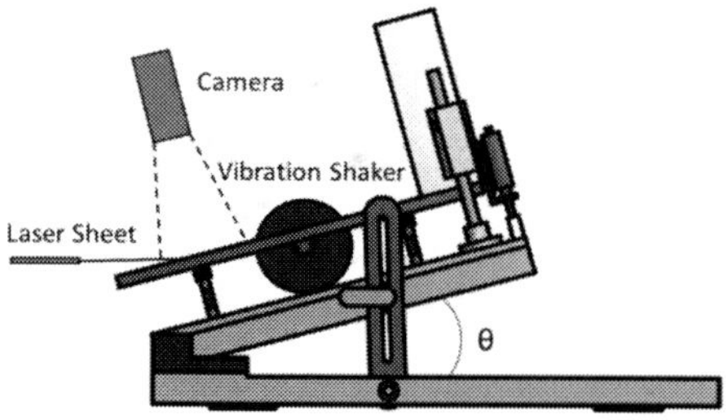

Figure 1. Experimental setup of glass beads (diameter d, $\rho \approx 2500\ kg/m^3$) flowing down a 40-cm-long and 10-cm-wide plane, covered with glass beads of average roughness λ, that can be inclined from horizontal up to $\theta = 35°$.

Table 1. The four systems of bulk and glued beads studied.

System	Bulk (d)	Rough surface (λ)
1	120 μm ±30	300 μm ±100
2	120 μm ±30	500 μm ±100
3	300 μm ±100	500 μm ±100
4	500 μm ±100	300 μm ±100

3 Experimental results and discussion

3.1 Existence of two regimes

Regarding the stationarity, the quantities h and u_s remain constant after a short transient regime of 20 s duration. After that, the flow is considered steady, and h and u_s relax to h_{stop} and zero value, respectively. The mean values $\bar{h}$ and $\bar{u}_s$ are calculated by averaging

DOI: 10.1051/epjconf/201714003051

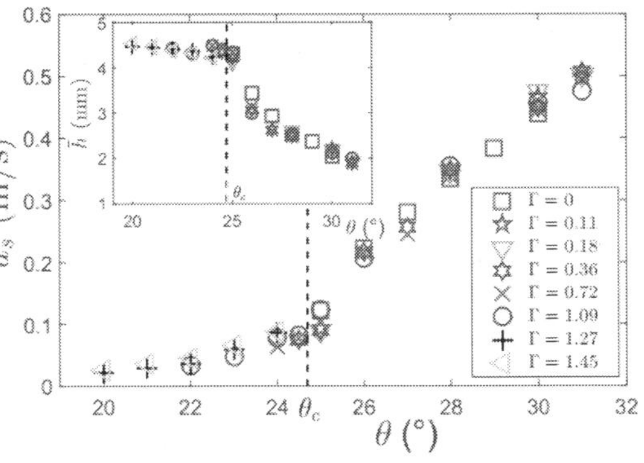

Figure 2. Evolution of $\bar{u}_s$ as a function of θ for the system 1 (δ = 5 mm) and different values of the Γ number. In insert, the variation of $\bar{h}$ with θ for the same parameters. Two regimes appear.

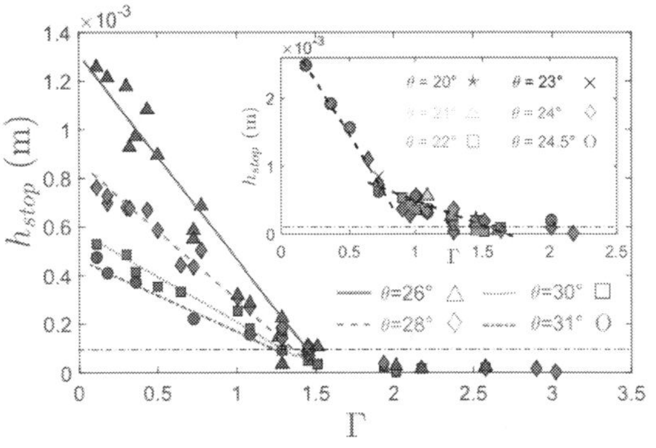

Figure 3. Evolution of h_{stop} as a function of the Γ number, for the system 1, for $\theta > \theta_c$. In insert, h_{stop} vs. Γ for $\theta < \theta_c$. The black dash-dot lines corresponds to the limit $h_{stop} = d$: below this, data values are similar to error measurements and are not taken into account.

these quantities over the steady-state regime. We define a critical angle θ_c, for which there is no flow in the absence of vibrations.

Evolution of $\bar{u}_s$ and $\bar{h}$ as functions of θ have been plotted in Fig. 2 for the system 1. The angle θ_c seems to be a transition between two distinct regimes. The fit of the variation of $\bar{u}_s$ with $\bar{h}$ leads to a power law behavior with an exponent $n = 3/2 \pm 0.08$ for $\theta > \theta_c$. This result is in agreement with previous experimental and numerical studies [5, 6]. For $\theta < \theta_c$, the exponent takes a smaller value of $n = 1/2 \pm 0.05$. It suggests that the flow dynamics flows down at small angles and confirms the existence of two regimes separated by a critical angle θ_c below which no flow occurs without vibrations:

- for $\theta > \theta_c$, $\bar{u}_s$ and $\bar{h}$ significantly vary when θ changes (high-angle regime),
- for $\theta < \theta_c$, $\bar{u}_s$ and $\bar{h}$ slightly vary when θ changes (low-angle regime).

We also study the flow structure along the plane (x-direction) in this steady-state regime. We observed that $\bar{h}$ gradually decreases along the plane and finally reaches a constant value for $\theta > \theta_c$. This behavior has already been observed in granular flows without vibrations [7, 8]. However, this phenomenon is nonexistent below the critical angle. A possible explanation is that flows are slow and homogenized by the vibrations for $\theta < \theta_c$, whereas they are heterogeneous due to gravity for $\theta > \theta_c$. To go through this limitation, a special care was taken to analyze the flow properties at a constant distance of 25 cm from the gate.

3.2 Study of arrested flow

The study of the variation of the deposits with the vibration intensity showed that $h_{stop}(\theta)$ is well controlled by Γ for all $\theta < \theta_c$ and for a given angle when $\theta > \theta_c$ (Fig. 3). The value of h_{stop} linearly decreases with Γ and vanishes at a critical value Γ_c which appears to be quasi-constant for all θ, in the range of values studied. This behavior has already been observed for the flow of a single bead down a transverse vibrated plane by Benedetti *et al.* [1]. We define Γ_c as the critical relative acceleration for which h_{stop} vanishes. To describe this dependency, we consider the flow as a three-layer flow: $h_{stop_{NV}}$ is the thickness of the deposits with no applied vibration, h_{stop} the thickness of the deposits and h the thickness of the flow. We define as $\left(h_{stop_{NV}} - h_{stop}\right)$ the fluid thickness that flows due to vibrations. By analogy with the boundary layer for a Newtonian fluid [9] and by dimensional analysis, we consider $h_{stop_{NV}}$ as the penetration depth affected by vibrations and hypothesize that:

$$h_{stop_{NV}} = \kappa \times \sqrt{\frac{\mu_{app}}{\rho\omega}} \quad (1)$$

here $\nu_{app} = \mu_{app}/\rho$ is the apparent viscosity (containing the dependence on d/λ), κ is a coefficient with a value of 4.64 in the Newtonian case without vibrations, and $\omega = 2\pi f$. In that case, the stress can be written as $\tau_s = \rho g h_{stop_{NV}}$. We suppose that τ_s scales, at the first order, as:

$$\tau_s = \mu_{app} \frac{A\omega}{h_{stop_{NV}} - h_{stop}} \quad (2)$$

Combining these hypothesis, we can predict how vibrations modify the values of the deposits height:

$$h_{stop}(\theta, \Gamma) = h_{stop_{NV}}(\theta)\left(1 - \frac{\Gamma}{\kappa^2}\right) \quad (3)$$

with $\kappa^2 = \Gamma_c$ the critical value of the relative acceleration. The value of κ is fixed at $1.20 \pm 0.2 < 4.64$ in experiments for the system 1. This suggests that the value of Γ_c is mainly controlled by rolling effects of the particles on the rough plane.

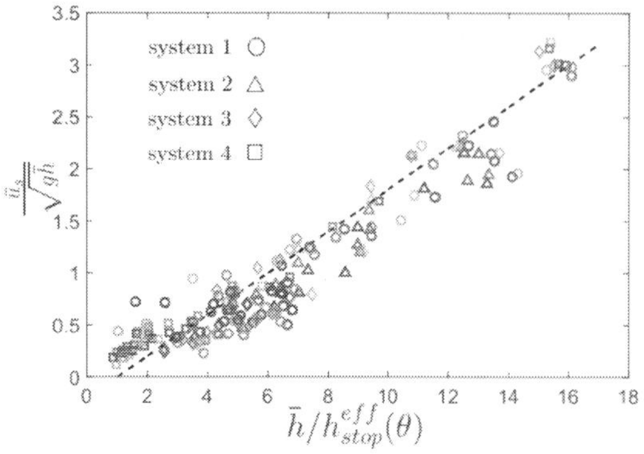

Figure 4. Evolution of the dimensionless velocity $\bar{u}_s/\sqrt{g\bar{h}}$ as a function of the rescaled thickness $\bar{h}/h_{stop}^{eff}(\theta)$ for the four systems studied (Tab. 1). Data are well fitted by this equation dotted line $\bar{u}_s/\sqrt{g\bar{h}} = 0.2\left(\bar{h}/h_{stop}^{eff}(\theta) - 1\right)$.

3.3 Study of high-angle regime

Previous studies have shown that the depth averaged velocity $\bar{u}$ is related to the thickness of the flow $\bar{h}$ in the steady-state regime by the flow rule [10–14]:

$$\frac{\bar{u}}{\sqrt{g\bar{h}}} = \alpha + \beta\frac{\bar{h}}{h_{stop}(\theta)} \tag{4}$$

where g is gravity. This scaling law is valid when there is no vibrations and $h_{stop}(\theta)$ contains all the dependence in θ and d/λ. In the high-angle regime, measurements showed a proportionality of $\bar{u}$ with $\bar{u}_s$ such as $\bar{u}/\bar{u}_s = 1/2 \pm 0.08$. This is not a Bagnold profile, suggesting that nonlocal effects are at play in our system [11, 12]. The quantities $\bar{u}_s$ and $\bar{h}$ are not affected by vibrations for a fixed angle (Fig. 2). Thus, we replace $h_{stop}(\theta)$ with $h_{stop_{NV}}(\theta)$ in Eq. 4. The data then collapse into four different straight lines corresponding to the four systems: each of them is a particular relative roughness. In this study, the function $h_{stop_{NV}}(\theta)$ does not take into account the influence of d/λ. Therefore we propose to define a new stopping height:

$$h_{stop}^{eff}(\theta) = h_{stop_{NV}}(\theta)\frac{d}{\lambda} \tag{5}$$

The data collapse into the same straight line (Fig. 4), the following scaling law seems to be suitable:

$$Fr = \frac{\bar{u}}{\sqrt{g\bar{h}}} = \tilde{\alpha}\left(\frac{\bar{h}}{h_{stop}^{eff}}(\theta) - 1\right) \tag{6}$$

where the influence of the bead size d, the roughness condition λ and the inclination angle θ is included in the single function $h_{stop}^{eff}(\theta) = h_{stop_{NV}}(\theta)\frac{d}{\lambda}$, and $\tilde{\alpha} = 0.1$ assuming $\bar{u}/\bar{u}_s = 1/2$.

3.4 Study of low-angle regime

In the low-angle regime, flows do not occur without vibrations and h_{stop} then diverges: the term $\bar{h}/h_{stop}(\theta)$

Table 2. The four systems studied and the corresponding values of the activation Γ_a number.

System	Relative roughness	Γ_a
1	1/3	0.25 ± 0.02
2	1/5	0.35 ± 0.03
3	3/5	0.12 ± 0.03
4	5/3	0.08 ± 0.03

in Eq. 4 tends to 0. The only remaining contribution is the one: $\bar{u} = \tilde{\gamma}\sqrt{\bar{h}}$, with $\tilde{\gamma} = \alpha\sqrt{g}$. This is consistent with our experimental results that lead to a power-law evolution of $\bar{u}_s(\bar{h})$ with an exponent $n = 1/2$ (see Sec. 3.1) and a ratio $\bar{u}/\bar{u}_s = 2/3 \pm 0.1$. A minimum Γ number, named Γ_a, is necessary to trigger the flow: for a given system, this value increases when θ decreases. The system needs more energy to break links between beads when the inclination decreases because the influence of gravity becomes lower. In the law-angle regime, vibrations impose the values of $\bar{u}_s$ and $h_{stop}(\theta)$. As expected, the scaling law (Eq. 4) does not describe the flow in our experiments. However, further investigations concerning the dependence of $\bar{u}_s$ on h_{stop} reveal that $\bar{u}_s \propto \sqrt{\bar{h}}/h_{stop}^2(\theta)$ better rationalizes our experimental data. For a given Γ, the surface velocity linearly varies with $\sqrt{\bar{h}}/h_{stop}^2(\theta)$ (Fig. 5, in insert). We expect a dependence of $\bar{u}_s$ with $\exp\left(-\frac{\Gamma}{\Gamma_a}\right)$, with Γ_a the minimum relative acceleration necessary to trigger the flow. We reasonably assume that the flow can be described by a vibration induced activated process for which the dimensionless equation can be written as:

$$\frac{\bar{u}_s}{\sqrt{g\bar{h}}}\left(\frac{h_{stop}}{d}\right)^2 \propto exp\left(-\frac{\Gamma}{\Gamma_a}\right) \tag{7}$$

The semilog plot of $\bar{u}_s/\sqrt{g\bar{h}}$ as a function of Γ provides us the value of Γ_a for each system, using the slope value of Γ_a (Tab. 2). The value of Γ_a increases when d/λ decreases because of the friction between beads [13]. Considering these values, Fig. 5 suggests that the scaling law (Eq. 7) correctly describes the flow in the low-angle regime. This exponential dependency shows that applying strong enough mechanical vibrations destabilizes the bulk structure. The probability for a grain to escape from a local trap raises with Γ and θ.

4 Conclusions

We studied granular avalanches down inclined and vibrated planes. Three results are highlighted:

(a) The existence of two flow regimes: a gravity-driven regime for $\theta > \theta_c$ and a vibration-driven regime for $\theta < \theta_c$.

(b) In the gravity-driven regime, vibrations only affect h_{stop} and so the apparent flow rule Fr=

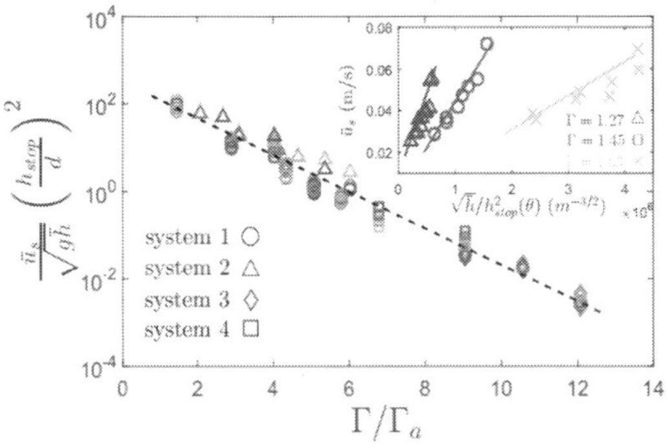

Figure 5. Semilog plot of $\bar{u}_s/\sqrt{g\bar{h}}$ as a function of Γ/Γ_a for the four systems studied (Tab. 1). In insert, $\bar{u}_s$ vs. $\sqrt{\bar{h}}/h^2_{stop}(\theta)$ for the system 2 and different values of Γ.

$\tilde{\alpha}\left(\bar{h}/h^{eff}_{stop}(\theta) - 1\right)$. We propose a linear dependency of h_{stop} with Γ such as $h_{stop}(\theta, \Gamma) = h_{stop_{NV}}(\theta)\,(1 - \Gamma/\Gamma_c)$.

(c) In the vibration-driven regime, flows occur due to vibrations. Increasing the relative roughness allow the minimum relative acceleration Γ_a to decrease. Flows can be finely tuned by applying external vibrations. The scaling law in this regime is given by $\frac{\bar{u}_s}{\sqrt{g\bar{h}}}\left(\frac{h_{stop}}{d}\right)^2 \propto exp\left(-\frac{\Gamma}{\Gamma_a}\right)$.

Acknowledgements

The authors thank S. Becker, J. Bianchin, F. Demeurie, J.-Y. Morel for technical assistance and the students A. Pascot, M. Sandoungout, M. Stiti for their work.

This work is funded by European Founds Interreg VA (Project PowderReg).

References

[1] A. Benedetti, P. Sornay, B. Dalloz, M. Nicolas, Phys. Rev. E **85**, 011307 (2012)
[2] C. Hanotin, S. Kiesgen de Richter, P. Marchal, L.J. Michot, C. Baravian, Phys. Rev. Lett. **108**, 198301 (2012)
[3] C. Hanotin, S. Kiesgen de Richter, L.J. Michot, P. Marchal, Journal of Rheology **59**, 253 (2015)
[4] C. Hanotin, P. Marchal, L.J. Michot, C. Baravian, S. Kiesgen de Richter, Soft Matter **9**, 9352 (2013)
[5] J. Vallance, Ph.D. thesis, Michigan Technological University (1994)
[6] E. Azanza, Ph.D. thesis, Ecole Nationale des Ponts et Chaussees (1998)
[7] Y. Forterre, Ph.D. thesis, Universite de Provence - Aix-Marseille I (2002)
[8] C. Ancey, Physical Review E **65** (2001)
[9] H. Schlichting, K. Gersten, *Boundary-layer theory* (Springer Science & Business Media, 2003)
[10] O. Pouliquen, Physics of Fluids **11**, 542 (1999)
[11] S. Deboeuf, E. Lajeunesse, O. Dauchot, B. Andreotti, Physical review letters **97**, 158303 (2006)
[12] F. Malloggi, B. Andreotti, E. Clement, Physical Review E **91** (2015)
[13] C. Goujon, Ph.D. thesis, Universite de Provence (2004)
[14] L.E. Silbert, J.W. Landry, G.S. Grest, Physics of Fluids **15**, 1 (2003)

Effect of Structure on Strength of Agglomerates using Distinct Element Method

Tina Bonakdar* and *Mojtaba* Ghadiri

Institute of Particle Science & Engineering, University of Leeds, Leeds, UK

Abstract. Knowledge of agglomerate strength is highly desirable for compression and tableting, dissolution and dispersion and mitigation of dust formation. The behaviour of agglomerates is affected by parameters such as density, agglomerate size, primary particle size, and interparticle bond strength. The method of agglomeration influences the evolution of structure, and this in turn affects its strength. Furthermore, the methods of strength characterisation, i.e. quasi-static or impact produce different results. To understand the role of structure and the influence of test method, agglomerate failure behaviour has been analysed by the use of the Distinct Element Method (DEM). We report on our work on the simulation of the breakage of the agglomerates for different porosities and impact conditions, where the role of impact speed and angle and type of contact bonding model have been evaluated. The adhesive contact model of JKR is used to form an agglomerate. The effect of the bonding level on the strength and size distribution of the clusters released as a result of failure has been investigated. This work also evaluates the effect of structure (porosity) on the strength of the agglomerates.

1 Introduction

Agglomerates may be formed in various ways and their mechanical strength depends on many factors due to the degree of freedom afforded by the primary particles. Weak agglomerates formed for example by spray-drying are prone to attrition, which has adverse effects on product quality, ability to process, dust formation and explosion hazards [1]. Therefore it is highly desirable to develop predictive ability for agglomerate breakage. One of the important factors, which has a strong influence on strength, is the agglomerate structure, yet the current modelling capability is not predictive. Quasi-static and impact behaviour differ markedly, and for the latter the velocity and impact angle dependence are not well understood [2]. For some structures, the impact strength depends mainly on the normal component of the impact speed, whilst for other structures the tangential component is also influential, but the factors affecting the difference are unknown [3]. In addition the functional dependence of the extent of breakage on impact speed is often a square power law, but this is not universal, again the source of difference not understood [4].

In this work we use the Distinct Element Method (DEM) to analyse the effect of agglomerate structure on the impact strength of agglomerates of elastic spheres that are bonded together by a contact adhesive force, following the JKR model [5]. A sensitivity analysis is carried out addressing the effects of the single particle properties (adhesion) and agglomerate structure (porosity) on the impact strength and size distribution of the clusters that are formed as a result of impact. This enables the underlying causes of variations in the dependency of impact speed and angle to be elucidated.

2 Methodology

Simulations are done using EDEM software, DEM Solutions, Edinburgh, UK. The primary particles are generated using six different particle types. The properties of all the particle types are the same, however they vary in particle size and number. The use of particle type makes it possible to vary porosity by removing some types from the structure. Particles are generated in a large space and a centripetal force is applied to bring them together to form an agglomerate. A relaxation step is applied after agglomerate formation in order to minimise the residual stresses within the agglomerate to avoid their possible effects on the breakage. The particles are then bonded together using the JKR contact adhesion model. Six different surface energies are used in order to make the agglomerates (0.5, 1, 1.5, 2, 3 and 5 J/m^2). The agglomerates are then impacted at different impact velocities (0.2, 0.5, 1, 5, 10 and 20 m/s). Two impact angles are used (normal and inclined). The same impact tests are repeated for agglomerates with three different levels of porosity. The impact reuslts are then analysed using the concept of damage ratio [6, 7], which is defined as the ratio of number of broken contacts to the initial number of contacts, to explore the effects of above-mentioned parameters on the impact strength of agglomerates.

* Corresponding author: t.bonakdar@leeds.ac.uk

3 Results and Discussion

Six agglomerates with different surface energies have been impacted at a range of impact velocities. The damage ratio as a function of impact speed is shown in Fig. 1.

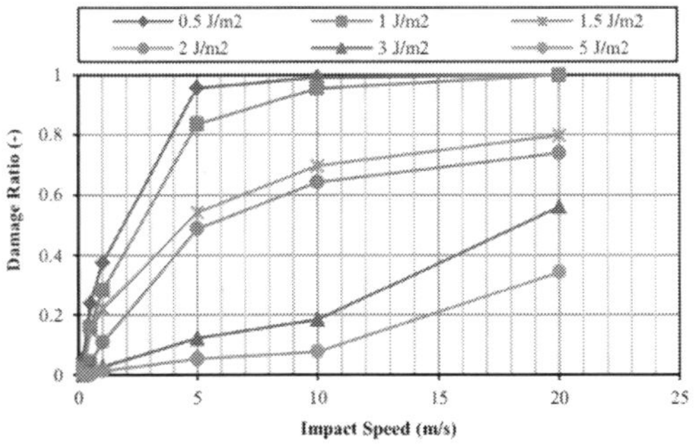

Fig. 1. Damage ratio as a function of impact speed for agglomerates with different cohesive interparticle forces

For a given surface energy, as the impact speed is increased, the damage ratio increases. At a given impact speed, the agglomerate with a lower surface energy breaks to a larger extent compared to the one with higher surface energy, which is as expected. However, three different trends are clearly observed, which are attributed to the pattern of breakage.

It is expected that the damage ratio follows a square of velocity relationship. Therefore the damage ratio has been plotted as a function of V^2 to see if such a dependency holds. Based on the results, a linear relationship is only observed for impact results of high surface energy agglomerates, but there is no such a dependency for low surface energy case (not shown here).

To explore the effect of impact angle on the extent of breakage, the agglomerates have been impacted at 60° and 90° (the impact angle is defined as the angle between agglomerate moving direction and the surface of the impact target). As an example the damage ratio as a function of impact speed is shown in Fig. 2 for impacting the agglomerate with 3 J/m^2 surface energy at these two angles.

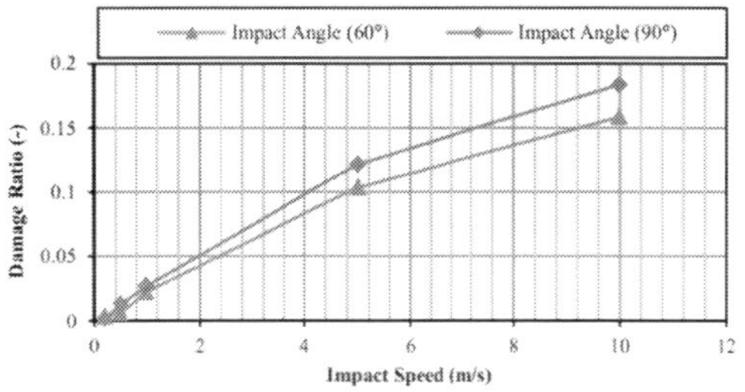

Fig. 2. Damage ratio as a function of impact speed for the same agglomerates (surface energy of 3 J/m^2) at different impact angles

Normal impacts cause more damage as compared to inclined ones. However, if the normal velocity component of the inclined impact is taken into account, then the data overlap, indicating that the agglomerate breakage is mainly dependent on the normal component (as shown in Fig. 3).

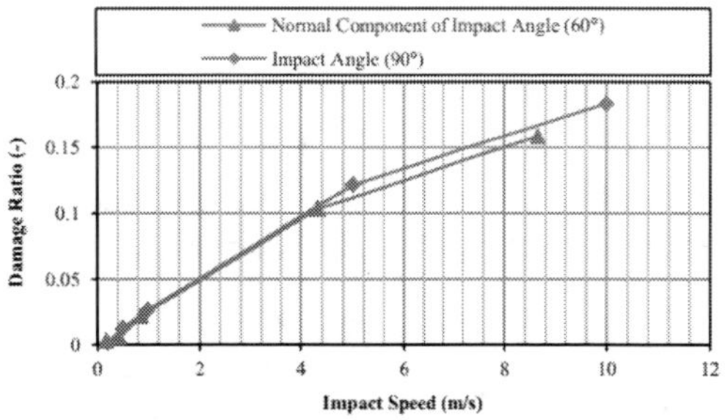

Fig. 3. Damage ratio as a function of normal component of the impact speed for the same agglomerates (surface energy of 3 J/m^2) at different impact angles

The effect of porosity on impact breakage has been studied by removing some particle types to make cavities inside the agglomerate. The impact breakage results of three cases, reference agglomerate (referring to the initial agglomerate) and more porous agglomerates (51% and 61% porosity) are shown in Fig. 4.

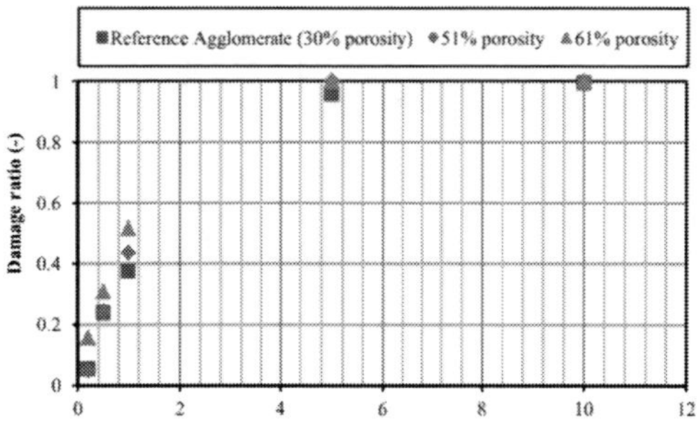

Fig. 4. Damage ratio as a function of impact speed for three agglomerates with different levels of porosity and surface energy of 0.5 J/m^2

For the three porosity levels analysed here, there are no notable differences in the damage ratio for all the impact velocities. Interestingly, applying the same analysis to the agglomerates with high surface energy shows a huge difference in damage ratio for three different porosities (as shown in Fig. 5). More work is needed to map out the breakage patterns in terms of these two factors.

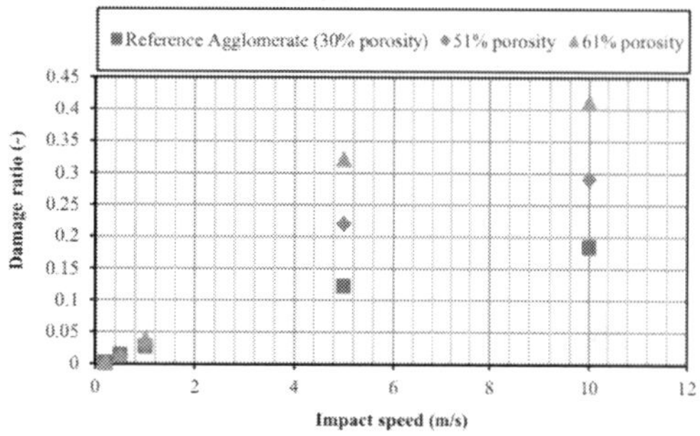

Fig. 5. Damage ratio as a function of impact speed for three agglomerates with different levels of porosity and surface energy of 3 J/m^2

The damage ratio for the above conditions was analysed as a function of impact speed. However, the breakage of contacts can also be related to Weber Number (We). Previously Kafui and Thornton [6] used this approach to analyse the agglomerate breakage. Weber number is defined based on the equation below,

$$We = \frac{\rho D V^2}{\Gamma} \tag{1}$$

where D and ρ are particle diameter and density of the particles, respectively. V is the impact speed. However, the modified version of the Weber number (We') as proposed by Thornton et al. [7] is more accurate, as it considers the speed below which there is no breakage (V_0).

$$We' = \frac{\rho D (V - V_0)^2}{\Gamma} \tag{2}$$

Moreno-Atanasio and Ghadiri [8] did an extensive piece of work on impact breakage of spherical agglomerate using DEM simulation. Based on their model, the number of broken contacts is proportional to the incident kinetic energy. By using an energy balance between the work required to break interparticle contacts and incident kinetic energy, they proposed the following relationship between damage ratio (Δ) and the surface energy,

$$\Delta \propto We' (\frac{ED}{\Gamma})^{2/3} \tag{3}$$

where E is the primary particle elastic modulus. Therefore, the simulation data have been analysed using Eq. (3) to explore the dependency of damage ratio on the surface energy as shown in Fig. 6. Based on their work, a unification is expected when the damage ratio is plotted as a Weber-dependant group. However, this is not the case. The trend observed in Fig. 6 is similar to that of Fig. 1. In both Figures, there are three clear trends. Each trend is attributed to a different pattern of breakage. Based on the observation of the agglomerates after impact, impacting the agglomerates with low surface energies (0.5 and 1 J/m^2) would result in disintegration, agglomerates with surface energies of 1.5 and 2 J/m^2 would result in fragmentation whilst for agglomerates with high surface energies (3 and 5 J/m^2) the damage is localised and chipping is the main pattern of breakage for these range of velocities. As an example the force distribution after impact at 0.5 m/s is shown in Figs 7 and 8 for agglomerates with 0.5 and 3 J/m^2 surface energy, respectively. Here the lines connecting the spheres are shown with their colours representing the magnitude of the contact force.

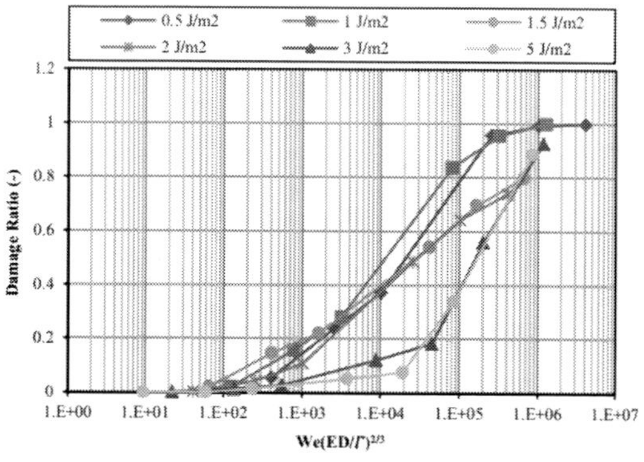

Fig. 6. Damage ratio as a function of Γ Eq. (7-8) for agglomerates with different surface energies

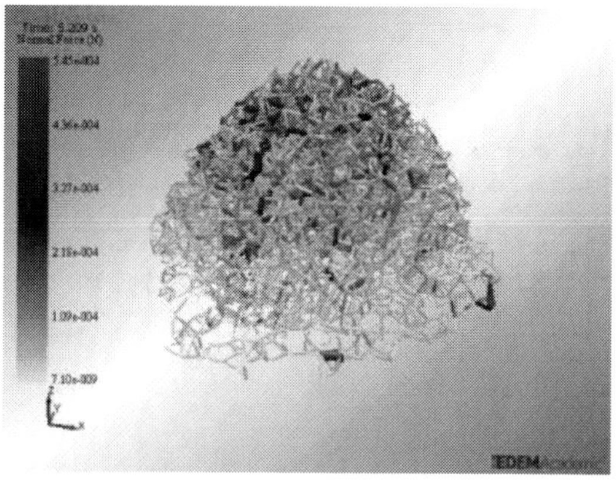

Fig. 7. Side view of force distribution after impacting at 0.5 m/s for agglomerates with 0.5 J/m^2 surface energy

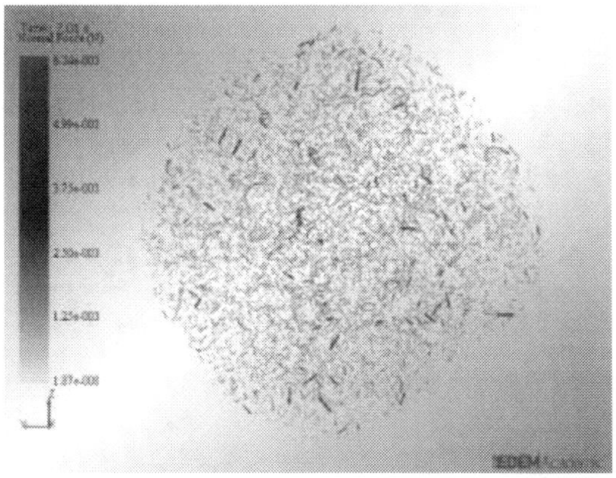

Fig. 8. Side view of force distribution after impacting at 0.5 m/s for agglomerates with 3 J/m^2 surface energy

The force distributions are exported after impact at 0.5 m/s. Side views of agglomerates with the surface energy of 0.5 J/m^2 and 3 J/m^2 are shown in Figs 7 and 8, respectively, where the impact side is at the base of the agglomerate. However, the data for both agglomerates have been exported at 0.01 s after impact, and the difference in the time scale shown in Fig. 7 is due to the time taken for agglomerate generation and applying surface energy at previous steps before impact. The disintegration of the agglomerate with 0.5 J/m^2 surface energy is clearly seen, as the bottom part of the agglomerate (close to the impact zone) is completely deformed and contacts broken in the impact zone and beyond. The force has also propagated within the agglomerate and reached to the top part, as shown in Fig. 7, causing breakage of the contacts. A different scenario is observed in Fig. 8 for 3 J/m^2 surface energy. There is no clear deformation at the impact zone (bottom part of the agglomerate) with little/ no contact breakage, and the force distribution has a different pattern as compared to Fig. 7.

Based on these observations, the lack of unification of breakage data in Fig. 6 is presumably due to different patterns of breakage. For tested agglomerates by Moreno-Atanasio, all of them failed through the same pattern of breakage, for which a good unification was obtained, as the damage was ratio plotted as a function of $We'(\frac{ED}{r})^{2/3}$. However, more probing of the differences in the method of agglomeration preparation is required to explore the underlying differences in the breakage behaviour of the two systems.

4 Conclusions

Agglomerate structure affects its strength. Impacting the agglomerates with different surface energies for a range of impact velocities shows three different trends of breakage, which are attributed to different patterns of breakage. The agglomerates with higher surface energy undergo less breakage compared to the agglomerates with lower surface energy, as intuitively expected. Impacting the agglomerates at the normal angle would result in a larger number of broken contacts compared to impact at inclined angles. It has been found, for the range of conditions analysed here, that the agglomerate failure is mainly dependent on the normal component of the impact speed. Three different levels of porosity have been used to explore the effect of porosity on the impact strength of agglomerates. Generally, more porous agglomerates break to a higher extent compared to less porous agglomerates, the difference being more pronounced for agglomerates with high surface energy.
As an ongoing work, the newly developed bonding model of Brown et al. [9] is used to simulate the impact breakage of agglomerates. The model is based on the Timoshenko Beam Bond Model (TBBM). This provides a more realistic representation of granules made using a liquid binder which on drying forms a solid bond.

References

1. T. Bonakdar, M. Ghadiri, H. Ahmadian, L. Martin de Juan, H. Tantawy, H. D, Smith, Impact attrition of spray-dried burkeite particles. *Powder Technol.* **304**. 1-7 (2016)
2. A. Samimi, R. Moreno, M. Ghadiri, M. Analysis of impact damage of agglomerates: Effect of impact angle. *Powder Technol.* **143-144**, 97–109 (2004)
3. A. D. Salman, M. J. Hounslow, J. P. K. Seville. *Granulation handbook*. **11** (2007)
4. J. Subero, M. Ghadiri, Breakage patterns of agglomerates. *Powder Technol.* **120,** 3 (2001)
5. K. L. Johnson, K. Kendall, A.D. Roberts, Surface energy and the contact of elastic solids. *Proceedings of the Royal Society of London A.*, **324**, 301-313 (1971)
6. K. Kafui, C. Thornton, Computer simulated impact of agglomerates. *Powders and Grains* (1993)
7. C. Thornton, K. Yin, M. Adams, 1996. Numerical simulation of the impact fracture and fragmentation of agglomerates. *J. Phys. D Appl. Phys*. **29**. 424-435 (1996)
8. R. Moreno-Atanasio, M. Ghadiri, M. 2006. Mechanistic analysis and computer simulation of impact breakage of agglomerates: Effect of surface energy. *Chem. Eng. Sci.* **61**, 8 (2006)
9. N. J. Brown, J. F. Chen, J. Y. and Ooi, J.Y. 2014. A bond model for DEM simulation of cementitious materials and deformable structures. *Granular Matter*. **16**, 3 (2014)

DOI: 10.1051/epjconf/201714002012

Assessing contact forces in granular materials from experimental measurements of kinematics

Mathias Tolomeo[1,*], *Vincent* Richefeu[1], *Gaël* Combe[1], *Jean-Noël* Roux[2], and *Gioacchino* Viggiani[1]

[1] *Univ. Grenoble Alpes, CNRS, Grenoble INP **, 3SR, F-38000 Grenoble, France*
[2] *Université Paris-Est, Laboratoire Navier, 2 Allée Kepler, Cité Descartes, 77420 Champs-sur-Marne, France*

Abstract. We propose a new approach to assess interparticle forces in granular materials, based on the combination of experimental measurements and numerical techniques. Experimental measurements mainly consist in the assessment of the full kinematics of grains in a quasi-statically loaded assembly; such measurements are performed by applying Digital Image Correlation (DIC) on high-resolution photographs taken during the tests. This experimental data is then combined with the application of numerical methods which account for grain equilibrium. Two different approaches are adopted for this purpose: a Static ElastoPlastic Computation (SEPC) and the Non-Smooth Contact Dynamics (NSCD). Both methods aim to find a set of contact forces that is mechanically admissible, based on kinematics measured from experiments. In this way, we aim to make an assessment of forces that is experimentally based, as it comes from experimental measurements, and realistic, in the sense of being mechanically admissible.

1 Introduction

The full understanding of the mechanical behaviour of granular materials needs a complete description of what occurs at the grain scale. In particular, it is necessary to focus on the complex mechanisms of force transmission, whose peculiar nature is well known, but not widely investigated yet. The transmission of forces between grains in contact plays a key role in the microscale mechanics of an assembly of particles; its study can prove essential also for the comprehension of phenomena occurring at the local scale, such as grain breakage.

Nowadays, thanks to numerical methods, we can easily reproduce and investigate such microscale behaviour; on the other hand, it is still not easy to make an experimental assessment of contact forces. Photoelasticity [1] is commonly used for this purpose; anyway, the use of this technique is limited by some constraints. Another solution, proposed by Andrade and co-workers [2, 3], is the *Granular Element Method* (GEM), which aims to analytically determine the contact forces acting on a volume on the basis of the stress field inside the volume and of (at least) one of the forces on its boundary. The stress field should be derived, through the definition of an appropriate constitutive law, from the strain field, which can be determined by means of Digital Image Correlation (DIC) measurements. However, this method sets constraints to the grain size and consequently to the number of grains, since each grain surface has to be large enough to be sufficiently discretised to allow strain measurements. Recently, a new approach has been proposed by Hurley et al. [4], representing the first attempt to measure interparticle forces in a non-idealised 3D granular material. This technique combines 3D x-ray diffraction (to assess grain strain, and derive grain stress tensors from it) with a numerical optimisation algorithm which infers the interparticle forces on the basis of these measurements and of grain equilibrium.

The method we propose is a combination of experimental measurements and a numerical approach. The experimental measurements consist in the determination of the full microscale kinematics of a 2D granular assembly, *i.e.*, grain motion (displacement and rotation). Two numerical techniques are used: a *Static ElastoPlastic Computation* (SEPC), which views the assembly as a system of nodes, connections, and supports, and solves the problem as a conventional structural mechanics one, by looking at successive states of equilibrium during a test; the *Non-Smooth Contact Dynamics* (NSCD) [5], which is used here as an indirect method to determine contact forces based on contact velocities. Despite the different approach, both these techniques aim to find a set of contact forces which is mechanically admissible and compatible with the experimental information (grain kinematics, macroscopic loading). This indirect use of NSCD was already proposed in [6]; however, some issues of non-uniqueness of the solutions were observed. In our application, it might prove useful to use some experimental information such as the status of a contact (*i.e.*, whether it is sliding or not), in order to overcome such issues.

*e-mail: mathias.tolomeo@3sr-grenoble.fr
**Institute of Engineering Univ. Grenoble Alpes

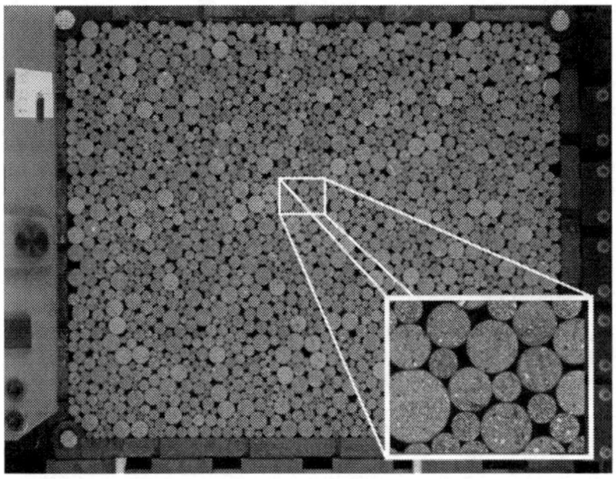

Figure 1: Image of an isotropically compressed sample in the $1\gamma2\varepsilon$ device, with zoom on a small amount of grains to show the speckle pattern on their visible face. The frame is 599.4 mm × 444.7 mm.

2 Experimental approach

2.1 Experimental device

The $1\gamma2\varepsilon$ apparatus, described in detail elsewhere [7, 8], allows to quasi-statically strain a 2D assembly of rollers (Schneebeli rods), with independent control on the macroscopic deformation components (vertical and horizontal strain, shear strain). Different kinds of tests were performed (isotropic compression, biaxial compression – vertical and horizontal –, simple shear). However, in this study, the results being discussed are mainly from the biaxial vertical compression test. Such test was carried out by first applying an isotropic loading with a mean stress of 100 kPa, and then increasing the vertical stress while keeping the horizontal one constant. The 2D material used for these tests is an assembly of around 1850 6 cm long wooden rods (Fig. 1), with four different diameters (8, 12, 14 and 20 mm).

2.2 Kinematic field measurements

During the tests, 80 Mpixel digital pictures of the sample were shot with a Phase One IQ180 camera, with a frequency of $5s$, corresponding to a strain increment in the vertical direction $\Delta\varepsilon_y \approx 0.01\%$. Each grain's visible surface was painted with a speckle pattern (Fig. 1) in order to track its displacement and rotation. The tracking procedure was carried out by means of a particular technique based on Digital Image Correlation, optimised for the case of discrete materials, for which the motion can be erratic; we refer to this technique as Particle Image Tracking (PIT) [9, 10].

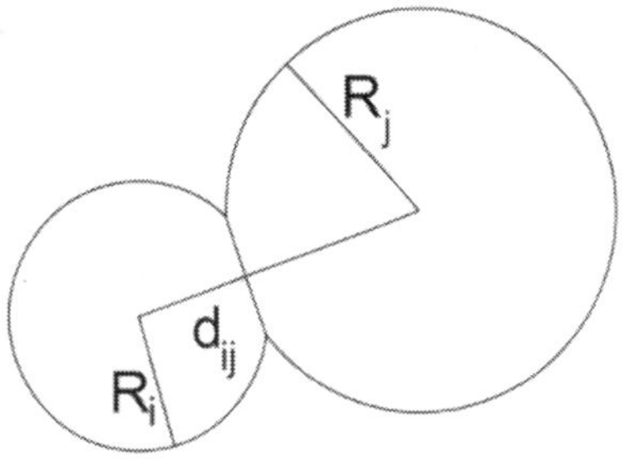

Figure 2: Sketch of a contact between two grains. The "fictitious" overlap is defined by Eq. (1). We assume the radii to be constant, while some deformation can occur in a small area around the contact point. With these assumptions, it is straightforward that x_{ij} can be negative when the grains in contact are compressed.

3 Contact forces assessment

3.1 Direct method

A first, rough assessment of contact forces was performed with a method which we refer to as "direct" or "DEM-like". This method aims to find a direct relationship between the contact force and some microscale kinematic indicator derived from the measurements. For this purpose, we introduced a microscale kinematic parameter which is defined as a "fictitious" overlap, following the overlap definition in common DEM. This definition is based on two assumptions: the grain is treated as a rigid body, so its radius stays constant; a very small deformation occurs in a narrow area around the contact point, when two grains in contact undergo compression. With these two assumptions, the definition of "fictitious" overlap comes straightforward (see sketch in Fig. 2):

$$x_{ij} = d_{ij} - R_i - R_j \,, \qquad (1)$$

Then, a simple linear contact law was assumed between the overlap and the normal contact force (in compression), also based on the results of a specific compression test which was performed on a single contact.

Having defined the normal contact forces in this way, it was possible to make a first assessment of their distributions in our assembly. Of course this method was not intended to give a precise evaluation of the exact value of the forces one-by-one, due to some unavoidable uncertainties in the determination of the overlap (coming mainly from the non perfect roundness of the surface of rods), but still some features of interest could be found: the statistics of normal forces in the isotropic compression test reproduce very well those obtained in similar DEM simulations in [11] (see Fig. 3); preferential contact orientations, with respect to those contacts subjected to the highest extent of compression, are aligned with the direction of maximum compression for each test respectively (biaxial vertical compression, horizontal compression, simple shear), thus showing a good correspondence between the microscopic and macroscopic behaviour of the assembly.

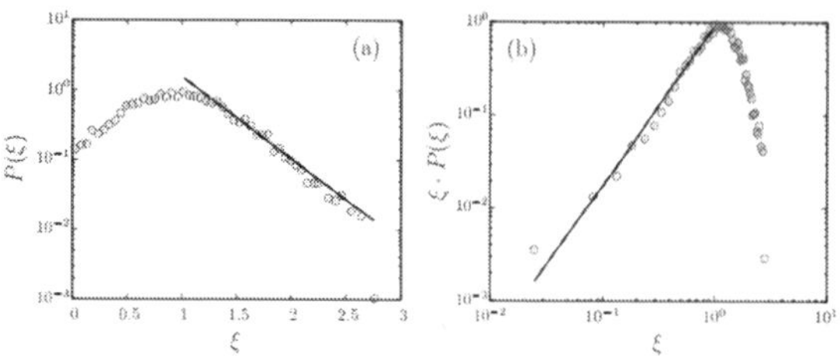

Figure 3: Semilogarithmic plot (a) and log-log plot (b) of the probability distributions of normalised "fictitious" overlaps $\xi = x/\langle x \rangle$. The linear fit of weak and strong overlaps, in the semilogarithmic and log-log plot respectively, show that the trend can be modelled through a power-law decay for weak overlaps and an exponential one for the strong overlaps.

Except for the previously mentioned sources of inaccuracy in the measurements, there are mainly two elements that this method is actually lacking in order to give a correct estimation of the whole set of contact forces. One is the history of tangential displacements, essential for a correct assessment of tangential forces; since we only have measurements in discrete time steps, we are missing a lot of information in terms of the tangential displacements occurring between two consecutive steps, and consequently in terms of the corresponding forces. The other missing feature is the mechanical balance of the grains: as it is not taken into account here, the set of forces obtained is not, in general, statically admissible. Introducing this concept, by means of the two numerical tools presented in the following sections, can lead to a more realistic solution.

The output of this direct method, although trivial and rough, is yet not to be discarded: using it as a starting point (guess solution) for the numerical tools that we propose can help to reach a solution, and in particular to orient the solution towards the "real" one, in case of non-uniqueness issues (as in [6]).

3.2 Static ElastoPlastic Computation

In the modelling of granular materials, quasistatic methods represent an important alternative to the most common numerical methods, which account for inertial effects (such as Molecular Dynamics and Contact Dynamics). Their peculiarity is to view the system, under a variable loading, as a succession of equilibrium states [12]. In this way, the assembly can be considered equivalent to a system of nodes, connections, and supports, and the problem can be studied as a conventional structural mechanics one [13]. Such methods are based on building a *rigidity matrix* $\underline{\underline{\mathbf{G}}}$ which describes the *granular structure* of the assembly, *i.e.*, it links grain-related quantities (forces and moments acting in correspondence of each degree of freedom and grouped in the vector $\mathbf{F}^{ext}$) to contact-related ones (contact forces grouped in the vector $\mathbf{f}$), from both a kinematic point of view and a static one, such as in the generalised equation of force equilibrium

$$\mathbf{F}^{ext} = \underline{\underline{\mathbf{H}}} \cdot \mathbf{f}\,, \qquad (2)$$

where $\underline{\underline{\mathbf{H}}} = \underline{\underline{\mathbf{G}}}^t$.

The method presented here, which will be referred to as Static ElastoPlastic Computation (SEPC), is meant to find a set of contact forces which is, at the same time, *statically admissible* (*i.e.*, in equilibrium with the external loading) and *plastically admissible* (*i.e.*, fulfilling the adopted friction law). This is done by means of an iterative procedure of double projection, of an initial set of forces (guess solution), respectively on the subspace of statically admissible solutions and on that of plastically admissible one. In particular, since we assume a Coulomb friction law, the boundary of the subspace of plastically admissible solutions corresponds to the Coulomb cone. This computation algorithm is fully detailed in [12].

3.2.1 Validation of the method

A first validation of the method was performed on synthetic data produced by Molecular Dynamics simulations of biaxial vertical compression tests, on an assembly which resembles a $1\gamma2\varepsilon$ sample in terms of number of particles and size distribution. The purpose of this validation was to test the ability of the method to determine a set of contact forces, given an initial geometrical configuration and an external loading, which fulfils the two conditions of *static admissibility* and *plastic admissibility*.

Having applied the method on some thousands of intermediate steps of the simulations, the first outcome is that in most cases it is possible to find a solution (Fig. 4), which, for strong normal forces, matches very well the MD forces (with a maximum relative error of around 10% [1]). However, some issues have been encountered, leading to an impossibility of finding a statically and plastically admissible solution in some of the intermediate steps analysed. In some situations, the problem was simply overcome by initialising the set of forces to a guess solution which is close to the expected one, *e.g.*, using the set of forces from the previous step, instead of starting with all zero forces; this shows that the success of the method is affected by the choice of the initial guess solution, as introduced in Sec. 3.1.

Yet this was not enough for some other situations, for which other issues were encountered. One typical issue, especially when dealing with MD simulations, is the occurrence of highly dynamic effects: in such situations, no static equilibrium could be reached, but the set of forces obtained was surprisingly consistent with the MD one (still with a maximum relative error for strong normal forces of about 10%), as in Fig. 4. Another typical issue of quasi-static methods, that should be taken into account when applying SEPC, is the stability of an equilibrium state, which can be studied through the condition that the *second-order work* is positive [12, 13].

[1]The maximum relative error is computed as $\max\left(\frac{\|f^N_{MD} - f^N_{SEPC}\|}{f^N_{MD}}\right)$

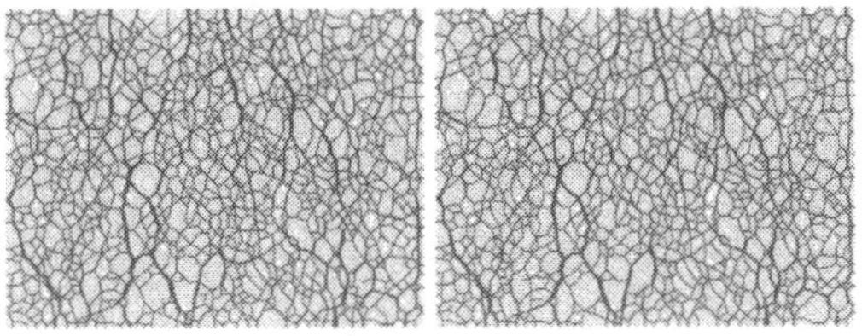

Figure 4: Maps of normal contact forces obtained with SEPC for an intermediate step (vertical strain $\varepsilon_y \approx 5\%$ for both – left and right – configurations) of a biaxial vertical compression MD simulation, where the thickness of each line is proportional to the normal force magnitude. The map on the left corresponds to a configuration for which the solution obtained is statically admissible; the map on the right, on the other hand, corresponds to a solution that is not statically admissible. In both cases, *force chains* clearly appear. Despite being not equilibrated, the set of forces in the configuration on the right is still clearly consistent with the external loading applied.

3.3 Non-Smooth Contact Dynamics

The application of Non-Smooth Contact Dynamics as an indirect method to assess contact forces from experimental kinematic data was introduced, at a preliminary stage, by [6]. The idea is to apply NSCD to find a set of contact forces which is mechanically admissible, *i.e.*, compatible with both the grain kinematics and the external loading applied. The kinematic information required by this method mainly consists in the relative velocities at the contact between each couple of grains; such information can be easily extracted from the grain kinematics assessment performed by means of the PIT technique.

This method might result complementary to the one in Sec. 3.2, as it can prove particularly useful for those situations in which the system undergoes dynamic events, and it is not possible to assume a static equilibrium.

The main issue concerning this method, as it was observed in [6], is the non-uniqueness of the solution, which sometimes can lead to a set of forces which might be mechanically compatible with the overall loading, but totally inconsistent with the kinematics, due to possible inaccurate inputs. To overcome such issues, some extra information can be exploited, *e.g.*, the status of a contact (*i.e.*, if it is a *sliding* contact or not). Another way to orient the method towards a "good" solution is to initialise the set of forces to a guess solution which should be close to the expected one, in the same way as in Sec. 3.2.1. To this purpose, a possible initial set of forces can be represented by the one derived from the "direct" method in Sec. 3.1.

4 Summary and outlook

In this study, we have presented the preliminary results of a combined experimental-numerical approach for contact forces assessment in granular materials. The experimental data comes from kinematic measurements performed by means of the PIT technique on a 2D rods assembly which behaves equivalently to a granular material. Such information is then combined with two different numerical methods, a quasi-static approach (SEPC) and the Non-Smooth Contact Dynamics, which, by injecting grain equilibrium, aim to find a set of forces which is compatible with the external loading and the grain kinematics. Some testing of the first method has already been performed on synthetic data from MD simulations, as a validation, proving the reliability of this method, and its robustness with respect to some inaccuracies in the inputs which may represent an issue when dealing with experimental measurements. In a next future, the applicability of both methods to experimental data will be addressed.

Acknowledgements

The first author thanks the LabEx Tec 21 (Investissements d'Avenir, Grant Agreement No. ANR-11-LABX-0030) for full financial support of his PhD, which this work is part of.

References

[1] T.S. Majmudar, R.P. Behringer, Nature (London) **435**, 1079 (2005)
[2] J.E. Andrade, C.F. Avila, Granular Matter **14**, 51 (2012)
[3] R.C. Hurley, E. Marteau, G. Ravichandran, J.E. Andrade, J Mech Phys Solids **63**, 154 (2014)
[4] R.C. Hurley, S.A. Hall, J.E. Andrade, J. Wright, Phys. Rev. Lett **117**, 098005 (2016)
[5] F. Radjaï, V. Richefeu, Mech Mater **41**, 715 (2009)
[6] V. Richefeu, G. Combe, R. Maurin, AIP Conference Proceedings **1542**, 429 (2013)
[7] F. Calvetti, G. Combe, J. Lanier, Mech. Cohes.-Frict. Mater **2**, 121 (1997)
[8] H. Joer, J. Lanier, J. Desrues, E. Flavigny, Geotech. Test. J. **15**, 129 (1992)
[9] G. Combe, V. Richefeu, AIP Conf. Proc. **1542**, 461 (2013)
[10] V. Richefeu, G. Combe, G. Viggiani, Géotechnique Letters **2**, 113 (2012)
[11] F. Radjaï, M. Jean, J.J. Moreau, S. Roux, Phys. Rev. Lett **77**, 274 (1996)
[12] J.N. Roux, G. Combe, in *Discrete-element Modeling of Granular Materials*, edited by F. Radjaï, F. Dubois (Wiley-Iste, 2011), chap. 3, pp. 83–116
[13] M.R. Kuhn, C.S. Chang, Int J Solids Struct **43**, 6026–6051 (2006)

AUTHOR INDEX

AUTHOR INDEX

AUTHOR INDEX

AUTHOR INDEX

AUTHOR INDEX

AUTHOR INDEX

AUTHOR INDEX

AUTHOR INDEX

AUTHOR INDEX

AUTHOR INDEX

Powders and Grains 2017 – 8th International Conference on Micromechanics on Granular Media

EPJ Web of Conferences Volume 140 (2017)

Montpellier, France
3 - 7 July 2017

Part 2 of 2

Editors:

Farhang Radjai
Saeid Nezamabadi
Stefan Luding
Jean-Yves Delenne

ISBN: 978-1-5108-4434-6

Printed from e-media with permission by:

Curran Associates, Inc.
57 Morehouse Lane
Red Hook, NY 12571

Some format issues inherent in the e-media version may also appear in this print version.

Printed by Curran Associates, Inc. (2017)

For additional information, please contact EDP Sciences – Web of Conferences at the address below.

EDP Sciences – Web of Conferences
17, Avenue du Hoggar
Parc d'Activité de Courtabœuf
BP 112
F-91944 Les Ulis Cedex A
France

Phone: +33 (0) 1 69 18 75 75
Fax: +33 (0) 1 69 28 84 91

contact-edps@webofconferences.org

Additional copies of this publication are available from:

Curran Associates, Inc.
57 Morehouse Lane
Red Hook, NY 12571 USA
Phone: 845-758-0400
Fax: 845-758-2633
Email: curran@proceedings.com
Web: www.proceedings.com

TABLE OF CONTENTS

PART 1

PART 2

Some statistics about research on granular materials during the last decade

William-Fernando Oquendo[1,*] and *Nicolas* Estrada[2,**]

[1] *Departamento de Ingeniería Industrial, Universidad de la Sabana, Chía, Colombia*
[2] *Departamento de Ingeniería Civil y Ambiental, Universidad de los Andes, Bogotá, Colombia*

Abstract. In this work we present several statistical analyses about research on the subject of granular materials during the last ten years. These statistics are based on the set of articles published in the Granular Materials section of *Physical Review E* during this period, comprising roughly 1000 documents (including rapid communications). We estimate the degree of contribution of countries, academic institutions, authors, and articles, in terms of the frequency of appearance on the articles' affiliations and the number of citations. On the first hand, these analyses serve as a recognition to researchers and research communities that have notably contributed to the development of knowledge on granular materials. On the second hand, the presented statistics allow for identifying countries and institutions where research in granular materials is the most developed, and also places where it can be developed in the future. Finally, these analyses allow for highlighting which cooperations amongst countries have been the most active during the last decade, which in turn allows for identifying potential links to be created and developed across this research network.

1 Description of the data set of articles

The set of articles that was analyzed was composed of all papers published in the Granular Materials section of *Physical Review E*, which was chosen as one of the possible reference titles in the research field of granular materials. The set comprised 951 articles, published between January 27 2006 (i.e., Vol. 73, Issue 1) and September 2 2016 (i.e., Vol. 94 Issue 3), with more than 1600 authors in total. From the results openly available on the internet, the following information was recorded for each article: publication date, volume, issue, title, DOI, number of citations, abstract, authors, and affiliations.

2 Countries and international cooperation

Figure 1 shows a world map in which countries have been assigned a color scale indicating the number of times each country appears on the authors' affiliations [1]. It can be seen that the number of appearances for each country varies strongly, from one to several hundreds. A strong concentration of appearances can be observed in a few countries. Table 1 shows the twenty top countries, ranked according to the number N of appearances. It can be seen that more than 50% of appearances correspond to the three countries on top of the list.

Table 1. Top twenty countries, ranked according to the number N of times that each country appears on the authors' affiliations.

Country	N (-)	(%)
USA	851	27.2
France	437	14.0
Germany	271	8.7
Spain	191	6.1
China	162	5.2
United Kingdom	147	4.7
Netherlands	146	4.7
Australia	87	2.8
Switzerland	69	2.2
Norway	64	2.0
India	63	2.0
Chile	61	1.9
Japan	60	1.9
Belgium	55	1.7
Taiwan	54	1.7
Italy	49	1.6
Argentina	45	1.4
Mexico	45	1.4
Israel	40	1.3
Hungary	32	1.0

Figure 2 shows the network of scientific cooperation links found in the set. Each time that two countries appeared on the authors' affiliations of a given article, a link was counted. Again, it can be seen that the number of links in which each country takes part varies strongly, from zero to more than a hundred. A strong concentration of link ends can be observed in a few countries. Table 2 shows

*e-mail: william.oquendo@unisabana.edu.co
**e-mail: n.estrada22@uniandes.edu.co

[1] It must be noted that the numbers reported on Fig. 1 do not represent the number of papers in which at least one of the authors is affiliated to an institution in the corresponding country, since a given country can appear several times in a single article

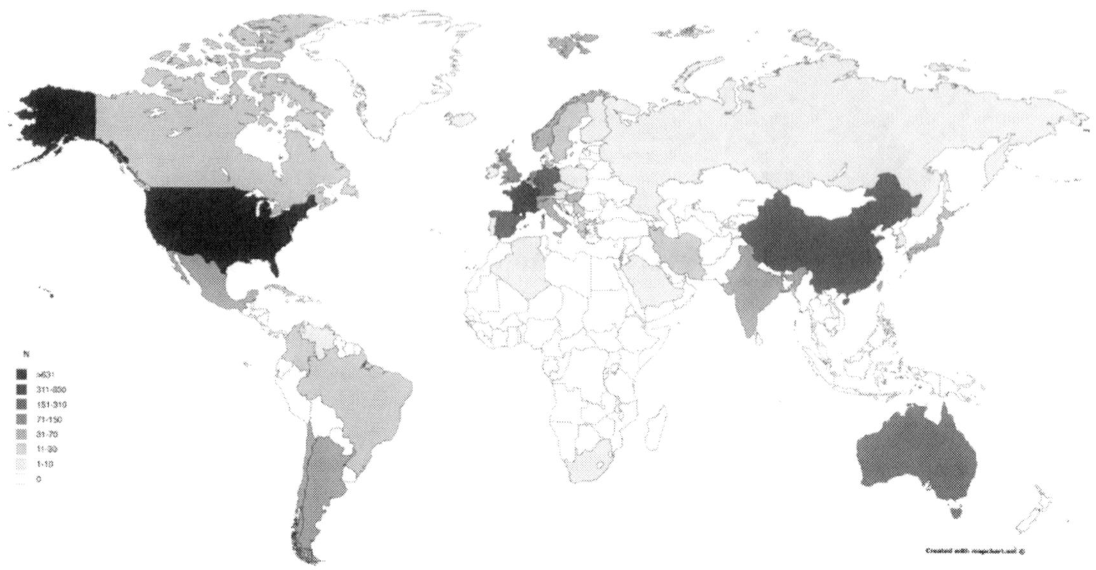

Figure 1. World map in which countries have been assigned a color scale indicating the number of times each country appears on the authors' affiliations.

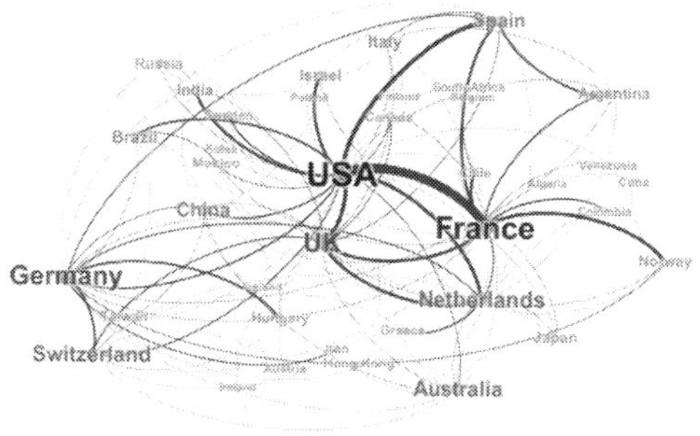

Figure 2. Network of scientific cooperation links found in the set. Thicker links represent more collaborations established.

the twenty top countries, ranked according to the number N of links in which each of them takes part. It can be seen that the five countries on top of the list take part in more than 50% of the total number of links.

3 Research institutions

The data set was used to assess the degree of contribution of research institutions, quantified by means of the number of times that each institution appears on the authors' affiliations [2] [3]. Table 3 shows the twenty top institutions, ranked according to the number N of appearances.

Table 2. Top twenty countries, ranked according to the number N of times that each country takes part in an international cooperation link.

Country	N (-)	(%)
USA	116	18.4
France	87	13.8
United Kingdom	54	8.6
Germany	48	7.6
Netherlands	40	6.4
Spain	34	5.4
Switzerland	23	3.7
China	19	3.0
Argentina	18	2.9
Australia	17	2.7
Norway	16	2.5
Brazil	15	2.4
India	15	2.4
Israel	14	2.2
Italy	12	1.9
Chile	11	1.7
Hungary	10	1.6
Mexico	9	1.4
Canada	8	1.3
Greece	7	1.1

[2] Again, it must be noted that these frequencies do not represent the number of papers on which at least one of the authors is affiliated to the corresponding institution, since a given institution can appear several times in a single article

[3] In our analyses, when a given author reported to be affiliated to several institutions, only the first of them was taken into account.

Table 3. Top twenty research institutions, ranked according to the number N of times that each institution appears on the authors' affiliations.

Institution	N (-)	(%)
Universidad de Sevilla (Spain)	72	4.8
University of Twente (Netherlands)	68	4.5
Université de Montpellier (France)	65	4.3
Princeton University (USA)	55	3.7
Université de Liège (Belgium)	54	3.6
University of Oslo (Norway)	52	3.5
Northwestern University (USA)	51	3.4
University of California (USA)	48	3.2
Universidad de Navarra (Spain)	46	3.1
Friedrich-Alexander Universität Erlangen-Nürenberg (Germany)	45	3.0
University of Pennsylvania (USA)	42	2.8
Université de Lyon (France)	41	2.7
Universidad de Extremadura (Spain)	41	2.7
Max Planck Institute for Dynamics and Self Organization (Germany)	40	2.7
Yale University (USA)	39	2.6
University of Chicago (USA)	39	2.6
Université de Rennes (France)	38	2.5
Universität Bayreuth (Germany)	37	2.5
Université Pierre et Marie Curie (France)	34	2.3
Université Paris Diderot (France)	31	2.1

4 Researchers

The data set was used to assess the degree of contribution of researchers, in terms of two indicators: the number of articles that each researcher authored, and the number of citations to articles that the researcher authored. Table 4 shows the twenty one top researchers, ranked according to the number N_a of articles that the researcher authored. This measure can be compared to the mean value for all authors in the data set: $\langle N_a \rangle = 2.1$.

Table 5 shows the twenty top researchers, ranked according to the number N_c of citations to articles that the researcher authored. This measure can be compared to the mean value for all authors in the data set: $\langle N_c \rangle = 17.2$.

5 Articles

Finally, the data set was used to asses the relevance of the articles, in terms of the number N_c of citations to these articles. Table 6 shows the twenty top articles, ranked according to N_c, as well as the citation rate $\dot{N}_c$, estimated as the number of citations divided by the article's age in years. These measures can be compared to the mean values for all articles in the data set: $\langle N_c \rangle = 8.6$ and $\langle \dot{N}_c \rangle = 1.4$ yr^{-1}.

6 Final comments

In this work we have performed several statistical analyses related to the research of granular materials for the last ten

Table 4. Top twenty one researchers, ranked according to the number N_a of articles that the researcher authored.

Researcher	N_a (-)
J. Javier Brey	24
Farhang Radjaï	21
Vicente Garzó	19
Salvatore Torquato	17
Hans J. Herrmann	16
Thorsten Pöschel	16
Nicolas Vandewalle	16
Iker Zuriguel	16
Emilien Azéma	14
Julio M. Ottino	14
Luis A. Pugnaloni	14
Douglas J. Durian	13
Jean-Christophe Géminard	13
Robert P. Berhinger	12
María Isabel García de Soria	12
Hisao Hayakawa	12
Richard M. Lueptow	12
Pablo Maynar	12
Diego Maza	12
Jean-Noël Roux	12
Renaud Toussaint	12
Average for the data set: $\langle N_a \rangle = 2.1$	

Table 5. Top twenty researchers, ranked according to the number N_c of citations to articles that the researcher authored.

Researcher	N_c (-)
Farhang Radjaï	384
Salvatore Torquato	380
Jean-Noël Roux	282
Martin van Hecke	256
Frank H. Stillinger	245
Emilien Azéma	219
Yang Jiao	211
Hans J. Herrmann	203
Vicente Garzó	189
Iker Zuriguel	182
Luis A. Pugnaloni	181
Julio M. Ottino	171
Wim van Saarloos	169
J. Javier Brey	162
Gary S. Grest	162
Diego Maza	158
Sidney R. Nagel	154
Thorsten Pöschel	151
Andrea J. Liu	149
Leonardo E. Silbert	145
Average for the data set: $\langle N_c \rangle = 17.2$	

Table 6. Top twenty articles from *Physical Review E*, ranked according to the number N_c of citations.

Article	N_c (-)	$\dot{N}_c$ (yr^{-1})
- C.H. Rycroft, G.S. Grest, J.W. Landry, and M.Z. Bazant. *Analysis of granular flow in a pebble-bed nuclear reactor*. 74(2), 2006.	96	9.5
- A. Donev, R. Connelly, F.H. Stillinger, and S. Torquato. *Underconstrained jammed packings of nonspherical hard particles: Ellipses and ellipsoids*. 75(5), 2007.	88	9.4
- A. Tordesillas, D.M. Walker, and Q. Lin. *Force cycles and force chains*. 81(1), 2010.	72	10.8
- E. Somfai, M. van Hecke, W.G. Ellenbroek, K. Shundyak, and W. van Saarloos. *Critical and noncritical jamming of frictional grains*. 75(2), 2007.	71	7.4
- S. Henkes, Y. Fily, and M.C. Marchetti. *Active jamming: Self-propelled soft particles at high density*. 84(4), 2011.	68	13.8
- M. Schröter, S. Ulrich, J. Kreft, J.B. Swift, and H.L. Swinney. *Mechanisms in the size segregation of a binary granular mixture*. 74(1), 2006.	68	6.7
- L.E. Silbert, A.J. Liu, and S.R. Nagel. *Structural signatures of the unjamming transition at zero temperature*. 73(4), 2006.	66	6.3
- V. Richefeu, M.S. El Youssoufi, and F. Radjaï. *Shear strength properties of wet granular materials*. 73(5), 2006.	64	6.2
- F. Alonso-Marroquín, I. Vardoulakis, H. J. Herrmann, D. Weatherley, and P. Mora. *Effect of rolling on dissipation in fault gouges*. 74(3), 2006.	62	6.2
- D.I. Goldman and P. Umbanhowar. *Scaling and dynamics of sphere and disk impact into granular media*. 77(2), 2008.	60	7.0
- T. Aste and T. Di Matteo. *Emergence of Gamma distributions in granular materials and packing models*. 77(2), 2008.	58	6.8
- A.R. Abate and D.J. Durian. *Approach to jamming in an air-fluidized granular bed*. 74(3), 2006.	57	5.7
- I. Agnolin and J.-N. Roux. *Internal states of model isotropic granular packings. I. Assembling process, geometry, and contact networks*. 76(6), 2007.	54	6.2
- Y. Jiao, F.H. Stillinger, and S. Torquato. *Optimal packings of superballs*. 79(4), 2009.	53	7.1
- P.-E. Peyneau and J.-N. Roux. *Frictionless bead packs have macroscopic friction, but no dilatancy*. 78(1), 2008.	53	6.5
- N.V. Brilliantov, N. Albers, F. Spahn, and T. Pöschel. *Collision dynamics of granular particles with adhesion*. 76(5), 2007.	51	5.8
- E. Azéma, F. Radjaï, R. Peyroux, and G. Saussine. *Force transmission in a packing of pentagonal particles*. 76(1), 2007.	49	5.3
- V. Vitelli, N. Xu, M. Wyart, A.J. Liu, and S.R. Nagel. *Heat transport in model jammed solids*. 81(2), 2010.	46	7.0
- V. Garzó, J.W. Dufty, and C.M. Hrenya. *Enskog theory for polydisperse granular mixtures. I. Navier-Stokes order transport*. 76(3), 2007.	46	5.1
- S. Henkes and B. Chakraborty. *Statistical mechanics framework for static granular matter*. 79(6), 2009.	45	6.2
Averages for the data set: $\langle N_c \rangle = 8.6$ and $\langle \dot{N}_c \rangle = 1.4$ yr^{-1}		

years. The data set was obtained from the openly available data of the journal *Physical Review E*, section Granular Physics (including Rapid Communications), which includes roughly 1000 documents and more than 1600 authors.

On the one hand, we found that the top five countries, ranked by the total number of articles published inside the time window, make more than 50% of the total of publications. This reflects a strong research network centered around these countries, something also reflected on the international collaborations.

On the other hand, the most published and cited authors show that the top authors publish and/or get cited much more than the average on each case. Some of the most published authors are also amongst the most cited authors, highlighting the relevance and broad impact of their research.

Finally, it should be noted that the presented analysis does not pretend to be an objective and complete overview of research on the field of granular materials, since it only included one section of one journal. This analysis should be understood as a proposal to look at our research community from a new perspective. In the future, it would be great to extend these analysis including journals such as *Physical Review Letters*, *Granular Matter*, *Europhysics Letters*, etc., or even the set of researchers that have attended the Powders and Grains conference.

Analysis of Minor Component Segregation in Ternary Powder Mixtures

Maryam Asachi, *Ali* Hassanpour*, *Mojtaba* Ghadiri, and *Andrew* Bayly

School of Chemical and Process Engineering, University of Leeds, Leeds, LS2 9JT, UK

Abstract. In many powder handling operations, inhomogeneity in powder mixtures caused by segregation could have significant adverse impact on the quality as well as economics of the production. Segregation of a minor component of a highly active substance could have serious deleterious effects, an example is the segregation of enzyme granules in detergent powders. In this study, the effects of particle properties and bulk cohesion on the segregation tendency of minor component are analysed. The minor component is made sticky while not adversely affecting the flowability of samples. The segregation extent is evaluated using image processing of the photographic records taken from the front face of the heap after the pouring process. The optimum average sieve cut size of components for which segregation could be reduced is reported. It is also shown that the extent of segregation is significantly reduced by applying a thin layer of liquid to the surfaces of minor component, promoting an ordered mixture.

1 Introduction

In many industries dealing with dry powder handling operation, such as pharmaceutical, food and agriculture segregation is encountered, particularly during handling, transportation and storage of the product. Segregation could have significant impact on the quality as well as economics of the production. This phenomenon mainly happens due to difference in particles physical properties such as size, density and shape [1, 2]. Some common examples of the segregation phenomenon that could cause problem are excessive concentration of the active ingredient in the mixture, such as insufficient bleach in a package of laundry detergent, taste variations in a powdered drink mixtures and variation of active pharmaceutical ingredients (API) content from tablet to tablet. Therefore, it is desirable to achieve a homogenous mixture [3].

One of the important factors affecting segregation phenomenon is variation in material properties. Particles with similar properties do not segregate, but segregation may take place as soon as there is variation in particles properties. Size distribution of particles is one of the main factors affecting this phenomenon. Usually a wider size distribution would result in more segregation [4]. Density variation of particles is another significant factor affecting segregation. Denser particles accumulate more in the centre of the powder mixtures like fine particles. Particle shape is also important for segregation. Spherical particles mostly behave like large and light particles and roll down the surfaces in a heap. In contrast, non-spherical particles almost collect closer to the centre of a heap similar to heavy particles [5]. Different segregation mechanisms could be seen as a result of particle interactions or influence of external forces causing the movement of particles in different directions as summarized by Bate [6].

Segregation of minor components such as expensive enzyme granules in detergent formulation could have significant economic impact on the production. By manipulating material properties, handling equipment, operational parameters and/or environmental conditions such as changing the bulk cohesion, it is possible to reduce segregation [4]. In this study, the segregation tendency of minor component (less than 3 % weight ratio), such as enzyme placebo granules in a ternary mixture has been assessed. Desired properties for segregation reduction of enzyme placebo granules have been evaluated by finding the optimum material properties and/or controlling the particle cohesion. Effect of bulk cohesion on segregation reduction of enzyme placebo granules was studied by applying a thin layer of liquid to the surfaces of placebo granules. Different coating percentages of the coating material were assessed for the reduction of segregation. The optimum coating percentage is reported according the results of flowability of both bulk as well as enzyme placebo granules.

2 Materials and Methodology

As a model material system, a ternary washing powder mixture, comprising blue Tetra Acetyl Ethylene Diamine (TAED), white spray-dried synthetic detergent powder (Arial LS Diamond), known as Blown Powder (BP), and red enzyme placebo granules, is used. All the materials

* Corresponding author: A.Hassanpour@leeds.ac.uk

used in this investigation were obtained from Procter and Gamble (P&G), Newcastle Innovation Centre, UK. To assess segregation, mixture of components (40 *g)* was poured in a box (dimension of 0.2×0.2×0.015 *m*) to make a heap of washing powders. Segregation of components was evaluated using image processing of the picture captured from the front face of the box using Photoshop7 software. The experimental set-up used here mainly represents the granular piles building up in detergent industry, which are exposed to a relatively high extent of segregation. For the measurement of segregation index of component, the particle area was first determined, the pixels of the defined area was measured and converted to the fraction of components. Coefficient of variation (COV) concept was used as the segregation index in this study (Eq. 1).

$$Segregation\ index = COV = \frac{\sigma}{\bar{x}_{Enzymeplacebo}} \quad (1)$$

where σ is the standard deviation and $\bar{x}_{Enzymeplacebo}$ is the mean number fraction of the enzyme placebo granules in a defined set of grids within the whole area of the mixture. Shape characteristic of individual particles was determined using Scanning Electron Microscope (SEM) by Hitachi TM3030 Bench Top SEM. Coating of enzyme granules by polyethylene glycol (PEG, Pluriol E 600) was performed in a mixer. PEG was used as coating material due to its compatibility with detergent formulation. Flowability of the coated enzyme placebo granules as well as the bulk of powders was measured using Schulze shear cell device. Bulk tapped density measurement was also done by pouring the mixture of powders in a graduated cylinder and tapping the poured powders until the pack reached equilibrium. The mass and volume of the powder were then recorded to estimate the bulk tapped density.

3 Results and Discussion

3.1. Optimum average sieve cut size for a well-mixed ternary mixture of laundry detergent powders

The bulk tapped density of the mode size of enzyme placebo granules is measured as 1450 $kg.m^{-3}$ which is higher than 530 $kg.m^{-3}$ and 400 $kg.m^{-3}$ measured for TAED and BP, respectively (sieve cut sizes between 300-600 *μm*) suggesting enzyme granules are denser than BP and TAED particles.

To find a well-mixed ternary mixture, first 50/50 weight ratio binary mixtures with the mode size of distribution of enzyme placebo granules (655 *μm*, obtained by British standard sieve analysis presented in Fig. 1) and different average sieve cut sizes of BP and TAED (sieve cut sizes between 300-500 *μm*) were separately prepared. The mixtures were then poured through a funnel fixed to the top side of the heap box to make heaps of binary mixture.

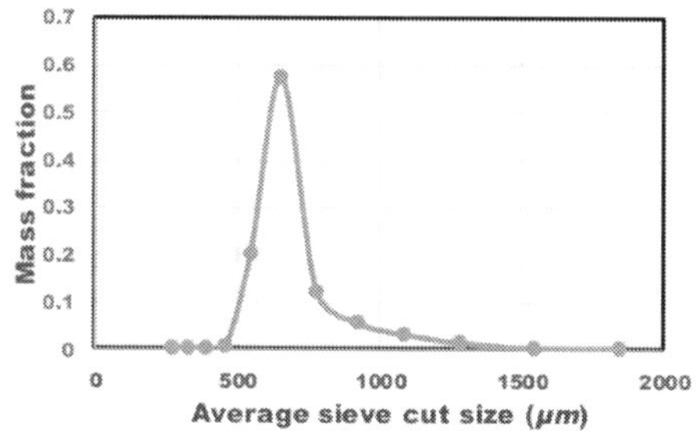

Fig. 1. Size distribution of enzyme placebo granules.

Figs. 2-b and 2-e demonstrate a relatively mixed condition of BP-placebo and TAED-placebo binary mixtures (390 *μm* – 655 *μm*). However, the binary mixtures of BP-placebo and TAED-placebo (328 *μm* – 655 *μm)* as shown in Figs. 2-a and 2-d, respectively undergone more segregation as compared to the previous cases. Size segregation of BP and TAED particles (328 *μm)* could dominate the density segregation of enzyme placebo granules (due to push-away effect) as shown in Figs. 2-a and 2-d. On the other hand, density segregation of enzyme placebo granules may dominate the size segregation of BP and TAED particles as shown in Figs. 2-c and 2-f.

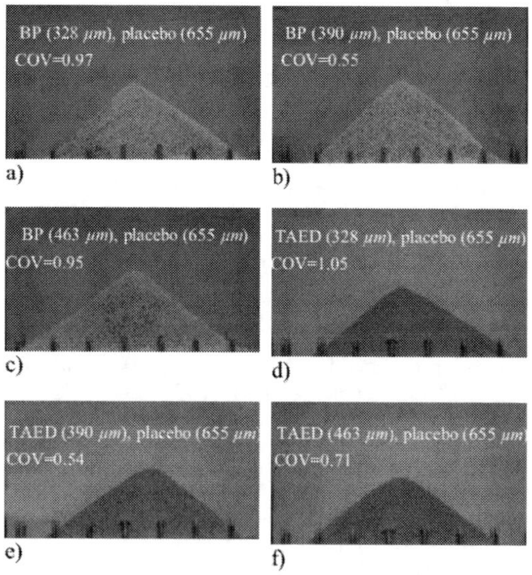

Fig. 2. Segregation of binary mixtures of, a-c) BP-placebo, and d-f) TAED-placebo, having different average sieve cut sizes.

Two different ternary mixtures have been prepared with full size distribution of components (Fig. 3-a) and with the optimum sizes suggested earlier (Fig. 3-b) both with 92.6/5.55/1.85 weight ratio of BP, TAED and enzyme placebo granules. The segregation index of enzyme placebo granules, obtained by the image analysis as described earlier, decreased from 1.37 (full size distribution) to 0.76 by applying the optimum average sieve sizes for a mixed ternary mixture.

a) 

b) 

Fig. 3. a) Ternary mixtures containing full size distribution of BP, TAED and enzyme placebo granules with 92.6/5.55/1.85 weight ratio, b) mixed ternary mixture with the optimum sizes.

SEM pictures of components with the optimum average sieve cut sizes of particles described earlier is presented in Fig. 4. It can be seen that the enzyme placebo granules are more spherical than BP and TAED particles. It has been reported that spherical particles act like light, big and free flowing particles in mixtures of spherical and irregular particles and they tend to roll down the surface of the heap [5]. In contrast, irregular particles behave like heavy and fine particles building up into the centre of the heap [5]. However in this case, shape segregation of enzyme placebo granules could be balanced with other material properties of components, notably size and density.

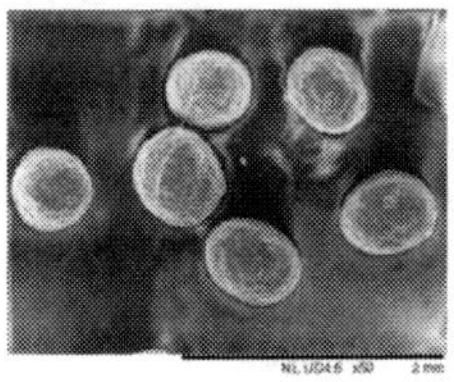

a)

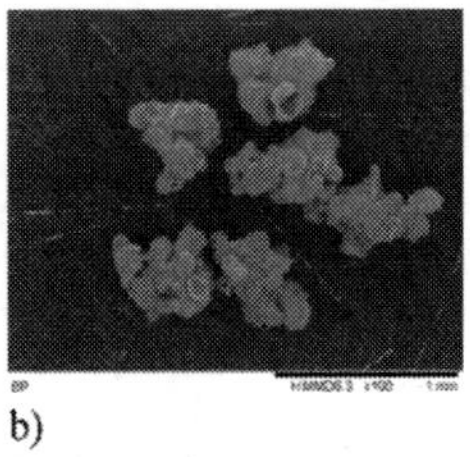

b)

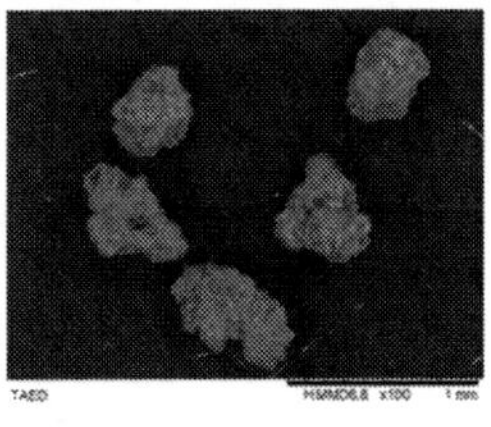

c)

Fig. 4. SEM pictures of the optimum average sieve cut size for a relatively mixed ternary mixture condition, a) enzyme placebo granules (655 μm), b) BP (390 μm), c) TAED (390 μm).

With the results shown earlier, it could be concluded that a relatively mixed ternary mixture could be generated by the addition of denser enzyme placebo particles with its mode size (655 μm) into the mixture of BP and TAED with average sieve cut size of 390 μm, balancing the effect of different material properties of particles. However, it may not be always practical to achieve the optimum average sieve cut sizes for a mixed ternary mixture to minimise segregation, particularly for industrial applications. Hence in this work, the effect of particle coating on the segregation reduction of enzyme placebo granules is investigated and reported in the next section.

3.2 Effect of coating on the reduction of segregation of enzyme placebo granules

To study the effect of coating on the reduction of segregation of enzyme placebo granules, different coating percentages of PEG (0.5 to 3.5% with the increment of 1%) have been applied to placebo granules as described earlier. The coated enzyme placebo granules were added to the mixture of full size distribution BP and TAED particles to form a heap of ternary mixture with 92.6/5.55/1.85 weight ratio of BP, TAED and enzyme placebo granules. The effect of coating on the segregation reduction of enzyme placebo granules is shown in Fig. 5-a. It could be observed that segregation of enzyme placebo granules decreased as coating level increased. However, the optimum coating level for a reasonable flowability of the particles should also be investigated. The flowability of both bulk mixture and coated enzyme placebo granules was analysed by the Schulze shear cell for which the results are shown in Fig. 5-b. It can be observed that the bulk flowability does not change significantly with the coating levels of 1.5 to 3.5 % (remains in easy-flowing regime). However, the behaviour of enzyme placebo granules shifts to the cohesive regime, particularly at lower pre-shear with 3.5% coating level. According to the Figs. 5-a and 5-b, 2.5% coating level could be enough to reduce segregation without compromising the flowability of both bulk and enzyme placebo granules. Ternary mixture with full size distribution of components after 2.5% coating level for enzyme placebo granules is shown in Fig. 6-a. It can be seen that some enzyme placebo particles (red particles) have reached to the surface of the heap, decreasing the overall segregation extent of enzyme placebo granules.

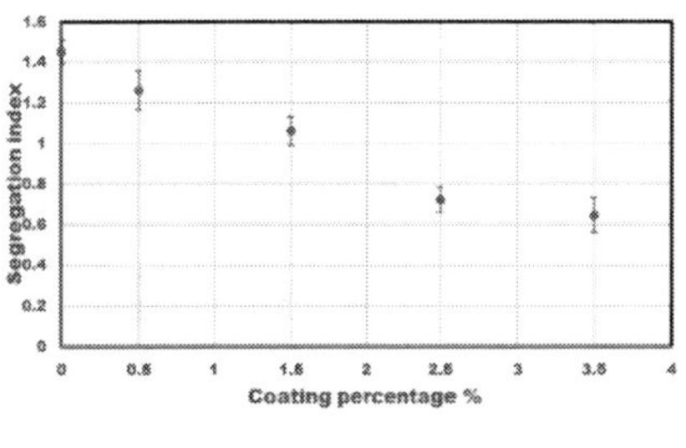

a)

DOI: 10.1051/epjconf/201714013013

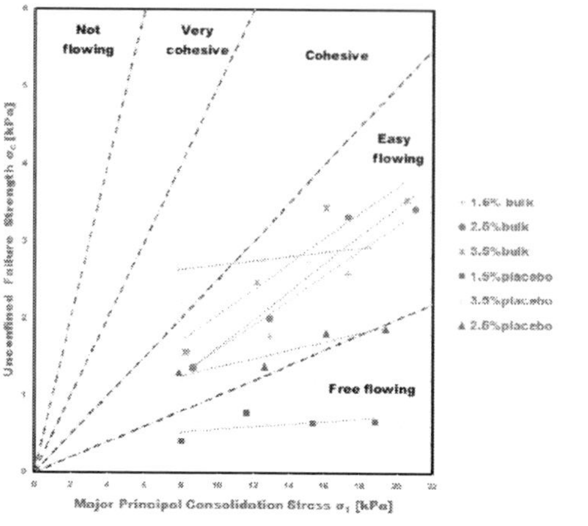

b)

Fig. 5. a) Segregation index versus coating percentage of PEG, b) flowability results of coated enzyme granules and bulk of detergent powders after coating.

The SEM picture of enzyme placebo granules after 2.5% coating percentage and after mixing with BP and TAED is illustrated in Fig. 6-b. It has been observed that BP particles adhere to the surface of coated enzyme placebo granules dispersing them within the bulk of mixture. Only fine BP particles adhere to the coated enzyme placebo granules and there is no evidence of TAED adherence. Adherence of fine BP particles onto the surfaces of the coated enzyme placebo granules modifies their shape and creates much rougher surfaces which further reduces their free movement in the mixture as compared to the case of free-flowing round enzyme placebo granules. This could lead to a reduction of the segregation of this component. Also, interlocking between particles could reduce the flowability of enzyme granules by the formation of liquid bridge between particles. It is recommended to test these hypotheses as part of future work. For instance, it would be useful to change the roughness of enzyme granules by other means to investigate its effect on the segregation.

a)

b)

Fig. 6. a) Full size distribution ternary mixture after 2.5% coating of enzyme granules, b) SEM picture of coated enzyme placebo after mixing with BP and TAED.

In the literature, Discrete Element Method (DEM) simulation has been widely applied to study the kinematics of the flow in multi-component mixture of granular materials [6, 7]. The dynamics of segregation of the ternary mixture studied in the current research can also be investigated using DEM technique, where effects of particle shape and surface roughness on the segregation tendency can be studied.

4 Conclusions

Segregation extent of minor component enzyme placebo granules has been investigated in this work. It is found that for enzyme placebo granules with the mode size of 655 μm, segregation can be reduced by using narrow sieve cuts of BP and TAED, with an average sieve cut size of 390 μm in a ternary mixture of laundry detergent powders, as compared to the case of using their full size distribution. The effect of coating of enzyme placebo granules with a thin layer of sticky liquid on the segregation reduction of this component is also studied. From flowability test results, it is shown that 2.5% coating by PEG can adequately reduce the segregation of enzyme placebo granules without compromising the flowability of bulk and enzyme placebo granules itself.

Acknowledgments

We would like to acknowledge the financial support from AMSCI (The Advance Manufacturing Supply Chain Initiative is a government supply chain fund which is helping to rebuild British manufacturing prowess). In addition, the authors would like to thank Ms Claire Duckitt (Procter and Gamble, Newcastle Innovation Centre, UK) for project coordination.

References

1. J.K. Prescott, T.P. Garcia, Pharm. Technol. **25**, 68 (2001).
2. X. He, X. Han, N. Ladyzhynsky, R. Deanne, Powder Technol. **236,** 85 (2013).
3. A.N. Huang, H.P. Kuo, Adv. Powder Technol. **25,** 163 (2014).
4. P. Tang, V.M. Puri, Part. Sci. Technol. **22**, 321 (2004).
5. G.G. Enstad, Handbook of Conveying and Handling of Particulate Solids. **10**, 589 (2001).
6. L. Bates, User guide to segregation. First edition, British Materials Handling Board, 1997.
7. B. Marks, P. Rognon, I. Einav, J. Fluid Mech. **690,** 499 (2012).
8. J. M. N. T. Gray and C. Ancey, J. Fluid Mech. **678**, 535 (2011).

DOI: 10.1051/epjconf/201714001004

Dense, collisional, shearing flows of compliant spheres

James Jenkins[1,*] and *Diego* Berzi[2]

[1]School of Civil and Environmental Engineering, Cornell University, Ithaca, NY 14853, USA
[2]Department of Civil and Environmental Engineering, Politecnico di Milano, 20133 Milano, Italy

Abstract. We outline the development of theory to describe, dense, collisional shearing flows of identical compliant spheres. We begin with two simple theories: one for rigid, nearly elastic spheres that interact through instantaneous, binary collisions; the other for compliant spheres that interact through multiple, enduring contacts. We then join the two extremes by adding compliance to the collisions and collisions to the spheres in enduring contact. Finally, we compare the predictions of the resulting theory with the results of discrete numerical simulations of steady, homogeneous shearing of compliant frictional spheres.

1 Introduction

In steady, homogeneous shearing flows of identical spheres at volume fractions less than about 0.49, spheres interact through collisions that can be regarded as instantaneous, binary, and uncorrelated. In this case, methods from the kinetic theory of dense gases that take into account the energy loss in a collision [1-3] can be used to predict the relationship between the forces necessary to maintain the flow and the rate of shearing. [4-6]

Above a volume fraction of 0.49, at which a first-order phase transition between disordered and ordered states in an equilibrated system of colliding hard spheres is first possible [7,8], molecular dynamics simulations show that correlations between collisions begin to influence the relationship between the components of the stress and the shear rate [9-11]. The introduction of an additional length scale in the relation for the rate of collisional dissipation of fluctuation energy associated with the size of clusters of interacting spheres modifies the stress relations in an appropriate way [12-15]. The length scale is determined by the competition between the orienting influence of the flow and the randomizing influence of the collisions, using a local balance between the rates of production and dissipation of fluctuation energy. This approach has been tested against discrete element simulations of steady flows in a variety of flow configurations [16–19].

These extended stress relations apply until the mean separation distance between the edges of the spheres vanishes, at least along the direction of principal compression, at which point, the stresses for hard spheres become singular [19]. The volume fraction at which this occurs is less than that for random close packing, $\phi_c = 0.64$, and is seen in numerical simulation to depend on the coefficient of sliding friction [20].

Replacing a rigid contact with a compliant contact has three consequences: it permits the introduction of a time associated with the duration of a collision before the hard-sphere singularity [21]; it relaxes the singularity of the stresses; and it allows prediction of the stresses at volume fractions greater than that at the singularity. The stresses at volume fractions beyond the hard-sphere singularity have parts that depend on the shear rate and parts that depend on the deformation of the contact. We refer to these as the rate-dependent and rate-independent parts of the stress [22,23].

Here, we focus on continuous, steady, homogeneous shearing of a dense aggregate of identical spheres over a range of solid volume fraction, ϕ, above 0.49. We first consider theories that are appropriate in limits on either side of, but distant from, the critical volume fraction ϕ_c at which chains of spheres first span the flow.

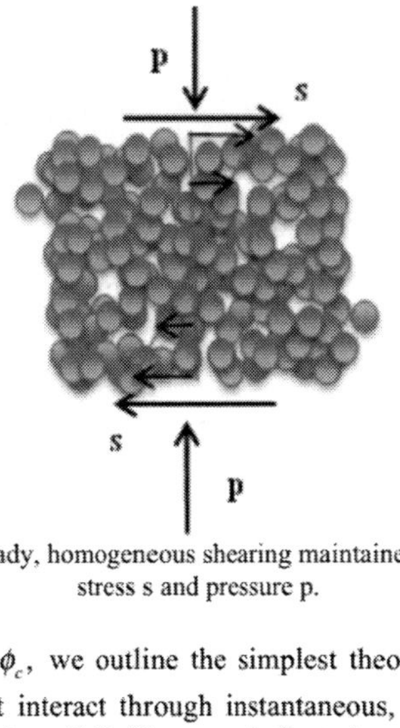

Fig. 1: Steady, homogeneous shearing maintained by shear stress s and pressure p.

For $\phi < \phi_c$, we outline the simplest theory for rigid spheres that interact through instantaneous, frictionless binary collisions [24]. Then, we indicate how this theory must be modified to apply to realistic situations. For

* Corresponding author: jtj2@cornell.edu

$\phi > \phi_c$, we treat deformable spheres that interact through a constantly changing network of longer lasting contacts with multiple neighbors [25]. Finally, we join the two extremes over the entire range of dense volume fractions by incorporating contact deformation in the collisional regime and impulsive momentum transfer in the deformational regime [26].

We treat spheres with a diameter, σ, mass, m, Young's modulus, E, Poisson's ratio, ν, friction coefficient: μ, Normal restitution, e, and tangential restitution, β. The aggregate has a number density $n = \phi / (\pi\sigma^3/6)$, mass density $\rho = mn = \rho_p \phi$, where ρ_p is the mass density of the material of the spheres, and is sheared at a rate u'.

2 Simple kinetic theory

The simplest kinetic theory [24] assumes instantaneous binary collisions, between a typical pair of frictionless, rigid spheres, with an impulse **J** exerted by sphere 2 upon sphere 1, and the change in velocities of the two spheres in a collision given by

$$\mathbf{c}_1' = \mathbf{c}_1 + \mathbf{J}/\mathrm{m} \text{ and } \mathbf{c}_2' \equiv \mathbf{c}_2' - \mathbf{J}/\mathrm{m} \quad (1)$$

where the prime denotes values after the collision. The component of the relative velocity $\mathbf{c}_{12}' \equiv \mathbf{c}_1' - \mathbf{c}_2'$, along the unit vector **k** directed from the center of sphere 1 to that of 2 is related to that before the collision by a coefficient of restitution e, with $0 \le e < 1$:

$$\mathbf{c}_{12}' \cdot \mathbf{k} = \ e(\mathbf{c}_{12}\ \ \mathbf{k}). \quad (2)$$

Then, the impulse may be determined as

$$\mathbf{J} = \frac{1}{2}\mathrm{m}(1+\mathrm{e})(\mathbf{k}\ \ \mathbf{c}_{12})\mathbf{k}. \quad (3)$$

The impulse figures in the calculation of the shear stress and pressure. The kinetic energy of the pair,

$$\mathrm{E} \equiv \frac{1}{2}\mathrm{m}\left(\mathrm{c}_1^2 + \mathrm{c}_2^2\right), \quad (4)$$

experiences the change

$$\Delta\mathrm{E} \equiv \mathrm{E}' - \mathrm{E} = \ \ \frac{1}{4}\mathrm{m}\left(1\ \ \mathrm{e}^2\right)\left(\mathrm{k}\ \ \mathrm{c}_{12}\right)^2. \quad (5)$$

The change in kinetic energy is important when calculating the rate of collisional dissipation.

Averages over the exchanges of momentum and energy in collisions are carried out by integrating over a complete pair particle distribution that governs the likelihood of collisions between pairs of spheres. In the simplest dense theory, the positions and velocities of colliding pairs are assumed to be completely uncorrelated - the assumption of Molecular Chaos - and the distribution is assumed to be expressed as the product of single particle velocity distributions and a correction to the collision frequency that incorporates the distance between edges, rather than centers.

The single particle velocity distribution is

$$f^{(1)}(\mathbf{c},\mathbf{r}) = \frac{n}{(2\pi T)^{3/2}} \exp\left(-\frac{C^2}{2T}\right), \quad (6)$$

where the velocity fluctuation $\mathbf{C}(\mathbf{r}) \equiv \mathbf{c} - \mathbf{u}(\mathbf{r})$ is given in terms of the local average velocity $\mathbf{u}(\mathbf{r})$, $C^2 \equiv \mathbf{C}\cdot\mathbf{C}$, at the point of contact $\mathbf{r}$ and the measure $T \equiv \langle C^2 \rangle / 3$ of the strength of the velocity fluctuations is often referred to as the granular temperature. The frequency correction is given in terms of the volume fraction dependence of the radial distribution function for a contacting pair in a dense aggregate [8]:

$$g_0(\phi) = \frac{0.85}{0.64 - \phi}. \quad (7)$$

With these, the averages are carried out using the distribution function

$$g_0(\mathbf{r}) f^{(1)}(\mathbf{c}_1, \mathbf{r} - \sigma\mathbf{k}/2) f^{(1)}(\mathbf{c}_2, \mathbf{r} + \sigma\mathbf{k}/2), \quad (8)$$

in which care is taken to distinguish between the locations of the enters of the colliding pairs and, in the simplest theory, spatial gradients of the average fields enter the theory through series expansions of these about the point of contact.

Integration yields expressions for the pressure

$$p = 2(1+e)\rho G T; \quad (9)$$

where $G \equiv \nu g_0$, the shear stress

$$s = \frac{4(1+e)}{5\pi^2} \rho G \sigma T^{1/2} u'; \quad (10)$$

and the rate of collisional dissipation

$$\gamma = \frac{12}{\pi^{1/2}} \rho G \frac{1-e^2}{\sigma} T^{3/2}. \quad (11)$$

The last of these may be written as

$$\gamma = n\ \left(1 - e^2\right) \frac{1}{2} mT\ \ \frac{24}{\pi^{1/2}} \frac{GT^{1/2}}{\sigma}, \quad (12)$$

to emphasizes that it is the product of the number of spheres per volume, the energy lost per collision, and the collision frequency. The inverse of the collision frequency is the average time of flight, τ_f, between collisions:

$$\tau_f = \frac{\pi^{1/2}}{24}\frac{\sigma}{GT^{1/2}}. \quad (13)$$

In a steady, homogeneous shearing flow, the granular temperature is determined through the energy balance:

$$su' - \gamma = 0. \quad (14)$$

When the expressions for shear stress and the rate of dissipation are employed in this, it may be solved for T:

$$T = \frac{2}{15}\frac{(\sigma u')^2}{(1-e^2)}. \quad (15)$$

Then, the shear stress and the pressure can be expressed in terms of the shear rate:

$$s = \frac{4(1+e)}{5\pi^{1/2}}\left(\frac{2}{15}\right)^{1/2}\rho G\frac{(\sigma u')^2}{(1-e^2)^{1/2}} \quad (16)$$

and

$$p = \frac{4(1+e)}{15}\rho G\frac{(\sigma u')^2}{(1-e^2)}. \quad (17)$$

We note for later comparison that the ratio s/p is independent of ϕ in the simplest theory.

3 Simple deformation theory

Pairs of spheres interact through contact forces that have normal and tangential components, P and $\mathbf{T}$, parallel and perpendicular to their line of centers. In the simplest linear theory these are, respectively, proportional to the normal and tangential displacements δ and $\mathbf{s}$ of the contact. The normal force exerted by sphere 2 on sphere 1 is

$$P = \frac{\pi}{4}E\sigma\delta, \quad (18)$$

while the tangential force is

$$|\mathbf{T}| = \begin{cases} \frac{\pi}{4}E\sigma s, & \text{if } |\mathbf{T}| < \mu P \\ \mu P, & \text{otherwise.} \end{cases} \quad (19)$$

The distribution of the magnitudes of the contact forces is roughly exponential [27]; so, for example,

$$f(P) = \frac{1}{\bar{P}}\exp\left(-\frac{P}{\bar{P}}\right), \quad (20)$$

where $\bar{P}$ is the average normal force. Discrete numerical simulations and physical experiments on photo-elastic disks indicate both the anisotropy and inhomogeneity of the distribution of contact forces in a sheared aggregate.

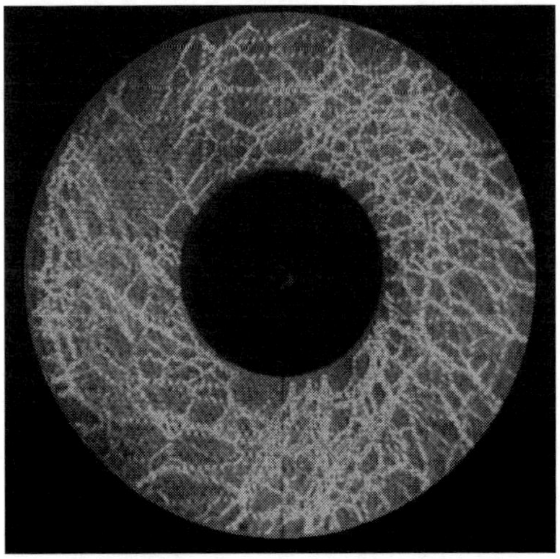

Fig. 2: Shearing of a planar aggregate of circular photo-elastic disks in the Behringer laboratory at Duke University, http://behringer.phy.duke.edu/

The orientational distribution of contacts is described by a function $A(\mathbf{k})$ of orientation k defined so that $A(\mathbf{k})d\Omega$ is the average number of contacts per particle within an element $d\Omega$ of solid angle centered at $\mathbf{k}$ [25]. The average total number of contacts per particle Z is called the coordination number and is given in terms of $A(\mathbf{k})$ by

$$Z = \iint_\Omega A(\mathbf{k})d\Omega. \quad (21)$$

When the contact distribution is isotropic,

$$A(\mathbf{k}) = \frac{Z}{4\pi}. \quad (22)$$

Anisotropic distributions of contacts are often described using a contact fabric tensor, $\mathbf{a}$:

$$A(\mathbf{k}) = \frac{Z}{4\pi}(1 + a_{ij}k_i k_j) \quad (23)$$

with $a_{ii} = 0$. Then,

$$a_{ij} = \frac{15}{2}\frac{1}{Z}\iint_\Omega A(\boldsymbol{k})\left(k_i k_j - \frac{1}{3}\delta_{ij}\right)d\Omega. \quad (24)$$

The stress, $\mathbf{t}$, is defined in terms of the force transmitted across area element $\Delta\Sigma$ with unit normal $\mathbf{N}$ by contacting pairs with orientations $\mathbf{k}$ within solid angle $d\Omega$ [28]:

$$nA(\mathbf{k})d\Omega\mathbf{F}(\sigma\mathbf{k}\cdot\mathbf{N})\Delta\Sigma, \qquad (25)$$

for $\mathbf{k}\cdot\mathbf{N}>0$. Similarly, for $\mathbf{k}\cdot\mathbf{N}<0$. Taking half the sum and integrating over all orientations **k** gives

$$t_{ij}=\frac{n\sigma}{2}\iint_{\Omega}A(\mathbf{k})F_i k_j d\Omega \qquad (26)$$

For example, for an isotropic contact distribution, an average volume strain, Δ, positive in compression, and the contact displacement assumed to be given by the average strain, $\delta=\sigma\Delta/3$, [25] the pressure is

$$p=\frac{1}{12}Z\phi E\Delta. \qquad (27)$$

In applying this result to continued shearing beyond the critical volume fraction, it's natural to relate the average volume strain to the excess of the volume fraction above the critical:

$$p=\frac{1}{12}Z\phi E\left(\phi-\phi_c\right). \qquad (28)$$

Numerical simulations of very slow, continuous, homogeneous shearing [29] indicate that the stress is given by

$$t_{ij}=-p\delta_{ij} \quad cpa_{ij}. \qquad (29)$$

Because only a_{xy} is nonzero,

$$s\equiv t_{xy}=\eta p, \qquad (30)$$

With

$$p=\frac{1}{12}Z\phi E\left(\phi-\phi_c\right), \qquad (31)$$

and η yet to be determined.

4 Corrections to simple kinetic theory

In order to make realistic, quantitative predictions for dense, dissipative, shearing flows the simple kinetic theory whose derivation was outlined above has to be modified in a number of ways.

First, a more realistic complete pair distribution function must be derived and employed. This has been done for elastic spheres [30]. The resulting modification to the theory for steady, homogeneous shearing outlined above for dissipative spheres [2] requires that in the shear stress (10), the factor 1+*e* be replaced by the scalar factor *J*:

$$J\equiv 1+e+\frac{2\pi(1+e)^2(3e-1)}{96-24(1-e)^2-20\left(1-e^2\right)}. \qquad (32)$$

Second, we must take into account the rotational degrees of freedom, and the additional loss of translational fluctuation energy to sliding, tangential restitution, and rotational fluctuation energy. This can be done by employing an effective normal coefficient of translation restitution, ε, that is derived using the balance of rotational fluctuation energy to determine the partition between rotational and translation energy in a steady, homogeneous flow. [31,32] The effective coefficient of restitution is simply expressed in terms of the coefficients of sliding friction and tangential restitution in two limits: when all contacts slide in a collision and μ is about 0. 1,

$$\varepsilon=e-\frac{\pi}{2}\mu \quad \frac{9}{2}\mu^2, \qquad (33)$$

and when all contacts come to rest at some time during a collision and μ is about 1,

$$\varepsilon^2=e^2-4\frac{1+\beta}{7} \quad 4\left(\frac{1+\beta}{7}\right)^2\left[1 \quad \frac{5(1+\beta)}{9-5\beta}\right]. \qquad (34)$$

Third, we must include the dependence seen in discrete numerical simulations of the critical volume fraction on friction. To do this, in g_0 of Eq. (7), we replace 0.64 by $\phi_c(\mu)$, where the dependence of ϕ_c on μ is taken from the simulations [20].

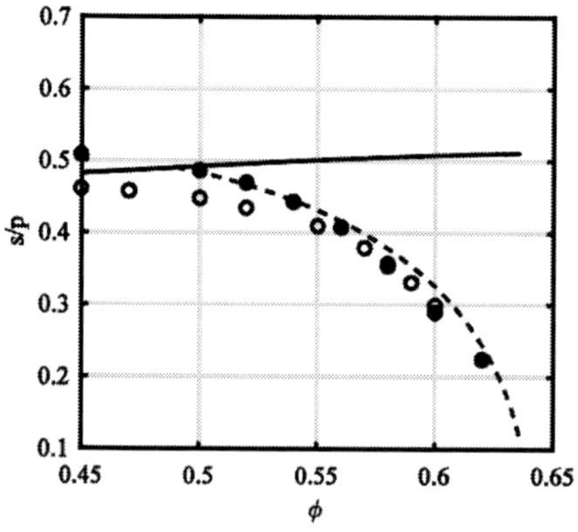

Fig. 3: The stress ratio versus volume fraction in discrete numerical simulations, for frictionless particles, when e = 0.7: event driven simulations (hollow circles [10]); soft particle simulations (filled circles [33]). The solid and dashed lines are the predictions of simple and extended kinetic theory.

Finally, we must incorporate the influence of velocity correlations that occur above $\phi=0.49$. At greater volume fractions, the assumption of molecular chaos breaks down and correlated velocities occur in clusters of size *L*, oriented along the axis of compression. The need for making such a correction is shown in Fig. 3, in which predictions and measurements of the ratio of shear stress to pressure versus volume fraction are shown for several typical systems.

A simple means of doing this is to introduce the cluster size in place of the particle diameter in the rate of collisional dissipation [12-14]

$$\gamma = \frac{12}{\pi^{1/2}} \frac{\rho G}{L} \left(1-\varepsilon^2\right) T^{3/2}. \quad (35)$$

The size of the cluster is determined by a balance between the orienting influence of the flow and the randomizing influence of the collisions:

$$\frac{L}{\sigma} = F(\phi) \frac{\sigma u'}{T^{1/2}}, \quad (36)$$

where the $F(\phi)$ is proportional to $\left(0.64-\phi\right)^{-1}$ [19]. That is, it is singular at random close-packing, not at the critical volume fraction. Then,

$$\frac{s}{p} = \frac{4J}{5\pi^{1/2}} \left[\frac{\left(1-\varepsilon^2\right)}{2JF} \right]^{1/3}, \quad (37)$$

where, in contrast to the ratio given by Eq. (16) and (17), there is a dependence on the volume fraction This relation is one of those plotted in Fig. 3, where it is referred to as that of the extended kinetic theory..

5 Deformation in collisions and collisions in deformation

We next incorporate the deformation of the contacts during a collision into the collisional theory. [21] This provides a way to regularize the singularity in the collisional constitutive relations that without deformation occurs at the critical volume fraction.

When the deformation of a contact is linear, the duration, τ_c, of a contact is

$$\tau_c = \frac{\sigma}{5} \left(\frac{\rho_p}{E} \right)^{1/2}, \quad (38)$$

while we have seen that the time between collisions is

$$\tau_f = \frac{\pi^{1/2}}{24} \frac{\sigma}{G T^{1/2}}. \quad (39)$$

Consequently, if we take into account the duration of a collision, the time between collisions increases from τ_f to $\tau_{f+}\tau_f$. In this case,

$$p = 2\rho(1+\varepsilon)GT = \rho(1+\varepsilon)\frac{\pi^{1/2}}{12}\frac{\sigma T^{1/2}}{\tau_f} \quad (40)$$

becomes

$$p = \rho(1+\varepsilon)\frac{\pi^{1/2}}{12}\frac{\sigma T^{1/2}}{\tau_f + \tau_c}. \quad (41)$$

Similarly,

$$s = \frac{J}{15} \rho \frac{\sigma^2}{\tau_f + \tau_c} u' \quad (42)$$

and

$$\gamma = \frac{1}{2} \rho \left(1-\varepsilon^2\right) \frac{\sigma}{L} \frac{T}{\tau_f + \tau_c}. \quad (43)$$

With the incorporation of deformation into the collisions, the collisional stresses and the rate of dissipation are no longer singular at $\phi = \phi_c$.

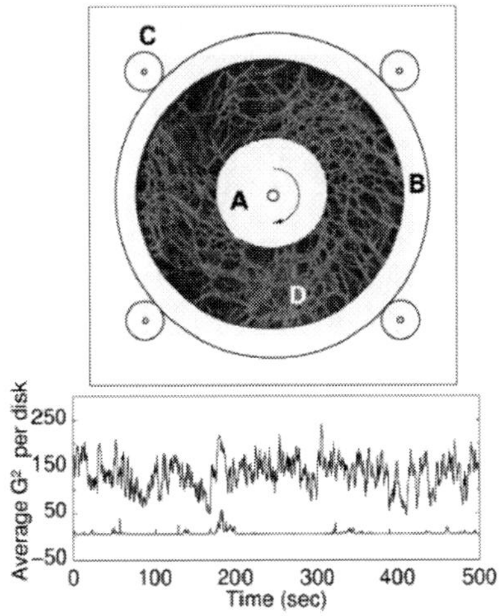

Fig. 4: Stress fluctuations below and above ϕ_c measured during shearing in the circular annulus in the Behringer laboratory, http://behringer.phy.duke.edu/.

The need to incorporate dynamics in the range of volume fraction above the critical can be seen in the results of experiments carried out on photo-elastic disks. Here, a dramatic difference in the strength of the fluctuations in stress is seen in Fig. 4 below and above an area fraction that we identify with the critical. Consequently, for $\phi \geq \phi_c$, there is still a collis-ional component of the exchange of momentum associated with the energy released by the breaking of chains. We assume that the frequency of these collisions is inversely proportional to the duration of a contact and evaluate the momentum transfer at the critical volume fraction.

We take the pressure in this regime to be the sum of contributions from collisions and deformation - that is, the sum of rate-dependent and rate-independent parts:

DOI: 10.1051/epjconf/201714001004

$$p=\rho_p\phi_c(1+\varepsilon)\frac{\pi^{1/2}}{12}\frac{\sigma T^{1/2}}{\tau_c}+\frac{\pi}{8}E(\phi-\phi_c). \quad (44)$$

We treat the shear stress in a similar way, taking into account the relation (30) between the shear stress and pressure when deformations alone occur:

$$s=\frac{J}{15}\rho_p\phi_c\frac{\sigma^2}{\tau_c}u'+\eta\frac{\pi}{8}E(\phi \quad \phi_c). \quad (45)$$

In this regime, we assume that the anisotropy of the collisions is the same as that of the deformation and take η to be the ratio of the collisional shear stress to the collisional pressure at ϕ_c:

$$\eta=\frac{4J}{5\pi^{1/2}}\frac{1}{(1+e)}\frac{\sigma u'}{T^{1/2}}. \quad (46)$$

Finally, in order to enable the determination of the strength of the velocity fluctuations at volume fractions greater than the critical, we adopt similar assumptions for the frequency of collision and the volume fraction in the rate of collisional dissipation:

$$\gamma=\frac{1}{2}\rho_p\phi_c\left(1-\varepsilon^2\right)\frac{\sigma}{L}\frac{T}{\tau_c}. \quad (47)$$

Then, in simple shearing flows, both below and above the critical volume fraction, T may be obtained as a solution to the energy balance (14). In this energy balance, the elastic part of the shear stress above the critical volume fraction is assumed to be recoverable and not to contribute to the rate of working.

6 Results

With T determined on either side of the critical volume fraction, the stress may be calculated and compared to the results of discrete numerical simulations [20,33,34]. The comparison for the pressure and shear stress, normalized by $\rho_p\left(\sigma u'\right)^2$, are shown in Figs. 5 and 6, respectively, for a wide range of contact stiffness, $\kappa\equiv(\pi/4)E\sigma$, made dimensionless by $\rho_p\sigma^3\left(u'\right)^2$. The predictions are, in general, good and provide an indication that at least some of the numerous assumptions made to obtain them have a basis in the physics.

A more stringent test results from plotting the ratio of the shear stress to the pressure. In this case, because he stiffness is assumed to enter into both quantities in the same way, the ratio is independent of stiffness. In contrast. the measured data is sensitive to the stiffness, especially for the more compliant spheres. The agreement for the stiffer spheres is relatively good.

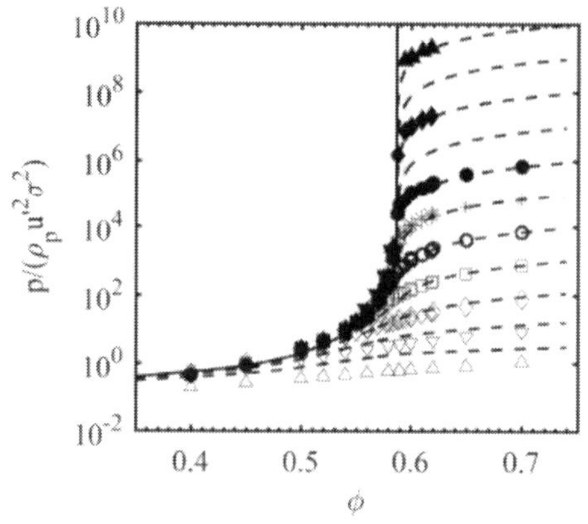

Fig 5: Predictions of the normalized pressure versus volume fraction for e = 0.7, μ = 0.5, values of the dimensionless contact stiffness ranging from 10 to 10^{11} (dashed curves) and data from the discrete numerical simulations [20,33,34].

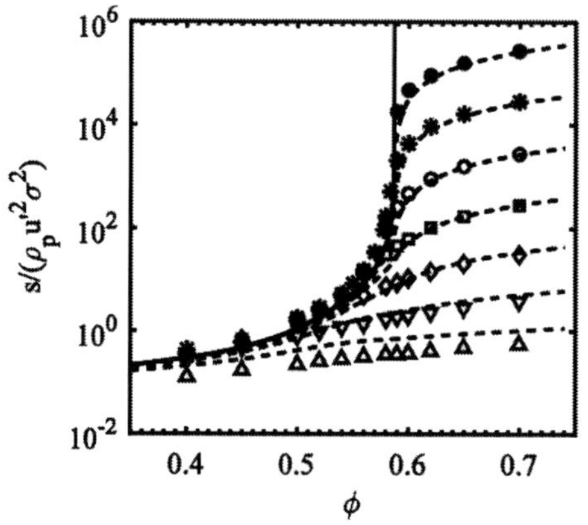

Fig 6: Predictions of the normalized shear stress versus volume fraction for e = 0.7, μ= 0.5, values of the dimensionless contact stiffness ranging from 10 to 10^7 (dashed curves) and data from the discrete numerical simulations [20,33,34].

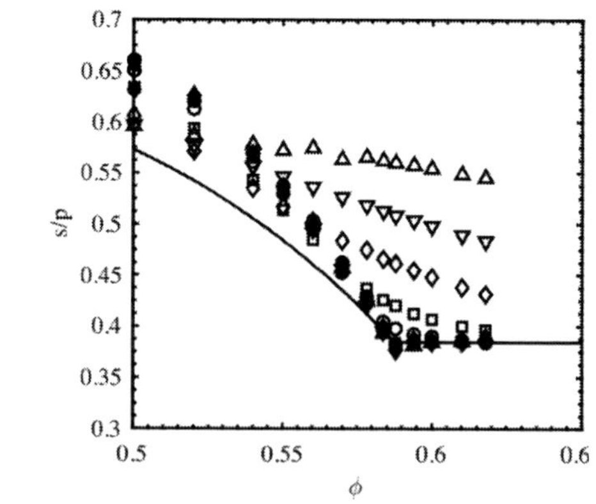

Fig. 7. Predictions of the ratio of shear stress to pressure versus volume fraction (solid curve) for e = 0.7 and μ= 0.5 and data from the discrete numerical simulations [20,33,34] for values of the dimensionless contact stiffness ranging from 10 to 10^7 (symbols).

7 Conclusions

We have outlined the derivation of a theory for steady, homogeneous shearing flows of dense aggregates of deformable, dissipative spheres. The incorporation of contact deformation in the collisional regime and impulsive momentum transfer in the deformational regime permits a smooth transition through the critical volume fraction and relatively good descriptions of the stresses on either side of it. The predictions compare well over the range of volume fraction before and after the hard sphere singularity and over nine orders of magnitude of the contact stiffness. The model requires no parameters other than the contact stiffness, and the coefficients of restitution and sliding friction of the spheres.

References

1. J.T. Jenkins, S.B. Savage, J. Fluid Mech. **130**, 187–202 (1983)
2. V. Garzo, J.W. Dufty, Phys. Rev. E **59**, 5895–5911 (1999)
3. I. Goldhirsch, Annu. Rev. Fluid Mech. **35**, 267–293 (2003)
4. E. Azanza, F. Chevoir, P. Moucheront, J. Fluid Mech. **400**, 199–227 (1999)
5. Y. Forterre, O. Pouliquen, Phys. Rev. Lett. **86**, 5886 (2001)
6. H. Xu, M. Louge, A. Reeves, Continuum Mech. Thermodyn. **15**, 321–349 (2003)
7. B.J. Alder, T.E. Wainright, J. Chem. Phys. **27**, 1208-1209 (1957)
8. S. Torquato, Phys. Rev. E **51**, 3170–3182 (1995)
9. N. Mitarai, H. Nakanishi, Phys. Rev. Lett. **94**, 128001 (2005)
10. N. Mitarai, H. Nakanishi, Phys. Rev. E **75**, 031305 (2007)
11. V. Kumaran, J. Fluid Mech, **632**, 145–198 (2009)
12. J.T. Jenkins, Phys. Fluids **18**, 103307 (2006)
13. J.T. Jenkins, Granular Matter **10**, 47–52 (2007)
14. J.T. Jenkins, D. Berzi, Granular Matter **12**, 151–158 (2010)
15. D. Berzi, Acta Mech. **225**, 2191–2198 (2014)
16. D. Berzi, J.T. Jenkins, Phys. Fluids **23**, 013303 (2011)
17. J.T. Jenkins, D. Berzi, Granular Matter **14**, 79–84 (2012)
18. D. Vescovi, D. Berzi, P. Richard, N. Brodu, Phys. Fluids **26**, 053305 (2014)
19. D. Berzi, D. Vescovi, Phys. Fluids **27**, 013302 (2015)
20. S. Chialvo, J. Sun, S. Sundaresan, Phys. Rev. E **85**, 021305 (2012)
21. H. Hwang, K. Hutter, Continuum Mech. Thermodyn. **7**, 357–384 (1995)
22. P. Johnson, P. Nott, R. Jackson, J. Fluid Mech. **210**, 501–535 (1990)
23. D. Berzi, C. di Prisco, D. Vescovi, Phys. Rev. E **84**, 1-6 (2011)
24. J.T. Jenkins, S.B. Savage, J. Fluid Mech. **130**, 187- 202, (1983)
25. J.T. Jenkins, O.D.L. Strack, Mech. Mater. **16**, 25–33 (1993)
26. D. Berzi, J.T. Jenkins, Soft Matter **11**, 4799-4808 188303 (2015)
27. C. Thornton, KONA **15**, 81-89 (1997)
28. J.T. Jenkins, in *Modern Theory of Anisotropy Elasticity and its Applications*, 368-377, SIAM (1990)
29. J. Sun, S. Sundaresan, J. Fluid Mech. **682**, 590–616 (2011)
30. S. Chapman, T.G. Cowling, *The Mathematical Theory of Non-Uniform Gases*, Cambridge University Press, 3rd Ed. (1970)
31. J.T. Jenkins, C. Zhang, Phys. Fluids **14**, 1228 (2002)
32. M. Larcher, J.T. Jenkins, Phys. Fluids **25**, 113301 (2013)
33. S. Chialvo, S. Sundaresan, Phys. Fluids **25**, 070603 (2013)
34. S. Ji, H.H. Shen, J. Rheol. **52**, 87–103 (2008)

Discrete Modelling of Compaction of Non-spherical Particles

Yi He[1], *Tim J.* Evans[2], *Aibing* Yu[1,3], *Runyu* Yang[1,*]

[1] School of Materials Science and Engineering, University of New South Wales, Sydney, 2052, Australia
[2] Rio Tinto Iron Ore Group, Australia
[3] Department of Chemical Engineering, Monash University, Clayton, 3900, Australia

Abstract. Compaction behaviour and mechanical response of a compact show strong dependence on particle shape. In this study, a numerical model based on the discrete element method (DEM) was developed to study the compaction behaviour of spheroidal particles. In the model, particle shape was approximated by gluing multiple spheres together. A bonded particle model was adopted to describe interparticle bonding force. The DEM model was first validated by comparing the properties of packing of spheroids (packing density, coordination number) with literature data and then applied to both die compaction and unconfined compression. In die compaction, the effect of aspect ratio on the densification was mainly due to the difference in the initial packing. In unconfined compression, the increase in compressive strength with increasing aspect ratio was attributed to the increase in the number of interparticle bonding. The findings facilitate a better understanding of the relation of particle shape to the compaction behaviour and compact strength.

1 INTRODUCTION

Compaction of particles has wide applications in industries. The physico-mechanical properties of a compact are critical to the performance of the compact in the after-compaction operations, such as storage, transportation and sintering. It is therefore important to understand the relationship between consolidation pressure and compact properties, including compact density and compact strength [1].

The macroscopic property of a compact is a collective outcome of the microscopic interactions between particles. For this purpose, the discrete element method (DEM) offers an efficient way to obtain micro-mechanical insight into its behaviour as it treats particles individually and explicitly considers the particle characteristics, material properties and the inter-particle forces. DEM has been adopted to investigate particle compaction, such as the effect of mechanical properties of particles [2, 3], particle-wall friction induced inhomogeneity [4], evolution of compact structure [5] and the effects of moisture, particle shape and particle size [6-8]. Recently we conducted a DEM study of the compressive strength of iron ore compacts using a bonded particle model [16]. The simulated stress-strain responses were comparable with those observed from the physical experiments.

Particles in real processes, however, are often non-spherical. It has been demonstrated that particle shape has a strong influence on the mechanical behaviour particle packing [9, 10] and flow [11]. For example, in the biaxial compression of elliptical particles, much larger strength was observed compared to that of disk-shaped particles [12]. Ting *et al.* [13] observed that the shear strength of a 2D assembly of ellipse shaped particles increased with particle angularity. The enhanced shear strength was attributed to the rolling resistance provided by particle interlocking. Chung and Ooi [14] obtained a better agreement with experimental results in the bulk response when the particle shape was considered under confined compression. Recently, Wiacek et al. [15] observed that the vertical pressure transfer was sensitive to the aspect ratio of elongated particles in the uniaxial compression. However, little effort has been devoted to investigating the mechanical response and strength of non-spherical particle compacts under the unconfined axial compression.

This study is to develop a DEM model to study the compaction of spheroid particles, mainly focusing on the effect of particle shape on compact density and strength. Particle geometry is modelled using a multi-sphere approach. The model is firstly validated against the packings of analytical shaped spheroids and then extended to study their behaviour in the die and unconfined compactions. Both the bulk behaviour and microstructure will be analysed.

2 Model description and simulation conditions

2.1 DEM model

* Corresponding author: r.yang@unsw.edu.au

For a non-spherical particle i of mass m_i and inertial tensor $\mathbf{I}_i$, its translational motion is defined based on its mass centre and its rotational motion obeys the Euler equations in body-fixed coordination system, given by,

$$m_i \frac{d\mathbf{v}_i}{dt} = \mathbf{F}_i + m_i\mathbf{g} \quad (1)$$

$$\mathbf{I}_i \frac{d\boldsymbol{\omega}_i}{dt} + \boldsymbol{\omega}_i \times \left(\mathbf{I}_i \times \boldsymbol{\omega}_i\right) = \mathbf{M}_i \quad (2)$$

where $\mathbf{v}_i$ and $\boldsymbol{\omega}_i$ are, respectively, the translational and angular velocity with respect to the mass centre. $\mathbf{F}_i$ is the sum of all externally applied forces acting on the particle, which include in this work the normal contact force $\mathbf{F}_{ij}^n$, tangential contact force $\mathbf{F}_{ij}^t$, capillary force $\mathbf{F}_{ij}^{cap}$ (for wet particles) and the bonding force $\mathbf{F}_{ij}^b$ (when a bond is present between particles). $\mathbf{g}$ is the gravity acceleration. $\mathbf{M}_i$ is the total moment on the particle, including the moment caused by tangential force and the moment induced by the tangential bonding force and bond bending/torsion.

Table 1 lists the force equations used in this work. The details can be found in our previous study [16].

Table 1 Equations used to calculate forces in this work.

Force	Equation
Normal contact force $\mathbf{F}_{ij}^n$	$\begin{cases} 4/3E^*R^{*1/2}\delta^{3/2} & (\delta<\delta_y) \\ F_y+\pi p_y R^*(\delta-\delta_y) & (\delta\ge\delta_y) \end{cases}$
Tangential force $\mathbf{F}_{ij}^t$	$\mu_s F_{ij}^n\left[1-\left(1-\frac{\min(\xi_s,\xi_{s,\max})}{\xi_{s,\max}}\right)^{1.5}\right]$
Capillary force $\mathbf{F}_{ij}^{cap}$	$\begin{cases} \frac{2\pi R^*\gamma\cos\theta_c}{1+1/[\sqrt{1+2V_L/(\pi R^*S^2)}-S]} & (P-P) \\ \frac{4\pi R^*\gamma\cos\theta_c}{1+S/\sqrt{\pi R/V_L}} & (P-W) \end{cases}$
Particle bonding force $\mathbf{F}_{ij}^b$	$\{\Delta\mathbf{F}\}=-\{\mathbf{K}\}\{\Delta\mathbf{x}\}$

2.2 Simulation conditions

In this work, the packing and compaction of ellipsoidal particles were simulated.

A simulation started with the random generation of particles without overlap inside a cylindrical container. Random orientations of the particles were introduced by assigning different initial values to their Euler angles. After the packing was formed, the upper punch then moved downward at a constant velocity to compress the particles till the compact reached the desired packing density. It then moved upward till it was separated from the compact top surface. The bonding forces were introduced between the contacting particles after a short period of relaxation (1s). The unconfined compression was conducted by removing the cylindrical wall and move the top punch downward.

Table 2 shows the properties of particle and bond used in present work. The values of the properties were determined by calibrating the simulation results with experiments of compaction of iron ore powders [16]. With these values, the simulations were able to reproduce similar compaction curves in the die and unconfined compaction to those observed in the experiments (Fig. 1).

Table 2 Parameters used for simulation.

Parameter	Value
Container diameter, D (m)	0.5
Particle number, N	30000
Equivalent diameter, d (m)	0.01
Particle density, ρ (kg·m^{-3})	5.17×10^3
Young's modulus, Y (GPa)	9
Yield pressure, p_y (MPa)	300
Poisson ratio, v	0.29
Interparticle friction coefficient, μ_{pp}	0.3
Particle-wall friction coefficient, μ_{pw}	0.1
Bond-area multiplier, λ	0.68
Young's modulus of bond, Y_b(MPa)	200
Bond stiffness ratio, k^n/k^t	1.0
Bond strength, σ_b(MPa)	15

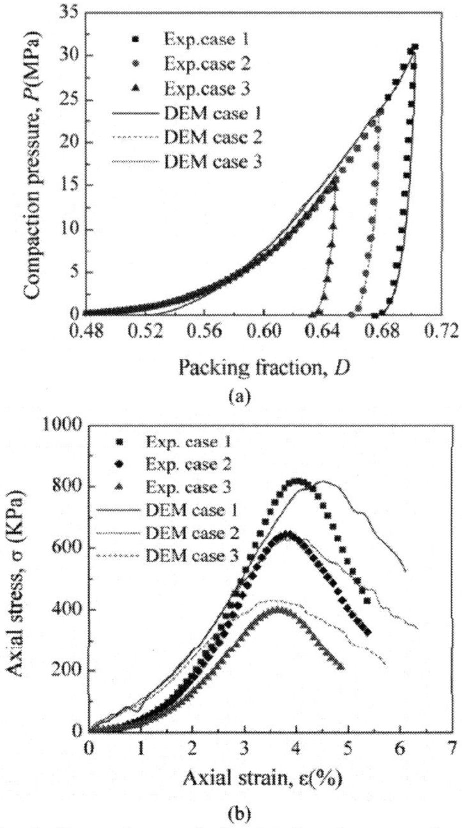

Fig. 1 Comparisons of simulated and measured axial pressure with packing fraction in the die and unconfined compactions [16].

A multi-sphere approach [17] was adopted to approximate the geometry of the spheroids. In this approach, a spheroidal particle was approximated by combining a number of overlapping primary spheres of different sizes. The spheroids have various aspect ratios from 0.2 to 3.0 to cover spheroids from oblate to prolate. The shapes were selected such that the comparison can be performed with a previous study using [18]. Fig. 2 shows the representative particle shapes in the study.

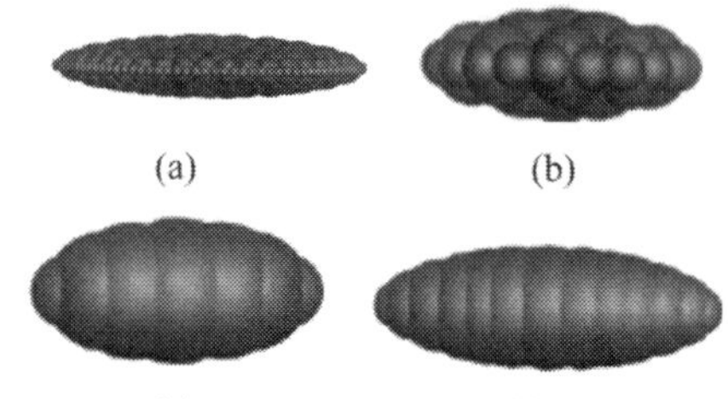

Fig. 2 Representative ellipsoidal particles of different aspect ratios: (a), 0.2; (b), 0.4; (c), 2.0; and (d), 3.0.

3 Results and Discussion

3.1 Particle packing and model validation

Fig. 3 shows the packed beds for three different aspect ratios of spheroids. The validity of the model was tested against a previous DEM simulation of ellipsoids packing with analytical shape [18].

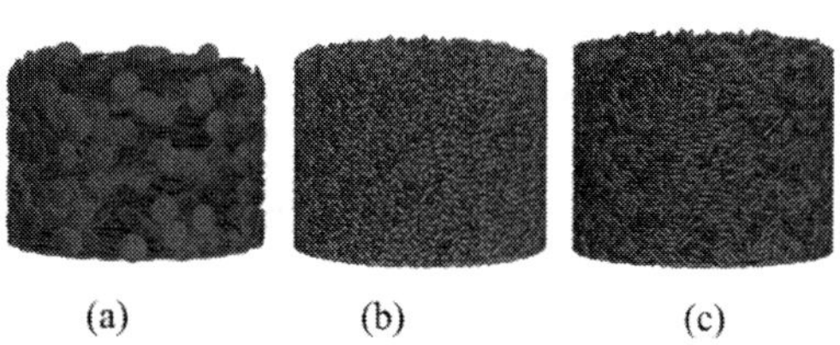

Fig. 3. Final packing of ellipsoids with aspect ratios of: (a) 0.2; (b) 2.0; and (c) 3.0.

Fig. 4(a) shows that variations of packing density with the aspect ratio. Here, the packing density was calculated from the region 3*d* away from the wall to minimise the wall effect. The figure shows that packing density increases with the aspect ratio and reaches a maximum at a ratio of 0.75. It then decreases with increasing aspect ratio and has a minimum when the ratio is 1 (packing of spheres). There is another peak for the prolate particles at an aspect ratio of 2. The comparison shows that the current model is able to reproduce results consistent to the previous study [18].

Fig. 4(b) shows the variation of mean coordination number (CN) with the aspect ratio. In this work only the contacted particles are counted. The mean CN appears to be the lowest for spherical particles and rises sharply with small deviations. Again, the general trends of the mean CN evolving with the aspect ratio is consistent with those reported earlier [18], thus validating the current model.

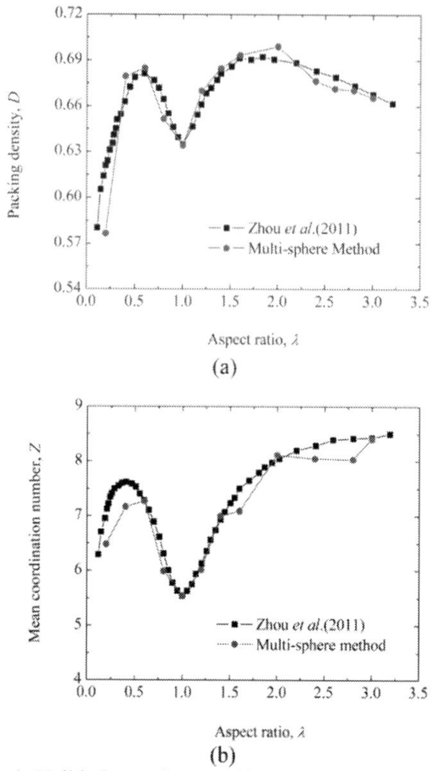

Fig. 4. Validations of (a) packing density and (b) mean coordination number with respect to aspect ratio.

3.2 Die compaction

Fig. 5 plots the variation of consolidation pressure for die compaction of different shapes of particles, showing a strong dependence on particle aspect ratio. The difference in the slope of the loading curves is related to the particle orientation which makes the vertical fore transmission more efficient for particles of larger aspect ratios [18]. In particular, the effect on initial packing density is more pronounced than that of the overall deformation. So the variation in pressure is due to the difference in the initial packing density.

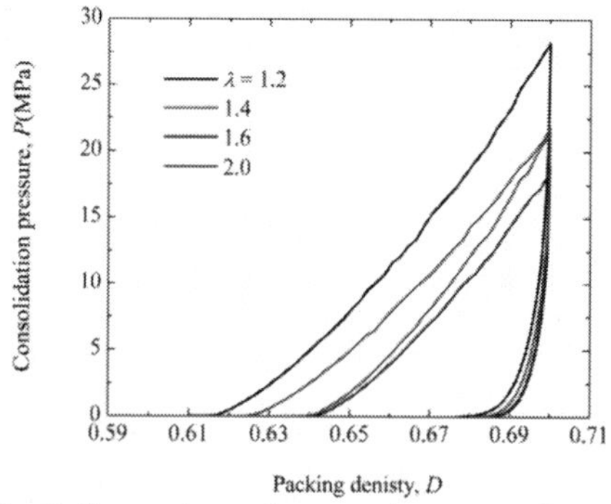

Fig. 5. Comparison of compaction curves for ellipsoidal particles with different aspect ratios.

In Fig. 6, both initial packing density and compact density after unloading are plotted against the aspect ratio. The general trend for the initial packing density is very similar to those obtained during packing as shown in Fig. 5 except for a relatively lower magnitude due to the boundary effect [19]. The overall trend of density variation persists after the unloading, which further demonstrates that difference in the overall compressibility is mainly due to the difference in initial packing density.

Fig. 6 also illustrates the consolidation pressure achieved by compressing the samples to a density of 0.7. As the particle aspect ratio increases, the consolidation pressure first decreases to the minimum at the aspect ratio of 1.6 and then increases with further increase in the particle aspect ratio.

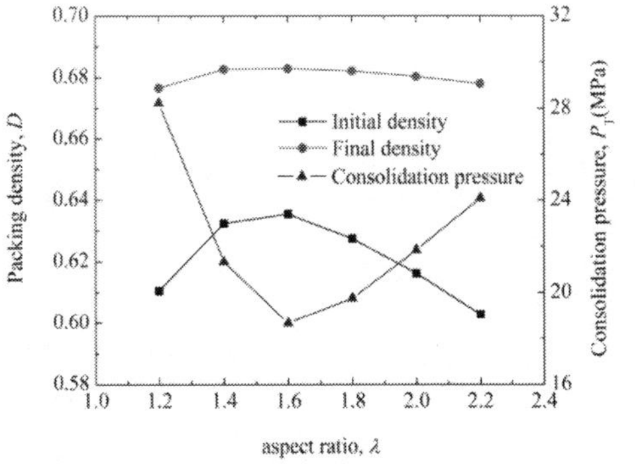

Fig. 6. Packing density variation with the aspect ratio.

The angular distribution of the force illustrates the anisotropy of the compact. Recent studies have shown that the spatial distribution of the normal contact force can be divided into two subnetworks: large and small forces [20]. As the large forces are mainly contribute to the compact stiffness, so it would be interesting to characterise the angular distribution of the large forces within the compact. Fig. 7 shows the angular distributions of all and large normal contact forces at the end of die compaction. Here the large forces are defined as those with magnitudes 3 times the mean force.

The angular distribution of all the normal contact forces (Fig. 7(a)) shows a peak between 40° and 70° for all the cases. It demonstrates the anisotropic nature of the force network. As the particle aspect ratio increases, the peak of the distribution gradually shifts left. Also, the proportion of the horizontally aligned forces decreases with increasing particle aspect ratio. Therefore, the anisotropy of the force network increases as the aspect ratio increases.

Fig. 7(b) shows the angular distribution of the large forces, which is clearly different from those of all the contact forces. The large forces tend to align along the vertical direction (compression direction). Furthermore, the influence of the compression on the alignment of the large forces become more significant as the particle aspect ratio is increases as reflected by the left-shifting and increasing peak value of the angular distributions.

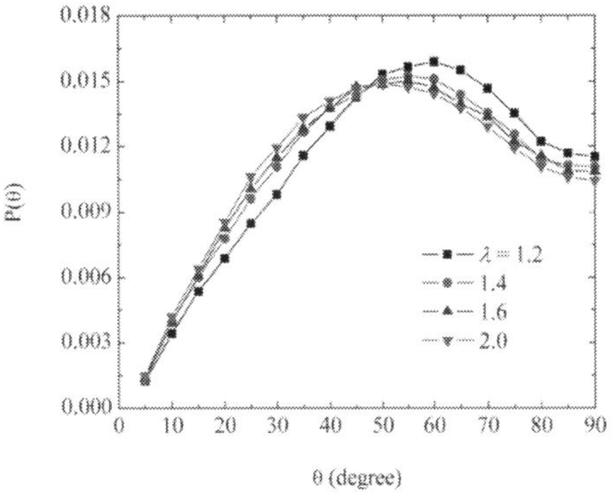

(a)

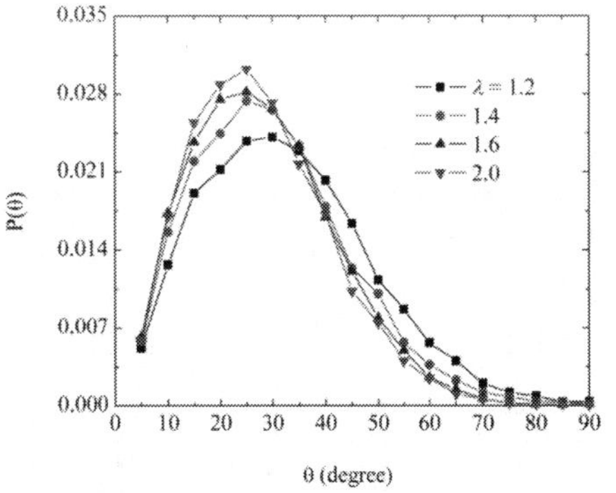

(b)

Fig. 7. Angular distributions of (a), all normal contact force; and (b), large forces with $f > 3.0$.

3.3 Unconfined compression

Fig. 8 shows the evolution of compressive stress plotted against the axial strain for the compacts subject to the unconfined compression. It can be seen that all the cases produced a pronounced peak in the compressive stress, followed by stress softening during compression. It is evident that particle aspect ratio affects the macroscopic behaviour of compact significantly as reflected by the loading stiffness as well as the axial strain at the peak stress. In contrast, the post-peak stress response is more or less independent on the variation in particle aspect ratio as indicated by the largely paralleled stress curve after peak stress.

A major advantage of discrete modelling is that the failure process can be readily accessed. Fig. 9 compares the fully developed failure patterns of three aspect ratios, 1.2, 1.6 and 2.0. Previous studies have linked the initiation and propagation of shear bands with particle rotation [21-24]. Shear banding is often associated with high gradients of particle rotation and large voids. Rolling resistance strongly influences in the orientation, thickness and the initiation of shear banding [25]. Here, the failure patterns are color-coded by particle angular velocity. It is obvious that an aspect ratio close to 1.0 results into a clear and single crack cutting through the compact without changing much the geometrical integrity of the rest part. In contrast, more complicated and dispersed failure patterns are observed for larger aspect ratios. This variation maybe related to the increased rotational resistance, which reduces the likelihood of shear-induced bond breakage during unconfined compression.

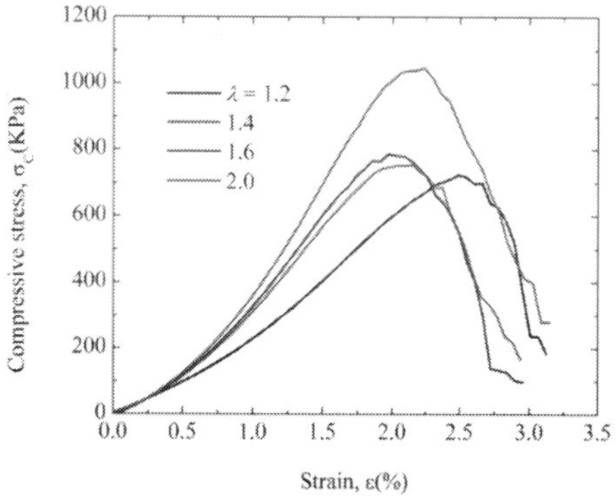

Fig. 8. Comparison of stress-strain relationships during unconfined compression test for different particle aspect ratios.

Fig. 9. Failure patterns of compacts of particles with different aspect ratios: (a), 1.2; (b), 1.6; and (c), 2.0. The particles are colour-coded by their angular velocities.

Fig. 10 shows the compressive strength, the peak stress during unconfined compression, as a function of particle aspect ratio. Despite of a non-monotonic relationship between packing density and aspect ratio, compressive strength is found to increase rapidly as the aspect ratio is increased above the value of 1.6.

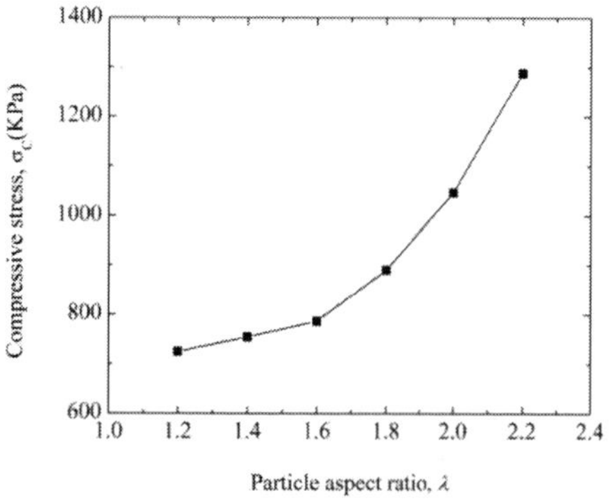

Fig. 10. Compressive strength as a function of particle aspect ratio.

4. Conclusions

The multi-sphere method was incorporated into the DEM to study the packing and compaction process of ellipsoidal particles. Validation was carried out based on packing of spheroidal particles. The developed model was further applied to die compaction and unconfined compression on the spheroidal particles. The main results are summaries below,

- The relationship between packing properties and aspect ratio is consistent with the DEM results of analytical shape representation.
- For ellipsoid compaction, difference in overall compressibility is mainly due to the difference in initial packing density. As the aspect ratio increases, consolidation pressure first decreases to a minimum at the aspect ratio of 1.6 and then increases.
- For unconfined compression of ellipsoidal particles, the mechanical response shows strong dependence on particle aspect ratio. Compressive strength increases rapidly as the aspect ratio above the value of 1.6.

The authors are grateful to the Australian Research Council and Rio Tinto Iron Ore group for financial support of this work.

References

1. M. Abdel-Ghani, J. G. Petrie, J. P. K. Seville, R. Clift and M. J. Adams, Powder Technology **65** (1-3), 113-123 (1991).
2. A. Samimi, A. Hassanpour and A. Ghadiri, Chem Eng Sci **60** (14), 3993-4004 (2005).
3. A. Hassanpour and M. Ghadiri, Powder Technology **141** (3), 251-261 (2004).
4. Y. Y. Foo, Y. Sheng and B. J. Briscoe, International Journal Of Solids And Structures **41** (21), 5929-5943 (2004).
5. Y. Sheng, C. J. Lawrence, B. J. Briscoe and C. Thornton, Eng Computation **21** (2-4), 304-317 (2004).
6. O. Skrinjar and P. L. Larsson, Particul Sci Technol **30** (4), 364-377 (2012).
7. O. Skrinjar and P. L. Larsson, Comput. Mater. Sci. **31** (1-2), 131-146 (2004).
8. S. C. Thakur, H. Ahmadian, J. Sun and J. Y. Ooi, Particuology **12**, 2-12 (2014).
9. F. X. Villarruel, B. E. Lauderdale, D. M. Mueth and H. M. Jaeger, Physical Review E **61** (6), 6914-6921 (2000).
10. A. Donev, I. Cisse, D. Sachs, E. Variano, F. H. Stillinger, R. Connelly, S. Torquato and P. M. Chaikin, Science **303** (5660), 990-993 (2004).
11. P. W. Cleary and M. L. Sawley, Appl. Math. Model. **26** (2), 89-111 (2002).
12. L. Rothenburg and R. J. Bathurst, Geotechnique **42** (1), 79-95 (1992).
13. J. M. Ting, L. Meachum and J. D. Rowell, Eng Computation **12** (2), 99-108 (1995).
14. Y. C. Chung and J. Y. Ooi, Modern Trends in Geomechanics **103**, 223-239 (2006).

15. J. Wiacek, M. Molenda, J. Horabik and J. Y. Ooi, Powder Technol **217**, 435-442 (2012).
16. Y. He, Z. Wang, T. J. Evans, A. B. Yu and R. Y. Yang, Int J Miner Process (2015).
17. J. F. Favier, M. H. Abbaspour-Fard, M. Kremmer and A. O. Raji, Engineering Computations **16** (4), 467-480 (1999).
18. Z. Y. Zhou, R. P. Zou, D. Pinson and A. B. Yu, Ind Eng Chem Res **50** (16), 9787-9798 (2011).
19. R. P. Zou and A. B. Yu, Chem Eng Sci **50** (9), 1504-1507 (1995).
20. F. Radjai, D. E. Wolf, M. Jean and J. J. Moreau, Phys Rev Lett **80** (1), 61-64 (1998).
21. J. P. Bardet and J. Proubet, Geotechnique **41** (4), 599-613 (1991).
22. N. Hu and J. F. Molinari, J. Mech. Phys. Solids **52** (3), 499-531 (2004).
23. C. Thornton and L. Zhang, Philos Mag **86** (21-22), 3425-3452 (2006).
24. A. Mohamed and M. Gutierrez, Granul Matter **12** (5), 527-541 (2010).
25. K. Iwashita and M. Oda, J Eng Mech-Asce **124** (3), 285-292 (1998).

 DOI: 10.1051/epjconf/201714006013

Simulation research of the buffering capacity of non-convex particle system under impact loading

Qi Zhang[1,*], *Siqiang* Wang[1], and *Meiying* Hou[2]

[1]Department of Mechanics, Taiyuan University of Technology, Taiyuan, China
[2]Key Laboratory of Soft Matter Physics, Institute of Physics, Chinese Academy of Science,Beijing, China

Abstract. The behavior of granular materials under the action of external force is a fundamental problem for the dynamics of granular media. This paper investigates the buffering capacity of three-dimensional non-convex granular beds subjected to vertical impact using the discrete element methods. The results display the particle shape significantly affects the buffering capacity of the equal mass granular bed. And the influence of aspect ratio of non-convex particles on buffering capacity exhibits an obvious non-monotonicity. The functional mechanism of inter-particle friction is discussed. It is showed that the shape of the non-convex particles has a great effect on the contact number of granular system and probability density functions of inter-particle forces.

1 Introduction

The behavior of granular systems is rather complex due to their dissipation and many-body nature[1-2]. At present, the most research of granular materials focus on the spherical particles system, while particles in nature are generally non-spherical[3]. Compared to spherical particles, the non-spherical particles are significant differences in the rheological behavior and structural properties like viscosity, yield stress and packing density, especially for non-convex particles system which could exhibit an effective cohesion through the interpenetration between particles[4-6].

In recent years, the impact problems of granular matter become particularly important in the field of materials engineering, soil mechanics and fracture mechanics[7-10]. Granular materials demonstrate many unique properties under external impact loading by means of energy dissipation. Particles could extend the time of impact by breaking and recombination of internal force chain network, thereby reducing the impact strength. Particle damper composed of spherical particles is widely used in engineering and industrial application. While dolosse made from concrete with complex non-convex geometric shape are used to protect harbour walls, breakwaters and shore earthworks. So, it is scientific significance and application prospects to study mechanical properties of granular materials under impact loading.

Especially, the buffering capacity is a fundamental issue for the dynamics of granular media[11]. However the energy dissipation and dynamic response during impact process of non-convex granular matter are still not well understood in both experiment and simulation. In this paper, based on the EDEM software platform, simulation of U-shaped granular bed under vertical impact loading is performed. The buffering capacity of U-shaped particles are analyzed by simulating different aspect ratios of particles.

2 Numerical simulation

To study the influences of non-convex particles shape on buffering capacity of granular system under impact load, three-dimensional discrete element method (DEM) with the soft contact approach is carried out to model the impacting process of spherical projectile on non-convex particle beds. Simulations in this study are performed by commercial software EDEM, the engineering simulation platform developed by DEM-Solutions Company. The motion of particles follows Newton's second law. The simplified Hertz-Mindlin model (Hertz-Mindlin no-slip model) which could make a good representation of the particle behavior during collisions, is employed to calculate the contact force between particles. Normal force and tangential force can be described in the form of nonlinear spring combined with dashpot, however the viscous forces such as liquid force were not considered here.

The multi-element method is used to represent the U-shaped particles as non-convex particles similar with previous article[6]. It is composed of several identically sized spherical elements overlapping each other as

* Corresponding author: zq_2012@126.com

shown in Fig.1. The diameter of a single unit is 2.8mm and the overlap is 1.4mm. There are 9 different proportions of length to width for U-shaped particle, which is ranged from 0.25:1 to 1.417:1 in this work. In the multi-element method, contact detection between particles is sphere-based, which leads to the figuring of contact forces between U-shaped particles can be conducted with a standard discrete element algorithm for spherical particle.

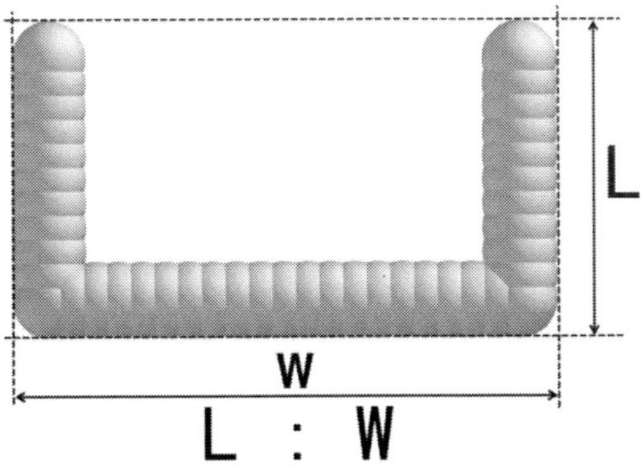

(a) U-shaped particle

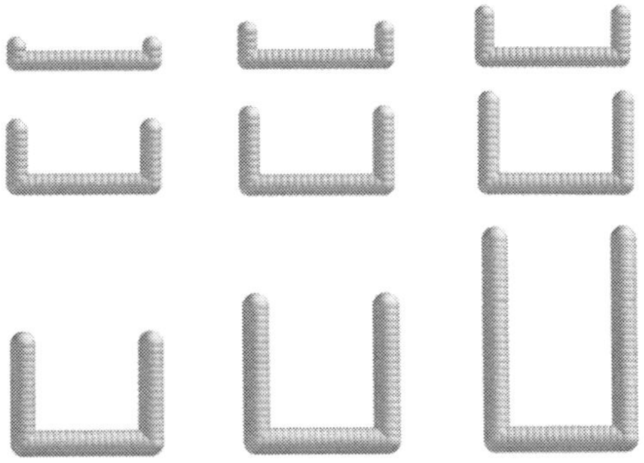

(b) different aspect ratio L/W

Fig. 1. U-shaped particles with different aspect ratio L/W generated by multi-element method

The cylindrical container is 80mm in diameter and 200mm in height. The main computational parameters in the DEM simulation are listed in Table 1. Firstly, U-shaped particles are generated randomly at the top of the cylindrical container. Secondly, all the beads fall into the container by rain-fall method. Then a rigid circular board appears above the surface of the granular bed closely and moves downwards 1 mm to flat the surface. When the velocity of particles is observably small, we remove the board. At last a spherical projectile with diameter of 14mm generates above the surface of granular bed and impacts the bed with constant velocity 1m/s as shown in fig.2. For different aspect ratio of U-shaped particles, the total mass of the granular bed is constant.

Table 1. Main computational parameters in DEM simulation

Parameters	Unit	Value
Projectile poisson ratio		0.31
Projectile shear modulus	GPa	4
Projectile density	g/cm^3	7.85
Projectile diameter	cm	14
U-shaped particle poisson ratio		0.215
U-shaped particle shear modulus	GPa	0.1
U-shaped particle density	g/cm^3	2.65
Coefficient of friction		0.5
Coefficient of restitution		0.5
Time step	s	1e-6

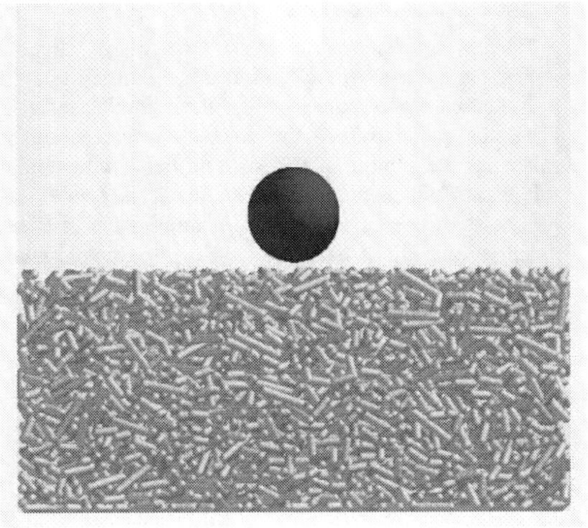

Fig. 2. The sketch map of simulation.

3 Numerical results

The first peak of impact force directly reflects the impact absorbing effect of granular materials. Therefore the compressive force on the base plate of the container is utilized to quantify the buffering capacity of non-convex particles as shown in Figure.3. The impact force of the base plate is obtained by subtracting the gravity of the granular bed from the total pressure on the base plate. The curve usually contains two parts, a large peak at beginning and a tail of small fluctuation wave continuing several milliseconds. The tail is not shown here. The results display that the impact load peaks decrease obviously with increase of particle aspect ratio from 0.125 to 0.792. Meanwhile the duration time grows with the increase of aspect ratio. There is no significant difference between granular beds in which aspect ratio is larger than 0.8. The negative value in the curve indicates that the particle system will oscillate in the process of impact. Then the oscillation results in a certain vibration

 DOI: 10.1051/epjconf/201714006013

of the base plate of the container so that sometimes the compressive force could go negative in Fig. 3. The simulation results are consistent with the reported experimental result[11].

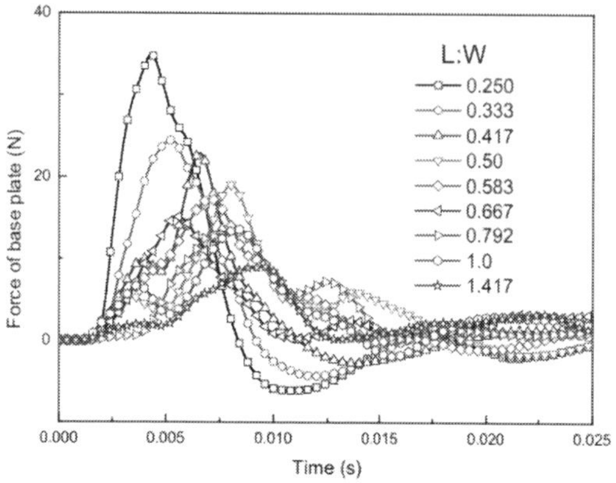

Fig. 3. Compressive force of base plate versus time under different granular aspect ratio.

The relationship between the compressive force peak and granular aspect ratio is plotted in Figure 4. Each line is obtained by average of five times simulation results. The dot line expresses the compressive force on base plate for the sample of spherical particles with diameter of 2.8mm which total mass of granular bed is equal to the mass of U-shaped particles. It implies that the buffer capacity of high aspect ratio U-shaped particles is better than spherical particle with equal mass.

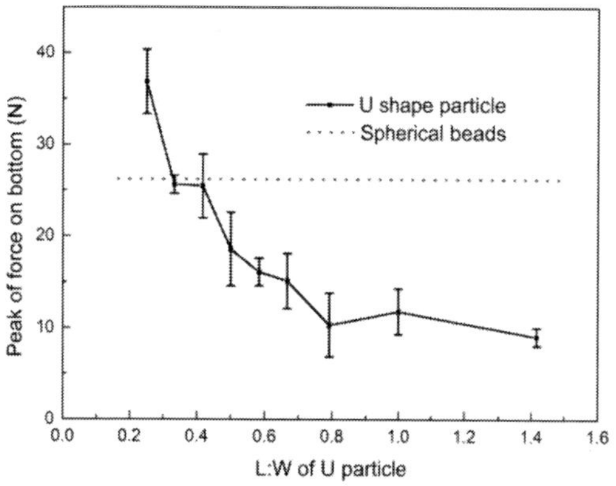

Fig. 4. The relationship between the compressive force peak and granular aspect ratio (the dot line represents compressive force peak of spherical beads)

4 Discussion

It is well known that sliding friction between particles is the main mechanism to dissipate energy of granular system. Prior study[12] has suggested non-spherical particles possess a higher effective friction, in comparison to spherical particles with the same inter-particle friction coefficient, due to suppression of rolling. On account of the entanglement between particles and the self-locking effect, the mechanism of the friction between non-convex particles is still not clear. We try to explain the influence of the non-convex particle shape on the cushioning properties of the particle system from the view of inter-friction.

Figure. 5 shows the variation of the average contact number during the impact process for different aspect ratio of U-shaped particles. The contact judgment algorithm of non-convex particles is the same as that of spherical particles. When two particles overlap, a contact is created. Once a contact has been detected, contact properties (such as particle prototype, size, and relative velocity) are passed to the contact model to calculate the force. It shows that for the low aspect ratio samples, the average contact number of granular bed reduces by nearly thirty percent in the duration of impact process; while for the high aspect ratio samples, the average contact number only reduces a small amount. Such behavior suggests that the entanglement between the non-convex particles and the inter-locking effect enable the system maintain relatively stable contact even suffering strong external disturbance.

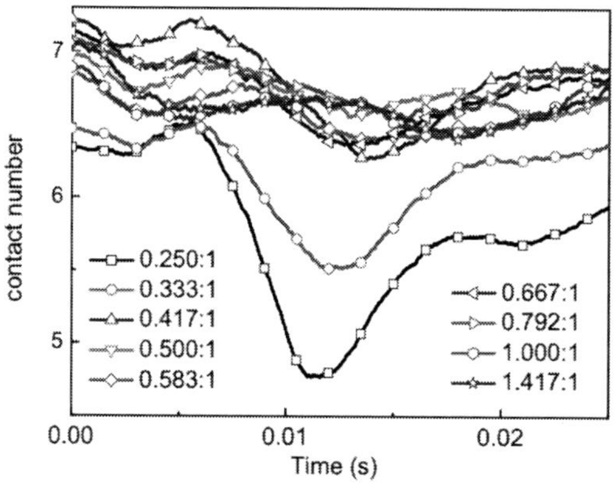

Fig. 5. Variation of the average coordination number during the impact process.

The entanglement of the non-convex particles and the inter-locking effect induce that the distribution of the internal non-uniform force chain is different from the spherical particles. The probability density function of the normal contact force between particles at the moment of collision is plotted in Fig.6. With the increasing of the aspect ratio of the sample, the probability of weak force is gradually decreasing, while the probability of the strong force appears gradually increasing. This feature of probability density function of normal force in the process of impact is similar. Such results imply that more stable contact and relative larger normal force lead to better energy dissipation ability for high aspect ratio non-convex particles.

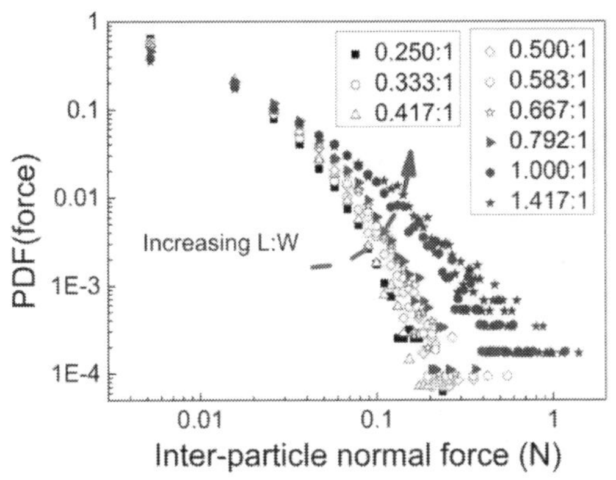

Fig. 6. Distribution of normal contact force for various aspect ratio particles

5 Conclusion

Granular material exhibits signally buffering capacity due to its dissipative nature. Numerical simulations show that the shape of non-convex particles notably influences the buffering capacity of granular bed. The buffering ability of high aspect ratio non-convex particles could be better than the spherical particles with equal mass. The shape of the non-convex particles significantly affects the contact number of granular system and probability density functions of inter-particle forces. We except this result could be helpful to study the ability of aleatory architectures[13] to withstand external disturbance.

We gratefully acknowledge the constructive suggestion of Prof. Shunying Ji (Dalian University of Technology) and Prof. Qicheng sun (Tsinghua University).

References

1. J.C. Ruiz-Suarez, Rep.Prog.Phys **76**, 066601 (2013).
2. K. Van der Weele, Contemporary Physics **49**, 157-178 (2008).
3. M.Z. Miskin, H.M. Jaeger, Soft Matter **10**, 3708-3715 (2014).
4. S. van den Wildenberg, Y. Yang, X. Jia, Granular Matter **17**, 419-426 (2015).
5. E. Brown, A. Nasto, A.G. Athanassiadis, H.M. Jaeger, Phys.Rev.Lett **108**, 108302 (2012).
6. N. Gravish, S.V. Franklin, D.L. Hu, D.I. Goldman, Phys.Rev.Lett **108**, 208001 (2012).
7. A. Seguin, Y. Bertho, P. Gondret.J.Crassous.EPL **88**, 44002 (2009).
8. M. Nishida, Y. Tanaka, Granular Matter **12**, 357-368 (2010).
9. J. Huang, S. Xu, S. Hu, INT.J.IMPACT.ENG **59**, 1-10 (2013).
10. M. Omidvar, M. Iskander, S. Bless, INT.J.IMPACT. ENG **93**, 211-221 (2016).
11. S. Ji, X. Chen, P. Li,Y. Yan, *Powders and Grains 2013*, 401-404 (2013).
12. H. Chen, Y.L. Liu, X.Q. Zhao, Y.G. Xiao,Y. Liu, Powder Technology **283**, 607-617 (2015).
13. S. Kellerl, H.M. Jaeger, Granular Matter **18**, 29(2016)

Discharge flow of granular media from silos with a lateral orifice and injection of air

Pascale Aussillous[1,⋆], *Yixian* Zhou[2,1], *Pierre* Ruyer[2], and *Pierre-Yves* Lagrée[3]

[1] *Aix-Marseille Univ., CNRS, IUSTI UMR 7343, 13013 Marseille, France*
[2] *Institut de Radioprotection et de Sûreté Nucléaire (IRSN), PSN-RES, SEMIA, LIMAR, Cadarache, St Paul-Lez-Durance, 13115, France*
[3] *Sorbonne Universités, UPMC Univ Paris 06, CNRS UMR 7190, Institut ∂'Alembert, F-75005 Paris, France*

Abstract. Few studies concern the prediction of the mass flow rate of a granular media discharged from a silo with a lateral orifice. However, this situation can have pratical interest considering a tank of granular material with a leak on its side. We studied experimentally the discharge of a vertical silo filled by spherical glass beads. We consider rectangular silos with a rectangular orifice. The impact of size, aspect ratio and position of the orifice and the effect of an additional air flow were studied. The measured parameters are the mass flow rate and the pressure along the silo, whereas the controlled parameters are the size of particles, and the flow rate of air. We identified two regimes of discharge according to the aspect ratio (of width to height) of the rectangular orifice. Increasing the air flow rate induces an increase of the granular media flow rate. Using a simple physical model to describe the grains and gas flow, we put in evidence the role played by the air pressure gradient at the outlet. Then we compared the experimental results with continuum Navier-Stokes simulations with the granular μ(I)-rheology. We showed that the continuum μ(I)-rheology describes well our discharge flow of granular media from silos, taking into account the effect of the position of the orifice as well as the coupling with the gas flow.

1 Introduction

Among the wide spectrum of granular flows, we focused on a rather specific configuration: a relatively narrow elongated vertical tube, with an opening located on its lateral wall, filled with spherical glass beads, discharges its contents in a constant pressure environment, a gas flowing throughout the tube. This case is of particular interest to understand conditions under which a set of nuclear solid fuel particles inside a typical PWR fuel rod, whose cladding would have failed under hypothetical accidental conditions, could disperse out of this rod. The fuel particles are generated from an initially cylindrical pellet (scale of 1cm) that can be fragmented (the smallest size being of the order of 10 μm) due to the irradiation process (burn-up of the fuel) or accidental conditions (large pressure and temperature variations inside the rod).

The flow rate of a granular medium from of a confined tank to an open space through a narrow orifice has clearly to be related to the classical hourglasses or hoppers flow rates, that have been studied since centuries. In those latter cases, the Beverloo relation [1], provides an universal scaling for the flow rate but requires some empirical constants to be applied to a specific configuration. Extension of the application range of this law to air-injection assisted [2–4], or inclined tanks have been already considered [5, 6]. The physical interpretation of the scaling is still an active research area: as an illustrative example, the widely known concept of a free-fall arch above the orifice has been recently invalidated as a correct idealization of the near orifice granular flow [7].

This bibliographical review motivates the need for a specific investigation of the granular flow of interest. This paper presents the main results of this study. First, we present the analysis of experiments we performed thanks to a dedicated device that simulates the main aspects of such a flow configuration. Then, we considered the potential of the model of the granular medium as an equivalent continuum material with a specific rheology to correctly reproduce the experimentally observed phenomena. For this purpose, we performed numerical simulations. Using the *Basilisk* program, we solved the Euler-Euler two-phase mass and momentum balance equations using, (i) a simplified incompressible Darcy flow model for the gas phase and, (ii) an effective viscosity formulation to model the so-called μ(I) rheology with uniform compacity for the granular phase.

2 Experimental study

2.1 Experimental device

The silo of granular materials

Several silos have been built for this study and we will focus on the results obtained in rectangular shape tanks as

⋆e-mail: pascale.aussillous@univ-amu.fr

DOI: 10.1051/epjconf/201714003052

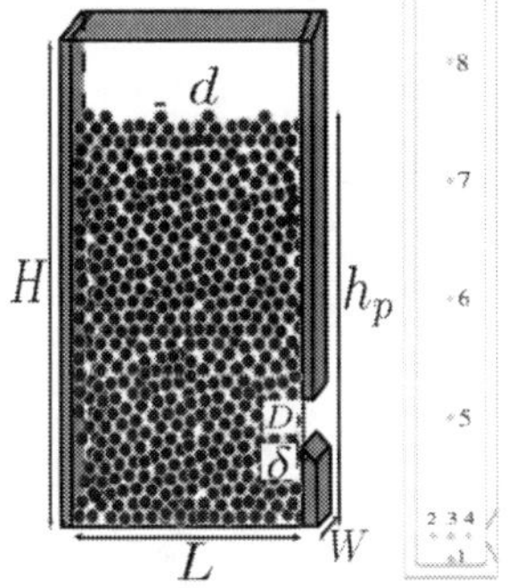

Figure 1. Schematic illustrations of the silo. Left hand side picture: main geometrical dimensions. The wall near the orifice, of thickness δ has been bevelled till no influence of δ on the flow rate can be observed. Right hand side picture: Pressure sensors locations.

Table 1. Dimension of the silo and size of the particles

D (mm)	W (mm)	d (μm)
[2.7, 5.4, 10, 15,; 20, 25, 30, 35]	[3.5, 5, 10, 20, 30, 40]	[124, 190, 375, 538, 762, 1129, 1347]

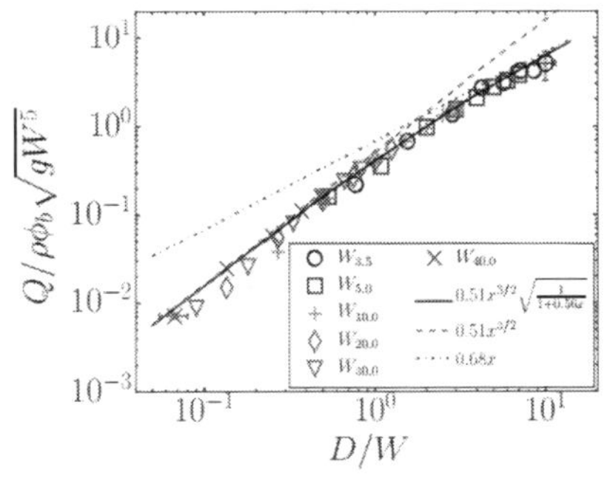

Figure 2. Experiments in the rectangular silo for a given particle diameter $d = 190\mu m$ and various silo widths, W, the solid line represents equation 1, $\sqrt{\kappa}c_D = 0.51$, $\sqrt{\kappa}c_W = 0.68$. Dashed, resp. dashed dotted lines, represents the asymptotics for $D \ll W$, resp. $W \ll D$.

illustrated in figure 1. The aperture on the lateral sidewall has a rectangular shape with a height D and a horizontal length W that corresponds to the width of the silo. The silo is initially filled with an height h_p of monodisperse in size d spherical glass beads of ρ_p density, the initial bulk volume fraction ϕ_b being estimated. The set of dimensions of the silo and of the particles size is given in table 1, the lateral extent, resp. the height, of the silo being fixed to L = 60mm, resp. H = 500mm. More details concerning the experimental device can be found in [8].

During the discharge of the silo, the beads are collected within a recipient, its mass being measured using an electronical balance (Mettler Toldeo 6002S) with an accuracy of 0.1g and a frequency of 20Hz. The flow rate is estimated from the time evolution of this mass and, in every case, a steady state regime is observed, the corresponding flow rate value being denoted Q (kg.s^{-1}) in the following.

The air injection system

For all the tests concerning the flow rate dependence to the geometry, the top of the silo is open to the atmosphere. In the tests related to the impact of an air flow on the discharge rate of granulars, this top is closed and the silo is airtight except the aperutre. An air injection system is then connected to the top of the rectangular silo. It provides a constant flow rate of air Q_{air} (m^3.s^{-1}) during the discharge. Pressure has been measured at different locations within the silo along the back wall, mainly along the median vertical axis and in the near orifice region, see figure 1. Similarly to the tests with open top silo, the granular flow rate is measured and a steady state value is observed. The experiments with an air flow have been performed only for a reduced set of geometries corresponding to the underlined D and W values in table 1 but for the whole set of d values.

2.2 Granular flow rate as a function of geometry

Let us consider the data collected for a given size of particles, say $d = 190\mu m$ and for the whole set of D and W values. Let us remark that, since, for this specific case, d is always at least one order of magnitude smaller than the dimensions D or W of the aperture, one can reasonnably neglect its impact on the flow rate as suggested both by the Beverloo law or the Janda model [9]. As a preliminary study, we considered the case of discharge flow rate of similar rectangular silos, the opening being at the bottom of the silo. In the latter case, the flow rate, say Q_b, follows a single Beverloo law $Q_b = c_D DW\sqrt{gD}$. For the case of a lateral opening of interest for this study, the dimensionless flow rate $Q/(\rho\phi_b\sqrt{gW^5})$ has been plotted for the whole set of D and W values on figure 2. It is worth pointing out that the data can not be reduced to a single power law function. Actually, two clearly distinct flow regimes can be distinguished according to the D/W aspect ratio of the aperture. For smallest value of this aspect ratio, *i.e.* when D is the smallest extension of the aperture, the flow rate scales as $\rho\phi_b DW\sqrt{gD}$. On the other side, when W is the smallest extension of the aperture, the flow rate scales as $\rho\phi_b DW\sqrt{gW}$. Therefore, we clearly identified two regimes according to the value of D/W, a behaviour that can be correctly modeled by the following relation that is illustrated on figure 2,

$$\frac{Q}{\rho\phi_b\sqrt{\kappa g W^5}} = c_D\left(\frac{D}{W}\right)^{3/2}\sqrt{\frac{1}{1+(c_D/c_W)^2 D/W}}, \quad (1)$$

where $0 \leq \kappa \leq 1$ corresponds to an empirical factor that reduces the driving force for the flow rate from g in the classical case of a bottom orifice to κg in the present case of a lateral orifice. For the larger particles size, we observed that the particles size effect can be well modeled by a dilation function that depends to d/D as proposed by Janda et *al.* [9]. We recovered the same behaviour for cylindrical silos as well. Additional results can be found in [10].

2.3 Granular flow rate as a function of air flow

The pressure space and time evolution is as follows. The pressure remains constant while the granular materials is

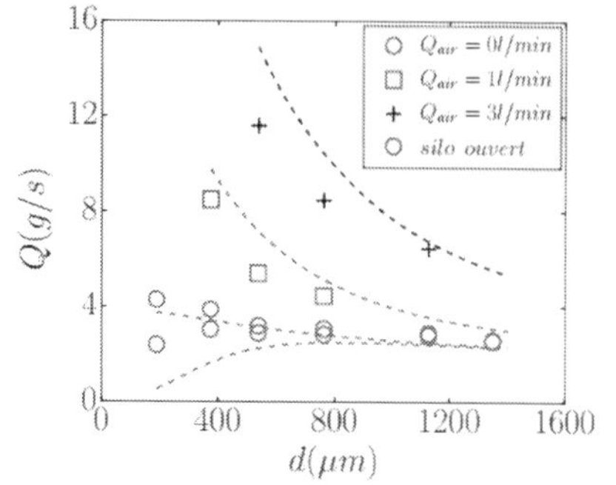

Figure 3. Rectangular silo with D = 10mm: steady-state flow rate of particles as a function of particles size for different air flow rates. The solid line represents equation 3.

above the sensor location. The pressure difference between two locations in the silo can be then successfully predicted by the Ergun model [11], for gas flow through porous media, the permeability coefficient depending on d and ϕ_b and considering the density ρ_f and the viscosity η of the air. As soon as the sensor is located above the granular materials, the pressure gradient within the air flow is negligible and its time evolution is governed by the Ergun predicted pressure difference between the top level of the granular materials and the aperture. It is worth pointing out that the pressure gradient is larger near the orifice, that can be simply related to the acceleration of the gas due to the reduction of the flow cross section of a factor D/L. The permeability in such a porous medium strongly depends on the particles size. Therefore, for the same air flow rate through the beads, the pressure gradient is larger for small particles.

The granular flow rate is affected by the gas flow rate and increases with it as illustrated in figure 3. In the specific case of no air flow rate at the top of the silo, the granular flow rate is lower than for the open top case when the particles size is small. Moreover, the granular flow rate that only slightly depends on the particles size in the open top silo case, becomes all the more sensitive to the air flow rate than the particles size is small. The air flow acts clearly as an additionnal driving force for the granular flow and its impact on it follows a similar tendency than the pressure gradient throughout the granular materials.

We observed that the granular flow rate was highly correlated with the pressure gradient in a region close to the aperture of the silo that acts as a driving force in addition to gravity. Let us recall that we deduced from the open top silo experiments that the driving force for a lateral aperture was scaling the flow rate as $\sqrt{\kappa g}$. Therefore, following the models proposed in [2–4], we consider that the variation of the granular flow due to air flow can be modeled as

$$Q = Q_0 \left(1 + \frac{1}{\kappa g \phi_b \rho_p} \frac{\partial p^f}{\partial x}|_{x=x_o}\right)^{1/2}, \qquad (2)$$

where Q_0 is the open top silo case flow rate and $\frac{\partial p^f}{\partial x}|_{x=x_o}$ the pressure gradient near the orifice, that can be estimated thanks to the viscous $\beta_v = 150\phi_b^2/(1-\phi_b)^3 d^2$ and inertial $\beta_i = 1.75\phi_b/(1-\phi_b)^3 d^2$ terms of the permeability as predicted by Ergun's model. This leads to a rather simple analytical model to relate the non-dimensional granular flow rate $\bar{Q} \hat{=} \frac{Q}{Q_0}$ to the non-dimensional air flow rate $\tilde{Q} = \frac{\phi_b \rho_p Q_{air}}{Q_0}$, given the bulk volume fraction ϕ_b and the particles size d

$$\bar{Q} = \frac{\left(-2N_i\tilde{Q} - N_v\right) + \sqrt{\Delta}}{2\left(1 - N_i\right)}$$

$$\text{with } \Delta = N_v^2 + 4\left(1 + N_v\tilde{Q}\right) + 4N_v\left(\tilde{Q}^2 - 1\right),$$

$$N_v = \frac{\eta\beta_v(\phi_b, d)Q_0}{\phi_b^2\rho_p^2\kappa g D W}, \text{ and } N_i = \frac{\rho_f d\beta_i(\phi_b, d)Q_0^2}{\phi_b^3\rho_p^3\kappa g D^2 W^2} \qquad (3)$$

The predicted variation of this analytical model is drawn with dashed lines on the bottom graph of figure 3. It is worth pointing out that, similarly to the cases with open top silo, we observed a deviation from the predicted results when the number of beads across the aperture (D/d) is low. Once again, this behaviour can be modeled thanks to a dilation of the granular flow thanks to a geometrical function as proposed by Janda: in presence of air flow, this dilation also impacts the permeability of the granular phase in the near orifice region, an effect that can then be taken into account in our model by considering the volumetric fraction near the orifice, say ϕ_o rather than in the bulk in the hereinabove model. More detailed informations can be found in [8].

3 Numerical simulations

3.1 Flow modeling and system of governing equations

We use the classical two-phase equations (see e.g. Jackson [12] or [13]) for gaz and granular phases (with the volume average velocity ponderation of the particules and fluid velocities). This gives for each phase mass and momentum balance equations. Flows are assumed to be incompressible and steady, inertia is neglected in the momentum balance equation of the air. This leads to a Darcy flow of air through the beads. The associated pressure gradient couples both momentum balance equations. The effective viscosity formulation of the μ(I) rheological law with uniform volume fraction ϕ_b is used for the momentum balance equation of the grains. To solve these continuous coupled equations we use the code *Basilisk* an extension of the code used in [14] for dry granular flows. The rectangular silo is simulated in a $2D$ computational domain, the silo being a $L \times H$ rectangular domain with a constant granular pressure boundary condition on the lateral orifice of size D. Steady air flow rate is imposed at the top of the silo.

3.2 Taking into account the wall friction

Let us first consider the case without air flow. To take into account the lateral friction, we average the momen-

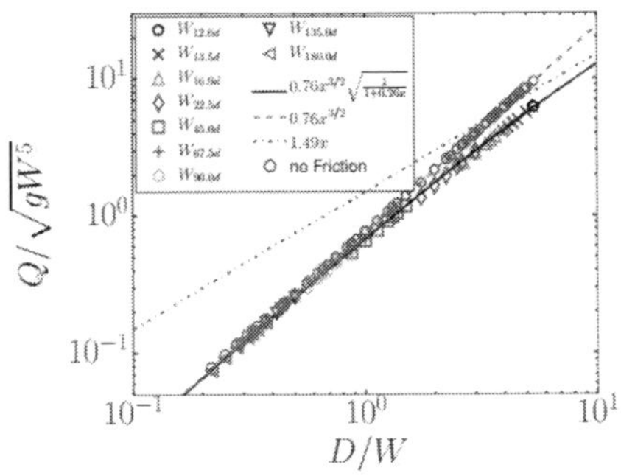

Figure 4. Numerical granular flow rate for various D and W values. Red circles correspond to numerical solutions of pure 2D flow, other symbols (corresponding to a given value for W) correspond to the averaged equations with the wall friction term. Plain and dashed lines correspond to the scaling laws.

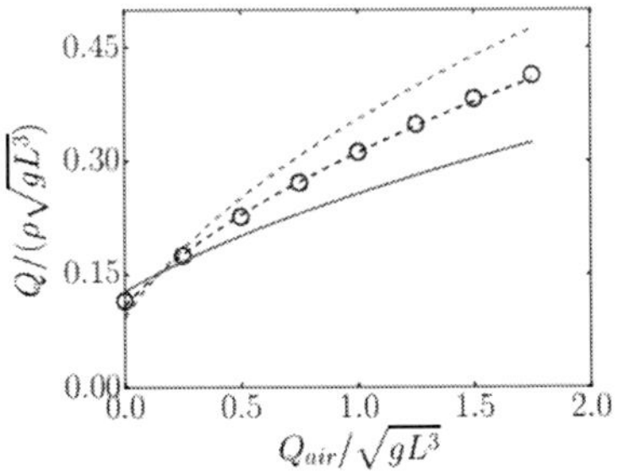

Figure 5. $Q/\rho\sqrt{gL^3}$ as a function of $Q_{air}/\sqrt{gL^3}$ for ($D = 0.40625L$, $W = 0.25L$). Symbols correspond to numerical simulation. The analytical fitting use equation 3 with (–) $A = 1$ and $\kappa = 1$, (– –) $A = 1$ and $\kappa = 0.36$, (– –) $A = 0.67$ and $\kappa = 0.36$.

tum equation across the width of the silo (in the Hele-Shaw spirit, [15]). This adds an averaged additional force from the front and back sidewalls in the momentum equation. This force reads $(2\mu_w/W)\mathbf{u}/||\mathbf{u}||$, this term being 0 for a 2D case ($W = \infty$). On figure 4, the granular flow rate for various D and W values has been plotted. In the $W = \infty$ case, we recover the scaling of the flow rate in $D^{3/2}$, whereas for the larger D/W values, the flow rate deviates from this law and tends to the experimentally observed scaling in $DW^{1/2}$. This result, further discussed in [10] clearly indicates that the wall friction plays a major role in the discharge flow rate from a silo having such an elongated geometry. It also extend the range of applicability of the μ(I) law. As a perspective, we will consider taking into account the dilation effect observed for small number of beads in the aperture D/d values by solving the additional ϕ(I) rheological law.

3.3 Air flow related additional driving force

The symbols in the curve of figure 5 illustrates that, as observed experimentally, the numerical simulation predicts an increase discharge flow rate when air flows through the aperture. This numerical behaviour has been correctly fitted by our simple analytical model (with $N_i = 0$ for this Darcy flow) considering the reduced κg driving force for the open-top case as well as a factor $A \leq 1$ in front of the fluid pressure gradient evaluated numerically at the orifice location. This latter factor is interpreted as follows: the granular flow is accelerated by the pressure gradient over a domain of D extent upstream the orifice location, rather than by the pressure gradient at the orifice location, where it is maximal ($A = 1$). The ability to fit the numerical result with the model used to interpret the experimental data is an additional evidence of the actual coupling between granular and air flows during such a discharge.

4 Conclusions

Using experiments and continuous simulation with a granular rheology we have proposed a model which predicts the flow rate of a granular media discharging from silos with a lateral orifice in the presence of an air flow rate.

References

[1] W.A. Beverloo, H.A. Leniger, J.V. de Velde, Chem. Eng. Sci. **15**, 260 (1961)
[2] J. de Jong, Q.E.J.J.M. Hoelen, Powder Technol. **12**, 201 (1975)
[3] J. de Jong, Powder Technol. **3**, 279 (1969)
[4] R. Nedderman, U. Tüzün, R. Thorpe, Powder Technology **35**, 69 (1983)
[5] A. Medina, D. Cabrera, A. López-Villa, M. Pliego, Powder Technology **253**, 270 (2014)
[6] D. Serrano, A. Medina, G.R. Chavarria, M. Pliego, J. Klapp, Powder Technology **286**, 438 (2015)
[7] S. Rubio-Largo, A. Janda, D. Maza, I. Zuriguel, R. Hidalgo, Phys. Rev. Lett. **114**, 238002 (2015)
[8] Y. Zhou, Ph.D. thesis, Aix-Marseille Univ. (2016)
[9] A. Janda, I. Zuriguel, D. Maza, Phys. Rev. Lett. **108**, 248001 (2012)
[10] Y. Zhou, P.Y. Lagrée, S. Popinet, P. Ruyer, P. Aussillous, submitted to J. Fluid Mech.
[11] S. Ergun, Chemical Engineering Progress **48(2)**, 89 (1952)
[12] R. Jackson, *The Dynamics of Fluidized Particles*, Cambridge Monographs on Mechanics (Cambridge University Press, 2000), ISBN 9780521781220
[13] P. Aussillous, J. Chauchat, M. Pailha, M. Médale, E. Guazzelli, J. Fluid Mech. **736**, 594–615 (2013)
[14] L. Staron, P.Y. Lagrée, S. Popinet, The European Physical Journal E **37**, 1 (2014)
[15] P. Jop, Y. Forterre, O. Pouliquen, J. Fluid Mech. **541**, 167–192 (2005)

 DOI: 10.1051/epjconf/201714007010

Effect of contact anisotropy on the crushing strength of aggregates

Aurélien Neveu[1,*], *Riccardo* Artoni[1,**], *Yannick* Descantes[1,***], and *Patrick* Richard[1,****]

[1]*LUNAM Université, IFSTTAR, MAST, GPEM, F-44340 Bouguenais, France*

Abstract. This work deals with the effect of the contact orientation distribution on the crushing of granular materials. At first, a simple drop weight experiment was designed in order to study the effect of the location of three contact edges on the fracture pattern and the strength of a model cylindrical particle. The sample was placed on two bottom contact edges symmetrically distributed with reference to the vertical symmetry plane of the particle and subjected to an impact at the top. Angle α between the plane connecting a bottom contact edge to the centerline of the cylinder and a vertical plane was varied. The energy required to fracture the particle was shown to be an increasing function of angle α. Peculiar crack patterns were also observed. Then, we present a discrete model of grain fracture [1] and employ it for a numerical analysis of the problem. The cylindrical particle is discretized by means of a space filling Voronoi tessellation, and submitted to a compression test for different values of angle α. In agreement with experiments, simulations predict a strong effect of the contact orientation on the strength of the particle as well as similar fracture patterns. The proposed numerical model is therefore an interesting tool for understanding fracture mechanisms with the purpose of optimizing the crushing process.

1 Introduction

Studying the fragmentation of cohesive materials is of importance for a wide range of natural and industrial processes. In the production of aggregates, rock blocks are crushed and the resulting fragments are required to meet high standards namely in terms of size and shape. Successive crushing steps are usually carried out to achieve the requested aggregate characteristics, leading to a waste of good quality raw materials and a high energy cost that could both be mitigated upon improving the crushing efficiency.

An important aspect of the crushing process is that loads are applied on a collection of particles, and therefore stresses are transmitted through particle contacts. In this perspective, some authors have proposed failure criteria for tensile [2] and plane shear [3] fracture modes taking into account the effect of the coordination number. However, the effect of anisotropy of the contact points location on the strength of the particle has not been deeply investigated.

In order to understand fracture mechanisms with the purpose of optimizing the crushing process, discrete numerical simulation is an important tool. In this work, we first examine from simple drop weight crushing tests how the contact orientation distribution on a cylindrical particle affects the particle strength and the fracture pattern. Then we employ a discrete model of cohesive material to the study of the effect of contact anisotropy on the shear strength. First the model is introduced, then numerical simulations of compression of cylindrical particles are presented.

*e-mail: aurelien.neveu@ifsttar.fr
**e-mail: riccardo.artoni@ifsttar.fr
***e-mail: yannick.descantes@ifsttar.fr
****e-mail: patrick.richard@ifsttar.fr

2 Experiments: Drop weight tests

In order to gather information on the effect of contact distribution on particle fracture and strength, a simple experiment was designed. Cylindrical mortar samples (D = 50 mm, H = 25 mm) were tested using a drop weight setup which allows varying the location of contacts at the bottom (see Fig. 1a). A 2kg mass held by an electro-magnet is dropped from variable height. The shock energy transmitted to the sample is assessed from the dropping height whereas a high speed video camera records the sample behaviour. For different values of the angle α, the probability of breaking the sample is displayed in Figure 1b as a function of the input energy. 20 to 30 tests were performed for each energy value. The figure shows that the energy required to crush the particle is an increasing function of angle α, at least in the range considered. This is even clearer when looking at the inset in Figure 1b, where the interpolated shock energy required to achieve 50% chance of breaking a sample is plotted as a function of angle α. The effect of α is very strong, given that the energy required for crushing at $\alpha = \pi/3$ is approximately four times larger than that at $\alpha = 0$.

The fracture pattern of the sample for two contact configurations ($\alpha = 0$ and $\alpha = \pi/6$) is also shown in Fig. 1c-d. It is clear that in the $\alpha = 0$ case, a classical fracture pat-

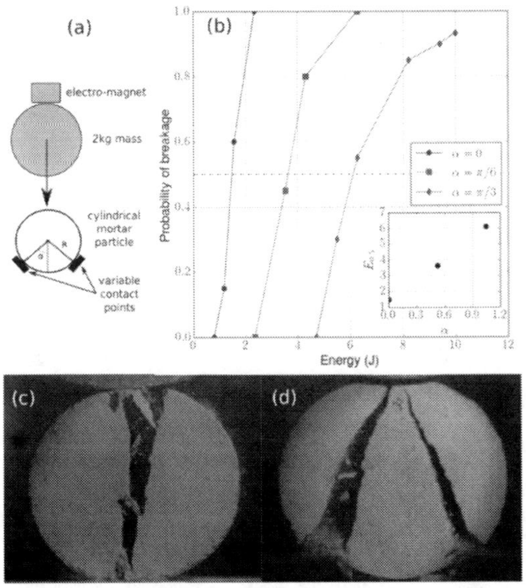

Figure 1. (a) Sketch of the experimental setup. A 2kg mass was dropped from an electro-magnet from a variable height on a cylindrical mortar sample; (b) probability of breakage of the sample as a function of the impact energy and the angle between base contact points. Fracture pattern of the sample from high speed video recordings for (c) $\alpha = 0$ and (d) $\alpha = \pi/6$.

tern is observed, with the development of a vertical crack between the two contact edges. On the other hand, when three contact edges are present, two cracks develop between the impact edge and the two bottom contact edges. Experiments prove that the particle strength and the fracture pattern are strongly influenced by the contacts configuration.

3 Cohesive interaction model

Numerical simulations using discrete Element Methods (DEM) have already been successfully used to describe the elastic behavior and rupture mechanism of a rock piece [4–6]. In these methods, a grain is represented by a collection of particles with contact bonds to model inter-particles cohesion. In a recent work [1], we introduced a cohesive interaction model for Discrete Element Methods which allows to model cohesion between contacting and non-contacting particles, suitable for any kind of particle shape. The idea behind the model is to treat independently contacts and cohesive interactions. Contacts may be modelled in different ways, for example using unilateral contact laws. Cohesion is set inside the modelled material by applying forces which oppose relative motion between particles. This relative displacement is computed between two cohesion points, each located on one of the two interacting particles. Cohesive interactions consist of a spring and a damper for both normal and tangential components. As cohesion is treated separately from contact, this allows to apply cohesive forces even if particles are not touching each other, whatever their shape. More than one cohesive interaction may be set between interacting particles to resist rotation, bending and torsion. In this work we study the behavior of a solid grain discretized by a space-filling tessellation (Voronoi). For the sake of simplicity, only one cohesive interaction is set between the centers of the Voronoi cells which share a face. The expression of the cohesive force in the reference frame of the cohesive interaction plane writes:

$$\boldsymbol{F} = F_n \cdot \boldsymbol{n} + F_t \cdot \boldsymbol{t} + F_s \cdot \boldsymbol{s}, \qquad (1)$$

with the normal (n) and tangential (t, s) components given by:

$$F_i = -k_i^s \Delta_i - k_i^v U_i, \text{ with } i = n, t, s \qquad (2)$$

where the Δ_i are the components of the relative displacements in the cohesive interaction plane reference frame, and the k_i^s are the corresponding stiffnesses. Energy dissipation is accounted for through viscous damping: U_i denotes the relative velocity of the interaction points along direction i and k_i^v the corresponding damping coefficients.

In order to determine the stiffnesses, an analogy is made with the continuum elastic modulus of a solid. Thus, the normal spring stiffness k_n^s is derived from the Young's modulus E_μ of an elastic body as :

$$k_n^s = \frac{E_\mu S}{l_0}, \qquad (3)$$

where S is the cohesive interaction surface, taken here equal to the area of the face shared by the Voronoi cells in contact. The initial length of a cohesive contact l_0 is the distance between the two cohesion points when the cohesive interactions are set. Since the cohesive interaction surface S and length l_0 differ for each pair of particles, a microscopic stiffness heterogeneity is thus introduced. The tangential spring stiffness is determined from the shear modulus, which corresponds to the shear stress to shear strain ratio. As it was shown in [1], the ratio of the normal to tangential stiffnesses may be expressed as follows:

$$\frac{k_n^s}{k_t^s} = \frac{k_n^s}{k_s^s} = 2(1 + \nu_\mu). \qquad (4)$$

The damping coefficients k_i^v are calculated to obtain a critical damping which prevents oscillations at contact. Cohesive interaction may break in tension or shear when the relative displacement reaches one of the following threshold values:

$$\Delta_n^{max} = \frac{l_0}{E_\mu} . \sigma_r^n, \qquad (5)$$

$$\Delta_{t,s}^{max} = \frac{2(1 + \nu_\mu)\, l_0}{E_\mu} . \sigma_r^t, \qquad (6)$$

with σ_r^n and σ_r^t the microscopic strength in tension and shear which are input parameters of the model, taken identical for all the interactions. The threshold displacements

 DOI: 10.1051/epjconf/201714007010

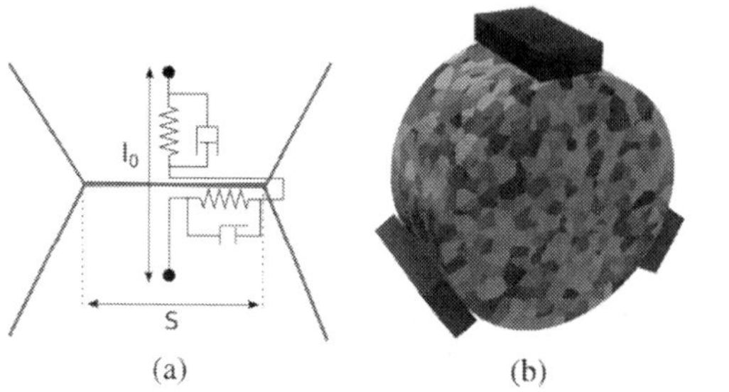

Figure 2. (a) 2D sketch of the cohesive interaction set between the centers of two Voronoi cells, with a spring-dashpot system along the normal and tangential components, (b) face view of the 3d Voronoi tessellation generated with Neper [7] (different colors for visualization purposes only).

have been chosen identical in tension and shear, which implies that:

$$\sigma_r^t = \frac{\sigma_r^n}{2(1+\nu_\mu)} \tag{7}$$

In the following, the rupture threshold will be referred as σ_r, with $\sigma_r^n = \sigma_r$ and σ_r^t given by Eq. (7).

4 Numerical simulations

The model described in the previous section has been implemented in the Contact Dynamics framework LMGC90 [8]. It is applied here to study the crushing of a cylindrical particle subjected to a three-edges loading.

The simulated particle is a cylinder of diameter d and thickness $0.5d$, which was discretized by a 3d Voronoi tessellation built by means of the Neper software [7]. Each cell of the tessellation was imported as a convex polyhedron for the simulation, for a total of 5000 polyhedra. An example of a sample generated from a Voronoi tessellation is shown in Fig. 2b.

Three walls are placed in contact with the sample. Two fixed walls are positioned symmetrically at the bottom, with their normal pointing towards the centerline of the cylinder; at the top, a wall is moved at constant velocity to apply the load on the body. The velocity is chosen of the same order of magnitude as in the impact test experiments. As in the experiments, the effect of angle α value on particle strength is investigated. Angle α (as defined in Fig. 1a) is varied between 0 and $\pi/3$. 6 repetitions were made for each angle. Simulations were performed without gravity. The stress scale is given by the microscopic Young's modulus E_μ, the mass scale by the mass of the cylinder m, the length scale by the cylinder diameter d, and the time scale by $\sqrt{\frac{m}{E_\mu d}}$. Several parameters were varied in the present study; the results discussed here were obtained for $\nu_\mu = 0.25$, $\sigma_r = 10^{-3}$, and a friction coefficient $\mu = 0.3$ between polyhedra.

The force-displacement curves obtained from the simulations are plotted in Fig. 3a. We can see that each curve displays a linear, elastic part, then reaches a maximum and decreases gently, suggesting the occurence of progressive damage of the grain. This behavior is confirmed by the evolution of the number of broken cohesive interactions (Fig. 3b): for small displacements no breakage of the cohesive interactions occurs, then the sample is progressively damaged.

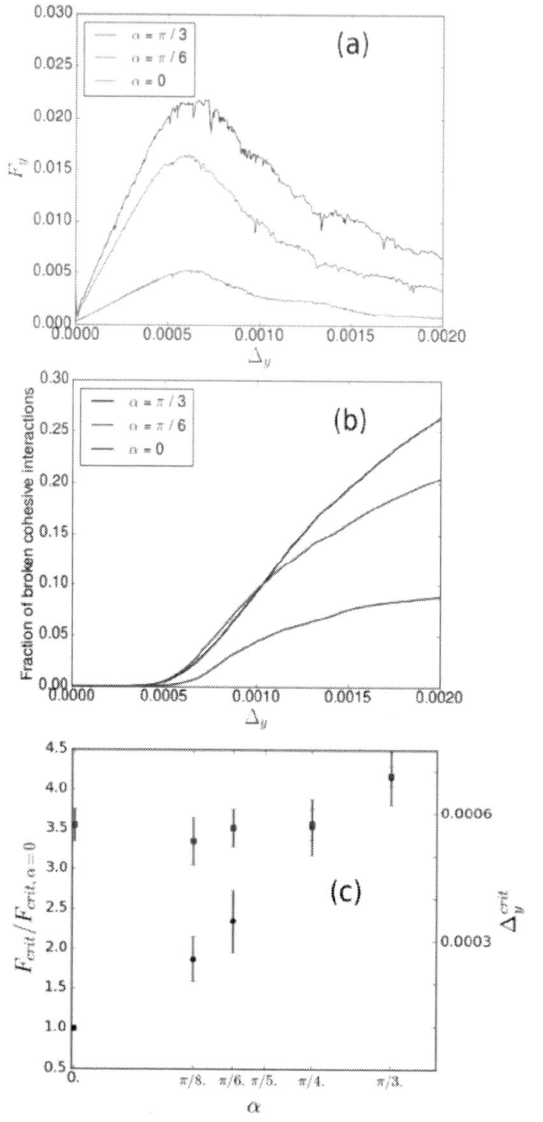

Figure 3. (a) Force-displacement curves from numerical simulations for three values of angle α; (b) fraction of broken cohesive interactions as a function of top wall displacement; (c) ratio between force maximum for a given value of α and force maximum for $\alpha = 0$ (•), critical displacement at force maximum (■).

The contact distribution strongly influences the slope of the force-displacement curve as well as the maximum of the force. The effect of angle α on the maximum of the force is shown in detail in Figure 3c, where the force is normalized by its value for $\alpha = 0$. The particle strength

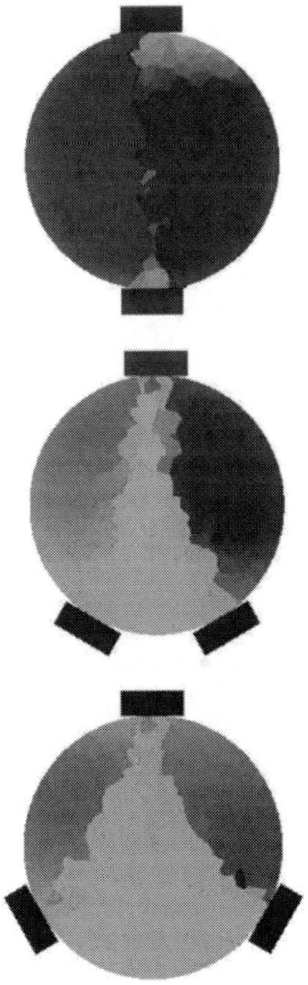

Figure 4. Fracture pattern for the numerical simulations for three values of angle α : from top to bottom, $\alpha = 0, \pi/6, \pi/3$. Colors correspond to the total displacement in the horizontal direction.

seems to increase fairly linearly with α, and the maximum force at $\alpha = \pi/3$ is approximately four times its value at $\alpha = 0$. This is in good agreement with the experimental data on impact crushing (see inset of Fig.2). By contrast, the displacements corresponding to the force maxima are similar, possibly because displacements in our Voronoi tessellation made of non-deformable particles concentrate in the fractures, hence they scale with the maximum displacement permitted in each cohesive bond crossed by the fracture.

Figure 4 displays the fracture patterns obtained for three values of angle α. As in the experiments, for $\alpha = 0$ a vertical crack is formed, and for three contact edges two cracks form between the impact and bottom contact edges.

5 Conclusions

The effect of the contact orientation distribution on the crushing of granular materials was studied by means of experimental and numerical tools. At first, a simple dynamic experiment was presented to highlight the effect of the location of three contact edges on the fracture pattern and the strength of a model cylindrical grain. The sample was placed on two bottom contact edges and received an impact by a falling weight at the top. The energy required to fracture the particle was shown to be an increasing function of the angle between the plane connecting a bottom contact edge to the centerline of the cylinder and a vertical plane. Peculiar crack patterns were also observed, connecting the impact edge to the other contact edges. A discrete model of grain fracture [1] was then employed for a numerical analysis of the effect of contact anisotropy. The cylindrical grain was discretized by means of a space filling Voronoi tessellation, and submitted to a compression test for different values of angle α. In agreement with experiments, simulations predict a strong effect of the contact orientation on the strength of the particle as well as similar fracture patterns. In particular, a linear reversible elastic behavior was found for small displacements of the top boundary, and a progressive damage was experienced for larger displacements. The proposed numerical model appears therefore a promising tool for understanding fracture mechanisms at microscale with the purpose of optimizing the crushing process. Future work will deal with more complex grain shapes and configuration of contacts.

References

[1] A. Neveu, R. Artoni, P. Richard, Y. Descantes, Journal of the Mechanics and Physics of Solids **95**, 308 (2016)

[2] O. Tsoungui, D. Vallet, J.C. Charmet, Powder Technology **105**, 190 (1999)

[3] O. Ben-Nun, I. Einav, Philosophical Transactions of the Royal Society of London A: Mathematical, Physical and Engineering Sciences **368**, 231 (2009)

[4] D. Potyondy, P. Cundall, International Journal of Rock Mechanics and Mining Sciences **41**, 1329 (2004)

[5] N. Weerasekara, M. Powell, P. Cleary, L. Tavares, M. Evertsson, R. Morrison, J. Quist, R. Carvalho, Powder Technology **248**, 3 (2013), discrete Element Modelling

[6] D. André, I. Iordanoff, J. luc Charles, J. Néauport, Computer Methods in Applied Mechanics and Engineering **213–216**, 113 (2012)

[7] R. Quey, P. Dawson, F. Barbe, Computer Methods in Applied Mechanics and Engineering **200**, 1729 (2011)

[8] M. Renouf, F. Dubois, P. Alart, Journal of Computational and Applied Mathematics **168**, 375 (2004)

EPJ Web of Conferences 140, 11008 (2017) DOI: 10.1051/epjconf/201714011008
Powders & Grains 2017

A hydrodynamic model for granular material flows including segregation effects

Dominik Gilberg[1,2,⋆], *Axel* Klar[2], and *Konrad* Steiner[1]

[1] *Fraunhofer ITWM, Germany*
[2] *TU Kaiserslautern, Germany*

Abstract. The simulation of granular flows including segregation effects in large industrial processes using particle methods is accurate, but very time-consuming. To overcome the long computation times a macroscopic model is a natural choice. Therefore, we couple a mixture theory based segregation model to a hydrodynamic model of Navier-Stokes-type, describing the flow behavior of the granular material. The granular flow model is a hybrid model derived from kinetic theory and a soil mechanical approach to cover the regime of fast dilute flow, as well as slow dense flow, where the density of the granular material is close to the maximum packing density. Originally, the segregation model has been formulated by Thornton and Gray for idealized avalanches. It is modified and adapted to be in the preferred form for the coupling. In the final coupled model the segregation process depends on the local state of the granular system. On the other hand, the granular system changes as differently mixed regions of the granular material differ i.e. in the packing density. For the modeling process the focus lies on dry granular material flows of two particle types differing only in size but can be easily extended to arbitrary granular mixtures of different particle size and density. To solve the coupled system a finite volume approach is used. To test the model the rotational mixing of small and large particles in a tumbler is simulated.

1 Introduction

Granular materials are the subject of intensive research for decades already, because they are a part of humans daily lives. Not only in geophysics *(dunes and avalanches)* or agriculture *(corn silos)*, but especially many different industrial processes deal with granular materials. The nature of the granular particles can strongly influence the behavior of the material and can lead to different phenomena. One of those phenomena is the segregation effect of particles due to different size, density or shape. Especially, in the food and pharmaceutical industry segregation leads to problems, as it counteracts the wanted mixing processes. This can worsen the quality of the products. Therefore, it is helpful to find a model that can describe the flow behavior of the granular material and additionally considers the effect of segregation.

This can be done using the discrete element method (DEM). But due to the long computation times of the DEM for large particle systems, a macroscopic model is a natural choice to simulate large industrial processes. Such continuum models already have been found to describe the flow behavior of granular materials, like the hydrodynamic model of Latz and Schmidt [1], which we will present and modify in the next section. Unfortunately, all these models need to be fitted to experimental data and are not able to additionally describe the segregation effect. Also there are promising continuum models that describe the effect of size-segregation. One of the first models of this kind was proposed by Bridgwater, Foo and Stephens in 1985 [2]. Later on models of similar structure were derived by Savage and Lun [3] or Gray and Thornton [4]. Larcher and Jenkins [5] extended kinetic theory for binary mixtures of nearly elastic spheres to predict the evolution of the concentration fractions of the two species. The modeling steps performed by Gray and Thornton are the basis of the segregation model we present in the upcoming section. In the last years extensions to their model were derived including size and density segregation [6],[7], but they are all modeled for given flow fields of idealized avalanches.

In the next section we present the mathematical models for the flow behavior and the segregation. We combine them to derive a model that shows the segregation effect in quite arbitrary flow fields. In Section 3 we present first simulation results gained by the model.

2 The mathematical models

The mathematical model consists of a hydrodynamic model of Navier-Stokes-Type describing the behavior of granular flows, and of a segregation equation, based on mixture theory. It is assumed that the granular mixture consists of small and large particles with equal material densities $\rho^{s*} = \rho^{l*} = const.$ In mixture theory one assumes that each constituent ν is present everywhere. Its percentage of the local volume is given by the volume fraction ϕ^{ν}.

⋆e-mail: dominik.gilberg@itwm.fraunhofer.de

Hence, in every point it holds

$$\phi^s + \phi^l + \phi^a = 1, \tag{1}$$

where ϕ^s, ϕ^l is the volume fraction of small and large particles, respectively. The interstitial pore space filled with air is denoted by ϕ^a. Further, the volume fraction of the whole granular material is given by

$$\Phi = \phi^s + \phi^l. \tag{2}$$

For the upcoming modeling process we do not ignore the volume occupied by air, as particles are not able to fill the whole space due to their shape. Nevertheless, we assume that the air has a passive role, where the pressure, and so the interaction on the particles is negligible small.

Classically, equations are formulated in terms of partial densities. They are given by a linear volume fraction scaling

$$\rho^v = \phi^v \rho^{v*}, \tag{3}$$

that equally holds for the granular mixture

$$\rho = \Phi\rho^*. \tag{4}$$

To avoid further variable changes all equations are directly given in terms of the respective volume fractions, as all upcoming terms are linear in the density, and $\rho^* = \rho^{s*} = \rho^{l*} = const.$

The final model includes solely an equation for the behavior of the small particles as the large particles are given through the saturation condition (2). To simplify the notation in the final model, we write ϕ instead of ϕ^s.

2.1 The granular flow model

The origin of the granular flow model is a set of hydrodynamic equations derived by Latz and Schmidt [1]. To reproduce the dilute flow regimes the model is based on kinetic theory. As granular materials come to rest they behave differently than in the fast dilute flow regimes. Kinetic theory is not able to reproduce this behavior correctly any more. Therefore, they modified the model using a theory from soil mechanics to extend its applicability to arresting granular flows. Their final set of hydrodynamic equations is a hybrid model of kinetic theory and the soil mechanical approach. We use a similar model for the granular flow as derived in [1]. As the original model does not take into account particles of different size, we apply some further modifications.

The granular system is given by the mass balance (6) and the momentum balance (7), that we solve for the granular volume fraction Φ and the mean granular velocity $\mathbf{u}$. One should keep in mind that the granular volume fraction Φ is bounded by a maximum volume fraction

$$0 \leq \Phi \leq \Phi^{max}(\phi) < 1, \tag{5}$$

due to the interstitial pore space between the particles. For particles of equal size, as in the original model, Φ_{max} is a constant. But for a system of small and large particles it is a function that depends on the local amount of small and large particles.

In addition to equation (6) and (7) we solve equation (8) for the granular temperature T, that accounts for the mean fluctuating kinetic energy in the system.

$$\partial_t \Phi + \nabla \cdot (\Phi \mathbf{u}) = 0 \tag{6}$$

$$\partial_t (\Phi \mathbf{u}) + \nabla \cdot (\Phi \mathbf{u} \otimes \mathbf{u}) = \nabla \cdot \bar{\sigma} + \Phi \mathbf{g}, \tag{7}$$

$$\partial_t (\Phi T) + \nabla \cdot (\Phi T \mathbf{u}) = \frac{3}{2}(\eta\kappa : \kappa - \nabla \cdot \mathbf{q}) - \varepsilon \Phi T \tag{8}$$

The stress tensor in equation (7) is given by

$$\bar{\sigma} = \eta\kappa - p\mathbf{I}, \quad \kappa = \nabla \mathbf{u}, \tag{9}$$

where η defines the granular viscosity and p the pressure. The gravitational acceleration is denoted by $\mathbf{g}$. The temperature equation (8) is similar to a heat equation. The Term $\eta\kappa : \kappa$ is called viscous heating and acts as a source term for the granular temperature. Whereas, the last term on the right hand side is a sink, which accounts for the loss of energy due to inelastic collisions. The parameter ε is the energy dissipation rate. If there are no forces acting on the granular material, a moving system will rapidly come to rest due to this term. The granular heat flux is given by

$$\mathbf{q} = -\lambda \nabla T, \tag{10}$$

where λ denotes the temperature viscosity.

Equations of state can be derived from kinetic theory to close the system [8]. The kinetic expressions are denoted by the subscript k. Additionally, so-called yield expressions are derived, that are solely active in the dense flow regime $\Phi > \Phi_0$. Accordingly, we use the subscript y. In the resulting hybrid model the pressure p can be split into a kinetic and a yield part

$$p = p_k + p_y \tag{11}$$

with

$$p_k = \Phi T g(\Phi, \phi) \tag{12}$$

$$p_y = \Theta(\Phi - \Phi_0) T_0 (\Phi - \Phi_0) g(\Phi, \phi), \tag{13}$$

where $\Theta(\cdot)$ is the Heaviside step function and

$$g(\Phi, \phi) := \left(1 - \frac{\Phi}{\Phi^{max}(\phi)}\right)^{-1} \tag{14}$$

the compressibility factor. If the local amount of granular material is in a physically valid state $\Phi < \Phi^{max}(\phi)$, the compressibility factor helps to ensure that the solution of the system stays valid. Similar to the pressure all transport coefficients in the system have a kinetic and an yield expression.

$$\eta = \eta_k + \eta_y, \quad \eta_k = \eta_0 \sqrt{T}\Phi g(\Phi, \phi), \quad \eta_y = \eta_k \frac{p_y}{p_k} \tag{15}$$

$$\varepsilon = \varepsilon_k + \varepsilon_y, \quad \varepsilon_k = \varepsilon_0 \sqrt{T} g(\Phi, \phi), \quad \varepsilon_y = \varepsilon_k \frac{p_y}{p_k} \tag{16}$$

$$\lambda = \lambda_k + \lambda_y, \quad \lambda_k = \lambda_0 \sqrt{T}\Phi g(\Phi, \phi), \quad \lambda_y = \lambda_k \frac{p_y}{p_k}. \tag{17}$$

In the dilute regime the yield pressure p_Y is zero and so the yield part of each transport coefficient is zero, too. In this case all coefficients reduce to their kinetic expression. In the dense flow regime p_Y is non zero. All parameters subscribed with 0 are constants, that either have to be looked up in literature or validated by experiments. A more detailed derivation of the model is given in [1].

2.2 The segregation equation

We assume that each constituent of the mixture can be described as a continuum. Additionally, we exclude agglomeration effects or breakage of particles. Therefore, they must fulfill the following mass and momentum balances:

$$\partial_t \phi^\nu + \nabla \cdot (\phi^\nu \mathbf{u}^\nu) = 0 \tag{18}$$

$$\partial_t(\phi^\nu \mathbf{u}^\nu) + \nabla \cdot (\phi^\nu \mathbf{u}^\nu \otimes \mathbf{u}^\nu) = \nabla \cdot \bar{\sigma}^\nu + \phi^\nu \mathbf{g} + \boldsymbol{\beta}^\nu. \tag{19}$$

The interaction term $\boldsymbol{\beta}^\nu$ in equation (19) describes the force on particle phase ν due to the other particles.

To derive an equation for particle phase ν, we start from the mass balance (18). Introducing the relative velocity of constituent ν to the bulk

$$\mathbf{u}^{\nu m} = \mathbf{u}^\nu - \mathbf{u} \tag{20}$$

we obtain the following form for the mass balance:

$$\partial_t \phi^\nu + \nabla \cdot (\phi^\nu \mathbf{u}) = -\nabla \cdot (\phi^\nu \mathbf{u}^{\nu m}). \tag{21}$$

The left hand side of equation (21) describes the transport with the bulk velocity $\mathbf{u}$, which is given by the granular flow model. The right hand side represents the segregation as it is a relative motion to the bulk.

Segregation happens in the direction of the gravitational force. In flowing granular materials shear forces open small voids in the granular system. The small particles are more likely to fall into these empty spaces and percolate through the granular matrix. As a consequence the large particles end up at the top of the system.

To find an expression for the relative velocity $\mathbf{u}^{\nu m}$ we follow the idea of Gray and Thornton [4] and introduce a pressure scaling

$$p^\nu = f^\nu p \quad \forall \nu. \tag{22}$$

While the small particles percolate through the granular system they carry less of the overburden pressure than the large ones. Hence, f^ν is the proportion of the load carried by phase ν, that must fulfill

$$f^s + f^l = 1. \tag{23}$$

We choose the load functions in a similar way to [4]

$$f^\nu = \frac{1}{\Phi}(\phi^\nu \pm \frac{B}{\Phi}\phi^s \phi^l), \tag{24}$$

where B is a non-dimensional constant. The equation with " + " stands for large particles and with " − " for small particles. The scalings by Φ are necessary to fulfill (23) and ensure that the effective segregation velocity is independent of Φ, as we do not work with volume fractions relative to the granular volume.

For the derivation of the relative velocity we look at the momentum balance (19), as it contains the information of pressure and particle interactions. The expression of the interaction force $\boldsymbol{\beta}^\nu$ must fulfill Newtons third law, which reduces to

$$\boldsymbol{\beta}^s + \boldsymbol{\beta}^l = 0, \tag{25}$$

as the air phase has no influence on the granular system. The interaction force $\boldsymbol{\beta}^\nu$ can be written as

$$\boldsymbol{\beta}^\nu = p \nabla f^\nu - \phi^\nu \frac{\mu}{\dot{\gamma}} \underbrace{(\mathbf{u}^\nu - \mathbf{u})}_{\mathbf{u}^{\nu m}} \quad \forall \nu. \tag{26}$$

The first term ensures that the segregation process is driven by intrinsic rather than partial pressure gradients [4]. The second term is a drag term of Darcy type as proposed in [9], where μ is the coefficient of inter particle drag. It is inversely proportional to the shear rate $\dot{\gamma}$. Consequently, the drag reduces for an increasing shear rate [6],[10].

The shear rate can be approximated by the granular temperature. In a steady state the equation for the granular temperature (8) can be written as

$$\frac{3}{2}\eta \dot{\gamma}^2 = \varepsilon \Phi T. \tag{27}$$

With the expressions (15) and (16) the shear rate is proportional to the square root of the granular temperature

$$\dot{\gamma} = \sqrt{\frac{2}{3}\frac{\varepsilon_0}{\eta_0} T}. \tag{28}$$

The expression for the relative velocity $\mathbf{u}^{\nu m}$ is a direct consequence of the previous assumptions. Neglecting the acceleration terms in the momentum equation (19), assuming that the stress is given by a lithostatic pressure field [4] and combining the previous relations (22), (23), (26) and (28), the relative velocity for the small particles is given by

$$\mathbf{u}^{sm} = \mathbf{S}^{sl} \frac{\phi^l}{\Phi} \tag{29}$$

The mean segregation velocity

$$\mathbf{S}^{sl} = \underbrace{-\frac{B}{\mu}\sqrt{\frac{2}{3}\frac{\varepsilon_0}{\eta_0}}}_{\mathbf{S}_0^{sl}} \mathbf{g}\sqrt{T} \tag{30}$$

is zero for all directions orthogonal to the gravitational force. Further, the relative velocity vanishes, when the system reaches a static state as the granular temperature decreases. The factor $\mathbf{S}_0^{sl}$ needs to be validated by experiments or DEM simulations. With the help of the saturation condition (2) the final segregation equation for the small particle phase is given by

$$\partial_t \phi + \nabla \cdot (\phi \mathbf{u}) = -\nabla \cdot (\mathbf{S}^{sl} \frac{\Phi - \phi}{\Phi} \phi) \tag{31}$$

with the temperature and gravity dependent segregation velocity

$$\mathbf{S}^{sl} = \mathbf{S}_0^{sl} \mathbf{g} \sqrt{T}. \tag{32}$$

The final system is given by the equations (6), (7), (8) and (31).

3 Simulation results

To demonstrate the behavior of the combined model, we simulate a plain rotating tumbler filled to a height of 75% with glass beads of different sizes with radius ratio $\frac{r_l}{r_s} = 2$. The material parameters for the glass beads are given in [11]. For the segregation strength, we choose $S_0^{sl} = 1.5$, which is the same order as used in [4]. The setting is sensible as the processes in such a tumbler are already well known for two particle mixtures. There are flow regimes including segregation effects as well as static regimes. For the performed simulation the tumbler rotates in a clockwise direction with a speed of 0.05 rounds per second (rps). Initially, the system is perfectly mixed with equal volume fractions for small and large particles. For this setup a special pattern should form due to the rotation. Near the surface the particles are in motion. Except of the surface layers the granular material is in a static state. In the shearing top layers the segregation happens. Here the small particles segregate, but they do not even reach the core of the granular material, where the system is static already. In the static regime the material rotates with the tumbler without further segregation. Finally, a ring of small particles forms around a completely unchanged core. The large particles accumulate next to the walls [12].

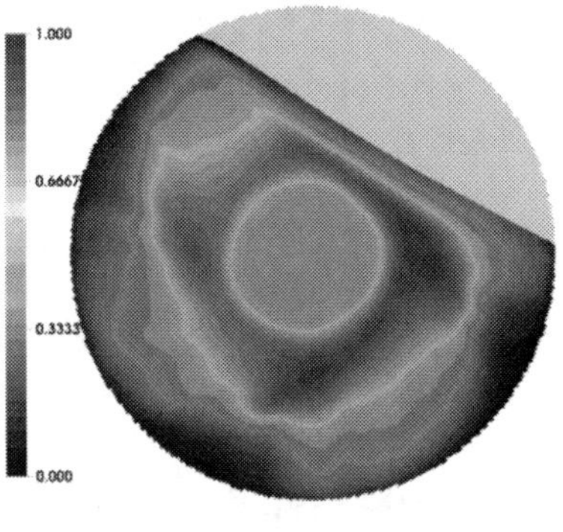

Figure 1. Volume fraction of the small particles relative to granular volume. The rotational velocity is 0.05 rps. The color table is fixed between 0 *(blue)* and 1 *(red)*.

Figure 1 shows the cross-section of the tumbler. The volume fraction of small particles relative to the granular volume is plotted after 30 seconds of rotation. One can see the expected pattern, where the small particles build a ringlike formation. The effect of the segregation process to the granular system can be seen in Figure 2. It shows the same state as Figure 1 for the granular volume fraction Φ in a rage of 0.56 and 0.66. In a mono-disperse system the packing profile is lithostatic. In a real application the maximum packing fraction for glass beads reaches a value of 0.632. For our poly-disperse system this value can grow up to 0.68, in regions with enough pressure and for $\phi^s \approx 0.3$. Our result shows the lithostatic profile superimposed by the segregation pattern. In the region of mainly one particle type the volume fraction fits to the values in mono-disperse system, whereas in the mixed regions the maximum packing fraction is bigger than 0.632.

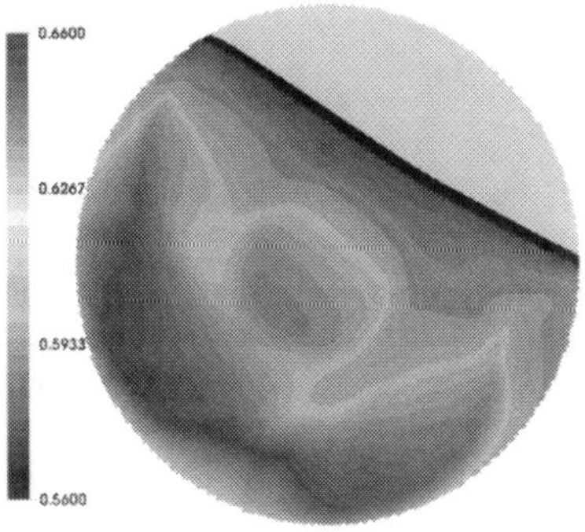

Figure 2. Volume fraction of the granular material. The rotational velocity is 0.05 rps. The color table is fixed between 0.6 *(blue)* and 0.66 *(red)*.

4 Conclusion

We modified and combined two models that are able to reproduce the radial segregation effect occurring in a rotating tumbler. The granular temperature dependence in the segregation equation regulates the segregation in the different regimes. Even if the simulation qualitatively leads to good results, one should keep in mind, that the applicability of the model to other processes still needs to be tested and validated by experiments or DEM simulations.

References

[1] A. Latz, S. Schmidt, Granular Matter **12**, 387 (2010)
[2] J. Bridgwater, W.S. Foo, D.J. Stephens, Powder Technology **41**, 147 (1985)
[3] S.B. Savage, C.K.K. Lun, Journal of Fluid Mechanics **189**, 311 (1988)
[4] Gray, J. M. N. T., A.R. Thornton, Proceedings of the Royal Society A: Mathematical, Physical and Engineering Sciences **461**, 1447 (2005)
[5] M. Larcher, J. Jenkins, Journal of Fluid Mechanics **782**, 405 (2015)
[6] D.R. Tunuguntla, O. Bokhove, A.R. Thornton, Journal of Fluid Mechanics **749**, 99 (2014)
[7] Gray, J. M. N. T., C. Ancey, Journal of Fluid Mechanics **779**, 622 (2015)
[8] N. Brilliantov, T. Pöschel, *Kinetic theory of granular gases*, Oxford graduate texts (Oxford University Press, Oxford, 2004), ISBN 0198530382
[9] L.W. Morland, Surv. Geophys. (Surveys in Geophysics) **13**, 209 (1992)
[10] B. Marks, P. Rognon, I. Einav, Journal of Fluid Mechanics **690**, 499 (2012)
[11] D. Niedziela, S. Rau, K. Steiner, S.d. Vita, M. Lutsche, M. Richter, M. Schmidt, To appear in: Chemical Engineering & Technology (2017)
[12] I.S. Aranson, L.S. Tsimring, Reviews of Modern Physics **78**, 641 (2006)

EPJ Web of Conferences **140**, 15016 (2017)
Powders & Grains 2017
DOI: 10.1051/epjconf/201714015016

Calibration of DEM parameters on shear test experiments using Kriging method

Bednarek Xavier[1]*, *Martin* Sylvain[1], *Ndiaye* Abibatou[2], *Peres* Véronique[1], and *Bonnefoy* Olivier[1]

[1]*Centre SPIN, LGF UMR CNRS 5307, École Nationale Supérieure des Mines de Saint-Étienne, 158 Cours Fauriel, F-42023 Saint-Étienne Cedex 2, France*
[2]*Areva-Melox, B.P. 93124, 30203 Bagnols sur Cèze Cedex, France*

Abstract. Calibration of powder mixing simulation using Discrete-Element-Method is still an issue. Achieving good agreement with experimental results is difficult because time-efficient use of DEM involves strong assumptions. This work presents a methodology to calibrate DEM parameters using Efficient Global Optimization (EGO) algorithm based on Kriging interpolation method. Classical shear test experiments are used as calibration experiments. The calibration is made on two parameters - Young modulus and friction coefficient. The determination of the minimal number of grains that has to be used is a critical step. Simulations of a too small amount of grains would indeed not represent the realistic behavior of powder when using huge amout of grains will be strongly time consuming. The optimization goal is the minimization of the objective function which is the distance between simulated and measured behaviors. The EGO algorithm uses the maximization of the Expected Improvement criterion to find next point that has to be simulated. This stochastic criterion handles with the two interpolations made by the Kriging method : prediction of the objective function and estimation of the error made. It is thus able to quantify the improvement in the minimization that new simulations at specified DEM parameters would lead to.

1 Introduction

Mixing of powders is an important process unit in several industries e.g. pharmaceuytical, food and nuclear industry. The Discrete-Element-Method (DEM) [1] can simulate such processes and help their understanding and optimization. However the calibration of micromechanical parameters remains the most sensitive issue of the modeling process.

There is currently no clear link between micromecanical parameters used in DEM and macromechanical measured ones. It appears that mixing behaviors of different powders can be related to different behaviors in shear testers as the Freeman Technology FT4 rheometer. The choice of shear test as calibration experiment seems thus suitable. It appears that DEM simulations of shear testers are mainly influenced by the Young's modulus and the static friction coefficient [2–4].

The aim of this paper is to present a calibration methodology using Efficient Global Optimization (EGO) algorithm based on Kriging interpolation method. The methodology is applied on the couple of DEM parameters Young 's Modulus and static friction coefficient.

Particular importance is given in this work to determine the number of particles to be used in the DEM simulation of the shear tester.

*e-mail: xavier.bednarek@emse.fr

2 Theoretical background & methodology

2.1 Discrete element modeling method

DEM describes the motion of spherical particles by integrating classical Newton's law in consecutive time steps. Positions, velocities and forces are updated at each time step. Contact-force models have been largely described in literature [5–7]. Here we use the most classical ones : the non-linear damped Hertz-Mindlin model for normal and tangential forces. Model of rolling friction [8] is used to represent the non-sphericity of particles. Viscous damping part of torque produced by rolling friction is negected. No cohesive forces are considered, but gravity is taken into account. Simulations were conducted with the open source software LIGGGHTS.

Choice of suitable time-step to integrate the motions is critical. The Rayleigh time step is often used when DEM simulations of dense regime are made. It represents the propagation time of a Rayleigh surface wave *around* a particle. As it is impossible in DEM paradigm to model such wave - the surface of an unique particle does not exist - we consider here that the validity of such a time step is a questionable. We thus use the critical time step given in (1).

$$\Delta t_{\text{crit}} = R\left(\frac{4}{3}\right)^{\frac{2}{3}} \pi^{\frac{1}{3}} \sqrt{\rho}\, \sigma_0^{-\frac{1}{6}} \tilde{E}^{-\frac{1}{3}} \qquad (1)$$

$\tilde{E} = E/(1 - \nu^2)$ being the reduced Young's modulus. The other symbols meaning can be found in table 1. Calculation of this time step is based on [9]. It represents the time needed for a perturbation of force propagating through a chain of identical particles in - Hertz - contact to cross a single one. It has the advantage of showing dependence on the compression σ_0 [Pa] of the chain. The resolution time step used here is about 5% of this critical time step.

As the size of the mixer - used in the target simulation - is fixed and as the computer/software performances limit the number of particles as about a million, the particle radius is fixed. This means that it is not conceivable to simulate one single real grain by one single numerical particle. Thus, the main challenge is to represent the rheology of a set of small grains with a set of larger numerical particles. It so is a critical point of first determine how many particles are needed to catch a *powder behavior.* Hence calibration experiments simulations are made with different number of particles, keeping constant the ratio between height and diameter of the shear cell and all DEM micromechanical parameters. From a certain number of particles, the result of the simulation no longer varies. The minimum number of particules to be used is therefore reached.

2.2 The standard shear test using the FT4 Freeman Technology rheometer

The standard shear cell technique for powder is based on the Jenike ring shear tester[10]. However unlike [3, 4], it is the cylindrical shape that is used here - as represented on figure 1 - particularly because the Freeman Technology FT4 rheometer is largely used in industrial application. Powder is placed into the shear cell and before shear a vertical stress σ is applied by the shear head to pre-compress the material sample. The shear head rotate at constant velocity, maintaining normal stress constant, and the shear stress τ is measured. Four shear steps are made with four differents vertical stresses : successively 9 kPa, 8kPa, 7 kPa and 6kPa. Before each shear step, a preshear step is made at σ =15 kPa.

(a) FT4 rheometer (b) DEM simulation

Figure 1. Shear tester used as calibration experiment

The effective measure is the yield locus of the powder, associated with the level of pre-compression. Indeed, each shear step should exhibit a clear maximum of measured shear stress before a strong collapse. This so called incipient faillure point corresponds to the critical point when the powder starts to flow under the shear stress applied. These four maximum shear stresses form the so called yield locus. Plot of these values versus the corresponding normal stress applied shows usually a linear trend. The slope is known as the angle of internal friction (AIF [°]) and the Y-intercept is reffered at cohesion C [Pa]. This last number is the extrapolated shear stress resitance that the powder should produce when sheared at zero normal stress. These two numbers describing the yield locus are then used to identify the appropriate couple of DEM parameters previously mentionned.

2.3 Efficient Global Optimization using Kriging

The goal of the calibration process is to find the couple of DEM parameters - Young's modulus and static coefficient of friction - that would minimize the distance between the simulated yield locus and the measured one. This distance can be viewed easily in the 2D-space C *vs* AIF as shown below on figure 4. As DEM simulations can be strongly time consuming, it is of great importance to have to do the least possible simulations to converge to the optimum.

The Efficient Global Optimization [11] is an optimization technique that is capable of choosing accuratelly new simulation points - new couples (E,μ_s) - to converge rapidly to the best solution. It can be viewed as a method to build progressively efficent design of experiments (DOE). The core of the method is to use Kriging interpolation [12] to predict the value of the objective function that has to be minimized - here the distance between simulated and experimental yield locus - and also the error made on this prediction. The smaller the prediction of this distance for a couple of DEM parameters and the the error on this prediction are, the bigger the chance of this couple to be choosen by the EGO algorithm to be the next point to be simulated is. In fact, it is the maximizatio of the Expected Improvment (EI) criterion (eq. 2) that allows the EGO algorithm to find this point.

$$\mathrm{EI}(\mathbf{x}) = (\min(\mathbf{Y}) - m_{\mathrm{K}}(\mathbf{x}))\Phi\left(\frac{\min(\mathbf{Y}) - m_{\mathrm{K}}(\mathbf{x})}{s_{\mathrm{K}}(\mathbf{x})}\right) + s_{\mathrm{K}}(\mathbf{x})\phi\left(\frac{\min(\mathbf{Y}) - m_{\mathrm{K}}(\mathbf{x})}{s_{\mathrm{K}}(\mathbf{x})}\right) \quad (2)$$

In equation (2), $\mathbf{x}$ is the couple (E,μ_s) on which the EI is calculated, $\mathbf{Y}$ is the vector of all the distances between simulations and experiment that have already been calculated, $m_{\mathrm{K}}(\mathbf{x})$ and $s_{\mathrm{K}}(\mathbf{x})$ are respectivelly the prediction of the distance for the point $\mathbf{x}$ and the error on this prediction and Φ and ϕ are respectivelly the standard normal distribution function and its probability density function.

In fact, the EGO algorithm used here works on the minimization of the 2-points-EI criterion. This criterion is similar to the one presented above but it allows the selection of two new points to be calculated simulateously. It is thus possible to take advantage of using a cluster of computers to conduct the calibration/optimization. The algorithm used is implemented in the R-package DiceOptim [13].

 DOI: 10.1051/epjconf/201714015016

3 Results and discussion

3.1 How many grains in powder?

Four simulations have been conducted with 12k, 31k, 78k and 200k particles, that respectivelly correspond to 25, 34, 46 and 63 particles per cell diameter. DEM parameter of these simulations are presented in table 1. Corresponding yield loci shows on figure 2 a clear convergence. It appears that beyond 31k particles, the measured behavior does not change. This can be explained by the difference of compactness induced by wall effet. Compacity before shear steps are indeed larger for small number of particles per cell diameter. The nearer the walls are, the smaller the compacity is. And a compacter granular medium is more difficult to be sheared.

Table 1. DEM parameters values

Parameter	Symbol	Value	Unit
Young's Modulus	E	25	GPa
Static Friction Coef.	μ_s	0.5	-
Poisson's Ratio	ν	0.25	-
Restitution Coef.	e	0.3	-
Rolling Friction Coef.	μ_R	0.5	-
Density	ρ	3000	kg.m^{-3}
Particle Radius	R	0.9	mm

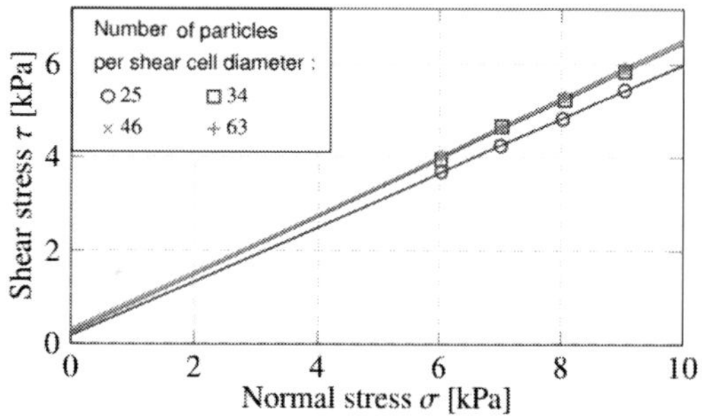

Figure 2. Yield loci of simulations with different numbers of particles

Figure 3 shows regular and smooth trends for the shear stresses measured. This is due to the non-friction wall hypothesis.

3.2 Comparison with experimental measurements

Given what has been observed above, calibration of DEM parameters has been started with 31k particles. Four simulations have been made to start the EGO algorithm. Figure 4 shows the result of these four simulations compared to experimental measurements made on two powders : Al_2O_3 with mean granulometry 88 μm and B_4C with mean granulometry 105 μm. Error bars for experiments are calculated from 3 repetitions of the full shear test and the ones for simulations are derived from classical statistics of linear regression.

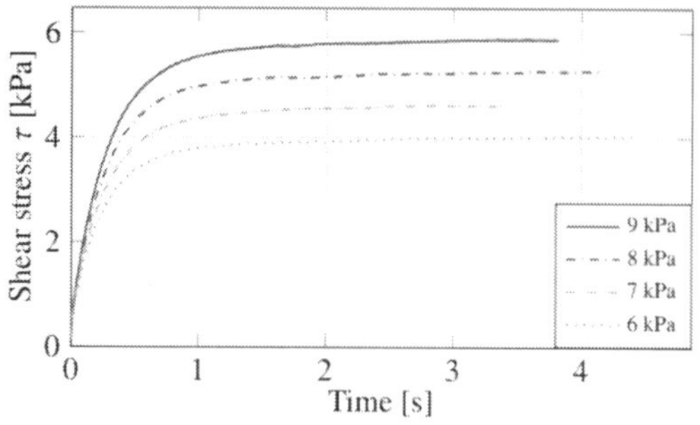

Figure 3. Shear steps for simulation with 78k particles

To find DEM parameters that lead to good calibration for B_4C, the two news points choosen by the EGO algorithm are (E=33GPa,μ_s=0.29) and (E=33GPa,μ_s=0.21). This choice can be easily comprehensible regarding figure 4 : B_4C point is obviously above the iso-Young's modulus line at 25 GPa, and near the iso-static friction line 0.25. As B_4C is very close to this line, the algorithm tends to look around.

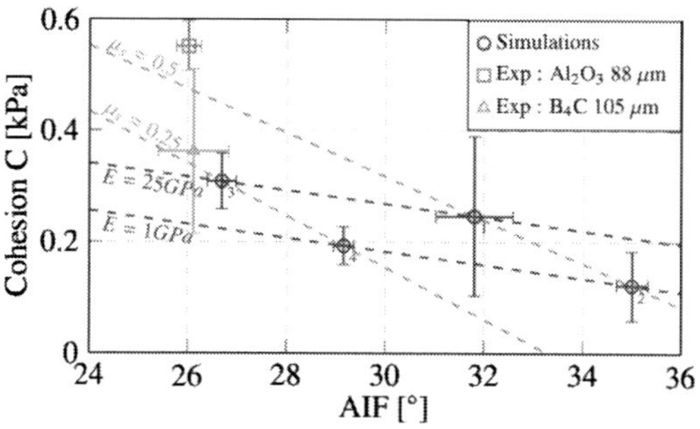

Figure 4. Cartography of the optimization

Figure 5 shows shear stress simulated at 9 kPa normal stress for both these new points compared to experimental ones. However it should be noted that these curves are not the result of a calibration that would have converged. Order of magnitude are reached, but the shape of curves are not very similar. Experimental curves for B_4C and Al_2O_3 show linear tends at the beginning, although simulations present much more concave shape. Experiment with glass beads shows also such a concave shape at the beginning. The concave shape can thus be explained by the mono-sized or quasi-mono-sized nature of the granulometry used. Indeed, the more extended granulometry of B_4C and Al_2O_3 allows much more granular rearrangement, retarding the rise of the shear stress.

Beyond the considerations about the number of particles made previously, the particle size distribution (PSD) seems to be a critical point. Indeed, catching the so called *powder behavior* might also need to take into account more precisely the PSD. It seems not sufficient to get enough particles, but it seems also important to reproduce the realistic partial coordination numbers of the packing

- ie the number of contact between particles of different sizes. The shear test made for the glass beads - that show a narrow PSD - is closer to the numerical results presented here, probably because of this.

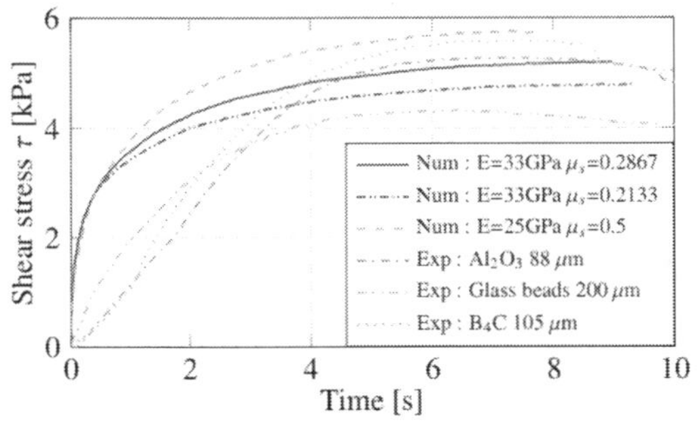

Figure 5. Comparison with experimental data at $\sigma = 9$ kPa

Moreover, the clear maximum of shear stress observed experimentaly is not reproduced by simulation. Here again this could be due to the mono-sized granulometry but recent simulations seems to show that there is a greater chance that this is due to the hypothesis of zero-friction at the walls of the shear cell. The work is currently moving in this direction.

3.3 Conclusion

Calibration of DEM parameters using the EGO algorithm seems promizing. The yield locus experimentally measured on powder sample can be approched by simulations. However, some assumptions made seem to be too restrictive to allow very good fitting of experimental curves : no friction at the wall and mono-sized granulometry. Future works must tackle these issues before continuing the proper use of the EGO algorithm.

Acknowledgements

This research was supported by Areva-Melox and Armines. This work was granted access to the HPC resources of IDRIS under the allocation 2016 - i2016097667 made by GENCI.

References

[1] P.A. Cundall, O.D. Strack, Geotechnique **29**, 47 (1979)
[2] C. Coetzee, D. Els, Computers and electronics in agriculture **65**, 198 (2009)
[3] T.A. Simons, R. Weiler, S. Strege, S. Bensmann, M. Schilling, A. Kwade, Procedia Engineering **102**, 741 (2015)
[4] I. Keppler, F. Safranyik, I. Oldal, Engineering Computations **33**, 742 (2016)
[5] J. Shäfer, S. Dippel, D. Wolf, Journal de physique I **6**, 5 (1996)
[6] H. Zhu, Z. Zhou, R. Yang, A. Yu, Chemical Engineering Science **62**, 3378 (2007)
[7] G.K. Barrios, R.M. de Carvalho, A. Kwade, L.M. Tavares, Powder Technology **248**, 84 (2013)
[8] J. Ai, J.F. Chen, J.M. Rotter, J.Y. Ooi, Powder Technology **206**, 269 (2011)
[9] C. Coste, E. Falcon, S. Fauve, Physical review E **56**, 6104 (1997)
[10] A.W. Jenike, Bulletin of the University of Utah **53** (1964)
[11] D. Ginsbourger, Ph.D. thesis, École des Mines de Saint-Étienne (2009)
[12] G. Matheron, Les cahiers du centre de morphologie mathématique de fontainebleau (1969)
[13] O. Roustant, D. Ginsbourger, Y. Deville, Journal of Statistical Software **51**, 54p (2012)

Vertical drag force acting on intruders of different shapes in granular media

Ali Abbas Zaidi[1], *Christoph* Müller[1,*]

[1]Laboratory of Energy Science and Engineering, Department of Mechanical and Process Engineering, ETH Zürich, 8092 Zürich, Switzerland

Abstract. The penetration of large objects into granular media is encountered commonly both in nature (e.g. impacts of meteors and projectiles) and engineering applications (e.g. insertion of tractor blades into sand). The motion of the impacting intruder in granular media is resisted by a granular drag force. In this work, we assess the effect of intruder shape on the granular drag force using discrete element modelling (DEM). The following intruder shapes were modelled: spherical, conical, cylindrical and cubical. We observed that the drag force can be described well by a power-law relationship with intrusion depth, independent of the intruder shape. However, the exponent of the power-law expression increases with increasing "flatness" of the intruder's impacting surface due to an increasing fraction of the granular media affected by the impact of the intruder.

1 Introduction

The insertion and impact of a solid object or intruder into a granular material is of both academic and industrial relevance. Typical examples include the formation of craters by the impact of meteors or projectiles, motion of wheels on gravel or sand , insertion of tractor blades or the penetration of a bullet into a bag of sand [1-5]. Previous work in this area has indicated that the granular drag force that acts on the intruder can be expressed as:

$$F(z,v) = F(z) + F(v) + F_0 \tag{1}$$

where z is the intrusion depth and v is the velocity of the intruder [6-10]. In Eq. (1), F(z) is the depth dependent hydrostatic drag force, F(v) is the velocity-dependent viscous drag force and F_0 represents a friction-like force acting on the intruder [11]. In some work the frictional component of the drag force is included in the hydrostatic term [9]. Using experiments, Peng et al. [12] and Hill et al. [13] proposed a power law expression for F(z), viz. $F(z) = kz^c$. In this work, we focused on low speed intruder dynamics, hence F(v) is neglected (generally F(v) can be neglected for intruder velocities less than about 0.8 m/s).

So far, most reports concerning the drag force acting on an intruder in a granular media, have studied spherical intruders. There is very little information available on how intruder shape affects the drag force. Hence, this work assesses the effect of intruder shape on the hydrostatic component of the granular drag force, F(z), using discrete element model (DEM) simulations.

2 Formulations

The DEM implemented in this work uses a soft sphere model [14]. The inter-particle collisions are modelled by a combination of a linear spring, a dashpot and a friction slider. The normal, F_n, and tangential component, F_t, of the contact force are calculated according to:

$$\boldsymbol{F}_n = (-K_n\delta_n - \eta_n \boldsymbol{G}\cdot\boldsymbol{n})\boldsymbol{n} \tag{2}$$
$$\boldsymbol{F}_t = -K_t\boldsymbol{\delta}_t - \eta_t\boldsymbol{G}_t \tag{3}$$
$$\text{If } |\boldsymbol{F}_t| > f|\boldsymbol{F}_n| \tag{4}$$
$$\boldsymbol{F}_t = -f|\boldsymbol{F}_n|\boldsymbol{t} \tag{5}$$

where K is the spring constant, η is the damping coefficient, $\boldsymbol{G}$ is the vector describing the relative velocity between the interacting particles, $\mathbf{G}_t$ is the slip velocity of the contact point and f is the coefficient of friction. The subscripts n and t refer to the normal and tangential direction of the contact, respectively, and $\boldsymbol{n}$ and $\boldsymbol{t}$ are the unit vectors in the normal and tangential direction of the particle-particle contact, respectively. Contact detection for non-spherical intruders is performed by discretizing the surface of the intruder into small triangular elements. The normal and tangential force are calculated by determining the overlap between the particle and the triangular element as shown in Fig. 1. Once the overlap (δ), has been determined the contact forces are calculated according to Eqs (2-5)

* Corresponding author: muelchri@ethz.ch

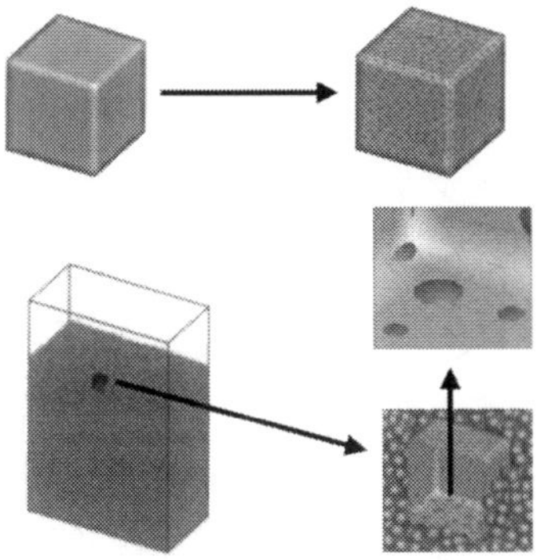

Fig. 1. Contact detection between the non-spherical intruder and particles.

The total force acing on an intruder is determined by summing up all forces acting on the elements of the intruder surface. Our numerical model was validated by MRI experiments and DEM simulations of spherical shaped intruders. In the latter validation case, a spherical intruder was pulled through a granular bed. The granular drag force calculated using the discretization scheme described above and using a conventional spherical DEM approach were compared. Good agreement between the different DEM schemes was observed.

3 Simulation Setup

To determine the drag force acting on non-spherical intruders in granular media, the following simulation set-up was used. The granular bed was contained in a rectangular container. The width, depth and height of the granular bed was 30×30×180 times the particle diameter (d_p). Particles were allowed to settle under gravity. The properties of particles used in the simulations are given in Table 1.

Table 1. Properties of the particles used.

d_p.	1.2 mm
Restitution coefficient	0.9
Friction coefficient	0.3
Elastic modulus	80000 N/m

The non-spherical intruders were pulled through the bed with a constant velocity of 0.2 m/s (hence, in a very good approximation, F(z) is the only component of the granular drag force that has to be considered). The shapes and sizes of the intruders simulated are summarized in Table 2. The small, medium and large sized intruders are referred to as, respectively, Case 1, Case 2 and Case 3 in the manuscript.

Table 2. Shapes and sizes of the intruder studied.

Intruder shape and direction of intrusion	Intruder dimensions
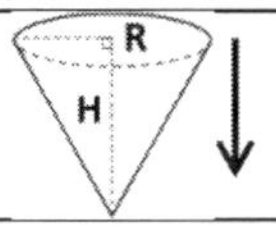	$H \times R = \begin{cases} 6d_p \times 3d_p \\ 12d_p \times 6d_p \\ 18d_p \times 9d_p \end{cases}$
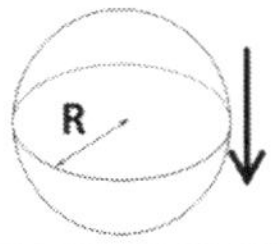	$R = 3d_p, 6d_p, 9d_p$
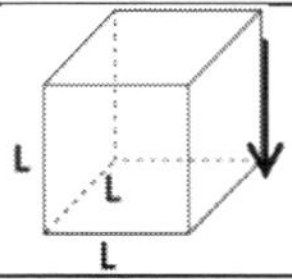	$L \times L \times L = \begin{cases} 6d_p \times 6d_p \times 6d_p \\ 12d_p \times 12d_p \times 12d_p \\ 18d_p \times 18d_p \times 18d_p \end{cases}$
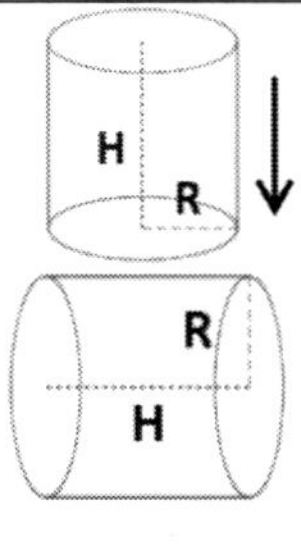	$H \times R = \begin{cases} 6d_p \times 3d_p \\ 12d_p \times 6d_p \\ 18d_p \times 9d_p \end{cases}$

In Table 2 the arrow indicates the direction of intrusion. For a cylindrical intruder two intrusion directions were studied.

4 Results & Discussions

Fig. 2 plots the granular drag force (divided by the intruder volume) as a function of the dimensionless intrusion depth (normalized by the height of the intruder) for different intruder shapes. The height of the granular bed was kept constant in all of the cases studied. We observe that independent of the intruder shape the drag force acting on the intruder follows a power law with intrusion depth *z*.

Similar to spherical intruders, dividing the drag force by the intruder volume leads to a size-independent power law expression for the drag force for all of particle shapes studied. We also observe that k $\approx 1.3 \times 10^{-4}$ independent of particle shape. However, the power-law coefficient c is a function of intruder shape and equals to 1.19, 1.3, 1.35, 1.38 and 1.41 for conical, spherical, cylindrical H, cylindrical V and cubic shapes, respectively (Table 3).

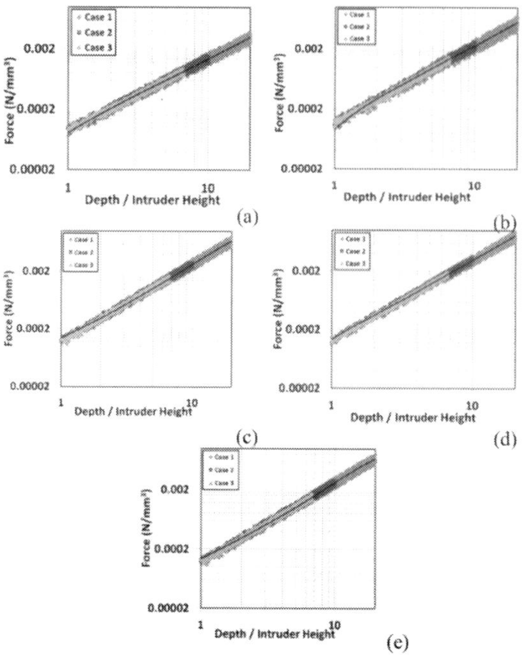

Fig. 2. Granular drag force acting on an intruder as a function of intrusion depth: (a) cone (b) sphere (c) cylinder configuration H (d) cylinder configuration V and (e) cube

Table 3. Expressions for the drag force for different intruder shapes.

Shape	**Power Law** $F(z) = kz^c$ $\left(k = 1.3 \times 10^{-4} \frac{\text{N}}{\text{mm}^{(c)}}\right)$
Cone	$F(z) = kz^{1.19}$
Sphere	$F(z) = kz^{1.3}$
Cylinder H	$F(z) = kz^{1.35}$
Cylinder V	$F(z) = kz^{1.38}$
Cube	$F(z) = kz^{1.41}$

It is worth noting that the value of c obtained for spheres is in agreement with previously reported experimental work [13]. Since for a given intrusion depth, cylinders and cubes experience a larger drag force than spheres or cones, it seems that the granular drag force increases with increasing "flatness" or "bluntness" of the impacting intruder surface.

To understand better this behaviour, we analysed in more detail the motion of particles around impacting intruders (Figs. 3 and 4). During the motion of the intruder, we determine the fraction of particles that have changed their original position (i.e. originally located in the purple box highlighted in Fig. 3).

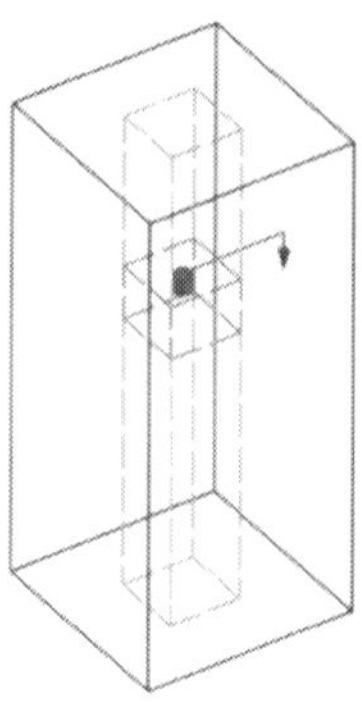

Fig. 3. Schematic diagram of a box that moves with the intruder and is used for data analysis. The dashed green lines highlight the path of the box during intruder motion

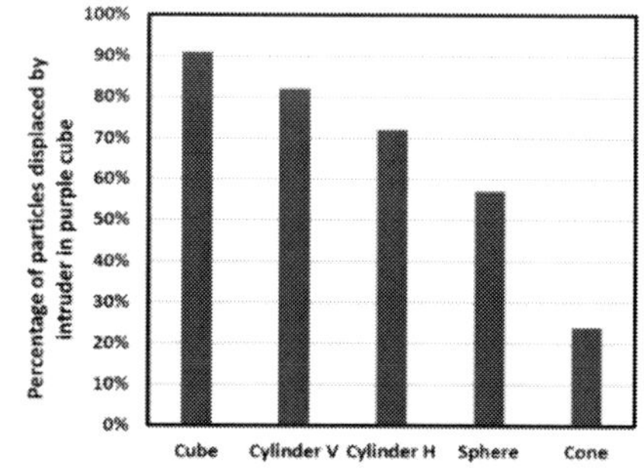

Fig. 4. Percentage of particles originally located in a box around the intruder that are affected by the motion of the intruder.

From Fig 4 we observe that the motion of a cubic intruder affects more particles than the motion of a conical intruder. The increase of the granular drag force (and the exponent of the power law) with an increase in intruder flatness hence, may be due to a higher energy loss due to a larger number of particle collisions.

5 Conclusions

The granular drag force acting on differently shaped intruders that are pulled with a constant velocity through a granular bed were studied using DEM. The following conclusions are drawn from this work:

(1) The hydrostatic component of the granular drag force obeys a power-law relationship with intrusion depth, independent of intruder shape.

(2) The granular drag force at a given intrusion depth increases with increasing"flatness" of the impacting intruder surface.

(3) The increasing exponent of the power-law relationship with increasing flatness of the intruder may be explained by an increasing number of particle-particle and particle-intruder collisions for "flatter" geometries.

Acknowledgement

The authors are grateful to the Swiss National Science Foundation (200021_153290) for financial support of this work.

References

1. R.A.F. Grieve, Nature **340**, 109 (1989)
2. H. Katsuragi, Lect. Notes Phys. **910**, 1 (2016)
3. A. H. Clark, L. Kondic, R.P. Behringer, Phys. Rev. Lett. **109**, 238302 (2012)
4. J. Crassous, D. Beladjine, A. Valance, Phys. Rev. Lett. **99**, 248001 (2007)
5. J. S. Uehara, M. A. Ambroso, R. P. Ojha, D. J. Durian, Phys. Rev. Lett **90**, 194301 (2003)
6. W. A. Allen, E. B. Mayfield, H. L. Morrison, J. Appl. Phys. **28**, 370 (1957)
7. M. J. Forrestal, V. K. Luk, Int. J. Impact Eng. **12**, 427 (1992)
8. D. I. Goldman, P. Umbanhowar, Phys. Rev. E **77**, 020101 (2008)
9. H. Katsuragi, D. J. Durian, Nat. Phys. **3**, 420 (2007)
10. P. Umbandowar, D. I. Goldman, Phys. Rev. E **82**, 010301 (2010)
11. M. Hou, Z. Peng, R. Liu, Y. Wu, Y. Tian, K. Lu, C. K. Chan, Sci. Technology Adv. Mat. **6**, 855 (2005)
12. Z. Peng, X. T. Xu, K. Q. Lu, M. Y. Hou, Phys. Rev. E **80**, 020101(R) (2009)
13. G. Hill, S, Yeung, S, A, Koehler, Europhys. Lett. **72**, 137 (2005)
14. P. A. Cundall, O. D. L. Strack, Geotechnique **29**, 47 (1979)

EPJ Web of Conferences **140**, 11009 (2017) DOI: 10.1051/epjconf/201714011009
Powders & Grains 2017

Extended granular micromechanics

Pasquale Giovine[1,⋆]

[1] *Department of Civil Engineering, Energy, Environment and Materials (DICEAM), 'Mediterranean' University of Reggio Calabria Via Graziella, 1 - Locality Feo di Vito, I-89122 Reggio Calabria - Italy*

Abstract. We analyze an extended mechanical model in the conservative case in order to describe a dilatant granular material with rotating grains for which the kinetic energy, in addition to the usual translational one, consists of three terms owing to microstructural motions: in particular, it includes the rotation of the grains, the dilatational expansion and contraction of the individual granules and of the granules relative to one another. Hence, we model the body as a continuum with a peculiar microstructure; after we follow classical procedures and define a variational principle of local type for a perfect fluid with microstructure, in accordance with the fluid-like behavior of granular materials: the motion equations are in good agreement with those obtained by other authors. At the end the particular case of a suspension of rotating rigid granules puts in evidence the possibility for granular materials to support shear stresses through the generation of micro-rotational gradients.

1 Preamble

A thermodynamic theory for the study of dilatant granular materials with rotating compressible granules was presented in [10] (see, also, [1] and [20]), a theory that generalized the model of distributed bodies in [16], which was widely used to study the slow flows of granular materials.

The material was assumed to consist of a suspension of dry cohesionless compressible spheres of uniform size in a compressible gas of negligible mass; the flow behavior has required a combination of suggestions from both fluid and solid mechanics owing to the fact that the material has an essentially fluid–like behavior, but it can also be heaped and, moreover, its bulk compressibility depends on the initial voids distribution in the reference placement (see the experimental results in [2]); we also assume a volume concentration close to that of packed particles, so that the mean free path of the particles is very short in comparison to the size of the particles themselves (as it is the case of cohesionless soil or sand with rough surface grains).

Additional balance equations for the microinertia was needed for new independent kinematical variables introduced in [10]: the volume fraction of the grains which describes the local arrangements of grains themselves and the micro-rotations of granules; so the granular material is a special case of perfect fluid with microstructure [5, 19].

In this work we analyze the conservative case for the model of granular materials proposed in [10] following the procedure used in [12]: an approach which was already successfully applied to a model of compressible immiscible mixtures of two perfect fluids [11] and of liquids with diffusing gas bubbles [13]. Also, an interesting application of the granular theory [12] was investigated in [14] for the study of seismic waves propagating through a sediment filled basin in the case of rigid grains; one advantage of our model, with respect to purely propagative models, was the reproduction of a nonlinear effect experimentally observed for real seismic waves: site amplification decreases as the amplitude of the incident wave increases.

The dynamic equations of motion are so obtained from a Hamiltonian variational principle of local type for a perfect fluid with microstructure, in accordance with the fluid–like behavior of granular materials, and then compared with those previously obtained by other authors.

At last, we consider the constrained case of a suspension of rotating rigid granules in a fluid matrix and notice that the microstructure behaves as that of a microrigid Cosserat's continua. By considering possible rotations of grains during the motion, we also show that, even when the volume fraction of the grain distribution is constant, the model predicts the possibility of supporting shear stress through the generation of microrotational gradients.

2 The model

The continuum model for a dry cohesionless dilatant granular materials with rotating grains here proposed is directly referred to the model proposed in [4], as specialized in [1, 7]. A material element of the body $\mathcal{B}$ is a sort of quasi–particle which consists of a grain and its immediate neighbours and that is called a 'chunk' of material; the chunks are thought of as envelopes which fill the body without voids between them and without diffusion of the grains through the envelope itself. The quasi-static motion of the body admits only dilatations, or contractions, of the individual (compressible) grain and of the grains relative to one another as well as rotations of the granule itself; there-

⋆e-mail: giovine@unirc.it

fore the continuum model depict the granular material as a medium with constrained affine microstructure [5].

The total kinetic energy per unit mass κ of the medium is then the sum of four components: a) the usual translational kinetic energy $\kappa_t := 2^{-1}\mathbf{v}^2$ related to the velocity $\mathbf{v}$ of the centre of mass $\mathbf{x}$ of the material element; b) the 'fluctuation' kinetic energy κ_f associated with the dilatancy of the chunk, that is the homogeneous expansions of it represented by the motion of individual grains relative to the centre of mass and expressed in terms of the rate of change of the chunk mass density ρ by $2^{-1}\gamma(\rho)\dot{\rho}^2$; the dot denotes the material time derivative, *i.e.*, $\dot{(\cdot)} := \frac{\partial(\cdot)}{\partial\tau} + \mathbf{v}\cdot\operatorname{grad}(\cdot)$, with τ the time. We observe that the chunk mass density ρ is equal to ν times the proper mass density ρ_m: $\rho = \rho_m\nu$, where $\nu \in (0,1)$ is the volume fraction of the grains; c) the dilatational kinetic energy κ_d related to local expansions (or contractions) of the compressible inclusions in the chunk and written in terms of the rate of change of the proper mass density of the grains ρ_m by $2^{-1}\alpha(\rho_m)\dot{\rho}_m^2$; d) the rotational kinetic energy $\kappa_r := 2^{-1}3\rho_m^2\alpha(\rho_m)\dot{\mathbf{R}}\cdot\dot{\mathbf{R}}$ due to grain rotations, with $\mathbf{R}$ a proper orthogonal tensor.

The form of constitutive coefficients $\gamma(\rho)$ and $\alpha(\rho_m)$ can be very general as they depend on geometric configurations of the grains and of the chunks and on the admissible micro-motions (see §2 of [4]). In [10, 12] simple microstructural motions and peculiar geometrical shapes for chunks and granules were assumed in order to compute them explicitly. In a mental magnification, one can imagine the chunk consisting of a spherical grain and its immediate spherical neighbors, and the envelope of the chunk as a spherical surface of variable radius ς containing all these spherical compressible inclusions of variable radius ϑ, with interstices filled with a fluid of negligible mass; the chunks and the compressible grains are also supposed to expand or contract homogeneously with independent purely radial pulsations, while rotations of single grains are not necessarily related to macro-rotations of the own chunk. Therefore, if the envelopes and the grains have the radius ς_* and ϑ_*, respectively, in a reference placement $\mathcal{B}_*$ of the material, one can calculate analytically the following expressions for functions γ and α:

$$\gamma(\rho) = \gamma_*\rho^{-\frac{8}{3}}, \quad \alpha(\rho_m) = \alpha_*\rho_m^{-\frac{8}{3}}, \tag{1}$$

with $\gamma_* = (16/351)\rho_*^{\frac{2}{3}}\varsigma_*^2$ and $\alpha_* = 15^{-1}\rho_{m*}^{\frac{2}{3}}\vartheta_*^2$.

The density per unit mass of the total kinetic energy of the material turns out to be

$$\kappa = 2^{-1}\mathbf{v}^2 + 2^{-1}\gamma(\rho)\dot{\rho}^2 + 2^{-1}\alpha(\rho_m)\left(\dot{\rho}_m^2 + 3\rho_m^2\,\dot{\mathbf{R}}\cdot\dot{\mathbf{R}}\right). \tag{2}$$

In order to reduce the developments to the essential things, we consider the case when the suspension is contained in a fixed rigid vessel, very large with respect to grains and to a single element of the medium. If $\mathcal{B}$ is the region of space delimited by the vessel and $\partial\mathcal{B}$ its boundary, kinematic compatibility imposes the condition

$$\mathbf{v}\cdot\hat{\mathbf{n}} = 0 \quad \text{on } \partial\mathcal{B}, \tag{3}$$

where $\hat{\mathbf{n}}$ is the unit exterior vector normal to $\partial\mathcal{B}$; moreover, the conservation of mass requires

$$\dot{\rho} + \rho\operatorname{div}\mathbf{v} = 0 \quad \text{in } \mathcal{B}. \tag{4}$$

3 The Hamiltonian variational principle

In order to obtain the dynamical balance equations of motion we use a Hamiltonian variational principle of local type in the conservative case, even if, in general, the viscous effects could not be ignored under dynamic conditions. The choice of a Eulerian variational principle, rather than Lagrangian, is not in contrast with the previous appeal to a reference placement; in fact the bulk compressibility of a granular material depends on the initial porosity (see [2]) and therefore the material has a preferred reference placement with respect to volume distribution, thus the choice for the former or the latter formulation is not so peremptory for such materials (see [3]).

So we suppose that the internal actions on $\mathcal{B}$ derive from a potential energy φ per unit mass; since the material is essentially thought as a perfect fluid in its behavior, we assume φ to be a function of the state of $\mathcal{B}$ through ρ, $\operatorname{grad}\rho$, ρ_m, $\operatorname{grad}\rho_m$, $\mathbf{R}$ and $\operatorname{grad}\mathbf{R}$.

When a variation $\delta\mathbf{x}(\mathbf{x},\tau)$, with $\mathbf{x}$ in $\mathcal{B}$ and the time τ in the interval $[\tau_0,\tau_1]$ ($\tau_1 > \tau_0$) of duration of the mechanical process, is added to the motion $\mathbf{x}$, a corresponding variation of the domain $\mathcal{B}$ and of the fields $\delta\rho_m$ and $\delta\mathbf{R}$ defined on $\mathcal{B}\times[\tau_0,\tau_1]$ is implied. Moreover, these fields are subject to their arbitrary variation in addition to that induced by $\delta\mathbf{x}$: for instance, the total variation of ρ_m is defined as

$$\hat{\delta}\rho_m := \delta\rho_m + (\operatorname{grad}\rho_m)\cdot\delta\mathbf{x}. \tag{5}$$

The condition (3) and the conservation of mass (4) require restrictions on the variation $\delta\mathbf{x}$, which must be tangent to the boundary, and the total variation of $\hat{\delta}\rho$:

$$\delta\mathbf{x}\cdot\hat{\mathbf{n}} = 0 \quad \text{on } \partial\mathcal{B}, \tag{6}$$

$$\hat{\delta}\rho + \rho\operatorname{div}\delta\mathbf{x} = 0 \quad \text{in } \mathcal{B}; \tag{7}$$

besides, we consider only variations $\delta\mathbf{x}$, $\delta\rho_m$ and $\delta\mathbf{R}$ which vanish at the extremes of the interval $[\tau_0,\tau_1]$ and on $\partial\mathcal{B}$.

In view of condition (6), we can define the global virtual work of external actions consequent to the virtual displacement $\delta\mathbf{x}$, to a change $\delta\rho_m$ of the proper density of the matrix and $\delta\mathbf{R}$ of the rotation of grains, as the quantity

$$\delta\mathcal{L} := \int_{\mathcal{B}} \rho\left(\mathbf{f}\cdot\delta\mathbf{x} + \beta\,\hat{\delta}\rho_m + \mathbf{O}\cdot\hat{\delta}\mathbf{R}\right)d\mathcal{B}, \tag{8}$$

where $\mathbf{f}$, β and $\mathbf{O}$ are the densities for unit mass of the external body, dilatational and tensor moment forces, respectively. Then the Hamiltonian principle asserts that, during the natural motion of the body, the equality

$$\hat{\delta}\int_{\tau_0}^{\tau_1} d\tau \int_{\mathcal{B}} \rho(\kappa-\varphi)d\mathcal{B} + \int_{\tau_0}^{\tau_1} \delta\mathcal{L}d\tau = 0 \tag{9}$$

holds for all virtual processes, where the applied operator $\hat{\delta}$ means the variation of the next integral functional defined on a variable region (see §37, ch. 7, of [8]).

The use of the transport theorem leads to the equality

$$\hat{\delta}\int_{\tau_0}^{\tau_1} d\tau \int_{\mathcal{B}} \rho(\kappa-\varphi)d\mathcal{B} = \int_{\tau_0}^{\tau_1} d\tau \int_{\mathcal{B}} \rho\hat{\delta}(\kappa-\varphi)\,d\mathcal{B}, \tag{10}$$

so we can write the variational principle (9) as follows

$$\begin{aligned}&\int_{\tau_0}^{\tau_1} d\tau \int_{\mathcal{B}} \{\rho\left[\mathbf{v}\cdot\hat{\delta}\mathbf{v} + 2^{-1}\gamma'(\rho)\dot{\rho}^2\hat{\delta}\rho + \gamma(\rho)\dot{\rho}\,\hat{\delta}(\dot{\rho})\right.\\&+\alpha(\rho_m)\left(\dot{\rho}_m\hat{\delta}(\dot{\rho}_m) + 3\rho_m\dot{\mathbf{R}}\cdot\dot{\mathbf{R}}\,\hat{\delta}\rho_m + 3\rho_m^2\dot{\mathbf{R}}\cdot\hat{\delta}(\dot{\mathbf{R}})\right)\\&\left.+2^{-1}\alpha'(\rho_m)\left(\dot{\rho}_m^2 + 3\rho_m^2\dot{\mathbf{R}}\cdot\dot{\mathbf{R}}\right)\hat{\delta}\rho_m\right] - \frac{\pi}{\rho}\hat{\delta}\rho - \zeta\hat{\delta}\rho_m\\&-\mathbf{r}\cdot\hat{\delta}(\operatorname{grad}\rho) - \mathbf{s}\cdot\hat{\delta}(\operatorname{grad}\rho_m) - \mathbf{\Sigma}\cdot\hat{\delta}(\operatorname{grad}\mathbf{R})\\&-\mathbf{N}\cdot\hat{\delta}\mathbf{R} + \rho(\mathbf{f}\cdot\delta\mathbf{x} + \beta\hat{\delta}\rho_m + \mathbf{O}\cdot\hat{\delta}\mathbf{R})\}\,d\mathcal{B} = 0,\end{aligned} \tag{11}$$

where the prime $'$ denotes derivation with respect to the argument and we have introduced the notations

$$\begin{aligned}&\pi := \rho^2\frac{\partial\varphi}{\partial\rho}, \quad \mathbf{r} := \rho\frac{\partial\varphi}{\partial(\operatorname{grad}\rho)}, \quad \zeta := \rho\frac{\partial\varphi}{\partial\rho_m},\\&\mathbf{s} := \rho\frac{\partial\varphi}{\partial(\operatorname{grad}\rho_m)}, \ \mathbf{N} := \rho\frac{\partial\varphi}{\partial\mathbf{R}}, \ \mathbf{\Sigma} := \rho\frac{\partial\varphi}{\partial(\operatorname{grad}\mathbf{R})}.\end{aligned} \tag{12}$$

The velocity variation $\hat{\delta}\mathbf{v}$ corresponding to $\delta\mathbf{x}$ is

$$\hat{\delta}\mathbf{v} = (\delta\mathbf{x})^{\cdot}, \tag{13}$$

while we can obtain the relations

$$\hat{\delta}(\dot{\rho}) = (\hat{\delta}\rho)^{\cdot}, \ \hat{\delta}(\operatorname{grad}\rho) = \operatorname{grad}\hat{\delta}\rho - (\operatorname{grad}\delta\mathbf{x})^T\operatorname{grad}\rho, \tag{14}$$

for the total variations of, *e.g.*, $\dot{\rho}$ and $\operatorname{grad}\rho$, from definitions of the time derivative and (5); ^{T} means transposition.

Usual developments (as the insertion of (13) and (14) into (11), the integration by parts, as well as the application of divergence and transport theorems with conditions (3) and (6), besides the vanishing of variations on the boundary $\partial\mathcal{B}$ and in τ_0 and τ_1) lead us to write (11) as follows:

$$\begin{aligned}&\int_{\tau_0}^{\tau_1} d\tau \int_{\mathcal{B}} \{[\rho(\mathbf{f}-\dot{\mathbf{v}}) - \operatorname{div}(\operatorname{grad}\rho\otimes\mathbf{r} + \operatorname{grad}\rho_m\otimes\mathbf{s}\\&+(\operatorname{grad}\mathbf{R})^T\odot\mathbf{\Sigma}^T)]\cdot\delta\mathbf{x} + \{\rho\beta - \zeta + \operatorname{div}\mathbf{s}\\&-\rho\left[\alpha(\rho_m)\left(\ddot{\rho}_m + \rho_m\dot{\mathbf{R}}\cdot\dot{\mathbf{R}}\right) + 2^{-1}\alpha'(\rho_m)\dot{\rho}_m^2\right]\}\hat{\delta}\rho_m\\&-\left[\rho\left(\gamma(\rho)\ddot{\rho} + 2^{-1}\gamma'(\rho)\dot{\rho}^2\right) + \frac{\pi}{\rho} - \operatorname{div}\mathbf{r}\right]\hat{\delta}\rho\\&+\left[\rho\left(\mathbf{O} - 3\left(\rho_m^2\alpha(\rho_m)\dot{\mathbf{R}}\right)^{\cdot}\right) - \mathbf{N} + \operatorname{div}\mathbf{\Sigma}\right]\cdot\hat{\delta}\mathbf{R}\}\,d\mathcal{B} = 0.\end{aligned} \tag{15}$$

where we used the expression $(1)_2$ of $\alpha(\rho_m)$; the tensor product $\odot$ between third–order tensors is so defined: $(\mathbf{\Gamma}\odot\mathbf{\Sigma})_{ij} := \mathbf{\Gamma}_{ihk}\mathbf{\Sigma}_{jhk}$.

By using relation (7) and other elementary calculations, we can express the integral in (15) as a linear functional of $\delta\mathbf{x}$, $\hat{\delta}\rho_m$ and $\hat{\delta}\mathbf{R}$; so, by requiring the integral to vanish for any values of them, we obtain the motion equations for dilatant granular materials with rotating grains

$$\begin{aligned}\rho\dot{\mathbf{v}} = \rho\mathbf{f} &- \operatorname{grad}\left[\pi - \rho\operatorname{div}\mathbf{r} + \rho^2\left(\gamma(\rho)\ddot{\rho} + 2^{-1}\gamma'(\rho)\dot{\rho}^2\right)\right]\\&- \operatorname{div}\left[\operatorname{grad}\rho\otimes\mathbf{r} + \operatorname{grad}\rho_m\otimes\mathbf{s} + (\operatorname{grad}\mathbf{R})^T\odot\mathbf{\Sigma}^T\right],\end{aligned} \tag{16}$$

$$\rho\left[\alpha(\rho_m)\left(\ddot{\rho}_m + \rho_m\dot{\mathbf{R}}\cdot\dot{\mathbf{R}}\right) + 2^{-1}\alpha'(\rho_m)\dot{\rho}_m^2\right] = \rho\beta - \zeta + \operatorname{div}\mathbf{s}, \tag{17}$$

$$\rho\rho_m\,\alpha(\rho_m)\left(3\rho_m\ddot{\mathbf{R}} - 2\dot{\rho}_m\dot{\mathbf{R}}\right) = \rho\mathbf{O} - \mathbf{N} + \operatorname{div}\mathbf{\Sigma}. \tag{18}$$

The balance equation for the moment of momentum is directly recovered when we impose the principle of material objectivity to the scalar potential φ, *i.e.*, its value is invariant if the observer changes, although its constitutive variables $\operatorname{grad}\rho$, $\operatorname{grad}\rho_m$, $\mathbf{R}$ and $\operatorname{grad}\mathbf{R}$ vary (the measures of ρ and ρ_m are not affected by changes of frame). This condition coincides with the following relation:

$$\mathcal{E}\left[\operatorname{grad}\rho\otimes\mathbf{r} + \operatorname{grad}\rho_m\otimes\mathbf{s} + (\operatorname{grad}\mathbf{R})^T\odot\mathbf{\Sigma}^T\right] = 0, \tag{19}$$

with $\mathbf{r}$, $\mathbf{s}$ and $\mathbf{\Sigma}$ given by $(12)_{2,4,6}$ and $\mathcal{E}$ being the Ricci's permutation tensor: (19) coincides with the symmetry of the tensor in the square parentheses and, consequently, of the Cauchy's stress tensor $\mathbf{T}$. Indeed, the balance (16) reduces, at least formally, to the Cauchy's equation

$$\rho\dot{\mathbf{v}} = \rho\mathbf{f} + \operatorname{div}\mathbf{T}, \tag{20}$$

if the symmetric Cauchy's tensor is defined as follows:

$$\begin{aligned}\mathbf{T} = &\left[\rho\operatorname{div}\mathbf{r} - \pi - \rho^2\left(\gamma(\rho)\ddot{\rho} + 2^{-1}\gamma'(\rho)\dot{\rho}^2\right)\right]\mathbf{I}-\\&-\operatorname{sym}\left[\operatorname{grad}\rho\otimes\mathbf{r} + \operatorname{grad}\rho_m\otimes\mathbf{s} + (\operatorname{grad}\mathbf{R})^T\odot\mathbf{\Sigma}^T\right],\end{aligned} \tag{21}$$

where $\operatorname{sym}(\cdot)$ is the symmetric part of the tensor $(\cdot)$.

Relations (17) and (18) are the balance equations for the micromomentum with ζ and $\mathbf{N}$ the densities per unit volume of the internal forces of dilatancy and of rotation, while $\mathbf{s}$ and $\mathbf{\Sigma}$ are the dilating microstress vector and the third order spinning hyperstress tensor, respectively. Here, we can recognize the dynamics equations of §8 of [5] that describe the motion of continua with microstructure, even if the constitutive relation (21) for the stress tensor $\mathbf{T}$ generalizes the analogous one for granular materials in §17 of [5] where less complex physical phenomena are studied.

The four basic fields ρ, $\mathbf{v}$, ρ_m and $\mathbf{R}$ are determined by the four equations (4), (17), (18) and (20), so the additional differential equation (3.16) for ρ which appears in [4] arouses some perplexity. However, it is possible to eliminate the Lagrange multiplier μ in the (3.15) of [4] by using (3.16) and reduce equations in agreement with us.

At the end, a comparison with equations $(69)_2$ and (85) of [10] that rule the motion of continua with spherical microstructure (see also [6]) is possible only if we insert the skew microspin tensor $\mathbf{Y} := -\dot{\mathbf{R}}\mathbf{R}^T$ in (17) and (18), so

$$\rho\left[\alpha(\rho_m)(\ddot{\rho}_m + \rho_m\mathbf{Y}\cdot\mathbf{Y}) + 2^{-1}\alpha'(\rho_m)\dot{\rho}_m^2\right] = \rho\beta - \zeta + \operatorname{div}\mathbf{s}, \tag{22}$$

$$\rho\left(\rho_m^2\,\alpha(\rho_m)\mathbf{Y}\right)^{\cdot} + \rho\rho_m^2\alpha(\rho_m)\mathbf{Y}^2 = 3^{-1}\mathbf{R}(\rho\mathbf{O} - \mathbf{N} + \operatorname{div}\mathbf{\Sigma})^T. \tag{23}$$

After we can take the deviatoric and the skew part of (23)

$$\begin{aligned}&\operatorname{dev}\left[3\rho\rho_m^2\alpha(\rho_m)\mathbf{Y}^2 - \mathbf{R}(\rho\mathbf{O} - \mathbf{N} + \operatorname{div}\mathbf{\Sigma})^T\right] = \mathbf{0},\\&\rho\left(\rho_m^2\,\alpha(\rho_m)\mathbf{Y}\right)^{\cdot} = 3^{-1}\operatorname{skw}\left[\mathbf{R}(\rho\mathbf{O} - \mathbf{N} + \operatorname{div}\mathbf{\Sigma})^T\right],\end{aligned} \tag{24}$$

to have, with (22), the requested equations.

4 The case of rigid grains

In this section we consider the flow of a large number of discrete inelastic particles at relatively high concentrations and with interstices filled by a fluid or a gas of negligible mass (as it is the case of cohesionless soil, such as sand with rough surface grains, or of fluidized particulate beds). Hence, we have to assume that the granules are rigid and the proper mass density ρ_m is constant, thus, for the constitutive hypothesis b) of §2, the chunk mass density ρ comes down to be proportional to the volume fraction ν of grains ($\rho = \rho_{m*}\nu$) and the conservation of mass (4) reduces to

$$\dot{\nu} + \nu \operatorname{div} \mathbf{v} = 0. \tag{25}$$

Moreover, the expression of kinetic energy changes, owing to such constraint; in particular we have from (2) and (1)

$$\hat{\kappa} = 2^{-1}\mathbf{v}^2 + 2^{-1}\hat{\gamma}(\nu)\dot{\nu}^2 + 2^{-1}\mu_*\dot{\mathbf{R}}\cdot\dot{\mathbf{R}}, \tag{26}$$

with $\hat{\gamma}(\nu) = (16/351)\,\nu_*^{\frac{2}{3}}\varsigma_*^2\nu^{-\frac{8}{3}}$, and $\mu_* = 5^{-1}\,\vartheta_*^2$.

The potential energy ψ is now a function of the state of $\mathcal{B}$ only through ν, $\mathbf{R}$, grad ν and grad $\mathbf{R}$; we also suppose that the external actions derive from a potential ω, which is a function of the place $\mathbf{x}$ in $\mathcal{B}$, of the volume fraction ν and of the rotation $\mathbf{R}$ (see [18] and [15]). Besides the relation (7) binds the fraction volume to vary as follows

$$\hat{\delta}\nu = -\nu \operatorname{div}\delta\mathbf{x} \quad \text{in } \mathcal{B}. \tag{27}$$

Therefore, now, the Hamiltonian principle is

$$\hat{\delta}\int_{\tau_0}^{\tau_1} d\tau \int_{\mathcal{B}} \rho(\hat{\kappa} - \psi - \omega)d\mathcal{B} = 0; \tag{28}$$

by going on as in the previous section we obtain equations

$$\begin{aligned} \nu\dot{\mathbf{v}} &= \nu\hat{\mathbf{f}} - \operatorname{div}\left[\operatorname{grad}\nu\otimes\hat{\mathbf{r}} + (\operatorname{grad}\mathbf{R})^T\odot\hat{\mathbf{\Sigma}}^T\right] \\ &- \operatorname{grad}\left[\hat{\pi} + \nu^2\left(\hat{\gamma}(\nu)\ddot{\nu} + 2^{-1}\hat{\gamma}'(\nu)\dot{\nu}^2 - \hat{\beta}\right) - \nu\operatorname{div}\hat{\mathbf{r}}\right], \end{aligned} \tag{29}$$

$$\mu_*\,\nu\ddot{\mathbf{R}} = \nu\hat{\mathbf{O}} - \hat{\mathbf{N}} + \operatorname{div}\hat{\mathbf{\Sigma}}, \tag{30}$$

where we introduced the notations

$$\begin{aligned} &\hat{\pi} := \nu^2\frac{\partial\psi}{\partial\nu},\ \hat{\mathbf{r}} := \nu\frac{\partial\psi}{\partial(\operatorname{grad}\nu)},\ \hat{\mathbf{\Sigma}} := \nu\frac{\partial\psi}{\partial(\operatorname{grad}\mathbf{R})}, \\ &\hat{\mathbf{N}} := \nu\frac{\partial\psi}{\partial\mathbf{R}},\ \hat{\mathbf{f}} := -\frac{\partial\omega}{\partial\mathbf{x}},\ \hat{\beta} := -\frac{\partial\omega}{\partial\nu},\ \hat{\mathbf{O}} := -\frac{\partial\omega}{\partial\mathbf{R}}. \end{aligned} \tag{31}$$

In (29) the usual pressure for fluids $\hat{\pi}$ is related to the compressibility of chunks; the spherical part of a Reynolds' stress tensor of the turbulence theory $-\nu^2\left(\hat{\gamma}(\nu)\ddot{\nu} + 2^{-1}\hat{\gamma}'(\nu)\dot{\nu}^2\right)\mathbf{I}$ evaluates the agitation within a chunk of material; finally, the stress of Ericksen's type $-\left[\operatorname{grad}\nu\otimes\hat{\mathbf{r}} + (\operatorname{grad}\mathbf{R})^T\odot\hat{\mathbf{\Sigma}}^T\right]$ measures the ability of rigid granular continua to support shear stresses in equilibrium, by inducing the generation of micro-rotation gradients, even when the proper mass density and the volume fraction of the grain distribution are constant.

Again, if we insert the microspin tensor $\mathbf{Y}$ into the balance (30) and apply the skew and symmetric operators, we calculate equations which govern the micromotion for the microrigid Cosserat's continua:

$$\begin{aligned} &\nu\dot{\mathbf{Y}} = \operatorname{skw}\left[\mu_*^{-1}\mathbf{R}\left(\nu\hat{\mathbf{O}} - \hat{\mathbf{N}} + \operatorname{div}\hat{\mathbf{\Sigma}}\right)^T\right], \\ &\operatorname{sym}\left[\mathbf{R}\left(\hat{\mathbf{N}} - \operatorname{div}\hat{\mathbf{\Sigma}}\right)^T\right] = \nu\left[\operatorname{sym}\left(\mathbf{R}\hat{\mathbf{O}}^T\right) - \mu_*\mathbf{Y}^2\right]. \end{aligned} \tag{32}$$

In particular, the first one represents the balance of rotational micromomentum (see equation (63) of [17])

$$\mu_*\nu\dot{\mathbf{Y}} = \nu\bar{\mathbf{O}} - \bar{\mathbf{N}} + \operatorname{div}\bar{\mathbf{\Sigma}}, \tag{33}$$

where $\bar{\mathbf{O}} := \operatorname{skw}(\mathbf{R}\hat{\mathbf{O}}^T)$ ($\bar{\mathbf{N}} := \operatorname{skw}(\mathbf{R}\hat{\mathbf{N}}^T)$) is the external (internal) rotational tensor moment and $\bar{\mathbf{\Sigma}}$, such that $\bar{\mathbf{\Sigma}}\mathbf{w} := \operatorname{skw}[\mathbf{R}(\hat{\mathbf{\Sigma}}\mathbf{w})^T]$, $\forall\mathbf{w}$, the third-order spinning hyperstress tensor; the second one is the equation for the reactions to the constraint ρ_{m*}=const.: indeed, if we consider the material expression of (27): $\nu = \nu_*\,\iota^{-1}$, with ι the determinant of deformation gradient, we observe that the volume fraction ν of grains is bound to the macrodeformation.

Remark: When the effects of rotations of chunks and of grains are also negligible ($\mathbf{R}$=$\mathbf{I}$), we recover the essence of the theory in [9]: in particular, the Coulomb's model for the stress at equilibrium in a granular material with incompressible grains: $\mathbf{T}_C = (\beta_0 - \beta_1\nu^2 + \beta_2\operatorname{grad}\nu\cdot\operatorname{grad}\nu + 2\beta_3\,\nu\,\Delta\nu)\mathbf{I} - 2\beta_4\operatorname{grad}\nu\otimes\operatorname{grad}\nu$ (β_i material constants) is obtained as a peculiar example (see, also, equation (9.1) of [16]). Alternatively, the complementary case in which the grains are elastic, is studied in [12].

Acknowledgements

The support of the "Gruppo Nazionale di Fisica Matematica (GNFM)" of the "Istituto Nazionale di Alta Matematica 'F. Severi' (INDAM)" is acknowledged.

References

[1] Ahmadi G, Int. J. Non–Linear Mech. **17**, 21-33 (1982)
[2] Arthur JR, Menzies BK, Geotech. **22**, 115-129 (1972)
[3] Bampi F, Morro A, J.Math.Phys.**25**, 2418-2421 (1984)
[4] Bedford A, Drumheller DS, Acta Mechanica **48**, 173-184 (1983)
[5] Capriz G, *Continua with Microstructure*, Springer Tracts in Nat. Phil. **35**, (Springer-Verlag, Berlin, 1989)
[6] Capriz G, Podio-Guidugli P, Arch. Rational Mech. Anal. **75**, 269-279 (1981)
[7] Chen K-C, Lan J-Y, Tai Y-C, Int. J. Solids Struct. **46**, 3882-3893 (2009)
[8] Gelfand IM, Fomin SV, *Calculus of Variations* (Prentice-Hall, Englewood Cliffs, New Jersey, 1963)
[9] Giovine P, Math. Phys. An. Geom. **2**, 179-196 (1999)
[10] Giovine P, in Capriz G, Giovine P, Mariano PM (Eds.): *Math. Mod. Gran. Matter*, Lect. Notes Appl. Math.S. **1937** (Springer, New-York, 2008) 167-192
[11] Giovine P, Oliveri F, Acta Mechan. **96** 85-96 (1993)
[12] Giovine P, Oliveri F, Meccanica **30**, 341-357 (1995)
[13] Giovine P, Archives Mechanics **58**, 363-380 (2006)
[14] Godano C, Oliveri F, Int. J. Non-linear Mech. **34**, 457-468 (1999)
[15] Goodman M, Cowin SC, ZAMM **56**, 281-286 (1976)
[16] Goodman M, Cowin SC, Arch. Rational Mech. Anal. **44**, 249-266 (1972)
[17] Grioli G, Cont. Mech. Thermod. **15**, 441-450 (2003)
[18] Jenkins JT, J. Applied Mech. **42**, 603-606 (1975)
[19] Jenkins JT, La Ragione L, in Giovine P, Goddard JD, Jenkins JT (Eds.): *IUTAM-ISIMM Symp. Math. Modeling Phys. Instances Granular Flows*, Conf.Proc.Series **1227** (AIP, Melville, New York, 2010) 31-40
[20] Kanatani KI, Int. J. Engng. Sci. **17**, 419-432 (1979)

Oblique shock waves in granular flows over bluff bodies

Nandu Gopan[1,*] and *Meheboob* Alam[1]

[1]*Jawaharlal Nehru Centre for Advanced Scientific Research, Jakkur PO, Bangalore 560064, INDIA*

Abstract. Granular flows around an object have been the focus of numerous analytical, experimental and simulation studies. The structure and nature of the oblique shock wave developed when a quasi-two dimensional flow of spherical granular particles streams past a fixed cylindrical obstacle forms the focus of this study. The binary granular mixture, consisting of particles of the same diameter but different material properties, is investigated by using a modified LIGGGHTS package as the simulation engine. Variations of the solid fraction and granular temperature are analysed and the shock-front is identified. The local Mach number is calculated to distinguish between the subsonic and the supersonic regions of the bow shock.

1 Introduction

The out of equilibrium nature of granular matter leads to dynamically complex flows. The flow of granular matter around an immersed object is a subject of considerable interest given the wide range of scenarios where such configurations are found– a better understanding of the flow dynamics would be essential not just to design better equipment for granular particle processing and handling but also to better grasp the dynamics of flows where thermal agitation plays no role.

Granular flows around an object have been the focus of numerous analytical [1–3], experimental [1, 2] and simulation studies [1, 4].The formation of plane shock waves [5], when a granular medium is perturbed faster than the velocity of sound, has been reported by previous studies. Experimental studies [6, 7] have shown that a stagnant zone formation, similar to the classical "bow-shock" detachment (see Fig. 1), is possible in granular flows. Analytical relations [5] for granular shocks have been developed and Discrete Element Modelling (DEM) simulation studies [4, 8] of single component granular flow past obstacles have been reported in literature.

The present study analyses results from soft particle DEM simulations of two component granular flows over a bluff body. The structure and nature of the oblique shock wave developed when a quasi-two dimensional flow of spherical granular particles streams past an immersed, fixed cylindrical obstacle forms the focus of this study. The aim is to analyse the bow-shock structure in a binary granular mixture and contrast it with well-known predictions for related molecular gas flows.

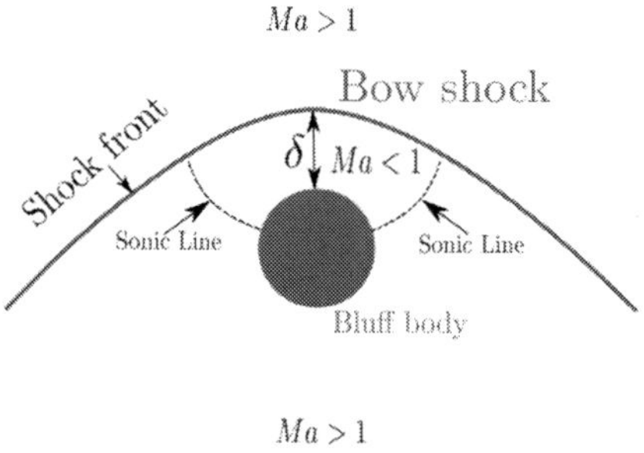

Figure 1: Schematic of bow-shock structure in molecular gas flow over a bluff body. The sonic line demarcates the subsonic and the supersonic regions, while δ denotes the detachment length.

Table 1: Simulation details

Simulation Parameter	Value
Particle diameter, d (m)	0.001
Time-step (s), Δt	5×10^{-7}
Upstream solid fraction, ϕ_∞	0.1
Obstacle diameter, D	$10d/50d/100d$
Upstream Velocity, $U_\infty(\frac{m}{s})$	0.475/1/2.375
Simulation domain size, $x \times y$	$500d \times 500d$
Upstream Mach Number, Ma_∞	2.64/5.9/13.7

2 Configuration and Numerical Method

The flow configuration is shown in Fig. 2 and the simulation details are given in Table 1.The DEM simulations are carried out by employing a modified version of the open-source software LIGGGHTS [9]. The granular flow consists of a binary mixture of steel and aluminium particles which flow over a cylindrical shaped obstacle, of length

*e-mail: nandug@jncasr.ac.in

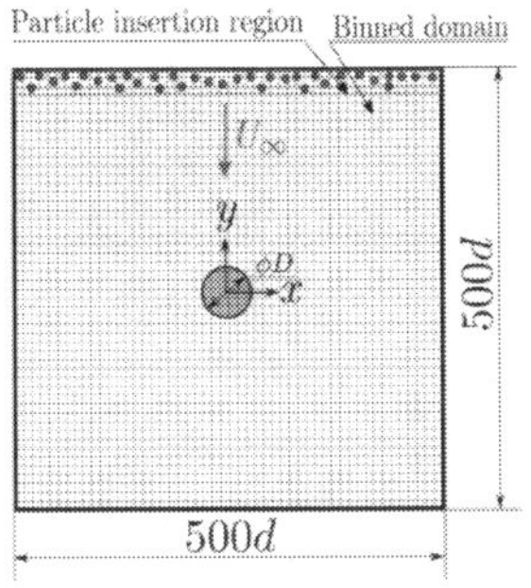

Figure 2: Schematic of the simulation domain. D and d denote the diameters of the cylindrical obstacle and granular particle, respectively, while x and y represent the cross-wise and streamwise directions, respectively. The particle insertion region spans $2d$ along the streamwise direction and $500d$ along the crosswise direction. The simulation domain is binned into square regions of side $3d$ for the calculation of macroscopic fields.

Table 2: Hertzian contact force model

Coefficient	Particle expression
β	$\frac{\ln(e)}{\sqrt{\ln(e)^2+\pi^2}}$
$k_n^* = \frac{k_n}{\sqrt{\delta_n}}$	$\frac{E\sqrt{d}}{3(1-\nu^2)}$
$k_t^* = \frac{k_t}{\sqrt{\delta_n}}$	$3\frac{1-\nu}{2-\nu}k_n^*$
$\gamma_n^* = \frac{\gamma_n}{\sqrt[4]{\delta_n}}$	$\beta\sqrt{\frac{5m_{eff}k_n^*}{2}}$
$\gamma_t^* = \frac{\gamma_t}{\sqrt[4]{\delta_t}}$	$\beta\sqrt{\frac{5m_{eff}k_t^*}{3}}$

d, modelled as aluminium. The Young's modulus (E) of steel and aluminium are taken as 210 GPa and 69 GPa, respectively, while the corresponding Poisson's ratios (ν) are taken to be 0.303 and 0.33; the material density of steel is taken as $\rho_{Steel} = 7850\ kg/m^3$, with $\rho_{Steel}/\rho_{Al} = 2.907$; the effective mass is $m_{eff} = m_{Steel}m_{Al}/(m_{Steel} + m_{Al})$. The coefficient of restitution (e) of all particles is set to 0.5. The particles have equal contribution to the upstream solid fraction of 0.1, representing an equimolar binary mixture.

The normal and tangential forces between particles are calculated, respectively, as:

$$F_n = k_n\delta_n + \gamma_n v_n^{rel} \tag{1}$$

$$F_t = \begin{cases} k_t\delta_t + \gamma_t v_t^{rel}, & \text{if } |F_t| \le \mu_f|F_n| \\ \mu_f|F_n|, & \text{otherwise} \end{cases} \tag{2}$$

Here, μ_f is the coefficient of Coulomb friction while subscripts n and t denote the normal and tangential components, respectively. The parameters k_n, k_t, γ_n and γ_t are calculated from the Young's modulus (E), Poisson's ratio (ν) and the coefficient of restitution (e) as given in Table 2. For all results presented in this study, μ_f is set to zero.

The total simulation time consists of 5 seconds of real time, of which the production run lasts for 4 seconds. The positions, velocities and the forces on each particle are noted and the macroscopic properties of interest are calculated by post-processing the particle trajectory data using an in-house post-processing code; the simulation domain is divided into a number of square bins of side $3d$ (see Fig. 2) and the macroscopic fields (density, granular temperature, Mach number, etc) in each bin are calculated by averaging over the entire production run.

In each bin of volume V_{bin} , the number-density of species α (steel or aluminium) is calculated from $n_\alpha = N_\alpha^{bin}/V_{bin}$ where N_α^{bin} is the 'average' number of particles of type α in that bin (averaged over time), and the local mass-density of species α is defined as

$$\varrho_\alpha = m_\alpha n_\alpha \equiv \rho_\alpha\phi_\alpha \tag{3}$$

where ρ_α is the "material" density of species α and ϕ_α is its volume fraction in a particular bin. The mixture/total mass-density is $\varrho = \sum_\alpha \varrho_\alpha = \varrho_{Steel} + \varrho_{Al}$. The mean velocity of each species is calculated from $\mathbf{u}_\alpha = \langle c_\alpha \rangle$, where $\mathbf{c}_\alpha$ is the instantaneous velocity of species α, with the angular bracket denoting averaging over time. The "mixture" velocity is defined as the mass-averaged velocity $\mathbf{u} = \sum_\alpha \varrho_\alpha \mathbf{u}_\alpha/\varrho$. The "species" granular temperature is calculated from

$$T_\alpha = \frac{1}{\mathcal{D}}\langle m_\alpha(\mathbf{c}_\alpha - \mathbf{u})^2\rangle, \tag{4}$$

and the "mixture" granular temperature is $T = \sum_\alpha \xi_\alpha T_\alpha$, where $\xi_\alpha = n_\alpha/n$ is the "local" (binwise) number fraction of species α.

Given the dilute nature of the flows considered, the flow regime can be understood to be inertial. The upstream Mach number, Ma_∞ is calculated via

$$Ma_\infty = \frac{U_\infty}{a_\infty}, \tag{5}$$

where the adiabatic sound speed, a, is given by [10]:

$$a = \sqrt{\left(\frac{\partial P}{\partial \rho}\right)_s} = \left[\frac{T}{m}\left(1 + \frac{2}{\mathcal{D}}F(\phi) + \frac{\phi}{F}\frac{dF}{d\phi}\right)F\right]^{1/2} \tag{6}$$

where $\mathcal{D}$ is the dimensionality of the system and

$$F(\phi) = 1 + (\mathcal{D} - 1)(1 + e)\phi g(\phi), \quad m = \sum_\alpha m_\alpha \tag{7}$$

and $g(\phi)$ is the radial distribution function

$$g(\phi) = \frac{16 - 7\phi}{16(1 - \phi)^2}. \tag{8}$$

Since the granular temperature varies along the streamwise direction, we have calculated the upstream sound-speed a_∞ by using T and ϕ averaged over two-bins from the inlet.

3 Results and Discussion

The simulations with (i) three different obstacle diameters (D = 10d, 50d and 100d) and (ii) three upstream Mach numbers (M_∞ = 2.64, 5.9 and 13.7) have been analysed. Note that the upstream Mach number Ma_∞ has been varied by varying the upstream inflow velocity U_∞ (see Table 1).

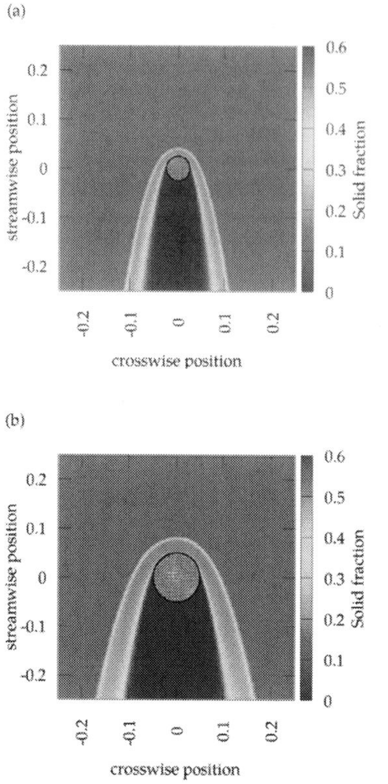

Figure 3: Solid fraction (ϕ) field calculated by bin-averaging over the entire production run, for (a) $D = 50d$ and (b) $D = 100d$. Both cases are for $Ma_\infty = 5.9$.

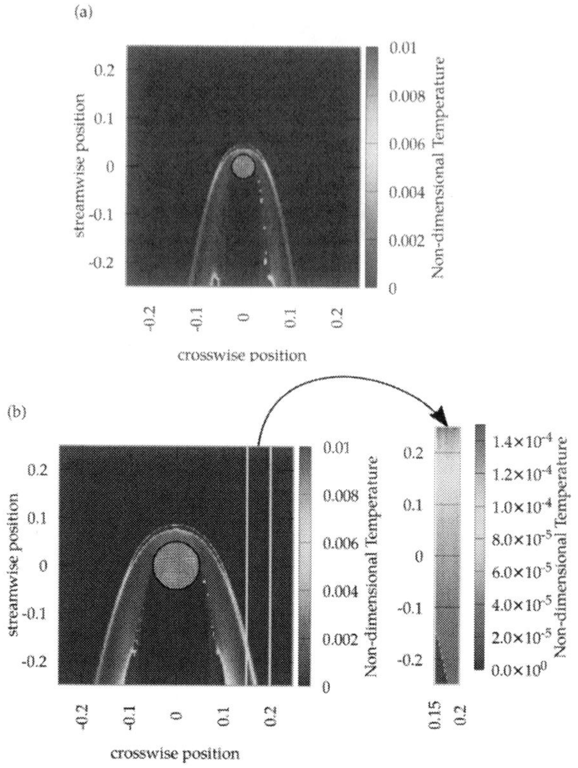

Figure 4: Granular temperature ($T/m_{av}U_\infty^2$) field, with $m_{av} = \sum_\alpha m_\alpha/2$; the parameter values are same as in Fig. 3. The right panel in (b) displays a vertical-strip of the simulation box (enclosed by two yellow lines in the main panel) with a lower range of temperatures.

3.1 Identification of Shock Front

The visualisation of particle trajectories at the end of the production run clearly shows a region of higher density around the cylindrical obstacle, reminiscent of a "bow shock", which is evident from the color-map of solid fraction in Fig. 3. Interestingly, the structure of the bow shock formed is qualitatively different from that known of a molecular gas (viz. Fig. 1) such that the region behind the obstacle (along the streamwise direction) is almost devoid of particles (Fig. 3) in the case of granular flows.

Analysing the dimensionless temperature field in Fig. 4, we find that a region of high temperature occurs between the shock wave front and the cylindrical obstacle. This seems counter-intuitive given the fact that one would expect the solid fraction and the granular temperature to be inversely dependent. However, this relationship is valid only for dilute granular flows. As the solid fraction fields (Fig. 3) have revealed, the solid fraction in the region between the shock front and the cylindrical obstacle is nearly 0.5, which cannot be considered dilute. This calls for an analysis of the stress-field behind the shock front, which can be taken up in future.

3.2 Detachment Region and the Sonic Line

To quantify the shock structure, the local Mach numbers (Ma) are calculated, see Fig. 5. An $Ma = 1$ iso-Mach line (i.e. the sonic line) is computed from the data which is zoomed as an inset panel in Fig. 5. The sonic-line (located just behind the shock-front) demarcates the subsonic and the supersonic regions of the flow – the shape of the sonic-line looks much different from its molecular analog (see the schematic in Fig. 1). Comparing the Mach-number fields in Fig. 5 with the corresponding temperature-fields in Fig. 4, we find that the high temperature region is a region of low Ma, representing a sub-sonic flow (since this region is clearly enveloped by the sonic-line). It has been verified that the high temperature region is a region of low velocities but of high velocity fluctuations (i.e. of higher values of granular temperature and sound speed) - this observation is in qualitative agreement with experimental results [6].

It must be pointed out that, for dissipative granular flows, the upstream Mach number Ma_∞ can be quite misleading. Due to inelastic collisions, the temperature (T)

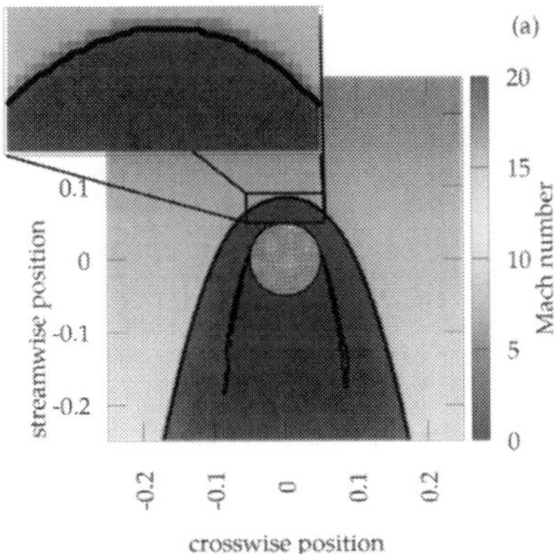

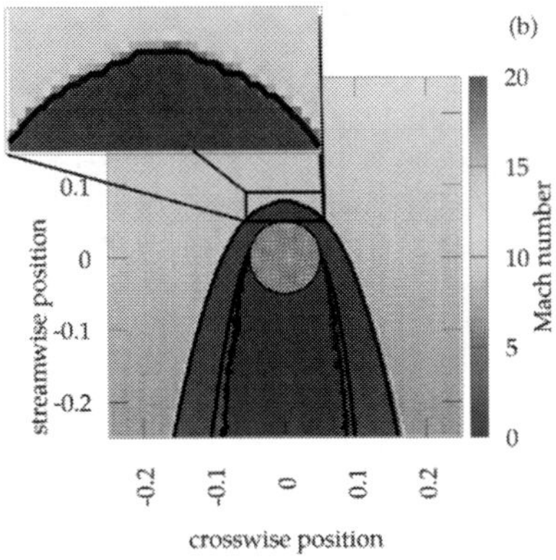

Figure 5: Local Mach number field for (a) Ma_∞ = 5.9 and (b) 13.7 with D = 100d. The inset is a magnified view of the detachment region, showing the sonic-line (Ma = 1).

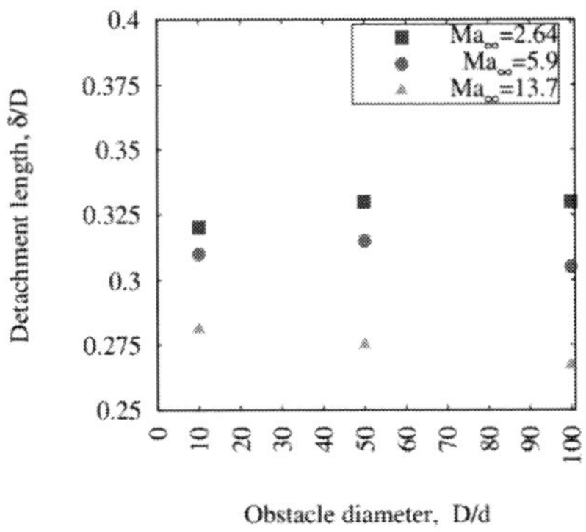

Figure 6: Non-dimensional detachment length, δ/D, plotted against obstacle diameter, D/d, for different Ma_∞.

(the right panel in Fig. 4b) and consequently the sound speed (a) decays downstream – this in turn increases the "local" Mach number as the particles progress downstream. The upstream Mach number can thus be a misleading metric to characterize the "granular" shock since the obstacle would see a different (and in-fact larger) Mach number upstream from the point of collision with granular particles. A similar conclusion has recently been noted in the case of plane granular shock waves [5].

The non-dimensional detachment length, δ/D, (see Fig. 1 for its definition) is plotted against D/d in Fig. 6. It is seen that δ/D decreases with increasing Ma but remains nearly constant, with varying D/d, implying that the detachment length (δ) is a function of obstacle diameter (D) for a given upstream Mach number, at least for the range of Ma_∞ considered in this study.

4 Summary and Outlook

DEM simulations of quasi-two dimensional dilute flows of bi-disperse spherical granular particles past a fixed cylindrical object were carried out. A high temperature region is observed in the "subsonic" detachment region, between the bluff body and the bow-shock front. It is found that inelastic collisions result in a decay of the temperature field along the flow direction, resulting in varying downstream Mach numbers (greater than Ma_∞). The detachment length (δ) is found to be a function of obstacle diameter (D) for the range of Ma_∞ studied. The future work should focus on a better understanding of the bow-shock and the shape of the sonic line by quantifying the effects of (i) the mean density, (ii) the molar-fraction of binary mixture, (iii) the material parameters and (iv) the upstream Mach number.

Acknowledgement

Nandu Gopan would like to gratefully acknowledge the Science and Engineering Research Board (SERB) for support under the National Post-Doctoral Fellowship (PDF/2015/001068).

References

[1] E.C. Rericha, C. Bizon, M.D. Shattuck, H.L. Swinney, Phys. Rev. Lett. **88**, 014302 (2001)
[2] T. Faug, P. Childs, E. Wyburn, I. Einav, Phys. Fluids **27**, 073304 (2015)
[3] X. Cui, J. Gray, J. Fluid Mech. **720**, 314 (2013)
[4] C.R. Wassgren, J.A. Cordova, R. Zenit, A. Karion, Phys. Fluids **15**, 3318 (2003)
[5] M.H.L. Reddy, M. Alam, J. Fluid Mech. **779**, R2 (2015)
[6] Y. Amarouchene, J.F. Boudet, H. Kellay, Phys. Rev. Lett. **86**, 4286 (2001)
[7] T. Atkinson, J. Butcher, M. Izard, R. Nedderman, Chem. Engg. Sci. **38**, 91 (1983)
[8] P.A. Cundall, O.D.L. Strack, Géotechnique **29**, 47 (1979)
[9] C. Kloss, C. Goniva, A. Hager, S. Amberger, S. Pirker, Prog. Comput. Fluid Dyn. **12**, 140 (2012)
[10] S. Savage, J. Fluid Mech. **194**, 457 (1988)

EPJ Web of Conferences 140, 07011 (2017) DOI: 10.1051/epjconf/201714007011
Powders & Grains 2017

The significance of relative density for particle damage in loaded and sheared gravels

Stephen Fityus[1] and *Emőke* Imre[2]

[1] The School of Engineering, University of Newcastle, University Drive, Callaghan, Australia, 2308.
[2] Kandó Kálmán Electrical Engineering Faculty, Obuda University BMF, Bp., Hungary

Abstract. For granular assemblages of strong particles, an increase in the relative density usually leads to a significant increase in shear strength, which is evident as a peak strength, accompanied by significant dilation as the peak strength is attained. This paper describes an experimental study of shearing in assemblages of weak particles, where particle breakage offsets dilation for all but the lowest of confining stresses. In such materials, prone to particle breakage, the shear strengths of loose and dense assemblages rapidly converge to similar values as confining stress increases, and any benefit of greater relative density is lost. This is attributed to the densification effect associated with the loading under a high stress prior to shearing, which is characterised by widespread particle breakage and the formation of smaller particles to occupy space between coarser ones. Interestingly, under both low and high stresses, there was a tendency for greater particle breakage in the loose samples, as a result of both shearing and compression. This result suggests that, despite the denser assemblage having its particles more rigidly constrained and less able to rearrange to avoid direct loading, the influence of greater load-spreading capacity afforded by an increased number of particle contacts in a denser sample, is more dominant in controlling breakage.

1 Introduction

In Geomechanics, the shearing of granular materials is usually studied in the context of soils in which the particles are small and strong relative to the stresses they are likely to encounter under working conditions [1-3].

Rocky materials (waste rock and rock fills) have received considerably less attention. It is generally considered that the strength of smaller particles is greater than the strength of larger particles of the same material, since the likelihood of larger particles containing flaws is inherently greater [4, 5]. Furthermore, structures composed of rock fill (waste rock dumps and rockfill dams) are typically large or very large [6], so the likelihood of particle breakage/crushing is much greater. Hence, the significance of breakage for the behaviour of coarse materials is potentially greater.

It is well understood that breakage in granular materials is related to the particular granular mechanisms which are active during shearing, such as particle rubbing, sliding and rolling, or particle breakage, which may involve comminution of thin corners/edges or complete particle failure [7]. Generally, rubbing, sliding and rolling dominate under lower stresses, whereas breakage will become more prevalent as the applied stresses increase and dilation is suppressed [8]. As a result, it is generally believed that the shear strength envelope is non-linear [9, 10].

Shearing mechanisms and non-linear shear strength envelopes have been linked through the phenomenon of dilation, (sample expansion during shear). The influence of the stress environment (confinement) on dilation has been well studied [1, 7], but the transition from dilation and particle sliding to compression and grain crushing is also affected by the relative density of the granular assemblage, and this has received considerably less attention.

In this paper, the significance of relative density for the dilation response of sheared gravels, and its consequences for the onset and extent of particle breakage are considered.

2 Methods

The material used in this study was mine waste consisting primarily of a siltstone which was obtained from Rolleston Coal mine in the Bowen Basin in Australia. The unconfined compressive strength (UCS) of rocks from which the sample was derived was of the order of 5 MPa.

* Corresponding author: stephen.fityus@newcastle.edu.au

For the purposes of this study, the samples were prepared so that all particles were larger than 4.75mm, and particle damage was characterised by the production of fines, where fines were defined as any particles passing the 4.75mm sieve after testing. So as to avoid issues with possible size effects [5, 6], each sample was prepared to have the same range of initial particle sizes, ranging between 4.75 and 19.6mm.

Tests were performed using a 300 x 300mm Prolab Direct Shear Machine (DSM) with a 1MPa capacity (Fig 1a). Tests were carried out with normal stresses ranging from 9 kPa to 3.5 MPa. To achieve stresses greater than 1 MPa, concrete and steel inserts (Fig 1b) with a 167mm circular sample space were manufactured, that could be inserted into the 300mm shear boxes, reducing the sample area by about 75%. For stresses less than 100 kPa, the mass of the box and insert became significant, and hydraulic control of the DSM (Fig 1d) was unable to apply and maintain the normal stress to sufficient accuracy. So, for the 9 and 48 kPa tests, a special lightweight top box was fabricated (Fig 1c) and the normal load was applied using deadweights (Fig 1 e, f).

Samples were carefully prepared by sieving the required fractions from bulk samples. A fraction mixture with roughly ¼ of each size was selected: 21% 4.75-6.7mm; 23% 6.7-9.5mm; 23% 9.5-13.2mm; 32% 13.2-19mm. After several bench trials, a target mass was determined to allow samples to be prepared to the same consistent, loose initial density for each test: 1126 kg/m3; corresponding to a relative density of 0%.

a) Prolab direct shear machine, with inserts installed

b) inserts for area reduction c) lightweight top box

d) hydraulic loading e) 9kPa loading e) 48kPa loading

Fig. 1. Features of the testing arrangement.

Dense specimens were created by dry pluviation, in which the sample was "rained" by delivering it slowly through a 1m long, 150mm diameter PVC tube, fitted with a 19mm grid mesh at half height. This resulted in the same consistent, dense initial density for each test of 1299 kg/m3; corresponding to a relative density of 71%.

Acknowledging that significant particle breakage may occur in the application of the normal compressive stress, even before the sample is sheared, particle breakage during shear was estimated as follows:

1. Prepare a new sample, apply compression, determine fines.
2. Prepare a new sample, apply compression, shear, determine fines.
3. Estimate fines due to shear as the difference between samples tested by compression and shear (2.), and compression only (1.).

Determination of fines after compression and/or shearing involved careful recovery of all the tested material and a sieve analysis. After testing under 3.5 MPa, the compression of the siltstone was so substantial that the recovered sample was effectively a monolith. Before it could be sieved, it had to be broken up. This was achieved very carefully by gently teasing out crushed and recompacted material that had clearly been reconstituted by the test process, whilst not breaking down primary particles that had not been damaged.

3 Results

Figure 2 shows the shear stress/shear displacement data for low and high density tests, plotted in terms of the secant friction angle (tan $^{-1}$ (shear stress/normal stress)). At low normal stresses, the shear strength of dense samples exceeds the shear strength of the loose samples, but with increasing normal stress, the difference reduces. For normal stresses of 250 kPa or greater, there is no significant difference in the peak shear strength, although the shape of the shear strength response differs.

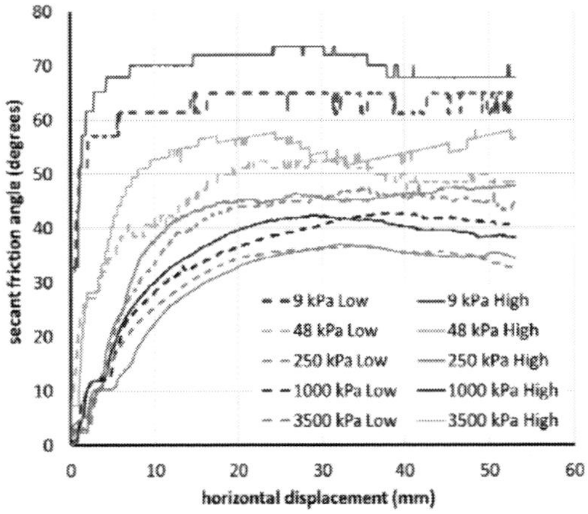

Fig. 2. Shear strength (secant friction angle) vs horizontal displacement for low density and high density tests.

Figures 3a and b show the peak secant friction angle for the low and high density samples. The severe and rapid reduction in the peak strength with increasing stress on a linear scale in Fig 3a) is consistent with the general expectation of shear strength envelope non-linearity. Whilst the non-linearity is strong for both the loose and dense samples, the friction angle, and its corresponding decrease with increasing normal stress, are greater for the denser samples.

However, by the time the normal stress reaches 250 kPa, the peak friction angle is similar for loose and dense samples, and it remains so as the normal stress becomes very large. The convergence of the loose and dense peak friction angles, approaching a normal stress of 250kPa, is clearly evident when the data is considered on a log normal stress scale in Fig 3b).

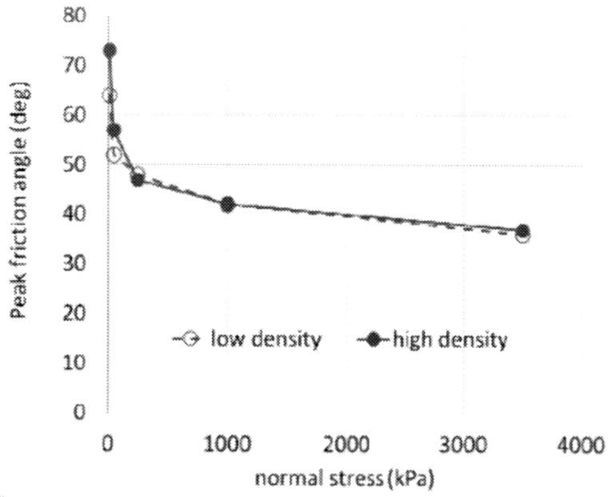

a) normal stress on a linear scale

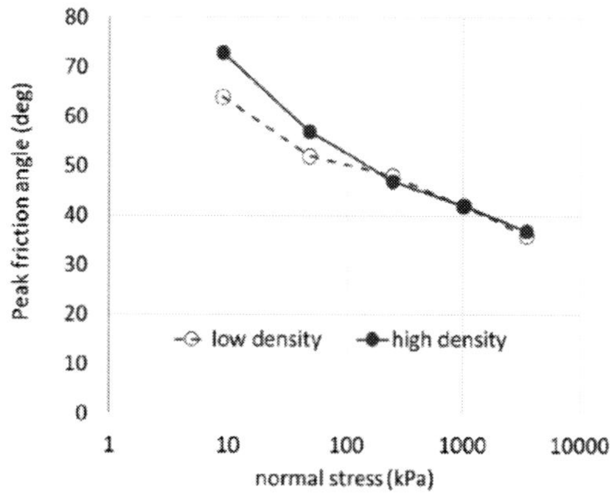

b) normal stress on a log scale

Fig. 3. Peak friction angle as a function of normal stress

Figure 4 shows the average vertical displacement (independent of top cap rotation) measured for all of the loose and dense sample tests. At very low normal stresses, the samples exhibit some compression before they begin to dilate; at higher normal stresses, they exhibit only compression. As expected, for the same normal stress, dilation is greater and compression is less in the denser samples than in the looser samples.

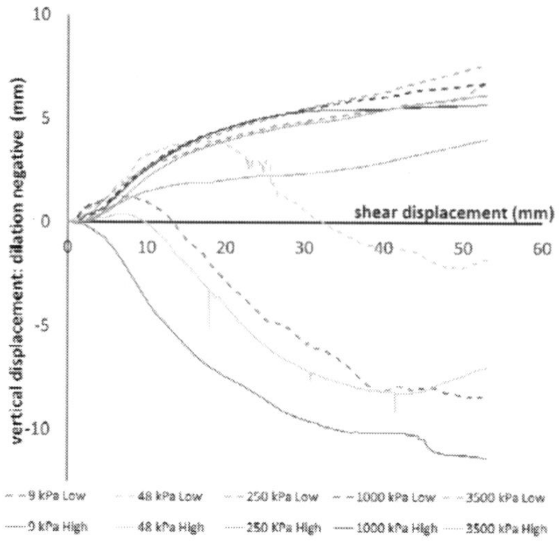

Fig. 4. Vertical displacement (dilation) vs shear displacement for loose and dense samples.

Figure 5 shows the effect of sample density on the extent of dilation/compression for different values of normal stress. The results suggest that dilation ceases at a normal stress of around 70 kPa for the loose samples, but persists up to normal stresses of around 150 kPa for denser samples. It would seem that in denser arrangements, the particle skeleton resists compression, and dilation is able to persist, beyond stresses that could force sample rearrangements by particle repacking, up to stresses which would instead cause particle breakage.

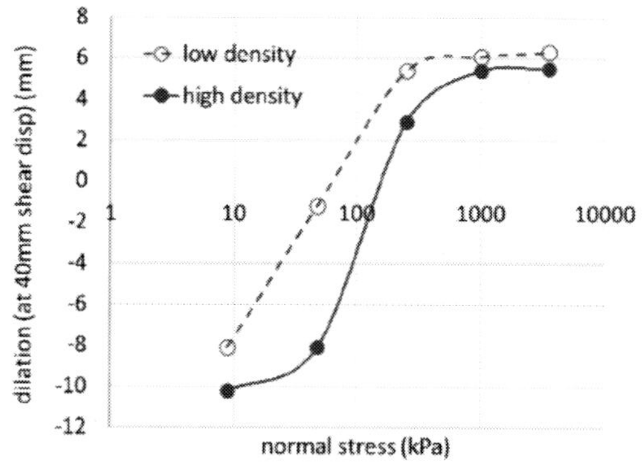

Fig. 5. Dilation (negative)/compression (positive) vs normal stress (values at 40mm shear displacement).

To explore this idea further, it is necessary to consider the breakage behaviour as a function of the vertical stresses applied. Figure 6 shows the breakage due to compression (application of the normal stress), breakage due to compression and shearing, and by difference, the breakage due to shear alone.

Figure 6 suggests that particle breakage due to compression is relatively low until a normal stress of 250 kPa is reached. Thereafter, it continues to increase steadily as the (log) normal stress increases.

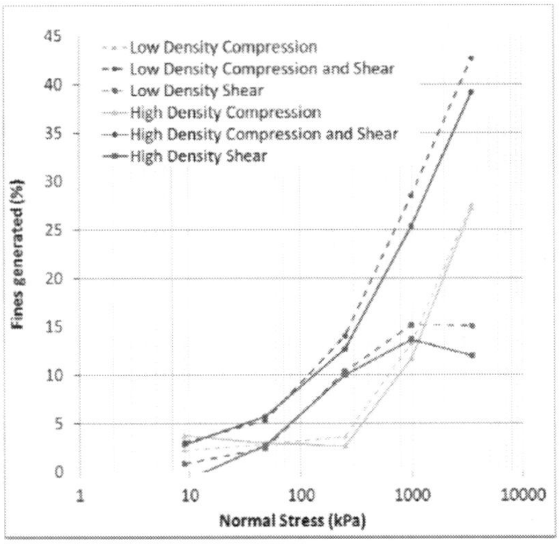

Fig. 6. Particle breakage due to compression, compression and shearing, and shearing in loose and dense samples.

Breakage due to shearing increases from the lowest of stresses, but increases more rapidly between normal stresses of 50 and 250 kPa, after which, the rate of breakage decreases, until reaching a maximum value at a normal stress of 1 MPa. At higher stresses, the high density sample actually experiences less breakage due to shear.

Other than for the high density sample at 9 kPa (where the negative breakage value indicates an obvious error in very small masses), the loose samples tend to experience similar breakage to dense samples under low stresses, and more breakage than dense samples at higher stresses. This suggests that dense samples, despite being stiffer under low stresses (less able to compress due to particle repacking) are actually less prone to particle damage under high stresses. The reason for this is likely to be that denser packing leads to an increased number of particle contacts, allowing the particle point pressures to be reduced, hence causing less damage when loaded under a given high stress. The consequence of this, however, is that the loose samples, in experiencing greater damage, will attain a more efficient compressed structure and greater densification, causing them to become more like the dense samples under high normal stresses.

4 Conclusions

In light of the above observations, the following behavioural model is proposed for the significance of relative density for coarse granular materials with weak particles.

Under low normal stresses (ie, those insufficient to cause significant particle breakage) dense materials will record higher strengths than loose materials, because loose materials will tend to be repacked during shear to become denser, whereas already-dense materials will dilate to maximise their mobilised grain-sliding resistance.

Under high normal stresses (ie, those sufficient to cause significant particle breakage) loose materials will undergo greater particle crushing during normal load compression than dense materials, allowing them to attain denser structures which, consequently, result in them attaining similar strengths as denser materials when sheared under the same high stresses.

The authors would like to acknowledge the excellent work of Mr Jarrad Phelps in obtaining the experimental results, and the funding support of Australian Coal Research Limited.

References

1. P. Lade, J. Yamamuro, and P. Bopp, (1996) Significance of particle crushing in granular materials. Journal of Geotech. Eng. 122 pp 309-316.
2. Y. Nakata, Y. Kato, M. Hyodo, A.F.L. Hyde and H. Murata (2001) One-dimensional compression behaviour of uniformly graded sand related to single particle crushing strength. Soils and Foundations. 41, pp 39–51.
3. M. Bolton (1986). The strength and dilatancy of sands. Géotechnique 36, No. 1, pp 65-78
4. C. Ovalle, C. Dano and P.-Y.,Hicher (2013). Experimental data highlighting the role of surface fracture energy in quasi-static confined comminution. International Journal of Fracture, Vol. 182, pp 123-130.
5. W. Weibull (1939) A statistical theory of the strength of materials. Proceedings of Royal Swedish Institute of Engineering. Res. 151
6. S. Linero, C. Palma and R. Apablaza, (2007) Geotechnical characterization of waste material in very high dumps with large scale triaxial testing. Proceedings of International Symposium on Rock Slope Stability in Open Pit Mining and Civil Engineering, Perth, Australia,.pp 59-75 (ed.: Potvin, Australian Centre for Geomechanics).
7. C. Ovalle, C. Dano, P.-Y.,Hicher and M. Cisternas, (2015) Experimental framework for evaluating the mechanical behavior of dry and wet crushable granular materials based on the particle breakage ratio. Canadian Geotechnical Journal 52. pp 587-598.
8. R. Marsal (1973) Mechanical properties of rockfill dams in Embankment dam engineering: Casagrande Volume., eds. R.C. Hirschfeld, , S.J. Poulos, Wiley, New York, pp 109-200.
9. T.M. Leps (1970) Review of shearing strength of Rockfill. Journal of Soil Mechanics and. Foundations Div. ASCE 96, pp1159-1170.
10. B. Indraratna, L. Wijewardena, and A. Balasubramaniam (1993) Large-scale triaxial testing of greywacke rockfill. Géotechnique 43, pp 539-543.

Coupling multibody system and granular dynamics

application to a 2D benchmark

Olivier Lantsoght[1,⋆], *Paul* Fisette[1,⋆⋆], *Frédéric* Dubois[2,⋆⋆⋆], *Olivier* Brüls[3,⋆⋆⋆⋆], and *Nicolas* Docquier[1,†]

[1] *Université Catholique de Louvain, Belgique*
[2] *Université de Montpellier, France*
[3] *Université de Liège, Belgique*

Abstract. We present a benchmark to experimentally validate the method we have developed to simulate mechanical devices interacting with granular media. Consequently the method we develop is able to solve problems involving bilateral and unilateral constraints. The mechanical device is modelled with the multibody system approach and is described by the use of relative coordinates. With these coordinates, the bilateral constraints are mainly due to loop-closure conditions (or specific bilateral constraints such as imposed screw-joint motion). The bilateral constraints are solved thanks to the coordinate partitioning method. The problem of contact dynamics, which introduces the unilateral constraints, is solved by a non-linear Gauss-Seidel procedure applied in the contacts coordinates. The procedure accounts for the constrained motion of the mechanism because the coordinate partitioning is also applied to the mapping between the contacts coordinates and the generalized coordinates. Consequently the solution of the unilateral constraints problem is locally compatible with the set of bilateral constraints coming from the mechanical system. The proposed planar benchmark consists of a modified four-bar mechanism blending a set of disks contained in a vibrating box. The numerical results reveals the influence of the vibration frequency on the granular compactness and thus the mechanism motion.

1 Introduction

Our research project deals with the tamping process of railway tracks. This process is part of the maintenance operation done on ballasted railway tracks. This particular operation restores the track geometry after some degradation. During the process the ballast grains are shacked at certain frequency range and then squeezed under the sleepers. Initially the ballast is a compact and rigid assembly of stones, when it is vibrated the compactness goes down which allows an easier squeezing. This application is an example of an articulated mechanism interacting with a granular media which takes advantage of the different state of behaviour of the ballast. In our work we want to simulate such process. To this purpose we suggest to couple the multibody systems dynamics (MBS) approach with the discrete element method (DEM) using non-smooth contact dynamics.

The objective of this paper is firstly to explain how we coupled MBS and DEM and secondly to present the experimental benchmark we are building to validate our methodology. The contact can be modelled either by the molecular dynamics (regularized contact law) or the non-smooth contact dynamics (non-regularized contact law). The contact dynamics has been chosen due to the high contact stiffness of the ballast and because it has proven effective to deal with railway ballast application. Previous works have been conducted on the coupling between MBS and non-smooth contact dynamic. For example Flores et al. [1] extend the approach of Jean [2] to unconstrained MBS. In Ref [3], Brüls and al. present the non-smooth generalized-α scheme which enables to enforce the unilateral constraints both at position and velocity levels for frictionless contact. This approach allows to solve the drift-off effect which is observed when the constraints are solved at velocity level only. The particularity of our work is the capacity of modelling granular media involving a lot of contacts and the ability of solving three-dimensional problem thanks to the NLGS procedure.

To validate our methodology, we are developing a planar benchmark that can be experimentally constructed and analyzed. The article will firstly describe this benchmark. Then the benchmark will be used to present the equation of motion of the MBS that will be coupled with the DEM equations of motion and contact equations. We will quickly describe the procedure used to solve the contact dynamics problem. Finally we will introduce some of our numerical results that were obtained in order to design the experimental benchmark. In near future, the benchmark will be build to compare experimental data's with numerical results.

⋆e-mail: Olivier.Lantsoght@uclouvain.be
⋆⋆e-mail: Paul.Fisette@uclouvain.be
⋆⋆⋆e-mail: Frederic.Dubois@umontpellier.fr
⋆⋆⋆⋆e-mail: O.Bruls@ulg.ac.be
†e-mail: Nicolas.Docquier@uclouvain.be

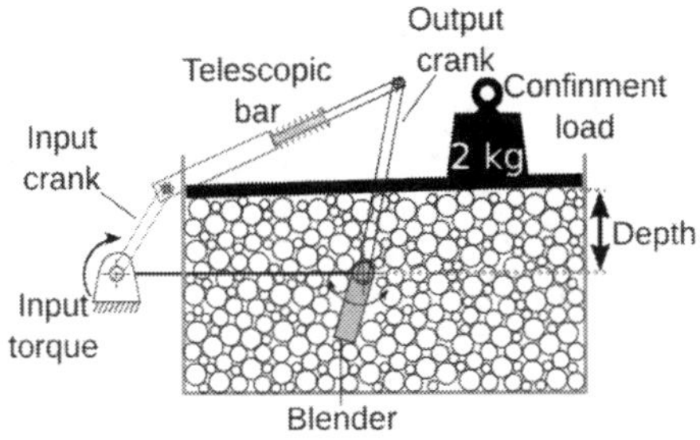

Figure 1. Sketch of the benchmark

2 Benchmark

The goal of the experimental benchmark is the consolidation of our development by experimental measurements. On one hand the benchmark has to represent a typical MBS that means containing at least one kinematic loop. On the other hand it has to be representative of the specific behaviour of granular media such as the fluidization or the jamming phenomena. To this purpose we suggest a benchmark (figure 1), composed of three parts:

1. The mechanical device is a **four-bar mechanism** i.e. a system composed of a floating rod connected to an input and output crank. Some compliance is introduced by using a telescopic bar instead of one arm. The position of the telescopic bar in the system (at the input, middle or output crank) is a design parameter to be investigated. The movement of the device is induced by a motor linked to the input crank. The last crank of the device is connected to a blender that will interact with the granular media.
2. The **granular media** is composed of disks of different radii. The disks are located in a box, with transparent lateral panels. This allows us to track the motion of the disks and observe the force network if they are made of a photoelastic material. The granular media is excited by the movement of the blender and also by a vertical movement of the box.
3. The **box**, containing the disks, can move vertically in order to introduce a vibration in the granular media. The upper part of the box is open in order to apply a confinement load on the disks.

Depending on the length of the different bars and on the position of the telescopic bar we may have a simple oscillation of the blender (as illustrated in figure 1) or a more complex movement including a full rotation. The motion of the mechanism, such as the compression of the telescopic bar or the movement of output crank, will be impacted by the compactness of the granular media. This compactness will be affected by the vibration generated by the box.

3 Equations of motion

Modelling this benchmark requires to couple two different techniques. On one hand the four-bar mechanism is well described by the MBS dynamics. On the other hand the DEM is well suited to describe the granular dynamics. It is the contact interaction that causes the coupling between the two systems.

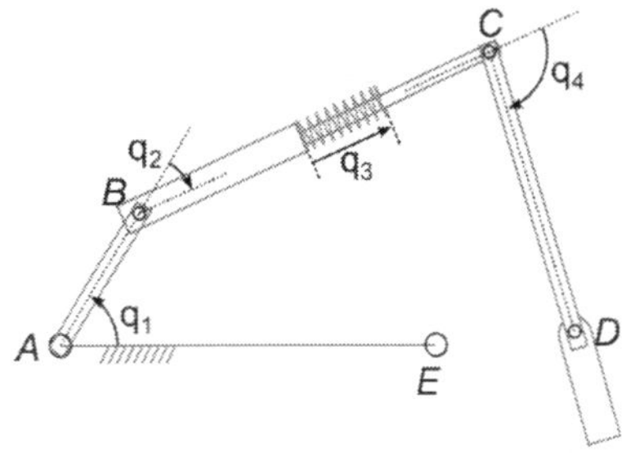

Figure 2. Kinematic description of the benchmark

3.1 Multibody system

The MBS formalism is used to model the mechanical device of the experiment, i.e. the four-bar mechanism. The kinematics of the system is characterized thanks to the relative coordinates ($\mathbf{q}_{mbs}$). This means that the position and orientation of a body is defined with respect to the previous body, the first body of the chain being the ground. The relative coordinates system implies that the assembly of tree-like mechanisms at the articulations is automatically satisfied. When a mechanism contains kinematic loops the generalized coordinates are completed by the following equations of constraints:

$$\begin{aligned} \mathbf{h}(\mathbf{q}_{mbs}) &= \mathbf{0} \\ \dot{\mathbf{h}}(\mathbf{q}_{mbs}, \dot{\mathbf{q}}_{mbs}) &= \mathbf{J}(\mathbf{q}_{mbs})\,\dot{\mathbf{q}}_{mbs} \\ \ddot{\mathbf{h}}(\mathbf{q}_{mbs}, \dot{\mathbf{q}}_{mbs}, \ddot{\mathbf{q}}_{mbs}) &= \mathbf{J}(\mathbf{q}_{mbs})\,\ddot{\mathbf{q}}_{mbs} + \dot{\mathbf{J}}(\mathbf{q}_{mbs}, \dot{\mathbf{q}}_{mbs})\,\dot{\mathbf{q}}_{mbs} \end{aligned} \quad (1)$$

with $\mathbf{J}(\mathbf{q}_{mbs}) = \frac{\partial \mathbf{h}(\mathbf{q}_{mbs})}{\partial \mathbf{q}^T_{mbs}}$. The kinematics of our four-bar mechanism is fully described by a set of 4 coordinates (figure 2) completed by a set of two constraints. In this example the set of constraints ($\mathbf{h}(\mathbf{q}_{mbs})$) imposes that the point D on the output bar is located at the position of the fixed point E on the ground. Such a constraint will only allow a rotation of the output bar about the point E. The dynamic equations of constrained MBS are:

$$\mathbf{M}_{mbs}(\mathbf{q}_{mbs})\ddot{\mathbf{q}}_{mbs} + \mathbf{c}(\mathbf{q}_{mbs}, \dot{\mathbf{q}}_{mbs}, \mathbf{frc}, \mathbf{trq}, t) = \mathbf{Q}(\mathbf{q}_{mbs}, \dot{\mathbf{q}}_{mbs}) + \mathbf{J}^T\lambda \quad (2)$$

with λ being the Lagrange multipliers representing the forces and torques due to the constraints, $\mathbf{Q}$ is the joint forces, $\mathbf{M}_{mbs}$ is the mass matrix of the system and $\mathbf{c}$ is the non linear dynamical vector that includes the effect of external forces **frc** and torques **trq**. The set of equations composed by equations 2 and 1 form a set of differential algebraic equation. This set of equations is reduced to a set of ordinary differential equation thanks to the coordinate partitioning method. Full details about the method are given in [4]. The main idea is to split the set of coordinates in two subsets. The first one contains the independent coordinates ($\mathbf{q_u}$) that will be time integrated while the second

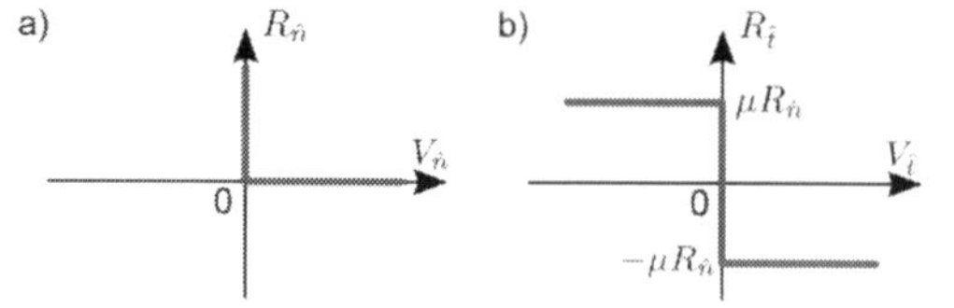

Figure 3. Contact force: a) Signorini's condition in normal direction b) Coulomb's condition in tangential direction

set contains the dependent coordinates ($\mathbf{q_v}$) that will be determined thanks to the equation of constraints. The equation of motion for independant coordinates is then written in the reduced form:

$$\mathbf{M_r}(\mathbf{q_u})\ddot{\mathbf{q}}_\mathbf{u} = \mathbf{F_r}(\mathbf{q_u}, \dot{\mathbf{q}}_\mathbf{u}) \tag{3}$$

3.2 Discrete Element Method

The DEM consists in analysing the granular media behaviour by computing the evolution of each particle of the media and its interaction with other particles. The position of each particles is described using the absolute coordinates (i.e. with respect to the ground fixed frame). Expressing and integrating the angular velocities in each body fixed frame leads to a constant mass matrix $\mathbf{M}$:

$$\mathbf{M}_{dem}\ddot{\mathbf{q}}_{dem} = \mathbf{F}_{dem}(\mathbf{q}_{dem}, \dot{\mathbf{q}}_{dem}) \tag{4}$$

with $\mathbf{F}_{dem}$ being the non linear dynamical vector including the external forces and torques. This equation must be completed to account for the interaction laws between grains and with the MBS, as explained in the next section.

3.3 Contact dynamics

To model the contact between bodies we choose the approach of non-smooth contact law which prevent the bodies from interpenetration. The equation in the normal direction is:

$$\text{if } g \leq 0 \text{ then } V_{\hat{n}} \geq 0,\ R_{\hat{n}} \geq 0,\ V_{\hat{n}}\, R_{\hat{n}} = 0 \tag{5}$$

where g is the gap, $V_{\hat{n}}$ and $R_{\hat{n}}$ are the normal component of the velocities and reaction force between the bodies. In the tangential direction a coulomb friction law is used:

$$\begin{array}{l} \text{if } g \leq 0 \text{ then } \|\mathbf{R}_{\hat{t}}\| \leq \mu R_{\hat{n}} \\ \text{and if } \|\mathbf{V}_{\hat{t}}\| \neq 0 \text{ then } \mathbf{R}_{\hat{t}} = -\mu R_{\hat{n}} \frac{\mathbf{V}_{\hat{t}}}{\|\mathbf{V}_{\hat{t}}\|} \end{array} \tag{6}$$

where μ is the friction coefficient, $\mathbf{V}_{\hat{t}}$ and $\mathbf{R}_{\hat{t}}$ are the tangential vector of the velocities and reaction force between the bodies. The equations 5 and 6 illustrated in figure 3 are referred as the "Signorini-Coulomb" condition.

We use the Newton impact law $V_{\hat{n}}^{+} = -eV_{\hat{n}}^{-}$, to describe the dynamics of contact. In this equation $e \in [0, 1]$ is the coefficient of restitution and the superscript $-$ and $+$ indicate the pre- and post-impact velocity. This law introduces discontinuities in the velocity field of the bodies with consequence that the equations of motion are no more integrable using standard methods based on the evaluation of the accelerations.

3.4 Coupled equations of motion

The coupling between MBS and DEM sets of equations is achieved by the contact conditions. This is taken into account by adding the unilateral constraints in the equations of motion of both system (eq. 3 & 4). These coupled equations:

$$\mathbf{M_r}(\mathbf{q_u})d\dot{\mathbf{q}}_\mathbf{u} = \mathbf{F_r}(\mathbf{q_u}, \dot{\mathbf{q}}_\mathbf{u})dt + \mathbf{H}_{\mathbf{r}_{mbs}}(\mathbf{q_u}, \mathbf{q}_{dem})\, d\mathbf{I}_{\mathcal{U}} \tag{7}$$

$$\mathbf{M}_{dem}d\dot{\mathbf{q}}_{dem} = \mathbf{F}(\mathbf{q}_{dem}, \dot{\mathbf{q}}_{dem})dt + \mathbf{H}_{dem}(\mathbf{q_u}, \mathbf{q}_{dem})\, d\mathbf{I}_{\mathcal{U}} \tag{8}$$

completed by the contacts equations (5 and 6) are written in terms of differential measures of the velocities $d\dot{\mathbf{q}}_\mathbf{u}$ and $d\dot{\mathbf{q}}_{dem}$. $\mathbf{dI}_{\mathcal{U}}$ are the differential measures of the contact reactions over the time step. The matrices $\mathbf{H}_{\mathbf{r}_{mbs}}$ and $\mathbf{H}_{dem}$ are used to project the quantities expressed in the contacts frame to the coordinate system of each domain: $\mathbf{V} = \mathbf{H}^T\dot{\mathbf{q}}$ where $\mathbf{H}$ is the concatenation of $\mathbf{H}_{\mathbf{r}_{mbs}}$ with $\mathbf{H}_{dem}$, $\dot{\mathbf{q}}$ the concatenation of $\dot{\mathbf{q}}_\mathbf{u}$ with $\dot{\mathbf{q}}_{dem}$ and $\mathbf{V}$ the velocities in the contacts (concatenation of $\mathbf{V}_{\hat{\mathbf{n}}}$ with $\mathbf{V}_{\hat{\mathbf{t}}}$ for each contact).

4 Resolution of the dynamics

We applied to our coupled equations of motion a time discretization which is close to the time stepping scheme of the Non-Smooth Contact Dynamics submitted by Moreau and Jean [2]. In this scheme the problem of the contact dynamics is written in the contact frame:

$$\begin{array}{ll} \mathbf{V}_{n+1} = & \mathbf{V}_{free} + \mathbf{W}\,\mathbf{I}_{\mathcal{U}} \\ \text{with :} & \mathbf{V}_{free} = \mathbf{H}^T_{(\mathbf{q}_m)}\left(\dot{\mathbf{q}}_n + \mathbf{M}^{-1}_{(\mathbf{q}_m)} h\mathbf{F}_{(\mathbf{q}_m, \dot{\mathbf{q}}_n, t)}\right) \\ & \mathbf{W} = \mathbf{H}^T_{(\mathbf{q}_m)}\mathbf{M}_{(\mathbf{q}_m)}\mathbf{H}_{(\mathbf{q}_m)} \\ & \mathbf{q}_m = \mathbf{q}_n + (1-\theta)\, h\dot{\mathbf{q}}_n \end{array} \tag{9}$$

where $\mathbf{M}$ and $\mathbf{F}$ groups the mass matrix, respectively the force vector, of MBS and DEM. h is the time step and θ a parameter of the time integrator. This problem is solved in agreement with the contact law (eqs. 5 and 6) with a Non-Linear Gauss-Seidel (NLGS) procedure. Full resolution of the contact dynamics is available in [2] and is no more detailed here. The NLGS procedure used to solve the contact problem enable us to deal with 3D problem.

5 Numerical analysis of the benchmark

We have applied the methodology described above to simulate some possible configuration of our four-bar mechanism. The results for the two configurations are summarized in the table 1. The radius of disks were randomly chosen between 5 equally spaced values between 6 and 10 mm. For physical data's (mass, inertia) the density of aluminium (2710 kg/m^3) and a width of 10 mm were chosen. The motor have an imposed operating speed of 15 rpm. The stiffness of the telescopic bar was chosen at 500 N/m. The amplitude of the box vertical motion is set at 4 mm. The confinement load is applied to increase the resistance of the media toward the blender motion and increase the amplitude of telescopic bar motion.

On the first configuration the complete rotation of the input crank will produce an oscillation on the output crank.

Table 1. Configuration of the benchmark (fig. 2), final dimensions may differ due to design constraints.

	Config. 1	Config. 2
Telescopic bar	*BC*	*AB*
Length $\|\|AB\|\|$	150 *mm*	200 *mm*
Length $\|\|BC\|\|$	575 *mm*	200 *mm*
Length $\|\|CD\|\|$	450 *mm*	200 *mm*
Length $\|\|AE\|\|$	450 *mm*	150 *mm*
Output motion	Oscillation	Rotation
Confinement load	2 *kg*	4 *kg*

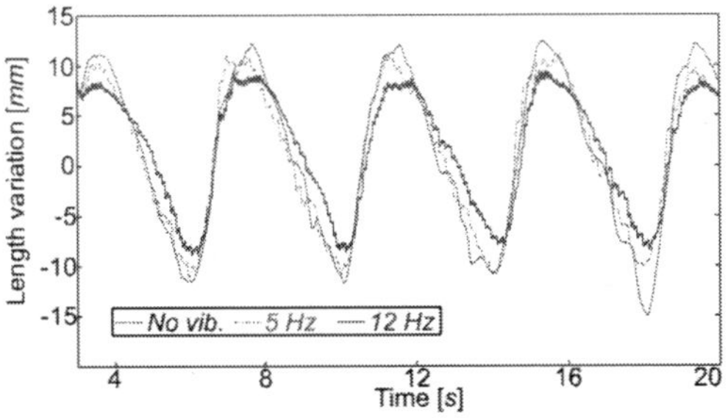

Figure 4. Variation of telescopic bar's length in configuration 1

The figure 4 illustrates the length variation of the telescopic bar for different vibration frequencies of the box. We can observe at each rotation of the input crank that the solicitation of the telescopic bar is higher in absence of vibrations. Moreover a low frequency vibration (5 *Hz*) leads to similar results as the case without vibration. If we apply a higher vibration than 12 *Hz*, we have observed (for example with 25 *Hz*, not traced for readability reason) that the solicitation of the telescopic bar is not reduced any more.

On the second configuration a complete rotation of the input crank, which is the telescopic bar, will produce a complete rotation of the output bar. However if the length variation exceeds a certain value an oscillation will be produced at the output crank. The figure 5 illustrates the length variation of the telescopic bar for different vibration of the box. With this configuration the difference of solicitations for different vibration frequencies is not systematically observed at each turn of the input crank. On average we can still observe a higher solicitation in the absence of vibration or with a low frequency in comparison with the vibration of 12 *Hz*.

The solicitation of the telescopic bar is due to the resistance of the granular media. This resistance is an image of the media compactness. We expect a sudden decrease of the compactness when the vibration have induced a fluidized state of the granular media. As first approximation it has been considered that this state was reached when the maximal acceleration due to the box vibration matches the gravity acceleration. This corresponds to a frequency of 8 *Hz* with the amplitude of 4 *mm*. We can observe, on figures 4 and 5, that the solicitation of the telescopic bar is reduced when the frequency is over 8 *Hz* in comparison of a lower frequency. Moreover once the state of fluidization is reached, the solicitation does not decrease any more when the vibration frequency is increased.

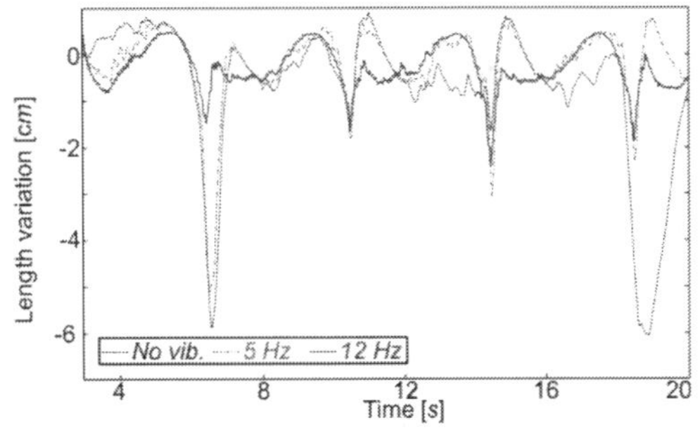

Figure 5. Variation of telescopic bar's length in configuration 2

6 Conclusion

We developed and applied a new methodology to solve problems involving constrained mechanical devices interacting with granular media. We modelled the fluidization of a granular media and observe the impact on the mechanical device. The results of the simulation were coherent with our expectations. We are now designing the experimental version of the benchmark in order to compare in a near future experimental data with numerical results.

In the future, we will challenge the time-stepping method for a better consideration of the non-linearity in the mechanical devices. A possible approach is to use a specific integrator for the smooth motion and then solving the non smooth part at each step as proposed in [5] and [3].

Acknowledgements

This work is supported by a FRIA grant from the Fonds National de la Recherche Scientifique (FNRS) of Belgium.

References

[1] P. Flores, R. Leine, C. Glocker, Multibody System Dynamics **23**, 165 (2010)

[2] M. Jean, Computer Methods in Applied Mechanics and Engineering **177**, 235 (1999)

[3] O. Brüls, V. Acary, A. Cardona, Computer Methods in Applied Mechanics and Engineering **281**, 131 (2014)

[4] N. Docquier, A. Poncelet, P. Fisette, Computer Methods in Applied Mechanics and Engineering **4**, 199 (2013)

[5] Q.z. Chen, V. Acary, G. Virlez, O. Brüls, International Journal for Numerical Methods in Engineering **96**, 487 (2013)

Dry granular avalanche impact force on a rigid wall of semi-infinite height

Adel Albaba[1,2,*], *Stéphane* Lambert[1,**], and *Thierry* Faug[1,***]

[1] *Univ. Grenoble-Alpes, Irstea, ETGR, 38402 St-Martin d'Hères, France*
[2] *Berne University of Applied Sciences, School of Agricultural, Forest and Food Science HAFL, Länggasse 85, CH-3052 Zollikofen, Switzerland*

Abstract. The present paper tackles the problem of the impact of a dry granular avalanche-flow on a rigid wall of semi-infinite height. An analytic force model based on depth-averaged shock theory is proposed to describe the flow-wall interaction and the resulting impact force on the wall. Provided that the analytic force model is fed with the incoming flow conditions regarding thickness, velocity and density, all averaged over a certain distance downstream of the undisturbed incoming flow, it reproduces very well the time history of the impact force actually measured by detailed discrete element simulations, for a wide range of slope angles.

1 Introduction

The mitigation of natural hazards related to landslides and avalanche requires a thorough understanding of the interaction between large-scale granular flows and civil engineering structures. This includes the development of new knowledge on the kinematics of free-surface granular flows—down inclines—disturbed by the presence of obstacles, as well as on the force imparted to the obstacle. One interesting approach is to develop depth-averaged analytic solutions, useful for engineers in charge of the design of structures subjected to landslides and avalanches, as recently reviewed in [1]. The present paper describes a contribution to this topic, showing how an analytic model based on depth-averaged shock theory can capture the impact force of a granular avalanche-flow on a rigid wall which cannot be overtopped (wall of semi-infinite height), provided that the relevant depth-averaged properties of the incoming flow are used as input parameters.

Numerical simulations using the discrete element method (DEM) to mimic the impact of a granular avalanche-flow on a rigid wall are briefly described in section 2. An analytic force model based on depth-averaged shock theory is presented in section 3. Section 4 addresses the main outcomes of our study regarding the cross-comparison between the depth-averaged analytic model and the results from DEM simulations. Finally, section 5 concludes on the need for future research.

2 DEM simulations

The impact of the granular avalanche-flow on the rigid wall is simulated with the help of DEM, using the open-source YADE software [2]. Samples were prepared using clumped particles ranging from 12.5 to 25 mm in diameter, interacting through a linear-dashpot model with a Mohr-Coulomb failure criterion. Particles were flowing in an inclined chute with different angles ranging from 30 to 55 degrees which is bound by a rigid wall at the bottom. Much more detail on the parameters used in the DEM simulations, as well as on the validation of the DEM model on existing laboratory tests on dry granular avalanche impact on a rigid wall, can be found in [3].

Figure 1 shows series of snapshots of the granular flow evolution. It starts with the impact of a very dilute front that is too discrete in nature and can not be represented by a continuum model. Afterwards, a more dense incoming flow arrives and a distinguished dead zone (gray) can be observed. That zone is followed by a green zone representing the part where the incoming flow conditions (height, velocity ..etc) are averaged and then used in the analytic model as inputs. Different inclination angles (Fig.1 top and bottom) leads to different evolving and final dead zone shape, basal friction contribution and propagating wave speed. The objective of the depth-averaged analytic model is to determine the evolution of impact force with time including all the complexities arising from different inclination angles.

3 Depth-averaged analytic model

The flow of granular materials down the incline, of slope α and constant width w, can be described by its thickness h_1, and its depth-averaged quantities, namely the velocity $\bar{u}_1 = \frac{1}{h}\int_0^{h_1} u(z)dz$ and the volume fraction ϕ_1. The avalanche-flow then impacts a rigid wall normal to the bottom and of semi-infinite height, thus preventing overflow downstream of the wall. As a result, the incoming flow of granular materials is pushed upstream the wall, which

*e-mail: adel.albaba@bfh.ch
**e-mail: stephane.lambert@irstea.fr
***e-mail: thierry.faug@irstea.fr

DOI: 10.1051/epjconf/201714003054

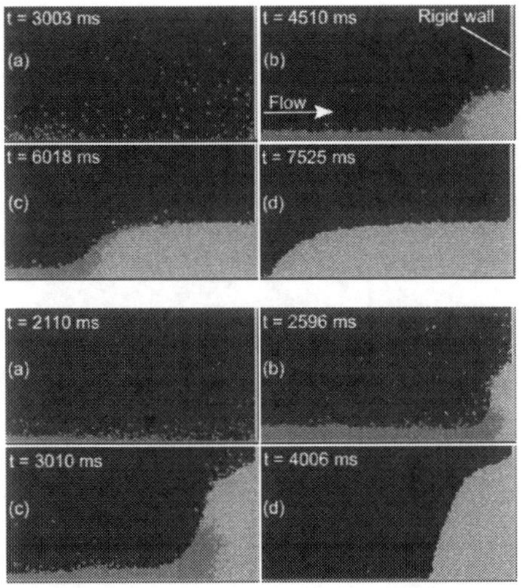

Figure 1. Impact of the avalanche-flow with the wall (a), formation (b), propagation of the granular bore (c), and final arrest of the flow (d), as simulated by DEM for two slopes angles $\alpha = 30°$ (top panel) and $\alpha = 42.5°$ (bottom panel). Orange: flowing particles, green: flowing parameters sampling zone, gray: dead zone particles

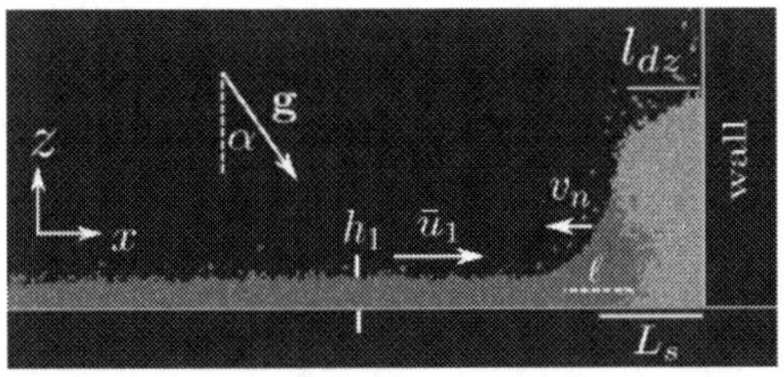

Figure 2. Granular bore forming upstream of a rigid wall impacted by a flow of granular materials: DEM simulations (example for $\alpha = 40°$) and notation

produces a granular shock-wave propagating upstream, the grains trapped between the wave and the wall forming a stagnant zone. A general picture of this granular flow-wall interaction is displayed in Fig. 2 showing a snapshot extracted from the DEM simulations presented in section 2.

3.1 Simple description of the granular bore

We propose to describe the interface between the incoming granular flow and the stagnant zone formed upstream of the wall using depth-averaged shock theory. As depicted in the sketch in Fig. 3, we define the following variables: the filling height h_{dz} against the wall, the length l_{dz} of the dead zone, the jump height h_2, the mean velocity $\bar{u}_2$ within the jump, the volume fraction ϕ_2 and v_n the propagation speed of the wave. The dead zone is defined by a volume fraction ϕ_{dz} corresponding to static close packing ($\phi_{dz} \sim 0.64$) and a mean velocity $u_{dz} = 0$. Also we consider that ϕ_2 (volume fraction within the shock) is equal to ϕ_{dz} (mean volume fraction inside the dead zone). Because we assume

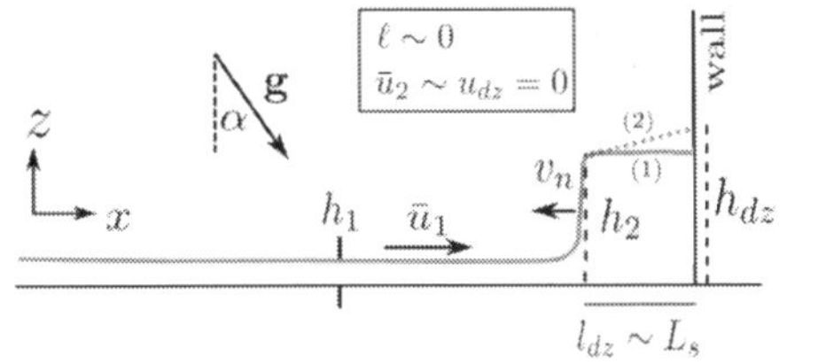

Figure 3. Sketch of a granular bore for the derivation of the depth-averaged shock model

that $\ell \sim 0$, we do not describe the variation of ϕ from ϕ_1 to ϕ_2 but rather consider a strong shock-wave in density.

The integral forms of mass and momentum conservation equations across the compressible jump are given by the following relations (see more detail in [4]):

$$[[\rho_P \phi h(\mathbf{u} - \mathbf{v_n}) \cdot \mathbf{n}]]w = 0, \tag{1}$$

$$[[\rho_P \phi h \mathbf{u}(\mathbf{u} - \mathbf{v_n}) \cdot \mathbf{n} + \frac{1}{2}\rho_P \phi g h^2 \cos\alpha \mathbf{n}]]w = \mathbf{S}. \tag{2}$$

The jump bracket $[[f]] = f_1 - f_2$ denotes the difference between the enclosed function f on the forward and rearward sides of the singular surface, following the notation commonly used in the literature about granular shocks [5, 6]. Hydrostatic pressure distribution is assumed at each side of the jump ($k = \sigma_{xx}/\sigma_{zz} = 1$ with $\sigma_{zz} = \rho g(h - z)\cos\alpha$), as well as uniform velocity profiles ($\beta = \bar{u^2}/\bar{u}^2 \sim 1$). The density of the granular fluid is $\rho_P \phi$, where ρ_P is the particle density. A detailed discussion on considering values of k and β slightly different from 1 (either upstream or downstream the shock) is given in [1, 4] and some other references therein. In the momentum equation, assuming that the jump volume shrinks into a singular surface ($\ell \sim 0$) yields $\mathbf{S} = 0$. In other words, the volume and surface forces acting over the finite length of the jump are neglected here, though a recent experimental study dedicated to standing jumps formed in granular flows revealed the importance of those forces under certain circumstances when the jumps become diffuse [4]. This crucial assumption will be further discussed while concluding the present paper.

In the problem of granular impact against a wall of semi-infinite height, the normal velocity at the outgoing section of the jump has to satisfy the velocity continuity, which means that $\bar{\mathbf{u}}_2 \cdot \mathbf{n} = u_{dz} = 0$. From mass conservation, it yields an expression for the velocity v_n at which the granular jump propagates upstream:

$$v_n = -\frac{(\bar{\mathbf{u}}_1 \cdot \mathbf{n})\mathbf{n}}{\frac{\phi_2}{\phi_1}\frac{h_2}{h_1} - 1}. \tag{3}$$

Note that the jump height h_2 is by construction smaller or equal to h_{dz}, as drawn in Fig. 3. h_2 can be predicted by injecting the relation for v_n in the momentum conservation equation across the shock (Eq. 2), which after some rearrangements takes the following form:

$$\left(\frac{h_2}{h_1}\right)^3 + A\left(\frac{h_2}{h_1}\right)^2 + B\left(\frac{h_2}{h_1}\right) + C = 0, \tag{4}$$

where $A = 1/\lambda_\phi$, $B = -(1 + 2Fr_1^2)/\lambda_\phi$, and $C = 1/\lambda_\phi^2$. We have defined here the density ratio $\lambda_\phi = \phi_2/\phi_1$ and the Froude number of the incoming flow $Fr_1 = \bar{u}_1/\sqrt{gh_1}\cos\alpha$. This equation is a cubic equation that can be solved by Cardano's method.

3.2 Impact force imparted to the rigid wall

Once the grains start piling-up against the wall, the force imparted to the rigid wall is given by the apparent weight of the grains squeezed between the incoming flow and the wall. The interface between the incoming flow and the grains trapped against the wall is modeled by a strong compressible jump (see Sec. 3.1). Let's write the static equilibrium for the dead zone:

$$\Sigma F_{ext} = F_s + W_{dz}\sin\alpha - \mu_{dz}W_{dz}\cos\alpha + R = 0. \quad (5)$$

The following forces are defined: F_s is the force of the strong jump exerted on the dead zone, $W_{dz}\sin\alpha$ is the weight of the dead zone along the slope, $\mu_{dz}W_{dz}\cos\alpha$ is the resistive friction force (where μ_{dz} denotes the Coulomb friction coefficient) and R is the reaction of the wall. The total force on the wall is given by the following expression:

$$F_n = -R = F_s + W_{dz}\cos\alpha(\tan\alpha - \mu_{dz}). \quad (6)$$

The weight of the dead zone is:

$$W_{dz} = \rho_P\phi_F g c_{dz} l_{dz} h_{dz} w, \quad (7)$$

where l_{dz} is the dead zone length and c_{dz} a shape coefficient. The distance L_s at which the granular wave is located at time $t + dt$ can be estimated from the value at time t using $L_s(t + dt) = L_s(t) + v_n dt$. In practice, it is mostly likely that the propagation speed is actually smaller than the theoretical value for v_n because of some dissipation (by friction and collisions between grains). However, if we assume that v_n matches the speed at which the material piles-up against the wall, we have: $l_{dz} \sim L_s$. L_s should then be always an overestimation of l_{dz}.

Moreover, it is worthwhile to stress that the jump height h_2 is not the run-up height h_{dz} on the wall. h_{dz} takes the following form: $h_{dz}(t + dt) = max(h_2(t), h_2(t + dt)) + h_+$, where h_+ is an additional contribution to the run-up, as shown in the sketch on the bottom panel of Fig. 3. Predicting h_+ remains a challenging question. There exists a time beyond which the jump height h_2, derived from shock theory, starts vanishing once the incoming flow starts dying. We then consider the maximum between $h_2(t + dt)$ and $h_2(t)$ because the filling height $h_{dz} = h_2 + h_+$ cannot decrease. Note that $c_{dz} > 2$ if $h_+ = h_{dz} - h_2$ is not nil, while $c_{dz} = 2$ if $h_{dz} = h_2$. The latter situation corresponds to a dead zone that is nearly rectangular—see case (1) drawn in the sketch on bottom panel of Fig. 3, and may happen if the slope α is lower than a typical friction angle φ associated with quasi-static deformations. The former situation holds for a dead zone being trapezoidal—see case (2) in the sketch on bottom panel of Fig. 3, and may occur when the slope α is greater than φ. In this case, the slope of the deposit is nearly equal to φ.

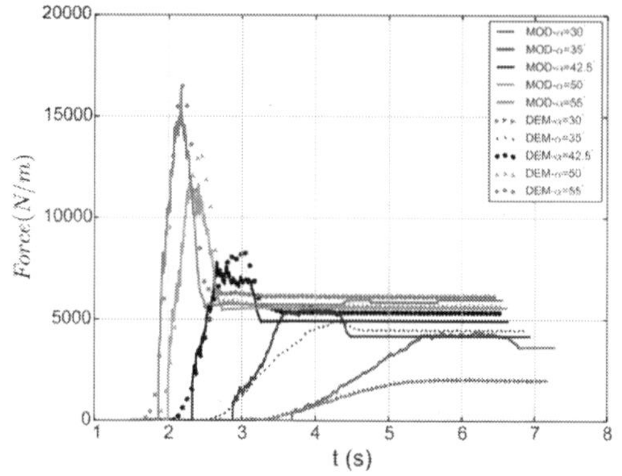

Figure 4. Temporal force signal on the rigid wall: prediction of the depth-averaged shock model (Eq. 6 fed with Eqs. 7 and 9, continuous lines) versus the force actually measured in DEM simulations (dots), for different slope angles α

Finally, in the DEM simulations described at Sec. 2 a finite-sized mass m_{tot} of material is initially released, which means that the mass of the dead zone is limited to this value m_{tot}: $\rho_P\phi_{dz}c_{dz}l_{dz}h_{dz}w < m_{tot}$. For given values of h_{dz}, ϕ_{dz} and c_{dz}, this yields a limit value for l_{dz}, itself approximated by L_s. In other words, the weight of the dead zone is restricted to the maximum value $m_{tot}g$.

The force of the incoming flow on the dead zone can be predicted from the integral form of momentum conservation across the shock formed between the incoming flow and the dead zone (the forces acting over the finite length ℓ of the dead zone are still neglected here):

$$\left[[[\rho_P\phi h\mathbf{u}(\mathbf{u} - \mathbf{v_n})\cdot\mathbf{n}]] - [[\int_0^h \sigma_x]]\mathbf{n}\right] w = 0. \quad (8)$$

Under the assumption of hydrostatic pressure distribution for the incoming granular flow, this yields:

$$F_s = \rho_P\phi_1 u_1^2 h_1 w\left(1 + \frac{1}{\frac{\phi_2}{\phi_1}\frac{h_2}{h_1} - 1} + \frac{1}{2Fr_1^2}\right) \quad (9)$$

Equation 6, fed with Eqs. 7 and 9, enables the estimation of the total force on the rigid wall, which is compared to the impact force actually measured in the DEM simulations in the next section.

4 Depth-averaged model versus DEM

A number of calculations were done to test the depth-averaged analytic force model presented in Sec. 3 and the various assumptions made. A full description of these tests will be given in a forthcoming paper. A key result is that the model was not able to capture the impact force actually measured in the DEM simulations if the properties of the incoming flow, taken at a singular surface before the jump, were used to feed the model. However, a good agreement could be found—over the whole range of slope

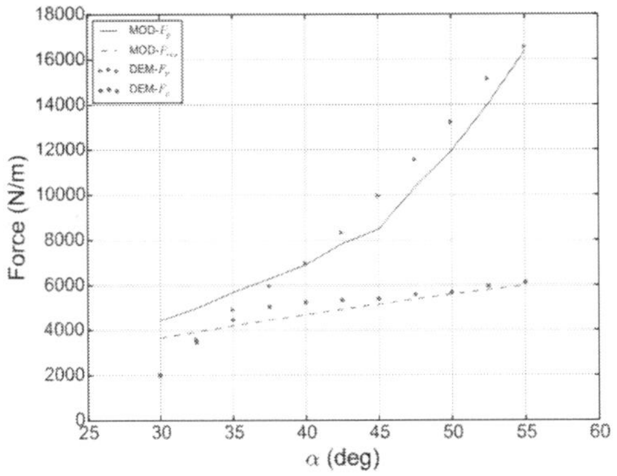

Figure 5. Peak and residual forces versus α: comparison between the shock model and the force measured in DEM

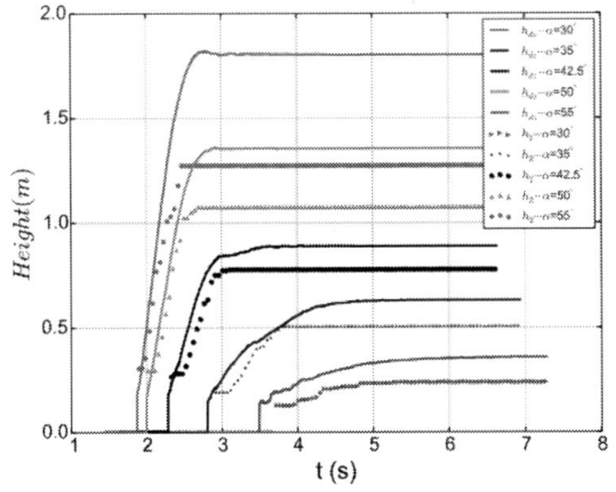

Figure 6. Evolution over time of the jump height h_2 predicted by the shock model (dots) and of the run-up h_{dz} measured in DEM (continuous line)

angles tested, by feeding the force model with the incoming flow properties averaged over a certain region, starting from the beginning of the jump and stopping after a distance $15d_{50}$ along the slope (see the green-colored regions on Fig. 1). These results are shown in Fig. 4.

The agreement is nearly perfect for the highest slope angles α but becomes less good while decreasing α. Note that there always exists—whatever the slope—a time window (until and even after the peak for the highest α) during which the increase of F with time is very well captured by the analytic force model. The agreement is less good after this time window because the strong compressible shock does not longer exist but is largely damped by energy dissipation caused by friction and taking place along the slope. There is then a transition towards a much more diffuse jump, leading to the formation of a final deposit of characteristic shape. Of course this complicated physics related to a fluid-to-solid transition is not considered in the shock model and prevails at the lowest α. As a result, the peak and residual forces cannot be well captured at the lowest α, as depicted in Fig. 5 that shows the peak force (F_{peak}) and the residual force (F_{res}) extracted from the temporal signals displayed in Fig. 5.

As already discussed in section 3.1, the (saturated) height of the granular bore is—by construction—an underestimate of the run-up height h_{dz} on the wall. This is confirmed by Fig. 6 showing the h_{dz} actually in DEM and the jump height h_2 predicted by the shock model. Interestingly, the depth-gradients $\partial h_{dz}/\partial t$ and $\partial h_2/\partial t$ are very similar, thus probing the relevance of describing the impact by the shock model presented in Sec. 3.

5 Discussion and conclusion

The present paper briefly presented DEM simulations of the impact of a dry granular avalanche-flow on a rigid wall. An analytic force model based on depth-averaged shock theory for compressible fluids was derived. This model was able to capture the evolution of the impact force with time actually measured in DEM if it was fed with the incoming flow properties averaged over a certain distance (equal to $15d_{50}$) independent of the slope angle. This result shows that there exists a relevant length-scale, namely the jump length along the slope, relevant to the problem. The assumption $\ell \sim 0$ appears to be too strong and will need further investigation in the future. A detailed description of both the DEM simulations and the depth-averaged force model, as well as an extended description of the tests made to compare the analytic force model to the DEM results, are addressed in a paper currently under preparation.

Acknowledgments

TF is grateful to funding by the People Programme (MC Actions) of the EU 7th FP under REA Grant Agreement No. 622899 (FP7-PEOPLE-2013-IOF, GRAINPACT).

References

[1] Faug T., Physical Review E **92**, 062310 (2015)
[2] https://yade-dem.org/doc/
[3] A. Albaba, S. Lambert, F. Nicot, B. Chareyre, Granular Matter **17**(5), 603 (2015)
[4] T. Faug, P. Childs, E. Wyburn, I. Einav, Physics of Fluids **27**(7), 073304 (2015)
[5] K.M. Hákonardóttir, A.J. Hogg, Physics of Fluids **17**(7), 077101 (2005)
[6] J.M.N.T. Gray, X. Cui, Journal of Fluid Mechanics **579**, 113 (2007)

Simulation of size segregation in granular flow with material point method

Minglong Fei[1,5,*], *Qicheng* Sun[1,**], *Kimberly* Hill[2,3,***], and *Gordon G. D.* Zhou[4,****]

[1] *State Key Laboratory of Hydroscience and Engineering, Tsinghua University, Beijing, 100084, China*
[2] *Department of Civil, Environmental, and Geo- Engineering, University of Minnesota, Minneapolis, MN 55455, U.S.A*
[3] *St. Anthony Falls Laboratory, University of Minnesota, Minneapolis, MN 55414, U.S.A*
[4] *Institute of Mountain Hazards and Environment, Chinese Academy of Sciences, Chengdu, 610041, China*
[5] *Beijing Municipal Institute of City Planning and Design, Beijing, 100045, China*

Abstract. Segregation is common in granular flows consisting of mixtures of particles differing in size or density. In gravity-driven flows, both gradients in total pressure (induced by gravity) and gradients in velocity fluctuation fields (often associated with shear rate gradients) work together to govern the evolution of segregation. Since the local shear rate and velocity fluctuations are dependent on the local concentration of the components, understanding the co-evolution of segregation and flow is critical for understanding and predicting flows where there can be a variety of particle sizes and densities, such as in nature and industry. Kinetic theory has proven to be a robust framework for predicting this simultaneous evolution but has a limit in its applicability to dense systems where collisions are highly correlated. In this paper, we introduce a model that captures the coevolution of these evolving dynamics for high density gravity driven granular mixtures. For the segregation dynamics we use a recently developed mixture theory (Fan & Hill 2011, New J. Phys; Hill & Tan 2014, J. Fluid Mech.) which captures the combined effects of gravity and fluctuation fields on segregation evolution in high density granular flows. For the mixture flow dynamics, we use a recently proposed viscous-elastic-plastic constitutive model, which can describe the multi-state behaviors of granular materials, i.e. the granular solid, granular liquid and granular gas mechanical states (Fei et al. 2016, Powder Technol.). The platform we use for implementing this model is a modified Material Point Method (MPM), and we use discrete element method simulations of gravity-driven flow in an inclined channel to demonstrate that this new MPM model can predict the final segregation distribution as well as flow velocity profile well. We then discuss ongoing work where we are using this platform to test the effectiveness of particular segregation models under different boundary conditions.

1 Introduction

When mixtures of particles with different size or density flowing on an incline under gravity, segregation would happen due to the gradients in total pressure and fluctuation fields. From different perspective of views, there are many effective models that can be used to describe segregation in granular flow. By considering the micro-scale collision possibility between particles, kinetic theory offers a robust framework to explain the mechanism of granular segregation [1, 2] ; on the opposite, the percolation-diffusion model [3, 4] mainly focuses on the macro-scale description of segregation process by balancing the percolation and diffusion effects in granular flow. The mixture theory, in-between the former two types of model, describes granular segregation with a macro-scale frame work but also offers some particle interaction mechanisms [5, 6]. In this paper, the mixture theory model proposed by Fan et al [6, 7] is employed, in which particle segregation is driven by the space gradient of gravity, density and granular temperature.

Granular materials may behave like solid, liquid or gas when they are under different load and/or dense conditions [8, 9], thus a unified constitutive model that can represent the mechanical behaviors of granular materials in different states is of importance in simulating granular flow segregation. Energy analysis of granular materials could offer some hints to develop such a model. The internal energy of particles inside granular materials can be divided as elastic energy and irreversible heat, the elastic energy comes from the reversible elastic contacts between particles, while the later one represents the irreversible dissipation that accompany particle interactions, such as irreversible deformation of a single particle, frictional sliding and structure rearrangement between particles. A non-linear viscous-elastic-plastic constitutive model that based on this analysis [10] is proposed to capture the dynamic of granular materials in this work.

The multi-state behaviors of granular materials may lead to the co-exist of large and small deformations in a same system, making it another challenge to simulate

*e-mail: mfei16@163.com
**e-mail: qcsun@mail.tsinghua.edu.cn
***e-mail: kmhill@umn.edu
****e-mail: gordon@imde.ac.cn

granular flow segregation with traditional Lagrangian and Eulerian methods. As an effective numerical method that mixed the advantage of Lagrangian and Euler schemes, material point method offers a good platform for that task. In MPM, granular media is divided as a set of material points that store all the neighbor material information around them. Meanwhile, momentum conservation equations are solved on Euler background gird nodes. During the calculation, material information is transferred between Lagrangian material points and Euler background gird nodes with the help of shape function. The Lagrangian material points make it simpler to trace the deformation and movement of materials when compared to traditional Eulerian method; meanwhile, in the end of each calculation circle, Euler background grids would be reset, accuracy loss caused by large mesh deformation in Lagrangian method is avoid [18]. Thus, MPM has the benefit of flexibility in tracking evolution of densities, velocities, and stress fields for different mechanical states of granular materials

2 Models

2.1 Mixture theory segregation model

For a binary size, dense granular mixture flowing on an incline with a slope angel of θ, the segregation flux Ψ_T^i of size specie i ($i = l, s$ for large and small particles, respectively) along segregation direction y (normal to the incline) in steady flow state can be expressed as [7]

$$\Psi_T^i = f^i(v^i - v)$$
$$= f\frac{(\psi^{c,i} - \psi^{k,i})}{c_D}\frac{1}{\rho}\frac{\partial\sigma_{yy}^k}{\partial y} + f\frac{(\phi^i - \psi^{c,i})}{c_D}g\cos\theta - f\frac{\eta}{c_D}\frac{\partial\phi^i}{\partial y} \quad (1)$$

where f denotes the volume fraction of granular solid, v^i and v are the flow velocity of specie i and the mixture along y direction. ρ is the density of granular mixture, $g\cos\theta$ denotes the component of gravity acceleration g normal to the incline. c_D and η are the linear drag and diffusion coefficients, and $\sigma_{yy}^k = \rho\overline{v'v'}$ is the kinetic stress of granular material, with v' the peculiar velocity of the mixture. Kinetic stress reflects the instant interaction between particles, compared to that, the enduring contact stress between particles is expressed as contact stress σ_{yy}^c. $\psi^{c,i} = \sigma_{yy}^{c,i}/\sigma_{yy}^c$ and $\psi^{k,i} = \sigma_{yy}^{k,i}/\sigma_{yy}^k$ are the stress partition coefficients of contact and kinetic stress, which show the weight of specie stresses in the mixture. $\phi^i = f^i/f$ denotes the concentration of specie i.

It shows clearly in Eq.(1) that segregation is conducted by the gradients of the fluctuation of flow, the gravity and the density; the first term reflects the effect of shear in granular segregation explicitly. However, velocity and stress are just input parameters in this model. In reference [6, 7], these parameters are measured directly from DEM simulation. To get the flow dynamic information of granular mixture, a constitutive model is required.

2.2 Non-linear viscous-elastic-plastic constitutive model

Based on the energy analysis mentioned ahead, mechanical behaviors inside granular materials are divided into the reversible elastic effect results from contacts between particles, and the irreversible dissipation caused by the frictional sliding and structure rearrangement inside granular materials. Respectively, the total stress of granular material σ_{ij} is divided into two terms, the elastic stress σ_{ij}^e that reflects the elastic energy stored inside granular material, and the dissipative stress σ_{ij}^d that describes the irreversible energy dissipation of granular materials during the momentum transfer caused by the velocity gradient.

Elastic stress is deduced from the derivation of elastic energy density w_e. A non-linear elastic energy density model first proposed by Jiang & Liu [11] and further improved by Sun et al. [12] is used,

$$w_e = AI_1^{2.5} + CI_1^{0.5}I_2 \quad (2)$$

thus elastic stress σ_{ij}^e can be expressed as

$$\sigma_{ij}^e = \frac{\partial w_e}{\partial\varepsilon_{ij}^e} = \left(2.5AI_1^{0.5} + 0.5CI_1^{0.5}\frac{I_2}{I_1^2}\right)(-\delta_{ij})I_1 + CI_1^{0.5}\varepsilon_{ij}^{e,0} \quad (3)$$

with A, C the elastic modulus related parameters, which are constant in this model, $C = 5q_\phi^2 A$, q_ϕ is the internal friction coefficient of granular material. $I_1 = -\varepsilon_{ii}^e$, $I_2 = \varepsilon_{ij}^{e,0}\varepsilon_{ji}^{e,0}$ are invariants of the elastic strain tensor $\varepsilon_{ij}^e\mathbf{e}_i\mathbf{e}_j$, with $\varepsilon_{ij}^{e,0} = \varepsilon_{ij}^e + I_1/3$. For dense granular flows, we have $I_1 > 0$.

To reflect the dissipative nature of granular materials, we refer the so called $\mu(I)$ rheology [13, 14] to express the dissipative stress σ_{ij}^d,

$$\sigma_{ij}^d = \mu(I)\sigma_m^e = -aI^b\sigma_m^e D_{ij} \quad (4)$$

with $D_{ij} = (\dot\gamma_{ij} - \delta_{ij}\dot\gamma_{kk}/3)/\dot\gamma$, $\dot\gamma_{ij} = d\varepsilon_{ij}/dt$ is the shear rate, $\dot\gamma = \sqrt{\dot\gamma_{ij}\dot\gamma_{ji}/2}$. $I = \dot\gamma d/\sqrt{\sigma_m^e/\rho^m}$ is the inertial number, ρ^m and d denote the material density and diameter of particle, respectively. a and b are material property parameters that is related to the dissipative nature of granular materials.

According to Eqs. (3) and (4), the total stress of granular materials can be expressed as

$$\sigma_{ij} = \sigma_{ij}^e H(I_1) + \sigma_{ij}^d$$
$$= \left[\left(2.5AI_1^{0.5} + 0.5CI_1^{0.5}\frac{I_2}{I_1^2}\right)\left(-\delta_{ij} - aI^bD_{ij}\right)I_1 + CI_1^{0.5}\varepsilon_{ij}^{e,0}\right]H(I_1) \quad (5)$$

with $H(I_1)$ the Heaviside function. When $I_1 \le 0$, $H(I_1) = 0$, which means particles lose contacts with each other, and the system would be in gaseous state.

2.3 Combined flow-segregation model

In the segregation model of Eq. (1), segregation flux is determined by the spatial distribution of stresses and their partition coefficients, but stress information is not supported by the model itself. In this work, however, stress profiles are achieved at the same time of segregation calculation by combing Eqs. (1) and (5). Considering that kinetic stress σ^k_{yy} represents the irreversible fluctuation of particles while σ^c_{yy} is the enduring contact between particles, their physical meaning are similar to that of σ^d and σ^e in Eq. (5), σ^k_{yy} and σ^c_{yy} in Eq. (1) are replaced by Eq. (4) and Eq. (3). What's more, along the segregation direction, elastic stress is balanced by gravity component in steady flow, and $\psi^{d,i} = \psi^{d,i}(\phi^i)$ can be expressed as a function of concentration ϕ^i [7, 17], thus we have

$$\Psi^i_T = f\frac{(\phi^i - \psi^{d,i}(\phi^i))}{c_D}\frac{1}{\rho}\frac{\partial\sigma^d_{yy}}{\partial y} - f\frac{\eta}{c_D}\frac{\partial\phi^i}{\partial y} \quad (6)$$

with $\psi^{d,i}(\phi^i) = \sigma^{d,i}_{yy}/\sigma^d_{yy} = \phi^i + b^{k,i}\phi^l\phi^s$, $b^{k,i}$ is a material parameter. In Eq. (6), ϕ^i is the target valuable that is to be solved, if the initial profile of concentration ϕ^i and the evolution of kinetic stress profile $\sigma^{d,i}$ are knew, the evolution of segregation in steady granular flow can be achieved.

3 Material point method for granular flow segregation

Like many other MPM simulations of two phase flow , two sets of materials could be used for binary mixture granular flow, with each set represents a separate size specie. However, taking advantage of the rheology of binary granular mixture can make the MPM simulation simpler. Even though components of the mixture differ in size, they obey the same rheology rule, i.e. Eq. (5); what's more, the flow velocity profiles of species in binary mixture of granular materials are the same to that of the mixture, and the mixture velocity profile is determined by applying specie average density and particle size $\bar{\rho} = \sum_{i=1}^{2}\phi^i\rho^i$, $\bar{d} = \sum_{i=1}^{2}\phi^i d^i$ in Eq. (4) [15, 16], then the space profile of σ^d_{yy} can be determined by Eq. (4). Since concentration profile ϕ^i only depends on σ^d_{yy} (Eq. (6)), segregation evolution could be achieved. Thus, in this work, just one set of material points is used to represent the mixture, and material parameters for the mixture are averaged from binary species. Note that the ordinary MPM is space-averaged over each Euler background grid cell, while segregation leads to space gradient of concentration inside each grid cell, thus calculating segregation with the ordinary MPM may lead to numerical viscus on the spacial evolution of specie concentration. To avoid the effect of numerical viscus on segregation result, a modified MPM frame with a extra Euler grid is used, granular flow is simulated on the original MPM grids, just like the ordinary MPM, while evolution of specie concentration is solved on the extra Euler grid nodes, as shown in figure 1.

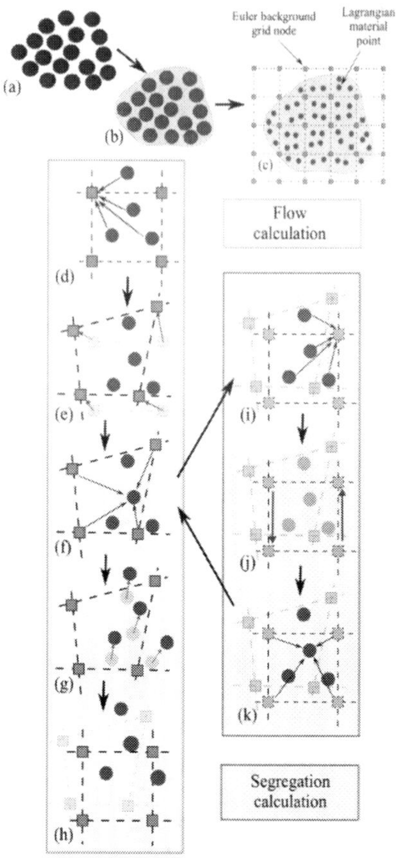

Figure 1. Calculation process of Material Point Method for binary mixture granular flow. (a) discrete granular mixture; (b) continuum granular mixture; (c) divide continuum granular mixture into material points (MP), background grid nodes (GN) are used as assists to solve field equations. A calculation cycle contains two parts, the flow calculation (the ordinary MPM [19]) (d)~(h) and the segregation calculation (i)~(k). (d) mapping MP information to GN; (e) solving momentum equation on GN; (f) mapping information back MP from GN; (g) updating MP information (h) resetting deformed grid. Segregation calculation begin when flow calculation is at step (f). (i) mapping specie concentration and mixture dynamic information from MP to extra Euler grid node; (j)transfer specie concentration between extra Euler grid nodes according to Eq. (6); (k) specie concentration are mapped back to MP.

4 Results and Discussion

An equal-volume mixture of two size granite particles (2 mm and 3 mm) flowing on an open channel is simulated with the combine model Eqs. (3), (4) and (6) by MPM, and results are compared with DEM simulation. In both MPM and DEM simulations, the channel is 22mm wide, with a slope angle of 26°, periodic boundaries are used in the front and rear of the channel, and the channel length is 88 mm, bottom is set as non-slip boundary. Sidewalls are vertical to the bottom incline. In the DEM simulation, 1841 large particles and 5168 small particles are mixed randomly and then deposited on the channel by gravity, 310 particles are glued to the lower incline to form a non-slip boundary. Frictional sidewalls are used, friction coefficient between sidewall and particles is $\mu_w = 0.3$, be-

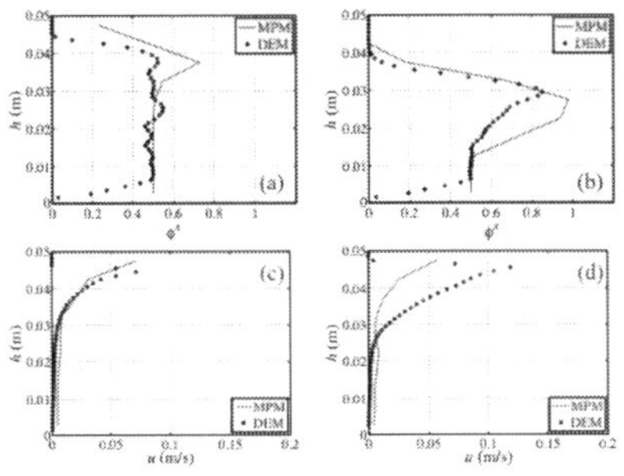

Figure 2. Results for MPM (red line) and DEM (black circle) at 1s (left) and 20s (right) of the simulation. (a), (b) are concentration profile of small particles; (c), (d) are velocity profiles of the mixture.

tween particles is $\mu_p = 0.4$. Elastic modulus of particle is 29 GPa, particle density is 2650 kg/m^3 [7]. In MPM simulation, 7200 material points are used to represent the mixture, the spacing between points is 0.25mm; other material parameters $A = 1.79 \times 10^9$, $q_\phi = 0.38$, $a = 0.46$, $b = 1.0$, $\phi = 0.6$ are used for flow dynamic calculation [10], and $\eta = 1.11 \times 10^{-5} m^2 s^{-2}$, $c_D = 6.8 s^{-1} s^{-1}$, $b^{k,s} = 0.39$ for segregation calculation [7].

Simulation results are shown in figure 2. As we can see, both the concentration profiles of small particle (figure 2-a, 2-b) and the mixture velocity profile (figure 2-c, 2-d) of MPM simulation can match well with those of DEM; small particles are transferred from top to bottom during the flow. Due to the frictional sidewall, particles near the bottom are almost still during the flow process, they form the erodible bed and change the mixture velocity profile from Bagnold type to exponent type profile [14, 20]; accordingly, there is almost no segregation near the bottom, concentration of small particle in this area keeps the original value of 0.5. Simulation results suggest that, the combine model Eq. (6) can represent the flow dynamics as well as segregation evolution of binary size granular mixture flow well, and material point method turns out to be a reliable numerical scheme to simulate granular flow. We also notice some deviations of velocity profile between DEM simulation and our model in figure 2-d. These deviation may be caused by the erodible bed near the bottom. Actually the erodible bed plays an important role on the rheology of granular flow [21], and it also have a significant effect on segregation results [22]. At the stage, our ongoing work is focusing on the relation between erodible bed and granular segregation. By applying kinetic theory and mixture theory in granular flows with different sidewall conditions, and comparing the segregation prediction between these two types of model, more hints about the relation between erodible bed, granular mixture rheology and segregation results may emerge.

Acknowledgements

The authors thank Prof. X. Zhang of Tsinghua Univ. for sharing the open source MPM code MPM3D [19]. This work has been supported by the National Natural Science Foundation of China (11572178,91634202), and the fund of State Key Laboratory of Geohazard Prevention and Geoenvironment Protection, Chengdu Univ. of Tech..

References

[1] J. T. Jenkins and Mancini, Phys. Fluids **1** (12), 2050 (1989).

[2] M. Larcher and J. T. Jenkins, J. of Fluid Mech. **782**, 405 (2015).

[3] J. Bridgwater, W. S. Foo and D. J. Stephens, Powder Technol. **41** (2), 147 (1985).

[4] Y. Fan, C. P. Schlick, P. B. Umbanhowar, J. M. Ottino, and R. M. Lueptow, J. of Fluid Mech. **741**, 252 (2014).

[5] J. Gray and A. R. Thornton, Proc. Roy. Soc. A **461** (2057), 1447 (2005).

[6] Y. Fan and K. M. Hill, New J. of Phys. **13** (095009) (2011).

[7] K. Hill and D. Tan, J. of Fluid Mech. **756**, 54 (2014).

[8] H. M. Jaeger, S. R. Nagel and R. P. Behringer, Rev. Mod. Phys. **68** (4), 1259 (1996).

[9] Y. Forterre and O. Pouliquen, Ann. Rev. Fluid Mech. **40** (1), 1 (2008).

[10] M. Fei, Q. Sun, X. Xu, F. Jin, and G. G. D. Zhou, Powder Technol **301**, 1092 (2016).

[11] Y. Jiang and M. Liu, Phys. Rev. Lett. **91** (14), 144301 (2003).

[12] Q. Sun, F. Jin, G. Wang, S. Song, and G. Zhang, Sci. Rep. **5**, 9652 (2015).

[13] F. da Cruz, S. Emam, M. Prochnow, J. N. Roux, and F. Chevoir, Phys. Rev. E **72** (2 Pt 1), 21309 (2005).

[14] P. Jop, Y. Forterre and O. Pouliquen, Nature **441** (7094), 727 (2006).

[15] A. Tripathi and D. V. Khakhar, Phys. Fluids **23** (11), 113302 (2011).

[16] B. Yohannes and K. M. Hill, Phys. Rev. E **82** (6), 61301 (2010).

[17] T. Weinhart, S. Luding and A. R. Thornton, AIP Conf. Proc. **1542** (1), 1202 (2013).

[18] Z. T. Ma, X. Zhang and P. Huang, Comp. Model. Eng. **55** (1), 61 (2010).

[19] X. Zhang, Mpm3d, http://www.mpm3d.com.

[20] GDR Midi, Eur. Phys. J. E **14** (4), 341 (2004).

[21] K. Kamrin and G. Koval, Phys. Rev. Lett. **108** (17), 178301 (2012).

[22] Y. Fan and K. Hill, Phys. Rev. Lett. **106** (21), 218301 (2011).

EPJ Web of Conferences **140**, 09027 (2017) DOI: 10.1051/epjconf/201714009027
Powders & Grains 2017

Toward multiscale modelings of grain-fluid systems

Bruno Chareyre[1,*], *Chao* Yuan[1], *Eduard* P. Montella[1], and *Simon* Salager[1]

[1] *Université Grenoble-Alpes, 3SR lab., Grenoble, France*

Abstract. Computationally efficient methods have been developed for simulating partially saturated granular materials in the pendular regime. In contrast, one hardly avoid expensive direct resolutions of 2-phase fluid dynamics problem for mixed pendular-funicular situations or even saturated regimes. Following previous developments for single-phase flow, a pore-network approach of the coupling problems is described. The geometry and movements of phases and interfaces are described on the basis of a tetrahedrization of the pore space, introducing elementary objects such as bridge, meniscus, pore body and pore throat, together with local rules of evolution. As firmly established local rules are still missing on some aspects (entry capillary pressure and pore-scale pressure-saturation relations, forces on the grains, or kinetics of transfers in mixed situations) a multi-scale numerical framework is introduced, enhancing the pore-network approach with the help of direct simulations. Small subsets of a granular system are extracted, in which multiphase scenario are solved using the Lattice-Boltzman method (LBM). In turns, a global problem is assembled and solved at the network scale, as illustrated by a simulated primary drainage.

1 Introduction

The simulation of wet granular materials with pendular bridges between pairs of solid grains has been a popular application of discrete element models (DEM). The study of mixed situations in which pendular bridges and larger clusters of wetting phase co-exist (the typical situation), eventually reaching one single cluster in the saturated regime, has received much less attention. This is certainly due to the modeling and computational difficulties more than the lack of scientific an practical interest. An hybrid modeling framework taking advantage of both the efficiency of the pore-network approach and the accuracy of direct fluid dynamics solutions at the microscale is proposed in this paper. In the first section physical processes are associated to the basic geometrical entities of a triangulation and the main governing equations of fluid-grain mixtures are listed. The saturated case is included since one-phase flow in large single-phase clusters is intrinsically linked to the more general multiphase flow problem. The last section describes how improved accuracy is gained by embedding direct simulations of multiphase events into the pore-network model.

2 Pore-scale modeling of fluid phases

2.1 Topology

The network representation of the pore space in three dimensions is obtained by Regular Triangulation [1] of the sphere assembly. The algorithm follows [2] and the implementation is based on the open source library CGAL [3, 4].

*e-mail: bruno.chareyre_a_3sr-grenoble.fr

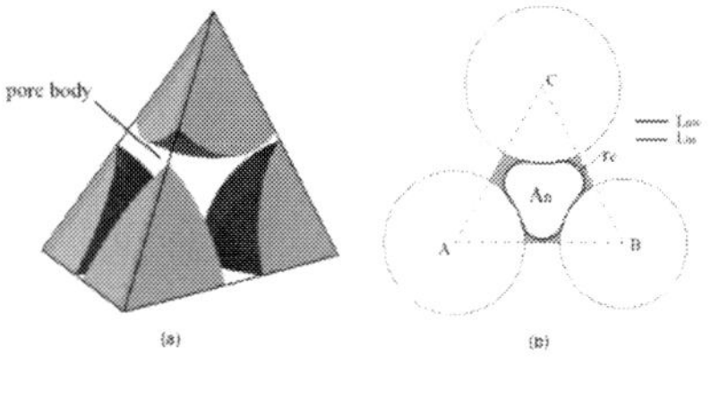

Figure 1. Pore geometry. (a) A pore defined by a tetrahedral element. (b) Geometry of a pore throat with interfaces in the MSP method.

In the triangulated geometry, a *pore body* is defined as the irregular cavity within a (3-D) tetrahedron, surrounded by four solid spheres (whose centers are the vertices of the corresponding tetrahedron, see Fig.1(a)). A *pore throat* is defined by the cross sectional area extending within a tetrahedral (2-D) facet (Fig.1(b)). Pair interactions (including solid-solid contacts) act along the (1-D) edges of the triangulation. Finally the 0-D points are attached to the solid objects to which total forces are assigned (Fig. 2). This Lagrangian mesh follows the solid grains.

2.2 One-phase flow

The pressure of a fluid flowing within a granular medium may not immediately appear as a discontinuous problem yet the changes of pore pressure are strongly localized near the pore throats (i.e. the facets of the triangulation), as shown in Fig. 3. This feature has been exploited by the

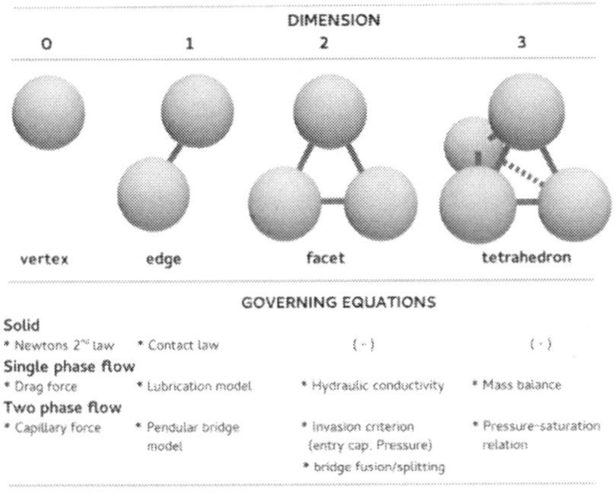

Figure 2. Simplices of a triangulation and the associated constitutive relations.

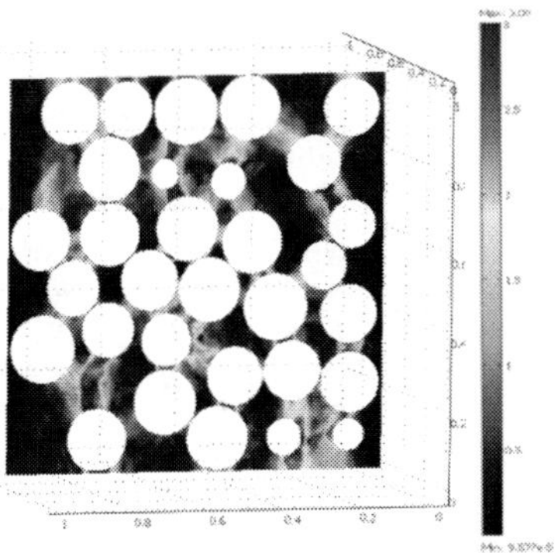

Figure 3. Gradient of fluid pressure in the Stokesian regime (FEM solution for a 3D random packing of 200 spheres); the values are normalized by the macroscale pressure, values equal to 3 or higher are in red color.

DEM-PFV scheme [2, 5]. In this method the fluid mass balance in one pore links interface fluxes with the rate of change $\dot{V}_i$ of the volume of pore i (a function of particle velocities). In addition the linearity of the flux-pressure relations (Stokesian regime) leads to the discrete form of a Poisson problem with the $\dot{V}_i$ as source terms, for i iterating over the pores

$$\dot{V}_i = \sum_{j=1}^{j=4} K_{ij}\,(p_j - p_i). \tag{1}$$

K_{ij} is the local hydraulic conductivity reflecting the local pore geometry [6]. Solving this linear system gives the p_i, from which drag forces on the grains are evaluated and added to the total forces (strong two-way coupling).

The lubrication forces, short-range hydrodynamic interactions between immersed particles, are not captured by the pore-scale mass balance. They should be added independently. The normal and shear components of the lubrication force read [7, 8] (the rotational terms are omitted

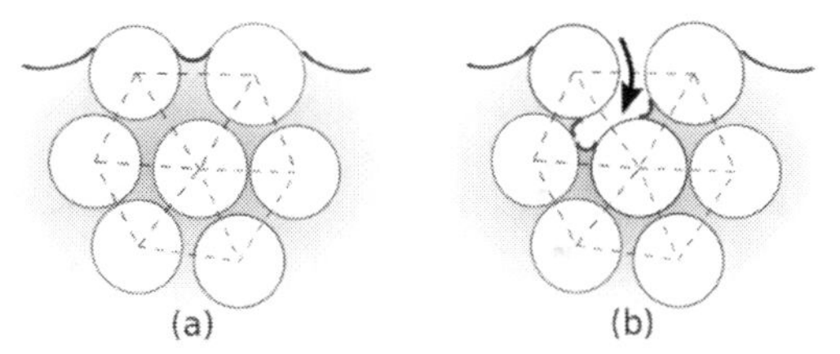

Figure 4. Elementary process of interface displacement during drainage.

for brevity, see [9])

$$F_n^L = \frac{3}{2}\,\pi\,\mu\,\frac{a^2}{h}\,\mathbf{v}_n \tag{2}$$

$$F_s^L = \frac{\pi\mu}{2}\,(-2a + (2a+h)\,ln(\frac{2a+h}{h}))\,\mathbf{v}_t \tag{3}$$

where $\mathbf{v}_n$ and $\mathbf{v}_t$ are the normal and tangential relative velocity, μ is the fluid viscosity, and a is the radius of the grains.

2.3 Immiscible fluids and interfaces

We now consider the case of two fluid phases in the pore space: a wetting phase and a non-wetting phase (hereafter w-phase and n-phase respectively) such as water and air, typically. For the sake of simplicity we admit that the particles are sufficiently small to make the gravitational effects negligible compared to surface tension effects (small values of the *Bond number* B_o). The geometry of the nw-interfaces follows the the Laplace-Young relation between the trace C of the curvature tensor and the capillary pressure p_c (pressure difference between the n-phase and the w-phase locally): $p_c = C\sigma_{wn}$ with σ_{wn} is the interfacial tension. The simplest shape formed by immiscible fluids is the axisymmetric wetting bridge connecting two spheres (edges of the triangulation), for which Laplace-Young equation leads to a one dimensional differential equation tractable with conventional methods [10]. Incorporating pendular bridges in DEM boils down to calculating the traction force resulting from a particular state of the bridge

$$f_{cap} = f_{cap}(u_n, p_c, \sigma_{wn}, R_1, R_2) \text{ for } \{u_n, ..., R_2\} \in \mathcal{D} \tag{4}$$

where u_n is the distance between spheres of radii R_1 and R_2. The bridges exist only in a bounded region $\mathcal{D}$ of the parameter space. Alternatively, many empirical relations can be found [11], determining f_{cap}, volume V, and the bounds of $\mathcal{D}$. Beyond pendular bridges, the interfaces in a porous medium frequently move in sudden piston-like steps (the so-called Haines jumps) from one equilibrium state to another. The typical equilibrium states have interfaces in the form of meniscii near the pore throats (Fig.4). A meniscus is stable as long as

$$p_c \leq p_c^e \tag{5}$$

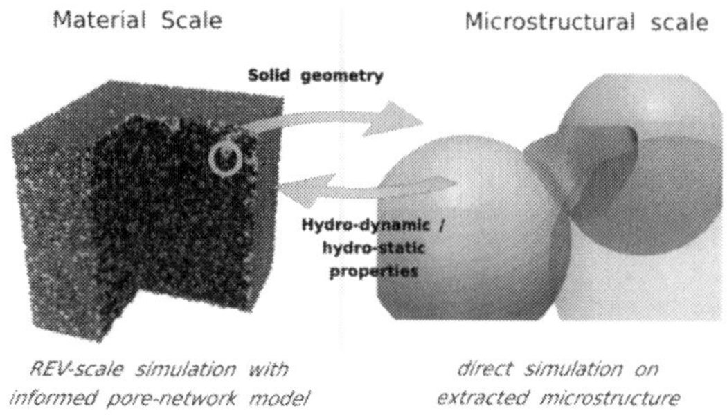

Figure 5. Multiscale coupling: pore network simulation of fluid(s) flow commanding and incorporating data from direct resolution at the micro-scale.

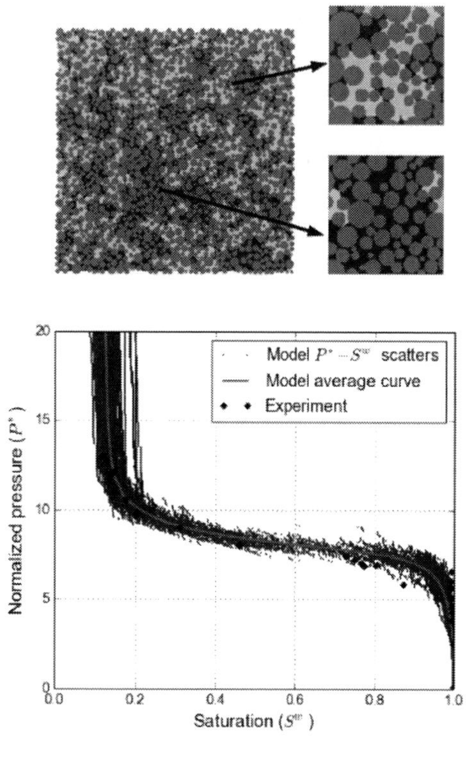

Figure 6. Network simulation of primary drainage in a packing of 40,000 spheres; (top) detail of phase distribution at average saturation S_r = 0.4, (bottom) water retention curves from repeated simulation compared to published data [15].

where p_c is the local capillary pressure and p_c^e is the throat *entry capillary pressure*. p_c^e may be approximated by the inscribed sphere method [12] or the Mayer-Stowe-Princen (MSP) method [13, 14] (Fig. 1.b), or by direct simulations as in the last section. The MSP method has been found to predict rather accurately measured drainage curves (see Fig. 6, based on [14]). The imbibition events are more complex. Piston-type invasion by the w−phase is possible depending on a new threshold value p_c^f (the previous drainage event reversed). The interface stability inside a pore needs

$$p_c \geq p_c^f. \tag{6}$$

Generally $p_c^f < p_c^e$, which is a major cause of hysteretic responses in multiphase systems. Alternatively, imbibition can proceed by progressive coalescence of pendular bridges, i.e. by first filling edges and facets instead of pore bodies; possibly leading to *snap-off*: the disconnection of a bubble of n-phase trapped in a pore after all the edges/faces have coalesced. Local rules driving such complex events can be found in [16]. Note that the accurate geometry of interfaces is only known for the pendular bridges [17] (unless direct simulation of the microstructure is performed - see last section). Due to this limitation the interfacial area in the funicular regions is only roughly approximated, which is detrimental in view of thermodynamic analysis [18, 19].

In addition to the movements of interfaces each phase can flow within a connected domain just like it does in fully saturated cases - this is handled by the single-phase PFV scheme. Transfers may also occur by *film flow* in the thin layer of water at the surface of the grains [20, 21]. A third type of transfer occurs by evaporation/condensation and vapor advection-diffusion [22]. The last two mechanisms tend to equilibrate capillary pressure all over the pore space after enough time, even between distant, apparently disconnected, clusters. Depending on flow rate and viscosity of the fluids, interface movements can be coupled to viscous dissipation (that is, when the capillary number C_a is too small for the flow to be quasi-static), leading to secondary pore imbibition in drainage as well as various patterns of fingering [23]. Beyond the pendular regime, very few authors derived analytical expressions of capillary forces in partial saturation [22, 24]. A general expression of the force is

$$\mathbf{f}_{cap} = \int_{\partial\Omega} (p(x)\mathbf{n} + \delta_L(x)\sigma_{wn}\mathbf{t})ds \tag{7}$$

where p is the pressure of the fluid phase present at point x and $\delta_L(x)$ is a line Dirac delta reflecting the triple wns-line. Obviously the evaluation of this integral needs accurate knowledge of the geometry of the phases around the particle, which is usually lacking.

3 Multiscale coupling

Nearly every equation in the previous section can be calibrated with the help of direct fluid dynamics (CFD) simulations of the actual microstructure (Fig. 5). For this purpose we employ a multicomponent Shan & Chen [25] Lattice Boltzmann method (LBM) implemented in Palabos [26]. Here primary drainage is shown for a proof of concept (the reader is referred to the oral or poster presentation for more examples). Primary drainage of an initially saturated material depends, mainly, on the distribution of the entry capillary pressure p_c^e (Eq. 6) in the pore network. The elementary problem that is solved with the LBM to determine p_c^e is shown in Fig.7 (note the pendular bridges

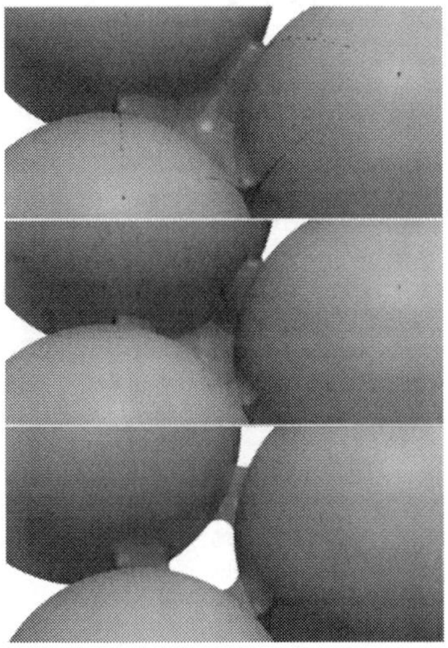

Figure 7. Progressive curvature and splitting of a meniscus within a pore throat at increasing capillary pressure; simulated using a multicomponent Shan & Chen LBM scheme and symmetry conditions on the sides of the throat.

staying behind the advancing main meniscus). The geometry of this throat-scale problem is defined by the positions of three spheres taken from a large (40,000 sphere) packing. The number of LBM problems to be solved is virtually the same as the number of facets in the triangulation (approximately 85,000). A dynamic selection of the potentially invaded throats is done at runtime to eliminate some of them. The task fits very well in massively parallel architectures where many throats can be simulated simultaneously. After determining p_c^e for every relevant throat the master (pore-network) process assign the p_c^e values to the network. In turns, invasion by the n-phase in the boundary value problem is simulated at the network scale. It couples capillary effects (Eq. 6 with updated p_c^e) and viscous flow of the receding w−phase (Eq. 1).

This multiscale coupling may seem to be merely one very large LBM problem decomposed in a series a subproblems with nearly the same number of LBM nodes in total - more or less what blind parallelization would achieve. However we argue that the computation time is decreased by order of magnitudes practically. Besides the fact not every throats have to be analyzed, this is mainly because the flow of the receding phase is solved as a pore-network problem. In fully resolved explicit methods like the LBM, the long range interactions associated to incompressible flow within large w-clusters introduce additional constrained and cause the dominant part of the increased computational cost. In contrast, multiscale coupling enable flexible combinations where the Pore-Network and CFD methods are used for what they do best.

References

[1] H. Edelsbrunner, N.R. Shah, Algorithmica **15**, 223 (1996)

[2] B. Chareyre, A. Cortis, E. Catalano, E. Barthélemy, Transport in Porous Media **92**, 473 (2012)

[3] J.D. Boissonnat, O. Devillers, S. Pion, M. Teillaud, M. Yvinec, Computational Geometry **22**, 5 (2002)

[4] C. Jamin, S. Pion, M. Teillaud, in *CGAL User and Reference Manual* (CGAL Editorial Board, 2016), 4.9 edn.

[5] E. Catalano, B. Chareyre, E. Barthélemy, Int. J. Numer. Anal. Met **38**, 51 (2014)

[6] A.T. Tong, E. Catalano, B. Chareyre, Oil and Gas Science and Technology-Revue d'IFP Energies Nouvelles **67**, 743 (2012)

[7] N. Frankel, A. Acrivos, Chemical Engineering Science **22**, 847 (1967)

[8] B. Van den Brule, R. Jongschaap, Journal of statistical physics **62**, 1225 (1991)

[9] D. Marzougui, B. Chareyre, J. Chauchat, Granular Matter **17**, 297 (2015)

[10] G. Lian, C. Thornton, M.J. Adams, Journal of colloid and interface science **161**, 138 (1993)

[11] A. Gladky, R. Schwarze, Granular Matter **16**, 911 (2014)

[12] T. Sweijen, E. Nikooee, S.M. Hassanizadeh, B. Chareyre, Transport in Porous Media **113**, 207 (2016)

[13] V. Joekar-Niasar, M. Prodanovic, D. Wildenschild, S. Hassanizadeh, Water Resources Research **46**, W06526 (2010)

[14] C. Yuan, B. Chareyre, F. Darve, Advances in Water Resources **95**, 109 (2016)

[15] K. Culligan, D. Wildenschild, B. Christensen, W. Gray, M. Rivers, A. Tompson, Water Resour. Res. **40** (2004)

[16] K. Melnikov, R. Mani, F. Wittel, M. Thielmann, H. Herrmann, Phys. Rev. E **92**, 022206 (2015)

[17] C. Chalak, B. Chareyre, E. Nikooee, F. Darve, European Journal of Environmental and Civil Engineering (2016)

[18] S.M. Hassanizadeh, W.G. Gray, Advances in Water Resources **16**, 53 (1993)

[19] R.J. Held, M.A. Celia, Advances in Water Resources **24**, 325 (2001)

[20] R. Mani, D. Kadau, D. Or, H.J. Herrmann, Physical review letters **109**, 248001 (2012)

[21] K. Melnikov, F. Wittel, H. Herrmann, Acta Geotechnica **11**, 539 (2016)

[22] A. Kharaghani, T. Metzger, E. Tsotsas, Chemical Engineering Science **75**, 267 (2012)

[23] R.G. Hughes, M.J. Blunt, Transport in Porous Media **40**, 295 (2000)

[24] C. Yuan, B. Chareyre, Comput Methods Appl Mech Eng. doi:10.1016/j.cma.2017.02.024 (2017)

[25] X. Shan, H. Chen, Physical Review E **49**, 2941 (1994)

[26] *Palabos Documentation v1.5* (retrieved from http://www.palabos.org, 2015)

DOI: 10.1051/epjconf/201714008013

Numerical modeling of the tensile strength of a biological granular aggregate: Effect of the particle size distribution

Karsta Heinze[1,2,⋆], *Xavier* Frank[2], *Valérie* Lullien-Pellerin[2], *Matthieu* George[1], *Farhang* Radjai[3,4], and *Jean-Yves* Delenne[2]

[1] *Laboratoire Charles Coulomb, UMR 5221, CNRS-Université de Montpellier, 163 Rue Auguste Broussonet, Montpellier, France*
[2] *UMR IATE, CIRAD, INRA, Montpellier SupAgro, Université de Montpellier, Montpellier, France*
[3] *LMGC, UMR 5508, CNRS-Université de Montpellier, 163 Rue Auguste Broussonet, Montpellier, France*
[4] *MSE2, UMI 3466, CNRS-MIT, DCEE, Massachusetts Institute of Technology, 77 Massachusetts Avenue, Cambridge, USA*

Abstract. Wheat grains can be considered as a natural cemented granular material. They are milled under high forces to produce food products such as flour. The major part of the grain is the so-called starchy endosperm. It contains stiff starch granules, which show a multi-modal size distribution, and a softer protein matrix that surrounds the granules. Experimental milling studies and numerical simulations are going hand in hand to better understand the fragmentation behavior of this biological material and to improve milling performance. We present a numerical study of the effect of granule size distribution on the strength of such a cemented granular material. Samples of bi-modal starch granule size distribution were created and submitted to uniaxial tension, using a peridynamics method. We show that, when compared to the effects of starch-protein interface adhesion and voids, the granule size distribution has a limited effect on the samples' yield stress.

1 Introduction

Wheat is one of the three most important cereal crops worldwide. The major part of the grain is the starchy endosperm, which consists of a multi-modal distribution of stiff starch granules [1, 2], that are embedded in a softer protein matrix (see Fig.1). Therefore, the endosperm can be considered as a cemented granular material.
The milling of the grains is a crucial step in the processing of wheat. It separates the different grain parts and results in a transformation of the grains into flour. The milling performance is determined by endosperm texture, which depends on genetics and environment. The genetic component influences grain hardness, which was suggested to result from the starch-protein adhesion strength. The environmental conditions impact the porosity of the protein matrix, commonly measured as endosperm vitreousness [3]. Both components have been studied by experimental and numerical approaches [4–7], while other possible sources for differences in milling behavior, such as the starch granule size distribution (SGSD), are still less well understood. Experimental studies on the connection of SGSD and endosperm texture are not numerous and even not convergent [8, 9]. Studies on the milling performance as a function of SGSD are contradicting, too [2, 10]. However, one point of general agreement is that SGSD is genetically controlled [8]. Therefore, the study of SGSD using numerical modeling appears essential to understand its specific effect on the fragmentation behavior and they also can help the selection of wheat varieties for breeding.
The fragmentation behavior of wheat endosperm has already been the focus of DEM or LEM studies [6, 7, 11], but with a focus on starch-protein adhesion and the volume fractions of the different material phases. We use a peridynamic approach [12, 13] to study how the variation of the starch granules polydispersity affects the yield stress and failure properties. Here we present the methodology in detail along with some first results.

⋆e-mail: kmheinze@um2.fr

2 Methods

2.1 Sample construction

Dense and homogeneous cemented granular samples were created, consisting of the three bulk phases: starch ρ_s in the form of discs, a matrix ρ_p, representing the embedding protein network and a void phase ρ_v, if matrix porosity is implemented. The condition $\rho_s+\rho_p+\rho_v = 1$ is fulfilled for all samples. For most samples ρ_v was chosen to be 0 and therefore all non-starch spaces were filled with the protein phase.
Square volumes were filled with discs of a bi-modal distribution of larger A-type discs and smaller B-type discs, embedded in a matrix phase. Two distributions of disc radii were implied: a normal distribution around a constant radius $r_A = 5 \cdot 10^{-3}$ (standard deviation $\sigma_A = 0.25 r_A$) for the A-type discs and a normal distribution around radius $r_B \leq r_A$ ($\sigma_B = 0.25 r_B$) for the B-type discs, which was varied. Due to the finite mesh size, a cutoff of

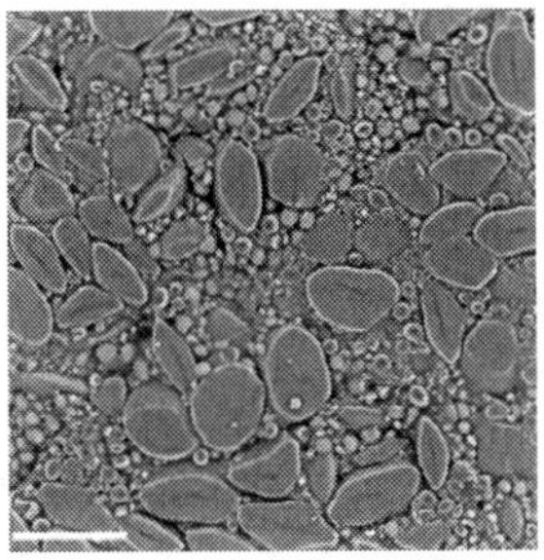

Figure 1. SEM image of a sectioned wheat endosperm, showing the dense packing of starch granules. Scale bar: $30\mu m$

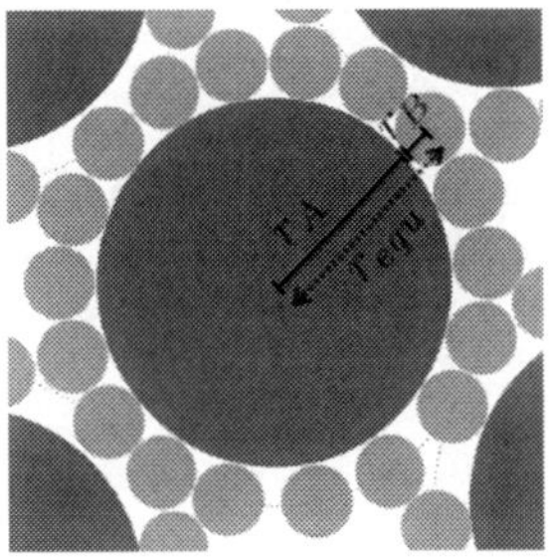

Figure 2. The schematic of an A-type disc of radius r_A surrounded by B-type discs of radius r_B illustrates the calculation of the area ratio to be $\mathcal{A} = 1.59$.

$c^{+,-}_{A,B} = r_{A,B} \pm 1.5\sigma_{A,B}$ had to be applied to avoid the occurrence of very small discs, which could not be meshed sufficiently. Consequently, the smallest allowed grain would be meshed by about 10 nodes along the diameter.

In the first step, a square box of the area $A_{box} = 900 * (\pi r_A{}^2)$ and length $L = \sqrt{A_{box}}$ was filled with A-type discs using a Fast Poisson Disc Sampling (FPDS) method [14]. The first disc with a random radius within the normal distribution around r_A was placed at a random position within the box. Then a high number of preliminary discs with random radii within the distribution around r_A were randomly placed in the periphery of the already existing disc. All preliminary discs were tested consecutively for overlap with the existing discs within a limited distance. If no such overlap exists, the preliminary disc was permanently placed into the sample volume. Additionally, a minimum distance d_{min} between the discs was required. The steps of randomly selecting a disc from the sample box, creating preliminary discs and checking for overlap were repeated. If after a finite number of iterations, no more discs could be permanently placed, the box was regarded as being filled.

In the second step, smaller discs with a random radius following the distribution around radius r_B were placed into the free spaces remaining in-between the bigger discs. The same FPDS approach as for the A-type discs was used, only that no minimum distance between discs was required and which therefore enabled a dense packing of discs.

Samples with fixed radius r_A and different radii r_B were created to investigate the effect of varying B-type disc radius on the mechanical properties. The radius r_B was varied from $r_B = 0.2r_A$ to $r_B = r_A$ in steps of $0.1r_A$. The resulting nine different radius ratios $\mathcal{R} = r_B/r_A$ were therefore [0.2, 0.3, 0.4, 0.5, 0.6, 0.7, 0.8, 0.9, 1.0]. For every radius ratio $\mathcal{R}$ five independent samples were created.

An increase in r_B inevitably resulted in a different disc packing, including a significant change in the number of A- and B-type discs and the total number of discs. Therefore, the sole parameter that was chosen to be kept constant between samples was the ratio of total area occupied by A-type discs to total area occupied by B-type discs:

$$\mathcal{A} = \sum_{ng_A}(\pi r_A^2) \Big/ \sum_{ng_B}(\pi r_B^2) \qquad (1)$$

$ng_{A,B}$ being the respective total number of discs per type. To define the constant $\mathcal{A}$ we assumed a coverage with B-type discs of 83% of a perimeter with the radius $r_{equ} = r_A + r_B$ around one A-type disc for $\mathcal{R} = 0.2$ in two dimensions. An illustrating scheme is shown in Fig.2. This coverage allowed some "sharing" of B-type discs between different A-type discs. The number of B-type granules n_B that cover 83% of the perimeter was estimated by

$$n_B = 0.83\frac{2\pi r_{equ}}{2r_B} = 0.83\pi\left(1 + \frac{r_A}{r_B}\right) \qquad (2)$$

From equation (1) and (2) it follows that

$$\mathcal{A} = \frac{\pi r_A^2}{n_B \pi r_B^2} = \frac{1}{0.83\pi}\frac{r_A^2}{(r_A r_B + r_B^2)} \qquad (3)$$

For $\mathcal{R} = r_B/r_A = 0.2$ it follows that $\mathcal{A} = 1.59$. This value was kept constant for all samples, while r_B was varied. To guarantee homogeneous packing along with a constant $\mathcal{A}$ a sufficient number of B-type discs was needed to fit in the spaces between the A-type discs. Therefore, the minimum distance d_{min} between A-type discs was increased along with an increasing r_B. Iterations of a step-wise increase of d_{min} were implied in the disc packing computations, until the required area ratio $\mathcal{A} = 1.59$ was met for every $\mathcal{R}$.

With a limited computation power and time, FPDS was able to achieve dense, but not the densest possible packing of discs. Therefore, a classical Discrete Element Methods (DEM) approach [11] was used to apply a slight compression over 100.000 time steps of $10^{-4}s$. The contact law of a classical linear spring-dash pot model was used with the relation of normal stiffness of the granules k_n to the confining pressure P being $\frac{k_n}{P} \approx 10^4$. As a consequence the overlap of the rigid granules during the compression was smaller than the size of the applied mesh [15]. Coulombs law of friction was applied with the sliding friction coefficient $\mu = 0.3$ to avoid segmentation of the small granules towards the lower sample border. The confining pressure was set to achieve maximum dense packing and the number of time steps was chosen sufficiently high to result in a mechanical equilibrium. Finally, all free spaces that were not occupied by discs of type A or B, were considered to be filled with matrix ($\rho_v = 0$). A square box of the length

$l_{x,y} = 24\,r_A = 0.12$ was cropped out of the center of the sample box to avoid boundary effects. Examples for five different samples with increasing $\mathcal{R}$ are shown in Fig.3.

For $\mathcal{R} = 0.2$ a sub-series of tests was carried out on microstructures containing void spaces ($\rho_v > 0$). The matrix filling was computed following the procedure described by Topin et al. [6] and resulted in samples with an average void volume of 3.5%.

2.2 Peridynamic mechanical simulations

The peridynamic approach is an alternative to continuum mechanics. It is based on integral equations and is more accurate than classical LEM [6, 7]. Peridynamics allows taking into account heterogeneous distributions of mechanical properties using bond-based schemes with brittle linear springs. The stress and strain tensors are computed on a rectilinear grid by integrating long-range interactions over a limited neighborhood. As the algorithm is based on non-local interactions, failure occurs in the form of damaged zones with a characteristic thickness.

The square sample volumes were meshed onto a rectilinear grid of 1024^2 nodes. Starch discs were considered as distinct elements, whereas the protein matrix was considered as one single element. Nodes were connected via brittle linear springs within a certain horizon, which was set to be three nodes in each direction. The bonds were classified to three cases: (I) connecting two nodes within the same material phase in the same element (node in ρ_s^{discX} to node in ρ_s^{discX}, all nodes within ρ_p to each other), (II) connecting two nodes within the same material phase in different elements (node in ρ_s^{discX} to node in ρ_s^{discY}) or (III) connecting nodes of different material phases (node in ρ_s to node in ρ_p). Depending on these cases, the bonds were characterized by an elastic modulus $E_{s,p,i}$ and toughness $K_{s,p,i}$. For simplification in this first study, granules of type A and B were assumed to have the same elastic and toughness properties. All bonds of case (II) and (III) were characterized by interface properties. Therefore a small interface layer was considered to be present around all starch discs, even if two discs are in contact with each other.

Two distinct series with the same sample micro-structures but different interface properties were investigated. The parameter values for elasticity and toughness of the different phases and series are summarized in Tab.1. Series 1 represents the case of the strongest possible adhesion between starch and protein, where protein matrix and interface have the same toughness ($K_p = K_i$). In Series 2 the toughness of the interface K_i was decreased by 50%. The parameters were chosen similar as was reported in [7], taking into account nano-mechanical measurements of starch and protein inside wheat endosperm by atomic force microscopy[16]. Samples were submitted to quasi-static uniaxial tension in the y-direction, with the lower boundary being fixed, until integrity failure occurred.

3 Results and Discussion

The increase of $\mathcal{R}$ lead to a less dense packing of discs and therefore more free spaces in between the discs, which

Table 1. The model parameters elasticity E in $[Pa]$ and toughness K in $[Jm^{-3}10^{-4}]$ are summarized for both series.

Series	Protein	Starch	Interface
S1	E_p=1 $K_p=1\cdot10^{-3}$	$E_s=4E_p$ $K_s=3.5K_p$	$E_i=E_p$ $\mathbf{K_i=K_p}$
S2			$E_i=E_p$ $\mathbf{K_i=0.5K_p}$

could be occupied by the matrix material. For the selected variations of $\mathcal{R}$ a non-negligible, non-linear increase of the protein matrix volume fraction ρ_p from 13% ($\mathcal{R}^{min} = 0.2$) to 19% ($\mathcal{R}^{max} = 1.0$) was observed, the starch volume fraction decreasing accordingly. In a cemented granular material, such as the endosperm, it is not only the SGSD that determines the amount of protein fraction, but also other factors such as packing density and presence of voids. Experimental studies on wheat are inconclusive whether or not SGSD is correlated with grain protein content [8, 17]. Though our first numerical samples could mimic the endosperm of vitreous wheat grains, which have low porosity, a direct comparison with the experimental results is unfeasible at this point. Future numerical simulations, which could investigate the relation of SGSD and protein content further, would need to take into account more parameters, such as void spaces and packing density.

All tested sample volumes showed a brittle behavior. Yield stress ρ^Y is given as normalized by the yield stress of pure matrix ($\rho^{[m]}$): $\rho^Y = \rho_y/\rho^{[m]}$; with ρ_y being the absolute yield stress. An increase of $\mathcal{R}$ lead to a decrease of the yield stress (Fig.4). The lowest yield stress was obtained for $\mathcal{R} = 0.9$, which was decreased by about 10% compared to the maximum yield stress obtained for $\mathcal{R} = 0.2$, for both interface toughness values respectively.

When comparing samples with the same micro-structures, but with a differing interface toughness K_i, a more drastic decrease of the yield stress by 50% is observed. In the range of the observed variation, which cover the range of biological micro-structures, the effect of the SGSD is of secondary order compared to the effect of the interface. The high significance of the interface properties has already been established by numerical simulations[6, 7] and our results are consistent with this. When yield stress is additionally normalized by the interface toughness K_i, the evolution of yield stress with $\mathcal{R}$ merges for Series 1 and 2 (see Fig.4b). The effect of $\mathcal{R}$ on yield stress is therefore independent of the specific interface properties.

As mentioned before, changes in $\mathcal{R}$ are inevitably accompanied by changes of the volume fraction of ρ_p, if ρ_v=0. Yield stress was found to decrease with increasing protein volume (data not shown). Because the increase in matrix volume fraction is the consequence of significant changes in the micro-structures, $\mathcal{R}$ and matrix volume fraction have a joint effect on the yield stress, which cannot easily be separated for the presented sample selection.

The introduction of voids into the samples with $\mathcal{R}$=0.2 (ρ_v=3.5%, ρ_p=8.6%, data not shown) had a significant effect on the yield stress in the same range as was observed for the 50% decrease of K_i. Samples with varying

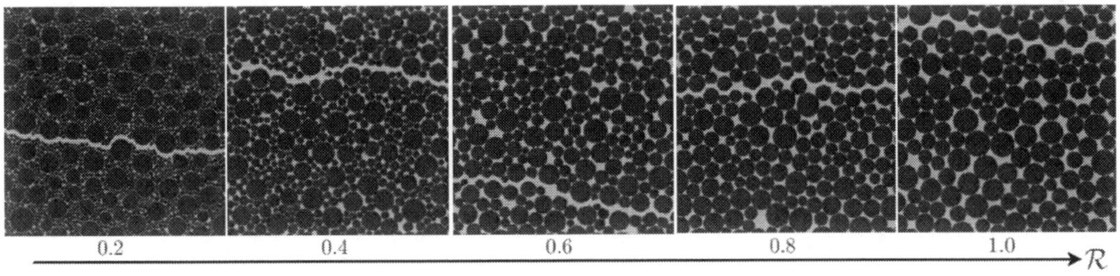

Figure 3. Five examples of square, bi-modal granular samples with increasing $\mathcal{R}$ are shown after integrity failure due to uniaxial tension. Starch discs are shown in blue. All voids are filled with protein matrix (yellow). The interface around the starch granules is highlighted in red. The fracture pathway, signaled by broken bonds, is shown in light blue.

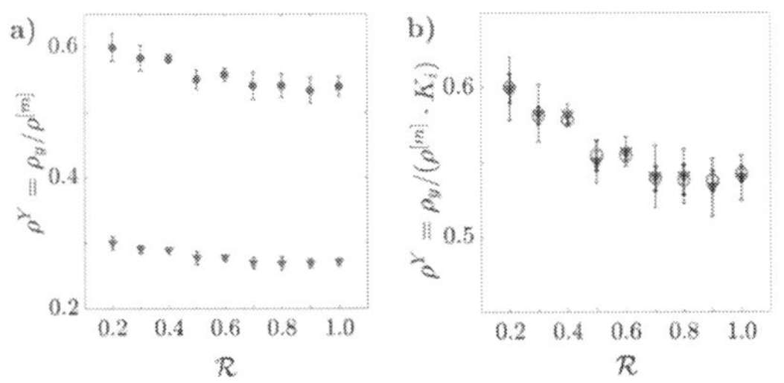

Figure 4. a: The evolution of yield stress ρ^Y with increasing $\mathcal{R}$ is shown for two series with differing interface toughness: K_i= 1.0 (circles, S1) and K_i=0.5 (triangles, S2). Yield stress was normalized by the yield stress of pure matrix. **b**: Both series merge, when yield stress is normalized by K_i.

$\mathcal{R}$, but fixed ρ_p are difficult to construct and it can be hypothesized that the necessary introduction of voids would strongly dominate over the effect of a changing $\mathcal{R}$.
It is visible in Fig. 3 that the fracture does not propagate through starch granules. This is a direct result of the value chosen for K_s, which affects the resulting amount of damage to the starch granules, but not ρ^Y (Data not shown).

4 Conclusion

We found that, when compared to the effects of interface adhesion and void inclusions, the SGSD presents a limited effect on the yield stress in tension. The material is strongest for the smallest size ratio (small/large) between the two granule populations. The effect of SGSD is coupled to changes in the volume fractions of the starch and protein phase, but is fully independent of the interface properties. An ongoing study investigates the effect of interface adhesion and void occurrence on the tensile properties for different volume ratios of A- and B-type granules. Additionally, further studies will investigate fracture under dynamical conditions. This is possible with the presented peridynamic assay, which enables simulations under conditions closer to the milling process than quasi-static tests.

Acknowledgment

K. Heinze received a French grant from INRA (CEPIA dept.) and Labex Numev as a PhD student. We thank the bioinformatics platform Toulouse Midi-Pyrenees (Bioinfo Genotoul) for providing computing resources.

References

[1] D.B. Bechtel, I. Zayas, L. Kaleikau, Y. Pomeranz, Cereal Chem **67**, 59 (1990)
[2] M. Edwards, B. Osborne, R. Henry, J Cereal Sci **48**, 180 (2008)
[3] F.X. Oury, P. Lasme, C. Michelet, M. Rousset, J. Abecassis, V. Lullien-Pellerin, Theoretical and Applied Genetics **128**, 913 (2015)
[4] V. Greffeuille, J. Abecassis, M. Rousset, F.X. Oury, A. Faye, C.B. L'Helgouac'h, V. Lullien-Pellerin, Theor Appl Genet **114**, 1 (2006)
[5] Y. Haddad, F. Mabille, A. Mermet, J. Abecassis, J. Benet, Powder Technol **105**, 89 (1999)
[6] V. Topin, J.Y. Delenne, F. Radjai, L. Brendel, F. Mabille, Eur Phys J E **23**, 413 (2007)
[7] E. Chichti, V. Lullien-Pellerin, M. George, F. Radjai, R. Affès, J.Y. Delenne, J of Food Eng **190**, 40 (2016)
[8] G. Igrejas, B. Faucher, D. Bertrand, D. Guibert, P. Leroy, G. Branlard, J Cereal Sci **35**, 103 (2002)
[9] D.B. Bechtel, I. Zayas, R. Dempster, J.D. Wilson, Cereal Chem **70**, 238 (1993)
[10] M. Edwards, B. Osborne, R. Henry, J Cereal Sci **52**, 314 (2010)
[11] J. Delenne, Y. Haddad, J. Bénet, J. Abecassis, J Cereal Sci **47**, 438 (2008)
[12] S. Silling, J Mech Phys Solids **48**, 175 (2000)
[13] P. Perré, G. Almeida, M. Ayouz, X. Frank, Ann For Sci **73**, 147 (2016)
[14] R. Bridson, *Fast Poisson Disk Sampling in Arbitrary Dimensions*, in *ACM SIGGRAPH 2007 Sketches* (ACM, New York, NY, USA, 2007), SIGGRAPH '07
[15] J.N. Roux, G. Combe, C R Phys **3**, 131 (2002)
[16] E. Chichti, M. George, J.Y. Delenne, F. Radjai, V. Lullien-Pellerin, Eur Polym J **49**, 3788 (2013)
[17] S.H. Park, J.D. Wilson, B.W. Seabourn, J Cereal Sci **49**, 98 (2009)

DOI: 10.1051/epjconf/201714014007

Experimental study of wave generation by a granular collapse

Manon Robbe-Saule[1], *Cyprien* Morize[1,*], *Yann* Bertho[1], *Alban* Sauret[1,2], and *Philippe* Gondret[1]

[1] *Laboratoire FAST, Univ. Paris-Sud, CNRS, Université Paris-Saclay, F-91405, Orsay, France*
[2] *Laboratoire SVI, CNRS, Saint-Gobain, 39 quai Lucien Lefranc, F-93303 Aubervilliers, France*

Abstract. We consider experimentally the wave generation by the collapse of a dry granular column into water. Our results show that for a given aspect ratio of the initial granular column, the amplitude of the generated wave depends on the column mass and for a given mass of the granular column, two distinct regimes are observed depending on the aspect ratio of the column a: for $a < 3.5$, the volume of the generated wave is directly proportional to a whereas for thinner column, $a > 3.5$, there is a saturation of the amplitude of the wave suggesting that the leading wave is generated before the collapse is fully completed.

1 Introduction

A tsunami is an ocean wave of very large wavelength than can propagate over large distances, typically few thousands of kilometers, and turns into a breaking wave of large amplitude which can cause considerable damages to infrastructures and population when reaching the coasts. Although many tsunamis arise from tectonic events, various past geological events [1] have shown that landslides near coastlines can lead to tsunami waves of significant amplitude. Despite this important hazard, tsunamis generated by landslides remain poorly understood and difficult to model, leading to a very approximated estimate of the associated hazards.

While the generation of tsunamis by a solid block has been widely studied since the 1970s [2], more realistic modeling requires taking into account the granular nature of the landslide. The collapse of dry or immersed granular media has been widely studied during last decades in various situations [3–7]. However, the collapse of dry granular material into water complicates drastically the situation. This complexity mainly arises from the interaction of a two-phases landslide (*i.e.* the solid beads and the air) into a third phase (*i.e.* water) and to the inaccurate modeling of the generation of the initial wave. In particular, it is necessary to accurately describe the entry of the granular material into water and the interactions between the grains and the liquid that lead to the generation of the impulse wave [8–10].

In this paper, we consider a model situation consisting in the collapse of a dry granular column into shallow water, leading to the generation of an impulse wave. We focus our attention on the generation mechanisms of the first wave and not on its later propagation.

*e-mail: morize@fast.u-psud.fr

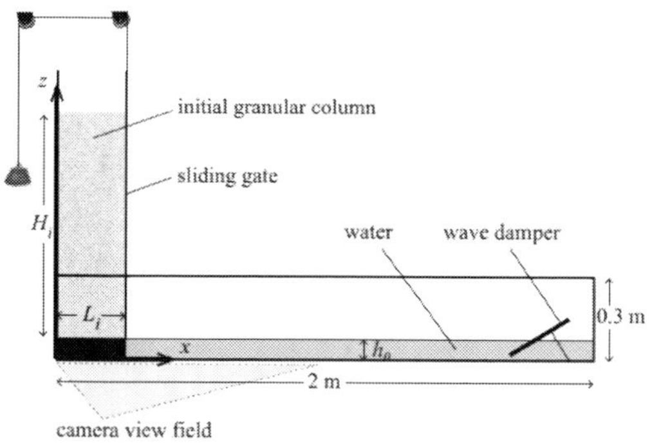

Figure 1. Schematic of the experimental setup.

2 Experimental setup

The experimental setup consists of a rectangular tank (2 m long, 0.3 m high and W=0.15 m wide) in which a sliding gate maintains a column of monodisperse dry grains of height H_i and length L_i, as shown in Fig. 1. The grains are located just above an undisturbed free surface of water of depth $h_0 = 55$ mm. Grains are glass beads of density $\rho = 2500$ kg m^{-3} whose diameter d is 5 mm. This range of diameters ensures that capillary effects can be neglected as the capillary length for water is of the order of 2 mm. In the following, we will note $a = H_i/L_i$ and $M = \rho\phi H_i L_i W$, the aspect ratio and the mass of the column respectively, where $\phi \simeq 0.64$ is the packing fraction of the granular column. Upon opening the sliding vertical door, the subaerial granular column collapses with a typical time $\tau^* = \sqrt{H_i/g}$ and impacts the free surface of water at the typical velocity $u^* = \sqrt{gH_i}$. This granular collapse generates an impulse wave that propagates at the typical velocity $c = \sqrt{gh_0}$ in shallow water conditions. The Froude number is then defined as $Fr = u^*/c = \sqrt{H_i/h_0}$. Note that the gate is opened fast enough to ensure that it does not affect the dynamics of the granular collapse. The aspect ratio of the tank ensures that the motion of the granular medium and of the generated impulse wave takes place along x-direction such that the situation remains two-dimensional. Glass beads of diameter 3 mm are glued on the bottom of the tank to prevent the grains from sliding over the surface. Finally, an inclined perforated plexiglas plate is located at the end of the tank to absorb the incoming wave and reduce the amplitude of the reflected wave.

The dynamics of the granular collapse and the wave generation are recorded by means of a video camera from the front side of the tank. The processing of the images is realized by using a custom-made Matlab routine based on thresholding method that allows to determine both the surface of the granular collapse and the free-surface $h(x, t)$ of the water during the wave generation. This procedure allows one to extract both the instantaneous amplitude $A(t) = h(x, t) - h_0$ and the length $\lambda(t)$ of the impulse wave as the grains collapse, with a resolution of about 0.7 mm. The dynamics of the collapse, the final shape of the deposit and the generated wave are then studied as a function of the aspect ratio of the granular column a, the total mass of grains M and their diameter d.

3 Phenomenology

Figures 2 and 3 show typical examples of a granular collapse of 5 mm glass beads into water, for two different initial masses M and aspect ratios a. Impulse waves, whose dynamics depends on the initial granular column, are observed when the grains impact the free surface of water. Indeed, both the amplitude A and the width λ of the generated wave are more important for larger volume of grains, leading to a breaking wave for the largest one (Fig. 3). Note also that in both situations a secondary wave of weaker amplitude is generated after the leading one, but this study goes beyond the scope of this paper.

These two examples highlight the fact that the penetration of a granular column into water is a complex processus involving three different phases: the water, the grains and the air between the grains. Indeed, the wetting of the dry grains is not instantaneous since a front of dry beads clearly appears at short times, below the initial water free surface. This dry/wet front rises progressively during the collapse in a typical timescale of imbibition of an equivalent porous medium. As a result, a volume of air is trapped between the grains in the early stage of the collapse and contributes to displace a given volume of water, and thus partially leads to the wave generation. Finally the air trapped in the granular medium escapes, and bubbles rise to the surface.

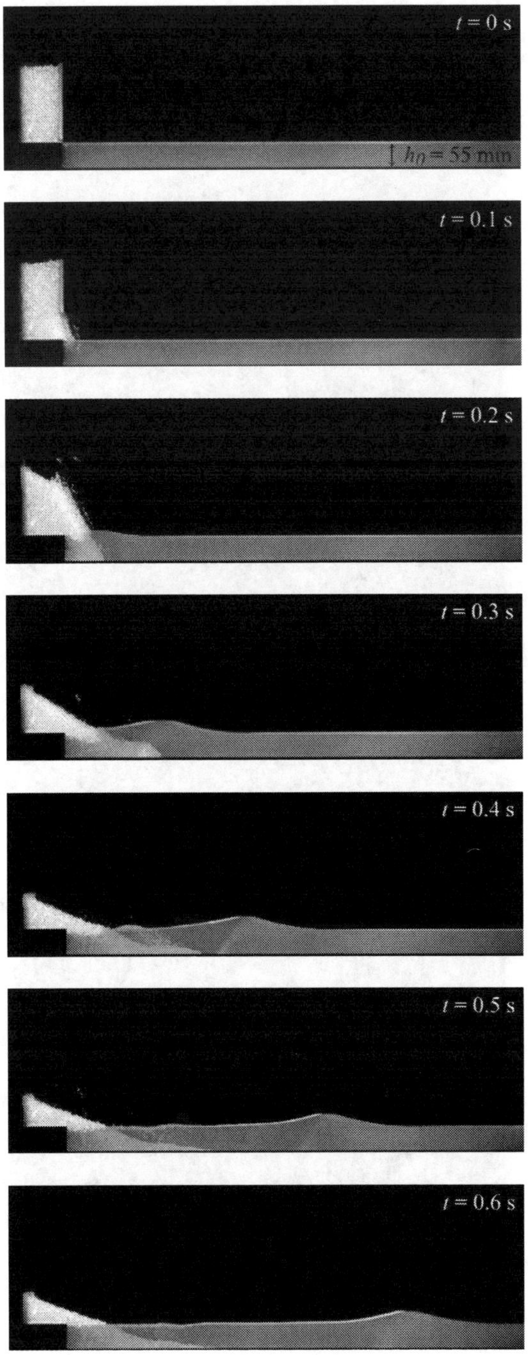

Figure 2. Image sequence showing the propagation of a non-breaking wave due to the collapse of a granular column (mass M = 2.8 kg, aspect ratio a = 1.8) of 5 mm glass beads into h_0 = 55 mm of water.

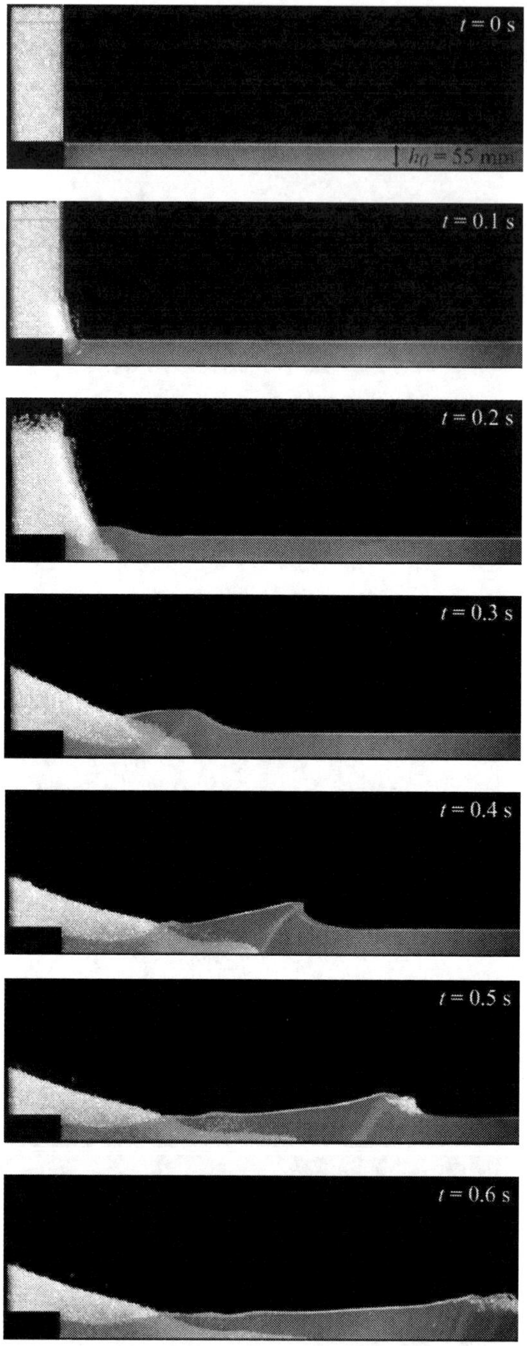

Figure 3. Image sequence showing the propagation of a breaking wave due to the collapse of a granular column (mass M = 6.5 kg, aspect ratio a = 2.7) of 5 mm glass beads into h_0 = 55 mm of water.

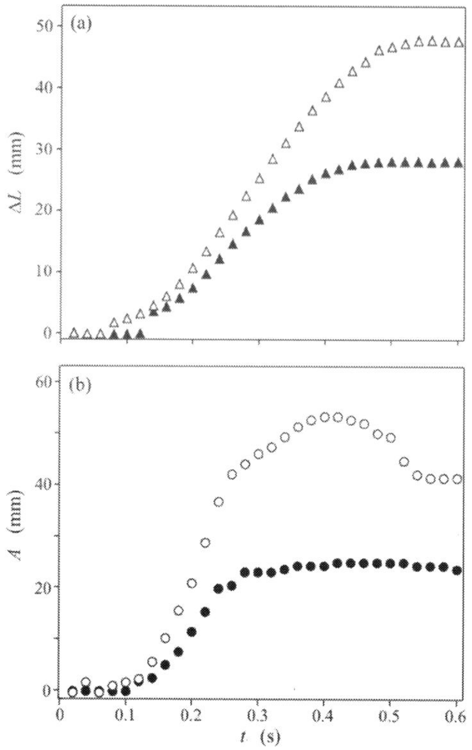

Figure 4. Time evolution of (a) the runout length ΔL and (b) the amplitude A of leading wave, for the (▲, •) non-breaking and (△, ○) breaking waves presented in Figs. 2 and 3 respectively.

4 Experimental results

The spread of the collapse of the granular column is extracted by measuring the runout distance, in the bottom of the tank at z=0, as a function of time [Fig.4(a)]. Once the collapse of the granular column is achieved, the runout distance is comparable to dry collapses observed in the literature [4].

Figure 4(b) displays the time evolution of the amplitude A of the leading wave generated by the granular collapse for the two experiments presented in Figs. 2 and 3. In both situations, one observes first that this amplitude grows strongly, in a similar way, as the grains collapse into water. Then, for the non-breaking wave, the amplitude reaches a maximum value $A_{max} \simeq 25$ mm. Even if the amplitude of the wave seems to be roughly constant over the small observation window available here, a decay as $x^{-1/3}$ is expected, where x is the distance from the point where the wave is generated, in agreement with the weakly dispersive linear shallow water approximation [8]. For larger initial volumes of grains, the amplitude grows until it reaches a larger maximum value $A_{max} \simeq 54$ mm. Once the wave breaks, its amplitude suddenly decreases because of energy dissipation. Thus, the characteristics of the wave generated by a granular collapse highly depend on initial conditions, especially the mass and the initial shape of the granular column impacting the liquid.

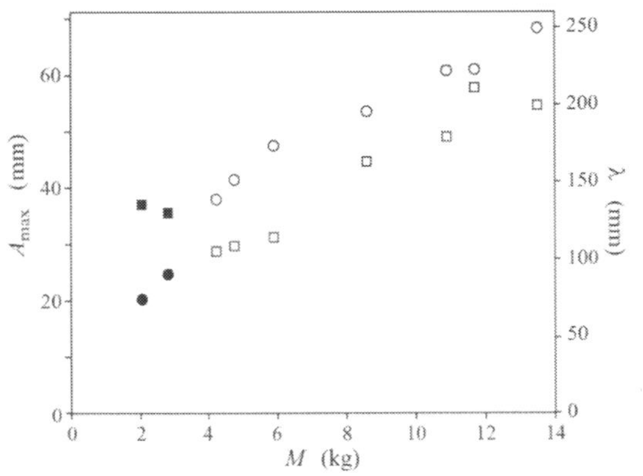

Figure 5. (○, •) Maximum amplitude A_{max} and (□, ■) mid-height width λ of the leading wave as a function of the mass of the initial column M, for a granular column of aspect ratio $a = 2 \pm 0.2$, and of grain diameters $d = 5$ mm. Empty and filles symbols correspond respectively to breaking and non-breaking waves.

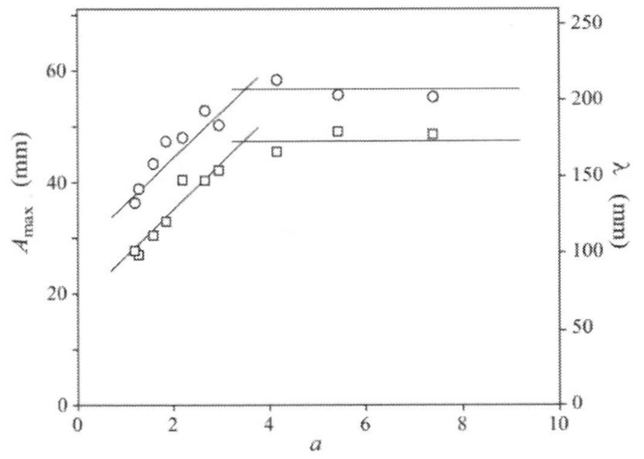

Figure 6. (○) Maximum amplitude A_{max} and (□) mid-height width λ of the leading wave as a function of the aspect ratio a of an initial column of mass $M = 6.5$ kg and of grain diameters $d = 5$ mm. All the data correspond to breaking waves. (—) Guidelines for the eyes.

Figure 5 represents the maximum amplitude A_{max} and the mid-height width λ of the leading wave as a function of the mass M of the initial column, for an initial granular column of aspect ration $a \simeq 2$. Note that λ is measured at the mid-height of the leading wave at $t = t_{max}$ corresponding to the time when the wave amplitude is maximum. Except for the non-breaking waves, one observes that the amplitude and the width of the leading waves increase approximatively linearly with the mass of the granular collapse. Moreover, one can note that the width of the waves is minimum at the transition between non-breaking and breaking waves. This is probably due to the aspect ratio of the breaking waves that have a larger height to width ratio in contrast with non-breaking waves.

Figure 6 represents the maximum amplitude A_{max} and the mid-height width λ of the leading wave as a function of the aspect ratio of the initial granular column a, for the same total mass of grains $M = 6.5$ kg. The amplitude as well as the mid-height width of the generated waves are observed to increase linearly with a for $a \lesssim 3.5$, while they reach a constant value for $a \gtrsim 3.5$. This saturation suggests that for high enough granular columns, the leading wave is generated before the collapse is fully completed. In other words, the grains located at the top of the column are not involved in the wave generation. An estimate of the characteristic times for the generation of the wave and for the granular collapse will help us to understand how the grain column contributes or not to the generated wave.

5 Conclusion

By using a small scale laboratory experiment, we investigated, at a very initial state, the generation of waves by the collapse of a dry granular material into water. These experiments show that the amplitude and the width of the generated waves depend on both the initial mass of grains impacting the water surface and the aspect ratio of the granular column. Both the amplitude and the width of the leading waves are found to be proportional to the initial mass of grains for small aspect ratio of the column while they both saturate for larger aspect ratio.

Acknowledgements

This work is supported by CNRS through its multidisciplinary program "Défi Littoral" via the projects SlideWave and SlideWave2. We are grateful to A. Hildenbrand and S. Viroulet for fruitful discussions. We thank J. Amarni, A. Aubertin, L. Auffray and R. Pidoux for their contribution to the development of the experimental setup.

References

[1] A. Hildenbrand, P. Y. Gillot, A. Bonneville, Geochem. Geophys. Geosyst. 7 (2006).
[2] P.L.-F. Liu, T.R. Wu, F. Raichlen, C.E. Synolakis, J.C. Borrero, J. Fluid Mech. **536**, 107 (2005).
[3] G. Lube, H. E. Huppert, R. S. J. Sparks, M. A. Hallworth. J. Fluid Mech. **508**, 175 (2004).
[4] E. Lajeunesse, J.B. Monnier, G.M. Homsy, Phys. Fluids **17**, 103302 (2005).
[5] N. J. Balmforth, R. R. Kerswell, J. Fluid Mech. **538**, 399 (2005).
[6] L. Rondon, O. Pouliquen, P. Aussillous. Phys. Fluids **23**(7), 073301 (2011).
[7] V. Topin, Y. Monerie, F. Perales, F. Radjai, Phys. Rev. Lett. **109**, 188001 (2012).
[8] S. Viroulet, A. Sauret, O. Kimmoun, EPL **105**, 34004 (2014).
[9] H. M. Fritz, W. H. Hager, H. E. Minor, Exp. Fluids **35**, 505 (2003).
[10] M. Cremonesi, A. Frangi, U. Perego, Comput Struct **89**, 1086 (2011).

Rheology of vibrated granular suspensions

Sebastien Kiesgen de Richter[1,★], *Caroline* Hanotin[2], *Naima* Gaudel[1], *Nicolas* Louvet[1], *Philippe* Marchal[2], and *Mathieu* Jenny[1]

[1] *Université de Lorraine, LEMTA, CNRS UMR 7563, Vandoeuvre-lès-Nancy, F-54500, France*
[2] *Université de Lorraine, LRGP, CNRS UMR 7274, Nancy, F-54000, France*

Abstract. In this work we investigate in details the flow behaviour of dense vibrated gravitational suspensions. We study the rheology in the stationary state by using a stress imposed rheometer (spectroscopy mechanics) coupled with a vibration cell, we show that applying well-controlled mechanical vibrations allows the control of the suspension viscosity by suppressing the apparent yield stress which is largely the cause of flow jamming. We show that the rheology in the stationary state is controlled by the competition between the reorganization time induced by the flow and the internal reorganization time induced by vibrations. We discuss the influence of particles size, suspending fluid viscosity and vibration parameters and demonstrate that the grains dynamics is controlled by the ratio between the lubrication stress and the granular pressure. This work evidences the major role played by the vibration induced lubrication stress on the liquefaction of vibrated granular suspensions.

1 Introduction

Industries are looking for tests, correlations or even rheological models for the flow of granular dispersions during transport in relation with some measurable physical features. These correlations and models should, notably, account for the impact of vibrations (amplitude, frequency) on their rheological behavior and structure (rupture, granulation). In practice, vibrations avoid the creation of localized blockages due to the frictional forces between particles and/or with the walls which might result in halting the transport. It is crucial to understand the underlying mechanisms of the blockage and the local effects of applied vibrations to optimize transport. It requires a better understanding of the rheological properties of vibrated granular dispersions, in order to identify the influence of a large number of parameters (size and morphology of particles, viscosity of the interstitial fluid, volume fraction, influence of the applied vibrations, geometrical and confinement effects) on the response of the system. From a more fundamental point of view, studying the influence of vibrations on the flowing behavior of granular dispersions represents a fascinating interplay between the jamming transition in concentrated suspensions and rheology. For instance, understanding the way the particles are organized together and the structure they adopt when vibrations are applied is fundamental to optimize the flow property of this materials. Some crucial common features between dry and fluid saturated monodisperse granular suspensions have recently been evidenced. It has been shown that in both case a visco-plastic $\mu(I)$ rheology describes accurately the flow of such dispersions[1]. The dynamical friction coefficient and therefore the yield stress depends on the dimensionless number I, which is the ratio of the internal reorganization time of the dispersion over the typical timescale of the flow. In this respect, the expression of I differs for dry and fluid-saturated flows , and can be interpreted in terms of the relevant time scales controlling grain motions. When vibrations are applied, granular dispersions rheology changes in a drastic way [2–9]. In particular, it modifies the rheology by changing the conditions for flow arrest. One challenge is to generalize the $\mu(I)$ approach when external vibrations control grains reorganizations. In that context, we show in this article that small amplitude vibrations strongly affect the rheology of dense gravitational suspensions. We evidence that the typical reorganization time extracted from mechanical spectroscopy is related to the dynamics of the medium at the grain scale induced by vibrations. This typical time depends on the lubrication Peclet number, Pe_{lub} defines as the ratio between the lubrication and the frictional stresses.

★e-mail: sebastien.kiesgen@univ-lorraine.fr

2 Mechanical spectroscopy

2.1 Materials and methods

We study the viscoelastic behavior of dense gravitational suspensions made of monodisperse glass beads (ρ_p = 2400 $Kg.m^{-3}$) immersed in a Newtonian fluid (Water/Polyalkylene glycol mixture, ρ_f = 1032 $Kg.m^{-3}$) when both shear and vibrations are applied. In this article, we present results obtained for three different particle diameter d = 49 ± 3 μm, 100 ± 8 μm, 335 ± 15 μm and interstitial fluid viscosities ranging from 40 to 160 $mPa.s$. We impose sinusoidal stresses (rheometer (AR 2000, TA

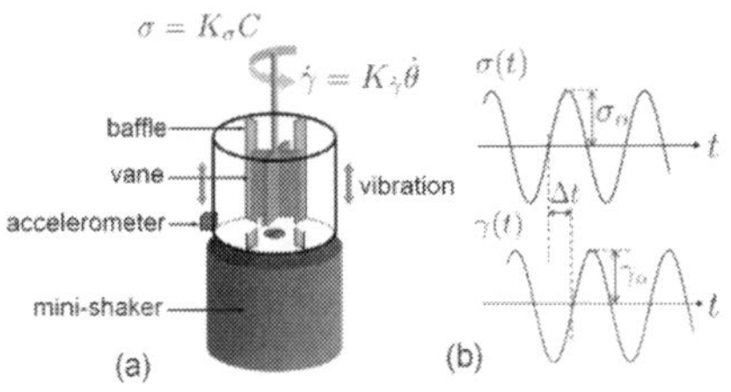

Figure 1. (a) sketch of the experimental setup used for mechanical spectroscopy experiments. (b) Applied sinusoidal shear stess and corresponding measured deformation.

Instruments)) defined by $\sigma(t)=\sigma_o cos(\omega t)$ with σ_o, the oscillation amplitude and ω, the pulsation. The sample was placed into a "powder cell", composed of a cylindrical baffled cup (25 mm radius; 5 mm baffle width) connected to a vibration shaker(Fig. 1 (a)). Measurements were carried out using a four blade vane device (10 mm radius; 30 mm length)(see hanotin et al. for details). Vertical sinusoidal vibrations are applied with controlled frequency and amplitude. In the following, we define a vibration stress σ_v defined as the mechanical energy supplied to the sample per unit volume, hence $\sigma_v = 1/2\rho A^2(2\pi f)^2$ with ρ the density of the sample, A and f the amplitude and frequency of vibrations, respectively. The vibration amplitudes and frequencies varies from 30 to 1075 μm and from 30 to 60 Hz, respectively.

First, we carried out a stress sweep experiment to ensure experiments are performed in the linear viscoelastic domain. The oscillation amplitude σ_o is increased from 1 to 1000 Pa, at a fixed frequency ω = 1 rad.s^{-1}. Moduli are independent of the stress for $\sigma_o \leq 50$ Pa, and the linear viscoelasticity range is then defined for all σ_o values lower than 30 Pa. This value changes slightly depending on the experimental parameters, i.e. bead size (d), fluid viscosity (η_f) and vibration stress (σ_v), but in any case for all samples, a value of $\sigma_o = 5$ Pa is always located in the linear viscoelasticity domain. For that reason, all frequency sweep tests (variation of ω between 0.01 and 100 rad.s^{-1}, at fixed σ_o) were carried out at $\sigma_o = 5$ Pa.

2.2 Results

Figure 2 shows the evolution of $G'(\omega)$ and $G''(\omega)$ for a granular suspension made up of 100 μm glass beads for η_f =40 mPa.s , with a vibration stress σ_v = 12 Pa (f = 40 Hz, A = 627 μm). The mechanical spectrum clearly exhibits two distinct regimes. At low frequency, the elastic and viscous moduli display a Maxwell behavior with $G' \sim \omega^2$ and $G'' \sim \omega$. In this region, a viscous Newtonian regime is observed in agreement with Hanotin et al(citation). For higher frequencies, the two moduli are equal and increase with $\omega^{1/2}$. The transition between both regimes occurs at a critical frequency ω_R corresponding to a critical time $\tau_R = 1/\omega_R$. We evaluate τ_R by fitting the terminal zone on the basis of a Maxwell model with [10]:

$$G' = G\frac{\omega^2\tau_R^2}{1+\omega^2\tau_R^2} \text{ and } G'' = G\frac{\omega\tau_R}{1+\omega^2\tau_R^2} \quad (1)$$

Where G is the shear modulus (modulus of rigidity).

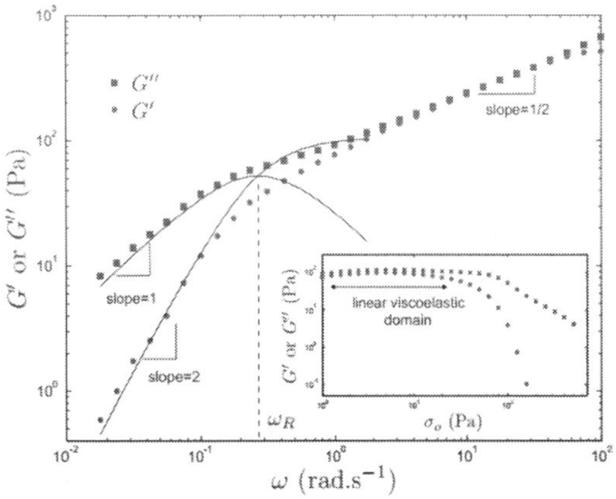

Figure 2. (Color online) Evolution of G' and G'' at a fixed stress amplitude σ_o = 5 Pa, for 100 μm glass beads immersed in a 40 mPa.s Emkarox-water solution (volume fraction $\phi \approx 0.61$) and subjected to a vibration stress σ_v=12 Pa (f = 40 Hz, A = 627 μm). The solid lines correspond to the fit of the terminal zone by Eq. (1) (Maxwell model), with $\tau_R = \omega_R^{-1}$ the relaxation time. Inset: Stress dependence of the elastic (G') and viscous moduli (G''), at a fixed pulsation ω = 1 rad.s^{-1}

From data of Fig. 2, the fit yields G = 105 Pa and $\tau_R = 3.7$ s $\Rightarrow \omega_R = 0.27$ rad.s^{-1}. Figure 3 displays the evolution of τ_R as a function of the vibration stress σ_v for various granular suspensions with three different bead size and fluid viscosity. For any given bead size and fluid viscosity, τ_R decreases when σ_v increases, and τ_R scales as $\sigma_v^{-1/2}$. Figure 3 also reveals that τ_R decreases with decreasing bead size, which suggests that rearrangements time are smaller for smaller beads. This is in aggreement with results of [7] which show that the viscosity is controlled by the lubrication Peclet number $Pe_{lub} = \sigma_{lub}/\sigma_f$, such as $\tau_R \propto Pe_{lub}^{-1} \propto (\eta_f A(2\pi f)/\sigma_f d)^{-1}$ where σ_f is the frictional stress (see [7]). According to the same interpretation τ_R should decrease when η_f increases, which is experimentally confirmed in Fig. 3.

We see from figure 2 that the Maxwell model fails to capture the evolution of the viscoelastic moduli at high frequencies ($G' \sim G'' \sim \omega^{1/2}$) which suggests the existence of a broad spectrum of relaxation times.

Assuming that the complex modulus G^* is the mean of Maxwell modes related to a power law distribution of relaxation times $\rho(\tau)$, it comes that $G^* =< \frac{i\omega\tau}{1+i\omega\tau} >$. This relation leads to $G' = G''$ when the spectrum of relaxation times takes the asymptotic behavior $\rho(\tau) \sim \tau^{-3/2}$.

This indicates an inverse square root dependence on frequency similar to that for the anomalous viscous loss

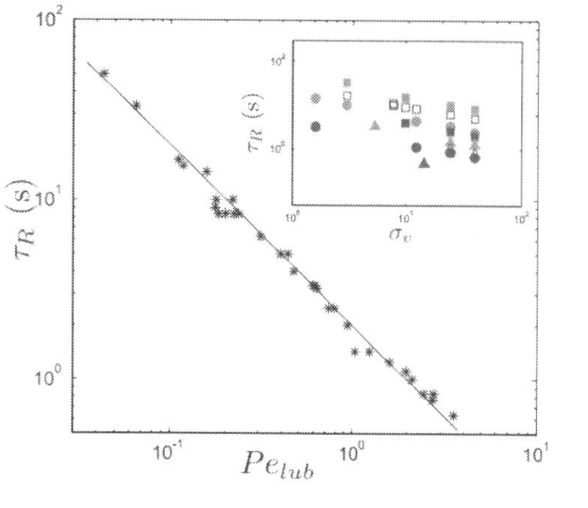

Figure 3. (Color online) (a) Evolution with Peclet number ($Pe_{lub} = \frac{\sigma_{lub}}{\sigma_f}$ of the viscoelestic time (τ_R) extracted from mechanical spectroscopy measurements. Inset: Viscoelastic time (τ_R) versus vibration stress (σ_v), for various values of bead size with diameters d= 335 μm (squares), d= 100 μm (circles) and d=49 μm (triangles) and for various Emkarox water mixtures with fluid viscosities η_f= 40 mPa.s (green), η_f= 70 mPa.s (red) and η_f= 160 mPa.s (blue).

in emulsions which suggests that the shear strain does not produce affine deformations but some regions slip instead of deforming due to the disorder of the packing.

If the the direction $\cos(\theta)$ of local weak planes with the shear gradient direction varies randomly throughout the material then a broad range of relaxation rates appear with $\tau \sim \tau_0 cos^{-2}(\theta)$ which leads to $G^* \sim (i\omega)^{1/2}$. Our results suggest that the model proposed by Liu et al.[11] is applicable to report the high frequency behavior of the dynamic shear modulus in vibrated granular suspensions.

3 Diffusing wave spectroscopy

3.1 Materials and methods

To probe the dynamics at the microscopic scale, we use MSDWS well adapted for sub-microns displacements [12]. The sample is poured into a cylindrical Plexiglas vessel (10 mm diameter; 1mm thickness; 50 mm height), at a height h = 35 mm. l^*,the average distance traveled by a photon in the sample before its direction becomes randomized approximately equals to $6d$ in our case with d the bead diameter. We use a He-Ne polarized laser spot of wavelength λ = 632.8 nm. The backscattered light is detected by a CCD camera, placed 10 cm behind the diaphragm. The camera records the different speckle images in 16 bits, with a size of 1392 × 1040 pixels. To probe dynamics at long time scales, the amplitude and frequency of vibrations as well as the aperture time of the camera have to be chosen to avoid fast decorrelations of the intensity due to cell vibrations. This condition are achieved by using amplitudes that are small compared to the grain diameters and vibration periods that are short (at least four times lower) with regard to the exposure times of the camera). The technique probes events occurring at long times (typically > 0.1 s) by integrating fast variations at short time scales. In our experiments, as the frequency and amplitude used varied from 60 to 160 Hz and from 4 to 15 μm, respectively, the samples were subjected to vibration stresses σ_v lower than 0.1 Pa. Before any data collection, the sample is first vibrated at f = 50 Hz and A = 400 μm during 10 minutes for removing air bubbles. It is then conditioned during five minutes at the desired vibration amplitude and frequency to ensure a "stationary state" where the volume fraction does not change anymore. We measure the intensity correlation function of speckle images by calculating:

$$\overline{g_2(t_0,t)} - 1 = \overline{\frac{\langle I(t_0+t).I(t_0)\rangle_{spkl} - \langle I(t_0)\rangle^2{}_{spkl}}{\langle I(t_0)^2\rangle_{spkl} - \langle I(t_0)\rangle^2{}_{spkl}}} \quad (2)$$

Where $I(t_0)$ and $I(t_0+t)$ are the intensities of the backscattered light respectively at the reference time t_0 and at the time t_0+t with t the delay time, $\langle\bullet\rangle_{spkl}$ is the average over different speckles, $\overline{\bullet}$ is the average over the time origin t_0. The expression of $g_2(t_0,t)-1$ is the one used by Kabla et al. [12] or by Caballero et al. [13].

3.2 Results

Figure 4 shows four correlations functions obtained with the same suspension but for different vibration stresses. These function are well fitted by exponential:

$$\overline{g_2(t_0,t)} - 1 = \beta\exp(-2\gamma\sqrt{6t/\tau_D}) \quad (3)$$

with $\gamma = 1.3 \pm 0.3$.

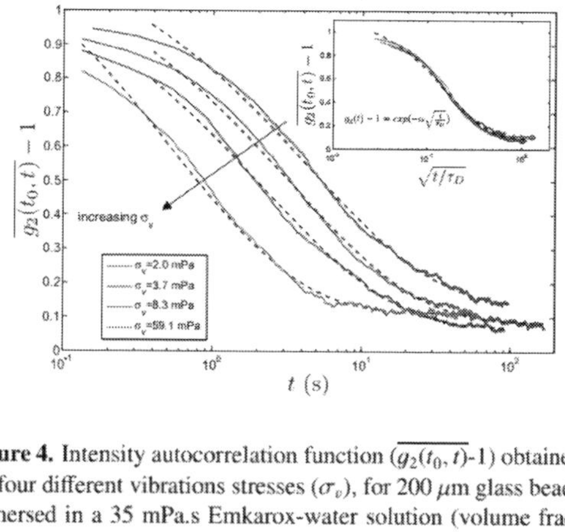

Figure 4. Intensity autocorrelation function ($\overline{g_2(t_0,t)}$-1) obtained for four different vibrations stresses (σ_v), for 200 μm glass beads immersed in a 35 mPa.s Emkarox-water solution (volume fraction $\phi \cong 0.61$). The solid lines represent the experimental data and the dashed lines materialize the fit given by Eq. (3). In the inset, the same data are represented as a function of $\sqrt{t/\tau_D}$ with τ_D the decorrelation time extracted from the fit with $\alpha \approx 6$.

Furthermore, the inset of Fig. 4 reveals that all the functions $\overline{g_2(t_0,t)}$-1 fall onto a master curve when plotted as a function of the reduced time $\sqrt{t/\tau_D}$. It then appears justified to use Eq. (3) for fitting our experimental data to extract the decorrelation time τ_D. Contrary to Brownian suspensions, the relaxation only results here from externally applied vibrations which induce the rupture of some contacts. The MSDWS technique used in this study probes the frequency of this slipping events (yielding frequency)which cause the displacement of grains and a reorganization of the contact network. Using a phenomenological two-state approach related to the inherent bimodal behavior of chain forces in granular packing ([8]), a given grain can be in two distinct states: a consolidated state(C) that ensures momentum (strong contact network)transport and a mobile state (M) disconnected (weak contact network). The granular medium thus reduces to a two-level system and the typical time extracted from MSDWS measurement can be interpreted as the transition time between states $C \longleftrightarrow M$ induced by external vibrations. Consequently, it is particularly relevant to compare τ_D with τ_R derived from rheological measurements. It clearly appears in Fig. 5 that τ_D, as τ_R, is controlled by Pe_{lub} and that both times evolve as $1/Pe_{lub}$. Moreover, the values of τ_D and τ_R overlap over values of Pe_{lub} extending over more than one decade, which means that τ_R completely match with the microscopic measurement of τ_D. These results show that the newtonian behavior observed at low shear rates in [7] is direcly related to the internal reorganization of grains due to vibrations. τ_R is linked to the newtonian plateau at low shear viscosity $\eta_0 = \tau_R\sigma_f/\gamma_c$ which results from the internal dynam- ics of the system induced by vibrations (frequency of yielding events). The critical deformation $\gamma_c(\sim 0.82)$ is then related to the breaking of contacts. This approach is in agreement with the recent numerical work of Lin et al. [14] who show that lubrication effects induced by vibrations in dense suspensions can open gaps between particles along chain forces and tunning shear thickening properties.

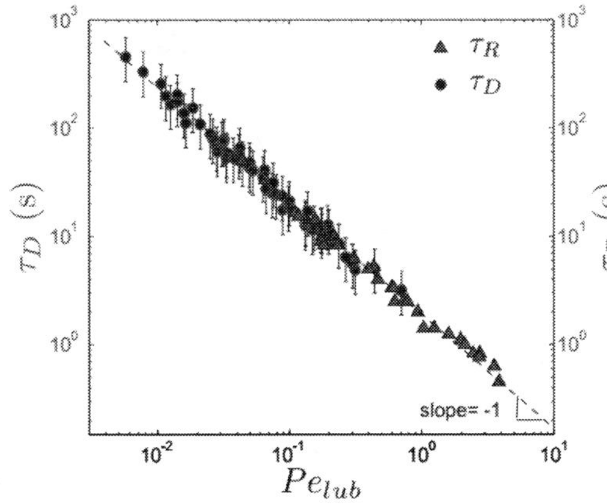

Figure 5. τ_D and τ_R as a function of the Peclet number ($Pe_{lub} = \frac{\sigma_{lub}}{\sigma_f} \propto \frac{\eta_f A(2\pi f)}{d\mu\Delta\rho\phi g z}$).

4 Conclusion

In this paper, we investigate the dynamics of vibrated granular suspensions by mechanical spectroscopy and multi-speckle diffusing wave spectroscopy (MSDWS), with the aim of relating microscopic dynamical mechanisms, at the grain scale, to the rheological behavior of the samples. Results shows that the viscosity of vibrated granular suspensions is controlled by rearrangements dynamics at large time scale related to contact breaking mechanisms. The mean life time of contact τ_D extracted from MSDWS is direcly related to the Maxwell time τ_R obtained by rheological measurements. viscosity are controlled by one relaxation time (larger than the vibration period) related to the flowing behavior of the medium when vibrations are applied which scales as Pe_{lub}^{-1}. In the near future, it would be of interest to probe the microscopic dynamics of these systems to visualize in detail how vibrations affect the particle motion at the grain scale. This could be for example achieved by carrying out fast tomography experiments such as those recently developed in synchrotron facilities or by using refractive index matching technics

References

[1] F. Boyer, E. Guazzelli, O. Pouliquen, Phys. Rev. Lett. **107**, 188301 (2011)
[2] G. D'Anna, G. Gremaud, Nature **413**, 407 (2001)
[3] G. D'Anna, P. Mayor, A. Barrat, V. Loreto, F. Nori, Nature **424**, 909 (2003)
[4] C. Hanotin, P. Marchal, L.J. Michot, C. Baravian, S.K.d. Richter, Soft Matter **9**, 9352 (2013)
[5] S.K. de Richter, G. Le Caër, R. Delannay, EPL (Europhysics Letters) **85**, 58004 (2009)
[6] S.K. de Richter, C. Hanotin, P. Marchal, S. Leclerc, F. Demeurie, N. Louvet, The European Physical Journal E **38**, 1 (2015)
[7] C. Hanotin, S.K. de Richter, P. Marchal, L.J. Michot, C. Baravian, Phys. Rev. Lett. **108**, 198301 (2012)
[8] C. Hanotin, S.K.d. Richter, L.J. Michot, P. Marchal, Journal of Rheology (1978-present) **59**, 253 (2015)
[9] P. Marchal, L. Choplin, N. Smirani, Journal of Rheology **53**, 1 (2009)
[10] J.D. Ferry, *Viscoelastic properties of polymers* (Wiley New York, 1980)
[11] A.J. Liu, S. Ramaswamy, T. Mason, H. Gang, D. Weitz, Physical review letters **76**, 3017 (1996)
[12] A. Kabla, G. Debregeas, Phys. Rev. E. **92**, 035501 (2004)
[13] G. Caballero, E. Kolb, A. Lindner, J. Lanuza, E. Clement, J. Phys.: Condens. Matter **17**, S2503 (2005)
[14] N.Y. Lin, C. Ness, M.E. Cates, J. Sun, I. Cohen, Proceedings of the National Academy of Sciences p. 201608348 (2016)

Response to a localized stress perturbation during a biaxial test

Sean McNamara[1,⋆], *Axelle* Amon[1,⋆⋆], and *Jérôme* Crassous[1,⋆⋆⋆]

[1] *Université de Rennes 1, Institut de Physique de Rennes (UMR UR1-CNRS 6251), Bât. 11A, 263 av. Général Leclerc, F-35042 Rennes cedex, France*

Abstract. Numerical simulations performed to complement experimental studies of a biaxial test show that stress and strain are both structured at the mesoscopic scale. Certain features can be explained by a calculation from linear elasticity.

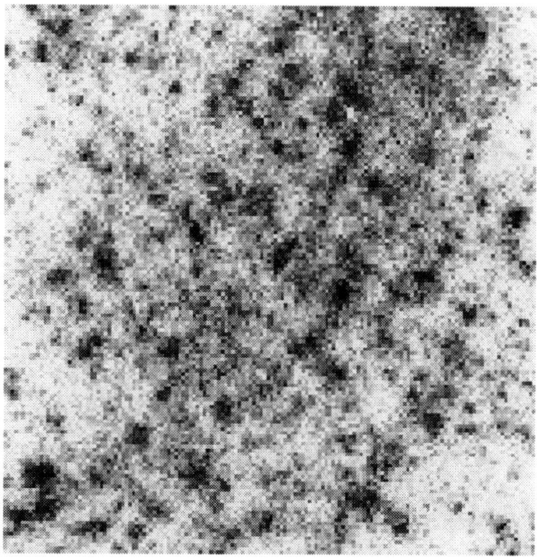

Figure 1. Experimental image showing strain inhomogeneities. The material is being compressed along the vertical axis, while a constant stress is applied at the horizontal boundaries.

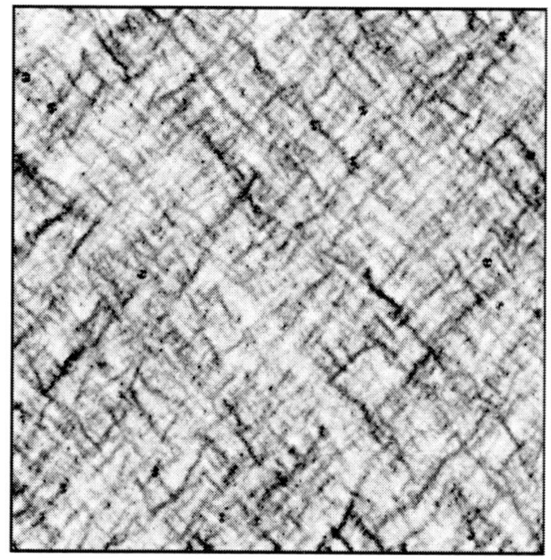

Figure 2. Simulation of a two-dimensional biaxial test with $N = 65536 = 256^2$ disks with friction coefficient $\mu = 0.2$. Imposed strain is 0.1%.

1 Introduction

The biaxial and triaxial tests are often presented as ways of imposing uniform stress and strain on material samples. Optical techniques based on diffusive-wave spectroscopy applied to a biaxial test [1], have been used to show that strain in granular materials is never uniform, even at the beginning of the test. In Fig. 1, we show a typical experimental image. The color code indicates a local measure of strain. with strain being concentrated in dark patches. These patches are themselves embedded in a diffuse irregular network of diagonal bands. It is important to note that these features appear long before failure, and seem to be independent of the formation of the shear band.

When numerical simulations with the similar boundary conditions (Fig. 2) are visualized in an analogous way, the same features appear, with the bands being more prominent [2].

One (yet to be verified) explanation of the diagonal bands is stress redistribution following localized plastic events. As shown by Eshelby [6], such events increase stress in four diagonally oriented lobes, while relaxing stress between these lobes. The increased stress could provoke new events along these directions. After many

⋆e-mail: sean.mcnamara@univ-rennes1.fr
⋆⋆e-mail: axelle.amon@univ-rennes1.fr
⋆⋆⋆e-mail: jerome.crassous@univ-rennes1.fr

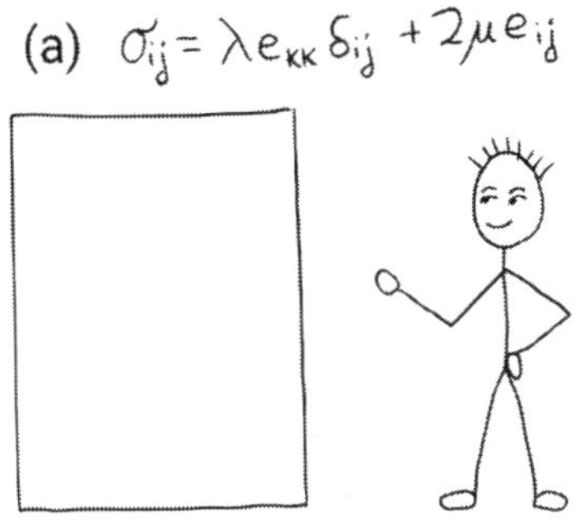

(a) An isotropic elastic medium!

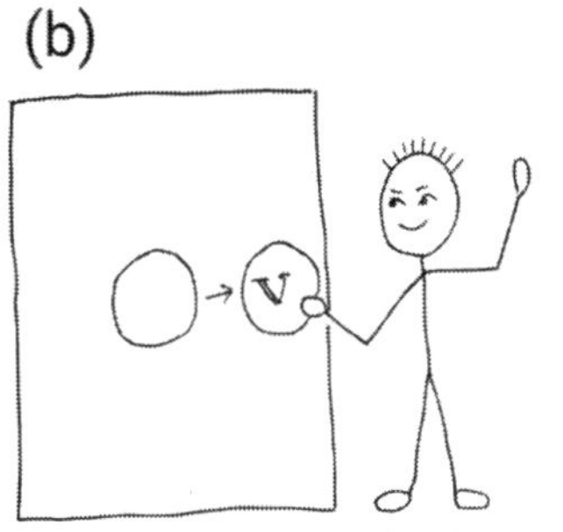

(b) Remove a small region 𝕍.

(c) 𝕍 changes shape!

(d) The stress σ^* returns 𝕍 to its original form.

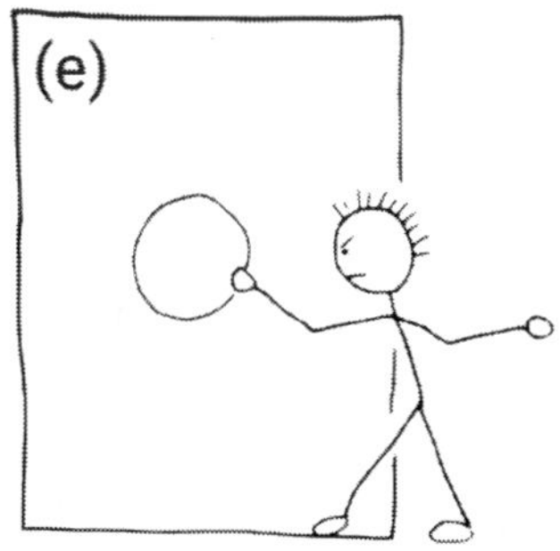

(e) Put 𝕍 back into the medium.

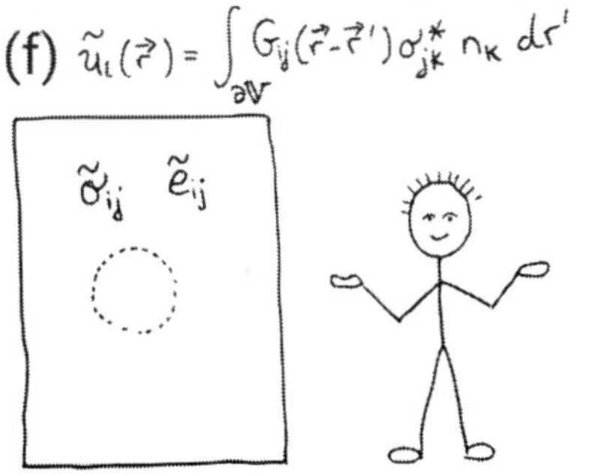

(f) The material adjusts to the new state of the inclusion.

Figure 3. The Eshelby calculation. A plastic event is modeled as a change of equilibrium configuration of a region 𝕍. The consequences are propagated by a Green function. See text for more details.

events, stress will be concentrated in a network of diagonal bands, like those seen in Figs. 1 and 2.

In this paper, we present evidence supporting this explanation by showing that a local perturbation does indeed redistribute stress as predicted by Eshelby. This is nontrivial, because Eshelby's calculation assumes a linear isotropic elastic medium, whereas a loaded granular material is nonlinear, anisotropic, and contains many sliding (plastic) contacts.

2 Boundary conditions

The experiments of Fig 1 and the simulations of Fig. 2 have similar boundary conditions. A granular sample of initial height H and initial length L is contained by four straight walls. The bottom wall is fixed, while the upper wall descends at a small, constant velocity: thus the change in height ΔH decreases from zero at an imposed constant rate. On the contrary, ΔL is not imposed, but adjusted so as to maintain a constant confining pressure p_0.

Of course, there are also many differences: The simulations are strictly two-dimensional whereas the experiments are done in a three-dimensional plane strain configuration. In the simulations, the walls are perfectly straight and smooth; in the experiments they are composed of a rubber membrane.

Frictionless walls minimize boundary effects, leading to mesoscopic strains that resemble those seen in the center of the experimental sample. Other features of the experiment, such as the formation and persistence of shear bands crossing the entire sample, are not well reproduced by the simulations, suggesting that boundary conditions play an important role in the formation of these features.

3 Macro-, meso- and microscales

To think clearly about these questions, we distinguish between three different length scales:

1. The *macroscopic* length scale is the size of the sample. The macroscopic stress and strain tensors are

$$\sigma = \begin{pmatrix} -p_0 & 0 \\ 0 & -\sigma_y \end{pmatrix} \quad \varepsilon = \begin{pmatrix} \Delta L/L & 0 \\ 0 & \Delta H/H \end{pmatrix}, \quad (1)$$

   where σ_y is the average normal stress on the upper wall. These tensors are functions of time, but not of space.

2. The *microscopic* scale is the grain diameter or smaller: microscopic quantities describe what happens at individual contacts or grains. Examples of microscopic quantities are contact forces and particle displacements.

3. Finally, between these two scales is the *mesoscopic* scale, where structures arise from the interaction

Figure 4. Stress perturbations arising from a perturbing stress applied in the center of the sample. Grayscale indicates the values of $\tilde{\sigma}_{xx} - \tilde{\sigma}_{yy}$ of each grain. Black indicates grains where $\tilde{\sigma}_{xx} - \tilde{\sigma}_{yy}$ increases the load, while where the load is decreased.

of several grains. Examples include force chains and non-affine motions. The strain inhomogeneities shown in Figs. 1 and 2 are mesoscopic. At these length scales, stress and strain can depend on position as well as time. A central question in this article is the interaction between macroscopic and mesoscopic stresses.

4 Elastic theory

In this section, we sketch the calculation of Eshelby [6], who showed how to calculate the effect of local plastic events occurring within an elastic medium. We begin by considering an isotropic elastic medium: Fig 3a. We then imagine isolating the region $\mathbb{V}$ where the event will occur: Fig. 3b.

The plastic event is modeled as a change of the zero stress configuration of $\mathbb{V}$. Instead of zero stress occurring at $\boldsymbol{\varepsilon} = 0$, it occurs at $\boldsymbol{\varepsilon} = \boldsymbol{\varepsilon}^*$. If $\mathbb{V}$ were circular and isolated from the rest of the elastic medium, it would adopt an elliptical form given by $\boldsymbol{\varepsilon}^*$, as shown in Fig. 3c.

It remains to calculate the effect of this new zero-stress configuration on the rest of the elastic medium. To do this, we first use the elastic stress-strain relation to calculate the stress $\boldsymbol{\sigma}^*$ needed to return $\mathbb{V}$ to its original form, see Fig. 3d.

Then $\mathbb{V}$ can be put back (mentally) into the elastic medium - see Fig. 3e. It will exert an additional stress $\boldsymbol{\sigma}^*$ on the boundary of $\mathbb{V}$. Using a Green function, the displacements $\tilde{\boldsymbol{u}}$ provoked by this additional stress are calculated everywhere in the medium (Fig. 3f). From these displacements, and additional strain $\tilde{\boldsymbol{\varepsilon}}$ can be found. Finally, the elastic stress-strain relation can be used again to find the corresponding stress $\tilde{\boldsymbol{\sigma}}$.

We will consider $\boldsymbol{\varepsilon}^*$ having the same directions as the macroscopic stress, that is, that height of $\mathbb{V}$ decreases while its length increases:

$$\boldsymbol{\varepsilon}^* = \begin{pmatrix} \varepsilon_1^* & 0 \\ 0 & \varepsilon_2^* \end{pmatrix}, \tag{2}$$

with $\varepsilon_1^* > 0$ and $\varepsilon_2^* < 0$.

One can then ask where $\tilde{\boldsymbol{\sigma}}^*$ will add to the macroscopic stress, possibly provoking new plastic events, and where, on the contrary, it will cancel the macroscopic stress, stabilizing the material. In the case of the biaxial test, the shear stresses are given by $\sigma_y - p_0$. We therefore calculate $\tilde{\sigma}_{yy} - \tilde{\sigma}_{xx}$. This quantity has a positive sign in four diagonal lobes radiating from $\mathbb{V}$. In these regions, $\tilde{\boldsymbol{\sigma}}$ adds to the macroscopic stress, possibly triggering new plastic events that concentrate further stress along other diagonal lobes. Superimposing many such events leads to concentrations of strain in diagonal bands, such as those seen in Fig. 1 and 2.

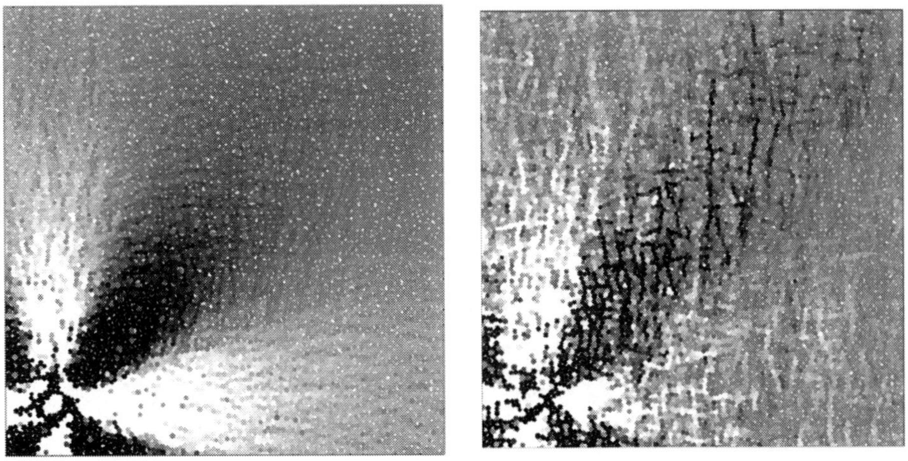

Figure 5. Magnified versions of the first and last panels of Fig. 4. The thresholds for the first panel were reduced to $3 \times 10^{-5} p_0$. One sees that the smooth elastic response co-exists peacefully with the granular microsctructure and the force chain meso-structure.

5 Numerical study of stress redistribution

The explanation advanced above explains the distribution of plastic events using a theory that reposes exclusively on calculations using elasticity. Several objections are possible:

1. We use only elasticity to explain plasticity,
2. An elastic medium has an extremely simple microscopic structure, but granular materials have complicated meso- and microscopic physics,
3. In the theory, the medium is isotropic, but a loaded granular material in a biaxial test is strongly anisotropic.

To investigate these questions, we apply a stress $\boldsymbol{\sigma}^*$ to the boundary of a small circular region $\mathbb{V}$ at the center of a numerically simulated granular material undergoing a biaxial test [3]. This stress can be applied by adjusting the forces of those contacts that cross the boundary of $\mathbb{V}$. We then examine the resulting change in stress $\tilde{\sigma}_{yy} - \tilde{\sigma}_{xx}$ provoked by the perturbing stress $\boldsymbol{\sigma}^*$. The stress on each grain is estimated by

$$\sigma_{ij} = \frac{1}{\pi R^2} \sum_{\alpha} F_i^{(\alpha)} r_j^{(\alpha)}, \quad (3)$$

where α labels the contacts. $\mathbf{F}^{(\alpha)}$ is the force transmitted by contact α, and $\mathbf{r}^{(\alpha)}$ is a vector pointing from the point of contact to the grain center, and R is the particle radius. The sum runs over all contacts of the grain.

In Fig. 4, we use a gray-scale to show the change in stress on each grain. If the stress change exceed $10^{-4} p_0$, the grain is black; it it is less than $-10^{-4} p_0$, the particle is white. If the stress lies between these two values, then a shade of gray is selected to indicate the stress level.

Fig. 4 confirms that the response of the material is indeed as predicted by the prediction of elastic theory: stress is increased in lobes oriented near 45°, and decreased above, below, and on each side of the inclusion.

Another change is also visible: as the test progresses, the texture of the lobes changes. To show this more clearly, in Fig. 5, we show enlargements of one lobe for the first and last images of Fig. 4. These pictures show that the lobes are not uniform. Instead, force chains can be discerned in both cases. But as the simulation progresses, the force chains become more widely separated, with a greater contrast with the rest of the grains. The elastic solution thus coexists with a changing mesoscale force chain structure.

References

[1] Le Bouil, A. *et al.* Gran. Matt. **16**, 1 (2014).

[2] Le Bouil, A., Amon, A., McNamara, S., and Crassous, J., Phys. Rev. Lett. **112**, 246001 (2014).

[3] S. McNamara, J. Crassous, A. Amon, Phys. Rev. E **94** 022907 (2016).

[4] F. Radjai, D.E. Wolf, M. Jean, J.-J. Moreau, *Phys Rev Lett* **80** 61 (1997)

[5] J.R. Williams and N. Rege, *Mech of Cohesive-Frictional mat* **2** 223 (1997)

[6] J. D. Eshelby, Proc. R. Soc. Lond. A **241**, 376-396 (1957).

Relaxation processes in Aeolian transport

Houssem Selmani[1,4,⋆], *Alexandre* Valance[1,⋆⋆], *Ahmed* Ould El Moctar[2], *Pascal* Dupont[3], and *Rabah* Zegadi[4]

[1] *Institut de Physique de Rennes, UMR UR1-CNRS 6251, Université de Rennes 1, 35042 Rennes cedex, France.*
[2] *Laboratoire de Thermocinétique, Polytech. Nantes, CNRS UMR 6607, 44306 Nantes, France.*
[3] *GCGM, INSA de Rennes, Campus Beaulieu, 35043 Rennes, France.*
[4] *Laboratoire de Mécanique de Précision Appliquée, Université Ferhat Abbas Sétif 1, Algeria.*

Abstract.
We investigate experimentally the relaxation process toward the equilibrium regime of saltation transport in the context of spatial inhomogeneous conditions. The relaxation length associated to this process is an important length in aeolian transport. This length stands for the distance needed for the particle flux to adapt to a change in flow conditions or in the boundary conditions at the bed. Predicting the value of this length under given conditions of transport remains an open and important issue. We conducted wind tunnel experiments to document the influence of the upstream particle flux and wind speed on the relaxation process toward the saturated transport state. In the absence of upstream particle flux, data show that the relaxation length is independent of the wind strength (except close to the threshold of transport). In contrast, in the case of a finite upstream flux, the relaxation length exhibits a clear increase with increasing air flow velocity. Moreover, in the latter the relaxation is clearly non-monotonic and presents an overshoot.

1 Introduction

Saltation is known as the primary mode of sediment transport in air and characterizes the movement of the particles jumping along the sand surface in ballistic trajectories.[1–4]. When a steady wind blows over a flat erodible bed, an equilibrium between the flow and the transported particles is achieved and the mass flux, Q, reaches a saturated value $Q = Q_{sat}$. Now, if we consider an unsteady or inhomogeneous situation, in which the wind velocity is subjected to a temporal or spatial change, the relaxation process towards the new saturated state takes a certain characteristic time or length [5, 6]. By linearizing the problem around the saturated state, one can describe the relaxation process by a first order differential which takes the following form for spatial inhomogeneous situations:

$$\frac{\partial Q}{\partial x} = -\frac{Q - Q_{sat}}{L_{sat}} \tag{1}$$

where L_{sat} is the characteristic length scale of the relaxation process and Q_{sat} is the saturated state corresponding to the new equilibrium transport regime. L_{sat} is usually referred to as the saturation length (L_{sat}) [5–8]. A similar first order differential equation can be written for unsteady situations where the space variable is replaced by the time and the saturation length by the saturation time τ_{sat}.

The relaxation process is driven by several intricate physical processes. If the wind speed is suddenly increased, the saltating particles will be accelerated but within a certain time delay due to particle inertia. New grains will be extracted from the bed either by aerodynamic or impact entrainment but at a certain rate depending on the efficiency of the particle pick-up process. Then, in response to the increase of the concentration of the saltating particles, the wind speed within the transport layer is lowered due to momentum extraction by the particles until a new equilibrium is reached. It turns out that several time or spatial scales may be a good candidate for the saturation time or length, respectively: the time needed for the negative feedback of transport on the wind to take place [9], the time needed to expel new grains from the sand bed [5], and the time needed to accelerate the grains to their new equilibrium velocity. It is generally argued that the latter time scale, usually referred to as the drag time τ_{drag} (or the drag length l_{drag} for the corresponding length scale) [10] is associated with the slowest of these processes and is therefore expected to control the relaxation process. The drag length is the distance for a particle initially at rest to equilibrate its velocity with that of the fluid. It is controlled by the particle inertia and is given by $l_{drag} = (\rho_p/\rho_a)d$. Experimental estimations of the saturation length by [6] indicate that the saturation length is independent of the wind strength and is about twice the drag length: $L_{sat} = 2l_{drag}$ (for $d = 200\ \mu m$, the saturation length is on the order of a meter). Eq. (1) tells us that the mass flux relaxes monotonously towards the equilibrium saturated state via a simple exponential behavior:

$$Q(x) = Q_{sat} + (Q_0 - Q_{sat})\exp(-\frac{x - x_0}{L_{sat}}) \tag{2}$$

⋆e-mail: selmani.h@univ-setif.dz
⋆⋆e-mail: alexandre.valance@univ-rennes1.fr

 DOI: 10.1051/epjconf/201714003055

Table 1: Values of the fitted parameters obtained from the data of Andreotti et al. [6].

$u^\star(m/s)$	$Q_{sat}(mm^2/s)$	$x_0(m)$	$L_{sat}(m)$
0.26	12.3555	0.5721	0.4607
0.33	21.0209	0.3151	0.3932
0.48	41.5917	0.1192	0.5160

where Q_0 is the upstream mass flux at the position $x = x_0$. This simple exponential behavior is expected to provide a good description of the relaxation process as long as the upstream mass flux Q_0 is not too far from the equilibrium saturated value Q_{sat}. Andreotti et al. [6] reported wind-tunnel experiments concerning the variation of the mass flux along the downstream direction and showed the last stage of the relaxation (i.e., for $Q > Q_{sat}/4$) is well described by Eq. (2) (see Fig. 1). The first stage of the relaxation process (i.e., for $0 < Q < Q_{sat}/4$) is however not captured by Eq. (2).

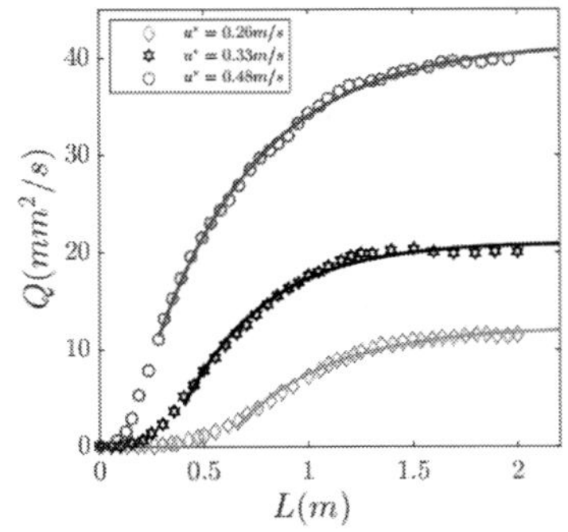

Figure 1: Variation of the mass flux along the downstream distance x for three different wind friction velocities: $u^\star$=0.26, 0.33 and 0.48 m/s (Wind-tunnel data from Andreotti et al [6]). The upstream mass flux for the three runs is $Q_0 = 0$. The solid lines represent the best fit using eq. (2) starting from $x = x_0$ where x_0 is defined such $Q = Q_{sat}/4$. See table 1 for the values of the fitted parameters x_0, L_{sat} and Q_{sat}.

Other experimental results indicate more complex relaxation processes with an overshoot which is accompanied by an oscillating behavior. In particular, [1] reported wind-tunnel experiments showing that this type of complex relaxation occurs in the case where the upstream mass flux Q_0 is non-zero and controlled independently from the wind speed with a hopper device (see Fig. 2). Eq. (2) is again too simple to capture the experimental data from Bagnold.

The above experimental results illustrate the complexity of the relaxation process and its sensitivity to the out-of-equilibrium situation which is investigated. The prediction and description of the relaxation processes and their characteristic length scale is an important issue in the context of the formation and dynamics of aeolian sand dunes.

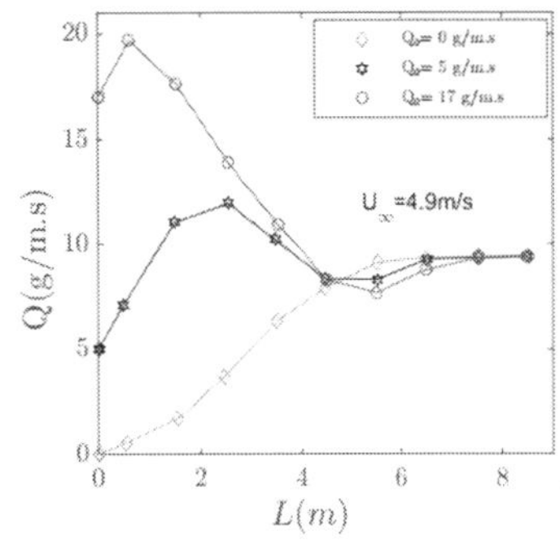

Figure 2: Variation of the mass flux along the downstream distance x (Data from Bagnold [1]) for three different upstream flux conditions: Q_0=0, 5 and 17 $g/m.s$. The nominal velocity of the air flow is $U_\infty = 4.9\ m/s$ for three runs.

In particular, the saturation length controls the length of the smallest dunes emerging from a flat sand bed [11, 12] and is therefore a crucial parameter in morpho-dynamic models of Earth's sandy landscapes [13–15].

Our aim here is to investigate through wind-tunnel experiments the relaxation process of the mass flux Q towards its saturated value Q_{sat} for different out-of-equilibrium situations. Inspired by Bagnold experiments, we conducted a series of wind-tunnel experiments varying systematically the upstream flux and wind conditions in order to document the relaxation process.

The paper is organized as follows. First, the wind tunnel facilities and protocols are described in Section 2, while results and analyses of the relaxation process are presented in Section 3. Finally, section 4 provide conclusion and outlook.

2 Experiments

The experiments reported here were performed in a 6.6 m long wind tunnel with a cross-section of dimensions 0.27 $m \times$ 0.27 m where the nominal air velocity U_∞ (i.e., the air speed outside from the boundary layer) can be varied between 0 to 20 m/s (see [16, 17] for further details). The floor of the tunnel is covered with a sand layer whose stream-wise length can be varied from 0.2 to 6 m. At the upstream edge of the sand bed, a hopper is installed on the roof of the wind-tunnel and allows to deliver a constant and prescribed upstream sand flux Q_0 (see Fig. 3). We used natural sand with a median diameter $d = 200\ \mu m$ and density $\rho_p = 2650\ kg/m^3$. The air speed is measured by using a Pitot tube located at a distance of 6 m from the beginning of the tunnel and set at a height of 130 mm from the floor. The Pitot tube is outside from the boundary layer and is used to assess the nominal air velocity U_∞.

We performed experiments with increasing bed length L from 0.2 to 6 m (with an increment of 0.5 m) for various air speeds (from 6 to 10 m/s) and three different upstream sand fluxes Q_0 ($Q_0 = 0, 35.6$ and 69.5 $(g/m.s)$; cf

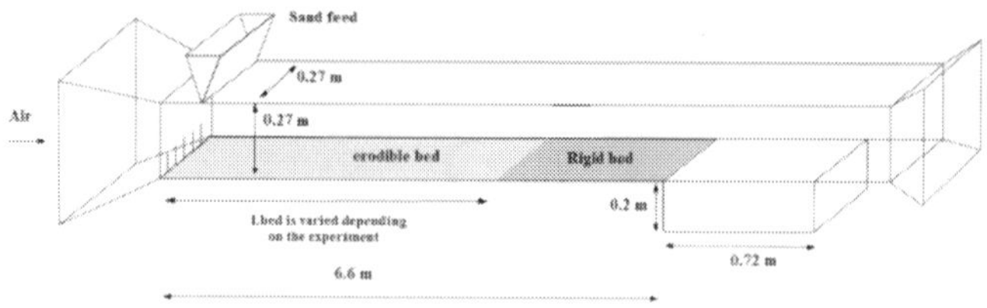

Figure 3: Schematic view of the wind tunnel.

Table 2: Air speeds U_∞ and upstream fluxes Q_0 used in the reported experiments.

$Q_0(g/m.s)$	0	35.6	69.5
$U_\infty(m/s)$	6.09	6.06	6.05
	7.06	7.05	7.07
	8.05	8.05	8.05
	9.07	9.04	9.05
	10.04	10.04	10.06

Table 2 for further details). The non-zero values of upstream fluxes correspond to saturated values of saltation transport obtained at U_∞ = 7.53 and 9.23 m/s, respectively.

Each run follows the same protocol. The sand bed is prepared with a uniform thickness of 2 cm and a given stream-wise bed length L. The sand bed surface is flattened before each run. The fan of the tunnel is tuned at a given wind speed while the hopper is opened at the same time delivering a prescribed upstream flux. A run lasts a few minutes (typically between 1 and 20 minutes according to the wind strength) in order to collect a measurable amount of sand at the exit of the tunnel. After each run, the collected sand is weighed and the resulting outgoing flux $Q(L)$ is determined as follows: $Q(L) = M/T \times W$, where M is the mass of the collected sand, T the duration of the run and W the wind-tunnel width.

These experiments allowed to document the relaxation process and estimate the characteristic relaxation length needed to reach the saturated state of transport starting from different out-of-equilibrium situations.

3 Results and discussion

The evolution of the mass flux with respect to the bed length for different upstream mass fluxes and air velocities is presented in Fig. 4. In the case of a zero upstream flux, the downstream mass flux increases monotonously with increasing bed length until an asymptotic value corresponding to the equilibrium saturated value is reached. The characteristic relaxation length of the whole process decreases within increasing wind speed but saturates to a constant value of between 1 m and 2 m for nominal air velocities greater than 8 m/s. In the case of a non-zero upstream flux, the evolution of mass flux with the bed length exhibits a peak value that is greater than the equilibrium saturated flux Q_{sat}. This overshoot effect is caused by the process of impact entrainment triggered by the particles launched upstream. The peak value of the mass flux Q increases with increasing both the upstream mass flux and air flow velocity. These data confirm the trends observed in Bagnold experiments [1]. However, unlike Bagnold observations, we do not see a clear oscillating behavior during the relaxation towards the saturated value.

Our data provide additional information about the characteristic relaxation length in case of non-zero upstream flux. Indeed, the latter increases with increasing wind strength. The influence of the upstream mass flux on the relaxation length is however not evidenced. Additional experiments would be needed to clarify this issue.

4 Conclusion

We carried out wind tunnel experiments to document the relaxation process towards the equilibrium regime of aeolian sand transport in the context of spatial inhomogeneous configuration by varying the upstream particle flux and wind speed. For zero upstream mass flux, the relaxation length characterizing the whole relaxation process decreases with increasing wind speed but reach an asymptotic value at large wind speed. These results confirm the measurements of Andreotti et al. showing the relaxation length (including the amplification process and the saturation one) diverge at the transport threshold. In contrast, for non-zero upstream mass flux, the relaxation length shows a clear increase with the wind strength. In addition, the relaxation exhibits, in this case, an overshoot already mentioned by Bagnold [1].

In the light of these results, it becomes clear that the simple relaxation equation (cf Eq. 1) is not able to capture the amplification process (including the overshoot) and the saturation process. It is therefore needed to model in details all the physical mechanisms involved in the relaxation process. Continuum models for aeolian sand transport have already been developed in the context of unsteady or spatial inhomogeneous conditions such as the one proposed by Sauermann et al. [5]. The next step is to make quantitative comparison with the predictions of these models and check their pertinence and validity.

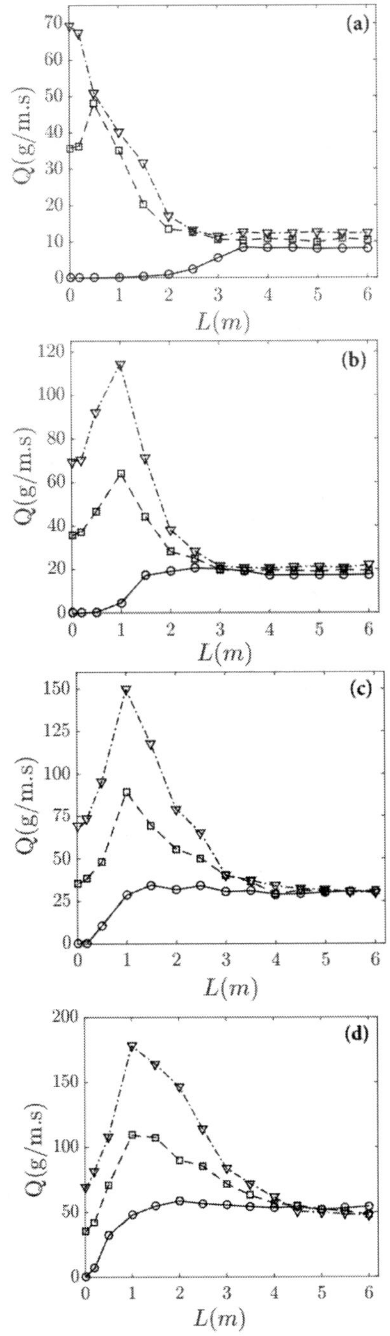

Figure 4: (a-d) The evolution of the sand mass flux as a function of the bed length L for 4 different air velocities respectively $U_\infty = 6, 7, 8$ and $9\ m/s$. For each air speed, three upstream fluxes were investigated $Q_0 =$ 0, 35.6, 69.5 $g/m.s$ represented respectively by (○), (□), (▽).

Acknowledgments

The authors would like to thank the Algerian Ministry of Higher Education and Scientific Research for the financial support in the framework of program PNE 2015-2016.

References

[1] R.A. Bagnold, Methuen, New York (1941)
[2] O. Durán, P. Claudin, B. Andreotti, Aoelian Research **3**, 243 (2011)
[3] J.F. Kok, E. Partel, T. Michael, D. Bou Karam, Report on progress in Physics **75**, 106901 (2012)
[4] A. Valance, K.R. Rasmussen, A. Ould El Moctar, P. Dupont, Comptes Rendus Physique **16**, 105 (2015)
[5] G. Sauermann, K. Kroy, H.J. Herrmann, Physical Review E. **64**, 031305 (2001)
[6] B. Andreotti, P. Claudin, O. Pouliquen, Geomorphology **123**, 343 (2010)
[7] T. Pähtz, J.F. Kok, E.J.R. Parteli, H.J. Herrmann, Physical Review Letters **111**, 218002 (2013)
[8] T. Pähtz, A. Omeradžić, M.V. Carneiro, N.A.M. Araújo, H.J. Herrmann, Geophysical Research Letters **42**, 2063 (2015)
[9] B. Andreotti, Journal of Fluid Mechanics **510**, 47 (2004)
[10] P. Hersen, S. Douady, B. Andreotti, Physical Review Letters **89**, 264 (2002)
[11] P. Claudin, B. Andreotti, Earth and Planetary Science Letters **252**, 30 (2006)
[12] E.J.R. Parteli, O. Durán, H.J. Hermann, Physical Review E **75**, 011301 (2007)
[13] C. Narteau, D. Zhang, O. Rozier, P. Claudin, Journal of Geophysical Research: Earth Surface **114**, F03006 (2009)
[14] E.J.R. Parteli, O. Durán, H. Toar, V. Schwammle, H.J. Hermann, Proceedings of the National Academy of Science **106**, 22085 (2009)
[15] E.J.R. Parteli, K. Kroy, H. Tsoar, J.S. Andrade, T. Poschel, Review and future plans, The European Physical Journal Special Topics **223**, 2269 (2014)
[16] T.D. Ho, A. Valance, P. Dupont, A. Ould El Moctar, Physical Review Letters **106**, 094501 (2011)
[17] T.D. Ho, A. Valance, P. Dupont, A. Ould El Moctar, Aeolian Research **223**, pp. 65–74 (2014)

DOI: 10.1051/epjconf/201714003056

Effect of geometric base roughness on size segregation

L. Jing[1], C. Y. Kwok[1,*], Y. F. Leung[2], and Y. D. Sobral[3]

[1]Department of Civil Engineering, The University of Hong Kong, Pokfulam Road, Hong Kong
[2]Department of Civil and Environmental Engineering, The Hong Kong Polytechnic University, Hong Kong
[3]Departamento de Matemática, Universidade de Brasília, Campus Universitário Darcy Ribeiro, 70910-900 Brasília, DF, Brazil

Abstract. The geometric roughness at boundaries has a profound impact on the dynamics of granular flows. For a bumpy base made of fixed particles, two major factors have been separately studied in the literature, namely, the size and spatial distribution of base particles. A recent work (Jing *et al.* 2016) has proposed a roughness indicator R_a, which combines both factors for any arbitrary bumpy base comprising equally-sized spheres. It is shown in mono-disperse flows that as R_a increases, a transition occurs from slip ($R_a < 0.51$) to non-slip ($R_a > 0.62$) conditions. This work focuses on such a phase transition in bi-disperse flows, in which R_a can be a function of time. As size segregation takes place, large particles migrate away from the bottom, leading to a variation of size ratio between flow- and base-particles. As a result, base roughness R_a evolves with the progress of segregation. Consistent with the slip/non-slip transition in mono-disperse flows, basal sliding arises at low values of R_a and the development of segregation might be affected; when R_a increases to a certain level ($R_a > 0.62$), non-slip condition is respected. This work extends the validity of R_a to bi-disperse flows, which can be used to understand the geometric boundary effect during segregation.

1 Introduction

Base roughness plays an important role in the flow of granular materials. For instance, the minimum thickness necessary to sustain a steady flow down incline is a function of base roughness [1]; the fingering instability developed at the front of granular avalanches is due to the change of basal resistance upon segregation [2–4]; an inadequate base roughness may be responsible for the unsteady flow regime and crystallisation in chute flows [5–7]; in small-scale experimental debris flows, base roughness is the key to produce a realistic deposition with levee formation [8].

To achieve sufficiently rough (i.e. non-slip) base conditions, geometric roughness is usually provided by gluing particles at the boundary surface. By contrast, a flat surface with only mechanical friction usually leads to a shear layer near the base, above which plug flows may occur with poorly-developed shear [5, 9, 10]. In discrete element modelling (DEM), one can construct a rough base by fixing a layer of particles with a regular/random packing [5, 9, 11–21]. For such a bumpy base, both the size and spatial distribution of the fixed particles are influential factors of the geometric roughness. To better understand the basal effect on flow dynamics, it is favorable to characterise geometric rougness in terms of these two factors. A recent work by Jing *et al.* [20] proposes an alternative indicator of base roughness, R_a. It is shown that in mono-disperse flows R_a well describes the geometric roughness of any arbitrary base composing equally-sized spheres; as R_a increases, a phase transition occurs from slip ($R_a < 0.51$) to non-slip ($R_a > 0.62$) regimes, with an intermediate regime of $R_a \in [0.51, 0.62]$.

In this paper, we adopt R_a to study the basal effects in both mono- and bi-disperse flows. In bi-disperse flows, one issue of interest is that size segregation leads to a variation of base roughness, thus affecting flow kinetics [19, 22]. When larger particles are initially in contact with the base (which is made of small particles), the relative roughness is low. The occurrence of size segregation and the subsequent upward migration of large particles change the relative roughness. As a result, the boundary value of velocity evolves with the progress of segregation.

The rest of this paper is organised as follows: In Sect. 2 we briefly introduce the simulation set-up, the strategy of base generation, and the characterisation of base roughness. Section 3 presents the effect of base roughness in both mono- and bi-disperse flows, with emphasis on the variation of geometric roughness during segregation and the consequence of such a variation. In Sect. 4 we conclude our findings.

2 Characterisation of base roughness

2.1 Case set-up

In our DEM simulations, periodic boundaries are imposed to both the flow (x) and vorticity (z) directions, and the top surface is free of constraint [Fig. 1(a)]. The base is roughened by randomly-packed particles [Fig. 1(b)]. At the initial state, a layer of small particles (diameter d_s) is placed

*e-mail: fiona.kwok@hku.hk

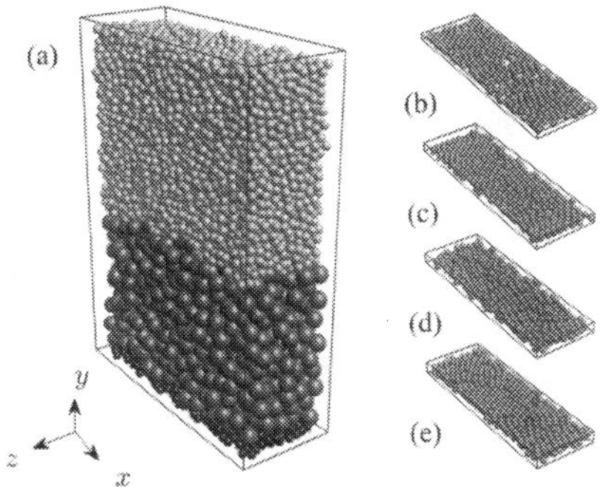

Figure 1. Simulation set-up. (a) A bi-disperse flow with periodic boundaries. (b–e) Bases constructed with $H_b/d_b = 1.0, 1.6, 1.8, 2.0$, respectively.

on top of a bed of large particles (diameter $d_l = S_r d_s$), in which S_r denotes the size ratio of the binary mixture. To initiate the flow, gravity is tilted in the xy plane to a desired inclination θ. The flow thickness H is a function of time t after the flow is initiated. Unless otherwise specified, we consistently use $d_s = 0.005$ m, $H = 40d_s$ at $t = 0$, and $\theta = 25°$; $S_r = 1$ and 2 are specified for mono- and bi-disperse flows, respectively; particle properties include density $\rho = 2650$ kg/m^3, Young's Modulus $E = 5 \times 10^6$ Pa and Poisson's ratio $\nu = 0.35$. The Hertz contact model is used, in which the coefficients of restitution and friction are $e = 0.5$ and $\mu = 0.5$, respectively. The two contact properties are not varied since we focus on geometric roughness in the present work. The effect of these parameters are discussed in [21].

2.2 Base construction

In the construction of bumpy bases, a variety of particle sizes and packings are adopted. The size ratio between flow particles (diameter d) and base particles (d_b) is denoted by $\Phi = d/d_b$, in which d is the average diameter of flow particles contacting the base. The (mean) spacing measured over a layer is noted as $\varepsilon_k d_b$, where ε is spacing and $k = 1, 2, 3$ represents the three sides [Fig. 2(a)]. A base can comprise multiple layers with thickness H_b [Fig. 1(b)–1(e)]. Let $y = 0$ be the top of a base, we firstly place a wall at $y = -H_b$, then pour base particles onto the wall, and finally trim particles beyond $y = 0$. By varying H_b/d_b, different distributions of base particles are produced. As reported in [20], $H_b/d_b = 1.8$ generally yields the roughest base for a given Φ in mono-disperse flows.

2.3 Base roughness

Here we present briefly the characterisation of geometric roughness. Consult Jing *et al.* 2016 [20] for more details.

Given a bumpy base made of particles, such as those in Figs. 1(b)–1(e), we assume that within each layer particles are centred at one common plane. We can discretise

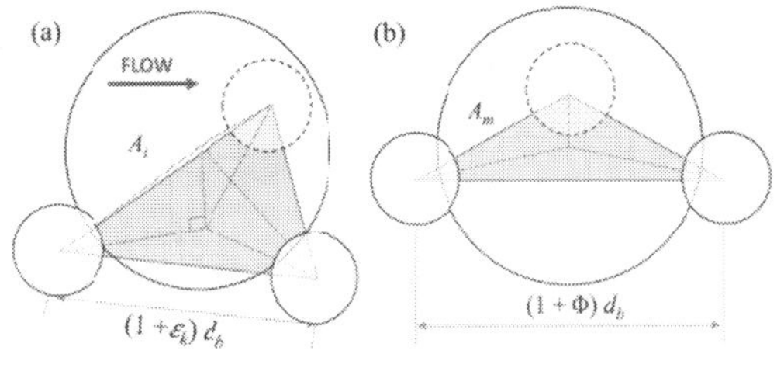

Figure 2. Definition of base roughness. The larger ball represent a flowing particle and smaller balls the fixed base particles. (a) An arbitrary void with area A_i. (b) An equilateral void with area A_m (the most stable with respect to flow).

this plane into an assembly of triangles using the Delaunay triangulation scheme [23]. The ith triangle represents a local void among three base particles (whose centroids are the vertices of the triangle), with area A_i [Fig. 2(a)]. We interpret roughness as the resistance to motion when a particle flows over the local void. The maximum roughness is achieved when the flowing particle is coplanar with the three base particles, in which case we denote the void area as A_m [Fig. 2(b)]. For a layer, we define its (mean) roughness as

$$R_a = \frac{1}{N}\sum_{i=1}^{N}\frac{A_i}{A_m} \tag{1}$$

where R_a stands for the roughness defined by area ratio, and N is the number of triangular voids. This definition combines two key steps shown in [20].

As seen in Fig. 2, A_i can be expressed in terms of ε_k

$$A_i = d_b^2\sqrt{\frac{\sum_k(1+\varepsilon_k)}{2}\prod_k\left[\frac{\sum_k(1+\varepsilon_k)}{2} - (1+\varepsilon_k)\right]} \tag{2}$$

and $A_m = \frac{\sqrt{3}}{4}(1+\Phi)^2 d_b^2$. Note that when A_i exceeds a critical value, the layer underneath void i also contributes to the base roughness. In this case, A_i/A_m is truncated such that $R_a \leqslant 1$, and a non-zero weight is assigned to the underneath layer. This process is referred to as *multi-layer composition* and is elaborated in [20]. Clearly, R_a is a function of spacing ε_k and size ratio Φ, with limits of zero and unity; for each layer, a wider spacing or a larger size of base particles tends to enhance the geometric roughness.

3 Effect of base roughness

3.1 Mono-disperse flow

Both the size and packing of base particles are influential to the flow kinetics [20]. As seen in Fig. 3, for the same packing ($H_b/d_b = 1.0$) but different size ratios ($\Phi = 0.5, 1.0, 2.0$), the velocity profiles show a significant difference. Larger base particles ($\Phi = 0.5$) provide larger voids to trap the flowing particles near the base [Fig. 3(b)], while smaller base particles ($\Phi = 1.0, 2.0$) produce sliding at the base. Notably, the velocity profiles are mainly

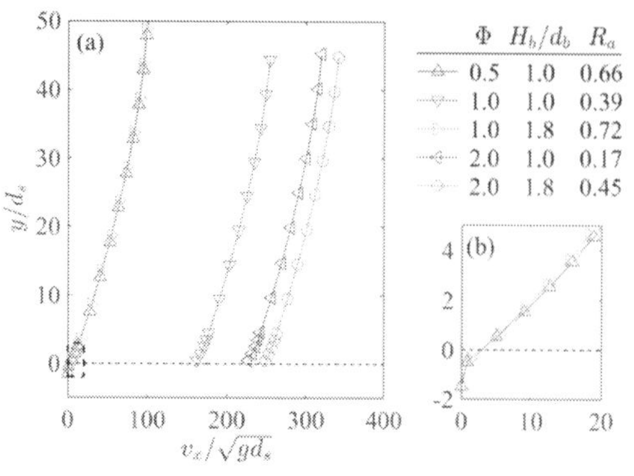

Figure 3. Basal effect on mono-disperse flows ($\theta = 25°$). (a) Velocity profiles at steady state. (b) Details near the base surface, i.e. the region with dashed borders in (a).

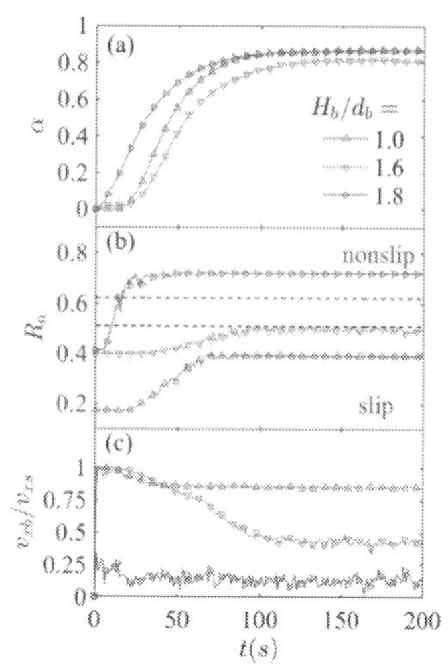

Figure 4. Basal effect on bi-disperse flows ($d_b = d_s$, $d_l = 2d_s$, $\theta = 25°$). Evolution of (a) degree of segregation α, (b) base roughness R_a, (c) the normalised base velocity v_{xb}/v_{xs}. Two dashed lines in (b) mark $R_a = 0.51$ and 0.62, respectively.

shifted horizontally, but their shapes remain similar. For size ratio $\Phi = 1.0$, $H_b/d_b = 1.8$ ensures non-slip condition to the flow because more bumps are generated [Fig. 1(d)]. However, when base particles are half the size of the flowing particles, i.e. $\Phi = 2.0$, the sliding can not be avoided by varying base constructions. In fact, as [20] pointed out, neither the size nor the packing is a sole indicator of the slip/non-slip regimes. With the base roughness R_a defined in Eq. 1, a transition from slip to nonslip is observed. In particular, an intermediate range of R_a is found connecting the regime of sliding ($R_a < 0.51$) and nonslip ($R_a > 0.62$). In the intermediate range, either slip or nonslip can occur [20]. Such a phase change is confirmed in Fig. 3, where a non-slip condition is achieved when $R_a = 0.66$ and 0.72, regardless of the values of Φ and H_b/d_b.

3.2 Bi-disperse flow

In bi-disperse flows, the effect of base roughness is more complicated due to the occurrence of segregation. Initially, large particles are in contact with the base, giving rise to a relatively larger size ratio ($\Phi = d_l/d_b$) than that after segregation ($\Phi = d_s/d_b$). If we consistently use $d_l = 2d_s$ and $d_b = d_s$, the actual size ratio between flowing particles (near the base) and base particles will vary from 2 to 1 during segregation. As a result, the geometric roughness R_a becomes a function of time, with which the basal condition evolves. In Fig. 4, the same bi-disperse mixture is flowing on three bases with different constructions, i.e. $H_b/d_b = 1.0, 1.6, 1.8$ [see Figs. 1(b)–1(d)], and the evolution of base roughness (thus basal velocity) is dependent on the progress of segregation.

To monitor the degree of segregation, we follow [19] to define a state parameter $\alpha = (1 - \Delta c/\Delta c_0)/2$, where $\Delta c = c_s - c_l$ is the distance between the centers of mass of small (c_s) and large (c_l) particles, and Δc_0 is the initial distance; $\alpha = 0$ is the initial state and $\alpha = 1$ means perfect segregation. Figure 4(a) presents the temporal development of α, which exhibits an exponential growth [15, 19]. The initial base roughness is generally low ($R_a \leqslant 0.4$) because the base is made of small particles; as segregation occurs and smaller particles reach the base, R_a increases until the maximum value is reached [Fig. 4(b)]. The values of R_a can be found in three zones: slip, nonslip and the transition in-between. The dashed lines in Fig. 4 mark $R_a > 0.62$ for the slip zone and $R_a < 0.51$ for the non-slip zone. Note that the three zones are established in mono-disperse flows [20], which still predict well the basal velocity conditions in the current segregation configurations. In Fig. 4(c), we plot the evolution of the normalised base velocity, v_{xb}/v_{xs}, where v_{xb} and v_{xs} are velocities at the base and surface layers, respectively. It can be observed that when $H_b/d_b = 1.0$, the varying R_a remains in the slip zone, and the base velocity of the flow is high ($v_{xb}/v_{xs} = 1$, indicating no shear); when $H_b/d_b = 1.6$, R_a evolves from slip zone to the transition zone, which corresponds to the drop of v_{xb}/v_{xs} from 1 to ~ 0.4; when $H_b/d_b = 1.8$, non-slip condition is respected for the whole duration (except the initial short time) and v_{xb}/v_{xs} remains small. Another interesting observation is that as basal sliding takes place, the process of segregation is delayed (see $t < 20s$ in Fig. 4). It is reported in [19, 21] that the sliding near the base may result in crystallisation in the lower layers of the flow, which hinders the development of shear and hence the observed delay of segregation. As small particles percolate towards the base, the crystallised structure can not sustain and then the degree of segregation grows rapidly.

After segregation, a steady state is reached in all presented bi-disperse flows. As shown in Fig. 5(a), the velocity profile for $H_b/d_b = 1.8$ obeys Bagnold's scaling [1, 11, 12, 21], with no sliding at the base. For $H_b/d_b = 1.0$ and 1.6, basal sliding arises and leads to a shift of the velocity profiles. Figure 5(b) shows the profiles of shear rate

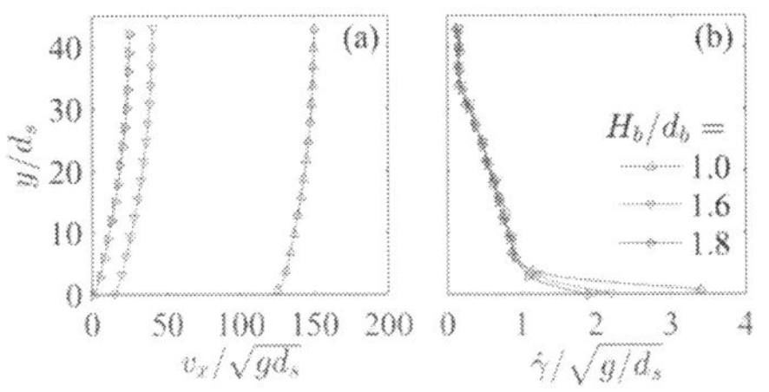

Figure 5. Basal effect on the steady state of bi-disperse flows ($d_b = d_s$, $d_l = 2d_s$, $\theta = 25°$). (a) Velocity profile. (b) Shear rate profile.

($\dot{\gamma}$) for all three cases; the velocity gradients are identical in the bulk of the flows. This indicates that although basal sliding may delay segregation and significantly change the overall speed of the flow, the development of shear is marginally affected provided that a steady state is reached. Note that if basal sliding happens in a perfectly mono-disperse flow, it is possible that crystallisation sustains in the flow, which might lead to an unsteady or plug-like flow where shear is less efficient [5, 20, 21].

4 Conclusion

This work presents the effect of geometric base roughness in granular chute flows. A roughness indicator R_a is used to represent a variety of roughness values due to different sizes and distributions of base particles. A phase transition from slip ($R_a < 0.51$) to nonslip ($R_a > 0.62$) regimes [20] is validated for both mono- and bi-disperse flows. Basal sliding occurs at low values of R_a, which mainly shifts the flow velocity to a higher magnitude but has a marginal influence on the shape of velocity profiles. In bi-disperse flows, the basal condition is a function of time because segregation takes place and R_a evolves with the progress of segregation. When large particles contact the base at the initial stage, significant sliding may arise and the development of segregation can be delayed. As small particles percolate towards the bottom, non-slip condition can be respected if R_a increases beyond 0.62.

In the future, we expect the roughness indicator R_a to be extended to more configurations, such as rotating drum, circular shear apparatus, and curved incline. It would facilitate the characterisation of geometric roughness and the interpretation of the boundary effect on mixing, segregation, and other granular pattern problems.

Acknowledgements

The work was supported by Research Grants Council of Hong Kong (Grant No. RGC/GRF 17203614), and the Research Institute for Sustainable Urban Development at The Hong Kong Polytechnic University, and FAP-DF, Brazil. This research is conducted in part using the research computing facilities and advisory services offered, by Information Technology Services, the University of Hong Kong.

References

[1] O. Pouliquen, Physics of Fluids **11**, 542 (1999)
[2] O. Pouliquen, J. Delour, S.B. Savage, Nature **386**, 816 (1997)
[3] M.J. Woodhouse, A.R. Thornton, C.G. Johnson, B.P. Kokelaar, J.M.N.T. Gray, Journal of Fluid Mechanics **709**, 543 (2012)
[4] J.L. Baker, C.G. Johnson, J.M.N.T. Gray, Journal of Fluid Mechanics **809**, 168 (2016)
[5] L.E. Silbert, G.S. Grest, S.J. Plimpton, D. Levine, Physics of Fluids **14**, 2637 (2002)
[6] V. Kumaran, S. Maheshwari, Physics of Fluids **24**, 053302 (2012)
[7] V. Kumaran, S. Bharathraj, Physics of Fluids **25**, 070604 (2013)
[8] T. de Haas, L. Braat, J.R.F.W. Leuven, I.R. Lokhorst, M.G. Kleinhans, Journal of Geophysical Research: Earth Surface **120**, 1949 (2015)
[9] R. Delannay, M. Louge, P. Richard, N. Taberlet, A. Valance, Nature Materials **6**, 99 (2007)
[10] W. Zhou, Z. Lai, G. Ma, L. Yang, Y. Chen, Granular Matter **18** (2016)
[11] L.E. Silbert, D. Ertaş, G.S. Grest, T.C. Halsey, D. Levine, S.J. Plimpton, Physical Review E **64**, 051302 (2001)
[12] A. Tripathi, D.V. Khakhar, Physics of Fluids **23**, 113302 (2011)
[13] T. Weinhart, A.R. Thornton, S. Luding, O. Bokhove, Granular Matter **14**, 531 (2012)
[14] A.R. Thornton, T. Weinhart, S. Luding, O. Bokhove, The European Physical Journal E **35** (2012)
[15] L. Staron, J.C. Phillips, Physics of Fluids **26**, 033302 (2014)
[16] D.R. Tunuguntla, O. Bokhove, A.R. Thornton, Journal of Fluid Mechanics **749**, 99 (2014)
[17] D.R. Tunuguntla, A.R. Thornton, T. Weinhart, Computational Particle Mechanics **3**, 349 (2016)
[18] U. D'Ortona, N. Thomas, R.M. Lueptow, Physical Review E **93**, 022906 (2016)
[19] L. Jing, C.Y. Kwok, Y.F. Leung, *Discrete element modelling of grain size segregation in bi-disperse granular flows down chute*, in *International Conference on Particle-Based Methods, Fundamentals and Applications* (Barcelona, Spain, 2015), `http://hub.hku.hk/handle/10722/213642`
[20] L. Jing, C.Y. Kwok, Y.F. Leung, Y.D. Sobral, Physical Review E **94**, 052901 (2016)
[21] L. Jing, C.Y. Kwok, Y.F. Leung, Y.D. Sobral, *Basal Effect in Mono- and Bi-Disperse Chute Flows*, in *7th International Conference on Discrete Element Methods* (Dalian, China, 2016)
[22] L. Staron, J.C. Phillips, Physical Review E **92**, 022210 (2015)
[23] *MATLAB Release 2016b*, The MathWorks, Inc., Natick, Massachusetts, United States (2016)

Physics of soil erosion at the microscale

Pierre Philippe[1,*], *Pablo* Cuéllar[1,2], *Florian* Brunier-Coulin[1], *Li-Hua* Luu[1], *Nadia* Benahmed[1], *Stéphane* Bonelli[1], and *Jean-Yves* Delenne[3]

[1] *IRSTEA, UR RECOVER, 3275 route de Cézanne, 13182 Aix-en-Provence, France*
[2] *BAM, Division 7.2 Buildings and Structures, Unter den Eichen 87, 12205 Berlin, Germany*
[3] *UMR IATE, INRA/CIRAD/Montpellier SupAgro, 2 place Pierre Viala, 34060 Montpellier, France*

Abstract. We focus here on the major and still relevant issue of soil erosion by fluid flows, and more specifically on the determination of both a critical threshold for erosion occurrence and a kinetics that specifies the rate of eroded matter entrainment. A state-of-the-art is first proposed with a critical view on the most commonly used methods and erosion models. It is then discussed an alternative strategy, promoting the use of model materials that allow systematic parametric investigations with the purpose of identifying more precisely the local mechanisms responsible for soil particle erosion and ultimately quantifying both critical onsets and kinetics, possibly through existing or novel empirical erosion laws. Finally, we present and discuss several examples following this methodology, implemented either by means of experiments or numerical simulations, and coupling erosion tests in several particular hydrodynamical configurations with wisely selected mechanical tests.

1 Introduction

Erosion and transport of cohesive soil sediments by fluid flows are of primary concern in many processes, either in natural or manmade environments, as fluvial morphological changes, spreading of pollutant or radionucleides in river sediments, impact of agricultural soil erosion on the global carbon cycle, local scour at bridge piers, external and internal erosion of earthen hydraulic structures [1].

The present contribution focuses on two related fundamental issues: (i) How to properly predict a critical condition, if any, for occurrence of erosion? (ii) How to accurately quantify the amount of eroded soil particles? It is first presented here a brief synthesis of the most usual methods used to tackle these issues with particular attention to what are the major remaining problems or contradictions which could limit the general scope, or even the validity, of these previous approaches. As a consequence, an alternative methodology is proposed, based on systematic investigations, either experimental or by means of numerical models, with artificial cohesive materials and aims to combine erosion tests in some specific fluid flow configurations with better targeted, or even specially developed, mechanical characterisations of our model systems. As illustrated by several preliminary tests following the previous strategy, a specific scope is put on the local processes of matter entrainment by a fluid flow inducing what is more globally called *soil's erosion* at a more wider scale, both in space and time.

*e-mail: pierre.philippe@irstea.fr

2 Quantification of soil erosion and related issues

Being able to quantify erosion of a soil by a neighboring fluid flow requires first to select both space and time scales over which the amount of eroded matter and the hydrodynamical load exerted by the fluid flow should be considered and quantified. Owing to the large variety of practical issues, the space scale can span from kilometer squares to microns, including for instance watersheds, riverbanks, soil specimens in the lab, or only few sediment layers, while the duration under consideration can go from seconds to months, or even more in some geological phenomena. This first choice prescribes the type of erosion models needed to achieve a consistent prediction, in particular regarding the quantities that can be used to account for the action of the eroding fluid flow. In a soil mechanics perspective as will be considered in the following, the size of a soil sample under study is typically equal to some centimeters, or tens of centimeters, and the rates of erosion are estimated over time periods of seconds to minutes. Regarding the flow regime which is turbulent in most applications, an hydrodynamical indicator, able to account for the hydrodynamic force exerted by the flow on the soil surface, is not easy to quantify since the fluid flow highly fluctuates in time and also, most often, in space. The simplest and most commonly followed solution consists in a crude averaging of the chosen quantity. Alternative approaches can be developed, mostly based on statistical modeling, but might be hard to apply successfully in practice since a realistic probability density function is needed. The same uncertainty about the use of average quantities arises considering the mechanical resistance of a soil to erosion. Here again, it is likely that the sensitivity to erosion varies from location to location within a soil, reflecting somehow the inherent heterogeneous nature of the material. Note also that in the particular case of a sandy soil, following the pioneer studies by Shields on fluvial sediment transport, the resistance to erosion can be expressed through a critical dimensionless shear-stress whose value can be approximated from simplified mechanistic models describing the equilibrium of a single particle and accounting for drag force, buoyant weight and friction. The more usual situation of a cohesive soil is far much complicated and knowledge is lacking on local processes of erosion at a natural soil surface to consider developing similar mechanistic models.

In practice, the very large majority of previous and actual studies seeking to assess the sensitivity of natural soils to fluid flow erosion have based their analysis assuming the following expression of a local erosion law that accounts for a threshold and relates $\dot{m}$, the rate of mass removal by units of time and surface, to τ, the shear-stress exerted by the fluid flow at the soil's surface [2, 4]:

$$\dot{m} = k_{er}(\tau - \tau_c) \text{ if } \tau > \tau_c, \quad (1)$$

$$\dot{m} = 0 \text{ otherwise.} \quad (2)$$

In this purely empirical modeling, τ_c (in Pa) is a critical shear-stress and k_{er} (in $s.m^{-1}$) is a kinetic erosion coefficient. Both parameters are sometimes called *soil's erodibility*. Several erosion tests have been developed to provide direct measurements or (sometimes crude) estimates of eroded mass and hydrodynamical shear-stress. Based on the local erosion law in Eq. 1 (or a slightly modified version), a fitting procedure ultimately provides the erodibility parameters. Correlations between those parameters and more common soil's properties are still being explored. If rather accurate trends are observed for a same soil, very different and hardly predictable values of erodibility can be found from one soil to another and also possibly from one erosion device to another. The latter result raises in particular questions regarding the postulated intrinsic nature of erodibility. The many different, and sometimes contradictory, empirical relationships proposed in the literature are far from providing a definitive answer. Another general issue concerns the use of average quantities, especially when describing the fluid flow, which appears in serious contradiction with a process as erosion which assumes the existence of an inherent soil's threshold. Indeed, the local hydrodynamic stress values are naturally distributed (fluctuations, turbulence) and the highest portion of this distribution might exceed the soil's critical threshold τ_c while the average stress remains below it, generating intermittent particle extraction and consequently sustaining over time a non-zero rate of erosion.

3 Methodology

Regarding the related literature, very little is known about the real processes at work at particle scale and which commonly hide behind the general term of erosion. Furthermore, as discussed previously, no clear picture has emerged from the most popular route based on direct empirical correlations between resistance to erosion and many different usual material properties obtained for a large variety of natural soils. This approach possibly reaches its limits, facing a rather complex phenomenon and the present contribution proposes to reverse the methodology by using purely artificial materials which can advantageously allow for both a much more accurate control of some key properties (as cohesion, particle size, etc) and, whenever possible, for a direct visualization of the basic local processes of solid matter removal by a fluid flow. Our goal is first to gain insights from the fundamental scale in order to reach a better understanding of the microstructural scenarios of soil's erosion. Second, to identify and rationalize these local mechanisms would constitute an important step towards clearer relationships between some pertinent usual properties of a material, whose characterization might requires the development of specific mechanical tests, and some more application-oriented empirical characterizations as the previous erodibility parameters k_{er} and τ_c, or possibly some new ones related to alternative local erosion laws to be proposed. Such a breakthrough could considerably supplement the usual approach for erosion risk assessment and accordingly motivates this work.

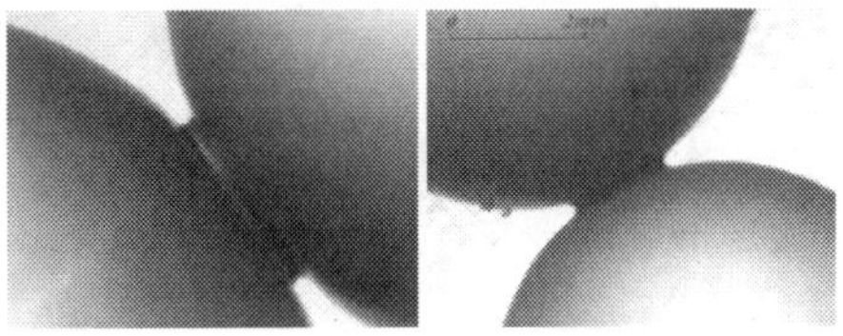

Figure 1. (Left) Capillary bridge of Ucon Oil between $3mm$ glass beads. (Right) Solid bridge of paraffin between $7mm$ glass beads after solidification of the original capillary bridge of liquid paraffin.

3.1 Model materials

For the present contribution, three different types of cohesive soils have been considered. The first type of materials that can be rather simply prepared and studied is made of large granular particles with adhesive bonds at inter-particle contacts. Such contact bonds can be either liquid or solid as illustrated in Figure 1. It is to be noticed the main differences between liquid and solid bonds: solid bridges are usually far much resistant than liquid ones but the latter can be broken and then created back with strong hysteresis while the rupture of a solid bond is permanent and irreversible. Regarding the grains, they can come from either a natural sandy material or an artificial one as the previous glass beads in Fig. 1 which can advantageously enable the additional implementation of two coupled optical techniques, Refractive Index Matching and Planar Laser Induced Fluorescence, to access to a direct visual-

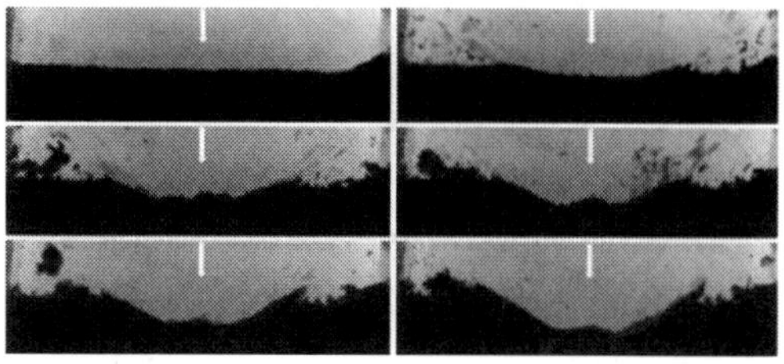

Figure 2. Typical sequence (from left to right and top to bottom) of scour erosion of a magic sand during a 2D water jet test. The arrow indicates the location of the impinging water jet.

ization of the local erosion process provided that the eroding fluid is an index-matched mixture [5]. The behaviour of such a cohesive granular material with solid bonds can also be investigated by means of the coupling between the Discrete Element Method (DEM) and the Lattice Boltzmann Method (LBM), which is certainly today one of the numerical methods being the most increasingly used for the modeling of fluid-grain systems. An attractive bond at each grain contact is implemented thanks to an additional cohesion model between the solid particles which is based on a relatively straightforward rheology (i.e. viscoelastic-perfectly-plastic) and on the de-bonding limit (yield) conditions introduced by Delenne and coworkers [3] which permit the consideration of bond ruptures in tension, shear and bending conditions.

The second type of model cohesive soils that can be easily studied through systematic investigations are chemically-treated sands, as *magic sand* and *kinetic sand*. The first one is a building toy made of hydrophobic sand grains and can consequently be viewed as a granular material with an effective "skin" cohesion arising from the Laplace over-pressure at the interface between air and water. Kinetic sand , made of 98% sand and 2% polydimethylsiloxane (a type of silicone oil), is another toy material specially developed to mimic the physical behaviour of wet sand with reversible adhesion between grains.

The third kind of artificial cohesive soils corresponds to homogeneous materials, modeled in a continuous medium perspective and able to mimic loamy soils made of very small particles with strong attractive interactions. A synthetic clay as *Laponite* appears as a sound choice since its yield-stress rheology can be systematically varied and because it also has the great advantage to be transparent and thus to allow here again for direct visualization of solid matter retrieval during an erosion test.

3.2 Coupling hydrodynamic and mechanical tests

As an integral part of our methodology, erosion and mechanical tests must be implemented in parallel. The choice of the hydrodynamical configuration inducing erosion determines the ability to measure and/or model accurately some key quantities as shear-stress or velocity and to reach sufficient intensity level for erosion to occur. In the present paper, some already existing configurations, as jet erosion

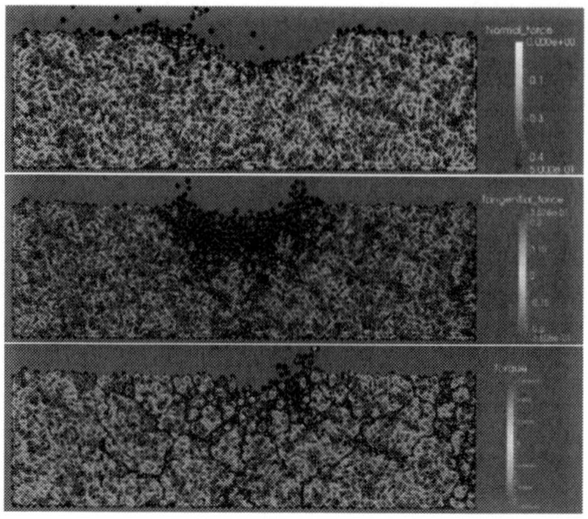

Figure 3. Jet erosion tests on a soil where the cohesion model at particle contacts presents a small yield value under tension (top), under shear (middle), and under bending (bottom), respectively.

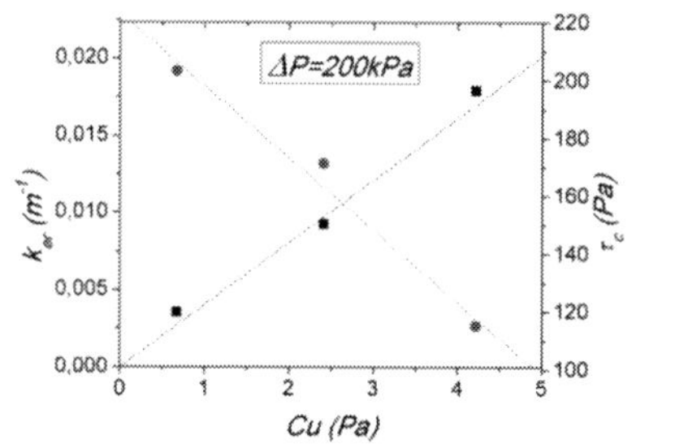

Figure 4. Linear correlations between the erodibility parameters τ_c (black squares) and k_{er} (blue circles) of a kinetic sand, deduced from the classical interpretation model with a 2D pressurised-air jet test (differential pressure $\Delta P = 200Pa$), and the undrained cohesion Cu, obtained from a fall-cone test.

test and hole erosion test, were used but sometimes with specific adaptations (fluid, geometry, optical techniques) enabling direct visualization of the local process. By way of example, Figure 2 shows a jet erosion test on a magic sand where the geometry was restricted to 2D for observation purpose. Some rather usual mechanical tests were carried out with our artificial materials, as fall-cone test, while several other ones were specifically developed for instance to measure tensile strength both at contact and sample scales [7].

4 Some preliminary results

Quite an extensive study of our first model material (i.e. glass beads bonded by solid bridges), subjected to erosion by impinging of an immersed jet, has already been carried out by part of the present authors and will be presented elsewhere [7]. However, several recent and original results can be presented, even if they are still preliminary

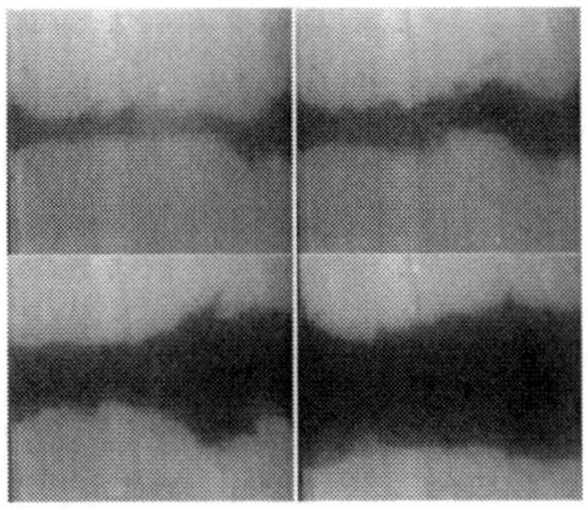

Figure 5. Time evolution of an initially cylindrical pipe in a transparent synthetic clay sample during a Hole Erosion Test with water flow from left to right. The flow rate is progressively increased and the pictures correspond (from left to right and top to bottom) to successive times $t = 94s$, $t = 200s$, $t = 350s$, and $t = 800s$. The dimension of the pictures is $70mm \times 58mm$.

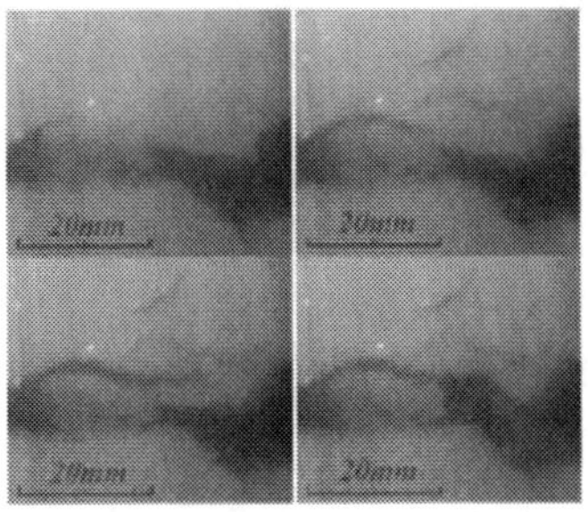

Figure 6. Close-up from the previous test (Fig. 5) where a crack propagation is clearly identified above the pipe and induces the final pull-out of a large aggregate.

at this stage. To start with, our first type of model material has also been studied in a jet erosion configuration but this time by means of a numerical DEM-LBM modeling as previously mentioned. More precisely, the influence of the cohesion model at particle contacts was tested through deliberately exaggerated soil's behaviours by imposing a very weak value of the solid bond yield alternatively under tension, shear and bending loading. As can be seen on Figure 3, widely different erosion processes are observed in the three distinct cases: (i) almost usual grain-by-grain erosion, (ii) remote debonding to a rather long extent within the sample, (iii) in-depth fracturing creating large aggregates.

Another example deals with the same kind of jet erosion test but performed this time on one of our second model materials, namely kinetic sand, and with pressurised air as eroding fluid. Changing the initial density of the kinetic sand allows to control quite accurately the undrained cohesion Cu estimated via a cone-fall test (by calibration from penetration depth of a falling cone, released from a given height, within a soil's specimen). During a jet test, a classical fitting procedure of the scour crater depth versus time using Eq. 1 provides the erodibility parameters τ_c and k_{er} which, as underlined in Figure 4, are nicely and almost linearly correlated to Cu.

Finally, a last preliminary study has focused on hole erosion test performed on a Laponite synthetic clay, our third model material of cohesive soils, using the classical Hole Erosion test (HET) device [1] where a controlled flow is generated along a preformed axial hole within the soil sample to determine a critical value of hydraulic shear stress. Figure 5 gives a general overview of the pipe enlargement versus time while Figure 6 is a close-up where crack initiation and propagation can be observed till a large aggregate is ultimately washed away.

5 Conclusion

The starting point of the present contribution was the identified limitations and remaining issues regarding theoretical and experimental state-of-the-art in quantifying erosion of natural soils. It was then proposed to reverse the classical methodology, substituting real soils by artificial materials, in both experimental and numerical perspectives, with the purpose to better control key properties of the materials so as to implement systematic analysis of sensitivity to erosion, including whenever possible direct visualization of the basic local processes of solid matter removal by a fluid flow as well as specifically selected, or developed, mechanical tests. Several examples of such model materials were first described before some encouraging preliminary results were presented and discussed. Much more work is now required to achieve comprehensive results continuing this strategy.

Acknowledgements

The authors acknowledge a financial support from the region PACA (France) through the APEX-ESCAPE project.

References

[1] S. Bonelli (Ed.). Erosion in Geomechanics Applied to Dams and Levees. Wiley-ISTE (2013)

[2] E. Partheniades. J. Hydr. Div. ASCE **91**(HY1) 105-139 (1965)

[3] J.-Y. Delenne, M.S. El Youssoufi, F. Cherblanc, J.-C. Bénet. Int. J. Num. Anal. Meth. Geomech. **28** 1577-1594 (2004)

[4] A. Knapen, J. Poesen, G. Govers, G. Gyssels, J. Nachtergael. Earth-Sciences Reviews **80**(1-2) 75-109 (2007)

[5] P. Philippe, M. Badiane, Phys. Rev. E. **87**(4) 042206 (2013)

[6] P. Cuéllar, P. Philippe, S. Bonelli, N. Benahmed, F. Brunier-Coulin, J. Ngoma, J.-Y. Delenne, F. Radjaï. Proceedings of the 4th International Conference on Particle-based Methods – Fundamentals and Applications, PARTICLES 2015. E. Oñate, M. Bischoff, D.R.J. Owen, P. Wriggers & T. Zohdi (Eds).

[7] F. Brunier-Coulin, J.-L. Sarrat, P. Cuéllar, P. Philippe. (Accepted for P&G2017).

 DOI: 10.1051/epjconf/201714009029

Volume fraction instability in an oscillating non-Brownian iso-dense suspension

Y.L. Roht[2,1,*], G. Gauthier[1,**], J.P. Hulin[1,***], D. Salin[1,****], R. Chertcoff[2,†], H. Auradou[1,‡], and I. Ippolito[2,§]

[1] Laboratoire FAST, Univ. Paris Sud, CNRS, Université Paris-Saclay, F-91405, Orsay, France
[2] Universidad de Buenos-Aires, Facultad de Ingeniería, Grupo de Medios Porosos, CONICET, Paseo Colón 850, 1063, Buenos Aires (Argentina).

Abstract. The instability of an iso-dense non-Brownian suspension of polystyrene beads of diameter 40 μm dispersed in a water-glycerol mixture submitted to a periodic square wave oscillating flow in a Hele-Shaw cell is studied experimentally. The instability gives rise to stationary bead concentration waves transverse to the flow. It has been observed for average particle volume fractions between 0.25 and 0.4, for periods of the square wave flow variation between 0.4 and 10 s and in finite intervals of the amplitude of the fluid displacement. The study shows that the wavelength λ increases roughly linearly with the amplitude of the oscillatory flow; on the other hand, λ is independent of the particle concentration and of the period of oscillation of the flow although the minimum threshold amplitude for observing the instability increases with the period.

1 Introduction

Dense suspension flows are encountered in various industrial processes (food processing, nuclear waste reprocessing, manufacturing of concrete or reinforced plastics ...), in biological (blood tests) or environmental engineering (wastewater treatment). This large area of applications gave rise to numerous studies both theoretical, numerical or experimental. However, some properties of suspension flows are not yet fully understood such as shear induced migration [1] and more precisely the dynamics of this migration in axisymmetric Poiseuille flows at moderate volume fractions ($\phi \leq 0.3$) [2]. Moreover recent numerical simulations [3] point out the strong influence of confinement on the suspension rheology. On the other hand, several studies of dry granular media demonstrate that vibration may induce volume fraction instabilities [4] and the build-up of patterns has already been observed in oscillating suspensions, both experimentally [5, 6] and in numerical simulations [7]. These latter studies dealt with dilute suspensions (volume fraction $\phi \leq 10\%$) of buoyant beads ($\Delta\rho \neq 0$) oscillating in bulk at moderate frequencies ($f \geq 6$ Hz).

The present study deals with the different case of viscous iso-dense suspensions of moderate volume fraction ($\phi \geq 0.3$) oscillating at low frequencies ($f \leq 2$ Hz): these suspensions are confined inside a Hele-Shaw cell of aperture $H = 0.4$ mm while the bead diameter is equal to 40 μm. The study is focused on the instability which develops in a finite range of periods and amplitudes and is marked by the appearance of periodic variations of the volume fraction ϕ of the suspension in the direction of the flow while it remains constant in the perpendicular direction. We shall study particularly the dependence of the wavelength of these structures on the period and amplitude of the flow, as well as on the concentration of the suspension.

2 Experimental set-up and data processing

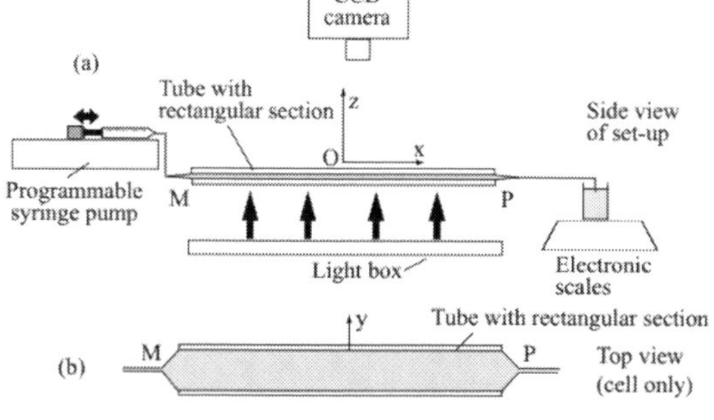

Figure 1. Schematic view of the experimental set-up

The experimental set-up is shown in Fig. 1. The suspension is contained in a Hele-Shaw cell of length

*e-mail: lucreroht@gmail.com
**e-mail: georges.gauthier@u-psud.fr
***e-mail: hulin@fast.u-psud.fr
****e-mail: salin@fast.u-psud.fr
†e-mail: ricardochertcoff@gmail.com
‡e-mail: auradou@fast.u-psud.fr
§e-mail: iippoli@fi.uba.ar

$L = 100\,\text{mm}$, thickness $H = 0.4\,\text{mm}$ and aspect ratio $W/H = 20$. The suspension is made of polystyrene beads of diameter $2a = 40\,\mu\text{m}$ immersed in a water glycerol mixture (mass concentration of glycerol $c = 21\%$) of viscosity $\eta = 1.8\,\text{mPa}\cdot\text{s}$ and with a density $\rho = 1050\,\text{kg}\cdot\text{m}^{-3}$ matching that of the particles at a temperature of 23° C. The oscillating flow of the suspension is induced by a programmable syringe pump fitted with 2.5 ml glass syringes: the variation of the flow rate with time is a symmetrical square wave of period $T \geq 0.4\,\text{s}$ adjustable by steps of 0.1 s. The peak-to-peak amplitude A of the mean displacement of the fluid in the cell has been varied between $A = 0.2\,\text{mm}$ and $A = 5\,\text{mm}$. The cell is horizontal (with its thickness parallel to the direction z of gravity) and illuminated from below. Images of the patterns induced by the instability are acquired by a Nikon D300s camera located at 10 cm above the cell and used in the movie mode: it captures 25 frames per second with a resolution of 1920×1080 pixels.

Fig. 2a displays a typical view from above of a part of the cell at a time such that a stationary regime of the instability generated by the oscillations of the flow along x has been reached: the instability leads to the appearance of periodic bands transverse to the flow. These bands are detected from variations of the transmitted light intensity which indicates that they correspond to variations of the particle concentration. The image has been enhanced by applying a low pass filter (smoothing out variations of the optical density in the x direction due to the bands) and by subtracting the filtered image from the original one: this eliminates low frequency variations and enhances the contrast of the bands.

As one can see on Fig. 2a, the bands are more visible and straighter in the region of the axis of symmetry: in order to reduce the influence of the noise of the image, we average therefore, for a given value of x, the grey levels of all pixels located within a range of distances y of width $\delta y \sim 0.1W$ close to the axis. The averages $I(x,t)$ obtained for the different distances x and times t are plotted as grey levels in the spatiotemporal diagram of Fig. 2b. The zigzag shape of the bands visible in the diagram reflects the periodic displacement along x of the volume fraction variations induced by the periodic flow: the peak-to-peak amplitude A of the displacements of the bands is used to characterize the displacement of the fluid imposed by the oscillating flow.

Finally, we compute, for all times t, the autocorrelation function:

$$F(\delta x, t) = \int I(x,t) \times I(x - \delta x, t)\,\mathrm{d}x. \qquad (1)$$

The function $F(\delta x, t)$ is plotted as grey levels in the spatiotemporal diagram of Fig. 2c as a function of time and of the interval δx. $F(\delta x, t)$ is symmetrical with respect to $\delta x = 0$ so that only the right part of the diagram has been plotted. The diagram displays alternate light and dark vertical bands: their period corresponds to the wavelength λ of the variations of the volume fraction induced by the instability. The visibility of the bands decreases with δx

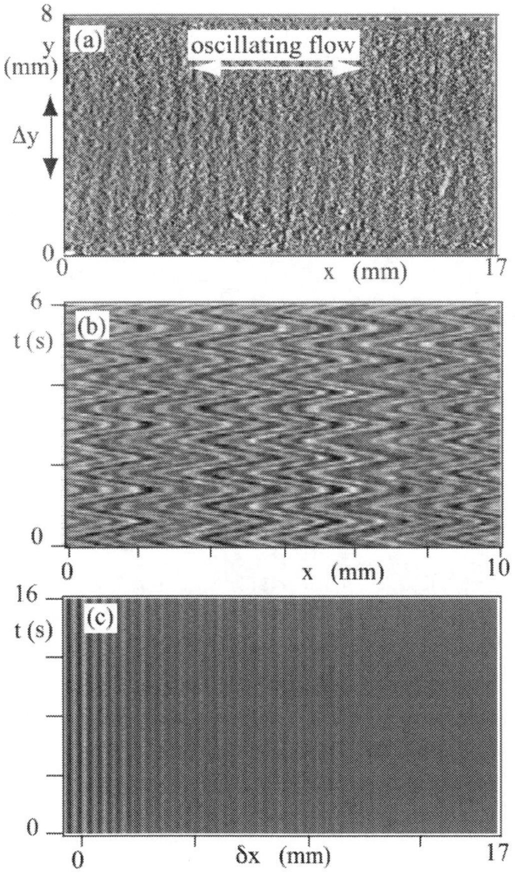

Figure 2. a) View from above of the instantaneous transmitted light pattern produced by the volume fraction instability of a suspension of polystyrene beads ($\phi = 0.35$) submitted to an oscillatory flow ($T = 0.8\,\text{s}$, $A = 2.34\,\text{mm}$) parallel to the direction x. Grey level variations correspond to those of the volume fraction. b) Spatiotemporal diagram of the local transmitted light intensities averaged over an interval $\Delta y = 0.1W$ in the transverse direction (grey level); vertical scale: time; horizontal scale: coordinate x along the flow direction. c) Spatiotemporal diagram of the autocorrelation function (grey levels) of the profiles of Fig. b; vertical scale: time; horizontal scale: interval δx (Eq. 1).

over a distance which corresponds to the spatial correlation length of the instability: depending on the experiment it may range from 5λ to 30λ. Practically, λ is determined by plotting the distances δx corresponding to the different maxima of F as a function of their number and performing a linear regression. In some of the diagrams of the type of Fig. 2c there is a small undulation of the vertical bands at large values of δx: this corresponds to a weak modulation of the wavelength by the oscillating flow.

3 Experimental results

Periodic bands transverse to the flow marking the occurrence of the instability were observed only for suspen-

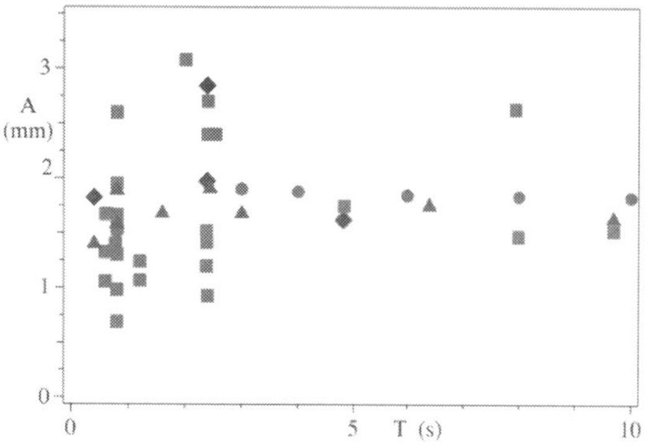

Figure 3. Map of the pairs of values of the amplitude A of the instability pattern oscillations and of the period T for which the instability has been observed in this study. Flowing fluid: $40\,\mu$m polystyrene particles in a 21% water-glycerold solution; particle fractions: $\phi = 0.25$ (♦), $\phi = 0.3$ (•), $\phi = 0.35$ (■), $\phi = 0.37$ (▲). Cell dimensions: $W = 8$ mm, $H = 0.4$ mm, $L = 100$ mm.

sions of large enough volume fraction $\phi \geq 0.25$ and up to $\phi = 0.37$. This latter upper limit of ϕ may actually reflect other effects such as the blockage of the suspension in the cell, at the outlet of the syringe or in the connecting tubes. Moreover, the intensity of the transmitted light decreases with the concentration which makes the patterns less visible and may account in part for their apparent disappearance.

We have studied the domain of existence of the instability which is displayed on Fig. 3. For an average volume fraction $\phi = 0.35$ ((■) symbol on Fig 3) and for the shortest imposed period $T = 0.4$ s, the instability appears above a threshold amplitude $A_c = 0.8$ mm. The instability pattern remains visible when A increases up to an upper limit $A_l \simeq 3$ mm above which it disappears and the suspension is no longer organized. Moreover, the lower threshold A_c increases slowly with the period T and reaches a plateau value $A_c \simeq 1.4$ mm for $T \gtrsim 5$ s. For lower volume fractions ϕ, the minimum amplitude A required for observing the instability increases: for $\phi = 0.25$ and $T = 0.4$ s, for instance, $A_c = 1.3$ mm.

As mentioned above, the wavelength λ of the sequence of bands at a given time is determined from the interval between the values of x corresponding to the maxima of the autocorrelation function $F(\delta x, t)$ defined by Eq. 1. A stationary flow and volume fraction variation regime may be assumed to be reached after $\sim 10 - 20$ oscillations and λ remains constant thereafter within some cases, a small modulation by $1 - 2\%$ of same period as that of the flow.

Figures 4a-b display respectively for several average concentrations ($0.25 \leq \phi \leq 0.37$) the variation of the wavelength λ as a function of either the period T or the measured amplitude A.

The amplitude A is approximately the same ($A = 1.75 \pm 0.25$ mm) for all data points of Graph 4a: we observe that λ is nearly constant with the period T for all three particle fractions ($\phi = 0.3, 0, 35$ and 0.37). More-

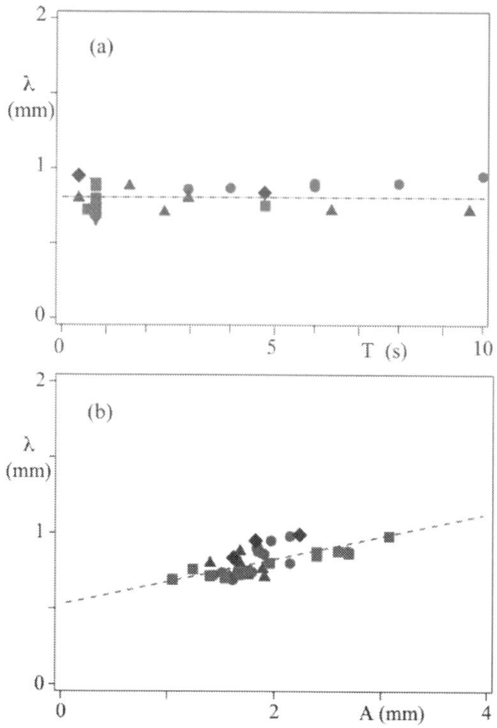

Figure 4. Variation of the average wavelength λ (mm) of the instability as a function of: (a) the period T of the flow ($A = 1.75 \pm 0.25$ mm) and: (b) the peak-to-peak amplitude A of the displacement of the instability pattern ($0.4 \leq T \leq 10$ s). The flowing solutions, the cell and the meaning of the symbols are the same as in Fig. 3.

over, there does not seem to be any systematic variation of λ with ϕ.

Since Graph 4a indicates that λ is independent of T, the data points corresponding to all values of T have been superimposed in Graph 4b as a function of A: the good collapse of the data points confirms this lack of dependence of λ on T. This graph shows that λ increases instead roughly linearly by about 25% when A increases from 1 to 3 mm with: $\lambda \simeq 0.54 + 0.14 \times A$ mm. It also confirms that there is no clear variation of λ with the particle volume fraction ϕ for a given amplitude A.

4 Discussion and conclusion

We have identified and analyzed an instability of iso-dense non Brownian suspensions in a Hele-Shaw cell submitted to an oscillatory flow velocity with a square wave variation at a low frequency. This instability occurs only for large enough mean volume fractions of the particles ($phi \geq 0.25$). On the one hand, the wavelength λ of this instability is independent of the period of the oscillations and of the particle volume fraction; on the other hand, the minimum threshold amplitude for observing the bands increases with the period T while it decreases at high volume fractions.

This instability of an oscillating suspension differs from those reported by other authors in one or several of its characteristics: lower frequency, larger volume fraction, $1D$ instability pattern with straight parallel bands. The periodic patterns studied in the present work are also different in nature from those resulting from an oscillating fluid shear flow above a granular sediment [8, 9]: such sediments might indeed appear due to sedimentation induced by a small residual density difference of the beads and the fluid or to the suspension getting structured with a sediment layer close to the wall and clear fluid between them [10, 11].

In order to eliminate the possible effect of gravity, several experiments have been performed, holding the cell with the y-axis vertical (Fig. 1: band patterns transverse to the horizontal flow and with the same wavelength as for horizontal cells were observed like in the latter.

Also, realizing experiments with a non-zero density contrast between the beads and the fluid hinders the appearance of the instability rather than fostering it. For a relative density contrast $\Delta\rho/\rho = 1.5\%$, A_c is significantly increased; for larger $\Delta\rho/\rho$ values, the periodic pattern does not appear any more.

Finally, a few preliminary visualizations of the distribution of the particles in the gap of the cell have been performed by means of a fluorescent dye dissolved in the fluid and of a laser sheet perpendicular to the cell: the small variations of the volume fraction $\phi(x,t)$ due to the instability resulted in variations of the intensity of the fluorescence light. These variations spanned the whole thickness of the cell and no layering effect was observed. The instability appears therefore as a succession of particles plugs that might result from shear induced migration: this implies that the important control parameter of the process would be the ratio A/H [12]. In order to test this hypothesis, further experiments will have to be performed with cells of different thicknesses. More information on the parameters determining the domain of existence of the instability and its wavelength is also needed: it will be obtained by varying the size and density of the particles as well as the density and viscosity of the fluid and the geometry of the flow cell (thickness, length and width). A more local analysis of the distribution of the particles will also be necessary to understand the processes involved.

Acknowledgements

We acknowledge support by the RTRA Triangle de la Physique, by the LIA PMF-FMF (Franco-Argentinian International Associated Laboratory in the Physics and Mechanics of Fluids) and from UBACyT 20020130100570BA. The thesis of Y.L. Roht is supported by a fellowship from the Peruilh foundation of the Faculty of Engineering of the Buenos-Aires University, by a Bec.Ar fellowship and by an Eiffel fellowship. The authors want to thank A. Aubertin, L. Auffray and R. Pidoux for their help in the design and realization of the experimental set-up and J.E. Wesfreid for helpful discussions.

References

[1] D. Leighton, A. Acrivos, Journal of Fluid Mechanics **181**, 415 (1987)

[2] B. Snook, J.E. Butler, É. Guazzelli, Journal of Fluid Mechanics **786**, 128 (2016)

[3] S. Gallier, E. Lemaire, F. Peters, L. Lobry, Phys. Rev. E **92**, 020301 (2015)

[4] D. Krengel, S. Strobl, A. Sack, M. Heckel, T. Pöschel, Granular Matter **15**, 377 (2013)

[5] P. Gondret, L. Petit, Physics of Fluids **8**, 2284 (1996)

[6] R. Moosavi, M. Maleki, M.R. Shaebani, J.C. Ruiz-Suárez, E. Clément, EPL (Europhysics Letters) **107**, 34006 (2014)

[7] V. Loisel, M. Abbas, O. Masbernat, E. Climent, Physics of Fluids **27**, 123304 (2015)

[8] G. Rousseaux, J. Kruithof, P. Jenffer, J.E. Wesfreid, Physical Review E **78**, 016302 (2008)

[9] M. Abedi, M.A. Jalali, M. Maleki, Journal of Fluid Mechanics **758**, 312 (2014)

[10] J.E. Butler, P.D. Majors, R.T. Bonnecaze, Physics of Fluids **11**, 2865 (1999)

[11] J.F. Morris, Physics of Fluids **13**, 2457 (2001)

[12] P. Nott and J. Brady, Journal of Fluid Mechanics **275**, 157 (1994)

EPJ Web of Conferences 140, 03057 (2017) DOI: 10.1051/epjconf/201714003057
Powders & Grains 2017

Experimental investigation of high speed granular flows down inclines

Joris Heyman[1,2], *Philippe* Boltenhagen[1], *Renaud* Delannay[1], and *Alexandre* Valance[1,*]

[1] *Institut de Physique de Rennes, CNRS UMR 6251, University of Rennes 1, 35 042 Rennes Cedex, France*
[2] *Géosciences Rennes, CNRS UMR 6118, University of Rennes 1, 35 042 Rennes Cedex, France*

Abstract. We report on laterally confined granular flow experiments on steep slopes. We provide evidences for the existence of different flow regimes with secondary flows. At moderate mass flow, we observe a first flow regime with a pair of longitudinal vortices which are localized close to the lateral walls and span progressively over the whole flow width with increasing flow rate. They are counter-rotative and induce a vertical upward motion of the grains at the wall. Upon a further increase of the mass flow rate, a transition is evidenced by a reversal of the rotation direction of the vortices which trigger in contrast a downward motion of the grains close to the lateral walls and an upward motion at the center of the channel. We argue that these flows bear some resemblance with the flow regimes reported recently in discrete element simulations.

1 Introduction

The inclined plane geometry forms particularly interesting boundary conditions for granular flows since, in addition to being comparable to several industrial and natural flows, it may be seen as a rheological test with constant friction. To date, experiments and simulations have focused mainly on mildly sloping planes, leading to fairly well understood flow regimes. More complex flows, including spanwise vortices [1–3], were obtained at slightly higher angles suggesting that upon further steepening, granular flows may reveal original features.

Recent discrete element method simulations of granular flows down steep and flat inclines [4, 5] revealed the existence of new flow regimes characterized by complex internal structures: secondary flows, heterogeneous particle volume fraction, symmetry breaking and dynamically maintained order. One of these regimes, referred to as "supported flow", is particularly interesting: it exhibits an uncommonly high bulk velocity, the granular flow being "supported" on a small collisional gaseous layer of highly agitated grains. Similarly to an air-cushion suspension, this layer reduces the effective wall friction and increases significantly the bulk velocity. These "supported" flows have extremely interesting potential applications in geophysics. The reduction in the effective friction due to the gaseous granular layer could indeed explain unexpected long run-out distances of large granular avalanches. However, although this hypothesis was suggested a long time ago by Campbell [6], no experimental evidences of these peculiar flow regimes have been constituted so far.

We present in this article preliminary experimental results on high speed granular flows. We provide evidences for the existence of different types of flow regimes with secondary flows which bear some similarities with the supported regimes recently reported in numerical simulations [5].

* e-mail: alexandre.valance@uni-rennes1.fr

2 Experimental set-up

The experiments were achieved using a 1.5 m long granular chute flow facility with lateral glass walls. The inclination θ of the chute can be varied from a few degrees to 50° and the spacing W between the lateral walls is also tunable from a few centimeters to a few decimeters. The chute facility is equipped with an automatic particle conveyor allowing to reach high flow discharges in steady configuration. The granular material used in the experiments is spherical glass beads of diameter $d = 1\ mm$ and density $\rho_p = 2650\ kg/m^3$.

The presence of lateral walls plays a major role for the observation of steady and fully developed flow regimes on relative short distances even for steep slopes. In fact, the additional lateral friction ensures the existence of stable and stationary flow regimes even for steep slopes as shown in recent simulations [5]. Moreover, the higher the lateral confinement is, the shorter is the transient to reach the steady regime. We thus carried out granular flow experiments on steep slopes with a rather strong confinement (i.e., small wall spacing) but nevertheless not too strong to allow for the apparition of secondary flow regimes. According to numerical simulations [7], a good compromise for the wall spacing is $W = 44d$ where d is the grain diameter.

The base of the chute consists of a flat and smooth aluminum substrate. The flow rate is accurately controlled by a hopper (see Fig. 1) installed at the entrance of the chute and set at 10 cm above the chute base. The speed of conveyor belt is tuned such that the hopper is always filled

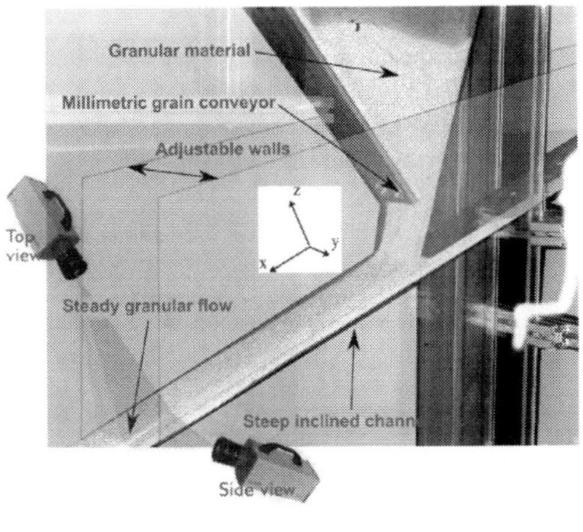

Figure 1. Photography of the granular chute flow set-up: the flow runs over a smooth inclined base and is confined between two lateral walls. The x direction corresponds to the main flow direction, the y direction is perpendicular to the lateral side wall and the z direction is normal to the flow base.

up with grains. This system ensures a steady flow rate for indefinite time.

The granular flow was characterized using a high speed video camera (APX RS photron) located at the end of the chute where the flow is expected to be fully developed. We recorded videos at a frame rate of 10000 frames per second from the transparent side walls but also from above. Using a self-developed particle tracking velocimetry algorithm, we were able to extract particle velocity and particle velocity fluctuations at the lateral wall and at the free surface (i.e., particle-air interface). In addition, we were able to assess the particle concentration at the wall.

3 Experimental Results

We report below the features of steady and fully developed granular flows obtained at an inclination angle $\theta = 30°$. Fig 2 presents the transverse profiles of the two components u and v (along the x and y direction, respectively) of the averaged particle velocity measured at the free surface. The longitudinal velocity u exhibits a parabolic-like profile at moderate flow rate but changes to a plug flow (flat profile) at higher flow rate (see Fig. 2.a).

Finite transverse velocity v informs us of the presence of secondary flows. At moderate flow rate, we observe two symmetrical lobes in the transverse velocity profile close to the lateral walls (see Fig. 2.b). These lobes are the signature of the presence of a pair of vortices. The positive and negative sign of the transverse velocity close to the left-hand side and right-hand side wall, respectively, indicate that the particles are moving away from the walls. The vortices are localized first close to the walls and span progressively over the whole flow width with increasing flow rate. At higher flow rate corresponding to a mass hold-up $H \approx 8d$[1], there is a sudden reversal of the rotation direction of the vortices. The vortices still appear as a pair. At the highest mass flow rate investigated so far, the transverse velocity exhibits a quasi-linear variation with a zero value at the middle of the channel and a maximum value at five grains diameters from the lateral wall.

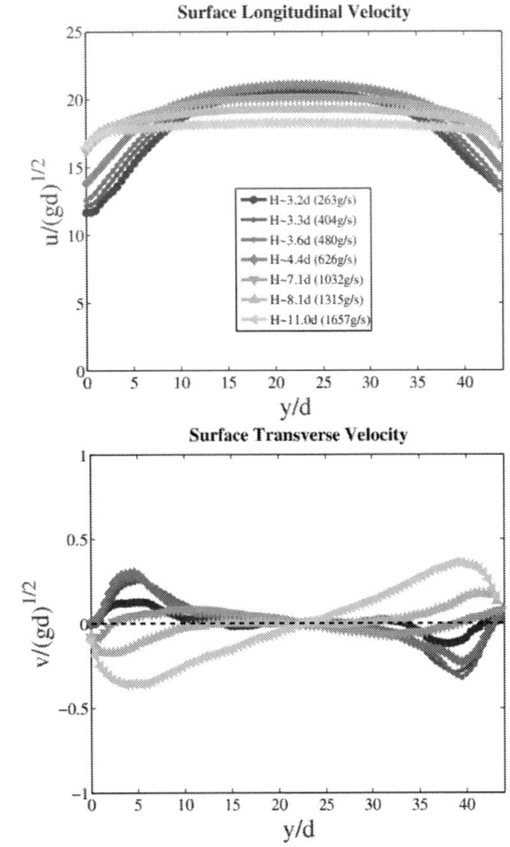

Figure 2. a) Top: transverse profiles of the longitudinal velocity (i.e, $u(y)$) at the free surface for increasing mass flow rate; b) Bottom: transverse profiles of the transverse velocity (i.e., $v(y)$) at the free surface for increasing mass flow rate; Chute flow configuration: $W = 44d$ and $\theta = 30°$.

In Fig 3, we present the corresponding vertical profile of the two components u and w (along the x and z direction, respectively) of the averaged particle velocity measured at the wall. The vertical profile of the longitudinal velocity $u(z)$ indicate a strong slip at the base which decreases with increasing flow rate. At moderate flow rate, the longitudinal velocity exhibits a Bagnold profile (convex curvature), while at higher flow rate, the profile is almost linear with a finite slip velocity (see Fig. 3.a).

The vertical velocity $w(z)$ reveals interesting features which are related with the finite transverse velocity mea-

[1]the mass hold-up corresponds to the depth-integrated volume fraction: $H = (1/W)\int_0^W dy \int dz\,\phi(y,z)$, where $\phi(y,z)$ is the particle volume fraction at position (y,z)

DOI: 10.1051/epjconf/201714003057

sured at the free surface. The finite vertical velocity confirms the existence of vortices (see Fig. 3.b). At low flow rate, the positive vertical velocity indicates that the particles are moving upwards and the parabolic shape of the profile informs us that the vortices spread over the whole depth of the flow. In contrast, at higher flow rate, there is a clear reversal of the sign of the vertical velocity in the upper part of the flow, indicating that the particles close to the wall have a downward vertical motion. Furthermore, we clearly see that the vortices observed from the free surface spread only over the upper three quarter of the flow depth. In addition, the S-shape of the velocity profile may suggest the possible emergence of an other vortex with an opposite direction of rotation in the lower quarter of the flow, although the data do not confirm this (since the vertical velocity remains negative). Further experiments are required to confirm this conjecture.

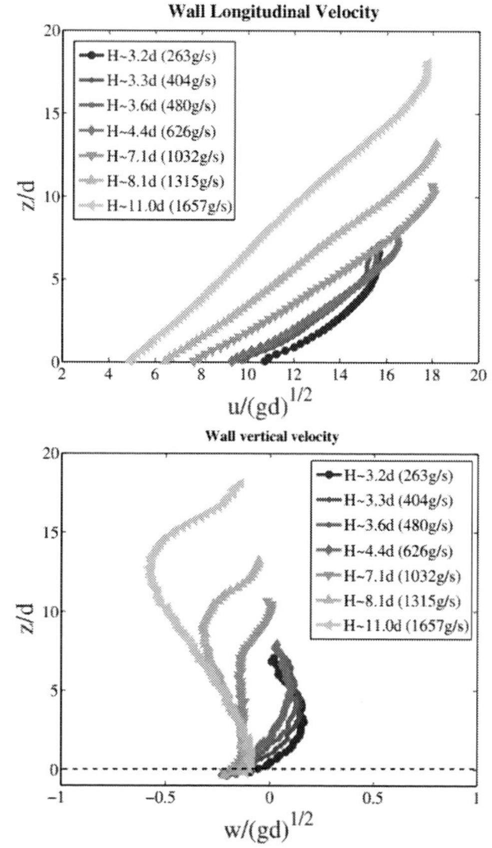

Figure 3. a) Top: vertical profile of the longitudinal velocity (i.e., $u(z)$) at the lateral wall; (b) Bottom: vertical profile of the vertical velocity (i.e., $w(z)$) at the lateral wall. Chute flow configuration: $W = 44d$ and $\theta = 30°$.

It is worthwhile to note that the existence of two superposed rows of vortices is seen in numerical simulations for high mass hold-up[5]. The apparition of an additional pair of vortices is actually characterized by a reversal of the rotation direction of the upper vortices in comparison with the lower ones.

Although, we do see the interior of the flow, the combination of velocity profiles measured at the free surface and at the lateral walls provides us with a rather good description of the secondary flow structure. The structure of the two different flow regimes identified from the velocity measurements is illustrated in Fig. 4. At low flow rate, we have two longitudinal vortices which induce an upward particle motion at the walls and a shift of the particles towards the center of the flow at the free surface. At higher flow rate, we observe a reversal of the rotation direction of the vortex. In the latter case, the vortices do not spread over the whole depth of the flow which may suggest the presence of an other pair of vortices.

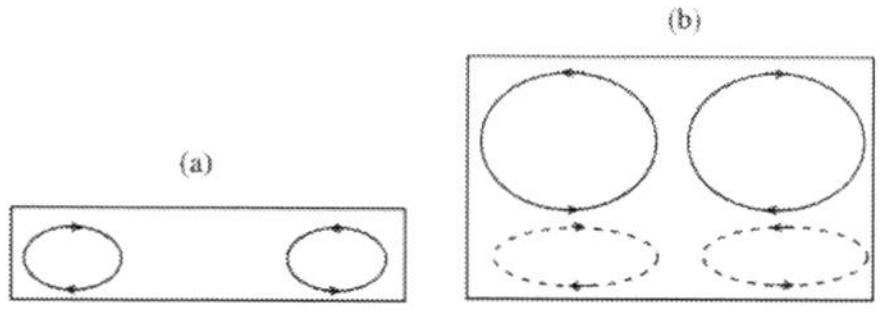

Figure 4. Sketch of the structure of the flow regimes obtained in the experiments: (a) Low flow rate regimes with longitudinal vortices inducing an upward motion of the particles close to the walls ($3 < H < 8$); (b) High flow rate regimes with vortices spinning in the reverse direction ($8 < H < 12$). The presence of an additional pair of vortices is a conjecture to be checked experimentally.

The last feature extracted from the video recording is the vertical profile of the particle concentration at the wall (see Fig. 5). The 2D image analysis does not provide us with a true measurement of the particle volume fraction but give us a proxy. Stereo imaging would be required to assess accurately the volume fraction. At moderate flow rate, the concentration exhibits a parabolic shape where the bulk flow is denser than the lower and upper layer. The density of the bulk increases with increasing flow rate. At higher flow rate, the concentration is much more uniform within the bulk (flat profile). The other salient feature is the small decrease of the bulk concentration when the mass hold-up is increased from $H = 7.1d$ to $H = 8.1d$. This is a signature of a small depletion of particles close to the walls but we can not rule out the possibility of an artefact of the tracking algorithm.

These preliminary measurements on high speed granular flows do not provide us with a full description of their structure since we can not probe the interior of the flow. On the basis of this partial description, we can not claim with certainty that the observed regimes are similar to the supported flows recently reported in discrete numerical simulations [5]. The latter indicate that the supported flows appear for inclination angle equal to or larger than 30° and are characterized by a highly concentrated core surrounded by an extremely dilute layer of highly agitated particles [5]. This highly depleted layer is located at the lateral walls and at the flow base and acts similar to a air-

DOI: 10.1051/epjconf/201714003057

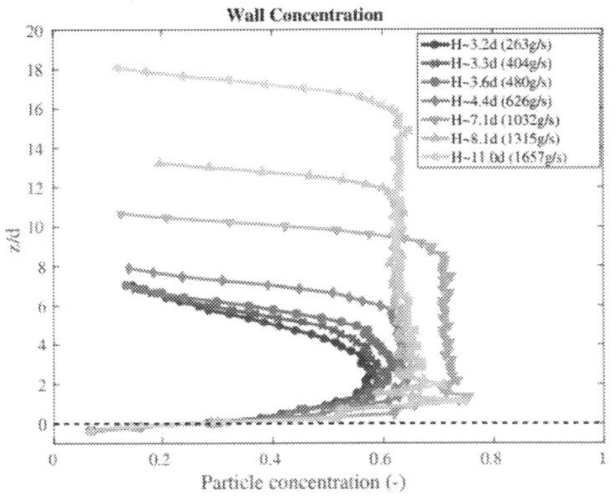

Figure 5. Vertical profile of the particle concentration at the lateral wall. Chute flow configuration: $W = 44d$ and $\theta = 30°$.

cushion suspension reducing the friction at the walls. At the moment, we do not have clear experimental evidences of the existence of a highly depleted layer at the later wall and at the base. We may suspect that the transition towards the supported flow regime is shifted to higher values of inclination angle when the lateral confinement is increased (i.e., for smaller wall spacing W) [7]. Actually the simulations of Brodu et al. [5] were performed for a flow width $W = 68d$ whereas the experiments were conducted with a wall spacing $W = 44d$.

4 Conclusion

We reported preliminary measurements on granular flow regimes on a steep slope which reveal complex structures with secondary flows as observed in recent numerical simulations [5]. Additional measurements are required to better document these high flow regimes. We plan in the near future to investigate these flows within a broader range of inclination angle and mass flow rate with the aim to provide clear evidences of the existence of the supported flow regime.

Besides, an effort should be made in the development of new experimental techniques to probe the interior of the flow because the classical imaging techniques are limited to the characterization of what happens at the wall and the free surface.

Acknowledgments

We acknowledge the support of the French "Institut des Sciences de l'Univers" (INSU) through the national research program "Aléas et Risques".

References

[1] Y. Forterre and O. Pouliquen, Phys. Rev. Lett. **86**, 5886 (2001).

[2] Y. Forterre and O. Pouliquen, J. Fluid Mech. **467**, 361–387 (2002).

[3] T. Börzsönyi, R.E. Ecke and J.N. McElwaine, Phys. Rev. Lett. **103**, 178302 (2009).

[4] N. Brodu, P. Richard and R. Delannay, Phys. Rev. E **87**, 022202 (2013)

[5] N. Brodu, R. Delannay , A. Valance and P. Richard, J. Fluid Mech. **769**, 218–228 (2015).

[6] C.S. Campbell, J. Geol. **97** , 653–665 (1989).

[7] J.-L. Ralaiarisoa., N. Brodu, R. Delannay and A. Valance, in preparation (2016).

Non-Dissipative Structural Evolutions in Granular Materials

Mehdi Pouragha[1,*] and *Richard* Wan[1,**]

[1] *Civil Engineering Department, University of Calgary, Canada*

Abstract. The structure of the contact network in granular assemblies can evolve due to either dissipative mechanisms such as sliding at contact points, or non-dissipative mechanisms through the phenomenon of contact gain and loss. Being associated with negligible deformations, non-dissipative mechanisms is actually active even in the small strain range of $\sim 10^{-3}$, especially in the case of densely packed assemblies. Hence, from a constitutive modelling point of view, it is crucial to be able to estimate such non-dissipative evolutions since both elastic and plastic properties of granular assemblies highly depend on contact network characteristics. The current study proposes an analytical scheme that allows us to estimate the non-dissipative contact gain/loss regime in terms of directional changes in the average contact force. The probability distribution of contact forces is used to compute the number of lost contact for each direction. Similarly, the number of newly formed contacts is estimated by considering the probability distribution of the gap between neighbouring particles. Based on the directional contact gain/loss computed, the changes in coordination number and fabric anisotropy can be found which, together with statistical treatments of Love-Weber stress expression, form a complete system of equations describing the evolution of other controlling microvariables. Finally, the results of the calculations have been compared with DEM simulations which verify the accuracy of the proposed scheme.

1 Introduction

The internal structure of jammed granular materials can change under external loading according to various mechanisms that can be classified into three broad categories:

- **Non-dissipative- reversible** (e): which refers to the so-called elastic deformation at interparticle contact points. Such processes conserve energy, contribute to overall strain, and are recoverable upon load reversal.
- **Dissipative- irreversible** (p): which includes the plastic deformation associated with sliding at contact points. Such processes dissipate energy, contribute to overall strain, and are, by nature, irrecoverable upon load reversal.
- **Non-dissipative- irreversible** (δ): which includes the changes in contact network topology due to contact gain and loss events. These events do not contribute to overall strain, nor do they dissipate energy. However, such structural alterations are irreversible and the previous structure cannot be recovered upon load reversal.

Unlike the common elastic/plastic dichotomy of mechanisms in continuum mechanics, the presence of non-dissipative/irreversible events in granular media has given rise to many peculiarities in granular media. Being associated with no deformation, such events are capable of altering the contact structure within minute strain ranges which has prompted the rapid adaptability of the internal structure with respect to the external loading observed in initially dense granular assemblies[1, 2]. Moreover, being non-dissipative by nature, such contact loss/gain mechanisms can theoretically coexist with purely elastic deformations of the domain which leads to a non-reversible elasticity and the concept of locked energy in closed stress loading paths[3]. This is in stark contrast with common beliefs that elastic response of granular materials should, by definition, preclude any evolution of the contact network[4, 5].

From a constitutive modelling point of view, it is crucial to be able to calculate the evolution of internal structure mainly because almost all micromechanical models (including models for elasticity of granular materials) heavily depends on the structure of contacts often expressed in terms of coordination number and fabric anisotropy. However, there have been very few studies in the literature on the analytical calculation of the contact gain/loss regimes and their contribution to ensuing structural evolutions[6].

As such, the current study proposes an analytical scheme to compute the change in fabric anisotropy and coordination number due to non-dissipative/irreversible mechanisms. Starting from a generic contact structure, the directional change in average contacts forces is calculated which, together with the directional distributions of contact forces and interparticle gaps, are used to estimate contact loss and gain regimes through a probabilistic approach.

*e-mail: mpouragh@ucalgary.ca
**e-mail: wan@ucalgary.ca

DOI: 10.1051/epjconf/201714002014

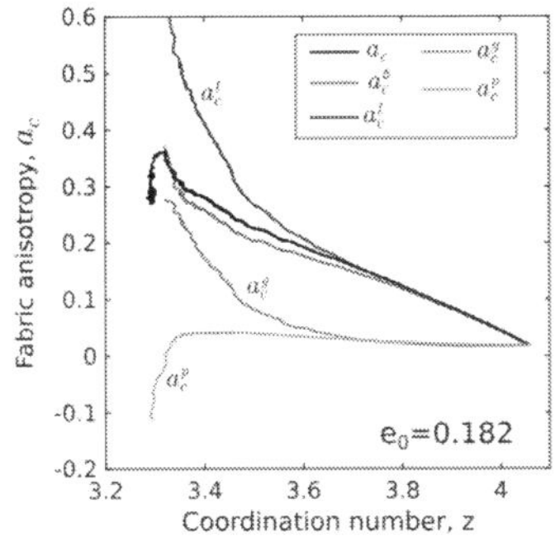

Figure 1. Decomposition of fabric anisotropy into contribution from different mechanisms.

2 Decomposing Fabric Evolution

The second-order fabric tensor describing the directional distribution of contact normals is classically defined as follows:

$$F_{ij} = \frac{1}{N_c}\sum_{\alpha=1}^{N_c} n_i^\alpha n_j^\alpha, \quad a_c = 2(F_1 - F_2) \tag{1}$$

where N_c is the total number of contacts, n_i is the contact normal vector, and a_c is fabric anisotropy. The directional probability density function of normal contacts can also be expressed in terms of fabric anisotropy as:

$$P(\theta) = \frac{1}{2\pi}(1 + a_c \cos 2\theta) \tag{2}$$

where the angle θ is measured from the major principal direction of F_{ij}. The incremental change in fabric tensor can be decomposed into contributions from lost (a_c^l), gained (a_c^g), and persisting contacts (a_c^p), with $a_c^\delta = a_c^g - a_c^l$ being the contributions of δ mechanism, and a_c^p of the p mechanism. The outcome of such decomposition in a typical 2D, biaxial DEM test on an initially dense granular assembly is given in Fig. 1. Similar results have been reported in [2]. It can be seen that δ mechanism dominates the overall structural evolution during the initial stages of deviatoric loading. The dissipative mechanism (p) becomes significant only after a threshold for coordination number is reached, as argued in [1].

3 Analytical Modelling

This section deals with calculating the directional distribution of lost and gained contacts through probabilistic treatments. Interested readers are referred to [7] for more details. We start by calculating the change in average normal contact force for each direction in terms of contact and force anisotropies. Next, assuming the probability distribution function (PDF) for magnitude of normal contact forces to be known for each direction, the number of lost contacts is calculated by integrating over contact forces with magnitudes smaller than the change in average normal force. A similar procedure is followed for gained contact with the integration being over the interparticle gaps smaller than the average deformation for each direction. It should be mentioned that, for the sake of simplicity, all the calculations are presented for 2D condition with no principal stress rotation. The calculation procedure, however, can be easily extended to be applicable to more general cases.

3.1 Average Normal Contact Force

The average normal contact force for direction θ is defined as $\bar{f}_n(\theta) = \left(\sum_{\alpha=1}^{N_c(\theta)} f_n^\alpha(\theta)\right)/N_c(\theta)$ which can change as either the sum of the forces or the number of contacts along θ changes. However, in order for the variables to be comparable at the initial state, here we are only interested in the change in average normal force due to the former effect. As such, the change in average normal force in the absence of contact gain/loss, $d\bar{f}_n^*(\theta)$, can be calculated as:

$$d\bar{f}_n^*(\theta) = \frac{1}{N_c(\theta)}\sum_{\alpha=1}^{N_c(\theta)} df_n^\alpha(\theta) = d\bar{f}_n(\theta) + \bar{f}_n(\theta)\frac{dN_c(\theta)}{N_c(\theta)} \tag{3}$$

Adopting the notation and anisotropy parameters introduced in [8], the directional distribution of average normal contact force can be approximated by $\bar{f}_n(\theta) = \bar{f}_n(1 + a_n\cos 2\theta)$ with $\bar{f}_n$ and a_n being the average normal force and normal force anisotropy, respectively. The change in average normal force can be written as:

$$d\bar{f}_n(\theta) = d\bar{f}_n(1 + a_n\cos 2\theta) + \bar{f}_n\, da_n\cos 2\theta \tag{4}$$

Moreover, recalling that $N_c(\theta) = \frac{N_{ct}}{2\pi}(1 + a_c\cos 2\theta)$ from fabric anisotropy relation, the relative change in number of contact along a given direction can be calculated as follows:

$$\frac{dN_c(\theta)}{N_c(\theta)} = \frac{dN_{ct}}{N_{ct}} + \frac{\cos 2\theta\, da_c}{1 + a_c\cos 2\theta} \tag{5}$$

with $dN_{ct}/N_{ct} = dz/z$, z being coordination number defined as average number of contacts per particle. By inserting Eqs. (4) and (5) into (3) and truncating second and higher orders of anisotropy parameters, the change in average normal force for a given direction can be expressed in terms of the change in prominent micro-variables, i.e.

$$d\bar{f}_n^*(\theta) = \bar{f}_n\left[\frac{d\bar{f}_n z + \bar{f}_n dz}{\bar{f}_n z} + \cos 2\theta\left(a_n\frac{d\bar{f}_n z + \bar{f}_n dz}{\bar{f}_n z} + (da_n + da_c)\right)\right] \tag{6}$$

On the other hand, based on expressions in [8], the stress increments can be written in terms of microvariables:

$$\frac{dp}{p} = \frac{d\bar{f}_n z + \bar{f}_n dz}{\bar{f}_n z}$$
$$\frac{dq}{p} - \frac{q\,dp}{p^2} = \frac{1}{2}(da_c + da_n + da_t) \tag{7}$$

As a further simplification, it is temporarily assumed that the contribution of da_t (anisotropy of tangential forces) is negligible, and as such, Eq. (7) can be substituted into (6) to replace the microvariables with boundary stresses, i.e.

$$d\bar{f}_n^*(\theta) = \bar{f}_n\frac{dp}{p}\left[1 + \cos 2\theta\left(a_n + 2 - 2\frac{q}{p}\right)\right] \tag{8}$$

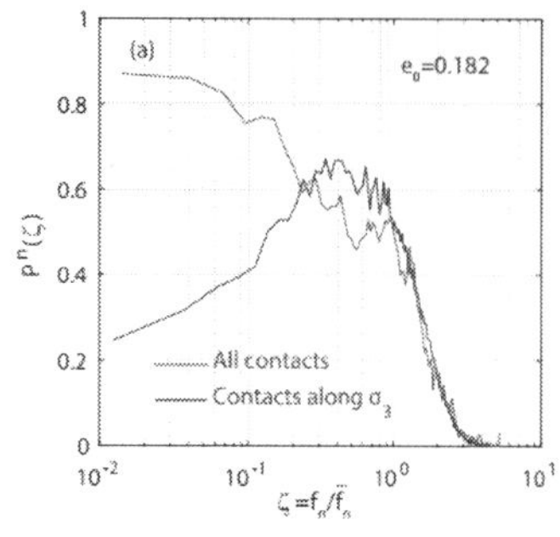

Figure 2. PDF of normal contact force magnitude for all the contacts, and contacts along the minor principal stress direction.

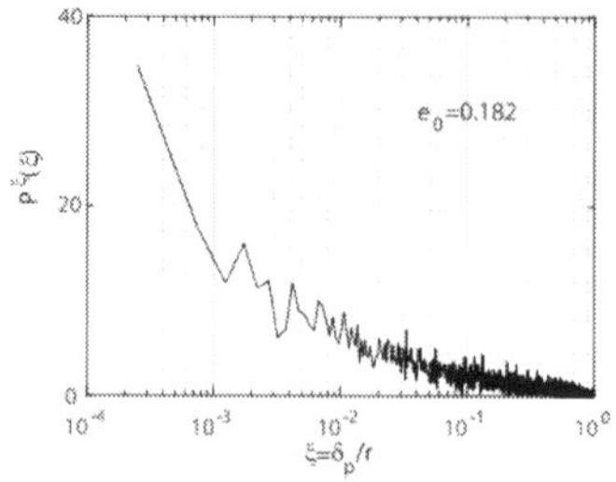

Figure 3. PDF of interparticle gaps for a 2D assembly.

The loss of contacts is assumed to happen in a region $\Omega^- = [pi/2-\Delta\theta, \pi/2+\Delta\theta]$ within which $d\bar{f}_n^*(\theta) < 0$. Similarly, new contacts are assumed to be gained within complementary interval $\Omega^+ = [-\Delta\theta, \Delta\theta]$ where $d\bar{f}_n^*(\theta) > 0$. The value of $\Delta\theta$ can be calculated by setting Eq. (8) to zero. Thus,

$$d\bar{f}_n^*(\theta) = 0 \xrightarrow{\text{Eq. 8}} \Delta\theta = \frac{1}{2}\arccos\left[\frac{1}{2(\frac{dq}{dp} - \frac{q}{p}) + a_n}\right] \tag{9}$$

3.2 Lost and Gained Contacts

The PDF of the magnitude of contact forces along a direction θ is assumed to be given by $P^n(f_n/\bar{f}_n(\theta))$ whose form does not change with direction. A typical form of such PDF is given in Fig. 2 where the PDF is shown to differ for all the contacts and contacts along minor stress direction. However, the lost contacts are mainly oriented close to minor stress direction and the dependency of PDF on the direction can be arguably ignored. The number of lost contacts along a direction θ can now be calculated by integrating over the normal forces whose magnitude is smaller than $d\bar{f}_n^*(\theta)$:

$$dN_l(\theta) = N_c(\theta)\int_0^{d\bar{f}_n^*(\theta)} P^n(\zeta \to 0^+, \theta)d\zeta \qquad \zeta = f_n/\bar{f}_n(\theta) \tag{10}$$

It should be noticed that the integration is performed only over the minute part of force PDF, $f_n = [0, d\bar{f}_n^*(\theta)]$, within which the distribution can be accurately replaced by an straight line. The total number of lost contacts can be calculated by integrating Eq.(10) over θ:

$$dN_{lt} = N_{ct}\frac{dz^l}{z} = 2\int_{\pi/2-\Delta\theta}^{\pi/2+\Delta\theta} dN_l(\theta)\, d\theta \tag{11}$$

The contribution of such a contact loss regime to fabric anisotropy change can also be calculated as follows:

$$da_c^l = 2\int_{\pi/2-\Delta\theta}^{\pi/2+\Delta\theta} \frac{dN_l(\theta)}{N_{ct}}\left(\cos^2\theta - \sin^2\theta\right) d\theta \tag{12}$$

Turning to gained contacts, the number of newly formed contacts along θ is calculated by integration over particle pairs with interparticle gaps smaller than the average elastic deformation of contacts along direction θ. Such an average deformation is approximated by $d\bar{f}_n^*(\theta)/k_n$ with k_n being normal contact stiffness. Compared to the procedure for calculating lost contacts, the normal force PDF is replaced here by PDF of interparticle gaps, $P^\xi(\delta_p/2\bar{r})$, with δ_p being face-to-face distance between neighbouring particles, and $\bar{r}$ the average radius of particles. Figure. 3 shows the form of P^ξ obtained from the 2D biaxial DEM simulation. Similar to the lost contacts calculation, the directional variation of P^ξ has been ignored. The number of gained contact along direction θ can now be calculated as follows:

$$dN_g(\theta) = \beta_{nc}N_c(\theta)\int_0^{d\bar{f}_n^*(\theta)/k_n} P^\xi(\xi \to 0^+)d\xi, \quad \xi = \frac{\delta_p}{\bar{r}} \tag{13}$$

The parameter β_{nc} determines the ratio of non-contacting neighbours[1] to the total number of contacts, which can be expressed in terms of coordination number assuming that the total number of neighbours (contacting and non-contacting) remains constant[7]. The contribution of gained contacts to coordination number (dz^g) and fabric anisotropy (da_c^g) changes can also be calculated similar to lost contacts, with the directional integration performed over Ω^+.

3.3 System of Equations

The above analyses involve five microvariables describing the statistics of contact and force network; z, a_c, $\bar{f}_n$, a_n, and a_t. Based on the decomposition given in Fig. 1, we can safely assume that microstructure of granular material changes solely due to non-dissipative δ mechanisms, and hence, the net changes in fabric anisotropy and coordination number calculated from contact gain and loss can be assumed to represent the total changes: $dz = dz^g - dz^l$ and $da_c = da_c^g - da_c^l$. Moreover, given the stress increments, Eqs.(7) provide us with two more equations. The last equation for closing the system of equations comes from the observation that the parameter a_n is strictly proportional to the deviatoric stress ratio q/p at the initial stages, with the proportionality coefficient remaining the same irrespective of the loading path[1, 7]. The five

[1]The term "neighbours" is referred here to particles which share a cell boundary in Dirichlet tessellation network

above-mentioned relationships can now be solved in terms of stress increments to find the evolution of the five microvariables.

4 Model Verification

The model's predictions of the microvariables evolution have been compared with DEM simulation of 2D biaxial tests on an initially dense assembly. The simulation parameters are: number of particles $\simeq 20,000$, particle size distribution $D_{max}/D_{min} = 1.25$, interparticle friction $\mu = 0.5$, initial void ratio $e_0 = 0.182$, initial coordination number $z_0 = 4.06$, and contact stiffness $k/p = 10^4$ which has been chosen to be the same for normal and tangential directions. The simulation results have been used to find the PDF's of force and interparticle gaps which enter the calculation procedure as an input. Fig. 4 shows the comparison between model predictions and DEM results where a satisfactory agreement is observed especially for the initial part of the loading. As anticipated, the simulation results diverge from the predictions as the deviatoric load becomes large enough, which indicates the onset for prevalence of dissipative p mechanisms.

5 Conclusion

The role of non-dissipative/irreversible mechanisms such as contact gain and loss events in altering the internal structure of granular media has been investigated. It has been shown, through decomposition of fabric anisotropy, that such mechanisms are the main cause behind the rapid adaptation of the granular microstructure with respect to external stresses, especially at the initial stages of deviatoric loading. The irrecoverable fabric realignment associated with non-dissipative/irreversible mechanisms is found to coexist with elastic mechanisms at strain levels as small as 10^{-2}, which compromises the possibility of deriving elastic deformational characteristics of granular materials from a energy potential. The study also puts forward an analytical procedure that relates the directional distribution of contact gain and loss regimes to the probability distribution functions of contact forces and interparticle gaps. The model has been validated by comparing the prediction of prominent microvariables to 2D DEM simulations.

Acknowledgements

This work was funded by the Natural Sciences and Engineering Research Council of Canada and Foundation Computer Modelling Group.

References

[1] M. Pouragha, R. Wan, Granular Matter **18**, 1 (2016)
[2] N. Kruyt, Mechanics of Materials **44**, 120 (2012)
[3] I.F. Collins, Geotechnique **55**, 373 (2005)
[4] C.S. Chang, S.J. Chao, Y. Chang, International journal of solids and structures **32**, 1989 (1995)
[5] B. Cambou, P. Dubujet, C. Nouguier-Lehon, Mechanics of Materials **36**, 1185 (2004)
[6] J.T. Jenkins, O.D. Strack, Mechanics of Materials **16**, 25 (1993)
[7] M. Pouragha, R. Wan, Non-Dissipative Fabric Evolution in Granular Materials, in preparation (2017)
[8] L. Rothenburg, R. Bathurst, Geotechnique **39**, 601 (1989)

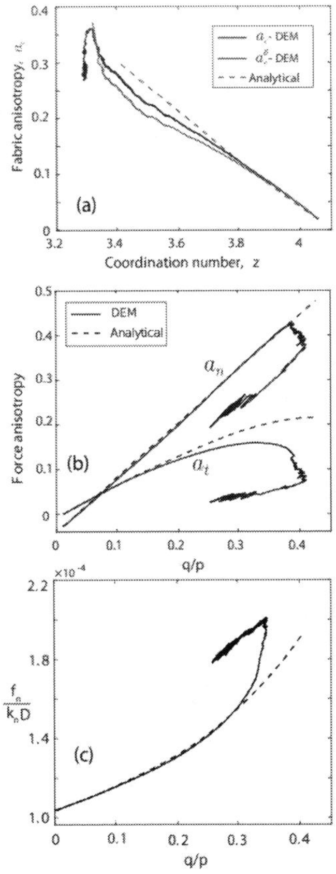

Figure 4. Comparison between model prediction and DEM simulations for (a) fabric anisotropy and coordination number, (b) force anisotropies, and (c) average contact forces.

An experimental study of the transient regime to fluidized chimney in a granular medium

Pierre Philippe[1,⋆], *Sarah* Mena[2], *Florian* Brunier-Coulin[1], and *Jennifer* Curtis[2,3]

[1] *IRSTEA, UR RECOVER, 3275 route de Cézanne, 13182 Aix-en-Provence, France.*
[2] *Department of Chemical Engineering, University of Florida, Gainesville, FL 32611, Florida, US*
[3] *College of Engineering, Department of Chemical Engineering, Kemper Hall, University of California, Davis, CA 95616, California, US*

Abstract. Localized fluidization within a granular packing along an almost cylindrical chimney is observed when an upward fluid-flow, injected through a small port diameter, exceeds a critical flow-rate. Once this threshold reached, a fluidized area is first initiated in the close vicinity of the injection hole before gradually growing upward to the top surface of the granular layer. In this work, we present an experimental investigation specifically dedicated to the kinetics of chimney fluidization in an immersed granular bed. Two different transient regimes are identified depending on wether the expansion of the fluidized area is rather fast and regular, reaching the final chimney state typically in less than 10 seconds, or, on the contrary, slow and very progressively accelerated, giving rise to transient duration up to 1 hour or even more. Some systematic investigations allow to propose several empirical scaling relations for the kinetics of chimney fluidization in the fast regular regime.

1 Introduction

Many industrial processes rely on particle bed fluidization, induced in most cases by a uniform fluid flow [1]. Much less often, the fluid injection can be restricted to a small area compared with the dimensions of the granular bed. Very little attention has been paid to this specific situation which generates only a partial fluidization of the system, through a vertical fluidized chimney, and defines lateral boundaries between fluidized and static zones. However, localized fluidization is relevant in some industrial applications, as spouted beds, but also in the particular context of embankment dam safety where such local structures, observed at the downstream foot of a levee and most commonly referred to as "sandboil", may indicate the presence, at an earlier stage, of an erosion process called backward erosion, listed as one of the four generic mechanisms of internal erosion [2]. Few experimental investigations have been conducted on steady-state localized fluidization, except the previous studies by Zoueshtiagh *et al.* [3] and by Philippe *et al.* [4], and even less regarding the transient upward expansion of the fluidized zone from an original cavity to a complete chimney [4].

Towards this end and for the purpose of expanding our previous preliminary study [4], an experimental study was carried out to investigate specifically this transient behaviour at the early stages of local fluidization. Our experimental procedure relies on the combination of Refractive Index Matching and Planar Laser Induced Fluorescence [4] allowing a direct visualization trough the granular material made of glass spheres immersed in an oil mixture. In addition to the flow rate at injection, the other parameters under study were the particle diameter (from $3mm$ to $10mm$), the injection diameter (from $10mm$ to $40mm$), and the initial granular bed height ($H_0 = 80mm$ or $125mm$). A systematic analysis of the fluidized area expansion with time was then undertaken, based on the measurement of the total duration of the transient phase which allows distinguishing between a slow and very progressively accelerated transient regime and a faster and regular one. In this later regime, a more systematic investigation was carried out to study the kinetics of fluidized area expansion, including the dependencies with the different control parameters, and to propose some empirical scaling relations.

⋆e-mail: pierre.philippe@irstea.fr

2 Experiments

The experimental investigation rely on the same optical techniques than used in our previous study [4]. These include Planar Laser Induced Fluorescence (PLIF) together with Refractive Index matching (RIM) [5]. RIM technique is based on the use of a solid phase and a liquid phase sharing the same refractive index and consequently giving rise to a translucent medium. After addition of a very small amount of fluorescent dye in the liquid, PLIF consists in illuminating the particulate system with a planar laser at a wavelength inside the fluorescence spectrum of the dye. Then an optical high-pass filter allows to recover a picture

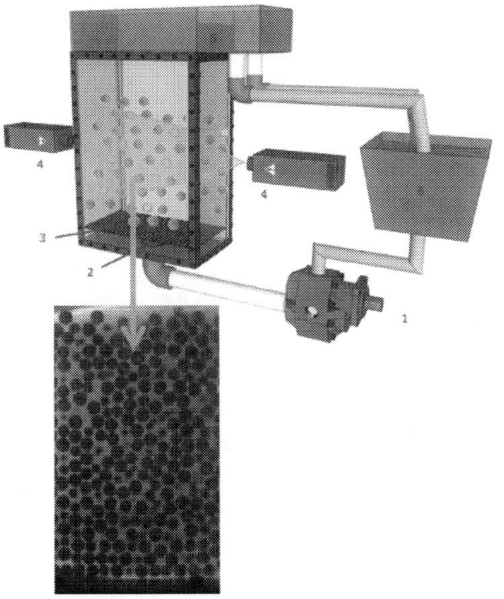

Figure 1. Experimental device (1: Gear pump; 2: Injection port; 3: Metallic mesh plate; 4. Planar lasers; 5: Over-flow set-up; 6: Reservoir) and typical picture of the granular bed obtained by coupling RIM and PLIF techniques.

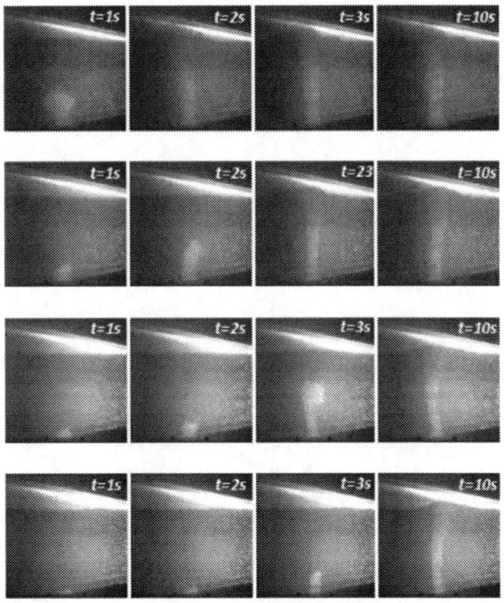

Figure 2. Sequences of transient fluidized area expansion ($d = 3mm$, $D = 20mm$, $H_0 = 125mm$) at successive time steps $t = 1s$, $t = 2s$, $t = 5s$, and $t = 10s$, with, from top to bottom, injection flow rates Q and transient duration T_0 equal to $Q = 117.5cm^3.s^{-1}$ and $T_0 = 1.80s$, $Q = 68.1cm^3.s^{-1}$ and $T_0 = 2.72s$, $Q = 50.4cm^3.s^{-1}$ and $T_0 = 4.65s$, $Q = 44.6cm^3.s^{-1}$ and $T_0 = 8.30s$.

where the microstructure is directly observed as illustrated in Figure 1.

In practice here, the solid phase is made of borosilicate glass beads (spherical shape) of density $\rho_s = 2230kg.m^{-3}$ and with particle diameter $d = 3, 5, 7$ or $10mm$. The liquid phase is a mixture with 90% (in mass) of light mineral oil and 10% of immersion oil. The liquid density is $\rho_f = 840kg.m^{-3}$ and the dynamical viscosity is approximately equal to $(28 \pm 2) \times 10^{-3}kg.m^{-1}.s^{-1}$ in the range of temperature experienced during this study. The refractive index of the liquid, which also approximately the one of the beads, was measured equal to 1.472 at 23°c.

During an experiment, images similar the one in Figure 1 are recorded by a camera positioned perpendicularly to the laser plane. Two different cameras have been used with frame rate varying from 1 to $125fps$. A sketch of the experimental set-up is shown in Figure 1. It consists in a rectangular box (width= $10cm$, height= $30cm$, length= $20cm$) filled by the beads sample on top of a horizontal metallic 2 mm mesh plate and over an injection port with changeable diameter $D = 10mm$, $20mm$, $30mm$, or $40mm$. The liquid is supplied at a slow rate by a gear pump from a reservoir to saturate the granular medium in the cell and till the over-flow condition at the top of the device is reached. Next the pump is stopped to prepare the initial bed state. For this purpose, a metallic rod is manually and systematically stirred around within the granular bed both to ensure a good reproducibility and to eliminate any trapped bubbles. The global solid volume fraction obtained with this preparation method is approximately $\Phi = 0.61 \pm 0.01$ for all bead diameters as estimated from the measurement of the initial bed's height H_0 which, in the present study, is chosen equal to either $H_0 = 80mm$ or $H_0 = 125mm$.

Starting from this reproducible initial sample, each experiment consists in imposing with the gear pump a prescribed constant flow-rate Q. The time needed for the pump to reach the selected flow-rate is small enough to be negligible compared to the transient time for fluidized chimney development. The later time, noted T_0, is highly sensitive to the flow-rate Q as discussed in more details in the following section.

3 Results

3.1 Transient regimes of chimney fluidization

As shown in the pictures of Figure 2, the duration of the transient fluidized area expansion increases more and more markedly as the injection flow rate is reduced. For transient time typically less than 10 to 20 seconds, this expansion is rather regular with a rate of expansion which increases by a factor no more than 10 from the beginning of the process to the final chimney steady-state. In this regime, the relationship between T_0 and Q is compatible with a power law divergence as shown in Figure 3:

$$T_0 \propto \left(\frac{Q}{Q_c} - 1\right)^{-n} \tag{1}$$

where Q_c appears as a critical flow rate while the exponent n remains between 1 and 0.5.

 DOI: 10.1051/epjconf/201714009030

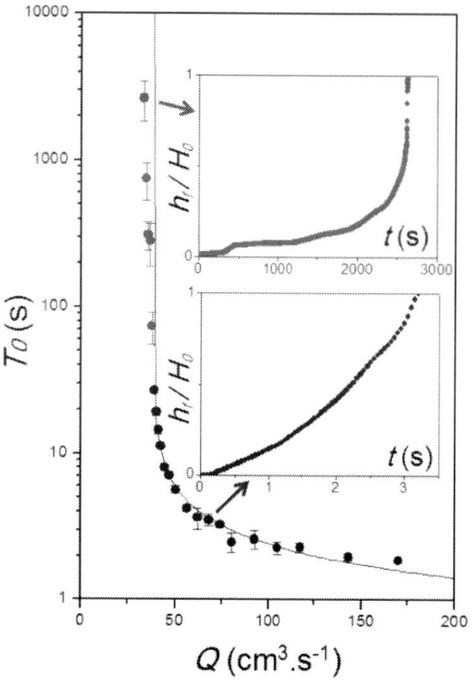

Figure 3. Transient duration T_0 versus injection flow rate Q for d = 3*mm*, D = 30*mm*, and H_0 = 125*mm*. The blue symbols account for the regular regime where a power law divergence is observed (Eq. 3.1 represented by the blue solid line) while the red symbols correspond to the ultra-slow regime as discussed in the text. Insets: typical time evolution of the fluidized height h_f in the ultra-slow regime (top and red symbols) and in the regular regime (bottom and blue symbols).

According to previous observations by Philippe *et al.*, it is likely that this critical flow rate demarcates the fluidized chimney regime ($Q > Q_c$) and the fluidized cavity regime ($Q < Q_c$). However, in this later regime, a fluidized chimney can be eventually obtained provided that one waits a sufficient time. This new regime is different from the regular one ($Q > Q_c$) regarding the transient time T_0 which increases by several orders of magnitude as can be seen in Figure 3. Analyzing the time evolution of the fluidized area height h_f obtained by some image processing, it is observed in Figure 3that, in this ultra-slow regime, the upward expansion is initially almost frozen, and consequently hardly to identify, and accelerate very progressively over about 4 to 5 decades in terms of expansion rate. This regime is also characterized by a bad reproducibility and a sharp increase of the standard deviation obtained for T_0. By comparison, the kinetics of fluidized expansion is faster and much better reproducible but also quite more regular with only a limited variation of acceleration during the transient phase as seen in Figure 3. Finally, as a consequence arising from the identification of this new regime, the real existence of a steady fluidized cavity regime is questionable, such situations being potentially

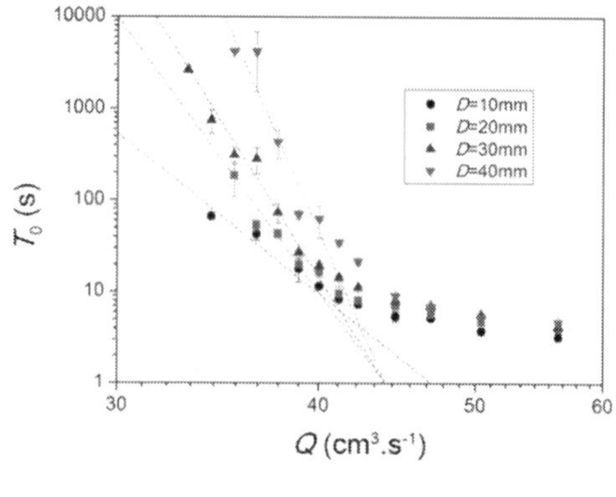

Figure 4. Log-log representation of the transient duration T_0 versus injection flow rate Q for d = 3*mm*, H_0 = 125*mm*, and for D = 10*mm*, D = 20*mm*, D = 30*mm*, and D = 40*mm*, respectively. The dashed lines are only guides for the eyes suggesting a linear relation between ln T_0 and ln Q.

a nearly frozen intermediate step of a complete fluidized chimney expansion. However, it has been observed that a chimney can come back to a cavity state when the flow rate is decreased [4] suggesting that a stationary fluidized cavity truly exists. In practice, the time range needed to decide on this issue is unfortunately out of any realistic experiments capability.

3.2 Influence of the control parameters

Regarding the ultra-slow regime, the transient duration seems to increases almost exponentially with the flow rate as suggested by the roughly linear trend in the semi-logarithmic representation in Figure 4: $\ln T_0 \propto \ln Q$. Several experimental series carried out with the four different injection diameter D have clearly underlined a strong influence of D on T_0. Indeed, there is a sharp increase of the slope between $\ln T_0$ and $\ln Q$ when the injection diameter gets wider, giving rise, for a same flow rate, to a significantly longer transient phase for a larger injection width. This effect is most probably due to a smaller velocity at injection when the diameter is enlarged that considerably reduces the expansion rate at the beginning of the fluidization. The end of the expansion is likely to be quite less dependent on the injection diameter D since, far enough from injection, the size of the chimney gets almost the same whatever the value of D [4].

A more systematic study was conducted in the regular regime, relying in particular on the divergence law in Eq. (1). It can first be established a weak influence of the injection diameter D on the apparent critical flow-rate Q_c: the later indeed slightly increases with D. Here again, this is likely due to a more moderate flow at injection for a larger injection width, which may slow down the beginning of the expansion kinetics. The transient duration is consequently increased, giving rise a slightly anticipated apparent divergence. A noticeable impact of D was also

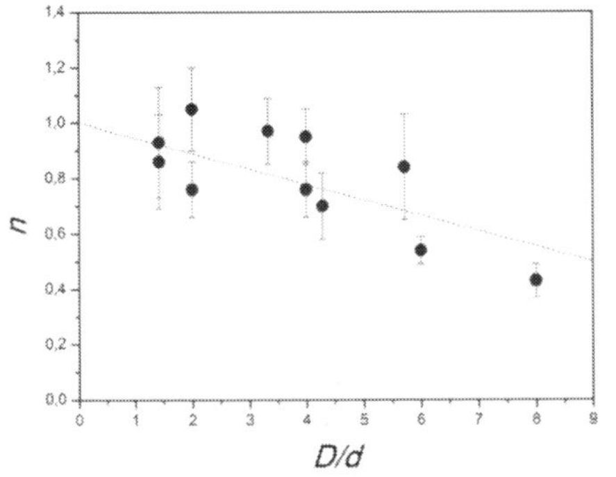

Figure 5. Plot of the exponent n of the divergence law in Eq.(1) versus the ratio $\frac{D}{d}$ of the injection width to the bead diameter. The dashed blue line stands for the linear relation given in Eq.(2).

found regarding the exponent n of the divergence law. This exponent actually decreases monotonously with D, varying for instance from $n \approx \frac{2}{3}$ for $D = 10mm$ to $n \approx \frac{1}{3}$ for $D = 40mm$ in the case of an initial sample of height $H_0 = 125mm$ with bead diameter $d = 3mm$. Considering now the influence of the bead diameter d, it was observed that larger values of n are obtained for larger bead diameter. Thus, for a same sample height but with beads of diameter $d = 5mm$, the range of variation for n is now from $n \approx 1$ for $D = 10mm$ to $n \approx \frac{2}{5}$ for $D = 40mm$. Practically, a convenient way to satisfactory rationalize both dependencies of n with D and d respectively can be proposed through the following relation:

$$n \approx 1 - 0.056\frac{D}{d} \qquad (2)$$

As can be seen in Figure 5, a rough agreement is obtained when testing the previous equation with the whole set of data.

Finally, a quadratic relation between the apparent divergence flow-rate Q_c and the bead diameter d was naturally recovered. This is simply the consequence of the fact that fluidization onset is not controlled by the flow rate transiting through the granular bed but by the pressure gradient induced by this flow. According to Darcy law (which is almost valid in the whole set of experiments since the particle Reynolds does not exceed 3 [6]), flow rate and pressure gradient are directly related through the permeability which is indeed proportional to the square of the bead diameter (see for instance Carman-Kozeny relation in [6]). Note that a complete dimensional analysis of localized fluidization accounting notably for this effect has been proposed more in details in [7].

4 Conclusion

By way of summary, several important findings can be underlined. First, a new and original transient regime to fluidized chimney has been highlighted and was characterized by an expansion of the fluidized area at injection that initially grows extremely slowly, in a quasi-frozen fluidized cavity configuration, before the kinetics accelerates very progressively till a final burst comparable to the more usual transient process observed in the already identified regular regime where the transient duration is related to the injection flow rate by a divergence law. On the contrary, this new "ultra-slow" regime is badly reproducible in terms of transient duration and the whole expansion process can last up to one hour or more which could explain why it has not been identified before [4]. A second significant result concerns the influence of both the injection width D and the bead diameter d on the transient kinetics. The injection diameter D was proved to have a huge influence on the duration of the transient phase in the "ultra-slow" regime which increases dramatically with D, probably because, for a same flow-rate Q, the local velocity at injection is significantly smaller and much less efficient in fluidizing the particles. In the regular expansion regime, based on a systematic use of the divergence law of duration time versus flow-rate, it can be reported that the apparent critical flow-rate Q_c scales with d^2, as an obvious impact of permeability, and increases slightly with injection port width D, and that the exponent n decreases almost linearly with the ratio D/d.

Acknowledgements

Fundings from NSF OISE 0968313 and from the region PACA (Provence-Alpes-Côte d'Azur, France) are gratefully acknowledged.

References

[1] M.J. Rhodes. Introduction to particle technology. New York: Wiley (1998).

[2] S. Bonelli (Ed.), Erosion in Geomechanics Applied to Dams and Levees. Wiley-ISTE (2013)

[3] F. Zoueshtiagh, A. merlen, Phys. Rev. E **75**(5) 053613 (2007)

[4] P. Philippe, M. Badiane, Phys. Rev. E. **87**(4) 042206 (2013)

[5] J.A. Dijksman, F. Rietz, K.A. Lorincz, M. van Hecke, W. Losert. Rev. Sci. Instrum. **83** 011301 (2013).

[6] J. Bear. Dynamics of Fluids in Porous Media. Dover (1988).

[7] S. Mena, L.-H. Luu, P. Cuéllar, P. Philippe, J. S. Curtis. (Submitted).

Fluid-solid transition in unsteady shearing flows

Dalila Vescovi[1,*], *Diego* Berzi[1], and *Claudio* di Prisco[1]

[1] *Department of Civil and Environmental Engineering, Politecnico di Milano, 20133 Milano, Italy*

Abstract. This paper focuses on the mechanical behaviour of granular systems under shearing, unsteady conditions. The results of numerical simulations of time evolving, homogeneous, shear flows of an assembly of frictional spheres, under constant volume conditions are illustrated. Simulations have been performed considering three volume fractions corresponding to fluid, solid and near-to-critical conditions at steady state. The three systems follow very different evolutionary paths, in terms of pressure, coordination number and stress ratio. Fluid-like and solid-like systems exhibit large and small fluctuations, respectively, in those quantities. A critical value of the coordination number seems to govern the transition from fluid to solid.

1 Introduction

The discontinuous and inhomogeneous nature of granular materials leads to complex mechanical behaviours, even in case of simple flow conditions. In particular, granular systems can behave like either fluids, meaning that they yield under shear stress, or like solids able to resist applied stresses without deforming [1]. If grains are widely spaced and free to move in any direction, interacting only through collisions, the medium behaves like a fluid and the stresses are proportional to the square of the strain rate under simple shearing (Bagnold scaling [2]). On the other hand, when particles are densely packed, a network of persistent contacts develops within the medium, and the granular material shows a solid-like, rate-independent behaviour. The mechanical response of the system during the solid-fluid transition, is still an open question, [3] although several constitutive models have been proposed to this purpose in the literature [4–9].
Whereas several numerical results have been obtained in the literature concerning steady, shearing granular flows, unsteady conditions have been less investigated. In this work, we investigate the fluid-solid transition in unsteady, homogeneous, shear flows of a collection of identical, frictional spheres.

2 DEM numerical simulations

The authors have performed DEM numerical simulations of unsteady simple shear flows of frictional spheres, using a linear spring-dashpot interaction model [10, 11], implemented in the open-source code Mercury-DPM (www.mercurydpm.org) [12, 13]. All the simulations have been carried out under constant volume in a cubic box of dimension L (figure 1). 2000 particles of diameter d, density ρ_p, i.e., mass $m_p = \rho_p \pi d^3/6$, were used. The size L of the domain has been chosen once the volume fraction ν is fixed (where, for a granular material composed of identical particles, the local volume fraction ν is defined as the ratio of the local material density to the particle mass density). We set the normal coefficient of restitution $e_n = 0.7$, the normal spring stiffness $k_n = 10^7$ Pa m, the tangential spring stiffness $k_t = 2/7\ k_n$ (corresponding to a tangential coefficient of restitution $e_t = e_n = 0.7$) and the inter-particle friction coefficient $\mu = 0.3$. In order to keep the system homogeneous in space, Lees-Edwards [14] periodic boundary conditions in the z direction and periodic boundary conditions in the x and y directions have been employed. The system is sheared by moving the boundaries in opposite directions at constant horizontal velocity V, and setting the global shear rate $\hat{\dot{\gamma}} = 2V/L$ equal to 0.1 s^{-1}. The instantaneous application of a relative velocity between the boundaries is not immediately transmitted within the entire granular material, given that particles in the core of the domain cannot initially feel the boundaries motion. As a consequence, the velocity profiles are not linear, instead they exhibit some curvature; the local shear rate $\dot{\gamma}$ (velocity gradient) is not homogeneous, but is maximum at the boundaries and minimum (close to zero) in the core. After a short time interval ($0 < t < 0.005$ s) the non-homogeneity is dissipated and $\dot{\gamma}$ becomes homogeneous in space and almost constant in time, equal to the global shear rate $\hat{\dot{\gamma}} = 0.1 s^{-1}$.
Three values of the constant volume fraction have been taken into consideration: 0.59, 0.60 and 0.62. For steady, shearing flows, solid volume fraction larger (lower) than a critical, ν_c, indicates that a granular system is solid-like (fluid-like), i.e., rate-independent components of the stresses are present (absent). ν_c is the largest volume fraction at which a randomly collisional granular material can be sheared without force chains spanning the entire

*e-mail: dalila.vescovi@polimi.it

 DOI: 10.1051/epjconf/201714003058

domain [15]. The critical volume fraction depends on the granular material, in particular is affected by the polydispersity of the system [16, 17] and the inter-particle friction coefficient [1]. Chialvo et al. [6] have measured that for identical particles with $\mu = 0.3$, ν_c is 0.598. Then, our three samples at steady state should be in fluid ($\nu = 0.59$), solid ($\nu = 0.62$) and near-to-critical ($\nu = 0.60$) conditions. At steady state, there is a univocal relation between the critical volume fraction and the corresponding critical coordination number C_c (where the coordination number C is defined as the average number of contacts between all particles), independent of the shear rate [9]. The coordination number is an important parameter for describing the interaction among grains and the characteristic time scale of the force chains [18]. Note that in this paper we are not disregarding particles with zero (floaters) or one contact (rattlers) even if they do not participate to the contact network. The "corrected" coordination number C^* has been introduced by neglecting floaters and rattlers [17, 19], and is related to the standard coordination number as $C = C^*(1 - \phi_r)$ [20], where ϕ_r is the rattler fraction. In this work, we assume the system to experience a transition from fluid to solid states, under unsteady conditions, independently of the volume fraction, when the coordination number reaches its critical value C_c. The corrected coordination number at the transition, C_c^*, has been measured by Sun and Sundaresan [21], and, for $\mu = 0.3$, $C_c^* = 4.6$. By assuming $\phi_r = 0.13 \pm 0.03$ as measured for frictionless particles [20], the critical coordination number is $C_c \approx 4$.

Figure 1. Sketch of the flow configuration. A granular material composed of identical, frictional spheres is homogeneously sheared at constant global shear rate $\dot{\gamma}$, with flow in horizontal direction (boundaries move in opposite horizontal directions with velocity $V = \dot{\gamma}L/2$), by using Lees Edwards boundary conditions. Colours indicate speed, from blue (zero velocity of the particles in the core of the domain) to red (maximum velocity of the particles at the boundaries)

The preparation of the system followed a standard three-step procedure [17, 22]: (i) frictionless spherical particles are randomly generated in the 3D box with low volume fraction and random velocities, such that they have sufficient space and time to exchange places and to randomize themselves; (ii) the inter-particle friction coefficient is activated to a given value μ_0 and the granular material is isotropically compressed to the target volume fraction ν; (iii) the particles are allowed to relax and dissipate their kinetic energy before either a further preparation or the actual element test is initiation.

At the end of the three stages, a static initial granular sample of volume fraction ν has been generated, with a certain coordination number C_0 and a defined isotropic pressure p_0. By performing steps (ii) and (iii) with different values of the inter-particle friction leads to different isotropic samples (i.e, different values of C_0 and p_0). For all the target volume fraction considered (0.59, 0.60 and 0.62), we have performed step (ii) and (iii) by imposing $\mu_0 = 0$ in order to attain initial fluid-like configurations ($p_0 = 0$ Pa and $C_0 = 0$). Finally, the inter-particle friction $\mu = 0.3$ is assigned, and the boundaries velocity are switched on with global shear rate $\dot{\gamma}$, constant in time.

Here and in the following, the variables are made dimensionless using the particle diameter d, density ρ_p and stiffness k_n. Then, the scaled time t, pressure p, shear stress s and shear rate $\dot{\gamma}$ are given, respectively, as: $t\sqrt{k_n/(\rho_p d^3)}$, pd/k_n, sd/k_n and $\dot{\gamma}\sqrt{\rho_p d^3/k_n}$.

3 Results

The evolution of the dimensionless pressure pd/k_n and coordination number C with the accumulated shear strain $\gamma = \dot{\gamma}t$ are illustrated in figure 2(a) and (b), respectively.

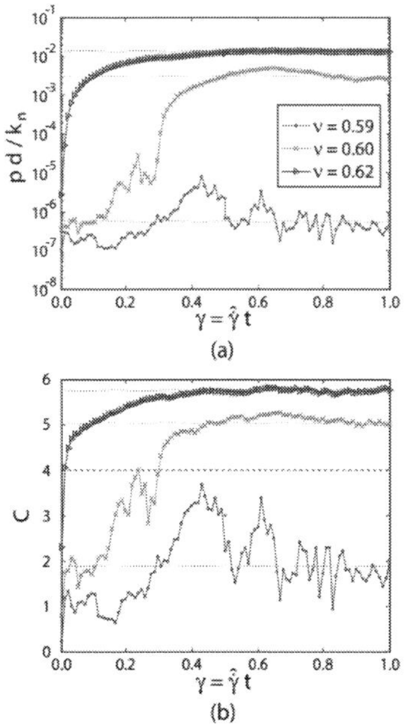

Figure 2. Evolution of (a) scaled pressure and (b) coordination number for different values of the volume fraction. Steady state values are denoted with dotted lines; the dashed line represents the critical coordination number

In all the systems, as was already mentioned, both the initial pressure p_0 and the coordination number C_0 are zero. At $t = 0$ ($\gamma = 0$), a dimensionless global shear rate $\dot{\gamma}\sqrt{\rho_p d^3/k_n} = 3.16 \cdot 10^{-5}$ is instantly applied. Pressure and coordination number monotonically increase with time until the steady state is reached, by following similar patterns, as was previously observed by Sun and Sundaresan [21]. The dotted lines in figure 2(a) and (b) represent the final steady state values of p and C, obtained by averaging in time over at least 1000 time steps. For $\nu = 0.59$, the steady state is reached in a very short time interval, with large fluctuations for both p and C. We take this as a signal that the system is fluid during the simulation; indeed, the coordination number is always less than the critical value 4 (dashed line in figure 2b). In the case $\nu = 0.62$, the system turns rapidly solid, with small fluctuations in both p and C, with the former one increasing by several orders of magnitude.
Finally, in the case $\nu = 0.60$ the behaviour of C and p is initially very similar to the fluid case with large oscillations and significant increase in the coordination number. Then, the granular material experiences a fluid-solid transition, and the fluctuations dramatically decrease. In this second stage, the pressure increases of almost 4 orders of magnitude.
The three cases of volume fraction investigated show how much the micro-structure of the system influences the time evolution of the variables. Starting from fluid conditions, that is in absence of force chains, systems at less than critical volume fraction ($\nu = 0.59$) do not require a long transient regime to reach the steady state, since no persistent chains of contacts among particles form during shear. Although the particles are quite largely spaced and interact by mean of collisions, the coordination number is never zero due to the presence of clusters of particles which do not span the entire domain. The continuous destruction and re-building of clusters, together with the free motion of single particles, gives rise to the observed large fluctuations in both coordination number and pressure. At volume fraction much larger than the critical ($\nu = 0.62$), the system quickly solidifies during shear. The chains of contacts that span the entire domain require time to develop within the sample and, as a consequence, a clear transient regime is shown in terms of both C and p. Once a contact network has been developed, the very small fluctuations of C and p are due to micro-structural rearrangements during shear. The larger the volume fraction, the smaller the fluctuations, since particles are more compacted and cannot easily abandon the own force chain. The transition from fluid to solid is evident in the case $\nu = 0.60$, where the volume fraction is very close to the critical value, and the particles interact through both force chains and collisions. The critical value $C = 4$ distinguishes the fluid-like (large fluctuations) from the solid -like (small fluctuations) behaviour (figure 2b).

Finally, the evolution of the stress ratio s/p with the shear strain γ is depicted in figure 3(a). Analogously to the other variables, smaller volume fractions are associated with larger oscillations of s/p. However, the three

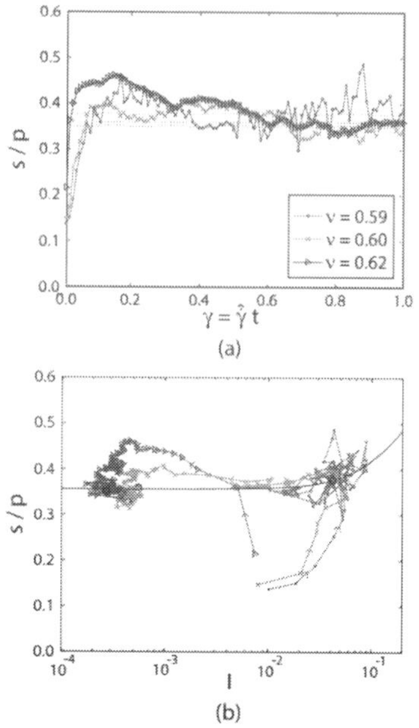

Figure 3. Stress ratio s/p plotted against (a) strain and (b) inertial number I for different values of the volume fraction. Steady state values of s/p are denoted with dotted line in (a) and filled big circles in (b). The solid line in (b) represents Equation (1) and arrows indicate the direction of increasing γ

patterns are pretty similar, with an increase at the initial stages, until $\gamma = 0.1 \div 0.2$, and then a slight softening to the final steady value.
It is also interesting to plot the stress ratio data measured in the transient against the inertial number: $I = \dot{\gamma} d\sqrt{\rho_p/p}$, introduced in Ref. [23–25]. In the case of steady, homogeneous shear flows of rigid particles, the French research group GDR-MiDi [23] has introduced a constitutive relation describing a wide variety of steady flows, at various strain rates. Chialvo et al. [6] have slightly revised that equation, providing a better fit of steady, simple shearing, especially at large I:

$$\frac{s}{p} = \eta_0 + \frac{\alpha_1}{(I_0/I)^{\beta_1} + 1} \tag{1}$$

with $\alpha_1 = 0.667$, $I_0 = 0.25$, $\beta_0 = 1.5$ and $\eta_0 = 0.357$ for $\mu = 0.3$. In figure 3(b) our unsteady numerical data (lines with symbols) are compared with Equation (1) (black solid line). During the transient phase, coordination number, pressure and shear stress are homogeneous in the domain, at each time step, whereas, as it was previously observed, the velocity profiles are initially not linear (for $\gamma < 0.05$) as, instead, is in the case of simple shear flows. This is due to instantaneous application of a relative velocity between the boundaries not immediately affecting the entire granular material. In order to disregard non local effects in this first time periods, we consider the bulk shear rate, averaged over the central region, discarding the regions in proximity of the boundaries. Nevertheless, the time interval during which the non-homogeneity is dissipated is very short, $0 < \gamma < 0.05$, for any volume fraction considered here, and for $\gamma > 0.05$, $\dot{\gamma}$ is homogeneous in space and almost constant in time, equal to the imposed global shear rate $\hat{\dot{\gamma}}$. In figure 3(b), the final steady state values of the three simulations, represented with big filled circles, belong to the curve defined by Equation (1). For $\nu = 0.59$, data taken at different times in the transient phase are characterized by large inertial number $10^{-2} < I < 10^{-1}$. Except a few data which lie below the steady state curve, corresponding to the first time steps of the simulation ($0 < \gamma < 0.1$), the numerical measurements seem to follow the curve pretty well. The outliers are due to large fluctuations characterizing fluid-like systems, as was previously observed for all the variables. On the other hand, for $\nu = 0.62$, differently from what observed in Ref. [26] for constant pressure simple shearing, Equation (1), nor any linearization of it, cannot predict the behaviour of solid-like, constant volume simple shear flows in the transient regime. The transition from fluid to solid conditions observed for $\nu = 0.60$ in terms of C, is evident also in figure 3(b). After a few time steps ($0 < \gamma < 0.1$) the data reach the steady state curve at $I \approx 5 \cdot 10^{-2}$, then follow it while I decreases till $I \approx 5 \cdot 10^{-3}$. This behaviour is similar to the fluid case $\nu = 0.59$, with less fluctuations. After that, the data detach from the curve and move above it, denoting the phase transition. Finally, the steady state is attained and just small fluctuations are observed.

4 Conclusions

The fluid-solid transition in granular materials has been investigated by performing discrete numerical simulations of unsteady, homogeneous, shear flows. The fluid-like behaviour has been shown to be characterized by large fluctuations in stresses and coordination number. Conversely, the fluctuations are much smaller when a contact network spanning the entire domain develops in the granular material (solid-like behaviour). The fluid-solid transition is characterized by a critical value of the coordination number, independent of the volume fraction. Finally, we have shown that, contrary to what has been observed for unsteady, homogeneous shearing at constant pressure, the relation between the stress ratio and the inertial number, valid for homogeneous steady flows, cannot reproduce the behaviour of unsteady homogeneous shearing at constant volume.

Acknowledgements

The authors express their special thanks to Professor Stefan Luding for valuable scientific discussions. Dalila Vescovi is supported by a fellowship from Fondazione Fratelli Confalonieri.

References

[1] L. Silbert, Soft Matter **6**, 2918 (2010)
[2] R. Bagnold, Proc. Royal Society A **255**, 49 (1954)
[3] S. Luding, Nature Phys. **12**, 531 (2016)
[4] P.C. Johnson, R. Jackson, J. Fluid Mech. **176**, 67 (1987)
[5] S.B. Savage, J. Fluid Mech. **377**, 1 (1998)
[6] S. Chialvo, J. Sun, S. Sundaresan, Phys. Rev. E **85**, 021305 (2012)
[7] D. Vescovi, C. di Prisco, D. Berzi, Int. J. Numer. Anal. Methods Geomech. **37**, 2937 (2013)
[8] D. Berzi, J. Jenkins, Soft Matter **11**, 4799 (2015)
[9] D. Vescovi, S. Luding, Soft Matter **12**, 8616 (2016)
[10] S. Luding, S. McNamara, Granular Matter **1**, 113 (1998)
[11] S. Luding, Eur. J. Environ. Civil Eng. **12**, 785 (2008)
[12] A. Thornton, T. Weinhart, S. Luding, O. Bokhove, Int. J. Modern Phys. C **23**, 1240014 (2012)
[13] T. Weinhart, A. Thornton, S. Luding, O. Bokhove, Granular Matter **14**, 289 (2012)
[14] A. Lees, S.F. Edwards, J. Phys. C: Solid State Phys. **5**, 1921 (1972)
[15] D. Berzi, D. Vescovi, Phys. Fluids **27**, 013302 (2015)
[16] V. Ogarko, S. Luding, Soft Matter **9**, 9530 (2013)
[17] N. Kumar, O.I. Imole, V. Magnanimo, S. Luding, Particuology **12**, 64 (2014)
[18] S. Ji, H. Shen, Chin. Sci. Bull. **51**, 646 (2006)
[19] F. Göncü, O. Duran, S. Luding, *Jamming in frictionless packings of spheres: determination of the critical volume fraction*, in *Powders and Grains 2009, AIP Conference Proc.* (M. Nakagawa and S. Luding (Eds.), 2009), Vol. 1145, pp. 531–534
[20] F. Göncü, S. Luding, *From "soft" particle simulations to macroscopic constitutive relations*, in *PARTEC Proc.* (Nurnberg Messe GmbH, 2010)
[21] J. Sun, S. Sundaresan, J. Fluid Mech. **682**, 590 (2011)
[22] O. Imole, N. Kumar, V. Magnanimo, S. Luding, KONA **30**, 84 (2013)
[23] GDR-MiDi, Eur. Phys. J. E **14**, 341 (2004)
[24] F. da Cruz, S. Emam, M. Prochnow, J. Roux, F. Chevoir, Phys. Rev. E **72**, 021309 (2005)
[25] P. Jop, Y. Forterre, O. Pouliquen, Nature **441**, 727 (2006)
[26] A. Ries, L.B.D. Wolf, Comp. Part. Mech. **3**, 303 (2016)

Calibration of micromechanical parameters for DEM simulations by using the particle filter

Hongyang Cheng[1,4,⋆], *Takayuki* Shuku[2], *Klaus* Thoeni[3], and *Haruyuki* Yamamoto[4]

[1] *Multi Scale Mechanics (MSM), Faculty of Engineering Technology, MESA+, University of Twente, The Netherlands*
[2] *Graduate School of Environmental and Life Science, Okayama University, Japan*
[3] *School of Engineering, The University of Newcastle, Australia*
[4] *Graduate School for International Development and Cooperation, Hiroshima University, Japan*

Abstract. The calibration of DEM models is typically accomplished by trail and error. However, the procedure lacks of objectivity and has several uncertainties. To deal with these issues, the particle filter is employed as a novel approach to calibrate DEM models of granular soils. The posterior probability distribution of the micro-parameters that give numerical results in good agreement with the experimental response of a Toyoura sand specimen is approximated by independent model trajectories, referred as 'particles', based on Monte Carlo sampling. The soil specimen is modeled by polydisperse packings with different numbers of spherical grains. Prepared in 'stress-free' states, the packings are subjected to triaxial quasistatic loading. Given the experimental data, the posterior probability distribution is incrementally updated, until convergence is reached. The resulting 'particles' with higher weights are identified as the calibration results. The evolutions of the weighted averages and posterior probability distribution of the micro-parameters are plotted to show the advantage of using a particle filter, i.e., multiple solutions are identified for each parameter with known probabilities of reproducing the experimental response.

1 Introduction

The Discrete Element Method (DEM) often brings forth new understanding that are difficult to acquire from sophisticated experiments [1]. Notwithstanding the versatility, the DEM model of a granular material with appropriate micromechanical parameters generally requires a 'calibration' against the macroscopic experimental response. The conventional calibration employs 'one at a time' sensitivity analyses of the parameters to derive the micro–macro relations. Depending on the contact laws and material types, various micro–macro relations are obtained. Among a handful of *systematic* calibration approaches, the design of experiments (DOE) methods are efficient in searching possible solutions in the multi-dimensional parameter space with a manageable number of DEM simulations and optimized outcomes. Hanley [2] applied the DOE to calibrate the DEM models of crushable agglomerates. The interaction between key parameters were effectively considered by the orthogonal arrays designed with the Taguchi methods. Nevertheless, the DOE requires the knowledge of interaction between parameters, which is generally unavailable for various types of granular materials. A significant limitation of the aforementioned approaches is that the calibration can only be conducted with instantaneous bulk behaviors (e.g., Young's modulus) rather than a complete history of stress–strain response. To deal with the above-mentioned difficulties, the particle filter[1] (PF) and sequential importance sampling (SIS), which can jointly account for loading history, are selected for the current problem. The PF applies the recursive formula of the sequential Bayesian estimation and approximates the posterior probability density function (PDF) with the SIS algorithm. The proposed calibration approach for DEM simulations is expedient, because the PF is well-justified for nonlinear and non-Gaussian geotechnical problems. Moreover, both the PF and SIS can be easily implemented in the open-source DEM package YADE [3], where parallel DEM simulations are run as model trajectories for the Monte Carlo sampling.

2 DEM simulations of granular materials

The parameter values for conducting three-dimensional DEM simulations are sampled from the posterior probability distribution, given the sequentially obtained experimental data. YADE [3] models granular materials as packings of solid grains with simplified geometries and vanishingly small intergranular overlaps. It tracks the kinematics of each grain within the explicit time integration scheme, based upon the net force resulted from the intergranular forces. In the present work, the robustness of the

⋆e-mail: h.cheng@utwente.nl

[1] In this context, the word 'particle' exclusively refers to a 'sample' for approximating posterior probability distribution functions and it should not be comprehended as a grain in the physical tests and numerical simulations.

PF for calibrating DEM models is examined by modeling a Toyoura sand specimen (50 × 50 × 100 mm) with a void ratio of 0.68 as a dense packing of polydisperse spherical grains. Among all the randomly generated dense packings, three (N_g = 1000, 2000 and 5000) are selected, because of their relatively smooth and spherical contact orientation diagrams after the initial isotropic compaction. The Hertz-Mindlin (HM) and classical linear (CL) contact laws are employed to govern the intergranular mechanical behaviors along the normal and tangential directions. Rolling/twisting moment transfer is allowed between adjoining spheres to account for the effect of grain roughness. Note that these non-physical parameters are necessary to obtain bulk friction angles which are close to experimental values of dense sands, without extra computational effort to capture the kinematics related to irregular shapes. Both intergranular tangential forces and moments are assumed to be bounded by Coulomb type plastic limits. Both contact laws have five micromechanical parameters, namely, the Young's modulus E_c and Poisson's ratio υ_c of contacting grains, factors β_m and η_m which define rolling/bending stiffnesses and maximum moments, and intergranular friction angle μ. The DEM simulations of sand specimens under drained triaxial compression (DTC) are performed in a strain-controlled quasistatic manner.

3 Data assimilation for granular materials with the particle filter

3.1 State space model and state estimation

Given a specimen of granular soil and its equivalent DEM packing, the physical measurements on the specimen and the prediction resulting from the DEM packing can be described by a nonlinear and non-Gaussian state space model:

$$x_t = f_t(x_{t-1}) + \upsilon_t \qquad (1)$$

$$y_t = h_t(x_t) + \omega_t \qquad (2)$$

where the state vector x_t consists of stress ratio σ_a/σ_r, radial strain ε_r and volumetric strain ε_v at a discrete loading step t, whereas the observation vector y_t was directly measured in the DTC experiments [4]; υ_t and ω_t are the system and observation errors whose PDFs follow normal distributions with zero means. f_t denotes the operator of the nonlinear dynamic model (i.e., DEM model), whose current state depends on the entire loading history. In the current problem, the nonlinear observation operator h_t is reduced to an identity matrix of size three.

3.2 Particle filter

The PF approximates PDFs via a set of particles (referred as an ensemble) and the associated weights. From the ensemble $\{x^{(1)}_{t-1|t-1}, x^{(2)}_{t-1|t-1}, ..., x^{(N)}_{t-1|t-1}\}$ and the weights $\{w^{(1)}_{t-1}, w^{(2)}_{t-1}, ..., w^{(N)}_{t-1}\}$, the filtered distribution $p(x_{t-1}|y_{1:t-1})$ at time $t-1$ is approximated as

$$p(x_{t-1}|y_{1:t-1}) \approx \frac{1}{N}\sum_{i=1}^{N} w^{(i)}_{t-1}\delta(x_{t-1} - x^{(i)}_{t-1|t-1}) \qquad (3)$$

where N is the number of particles, δ is the Dirac delta function, and the superscript (i) indicates the ith particle $x^{(i)}_{t-1|t-1}$ and its associated weight $w^{(i)}_{t-1}$. It should be noted that all weights are greater or equal to zero and the summation of them must be one. With the ensemble sampled using the Monte Carlo Method, the recursive formulas for approximating the one-step-ahead forecast and filtered distributions $p(x_t|y_{1:t-1})$ and $p(x_t|y_{1:t})$ are simplified as follows:

One-step-ahead forecasting

$$\begin{aligned} p(x_t|y_{1:t-1}) &= \int_{-\infty}^{\infty} p(x_t|x_{t-1})p(x_{t-1}|y_{1:t-1})dx_{t-1} \\ &\approx \sum_{i=1}^{N} w^{(i)}_{t-1}\delta(x_t - x^{(i)}_{t|t-1}) \end{aligned} \qquad (4)$$

Filtering

$$\begin{aligned} p(x_t|y_{1:t}) &= \frac{p(y_t|x_t)p(x_t|y_{1:t-1})}{p(y_t|y_{1:t-1})} \\ &= \sum_{i=1}^{N} \tilde{w}^{(i)}_t w^{(i)}_{t-1}\delta(x_t - x^{(i)}_{t|t-1}) \\ &= \sum_{i=1}^{N} w^{(i)}_t\delta(x_t - x^{(i)}_{t|t-1}) \end{aligned} \qquad (5)$$

with

$$\tilde{w}^{(i)}_t = \frac{p(y_t|x^{(i)}_{t|t-1})}{\sum_j p(y_t|x^{(j)}_{t|t-1})w^{(j)}_{t-1}} \qquad (6)$$

If the observation system is linear and $p(y_t|x_{t|t-1})$ of the ith particle reads

$$p(y_t|x_{t|t-1}) = \frac{1}{(2\pi)^{m/2}|R_t|} \times \exp\{-\frac{[y_t - H_t(x^{(i)}_{t|t-1})]^T R_t^{-1}[y_t - H_t(x^{(i)}_{t|t-1})]}{2}\} \qquad (7)$$

where m is the dimension of x_t, R_t is a predetermined covariance matrix for the observation error, and H_t is the observation operator in matrix form. Given the previous weight $w^{(i)}_{t-1}$ and current regulated weight $\tilde{w}^{(i)}_t$ of each particle (Equation 6), new weight $w^{(i)}_t$ can be updated as

$$w^{(i)}_t = \tilde{w}^{(i)}_t w^{(i)}_{t-1} \qquad (8)$$

3.3 Sampling and filtering procedure

The loading step t ranges from one to one hundred. For the current simple problem with few parameters, a total number of 2000 particles N_p is sufficient to achieve good accuracy. The readers are referred to [5] for further details. The proposed calibration procedure is summarized as follows:

1. Initialization:
   Generate particles $\{x^{(1)}_0, x^{(2)}_0, ..., x^{(N)}_0\}$ from the initial distribution $p(x_0)$ by initializing parallel DEM simulations with a low-discrepancy sequence in $\mathbf{R}^n$ (n is the number of parameters to be identified).

Table 1. Means and standard deviations of the particles.

	$\log(E_c)$	υ_c	β_m	η_m	μ (°)
MEAN	9	0.25	0.5	0.5	30
STDEV	2	0.25	0.5	0.5	30

2. Prediction:
Predict the state of each particle $x_{t-1}^{(i)}$ based on the parallel DEM simulations run for the $t-1$ loading step, at which the observation data y_{t-1} are obtained.

3. Filtering:
Calculate the weight of each particle $w_t^{(i)}$ that expresses the 'fitness' of the prior particles to the observation data y_t with Equation 8 and Equation 6.

4. Weight updating:
The filtered distribution $p(x_t|y_{1:t})$ is approximated by the particles and the updated values of the associated weights. Return to Step 2 with $p(x_t|y_{1:t})$, $\{x_t^{(1)}, x_t^{(2)}, ..., x_t^{(N)}\}$ and $\{w_t^{(1)}, w_t^{(2)}, ..., w_t^{(N)}\}$ for the prediction and filtering at the $t+1$ loading step.

4 Calibration of micro-parameters for DEM simulations

A Halton sequence generated with the means and standard deviations in Table 1 is used as the particles (N_p = 2000). Each particle, which represents a specific combination of micro-parameters, is assigned to the contact law (HM or CL) for the three initially prepared 'stress-free' packings (N_g = 1000, 2000 and 5000). 2000 DEM simulations of DTC tests are then conducted in a two-phase loading program: the packings are first isotropically compressed to a prescribed confining stress (σ_c = 0.2 MPa), and then sheared under triaxial compression in a quasistatic manner. The state variables are extracted at each loading step where the measurements were made. In what follows, the respective evolutions of the weighted averages of each micro-parameter obtained from different contact laws and DEM packings are first shown to examine the efficiency of the PF. The posterior PDF at different simulation steps is also plotted to further demonstrate the features of the PF.

4.1 Weighted averages of micro-parameters

The values randomly generated for each parameter are weighted by the approximate posterior PDF at each loading step. The resulted evolutions of the weighted averages of the micro-parameters are illustrated in Figure 1. While updating the weights through the entire loading history, the PF eventually results in less pronounced variation of the posterior PDF, which can also be understood from the weighted averages converging into the true values. The combinations of micro-parameter values that are associated with higher weights at the converged steps are identified as the calibration results. Although, convergence is reached for both contact laws and all three DEM packings,

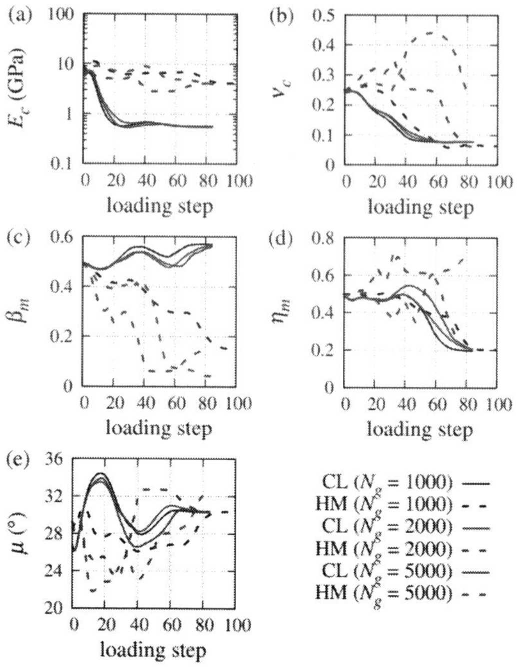

Figure 1. Evolution of weighted averages of the micro-parameters versus loading step.

the convergence rate appears to be lower with larger fluctuations in the cases of DEM models governed by the HM law (dashed curves). Note that the true values which the weighted averages converge to are very close for the three packings governed by the same contact law. The agreement is almost perfect in the cases where the CL law governs intergranular behaviors (solid curves). The converged weighted averages of μ and η_m of the HM and CL laws (with moment transfer) differ slightly, whereas the differences between the calibration results for E_c, υ_c and β_m of the two contact laws are more evident. This is because the same Coulomb type friction is applied on tangential forces and contact moments, despite that the normal and tangential stiffnesses are calculated from different contact laws.

4.2 Posterior PDFs of micro-parameters

The advantage of the PF in a parameter identification problem is the ability to approximate the posterior PDF in a multi-dimensional parameter space. By taking advantage of the PF and SIS, the updated distribution can be understood at each loading step, as shown in Figure 2. Subplots (a)–(e) and (f)–(j) show respectively the posterior PDF of the five micro-parameters of the HM and CL laws at five consecutive loading steps with an increment of 20 steps, given the triaxial response of Toyoura sand.

It can be seen that the initial distribution along the E_c and μ axes gradually shifts from a uniform to a Gaussian-like distribution. It is surprising to find that the preferable values for E_c and μ are already identifiable within the first

20 loading steps (ε_a = 2%). The distribution associated with υ_c appears to exhibit a bimodal character at the 40th loading step (ε_a = 4%) which corresponds to the peak stress state. However, the weights are only assigned to few particles after the 60th steps. A further refinement of the range of υ_c would help to better represent the bimodal distribution. Although suitable values for β_m and η_m are identified, the associated posterior distributions appear to be more scattered and do not evolve consistently, even in the early stage of the calibration process. The fact that the distributions can hardly be described by a Gaussian or mixtures of Gaussians might be due to the non-physical nature of these rolling/twisting parameters.

Similar to the converging weighted averages of the micro-parameters as shown in Figure 1, the posterior PDFs collapse for the DEM packings with different numbers of grains (N_g = 1000, 2000 and 5000), governed by the same contact law. This finding verifies a scaling rule which facilitates the calibration of DEM models, that is, a minimal number of DEM grains can be used for fast calibration. For DEM simulations which require a large number of grains, same parameters can be used, as long as the initial void ratio and stress state are the same as in the simulations run for the calibration. For the current parameter identification problem, 2000 simulations were conducted for each DEM packing. It took less than two days to run 1000-grain quasistatic DEM simulations on a 16-core workstation.

4.3 Simulation results versus measurement data

The simulation results reproduced with the identified parameters (particles with the highest weights) are plotted with the experimental data in Figure 3. Good agreements are achieved, regardless of the different contact laws and numbers of grains applied in the DEM models. While the PF is proved to be an effective tool in calibrating the DEM models, further study is needed to investigate the capacity of this method on the DEM modeling of granular materials at various stress states and more complex situations.

5 Conclusions

A new calibration method is proposed for the DEM simulations of granular soils. The PF and SIS algorithms are implemented to work with the open-source DEM package YADE. The micro-parameters for the DEM simulations of granular soils under triaxial compression are successfully calibrated with the proposed method. The posterior probability distribution is an important feature which is approximated by the PF. It helps to gain insights into parameter identification problem. It is found that the converged posterior PDFs collapse for different DEM simulations governed by the same contact law. A scaling rule is verified and can be adopted as a guideline for calibrating DEM models in a cost-effective manner.

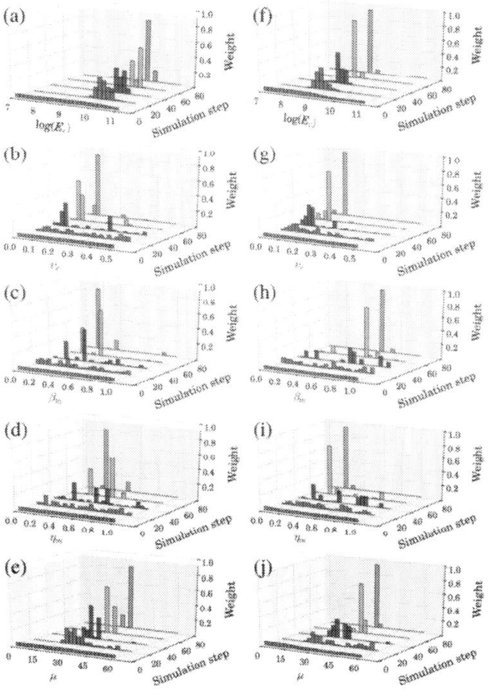

Figure 2. Evolution of posterior probability distribution associated with each micro-parameter. Subplots (a–e) are made with the HM law and subplots (f–j) with the CL law.

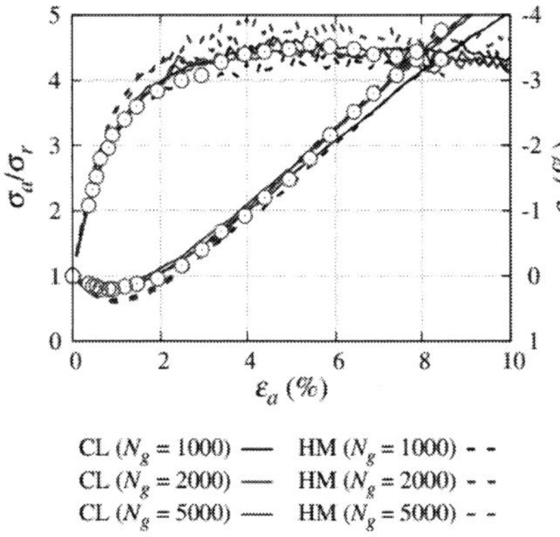

Figure 3. Comparison of experimental data and DEM simulation results reproduced with the identified parameters.

References

[1] A. Singh, V. Magnanimo, K. Saitoh, S. Luding, Phys. Rev. E **90**, 022202 (2014)

[2] K.J. Hanley, C. O'Sullivan, J.C. Oliveira, K. Cronin, E.P. Byrne, Powder Technology **210**, 230 (2011)

[3] V. Šmilauer, B. Chareyre, J. Duriez, A. Eulitz, A. Gladky, N. Guo, C. Jakob, J. Kozicki, in *Yade Documentation*, edited by T.Y. Project (2015), 2nd edn.

[4] D. Sun, W. Huang, D. Sheng, H. Yamamoto, Key Eng. Mater. **340-341**, 1273 (2007)

[5] T. Shuku, A. Murakami, S.i. Nishimura, K. Fujisawa, K. Nakamura, Soils and Foundations **52**, 279 (2012)

DOI: 10.1051/epjconf/201714002015

Heat transfer rate within non-spherical thick grains

Florian Huchet[1,*], *Patrick* Richard[1], *Jules* Joniot[1], *and Laurédan* Le Guen[1]

[1]LUNAM, Université de Nantes, IFSTTAR, MAST dpt, GPEM, CS4, 44344 Bouguenais, France

Abstract. The prediction of the internal heat conduction into non-spherical thick grains constitutes a significant issue for physical modeling of a large variety of application involving convective exchanges between fluid and grains. In that context, the present paper deals with heat rate measurements of various sizes of particles, the thermal sensors being located at the interface fluid/grain and into the granular materials. Their shape is designed as cuboid in order to control the surface exchanges. In enclosed cone-shaped apparatus, a sharp temperature gradient is ensured from a hot source releasing the air stream temperature equal to about 400°C. Two orientations of grain related to the air stream are considered: diagonally and straight arrangements. The thermal diffusivity of the grains and the Biot numbers are estimated from an analytical solution established for slab. The thermal kinetics evolution is correlated to the sample granular mass and its orientation dependency is demonstrated. Consequently, a generalized scaling law is proposed which is funded from the effective area of the heat transfer at the grain-scale, the dimensionless time being defined from the calculated diffusional coefficients.

1 Introduction

The response to the temperature variation of the granular assembly surrounded by hot fluid stream is currently encountered in the multiphase flows. Indeed, many applications involving different size and geometry of grains submitted to a convective transfer are emphasized in chemical and materials science or civil engineering domains dealing respectively with various types of gas/solid interaction such as the packed bed [1], the vibrofluidized bed [2] or the rotor-stator system [3].

Discrete element simulations are undoubtedly a relevant tool to study such systems. A common assumption is to consider that grains have a uniform internal temperature, *i.e.* their Biot number, defined as $Bi = hd/2k_s$ (where h is the convective heat transfer, k_s the thermal conductivity and d the grains size) is much lower than 1. Thus, the Bi number informs about the use of the lumped capacitance assumption, these latter hypothesis being applicable to the lowest values corresponding to small particle size (Bi<<1). In that latter case, heat flux is transferred instantaneously from the external layer of the particles to the bulk. The sense of the heat transfer is inverted for hot particles submerged into a fresh fluid. One can refer to the examples of the combustion of char [4] or lime shaft kiln simulation [5] where this assumption can be not relevant. Although it can be justified in many practical cases, this assumption remains invalid in the case of large grains. In that context, the present paper deals with experimental study of the unsteady conduction of non-spherical thick grain submitted to intense temperature gradient in two orientations related to the fluid stream (see fig. 1). The establishment of a simple thermal law devoted to non-spherical particle has a great potential for its implementation in Euler-Lagrange code.

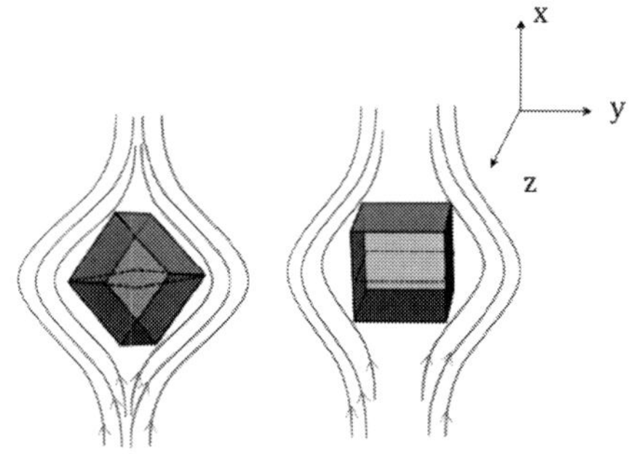

Fig. 1. Two geometrical arrangements perpendicularly to the heat stream are tested, one in diagonal offering an augmented exchange surface while the second is set in the straight configuration offering only one face to the hot stream. The heat transfer area being represented in solid line.

2 Experimental set-up

Rock material (gneiss) is chosen as a model of granular materials. Carefully prepared from milling process in order to be designed as cuboid shape, four samples of different size are made (see picture in fig. 2.a. and table 1) to control the surface exchanges. The experimental apparatus is composed of an enclosed cone-shaped system, which controls the air hot stream in upward direction from a lab-scale boiler located below the granular sample. A grid composed of ultra-fin steal wires (1 mm in diameter) is disposed at the interface between

* Corresponding author: florian.huchet @ifsttar.fr

the hot source and the measurement area reducing any disturbance between the grain and the hot stream ($d_{material}$>>$d_{steal\ wire}$). The heat flux is controlled from a burner at zero-pressure gradient (*i.e.* no imposed flow rate), only the buoyancy is responsible of an upward heat flux toward the granular sample. According to [6, 7], the stream velocity in the vicinity of the rock is estimated to 4.3 m/s.

a

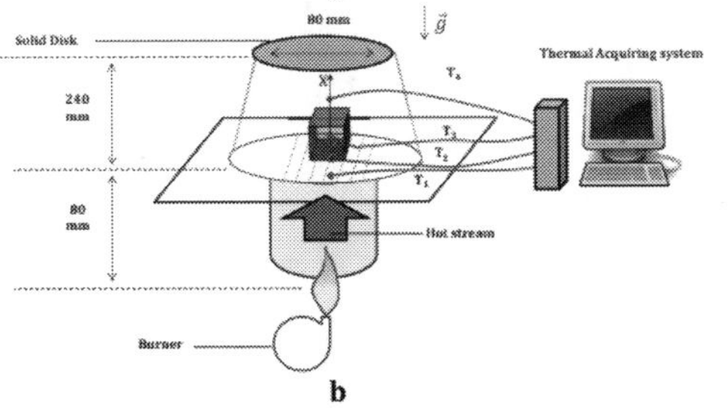

b

Fig. 2a. Rocks materials with four different sizes are milled while holes are bored at the center of the surface samples.
Fig. 2b. The granular sample is composed of two holes. One is accomplished in the near-wall surface to set one K-thermal probe (T_2) at l =1 mm. The second is drilled in the heart of the granular medium close to the central location (T_3). The rest of the thermal probes are located above (T_4) and below the grain (T_1). The thermoregulation is ensured from a burner from the thermal probe, T_1, ranged between 350°C and 450 °C.

The surface of the granular matter « seen » by the hot fluid is determinant to characterize the interaction between fluid and grain, especially for non-spherical particles. The state of art related to that topic remains confused between the works established in the case of the forced convection [8, 9] or the free convection [10, 11]. In that context, two orientations were investigated, one located diagonally with respect to the fluid stream while one is straight with respect to the fluid stream. An effective surface, S_{eff}, is required to give an insight of the shape factor. Defined as the projected area for the diagonal orientation, one face as the inner heat transfer area is considered for the straight orientation case (see sketch in figure 2.b.). Note that the aspect ratio between the two orientations is $\sqrt{2}$. The table 1 recaps for each sample the ratio between the effective surface and the overall area of transfer, S, the mass of the grain, and the characteristics length of the thermal diffusivity between the inner face and the center of the sample. Afterward a particular attention of the positioning of the probes, the cone-shape is systematically contained in closed-volume, from a solid disk, before the beginning of the experiments. Then, all the experiments were conducted in confined media with a temperature regulation of the hot stream between 350°C and 450°C controlled from the first probes, T_1.

Table 1. Physical characteristics of each cuboid granular sample.

Granular sample	S_1	S_2	S_3	S_4
Mass (g)	287	288	91.5	50
Area of transfer, S (cm²)	127	138.5	58	39.5
Diffusional tickness, L (cm)	2.30	2.40	1.55	1.28
S_{eff}/S Diagonal orientation	0.264	0.235	0.262	0.277
S_{eff}/S Straight orientation,	0.187	0.166	0.185	0.196

For each experiment, four thermocouples are positioned (see fig. 2.b) such as: (*i*) Below the grain to measure the hot stream, T_1, (*ii*) Through a thin hole, l=1 mm in depth at the inner face to measure the wall temperature, T_2 (*iii*) Into a deeper hole until the center of the sample to measure the bulk temperature, T_3, corresponding to the diffusion length, L, (*iv*) Above the sample to measure the gas temperature, T_4. Note that the volume devoted to the thermocouples is small (less than 2% of the total volume). Consequently we assume that its effect on the conductivity is negligible.

As aforementioned, the thermal apparatus is composed of four thermocouples of type K connected to an acquiring system. The sampling frequency is equal to 1 hertz and the experiment is performed during 10 minutes. An example of the temperature time-traces is presented in the figure 3 for each probe location. A sharp temperature increase is noticeable for T_1 probe. This shows the beginning of the experiment. One can see the heat transfer competition between convection and radiation modes. Indeed, T_1 and T_2 probes locations are close to the hot stream compared to T_4 probe location into the fluid flow.

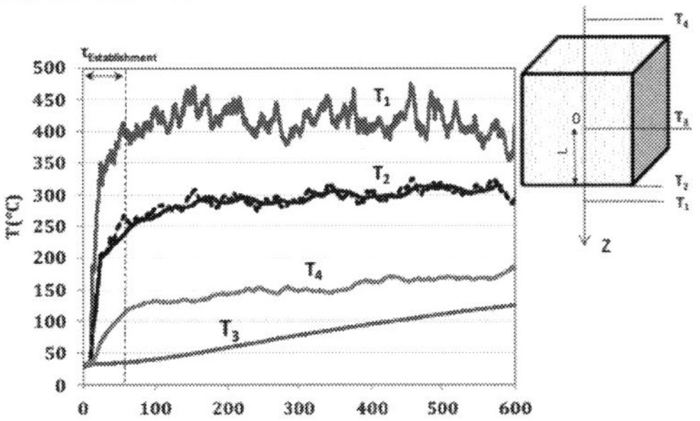

Fig. 3. Experimental apparatus controlled by a burner whose the first thermal probe T_1 is adjusted around a value ranged between 350 and 400 °C {T_1>T_2>T_4>T_3}. The signals measured from probes close to the heat flux are subject of noise imputed to the intense thermal gradient. A sliding-average window is able to filter the fluctuations (see continuous black line).

The time of thermal establishment, $\tau_E = 60\ s$ is necessary to reach a constant temperature T_1 close to 400 °C. T_3 corresponds to the bulk temperature of the granular material; its variation is slow and undisturbed since the kinetic is governed by the thermal diffusion. The elevated temperature fluctuations are filtered in order to obtain a suitable signal to process of the data analysis. Then, according to the temperature variation, different sliding-average windows are applied in order to filter the fluctuations of T_2 (t) as seen in fig. 3.

3 Results and discussion

3.1 Heat Transfer properties

The unsteady heat conduction equation in anisotropic medium can be expressed in dimensionless form as:

$$\frac{\partial T}{\partial t} = a\left(\frac{\partial^2 T}{\partial r^2} + \frac{n}{r}\frac{\partial T}{\partial r}\right) \qquad (10)$$

With n=0, 1 and 2 for slab, infinite cylinder and sphere respectively, and a the thermal diffusivity.

The solution for a slab (with r=X) is solved and given in function of the Fourier number ($Fo = t/aL^2$) such as:

$$T(X,t) = \frac{T_1 - T_X}{T_1 - T_{X,t=0}} = \sum_{i=1}^{n} \frac{2sin\beta_n \cos(\beta_n X/L)}{(\beta_n + sin\beta_n cos\beta_n)} \exp(-B_n{}^2 Fo) \qquad (11)$$

Where B_n is the nth positive root of:

$$\beta \tan \beta = Bi \qquad (12)$$

Consequently, the thermal diffusivity, a, can be assessed by minimizing the relative error between the analytical model calculated from the equation (11) and the temperature measurements (with $T_{X=L} = T_3$) performed for each sample at a given orientation. The average thermal diffusivity of our materials is found equal to 2.90×10^{-7} m²/s with a relative error equal to 6.5 % calculated at Fo>0.2.

Thus, the ratio method has been applied to assess the Biot number of each sample. As reported by [12], it is based on the evaluation of $T(X,t)$ at any location in such way that the equation (11) can be reduced at the first order for the Fourier number superior to 0.2 :

$$\forall X : \frac{T_1 - T_X}{T_1 - T_{X,t=0}} = \frac{2sin\beta_n \cos(\beta_n X/L)}{(\beta_n + sin\beta_n cos\beta_n)} \exp(-\beta_n{}^2 Fo) \qquad (13)$$

By taking into account two positions of measurement into the grains, it comes:

$$\begin{cases} \frac{T_1 - T_{X=0}}{T_1 - T_{X,t=0}} = \frac{2sin\beta_n}{(\beta_n + sin\beta_n cos\beta_n)} \exp(-\beta_n{}^2 Fo) \\ \frac{T_1 - T_{X=L-l}}{T_1 - T_{X,t=0}} = \frac{2sin\beta_n \cos(\beta_n (L-l)/L)}{(\beta_n + sin\beta_n cos\beta_n)} \exp(-\beta_n{}^2 Fo) \end{cases} \qquad (14, 15)$$

With T_1, the inlet temperature condition, $T_{X=0} = T_3$ and $T_{X=L-l} = T_2$

The ratio between (14) and (15) defines:

$$Z = \frac{T_1 - T_2}{T_1 - T_3} = \cos(\beta(L-l)/L) \qquad (16)$$

From a geometrical identification showed in figure 4.a, β is obtained from the equation (16) and the Biot number is approximated from the equation (12). As expected, Bi is found superior to 1 with the major values imputed to the straight orientation. This latter result corresponds slightly to a heat transfer augmentation related to its geometric configuration with respect to the heat flux.

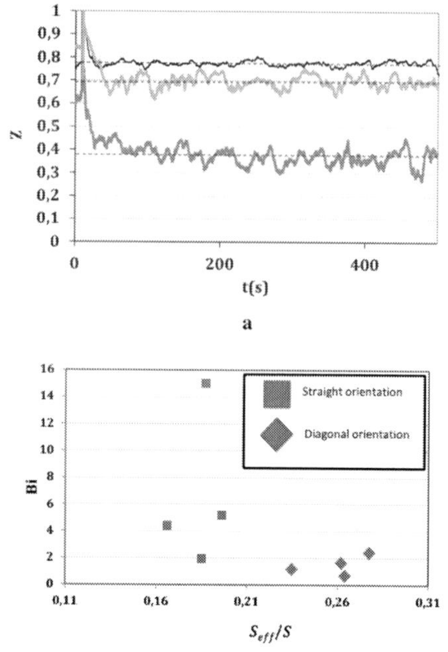

Fig. 4 a. Left: Examples of experimental Z values. **b.** Biot number assessed for each sample.

3.2 Scaling based on the dimensionless surface

To damp the anisotropy effect on the Biot numbers results, a dimensionless criterion is necessary. Thus, the product of the dimensionless temperature T^+ and the shape factor, S^+ is, here, proposed such as:

$$S^+T^+ = \frac{S_{eff}}{S}\frac{|T_{bulk} - T_{wall}|}{T_{wall}} = \frac{S_{eff}}{S}\frac{|T_3 - T_2|}{T_2} \qquad (17)$$

Such dimensionless parameters plotted *versus* time give still large discrepancies (not shown here). Besides, the time must be converted into a dimensionless number. Inspired to the Fourier number, this is defined from the diffusivity coefficient $\{a = \rho c_p/\lambda$ en m²/s$\}$ and the characteristics length, L, corresponding to thermal thickness of diffusion (see table 1):

$$t^+ = \frac{t}{L^2/a} \qquad (18)$$

Consequently, the overall curves $T^+ = f(t^+ = Fo)$ are well-scaled from the Fourier Number as seen in the figure 5.a. Two relationships are proposed to correlate the experimental data respectively at the lowest and the highest Fo numbers:

- At Fo < 0.02 :

$$T^+S^+ = \frac{1}{\alpha} e^{\left[\ln(\alpha.Fo)\frac{\beta}{\alpha^2}\right]^{\beta}} \qquad (19)$$

with $\beta = 5$, $\alpha = 10$ for $0.01 < \mathrm{Fo} < 0.02$, and $4 \leq \alpha < 10$ for $\mathrm{Fo} < 0.01$;

- At Fo > 0.02 :

$$T^+S^+ = \gamma e^{-\delta Fo} \qquad (20)$$

with $0.14 < \gamma < 0.15$, $1 < \delta < 1.5$.

Note that at the lowest Fourier numbers inferior to 0.01, the physical scaling is not representative of the overall experimental data. In this range of the *Fo* number, the transverse diffusional is more sensitive and a 2-D heat conduction model seems required. Above a *Fo* number equal to 0.015, it seems quite reasonable to keep a 1-D model as the equation (10).

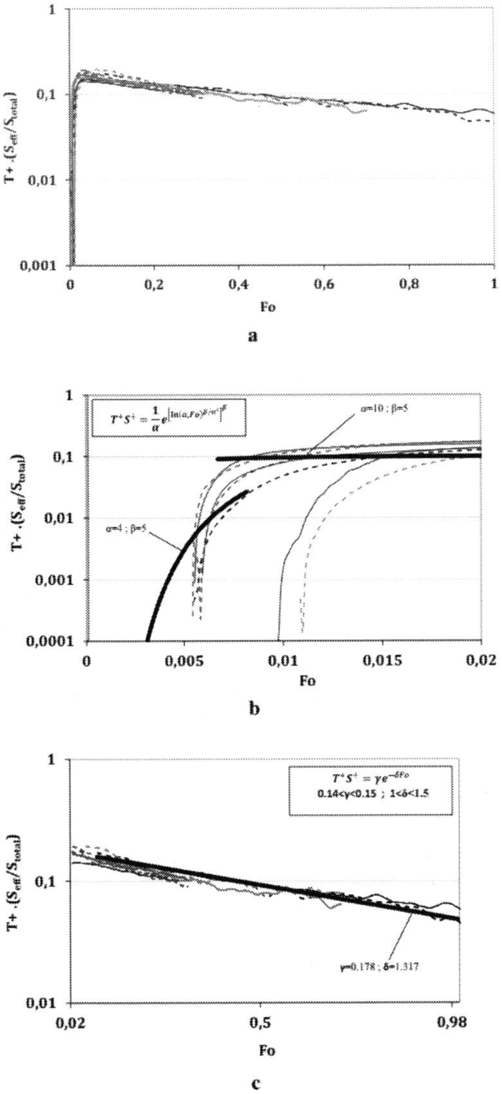

Fig.5.a. Dimensionless representation of the conduction phenomena into non-spherical grain with various orientations. **b.** Focus at the low Fourier number **c.** Focus at the high Fourier number.

4 Conclusions

This work is devoted to the unsteady conduction of cuboid rock materials submitted to an imposed upward heat flux. Two fluid flow orientations have been explored such as the grains, from the same origin, are oriented diagonally or straight with respect to the heat flux. The experimental results show that the *Bi* number is, as expected, superior to 1. Ranged between 0.7 and 15, the highest values of *Bi* are connected to the straight orientation. To damp the geometrical configuration upon the heat transfer coefficients, our results are scaled from new original geometrical parameters based on the effective area of transfer as shape factor, which is found representative of the overall measured materials temperature in terms of $T^+S^+=f(Fo)$. Two correlations are expressed in the range of the *Fo* numbers such as:

- At $Fo < 0.02$:

$$T^+S^+ = \frac{1}{\alpha} e^{\left[\ln(\alpha.\mathrm{Fo})\alpha^{\frac{\beta}{\alpha^2}}\right]^{\beta}} \quad (19)$$

- At $Fo > 0.02$:

$$T^+S^+ = \gamma e^{-\delta \mathrm{Fo}} \quad (20)$$

Note that at the lowest *Fo* number, the fit is much less representative of the experimental data. This difference is imputed to the transversal diffusion, thus the problem cannot be reduced to one-dimensional model.

References

1. Yu Y.S., Liu W.Q., An H., Yang F.S., Wang G.X., Feng B., Zhang Z.X., Rudolph V., Int. J. Greenhouse Gas C., **10**, 510-519 (2012)
2. Raison C.E., Thomas B., Squires A.M., Powder Tech., **169**, 136-142 (2006)
3. Le Guen L., Huchet F., Dumoulin J., Baudru Y., Tamagny P., Appl. Therm. Sci., **54**, 131-139 (2013)
4. Zhou H., Flamant G., Gauthier D., Chem. Eng. Sci., **59**, 4205-4215 (2004).
5. Bluhm-Drenhaus T., Simsek E., Wirtz S., Scherer V., Chem. Eng. Sci., **65**, 2821-2834 (2010)
6. Yan Z., PhD thesis, Université de Poitiers, France (2014)
7. Rouse H., Yih C S., Humphreys H W., "Gravitational convection from boundary source." Tellus **4**, 201-210 (1952).
8. Duan Z., He B., Duan Y., Scientific report **5**, 12304 doi: 10.1038/srep12304 (2015)
9. Richter A., Nikrityuk P. A., Int. J. Heat Mass Transfer **55**, 1343–1354 (2012).
10. Hassani A.V., Hollands K.G.T., J. Heat Transfer **111**, 363-371 (1989)
11. Yovanovich, M.M., ASME HTD **82**, 121-129 (1987).
12. Awuah G.B., Ramaswamy H.S., Simpson B.K., Food Res. Int., 28 **3**, 261-271 (1995)

Jamming by compressing a system of granular crosses

Hu Zheng[1,2,*] *Dong* Wang[2,**] *Jonathan* Barés[2,3,***] *Robert* Behringer[2,****]

[1]*School of Earth Science and Engineering, Hohai University, Nanjing 210098, China*
[2]*Department of Physics, Center for Nonlinear and Complex Systems, Duke University, Durham, NC, USA*
[3]*Laboratoire de Mécanique et de Génie Civil de Montpellier, Université de Montpellier, France*

Abstract.
A disordered stress-free granular packing can be jammed, transformed into a mechanically rigid structure, by increasing the density of particles or by applying shear deformation. The jamming behavior of systems made of 2D circular discs has been investigated in detail, but very little is known about jamming for non-spherical particles, and particularly, non-convex particles. Here, we perform an experimental study on jamming by compression of a system of quasi-2D granular crosses made of photo-elastic crosses. We measure the pressure evolution during cyclic compression and decompression. The Jamming packing fraction of these quasi-2D granular crosses is $\phi_J \simeq 0.475$, which is much smaller than the value $\phi_J \simeq 0.84$ for-2D granular disks. The packing fraction shifts systematically to higher values under compressive cycling, corresponding to systematic shifts in the stress-strain response curves. Associated with these shifts are rotations of the crosses, with minimal changes in their centers of mass.

1 Introduction

Granular materials can be easily found in nature and in industrial applications. They have gas-like, liquid-like or solid-like properties depending on their density or shear stress [1–4]. Many industrial devices exploit these state properties, and a number of geo-hazards involve uncontrolled state changes. For example, in controlled flow in an industrial process, such as in a hopper, gas-like or liquid-like states are desirable [5]. On the other hand, most geo-hazards, like landslide or debris flow occur when a jammed solid state transitions to a liquid-like state [6, 7]. Hence, it is very important to understand the granular phase properties, and the transitions between fluid and solid, *i.e.* unjammed or jammed, states.

Characterizing the jamming transition in disordered granular systems has attracted significant interest [2, 8–12]. Liu and Nagel [8] hypothesized a zero temperature jamming phase diagram. In a space of shear stress, τ and packing fraction, ϕ, they proposed jammed states for $\tau < \tau_Y(\phi)$, where the yield stress curve, $\tau_Y(\phi)$, is a monotonically increasing curve that intersects the ϕ axis at ϕ_J. This proposal was substantiated by O'Hern et al. using numerical simulations of frictionless discs. These studies used protocols that produced macroscopically isotropic states. Recently, Bi et al. [2] showed that for frictional disks, jammed states occur below the frictionless isotropic ϕ_J. They used a protocol consisting of shear strain applied to an initially unjammed (stress-free) isotropic state. In a range $\phi_S < \phi < \phi_J$, they found that increasing shear strain led from unjammed states to fragile and then robustly shear jammed states, where the last two states are characterized by non-zero shear stress. The isotropic jamming point of a two dimensional frictionless disc granular system is $\phi \simeq 0.84$ [9, 13].

However, most of the experiments and simulations of jamming have focused on simple geometry particles like discs and spheres or at least on convex particles. Hence the focus of the present experiments is on non-convex cross-shaped particles. Role of particle shape has been recognized as significant in controlling the microstucture of various granular materials [1, 14–16]. For instance, Athanassidis et al. [17] have shown that the macroscopic mechanical properties of a granular material can be controlled by tuning the grain shape via a 3D printing approach to make different shape particles including tetrahedra, octahedra, and dodecahedra[17]. And several authors have shown that 3D systems of rods [18] or of non-convex particles, such as staples, or star-shaped particles have interlocking or entangled states that are mechanically stable [19–21]. For instance, star-shaped non-convex particles, *i.e.* hexapods are interesting to both granular science and architecture design. These particles form stable granular aggregates by pouring [22]. Zhao et al. [21] conducted laboratory experiments to show the stability and structure of poured hexapod packings by varying the arm length and coefficient of friction between particles. However, experimental methods generally cannot detect the internal details of such packings, or the forces between particles, which

*e-mail: hz64@phy.duke.edu
**e-mail: dw148@phy.edu.ed
***e-mail: jb@jonathan-bares.eu
****e-mail: bob@phy.duke.edu

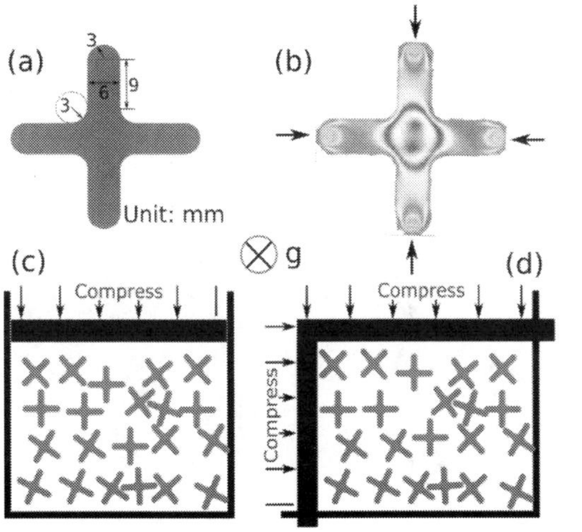

Figure 1. (a): Sketch of cross particle geometry. (b): Photo-elastic response generated by compressing a cross particle along each arm. (c): Schematic of uniaxial compression, using the bi-axial apparatus. (d): Sketch of biaxial compression. The gravity direction is given by **g**.

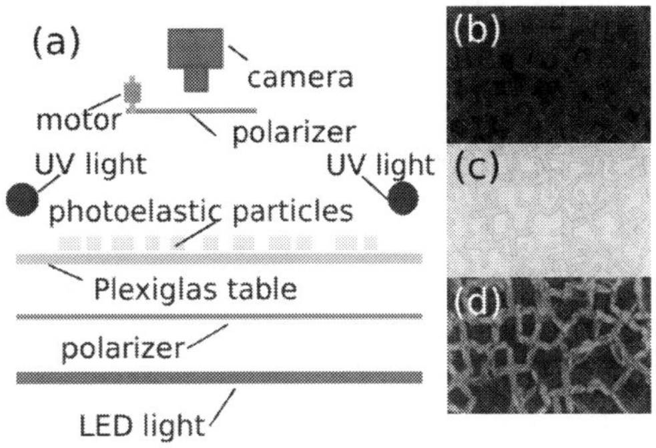

Figure 2. Experimental apparatus and typical experimental image data. (a): Schematic of the experimental apparatus, side view. Light from an LED source passes through a first circular polarizer, then through the transparent base of the apparatus and the particles. A stepper motor rotates a polarizer into or out of the line of sight of the camera, yielding polarized or non-polarized images. (b): Image with only UV light from above. Fluorescent bars on the arms yield the cross orientation. (c): Image with normal light and without the second polarizer. These images are used to detect particle positions. (d): Polarized image used to measure the pressure inside of the each particle.

are important for understanding their mechanical properties.

The photo-elastic method, used here, is an effective experimental technique to study 2D granular materials [10, 23–26] because it provides access to the forces, and hence local *pressure* on each particle. We combine this technique with direct visualiation of particle displacements and rotations. The latter is particulary important for crosses as they gradually evolve to lower stress states at a given ϕ under cyclic compression/decompression. We apply strain with a bi-axial apparatus [2, 27] which allows controlled strains in two directions, as in Fig. 1(c) and (d).

2 Experimental techniques and protocols

2.1 Experimental techniques

Specifically, the experiments are conducted using photo-elastic crosses in a two dimensional bi-axial apparatus. We cyclically compress this system through the isotropic jamming point, and then decompress the system to the original boundary configuration. The geometry of the crosses, cut out of a 6.35mm thick photo-elastic sheet by water-jet cutting, is shown in Fig. 1 (a). The length of each arm is 18*cm*, from the center of cross to the edge. The end of each arm is a semicircle and the corner between two neighbor arms is arc-shaped to avoid force concentration. The radii of the semicircle and arc is 3cm. The friction coefficient between crosses is ~ 0.7.

The particle photo-elastic property allows us to measure the forces acting on them via digital images of the fringe patterns, as typified by Fig. 1 (b). In particular, the concentrated intensity variation close to a contact point gives a measure of the force at a contact, and helps determine the existence of contacts. The density of fringes corresponds to the magnitude of the force: the more fringes, the bigger the force. Fig. 1 (b) shows a photo-elastic image of a cross particle subjected to compression in all four directions, as indicated by the arrows. We use the 'G^2' method [23] as measure of the fringe density, and hence of the forces acting on a particle. The G^2 value is defined as the pixel-wise intensity gradient in the photo-elastic digital image. Moreover, the actual contact force can also be calculated in principle based on a complex fringe pattern-fitting algorithm pioneered by Majmudar [27]. In this paper, we only use G^2 as a measure of the pressure of each particle, which has been demonstrated to be a one-to-one function of pressure at the particle scale [28].

2.2 Experimental apparatus

The experimental apparatus performed bi-axial and uni-axial compression. An illustration of the system is shown in Fig 2 (a). The photo-elastic cross particles are placed on a transparent Plexiglas table in a horizontal plane, such that gravity plays no role in the granular behavior, other than through weak friction with the base (0.36 ± 0.04). Four UV tube lights illuminate the UV-fluorescent ink on the top of each cross (Fig. 2 (b)), allowing us to track the position and orientation of each cross. An LED array provides homogeneous light from below. A circular polarizer sheet on the top of the LED array polarizes the light. A second crossed polarizer is rotated into or out of position in front of a camera, providing normal and polarized images (Fig. 2 (c) and (d)). The camera (Canon® 70D) has a resolution of 3648 × 5472 pixels, corresponding to ~ 31,000 pix^2 per cross.

2.3 Protocol

We compressed the system quasi-statically in steps of 1mm. After each step, we obtained three images, one with only UV light, one polarized and a normal image with light from the LED array. The initial size of the shear box was 47.2 × 39.3cm^2. 270 crosses were placed in the shear box. Before compressing the system, we ensured that there was no initial stress in the system, which meant that no fringes were visible in the polarized image of the initial state.

3 Results and discussion

We focus on the jamming transition where we contrast system properties that were achieved by compressing either uniaxially or isotropically, *i.e.* biaxially. Below, we distinguish these two protocols by unicompression or bicompression. For these two protocols, we determined the contact number, the force chain network, and the system pressure, and used these properties to characterize the jamming transition.

The system pressure (G^2) evolution as a function of the packing fraction for multiple compression cycles is shown in Fig. 3. The blue line corresponds to unicompression and the red one to bicompression. The green region indicates the transition area of the granular crosses from an unjammed to a jammed state. Manifestly, the jamming point is not a unique function of packing fraction, and it also depends on the compression protocol. The difference between unicompression and bicompression can be seen from the inset images in Fig. 3. The force chain network generated by unicompressing is more anisotropic than the bicompression case, reflecting the more anisotropic character of unicompression. The difficulty of rearrangement in the cross packing also caused a relative fast G^2 development by unicompressing with a slope of about 7.8. For the bicompression system, the slope of G^2 is about 7.2, and the system evolution is smoother.

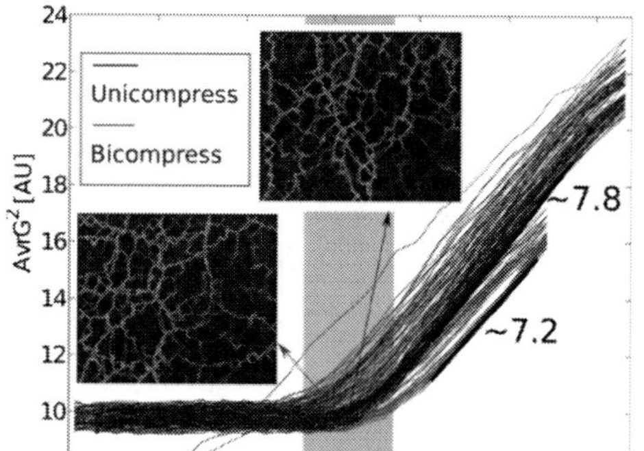

Figure 3. System pressure, expressed by G^2 in arbitrary units, *vs.* compressive strain for multiple cycles. Insets: Image indicated by red arrow is a photo-elastic image at the jamming point for unicompressing while the one indicated by blue arrow is for bicompression. The force chain networks of the two images indicate that a more isotropic network occurs for the bicompression case. The green area shows the jamming transition zone.

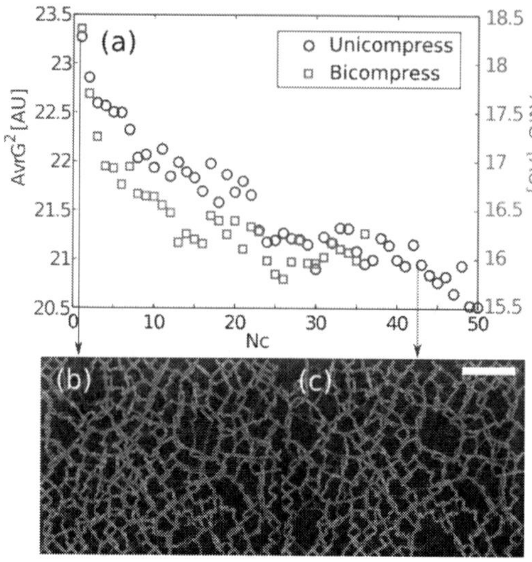

Figure 4. (a): Maximum G^2 at the highest packing fraction of each compression circle. Note that the maximum compressions and vertical scales are different for the two compression protocols. The maximum G^2 decays with N_c, the number of compression cycles for both experimental methods. Alternatively, this can be seen as an increase in ϕ_J as a function of N_c. (b): Photo-elastic image when the system firstly reached the maximum packing fraction, ϕ_{max} = 48%. (c): Photo-elastic image at the maximum packing fraction after 43 shearing cycles. The white bar in (b) indicates 10cm. (b) and (c) share the same scale.

The crosses move little during cyclic compression, G^2 decays over multiple cycles, as seen in 2D disc system [13] and in simulations of frictionless spheres. However, G^2 appears to reach a plateau, within moderate fluctuations, for large enough N_c for bicompression, although not necessarily for unicompressions. Fig. 4 (b) and (c) show the force chain networks at the packing fraction, ϕ = 48%. The photo-elastic response at cycle 43 is slightly weaker than after the first cycle.

We compared the position and orientation of each cross in the unicompression system, as shown in Fig. 5, to see that the position of each particle changes very little, even after 43 compression cycles; consequently, the coarse-grain density also changes little over the 43 cycles. However, the particle orientations change substantially. It is reasonable to conclude that the pressure decay at the maximum packing fraction is strongly correlated with changes in the particle orientations.

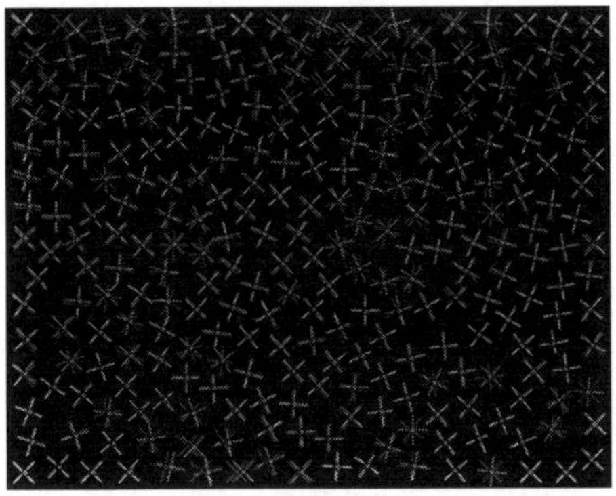

Figure 5. Positions and orientations of the crosses for unicompression initially (red crosses) and after 43 compression cycles (green crosses).

4 Conclusion

We have experimentally investigated the jamming transition of a packing formed from 2D photoelastic crosses subject to multiple cycles of compression that is either uniaxial or biaxial. The jamming point of this system is between 47.5% and 48.0%, and the specific jamming density is sensitive to details of the compression protocol, including the cycle number and whether the compression is unixial of biaxial. The rate of increase of G^2 with ϕ is higher for unicompression than for bicompression. Several measures are sensitive to the cycle number, including the jamming point and the pressure at a given packing fraction – two properties that are correlated. There is little change in the positions of individual particles, but there is significant rotation of crosses during the course of multiple cycles, and this rotation provides a mechanism for stress relaxation.

Acknowledgements

This work is supported by NSF Grant No.DMR1206351, NASA Grant No. NNX15AD38G, the W.M. Keck Foundation and a RT-MRSEC fellowship. HZ thanks NSFC Grant No. 41672256 and the NSF(Jiangsu, China) Grant No. BK20140845 for financial support.

References

[1] H.M. Jaeger, S.R. Nagel, R.P. Behringer, Reviews of Modern Physics **68**, 1259 (1996)

[2] D. Bi, J. Zhang, B. Chakraborty, R.P. Behringer, Nature **480**, 355 (2011)

[3] M. van Hecke, Journal of Physics: Condensed Matter **22**, 033101 (2010)

[4] Alexander O.N. Siemens, Martin van Hecke, Physica A **389**, 4255 (2010)

[5] J. Tang, R.P. Behringer, Europhysics Letters **114**, 34002 (2016)

[6] G.Q. Chen, R.Q. Huang, Q. Xu, T. Li, M. Zhu, Journal of Mountain Science **10**, 532 (2013)

[7] H. Zheng, Y. Huang, Environmental Earth Sciences **73**, 4853 (2015)

[8] A.J. Liu, S.R. Nagel, Nature **396**, 21 (1998)

[9] C.S. O'Hern, L.E. Silbert, A.J. Liu, S.R. Nagel, PHYSICAL REVIEW E **68**, 011306 (2003)

[10] R. Behringer, D. Bi, B. Chakraborty, S. Henkes, R.R.Hartley, Physical Review Letters **101**, 268301 (2008)

[11] J. Ren, J.A. Dijksman, R.P. Behringer, Physical Review Letters **110**, 018302 (2013)

[12] H. Zheng, J.A. Dijksman, R.P. Behringer, Europhysics Letters **107**, 34005 (2014)

[13] T. Majmudar, M. Sperl, S. Luding, R. Behringer, Physical review letters **95**, 058001 (2007)

[14] J. Duran, *Sands, powdersm and Grains: An Introduction to Physics of Granular Materials* (Springer, Berlin, 1999)

[15] D.M. Wood, *Soil Behavior and Critical State Soil Mechanics* (Cambridge University Press, Cambridge, 1990)

[16] E. Azéma, F. Radjaï, R. Peyroux, G. Saussine, Physical Review E **76**, 011301 (2007)

[17] A. Athanassiadis, M. Miskin, P. Kaplan, N. Rodenberg, S. Lee, J. Merritt, E. Brown, J. Amend, H. Lipson, H. Jaeger, Soft Matter **10**, 48 (201)

[18] K. Desmond, S.V. Franklin, PHYSICAL REVIEW E **73**, 031306 (2006)

[19] C. Schreck, N. Xu, O'HernaC.S., Soft Matter **6**, 2960 (2010)

[20] T. Marschall, S. Franklin, S. Teitel, Granular Matter **17**, 121 (2015)

[21] Y. Zhao, Kevin Liu, Matthew Zheng, Jonathan Barés, Karola Dierichs, Achim Menges, Robert P. Behringer, Granular Matter **18**, 1 (2016)

[22] K. Dierichs, A. Menges, Architectural Design **85**, 86 (2015)

[23] D. Howell, R.P. Behringer, C. Veje, Physical Review Letter **82**, 5241 (1999)

[24] J. Geng, D. Howell, E. Longhi, R.P. Behringer, G. Reydellet, L. Vanel, E. Clément, S. Luding, Physical Review Letter **87**, 035506 (2001)

[25] A.H. Clark, R.P. Behringer, Europhysics Letters **101**, 64001 (2013)

[26] A.H. Clark, A.J. Petersen, R.P. Behringer, Physical Review E **89**, 012201 (2014)

[27] T.S. Majmudar, R.P. Behringer, Nature **435**, 1079 (2005)

[28] M. Cox, D. Wang, J. Barés, R.P. Behringer, Europhysics Letters **115**, 64003 (2016)

EPJ Web of Conferences 140, 09045 (2017) DOI: 10.1051/epjconf/201714009045
Powders & Grains 2017

Shear-induced organization of forces in dense suspensions: signatures of discontinuous shear thickening

Sumantra Sarkar[1,*], *Elan* Shatoff[2], *Kabir* Ramola[2], *Romain* Mari[3], *Jeffrey* Morris[4,5] and *Bulbul* Chakraborty[2,**]

[1] *Physics of Living Systems, MIT, 400 Technology Square, Cambridge, MA 02139, USA*
[2] *Martin Fisher School of Physics, Brandeis University, Waltham, MA 02454, USA*
[3] *Department of Applied Mathematics and Theoretical Physics, Centre for Mathematical Sciences, University of Cambridge, Cambridge CB3 0WA, UK*
[4] *Benjamin Levich Institute, City College of New York, New York, NY 10031, USA*
[5] *Department of Chemical Engineering, City College of New York, New York, NY 10031, USA*

Abstract. Dense suspensions can exhibit an abrupt change in their viscosity in response to increasing shear rate. The origin of this discontinuous shear thickening (DST) has been ascribed to the transformation of lubricated contacts to frictional, particle-on-particle contacts. Recent research on the flowing and jamming behavior of dense suspensions has explored the intersection of ideas from granular physics and Stokesian fluid dynamics to better understand this transition from lubricated to frictional rheology. DST is reminiscent of classical phase transitions, and a key question is how interactions between the microscopic constituents give rise to a macroscopic transition. In this paper, we extend a formalism that has proven to be successful in understanding shear jamming of dry grains to dense suspensions. Quantitative analysis of the collective evolution of the contact-force network accompanying the DST transition demonstrates clear changes in the distribution of microscopic variables, and leads to the identification of an "order parameter" characterizing DST.

1 Introduction

A remarkable property of dense stabilized suspensions of particles in the tens of nanometers to tens of micrometer size range is that they can abruptly transform from a low to a high viscosity phase (or even a solid-like phase) with increasing applied stress [1–5]. Under steady shear, these suspensions undergo a discontinuous shear thickening (DST) transition [1–3]. A confluence of ideas from the granular and fluid dynamics communities has led to new understanding and a new set of questions regarding the flowing and jamming behavior of dense suspensions [3, 6, 7]. In this paper, we analyze the DST transition using a force-space representation [8] that was originally developed for granular systems. We extend the formalism to suspensions and identify distinct, quantitative signatures of the DST transition in this representation.

There is developing consensus that particle contact and friction play a crucial role in dense suspension rheology [9, 10]. Combining fluid mechanical interactions with contact friction between particles has been shown to capture critical features of both DST [11–14] and shear induced jamming (SJ) [15]. Furthermore, clear connections have been made between a dynamic shear jamming front and impact-driven solidification [5]. The appearance of non-monotonic flow curves in numerical models of DST [11, 12] is reminiscent of classical phase transitions. As in classical phase transitions, the key question is how interactions between the microscopic constituents give rise to a macroscopic transition.

The essential idea underlying the jamming-based rheology model [3, 6] for DST is that it requires a shear-stress driven transition from the lubricated to the frictional branch of the viscosity. Reduction of the jamming packing fraction through the creation of frictional contacts [7] yields a higher viscosity at any ϕ. If shearing changes the relative fractions of frictional and lubricated contacts, the suspension can transition to the frictional branch. We are interested in understanding the nature of the collective reorganization of particles that gives rise to this change in rheology. Simulations show that DST is accompanied by significant changes in the network of frictional contacts and contact forces with minimal changes in structure factors and pair correlation functions [2]. We present a quantitative analysis of changes in the force network by studying the organization in "force space", which is dual to the positional network of grains in a sense to be defined below [8, 16]. The density of points in this space naturally decreases as the stress increases since the number of contacts is roughly constant, but more importantly we show that the rheological behavior characterizing DST is accompanied by changes in the form of the density distribution of points in this space. The change in distribution is a consequence of the constraints of local mechanical equi-

*e-mail: sumantra@mit.edu
**e-mail: bulbul@brandeis.edu

librium applied to grains as contacts change from lubricated to frictional.

2 Model

The combination of frictional and viscous interactions between surfaces of suspended particles has recently been addressed through simulations [11–14] that use a hybridization [11, 12] of Stokesian Dynamics (SD) with discrete-element modeling (DEM) [17]. The SD method is simplified by only considering the near-field, or lubrication, hydrodynamic interactions. To allow contact, the lubrication resistance singularity is cut off at a distance h_c between particle surfaces. The DEM approach uses a Coulomb friction law: for $F_{\tan} \leq \mu F_{norm}$, where μ is the interparticle friction coefficient, there is no slipping of the contact, while there is slip when the friction is fully mobilized for $F_{\tan} = \mu F_{norm}$. To model the transition from lubricated to frictional contacts, a repulsive force, representative of electrostatic repulsion or steric hindrance due to a grafted layer, is used. For low stress or low shear rate, the repulsive force maintains surfaces separated at $h > h_c$ and thus interactions are lubricated, as if $\mu = 0$, while at large enough stress to push particles to $h < h_c$, frictional contacts are activated. 2D simulations of this model capture the progressively steeper shear thickening with increasing ϕ (Fig. 2a) seen in experiments.

3 Force Balance in 2D Suspensions

Owing to the size ($\leq 10\mu m$) of the particles and the shear rates at which DST is observed (usually around $1s^{-1}$), it is a very low Stokes number phenomenon, and particle inertia can be neglected. As a consequence, the constraints of mechanical equilibrium are strictly obeyed at both the local and global levels. We analyze data obtained from stress-controlled simulations of a bidisperse suspension of particles using the hybrid SD-DEM model [11, 12] in 2D. Each disk in the suspension experiences a hydrodynamic force, $\vec{F}_H$, a short range repulsive force, $\vec{F}_R$, and frictional contact force, $\vec{F}_C$. The hydrodynamic component includes a non pairwise force (Stokes drag) and pairwise forces (lubrication), whereas $\vec{F}_R$ and $\vec{F}_C$ are pairwise interactions. The force balance equation for each disk is $\vec{F}_H + \vec{F}_R + \vec{F}_C = 0$. The repulsive force is modeled as a short range electrostatic repulsion,

$$\vec{F}_R^{(i,j)} \propto \begin{cases} -\frac{a_i a_j}{a_i+a_j} e^{-\kappa h^{(i,j)}} & h^{(i,j)} \geq 0 \\ -\frac{a_i a_j}{a_i+a_j} & h^{(i,j)} \leq 0\,, \end{cases} \quad (1)$$

with Debye length, $1/\kappa$, much smaller than the radius of either particle, $a = (a_i$ or $a_j)$. Adjacent grains that overcome this repulsive force form frictional contacts and contribute to the $\vec{F}_C$. Thus, $|F_R|$ sets a stress scale, $\sigma_0 \approx \frac{|F_R|}{a^2}$, whose interplay with the hydrodynamic stress scale determines the stress (shear rate) dependent rheology.

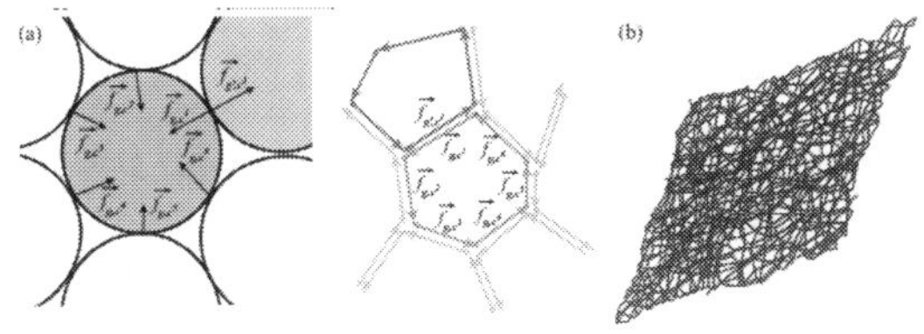

Figure 1. (a) Construction of force tiles: The vectors $\vec{f}_{g,c}$ form a closed polygon (light brown) (Eq. 2), and there is strict edge sharing between grains g and g' (Eq. 3). (b) A force tiling constructed from a configuration at $\sigma = \sigma_{xy}/\sigma_0 = 0.1$. The shape and size of the bounding parallelogram is determined by the components of the imposed stress tensor: σ controls the opening angle. The number of tiles is equal to the number of particles in the force-bearing network

3.1 Force tilings in granular materials

In packings of dry grains, labeling the contacts of grain g by (g, c) (Fig. 1), the force balance condition is:

$$\vec{F}_C \equiv \sum_c \vec{f}_{g,c} = 0, \quad (2)$$

where the sum is taken over all the contacts $\{c\}$ for a given grain g. In addition, Newton's *third law* states:

$$\vec{f}_{g,c} = -\vec{f}_{g',c}. \quad (3)$$

These constraints can be used to construct a representation known as the "force tile" representation or Maxwell-Cremona tiles [18]. Eq. (2) can be represented as a *closed polygon*, if the sum is taken *cyclically* over the contacts for each grain. Eq. 3 dictates that the tiles of touching grains have to share an edge. We can, therefore, construct a *force tiling*, as illustrated in Fig. 1. The vertices of the force tiling form a point pattern. In previous studies of the shear-jamming process in dry granular systems, it has been shown that the onset of shear-jamming is marked by distinct changes in this point pattern [8, 16].

3.2 Generalization to Suspensions

The Stokes drag is a non-pairwise force that prevents individual force tiles from closing: $\sum_c \vec{f}_{g,c} = \vec{f}_{Stokes}$. In addition, suspended grains experience pairwise, non-contact forces. The latter can be included by extending the definition of the contact network, $\{g, c\}$, and $\{\vec{f}_{g,c}\}$ to include $\vec{F}_R$, $\vec{F}_c$ and the pairwise lubrication force. Given these, and $\vec{f}_{Stokes}$, there is a unique way of obtaining effective contact forces such that all force tiles close [19]. We use an iterative algorithm [8] in which the contact forces are modified to ensure that Eqs. 2 and 3 are satisfied at every iteration. The final solution is unique up to global translation.

4 Results

We construct force tilings for each member of the ensemble of steady states over a range of shear stresses,

$\sigma \equiv \sigma_{xy}/\sigma_0$ and at packing fractions, $\phi = 0.76$ and 0.78. The point patterns of force-tiling vertices from three representative configuration at σ below and above the DST transition for $\phi = 0.78$ are shown in Fig. 2. The lengths of the boundaries increase with increasing σ, and they have been scaled by σ in order to aid in the visualization of the patterns. It should be pointed out that the shape of the bounding boxes fluctuate even at a given σ, which fixes only the opening angle. Fig. 2 (b) illustrates the effect of increasing σ on the density of vertices. This density is observed to follow $\rho_v \propto 1/\sigma^2$ for up to six decades, and occurs as ϕ is held constant with no change in the real-space density. Given the dramatic change in ρ_v across the DST, one could expect significant changes in the *patterns* of vertices. We have employed a clustering analysis of these point patterns based on purely local densities. In the next section we demonstrate that there is a clear signature of the DST transition in the clustering properties of the vertices.

4.1 Point Patterns of Vertices

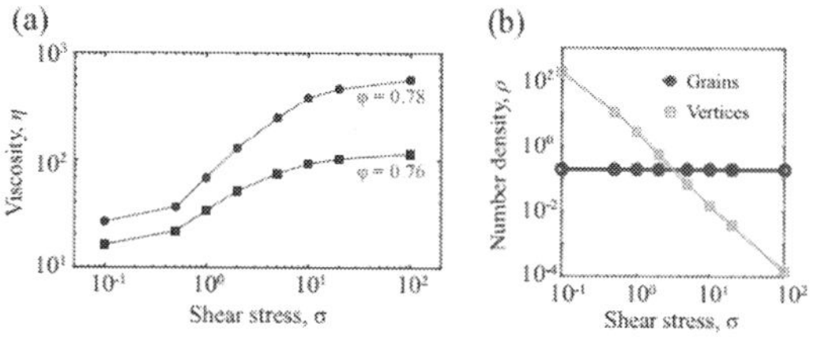

Figure 2. (a) Viscosity as a function of shear stress in stress-controlled simulations of a non-Brownian suspension. (b) Density of grains and vertices as a function of shear stress.

4.1.1 Density based clustering analysis

If the repulsive force between two contacts is overcome to create a frictional contact, the constraints of mechanical equilibrium will necessarily induce a change in the shape of the force tiles, and hence a collective reorganization of the point pattern of vertices. We performed a density based clustering analysis of the vertices of force tiles using the DBSCAN algorithm [20] to detect these changes. In DBSCAN, two points belong to the same cluster if their distance is less than a probing length scale, l (which here will dimensionally be a stress scale). The connected set of all such points define a unique cluster, which are separated from each other by distances greater than l. In usual implementations of DBSCAN, clusters are determined using an optimum probing radius.

We use DBSCAN to analyze how the clustering pattern changes as the probing length scale is varied. For a large enough length scale, all points will belong to one cluster and for small enough length scales, each point will be its own cluster. Our algorithm probes the density distribution of point patterns at different length scales by monitoring the number of clusters, $N_c(l)$, as a function of the probe length, l. For a point pattern with uniform density, $N_c(l)$ decreases continuously with l. In a periodic lattice, where the distance distribution of nearest neighbors is a delta function, $N_c(l)$ exhibits a jump discontinuity at the lattice spacing. For a complex pattern, we expect $N_c(l)$ to show significant changes in its derivatives at length scales where the distance distribution has structure.

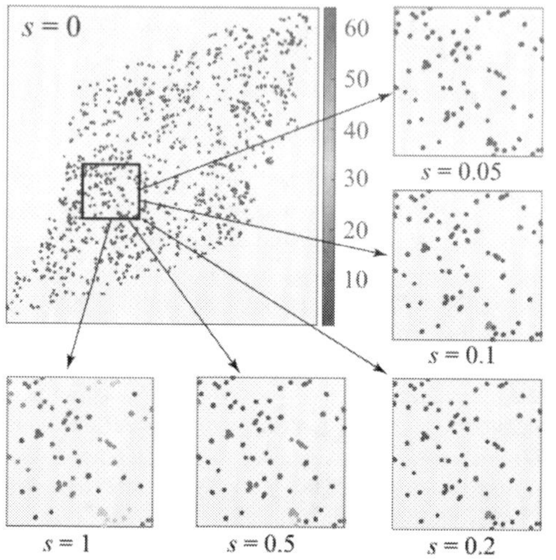

Figure 3. Illustration of DBSCAN analysis. Left: Sample tiling at $\phi = 0.78$, $\sigma = 100$ at $s = 0$: all clusters are of size 1 and $N_c(s = 0) = N_v$, the number of vertices. Right panel: Zoomed in point pattern analyzed with increasing probe length s. The clusters are colored by their size as indicated by the color bar.

In applying the clustering analysis to point patterns in force tilings, the probing "length scale" is to be interpreted as an isotropic probing stress scale, σ_{probe}. We measure the variation in the number of clusters $N_c(s)$ as the dimensionless probe stress, $s = \sigma_{probe}/\sigma$, is increased (Fig. 3). Identifying significant variations in $N_c(s)$ yields characteristic stress scales in the patterns, just as in real-space.

4.1.2 Clustering Analysis of DST

We calculate $N_c(s)$ for each force tiling and then ensemble average over tilings sampled at a given ϕ and σ. Fig. 4 (a) shows that at $\phi = 0.78$, and σ below the DST transition, $N_c(s)$ decays continuously with s, indicating the lack of any characteristic scale in the distribution of distances between the vertices. We note that vertices that are close to each other are not necessarily connected by an edge in the tiling. Therefore, the distance distribution is not equivalent to the contact force distribution.

Above the DST transition, we can clearly identify three different decay regimes in $N_c(s)$, with a plateau-like structure clearly identifiable in the derivative, $dN_c(s)/ds$ shown in Fig. 4 (b). As seen from Fig. 4 (c), this structure in $dN_c(s)/ds$ is much more pronounced for $\phi = 0.78$ than it is for $\phi = 0.76$. The observed structure in $N_c(s)$ indicates that there are density inhomogeneities in the pattern of the vertices characterized by both s^*, marking the onset of the plateau, and s^{**}, the edge of the plateau. The abrupt change in $dN_c(s)/ds$ to a much smaller value at s^* implies that incremental changes in s beyond s^* incorporate the "noise points": isolated vertices that lie in low-density regions. Beyond s^{**}, $dN_c(s)/ds$ starts decreasing as s increases indicating that clusters separated by $s \geq s^{**}$ are beginning to merge. Fig. 3 illustrates how the cluster pattern changes with s. The difference $\Delta s = s^{**} - s^*$ provides a measure of the separation of high density regions in force tilings: the more pronounced the plateau the sharper is the distinction between high and low density regions.

Fig. 4 (d) demonstrates that Δs increases sharply at a characteristic value of σ that depends on ϕ. Moreover the saturation value of Δs decreases with decreasing ϕ, suggesting that Δs can be considered as an "order parameter" characterizing DST.

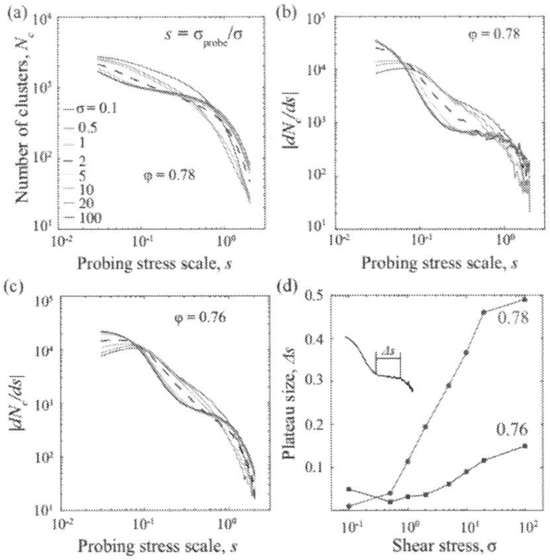

Figure 4. Top: Cluster analysis at $\phi = 0.78$: Number of clusters, N_c, versus $s = \sigma_{\text{probe}}/\sigma$, at different values of σ. b) dN_c/ds vs s. Each curve is obtained by averaging over fifty steady state configurations. The black dashed lines highlight the behavior change at $\sigma \approx 2$. Bottom: (c) dN_c/ds vs s for $\sigma = 0.1$ and 100 at $\phi = 0.76$ compared to $\phi = 0.78$. (d) Plateau size Δs (See inset for the definition) as a function of the control shear stress σ for packing fractions 0.76 and 0.78 .

5 Discussion

Using a clustering analysis, we have demonstrated that the rheological changes at DST are accompanied by a collective reorganization in the space of forces. This collective response is necessitated by the constraints of mechanical equilibrium, applied at the local grain level, as lubrication forces change to frictional contact forces. We identify a characteristic stress scale in the pattern, Δs, from the plateau in $dN_c(s)/ds$. This plateau suggests that the vertices are clustered into clumps with an approximately uniform density of points with the clumps separated by $\approx \Delta s$. Equilibrium clumped phases are observed in systems with ultrasoft interaction potentials [21, 22]. Indeed, our preliminary DBSCAN analysis of the clumped phase shows a plateau structure remarkably similar to Fig. 4 (b). We are currently exploring this connection further.

Acknowledgements

The authors acknowledge useful conversations with Jetin Thomas and Abhinendra Singh. SS, KR and BC acknowledge the support of NSF-DMR 1409093 and the W. M. Keck Foundation. JM and BC acknowledge NSF-CBET 1605283.

References

[1] E. Brown, H.M. Jaeger, Rep Prog Phys **77**, 046602 (2014)
[2] R. Mari, et. al., Journal of Rheology **58**, 1693 (2014)
[3] M. Wyart, M. Cates, Phys Rev Lett **111**, 218301 (2013)
[4] S.R. Waitukaitis, H.M. Jaeger, Nature **487**, 205 (2012)
[5] E. Han, I.R. Peters, H.M. Jaeger, Nat Commun **7**, 12243 (2016)
[6] M.E. Cates, M. Wyart, Rheologica Acta pp. 1–10 (2014)
[7] E. Guazzelli, J.F. Morris, *A Physical Introduction to Suspension Dynamics* (Cambridge University Press, 2011)
[8] S. Sarkar, et. al., Phys Rev E **93**, 042901 (2016)
[9] E. Brown, H.M. Jaeger, J. Rheol. **56**, 875 (2012)
[10] D. Lootens, et. al., Phys. Rev. Lett. **95**, 268302 (2005)
[11] R. Seto, et. al., Phys. Rev. Lett. **111**, 218301 (2013)
[12] R. Mari, et. al., J. Rheol. **58**, 1693 (2014)
[13] C. Heussinger, Phys. Rev. E **88**, 050201(R) (2013)
[14] N. Fernandez, et. al., Phys. Rev. Lett. **111**, 108301 (2013)
[15] R. Mari, et. al., Phys. Rev. E **91**, 052302 (2015)
[16] S. Sarkar, et. al., Phys. Rev. Lett. **111**, 068301 (2013)
[17] P.A. Cundall, O.D.L. Strack, Geotechnique **29**, 47 (1979)
[18] B.P. Tighe, T.J.H. Vlugt, Journal of Statistical Mechanics: Theory and Experiment **2011**, P04002 (2011)
[19] K. Ramola, B. Chakraborty, submitted to Journal of Statistical Physics
[20] M. Ester,et. al., , AAAI Press, 226, (1996)
[21] J.B. Delfau, et. al., Phys. Rev. E **94**, 042120 (2016)
[22] W. Klein, et. al., Physica A **205**, 738 (1994)

 DOI: 10.1051/epjconf/201714015017

Numerical insight into the micromechanics of jet erosion of a cohesive granular material

Pablo Cuéllar[1,2*], *Zeyd* Benseghier[1], *Li-Hua* Luu[1], *Stéphane* Bonelli[1], *Jean-Yves* Delenne[3], *Farhang* Radjaï[4,5], and *Pierre* Philippe[1]

[1]IRSTEA, UR RECOVER, 3275 route de Cézanne, 13182 Aix-en-Provence Cedex 5, France
[2]BAM, Division 7.2 Buildings and Structures, Unter den Eichen 87, 12205 Berlin, Germany
[3]UMR IATE, INRA – CIRAD - Montpellier Supagro, 2 place Pierre Viala, 34060 Montpellier, France
[4]LMGC, CNRS - Université Montpellier 2, Place Eugène Bataillon, 34095 Montpellier, France
[5]MSE2, UMI 3466 CNRS-MIT, CEE, MIT, 77 Massachusetts Avenue, Cambridge CA 02139, USA

Abstract. Here we investigate the physical mechanisms behind the surface erosion of a cohesive granular soil induced by an impinging jet by means of numerical simulations coupling fluid and grains at the microscale. The 2D numerical model combines the Discrete Element and Lattice Boltzmann methods (DEM-LBM) and accounts for the granular cohesion with a contact model featuring a paraboloidal yield surface. Here we review first the hydrodynamical conditions imposed by the fluid jet on a solid granular packing, turning then the attention to the impact of cohesion on the erosion kinetics. Finally, the use of an additional subcritical debonding damage model based on the work of Silvani and co-workers provides a novel insight into the internal solicitation of the cohesive granular sample by the impinging jet.

1 Introduction

The physical phenomenon of surface erosion afflicts often the earthen hydraulic constructions such as earth-dams and levees [1]. In this context, the Jet Erosion Test (JET) has proved useful for the assessment of the sensitivity of soils to the occurrence of surface erosion. The interpretation of results can be made in terms of the critical threshold of hydrodynamic shear stress τ_c and the erosion modulus k_d that quantifies the erosion kinetics [2]. A simple erosion law is thereby commonly adopted as $E = k_d(\tau - \tau_c)$ where E is the erosion rate and τ is the hydraulic shear stress. This implies a description of the erosion evolution with one single variable, the shear stress, which is averaged over time and space aiming to represent the hydrodynamic conditions at the fluid-solid interface. However, this is a rough simplification of the complex conditions at the surface under an impinging jet, where actually the shear stress should be zero right at the impingement point.

This paper aims to provide a micromechanical insight into the mechanisms taking place during the jet erosion of a cohesive granular material. Firstly, we introduce a numerical model considering on the one hand a model for cohesive intergranular bonds and then a suitable extension for transient subcritical debonding processes (i.e. damage). Then, the results from a parametric study varying the bond strength are put forward. Finally, the erosion mechanisms within the granular sample at the onset of erosion are briefly discussed in the light of preliminary results with the extended damage model.

2 Numerical methods

The analysis of fluid-solid interactions at the micro-scale is performed here numerically, combining the Lattice Boltzmann Method (LBM) for the fluid analysis with the Discrete Element Method (DEM) for the solid particles (see further applications for instance in [3,4]).

2.1 Solid mechanics

A Molecular Dynamics method has been used here to describe the granular soil as a two-dimensional assembly of round particles whose trajectories are governed by Newton's equations of motion. The interaction between particles for the case of purely frictional contacts is formulated in terms of an interaction force F with normal and shear components and an interaction moment M applied at the common contact point. The normal force F_n depends on the local interpenetration δ_n through a viscoelastic relationship featuring a normal stiffness k_n and damping coefficient η_n. On the other hand, a viscous-regularized form of Coulomb's law is used here to compute the shear force arising at a frictional contact in dependence of the sliding velocity $\dot{\delta}_s$ by means of a static friction coefficient μ and the viscous coefficient of shear regularization k_s. Finally, the interaction moment is defined by the shear force with the particle's radius as lever arm and a rolling friction component that depends on the relative velocity of rotation through a rolling friction coefficient μ_ω and a coefficient of regularization k_ω.

* Corresponding author: pablo.cuellar.moro@gmail.com

2.1.1 Intergranular cohesion

For the analysis of cohesive granular samples we consider that all particles initially at contact are bound by a solid bridge with an elastic rheology characterized by the normal and shear bond stiffnesses $k_{n,b}$ and $k_{s,b}$. An elastoplastic model with paraboloidal yield surface g_u in the space of contact forces (a three-dimensional space in terms of F_n, F_s and M) provides here the limits of cohesion allowing for tensile normal forces as long as they remain in the interior of the yield surface (Figure 1). Whenever the contact forces reach or trespass the yield surface, the cohesive bond is broken and the contact becomes purely frictional. For convenience, the single thresholds for normal, shear and moment interactions C_n, C_s and M_b have all been set to depend here only on a single parameter $C = C_n = 2C_s = M_b/(2D_{mean})$, thereby fixing the shape of the paraboloid for different degrees of cohesion. This way, the parameter C represents the strength, or degree of cementation, of the solid bond and permits the characterization of the relative bond strength in a polydisperse assembly by defining a dimensionless number $B = C/(\Delta\rho\ g\ V)$ as the ratio of the bond cohesion C to the particle's own buoyant weight.

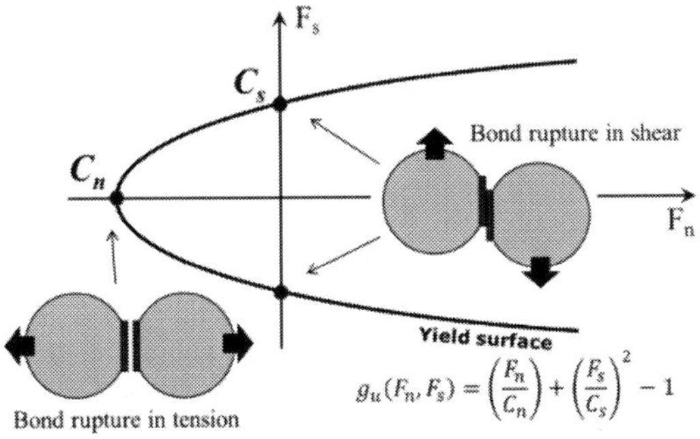

Fig. 1. Section of the yield surface of cohesive bonds in the plane of normal and shear interaction forces, after [5]. The third dimension of interaction moments is not depicted here.

In general, it can be noted that $B \geq 0$ and that solid bonds with $B < 1$ tend to be unstable and short-lived since any slight rearrangement of the assembly under its own weight has the potential to cause bond rupture. Here it was further observed that the granular assemblies remained completely bonded under gravity whenever $B \geq 3$ for all particles in the sample.

2.1.2 Damage model. Subcritical debonding

The cohesion model presented so far is suitable to reproduce the cohesive response to instantaneous solicitations such as dynamic impacts and transient peaks of applied forces for instance, but it lacks itself of transience beyond the dual state-disjunctive (from intact to broken bond state). A characteristic time can however be introduced for instance by means of the subcritical debonding concept (see e.g. [6]) which permits the possibility of a progressive degradation of the cohesive strength for solicitations contained within the yield surface (i.e. subcritical solicitations). To this end, an additional surface, the damage surface, is defined in the interior of the failure surface discriminating the bond solicitations that induce damage (space of interaction forces bounded by the yield and damage surfaces, depicted in pink colour in Figure 2) from those that cause no bond degradation (space contained within the damage surface, shown in green colour). All other solicitations, outside the yield surface, cause an immediate rupture of the cohesive bond, as in the original model.

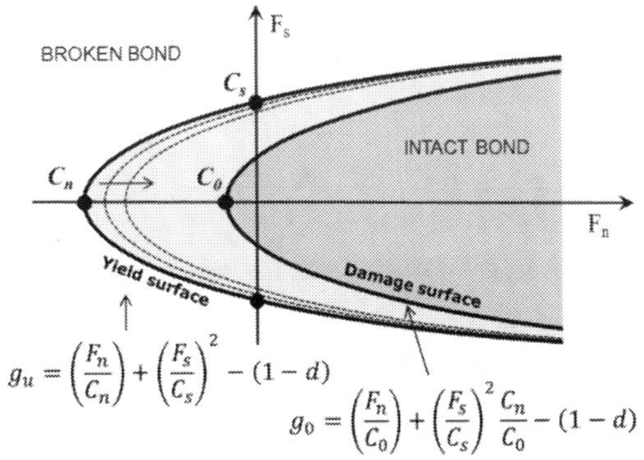

Fig. 2. Application of the subcritical debonding concept after [6] to the cohesion model of Delenne and coworkers [5]. The third dimension (moments) has been omitted in the figure.

The transience can then be introduced into the model by any suitable definition of a damage variable d and its time derivative, for instance as follows:

$$\dot{d} = \frac{\langle g_0(F_n, F_s, M, d)\rangle}{\eta C_0} \quad (1)$$

$$0 \leq d \leq d_c(F_n, F_s, M) \leq 1 \quad (2)$$

where $\langle\cdot\rangle$ denotes the MacCauley brackets ($\langle x\rangle = x$ if $x \geq 0$; $\langle x\rangle = 0$ if $x < 0$), η is a characteristic time, C_0 stands for the initial damage threshold under pure tensile forces, $g_0(F_n, F_s, M, d)$ is the damage criterion and d_c is the ultimate value of damage, which depends both on the material and loading parameters [6]. It can be noted that, since the yield and damage surfaces g_u and g_0 include now a negative dependency on d, this formulation implies that both failure and damage surfaces are actually displaced towards the origin as the damage variable grows, thus increasing the susceptibility of the bond to further damage or rupture.

2.2 Fluid dynamics

The D2Q9 scheme of the lattice Boltzmann method including a multiple relaxation time (MRT) is employed here to model the transient flow of the fluid phase [7]. The fluid dynamics are this way computed basically in just two steps for the collision and advection of the fluid particles. A set of rules ensuring the conservation of mass, momentum and energy lead to the retrieval of the incompressible Navier-Stokes equations in conditions of low Mach numbers. The fluid phase is thus fully defined by specifying the lattice grid size Δx and lattice speed c_s, the components s_α of the diagonal relaxation matrix S (inverse of the different relaxation times) and the fluid material parameters of density ρ_f and kinematic viscosity ν. Further details can be found in [3].

The method for momentum exchange by Bouzidi and coworkers [8] provides here the coupling between the fluid and solid phases and permits the computation of the hydrodynamic forces on the discrete particles. Since the size of time steps required for the fluid computation is larger than for the discrete solid phase, a sub-cycling time integration technique has been employed, performing two DEM subcycles for each LBM step [9]. In order to retrieve a non-zero permeability through the two-dimensional assembly of solid grains, a reduced "hydraulic" radius is introduced [10].

3 Jet erosion

3.1 Jet hydrodynamics at the soil surface

Figure 3 illustrates the hydrodynamic conditions derived by the consideration of a fluid jet with prescribed velocity at the injection nozzle (nozzle diameter b) which impinges orthogonally on the surface of a granular packing located at a distance H from the nozzle, as described more thoroughly in [11].

The geometrical, material and rheological parameters employed for the simulation have been chosen here either for convenience or based on usual values from the literature (see e.g. [12]) and are summarized in Table 1. The flow may be described here as an inertial laminar one (jet's Reynolds number $Re_j \sim 50$ to 200) in the transition from a laminar to a turbulent regime. The flow is confined between the solid surface and the upper, closed boundary, while the lateral boundaries have been left open, and so the appearance of two convective cells can be observed. The profiles of fluid velocity, pressure and shear stress right over the solid surface (Figure 4) show the characteristic "M"-shape profile of fluid velocity with the stagnation point right under the jet's axis and its complementary maximum of fluid pressure. However, the maxima of shear stress, as well as the highest pressure gradients, are located actually right on the spots of the most prominent grains (the most exposed ones) of the irregular granular surface.

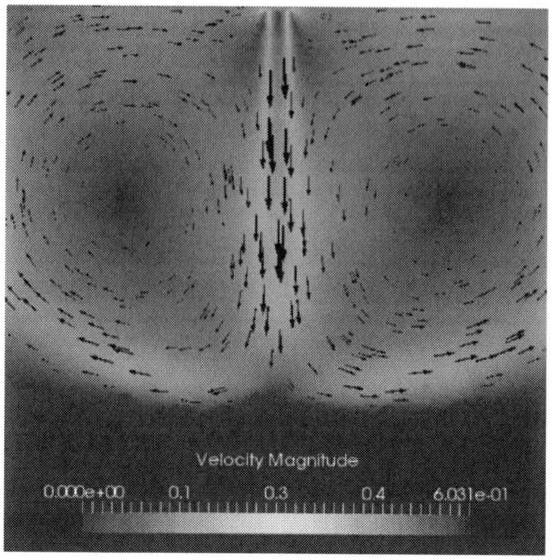

Fig. 3. Fluid velocity field of the jet's impingement on a fixed irregular surface. Arrow marker size is proportional to the velocity magnitude, shown in units of [m/s], at each point.

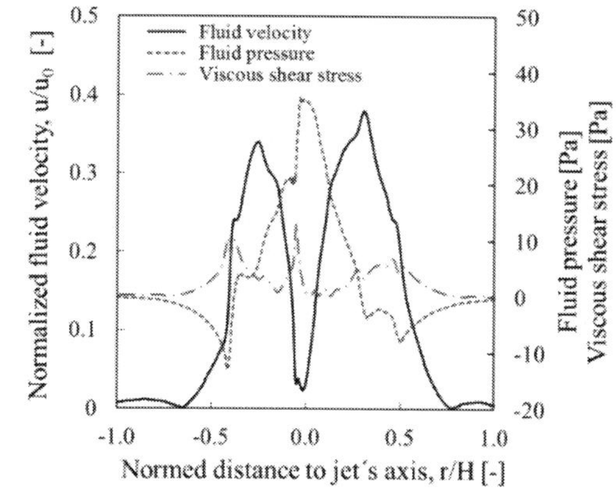

Fig. 4. Hydrodynamic profiles right above the surface of the granular packing.

This seems to imply that the topology of erosion, at least at its onset, can be dictated by the irregularities of the surface in the relevant impingement area, approximately in the range (-H, H) of radial distance from the jet's axis.

Table 1: Geometrical, material and model parameters for the presented simulations

Solid phase		Fluid phase	
Particle mean size, D_{mean}	3×10^{-3} m	Jet's distance to surface, H	7×10^{-2} m
Polydispersity, D_{max} / D_{min}	1.5	Jet's nozzle size, b	5×10^{-3} m
Length of granular sample, L	2.63×10^{-1} m	Kinematic fluid viscosity, ν	4 to 5×10^{-5} m^2/s
Height of granular sample, H_s	8×10^{-2} m	Fluid density, ρ_f	847 kg/m^3
Particle density, ρ_s	2230 kg/m^3	Lattice grid size, Δx	2.3×10^{-4} m
Normal contact stiffness, k_n	1.1×10^5 N/m	Lattice speed, c_s	10 m /s
Shear contact stiffness, k_s	1.1×10^5 N/m	Hydraulic radius factor, R_h	0.8
Rolling stiffness, k_ω	$0.1 \times k_n$	Inlet fluid velocity, u_0	0.45 to 1.5 m/s
Friction coefficients, $\mu=3 \cdot \mu_\omega$	0.3		

3.2 Soil erosion. Simple cohesion

When the soil grains are allowed to move under the action of the fluid jet with a velocity above the erosion critical threshold, a scouring crater will generally form. To quantify the erosion kinetics, it is necessary to specify an erosion criterion filtering the eroded from the non-eroded particles. Here, we classify a particle as eroded if it attains at any moment a kinetic energy above $2x10^{-5}$ Joules [11]. This was useful to discriminate the grains at the debonding front separating the cohesive assembly from the detached and re-settled particles.

Figure 5 shows the time evolution of the relative eroded mass (i.e. the proportion of eroded particles compared to the total mass of the granular assembly) for different strengths of the cohesive bonds and a jet velocity of 1.5 m/s. The purely frictional sample ($B = 0$) shows a sharp increase of eroded mass when the fluid jet reaches the soil surface and, after 5 seconds, half of the assembly has already been eroded. This proportion is significantly reduced by the introduction of cohesion, while the complete absence of erosion is achieved when the bond strength is $B = 25$, indicating the critical cementation degree for the beginning of erosion.

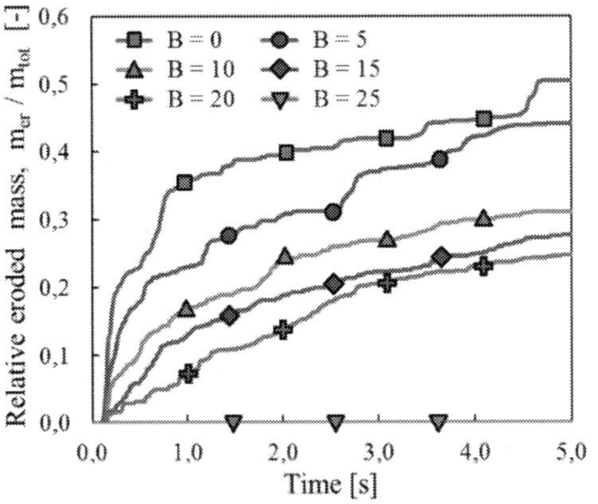

Fig. 5. Erosion kinetics in dependence of the bond strength.

3.3 Bond damage caused by an impinging jet

The extended subcritical debonding model can provide an insight into the micromechanics of erosion even before the erosion onset. Figure 6 shows the network of cohesive bonds among the soil particles ($D_{mean} = 1$mm) in the impingement area after 5 seconds of simulated jet flow with $C = 1$N, $C_0 = 0.01$N and $\eta = 0.1$s.

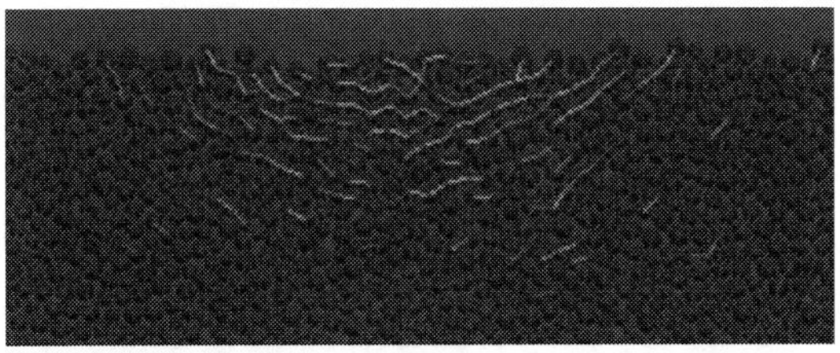

Fig. 6. Network of damaged cohesive bonds at simulated time $t = 5$ s. Colour scale from dark blue for $d = 0$ to red for $d = 1$.

The figure appears to show a pattern of damage with a clear preferential directionality along a tangential polar coordinate centred on the impingement point. A closer inspection of the damaged bonds (Fig. 7) shows as well a second preferential damage direction deviating 30° from the tangential direction. These results, along with the fact that the most damaged bonds are located below the exposed surface, appear to suggest that the fluid pressure gradients are at least a major agent (if not the driving one) behind the jet erosion.

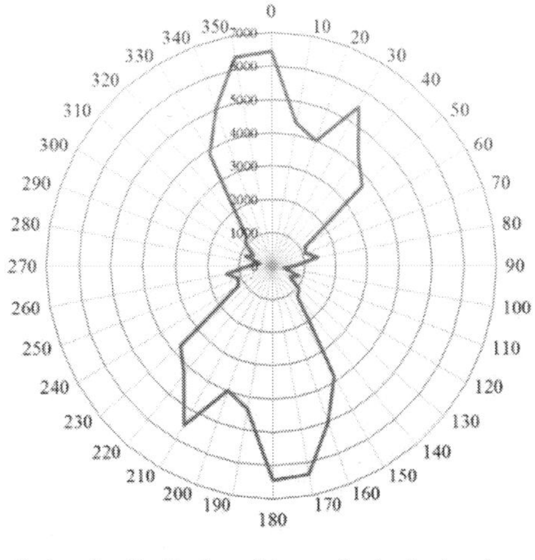

Fig. 7. Angular distribution of damaged cohesive bonds. Angular deflections from polar tangent centred on the jet's impingement point.

The French regional administration Provence-Alpes-Côte d'Azur has kindly provided the funding for these investigations in the frame of the ESCAPE research project.

References

1. M. Foster, R. Fell, M. Spannagle, Can. Geotech. J. **37** (2000)
2. G.J. Hanson, K.R. Cook, J. Applied Engineering in Agriculture. **20**,4 (2004)
3. J. Ngoma, PhD thesis, Université d'Aix-Marseille (2015)
4. L.-H. Luu, *et al.*, Proc. 8th Int. Conf. on Micromech. of Granular Media, Powders & Grains (2017)
5. J.-Y. Delenne, M.S. El Youssoufi, F. Cherblanc, J.-C. Bénet, Int. J. Numer. Anal. Meth. Geomech., **28** (2004)
6. C. Silvani, T. Désoyer, S. Bonelli, Int. J. Numer. Anal. Meth. Geomech. **33** (2009)
7. P. Lallemand, L.-S. Luo, J. Comp. Phys. **184** (2003)
8. M. Bouzidi, M. Firdaouss, P. Lallemand, Phys. Fluids, **13**,11 (2001)
9. Y.T. Feng, K. Han, D.R.J. Owen, Int. J. Numer. Meth. Engng **72** (2007)
10. D.F. Boutt, B.K. Cook, J.R. Williams, Int. J. Numer. Anal. Meth. Geomech. **35** (2011).
11. P. Cuéllar *et al.*, Proc. 8th Int. Conf. on Scour and Erosion, ICSE (2016)
12. S. Luding, Granular Matter, **10** (2008)

Comparison of multi-sphere and superquadric particle representation for modelling shearing and flow characteristics of granular assemblies

Behzad Soltanbeigi[1,*], *Alexander* Podlozhnyuk[2,**], *Jin Y.* Ooi[1], *Christoph* Kloss[2,3], and *Stefanos-Aldo* Papanicolopulos[1]

[1] *Institute for Infrastructure and Environment, School of Engineering, The University of Edinburgh, EH9 3JL, Edinburgh, UK*
[2] *DCS Computing GmbH, Industriezeile 35, 4020, Linz, Austria*
[3] *CFDEMresearch GmbH, Industriezeile 35, 4020, Linz, Austria*

Abstract. In the current study, complex-shaped particles are simulated with the Discrete Element Method (DEM) using two different approaches, namely Multi-spheres (MS) and Superquadrics (SQ). Both methods have been used by researchers to represent the shape of real particles. However, despite the growing popularity of utilizing MS and SQ particles in DEM simulations, few insights have been given on the comparison of the macro scale characteristics arising from the two methods. In this respect, initially the characteristics of the two shape representation methods are evaluated in a direct shear test simulation. The results suggest that controlling the sharpness of the edges for SQ particles can lead to a good agreement with the results of MS particles. This way, a set of SQ and MS particles, which are numerically calibrated in the shear tester, are obtained. Furthermore, the macro-scale responses of the numerically calibrated particles are assessed during a slow shearing scenario, which is achieved through simulating quasi-static flow of the particles from a flat-bottom silo. The results for mass discharge, flow profile and wall pressure show a good quantitative agreement. These findings suggest that the numerically calibrated MS and SQ particles in the shear tester can provide similar bulk-scale flow properties. Moreover, the results highlight that surface bumpiness for MS particles and corner sharpness for SQ particles change the characteristics of particles and play a significant role in the shear strength of the material composed of these particles.

1 Introduction

Most DEM codes use spherical particles to reduce the computational cost of simulations, although in reality particles are mostly of irregular shape. Several non-spherical shape descriptions have been proposed in the literature, the most popular approach in DEM being the multi-sphere approach (MS) [1]. In this description, spheres are allowed to overlap and be glued together to approximate an arbitrarily shaped particle. Another shape description, recently implemented in the DEM package LIGGGHTS [2], is the superquadric (SQ) [3, 4] that is an extension of spheres and ellipsoids.

Several researchers have studied the characteristics of different shape representation methods and compared them at the micro and macro level. Cleary and Sawley [5] compared the discharge of SQ and spherical particles from a hopper and showed that the non-sphericity causes a slower flow up to 30% and also changes the flow kinematics. Fraige et al. [6] simulated spherical and cubical particles in a flat-bottom silo and concluded that cubic-shape particles provide a packing with lower porosity and increased resistance to flow compared to spheres. Tao et al. [7] used the MS approach to represent corn-shape particles and compared the flow properties with spherical particles. They showed that the downward velocity of the clusters shows higher variation, compared to spheres, along the width of the silo (from centre toward walls). Furthermore, they observed that the mean voidage of packings for non-spherical particles is smaller than that of the spherical particles. Markauskas et al. [8] evaluated the capability of MS to provide ellipsoidal particles, which can replace the perfectly smooth ellipsoids generated using the SQ technique. Assigning the variable parameter as the number of sub-spheres, the MS particles were characterized through studying the angle of repose, porosity and coordination numbers. It was pointed out that increasing the number of sub-spheres exhibits a clear tendency to mimic macroscopic parameters of a smooth ellipsoid.

Based on the above studies, it is clear that several attempts have been made to understand the characteristics of different shape approximation methods. However, there is still a lack of a comprehensive study that investigates the bulk response of the SQ and MS particles under various compression and shearing conditions. Accordingly, this paper aims to provide a better understanding of the micro/macro properties of the MS and SQ particles and also investigates the potential similarities. In this respect, a series of shearing tests, with MS and SQ particles in a

*e-mail: b.soltanbeigi@ed.ac.uk
**e-mail: alexander.podlozhnyuk@dcs-computing.com

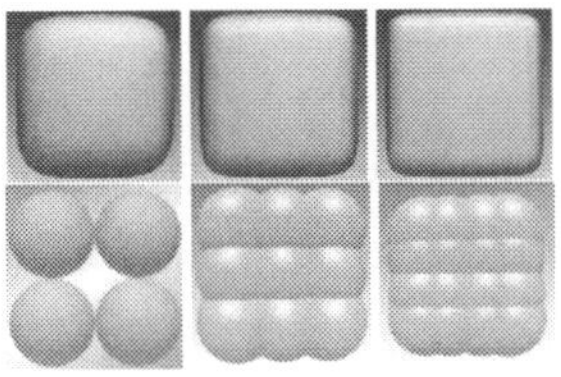

Figure 1. Particle shapes used. SQ_{N4},SQ_{N6} and SQ_{N8}, top, from left to right. And MS_8, MS_{27} and MS_{64}, bottom, from left to right

Jenike shear tester, are conducted to determine the role of edge sharpness in SQ and number of sub-spheres (surface bumpiness) in MS particles. Subsequently, the flow properties of SQ and MS particles with similar macroscopic shearing response, are evaluated through comparing various bulk-scale features during a silo discharge.

2 Methodology

This section provides information regarding material properties and testing procedures that have been followed. The considered particles have cubical shapes with edge length of d = 2mm. Cubes were approximated by SQ particles in LIGGGHTS and by MS particles in EDEM [9] (Fig.1). The considered material is selected from the Generic EDEM Material Model (GEMM) Database, which provides suitable material inputs for accurate representation of bulk materials. Table 1 shows the material properties for particles and geometry.

The equation that governs the shape of SQ particles, given by Barr [10], is as follows:

$$f(x, y, z) \equiv (|x/a|^{n_2} + |y/b|^{n_2})^{n_1/n_2} + |z/c|^{n_1} - 1 = 0.$$

A detailed description of a contact detection and a contact force algorithm for two SQ particles implemented in LIGGGHTS can be found in [4]. Cubical particles can be modelled by superquadrics taking $a = b = c = d/2$ and taking $n_1 = n_2 = N > 2$, where N controls the level of edge sharpness/blockiness. Different levels of edge sharpness (N = 4, 6 and 8, denoted as SQ_{N4}, SQ_{N6} and SQ_{N8} correspondingly, see Fig.1) were used in this paper.

Multispheres, which approximate the shape of particles by overlapping or touching of spheres, are used as an approximation of the real shape irregularities [1, 11]. Cubes were modelled in EDEM by multi-spheres using 8, 27 and 64 sub-spheres of equal radius (denoted as MS_8, MS_{27} and MS_{64} correspondingly, see Fig. 1).

The Jenike shear tester is widely used for measuring flow properties of particulate solids, [12]. In this test the granular material is placed in a split cylindrical box. Then, the material is consolidated by applying a constant vertical load σ_v to the lid section. Later, the top half of the cylinder (ring) is moved at a constant translational velocity (Fig.2). The measured quantity is the force required for this movement that can be converted to an average shear stress τ. The quasi-static flow of SQ and MS particles is further assessed in silo discharge. The flat-bottom silo has dimensions $50d \times 5d \times 100d$ in x, y, z directions correspondingly. Periodic boundary conditions are applied in the y-direction. The orifice dimension is $10d \times 5d$.

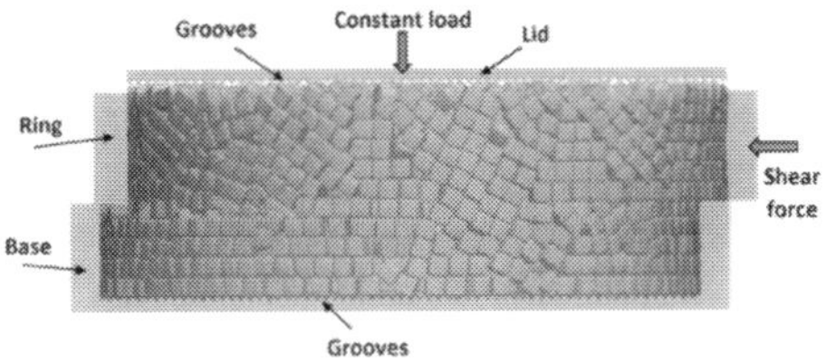

Figure 2. Jenike shear tester filled with superquadric particles

Table 1. DEM material properties

Parameter	Value
Density (particle) ρ [kg/m^3]	4100
Coefficient of particle-particle friction μ^s_{pp}	0.56
Coefficient of particle-wall friction μ^s_{pw}	0.45
Coefficient of restitution (particles), e_p	0.15
Coefficient of restitution (wall), e_w	0.5
Poisson ratio (particles), ν_p	0.25
Poisson ratio (wall), ν_w	0.25
Shear modulus (particles), G_p[Pa]	10^7
Shear modulus (wall), G_w[Pa]	10^{10}

3 Results

3.1 Jenike Shear Test

Before comparing MS and SQ particles, simulations with mono-sized spherical particles were conducted to prove equivalence of the setups in the LIGGGHTS and EDEM codes. Results showed a reasonably good agreement with a maximum difference of 3 %.This discrepancy can be explained by the difference in initial particle configuration within the generated packings.

The DEM time-step was chosen as $\Delta t = 2 \cdot 10^{-6}$ s (5 % of Rayleigh time) in all simulations. Two types of packing, using MS and SQ particles, were generated to assess the dependency of the results on the density of the initial packing. In the dense packing, the particle-particle friction coefficient μ^s_{pp} was set to zero at the filling stage and changed back to $\mu^s_{pp} = 0.56$ before applying σ_v. On the contrary, the loose packing had $\mu^s_{pp} = 0.56$ during the whole simulation .

For all simulations, material properties were kept identical and packings were generated using SQ_{N4}, SQ_{N6}, SQ_{N8} and MS_8, MS_{27}, MS_{64} particles. Figure 3 presents the information on the volume of single MS and SQ particles (V_p) and the solid fraction (ϕ) of the packings for the shear tester. The results for ϕ are plotted against the ratio of V_p to the volume of the perfect cube (V_c = 8mm^3). It is observed that for MS particles, MS_8 has the lowest value of V_p/V_c, while the difference between MS_{27} and MS_{64} is much smaller. Additionally, the V_p/V_c value has a direct relation with N for SQ particles and $N = 8$ has the maximum value accordingly. For the solid fraction, it can be seen that MS_8 provides the packing with the largest ϕ. Furthermore, it is clear that for both loose and dense packings, SQ particles show an approximately constant value for ϕ (independent of the N value). MS_{27} and MS_{64} also follow the same trend for ϕ as SQ particles.

The corresponding shear stress curves for SQ and MS particles as a function of shear displacement (D) are repre-

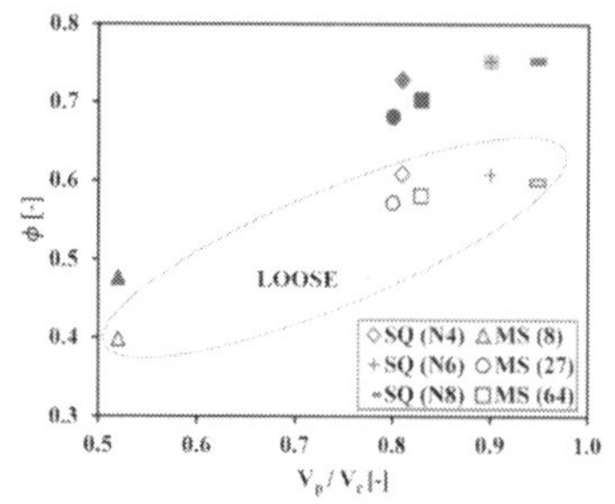

Figure 3. Solid fraction of the packings with different particle shapes (the filled markers are for dense packing)

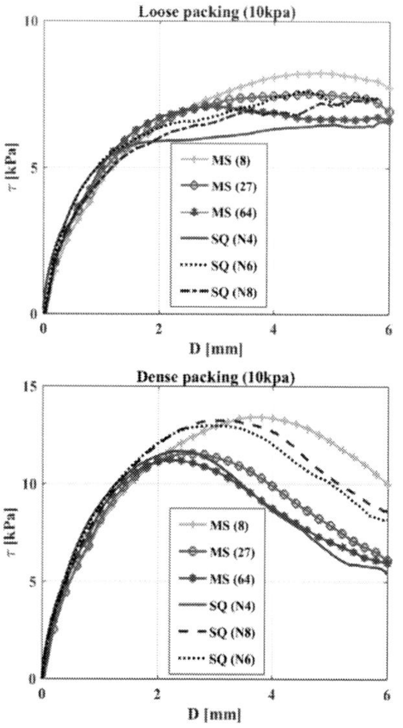

Figure 4. Jenike direct shear simulations considering SQ and MS particles a)loose packings b) dense packings.

sented in Fig. 4. It can be seen that, in loose packing, SQ_{N4} has the lowest strength during shearing. However, it can be noticed that after increasing the blockiness to $N = 6$, the material shows a higher strength, but a further blockiness (i.e. $N = 8$) plays no significant role in the shearing response of the SQ particles.

For the loose packings of the MS particles, the MS_{27} and MS_{64} show a similar peak strength, while a slightly lower residual strength can be seen for the MS_{64}. The MS_8 is providing a peak and residual strength larger than for MS_{27} and MS_{64}, which is an indication of increased interlocking among the MS_8 particles. Additionally, for the dense packing, increasing blockiness for SQ particles results in higher shear strength. A similar response is seen for the MS particles when the surface bumpiness is increased (i.e. the number of sub-spheres is reduced).

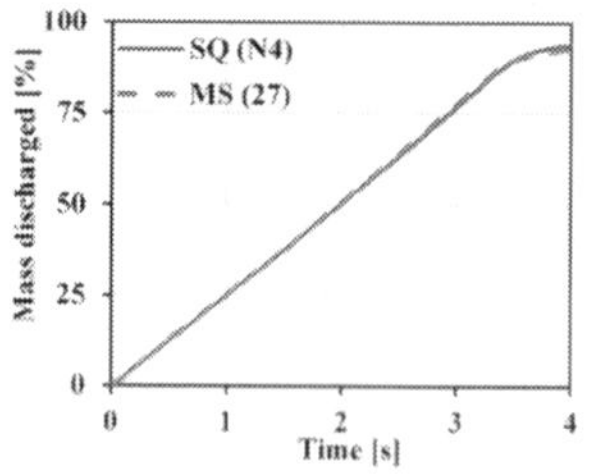

Figure 5. Cumulative discharged mass

It is also observed that SQ_{N4} is in good agreement with MS_{27} and MS_{64} and show a similar peak strength for the dense sample. Although in the loose samples the peak strength for SQ_{N4} stays approximately 15 % lower than MS_{27} and MS_{64}, the residual strength is relatively comparable. Accordingly, SQ_{N4} and MS_{27} and MS_{64} can be considered as numerically calibrated materials in a Jenike shear tester. The next section provides comparison of the macro-scale features of calibrated particles during silo flow. Due to the very high computational time for MS_{64}, the silo discharge is just simulated with MS_{27} particles and compared with results of SQ_{N4}.

3.2 Flat-bottom silo

Using the calibrated MS and SQ particles (MS_{27} and SQ_{N4}), a total of about 20 000 particles were generated in each software to model the filling and discharge of the flat-bottom silo. Figure 5 shows that the mass discharged is independent of the shape representation technique for the calibrated particles.

Snap-shots have been taken from different instances of the discharged mass (M_D) to provide an insight into flow profiles for both particle types. Figure 6 shows the particles inside the silo at 10, 30 and 60 % of the discharge. It can be seen that flow profiles are similar both in funnel and mass flow sections. This confirms the earlier results shown in Figure 5.

The wall pressure during discharge is also compared for both particle types. In this respect, the height of the wall is divided into equal-size bins and total normal contact forces are computed by using spatial averaging and plotted in Figure 7. Results are time-averaged over 0.05 seconds of the discharge to reduce the fluctuations of instantaneous values. Results show that there is significant difference between the two particles types at 10% of the discharge. This might be due to inappropriate spatial and temporal averaging windows or a result of the different initial particle configuration. However, it is clear that in the steady state stage of the discharge, the wall pressure distributions for the two particle types are in good agreement.

4 Conclusion

In this work, the behavior of SQ particles, at different edge sharpness levels, and MS particles, with different numbers of sub-spheres, have been evaluated in a Jenike shear test and the following conclusions are drawn:

• For the loose packings, the change in edge/corner sharpness (from $N = 4$ to $N = 6$) increases the strength o the

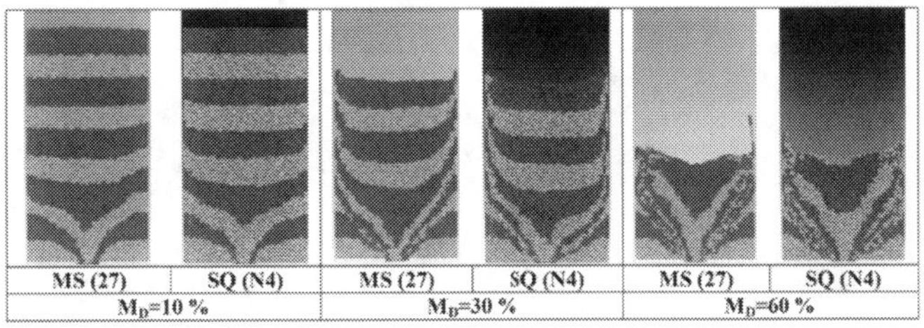

Figure 6. Flow profiles for both SQ and MS particles at $M_D = 10$, 30 and 60 %

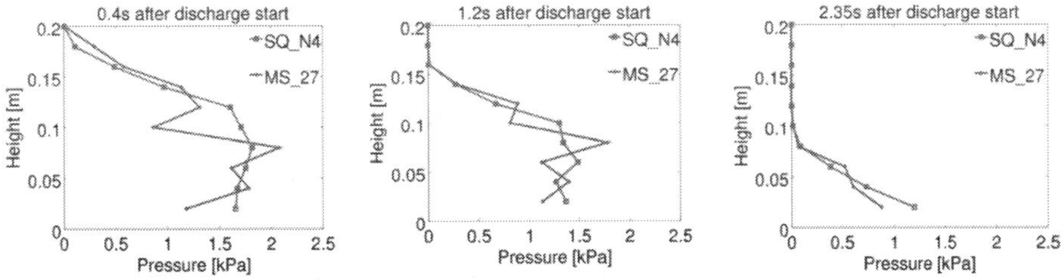

Figure 7. Normal wall pressure distribution along the height of the wall at 10%, 30% and 60% of the discharge (from left to right).

material. However, further increase of N plays no major role in the shearing response of SQ particles. For dense packing, increasing N leads to higher shearing resistance respectively for SQ particles.

• The bulk response of MS cubes is governed by the bumpiness of the surface in both loose and dense samples; the higher bumpiness, which means lower number of spheres contribute in building the surface of the single MS particle, increases the resistance of the material against shearing.

• It is also observed that the shearing result of SQ_{N4} is in good agreement with MS_{27} and MS_{64} and show a similar peak and residual strength in dense samples. For the loose samples, although the peak strength for SQ_{N4} is approximately 15 % lower than MS_{27} and MS_{64}, the residual strength is relatively comparable for these pair of particles (assumed as numerically calibrated particles).

The behaviour of the calibrated particles is also assessed during silo discharge and the key findings are:

• The discharge rate is matching well for MS_{27} and SQ_{N4} particles.

• Flow profiles are similar in both funnel and mass flow sections at various stages of the discharging process.

• The wall pressure distribution differs during the initial states of the discharge, however during steady state there is a good agreement between results for SQ and MS.

To sum up, by using a proper numerical calibration procedure (Jenike tester simulation), it is possible to select roughly equivalent SQ and MS representations that indeed produce very similar results in a silo flow simulation. However, further investigations are essential for generalization of the findings for various particle shapes.

Acknowledgments

This work has been carried out as a part of the T-MAPPP project, an EU-funded Framework 7 Marie Curie Initial Training Network. The financial support provided by the European Commission is gratefully acknowledged. The assistance received from Dr J. P. Morrissey is also acknowledged.

References

[1] J. Favier, M. Abbaspour-Fard, M. Kremmer, A. Raji, Engineering Computations **16**, 467 (1999)

[2] C. Kloss, C. Goniva, A. Hager, S. Amberger, S. Pirker, Progress in Computational Fluid Dynamics, An International Journal **12**, 140 (2012)

[3] P.W. Cleary, Engineering Computations **21**, 169 (2004)

[4] A. Podlozhnyuk, S. Pirker, C. Kloss, Computational Particle Mechanics (2016)

[5] P.W. Cleary, M.L. Sawley, Applied Mathematical Modelling **26**, 89 (2002)

[6] F.Y. Fraige, P.A. Langston, G.Z. Chen, Powder Technology **186**, 224 (2008)

[7] H. Tao, B. Jin, W. Zhong, X. Wang, B. Ren, Y. Zhang, R. Xiao, Chemical Engineering and Processing: Process Intensification **49**, 151 (2010)

[8] D. Markauskas, R. Kačianauskas, a. Džiugys, R. Navakas, Granular Matter **12**, 107 (2009)

[9] D. Solutions, Edinburgh, UK (2014)

[10] A.H. Barr, IEEE Computer Graphics and Applications **1**, 11 (1981)

[11] C. Zhou, J.Y. Ooi, Mechanics of Materials **41**, 707 (2009)

[12] J. Härtl, J.Y. Ooi, Powder Technology **212**, 231 (2011)

The distribution of saturated clusters in wetted granular materials

Shuoqi Li[1], *Dorian Hanaor*[2] and *Yixiang Gan*[1*]

[1] School of Civil Engineering, The University of Sydney, Sydney, NSW 2006, Australia

[2] Institute for Materials Science and Technology, TU Berlin, Berlin, Germany

Abstract. The hydro-mechanical behaviour of partially saturated granular materials is greatly influenced by the spatial and temporal distribution of liquid within the media. The aim of this paper is to characterise the distribution of saturated clusters in granular materials using an optical imaging method under different water drainage conditions. A saturated cluster is formed when a liquid phase fully occupies the pore space between solid grains in a localized region. The samples considered here were prepared by vibrating mono-sized glass beads to form closely packed assemblies in a rectangular container. A range of drainage conditions were applied to the specimen by tilting the container and employing different flow rates, and the liquid pressure was recorded at different positions in the experimental cell. The formation of saturated clusters during the liquid withdrawal processes is governed by three competing mechanisms arising from viscous, capillary, and gravitational forces. When the flow rate is sufficiently large and the gravity component is sufficiently small, the viscous force tends to destabilize the liquid front leading to the formation of narrow fingers of saturated material. As the water channels along these liquid fingers break, saturated clusters are formed inside the specimen. Subsequently, a spatial and temporal distribution of saturated clusters can be observed. We investigated the resulting saturated cluster distribution as a function of flow rate and gravity to achieve a fundamental understanding of the formation and evolution of such clusters in partially saturated granular materials. This study serves as a bridge between pore-scale behavior and the overall hydro-mechanical characteristics in partially saturated soils.

1 Introduction

In soil mechanics, multiphase fluid flows in porous medium have been subjected to a considerable number of studies over the past decades since it is important to a broad range of applications, including oil recovery, risk assessment of landslides, and groundwater hydrology [1,2]. Strong variations exist during mixing two or more immiscible fluids in porous medium, depending on two dimensionless numbers, *i.e.*, the Capillary number (Ca) and Bond number (Bo) [3,4]. Experimental and numerical studies concerning drainage processes in porous media often tend to focus on the evolution of viscous fingering at various flow rates [5,6]. However, studies on the morphological characteristics of liquid phase entrapment behind the front (interface between wetting and non-wetting fluids) remain relatively scarce.

This study aims to find a link between the two aforementioned dimensionless numbers and saturated cluster size distribution by establishing an experimental system where capillary, gravity and viscous forces can be examined. Here in, we focused on saturated clusters, which were formed when a liquid phase fully occupied the pore space between solid grains in a localised region. We have studied the saturated clusters obtained by draining water from a porous medium. To tune the gravitational and viscous forces, different tilting angles and drainage rates were employed on the system. The temporal and spatial information of the porous medium was characterised by means of a digital camera. The size distribution of saturated clusters were extracted from a series of images and statistically analysed by employing the maximum likelihood estimation [7].

The present paper is organised as follows. The experimental setup and methods for data analyses are described in Section 2. In Section 3, the saturated cluster distributions under various loading scenarios are presented and discussed. Finally, conclusions are presented in Section 4.

2 Experimental setup and methods

The experimental system was designed for temporally and spatially resolved imaging of interface motions in micro-models consisting of glass beads. The key elements of the experimental setup are shown in Fig. 1(a). A two-dimensional porous medium was constructed by a single layer of glass beads with a diameter of 2 mm between two parallel glass plates. The glass plates and beads were then slightly compressed by

* Corresponding author: yixiang.gan@sydney.edu.au

Table 1. Evaluative parameters mean front velocity and generalized bond number (Bo*) are given.

Tilt Angle (°)	0					5			10		15
Drainage Rate (mL/min)	1	2	5	10	20	1	5	20	5	20	10
Mean Front Velocity (mm/s)	0.116	0.231	0.579	1.157	2.315	0.116	0.579	2.315	0.579	2.315	1.157
Bo* (-)	-1.444 $\times10^{-4}$	-2.889 $\times10^{-4}$	-7.222 $\times10^{-4}$	-0.001	-0.003	0.005	0.004	0.002	0.010	0.007	0.014

a few clamps, and silicone glue were applied on the bottom and side edges, leaving the top side connected to the atmosphere. The observable area was 180 mm×270 mm, and the resulting porosity was about $n \approx 0.4$. After the model was installed, the outlet channel was connected to a dispensing pump (Masterflex 77200-62, USA) that controlled a stable volumetric injection and withdrawal flow rate. A digital camera (Nikon D3300) controlled by a computer over digiCamControl software was used to record the experiments at given time eclipses based on different flow rates. Each image contains 1000×1500 pixels, which corresponds to a spatial resolution of 30 pixels per mm^2. An electroluminescent panel provided sufficient background illumination for different shutter speeds. Both the light source and model could be tilted spontaneously to various inclination angles to vary the gravitational component on the fluid flows in the porous medium. The component of gravity along the model g' is given by $g\ \sin(\alpha)$, where g is the gravitational acceleration and α is the inclination angle.

Water was used as the wetting fluid, dyed with dye tracer (Cole-Parmer 00298-06), and air was the non-wetting fluid invading from the open top edge. Experimental values of the wetting fluid are calculated by assuming room temperature at 25°C. The surface tension between two fluids γ is 72.75 mN/m. The wetting fluid has a viscosity μ of 1.00 mPa.s and a density ρ of 998 kg/m3. The permeability of the medium k was estimated from the Kozeny-Carman relation as 4.89×10-9 m^2 [8]. The applied flow rates ranged from 1 ml/min to 20 ml/min for different tilt angles. These resulted in the mean front velocity determined from volumetric flow rates at the cross-section area taking into account the porosity (shown in Table 1).

To quantify magnitudes of viscous or gravitational forces relative to capillary forces in the following experiments and analyses, we employed dimensionless Capillary (Ca) and Bond (Bo) numbers which are defined as follows [9]:

$$\mathrm{Bo}=\frac{\rho g' a^2}{\gamma} \quad (1)$$

$$\mathrm{Ca}=\frac{\mu v a^2}{\gamma k} \quad (2)$$

with the wetting liquid density (ρ), the component of gravitational acceleration (g'), typical pore size (a) assumed to be 1/3 of bead diameter (more details can be found in [10]), viscosity of the wetting fluid (μ), surface tension (γ), the mean front velocity (v), permeability of the medium (k).

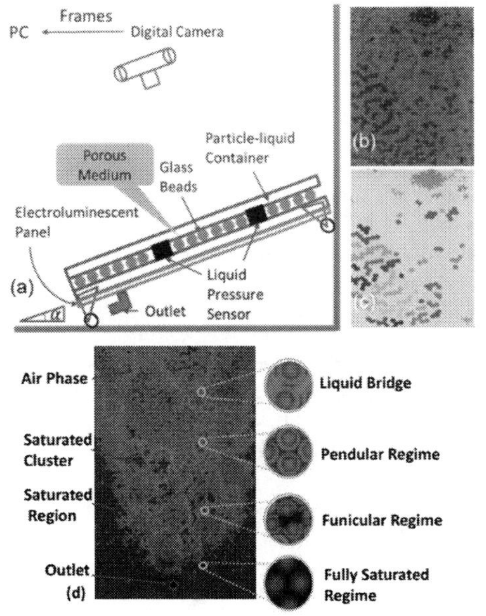

Fig. 1. Diagram of the experimental setup. (a) The camera, electroluminescent panel and cell are all connected to a common frame that can be tilted by an angle. The liquid outlet on the bottom is connected to a syringe pump that injects and withdraws the liquid. (b) A detailed zoom-in section from the raw image extracted from the camera showing the structure of the porous network, the non-wetting phase (green) and wetting phase (red). (c) The same section after applying image processing. Individual saturated clusters are coloured randomly to allow the visualisation of separated clusters for different sizes. (d) An example experimental image revealing different saturation regimes.

We also define a generalised Bond number [9,11]

$$\mathrm{Bo}^{*}=\mathrm{Bo}-\mathrm{Ca}. \quad (3)$$

In previous studies on the liquid front, it is found that the transition from stable to unstable front morphology conditions occurs at Bo*=0. The unstable front morphology exhibits capillary or viscous fingering during drainage stages. If the generalised Bond number is larger than zero, the gravitational force will overcome the viscous effects and lead to a flat plateau of the liquid front. However, when it is less than zero, the geometry of the front converts into unstable patterns.

The grayscale pixel-intensity histogram of raw images have two distinguishable peaks corresponding to the non-wetting and wetting fluids. In order to separate the air-filled region from the saturated region, all grey levels under a given threshold were set to black ('wet') while the others were white ('dry'). The thresholds varied from images to images to account for the frequency changes of the electroluminescent panel. After the thresholding process, the whole observation area was separated into different regions with glass beads determined to be part of wet or dry region. This treatment did not affect the following analyses since only the saturated regions that are larger than the projection area of a typical glass bead (π mm^2 in this study) were considered. After removal of trapped liquid volume with the projection area smaller than a single glass bead, the resulting black and white picture could provide the required geometrical information. To allow visualising separate clusters of different sizes, analyses of morphological components have been carried out and colours were assigned to localised saturated clusters, as shown Fig. 1(c). Fig. 1(b) and (c) show that crystallisation indeed exists in our observation. It is foreseen that this localised structure indeed correlates to the resulting saturation states. Further correlation analysis between the local crystallised region and saturation stage is being carried out. Liquid bridges which are observed as ring-like structures as shown in Fig. 1(d) can be of interest in future study. However, at the present stage, only saturated regions larger than π mm^2 are taken into account.

3 Results and discussion

In order to quantify the magnitudes of drainage-injection curves, liquid saturation (Sr) is defined as

$$Sr = \frac{V_W}{V_V}, \tag{4}$$

where V_w and V_V are the total volume of water and voids, respectively.

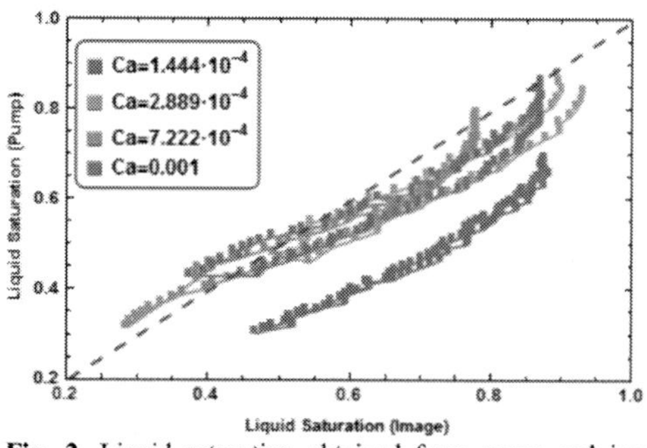

Fig. 2. Liquid saturation obtained from pump and image separately during drainage-injection cycles for different drainage rates (1 ml/min, 2 ml/min, 5 ml/min, 10 ml/min). The experimental cell was horizontally placed, i.e. Bo*= -Ca. Each dot is an individual image of the liquid content profile. The green dashed line indicates the points for these two saturation levels equal with each other.

Fig. 2 shows the liquid saturation calculated from the pump in comparison to the liquid saturation obtained from image-based processing for the entire experimental region for the drainage-injection cycles with different capillary number, Ca. The upper branch of each hysteresis curve is the injection section and the lower one is the drainage section. Due to the removal of trapped liquid volume, usually existing as liquid bridges between grains that can only be observed during drainage processes, the accumulated saturated region of drainage process is expected to be smaller than the accumulated area in injection process. The impact of Ca on different drainage-injection curves is driven by the contribution of drainage rates. When Ca (or the drainage rate) is extremely small, i.e. Ca $= 1.44 \times 10^{-4}$, due to the slow flow rate, more water is trapped in the form of liquid bridges, the blue curve shows a noticeable difference compared with other cases. As Ca is increased, the curves are overlaid with each other and shifted closer to the green dashed line (points with liquid saturation extracted from pump and image equal to each other), the liquid saturation of drainage-injection range stays constant at around 0.25-0.85.

The saturated clusters are statistically analysed in terms of geometrical properties. The largest saturated cluster in each image, corresponding to the saturated region behind the continuous liquid front, is excluded from the statistical analysis of cluster sizes. Here, we first look at the effective size of the saturated clusters, defined as $d_i = \sqrt{A_i}$, where A_i denotes the area of the cluster (area of localised saturated region). By rearranging the experimental data into multiple sub lists by a given bin size (0.91 mm), the size distributions of saturated clusters for drainage under different Bo* are depicted in Fig. 3. The probability density tends to the peak value around 2-3 mm. The decrease of probability is then generated by the sizes of clusters in the log-linear scale. A lognormal distribution function was selected here to represent the experimental data.

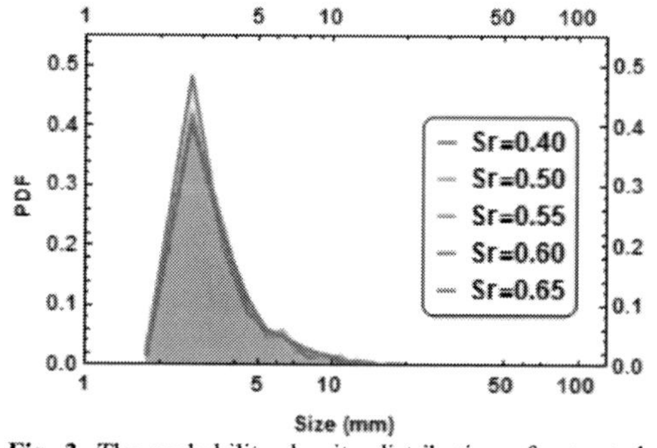

Fig. 3. The probability density distribution of saturated cluster sizes for drainage process for different liquid saturation level with Bo*=-1.444×10^{-4}.

The experimental probabilities of the area distributions are fitted using the lognormal distribution function as (see Fig.4)

$$f(X|\mu,\sigma^2) = \frac{1}{\sqrt{(2\pi\sigma^2)}X} \exp\left[-\frac{(\ln X-\mu)^2}{2\sigma^2}\right] \tag{5}$$

where μ and σ are the mean and the standard deviation of the logarithm of random variables, respectively.

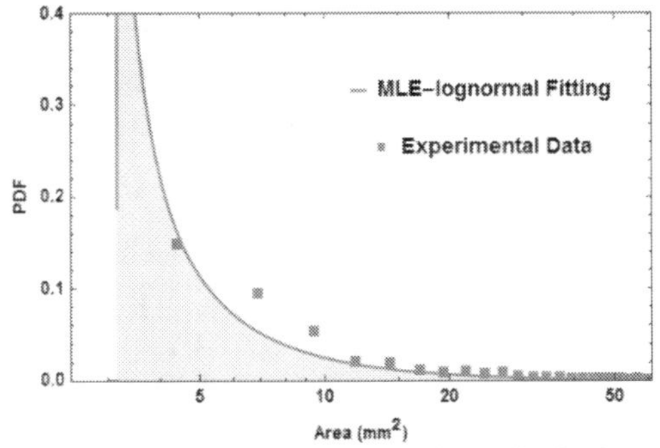

Fig. 4. Log-linear plot of saturated cluster distribution at the same saturation (Sr=0.55) for drainage process with Bo*=0.002 as a function of area. The dots correspond to the experimental data and the continuous line is obtained by applying all experimental data using maximum likelihood estimation method based on Eq. (5).

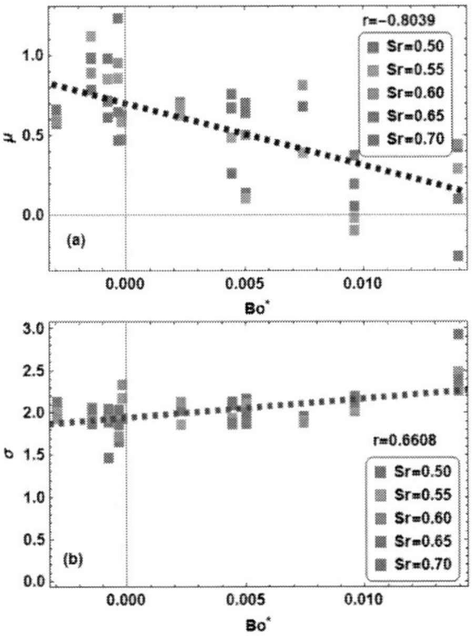

Fig. 5. Plot of lognormal distribution parameters: (a) scale parameter μ and (b) shape parameter σ as a function of generalised Bond number, Bo*, for different saturation levels, correlation coefficient r is also provided.

Fig. 5 shows the relationship between lognormal distribution parameters and generalised Bond number Bo*. Since for both subfigures in Fig. 5 the absolute value of correlation coefficient r is larger than 0.65, a relatively strong linear relationship lays in between these two parameters and Bo*. It is found that the scale parameter μ is inversely correlated to Bo*. During the drainage stage, it has been observed experimentally that at a large positive generalised Bond number, the liquid front quickly grows to a plateau, with small amount of clusters left behind. As the generalised Bond number decreases, narrowed branched fingers appear. For large negative values of Bo*, some large sized saturated clusters are trapped in the non-wetting phase due to the instability induced by viscous forces.

4. Conclusions

In this study we experimentally evaluated the saturated cluster size distribution and its evolution during drainage processes of wetted granular materials. The distribution is found to follow a lognormal distribution. It is concluded that the scale parameter μ and shape parameter σ negatively and positively correlate with Bo*, respectively. Due to the fact that a strong linear dependence exists between lognormal distribution parameters and generalised Bond number, Bo*, it is possible to generate a function containing μ, σ and Bo* for describing the cluster distribution at various saturation levels. Knowing this, combined with the total surface energy calculated by the area of interfaces, it is possible to include this additional grain-scale information in constitutive modelling of unsaturated soils using both the degree of saturation and generalised Bond number, Bo*.

Acknowledgments

Financial support for this research from the Australian Research Council through grants DE130101639 is greatly appreciated.

References

1 J. Bear, *Dynamics of fluids in porous media*. (Courier Corporation, 2013).
2 Y. Gan, F. Maggi, G. Buscarnera, and I. Einav, Géotechnique Letters **3** (4), 152 (2013).
3 R. Lenormand, E. Touboul, and C. Zarcone, Journal of Fluid Mechanics **189**, 165 (1988).
4 D. G. Avraam and A. C. Payatakes, Journal of Fluid Mechanics **293**, 207 (1995).
5 R. Lenormand and C. Zarcone, Physical Review Letters **54** (20), 2226 (1985).
6 R. Chandler, J. Koplik, K. Lerman, and J. F. Willemsen, Journal of Fluid Mechanics **119**, 249 (1982).
7 I. J. Myung, Journal of mathematical Psychology **47** (1), 90 (2003).
8 A. E. Scheidegger, Soil Sci. **86** (6), 355 (1958).
9 Y. Méheust, G. Løvoll, K. J. Måløy, and J. Schmittbuhl, Physical Review E **66** (5), 051603 (2002).
10 D. J. Crisp and R. Williams, Marine Biology **10** (3), 214 (1971).
11 F. Moebius and D. Or, Water Resources Research **50** (11), 8441 (2014).

 DOI: 10.1051/epjconf/201714002016

Measurement of effective thermal conductivity of compacted granular media by the transient plane source technique

Weijing Dai[1], *Yixiang* Gan[1,*]

[1]School of Civil Engineering, the University of Sydney, Darlington, Sydney, Australia

Abstract. To successfully realise industrial applications handling granular media, especially those involving heating and cooling processes, the temperature fields must be properly evaluated according to the accurate thermal properties of the media. The knowledge the effective thermal conductivity is regarded as one of the fundamental aspects. However, due to the complicated relations between the effective thermal conductivity and the heterogeneity and complexity in the structures and composition of the granular media, the quantitative prediction of the conductivity is challenging. Therefore, experimental investigation of the effective thermal conductivity becomes desired and this can provide first-hand data for industrial reference and serve as the benchmark for the theoretical analysis. In this study, the transient plane source technique is employed to investigate the effective thermal conductivity of compacted granular beds by the application of the commercially available Hot Disk system. The granular beds of different particle size ranges are characterised under different mechanical loading conditions by different sensors. Experimental results are discussed and suggestion to achieve reliable experimental designs is provided.

1 Introduction

Granular media, for a long time, is one of the most important material systems widely used in industry including chemical synthesis, food production, thermal insulation, additive manufacturing and other application consisting of granular beds. Efforts have been devoted to investigate the properties and features of granular media to provide knowledge for the realisation and optimisation of those applications. One step further into them, it is clear that the thermal transport is a crucial physical phenomenon determining the capability of not only the high-end technologies but also the conventional industries. For example, powder beds based additive manufacturing, promising in fabricating products of complex geometries, requires precise control of the thermal energy input into the processed beds to selectively melt a target region [1]; thermal management techniques of lithium battery systems is critical due to the safety concern in the application of high power and fast charging [2]. A growing demand on the appropriate knowledge about the effective thermal conductivity of granular media rises because the effective thermal conductivity of the media is one of the key properties to evaluate the thermal transport phenomenon [3]. With knowing the effective thermal conductivity of beds, numerical simulations can be effectively employed to investigate the potential issues in real scenarios without costly actual implementation [4,5]. Besides, the micro-mechanisms that are difficult to detect in-situ can be validated by investigating the correlation between the mechanisms and the accessible effective thermal conductivity [6]. Considerable progresses have been made in the field of the characterisation methods for the effective thermal conductivity of granular media and catalysed the development of the theories to analyse the measured data.

Theoretically, the Zehner-Schlunder-Bauer model used a unit cell constructed by two particle halves and filling fluid phase to represent a granular media according to the packing properties of the media. The effective thermal conductivity of granular media is derived by the parallelisation rule of the thermal resistance of the thermal transport components of the unit cell [7,8]. This model has been later modified [9] and adopted [6] widely by other researchers to study related problems. Besides the evolution of theories, experimental research about a variety of granular media is of equal importance. Unsaturated sand and soil have been measured by the steady-state probe method [10,11], but the attributes of those granular media is hard to interpret, limiting the understanding of the thermal transport mechanisms involved. Hot wire method was employed to measure the effective thermal conductivity of glass beds at different ambient pressured [12] and it is observed the effective thermal conductivity increased with the increasing ambient pressure. The effect of applied mechanical loading onto granular beds has been studied by Tehranian, *et al* [13] and Widenfeld, *et al* [14], applying the steady-state measurement techniques. Recently, the transient plane source technique has been developed to measure the effective thermal conductivity

* Corresponding author: yixiang.gan@sydney.edu.au

of particulate materials. Anisotropic effect on the effect thermal conductivity of granular beds introduced by the uniaxial compression was observed by Mo *et al* [15]. Using a similar technique, Liang *et al* [16] found a sensitive dependency of the effective thermal conductivity of sands on the amount of hydrates. Since this transient plane source technique attracts more attention in the community working with granular media, some questions that may be encountered in the practise of this technique remain unclear and have not been discussed previously by the researchers. In general, this technique comes with different sized sensors. The size contrast between the sensor and the particles has potential influence on the measurement. In this study glass beads of relative low thermal conductivity and steel ball bearings of relative high conductivity are utilised to investigate the dependency of the effective thermal conductivity upon the particle size. The response of effective thermal conductivity of these granular materials subjected to mechanical loading is tested. Finally, we will address the influence of the sensor size and comment on the selection of the sensor according to our measurement.

2 Experimental setup

2.1 Transient plane source technique

The transient plane source technique utilises a thin metal stripe or a metal disk to act as a continuous plane heat source and at the same time to serve as a resistive temperature sensor [17]. In principle, the temperature profile during the heating period of the sensor is sensitive to the surrounding materials. Therefore, by monitoring the resistance of the sensor which reflects the temperature change over the heating period, the thermal properties of the materials surrounding the metal film can be extracted. The advantages of this transient measurement technique over steady-state measurement technique are: (1) short measurement time, (2) little restriction of the geometries of samples, (3) non-invasive requirement on the sample structure, (4) no dependency on the measured temperature gradient ∇T. Based on this transient technique, the hot disk sensor was introduced [18] to measure thermal conductivity, thermal diffusivity and specific heat of different kinds of materials. This sensor adopts a double spiral nickel ring for spontaneously sensing and heating. This nickel ring is sealed inside an insulated film to prevent the electrical leakage and the potential damage to the metal. After that, the Hot Disk systems integrating the sensor with the electrical measurement equipment gains popularity in the thermal properties measurement. Various kinds of materials, like solid, liquid, gel [19], and powder beds [15], have been measured by the Hot Disk systems.

2.2 Measurement kit for granular beds

A measurement kit is designed and fabricated to facilitate the experiments on granular beds. As illustrated in Figure 1, two stainless steel rings of 2 cm height form a cylindrical sample holder with an inner diameter of 37 cm. The sensor is placed between two rings with the sensor plane parallel to the cross-section of the rings. The granular materials are poured into the lower ring first, and then the surface of the particle bed is flattened to ensure the horizontal placement of the flexible sensor. The other ring is stacked onto the top of lower ring. The inner space of the ring is filled with the particles and the top surface is flattened as well. The mass of the filling granules is recorded. With the density of the bulk material known, the packing factor of the granular beds can be obtained by calculating the ratio between the solid volume of particles and the total volume of the inner space of two rings. A short rod with diameter slightly being smaller than the inner diameter of rings can be used to apply the external mechanical loading onto the top surface of the granular bed. Currently, the increase of the mechanical loading is realised by stacking weights onto the rod.

Figure 1. The simple measurement kit including two rings (a) and (b) and a short steel rod to support weights (c).

2.3 Types of granular beds

In this study, six glass beds of different diameters, e.g.. (1) 0.1~0.2 mm, (2) 0.25~0.5 mm, (3) 0.5~0.75 mm, (4) 0.75~1.0 mm, (5) 1.0~1.25 mm and 6) 1.25~1.5 mm, are measured by using the same testing procedures. Optimal procedures are selected for the following study to produce reliable measurement. Hot disk sensors of two sizes are used to test the potential variations caused by the sensor size. The change in the effective thermal conductivity of glass beds subjected to external mechanical loadings is investigated by stacking different weights. Moreover, granular beds of different sized (1 mm and 2 mm) stainless steel balls are measured following the similar procedures.

3 Experimental result and discussion

The assembly of the measurement kit is shown in Figure 2 with half of the cylindrical sample holder filled by the glass beads of diameters ranging from 0.75 mm to 1.00 mm. Two sensors #5465 and #5501 of different sizes are shown in the Figure 2 (a) and (b) respectively, and the size contrast can be seen by the comparison between the sensor and the rings.

 DOI: 10.1051/epjconf/201714002016

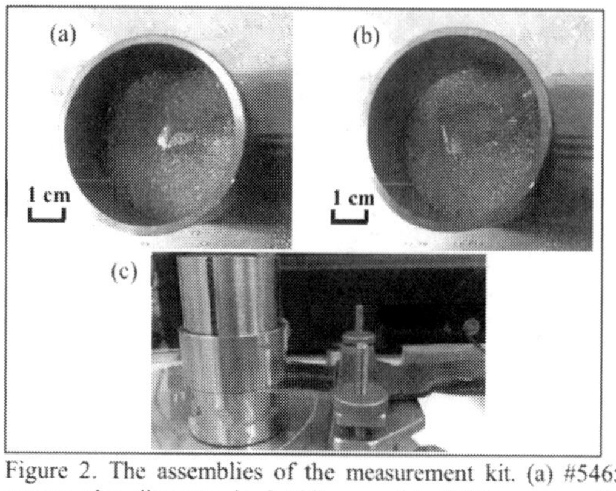

Figure 2. The assemblies of the measurement kit. (a) #5465 sensor, the diameter is 3.189 mm; (b) #5501 sensor, the diameter is 6.403 mm; (c) the assembly for applying mechanical loading.

The effective thermal conductivity with respect to the granule size is measured first. It can be seen from Figure 3 that the effective thermal conductivity increases with the increasing of the granule size. This increasing trend is consistent with the result shown by our numerical study [20] and this phenomenon can be explained by the Smoluschowski effect which is originally observed as the decrease of the effective thermal conductivity of the porous media during evacuation of gases from the media. Generally, this effect is introduced by the increase of the Knudsen number, the ratio of the mean free path of gas molecules over the dimension of a space filled by the gas. The studies of the variable ambient pressure proved the effective thermal conductivity of granular materials is reduced by the increase of the mean free path of the gas molecules [12,21]. In our study, although the gas pressure of the beds is the nearly constant air atmosphere pressure, the Smoluschowski effect is induced by the decrease of the pore space due to the reduction of the granule size. This decrease increases the Knudsen number, as is achieved by the decrease of the gas pressure in other studies. One thing shown by the measurement result is worth to notice that the small sensor tends to give a larger effective thermal conductivity. This indicates that the measured value may depend on the size of the sensor used.

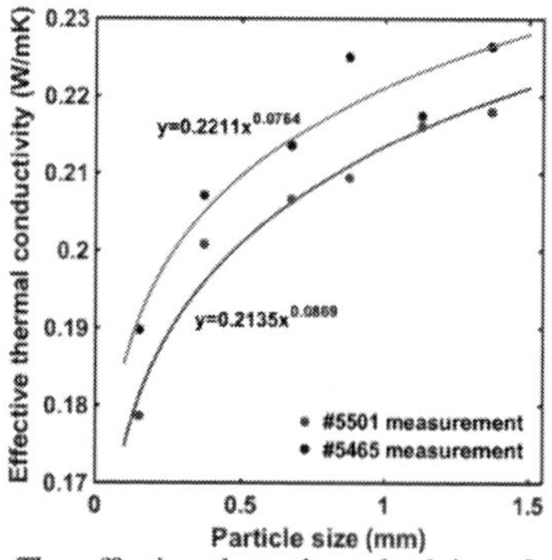

Figure 3. The effective thermal conductivity of glass beds consisting of beads in different diameter ranges. The horizontal coordinates of the points corresponding to the mid-point of the size range.

The effective thermal conductivity versus packing factor of the beds in this study is plotted in Figure 4. The material properties of the glass beds were input into the Zehner-Schlunder-Bauer model to calculate the effective thermal conductivity. The calculated conductivity versus packing factor is also plotted in the Figure 4. A higher packing factor generally means a higher effective thermal conductivity. The measured result deviates from the model prediction when the packing factor over 0.6. Because the Zehner-Schlunder-Bauer model does not take the Smoluschowski effect into account, it may not be able to fully capture the scenario of this study, in particular for these beds with smaller granules. However, as the packing factor has not been well controlled in current step of this study, the scattering of effective thermal conductivity in the packing factor range 0.60~0.63 require further work to interpret.

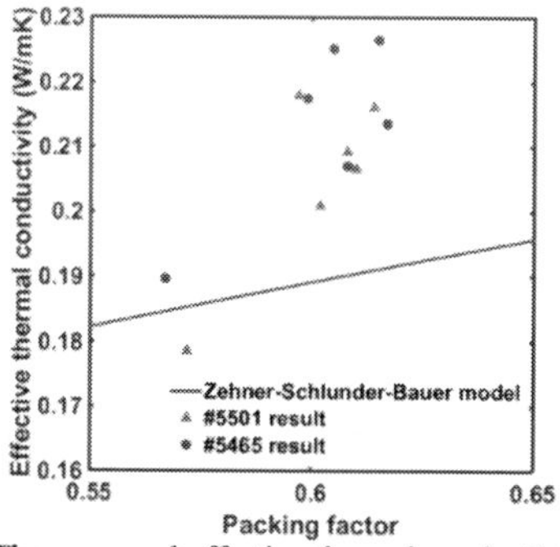

Figure 4. The measured effective thermal conductivity of beds versus packing factors. The solid line shows the calculated Zehner-Schlunder-Bauer conductivity of the same beds.

Then three kinds of glass beds (small--0.1~0.2 mm, medium--0.5~0.75 mm, large--1.25~1.5 mm) were selected to investigate the influence of the mechanical loading. The effective thermal conductivity of glass beds shows a slight increment with increasing the mechanical loading in the tested range. This increase is relative obvious when the loading was just increased from 0 to about 20 kPa. Further increase shows a minor effect. Again the granule size effect can be observed.

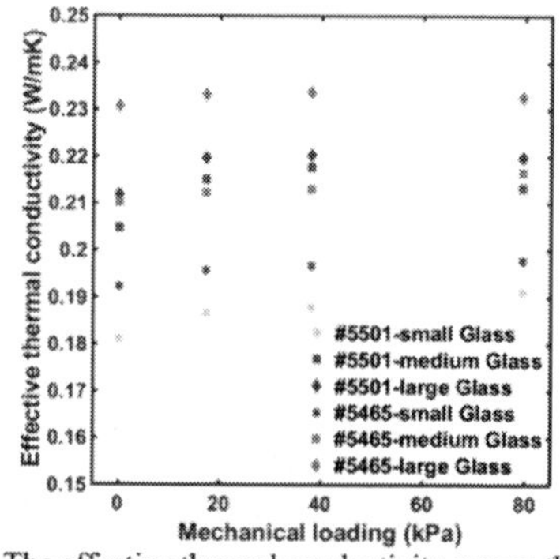

Figure 5. The effective thermal conductivity versus the external mechanical loadings for glass beds.

In parallel, the steel beds were studies in the same way as glass beds. In contrast to the result from the glass beds, the enhancement in the effective thermal conductivity occurs in the entire mechanical loading range. This continuous rise in the conductivity can be resulted from the relatively high thermal conductivity of steel compared to the glass. The higher loading pressure is supposed to enlarge the contact area between particles, giving rise to the conduction through the solid phase. Similar to the glass beds, a larger particle size shows a higher effective thermal conductivity.

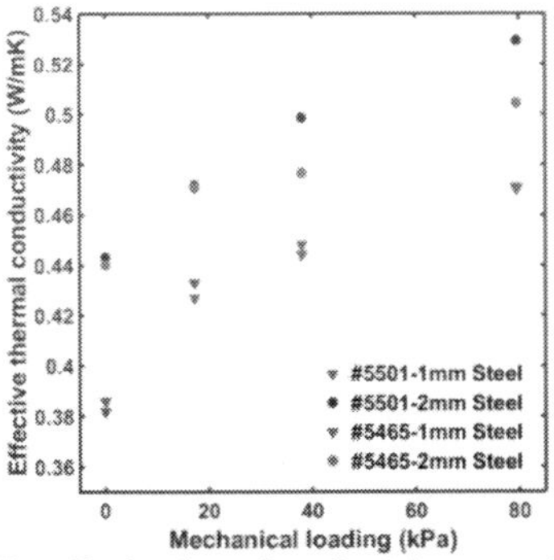

Figure 6. The effective thermal conductivity versus the external mechanical loadings for steel beds.

In terms of the influence of the sensor size, it is obvious in the glass beds measurement, but looks like playing no significant role in the steel beds measurement. It may be possible that the low thermal conductivity of the glass, producing a localised heating profile, is responsible for this observation. However, conclusive comments are hard to given in this stage. Further experimental study combined with numerical simulation method is required to clarify this influence so that the measured results can be better understood.

4 Conclusion

The effective thermal conductivity of glass beds as well as steel beds has been measured by the Hot Disk system based on the transient plane source technique. It is clearly demonstrated that the effective thermal conductivity of the granular beds is influenced by the granule size and it increases with the increasing granule size. This phenomenon can be explained by the Smoluschowski effect on the thermal conductivity of interstitial gases. The lowest thermal conductivity of 0.1~0.2 mm glass beds can also be attributed to the lowest packing factor. It is worth to study how the packing factor varies the effective thermal conductivity in a more controllable experiment procedure. The mechanical loading plays a positive role in enhancing the thermal transport process in the measured beds. However, such a positive effect is more significant in the steel beds compared to the glass beds because the intrinsic thermal conductivity of steel is much higher than the glass.

The variation in the measured effective thermal conductivity by using sensors of different sizes is observed in this study, in particular for beds with intrinsic materials having lower conductivity, subjected to the measurement conductivity by the Hot Disk system. Further research is needed to provide a sound theoretical understanding and fundamental explanation to this phenomenon.

Acknowledgements: Financial support for this research from the Australia Research Council through grant DE130101639 is greatly appreciated.

References

1 M. Markl and C. Körner, Annual Review of Materials Research **46** (1), 93 (2016).
2 K. Zhao, M. Pharr, J. J. Vlassak, and Z. Suo, Journal of Applied Physics **108** (7), 073517 (2010).
3 Y. Gan and M. Kamlah, Journal of the Mechanics and Physics of Solids **58** (2), 129 (2010).
4 A. Bauereiß, T. Scharowsky, and C. Körner, Journal of Materials Processing Technology **214** (11), 2522 (2014).
5 Y. Gan, Francisco Hernandez, Dorian Hanaor, Ratna Annabattula, Marc Kamlah and Pavel Pereslavtsev, Fusion Science & Technology **66** (1), 83 (2014).
6 M. Bahrami, M. M. Yovanovich, and J. R. Culham, International Journal of Heat and Mass Transfer **49** (19-20), 3691 (2006).
7 P. Zehner and S. EU, Chemie Ingenieur Technik **42** (14), 933 (1970).
8 R. Bauer and E. Schlunder, International Chemical Engineering **18** (2), 189 (1978).
9 C. T. Hsu, P. Cheng, and I. W. Wong, International Journal of Heat and Mass Transfer **37** (17), 2751 (1994).
10 S. Yagi and D. Kunii, AIChE Journal **3** (3), 373 (1957).
11 W. Woodside and J. H. Messmer, Journal of Applied Physics **32** (9), 1688 (1961).
12 M. A. Presley and P. R. Christensen, Journal of Geophysical Research: Planets **102** (E3), 6551 (1997).
13 Fatollah Tehranian and M. A. Abdou, Fusion Science & Technology **27** (1995).
14 G. Widenfeld, Y. Weiss, and H. Kalman, Powder Technology **133** (1–3), 15 (2003).
15 J. Mo, D. Garrett, and H. Ban, International Journal of Thermophysics (2015).
16 D. Li and D. Liang, International Journal of Heat and Mass Transfer **92**, 8 (2016).
17 S. E. Gustafsson, Review of Scientific Instruments **62** (3), 797 (1991).
18 M. Gustavsson, E. Karawacki, and S. E. Gustafsson, Review of Scientific Instruments **65** (12), 3856 (1994).
19 Y. He, Thermochimica Acta **436** (1-2), 122 (2005).
20 W. Dai, Y. Gan, and D. Hanaor, Submitted.
21 A. V. Gusarov and E. P. Kovalev, Physical Review B **80** (2), 024202(1) (2009).

EPJ Web of Conferences **140**, 02017 (2017)
Powders & Grains 2017
DOI: 10.1051/epjconf/201714002017

FEM modeling of the sandpile dip effect

Isadora Cota Carvalho[1,*], *Diego* Santos[1,**], and *A. P. F.* Atman[2,3,***]

[1]*Materials Engineering Department, Centro Federal de Educação Tecnológica de Minas Gerais, CEFET-MG, Av. Amazonas 5651, Belo Horizonte, MG, Brazil. CEFET-MG.*
[2]*Departamento de Física e Matemática, Centro Federal de Educação Tecnológica de Minas Gerais, CEFET-MG, Av. Amazonas 7675, Belo Horizonte, MG, Brazil.*
[3]*National Institute of Science and Technology for Complex Systems, INCT-SC – CEFET-MG.*

Abstract. Previously, it was shown that the pressure "dip" in the stress profile observed at the bottom of sandpiles prepared by successive avalanches can be related with the skewness effect observed inthe stress response of sheared granular layers to a localized vertical overload. Here, we present a FEM study of the sandpile dip effect using an alternative geometry ("wagon") to build the pile, considering a simplified anisotropic elastic model, with an orthotropy angle. We explore the effect of different boundary conditions and grid coarsening on numerical stress response profiles. Our results indicate that the wagon geometry combined with partially fixed boundary condition may describe better experimental results than the conical geometry, a remark which can contribute for better comprehension of this effect.

1 Introduction

One of paradigmatic features in the physics of static granular materials is the sandpile stress dip effect [1–4]. Following an intense debate in the literature along several years, first, to recognize this effect [5–7] and, later, to furnish a fair theoretical description [8–10]. Nowadays, it is well established that this dip effect is due to the preparation history of the sandpile [11–16] which lead to a spontaneous pattern formation on the force chains along the layer. Aside the debate along the mechanical origins for the dip, a rich discussion concerning how forces are transmitted along granular materials have emerged in this context, forming a theoretic panorama for stress propagation in granular layers: hyperbolic partial differential equations (PDE) are associated with the stress propagation in ordered materials, while elliptic PDE are associated with the stress propagation in disordered assemblies [17], but transition between these behaviors can occur under given conditions of friction or force magnitude [18, 19].

In this context, a tool initially designed to distinguish between the very different mathematics of hyperbolic and elliptic PDEs was proposed: the stress response function (SRF) [8, 20]. It consists basically to measure the pressure profile at the bottom of a layer in response to a localized overload at its top surface. Elliptic PDEs predict a single broad peak for this stress response while hyperbolic models would yield two peaks in 2D or a ring in 3D. Several experiments [11, 14, 21–23] and simulation studies [15, 16, 24–26] have addressed this issue and the above scenario for the stress propagation in granular materials is well established today and fairly verified [4, 27].

Nevertheless, there are still several unresolved questions. For instance, Atman *et al.* [15] have proposed a relation of the stress dip observed in sandpiles prepared with successive avalanches with the skewness angle observed in SRF profiles of sheared granular slabs[14], a remarkable example of strain induced anisotropy. However, the comparison of experimental results and numerical data did not presented a good match, except when one of the control parameter was set to infinity [15]. In that paper, the authors have suggested that the geometry used in the construction of the 3-D conical piles maybe could have some influence in the disagreement observed in the results, since they considered a 3-D conical pile produced by rotation of a triangular element around the vertical axis. This geometry led to a sharp tip, a point in the mathematical sense, where the overload was applied on the top surface of the sandpile. Here, we present an extension of this work considering a trapezoidal geometry to build the pile – Figure 1, and study the dependence of the skewness angle β with the orthotropy angle τ of the layer. We also study the effect of using different boundary conditions for the substrate in the SRF: completely fixed or partially fixed, and also different grid coarsening.

The paper is structured as following: the next section is devoted to present the basic methodology used to build the trapezoidal layer ("wagon") in CAST3M^{TM} [28] language, an open source for Finite Element Method (FEM). We also present the analytical treatment used to obtain the SRF function with this geometry and the details of the numerical model. Next, the results obtained for the range

*e-mail: isadora.cota@gmail.com
**e-mail: diego.so2@hotmail.com
***e-mail: atman@dppg.cefetmg.br

DOI: 10.1051/epjconf/201714002017

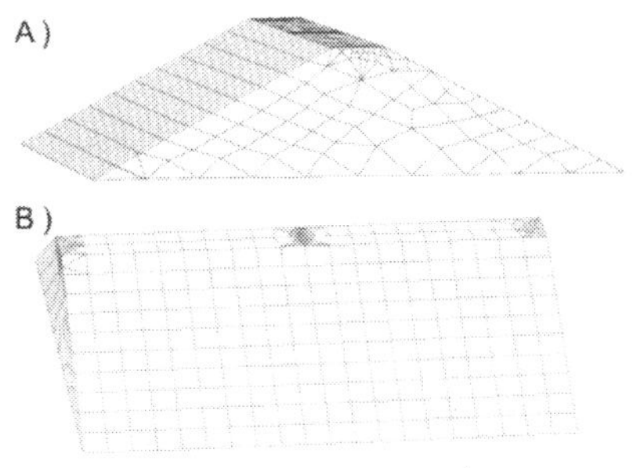

Figure 1. The "wagon" geometry. We show two different side views of the prismatic pile considered to measure the SRF profiles. Above: $x - z$ plane. The SRF profiles are measured along the x axis at the bottom of the pile. Below: $y - z$ plane. Note the translational symmetry along the y direction and the increasing of resolution (finer grid coarsening) at the center of the pile, where the SRF is measured.

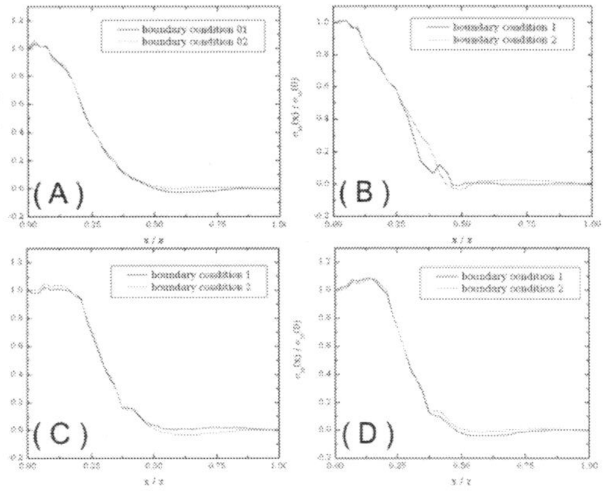

Figure 2. Comparison of stress response profiles for different boundary conditions. Panels a), b) and c) show results for Model 1, Model 2 and Model 3, respectively, with $\tau = 0^o$. Panel d) show the result for Model 3 and $\tau = 15^o$.

of parameters and boundary conditions tested are shown and discussed. Finally, we draw some conclusions and perspectives to future investigations and our acknowledgements.

2 Methods

Here we present briefly the elastic calculations considered to build the model and obtain the stress response function profiles. Due to didactic purposes, we first will present the 2D case before to generalize the result for the 3D numerical model. In both 2D and 3D, isotropic elasticity describes a given material by means of two constitutive parameters, the Young modulus E and the Poisson ratio ν. A third constitutive parameter, the shear modulus G, can be written in function of E and ν in the isotropic case. The anisotropic case is a bit more complicated, since the number of parameter grows fast with dimension. Here, we will consider an anisotropic elastic model with an orthotropy angle, which leads to five independent parameters in 2D [17, 22], as shown below. A meaningful fitting study of stress response profiles with this model was published recently [26].

The motivation to use an anisotropic model with an orthotropy angle is the possibility to reproduce a well known experimental phenomena of granular assemblies submitted to plane shear, the tilting of chain forces at an angle of 45° with respect to the shear direction. This rearrangement of force chains lead to a reinforcement of contacts in that direction and, consequently, a softening of those contacts in the perpendicular directions. Thus, the Simple Elastic Model (SEM) in 2D [16] considers a stiff direction (direction 1) which makes an orthotropy angle τ relative to the vertical axis z (x is the horizontal direction), and a softer direction (direction 2) with a smaller Young modulus $E_2 < E_1$. Along with two Poisson coefficients, ν_{12} and ν_{21}, and the shear modulus G, we can write the strain-stress relation in the matrix form, expressed in these principal stress directions:

$$\begin{pmatrix} u_{11} \\ u_{22} \\ u_{12} \end{pmatrix} = \begin{pmatrix} \frac{1}{E_1} & -\frac{\nu_{21}}{E_2} & 0 \\ -\frac{\nu_{12}}{E_1} & \frac{1}{E_2} & 0 \\ 0 & 0 & \frac{1}{2G} \end{pmatrix} \begin{pmatrix} \sigma_{11} \\ \sigma_{22} \\ \sigma_{12} \end{pmatrix}, \quad (1)$$

The symmetry properties of the compliance matrix leads to the following $E_2/E_1 = \nu_{21}/\nu_{12}$, and the requirements for a proper definition of the elastic free energy are: $E_1, E_2, G > 0$ and $1 - \nu_{12}\nu_{21} > 0$. The isotropic limit is obtained with $E_1 = E_2 = E$, $\nu_{12} = \nu_{21} = \nu$ and $G = \frac{E}{2(1+\nu)}$.

Otto *et al* [17], have studied analytically the stress tensor components for the SEM model in the case of a semi-infinite medium ($z > 0$) loaded with a unitary point force at the origin $x = z = 0$. For a given anisotropy angle τ, they have shown that the SRF cna be described with only two parameters:

$$t = E_2/E_1 = \nu_{21}/\nu_{12}, \quad (2)$$

$$r = \frac{1}{2} E_2 \left(\frac{1}{G} - \frac{\nu_{12}}{E_1} - \frac{\nu_{21}}{E_2} \right). \quad (3)$$

For a layer of finite thickness, an additional parameter is needed to be describe the profiles [15, 22]. The extension of the model to consider an arbitrary orthotropy angle τ and finite thickness for the layers was presented recently [25, 26]. To recover the stress components at the normal and tangential coordinates, a rotation matrix should be applied,

$$Q = \begin{pmatrix} \cos^2\tau & \sin^2\tau & 2\cos\tau\sin\tau \\ \sin^2\tau & \cos^2\tau & -2\cos\tau\sin\tau \\ -\cos\tau\sin\tau & \cos\tau\sin\tau & \cos^2\tau - \sin^2\tau \end{pmatrix}. \quad (4)$$

Considering that

$$\begin{pmatrix} u_{nn} \\ u_{tt} \\ u_{tn} \end{pmatrix} = \mathcal{W}_\tau \begin{pmatrix} \sigma_{nn} \\ \sigma_{tt} \\ \sigma_{tn} \end{pmatrix} \quad \text{with} \quad \mathcal{W}_\tau = Q^{-1}\mathcal{W}_\dagger Q\,, \quad (5)$$

where $\mathcal{W}_\dagger$ is the compliance matrix. The explicit form of $\mathcal{W}$ can be found in [26].

The generalization of this approach for $3D$ is immediate, considering two axes (directions 2 and 3) orthogonal to direction 1 as isotropic and soft directions, in the principal stress reference system. In the real space, besides the horizontal direction x along which the shearing is applied, and the vertical direction z, a third direction, y, perpendicular to these two, is now considered. The 3D equivalent of the compliance matrix has a similar structure to the 2D

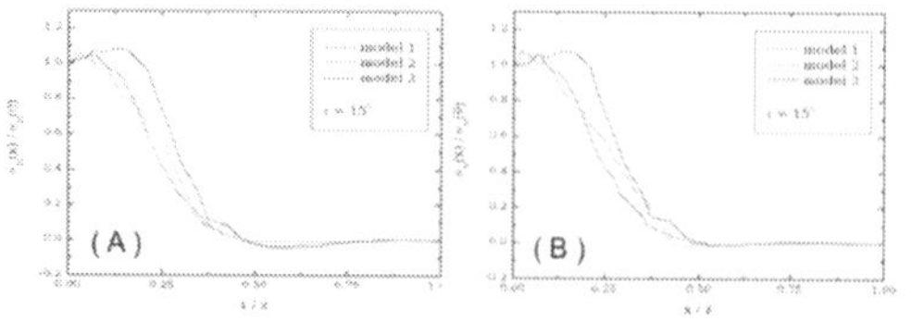

Figure 3. Comparison of the SRF considering different boundary conditions. Left panel show results for the three models considered and fixed boundaries. Right panel show the results considering partially fixed boundaries.

case, but with three Young moduli E_i, three shear moduli G_i and six Poisson ratios ν_{ij} (i and j must be different). Again, the symmetry of the matrix gives three relations of the form $\nu_{ij}/E_i = \nu_{ji}/E_j$. In the final analysis, there are 9 independent coefficients besides the anisotropy angle τ. Stability now requires that $E_i, G_i > 0$, $1 - \nu_{ij}\nu_{ji} > 0$ and $1 - \sum_{i \neq j} \nu_{ij}\nu_{ji} - \nu_{12}\nu_{23}\nu_{31} - \nu_{21}\nu_{32}\nu_{13} > 0$.

To our knowledge, there is no closed analytical expression for the stress response profiles in 3D up to now, and we can use a Finite Element open source code developed in soil mechanics called CAST3M^{TM} [28] to compute the stress components. We test three grid coarsening values to observe the sensibility of results with the grid resolution: "Model 1" ($41 \times 13 \times 13$), "Model 2" ($41 \times 15 \times 15$) and "Model 3" ($43 \times 13 \times 13$)cells in the x, y and z directions respectively. We also considered two boundary conditions for the substrate: a completely blocked condition – "Condition 1", where no displacement is allowed in any direction ($U_z = Ux = U_y = 0$); and a partially blocked condition – "Condition 2", where only the corners of the substrate and the vertical direction ($U_z = 0$) are blocked, while in (x, y) directions small displacements are allowed.

Following the reasoning of Atman *et al.*, we limit the parameter space to better understand the role of each player: the average stiffness $\overline{E} = (E_1+E_2+E_3)/3$, the shear modulus G and τ, the orthotropy angle, measured with respect to the vertical direction. Thus, we fixed the same value of G for all directions - $G_1 = G_2 = G_3 = G = \frac{\overline{E}}{2(1+\nu)}$), with $\nu_{ij} = 0.3$ and $\overline{E} = 150MPa$. In the spirit of the $2D$ analytical results, we use two rescaled parameters to present our data, i.e. the stiffness ratio $t = E_2/E_1$, and the shearing ratio $u_G = E_2/G$. This last parameter is analogous to the parameter r in $2D$ (see equation 2). For the anisotropic samples, we have considered $t = 1.5$, and keep all other parameters fixed.

3 Results

An essential result from shearing experiments in $2D$ and in $3D$ was that the orthotropic direction is tilted with an angle $\tau = 45^o$ with the shear direction, leading to a skewness angle of $\sim 22o$ in 2D and around 8° in $3D$. By means of a fitting procedure Atman *et al.* have obtained $t = 1.5$ and $u_G = 2.14$ for the best fit of SRF in shearing experiments with sandbox. When applying these values in a conical pile numerical finite element model, build a pile from the rotation or a triangular around the z axis, the authors did not found a good fit (skewness angle $\eta = 3^o$). The best fit for the SRF from sandpile experiments seems to correspond to the limit $u_G \to \infty$ [15], a result that still is

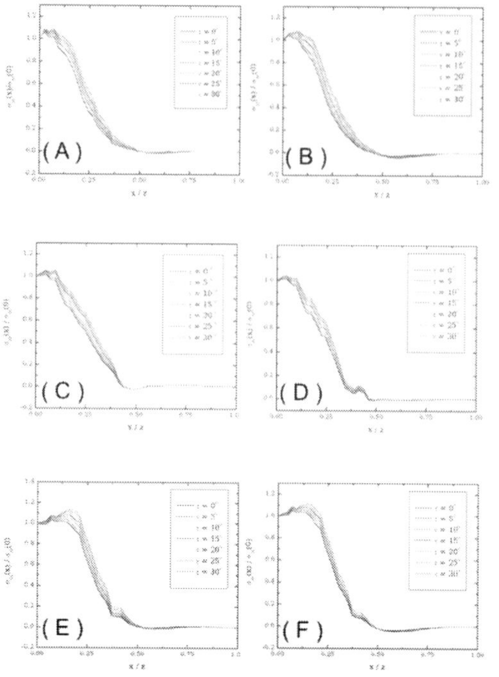

Figure 4. Dependence of the SRF with the orthotropy angle. a) Model 1, boundary condition 1. b) Model 1, b.c. 2. c) Model 2, b.c. 1. d) Model 2, b.c. 2. e) Model 3, b.c. 1. f) Model 3, b.c. 2.

obscure. Here, we test a different geometry, that we called "wagon", in the hope to contribute to the understanding of the reasons for this mismatch.

In Figure 1 we show the "wagon" geometry that we have considered to build the pile, and in Figure 2 the corresponding SRF profiles. These results were obtained considering the SEM model with $t = 1.5$ and $u_G = 2.08$, but $\tau = 0^o$ to test only the influence of the grid coarsening and boundary conditions. It is clear that the coarse-graining has a small influence on the SRF profiles, but the boundary conditions plays a less important role. The partially blocked condition furnished results which seem to better reproduce the profiles obtained in experiments or simulations. The comparison of the results for the different grid coarsening is shown in Figure 3.

The dependence of the SRF profiles with the orthotropic angle is shown in Figure 4 for each grid coarsening considered and different boundary conditions. Figure 5 show the summary of the results obtained, displaying the dependence of the skewness angle with the control parameters considered here. The main result is that the skewness angle can be considered enhanced with a proper combination of boundary conditions and orthotropy angle. The grid coarsening plays a role on the stress profiles, it is related mainly to the steepness of the response peak. This result indicates that the mathematical feature of the conical pile, with a perfect sharp tip maybe was the cause for the mismatch of the results. However, we believe that we have to better improve the aspect ratio of the wagon geometry in order to reproduce better the experimental SRF profiles of avalanched sandpiles. Moreover, we hope to test other values for t and u_G in order to understand better

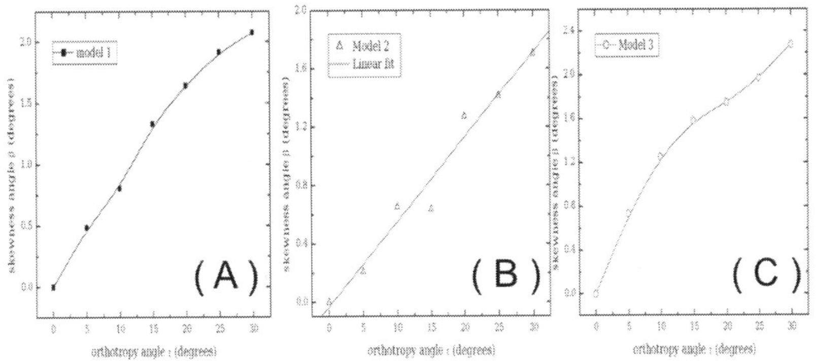

Figure 5. Skewness angle variation in function of the orthotropy angle for the different models considered, as indicated in the insets.

what is the reason for the mismatch in the skewness angle, which still is verified here.

4 Conclusion

We present an extension of previous study of the FEM modeling of stress dip effect in sandpiles prepared by successive avalanches. We test different boundary conditions and grid coarsening combinations to study their influence on the stress response profiles. Our results indicate that a proper combination of boundary condition, grid coarsening and elastic parameters could furnish results which could fit better the experimental and simulation results. We hope to explore the parameters space and also test other grid coarsening combinations to understand better the reason for the mismatch in the skewness angle which persists with the wagon geometry.

Acknowledgements

We thank Prof. Philippe Claudin and Éric Clément for illuminating discussions. ICC thanks CEFET-MG, PMMH/ESPCI and P. Claudin for the scientific interchange during her graduation.

References

[1] *Physics of Dry Granular Media*, H.J. Herrmann, J.-P. Hovi, and S. Luding editors, NATO ASI Series, Kluwer, (1997);

[2] R.P. Behringer and J.T. Jenkins editors, *Powders and Grains 97*, Balkema, (1997);

[3] H.M. Jaeger, S.R. Nagel, and R.P. Behringer, Rev.Mod. Phys. **68**, 1259 (1996).

[4] A. Mehta, *Granular Physics*, Cambridge Univ. Press, (2007).

[5] J.Šmíd, and J. Novosad, Proc. Powtech. Conference 1981, Ind. Chem. Eng. Symp. **63**, D3V 1 (1981).

[6] R. Brockbank , J.M. Huntley, and R.C. Ball, J. Phys. (France) II **7**, 1521 (1997).

[7] L. Vanel, D.W. Howell, D. Clark, R.P. Behringer and E. Clément, Phys. Rev. E **60**, R5040 (1999).

[8] J.-P. Bouchaud, M.E. Cates and P. Claudin, J. Phys. (France) I **5**, 639 (1995).

[9] J.P. Wittmer, M.E. Cates and P. Claudin, J. Phys. (France) I **7**, 39 (1997).

[10] P. Claudin, Ph.D. thesis *La physique des tas de sable*, Annales de Physique **24**, n°2, pp 1-210 (1999).

[11] J. Geng, D. Howell, E. Longhi, R. P. Behringer, G. Reydellet, L. Vanel, E. Clément and S. Luding, Phys. Rev. Lett. **87**, 035506 (2001).

[12] J.-J. Moreau, in the proceedings of the *colloque 'Physique et mécanique des matériaux granulaires'*, LCPC Editor, Champs-sur-Marne (France), 199 (2000).

[13] S. Luding, Phys. Rev. E, 55 4720 (1997).

[14] J. Geng, G. Reydellet, E. Clément and R.P. Behringer, Physica D **182**, 274 (2003).

[15] Atman A.P.F., Brunet P., Geng J., Reydellet G., Claudin P., Behringer R.P., Clément E.: Eur. Phys. J. E **17**, 93 (2005).

[16] Atman A.P.F., Brunet P., Geng J., Reydellet G., Combe G., Claudin P., Behringer R.P., Clément E.: J. Phys. Cond. Mat. **17**, S2391 (2005).

[17] M. Otto, J.-P. Bouchaud, P. Claudin and J.E.S. Socolar, Phys. Rev. E **67**, 031302 (2003).

[18] Atman A.P.F., Claudin P., Combe G.: Comput. Phys. Comm. **180**, 612 (2009).

[19] Goldhirsch I., Goldenberg C.: Nature **435**, 188 (2005).

[20] P.-G. de Gennes, Physica A **261**, 267 (1998); Rev. Mod. Phys. **71**, S374 (1999).

[21] G. Reydellet and E.Clément, Phys.Rev.Lett. **86**, 3308 (2001)

[22] D. Serero, G. Reydellet, P. Claudin, E. Clément and D. Levine, Eur. Phys. J E **6**, 169 (2001).

[23] N.W. Mueggenburg, H.M. Jaeger and S.R. Nagel, Phys. Rev. E **66**, 031304 (2002).

[24] Goldenberg C., Atman A.P.F., Claudin P., Combe G., Goldhirsch I.: Phys. Rev. Lett. **96**, 168001 (2006).

[25] Atman A.P.F., Claudin P., Combe G., Mari R.: Europhys. Lett. **101**, 44006 (2013).

[26] A.P.F. Atman, P. Claudin, G. Combe, G.H.B. Martins, Granular Matter, **16**:2, 193 (2014).

[27] B. Andreotti, Y. Forterre, Y. and O. Pouliquen, *Granular Media: Between Fluid and Solid*, 2013, Cambridge University Press.

[28] with CAST3M, http://www-cast3m.cea.fr/

DOI: 10.1051/epjconf/201714009032

Influence of granulometry in the Hurst exponent of air liquid interfaces formed during capillary rising in a granular media

Guilherme L. Gontijo[1,*], *Flávia B.* Souza[1], *Rafael M. L.* Braga[2], *Pedro H. E.* Silva[2], *Maury D.* Correia[5,**], and *A. P. F.* Atman[3,4,***]

[1] *Civil Engineering Department, Centro Federal de Educação Tecnológica de Minas Gerais, CEFET-MG, Av. Amazonas 7675, Belo Horizonte, MG, Brazil.*
[2] *Materials Engineering Department, CEFET-MG, Av. Amazonas 5253, Belo Horizonte, Brazil,*
[3] *Departamento de Física e Matemática, CEFET-MG, Av. Amazonas 7675, Belo Horizonte, MG, Brazil.*
[4] *National Institute of Science and Technology for Complex Systems, INCT-SC – CEFET-MG.*
[5] *Centro Brasileiro de Pesquisas Físicas, CBPF. R. Xavier Sigaud, 150, Rio de Janeiro, RJ, Brazil.*

Abstract. We report results concerning the fractal dimension of a air/fluid interface formed during the capillary rising of a fluid into a dense granular media. The system consists in a modified Hele-Shaw cell filled with grains at different granulometries and confined in a narrow gap between the glass plates. The system is then placed onto a water reservoir, and the liquid penetrates the medium due to capillary forces. We measure the Hurst exponent of the liquid/air interface with help of image processing, and follow the temporal evolution of the profiles. We observe that the Hurst exponent can be related with the granulometry, but the range of values are odd to the predicted values from models or theory.

1 Introduction

Fluid flow in heterogeneous media is a wide spreading phenomenon found in several natural and industrial processes, such as groundwater formation, percolation, oil extraction, chromatography, crack formation, etc. [1–3]. Due this feature, researchers from various fields, such as chemistry, physics and engineering in general, became interested on it, and several approaches combining theory, experiment and numerical modeling have been applied [4–6].

A key concept to understand this phenomena is pattern formation, particularly fractal geometry and analysis [4], and despite several efforts devoted to understand the fluid flow in granular media, a unified description for this phenomena still lacking [7–11]. Previously, Atman *et al.* have studied the air/liquid interface formed during the capillary rising of water into a granular media [12], but the results did not fit to the expected theoretic models. Here, we present an extension of that work presenting new results for the Hurst exponent of the air/liquid interface now considering granular media with three different granulometries. We also considered different widths for the gap between the plates of the Hele-Shaw cell used in the experiments, and compare the results with previous experiments, simulations and the predicted values of the Depinning by Directed Percolation model (DPD) [4, 5].

*e-mail: glgontijo@gmail.com
**e-mail: maury@cbpf.br
***e-mail: atman@dppg.cefetmg.br

The paper is structured as follows: after this brief introduction, we present the experimental setup and methodology in the next section. After, we discuss the results obtained and close with our conclusions and perspectives.

2 Experimental Setup

We consider a Hele-Shaw cell filled with a granular material (sand) with different gradings (granulometries), assembled to form an apparatus which can be placed over an water reservoir – Figure 1. The glass plates have dimensions $(400 \times 300 \times 4)mm$, put apart by spacers with 4 mm each, placed at the four corners in the gap between the plates. We also run experiments with 2 and 3 spacers for a given grading. To fill the apparatus with sand, a special bond tape is placed at lateral borders to block the water passage, and also help to hold the plates together. The gap at bottom of the apparatus is filled with a fine polystyrene foam to avoid sand evasion but allowing the water passage. The granular media used was a white quartz sand, with different gradings: $45-90$ mm, $70-120$ mm and $125-600$ mm, which means that the grains have 95% of their average radii within that range. Grain geometries are very different, mostly consisting of polyhedrons with several faces, but which could be approximated by a sphere with a mean radius r.

The cell is filled with sand, gently poured atop the cell, with help of a sheet of paper. The top side of the cell is left open and the sand is poured until the gap is completely

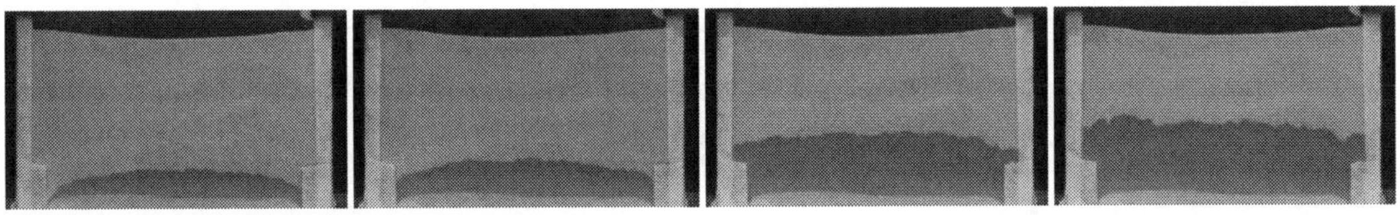

Figure 1. Fluid displacement in heterogeneous media. From left to right, snapshots of the experimental assembly showing the capillary rising of the fluid into the granular media, evincing the high contrast between the wet and dry phases.

filled. During this process, sometimes we tapped the system to compact and homogenize the media, but some degree of segregation is unavoidable. The granular media formed is very heterogeneous, characterized as a highly connected porous media with strong capillary pressure.

Next, the apparatus is fixed at the four corners with clamps, and placed vertically over a water reservoir, partially filled at that moment. A digital video camcorder Canon^{TM} Full HD Vixia HF G30, with (1920x1080) pixels, was placed at $1.5m$ in front of the assembly. The level of water in reservoir is then slowly raised until it forms a flat interface with the glass bottom. Just before the level of the water in the reservoir reaches the bottom of the plates, the camera is turned on. A complete record of an experiment typically takes around 15 minutes, thus we can disregard evaporation effects and consider the water-glass interface angle as constant - the volume of water of the reservoir ($40L$) is much bigger than the water volume inside the cell ($100mL$).

As soon the water touch the granular media, a liquid/air interfaced is formed, altering the contrast of wet area, allowing the identification of the interface after image treatment. The images were treated with ImageJ^{TM} software with a sequence of numerical operations [6], as detailed in Figure 2. As resutl, a single valued function corresponding to the profile of the liquid/air interface, $h(x.t)$, where x indicates the horizontal position, can be extracted.

We have performed five runs considering exactly the same procedure described above, for each set of the tested parameters. For each run, we and took snapshots of the cell along all the experiment, at predetermined times. We calculate the Hurst exponent H, for each interface considered, using detrended fluctuation analysis (DFA) [13, 14]. Then, the average value of the Hurst exponents at a given time, $H(t)$, can be plotted against time – Figure 3.

The roughness of the interface is defined as the variance of the function $h(x,t)$:

$$w(t) = \sqrt{\frac{1}{2N}\sum_{i}^{N}\left(h_i - \overline{h}\right)^2} \tag{1}$$

where h_i is the height of the interface at the column i, and i is numbered in function of the pixel resolution of the camera. DFA considers the roughness in a local scale ϵ regarding to the best linear fit for the data in this scale:

$$w(\epsilon) = \sqrt{\frac{1}{1+2\epsilon}\sum_{i-\epsilon}^{i+\epsilon}(h_i - (\alpha(\epsilon)x_i + \beta(\epsilon)))^2}\ , \tag{2}$$

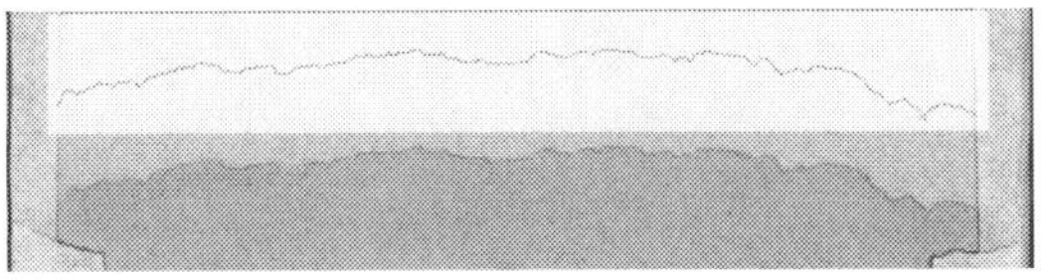

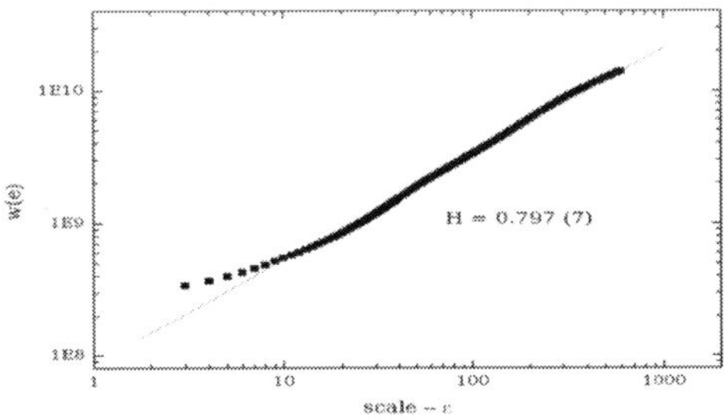

Figure 2. Image treatment of the air/liquid interface. The top panel shows the image captured from the digital camera, which was treated with ImageJ^{TM} in order to obtain a single valued function, shown above the treated image. The bottom panel show the methodology to calculate the Hurst exponent, eq. 3. The continuous line indicates a power law fit giving the exponent $H = 0.797(7)$.

where $\alpha(\epsilon)$ and $\beta(\epsilon)$ are, respectively, the angular and the linear fit coefficients at the local scale ϵ.

Then, we can obtain the Hurst exponent from the relation:

$$w(\epsilon) \sim \epsilon^H\ , \tag{3}$$

as shown in Figure 2. In a log × log plot, it corresponds to the angular coefficient of the curve. The linear behavior of $w(\epsilon)$ is a sign of the self-affinity of the function considered, confirming its fractal feature.

3 Results and Discussion

We have performed five runs for each set of experimental parameters, and the results shown correspond to the average values. Figure 3 show the results for different spacers and (45 – 90 grading. We can observe that as the gap width increases, the results become more noisy, and fluctuates considerably. This is consistent with a layer-by-layer forming interface, and could be related with the three dimensional feature of the media as the gap width increases. Considering the Hurst exponent, the stationary results shown a tendency for a decreasing of its value as

DOI: 10.1051/epjconf/201714009032

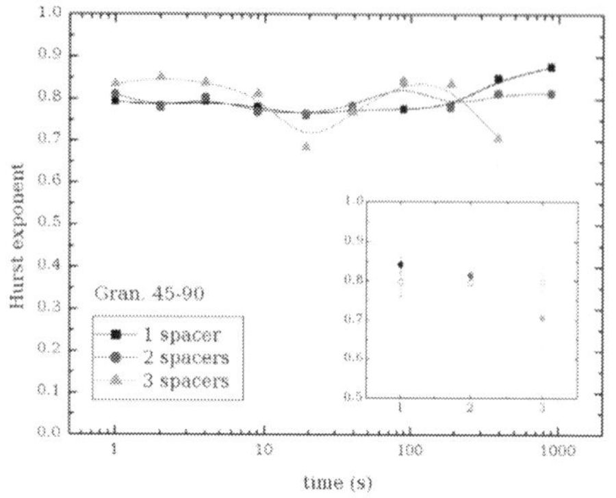

Figure 3. Temporal evolution of the Hurst exponent considering different gap widths. The grading considered was 45 − 90 *mm* and we show results for 1, 2 and 3 spacers placed in the gap. Despite the curves are pretty close to each other, in the inset its possible to verify a tendency on the Hurst exponent value to decrease as the gap width increases. In the inset, open symbols correspond to the temporal average of the Hurst exponent, and the filled ones correspond to stationary values.

the spacing increases, a results which indicates that the interface become more flat with. Again, this result is compatible with the three-dimensional feature of the interface as the gap width increases.

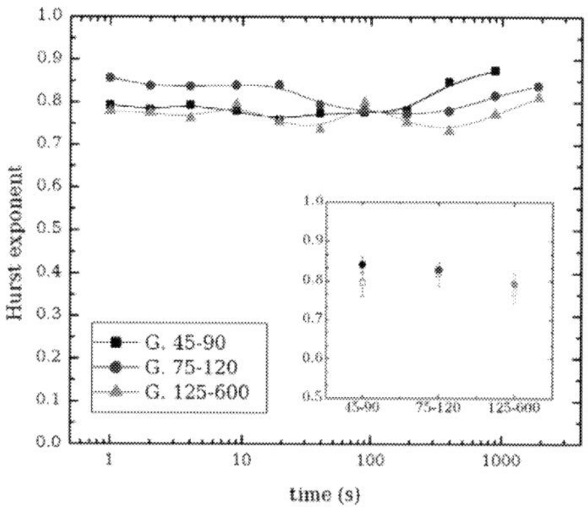

Figure 4. Time evolution of the Hurst exponent for the air/liquid interface for different sand gradings. Again, the curves are quite close to each other, but we can observe a tendency of smaller values of Hurst exponent when the grading is larger, an observation which agrees with our expectation since the media became more connected as the granulometry increases.

In Figure 4 we shown the results for the Hurst exponent obtained for the same gap width (single spacer) but with different gradings. Surprisingly, the results do not shown great dispersion. Only a slight tendency to decreasing of the Hurst exponent value was observed for larger values of gradings. All results for the Hurst exponent fluctuates around $H \sim 0.8$. This value do not agree with none model or other experiment that we are aware, except [?]. The conclusion is that the high connectivity of the granular media implies to smoother surfaces, rising the value of the Hurst exponent above to described in literature. We have developed a model to study this problem, but for the moment our results do not fit well the experiments.

4 Conclusions and Perspectives

we have studied the influence of the gap width and the sand grading on the value of the Hurst exponent of the air/liquid interface formed during the displacement of air by water into a heterogeneous granular media. Our results indicates that an average value of $H \sim 0.8$ for all experiments, a result which was quite surprisingly for us. A slight tendency of smoother surfaces (smaller H values) was observed for larger gaps and increasing the grading. We hope to extend this work using grading ranges more narrow and also testing thicker and narrower gap widths.

Acknowledgements

PHES and FBS thank CNPq for financial support. APFA and RMB thank CEFET and FAPEMIG for financial support.

References

[1] Y. Shikhmurzaev, *Capillary Flows With Forming Interfaces* (Chapman & Hall/CRC, 2008), ISBN 9781584887485

[2] M. Sahimi, Review Modern Physics **65**, 1393 (1993)

[3] D. Dalmas, E. Barthel, D. Vandembroucq, Journal of the Mechanics and Physics of Solids **57**, 446 (2009)

[4] A.L. Barabási, H.E. Stanley, *Fractal Concepts in Surface Growth*, 1st edn. (Cambridge University Press, 1995)

[5] L.A.N. Amaral, A.L. Barabási, S.V. Buldyrev, H.A. Makse, H.E. Stanley, Physical Review E **52**, 4087 (1995)

[6] A. Atman, *Crescimento de interfaces em meio poroso* (1998), also available at `http://www13.fisica.ufmg.br/ posgrad/` (in Portuguese)

[7] P.B. Umbanhowar, F. Melo, H.L. Swinney, Nature **382**, 793 (1996)

[8] S.F. Pinto, M.S. Couto, A.P.F. Atman, S.G. Alves, A.T. Bernardes, H.F.V. de Resende, E.C. Souza, Physical Review Letters **99** (2007)

[9] C. Chevalier, A. Lindner, E. Clément, Physical Review Letters **99** (2007)

[10] A. Lindner, D. Bonn, E.C. Poiré, M.B. Amar, J. Meunier, Journal of Fluid Mechanics **469**, 237 (2002)

[11] J. Nase, A. Lindner, C. Creton, Physical review letters **101**, 074503 (2008)

[12] A. Atman, G. Combe, T. Ferreira, J. Barros, *Pattern formation during capillary rising of a fluid front into a granular media*, in *POWDERS AND GRAINS 2013: Proceedings of the 7th international conference on micromechanics of granular media, AIP Conference Proceedings* (2013), Vol. 1542, pp. 1075–1078

[13] C.K. Peng, S.V. Buldyrev, S. Havlin, M. Simons, H.E. Stanley, A.L. Goldberger, Physical Review E **49**, 1685 (1994)

[14] J.G. Moreira, J.K.L. da Silva, S.O. Kamphorst, Journal of Physics A: Mathematical and General **27**, 8079 (1994)

Modelling of Coke Layer Collapse during Ore Charging in Ironmaking Blast Furnace by DEM

Yoichi Narita[1,*], *Hiroshi* Mio[1], *Takashi* Orimoto[1] and *Seiji* Nomura[1]

[1] Process Research Laboratories, Nippon Steel and Sumitomo Metal Corporation. Futtsu, Chiba 293-8511, Japan

Abstract. A technical issue in an ironmaking blast furnace operation is to realize the optimum layer thickness and the radial distribution of burden (ore and coke) to enhance its efficiency and productivity. When ore particles are charged onto the already-embedded coke layer, the coke layer-collapse phenomenon occurs. The coke layer-collapse phenomenon has a significant effect on the distribution of ore and coke layer thickness in the radial direction. In this paper, the mechanical properties of coke packed bed under ore charging were investigated by the impact-loading test and the large-scale direct shear test. Experimental results show that the coke particle is broken by the impact force of ore charging, and the particle breakage leads to weaken of coke-layer strength. The expression of contact force for coke in Discrete Element Method (DEM) was modified based on the measured data, and it followed by the 1/3-scaled experiment on coke's collapse phenomena. Comparing a simulation by modified model to the 1/3-scaled experiment, they agreed well in the burden distribution.

1 Introduction

A blast furnace is a huge reactor used for reducing and smelling iron ore to produce pig iron. Coke and ore particles are supplied through the top of the furnace and stacked alternately in layers. A hot air is blown into the lower section of the furnace, reacts with coke to generate carbon monoxide, which reduces iron ore. Thus, the gas flow distribution has strong influence to efficiency and productivity of blast furnace. To control the gas flow, optimizing the burden distribution is very important. However, a coke layer-collapse occurs during ore charging onto the inclined coke bed because of the density difference between them. The coke particles have about three times less density than that of ore particles. This coke layer-collapse affects to the burden distribution and ore to coke layer thickness ratio. Furthermore, in the actual blast furnace, the coke layer thickness after collapse cannot be measured because the coke layer is covered with ore layer. Since many experimental studies have been carried out to investigate the coke collapse and clarify dominant factors [1-6]. Some studies have used numerical method like Discrete Element Method (DEM)[7] for better understanding of the phenomena involved. DEM can provide deeper insight on the particle flow [8], but the mechanics of coke packed bed has not been fully analyzed yet, because the bed has peculiar mechanical properties. A large number of investigations have been published on the mechanical properties of geomaterials, which consist of various shape of particles such as rounded rock, crushed stone, sea sand and so on. Especially, packed bed of volcanic coarse-grained soil, which is mainly formed from pumice particles, has peculiar mechanical properties. The shear resistance angle of packed bed is significantly reduced with the increase of isotropic consolidation stress due to the particle breakage [9-10].

In this study, three kinds of experiment were carried out to investigate the influence of particle breakage on coke collapse. First, the particle breakage and stress on its packed bed under ore charging were measured by an impact loading test. Second, the shear resistance angle under compressive stress was measured by a large-scale direct shear test. Third, a scaled model experiment of ore charging onto blast furnace was carried out to measure the coke collapse. The characteristic of coke particle and coke packed bed under stress were discussed. The contact model of DEM was modified on the basis of the experimental result to improve estimation accuracy in coke collapse.

2 Experiments

2.1 Impact-loading test

Fig.1 shows a schematic illustration of the impact-loading test. 10 kg of coke particles having 15-20mm diameter was put into a sampling box. The sampling box has a cylindrical shape of 600 mm diameter. 75kg of sinter particle having 10-15mm was charged from a hopper, which is 6300mm above the bottom of the sampling box. A pressure measurement film (Prescale LW, Fujifilm Business Supply Co.) was stuck to the inner bottom of the sampling box to measure the impact force acting on coke packed bed bottom. After the test,

* Corresponding author: narita.58d.yoichi@jp.nssmc.com

coke and ore were taken out from the sampling box. The particle size distribution was measured, and mass fraction of broken coke particle (<15mm) was obtained. The pressure measurement film was analyzed by the digital analysis system (FPD-9210, Fujifilm Business Supply Co.) to calculate the area receiving a stress of 4MPa, tensile strength of coke[11], or more.

2.2 Large-scale direct shear test

Fig.2 shows a schematic illustration of the large-scale direct shear test apparatus. The shear box is 400mm cube. The experimental conditions are tabulated in Table 1. Coke particles having 19-75mm in diameter was poured into the shear box. The shear box can be divided into upper and lower parts. The upper part is free to move in the vertical direction, and the lower part is free to move in the horizontal direction. The top plate was placed on the sample. When a normal load was applied vertically to the top plate, the particles were constricted and the top plate sank gradually. After the constriction was settled, a shear load was applied. The shear load was controlled to retain the displacing speed of lower part at a target speed. The vertical displacement of the top plate and the shear load was recorded during shearing. The shear resistance angle ϕ_r was calculated by the following equation.

$$\phi_r = \sigma_{s,max} / \sigma_n \qquad (1)$$

where $\sigma_{s,max}$ and σ_n are the maximum shear stress and the normal stress, respectively. After the shear test, coke was taken out from the shear box. The particle size distribution was measured, and the mass fraction of small particle (<19mm) was obtained

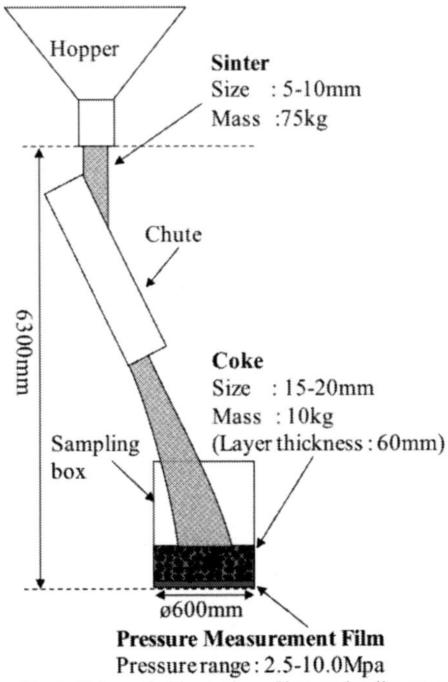

Fig. 1. Schematic illustration of impact loading test

2.3 Scaled model experiment of ore charging

The 1/3 scale burden distribution model was used for evaluating the coke collapse(Fig.3). First, the burden was charged to the small bell hopper. Second, the small bell was lowered, and the burden in the small bell hopper dropped on the large bell and stored in the large bell hopper. Third, the large bell was lowered and the burden was discharged into the furnace. By means of repeating this procedure, coke and ore are stacked alternately in layers.

In this study, 1.0 ton of coke having a diameter of 25mm or less was stacked in advance. 2.6 ton of ore particle having 3-15mm diameter was charged onto coke layer. After charging, the burden in the furnace was collected. The relative thickness of coke was calculated from mass fraction of coke.

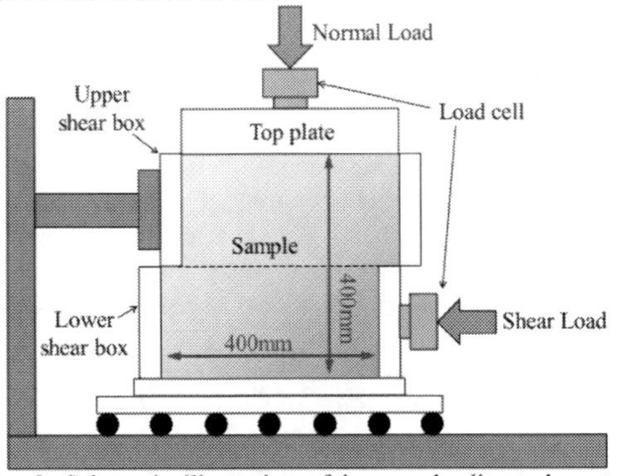

Fig. 2. Schematic illustration of large-scale direct shear test apparatus

Table 1. Experimental conditions.
(Large-scale direct shear test)

Material	Coke
Size[mm]	19-75
Samplebox size[mm]	400×400×400
Shear displacement [mm]	80
Normal stress [kPa]	50,100,150

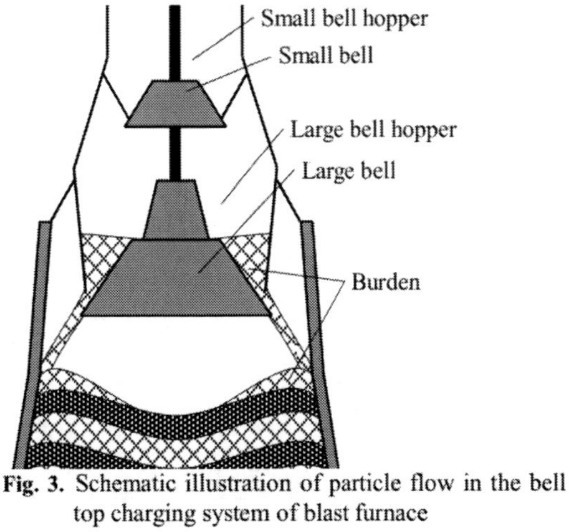

Fig. 3. Schematic illustration of particle flow in the bell top charging system of blast furnace

3 Simulation

3.1 Discreet Element Method

DEM is one of the most famous numerical methods for simulating the motion of a large number of particles. Basically, particles are dealt as spheres. The contact forces, which act on each particle, are computed, and then particles' velocity and position are updated within a small time-step. Unlike a continuum analysis of the finite element method, DEM is suitable for analyzing the dynamic behaviour of the individual granular flow. Generally, the simple contact models (Voigt model and Coulomb's friction rule) are used to compute a normal contact force, $\boldsymbol{F}_{\mathrm{n}}$, and a tangential contact Force, $\boldsymbol{F}_{\mathrm{t}}$, as follows.

$$\boldsymbol{F}_{\mathrm{n,ij}} = \left(K_{\mathrm{n}}\Delta u_{\mathrm{n,ij}} + \eta_{\mathrm{n}}\frac{\Delta u_{\mathrm{n,ij}}}{\Delta t}\right)\boldsymbol{n}_{\mathrm{ij}} \tag{2}$$

$$\boldsymbol{F}_{\mathrm{t,ij}} = min\left\{\begin{matrix}\mu\left|\boldsymbol{F}_{\mathrm{n,ij}}\right|\boldsymbol{t}_{\mathrm{ij}}, \\ \left|\left[K_{\mathrm{t}}\left(\Delta\boldsymbol{u}_{\mathrm{t,ij}} + \Delta\boldsymbol{\phi}_{\mathrm{ij}}\right) + \eta_{\mathrm{t}}\left(\frac{\Delta\boldsymbol{u}_{\mathrm{t,ij}} + \Delta\boldsymbol{\phi}_{\mathrm{ij}}}{\Delta t}\right)\right]\right|\boldsymbol{t}_{\mathrm{ij}}\end{matrix}\right\} \tag{3}$$

where, K and η are the spring and the damping coefficients. $\Delta\boldsymbol{u}$ and $\Delta\boldsymbol{\phi}$ are a relative translational displacement of the gravitational center between two particles and a relative displacement at the contact point caused by particle rotation. μ is the frictional coefficient. $\boldsymbol{n}_{\mathrm{ij}}$ and $\boldsymbol{t}_{\mathrm{ij}}$ denote the unit vector from i-th particle to j-th one in the normal and the tangential components. The subscripts n and t also denote the normal and the tangential components.

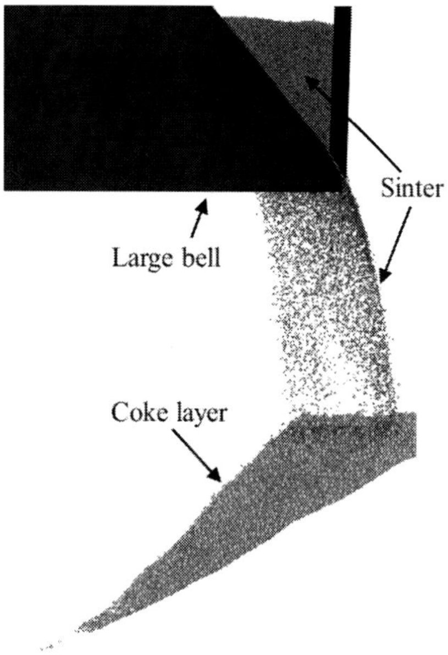

Fig. 4. Simulated charging system.

Table 2. Particle properties used in DEM

	Coke	Ore
Diameter [mm]	8.5 - 22.5	4.0 - 12.5
Young's modulus [GPa]	0.54	3.50
Poisson's ratio [-]	0.22	0.25
Particle density [kg/m³]	1050	3300

3.2 Simulation Setup

Ore charging behaviour into the 1/3 scale experimental model was simulated by DEM(Fig. 4). The particle properties are tabulated in Table 2. To reduce the computational load, 45° section of the furnace was simulated and the periodic boundaries were used to enforce the conditions for axi-symmetry in the model.

4. Results and discussion

4.1. Mechanical properties of coke packed bed

The results of the impact loading test are tabulated in Table 3. The coke particle was broken by the impact stress of ore charging. Pressure measurement film shows that the stress exceeding tensile strength of coke acts on the bottom of coke packed bed.

Fig.5 shows the mass fraction of small particle (<19mm) and the shear resistance angle obtained by the large-scale direct shear test. In the case of the normal stress of 50kPa, the coke packed bed has a higher shear resistance angle of 60deg. It indicates that the coke particle is easy to interlock each other and its packed bed has a high shear strength. However, increase of normal steress leads to more breakage of coke particles, and reduction of the shear resistance angle.

Therefore, the impact stress of ore charging lead to coke particle breakage and weaken of shear strength of coke packed bed. Interlocking and breakage of coke particle have a strong influence on the mechanical properties of coke packed bed.

Table 3. Strong stress area on the bottom of sampling box and mass fraction of broken coke particle (impact-loading test).

Mass fraction of broken particle [%]	15
>4MPa Stress area [mm²]	169

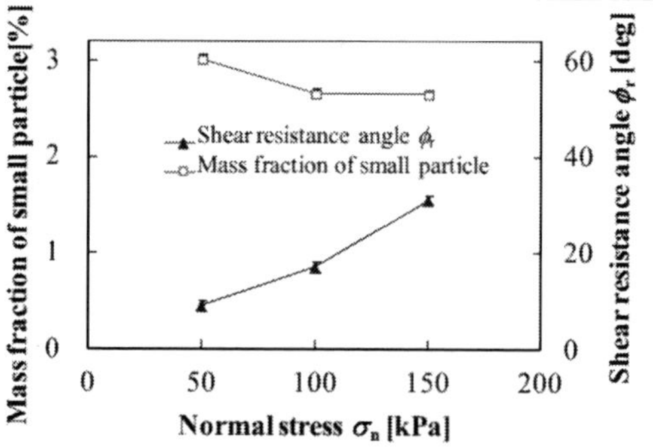

Fig. 5. Increment of fines content and shear resistance angle obtained by large-scale direct shear test

4.2. Modification of contact model of DEM

On the basis of the experimental results, the contact model of DEM was modified to simulate the breakage of particles. In order to simulate weaken of the coke-layer strength under high normal stress, the tangential contact force $\boldsymbol{F}_t$ is modified as follows,

$$\boldsymbol{F}_{t,ij} = min \left\{ \begin{array}{c} \mu F_{n,max}\, \boldsymbol{t}_{ij} \\ \mu |\boldsymbol{F}_{n,ij}| \boldsymbol{t}_{ij} \\ \left[K_t \left(\Delta \boldsymbol{u}_{t,ij} + \Delta \boldsymbol{\phi}_{ij} \right) + \eta_t \left(\dfrac{\Delta \boldsymbol{u}_{t,ij} + \Delta \boldsymbol{\phi}_{ij}}{\Delta t} \right) \right] \boldsymbol{t}_{ij} \end{array} \right\} \quad (4)$$

where, $F_{n,max}$ is the threshold value of normal contact force which was associated with the strength of coke particle. In this study, $F_{n,max}$ is set at 3N/contact. As the result of this modification, the effective friction coefficient μ_{eff} (=$|\boldsymbol{F}_t|/|\boldsymbol{F}_n|$) is decreased under high normal stress as shown in Fig. 6. The frictional coefficient μ is set at higher value than that of original model in order to simulate the high shear resistance angle caused by the interlocking of particle.

Ore charging behaviour into the blast furnace was simulated by the modified model and the burden distribution was compared with the experimental result. Fig.8 shows the relative coke thickness in radical direction. In comparison with the original, the modified model simulated the thin coke layer near the wall and the thick coke layer in the center of furnace. That indicates the increasing of coke collapse. It is found that the modified model correlated with the experimental result.

5 Conclusion

The impact-loading test and the large-scale direct shear test were carried out to investigate the mechanical properties of coke packed bed. Influence of interlocking and breakage on mechanical properties of the packed bed was discussed. The summaries are as follows:

(1) Under the normal stress of 50kPa or less, coke packed bed has higher shear strength because coke particles are easy to interlock each other.
(2) Increasing of normal stress leads to weaken of coke-layer strength because of particle breakage.
(3) Breakage of coke particle is also occurred by impact force of ore charging.

Based on the experimental results, the contact force model in DEM was modified to simulate weaken of friction coefficient under high normal stress. The modified DEM correlated with the experimental result.

References

1. Y. Kajiwara, T. Jimbo, T. Joko, Y. Aminaga and T. Inada: *Trans. Iron Steel Inst. Jpn.*, **24** (1984), 799.
2. Y. Okuno, K. Kunitomo, T. Irita and S. Matsuzaki, *Tetsu-to-Hagane* **72** (1986) P 783
3. K. Takeda, Y. Konishi, S. Taguchi, and T. Fukutake, *Tetsu-to-Hagane* **73** (1987) P 282
4. J. Jiminez, J. Mochon, A. Formoso, J. S. Ayala, *ISIJ Int.*, **40** (2000), 114.
5. T. Nouchi, K. Takeda, H. Itaya, *Tetsu-to-Hagane* **86** (2000) P 293
6. T. Mitra, H. Saxén, *ISIJ int.*, **56** (2016), 1570
7. P. A. Cundall, O. D. L. Strack, *Geotechnique*, **29** (1979), 47-65.
8. H. Mio, S.Komatsuki, M. Akashi, A. Shimosaka, Y. Shirakawa, J. Hidaka, M. Kadowaki, H. Yokoyama, S. Matsuzaki, K. Kunitomo, *ISIJ int.*, **50** (2010), 1000.
9. T. Nakata, S. Miura, *Proc. Jpn. Soc. Civ. Eng.*, **63** (2007), 224
10. S. Ohtsuka, Y. Konishi, K. Isobe, S. Endo, *Jpn. Geotech. J.* , **10** (2015), 381
11. J. W. Patrick, A. E. Stacey,. *Fuel*, **51** (1972), 81

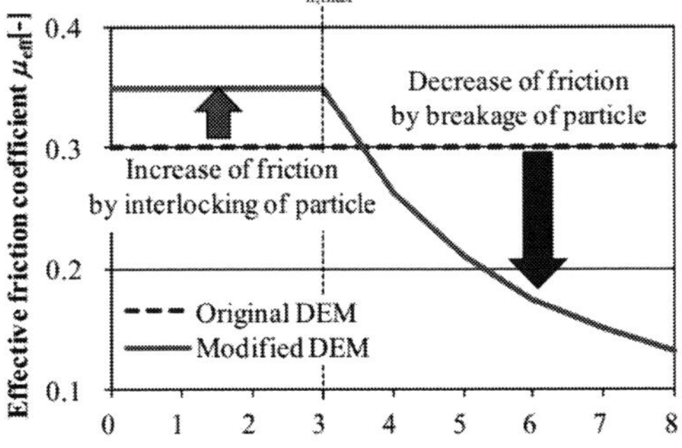

Fig. 6. Effective frictional coefficient of DEM

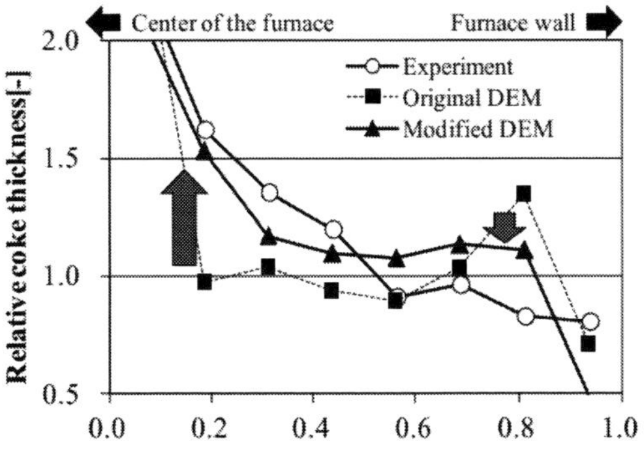

Fig. 7 Relative coke thickness in radical direction.

Experimental study of shear rate dependence in perpetually sheared granular matter

Sophie Yang Liu[1,*], *François* Guillard[1], *Benjy* Marks[1], *Pierre* Rognon[1,2], and *Itai* Einav[1,3,**]

[1] *Particles and grains laboratory, school of Civil Engineering, The University of Sydney, Sydney, NSW 2006, Australia*
[2] *IUSTI-CNRS UMR 7343, Aix-Marseille University, 13453 Marseille, France*
[3] *Department of Civil, Environmental and Geomatic Engineering, Faculty of Engineering Science, University College London, London WCIE 6BT, United Kingdom*

Abstract. We study the shear behaviour of various granular materials by conducting novel perpetual simple shear experiments over four orders of magnitude of relatively low shear rates. The newly developed experimental apparatus employed is called "3D Stadium Shear Device" which is an extended version of the 2D Stadium Shear Device [1]. This device is able to provide a non-radial dependent perpetual shear flow and a nearly linear velocity profile between two oppositely moving shear walls. Using this device, we are able to test a large variety of granular materials. Here, we demonstrate the applicability of the device on glass beads (diameter 1 mm, 3 mm, and 14 mm) and rice. We particularly focus on studying these materials at very low inertial number I ranging from 10^{-6} to 10^{-2}. We find that, within this range of I, the friction coefficient μ of glass beads has no shear rate dependence. A particularly appealing observation comes from testing rice, where the attainment of critical state develops under much longer duration than in other materials. Initially during shear we find a value of μ similar to that found for glass beads, but with time this value decreases gradually towards the asymptotic critical state value. The reason, we believe, lies in the fact that rice grains are strongly elongated; hence the time to achieve the stable μ is primarily controlled by the time for particles to align themselves with respect to the shear walls. Furthermore, the initial packing conditions of samples also plays a role in the evolution of μ when the shear strain is small, but that impact will eventually be erased after sufficient shear strain.

1 Introduction

The mechanics of granular flow have been widely studied experimentally. Three flow regimes of granular materials include the quasi-static, inertial and collisional regimes [2], which are classified using the dimensionless Inertial number I [3]. It is defined as the ratio of the confinement timescale and the typical time of deformation [4, 5]: $I = \dot{\gamma} d \sqrt{\frac{\rho}{\sigma_{yy}}}$, where $\dot{\gamma}$ is the shear rate, σ_{yy} is the normal stress, d and ρ are the diameter and density of the grain respectively. This number is treated as the index of inertial effects in granular flows at various shear rates. Collisional rapid flows has been described by kinetic theories which provide a set of constitutive equations connecting the mean density, the mean velocity, and the granular temperature to the fluctuational energies induced by the binary collisions between particles [6]. The behaviour of the dense confined granular assemblies under extremely low shear rates are usually captured by an elasto-plastic, rate-independent constitutive law that characterises the critical state by the internal friction angle ϕ and the critical solid fraction υ_c [7]. However, a comprehensive constitutive law is still lacking to fully capture the characteristics of granular flow in the intermediate inertial regime. Many studies [4, 5, 8, 9] have indicated a positive shear rate dependence of the friction coefficient μ, defined as shear stress over normal stress. In the study of da Cruz, et al. [5], the friction and dilatancy laws were proposed by conducting DEM simulation of 2D disc particles assembly in a plane shear configuration. It stated that the solid fraction υ linearly decreases, but the friction coefficient μ linearly increases with the increasing I over 10^{-4} to 10^{-1}. However, Kuwano, et al. [10] carried out annular shear cell testing with 270μm diameter glass beads, and they reported a crossover from negative to positive shear rate dependence of the friction coefficient over the range of inertial number I from 10^{-6} to 1.

In this study, we employed a novel experimental apparatus called the "Stadium Shear Device" (SSD) that enables testing granular materials under perpetual shear conditions, here for the first time in three dimensions (i.e. 3D SSD). Unlike the Couette geometry which can only produce a radial dependent stress field, our new device is able to produce a non-radial dependent stress field in the central region[1]. We conducted perpetual shear tests on a large variety of materials, and present results examining the shear rate dependency of friction coefficient μ and the evolution of μ with shear strain.

* e-mail: sophie.liu@sydney.edu.au
** e-mail: itai.einav@sdyney.edu.au

2 Experimental methodology

2.1 Introducing the 3D SSD

A schematic diagram of the 3D Stadium Shear Device is shown in Fig. 1(a). A belt customised from the belt designation Bando HTS 966-14M is centrally placed in a confining case (540 mm length ×250 mm width ×190 mm height). The belt has the same height as the confining case, and can be driven with shear rates in the range of $0.00645s^{-1} \sim 0.645s^{-1}$ from the power supplied by a motor and gear assembly installed on top of the driven sprocket. The granular material filled inside the belt is horizontally confined by two side plates of 175 mm length ×100 mm height, and vertically confined by a top and base plates as indicated in Fig. 1(a). Note that the sample has to be filled to a height above the side plate to accurately measure the normal stress using the two load cells (Mark-10 MR01-100#, one at front and one at back) attached on the side plates. Shear force is recorded by a torque transducer (Kistler 4520A050, range: 50 N.m) coaxially installed under the motor. There are 6 rows of 7 rollers bearings half-embedded on the surface of side plates to allow the belt run with minimal resistance. The friction between the running belt and the top plate are minimised using a brush-like draught excluder glued along the side of the top plate. In addition, the belt is sitting inside a U-shaped groove on the base plate with a Teflon sheet inserted to reduce the friction.

Fig. 1(b) schematically shows the stress state in the central confined region of the device, where σ_{zz} is the vertical normal stress applied through the top plate, and $\sigma_{yy} = \frac{F_N}{lh}$ the lateral stress, where F_N is the normal force measured by load cells, $l = 412$ mm is the projected longitudinal length of the belt and h ($h > 100$ mm) is the height of the sample. The in-plane shear stress $\tau_{xy} = \frac{S}{hp} = \frac{T}{rhp}$, where S is the shear force, T the torque applied by the motor and gear assembly, r the pitch radius of the sprocket, h the sample height and $p = 966$ mm the circumference of belt, respectively. The perpetual shear strain condition of the granular assembly within the belt is schematically illustrated in Fig. 1(c). Red thick arrows represent the directions of global shear direction and grey arrows indicate the localised shear directions within the sample. Due to the particular "stadium" shape of the belt, this device is able to perpetually shear particles and produce a continuous plane shear condition in the central region while re-circulating particles at either end without particle congestion. Although not shown here for brevity purposes, we used DEM, FEM and radiography studies to extract velocity profiles quite similar to the schematic one shown in Figure 1c.

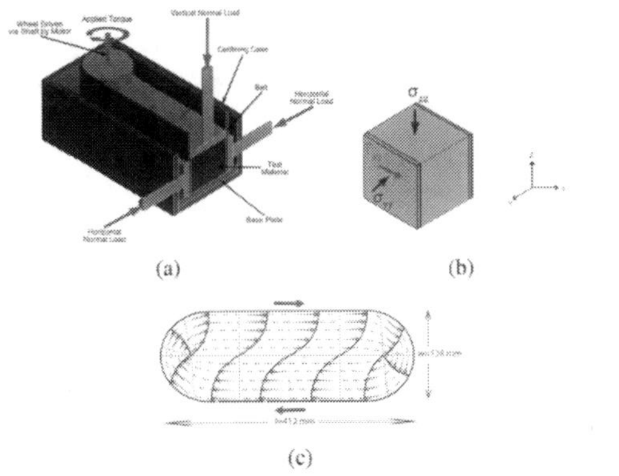

Figure 1. The 3D Stadium Shear Device (SSD). (a) An overview of the deive. (b) Schematic diagram of the granular assembly in the central region. (c) Schematics of velocity field. Red thick arrows show the global shear direction and grey thin arrows illustrate the flow.

2.2 Calibration procedure

A first calibration procedure was carried out to account for the forces induced by bending of the belt and frictions along interfaces between the metal plates and the belt. Before any experiment, the top plate was firstly kept at a certain height similar to the one it would have during the experiments (without material to be tested), and then the readings from both the load cells and the torque transducer were synchronously recorded when the belt was running at the velocity that would be operated during the experiment to be conducted later on. To avoid any possible hysteresis effects from the measuring meters, the same calibration procedure was repeated after every experiment. Since the calibration was made without the material, the forces measured were purely from the mechanical frictions mentioned above. The average force from two calibrations (before and after the experiment) was deducted from the corresponding shear and normal stresses during the experiments.

A second calibration procedure was conducted to examine the accuracy of the two load cells symmetrically attached to the adapt plates, which are connected to the side plates via bars on each side. This might lead to inaccurate readings of the normal force generated by shearing the sample. The calibration first involved tilting over the device in a horizontal direction, and then evenly placing over the side of the belt a bag of lead shot with known weight. This procedure was repeated many times with different weights. By comparing the readings of load cells with the actual weights, a factor of 1.71 with an error bar of 6% was adopted for the normal forces measured.

2.3 Experimental procedure

In the present study, we tested glass beads with three different diameters (1 mm, 3 mm and 14 mm), for which the density of grain is roughly 2400 ± 100 kg/m^3 with less than 10% polydispersity. The mean aspect ratio (ratio between the smallest diameter and the largest diameter orthogonal to it) is 0.926. Another material tested was rice with a mean equivalent diameter of 3.8 mm and density of 1300 kg/m^2. The mean aspect ratio of the rice measured by macrophotographic imaging was 0.304, and the mean circularity, defined as a function of perimeter P and area A, $\frac{4\pi A}{P^2}$ was found to be 0.542.

A set of tests with glass beads was conducted to investigate the relation between μ and inertial number I over the range of $10^{-6} \sim 10^{-2}$. The granular system was sheared to the same shear strain under different shear rates, and the normal and shear stress were measured only after the system achieved the critical state. Further discussion about the critical state is addressed in the results and discussions section. We firstly measured the shear and normal stresses under the lowest shear rate, and incrementally increased the shear rate to the highest possible. High reproducibility was indicated by repeating the same test from the highest shear rate to the lowest. The shear rate in our experiment is defined as velocity of the belt V divided by the width between two shearing walls w = 124 mm.

Furthermore, the evolution of μ with nominal shear strain $\epsilon = x/w$ (x being the distance travelled by the belt) was analysed with different particle shapes. In this test, the samples (3 mm glass beads and rice) were sheared to a shear strain of 240 at a constant shear rate with vertical stress σ_{zz} fixed. In addition, since the mechanical behaviour of a granular assembly is strongly dependent on its initial packing conditions, we compared the stress response and the critical states from dense and loose packed rice after more than 15000 shear strain. The loose sample was prepared by gently pouring a small amount of material into the device from minimal height until the target depth, while the dense one was prepared by compacting the sample layer by layer by hand.

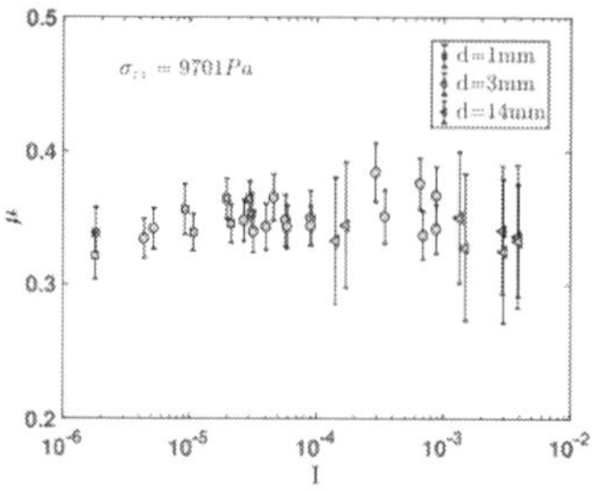

Figure 2. The shear rate dependence of friction coefficient μ for glass beads (initial void ratio is 0.645) with three diameters: 1 mm, 3 mm and 14 mm.

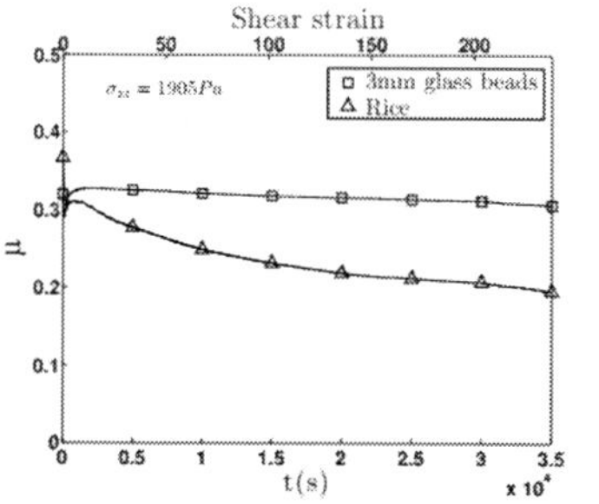

Figure 3. The friction coefficient μ over time and shear strain for shear rate of $0.00645s^{-1}$ on 3 mm glass beads and rice.

3 Results and discussions

In Fig. 2, the friction coefficient μ of glass beads at steady state is shown as a function of inertial number I over the range of $10^{-6} \sim 10^{-2}$. The low $I < 10^{-2}$ values correspond to granular flows in quasi-static regime. The symbols and error bars are the means and standard deviations of the stress readings over 10 belt turns, of which the accumulative shear strain is ϵ=79.2. By comparing the results from 14 mm glass beads with other smaller glass beads, one can find that the error bars are relatively large due to the smaller w/d ratio (w/d equal to 8.7, 40.7 and 122 for 14 mm, 3 mm and 1 mm glass beads respectively). The value of μ fluctuates slightly between $0.32 \sim 0.38$ and shows no dependence on the shear rate. This shear rate independence at the quasi-static regime is in agreement with previous studies. For example, da Cruz, et al. [11] and Iordanoff and Khonsari [12] identified shear rate independence in the quasi-static regime from discrete numerical simulations of 2D discs in a plane shear configuration, and the same characteristic was captured in [13] for polystyrene beads and glass beads in a Couette geometry. In particular, the value of μ from our experiments is very close to the μ_s that is an important parameter in the $\mu(I)$ rheology relationship [14]. This model assumes that the friction coefficient goes towards a minimum value μ_s at very low I, and μ_s is found to be equal to 0.38 for glass beads [15]. In [9], a weakly negative dependence of friction coefficient μ was found when $I < 10^{-3}$ and μ started increasing noticeably when $I > 10^{-2}$. However, no crossover from negative to positive shear rate dependence within this I range could be observed in our experiments. At this stage, we can not obtain data at I larger than 10^{-2} for glass beads because the velocity after solid-liquid transition is beyond the range covered by our device. In the future, further improvements will be implemented to increase the velocity range of the device.

To study the impact of particle shape on the friction coefficient μ, we compare μ as a function of shear strain for the tests of 3 mm glass beads and rice. As shown in Fig. 3, the rice exhibits a softening of friction coefficient with increasing shear strain, in contrast to the steady flow observed in glass beads. The value of μ decreases to two thirds of the initial value after shearing the rice to 240 shear strain. The same amount of friction coefficient reduction can be found in [16]. We believe that compared to spherical grains, elongated particles need a much longer shear duration to align themselves against the streamline, which increases the time to achieve the critical state. Further investigation into the alignment time and the orientation of the particles is out of the scope of the current paper.

The mechanical behaviour of a granular system is highly dependent on its initial preparation, and is expected to eventually reach a critical state that is independent of the initial packing conditions after sufficient shear strain [17]. In Fig. 4, we show the shear and normal stresses

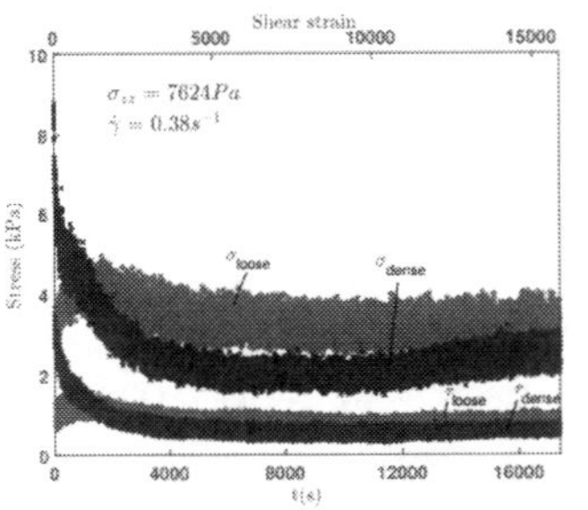

Figure 4. The normal and shear stress measured for dense and loose packed samples of rice.

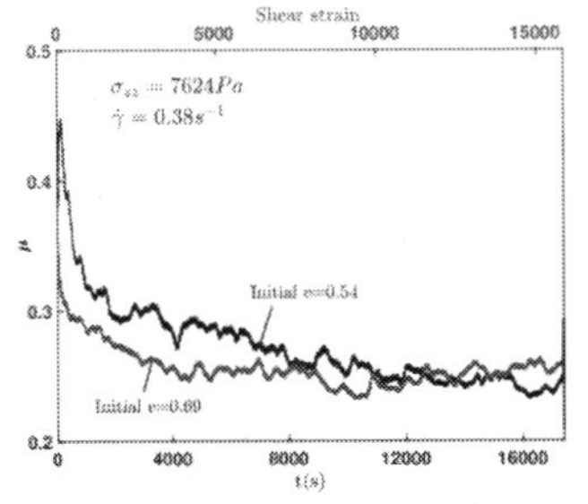

Figure 5. Friction coefficient μ over time and shear strain for the experiments shown in Fig.4. Note that the lines shown were smoothed over 160 s of data.

as a function of shear strain γ for both the initially dense (e=0.54) and loose (e=0.69) samples of rice. The samples are sheared at a shear rate of $0.38s^{-1}$ to a large deformation (15000 strain) with a constant vertical stress. The stress of the loose sample first reaches a peak at shear strain 600 before declining to a plateau at around 4000 shear strain, while the stress of the dense sample reduces from a higher initial value compared to the loose sample and may require more than 12000 shear strain to reach its critical state. Both systems converge to an identical value of 4 kPa for normal stress and 1 kPa for shear stress at the critical state. We also plot in Fig. 5 the friction coefficient μ as a function of shear strain and the same characteristics are observed: the value of μ converges to 0.25 after 16000 shear strain for dense system but only 4000 shear strain is required to reach the critical state for the loose one. Radjaï and Roux [7] also studied the critical state of initially dense and loose packed samples by conducting plane shear simulations with an assembly of rigid discs in two dimensions. Similar results were found in [7] except the peak of stress ratio was only observed in the dense packed system.

4 Conclusion

After conducting robust experiments with the 3D Stadium Shear Device, we found that the friction coefficient μ of glass beads is not dependent on the shear rate over four orders of magnitude at relatively low inertial numbers. The particle shape plays a role in the evolution of μ as a function of the shear strain. In particular, μ of granular assembly comprising elongated particles declines gradually to a plateau and reaches a critical state after very large deformation. By contrast, μ of spherical particles is fairly constant with increasing shear strain. The evolution of stress and μ at the beginning of shear is strongly influenced by the initial packing state but eventually collapse to the same value. Further research into the initial packing condition and its impacts on the stress responses and dilation curves will be an interesting topic to study. It would also be worthwhile to further examine (by numerical simulations) the properties of the device, for example, the shear strain condition and the stresses generated in the non-affine and rotational flows under the circular cap.

References

[1] T. Miller, P. Rognon, B. Metzger, I. Einav, Phys. Rev. Lett. **111**, 058002 (2013)
[2] C.S. Campbell, Journal of fluid mechanics **465**, 261 (2002)
[3] S.B. Savage, Advances in applied mechanics **24**, 289 (1984)
[4] G. MiDi, The European Physical Journal E **14**, 341 (2004)
[5] F. da Cruz, S. Emam, M. Prochnow, J.N. Roux, F. Chevoir, Physical Review E **72**, 021309 (2005)
[6] I. Goldhirsch, Annual review of fluid mechanics **35**, 267 (2003)
[7] F. Radjaï, S. Roux, The physics of granular media pp. 165–187 (2003)
[8] P. Jop, Y. Forterre, O. Pouliquen, Nature **441**, 727 (2006)
[9] T. Hatano, Physical Review E **75**, 060301 (2007)
[10] O. Kuwano, R. Ando, T. Hatano, Geophysical Research Letters **40**, 1295 (2013)
[11] F. Da Cruz, F. Chevoir, J. Roux, I. Iordanoff, Tribology series **43**, 53 (2003)
[12] I. Iordanoff, M. Khonsari, Journal of Tribology **126**, 137 (2004)
[13] F. Da Cruz, F. Chevoir, D. Bonn, P. Coussot, Physical Review E **66**, 051305 (2002)
[14] P. Jop, Y. Forterre, O. Pouliquen, Journal of Fluid Mechanics **541**, 167 (2005)
[15] Y. Forterre, O. Pouliquen, Journal of Fluid Mechanics **486**, 21 (2003)
[16] T. Börzsönyi, B. Szabó, G. Törös, S. Wegner, J. Török, E. Somfai, T. Bien, R. Stannarius, Phys. Rev. Lett. **108**, 228302 (2012)
[17] A. Schofield, P. Wroth (1968)

Partial jamming and non-locality in dense granular flows

Prashidha Kharel[1,*] and *Pierre* Rognon[1,**]

[1] *Particles and Grains Laboratory, School of Civil Engineering, The University of Sydney, 2006 NSW, Australia*

Abstract. Dense granular flows can exhibit non-local flow behaviours that cannot be predicted by local constitutive laws alone. Such behaviour is accompanied by the existence of diverging cooperativity length. Here we show that this length can be attributed to the development of transient clusters of jammed particles within the flow. By performing DEM simulation of dense granular flows, we directly measure the size of such clusters which scales with the inertial number with a power law. We then derive a general non-local relation based on kinematic compatibility for the existence of clusters in an arbitrary non-homogenous flow. The kinematic nature of this derivation suggests that non-locality should be expected in any material regardless of their local constitutive law, as long as transient clusters exist within the flow.

1 Introduction

The bulk rheology of dense granular flows can be described by a local constitutive law involving an effective viscosity and yield stress, both depending on the normal stress [1–3]. However, this local description fails to capture the behaviour of flows in heterogeneous configurations, in particular, flows near walls or when stress and shear strain-rate gradients are present in the flow. In such configurations, the influence on the flow at one point due to flows at its neighbouring points results in nontrivial flow features, e.g., S-shaped velocity profile in plane shear flow between two walls [4, 5] and wide shear bands observed in slit-bottom cell experiments [6].

Various non-local models have been introduced that successfully account for these situations [4, 5, 7–12]. In spite of the differences in their formulations, these non-local models are all based on the key assumption that there exists a meso-scale length ℓ which is larger than the grain size d and that it represents the extent of non-locality, i.e., the influence on the flow at one point due to flows at its neighbouring points. An association of this length with the critical stopping height of grains on inclined plane [13–15] has been made [16], which suggested possible link between jamming and non-locality. Here we show how this length representing non-locality can be attributed to the development of correlated structures of jammed particles within the flow.

*e-mail: prashidha.kharel@sydney.edu.au
**e-mail: pierre.rognon@sydney.edu.au

2 Multiple shear bands and clusters in homogeneous shear flow

In order to study the microstructure of the dense granular flows, we perform discrete element simulation, as in [5], of a two dimensional assembly of 10,000 grains sheared homogeneously with a shear strain-rate of $\dot{\gamma}$ under constant normal pressure P, as shown in Fig. 1. The grains have density ρ, mean diameter d and a polydispersity of $\pm 20\%$ in order to avoid crystallisation and any permanent shear banding. Bi-periodic Lees-Edwards boundary condition [17] is used to ensure the time averaged shear flow is homogeneous. Overlapping grains interact via normal inelastic forces and tangential frictional forces with Young's modulus E, coefficient of restitution $e = 0.5$, and grain-to-grain coefficient of friction $\mu_g = 0.5$. The typical overlap is of the order of $\kappa = P/E = 10^{-3}$. The system is initially set randomly with no contacts, then the pressure P and the shear strain rate $\dot{\gamma}$ are applied. All measurements reported here are after the system has reached a steady state.

As a way to identify correlated structures, triplets of neighbouring grains are identified using a Delaunay triangulation and a criteria is employed to determine which triplets are jammed. Amongst several possibilities, we used a criteria based on the maximum normal relative velocity between pair of grains in the triplet k: $|\Delta v| = max(|\Delta v^k_{12}|; |\Delta v^k_{13}|; |\Delta v^k_{23}|)$ (see Fig. 1(a)). The triplet is considered kinematically jammed if all of the three relative normal velocities are smaller than the average relative normal velocity $\langle|\Delta v|\rangle$ between all pair of neighbouring grains in the full system. The flow is composed of transient orthogonal shear bands that create a lattice of highly sheared zones [18, 19] delimiting the boundaries of cluster of grains subjected to little if any shear deformation, thus moving briefly like rigid bodies. We then count the num-

DOI: 10.1051/epjconf/201714003060

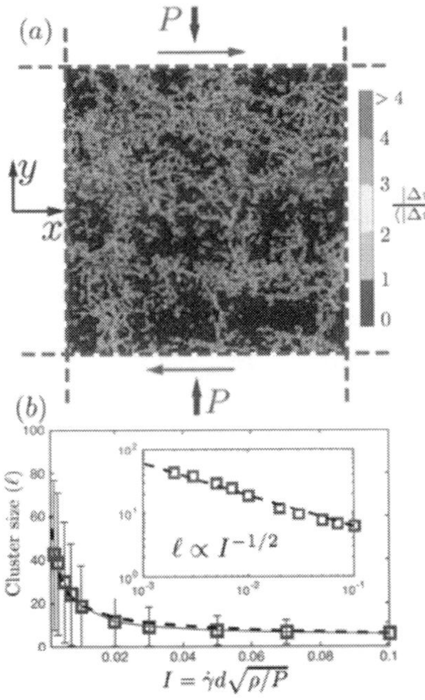

Figure 1. Clusters in sheared granular flows: (a) System of 10000 grains sheared homogeneously by using bi-periodic boundary condition. The colours represent a snapshot of relative velocities between the grains for inertial number $I = 0.001$. The partially jammed clusters are shown in dark blue (b) The average cluster size (squares) and standard deviation (vertical bars) for different inertial numbers. Inset: same plot in log-log axis. Dashed curves indicate a power law 1/2.

ber of grains, n, in each clusters and calculate its size as $\ell = \sqrt{n}d$. The average size of such jammed clusters scale with the inertial number with the following power law, as shown in Fig 1 (b):

$$\ell \propto \frac{d}{\sqrt{I}} \tag{1}$$

Similar scaling has been observed for other non-local length scales [5, 10, 11] measured indirectly in various systems, which suggests a possible relation between the clusters and non-locality.

3 Derivation of continuum non-local relation

Let us now seek to derive a general relation for non-local rheology by considering the existence of correlated structures in an arbitrary non-homogeneous flow. The first assumption is that, in the dense regime, the flow is comprised of transient clusters of grains that move like rigid bodies. Since, rigid body motion will result in constant vorticity throughout the cluster size, such clusters will redistribute the vorticity over their size. The local value of the vorticity within a flow should therefore be affected by the value of the vorticity in the surrounding.

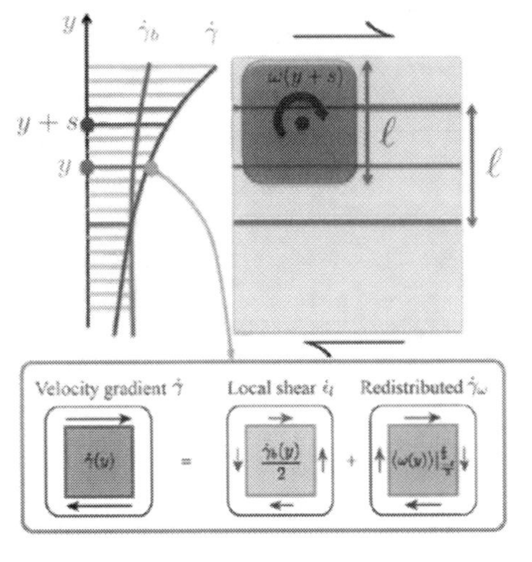

Figure 2. Illustration of the proposed mechanism at the origin of non-locallity. *(Top)* Actual shear rate profile $\dot\gamma(y)$ in a heterogeneous sheared layer compared to the shear rate profile $\dot\gamma_b$ predicted by a local constitutive law with no account for non-local effects; A clusters of jammed particles located at a position $y + s$ is represented, distributing its vorticity over its size ℓ. *(bottom)* The local velocity gradient is comprised of i) a local pure shear strain rate $\dot\epsilon = \dot\gamma_b/2$ governed by the local stresses via a local constitutive law, and a contribution $\dot\gamma_\omega = \langle\omega(y)\rangle|_{-\ell/2}^{\ell/2}$ coming from the vorticities of nearby clusters.

The second assumption is that the velocity gradient can be decomposed into a local shear strain deformation that is not redistributed, $\dot\epsilon_l$, and a shear rate due to redistribution of vorticity, $\dot\gamma_\omega$:

$$\dot\gamma(y) = \dot\epsilon_l(y) + \dot\gamma_\omega(y) \tag{2}$$

The local shear deformation is responsible for mechanical dissipation in localised shear bands, which shall follow the local constitutive law. On the other hand, the vorticity governs the cluster rotation rate which gets redistributed along the transverse direction.

As shown in figure 2, when a cluster of size ℓ develops at some position $y + s$, it redistributes the vorticity $\omega(y + s) = \dot\gamma(y + s)/2$ over the zone comprised over its size ℓ. Given that these clusters are transient and develop at different locations, the net shear rate at a point y due to redistribution of vorticity is given by the average of the vorticities of points between $y \pm \ell/2$, i.e.,

$$\dot\gamma_\omega(y) = \langle\omega(y)\rangle|_{-\ell/2}^{\ell/2} = \frac{1}{\ell}\int_{-\ell/2}^{\ell/2}(\dot\gamma(y + s)/2)ds$$

By contrast, local pure shear is not redistributed, which is entirely governed by the local stresses in the layer and is therefore the same as in the case of homogeneous shear flow, i.e., $\dot\epsilon_l(y) = \dot\epsilon_b(y) = \dot\gamma_b(y)/2$, where $\dot\gamma_b$ is the shear rate predicted by the local constitutive law. Substituting these two results in (2) leads to:

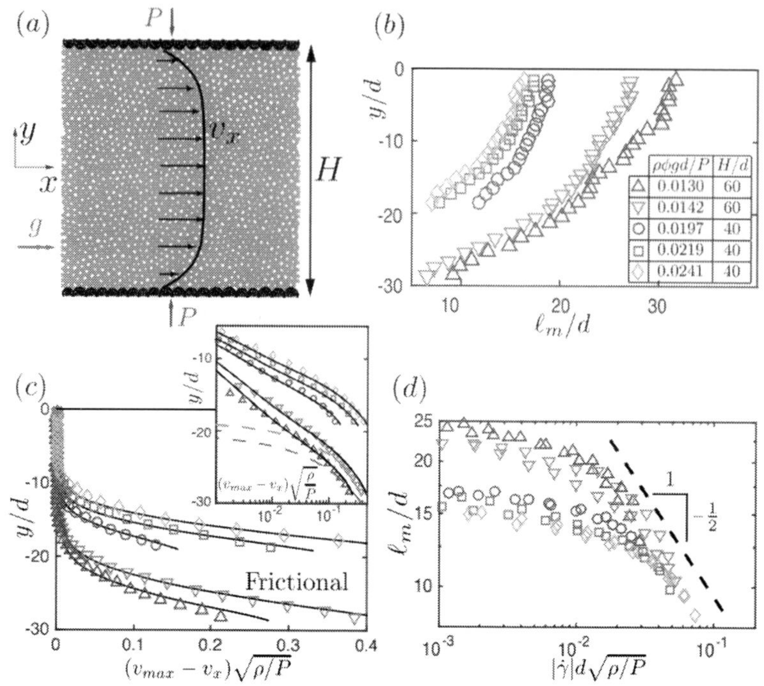

Figure 3. Non-locality and clusters in a non-homogeneous shear flow. (a) $H \times H$ Poiseuille flow geometry where wall grains (black) with average diameter $2d$ are stationary along x and subjected to a pressure P along y, whereas flowing grains (blue) with average diameter d are subjected to a constant force of $\pi d^2 \rho g/4$ along x. Periodic boundary condition is used along the x axis. (b) Measured cluster size profile, (c) velocity profile [inset: same in semi-log scale], and (d) measured cluster size versus inertial number within various Poiseuille flows. Symbols represent data from DEM simulations, solid black curves in (c) are solutions of eq. (4) and dashed lines in the inset of (c) are velocity predictions using local constitutive law for H/d=60.

$$2\dot{\gamma}(y) = \dot{\gamma}_b(y) + \frac{1}{\ell}\int_{-\ell/2}^{\ell/2} \dot{\gamma}(y+s)ds \quad (3)$$

By using a Taylor expansion of $\dot{\gamma}(y+s)$ with respect to s about $s = 0$, the integral in (3) becomes:

$$\frac{1}{\ell}\int_{-\ell/2}^{\ell/2}\left(\dot{\gamma}(y) + \frac{\partial\dot{\gamma}(y)}{\partial y}s + \frac{1}{2}\frac{\partial^2\dot{\gamma}(y)}{\partial y^2}s^2 + O\left(\frac{\partial^3\dot{\gamma}}{\partial y^3}\right)\right)ds$$

$$= \dot{\gamma}(y) + \frac{\ell^2}{24}\frac{\partial^2\dot{\gamma}(y)}{\partial y^2} + O\left(\frac{\partial^4\dot{\gamma}}{\partial y^4}\right)$$

Introducing the second order approximation of this expression into (3) leads to a non-local equation governing the shear rate $\dot{\gamma}(y)$:

$$\dot{\gamma}(y) - \dot{\gamma}_b(y) = \frac{\ell^2}{24}\frac{\partial^2\dot{\gamma}}{\partial y^2}. \quad (4)$$

This equation was derived solely based on kinematic compatibility, so it should be valid for any material regardless of their local constitutive law, as long as transient clusters of size ℓ exists in the flow. This differential equation is similar to the relation used in the Kinetic Elastoplastic (KEP) model [8–10], except that, here, the fluidity is replaced by shear strain-rate. In the quasi-static limit, the KEP model reduces to (4).

4 Non-locality in non-homogeneous shear flow

In order to assess the validity of the non-local model (4), we now consider a series of granular flows in a Poiseuille geometry (see Fig. 3(a)). This geometry involves a stress gradient, expressed as $\mu(y) = \tau(y)/P = \frac{\rho\phi g}{P}y$ for a solid fraction ϕ, and the presence of walls, which both induce non-local effects [5, 10]. Figure 3(b) shows the profile of cluster size $\ell_m(y)$, which is measured using a protocol similar to that in the homogeneous case, measuring the average relative velocity for each horizontal layer across the flow. Figure 3(c) shows that the model (4) is able to reproduce all the measured velocity profiles. To calculate the bulk shear rate $\dot{\gamma}_b$, we use the constitutive relation of the form $\mu = \mu_o + bI$, where $\mu_o = 0.266$, $b = 1.54$, fitted from the plane shear flow data. For the cluster size we used $\ell(y) = A\ell_m(y)$, where $A = 0.55$ is the single fitting constant which depends on the protocol used to measure the cluster size[1]. The two boundary conditions used are the values of the shear rate at both walls directly measured from the DEM simulations. Using Neumann boundary condi-

[1]In order to be consistent, we have used instantaneous velocities to measure cluster size. Using time averaged velocities does affect the absolute values of measured cluster size. However, the relative power law scaling in (1) is not affected.

tions, as in [10, 12], also produces similar result. Figure 3(d) shows that size of clusters and local inertial numbers follow a power law consistent with (1) for smaller cluster sizes. However, larger cluster sizes seemingly saturate to some finite size, which may be attributed to a finite size effect of the system [5].

5 Conclusions

Here we showed that in the dense regime, granular flow is composed of the correlated structures of jammed particles. The typical size of such correlated structure scales with the inertial number with a power law, which is consistent with the scaling of non-local length measured in various systems [5, 10, 11]. By considering the existence of such structures in an arbitrary non-homogeneous flows, we derived a general non-local relation which is similar in form to the KEP model. The fact that this relation was derived solely based on kinematics of the flow suggests that it should be valid for any material regardless of their constitutive law, as long as transient clusters exists in the flow.

More generally, correlated motion of particles [20–22], and avalanche-like rearrangements have been observed in various glassy materials like sand [23], granular materials [15, 24, 25], cohesive grains [26], foams [27], suspensions [8, 28], colloids [29, 30], and Lennard-Jones glass [31, 32]. These observations suggest that clusters of particles may exist in many glassy materials. Our derivation indicates that non-local behaviours should then be expected. Furthermore, the link made here between clusters and non-locality creates a promising opportunity to better interpret, analyse and unify the non-local behaviours of glassy materials through a thorough characterisation of cluster size and formation mechanisms for specific materials and flow configurations.

References

[1] G. MiDi, The European Physical Journal E **14**, 341 (2004)
[2] F. da Cruz, S. Emam, M. Prochnow, J.N. Roux, F. Chevoir, Physical Review E **72**, 021309 (2005)
[3] P. Jop, Y. Forterre, O. Pouliquen, Nature **441**, 727 (2006)
[4] T. Miller, P. Rognon, B. Metzger, I. Einav, Physical review letters **111**, 058002 (2013)
[5] P.G. Rognon, T. Miller, B. Metzger, I. Einav, Journal of Fluid Mechanics **764**, 171 (2015)
[6] D. Fenistein, M. van Hecke, Nature **425**, 256 (2003)
[7] O. Pouliquen, Y. Forterre, Philosophical Transactions of the Royal Society A: Mathematical, Physical and Engineering Sciences **367**, 5091 (2009)
[8] J. Goyon, A. Colin, G. Ovarlez, A. Ajdari, L. Bocquet, Nature **454**, 84 (2008)
[9] L. Bocquet, A. Colin, A. Ajdari, Physical review letters **103**, 036001 (2009)
[10] K. Kamrin, G. Koval, Physical Review Letters **108**, 178301 (2012)
[11] M. Bouzid, M. Trulsson, P. Claudin, E. Clément, B. Andreotti, Physical review letters **111**, 238301 (2013)
[12] D.L. Henann, K. Kamrin, Proceedings of the National Academy of Sciences **110**, 6730 (2013)
[13] A. Daerr, S. Douady, Nature **399**, 241 (1999)
[14] D. Ertaş, T.C. Halsey, EPL (Europhysics Letters) **60**, 931 (2002)
[15] O. Pouliquen, Physical Review Letters **93** (2004)
[16] K. Kamrin, D.L. Henann, Soft Matter **11**, 179 (2015)
[17] A. Lees, S. Edwards, Journal of Physics C: Solid State Physics **5**, 1921 (1972)
[18] M. Falk, J. Langer, Physical Review E **57**, 7192 (1998)
[19] M.L. Falk, J. Langer, Annual Review of Condensed Matter Physics **2**, 353 (2011)
[20] R. Khosropour, J. Zirinsky, H.K. Pak, R.P. Behringer, Physical Review E **56**, 4467 (1997)
[21] F. Radjai, S. Roux, Physical review letters **89**, 064302 (2002)
[22] P. Rognon, I. Einav, Physical Review Letters **105** (2010)
[23] S. Abedi, A.L. Rechenmacher, A.D. Orlando, Granular Matter **14**, 695 (2012)
[24] A.S. Keys, A.R. Abate, S.C. Glotzer, D.J. Durian, Nature Physics **3**, 260 (2007)
[25] S. Henkes, D.A. Quint, Y. Fily, J. Schwarz, Physical review letters **116**, 028301 (2016)
[26] P.G. Rognon, J.N. Roux, M. Naaim, F. Chevoir, Journal of Fluid Mechanics **596**, 21 (2008)
[27] D. Durian, Physical Review E **55**, 1739 (1997)
[28] G. Düring, E. Lerner, M. Wyart, Physical Review E **89** (2014)
[29] L. Berthier, G. Biroli, J.P. Bouchaud, L. Cipelletti, D.E. Masri, D. L'Hôte, F. Ladieu, M. Pierno, Science **310**, 1797 (2005)
[30] P. Ballesta, A. Duri, L. Cipelletti, Nature Physics **4**, 550 (2008)
[31] C. Donati, S.C. Glotzer, P.H. Poole, W. Kob, S.J. Plimpton, Phys. Rev. E **60**, 3107 (1999)
[32] A. Lemaître, C. Caroli, Physical review letters **103**, 065501 (2009)

Powder keg divisions in the critical state regime: transition from continuous to explosive percolation

Zongzheng Zhou and *Antoinette* Tordesillas*

School of Mathematics and Statistics, University of Melbourne, 3010, Australia

Abstract. The underlying microstructure and dynamics of a dense granular material as it evolves towards the "critical state", a limit state in which the system deforms with an essentially constant volume and stress ratio, remains widely debated in the micromechanics of granular media community. Strain localization, a common mechanism in the large strain regime, further complicates the characterization of this limit state. Here we revisit the evolution to this limit state within the framework of modern percolation theory. Attention is paid to motion transfer: in this context, percolation translates to the emergence of a large-scale connectivity in graphs that embody information on individual grain displacements. We construct each graph $G(r)$ by connecting nodes, representing the grains, within a distance r in the displacement-state-space. As r increases, we observe a percolation transition on $G(r)$. The size of the jump discontinuity increases in the lead up to failure, indicating that the nature of percolation transition changes from continuous to explosive. We attribute this to the emergence of collective motion, which manifests in increasingly isolated communities in $G(r)$. At the limit state, where the jump discontinuity is highest and invariant across the different unjamming cycles (drops in stress ratio), $G(r)$ encapsulates multiple kinematically distinct communities that are mediated by nodes corresponding to those grains in the shear band. This finding casts light on the dual and opposing roles of the shear band: a mechanism that creates powder keg divisions in the sample, while simultaneously acting as a mechanical link that transfers motion through such subdivisions moving in relative rigid-body motion.

1 Introduction

Percolation has profound implications for many disordered complex systems in social environments, Nature and technology: e.g., outbreak of an epidemic disease, social upheaval, spread of fire in a forest, computer virus or malware on computer networks, and power grid failure (see [1–4], and references therein). In multiphase granular and soft matter systems, percolation has been studied extensively in the context of fluid transport and applied in various industrial processes, including oil and gas recovery, dewatering, and in the manufacture of concrete, slurries and pastes [5, 6]. In this context, the study of percolation concerns itself with a nontrivial critical behavior or phase transition that manifests as a sudden onset of large-scale connectivity in the system.

In 2009, Achlioptas and co-workers introduced a new class of percolation phenomena, coined *explosive percolation* [7]. Compared with ordinary percolation where the transition is continuous, explosive percolation is marked by an abrupt jump or discontinuous transition to a large-scale connectivity through steps that are much smaller than the system size. Though still in its infancy [4], explosive percolation has already attracted enormous interest with some applications, albeit still few, to finite real-world networks. For example, it has been shown that monitoring the explosive percolation process on modular networks is useful for understanding the underlying community structure of proteins [2] and social networks [3]. The basic idea behind these real networks is that an initially disordered system can suddenly become ordered through clustering behavior; interacting units of a complex system, which behave in similar ways, form clusters or cliques. In the representative networks where the nodes are the units, such clustering behavior manifest as "network communities", which are partitions of graphs into groups of nodes with high levels of intra-connections and relatively sparse inter-connections.

Here, we search for similar signatures of critical behavior in the "critical state regime" where distinct patterns in motion are widely observed. Take, for instance, samples undergoing localized failure. In the transition to this limit state, the sample suddenly splits into subdomains that move in relative rigid-body motion, while mediated by shear bands rich in vortices [8]. Once in the limit state, these "powder keg" divisions in kinematics intermittently form across unjamming cycles (drops in stress ratio), in between periods of relatively quiescent jamming cycles (rises in stress ratio). Each such unjamming cycle is triggered by collective force chain failure by buckling inside the shear band, a mechanism that causes adjacent subdomains on either side of the band to unload elastically while undergoing relative rigid-body motion. These

*e-mail: atordesi@unimelb.edu.au

failure events, accompanied by bursts to a peak in dissipation rate, kinetic energy and nonaffine deformation at the macroscale, have been widely observed in discrete element simulations [9].

Given the rich mechanisms and patterns in the limit state, our objective in this study is twofold: (a) to uncover a possible percolation transition in the dynamics of a dense granular material in the limit state, and (b) to use this critical percolation (if it exists) to gain new insights into the significance of self-organized, collective motions that characterize this limit state. To this end, we examine the percolation process in graphs embodying information on individual grain motions. Our data set is from a discrete element simulation of a biaxial compression test in which the sample deforms in the presence of strain localization in the limit state. We study the emergence of a large-scale connectivity in graphs $G(r)$ where the nodes are the grains; nodes close to each other in the displacement-state-space (i.e., within a distance r) are preferentially connected by links. As we tune the parameter r from small to large, a percolation transition characterised by the size of the largest connected component can be observed. We study the nature of this percolation transition and attempt to relate the evolution of the critical point for percolation transition to the observed cycles of jamming-unjamming dynamics as evident in the rises and falls in the stress ratio.

The remainder of this paper is organized as follows. We describe the data set analysed in Section 2 and the method in Section 3. Key results are discussed in Section 4 before the concluding summary in Section 5.

2 Data

The data set comes from a well-studied simulation of 5098 spherical grains submitted to planar biaxial compression under constant confining pressure [9]. This simulation, which is from a model developed using the discrete element method (DEM), has been described elsewhere and many facets of it have been reported in the literature (e.g., [9, 10]). It incorporates a contact moment (or rolling resistance) to account for the effect of particle shape angularity, e.g. resistance to relative rotations due to particle interlocking, which is encountered in real irregularly shaped grains. We have DEM output data for every grain over 299 equilibrium states, spanning all regimes of the loading history that culminates in the limit state where the sample fails by strain localization. A single fully developed shear band forms, which is inclined approximately along the main forward diagonal (from bottom left to top right corner of each sample) with a global anticlockwise rotation. The shear band governs system dynamics in the limit state: distinct fluctuations manifest in the stress ratio, indicative of stick-slip dynamics. During stick or rise in stress ratio, the dominant mesoscale mechanism inside the shear band is jamming whereby clusters of particles form new stable, densely-connected configurations embodying force chains. During the ensuing slip or drop in stress ratio, the dominant mesoscale mechanism inside the shear band is unjamming whereby force chains fail by buckling. Force

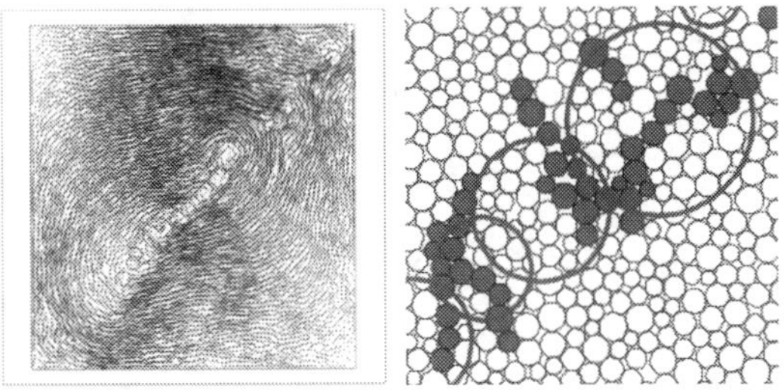

Figure 1. A representative displacement vector field during unjamming in the limit state showing the shear band (left); a close-up view of buckling force chains shown here by the red colored grains (right). Displacement vectors in vortices, identified and characterized in [8], are highlighted by the blue circles.

chain buckling is accompanied by, and induces, local grain rearrangements that are highly nonaffine, dilatant, and dissipative [9, 10]. These rearrangements manifest a pattern of motion that is rich in vortices — and mediates the relative rigid-body motion of the regions on either side of the shear band, as shown in Figure 1.

3 Methods

Let N be the total number of grains in the sample, and $(x_i(t), y_i(t), z_i(t))$ be the location of grain i, where $i = 1, 2, 3, ..., N$, at each equilibrium state t. For the data described in Section 2, $N = 5098$. The displacement of grain i is $d_i(t) = (\Delta x_i(t), \Delta y_i(t), \Delta z_i(t))$, where

$$\Delta x_i(t) = x_i(t) - x_i(t-1)$$
$$\Delta y_i(t) = y_i(t) - y_i(t-1)$$
$$\Delta z_i(t) = z_i(t) - z_i(t-1)$$

At each equilibrium state t, we construct a graph $G(r)$ in the displacement-state-space (DSS) where the nodes represent the grains. Two nodes, i and j, share a link if the difference in their displacements satisfies

$$|d_i(t) - d_j(t)| \le r. \quad (1)$$

Thus, two nodes are connected if their Euclidean distance in DSS is at most r, where $r > 0$ is a free parameter (Figure 2). We then study how $G(r)$ changes as r increases from 0 by a constant and small increment δ. When r is small, only a few links exist: G is sparsely connected and is a multi-component graph. As r increases, more and more nodes are connected, so that eventually G becomes a single-component graph. Let $S(r)$ be the size of the largest connected component in $G(r)$, the so-called giant component of $G(r)$. The size of S, given by the total number of nodes in S, increases from 0 to its maximum possible value of N. In Figure 2, $N = 27$, $G(r)$ is an 8-component graph (sizes 10, 7, 4, 2, 1,1,1,1) and $S(r)$ is size 10; but increasing r to $r + R$ (constant $R >> \delta$) would lead to $G(r) = S(r)$ of size 27 because of the bridge nodes.

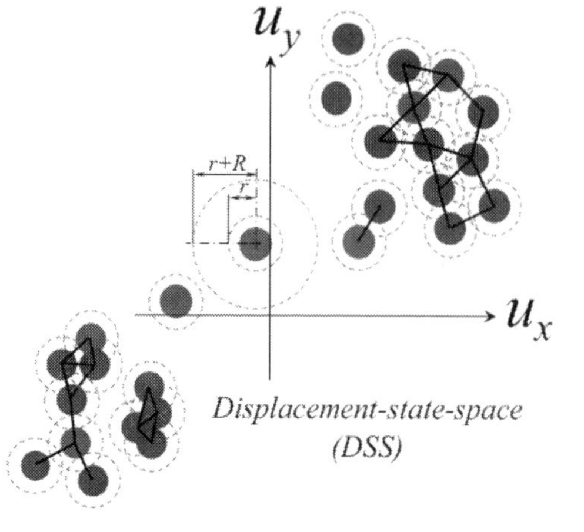

Figure 2. Schematic plot of the method for constructing $G(r)$ for $N = 27$ nodes. Each node is located at its corresponding grain displacement vector (u_x, u_y). For a given radius r, two nodes within a distance r (radius of dashed circle) in DSS are connected by a link. Red nodes act as a bridge between the two largest communities of sizes 10 (top-right) and 7 (bottom-left), respectively.

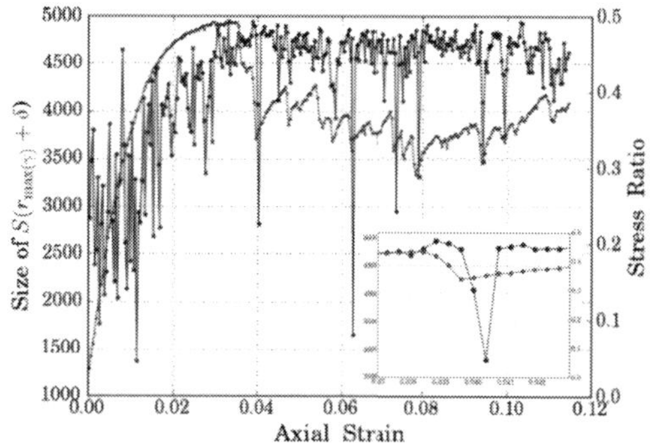

Figure 3. Evolution with strain of the size of $S(r_{max(\gamma)})$ and stress ratio. The significant drops in the critical state regime occur during jamming cycles: inset enlarges the rise in stress after ≈ 0.04 axial strain.

4 Results

In percolation theory, a key quantity is the order parameter, which is used to characterize the degree of order of the system. The order parameter is a normalized quantity between 0 and 1. A system is termed disordered (ordered) if the corresponding order parameter is close to 0 (1). Moreover, a percolation process is called continuous (explosive) if the order parameter is a continuous (discontinuous) function of the microscopic parameter r. Order parameters can be defined in various ways as needed. The most common order parameter in percolation theory is the probability that a selected node belongs to the giant component $S(r)$. Given that $r > 0$, this probability can be expressed as

$$p(r) = \frac{S(r)}{N}, \quad (2)$$

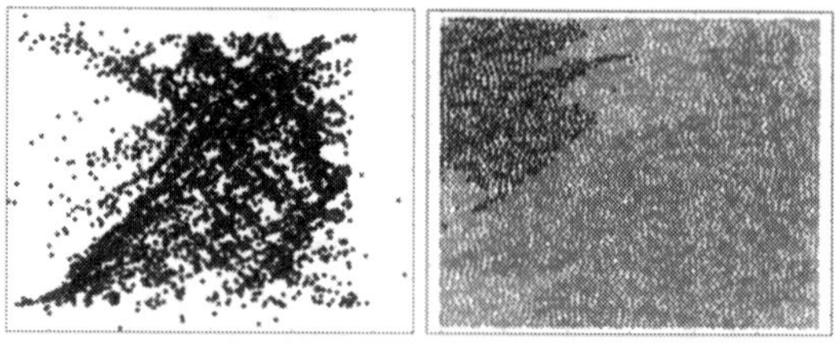

Figure 4. *Pre-failure regime.* Scatter plot of displacements in DSS at the axial strain interval of [0.0008,0.0012] (left). Member nodes of the different components in $G(r)$ just before percolation transition $S(r_{max(\gamma)} - \delta)$, mapped back into the physical domain of the sample: members of the largest and second largest connected components are colored red and blue, respectively, with the rest in grey (right).

with $S(r)$ the size of the largest connected component in $G(r)$. When r is small, $G(r)$ is sparsely connected: $p(r)$ is close to zero. When r is large, $G(r)$ is densely connected: $p(r)$ is close one.

For finite systems, the function $p(r)$ can contain distinct jump discontinuities, $\gamma(r) >> \gamma(r - \delta)$, as r is increased at small and fixed increment δ, where $\gamma(r) = [p(r) - p(r - \delta)]/\delta$. The evolution with strain of the size of $S(r_{max(\gamma)})$ where $r = r_{max(\gamma)}$ is where γ is maximum is shown in Figure 3. We see that $S(r_{max(\gamma)})$ increases initially but becomes almost invariant in the limit state. This trend is consistent with the distribution of displacements in the DSS as shown both prior to and during failure in Figures 4(left) and 5(left), respectively. The drops in the size of $S(r_{max(\gamma)})$ occur during jamming when grains lock up and their motions momentarily subdued.

The corresponding visualisations in Figures 4(right) and 5(right) show the locations of the member nodes of the different components in $G(r)$ just *before* the percolation transition, at $r = r_{max(\gamma)} - \delta$, in the physical domain of the sample. Two key features of the displacement-state-space in Figure 5 distinguish it from that in Figures 4. First, there are two distinct communities of nodes representing the two oppositely moving rigid-body parts of the sample, whenever the shear band unjams. Second, the nodes in the shear band mediate, and thus enables motion transfer, between the two communities.

Finally, we see from Figure 6 that the transition becomes explosive in the limit state — when a fully developed shear band inhabits the sample. Here we observe a prolonged period when the size of the giant component $S(r)$ is almost invariant as r is increased incrementally by an amount δ, where δ is very small relative to system size. This prolonged invariance of $S(r)$ is indicative of the wide separation of the largest components (network communities) in $G(r)$ in the limit state. The shear band, however, acts as bridge nodes: compare the depiction in Figure 2 and the distribution of the nodes in DSS in Figure 5. Due to the shear band nodes, we observe a sudden and large jump increase in $p(r)$ to just below 1, indicating an explo-

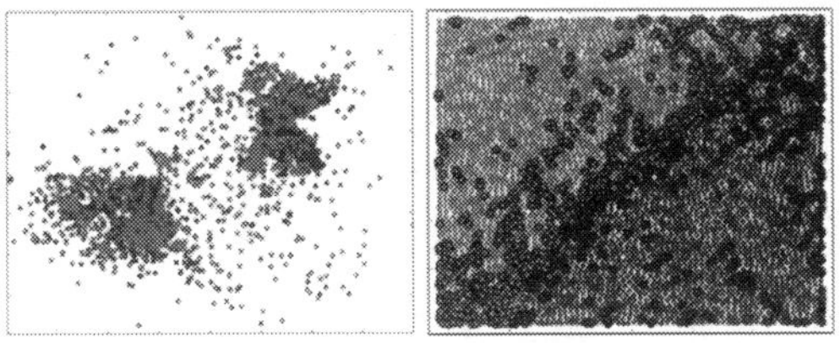

Figure 5. *Failure regime.* Scatter plot of displacements in DSS at the axial strain interval of [0.0388,0.0392] (left): here nodes highlighted in red correspond to those grains with the highest rotations (top 20%) which are mainly in the shear band. Member nodes of the different components in $G(r)$ just before explosive percolation $S(r_{\max(y)} - \delta)$, mapped back into the physical domain of the sample: members of the largest and second largest connected components are colored red and blue, respectively, with the rest in grey (right). Grains with the highest rotations are highlighted by the black outlined circles.

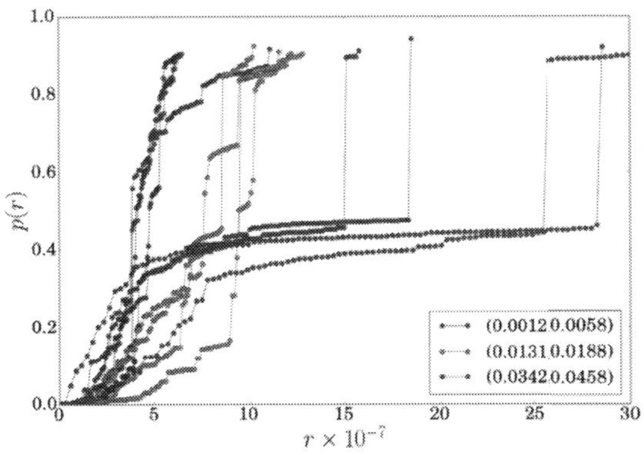

Figure 6. Evolution of $p(r)$ versus r in the lead up to and during the limit state.

sive growth in the size of $S(r)$ to near its maximum value just after the explosive percolation transition.

5 Conclusion

The process of percolation in graphs embodying information on individual grain kinematics is investigated to determine the existence of a possible critical transition in the underlying microstructure and dynamics of a dense granular material as it evolves towards the so-called critical state regime. We find that the presence of strain localization is precisely that which renders this limit state critical when viewed from the perspective of a topological-dynamical phase transition. Moreover the transition is consistent with that of an explosive percolation. We attribute this to the emergence of self-organized collective motion, which result in increasingly isolated kinematical communities that are mediated by the shear band. This finding casts light on the dual and opposing roles of the shear band: a mechanism that creates subdivisions in the sample during unjamming events, while simultaneously acting as a mechanical link that transfers motion through such subdivisions moving in relative rigid-body motion. A full analysis of the percolation properties of $G(r)$ in the lead up to and during the limit state, including the evolution of the critical percolation transition is reported in [11].

Acknowledgements

The authors acknowledge the support of the Australian Research Council (DP120104759), the U.S. Air Force (AFOSR 15IOA059), and the U.S. Army Research Office (W911NF-11-1-0175, W911NF-15-1-0527).

References

[1] A. Aharony, D. Stauffer, *Introduction To Percolation Theory* (Taylor & Francis, 2003)
[2] Rozenfeld, H. D., Gallos, L. K., Makse, H. A., Eur. Phys. J. B **75**, 305 (2010)
[3] R.K. Pan, M. Kivelä, J. Saramäki, K. Kaski, J. Kertész, Phys. Rev. E **83**, 046112 (2011)
[4] R.M. D'Souza, J. Nagler, Nature Physics **11**, 531 (2015)
[5] B. Brady, E. Brown, *Rock Mechanics: For underground mining* (Springer Netherlands, 2007)
[6] C. Hall, W. Hoff, *Water Transport in Brick, Stone and Concrete* (Taylor & Francis, 2002)
[7] D. Achlioptas, R.M. D'Souza, J. Spencer, Science **323**, 1453 (2009)
[8] A. Tordesillas, S. Pucilowski, Q. Lin, J.F. Peters, R.P. Behringer, Journal of the Mechanics and Physics of Solids **90**, 215 (2016)
[9] A. Tordesillas, Philosophical Magazine **87**, 4987 (2007)
[10] A. Tordesillas, S. Pucilowski, D.M. Walker, J.F. Peters, L.E. Walizer, International Journal for Numerical and Analytical Methods in Geomechanics (2014)
[11] Z. Zhou, A. Tordesillas, in preparation (2017)

DEM Simulation of Particle Stratification and Segregation in Stockpile Formation

Dizhe Zhang[1], *Zongyan* Zhou[1,*], and *David* Pinson[2]

[1]Laboratory for Simulation and Modelling of Particulate Systems (SIMPAS), Department of Chemical Engineering, Monash University, VIC 3800, Australia
[2]BlueScope Steel, Port Kembla, NSW 2505, Australia

Abstract. Granular stockpiles are commonly observed in nature and industry, and their formation has been extensively investigated experimentally and mathematically in the literature. One of the striking features affecting properties of stockpiles are the internal patterns formed by the stratification and segregation processes. In this work, we conduct a numerical study based on DEM (discrete element method) model to study the influencing factors and triggering mechanisms of these two phenomena. With the use of a previously developed mixing index, the effects of parameters including size ratio, injection height and mass ratio are investigated. We found that it is a void-filling mechanism that differentiates the motions of particles with different sizes. This mechanism drives the large particles to flow over the pile surface and segregate at the pile bottom, while it also pushes small particles to fill the voids between large particles, giving rise to separate layers. Consequently, this difference in motion will result in the observed stratification and segregation phenomena.

Keywords: stockpile, segregation, stratification, DEM

1 Introduction

The pattern formation of particle stockpiles is primarily a practical concern. During the stockpiling process, segregation and stratification happen spontaneously and directly lead to the de-mixing state of the particulate system. Therefore, in order to increase the stockpiles' mixing quality, it is of great significance to have a better understanding of these two phenomena.

It is already acknowledged that segregation and stratification are both induced by the difference of particle properties among granular mixtures. For segregation, significant investigation of its mechanisms has been undertaken. Williams [1-3], and Drahun and Bridgwater [4] experimentally studied the heap flow segregation using binary mixtures of spherical particles. They demonstrated a percolation mechanism where small particles tend to sink into the voids while large particles are able to float onto the surface and roll down to the pile bottom. For stratification, many researchers have reported this feature with the use of particles in different shapes. They also further proposed several mechanisms to explain this phenomenon. Makse et al. [5-7] believed that stratification is induced by a "kink" mechanism and the key parameter controlling the process is the repose angle of different species. Gray et al. [8-10] experimentally proved that surface avalanches are inducing the stratification. From their work, avalanches work as a kinetic sieve which triggers the particle layering, while dispersed shock waves can further bring the granular matters rapidly to rest. This theory was also extensively studied by Baxter et al. [11] with poured granular heaps. Different from the previous researchers, Benito et al. [12] experimentally demonstrated the existence of stratification using only spherical particles, and they further conducted parameter studies in an attempt to determine the parameter that initiates stratification.

In this article, we present a numerical study of spontaneous segregation and stratification of binary spherical particles when forming a stockpile. With the use of a discrete element model, both of these phenomena were observed and a comprehensive view of the particle movements was obtained. After a detailed parameter study of size ratio, injection height and mass ratio, Benito et al.'s findings [12, 13] on stratification were validated and further the effects of these parameters on stockpile mixing quality were demonstrated. Of the mechanisms dominating the pattern formation, we found that stratification is induced by the void-filling mechanism, while further pile avalanches and motion of large particles will result in segregation.

* Zongyan Zhou: zongyan.zhou@monash.edu

2 Simulation method and conditions

Simulations were conducted based on a DEM model previously developed by Zhou et al.[13]. The simulated problem is shown schematically in Fig. 1. The device includes two parts: a particle injection device and a receiving box ($200mm \times 100mm \times 5mm$). During the growth of the stockpile, the distance between injection surface and top of stockpile (injection height) was set constant by moving the injection surface vertically during the stockpile formation. Specifically for this work, the injection height is set as either 10mm or 50mm to study its effects on pattern formation.

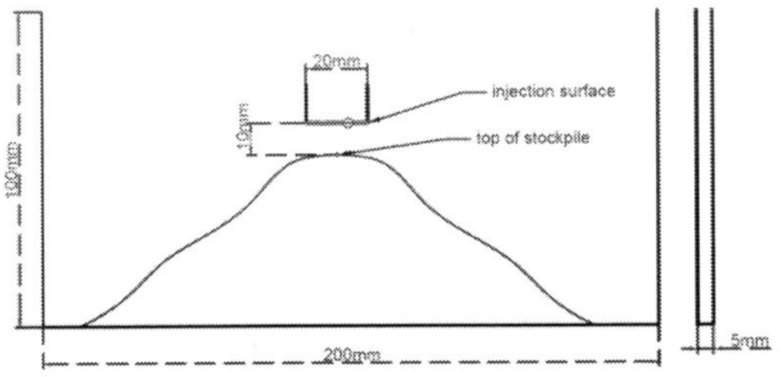

Fig. 1 Schematic diagram of the simulation conditions. (Injection surface: the surface where particles are generated and released; top of stockpile: the highest point of the formed stockpile)

The binary granular mixtures used in the simulations are formed by spherical particles of two different size ratios (1mm and 3mm, 2mm and 3mm). The mass ratio R (the total mass ratio of the two species which also represents the flow rate ratio of large particles versus small particles) was set to be 0.5 or 1. Other physical properties of the particles were set to those of glass beads. Details of the input variables and their values are shown in Table.1.

Table 1 Input variables and their values

Input variables	Value
Young's module	1×10^8 N/m^2
Poisson's ratio	0.3
Sliding coefficient (P-P)	0.4
Sliding coefficient (P-W)	0.4
Rolling coefficient (P-P)	0.3
Rolling coefficient (P-W)	0.3
Particle density	2.7×10^3 kg/m^3
Particle number	20,000
Flow rate	0.1g/s

3 Results and discussions

Fig. 2 shows some of the patterns obtained from the simulations. After pouring a well-mixed particle stream into the slot, segregation and stratification spontaneously occur within the stockpiles. These patterns vary when changing the parameters including size ratio, injection height and mass ratio. It is clearly observed in Fig. 2 that the presence of segregation is not strictly limited by the testing parameters. For example, while a size ratio of 3:2 is already large enough to trigger stratification, stratification is only observed in cases with size ratios exceeding 3:1 combined with a low injection height.
In the following sections, parameter effects in terms of the mixing quality and mechanisms of stratification and segregation will be discussed in detail.

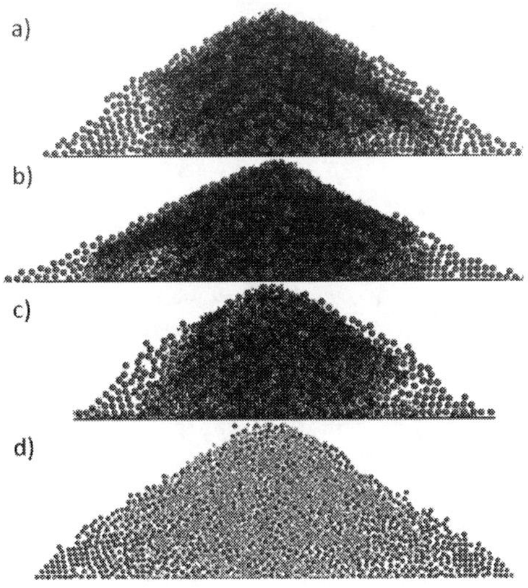

Fig. 2 Stockpile patterns of different cases : a) size ratio 3:1, injection height 10mm, mass ratio R=1; b) size ratio 3:1, injection height 10mm, mass ratio R=0.5; c) size ratio 3:1, injection height 50mm, mass ratio R=1; d) size ratio 3:2, injection height 10mm, mass ratio R=1. Particle sizes have been scaled in the figure (scale factor = 0.7).

3.1. Mixing quality

A common method to quantify the mixing quality of a granular mixture is to use the mixing index. In this work, the mixing index M, first introduced by Lacey [14] and further developed by Feng et al. [15] was adopted. This index is given by

$$M = \frac{S_0{}^2 - S^2}{S_0{}^2 - S_R{}^2} \quad (1)$$

where S^2 is the actual variance, while $S_0{}^2$ and $S_R{}^2$ are the variances of the completely segregated and completely mixed states, respectively. So a completely segregated mixture will have M as 0. More details about this equation can be found in Feng et al.'s work [15].

Fig. 3 demonstrates the comparisons of M among different cases. Firstly, the particle size ratio can effectively influence the pattern formation as large size ratio will induce stronger segregation [4]. Then, for the injection height, it will affect the impact velocity, thus influencing the further particle movement. As shown from Fig. 3, a higher injection point with larger impact velocity

 DOI: 10.1051/epjconf/201714015018

can slightly reduce the measured segregation. Finally, the mixing quality is also largely affected by R that larger R indicates lower M. These results can be explained by the void-filling mechanism as discussed in the following sections.

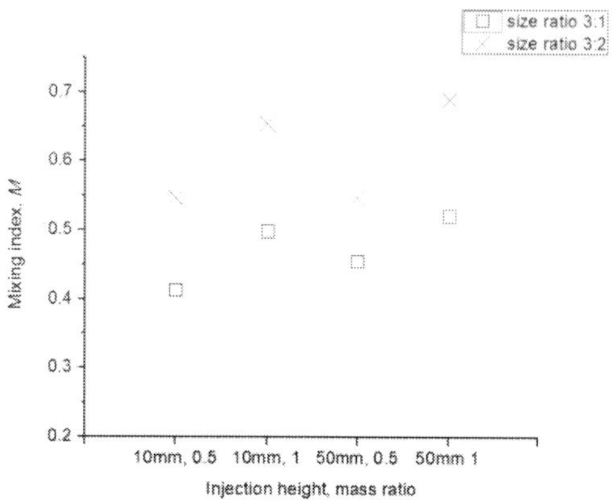

Fig. 3 Effects of size ratio, injection height and mass ratio on mixing quality in terms of mixing index, M

3.2. Stratification

From Fig. 2, particle stratification is easily observed in Fig. 2a) and Fig. 2b). This layering phenomenon can be explained by a void-filling mechanism which has been used to explain the situations in chute flow and vibrated granular beds [16, 17]. When particles of different sizes are mixed together within an active system, this mechanism presents the process that gravity and inertia which drives small particles to sink or percolate into the voids between large particles instead of flowing on the particle surface. Fig. 4 shows the specific situation for the stockpiling process. When pouring well mixed granular mixtures onto a flat surface to form a pile, the small particles percolate into the voids while large particles are able to flow on the top of the surface. This difference in motion will then cause the accumulation of small particles between large particles, thus divide the large particles into separate single layers. The lower layer remains motionless while the upper layer is free to slide until the particles reach a "block".

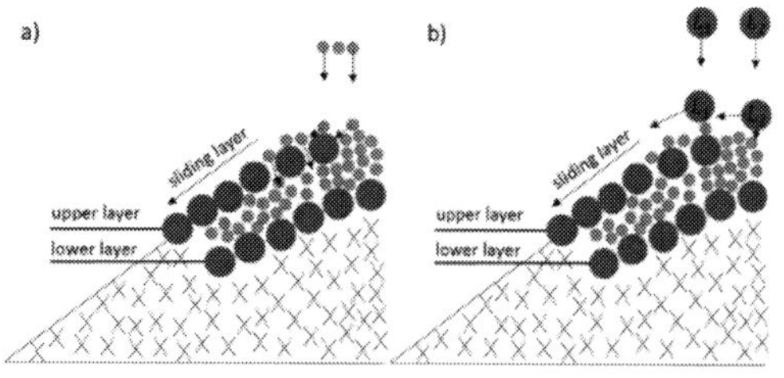

Fig. 4 Comparison of the motion of different particles: a) motion of small particles; b) motion of large particles

The "block" is a special area within the region where the void-filling process occurs. With the growth of the stockpile, available voids for small particles to fill into become fewer and the porosity of the area decreases. When some critical points have extremely low porosity, the void-filling process will then stop. The area with extremely low porosity has a high resistance to flow and represents the "block". Fig. 5 shows the "block" inside the stockpile identified from spatial distribution of porosity.

As the "block" stops the small particles' penetration motion, the outer layer of the stockpile becomes more unstable as additional particles drop onto the pile. This instability then causes further avalanche of the outer particles, especially for large particles. This avalanche behaviour consequently reforms the distribution of outer particles and removes the "block", which will then reinitiate the void-filling process and successively regenerate the stratification phenomenon leading to the regular layered stratification pattern.

The regular motion of large particles is shown schematically by particle L_1 in Fig. 4b). These large particles tend to flow onto the particle surface until reaching the pile bottom or forming the sliding layer until a "block" forms. However, there is still another type of particle motion that should be considered. Sometimes, for large particles like L_2, their inertia drives them to sink into the pile and subsequent burial by the later particles. Such particles remain motionless within the pile and their presence can be used to explain the irregular distribution of some large particles at the centre area of the stockpile.

The mechanisms discussed above can be used to explain the effects of different parameters on stratification. In Fig. 2, stratification is not observed for cases with large injection height and small size ratio. For large injection height, it will significantly influence the impact velocity when particles drop on the stockpile, as large particles with higher impact velocity will easily percolate into the pile instead of flowing on the particle surface. As mentioned above, this motion will then cause the irregular distribution of large particles within the stockpiles instead of the stratification. For size ratio, as stratification is mainly driven by the void-filling mechanism, available voids can be the most important factor for triggering this process. Lower size ratio will make the voids comparably smaller for small particles which increases the difficulty for them to enter into the available voids. This decrease of available voids will then weaken the void-filling process, thus decrease the stratification.

3.3. Segregation

Segregation specifically represents the phenomenon that large particles tend to segregate at the bottom of the stockpile. This phenomenon can be clearly observed from the patterns simulated in Fig. 2.

As shown in Fig. 4, the void-filling mechanism differentiates the motion of particles with different sizes. For large particles, gravity and inertia drive them to flow onto the particle surface, and consequently, large particles will segregate to the bottom of the pile. Similarly, when the particles avalanche following the formation of a "block", the outer particles also flow to the bottom. Such avalanches also result in the segregation of large particles as the outer particles are mostly large particles as the small particles are more likely to percolate inside. This finding is also consistent with Drahun and Bridgewater's experiments with free surface flow [4].

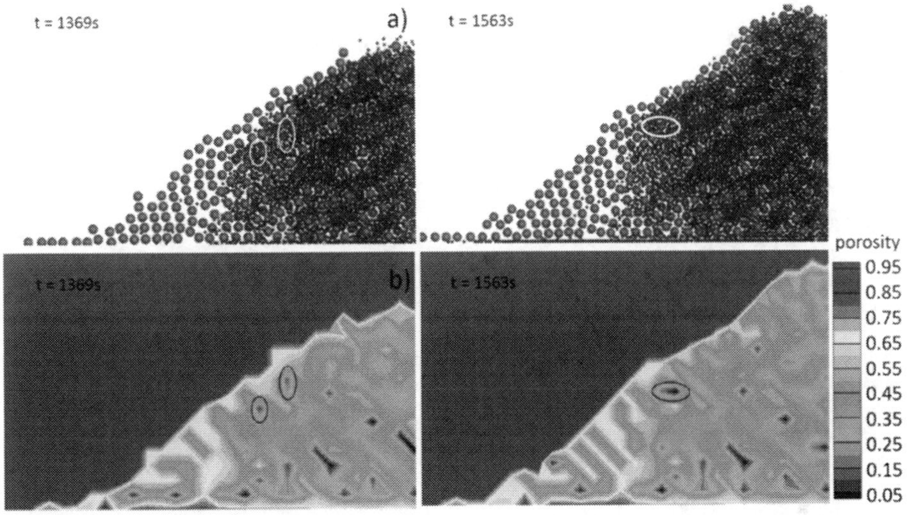

Fig. 5 a) Stockpile particle distribution patterns and b) porosity spatial distribution when the void-filling process stops. The figures are captured at 2 different times (t = 1369s and t = 1563s); the circled area is where "block" is located; particle sizes have been scaled in the figure (scale factor = 0.7).

4 Conclusions

A numerical study based on DEM simulations was carried out to study the stockpile formation in terms of the segregation and stratification processes. The effects of factors including size ratio, injection height, and mass ratio, *R* were investigated, and a relationship between the mixing quality and these factors was found. A larger size ratio, lower injection height and comparably smaller *R* can induce stronger segregation, which also represents a lower mixing quality. Furthermore, the stratification phenomenon was successfully simulated with the use of spherical particles, and the simulation conditions and results are consistent with Benito et al.'s physical experiments [12].

In terms of the underlying mechanism causing stratification and segregation, the void-filling mechanism dominates and its effects can explain the occurrence of these two phenomena. This mechanism drives particles of unequal size to move differently, which will then induce the specific phenomena.

Acknowledgements

This work was financially supported by Australian Research Council (ARC) and BlueScope Steel. Simulations were conducted with the support of NCI National Facility.

References

1. J. Williams, Fuel Soc. J, 1963. **14**: p. 29-34.
2. J. Williams, Powder Technol, 1976. **15**(2): p. 245-251.
3. J. Williams, Powder Technol, 1968. **2**(1): p. 13-20.
4. J. Drahun, J. Bridgwater, Powder Technol, 1983. **36**(1): p. 39-53.
5. P. Cizeau, H. Makse, H. Stanley, Phys Rev E, 1999. 59(4): p. 4408-4421.
6. H. Makse, P. Cizeau, H. Stanley, Phys. Rev. Lett. 1997. **78**(17): p. 3298-3301.
7. H. Makse, P. Cizeau, H. Stanley. Physica A, 1998. **249**(1–4): p. 391-396.
8. J. Gray, B. Kokelaar, J Fluid Mech, 2010. **652**: p. 105-137.
9. J. Gray, K. Hutter, Continuum Mech Therm, 1997. **9**(6): p. 341-345.
10. J. Gray, C. Ancey, J Fluid Mech, 2009. **629**: p. 387-423.
11. J. Baxter, U. Tuzun, D. Heyes, I. Heyati, P. Fredlund, Nature, 1998. **391**(6663): p. 136-136.
12. J. Benito, I. Ippolito, A. Vidales, Powder Technol, 2013. **234**: p. 123-131.
13. Y. Zhou, B. Wright, R. Yang, B. Xu, A. Yu, Physica A, 1999. **269**(2–4): p. 536-553.
14. P. Lacey, Chem Eng Res Des, 1997. **75**: p. S49-S55.
15. Y. Feng, B. Xu, S. Zhang, A. Yu, P. Zulli, AIChE J, 2004. **50**(8): p. 1713-1728.
16. S. Savage, C. Lun, J Fluid Mech, 1988. **189**: p. 311-335.
17. Arshad, K., Rep Prog Phys, 2004. **67**(3): p. 209.

Self-organization in the localised failure regime: metastable attractors and their implications on force chain functionality

Sebastian Pucilowski[1], *Antoinette* Tordesillas[1,*], and *Gary* Froyland[2]

[1] *School of Mathematics & Statistics, University of Melbourne, VIC 3010, Australia*
[2] *School of Mathematics & Statistics, University of New South Wales, NSW 2052, Australia*

Abstract. In transitive metastable chaotic dynamical systems, there are no invariant neighbourhoods in the phase space. The best that one can do is look for metastable or almost-invariant (AI) regions as a means to decompose the system into its basic self-organising building blocks. Here we study the metastable dynamics of a dense granular material embodying strain localization in 3D from the perspective of its conformational landscape: the state space of all observed conformations as defined by the local topology of individual grains relative to their first ring of contacting neighbors. We determine the metastable AI sets that divide this conformational landscape, such that grain rearrangements from one conformation to another conformation in the same AI set occurs with high probability: by contrast, grain rearrangements involving conformational transitions between AI sets are unlikely. The great majority of conformational transitions are identity transitions: grains rearrange and exchange contacts to preserve those topological properties with the greatest influence on cluster stability, namely, the number of contacts and 3-cycles. Force chains show a clear preference for that AI set with the most number of accessible and highly connected conformations. Here force chains continually explore the conformational landscape, wandering from one rarely inhabited conformation to another. As force chains become overloaded and buckle, the energy released enables member grains to overcome the high dynamical barriers that separate metastable regions and subsequently escape one region to enter another in the conformational landscape. Thus, compared to grains locked in stable force chains, those in buckling force chains, confined to the shear band, show a greater propensity for not only non-identity transitions within each metastable region but also inter-transitions between metastable regions.

1 Introduction

Grain rearrangements around force chains are indispensable for effective force transmission. Unlike their counterparts in architectural structures, columnar force chains are far from static. Grain motions in and around force chains continually occur, enabling force chains to explore a range of mutating conformations in a rich conformational and stability landscape [1–3]. Emerging evidence suggests that this landscape, populated by hills and valleys of varying depths and widths, shapes the rich ordered patterns of interaction in granular systems that emerge in the lead up to and during failure [1–4]. Here we focus on this process of self-organization in the localized failure regime and explore the organization of this landscape. We search for metastable attractor regions in this landscape to understand persistent or frequently occurring local grain arrangements and rearrangements in a dense granular dynamical system embodying shear bands in three dimensions. To understand force chain formation and evolution in this context, we go beyond 'static' information and determine the structure and dynamics underlying *conformational transitions* of member grains in force chains, as the system evolves from one equilibrium state to the next.

Specifically, building on [1, 2], we investigate the presence of metastable attractors which facilitate the creation and persistence of force chains in 3D frictional materials. The role of conformational transitions in force chains has been explored in 2D experiments on synthetic (photoelastic), as well as virtual, granular materials [1]. In these studies, the conformations are characterized by the topological configurations inhabited by a grain relative to its first ring of contacting neighbors. All such conformations observed in the system during loading define a conformational state space. This state space is then divided into regions, known as almost-invariant metastable sets, between which conformational transitions are unlikely: that is, there is a high probability that a grain undergoes a rearrangement from one conformation to another in the same metastable AI set. As to which of these sets force chains prefer to inhabit is crucial for the formulation of robust constitutive stress-strain relations for granular media [5, 6]. Equally important, especially for control of the underlying self-organization process, is the transition pathway followed by force chains in this conformational landscape under varying loading conditions. Such an exploration into 3D systems is challenging given the scarcity of data from physical experiments on interparticle contact forces for many

*e-mail: atordesi@unimelb.edu.au

DOI: 10.1051/epjconf/201714010007

Table 1. DEM model parameters and material properties.

Parameter	Value
Applied strain rate $\dot{\epsilon}_{yy}$	-5×10^{-2} s^{-1}
Confining pressure P	494.91 k N m^{-2}
Particles n	55100
Particle radii R	$(0.76 - 1.52) \times 10^{-3}$ m
Particle density ρ	2650 kg m^{-3}
Modulus of elasticity E_c	91.86 M Pa
Poisson's ratio ν_c	0.5
Rolling stiffness coefficient β	1.0
Inter-particle friction angle μ	0.7
Rolling coefficient $\eta = \mu^r$	0.02
Wall friction angle μ_{wall}	0.0 – 0.7

stages of a single test [2]. This challenge was recently overcome for near-frictionless hydrogel spheres [2]. However, transitions in the conformational landscape of 3D frictional systems representative of common real granular materials (e.g. sand) have not been explored and is the focus of this study. Given the absence of requisite data on interparticle contact forces in real granular systems, we use data from discrete element simulations.

Our analysis proceeds in two phases. In Phase 1, we establish the conformational landscape, a state space in which all observed conformations reside. In Phase 2, we divide this state space into the metastable AI sets. With attention paid to force chains, our objective here is to find the conformations and conformational transitions at the opposite extremes of the frequency spectrum: the most prevalent *intra*-transitions *within*, and the rare *inter*-transitions *between*, AI sets. A comprehensive analysis of the conformational landscape of dense granular systems in the limit state is reported in [4].

2 Data

The data set used here is from a DEM simulation of various assemblies of frictional 3D spherical particles submitted to triaxial compression under constant confining pressure. The simulation uses the YADE package [7]. Here we focus on a three-dimensional analogue of a well-studied two-dimensional system [3, 4, 8]. The same simulation and material properties are used as the two-dimensional system where possible to enable direct comparison with previously published results. A summary of the simulation and material parameters is provided in Table 1. Figure 1 (inset) shows the stress-strain response that is typical of densely packed granular systems [9].

3 Method

Analysis proceeds in two phases. In Phase 1, we establish the conformational state space for the triaxial test sample as it evolves from one equilibrium state to another in the failure or "persistent shear band" (PSB) regime where the underlying dynamics is assumed to be stationary. This begins with the contact network, in which the grains and their contacts are represented by nodes and links in a network [1, 2]. For the entire series of equilibrium states in the PSB regime, we enumerate all observed conformations of every node in the contact network and its local connectivity, d^t_s: d is the node degree (number of links), t is the number of 3-cycles, and s is the number of shared edges between the 3-cycles. For example, in the physical space of the sample, conformation 3^1_0 corresponds to a grain with three contacting neighbors and a single rotationally frustrated 3-cycle [3, 10].

In Phase 2, we partition the conformational state space into their almost-invariant sets, following the method in [11]. To do this, we perform a spectral analysis of the Markov transition matrix P that is constructed from the information gained in Phase 1. The conformations correspond to a collection of $S = \{1, \ldots, n\}$ states, where n is the total number of unique conformations observed in the PSB regime. We also track all the grain rearrangements between all consecutive pairs of equilibrium strain states and map these into P as a transition from conformation state i to conformation state j. This leads to the row-stochastic $n \times n$ transition matrix P, such that P_{ij} denotes the conditional probability of transitioning from state i to state j in one strain interval. The aim is to partition the collection of states S into subsets $A_k \subset S$, $k = 1, \ldots, K$, with $\bigcup_{k=1} A_k = S$ and $A_k \cap A_\ell = \emptyset$, $k \neq \ell$. The eigenvectors of P are used to partition the conformational state space S into metastable regions A_k known as almost-invariant (AI) sets, between which conformational transitions are unlikely. The elements A_k are then partitioned as follows: for each $k = 1, \ldots, K$, the probability that A_k at time t transitions to A_k at time $t + 1$ is as large as possible, while for $k \neq \ell$ the probability that A_k at time t transitions to A_ℓ at time $t + 1$ is as small as possible. Such a partition $\{A_k\}^K_{k=1}$ is called *almost-invariant* because each partition element acts approximately as its own dynamical process and is almost disconnected dynamically from the other partition elements.

4 Results

A total of 363 unique grain conformations are observed in the persistent shear band (PSB) regime, which leads to a stochastic transition matrix P of dimensions 363×363. However, the leading unit eigenvector of P is not unique. We correct this by interpreting P as an adjacency matrix of a graph G, and retaining the edges associated with the largest connected component of the graph G. This removes 17 rare conformations, corresponding to those with no information about their next transition: that is, transitions which when entered cannot be left, and isolated small components of G. This leads to a corrected stochastic transition matrix $\tilde{P}$ of size 346×346. From the leading unit eigenvector of $\tilde{P}$, we obtain the probability density distribution that quantifies the frequency that a given conformation is inhabited during the PSB regime (Figure 1). We found five metastable regions, AI sets $A - E$ (overlaid symbols in Figure 1), with the proportions of grains inhabiting

DOI: 10.1051/epjconf/201714010007

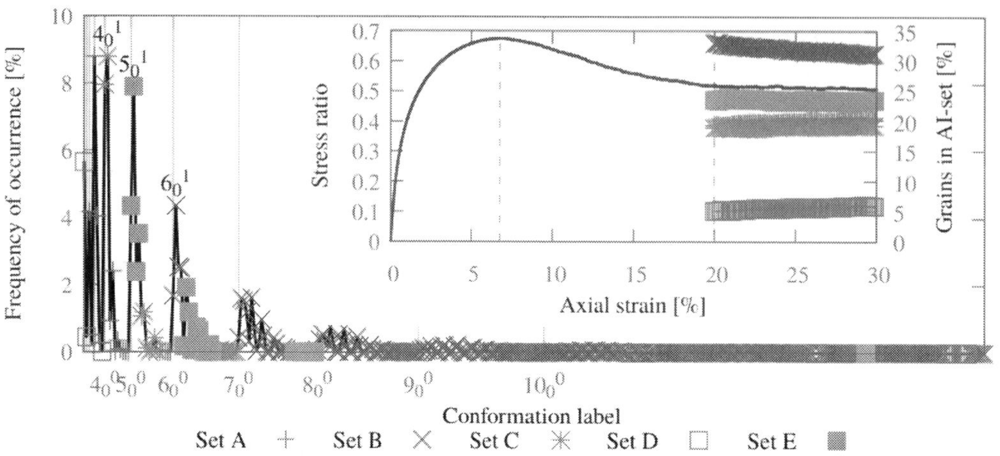

Figure 1. Frequency of occurrence of each conformation (black curve); overlaid is the classification of each conformation into one of the five almost-invariant set $A - E$ (symbols). (inset) Stress ratio and percentage of grains in each almost-invariant set.

each metastable region across the PSB regime shown in the inset. The near-invariant strain evolution of the population of grains in each region provides corroborating evidence to support past observations that the PSB regime exhibits stationary dynamics.

Prevalent conformations: We observe several prominent peaks in the probability density distribution, suggesting the presence of *preferred* self-assembled mesostructures. The top five peaks corresponding to the most preferred conformations, in decreasing order of occurrence, are: 4_0^1 (set C), 3_0^0 (set A), 5_0^1 (set E), 4_0^0 (set C) and 0_0^0 (i.e. a rattler, set D). Outside these peaks are infrequently inhabited grain conformations. As shown in Figure 1(inset), the majority of these are in set B, the most populated region with the largest number of accessible conformations that are densely connected (i.e. $d \geq 6$ and $s \geq 1$) yet rarely visited. The least populated region in the conformation state space is D, which consists of rattlers and poorly supported conformations (e.g. 0_0^0, 1_0^0 and 2_0^1).

Prevalent conformational transitions: Identity transitions comprise around 90% of all grain transitions (i.e. no change in local topology with respect to degree and 3-cycle membership and arrangement). The five most frequently occurring non-identity transitions are $3_0^0 \rightarrow 4_0^1$ (i.e. gain of a 3-cycle), $4_0^1 \rightarrow 3_0^0$ (i.e. loss of a 3-cycle), $3_0^0 \rightarrow 4_0^0$ (i.e. gain of a contact), $4_0^0 \rightarrow 3_0^0$ (i.e. loss of a contact), and $2_0^0 \rightarrow 3_0^0$. These preferred transitions are cyclic (e.g. Figure 2) and are characteristic of the underlying dynamics in the shear band where grain rearrangements are greatest.

Local arrangements around force chains: Grains found in force chains, classified according to their almost-invariant set membership, show a distinct preference for the densely connected conformations in set B and spend considerably less time visiting metastable regions with poorly supported conformations: A 6%, C 11%, D < 1%, E 24% versus B 58% (Figure 3(a)). Set B distinguishes itself from the other four regions of the conformational landscape by having the most number of accessible yet infrequently visited conformations. As the conformations in B have high d, resident grain clusters have more degrees of freedom to explore accessible sites in this region of the conformational state space. The prevalence of conformations with 3-cycles (increasing t and s) that frustrate grain rotations and provide truss-like support to force chains make B an ideal breeding ground for force chains [10].

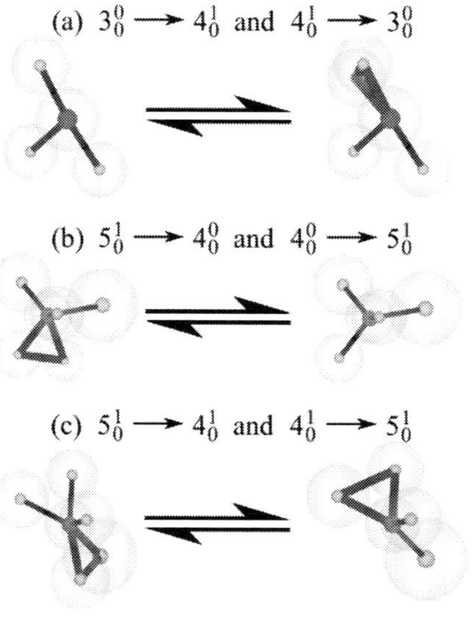

Figure 2. Top three most prevalent cyclic transitions involving a 3-cycle and their corresponding AI set (focal grain in red, 3-cycle in blue).

Local rearrangements in and around force chains: Rearrangements around force chains typically preserve structural stability, with 90% of conformational transi-

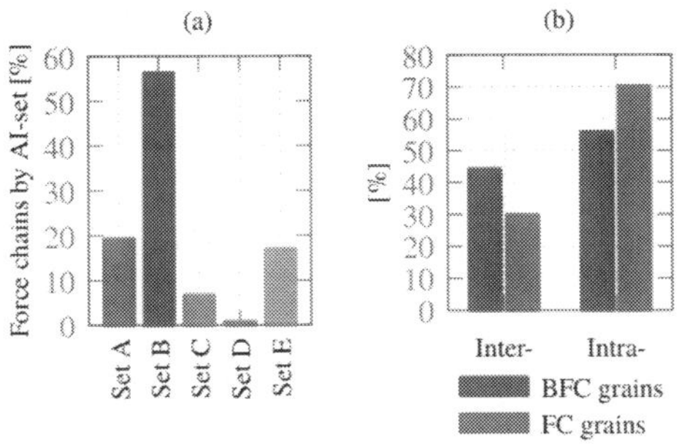

Figure 3. (a) Force chain grain conformations classified by their respective AI set. (b) Proportion of inter- and intra- transitions for grains found in and around force chains (FCs) and buckling force chains (BFCs).

tions maintaining the same number of contacts and 3-cycle topology (i.e. identity transitions). In stark contrast, BFCs are far more unstable, with only 56% of rearrangements being identity transitions. The non-identity transitions can be classed into two categories: intra-transitions and inter-transitions (Figure 3(b)). Compared to force chains which are locked in more stable configurations, grains in unstable BFCs are more likely to overcome the dynamical barriers that separate the metastable regions in the conformational landscape. Indeed, around 30% (44%) of grains in force chains (BFCs) undergo inter-transitions. Overcoming the dynamical barriers that separate the metastable regions in the conformational landscape requires an expenditure of energy. In this context, the propensity for BFCs to traverse a dynamical barrier, escaping a metastable region to inhabit another, is entirely consistent with past findings that showed these failure events are accompanied by bursts in kinetic energy, dissipation and mesoscale nonaffine deformation [1, 3, 4, 6, 8, 12].

5 Conclusion

We study the process of self-organization, on the scale of a grain and its first ring of contacting neighbors, during localised failure of a dense granular material submitted to triaxial compression. Attention is paid to those equilibrium states when the sample deforms in the presence of a fully developed persistent shear zone. The conformational landscape of the system embodies five metastable attractor regions, in each of which there is a high (very low) probability that a grain undergoes a rearrangement from one conformation to another in the same (different) region, across a single strain interval in the persistent shear band regime. The great majority of conformational transitions are identity transitions, implying that grains rearrange and exchange contacts to preserve the topological properties with the greatest influence on cluster stability: degree and the number of 3-cycles. Of the five metastable regions, force chains show a clear preference for that region with the most number of accessible and highly connected conformations. Here force chains continually explore state space, wandering from one rarely inhabited conformation to another. Force chain failure by buckling, confined to the shear band, involves grains that show a greater propensity to follow trajectories that involve non-identity transitions either within each metastable region or inter-transitions between metastable regions. The latter suggests that the stored energy released in force chains during buckling enables member grains to overcome the dynamical barriers that separate metastable regions and subsequently escape one region to enter another in the conformational landscape.

Acknowledgements

The authors acknowledge the support of the Australian Research Council (FT140100227, DP120104759), the U.S. Air Force (AFOSR 15IOA059), and the U.S. Army Research Office (W911NF-11-1-0175, W911NF-15-1-0527).

References

[1] A. Tordesillas, D.M. Walker, G. Froyland, J. Zhang, R.P. Behringer, Physical Review E **86** (2012)
[2] D.M. Walker, A. Tordesillas, N. Brodu, J.A. Dijksman, R.P. Behringer, G. Froyland, Soft Matter **11**, 2157 (2015)
[3] A. Tordesillas, Q. Lin, J. Zhang, R. Behringer, J. Shi, Journal of the Mechanics and Physics of Solids **59**, 265 (2011)
[4] S. Pucilowski, A. Tordesillas, G. Froyland, Powder Technology (2016), in preparation
[5] A. Tordesillas, M. Muthuswamy, Journal of the Mechanics and Physics of Solids **57**, 706 (2009)
[6] A. Tordesillas, S. Pucilowski, Q. Lin, J.F. Peters, R.P. Behringer, Journal of the Mechanics and Physics of Solids **90**, 215 (2016)
[7] V. Smilauer, E. Catalano, B. Chareyre, S. Dorofeenko, J. Duriez, N. Dyck, J. Elias, B. Er, A. Eulitz, A. Gladky et al., *Yade documentation 2nd ed* (2015)
[8] A. Tordesillas, Philosophical Magazine **87**, 4987 (2007)
[9] J. Kozicki, J. Tejchman, H.B. Mühlhaus, International Journal for Numerical and Analytical Methods in Geomechanics **38**, 1923 (2014)
[10] A. Tordesillas, D.M. Walker, Q. Lin, Physical Review E **81**, 011302 (2010)
[11] G. Froyland, Physica D: Nonlinear Phenomena **200**, 205 (2005)
[12] H. Zhu, G. Veylon, F. Nicot, F. Darve, European Journal of Environmental and Civil Engineering pp. 1–24 (2016)

Compaction dynamics of crunchy granular material

François Guillard[1,*], *Pouya* Golshan[2], *Luming* Shen[1], *Julio R.* Valdès[2], and *Itai* Einav[1,3]

[1] *Particles and Grains Laboratory, School of Civil Engineering, The University of Sydney, Sydney, New South Wales 2006, Australia*
[2] *Geo-Innovations Research Lab, Department of Civil, Construction, and Environmental Engineering, San Diego State University, San Diego, California 92182, USA.*
[3] *Department of Civil, Environmental & Geomatic Engineering, Faculty of Engineering Science, University College London, London WC1E 6BT, UK.*

Abstract. Compaction of brittle porous material leads to a wide variety of densification patterns. Static compaction bands occurs naturally in rocks or bones, and have important consequences in industry for the manufacturing of powder tablets or metallic foams for example. Recently, oscillatory compaction bands have been observed in brittle porous media like snow or cereals. We will discuss the great variety of densification patterns arising during the compaction of puffed rice, including erratic compaction at low velocity, one or several travelling compaction bands at medium velocity and homogeneous compaction at larger velocity. The conditions of existence of each pattern are studied thanks to a numerical spring lattice model undergoing breakage and is mapped to the phase diagram of the patterns based on dimensionless characteristic quantities. This also allows to rationalise the evolution of the compaction behaviour during a single test. Finally, the localisation of compaction bands is linked to the strain rate sensitivity of the material.

When a brittle material undergoes compaction, a wide variety of compaction patterns can be observed. Static compaction bands are observed very commonly in nature, for example in bones, rocks such as sandstone [1–3], or in meteoritic impacts [4]. In industrial processes, compaction of brittle porous media is involved during fabrication of pharmaceutical drugs [5], or when using metallic foams [6] or honeycomb structures [7]. The localisation of the compaction in dense bands leads to inhomogeneities in the material, modifying its properties, which can have important consequences on its mechanical or chemical response. Usually, the bands of lower porosity resulting from the loading of material remain static in the medium. However, it has been shown recently that moving compaction bands can also appear during the compression of brittle porous media like puffed rice [8] or snow [9]. In these cases, the compaction zone travels within the medium, leading to oscillating compaction bands. However, the precise mechanism leading to the localisation of the compaction bands, and how this compression behaviour can be generalised to other type of brittle porous media remain elusive.

In this paper, we will first show that a wide variety of compaction patterns can appear when compressing a medium made of brittle porous material (puffed rice), including periodic travelling densification bands. These different compaction patterns will be explained using a simple spring lattice model [10]. The condition of appearance of the different patterns will be related to the characteristic times involved in the physical processes, and the evolution of the compaction behaviour over time will be discussed. Finally, the localisation of the compaction bands will be related to the strain rate sensitivity of the medium.

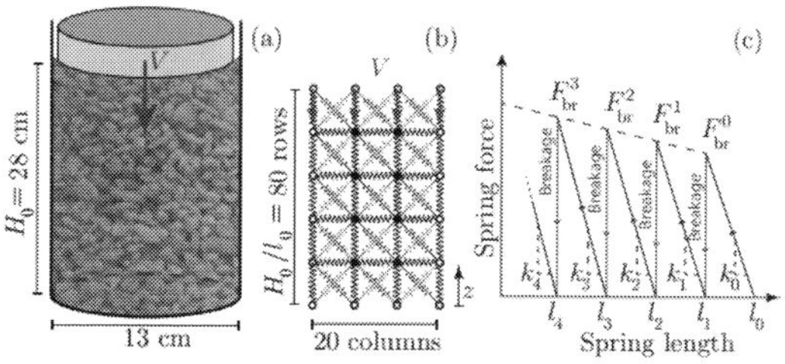

Figure 1. (a) Sketch of the experimental device for dry compaction of puffed rice at constant velocity. (b) Spring lattice geometry used in simulations. (c) Relationship between force and displacement for the simulated breakable springs.

Fig. 1a shows the experimental device used for the compression of a dry brittle porous material. The medium is made of puffed rice placed in a transparent cylindrical tank up to a height H_0, compressed from the top at constant velocity V. The side view of the tank is recorded during the compaction, allowing us to measure the velocity field over time using *Particle Image Velocimetry* (PIV). Since the experiment has an axial symmetry, all the velocities are vertical and averaged horizontally, and the velocity depends only on the vertical position z in the sam-

*e-mail: francois.guillard@sydney.edu.au

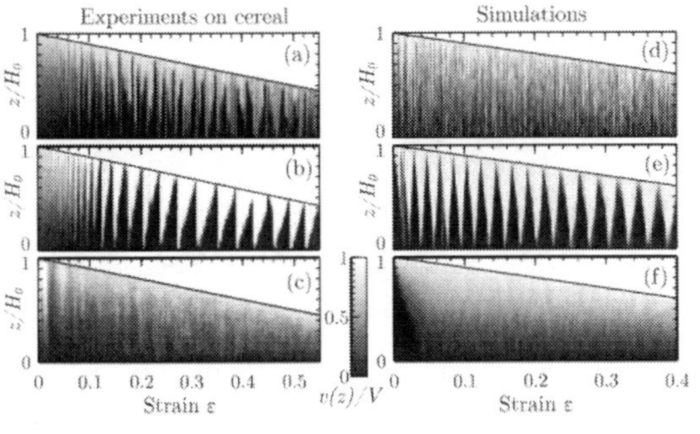

Figure 2. Heatmap of the medium velocity at a given depth v scaled by top velocity V as a function of scaled depth and strain, measured in experiments and simulations. (a,b,c) Experiments with $V = 0.013\,\mathrm{mm.s^{-1}}$, $1.3\,\mathrm{mm.s^{-1}}$ and $300\,\mathrm{mm.s^{-1}}$, respectively. (d,e,f) Model results with $V = 5\times10^{-6}$, 2×10^{-5} and 7×10^{-5}, respectively (with $k_0 = 2\times10^{-3}$, $F^0_{\mathrm{br}} = 2\times10^{-6}$, $\eta = 1\times10^{-3}$).

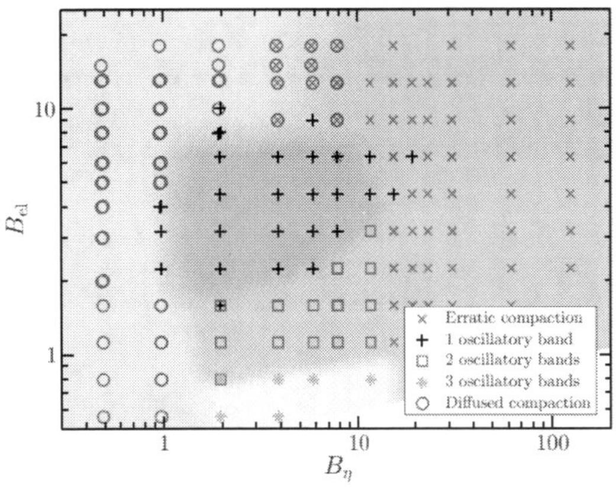

Figure 3. Compaction patterns observed in model simulations as a function of the dimensionless numbers B_η and B_{el}.

ple. Fig. 2a-c show the spatiotemporal velocity field for three different experiments with different compaction velocities. At low velocity, the compaction is mainly erratic, with different zones of the sample compacting at different time, without correlation between them. When the velocity is higher, the periodic compaction bands observed by Valdès et al. [8] are recovered. In that case the sample splits into two zones: a static zone at the bottom, and a zone moving at the velocity of the top compacting plate above. The boundary between these two zones moves upward over time. Finally, at the highest velocity, the whole sample compresses at the same time, leading to a homogeneous compaction, the velocity $v(z)$ at a given height z being approximately given by $v(z) = \frac{V}{H}z$.

To better understand the origin of these various compaction patterns, numerical simulations are performed using a spring lattice model [11, 12]. In this model, a lattice of nodes of mass m is created, each of the node being linked to his neighbours by springs that are able to undergo repetitive breakage events (Fig. 1b). The lattice is completely regular, and all the springs have same breakage law and parameters, so there is no disorder in the simulations. The equation of motion are solved for each of the nodes using a verlet algorithm with timestep Δt and the lattice is deformed accordingly. In addition to the force due to the springs, each node is also submitted to a global viscous damping force ηv_j, where η is the viscosity and v_j the velocity of node j. The springs in the lattice are able to break multiple time when they are compressed, following the force-displacement law of Fig. 1c. When the force in a spring overtake its strength F_{br}, it breaks, ie. its equilibrium length l is set to its current length and its stiffness k and strength F_{br} are increased. If i denotes the number of time a given spring already broke, the increase of stiffness and strength is set as $k_{i+1} = k_i + ak_0$ and $F^{i+1}_{\mathrm{br}} = F^i_{\mathrm{br}} + aF^0_{\mathrm{br}}$, where $a = 0.01$. During compaction of the lattice, each spring is therefore able to break as many time as needed, each breakage event releasing the force in the corresponding spring. The simulations are non-dimensional, with basic units of mass $[M] = m$, length $[L] = l_0$ and time $[T] = \Delta t$. Velocity field are coarse-grained from the node velocity [13].

When performing these simulations with different compaction velocity, compaction patterns similar to the ones appearing in experiment are observed. Fig. 2d-f shows that we recover the erratic compaction, single travelling compaction band and homogenous compaction observed in the experimental data (Fig. 2a-c).

Interestingly other compaction patterns are also observed in the simulations, corresponding to multiple compaction bands travelling through the sample at the same time. In fact, up to three oscillatory bands have been observed in the spring lattice model simulations. It remains unclear if these pattern also occur in the experiment as they may be too faint to be observed.

The model allows us to test the effect of the physical parameters on the observed compaction patterns. Three different characteristic times can be extracted from the simulation. The elastic time τ_{el} corresponds to the time needed for an elastic wave to travel the height of the sample. The viscous time τ_η corresponds to the slow down time of a node due to viscosity. Finally, the breakage time corresponds to the typical time between two successive breakage of a node given the strain rate that is applied. These times are defined from the simulation parameters with:

$$\tau_{\mathrm{br}} = \frac{F_{\mathrm{br}}}{kl}\frac{H}{V}; \quad \tau_{\mathrm{br}} = \frac{H}{l}\sqrt{\frac{m}{k}}; \quad \tau_\eta = \frac{m}{\eta} \qquad (1)$$

where the indices i have been removed for simplicity. Note that in principle these characteristic times evolve during a given simulation, so we start by focusing only on their initial value and on the first identifiable compaction pattern on a given test.

From these characteristic times two dimensionless groups can be derived, an 'elasto-breakage' number B_{el}

 DOI: 10.1051/epjconf/201714007012

and a 'visco-breakage' number B_η, defined as:

$$B_{el} = \frac{\tau_{br}}{\tau_{el}} = \frac{F_{br}}{V\sqrt{km}} \; ; \; B_\eta = \frac{\tau_{br}}{\tau_\eta} = \frac{F_{br}H\eta}{klVm} \qquad (2)$$

These two dimensionless numbers allow to clarify the range of parameters leading to the observed compaction patterns. Fig. 3 shows the patterns observed as a function of B_{el} and B_η. At high B_η, when the dissipation is high, the compaction is erratic since when breakage happens in the medium the elastic energy is rapidly dissipated and therefore there is no correlation between the breakage event in time. At low B_η on the contrary, there is very low dissipation and the medium is very weak so the whole sample breaks at the same time, leading to a homogeneous compaction. In between these two extremes, single of multiple oscillatory compaction band can appear, depending on B_{el}. Interestingly, the range of parameter leading to these bands gets smaller as the number of bands increases, similar to what is observed in the physics of bifurcation near phase transition. This limited range of parameters could also explain that these multiple oscillatory compaction bands have not been observed experimentally.

To see how general these results are for different type of materials, we also performed simulations where the dissipation mechanism is changed from global viscosity at the nodes to local viscosity inside each bond (Kelvin-Voigt model). The force in the spring is therefore not only computed from their extension, but also from the relative velocity of the nodes i and j it is attached to, adding a force $\eta(\mathbf{v}_j - \mathbf{v}_i)$ on node i. The characteristic viscous time is therefore modified to $\tau^m_{eta} = \eta/k$ to account for the change in the damping mechanism, and the 'visco-breakage' number to $B^m_\eta = \tau_{br}/\tau^m_\eta$. Fig. 4 shows the phase diagram for this model. Clearly the patterns observed in this case are very similar to the one observed with global damping, and the phase diagram is only marginally modified. Travelling compaction bands seems to be independent of the details of the breakage and dissipation mechanism, and therefore should be a general behaviour appearing in the compaction of brittle porous media for a certain range of parameters, which is also supported by their observation in snow [9] as well as puffed rice.

As mentioned earlier, the previous phase diagrams are based on the first compaction pattern observed in the simulations. However, B_{el} and B^m_η are evolving during a given simulation, due to the decrease of the height of the sample and the increase of the spring breakage inside the sample that modifies the average strength and stiffness of the springs. By taking these effects into account, one can rewrite B_{el} and B^m_η over time, with $\bar{\imath}$ the average number of breakage per spring, and t the simulation time:

$$B_{el} = \frac{F_0\sqrt{1+\bar{\imath}a}}{V}\sqrt{\frac{m}{k_0}} \; ; \; B^m_\eta = \frac{F_0(1+\bar{\imath}a)(H_0 - Vt)}{V\eta l}. \qquad (3)$$

Interestingly, the evolution of B_{el} and B^m_η over time implies that the same sample can experience different patterns of compaction over one simulation. Fig. 5 shows one particular simulation with bond viscosity where the material

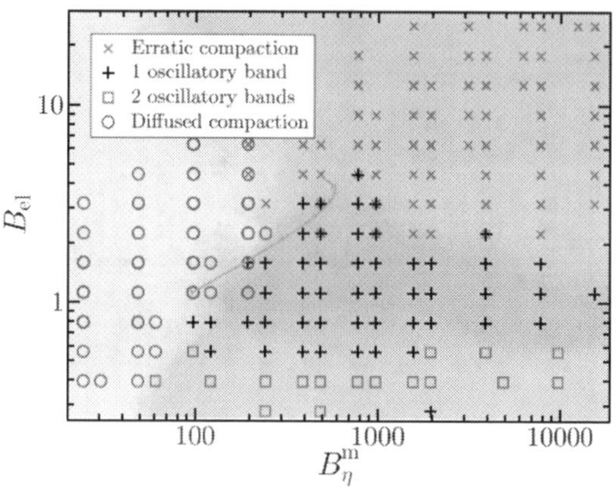

Figure 4. Compaction patterns observed in model simulations as a function of the dimensionless numbers B^m_η and B_{el} for bond viscosity. The continuous line is the path followed by the simulation presented in Fig. 5, the ticks on the line are equally spaced in time.

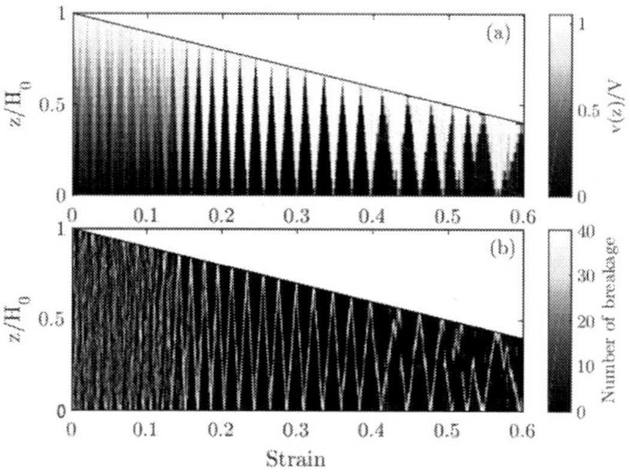

Figure 5. Velocity field (a) and total breakage events per row and per timestep (b) function of the strain for a given simulation. The simulation follows the path in the phase space indicated in figure 4. $V = 2 \times 10^{-5}$, $\eta = 0.02$, $k_0 = 5 \times 10^{-4}$, $F^0_{br} = 5 \times 10^{-7}$.

moves along the orange line in Fig. 4 during the experiments. The simulation starts with a diffuse compaction patterns. As the material breaks, both B_{el} and B^m_η increase, and the sample reaches the region where a travelling compaction band can localise and propagate into the medium. Toward the end of the simulation, the breakage rate decreases on average because the bonds that were initially inclined at 45° are now almost horizontal flat, and therefore the height dependency term in B^m_η takes over and reduces its value, while B_{el} keeps increasing slowly. The sample then starts to display more erratic compaction behaviour as it reaches this configuration. The same simulation can lead to multiple different compaction patterns over time because of the evolution of the geometry of the sample and of the internal bond properties due to breakage.

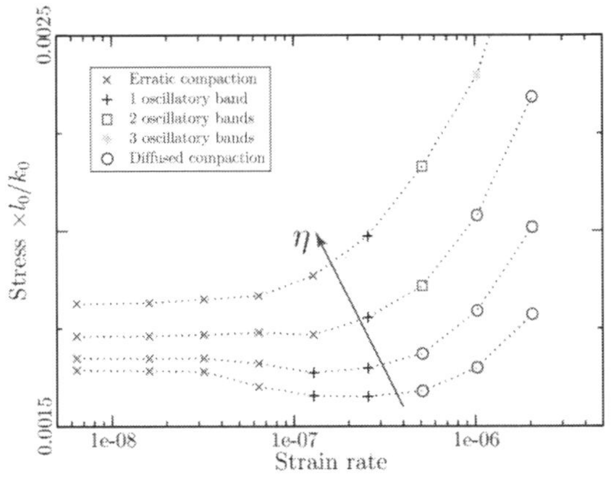

Figure 6. Stress as a function of compaction strain-rate for different simulations with global viscosity. Points on the same dotted line have the same viscosity, respectively $\eta = 2.5 \times 10^{-4}$, $\eta = 5 \times 10^{-4}$, $\eta = 1 \times 10^{-3}$, $\eta = 2 \times 10^{-3}$. $k_0 = 4 \times 10^{-3}$, $F^0_{\mathrm{br}} = 4 \times 10^{-6}$.

The phenomenon of localisation of compaction bands is reminiscent of similar behaviour observed in the extension of metallic alloys [14–16]. In this case, applying a tensile load to the material results in intermittent stress drops or 'jerky flow'. This instability is known to appear because of the strain-rate softening properties of the material, ie. the softening of the material with increasing pulling velocity, that leads to the instability of some strain rate states. It is relatively simple to check if the same kind of behaviour is observed in our model. To this end, the applied stress at the top is measured in simulation for different compacting velocity along with the identified compaction pattern (Fig. 6). It is clear that the oscillatory compaction bands are associated with the strain-rate softening of the material, at least when the dissipation is low. Interestingly, when η is increased, travelling compaction bands can still be observed for similar strain-rate range even if the medium does not seem to show macroscopic strain-rate softening. This is due to the fact that only bond level strain-rate softening is required to have localisation of a compaction band, whereas the bulk material can actually exhibit strain-rate hardening as a whole. This effect can have important consequences since its suggests that assessing the mechanical properties of a material at the macroscopic level may hide some important microscopic properties that can greatly modify the behaviour of the material.

In conclusion, a wide variety of compaction behaviour have been observed when brittle porous materials are compressed. In absence of water, when the grains breaks due to the applied stress only, compaction leads to either erratic behaviour at low velocity, single or multiple compaction bands oscillating in the medium when the velocity is increased, and homogeneous compaction at large velocity. These patterns have been observed experimentally and modelled numerically using a lattice of springs able to undergo multiple breakage events. The generality of the processes involved in the apparition of these different behaviours suggest that other brittle porous material should exhibit similar phenomena. This effect of localisation of compaction bands is related to the shear-rate sensitivity of the material at microscopic level, and can be linked to similar stress localisation effects observed in metallic alloys [16] and to the apparition of cnoidal or solitary wave in non-linear dynamic systems [17].

Acknowledgements

I.E. and L.S. thank the Australian Research Council for support through project DP130101291. J.R.V. thanks the USA National Science Foundation for support through Grant CBET1336952.

References

[1] W.A. Olsson, Mechanics of Materials **33**, 659 (2001)
[2] V. Vajdova, T.F. Wong, Geophysical Research Letters **30**, 1 (2003)
[3] P. Baud, E. Klein, T.F. Wong, Journal of Structural Geology **26**, 603 (2004)
[4] S.G. Love, F. Hörz, D.E. Brownlee, Icarus **105**, 216 (1993)
[5] C.Y. Wu, O.M. Ruddy, A.C. Bentham, B.C. Hancock, S.M. Best, J.A. Elliott, Powder Technology **152**, 107 (2005)
[6] A.F. Bastawros, H. Bart-Smith, A.G. Evans, Journal of the Mechanics and Physics of Solids **48**, 301 (1999)
[7] S. Papka, S. Kyriakides, Acta Materialia **46**, 2765 (1998)
[8] J.R. Valdes, F.L. Fernandes, I. Einav, Granular Matter **14**, 71 (2011)
[9] T.W. Barraclough, J.R. Blackford, S. Liebenstein, S. Sandfeld, T.J. Stratford, G. Weinländer, M. Zaiser, arXiv:1501.0218 (2015), `arXiv:1501.0218`
[10] F. Guillard, P. Golshan, L. Shen, J.R. Valdes, I. Einav, Nature Physics **11**, 835 (2015)
[11] R. Katsman, E. Aharonov, H. Scher, Mechanics of Materials **37**, 143 (2005)
[12] S. Kale, M. Ostoja-starzewski, Phys. Rev. E **042405**, 1 (2014)
[13] B.J. Glasser, I. Goldhirsch, Physics of Fluids **13**, 407 (2001)
[14] P. Penning, Acta Metallurgica **20**, 1169 (1972)
[15] L.P. Kubin, Y. Estrin, Acta Metallurgica **33**, 397 (1985)
[16] T.A. Lebedkina, M.A. Lebyodkin, Acta Materialia **56**, 5567 (2008)
[17] E. Veveakis, K. Regenauer-Lieb, Journal of the Mechanics and Physics of Solids **78**, 231 (2015)

The behaviour of free-flowing granular intruders

Edward Wyburn[1,*], *François* Guillard[1], *Benjy* Marks[1], and *Itai* Einav[1,2]

[1] *Particles and Grains Laboratory, School of Civil Engineering, The University of Sydney, Sydney, New South Wales 2006, Australia*
[2] *Department of Civil, Environmental & Geomatic Engineering, Faculty of Engineering Science, University College London, London WC1E 6BT, UK*

Abstract. Particle shape affects both the quasi-static and dynamic behaviour of granular media. There has been significant research devoted to the flowability of systems of irregularly shaped particles, as well as the flow of grains around fixed intruders, however the behaviour of free flowing intruders within granular flows remains comparatively unexplored. Here, the effect of the shape of these intruder particles is studied, looking at the kinematic behaviour of the intruders and in particular their tendency of orientation. Experiments are carried out within the Stadium Shear Device, which is a novel apparatus able to continuously apply simple shear conditions to two-dimensional grain analogues. It is found that the intruder shows different behaviour to that of the bulk flow, and that this behaviour is strongly shape dependent. These insights could lead to the development of admixtures that alter the flowability of granular materials.

1 Introduction

Granular materials exhibit complex behaviour that is not found in other solids or liquids. The properties of the granular mass, whether it be static strength or their subsequent flowing behaviour, are dependent on many factors, including: particle surface roughness, static electricity and cohesive forces, as well as particle size, density, stiffness and shape. Of these factors, shape is particularly important. The majority of granular materials which are produced, collected or conveyed are of an irregular shape. Many tests, both experimental and computational, have used spheres or cylinders due to the simplicity of those shapes [1] with glass beads often chosen for fundamental experiments. There have been studies conducted on systems entirely consisting of irregularly shaped or elongated particles [2–5], as well as the influence of mixtures of regular and elongated particles [6]. The behaviour of a single elongated free flowing intruder within a 2D layer of spheres has also been studied [7]. The elongation of particles in particular introduces many complexities such as the change in jamming transition, altering of shear strength parameters, the orientation of these particles with the principal stress region within sheared assemblies, and the strong dependence on initial conditions [5]. There exist no adequate models to explain these additional factors which are introduced into the rheology of granular materials by anisotropic shaped particles – even for relatively simple shapes such as ellipsoids or rods [5].

There exist many conventional ways of shear testing granular materials, however each of these tests has distinct limitations placed upon it by the geometry. Shear box tests impart simple shear on a sample, however they stipulate the failure plane and have a limited scope for deformation. Triaxial apparatus allow strict control of loading and drainage conditions yet these tests are only able to apply limited deformation and the geometry of the membrane alters as the specimen fails. Couette cells and ring shear are able to continuously shear the sample, however due to their circular geometry contain a radial stress gradient [8]. Preliminary results are provided in this paper which look at the behaviour of rod and teardrop shaped particles of different aspect ratios placed within the 2D Stadium Shear Device (2D SSD – discussed in the next section). The 2D SSD is similar in its loading conditions to the "$1\gamma 2\epsilon$" apparatus [9], which also imparts a shear displacement on 2D grain analogues, yet the 2D SSD is able to perpetually apply simple shear at a constant velocity due to the recirculation of the grains within the curved end zones. The 2D SSD has been used for experiments [10–12] and simulated using spherical and irregular particles [3, 13].

2 Method

A schematic of the 2D SSD is shown in Figure 1(a,b). The granular mass has shear imparted upon it by a belt which contains asperities at 14 mm. The rollers directing the belt have constant force springs attached which apply constant confining pressure boundary conditions. A great advantage of the device is the stadium shape which allows for perpetual shear achieved by the gap between the wheels and granular mass allowing for the recirculation of the grains underneath (Figure 1(b)). These tests were all conducted in fixed pressure loading conditions with the same applied pressure and shear rate, which produced an iner-

*e-mail: edward.wyburn@sydney.edu.au

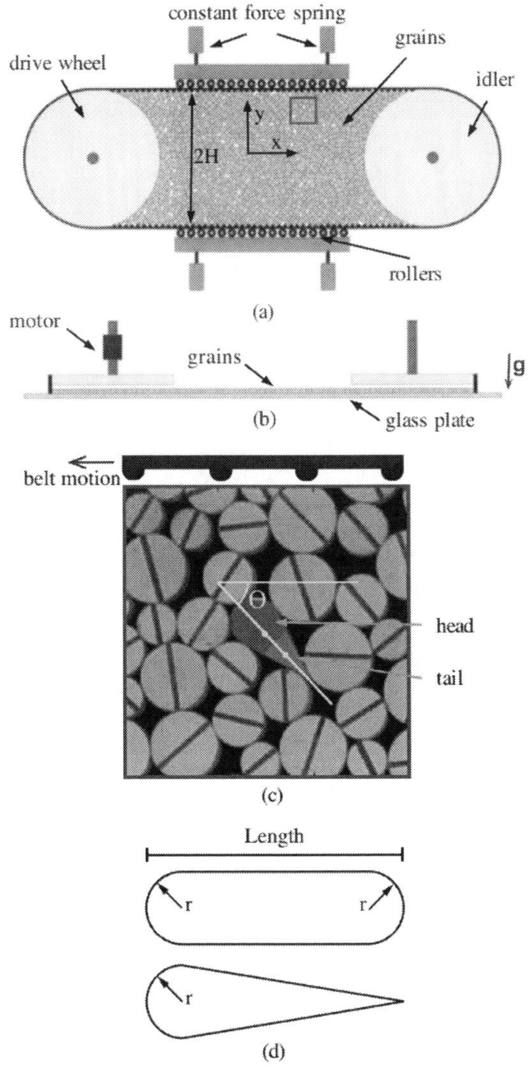

Figure 1. Diagram of the 2D Stadium Shear Device [12]. (a) top view of the device. (b) elevation view. (c) the long "teardrop" shaped particle sitting within the device surrounded by the typical cylinders. The measured angle is drawn with the yellow line and the two dots correspond to the centroid of the entire particle and the blue area respectively. (d) Diagram showing rod and teardrop intruders.

tial number of approximately 0.04 using an area-weighted average inertial number [14].

The grains within the 2D SSD are 10mm high nylon cylinders of three different diameters (12, 15 and 20 mm) in order to prevent crystallisation from occurring. These grains can be seen in Figure 1(c) along with an intruder particle. These nylon cylinders are durable and have a relatively low friction coefficient of 0.072 against glass [11]. A single 3D printed polylatic acid (PLA plastic) intruder has been placed within the matrix of cylinders for each

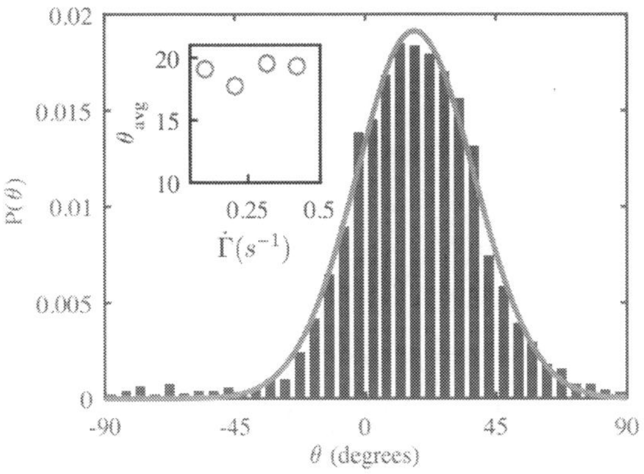

Figure 2. Determination of mean angle θ_{avg} for the long rod particle. Here $\dot{\Gamma}$ is the macroscopic shear rate of the device, taken as V_{belt}/H where V_{belt} is the belt velocity. Here the angle has been plotted in the interval $\theta \in [-90, 90)$ due to the symmetry of the rod shape. A gaussian is fitted to the data and θ_{avg} is taken as the value of the peak of the fitted distribution. Inset: θ_{avg} as a function of shear rate.

test. Intruders are grouped into two categories, rods and teardrops, and two different length shapes have been analysed for each category.v These intruders are also 10 mm high and their properties are summarised in Table 1, with aspect ratio defined as the ratio between maximal length and head diameter for all shapes. The first category of intruders can be called rods or capsules. Similar shapes are found widely in nature and industry and have previously received a lot of attention [6, 15, 16]. The second category of intruders, which we have named a teardrop shape, is analogous to a rod particle with one radius equal to zero. These teardrops were produced by taking the convex hull of a cylinder and a point at the desired distance.

Particle tracking velocimetry (PTV) has been used to quantify the full field data. This was performed on videos in 1920×1080 resolution with a nominal frame rate of 60 fps, which allowed for a displacement of less than 1/4 of the smallest grain diameter between frames. The camera was placed above the device and data was recorded in the central region where pressure is imparted by the springs and the flow is steady. In order to gain significant and smoother statistics for each intruder, since only one was within the mass, the device had to be sheared for at least 200 shear deformations which is many more than was found to be required for the bulk mass. Each intruder particle was tagged with blue tape in order to easily define the tail end and gather the orientation. For the intruders the angle θ was measured from the centroid of the entire object to the centroid of the tagged tail, which are the yellow dots in Figure 1(c), and gathered in the range of $\theta \in [-180, 180)$. For the rod intruders - due to the shape having two axes of symmetry - these angles are doubled by considering an interval of $\theta \in [-90, 90)$.

In Figure 2 an example of the method for determination of the mean angle, θ_{avg}, with a fitted gaussian is

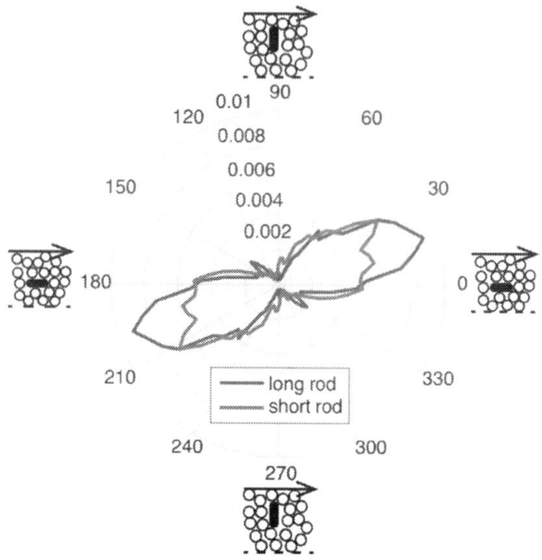

Figure 3. Probability density function P(θ) of the orientation of the major axis for the rod intruder particles. The radial distance is the probability, and the angle on the graph corresponds to the particle orientation relative to the belt flow. The data has been discretised into bins of 5 degree width to reduce noise. Due to the symmetry of the rod shape this data has been collected in the range of $\theta \in [-90, 90)$ and reflected by 180°. The diagrams surrounding the axes show the orientation of the particle with respect to the mean flow direction. A clockwise rotation on the plane corresponds to a clockwise rotation of the particle.

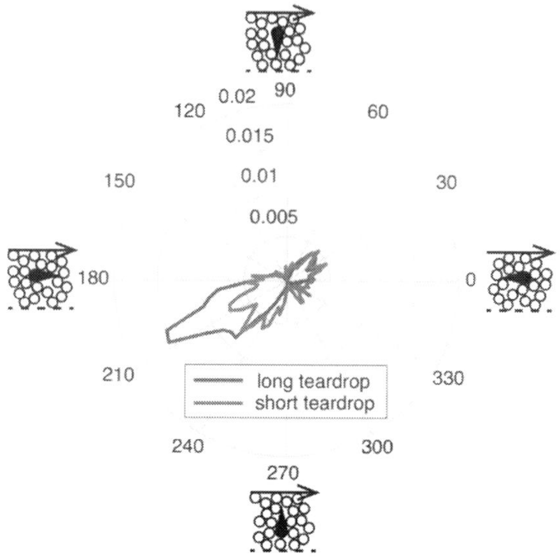

Figure 4. Probability density function P(θ) of the orientation of the major axis for the teardrop intruder particles. The radial distance is the probability, and the angle on the graph corresponds to the particle orientation measured from the head to the tail. The data has been discretised into bins of 5 degree width to reduce noise. The diagrams surrounding the axes show the orientation of the particle with respect to the mean flow direction at that angle on the graph. A clockwise rotation on the plane corresponds to a clockwise rotation of the particle.

shown. The inset shows tests at four different macroscopic shear rates, with the same confining pressure, which shows the mean orientation's independence from the shear rate – which has also been found in previous studies of masses of elongated particles [4–6]. The data presented in the results section was gathered at the highest shear rate shown here. A similar process to find the value of θ_{avg} was conducted on the teardrop shaped particles. For those particles instead of taking angles in an interval of $\theta \in [-90, 90)$ the range was taken as the 180 degrees of the probability density distribution which yielded the highest probability. For the long teardrop this came to be $\theta \in [142, 322)$. The same process was repeated for the short teardrop but the angle probability for this shape does not show a clear orientation preference and results in a poor fit (see Figure 4).

3 Results

The graphs in Figures 3 and 4 show the probability density $P(\theta)$ of the orientation of the intruder particles. The measured angles are relative to the belt motion of the half of the flow which the intruder is positioned in. Rod and teardrop shaped particles are both observed to align in a direction that is not parallel to the mean flow direction. The difference in preferred orientation of around 4 degrees between the short and long rod shape is due to a relatively small difference in elongation (0.5 rod diameters). When the intruder leaves the central zone and enters the areas of circulation below the spindles, mixing occurs due to the strong vortices within this section which often trap the intruder for some time. After leaving these vortices the intruder emerges at a new position and orientation.

Table 1. Intruder particle properties

Intruder	Head Diameter	Length	θ_{avg}
Rod - Short	10 mm	30 mm	23.1 ≅ 203.1°
Rod - Long	10 mm	35 mm	19.4 ≅ 199.4°
Teardrop - Short	15 mm	22 mm	215.5°
Teardrop - Long	15 mm	32 mm	207.2°

The teardrop particles have a strongly asymmetric distribution P(θ) due to the shape's single axis of symmetry. These particles interestingly have a tendency to "wedge" where the motion is lead by the tail end of these particles and this pointed end acts as the thin end of a wedge between the other grains. This wedging behaviour is shown by the value of θ_{avg} which corresponds to the teardrop shape's head moving within the slower grains and its tail moved by the grains in the swifter moving layer. The wedging mechanism agrees with the findings in [7], where intruder velocity was found to be almost entirely in the direction of the thin end of the intruder particle – however their motion was induced by out-of-plane vertical vibration, which is fundamentally different to shearing the granular mass. There exists a large reduction in wedg-

DOI: 10.1051/epjconf/201714006016

ing behaviour for the shorter tailed teardrop. The shorter tailed teardrop is relatively close to a circle, its circularity being 0.68 when defined as the ratio between the radius of the minimum circumscribing circle and the maximum inscribed circle. While there is somewhat of a tendency for wedging to occur with this shape, it does start to approach the behaviour of a circle where there would be expected to be no systematic value for θ_{avg}. It is uncertain whether a much longer teardrop shape would exhibit this same wedging behaviour, or if due to the larger aspect ratio the larger coordination number would mean that a full rotation of the particle becomes extremely unlikely and the initial condition of intruder orientation is of much more import. It is also uncertain if the mixing beneath the rotors would negate this effect.

The data used in the figures and tables presented has been taken from sections of the flow sufficiently far from the belt and the centre. This was achieved by splitting the flow across the y coordinate into seven portions and ignoring data gathered within the central portion and the portion nearest to each belt. Data near the belt has been ignored due to entrainment occurring within the asperities of the belt which would artificially increase the statistics for a particular angle as the intruder was dragged around at a constant orientation. Since the orientation is taken with respect to the mean flow direction and belt motion, when the intruder crosses the centreline this reverses the angle – which in the case of the teardrop, due to its singly-symmetric cross section and subsequent asymmetric P(θ) diagram, makes this data spurious. The reversed information around the centre is still valid for the doubly symmetric rod shapes but has not been included.

The values for θ_{avg} of the rods shown here are higher than reported for those of entirely elongated particles, about 10 degrees higher than previous studies where the average angle of rods of aspect ratio 3-3.5 within dynamic sheared assemblies are between approximately 5-10 degrees [5], yet still follow the same trend of increasing θ_{avg} with a reduction of aspect ratio. A number of reasons can explain this significantly higher value of θ_{avg} for the intruder, however the most likely cause is that the majority of these dynamic simulations or experiments are in 3D where the extra degree of freedom causes looser packings. In addition it is likely that θ_{avg} for the intruder is increased because it is a singular entity within a mass of cylinders which are analogous to rods/teardrops with an aspect ratio of 1. If the angle of mixtures follows the previously observed trends of inverse proportionality between aspect ratio and θ_{avg} [4–6] which was also observed within our tests, it would likely increase the value of θ_{avg}.

4 Conclusions

Preliminary results have been presented which show the behaviour of the intruder particles. Of particular interest was the "wedging" of the teardrop particles, where the thin tail lead the motion by squeezing between other grains. The behaviour of a single rod intruder follows the same trends as previous studies of elongated masses, where θ_{avg} is not in the principle flow direction. The presented results are promising and provide good reason to conduct more studies on the effect of elongation as well as the effects of inclusion of multiple intruders on both mean flow fields and the dynamics of the individual intruders.

Acknowledgements

The authors would like to acknowledge the financial support provided by the Australian Research Council through Discovery Project DP160104310. The authors would also like to thank Garry Towell and Ross Barker for their continued support, as well as Tom Miller for his pioneering work.

References

[1] GDR MiDi, The European Physical Journal E **14**, 341 (2004)

[2] G. Lumay, F. Boschini, K. Traina, S. Bontempi, J.C. Remy, R. Cloots, N. Vandewalle, Powder Technology **224**, 19 (2012)

[3] A.X. Jerves, R.Y. Kawamoto, J.E. Andrade, Acta Geotechnica **11**, 493 (2016)

[4] T. Börzsönyi, B. Szabó, S. Wegner, K. Harth, J. Török, E. Somfai, T. Bien, R. Stannarius, Phys. Rev. E **86**, 051304 (2012)

[5] T. Börzsönyi, R. Stannarius, Soft Matter **9**, 7401 (2013)

[6] E. Azéma, I. Preechawuttipong, F. Radjai, Phys. Rev. E **94**, 042901 (2016)

[7] N. Kumar, H. Soni, S. Ramaswamy, A.K. Sood, Phys. Rev. E **91**, 030102 (2015)

[8] D. Howell, R. Behringer, C. Veje, Physical Review Letters **82**, 5241 (1999)

[9] F. Calvetti, G. Combe, J. Lanier, Mechanics of Cohesive-frictional Materials **2**, 121 (1997)

[10] T. Miller, P. Rognon, B. Metzger, I. Einav, Phys. Rev. Lett. **111**, 058002 (2013)

[11] T. Miller, Ph.D. thesis, The University of Sydney (2014)

[12] P.G. Rognon, T. Miller, I. Einav, Granular Matter **17**, 177 (2015)

[13] P.G. Rognon, T. Miller, B. Metzger, I. Einav, Journal of Fluid Mechanics **764** (2015)

[14] P.G. Rognon, J.N. Roux, M. Naaïm, F. Chevoir, Physics of Fluids **19**, 058101 (2007)

[15] P.W. Cleary, M.L. Sawley, Applied Mathematical Modelling **26**, 89 (2002)

[16] P.W. Cleary, Powder Technology **179**, 144 (2008)

Experiments and DEM Simulations of Granular Ratcheting

Gianluca Zorzi[1,*], *Riccardo* Artoni[2,**], and *Fabio* Gabrieli[3,***]

[1] *GuD CONSULT, Darwinstraße 13, 10589, Berlin, Germany*
[2] *LUNAM Université, IFSTTAR, MAST, GPEM, F-44340 Bouguenais, France*
[3] *Icea Dept., University of Padova, Via Ognissanti 39, 35129 Padova, Italy*

Abstract. In this work we studied the effect of cyclic loading on a granular packing by means of numerical simulations and experiments. A confined packing of glass beads was prepared and one of the walls was moved cyclically with a prescribed amplitude of the order of the particle diameter. Different amplitudes were tested, and their effect on the free surface evolution, the force transmitted to the moving wall and the displacement patterns in the material was characterized. Discrete numerical simulations were also carried out with the specific purpose of evaluating the effect of the particle shape on the dynamics of the system. The displacement amplitude of the moving wall was shown to increase the maximum force experienced at the end of the compressive phase of the wall movement; the angularity of the particles had a similar effect. Force-wall displacement curves displayed a peculiar hysteretic behavior. The evolution of the system towards an asymptotic state was shown to be faster for spheres than for angular particles; the latter displayed an interesting long-time evolution of the force-displacement paths which deserves deeper investigations.

1 Introduction

Granular ratcheting is an interesting phenomenon which arises spontaneously when a granular material is subjected to cyclic loads or, more generally, to cyclic strains. It consists in the accumulation of irreversible strains in the granular mass and in a modification of its mechanical response. The effects of granular ratcheting has been shown in the form of convective flow [1, 2], densification, soil settlementand a stiffening of the soil-structure interaction (SSI) [3]. This evolution of the soil under cyclic loading is mainly due to the rearrangement of the initial packing and could be an important aspect to be considered for the vulnerability of the underground structures. From an engineering point of view it is important to understand and predict this change of the soil behaviour under cyclic loading especially in the long-term SSI problem. In this paper, in order to characterize the ratcheting phenomenon, joint laboratory tests and DEM simulations of a granular packing subject to the cyclic displacement of a lateral wall were carried out. We focused on the effect of the displacement amplitude and of the particle shape on forces transmitted to the moving wall, and on the macroscopic motion of the granular mass. In the following we first discuss experimental conditions and setup, and then introduce the numerical method.

*e-mail: zorzi@gudconsult.de
**e-mail: riccardo.artoni@ifsttar.fr
***e-mail: fabio.gabrieli@unipd.it

2 Experiments

2.1 Experimental set-up

Experimental tests were performed using a glass box of size $L = 150 \times W = 50 \times H = 90$ mm in which one of the lateral walls was cyclically moved with a prescribed amplitude (see Fig. 1 for a sketch of the set-up). Glass beads were poured in the box and then leveled to reach a material height of 90 mm.

All tests were conducted in quasi-static conditions with $(dx/dt)_{max} < 10^{-3}$ mm/s by means of a manual inbuilt micrometrical screw, equipped with precision bearings to avoid slack and a load cell in order to record the horizontal lateral force. A triangle wave horizontal displacement was imposed by the screw with different amplitudes. The materials used in the tests discussed here were glass beads (soda-lime silicaglass) with an average diameter of 1.5 mm. Results from other tests (for different materials and particle diameters), and more information regarding the preparation and conduction of the tests can be found in [2]. The present paper will focus on two amplitude/particle diameter ratios (named A/D in the following) equal to 1 and 0.50. For each A/D value, 50 wall displacement cycles were performed. A digital camera Canon EOS 400D equipped with EF-S 18-55 lens was placed perpendicular to the box in order to capture the grain motions behind the transparent wall. The image sequences were analysed with an image analysis module for MATLAB called GeoPIV_RG [4], which implements image cross-correlation algorithm to identify the motions of a grid of image patches.

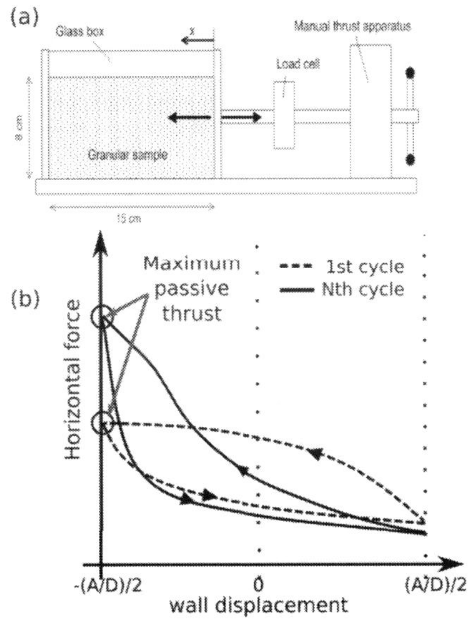

Figure 1. (a) Experimental device. A $L = 150 \times W = 50 \times H = 90$ mm rectangular cuboid is filled with a granular material (glass beads). One of the lateral walls is cyclically moved with a given amplitude in quasistatic conditions. (b) Typical profiles of the horizontal force exerted on the wall as an function of the wall displacement.

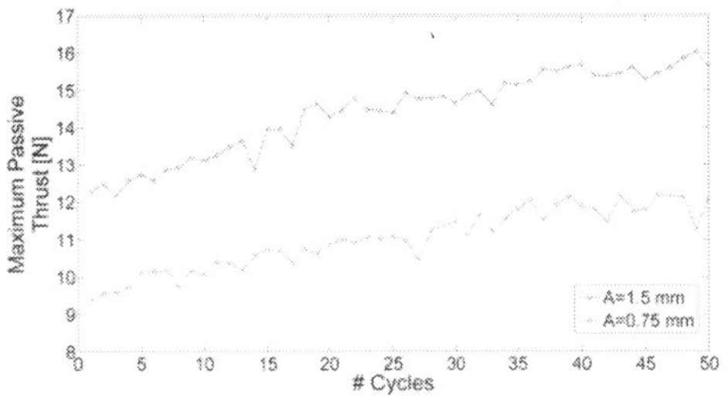

Figure 2. Evolution of the maximum passive thrust (i.e. the maximum force exerted during each compression cycle of the wall displacement) with the number of cycles.

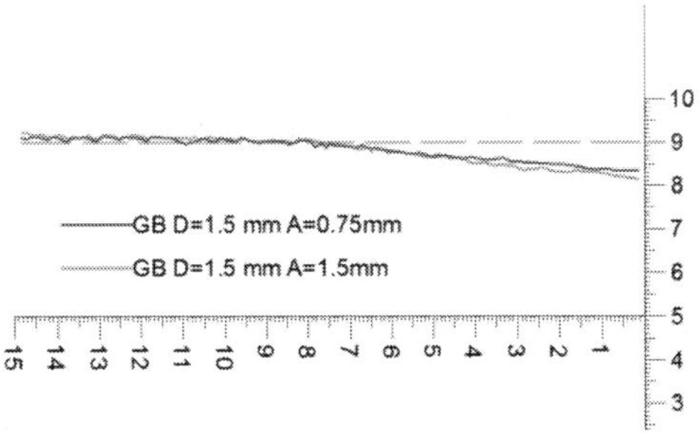

Figure 3. Free surface profiles after 50 cycles, for 1.5 mm glass beads with A/D = 1 (green line)and 0.50 (red line). The dashed line corresponds to the initial surface level.

2.2 Experimental results

Due to the periodic displacement of the wall, the force exerted by the material on the wall has a typical hysteretic evolution, which is sketched in Fig. 1. During the (passive) compression phase, the force increases progressively with the wall displacement, while in the (active) relaxation phase the system follows another path, characterized by lower values of the force. The hysteresis loop evolves during the cycles. Figure 2 displays the evolution of the maximum passive thrust (i.e. the maximum force exerted during each compression cycle of the wall displacement) with the number of cycles. It is evident that the maximum force increases with the number of cycles, and that an increase in the displacement amplitude yields an increase in the force. It is reasonable to expect that the maximum passive thrust will reach an asymptote for a sufficiently large number of cycles. In our experiments the maximum number of cycles was not sufficient to characterize this long time behavior of the ratcheting phenomenon.

Figure 3 shows the free surface profile after 50 cycles for the two values of A/D introduced above. It is clear that both tests show a settlement of the granular material close to the moving wall. This settlement is only slightly higher the larger the amplitude.

The convective flux induced by the cyclic displacement of the lateral wall can be appreciated by tracking the position of a grid of patches in the sequence of images taken when the screw passes at the zero position. This is shown in Figure 4. It is clear that the larger displacement amplitude displays a larger permanent displacement. Most of irreversible strains are just behind the moving wall where the particles move exhibiting a net vertical displacement, which is not due to a simple gravitational densification, since it persists at each cycle. Deeper particles near the wall deviates from the vertical, and the horizontal component results larger for larger displacement amplitude. In the middle region of the sample the material moves with different magnitude but approximately with the same direction. Far away from the wall the cumulative displacement is gentle and diagonal-upward, compatible with a passive stress condition. No closure of the convective flux (i.e. permanent displacement towards the wall) is detected near the ground surface during the tests. This is very likely consistent with the lack of an asymptotic behaviour in the maximum passive force trend.

3 Numerical simulations

3.1 Numerical method

Numerical simulations were performed using the non smooth contact dynamics method [5], as implemented in the LMGC90 open source framework [6]. Simulations were carried out in a rectangular cuboid geometry of dimensions $L = 30d \times W = 10d \times H = 20d$, where d is the particle characteristic length. Simulations were performed in dimensionless units with lengths rescaled by d,

 DOI: 10.1051/epjconf/201714003061

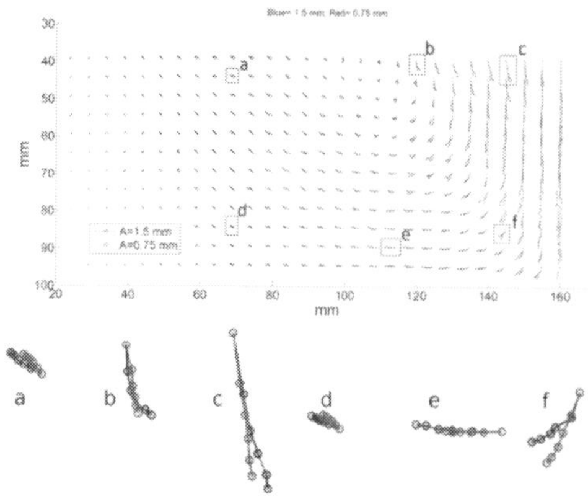

Figure 4. (top) Paths of the irreversible displacements after 50 cycles obtained by crosscorrelation PIV, for two values of the displacement amplitude. (bottom) Focus on some peculiar paths with the letters referring to the top picture.

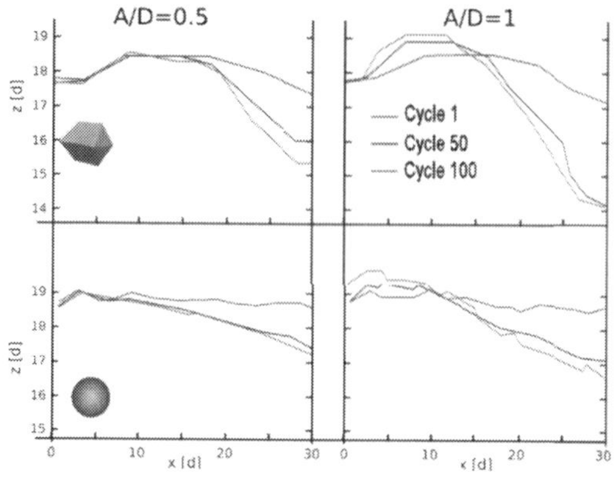

Figure 5. Free surface profiles for different cycle numbers for simulations with different particle shapes and displacement amplitudes.

time rescaled by $\sqrt{d/g}$, masses rescaled by ρd^3. In order to study the effect of the the particle shape on the ratcheting phenomenon, two types of monodisperse granular particles with different shapes were used in the simulations (spheres, and pinacoids, as sketched in Fig. 5). The pinacoid is an example of a very angular particle which was already discussed by Camenen et al. [7]. The one chosen in this work has all the three axis equal to d and a side angle of 60°. Contacts were modeled as inelastic with a particle-particle and particle-wall friction coefficient of $\mu_p = \mu_{pw} = 0.5$. The initial packing was obtained by a rainfall simulation with a stability criteria of total kinetic energy over total potential energy less than 10^{-8}. Once a stable packing was reached, the ratcheting simulation started. This paper will present two different normalized displacement amplitude $A/D = 0.5$ and 1 tested for each particle shape. The same displacement pattern starting from $(A/D)/2$, inverting the motion and approaching $-(A/D)/2$ was imposed for 100 cycles. Wall velocity was maintained constant and slow to simulate quasi-static conditions. In order to have approximately the same initial level independently of particle shape, the number of particles was different between spheres ($N = 5590$) and pinacoids ($N = 5990$).

3.2 Numerical results

The evolution of the free surface profiles with the number of cycles for different particle shapes and displacement amplitudes is shown in Figure 5. It is clear that the largest part of the evolution occurs between 1 and 50 cycles, and that particle shape strongly influences the final surface shape. Increasing the angularity of particles yields a more pronounced surface variation. Moreover, as it was shown in experiments, increasing the displacement amplitude yields a slightly more pronounced variation of the surface profile. The surface variation is more important than in the experiments; this is probably an effect of the difference in the ratio between particle size and system dimension.

In figure 6a, the maximum passive thrust obtained from the simulations is plotted as a function of the number of cycles. The force displays strong fluctuations from one cycle to the other, probably due to the relatively small system size in terms of number of particles. However, we can see that an initial increase of the force is followed by a nearly constant value for all the simulations after approximately 50 cycles. This is consistent with the fact that free surface displays only a small evolution after 50 cycles, as it was shown in Figure 5.

The force-wall displacement profiles for spheres (Fig. 6b) confirm this point. Between 50 and 100 cycles, the force-displacement curves follow nearly the same hysteresis loop. The system has reached an asymptotic state. As regards pinacoids (Fig. 6c), however, even if the free surface displays little evolution and the maximum force is relatively constant, the hysteresis loops definitely continue to evolve from 50 to 100 cycles. While displaying nearly the same final value, the force displacement curve during the compressive phase is characterized by an increasingly nonlinear path, with a steeper evolution. The total work needed to move the wall in a compressive cycle seems to decrease. As concerns the effect of displacement amplitude, Fig. 6a shows that the larger the A/D value the larger are the forces, as it was shown by the experimental results. Particle shape also affects the maximum force values, with the angular particles displaying larger values of the force.

4 Conclusions

In this work we studied the effect of cyclic loading on a granular packing by means of numerical simulations and experiments. A packing of glass beads confined between four lateral walls was prepared and one of the walls was moved cyclically with a prescribed amplitude of the order of the particle diameter. Different amplitudes were tested, and their effect on the free surface evolution, the

(a)

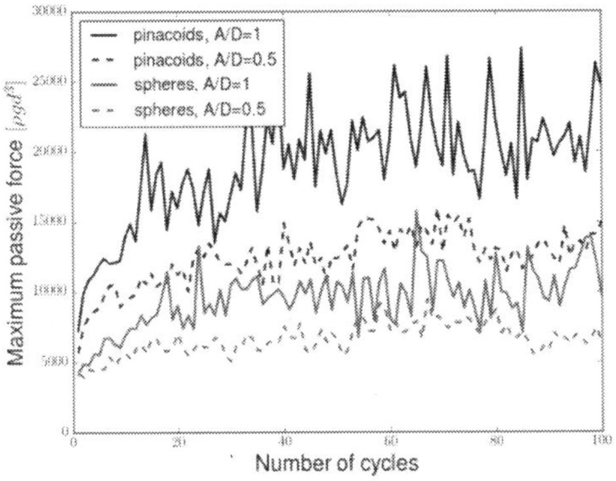

(b)

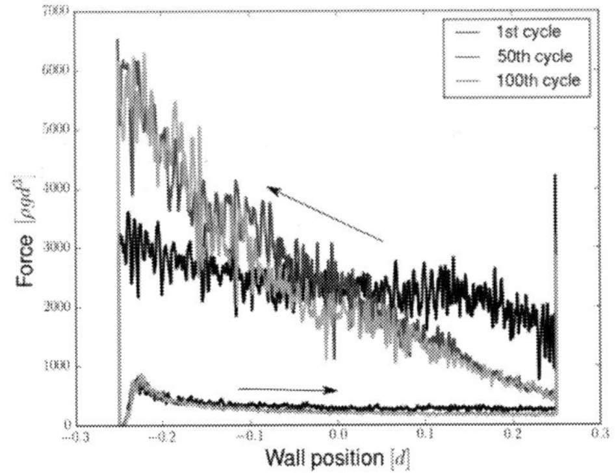

(c)

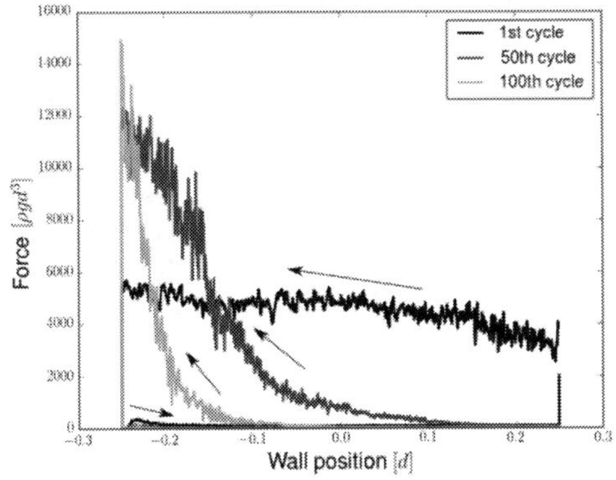

Figure 6. (a) Maximum passive thrust as a function of the number of cycles for spheres, pinacoids and two values of the displacement amplitude A/D. (b-c) Hysteresis loops for spheres (b) and pinacoids (c) with $A/D = 0.5$ at different cycles

force transmitted to the moving wall and the displacement patterns in the material was characterized. Numerical simulations by the nonsmooth contact dynamics method were also carried out with the specific purpose of evaluating the effect of the particle shape (spheres vs angular particles) on the dynamics of the system. The displacement amplitude of the moving wall was shown to increase the maximum force experienced at the end of the compressive phase of the wall movement; the angularity of the particles had the same effect. The evolution of the system towards an asymptotic state was shown to be faster for spheres than for angular particles; the latter displayed an interesting evolution of the force-displacement hysteresis loop which is probably related with microstructural changes and deserves deeper investigations. The effect of system size and other particle shapes will also be addressed in the future.

Acknowledgements

The numerical simulations were carried out at the CCIPL (Centre de Calcul Intensif des Pays de la Loire) under the project MTEEGD.

References

[1] F. Alonso-Marroquín, H.J. Herrmann, Phys. Rev. Lett. **92**, 054301 (2004)

[2] F. Gabrieli, G. Zorzi, R. Wan, in *Geomechanics from Micro to Macro* (CRC Press, 2014), pp. 601–606.

[3] P. Cuéllar, S. Georgi, M. Baeßler, W. Rücker, Granular Matter **14**, 11 (2012)

[4] D. White, W. Take, M. Bolton, Geotechnique **53**, 619 (2003)

[5] M. Jean, Computer Methods in Applied Mechanics and Engineering **177**, 235 (1999)

[6] M. Renouf, F. Dubois, P. Alart, Journal of Computational and Applied Mathematics **168**, 375 (2004)

[7] J.F. Camenen, Y. Descantes, P. Richard, Phys. Rev. E **86**, 061317 (2012)

Comparing numerically exact and modelled static friction

Dominik Krengel[1,*] and *Hans-Georg* Matuttis[1,**]

[1] *The Graduate School of the University of Electro-Communications, Department of mechanical engineering and intelligent systems, 1-5-1 Chofugaoka, Chofu, 182-8585 Tokyo,Japan*

Abstract. Currently there exists no mechanically consistent "numerically exact" implementation of static and dynamic Coulomb friction for general soft particle simulations with arbitrary contact situations in two or three dimension, but only along one dimension. We outline a differential-algebraic equation approach for a "numerically exact" computation of friction in two dimensions and compare its application to the Cundall-Strack model in some test cases.

1 Introduction

Coulomb sliding friction is a dissipative force, given by the equation

$$F_{\mathrm{Fr,dyn}} = -\mu F_{\mathrm{normal}} \mathrm{sign}(v), \tag{1}$$

where μ is the coefficient of friction, F_{normal} the normal force and v the velocity. Its implementation is computationally as simple as for viscous friction. Static Coulomb friction, on the other hand, is a constraint of motion

$$-\mu F_{\mathrm{normal}} \leq F_{\mathrm{Fr,dyn}} \leq \mu F_{\mathrm{normal}}, \tag{2}$$

which fixes two bodies tangentially to each other. Its implementation is algorithmically much more demanding: For zero tangential velocity, the tangential inter-particle force must compensate the external forces and fullfill inequality (2). Currently, there exists no mechanically consistent theory of Coulomb friction for two- or three dimensional particle aggregates for arbitrary contact situations. Existing exact algorithms are only suitable for one dimension, see Sec. 2. In this research, we outline the formalism for the numerically exact implementation of Coulomb friction in generalisation of the one-dimensional formalism and show the implementations for some test cases. Rolling friction is neglected, it is generally one or two orders of magnitude lower than sliding friction. The same friction coefficient is used for dynamic and static friction, measurable differences are usually due to time-dependant surface changes (e.g. accumulation of humidity in interstices) which cannot be absorbed into mere constants anyway, see Ref. [9].

2 Existing Algorithm

For the static case, Coulomb friction is a constraint of motion. Cundall and Strack [2] were the first to include a tangential force which was incremented proportional to the relative sliding distance of the contacting particles. The drawback is that this models a degree of freedom, which may release energy into the system when there is no preferred direction of motion (e.g. in vibrated systems), not a constraint force. Stewart [11] proposed a mathematical formulation where the contact situation is absorbed into a linear system which is solved to obtain the friction values. Nevertheless, only examples for contacts in one dimension were treated and it was left open how the inequality eq. (2) can be satisfied mathematically. While formalisms in convex analysis allow the description of tangential interparticle forces which fullfill the formal conditions, there is no guarantee that the variation during time evolution is physical or even smooth. Moreau et al. [8, 10] have treated the problem in their "contact mechanics" in a generalization of Newtonian kinematics which we would like to avoid (also for the sake of applicability of numerical integrators). Further, in contact mechanics the normal and tangential forces are evaluated simultaneously, while we would like to obtain friction forces as reactions to the existing normal forces. Finally, the approach used rigid particles, which makes the investigation of some problems unfeasible, be it because it is the limit of small strains on the one hand or because the sound velocity is infinity so that issues of shock propagation cannot be dealt with.

3 Linear oscillator with Coulomb friction

The ordinary differential equation (ODE) of a simple linear oscillator with dry friction

$$m\ddot{x} = -\mu F_N \,\mathrm{sign}(\dot{x}) - kx \tag{3}$$

is not continuous for sign changes of $\dot{x}$, which require a special treatment. The mathematical theory goes back to A.F. Filippov [3], while in the numerical treatment, we will follow Hairer and Wanner [4]. The jump in eq. (3)

*e-mail: krengel@matuttis.mce.uec.ac.jp
**e-mail: hg@mce.uec.ac.jp

is dealt with by separating the equations formally into

$$f = m\ddot{x} = \begin{cases} f_I & \text{for } \dot{x} > 0 \\ f_{II} & \text{for } \dot{x} < 0 \end{cases} \quad (4)$$

with region I for positive and region II for negative velocity. For $\dot{x} = 0$, additional formalism is necessary. The

$$f_I = f_{\text{ext}} - \mu F_N \quad (5)$$
$$f_{II} = f_{\text{ext}} + \mu F_N, \quad (6)$$

are the sums of all forces in region I, respectively II, where for eq. (3) we have used as external force $f_{\text{ext}} = -kx$. The velocity $\dot{x}$ is used as "indicator function" $g(x, \dot{x}) = \dot{x}$ to determine whether the system is in region I or II (Fig. 1). The solution at $\dot{x} = 0$ depends on the direction of the flow of the ODE, which is given by the (signed) scalar product between the velocity and the corresponding acceleration

$$a_I = \langle \nabla g, f_I/m \rangle, \quad \text{(region I)} \quad (7)$$
$$a_{II} = -\langle \nabla g, f_{II}/m \rangle, \quad \text{(region II).} \quad (8)$$

$\nabla = \nabla_{\dot{x}}$ is the directional derivative with respect to the velocity. There are three scenarios for the sign of the a_i's (compare Fig. 1):
1. $a_I \cdot a_{II} < 0$: the friction is dynamic, the flow crosses the x-axis at ($v = 0, |x|, x > 0.4$ in Fig. 1)
2. $a_I < 0, a_{II} < 0$: the flow pushes into the constraint at ($v = 0, |x|, x < 0.4$ in Fig. 1)
3. $a_I > 0, a_{II} > 0$: the flow pulls away from the constraint, which usually only occurs for an unfortunate choice of initial conditions (Painlevé paradox, see [7], section 3.3.1).
For case 2, to maintain the constraint, the right hand side of eqs. (7, 8) is neither f_I nor f_{II}, but

$$f(\dot{x}, \lambda) = (1 - \lambda) f_I + \lambda f_{II}, \quad (9)$$

a solution in the convex hull, where $0 \le \lambda \le 1$ is a Lagrange parameter. This means solving

$$\ddot{x} = f(\dot{x}, \lambda)/m, \quad (10)$$
$$0 = g(x, \dot{x}). \quad (11)$$

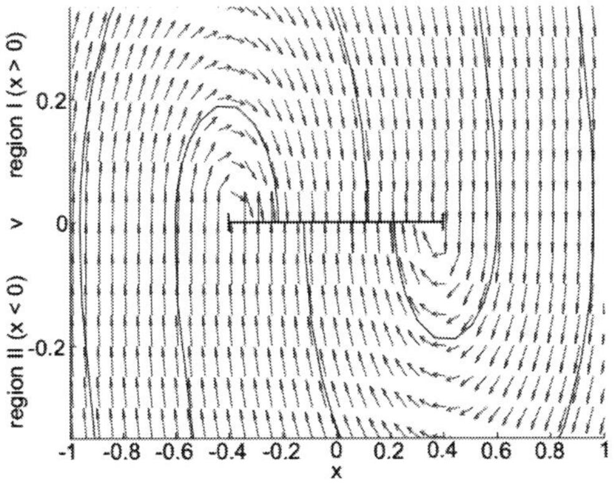

Figure 1. Phaseflow for a linear oscillator with dry friction eq. (3) with $m = 0.25$, $k = 1$ and $\mu = 0.4$. For static friction the flow of $\dot{x} < 0$ and $\dot{x} > 0$ pushes against each the singularity between $(-\mu, \dot{x} = 0)$ and $(+\mu, \dot{x} = 0)$. For dynamic friction ($|x| > \mu$) and $\dot{x} = 0$, the flow continues in the same direction.

Differentiating eq. (11) with respect to time allows to express λ in terms of a_I and a_{II},

$$\lambda = \frac{a_I}{a_I + a_{II}}. \quad (12)$$

Using eqs. (7, 8) and (in one dimension) $\nabla g = \pm 1$, the equation further simplifies to

$$\lambda = \frac{f_I}{f_I - f_{II}}. \quad (13)$$

This is a remarkable result, as the highly non-linear non-smooth static friction can be computed by solving a mere linear scalar equation. The linearity in the $f_{\dots}$ of eqs. (7,8) allows to write the force as

$$f(\dot{x}, \lambda) = f_{\text{ext}} - \left(\frac{f_{\text{ext}}}{\mu F_N}\right) \mu F_N. \quad (14)$$

That means that at $\dot{x} = 0$, static friction compensates the external forces numerically exact, $f(\dot{x} = 0, \lambda) = 0$ (within the error of the solver). The scalar products eqs. (7, 8) can be numerically approximated as

$$a_I \approx (\dot{x} + \delta_I f_I/m)/\delta_I, \quad (15)$$
$$a_{II} \approx -(\dot{x} + \delta_{II} f_{II}/m)/\delta_{II}. \quad (16)$$

The δ_I, δ_{II} have to be chosen small enough, so that the flow cannot cross the manifold, but large enough, so that the f_I, f_{II} are still taken into account. The δ_I, δ_{II} are given by

$$\delta_{I,II} = -0.9m \cdot \dot{x}/f_{I,II}. \quad (17)$$

4 Particle on an inclined slope

The most simple case for a two-dimensional particle system is a square block sliding down a slope. It is a single contact system, with gravity $g = 9.81[\text{m/sec}^2]$ as external force. We initialise the square particle orientated parallel to the slope without overlap. The slope has an inclination angle of $\alpha = \pi/6 = 30°$, slightly lower than the critical angle of $\approx 31°$ for our $\mu = 0.6$, so the block should come to rest. The particle and the slope are modelled as convex polygons. For the interaction we use the centroid of overlap region as the force-point, the normal direction of the contact is the weighted average of the connecting lines between the centroid and the intersection points. The elastic force is proportional to the particle overlap A and the Young's modulus Y and inverse proportional to the inverse characteristic length l. (see Ref. [7] for more details). The damping force is proportional to the change in overlap area

$$F_{\text{damp}} = -D\sqrt{Y_{2D} m_{\text{eff,tan}}} \frac{\dot{A}}{l}. \quad (18)$$

In the Cundall-Strack model for Coulomb friction, the tangential force is incremented by

$$dF = -k_t v \, dt = -k_t \, dx \quad (19)$$

and cut off at μF_{normal}. Without cutoff, it integrates out to an harmonic oscillator $F = -k_t x$. Accordingly, as long as

 DOI: 10.1051/epjconf/201714015020

Table 1. Simulation parameters for the inclined slope.

dt	ρ	Y	μ	γ	k_t	γ_t
$[s]$	[kg/m^2]	[kg/ms^2]				
$5 \cdot 10^{-6}$	5000	$5 \cdot 10^7$	0.6	1.5	$\frac{2}{7}Y$	$\frac{1}{4}\sqrt{\lvert k_t m_{\text{eff,tan}}\rvert}$

there is no dissipation, a particle on a slope $\ll \arctan(\mu)$ will oscillate around the equilibrium position. Adding

$$dF_{\text{damp}} = -\gamma_t a\,dt = -\gamma_t\,dv \tag{20}$$

as damping term to eq. (19) can lead to permanent slipping close to the critical angle, as the dv is opposite to v and therefore the grip of the tangential force is retarded. Instead, an instantaneous damping $\propto -v$ is conventionally introduced, which is nevertheless out of phase with eq. (19), so very small prefactors must be used. Physical parameters are given in Tab. 1, and remain the same unless specified otherwise. For CS, one can speed up the grip by using a larger prefactor in the incrementation: Nevertheless, this may make the use of a smaller timestep necessary. Our exact friction implementation (DAE) shows stronger drift far from the critical angle compared to the Cundall-Strack (CS) model (Fig. 2). At the critical angle, the CS model shows a later onset of static friction, i.e. a later "grip", compared to the DAE implementation (Fig. 3), in particular the damped DAE implementation has a significantly earlier onset than any of the other configurations. As can be seen in Fig. 4, the drift decreases with the timestep, a typical accuracy effect in the implementation of DAEs. The difficulty of implementing "numerically exact" static friction for discrete element simulations lies in the fact, that stationary configurations must be obtained by dynamic simulations in the presence of noise, which is a superposition of discretisation error, defective position information, as well as zero-point oscillations of particles around the equilibrium position. Obtaining static equilibria is a much greater challenge than computing mere dynamics: In dynamic computer simulations, bodies may move due to the noise amplitudes alone, while static equilibria have to be obtained within the numerical accuracy of the integrator. The drift increases with the timestep. The implicit Backward-Difference-Formulae (BDF, Gear Predictor-Corrector) show less drift than the explicit Heun-integrator ("2.nd order Runge-Kutta", implemented in predictor-corrector fashion, see [7], p. 80), with the second order BDF integrator being more accurate than the fifth order integrator. On the other hand, particle size has a much lower impact on the drift for implicit integrators. Fig. 5 shows that only larger particle sizes (>0.1m) cause an increase in the drift for the BDF-integrators. For the Heun-integrator, the drift shows significant increase both for very small and for very large particle sizes. This is consistent with the common practice in numerical analysis that implicit integrators are used for the solution of DAE's due to better stability compared to explicit integrators. If the drift is too large for the time scale considerations, it would be possible to "fix" the particle by velocity projection (see [5], p. 471). This would correspond to setting the velocity zero for a non-moving floor, and to project onto zero relative velocities for moving contacting particles.

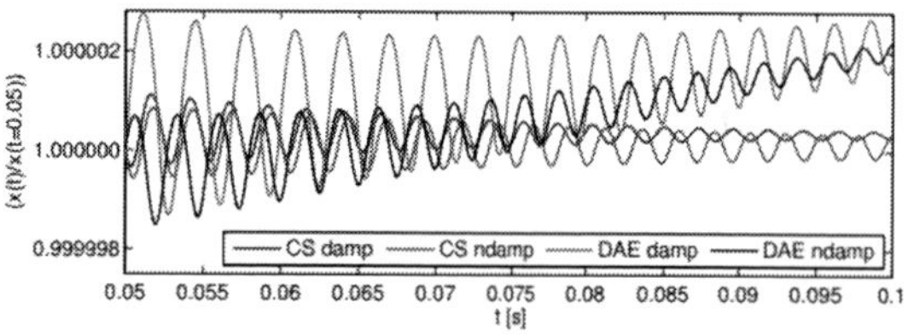

Figure 2. Horizontal position for a 1 cm particle on a slope with half the critical angle (15°) for CS and DAE friction implementations with and without damping.

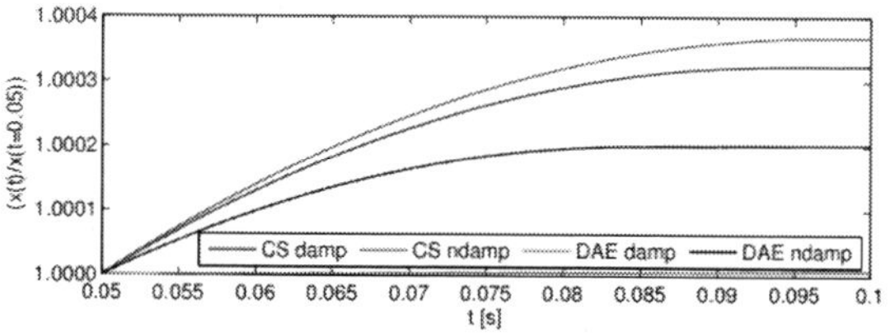

Figure 3. Horizontal position for a 1 cm particle on a slope with the critical angle (30°) for CS and DAE friction implementations with and without damping.

5 Many particle systems

We adapt the approach to many-body systems in two-dimension in the following way: The constraints are evaluated for each contact. Instead of the single particle mass, the reduced (tangential) mass

$$m_{\text{eff,tan}} = (1.0/m_1 + 1.0/m_2 + r_{1,\perp}^2/I_1 + r_{2,\perp}^2/I_2)^{-1} \tag{21}$$

is computed from the masses $m_{...}$ and the moments of inertia $I_{...}$ with the projection $r_{...,\perp}$ of the contact vectors (from the centres of mass to the contact point) onto the normal direction. The external forces f_{ext} in eqs. (5, 6) are replaced by the relative accelerations, scaled by the effective tangential mass,

$$f_{\text{ext}} \to (a_2 - a_1) \cdot m_{\text{eff,tan}}. \tag{22}$$

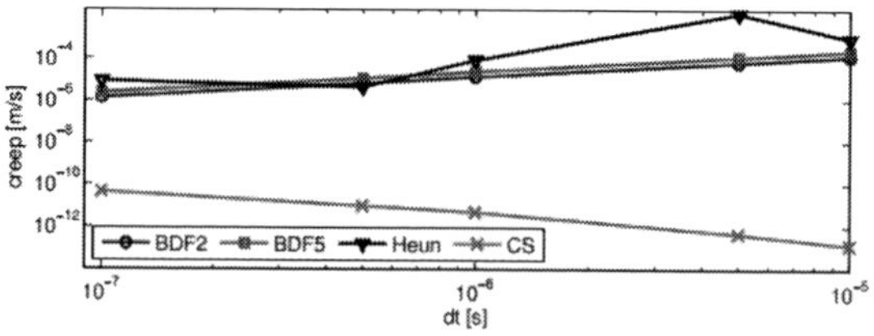

Figure 4. Numerical drift for a 1cm sized particle on a slope in theoretical force equilibrium, depending on the timestep dt.

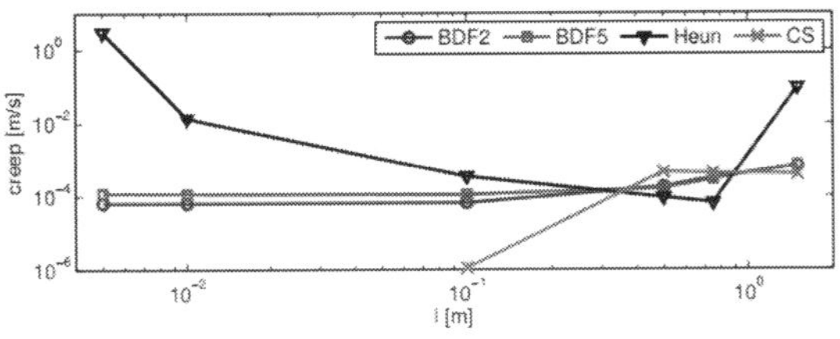

Figure 5. Dependence of the numerical drift on the particle size l for $dt = 5 \cdot 10^{-6}$s.

Like solids form in molecular dynamics through equilibration of forces for the inter-particle interactions, the equilibration of the tangential force for each contact leads to many-particle agglomerates in static equilibrium. As no matrix formulation is necessary, no condition number considerations are necessary and no instantaneous propagation of forces occurs. Mathematical convergence proofs are not possible, as the stability depends on the physical stability: even for experimental heaps, vibration can destroy the stability, so it is the entirety of parameters (initial velocities, friction coefficients, damping, state of motion of boundaries, ...) which determines whether e.g. a heap is stable or not. For the many particle system, we compare both friction approaches with via a granular heap of 1625 particles formed from a hopper, see Fig. 6). The particles have between 5 and 10 corners, distributed randomly on an ellipse with 2.5 mm mean radius and eccentricity ≈ 0.8 (for more details, see [7], chapter 7), and the same parameters as in section 4. For slopes of $\approx 30°$, the drift is shown in Fig. 7: For the CS-model (equivalent to Baumgarte-stabilization a DAE-problem[1], i.e. a parametrization of an oscillation around the constraint), the drift is considerably smaller, as it is , while the drift of the DAE-approach depends on the timestep.

Figure 6. Heap with particles dropped from a hopper. Some particles are still falling and appear airborne in the graphic.

6 Conclusion

We have implemented "numerically exact" Coulomb friction based on a DAE-formalism where the constraints are computed based on the interparticle forces for many body problems in two dimensions, which does not need a matrix formulation, by evaluating the frictional constraints. Compared to the Cundall-Strack model, the "grip" of the numerically exact friction is significantly faster and there is no internal degree of freedom which can release energy in the system. This makes the approach superior for granular systems where realistic damping is necessary, for fluidisation under vibration (as for earthquakes) or the onset of landslides. Depending on the particle size, the Cundall-Strack model is either close to optimal, as it is equivalent to a Baumgarte-stabilisation of a DAE-constraint, or it fails as the grip acts so slowly that no mechanical equilibrium is reached at all. While the weak point of the Cundall-Strack model is the parameter dependence, a weak point of the DAE-approach in many body configurations is the numerical drift due to the finite time discretiztion accuracy. This is a problem shared by all approaches which compute the friction as equilibrium force for DAE-solutions. Reducing the timestep considerably is not an option, as the simulation times will become too long. If necessary, the drift can be eliminated by stabilisation by projection (in particular velocity projection) [5, 6], which for single-particle problems corresponds to setting the velocity, while for many-body problems, the algorithm is more complicated.

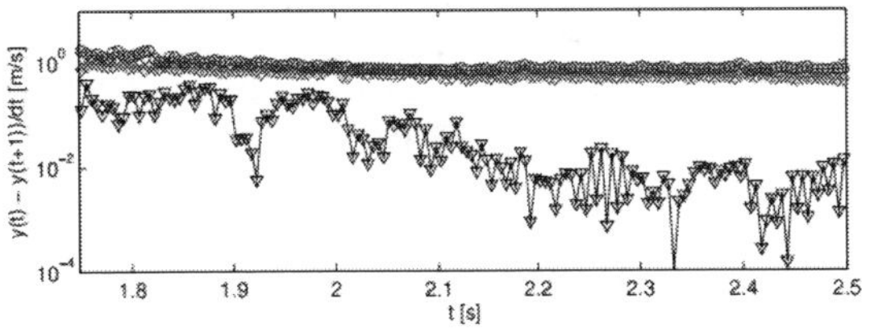

Figure 7. Drift of the centre of mass (□) for dt=10^{-5}s (○) and dt=$5 \cdot 10^{-6}$s (▽) for DAE as well for the CS with dt=10^{-5}s.

References

[1] J. Baumgarte, Comp. Meth. Appl. Mech. Eng., **1**, 11-16, (1972)

[2] P.A. Cundall, O.D.L. Strack, Géotechique, **29**, 47-65, (1979)

[3] A.F. Filippov, *Differential Equations with Discontinuous Righthand Sides* (Springer, 1988) 181

[4] E. Hairer, S.P. Nørsett, G. Wanner *Solving Ordinary Differential Equations, I. Nonstiff Problems*, (Springer 1987) 196

[5] E. Hairer, G. Wanner *Solving Ordinary Differential Equations, II. Stiff and Differential Algebraic Problems*, (Springer 1991)

[6] C. Lubich, BIT Numerical Mathematics **31**, 545-550 (1991)

[7] H.-G. Matuttis, J. Chen, Understanding the Discrete Element Method, Wiley (2014)

[8] J.J. Moreau. Unilateral contact and dry friction in finite freedom dynamics. Nonsmooth mechanics and applications, CISM Courses Lect. 302, 1-82 (1988)

[9] E.Rabinowicz. *Friction and Wear of Materials* (Wiley 1965)

[10] F.Radjai, P. Evesque, D. Bideau, S.Roux, Phys. Rev. E **52** 5555 (1995)

[11] D.E. Stewart, SIAM Review **42**, 3-39 (2000)

EPJ Web of Conferences **140**, 09033 (2017) DOI: 10.1051/epjconf/201714009033
Powders & Grains 2017

Simulation of Bridging at the Static Surface Filtration by CFD-DEM Coupling

David Hund[1,*], *Sergiy* Antonyuk[1] and *Siegfried* Ripperger[1,2]

[1]Institute of Particle Process Engineering, University of Kaiserslautern, 67663 Kaiserslautern, Germany

[2]IT for Engineering (it4e) GmbH, 67663 Kaiserslautern, Germany

Abstract. To achieve a better understanding of the bridging mechanism at the static surface filtration, simulations of particle deposition at a three-dimensional pore model of a filter were performed. The simulations are used to investigate the total filter resistance, which results from interferences between filter medium, particles and fluid. Thereby a CFD (Computational Fluid Dynamics) method for flow simulation was coupled with a Discrete Element Method (DEM) which calculates the particle interactions. The results show that a built particle bridge influences the flow through an unclosed pore. It was concluded that bridging of open pores is hindered at high velocities of approach caused by increasing percentage of closed pores. Moreover, it could be shown, that the total filter resistance depends on the approaching velocity and shape of particles.

1 Introduction

Solid-liquid filtration is the separation of dispersed particles from a fluid by means of filter media. At the static surface filtration the solid particles deposit on the surface of the filter medium. Further accumulation of particles lead to the filter cake build up. For the filtration processes, which are operating with short cycle times to increase the throughput, the filter resistance and the structure of the first particle layer are important characteristics. By the use of open-pore filter medium, the through-flow resistance can be reduced and the filtrate throughput can be increased. In such filtration processes the so called bridging can take place. Bridging means the pores are blocked by accumulated particles even though their diameters are smaller than the diameters of the pores of the filter medium (Fig. 1) [1, 2, 3].

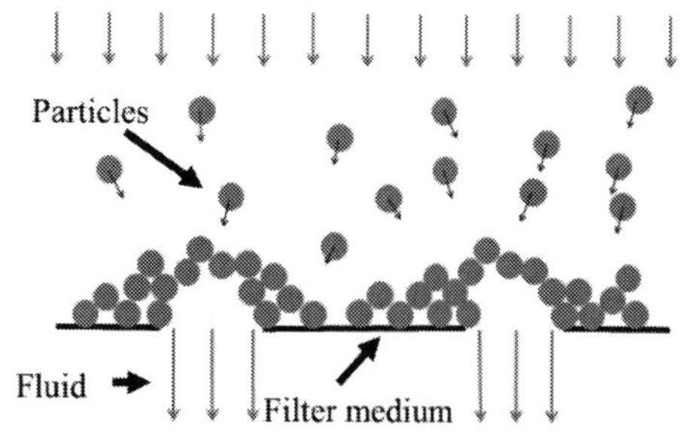

Fig. 1. Schematic illustration of the bridging mechanism

At the beginning of cake filtration processes, bridging is of particular importance, because it enables the cake building. Previous experimental studies have shown that the total filter medium resistance increases greatly as a result of interferences between the filter medium and the first particle layer [3]. In industrial filtration, this interference resistance can be 10 to 40 times the value of the unladen filter medium resistance. It contributes significantly to the total filter resistance [4].

In order to investigate the interference resistance and to understand the process of bridging, simulations for particle deposition were carried out. For this a simplified pore model was used considering particle-particle and particle-filter medium interactions. For this purpose, a CFD method to simulate the flow and a Discrete Element Method (DEM) to simulate the particle interactions were coupled.

2 Simulation Methods

2.1 Discrete Element Method (DEM)

The DEM, developed by Cundall and Strack in 1978 can be used to simulate the dynamics of particulate systems [5]. The method describes each particle in the system and its interactions with other particles or rigid walls (Fig. 2). Based on the force balance for particle i the equation of translational motion (1) is solved.

$$m_i \frac{d\boldsymbol{v}_i}{dt} = \sum_{j=1}^{k_i} \boldsymbol{F}_{c,ij} + \boldsymbol{F}_{g,i} + \boldsymbol{F}_{d,i} \qquad (1)$$

In the equations m_i and $\boldsymbol{v}_i$ are the mass and the velocity of particles i. $\boldsymbol{F}_{c,ij}$ are the contact forces between particles i and j and they are summed over the k_i particles in contact with particle i [5]. $\boldsymbol{F}_{g,i}$ is the gravitational force

* David Hund: david.hund@mv.uni-kl.de

and $F_{d,i}$ is the drag force acting on particle i. $F_{d,i}$ results from the interaction with the fluid.

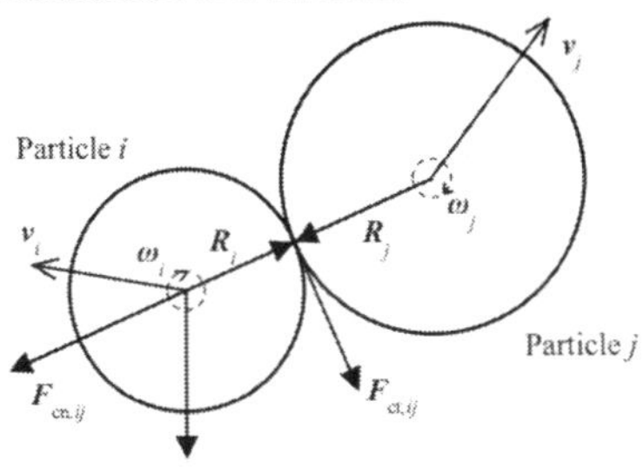

Fig. 2. Illustration of particle i in contact with particle j

The contact force $F_{c,ij}$ between two particles can be decomposed into a normal $F_{cn,ij}$ and a tangential force $F_{ct,ij}$.

$$\boldsymbol{F}_{c,ij} = \boldsymbol{F}_{cn,ij} + \boldsymbol{F}_{ct,ij} \quad (2)$$

For the calculation of $F_{cn,ij}$ a non-linear model, based on Hertz [6] combined with a viscous damping term according to Tsuji [7] and adhesion according to the Johnson-Kendall-Roberts (JKR) theory [8] was used. The damping parameter α (Table.1) of the damping term [7] was determined based on the coefficient of restitution (COR) $e = 0.5$. But the effective COR of the whole model depends on the impact velocity. $F_{ct,ij}$ is modelled based on the non-slip solution of the theory provided by Mindlin and Deresiewicz [9].
The contact forces $F_{c,ij}$ act at the contact point between particles i and j. The lines of action of the contact forces which do not pass the centre of the particle will generate a torque M_{ij}. It is given by:

$$\boldsymbol{M}_{ij} = \boldsymbol{R}_i \times \boldsymbol{F}_{ct,ij} \quad (3)$$

where R_i is a vector from the mass centre of the particle to the contact point [4]. The governing equation for the rotational motion of particle i with the moment of inertia J_i is written as:

$$J_i \frac{d\boldsymbol{\omega}_i}{dt} = \sum_{j=1}^{k_i} \boldsymbol{M}_{ij} \quad (4)$$

where ω_i is the angular velocity of particle i.

2.2 Governing equations in the CFD approach

To consider that particles influence the fluid, the description of the flow in this work is made via modified Navier-Stokes equations. The averaged mass (1) and momentum (2) conservation equations are written as:

$$\frac{\partial(\varepsilon\rho_F)}{\partial t} + \nabla\cdot(\varepsilon\rho_F\boldsymbol{u}) = 0 \quad (5)$$

$$\frac{\partial}{\partial t}(\rho_F\varepsilon\boldsymbol{u}) + \nabla\cdot(\rho_F\varepsilon\boldsymbol{u}\boldsymbol{u}) = -\nabla p + \nabla\cdot(\varepsilon\bar{\tau}) + \varepsilon\rho_F\mathbf{g} - \boldsymbol{S} \quad (6)$$

with the fluid volume fraction ε, the averaged velocity $\boldsymbol{u}$, the density ρ_F of the fluid, the gravity $\mathbf{g}$ and the viscous stress tensor $\bar{\tau}$. $\boldsymbol{S}$ is an external volumetric force which considers the drag force of the particles on the fluid.

2.3 CFD-DEM coupling

In this work, a four way coupling was used which means that particle-fluid, fluid-particle, particle-particle and particle-wall interactions were considered. In Fig. 2 is shown schematically the coupling of CFD and DEM. In the first step the conservation equations are solved by the CFD. Afterwards the flow parameters are transferred to DEM to calculate the hydrodynamic forces acting on the particles. Then the balances of forces for each particle are solved and the new velocity and the new position of the particles are calculated. Finally the fluid flow is computed again by CFD depending on the new fluid volume fraction of the mesh cell and the source term $\boldsymbol{S}$. In general the source term $\boldsymbol{S}$ considers all fluid forces acting on the particle. In the present work only the drag force of all particles $\boldsymbol{F}_d$ is considered.

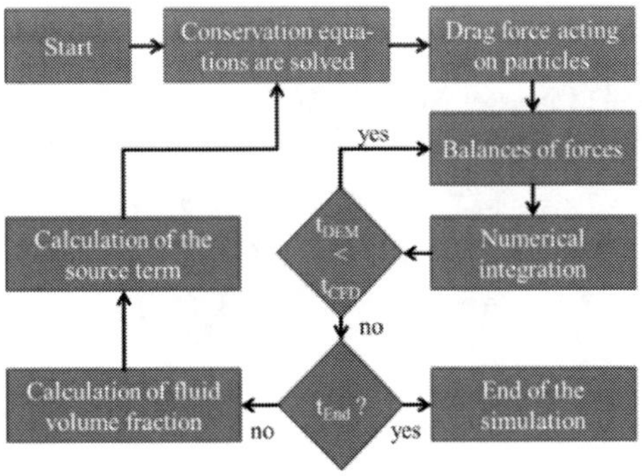

Fig. 3. Simulation flowchart of CFD-DEM coupling

The drag force on an isolated particle is calculated by the following equation [10]:

$$\boldsymbol{F}_d = \frac{\rho_F}{2} c_d(\mathrm{Re_P}) A_P (\boldsymbol{u}-\boldsymbol{v})|\boldsymbol{u}-\boldsymbol{v}| \quad (7)$$

where c_D is the drag coefficient and A_P the cross section of the particle. Depending on the particle Reynolds number $\mathrm{Re_P}$ c_D is defined according to Stokes, Schiller and Naumann [11] and Newton [12] respectively.

$$\mathrm{Re_P} = \frac{\rho_F\varepsilon|\boldsymbol{u}-\boldsymbol{v}|d_P}{\eta} \quad (8)$$

$$\text{if} \quad \mathrm{Re_P} \le 0.5 \quad c_d = \frac{24}{\mathrm{Re_P}} \quad (9)$$

$$\text{if} \quad 0.5 < \mathrm{Re_P} \le 1000 \quad c_d = \frac{24(1+0.15\mathrm{Re_P^{0,687}})}{\mathrm{Re_P}} \quad (10)$$

$$\text{if} \quad 10^5 > \mathrm{Re_P} > 1000 \quad c_d = 0.44 \quad (11)$$

3 Results

3.1 Pore model and simulation set up

The filter medium, as shown in Fig. 3, is represented by a simplified pore model with cylindrical pores and a pore diameter of $d_{Pore} = 20$ µm. The suspension flows through the filter medium perpendicular. Besides the bridging at a single pore, the influence of a blocked pore on the bridging at surrounding pores is considered. The pore model includes four pores. To exclude boundary effects periodic boundary conditions were used.

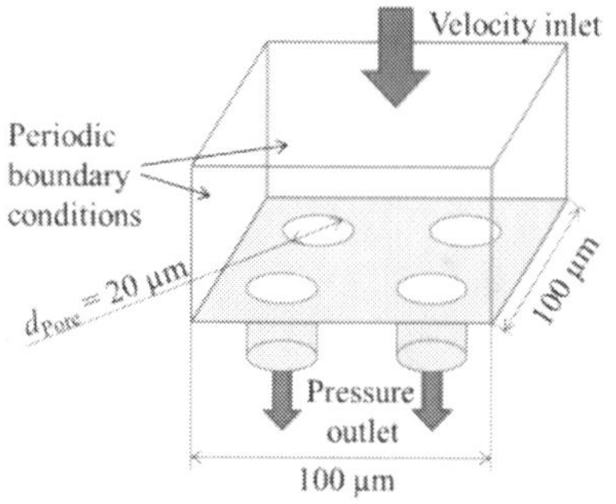

Fig. 4. Simulation model with fur cylindrical pores

In this work, the bridging formation on both spherical and cylindrical polyamide particles is investigated (Fig. 4). The two particle systems are identical in volume.

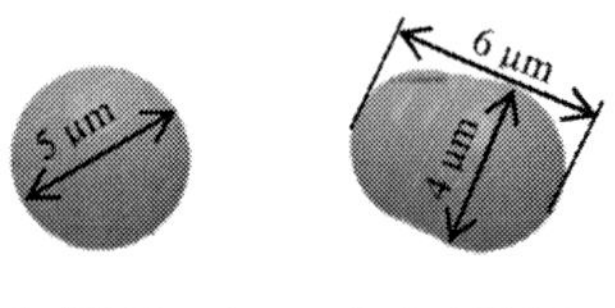

Fig. 5. Spherical (left) and non-spherical (right) particles

In DEM the non-spherical particles were modelled with the multi-sphere approach [12]. In this approach three smaller spherical particles were connected in order to form a cylindrical particle. In Table 1 all necessary simulation parameters are shown. Further information are given in the corresponding literature.

Table 1. Simulation setup

Pore Model	
Side length	L_S = 100 µm
Pore Diameter	D_{Pore} = 20 µm
Particle phase	
Particle size	
Spherical	d_P = 5 µm
Non-spherical	L_P = 6 µm, H_P = 4 µm
Particle density	ρ_P = 1180 kg/m^3
Shear modulus [6, 7]	G = 0.89 GPa
Young´s modulus [6, 7]	E = 2.5 GPa
Poisson ratio [6, 7]	υ = 0.4
Friction coefficient [9]	μ = 0.5
Damping parameter [7]	α = 1.728
Surface energy [8]	γ = 0.41 J/m^2
Fluid phase	
Density	ρ_F = 998.2 kg/m^3
Viscosity	η_F = 0.001 Pa·s

3.2 Bridging mechanism

The bridging simulations were performed in dead-end mode. Either a constant pressure difference or a constant specific filtrate volume flow was prescribed. The simulations were terminated if the pores were blogged by the particles. The bridge formation is initially discussed at a constant flow velocity of u = 0.5 m/s and at a volume concentration of the particles in the fluid of c_v = 0.1 %. In Fig. 5 both pictures (left and right) show a section through two of four pores. On the left side the volume fraction and on the right the associated particle velocity is illustrated. It is clearly indicated that the particles first deposited around the pores and that the process of bridging is still not completed.

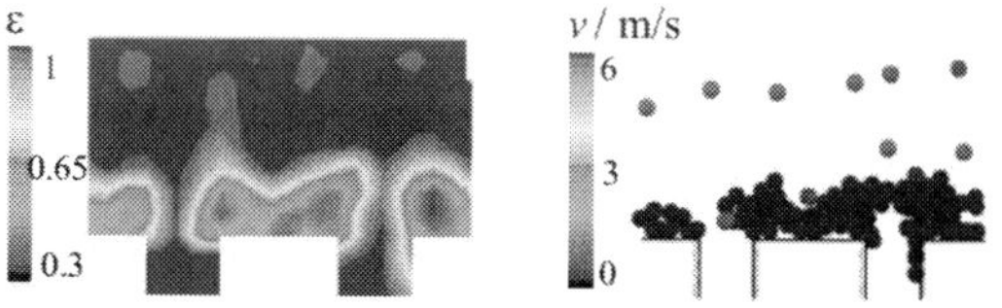

Fig. 6. Volume fraction of fluid ε while bridging (left) and the associated particle velocity v (right)

To figure out the influence of a clogged pore on the flow through the other pores the specific flow rate of each pore is presented in Fig. 6. First, the flow rate decreases in pore 3 because of deposited particles. As a consequence the averaged flow rates in the other pores increase due to the constant flow rate at the inlet. After t = 3 ms pores 1, 3 and 4 are blogged and only pore 2 is unclosed. After t = 4.3 ms even pore 2 is clogged and the specific flow rate in each pore is constant. Because a nearly built bridge over pore 2 collapses at t = 2.3 ms the flow rate in pore 2 increases abruptly.

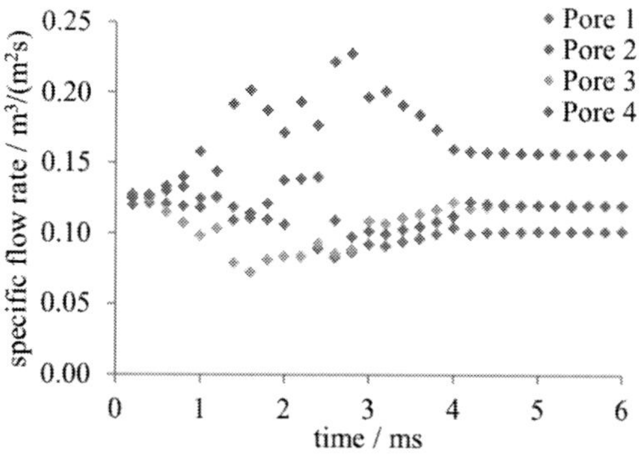

Fig. 7. Specific flow rate in each pore at a constant flow velocity at the inlet u = 0.5 m/s, c_v = 0.1 %, d_P = 5 µm

Fig. 7 shows the corresponding pressure drop across the filter medium over the time. At t = 0.1 ms the pressure drop results from the flow resistance of the unladen filter medium at a constant flow velocity u = 0.5 m/s. Afterwards the pressure drop increases because of deposited particles around the pores and clogging of pores.

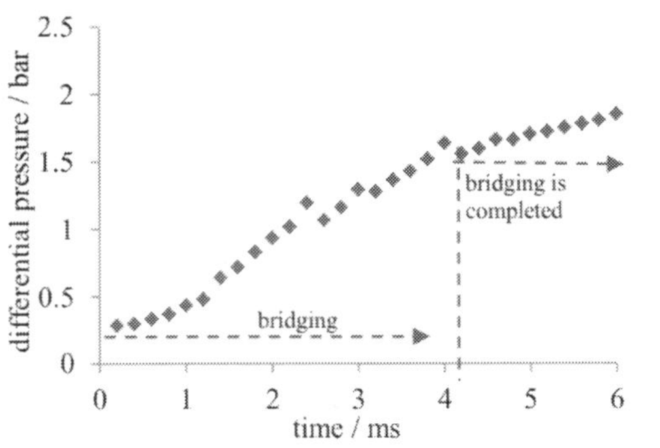

Fig. 8. Pressure drop across the filter medium at a constant flow velocity at the inlet u = 0.5 m/s, c_v = 0.1 %, d_P = 5 µm

As well as presented in Fig. 6 at t = 2.3 ms the pressure drop decreases because of a collapsed bridge. If all pores are clogged the filter cake build up begins and the pressure drop increases linearly.

Fig. 8 illustrates the filter cake height over time for different specific flow rates. If the filter cake height increases linearly with time bridging is finished. It can be seen that the higher the flow velocity the later bridging is completed. Because of high drag forces acting on particles, at a flow velocity of u = 2 m/s no bridging takes place.

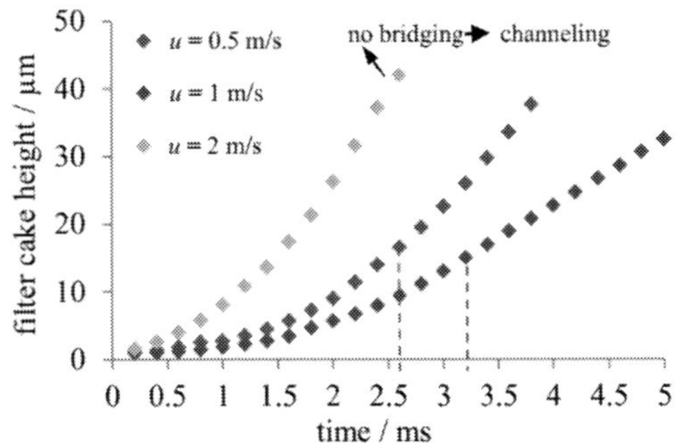

Fig. 9. Filter cake height over time for different specific flow rates u = 0.5 m/s, u = 1 m/s, u = 2 m/s, c_v = 0.1 %, d_P = 5 µm

In Fig. 9 the corresponded volume fractions of fluid above the pores are shown. As can be seen the higher the flow velocity the higher the built bridge and at a flow velocity of u = 2 m/s continuous channels are formed.

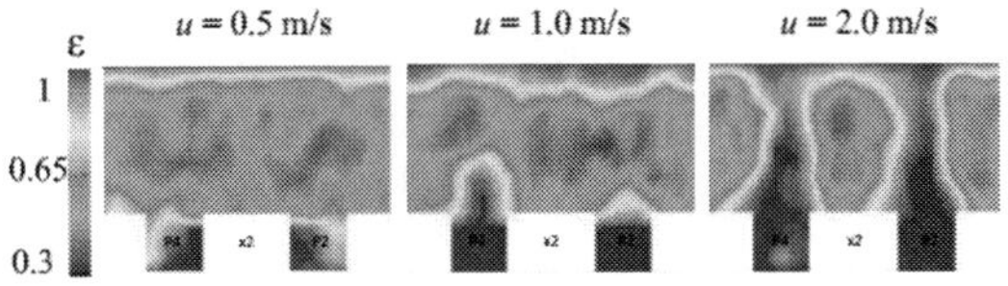

Fig. 10. Volume fraction ε of fluid at different flow velocities u = 0.5 m/s, u = 1 m/s, u = 2 m/s, c_v = 0.1 %, d_P = 5 µm

Finally the bridging mechanism with different shapes of particles (Fig. 4) at a constant pressure drop will be discussed. The filter cake height over time is presented in Fig. 10. The results show that bridging at non-spherical particles is completed earlier. The cylindrical particles are better placed together. Due to the higher porosity in the filter cake of the non-spherical particles, the slope in the linear region is higher. Thus the filter cake build up at non-spherical particles is faster than with spherical particles.

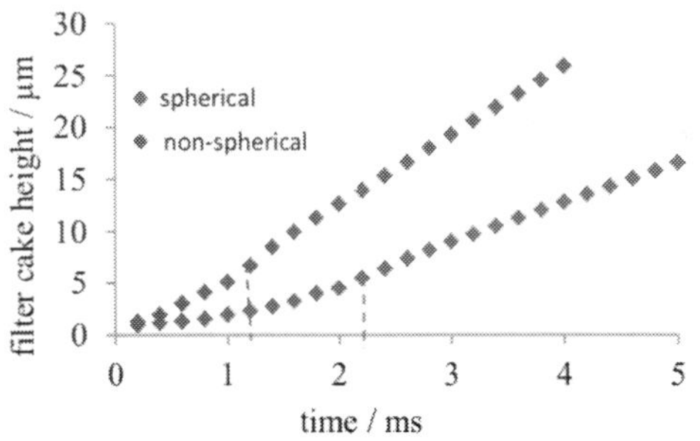

Fig. 11. Filter cake height over time for different particle shapes, Δp = 0.5 bar, c_v = 0.1 %

4 Conclusions

In this manuscript a CFD-DEM coupled simulation method was introduced. The method appears to describe and investigate the influence of built bridges on the flow through of unclosed pores. It was found out that bridging in open pores is hindered at high velocities of approach caused by increasing percentage of closed pores. Further could be shown that the shape of particles influences the time until the pores are clogged. Compared to the spherical particles the cylindrical particles which were modelled using the multi–sphere approach clog the pore first.

References

1. G. Agbangla, E. Climent, P. Baccin, Sep Purif Technol, **101**, 42-48 (2012)
2. C. Schnitzer, S. Ripperger, Filtr. und Sep. **2**, 58-60 (2009)
3. A. Rushton, The Chem. Eng., (1970)
4. J.W. Tichy, VCH-Verlag, (2007)
5. P.A. Cundall, O.D. Strack, Geotechnique **29**, 47-65 (1979)
6. H. Hertz, J reine angew Math, **92**, 156-171 (1882)
7. Y. Tsuji, T. Tanaka, T. Ishida, Powder Technol. **71**, 239-250 (1992)
8. K.L. Johnson, K. Kendall, D. Roberts, Proc R. Soc. Lond. 324, 301-313 (1971)
9. R. D. Mindlin, H. Deresiewicz, Adv energy syst div, 327-344 (1953)
10. Lord Rayleigh, C R, **159**, 109-110 (1913)
11. L. Schiller, A. Naumann, Z Ver Deutsch Ing **77**, (318-320) 1935
12. R. Johnson, Handbook of Fluid Dynamics, 2016
13. J. Favier et al., Eng. Comput. **16**, 467–480 (1999)

DOI: 10.1051/epjconf/201714003062

Effective friction of granular flows made of non-spherical particles

Ellák Somfai[1,⋆], *Dániel B.* Nagy[1], *Philippe* Claudin[2], *Adeline* Favier[2], *Dávid* Kálmán[1], and *Tamás* Börzsönyi[1]

[1] *Institute for Solid State Physics and Optics, Wigner Research Centre for Physics, Hungarian Academy of Sciences, P.O. Box 49, H-1525 Budapest, Hungary*
[2] *Physique et Mécanique des Milieux Hétérogènes, PMMH UMR 7636 ESPCI – CNRS – Univ. Paris-Diderot – Univ. P.M. Curie, 10 rue Vauquelin, 75005 Paris, France*

Abstract. Understanding the rheology of dense granular matter is a long standing problem and is important both from the fundamental and the applied point of view. As the basic building blocks of granular materials are macroscopic particles, the nature of both the response to deformations and the dissipation is very different from that of molecular materials. In the absence of large gradients, the best approach formulates the constitutive equation as an effective friction: for sheared granular matter the ratio of the off-diagonal and the diagonal elements of the stress tensor depends only on dynamical parameters, in particular the inertial number. In this work we employ numerical simulations to extend this formalism to granular packings made of frictionless elongated particles. We measured how the shape of the particles affects the effective friction, volume fraction and first normal stress difference, and compared it to the spherical particle case. We had to introduce polydispersity in particle size in order to keep the systems of the more elongated particles disordered.

1 Introduction

After a few decades of research, the physics of granular material has reached today a good level of general understanding [1]. In particular, it has been shown that the rheology of dense granular flows takes a simple form when working at controlled pressure P [2]. It this case, the constitutive laws have been formalized in terms of the so-called inertial number $I = \dot{\gamma} d / \sqrt{P/\rho}$, where $\dot{\gamma}$ is the shear rate and d and ρ are the grain size and density [3–6]. This dimensionless quantity compares the macroscopic shearing time scale $1/\dot{\gamma}$ to the microscopic time scale of local plastic events $d/\sqrt{P/\rho}$. In the limit of rigid grains, the effective friction coefficient μ, which compares the shear stress τ to the pressure P, as well as the volume fraction ϕ of the system must be functions of I, and one now refers to this framework as the $\mu(I)$-rheology for granular flows. The shape of the function $\mu(I)$ has first been determined by means of numerical simulations [4–6], also complemented by experimental measurements [7, 8]. Its behavior has been related to grain contact distribution, whose anisotropy increases with I [9]. This rheology is able to recover Bagnold scaling in flows down an incline plane [10]. More generally, this approach has been successfully applied in different geometries [2, 3], e.g. silo discharge [11], granular chute flows [12–15] and granular column collapse [16, 17], as well as to describe dynamic compressibility effects associated with spontaneous oscillatory motion [18, 19]. Interestingly, experiments and numerical simulations have recently provided evidence for limitations of the $\mu(I)$-rheology, when some non-local effects come into play (see e.g. [20] and references therein).

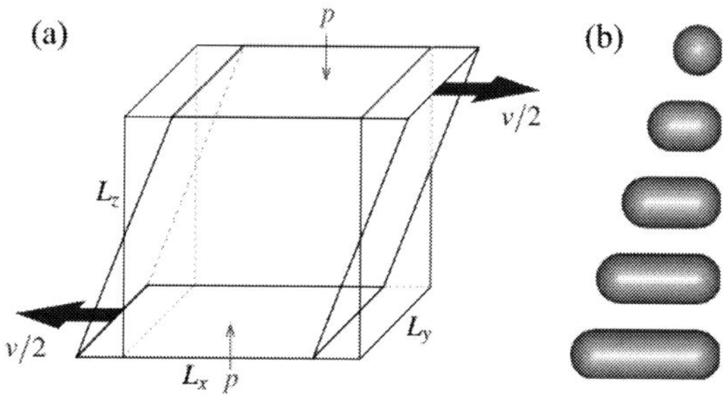

Figure 1. Numerical setup. (a) The simulation box. The velocity difference between the top and bottom sides is $v = \dot{\gamma} L_z$, where L_z is adjusted continuously as if a contstant pressure was applied to the top and bottom sides. (b) Particles used: sphere ($Q = 1$) and spherocylinders with aspect ratio $Q = 1.5$, 2, 2.5 and 3.

Many works are now interested in possible extensions of this rheology, when another effect is added, such as cohesion [21], particle softness [22] or activity [23]. This framework has also been extended to the case of granular suspensions [24, 25]. In this case, the dimensionless viscous number J, which compares the shear rate to the viscous frequency P/η_f, where η_f is the viscosity of the fluid, must be used instead of, or in complement to I. Interestingly, the case of Brownian suspensions can also be described in this framework [26]. Finally, these constitutive laws have been incorporated to models for a diphasic description of sediment transport [27–30].

⋆e-mail: somfai.ellak@wigner.mta.hu

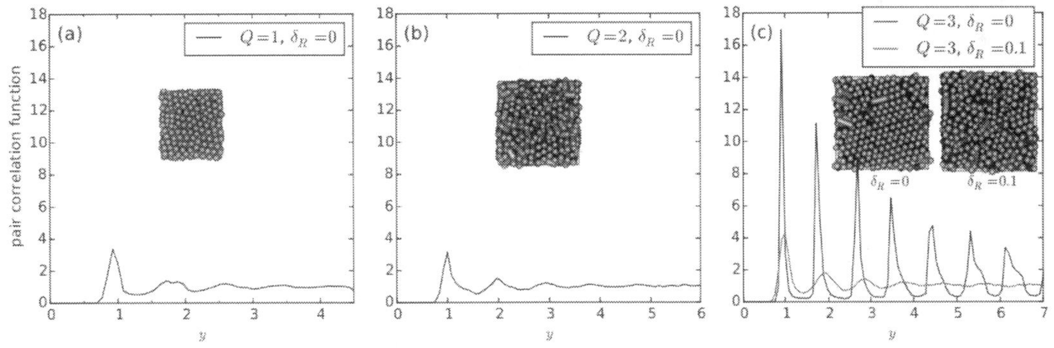

Figure 2. Crystallization. The pair correlation function is plotted in the neutral y direction for (a) spheres, (b) $Q = 2$ and (c) $Q = 3$ spherocylinders. Monodisperse spheres and $Q = 2$ spherocylinders show no crystallization: the pair correlation function rapidly converges to its asymptotic value of 1 after a few oscillations. However, for $Q = 3$ monodisperse spherocylinders the peaks are very sharp, and the oscillation is still very strong even at the 7th neighbor. Introducing polydispersity at $\delta_R = 0.1$ (red curve) restores disorder. The insets are side views of the simulation box taken from the x direction. The ordering for $Q = 3$, $\delta_R = 0$ is clearly visible.

Several recent studies investigated the behavior of elongated grains under shear [31–37]. These studies mostly focused on quasistatic flows [31, 32] or on the orientational distribution of particles as a function of grain elongation and shear rate [33–36]. Stresses were analyzed as a function of particle shape for glued-spheres and true cylinders [38]. In these systems, the shear induced alignment (average orientation of the grains) is not parallel to the streamlines but encloses a small angle with it. In the present work, we numerically consider the case of elongated grains shaped as spherocylinders, and investigate how the $\mu(I)$-rheology is modified with respect to the aspect ratio Q of the grains.

2 Numerical system

In the numerical measurements we used the 3-dimensional plane Couette geometry [Fig. 1(a)], where the boundary conditions were periodic in the flow (x) and neutral (y) directions, and Lees-Edwards in the velocity gradient (z) direction. The box was filled with elongated particles: we used spherocylinders with aspect ratio $Q = \ell/2R =$ 1.5, 2, 2.5 and 3, where the R is the radius of a particle (both of its spherical cap and of its cylindrical body), and ℓ is its total length. For reference we used spheres ($Q = 1$) as well. The interaction between the particles in this work was frictionless: the repulsive force was proportional to the overlap and contained an additional velocity-dependent dissipative term: $\mathbf{F}_{ij} = (-k\,\delta_{ij} + b\,\mathbf{v}_{c,ij} \cdot \hat{c}_{ij})\,\hat{c}_{ij}$, where $\mathbf{F}_{ij}$ is the force between particle i and j, the overlap is $\delta_{ij} = R_i + R_j - d_{ij}$; the distance d_{ij} and its unit length direction $\hat{c}_{ij}$ between the core line segments of the particles (connecting the centers of the spherical caps) is calculated by an efficient algorithm [39]. The velocity difference between particles at contact, $\mathbf{v}_{c,ij}$, is calculated using the velocity difference between the centers of the particles, their rotation angular velocities and the relative position of the contacts. The prefactor b of the dissipative term is set such that the restitution coefficient e takes a specified value, we used $e = 0.5$. The length unit was the average particle diameter (not length), the mass unit was set by specifying unit density for the particles (they were solid bodies), and fixing the stiffness to $k = 1$ defined the force scale and consequently the pressure and time scales.

The time evolution was calculated by a velocity-Verlet scheme, with particle rotations represented by quaternions [40]. The time step was taken to 1/100 of what would be the duration of a binary collision between two particles.

The initial conditions were prepared by a strongly overdamped dynamics, afterwards the system was sheared by a constant shear rate $\dot{\gamma}$. During shear the L_x and L_y sides of the box were kept fixed, but L_z was adjusted by a feedback loop such that the P_z component of the stress tensor, monitored continuously, fluctuated around a predefined value P; we used $P = 10^{-3}$ in our units.

To ensure stationary state, before making measurements we sheared the system for shear strain $\gamma = 25$ starting from a random initial condition, or for $\gamma = 10$ starting from a final state of the same system sheared with a different shear rate; all transients have decayed to below noise level at this deformation. After ensuring no crystallization (see next Section), we checked that in the stationary state the density and particle level quantities like the average alignment angle and nematic order parameter were homogenous in the simulation box. The velocity profile was linear, consistent with the prescribed $\dot{\gamma}$ – i.e. no shear banding was observed.

After the stationary state has been reached, we measured the particle volume fraction ϕ and the stress tensor σ at regular intervals and recorded their time average over strain scale $\gamma = 5$. The stress tensor was defined as

$$\sigma_{\alpha\beta} = \frac{1}{V}\sum_{<ij>}(\mathbf{r}_i - \mathbf{r}_j)_\alpha\,\mathbf{F}_{ij,\beta} - \frac{1}{V}\sum_i m_i \mathbf{v}^*_{i,\alpha}\mathbf{v}^*_{i,\beta}\,,$$

where $\mathbf{v}^*_i$ is the non-affine velocity of particle i in excess of the average shear velocity at that position; the second, kinetic term is negligible except at high shear rates. The most relevant elements of σ are the shear stress σ_{zx} and the diagonal elements $\sigma_{xx} = -P_x$, $\sigma_{yy} = -P_y$ and $\sigma_{zz} = -P_z$ (the compressive pressures are positive).

3 Crystallization

With our numerical simulation we aimed to model the behavior of real, disordered granular systems. However we observed that our numerical model produces configurations ordered spatially in the y and z directions, when the shear rate was not too large for sufficiently elongated particles.

According to common wisdom 2D packings of monodisperse disks crystallize easily, while in 3D this phenomenon is less pronounced. It has been observed, however, that in 3D numerical systems of identical spheres crystallization may happen after sufficiently long shear [41] or shaking [42]. Experiments of sheared identical spheres also displayed ordering, where chains of particles formed along the streamlines, and neighboring chains were arranged on a triangular lattice in the perpendicular plane [36].

We also observed in our simulations that small periodic packings (around 500 particles) crystallized easily unders shear, though that order was quickly destroyed by a few percent of polydispersity, slight deviation from spherical shape, and even by increasing the system size.

For the typical system size used in this work (1000 particles) no crystallization was observed for spheres and $Q = 2$ spherocylinders, but those with $Q = 3$ displayed strong spatial order. To quantify this we calculated the pair correlation function in the neutral y direction, the results are shown in Fig. 2. In order to destroy this order we introduced polydispersity: the particles were of identical aspect ratio, but their radii were distributed according to a uniform distribution, characterized by the variation coefficient δ_R (the ratio of standard deviation of radii to their average). At polydispersity $\delta_R = 0.1$ no spatial order has been observed for the $Q = 3$ spherocylinders; for consistency we used this value of polydispersity for all particle shapes in the following rheological measurements.

4 Rheological laws

Our main goal in this work was the measurement of rheological quantities as a function of the inertial number I for different particle shapes; the results are shown on Fig. 3.

The effective friction, defined as the ratio of the shear stress and the control pressure, $\mu = \sigma_{zx}/P_z$, is shown on Fig. 3(a). The general trend of the elongated particles is similar to that of spheres: μ tends to a nonzero value in the quasistatic ($I \to 0$) limit even for frictionless particles, and gradually increases with the inertial number. It is interesting to note that the dependence appears to be non-monotonic with Q, as in the quasistatic limit the effective friction for $Q = 1.5$ appears to be larger than for either spheres or more elongated particles; though this value needs to be extracted systematically for a sharp statement.

The particle volume fraction ϕ is plotted in Fig. 3(b). For spheres in the quasistatic limit it tends to a value close

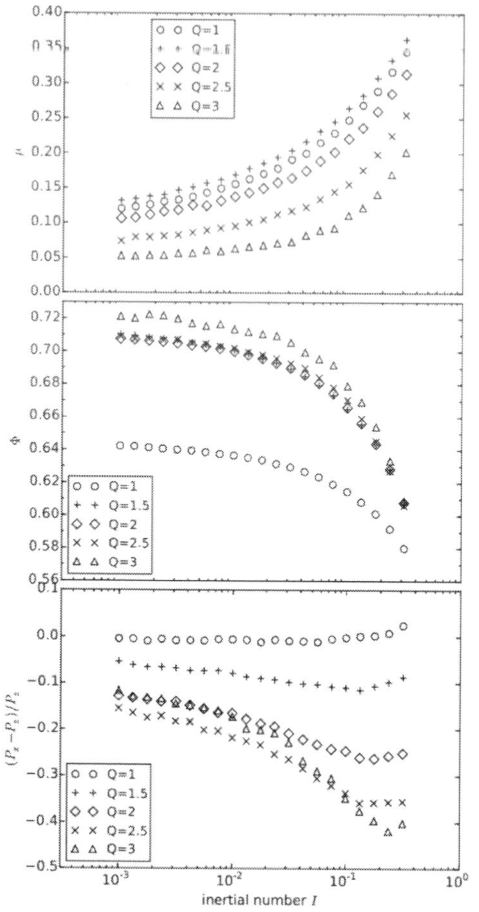

Figure 3. Rheological laws: (a) effective friction coefficient, (b) volume fraction and (c) normalized first normal stress difference as a function of I for five different particle shapes.

to that quoted for random close packing, though for comparing actual values one needs to take into account the polydispersity we used. For elongated particles the quasistatic volume fraction is higher, similarly to what has been observed for ellipsoids [43]. For increasing inertial numbers the constant pressure volume starts to increase as the collisions become increasingly violent, consequently the volume fraction starts to drop.

Finally the first normal stress difference was measured as well, which we normalize by the control pressure as $N_1/P_z = (P_x - P_z)/P_z$; see Fig. 3(c). As expected, for spheres this quantity is zero up to moderate inertial numbers, and starts to deviate from zero only in the collisional regime. For elongated particles the normal stress difference is nonzero even for small inertial numbers. The quasistatic limit needs to be determined carefully before a conclusive statement can be made about its value for spherocylinders. The second normal stress difference (not shown) is smaller in magnitude, its sign is positive, and the values for $Q > 1.5$ seem to collapse on each other.

DOI: 10.1051/epjconf/201714003062

In summary, we performed numerical simulations to determine the rheological properties of homogenously sheared granular packings of frictionless spherocylinders. We found that the qualitative behavior of the effective friction is similar to that of spheres: takes a nonzero value in the quasistatic limit, and increases with the inertial number; furthermore its aspect ratio dependence is non-monotonic. The volume fraction of sheared spherocylinders is higher than spheres for all aspect ratios we considered. The first normal stress difference for elongated particles deviates from zero even for small inertial numbers, unlike for spheres.

Our short term plans include extracting systematically the quasistatic limit of μ, ϕ and the normal stress differences, determining their aspect ratio dependence, and trying to establish a link between these macroscopic rheological properties and the particle level behavior like shear induced alignment and coordination of elongated particles.

Acknowledgements

We thank B. Andreotti, E. Clément and B. Szabó for useful discussions. This work was supported by the Hungarian-French bilateral TéT/Balaton exchange programme and by the Hungarian National Research, Development and Innovation Office NKFIH under grant OTKA K 116036.

References

[1] B. Andreotti, Y. Forterre and O. Pouliquen, *Granular Media, Between Fluid and Solid*, Cambridge University Press (2013).
[2] Y. Forterre and O. Pouliquen, Ann. Rev. Fluid Mech. **40**, 1-24 (2008).
[3] GDR MiDi, Eur. Phys. J. E **14**, 341-365 (2004).
[4] F. da Cruz , S. Emam, M. Prochnow , J.-N. Roux and F. Chevoir, Phys. Rev. E **72**, 021309 (2005).
[5] P. Jop, Y. Forterre and 0. Pouliquen, Nature **441**, 727-730 (2006).
[6] T. Hatano, Phys. Rev. E **75**, 060301(R) (2007).
[7] P. Jop, Y. Forterre and O. Pouliquen, J. Fluid Mech. **541**, 167-192 (2005).
[8] A. Fall, G. Ovarlez, D. Hautemayou, C. Mézière, J.-N. Roux and F. Chevoir, J. Rheology **59**, 1065-1080 (2015).
[9] E. Azéma and F. Radjaï, Phys. Rev. Lett. **112**, 078001 (2014).
[10] L.E. Silbert, D. Ertas, G.S. Grest, T.C. Halsey, D. Levine, and S.J. Plimpton, Phys. Rev. E **64**, 051302 (2001).
[11] L. Staron, P.-Y. Lagrée and S. Popinet, Eur. Phys. J. E **37**, 5 (2014).
[12] J.M.N.T. Gray and A.N. Edwards, J. Fluid Mech. **755**, 503-534 (2014).
[13] J. Baker, T. Barker and J.M.N.T. Gray, J. Fluid Mech. **787**, 367-395 (2015).
[14] T. Barker, D.G. Schaeffer, P. Bohorquez and J.M.N.T. Gray, J. Fluid Mech. **779**, 794-818 (2015).
[15] T. Börzsönyi, R.E. Ecke and J.N. McElwaine, Phys. Rev. Lett. **103**, 178302 (2009).
[16] P.-Y. Lagrée, L. Staron and S. Popinet, J. Fluid. Mech. **686**, 378-408 (2011).
[17] I. Ionescu, A. Mangeney, F. Bouchut and O. Roche, J. Non-Newton. Fluid. Mech. **219**, 1-18 (2015).
[18] L. Staron Phys. Rev. E **86**, 041307 (2012).
[19] M. Trulsson, M. Bouzid, P. Claudin and B. Andreotti, Europhys. Lett. **103**, 38002 (2013).
[20] M. Bouzid, A. Izzet, M. Trulsson, E. Clément, P. Claudin, B. Andreotti, Eur. Phys. J. E **38**, 125 (2015).
[21] P.G. Rognon, J.-N. Roux, D. Wolf, M. Naaim and F. Chevoir, Europhys. Lett. **74**, 644-650 (2006).
[22] A. Favier de Coulomb, M. Bouzid, P. Claudin, E. Clément and B. Andreotti, Subm. to Phys. Rev. Lett.
[23] A. Peshkov, P. Claudin, E. Clément and B. Andreotti, Europhys. Lett. **116**, 14001 (2016).
[24] F. Boyer, E. Guazzelli and O. Pouliquen, Phys. Rev. Lett. **107**, 188301 (2011).
[25] M. Trulsson, B. Andreotti and P. Claudin, Phys. Rev. Lett. **109**, 118305 (2012).
[26] M. Trulsson, M. Bouzid, J. Kurchan, E. Clément, P. Claudin and B. Andreotti, Europhys. Lett. **111**, 18001 (2015).
[27] M. Ouriemi, P. Aussillous and E. Guazzelli, J. Fluid Mech. **636**, 295-319 (2009).
[28] T. Revil-Baudard and J. Chauchat, J. Geophys. Res. Oceans **118**, 619-634 (2013).
[29] P. Aussillous, J. Chauchat, M. Pailha, M. Médale and E. Guazzelli, J. Fluid Mech. **736**, 594-615 (2013).
[30] F. Chiodi, P. Claudin and B. Andreotti, J. Fluid Mech. **755**, 561-581 (2014).
[31] E. Azéma and F. Radjaï, Phys. Rev. E **81**, 051304 (2010).
[32] E. Azéma and F. Radjaï, Phys. Rev. E **85**, 031303 (2012).
[33] C.S. Campbell, Phys. Fluids **23**, 013306 (2011).
[34] T. Börzsönyi, B. Szabó, G. Törös, S. Wegner, J. Török, E. Somfai, T. Bien and R. Stannarius, Phys. Rev. Lett. **108**, 228302 (2012).
[35] T. Börzsönyi, B. Szabó, S. Wegner, K. Harth, J. Török, E. Somfai, T. Bien and R. Stannarius, Phys. Rev. E **86**, 051304 (2012).
[36] S. Wegner, R. Stannarius, A. Boese, G. Rose, B. Szabó, E. Somfai, T. Börzsönyi, Soft Matter **10**, 5157-5167 (2014).
[37] G. Lu, J.R. Third, C.R. Müller, Chem. Eng. Sci. **127** 425 (2015).
[38] Y. Guo, C. Wassgren , W. Ketterhagen , B. Hancock , B. James and J. Curtis, J. Fluid Mech. **713**, 1 (2012).
[39] C. Vega, S. Lago, Comput. Chem. **18**, 55-59 (1994).
[40] I. P. Omelyan, Comput. Phys. **12**, 97-103 (1998).
[41] J. C. Tsai, G. A. Voth, J. P. Gollub, Phys. Rev. Lett. **91**, 064301 (2003).
[42] O. Pouliquen, M. Nicolas, P. D. Weidman, Phys. Rev. Lett. **79**, 3640-3643 (1997).
[43] A. Donev, *et.al.*, Science **303**, 990-993 (2004).

Velocity autocorrelation function in uniformly heated granular gas

Shikha Kumari[1,*] and *Syed Rashid* Ahmad[1,**]

[1] *Department of Physics, Jamia Millia Islamia, New Delhi 110025, India*

Abstract. In this paper we study aging of the velocity autocorrelation function of a uniformly heated granular gas using large scale event-driven molecular dynamics simulations in both 2 and 3 dimensional system. The system is heated by adding Gaussian white noise to each velocity component of all the particles. After a few collisions per particle, the system attains steady state. During early stages, the velocity autocorrelation function shows aging as there is explicit dependence of the function on τ_w but after steady state is reached, the velocity autocorrelation function become independent of τ_w and does not show any aging. Velocity correlations develop even after steady state is reached, these correlations are less pronounced in 3 dimensional system.

1 Introduction

A Granular material is a collection of mesoscopic particles called grains. These materials are abundant in the universe in different shapes with sizes ranging from powders to huge asteroids. Depending on the average kinetic energy of the individual grains, the system of granular matter can exhibit properties of solid, liquid or gas [1]. These materials are enriched with variety of phenomena when they are considered to be solid, liquid or gas [1, 2]. The most important feature of the granular matter is that the constituent grains lose energy upon collision i.e. collisions are inelastic, thus leading to dissipation of energy. This inelastic property of granular matter causes variety of interesting phenomena such as pattern formation, clustering, jamming, size separation, etc in the system [1–4]. Some examples of natural processes involving granular matter are formation of dunes and other patterns in deserts, earthquakes, landslides, erosion etc [2].
In this study, we are considering granular gas [5–7], i.e. a granular system in which grains move separately and are not in contact with each other. Technically, granular gas is a dilute system in which the mean free path of particles is much greater than average particle size. The grains interact when they collide or overlap and these interactions are dissipative because of which they lose energy. If no energy is injected into the system, the system continue to lose energy and settles in an inhomogeneous cooling state [7, 8] where particle-rich clusters emerge in the system and within the clusters particles move in parallel directions. For studies related to these phenomena see Ref. [3, 9–11]. If energy is injected into the system to compensate energy loss due to collisions, the system settles to a non-equilibrium steady state. Energy is injected into the system through standard geometries like horizontal or vertical vibrations [12, 13], rotation [14], shear [15], applying magnetic [16] or electrostatic [17] forces. For non-equilibrium systems, certain statistical quantities are history dependent i.e. they depend on both the time interval of measurement and the waiting time from where the measurement started. In case of granular gases, the velocity autocorrelation function exhibits this kind of explicit dependence on waiting time.
Earlier, a theoretical study of the velocity autocorrelation function of granular system was done in Ref.[18], Ahmad and Puri [8] have studied aging of the velocity autocorrelation function for a force free granular gas through numerical simulations. We aim at studying the velocity autocorrelation function for the forced case. In this work, we are considering uniformly heated granular gas. In previous work on heated systems, there have been studies on the velocity distributions in granular fluids [19], velocity correlations and distributions in granular gases [20, 21], velocity distributions in heated viscoelastic granular gas [22]. Till now no study has been done on the velocity autocorrelation function in heated systems. The velocity autocorrelation function is used to study history dependence in the system i.e. memory of previous behaviour. During each particle collision, local parallelization of particle velocities takes place i.e. correlations appear in velocity field. This correlated motion leads to formation of regions where particle density is more and also regions where density is less. The persistence of this domain growth is quantified by the velocity correlation and autocorrelation function. We study the behaviour of the velocity autocorrelation function in heated granular gas and compare the results with force free case. The system is subject to Gaussian white noise [19–21, 23, 24], where all particles are uniformly heated.
The paper is organised as follows. In Sec. 2, we briefly describe the system, i.e. we explain the equation of motion governing heated granular gas, how the particles veloci-

*e-mail: k.shikha.physics@gmail.com
**e-mail: srahmad@jmi.ac.in

ties update after interaction and what is the model used for heating the system. In Sec. 3, we report detailed simulation results from MD simulations and its discussion. Finally, in Sec. 4 we summarize and conclude the paper.

2 Description of System

The most important feature of granular matter is that the pairwise collisions are inelastic. The post-collision velocities are given in terms of pre-collision velocities as,

$$\mathbf{v}'_{i,j} = \mathbf{v}_{i,j} \mp \frac{1+\epsilon}{2} (\mathbf{v}_{i,j} \cdot \mathbf{e})\, \mathbf{e} \tag{1}$$

where $\mathbf{v}'$ is post-collision velocity and $\mathbf{v}$ is pre-collision velocity, $\mathbf{e}$ is unit vector joining center of ith and j^{th} particle at the time of collision and ϵ is called coefficient of restitution defined as,

$$\epsilon = \left| \frac{(\mathbf{v}'_{\mathbf{ij}} \cdot \mathbf{e})}{(\mathbf{v}_{\mathbf{ij}} \cdot \mathbf{e})} \right| \tag{2}$$

where v_{ij} and v'_{ij} are relative velocities of ith and j^{th} particles before and after collision respectively. The kinetic energy of the granular system is defined as $E = \frac{\langle v^2 \rangle}{d}$, where $\langle v^2 \rangle$ is the mean-squared velocity and d is the dimension of the system.

2.1 Uniformly Heated System

The stochastic equation of motion for granular system driven by a white noise thermostat is described as,

$$\frac{d\mathbf{v}_i}{dt} = \frac{\mathbf{F}_i^c}{m} + \frac{\mathbf{F}_i^t}{m} \tag{3}$$

where m is the mass of the particle, $\mathbf{F}_i^c$ is the force on the ith particle ($i = 1, 2, ...N$) due to pairwise collision and $\mathbf{F}_i^t$ is the external force which is considered as Gaussian white noise with zero mean and is uncorrelated for different particles i.e,

$$\begin{aligned} \langle\, F^t_{i,\alpha}(t)\, F^t_{j,\beta}(t')\,\rangle &= \xi_0^2\, \delta_{ij}\, \delta_{\alpha\beta}\, \delta(t - t') \\ \langle\, \mathbf{F}^t_i(t)\,\rangle &= 0 \end{aligned} \tag{4}$$

where $\alpha, \beta = \{x, y, z\}$, ξ_0 characterizes the strength of stochastic force, δ_{ij} and $\delta_{\alpha\beta}$ are Kronecker delta and $\delta(t-t')$ is the delta function.

2.2 The Velocity Autocorrelation Function

The velocity autocorrelation function is defined as,

$$A(\tau_w, \tau) = \frac{1}{N} \sum_{i=1}^{N} \langle \mathbf{v_i}(\tau_\mathbf{w}) \cdot \mathbf{v_i}(\tau) \rangle \tag{5}$$

It measures correlations in the velocity of a particle in time. It is used to study history dependence i.e. the memory of previous behaviour of particles in the system. The autocorrelation function is calculated at collision time τ with respect to reference or waiting time τ_w. For an equilibrium system autocorrelation function depends on time difference $\tau - \tau_w$, but for a non-equilibrium system it depends explicitly on both τ and τ_w rather than their difference. This dependence of autocorrelation function on τ_w is called aging or history dependency. A cooling granular gas is a non-equilibrium system and therefore the corresponding velocity autocorrelation function exhibits aging properties.

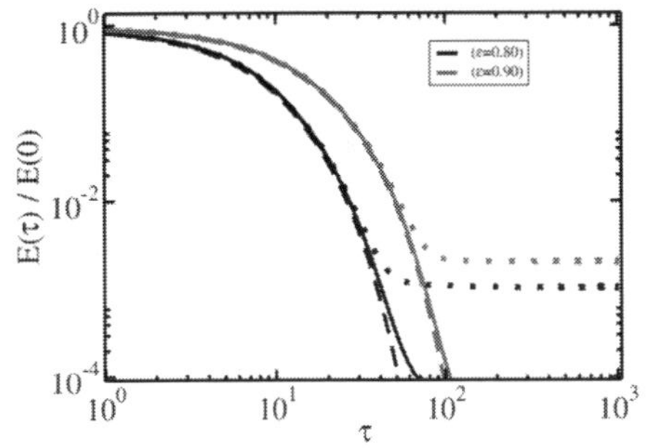

Figure 1. Time evolution of the normalised kinetic energy $E(\tau)/E(0)$ for force free and heated case for 2 dimensional system. Coefficient of restitution $\epsilon = 0.80$ and 0.90. Dotted lines represent simulation results for heated gas, solid lines represent simulation results for force free gas and dashed lines represent Haff's Law.

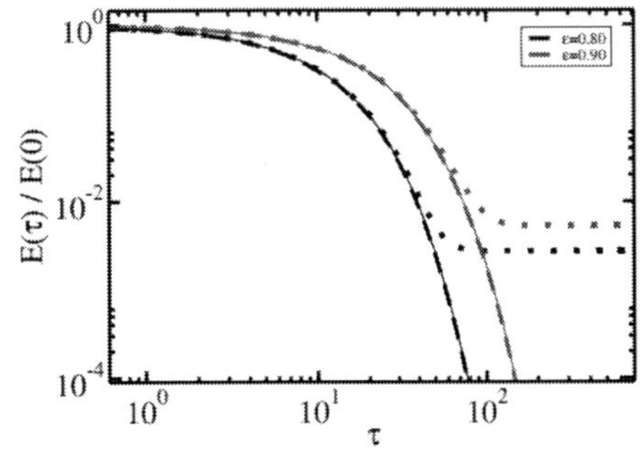

Figure 2. Time evolution of the normalised kinetic energy $E(\tau)/E(0)$ for heated case for 3 dimensional system. Coefficient of restitution $\epsilon = 0.80$ and 0.90. Dotted lines represent simulation results for heated gas, solid lines represent simulation results for force free gas and dashed lines represent Haff's Law.

2.3 Simulation Details

We use event driven molecular dynamics simulation [25] to study the heated granular system. The system consists of $N = 100000$ particles having unit mass and unit diameter. We have simulated these particles in both 2-dimensional and 3-dimensional box. Size of the box for both the cases is chosen such that the number density is 0.02 for both the cases and periodic boundary conditions are applied in all the directions. Since we are using event driven simulations the velocities of grains are updated according to Eq. (1), and during simulations we add heat

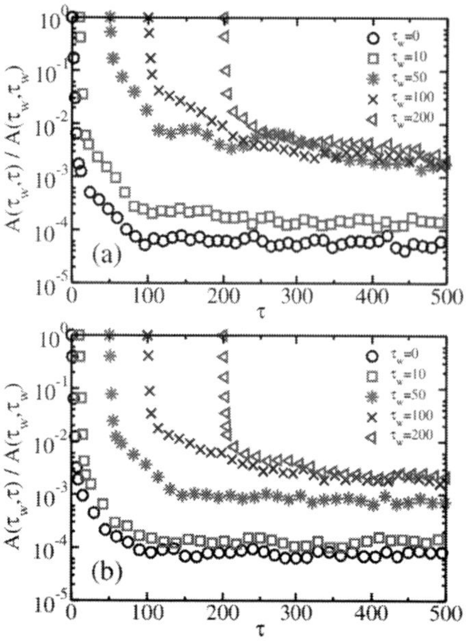

Figure 3. The velocity autocorrelation functions for 2 dimensional system with different reference time τ_w = $0, 10, 50, 100, 200$. Coefficient of restitution values taken are (a) $\epsilon = 0.80$ and (b) $\epsilon = 0.90$.

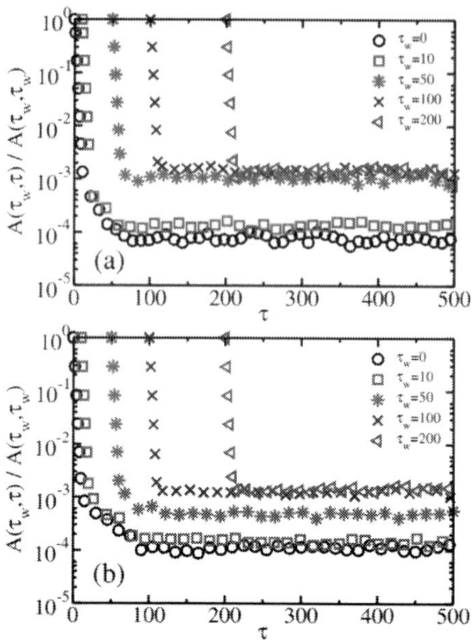

Figure 4. The velocity autocorrelation functions for 3 dimensional system with different reference time τ_w = $0, 10, 50, 100, 200$. Coefficient of restitution values taken are (a) $\epsilon = 0.80$ and (b) $\epsilon = 0.90$.

component to velocity of each particle after every time step dt using following equation,

$$v_i(t + dt) = v_i(t) + \sqrt{r}\sqrt{dt}\,\xi \tag{6}$$

where ξ is random variable which is uniformly distributed between $[-\frac{1}{2}, \frac{1}{2}]$, r is amplitude of noise. After adjusting the velocities, the system is shifted to center-of-mass frame,

$$\mathbf{v}_i = \mathbf{v}_i - \frac{1}{N}\sum_{i=1}^{N} \mathbf{v}_i \tag{7}$$

This ensures conservation of momentum and that the speed of the particles do not increase indefinitely. For convenience we have used a new time variable τ, which represents the average number of collisions a particle has suffered ($\tau = 1$ implies one collision per particle). In our simulations we have used $r = 0.1$ for 3-dimensional case and $r = 10$ for 2-dimensional case. A high value of the noise amplitude r is required in 2-dimensional case since, $r = 0.1$ was not sufficient to overcome dissipation in the system. The system is initialised by randomly placing particles in the box using random number generator. Then, velocities are assigned to each particle randomly in such a way that $\sum_i v_i = 0$. The kinetic energy E and the aging behaviour of heated system is studied for different values of coefficient of restitution ($\epsilon = 0.80, 0.85, 0.90$). For a particular value of ϵ the system is evolved till collision time $\tau = 1000$ and the velocity autocorrelation function is calculated according to Eq.(5) with respect to different reference times $\tau_w = 0, 10, 50, 100, 200$. All statistical quantities are averaged over 30 independent runs.

3 Results and Discussion

In Fig.1, time evolution of the normalised kinetic energy of granular gas is shown on log-log scale for a 2 dimensional system. For force free granular system, initially the energy of the system cools down but the density is uniform throughout. This stage is called Homogeneous Cooling State (HCS) [26]. But, this state is unstable and at later stages clusters develop and the system continues to lose energy in Inhomogeneous Cooling State (ICS) [7, 8]. For a heated granular system during early stages, for a particular value of coefficient of restitution for both force free and heated granular gas evolve in same fashion. But at later stages, the force free granular gas continues to cool, whereas heated granular gas settles to a steady state. During steady state, the kinetic energy of the system becomes independent of time. Further, it is clear that, as the value of ϵ is increased i.e. the system tends towards elastic limit, the value of steady state energy E_{ss} of the system increases. For $\epsilon = 0.80$, E_{ss} is approximately 1.03×10^{-3} and for $\epsilon = 0.90$, E_{ss} is approximately 2.02×10^{-3}. Also, the time at which the system reaches steady state also increases with increasing the value of ϵ. For $\epsilon = 0.80$, after approximately 50 collisions per particle have occurred steady state is reached and for $\epsilon = 0.90$, steady state is reached after approximately 100 collisions per particle have occurred. In Fig.2, time evolution of the normalised kinetic energy of granular gas is shown on log-log scale for 3 dimensional system. Similar results are observed for 3 dimensional system as well.

In Fig.3, we plot the normalised velocity autocorrelation function $\bar{A}(\tau_w, \tau) = A(\tau_w, \tau)/A(\tau_w, \tau_w)$ versus τ for $\epsilon = 0.80$ and $\epsilon = 0.90$ for 2-dimensional system on linear-log scale and in Fig.4 same is plotted for 3-dimensional system. A comparison study was done of these autocorrelation graphs with the normalised autocorrelation graphs obtained for force free granular gas Ref [8]. For force free gas, the decay of $\bar{A}(\tau_w, \tau)$ becomes slower for large values of waiting time which is referred to aging property of granular gas. Further, it is seen that the decay of $\bar{A}(\tau_w, \tau)$ is slower than exponential function in HCS which is due to correlations that build in the system prior to ICS. In the case of heated granular gas it is observed that at early stages when the system cools down like a force free gas, correlations build up in the system and $\bar{A}(\tau_w, \tau)$ decays slower than exponential. Also, there is an explicit dependence of $\bar{A}(\tau_w, \tau)$ on both τ and τ_w showing aging property. When steady state is reached, $\bar{A}(\tau_w, \tau)$ become independent of τ_w and does not show aging. However, the rate of decay of $\bar{A}(\tau_w, \tau)$ is slow showing presence of correlations in the system. On comparing Fig.3a and Fig.3b with Fig.4a and Fig.4b respectively it is observed that these velocity correlations are more pronounced in 2 dimensional system as compared to 3 dimensional system.

4 Conclusion

Aging behaviour of the velocity autocorrelation function is studied for uniformly heated granular gas. The results are compared with existing results for a force free granular gas. It is observed that initially $\bar{A}(\tau_w, \tau)$ depends on both τ and τ_w independently but after steady state is reached $\bar{A}(\tau_w, \tau)$ becomes independent of τ_w. Even in the steady state, the decay of $\bar{A}(\tau_w, \tau)$ is slower than exponential. This shows existence of velocity correlations in uniformly heated granular gas even after steady state is reached. These correlations are less pronounced in 3 dimensional system.

Acknowledgements

S.K. acknowledges the computational facilities provided for this work by Department of Physics, Jamia Millia Islamia, New Delhi, India. S.K. also acknowledges University Grants Commission, India for financial support.

References

[1] H. M. Jaeger, S. R. Nagel, and R. P. Behringer, Rev. Mod. Phys. **68**, 1259 (1996).
[2] I. S. Aranson and L. S. Tsimring, Rev. Mod. Phys. **78**, 641 (2006).
[3] A. Kudrolli, M. Wolpert, and J. P. Gollub, Phys. Rev. Lett. **78**, 1383 (1997).
[4] L. P. Kadanoff, Rev. Mod. Phys. **71**, 435 (1999).
[5] T. Pöschel and S. Luding, *Granular Gases, in: Lecture Notes in Physics Vol. 564* (Springer, Berlin, 2001).
[6] T. Pöschel and N. V. Brilliantov, *Granular Gas Dynamics, in: Lecture Notes in Physics Vol. 624* (Springer, Berlin, 2003).
[7] N. V. Brilliantov and T. Pöschel, *Kinetic Theory of Granular Gases* (Oxford University Press, Oxford, 2004).
[8] S. R. Ahmad and S. Puri, Phys. Rev. E **75**, 031302 (2007).
[9] I. Goldhirsch and G. Zanetti, Phys. Rev. Lett. **70**, 1619 (1993).
[10] S. Kumari and S. R. Ahmad, AIP Conference Proceedings **1728**, 020680 (2016).
[11] S. K. Das and S. Puri, EuroPhys. Lett. **61**, 749 (2003).
[12] R. D. Wildman and D. J. Parker, Phys. Rev. Lett. **88**, 064301 (2002).
[13] K. Feitosa and N. Menon, Phys. Rev. Lett. **88**, 198301 (2002).
[14] O. Zik, D. Levine, S. G. Lipson, S. Shtrikman and J. Stavans, Phys. Rev. Lett. **73**, 644 (1994).
[15] L. Bocquet, W. Losert, D. Schalk, T. C. Lubensky and J. P. Gollub , Phys. Rev. E **65**, 011307 (2001).
[16] A. Snezhko, I. S. Aranson and W. K. Kwok, Phys. Rev. Lett. **94**, 108002 (2005).
[17] I. S. Aranson and J. S. Olafsen, Phys. Rev. E **66**, 061302 (2002).
[18] E. Ben-Naim and P. L. Krapivsky, Phys. Rev. E **66**, 011309 (2002).
[19] T. P. C. van Noije and M. H. Ernst, Granular Matter **1**, 57 (1998).
[20] D. R. M. Williams and F. C. MacKintosh, Phys. Rev. E **54**, R9 (1996); D. R. M. Williams, Physica A **233**, 718 (1996).
[21] Sung Joon Moon, M. D. Shattuck, and J. B. Swift, Phys Rev. E **64**, 031303 (2001).
[22] Awadhesh Kumar Dubey, Anna Bodrova, Sanjay Puri and Nikolai Brilliantov, Phys Rev. E **87**, 062202 (2013).
[23] G. Peng and T. Ohta, Phys. Rev. E **58**, 4737 (1998).
[24] J. M. Montanero and A. Santos, Granular Matter **2**, 53 (2000).
[25] D. C. Rapaport, *The Art of Molecular Dynamics Simulation, Second edition* (Cambridge University Press, Cambridge, 2005).
[26] P. K. Haff, J. Fluid Mech. **134**, 401 (1983).

Multiscale modeling of transport of grains through granular assemblies

Ignacio G Tejada[1,*], *Luc* Sibille[2], *Bruno* Chareyre[2], *Chuheng* Zhong[3], and *Didier* Marot[3]

[1] *Universidad Politécnica de Madrid, ETSI de Caminos, Canales y Puertos, Madrid (Spain)*
[2] *Université Grenoble Alpes, CNRS, Lab. 3SR, Grenoble (France)*
[3] *Université de Nantes, CNRS, Institut GeM, Saint-Nazaire (France)*

Abstract. We investigate the transport of moderately large passive particles through granular assemblies caused by seeping flows. This process can only be described by highly nonlinear continuum models, since the local permeability, the advection and dispersion mechanisms are strongly determined by the concentration of transported particles. Particles may sometimes get temporally trapped and thus proper kinetic mass transfer models are required. The mass transfer depends on the complexity of the porous medium, the kind of interaction forces and the concentration of transported particles. We study these two issues by means of numerical and laboratory experiments. In the laboratory we use an oedo-permeameter to force sand grains to move through a gravel bed under conditions of constant hydraulic pressure drop. To understand the process, numerical experiments were performed to approach particle transport at the grain scale with a fully coupled method. The DEM-PFV combines the discrete element method with a pore scale finite volume formulation to solve the interstitial fluid flow and particle transport problems. These experiments help us to set up a continuum transport model that can be used in a boundary value problem.

1 Introduction

Particle transport in porous media is a widely studied problem with applications in many different contexts [1]. Even when the macroscopic conditions for the flow and transport problem are homogeneous, preferential paths for particle and fluid flows as well as localized filtration regions are often observed [2, 6]. We investigate how seeping flows convey passive particles through granular assemblies made of coarser particles (size ratio, transported to coarse particles, around 1/10). In such circumstances there is a significant relative motion between the transported particles and the carrier fluid, as shown by their average velocities or tortuosities (the length of a path connecting two points divided by the distance between them).

1.1 Continuum transport model

Under a continuum approach [3], the system can be viewed as a three-phase medium with a representative volume element V consisting of (i) a solid fraction of coarse and fixed particles (the coarse skeleton) V_c, (ii) a set of solid fine particles not attached to the coarse skeleton V_p and (iii) a fluid that fills the space left by solid particles V_w. The porosity of the clean coarse medium (i.e. without fine particles in the voids) is $\Phi_0 = (V - V_c)/V$ while the actual porosity is $\Phi = V_w/V$. The concentration of transported particles is defined as $C = V_p/V$, thus $\Phi = \Phi_0 - C$. In the case of moderately large particles, the particle flux, $\mathbf{q_p}$, is much smaller than the water flux, $\mathbf{q}_w = \Phi\mathbf{v_w}$ (where $\mathbf{v_w}$ is the velocity of water), so that the flow problem can be separately solved:

$$\nabla.\mathbf{q}_w = 0 . \quad (1)$$

Darcy's law states $\mathbf{q}_w = \frac{1}{\mu}\boldsymbol{\kappa}\gamma_w\nabla(h)$, where μ the viscosity of the fluid, $\boldsymbol{\kappa}$ the intrinsic permeability and $h = p/\gamma_w + z$ the hydraulic head (with z the elevation, γ_w the specific weight of water and p the fluid pressure). Some theoretical approaches aim at explaining the influence of the features of the coarse skeleton and the concentration on $\boldsymbol{\kappa}$. For example, in the Kozeny-Carman's [4, 5] model,

$$\kappa \propto \frac{1}{\mathcal{T}}\frac{\Phi^3}{(1-\Phi)^2} , \quad (2)$$

when the porous medium consists of a collection of curving pipes through which the water flows according to Poiseuille's law. The average tortuosity of the pipes is $\mathcal{T}$. We refer to $\boldsymbol{\kappa}_0$ as the permeability of the clean structure, which can be expressed in terms of the average tortuosity of the clean structure, $\mathcal{T}_0$. The inclusion of fine particles changes $\mathcal{T}$ and Φ and makes $\boldsymbol{\kappa}$ dependent on C.

Regarding the transport of passive particles, the advection - (normal) dispersion paradigm states that the evolution of C is given by:

$$\frac{\partial C}{\partial t} + \boldsymbol{\nabla}\cdot(\mathbf{v}_p C) - \boldsymbol{\nabla}\cdot(\boldsymbol{D}\boldsymbol{\nabla}C) = 0 , \quad (3)$$

where $\mathbf{v}_p$ is the advective velocity of the particles and $\boldsymbol{D}$ is the hydrodynamic dispersion tensor. The advective and

*e-mail: ignacio.gtejada@upm.es

dispersive fluxes of particles are, respectively $\mathbf{q}_{p,a} = \mathbf{v}_p C$ and $\mathbf{q}_{p,d} = -D\nabla C$ (Fick's law). The advective velocity is parallel to that of the fluid. The relative advective velocity of the particles is defined as $\lambda = \mathbf{v}_p/\mathbf{v}_w$ and depends on the coarse skeleton, the transported particles and C (as observed in interparticle percolation by gravity [9, 10]). The hydrodynamic dispersion depends on $\mathbf{v}_w$ (e.g., [7]) and can be expressed as $\boldsymbol{D} = |\mathbf{v}_w|\delta$, where δ is the dispersivity, to emphasize the fact that the dispersion is scaled by the water velocity.

Eq. 3 is valid when the number of transported particles is constant (they do not join the coarse skeleton and no new particles are eroded from the later). However, sometimes particles can be temporarily immobilized. For example, when they get trapped due to sticking at surfaces, crevices, caverns, etc. [8], or when they clog constrictions (individually or collectively). Then, kinetic mass transfer laws must be included in the right-hand side of Eq. 3. These laws depend on the complexity of the porous medium and interaction forces, but also often do on C. For example, in this case, the intermittent, purely mechanical, formation of blockages in the constrictions due to physical crowding can be properly reproduced by 1st-order kinetic mass transfer laws, and Eq. 3 becomes:

$$\frac{\partial C_f}{\partial t} + \nabla \cdot (\mathbf{v}_p C_f) - \nabla \cdot (\boldsymbol{D}\nabla C_f) = -\frac{C_f}{\tau_f} + \frac{C_t}{\tau_t}, \quad (4)$$

$$\frac{\partial C_t}{\partial t} = +\frac{C_f}{\tau_f} - \frac{C_t}{\tau_t}$$

where $C_{t(x,t)}$ and $C_{f(x,t)}$ are, respectively, the concentrations of trapped and free particles ($C_{(x,t)} = C_{t(x,t)} + C_{f(x,t)}$), τ_t is the mean lifetime of trapping events and τ_f is that of travelling periods. These mean lifetimes may change considerably with C. As κ (and hence $\mathbf{v}_w$), λ, τ_f and τ_s, and, to a lower extent, δ, are strongly determined by C, the transport of moderately large particles can only be described by a highly nonlinear continuum model. A micromechanical approach may help to get some insight on the effect of C on these parameters.

1.2 Micro-mechanical transport model

Seeping water conveys fine particles through the coarse skeleton by means of viscous and hydraulic pressure forces. The particles interact mechanically with each other and with the coarse skeleton. Under a micromechanical approach, their individual paths are considered.

From grain level considerations, Eqs. 1 and 4 can be derived. The computation of the $\mathbf{q}_w$ over a representative volume element is straightforward, so that if the hydraulic pressure drop is known, then κ is known for each given value of C. Regarding $\mathbf{q}_p$, individual particles can be tracked to generate an ensemble of trajectories over large times and distances. The center of mass of the particles is expected to move linearly with time (t) at the advective velocity: $\langle \mathbf{x} \rangle = \mathbf{v}_p t$.

The Fick's law requires a linear growth with time of the centered mean square displacement of transported particles ($\sigma^2 = \left\langle |\mathbf{x} - \mathbf{v}_p t|^2 \right\rangle \propto t$). In 2D, $\sigma_x^2 = 2D_x t$. However on very short time scales (e.g. below the typical time for interpore transit) the motion of particles involves time correlations and Fick's law does not applies. On timescales comparable to the lifetime of blockages or to the typical time between consecutive blockages [11], the total C can be splitted into trapped C_t and free fractions C_f and kinetic mass transfer laws can be used (as in Eq. 4). 1st-order kinetic mass transfer laws are justified by exponential statistical distributions of the duration and frequency of trapping events [11]. For a given C the probability of a particle to get trapped depends on the number of constrictions that it crosses per unit of time, so that τ_f should decrease almost linearly with the particle velocity. In other words, $\tau_f|\mathbf{v}_p| = l_f$, where l_f is the filtration length. In contrast, the duration of trapping periods, represented by τ_t, should be related to the hydraulic pressure gradient, so that $\tau_t = f(C)$. As a first approximation, $\tau_t|\mathbf{v}_p| = l_t$, where l_t is the length that the particle would have moved if it had not been trapped.

2 Methodology

We performed a laboratory experiment and many simulations of particle transport. The laboratory experiment allowed us to understand the transport of moderately large particles caused by water seepage whereas the numerical simulations helped us to set up a continuum transport model. At this stage of the research the simulations did not aim at reproducing the laboratory experiments.

The experimental device used at the laboratory was a large oedo-permeameter (see a detailed description in [12]). This device comprises of a cylindrical cell made of a rigid transparent acrylic tube of diameter 0.279 m. The granular specimen, depicted in Fig. 1, consisted of a binary mixture of sand and gravel (1:1) placed over a 0.200 m high bed of gravel. We used Fontainebleau sand (median diameter, $D_{50} = 0.207 \cdot 10^{-3}$ m, i.e. 50 % of the grains, by mass, is finer than $0.207 \cdot 10^{-3}$ m) and a gravel from Sablière Palvadeau ($D_{50} = 3.309 \cdot 10^{-3}$ m). The binary mixture was prepared with a water content of 3 % (by mass) and placed to fill the gap between the acrylic tube and an impervious internal plastic bucket with an average diameter of 0.209 m. The bucket left a 0.150 m high ring (see Fig. 1). This configuration allowed us to observe how the sand spread in the gravel bed as the water flowed. The whole sample was saturated first with CO_2 and then with water and subjected to a downward water seepage caused by a fixed hydraulic pressure drop. After some time, the water flow was stopped and the cell was gently drained. Some samples were collected at different sites and their sand content was measured (the ratio of the dry weight of sand to the dry weight of the sample). The fact that the specimen remained partially saturated, helped to keep sand particles where they were at the end of the seepage stage.

The numerical method we use is the DEM-PFV, which applies for incompressible and Newtonian fluids at low Reynolds numbers. The discrete element method (DEM) [16] is combined with a pore-scale finite volume

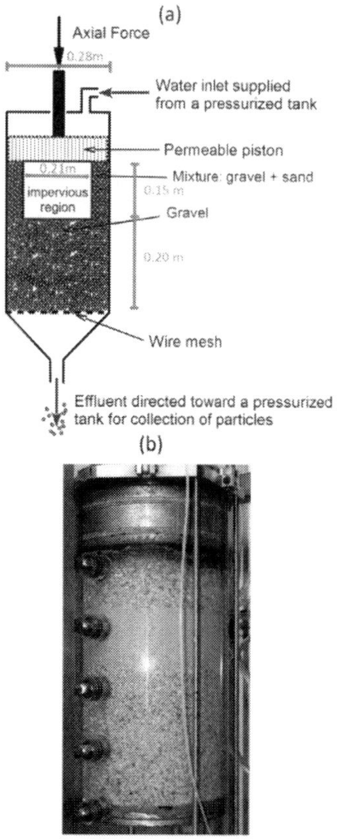

Figure 1. (a) Schematic and (b) picture of the laboratory test performed with an oedo-permeameter.

formulation (PFV) [13–15]. We use linear elastic contact laws between the frictionless particles. The finite volumes are geometrically defined after a tessellation of the granular assembly[1]: finite volumes are the tetrahedra whose apexes coincide with particle centers. When solid particles move, tetrahedra change, and the variations of volume must be equal to the net water fluxes over their four faces (mass balance and incompressibility conditions). Inter-pore fluxes are supposed to be proportional to the difference of hydraulic pressure between consecutive tetrahedra, the distance of their centers of mass and the hydraulic section of the shared face. This is congruent with Stokes flows [13]. The forces that the fluid flow exerts on solid particles are computed after the integration of fluid pressure and viscous stresses over the wet surfaces of solid particles and the hydraulic sections of tetrahedron faces. This ensures a full coupling.

In the simulations we used random, crystalline and quasi-crystalline[2] fixed granular assemblies of monodispersed spheres of size $d_c = 5 \cdot 10^{-3}$ m. These structures are

[1] Transported particles are also considered for the tessellation.

[2] Randomly generated from the corresponding crystalline structures by enlarging unit cells and slightly displacing particles from crystalline positions.

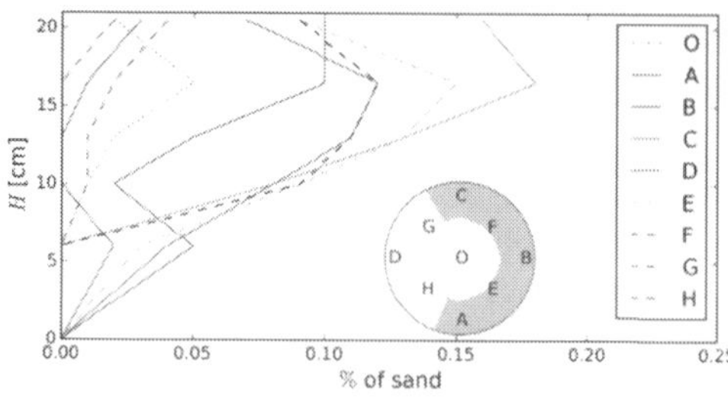

Figure 2. Vertical profiles of sand content after the water seepage. Each curve represents a vertical profile starting from the bottom specimen at a given point located in a cross section as depicted in the inset (gray shaded area approximately corresponds to the zone of localized erosion in the upper ring mixture).

included in parallelepiped periodic simulation boxes. Fine particles of size $d_f = d_c/8$ are randomly introduced within the voids (all over the simulation box) up to an average concentration C . The size ratio ensures that fine particles cannot get trapped alone but allows them to collectively form clogging arches in the constrictions that may cause larger blockages. For any coarse skeleton, blockages are more likely at high C. The simulations provide us a statistical sample of particle trajectories. The longitudinal velocity, the dispersion (in the direction of $-\nabla h$), τ_f and τ_t can be measured from the statistical sample.

3 Results and discussion

In the laboratory experiment, the ring containing the mixture was eroded in a few seconds under the action of the water seepage. The process was localized. The main water flow occurred through a way opened in the ring. The sand particles that had been eroded from the ring moved down and settled. No significant amount of sand was collected at the outlet of the cell after the experiment and the sand content at different sites within the specimen is shown in Fig. 2. Although the corresponding boundary value problem was supposed to be axisymmetric, the transport was not symmetric at all. This instability unveils a highly non-linear behavior: the water flows where there is less sand and the particle transport is stopped when water finds continuous paths to flow and deposits the sand aside. This emphasizes the importance of the relationship between κ and C.

The numerical experiments show (Fig. 3a) how κ changes with C. These values have been fitted with Kozeny-Carman models (Eq. 2). For each coarse skeleton, an expression of the form $\mathcal{T} = a\sqrt{C}$, for $C > C_0$, and $\mathcal{T} = \mathcal{T}_0$ otherwise, was used. Each coarse skeleton had a porosity and tortuosity that determined its permeability. In all the cases, v_p decreased with C, whereas λ increased with C (Fig. 3b). This is because with a higher number of fine particles makes them to move more evenly with the fluid. D is scaled by v_w, but δ remains considerably constant at different C levels (Fig. 3c) (i.e. the dispersivity mainly varies from one coarse skeleton to another).

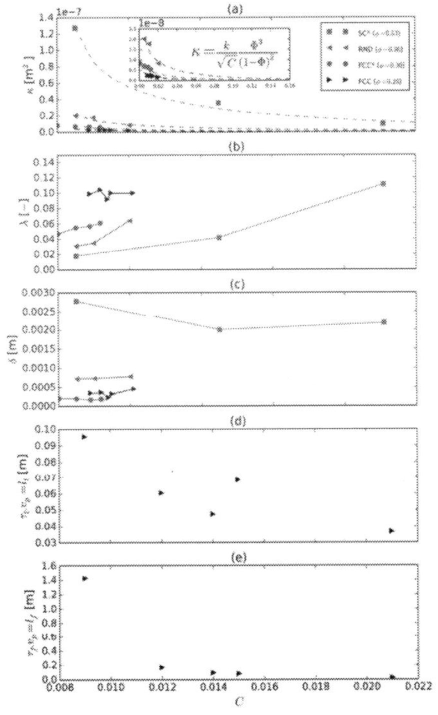

Figure 3. Effect of C on the parameters of Eq. 4 for different coarse skeletons (crystalline: face-centered cubic FCC, quasi-crystalline: face-centered cubic FCC* and simple cubic SC*, and random RND). (a) κ fitted with Kozeny-Carman like expression. (b) Relative advective velocity. (c) Dispersivity. (d) Waiting and (e) filtration lengths (only in FCC structures since trapping events are impossible or quite rare in other structures).

Finally, with respect to the formation of collective clogging arches and blockages in the constrictions (the only trapping mechanism that our model is able to reproduce; only for the FCC structure), at low C, the expected walking length can be very long while it decreases greatly with C (Fig. 3e). The lifetime of trapping events is less variable (Fig. 3d).

4 Concluding remarks

Laboratory and numerical micromechanical experiments show that the transport of moderately large passive particles caused by seeping flows requires nonlinear transport models which may lead to preferential paths, where water flows and particles are deposited aside. In these models the macroscopic permeability is strongly determined by the presence of particles. In consequence, the concentration scales advection, dispersion, filtration and erosion processes. Furthermore, the relative advective velocity of the particles, the filtration and erosion times and the dispersivity also depend on the concentration. The next challenge is to embed such a macroscopic model, including all these dependencies, in a continuum numerical method to tackle boundary value problems.

Acknowledgements

This work has been partially supported by the LabEx Tec 21 (Investissements d'Avenir - grant agreement n ANR-11-LABX-0030).

References

[1] L.M. McDowell-Boyer, J.R. Hunt, N. Sitar, Water Resources Research **22**, 1901 (1986)

[2] S.E. Silliman, E.S. Simpson, Water Resources Research **23**, 1667 (1987)

[3] E. Papamichos, I. Vardoulakis, Computers and Geotechnics **32**, 47 (2005)

[4] J. Kozeny, Sitzungs-ber. Akad. Wiss. Wien **136**, 271 (1927)

[5] P. Carman, Trans. Inst. Chem.Eng. **15**, 150 (1937)

[6] G. Gao, H. Zhan, S. Feng, G. Huang, X. Mao, Journal of Hydrology **377**, 391 (2009)

[7] J. Bear, Y. Bachmat, *Fundamentals of Transport Phenomena in Porous Media* (Springer Netherlands, Dordrecht, 1984), pp. 3–61

[8] J.P. Herzig, D.M. Leclerc, P.L. Goff, Industrial & Engineering Chemistry **62**, 8 (1970)

[9] I. Ippolito, L. Samson, S. Bo Journal E **3**, 227 (2000)

[10] F. Lominé, L. Oger, Phys. Rev. E **79**, 051307 (2009)

[11] I.G. Tejada, L. Sibille, B. Chareyre, EPL (Europhysics Letters) **115**, 54005 (2016)

[12] Y. Sail, D. Marot, L. Sibille, A. Alexis, European Journal of Environmental and Civil Engineering **15**, 799 (2011)

[13] B. Chareyre, A. Cortis, E. Catalano, E. Barthélemy, Transport in Porous Media **92**, 473 (2012)

[14] E. Catalano, B. Chareyre, E. Barthélémy, International Journal for Numerical and Analytical Methods in Geomechanics **38**, 51 (2014)

[15] V. Šmilauer et al., *Yade Documentation 2nd ed* (The Yade Project, 2015), http://yade-dem.org/doc/

[16] P.A. Cundall, O.D.L. Strack, Géotechnique **29**, 47 (1979)

Localization in an anisotropic planar aggregate of spheres

Luigi La Ragione[1,⋆], *Vincent C.* Prantil[2], and *James T.* Jenkins[3]

[1] *Dipartimento di Scienze dell'Ingegneria Civile e dell'Architettura, Politecnico di Bari*
[2] *Department of Mechanical Engineering, Milwaukee School of Engineering 53202 Milwaukee, WI, USA*
[3] *School of Civil and Environmental Engineering, Cornell University 14853 Ithaca, NY, USA*

Abstract. We present a micro-mechanical model that is able to predict localization in a sheared planar aggregate of spheres. We assume a non-linear contact law between interacting particles that deform differently from an affine deformation. Equilibrium determines this deviation. The aggregate is isotropically compressed and then sheared so anisotropy develops because of contacts deletion and a non-linear contact law. Because of anisotropy and fluctuations in particles deformation, the resulting macroscopic stiffness tensor, which relates increments in the average stress with increments in the average strain, is characterized by a lack of major symmetry, $\mathcal{A}_{ijkl} \neq \mathcal{A}_{klij}$. At given shear strain and coordination number it is possible to detemine a plane over which discontinuity in strain occurs; this is identified as localization.

1 Introduction

The authors have recently proposed a micro-mechanical model that highlights a possible mechanism for localization in granular materials [1]. Following previous work [2–4], we identify localization with the possibility of a discontinuity in a component of the incremental strain at a predicted value of the shear strain and a predicted orientation relative to the axis of greatest compression. This is equivalent to a solution for the zero eigenvalue of the acoustic tensor associated with the macroscopic stiffness tensor of the aggregate [5]. In general this condition is possible when the stiffness tensor, that relates increments in the average stress with increments in the average strain, exhibits a lack of major symmetry, $\mathcal{A}_{ijkl} \neq \mathcal{A}_{klij}$. This seems a natural state for a granular material and we recover this condition without any hypothesis a priori [6, 7]. We simply require that particles should move according to an average strain and a fluctuation that satisfy equilibrium, with the simultaneous presence of anisotropy. Then a macroscopic stiffness tensor results with no major symmetry. This occurs when the incremental response of the aggregate, at relatively high shear stress, is essentially frictionless and the incremental contribution associated with the tangential force is almost negligible [8].

2 Theory

In this paper, we consider a random aggregate of identical particles with diameter D that has been compressed and then sheared. We assume a non-linear contact law with a normal force that follows the Hertz relation between the overlapping and the contact, and a tangential force that is bilinear with an elastic resistance followed by frictional sliding. The aggregate is axially compressed at constant pressure and, as the shear strain increases, anisotropy develops because of contact deletion and because the normal component of the contact force varies with the orientation of the pair with respect to the direction of greatest compression. The deviatoric stress, q, also increases and both numerical simulations [8] and theoretical models [9] show that it is possible to distinguish a contribution to the deviatoric stress associated with the normal component of the contact force, q^N, from that associated with the tangential force, q^T. The result of both numerical simulations and theory is that, close to the stress peak, the deviatoric stress is prescribed almost entirely by q^N (e.g. [8], fig. 4). Therefore, in the regime of deformation that preceeds the peak, we assume the aggregate to be in an anisotropic stressed state, incrementally frictionless (because q^T does not changes) under a given coordination number, $\hat{k}$ (the average number of contacts per particle).

When an average deformation is applied over a random aggregate, a particle's motion needs to deviate from the average in order to achieve equilibrium. The result is a rather complex system in which particle's motion is described by an average and a fluctuation. Here we work in a simpler situation. We consider a typical pair of spheres A and B. Following [10, 11], the increment in the relative displacement, $\dot{\mathbf{u}}^{(BA)}$, between contacting spheres A and B is assumed to be the sum of the average and a fluctuation:

$$\dot{u}_i^{(BA)} = \dot{E}_{ij} d_j^{(BA)} + \dot{\Delta}_i^{(BA)}, \tag{1}$$

where $\dot{\mathbf{E}}$ is the increment in average strain and $\dot{\mathbf{\Delta}}^{(BA)}$ is the increment in the difference of the fluctuations in displacements. Unlike the pair $A-B$, the neighbor particles in contact with the pair $A-B$, say n, move according to the

⋆e-mail: luigi.laragione@poliba.it

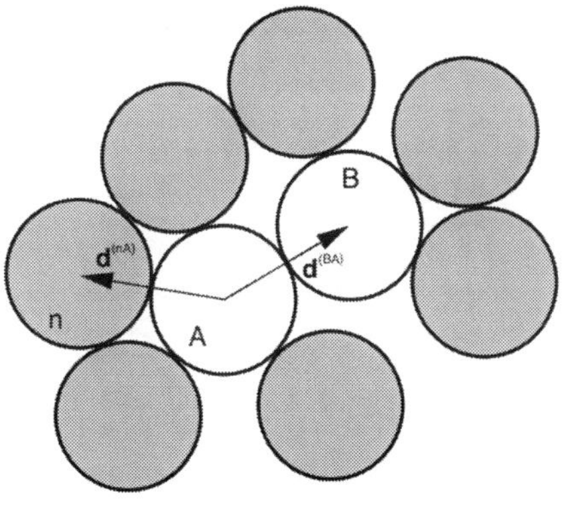

Figure 1. Pair $A - B$ and the neighbor particles n.

average strain (see Fig. 1):

$$\dot{u}_i^{(nA)} = \dot{E}_{ij} d_j^{(nA)} + \frac{1}{2}\dot{\Delta}_i^{(BA)} - \frac{1}{2}\dot{\Sigma}_i^{(BA)}, \qquad (2)$$

where $\dot{\Sigma}_j^{(BA)}$ is the increment in the sum of the fluctuations in displacements. Because the aggregate is seen incrementally frictionless, the incremental contact force between the two contacting spheres, A and B, is along their line of centers:

$$\dot{F}_i^{(BA)} = K_N^{(BA)} \hat{d}_i^{(BA)} \hat{d}_j^{(BA)} \dot{u}_j^{(BA)}, \qquad (3)$$

in which

$$K_N^{(BA)} = \frac{GD^{1/2}}{(1-\nu)} \left(\delta^{(BA)}\right)^{1/2}, \qquad (4)$$

where $\delta^{(BA)}$ is the inter-particle overlap, G and ν are, respectively, the shear modulus and the Poisson ratio of the material of the spheres, and $\hat{\mathbf{d}}^{(BA)} = (\sin\theta, \cos\theta)$ is the unit vector from the center of A to the center of B. We assume that the normal component of the contact displacement, $\delta^{(BA)}$, present in the contact stiffness, Eq. (4), depends only on the average strain [12] :

$$\delta^{(BA)} = -DE_{ij} \hat{d}_i^{(BA)} \hat{d}_j^{(BA)}. \qquad (5)$$

In the assumed biaxial compression, the area strain λ (taken positive for a decrease in area) and the shear strain γ are, respectively:

$$\lambda = -(e_{11} + e_{22}) \qquad (6)$$

and

$$\gamma = -\frac{1}{2}(e_{22} - e_{11}). \qquad (7)$$

Then, according to Eq. (5), the normal component of the contact displacement can be written as

$$\delta^{(BA)} = \frac{D}{2}\left(\lambda - 2\gamma + 4\gamma\cos^2\theta\right). \qquad (8)$$

2.1 Equilibrium

At given stressed state, the equations of force equilibrium for sphere A are [10]

$$\frac{1}{2}A_{ij}^{(BA)}\left(\dot{\Delta}_j^{(BA)} - \dot{\Sigma}_j^{(BA)}\right) + \frac{1}{2}K_N^{(BA)}\hat{d}_i^{(BA)}\hat{d}_j^{(BA)}\left(\dot{\Delta}_j^{(BA)} + \dot{\Sigma}_j^{(BA)}\right) = -DJ_{ijk}^{(BA)}\dot{E}_{jk}$$

with

$$A_{ij}^{(BA)} \equiv \sum_{n=1}^{N^{(A)}} K_N^{(nA)} \hat{d}_i^{(nA)} \hat{d}_j^{(nA)}, \qquad (9)$$

where $N^{(A)}$ is the number of spheres in contact with A, and

$$J_{ijk}^{(BA)} \equiv \sum_{n=1}^{N^{(A)}} K_N^{(nA)} \hat{d}_i^{(nA)} \hat{d}_j^{(nA)} \hat{d}_k^{(nA)}. \qquad (10)$$

We employ $\mathbf{d}^{(AB)} = -\mathbf{d}^{(BA)}$ and interchange A and B to obtain the force equilibrium equations for sphere B.

The solutions of these local equilibrium equations, up to an error of $1/\hat{k}$, are $\Delta^{(BA)}$ and $\Sigma^{(BA)}$. The former is our interest,

$$\dot{\Delta}_i^{(BA)} = -D\left[\left(A_{ij}^{(BA)}\right)^{-1} J_{jkm}^{(BA)} - \left(A_{ij}^{(AB)}\right)^{-1} J_{jkm}^{(AB)}\right]\dot{E}_{km}. \quad (11)$$

The local solutions depend upon the local neighborhood of each pair. Also, both $\mathbf{J}$ and $\mathbf{A}$ depend on the contact stiffness K_N through Eqs. (4) and (8).

Tensors $\mathbf{A}^{(BA)}$, $\mathbf{A}^{(AB)}$, $\mathbf{J}^{(BA)}$ and $\mathbf{J}^{(AB)}$ are based on the existence of the contact between A and B, and, as a consequence, they depend on $\mathbf{d}^{(AB)}$ or $\mathbf{d}^{(BA)}$. Average values of these sums and differences of the fluctuations result from averaging over the neighborhoods of pairs with nearly the same orientation. Because the averages depend upon the existence of a contact, we refer to them as conditional averages. We denote them with an over-bar and we have

$$\overline{\dot{\Delta}_j^{(BA)}} = -2d\overline{\left(A_{ji}^{(BA)}\right)^{-1} J_{imn}^{(BA)}}\dot{E}_{mn}, \qquad (12)$$

$$\overline{\dot{\Sigma}_j^{(BA)}} = 0. \qquad (13)$$

Again following [10, 11]

$$\overline{J_{ijk}^{(BA)}} \equiv K_N[\beta_1 \hat{d}_i^{(BA)} \hat{d}_j^{(BA)} \hat{d}_k^{(BA)} + \beta_2\left(\delta_{ik}\hat{d}_j^{(BA)} + \delta_{ij}\hat{d}_k^{(BA)} + \delta_{kj}\hat{d}_i^{(BA)}\right)], \qquad (14)$$

$$\overline{A_{ij}^{(BA)}} = \beta_3 K_N \delta_{ij} \qquad (15)$$

with

$$\beta_1 = 1 + \frac{2}{3}g - \rho\frac{2}{3};$$

$$\beta_2 = \rho\left(\frac{2}{3} - \frac{\sqrt{3}}{4}\right) - \frac{2}{3}g$$

with

$$\rho = \frac{3(k-2)}{2\pi} \text{ and } g = \frac{k}{2\pi},$$

while

$$\beta_3 = \frac{1}{3} + \rho\left(\frac{\sqrt{3}}{12} + \frac{\pi}{6}\right) + g\frac{\pi}{2}.$$

Making the simplifying assumption that the average of a product is the product of the averages, the average of the difference in fluctuations in translation is given by

$$\overline{\dot{\Delta}_i^{(BA)}} = DR_{ijk}\dot{E}_{jk}, \tag{16}$$

with

$$R_{ijk} = \alpha_1 \hat{d}_i^{(BA)}\hat{d}_j^{(BA)}\hat{d}_k^{(BA)}\dot{E}_{jk} + \alpha_2\left(\delta_{ij}\hat{d}_k^{(BA)} + \delta_{ik}\hat{d}_j^{(BA)} + \delta_{kj}\hat{d}_i^{(BA)}\right)\dot{E}_{jk}$$

where

$$\alpha_1 = -2\beta_3^{-1}\beta_1 \text{ and } \alpha_2 = -2\beta_3^{-1}\beta_2.$$

With Eq. (16), Eq. (1) can be written as

$$\overline{\dot{u}_i^{(BA)}} = D(1+2\alpha_2)\dot{E}_{ij}\hat{d}_j^{(BA)} + D\left(\alpha_1\hat{d}_j^{(BA)}\dot{E}_{jk}\hat{d}_k^{(BA)} + \alpha_2\dot{E}_{kk}\right)\hat{d}_i^{(BA)}. \tag{17}$$

3 Stiffness tensor

The incremental average stress is

$$\dot{\sigma}_{ij} = \frac{nD\hat{k}}{\pi}\int_0^{\theta_1} \dot{F}_i\hat{d}_j d\theta, \tag{18}$$

where the integration is limited to contacting pairs, the distribution function is assumed isotropic, $\hat{k}/(2\pi)$ and n is the number of particles per unit area. For symmetry condition we have restricted our attention to the range $0 \leq \theta \leq \theta_1$, with $\theta_1 \leq \pi/2$. The upper limit in the integral refers to the deletion angle, θ_1, the angle associated with $\delta^{(BA)} = 0$,

$$\theta_1 = \cos^{-1}\sqrt{\frac{2\gamma - \lambda}{4\gamma}}. \tag{19}$$

That is, the deletion angle is the orientation of the normal component of the contact force that relaxes to zero which corresponds to the vanishing of the contact stiffness.

Under isotropic pressure, fluctuations play a minor role [10]; therefore the condition of constant pressure is developed with the simple average strain hypothesis and we obtain for $\dot{p} = 0$ the following expression

$$\int_0^{\theta_1}\left(\lambda - 2\gamma + 4\gamma\cos^2\theta\right)^{1/2}\left(\dot{\lambda} - 2\dot{\gamma} + 4\dot{\gamma}\cos^2\theta\right)d\theta = 0. \tag{20}$$

We integrate with Simpson's rule to derive a relation between the area strain λ and the shear strain γ.

From Eq. (18), with the incremental contact force given by Eq. (3) and the incremental displacement given by eq. (17), we obtain

$$\begin{aligned}\dot{\sigma}_{ij} &= \omega(1+\alpha_1+2\alpha_2) \qquad (21)\\ &\times \int_0^{\theta_1}\left(\lambda - 2\gamma + 4\gamma\cos^2\theta\right)^{1/2}\hat{d}_k\hat{d}_m\hat{d}_i\hat{d}_j\dot{E}_{km}d\theta\\ &+ \omega\alpha_2\int_0^{\theta_1}\left(\lambda - 2\gamma + 4\gamma\cos^2\theta\right)^{1/2}\hat{d}_i\hat{d}_j\delta_{km}\dot{E}_{km}d\theta.\end{aligned}$$

where $\omega = nD^3\hat{k}G/[(1-\nu)\pi\sqrt{2}]$. Anisotropy is present in the stiffness and in the limit of integration where the integrals are extened to θ_1 which varies as the shear strain increases, Eq. (19). In a more compact form we have

$$\dot{\sigma}_{ij} = \mathcal{A}_{ijkm}\dot{E}_{km}, \tag{22}$$

where, with **h** the vertical axes of compression,

$$\begin{aligned}\mathcal{A}_{ijkm} &= \eta_1 h_m h_k h_j h_i + \eta_2\delta_{ij}\delta_{km} + \eta_3\left(\delta_{ik}\delta_{jm} + \delta_{im}\delta_{jk}\right)\\ &+ \eta_4\delta_{km}h_i h_j + \eta_5 h_m h_k\delta_{ij} \qquad (23)\\ &+ \eta_6\left(\delta_{ik}h_j h_m + \delta_{jm}h_k h_i + \delta_{im}h_k h_j + \delta_{jk}h_i h_m\right),\end{aligned}$$

with

$$\begin{aligned}\eta_1 &= \omega(1+\alpha_1+2\alpha_2)a_1,\\ \eta_2 &= \omega(1+\alpha_1+2\alpha_2)\frac{a_2}{3} + \omega\alpha_2 a_3\\ \eta_3 &= \omega(1+\alpha_1+2\alpha_2)\frac{a_2}{3},\\ \eta_4 &= \omega(1+\alpha_1+2\alpha_2)\left(a_4 - \frac{a_2}{3}\right) + \omega\alpha_2 a_5,\\ \eta_5 &= \omega(1+\alpha_1+2\alpha_2)\left(a_4 - \frac{a_2}{3}\right),\end{aligned}$$

and

$$\eta_6 = \omega(1+\alpha_1+2\alpha_2)\left(a_4 - \frac{a_2}{3}\right),$$

in which,

$$\begin{aligned}a_1 &\equiv \int_0^{\theta_1}\left(\lambda - 2\gamma + 4\gamma\cos^2\theta\right)^{1/2}\\ &\times \left(\cos^4\theta - 6\sin^2\theta\cos^2\theta + \sin^4\theta\right)d\theta,\\ a_2 &\equiv \int_0^{\theta_1}\left(\lambda - 2\gamma + 4\gamma\cos^2\theta\right)^{1/2}\sin^4\theta d\theta,\\ a_3 &\equiv \int_0^{\theta_1}\left(\lambda - 2\gamma + 4\gamma\cos^2\theta\right)^{1/2}\sin^2\theta d\theta,\\ a_4 &\equiv \int_0^{\theta_1}\left(\lambda - 2\gamma + 4\gamma\cos^2\theta\right)^{1/2}\sin^2\theta\cos^2\theta d\theta,\\ a_5 &\equiv \int_0^{\theta_1}\left(\lambda - 2\gamma + 4\gamma\cos^2\theta\right)^{1/2}\cos 2\theta d\theta.\end{aligned}$$

4 Localization condition

We investigate the possibility of a discontinuity in a component of the incremental strain over a band with unit normal **l**. Following [2, 3] we have

$$\det|l_i\mathcal{A}_{ijpq}l_q| = 0. \tag{24}$$

With $\mathbf{l} = (\cos\psi, \sin\psi)$, and the stiffness tensor given by Eq. (23), the localization condition corresponds to

$$0 = Z_1\cos^4\psi + Z_2\cos^2\psi + Z_3$$

with

$$\begin{aligned}Z_1 &= \mathcal{A}_{1212}(\mathcal{A}_{1122} + \mathcal{A}_{2211} + \mathcal{A}_{1111} + \mathcal{A}_{2222})\\ &+ \mathcal{A}_{1122}\mathcal{A}_{2211} - \mathcal{A}_{2222}\mathcal{A}_{1111},\\ Z_2 &= \mathcal{A}_{2222}\mathcal{A}_{1111} - \mathcal{A}_{1212}(\mathcal{A}_{1122} + \mathcal{A}_{2211} + 2\mathcal{A}_{2222})\\ &- 2\mathcal{A}_{1122}\mathcal{A}_{2211},\\ Z_3 &= \mathcal{A}_{2222}\mathcal{A}_{2112}.\end{aligned}$$

Table 1. Localization angle at given coordination number and normalized shear strain

$\hat{k}$	ψ (deg)	γ/λ_0
5	16.78	2.77
4.5	18.33	2.67
4	20.60	2.44
3.5	23.70	2.25
3	28.57	1.65
2.5	38.35	0.83

With $m = \cos^2\psi$ we have

$$m_{1/2} = \frac{-Z_2 \pm \sqrt{Z_2^2 - 4Z_3Z_1}}{2Z_1}. \tag{25}$$

The components of the stiffness tensor which involve the knowledge of integrals $a_1, a_2, ..a_5$ are determined by means of Simpson's rule. At lower shear strain the discrimant is negative in the range of coordination number between 6 and 2. Therefore we determine the value of the strain for which a real value of the angle is first possible; this occurs when first the discrimant $Z_2^2 - 4Z_1Z_3$ vanishes. There,

$$\cos\psi = \sqrt{\frac{-Z_2}{2Z_1}}. \tag{26}$$

Results are reported in Table 1.

5 Conclusion

While it is well-known that localization in granular materials is associated with the solution of the vanishing of the acoustic tensor related to the stiffness tensor of the aggregate, it is less obvious why the stiffness tensor should not be characterized by a major symmetry. In [2], this condition is physically justified by the non-associative flow rule for an elasto-plastic material with a yield surface characterized by a vertex structure. In [4], it is not assumed a priori and the numerical micro-mechanical approach seems to support this condition, although the authors refer again to the loss of the plastic normality. In [13], in the context of a micro-mechanical model, when particles are constrained to move with the average strain, it is possible to predict a potential function different from the yield function and so a non-associative flow. Nevertheless the same model, because of the absence of fluctuation, does not predict a loss of symmetry in the stiffness tensor and therefore no localization. Here we show that it is anisotropy and fluctuations, two key mechanisms in granular materials, that provide a structure to the macroscopic stiffness tensor that leads to the possibility of the vanishing of the acoustic tensor, for a given coordination number, at a given shear strain. Numerical simulations may be a tool to understand which quantities are crucial to obtain the loss of major symmetry in the stiffness tensor and how fluctuations, anisotropy are associated with plastic normality.

Acknowledgements

L. La Ragione is grateful to Office of Naval Research under the grant N62909-17-1-2048 and GNFM (Italy) to support his research.

References

[1] La Ragione L., Prantil, V.C., Jenkins J.T., Journal of the Mechanics of Physics and Solids, **83**, 46-159 (2015)
[2] Rudnicki, J.W. and Rice, J. R. Journal of the Mechanics and Physics of Solids **23** , 371-394 (1975).
[3] Vardoulakis, I., Mechanics Research Communication **3,** 209-214 (1976)
[4] Wren, J.R., Borja, R.I., Computer Methods in Applied Mechanics and Engineering **141**, 221-246 (1997)
[5] Hill, R., Journal of the Mechanics of Physics and Solids, **10**, 1-16 (1962)
[6] La Ragione, L., Oger, L., Recchia, G., Sollazzo, A., Proceedings of the Royal Society A: Mathematical, Physical and Engineering Sciences,**471**, 13-31 (2015)
[7] La Ragione, L., Journal of the Mechanics of Physics and Solids, **95**, 147-168 (2016)
[8] Thornton, C. and Antony, S.J., Philosophical Transactions of the Royal Society of London Series A **356**, 2763-2782 (1998)
[9] Jenkins, J. T. and Strack, O.D.L., Mechanics of Materials **16**, 25-33 (1993).
[10] Jenkins, J. T., Johnson, D. L., La Ragione, L. and Makse, H., Journal of the Mechanics and Physics of Solids **53**, 197-225 (2005)
[11] Jenkins, J.T., Agnolin, I., La Ragione, L., Mechanics of Materials **38**, 687-701 (2006)
[12] Kuhn, M., Mechanics of Materials, **42**, 827-840 (2010)
[13] La Ragione, L., Prantil, V.C. and Sharma, I., International Journal of Plasticity **24,** 168-189 (2008).

Elongated grains in a hopper

Tamás Börzsönyi[1,*], *Ellák* Somfai[1,**], *Balázs* Szabó[1,***], *Sandra* Wegner[2], *Ahmed* Ashour[2,3,****], and *Ralf* Stannarius[2,†]

[1] *Institute for Solid State Physics and Optics, Wigner Research Centre for Physics, Hungarian Academy of Sciences, P.O. Box 49, H-1525 Budapest, Hungary*

[2] *Institute of Experimental Physics, Otto-von-Guericke-University, Universitätsplatz 2, D-39106 Magdeburg, Germany*

[3] *Future University, End of 90th St., Fifth Settlement, New Cairo, Egypt*

Abstract. Flow and clogging of granular materials in a 3-dimensional hopper is investigated experimentally. We use X-ray tomography and optical methods to study this phenomenon for spherical and elongated particles. The X-ray tomograms provide information on the bulk of the hopper filling, and allow to determine the particle positions and orientations inside the silo, as well as spatial variations of the local packing density. We find that particles show a preferred orientation and thereby an enhanced order in the flowing zone of the silo. Similarly to simple shear flows, the average orientation of the particles is not parallel to the streamlines but encloses a certain angle with them. The clogged state is characterized by a dome, i. e. the geometry of the layer of grains blocking the outflow. The number of grains forming this blocking layer is larger for elongated grains compared to the case of spheres of the same volume.

1 Introduction

Flow and clogging of granular materials in hoppers still pose various questions which are intensively investigated nowadays [1–4]. This involves the statistics of clogging events [2, 5], the flow field and velocity fluctuations [3, 6, 7], possible precursors of clogging [7], and the microscopic features of the clogged state including the visualization and analysis of the blocking forces [1, 8, 9]. Simplified model systems using spherical (or circular) particles in a confined 2-dimensional (2D) geometry have been well investigated. On the basis of these studies there is growing interest for the characterization of more general 3D systems with more common (irregular, angular, elongated) grain shapes.

There are several unsolved problems regarding flow and clogging in 3D hoppers. One of these is the influence of grain shape. Even for simple elongated grains, such as spherocylinders or ellipsoids it is not clear whether particle elongation leads to increased or decreased flow rates. Such studies typically compare particles (spherical and elongated) with equal area (2D) or volume (3D). In 2D systems Cleary and Sawley found reduced flow rates for elongated particles with *nonzero* friction [10, 11], while Langston found increased flow rates for elongated particles with *zero* friction [12] compared to the case of circles. In 3D simulations Langston found the same flow rate for spheres and spherocylinders with *zero* friction [12], while Liu found reduced flow rates for both prolate and oblate ellipsoids with *nonzero* friction [13]. On the other hand, Li reported increased flow rates for oblate ellipsoids with *nonzero* friction [14]. The flow rate and clogging probability is strongly influenced by the orientation of the particles. In earlier studies, we systematically investigated particle rotation and average alignment in shear flows [15–20].

In the current study, we focus on the distribution of particle orientations in a 3D hopper for rod-like grains. We show that similarly to simple shear flows, the average orientation of the rods is not parallel with the streamlines, but encloses a certain angle with them. In the clogged state we charaterize the first layer of the grains above the orifice and show, that the number of grains in this shell is larger for elongated grains than for spheres of comparable volume.

2 Experimental methods and materials

An axisymmetric, nearly cylindrical container with a flat bottom (see figure 1a) was filled with a granular material. The diameter of the hopper was 19 cm, its height was 21.4 cm. At the bottom of the container, a removable plate was installed with an orifice with radius R. The orifice size was adjustabe by using plates with different R.

The granular materials investigated are shown in figures 1b-d. The elongation of the grains is characterized by the aspect ratio $Q = L/d$. We define the equivalent radius r_{eq} which is the radius of a sphere with the same volume as the particle. Three types of grains were used in the experiments presented here:

*e-mail: borzsonyi.tamas@wigner.mta.hu
**e-mail: somfai.ellak@wigner.mta.hu
***e-mail: szabo.balazs@wigner.mta.hu
****e-mail: ashour_55@yahoo.com
†e-mail: ralf.stannarius@uvgu.de

- wooden pegs with cylinder shape and tapered ends, diameter $d = 5$ mm, length $\ell = 25$ mm, aspect ratio $Q = 5$, equivalent radius $r_{eq} = 4.9$ mm
- peas, with small deviations from a perfect sphere, polydisperse with mean diameter of 7.6 mm, standard deviation 0.23 mm, $r_{eq} = 3.8$ mm
- monodisperse spherical plastic (airsoft) balls with diameter of 6 mm.

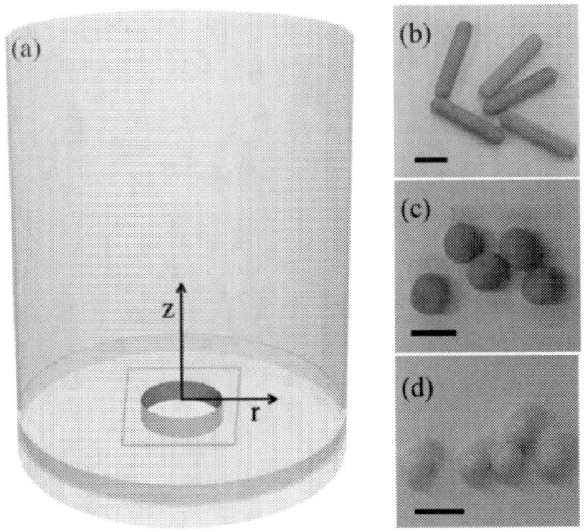

Figure 1. (a) Axisymmtertic hopper with flat bottom and circular orifice with radius R. (b-d) Granular materials (b) pegs with aspect ratio $L/d = 5$, (c) peas and (d) plastic (airsoft) balls. The scale bars correspond to 1cm.

The orifice size was in the range of $1 < R/r_{eq} < 5$, and the clogged configuration was characterized by two methods:

(i) An X-ray tomogram (CT) of the lower half of the silo was recorded using the robot-based flat panel X-ray C-arm system Siemens Artis zeego of the STIMULATE-lab, Otto von Guericke University, Magdeburg. The chosen spatial resolution was 2.03 pixel/mm, with recorded volumes of 25.2 cm × 25.2 cm × 19 cm. The 3D arrangement of the particles was then determined from the X-ray tomogram.

(ii) A digital camera was placed close to the orifice and the number of grains forming the last (clogging) layer was determined by image processing.

3 Results and discussion

For the experiments, the silo was filled with a closed orifice, then the orifice was opened and the granular flow was running until the system clogged. This clogged configuration was investigated (CT or digital camera), and thereafter the system was perturbed at the orifice to generate flow again. About 30 configurations were recorded for each material. First, we will show data obtained with CT for pegs, focusing on grain orientation, density distribution and ordering, then we will analyze data obtained by a digital camera for all three materials focusing on the number of grains which build up the last layer (dome) which blocked the flow.

Vertical cross sections (central slice) of two subsequent tomograms are presented for pegs in figure 2. The first tomogram was taken after several avalanches, thus the orientation and packing of the grains reflects the properties

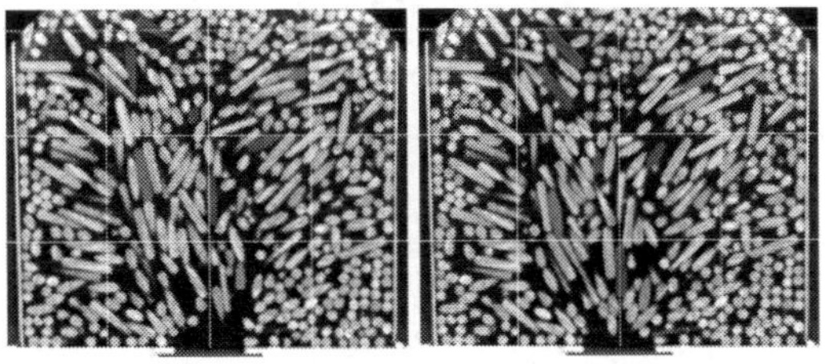

Figure 2. Vertical cross sections of two subsequent tomograms (two clogged states separated by a single avalanche) for pegs. In the flowing region particles are aligned. The pictures show the lower 16 cm of the nearly cylindrical hopper, a rectangular mesh of 5 cm × 5 cm has been superimposed to guide the eye. A few colored grains indicate the displacement of selected particles between the two subsequent clogged states. The diameter of the hopper outlet was 35 mm.

of the developed flow and alignment (initial conditions erased in the flowing regions). As explained above, these tomograms were taken in the clogged state. During the discharge between these two tomograms, approximately 60 particles left the hopper. The displacement of the material due to the flow is visualized by a few selected particles which are marked with colors in figure 2. It is clearly visible that the central region moves faster, especially near the outlet. Stagnant zones can be identified near the walls especially at the bottom of the hopper. In the flowing region particles are expected to be oriented nearly (but not exactly) parallel to the streamlines [15–17]. In order to quantify the ordering and alignment of the particles and to determine the flow field and the density distribution, tomograms from 26 subsequent clogged states have been analyzed. In figures 3(a)–(b), the color maps visualize the density distribution, and the average alignment angle (θ_{av}) of the particles in the $r - z$ plane. The alignment angle is measured with respect to the streamlines, which are indicated by white lines. The streamlines were calculated using the direction of average particle displacement between subsequent clogs. In order to better visualize the correlation between flow and orientation, on figure 3a we represented the average orientation of the particles by ellipses. The order parameter is represented by the flattening of the ellipses, S=1 (perfect order) would be represented by a thin line, and S=0 (isotropic state) by a circle [4].

Altogether, figures 3a-b show that in the flowing (sheared) regions we find well oriented rods due to the shear flow, while near the walls tangential alignment is preferred due to the boundary-induced ordering. We see, that in the shear flow the average orientation is not parallel to the streamlines, but there is a flow alignment angle as

DOI: 10.1051/epjconf/201714006017

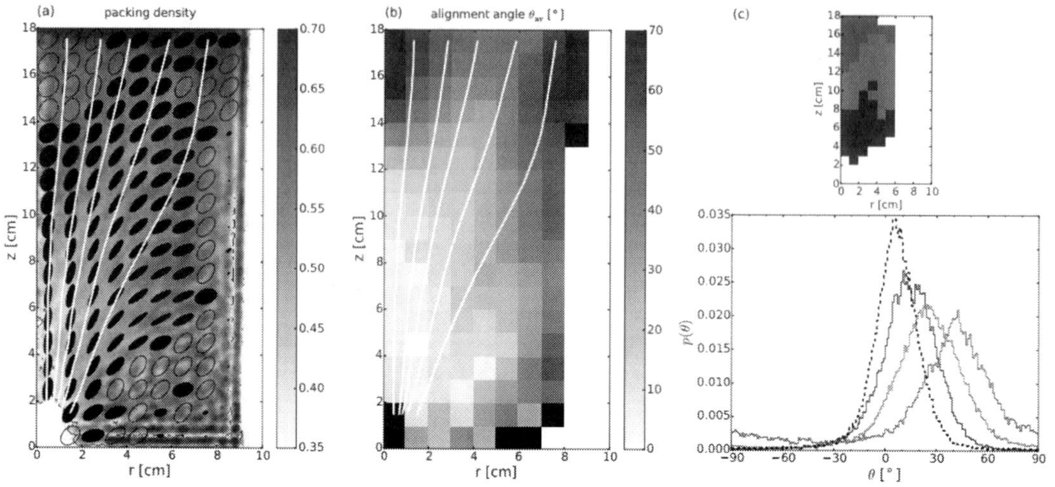

Figure 3. Distributions of (a) packing density, (b) and average alignment angle θ_{av} in the $r-z$ plane. Streamlines are determined from particle displacements between subsequent tomograms and are indicated by white lines. The alignment angle is measured with respect to the flow direction. On panel (a) the alignment and ordering of the particles is represented by ellipses (see text). Data have been calculated from 26 subsequent tomograms (examples shown in figure 1). (c) Distribution of particle orientations θ with respect to the streamlines in three regions as indicated in the top panel. The dashed line corresponds to a stationary distribution in a shear flow [16].

expected [15–17]. The sheared regions are characterized by lower packing fraction, consistent with our earlier observation in shear flow, where the shear induced dilation was found to be only partially compensated by a slight density increase due to the ordering of the grains [19].

In order to characterize the evolution of the grain alignment angle during flow, we plotted the distributions of θ_{av} in three different regions of the hopper in figure 3c. The three regions are marked with different colors (see top panel of figure 3c). The corresponding 3 curves show gradual narrowing of the angular distribution and decreasing of the average angle as the material is exposed to larger and larger deformation as it is displaced downwards in the hopper. In figure 3c we also show the distribution of angles of the same particles in stationary split bottom Couette flow (dashed line) [20]. We see that near the outlet (blue curve) the angular distribution almost reached the case of stationary shear.

In the following we characterize the arrangement of particles in the first layer of grains (above the orifice). In 2D a continuous force chain from one side of the orifice to the other can be used to identify the particles that form the blocking arch. The definition of the first layer is not straightforward in 3D, one can only roughly determine the blocking layer of particles. Of course, not only the grains in that first layer contribute to the clog, i. e. removal of other particles can also destabilize the blocked state, in 3D as well as in 2D. In our 3D experiments, we roughly define the blocking particles by selecting the minimal number of particles that form a closed shell above the outlet which cannot be penetrated by particles above that layer. A reasonable assumption for the mean number N of monodisperse spheres of radius $r = r_{eq}$ blocking an orifice with radius R can be estimated when one compares the area of a hemisphere (dome) of radius $R + r_{eq}$ with the cross section area of each grain, $\pi r_{eq}{}^2$. This yields $N \approx 2\gamma(R/r_{eq} + 1)^2$ The parameter γ is an area-filling factor which accounts for the fact that the grains can be displaced out of the hypothetical hemisphere. One may expect that γ is of the order of one, larger than the maximum packing fraction of circles in a plane, 0.906. The number of particles identified in the first layer of the clog are presented as a function of R in figure 4 for spherical particles and rods, respectively. Data for peas and monodisperse smooth plastic spheres collapse to the same curve with $\gamma \approx 1.03$. Thus for these samples differences in particle properties like detailed geometrical shape and friction play only an insignificant role. For the case of cylindrical particles (figure 4) the growth of N with increasing R is faster with $\gamma \approx 1.21$. This is a consequence of grain alignment, i.e. that grains forming the blocking shell are preferentially oriented such that longer axes are nearly perpendicular to the local shell surface. Thus, the shell is thicker (in units of the equivalent radius) for elongated grains, one needs more cylinders to form the blocking shell than spheres of the same equivalent radius.

We are currently working on the characterization of clogging statistics and hopper flow rate as a function of particle elongation for rod-like particles [21] in a 3D hopper. Characterization of the flow field as a function of grain shape in a quasi-2D hopper is also in progress. For elongated grains the profile of the vertical velocity clearly deviates from gaussian, and tunnel flow occurs [22].

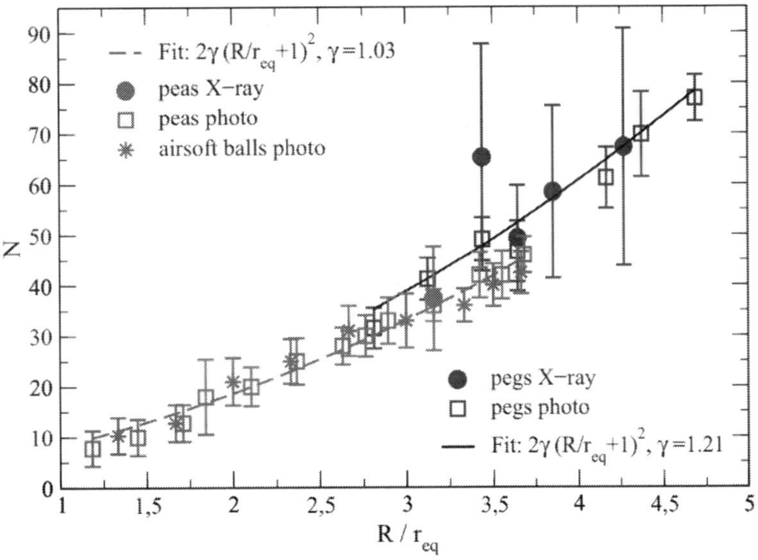

Figure 4. Number of particles forming the blocking dome structure as a function of the orifice radius for peas, airsoft balls and pegs. The fits with Eq. (1) result $\gamma \approx 1.03$ and $\gamma \approx 1.21$ for beads and rods, respectively. Each datum point is an average of 30 measurements.

Acknowledgements

We thank G. Rose and the STIMULATE-lab of Otto von Guericke University Magdeburg for the opportunity to use the X-ray CT facilities (funded by BMBF, reference No. 03IP710) and support. Financial support from the DAAD/MÖB researcher exchange program (Grants No. 57141022 and 64975), the Hungarian Scientific Research Fund (Grant No. OTKA NN 107737) and the János Bolyai Research Scholarship of the Hungarian Academy of Sciences are acknowledged.

References

[1] J. Tang and R.P. Behringer, EPL **114**, 34002 (2016)

[2] C. C. Thomas and D. J. Durian, Phys. Rev. Lett. **114**, 178001 (2015)

[3] S.M. Rubio-Largo, A. Janda, D. Maza, I. Zuriguel, and R.C. Hidalgo Phys. Rev. Lett. **114**, 238002 (2015)

[4] T. Börzsönyi, E. Somfai, B. Szabó, S. Wegner, P. Mier, G. Rose, and R. Stannarius, New J. Phys. **18**, 093017 (2016)

[5] I. Zuriguel, A. Garcimartín, D. Maza, L. A. Pugnaloni, and J. M. Pastor, Phys. Rev. E **71**, 051303 (2005)

[6] S. Tewari, M. Dichter and B. Chakraborty, Soft Matter **9**, 5016 (2013)

[7] C. C. Thomas and D. J. Durian, Phys. Rev. E **94**, 022901 (2016)

[8] J. Tang and R.P. Behringer, CHAOS **21**, 041107 (2011)

[9] R.C. Hidalgo, C. Lozano, I. Zuriguel, and A. Garcimartín, Granular Matter **15**, 841-848 (2013)

[10] P.W. Cleary. *Second International Conference on CFD in the Minerals and Process Industries*, pages 71–76, 1999.

[11] P.W. Cleary and M.L. Sawley. Applied Mathematical Modelling **26**, 89 (2002)

[12] P.A. Langston, M.A. Al-Awamleh, F.Y. Fraige, and B.N. Asmar. Chem. Eng. Sci. **59**, 425 (2004)

[13] S.D. Liu, Z.Y. Zhou, R.P. Zou, D. Pinson, and A.B. Yu. Powder Technology **253**, 70 (2014)

[14] J. Li, P.A. Langston, C. Webb, and T. Dyakowski. Chemical Engineering Science **59**, 5917 2004.

[15] T. Börzsönyi, B. Szabó, G. Törös, S. Wegner, J. Török, E. Somfai, T. Bien, and R. Stannarius, Phys. Rev. Lett. **108**, 228302 (2012)

[16] T. Börzsönyi, B. Szabó, S. Wegner, K. Harth, J. Török, E. Somfai, T. Bien, and R. Stannarius, Phys. Rev. E **86**, 051304 (2012)

[17] S. Wegner, T. Börzsönyi, T. Bien, G. Rose, and R. Stannarius, Soft Matter **8**, 10950–10958 (2012)

[18] T. Börzsönyi and R. Stannarius, Soft Matter **9**, 7401 (2013)

[19] S. Wegner, R. Stannarius, A. Böse, G. Rose, B. Szabó, E. Somfai, and T. Börzsönyi, Soft Matter **10**, 5157 (2014)

[20] B. Szabó, J. Török, E. Somfai, S. Wegner, R. Stannarius, A. Böse, G. Rose, F. Angenstein, and T. Börzsönyi, Phys. Rev. E **90**, 032205 (2014)

[21] A. Ashour, S. Wegner, T. Trittel, T. Börzsönyi, and R. Stannarius, Soft Matter **13**, 402 (2017)

[22] B. Szabó, E. Somfai, S. Wegner, A. Ashour, D. Fischer, R. Stannarius, and T. Börzsönyi, In preparation.

Evolution of particle breakage studied using x-ray tomography and the discrete element method

Zeynep Karatza[1,2,*], *Edward* Andò[2], *Stefanos-Aldo* Papanicolopulos[1], *Gioacchino* Viggiani[2], and *Jin Y.* Ooi[1]

[1]School of Engineering, The University of Edinburgh, Edinburgh EH9 3JL, U.K.
[2]Univ. Grenoble Alpes, CNRS, Grenoble INP*, 3SR, F-38000 Grenoble, France
* Institute of Engineering Univ. Grenoble Alpes

Abstract. Particle breakage can significantly change the fabric (size and shape of particles and contact network) of a granular material, affecting highly the material's macroscopic response. In this paper, oedometric compression tests are performed on zeolite specimens and x-ray computed micro-tomography is employed, to acquire high resolution 3D images of the specimens throughout the test. The images are processed, to describe breakage spatially and quantify it throughout the test and gain information about the mechanisms leading to particle breakage. In addition to the image processing, the discrete element method (DEM) is used to study the initiation and likelihood of particle breakage, by simulating the experimental test during the early stages of loading and using quantitative results from the images to inform and validate the DEM model. A discrete digital image correlation is used, in order to incrementally identify intact grains and simultaneously get results about the strain field within the specimen, as well as the kinematics of individual grains and fragments. In the initial stages of breakage, there is a clear boundary effect on the spatial distribution of breakage, as it is concentrated at the moving boundary (more than 90% of total breakage) and circumferentially (more than 70% of total breakage) close to the apparatus cell. The DEM model can reproduce the bulk response of the material until the point where substantial breakage governs the macroscopic response and it starts to soften. Additionally, there is an initial indication that the spatial distribution of the force network matches the localisation of breakage radially, but it does not seem to localise close to the loading platen. This analysis will enrich our understanding of the mechanisms and evolution of particle breakage.

1 Introduction

Particle breakage can significantly change the fabric (size and shape of particles and contact network) of a granular material, affecting highly the material's constitutive behaviour. The discrete element method (DEM) has been used to numerically model breakage and investigate the underlying micro-mechanics that govern the fracturing behaviour (for example, [1-3]). In many cases, the lack of experimental 3D micro-scale information regarding particle deformation, significantly limits the validation of numerical models. Additionally, the models might lack the necessary input to accurately reproduce breakage behaviour.

The main objective of this work is to gather quantitative information regarding breakage of a granular material (zeolite) subjected to strain-controlled oedometric compression, using x-ray computed micro-tomography (XCT). The use of 3D imaging with XCT to study particle scale mechanics has been increasing recently (for example, [4-6]), enabling a direct comparison of various micro-mechanisms with DEM models (for example, [7-9]). The high resolution of the images in this work allows the extraction of quantitative data about the fabric of the material and the visualisation of the spatial distribution of deformation.

2 Testing material and experimental campaign

A total of 6 tests were performed on zeolite specimens, all with similar initial grading (very uniform) and relative density (medium density), with a D_{50} of 1.36mm. The specimens were loaded until the majority of the particles break. Experiments were carried out with 3 different loading rates: 25μm/min, 50μm/min and 100μm/min, before the XCT was performed, and it was determined that the strain rate did not affect the response of the zeolite. The macroscopic response of the tests (Figure 1) shows a good match of the initial stiffness, which suggests the tests are reproducible, and a well-defined yielding. From the response it is expected that significant breakage would start to occur at an axial stress of 2MPa (Point LS3 in Figure 1). The different stages of loading (LS#), corresponding to the necessary pauses to perform

* Corresponding author: z.karatza@ed.ac.uk

the XCT, are indicated on Figure 1 for test OSC1 (tested at an axial loading rate of 50μm/min).

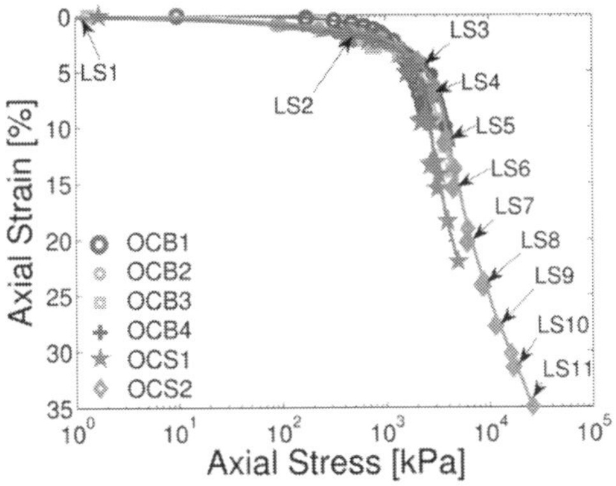

Figure 1. Macroscopic response of oedometric tests.

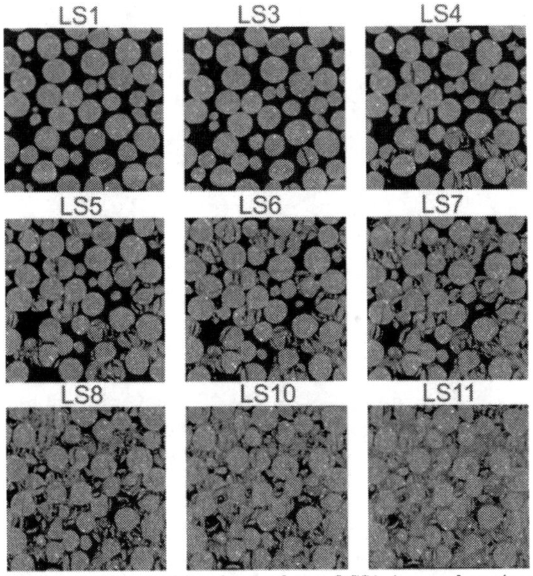

Figure 2. Extent of breakage of test OCS1 (zoomed section from XCT images).

The oedometer cell was custom-made out of PEEK (PolyEtherEtherKetone), specifically to enable XCT imaging. Two cells where built and the respective specimens have a ratio of diameter/height of 4.5/1.5 (OCB#) and 1.5/1.5 (OCS#). For brevity, in this work only results from test OCS1 are presented and compared to the DEM model. The mean diameter of each particle comprises of approximately 120 pixels (voxel size of 12μm), which provides plenty of detail to identify contacts and individual fragments. Figure 2 shows the evolution of breakage during testing of specimen OCS1. From a visual investigation, it is concluded that breakage develops irregularly in space and time, well after yielding (LS8), similarly to what was observed in [9]. Thereafter breakage seems to develop continuously and uniformly.

3 Discrete Digital Image Correlation

A discrete Digital Image Correlation (DIC) is employed, in order to incrementally identify intact and broken grains and simultaneously get results about the strain field within the specimen and the kinematics (translation and rotation) of individual grains. It is based on the code TomoWarp2, described in [10], where in this case the correlation window is adapted to each particle. Figure 3 lists the basic steps of the DIC algorithm and presents an example of a result. The grains that break are analysed regarding their spatial distribution, coordination number right before breakage, the size and shape of the intact grains and the fragments (whenever they could be identified), and the evolution of the grading of the whole specimen.

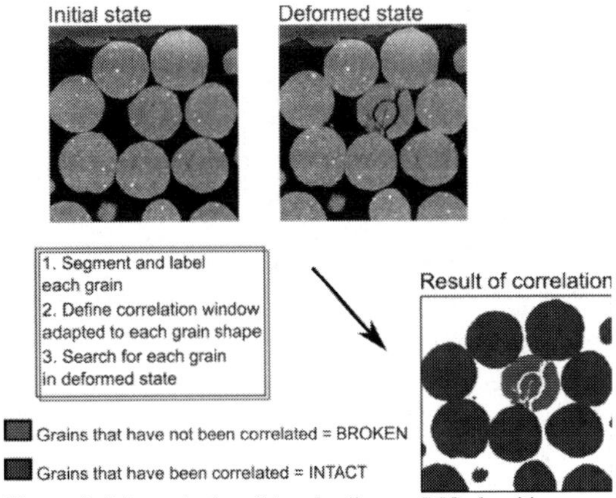

Figure 3. Schematic describing the discrete DIC algorithm.

4 Discrete Element Method

For the DEM simulations, a Hertz-Mindlin contact model was used to study the changes in the fabric of the specimen right before the majority of breakage occurs (Point LS4 in Figure 1). Right after this point, there is a substantial amount of softening in the macroscopic response caused by the extensive breakage that cannot be simulated by a simple Hertz-Mindlin contact law. This can also be seen in Figure 4 where after 7% of axial strain (LS4) the numerical and experimental results start to deviate.

Table 1. DEM particle parameters for test OCS1.

Parameter	Value	Units
Particle density, ρ	2180	kg/m^3
Poisson's ratio, ν	0.25	-
Young's modulus, E	4.5	GPa
Coef. of restitution, e	0.01	-
Coef. of static friction μ_s	0.2	-
Coef. of rolling friction, μ_r	0.05	-
Mean diameter, D_{50}	1.36	mm
St. Dev. of D_{50}	0.0322	-

In this work each particle is represented by a single rigid sphere in the DEM model. An assembly of 1056 grains is generated in a virtual space and left to randomly fill a cylinder that matches the dimensions of the oedometer cell, to simulate the actual specimen and experimental test. For the simulation some friction has been assigned to the boundary (PEEK was also chosen because of its low friction). Table 1 summarises the DEM parameters chosen for this simulation. The material properties (ρ, ν and E) are chosen by performing single-particle compression tests on real zeolite particles and comparing them with DEM simulations of the same test. In this first implementation the contact parameters (e, μ_s and μ_r), are assigned to match the packing density of the DEM to that of test OCS1, since the filling density will be governed by the frictional behaviour (sliding and rolling) of the particles. Finally, the physical parameters of the DEM model (D_{50} and St. Dev of D_{50}) are an exact match to what has been measured from the images for test OCS1. In future work, in order to simulate breakage a bonded contact model will be used, as described in [11], based on the Timoshenko beam theory which considers axial, shear and bending behaviour of the bond, to create multi-sphere particles to simulate each zeolite grain.

5 Results

The main focus of this ongoing work is to quantify the amount of breakage from the images and study the likelihood of breakage with DEM. Figure 4 shows the results from the DEM simulation in comparison to the stress strain response of test OCS1.

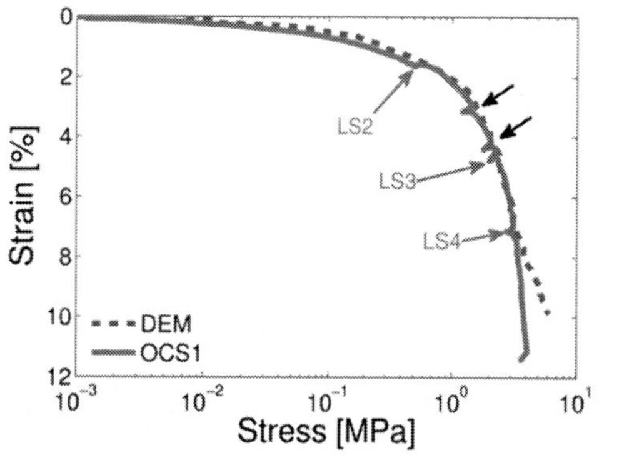

Figure 4. Comparison of early loading stages of DEM and test OCS1 macroscopic response.

There are various slight stress relaxations present in the experimental curve (the largest ones marked with the two black arrows in Figure 4), that are attributed to fragmentation of the micro-asperities of the particles and potentially some crack initiation. These asperities, create multiple contacts between two particles, whereas in DEM a contact occurs over an infinitesimal area and there can only be one contact point between two spherical particles; therefore, the overall DEM response is expected to be slightly stiffer than the experimental. This has been accounted for by assigning a rolling friction to the particles.

From the DIC results, it is found that no particles break until LS2, 4% of particles (42 grains) break between LS2 and LS3 and 14.5% (153 grains) between LS3 and LS4. At LS4, 18.5% of grains break since the beginning of the test. Figure 5 shows 3D images of all the intact particles that broke; it is clear that the breakage is concentrated close to the moving (bottom) boundary and the cell. Specifically, breakage initiates between LS2 and LS3 and in LS3 more than 90% of all breakage is concentrated in an area with height $3.8D_{50}$ (measured from the moving boundary) and radially more than 70% of the breakage is concentrated in a maximum distance of $3D_{50}$ from the cylindrical boundary.

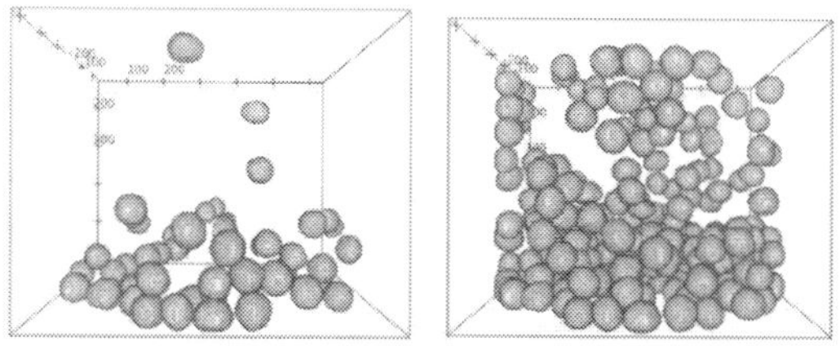

Figure 5. 3D images of the intact particles that have broken in LS3 (left) and LS4 (right), from the DIC analysis.

An initial approach is to investigate the distribution of forces, both normal and shear and how they compare to the spatial distribution of breakage. Figure 6 shows 3D images from the DEM results of the normal and tangential force network at LS3 and LS4, with respect to the average normal and tangential force at each loading stage (LS3: $F_{N,avg}$ = 3.64N and $F_{T,avg}$ = 0.35N; LS4: $F_{N,avg}$ = 4.56N and $F_{T,avg}$ = 0.43N). As it can be seen in Figure 6, the normal forces higher than $4F_{N,avg}$ develop radially, close to the cylindrical boundary and more than 85% of the forces higher than $3F_{N,avg}$ are lying within a $3D_{50}$ annular section from the cylindrical boundary. This correlates well with where the majority of breakage is concentrated. Regarding the higher shear forces they do seem to localise, however the pattern is not as clear as for the normal force network. The next step will be to study the kinematics and compare the shear and volumetric strains of the DIC analysis and the DEM results. We expect to see a localised compaction band at the initial stages of breakage close to the moving boundary as indicated by the localisation of breakage and the porosity maps created from the images. The coarse graining of the DEM results could be challenging due to the small number of grains (10 grains per oedometer diameter).

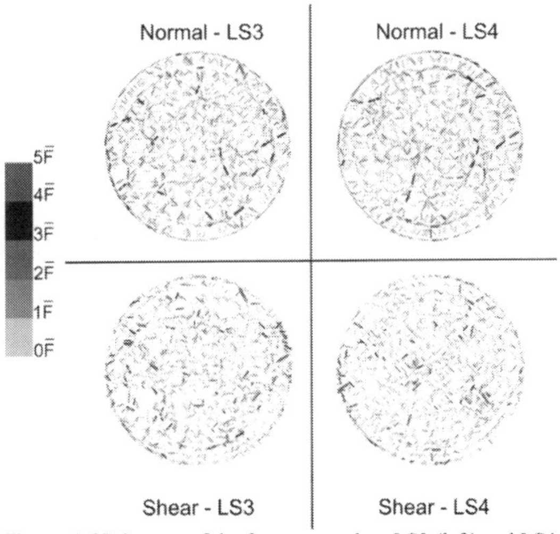

Figure 6. 3D images of the force network at LS3 (left) and LS4 (right), from the DEM analysis. [Top: normal and bottom: shear].

6 Conclusions

In summary, the recent use of XCT in geomechanics has allowed the extraction of valuable quantitative 3D information, locally or not, for entire specimens during loading. The increasing use of the DEM to study kinematics and deformation of granular materials, has led to a rising need for particle-scale quantitative experimental results (especially regarding the meso-scale) for validation. In this paper, an effort was made to interdependently validate the measurements of both the XCT and the DEM, as well as to quantitatively describe particle breakage. The main findings of this study are:

- The macroscopic response of the tests is reproducible and shows a well-defined yielding.
- Breakage initiates after 3% of axial strain (1MPa of axial stress) and it becomes substantial (18.5% total breakage) after 7% axial strain (8MPa axial stress).
- A discrete DIC is employed that facilitates the incremental study of grain breakage, in terms of fabric and spatial distribution. So far, the initial stages of loading are evaluated and it is found that there is a clear boundary effect on the spatial distribution of breakage, as it is concentrated at the moving boundary (more than 90% of total breakage) and circumferentially (more than 70% of total breakage).
- For the DEM model, initially a Hertz-Mindlin contact law is used to simulate the early stages of loading and the material particles are represented by spheres of the same diameter and overall grading.
- The DEM model can reproduce the bulk response of the material until the point where substantial breakage governs the macroscopic response and it starts to soften.
- From the results of the DEM model the force network shows some localisation of the large forces, more so of the normal forces than the shear.

In the future, breakage will be simulated with the DEM by using a bonded model to create clumps that represent each particle. The fines produced by the DEM simulations will be compared to the XCT fines, that will be measured with the technique described in [5], complemented by the discrete DIC presented in this paper. Also, the 3D images will be further on investigated in order to gain more information on the contact network and how that affects the particles that break. Finally, the kinematics (stress and strain field) will be investigated and compared for both the DEM and XCT results.

Acknowledgements

The authors would like to thank Pascal Charrier for his valuable help during the XCT, Laurent Debove for his help for the creation of the oedometer and Olumide Okubadejo for his work on the discrete DIC. Laboratoire 3SR is part of the LabEx Tec 21 (Investissements dAvenir - grant agreement nANR-11-LABX-0030). Z. Karatza would like to thank the UK Engineering and Physical Sciences Research Council (EPSRC) DTP PhD studentship and the International Fine Particles Research Institute (IFPRI) for funding her PhD research. S.-A. Papanicolopulos acknowledges the support received from the People Programme (Marie Curie Actions) of the European Union's Seventh Framework Programme (FP7/2007-2013) under REA grant agreement n 618096.

References

[1] D. Robertson, M. D. Bolton, Proc. of 4th international conference on micromechanics of granular media: powders and grains 623-626 (2001).
[2] O. Harireche, G. R. McDowell, Granular Matter **5**, 3 (2003).
[3] I. Laufer, Granular Matter **17**, 389 (2015).
[4] S. A. Hall, M. Bornet, J. Desrues, Y. Pannier, N. Lenoir, G. Viggiani, P. Bésuelle, Géotechnique **60**, 5 (2010).
[5] Z. Karatza, E. Andò, S.-A. Papanicolopulos, J. Y. Ooi, G. Viggiani, Géotechnique (accepted for publication in 2017).
[6] E. Andò, S. A. Hall, G. Viggiani, J. Desrues, P. Bésuelle, Acta Geotechnica **7**, 1 (2012).
[7] M. B. Cil, K. A. Alshibli, Géotechnique **64**, 5 (2014).
[8] M. B. Cil, K. A. Alshibli, Comptes Rendu Mechaniqué **343**, 2 (2015).
[9] F. Guillard, P. Golshan, L. Shen, J. R. Valdes, I. Einav, Nature Physics 11, 835 (2015).
[10] E. Tudisco, S. A. Hall, E. M. Charalampidou, N. Kardjilov, A. Hilger, H. Sone, Physics Procedia **69**, 509-515 (2015).
[11] N. J. Brown, J.-F. Chen, J. Y. Ooi, Granular Matter **16**, 299 (2014).

Dilute suspensions in annular shear flow under gravity: simulation and experiment

Kevin Schröer[1,*], *Patrick* Kurzeja[1], *Stephan* Schulz[1], *Philipp* Brockmann[2], *Jeanette* Hussong[2], *Peter* Janas[3], *Irenaeus* Wlokas[3], *Andreas* Kempf[3], and *Dietrich E.* Wolf[1]

[1] *Faculty of Physics and CENIDE, University of Duisburg-Essen, Duisburg, Germany*
[2] *Institute for Thermo and Fluid Dynamics - Hydraulic Fluid Machinery, Ruhr-University Bochum, Germany*
[3] *Institute for Combustion and Gas Dynamics - Fluid Dynamics and CENIDE, University of Duisburg-Essen, Duisburg, Germany*

Abstract. A dilute suspension in annular shear flow under gravity was simulated using multi-particle collision dynamics (MPC) and compared to experimental data. The focus of the analysis is the local particle velocity and density distribution under the influence of the rotational and gravitational forces. The results are further supported by a deterministic approximation of a single-particle trajectory and OpenFOAM CFD estimations of the overcritical frequency range. Good qualitative agreement is observed for single-particle trajectories between the statistical mean of MPC simulations and the deterministic approximation. Wall contact and detachment however occur earlier in the MPC simulation, which can be explained by the inherent thermal noise of the method. The multi-particle system is investigated at the point of highest particle accumulation that is found at 2/3 of the particle revolution, starting from the top of the annular gap. The combination of shear flow and a slowly rotating volumetric force leads to strong local accumulation in this section that increases the particle volume fraction from overall 0.7% to 4.7% at the outer boundary. MPC simulations and experimental observations agree well in terms of particle distribution and a close to linear velocity profile in radial direction.

1 Introduction & Goals

Particle-laden flows in narrow annular gaps of hydraulic fluid machinery may alter the systems' performance even at moderate seeding densities. However, the detailed particle behavior and underlying mechanisms such as local particle accumulation or particle segregation effects in Taylor-Couette flows of narrow gaps are still little understood up to date [1, 2].

Various numerical and experimental approaches have been developed to analyze complex suspensions, including combined discrete element and finite volume methods [3] or imaging techniques [4]. A young simulation technique emerged from the so-called multi-particle collision dynamics method (MPC) that mimics the fluid by coarse-grained particles and a simplified collision model [5]. The MPC technique fully incorporates hydrodynamic interactions as well as thermal fluctuations and benefits from a highly parallelized numerical implementation [6]. The present study aims to develop numerical and experimental complementary methods to study the particle dynamics of a dilute, sheared, rotating suspension under the influence of gravity. It focuses on the local particle velocity and density variations while verifying the MPC method as a predictive, stochastic tool by means of an especially adapted A-PTV method. A finite volume CFD simulation is additionally used to determine the threshold for overcritical laminar flow. After introducing the methodology, we investigate a single-particle simulation and compare it with a deterministic approximation. Afterwards, we compare a multi-particle simulation with experimental results at the position of highest local particle density. Special attention is given to the capabilities and limitations of the MPC method.

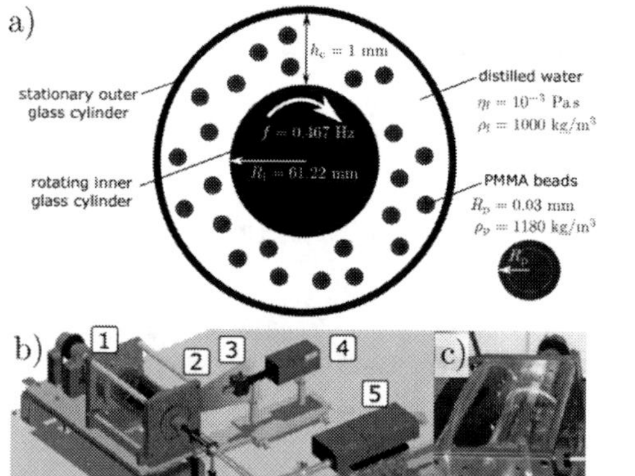

Figure 1. a) Sketch of the system investigated by experiment and simulations (not to scale); b) components of experimental setup: 1 Taylor-Couette cell, 2 microscope objective, 3 microscope tubus, 4 CCD camera, 5 Nd:YAG laser; c) close up of the Taylor-Couette cell.

*e-mail: kevin.schroeer@uni-due.de

DOI: 10.1051/epjconf/201714009034

2 Methods

2.1 System of investigation and dimensionless characterization

Our setup consists of two concentrically and horizontally aligned glass cylinders with 1 mm spacing that will be called the Taylor-Couette cell, see Fig. 1. Its inner cylinder rotates with $f = 0.467$ Hz ($\sim v_0 = 0.18$ m/s at its outer edge), while the outer cylinder is stationary. The annular gap between both cylinders is filled with distilled water and monodisperse PMMA beads of 60 μm diameter. The ratio of channel height ($h_c = 1$ mm) to the radius of the inner cylinder ($R_i = 61.22$ mm) is 0.016 and thus minimizes the influence of curvature when focusing on a local section.

The two dimensionless key parameters of the problem at hand are the Froude number $\mathrm{Fr} = v_0^2/(R_i\, g) = 0.055$, balancing the amplitudes of centrifugal and gravitational forces, and the dimensionless frequency $f^* = f\, h^2\, \rho_f/\eta_f = 4.5 \cdot 10^{-7}$, i.e. the ratio of viscous relaxation time to the rotation period (with $g = 9.81$ m/s^2, fluid density $\rho_f = 1000$ kg/m^3 and dynamic viscosity $\eta_f = 0.001$ Pa s). These parameters place the presented setup between pure vertically aligned sedimentation ($\mathrm{Fr} \to 0$) and radially oriented centrifugation ($\mathrm{Fr} \to \infty$). The particles are slightly denser compared to the fluid ($\rho_s/\rho_f = 1.18$). Very low characteristic Mach and channel Reynolds numbers, $\mathrm{Ma} = v_0/c_{\mathrm{sound}} = 1.25 \cdot 10^{-4}$ and $\mathrm{Re_{ch}} = v_0\, h_c\, \rho_f/\eta_f = 190$, support the assumption of incompressible, laminar flow. The particle Reynolds number based on the shearing rate is $\mathrm{Re_p} = (v_0/h_c)\, R_p^2\, \rho_f/\eta_f = 0.162$. Thermal fluctuations are furthermore negligible owing to the large particle Peclet number of $\mathrm{Pe} = (3\pi\, R_p^3\, \eta_f\, v_0)/(2h_c k_B T) = 2.26 \cdot 10^7$ at room temperature.

2.2 MPC Simulation

The MPC method mimics the fluid in a stochastically coarse-grained manner by mesoscopic fluid point particles [5]. These fluid particles alternately undergo a streaming and a collision step. During the streaming step, the fluid particles move independently, keeping their velocity. During the collision step, particles are sorted into a grid of collision cells, and their velocities relative to the cell's mean velocity are rotated about a rotation axis randomly generated for each cell and time step. This collision rule, known as stochastic rotation dynamics (SRD), ensures local momentum and energy conservation. In the current numerical setting, the collision-cell width is chosen to be $0.01\, h_c$ with 20 fluid particles per cell. The rotation angle of 100° and the mean free path of 0.32 lead to $\Delta t_{\mathrm{MPC}} = 11.24$ ms.

The suspended solid particles are integrated by the Velocity Verlet algorithm imposing $\Delta t_{\mathrm{MD}} = 0.1 \Delta t_{\mathrm{MPC}}$ and particle-particle interaction via a hard-sphere potential. Fluid-particle coupling is achieved by a bounce-back rule during the streaming step and the use of virtual particles during the collision step [7].

Instead of the full geometry, a cubic box of side length h_c is simulated, neglecting curvature of the boundaries. The simulation box is interpreted as a reference frame that is moving along the cylinder gap with mean fluid velocity $v_0/2$. The box boundaries are shearing rigid walls in radial direction and periodic boundaries in all other directions. In addition to the gravitational force, Coriolis and centrifugal forces are applied to the suspended colloids. Following the reference frame, the gravitational force is rotated accordingly in order to recover its varying influence at different angular positions along the cylinder. In the case of a single particle, rotation is adjusted by the global particle position. Because incompressibility in the MPC fluid is not assured, applying the above mentioned forces to the fluid particles would lead to non-physical artifacts. Instead, effective buoyancy is taken into account according to the colloid-fluid density mismatch.

2.3 Experiment

Centrifugal and gravity-driven particle dynamics were studied in an optically accessible Taylor-Couette setup by means of shadowgraphy imaging (see Fig. 1 b)). A frequency-doubled dual cavity Nd:YAG laser (New Wave Solo III-15) with 532 nm wavelength and 8 ns pulse length was used for volume illumination. The laser beam was guided into the Taylor-Couette setup to excite a red fluorescent color-coated plate. In this way, incoherent fluorescence light is emitted passing through inner and outer glass cylinders in positive radial direction to create a homogeneous, speckle-free background illumination. Images were taken with a 12 bit dual frame CCD-cameras (Imager Intense, LaVision) with 1376 x 1040 image size. Using a microscope objective of 5x magnification and 12 μm focal depth (CFI60 TU Plan (MUE12050), Nikon), an image resolution of 1.26 μm/pixel is reached, yielding a field of view of 1.76 mm x 1.32 mm. A suspension of PMMA particles (Microbeads AS; final diameter as provided by the manufacturer of 62.8 μm ± 1.2 % standard deviation) was injected in the inlet of the system briefly before the experiment was initiated. Fifty double-frame images were taken every 5 μm along the gap width. A 3D-PTV procedure was developed similar to Cierpka et al. [8] to achieve an out-of plane particle position reconstruction. Radial particle positions could be determined with an overall standard deviation of 34 μm accuracy. The reference position at the outer cylinder was determined by reducing invalid detections outside the channel to less than 4 % of all particles.

2.4 Finite Volume Simulation

As a second numerical approach, the 3D-CFD software OpenFOAM was used to estimate the transition range towards Taylor-Couette-instabilities [3]. The convective fluxes are discretized by a total variation diminishing scheme (TVD) with the Sweby limiter. A larger simulation box of 5 mm x 5 mm x 1 mm was chosen to account for the system's curvature and thus implicitly consider centrifugal and Coriolis forces. No-slip boundary conditions are applied to the cylinder walls and periodic boundaries are used in axial and circumferential direction.

2.5 Deterministic approximation of the single-particle trajectory

The steady-state trajectory of a single particle was theoretically approximated by Newton's second law and solved numerically using MATLAB's ode solver ode15s. The governing equations account for inertia, Stokes drag, gravitation, the Saffman lift force [9] and the virtual-mass force [10]. A stationary flow field, elastic slip motion at contact with the outer wall and negligible angular motion of the particle were assumed. Coriolis and Basset forces were neglected, because of small radial and relative velocities between particle and fluid.

3 Results and Discussion

3.1 Methodological peculiarities and limitations

We consider it helpful to discuss the major peculiarities and limitations of each method before comparing and evaluating the specific results. This clarification helps to justify the assumptions made and to highlight potentially deviating conditions beforehand.

Thermal noise incorporated in the MPC method reduces the Peclet number from $Pe = 2.26 \cdot 10^7$ in the experiment to 28 in the simulation. In the case of multiple particles, all particles are exposed to the same orientation of the gravity force that is determined by the location of the reference frame. This assumption is motivated by small curvature effects within the simulation box. Nevertheless, particles on an inner radial trajectory rotate faster and should experience a shorter force oscillation and vice versa. Using the mean fluid velocity for the reference frame is therefore assumed as a compromise.

The formation of a distinguished particle band perpendicular to the cylinder axis was observed in the experiment. This is a well-known phenomenon in particle-laden drum flows exceeding a particle-density threshold for intermediate frequencies [11]. In the present case, the band formation is 5 mm to 10 mm in thickness, depending on the circumferential position, and thus qualified as a homogeneous particle distribution on the length scale of 1 mm to be compatible with the assumptions of the numerical approaches. The measurements were hence performed in the centerline of the band to ensure the most periodic conditions possible.

Taylor-Couette-instabilities were prevented by keeping the channel Reynolds number $Re_{ch} = 190$ below the critical value of $Re_{ch,crit} \approx 350$, which was determined by a linear stability analysis. Accordingly, no Taylor vortices are observed for the operation point of the experiment. This was verified by the finite volume method via a parameter test that identified a transition at $Re_{ch,crit} \geq 3800$.

3.2 Single particle

The deterministic single-particle approximation predicts a long-term trajectory, along which single particles will move independently of their starting position as illustrated in Fig. 2. The MPC simulation resembles this characteristic trajectory, distinguishing a convection-dominated drift period from a contact period at the lower left wall. Averaged over six revolutions, the simulation predicts closest approach to the wall approximately 25° earlier. The individual retention of minimum particle-wall distance also fluctuates and is generally smaller in the MPC simulation. The comparably strong delay observed in the deterministic approximation may result from the simplifying assumptions yielding an underestimation of the total lift force [12]. A final quantitative evaluation exceeds the scope of this work, and the necessity of a more precise modeling of close contact will be discussed below against the background of multiple particles.

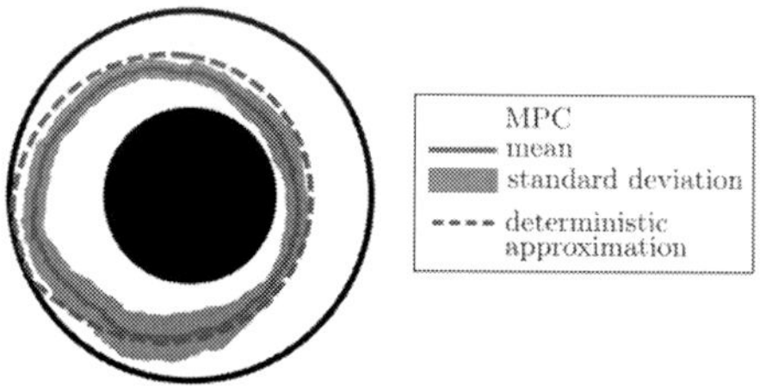

Figure 2. Comparison between steady-state deterministic approximation and MPC simulation of the trajectory of a single particle (not to scale, channel height is magnified).

3.3 Multiple particles

The study of multiple particles is focused on a section at 240° clockwise revolution, where the single-particle results predict strongest particle-wall interaction and therefore the highest particle accumulation, see Fig. 3 a). The corresponding particle distribution over the annular gap derived from experiments and simulations are in good agreement, as can be seen in Fig. 3 b). The simulations were performed with 63 particles per mm^3 corresponding to a mean particle volume fraction of 0.7% in this section as found in the experiment. However, most particles are located in an interval of approximately 15% gap-width centered around the maximum density at 1.5 particle diameters (i.e. 90 μm) outer wall distance, leading to a local volume fraction of 4.7%. In this region, the slope of the almost linear particle velocity profile is slightly reduced, see the left 180 μm interval in Fig. 3 c). This behavior can be qualitatively understood by particle-induced changes in viscosity, estimated by the Krieger model [13] to reach 13% deviation in this region of highest accumulation compared to only 2% for the section's mean volume fraction.

Particles do not adhere to the resting wall collectively. Their behaviour rather resembles the one of a single particle as observed from the MPC simulation. We assume that a wall-lift force prevents the particles from getting into direct contact with the wall [12]. Also direct contact or short-range hydrodynamic interaction between particles can enhance repulsion in the particle cloud.

We suggest that the accuracy of the present MPC simulation mainly depends on particle-fluid and particle-particle momentum exchange, because of the rare particle-wall contacts. This is supported by the similarity between the stochastic MPC approach and the experimentally observed attributes of the particle and velocity distributions. Possible sources for quantitative deviations are the finite Pe number as well as periodic boundary conditions. Slight layering is also observed in the simulation, see the local peaks approximately one particle diameter apart in Fig. 3. This can be explained by the constant rotation period assumed for the numerically implemented reference frame that allows for a development of particle layers before break-up. Future studies extending the periodically confined simulation box will allow for a more accurate implementation of the gravity force acting individually on each particle such that the time-dependent formation can be predicted more precisely.

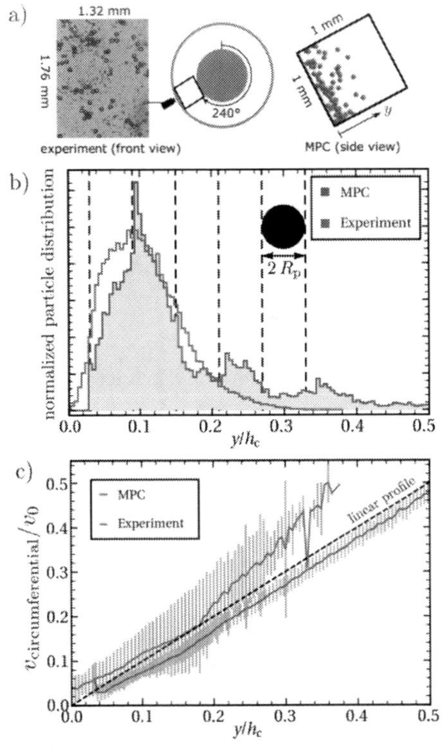

Figure 3. a) Position of the evaluation region at 240° for the multi-particle study with snapshots of experimental and MPC data; b) density profile of the particle distribution per 5 μm-interval for experiment and MPC simulation normalized by total particle number and plotted over normalized radial distance y from the outer cylinder; c) velocity profile for experiment and MPC simulation averaged over 5 μm-intervals with standard deviation as vertical lines plotted over normalized radial distance y from the outer cylinder.

4 Conclusion

We studied dilute suspensions in annular shear flow numerically and experimentally with a ratio of radial centrifugal forces to vertical gravity forces of 0.055. The single-particle trajectory predicted by the MPC simulation shows good agreement with the deterministic approximation during the distinct periods of free drift and wall contact. The closest approach to the wall and the subsequent release appeared earlier in the MPC simulation. This can be explained by thermal fluctuations that are inherent to the MPC approach and better reflect stochastic perturbations of the steady-state trajectory. The multi-particle system accordingly shows rare particle-wall contacts and a peak of the particle distribution about 1.5 diameters away from the outer wall, leading to a locally increased volume fraction (0.7% → 4.7%) and viscosity (2% → 13%). These characteristics are also well predicted by the MPC simulations; notwithstanding the algorithmic restrictions in terms of Peclet number, Mach number and periodic boundaries. Future studies will focus on the improvement of the MPC simulation in terms of larger Peclet numbers and a gravitational stimulation that acts individually on the time-dependent position of each particle.

Acknowledgements

Funding by the Mercator Research Center Ruhr (project 2015-0021) is gratefully acknowledged. We thank Godehard Sutmann at Jülich Supercomputing Centre for providing fruitful discussions and the basic algorithm of the MP^{2}C implementation. Patrick Kurzeja gratefully acknowledges financial support by DFG grant KU 3351/2-1.

References

[1] J. Lee, A.J. Ladd, Journal of Fluid Mechanics **577**, 183 (2007)
[2] G. D'Avino, P. Maffettone, Journal of Non-Newtonian Fluid Mechanics **215**, 80 (2015)
[3] C. Kloss, C. Goniva, A. Hager, S. Amberger, S. Pirker, Progress in Computational Fluid Dynamics, an International Journal **12**, 140 (2012)
[4] F. Blanc, F. Peters, E. Lemaire, Applied Rheology **21**, 23735 (2011)
[5] J. Padding, A. Louis, Physical Review E **74**, 031402 (2006)
[6] G. Gompper, T. Ihle, D. Kroll, R. Winkler, in *Advanced computer simulation approaches for soft matter sciences III* (Springer, 2009), pp. 1–87
[7] M. Hecht, J. Harting, T. Ihle, H.J. Herrmann, Physical Review E **72**, 011408 (2005)
[8] C. Cierpka, R. Segura, R. Hain, C. Kähler, Measurement Science and Technology **21**, 045401 (2010)
[9] P. Saffman, Journal of fluid mechanics **22**, 385 (1965)
[10] P. Michael, Physics of Fluids (1958-1988) **8**, 1263 (1965)
[11] W. Matson, M. Kalyankar, B.J. Ackerson, P. Tong, Physical Review E **71**, 031401 (2005)
[12] E.S. Asmolov, Journal of Fluid Mechanics **381**, 63 (1999)
[13] J.J. Stickel, R.L. Powell, Annu. Rev. Fluid Mech. **37**, 129 (2005)

Numerical investigation of compaction of deformable particles with bonded-particle model

Maksym Dosta[1,*], *Clara* Costa[2], and *Hazim Al-Qureshi*[2]

[1]Hamburg University of Technology, 21073 Hamburg, Germany
[2] Federal University of Santa Catarina, 89218-000 Joinville, Brazil

Abstract. In this contribution, a novel approach developed for the microscale modelling of particles which undergo large deformations is presented. The proposed method is based on the bonded-particle model (BPM) and multi-stage strategy to adjust material and model parameters. By the BPM, modelled objects are represented as agglomerates which consist of smaller ideally spherical particles and are connected with cylindrical solid bonds. Each bond is considered as a separate object and in each time step the forces and moments acting in them are calculated. The developed approach has been applied to simulate the compaction of elastomeric rubber particles as single particles or in a random packing. To describe the complex mechanical behaviour of the particles, the solid bonds were modelled as ideally elastic beams. The functional parameters of solid bonds as well as material parameters of bonds and primary particles were estimated based on the experimental data for rubber spheres. Obtained results for acting force and for particle deformations during uniaxial compression are in good agreement with experimental data at higher strains.

1 Introduction

1.1 State of the art

The process of powder compaction can be subdivided into the three main steps. In the first step, particles are subject to sliding and rearrangement without deformation and agglomerate breakage. The second step involves a great amount of plastic or elastic deformation and fragmentation at high pressures, filling the green body voids. Finally, in the last step the compact usually shows a near asymptotic behaviour, where the load transfers due to friction limits densification [1].
The particle behaviour during compaction is a collective result of the discrete interparticle interactions, being affected by individual particle properties, such as hardness, roughness and surface energy. In order to understand the mechanical properties of a compact and the process itself, better knowledge of relationships between individual properties and the macro behaviour is needed. Various approaches in the literature are based on continuum mechanics, as the model proposed by Al-Qureshi et al. [2,3], where large-deformation solution for the axisymmetric problem is proposed, even including work hardening of metallic powders. However, those models act only at the bulk scale, being phenomenological in nature. Those models cannot relate the macroscopic response of a compact to the microscopic inter-particle interaction information, such as particle rearrangement in the first step of compaction [4].
To describe a compaction process on the microscale the Discrete Element Method (DEM) can be used. This simulation method is extensively used in science and industry in a wide range of applications. In the DEM each particle is considered as a separate element and for each particle Newtonian equations of motion are solved. To calculate interparticle interactions hard and soft sphere types of models can be employed. In the case of the soft sphere model it is supposed that particles can undergo small virtual deformations. Based on these values and interparticle overlaps the contact forces are calculated. However, all stressed particles are still treated as ideally spherical objects without considering of shape modification. Moreover, most of contact models are based on the theory of Hertz, which can be applied for relative small deformations. Therefore, the DEM cannot be directly applied for modelling of second of third steps of compaction.
Many different methods have been proposed in the literature, such as Olson and Larsson [5], which use micromechanical tests and FEM to calibrate the particle properties and contact formulations. These were then implemented in the DEM. Hathong et al. [6] have used the Meshed Discrete Element Method (MDEM), in which elements are deformable for small-scale experiments, yielding effective properties used on a full-scale simulation with the DEM. Jerier et al. [7] modified the DEM at high densifications (>85%) during a powder

* Corresponding author: dosta@tuhh.de

compaction by using contact models derived from FEM simulations, using Voronoi cells. A similar approach was also used by Zhang [8], Frenning [9] and Güner et al. [10].

However, despite various research works performed in this area, there is no efficient way to perform simulation of compaction on the microscale. In this contribution the novel modelling approach based on the Bonded-Particle-Model (BPM) is presented and its advantages and disadvantages are discussed.

1.2 Bonded-Particle Model (BPM)

The BPM is an extension of the DEM, where each two primary particles can be connected by solid or liquid bonds [11-14]. This allows to create spherical or irregular-shaped objects as aggregates consisting of primary particles connected with bonds [13]. Moreover, it allows to simulate elasto-plastic deformation or even breakage of modelled objects.

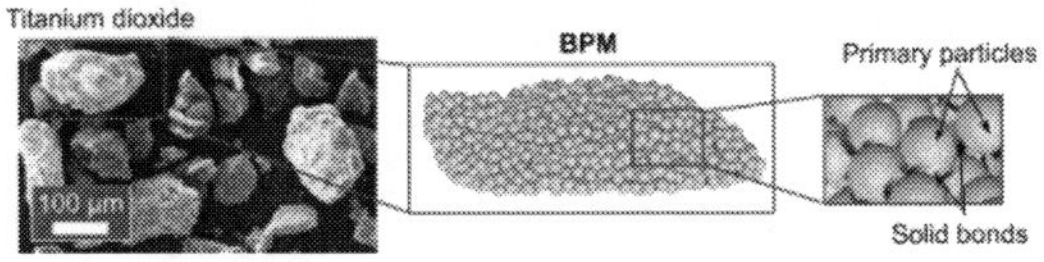

Fig. 1. Representation of titanium dioxide with BPM.

In this contribution the BPM was used to simulate behaviour of spherical rubber particles. Solid bonds between primary particles were described with elastic solid beam model [12]. In each time step, based on the deformations of bonds, the forces and moments acting on primary particles were estimated. As a result, the relative movement of primary particles and deformation of initial objects were calculated.

To perform a simulation of compaction, the BPM was implemented into component-based simulation framework MUSEN [14].

2 Parameter Estimation

The spherical rubber particles of diameter 2 cm have been used as model material in this contribution. The rubber materials are thermoset polymers known for their elastic behaviour and ability to withstand extensive deformation and to return to the original state once the applied force is removed.

Experimental data previously obtained by Shima et al. [15] have been used as reference values. In their experiments the spherical particles were uniaxial compressed and resulted compressive force and lateral deformation were measured, as it is schematically shown in Fig. 2. The force was measured by the load cell and the total displacement by a displacement meter. The starting compression point is visually detected when the rubber and the plate start touching each other. In the reference experiment the lateral extension (U(R)) for the diameter of the spheres was measured using photography at each displacement level, measuring at each force level the expended displacement of the horizontal diameter and taking half of the total as the lateral extension. During the experiment the radius of contact surface (a) between the rubber sphere and the plate was measured through a white paper on the bottom of the sample that was coloured during the compression due to a red ink on the rubber.

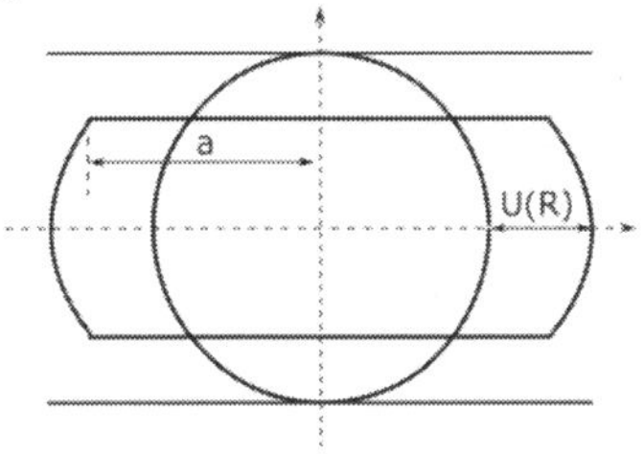

Fig. 2. Schematic representation of performed experiment.

In order to simulate complex behaviour of the rubber sphere a unimodal distribution of primary particles connected by solid bonds was chosen. To evaluate applicability of the proposed modelling approach, the two different shape discretizations were analysed (see Tab. 1). On the one hand a large number of primary particles allows to get better shape representation, on the other hand it leads to much higher computational effort.

Table 1. Representation of rubber sphere with BPM.

	Fine	Coarse
Primary particles	1171	93
Diameter of primary particle [mm]	1.6	3.5
Solid bonds	8135	303
Bond diameter [mm]	1.28	2.8

Material parameters like density, Poisson ratio or friction coefficients of primary particles and bonds was taken equal to material parameters of rubber. However, due to the modifications of a structural model (transition from solid rubber particle to agglomerate) Young's moduli of primary particles E_p and bonds E_b were used as adjustment parameters. For fine approximation were used: E_p=10.8 MPa and E_b=0.2 MPa, for coarse: E_p=17.6 MPa and E_b=0.37 MPa.

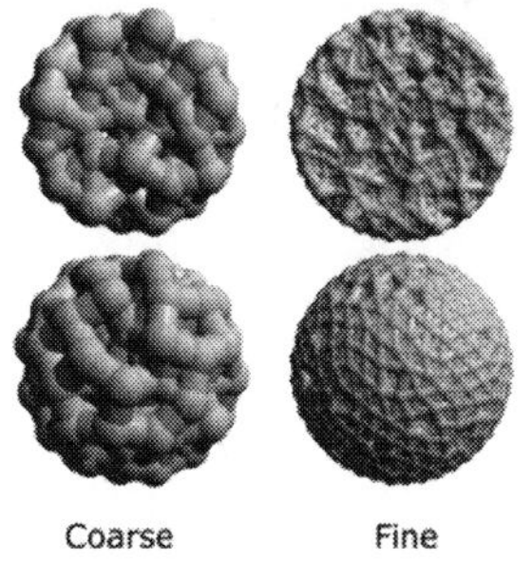

Fig. 3. Coarse and fine representation of rubber sphere.

Some materials such as elastomers have a different behaviour in tension or compression [16]. In order to reproduce such behaviour a solid bond model was extended with an additional parameter to describe bimodularity. This parameter defines the difference between effective Young's moduli of bond in tension and compression [16].

3 Simulation results

3.1 Single particle compression

In order to adjust model parameters and to validate the proposed approach, uniaxial compression of a single agglomerate was employed. The investigated aggregate was placed between the two walls, whereas the upper wall was moved with predefined velocity, as it is schematically shown in Fig. 4. To avoid oscillations caused in result of wave propagation in material, the slow compression velocity was chosen.

Fig. 4. Compression of single rubber particle.

In Fig. 5 and Fig. 6 main characteristics obtained from simulation studies for fine and coarse shape approximations are shown. In both figures the increase of the compression force F and radial elongation U(R) are represented. To see the influence of initial placement of primary particles several aggregates of each type were modelled. The deviations of simulation results are shown with error bars in Fig. 5 and Fig. 6. Moreover, it can be seen there exist differences between results obtained from coarse and fine representation, as well as coarse representation leads to larger deviations. This is caused due to higher material anisotropy caused due to representation of spherical shape only with 93 primary particles.

From the obtained results the conclusion can be drawn, that in compare to the Hertz model the BPM representation allows to predict correctly the force as well as the deformation of objects. Moreover, even a coarse description of an agglomerate shape allows to describe the deformation correctly. The Hertz model is not applicable for large deformations, therefore for strain greater than 10% it is visualized as dotted line.

It is to note, however, that the proposed model allows to describe only those mechanical deformation of spherical objects for strains that are smaller than 50%. For larger strains, the BPM model is not able to describe objects behaviour.

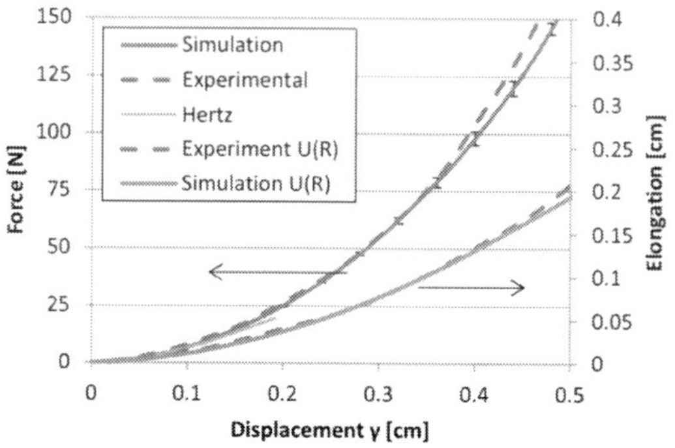

Fig. 5. Fine representation (1171 primary particles).

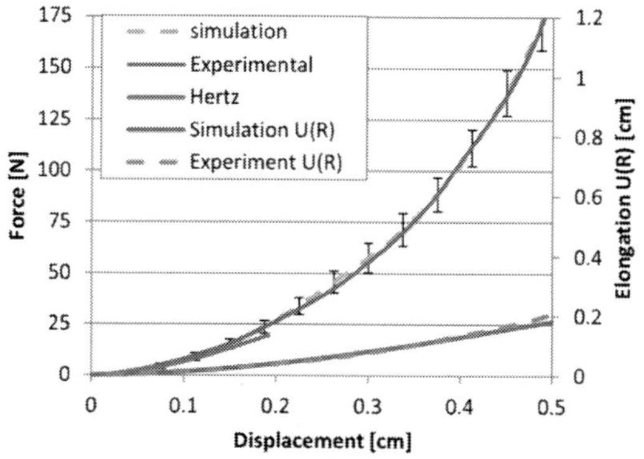

Fig. 6. Coarse representation (93 primary particles).

3.2 Compression of particle collective

After fitting of the parameters for the BPM, the packing consisting of about 2500 ideally spherical rubber particles was modelled without considering of particle deformation. For this purpose Hertz-Mindlin model was used. The particles were filled into cylindrical volume. This packing was uniaxially compressed and the obtained behaviour was compared to the behaviour of the similar packing consisting of coarse aggregates (deformable particles) as it is schematically shown in Fig 7.

In Fig. 8 the obtained force-displacement characteristics for both cases is shown. Here, the forces acting on upper and bottom plates are presented. It can be seen, that there is a decisive difference between simulation results, where the non-deformable spherical particles were modelled by using the Hertz-Mindlin model [12] and where deformable agglomerates were modelled by the BPM. Due to a set of simplifications in the case of spherical particles, like application of the Hertz theory and negligence of deformation, there is an underprediction of the force for the large strains.

However, it should be mentioned, that a coarse shape representation in the BPM, results in overpredicted initial packing density. This is caused due to penetration of primary particles of agglomerate into voids between primary particles of neighbour agglomerates. Therefore, for the detailed investigation of a compaction process a better shape representation and a higher amount of primary particles are needed.

Fig. 7. Compression of particle collective consisting of coarse aggregates (Right side: particles are coloured according to their velocity).

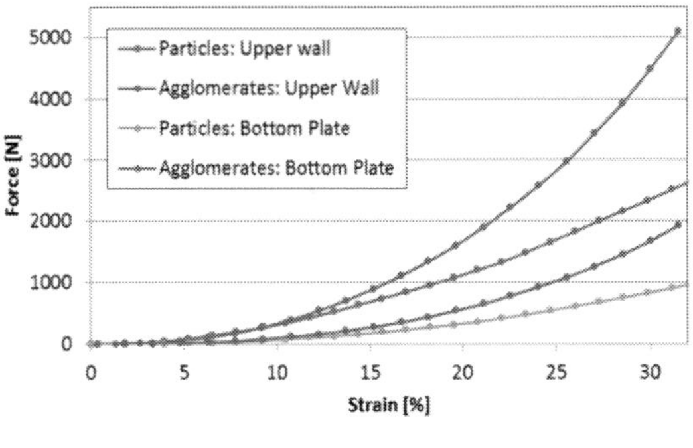

Fig. 8. Force-displacement characteristics obtained from compression of rubber particles.

4 Conclusions

In this contribution the application of the Bonded Particle Model for numerical investigation of compaction process was proposed. This technique was validated on experimental results obtained from uniaxial compression of a single rubber particle. From the results it is apparent that in compare to the convenient DEM, this technique is a very effective way to describe large deformations of objects caused in result of compressive stresses. However, for a precise description of particle shapes fine discretization should be employed.

Application of proposed technique for particle collective showed that considering the particle deformation leads, as expected, to higher compressive forces. However, for the further model additional validation experiments are needed and, therefore, should be performed in future.

References

1. A. Samimi, A. Hassanpour, M. Ghadiri. CES **60**, 3993-4004, (2005).
2. H.A. Al-Qureshi, M.R.F. Soares, D. Hotza, M.C. Alves, A.N. Klein. J. of Materials Processing Techn. **199**, 417-424, (2008).
3. H.A. Al-Qureshi, A. Galiotto, A.N. Klein. J. of Materials Processing Techn. **166**, 135-143, (2005).
4. H.P. Zhu, Z.Y. Zhou, R.Y. Yang, A.B. Yu. CES 63, 5728-5770, (2008).
5. E. Olsson, P.-L. Larsson. Powd. Techn. **243**, 71-78 (2013).
6. B. Hathong, J.-F. Jerier, P. Doremus, D. Imbault, F.-V. Donze. Int. J. of Solids and Structures **46**, 3357-3364, (2009).
7. J.-F. Jerier, B. Hathong, V. Richefeu, B. Chareyre, D. Imbault, F.-V. Donze, P. Doremus. Powd. Techn., **208**, 537-541, (2011).
8. J. Zhang. Composite Sci. and Techn., **62**, 2048-2053, (2009).
9. G. Frenning. CES **65**, 2464-2471, (2010).
10. F. Güner, Ö.N. Cora, H. Sofuoglu. Powd. Techn., **271**, 238-247, (2015).
11. D.O. Potyondy. Geosystem Eng. **18**, 1-28 (2015).
12. M. Dosta, S. Dale, C.R. Wassgren, S. Heinrich, J.D. Litster. Powd. Techn. **299**, 87-97 (2016).
13. S. Kozhar, M. Dosta, S. Antonyuk, S. Heinrich, L. Gilson, U. Bröckel. Adv. Powd. Techn. **26**, 767-777 (2015).
14. M. Dosta, S. Antonyuk, S. Heinrich. IECr **52**, 11275-11281 (2013).
15. S. Shima, Y. Tatara, M. Lio, J.C. Lucero. JSME Int. J. **36**, 197-205 (1993).
16. M. Destrade, J.G. Murphy, B. Rashid. Soft Cond. Matter. **1**, 29-37 (2009).

Slip-stick excitation and travelling waves excite silo honking

Katarzyna Warburton[1], *Elze* Porte[2,3], and *Nathalie* Vriend[1,*]

[1] *University of Cambridge, UK, Department of Applied Mathematics and Theoretical Physics*
[2] *Imperial College London, UK, Department of Mechanical Engineering*
[3] *University of Twente, Netherlands, Department of Mechanical Engineering*

Abstract. Silo honking is the harmonic sound generated by the discharge of a silo filled with a granular material. In industrial storage silos, the acoustic emission during discharge of PET-particles forms a nuisance for the environment and may ultimately result in structural failure. This work investigates the phenomenon experimentally using a laboratory-scale silo, and successfully correlates the frequency of the emitted sound with the periodicity of the mechanical motion of the grains. The key driver is the slip-stick interaction between the wall and the particles, characterized as a wave moving upwards through the silo. A quantitative correlation is established for the first time between the frequency of the sound, measured with an electret microphone, and the slip-frequency, measured with a high-speed camera. In the lower regions of the tube, both the slip-stick motion and the honking sound disappear.

1 Introduction

The discharge of industrial silos is known to produce low frequency resonance known as *silo honking* or *silo music* [1, 3–5]. The sound forms a nuisance for the environment when the structure starts to act as a loudspeaker. The acoustic emission consists of a main frequency (typically between 50 - 200 Hz) and a fixed set of harmonics.

We model silo honking in a laboratory set-up with a small diameter vertical tube filled with fine particles, reproducing the key features of silo music [1]. A typical spectrogram is shown in figure 1 and shows two distinct resonances contributing to the honk: a falling and a rising one. When the distinct resonances intersect in frequency at a given time, the amplitude temporary increases. At irregular times jumps towards a higher frequency occur, usually representing an alteration of approximately 5 Hz. A peculiar observation is that the duration of the acoustic emission is significantly shorter than the total duration of the discharge: the audible sound stops abruptly, well before the silo is emptied of particles.

The falling signal is the air resonance, not too dissimilar from an organ pipe, forming an open-closed tube system of the resonating air layer between the bulk surface and the top of the tube [6]:

$$f_n = \frac{c}{4H_a}(2n-1), \tag{1}$$

with air layer height H_a, air velocity c, and odd harmonics ($2n-1 = 1,3,5...$). The frequency scales inversely with the length, resulting in a decreasing frequency as the silo discharges and H_a increases. The structural resonance typically starts at a lower frequency, displays all integer harmonics (n = 1,2,3...) and the frequency increases slowly by a factor of around 1.5 during silo discharge. This structural resonance is not a result of the natural modes of the tube: the bending, radial-axial and longitudinal modes are excluded as the dominant mechanism for setting the structural frequency due to their incompatible harmonic structure and/or frequency range. It is the source of this structural resonance that we investigate in this paper.

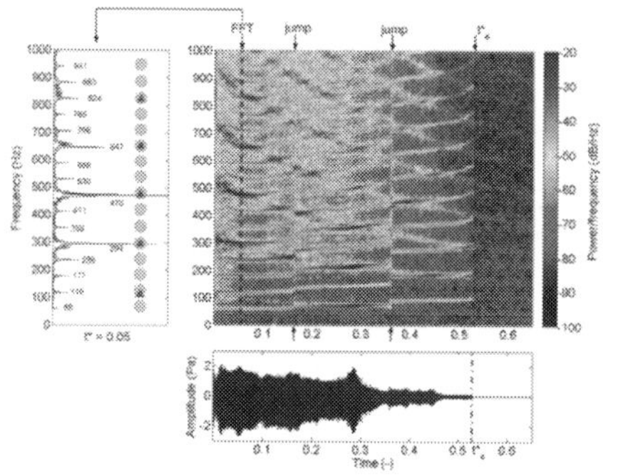

Figure 1. A typical spectrogram, signal and FFT plot [2], featuring the falling air and rising structural resonance and sudden jumps. Note that time has been non-dimensionalised by the total discharge time and time t_e^* denotes the abrupt cessation of the sound, just past halfway through the discharge.

*e-mail: n.m.vriend@damtp.cam.ac.uk

DOI: 10.1051/epjconf/201714010009

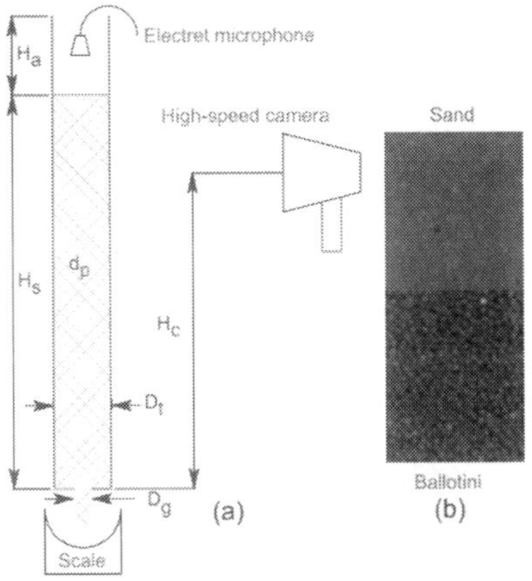

Figure 2. (a) A schematic of the experimental setup, and (b) images of grains taken by the camera (top, 0.15-0.25mm sand, below, 0.3-0.4mm ballotini).

2 Methods

2.1 Experimental setup

For our laboratory experiments, we model the silo as a $2m$ long vertical tube (figure 2), filled with a fixed mass of 4 kg to a height of approximately $H_s = 1.45m$, which leaves an air layer H_a. We vary the wall material (perspex and glass) while keeping the inner tube diameter essentially constant ($D_t = 50mm$ versus $D_t = 51mm$, respectively).

The tube is supported at the base by an MDF plate with outlet gaps of diameters D_g between $8mm$ and $14mm$, which rests on an aluminium frame. The sides are attached to a rail via a padded strapping, allowing the tube to vibrate without being rigidly fixed to the supporting structure. The tube is filled to H_s with granular material of different composition and size. In this work, we used sieved particles, of both silver sand ($d_p = 150 - 250\mu m$) and ballotini ($d_p = 90 - 150\mu m$, $150 - 250\mu m$, or $300 - 400\mu m$).

Above the grain layer, near the top of the tube, a small electret microphone with a sampling frequency of $44100Hz$ is suspended, acquiring acoustical data for the duration of the discharge. The rate of discharge is derived by tracking the discharged mass with a set of Ohaus Scout Pro electronic scales. Video footage, collected at various heights H_c, is acquired by up to two Photon FASTCAM SA1.1 high-speed cameras through the transparent tube walls.

2.2 Processing of camera footage

The camera is mounted at various heights H_c between 0.7m and 1.1m at 10cm intervals, to image the particle motion at the walls at different stages of the discharge. Visual data is post-processed with three different modes: (1) tracking of the bulk surface, (2) particle tracking velocimetry routines to measure particle displacements, and (3) particle image velocimetry routines to measure flow structures. For mode (1), a wide-angle 24 mm lens was used to obtain a visible 50 cm region of the tube at 60 fps, showing pale grains against a black background. This image was used as input in a Sobel edge detection filter, and produced position of the free surface as a function of time. For mode (2) and (3), a 60 mm lens collected data at a higher frame rate and focussing on a much smaller area (5 cm by 1 cm), to resolve particle motions and flow structure respectively. For mode (2), one camera at 2000 fps was used, and for mode (3), two cameras, placed 50 cm apart at $H_c = 0.65$m and $H_c = 1.15$m, were used at 5400 fps to measure the wave of motion by cross-correlation.

3 Results and discussion

3.1 Structural resonance

Typically, the first harmonic of the audio recording is initially in the range of $50-70Hz$, varying slightly depending on atmospheric conditions and packing fraction. However, during discharge this first harmonic steadily increases, almost doubling its initial value by the time that sound abruptly ceases. Interestingly, changing the wall material did not alter the initial frequency significantly, even though the frictional characteristics between the particles and the wall material are quite different. The particle choice did make a difference: main harmonic for sand was initially approximately 20% higher than that for ballotini.

The integer harmonics (n = 1,2,3,...) and slowly rising frequency of the structural resonance (figure 1) indicate a closed-closed system of decreasing length or increasing velocity. Several authors [4, 7] suggested that the structural resonance is at a natural eigenmode frequency of the silo. Our initial work [2] demonstrated this explanation is inadequate due to the harmonic structure and the frequency range. To support this statement, we now performed proof-of-concept experiments with a PCB 352C22 accelerometer attached to different positions on the tube during discharge. No radial vibrations could be recorded for the perspex tube and only low amplitude vibrations were recorded with the thinner glass tube. The axial modes are probed by closing the base of the tube with a flexible latex sheet and simply tapping the filled tube through the sheet. Motion measured by the accelerometer in axial direction matched the characteristics of the structural resonance, but the lowest natural resonant frequency of our tube was $80Hz$, which is too high to account for the silo honking.

3.2 Slip-stick frequency

As reported by Buick et al. [8] and Börzsönyi et al. [4], particles undergo periodic slip-stick motion down the tube, with extended sticking to the tube wall followed by acceleration at close to g. As seen from the displacement data (blue) in figure 3, each slip-stick cycle is extremely regular, and therefore it is possible to extract its recurrent frequency using temporal frequency analysis. Of note is the slight relaxation, resulting in a small upward displacement and small negative velocities, which occurs close to the bulk surface for early stages of the process for sand particles and the smaller ballotini sizes. In a novel approach,

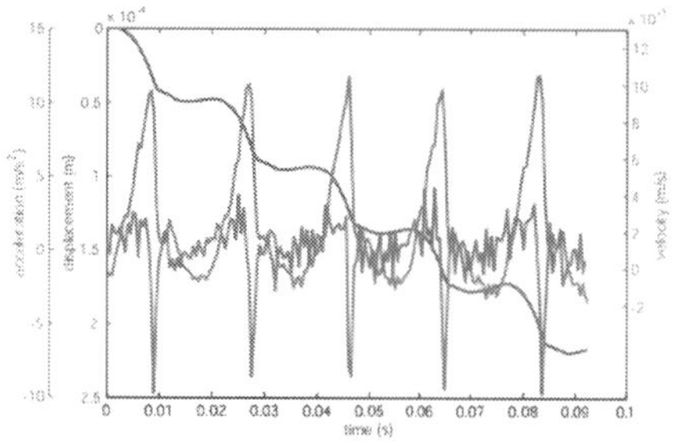

Figure 3. Average particle motion, velocity, and acceleration for ballotini particles in a glass tube at $H_c = 1.1$m. Each slip is around 0.05mm, with velocities reaching $10^{-3}ms^{-1}$, and deceleration around g.

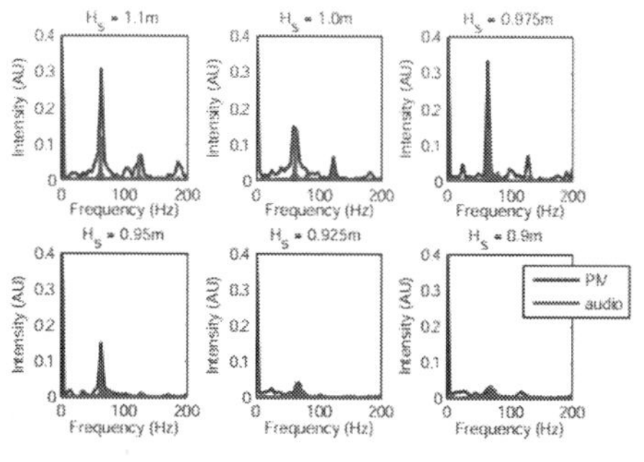

Figure 4. Comparison between the frequency of slip-stick particle motion and recorded audio, for ballotini particles in a perspex tube. The data was collected when the free surface of the sand moved from $H_s = 1.1$m to $H_s = 0.9$m, while the camera was located at $H_c = H_s - 0.05m$: the higher harmonics cut out first, and all sound and slip-stick motion ceases below 0.9m.

we capture simultaneously audio data and record particle motion to investigate the correlation between the two measurements. Figure 4 illustrates that the slip-stick frequency derived from particle motion and the audible structural resonance show an exact match within the available resolution of $0.5Hz$ for a variety of stages in the release. This surprising result has not been reported before: previous work identified slip-stick as a characteristic of silo honking, but a quantitative correlation has never been established. The correspondence between slip-stick was unaffected by flow rate, particle type, or height of recording, and seen in every single run. As the free surface of the sand moved below 0.9 m, both the slip-stick motion and the audible frequency were significantly reduced and disappeared.

3.3 Slip-stick steps

Important characteristics of slip-stick beyond the frequency are the step-size and the maximum and average velocity of motion. The rate of discharge through the base gap, measured with a scale, is not periodic but continuous. The slip-stick behavior gradually disappears as the free surface drops to heights below $H_s = 0.9$m, only occurs to a small extent below the critical height at which honking cuts out in early parts of the experiment and is always absent in the lower regions of the tube. This indicates that the upper portion of column of particles has the largest role in exciting silo honking.

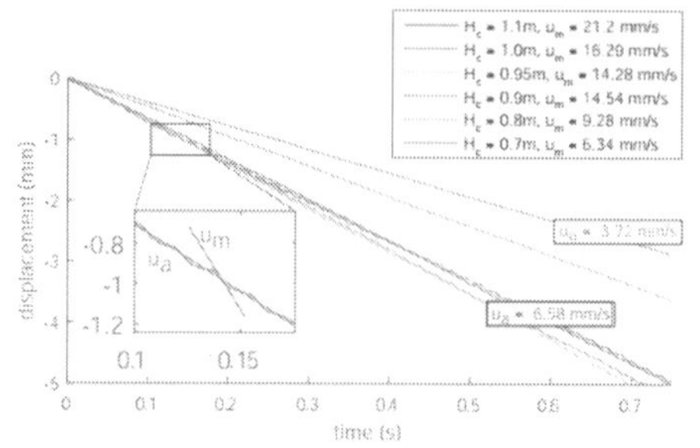

Figure 5. Initial particle motion for ballotini on perspex at different heights (plotted on the right hand axis). Little slip-stick motion is observed at 0.8m and below. Note also that particles at the top of the tube initially move faster, as seen in the varying gradient: $u_a = 6.58mms^{-1}$ for $H_c = 1.1$m to $u_a = 3.72mms^{-1}$ for $H_c = 0.7$m.

Figure 5 shows that average velocity of the grains at the side-wall, whether they undergo slip-stick or not, is initially approximately constant around $u_a = 7mms^{-1}$, but drops dramatically by a factor 2 for much lower heights of the free surface. As the overall discharge rate does not change, we attribute this to a different internal velocity profile, where in the lower regions of the tube larger continuous shear rates occur close to the wall, where in the upper regions these large shear rates are only intermittent. This assessment is further strengthened by the observation that the instantaneous maximum velocity during the slip-portion of the motion is very high (up to $u_m = 21mms^{-1}$) initially at $H_c = 1.1$m, but consistently decreases with height, by a factor of 3.5 at heights of $H_c = 0.7$m. This can be visualized by the fact that for the upper regions, the velocity fluctuations and maximum velocity peak are high, but spread over a small time step size, while for the lower regions, the maximum velocity is much lower and smeared out over time. This process is a continuous transition, to a point where the periodic structure of slip-stick is difficult to identify and the flow becomes continuous as indicated in figure 5 for $H_c = 0.7$m.

The amplitude of the sound emission, for both sand and ballotini particles, was higher in the glass tube compared with the perspex tube. An interesting observation was made with regards to the sound duration in the glass tube: the honking time for sand shortened, while it lasted longer for ballotini particles.

3.4 Travelling waves

Regions of slip-stick motion propagate upwards through the tube, in a motion resembling a wave as illustrated in figure 3. For a given height, there is a gradual velocity increase followed by a rapid deceleration. This can be explained by a gradual expansion of packing fraction, where particles fall into gaps left by the movement of lower particles, followed by a sudden stop. In figure 6 we show we

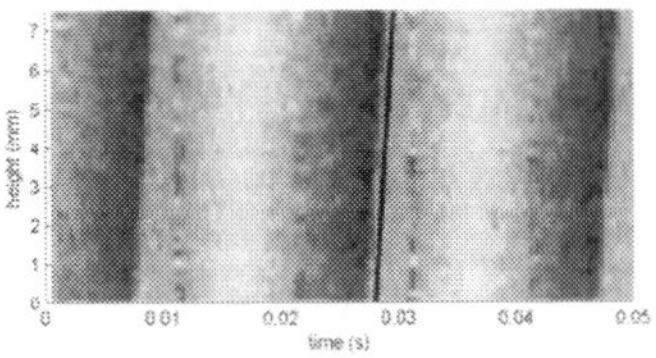

Figure 6. Waves of motion (red) propagate upwards through the imaged region extremely rapidly ($\approx 60ms^{-1}$), as seen from the steep gradient of these lines.

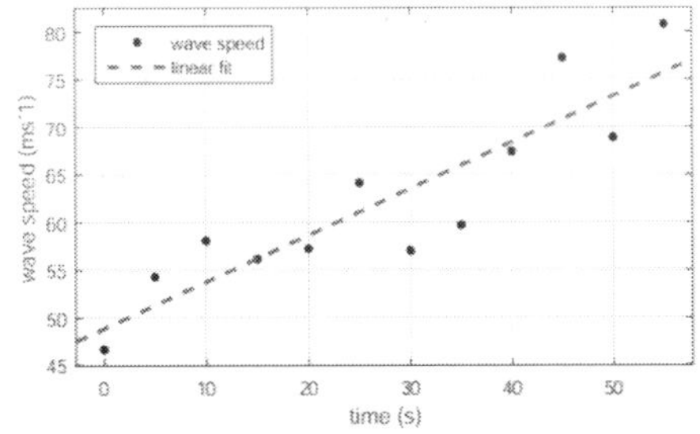

Figure 7. Wave speed c increases linearly with time t, $c = At + b$, with $A = 0.49ms^{-2}$, $b = 48.9ms^{-1}$, coefficient of determination $R^2 = 0.80$.

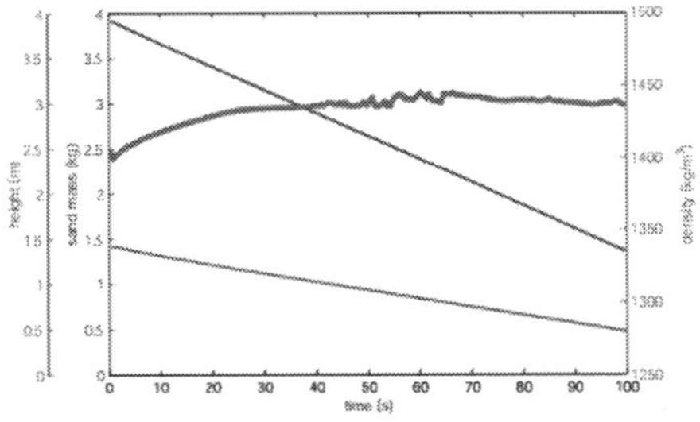

Figure 8. Mean surface height, particle mass, and density during the flow of sand particles in a perspex tube through a gap width of 10mm.

can estimate the speed of this slip-stick wave c, which is distinct from the particle speed u_a and u_m, by looking at the phase shift between particle velocity peaks at each end of the imaged region Using two cameras at a separation of 50cm, we obtain a more accurate measure of mean wave speed over the whole tube. Locally, wave speed is known to vary with height [4], but average speed provides a more important measurement for setting the frequency. For 0.3-0.4mm ballotini, the wave speed increased during the flow by a factor of 2, as shown in figure 7. During this time, the honking frequency also increases, but not as rapidly. A central question is therefore what determines the length scale of the slip-stick.

One unknown factor is the packing fraction, which changed significantly during the flow. The wave speed in granular materials scales with the pressure as $p^{1/6}$, with pressure depending on both density and overburden height. The flow rate at the outflow point is constant, as confirmed by measurements from the scale. The height of the bulk surface, however, does not decrease at a constant rate. Initially, the height drops more rapidly during the first third of the flow, after which the level drops at a constant rate as illustrated in figure 8. This corroborates evidence from particle tracking where the average particle velocity was higher at the top of the tube, implying a lower average density here. The changing height but constant outflow rate results in an increasing packing fraction and density during the first third of the flow duration, after which the packing fraction stabilizes, providing another factor for increasing wave speed and increasing resonant frequency over time.

4 Conclusion

We find that the slip-stick motion and the emitted acoustical frequency are strongly correlated with each other: the repetitive frequency of slip-stick is equal to the frequency of the sound produced by the structural resonance in the tube during discharge. The upward travelling waves, connected to a convective linear stability in [5], create regions where active slip-stick occurs. The key open question is what the effect is of the frictional characteristics between the different grains and the tube choices, and how this quantitatively sets the frequency of slip-stick motion – further work is necessary to answer this question.

We would like to acknowledge support by Dr. Marc Masen (Imperial College London) and Professor André de Boer (University of Twente) at earlier stages of this project, and the collaboration with Professor Jim Woodhouse (University of Cambridge) on the deployment of accelerometers. NMV would like to acknowledge financial support from the Royal Society Dorothy Hodgkin Research Fellowship RG70273 and KW support from a LMS summer bursary and the Bridgewater programme.

References

[1] K. Wilde, J. Tejchman, M. Rucka, and M. Niedostatkiewicz, Powder Technol. **198**, 38 (2010)

[2] E. Porte, M. Masen, N.M. Vriend and A. de Boer, APS-DFD conference abstract, BAPS.2015.DFD.L10.12, (2015)

[3] J. M. Buick, J. Chavez-Sagarnaga, Z. Zhong, J. Y. Ooi, Pankaj, D. M. Cambell, and C. A. Greated, J. Engrg. Mech. **131**, 299 (2005)

[4] T. Börzsönyi and Z. Kovacs, Phys. Rev. E **83**, 03230 (2011)

[5] L. Bonneau and T. Catelin-Jullien, B. Andreotti, Phys. Rev. E **82**, 011309 (2010)

[6] J.W.S. Rayleigh, *The Theory of Sound*, (Dover, London, 1945)

[7] K. Wilde, M. Rucka, and J. Tejchman, Powder Technol. **186**, 113 (2008)

[8] J. M. Buick, Pankaj, J. Y. Ooi, J. Chavez-Sagarnaga, A. Pearce, G. Houghton, J. Phys D: Appl. Phys. **37**, 2755-2760 (2004)

Approximation of mechanical properties of sintered materials with discrete element method

Maksym Dosta[1,*], *Robert* Besler[2], *Christian* Ziehdorn[1], *Rolf* Janßen[2], *Stefan* Heinrich[1]

[1]Institute of Solids Process Engineering and Particle Technology, Hamburg University of Technology, 21073 Hamburg, Germany

[2]Institute of Advanced Ceramics, Hamburg University of Technology, 21073 Hamburg, Germany

Abstract. Sintering process is a key step in ceramic processing, which has strong influence on quality of final product. The final shape, microstructure and mechanical properties, e.g. density, heat conductivity, strength and hardness are depending on the sintering process. In order to characterize mechanical properties of sintered materials, in this contribution we present a microscale modelling approach. This approach consists of three different stages: simulation of the sintering process, transition to final structure and modelling of mechanical behaviour of sintered material with discrete element method (DEM). To validate the proposed simulation approach and to investigate products with varied internal structures alumina powder has been experimentally sintered at different temperatures. The comparison has shown that simulation results are in a very good agreement with experimental data and that the novel strategy can be effectively used for modelling of sintering process.

1 Introduction

The sintering process is one of the most widely used heat treatment processes in the ceramic industry. During sintering densification of heated green body occurs and the free enthalpy, as well as, the surface and contact energies are minimized [1]. However, despite the wide applications of sintering process, until now there are no precise methods allowing to describe the process with high detailing grade. Nowadays, most of applied models are based on the continuum mechanical approach, where the material microstructure, anisotropy or high porosity cannot be effectively modeled [2].
In this contribution the application of the discrete element method (DEM) for modeling of sintering process is proposed. The DEM was initially formulated and applied to predict behavior of granular materials consisting of ideally spherical particles. However, in the recent years this method was extended with new contact models and calculations algorithms. One of the interaction laws was specially developed for modeling of material densification occurring during sintering process [3]. The main advantage of the DEM compared to experimental investigations is the detailed comprehensive information about trajectories and position of individual particles as well as forces acting on them. As an example Parhami and McMeeking [3] have proposed DEM model for description of initial stage sintering. This model was afterwards effectively applied by Martin et al. [4] for modelling of defects evolution. Based on this model Besler et al. [5] have investigated structural deformation of periodic macro porous alumina as well as densification of ceramic-metal composite materials with varying metal content [6].
Another extension of the DEM is a bonded particle model (BPM). By the BPM each two primary particles can be connected with solid or liquid bonds. Similarly to primary particles, all bonds are treated as separate objects. This allows to represent complex material structures with high porosities and to investigate their mechanical behaviour [7]. For example Kozhar et al. [8] have investigated breakage behaviour of micrometer-sized titania at compression with the BPM, Dosta et al. [9] applied this approach to predict strength of glass agglomerates.
In this contribution both extensions of the DEM which have been described above have been used for approximation of mechanical properties of sintered materials. For this purpose novel simulation strategy consisting of three stages has been proposed:

- first stage: modelling of the sintering process, where material densification is calculated using DEM;
- second stage: transition from a set of overlapped spheres to the bonded-particle model, consisting of non-overlapped spheres connected with solid bonds;
- third stage: investigation of mechanical material properties during three-points bending tests.

In this study, alumina powder (Al_2O_3) has been used as model material. Due to various chemical and physical properties, the processing of alumina has a set of advantages compared to other materials; therefore it is widely used for sintering.

* Corresponding author: dosta@tuhh.de

2 Experimental measurements

In order to validate model a set of experiments was performed to measure the mechanical properties of sintered material. In the first step, the green body was produced by material densification through uniaxial compression. For this purpose the press Model 10 of company Paul-Otto Weber GmbH with matrix dimensions of 470 x 45 x 620 mm was used.
Afterwards, the sample was pressed again through cold isostatic pressing with a pressure of 150 MPa. The resulting green body was analyzed in order to obtain information about its composition (e.g. density, dimensions).
Thereafter, all samples were sintered by varying temperatures between 1100 °C and 1250 °C in the chamber oven HT 04/17 of company Nabertherm GmbH. In the case of all studies the heating up and cooling down process was controlled by a temperature increase and decrease of 10 K·min^{-1}. The holdup time for each sintering temperature was equal to one hour.
Directly after the sintering process, all samples were measured again to obtain information about their new geometrical characteristics and density. In Fig. 1 dependency between sintering temperature and relative density is illustrated.

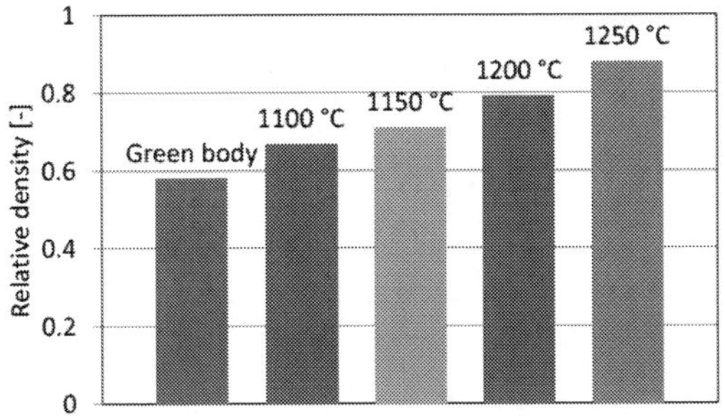

Fig. 1. Dependency of samples densities from sintering temperature.

Before proceeding with the bending tests, each specimen was polished coplanar to ensure the correct measurement of Young's modulus. All bending tests were accomplished by the three point bending process in a self-developed bending device at the Institute of Advanced Ceramics of Hamburg University of Technology. In order to minimize position effects, each specimen was loaded and unloaded for three times with a maximum loading force of 60 N. To receive the value of the slope of the force-displacement characteristic, only the last discharge cycle with values between 50 N and 20 N was taken into account.

3 Modelling of sintering process

For the modeling of sintering process the contact model proposed by Parhami and McMeeking [3] was implemented into the component-based simulation framework MUSEN [10]. According to this model, between each two contacting particles, the forces in normal and tangential direction are acting. The normal force is described with Eq. 1 and consists of two terms:

- viscous force, which describes resistance against relative motion;
- sintering force acting toward the contacting particles.

$$F_n = \pi v_{rel} a^4/(2\beta\Delta_b) - \alpha\pi R\gamma/\beta \qquad (1)$$

where v_{rel} is a relative velocity, a is a contact radius, α and β are parameters which define ratio between grain boundary to surface diffusion, R is a particle radius, Δ_b diffusion parameter.

The force in tangential direction depends on the relative velocity in tangential direction, the dimensionless coefficient of friction and the contact radius between two particles. Detailed description of the model and its application on alumina or on ceramic-metal composites can be found in Besler et al. [5], [6].

The main parameters which have been used for DEM simulation of sintering process are listed in Table 1. In order to decrease calculation time a scaling factor has been introduced. With help of scaling factor the mass of primary particles has been increased. Such scaling allow to increase calculation time step and as a consequence to decrease calculation time.

Table 1. Main simulation parameters.

Parameter	Value
Density [kg/m^3]	3950
Particle diameter [nm]	100
Surface energy [J/m^2]	1.1
Atomic volume [m^3]	8.47e-30
Particles number	99000
Scaling factor	1e+20

Contrary to the experiments, the simulation of sintering has been performed only for one specific value of temperature 1200 °C. As result, the time-dependent change of the relative density in the sample has been obtained. In Fig. 2 time-dependent propagation of the relative density is shown with solid line. It can be seen that with an increased sintering time the density increased from 0.58 (green body) up to 0.9.

In order to get direct correlation between simulation and experimental data four time points with porosities equal to the experimental obtained data (0.67, 0.71, 0.79, 0.88) have been selected. These time points have been used afterwards as initial data for the second modeling stage.

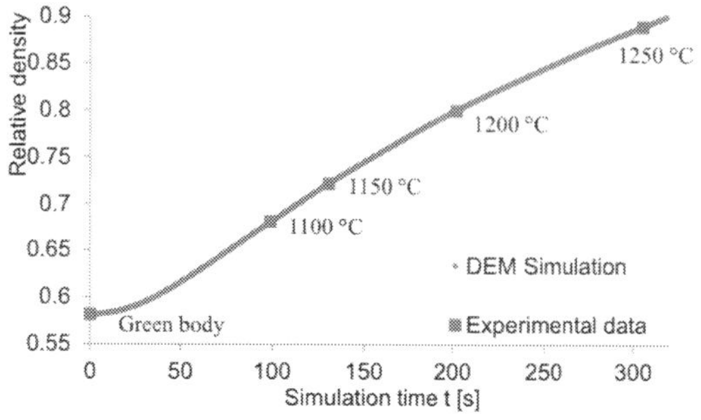

Fig. 2. Time-dependent propagation of relative density.

4 Transition to BPM

After modeling of sintering process the alumina sample was represented as a set of overlapped ideally spherical particles. In order to perform further calculations of the mechanical properties, the transition to the bonded-particle model has been done. During this transition, each two overlapped primary particles have been converted into the system of smaller non-overlapped spherical particles with cylindrical solid bond between them, as it is schematically illustrated in Fig. 3.

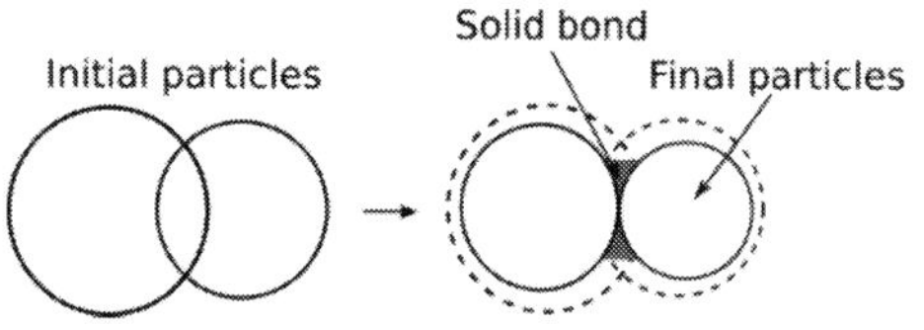

Fig. 3 Schematic representation of transition algorithm.

The transformation was performed as an iterative process. In the every iteration of algorithm, all pairs of the overlapped particles in the sample were identified. Afterwards, radii of overlapped particles R_n, R_m in each pair were decreased according to:

$$R_{n,i+1}=R_{n,i}-\Delta R;\ R_{m,i+1}=R_{m,i}-\Delta R \quad (2)$$

where i is the number of iteration and ΔR is the decrement which was equal to 0.0001 nm. To fulfill mass conservation, the volume of the corresponding solid bond between particles n and m (V_{nm}) was increased according to Eq. 3. Iterations have been repeated until there is no more pair of overlapped particles in the system.

$$V_{nm,i+1} = V_{nm,i}+4/3\pi(R^3_{m,i}+ R^3_{m,i} -R^3_{m,i+1}- R^3_{n,i+1}) \quad (3)$$

When the algorithm was finished the radii and lengths of all bonds were calculated based on their volumes. During calculation it was considered, that the length of the bond L_b between two particles with radii R_1 and R_2 is equal to:

$$L_b = L_{pp}-(R_1^2-R_b^2)^{0.5}-(R_2^2-R_b^2), \quad (4)$$

where L_{pp} is distance between particle centres and R_b is bond radius. In Fig. 4 the size distribution of bonds radii obtained for sintering temperature 1100 (porosity 0.67) is shown.

In Table 2 main properties of bonds for different case studies are listed. It can be seen, that increase of the sintering temperature leads to the significant increase of the solid bonds number. This is caused due to sample shrinkage and formation of new interparticulate contacts.

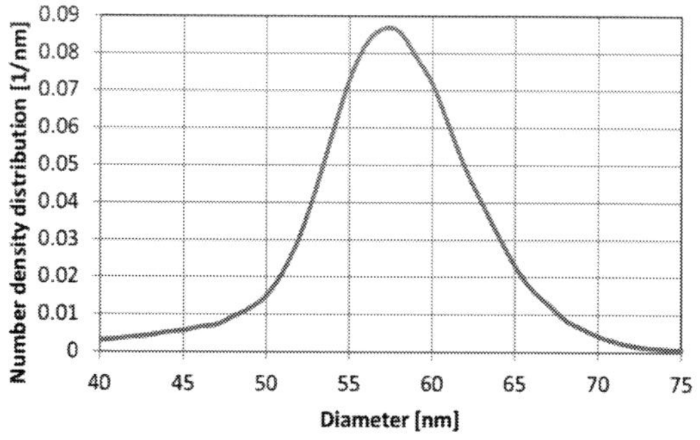

Fig. 4 Distribution of solid bonds according to their radii.

Table 2. Parameters of solid bonds.

Parameter	1100 °C	1150 °C	1200 °C	1250 °C
Average diameter [nm]	57.2	56.89	56.26	55.56
Average length [nm]	23.9	24.5	25.46	26.51
Bonds number [-]	270514	288655	321888	353318

5 Approximation of mechanical properties

In the final stage mechanical properties of material has been modeled with DEM. For this purpose three-point bending test has been used. Sample was placed between three cylindrical punches and upper cylinder was moved downwards with predefined velocity. For particle-particle or particle-wall interactions Hertz-Mindlin contact model was used. The solid bonds have been calculated according to the model of elastic solid beam [9].

Young's modulus and Poisson ratio of primary particles and bonds were equal to 400 GPa and 0.23. These are the parameters of alumina material which was used in the experimental measurements.

In Fig. 5 simulation results for one case study are shown. In this figure particles are hided and only solid bonds are shown. The bonds are colored according to the distribution of forces acting in them. It can be seen that in the upper part of sample mainly compression forces are acting on bonds and in the lower part the tension forces.

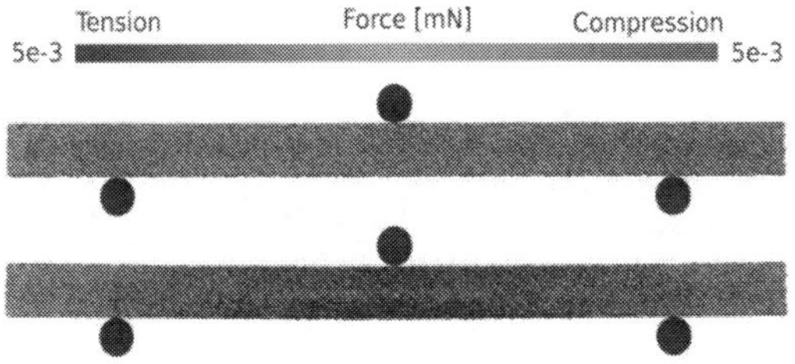

Fig. 5. Forces acting in solid bonds during three-points bending test.

Based on the simulation results the Young's moduli of the final samples have been approximated. This has been done based on the analysis of the obtained force-displacement characteristics. The numerically and experimentally obtained dependencies between sample porosity and Young's modulus are shown in Fig. 6. In the same figure the correlation proposed by Knudsen [11] is illustrated.

From the analysis of obtained results the conclusion can be drawn that proposed model is in a relative good agreement with the experimental results and is able to predict the stiffness of the sintered material for the relative densities in the range between 0.7 and 0.9. However, it should be mentioned, that the DEM model for calculation of sintering process does not contain terms which are able to describe grain growth. As a consequence this model cannot be applied for densities larger than 0.9.

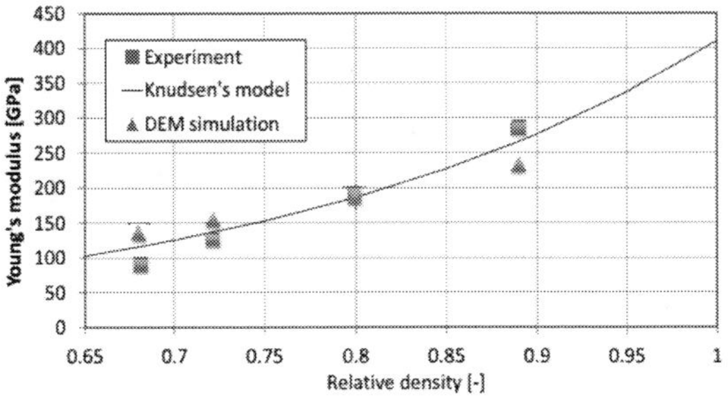

Fig. 6. Dependency of the Young's modulus on the relative density of sintered material.

6 Conclusion

In this contribution the microscale modeling approach to predict stiffness of sintered materials has been proposed. This approach based on the discrete element method and bonded-particle model and allows to investigate properties of sintered material. However, the developed model can be effectively applied only for relative porosities in the range between 0.7 and 0.9. In the further investigations it is planned to extend sintering model with terms to describe grain growth.

Acknowledgments

We gratefully acknowledge financial support from the German Research Foundation (DFG) via SFB 986 "M3", project A3 and C5.

References

1. H. Salmang, H. Scholze, R. Telle. *Keramik* (Springer Verl., 2007).
2. E.A. Olevsky. Mat. Sci. Eng. R. **2**, 41-100, (1998).
3. F. Parhami, R. McMeeking, Mech. Mat. **2**, 111-124 (1998).
4. C.L. Martin, H. Camacho-Montes, L. Olmos, D. Bouvard, R.K. Bordia, J. Am. Ceram. Soc. **7**, 1435-1441 (2009).
5. R. Besler, M. Rossetti, M. Dosta, S. Heinrich, R. Janssen, J. of the Amer. Ceram. Soc., **26**, 1021-1030 (2015).
6. R. Besler, M.R. da Silva, M. Dosta, S. Heinrich, R. Janssen, J. of the Europ. Ceram. Soc. **36**, 2245-2253 (2016).
7. D.O. Potyondy. Geosystem Eng. **18**, 1-28, 2015.
8. S. Kozhar, M. Dosta, S. Antonyuk, S. Heinrich, L. Gilson, U. Bröckel. Adv. Powd. Techn. **26**, 767-777 (2015).
9. M. Dosta, S. Dale, C.R. Wassgren, S. Heinrich, J.D. Litster, Powd. Techn. **299**, 87-97 (2016).
10. M. Dosta, S. Antonyuk, S. Heinrich. Ind. Eng. Chem. Research **52**, 11275-11281 (2013).
11. F.P. Knudsen, Journ. of Amer. Ceram. Soc. 45, 94-95, (1967).

Models of current sintering

Sebastian Angst[1,*], *Lukas* Engelke[1], *Markus* Winterer[2], and *Dietrich E.* Wolf[1,**]

[1]*Faculty of Physics and CENIDE, University of Duisburg-Essen, D-47057, Duisburg, Germany*
[2]*Nanoparticle Process Technology and CENIDE, University of Duisburg-Essen, D-47057, Duisburg, Germany*

Abstract. Densification of (semi-)conducting particle agglomerates with the help of an electrical current is much faster and more energy efficient than traditional thermal sintering or powder compression. Therefore, this method becomes more and more common among experimentalists, engineers, and in industry. The mechanisms at work at the particle scale are highly complex because of the mutual feedback between current and pore structure. This paper extends previous modelling approaches in order to study mixtures of particles of two different materials. In addition to the delivery of Joule heat throughout the sample, especially in current bottlenecks, thermoelectric effects must be taken into account. They lead to segregation or spatial correlations in the particle arrangement. Various model extensions are possible and will be discussed.

1 Introduction

Current-activated pressure-assisted densification (CAPAD) is used to rapidly densify (semi-)conducting powders with little grain growth, which enables an upscalable production of nanostructured bulk material [1, 2]. A low pressure in the range of 10-100 MPa is applied, while a kA current is flowing through the initially porous powder. The pressure alone is too weak for a significant densification. Dissipation of electrical energy is used to heat up the powder, which enables fast and almost complete densification.

The mechanisms underlying current "sintering" are very controversial. Therefore the experimental procedure is known under several different names: Field assisted sintering technology (FAST), spark plasma sintering (SPS), and current-activated pressure-assisted densification (CAPAD) [3], which we adopt in this paper. Especially during early stages of the process particle rearrangement is among the relevant densification mechanisms [2, 4, 5].

Several approaches to simulate CAPAD were developed in recent years, among them thermo-electrical-mechanical finite element studies [6, 7]. Furthermore, network models have been investigated, which focus on the complex feedback between microstructure, current flow and densification [8, 9], but do not address issues of ordinary thermal sintering such as solid neck growth at particle contacts, grain coarsening and pore shrinkage by grain boundary diffusion [10], which may also play a role in CAPAD.

Here, we report recent results obtained with an extension of this network approach to binary particle mixtures. We predict that the Peltier effect leads to correlated particle rearrangement. If the porosity is high enough, this may result in segregation of the two components. As an outlook we expand the view even further and discuss recent experiments, where phases of different conductivity are formed via current driven segregation and second phase formation during the densification process. Finally, we sketch, how we include heat transport through pores of the particle agglomerate.

*e-mail: sebastian.angst@uni-due.de
**e-mail: dietrich.wolf@uni-due.de

2 Model

The model is based on the Onsager-de Groot-Callen theory of thermoelectricity (e.g. [11]), which provides a description of the electrical current density $\mathbf{j}$ and the heat current density $\mathbf{j}_Q$:

$$\mathbf{j} = -\sigma\left(\nabla\mu/q + \alpha\nabla T\right) \tag{1}$$

$$\mathbf{j}_Q = \pi\mathbf{j} - \kappa\nabla T. \tag{2}$$

σ being the electrical conductivity, the first term on the right hand side of eq. (1) is Ohm's law, describing electrical current due to a gradient of the electrochemical potential μ. According to the Seebeck effect, a temperature gradient ∇T drives mobile charge carriers as well, which is expressed by the second term with the Seebeck coefficient α. The second term in eq. (2) is Fourier's law of heat conduction with the heat conductivity κ. The first term describes the Peltier effect that electrical currents transport also heat proportional to the Peltier coefficient π, which is related to the Seebeck coefficient via the Kelvin relation $\pi = \alpha T$. The time evolution of the local temperatures T follows from

$$c\frac{\partial T}{\partial t} + \nabla\cdot\mathbf{j}_Q = -\frac{\nabla\mu}{q}\cdot\mathbf{j}, \tag{3}$$

where c denotes the specific heat capacity times mass density. The right hand side is the electrical power density,

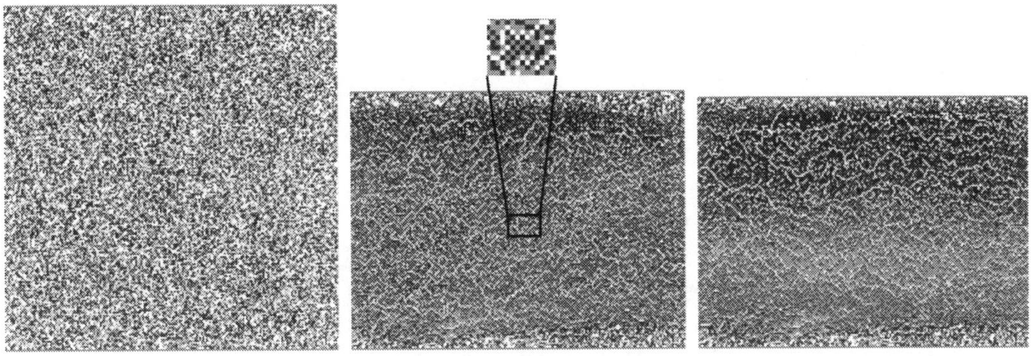

(a) $t = 0$ (b) $t = 500$ (c) $t = 2000$

Figure 1: Configurations of a CAPAD simulation with two different particle types, type B (blue) and type O (orange). We choose $\alpha_B = 1$ and $\alpha_O = 0$, while all other transport parameters are equal. The number of particles of type B and of type O are equal. The current I is flowing from the bottom to the top. Time is measured in units of the heat capacity C divided by the heat conductance K, which are the same for all particles in this simulation.

which contributes to the temperature evolution as a source term. Compared to heat diffusion, charges redistribute extremely fast. Therefore it is reasonable to assume that the local electrochemical potentials μ adapt instantaneously to the temperature change. They are obtained from Kirchhoff's first law,

$$\nabla \cdot \mathbf{j} = 0. \quad (4)$$

In this paper we present only two-dimensional simulations in order to explain the ideas most clearly. We use the network model defined more generally in [12] to discretize the above equations on a square lattice. Two particle types are randomly distributed on the lattice sites with an initial solid volume fraction $\rho = 0.6$. Both particle types occur with the same probability. At the top and the bottom, electrodes are located and kept at a constant heat bath temperature T_{HB}.

A particle i is characterized by the electrochemical potential μ_i, temperature T_i and heat capacity C_i, while the bonds carry the transport coefficients. Solving the discretized versions of eqs. (3) and (4), we obtain the time evolution of μ_i and T_i. If a particle's temperature exceeds a mobilization threshold T_m, reduced friction allows it to move into a neighboring void, i.e. to a randomly chosen empty neighbor site [8]. If several particles are mobilized at the same time, they move in random order. As a result of the particle motion the connection between the electrodes may be lost. In this case the upper electrode and all particles connected to it move downwards until percolation is reestablished.

It is assumed that the particle reorganization, once the reduced friction renders the system mechanically unstable, happens very fast compared to heat diffusion. Therefore in the present model the particle displacement to a neiboring empty site takes just one time step. A more accurate representation of the particle dynamics by discrete element methods was tested [9], but would be too time consuming for the exploratory study intended here.

In this work, the two different types of particles have identical properties except the Seebeck coefficient. As natural units we use the heat bath temperature T_{HB}, the electrical conductance of a bond G, the heat conductance K, the heat capacity of a site C and the particle diameter. We set $T_m = 1.1$ and find that $I = 1$ is a reasonable value for the electrical current, which is flowing from bottom to top electrode.

3 Segregation in binary mixtures

3.1 Time evolution of the structure

Figure 1 shows snapshots of a binary mixture on a 200×200 square lattice. B-particles marked blue have $\alpha_B = 1$ and O-particles marked orange are characterized by $\alpha_O = 0$. While in the initial configuration particles are homogeneously distributed, we recognize a slight segregation at $t = 500$ (see fig. 1b). In the mixed central region, strong correlations between B- and O-particles emerge. Some areas exhibit a checkerboard-like pattern (CBP), which is a lattice artifact, but indicates the spatial correlations that are expected to emerge also without a lattice. At $t = 2000$ the strength of segregation has increased even further and is clearly visible (see fig. 1c). Particles with the higher Peltier coefficient (blue) are concentrated above, those with low Peltier coefficient (orange) below. As the current flows from the bottom to the top, it follows that it carries less heat towards the center region than away from it. This explains that Peltier cooling occurs in between the segregated areas. We checked that these patterns are caused by the Peltier effect (see below).

The mechanism for the pattern formation becomes clear, when one considers the mobility of a local configuration, which requires that the mobilization temperature is exceeded and empty neighboring sites are available. Due to the Peltier effect, O-B-interfaces in current direction are cooled, while B-O-interfaces are heated. B-B- or O-O-interfaces do not experience any interface Peltier effect.

 DOI: 10.1051/epjconf/201714013014

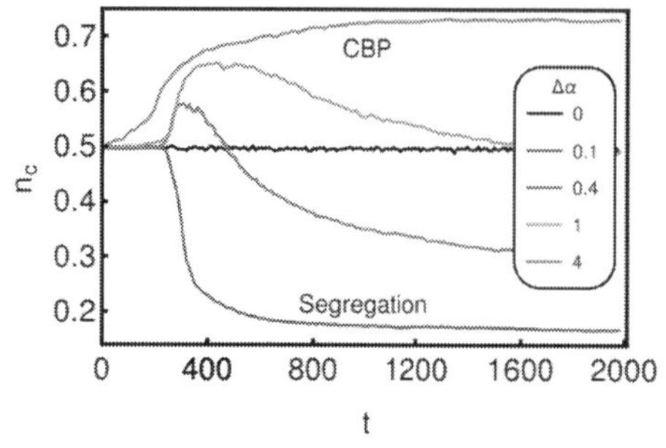

Figure 2: The relative amount of contacts between different types of particles over time. For large α_B the CBP occurs rapidly and is maintained for the whole simulation time, while for intermediate α_B the CBP vanishes with time. For small α_B particles segregate rapidly, while for large α_B the process is much slower.

As a consequence, it is more likely that particles having a B-O-interface become mobile, and their elimination leads to more stable configurations. However, in fig. 1c one still finds regions with CBP, which suggests that the CBP is stable as well. Considering a B-particle surrounded by O-particles (or vice versa), and assuming isothermal conditions, we deduce that due to current conservation the Peltier heating and cooling at the four interfaces neutralize each other. In general however, temperatures in neighboring particles differ, leading to a net cooling or heating of the center particle in a CBP. As we will show below, cooled configurations are more likely. Starting from the initial, random configuration, the creation of a CBP gets rid of Peltier heating.

3.2 Influence of the Seebeck coefficient difference

Now, we discuss the influence of the Seebeck coefficient difference $\Delta\alpha = \alpha_B - \alpha_O = \alpha_B$ and show that depending on $\Delta\alpha$ the self-organization leads to different configurations. Therefore we consider the amount of contacts between particles of different types C_{CB} divided by the total number of contacts C (see fig. 2), $n_c = C_{CB}/C$. In a perfect checkerboard pattern it reaches 1, while it approaches 0, if the distinct particle types arrange in two separate stripes. We verified that the stripes are oriented as indicated in fig. 1. From fig. 2 we recognize that in the initial configurations approximately half of the contacts are between equal and the other half between distinct particles. If both particles have the same Seebeck coefficient, here $\alpha_O = \alpha_B = 0$, none of the two patterns are created, and n_c fluctuates around 0.5. For $\Delta\alpha \neq 0$, n_c changes significantly at $t \approx 220$ for all α_B except $\alpha_B = 4$, where strong Peltier heating leads to a faster increase. Considering $\alpha_B < 4$, n_c grows at first, but decreases after reaching a maximum, which agrees well with the observations from the snapshots in fig. 1 ($\alpha_B = 1$). In contrast, for $\alpha_B = 4$ the CBP is preserved.

In order to understand the pattern evolution, it is instructive to consider the time dependence of solid volume fraction ρ (see fig. 3) and the amount of mobile particles n_m (see fig. 4). In the beginning, two different behaviors are present: one strongly affected by Joule ($\alpha_B \lessapprox 1$) and the other by Peltier heating ($\alpha_B \gtrapprox 1$). For small Seebeck coefficients, ρ and n_m behave similar to $\alpha_B = 0$. Although Joule heating dominates, the Peltier effect does have an impact on the pattern formation (see fig. 2) Joule heating leads to a large amount of mobile particles. Due to their random motion, it is improbable that the percolation is interrupted so that densification is inhibited. Segregation, on the other hand, is then promoted by the prevailing low density.

Considering large $\alpha_B = 4$, Peltier-driven compaction is observed. Particles are mobilized rapidly due to strong Peltier heating, leading to a strong and fast densification at the beginning of the process. Peltier heated connections are unstable and vanish, but Peltier cooled contacts remain withdrawing heat from the sample, which keeps the total amount of mobile particles relatively small.

In order to underline the role of Peltier cooling, we examine the mean heat per particle at $t = 2000$ distinguishing between average Peltier heat $\bar{P}_P$ and average Joule heat $\bar{P}_J$ (see inset of fig. 4). The configurations have evolved such that the Peltier heat averaged over all particles is negative,

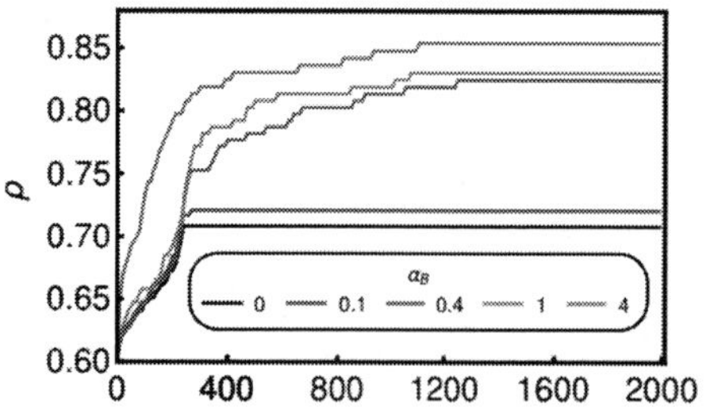

Figure 3: Solid volume fraction ρ in dependence of time for various α_B.

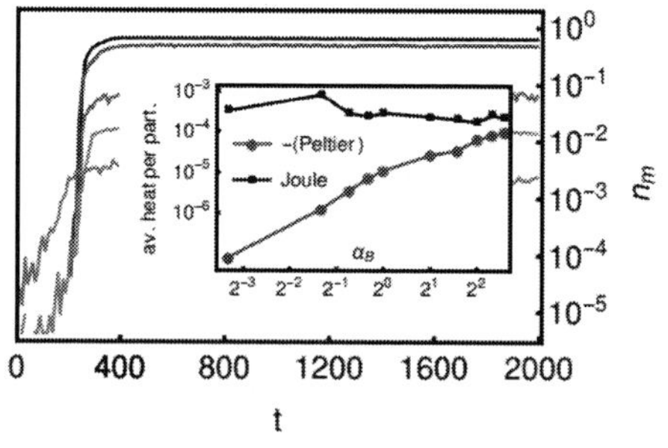

Figure 4: Relative amount of mobile particles n_m in dependence of time for various α_B. The inset shows the mean heat per particle distinguishing between Peltier and Joule heat. Color-coding of α_B as in fig. 3.

$\bar{P}_P < 0$. For large α_B, $\bar{P}_J$ and $-\bar{P}_P$ are of the same order of magnitude, which explains the low n_m. In combination with the high density, this prohibits further restructuring and CBP becomes a stationary pattern.

4 Conclusion and Outlook

By means of simulations we investigated current-activated pressure-assisted densification of binary particle mixtures. We discovered pattern formation due to a difference in the Peltier coefficients of both materials. Within our model, the Peltier heat created at particle interfaces determines the stability of a local configuration. If the Peltier effect predominantly cools a particle, its position is more likely to be stable and vice versa. This basic effect together with the total amount of mobile particles and the available void space determines, whether the random mixture of the two species becomes more uniform (here: a checkerboard-pattern) or segregates into separate stripes.

In conclusion, we observed pattern formation in current sintering, when a local property is switched by the electrical current and is influencing the sintering process. Therefore, we predict a large effect especially for systems where the local property is directly coupled to the electrical current, such as electrical conductivity. This should be the case, when highly doped Al:ZnO is treated by CAPAD [13]. In this case, the creation of a secondary phase is crucial to the process as will be explained in the following.

In bulk ZnO, Al is virtually insoluble up to high temperatures. However, this does not necessarily apply to ZnO nanoparticles where the solubility might be much higher. Alternatively, the high dopant concentration in the ZnO nanoparticles could be frozen in due to their synthesis. In any case, there is experimental evidence that the sintering process destabilizes the solid solution locally and leads to formation of a separate spinel phase. This has a twofold impact on the electrical transport. On the one hand, spinel is an insulator, and on the other hand, the Al concentration in the vicinity of the spinel phase is decreased reducing the electrical conductivity of the doped nanoparticles.

We already know that in the early stages of current sintering spatial variations of the electrical conductivity and the microstructure can lead to the occurrence of hot spots [8]. We expect in this case that the increased temperature facilitates spinel formation which in turn reduces the electrical conductivity of the cross section in which the spinel is located. Hence, we postulate that Joule heating of this cross section increases which causes the spinel to grow in bands perpendicular to the general direction of electrical current, as observed. If, furthermore, the spinel phase limits the mobility of the particles and grain growth, this would explain the observation that the spinel-rich bands retain a porous, particle-like structure. Once spinel has formed, Al ions diffuse towards the precipitate and get bound there, which depletes the Al concentration in vicinity of spinel.

For the purpose of checking these hypotheses, several extensions of the model discussed in this paper are needed: The Onsager network must not only deal with electrical and heat transport, but include also the transport of Al atoms. The coarsening of the particles must be implemented in a way that is not restricted to stress-free tree-like agglomerates [14]. The displacement dynamics of the particles should be simulated with discrete element methods as in [9].

Last but not least, heat transport by radiation and gas flow in the pores require a further extension of the model. For this purpose we spatially decomposed the system into simplices using a Delaunay tessellation of the particle centers. Each Delaunay simplex represents a discrete element of the pore volume and constitutes a node of the network. This node is connected to the nodes of the adjacent simplices and nodes at the corners of the simplex representing the particles. Conductances and heat capacities are determined geometrically. Our method has several advantages over conventional methods. Due to the properties of the Delaunay tessellation, we are able to resolve even small, permeable gaps between particles. Also, the total numbers of nodes and edges which determines the computing time is only defined by the number of particles and does not depend on the geometry.

Acknowledgment

This research has been supported by DFG grants WO 577/6 within the priority program SPP 1386 "Nanostructured Thermoelectric Materials: Theory, Model Systems and Controlled Synthesis", and by the grants WO 577/13 and WI 981/16-1 within the priority program SPP 1959 "Manipulation of matter controlled by electric and magnetic fields".

References

[1] Z. A. Munir et al., J. Mater. Sci. **41**, 763 (2006)
[2] O. Guillon et al., Adv. Eng. Mater. **16**, 830 (2014)
[3] J.E. Garay, Ann. Rev. Mater. Res. **40**, 445 (2010)
[4] R. Marder et al., Scr. Mater. **82**, 57 (2014)
[5] L. Wang et al., J. Asian Ceram. Soc. **3**, 183 (2015)
[6] D. Giuntini et al., Materials **6**, 2612 (2013)
[7] E. A. Olevsky, J. Am. Ceram. Soc. **95**, 2414 (2012)
[8] D. Schwesig et al., Nanotechnology **22**, 135601 (2011)
[9] S. Angst et al., in *AIP Conference Proceedings* (AIP Publishing, 2013), Vol. 1542, p. 593
[10] D. E. Wolf et al., in *Nanoparticles from the Gasphase*, edited by A. Lorke et al. (Springer, Berlin, 2012), p. 161
[11] C. Goupil et al., Entropy **13**, 1481 (2011)
[12] S. Angst et al., New J. Phys. **18**, 043004 (2016)
[13] D. Gautam et al., J. Mater. Chem. A **3**, 189 (2014)
[14] M.L. Eggersdorfer et al., Langmuir **27**, 6358 (2011)

Dynamics of wetting explored with inkjet printing

Simeon Völkel[1,*] and *Kai* Huang[1,**]

[1] *Experimentalphysik V, University of Bayreuth, 95440 Bayreuth, Germany*

Abstract. An inkjet printer head, which is capable of depositing liquid droplets with a resolution of 22 picoliters and high repeatability, is employed to investigate the wetting dynamics of drops printed on a horizontal plane as well as on a granular monolayer. For a sessile drop on a horizontal plane, we characterize the contact angle hysteresis, drop volume and contact line dynamics from side view images. We show that the evaporation rate scales with the dimension of the contact line instead of the surface area of the drop. We demonstrate that the system evolves into a closed cycle upon repeating the depositing-evaporating process, owing to the high repeatability of the printing facility. Finally, we extend the investigation to a granular monolayer in order to explore the interplay between liquid deposition and granular particles.

1 Introduction

Granular materials are ubiquitous in nature, industry and daily lives [1]. Despite intense investigations motivated by applications in geoscience, chemical and civil engineering, the physics of granular materials, are still far from being understood, particularly when they are partially wet [2–4]. The mechanical properties of a granular material change dramatically as a tiny amount wetting liquid is added, representing the formation of capillary bridges between adjacent particles [5, 6]. The enhancement of rigidity leads to an easily moldable material such as wet sand on the beach for sculpturing [7, 8]. In the pendular regime [3] with capillary bridges formed between adjacent particles, previous investigations showed that the collective behaviors of partially wet granular materials, such as clustering, phase transitions and pattern formation, are related to the capillary force induced by the wetting liquid [9–13]. However, as more liquid is added, it is still challenging to understand how a wetting liquid film advances inside a geometrically heterogeneous granular packing, as well as how the particles respond to the capillary interactions induced by the additional wetting liquid.

Wetting is also an ubiquitous phenomenon that has attracted interest from physics, chemistry and engineering communities over the past decades [14, 15]. More recently, there has been a growing interest in understanding the wetting dynamics on patterned and heterogeneous surfaces, motivated by a better control of wettability and liquid transport [16, 17]. Nevertheless, there are still open questions (e. g. contact angle hysteresis, contact line dynamics at the onset of the pinning-depinning transition, etc.) [18–20].

Here, we use a commercial inkjet printer head (HP 51645A) to explore the dynamics of wetting, because it enables a fine control of the drop volume with high repeatability. For both a horizontal plane and a granular monolayer, repeating the print-evaporate process leads to reproducible contact line dynamics.

2 Setup

Figure 1 shows a sketch of the experimental setup. An empty print catridge (HP C6125A) is filled with purified water (LaborStar TWF, surface tension $\gamma = 0.072\,\mathrm{N\,m^{-1}}$). Its printhead is pointing downwards in order to deposit droplets vertically onto the horizontal substrate [polytetrafluoroethylene (PTFE), and a monolayer of glass beads (SiLiBeads Type S)]. In order to control all nozzles (printing frequency 2.8 kHz) individually, we use a custom microcontroller board (Microchip PIC18F6722). Two types

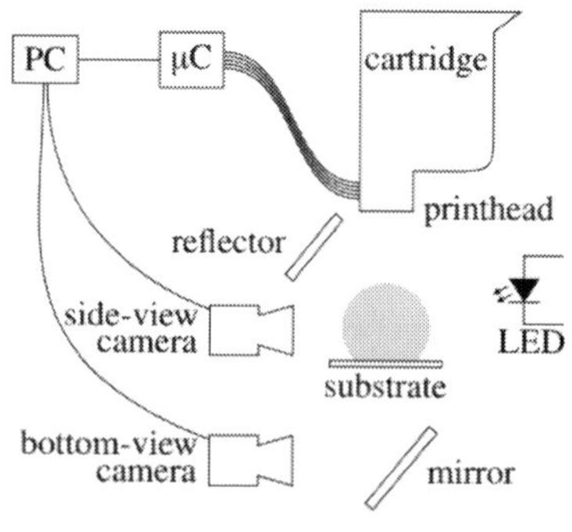

Figure 1: Schematic diagram of the experimental setup. μC stands for microcontroller. A reflector is used to provide back-light illumination for the bottom view camera.

*e-mail: simeon.voelkel@uni-bayreuth.de
**e-mail: kai.huang@uni-bayreuth.de

of cameras (Lumenera LU135M, Nikon D7000) are used to take side view and bottom view images.

A computer program (PC) triggers the image acquisition for one experimental cycle N_{print} times during the printing phase, each time after a specific volume V_{step} has been deposited, and N_{evap} times during evaporation phase, at a fixed interval T_{evap}. The ambient temperature is regulated to 20 °C. The relative humidity is in the range from 20 % to 40 %. The average droplet volume is determined to be 22.4(11) pL by image analysis of side view images.

3 Image analysis

Figure 2 summarizes the image analysis procedure to obtain the geometric properties of the sessile drop. The side view images are cropped such that the bottom of the rectangular region of interest (ROI) separates the drop from the substrate, and binarized to segment the image.

Figure 2a defines the quantities of the drop. The volume V is estimated assuming rotational symmetry along the vertical axis of the drop in each line of the image, as shown in Fig. 2b. The left and right contact angles ϕ_l and ϕ_r as well as the (horizontal) x-coordinates of the associated contact points $x_{c,l}$, $x_{c,r}$ are determined with subpixel resolution by fitting locally a parabola using the standard least square method to the drop contour. Short[1], adjacent segments of the contour are used, as indicated by Fig. 2c. As shown in Fig. 2d, the fit is not performed in the laboratory frame (x, y), but in the rotated (x', y') coordinate system set by the first (1) and last point (n) of the contour segment. This is to circumvent ambiguity problems with contact angles close to 90° as sketched in Fig. 2c and increase both robustness and accuracy. The diameter of the contact line d_c is estimated as the distance between the two contact points (cf. Fig. 2a).

[1]For estimating the (vertical) radius of curvature of a sessile drop close to the contact points, its contour in the side view can be approximated as a part of an ellipse. The short main axis gives a good estimate independent of the drop volume and contact angle. Here, a contour segment one order of magnitude shorter is used. Fig. 2c is not to scale.

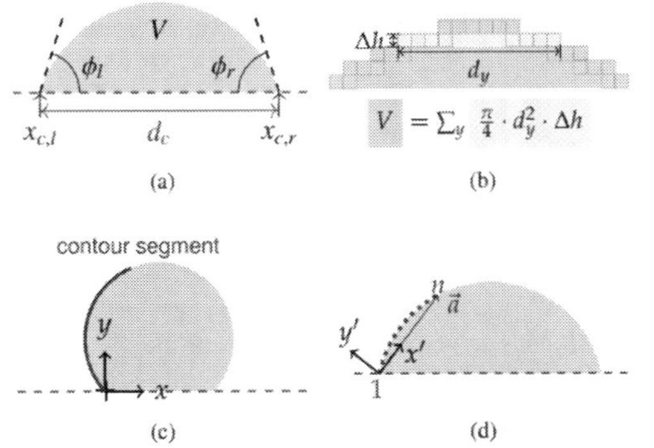

Figure 2: Image analysis procedure and definitions of drop quantities. The red dashed lines indicate the substrate.

4 Results and discussion

Figure 3 shows the development of d_c as a function of $V^{\frac{1}{3}}$ for six print-evaporate cycles. Starting with an initial drop (1), its volume is increased by adding water droplets from above using the inkjet printhead. At first, d_c remains constant until the advancing contact angle is reached (2). As further liquid is added, the drop expands laterally (3), (4). During evaporation, the drop again first keeps d_c constant until the receding contact angle is reached (5). Further evaporation lets the drop shrink laterally (6). When adding some liquid again, the same behaviour recurs. The overlapping curves for cycles 4–6 illustrate the excellent repeatability of the experimental setup and that the deposition-evaporation process can be driven into a closed cycle by just periodically adding a fixed volume with high precision.

In the limit of small drops (i. e. small Bond-number $\text{Bo} = \rho g V^{\frac{2}{3}} \gamma^{-1}$, where ρ: fluid density, g: gravitational acceleration), interfacial tension determines the equilibrium shape of a drop and all linear dimensions of the drop scale with $V^{\frac{1}{3}}$. With growing volume, the relative influence of gravity increases and finally flattens the drop [21]. The drops (2), (3) and (4) as well as (1), (5) and (6) have consistent shapes. It is worth noting that both lateral expansion and shrinking of the drop follow a sloped straight line in Fig. 3: The sessile drops ($\text{Bo} < 1.65$) are thus small. However, due to contact angle hysteresis, the shapes of an expanding and contracting drop differ, which results in two distinct slopes in Fig. 3.

Figure 4 shows the contact angle at the right edge ϕ_r plotted over d_c during the last three print-evaporate cycles. While the contact line is advancing or receding, the contact angle remains approximately constant. As the contact angle for the two directions of movement differs, contact angle hysteresis is clearly visible. In the inset images (1)–(6) of Fig. 3, the drop expands to his right side, keeping

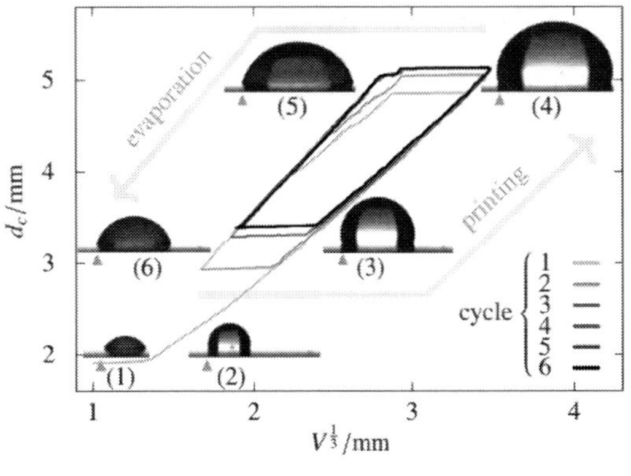

Figure 3: Scaling of the contact line diameter d_c with the drop volume V for water on PTFE. Six cycles of $N_{\text{print}} = 120$, $V_{\text{step}} = 0.34(2)\,\mu\text{L}$, $N_{\text{evap}} = 120$, $T_{\text{evap}} = 60\,\text{s}$ are shown. The triangles mark the same position on the substrate. See text for details.

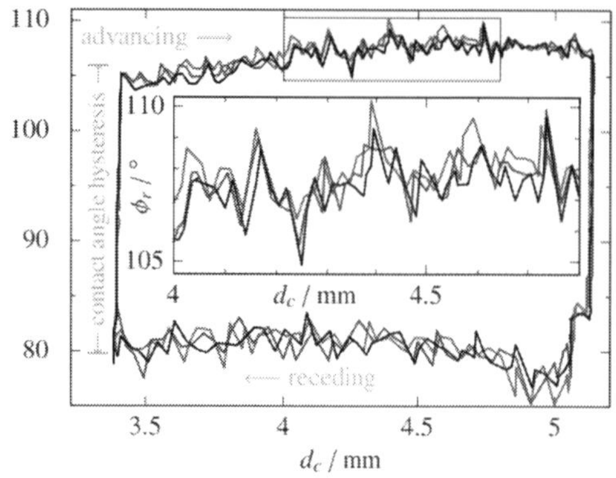

Figure 4: Contact angle ϕ_r of water on PTFE measured from side view images at the freely moving edge. Coloring corresponds to Fig. 3.

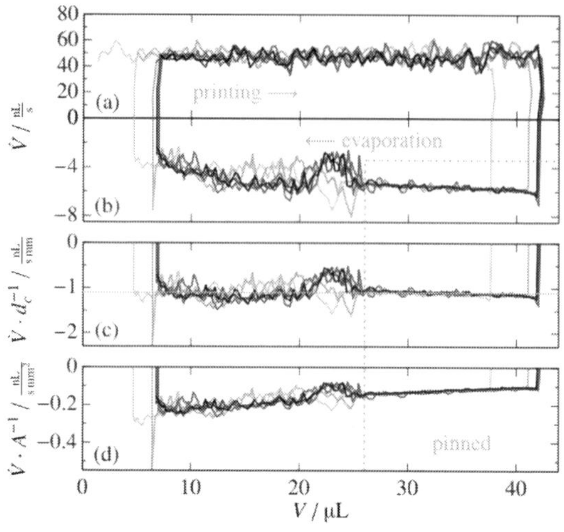

Figure 5: (a) and (b): Effective printing and evaporation rate estimated from side view images. (c) and (d): Comparison of two scaling laws for the evaporation rate. The blue line corresponds to $-1.1\,\mathrm{nL\,s^{-1}\,mm^{-1}}$ and is a guide to the eye. See text for details. Coloring corresponds to Fig. 3.

his left edge at a fixed position. This is due to inhomogenities on the substrate, resulting in a slightly larger contact angle hysteresis there than for the rest of the contact line. Thanks to the pinned left side in the experiments shown in Fig. 4, equal d_c results in coincident positions $x_{c,r}$ of the right edge of the drop. The inset in Fig. 4 magnifies and highlights a part of the graph. The strong correlation between subsequent runs reveals that the fluctuations in the measured contact angles come from the drop's exploration of the surface heterogenities. Figure 5 shows that the average deposition rate of the used printing-protocol exceeds the evaporation rate by an order of magnitude. While evaporation drives the recedence of the contact line, its influence on the advance is marginal. As a result, the advancing contact angle in Fig. 4 exhibits a more prominent correlation than the measurements during the receding phase.

As the blue horizontal line in Fig. 5c illustrates, the evaporation rate scales with the contact line diameter. Thus we conclude that the drop evaporates mainly through its three phase line (see [22]) instead of its surface area (see Fig. 5d for a comparison). The scaling of the evaporation with the drop geometry reveals that our sessile drop is also small in this regard.

Having illustrated the repeatability provided by the setup when using a flat substrate, we now extend our investigation to a granular monolayer. Concerning wetting in a granular system, one of the most important challenges is to understand the interplay between wetting liquid and granules. The particles are subject to capillary forces if they are not immersed in liquid. In a loose packing, a movement of the liquid interface can therefore lead to a movement of granules, if they are not fully constrained by their neighbours. The details of the advancing and receding of the liquid interface, the involved wetting and dewetting, as well as rupture events of capillary bridges are strongly dependend on the the configuration of neighboring particles [23].

For the following experiments, spherical glass beads of 0.3 mm to 0.4 mm diameter are put on top of a horizontal glass slide. By tapping they are brought into a two-dimensional random packing. Further beads are added until the bottom of a 25 × 25.5 mm rectangular enclosure is filled. After initial reorganization of some particles, the system again can apparently be driven into a closed cycle by periodically adding and evaporating water.

Figure 6 compares two subsequent printing-cycles on this three-dimensionally patterned substrate. After removing the background by division, the overlaid images are obtained by putting into their red and blue channels the image of one cycle and in their green channels the image of the subsequent one. Qualitatively, Fig. 6 provides the following features: (i) The contour of a sessile drop on a granular layer does not exhibit a circular contact line due to the prominent structure of the substrate. (ii) The deviation from a circle is more than the particle size, especially if the wettability of the substrate is high. This is partially due to the fact that the wetting of a neighbouring particle is a discrete event which leads to an abrupt change of the liquid interface position. (iii) While for drops on an inclined homogeneous plane the aspect ratio of the bottom view is a suitable parameter for describing, e. g., the onset of sliding [20], the irregularly jagged contact line shown here raises the question: Which shape descriptors are suitable for the case of a partially wetted granular layer? Finally, the tiny difference in the shape of the wetted area (see colored regions in Fig. 6) clearly illustrates that, even in this complex system, this setup offers a high repeability.

5 Summary

In this work, we use a commercial inkjet printer head, which provides a fine volume control of liquid deposition

 DOI: 10.1051/epjconf/201714009035

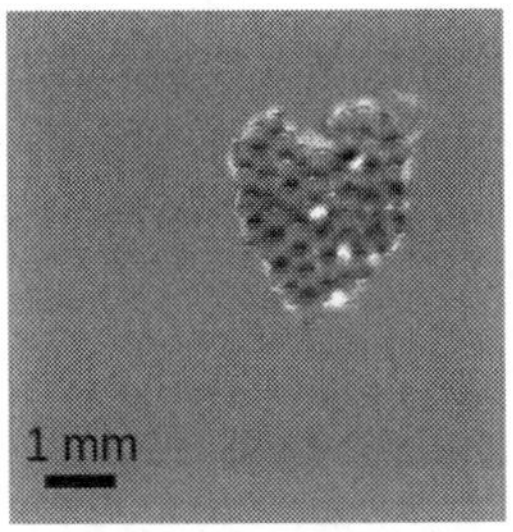

(a) $V = 0.56(3)\,\mu L$

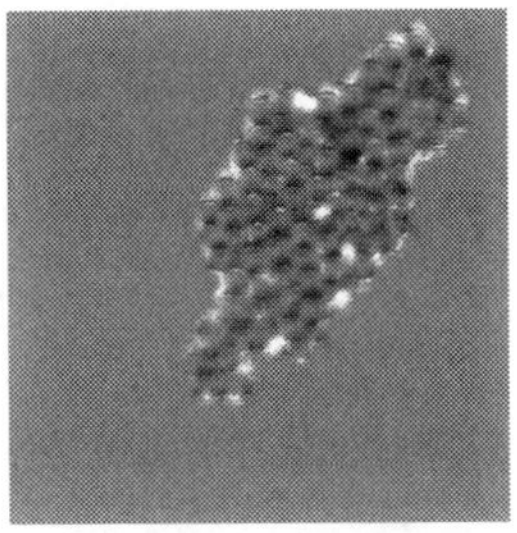

(b) $V = 1.22(6)\,\mu L$

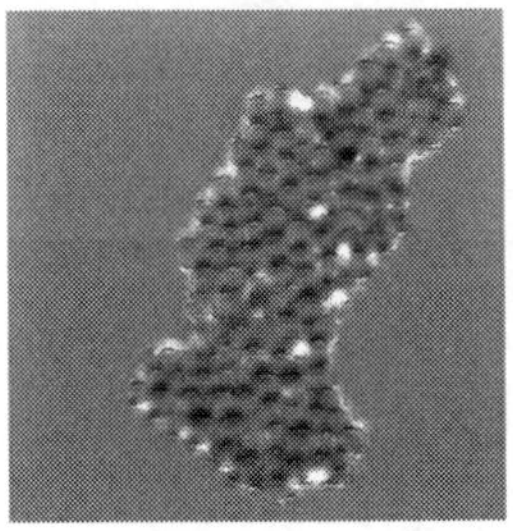

(c) $V = 1.68(9)\,\mu L$

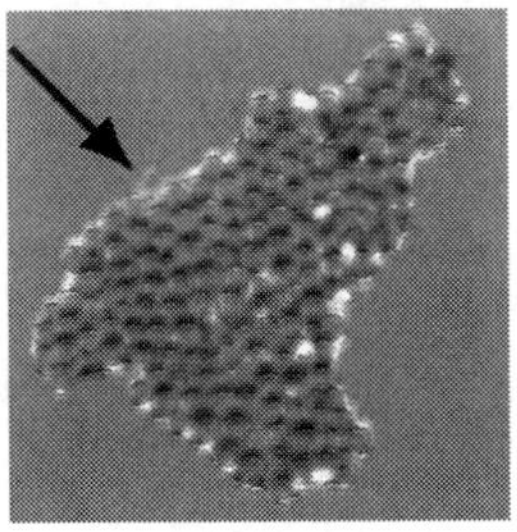

(d) $V = 2.24(11)\,\mu L$

Figure 6: Overlay of typical bottom view images taken in two subsequent printing cycles ($N_{\text{print}} = 100$, $V_{\text{step}} = 22.4(11)\,\text{nL}$, $N_{\text{evap}} = 40$, $T_{\text{evap}} = 30\,\text{s}$). Unsaturated pixels show regions of identical wetting while green and purple regions highlight the differences of the wetted area. See, for example, the region marked by the arrow in (d).

in a well controlled and cost-effective way, to explore the wetting dynamics of liquid drops on a plane and a granular monolayer.

For a sessile drop on a plane, we demonstrate the excellent repeatability of the setup through applying print-evaporate cycles and monitoring the drop geometry. Contact angle hyteresis is observed. The evaporation rate is found to scale with the diameter of the drop contact line, which is in agreement with a previous investigation [22]. Thanks to the high and adjustable printing rate ($63(3)\,\text{nL}\,\text{s}^{-1}$ per nozzle), water evaporation can be easily compensated by printing in typical experimental conditions.

Remarkably, we find that, even on a discrete substrate like a granular layer, this print-evaporate protocol leads to reproducible contour line dynamics. This feature facilitates further investigations on the interactions between a wetting liquid and a granular bed, in order to shed light on the advance of additive manufacturing as well as other drop-on-demand applications [24, 25].

References

[1] J. Duran, *Sands, Powders and Grains (An Introduction to the Physics of Granular Materials)*, 1st edn. (Springer-Verlag, New York, Inc., 2000)

[2] S. Herminghaus, Adv. Phys. **54**, 221 (2005)

[3] N. Mitarai, F. Nori, Adv. Phys. **55**, 1–45 (2006)

[4] K. Huang, Habilitation thesis, University of Bayreuth (2014)

[5] T. Halsey, A. Levine, Phys. Rev. Lett. **80**, 3141 (1998)

[6] K. Huang, M. Sohaili, M. Schröter, S. Herminghaus, Phys. Rev. E **79**, 010301 (2009)

[7] M. Scheel, R. Seemann, M. Brinkmann, M. Di Michiel, A. Sheppard, B. Breidenbach, S. Herminghaus, Nat. Mater. **7**, 189 (2008)

[8] M. Pakpour, M. Habibi, P. Møller, D. Bonn, Sci. Rep. **2** (2012)

[9] K. Huang, K. Roeller, S. Herminghaus, Eur. Phys. J. Spec. Top. **179**, 25 (2009)

[10] K. Huang, I. Rehberg, Phys. Rev. Lett. **107**, 028001 (2011)

[11] K. Huang, M. Brinkmann, S. Herminghaus, Soft Matter **8**, 11939 (2012)

[12] C. May, M. Wild, I. Rehberg, K. Huang, Phys. Rev. E **88**, 062201 (2013)

[13] L. Butzhammer, S. Völkel, I. Rehberg, K. Huang, Phys. Rev. E **92**, 012202 (2015)

[14] P.G. de Gennes, Rev. Mod. Phys. **57**, 827 (1985)

[15] D. Bonn, J. Eggers, J. Indekeu, J. Meunier, E. Rolley, Rev. Mod. Phys. **81**, 739 (2009)

[16] C. Semprebon, P. Forsberg, C. Priest, M. Brinkmann, Soft Matter **10**, 5739 (2014)

[17] M. Sbragaglia, L. Biferale, G. Amati, S. Varagnolo, D. Ferraro, G. Mistura, M. Pierno, Phys. Rev. E **89**, 012406 (2014)

[18] U. Thiele, E. Knobloch, Phys. Rev. Lett. **97** (2006)

[19] H.B. Eral, D.J.C.M. 't Mannetje, J.M. Oh, Colloid Polym. Sci. **291**, 247 (2013)

[20] C. Semprebon, M. Brinkmann, Soft Matter **10**, 3325 (2014)

[21] C.W. Extrand, S.I. Moon, Langmuir **26**, 11815 (2010)

[22] H. Hu, R.G. Larson, J. Phys. Chem. B **106**, 1334 (2002)

[23] C. Semprebon, M. Scheel, S. Herminghaus, R. Seemann, M. Brinkmann, Phys. Rev. E **94**, 012907 (2016)

[24] A. Yarin, Ann. Rev. Fluid Mech. **38**, 159 (2006)

[25] B. Derby, Ann. Rev. Mat. Res. **40**, 395 (2010)

DOI: 10.1051/epjconf/201714003063

A microscopic theory for discontinuous shear thickening of frictional granular materials

Kuniyasu Saitoh[1] and *Hisao* Hayakawa[2]

[1] *Advanced Institute for Materials Research, Tohoku University, 2-1-1 Katahira, Aoba-ku, Sendai 980-8577, Japan*
[2] *Yukawa Institute for Theoretical Physics, Kyoto University, Kyoto 606-8502, Japan*

Abstract. We extend a recent theory for the rheology of frictionless granular materials [1] to the case of frictional disks in two dimensions. Employing a frictional contact model for molecular dynamics simulations, we derive difference equations of the shear stress, the granular temperature, and the spin temperature from the generalized Green-Kubo formula, where all the terms are given by microscopic expressions. The numerical solutions of the difference equations not only describe the flow curve, but also reproduce the hysteresis of shear stress, which can be the signature of discontinuous shear thickening of frictional disks.

1 Introduction

Understanding of the rheology of granular materials is important for industrial applications so that its microscopic descriptions have been desired in science and technology [2]. It is now known that the rheology drastically changes if the fraction exceeds the jamming point, i.e. the shear stress changes from the liquid- to solid-branches of flow curves. Besides the strong dependence on the fraction, the friction between constituent particles also triggers the anomalous flow behavior, i.e. the *discontinuous shear thickening* (DST), where the shear stress discontinuously jumps from liquid- to solid-branches, and vice versa [3–5]. Recently, a microscopic theory for the rheology of frictionless granular materials have been developed by Suzuki and one of authors [1], where the critical scaling of shear viscosity and that of granular temperature below jamming were well predicted without any fitting parameters.

In this paper, we apply their theory to the case of frictional disks in two-dimension such that the DST observed in molecular dynamics (MD) simulations [3, 5] can be explained.

2 Theory

First, we describe translational motions of frictional disks by the SLLOD equation [6]. We assume that the system is homogeneously sheared along the x-axis under the Lees-Edwards boundary condition. Then, time derivatives of the position, $\mathbf{r}_i = (x_i, y_i)$, and *peculiar momentum*, $\mathbf{p}_i = (p_{ix}, p_{iy})$, of i-th particle ($i = 1, \ldots, N$) are given by

$$\dot{\mathbf{r}}_i = m_i^{-1}\mathbf{p}_i + \dot{\boldsymbol{\gamma}} : \mathbf{r}_i\,, \qquad \dot{\mathbf{p}}_i = \mathbf{F}_i - \dot{\boldsymbol{\gamma}} : \mathbf{p}_i\,, \tag{1}$$

respectively, where m_i and $\mathbf{F}_i = (F_{ix}, F_{iy})$ are the mass of i-th particle and the force acts on the i-th particle, respectively. On the right-hand-sides of Eq. (1), each element of the two-rank tensor, $\dot{\boldsymbol{\gamma}}$, is defined as $\dot{\gamma}_{\alpha\beta} = \dot{\gamma}\delta_{x\alpha}\delta_{y\beta}$ ($\alpha, \beta = x, y$) with the shear rate, $\dot{\gamma}$, such that $\dot{\boldsymbol{\gamma}} : \mathbf{r}_i = (\dot{\gamma}y_i, 0)$.

Next, rotational motions of the disks are described by the Euler equation of motions, where time derivatives of the angular position, $\boldsymbol{\varphi}_i$, and angular velocity, ω_i, of i-th particle are given by

$$\dot{\boldsymbol{\varphi}}_i = \omega_i \times \boldsymbol{\varphi}_i\,, \qquad I_i\dot{\omega}_i = \mathbf{M}_i\,, \tag{2}$$

respectively. In Eq. (2), I_i and $\mathbf{M}_i$ represent the moment of inertia and the torque acts on the i-th particle, respectively. From Eqs. (1) and (2), the *phase variables* are defined as $\boldsymbol{\Gamma} \equiv \{\mathbf{r}_i, \mathbf{p}_i, \boldsymbol{\varphi}_i, \omega_i\}$.

2.1 Frictional contact model

To study the DST of frictional disks, we employ a frictional contact model [7], where the total force acts on the i-th particle is divided into the normal and tangential directions as $\mathbf{F}_i = \mathbf{F}_{in} + \mathbf{F}_{it}$ (the subscripts, n and t, are, respectively, used for the normal and the tangential components). Here, both the normal and tangential forces are modeled by linear spring-dashpot and are further decomposed into (reversible) elastic and (irreversible) viscous parts as $\mathbf{F}_{in} \equiv \mathbf{F}_{in}^{(\mathrm{el})} + \mathbf{F}_{in}^{(\mathrm{vis})}$ and $\mathbf{F}_{it} \equiv \mathbf{F}_{it}^{(\mathrm{el})} + \mathbf{F}_{it}^{(\mathrm{vis})}$, respectively. For later use, we also introduce the elastic and viscous parts of the total force as $\mathbf{F}_i = \mathbf{F}_i^{(\mathrm{el})} + \mathbf{F}_i^{(\mathrm{vis})}$, where each part is decomposed into $\mathbf{F}_i^{(\mathrm{el})} \equiv \mathbf{F}_{in}^{(\mathrm{el})} + \mathbf{F}_{it}^{(\mathrm{el})}$ and $\mathbf{F}_i^{(\mathrm{vis})} \equiv \mathbf{F}_{in}^{(\mathrm{vis})} + \mathbf{F}_{it}^{(\mathrm{vis})}$. In the MD simulations of frictional disks, the DST is enhanced for strong friction (the friction coefficient, $\mu = 2$, is used for Coulomb's friction [3, 5]). Therefore, in this study, we neglect the switch from static friction, $\mathbf{F}_{it}$, to dynamical one, $\mu|\mathbf{F}_{in}|$, where all the disks in contact do not slip with each other.

In the normal force, $\mathbf{F}_{in}$, the elastic and viscous parts are given by $\mathbf{F}_{in}^{(\mathrm{el})} = k_n \sum_{j \neq i} \Theta(\xi_{ij})\xi_{ij}\mathbf{n}_{ij}$ and $\mathbf{F}_{in}^{(\mathrm{vis})} =$

$-\eta_n \sum_{j\neq i} \Theta(\xi_{ij})(\dot{\mathbf{r}}_{ij} \cdot \mathbf{n}_{ij})\mathbf{n}_{ij}$, respectively, where k_n and η_n are introduced as a spring constant and viscosity coefficient, respectively. Here, $\Theta(\xi_{ij})$ is the Heaviside step function (which is 1 for $\xi_{ij} > 0$ and 0 otherwise), where $\xi_{ij} \equiv R_i + R_j - r_{ij}$, R_i (or R_j), $r_{ij} \equiv |\mathbf{r}_i - \mathbf{r}_j|$, $\mathbf{n}_{ij} \equiv (\mathbf{r}_i - \mathbf{r}_j)/r_{ij}$, and $\dot{\mathbf{r}}_{ij} \equiv \dot{\mathbf{r}}_i - \dot{\mathbf{r}}_j$ represent an overlap between the disks (i and j), the radius of i-th (or j-th) disk, the distance between the disks, the unit vector parallel to the relative position, and the relative velocity, respectively.

In the tangential force, $\mathbf{F}_{it}$, the elastic and viscous parts are written as $\mathbf{F}_{it}^{(\mathrm{el})} = -k_t \sum_{j\neq i} \Theta(\xi_{ij}) \int_{\tilde{t}_{ij}}^{t} \mathbf{v}_{ijt}(s)ds$ and $\mathbf{F}_{it}^{(\mathrm{vis})} = -\eta_t \sum_{j\neq i} \Theta(\xi_{ij})\mathbf{v}_{ijt}$, respectively, where k_t and η_t are the spring constant and the viscosity coefficient, respectively, and the disks, i and j, get into contact at time $\tilde{t}_{ij}$. Here, $\mathbf{v}_{ijt} = \mathbf{n}_{ij} \times (R_i\omega_i + R_j\omega_j) + \dot{\mathbf{r}}_{ij} - (\dot{\mathbf{r}}_{ij} \cdot \mathbf{n}_{ij})\mathbf{n}_{ij}$ represents the tangential relative velocity at the contact point [7], which is perpendicular to the normal unit vector between the disks, i.e. $\mathbf{v}_{ijt} \cdot \mathbf{n}_{ij} = 0$. Then, the elastic part, $\mathbf{F}_{it}^{(\mathrm{el})}$, is proportional to the relative displacement, $\int_{\tilde{t}_{ij}}^{t} \mathbf{v}_{ijt}(s)ds$, which depends on the memory during a duration of contact, $\tau_{ij}(t) \equiv t - \tilde{t}_{ij}$.

The total torque acts on the i-th particle also consists of elastic and viscous parts as $\mathbf{M}_i = \mathbf{M}_i^{(\mathrm{el})} + \mathbf{M}_i^{(\mathrm{vis})}$. If we write the tangential forces as $\mathbf{F}_{it}^{(\mathrm{el})} = \sum_{j\neq i} \mathbf{f}_{ijt}^{(\mathrm{el})}$ and $\mathbf{F}_{it}^{(\mathrm{vis})} = \sum_{j\neq i} \mathbf{f}_{ijt}^{(\mathrm{vis})}$, each part of the torque is given by $\mathbf{M}_i^{(\mathrm{el})} = R_i \sum_{j\neq i} \mathbf{f}_{ijt}^{(\mathrm{el})} \times \mathbf{n}_{ij}$ and $\mathbf{M}_i^{(\mathrm{vis})} = R_i \sum_{j\neq i} \mathbf{f}_{ijt}^{(\mathrm{vis})} \times \mathbf{n}_{ij}$.

2.2 Hamiltonian and Liouville equation

In our system, the Hamiltonian is introduced as $\mathcal{H} = \sum_i (\mathbf{p}_i^2/2m_i + I_i\omega_i^2/2) + U_n + U_t$, where U_n and U_t are elastic potentials stored in the normal and tangential directions, respectively. Because the Hamiltonian depends on time through the phase variable, $\mathbf{\Gamma}$, its time derivative is given by $\dot{\mathcal{H}} = \sum_i (m_i^{-1}\mathbf{p}_i \cdot \dot{\mathbf{p}}_i + I_i\omega_i \cdot \dot{\omega}_i - \mathbf{F}_{in}^{(\mathrm{el})} \cdot \dot{\mathbf{r}}_i - \mathbf{F}_{it}^{(\mathrm{el})} \cdot \dot{\varphi}_i)$, where we have used the relations, $\mathbf{F}_{in}^{(\mathrm{el})} = -\partial U_n/\partial \mathbf{r}_i$ and $\mathbf{F}_{it}^{(\mathrm{el})} = -\partial U_t/\partial \varphi_i$. From Eqs. (1) and (2), the time derivative is rewritten as $\dot{\mathcal{H}} = -\dot{\gamma} S \sigma_{xy} - 2\mathcal{R} + \mathcal{J}$. Here,

$$\sigma_{xy} \equiv \frac{1}{S}\sum_i \left(\frac{p_{ix}p_{iy}}{m_i} + y_i F_{ix}\right), \tag{3}$$

$$\mathcal{R} \equiv -\frac{1}{2}\sum_i \left\{\mathbf{F}_i^{(\mathrm{vis})} \cdot \dot{\mathbf{r}}_i + \mathbf{M}_i^{(\mathrm{vis})} \cdot \omega_i\right\}, \tag{4}$$

$$\mathcal{J} \equiv \sum_i \left\{\mathbf{F}_{it}^{(\mathrm{el})} \cdot (\dot{\mathbf{r}}_i - \omega_i \times \varphi_i) + \mathbf{M}_i^{(\mathrm{el})} \cdot \omega_i\right\}, \tag{5}$$

are introduced as the shear stress [1], the *dissipation function*, and the rate of energy production by the *couple stress* [8], respectively, where S represents the system area.

If we introduce a *dynamical variable*, $A(t)$, as a function of phase variable, $\mathbf{\Gamma}$, its time derivative is given by

$$\frac{d}{dt}A(t) = \dot{\mathbf{\Gamma}} \cdot \frac{\partial}{\partial \mathbf{\Gamma}} A(t) \equiv i\mathcal{L}A(t), \tag{6}$$

where $i\mathcal{L} \equiv \dot{\mathbf{\Gamma}} \cdot (\partial/\partial\mathbf{\Gamma})$ is defined as the *Liouville operator* [6]. From Eqs. (1) and (2), the Liouville operator is decomposed as $i\mathcal{L} = i\mathcal{L}_{\mathrm{el}} + i\mathcal{L}_{\dot{\gamma}} + i\mathcal{L}_{\mathrm{vis}}$ [2].

On the other hand, the time development of N-body distribution function, $f(t)$, is described by the *Liouville equation* as

$$\frac{\partial}{\partial t} f(t) = -\left(\dot{\mathbf{\Gamma}} \cdot \frac{\partial}{\partial \mathbf{\Gamma}} + \frac{\partial}{\partial \mathbf{\Gamma}} \cdot \dot{\mathbf{\Gamma}}\right) f(t) \equiv -(i\mathcal{L} + \Lambda) f(t), \tag{7}$$

where $\Lambda \equiv (\partial/\partial\mathbf{\Gamma}) \cdot \dot{\mathbf{\Gamma}}$ is defined as the *compression factor* [6]. The Liouville equation is also written as $\partial f(t)/\partial t = -i\mathcal{L}^\dagger f(t)$, where $i\mathcal{L}^\dagger \equiv i\mathcal{L} + \Lambda$ is the adjoint of Liouville operator so that the Liouville operator is non-Hermitian, i.e. $i\mathcal{L}^\dagger \neq i\mathcal{L}$ [6]. From Eqs. (1) and (2), we find that the compression factor is given by

$$\Lambda = \sum_i \left\{\frac{\partial}{\partial \mathbf{p}_i} \cdot \mathbf{F}_i^{(\mathrm{vis})} + \frac{\partial}{\partial \omega_i} \cdot (I_i^{-1}\mathbf{M}_i^{(\mathrm{vis})})\right\} \tag{8}$$

which only consists of irreversible viscous parts, $\mathbf{F}_i^{(\mathrm{vis})}$ and $\mathbf{M}_i^{(\mathrm{vis})}$. Then, it is readily found that the first and second terms on the right-hand-side of Eq. (8) are reduced to $(\partial/\partial\mathbf{p}_i) \cdot \mathbf{F}_i^{(\mathrm{vis})} = -m_i^{-1}(\eta_n + \eta_t)z_i$ and $(\partial/\partial\omega_i) \cdot (I_i^{-1}\mathbf{M}_i^{(\mathrm{vis})}) = -I_i^{-1}R_i^2\eta_t z_i$, respectively, where $z_i \equiv \sum_{j\neq i}\Theta(\xi_{ij})$ is introduced as the coordination number of the i-th particle. If we use the moment of inertia for a two-dimensional disk, $I_i = m_i R_i^2/2$, and assume that every mass is identical, $m_i = m$, we find that the compression factor is proportional to the total number of contacts, i.e. $\Lambda = -(\eta_n + 3\eta_t)N_c/m$ with $N_c \equiv \sum_i z_i$.

2.3 Generalized Green-Kubo formula

To describe the shear stress of frictional disks, let us remind the measurement of *flow curves* [3]: After the system is equilibrated under shear, the shear rate is increased from $\dot{\gamma}$ to $\dot{\gamma} + \Delta\dot{\gamma}$ (at time $t = 0$) such that the shear stress relaxes from $\langle\sigma_{xy}\rangle_{\dot{\gamma}}$ to $\langle\sigma_{xy}\rangle_{\dot{\gamma}+\Delta\dot{\gamma}}$, where $\langle\ldots\rangle_{\dot{\gamma}}$ represents the average in a non-equilibrium steady state under shear with $\dot{\gamma}$. The *generalized Green-Kubo formula* around the steady state characterized by $\dot{\gamma}$ is expressed as

$$\langle\sigma_{xy}(t)\rangle_{\dot{\gamma}+\Delta\dot{\gamma}} \simeq \langle\sigma_{xy}\rangle_{\dot{\gamma}} + \int_0^t ds \langle\sigma_{xy}(s)\Omega\rangle_{\dot{\gamma}}, \tag{9}$$

where $\Omega \equiv \beta\dot{\mathcal{H}} - \Lambda = -\dot{\gamma}\beta S\sigma_{xy} - 2\beta\mathcal{R} + \beta\mathcal{J} - \Lambda$ with the inverse of granular temperature, $\beta \equiv 1/\langle T\rangle_{\dot{\gamma}}$, is the *work function* [6, 9] and $\langle\sigma_{xy}\rangle_{\dot{\gamma}} \neq 0$. Assuming an exponential decay of the time correlation function, $\langle\sigma_{xy}(s)\Omega\rangle_{\dot{\gamma}} \simeq \langle\sigma_{xy}\Omega\rangle_{\dot{\gamma}} e^{-s/\tau_{\mathrm{rel}}}$, with a relaxation time, τ_{rel}, and taking a long time limit ($t \to \infty$), we reduce Eq. (9) to $\langle\sigma_{xy}\rangle_{\dot{\gamma}+\Delta\dot{\gamma}} \simeq \langle\sigma_{xy}\rangle_{\dot{\gamma}} + \tau_{\mathrm{rel}}\langle\sigma_{xy}\Omega\rangle_{\dot{\gamma}}$, where we have rewritten the steady state value as $\langle\sigma_{xy}\rangle_{\dot{\gamma}+\Delta\dot{\gamma}} \equiv \langle\sigma_{xy}(\infty)\rangle_{\dot{\gamma}+\Delta\dot{\gamma}}$ (see Ref. [1] for more rigorous treatment). If we assume that the non-equilibrium steady state is not far from the equilibrium state, we can rewrite the correlation function as

[1] It is also written as $\sigma_{xy} = \sum_i m_i^{-1} p_{ix}p_{iy}/S + \sum_i \sum_{j\neq i}(y_i - y_j)f_{ijx}/2S$.

[2] Each part is explicitly written as $i\mathcal{L}_{\mathrm{el}} = \sum_i \{m_i^{-1}\mathbf{p}_i \cdot (\partial/\partial\mathbf{r}_i) + \mathbf{F}_i^{(\mathrm{el})} \cdot (\partial/\partial\mathbf{p}_i) + (\omega_i \times \varphi_i) \cdot (\partial/\partial\varphi_i) + I_i^{-1}\mathbf{M}_i^{(\mathrm{el})} \cdot (\partial/\partial\omega_i)\}$, $i\mathcal{L}_{\dot{\gamma}} = \dot{\gamma}\sum_i (y_i\partial/\partial x_i - p_{iy}\partial/\partial p_{ix})$, and $i\mathcal{L}_{\mathrm{vis}} = \sum_i \{\mathbf{F}_i^{(\mathrm{vis})} \cdot (\partial/\partial\mathbf{p}_i) + I_i^{-1}\mathbf{M}_i^{(\mathrm{vis})} \cdot (\partial/\partial\omega_i)\}$.

Table 1. Coefficients in the difference equations, Eqs. (10)-(12), where $n_{ij\alpha}$ ($\alpha = x, y$) represents the α-component of the normal unit vector between the particles (i and j), $\mathbf{n}_{ij}$.

$$
\begin{aligned}
a_0 &= 2\tau_{\rm rel}k_n^2 \sum_{\substack{i,j,k,l\\(i\neq j,k\neq l)}} \langle \Theta(\xi_{ij})\Theta(\xi_{kl})\xi_{ij}\xi_{kl}r_{ij}r_{kl}n_{ijx}n_{klx}n_{ijy}n_{kly}\rangle_{\dot\gamma}\\
a_1 &= (mS)^{-1}\tau_{\rm rel}\sum_{\substack{i,j,k\\(i\neq j,i\neq k)}} \langle 3\eta_n^2\Theta(\xi_{ij})^2 r_{ij}^2 n_{ijx}^2 n_{ijy}^2 + 3\eta_t^2\langle\Theta(\xi_{ij})^2 r_{ij}^2 n_{ijy}^4 + \eta_n^2\Theta(\xi_{ij})\Theta(\xi_{ik})\left(n_{ijx}n_{ikx}+n_{ijy}n_{iky}\right) r_{ij}r_{ik}n_{ijx}n_{ikx}n_{ijy}n_{iky}\\
&\quad +\eta_n\eta_t\Theta(\xi_{ij})\Theta(\xi_{ik})r_{ij}r_{ik}n_{ijx}n_{ijy}n_{iky}^2\left(n_{ijx}n_{iky}-n_{ijy}n_{ikx}\right) + \eta_t^2\Theta(\xi_{ij})\Theta(\xi_{ik})r_{ij}r_{ik}n_{ijy}^2 n_{iky}^2\left(n_{ijy}n_{iky}+n_{ijx}n_{ikx}\right)\rangle_{\dot\gamma}\\
a_2 &= (mS)^{-1}\tau_{\rm rel}\sum_{\substack{i,j,k\\(i\neq j,i\neq k)}} \langle 6\eta_t^2\Theta(\xi_{ij})^2 r_{ij}^2 n_{ijy}^4 + 2\eta_t^2\langle\Theta(\xi_{ij})\Theta(\xi_{ik})r_{ij}r_{ik}n_{ijy}^2 n_{iky}^2\rangle_{\dot\gamma}\\
b_0 &= 2\tau_{\rm rel}k_n^2 \sum_{\substack{i,j,k,l\\(i\neq j,k\neq l)}} \langle \Theta(\xi_{ij})\Theta(\xi_{kl})\xi_{ij}\xi_{kl}r_{ij}r_{kl}n_{ijx}n_{klx}n_{ijy}n_{kly}\rangle_{\dot\gamma}\\
b_1 &= (mS)^{-1}\tau_{\rm rel}\sum_{\substack{i,j,k\\(i\neq j,i\neq k)}} \langle 3\eta_n^2\Theta(\xi_{ij})^2 r_{ij}^2 n_{ijx}^2 n_{ijy}^2 + 3\eta_t^2\langle\Theta(\xi_{ij})^2 r_{ij}^2 n_{ijy}^4 + \eta_n^2\Theta(\xi_{ij})\Theta(\xi_{ik})\left(n_{ijx}n_{ikx}+n_{ijy}n_{iky}\right) r_{ij}r_{ik}n_{ijx}n_{ikx}n_{ijy}n_{iky}\\
&\quad +\eta_n\eta_t\Theta(\xi_{ij})\Theta(\xi_{ik})r_{ij}r_{ik}n_{ijx}n_{ijy}n_{iky}^2\left(n_{ijx}n_{iky}-n_{ijy}n_{ikx}\right) + \eta_t^2\Theta(\xi_{ij})\Theta(\xi_{ik})r_{ij}r_{ik}n_{ijy}^2 n_{iky}^2\left(n_{ijy}n_{iky}+n_{ijx}n_{ikx}\right)\rangle_{\dot\gamma}\\
b_2 &= (mS)^{-1}\tau_{\rm rel}\sum_{\substack{i,j,k\\(i\neq j,i\neq k)}} \langle 6\eta_t^2\Theta(\xi_{ij})^2 r_{ij}^2 n_{ijy}^4 + 2\eta_t^2\langle\Theta(\xi_{ij})\Theta(\xi_{ik})r_{ij}r_{ik}n_{ijy}^2 n_{iky}^2\rangle_{\dot\gamma}\\
c_0 &= 2\tau_{\rm rel}k_n^2 \sum_{\substack{i,j,k,l\\(i\neq j,k\neq l)}} \langle \Theta(\xi_{ij})\Theta(\xi_{kl})\xi_{ij}\xi_{kl}r_{ij}r_{kl}n_{ijx}n_{klx}n_{ijy}n_{kly}\rangle_{\dot\gamma}\\
c_1 &= (mS)^{-1}\tau_{\rm rel}\sum_{\substack{i,j,k\\(i\neq j,i\neq k)}} \langle 3\eta_n^2\Theta(\xi_{ij})^2 r_{ij}^2 n_{ijx}^2 n_{ijy}^2 + 3\eta_t^2\langle\Theta(\xi_{ij})^2 r_{ij}^2 n_{ijy}^4 + \eta_n^2\Theta(\xi_{ij})\Theta(\xi_{ik})\left(n_{ijx}n_{ikx}+n_{ijy}n_{iky}\right) r_{ij}r_{ik}n_{ijx}n_{ikx}n_{ijy}n_{iky}\\
&\quad +\eta_n\eta_t\Theta(\xi_{ij})\Theta(\xi_{ik})r_{ij}r_{ik}n_{ijx}n_{ijy}n_{iky}^2\left(n_{ijx}n_{iky}-n_{ijy}n_{ikx}\right) + \eta_t^2\Theta(\xi_{ij})\Theta(\xi_{ik})r_{ij}r_{ik}n_{ijy}^2 n_{iky}^2\left(n_{ijy}n_{iky}+n_{ijx}n_{ikx}\right)\rangle_{\dot\gamma}\\
c_2 &= (mS)^{-1}\tau_{\rm rel}\sum_{\substack{i,j,k\\(i\neq j,i\neq k)}} \langle 6\eta_t^2\Theta(\xi_{ij})^2 r_{ij}^2 n_{ijy}^4 + 2\eta_t^2\langle\Theta(\xi_{ij})\Theta(\xi_{ik})r_{ij}r_{ik}n_{ijy}^2 n_{iky}^2\rangle_{\dot\gamma}
\end{aligned}
$$

$\langle\sigma_{xy}\Omega\rangle_{\dot\gamma} \simeq \langle\sigma_{xy}\Omega\rangle_{\rm eq} + \tau_{\rm rel}\langle\sigma_{xy}\Omega^2\rangle_{\rm eq}$, where $\langle\dots\rangle_{\rm eq}$ represents the ensemble average in the equilibrium state. Then, we find that the shear stress is expanded into the series of relaxation time as $\langle\sigma_{xy}\rangle_{\dot\gamma+\Delta\dot\gamma} \simeq \langle\sigma_{xy}\rangle_{\dot\gamma} + \tau_{\rm rel}\langle\sigma_{xy}\Omega\rangle_{\rm eq} + \tau_{\rm rel}^2\langle\sigma_{xy}\Omega^2\rangle_{\rm eq}$. The *parity* of phase variables requires $\langle\sigma_{xy}\Omega^2\rangle_{\rm eq} = 0$ [1] and finally, the shear stress is given by $\langle\sigma_{xy}\rangle_{\dot\gamma+\Delta\dot\gamma} \simeq \langle\sigma_{xy}\rangle_{\dot\gamma} + \tau_{\rm rel}\langle\sigma_{xy}\Omega\rangle_{\rm eq}$, where the equilibrium averages, $\langle\sigma_{xy}\Omega\rangle_{\rm eq} = -\dot\gamma\beta S\langle\sigma_{xy}^2\rangle_{\rm eq} - 2\beta\langle\sigma_{xy}\mathcal{R}\rangle_{\rm eq} + \beta\langle\sigma_{xy}\mathcal{J}\rangle_{\rm eq} - \langle\sigma_{xy}\Lambda\rangle_{\rm eq}$, can be explicitly calculated from Eqs. (3)-(5) and (8). Applying the parallel procedure to the granular temperature, $\langle T\rangle_{\dot\gamma}$, and rotational temperature, $\langle T_t\rangle_{\dot\gamma}$ [3], we find difference equations as

$$\langle\sigma_{xy}\rangle_{\dot\gamma+\Delta\dot\gamma} = \langle\sigma_{xy}\rangle_{\dot\gamma} - a_0 - a_1\dot\gamma - \frac{\dot\gamma h(\dot\gamma) + a_2\dot\gamma\langle T_t\rangle_{\dot\gamma}}{\langle T\rangle_{\dot\gamma}}, \tag{10}$$

$$\langle T\rangle_{\dot\gamma+\Delta\dot\gamma} = (b_1+1)\langle T\rangle_{\dot\gamma} + b_0\dot\gamma^2 - b_2\langle T_t\rangle_{\dot\gamma}, \tag{11}$$

$$\langle T_t\rangle_{\dot\gamma+\Delta\dot\gamma} = (c_1+1)\langle T_t\rangle_{\dot\gamma} + (c_0\dot\gamma^2 - c_2\langle T_t\rangle_{\dot\gamma})\frac{\langle T_t\rangle_{\dot\gamma}}{\langle T\rangle_{\dot\gamma}}, \tag{12}$$

where the coefficients (a_0, a_1, a_2, b_0, b_1, b_2, c_0, c_1, and c_2) are listed in Table 1 [4].

To explicitly calculate the coefficients in Table 1, we need to calculate the three- and four-body correlations [1] and have to introduce the radial distribution function for frictional disks. In addition, it is necessary to determine the two time scales, i.e. the relaxation time, $\tau_{\rm rel}$, and duration of contact, τ_{ij}, which is far beyond the scope of this work. Therefore, in the following analysis, we simply regard the coefficients as fitting parameters.

[3]The granular temperature and rotational temperature are introduced as $T \equiv N^{-1}\sum_i \mathbf{p}_i^2/2m$ and $T_t \equiv N^{-1}\sum_i I_i\omega_i^2$, respectively.

[4]To calculate the correlation functions, the elastic part of tangential force is approximated to $\mathbf{F}_{it}^{(\rm el)} \simeq -k_t\sum_{j\neq i}\Theta(\xi_{ij})\tau_{ij}\mathbf{v}_{ijt}$.

2.4 Flow curves

To draw the flow curve of frictional disks, we numerically solve the difference equations (10)-(12). Here, we scale the mass, length, and time by m, d (the diameter of disk), and $t_0 \equiv \eta_n/k_n$ such that the shear stress, granular (rotational) temperature, and shear rate are scaled as $\langle\sigma_{xy}^*\rangle_{\dot\gamma} \equiv (t_0^2/m)\langle\sigma_{xy}\rangle_{\dot\gamma}$, $\langle T^*\rangle_{\dot\gamma} \equiv (t_0^2/md^2)\langle T\rangle_{\dot\gamma}$ ($\langle T_t^*\rangle_{\dot\gamma} \equiv (t_0^2/md^2)\langle T_t\rangle_{\dot\gamma}$), and $\dot\gamma^* \equiv t_0\dot\gamma$, respectively. In addition, the coefficients on the right-hand-sides of Eqs. (10)-(12) are nondimensionalized as $a_0^* \equiv (t_0^3/m)a_0$, $a_1^* \equiv (t_0^2/m)a_1$, $a_2^* \equiv (t_0^2/m)a_2$, $b_0^* \equiv (t_0/md^2)b_0$, $b_1^* \equiv t_0b_1$, $b_2^* \equiv t_0b_2$, $c_0^* \equiv (t_0/md^2)c_0$, $c_1^* \equiv t_0c_1$, and $c_2^* \equiv t_0c_2$, which are used as numerical parameters (see the caption of Fig. 1).

Figure 1 displays numerical solutions of scaled shear stress, where the shear rate increases (the open circles) and then decreases (the open squares) as shown by the two arrows. Increasing the shear rate, we find that the shear stress satisfies *Bagnold's scaling*, i.e. $\langle\sigma_{xy}^*\rangle_{\dot\gamma} \sim \dot\gamma^{*2}$ (the dotted line) in the slow flow regime ($\dot\gamma^* \lesssim 10^{-5}$), while it becomes linear, i.e. $\langle\sigma_{xy}^*\rangle_{\dot\gamma} \sim \dot\gamma^*$ (the inset) in the plastic flow regime ($\dot\gamma^* \gtrsim 10^{-4}$) as known by the *$\mu$-I rheology* [10]. On the other hand, if we decrease the shear rate, the shear stress exhibits irreversibility, where it becomes almost rate-independent, i.e. $\langle\sigma_{xy}^*\rangle_{\dot\gamma} \sim \sigma_Y$ with the *yield stress*, σ_Y, in the solid-like regime ($\dot\gamma^* \lesssim 10^{-4}$). Therefore, our model, Eqs. (10)-(12), well captures not only all the flow curve of frictional disks, but also the *hysteresis* of shear stress, which is the signature of the DST.

3 Summary

In this study, we have extended the previous microscopic theory for frictionless granular particles [1] to the case of two-dimensional frictional disks, where all the flow behavior, i.e. Bagnold's scaling, $\langle\sigma_{xy}^*\rangle_{\dot\gamma} \sim \dot\gamma^{*2}$, the plastic flow, $\langle\sigma_{xy}^*\rangle_{\dot\gamma} \sim \dot\gamma^*$, the rate-independent yielding stress,

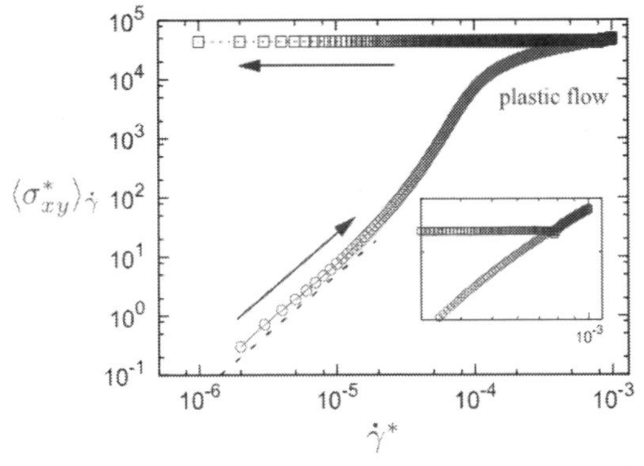

Figure 1. (Color online) Numerical results of scaled shear stress, $\langle\sigma^*_{xy}\rangle_{\dot\gamma}$. The scaled shear rate, $\dot\gamma^*$, is first increased (the open circles) and then decreased (the open squares) as shown by the two arrows, where the scaled increment is given by $\Delta\dot\gamma^* = 10^{-6}$. The dotted line represents Bagnold's scaling, $\langle\sigma^*_{xy}\rangle_{\dot\gamma} \sim \dot\gamma^{*2}$, and the inset shows the zoom-in to the plastic flow regime. Here, we choose numerical values of the coefficients, $a^*_0 = 2.0\times10^{-7}$, $a^*_1 = 4.0\times10^{-7}$, $a^*_2 = 1.0\times10^{-7}$, $b^*_0 = 3.0\times10^{-7}$, $b^*_1 = 1.6\times10^{-7}$, $b^*_2 = 1.0\times10^{-7}$, $c^*_0 = b^*_0$, $c^*_1 = b^*_1$, and $c^*_2 = b^*_2$.

$\langle\sigma^*_{xy}\rangle_{\dot\gamma} \sim \sigma_Y$, and the hysteresis of shear stress as the signature of DST, is well described by the difference equations (10)-(12). All the terms in Eqs. (10)-(12) including the coefficients have been derived from the generalized Green-Kubo formula, Eq. (9), where the canonical ensembles are assumed for the system under shear and the time correlation function between the shear stress and work function, i.e. $\langle\sigma_{xy}(s)\Omega\rangle_{\dot\gamma}$, is approximated by the exponential function with the relaxation time, $\tau_{\rm rel}$. Although it is needed to introduce the radial distribution function for frictional disks and determine the two time scales, i.e. the relaxation time and duration of contact, the coefficients in the difference equations have been provided by the microscopic expressions as listed in Table 1. However, we have used the coefficients as numerical parameters to draw the flow curves of frictional disks (Fig. 1). From our results, we can expect that the coupling with rotational temperature, which is absent in the case of frictionless particles [1], triggers both the transition from the liquid-like branch to solid-like one and the hysteresis behavior of shear stress.

Nevertheless, the transition from the liquid-like behavior (unjammed state) to the solid-like branch (jammed state) during the increase of shear rate (the open circles in Fig. 1) is much smoother than that observed in MD simulations [3, 5]. In addition, we have not observed a sudden drop from the solid-like branch to the liquid-like one (the open squares in Fig. 1). Therefore, we will need further investigations of our microscopic approach to the DST, where more rigorous treatments of the coefficients, time scales, and the radial distribution function might be crucial. In addition, it is important to implement the recent theory of non-Brownian suspensions, where the competing divergences of shear viscosity with the smooth shift of jamming point explain the discontinuous shear thickening [11].

In conclusion, we have developed a microscopic theory for the DST of two-dimensional frictional disks, where the difference equations of the shear stress, the granular temperature, and the rotational temperature derived from the generalized Green-Kubo formula well explain all the flow behavior including the hysteresis of the shear stress.

Acknowledgements

We thank K. Suzuki, S. Takada, M. Otsuki, and T. Kawasaki for fruitful discussions. This work was supported by the World Premier International Research Center Initiative (WPI), Ministry of Education, Culture, Sports, Science, and Technology, Japan (MEXT), Kawai Foundation for Sound Technology & Music, Grant-in-Aid for Scientific Research B (Grants No. 16H04025), and Grant-in-Aid for Research Activity Start-up (Grant No. 16H06628) from the Japan Society for the Promotion of Science (JSPS).

References

[1] K. Suzuki, H. Hayakawa, Phys. Rev. Lett. **115**, 098001 (2015)

[2] J. Lemaitre, J.L. Chaboche, *Mechanics of Solid Materials* (Cambridge University Press, Cambridge, UK, 1990)

[3] M. Otsuki, H. Hayakawa, Phys. Rev. E **83**, 051301 (2011)

[4] R. Seto, R. Mari, J.F. Morris, M.M. Denn, Phys. Rev. Lett. **111**, 218301 (2013)

[5] M. Grob, C. Heussinger, A. Zippelius, Phys. Rev. E **89**, 050201(R) (2014)

[6] D.J. Evans, G. Morriss, *Statistical Mechanics of Nonequilibrium Liquids*, 2nd edn. (Cambridge University Press, Cambridge, UK, 2008)

[7] S. Luding, J. Phys.: Condens. Matter **17**, S2623 (2005)

[8] N. Mitarai, H. Hayakawa, H. Nakanishi, Phys. Rev. Lett. **88**, 174301 (2002)

[9] H. Hayakawa, M. Otsuki, Phys. Rev. E **88**, 032117 (2013)

[10] G.D.R. MiDi, Eur. Phys. J. E **14**, 341 (2004)

[11] M. Wyart, M.E. Cates, Phys. Rev. Lett. **112**, 098302 (2014)

Transverse mixing of ellipsoidal particles in a rotating drum

Siyuan He[1], *Jieqing* Gan[1], *David* Pinson[2], and *Zongyan* Zhou[1,*]

[1]Department of Chemical Engineering, Monash University, Melbourne, VIC 3800, Australia
[2]BlueScope Steel, Port Kembla, NSW 2505, Australia

Abstract. Rotating drums are widely used in industry for mixing, milling, coating and drying processes. In the past decades, mixing of granular materials in rotating drums has been extensively investigated, but most of the studies are based on spherical particles. Particle shape has an influence on the flow behaviour and thus mixing behaviour, though the shape effect has as-yet received limited study. In this work, discrete element method (DEM) is employed to study the transverse mixing of ellipsoidal particles in a rotating drum. The effects of aspect ratio and rotating speed on mixing quality and mixing rate are investigated. The results show that mixing index increases exponentially with time for both spheres and ellipsoids. Particles with various aspect ratios are able to reach well-mixed states after sufficient revolutions in the rolling or cascading regime. Ellipsoids show higher mixing rate when rotational speed is set between 25 and 40 rpm. The relationship between mixing rate and aspect ratio of ellipsoids is established, demonstrating that, particles with aspect ratios of 0.5 and 2.0 achieve the highest mixing rates. Increasing rotating speed from 15 rpm to 40 rpm does not necessarily increase the mixing speed of spheres, while monotonous increase is observed for ellipsoids.

1 Introduction

Rotating drums are widely used in industry for granular mixing. The mixing process of spherical particles has been studied extensively in the past decades. For example, Henein et al. [1] studied the segregation in a rotary kiln experimentally. Alizadeh et al. [2] studied the mixing and segregation process in a rotating drum by particle tracking method. Liu et al [3] used Discrete Element Method (DEM) to simulate the transverse mixing of wet particles. They found that the mixing of dry particles was poorest at 64% filling level and increasing the cohesion at that level improved particle mixing. Similarly, Arntz et al [4] focused on the impact of rotational speed and filling level by DEM. They found that there was strong correlation between flow regime and segregation pattern and the most intense segregation occurred when the fill fraction is larger than 65%.

Though comprehensive study have been implemented on mixing of spherical particles, the mixing of non-spherical particles have received limited study yet. Particle shape is an important factor that affects the mixing behaviour and thus mixing rate. Geng et al [5] focused on the mixing rate of cylinder-shaped particles and found that the mixing rate of slender particles was faster than for spheres. Pereira et al [6] studied the streak patterns formation in binary granular mixture, considering both the size and shape effects. They proposed that the growth and stabilisation of a regular streak pattern required the two sets of particles to differ significantly in angle of repose. Though the efforts have been made on the mixing of non-spherical particles, e.g. slender particles or super-quadrics, the mixing process of ellipsoidal particles with different aspect ratios has not been studied yet. In this work, we focus on the mixing of ellipsoids with various aspect ratios in comparison with spheres. The mixing patterns for both ellipsoids and spheres are presented to examine the evolution of the structure of the beds. By introducing mixing index and mixing rate, the quality and speed of mixing process are quantified, respectively.

2 Simulation methods and conditions

According to the DEM proposed by Cundall and Strack [7], a particle can have two types of motion: translational and rotational, which are determined by Newton's second law of motion. DEM for ellipsoidal particles have been reported by Dziugys [8]. The DEM for ellipsoids was developed by the authors. The governing equations for the translational and rotational motion of particle i with radius R_i, mass m_i, and moment of inertia I_i can be written as:

$$m_i \frac{d\mathbf{v}_i}{dt} = \sum_{j=1}^{k_c}(\mathbf{f}_{c,ij} + \mathbf{f}_{d,ij} + \mathbf{f}_{v,ij}) + m_i\mathbf{g} \quad (1)$$

and

$$I_i \frac{d\boldsymbol{\omega}_i}{dt} = \sum_{j=1}^{k_c}(\mathbf{M}_{t,ij} + \mathbf{M}_{r,ij} + \mathbf{M}_{n,ij}) \quad (2)$$

where $\mathbf{v}_i$ and $\boldsymbol{\omega}_i$ are the translational and angular velocities of the particle, respectively, and k_c is the number of particles in interaction with the particle. As plotted in Fig.1, the forces involved are: the gravitational force $m_i\mathbf{g}$, and inter-particle forces between particles, which include elastic force $\mathbf{f}_{c,ij}$, and viscous damping force $\mathbf{f}_{d,ij}$. For fine particles, the van der Waals force $\mathbf{f}_{v,ij}$ needs to be considered. These inter-particle forces can be resolved

* Corresponding author: zongyan.zhou@monash.edu

into the normal and tangential components at a contact point. The torque acting on particle i by particle j includes two components: $\mathbf{M}_{t,ij}$ which is generated by the tangential force and causes particle i to rotate, and $\mathbf{M}_{r,ij}$, commonly known as the rolling friction torque, is generated by asymmetric normal forces and slows down the relative rotation between particles. For ellipsoids, addition torque $\mathbf{M}_{n,ij}$ should be added because the normal contact force and van der Waals force do not necessarily pass through the particle centre. Other critical features for ellipsoids are introduced briefly below:

- Determining the interaction forces by calculating the radius R*, which is strongly related to the radii of the curvature at a contact point;
- Using geometric potential algorithm to detect the contacts between ellipsoids; and
- Particle orientation is defined by three Euler angles (ϕ, Θ, Ψ) based on the quaternion method.

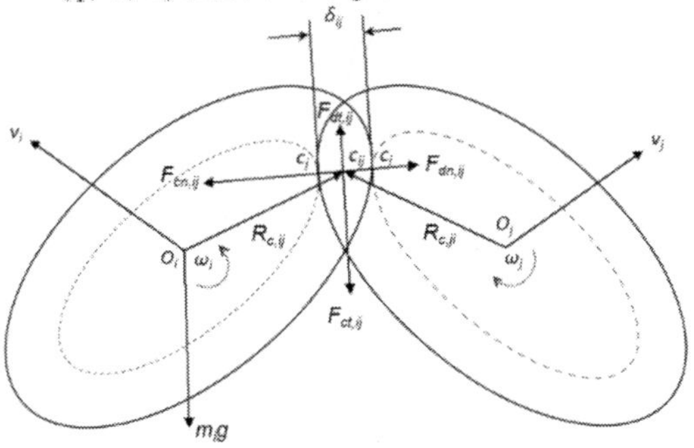

Fig.1. Two-dimensional diagram of forces acting on particle i contacting particles j

For the shape of ellipsoids, three principle diameters are used to describe: 2a, 2b and 2c, where a is the principal radius in the polar direction, b and c are principal radii in the equatorial plane. For spheroids used in this work, b = c which means that the equatorial plane is a circle. The parameter aspect ratio AR is defined as AR=a/c. Spheroids with aspect ratios varying from 0.25 to 3.0 are used in this work. For a prolate (a>b=c) spheroid, aspect ratio AR > 1; for a sphere (a=b=c), AR = 1; and for an oblate (a<b=c) spheroid, AR< 1.

In the current study, we quantify the extent of mixing by Lacey mixing index M given by [9]

$$M = \frac{\delta_0^2-\delta_t^2}{\delta_0^2-\delta_r^2} \quad (3)$$

where δ_t^2 is the variance of the current state of mixing. $\delta_0^2 = q(1-q)$ and $\delta_r^2 = q(1-q)/p$ are the variances of the in a completely separated system and in a well-mixed system, respectively. q is the total proportion of either type of particles in the mixture. p is the average number of particles in a cell. The size of cells are set as $4d_v$ in the current study, where d_v denotes the volume-equivalent diameter of the particles. The sliding friction coefficient of the end-walls are set as 0.04 in the simulation, which is small enough to ignore the effects of end-walls.

3 Results and discussion

3.1 Mixing patterns

As shown in Fig.2, different mixing patterns can be observed for ellipsoids and spheres. Evidently, the beds of ellipsoidal particles have reached cascading regime, which is characterised by the S-shaped surface. However, the flow surface for spheres remains flat, which means spheres are still in rolling regime. As the drum rotates, more avalanching occurs, and better mixing quality is achieved. Each time when red particles finish flowing downward the free surface, they will re-enter the particle mixture, move upward against the side wall, and will experience the avalanching of the next round. As the bed is turned over repeatedly, more red particles accumulate in the central part of the bed in a spiral shape and less red particles would take part in the avalanching motion. As a result, the bed become more hierarchical, indicating the progress of mixing. After rotating for 6 seconds, obvious stratification structures and large mixing interface can be observed for the beds of ellipsoids, while preliminary state of mixing for spherical particles. At the 8th second, red ellipsoidal particles have distributed in the bed homogeneously with the disappearance of the stratification structure, which indicates a perfectly mixed state of the bed. However, for spheres, stratification structure just appears. From the evolution of the mixing patterns of spheres and ellipsoids, ellipsoids are evidently mixed faster than spheres at the same rotating speed.

3.2 Effects of aspect ratio on mixing rate

Figs. 3 and 4 show the temporal evolutions of mixing index for oblate and prolate spheroids. The particles of different colours are completely separated initially, i.e. the mixing index is close to 0. After sufficient revolution of the drum, both spheres and ellipsoids are able to reach a well-mixed state eventually, i.e. the mixing index is close to 1. For spheres, the increase of mixing index can be fitted by an error function, which is given by

$$M(t) = M_\infty + (M_0 - M_\infty)\mathrm{erf}(Rt) \quad (4)$$

Where the error function $\mathrm{erf}(x) = 2/\sqrt{\pi}\int_0^x e^{-x^2}\,dx$. M_∞, M_0 and R are the final mixing index, initial mixing index and mixing rate respectively, which can be obtained from the fitting of simulation data. Mixing rate quantifies the speed of mixing of particles in different colours. Obviously, the same fitting function can be applied to the temporal growth of the mixing index for ellipsoids. Moreover, for both oblate and prolate spheroids, the mixing index become steady and close to 1.0 after rotating for around 10 seconds, while 30 seconds for spherical particles. Clearly, ellipsoids are mixed faster than spheres at the same rotating speed. For oblate spheroids, the speed of mixing is similar. The particles with aspect ratio of 0.50 are mixed faster than the ones with aspect ratio of 0.25. Similarly, slight difference of mixing speed is observed among prolate spheroids of different aspect ratio. The particles with aspect ratio of 2.0 are mixed fastest for prolate spheroids.

DOI: 10.1051/epjconf/201714006018

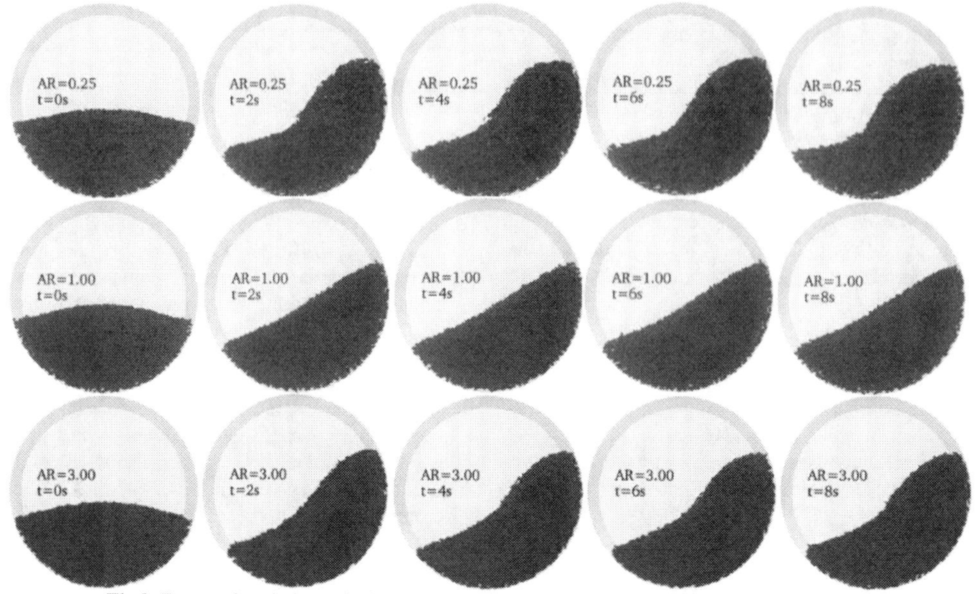

Fig.2. Temporal evolutions of mixing patterns for particles of different aspect ratios at 15 rpm.

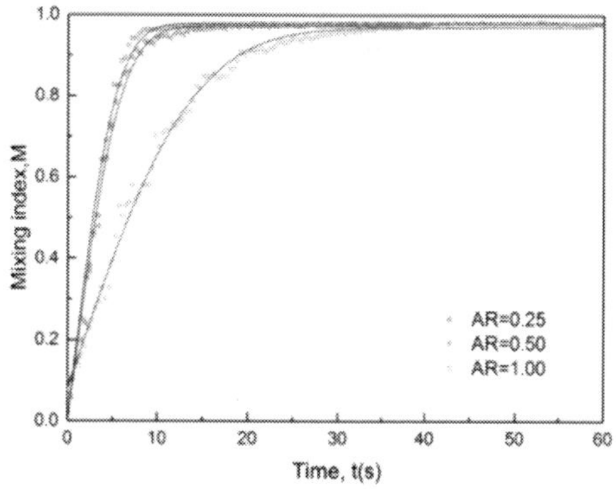

Fig. 3. Evolution of the mixing index of spheres and oblate spheroids with time at 30 rpm. The solid curves are the fitting results with Eq. (4).

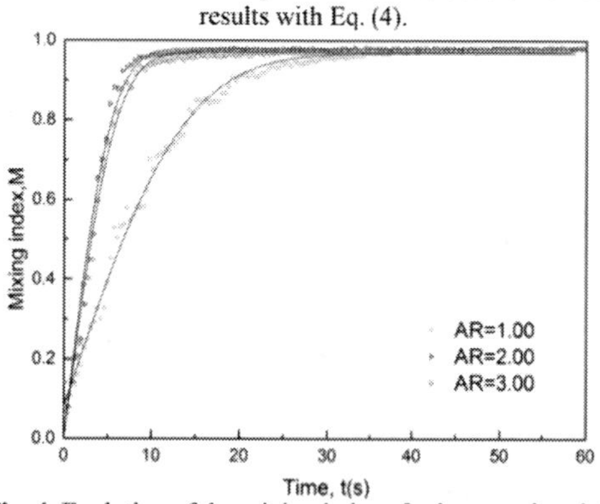

Fig. 4. Evolution of the mixing index of spheres and prolate spheroids with time at 30 rpm. The solid curves are the fitting results with Eq. (4).

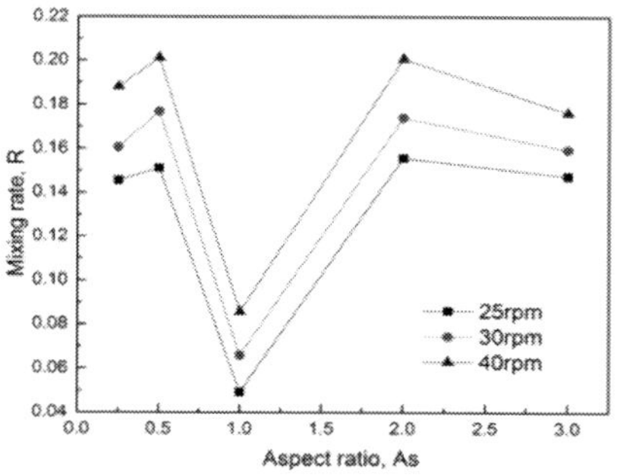

Fig.5. The relationship between mixing rate and aspect ratio for different rotating speeds.

Fig.5 shows the relationship between the mixing rate and aspect ratio under different rotating speeds. Obviously, for the same rotating speed, spheres are mixed much slower than ellipsoids, which is in consistent with the evolution of mixing patterns. However, for oblate or prolate spheroids, slight change on mixing rate can be observed when the aspect ratio varies. Prolate and oblate spheroids tend to have similar mixing rate when there is similar deviation of the aspect ratio from the unity. Though slight difference of mixing rate is observed between ellipsoidal particles, granules with aspect ratio of 0.50 and 2.0 have the largest mixing rate, which is approximately 3 times faster than spherical particles.

Additionally, when the rotating speed increases from 25rpm to 40rpm, the mixing rate increases for both spheres and ellipsoids. It is different from the results obtained by Liu et al. [3], where the mixing rate of spheres decreases when the rotating speed increases. It is because the mixing rate obtained in the current study is calculated by the evolution of mixing index with time other than revolutions. It means that, though the mixing efficiency for each revolution decreases with the increasing rotational speed [3], the mixing rate increases due to more revolutions are completed in a certain period of time.

3.3 Evolution of the mixing index in a wide range of rotating speed

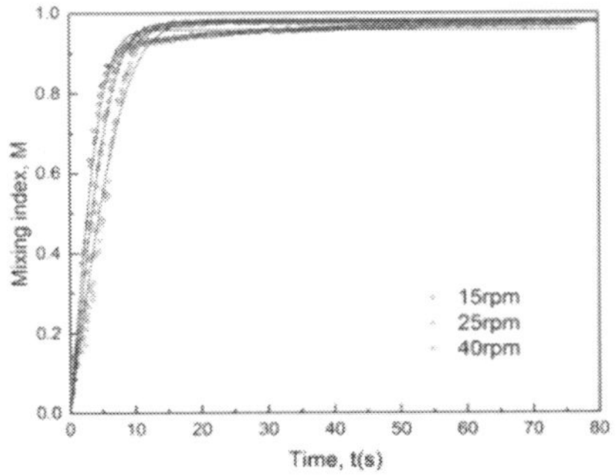

(a)

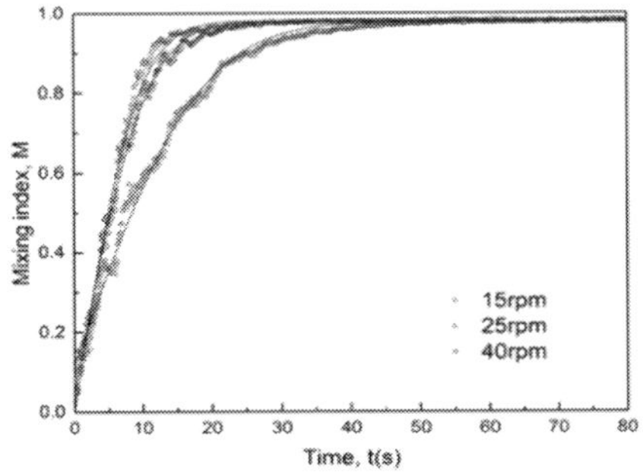

(b)

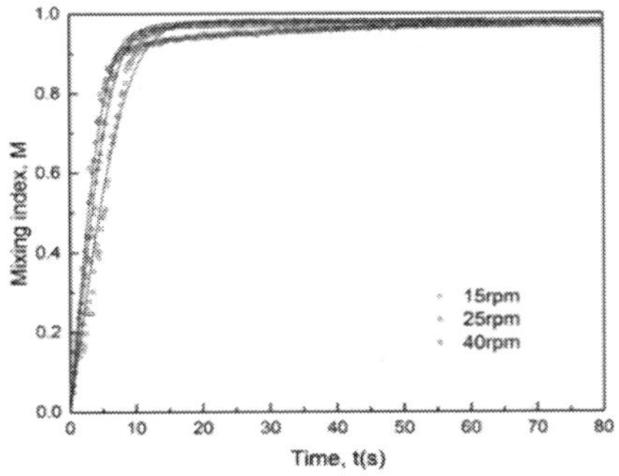

(c)

Fig.6. The temporal variation of mixing index for different rotating speeds: (a) AR=0.25; (b) AR=1.00; (c) AR=3.00.

Fig.6 shows the temporal evolutions of mixing index of ellipsoids and spheres for different rpm. The inclination of the curve denotes the speed of mixing. Though particle shape changes, an exponential increase of mixing index remains for ellipsoidal particles. For spheres, the speed of mixing is sensitive to the variation of rotational speed. From 15 rpm to 25 rpm, the speed of mixing decreases sharply. It means that increasing rotating speed leads to a significant decrease in "mixing efficiency", i.e., the number of revolutions required to turn over the granular bed, which is consistent with the previous study [10]. However, when the rotating speed increases from 25 rpm to 40 rpm, the mixing speed increases, which means the mixing speed is dominated by the rotating speed other than the "mixing efficiency".

However, this trend cannot be applied to ellipsoidal particles. Slight change is observed between ellipsoidal particles when rotating speed varies from 15rpm to 40 rpm. As the rotating speed increases, the mixing speed of ellipsoids increases as well. It means that, with the increasing rotational speed, the "mixing efficiency" does not change much so that the number of revolutions in a certain time dominates the speed of mixing.

4 Conclusions

DEM is used to examine the effect of particle shape on the mixing of particles in a rotating drum. It indicates that after sufficient time of revolution, both ellipsoids and spheres can reach well-mixed state. Similar to spherical particles, the mixing index of ellipsoidal particles increases exponentially with time. From 25 rpm to 40 rpm, ellipsoids tend to have larger mixing rate than spheres at the same rotating speed. Though slight difference of mixing rate is observed between ellipsoidal particles, granules with aspect ratio of 0.50 and 2.00 are mixed fastest. For a wider range of rotational speed, i.e. from 15rpm to 40 rpm, the speed of mixing decreases first and reaches a minimum at 25 rpm, then increases in the range of 25 rpm to 40 rpm. However, for ellipsoidal particles, monotonous increase of mixing speed is observed.

Acknowledgements

The authors are grateful to Australian Research Council and BlueScope Steel for their financial support of this work.

References

1. H. Henein, J.K. Brimacombe, A.P. Watkinson, Metall. Trans. B **16**, 4 (1985)
2. E. Alizadeh, O. Dubé, F. Betrand, J. Chaouki, AIChe J. **59**, 6 (2013)
3. P.Y. Liu, R.Y. Yang, A.B. Yu, Chem. Eng. Sci. **86** (2013)
4. M.M.H.D. Arntz, W.K. den Otter, W.J. Briels, P.J.T. Bussmann, H.H. Beeftink, AIChe J. **54**, 12 (2008)
5. F. Geng, Z.L. Yuan, Y.M. Yan, D.S. Luo, H.S. Wang, B. Li, D.Y. Xu, Powder Technol. **193**, 1 (2009)
6. G.G. Pereira, S. Pucilowski, K. Liffman, P.W. Cleary, Appl Math Model **35**, 4 (2011)
7. P.A. Cundall, O.D.L. Strack, Geotechnique **29**, 1 (1979)
8. A. Dziugys, B. Peters, Granul Matter **3**, 4 (2001)
9. P.M.C. Lacey, J. Appl. Chem. **4**, 5 (1954)
10. G.J. Finnie, N.P. Kruyt, M. Ye, C. Zeilstra, J.A.M.Kuipers, Chem. Eng. Sci. **60**, 15 (2005)

DOI: 10.1051/epjconf/201714003064

Kinetic theory for strong uniform shear flow of granular media at high density

W. Till Kranz[1,*] and Matthias Sperl[1,2]

[1] Institut für Materialphysik im Weltraum, Deutsches Zentrum für Luft- und Raumfahrt e.V., Linder Höhe, 51147 Köln, Germany
[2] Institut für Theoretische Physik, Universität zu Köln, Zülpicher Straße 77, 50937 Köln, Germany

Abstract. We discuss the uniform shear flow of a fluidized granular bed composed of monodisperse Hertzian spheres. Considering high densities around the glass transition density of inelastic Hertzian spheres, we report kinetic theory expressions for the Newtonian viscosity as well as the Bagnold coefficient. We discuss the dependence of the transport coefficients on density and coefficient of restitution.

1 Introduction

Flows of dense granular materials are ubiquitous in nature and industry [1–5]. For efficiency [6] and disaster prevention [7] a solid understanding of the rheology of these flows is crucial. Inclined plane flows both in industrial and geophysical settings have been successfully modeled by inelastic Hertzian spheres [5, 8–11] using molecular-dynamics simulations. Discrete element models are constrained, however, to relatively small system sizes due to computational demands.

For a continuum modeling which can be applied at very large scales and a theoretical analysis, it is essential to have access to the transport coefficients as functions of the system parameters. For low densities, the kinetic theory of granular flows is well developed based on the inelastic Boltzmann equation [12–14]. It yields predictions for, amongst others, the diffusivity and the shear viscosity. For high densities and especially away from the linear response regime, results derived from first principles are rare.

The Integration Through Transients (ITT) formalism has been very successful in describing shear in colloidal suspensions at densities around the glass transition and for finite shear rates [15–17]. A glass transition is observed [18] and predicted for fluidized granular beds also [19, 20]. This allowed us recently to generalize the ITT formalism to the far from equilibrium regime of a driven granular medium [21]. In this contribution we will present results for the viscosity and Bagnold coefficient. Hayakawa *et al.* have derived two related sets of equations (Refs. [22] and [23]) where only the second version has yielded numerical results for the transport coefficients so far.

The structure of the paper is as follows. In Sec. 2 we will define the model and control parameters. A brief sketch of the theory and the required inputs are given in Sec. 3 before we will discuss the transport coefficients for the limiting cases of weak (Sec. 4) and strong (Sec. 5) shear. We will close with a discussion in Sec. 6.

2 Model

We consider a monodisperse system of $N \to \infty$ inelastic Hertzian spheres [24, 25] of diameter d and mass $m = 1$ in a volume V such that the density $n = N/V$ remains finite. The overlap potential is given by

$$V(r_{ij}) = \Gamma T(1 - r_{ij}/d)^{5/2} \quad \text{for } r_{ij} < d \tag{1}$$

where Γ controls the strength of the repulsion and $T = \langle v^2 \rangle/3$ is the granular temperature. The distance between two particles is denoted by $r_{ij} = |\boldsymbol{r}_i - \boldsymbol{r}_j|$. Here we will focus on the practically relevant case of rather stiff particles, $\Gamma \gg 1$. Dissipation is quantified by a constant coefficient of restitution $0 \leq \epsilon < 1$ [12].

We consider the quiescent system to be in a fluidized state [18, 26, 27] and model the fluidization by a Gaussian random force $\boldsymbol{\xi}_i(t)$ with zero mean $\langle \boldsymbol{\xi}_i \rangle = 0$ acting on each particle individually. Its variance $\langle \xi_i^\alpha(t) \xi_j^\beta(t') \rangle = 2P_D \delta_{ij} \delta^{\alpha\beta} \delta(t - t')$ is characterized by the driving power P_D.

We imagine that at time $t = 0$ the system is in a stationary state where the dissipation is balanced by the driving force resulting in a stationary temperature T_0. At this point we switch on a uniform linear shear profile modeled by the SLLOD equations of motion [28]. We prescribe the shear rate $\dot{\gamma}$ and monitor the resulting shear stress σ. For finite shear rates the shear will contribute to the energy input by a term $\dot{\gamma}\sigma$.

If we wait long enough the system will reach a new stationary state with a new temperature T and constant mean shear stress σ. This is the state we want to consider in the following. For a given material of the particles prescribed by the pair (ϵ, Γ), the macroscopic state of the system is fully specified by the packing fraction $\varphi = \pi n d^3/6$, the temperature T_0 (or, equivalently, T) resulting from the energy balance, and the shear rate $\dot{\gamma}$.

* e-mail: till.kranz@dlr.de

DOI: 10.1051/epjconf/201714003064

3 Calculating the Transport Coefficients

The transport coefficients which have traditionally been calculated from the Boltzmann-Enskog equation using, e.g., the Chapmann-Enskog or related methods [29], are necessarily valid only to lowest order in the shear rate $\dot{\gamma}$ and the packing fraction φ. To go beyond the lowest order presents a considerable challenge.

For densities around the (dynamic) glass transition, the ITT formalism enables to calculate corrections beyond the linear order in packing fraction and for arbitrary shear rates. It relates expectation values in the stationary sheared state to the (equilibrium) unsheared state and captures the transient dynamics after the switch-on of shear in terms of time-dependent density correlation functions [17]. For the latter, mode-coupling theory (MCT) [30] affords a faithful description for the densities under consideration.

Granular MCT allows for a generalization of the ITT formalism for dissipative particles. A crucial input for both MCT and ITT are static structure factors, S_k, in the quiescent state. To this end we use numerical solutions of the Hypernetted-Chain-Equation [31] for Hertzian spheres [32].

For the low-density (bare) transport coefficients, we use the Enskog expressions provided by Garzó and Montanero [14]. Especially important are the shear viscosity η_0 and the related sound-damping constant D_S.

4 Linear Rheology

For small shear rate $\dot{\gamma} \to 0$ and packing fractions $\varphi < \varphi_c$ below the glass transition, we recover the linear response of a Newtonian fluid. We consider two values for the Hertzian stiffness $\Gamma = 1000$ and $\Gamma = 1500$ as well as several values for the coefficient of restitution ϵ. For the packing fraction we choose values high enough for the validity of our approach but well below the hard-sphere random close packing density $\varphi_J \approx 0.64$ [33] such that we avoid lasting contacts between particles.

The ITT formalism yields the following expression for the viscosity

$$\eta = \eta_0 + \frac{3}{10\pi^2} \times \frac{(1+\epsilon)^2}{4} \times \frac{T}{dD_S}\varphi^2\chi^2\tilde{\eta}[S](cd/D_S) \quad (2)$$

where χ is the pair correlation function at contact, c denotes the (long wavelength) speed of sound, and[1]

$$\tilde{\eta}[S](K) = \int_0^\infty dk \frac{K^2 + 2k^2}{K^2 + k^2} \times \frac{[j_0''(k)]^2}{S_k^2}. \quad (3)$$

We expect that generally the viscosity will be lower for more dissipative particles (smaller ϵ) but note that D_S strongly depends on ϵ especially for nearly elastic particles [14]. The influence of the particles stiffness Γ will enter through the static structure factor S_k.

For moderate densities (cf. Fig. 1), we observe that the viscosity increases with ϵ as expected. Note that we obtain values that are far above the Boltzmann-Enskog value

[1] Here $j_0(x)$ is the zeroth order spherical Bessel function.

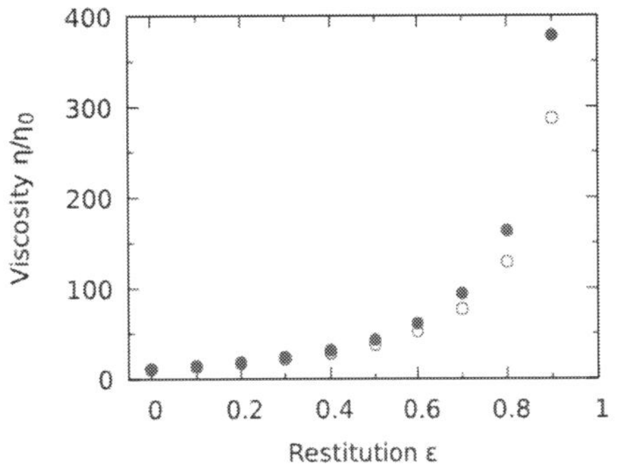

Figure 1. Viscosity η relative to its low density limit η_0 as a function of the coefficient of restitution ϵ for packing fraction $\varphi = 0.51$ and stiffness $\Gamma = 1000$ (open), and 1500 (filled).

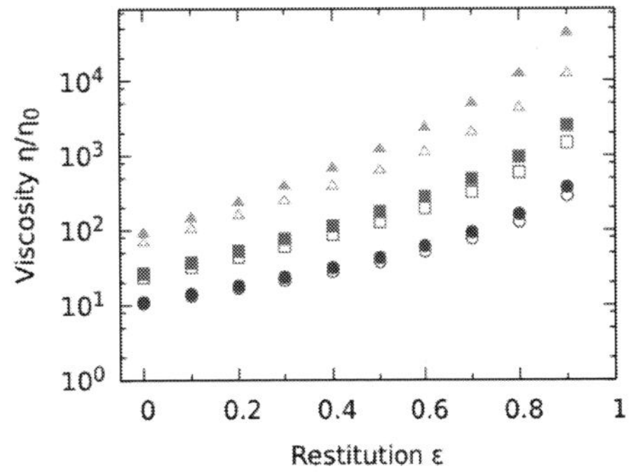

Figure 2. Viscosity η relative to its low density limit η_0 as a function of the coefficient of restitution ϵ for packing fractions $\varphi = 0.51$ (violet disks), 0.56 (green squares), and 0.61 (blue triangles) and stiffness $\Gamma = 1000$ (open), and 1500 (filled).

η_0 valid for vanishing density. Also the stiffer particles have a higher viscosity. This increase is explained by the approach to the glass transition $\varphi_c(\Gamma)$ which is lower for stiffer particles [32, 34].

In general, the viscosity increases over several orders of magnitude upon increasing the density (cf. Fig. 2. Note that we span more than two orders of magnitude in viscosity by varying the inelasticity ϵ at the highest packing fraction considered.

5 Nonlinear Rheology

For strong shear such that shear heating dominates the fluidization, $\sigma\dot{\gamma} \gg nP_D$, we obtain Bagnold scaling where the fluidized bed no longer behaves like a Newtonian fluid, $\sigma = \eta\dot{\gamma}$, but $\sigma = B\dot{\gamma}^2$. This peculiar behavior of granular fluids has first been observed and described by Bagnold [35]. Kinetic predictions for the generalized (Bagnold) viscosity B (which has dimensions of an inverse length) are sparse.

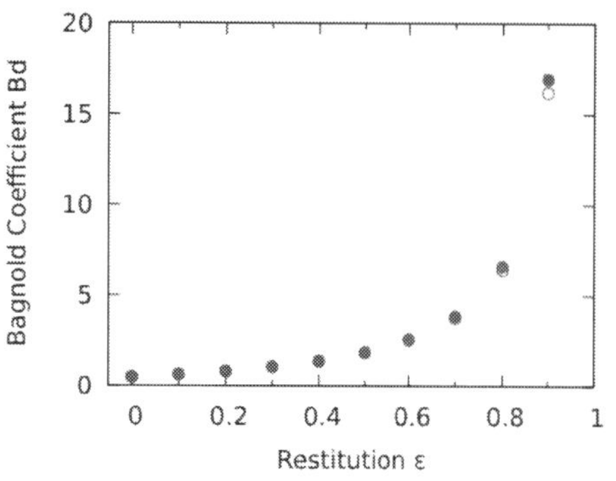

Figure 3. Bagnold coefficient $B \cdot d$ (where d is the particle diameter) as a function of the coefficient of restitution ϵ for packing fraction $\varphi = 0.51$ and two values of the stiffness $\Gamma = 1000$ (open), and 1500 (filled).

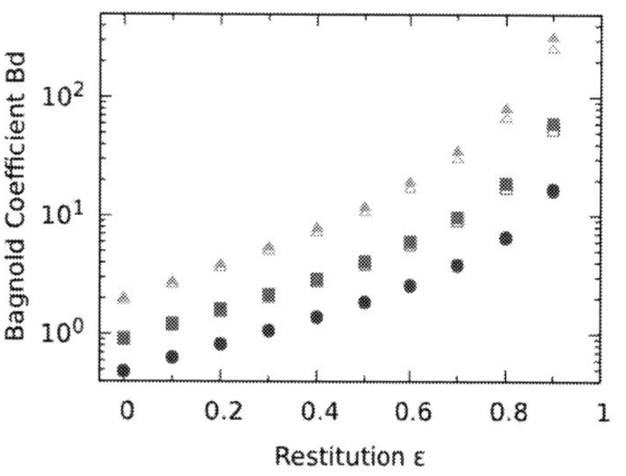

Figure 4. Bagnold coefficient $B \cdot d$ as a function of the coefficient of restitution ϵ for packing fractions $\varphi = 0.51$ (violet disks), 0.56 (green squares), and 0.61 (blue triangles) and two values of the stiffness $\Gamma = 1000$ (open), and 1500 (filled).

The ITT formalism sketched above yields an expression for the Bagnold coefficient [21] in terms of the coherent scattering function. Numerical evaluation of the resulting generalized Green-Kubo-integrals yields values for the Bagnold coefficient B.

We observe (cf. Fig. 3) a very weak dependence of B on the stiffness Γ but, just as for the viscosity, a strong increase with the coefficient of restitution ϵ. As is expected also the Bagnold viscosity increases with increasing density (cf. Fig. 4)

6 Discussion

We presented first results from an granular ITT formalism for strongly sheared dense granular fluids. It yields kinetic theory expressions for the viscosity beyond the lowest order in density. These allowed us to calculate viscosities well above the Boltzmann viscosity η_0 and to determine the dependence on the inelasticity (quantified by the coefficient of restitution ϵ) and on the parameter Γ controlling the stiffness of the Hertzian potential. The general result is that the more elastic and the stiffer the particles are, the higher is the viscosity of the granular fluid.

We also presented predictions for the Bagnold coefficient B which determines the non-Newtonian rheology of granular hard spheres at high shear rates. Like the Newtonian viscosity, it is a measure of the sluggishness of the fluid and thus increases towards the glass transition packing fraction of the quiescent fluid. We found that B is rather insensitive to the stiffness Γ.

In the future we would expect our constitutive relations to be used in the modeling of dense granular at practically relevant shear rates. Kumaran [36] recently presented work in this direction.

Acknowledgments

We acknowledge continuing discussions with Annette Zippelius, Matthias Fuchs, and Fabian Frahsa and thank Marco Heinen for the structure factor code. W. T. K. acknowledges financial support by FOR1394.

References

[1] K. Hutter, N. Kirchner, *Dynamic response of granular and porous materials under large and catastrophic deformations* (Springer, 2013)
[2] M. Ausloos, R. Lambiotte, K. Trojan, Z. Koza, M. Pe et al., Physica A **357**, 337 (2005)
[3] R.M. Iverson, Rev. Geophys. **35**, 245 (1997)
[4] J.J. Lissauer, Annu. Rev. Astron. Astrophys. **31**, 129 (1993)
[5] H.M. Jaeger, S.R. Nagel, Science **255**, 1523 (1992)
[6] D. Kunii, O. Levenspiel, *Fluidization engineering* (Elsevier, 1991)
[7] D. Brunsden, Geomorphology **30**, 13 (1999)
[8] C.S. Campbell, Powder Technology **162**, 208 (2006)
[9] L.E. Silbert, D. Ertaş, G.S. Grest, T.C. Halsey, D. Levine, S.J. Plimpton, Phys. Rev. E **64**, 051302 (2001)
[10] X. Cheng, J.B. Lechman, A. Fernandez-Barbero, G.S. Grest, H.M. Jaeger, G.S. Karczmar, M.E. Möbius, S.R. Nagel, Phys. Rev. Lett. **96**, 038001 (2006)
[11] C.H. Rycroft, G.S. Grest, J.W. Landry, M.Z. Bazant, Phys. Rev. E **74**, 021306 (2006)
[12] N.V. Brilliantov, T. Pöschel, *Kinetic theory of granular gases* (Oxford University Press, 2010)
[13] T. Pöschel, S. Luding, eds., *Granular gases*, Vol. 564 (Springer Science & Business Media, 2001)
[14] V. Garzó, J.M. Montanero, Physica A **313**, 336 (2002)
[15] M. Fuchs, M.E. Cates, Phys. Rev. Lett. **89**, 248304 (2002)
[16] J.M. Brader, M.E. Cates, M. Fuchs, Phys. Rev. Lett. **101**, 138301 (2008)
[17] M. Fuchs, M.E. Cates, J. Rheol. **53**, 957 (2009)
[18] A.R. Abate, D.J. Durian, Phys. Rev. E **74**, 031308 (2006)

[19] W.T. Kranz, M. Sperl, A. Zippelius, Phys. Rev. Lett. **104**, 225701 (2010)

[20] W.T. Kranz, M. Sperl, A. Zippelius, Phys. Rev. E **87**, 022207 (2013)

[21] W.T. Kranz, F. Frahsa, A. Zippelius, M. Fuchs, M. Sperl, in preparation

[22] H. Hayakawa, M. Otsuki, Prog. Theor. Phys. **119**, 381 (2008)

[23] K. Suzuki, H. Hayakawa, Phys. Rev. Lett. **115**, 098001 (2015)

[24] J.C. Pàmies, A. Cacciuto, D. Frenkel, J. Chem. Phys. **131**, 044514 (2009)

[25] L.D. Landau, E.M. Lifshitz, *Theory of elasticity* (Pergamon Press Oxford, 1986)

[26] R.P. Ojha, P.A. Lemieux, P.K. Dixon, A.J. Liu, D.J. Durian, Nature **427**, 521 (2004)

[27] M. Schröter, D.I. Goldman, H.L. Swinney, Phys. Rev. E **71**, 030301 (2005)

[28] G.P. Morriss, D.J. Evans, *Statistical Mechanics of Nonequilbrium Liquids* (ANU Press, 2007)

[29] S. Chapman, T.G. Cowling, *The mathematical theory of non-uniform gases: an account of the kinetic theory of viscosity, thermal conduction and diffusion in gases* (Cambridge University Press, 1970)

[30] W. Götze, *Complex dynamics of glass-forming liquids: A mode-coupling theory*, Vol. 143 (OUP Oxford, 2008)

[31] G.A. Martynov, G.N. Sarkisov, Mol. Phys. **49**, 1495 (1983)

[32] M. Heinen, M. Sperl, in preparation

[33] S. Torquato, F.H. Stillinger, Rev. Mod. Phys. **82**, 2633 (2010)

[34] L. Berthier, A.J. Moreno, G. Szamel, Phys. Rev. E **82**, 060501 (2010)

[35] R.A. Bagnold, Proc. Royal Soc. A **225**, 49 (1954)

[36] V. Kumaran, J. Fluid Mech. **756**, 555 (2014)

Cooling of 3D granular gases in microgravity experiments

Kirsten Harth[1,2,*], *Torsten* Trittel[1], *Sandra* Wegner[1], and *Ralf* Stannarius[1,**]

[1] *Institute of Experimental Physics and MARS, Otto von Guericke University, Magdeburg, Germany*
[2] *current address: Physics of Fluids, University of Twente, Enschede, The Netherlands*

Abstract. Granular gases are a most peculiar state, superficially similar to molecular gases: They are loose ensembles of moving grains, rarely interacting with each other and with container walls. The most investigated scenario is the "granular cooling", the collective loss of energy from an initially excited state. We present an experimental study of the cooling of a 3D granular gas of rodlike grains in micro-gravity. Driven steady states of non-spherical grains are characterized by a lack of energy equipartition between the degrees of freedom of translation and rotation. Excitation by vibrating walls additionally introduces strong gradients in the direction and magnitude of translational velocities. We show that the degrees of freedom equilibrate during granular cooling in the homogeneous cooling state. The energy loss follows a t^{-2} scaling. In addition, the alignment of the rod axes with the excitation direction and with the instantaneous velocities are altered during this process.

Excited granular materials are frequently encountered in nature, from dust storms to the Saturn rings. Many technological processes involve transport and handling of granular materials, often also moving loose or compact granular matter. However, an appropriate modeling remains challenging, not only due to insufficient computational resources to deal with most of these many-particle systems, but also due to the large range of scales involved in continuum models which have to capture the range of container to contact dimensions. Collisions, packing and jamming of the grains result in essential differences between granular and fluid flows, e.g. clogging and intermittent flows of hoppers or the occurrence of stick-slip events. The grain size distributions and shapes may influence the phenomena tremendously, e.g. via alignment [1] or secondary flow [2] of anisometric grains.

Ensembles of grains in which collisions almost exclusively involve pairs of particles are comparatively simple. In the extremely dilute case of granular gases with grain volume fractions of only few percent, individual collisions are directly accessible in experiments, and the influence of grain properties on the ensemble behavior can be studied. In addition, granular gases represent one of the simplest non-equilibrium many particle systems in statistical mechanics. The main difference to molecular gases lies in the dissipation of mechanical energy during collisions. This fact motivated various theoretical and numerical studies in the literature. Experimental validation is scarce, primarily due to the poor compatibility of favorable theoretical conditions (e.g. large number of particles, periodic boundary conditions, random energy supply in driven systems) with experimental realizability.

*e-mail: kirsten.harth@ovgu.de
**e-mail: ralf.stannarius@ovgu.de

One of the most intriguing and most addressed aspects of granular gas dynamics is the loss of kinetic energy from an initially homogeneous agitated state, the "granular cooling". Related to it are the formation of planetesimals, the evolution of planetary rings, or the dynamics of objects in the asteroid belt. Haff [3] described the energy loss in a spatially homogeneous granular gas of frictionless spherical grains: The kinetic energy decays with time t as

$$E_{\text{kin}}(t) = \frac{E_{\text{kin}}(t=0)}{(1+t/\tau)^2} \propto t^{-2} \quad (t \to \infty), \tag{1}$$

where the characteristic dissipation rate

$$1/\tau = 2(1-\epsilon^2)\cdot \overline{|\vec{v}|}/\lambda \tag{2}$$

depends on the normal coefficient of restitution ϵ, the mean free path λ and the mean absolute velocity $\overline{|\vec{v}|}$. All translational degrees of freedom (DOF) are treated equivalently there, rotations are not included. Haff simplifies in that the squared mean absolute velocity is set equal to the mean squared velocity. With proceeding energy decay, density inhomogeneities and clusters can develop in the system, first demonstrated in a 2D simulation [4] and observed in a small ensemble of levitated grains [5]. Depending on the grain shape [6, 7] or for finite system size [8], those clusters may be unstable, they can breakup and dissolve. Simulations of ensembles of elongated grains predict a loss of the equipartition of kinetic energies in general [6, 7, 9, 10]. It is *a priori* unclear whether Haff's prediction will also apply to a granular gas of (frictional) rodlike grains. The models in the literature do not provide unique answers [6, 10].

Experimental realization of a homogeneously cooling granular gas is challenging: The initial excitation of the grains generally requires vibrations of the container or its walls. An alternative magnetic excitation scheme was recently developed to provide contactless forcing in the bulk [11]. For two-dimensional granular gases, the interaction with the substrate must be taken into account [12]. One way out is levitation by electromagnetic fields [5, 13] or by flow [14, 15]. The realization of 3D granular gases in low external excitation requires microgravity. The first study by Falcon et al. investigated continuously driven ensembles of spheres and revealed dynamic clustering [16], which was investigated in detail and modeled in Refs. [17, 18]. Afterwards, mainly 2D systems of few particles were studied [19, 20]. Maaß et al. [5] and Weidling et al. [21] present results for small ensembles of highly dissipative grains during granular cooling, obtained from 2D perspective views.

A spatially homogeneous grain distribution in a steady state can be achieved with rodlike grains [22–24]. An advantage of rods over spheres is the reduction of the mean free path at equal volume fraction of particles and the trackability of rotations. In a steady excited state, we found an excess of kinetic energy of translational compared to rotational degrees of freedom from a 2D perspective evaluation of the 3D granular gas [23]. Velocity distributions in the coordinates $i \in \{x, y, z\}$ are non-Gaussian, rather $p(v_i) \propto \exp(-|v_i|^{1.5})$. The direction of the vibrating walls (x) is distinguished by the highest kinetic energy and by exponential high-velocity tails. We recently improved the setup and developed methods for preparing and obtaining 3D data of granular gases in few seconds. With a stereo camera setup, rods become trackable in 3D. This allows 3D experiments using drop tower shots at the ZARM Bremen [24], offering $\approx$ 9 s of excellent micro-gravity. Parabolic flight experiments are inappropriate for low-excitation experiments in general [24] due to the g-jitter, but may be applied for investigation at strong excitation [18] or of dense granular systems [25].

This presentation will provide the first detailed experimental study of free homogeneous cooling of a 3D granular gas. Our setup is similar to that of Refs. [22, 24]: It consists of a cuboid container of $\approx 11\times8\times8$ cm^3 with two movable side walls. Front and top walls are made of ITO coated acrylic glass, other walls are flat white painted aluminum. Front and top view are recorded using two GoPro cameras with c-mount objectives at 100 frames per second, 1280×960 pixel^2. Our granular gases consist of 374 colored insulated copper wire pieces with $\varnothing \approx 1.35$ mm, length $\ell = 10$ mm, corresponding to a volume fraction of 0.76% and an estimated mean free path of 2.2 mm [27]. Here we restrict to an average over 3 experiment runs, additional data are currently being evaluated and published elsewhere. The mass of a rod is $m = 0.0375$ g, and moments of inertia for a rotation axis parallel and perpendicular to the symmetry axis are $J_\parallel = 4.63\cdot10^{-12}$ kg m^2 and $J_\perp = 315\cdot10^{-12}$ kg m^2, respectively.

The initial state is prepared by injection of kinetic energy with the vibrating walls at an amplitude $A \approx 6$ mm

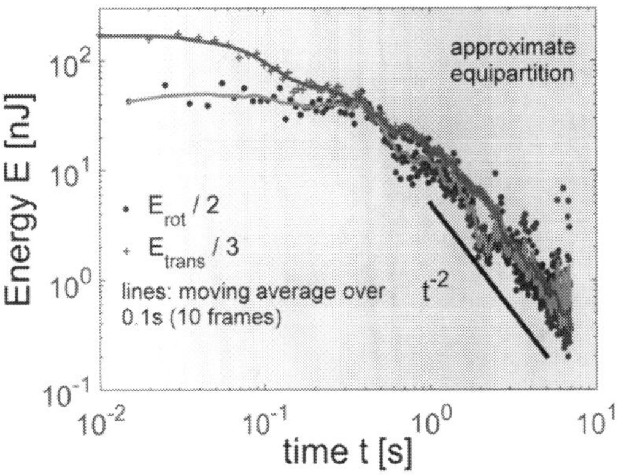

Figure 1. Decay of the translational and rotational kinetic energies per rod and DOF: The initial excess energy in translational over rotational DOF equilibrates within $\approx 2\ldots2.5$ s, to reach approximately equipartition of kinetic energy.

(peak-peak) and frequency $f = 30$ Hz for $1\ldots1.5$ s. Data are averaged over all experiment runs. Colored grains are tracked in each perspective by a semi-automatic custom routine. 3D data are calculated using the camera calibration toolbox [26] supplemented by a custom routine. The experimental uncertainty of positions and orientations depends on the grain orientation in both perspectives and the overlap with other rods, it is maximal e. g. when the rod makes only a small angle to the z-axis. We plot the time-dependent positions (center of mass) and angles to the coordinate axes for each rod and identify segments of undisturbed motion of each rod manually. The resulting segments of center of mass translations in each coordinate are fitted by linear functions. For rotations, we compute the reorientation angle in a time step from the vector product of initial and final orientation, then divide by the respective time step (may be larger than 1 frame). After that, we discard obvious detection errors and average all values of the rotational velocity obtained in a time sequence between rod collisions (as for the translational velocities). As the two degrees of freedom corresponding to rotations around axes perpendicular to the long rod axis cannot be distinguished, we consider only the mean squared rotational velocities of both degrees of freedom together, $\omega_\perp$. We expect the third rotational DOF (around the long particle axis) to be much less excited as it is only excited via friction during rod-rod collisions, but it has not been measured here quantitatively. Fitting is especially important during the second half of our measurement time and for the rotational velocities, to minimize the influence of detection noise. On average, the rod orientations possess larger errors than the center of mass positions. We excluded rods with positions less than 1 cm from the walls in this evaluation, they are assumed to be not equilibrated with the bulk due to interactions with the container walls.

Figure 1 shows the decay of the translational and the rotational energy per rod and degree of freedom over the complete accessible time of $\approx$ 7 s. In the driven

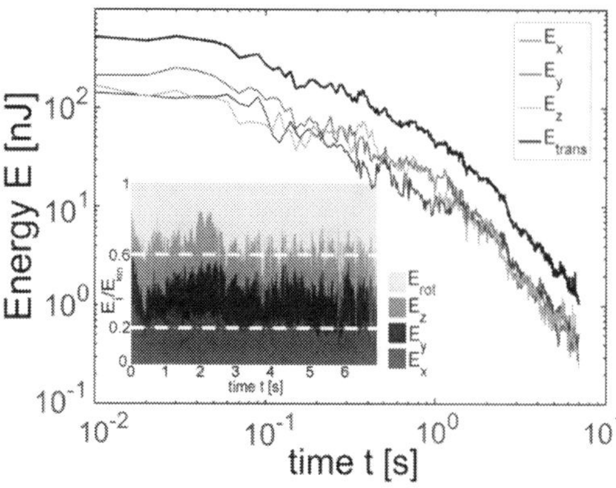

Figure 2. Translational kinetic energy per particle and DOF in the directions of the box coordinate system, where x is the direction enclosed by the vibrating walls, y and z by non-moving walls. Inset: Instantaneous fraction of the kinetic energy per DOF and the rotational energy (2 DOF). Dashed lines indicate $1/5\ E_{\mathrm{kin}}(t)$ and $3/5\ E_{\mathrm{kin}}(t)$.

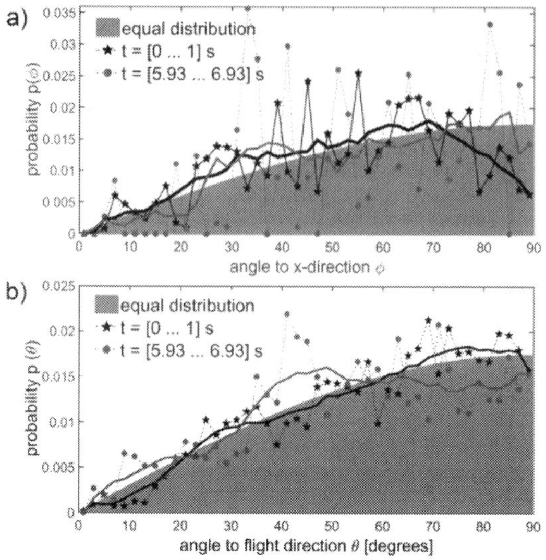

Figure 3. Distributions of the angles of the long axes of the rods respective to a) the direction of wall excitation x, and b) the current flight direction, during the initial and last second of the measurement. Grey areas represent equally distributed orientations.

steady state, equipartition is violated as expected from Refs. [9, 10, 23]: Initially, the mean translational energy per DOF is $\approx$ 180 nJ, whereas the rotational energy per DOF is $\approx$ 40 nJ, less than one quarter of the translational kinetic energy. Both values strongly approach each other during the first 0.2 s of cooling. Long-term deviations from an equilibrated state are observed until $t \approx 2\ldots2.5$ s. After this, the DOF equilibrate within experimental noise. Then, both translational and rotational kinetic energy decay $\propto t^{-2}$, as predicted by Haff [3] and in qualitative agreement with Refs. [10]. In particular, this is in contrast to the 2D simulations of elongated viscoelastic particles by Kanzaki et al. [6], who predicted differing power laws for translational and rotational DOF.

The initial bias in the energy partition must be attributed to the external driving of the system. The direction that is directly excited by shaking walls is distinguished by a higher initial kinetic energy than the indirectly excited DOF, see Fig. 2. The bias is larger than in our previous steady state experiments [23], probably due to the stronger excitation at approximately equal mean free paths. In particular, the velocity distributions in excitation direction x differ depending on the distance to the walls, as discussed in Refs. [20, 27]. Thus, the assumption of a spatially homogeneous initial state is not in accordance with the velocity field expected in this experiment. In the inset of Fig. 2, we display the fractions of the mean total energy per rod for the different DOF. We observe a pronounced transfer from all translational DOF to the rotational DOF during the first 0.5 s of cooling, followed b a slight growth in translational energy until $t \approx 2.5$ s. These transient deviations from equipartition are also clearly identified in Fig. 1. Equilibration (within the accuracy of the current data) of the three translational degrees of freedom occurs on a similar time scale as the equilibration of translational and transverse rotational degrees of freedom. Rotations around the symmetry axis are only excited via friction, and not measured here.

Finally, let us consider the distribution of orientations of the rods in order to characterize differences in the initial and final states in more detail. Fig. 3a) shows the distribution of the angles ϕ between of the rods' long axes and the excitation direction x, and b) shows the distribution of the angles θ of the long axes respective to the momentary flight direction. Comparing the first and the last seconds to a random distribution, one finds initially a decreased probability of rod motion perpendicular the x direction. This feature is caused by the strong excitation by the walls, it reflects the incomplete equilibration of the directions of motion via collisions in he excited state. Wright et al. [28] studied rods bouncing onto a vibrating wall. They reported consistently lower probabilities of impacts with the long axis parallel to the wall, but provided no corresponding data after rebound. Our recent results on energy input and distribution of rods rebounding from a sinusoidally vibrating wall reveal a strong excess of translational energy in direction of vibration over the injected rotational energy and translational kinetic energy of the other DOF [29]. This bias vanishes rapidly during granular cooling. Obviously, the collisions with the non-moving walls introduce only negligible bias (if any at all).

A related question is whether the permanent collisions in absence of external excitation influence the orientational distribution of the rods respective to their direction of motion. For rough spherical grains, correlations between the axis of rotations and translations were reported in the literature [30]. We find clear evidence of a preferential alignment in Fig. 3b: Initially, the rods translate in arbitrary direction respective to their long axis. However, as the excitation bias ends, partial alignment occurs, evident in the last second of our measurement. The scattering cross section of a rotating rod is maximized when it is directed perpendicular to its current direction of motion (propeller-like, $\vec{v} \parallel \vec{\omega}$). This results in higher collision rates for these rods compared to those with smaller angles θ between the rod symmetry axis and $\vec{v}$. Note the difference to Ref. [14], where the preferential motion along the rod axis is due to the high area fraction.

Summarizing, we present the first 3D experimental analysis of a freely cooling granular gas on the particle level. During the 7 s duration of the measurement, the kinetic energy decays by more than 2 orders of magnitude. We do not observe significant density inhomogeneities. While the grain distribution is initially close to homogeneous, the velocity fields include strong gradients especially in excitation direction, and energy equipartition is violated. After the excitation is stopped, we observe an equilibration of all DOF within statistical noise during the first $\approx$ 2.5 s of granular cooling. The full evaluated data sets will provide more detailed insights into the cooling process and its influence, e.g. on the shape of velocity distributions or spatial inhomogeneities. In direct consequence, it will be of interest to extend the observation time to search for a potential onset of spatial inhomogeneities or clustering of grains, a corresponding experiment is scheduled for a sounding rocket flight in 2017. Investigation of the energy partition between different components of particle mixtures provide a perspective to yield more insight into the ensemble dynamics of polydisperse or mixed shape systems.

Acknowledgements

The authors acknowledge funding by the German Aerospace Center DLR, projects 50WM1241 and 50WM1344, and by the German Science Foundation (DFG) Grant STA 425/34-1. We cordially thank the ZARM staff in Bremen for their excellent support in performing the drop tower experiments.

References

[1] e.g. T. Börzsönyi et al., Phys. Rev. Lett. **108** 228302 (2012); Phys. Rev. E **86** 051304 (2012); New J. Phys. **18** 093017 (2016)

[2] G. Wortel et al., Soft Matter **11** 2570 (2015); D. Fischer et al., New J. Phys. **18** 113006 (2016)

[3] P. K. Haff, J. Fluid Mech. **134** 401 (1983)

[4] I. Goldhirsch and G. Zanetti, Phys. Rev. Lett. **70** 1619 (1993)

[5] C. C. Maaß et al., Phys. Rev. Lett. **100** 248001 (2008)

[6] T. Kanzaki, R. C. Hidalgo, D. Maza, and I. Pagonabarraga, J. Stat. Mech. **2010** P06020 (2010)

[7] M. Huthmann, T. Aspelmeier, and A. Zippelius, Phys. Rev. E **60** 654 (1999)

[8] M. Hummel and M. G. Mazza, Phys. Rev. E **93** 022905 (2016)

[9] T. Aspelmeier, G. Giese, and A. Zippelius, PRE **57** 857 (1998); G. Costantini et al., J. Chem. Phys. **122** 164505 (2005)

[10] F. Villemot and J. Talbot, Gran. Matter **14** 91 (2012); S.M. Rubio-Largo, et al., Physica A **443** 477 (2016)

[11] M. Sperl, private communication, unpublished.

[12] J. S. van Zon et al., Phys. Rev. E **70** 040301 (2004)

[13] A. Snezhko, I. S. Aranson and W.-K. Kwok, Phys. Rev. Lett. **94** 108002 (2005); K. Kohlstedt et al., Phys. Rev. Lett. **95** 068001 (2005)

[14] L. J. Daniels et al., Phys. Rev. E **79** 041301 (2009)

[15] K. Nichol and K. E. Daniels, Phys. Rev. Lett. **108** 018001 (2012); J. C. Burton et al., Phys. Rev. E **88** 062204 (2013)

[16] E. Falcon et al., Phys. Rev. Lett. **83** 440 (1999)

[17] E. Opsomer, F. Ludewig, and N. Vandewalle, Phys. Rev. E **84** 051306 (2011)

[18] A. Sack et al., Phys. Rev. Lett. **111** 018001 (2013); J. Kollmer et al., Phys. Rev. Appl. **3** 024007 (2015); AIP Conf. Proc. **1542** 811 (2013)

[19] e.g. S. Tatsumi et al., J. Fluid Mech. **641** 521 (2009); Y. Grasselli, A. Bossis, and R. Morini, Eur. Phys. J. E **38** 8 (2015)

[20] C. Yanpei et al., J. Physique **327** (2011)

[21] R. Weidling, C. Güttler, and J. Blum, Icarus **218** 688 (2012)

[22] K. Harth et al., European Space Agency **SP-700**, 493 (2011)

[23] K. Harth et al., Phys. Rev. Lett. **110** 144102 (2012); AIP Conf. Proc. **1542** 807 (2013)

[24] K. Harth et al., Adv. Space Res. **55** 1901 (2015)

[25] P. D. Yu, S. Frank-Richter, and M. Sperl, Gran. Matter **16** 165 (2014); N. Murdoch et al., Granular Matt. **15** 129 (2013)

[26] Camera Calibration Toolbox by J.-Y. Bouguet, www.vision.caltech.edu/bouguetj/calib$_d$oc/

[27] K. Harth, T. Trittel and R. Stannarius, in preparation

[28] H. S. Wright, M. R. Swift, and P. J. King, Phys. Rev. E **74** 061309 (2006)

[29] T. Trittel, K. Harth, and R. Stannarius, under review (2017).

[30] M. Huthmann and A. Zippelius, Phys. Rev. E **56** R6275 (1997); N.V. Brilliantov et al., Phys. Rev. Lett. **98** 128001 (2007)

Submicron structure random field on granular soil material with retinex algorithm optimization

Yu Liang, Chenyuan Tao, Bingcheng Zhou, Shuai Huang, Linchong Huang*

School of Engineering, Sun Yat-sen University, Guangzhou, Guangdong, 510275, China

Abstract. In this paper, a Retinex scale optimized image enhancement algorithm is proposed, which can enhance the micro vision image and eliminate the influence of the uneven illumination. Based on that, a random geometric model of the microstructure of granular materials is established with Monte-Carlo method, the numerical simulation including consolidation process of granular materials is compared with the experimental data. The results have proved that the random field method with Retinex image enhancement algorithm is effective, the image of microstructure of granular materials becomes clear and the contrast ratio is improved, after using Retinex image enhancement algorithm to enhance the CT image. The fidelity of enhanced image is higher than that dealing with other method, which have explained that the algorithm can preserve the microstructure information of the image well. The result of numerical simulation is similar with the one obtained from conventional three axis consolidation test, which proves that the simulation result is reliable.

1 Introduction

How to establish the relationship between macro and micro is the key problem to be solved in the study of non-homogeneity of rock and soil materials across scales. In the research of multi-scale model, different macro and micro relationship rules are established based on different assumptions generally, to solve the problem of weak coupling or strong coupling scale (Landis E N. et al. 2003; Andrade J E. et al. 2006).

In this paper, the Monte Carlo method is adopted to randomly generate a polygonal element with a given size and surface features, to describe the various structural elements in the granular materials. According to the characteristics of the microstructure, the generated polygonal cells are composed of classic particle cell, confabulation unit and pore cell.

Based on the Retinex theory, this paper has defined the optimization criterion according to contrast statistics and proposed scale optimized Retinex algorithm. The image filtering technique and image enhancement method are used, to analyse the root cause of hindering micro visual images enhancement, aim at the problem of uneven illumination of the micro vision image, and enhance the micro vision image, improving the contrast and simultaneously eliminating the influence of the uneven illumination.

2 Retinex algorithm of Scale optimization

Land et al (1971,1977,1983) had proposed Retinex theory in the literature. Retinex is the compound word of retina and cortex, which represents a model of human vision perception of color and brightness. The model considers that the human eyes can stably perceive the object's colors and details when the same object is under different illuminations or in the shadow, without being effected by illumination change. According to the theory, the feeling caused by feature point in the human eye is determined by the reflection function of the object, this function is only related to the properties and the composition of the material, and it is not related to the distribution and transformation of the light intensity. So, in the computer vision image with uneven illumination, one or more paths can be taken around a pixel point to calculate the average value of the pixel path as the light component. The gray value of the pixel is divided by the light component to get the reflection component, that is for eliminating the effect of the light. The center / surround Retinex using Gauss form function can estimate the light component very well. The scale parameter of Gauss's function has the function of balancing and compressing the dynamic range and retaining the original image information. The larger the parameter, the more the gray dynamic range is compressed, and details will be affected accordingly; the smaller the parameter, the more the image is sharpening.

2.1 Retinex algorithm of Scale optimization design

Micro visual images which adopt coaxial illumination is enlarged by microscope magnification, making 68 μ m high, 85 μm width field to obtain the light illumination energy disperse to 1200 x 1500 pixels of image, and the

* Corresponding author: hlinch@mail.sysu.edu.cn

image intensity density becomes extremely weak. Due to the uneven distribution of light intensity of the light source coaxial, the gray value of pixels in the central region is different from the surrounding area. If the illumination is too large, the light intensity in the center region of the image will be saturated. So, a moderate intensity of light is needed. But this will lead to low overall image pixel gray value, and low contrast between metal lines corresponding to the pixel gray value and the substrate pixel gray value. In order to eliminate the influence of uneven illumination, the center / surround Retinex model is the best choice, the reflection component is obtained by estimating the light component to observe the morphology of the line.

According to this feature, a selection criterion of image enhancement scale optimization is proposed in this paper, the parameters of the selected Gauss scale parameters can be optimized and the estimated light component is the best based on this criterion. In the ideal reflection component image, the gray level's contrast of metal line and the pixel contrast of substrate is large and the difference between them is small, it actually calls for the mean value of the local contrast is large enough while the variance of contrast in different regions is as small as possible. Conversely, if the smooth transition would lead to the overlarge of the compressed dynamic range , It will lead to the combination of the pixels of different materials so that the details of microstructure will be damaged, the corresponding results include the contrast of the local area is not uniform and the variance of contrast becomes larger; If the smoothing is not enough, there is a bias in the estimation of the illumination component, the effect of uneven illumination is insufficient, still hinder the improvement of contrast, thus the average contrast is small (Elad 2005).

In order to achieve the goal of eliminating the effect of illumination uniformity and image fidelity, a selection of a reasonable scale parameter c is needed. The optimization of the standard is to make the reflection image component of the local area contrast biggest, and make the variance of contrast in different regions least at the same time. So:

$$r(x, y, c) = \exp\{\ln i(x, y) - \ln[i(x, y) * G(x, y, c)\} \quad (1)$$

Here, $r(x, y, c)$ represents the reflection component of the pixel (x, y) controlled by c . The enhanced reflection component image is denoted as $\mathrm{R}(c)$, in this component image, five windows are taken to calculate the local contrast. The windows' center pixels are distributed in the whole image evenly. This measurement is to avoid the point by point calculation of the neighborhood contrast of pixels in the whole image, so as to reducing the calculation amount. The selected window size is better as large as the size that can include the linear part and substrate part, according to the imaging system parameters disposed in this project, the window size is 120×120 pixels, and the five windows are distributed on the upper left, upper right, center, left and right of the image. The contrast m_k of the k - window can be calculated according to formula（2):

$$m_k = \frac{I_{k,\mathrm{Max}} - I_{k,\mathrm{Min}}}{I_{k,\mathrm{Max}} + I_{k,\mathrm{Min}}} \quad (2)$$

The calculating formula of the average window contrast $\overline{\mathrm{M}}[\bullet]$ and the variance $\sigma[\bullet]$ can be defined as follows:

$$\overline{M}[\bullet] = \frac{1}{5}\sum_{k=1}^{5} m_k \quad (3)$$

$$\sigma[\bullet] = \frac{1}{5}\sum_{k=1}^{5}(m_k - \overline{M})^2 \quad (4)$$

According to the analysis above, based on the optimal criterion of the average value of the window contrast and variance structure scale, the calculating formula can be defined as follows:

$$J(c) = \arg\min_c\{\frac{A_1}{1 + \overline{M}[\boldsymbol{W}(\boldsymbol{R}(c))]} + A_2 \times \sigma[\boldsymbol{W}(\boldsymbol{R}(c))]\} \quad (5)$$

The W[•] represents choosing uniformly distributed windows in the reflection component image $\mathrm{R}(c)$. The optimization criterion defined by the formula (5) is composed of two parts, its function are as follows: The first part is to maximize the overall contrast, plus 1 in the denominator, in order to avoid the window contrast of the blank image area becomes zero; The second part minimizes the variance of the contrast of all windows to preserve detail information. Positive constants are used to make the two parts A_1 and A_2 numerical equilibrium, its ratio is as same as the ratio of the mean of the contrast and the magnitude of the variance of enhanced image, in the practical application, the ratios mentioned above can be chosen according to experience. Finally, the optimal scale parameter c is substituted in formula (1) to calculate the reflection component image, the scale optimization of Retinex image enhancement can be obtained.

2.2 Algorithm simulation

In order to test the adaptability of the algorithm to the illumination change, the simulation experiment was carried out with the manual simulated image. The simulated image is composed of black and white stripes with a width of 15 pixels, its structure is similar to that of the photoresist mask, simulating the intensity distribution of coaxial light source in the form of Gauss's function, The light component model is shown in the formula (6):

$$l(x, y) = 1 + \frac{l}{2\pi\gamma}\exp(-\frac{x^2 + y^2}{\gamma^2}) \quad (6)$$

The γ is the coaxial halo size, l is the modulation ratio of coaxial light and ambient light. The halo scale parameters take 265, 270, 275,... 310, generating 10 images in the simulation. In this paper, the scale optimized Retinex algorithm is proposed to deal with the simulated image, the results obtained are in the normalized mean square error (NMSE) operation with the original simulated image, to measure the fidelity of the image before and after the enhancement. Treatment results are shown in Table 1, although the image light component changes, the reflection component of the smoothing function is unchanged, so Gauss scale estimation values are almost the same, and the estimated value of the scale is related to the number of fringe pixels. The simulation results show that the Retinex enhancement algorithm based on scale optimization can adapt to the changes of light intensity and scale.

Table 1 Simulation results of scale optimization retinex algorithm

γ	c	NMSE	γ	c	NMSE
265	16	0.0242	290	17	0.0194
270	14	0.0210	295	17	0.0211
275	14	0.0229	300	17	0.0241
280	14	0.0268	305	15	0.0214
285	14	0.0295	310	17	0.0265

The results are shown in FIG.1. It is obvious that the image which is enhanced in this paper eliminate the impact of uneven illumination, and the result of image segmentation is closer to the micro structure feature of the original image, the enhanced results of the histogram equalization method and the method of filtering can not eliminate the light component successfully.

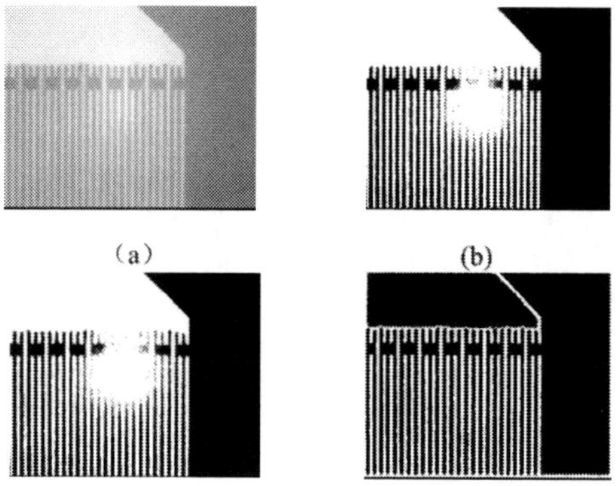

(a) (b) (c) (d)

Fig. 1 The comparison of enhanced micro-vision images for photomask. (a) Original picture; (b) Image enhanced by histogram equalization method; (c) Image enhanced by the method of the filter; (d) Image enhanced by algorithm proposed in this paper

3 Model calculation of geo-materials random field

In this section, we present a random field model of a typical granular materials of consolidation process in the Pearl River delta area of China.

As shown in Table 2, the established model of granular materials microstructure is relatively close to the actual soil sample statistically. FIG.2 is the contrast image of the soil microstructure random field model of soil sample and actual soil sample microstructure, where the black polygon represents the pore unit, polygons in other colors represent different kinds of structural units, and the remaining white space represents structure connection. In soil sample actual microstructure image, the black part represents the pore, and the white part represents the soil skeleton. The soil sample microstructure random field model and the soil sample actual microstructure image are relatively similar, which proves that the proposed modeling method is practical and effective (Fenton. et al. 2007).

Table 2. Microstructure geometric model component statistics of soil sample compared with the test results

Reduced Parameter	Test results	Model results
Void Ratio	1.026	0.998
Silt Content	37.2%	62.8%
Clay Content	37.15%	62.85%

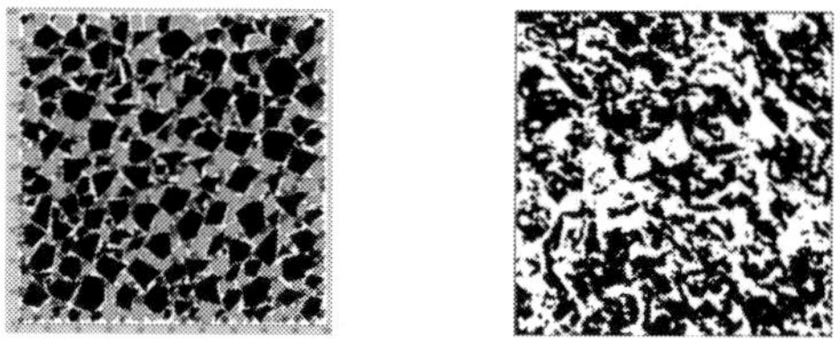

(a) (b)

FIG.2. (a) Soil microstructure random field geometry model & (b) Scanning electron microscopy images of microscopic structure

3.1 Numerical calculation of three axis consolidation of micro random field

The soil samples from an expressway in the Pearl River delta area of China are collected to conduct the indoor triaxial consolidation test. In this test, the curves of pore water pressure versus time of the soil samples are measured. Then the data of the indoor triaxial consolidation test will be used to conduct the finite element numerical simulation with the geometric model above. The parameters of random field geometric model will be given shown as Table 3.

Table 3. Parameters of triaxial consolidation test

Comparison	Width	Height	e	Particle	Clay Particle
Test results	195 μm	400μm	1.026	37.2%	62.8%
Computation results	195 μm	400μm	1.002	35.82%	64.18%

3.2 Calculation results and analysis

Intercept the 10, 50, 100, 200, 300, 500 sub steps of simulation data, count the boundary of each sub step model and the displacement of each node on the right boundary, average them respectively, the curves of average compression of the model's vertical and horizontal directions are therefore obtained. Comparing the body strain with the practical one (such as FIG.3), it is shown that the trend of simulate and practical curves are similar, which proves that the results of simulation are practical and reliable.

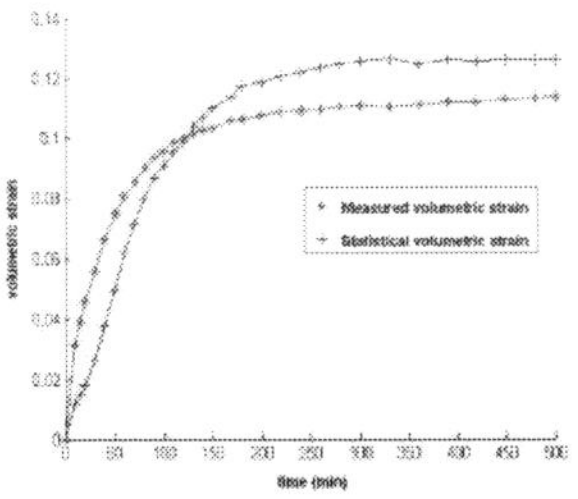

Fig.3 Volumetric strain comparison of model computation results with triaxial test results within each substep

4 Conclusion

(1) The Monte Carlo method is used to generate a certain amount of polygon unit in random shape, size and surface characteristics, which represented all kinds of structural units of granular materials, and these units were randomly distributed to the area of a given size, and the stochastic geometric model for the microstructure of granular materials is established.

(2) The algorithm is suitable for the coaxial light source with different scale and intensity, which can estimate the light component and the reflection component accurately. The effect of the enhanced image are better than that with the method of filtering and histogram equalization. It shows that this algorithm can preserve the microstructure information of the micro image while enhancing the image.

(3) The result of numerical simulation are similar with the one obtained from conventional three axis consolidation test, which proves that the results of simulation are practical and reliable. And the needed modeling parameters are few and easy to obtain from the indoor test.

Acknowledgments

This work has been supported by the National Natural Science Foundation of China (No.& 51678578), the Natural Science Foundation of Guangdong Province, China (No.2016A030313233), the Guangdong Provincial Science & Technology Program of China (No. 2015A020217004), and the Department of Communications of Guangdong Province of China (No. 2016-02-026), these supports are gratefully acknowledged.

References

1. G. A. Fenton , D. V. Griffiths. Probabilistic methods in geotechnical engineering, chapter The random finite element method (RFEM) in bearing capacity analysis. (Springer Wien New York, 2007).
2. R. Popescu, G. Deodatis, A. Nobahar. Effects of random heterogeneity of soil properties on bearing capacity. Probabilist. Eng. Mech., **20** (2005).
3. E. N. Landis, E. N. Nagy, D. T. Keane. Microstructure and fracture in three dimensions. Eng. Fract. Mech, **70,** 5 (2003).
4. J. E. Andrade, R. I. Borja. Capturing strain localization in dense sands with random density. Int. Numer. Meth. Eng., 2006, **67,** 1 (2006).
5. E. H. Land, J. Mccann. Lightness and retinex theory . J. Opt. Soc. Am., **61,** 1 (1971).
6. E. H. Land. The retinex theory of color vision. (Scientific America, 1977).
7. E. H. Land. An alternative technique for the computation of the designator in the retinex theory of color vision. P NATL A SCI INDIA B, **83,** 10 (1986).
8. M. Elad. Retinex by two bilateral filters. (Springer, 2005).

 DOI: 10.1051/epjconf/201714002019

Determination of the equivalent intergranular void ratio - Application to the instability and the critical state of silty sand

Trung-Kien Nguyen[1], *Nadia* Benahmed[2,*], and *Pierre-Yves* Hicher[3]

[1]Geotechnical Engineering Division, University of Transport Technology, Hanoi, Vietnam
[2]Irstea, RECOVER Research Unit, IRSTEA, Aix-en-Provence, France
[3]Ecole Centrale de Nantes, France

Abstract. This paper presents an experimental study of natural Camargue silty sand mechanical behavior. The analysis of test results was based on the use of the equivalent intergranular void ratio instead of the global void ratio. The calculation of equivalent intergranular void ratio requires the determination of parameter b which represents, physically, the fraction of active fines participating on the chain forces network, hence on the soil strength. A new formula for determining the parameter b using an approach based on the coordination number distribution and probability calculation is proposed. The validation of the developed relationship was done through back-analysis of published datasets in literature on the effect of fines content on silty sand behavior. It is shown that the equivalent intergranular void ratio calculated with the b value obtained by the new formula is able to provide a strong correlation, not only at the critical state, but also at the onset of instability of various silty sands, in different terms such as peak deviator stress, peak stress ratio or cyclic resistance. Therefore, it is suggested that the use of the equivalent void ratio concept and the new formula that we have put forward for b determination is highly desirable in predicting of silty sand behavior.

1 Introduction

The effect of fines content is an important parameter to take into account when evaluating the mechanical behavior of silty sand since it is widely encountered in nature. Several alternative density variables were proposed in order to better capture the mechanical response of this kind of material like, for example, the intergranular void ratio [1] and the specific relative density [2]. However, the analysis results adopting these density variables are still controversial.

In the sand-dominated sand-fine mixture like silty sand, according to Thevanaygam et al. [3], there are inactive fine particles that stay in the voids created by coarse grains and do not contribute to material resistance) and active fine particles that touch and separate the coarse grains. In order to take into account either active and inactive fine particles, a new density variable was proposed: the equivalent intergranular void ratio which is calculated as follows:

$$e^* = \frac{e + (1-b)f_c}{1-(1-b)f_c} \quad (1)$$

where e is the void ratio, f_c is fines content and b is the fraction of active fines.

Many authors afterward reported that the equivalent intergranular void ratio has strong correlation with many aspects of silty sand mechanical behavior [3]–[12]. The main purpose of this work is to verify the effectiveness of the equivalent intergranular void ratio concept in characterizing the mechanical behavior of natural silty sand.

2 Experimental procedure

The soil used in this study is poorly graded natural Camargue silty sand which was taken from flood protection dikes. The clean sand and clean fines were separated by sieving the silty sand in order to prepare different mixtures. The sand is uniform with angular and rounded grains, a uniformity coefficient C_U = 2.09 and an effective diameter D_{50} = 0.25 mm. Its minimum and maximum void ratio is equal to 0.61 and 1.10, respectively. The fines used are low plastic fines with a liquid limit W_L=32 and a plasticity index IP = 8. The host sand and host silt were blended together to obtain the different mixtures with the different desired target fines contents (by weight) varying from 0, 5, 10, 15 and 20%. The grain size distributions of all the materials tested in this study are presented in Figure 1.

A fully computed controlled Dynamic Triaxial System from Wykeham Farrance was used. The axial load was monitored by an internal submersible load transducer while the axial displacement was measured by an LVDT mounted on the top platen. The volume variation of the sample was determined by an automatic volume change device.

* Corresponding author: nadia.benahmed@irstea.fr

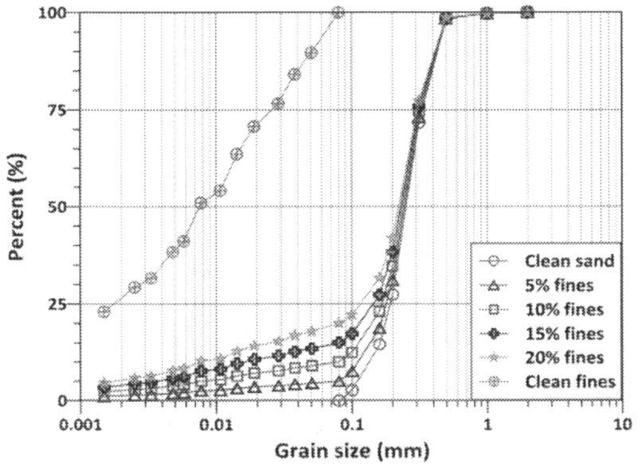

Fig. 1. Grain size distribution curves.

The soil samples of 140 mm height and 70 mm diameter were prepared by the moist tamping method. Particular care was taken to avoid any potential segregation, especially in the case of high fines content. To saturate the specimens, CO_2 was flushed under a pressure of 20 kPa during 30 minutes, followed by de-aired water of about three times the specimen volume. Back pressures ranging from 400 to 500 kPa were then applied to ensure full saturation of the specimens and to obtain a Skempton coefficient B higher than 0.96.

3 Experimental results

In order to investigate the effect of fines content on the behavior of Camargue silty sand, several triaxial tests were performed under both drained and undrained conditions. More details on the experimental program can be found in [13].

3.1 Instability analysis

Figure 2 shows the evolution of the deviator stress versus the axial strain in undrained tests of different mixtures. We note that all the curves show a very clear peak of resistance appearing at low axial deformation, followed by a sudden decrease until reaching a low (or null) residual shear strength.

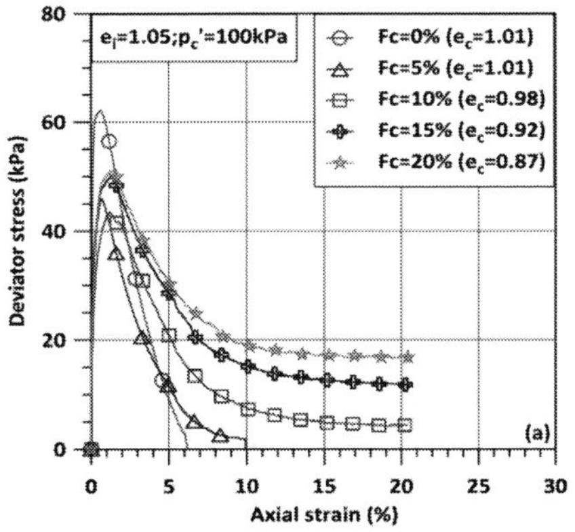

Fig. 2. Effects of fines content on undrained response of silty sand

In order to point out the effect of fines content on the onset of the instability, we reported in Figure 3 the peak deviator stress of all the tests against the post-consolidation void ratio. It is obvious that there is no unique relation between peak deviator stress and void ratio. Indeed, each mixture forms a different line. It is interesting to note that with the increasing of fines amount until Fc=20%, this line moves downward. In other words, for a given post-consolidation void ratio, more the fines content is high, more the peak deviator stress is small.

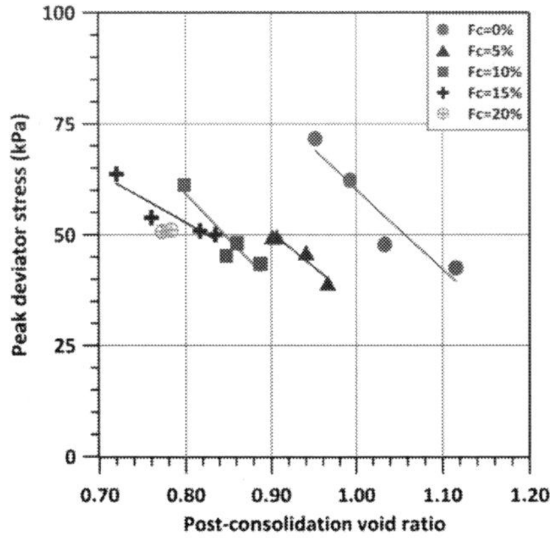

Fig. 3. Correlation between the peak deviator stress and the void ratio

3.2 Critical state

In this study, all the tests in drained conditions were conducted up to 30% of axial strain. Most of them have reached a stable state of both volumetric strain and deviator stress. We report in Figure 4 all points obtained at critical state for different mixtures as well as for the clean sand and the clean fines. It is interesting to note that each mixture has a different critical state line, which implies that the addition of fines particles creates different materials. The critical state line moved downward to a lower void ratio as the fines content increased up to 15%. Similar results were reported by other authors in literature.

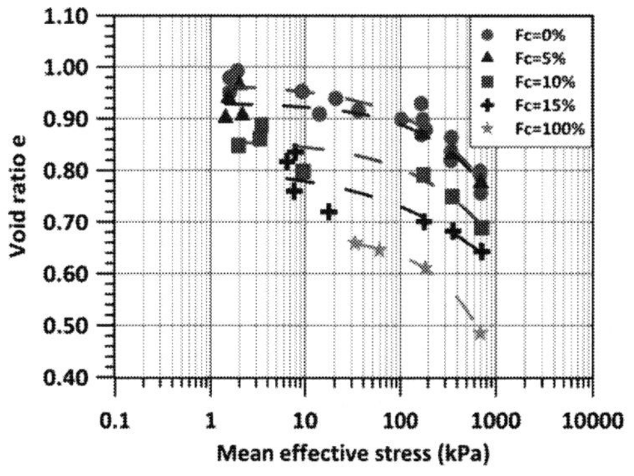

Fig. 4. Critical state lines in e - logp' space

4 Application of the equivalent intergranular void ratio

In this section, the equivalent intergranular void ratio was used to analyze the obtained experimental results. In order to calculate this density variable, it was necessary to determine the b parameter value. Some authors identified this value by back-analysis in order to obtain the best fit to their experimental data [4], [7]. This means that the value of b was chosen arbitrarily and it is a constant for a certain range of fines percentage. By using semi-empirical approach, Rahman et al. introduced the first formula to estimate the parameter b as follows [10]:

$$b = \left[1 - \exp\left(-2.5 f_c^{2} / k\right)\right]\left(r f_c / f_{thre}\right)^{r} \qquad (2)$$

where $r = d_{50}/D_{10}$ (d_{50} is mean diameter of fines particles whereas D_{10} is effective diameter of sand particles); $k = 1 - r^{0.25}$; f_{thre} is the threshold fines content (the fines content that separates the sand-dominated structure and the fines-dominated structure).

Mohammadi and Qadimi evaluated the formula (2) with more datasets and reported that this formula does not work for 9 datasets out of 17 [12]. In addition, the b parameter calculated by the equation (2) does not depend on the void ratio, which is not reasonable physically. Therefore, based on these observations, we developed a new formula to calculate the parameter b.

The basic idea of the new approach assumed that in three-dimensional space, a certain particle cannot move freely if it is "stucked" between at least four other particles. When the structure is loaded, such a particle could transmit forces. Thus, an active fine must have the coordination number greater than or equal to 4 and conversely, an inactive fine must have the coordination number smaller or equal to 3.

Since many researchers reported that the distribution of the coordination number can be represented by the normal distribution or Gaussian distribution [14]–[19], the fraction of inactive fine (1-b) can then be calculated as the integral of the probability density function of the normal distribution, from 0 to 3.5:

$$1 - b = \frac{1}{\sigma\sqrt{2\pi}} \int_{x=0}^{x=3.5} \exp\left(-\frac{(x - \overline{C_S})^2}{2\sigma^2}\right) dx \qquad (3)$$

where C_S and σ are respectively the mean value and the standard deviation of the coordination number distribution of the fine particles.

Then, it is necessary to relate these two parameters to the material parameters. Since the researches in literature reported generally the results on the mean coordination number of an homogenous mixture, we need to develop a physical model of the coordination number of the binary mixture, that give us the following relationship:

$$\overline{C_S} = 4.7\left(1 - f_C + 2 f_C X\right) / \left(X\left(e + f_C\right)\right) \qquad (4)$$

where $X = D_{50}/d_{50}$

A relation between σ and C_S can be obtained by approximating the experimental results reported by Oda as follows [17]:

$$\sigma = 2\overline{C}_S \qquad (5)$$

So, we obtain the final equation to calculate b parameter as follows:

$$b = 1.3 - \left[1 + \exp\left(0.85 - 0.63\frac{eX + f_c X}{1 - f_c + 2 f_c X}\right)\right]^{-1} \qquad (6)$$

In order to evaluate the applicability of the new proposed formula, apart from the results of the current study, we reanalyzed 14 others datasets published in the literature, divided into two groups. The first one related to the instability onset and the second one related to the critical state. These results were presented in original papers, for the first group, as different relations between the classical void ratio and one instability parameter (peak deviator stress, peak stress ratio, cyclic liquefaction resistance), whereas for the second group as different critical state lines with the fines contents.

The back-analysis procedure of one dataset is processed as follows: the original data on void ratio, fines content and diameter ratio were collected in order to calculate the corresponding equivalent intergranular void ratio using the new formula. Then, we use the latter instead of void ratio and verify if we can obtain a single correlation presenting all dataset points. For an objective assessment, we adopted the benchmark used by Rahman et al. [10], following that the data points can be represented satisfactorily by a single correlation if the value of the root mean square deviation (RMSD) is smaller than 0.043. In addition, we also reported the more popular coefficient of determination R^2 for each obtained trend curve.

Among 16 back-analyzed datasets, the equivalent intergranular void ratio concept calculated with the b parameter obtained from the new formula fails only in one case ([3]) and approaches the benchmark (RMSD=0.045) in three others cases ([9], [10], [20]). It could be explained by the fact that in these cases, the sand and fines used have a very different mineralogy, while the concept of the equivalent intergranular void considers implicitly that the mechanical contributions of the sand grains and of fine grains are equals. For all the others cases, the equivalent intergranular void ratio concept calculated with the b parameter obtained from the new formula can be applied successfully, as shown for example in the case of the current study in Figure 5 and Figure 6.

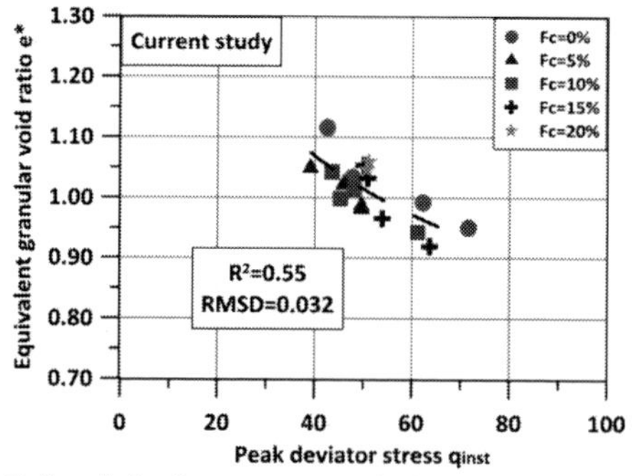

Fig. 5. Correlation between q_{peak} and e* (calculated with the b obtained by the new formula)

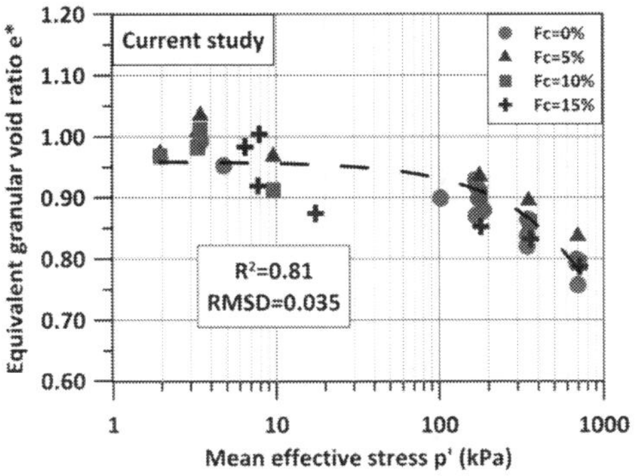

Fig. 6. Critical state line in e* - logp' space (e* calculated with the b obtained by the new formula)

5 Conclusion

The equivalent intergranular void ratio appears to be a pertinent density variable in analyzing the mechanical response of silty sand. This alternative density variable showed a strong correlation with both the peak deviator stress and the mean effective stress at the critical state.

In order to calculate the equivalent intergranular void ratio, we proposed in this paper a new formula to compute the fraction of active fines based on the distribution of the coordination number. This new formula takes into account the effect of the fine content, the void ratio and the diameter ratio of the large and small grains. The back-analysis of many different datasets published in the literature confirmed the effectiveness of this formula with different materials, especially when the large and small grains have a similar mineralogy. Therefore, it is highly recommended to use the equivalent intergranular void ratio as an analysis variable in silty sand mechanical behavior prediction.

References

1. R. H. Kuerbis, D. Negussey, and Y. P. Vaid, "Effect of gradation and fines content on the undrained response of sand," Geotech. Spec. Publ., no. **21**, pp. 330–345, (1988).
2. C. P. Polito and J. R. Martin II, "A reconciliation of the effects of non-plastic fines on the liquefaction resistance of sands reported in the literature," Earthq. Spectra, vol. **19**, no. 3, p. 635, (2003).
3. S. Thevanayagam, T. Shenthan, S. Mohan, and J. Liang, "Undrained fragility of clean sands, silty sands, and sandy silts," J. Geotech. Geoenvironmental Eng., vol. **128**, no. 10, pp. 849–859, (2002).
4. Q. Ni, T. S. Tan, G. R. Dasari, and D. W. Hight, "Contribution of fines to the compressive strength of mixed soils," Géotechnique, vol. **54**, no. 9, pp. 561–569, (2004).
5. S. A. Naeini and M. H. Baziar, "Effect of fines content on steady-state strength of mixed and layered samples of a sand," Soil Dyn. Earthq. Eng., vol. **24**, no. 3, pp. 181–187, (2004).
6. S. L. Yang, R. Sandven, and L. Grande, "Instability of sand-silt mixtures," Soil Dyn. Earthq. Eng., vol. **26**, no. 2–4 SPEC. ISS., pp. 183–190, (2006).
7. S. L. Yang, R. Sandven, and L. Grande, "Steady-state lines of sand-silt mixtures," Can. Geotech. J., vol. **43**, no. 11, pp. 1213–1219, (2006).
8. T. G. Murthy, D. Loukidis, J. A. H. Carraro, M. Prezzi, and R. Salgado, "Undrained monotonic response of clean and silty sands," Géotechnique, vol. **57**, no. 3, pp. 273–288, (2007).
9. A. Papadopoulou and T. Tika, "The effect of fines on critical state and liquefaction resistance characteristics of non-plastic silty sands," Soils Found., vol. **48**, no. 5, pp. 713–725, (2008).
10. M. M. Rahman, S.-C. R. Lo, and C. T. Gnanendran, "On equivalent granular void ratio and steady state behaviour of loose sand with fines," Can. Geotech. J., vol. **45**, no. 10, pp. 1439–1456, (2008).
11. C. Stamatopoulos, "An experimental study of the liquefaction strength of silty sands in terms of the state parameter," Soil Dyn. Earthq. Eng., vol. **30**, no. 8, pp. 662–678, Aug. (2010).
12. A. Mohammadi and A. Qadimi, "A simple critical state approach to predicting the cyclic and monotonic response of sands with different fines contents using the equivalent intergranular void ratio," Acta Geotech., May (2014).
13. N. Benahmed, T. K. Nguyen, P. Y. Hicher, and M. Nicolas, "An experimental investigation into the effects of low plastic fines content on the behaviour of sand/silt mixtures," Eur. J. Environ. Civ. Eng., vol. **18**, Jul. (2014).
14. W. O. Smith, P. D. Foote, and P. F. Busang, "Packing of homogeneous spheres," Phys. Rev., vol. **34**, no. 9, pp. 1271–1274, (1929).
15. J. D. Bernal and J. Mason, "Co-ordination of randomly packed spheres," Nature, vol. **188**, pp. 910–911, (1960).
16. M. Akarawa and M. Nishino, "Contact number and porosity in randomly packed sphere mixtures," J. Soc. Mater. Sci. Japan, vol. **22**, pp. 658–662, (1973).
17. M. Oda, "Co-ordination number and its relation to shear strength of granular material," Soils Found., vol. **17**, no. 2, pp. 29–42, (1977).
18. J. S. Goodling and M. S. Khader, "Co-ordination number cylindrical bed distribution of spherical particles in a packed," Powder Technol., vol. **44**, no. 1, pp. 53–55, (1985).
19. R. Y. Yang, R. P. Zou, and A. B. Yu, "Computer simulation of the packing of fine particles," Phys. Rev. E. Stat. Phys. Plasmas. Fluids. Relat. Interdiscip. Topics, vol. **62**, no. 3 Pt B, pp. 3900–3908, Sep. (2000).
20. C. P. Polito, "The effects of non-plastic and plastic fines on the liquefaction of sandy soils," Virginia Polytechnic Institute, (1999).

Vertical motion of particles in vibration-induced granular capillarity

Fengxian Fan[1,*], *Ju* Liu[1], *Eric J R* Parteli[2] and *Thorsten* Pöschel[3]

[1]Shanghai Key Laboratory of Multiphase Flow and Heat Transfer in Power Engineering, University of Shanghai for Science and Technology, 200093 Shanghai, China
[2]Department of Geosciences, University of Cologne, 50969 Cologne, Germany
[3]Institute for Multiscale Simulation, Universität Erlangen-Nürnberg, D-91052 Erlangen, Germany

Abstract. When a narrow tube inserted into a static container filled with particles is subjected to vertical vibration, the particles rise in the tube, much resembling the ascending motion of a liquid column in a capillary tube. To gain insights on the particle dynamics dictating this phenomenon – which we term **granular capillarity** – we numerically investigate the system using the Discrete Element Method (DEM). We reproduce the dynamical process of the granular capillarity and analyze the vertical motion of the individual particles in the tube, as well as the average vertical velocities of the particles. Our simulations show that the height of the granular column fluctuates in a periodic or period-doubling manner as the tube vibrates, until a steady-state (capillary) height is reached. Moreover, our results for the average vertical velocity of the particles in the tube at different radial positions suggest that granular convection is one major factor underlying the particle-based dynamics that lead to the granular capillarity phenomenon.

1 Introduction

When vertical vibration is introduced into granular systems, the granular materials can exhibit a diversity of interesting phenomena, such as size segregation, density waves, convection and heap formation [1]. A particularly interesting phenomenon is that when a narrow tube inserted into a static container filled with particles is subjected to vertical vibration, the particles rise in the tube. Because this process resembles the rise of a liquid in a capillary tube, we term it **granular capillarity**. This phenomenon has attracted the interest of many researchers since it offers a potential means to pump or lift granular systems [2].

Since the first experimental observations of granular capillarity by Liu et al. [2], much effort has been made to unravel and model its underlying mechanisms [3-5]. In particular, particle-based numerical models have been developed to simulate the particle dynamics due to the vibrating tube [4]. However, these simulations were performed for one oscillation cycle of the vertically vibrating tube only, and thus they could not reproduce the rising process of the granular column in the tube observed experimentally in the course of many oscillation cycles. Moreover, it remains uncertain which physical mechanisms dictate the ascending motion of the granular column in the tube [5].

Numerical simulations using the Discrete Element Method (DEM) [6] can offer a helpful tool in the investigation of the granular capillarity phenomenon. In the DEM, Newton's equations of translational and rotational motion for all particles are simultaneously solved by taking into account the forces and torques acting on each particle. Here we use three-dimensional (3D) DEM simulations to reproduce the experimentally observed granular capillarity phenomenon and investigate the particle-based processes that lead to referred phenomenon. We will show that the granular capillarity effect is associated with granular convection in the system, which is inherent to particulate materials subject to vibrations.

2 DEM model

According to the DEM, the equations for the translational and rotational motion of particle i with radius R_i, mass m_i and moment of inertia I_i , read

$$m_i \frac{\mathrm{d}\mathbf{v}_i}{\mathrm{d}t} = \sum_{j=1}^{n}\left(\mathbf{F}_{ij}^{n} + \mathbf{F}_{ij}^{t}\right) + m_i\,\mathbf{g} \tag{1}$$

$$I_i \frac{\mathrm{d}\omega_i}{\mathrm{d}t} = \sum_{j=1}^{n}\left(\mathbf{M}_{ij,t} + \mathbf{M}_{ij,r}\right) \tag{2}$$

where $\mathbf{v}_i$ and ω_i are the translational and angular velocities of particle i, t is the time, n is the number of particles in contact with particle i, $\mathbf{g}$ is the acceleration of gravity, while $\mathbf{F}_{ij}^{n}$ and $\mathbf{F}_{ij}^{t}$ are the interaction forces between particle i and particle j in normal and tangential directions, respectively. Moreover, the torques acting on particle i are $\mathbf{M}_{ij,t}$ due to the tangential force $\mathbf{F}_{ij}^{t}$, and $\mathbf{M}_{ij,r}$ due to the rolling resistance.

* Corresponding author: fanfengxian@hotmail.com

We assume viscoelastic interaction [8] for the normal force $\mathbf{F}_{ij}^n$ and apply a modified Cundall-Stract model [6] for the tangential force $\mathbf{F}_{ij}^t$. Hence the forces read

$$\mathbf{F}_{ij}^n = \min\left(0, -\rho\delta_{nij}^{3/2} - \frac{3}{2}A_n\rho\sqrt{\delta_{nij}}\dot{\delta}_{nij}\right)\mathbf{e}_n \quad (3)$$

$$\mathbf{F}_{ij}^t = -\min\left(\mu_s\left|\mathbf{F}_{ij}^n\right|, \int_{\text{path}}\frac{4G}{2-\nu}\sqrt{R_{\text{eff}}\delta_{nij}}ds + A_t\sqrt{R_{\text{eff}}\delta_{nij}}v_{tij}\right)\mathbf{e}_t \quad (4)$$

where $\delta_{nij} = R_i + R_j - |\mathbf{r}_i - \mathbf{r}_j|$ is the overlap distance between the particles at positions $\mathbf{r}_i$ and $\mathbf{r}_j$, $\mathbf{e}_n = (\mathbf{r}_j - \mathbf{r}_i)/(|\mathbf{r}_j - \mathbf{r}_i|)$ is the normal unit vector, ρ is the elastic parameter, which is a function of the Young's modulus, Y, the Poisson's ratio, ν, and the effective radius, $R_{\text{eff}} = R_iR_j/(R_i + R_j)$,

$$\rho = \frac{2Y}{3(1-\nu^2)}\sqrt{R_{\text{eff}}} \quad (5)$$

and A_n is the dissipative parameter, which depends not only on the Young's modulus and the Poisson's ratio, which are easily available for a variety of materials, but also on the material viscosities, which are not directly available. In order to determine A_n, we use an expression of the coefficient of restitution, ε, for the collision of the particles i and j [9]. Moreover, μ_s is the sliding friction coefficient, $G=Y/[2(1+\nu)]$ is the shear modus, v_{tij} is the relative tangential velocity at the point of contact, $A_t=A_nY/(1-\nu^2)$ is the tangential dissipative parameter [10] and $\mathbf{e}_t$ is the tangential unit vector. The integral in Eq. (4) is performed over the displacement of the particles at the point of contact for the duration of the contact.

According to the directional constant torque model [11], the torque associated with the rolling resistance, $\mathbf{M}_{ij,r}$, can be expressed as

$$\mathbf{M}_{ij,r} = -\mu_r\rho\delta_{nij}{}^{3/2}\frac{\omega_{ij}}{|\omega_{ij}|}R_{\text{eff}} \quad (6)$$

where μ_r is the coefficient of rolling resistance and ω_{ij} is the relative angular velocity between particle i and particle j.

In the DEM simulation, the time step Δt must be small enough to make sure that the Newton's equations are accurately solved. In this work, we take $\Delta t=t_{\text{col}}/30$, where t_{col} is the collision time, which can be estimated using the collision time for the center-to-center collision of frictionless elastic particles

$$t_{\text{col}} \approx 3.21\left(\frac{m_{\text{eff}}}{\rho}\right)^{2/5} v_{\text{imp}}^{-1/5} \quad (7)$$

where $m_{\text{eff}}=m_im_j/(m_i+m_j)$ is effective mass, and v_{imp} is the impact velocity. In this work, a reference impact velocity v_{imp}=1.0 m/s is adopted to compute t_{col}.

3 Numerical experiments

Our simulation system consists of N=200,000 mono-sized particles with diameter d_p=0.6 mm and density ρ_p=2500 kg/m^3. The particles are filled in a cylindrical silo with inner diameter D=28.2 mm, while the tube has inner diameter d=8 mm and thickness δ=0.6 mm. The tube is inserted into the center of the silo and filled up to the same level as the granular surface in the other volume. The lower end of the tube is initially at z=0 mm, and the bottom of the silo is at z=−30 mm, while the heights of the silo and the tube are large enough to contain all the particles during the simulation. The frictional wall boundary condition is applied for the walls of both silo and tube. The equations used for computing the forces between particles and the walls are the same used for modelling particle-particle collisions where one of the contact partners is of infinite mass and radius.

We vibrate the tube by applying a vertical displacement z which is a sinusoidal function of time,

$$z=A\sin(2\pi ft) \quad (8)$$

where A and f are the amplitude and the frequency, respectively. In this work, A=12.42 mm and f=10 Hz are applied. The physical properties of the particles used in the simulation are listed in Table 1. The simulations were performed with 4 processors in the workstation cluster equipped with Intel Core i7-4790 CPUs. It took about 45 hours to run a simulation case for the tube vibration time of 100T (T is the period of tube vibration).

Table 1. Physical properties of the particles.

Parameter	Symbol	Value
Young's modulus	Y	10 MPa
Poisson's ratio	ν	0.25
Coefficient of restitution	ε	0.5
Particle-particle sliding friction coefficient	$\mu_{s,pp}$	0.8
Particle-wall sliding friction coefficient	$\mu_{s,pw}$	0.58
Particle-particle rolling friction coefficient	$\mu_{r,pp}$	0.1
Particle-wall rolling friction coefficient	$\mu_{r,pw}$	0.05

4 Results and discussion

4.1 Phenomenon of granular capillarity

Fig. 1 shows snapshots of the simulation of the granular capillarity phenomenon. The snapshots show two-dimensional cuts of the granular system along the tube axis, the color of each particle indicates its initial position at t=0. Particles which, at t = 0, are within the cylindrical shell centered at the tube's central axis with radius equal to the tube's radius, are colored blue, while particles outside this shell at t = 0 are colored red. The particles keep this color code throughout the simulation, independently of how individual particle positions change with time. The simulation starts with the free surface of the particles in the tube and in the silo levelled. Due to the friction between the particles and the tube wall, the particles in the tube rise once the tube moves upwards, leaving a void on the bottom of the tube. This void is filled rapidly by the nearby particles in the silo, which enter the tube through its bottom end thereby increasing the height of the granular column in the tube. That is, the aforementioned void filling mechanism leads to an increase of the granular column height in the first oscillation cycles of the tube.

However, after several oscillation cycles (~ $70T$), H_c slowly achieves a steady-state value, as can be seen from Fig. 2 where we plot the granular column height as a function of time. We see from this figure that H_c fluctuates as the tube vibrates. The period and amplitude of the fluctuation change with the vibration time. It should be noted that during t=0~$2T$, the large fluctuation of H_c is caused by inertial effects. After the initial stage, the evolution of H_c shows three stages in the granular capillarity: the periodic rising stage, the period-doubling rising stage and the period-doubling terminal stage. In the periodic rising stage, H_c fluctuates with the same period as the tube vibration as shown during $2T<t<9T$. When $t>9T$, the development of the granular capillarity turns to a period-doubling stage, where H_c oscillates in a period of $2T$ and with a much larger amplitude compared with the periodic stage. As the time increases further to $70T$, H_c approaches a steady-state value. It has been observed that the appearance and duration of the above stages differs greatly with the change of the vibration condition and the sizes of the tube and the silo.

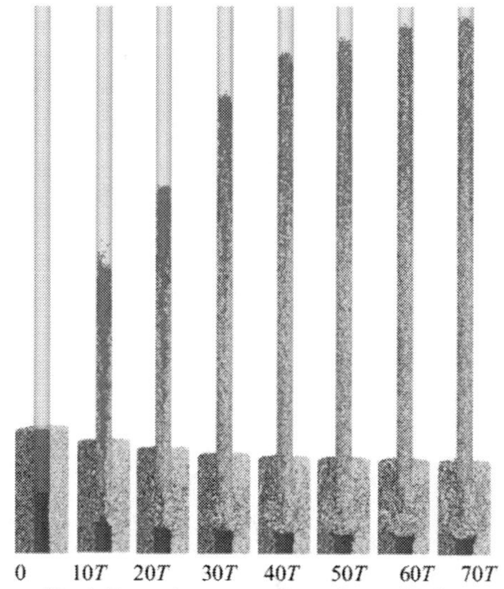

Fig. 1. Dynamic process of granular capillarity.

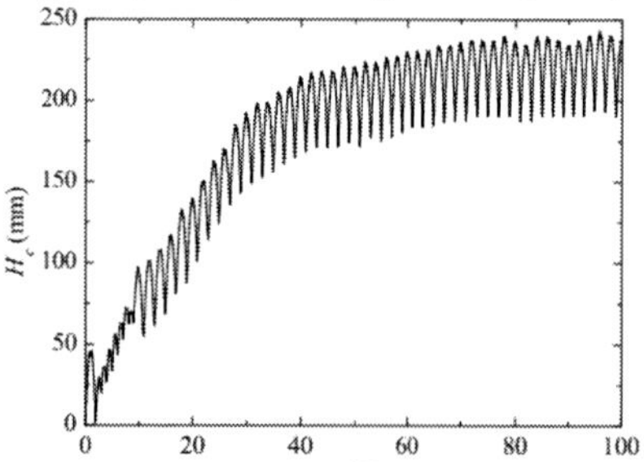

Fig. 2. Height of the granular column as a function of time.

4.2 Vertical motion of individual particles

We show the vertical displacements and velocities of three individual particles at different vertical positions along the tube in Fig. 3. It can be seen from Fig. 3a that, in the considered range $5 < t/T < 15$, all three particles a-c rise along the tube but their ascending motion is modulated by oscillations of different magnitudes. Indeed, we see that the higher the vertical position the larger the amplitude. We have found that the uppermost layers of the granular column have a smaller solid fraction compared to the center and bottom of the tube, which explains why the modulation amplitude for particle "c" is larger than for particles "b" and "a". However, we are currently investigating the particle-based processes underlying the rise of the granular column in more detail and the referred behavior still remains to be elucidated. Moreover, we note that there is a phase lag between the oscillation of the tube and the vertical particle trajectories for all particles. It is still uncertain how this phase lag is related to the transition of the periodic rising stage to the period-doubling rising stage mentioned in the previous section. However, it is observed that in the period-doubling stage, the particle in the top oscillates with a period of $2T$, whereas the particles in the bottom undergoes a transition from periodic to period-doubling oscillation.

Generally, once the tube starts to move upward from its neutral position with its bottom end at z=0, the particles experience decelerated rising, accelerated falling, decelerated falling, and accelerated rising in a cycle of tube vibration, as shown in Fig. 3(b). The downward acceleration during the particles' decelerated rising and accelerated falling is approximately g on average, indicating that the gravity is the dominant force. On the other hand, the average upward acceleration during the particles' decelerated falling and accelerated rising is approximately $7g$, which is larger than the maximum acceleration of the tube vibration ($A\omega^2 \approx 5g$). It indicates that the upward acceleration is caused not only by the friction but also by the collision between the particles.

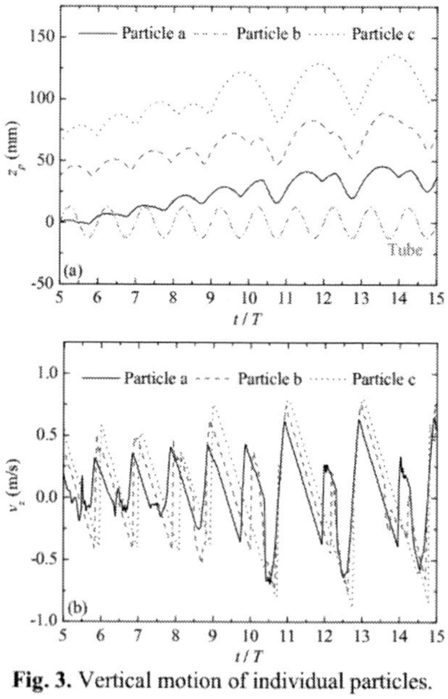

Fig. 3. Vertical motion of individual particles.

4.3 Average vertical velocity of particles

The average vertical velocity of the particles in the tube during different stages of the granular capillarity as a function of dimensionless radial position is presented in Fig. 4. In the periodic rising stage, the rising velocity of the particles near the wall is larger than in the center, while the rising velocity of the particles in the center is larger than near the wall in the period-doubling stage. Moreover, the rising velocity is smaller in the period-doubling rising stage compared with the periodic rising stage. During the periodic doubling steady-state stage, the particles in the tube center move upward whereas the particles near the tube wall move downward, indicating the occurrence of granular convection. Due to the convective motion of the particles in the tube, the net particle mass fluxes near the walls and in the center compensate each other, and thus the average height of the granular column achieves a stable value.

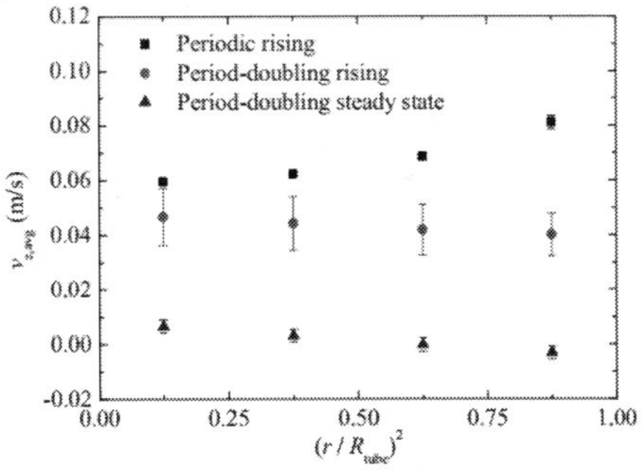

Fig. 4. Average vertical velocity of the particles in the tube as a function of dimensionless radial position $(r/R_{tube})^2$.

Therefore, our simulations show that the phenomenon of granular capillarity is related to the process of granular convection which is characteristic of granular systems subject to vibrations. We have seen that, as the tube vibrates, convection occurs within the granular material in the tube's outer volume (that is, in the silo). Indeed, we have performed simulations without frictional forces between particles and the silo's walls and found that, under this condition, the granular column does not rise along the tube. Moreover, we also found that simulations using periodic boundary conditions in the lateral directions also showed no granular capillary effect. These observations lead us to the conclusion that the interactions between the particles and the frictional walls of the silo play an important role for the capillary behavior of the granular column in the tube [12]. However, future work is needed to characterize the convection mechanism in the silo and the tube, as well as to explain whether and how this mechanism affects the capillary phenomenon.

5 Conclusions

In conclusion, we have presented the simulation of the experimentally reported granular capillarity phenomenon [2-5,14], which was caused by vertical vibrations of a vertical tube inserted in a container filled with particles. It is interesting to note that, in previous experiments, vertical vibrations were applied to the silo instead of the tube, and the phenomenon also occurred. Therefore, our simulations suggest that the phenomenon is robust upon whether the silo or the tube is subject to the vibrations.

It is worthy noting that the granular matter exhibits rich dynamic behavior under vertical vibration, depending on the system size, the vibration condition and the amount of particles. Future work is thus necessary to examine the effects of the parameters mentioned above. Moreover, it is still uncertain how the phenomenon of jamming, inherent to granular flows through pipes and narrow channels, may be influencing particle behavior and the capillary height. To clarify this role, more systematic investigation under different conditions of solid fraction, frictional properties and texture of the tube's inner-wall, as well as tube-to-particle diameter – for instance by extending the study of Ref. [13] for the system of silo and tube – is required.

The authors gratefully acknowledge the financial support provided by Natural Science Foundation of China (Grant No. 51206113, 51176128 and 51576130) and Shanghai Science and Technology Commission, China (Grant No. 13DZ2260900). We also thank the German Research Foundation (DFG) for funding through the Cluster of Excellence "Engineering of Advanced Materials", ZISC, FPS, the Collaborative Research Center SFB814, and grant PO472/20-2. We gratefully acknowledge the computing time granted by the John von Neumann Institute for Computing (NIC) and provided on the supercomputer JUROPA at Jülich Supercomputing Centre (JSC).

References

1. H.M. Jaeger, S.R. Nagel, R.P. Behringer, Phys. Today **49**, 32-38 (1996)
2. C.P. Liu, P. Wu, L. Wang, Soft Matter. **9**, 4762-4766 (2013)
3. C.P. Liu, F.W. Zhang, P. Wu, L. Wang, Powder Technol. **259**, 137-143 (2014)
4. F.W. Zhang, L. Wang, C.P. Liu, P. Wu, Acta Phys. Sin. **63**, 014501 (2014)
5. Y. Liu, J.H. Zhao, Chinese Phys. B **24**, 034502 (2015)
6. P.A. Cundall, O.D. Strack, Geotechnique **29**, 47-65 (1979)
7. C. Kloss, C. Goniva, A. Hager, S. Amberger, S. Pirker, Prog. Comput. Fluid Dy. **12**, 140-152 (2012)
8. N.V. Brilliantov, F. Spahn, J.M. Hertzsch, T. Pöschel, Phys. Rev. E **53**, 5382–5392 (1996)
9. P. Müller, T. Pöschel, Phys. Rev. E **84**, 021302 (2011)
10. C.H. Rycroft, A.V. Orpe, A. Kudrolli, Phys. Rev. E **80**, 031035 (2009)
11. J. Ai, J.F. Chen, J.M. Rotter, Powder Technol. **206**, 269–282 (2011)
12. F. Fan, E.J.R. Parteli, T. Pöschel, Phys. Rev. Lett. (submitted, 2016).
13. F. Verbücheln, E.J.R. Parteli, T. Pöschel, Soft Matter **11**, 4295-4305 (2015)
14. I. Sanchez, O. Pozo, J.R. Darias, Int. Conf. Agricultural Eng. CIGR-AgEng2012, Valencia, 8-12 July (2012)

Simulation study of the discharge characteristics of silos with cohesive particles

David Hund*, *Dominik* Weis, *Robert* Hesse and *Sergiy* Antonyuk

Institute of Particle Process Engineering, University of Kaiserslautern, 67663 Kaiserslautern, Germany

Abstract. In many industrial applications the silo for bulk materials is an important part of an overall process. Silos are used for instance to buffer intermediate products to ensure a continuous supply for the next process step. This study deals with the discharging behaviour of silos containing cohesive bulk solids with particle sizes in the range of 100-500 µm. In this contribution the TOMAS [1,2] model developed for stationary and non-stationary discharging of a convergent hopper is verified with experiments and simulations using the Discrete Element Method. Moreover the influence of the cohesion of the bulk solids on the discharge behaviour is analysed by the simulation. The simulation results showed a qualitative agreement with the analytical model of TOMAS.

1 Introduction

In industry silos are mainly used to store intermediate products in an overall process. In order to ensure a continuous supply for the next process step it is important to know the discharging behaviour of a silo. The discharging becomes more complex, when the silo contains cohesive bulk solids with fine particle sizes in the range about 100 µm or smaller. Since there are a large number of interfering factors, like the influence of the surounidng air on particles with d < 500 µm, affecting the discharging behavior of a silo an analytical calculation is difficult. However, there are many empirical approximations and models based on dimensional analysis. In this contribution the model of TOMAS [1,2] for stationary and non-stationary discharging velocity of a convergent hopper (1) is verified with simulations using the Discrete Element Method (DEM).

$$v(t) = \frac{g\left(1-\frac{b_{min}}{b}\right)\cdot\tanh\left(\frac{t}{t_{76}}\right)}{\left(\frac{9\cdot\eta_f}{\rho_s\cdot d_p^{\,2}\cdot\varepsilon^2}\right)\cdot\tanh\left(\frac{t}{t_{76}}\right)+\frac{1}{t_{76}}} \qquad (1)$$

In the model in Eq. (1) t_{76} is a time parameter, which describes the time needed to reach 76 % of the stationary outflow velocity and b_{min} is a geometrical parameter, which corresponds to the minimum width of the silo outlet to ensure that no bridging occurs. Both parameters can be estimated with the granular flow properties of the bulk solids as given in [2]. Moreover particle parameters, like particle diameter d_p and density ρ_s, the bulk porosity ε as well as the fluid viscosity η_f are included.

By using a contact model which includes also attractive forces in DEM the influence of the cohesion of the bulk solids on the discharging behaviour is analysed in the simulations. Thereto the hopper was completly filled at the beginning and discharged afterwards.

2 Simulation methods

The DEM, introduced firstly in 1979 by Cundall and Strack [3], can be used to simulate the dynamics of particulate systems. The method describes each particle in the system and its interactions with other particles or rigid walls via contact forces F_c. The contact forces can be devided in a normal component F_n and a tangential component F_t. In this study the particles are assumed to be spherical and indestructible. Based on the force balance for every single particle the equations of motion (2) and (3) are numerically solved.

$$m_p\frac{dv_p}{dt} = F_g + \sum_i^k\left(F_{c,i} + F_{coh,i}\right) + F_d \qquad (2)$$

$$J_p\frac{d\omega_p}{dt} = \sum_i^k M_{p,i} = \sum_i^k F_{t,i}\cdot r_{p,i} \qquad (3)$$

In the equations m_p, and J_p are the mass of a particle and its moment of inertia while v_p and w_p are the velocity and the angular velocity. The tangential force $F_{t,i}$ due to the interactions with contact partner i is acting at the contact point with the radius $r_{p,i}$ and causes the torque $M_{p,i}$. The contact forces need to be modeled with an appropriate contact model. For particles with a size less than 500 µm cohesive forces can become important. For that reason a model for the cohesion force $F_{coh,i}$ on the particles was included in the force balance. Also the drag force F_d caused by the interaction of the particles with the surrounding air and the gravitational force F_g is considered. The contact forces $F_{c,i}$ were modelled in DEM according to the soft sphere approach, which is based on the calculation of the forces depending on the overlap of the contact partners. For the calculation of the repulsive force in normal direction a non-linear elastic model, based on the Hertz theory [4], is used. To account for the energy dissipation during a collision viscous

* Corresponding author: david.hund@mv.uni-kl.de

damping terms for the normal and tangential contact were considered, taken from the work of Tsuji et al. [6].

$$F_n=\frac{4}{3}E^*\sqrt{R^*\cdot\delta_n}\cdot\delta_n-\alpha\sqrt{2\cdot E^*\sqrt{R^*\cdot\delta_n}\cdot m^*}\cdot\dot{\delta}_n \quad (4)$$

In the equation (4) for the normal force, E^* is the equivalent Young´s modulus, R^* the equivalent radius of the particles and δ_n the normal overlap. The damping term which considers the energy dissipation depends on the equivalent mass m^*, the relative velocity in normal direction $\dot{\delta}_n$ and a damping factor α, which depends on the restitution coefficient e [6]. The tangential force is modelled based on the non-slip solution of the theory provided by Mindlin and Deresiewicz [5] and is limited by Coulomb friction $\mu\cdot F_n$:

$$F_t=8\cdot G^*\sqrt{R^*\cdot\delta_n}\cdot\delta_t-\alpha\sqrt{8\cdot G^*\sqrt{R^*\cdot\delta_n}\cdot m^*}\cdot\dot{\delta}_t \quad (5)$$

Equation (5) contains the tangential overlap δ_t and the relative velocity in tangential direction $\dot{\delta}_t$. Also the shear modulus G^* is included. The attractive forces are modelled using a linear term as follows:

$$F_{coh,i}=k\cdot A \quad (6)$$

where k is the energy density, which contais all cohesive effects, and A is the contact area. In this approach the particle cohesion in a contact increases proportionally to the applied force and forming contact area. A behaviour like this was qualitatively confirmed for silica particles by atomic force instruments (AFM) in [7]. This effect will be dominant if the yield point of the particles is reached and the plastic contact area increases [8]. The influence of the cohesion on the force-displacement curve is shown in Figure 1.

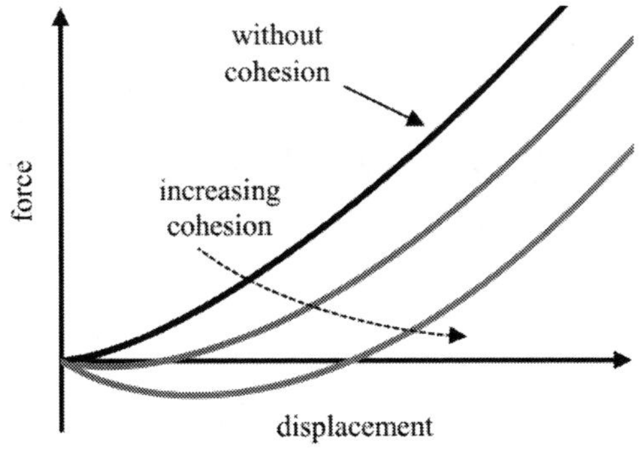

Fig. 1. Force-displacement curves with different cohesion

To account for the influence of the surrounding air on the particle motion, the drag force was calculated according to the work of Lord Rayleigh [9]:

$$F_D=\frac{\rho_F}{2}\cdot c_D(Re_P)\cdot A_P\cdot(v_F-v_P)^2 \quad (7)$$

where ρ_F and v_F are the fluid density and fluid velocity, respectively. A_p is the projection area of the particle and v_p its velocity. Based on the assumption that the particles do not affect the air, a constant fluid velocity of $v_F = 0$ is set. The drag coefficient c_D depends on the particle Reynolds number Re_p and can be calculated based on the equation of Kaskas [10]:

$$c_D=\frac{24}{Re_p}+\frac{4}{\sqrt{Re_p}}+0.4 \quad (8)$$

$$Re_P=\frac{\rho_F(v_F-v_P)d_P}{\mu_F} \quad (9)$$

As fluid properties the dynamic viscosity μ_F and the fluid density influence the particle Reynolds number.

3 Simulation results

For the simulations of the discharge behaviour of a convergent hopper, a monomodal particle system with a particle diameter of d = 500 µm was used. The energy density in the cohesion force acting on the particles was varied in the contact model.

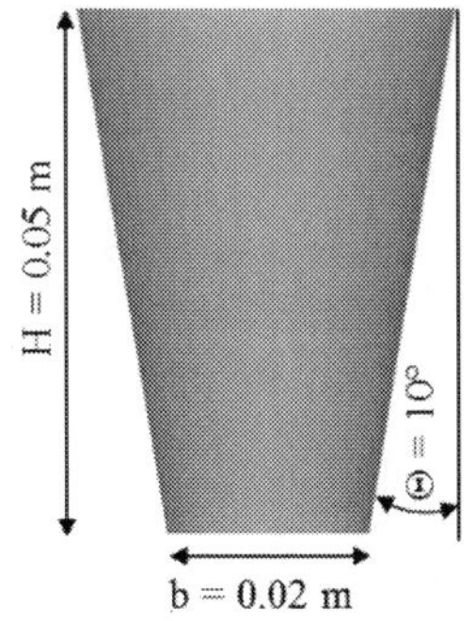

Fig. 2. Geometry of the simulated hopper

In Figure 2 the geometric dimensions of the convergent hopper are shown. The hopper has an outflow width of b = 20 mm and an angle Θ = 10°. The simulation paramters are given in Table 1.

Table 1. Simulation setup

hopper	
width at the outflow	b = 0.02 m
angle of the silo cone	Θ = 10°
particulate phase	
particle diameter	d_P = 0.0005 m
particle density	ρ_P = 1190 kg/m^3
shear modulus	G = 1064 GPa
young´s modulus	E = 3000 GPa
friction coefficient	µ = 0.5
restitution coefficient	e = 0.85
energy density	$k_1 = 0.2\cdot10^6$ J/m^3
	$k_2 = 1.0\cdot10^6$ J/m^3
	$k_3 = 3.3\cdot10^6$ J/m^3
time step	$\Delta t = 2\cdot10^{-10}$ s
fluid phase	
density	ρ_F = 1.29 kg/m^3
dynamic viscosity	μ_F = 17.1 µPa·s

To compare the results obtained by the simulations of the discharging behaviour with the empirical approximation in Eq. 1 the minimum outflow width b_{min} has to be known. b_{min} can be calculated according to Jenike model [11] with paramters which can be measured by a shear test. However, in this study simulations were performed to obtain b_{min}. The outlet diameter b of the geometry used in DEM was gradually reduced at a constant angle of the cone until no outflow of particles occured. This was determined by considering the average axial velocity of all particles. If the average velocity drops below a critical value for at least 10 seconds, the simulation was terminated. The results of b_{min} determined in this way are shown in Table 2.

Table 2. b_{min}/b for different energy densities determined by simulations

energy density	b_{min}/b
$k_1 = 0.2{\cdot}10^6$ J/m^3	0.075
$k_2 = 1.0{\cdot}10^6$ J/m^3	0.080
$k_3 = 3.3{\cdot}10^6$ J/m^3	0.730

Figure 3 exemplarily shows a snapshot of the particle velocities after a simulation time of t = 0.02 s for k_1. At the right side a rz – cross section is displayed at the top. The maximum velocities (red) are reached directly above the outlet, while particles in the upper regions close to the wall have still a velocity of v = 0 m/s (blue).

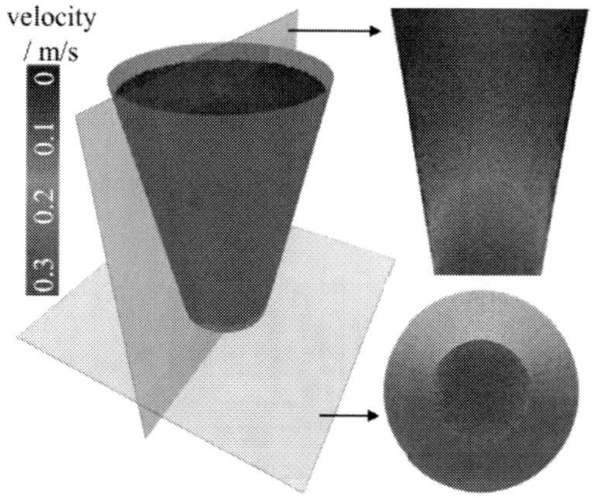

Fig. 3. Snapshot of the particle velocities after a simulation time of t = 0.02 s for k_1

The horizontal cross-section, displayed at the right side in Figure 3 at the bottom, shows that the particle velocity is minimal in the area of the wall (green) and increases towards the center.

In Figure 4 the averaged particle velocities at the outflow of the hopper are shown over the time. The data points show the results of the simulation while the lines are the solution of the empirical equation (1). The colours represent the different energy densities. The simulation results show a decreasing stationary outflow velocity with increasing cohesive forces. Comparing the course of the simulated velocities with the empirical results, it can be seen that the time needed to reach the stationary velocity is in a good agreement. Furthermore, it can be seen that the velocities obtained from the simulations are generally higher than the velocities calculated with Eq. (1). This deviation is in the same order of magnitude as the deviation showed by Tomas in [1] where the analytical results of his model for stationary velocity was compared with experiments.

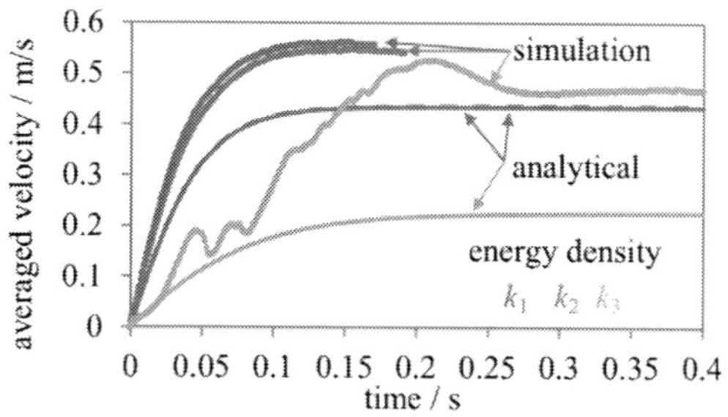

Fig. 4. Averaged particle velocity at the outflow over time for different cohesion forces

The high fluctuation of the velocity at k_3 (Fig. 4) is caused by agglomeration of particles to clumps in the silo outlet. This effect is illustrated in Figure 5 which shows the particle velocities (left) and the porosity (right) in rz – cross sections at t = 0.03 s. Due to bridging, agglomerates are breaking out of the bulk, what leads to an increase in the porosity between the agglomerate and the bulk. Figure 4 also shows, that for k_3 the maximal outflow velocity at t = 0.21 s is higher than the stationary outflow velocity. This effect is also a result of the breaking out agglomerates based on bridging at the beginning of the outflow. For this reason,

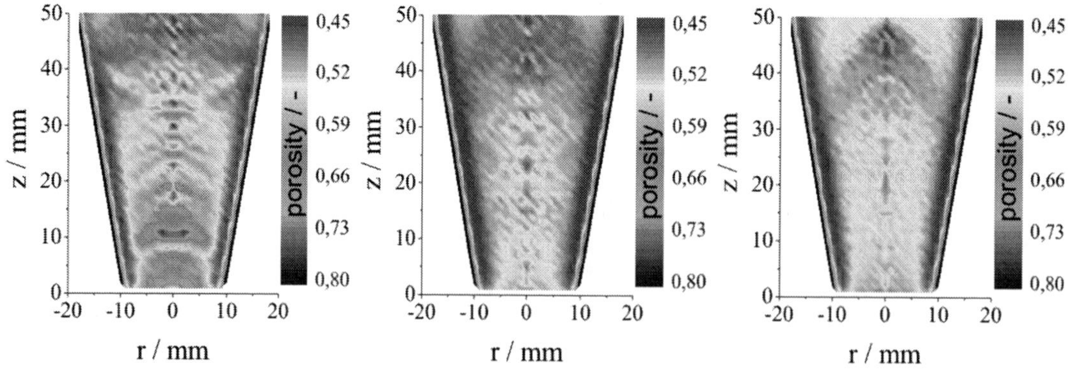

Fig. 6. Porosity for k_3 at simulation time of t = 0.18 s (left), t = 0.22 s (middle) and t = 0.27 s (right)

the porosity increases in the lower regions of the hopper (Figure 6, left). At a certain height of the hopper the width is getting big enough that briding not longer occurs. Because of the higher porosity in the lower regions of the hopper the particles achieve high velocities. This in turn leads to a decrease of the porosity at t = 0.22 s (Figure 6, middle) due to the compaction at the outlet. Finally, as can be seen in (Figure 6 right), at t = 0.27 s the porosity slightly increases again until the outflow is getting stationary.

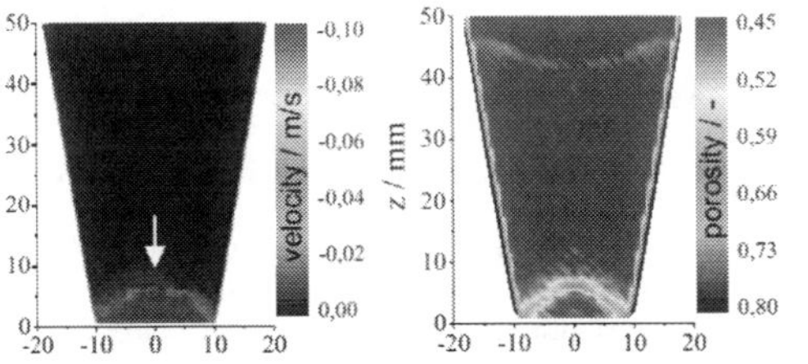

Fig. 5. Particle velocities (left) and the porosity (right)

The simulations also allow the determination of the velocity at the outflow as a function of the radial position of the particles (Fig. 7). The maximum of the velocity is reached in the centre of the hopper. With increasing radial position the velocity drops to smaller values. The velocity gradient is getting higher with decrasing distance to the wall at r = 10 mm. The curves for the different energy densities are qualitatively similar. However, with increasing energy density the values are shifted towards smaller velocities. At the high energy density k_3 in which b_{min} is in the range of b the averaged velocities are significantly lower then in the other cases.

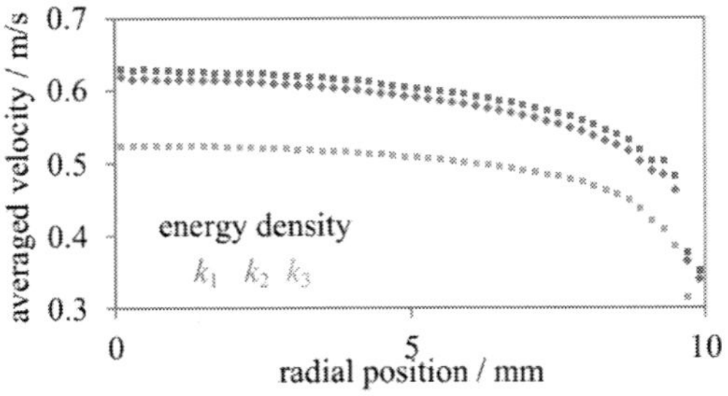

Fig. 7. Averaged particle velocity at the outflow over radial position for different energy densities in stanionary case

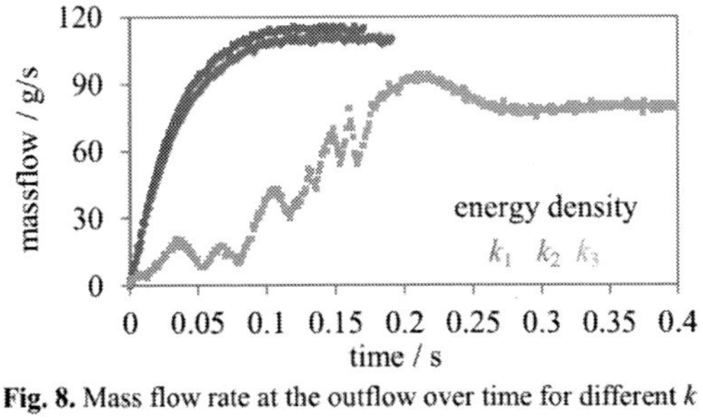

Fig. 8. Mass flow rate at the outflow over time for different k

Figure 8 shows a comparison of the mass flow rate at the outlet over time for different energy densities. It can be seen that the mass flow rate decreases with increasing values for the energy density. On the one hand this is a result of the decreasing outflow velocities, but on the other hand it is also caused by an increase in porosity.

4 Conclusions

In this work the discharge characteristics of a silo hopper which contains a cohesive bulk solids was investigated by DEM simulations. By varying the cohesive forces between the particles, the effect of the cohesion on the outflow was analysed. To model the contact force the Hertz-Mindlin model combined with a linear cohesion term was used. A comparison of the outflow velocities obtained in the simulations with velocities calculated with an empirical equation of TOMAS shows a qualitative agreement. Howeever, the simulated velocities are higher than the velocities of the empirical model. At a very high energy density in which b_{min} is in the range of b, the simulation showed that the outflow velocity does not increase steadily up to the stationary velocity. The sudden changes are caused by the breaking out of agglomerates from the bulk. Such phenomena can not be depicted with the equation of TOMAS. Furthermore, the simulation results showed, that the influence of the cohesion on the stationary massflow is bigger than on the stationary velocity because of the changing porosity.
In the next step the discharging of silos containing a highly cohesive bulk material will also be investigated in experiment. The velocities and mass flow rates which can be found in these experiments will also be compared. For this purpose, the flow properties of the bulk solid and particle contact properties were determined in shear tests using the methods described in [12].

References

1. J. Tomas, Habilitation Thesis, TU Freiberg (1991)
2. J. Tomas Chem. Technik **43**, 307-309 (1991)
3. P. A. Cundall, O. D. Strack, Geotechnique **29**, 47-65 (1979)
4. H. Hertz, Journal für die reine und angewandte Mathematik **92**, 156-171 (1882)
5. R. D. Mindlin, H. Deresiewicz, ASME, 327-344 (1953)
6. Y. Tsuji, T. Tanaka, T. Ishida, Powder Technol. **71**, 239-250 (1992)
7. R. Tykhoniuk et al., Chem. Eng. Sci. **66**, 2843-2864 (2007)
8. J. Tomas, Chem. Eng. Sci. **62**, 5925-5939 (2007)
9. Lord Rayleigh, Comptes rendus **159**, 109-110 (1913)
10. A. Kaskas, Diploma Thesis, TU Berlin (1964)
11. A.W. Jenike, Bulletin No. 123 of the Utah Engineering Experiment Station, 1970
12. S. Heinrich, M. Dosta, S. Antonyuk, Adv. Chem. Eng. **46**, 83-135 (2015)

DOI: 10.1051/epjconf/201714006019

Fluidization of spherocylindrical particles

Vinay V. Mahajan[1,2], *Tim* M.J. Nijssen[1], *Barry* W. Fitzgerald[1,2], *Jeroen* Hofman[1], *Hans* Kuipers[1], and *Johan* T. Padding[1,2,⋆]

[1] *Chemical Engineering department, TU Eindhoven, Netherlands*
[2] *Process and Energy, TU Delft, Netherlands*

Abstract. Multiphase (gas-solid) flows are encountered in numerous industrial applications such as pharmaceutical, food, agricultural processing and energy generation. A coupled computational fluid dynamics (CFD) and discrete element method (DEM) approach is a popular way to study such flows at a particle scale. However, most of these studies deal with spherical particles while in reality, the particles are rarely spherical. The particle shape can have significant effect on hydrodynamics in a fluidized bed. Moreover, most studies in literature use inaccurate drag laws because accurate laws are not readily available. The drag force acting on a non-spherical particle can vary considerably with particle shape, orientation with the flow, Reynolds number and packing fraction. In this work, the CFD-DEM approach is extended to model a laboratory scale fluidized bed of spherocylinder (rod-like) particles. These rod-like particles can be classified as Geldart D particles and have an aspect ratio of 4. Experiments are performed to study the particle flow behavior in a quasi-2D fluidized bed. Numerically obtained results for pressure drop and bed height are compared with experiments. The capability of CFD-DEM approach to efficiently describe the global bed dynamics for fluidized bed of rod-like particles is demonstrated.

1 Introduction

Non-spherical particles are frequently encountered in fluidisation processes such as in the production of biomass fuel in fluidised bed gasifiers [1]. Prior to fludisation, biomass particles are milled, resulting in particle shapes such as cylinders with an aspect ratio between 3 and 5. When fluidised, these particles experience anisotropic drag and lift conditions due to the surrounding fluid [2] and anisotropic interactions with adjacent particles and boundaries. High aspect ratio particles experience much larger drag torques in comparison to spherical particles of the same volume. The altered drag and collision characteristics directly affect the voids and heterogeneous structures, as well as mass and heat transfer rates. As such it is imperative to account for orientation, alignment and wall proximity effects in a description of particle collisions and the formulation of solid-fluid drag relations.

A number of established methods are available to simulate fluidised beds comprising of spherical particles [3]. At the macroscopic or industrial scale, the discrete bubble model (DBM) treats the gas or fluid phase in a Lagrangian manner while the solid particles are classified using an Eulerian approach [4]. At the microscopic scale, molecular dynamics (MD) accounts for both the particles and the fluid discretely, representing a Lagrangian-Lagrangian model. Between these limiting cases, other algorithms provide a more balanced interpretation of particle and gas/fluid dynamics. For example, the two-fluid model (TFM) approximates the gas/fluid and solid phases with generalised Navier-Stokes expressions and a description of solid phase rheological properties [3]. On the other hand, a coupled computational fluid dynamics and discrete element method (CFD-DEM) approach describes the gas phase as a continuum while the particles are represented as discrete entities [5, 6]. The CFD-DEM approach has been extensively applied to study particle scale flows for spherical particle beds. This is an idealised view, however, and in reality particles are non-spherical in nature. In this work we present a CFD-DEM model to study a laboratory scale fluidised bed consisting of spherocylinder or rod-like particles. We implement an accurate contact detection approach for spherocylinders while employing an appropriate drag coefficient expression for non-spherical particles [7]. We have also constructed an experimental apparatus to study laboratory scale fluidisation and compare experimental results on pressure drop and bed height with simulations.

2 Methods

2.1 Numerical Method

We have developed a CFD-DEM solver to simulate spherocylinder particles in fluidised beds that couples an OpenFOAM CFD solver and the DEM solver LIGGGHTS. The CFD-DEM model is developed on the Open Source coupling engine CFDEM, which allows for program execution for a user-prescribed number of time steps after which

⋆e-mail: j.t.padding@tudelft.nl

data is exchanged between the OpenFOAM fluid solver and the LIGGGHTS particle solver.

2.1.1 CFD

The fluid phase is described in an Eulerian manner by imposing a mesh of equal sized cells on the fluidised bed. The fluid velocity is spatially averaged at each cell. The motion of the fluid phase is calculated by solving the continuity equation (Eq. (1)) and the momentum conservation expression (Eq. (2))

$$\frac{\partial(\epsilon_f \rho_f)}{\partial t} + \nabla \cdot (\epsilon_f \rho_f \mathbf{u}_f) = 0 \quad (1)$$

$$\frac{\partial(\epsilon_f \rho_f \mathbf{u}_f)}{\partial t} + \nabla\cdot(\epsilon_f \rho_f \mathbf{u}_f \mathbf{u}_f) = -\epsilon_f \nabla p + \nabla\cdot(\epsilon_f \tau_f) + \mathbf{R}_{f,p} + \epsilon_f \rho_f \mathbf{g} \quad (2)$$

where ϵ_f is fluid volume fraction, ρ_f is fluid density, $\mathbf{u}_f$ is fluid velocity, τ_f is the stress tensor, $\mathbf{R}_{f,p}$ represents the momentum exchange between the fluid and particle phases and $\mathbf{g}$ is gravity.

As an exact drag force coefficient for the spherocylinders is unavailable, as an approximation, we use the drag force correlation formula for non-spherical particles from Hölzer and Sommerfeld [7]

$$C_D = \frac{8}{Re}\frac{1}{\sqrt{\phi_\perp}} + \frac{16}{Re}\frac{1}{\sqrt{\phi}} + \frac{3}{\sqrt{Re}}\frac{1}{\phi^{3/4}} + 0.4210^{0.4(-log\phi)^{0.2}}\frac{1}{\phi_\perp} \quad (3)$$

where Re, ϕ and $\phi_\perp$ are the Reynolds number, sphericity and crosswise sphericity respectively. To account for the effects of surrounding particles, the drag force scheme of Di Felice [8] is used, which is expressed as

$$\boldsymbol{F}_{pf,i} = \frac{1}{2} C_D \rho_f A_\perp \epsilon_f^{1-\chi} \left|\boldsymbol{u}_f - \boldsymbol{v}_i\right| (\boldsymbol{u}_f - \boldsymbol{v}_i) \quad (4)$$

where C_D is the drag coefficient (Eq. (3)), $A_\perp$ is the cross-sectional area perpendicular to the flow, χ is a correction factor accounting for crowding effects, $\boldsymbol{u}_f$ is the gas velocity interpolated to the location of particle i, and $\boldsymbol{v}_i$ is the velocity of particle i. This expression is commonly used for spherical particles but has recently been applied to non-spherical particles [9]. The correction factor χ is calculated as a function of the Reynolds number $Re = \epsilon_f \rho_f d_e \left|\boldsymbol{u}_f - \boldsymbol{v}_i\right| / \eta_f$ using the expression $\chi = 3.7 - 0.65\exp(-(1.5 - \log(Re))^2/2)$ where d_e is the equivalent diameter and η_f is the fluid viscosity. A pressure-based solver using "Pressure Implicit Split Operator (PISO)" pressure-velocity coupling is used to solve the fluid equations, where an implicit momentum predictor followed by a series of pressure solutions and explicit velocity corrections is performed.

2.1.2 DEM

For the DEM component, spherocylinders are depicted as particles with rotational and translational degrees of freedom. Particle trajectories are integrated using a Velocity Verlet method based on local contact forces and torques. The motion of a particle p can be calculated from Newton's second law of motion with the expression

$$m_p \frac{d\mathbf{v}_p}{dt} = \mathbf{F}_{p,n} + \mathbf{F}_{p,t} + \mathbf{F}_{p,f} + \mathbf{F}_{p,p} + \mathbf{F}_{p,b} \quad (5)$$

where $\boldsymbol{F}_{p,n}$ is the normal contact force, $\boldsymbol{F}_{p,t}$ is the tangential contact force, $\boldsymbol{F}_{p,f}$ is the drag force exerted by the fluid phase to the particles, $\boldsymbol{F}_{p,p}$ denotes the pressure gradient (or buoyancy) acting on the particles. Gravity is incorporated in the body force $\boldsymbol{F}_{p,b}$. The angular momentum on particle p is calculated with $\mathbf{I}_p \frac{d\omega_p}{dt} = \mathbf{T}_p$ where $\mathbf{I}_p$, ω_p and $\mathbf{T}_p$ are the moment of inertia, angular velocity and torque for particle p. Particle orientation is described by quaternions.

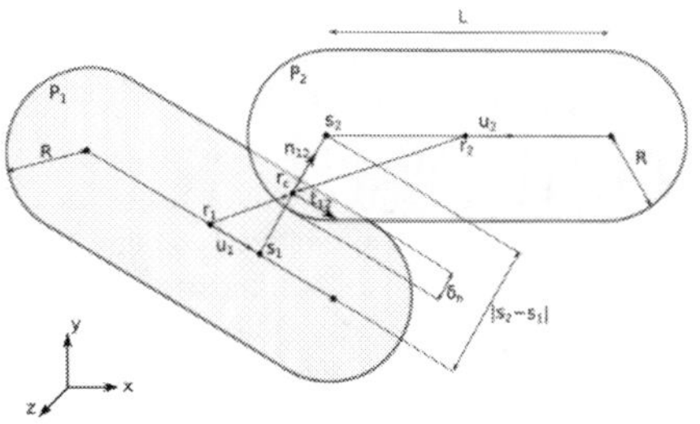

Figure 1: A contact between two spherocylinders.

The detection of contacts between spherocylinder particles is more involved than for the case of spherical particle contact detection. Figure 1 shows two spherocylindrical particles P_1 and P_2 experiencing a contact. For particle i, R is the characteristic radius, $\mathbf{r}_i$ is the centre of mass, L is the shaft length, $\mathbf{u}_i$ is the unit orientation vector, $\mathbf{v}_i$ is the translational velocity and ω_i is the angular velocity. The shortest distance between the particles is $|\mathbf{s}_2 - \mathbf{s}_1|$, where $\mathbf{s}_2$ and $\mathbf{s}_1$ are points on the central axes of the respective particles. $\mathbf{r}_c$ is the mid-point between $\mathbf{s}_2$ and $\mathbf{s}_1$ and δ_n is the degree of overlap between the particles with $\mathbf{t}_{12}$ the tangential unit vector and $\mathbf{n}_{12}$ the normal unit vector. The normal contact force exerted on P_1 by P_2 is approximated as a linear dashpot given by

$$\boldsymbol{F}_{12,n} = -k_n \delta_n \boldsymbol{n}_{12} + \eta_n \boldsymbol{v}_{12,n} \quad (6)$$

where k_n is the normal elastic constant, η_n is the normal damping coefficient and $\mathbf{v}_{12,n}$ is the normal relative velocity. A Coulomb-type friction law is used for the tangential forces such that

$$\boldsymbol{F}_{12,t} = min(-k_t|\boldsymbol{\delta}_t|\boldsymbol{t}_{12} + \eta_t \boldsymbol{v}_{12,t}, -\mu \left|\boldsymbol{F}_{12,n}\right| \boldsymbol{t}_{12}) \quad (7)$$

where k_t is the tangential elastic constant, $\boldsymbol{\delta}_t$ is the tangential overlap, which is defined as the time integral of the tangential relative velocity since initial particle contact, η_t is the tangential damping coefficient, μ is the friction coefficient and $\boldsymbol{v}_{12,t}$ is the tangential relative velocity.

In simulations, the CFD time step $\Delta t_{CFD} = 1 \times 10^{-4}$ s, the CFD cell size is 0.015 m, the DEM time step $\Delta t_{DEM} = 0.1\Delta t_{CFD}$, $\mu = 0.3$ and the number of particles is 32,400.

It is crucial that the fluid/solid void fraction is accurately calculated in the CDF-DEM model due to its strong influence on pressure drop in the particle bed. To enable calculation of the solid fraction contribution of any spherocylinder particle to grid cells, each particle is embedded with n_{sp} evenly spaced satellite points throughout its volume. Each satellite point within a particle carries an equal volume weight. When the parent grid cell for a particle is identified, the particle volume is assigned to the parent cell or neighbouring cells subject to the position of the satellite points. This method is flexible and easily optimised through variation of n_{sp}.

2.2 Experimental

2.2.1 Setup

The focus of this study is particle behaviour near walls. Therefore we have constructed a quasi-2D fluidised bed setup of width 0.3 m, depth 0.05 m and height 1 m. The side and back walls are made of anodized aluminium while the front wall is made of 10 mm thick tempered glass. The distributor plate is made of sintered stainless steel with a pore size of 100 microns. A differential pressure sensor NXP MPX5050DP with a range from 0 to 50 kPa (0 to 72.5 psi) and 0.2 to 4.7 V output is installed just above the distributor plate to measure the pressure drop.

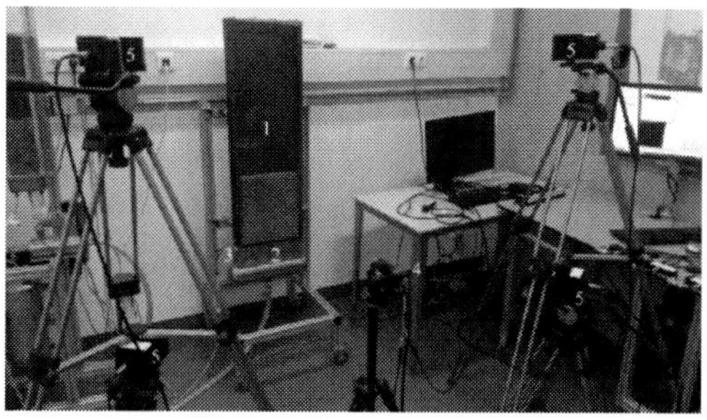

Figure 2: Snapshot of the experimental setup. (1) Fluidised bed, (2) Gas distributor, (3) Pressure sensor, (4) Camera, (5) Lights.

A Dantec Dynamics FlowSense EO 16M camera is used to capture images for Digital Image Analysis (DIA). The camera has a resolution of 4872 x 3248 pixels, 4.2 frames/s at full resolution, inter-frame time of 300 ns and pixel size of 7.4 μm. Four LED lights, arranged in a 2 by 2 grid were used to illuminate the bed. The setup is shown in Fig. 2. During experiments the density of air is 1.1 kg/m^3 while the viscosity is 1.88 x 10^{-5} Pa s.

2.2.2 Particles

The spherocylinder particles have been 3D printed using the selective laser sintering technique, a process that is cheap and highly precise. Made from a composite of aluminium and nylon dust, the particles have high strength and low density. Each particle has a diameter of 3 mm and length of 12 mm and have a density of 1395 kg/m^3. Thus they can be classified as Geldart D type particles Prior to the experiments, the particles were tested for attrition. At rest the bed height is approximately 0.3 m. The total weight of the bed in experiment and simulation is the same.

3 Results

Pressure Drop and Bed Height

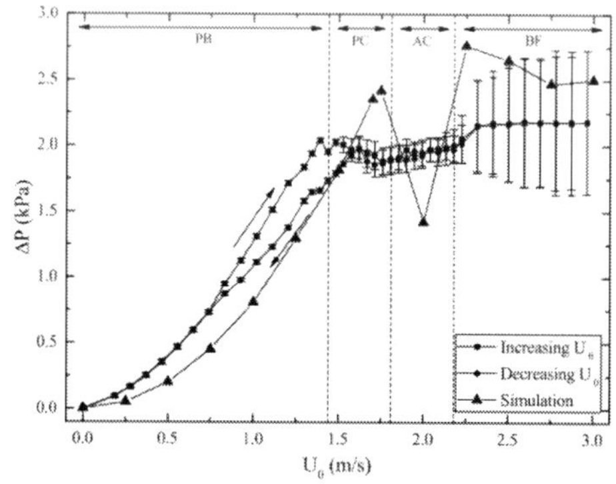

Figure 3: Variation of pressure drop ΔP with fluid velocity U_0. Error bars represent the standard deviation.

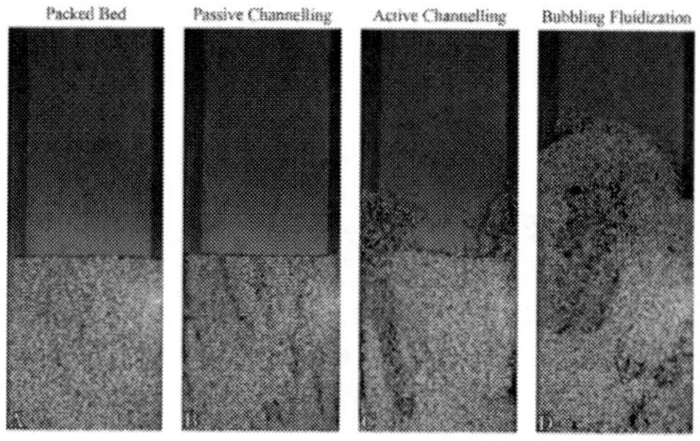

Figure 4: Examples of the four flow regimes observed in experiments (as indicated in Fig. 3).

Fig. 3 shows the pressure drop ΔP for varying fluid velocity U_0 from experiments and simulations. In the experiments, pressure measurements were made at 5 m^3/h intervals in gas flowrate by sampling at 100 Hz for 100 seconds for increasing and decreasing flowrates as indicated by the arrows and the trends in Fig. 3. Four flow regimes have been identified, as indicated on Fig. 3; namely packed bed (PB), passive channeling (PC), active channeling (AC) and bubbling fluidisation (BF). Snapshots of the bed in each regime are shown in Fig. 4. In the case of spherical particles with equivalent particle volume, the channeling regimes are not observed. This behaviour is specific

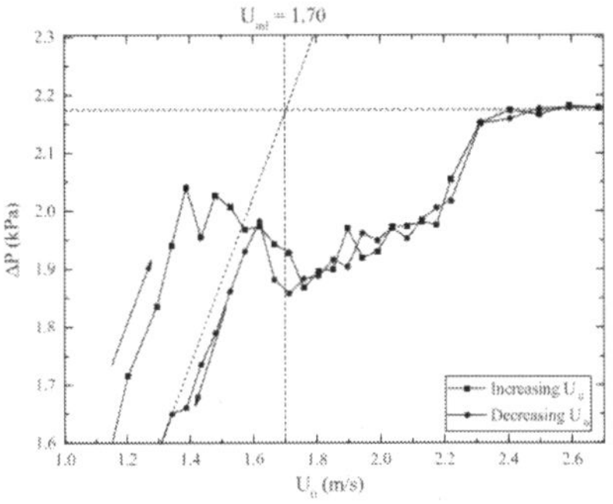

Figure 5: Close-up of pressure drop curve showing calculation of U_{mf} = 1.70 m/s from experiments.

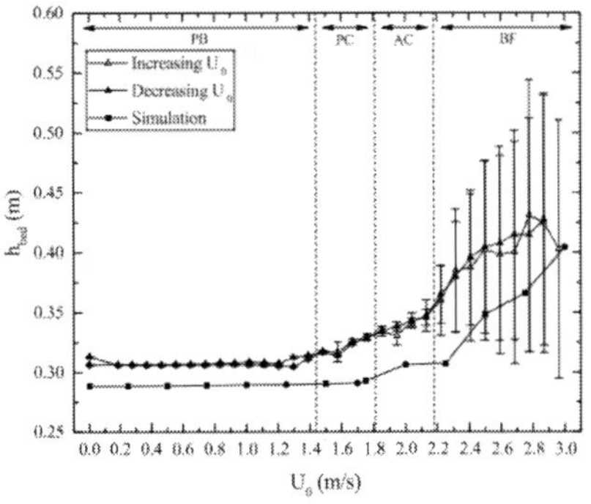

Figure 6: Variation of bed height (h_{bed}) with U_0 from experiments and simulations.

to extreme Geldart D particles i.e. with spherical diameter ≥ 3 mm. Notably the variation of ΔP here differs from spherical particles , where the pressure drop curve is smoother, since we observe the development of an initial plateau in the channeling regime followed by a step up to a second plateau in the BF regime. Our observation is in qualitative agreement with previous studies on rod-like particles [10]. In addition we observe a hysteretic response in ΔP. We estimate the minimum fluidisation velocity U_{mf} = 1.70 m/s, which marks the intersection of the maximum pressure drop or plateau in the BF regime with the ΔP curve for decreasing U_0 in the PC regime (Fig. 5).

Fig. 6 shows the variation of the bed height h_{bed} with U_0 from experiments and simulations. The bed height is extracted from experimental images using a maximum gradient routine implemented using MATLAB and the Image Processing Toolbox^{TM}. Over all regimes we find that h_{bed} calculated from simulations is less than the experimental value. This difference in the trends might be attributed to a lower value of μ that is used in the simulations in comparison to the coefficient of friction of the aluminium/nylon composite particles in experiments.

4 Conclusions and Outlook

We have presented results on the variation of pressure drop and bed height from CFD-DEM simulations and experiments of the fluidisation of spherocylinder particles. With varying flow velocity U_0, we have identified four distinct flow regimes. For low U_0, we have a packed bed regime while for high U_0 we observe the onset of intense bubbling dynamics. For intermediate flow velocity, we observe passive channelling, which is succeeded by active channelling with increasing U_0 (Fig. 4). Simulation results recover to good agreement both the variation in pressure drop and bed height from experiments.

As part of the future work, we will include drag force closures specific to spherocylinder particles. These closures can be defined using DNS simulations for particles oriented at different angles, under the effect of walls, at different Reynolds number and particle volume fraction. We will determine the exact coefficient of friction and restitution for the particles by experimentally studying the interaction of a single particle with differing surfaces made from the particle composite material, glass and aluminium. Given its importance for calculation of the pressure drop, we will also revise the approach for estimating the fluid/solid void fraction. For comparative purposes, we will also study fluidized spherical particles both experimentally and numerically.

Acknowledgements

We thank the European Research Council (ERC) for financial support, under its Consolidator Grant scheme, contract no. 615906 (nonSphereFlow). This work was carried out on the Dutch national e-infrastructure with the support of SURF Cooperative.

References

[1] H. Cui, J.R. Grace, Chem. Eng. Sci. **62**, 45 (2007)

[2] M. Zastawny, G. Mallouppas, F. Zhao, B. van Wachem, International Journal of Multiphase Flow **39**, 227 (2012)

[3] M. van der Hoef, M. van Sint Annaland, N. Deen, J. Kuipers, Annu. Rev. Fluid Mech. **40**, 47 (2008)

[4] D. Darmana, N. Deen, J. Kuipers, Chem. Eng. Sci. **60**, 3383 (2005)

[5] N. Deen, M.V.S. Annaland, M.V. der Hoef, J. Kuipers, Chem. Eng. Sci. **62**, 28 (2007)

[6] H. Zhu, Z. Zhou, R. Yang, A. Yu, Chem. Eng. Sci. **63**, 5728 (2008)

[7] A. Hölzer, M. Sommerfeld, Powder Technology **184**, 361 (2008)

[8] R.D. Felice, Int. J. Multiphase Flow **20**, 153 (1994)

[9] T. Oschmann, K. Vollmari, H. Kruggel-Emden, S. Wirtz, Procedia Engineering **102**, 976 (2015)

[10] K. Vollmari, R. Jasevicius, H. Kruggel-Emden, Powder Technology **291**, 506 (2016)

DOI: 10.1051/epjconf/201714003065

Effect of cohesion on local compaction and granulation of sheared soft granular materials

Sudeshna Roy[1,*], *Stefan* Luding[1,**], and *Thomas* Weinhart[1,***]

[1] *Multi Scale Mechanics (MSM), MESA+, CTW, University of Twente, PO Box 217, 7500 AE Enschede, The Netherlands*

Abstract. This paper results from an ongoing investigation of the effect of cohesion on the compaction of sheared soft wet granular materials. We compare dry non-cohesive and wet moderately-to-strongly cohesive soft almost frictionless granular materials and report the effect of cohesion between the grains on the local volume fraction. We study this in a three dimensional, unconfined, slowly sheared split-bottom ring shear cell, where materials while sheared are subject to compression under the confining weight of the material above. Our results show that inter-particle cohesion has a considerable impact on the compaction of soft materials. Cohesion causes additional stresses, due to capillary forces between particles, leading to an increase in volume fraction due to higher compaction. This effect is not visible in a system of infinitely stiff particles. In addition, acting oppositely, we observe a general decrease in volume fraction due to increased cohesion for frictional particle, which we attribute to the role of contact friction that enhances dilation.

1 Introduction

Unsaturated granular media of particles with interstitial liquid in the form of bridges between particle pair, display bulk cohesion, which can be tuned using different liquids with varying surface tension. Earlier studies have been on density correlation in terms of fractal structure of aggregates for dry cohesive powders [1]. Cohesive grains being sensitive to stress intensity as well as direction exhibit much larger variations in their equilibrium densities. Moreover, adhesion would enhance the role of sliding and rolling friction, because the limiting values for tangential contact forces and rolling moments are both proportional to the elastic repulsive part of the normal force. Fournier et al. [2] observed that wet granulates are significantly less densely packed than dry granular materials, but the packing densities in only weakly dependent on the amount of wetting liquid, an obvious reason being that the forces exerted by the liquid bridges are very weakly dependent on bridge volume [3]. At small liquid content and after sufficient equilibration, the interior of the wet granulate is expected to be characterized by a network of liquid bridges connecting adjacent grains. It is clear that the connectivity of this network of liquid bridges is of importance for the mechanics of the wet granular materials, be it directly due to the capillary forces itself or due to the enhancement of the mutual friction between the grains by the increased internal pressure. For the wet granular materials, this pressure increase is of the order $\Delta p \approx \sigma/r$ given by the Laplace-Young equation, where r is a typical radius of the grains and σ is the surface tension of liquid. The local volume fraction of the bulk on macro-scale is connected to the pressure gradient and is thus proportional to σ.

We study here the packing fraction in the critical state for non-cohesive to strongly cohesive systems by varying the surface tension of the liquid. Details of the simulation set-up is given in Sec. 2.1. Wet granular materials are cohesive and particles can stick together and form local agglomerates or granules, due to formation of clusters of particles for very cohesive systems, as shown in figure 1 and 4(c). Figure 1 shows the top view of the split bottom shear cell geometry with colors blue to red indicating z coordinate values of the particles. It is observed from the figure that some particles form clumps near the surface while the particles near the base are seen through the hollow region. Hence, the system is highly inhomogeneous in its spatial distribution.

2 Model System

2.1 Geometry

Split-Bottom Ring Shear Cell: We use MercuryDPM [4, 5], an open-source implementation of the Discrete Particle Method, to simulate a shear cell with annular geometry and a split bottom plate, as explained in [6]. Earlier studies used similar rotating set-ups, including [7, 8]. The geometry of the system consists of an outer cylinder (radius R_o = 110 mm) rotating around a fixed inner cylinder (radius R_i = 14.7 mm) with a rotation frequency of $\Omega = 2\pi f$ and f = 0.01 revolutions per second. The granular material is confined by gravity between the two concentric cylinders and the bottom plate, with a free top surface.

*e-mail: s.roy@utwente.nl
**e-mail: s.luding@utwente.nl
***e-mail: t.weinhart@utwente.nl

The bottom plate is split at radius R_s = 85 mm. Due to the split at the bottom, a narrow shear band is formed. It moves inwards and widens towards the flow surface. This set-up thus features a wide shear band away from the bottom and the wall which is free from boundary effects. The shear cell is filled up to a height of $H \approx 40$ mm under dry conditions, with particles of mean diameter $d_p = \langle d \rangle \approx 2.2$ mm, polydispersity $d_{\max}/d_{\min} = 2.0$ and gaussian distribution with width of $1 - \langle d \rangle / \langle d^2 \rangle = 0.0357$, where a is the particle radius. Thus, the shear band remains far away from the inner wall. The inter-particle friction coefficient is $\mu_p = 0.01$ and other parameters are detailed in [6].

In earlier studies [9], a quarter of this system ($0° \leq \phi \leq 90°$) was simulated using periodic boundary conditions. In order to save computation time, here we simulate only a smaller section of the system ($0° \leq \phi \leq 30°$) with appropriate periodic boundary conditions in the angular coordinate, unless specified otherwise. We have observed no noticeable effect on the macroscopic behavior in comparisons between simulations done with a smaller (30°) or larger (90°) opening angle. Note that for very strong attractive forces, the system becomes inhomogeneous and looses its radial symmetry: i.e. agglomeration of particles occurs. Then, particles interact on a larger length scale and thus the above statement is not true anymore.

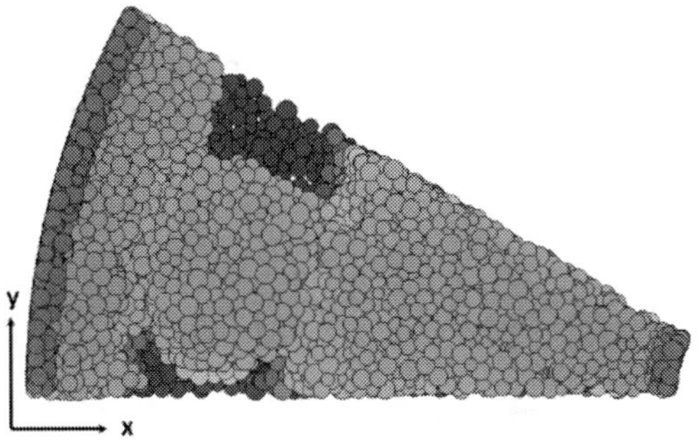

Figure 1: Top view of highly cohesive wet granular materials (Bo_g = 34.6). Different colors are the measure of low (blue), medium (green), high (yellow) and walls (red) z-coordinate of the particles.

2.2 Contact model and parameters

We use a linear visco-elastic frictional contact model in combination with Willet's capillary bridge model [3, 6, 10]. In order to see the effect of varying cohesive strength on the macroscopic rheology of wet materials, we vary the intensity of the maximum capillary force $f_{\rm cap}^{\max} = \pi d\sigma \cos\theta$, by varying the surface tension of the liquid σ, while keeping the volume of liquid bridges constant, (V_b = 75 nl), corresponding to a liquid saturation of 8% of the voidage. We compare here the volume fractions of non-cohesive to moderate to strongly cohesive granular materials, with surface tension of liquid ranging from $\sigma = 0$ Nm^{-1}, up to $\sigma = 5$ Nm^{-1} for strongly cohesive systems. The contact angle is fixed at $\theta = 20°$.

3 Dimensionless numbers

The effects of varying shear rate, pressure, stiffness and cohesion can be modelled using three dimensionless numbers, expressed as a ratio of time-scales as given in Tab. 1, where the subscripts $\dot\gamma, p, k$ and c denote strain-rate, pressure, stiffness and cohesion respectively [10]. In addition, we define the *global* Bond number as $Bo_g = f_{\rm cap}^{\max}/(p_{\rm mean} d_p^{\ 2})$, where $p_{\rm mean}$ is the mean pressure in the system (at about half filling height $H/2$).

Table 1: Dimensionless numbers for the model

Dimensionless number	Definition	Time scale ratios
Inertial number I	$\frac{\dot\gamma d_p}{\sqrt{p/\rho}}$	$t_p/t_{\dot\gamma}$
Softness p^*	$\frac{p d_p}{k}$	$(t_k/t_p)^2$
Local Bond number Bo	$\frac{f_{\rm cap}^{\max}}{p d_p^{\ 2}}$	$(t_p/t_c)^2$

4 Rheological model

The macroscopic quantities are obtained by spatial coarse graining with temporal averaging of the system in steady state as detailed in [6, 10]. We study the effect of the above mentioned dimensionless numbers on the local volume fraction ϕ in the critical state [10]. The local volume fraction is expected to be dependent on various factors like the dilation, compression and the inter-particle friction. In the following sections, we discuss more of the effect of each dimensionless number on the local volume fraction.

4.1 Non-cohesive granular materials

For dry granular materials, $Bo = 0$, the rheology only depends on p^* and I. The dependence of the macroscopic friction coefficient $\mu = \tau/p$ on p^* and I has been studied in [9, 10]. In order to complete the rheology for soft, compressible particles, a relation for the solid volume fraction (packing fraction) as function of pressure and shear rate is missing for dry non-cohesive materials. In [11], the following dependency was observed:

$$\phi(I, p^*) = \phi_o f_I(I) f_p(p^*) \qquad (1)$$

with the critical or steady state density under shear, first order i.e., in the limit of vanishing pressure and inertial number, $\phi_o = 0.64$, $f_p(p^*) = (1 + p^*/p_o^*)$, $f_I(I) = (1 - I/I_o)$. The typical strain rate for which dilation would turn to fluidization is $I_o = 0.85$, and the typical pressure level for which softness leads to huge densities is $p_o^* = 0.33$ [11]. Note that both correction functions are first order, i.e. they are valid only for sufficiently small arguments. Because of slow quasistatic flows in our simulations, no strong dilation is observed, i.e., no strong dependence of ϕ on the local shear rate. On the other hand, too large inertial numbers would fully fluidize the system so that the rheology should be that of a granular fluid, for which kinetic theory applies, while too large pressure would lead to enormous overlaps, for which the contact model and the particle simulation with pair forces become questionable.

 DOI: 10.1051/epjconf/201714003065

4.2 Cohesive granular materials

In cohesive flows, attractive forces enhance the local stresses acting on the particles and adds an attractive force which acts as an effective confinement and increased compression, i.e. increased volume fraction (at least for our nearly-frictionless spheres). Rough and frictional particles should display stronger dilatancy, a reduced volume fraction under shear. Having low inter-particle friction, we observe an overall increase in local volume fraction with Bo (plot not shown here). This overall effect is distinguished as contribution from compression of soft particles and due to structural changes in presence of friction as discussed in the following subsections.

4.2.1 Effect of cohesion on soft particles

In cohesive flows, cohesion enhances the compressive pressure acting on the particles. This can be quantified as follows: we split the net pressure into two components, $p = p_{\text{rep}} - p_{\text{cap}}$, denoting the respective contributions of repulsive and cohesive contact forces. The ratio between the total cohesive contribution and the total pressure is given by the local bond number, $Bo = p_{\text{cap}}/p$, and thus $p_{\text{rep}} = (1 + Bo)p$. As the geometrical compression (deformation at each contact) is related to the *repulsive* stress, it is the compressive pressure p_{rep} that has to be considered in the softness factor f_p. Thus, the modified local softness correction for cohesive systems is given as:

$$f_p(p^*_{\text{rep}}) = f_p((1 + Bo)p^*) \tag{2}$$

For dry non-cohesive system when $Bo = 0$, one has $f_p((1 + Bo)p^*) = f_p(p^*)$. This is similar in spirit with the modified inertial number as presented in [12], which takes into account the cohesive contribution in effective stress. A similar modification in the inertial number is also required in the inertial regime weakly effective in the quasistatic regime.

4.2.2 Generalized effect of cohesion

In addition to the effect of cohesion on the softness of particles, cohesion also changes the microstructure [1, 2] in presence of rolling and sliding friction. To study this additional effect, we analyse the local packing fraction ϕ scaled by f_I and f_p as a function of the local Bond number Bo as shown in figure 2. All the data shown in figure 2 correspond to the critical state, though a shear band is not clearly defined in strongly cohesive systems ($Bo_g > 3.46$). It is observed from figure 2 that all data for different Bo_g collapse, following a trend and the dependence is given by the solid line in figure 2 as:

$$f_c(Bo) = \left[1 - \left(\frac{Bo}{Bo_c}\right)^{\alpha}\right] \tag{3}$$

where, $\phi_o = 0.65, \alpha = 0.57$ and $Bo_c = 729$ denotes the limiting Bond number above which the correction is not applicable anymore. The dash-dotted line represents a linear fit $f_c^{\text{lin}}(Bo)$ that ignores the larger Bo data, with $\alpha = 1$. Note that due to different friction $\mu_p = 0.5$, $\phi_o = 0.60, \alpha = 0.71$ and $Bo_c = 42.63$ for the dashed fit in figure 2 corresponding to the data from [13] for frictional monodisperse particles.

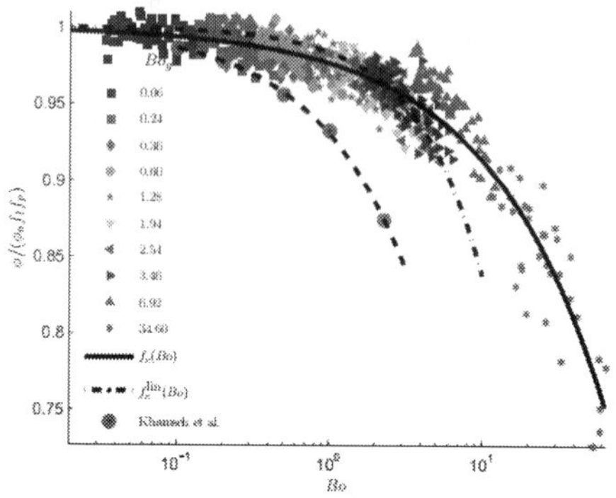

Figure 2: Scaled local packing fraction $\phi/(\phi_o f_I f_p)$ as a function of the local Bond number Bo showing the general decrease in scaled local volume fraction with cohesion. The solid line is given by Eq. (3) with parameters given in Tab. 2. The dash-dotted line corresponds to a simple linear fit $f_c^{\text{lin}}(Bo) = 1 - Bo/62$ corresponding to the data with $Bo < 3$. The dashed line corresponds to the fit for the data from [13].

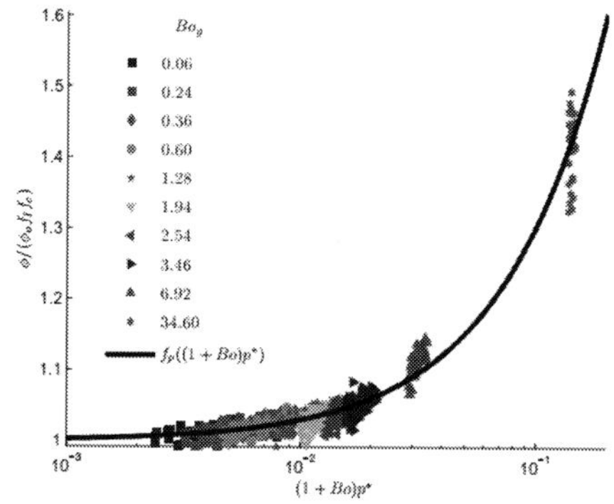

Figure 3: Scaled local packing fraction $\phi/(\phi_o f_I f_c)$ as a function of $(1 + Bo)p^*$ showing the effect of compaction of soft particles with increasing cohesion.

To confirm the effect of cohesion on the compressibility of soft particles, we re-plot figure 2 in another possible way, i.e., $\phi/(\phi_o f_I f_c)$ as a function of $(1 + Bo)p^*$ as shown in figure 3. In a way, this is the effect of compression of soft particles with increasing cohesion which is given by the function $f_p((1 + Bo)p^*)$ as shown by the solid line in the figure.

Thus the complete rheology for the local volume fraction is given as:

$$\phi(I, p^*, Bo) = \phi_o f_I(I) f_p((1 + Bo)p^*) f_c(Bo) \tag{4}$$

In the case of rigid particles, $p^* \to 0$ so that Eq. 4 reduces to $\phi_{\text{stiff}} = \phi(I, 0, Bo) = \phi_o f_I f_c$.

DOI: 10.1051/epjconf/201714003065

Table 2: Coefficients for the model

Dimensionless number	Corrections	Coefficients
Critical volume fraction (ϕ_o)		$\phi_o = 0.65$
Inertial number (I)	$f_I = \left(1 - \frac{I}{I_o}\right)$	$I_o = 0.85$
Softness (p^*_{rep})	$f_p = \left[1 + \frac{p^*_{rep}}{p_o^*}\right]$	$p_o{}^* = 0.33$
Cohesion (Bo)	$f_c = \left[1 - \left(\frac{Bo}{Bo_c}\right)^{\alpha}\right]$	See Sec. 4.2.2.

4.2.3 Discussion

Cohesion can either contribute to a decrease or an increase in the local volume fraction of sheared materials, depending on the inter-particle friction and the softness of materials. Note that compression is prevailing for soft particles but is negligible when $p^* > 0$ in the limit of infinite stiffness, when the local volume fraction is expected to increase with Bo. Berger et al. [12] uses contact dynamics for simulation which assumes that particles are infinitely stiff. Khamseh et al. [13] also shows that the volume fraction decreases for higher cohesion where the simulations are done using DEM with very stiff glass beads of Young modulus $E = 70$ GPa, so that for both, the softness effect is negligible.

Figure 4 (a), (b) and (c) show the contour plot of the spatial distribution of local packing fraction with the magnitude given by the color map for different $Bo_g = 0, 1.94$ and 34.6, respectively. Focusing on the shear band center, the mean volume fraction is close to 0.65 for non-cohesive materials as shown in figure 4(a). In comparison, the mean volume fraction of the strongly cohesive materials is 0.75 in figure 4(c). The vertical center of mass of the materials decreases by 25% from non-cohesive (a) to strongly cohesive (c) materials.

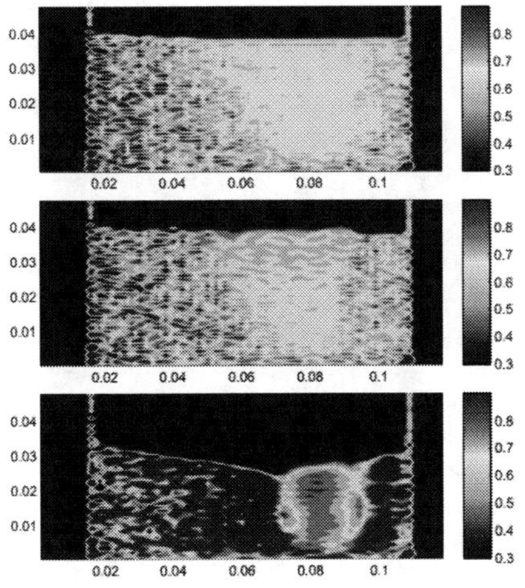

Figure 4: Contour plot of volume fraction for (a) $Bo_g = 0.0$, (b) $Bo_g = 1.94$ and (c) $Bo_g = 34.6$ in the $r - z$ plane. Both (a) and (b) are homogeneous in cylindrical direction, while (c) displays granule formation as shown in figure 1.

5 Conclusion

We studied the local packing fraction of dry and wet granular materials as a function of dimensionless numbers, namely, the inertial number I, the softness p^* and the Bond number Bo. Focus is on the effect of cohesion (quantified by the Bond number). Earlier studies have shown that the packing fraction of dry granular materials is to the first order linearly dependent on I and p^*.

We observe that the cohesive stress contributes to the softness for wet granular materials. This leads to an additional linear increment of local volume fraction with Bo. This effect is prevailing for soft particles and becomes negligible in the limits of stiff particles. Additionally, the local volume fraction decreases as a power law dependence on the local Bond number Bo. This is a more general effect of Bo on the local volume fraction related to the structural changes and increasing friction force of the materials.

Our results show that for strongly cohesive systems, our almost frictionless materials are overall densely compacted and center of mass of the bed drops by approximately 25%. The coordination number increases with local volume fraction. However, the trends are not clear enough and to have a complete understanding of the rheology, the dependence of the coordination number and the granular temperature on p^*, I and Bo should be studied.

Acknowledgements

We acknowledge our financial support through Technologiestichting STW Project 12272.

References

[1] F.A. Gilabert, J.N. Roux, A. Castellanos, Phys. Rev. E **75**, 11303 (2007)

[2] Z. Fournier, D. Geromichalos, S. Herminghaus, M.M. Kohonen, F. Mugele, M. Scheel, M. Schulz, B. Schulz, C. Schier, R. Seemann et al., J. Phys.: Condens. Matter **17**, S477 (2005)

[3] C.D. Willett, M.J. Adams, S.A. Johnson, J. Seville, Langmuir **16**, 9396 (2000)

[4] A. Thornton, T. Weinhart, S. Luding, O. Bokhove, Int. J. Mod. Phys. C **23, 124001** (2012)

[5] T. Weinhart, A. Thornton, S. Luding, O. Bokhove, Granul. Matter. **14**, 289 (2012)

[6] S. Roy, A. Singh, S. Luding, T. Weinhart, Computational Particle Mechanics **3**, 449–462 (2016)

[7] X. Wang, H.P. Zhu, A.B. Yu, Granular Matter **14**, 411 (2012)

[8] E. Woldhuis, B.P. Tighe, W. Saarloos, The European Physical Journal E: Soft Matter and Biological Physics **28**, 73 (2009)

[9] A. Singh, V. Magnanimo, K. Saitoh, S. Luding, New J. Phys. **17**, 043028 (2015)

[10] S. Roy, S. Luding, T. Weinhart, New Journal of Physics (accepted) (2017)

[11] S. Luding, A. Singh, S. Roy, D. Vescovi, T. Weinhart, V. Magnanimo, The 7th International Conference on Discrete Element Methods. (2016)

[12] N. Berger, E. Azéma, J.F. Douce, F. Radjai, EPL (Europhysics Letters) **112**, 64004 (2016)

[13] S. Khamseh, J.N. Roux, F. Chevoir, Physical Review E **92**, 022201 (2015)

Time-resolved granular dynamics of a rotating drum in a slumping regime as revealed by speckle visibility spectroscopy

V. Zivkovic[1]*, *H. Yang*[2], *G. Zheng*[2], *M. Biggs*[3]

[1]School of Chemical Engineering and Advanced Materials, Newcastle University, NE1 7RU, United Kingdom
[2]a. School of Optical-Electrical and Computer Engineering, b. school of medical instrument and food engineering, c. Shanghai Key Lab of Modern Optical System, d. Engineering Research Center of Optical Instrument and System, Ministry of Education, University of Shanghai for Science and Technology, Shanghai, 200093, China
[3]School of Science, Loughborough University, LE11 3TU, United Kingdom

Abstract. Granular materials in rotating drums are of wide interest not only because of their extensive use in the industrial contexts, but also as model systems in the study of natural disasters, such as avalanches or landslides. Most of available experimental methods are restricted to surface layer flows and dilute systems whilst the remainder can only resolve the granular dynamics to a fine scale with relatively poor temporal resolution or vice versa. In contrast, speckle visibility spectroscopy (SVS) is able to resolve the average of the three components of motion of grains in dense systems in small volume of granular media several layer deep with spatio-temporal resolutions that allow the probing of the granular micro-dynamics. We have used this technique to study granular dynamics of surface avalanche flow in the slumping regime using both spherical glass and irregular sand particles. Although results are very similar, we determined that visually observed compaction at the beginning of avalanche process for irregular sand particles influence time evolution of the particle fluctuation velocity during avalanches.

1 Introduction

Granular flows in rotating drums are of wide interest not only as model systems in the study of the physics of granular media, but also because of their extensive use in the chemical, minerals, pharmaceutical and food industry where they are employed to effect process as diverse as mixing and granulation. It is well known that as the drum rotational speed increases, the granular flow takes on one of a number of regimes [1], including slumping, which occurs at lower rotational speeds with discrete avalanches and is of wide relevance to industrial processes as well as environmental problems such as dune migration, sediment transport, landslides and natural avalanches in general [2,3]. However, the overwhelming majority of the studies of rotating drum systems have focussed on special case of steady continuous surface flow at constant dynamic angle (rolling and cascading regime) mostly due to their industrial importance. Although non-spherical particles are rather common both in industrial and environmental context, experimental investigations of non-spherical and irregular particles behaviour in rotating drums are scarce and rather limited in scope. For example, there is no experimental measurement of time resolved particle dynamics of transient flowing layer during discrete avalanches of slumping regime for non-spherical particles and even only few studies for spherical [4,5].

There are a range of experimental methods available for the study of the granular dynamics in rotating drums, including particle image velocimetry (PIV) [4,6-8], particle tracking velocimetry (PTV) [9,10], PEPT [11,12], magnetic resonance imaging [13,14], and x-ray microtomography [15]. Some of these are restricted to two components of motion (PIV and PTV) whilst the remainder can only resolve the granular dynamics to a fine scale with relatively poor temporal resolution or *vice versa*. In contrast, diffusing-wave spectroscopy [16], a dynamic light scattering technique, is able to resolve the average of the three components of motion of grains in dense systems with very fine spatiotemporal resolutions [17,18] that allow the probing of the dynamics of avalanches. However, it is based on temporal correlation functions calculated as a time average, so it is not appropriate for systems with transient dynamics. In such cases, the related method of speckle visibility spectroscopy (SVS) [5,19-21] can be used.

In this paper, we report the SVS-based study of the avalanching of particles in a rotating drum under the slumping regime. After introducing the SVS method and the rotating granular drum flow system to which it has been applied here, we report time resolved avalanche dynamics results for spherical glass and non-spherical sand particles followed by their qualitative comparison.

2 Experimental details

2.1 Experimental setup

The results reported here were obtained in a cylindrical drum, Fig. 1, whose inner diameter, D, and length, L, are

* Corresponding author: vladimir.zivkovic@ncl.ac.uk

142 mm and 200 mm respectively. The rotating drum, which is made of clear Plexiglas to permit optical access, was filled with granular material. The drum was placed on a pair of rollers turned by a DC motor in 0.1-2.3 revolutions per minute (RPM). Points on the free surface of the granular material were studied here by laser illumination as shown in Fig. 1, some at the top of the inclined free surface, whilst others were located at the bottom of this surface. Two granular material were considered: spherical glass particles (500-600 μm diameter range) and glass sand grains (1.0-1.5 mm). Based on shape descriptors obtained through image analysis, the sand grains would be categorised as subangular particles in the qualitative shape measure classification [22].

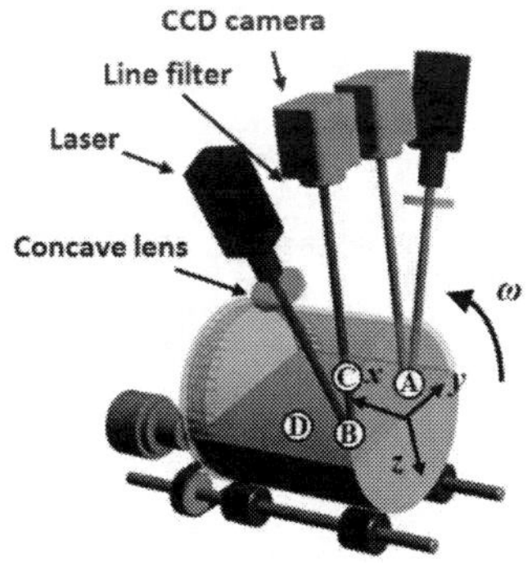

Fig. 1. Schematic of the experimental setup.

2.2 SVS methodology

The speckle visibility spectroscopy involves illuminating the avalanching free surface spots of the granular materials with lasers. In case of sand particles we have used synchronized double speckle visibility spectroscopy [23] using two monochromatic lasers of wavelengths 671 nm and 532 nm while only 532 nm laser was used for glass particles study. The light beams were passed through a concave lens before normally illuminating a spots of around 10 mm diameter on the inclined free surface of the granular material. The photons that emerged from the granular material after diffusing within it form a speckle pattern that was detected using high speed line scan CCD cameras (up to 50 kHz). The cameras were placed with its optical axis normal to the drum and some 350 to 400 mm from it, such that the ratio of pixel to speckle size was about 0.5. In the absence of any motion of the particles, the scattered light is characterised by a constant intensity as illustrated by the example in Fig.2(a). Collective motion of particles as a rigid body leads to the spatial and temporal fluctuations of the scattered light resulting in random intensity of lights across the pixels, as illustrated by the example in Fig. 2(b). However, only the relative motion of particles between each other's results in blurring of captured images and for a given exposure time, the faster the dynamics of the grains, the more the speckle image is blurred and the lower the contrast – this enables the capture of rapid changes in the granular material with time such as that which occurs in an avalanche as illustrated in the example of Fig. 2(c). This variation in intensity can be quantified by the variance of the intensity [19]

$$V_2(T) \propto \langle I^2 \rangle_T - \langle I \rangle^2 \qquad (1)$$

where $\langle \cdots \rangle_T$ denotes the average over pixels exposed for a duration T. The proportionality constant of $V_2(T)$ is set by the laser intensity and the ratio of speckle to pixel size (*i.e.* it is set-up dependent). It can, however, be eliminated by considering the variance ratio $V_2(2T)/V_2(T)$ [19]. For diffusely backscattered light from particles moving with a random ballistic motion, where the power spectrum is Lorentzian [24], the theory of SVS [19] gives the variance ratio as

$$\frac{V_2(2T)}{V_2(T)} = \frac{e^{-4\Gamma T} - 1 + 4\Gamma T}{4(e^{-2\Gamma T} - 1 + 2\Gamma T)} \qquad (2)$$

where $\Gamma = 4\pi\delta v/\lambda$. The mean fluctuation in the speed of the particles, $\langle \delta v^2 \rangle$, which is equal to the collision velocity and directly related to the so-called granular temperature [36], can be obtained by inverting Eq. (2), as given in Fig. 2(d) which was obtained from the raw data shown in Fig 2(c). Comparison of these two subfigures clearly shows that only the relative motion of particles during avalanching is detected while the rigid body motion before and after avalanching event is shown with the near zero collisional velocity.

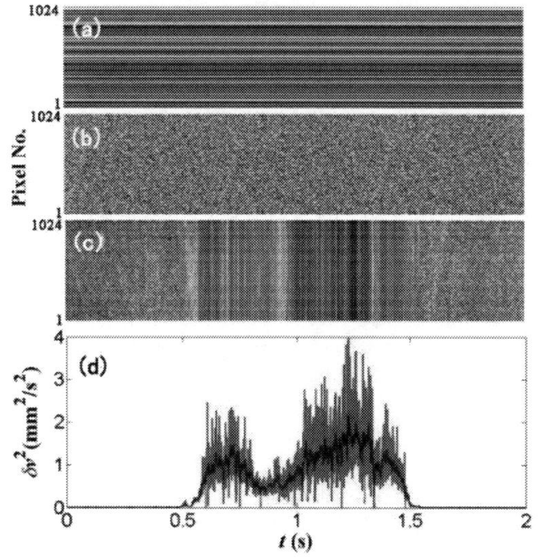

Fig. 2. Example of SVS results and analysis for sand particles in the drum: (a) intensity across the CCD pixels *vs.* time when the drum is stationary; (b) intensity across the CCD pixels *vs.* time when the drum is rotating at a speed of 1.53 RPM and no avalanche event occurs; (c) the same as (b) except an avalanche event has occurred in the period of around 0.5-1.5 s; (d) The grey (red online) line represents the raw particle fluctuation velocity (δv^2) obtained from the data in (c), whilst the black line represents the data after being subject to a Butterworth low pass filter to reduce the noise.

3 Results and discussion

3.1 Avalanche statistics

Fig. 3 shows the typical SVS-derived time trace of the granular temperature, $\langle \delta v^2 \rangle$, over several avalanche events for both spherical glass and irregular sand particles. Results for the glass particles, Fig. 3(a), are from the top point on the inclined surface at a drum at rotational speed of 0.15 RPM along with a filtered version thereof, which removes CCD camera related noise. Looking at a single avalanche event shows that the particle velocity fluctuation accelerates progressively to a maximum before similarly decelerating. The avalanche duration, t_d, rest time, t_r, and the peak of the particle velocity fluctuation, (δv_p^2), can be straightforwardly measured from the filtered trace for many slumping cycles to enable good statistics (at least 100 avalanches analyzed).

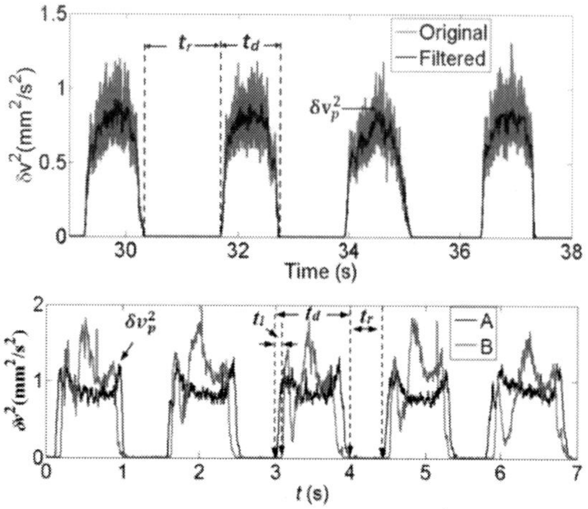

Fig. 3. Time trace of granular temperature, (δv^2) over several avalanches for (a) glass particles in a drum rotating at a speed of 0.15 RPM and (b) sand grains in drum rotating at 1.53 RPM.

The synchronized granular temperature time trace of sand grains over five slumping cycles at a drum speed of 1.53 RPM from the top, point A, and bottom, point B, of the inclined surface is given in Fig. 3(b). Looking at a single avalanche event for the point A shows that the particle fluctuation velocity accelerated quickly then reaching plateau for some time before decelerating quickly again in contrast to spherical glass particle experiments where this is more progressively transition. This is probably related to observed compaction of the drum at the beginning of avalanching process followed by expansion afterwards as indicated in plot of relative area of the drum during typical avalanche, Fig. 4. The compaction was not detected in case of spherical glass beads confirming that the particle shape is important for granular compaction [25].

Another difference are two peaks of the SVS signal at the beginning and the end of the avalanching period for point B albeit sometimes this is not very clear. Again visual observation (not shown here) give possible explanation as a small mound forms nearly at the centre of the avalanching surface approximately halfway through the avalanche event. Therefore, the bottom point (B) granular temperature shows some peaks and drops midway through the avalanche period due to the passing of this mound, while the top point A is not as influenced and has more stable signal.

Avalanching statistics in the middle of drum (not shown here) is also more stable and the granular temperatures are lower in comparison with near wall measurements shown in Fig. 3(b). This is in line with the previous reports that the flow velocity of the grains is lower in the middle than that near the front wall of the drum for the wall effect [26,27]. The stability of the signals would indicate that the visually observed mound is appearing only near the end walls and does not influence the avalanche dynamics in the middle of the drum.

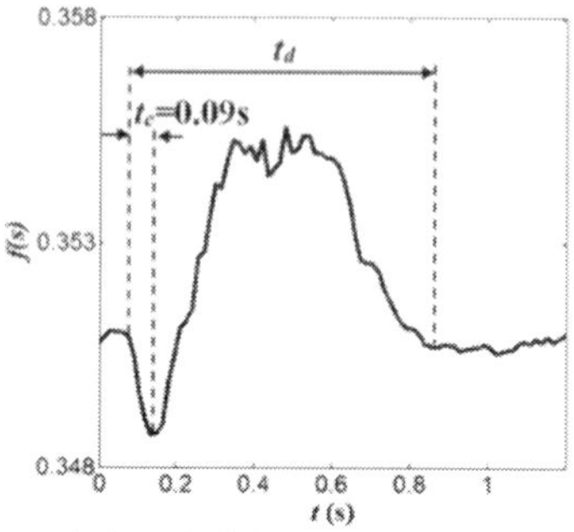

Fig. 4. The evolution of relative area, $f(s)$, during avalanching from the movies of the front face at a drum speed of 1.53 RPM.

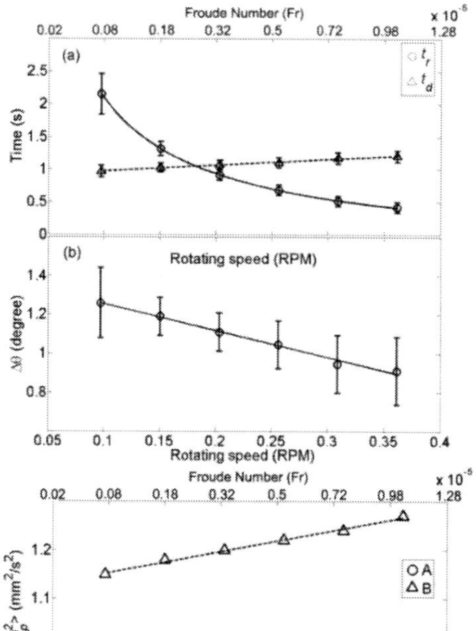

Fig. 5. (a) The average rest time, t_r, and the average avalanche duration time, t_d; (b) $\Delta\theta$by obtained by SVS analysis and (c) the average of the peak of particle fluctuation speed, (δv_p^2), of the avalanches for the point A (top) and B (bottom) as a function of the rotating speeds and Froude number for spherical glass particles.

3.2 Variation of slumping features with drum speed

Figures 5 and 6 shows variation of various avalanche statistics with drum speed and corresponding Froude number for glass and sand particles. The trends are quite similar but with some differences in the values. Obvious one is that drum with sand particles is operating at higher drum speed and consequently Froude number which in in line with previous studies that transition to rolling regime is highly influence by particle shape [3]. Another is that the difference between upper and lower angle of repose is almost 5 times bigger for sand particles and consequently the rest time between avalanches is also approximately 5 times bigger. However, the trends of various avalanche characteristics with the drum speed are very similar for both particles such as constant avalanche duration, decrease of the difference between upper and lower angle of repose and linear increase of granular temperature with the drum speed.

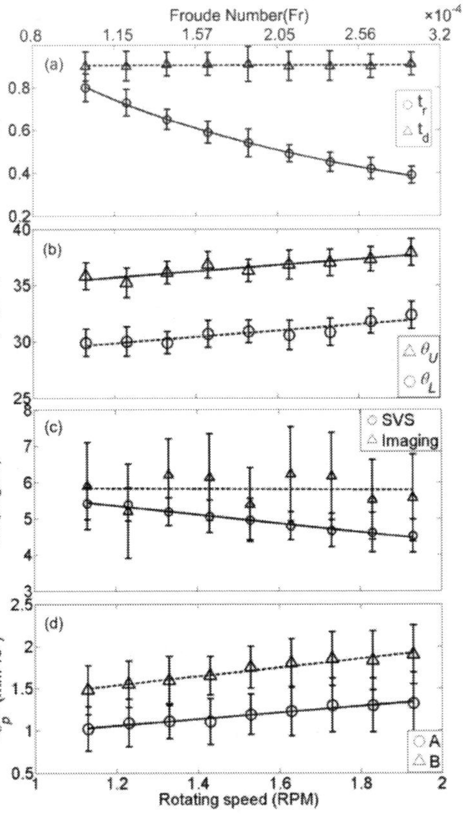

Fig. 6 (a) The average rest time, t_r, and the average avalanche duration time, t_d,; (b) upper and lower angle of repose from image analysis; (c) $\Delta\theta$ by SVS (open circle) and imaging technique (triangle) and (d) the average of the peak of particle fluctuation speed, $\langle\delta v_p^2\rangle$, of the avalanches for the point A (top) and B (bottom) as a function of the rotating speeds and Froude number for sand grains.

4 Conclusion

Our study showed that the speckle visibility spectroscopy is able to measure time resolved granular micro-dynamics of intermittent surface avalanche flow inside rotating drum operating in a slumping regime. There are some differences in time evolution of granular temperature between glass and sand particles probably due to visually observed compaction and consequent formation of the mound as a result of irregular shape of sand grains.

References

[1] H. Henein, J. K. Brimacombe, and A. P. Watkinson, Metall. Trans. B **14**, 191 (1983).
[2] G. Seiden and P. J. Thomas, Rev. Mod. Phys. **83**, 1323 (2011).
[3] J. Mellmann, Powder Technol. **118**, 251 (2001).
[4] S. C. du Pont, R. Fischer, P. Gondret, B. Perrin, and M. Rabaud, Phys. Rev. Lett. **94**, 048003 (2005).
[5] H. Yang, R. Li, P. Kong, Q. C. Sun, M. J. Biggs, and V. Zivkovic, Phys. Rev. E **91**, 042206 (2015).
[6] S. H. Chou and S. S. Hsiau, Powder Technol. **226**, 99 (2012).
[7] H. Komossa, S. Wirtz, V. Scherer, F. Herz, and E. Specht, Powder Technol. **264**, 96 (2014).
[8] H. T. Chou, S. H. Chou, and S. S. Hsiau, Powder Technol. **252**, 42 (2014).
[9] R. Mueller and P. Kleinebudde, Powder Technol. **173**, 51 (2007).
[10] N. Jain, J. M. Ottino, and R. M. Lueptow, Phys. Fluids **14**, 572 (2002).
[11] D. J. Parker, A. E. Dijkstra, T. W. Martin, and J. P. K. Seville, Chem. Eng. Sci. **52**, 2011 (1997).
[12] S. Y. Lim, J. F. Davidson, R. N. Forster, D. J. Parker, D. M. Scott, and J. P. K. Seville, Powder Technol. **138**, 25 (2003).
[13] K. Yamane, M. Nakagawa, S. A. Altobelli, T. Tanaka, and Y. Tsuji, Phys. Flluids **10**, 1419 (1998).
[14] C. R. Müller, D. J. Holland, A. J. Sederman, M. D. Mantle, L. F. Gladden, and J. F. Davidson, Powder Technol. **183**, 53 (2008).
[15] G. T. Seidler, G. Martinez, L. H. Seeley, K. H. Kim, E. A. Behne, S. Zaranek, B. D. Chapman, S. M. Heald, and D. L. Brewe, Phys. Rev. E **62**, 8175 (2000).
[16] D. J. Pine, D. A. Weitz, P. M. Chaikin, and E. Herbolzheimer, Phys. Rev. Lett. **60**, 1134 (1988).
[17] V. Zivkovic, M. J. Biggs, D. H. Glass, and L. Xie, Adv. Powder Technol. **20**, 227 (2009).
[18] V. Zivkovic, K. Berry, D. H. Glass, and M. J. Biggs, in *POWDERS AND GRAINS 2013: Proceedings of the 7th International Conference on Micromechanics of Granular Media* (AIP Publishing, 2013), pp. 1266.
[19] R. Bandyopadhyay, A. S. Gittings, S. S. Suh, P. K. Dixon, and D. J. Durian, Rev. Sci. Instrum. **76**, 093110 (2005).
[20] H. Katsuragi, A. R. Abate, and D. J. Durian, Soft Matter **6**, 3023 (2010).
[21] H. Yang, G. L. Jiang, H. Y. Saw, C. Davies, M. J. Biggs, and V. Zivkovic, Chem. Eng. Sci. **146**, 1 (2016).
[22] F. Altuhafi, C. O'sullivan, and I. Cavarretta, J Geotech. Geoenviron. **139**, 1290 (2012).
[23] R. Li, H. Yang, G. Zheng, B. F. Zhang, M. L. Fei, and Q. C. Sun, Powder Technol. **295**, 167 (2016).
[24] P. A. Lemieux and D. J. Durian, JOSA A **16**, 1651 (1999).
[25] N. Vandewalle, G. Lumay, O. Gerasimov, and F. Ludewig, Eur. Phys. J. E **22**, 241 (2007).
[26] N. A. Pohlman, J. M. Ottino, and R. M. Lueptow, Phys. Rev. E **74**, 031305 (2006).
[27] M. Arntz, W. K. Den Otter, H. H. Beeftink, R. M. Boom, and W. J. Briels, Granular matter **15**, 25 (2013).

 DOI: 10.1051/epjconf/201714003066

Increasing of horizontal velocity of particles leaving a belt conveyor

Abraão Tavares[1,*] and Allbens Faria[1,**]

[1] Departamento de Física e Matemática, Centro Federal de Educação Tecnológica de Minas Gerais, Av. Amazonas, 7675, Belo Horizonte, Brazil.

Abstract. We investigate the transport of granular materials by a conveyor belt via numerical simulations. We report an unusual increasing of particles horizontal velocity when they leave the belt and initiate free-fall. Using Discrete Elements Method, the mechanism underlying this phenomenon were investigated, and a study on how particle and system properties influences this effect were conducted.

1 Introduction

Granular materials play a fundamental role in a wide range of applications, from grain storage in the food industry to space exploration. However, even seemingly trivial questions about the behavior of these materials do not have a clear answer. [1]

A variety of different granular flows has been studied by GDR-MiDi in [2], where different flows in free fall (where the driving force is gravity) systems were characterized. Here we focus on a flow guided by a belt conveyor.

Belt conveyors are key components in any mine around the world. The velocity profile of the material being placed, transported or dropped from the belt plays an important role on the durability of mechanical components as well in the energy consumption of conveyor's drives. It is directly correlated to equipment wear and lifetime.

Via numerical simulations using molecular dynamics technique (also know as Discrete Element Method, or DEM [3]), we can evaluate and study the spatiotemporal behavior of many properties of a given granular material. We have detected a consistent increase in horizontal velocity of particles at the end of a conveyor. In this paper, we use 2D DEM simulations to demonstrate the existence of this phenomena and investigate the influence of some system parameters.

2 Methodology

In our studies, we use the molecular dynamics technique [4] to simulate a 2D flow of particles over a belt conveyor. The simulation model is as described in [5] and [6] adapted to incorporate objects other than particles (walls and belt), and a constant stream of particles. The contact model used for particle-object contact is the same used for particle-particle contacts.

As illustrated in Figure 1, the simulated base system consists of monodisperse material being transported by a conveyor-like geometry, with a 3.5m long belt, an inbox for particle entrance of 1.5m wide and width of 1.5m, with an opening for material flow of 0.5m. Belt velocity is constant at 3m/s in all simulations, this velocity is transmitted to particles without an actual movement of the geometry. An object that transmits movement without geometry displacement is a common technique used in many DEM software packages. The material being transported on this initial simulation is monodisperse (all particles have the same diameter).

For data collection we have defined 3 regions of interest, *Before*, *During* and *After* the belt endpoint, which is located at coordinates (0,0). On a real conveyor, the belt endpoint is actually where the belt touches the pulley and particles start a ballistic movement. Results shown in this paper are based on data from particles inside these regions of interest.

On this study, we will call the velocity modulus as V, and it's x and y components as V_x and V_y, respectively.

Particle size was chosen to balance detail level and processing time, in monodisperse simulations all particles have a radius of 0.025m and between 0.023m – 0.027m for polydisperse cases. Other material parameters used in this study include a normal and tangential stiffness(k_n and k_t) of 10E8 and dumping factor (gn) of 100. Particles, walls, and belt are considered to be of the same material.

3 Results

The simplest evidence of the effect existence can be seen in the visualization of longitudinal component (V_x) of the velocities of particles. On Figure 1(Right) we can observe V_x of particles on the conveyor are equal to the belt velocity, just as expected. As particles leave the conveyor and start ballistic movement their V_x spontaneously increases, proportional to their height on material layers, creating a

*e-mail: abraaomt@gmail.com
**e-mail: atman@dppg.cefetmg.br

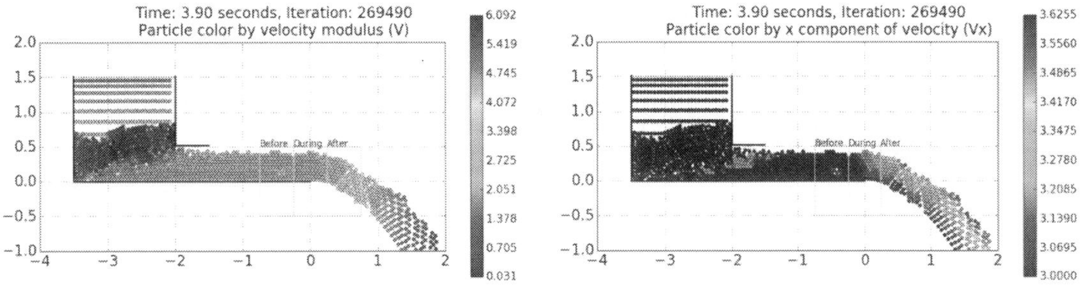

Figure 1. The simulated system: A basic conveyor with a particle inlet at the top left, a moving belt with constant velocity from left to right and three regions of interest (named "Before", "During" and "After") for data collecting. **Left:** Particles colored by they velocity modulus V, color scale automatically adjusted from minimum do maximum values in that frame. **Right:** Particles colored by they V_x, color scale set for range 3.0m/s(belt velocity) to 3.7m/s, for detail enhancement and better visualization of the effect. Note the V_x increasing that appears during the transition from forced flow to free fall, as well a velocity gradient from bottom to top is visible.

velocity distribution where particles next to the belt maintain their V_x close to 3.0m/s and particle far from the belt increases their V_x up to 3.8m/s(27%) in whole simulation time (See Figure 2, V_x after belt's end, between 6 and 7 seconds)

On Figure 2 we show the temporal evolution of V_x in the 3 regions of interest. We show the maximum, minimum and mean values of all particles inside each region, during 7 seconds of simulation. Please note that in $t = 0$ the particles start to fall from the inlet and takes about 1.0 – 1.5 seconds to reach the regions of interest.

From Figure 2, V_x before belt's end is, as expected, equal to the belt velocity with small oscillations. After belt endpoint the $V_x mean$ becomes higher than 3.2m/s and $V_x max$ oscillates around 3.6m/s.

3.1 Parameter influence

In our previous work [7], we have detected the existence of this phenomena in 3D simulations, using commercial simulation package Rocky[8]. A experiment have been designed to understand which of the three parameters *height of material*, *friction (between particles)* and *adhesion (between particles)* have more influence in the effect (more influence = higher V_x). As a result, the combination of parameters *height of material* and *friction* induced higher values of V_x. Since the company refuses to give information about the method used for they software, further investigations have to be conducted with our code.

Once demonstrated the effect existence by previous results, we can explore the influence of system parameters on it. Particle size distribution has been tested, as well material height and belt slope. Although the same overall gradient is clearly observable in all simulated systems, quantitative analyses show detailed behavior differences.

Particle size distribution: While we have used monodisperse material (all particles with the same diameter) in our base system, here the material is polydisperse, with particles diameters varying from 0.023m – 0.027m chosen from a uniform distribution. All other parameters are the same as in base system.

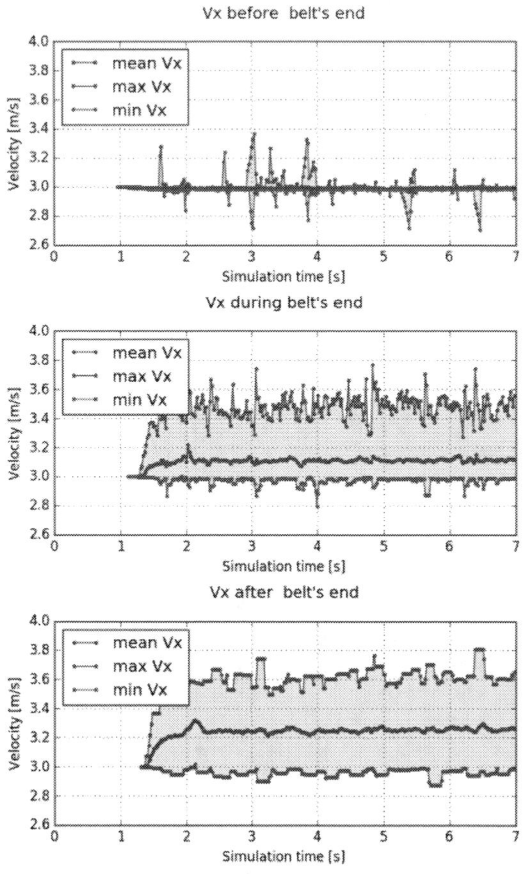

Figure 2. $V_x max, min, mean$ for all particles in each region of interest, in the base system (monodisperse material). It is clear the increase of maximum and mean V_x during and after the end of the conveyor. A opposite behavior can be observed for the minimum V_x, its value became <3.0 m/s more frequently *during* the transition in the end of conveyor and always < 3.0m/s *after* the transition

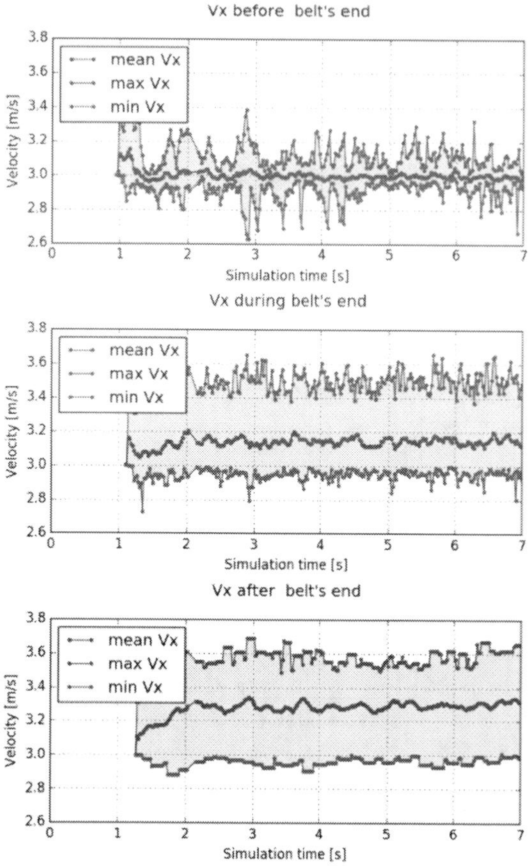

Figure 3. $V_x max, min, mean$ for all particles in each region of interest, in a polydisperse system. In this system, we can observe a much higher variability in V_x before the endpoint, although the mean value stays around 3.0m/s as expected. Maximum V_x after the transition is slightly less than the values obtained for the monodisperse system.

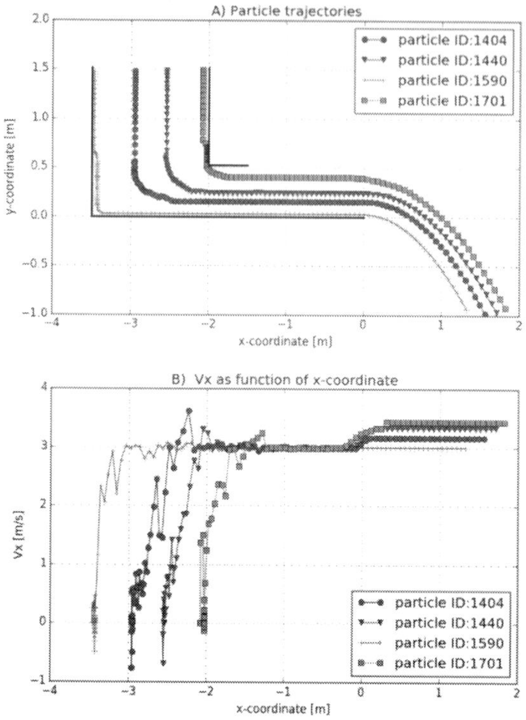

Figure 4. Influence of layer height on V_x increase. Tracking position and V_x of 4 particles in different layer heights, we can see the higher layers give a higher and earlier increase of V_x

In Figure 3 is visible a much higher oscillation of maximum and minimum values on *before* region, and a slightly higher mean value in *during* and *after* regions. A full explanation for this observed behavior is a work in progress.

Height of the material layer: As can be inferred from the velocity gradient in Figure 1(Right), material layer height directly influences the effect. Here the test consists of tracking particles that flow in different layers and analyze the behavior of its V_x along its trajectory.

Figure 4A shows the x, y position of 4 chosen particles along their lifespan in the base system, for each of these particles Figure 4B shows its V_x as function of x coordinate. We can see that higher the material layer, higher and earlier is the increase of V_x.

Belt slope Belt conveyors are frequently designed to operate with a small slope, limited by material properties. This slope cannot be too high to avoid material rolling. Two values for belt slope (+5° and -5°) were tested transporting monodisperse material and compared to horizontal belt position (0°). The results in Figure 5 are obtained calculating the temporal mean of $maxV_x$, $minV_x$ and $meanV_x$ from simulations with belt at above mentioned angles. Values before 3.0 seconds were discarded in order to consider only the flow in stationary regime. As V_x is measured in global X direction, is expected a reduction in observed V_x in inclined systems. A particle traveling at 3.0m/s in a linear trajectory at +-5° has a V_x of $3 \times \cos(5°) = 2,9885 m/s$.

A positive belt angle generates an upstream flow that reduces more significantly the maximum and mean values V_x but not the minimum V_x. In another hand, the negative slope generates a downstream flow that significantly reduces the minimum of V_x

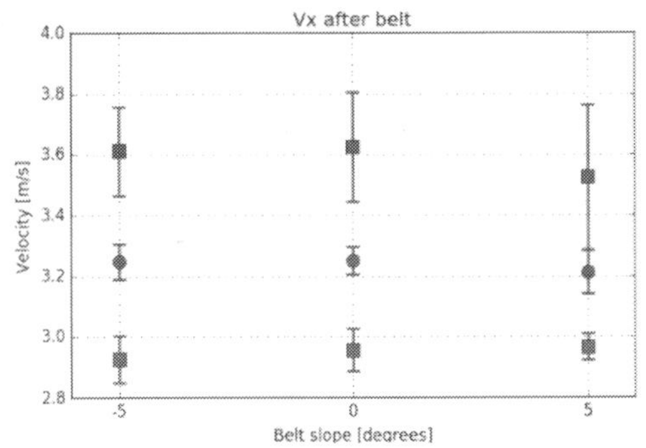

Figure 5. Results for inclining the belt. Here we show the mean values for V_x maximum (upper squares), mean (discs), and minimum (bottom squares) averaged in time for 3 belt slope angles. Angles are measured clockwise and 0 means the base state were belt is horizontal.

4 Conclusion

From the moment particles being transported by a belt conveyor leave the belt, is expected that only vertical component of its velocity (V_y) increases since gravity is the only force acting on particles.

However, we have verified via numerical simulations that the horizontal velocity of particles does increase after the belt endpoint and this effect is directly proportional to the height of material layer and influenced by particle size distribution, belt slope, friction and adhesion [7].

One possible explanation for the observed phenomena is the conservation of angular moment. Despite the absence of adhesion between particles, the natural packing and friction in the material being transported make it behave like a solid, at the transition at the end of the belt, these solid like properties are still present and the stacks of particles appear to rotate, accelerating the external ones.

Further investigation is necessary and the results presented here can guide on this path. Theoretically, measurements about this effect on real systems could be obtained by computer vision, filming the external upper and bottom layers of material on a conveyor discharge, and estimating velocities differences via optical flow technique.

References

[1] J. Duran, *Sands, Powders, and Grains*, Vol. 49 of *Partially Ordered Systems* (Springer New York, New York, NY, 2000), ISBN 978-1-4612-6790-4
[2] G.D.R. MiDi, GDR Midi, European Physical Journal E **14**, 314 (2004), 0312502
[3] S. Luding, The Physics of Granular Media **3**, 297 (2005)
[4] M. Allen, D. Tildesley, *Computer simulation of liquids* (1987)
[5] A.P.F. Atman, P. Brunet, J. Geng, G. Reydellet, G. Combe, P. Claudin, R.P. Behringer, E. Clément, Journal of Physics: Condensed Matter **17**, S2391 (2005), 0411734
[6] A.P.F. Atman, P. Claudin, G. Combe, Computer Physics Communications **180**, 612 (2009)
[7] A.M. Tavares, A.P.F. Atman, *Estudo do aumento espontâneo na velocidade de partículas em transportadores de correia*, in *Proceeding Series of the Brazilian Society of Applied and Computational Mathematics*, edited by M. Bernardes (SBMAC, São Carlos, SP, Brazil, 2016)
[8] ESSS, *Rocky DEM*, http://www.rocky-dem.com, accessed: 2016-10-15

A comparison between DEM and MPM for the modeling of unsteady flow

Fabio Gracia[1,2,⋆], *Pascal* Villard[1], and *Vincent* Richefeu[1]

[1] *Univ. Grenoble Alpes, 3SR, F-38000, Grenoble, France*
[2] *IMSRN, Montbonnot, F-38330, France*

Abstract. In order to provide a comprehensive comparison between two current numerical methods employed in the modeling of rock avalanches, the Discrete Element Method (DEM) [3] and the Material Point Method (MPM) [1] were used to simulate the mass propagation along a 45° plane transitioning to an horizontal plane. When using the DEM, a 3D code using tetrahedral elements was used and the flow was channelized by means of frictionless walls. For the MPM simulations, a 2D code was developed and plane strain simulations were run. Comparisons were made in terms of run-out distance and energy dissipated. Influence of parameters such as initial sample geometry, basal friction coefficient and shape of blocks composing the sample was studied.

1 Introduction

Numerical studies involving DEM and MPM were carried out in order to test the ability and the relevance of two kinds of numerical methods (discrete and continuous methods) to reproduce the kinematics of granular avalanches (propagation distances and amounts of dissipated energies). In the field of rock mechanics, the continuum approaches are generally considered when large volumes (*i.e.* large number of blocks) are involved. For this study, the MPM was chosen over the continuum-based methods because of its hybrid Lagrangian and Eulerian descriptions, which gives it the ability to manage naturally large deformations. This makes it an ideal method for the modeling of granular avalanches involving large masses. However, the continuity assumption may also limit the domain of validity of continuum approaches, specifically when dealing with rock flows. Additionally, the computational times required by the MPM simulations are significantly low, but the dissipations means and more generally the rheologies have to be given explicitly by macroscopic constitutive laws with, unavoidably, the continuity of the medium in mind.

On the other hand, the DEM requires an accurate definition of the initial condition for the terrain, block shapes and families of discontinuities, together with prohibitive computation times. As a consequence DEM turns out to be more appropriate for medium sized volume (*i.e.* few hundred blocks). Moreover, the complex rheologies inherent in granular materials, are naturally well captured by the DEM – provided that a minimum set of features are taken into account. In our case, these features are the block shapes and adequate dissipative contact models. The price to pay is however long computational times.

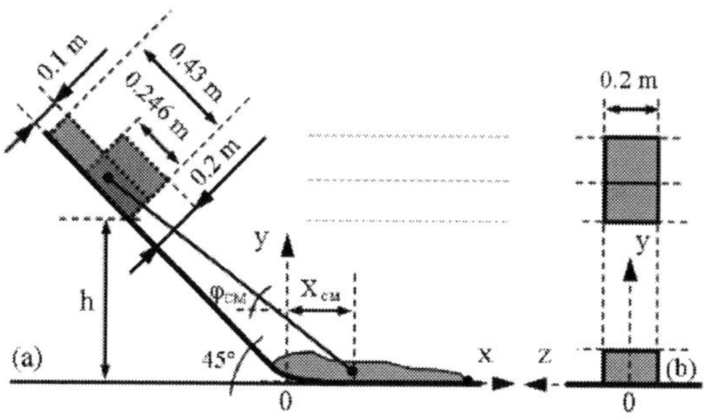

Figure 1. Geometry of the simulation setup

The aim of this paper is to compare the two types of methods on the base of numerical simulations that consist, first, in the release of randomly packed bricks within a parallelepiped box for the DEM, and second, in a continuum mass governed by an elasto-plastic law for the MPM. The setup is depicted in Figure 1.

A total of 27 configurations were setup using both numerical methods in order to study the influence of parameters such as initial sample geometry, basal friction coefficient and shape of blocks composing the sample (the latter affecting DEM only).

2 Numerical setup

For all the configurations, randomly packed blocks (cubes of $0.01 \times 0.01 \times 0.01$ m^3 or cuboid elements of $0.02 \times 0.01 \times 0.005$ m^3) were used for the DEM simulations while a continuum mass governed by an elasto-plastic law was considered for the MPM simulations. The container was positioned on an inclined plane at a given height, which is

⋆e-mail: fabio.gracia@3sr-grenoble.fr

one of the parameters to be studied. Between the slope and the horizontal area, a smooth transition was set by means of a curvature with 0.3 m as radius. DEM simulations were performed in a three-dimensional configuration; the flow was thus canalized to make the comparison possible with the two-dimensional MPM simulations. The canal width was set to 0.2 m to make the flowing truly three-dimensional while limiting the number of elements. This width was chosen so that the ratio between the width and the size of a DEM particle was at least 10. Two different types of containers with varying aspect ratios were used as can be seen in Figure 1. The plane inclination angle was fixed at 45°, while the height at which the block was placed (which refers to the height of the corner closer to the horizontal plane) had three different values: 0.4, 0.8 and 1.6 m.

3 MPM and DEM laws

As mentioned above, simulations based on the DEM were run using a 3D code while simulations based on the MPM were run using a 2D code. In order to compare the results coming from both methods in an appropriate manner, considerations such as using plane strain when running the MPM simulations and adjusting the MPM material bulk density had to be taken into account to deal with the same mass of material. A classical Mohr-Coulomb elasto-plastic law was used to describe the granular material mechanical behavior. A constant dilatancy angle, close to zero, was used in order to prevent volume increase of the sample. MPM elasto-plastic parameters can be found in Table 1. Within the mass the energy dissipation $\mathcal{W}^\star$ is given by:

$$\mathcal{W}^\star(t) = \sum_p \int_0^t V_p(\sigma : \dot{\varepsilon})\mathrm{d}t \qquad (1)$$

where V_p, σ and $\dot{\varepsilon}$ are respectively the volume, the stress and the strain increment held by each material point p. Note that, the index p is not used here for the total stress and total strain to avoid any confusion with plastic terms. Note also that, here and in the following, a parameter super-scripted with a star ($\star$) relates to the bulk material. Otherwise the parameter concerns the interface between the flowing material and the slope.

For the DEM, the interactions between the blocks are governed by a specific contact law established at each contact point, allowing dissipation in the normal and tangential direction of the contact [2][4]. In the perpendicular direction of the contact a linear elastic law with two different normal stiffnesses (k_n), depending on whether it is loading or unloading, is used. The two stiffnesses relate by using a restitution parameter e_n^2 with a value between 0 and 1 which allows for normal energy dissipation. In the tangential direction, a Coulomb frictional model was used, while using k_t as tangential stiffness and μ as friction parameter defined as $\mu = \tan\phi$ with ϕ being the friction angle. The contact parameters governing the mechanical behavior of the DEM elements are given in Table 2.

Table 1. MPM Elasto-plastic properties

Young Modulus (MPa)	0.8
Poisson ratio	0.42
Friction angle (deg)	28.6
Cohesion (MPa)	0
Dilatancy angle (deg)	~ 0

Table 2. DEM block-block contact parameters

$k_n^\star$ (N/m)	10^4
$k_t^\star/k_n^\star$	0.4
$\mu^\star$	0.7
$(e_n^\star)^2$	0.2

Table 3. DEM and MPM contact parameters between mass (or blocks) and boundaries at the bottom

k_n (N/m)	10^4
k_t/k_n	0.4
e_n^2	0.2

The elastic contact law and Coulomb frictional model that were used for the normal and tangential contact between particles in the DEM was used to model the boundary conditions (slope, transition and horizontal plane) in both MPM and DEM. The parameters can be found in Table 3. This external force will be added directly to the particles in DEM, while in MPM it will be included in the body forces term involved in the conservation of linear momentum equation.

Concerning the energy dissipation between the blocks (DEM) and the boundary conditions (DEM and MPM), the overall cumulated dissipation was split into 3 contributions: (1) the cumulated work $\mathcal{W}_n$ of the normal forces (contacts and collisions) at the base of the flow, (2) the cumulated work $\mathcal{W}_t$ of the tangential forces (friction) at the base of the flow, and (3) the work $\mathcal{W}^\star$ of all internal action in the flowing mass. For the DEM, the dissipation within the mass is simply obtained as the sum of cumulated works of contact forces between the blocks in the normal and tangential directions:

$$\mathcal{W}^\star(t) = \mathcal{W}_\mathrm{n}^\star(t) + \mathcal{W}_\mathrm{t}^\star(t) \qquad (2)$$

4 Selected results

The numerical configurations were defined by varying the height of release h, the shape of the blocks and the shape of the container (A, B, C), and the basal friction coefficient μ. These configurations are summarized in Table 4.

We compared first (Figure 2) the kinematics of the flow for DEM and MPM in a particular case (simulation #C1, $\mu = 0.3$). We then compared (Tables 5 and 6) the position of the center of mass X_{CM} (called *propagation distance* here) once the mass had stopped, as well as the distribution of the mass around it. Results were grouped using the basal friction, which is one of the driving parameters in our study. The distribution of the mass was quantified using the standard deviation σ_X of the positions of all blocks

Table 4. Configurations defined by changing the release height, the shape of the blocks, and the shape of the container

	Height (m)	Block Shape	Container shape
#A1	0.8	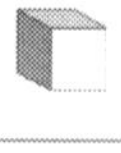	
#A2	1.6		
#A3	0.4		
#B1	0.8	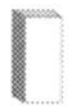	
#B2	1.6		
#B3	0.4		
#C1	0.8	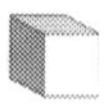	
#C2	1.6		
#C3	0.4		

Table 5. MPM and DEM run-out and energy dissipation results with a basal friction coefficient of 0.3

► **MPM,** $\mu = 0.3$

	X_{CM} (m)	σ_X (m)	$\mathcal{W}_n$ (J)	$\mathcal{W}_t$ (J)	$\mathcal{W}^*$ (J)
#A1	1.32	0.44	0.15	151.89	15.49
#A2	2.48	0.55	0.38	293.45	18.12
#A3	0.74	0.37	0.10	80.17	14.65
#B1	1.42	0.44	0.09	111.23	6.70
#B2	2.57	0.57	0.19	205.70	8.80
#B3	0.84	0.35	0.05	63.99	5.37
#C1	1.40	0.42	0.09	112.55	6.65
#C2	2.56	0.55	0.20	209.38	8.76
#C3	0.82	0.33	0.05	64.08	5.39

► **DEM,** $\mu = 0.3$

	X_{CM} (m)	σ_X (m)	$\mathcal{W}_n$ (J)	$\mathcal{W}_t$ (J)	$\mathcal{W}^*$ (J)
#A1	1.30	0.39	0.98	135.40	14.34
#A2	2.46	0.55	1.51	261.29	17.72
#A3	0.71	0.31	0.79	71.18	13.29
#B1	1.40	0.33	0.16	121.82	6.97
#B2	2.55	0.41	0.26	217.34	9.86
#B3	0.83	0.28	0.11	70.57	5.14
#C1	1.39	0.39	0.60	122.04	7.51
#C2	2.54	0.51	0.99	205.51	10.21
#C3	0.81	0.30	0.39	69.51	5.89

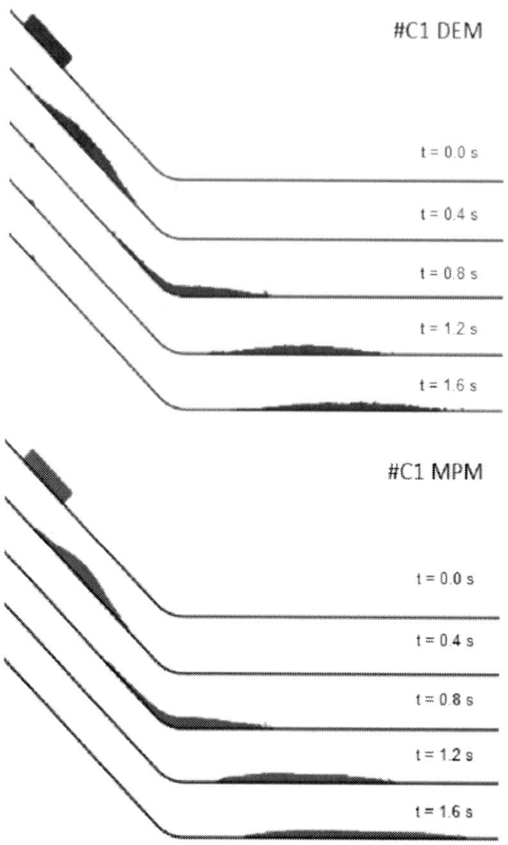

Figure 2. Simulation #C1. DEM (top) and MPM (bottom). The friction coefficient at the base is $\mu = 0.3$

or material points depending on the method. In Figure 2 (the figure at the top corresponds to the DEM simulations, while the one at the bottom refers to the MPM ones) we see that the flow evolution from the beginning through the very end correlates properly. Some minor discrepancies can be seen in the shape of the final deposit, as the MPM final deposit spreads a little further. The few blocks that escaped from the mass (spotted at the front of the DEM deposit) are completely absent in the MPM deposit because of its continuous nature.

However, even if in some cases there are outstanding similarities, when taking a closer look to the position of the center of mass and the standard deviation around it, the shape of the final deposits are not exactly the same. In particular, the curvature near the slope transition differs between MPM and DEM simulations, and the scattering of the blocks in front of the deposit is completely missed by the MPM. Apart from that, the main mismatch seems to appear at the very end, when the final heap is forming.

All configurations were first tested with a relatively low friction coefficient at the base: the friction angle was set to 16.7°, which is considerably lower than the internal friction angle of the granular material of 28.6°. The results obtained in terms of propagation and spreading of the mass, and total dissipation modes are summarized in Table 5 for both MPM and DEM simulations. Close correlations between the two methods can be seen.

In regard to the center of mass, X_{CM}, (see Table 5), the value reached by both methods is very similar in most of the simulations. This first observation is rather surprising when one realizes that the constitutive model used in the MPM is rather simple.

In terms of energy dissipation, Table 5 clearly shows that the driving means of dissipation is the friction at the base for both DEM and MPM simulations. This is an expected result since the internal friction angle (*i.e.*, an input parameter of the constitutive model for the MPM, and a parameter mainly related to the friction between blocks in the DEM) is higher than the basal friction coefficient. Internal dissipation achieves nearly the same values in both MPM and DEM. It is however slightly greater for higher container (cases #A). The cases correlating the least (although not largely different) happen to be the ones with higher release heights (cases #B2 and #C2) for which the constitutive model has a larger influence in the stretching of flowing mass.

Table 6. MPM and DEM run-out and energy dissipation results with a basal friction coefficient of 0.5

▸ **MPM,** μ = 0.5

	X_{CM} (m)	σ_X (m)	$\mathcal{W}_n$ (J)	$\mathcal{W}_t$ (J)	$\mathcal{W}^*$ (J)
#A1	0.45	0.32	0.27	90.71	67.86
#A2	0.76	0.40	0.37	229.41	73.64
#A3	0.29	0.27	0.25	33.42	53.29
#B1	0.47	0.29	0.9	78.40	35.14
#B2	0.82	0.38	0.18	185.74	27.02
#B3	0.30	0.24	0.09	37.16	27.61
#C1	0.46	0.29	0.10	80.29	34.56
#C2	0.81	0.36	0.15	174.32	39.42
#C3	0.30	0.23	0.09	36.96	27.96

▸ **DEM,** μ = 0.5

	X_{CM} (m)	σ_X (m)	$\mathcal{W}_n$ (J)	$\mathcal{W}_t$ (J)	$\mathcal{W}^*$ (J)
#A1	0.42	0.28	4.10	99.82	44.18
#A2	0.77	0.41	7.20	208.90	62.37
#A3	0.25	0.22	2.68	46.31	32.49
#B1	0.45	0.24	0.28	114.32	13.46
#B2	0.81	0.32	0.41	214.78	18.82
#B3	0.26	0.20	0.19	64.12	9.61
#C1	0.45	0.25	3.21	94.23	30.93
#C2	0.80	0.37	5.79	186.45	44.13
#C3	0.27	0.19	1.81	50.65	20.97

As for #C1, (shown in Figure 2), results correlated properly for the other simulations (from #A1 to #C3). Nonetheless, when using bricks instead of blocks in the DEM simulations, the spreading (σ_X) of the final deposit was reduced; This reduced spreading is an expected result since the rotation of single elements is greatly dependent on its shape, in this case being hindered by the use of bricks. The continuum approach however is not naturally able to simulate it in the way the DEM does.

The container shape becomes important, as well as the block shape, when analyzing the amount of energy dissipated within the granular mass. When using the square container, the energy dissipated in the bulk is larger. These differences might be attributed to the constitutive model itself or to the parameters used since calibration at a very early stage was performed. In order to give a stronger role to the constitutive model, a second series of simulations were carried out with a higher friction at the base.

The friction coefficient at the base was then set to $\mu = 0.5$ corresponding to an angle of 26.6°, which is still below but close to the internal angle of friction of 28.6°. The Table 6 summarizes the results obtained. At first glance it can be seen that similar results are found when comparing DEM and MPM in terms of propagation distance and spreading of the deposit, regardless of the aspect ratio of the starting box, release height or shape of blocks.

When compared with the results obtained with the low basal friction, $\mu = 0.3$, the propagation distances and spreading of the deposit were found to be smaller for all configurations. The observations made with low basal friction are still valid with a basal friction angle close to the internal friction angle.

In regard to the internal dissipation due to plastic straining, an increase can be seen. It is interesting to notice that the better matches are obtained when using a shallow container and cubic blocks for which the assumption of continuity is somehow more appropriate. In other configurations (in particular with bricks), the difference of internal dissipation is more pronounced because of the simplicity of the constitutive model. Despite this, the total amount of dissipated energy is rebalanced thanks to an increase of the friction dissipation at the base.

5 Conclusions

The use of the material point method has been evaluated in the case of a flow in the transient regime that involves finite strains. In the comparisons presented, the DEM results were taken as the ground truth. Although an extensive simplification has been considered in the continuous model, the MPM has proven to be of interest in the modeling of large size rock avalanches or landslides. In terms of energy dissipated, with the assumptions made for the constitutive law of the continuum model (Mohr-Coulomb and values of parameters) some discrepancies started to become evident as the basal friction coefficient was increased (which involves shearing within the granular mass), but the results were still related.

In the field of soil (hydro-)mechanics or computational fluid dynamics, many constitutive models exist. The MPM provides a framework that is not more restrictive than all approaches based of the continuum assumption. Accordingly, some events such as debris flows, mud flows or snow avalanches – deemed more complex to model – can be dealt with by implementing the best suited constitutive models. A more ambitious project is to make use of two-scale modeling by replacing the constitutive law by DEM simulations (representative to the material) for each material point.

Acknowledgements

The Laboratoire 3SR is part of the LabEx Tec 21 (Investissements d'Avenir, Grant Agreement No. ANR-11-LABX-0030).

References

[1] D. Sulsky, Z. Chen, H. L. Schreyer, Computer Methods in Applied Mechanics and Engineering **118**, pp. 179-196 (1994)

[2] G. Mollon, V. Richefeu, P. Villard, D. Daudon, Journal of Geophysical Research – Earth Surface **117** (2012)

[3] P. A. Cundall, O. D. L. Strack, Géotechnique **29**, pp. 47-65 (1979)

[4] V. Richefeu, G. Mollon, D. Daudon, P. Villard, Engineering Geology **149**, pp. 78-92 (2012)

[5] V. Richefeu and P. Villard , *Modeling Gravity Hazards from Rockfalls to Landslides* (ISTE Press Ltd, Elvesier Ltd, UK, 2016) 161 pp

Micromechanics of non-active clays in saturated state and DEM modelling

Arianna Gea Pagano[1,*], *Alessandro* Tarantino[1], *Matteo* Pedrotti[1], *Vanessa* Magnanimo[2], *Kit* Windows-Yule[2] and *Thomas* Weinhart[2]

[1] University of Strathclyde, Department of Civil and Environmental Engineering, Glasgow, UK
[2] University of Twente, Faculty of Engineering Technology, Multi-Scale Mechanics (MSM), Enschede, Netherlands

Abstract. The paper presents a conceptual micromechanical model for 1-D compression behaviour of non-active clays in saturated state. An experimental investigation was carried out on kaolin clay samples saturated with fluids of different pH and dielectric permittivity. The effect of pore fluid characteristics on one-dimensional compressibility behaviour of kaolin was investigated. A three dimensional Discrete Element Method (DEM) was implemented in order to simulate the response of saturated kaolin observed during the experiments. A complex contact model was introduced, considering both the mechanical and physico-chemical microscopic interactions between clay particles. A simple analysis with spherical particles only was performed as a preliminary step in the DEM study in the elastic regime.

1 Introduction

Microstructure of discrete materials such as soils affects the macroscopic behaviour of the material itself. The Discrete Element Method (DEM), first proposed by Cundall and Strack ([1]) is a numerical technique that can be successfully used in order to simulate those aspects of soil behaviour that cannot be anticipated at the scale of the sample, although derivable from the particle scale. Two of the many issues that need to be addressed in the application of the DEM for clayey geomaterials are ([2]): a) physico-chemical microscopic interactions between particles have to be taken into account as well as conventional mechanical interactions and b) the plate-like (3D) or rod-like (2D) shape of clay particles plays a key role in the interpretation of soil macroscopic behaviour and cannot be ignored during the analysis.

The paper presents an experimental investigation carried out on saturated kaolin clay samples in order to assess the effect of the pore fluid characteristics affecting kaolin's microstructure on the one-dimensional compressibility of the soil. A micro-mechanical conceptual model was estimated from the experimental observations and used as a basis for a three dimensional DEM analysis, considering both mechanical and physico-chemical microscopic interactions. A simple preliminary analysis with spherical particles only was performed in order to reproduce qualitatively some of the microscopic mechanisms concerning the elastic behaviour of the sample. Further simulations can be performed using rod-like and plate-like particles, already implemented in the model, in order to explore the effect of particles shape on the compressibility behaviour in the plastic regime.

2 Experimental investigation

2.1. Background

Electrochemical interactions existing between kaolin clay particles (mainly Coulomb and van der Waals interactions) are known to control the macroscopic behaviour of non-active clays. Such interactions affect particles arrangement, and are affected by both pH and dielectric permittivity.

Pore fluid pH has been proven to affect the distribution of charges on kaolin particles edges. While particle surfaces are assumed to be permanently negatively charged due to isomorphic substitution of cations with negative charges ([3], [4]), particles edges can be either positively or negatively charged due to protonation/deprotonation of broken edges induced by changes in pH ([3], [5], [6]). For low pH values (acid solutions), the face-to-face interaction is repulsive while the edge-to-face interaction is attractive. This charge distribution leads to open structures. For high pH values (alkaline solutions), both face-to-face and edge-to-face interactions are repulsive. This charge distribution leads to close structures.

On the other hand, dielectric permittivity is assumed to affect the extent of Coulomb repulsive forces existing between particle faces. By considering two facing particles as two circular plates, the repulsive force F [N] can be calculated as ([7]):

$$F = \frac{\pi a^2}{8\varepsilon}\rho^2\left(\frac{d}{\sqrt{a^2+d^2}} - 1\right) \quad (1)$$

* Corresponding author: arianna.pagano@strath.ac.uk

where a is the radius of the plates [m], d is the plates distance [m], ρ is the electrical charge distribution on the two plates [C/m^2] and ε is the dielectric permittivity of the media between the two plates [F/m], defined as the vacuum permittivity times the relative permittivity.

2.2. Experimental setup

Four cylindrical Speswhite kaolin samples of 100 mm diameter and different heights were tested in a modified triaxial cell apparatus. A cylindrical splitting mould was used to give the required shape to the samples and avoid lateral deformations during the tests. Each sample was saturated with a different pore fluid prior to the test, in order to assess the effect of pore fluid pH and dielectric permittivity on the one dimensional compressibility behaviour. The properties of each sample are shown in Tables 1 and 2.

A loading frame was used to load/unload the samples in oedometric loading condition. A constant strain rate, slow enough to guarantee drained conditions during the test, was applied continuously at the bottom of the cell. A load cell was used to measure the applied force.

Table 1. Samples tested to assess the effect of permittivity.

Test n°	Height [mm]	Pore fluid	Pore fluid content [%]	Dielectric permittivity
1	74.9	Air (dry sample)	100	1
2	83.3	Ethanol	100	25
3	84.3	Deionised water	100	80

Table 2. Samples tested to assess the effect of pH.

Test n°	Height [mm]	Pore fluid	Pore fluid content [%]	pH
3	84.3	Deionised water	100	4
4	91.9	Water + dispersant	55	9

2.3. Experimental results

Figure 1 shows the one dimensional compressibility curves obtained for the four reconstituted samples. By comparing the results of tests 1-2-3 and tests 3-4 the effect of dielectric permittivity and pH of the pore fluid respectively can be investigated.

2.3.1 Effect of dielectric permittivity

The compressibility curves for tests 1-2-3 show a different behaviour during loading and unloading-reloading cycles. Samples having lower values of dielectric permittivity show higher void ratio values e, defined as the ratio between the volume of voids and the volume of solids, for the same vertical effective stress, σ', as a consequence of the higher repulsive force developed between particles. However, upon loading (irreversible volume change), the lower the permittivity, the more compressible is the sample. Upon unloading-reloading (reversible volume change), this trend is reversed: samples having low dielectric permittivity swell less than samples with high dielectric permittivity and, hence, are less compressible upon reloading.

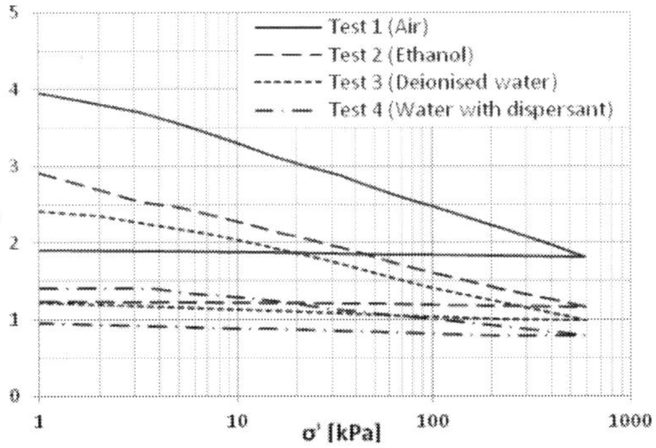

Fig. 1. One dimensional compressibility curves

2.3.2 Effect of pH

The compressibility curves for tests 3-4 show the effect of pH during loading and unloading-reloading cycles. Upon loading, the sample saturated with demineralised water at pH4 (test 3) exhibits higher values of void ratio and a much more compressible behaviour than the sample saturated with water and dispersant at pH9 (test 4). Upon unloading, the swelling of the two samples appears to be very similar, although there is still a difference in void ratio.

2.4. Micro-mechanical conceptual model as a basis for the DEM analysis

For both pH-induced configurations (edge-to-face and face-to-face), the lower the pore fluid dielectric permittivity the "stiffer" is the microscopic contact between particles, i.e. the higher is the repulsive force developed for the same particles distance. This affects the initial macroscopic configuration of the soil: the lower the dielectric permittivity, the more open is the structure and the higher is the void ratio. However, samples with open microscopic structures exhibit a more compressible behaviour at the macroscopic scale in the plastic regime (first loading). Loading can cause the destruction of some of the edge-to-face contacts between particles: the more open is the structure, the higher is the loss in terms of void ratio every time an edge-to-face contact is permanently lost upon first loading, leading to a faster drop of void ratio with applied load. Vice versa, in the elastic regime (unloading-reloading), the swelling-compressibility behaviour is affected by the "stiffness" of the particles contact only: highly repulsive contacts (low dielectric permittivity) will reduce swelling upon unloading and compressibility upon reloading.

3 DEM analysis

3.1. DEM implemented model

The micro-mechanical conceptual model previously introduced was implemented in a three dimensional DEM analysis.

A modified contact model was introduced in order to take into account both mechanical and physico-chemical microscopic interactions (Figure 2). The first interaction is conventionally defined in DEM analyses as the relationship between the overlap of particles in contact, δ_{ij}^n, and the repulsive contact force in the normal direction, f_{ij}^n, developing when overlapping occurs $(\delta_{ij}^n > 0)$. The second interaction has to be defined as the relationship between the particles distance $(\delta_{ij}^n < 0)$ and the attractive/repulsive force developing before the overlap occurs. These "long range" forces will be used to mimic the effect of Coulomb interaction, either repulsive or attractive, and van der Waals interaction, generally assumed to be attractive for clay particles. The total force acting on the particles is obtained by adding together these three contributions, calculated as follows:

$$f_{ij}^n = \begin{cases} 0, & \delta_{ij}^n < 0 \\ k_n \delta_{ij}^n, & \delta_{ij}^n > 0 \end{cases} \quad (2)$$

$$f_{ij,Coul}^n = \begin{cases} 0, & \delta_{ij}^n < \delta_{ij}^{n,1} \\ k_{Coul}(\delta_{ij}^n - \delta_{ij}^{n,1}), & \delta_{ij}^{n,1} < \delta_{ij}^n < \delta_{ij}^{n,2} \\ k_{Coul}(\delta_{ij}^{n,2} - \delta_{ij}^{n,1}), & \delta_{ij}^n > \delta_{ij}^{n,2} \end{cases} \quad (3)$$

$$f_{ij,vdW}^n = \begin{cases} 0, & \delta_{ij}^n < \delta_{ij}^{n,2} \\ k_{vdW}(\delta_{ij}^n - \delta_{ij}^{n,2}), & \delta_{ij}^{n,2} < \delta_{ij}^n < 0 \\ k_{vdW}(0 - \delta_{ij}^{n,2}), & \delta_{ij}^n > 0 \end{cases} \quad (4)$$

where k_n is the stiffness of the particles themselves (always positive), k_{Coul} is the stiffness of Coulomb interaction $f_{ij,Coul}^n$ (either positive or negative), k_{vdW} is the stiffness of van der Waals interaction $f_{ij,vdW}^n$ (assumed to be negative), and $\delta_{ij}^{n,1}$ and $\delta_{ij}^{n,2}$ are the lower limits of Coulomb and van der Waals interaction range respectively.

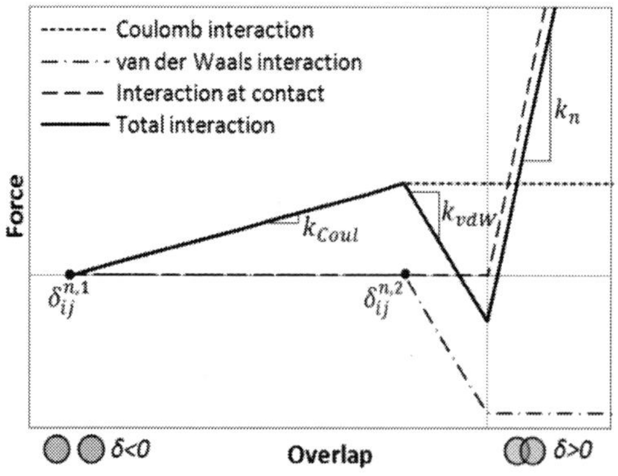

Fig. 2. Qualitative example of the implemented contact model

As regards the particles shape, three possible solutions have been implemented: spherical, rod-like and plate-like particles (Figure 3). For rods and platelets, each particle interacts with the surrounding particles with the previously mentioned contact law, plus an additional attractive interaction that glues together particles belonging to the same rod or platelet.

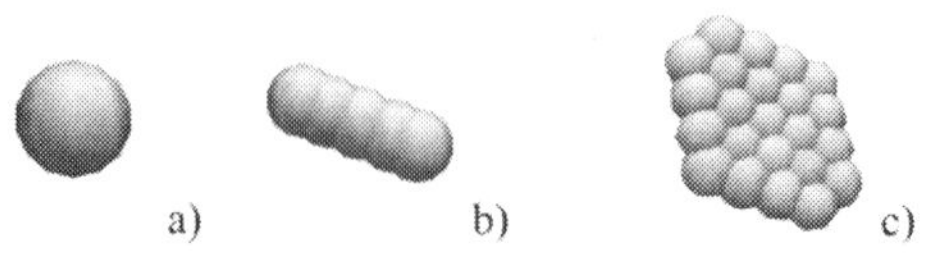

Fig. 3. Particle shape: a) spherical, b) rod-like and c) plate-like

3.2. DEM simulations: setup

Three preliminary simulations were performed using 3D assemblies of spherical particles only. Each simulation consists in an initial isotropic compression from gas state, followed by a sample relaxation after a solid state condition is reached and a 1D compression in the *z* direction. In order to guarantee the homogeneity of the samples, the simulations were performed using periodic boundary conditions and a constant strain rate during each stage.

The simulations were performed using assemblies of 400 particles. Particle diameters were randomly chosen between 0.5 and 1.2×0.5 m. For sake of simplicity, van der Waals interactions were ignored during the simulations and only repulsive (positive) Coulomb interactions were considered for negative δ_{ij}^n values. The parameters defining the contact law for each simulation are shown in Table 3.

Table 3. Contact model parameters.

Simulation n°	k_n [N/m]	k_{Coul} [N/m]	k_{vdW} [N/m]	$\delta_{ij}^{n,1}$ [m]	$\delta_{ij}^{n,2}$ [m]
1	1×10^6	0	0	0	0
3	1×10^6	100	0	1	0
4	1×10^6	500	0	1	0

3.3. DEM simulations: results and discussion

3.3.1 Preparation and relaxation

The three samples were isotropically compressed up to the same target value of elastic energy (energetically defined gas/solid transition criterion) and then relaxed. As expected, the higher the repulsive stiffness k_{Coul}, the sooner the target value was reached (Figure 4). After preparation, a relaxation stage was performed in order to reach a stable, quasi-static condition. The quasi-static condition was identified and guaranteed energetically by relaxing the sample until a target value of elastic energy to kinetic energy ratio was reached.

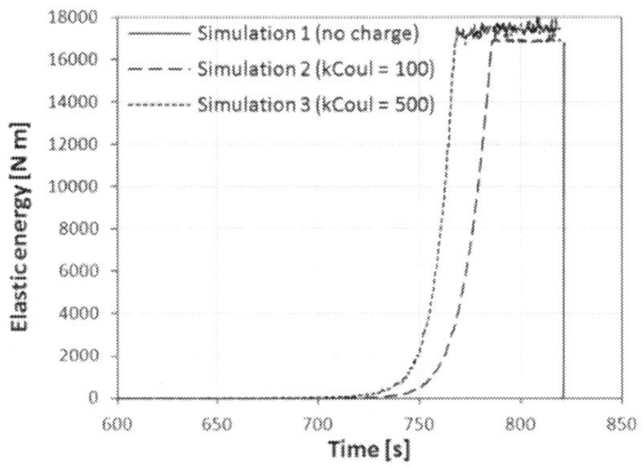

Fig. 4. Elastic energy versus time during preparation/relaxation

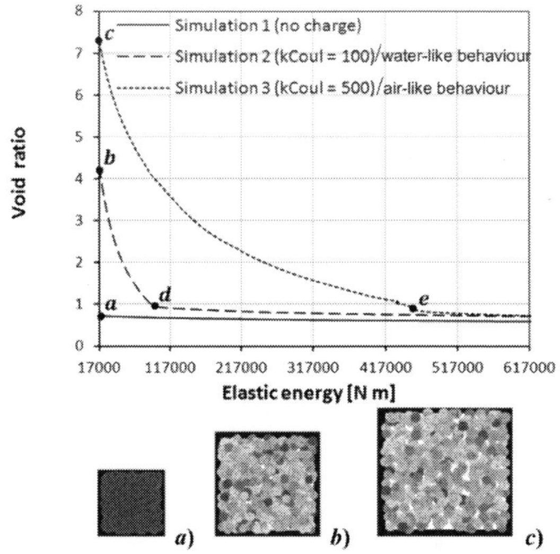

Fig. 5. Void ratio versus elastic energy during 1D compression

3.3.2 One dimensional compression

Figure 5 shows the evolution of void ratio with elastic energy during vertical 1D compression for the three simulations. The extent of the repulsion affects both the sample configuration and the energetic response during compression. The initial void ratio, corresponding to the last step of the relaxation stage, increases significantly with k_{Coul}: highly repulsive particles reach the gas/solid transition stage arranging themselves in a more open structure than weakly repulsive or non repulsive particles (Figure 5 a, b, c). Upon loading, the sample with no Coulomb interactions (simulation 1) shows the least compressible behaviour, since its compressibility depends on the value of k_n only. For simulations 2 and 3, the higher the k_{Coul} the less compressible is the sample. After reaching a threshold void ratio value, the slope of the compressibility curve becomes almost parallel to the one obtained for simulation 1 (points d, e in Figure 5). This change in trend occurs when charged particles, initially interacting only in the $\delta_{ij}^{n}<0$ range, come into contact: at this point, the compressibility of the whole system will be mainly depending on k_n, whose order of magnitude is much higher than the one of k_{Coul}.

The void ratio versus elastic energy curve can be qualitatively compared to the compressibility curves obtained experimentally. Elastic energy gives a measure of the confining pressure acting on the particles in the main direction of loading and, hence, can give us an idea of the effective vertical stress existing in the sample. Since spherical particles are not adequate to predict any of the edge-to-face and face-to-face configuration mechanisms occurring in kaolin samples during first loading, conclusions can be drawn about the elastic behaviour only. The results obtained in the simulations confirm the trend observed during the experiments when assessing the effect of dielectric permittivity: the higher the repulsion, i.e. the lower the dielectric permittivity, the lower the compressibility in the elastic regime.

4 Conclusions

An experimental investigation was carried out in order to assess the effect of pH and dielectric permittivity on the one dimensional compressibility of kaolin clay samples. Two different microscopic mechanisms have been assumed to take place during the tests: a configuration-induced effect, occurring during first loading, and a charge-induced effect, occurring during unloading-reloading cycles.

A simple preliminary set of DEM simulations with spherical particles only and a suitable modified contact model was already able to simulate the microscopic mechanism occurring during unloading-reloading cycles. Further simulations need to be performed using rod-like and plate-like particles, in order to explore the effect of particles shape on the compressibility behaviour in the plastic regime.

References

1. P. A. Cundall, O. D. Strack, Geotechnique, **29**, 47-65 (1979)
2. M. Yao, A. Anandarajah, J. Of Eng. Mech., **129**, 585-596 (2003)
3. J. Mitchell, K. Soga, *Fundamentals of Soil Behavior* (John Wiley and Sons Inc. Hoboken, NJ, 2005)
4. Y. H. Wang, W. K. Siu, Can. Geot. J., **43**, 587-600 (2006)
5. B. Rand, I. E. Melton, J. of Coll. and Int. Sc., **60**, 308-320 (1977)
6. H. Van Olphen, *Clay Colloid Chemistry: For Clay Technologists, Geologists and Soil Scientists* (John Wiley, 1977)
7. M. Pedrotti, PhD thesis, University of Strathclyde (2015)

Experimental measurement of granular fabric and its evolution under shearing

Max Wiebicke[1,2,3,⋆], *Edward* Andò[1,2], *Erminio* Salvatore[1,2,4], *Gioacchino* Viggiani[1,2], and *Ivo* Herle[3]

[1] *Univ. Grenoble Alpes, 3SR, F-38000 Grenoble, France*
[2] *CNRS, 3SR, F-38000 Grenoble, France*
[3] *Technische Universität Dresden, Institute of Geotechnical Engineering, Germany*
[4] *University of Cassino, Cassino, 03043, Italy*

Abstract. X-ray μ computed tomography (CT) made three-dimensional in-situ imaging of natural granular materials possible. Previous work using x-ray μCT and triaxial compression tests has studied the 3D kinematics of individual grains during shear banding [1]. This works aims to supplement these measurements of kinematics with the measurement of different fabric entities, such as particle or contact orientations. It was found that the individual orientations of the different fabric entities pick up on the forming and the direction of the evolving shear band. The evolution of the anisotropy of the bulk of orientations corresponds to the macroscopic behaviour during the shearing test.

1 Introduction

Extracting fabric in natural granular materials has been the subject of many studies since a long time [2]. However, due to technical limitations these studies were either only possible on perfect model materials, post-mortem specimen or in numerical simulations. With the rise of x-ray μ computed tomography (CT) these limitations have been pushed to new horizons. X-ray μCT enables the acquisition of three-dimensional images of intact specimens at certain resolutions. The accuracy of the measurement of different fabric entities, such as contact and particle orientations, is increased with increasing resolution of these images. However, a trade-off between the resolution and the size of the specimen has to be made, in order to stay mechanically representative [3]. Due to these limited resolutions, some problems e.g. in the detection of contacts and the determination of their orientation arise [4, 5]. Using standard approaches of treating such images yielded several problems such as e.g. the systematic over-detection of contacts or a strong bias of the determined contact orientations. Improvements of these approaches have been suggested [5].

In this study, the fabric is extracted from tomographic images of an experiment on a real natural granular material using improved image analysis tools and it is related to the emergence of a shear band.

2 The experiment

The experiment in this study is a triaxial compression test on a dry Hostun sand at a confining pressure of 100 kPa.

⋆e-mail: max.wiebicke@tu-dresden.de

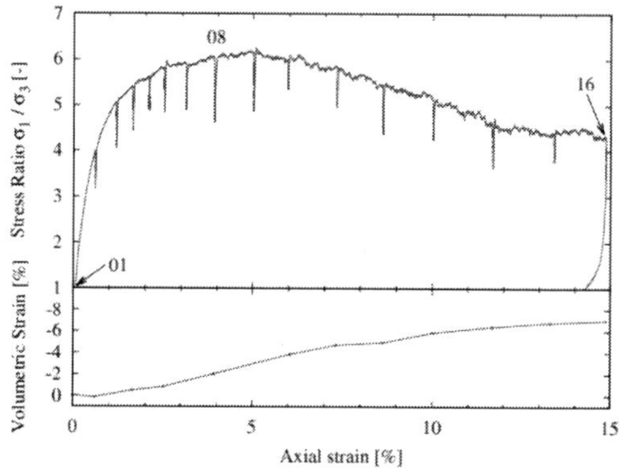

Figure 1. Macroscopic results of the triaxial compression test

The specimen was prepared by pluviation with a resulting initial relative density of 83.2%. More information on the experiment can be found in [1, 6]. In order to be able to identify individual grains in the tomographic images acquired throughout the test, the size of the specimen was chosen to be relatively small (height = 24.31 mm, diameter = 10.70 mm) with the images being acquired at a resolution of 15 μm/pixel.

The macroscopic response of the specimen is plotted in Figure 1. Images were acquired throughout the experiment with x-ray μ-CT; the relaxation parts of the stress-strain behaviour correspond to the states at which the experiment was paused and the images were acquired.

 DOI: 10.1051/epjconf/201714002020

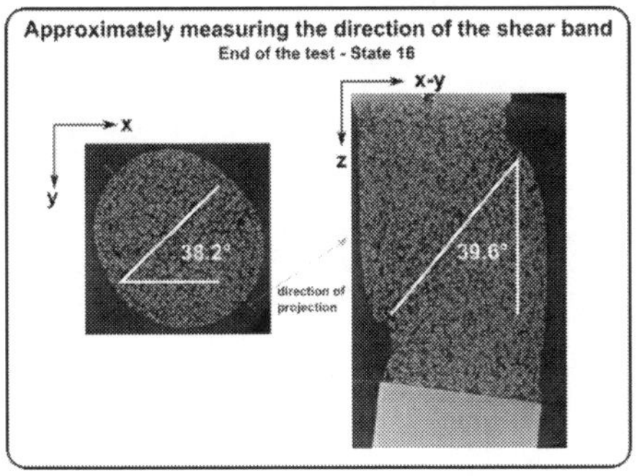

Figure 2. Measuring the orientation of the shear band in spherical coordinates

A single shear band formed during the compression test. Its orientation in the final state (16) in spherical coordinates was measured to be $\phi \approx 51.8°$ and $\theta \approx 50.4°$ with ϕ being the azimuthal angle and θ being the polar angle, see Figure 2.

3 Measuring fabric

In order to extract the two chosen fabric entities, contact and particle orientations, different approaches were chosen and will be explained in the corresponding sections. The images published in [1] have been used as input for the image processing. As described in [1], the initial grey-scale image (with the grey-values corresponding to x-ray attenuation and thus, mass density) is first binarised based on a physical threshold. The particles in this binary image are then separated using the high speed 3D Watershed algorithm mentioned in [1] and labelled. Using the labelled image, a connectivity, describing which particles are in contact with each other, can be determined; this procedure is outlined in [6].

Only two subsets of the sample are considered in the present study. One subset is located to capture a part of the evolving shear band, the other is well outside the shear band, close to the top platen of the specimen. Each subset has a size of 300x300x300 pixels, corresponding to 4.5x4.5x4.5 mm, and containing 1230 to 1743 particles at different stages of the test. To follow the evolution of fabric in both subsets throughout the test, three different states are chosen: 01 - the initial state, 08 - peak stress state, 16 - the end of the experiment.

In order to statistically express and gather the individual orientations, a second order fabric tensor [7] is used:

$$\mathbf{F} = \frac{1}{N} \sum_{i \in N} \mathbf{o}_i \otimes \mathbf{o}_i \tag{1}$$

with $\mathbf{o}_i$ being an individual orientation and N being the number of orientations. The difference between the first and third eigenvalue of the tensor $\mathbf{F}$ was chosen to be a measure for the anisotropy and a way to follow the evolution throughout the test.

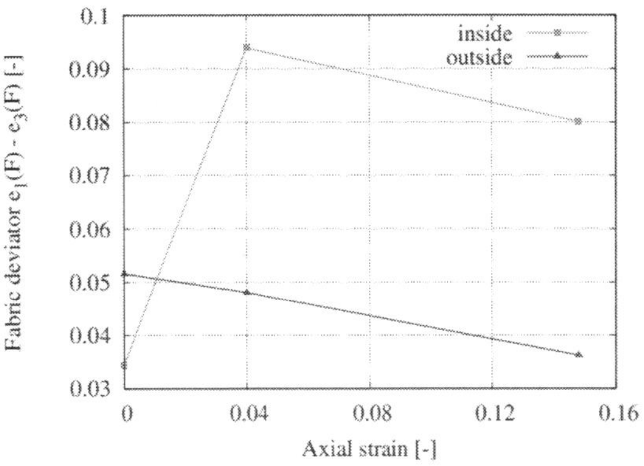

Figure 3. Evolution of the deviatoric contact fabric

3.1 Contact orientations

In order to deal with the systematic over-detection of contacts when using standard approaches as the one described above, a local refinement is applied on the contacts that were detected [5]. Within this approach, a local threshold (that is higher than the global one) is applied on every contact in the connectivity for improvement, i.e. if the corresponding particles do not appear to be in contact after applying the local threshold, the contact is deleted from the connectivity. The orientation of the contacts, which is defined as a normal to the contact plane, is then assessed using the random walker, an advanced watershed algorithm [4, 8].

The individual contact orientations of both subsets in the three chosen states are plotted in Figure 4. The contact orientations in the initial state appear to be either mostly vertical or horizontal, which corresponds to the center and the outer ring in the plots. Only some contacts are oriented in mixed directions (horizontal and vertical). That holds for the orientations in both subsets as the initial state seems to be homogeneous. In both following states, the peak (08) and the end of the test (16), the orientations correspond to the direction of the shear band. The normal to the shear band as well as the directions of the shear band are marked in Figure 4. The contact fabric outside of the shear band undergoes only minor changes with no preferable orientations.

The evolution of the deviatoric contact fabric with the macroscopic loading in the subsets is shown in Figure 3. Outside the shear band the anisotropy of the contact orientations is decreasing with increasing axial strain. In the subset inside the shear band the anisotropy is increased at the peak stress state (08) with respect to the initial state. The deviatoric contact fabric at the end of the test (16) shows a slightly lower anisotropy than at the peak stress state and it is higher than at the initial state. This evolution corresponds to the evolution of the stress ratio during shearing which is increasing until the peak state and then slightly decreasing to the end of the test, compare with Figure 1.

 DOI: 10.1051/epjconf/201714002020

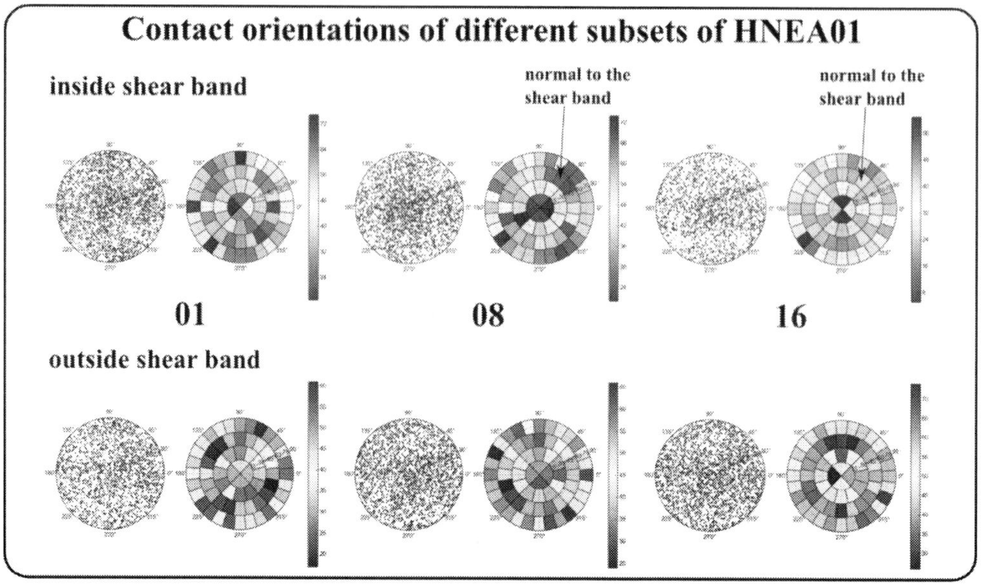

Figure 4. Plots of projected contact orientations - chosen projection: Lambert azimuthal equal area projection. Individual and binned orientations.

3.2 Particle orientations

The moment of inertia tensor was chosen for the characterisation of the particle orientations. Its eigenvectors correspond to the principal particle orientations: longest, intermediate and shortest axis. A study on the particle orientations using a high resolution scan of an individual Hostun sand grain [5] showed that in case of non-spherical particles there is usually one very clearly defined axis and two less clearly defined axes. Depending on the shape the most clearly defined axis is either the long or the short axis. In this case and in general for Hostun sand, the long axis, i.e. the eigenvector of the moment of inertia tensor corresponding to the smallest eigenvalue, is the most clearly defined axis of most particles. Thus, for consistency, this orientation is defined as the principal particle orientation for each particle.

The individual particle orientations of both subsets in the three chosen states are plotted in Figure 6. The initial states of both subsets are again very similar, with the longest axis of the particles being orientated strongly horizontal. The same still holds for the peak stress state (08) with only minor changes of the orientations. At the end of the test (16), the particles inside the shear band appear to reorientate themselves in the plane of the shear band. That however happens again at the horizontal plane with only minor vertical components of the orientations. This alignment is marked in Figure 6 as the opposite to the normal of the shear band. Outside of the shear band the particle orientations undergo only minor changes at the end of the test.

The evolution of the deviatoric particle fabric with the macroscopic loading in the subsets is shown in Figure 5.

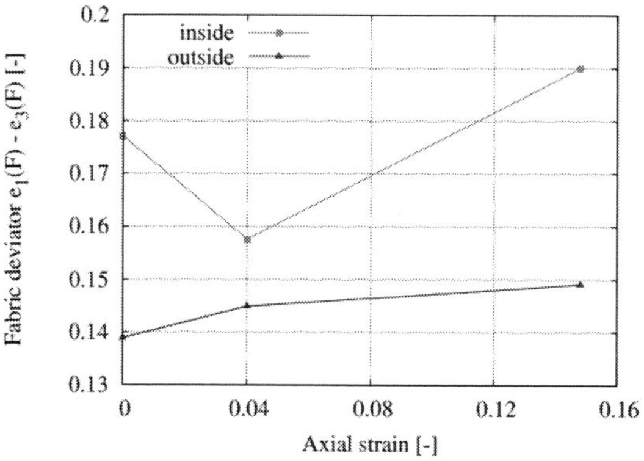

Figure 5. Evolution of the deviatoric particle fabric

The devitoric fabric outside the shear band undergoes only minor changes. Inside the shear band the anisotropy of the particle orientations at the peak stress state (08) is lower than the initial anisotropy. At the end of the test, however, the anisotropy has increased to a higher value than at the initial state. This corresponds to the observation on the individual orientations: the particle orientations do not clearly respond to the onset of the localisation at the peak state, but a tendency can be observed later at the end of the test, see Figure 6.

4 Conclusions

Different fabric entities were extracted from tomographies that were acquired throughout a triaxial compression test

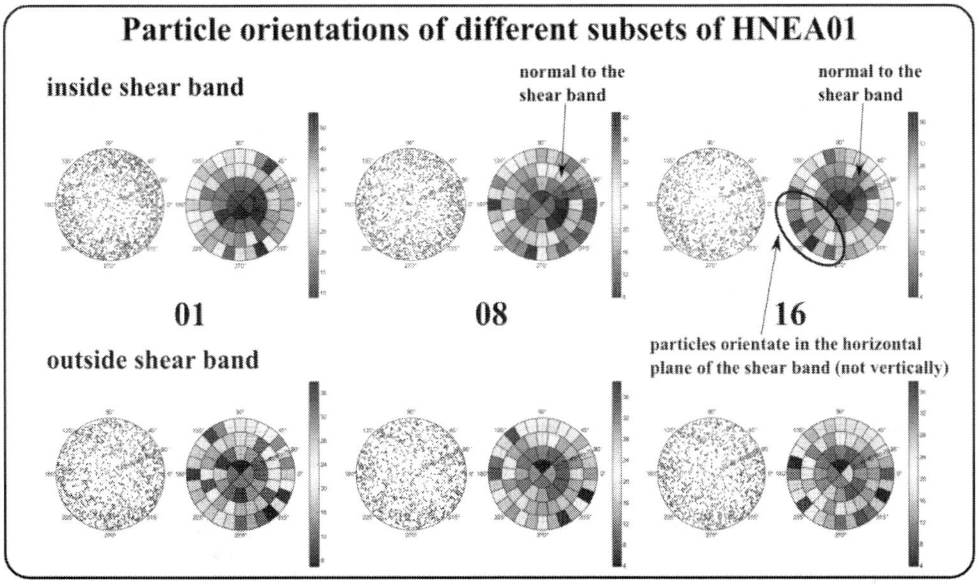

Figure 6. Plots of projected particle orientations - chosen projection: Lambert azimuthal equal area projection. Individual and binned orientations.

on Hostun sand. Using improved image analysis tools, individual contact as well as particle orientations were determined on subsets of the specimen that are located inside and outside an evolving shear band. Interparticle contacts and particle orientations align in the plane of the shear band with ongoing macroscopic shearing. Contact orientations reflected the shear band already at the peak stress state whereas particle orientations appear to pick up the shear band with a delay.

The evolution of anisotropy of the contact fabric inside the shear band corresponds to the evolution of stress, increasing up to the peak stress and decreasing slightly afterwards. The particle fabric picked up the later alignment with the shear plane but showed no clear evolution.

However, the second order approximation of the fabric tensor is limited as shown in [7] and a further study will investigate the need for higher order approximations. Another limiting factor might be the choice of the measure of anisotropy. Different invariants of the fabric tensor are subject of another study. Furthermore, the whole specimen with all its particles and contacts will be investigated by the same means.

Acknowledgements

Laboratoire 3SR is part of the LabEx Tec 21 (Investissements d'Avenir - grant agreement n°ANR-11-LABX-0030). The research leading to these results has received funding from the European Research Council under the European Union's Seventh Framework Program FP7-ERC-IDEAS Advanced Grant Agreement n° 290963 (SOMEF) and from the German Research Foundation (DFG) n° HE2933/8-1.

References

[1] E. Andò, S.A. Hall, G. Viggiani, J. Desrues, P. Bésuelle, Acta Geotechnica (2012)

[2] M. Oda, Soils and foundations **12**, 1 (1972)

[3] E. Andò, G. Viggiani, S.a. Hall, J. Desrues, Géotechnique Letters **3**, 142 (2013)

[4] C. Jaquet, E. Andó, G. Viggiani, H. Talbot, *Estimation of Separating Planes between Touching 3D Objects Using Power Watershed*, in *International Symposium on Mathematical Morphology* (2013), Vol. 11, pp. 452–463

[5] M. Wiebicke, E. Andó, G. Viggiani, I. Herle, *Towards the measurement of fabric in granular materials with x-ray tomography*, in *Proceedings of the 6th international Symposium on Deformation Characteristics* (2015)

[6] E. Andò, Ph.D. thesis, Université de Grenoble (2013)

[7] K.I. Kanatani, International Journal of Engineering Science **22**, 149 (1984)

[8] L. Grady, *Random walks for image segmentation*, in *IEEE Transactions on Pattern Analysis and Machine Intelligence* (2006), Vol. 28

EPJ Web of Conferences **140**, 13011 (2017) DOI: 10.1051/epjconf/201714013011
Powders & Grains 2017

Dynamic interactions of polystyrene particles with microstructured surface manufactured by cold spray

Fabian Krull[1,*], *Paul* Breuninger[1] and *Sergiy* Antonyuk[1]

[1] *Institute of Particle Process Engineering, University of Kaiserslautern, Germany*

Abstract. Dynamic interactions of particles with component surfaces often occur during production, processing and transport of granular materials. The morphology of component surfaces has a significant effect on the impact behaviour of particles. The micro processes during an impact (e.g. rebound, rolling, sliding and sticking) can be changed by microstructuring of the surface. This paper deals with the estimation of the energy dissipation processes of polystyrene particles (400 – 700 μm) impacting on a stainless steel surface. The surface was structured with stainless steel particles (mean particle size 3.5 μm) via a cold spray process. Impact experiments were performed and restitution coefficient was measured. The proportion of the energy dissipation by impact on structured surface was 3.87–10.92 times higher than the dissipation on a polished surface. The additional energy absorption can be explained by multiple particle contacts with the steel asperities and additional friction due to the indentation of these asperities in the polystyrene particles. The particle size did not show any influence on the average energy absorption by collision on the microstructured surface. Only the coefficient of variation got higher, because of the inhomogeneous particle distribution.

1 Introduction

During production, processing and transport particles often collide with the component surfaces. Due to contact stresses and adhesion, the particle sticking, breakage or attrition or surface erosion can take place. Therefore the particle impact was investigated widely. Many experiments with different impact scenarios were performed. Brach et al. describe the particle impact with flat surfaces and consider the rotational dissipation [1,2]. Antonyuk et al. investigated the breakage of granules at high impact velocities [3]. The influence of wet surfaces on the energy dissipation was investigated by Crueger et al. [4].

Besides the materials properties the morphology of surfaces has also a significant effect on the micro processes of the particle impact. Therefore a detailed investigation on the surface morphology is needed. This paper deals with the influence of an inhomogeneous microstructure on the energy dissipation of fine, elastic and soft particles during impact. A stainless steel surface was structured with stainless steel particles (mean diameter of 3.5 μm) via a cold spray process. Polystyrene particles were chosen as contact partner for impact tests. The diameter of these particles is larger than the asperities height of the surface by a factor of 100, so multiple contacts between particle and surface were possible.

*e-mail: Fabian.Krull@mv.uni-kl.de

2 Materials and methods

Single particle tests were carried out by a free fall setup (figure 1). Spherical polystyrene particles from Microparticles GmbH in a size range of 400 – 700 μm were used. The experiments were performed with a microstructured and polished surface of the same material. The manufacturing of the microstructured surface is explained in section 2.2. Each experiment was performed 50 times to obtain the influence of particle and surface roughness. The coefficient of restitution (CoR) e was used for the evaluation of the energy absorption during the particle impact. The energy absorption is equal to the energy dissipation by the particle. This coefficient depends on the ratio between kinetic energies of the translational movement after ($E_{kin,rebound}$) and before ($E_{kin,impact}$) particle impact at the point of contact and is defined as:

$$e = \sqrt{\frac{E_{kin,rebound}}{E_{kin,impact}}} = \sqrt{1 - \frac{E_{abs}}{E_{kin,impact}}} = \frac{v_{rebound}}{v_{impact}} \quad (1)$$

where E_{abs} describes the energy absorption. v_{impact} and $v_{rebound}$ are the total velocities of the particle before, respectively after impact. Furthermore is the CoR an important parameter for contact models in discrete element method (DEM) simulations.

2.1 Experimental setup for impact tests with microparticles

The particles were fixed by a vacuum nozzle and drop under gravity onto the surface. The impact velocity was adjusted by the height of fall of the particle. The vacuuum nozzle consists of a dosage needle and a vacuum pump. The smallest inner diameter of the needle is 60 μm. Thus particles > 60 μm can be investigated. The particle were released by switching off the pump or opening a valve to execute a shock pressure to overcome van der Waals forces. The experiments took place in ambient air. The impact was recorded with three high speed cameras. Two cameras were arranged at an angle of 90° to gain a three dimensional measurement method for the particle trajectory and velocity. The velocity and trajectory can be obtained for particle diameter down to 50 μm. The frame rate of the recording of the first two cameras depends on the particle size. The minimum frame rate is 4,200 fps with a resolution of 1920×1280 px^2. With smaller particles the frame rate can increase up to 20,000 fps and higher. In figure 2 the recorded velocity profile of a polystyrene particle is shown. With this profile the impact velocity can be determined directly. The micro processes of the impact were recorded by a third high speed camera with a microscope objective. This camera has a minimum frame rate of 9,300 fps with a resolution of 1600×1200 px^2 and can record particles with a diameter down to 10 μm.

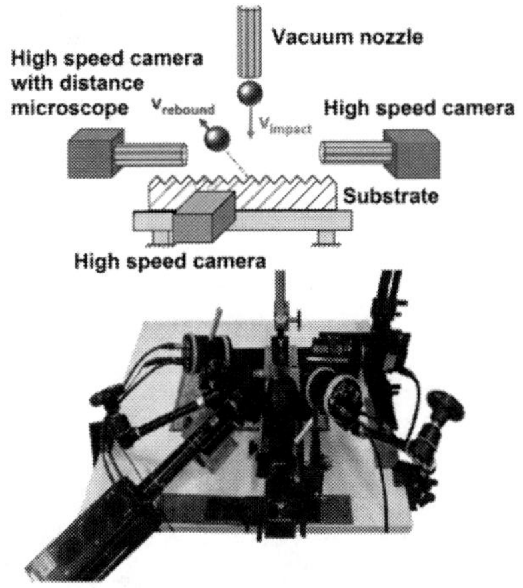

Figure 1. Sketch and image of the experimental setup for particle impact testing

2.2 Cold spray process

The surface was structured by a cold spray method [5]. Single stainless steel particles were dispersed in a nitrogen gas stream, heated up to 500°C and accelerated through a Laval nozzle up to supersonic velocities. Particles impacted on the surface and are fixed on the surface since the kinetic energy is transferred into heat in the contact zone. Nevertheless some particles with a small kinetic energy rebound from the surface and leave craters (figure 3) in the contact zone. Due to a small particle concentration in the sprayed aerosol and fast movement of the substrate vertical to the spray direction a statistically structure of the surface was created. In figure 3 the structured surface is shown. To compare the structured and original polished surfaces only a part of the steel substrate was cold sprayed by steel particles.

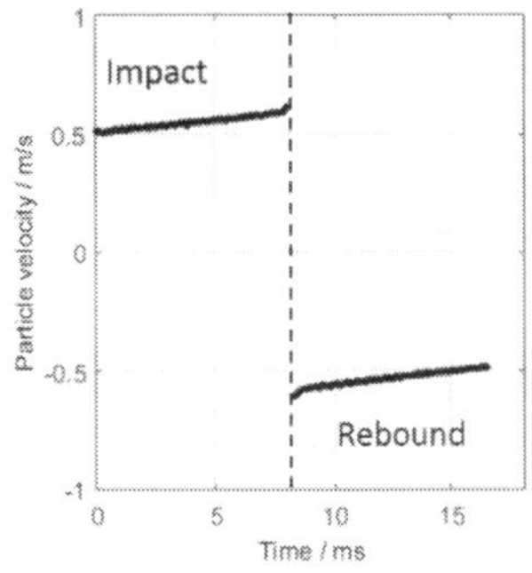

Figure 2. Velocity profile of 665 μm polystyrene particle before and after impact

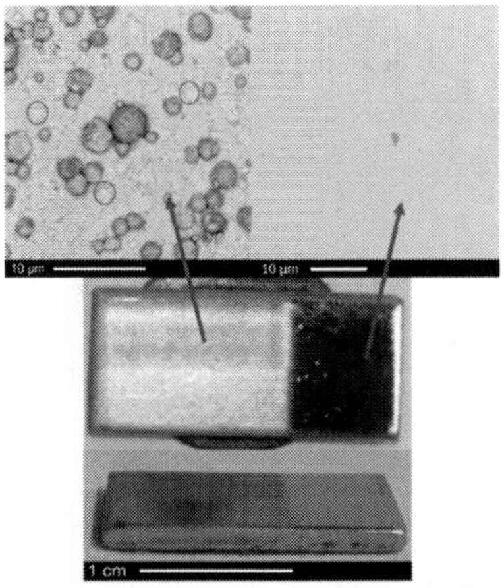

Figure 3. Top: SEM images of cold sprayed and polished surface. Examples of craters are within the red circles. Bottom: Image of the steel substrate for experiments

2.3 Characterisation of the surface morphology

The topography of the surfaces was measured via scanning probe microscopy (SPM). With a triboindenter from Hysitron the surfaces were scanned with a Berkovich tip, which is used for nanoindentation. This measurement was performed within an area of 90 μm^2 for cold sprayed surface and an area of 40 μm^2 for polished surface (figure 4). With the data of the SPM the maximum height, average height and the roughness were determined (table 1). The maximum height of the microstructured surface corresponds roughly to the mean radius of the steel particles. However the average height is 646 nm, because of the plastic crater-shaped pits remained due to rebound of some particles and inhomogeneous particle distribution on the surface. The surface is not completely covered with particles (figure 3). Therefore there are many areas on the surface without particles which decrease the average height.

Compared to the microstructured surface the polished surface is very smooth. The SPM measurements revealed an average height of 15 nm and a root mean squared roughness of 5 nm. The asperity peak (figure 4) of 107 nm can be explained by impurities on the surface.

Figure 4. Topography of the surfaces: (top) cold sprayed surface; (bottom) polished surface

Table 1. Topography of polished and microstructured surfaces

	polished	structured
maximum height	107 nm	2 µm
average height	15 nm	646 nm
root mean squared roughness	5 nm	311 nm

3 Results and Discussion

The experiments were performed with two different particle sizes (400 and 665 µm) and velocities (0.3 and 0.6 m/s). The CoR was obtained for a polished and microstructured surface to determine the influence of the microstructure on the energy dissipation. The ratio $f_i = \frac{E_{abs,i}}{E_{kin,impact,i}}$ were used to determine the energy absorption. The comparision between the different scenarios were executed by the ratio $\frac{f_a}{f_b}$ where a and b stand for the different impact scenarios.

3.1 Influence of microstructuring on energy dissipation

In figure 5 the CoR is shown for 655 µm polystyrene particles by different impact velocities. The proportion of the energy absorption due to the microstructuring was by a factor of 3.86 higher than the polished surface at 0.3 m/s. The ratio of the energy absorption due to the microstructuring increased with increasing impact velocity. At 0.6 m/s the ratio between structured and polished surfaces was 10.92.

A comparison of the influence of the impact velocity, for the same surface, shows that a reduction of the velocity led to an extra energy dissipation. The proportion of the particle energy loss was 3.37 times higher at the polished and 1.19 times at the structured surface.

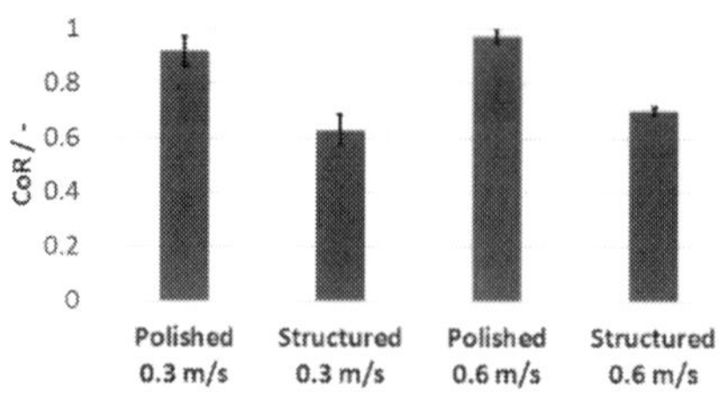

Figure 5. Comparison of the coefficient of restitution (CoR) for polished and microstructured surfaces. The polystyrene particle had a diameter of 655 µm

The additional energy dissipation due to produced microstructure can be explained with the sketch in figure 6. During the particle impact on the microstructured surface several local contact points with stainless steel asperities are formed. Due to the multiple contacts no continuous contact area, as in the case of smooth contact surfaces, is created. The contact force is distributed in the contact points of asperities with the particle. It is evident that these local forces are acting not normal to the impact direction that can lead to the additional energy dissipation due to local stressing in the tangential direction and to an oblique rebound of the particle. Furthermore the steel asperities can indent into the polystyrene particle, because the polymer is much softer than steel. The indentation causes friction during impact and rebound of the particle. This process results in an additional dissipation of the energy. A plastic deformation of the steel asperities can be excluded, because the high stiffness of asperities and high softness of the polymer particles. Also the kinetic energy of the particle before impact was not high enough to cause plastic deformation. The polystyrene particles can deform plastically due to the local forces and stressing.

The additional energy absorption by microstructuring with increasing impact velocity can be explained with the contact area of the particle and surface. These area increased with the velocity. Therefore more asperities of the microstructuring surface can indent into the particle.

The higher energy dissipation at lower impact velocities can be explained with the electrostatic adhesion and van der Waals forces. They contribution to the energy loss is increased with decreasing the particle velocity because of the decrease of kinetic energy of the particle. Compared to the polished surface the contact area and thereby van der Waals forces of the microstructured surface were smaller. Therefore the proportion of the energy dissipation increased slightly.

3.2 Influence of particle size on energy dissipation on microstructured surface

The influence of the particle size on the energy dissipation processes was determined on the microstructured surface with an impact velocity of 0.3 m/s. In figure 7 the results of the experiments are shown. The particle size did not have an influence on the average energy dissipation. The CoR

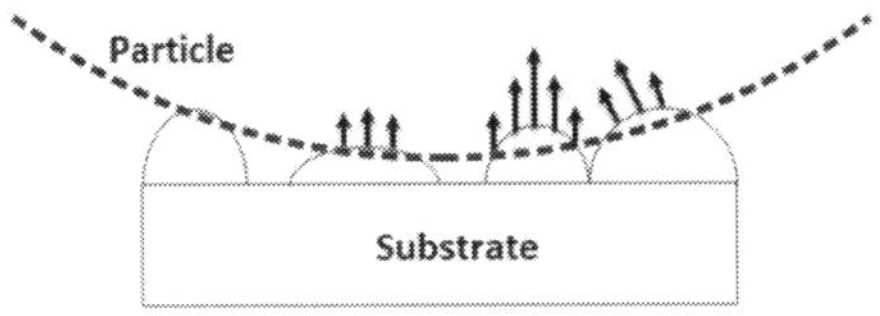

Figure 6. Sketch of asperities indentation into particle

was obtained by around 0.6 for both particle sizes. The ratio between the 400 μm and 665 μm was 1.05. Nevertheless the coefficient of variation of the CoR for the smaller particles is higher (12% to 6%) compared to the 665 μm particles. This can be explained with the inhomogeneous stainless steel particle distribution on the microstructured surface. The SEM image (figure 3) of the surface shows places of original surface and small craters between the cold sprayed steel particles. Smaller polystyrene particles had a higher probability to collide with this particle free, flat part of the surface which led to less energy dissipation. Also the probability to collide only with steel particles could be higher which led to more energy dissipation. Montaine et al. and Gunkelmann et al. investigated that small asperities can cause transfer of linear momentum into rotations which lead also to an energy dissipation. The scatter of the coefficient of variation was also caused by the apserities density [6,7]. Therefore the coefficient of variation of the impact tests of the CoR was at 400 μm particles twice as big as the deviation of 665 μm particles.

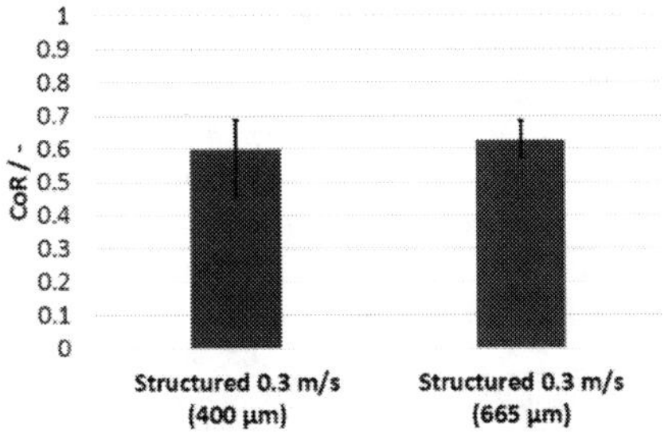

Figure 7. Comparision of the coefficient of restitution (CoR) for 400 μm and 665 μm particles. The particle impacts were performed with the structured surface at an impact velocity of 0.3 m/s

4 Conclusion

In this paper the influence of a statistically microstructured stainless steel surface on the energy dissipation during particle collision was investigated. Experiments were performed with polystyrene particles in the size range of 400 – 700 μm at different impact velocities (0.3 – 0.6 m/s). The experiments have shown that the energy dissipation depends on the microstructure and impact velocity. The microstructured surface absorbed relative to the kinetic energy at impact around 3.87 – 10.92 times more energy compared to a polished surface due to an inhomogeneous stress distribution at the particle contact with asperities. Furthermore, an additional dissipation is caused by friction during the indentation of the steel particles into the polystyrene particles. A variation of the particle size did not show any effects on the average energy dissipation for the structured surfaces at an impact velocity of 0.3 m/s. However the coefficient of variation of the 400 μm was higher compared to the 665 μm particles. This is caused by the inhomogeneous particle distribution of the cold sprayed particles on the microstructured surface. Because of the smaller diameter, there is a higher chance for the 400 μm particles to collide with a polished area, so that no additional friction processes occurs.

Acknowledgements

This research was funded by the German Research Foundation (DFG) within the CRC 926 “Microscale Morphology of Component Surfaces”, subprojects A08 and B03.

References

[1] Raymond M. Brach , Patrick F. Dunn and Xinyu Li, Journal of Adhesion **74**, 227-282 (2000)

[2] R. M. Brach, P. F. Dunn and W. Cheng, Journal of Aerosol Science and Technology **30**, 1321-1329 (1999)

[3] Sergiy Antonyuk, Manoj Khanal, Juergen Tomas, Stefan Heinrich, Lothar Moerl, Chemical Engineering and Processing **45**, 838-856 (2006)

[4] B. Crüger, S. Heinrich, S. Antonyuk, N. G. Deen, J.A.M. Kuipers, Chemical Engineering Research and Design **110**, 209-219 (2016)

[5] K. Schmidt,S. Buhl, N. Davoudi, C. Godard, R. Merz, I. Raid, E. Kerscher, M. Kopnarski, C. Müller-Renno, S. Ripperger, J. Seewig, C. Ziegler, and S. Antonyuk: Ti Surface Modification by Cold Spraying with TiO2 Microparticles, Surface and Coatings Technology **309**, 749-758 (2017)

[6] Gunkelmann et al. "Stochastic behavior of the coefficient of normal restitution" Phys. Rev. E 89, 022205 (2014)

[7] Montaine et al. "Coefficient of restitution as a fluctuating quantity" Physical Review E 84, 041306 (2011)

Effects of a large number of cycles on pile shaft resistance analyzed at the grain scale using x-ray tomography

Jeanne Doreau–Malioche[1,*], *Gaël* Combe[1], *Jean-Benoît* Toni[1], and *Gioacchino* Viggiani[1]

[1] *Univ. Grenoble Alpes, CNRS, Grenoble INP*[**], *3SR, F-38000 Grenoble, France*

Abstract. This study presents the results of laboratory-scale cyclic loading tests performed on an instrumented pile in sand using x-ray micro tomography and three-dimensional (3D) image analysis techniques. The macroscopic behaviour of sand-pile interface shows a two–phases evolution during cycles with a non negligible increase of shaft friction in the second phase. A discrete version of Digital Image Correlation (DIC) is employed to analyze quantitatively the mechanisms occurring at the grain scale. Using segmented images this method is able to correlate and follow each grain individually between two configurations. Displacement fields are measured and compared for different amounts of cycles. Grain breakage and density evolution are investigated using grey level measurements. The results provide a better understanding of the phenomena observed at the macroscale for a high number of load cycles.

1 Introduction

Foundations of offshore platforms, wind or tidal turbines, are submitted to a significant number of load cycles due to specific environmental conditions. High cyclic loadings involving a complex response of the soil were shown to give rise to abrupt changes mainly in shaft resistance. Although potentially important, such effect lacks a micromechanical understanding and remains rarely addressed in current pile design criteria. Thus, mechanisms controlling the macroscopic behaviour of sand-pile interface must be investigated at the grain-scale.

From this point of view, literature gathers numerous experimental studies focused on the interface conducted at different scales on laboratory models. Links between sand kinematics, local variations in sand density, grain crushing and the macroscopic response were proposed. Yang et al. [1] studying model piles installed in a uniform pressurized sand and parallel interface ring shear tests concluded on the potential influence of particle crushing, interface abrasion, shear banding and fines migration on the behaviour of displacement piles in sand during the installation and following cycles. Similar results were observed using DIC and postmortem images for cone penetration tests performed in a half-circular chamber [2].

Using laboratory model piles installed in a large calibration chamber Silva [3] investigated the soil–pile interface response during high cyclic loading tests. Local stress measurements showed a radial contraction of the soil close to the pile. Preliminary postmortem observations on grain breakage and local density changes at the interface were linked to shaft friction evolution during cycles. After a degradation phase observed by several authors, an increase in skin friction was measured by Bekki et al.[4] for high number of cycles (between 100 and 100 000 cycles). These phenomena were attributed to the formation of a shear band of crushed grains around the shaft and a sand densification around the probe.

However, these mechanisms were observed using various optical methods either postmortem on sections of the tested specimen or with plane strain devices mainly during pile installation. The present work aims at establishing a direct link between the macroscopic and microscopic behaviour of the sand–pile interface during cycles. This paper presents the first results of *in operando* cyclic loading tests using x–ray tomography for the 3D–analysis of the deformation mechanisms at the interface.

2 Methodology

2.1 Experimental setup

The tests were conducted in a mini-calibration chamber in the tomograph of Laboratoire 3SR, in Grenoble, France. An instrumented 14mm diameter cone–ended aluminium model pile was first installed by monotonic loading in a dry calcareous sand sample (Glageon sand is derived from a carbonate rock crushed in Bocahut quarry, France) under an isotropic stress of 100kPa. Following the installation phase, the model pile was submitted to a thousand axial displacement-controlled cyclic loadings.

The samples (70 diameter and 120mm height) were prepared by deposition of consecutive layers of 20mm thickness and a density of $1.5 g.cm^{-3}$ (loose samples). The material was chosen for its mean grain size (D_{50} =1mm),

*e-mail: jeanne.doreau@3sr-grenoble.fr
**Institute of Engineering Univ. Grenoble Alpes

its uniform grading and its low resistance to breakage. The size of the sample and the mean grain size are dictated by the scanning conditions of the tomograph.

The minicalibration chamber consists in a pseudo-triaxial cell transparent to x−rays. The confining pressure was maintained constant during the test. For practical reasons, the pile was installed by jacking from the bottom of the cell at a rate of 0.06mm/s with a pile embedded length of 45mm. The cycles were performed at the same loading rate with an amplitude of ±0.5mm around the pile tip reference position, alternating between compression and tension phases (two−ways cycles). The model pile is instrumented with a load sensor at its end to measure the tip load. By subtracting the total load applied on the pile, its shaft capacity was also calculated and followed during cycles.

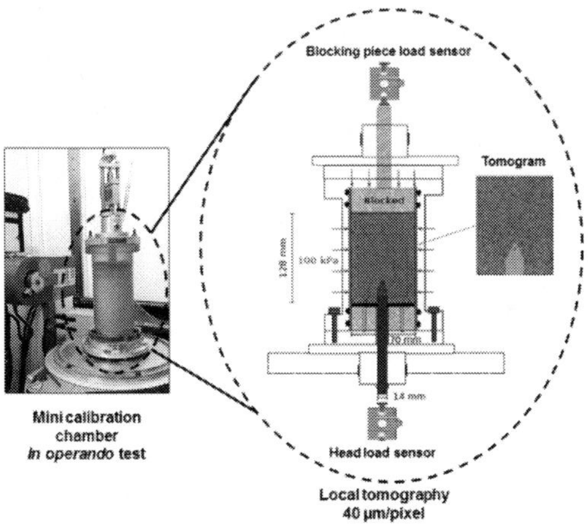

Figure 1. Experimental setup during x-ray micro tomography.

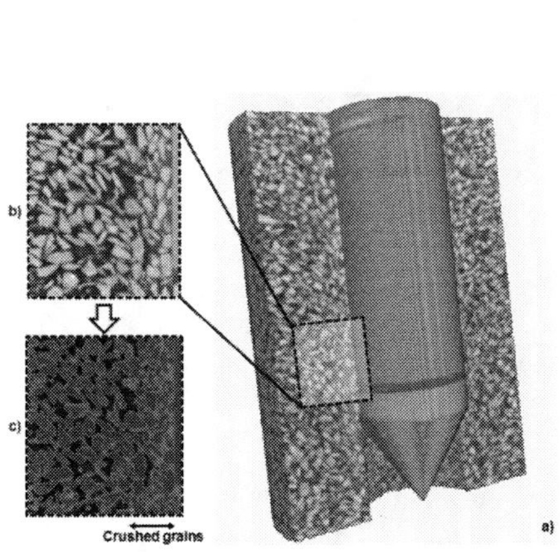

Figure 2. a) 3D reconstructed volume after pile installation in local tomography. b) Zoom on the interface. c) Detection of different phases (red:grains and pile, blue:air and green:fines).

2.2 Microscale image analysis

The loading on the model pile was applied progressively, stopping at defined amount of cycles for 3D scanning (1, 10, 50, 100, 500, 1000 cycles). Scans were taken in local tomography, with a field of view focused on the tip and the shaft, for a pixel size of 40μm. Scans were settled to a tension of 150kV and a current of 200μA. For each tomography, 1024 radiographies were made of the sand samples for the reconstruction of the entire volume.

To extract information at the microscale, different image analysis techniques were used. Fields of displacement and strain can be calculated using the 3D DIC code, TomoWarp2, by assuming a continuous displacement field, at least within each subset (containing about 8 grains) and, locally, the transformation to be a rigid translation. In this study, a discrete version of TomoWarp2 [5], recently implemented, able to correlate each grain individually was employed to take into account the granular aspect of the sand and of its mechanical response. In this approach, the correlation window is centred on individual grains and follows the actual shape of the grain. Full kinematics of each particle were then measured for different loading stages.

To characterize the state of the granular assembly during cycles, x-ray tomograms were analysed based on grey intensity changes. Density variation was followed at the interface for different amount of cycles as well as the quantity of fines produced by grain breakage. The analyses were conducted using the free and open−source image processing software Fiji.

3 Results

3.1 Friction force at the grain−pile interface

The measurement of the force induced by friction at the sand-pile interface during cyclic loadings indicates two separated regimes. For the first hundred cycles, shaft friction does not vary significantly, whereas a constant increase is observed for the following cycles (see Figure 3a). Figure 3b represents the maximum friction measured when the tip displacement reaches ±0.5mm for various two-ways displacement−controlled cyclic tests performed on Glageon sand, with identical testing parameters. A similar trend showing the reproducibility of the results is observed. Comparing the maximum friction obtained before and after cycles, the friction force has doubled, with a sharp transition between the two regimes. It is important to note the difference in friction intensity between the tests GLAG−C3, GLAG−C2 and GLAG−C1. Despite the smoothing with a very fine sandpaper before each experiment, a slight change in pile surface roughness is expected.

The above results are consistent with those previously obtained in the large calibration chamber of 3SR for a higher number of cycles [3]. In the same testing conditions, the shaft friction was shown to increase after 10000 cycles. The difference in the number of cycles necessary to reach the same trend can be explained by two different scale parameters: the initial density of the tested specimen and the aspect ratio between the model pile diameter and the mean grain size (14 in the present study against 166 for the large calibration chamber).

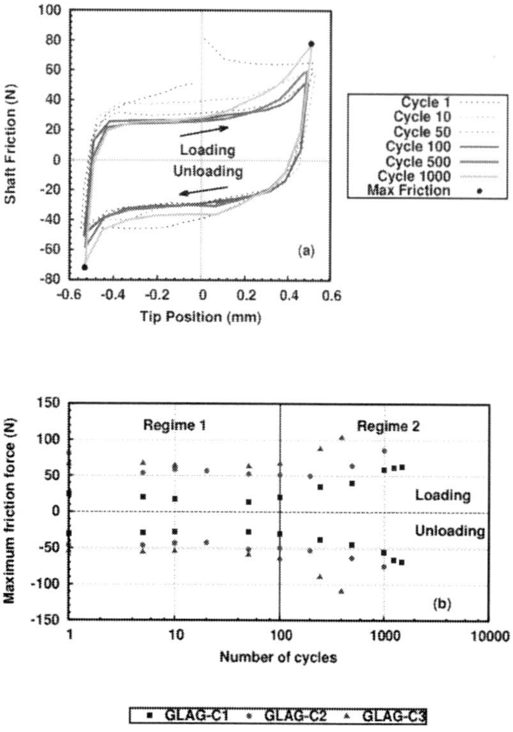

Figure 3. a) Shaft friction evolution during cycles for GLAG-C2 (the black arrows indicate the loading and unloading). b) Identification of two regimes in shaft friction evolution (change in trend at 100 cycles noted by the black vertical line).

3.2 Granular kinematics

3D–DIC was successfully used to study displacement and strain fields during the installation phase of a pile ([6], [7] and [8]). In the present case, the effects of cycles were analysed at the grain scale. Typical results from the discrete approach of 3D–DIC for two loading steps are presented : between 10–50 cycles and 500–1000 cycles. Displacements are plotted for each individual grain in a vertical plane passing through the pile axis.

Although the main granular rearrangement is occurring during the pile penetration, cyclic loading induces significant radial displacements. The results show a global movement of the grains towards the pile shaft for both loading steps (see Figure 4b). However, the sand mass undergoes tiny vertical displacements (see Figure 4a). Due to the low roughness of pile surface, grains tend to slide alongside the shaft. By comparing vertical and horizontal displacement fields after 50 and 1000 cycles, two different behaviours can be observed. The first tens of cycles cause higher grain movements within a larger area around the shaft.

Besides grain kinematics, discrete DIC proves to be a powerful tool offering a wide range of further possible analyses in terms of grain orientations, grain shape evolution and grain breakage quantification.

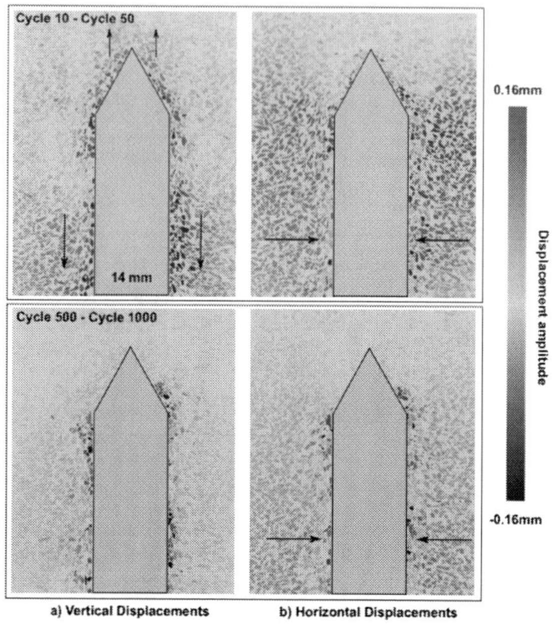

Figure 4. Individual grain displacements from discrete DIC for two loading stages: black arrows in the figure show global directions (GLAG–C2).

3.3 Local densification at the interface

Preliminary observations of raw x-ray tomograms show a clear evolution in density at the interface during cycles. 3D images were then further analysed looking at grey level in order to assess quantitatively the effect of cycles. Three phases were first identified in the images: pores (low grey level), intact grains (high grey level) and fines produced by crushing. For the given spatial resolution, the powder produced by grain breakage can not be clearly observed but by partially filling voxels, it can be associated to an intermediate mean grey level. However, fine particles and the edges of intact grains present similar x-ray absorption, leading to possible misinterpretation of particle boundaries. Choosing an appropriate threshold, the percentage of voxels belonging to each phase was estimated within 5–voxels–thickness hollow cylinders centred on the pile axis. Figure 5a shows the evolution of the density of each phase plotted in terms of distance from the shaft normalized by the mean grain size and the amount of applied cycles.

For cycle 1, each point represents one measurement at a given distance from the shaft. The two first data points should not be taken into account due to boundary effects at the sand–pile border on the images. The results show a significant increase of the percentage of voxels associated to intact grains (without taking fines into account) during cycles. Close to the interface, the quantity of grains is 20% higher after 1000 cycles, which indicates a local densification at the interface induced by cyclic loadings. It can also be observed that the region affected by the loadings is about $2{\cdot}D_{50}$ for cycle 1 and $4{\cdot}D_{50}$ for cycle 1000, which agrees with results given in [1] . In the rest of sand mass, the proportion of grains and pores remains constant (neglecting intensity changes between two scans).

In terms of fines production (see Figure 5b), only a slight variation (2%) is observed between the first and second loading stages. Later on, the percentage of fines is stable, meaning that most of particle crushing occurred during pile installation. A rearrangement around the shaft of crushed grains produced at the pile tip may be induced due to their own weight during the first loading and unloading steps, which might explain the apparent increase of fines quantity shown in figure 5b.

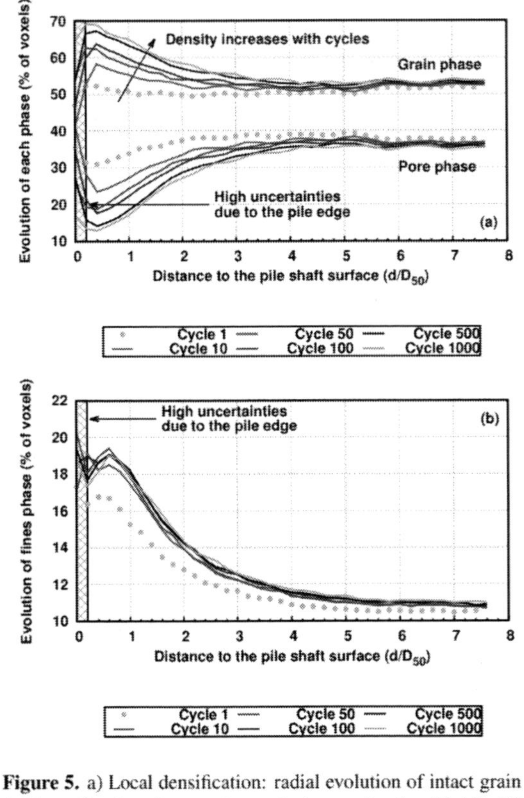

Figure 5. a) Local densification: radial evolution of intact grain (without fines) and pore phases during cycles for GLAG−C2 (scale: $\frac{7*d}{D_{50}} = 7mm = 0.5*$pile diameter. b) Evolution of fines quantity during cycles (GLAG−C2).

4 Conclusions

X-ray tomography and 3D image analysis were used to investigate cyclic load effects on sand-pile interface at the grain scale. Two distinct regimes were identified in shaft friction evolution according to the number of applied cycles. For these two phases, the measurement of grain kinematics revealed a different behaviour of the sand mass at the interface. In addition, a quantitative study of density evolution showed a significant densification during cycles. These results suggest that around 100 cycles are needed for the sand to rearrange and reach a threshold density for which the friction starts increasing substantially.

5 Perspectives

Further image processing should lead to a better understanding of the link between grain kinematics, density changes and the different regimes in shaft friction evolution. A new technique using discrete DIC results is under development in 3SR in order to assess quantitatively grain breakage. This tool can be used to follow the evolution of grain size and shape during both monotonic and cyclic loading. Other experiments will be performed to explore the influence of various parameters: grain size with respect to pile diameter, grain crushability, amplitude of cycles and pile roughness.

Acknowledgements

The authors gratefully acknowledge LabEx Tec21 (Investissements d'Avenir, Grant Agreement No. ANR-11-LABX-0030) for its financial support. The authors also extend their thanks to Pascal Charrier, Christophe Dano, Matías Silva and Edward Andò for their contributions.

References

[1] Yang, Z. X., Jardine, R. J., Zhu, B. T., Foray, P. & Tsuha, C. H. C., Géotechnique **60**, No. 6, 469-482 (2010)

[2] Arshad, M. I., Tehrani, F. S., Prezzi, M. & Salgado, R., Géotechnique **64**, No. 7, 551-569 (2014)

[3] Silva, I. M. F., Université Grenoble Alpes, https://tel.archives-ouvertes.fr/tel-01204765 (2014)

[4] Bekki, H., Tali, B., Canou, J., Dupla, J. C. & Bouafia, A., J. Appl. Eng. Sci. Technol. **2**, No. 2, 51-55 (2016)

[5] Tudisco, E., Andò, E., Cailletaud, R. & Hall, S. A., TomoWarp2: a local Digital Volume Correlation code, submitted to Software XC, (2016)

[6] Silva, M. & Combe, G., Sand displacement field analysis during pile installation using x-ray tomography and digital image correlation. International Symposium on Geomechanics from Micro to Macro. Sept. 1-3, Cambridge, UK, 1599-1603, (2014)

[7] Silva, M. & Combe, G., Soil-Pile interaction during pile installation. Observations from X-Ray tomography and 3D-DIC. 15th PCSMGE, 8th SCRM, 6th IS-BA, 2015, Buenos Aires, Argentina, 600-607, (2015)

[8] Silva, M., Doreau-Malioche, J. & Combe, G., Champs cinématiques dans un sable lors de l'enfoncement d'un pieu par tomographie RX: comparaison des corrélations numérique continue et discrète. 22ème Congrès Français de Mécanique, 2015, Lyon, France, (2015)

Stochastic model for the micromechanics of jammed granular materials: experimental studies and numerical simulations

Mathias Tolomeo[1,⋆], *Kuniyasu* Saitoh[2,⋆⋆], *Gaël* Combe[1], *Stefan* Luding[3], *Vanessa* Magnanimo[3], *Vincent* Richefeu[1], and *Gioacchino* Viggiani[1]

[1] *Univ. Grenoble Alpes, CNRS, Grenoble INP ⋆⋆⋆, 3SR, F-38000 Grenoble, France*
[2] *WPI Advanced Institute for Materials Research, Tohoku University, 2-1-1 Katahira, Aoba-ku, Sendai 980-8577, Japan*
[3] *Faculty of Engineering Technology, MESA+, University of Twente, Drienerlolaan 5, 7522NB, Enschede, The Netherlands*

Abstract. To describe statistical properties of complicated restructuring of the force network under isotropic compression, we measure the conditional probability distributions (CPDs) of changes of overlaps and gaps between neighboring particles by experiments and numerical simulations. We find that the CPDs obtained from experiments on a 2D granular material exhibit non-Gaussian behavior, which indicates strong correlations between the configurations of overlaps and gaps before and after deformation. We also observe the non-Gaussian CPDs by molecular dynamics simulations of frictional disks in two dimensions. In addition, the numerically calculated CPDs are well described by q-Gaussian distribution functions, where the q-indices agree well with those from experiments.

1 Introduction

The mechanics of granular materials has been widely investigated because of their importance in industry and science. However, mechanical properties of *jammed* granular materials are still not fully understood due to the disordered configurations of the constituent particles [1]. At the microscopic scale, the mechanical response to quasi-static deformations manifests as changes of the contact and force network [2], where complicated rearrangements of the constituents cause irreversible restructuring. This evolution under quasi-static deformation is described by the change of the probability distribution function (PDF) of these forces, where many theoretical studies [3, 4] have been devoted to determine their functional forms observed in experiments [5] and numerical simulations [6]. Macroscopic quantities are then defined as statistical averages over the contact forces.

Recently, we have proposed a stochastic approach to the irreversible restructuring of the force network in bidisperse frictionless [7] and polydisperse frictional [8] particles, where the change of the PDF of both forces and interparticle "gaps" was predicted by a master equation. The master equation encompassed the mechanical response to compression (or decompression), where transition rates, or conditional probability distributions (CPDs), fully determine the statistics of the micromechanic evolution of jammed granular materials.

⋆e-mail: mathias.tolomeo@3sr-grenoble.fr
⋆⋆e-mail: kuniyasu.saitoh.c6@tohoku.ac.jp
⋆⋆⋆Institute of Engineering Univ. Grenoble Alpes

In this paper, we measure the CPDs for the master equation by experiments on wooden cylinders [9] and compare our experimental results with those of numerical simulations.

2 Theoretical background

2.1 Microscopic responses of overlaps

When granular materials are deformed, the force network is restructured because of complicated rearrangements of the constituent particles (*i.e.*, *non-affine deformation*), see Fig. 1. To detect this restructuring of the contact and force network (including *opening and closing contacts*), we introduce the Delaunay triangulation (DT) (Fig. 2), where not only the particles in contact, but also the nearest neighbors without contact, *i.e.*, *virtual contact*, are connected by Delaunay edges. Then, we define a generalized overlap between the particles, i and j, as

$$x_{ij} \equiv R_i + R_j - D_{ij}\ , \qquad (1)$$

where R_i (R_j) and D_{ij} represent the radius of i-th (j-th) particle and the length of the Delaunay edge connecting the particles, i and j, respectively. Note that the generalized overlap is positive if the particles are in contact, while it becomes negative if the particles are not in contact (hereinafter, *virtual* contact), see Fig. 1.

The *affine response* of generalized overlap to isotropic compression is given, at first order, by $x_{ij}^{\mathrm{affine}} = x_{ij} + (D_{ij}\delta\phi/2\phi)$, where the area fraction of constituent particles increases from ϕ to $\phi + \delta\phi$ with the (small) increment,

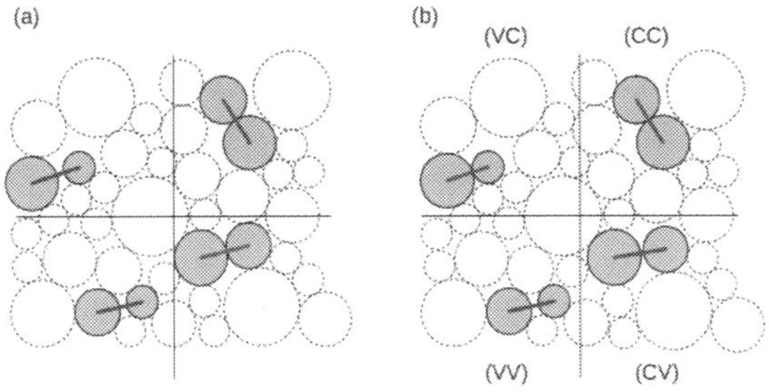

Figure 1: (Colour online) Schematic pictures of jammed packings (a) before and (b) after quasi-static deformation, where the red and blue solid lines represent contact status (red for grains in contact, blue for virtual contacts). The four kinds of transitions, (CC) contact-to-contact, (VV) virtual-to-virtual, (CV) contact-to-virtual, and (VC) virtual-to-contact, are displayed.

$\delta\phi$. In granular materials, however, constituent particles are randomly arranged and force balance is violated by compression. Then, the particles exhibit a deviation from affine behaviour and complicated rearrangements such that the force balance is restated. The system relaxes to another mechanical equilibrium and the generalized overlap changes to a new value, $x'_{ij} \neq x^{\text{affine}}_{ij}$, that is the *non-affine response* of the generalized overlaps. As shown in Fig. 1, there are only four kinds of transitions from x_{ij} to x'_{ij}: both overlaps are positive ($x_{ij}, x'_{ij} > 0$) or negative ($x_{ij}, x'_{ij} < 0$), if they do not change sign and thus (virtual) contacts are neither generated nor broken. We call these transitions *contact-to-contact* (CC) and *virtual-to-virtual* (VV), respectively. On the other hand, positive overlaps can change to negative ($x_{ij} > 0$ and $x'_{ij} < 0$) and negative ones can become positive ($x_{ij} < 0$ and $x'_{ij} > 0$), where existing contacts are broken and new contacts are generated, respectively. We call these transitions *contact-to-virtual* (CV) and *virtual-to-contact* (VC), corresponding to *opening and closing contacts*, respectively.

In the following, we scale the generalized overlaps by the averaged overlap before compression, $\bar{x}$, and introduce the scaled overlaps before deformation and after relaxation as $\xi \equiv x_{ij}/\bar{x}$ and $\xi' \equiv x'_{ij}/\bar{x}$, respectively, where we omit the subscript, ij, after the scaling.

2.2 A master equation for the PDFs of overlaps

The restructuring of the force network attributed to the transitions (CC, VV, CV, and VC) is encompassed by the evolution of the PDF of scaled overlaps, $P_\phi(\xi)$, where the subscript, ϕ, indicates the area fraction of the system. The PDF changes to $P_{\phi+\delta\phi}(\xi')$ under quasi-static isotropic compression [1]. However, it is difficult to predict the change from $P_\phi(\xi)$ to $P_{\phi+\delta\phi}(\xi')$ because the change from ξ to ξ' is *stochastic* rather than deterministic.

[1]The total number of Delaunay edges is conserved and the PDF is normalized as $\int_{-\infty}^{\infty} P_\phi(\xi)d\xi = \int_{-\infty}^{\infty} P_{\phi+\delta\phi}(\xi)d\xi = 1$.

To describe this stochastic evolution of the force network (or the DT), we connect the PDF after relaxation to that before compression through the *Chapman-Kolmogorov equation* [10],

$$P_{\phi+\delta\phi}(\xi') = \int_{-\infty}^{\infty} W(\xi'|\xi)P_\phi(\xi)d\xi\,, \tag{2}$$

where the transition from ξ to ξ' is assumed to be a *Markov process*. On the right-hand-side of Eq. (2), the CPD of scaled overlaps has been introduced as $W(\xi'|\xi)$ [2]. Then, a *master equation* for the infinitesimal evolution of the PDF can be derived from Eq. (2) as [10]:

$$\frac{\partial}{\partial\phi}P_\phi(\xi') = \int_{-\infty}^{\infty}\Big[T(\xi'|\xi)P_\phi(\xi) - T(\xi|\xi')P_\phi(\xi')\Big]d\xi\,, \tag{3}$$

where $T(\xi'|\xi) = \lim_{\delta\phi\to 0} W(\xi'|\xi)/\delta\phi$ has been introduced as the transition rate. The first and second terms on the right-hand-side of Eq. (3) represent the gain and loss of overlaps after deformation, ξ', respectively. Therefore, the transition rates or the CPDs describe the statistical properties of the micromechanic restructuring of the force network.

3 Experimental apparatus

In order to experimentally validate the stochastic approach presented in Sec. 2, we performed isotropic compression tests with the "1γ2ϵ" apparatus described in [9, 11]. To reproduce the behaviour of a granular assembly in 2D, Schneebeli rods (*i.e.*, 6 cm long cylindrical rollers) were used. The assembly, consisting of around 1850 rollers, is polydisperse (diameters from 8 to 20 mm). It is enclosed by a rectangular frame, see Fig. 2. Wooden rods (ash wood) were employed in order to take advantage of their quasi-elastic behaviour within the stress range applied (up to 250 kPa) at the boundaries of the sample.

The grain kinematics are assessed by applying Digital Image Correlation (DIC) on a set of 80 Mpixel digital images. In particular, the technique applied, called Particle Image Tracking, is an optimisation of DIC for the case of discrete materials, for which grain motion can be erratic [9, 12].

Assuming grains as 2D circles, contact deformation is modelled through the definition of a particle overlap, as in DEM, which is computed after detection of particle positions, following the definition in Eq. (1).

4 Numerical simulations

To model our experiments by numerical calculations, we employ molecular dynamics (MD) simulations of disks. The number of disks and the size-distribution are resembling those in our experiments, as well as the stiffness level κ, which allows us to assume that the same level of contact deformation is attained in the two cases.

The normal force between the particles in contact is modeled by a linear spring-and-dashpot, *i.e.*, $f_{nij} = k_n y_{ij} -$

[2]By definition, the CPD is normalized as $\int_{-\infty}^{\infty} W(\xi'|\xi)d\xi' = 1$.

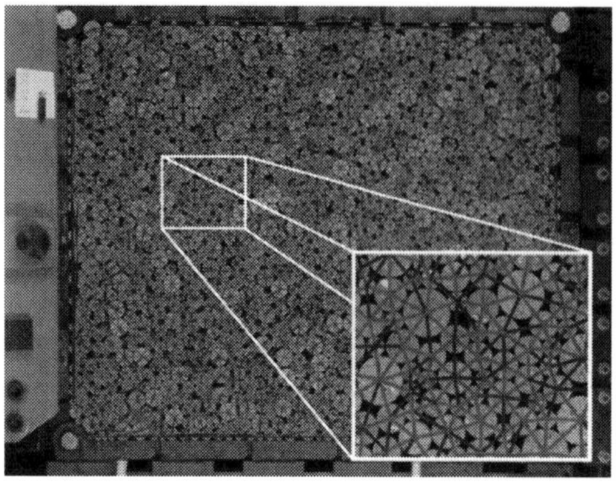

Figure 2: (Colour online) Image of the initial configuration of a sample under an isotropic stress loading of 20 kPa in the "$1\gamma2\epsilon$" device. The frame is 599.4 × 444.7 mm. The solid lines represent the edges obtained with a Delaunay Triangulation: red lines are for real contacts, while virtual contacts (*i.e.*, neighbours) are in blue.

$\eta_n \dot{y}_{ij}$, for $y_{ij} > 0$, (having assumed a unit mass for all disks, despite the different sizes), where k_n and η_n are a normal stiffness and normal viscosity coefficient, respectively, and y_{ij} is the overlap. Thus, the relative speed in the normal direction is given by its time derivative, $\dot{y}_{ij}$. The tangential force is also introduced as $f_{tij} = k_t z_{ij} - \eta_t \dot{z}_{ij}$, which is switched to the sliding friction or *Coulomb's friction*, $\mu|f_{nij}|$, when it exceeds this threshold. Here, we introduced $k_t = k_n/2$, $\eta_t = \eta_n/4$, and $\mu = 0.5$ (consistently with the experimental value) as tangential stiffness, tangential viscosity coefficient, and friction coefficient, respectively. In addition, z_{ij} and $\dot{z}_{ij}$ represent relative displacement and relative speed in tangential direction, respectively [13].

We first generate a dense packing of the disks in a $L \times L$ square periodic box as described in [7]. Then, we apply isotropic compression to the system by rescaling every radius proportionally to the area fraction increment $\delta\phi$ and to the radius itself. After compression, we relax the system to mechanical equilibrium. For later analyses, we introduce a *scaled strain increment* as $\gamma \equiv \delta\phi/(\phi - \phi_J)$, with the so-called jamming density, $\phi_J \simeq 0.8458$ [14], where we have confirmed in our previous study of bidisperse frictionless particles [7] that the widths of the CPDs were proportional to γ.

5 Results

Fig. 3 shows experimental results of the CPDs for CC transitions, where the scaled strain increment was $\gamma \simeq 0.074$. In this figure, the CPDs are nicely *symmetric* and are well described by $\gamma W_{CC}(\xi'|\xi) = f_c(\Xi_c/\gamma)$ (the solid lines), where $\Xi_c \equiv \xi' - m_c(\xi)$ is the distance from the mean value, $m_c(\xi)$, and

$$f_c(x) = \frac{1}{c(q_c)}\left[1 + \frac{x^2}{n(q_c)V_c^2}\right]^{\frac{1}{1-q_c}} \quad (4)$$

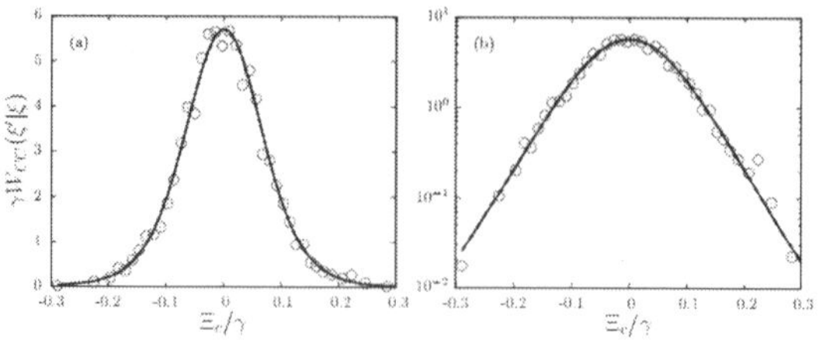

Figure 3: The CPDs (a) for CC transitions multiplied by the scaled strain increment, γ, and plotted against scaled distances from the mean values, Ξ_c/γ, with its semi-logarithmic plot (b), in experiments. The plot refers to one single loading step. Scaling of overlaps by the strain increment, γ, was applied. The solid lines are q-Gaussian distributions with $q_c \simeq 1.21$.

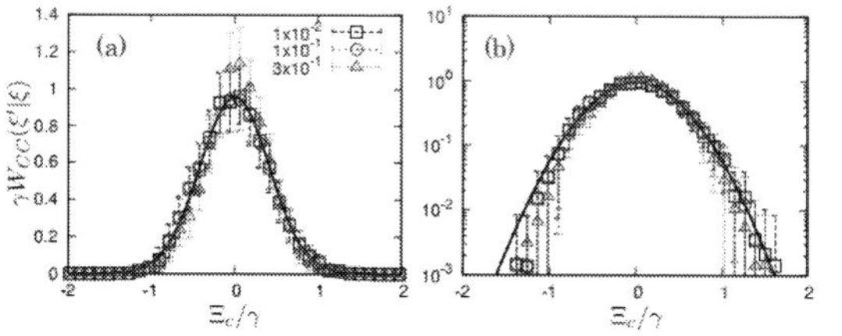

Figure 4: The CPD (a) for CC transitions with its semi-logarithmic plot (b) from numerical simulations. The different symbols represent different values of the scaled strain increment, γ, as listed in the legend of (a). The solid lines are q-Gaussian distributions with $q_c \simeq 1.19$.

is the *q-Gaussian distribution function* [15] with:

$$\begin{aligned} n(t) &= (t-3)/(1-t) \\ c(t) &= V_c\sqrt{n(t)}B(1/2, n(t)/2)\ , \end{aligned} \quad (5)$$

where $B(x, y)$ is the beta function. Then, the shape of the CPD is characterized by the *q-index*, q_c, defined in the range between $1 < q_c < 3$, where we find $q_c \simeq 1.21$ from the experiments. Therefore, the CPD for CC transitions deviates from the Gaussian distribution (corresponding to the limit $q_c \to 1$), indicating some moderate correlations between the changes of overlaps.

Fig. 4 shows our numerical results of the CPDs for CC transitions, where the different symbols represent different values of the scaled strain increment, γ. In this figure, all the data well collapse if we multiply the CPDs, $W_{CC}(\xi'|\xi)$, and distances from the mean value, Ξ_c, by γ and $1/\gamma$, respectively, which means that the CPDs are *self-similar* against the change of scaled strain increment γ, for small γ [7]. The solid lines are the q-Gaussian distributions, Eq. (5), where the q-index is given by $q_c \simeq 1.19$ which is very close to our experimental result.

Fig. 5 displays experimental results of the CPDs for VV transitions, where $\gamma \simeq 0.074$. The solid lines represent q-Gaussian fits, where the q-index is fitted as $q_v \simeq 1.31$.

Fig. 6 shows our numerical calculations of the CPDs for VV transitions, where the different symbols represent

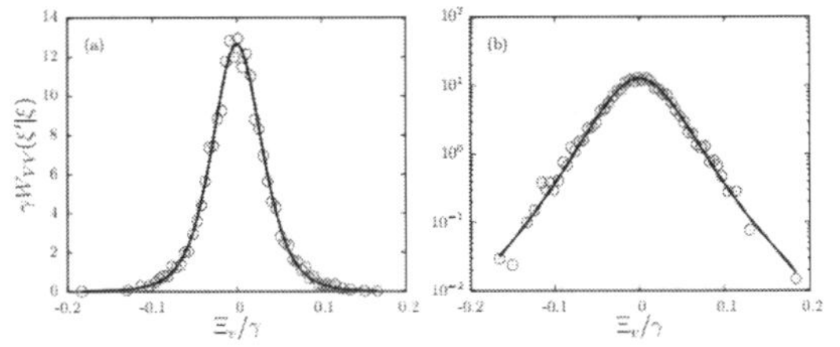

Figure 5: The CPD (a) for VV transitions with its semi-logarithmic plot (b) in the experimental case, as in Fig. 3. The solid lines are q-Gaussian distributions with $q_v \simeq 1.31$.

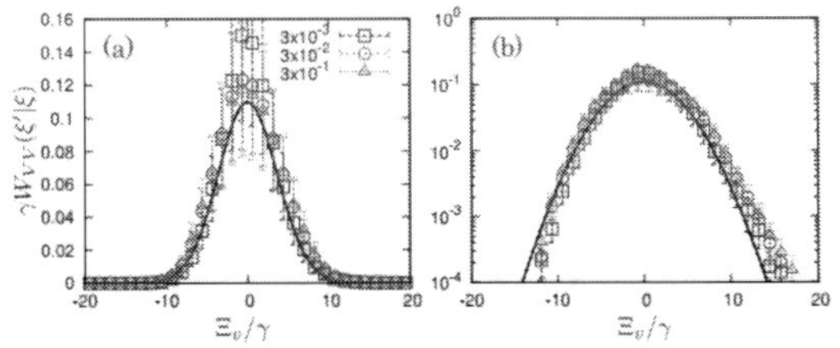

Figure 6: The CPD (a) for VV transitions with its semi-logarithmic plot (b), from numerical simulations, as in Fig. 4. The solid lines are q-Gaussian distributions with $q_v \simeq 1.09$.

different values of γ. Here, we also confirm the self-similarity of the CPDs, where they are well collapsed by the same scaling used in Fig. 4. In this figure, the solid lines are q-Gaussian fits with the q-index, $q_v \simeq 1.09$, which turns out to be considerably smaller than for the experiments and for frictionless data [7, 8].

6 Summary and outlook

In this study, we have experimentally and numerically investigated the CPDs of overlaps and gaps between neighboring particles in jammed granular materials. From our experimental tests of isotropic compression, we have found that the changes of overlaps and gaps are strongly correlated such that the CPDs exhibited the non-Gaussian behavior. Assuming Tsallis' q-Gaussian statistics for the complicated restructuring, we have described the CPDs by the q-Gaussian distribution functions, where the experimentally obtained q-indices well agreed with those obtained from our MD simulations of frictional disks.

Similar q-indices are related to similar shapes of the curves in Figs. 3-4; there are several possible interpretations for this similarity, as it can be related to long-range correlations (meaning that the same correlation length is found in experiments and simulations), but also to long-time correlations, memory and other alternative physical explanations. Apart from the exact interpretation, the most relevant feature to highlight is the reproducibility of some aspects between experiments and simulations.

In future, we plan to extend our study to the case of simple shear deformations, where more drastic restructuring and "recombinations" of the force network can be expected. For completeness, we also need to study the CPDs for contact-to-virtual (CV) and virtual-to-contact (VC) transitions, as in Ref. [7].

Acknowledgements

This work was financially supported by the NWO-STW VICI grant 10828. K.S. was also supported by the World Premier International Research Center Initiative (WPI), Ministry of Education, Culture, Sports, Science, and Technology, Japan (MEXT), Kawai Foundation for Sound Technology & Music, Grant-in-Aid for Scientific Research B (Grants No. 16H04025 and No. 26310205), and Grant-in-Aid for Research Activity Start-up (Grant No. 16H06628) from the Japan Society for the Promotion of Science (JSPS). A part of numerical computations has been carried out at the Yukawa Institute Computer Facility, Kyoto, Japan.
Laboratoire 3SR is part of the LabEx Tec 21 (Investissements d'Avenir, Grant Agreement No. ANR-11-LABX-0030).

References

[1] J. Lemaitre, J.L. Chaboche, *Mechanics of Solid Materials* (Cambridge University Press, Cambridge, UK, 1990)
[2] T.S. Majmudar, M. Sperl, S. Luding, R.P. Behringer, Phys. Rev. Lett. **98**, 058001 (2007)
[3] S. Henkes, B. Chakraborty, Phys. Rev. E **79**, 061301 (2009)
[4] J.H. Snoeijer, T.J.H. Vlugt, M. van Hecke, W. van Saarloos, Phys. Rev. Lett. **92**, 054302 (2004)
[5] D.M. Mueth, H.M. Jaeger, S.R. Nagel, Phys. Rev. E **57**, 3164 (1998)
[6] L.E. Silbert, G.S. Grest, J.W. Landry, Phys. Rev. E **66**, 061303 (2002)
[7] K. Saitoh, V. Magnanimo, S. Luding, Soft Matter **11**, 1253 (2015)
[8] K. Saitoh, V. Magnanimo, S. Luding, Comp. Part. Mech. (2016), doi: 10.1007/s40571-016-0138-z
[9] G. Combe, V. Richefeu, M. Stasiak, A.P. Atman, Phys. Rev. Lett. **115**, 238301 (2015)
[10] N.G. van Kampen, *Stochastic Processes in Physics and Chemistry, 3rd edition* (Elsevier B. V. Amsterdam, The Netherlands, 2007)
[11] F. Calvetti, G. Combe, J. Lanier, Mech. Cohes.-Frict. Mater **2**, 121 (1997)
[12] G. Combe, V. Richefeu, AIP Conf. Proc. **1542**, 461 (2013)
[13] S. Luding, J. Phys.: Condens. Matter **17**, S2623 (2005)
[14] M. van Hecke, J. Phys.: Condens. Matter **22**, 033101 (2010)
[15] C. Tsallis, J. Stat. Phys. **52**, 479 (1988)

Stochastic heterogeneous material modeling for wave propagation in a ballast layer

Lúcio de Abreu Corrêa[1,*], *Régis* Cottereau[1,**], *Charles* Voivret[2,***], *Sofia* Costa D'Aguiar[2,****], *Estelle* Bongini[2,†], and *Baldrick* Faure[2,‡]

[1] *MSSMat, CNRS, CentraleSupélec, Université Paris-Saclay, 92290 Châtenay-Malabry, France*
[2] *I&R, SNCF, 75012, Paris, France*

Abstract. This paper discusses the dynamical behavior of a randomly-fluctuating heterogeneous continuum model of the ballast. The Young's modulus is modeled as a random field parameterized by its average, its variance and a correlation model. A numerical model of the ballast and the surrounding soil is then constructed based on an explicit spectral element solver. This model allows to describe numerically the wave field generated, as well as to construct dispersion curves for the ballast-soil model. The influence of heterogeneity is discussed by comparison with a similar model where the ballast is assumed homogeneous.

1 Introduction

The competition with other means of transportation has increased the demand for performance in the railway industry. The dynamical loads caused by the passage of high-speed trains accelerate track deterioration and damage neighbor buildings [1]. In order to study this problem two classes of numerical models are used to try to predict this behavior: (1) discrete approaches, in which each grain of the ballast is represented by a rigid body and interacts with its neighbors through nonlinear contact forces (using molecular dynamics or non-smooth contact dynamics); and (2) continuum approaches, in which the ballast is replaced by a homogenized continuum and the classical Finite Element Method (FEM, or similar) can be used to solve the problem. The discrete approaches are today capable of solving only a few meters-length of ballast. The absence of an efficient scheme for parallelization over large clusters of computers [2] and the coupling between the grains and the soil [3] remains open problems. On the other hand, the methods based on continuum mechanics (like FEM) can solve very large models thanks to efficient parallelization schemes. However, a homogeneous continuum cannot represent the complex fields of stress and strain present in the ballast layer.

The dynamical behavior of ballast can be completely described by its dispersion curve. The numerical simulation of dispersion curves in railway tracks is usually performed on homogeneous media [4–6]. The most advanced models consider 3D finite elements and model reduction [7]. Unfortunately, these homogeneous models cannot reproduce accurately the experimental observations [8]. The discrete nature of the material and the complex interaction between the grains produce a "non trivial" mechanical behavior. It is necessary to consider the granular character of the material. The simulation of wave propagation in granular media for simple geometries has been treated using discrete models, and has succeeded in reproducing a wide range of experiments [9, 10]. The influence of pre-stress on the wave velocity in particular can be simulated [11] as well as the influence of the pressure in the grain network [12]. The simple addition of random "heterogeneity" in a mass-spring network allows to reproduce the complex frequency response seen in measurements of sound propagation in a granular system [13]. Note that [9] observed the coexistence of an attenuated coherent signal and the "coda", which is a classical feature of seismic signals [14].

The objective of this work is to use a statistical approach to model the mechanical properties of the ballast layer. This statistical approach aims to reproduce in a continuum model the heterogeneity of the stress field in a discrete granular model. The mechanical properties (in practice the Young's modulus) are modeled as stochastic random field. This approach was introduced in [15] and we apply it here to the construction of dispersion curves of ballasted railway tracks.

2 Numerical Framework

The key point of this model is the description of the mechanical properties. To build this stochastic continuum heterogeneous medium we need to define two things:

*e-mail: lucio.de-abreu-correa@centralesupelec.fr
**e-mail: regis.cottereau.centralesupelec.fr
***e-mail: charles.voivret@sncf.fr
****e-mail: sofia.costadaguiar@sncf.fr
†e-mail: estelle.bongini@sncf.fr
‡e-mail: baldrick.faure@sncf.fr

Figure 1. Example of one of the 29 samples of granular medium simulated with the NSCD.

(1)first-order marginal law; and (2) a correlation model. The first-order marginal law for the Young Modulus was identified using a optimization process when one discrete model was compared with one stochastic continuous one. The correlation model chosen here is based in the literature.

The second point is the wave propagation in this stochastic heterogeneous model. The construction of the dispersion diagrams was made using 3 different models, they were described in the Sec. 2.2. The methodology adopted was based on a 2D-FFT. We applied the procedure explained by Alleyne [16] to recover the dispersion curves with time-displacement signals.

2.1 Description of the mechanical properties

2.1.1 Discrete Model

The model consists of 29 cylinders with a radius of 35 cm and a height of 39 cm. Samples were taken from this cylinders as 48X48X32 cm^3 orthogonal parallelepiped. The total number of particles in each cylinder is around 2700. The particles are convex polyhedra with diameters between 2.5 cm and 5 cm, whose shapes have been obtained by digitalization from real ballast, one example was presented in the Fig. 1. An isotropic confinement pressure of 60 kPa is applied on all samples, a vertical load on the upper face of 63 kN, and the gravity is considered within all samples. The model was calculated using non-smooth contact dynamics code [17].

2.1.2 Continuum Model

The continuum model has one cubic sample of 48X48X32 cm^3, taken from cylinders with radius of 35 cm and 39 cm high. A radial pressure of 60 kPa is applied, a pressure of 163 kPa was applied on the upper face. Gravity is considered, the density was assumed constant, 1500 kg/m^3. In terms of first-order marginal law, we will model Young's modulus field with a log-normal distribution. This model requires the definition of a mean value $\bar{E}$, a variance σ^2, and a correlation model. We took $\bar{E}$ = 80 MPa given by the INNOTRACK Project report [18], variance (estimate below) and correlation length. A theoretical correlation model was used here to estimate the correlation function, based in the Percus-Yevick model [19] , the characteristic diameter was took as 3.7 cm. The value of the Poisson ratio was taken from the report [18], as $\nu = 0.23$.

2.1.3 Identification Process

This task requires the comparisons between the discrete and the continuum media. We choose the stress (σ_{zz}) as comparison variable. This choice try to represent the same statistical behavior between the equivalent stress in discrete medium and the continuum one.

It is interesting to define a generalization of the notion of stress tensor for these granular media. Several proposals have been made by [20–22] for the static regime. Given a network of contact forces f_α^c (at contact points c and with coordinates α) in a granular medium, the equivalent stress field were used by [23, 24], and can be defined as:

$$\sigma_{\alpha\beta}^V = \frac{1}{V}\sum_{c=1}^{N_c} f_\alpha^c \ell_\beta^c, \qquad (1)$$

where ℓ^c is the vector linking the centers of the two particles in contact at c, the sum is on the N_c contact points in the averaging volume V, see [25] for more details. Obviously, this quantity depends on the size of the averaging volume (with more variability over the whole sample for smaller volumes).

One realization of the random Young Modulus was inserted in a classical FEM model. The stress distribution found in this model is compared using the L2 norm with the discrete one. At this moment the optimization process take place. We used a Nelder-Mead simplex algorithm [26]. This process stopped when the tolerance was reached. The process found the minimum distance is at $\sigma_E^2/\bar{E}^2 \approx 15.6$. For this variance a comparison of the stress distribution in log-log scale was presented in the Fig. 2, where a good approximation for the values around the mean was noted, the exponential behavior was well fitted, however the match in the tails was not perfect.

A small remark must be made here: This approach can be applied for others discrete models and contact models. Models based on molecular dynamics [27], for example, can add the influence of the grain stiffness. Other contact mechanics, like the inclusion of the rolling contact [28, 29] adding rolling resistance, they will change the equivalent stress distribution. This modifications would be captures by the statistical properties of the continuum model.

2.2 Wave propagation model

2.2.1 Geometry and boundary conditions

Three models were built, one of those models can be seen in the Fig.3, it was based in a 250000 elements, with 375 degree of freedoms in each element, given $\approx 94\text{x}10^6$ degrees of freedom. This model is 32 meters long. One prescribed point force was impose as excitation. A rickers

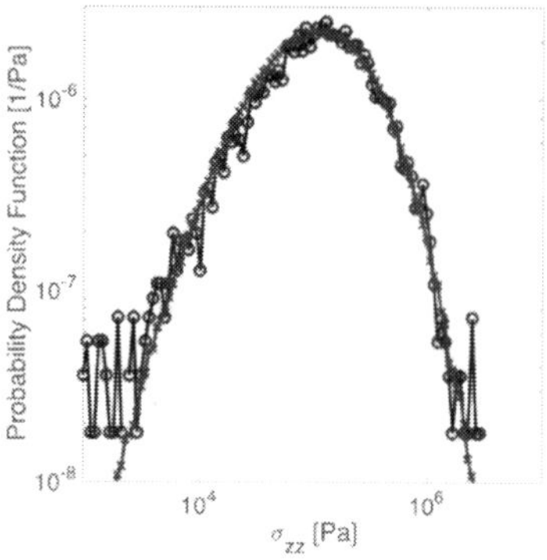

Figure 2. Relative distance between the probability distribution function σ_{zz} with $V = 8^3 cm^3$. Comparison of the probability distribution function with normalized variance $\sigma_E^2/\bar{E}^2 \approx 15.6$. The black line with circles is the equivalent stress distribution in the granular media and the blue line with cross is the stress distribution in the FEM model.

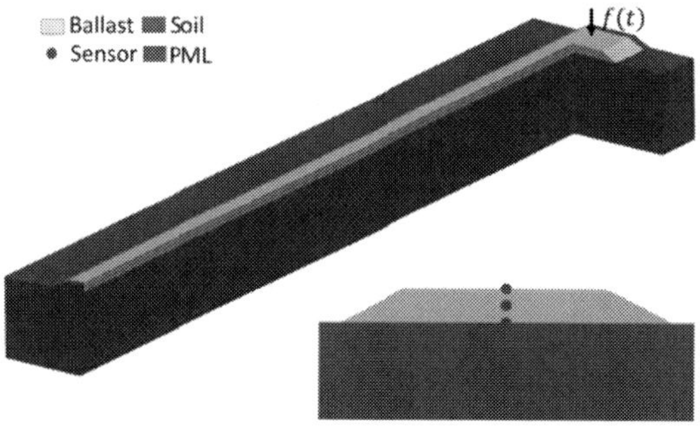

Figure 3. Cut view of the geometry and position of the captures lines. The red lines corresponds to the position of the captures lines. Three different sets where plotted at the top of the ballast, in the half of the layer, and on the interface between soil and ballast. The force is represented by the black solid arrow.

pulse with amplitude 1 and central frequency 210 Hz, for the other models the central frequency were modified. The Fig. 3 show also the region where vertical load is applied in the top of the ballast layer. The other two models had 320000 and 380000 elements with $120x10^6$ and $142x10^6$ degrees of freedom. They only difference is the length of the railway, one is 40 and the other 48 meters long. The excitation for those models were 100 and 30Hz. Due the large number of degree of freedoms present in the model we choose a parallelizable and scalable code. This code, called SEM3D, it solve the dynamic equation in a time domain, using a explicit time integration [30, 31].

The Fig.3 show these position of the captures used to record the displacement field in the model. Three rows of captures were put inside of the ballast co-linear to the longitudinal axis, placed in the top, half, and the bottom of the ballast layer.

The mechanical properties used in the soil are the same of the ballast the follow: V_s = 150m/s, V_p = 380m/s, and ρ = 1500kg/m^3. Two kind of simulations were performed in each model, homogeneous and heterogeneous for the ballast material. The statistical mean of the mechanical properties was keep the same in this two models.

3 Influence of the heterogeneity

We start with the homogeneous model of the ballast. The dispersion curves are plotted in Fig. 4, here we present just the result of the top line, the other had similar results. As expected, we retrieve approximately the dispersion relation for a Rayleigh wave in a homogeneous medium, with a velocity close to that of a shear wave (indicated by the black dotted line on Fig. 4). Although most of the displacement field corresponds to shear along the line that we are measuring, a small amount of P wave is also encountered at extremely low frequency, probably due a coupling between P and S waves in the low frequency.

The dispersion relations corresponding to the heterogeneous ballast are plotted in Fig. **??**. Two important effects can be observed: (i) the propagating behavior is dispersive in the low frequency range, with a very strong slowing of the wave velocity with increasing frequency, similar to that observed in the experimental work of [32] on unconsolidated granular packings, and (ii) the dispersion relation vanishes at higher frequencies. Note that it does not mean that there is no energy in the system, but only that the energy does not propagate. We also observed some energy along the P-wave dispersion curve, which is probably due to a strong coupling of P and S wave on the heterogeneities of the ballast.

4 Conclusions

In this paper, we constructed large scale randomly-fluctuating continuum models of ballasted railway tracks. These models allowed to construct dispersion curves for the ballasted layers. The importance of the heterogeneity was stressed very strongly. In particular, it was shown to create relatively slow waves at medium frequencies and a filter effect in the higher frequencies.

Because of the geometry of the ballast, and its heterogeneity, it is absolutely impossible to solve the dynamical equations analytically. Computer simulations are therefore an irreplaceable and unavoidable tool to understand the dynamical behavior of the soil-ballast system.

References

[1] D.P. Connolly, G. Kouroussis, O. Laghrouche, C.L. Ho, M.C. Forde, Construction and Building Materials (2014)

[2] T.M.P. Hoang, G. Saussine, D. Dureisseix, P. Alart, European Journal of Computational Mechanics **21**, 242 (2012)

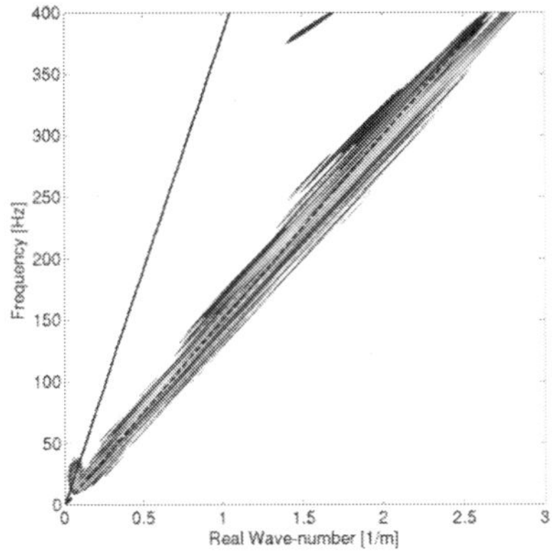

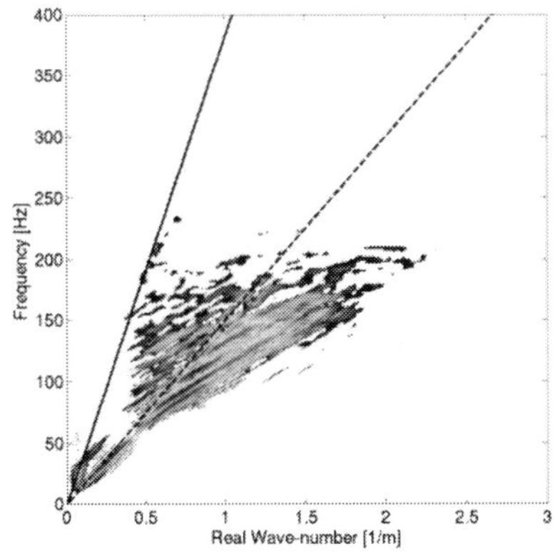

Figure 4. Dispersion curves for the homogeneous ballast resting on soil, upper figure, and dispersion curves for the heterogeneous ballast, lower figure. Dispersion relations for the shear wave (black dotted line) and pressure wave (black solid line) in the ballast are also represented.

[3] C. Wellmann, P. Wriggers, Computer Methods in Applied Mechanics and Engineering **205-208**, 46 (2012)
[4] X. Sheng, C. Jones, D. Thompson, **272**, 909 (2004)
[5] J.P. Freitas da Cunha, Ph.D. thesis (2013)
[6] X. Bian, C. Cheng, J. Jiang, R. Chen, Y. Chen, Acta Geotechnica pp. 1–14 (2015)
[7] E. Arlaud, S. D'Aguiar, E. Balmes, Computer Methods and Recent Advances in Geomechanics pp. 1193–1198 (2014)
[8] P. Anbazhagan, B.N. Indraratna, G. Amarajeevi, Faculty of Engineering-Papers (Archive) pp. 1–11 (2011)
[9] X. Jia, C. Caroli, B. Velicky, Physical Review Letters **82**, 1863 (1999)
[10] B.P. Lawney, S. Luding, Acta mechanica **225**, 2385 (2014)
[11] H. Shin, J.C. Santamarina, Journal of Geotechnical and Geoenvironmental Engineering **139**, 483 (2012)
[12] J.D. Goddard, Proceedings of the Royal Society A: Mathematical, Physical and Engineering Sciences **430**, 105 (1990)
[13] M. Leibig, Physical Review E **49**, 1647 (1994), 9301032
[14] K. Aki, J. Geophys. Res. **74**, 615 (1969)
[15] L. de Abreu Corrêa, J.C. Quezada, R. Cottereau, S. Costa d'Aguiar, C. Voivret (2017), unpublished
[16] D. Alleyne, P. Cawley, Journal of Acoustical Society of America **89**, 1159 (1991)
[17] LMGC, *Software for contact mechanics*, https://subver.lmgc.univ-montp2.fr/trac_LMGC90v2/
[18] INNOTRACK D2.1.3, Tech. Rep. Project no. tip5 ct-2006-031415, European commission - sixth framework program (2009)
[19] J. Quintanilla, Phys. Rev. E **60**, 5788 (1999)
[20] A. Drescher, G. de Josselin de Jong, **20**, 337 (1972)
[21] L. Rothenburg, R.J. Bathurst, **39**, 601 (1989)
[22] J.J. Moreau, *Numerical investigation of shear zones in granular materials*, in *Proceedings of HLRZ-workshop on friction, arching, contact dynamics* (1997)
[23] A.E.H. Love, *A Treatise on the Mathematical Theory of Elasticity*, Number v. 4 in A Treatise on the Mathematical Theory of Elasticity (Cambridge University Press, 1927)
[24] C. Voivret, Ph.D. thesis, Université Montpellier 2 (2008)
[25] J. Weber, Bulletin de liaison des Ponts et Chaussées pp. 1–20 (1966)
[26] J.C. Lagarias, J.A. Reeds, M.H. Wright, P.E. Wright, SIAM Journal on optimization **9**, 112 (1998)
[27] S.A. Galindo-Torres, D.M. Pedroso, Physical Review E - Statistical, Nonlinear, and Soft Matter Physics **81**, 1 (2010)
[28] J.M. J., Y.H. S., H. D., Computers and Geotechnics **32**, 340 (2005)
[29] J.M. J., S.Z. F., W.J. F., Computers and Geotechnics **65**, 147 (2015)
[30] D. Komatitsch, J. Ritsema, J. Tromp, Science (New York, N.Y.) **298**, 1737 (2002)
[31] D. Komatitsch, Geophysical Monograph Series **157**, 205 (2005)
[32] X. Jacob, V. Aleshin, V. Tournat, P. Leclaire, W. Lauriks, V.E. Gusev, Phys. Rev. Lett. **100**, 1 (2008)

Experimental characterization of mechanical properties of the cement paste-aggregate interface in concrete

M. Jebli[1,3,⋆], *F.* Jamin[1,3], *E.* Malachanne[2], *E.* Garcia-Diaz[2], and *M.S.* El Youssoufi[1,3]

[1]*LMGC, Univ. Montpellier, CNRS, Montpellier, France*
[2]*C2MA, Ecole des mines d'Alès, Alès, France*
[3]*MIST Lab., IRSN, CNRS, Univ. Montpellier, France*

Abstract. Granular materials are extensively used in the field of civil engineering. These materials are either used in their dry state, or mixed with water or with a binder. In the case of concrete, the binder could be cement or mortar. For ordinary concretes, it is generally admitted that there is a thin heterogeneous zone of paste, with a thickness of about 15-60 μm, surrounding the aggregates surface. This zone, commonly named the Interfacial Transition Zone (ITZ), is characterized by a higher porosity than the bulk paste and a high concentration of the portlandite crystals. Some of these crystals react with the aggregates' surface (limestone aggregates), leading to a good adhesion. In this work, the mechanical properties of the cement paste and of the cement-aggregate interface are experimentally analyzed. Experimental tensile and shear tests are performed on parallelipipedic samples. These samples are made by linking limestone aggregates with Portland cement paste using a water / cement ratio of 0.5. The results show that the cement-aggregate interface is the weak zone in the composite.

1 Introduction

Concrete is a composite material of aggregate particles and cement paste. The mechanical properties of concrete depend on the properties of the cement paste, the aggregates and the cement paste-aggregate interface. The cement-aggregate interface is characterized by a higher porosity than that found in the bulk paste. For this reason, the mechanical properties of this zone are considered different than those of the cement paste.

There is ample evidence that the properties of concrete, such as the tensile strength, modes of failure and the permeability; are significantly influenced by the nature of the existing interfaces between the cement paste and the aggregates. However, due to lack of understanding about these interfaces, very little attempts have been made to characterize the mechanical properties of this zone. Some studies have been devoted to the direct characterization of the mechanical properties of the interface between aggregates and cement paste, by using tensile and shear tests.

The literature of bond strength prior to 1965 has been admirably reviewed by Alexander et al. [1] and by Shah and Slate [2]. Both papers support the finding that the tensile strength of the cement-aggregate bond is lower than that of cement paste. Recently, Rao and Prasad [3] and Gu et al. [4] found that the tensile strength of the interface between aggregate and mortar was about one third to one half of the tensile strength of the mortar. Mielniczuk et al. [5] and Jebli et al. [6] have characterized the behavior of a composite with spherical aggregate using compression and tensile tests. They showed, at this scale, that the proposed models on the macroscopic scale by Eurocode 2 are confirmed to predict behavior at rupture.

In conclusion, all experimental studies to characterize the interface between cement paste and aggregate remain insufficient to assess the behavior law at local scale. In this context, the objective of this study is to determine the mechanical properties of the interface between the aggregate and the cement paste composite by direct tensile and shear tests.

⋆e-mail: mouad.jebli@umontpellier.fr

2 Experimental program

The selected cement is a Portland cement CEM II 32.5 CP2 NF used to prepare all composites with a water/cement ratio of 0.5.

2.1 Preparation for samples

Natural limestone parallelepipedic aggregates were made with dimensions of 10x10x15 mm^3.
In this experimental study, two types of samples whose dimensions are shown in Figure 1, were made:

- For tensile tests, the samples have a thin cement paste layer of 2 mm between two limestone aggregates (Fig. 1a).
- For direct shear tests, the samples are composed of a limestone aggregate and cement paste (Fig.1b).

The samples were prepared using specific molds adapted to the desired dimensions of the composite. All samples

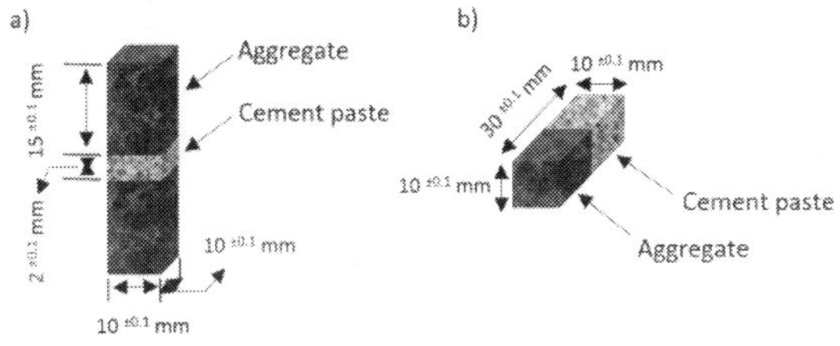

Figure 1. Sample sizes for direct tensile test a) and shear test b)

are demolded after 24 hours of hydration and are immediately immersed in water saturated with lime at a temperature of 21 °C to prevent the leaching and to decrease concrete shrinkage.
At different times of hydration (2, 7, 15, 28 and 90 days), we examine both the direct tensile and the shear bond strengths of the composite by means of specific devices. The shear tests are conducted for different normal stresses. All the tests are designed to characterize mechanical interaction laws at the local scale of the cement paste-aggregate interface.
To obtain the mechanical properties of the cement paste, mechanical tests were also carried out on cement paste samples of dimensions equal to those of composites.

2.2 Direct tensile test

To perform mechanical tests of direct tensile, experimental devices were designed and manufactured to be adapted to the size and shape of the desired samples. The principle of these tests is simple, but the preparation and implementation should be made carefully. Indeed, each mis-alignment of the supports or in the tilt sample induced parasitic bending which changes the results. Four metal shims with holes are first fixed onto the sample as shown in Figure 2. To perform the test, the sample is fixed by placing the metal rods in the holes of shims and of the U-shaped support set on the machine.

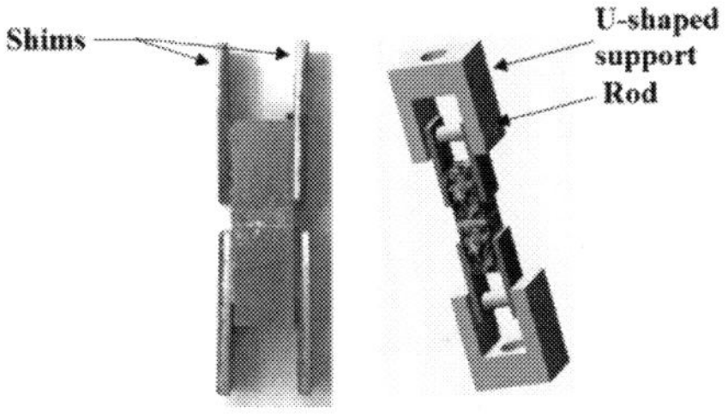

Figure 2. Experimental device for direct tensile test

2.3 Direct shear test

This test is among the most difficult tests to be performed on cementitious materials, especially to characterize the cement paste-aggregate interface due to the lack of appropriate experimental tools. The objective of our original approach is to access the Mohr-Coulomb law of cement paste-aggregate interface (local cohesion and friction angle). To perform this test, experimental devices were designed and manufactured in a way that the rupture is located at the cement paste-aggregate interface. The device consists of two half-boxes, the distance between these two half-boxes is around 0.5 mm to minimize the sample bending. The interface is fixed in the middle of the device with screws. These crews are also used to apply a normal force. Figure 3 illustrates the experimental apparatus in which the sample is placed inside the two half-boxes.

Figure 3. Experimental device for direct shear test

All tests were performed on a "Mechanical Testing System" by imposing a constant loading speed of 0.01 mm/s. Mechanical tests were filmed with a fast camera Vision Research Phantom V12 (acquisition speed of 10 to 40 thousand pictures per second), to correct the displacement during the tests by image correlation.

3 Experimental results

It is proposed to present the experimental results obtained from the direct tensile and shear tests, on cement paste and the composite. For all the tensile and shear tests, it is observed that all the breaks on the composite occur at the cement-aggregate interface (Fig. 4), which represents the weakest zone.

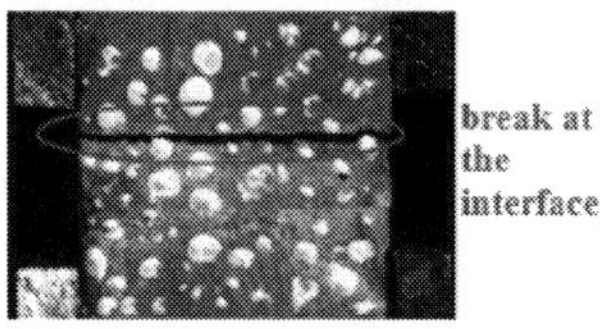

Figure 4. Crack growth at interface during tensile test

3.1 Direct tensile tests

Figure 5 shows the stress evolution as a function of the strain respectively for the composite "cement paste-aggregate" (Fig. 5a) and for pure cement paste (Fig. 5b). Significant increase in strength $\sigma_r(j)$ at the beginning of hydration, and a progressive stabilization after three weeks of hydration for the composite and cement paste is observed.

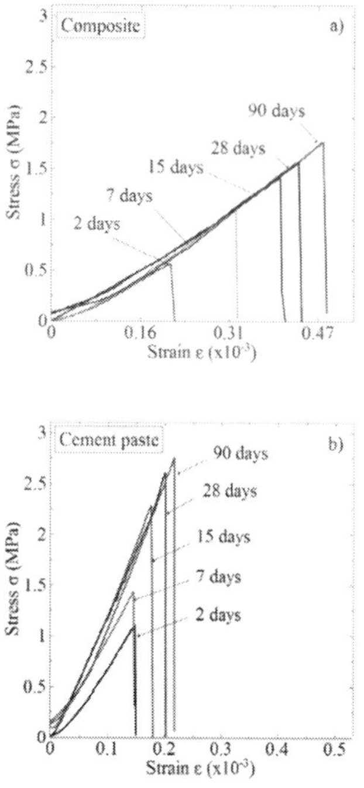

Figure 5. Evolution of the stress vs strain during tensile tests for composite a) and cement paste b)

Figure 6 shows the evolution of the strength $\sigma_r(j)$ in function of hydration time j during tensile tests for cement paste and composite. The results show a similar evolution of the strength, but at each hydration time the cement paste has higher characteristics than the composite. In addition, ratio between composite and cement paste is about 2/3 for the tensile strength. These results tend to confirm that the cement paste-aggregate interface is the weakest zone in composite. An increase of the strain values is also observed with curing, with higher values for the composite than that for the cement paste. Indeed this shows the more "ductile" character of the composite than of the cement paste.

3.2 Direct shear tests

The curves of stress - strain, obtained during direct shear tests (without normal stress), are presented in Figure 7 at different hydration times during the first 40 days for the cement paste and composite. These figures show a fast increase in the shear strength and stiffness of the samples at the beginning of hydration. We noticed a decrease in the evolution speed of the shear strength after 28 days. With both the shear test and the tensile test, a rapid drop in stress is observed just after it reached its peak. So, an absence of the softening part of the stress-strain curve is observed in macroscopic conventional mechanical behavior under shear stress. An increase of the strain values is

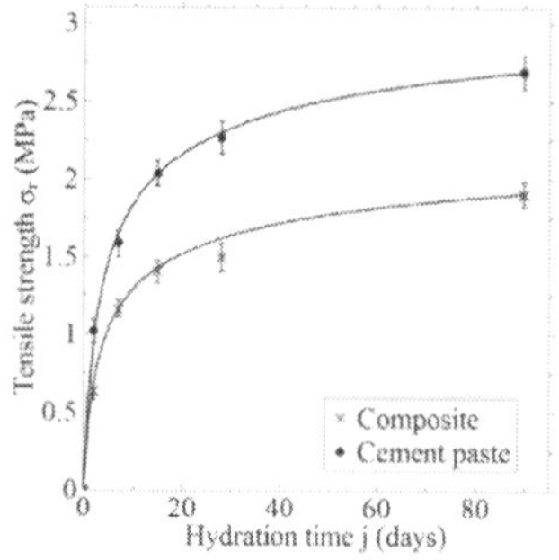

Figure 6. Evolution of tensile strength as a function of hydration time during tensile tests for composite and cement paste

also observed in the case of the composite with curing, with lightly higher values of the composite beyond 7 days than that of the cement paste. In this case the crack spreads quickly in the cement-aggregate interface. For the cement paste, breaking is conducted in the same strain value whatever the time of hydration.

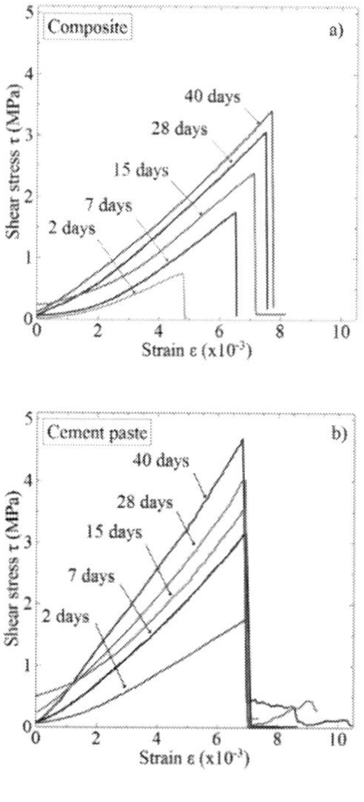

Figure 7. Shear stress-strain curves for composite a) and cement paste b)

The values of strength $\tau_r(j)$ (MPa) as a function of the hydration time j (in days), for the shear test for the cement paste and the composite, are shown on Figure 8. This figure shows a similar increase in strength during hydration time for both types of samples. After the characterization of the effect of hydration time on the shear strength, respectively for pure cement paste and for the cement paste-aggregate interface, we focus on the influence of normal

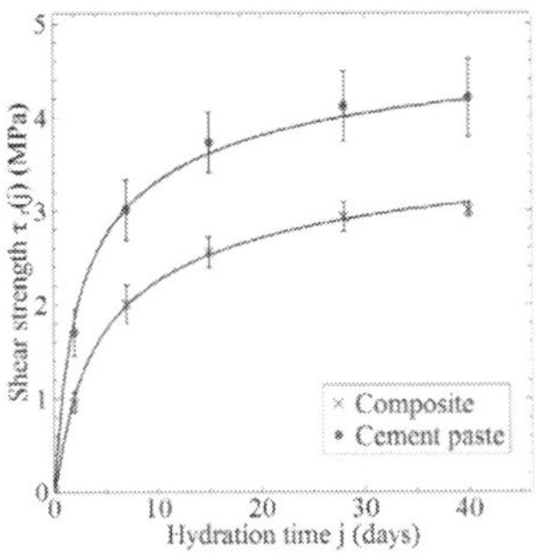

Figure 8. Evolution of the shear strength as a function of hydration time for the composite and the cement paste

stress on the shear strengths. The samples, at a fixed hydration time (40 days), are subjected to direct shear tests at different normal stresses. The obtained stress-strain curves are shown in Figure 9 for different normal stresses (0, 0.5, 1, 1.5 MPa).

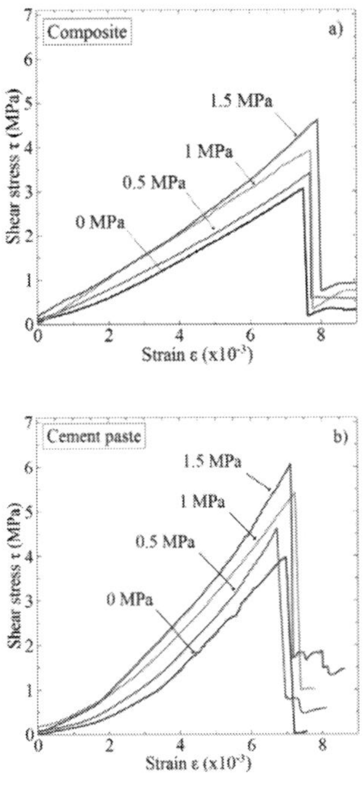

Figure 9. stress-strain curves during shear tests for composite a) and cement paste b) at different normal stresses and at 40 days

We noticed that there is a strong dependence of mechanical behavior of composite and cement paste on normal stress. For both samples, an increase in shear strength due to application of normal stress is noted. The initial tangential modulus also increases in the same way. This evolution of shear strength according to normal stress (Fig. 10) is used to calculate the local friction angle and cohesion according to the Mohr-Coulomb criterion. We noticed a linear evolution of the shear strength with the increase of normal stress, for both cement paste and composite. The results, presented in Table 1, show that the local cohesion c and the local friction angle Φ of the paste are greater than that of the composite.

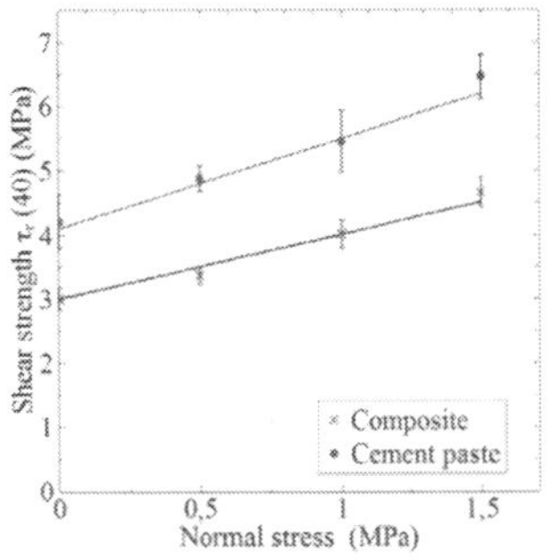

Figure 10. Mohr - Coulomb criterion for the composite and the cement paste at 40 days of hydration

Table 1. Local cohesion c and friction angle Φ of the composite and the cement paste at 40 days of hydration

	Cement paste	Composite
Local cohesion c (MPa)	4.1	2.9
Local friction angle Φ (°)	54	47

4 Conclusion

These results are a first approach to characterize the mechanical properties of the interface with the original experimental tools. The experimental results showed that the cement-aggregate interface is the weakest zone in composite. For both tests the strain values of the composite are higher than those of cement paste. Indeed, the more "ductile character" of the composite compared to the cement paste was more marked for the tensile test than for the shear test. These results at the local scale will improve numerical models (eg CZM) by identifying the parameters to be included in the calculation codes.

References

[1] K.M. Alexander, D.J. Gilbert, J. Wardlaw (1968)
[2] S.P. Shah, F.O. Slate, Cement and Concrete Assoc, London UK (1968)
[3] G.A. Rao, B.R. Prasad, Cement and Concrete Research **32**, 253 (2002)
[4] X. Gu, L. Hong, Z. Wang, F. Lin, Construction and Building Materials **46**, 156 (2013)
[5] B. Mielniczuk, M. Jebli, F. Jamin, M.S. El Youssoufi, C. Pelissou, Y. Monerie, Cement and Concrete Research **79**, 235 (2016)
[6] M. Jebli, F. Jamin, E. Garcia-Diaz, M. El Omari, M.S. El Youssoufi, Cement and Concrete Composites **12**, 241 (2016)

DOI: 10.1051/epjconf/201714015024

Methods of parallel computation applied on granular simulations

Gustavo H. B. Martins[1,*] and *Allbens P. F.* Atman[1,2,3,**]

[1]*Programa de Pós-Graduação em Modelagem Matemática e Computacional - PPG-MMC, Centro Federal de Educação Tecnológica de Minas Gerais, CEFET–MG, Av. Amazonas 7675, 30510-000, Belo Horizonte, MG, Brazil.*
[2]*Physics and Mathematics Departament, Centro Federal de Educação Tecnológica de Minas Gerais, CEFET–MG, Av. Amazonas 7675, 30510-000, Belo Horizonte, MG, Brazil.*
[3]*Instituto Nacional de Ciência e Tecnologia de Sistemas Complexos (INCT-SC).*

Abstract. Every year, parallel computing has becoming cheaper and more accessible. As consequence, applications were spreading over all research areas. Granular materials is a promising area for parallel computing. To prove this statement we study the impact of parallel computing in simulations of the BNE (Brazil Nut Effect). This property is due the remarkable arising of an intruder confined to a granular media when vertically shaken against gravity. By means of DEM (Discrete Element Methods) simulations, we study the code performance testing different methods to improve clock time. A comparison between serial and parallel algorithms, using OpenMP® is also shown. The best improvement was obtained by optimizing the function that find contacts using Verlet's cells.

1 Introduction

Granular materials have high relevance in Nature as well in humans activities like in mining, in food industry, in construction technology [3–9]. Basically, granular materials are everywhere, from sand to snow, from iron ore pellets to corn grains, from dust to stones. This ubiquity justify the high importance to known their behavior in order to use and manipulate them under different situations.

Several phenomena displayed by granular materials are quite surprising, and in general is a very difficult task to predict the behavior of a substantial quantity of granular material subjected to external loading, a very common situation. Perform experiments under controlled environment is the most desired approach to acquire knowledge about these systems, but most of time only with a detailed modeling of the system is possible to verify theoretical hypothesis. Thus, it is essential to develop a reliable computational model able to predict and reproduce granular behavior under different situations, and increasing system sizes.

Given the importance and presence of granular materials in the world, simulations offer lots of advantages for testing possible characteristic states, to predict behaviors, to save money in execution of projects, and to help plan next moves in engineering for instance. But some disadvantages may be present also, as the increasing of computation time for large systems sizes. With the increasing number of agents in simulation, higher is the computation time. Most cases, this increasing relation is quadratic in time because the nature of operations.

To improve simulations, and save time, more than one machine can execute the tasks at same time. Parallelization can solve single parts of the problem separately and get them together as part of the final solution, saving total operation time. Also, many researchers are using parallel computation resources in scientific programs to simulate DEM in particles systems [1, 2]. These DEMs are commonly used to simulate different kind of systems, like dense granular materials.

Is possible to perform parallelization in clusters of pc's or multicore computers, since they have more than one CPU (Central Processing Unit) available, that exchange information between them. In the model we use, some operations can be done separately, each one in one processor. More processors used, generally result in lower total execution time from a given job. The increasing number of cores in processors nowadays has turned this possibility very accessible. Also, GPUs (Graphics Processing Units) are becoming a cheaper option to improve a lot computational power.

Given all these possibilities to improve simulations, we propose to measure different techniques to simulate a paradigmatic example of granular system, the Brazil Nut Effect (BNE), and compare the performance of the algorithms. The BNE is a typical segregation phenomena observed in granular materials, which still have several open issues to be explored by simulations. We present results of BNE with frictional and frictionless walls.

In section 2, the methodology to simulate the system is shown, with different proposals to improve the time spent

*e-mail: gmartins@adm.cefetmg.br
**e-mail: atman@dppg.cefetmg.br

DOI: 10.1051/epjconf/201714015024

on serial and parallel algorithms, and some results. In section 3, the BNE results using these codes. In section 4, the main conclusions we got from our experiments.

2 Methods to save time in simulations

To simulate granular systems computationally, some models can be used [10]. One of this models considers rigid grains which are allowed to display some degree of interpenetrations, which are used in Kelvin-Voight rheology model, Cundal-Strack elastic contact in normal directions, and Coulomb friction in tangent direction, as shown in figure 1. By this, MD (Molecular Dynamics, that is a DEM) can be used [10–12] to integrate the motion equations of the grains directly from Newtons' laws. By choosing an appropriated time step, several different methods can be used to integrate these equations.

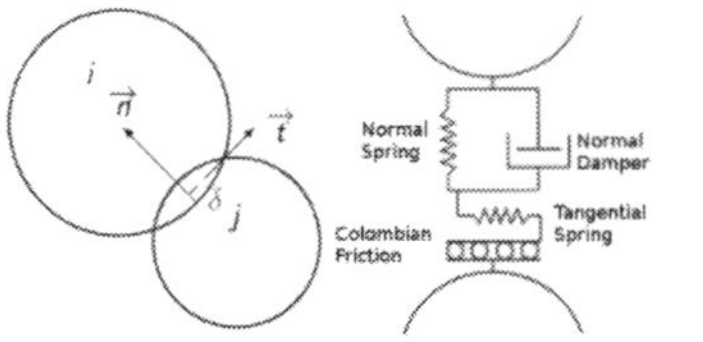

Figure 1. Force model between grain's interactions. i and j are particles that have contact each other. δ is the interpenetration value between the contact ij. **n** is the direction of the normal contact force. **t** is the direction of the tangential friction force.

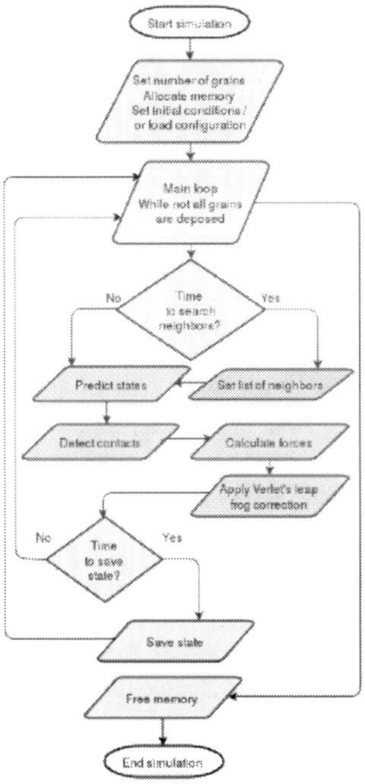

Figure 2. Flowchart of the simulation to simulate granular system using MD. Yellow operations can't be parallelized. Blue operations can be parallelized using CPUs. Green operation is the function that consume most processing, and can be parallelized based onto the type of implementation: *Set list of neighbors* as Verlet's list is the first improvement to reduce time in simulations. *Set list of neighbors* as Verlet's cells is the second improvement to reduce time in simulations.

In MD, a predictor-corrector method is used, and its scheme can be seen in Figure 2. The routine *Predict states* solves the kinematic equations over all grains over a single step. The routine *Detect contacts* find grains in contacts each other and store into a list. For the implementation without routine *Set list of neighbors*, *Detect contacts* look for all possible pair of grains for contacts, at every time step, and requires two loops over all grains. Otherwise, *Detect contacts* focus the search to the list of neighbor grains, and require one loop over the neighbor list. Both cases, it is produced a list of contacts, with the predicted states. The routine *Calculate forces* calculate the contact forces and also gravitational, requiring one loop over all particles and one loop over the list of contacts. The routine *Correct predicted states* corrects the predicted states and requires one loop over all particles. If the routine *Set list of neighbors* have Verlet's list implemented, it will search all possible pairs of neighboring grains, and requires two loops over all grains with a given frequency. If the routine *Set list of neighbors* have Verlet's cells implemented, it will search all neighbors grains' dividing the space in regions and look for possobli pair of contact grains only inside a cell and its neighboring cells. It requires one loop over all grains and other over all grains in the cells'neighbors.

To parallelize the simulation, all barriers should be identified to do a proper implementation. The first barrier is a temporal barrier, and happens in simulation each step. This first barrier can't be jumped by the temporal dependency of the system, as future depends on all past calculation done. Each particle can be treated independently, and at that point, the following routines *Predict states*, *Detect contacts*, *Correct predicted states* and *Set list of neighbors*, using Verlet's list, have no barrier to parallelize, so all loops were done to distribute all calculations to each processor available independently to read and write on memory. In the routine *Set list of neighbors*, using none of Verlet's improvement, the *Set list of neighbors* have no function, once all time steps the contact is calculated inside *Detect contacts*, and no barrier is present in parallelization. If the routine *Set list of neighbors* is using Verlet's cells, one barrier happens to insert grains in the cell it belongs, because to write on one same cell, two processors may try to do at same time. In the routine *Calculate forces* a barrier may happen, depending on implementation. If the 3rd Newton's law is applied on both bodies at same computational sequence when one contact happen, a barrier should be present to prevent memory concurrency by two or more contacts in a particle. This implementation gives an advantage of computation in serial execution that reduces the calculation by half, but a disadvantage in parallel execution by one barrier. The way done here is to calculate the pair of action and reaction separately, in each particle's loop.

The most expansive routine is detect contacts that searches all grains in contact each other, and its computational complexity can be written in function of the number of grains (n), as $O(n^2)$ with Verlet's list, and $O(n \log n)$ with Verlet's cells. Other routines have computational complexity $O(n)$. So, the total complexity of this algorithm can be written as $O(n^2)$ for Verlet's list, and $O(n \log n)$ for Verlet's cells.

2.1 Time spend

The clock time to simulate the system depends on the number of particles, the number of steps the system is simulated, the dynamic of the grains, the number of process running on the machine or cluster, the machine or cluster itself, the method used to implement the solution and many other controllable and uncontrollable variables.

To have an optimized parallel code, one need to know how much time serial functions costs, and their parallelizable portion. The results of time spent by each routine, in seconds, can be found on table 1, table 2 and table 3. The function *Detect contact* is the one which displays best results for application of parallization algorithm. In fact, Verlet's list and Verlet's cells are optimizations for the search algorithm. Both saves computational time and store closer grains into one list. The system do not update the neighbor list all steps of simulation, but only at a given frequency.

Table 1 shown the profile of serial runs of the code with 10^3 grains. For this number of grains, Verlet's cells and Verlet's list have little impact on execution times, but the simple search is the slowest, as expected. Table 2 have the profile of serial running of the code for 10^4 grains, and at this number of grains, Verlet's cells are faster than Verlet's list. Table 3 have the profile of serial running of the code for 10^5 grains, only for Verlet's list and Verlet's cells, because the simplest implementation got no result running after one week with this number of grains, and Verlet's cells are still faster than Verlet's list.

To compare and understand the parallelization, some metrics are defined. One of them is the performance of the system, that is the clock time measured by running that code. Faster codes gives lower results. Our results of performance can be found in figure 3, and for simulations with number of grains higher than 10^3, Verlet's cells gives best results.

The speedup is the comparison between clock time of execution in serial and the clock time of execution in parallel of same algorithm. It gives the tendency of the parallelization of the code. Best results gives higher curves. Our results of speedup, shown in figure 4, evince that parallelization decreased clock time to perform simulations aboce than 10^3 particles in the system. We can also conclude that *Detect contacts* without Verlet's implementations and Verlet's list are much more parallelizable than Verlet's list.

The efficiency is the comparison between speedup and the number of processors used. It gives the average utilization of each processor in the algorithm. Best results tends

Table 1. Clock time spent, in seconds, on each routine of serial running of the code for 10^3 grains for 10^3 time steps.

Function	Simplest	List	Cells
Predict states	0.01	0.03	0.03
Detect contact	22.47	0.15	0.18
Calculate forces	0.01	0.01	0.01
Correct states	0.01	0.04	0.02
List of neighbors	- -	0.22	0.01

Table 2. Clock time spent, in seconds, on each routine of serial running of the code for 10^4 grains for 10^3 time steps.

Function	Simplest	List	Cells
Predict states	0.27	0.19	0.21
Detect contact	2048	1.98	1.84
Calculate forces	0.09	0.06	0.05
Correct states	0.27	0.34	0.30
List of neighbors	- -	23.18	0.05

Table 3. Clock time spent, in seconds, on each routine of serial running of the code for 10^5 grains for 10^3 time steps.

Function	Simplest	List	Cells
Predict states	- -	2.93	2.87
Detect contact	- -	19.87	19.77
Calculate forces	- -	0.93	0.86
Correct states	- -	3.87	3.73
List of neighbors	- -	2,322.96	0.59

to have value 1, the full use of each processor. Figure 4 shown us that *Detect contacts* without Verlet's implementations and Verlet's list uses more the computational resources at same time than Verlet's cells implementation. For higher number of grains, they occupy more the computational resources, justifying the good use of parallelization with large systems.

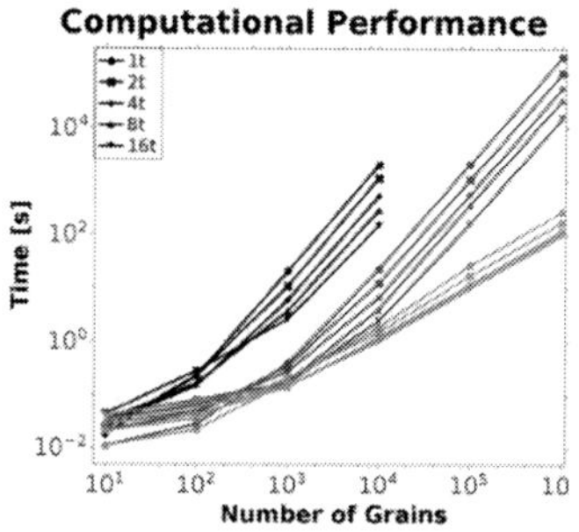

Figure 3. Performance of the system. This compares the results without Verlet's methods (in black), Verlet's list (in red) and Verlet's cells (in green). The number of processors used to simulate varies from 1 to 16, indicated on the legend.

3 Brazil Nut Effect (BNE)

The system has been validated comparing results with the theory of BNE [8, 9, 13, 14]. BNE is a segregation phe-

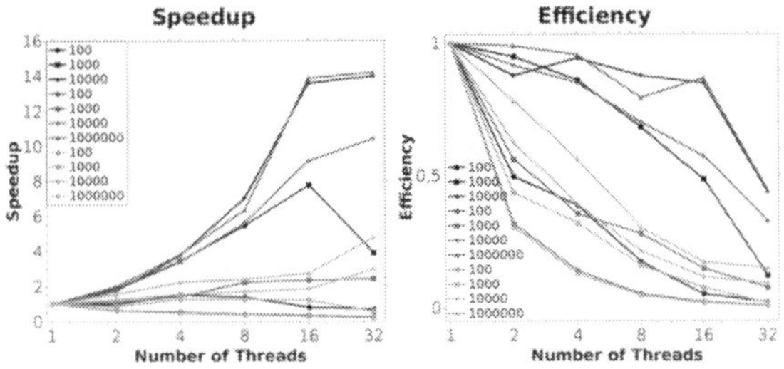

Figure 4. Speedup and Efficiency of the system. This compares the results without Verlet's methods (in black), Verlet's list (in red) and Verlet's cells (in green). The number of grains used to simulate varies from 100 to 1000000, indicated on the legend.

nomena which occurs when a system is shaken and a grain which is larger than the other grains in the media, rises to the surface. Friction is also an important influence to BNE, as can be seen in Figure 5, the larger grain rises in both cases for dimensionless accelerations higher than 1.00, but comparing friction walls with frictionless walls, the intruder with friction walls rises much more than the one without friction.

The parameters of the simulated system were: 2500 grains, uniform polydispersion of the radius of grains around 5%, friction coefficient of 0.5 between grains, and between grains and walls, normal spring stiffness of 1000 adimensional unities [3], tangential spring stiffness of 750 a.u., the radius of the intruder equals 3 times the radius of the larger grains in the media, density of the system equals to π^{-1} in a. u., critical damping coefficient in normal direction, time step of one tenth of the typical colision time for the smallest grains in the system. An imposed sinusoidal force is submitted to the box at 2500 a.u. and the dimensionless acceleration, $\Gamma = A\omega^2/g$, is varying between 0.85 to 1.5.

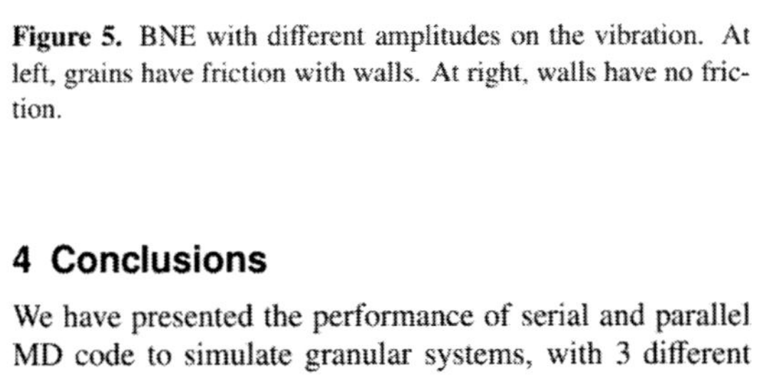

Figure 5. BNE with different amplitudes on the vibration. At left, grains have friction with walls. At right, walls have no friction.

4 Conclusions

We have presented the performance of serial and parallel MD code to simulate granular systems, with 3 different implementations on the computational function that costs most. The implementation that shown best performance was Verlet's cells, and parallelization present higher gain with increasing number of grains. The BNE simulations evince that intruder's ascension is faster and get higher stationary positions when friction is present at grains and walls.

Acknowledgements

We thanks to CEFET-MG that gives us materials and conditions to develop our research. Special thanks to CEFET-MG, FAPEMIG, CAPES and CNPq for funding the research.

References

[1] Radeke, Charles A. and Glasser, Benjamin J. and Khinast, Johannes G., Chemical Engineering Science **65**, 6435-6442 (2010).

[2] Cruz-Hidalgo, Raúl and Kanzaki, T. and Alonso-Marroquin, F. and Luding, S., Powders and Grains 2013 AIP Conference Proceedings **1542**, 169-172 (2013).

[3] Atman, A. P. F. and Claudin, P. and Combe, G., Computer Physics Communications **180**, 612–615 (2009).

[4] Andreotti, B. and Forterre Y. and Pouliquen O., *Granular Media: Between Fluid and Solid* (Cambridge University Press, São Paulo, 2013).

[5] Atman, A. P. F. and Claudin, P. and Combe, G. and Martins, G. H. B., Granular Matter **16** 193-201 (2014).

[6] J. Duran, A. Reisinger, and P. de Gennes, *Sands, Powders, and Grains: An Introduction to the Physics of Granular Materials* (Springer, New York, 1999).

[7] Atman, A. P. F. and Kolb, E. and Combe, G. and Paiva, H. A. and Martins, G. H. B., Powders and Grains 2013 AIP Conference Proceedings **1542**, 381-384 (2013).

[8] Mehta, A., *Granular Physics* (Cambridge University Press, São Paulo, 2007).

[9] Herrmann, H. and Hovi J. and Luding S., *Physics of Dry Granular Media* (Springer, Netherlands, 2013).

[10] Pöschel, T. and Schwager, T., *Computational Granular Dynamics: Models and Algorithms* (Springer, Berlin Heidelberg, 2005).

[11] Allen, M. P. and Tildesley, D. J., *Computer Simulation of Liquids* (Clarendon Press, Oxford, 1991).

[12] Hedman, Fredrik, *Algorithms for Molecular Dynamics Simulations* (Institutionen för fysikalisk kemi, oorganisk kemi och strukturkemi, 2006).

[13] Gutiérrez, G. and Pozo, O. and Reyes, L. I. and Drake, J. F. and Ott, E. and others, Europhysics Letters **67**, 369-372 (2004).

[14] Brito, R. and Soto, R., The European Physical Journal Special Topics **179**, 207-219 (2009).

Confined granular flows on a heap: from simulations to experiments

Patrick RICHARD[1,*], *Alexandre* VALANCE[2,**], *Renaud* DELANNAY[2,***], and *Philippe* BOLTENHAGEN[2,****]

[1] *LUNAM Université, IFSTTAR, MAST, GPEM, F-44340 Bouguenais, France*
[2] *Institut de Physique de Rennes, Université Rennes 1, UMR CNRS 6251, F-35042 Rennes, France*

Abstract. Surface granular flows are studied by means of discrete element simulations. The investigation of their kinematic properties suggests the existence of two regimes: one close to the jamming, for which the shear rate increases with the angle of the flow, and another one, far from jamming, for which the shear remains constant while the flow rate increases. Although the simulations reproduce experimental results for a given gap between sidewalls, they fail to capture the effect of confinement on the flows' properties (shear rate, flow height...).

1 Introduction

Significant progress has been made during the last decades in understanding gravity driven granular flows; nevertheless a full understanding of their behavior is far from being attained and their study remains an active field of research [1]. Here, we focus on steady and fully developed surface flows confined between two vertical sidewalls, *i.e.* systems for which the flowing particles appear to be limited to a surface layer above a "frozen" bulk region. In such systems, there is a critical flow rate above which the flow occurs on a static heap which forms along the base [2]. Its angle, θ, is determined by the flow rate, the internal dissipation of the flow and its friction with sidewalls. Using a balance of momentum for the flowing layer, one can derive an affine scaling law linking θ, the height of the layer, h, and the width of the channel, W:

$$\tan\theta = \mu_I + \mu_W h/W, \tag{1}$$

where μ_I and μ_W are effective friction coefficients [2]. Flows atop this sidewall-stabilized heap (SSH) differ fundamentally from flows on bumpy bases as they occur over erodible bases. However they still display steady and fully developed (SFD) features. The sidewall effect is not limited to important confinements (i.e. small gaps between sidewalls) since it has been experimentally quantified for flows on top of a static pile in wide channels [3]. We have performed a systematic, large-scale simulation study of SFD confined surface flows of cohesionless granular media in three dimensions. In agreement with the literature, we show that SSH flows can be attained for a wide range of angles. Interestingly at large angles the solidlike, liquidlike and gaslike behaviors coexist. This type of flows is therefore a relevant choice to test theories aiming to describe this coexistence.

The outline of this paper is the following. The next section is devoted to a short description of the numerical methods. Then, in section 3, we will study the relation between the angle of the flow and the flow rate and how it is influenced by confinement and by a modification of the micromechanical parameters (coefficients of restitution and friction). The kinematic properties of the flow will be discussed in section 4. Finally, in section 5, we compare our numerical results with experimental ones. In particular we show that even if we adjust the micromechanical parameters to match the experimental results for a given gap between sidewalls, W, our simulations are unable to capture the experimental dependence of the flow properties with W. This result questions the classical implementation of contact laws in soft sphere Discrete Element Method (DEM) simulations.

2 Numerical methods

To carry out our simulations we use our own implementation [4] of the classical DEM where Newton's equations of motion for a system of "soft" but stiff grains are integrated. It includes friction and rotation. Since the DEM is well known and can be found in many papers (*e.g.* among many others [5]), we just present here the forces used in this work. For the normal force between two overlapping spheres we use a standard linear spring-dashpot interaction model: $\mathbf{f}_n = k_n\boldsymbol{\delta}_n - \gamma_n\mathbf{v}_n$, where $\boldsymbol{\delta}_n$ is the normal overlap, k_n is the spring constant, γ_n the damping coefficient and $\mathbf{v}_n$ the normal relative velocity. Likewise we model the tangential force (which induces rotation for frictional grains) as a linear elastic and linear dissipative force in the tangential direction: $\mathbf{f}_t = -k_t\boldsymbol{\delta}_t - \gamma_t\mathbf{v}_t$, where k_t is the tangential spring constant, $\boldsymbol{\delta}_t$ the tangential overlap (obtained

*e-mail: patrick.richard@ifsttar.fr
**e-mail: alexandre.valance@univ-rennes1.fr
***e-mail: renaud.delannay@univ-rennes1.fr
****e-mail: philippe.boltenhagen@univ-rennes1.fr

by integrating surface relative velocities during elastic deformation of the contact), γ_t the tangential damping and $\mathbf{v}_t$ the tangential velocity at the contact point. The magnitude of $\boldsymbol{\delta}_t$ is truncated as necessary [6] to satisfy the Coulomb law, $|\mathbf{f}_t| \leq \mu_{gg} |\mathbf{f}_n|$, where μ_{gg} is the grain-grain friction coefficient. The granular material is an assembly of N dissipative spheres (average diameter $d = 2R$ and average mass m) submitted to gravity $\mathbf{g}$. A small polydispersity of $\pm 20\%$ is considered to prevent crystallization. The following values of the parameters are used: $k_n/mg = 5.6 \times 10^6$, $k_t = 2k_n/7$, $\gamma_t = 0$. The value of γ_n is adjusted to obtain the desired value of the normal restitution coefficient e_n [7]. The simulation chute (see figure 1) consists of a $3D$ cell which can be inclined relatively to the horizontal by an angle θ. Its size in $x-$ direction (main flow direction) is set to $25.3d$. In z direction (*i.e.* normal to the free surface of the flow) the size of the cell is large and considered as infinite. Frictional sidewalls are located at positions $y = 0$

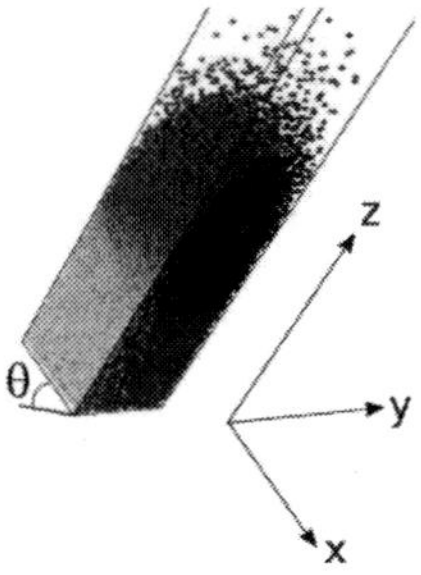

Figure 1. Typical 3D snapshot for chute flow: $W/d = 20$, $N = 24,000$. The angle of the flow is 45°, the coefficient of restitution e_n is equal to 0.88, and the friction coefficients μ_{gg} and μ_{gw} are set to 0.5. Flow is directed down the incline.

and $y = W$. They are treated as grains of infinite mass and radius and characterized by the grain-sidewall friction coefficient μ_{gw}. The bottom of the cell is made bumpy by pouring under gravity $\mathbf{g}$ a large number of grains in the cell and by gluing (*i.e.* setting their mass to infinity) those that are in contact with the plane $z = 0$. The initial state is a randomly diluted hexagonal lattice with disorder both on grain positions and velocities. Initially the angle was set to a large angle $\theta \approx 70°$ for a significant duration to remove any influence of the initial state and then decreased to the desired value. Our results are given in nondimensional quantities by using d, $\sqrt{d/g}$ and m as units of length, time and mass respectively. The flow domain is partitioned in horizontal strips whose height is equal to the average grain size. The average quantities corresponding to macroscopic field variables are determined for each strip by averaging over time the properties of the grains whose center is located in the considered strip. Although such an averaging method may induce some artefacts when computing properties involving fluctuations (*e.g.* granular temperature [8, 9]), it is not the case for the properties reported here (volume fraction and velocity).

3 Flow rate

Our initial state being very energetic, the mass of particles flowing out of the cell of the simulation per unit time decreases with time towards a constant value corresponding to a SFD flow which establishes throughout the simulation cell. Within the control parameters investigated so far, we obtain either a SFD flow or a completely jammed state, *i.e.*, no flow. For each cell width W we can therefore define an angle θ_1 below which the flow jams. This angle decreases with increasing W and reaches θ_1^∞ the "starting angle" classically obtained in case of laterally unbounded flows. Note also that, in vicinity of θ_1, the temporal evolution of the flow rate may be intermittent in agreement with experiments [10]. This intermittency induces a significant uncertainty on the determination of θ_1. Indeed, an intermittent behavior may be observed for a long period of time before the system comes to rest. It is also important to emphasize here that the presence of flat frictional walls allows the existence of steady flows at much higher angle than θ_2^∞, angle defined as the limit angle above which unconfined flows are accelerated [11]. Besides, even at very steep angles, the flow reaches a steady state. It seems that there is no upper limit angle for the existence of steady states for confined flows. These results point out the crucial importance of sidewalls on the behavior of granular flows. Additionally, as already been reported in the literature (*e.g.* [12, 13]), the angle θ_1 depends on the lateral width W: the wider the gap is the smaller is θ_1 [14]. We now focus on the flow rate Q and

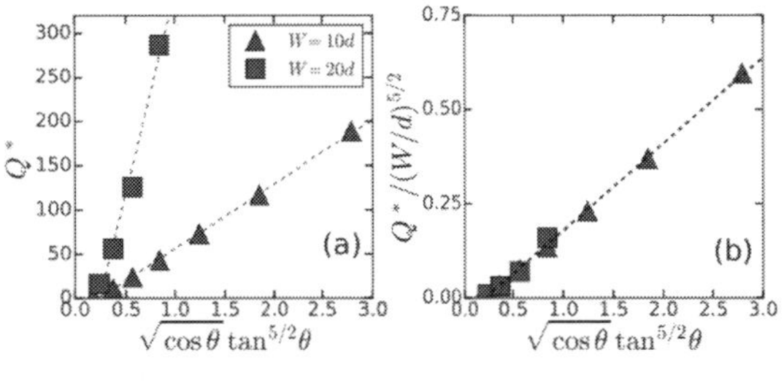

Figure 2. (a) Flow rate per unit width $Q^\star$ versus $\sqrt{\cos\theta}\tan^{5/2}\theta$ for different widths W. (b) Rescaled flow rate versus $\sqrt{\cos\theta}\tan^{5/2}\theta$.

more precisely on the flow rate per unit width, $Q^\star = Q/W$ which is an increasing function of the gap width W (figure 2(a)), as a consequence of the declining influence of the wall friction [2, 3]. In agreement with [3] we found that $Q^\star$ scales with $W^{5/2}$ (figure 2(b)). Our results are also in reasonable agreement with the analysis reported in [15] where the authors predict a dependance of the flow rate with $\sqrt{\cos\theta}h^{5/2}$ *i.e.* in our case $\sqrt{\cos\theta}\tan^{5/2}\theta$ (see equation 1). We investigated the influence of the microscopic parameters of particle/particle interactions. We varied three parameters (the particle/particle restitution coefficient e, the grain/grain Coulomb friction μ_{gg} and the grain/wall friction μ_{gw}) and analyzed their respective influence on the flow rate. It turns out that the flow rate is weakly sensitive to the restitution coefficient e, except when the elastic limit is approached (i.e., $0.9 < e < 1$). In contrast, the friction coefficients μ_{gg} and μ_{gw} have a much more significant impact on the flow rate. For a given μ_{gg} coefficient, the flow rate exhibits a strong monotonic decrease with increasing grain-wall friction. When μ_{gw} is increased from 0.1 to 0.3, the flow rate is decreased by one order of magnitude. Identically, the flow rate decreases with increasing grain-grain friction but at a smaller rate and seems to reach an asymptotic lower limit when μ_{gg} exceeds μ_{gw}.

4 Kinematic properties

In agreement to what has been reported in [4] the vertical profile of the volume fraction (not shown) is found to collapse on a single master curve

$$\nu(z) = \frac{\nu_0}{2}\left[1 + \tanh\left(z/l_\nu\right)\right], \qquad (2)$$

with origin such that $\nu = \nu_0/2$ at $y = 0$. In the static pile where $y \gg l_\nu$, $\nu \to \nu_0 = 0.6$. The characteristic length of the variation of ν, l_ν, scales with W and increases with inclination, $l_\nu/W \simeq \eta\times[\tan\theta - \tan\theta_0]$, where $\theta_0 \approx 17°$ and $\eta \approx 1.2$, a relation that we established for $W/d = 10$ and 20 (figure 3). Note that such a relation has been successfully tested for other confinements ($W/d = 5$ and $W/d = 30$) [14]. Of course, the quantities θ_0 and η change with μ_{gg} and μ_{gw}. Two remarks deserve to be mentioned here. First, the variation of the characteristic l_ν with $\tan\theta$ deviates from an affine law when approaching the jamming transition. Second, the characteristic length l_ν can be used to define unambiguously the height of the of the flowing layer : $h = 2l_\nu$ which also scales as $W\tan\theta$ as reported in [4].

As expected, the vertical profiles of the velocity are localized at the free surface, with a linear profile at the top followed by an exponential tail. In case of high flow angles a decrease of the velocity in the gaseous layer is also observed (figure 4(a)). Interestingly, these profiles can also be scaled by a characteristic speed which depends on l_ν : $V_0 \sim (l_\nu/d)\sqrt{gW}$ except close to jamming (figure 4(b)). This suggests the existence of two flow regimes, one far from jamming characterized by the length scale l_ν and a shear rate $\dot\gamma = \partial_y v_x$ which does not depend on θ and the other close to jamming for which the shear rate increases with θ.

5 Comparison with experiments

We carried out experiments with the aim to validate our numerical approach. Glass beads of diameter $d = (500 \pm 100)\,\mu m$ are poured between two parallel, vertical glass plates separated by W from a hopper orifice controlling the flow rate, thus fixing h and θ. A Photron APX RS camera of 1024×1024 resolution for 50−30000 Hz frame rate and a particle tracking algorithm are used to track rapid grains

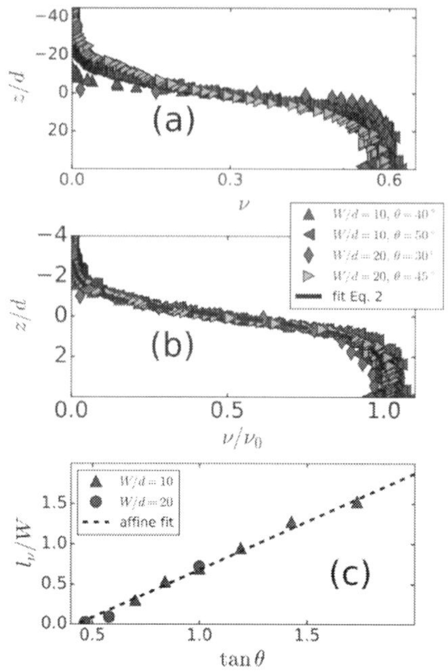

Figure 3. The profiles of volume fraction ν at inclinations and gap between sidewalls shown (a) are well fitted by Eq. 2 and can be rescaled into a single curve when plotting ν/ν_0 versus z/l_ν (b), where l_ν is the characteristic length of the variation of ν, which scales with W and is affine with $\tan\theta$ (c).

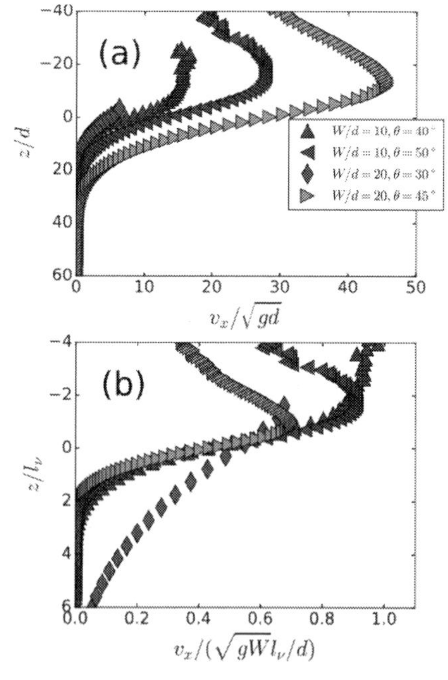

Figure 4. Vertical profiles of the horizontal particle velocity for various inclination angles θ and widths W.

at the sidewalls. We first adjust the value of μ_{gg} and μ_{gw} in our simulations to fit the experimental dependance of the evolution of the flow rate with the angle of the flow θ (the effect of e is weak). For $W = 10d$, we obtain a good agreement between experiments and numerical simulations by setting the friction coefficients to $\mu_{gg} = 0.2$ and $\mu_{gw} = 0.48$ (see figure 5(a)). Interestingly, the same values also give

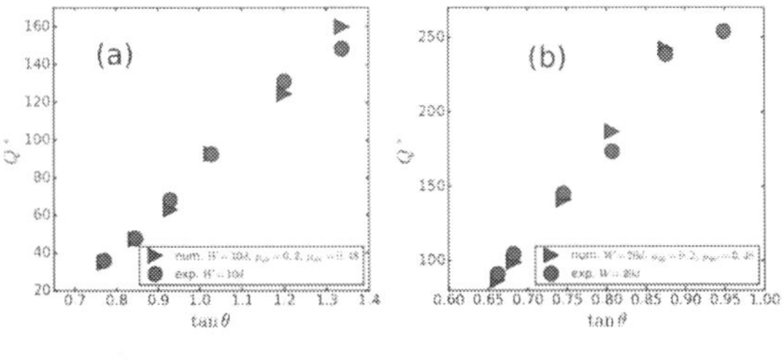

Figure 5. The simulations with $\mu_{gg} = 0.2$ and $\mu_{gw} = 0.48$ are capable to reproduce the experimental evolution of the flow rate per unit width Q^* with the angle of the flow θ both for $W = 10d$ and $W = 20d$.

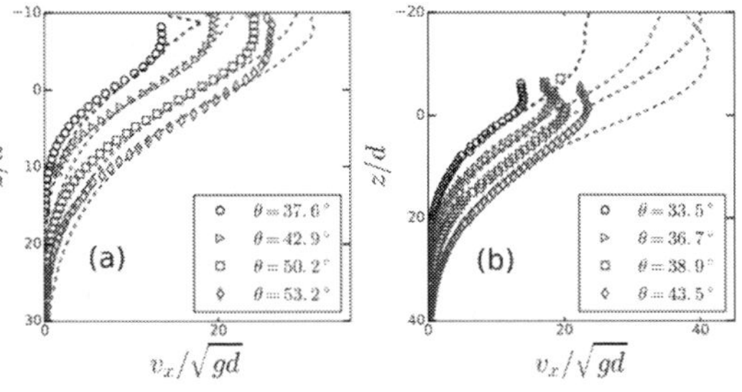

Figure 6. Vertical profiles of the horizontal particle velocity for various inclination angles θ and for $W = 10d$ and $W = 20d$.

a good fit of the flow rate for $W = 20d$ (figure 5(b)). However, we have to keep in mind that the flow rate is a macroscopic quantity. Consequently we also have to verify that the same friction coefficients allow to reproduce microscopic quantities like velocity profiles. Figure 6(a) shows that the numerical velocity profiles obtained for $W = 10d$ matches reasonably the experimental one. The height of the numerical seems to a be a little bit higher than the experimental one but it might be a consequence of the very low volume fraction at the top of the flow which leads to an underestimation of the height in experiments. On the contrary, for $W = 20d$ (figure 6(b)) the numerical flows have a much greater height than the experimental ones, which cannot be explained by the low density region at the top of the flow. Consequently, this result questions the relevancy of such numerical methods to model quantitatively granular flows: our simulations are capable to reproduce the experimental results for a given gap between sidewalls, W, –at least qualitatively– but not the experimental dependence of the flow properties with W.

6 Conclusion

Our study showed that although our simulations reproduce reasonably experiments for a given confinement, the experimental dependence of the flow properties with W is not captured by our simulations. This result questions the way that contact laws are classically implemented in soft sphere DEM simulations. As noticed previously in several works (*e.g.* [16]) the dissipation of multiparticle contacts is probably much too low. Thus, our work highlights the importance of dissipative force modeling in DEM.

Acknowledgements

We thank the Centre de Calcul Intensif des Pays de la Loire (CCIPL) for the use of their facilities (project MTEEGD).

References

[1] R. Delannay, M. Louge, P. Richard, N. Taberlet, A. Valance, Nature Materials **6**, 99 (2007)
[2] N. Taberlet, P. Richard, A. Valance, W. Losert, J.M. Pasini, J.T. Jenkins, R. Delannay, Phys. Rev. Lett. **91**, 264301 (2003)
[3] P. Jop, Y. Forterre, O. Pouliquen, Journal of Fluid Mechanics **541**, 167 (2005)
[4] P. Richard, A. Valance, J.F. Métayer, P. Sanchez, J. Crassous, M. Louge, R. Delannay, Physical Review Letters **101**, 248002 (4) (2008)
[5] L.E. Silbert, D. Ertaş, G.S. Grest, T.C. Halsey, D. Levine, S.J. Plimpton, Phys. Rev. E **64**, 051302 (2001)
[6] L. Brendel, S. Dippel, *Lasting Contacts In Molecular Dynamics Simulations*, in *Physics of Dry Granular Media* (Kluwer Academic Publishers, 1998), pp. 313–318
[7] S. Luding, European Journal of Environmental and Civil Engineering - EJECE 12 **12**, 785 (2008)
[8] H. Xu, A.P. Reeves, M.Y. Louge, Review of Scientific Instruments **75**, 811 (2004)
[9] R. Artoni, P. Richard, Phys. Rev. E **91**, 032202 (2015)
[10] R. Fischer, P. Gondret, M. Rabaud, Phys. Rev. Lett. **103**, 128002 (2009)
[11] GDR-MiDi, Eur. Phys. J. E **14**, 341 (2004)
[12] C.h. Liu, H.M. Jaeger, S.R. Nagel, Phys. Rev. A **43**, 7091 (1991)
[13] S. Courrech du Pont, P. Gondret, B. Perrin, M. Rabaud, EPL (Europhysics Letters) **61**, 492 (2003)
[14] P. Richard, A. Valance, R. Delannay, P. Botltenhagen, in preparation (2016)
[15] J.L. Baker, T. Barker, J.M.N.T. Gray, Journal of Fluid Mechanics **787**, 367 (2016)
[16] L. Pournin, T.M. Liebling, A. Mocellin, Phys. Rev. E **65**, 011302 (2001)

DOI: 10.1051/epjconf/201714003068

Flowability of granular materials with industrial applications

An experimental approach

Joel Torres-Serra[1,2,*], *Enrique* Romero[2], *Antonio* Rodríguez-Ferran[2], *Joan* Caba[1], *Xavier* Arderiu[1], *Josep-Manel* Padullés[1], and *Juanjo* González[1]

[1] *Técnicas Mecánicas Ilerdenses, S.L.*
[2] *Universitat Politècnica de Catalunya*

Abstract. Designing bulk material handling equipment requires a thorough understanding of the mechanical behaviour of powders and grains. Experimental characterization of granular materials is introduced focusing on flowability. A new prototype is presented which performs granular column collapse tests. The device consists of a channel whose design accounts for test inspection using visualization techniques and load measurements. A reservoir is attached where packing state of the granular material can be adjusted before run-off to simulate actual handling conditions by fluidisation and deaeration of the pile. Bulk materials on the market, with a wide range of particle sizes, can be tested with the prototype and the results used for classification in terms of flowability to improve industrial equipment selection processes.

1 Introduction

Within the industry of bulk material handling, bagging machinery is designed and manufactured to store and transport granular materials. This equipment can carry out dosing of powders and grains through gravimetric feeder systems among others [1], which consist of a batch weighing device that has to be fed on an adjustable flow of granular matter downstream a reception hopper.

The feeder selection process is still a challenging issue nowadays. Different types of feeding solutions exist, which mainly include:

- free discharge by gravity feeders or rotary valves;
- positive displacement by either belt, screw or vibratory tray feeders;
- pneumatic transport using fluidisation chambers.

The output rates of the feeder systems are affected by irregularities in dry granular flows, as well as the quality of the produced bags and the overall performance of the lines, evidenced through the formation of jamming processes even leading to stable arching or ratholing structures [2–4].

In turn, distinct material properties are shown by the diversity of products on the market and which are crucial in establishing the suitability of each bulk handling equipment. The lack of a standard framework for the industry as per flowability assessment of powders and grains demonstrates a need for experimental methodologies that are able to simulate actual handling conditions [5, 6].

*e-mail: j.torres@tmipal.com

Figure 1. Retained mass fractions of a refractory mortar sample after sieve analysis with sieves of mesh sizes 500, 200, 140, 100 and 71 μm (ISO 3310-1:2000 Series)

2 Methods

2.1 Testing techniques

The behaviour of granular materials can be described by means of the effect of both micro-scale and macro-scale properties on the mechanics of flow regimes [7]. At the particle level, the effect of parameters such as size, shape and density has been reported [8–10]. Complexity of the materials with actual industrial application is illustrated in figure 1, showing grain size dispersity of a refractory mortar sample by sieve analysis. Also, the granular shape of a food additive obtained by optical microscopy can be seen in figure 2.

At the bulk scale, robust characterization measures include compressibility in terms of uniaxial compression stress or flow factors obtained from shear testing in the quasi-static regime [11]. Rheological properties of granular materials are investigated to describe flowability of dense phases, including mixing and avalanching procedures [12, 13]. Also, indentation or penetrometry tests are

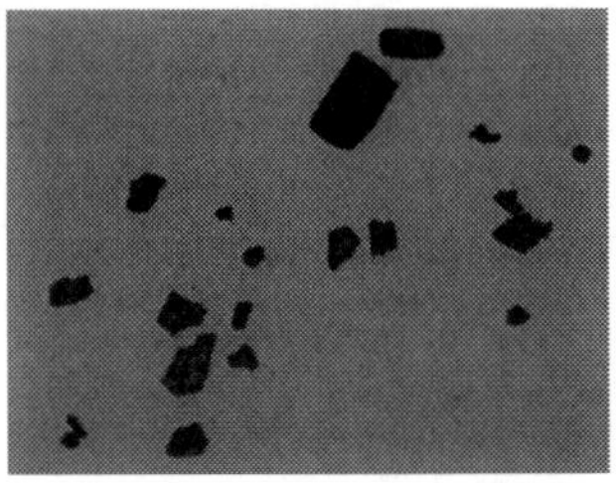

Figure 2. Projected particle shapes for image analysis of a food additive sample (d_{50} = 231 μm) mounted on a digital microscope at 4 optical magnifications

used to describe cohesive behaviour [14]. Furthermore, methodologies with a more indirect interpretation exist despite involving traditionally used parameters for instance discharging rates, compressibility in terms of tapping densities, angles of repose or dustiness [15–18], the latter being representative of dilute flows. Analysis of the aerated flow regime is also conducted via bulk permeability and bed expansion tests [19, 20].

In addition, the environmental conditions of granular flows have an effect on their behaviour. Hygroscopicity of bulk materials, essentially governed by the surface condition of constituent particles [21], is measured through relative humidity of surrounding air, water activity of the particles and equilibrium moisture content of the substance [22]. Some attempts have been made to incorporate temperature monitoring into flow testers [23]. What is more, exposure time to humidity and temperature conditions, as well as sampling of the materials, control among others caking processes, which alter flowability [24].

2.2 Granular column collapse

In recent years, granular column collapse experiments have been the object of study in a strive to better understand dry granular flows with implications, not only to earth sciences, but also to industry [25].

Figure 3 shows the test set-up, consisting of a container where a particle assembly is held. Once the granular column is formed, the container is instantaneously removed and the granular material is let spread out driven by gravity. Flow dynamics is mostly governed by interparticle contacts over particle-wall interactions [26].

The test yields direct observation of the kinematics of generated mass flows, thus focusing on the overall characterization of dense and dilute flow regimes beyond isolated measurements of bulk material and particle properties.

The initial configuration of the pile is adjusted regarding two main aspects:

- Geometry, considering quasi-two-dimensional and axisymmetric layouts [27, 28]. Inspection in the cylindrical case requires cutting the pile vertically to observe the effect of collapse on the final deposit, while the bulk remains unobservable during run-off. Semiaxisymmetric geometries have been explored to circumvent such drawbacks [29].
- Packing state, considering air fluidisation of the samples before release of the unsteady flow of powders and bulk solids [30].

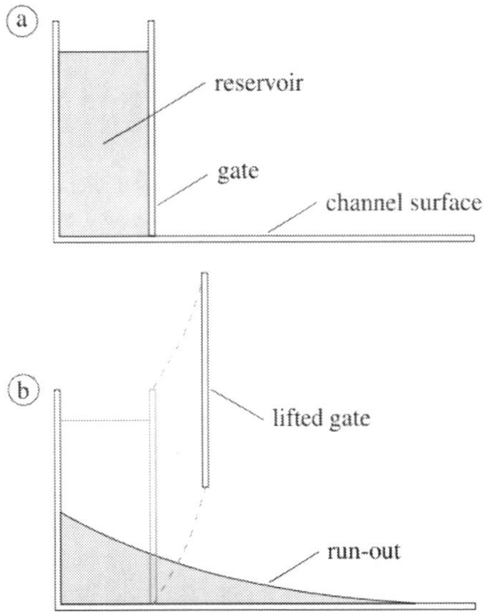

Figure 3. Schematic of a quasi-two-dimensional granular collapse test set-up: a) Initial configuration; b) Final deposit.

Further exploration of this methodology includes analysis of the effect of moisture content [31] or testing collapse on inclined planes [32].

3 Prototype

3.1 Proposal

A new prototype has been developed to provide insight into the mechanics of industrially produced powders and bulk solids based on run-out experiments. Various existing apparatus on the market aimed at the characterization of different flowability states are compared and their main features shown in table 1, in contrast with the capabilities of the new prototype. Traditional testers are not taken into account for comparison, despite being commercially distributed [33]. The objective is to reproduce the dynamic behaviour of granular samples in actual handling processes involving the release of material from a container at different packing conditions determined by the feeding systems in use.

Table 1. Comparison between proposed flowability prototype and commercial devices

	Prototype	Static testers [34]	Rheometers [35]	Dynamic testers [36]
Shear properties		•	•	
Compressibility	•	•	•	
Permeability	•		•	
Dynamic energy profile			•	
Dynamic morphology	•			•
Dynamic load pattern	•			•
Characteristic velocities	•			

 DOI: 10.1051/epjconf/201714003068

3.2 Description

Figure 4 shows the assembly of the prototype. A quasi-two-dimensional arrangement is considered for better inspection of the tests regarding bulk velocity fields through lateral walls and longitudinal morphology and load distribution during run-off. Therefore, the device consists of a 2000 mm long, 160 mm wide horizontal channel with and anodized aluminium plate at the base and vertical glass walls with a height of between 350 and 150 mm, accounting for flow front height decrease due to propagation as well as for accessibility purposes during operation. A reservoir of a square 150 mm sided base and 350 mm wall height is attached to the channel in order to contain granular columns with aspect ratios going from 1 to 2 times the initial height with respect to the base length of the reservoir.

A lifting gate allows for instantaneous material release from the reservoir into the channel. It is made of carbon fibre reinforced polymer components held together by a 3D-printed polymeric frame as a means to lighten the mechanism. Motion of the gate is controlled by pneumatic valve actuators in two phases. First, fast horizontal separation from the vertical free surface of the column allows for the onset of flow; second, the gate is slowly withdrawn out of the reservoir and channel to clear vertical space for run-out visualization.

At the reservoir base, an air chamber allows for both air injection and suction for fluidisation and deaeration of the pile. The aerated state is representative of industrial processes such as pneumatic conveying and fluidisation chamber feeding. In this sense, previous reported experience has addressed quasi-static characterization [37]. The testing procedure considers gravity-driven flows at three different initial packing states:

- poured random packing as a result of free fall of granular material into the reservoir;
- loose packing obtained from poured packing by an imposed airflow —between minimum fluidisation and bubbling pressure [38]—into the poured material to observe aeration behaviour prior to the gate activation;
- dense packing resulting from deaeration of initially poured packing so as to evaluate air permeability of the different materials, which is a common technique applied in bagging systems.

To this end, injected air flows come out of an air compressor and into an air dryer, from where they are transported and regulated by a flowmeter, before being delivered to the air chamber, which in turn includes a manometer; air suction is imposed by inverting the air flow sense throughout the system. A permeable polyester needle-punched felt layer prevents particles of sampled materials from entering the air chamber, as used in actual industrial equipment.

At the channel base, logarithmically-spaced beam load cells (Modelo 104, UTILCELL, Spain) are installed to register dynamic load distributions to track the evolution of the stress state of the bulk at the channel surface. Power supply of the force transducers includes signal conditioning (SWIFT PANEL, UTILCELL, Spain).

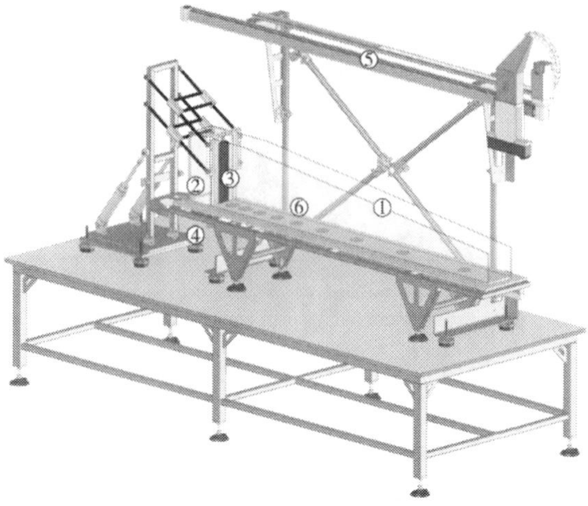

Figure 4. View of the prototype with numbered parts: 1) Channel; 2) Reservoir; 3) Lifting gate; 4) Air chamber; 5) Linear guide and laser scanner mount; 6) Membranes and load cell mounts.

Moulded silicone membranes placed at the channel base undergo small elastic deformations during run-off and, consequently, transmit dynamic reaction forces to the load cells.

Moreover, a 2D/3D laser scanner (GOCATOR 2150, LMI Technologies Inc, Canada) is mounted on a linear guide aligned with the channel and with positioning control by servomotor. The profile sensor provides three-dimensional maps of the final configuration for morphology analysis of the run-out. On the one hand, the quasi-two-dimensional assumption of the experiment can be checked. On the other hand, different shape patterns of the resulting deposits related to intrinsic bulk properties can be extracted, including measurement of traditional parameters such as the angle of repose [39].

Finally, a high speed video camera (PXW-FS5, SONY Corporation, Japan) is incorporated into the assembly of the prototype. The use of high speed digital video recording systems and image analysis post-processing techniques, such as the Particle Image Velocimetry [40], enables the extraction of relevant properties from the visualization of the established two-dimensional wall velocity field and flow patterns during run-off. Additionally, image analysis through column height monitoring is also used for measurements of compressibility by uniaxial compression using weights and bed expansion by air fluidisation.

4 Concluding remarks

Few previous experiences with granular column collapse of fine or cohesive powders are reported in existing research literature, typically resorting to monodisperse glass beads. A wide range of powders and bulk solids are dealt with in the packing industry, produced by several industrial sectors, namely the food, agri-food, construction and mining, chemical and petrochemical, and recycling industries.

Decision making strategies towards selecting the adequate machinery often rely on qualitative estimations of the flowability of granular materials, thus lacking a robust physical approach to mass flow phenomena. The presented prototype is intended to test bulk materials with particle sizes from the scale of micrometres to millimetres to provide a new flowability assessment methodology and, at the same time, allow the effect both particle and bulk properties have on the observed behaviour to be identified.

Results coming from prototype testing are to be used for defining a classification tool for any granular material of interest to the packing industry, based on establishing flowability clusters from properties that control regularity of product feed rates. Thus, a new methodology is to be introduced, leading to the choice of the most adequate feeder system for each material among the existing solutions presently in use.

References

[1] S. Nowak, Powder and Bulk Engineering **20**, 25 (2015)
[2] M.E. Cates, J. Wittmer, J.P. Bouchaud, P. Claudin, Physical Review Letters **81**, 1841 (1998)
[3] A.J. Matchett, Powder Technology **171**, 133 (2007)
[4] K. Johanson, Powder Technology **141**, 161 (2004)
[5] J.A.H. de Jong, A.C. Hoffmann, H.J. Finkers, Chemical Engineering Progress **95**, 25 (1999)
[6] M. Krantz, H. Zhang, J. Zhu, Powder Technology **194**, 239 (2009)
[7] Y. Forterre, O. Pouliquen, Annual Review of Fluid Mechanics **40**, 1 (2008)
[8] S. Berretta, O. Ghita, K.E. Evans, A. Anderson, C. Newman, *Size, shape and flow of powders for use in Selective Laser Sintering (SLS)*, in *Proceedings of VRAP 2013, 6th International Conference on Advanced Research in Virtual and Rapid Prototyping* (2013), pp. 49–54
[9] T. Horio, M. Yasuda, S. Matsusaka, International Journal of Pharmaceutics **473**, 572 (2014)
[10] A. Santomaso, P. Lazzaro, P. Canu, Chemical Engineering Science **58**, 2857 (2003)
[11] J. Schwedes, Granular Matter **5**, 1 (2003)
[12] S. Forrest, J. Bridgwater, P.R. Mort, J. Litster, D.J. Parker, Powder Technology **130**, 91 (2003)
[13] X.Y. Liu, E. Specht, J. Mellmann, Powder Technology **154**, 125 (2005)
[14] A. Hassanpour, M. Ghadiri, Particle & Particle Systems Characterization **24**, 117 (2007)
[15] N. Vlachos, I.T. Chang, Powder Technology **205**, 71 (2011)
[16] G. Lumay, F. Boschini, K. Traina, S. Bontempi, J.C. Remy, R. Cloots, N. Vandewalle, Powder Technology **224**, 19 (2012)
[17] D. Geldart, E. Abdullah, A. Hassanpour, L. Nwoke, I. Wouters, China Particuology **4**, 104 (2006)
[18] E.A. Ural, *Ad Hoc Methods to Characterize Material Dustiness and Entrainability* (Springer-Verlag New York, 2011), pp. 51–70
[19] V. Le, E. Robins, M. Flament, European Journal of Pharmaceutics and Biopharmaceutics **76**, 464 (2010)
[20] F. Yazdanpanah, S. Sokhansanj, A. Lau, C. Lim, X. Bi, S. Melin, Biomass and Bioenergy **35**, 1960 (2011)
[21] A. Nokhodchi, Pharamaceutical Technology **29**, 46 (2005)
[22] V. Murikipudi, P. Gupta, V. Sihorkar, Pharmaceutical Development and Technology **18**, 348 (2013)
[23] I. Tomasetta, D. Barletta, M. Poletto, Advanced Powder Technology **24**, 609 (2013)
[24] E. Teunou, J.J. Fitzpatrick, E.C. Synnott, Journal of Food Engineering **39**, 31 (1999)
[25] G. Lube, H.E. Huppert, R.S.J. Sparks, M.A. Hallworth, Journal of Fluid Mechanics **508**, 175 (2004)
[26] E. Lajeunesse, A. Mangeney-Castelnau, J.P. Vilotte, Physics of Fluids **16**, 2371 (2004)
[27] C. Mériaux, T. Triantafillou, Physics of Fluids **20**, 033301 (2008)
[28] J.M. Warnett, P. Denissenko, P.J. Thomas, E. Kiraci, M.A. Williams, Granular Matter **16**, 115 (2014)
[29] E. Lajeunesse, J.B. Monnier, G.M. Homsy, Physics of Fluids **17**, 103302 (2005)
[30] O. Roche, S. Montserrat, Y. Niño, A. Tamburrino, Journal of Geophysical Research: Solid Earth **115**, B09206 (2010)
[31] F. Gabrieli, R. Artoni, A. Santomaso, S. Cola, Physics of Fluids **25**, 103303 (2013)
[32] F. Maeno, A.J. Hogg, R.S.J. Sparks, G.P. Matson, Physics of Fluids **25**, 023302 (2013)
[33] E. Abdullah, D. Geldart, Powder Technology **102**, 151 (1999)
[34] L. Parrella, D. Barletta, R. Boerefijn, M. Poletto, KONA Powder and Particle Journal **26**, 178 (2008)
[35] R. Freeman, Powder Technology **174**, 25 (2007)
[36] B.C. Hancock, K.E. Vukovinsky, B. Brolley, I. Grimsey, D. Hedden, A. Olsofsky, R.A. Doherty, Journal of Pharmaceutical and Biomedical Analysis **35**, 979 (2004)
[37] A. Castellanos, J. Valverde, M. Quintanilla, KONA Powder and Particle Journal **22**, 66 (2004)
[38] D. Geldart, *The Characterisation of Bulk Powders* (Blackwell Publishing Ltd., 2009), pp. 132–150
[39] K.A. Holsapple, Planetary and Space Science **82–83**, 11 (2013)
[40] W. Thielicke, E.J. Stamhuis, Journal of Open Research Software **2**, e30 (2014)

Micromechanical formulation of the mobilized stress ratio under principal stress rotation in granular materials

Homayoun Shaverdi[1], *Farzin* Kalantary[2], *Mohd R.* Taha[3]

[1]Civil Engineering Department, Ilam University, Iran
[2]Civil Engineering Department, K.N. Toosi University of Technology, Iran
[3]Civil and Structural Engineering Department, Universiti Kebangsaan Malaysia(UKM), Malaysia

Abstract. In this paper the mobilized stress ratio is investigated by using the static equilibrium at the micro scale level analysis. The mobilized stress ratio is obtained in the distribution of the contact normal, $E(\beta)$, and the inter-particle mobilized friction angle, ϕ_μ. The second invariant of the fabric tensor, α is used to account for fabric and its evolution. There is a difference between the theoretical rotation of principal stress axes that is obtained by the classical equations of the mechanics of materials and the experimental rotation of principal stress axes. The experimental rotation is related to the anisotropic parameter, α. The non-coincidence between the theoretical and experimental is related to the relative direction of the mobilized plane and the bedding plane. A comparison with experimental tests demonstrates the validity of this formulation.

1 Introduction

Experimental tests have shown that the gradual rotation of the principal stress axes during the progressive shearing of granular materials cause the gradual rotation of the preferred direction of the contact normal (Oda & Konishi, 1974, Oda, 1975, Arthur et al., 1977, Miura et al., 1986, Yang et al., 2016). Roscoe et al., (1967) showed that the rotation of the principal stress axes happens during simple shear of sand. They found that the rotation of the principal axes occurs when the shear stress τ increases and they also found that the inclination angle, ψ, of the maximum principal stress axis with respect to the vertical direction tends to approach a certain asymptotic value. There is the non-coincidence between the principal stress direction and the principal strain rate direction when the applied shear loads are rotating (Arthur et al., 1977, Wong & Arthur, 1986, Gutierrez et al., 1991, Yang et al., 2016). At the start of shearing there is a non-coaxiality between stress and fabric, and also between strain rate and fabric. Shaverdi et al. (2014) showed that fabric has a bridge-like role in the shear deformation. The contact normal tends to be coaxial with the principal stress direction during shear, while the principal strain increment also tends to be coaxial but at a different rate. At large shear strain, stress, strain and fabric will coincide. The contact normal are the linkage between stress and strain.

Oda (1972) and Oda & Konishi (1974) observed that the intensity of the preferred orientation of the contact normal (fabric anisotropy). In other words, fabric anisotropy has a major effect on the shear strength of granular materials in the presence of the rotation of the principal stress axes (Roscoe et al., 1970, Oda, 1972, Cai et al., 2013, Yang et al., 2016). The stress and strain are linked via the contact normal. The major direction of fabric anisotropy always lays between stress and strain principal directions. This non-coaxiality between stress and strain rate directions result in the difference between the real direction of the major principal stress and the theoretical direction. By micromechanical analysis Oda & Konishi (1974) and Yang et al. (2016) among others showed that the direction of the contact forces are different from the direction of the contact normal. Many researchers have also shown that the rotation of principal stress directions have to be related to the fabric anisotropy parameters. Oda & Konishi (1974) by considering the static equilibrium at the micro scale analysis proposed the following equation:

$$\left\{J_1\tan^2(45+\frac{\phi_\mu}{2})-J_2\right\}\tan^2\psi-(J_3-1)\left\{\tan^2(45+\frac{\phi_\mu}{2})-1\right\}\tan\psi+\left\{J_1-J_3\tan^2(45+\frac{\phi_\mu}{2})\right\}=0 \quad (1)$$

$$\text{where: } J_1=\frac{\int_{-\pi/2}^{\pi/2}2E(\beta)\sin\beta\,d\beta}{\int_0^{\pi}2E(\beta)\sin\beta\,d\beta} \quad (2)$$

* Corresponding author: h.shaverdi@ilam.ac.ir, homayoun_shaverdi@yahoo.com, [2] fz_kalantary@kntu.ac.ir, [3] profraihan@gmail.com

$$J_2 = \frac{\int_0^{\pi} 2E(\beta)\cos\beta \,d\beta}{\int_0^{\pi} 2E(\beta)\sin\beta \,d\beta}, \quad J_3 = \frac{\int_{-\pi/2}^{\pi/2} 2E(\beta)\cos\beta \,d\beta}{\int_0^{\pi} 2E(\beta)\sin\beta \,d\beta}$$

where $E(\beta)$ is the distribution of the contact normal, β is the direction of the contact normal with respect to the principal stress direction, ψ is the direction of the principal stress axes with the vertical direction. This equation cannot easily be applied to determine the direction of the major principal stress. In order to obtain and develop a practicable equation to calculate the direction of the major principal stress, the micro-level parameters that have been developed by Shaverdi & Taha (2013) and Shaverdi et al., (2014) are used. These parameters are related to the macro-level variables such as the shear stress, the confining pressure and the mobilized stress ratio.

2 Theoretical considerations

Rothenburg & Bathurst (1989) showed that the distribution of the contact normal can be represented by the following equation:

$$E(\beta) = \frac{1}{2\pi}(1+\alpha\cos 2(\beta-\beta_f)) \tag{3}$$

where α is the degree of anisotropy and β_f is the direction of the major principal fabric.
Putting Eq.(3) into Eq.(2) and some manipulation we will find:

$$J_1 = J_2 = \frac{2\alpha\sin 2\beta_f}{3-\alpha\cos 2\beta_f}, \quad J_3 = \frac{3+\alpha\cos 2\beta_f}{3-\alpha\cos 2\beta_f} \tag{4}$$

Incorporating Eq.(4) into Eq.(1) and solve the second order equation to obtain $\tan\psi$, the essential equation to calculate the rotation of principal stress will be obtain as follows:

$$\tan\theta_{\sigma r} = \tan\psi = \frac{(1-J_3)\pm\sqrt{(J_3-1)^2+4J_1}}{2J_1} \tag{5}$$

It is obvious that the real rotation of the principal stress is dependent on the contact normal distribution functions (fabric anisotropy). Shaverdi et al. (2014) and Taha & Shaverdi (2013) proposed some equations to calculate the degree of anisotropy α and the direction of the fabric tensor during shear process, β_f. As mentioned already, the rotation of the principal stress axes have to be related to the fabric anisotropy parameters only. Eq. (5) can micromechanically predict the variation of the principal stress axes by considering the fabric anisotropy parameters (α,β_f). The required equation is developed to consider the direction of the major principal stress direction by micro-level parameters. In the shearing process, these parameters (α,β_f) can be obtained by macro-level quantities such as the shear stress and the confining pressure. Unlike the Oda & Konishi's (1974) equation, that required the distribution of the contact normals (J_1,J_2,J_3) and the internal friction angle (ϕ_μ), the developed equation can calculate the direction of the major principal stress by using the macro quantities directly (because α and β_f can be obtained from the ratio of the shear to normal stresses and the variation of the shear stress).

3 Mobilized stress ratio in simple shear test

Contact force transmitted through the contact surface from the grain P_i to the grain P_i' can be resolved in the principal directions X, Y. As shown in Fig.1, the interparticle sliding have to occur when the following equation is satisfied:

$$\frac{F_{y'i}}{F_{x'i}} = \tan(\frac{\pi}{2}+\phi_\mu-\beta') \tag{6}$$

Considering the distribution low of contact forces by Oda (1972) and Horne (1956), the contact forces can be resolved in the principal stress axes X' and Y'. Finally, we will obtain (due to the limitation of space we cannot explanation other steps)(Oda & Konishi, 1974):

$$\frac{\tau}{\sigma} = \frac{\left\{1+\tan^2(45+\frac{\phi_\mu}{2})\right\}K + \left\{\tan^2(45+\frac{\phi_\mu}{2})-1\right\}\tan\psi}{\tan^2\psi+\tan^2(45+\frac{\phi_\mu}{2})+\left\{\tan^2(45+\frac{\phi_\mu}{2})-1\right\}K\tan\psi} \tag{7}$$

$$\text{where:} \quad K = \frac{4}{3}\frac{\alpha\sin 2\beta_f}{(-2+\alpha\cos 2\beta_f)} \tag{8}$$

Eqs.(7) and (8), show that the mobilized stress ratio is related to the fabric anisotropy (α,β_f) and the interparticle friction angle (ϕ_μ). The anisotropic parameters and their evolution can be obtained by the relationships that have been developed by Shaverdi et al. (2014). The theoretical mobilized stress ratio can be compared with the experimental mobilized stress ratio (Eqs. (7) and (8)), as shown in Fig.2.

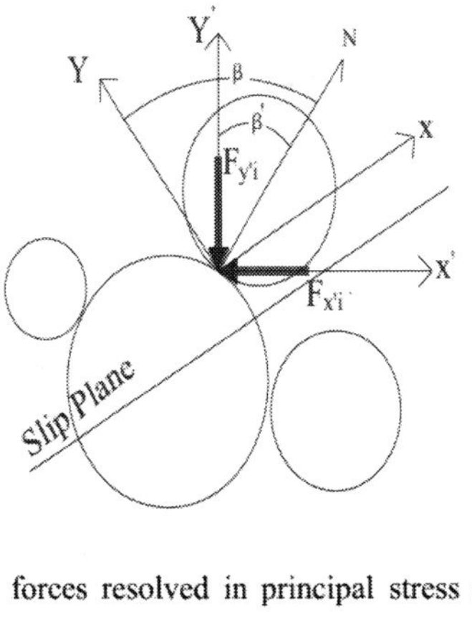

Fig. 1. Contac forces resolved in principal stress axes X' and Y'

DOI: 10.1051/epjconf/201714015025

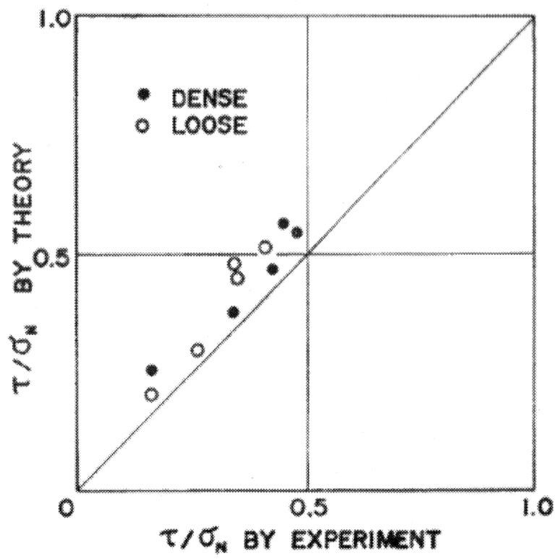

Fig. 2. Relationship between experimentally determined τ/σ_N values and calculated ones of two dimensional granular model deformed under simple shear (after Oda & Konishi 1974)

4 Verification with the experimental tests

The theoretical rotation of the principal stress and the direction of the major principal stress can be obtained as follows:

$$\tan 2\theta_{\sigma t} = \frac{2\tau_{xy}}{\sigma_x - \sigma_y} \tag{9}$$

The proposed equation (Eq.(5)) can be compared with the theoretical equation (Eq.(9)), as shown in Fig.3. It can be seen that this equation can well predict the rotation of principal stress.

Oda & Konishi (1974) conducted some experimental tests by using photo elastic materials. They were obtained the real direction of the major principal stress. By the methods developed by Budhu (1985) and Ochiai (1975) to calculate the theoretical rotation of the principal stress axes (Eq.(9)) we found that there is a difference between the real direction of the major principal stress and the theoretical of it. The actual direction of the major principal stress is obtained from the equation proposed by the authors that corresponds exactly to the direction proposed by Oda & Konishi (1974). In the other words, the comparisons Figs. 3a & 3b, the actual direction of principal stress is obtained from Eq. (5) and the theoretical direction is obtained from Eq. (9).As proposed by Oda & Konish (1974), the interparticle friction angle for the dense specimens (D) is $\phi_\mu = 26^o$ and for the loose ones (L) is $\phi_\mu = 22^o$. the difference between the theoretical and the real principal stress rotation decreased by shearing process, as shown in Fig. 3.

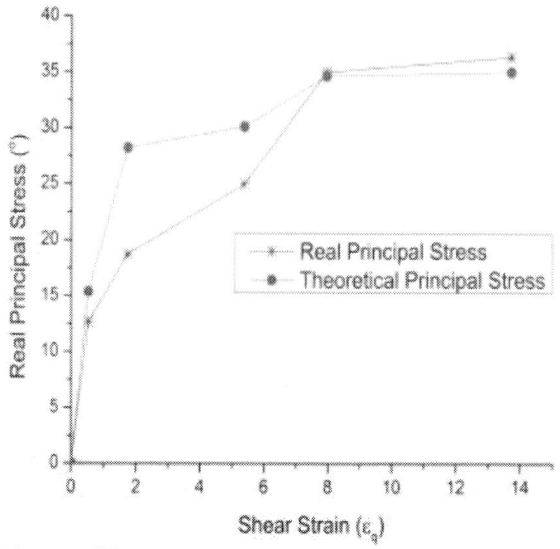

Fig. 3a Difference between the real and the theoretical principal stress rotation for the loose specimen

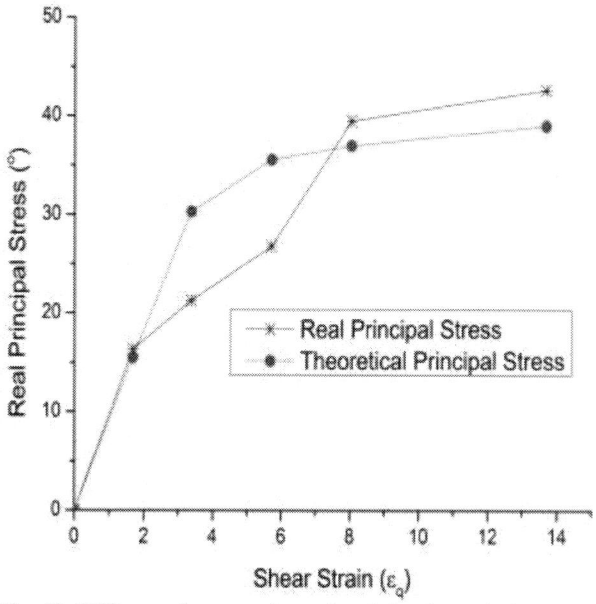

Fig. 3b Difference between the real and the theoretical principal stress rotation for the dense specimen

5 Conclusion

Using micromechanical analysis it has been shown that the mobilized stress ratio can be attributed to the interparticle friction angle, ϕ_μ and the anisotropic parameters, α, β_f . The Oda & Konishi's (1974) equation is developed to obtain the direction of the major principal stress by using the macro-level quantities. Micro-level parameters that have been proposed by Oda

& Konishi (1974) are not easily obtainable in classical laboratory tests. In this research, however, the suitable micro-level parameters have been chosen to calculate the direction of the major principal stress. The equation was proposed to obtain the real direction of the major principal stress during shear process in the simple shear test. The experimental tests reveal the accuracy of the proposed equations.

References

1. M. Oda, J. Konishi, Soils & Found., **14,** 39-53(1974)
2. M. Oda, Soils & Found., **15,** 17-29 (1975)
3. J.R.F. Arthur, K.S. Chua, T. Dunstan, Geotechnique, **27,** 13-30 (1977)
4. K. Miura, S. Miura, S. Toki, Soils & Found., **26,** 36-52 (1986)
5. L.T. Yang, X. Li, H.S. Yu, D. Wanatowski, Acta Geotech., (2016)
6. K.H. Roscoe, R.H. Bassett, E.R.L. Cole, *Proc. Geotech, Conf.*, Olso, 231-237 (1967)
7. Wong, J.R.F. Athur, Geotechnique, **36**, 215-226 (1986)
8. M. Gutierrez, K. Ishihara, I. Towhata, Soils Found., **31**, 121-131 (1991)
9. H. Shavedri, M.R. Taha, F. Kalantary, Gran. Matter, **16**, 675-685 (2014)
10. M. Oda, Soils Found., **12**, 17-36 (1972)
11. Cai, H.S. Yu, D. Wanatowski, X. Li, J. Geotech. Geoenviron. Eng., **139**, 1381-1395 (2013)
12. L. Rothenburg, J. Bathurst, Geotechnique, **39**, 601-614 (1989)
13. M.R. Taha, H. Shaverdi, Mech. Mater., **69**, 173-184 (2013)
14. M.R. Horne, *Proc. Roy. A.*, **286**, 79-97 (1956)
15. M. Budhu, Geotech. Geoenviron. Eng., **111**, 698-711 (1985)
16. Ochiai, Soils Found., **15**, 113-118 (1975)

EPJ Web of Conferences **140**, 12015 (2017) DOI: 10.1051/epjconf/201714012015
Powders & Grains 2017

Sand – rubber mixtures submitted to isotropic loading: a minimal model

Auriane PLATZER[1,2,*], *Salman* ROUHANIFAR[3,**], *Patrick* RICHARD[1,***], *Bogdan* CAZACLIU[1,****], and *Erdin* IBRAIM[3,†]

[1] *LUNAM Université, IFSTTAR, MAST, GPEM, F-44340 Bouguenais, France*
[2] *GeM (Institut de recherche en génie civil et mécanique), CNRS UMR 6183, École Centrale de Nantes, F- 44321 Nantes, France*
[3] *Department of Civil Engineering, University of Bristol. Bristol, BS8 1TR, UK*

Abstract. The volume of scrap tyres, an undesired urban waste, is increasing rapidly in every country. Mixing sand and rubber particles as a lightweight backfill is one of the possible alternatives to avoid stockpiling them in the environment. This paper presents a minimal model aiming to capture the evolution of the void ratio of sand-rubber mixtures undergoing an isotropic compression loading. It is based on the idea that, submitted to a pressure, the rubber chips deform and partially fill the porous space of the system, leading to a decrease of the void ratio with increasing pressure. Our simple approach is capable of reproducing experimental data for two types of sand (a rounded one and a sub-angular one) and up to mixtures composed of 50% of rubber.

1 Introduction

The number of scrap tyres is increasing rapidly in both developed and developing countries. Consequently, the accumulation of used tyres is becoming a real societal problem. In this context, the possibility of recycling and mixing rubber chips derivatives with granular soil particles can be a solution for some geotechnical applications like backfilling for retaining structures, slope and highway embankment stabilization, road constructions, soil erosion prevention and seismic isolation of foundations [1]. In addition to their remarkable mechanical properties [2], such soil-rubber chip composite mixtures have interesting acoustical and drainage properties too [3].

While their potential range of applications is wide, a full understanding of their behaviour, including internal interaction mechanisms resulting from the combination of two particular materials, one soft, tyre chip rubber, and one rigid, granular soil, deserves further studies despite recent interesting works [5–7]. Here, we present a theoretical model aiming to capture the evolution of the void ratio of sand-rubber mixtures of various volume ratios of rubber undergoing an isotropic compression loading. For the sake of simplicity, we focus on mixtures composed of rubber and sand particles of similar particle size distributions. Our model is based on the idea that, submitted to a pressure, the rubber chips deform and partially fill the porous space of the system leading to a void ratio which decreases with increasing pressure. Assuming the sand particles as infinitely rigid and the rubber particles as incompressible, our approach –similar to mean-field approaches– leads to an expression of the void ratio of the mixture as a function of: (i) the fraction of rubber that deforms and fills the porous space, (ii) the void ratio for the soil without rubber chips and (iii) the volume ratio of rubber. Assuming (i) varies exponentially with the applied pressure, we compare the predictions of our model with experimental test results and discuss its limits. Despite its simplicity, our approach successfully reproduces experimental data for two types of sand (a rounded one and a sub-angular one) and up to mixtures composed of 50% of rubber.

2 Model

To derive our model we consider granular packings made of sand grains and rubber chips. For the sake of simplicity, we focus here on packings for which all the particles have the same size distribution independently of their type (sand or rubber). We also consider an isotropic compression of the materials and the evolution with the pressure of the void ratio *i.e.* the ratio between volume of the porous space and volume of the solids. Intuitively, a simple linear combination of the respective void ratios according to the proportion of rubber is not relevant. Indeed, such an approach assumes the absence of correlation between the two phases. This is clearly not the case here, due to the potential important deformation of the rubber particles which affects the structure of the whole system. The volume ratio of rubber, x_R is defined as the ratio $V_R/(V_R+V_S)$ where V_R and V_S are respectively the volume of rubber and that of sand. Our model is based on the following assumptions:

1. sand grains are infinitely rigid,
2. rubber chips are incompressible,

*e-mail: auriane.platzer@eleves.ec-nantes.fr
**e-mail: salman.rouhanifar@bristol.ac.uk
***e-mail: patrick.richard@ifsttar.fr
****e-mail: bogdan.cazacliu@ifsttar.fr
†e-mail: erdin.ibraim@bristol.ac.uk

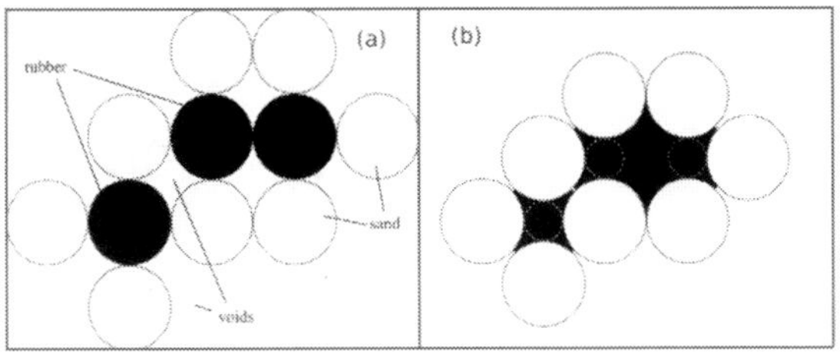

Figure 1. Sketch explaining our fourth assumption (see text). We assume that, due to the deformation of rubber particles, the porous space of an initially undeformed packing (a) is partially filled by the rubber (b).

3. the repartition of rubber chips is spatially homogenous,

4. we neglect particle rearrangements, *i.e.* the particles' neighborhood do not change. We just consider that the rubber chips undergo deformation due to pressure and fill the porous space (see figure 1). Such an assumption is fully consistent with the work of Feng and Sutter [4], in which they treat the volume of rubber chips as voids to estimate the maximum shear modulus.

5. Finally, we restrict ourselves to pressures important enough ($p > p^* = 100$ kPa) to probe the properties of the granular material, not those of the grain surface. To estimate this threshold, we considered two D-diameter spherical sand grains interacting through the Hertz law. The macroscopic pressure necessary to achieve a deformation of the order of magnitude of the surface's asperities ($\delta \approx 100$ nm) is $p^* = \kappa\delta^{3/2}/D^{3/2}$, where κ is a constant depending on the Young modulus E and on the Poisson ratio ν through $\kappa = 4E/\left[3\pi\left(1-\nu^2\right)\right]$. Using $E = 90$ GPa, $\nu = 0.4$ leads to $p^* \approx 100$ kPa for D of a few tenths of millimeter.

Note that, due to the two first assumptions, the volume ratio of rubber, x_R, is constant whatever the loading conditions. Let us now define a volumetric function f which relates the total volume of rubber that fills the porous space to the total volume of rubber. This fraction, called the deformed fraction of rubber, depends on x_R and on the pressure p. In the following we will derive an expression of f relating this quantity to x_R and p. For a given x_R, the evolution of deformed fraction of rubber, f, with the pressure is intuitive : at the early stages of the compression process the pores are large and easy to fill. Consequently, f increases with the pressure. As the pores get filled, it becomes more and more difficult to fill the remaining pores and f reaches a constant value. The simplest way to model such an increase towards a saturation value is to assume that f obeys the following first order differential equation

$$p_0(x_R)\left.\frac{\partial f}{\partial p}\right|_{x_R} + f(x_R,p) = F(x_R), \qquad (1)$$

where $p_0(x_R)$ is a characteristic pressure and $F(x_R)$ the maximum deformed fraction of rubber for an infinite pressure, *i.e* $f(x_R, p \to \infty)$. Such kind of equation are classically used to model systems which involve a driving force and viscous-like force. For example the fall of an object in a viscous fluid (the system is driven by gravity and slowed down by viscosity) or the loading of a capacitor (the system is driven by an electromotive force and slowed down by the Ohm law). Here, the system is driven by the compression and slowed down by the ability of the porous space to resist to the rubber filling (a permeability-like quantity). The solution of the preceding equation is

$$f(x_R,p) = f^* + (F(x_R) - f^*)\left[1 - \exp\left(-\frac{p-p^*}{p_0(x_R)}\right)\right], \qquad (2)$$

where $f^* = f(x_R, p = p^*)$. Since we restrict ourselves to pressures greater than p^*, it is more natural to use $\tilde{f} = f - f^*$ and $\tilde{p} = p - p^*$ instead of respectively f and p. Then:

$$\tilde{f}(x_R,\tilde{p}) = (F(x_R) - f^*)\left[1 - \exp\left(-\frac{\tilde{p}}{p_0(x_R)}\right)\right]. \qquad (3)$$

The parameters $p_0(x_R)$ and $F(x_R)$ are unknown functions of x_R that can be determined from experiments. Indeed, $\partial\tilde{f}/\partial\tilde{p}$ is an affine function of the pressure $\tilde{p}$ whose intercept and slope are respectively given by $\ln\left[(F(x_R) - f^*)/p_0(x_R)\right]$ and $-1/p_0$. Although it is probably possible to measure $f(x_R, p)$ using X-ray tomography at several stages of the compression process such an experimental campaign would be long and fastidious. Yet, it is possible to test our model by studying the evolution of the void ratio with pressure. As mentioned above, the void ratio is classically defined as $e = V_V/V_{sol}$, where V_V and V_{sol} are respectively the volume of the porous space and the volume of solids (*i.e.* the volume of the rubber chips and of the sand grains). Contrary to the former volume, the latter depends neither on the volume ratio of rubber, nor on the pressure. The volume of the porous space decreases with increasing pressure due to the gradual filling of the pores by the deformed rubber following the equation

$$V_V(x_R,p) = V_V(x_R = 0,p) - f(x_R,p)\,x_R\,V_{sol}. \qquad (4)$$

Equivalently, the void ratio is given by

$$e(x_R,p) = e(x_R = 0,p) - f(x_R,p)\,x_R. \qquad (5)$$

Therefore, our model predicts that the void ratio decreases with the pressure according to the following equation:

$$e(x_R,p) = e(x_R = 0,p) - x_R f^* - x_R\,(F(x_R) - f^*)\left[1 - \exp\left(-\frac{p-p^*}{p_0(x_R)}\right)\right]. \qquad (6)$$

This prediction requires the knowledge of (i) the evolution of the void ratio versus the pressure for a 100% sand sample (ii) the characteristic pressure, p_0, and (iii) the maximal deformed fraction of rubber, F. Since these three prerequisites depend on the type of sand used, it is necessary to test our model using at least two types of sand.

3 Comparison with experiments

Two silica based uniform sand materials, Leighton Buzzard fraction A characterised as a coarse material (mean grain diameter, D_{50} = 1.6mm) and Hostun RF characterised as fine material (D_{50} = 0.38mm), have been chosen for this study. The former sand is natural, the latter is industrially produced by crushing. The rubber particles are obtained from the shredding of used lorry tires and consist of polymer, acetone, carbon black, ash and Sulphur. Their Young modulus is approximately 3.5MPa. Following a tedious process that involved washing, drying, sorting and sieving, equivalent rubber granular materials have been engineered to match the particle size distributions of the two sands. The fabrication process of all samples of cylindrical shape (70 mm diameter and 70 mm height) is done using the moist tamping. The method implies the succession of three stages: mixing, deposition and compaction. Sand and rubber particles are mixed by adding 10% water content and the mixture is placed into the triaxial cylindrical mould in three successive layers. Each layer is compacted up to a pre-definite height using a circular tamper with a diameter being half of the sample diameter. While this technique appears to be more effective in discouraging segregation of the composite constituents, providing good control of sample density and homogeneous distribution of rubber, it also produces a soil-rubber fabric which may correspond to that obtained in rolled-compacted construction fills. Once the sample is fabricated, a top cap together with a lateral latex membrane seal the sample and a small vacuum of 20kPa is applied. This vacuum is necessary for the protection of sample's fabric and stability during removal of the cylindrical mould. At this stage, the mixtures have initial void ratios which are reported in table 1 for various volume ratios of rubber x_R = 0.1, 0.2, 0.3, 0.5 and 1. The triaxial cell containing the cylindrical sample is then filled with water and pressurized to a pressure of 30kPa while the vacuum is released. All samples are saturated by applying CO_2 for around 45 minutes followed by flushing de-aired water to obtain Skempton coefficient (B) higher than 95%. A backpressure of 100kPa is normally used. Both the back pressure and cell (total pressure) pressure are measured by pressure transducers. The samples are then loaded isotropically up to the maximum pressure of 550kPa at which point, an unloading stage is applied. During the isotropic compression, the sample volumetric changes are measured by a volume change measurement device accounting for the amount of water that is expelled in the process.

Table 1. Initial void ratios of the samples

Leighton Buzzard Sand-Rubber Mixtures		Hostun Sand-Rubber Mixtures	
x_R	void ratio	x_R	void ratio
0.	0.630	0.	0.746
0.05	0.627	0.05	0.741
0.1	0.625	0.1	0.739
0.2	0.620	0.2	0.736
0.3	0.618	0.3	0.736
0.5	0.604	0.5	0.719
1.	0.602	1.	0.707

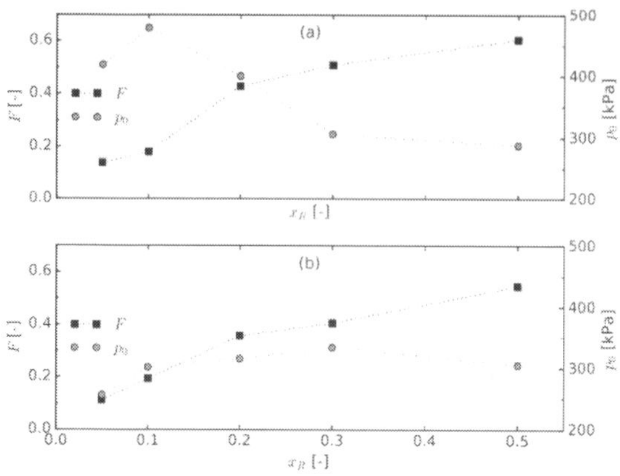

Figure 2. The parameters F and p_0 ((a): Hotsun sand (b) : Leighton Buzzard sand) are obtained by fitting experimental data

We have reported in figure 2 the values of the parameters p_0 and F versus the volume ratio of rubber for mixture made of rubber and Hotsun sand (a) and of rubber and Leighton Buzzard sand (b). Both quantities are obtained by fitting $\ln \partial \tilde{f}/\partial \bar{p}$ by an affine function (see section 2). The characteristic pressure p_0 depends on the type of sand. Its evolution with x_R is not clear since it is not monotonic. On the contrary, F (the maximum fraction of deformed rubber) increases with x_R but seems to reach a plateau.

To compare directly the experimental results with the prediction of our model we report in figure 3 (for rubber-Hotsun sand mixtures) and in figure 4 (for rubber-Leighton Buzzard mixtures) the experimental void ratios versus the applied pressure as well as $\tilde{e}$, the void ratios determined by our theoretical predictions *i.e.*

$$\tilde{e}(x_R, \bar{p}) = e(x_R = 0, \bar{p}) - f(x_R, \bar{p})\, x_R. \qquad (7)$$

A good agreement between our minimal model and experimental data is found for the two types of sand. This demonstrates that the main idea of our model *i.e.* the pores are partially filled by the deformed fraction of the rubber particles is reasonable, while neglecting the particle rearrangement effect. However, a full validation of our model is still necessary. In particular it would be interesting to calibrate p_0 and F using other loading conditions.

4 Discussion

As mentioned above, our model is based on the idea that due to compression the rubber particles deform and partially fill the porous space of the system. Although the comparison made with experimental results seems to validate this idea, the study of the deformation of the rubber particles during the compression process (*e.g.* by X-ray tomography or by numerical simulation coupling Discrete Element Method to model sand and finite element

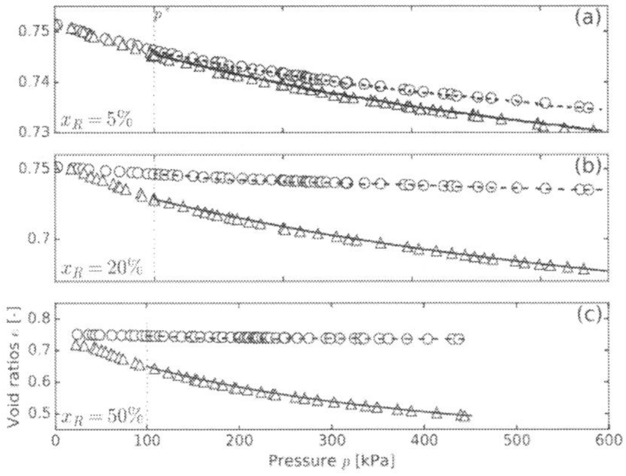

Figure 3. Comparison between experimental void ratios and those obtained by our model for several volume ratios of rubber and the Hotsun sand. Triangles (resp. circles) correspond to the sample with rubber ($e(x_R, p)$) (resp. without rubber $e(x_R = 0, p)$ and the solid line (resp. dotted line) to the prediction of our model $\tilde{e}(x_R, p)$ (resp. $\tilde{e}(x_R = 0, p)$).

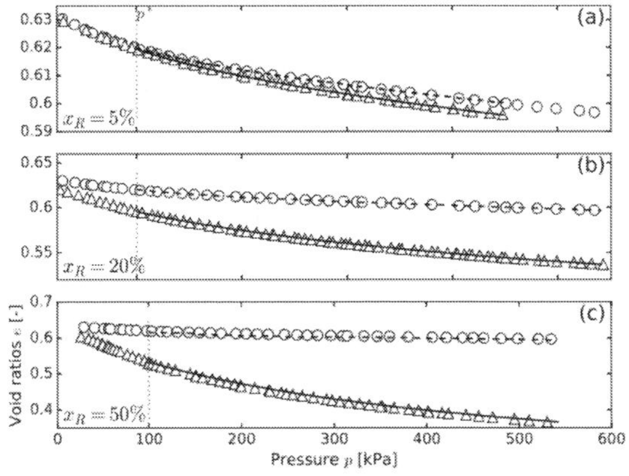

Figure 4. Comparison between experimental void ratios and those obtained by our model for several volume ratios of rubber and the Leighton Buzzard sand. Triangles (resp. circles) correspond to the sample with rubber ($e(x_R, p)$) (resp. without rubber $e(x_R = 0, p)$ and the solid line (resp. dotted line) to the prediction of our model $\tilde{e}(x_R, p)$ (resp. $\tilde{e}(x_R = 0, p)$).

modelling of rubber) is necessary to fully validate our approach. In the absence of such a study we can discuss the assumptions we used and their validity. Among them, some are fully justified. For example the assumption that sand is infinitely rigid is justified by the value of the Young modulus of the sand which is three order of magnitude greater than that of rubber. Similarly, assuming the incompressibility of rubber is justified by its Poisson coefficient whose value is close to 0.5. Other assumptions probably deserve deeper justifications. Firstly, the range of pressure studied. We indeed assume that our model is only valid for pressures greater than a threshold p^* below which the deformation is too weak to neglect the importance of the surface properties of the grains. Relaxing this assumption would require to model the surface properties of both the Hotsun sand and the Leighton Buzzard sand and include them in the model, which is out of the scope of this paper. Secondly, we have neglected particle rearrangements (each particle keeps its neighbourhood). Although reasonable in the case of packing submitted to compression this is probably no more the case when the packings are submitted to shear. Thirdly, we have assumed that the rubber grains are homogeneously distributed within the sample. In other words, every rubber grain has, statistically, the same neighbourhood. Clusters are allowed but their probability of existence have to be independent of their position.

Finally, it is worth mentioning three main perspectives of this work. First, what happens for volume ratios of rubber greater than 0.5? Is our model still valid? Second, how can we adapt our model for packings where sand and rubber particles have different size distributions? Third, as mentioned above, a direct measurement of the deformed fraction of rubber f would be an ultimate test for our model.

5 Conclusions

We derived a model aiming to predict the behaviour of mixtures of sand grains and rubber chips undergoing isotropic compression. We restricted ourselves to mixtures for which the size distributions of the sand and rubber particles are similar. Our model assumes that the deformed fraction of rubber, which increases with the pressure, fills the porous space and that its evolution with the pressure obeys a first order equation whose parameters can be determined experimentally. The experimental results are captured in a satisfactory way. The next step of this study is to adapt our model to the cases where the sand particles and the rubber chips significantly differ in size.

References

[1] Lee, J. H.; Salgado, R.; Bernal, A. and Lovell, C. W., Journal of Geotechnical and Geoenvironmental Engineering,**125**, 132-141 (1999)

[2] Youwai, S. and Bergado, D. T., Canadian Geotechnical Journal, **40**, 254-264 (2003)

[3] Valdes, J. R. and Evans, T. M., Canadian Geotechnical Journal, **45**, 588-595, (2008)

[4] Feng, Z. and Sutter, K., Geotechnical Testing Journal, **23**, 338-344 (2000)

[5] Eidgahee, D. R. and Hosseininia, E. S., AIP Conference Proceedings, **1542**, 269-272 (2013)

[6] Perez, J. L.; Kwok, C. and Senetakis, K., Computers and Geotechnics , **80**, 199 - 214, (2016)

[7] Lee, C.; Shin, H. and Lee, J.-S., International Journal for Numerical and Analytical Methods in Geomechanics, **38**, 1651-1663 (2014)

A CFD-DEM study of single bubble formation in gas fluidization of spherical and non-spherical particles

Siddhartha Shrestha and *Zongyan* Zhou*

Laboratory for Simulation and Modelling of Particulate Systems, Department of Chemical Engineering, Monash University, Victoria 3800, Australia

Abstract. Bubble dynamics significantly affect the hydrodynamics of gas-solid fluidized bed since they influence the gas-solid mixing. In this study, simulations using CFD-DEM were carried out to characterize the bubble size and shape for a bubble formed at a single orifice in gas-solid fluidized bed. Impact of parameters such as jet velocity, orifice size and particle shape on bubble equivalent diameter and bubble aspect ratio were analysed and discussed. Bubble equivalent diameter was found to increase with increasing jet velocity, decreasing bed width to orifice width ratio, and particle shape deviating from spherical. The bubble shape illustrated by aspect ratio, was found to elongate more as it rise through the bed and then commence to expand horizontally after it was detached from the orifice. Aspect ratio was found to be closer to a circle for the bubble at higher jet velocity, lower orifice width to bed ratio and for non-spherical particles.

1 Introduction

Fluidization technology has been widely used in many industrial applications such as food, pharmaceuticals, energy and environment since they are associated with good heat and mass transfer. The flow behaviour in gas-solid fluidized beds are very complex because of the fluid-particle and particle-particle interactions. Bubbles often exist in fluidized beds and are termed as "motor of fluidization"[1]. They govern hydrodynamics of fluidized beds, induce gas-solid mixing and promote heat and mass transfer [2]. Thus a good understanding of bubble formation at an orifice and its dynamics is essential to aid in design and development of gas-solid fluidized beds.

Various experimental studies have been carried out to portray bubble formation at an orifice [3] and to establish theoretical models [4, 5]. In recent years numerical techniques have been widely used to study gas-solid fluidized beds and also have become a significant approach to characterize bubble behaviour. Numerical techniques for gas-solid systems can be either continuum- or discrete-based with respect to the solid phase. The former is called two-fluid model (TFM), where both gas and solid phases are considered continuous and fully interpenetrating. The latter is referred as discrete element model (DEM) where particle phase is modelled as discrete phase and gas phase as continuous. Several authors have used TFM to study bubble formation at a single orifice and its dynamics [6, 7]. However, DEM simulations have advantage over TFM simulations because the solid phase is modelled much more accurately, in particular with respect to friction [8-10]. Olafe et al. [9] found that the bubble size and detachment times were sensitive to gas injection velocity, particle sizes and bed size. Rong et al. [11] studied the influence of particle properties (size and density), jet velocity and bed width on bubble characteristics. These parameters were found to affect the bubble dynamics. However, effect of particle shape and orifice size which can influence the bubble dynamics has not been previously investigated.

In this study, CFD-DEM is used to study the bubble formation at a single orifice and depict its size and shape with respect to jet velocity, orifice size and particle shape.

2 Models and simulation conditions

In CFD-DEM, the solid phase is based on the DEM and gas phase is modelled as TFM. The governing equations for particle i with mass m_i and moment of inertia I_i can be written as

$$m_i \frac{d\mathbf{v}_i}{dt} = \mathbf{f}_{pf,i} + \sum_{j=1}^{k_i}(\mathbf{f}_{c,ij} + \mathbf{f}_{d,ij}) + m_i\mathbf{g} \quad (1)$$

$$I_i \frac{d\omega_i}{dt} = \sum_{j=1}^{k_i}(\mathbf{M}_{t,ij} + \mathbf{M}_{r,ij} + \mathbf{M}_{n,ij}) \quad (2)$$

where $\mathbf{v}_i$ and $\boldsymbol{\omega}_i$ are the translational and angular velocities of the particle, respectively, and k_i is the number of particles interacting with the particle. The forces involved are: particle fluid interaction force $\mathbf{f}_{pf,i}$, the gravitational force $m_i\mathbf{g}$, and the inter-particle forces between the particles, which include elastic force $\mathbf{f}_{c,ij}$ and viscous damping force $\mathbf{f}_{d,ij}$. The torques acting on particle i by particle j include: $\mathbf{M}_{t,ij}$ generated by the tangential force, $\mathbf{M}_{r,ij}$ commonly known as the rolling friction torque, and also the torque $\mathbf{M}_{n,ij}$ generated by the normal force when the normal force does not pass through the particle centre. The governing equations for gas phases are given as,

$$\frac{\partial \varepsilon_f}{\partial t} + \nabla \cdot (\varepsilon_f \mathbf{u}) = 0 \quad (3)$$

$$\frac{\partial(\rho_f \varepsilon_f \mathbf{u})}{\partial t} + \nabla \cdot (\rho_f \varepsilon_f \mathbf{u}\mathbf{u}) = -\nabla p - \mathbf{F}_{pf} + \nabla \cdot (\varepsilon_f \boldsymbol{\tau}) + \rho_f \varepsilon_f \mathbf{g} \quad (4)$$

* Corresponding author: zongyan.zhou@monash.edu

where $\mathbf{u}$, ρ_f and p are the fluid velocity, density and pressure, respectively. $\mathbf{F}_{pf}$ is the volumetric particle-fluid interaction force in a computational CFD cell of volume, ΔV. $\boldsymbol{\tau}$ and ε_f are the fluid viscous stress tensor and local porosity.

For details regarding the governing equations and the equations to calculate the interaction forces and torques, refer to the literatures [12, 13]. The bed geometry used in this work is a slot model with periodic boundary conditions applied to the front and rear direction. For such a geometry, two-dimensional CFD and three dimensional DEM are employed as used elsewhere [12]. The bed is produced by generating particles randomly. Particles are generated randomly at different heights and allowed to come to rest to form a bed [13]. The bed is then fluidized by gas from central orifice at the bottom which is of size 4 CFD cells (0.001 m). Table 1 lists the parameters including the bed geometry, physical properties of particles and fluid used in the simulation.

Table 1. Various parameters used in simulation

Variable	Values
Bed geometry	
Bed width (x)	0.02 m
Bed height (z)	0.08 m
Bed thickness (y)	0.0004 m
Orifice width	4 CFD cells
CFD cells (x, z)	320 x 80 cells
CFD cell size (Δx , Δz)	0.00025 m
Particle Properties	
Shape	Oblate, spherical, prolate
Aspect ratio, η	0.75, 0.85, 1, 1.15, 1.30
U_{mf}, m/s	0.003, 0.0035, 0.005, 0.004, 0.003
Number of Particles, N_p	100,000
Particle size, d_p	100 μm
Particle density, ρ_p	1450 kg/m^3
Particle-particle sliding friction, μ_s	0.4
Particle-wall sliding friction, μ_r	0.4
Normal damping co-efficient, c_n	0.3
Young's modulus, E	1 x 10^7 Pa
Poisson ratio, ν	0.3
Time step, Δt	1.6 x 10^{-6}
Gas properties	
Density	1.205 kg/m^3
Viscosity	1.8 x 10^{-5} Pa s

3 Results and discussion

In order to extract the bubble characteristics, bubble boundary has to be defined which was evaluated from the simulated void fraction data. The threshold void fraction was taken as 0.8, and the regions above $\varepsilon = 0.8$ were defined as bubbles. Assuming bubbles to be two dimensional, the bubble equivalent diameter was defined as follows:

$$D_e = \sqrt{4A/\pi} \tag{5}$$

where A is the bubble area. The aspect ratio of a bubble describes the closeness of bubble shape to circle and is defined as:

$$\sigma = D_v/D_h \tag{6}$$

where D_v and D_h represents the vertical and horizontal diameter through bubble centre respectively.

3.1 Influence of Jet velocity

Different jet velocities were injected from the centre of bed with orifice size equal to 4 CFD cells (i.e. 0.001 m of orifice width) to examine its effect. Fig. 1 shows formation of DEM bubbles at different times after injection of jet velocity of 0.1 (a-e), 0.15 (f-j) and 0.25 (k-o) m/s. For all the injection velocities applied, the bubbles detach the distributor around 0.046s. The bubble detachment time is unaffected by the injection velocity. Similar conclusion was shown in previous study [10]. However the rise of the bubble, size and shape are significantly affected by the jet velocity. It can be seen from Fig. 1 (c, h and m) that at the same time the bubble for the higher velocity is at much higher level in the bed than for the lower velocity. Moreover, at 0.11s (Fig. 1 d, i and n), the bubble for U = 0.25 m/s is at the top of the bed while the bubble for U= 0.1 m/s is at the middle of the bed. This might be because the bubble rise velocity is greatly affected by the size of the bubble which is generally expressed as the bubble equivalent diameter. In literature, most of the correlations developed to calculate the single bubble rise velocity is correlated with the bubble equivalent diameter [14]. However, in this paper we do not intend to investigate the single bubble rise velocity but rather limit our investigation to the size and shape of the first bubble formed and the effect of several parameters such as jet velocity, orifice size and particle shape.

Fig. 2 shows the effect of jet velocity on bubble equivalent diameter for the formation of the first bubble. The bubble size grows with time when it rises through the bed to a maximum value and then remains approximately constant. At the same time the bubbles formed for large velocity is larger. This result is consistent with the expression for bubble mass balance and literature results [3, 9]. Fig. 3 shows the calculated bubble aspect ratio during the bubble growth for different jet velocity. For all jet velocity, as the bubble rises through the bed it expands in the vertical direction thereby becomes more and more elongated. However, for lower jet velocity, the elongation is more significant and existing for long time. After the bubble is detached, the bubble stretches in the horizontal direction due to development of wake which results in the decrease of the aspect ratio. It can be established that for the higher velocity, the bubble tends to be more spherical when it is rising through the bed.

3.2. Effect of orifice size

In order to assess the effect of orifice size on the bubble characteristics, the CFD cell size in the centre of

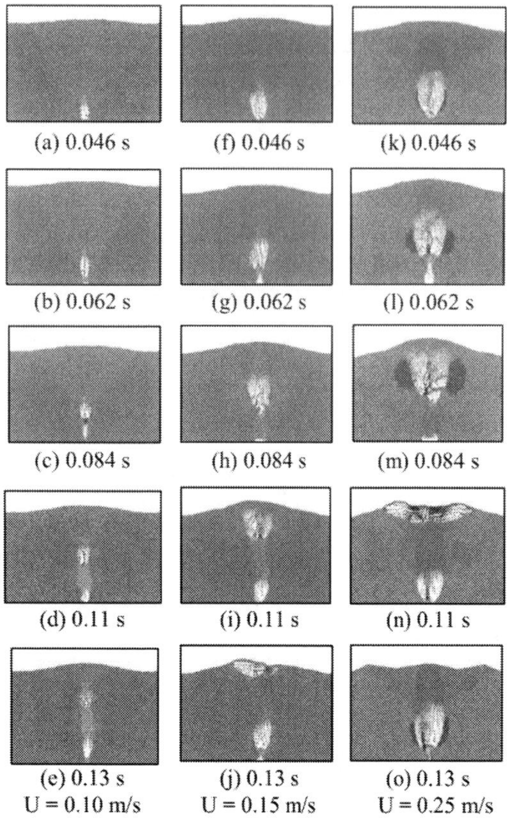

Fig. 1. DEM bubbles at different times after injection of jet velocity of U= 0.10 m/s (a-e), U = 0.15 m/s (f-j) and U = 0.25 m/s (k-o)

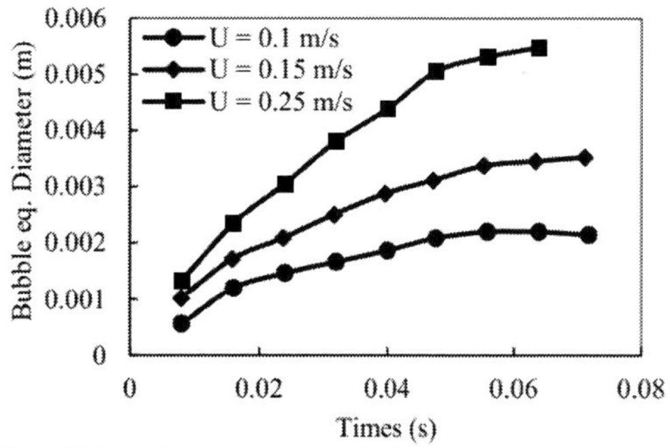

Fig. 2. Effect of jet velocity on bubble equivalent diameter.

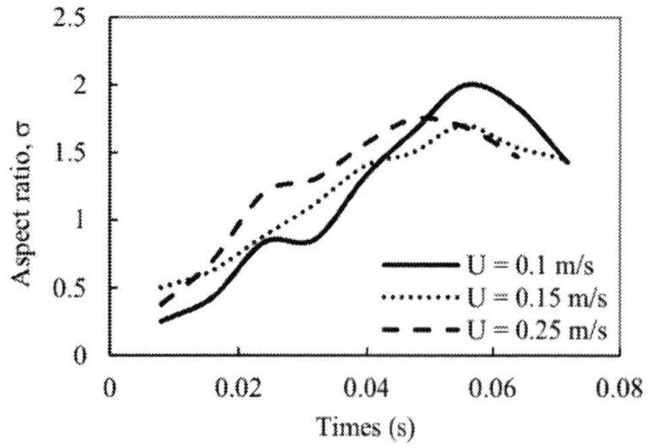

Fig. 3. Bubble aspect ratio for different jet velocity.

the bed which represented the orifice size in the simulations were varied while jet velocity was maintained at 0.1 m/s. The base case was 4 CFD cells which is equal to 0.001m orifice width and bed width to orifice width ratio of 20. CFD cell size was increased from 4 to 6 and 8 which is equal to 0.0015m and 0.002 m, and bed width to orifice width ratio of 13.3 and 10, respectively. Fig. 4 shows the bubbles at 0.046 s for different orifice size for jet velocity of 0.1 m/s, indicating that the bubbles formed for larger orifice size is larger. Similar result is also illustrated by Fig. 5 which shows the effect of increasing orifice size on the growth of bubble equivalent diameter. Increasing the orifice size increases the bubble size. Figs. 4 and 5 shows that as the bed width to orifice width ratio decreases, the size of the bubble increases. A case in which orifice size was decreased to that of base case, i.e. with orifice size equal to 2 CFD cells was also simulated, the results are not shown here. However, in this case no bubbles were formed.

Fig. 6 shows the effect of orifice size on the bubble aspect ratio at jet velocity of 0.1 m/s. As the bubble rises through the bed, the bubble aspect ratio keeps increasing, and indicates that the bubble is growing in the vertical direction. This is similar to previous case for all orifice sizes. But the shape of the bubble formed with increasing the size of the orifice or decreasing the bed width to orifice width ratio tend to be more spherical. Bubble formed after the detachment from the orifice for orifice size of 8 CFD cell was the closest to spherical with bubble aspect ratio of 1.25.

3.3 Effect of particle shape

To date, simulation studies related to the formation of bubble at single orifice are carried out with spherical particles. This cannot fully represent the reality since in practice the particles are mostly non-spherical, of either regular or irregular shapes [12]. Particle shape is one of the most important particle properties which can affect the flow dynamics and influence the size and shape of the bubble. To investigate the effect of particle shape, spheroids are considered, with its aspect ratio (AR) varying from 0.75 to 1.30. The aspect ratio of spheroids/ellipsoids is defined as the ratio of the polar to equatorial lengths. DEM study of ellipsoidal particles have been well developed [12]. In this study, bed of spheroid particles with AR of 0.75, 0.85, 1.15 and 1.30 were generated and then gas was injected at jet velocity of 0.1 m/s from the centre of the bed with orifice width of 4 CFD cells.

Fig. 7 shows the effect of different particle shape (characterized by AR) on the bubble size. The bubble equivalent diameter is the smallest for spherical particles. As the AR deviates from 1 (either greater or smaller than 1), the size of the bubble increases. This might be because when the AR is deviating from 1, the particle-particle interaction and fluid drag becomes relatively weak. In addition, this can also be related to the effect of minimum fluidization velocity. Particles with lower minimum fluidization velocity have been shown to generate larger bubbles [7, 11]. Decreasing AR from 1 up to 0.25 and increasing AR from 1 up to 2 have shown to decrease the minimum fluidization velocity [13]. Bubbles for AR = 0.75 and AR = 1.30 have the highest bubble equivalent diameter since they have the lowest minimum fluidization velocity (refer Table 1).

Fig. 8 shows the effect of particle shape on bubble aspect ratio. It can be seen that the bubble aspect ratio grows in similar way for all the particles, however for ellipsoidal particles the decrease in aspect ratio is not as sharp as for the spherical particles. Bubble for particles deviating from AR = 1 tend to be closer to spherical than bubble for AR = 1.

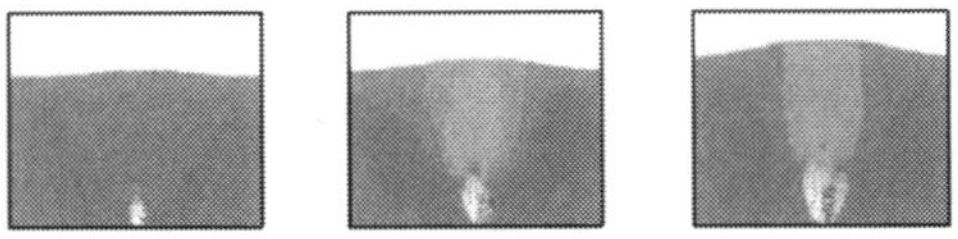

(a) 4 CFD cells (b) 6 CFD Cells (c) 8 CFD Cells

Fig. 4. DEM bubbles at 0.046 s for different orifice size at jet velocity of 0.10 m/s.

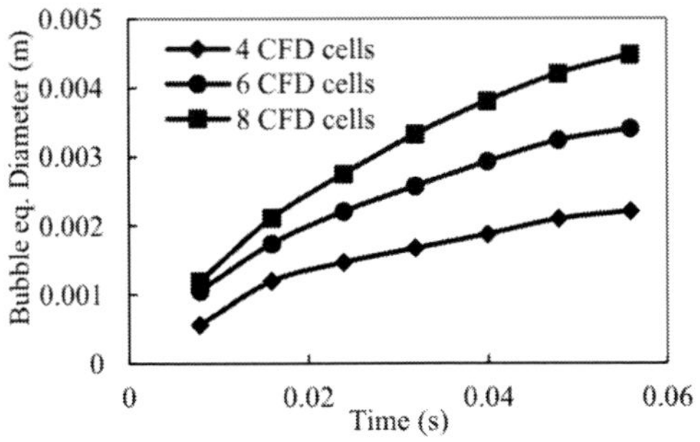

Fig. 5. Effect of orifice size on bubble equivalent diameter, when jet velocity = 0.10 m/s.

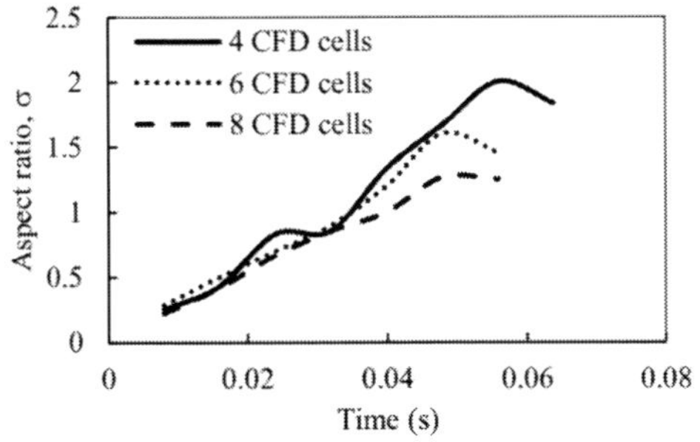

Fig. 6. Effect of orifice size on bubble aspect ratio, when jet velocity = 0.10 m/s.

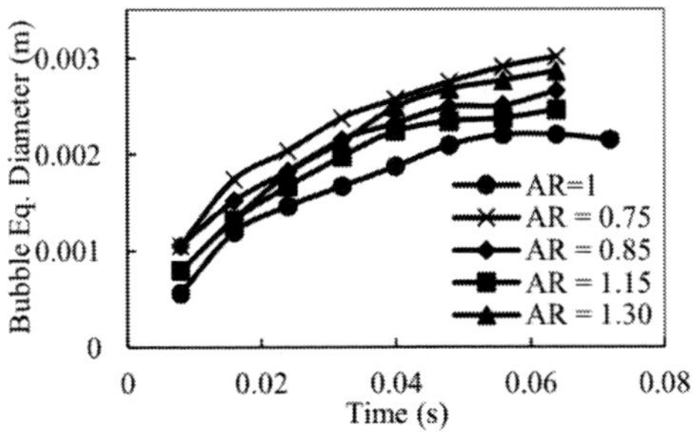

Fig. 7. Effect of particle shape on bubble equivalent diameter, when jet velocity = 0.10 m/s.

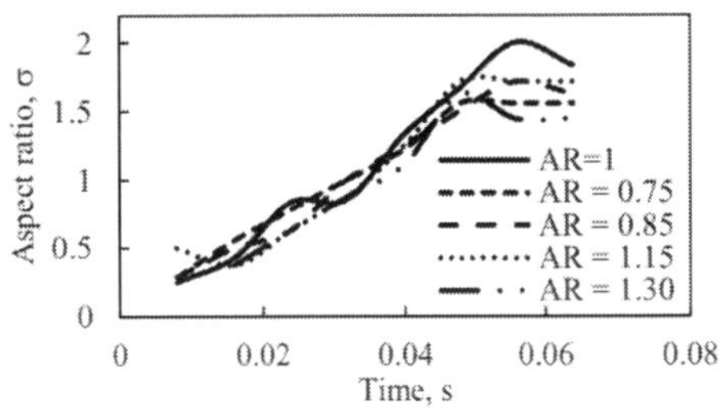

Fig. 8. Effect of particle shape on bubble aspect ratio, when jet velocity = 0.10 m/s.

4. Conclusion

Formation of a bubble at a single orifice was investigated using CFD-DEM to study the bubble characteristics such as bubble equivalent diameter and bubble aspect ratio. Effect of different parameters such as jet velocity, orifice size and particle shape on bubble characteristics were explored. The bubble equivalent diameter which delineates the size of the bubble was found to increase with increasing jet velocity and orifice width. It was also found to be greater for the non-spherical particles whose aspect ratio was deviating from 1. The bubble aspect ratio was found to elongate in the vertical direction as the bubble rose through the bed, and it expanded in the horizontal direction when it was detached from the orifice. The bubbles for higher jet velocity, large orifice size and for non-spherical particles tend to be more spherical.

References

[1] M. Rüdisüli, PhD. Thesis **ETH ZURICH** (2012).
[2] A. Busciglio, G. Vella, G. Micale, L. Rizzuti, Chem. Eng, J., **140** (2008) 398-413.
[3] A.D. Lotfabad, S. Movahedirad, M. Sadeghi, Iran. J. Chem. Eng. (IJChE), **13** (2016) 61.
[4] D. Harrison, L. Leung, Chem. Eng. Res. Des., **39** (1961) 409-414.
[5] H.S. Caram, K.-K. Hsu, Chem. Eng. Sci., **41** (1986) 1445-1453.
[6] J. Kuipers, W. Prins, W. Van Swaaij, Chem. Eng. Sci., **46** (1991) 2881-2894.
[7] J. Nieuwland, M. Veenendaal, J. Kuipers, W. Van Swaaij, Chem. Eng. Sci., **51** (1996) 4087-4102.
[8] D. Gera, M. Gautam, Y. Tsuji, T. Kawaguchi, T. Tanaka, Powder Technol., **98** (1998) 38-47.
[9] O.O. Olaofe, M.A. van der Hoef, J.A.M. Kuipers, Chem. Eng. Sci., **66** (2011) 2764-2773.
[10] J.K. Pandit, X. Wang, M. Rhodes, Chem. Eng. Sci., **62** (2007) 159-166.
[11] L.-w. Rong, J.-m. Zhan, C.-l. Wu, Adv. Powder Technol., **23** (2012) 398-405.
[12] Z. Zhou, D. Pinson, R. Zou, A. Yu, Chem. Eng. Sci., **66** (2011) 6128-6145.
[13] J. Gan, Z. Zhou, A. Yu, AIChE Journal, **62** (2016) 62-77.
[14] S. Karimipour, T. Pugsley, Powder Technol., **205** (2011) 1-14.

 DOI: 10.1051/epjconf/201714003069

Homogenization of granular pipe flow by means of helical inner-wall texture

Felix Verbücheln[1,*], *Eric J. R.* Parteli[2,**], and *Thorsten* Pöschel[1,***]

[1] *University of Erlangen-Nuremberg, Erlangen, Germany*
[2] *University of Cologne, Cologne, Germany*

Abstract. The homogenization of granular flows through narrow pipes is important for a broad range of technological and industrial applications. Here we show, by means of molecular dynamics simulations, that such homogenization can be achieved by adding a helical inner-wall texture to the pipe, without the need for energy input from any external source. By using such a texture, jamming is prevented and the granular flux can be predicted using a modified Beverloo equation that accounts for the wavelength of the helical texture.

1 Introduction

Granular pipe flows are characterized by intermittent behavior and large, potentially destructive solid fraction variations in the transport direction. Although the dynamics of density waves in granular pipe flows have been studied extensively in the past [1–5], it is still a challenging problem to control the mass flux of the granular material flowing through a pipe (see e.g. Ref. [6]).

Most of the proposed strategies to control the transport along the pipe involve energy input from an external source, e.g. through application of electric fields [7] or mechanical perturbations [8]. One exception is the work of Zuriguel *et al.* [9], who showed how insertion of an obstacle just above the outlet of a silo can significantly reduce the probability that the granular flow is arrested due to the formation of an arch blocking the silo's outlet [9].

Here we demonstrate a method to homogeneize the mass flux and avoid flow blockage in granular pipe flows without necessity of applying any external source of energy to the system. By means of particle-based numerical experiments, we will show that it is possible to achieve flows with prescribed characteristics regarding the particle distribution within the pipe and the mass flow rate of the granular material by adjusting the geometric properties of the helix-shaped texture.

2 Numerical experiments

We simulate the process using the Discrete Element Method (DEM), that is, simultaneously solving Newton's equations of translational and rotational motion for all particles. We assume viscoelastic interaction in normal direction [10, 11] and apply a modified Cundall-Strack model [12] for the tangential direction [13]. The integration was performed using LIGGGHTS [13, 14], while the values of the model parameters are listed in Tab. 1. While a detailed description of the model has been presented before [14–18], in the following we focus on the description of the numerical experiments of granular pipe flow and, in the subsequent section, the model of the helix-shaped texture to homogenize the particle transport [14].

The pipe has a circular cross-section of diameter D mm and its length is $L = 1$ m. The inner-wall of the pipe is frictional, though it is modelled as a flat surface, i.e. the pipe is not built with particles (see Ref. [14]). At time $t = 0$, N particles are placed at random positions within the pipe, where N is chosen such that the sum of the volumes of all particles amounts to a prescribed fraction V_ϕ of the pipe volume. The initial velocity of the particles in radial direction is chosen randomly between $-v_r$ and v_r, with $v_r = 0.01$ m/s, while the particles have vanishing initial velocity in the axial direction ($v_z = 0$). Periodic boundary conditions are applied in the vertical (z) direction.

Table 1. Numerical values of the parameters used in the simulations.

parameter	symbol	value
particle material density	ρ_p	2650 kg/m^3
particle diameter	d	1.2 mm
Young's modulus	Y	10^8 Pa
Poisson's ratio	ν	0.24
Coulomb's friction coefficient	μ	0.5
pipe length	L	1 m
timestep	Δt	6×10^{-7}s

3 Results and Discussion

We perform simulations using a constant particle diameter d as specified in Table 1 and different values of pipe diameter D. We first investigate the total kinetic energy

*e-mail: verbuechelnf@googlemail.com
**e-mail: eric.parteli@uni-koeln.de
***e-mail: thorsten.poeschel@fau.de

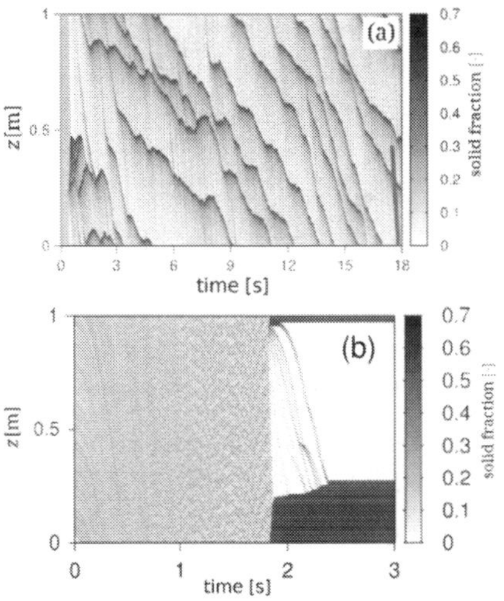

Figure 1. Spatio-temporal images of the packing fraction along the tube for a pipe to particle diameter $D/d = 4$ (a) and $D/d = 2.5$ (b). The slope of the line at the lower right corner in (a) indicates the average axial velocity of the particles.

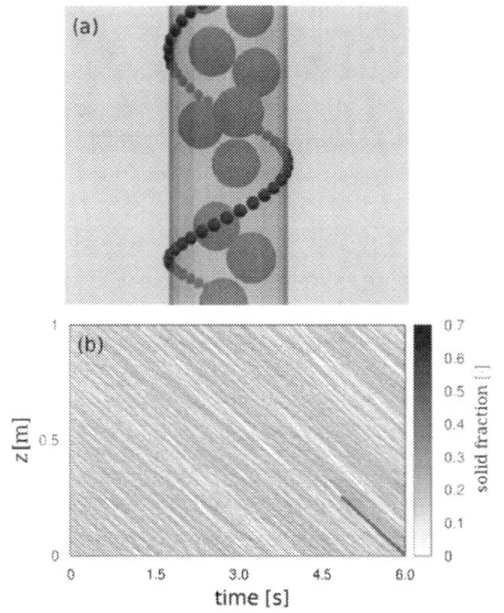

Figure 2. Flow of granular particles (blue) through a vertical pipe with $D/d = 2.5$ to which the helical inner-wall texture (constituted by the red beads) is applied; (b) corresponding spatio-temporal diagram of the solid fraction along the axial position. The slope of the line at the lower right corner indicates the average axial velocity of the particles.

of particles flowing through a vertical pipe with $D/d = 4$. Collisions between the particles and between the particles and the wall lead, then, to deceleration of the particles and a decrease in the kinetic energy of the system. After a transient time, the energy gain of the particles due to gravity is nearly compensated by the energy dissipation due to collisions such that the total kinetic energy fluctuates around a constant value [1, 14].

The spatio-temporal image of the volume fraction along the tube (Fig. 1a) shows the emergence of density waves [1, 14], with the development of recurrent *clogging* characterized by the formation of *plugs*, that is regions with high packing fraction (thick dark lines in Fig. 1a) which can either converge or diverge with time. The flow is inhomogeneous and associated with strong fluctuations of particle average velocities and solid fraction both in time and position along the pipe.

The average flow rate is dictated by the behavior of grains within the plug zones [19]. Each plug zone is associated with the formation of an unstable arch, the height of which should scale with D. Particles falling after the break of an arch can accelerate freely due to the action of gravity thus leading to a scaling of the average vertical velocity of the grains, $\langle v_z \rangle$, with $\sqrt{D}$ [19]. Therefore, a scaling of the average mass flux exiting the pipe through its bottom end per unit time, Q_0, with $D^{5/2}$ is expected. This scaling indeed governs the mass flux of a granular material flowing out a silo through an orifice of diameter D, which follows the well-known Beverloo equation [6, 9, 19, 20],

$$Q_0 = A\rho_p \sqrt{g} \cdot (D - kd)^{2.5}, \quad (1)$$

where the coefficients A and k must be determined from the fit to the data. In Ref. [14] we have shown that such a scaling captures well the dependence of the mass flux on the pipe diameter for $D/d > 3$.

Moreover, we find that, for $D/d \lesssim 3$, *jamming* occurs thus leading to complete blockage of the granular flow. As an example of the flow in the jamming regime, we show in Fig. 1b the spatio-temporal diagram of the packing fraction for $D/d = 2.5$. As we can see in this figure, after about 1.8 s, a large, stable plug is formed, which corresponds to the dark horizontal lines in the spatio-temporal diagram. Although we indeed did not observe complete flow blockage for $D/d > 3.0$, the flow in this regime is intermittent and characterized by density waves as discussed in detail in Ref. [14] and demonstrated previously [1, 2]. Therefore, in the following our aim is to develop a method to homogeneize the flow thus avoiding the formation of density waves that lead to jamming in granular pipe flows.

We investigate the vertical flow of a granular material down a pipe of circular cross-section that has a helical inner-wall texture as depicted in Fig. 2a. This texture is constituted of small beads of diameter $d_s = D/10$, which are fixed to the inner-wall of the pipe and have the material properties listed in Table 1. Each constituent bead is fixed to the wall at its center such that one hemisphere of each bead is within the inner volume of the pipe. The diameter of the pipe in the presence of the helix is adjusted such that the total volume within the pipe is the same as in the simulations without the texture elements.

As we can see jamming does not occur in such a pipe in the presence of the helical texture. A steady downward flow of the granular material is observed, whereas the particles are more homogeneously distributed throughout the pipe compared to the simulations without the helical tex-

DOI: 10.1051/epjconf/201714003069

ture. This result can be understood by noting that, as the particles collide with the beads fixed on the inner-wall, they are deflected to the center of the pipe, thus reducing the probability for the formation of stable archs. On the other hand, due to the collisions of the particles with the inner-wall beads, the average axial velocity of the particles is reduced, thus causing a decrease in the average mass flux in the axial direction [14]. Therefore, by introducing the helical texture of Fig. 2a, jamming can be prevented at the cost of a lower rate of particle flux down the pipe.

We find that the average mass flux depends significantly on the wavelength of the helix, λ. In order to quantitatively describe the effect of the helix wavelength on the mass flux, we first note that in the steady-state, the gain in momentum of the particles due to gravity is fully compensated by the momentum loss due to collisions with the pipe's inner-wall as well as with the other particles. In the presence of the helix, there is an additional contribution to the rate with which axial momentum is dissipated. This contribution, $\dot{p}_{\rm diss,helix}$, is due to collisions of the particles with the beads composing the helix. It is reflected by a relative decrease in the magnitude of the steady-state axial velocity of the particles, compared to the value without helix [14]. Therefore, the steady-state flux Q in the presence of the helix relates to the flux Q_0 without helix through the expression

$$Q = Q_0 - Q_{\rm diss,helix}, \tag{2}$$

where $Q_{\rm diss,helix}$ is the amount by which the steady-state axial mass flux is reduced when the helical texture is present.

In the small pipe to particle diameter ratio investigated here, it is reasonable to assume that the average mass flux in the axial direction is proportional to the average axial momentum of the particles, and that $Q_{\rm diss,helix}$ is nearly proportional to $\dot{p}_{\rm diss,helix}$. As a result of particle-helix collisions along the pipe, the particles are deflated to the pipe's central axis which is why there is an increase in the average radial momentum when the helix is present [14]. However, we note that due to symmetry, for any radial direction, the contribution of the collisions to increasing the radial momentum at the upper half of one helix wavelength is the same as at the lower half. For a given pipe diameter D and an helix wavelength λ made up of n_s beads, the collisions between particles with $n_s/2$ beads from the upper and lower hemispheres (half-wavelengths) deflate the particles in *opposite* directions. We thus need to calculate the rate with each the particles gain radial momentum due to collisions with one single hemisphere. The total rate is then twice the contribution from one half-hemisphere.

Fig. 3 represents both hemispheres unravelled from the inner-wall as two right triangles, each with legs D and $\lambda/2$. Note that $\lambda/2$ is the length corresponding to $n_s/2$ beads, which is the contribution of each hemisphere to increase the radial momentum along the pipe's cross section. From the figure, we see that the rate of axial momentum dissipation due to each hemisphere must scale with $p_\perp$, the component of the particles' average axial momentum p_0 perpendicular to the hypothenuse of each triangle. Therefore, $\dot{p}_{\rm diss,helix} \propto 2\,p_\perp = 2\,p_0 \sin\left[\Theta_{D,\lambda/2}\right]$, where $\Theta_{D,\lambda/2} = \arctan\left[\pi D/(\lambda/2)\right]$ (see Fig. 3).

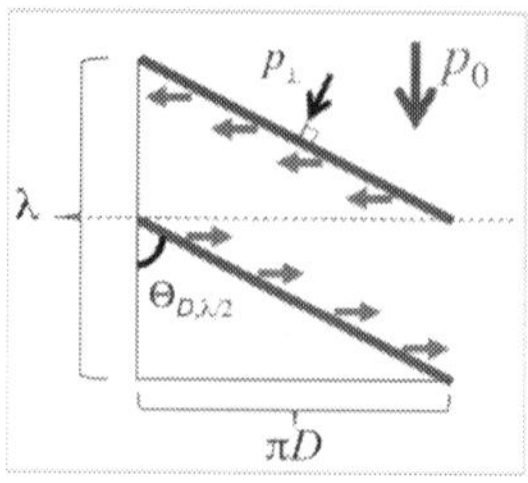

Figure 3. Schematic diagram displaying the average axial momentum p_0 and the equivalent angle $\Theta_{D,\lambda/2} \equiv \arctan[\pi D/(\lambda/2)]$, which dictates the energy dissipation over one helix wavelength due to collisions with the $n_s/2$ beads composing one half-wavelength of the helix. The component $p_\perp = p_0\sin[\arctan[\pi D/(\lambda/2)]]$ is also indicated. The horizontal arrows along the helix indicate the direction from the helix to the central axis of the pipe.

Following our assumption that $Q_{\rm diss,helix}$ should be proportional to $\dot{p}_{\rm diss,helix}$, we can write $Q_{\rm diss,helix} \approx 2bQ_0\sin\left[\arctan\left[\pi D/(\lambda/2)\right]\right] = 2bQ_0 \cdot \left[2\pi D/\sqrt{\lambda^2-(4\pi D)^2}\right]$, where b is a parameter that encodes information on the dissipative properties of the collisions. Thus, from Eq. (2),

$$\frac{Q}{Q_0} = 1 - B\left[\frac{2\pi D}{\sqrt{\lambda^2-(2\pi D)^2}}\right], \tag{3}$$

where Q_0 is the flux without helix and the constant $B = 2b$ encodes information on energy dissipation due to collisions with the helix. Thus, the value of B should depend on the material properties and particle diameter. To verify Eq. (3), we compute the expectation value of the flux, $Q = \int q f_\lambda(q)dq$ as a function of λ, where $f_\lambda(q)$ is the probability density function associated with the wavelength λ. The result of this calculation is denoted by the symbols in Fig. 4b, in which Q appears rescaled with Q_0. The continuous line denotes the best fit using Eq. (3), which gives $B \approx 0.74$. As we can see from this figure, the agreement between model and simulation results is excellent.

The inset of Fig. 4b shows the rescaled standard deviation of the flux, σ/σ_0, as a function of $1/\lambda$, where σ_0 is the standard deviation obtained in the simulations without helix. The value of σ/σ_0 gives a measure of the homogeneity of the flux along the pipe — the smaller σ the more homogeneous the flux. By fitting the simulation data using the equation

$$\frac{\sigma}{\sigma_0} = 1 - C\cdot\left[\frac{2\pi D}{\sqrt{\lambda^2-(2\pi D)^2}}\right], \tag{4}$$

we obtain $C \approx 0.9$. In conclusion, both the flux and its standard deviation can be obtained from the geometric parameters of the helix using Eqs. (3) and (4), respectively.

However, obviously Eq. (3) can only be valid for the regime of small pipe to particle diameter ratios investigated here. In particular, it can be shown that the model predicts negative flux values for $D/d \gtrsim 10$ [14]. Moreover,

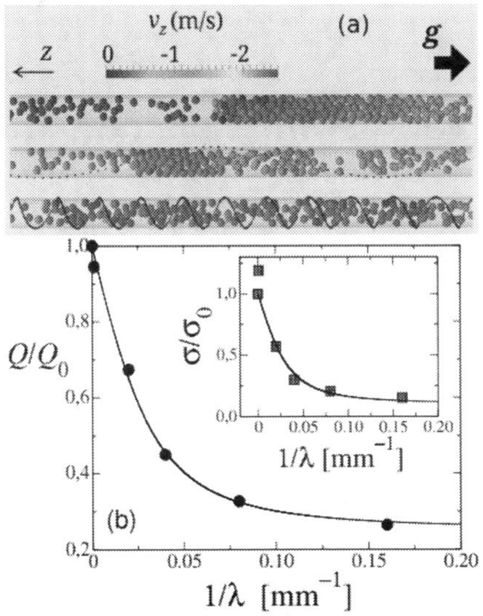

Figure 4. (a) Simulations with no helix and with helix of wavelength λ = 25 mm and 6.25 mm (from top to bottom); (b) expectation value of the rescaled mass flux Q/Q_0 (main plot) and non-dimensional standard deviation σ/σ_0 (inset) as a function of $1/\lambda$. Symbols denote simulation results, continuous lines are best fits to the model described in the text, while D/d = 3.5, $Q_0 \approx 0.0074\,\mathrm{kg/m^3}$ and $\sigma_0 \approx 0.00025\,\mathrm{kg/m^3}$.

as $\lambda \to 0$ the flux through the pipe must follow the original Beverloo equation with Q_0 corresponding to a pipe with no helix and diameter $D - d_s$, where d_s is the size of the spheres constituting the helix. Further work is thus needed in order to elucidate the dependence of model fit parameters on the number of helix beads per wavelength λ, as well as on the particles' frictional and dissipation properties.

Conclusions

In conclusion, we have presented a method to obtain steady flows of granular materials through narrow pipes, which consists of applying a helical texture to the pipe's inner wall. The helix texture prevents the formation of stable plugs and avoids jamming of the granular flow. We believe that application of the helical texture presented here could be used to enhance not only gravity-driven pipe flows but also fluid-driven particle transport through both through vertical and horizontal pipes, which remains to be investigated in the future. Moreover, the influence of wear should be also investigated.

Acknowledgements

We thank the German Research Foundation (DFG) for funding through the Cluster of Excellence "Engineering of Advanced Materials", ZISC, FPS, the Collaborative Research Center SFB814, and grant PO472/20-2. We gratefully acknowledge the computing time granted by the John von Neumann Institute for Computing (NIC) and provided on the supercomputer JUROPA at Jülich Supercomputing Centre (JSC).

References

[1] T. Pöschel, J. Phys. I France **4**, 499-506 (1994).
[2] T. Riethmüller, L. Schimansky-Geier, D. Rosenkranz and T. Pöschel, J. Stat. Phys. **86**, 421-430 (1997).
[3] J. L. Aider, N. Sommier, T. Raafat and J. P. Hulin, Phys. Rev. E **59**, 778-786 (1999).
[4] Y. Bertho, F. Giorgiutti-Dauphiné, T. Raafat, E. J. Hinch, H. J. Herrmann and J. P. Hulin, J. Fluid Mech. **459**, 317-345 (2002).
[5] I. Bratberg, F. Radjai and A. Hansen, Phys. Rev. E **71**, 011301 (2005).
[6] A. Janda, I. Zuriguel, J. Bienzobas, A. Garcimartín and D. Maza, AIP Conf. Proc. **1542**, 710 (2013).
[7] W. Chen, M. Hou, K. Lu, Z. Jiang and L. Lam, Phys. Rev. E **64**, 061305 (2001).
[8] M. L. Hunt, R. C. Weathers, A. T. Lee, C. E. Brennen and C. R. Wassgren, Physics of Fluids **11**, 68-75 (1999).
[9] I. Zuriguel, A. Janda, A. Garcimartín, C. Lozano, R. Arévalo and D. Maza, Phys. Rev. Lett. **107**, 278001 (2011).
[10] N. V. Brilliantov, F Spahn, J. M. Hertzsch, T. Pöschel, Physical review E **53**, 5382 (1996).
[11] T. Pöschel and T. Schwager, *Computational Granular Dynamics* (Springer, Heidelberg, 2005).
[12] P.A. Cundall, O.D.L. Strack, Geotechnique **29**, 47 (1979).
[13] C. Kloss, C. Goniva, A. Hager, S. Amberger and S. Pirker, Prog. Comput. Fluid Dy. **12**, 264 (2012). www.liggghts.com
[14] F. Verbücheln, E. J. R. Parteli and T. Pöschel, Soft Matter **11**, 4295-4305 (2015).
[15] E. J. R. Parteli, AIP Conf. Proc. **1542**, 185-188 (2013).
[16] H. Pourtavakoli, E. J. R. Parteli and T. Pöschel, New J. Phys. **18**, 073048 (2016).
[17] E. J. R. Parteli and T. Pöschel, Powder Technology **288**, 96-102 (2016).
[18] E. J. R. Parteli, J. Schmidt, C. Blümel, K.-E. Wirth, W. Peukert and T. Pöschel, Scientific Reports **4**, 6227 (2014).
[19] C. Mankoc, A. Janda, R. Arévalo, J. M. Pastor, I. Zuriguel, A. Garcimartín and D. Maza, Granular Matter **9**, 407 (2007).
[20] W. A. Beverloo, H. A. Leniger and J. van de Velde, Chemical Engineering Science **15**, 260-269 (1961).

Active earth pressure model tests versus finite element analysis

Magdalena Pietrzak*

Koszalin University of Technology, Department of Civil, Geodetic and Environmental Science, 74-453 Koszalin, Poland

Abstract. The purpose of the paper is to compare failure mechanisms observed in small scale model tests on granular sample in active state, and simulated by finite element method (FEM) using Plaxis 2D software. Small scale model tests were performed on rectangular granular sample retained by a rigid wall. Deformation of the sample resulted from simple wall translation in the direction 'from the soil" (active earth pressure state. Simple Coulomb-Mohr model for soil can be helpful in interpreting experimental findings in case of granular materials. It was found that the general alignment of strain localization pattern (failure mechanism) may belong to macro scale features and be dominated by a test boundary conditions rather than the nature of the granular sample.

1 Introduction

The purpose of the paper is to compare failure mechanisms observed in small scale model tests on granular sample in active state, and simulated by finite element method (FEM) using Plaxis 2D software.

Experimental failure mechanisms were obtained by image analysis (particle image velocimetry PIV, geoPIV_RG software). Failure mechanism in a granular material consists of a system of localised deformation (shear bands) best visible on shear strain maps which can be calculated from displacement fields (Fig.2A, Fig.3 Experiment).

The corresponding mechanisms produced by numerical simulation (FEM) are best visible on maps of shear strain increments (Fig.2B, C, Fig.3 Calculations).

1.1 PIV method

Particle image velocimetry is a well-known velocity measuring non-invasive procedure, originally developed for fluid mechanics and used for the analysis of displacements in tests on soil models.

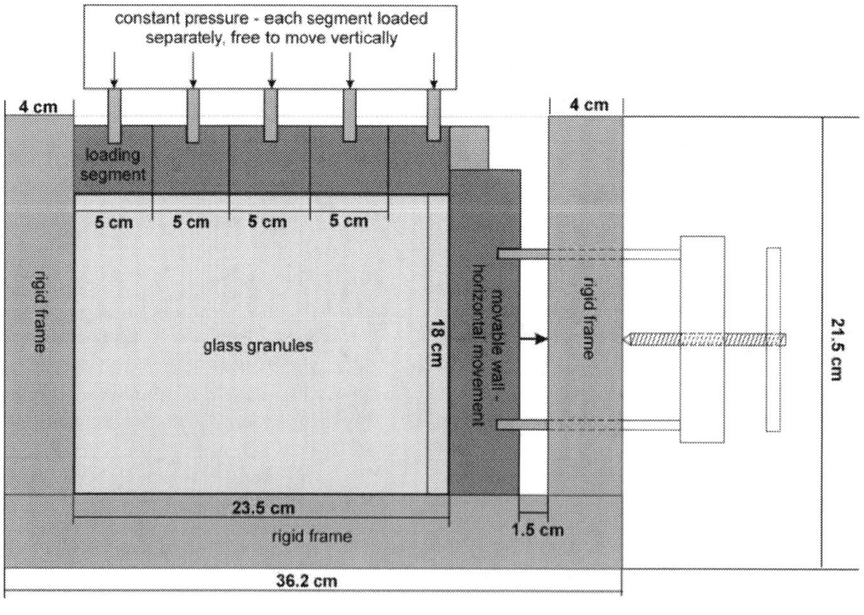

Fig. 1. Experimental setup.

* Corresponding author: magdalena.pietrzak@tu.koszalin.pl

 DOI: 10.1051/epjconf/201714002022

It operates by tracking spatial variations of brightness within an image (divided into a mesh of PIV patches) by comparing successive images so that displacement data can be extracted from sequences of images, and strains then calculated from gradients of measured displacements, [1-3].

The geoPIV_RG software developed by White and Take [1-3] for granular materials was employed in this study.

The first step in geoPIV_RG application is to divide the digital image into square patches of pixels forming a regular mesh – the extremes of the mesh coarseness were discussed by Leśniewska and Muir Wood in [4].

According to White et al. [3], the precision of PIV is a strong function of the patch size that is to be compared from image to image and is also influenced by the image content. If the patch is too small, then the software may not be able to recognize the displaced patch. A finer grid provides more details but also a greater potential for random data fluctuations. A trial and error method indicated that a reliable result (a displacement field containing no spurious vectors) could be obtained by using the mesh of square 40 x 40 pixel patches.

1.2 Finite element method – Plaxis software

Finite element method is based on formalism of the mechanics of continuous media [5]. Plaxis 2D is mainly used for analysis of boundary problems in the field of soil mechanics and geotechnics.

Numerical analysis by Plaxis in this paper consisted of simulation of the model tests' initial geometry and then the granular sample deformation under applied loads.

15 nodes triangular elements were employed in the analysis, containing 3 external and 12 internal nodes.

Results of finite element analysis depend on many factors, such as adopted constitutive models of materials, values of material parameters, boundary conditions. Plaxis contains a number of constitutive models, from which the linear elastic model is most often applied for structural elements, and elasto-plastic Mohr-Coulomb model and its extensions for soils.

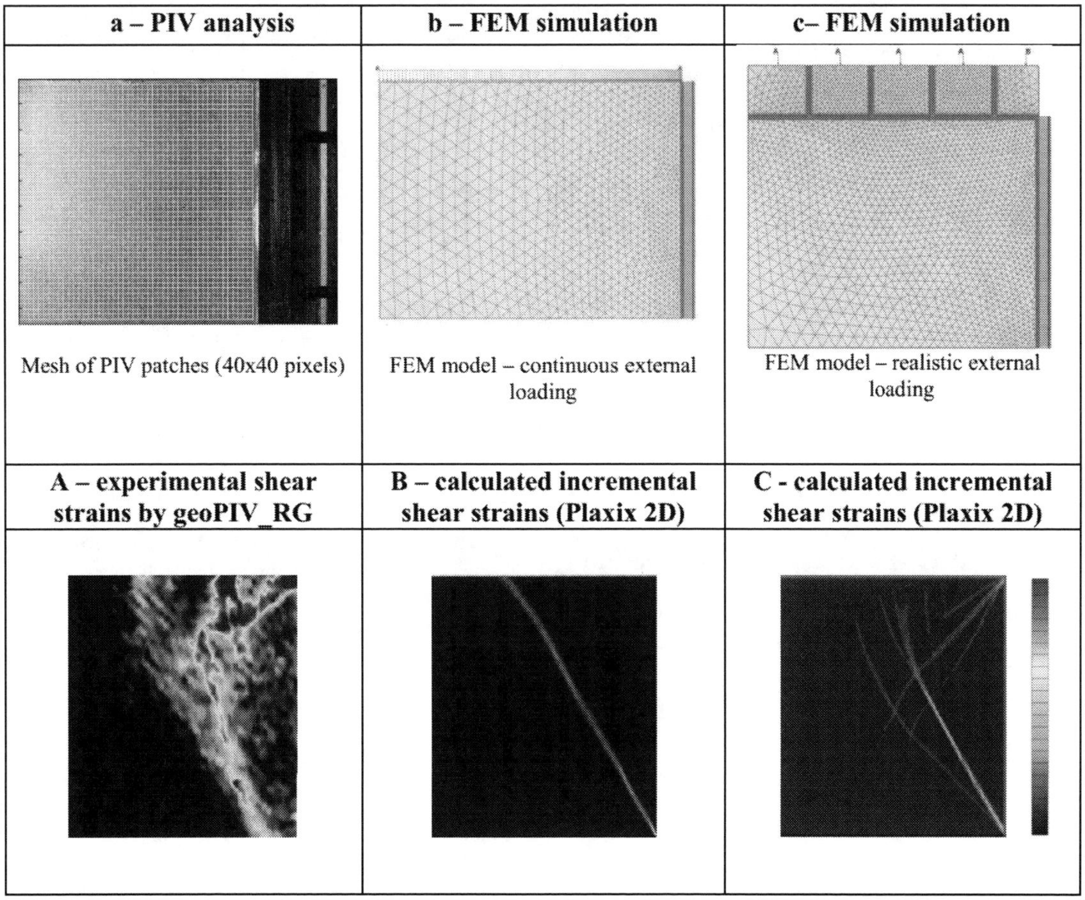

Fig. 2. Comparison of experimental (column A) and calculated (column B and C) shear strain fields evolved due to a vertical wall (right boundary) incremental horizontal displacement.

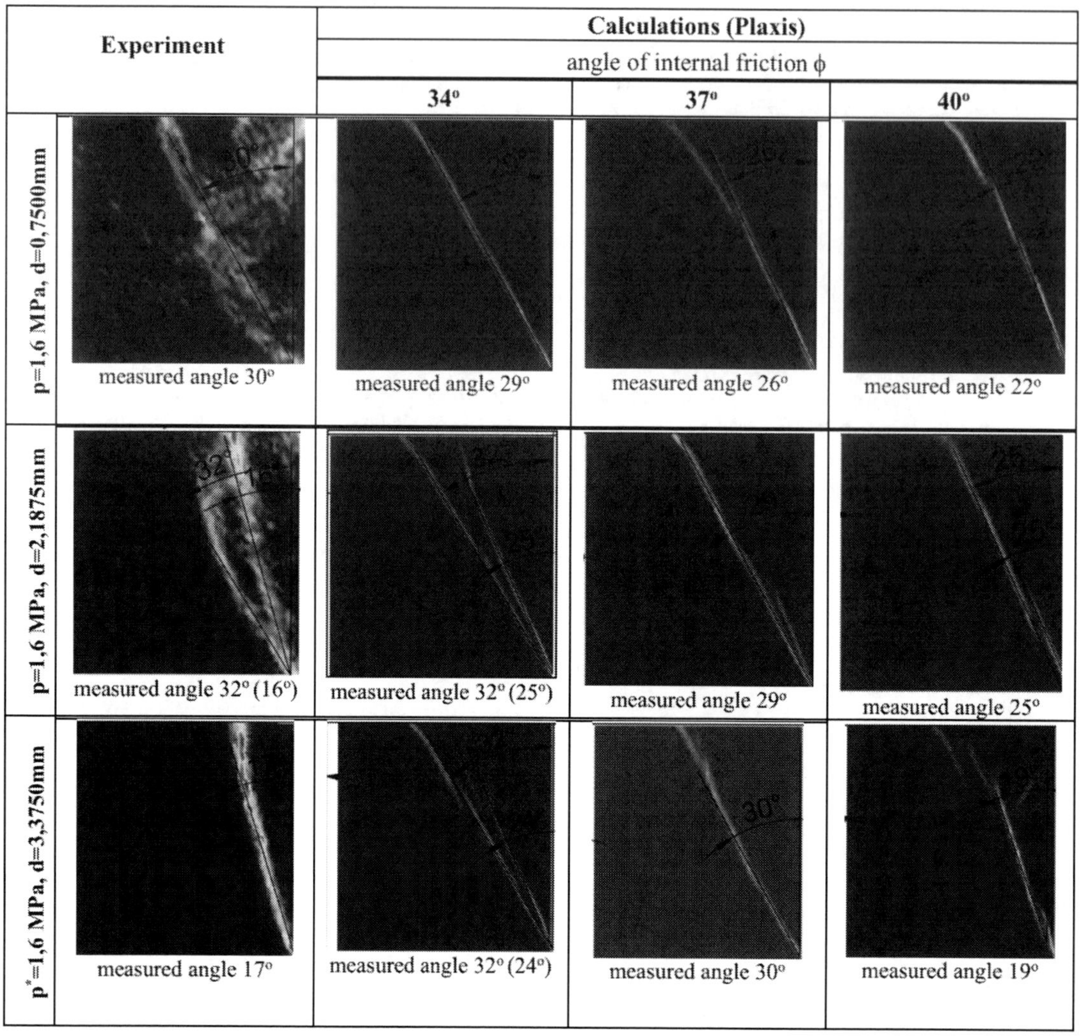

Fig. 3. Average inclination of strain localisations: experiment and FEM simulation. * - p – load, d - displacement

1.3 Materials

Starlitbeads1000 spherical glass granules were used to form a granular sample (d_{50} = 1.1 mm). Only dense samples were investigated. The glass granules were selected to represent soil due to their transparency.

The tests were registered both in ordinary and polarized light, to give not only strain, but also stress information based on photo-elasticity, but the latter is not a subject of this paper. Due to the demands of photo-elastic method the granular specimen was saturated with clove oil, having the same refraction index as glass [4, 6-11]. Volume weight of granules was 16,80 kN/m3.

1.4 Model tests

Small scale model tests were performed on rectangular granular sample retained by a rigid wall. Deformation of the sample resulted from simple wall translation in the direction 'from the soil" (active earth pressure state).

The experimental setup and the techniques applied in this research were described earlier [4, 6-11].

The granular sample consisted of about 10^6 spherical glass particles (~1mm in diameter) – a substitute material for sand.

The sample was tested in a plane strain apparatus, equipped with a loading system and rigid model wall able to translate horizontally. The granular material, placed in the box by pouring from a hopper, was statically loaded on the top surface and deformed due to incremental horizontal translation of right boundary. Model dimensions and arrangement are given in Fig.1.

Small increment of the boundary translation was applied (less than 0.25 grain diameter) to capture momentary changes within strain fields.

All the images used in this research were taken by an industrial grade camera.

2 Numerical simulations by FEM

The value of the stiffness modulus of glass granules was adopted based on triaxial compression test and was equal to 150 000 kPa. For loading segments an elastic model of material was adopted and parameters for steel were accepted, from which they were made.

Maps of shear strain increments were selected to be presented, because they indicate best the locations of strains. Examples of such maps obtained for both types of loading are shown on Fig. 2B and 2C.

Since experimentally approximated internal friction angles ranged from 27° to 34°, calculations were carried out for the slightly wider range 25° – 40°.

Specific boundary conditions – external load distributed on the sample through the set of five metal segments, Fig.1, [4, 6-11] arose the question of their influence on the deformation field. Therefore, a simulation of model test was carried out for two types of load distribution: direct (continuous load Fig. 2b) and applied through the five rigid segments, like it was in reality (Fig. 2c).

Figs 2b and 2c present the results of simulations for both versions of top boundary loading. It is clear that segmented loading (Fig.1) may promote multiple strain localisation. The arrangement of shear bands shown in Fig.2C do not depend on the mesh refinement (the width of shear bands does) and is very close to what was observed in some tests (Fig.2A).

The other issue addressed by numerical analysis was the inclination of strain localisations observed in model tests – it varied significantly as Fig.3 (Experiment) shows.

It can be seen from Fig.3 (Calculations) that the inclination of strain localisations depends on the value of internal friction angle in a way explaining any difference in shear bands inclination, observed experimentally.

3 Conclusions

The mechanisms of a granular material deformation are still not sufficiently recognized, even in case of such a simple test arrangement, like vertical retaining wall moving horizontally. Even simple Coulomb-Mohr model can be helpful in interpreting experimental findings in case of granular materials.

It was found that the general alignment of strain localization pattern (failure mechanism) may belong to macro scale features rather and be dominated by a test boundary conditions rather than the nature of the granular sample.

Acknowledgments

This work was done within the frame of the project nr.2011/03/B/ST8/05865, Polish National Science Centre.

References

1. D. J. White, W. A. Take, Technical Report D-SOILS-TR322. Cambridge University Engineering Department (2002)
2. D. J White, D. J., W. A. Take, M. D. Bolton, Geotechnique 53, 619–631 (2003)
3. D. J White, M. Randolph, B. Thompson, Int. J. Phys. Model. Geotech. 3, 1–12 (2005)
4. D. Lesniewska, D. Muir Wood, Geotechnique, 60, doi: 10.1680/geot.2010.60.00.1 (2010)
5. O. C. Zienkiewicz, M. Huang, M. Pastor, Int. Jnl. For Numerical and Analytical Methods in Geomechanics, 19, 127-148 (1995)
6. D. Lesniewska, D. Muir Wood, J Eng Mech 135:1038–1054 (2009)
7. D. Muir Wood, D. Lesniewska, Granul Matt 13:395–415 (2011)
8. M. Pietrzak, D. Lesniewska, Studia Geotechnica et Mechanika, XXXIV, 4 : 69-77 (2012)
9. D. Lesniewska, M. Pietrzak, AIP Conf. Proc. 1542, 425-428 (2013)
10. D. Lesniewska, M. Pietrzak, Springer Series in Geomechanics and Geoengineering, 273-278 (2015)
11. D. Lesniewska, M. Pietrzak, J. Tejchman, Geomechanics from Micro to Macro, 1-2, 1025-1030 (2015)

Wave propagation of spectral energy content in a granular chain

Rohit Kumar Shrivastava[1,*] and *Stefan* Luding[1,**]

[1] *Multiscale Mechanics (MSM), MESA+, Faculty of Engineering Technology (CTW), PO Box 217, 7500 AE Enschede, Netherlands.*

Abstract. A mechanical wave is propagation of vibration with transfer of energy and momentum. Understanding the spectral energy characteristics of a propagating wave through disordered granular media can assist in understanding the overall properties of wave propagation through inhomogeneous materials like soil. The study of these properties is aimed at modeling wave propagation for oil, mineral or gas exploration (seismic prospecting) or non-destructive testing of the internal structure of solids. The focus is on the total energy content of a pulse propagating through an idealized one-dimensional discrete particle system like a mass disordered granular chain, which allows understanding the energy attenuation due to disorder since it isolates the longitudinal P-wave from shear or rotational modes. It is observed from the signal that stronger disorder leads to faster attenuation of the signal. An ordered granular chain exhibits ballistic propagation of energy whereas, a disordered granular chain exhibits more diffusive like propagation, which eventually becomes localized at long time periods. For obtaining mean-field macroscopic/continuum properties, ensemble averaging has been used, however, such an ensemble averaged spectral energy response does not resolve multiple scattering, leading to loss of information, indicating the need for a different framework for micro-macro averaging.

1 Introduction

Disorder in granular materials (like soil) can manifest in many ways (contact disorder, geometrical disorder, asphericity, layering, etc.) from grain level to system level. All may have an effect on the mechanical wave transmission through the granular material in it's own unique way (for instance, contact disorder due to tiny polydispersity can reduce the mechanical wave speed [9]). Knowing this effect can aid us in numerous ways for subsurface exploration or for non-destructive testing. Thus, there is a need to study the effects of disorder individually and hence the focus in this article will only be on mass disorder, for which 1D granular chain has been chosen so that the P-wave mode is isolated from shear or rotational modes. A mechanical wave propagating through this 1D granular chain is bound to suffer from multiple scattering. However, regardless of scattering, linear waves preserve some coherence and that manifests as intensity correlations ([8]). The results obtained from the chain also represent attributes of both longitudinal P-waves (compressional) and S-waves (shear) in a 3D system as stated in [10]; The frequency filtering effects are similar to those in a 3D system, as observed in [9]. All the more reason to study energy content and spectral energy response of the propagating wave.

*e-mail: r.k.shrivastava@utwente.nl
**e-mail: s.luding@utwente.nl

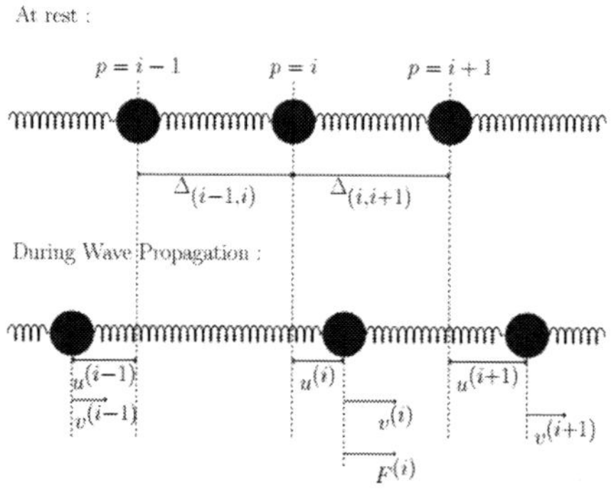

Figure 1. Chain of granular elements during dynamic wave propagation with length scaled by the characteristic equilibrium distance $\tilde{\Delta}_o$.

2 Micromechanical model of 1d chain

A one dimensional granular chain of mass disordered mesoscopic particles is modeled using only a repulsive interaction force,

$$F_{(i,i+1)} = \kappa_{(i,i+1)}\delta_{(i,i+1)} = \kappa_{(i,i+1)}(\Delta_{(i,i+1)} - (u^{(i)} - u^{(i+1)})) \quad (1)$$

The granular chain has a high pre-confining force which does not allow opening and closing of contacts (no occurrence of Nesterenko's sonic vacuum [2, 11]). Hence, a linear contact model is valid to compute the repulsive

Table 1. Dimensional, scaling and hence, the non-dimensionalized parameters.

Dimensional	Scaling	Non-dimensional
$\tilde{m}^{(i)}$	$\tilde{m}_o$	$b^{(i)} = \tilde{m}^{(i)}/\tilde{m}_o$
$\tilde{\kappa}_{(i,j)}$	$\tilde{\kappa}_o$	$\kappa_{(i,j)} = \tilde{\kappa}_{(i,j)}/\tilde{\kappa}_o$
$\tilde{\Delta}_{(i,j)}$	$\tilde{\Delta}_o$	$\Delta_{(i,j)} = \tilde{\Delta}_{(i,j)}/\tilde{\Delta}_o$

interaction forces from the overlap ($\delta_{(i,i+1)}$) [4] for small amplitudes. Table 1 contains the scaling parameters required to arrive at the non-dimensional equation of motion of particle i. $b^{(i)}$ is the non-dimensional mass of particle i and $\tilde{m}_o$ is the average mass of the particles. $\kappa_{(i,j)}$ and $\Delta_{(i,j)}$ are the non-dimensionalized stiffness and overlap between particles i and j, respectively, scaled by the characteristic stiffness ($\tilde{\kappa}_o$) and the corresponding characteristic overlap ($\tilde{\Delta}_o$). Time (t) is scaled by $\sqrt{\frac{\tilde{m}_o}{\tilde{\kappa}_o}}$ (natural frequency of the chain) to give non-dimensional time (τ). The equation of motion of a general particle i is given by,

$$\begin{aligned} F^{(i)}(\tau) = b^{(i)}\frac{\mathrm{d}^2u^{(i)}}{\mathrm{d}\tau^2} &= \kappa_{(i-1,i)}\left[\Delta_{(i-1,i)} - (u^{(i)} - u^{(i-1)})\right] \\ &- \kappa_{(i,i+1)}\left[\Delta_{(i,i+1)} - (u^{(i+1)} - u^{(i)})\right] \\ &= -\left[2u^{(i)} - u^{(i-1)} - u^{(i+1)}\right]. \end{aligned} \quad (2)$$

where $u^{(i)}$ is the non-dimensionalized displacement of particle i, scaled by characteristic overlap ($\tilde{\Delta}_o$). Since, the focus of the analyses here has been only on the effects of mass disorder, it makes us take all interparticle stiffnesses equal to the characteristic stiffness ($\tilde{\kappa}_o$) which makes the non-dimensionalized interparticle stiffness, $\kappa_{(i,j)} = 1$ and implicitly the interparticle overlap at rest becomes, $\Delta_{(i,j)} = 1$. Using $\kappa_{(i,j)} = 1$, $\Delta_{(i,j)} = 1$ and assembling equations (2) in the granular chain, we arrive at a matrix form of these equations given by,

$$\mathbf{M}\frac{\mathrm{d}^2\mathbf{u}}{\mathrm{d}\tau^2} = \mathbf{K}\mathbf{u}, \quad (3)$$

where $\mathbf{M}$ is a diagonal matrix containing $b^{(1)}, b^{(2)}, ... b^{(N)}$ (masses of the particles) along the diagonal. $\mathbf{K}$ contains -2 along the diagonal, +1 along sub-diagonal and super-diagonal elements and 0 elsewhere. Defining $\mathbf{A} = \mathbf{M}^{-1}\mathbf{K}$ and assuming normal mode motion of particles, equation (3) becomes $\mathbf{A}\mathbf{u} = \omega^2\mathbf{u}$ with $\omega^2_{(j)}$ as eigenvalues of the matrix $\mathbf{A}$ and $\omega_{(j)}$ as eigenfrequencies. Defining $(N \times N)$ matrix $\mathbf{S}$, containing eigenvectors of $\mathbf{A}$, $\mathbf{s}_{(j)}$ as the columns and arranged in increasing order with respect to their corresponding eigenfrequencies. $\mathbf{S}$ is further orthonormalized to satisfy the relation $\mathbf{S^T M S} = \mathbf{I}$ (Identity matrix). Using initial conditions $\mathbf{v}_o = [v_0\ 0\ 0\ ...\ 0]$ (initial velocity) and $\mathbf{u}_o = 0$ (initial displacement) as impulse boundary conditions, we get the displacement and velocity history of particles as,

$$\begin{aligned} u^{(p)}(\tau) &= v_o \sum_{j=1}^{N} \frac{S_{pj}S_{1j}\sin(\omega_{(j)}\tau)}{\omega_{(j)}}, \\ v^{(p)}(\tau) &= v_o \sum_{j=1}^{N} S_{pj}S_{1j}\cos(\omega_{(j)}\tau). \end{aligned} \quad (4)$$

2.1 Mass disorder and disorder parameter

The diagonal elements of the mass matrix $\mathbf{M}$, $b^{(1)}$, $b^{(2)}$, ... $b^{(N)}$ (masses of the particles) have been selected from a normal distribution $f^{(n)}(b) = \frac{1}{\xi\sqrt{2\pi}}e^{-\frac{(b-1)^2}{2\xi^2}}$ whose **standard deviation** (ξ) quantifies the **disorder** of the granular chain. The scaled average of the distribution is 1. The aforementioned model has been used previously in [10] for a frequency filtering study of the displacement wave (mass disorder dependent frequency filtering and attenuation) and to show that the shape of disorder probability (binary, uniform, normal or any other distribution) produce quantitatively similar frequency filtering, up to a certain strength of disorder.

2.2 Total energy signal from displacement and velocity histories

We go one step further and determine the total energy signal per particle which is given by,

$$E^{(p)}(\tau) = \underbrace{\frac{1}{2}b^{(p)}(v^{(p)}(\tau))^2}_{\text{Kinetic Energy}} - \underbrace{\frac{1}{2}F^{(p)}(\tau)\left(u^{(p)}(\tau)\right)}_{\text{Potential Energy}}. \quad (5)$$

The position of the center of total energy in the granular chain is given by [5],

$$R(\tau) = \frac{1}{E_{tot}}\sum_{p=1}^{N} pE^{(p)}(\tau), \text{ with, } E_{tot} = \sum_{p=1}^{N} E^{(p)}(\tau). \quad (6)$$

The squared width of the propagating pulse is,

$$r^2(\tau) = \frac{1}{E_{tot}}\sum_{p=1}^{N}(p - R(\tau))^2 E^{(p)}(\tau), \quad (7)$$

2.3 Model system and ⟨ensemble⟩ averaging

A 256 element long granular chain has been used with disorder parameters $\xi = 0.0$, $\xi = 0.1$, $\xi = 0.2$, $\xi = 0.35$, $\xi = 0.5$ and $\xi = 0.8$ for the mass distribution of the chain. The time duration for the computation was 256 (time duration till which the pulse has propagated to about the last particle in the granular chain) and the time step utilized for the data output was 0.032. 500 ensembles/realizations of granular chains have been utilized and displacement, velocity, force and hence, total energy histories were obtained and ensemble averaged for different disorder parameters. From here onwards angular brackets (⟨⟩) will denote ensemble averaging.

3 Results

$\langle R(\tau)\rangle$, $\langle r^2(\tau)\rangle$, $\langle E^{(p)}(\tau)\rangle$ and the Fourier transform of $\langle E^{(p)}(\tau)\rangle$ ($\langle \mathcal{E}^{(p)}(\omega)\rangle$) are discussed in this Section.

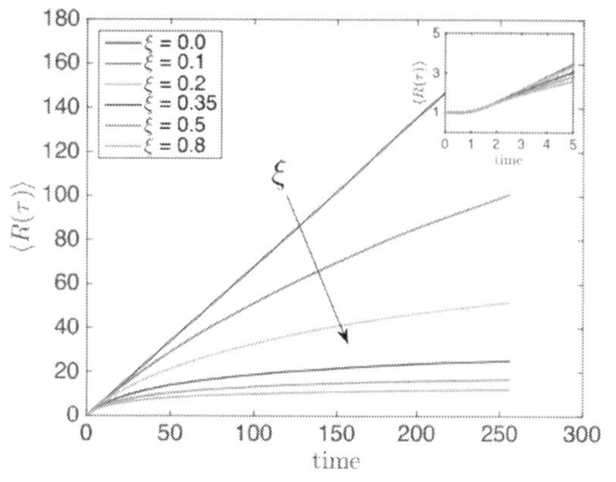

Figure 2. Propagation of center of energy $\langle R(\tau)\rangle$ in different disordered granular chains with the arrow indicating increasing disorder.

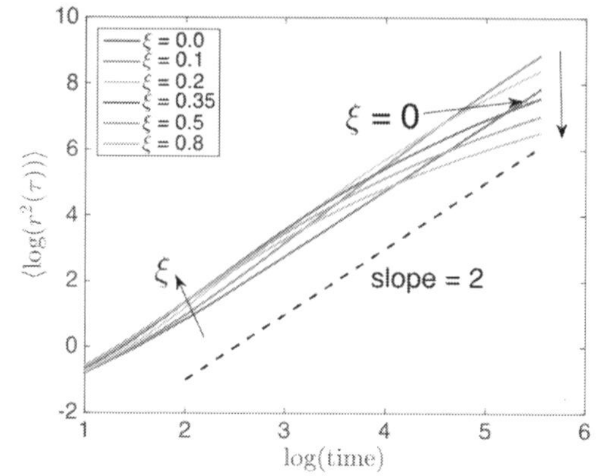

Figure 3. Squared width of the pulse with respect to time (after self-demodulation) for different parameters in log-log scale. Line is a guide to the eye. Arrows indicates the trend for $\xi > 0$.

3.1 Propagation of center of energy

In order to monitor the propagation of the center of energy during propagation of a pulse in a disordered granular chain. $R(\tau)$ from equation (6) has been computed. In Figure 2, the center of energy of propagating pulses for 500 realizations is plotted against time for different disorders. Initially the center of energy does not propagate (as shown in the inset figure), this is the time during which the initial high frequency impulse is self-demodulated [6] by the granular chain (in contrast to a Gaussian pulse [5]) and after this short time interval, for a short time, the center of energy propagates with the same speed for different disorder parameters. $\xi = 0.0$ has linear (ballistic) propagation of center of energy whereas, $\xi > 0$ displays nonlinear propagation of the center of energy with propagation speed decreasing with increasing time. Stronger disorders have stronger decrease in propagation speed with increase in time. Unlike $\xi = 0.0$, higher ξ have the center of energy becoming confined in a finite space and this confinement space is smaller for stronger disorder. Such confinement of energy of mechanical waves has often been attributed to Anderson's localization [5], when elastic waves get trapped in a confined space due to the disorder in the medium [1] which also becomes more enhanced with increase in disorder.

3.2 Squared width of propagating pulse

The ensemble averaged, squared width of the pulse obtained from equation (7) has been plotted in Figure 3 against time in log-log scale to see whether the energy dispersion is diffusive in nature. It is observed that for an impulse response, after the self-demodulation time interval (recognized from the inset in Figure 2), the energy propagation is slightly superballistic for low disorder parameters (eg. $\xi = 0.1$) and gets nonlinear towards diffusive for high disorder parameters (eg. $\xi = 0.8$).

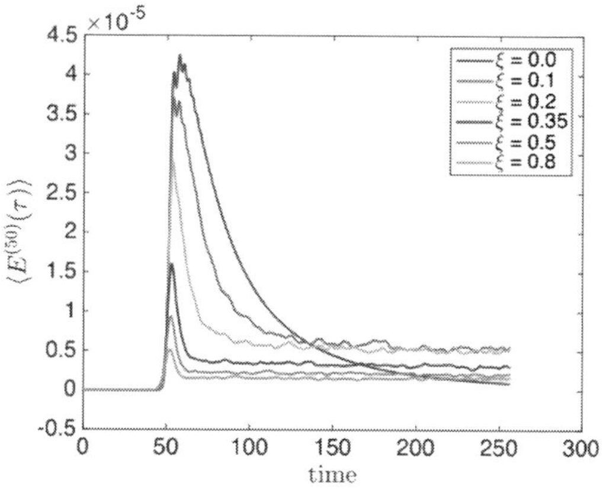

Figure 4. Averaged Energy history of 50th particle in a 256 particles long granular chain for different disorder parameters.

3.3 Spectral energy analysis

The ensemble averaged total energy signal $\langle E^{(p)}(\tau)\rangle$ for the 50th particle in a 256 long chain for different disorder parameters is plotted in Figure 4. We can observe that the initial coherent wavefront is attenuating faster in stronger disordered media as indicated in [10], the peak of this coherent wavefront gives the maximum **peak** energy (E_P) attained by the particle during wave propagation which has been plotted in Figure 5(a) for different disorder parameters (ξ). Due to ensemble averaging, the multiply scattered incoherent part [7] has attained a constant value (with very minor fluctuations) giving the **residual** energy (E_R) contained in the particle after the initial pulse has passed across the particle, this residual energy at $\tau = 256$ has also been plotted in Figure 5(a) for different ξ, it shows that the residual energy contained in the particle decreases with increasing disorder. The ratio between the **residual** energy and the **peak** energy shows an increasing trend as shown

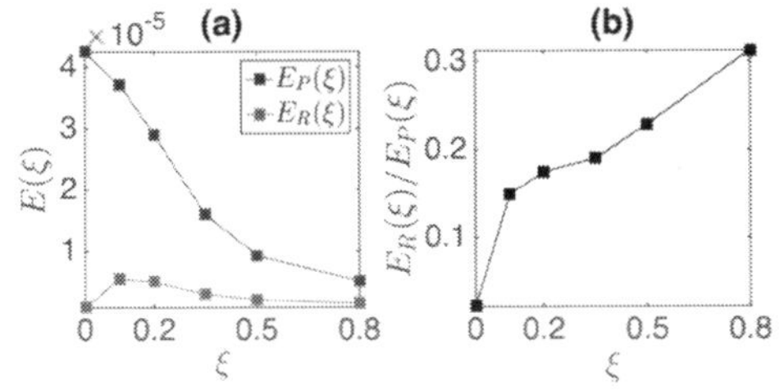

Figure 5. (a) Peak energy & Residual energy and (b) their ratio at the 50th particle for different ξ.

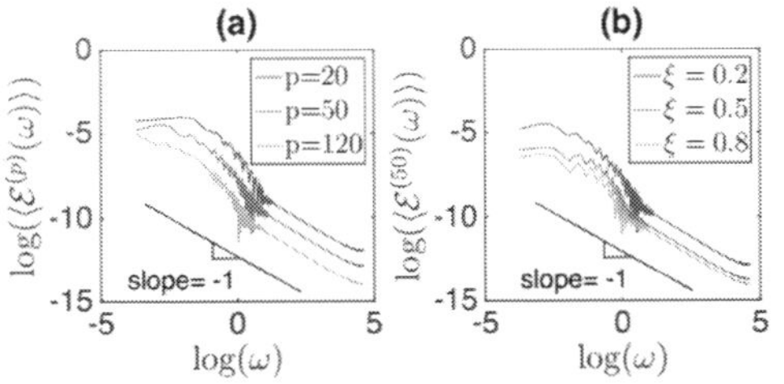

Figure 6. Ensemble averaged spectral energy response of (a) 20th, 50th and 120th particle for $\xi = 0.2$ and (b) $\xi = 0.1, 0.2$ and 0.5 for 50th particle in log-log scale. The FFT-time window is 0 to 256 time units and time step is $\Delta t = 0.0313$.

in 5(b). However, the spectral energy response for any particular particle remains qualitatively the same (Figure 6 (b)) irrespective of disorder, exhibiting only decrease in amplitude of the Fourier spectra and a slight shift towards lower frequency, resulting in loss of information contained in the multiple scattering part.

This suggests adoption of an alternative approach/model for the micro-macro transition rather than relying only on ensemble averaging. Distance and disorder (Figure 6(a) and Figure 6(b)) have no effect on this power law relationship, only the amplitude decreases with the position of the particle and disorder of the granular chain. This ω^{-1} power-law observation is not new as a similar relationship was observed with in Ref. [3].

4 Conclusion

It was observed in Section 3.1 that an ordered granular chain system has ballistic wave propagation of center of energy and a disordered system has a diffusive like propagation at larger times, when the energy becomes localized in space. For stronger disorder, this effect is enhanced. Figures 4 and 5 confirm, as mentioned in [10], that stronger disorder causes faster attenuation of signals. The spectral energy response exhibits a power-law relationship (Figure 6). A main cause of concern is that ensemble averaging causes the spectral energy response ($\langle \mathcal{E}^{(p)}(\omega) \rangle$) to become qualitatively similar for a particular particle irrespective of disorder. This results from the fact that averaging causes the multiply scattered part of the signal to cancel itself as it is not self-averaging hence, leading to loss of information. Therefore, a different micro-macro transition procedure needs to be developed.

Acknowledgements

This work is part of the Industrial Partnership Programme (IPP) 'Computational sciences for energy research' of the Foundation for Fundamental Research on Matter (FOM), which is part of the Netherlands Organisation for Scientific Research (NWO). This research programme is co-financed by Shell Global Solutions International B.V.

References

[1] P. W. Anderson, Phys. Rev. E **109**, 1492-1505 (1958).
[2] V. F. Nesterenko, J. Appl. Mech. Tech. Phys. **24**, 733-743 (1983).
[3] Chu-heng Liu and R. Sidney Nagel, Phys. Rev. Lett. **68**, 2301-2304 (1992).
[4] R. S. Sinkovits and S. Sen, Phys. Rev. E **54**, 6857-6865 (1996).
[5] P. B. Allen and J. Kelner, American Journal of Physics **66**, 497-506 (1998).
[6] V. Tournat, V. E. Gusev and Castagnède, Phys. Rev. E **70**, 056603 (2004)
[7] X. Jia, Phys. Rev. Lett. **93**, 154303 (2004)
[8] R. L. Weaver, Science **307**, 1568-1569 (2005)
[9] O. Mouraille and S. Luding, Ultrasonics **48**, 498-505 (2008)
[10] B. P. Lawney and S. Luding, Acta Mechanica **225**, 2385-2407 (2014).
[11] N. Upadhyaya, L.R. Gómez and V. Vitelli, Phys. Rev. X **4**, 011045 (2014)
[12] J. O'Donovan, E. Ibraim, C. O'Sullivan, S. Hamlin, D. Muir Wood, G. Marketos, Granular Matter **18**, 56 (2016)
[13] R. K. Shrivastava and S. Luding, Nonlin. Processes Geophys., submitted (2017).

Initial stage sintering of polymer particles – Experiments and modelling of size-, temperature- and time-dependent contacts

Regina Fuchs[1,*], *Thomas* Weinhart[2,**], *Ming* Ye[1], *Stefan* Luding[2], *Hans-Juergen* Butt[1], and *Michael* Kappl[1,***]

[1] *Max–Planck-Institute for Polymer Research, Physics at Interfaces, Ackermannweg 10, 55218 Mainz, Germany*
[2] *Multiscale Mechanics, Engineering Technology, MESA+, University of Twente, PO Box 217, 7500 AE Enschede, The Netherlands*

Abstract.
The early-stage sintering of thin layers of micron-sized polystyrene (PS) particles, at sintering temperatures near and above the glass transition temperature T_g ($\sim 100°$C), is studied utilizing 3D tomography, nanoindentation and confocal microscopy. Our experimental results confirm the existence of a critical particle radius ($r_{crit} \sim 1\,\mu$m) below which surface forces need to be considered as additional driving force, on top of the usual surface-tension driven viscous flow sintering mechanism. Both sintering kinetics and mechanical properties of particles smaller than r_{crit} are dominated by contact deformation due to surface forces, so that sintering of larger particles is generally characterized by viscous flow. Consequently, smaller particles require shorter sintering. These experimental observations are supported by discrete particle simulations that are based on analytical models: for small particles, if only viscous sintering is considered, the model under-predicts the neck radius during early stage sintering, which confirms the need for an additional driving mechanism like elastic-plastic repulsion and surface forces that are both added to the DEM model.

1 Introduction

Recent developments in 3D printing and Additive Manufacturing [1] enable the fabrication of individualized serial products based on powders and grains. To date, none of the conventional techniques like, e.g., selective laser sintering allow the production of interconnected porous polymer scaffolds with a variety of pore sizes and reproducible morphology, which are commonly used to stimulate the formation of new tissue [2]. This forces the attention of scientists to the initial stage of sintering, where the porous structure is characterized by the formation of necks between individual particles. Achieving the desired end-product by initial stage sintering of polymer particles of various sizes without a significant temperature increase in the bulk of each particle requires a fundamental understanding of sintering under varying temperature–, pressure– and time–conditions. In the present work, initial stage sintering of polystyrene (PS) particle layers (< 30 particle diameters) with radii of $0.25 - 2\,\mu$m was analysed utilizing 3D tomography (FIB/SEM), nanoindentation and confocal microscopy. The mechanical properties as well as the real-time topography deformation within the sintered particle layers during nanoindentation are investigated. Moreover, the experimental results are correlated with Discrete Element Method (DEM) simulations to calibrate a temperature– and pressure–dependent sintering model that includes the contribution of surface forces.

*e-mail: fuchsr@mpip-mainz.mpg.de
**e-mail: t.weinhart@utwente.nl
***e-mail: kappl@mpip-mainz.mpg.de

2 Experimental details

2.1 Preparation of particle and samples

PS spheres featuring nominal particle radii of $0.25\,\mu$m, $0.5\,\mu$m, $0.75\,\mu$m and $2\,\mu$m, molecular weights (M_w) of 110 709–312 010 g/mol and a glass transition temperature T_g of 94 – 99°C were synthesized by dispersion polymerization as reported in Zhang et al. [3] and stored in aqueous solution. For each particle radius, multi-layer films were realized by placing $10\,\mu$l of a particle dispersion (1:1 volume ratio, PS particle:ethanol) on an oxygen–plasma hydrophilised glass substrate and dried for at least 24 h in air. Sintering was carried out in air on a heating stage at temperatures of 90°, 95°, 100°, 105° and 110°C. The rate of temperature change was set to be 15°C/min during heating. Holding periods of 20, 40, 60, 90, 120 and 180 s were set at each temperature for each particle radius, respectively. After sintering, the samples were quickly cooled down to room temperature on a metal platform. The average layer thickness was reduced by up to max. 50% as the sintering temperature increases from room temperature to 110°C. Fluorescently labeled PS particles with radii of $2\,\mu$m were prepared by staining the PS shell of the particles with a solution of Nile Red and Xylol for 24 h. Afterwards, the solvent was removed under vacuum.

2.2 Characterization of microstructure with FIB/SEM reconstruction

The sintered PS particle films were coated with 100 nm Pt layer on top. A focused–ion beam instrument com-

DOI: 10.1051/epjconf/201714013012

bined with a scanning electron microscope (FIB/SEM), (Nova 600 Nanolab, FEI, USA), was used for sectioning and imaging sequential 2D cross-sectional surface images. A protective Pt layer of 1 μm thickness and $12 \times 8\ \mu m^2$ area was deposited on the sample. A volume of $12 \times 8 \times 6\ \mu m^3$ was milled slice–by–slice with 20 nm distance between two consecutive images. The actual 3D volumes of the sintered films were reconstructed based on the as-recorded stacks of images using Amira 4.1 (Visage Imaging, San Diego, USA). The reconstruction process involved i) alignment of the images, ii) resampling and iii) segmentation. The radius of at least 12 particles as well as the mean neck radius x was determined for each sintered sample.

2.3 Characterization of mechanical properties with Nanoindentation and Confocal Microscopy

Nanoindentation measurements were performed with a MFP NanoIndenter (Asylum Research, Santa Barbara, CA, spring constant k=2390 N/m) equipped with a spherical ruby indenter (d=127 μm). Indentations were performed in load–controlled mode. The applied load varies between 1 and 4 mN with loading rates between 200 μN/s and 800 μN/s. In order to get an estimate of the standard deviation of the mechanical properties, each sample was tested at least at 12 individual positions separated by at least 100 μm. The reduced elastic modulus (E_{red}) for each sample was obtained from the unloading portion of the load-displacement curve using the Oliver and Pharr method with a spherical area function [4]. Additionally, the nanoindenter was placed on the sample stage of a custom-built laser scanning confocal microscope (LSCM), which has the capability to measure in–situ the real–time deformation within the sintered particle layers during nanoindentation. The scan rate of the LSCM was one 2D frame per second with a sampling resolution of 0.235 μm/pixel in x– and y–direction and 0.570 μm/pixel in z–direction. The ruby sphere was indented into the sintered fluorescently labelled PS particle film in displacement–controlled mode with a maximum displacement of 5 μm, while the structure was imaged by LSCM. For the LSCM imaging, the sintered films needed to be infiltrated with a liquid that matches the refractive index of the PS (Cargille Laboratories, USA).

3 Contact mechanics

To simulate the sinter and indenting process, we model adhesive, elastoplastic, and dissipative contact forces, with the normal force between two overlapping spherical particles given by

$$f_n = -f_a + k(\delta - \delta_p) + \gamma\dot{\delta} \quad \text{if} \quad \delta > \delta_p. \tag{1}$$

where δ denotes the total particle overlap and δ_p the *plastic deformation*. The adhesion force f_a is assumed constant, while the elastoplastic and dissipative forces are linear functions of the *effective, elastic overlap* $(\delta - \delta_p)$ and normal velocity $\dot{\delta}$, respectively, with k denoting the contact stiffness and γ the dissipation coefficient, see also [5].

We further assume that the stiffness increases with the amount of plastic deformation, due to the increased contact surface, with $k = k_l$ for zero plastic overlap up to a maximum of $k = k_{crit}$ at maximum plastic overlap δ_p^{crit}, which is set to avoid solid volume fractions above one (fully sintered) [5]:

$$k = k_l + (k_{crit} - k_l)(\delta_p/\delta_p^{crit})^{\psi}. \tag{2}$$

The power ψ can be set to $\psi = 1/2$ to reproduce Hertz-like behavior, or to any other value including $\psi = 0$ to deactivate this non-linearity, or to a linear interpolation with $\psi = 1$, as done in the following. Below $\delta_p = \delta_p^{crit}$, particles are soft, and thus deform, proportionally to the overlap, when the overlap between the particles reaches a new maximum (i.e. $k(\delta - \delta_p) > k_l\delta$).

$$\delta_p = \begin{cases} \frac{k-k_l}{k}\delta & \text{if } k(\delta - \delta_p) > k_l\delta. \\ \frac{k+k_c}{k}\delta & \text{if } k(\delta - \delta_p) < -k_c\delta. \end{cases} \tag{3}$$

The reduction in plastic overlap during unloading is introduced to model contact cohesion. This causes the loading-unloading-reloading behavior shown in Figure 1.

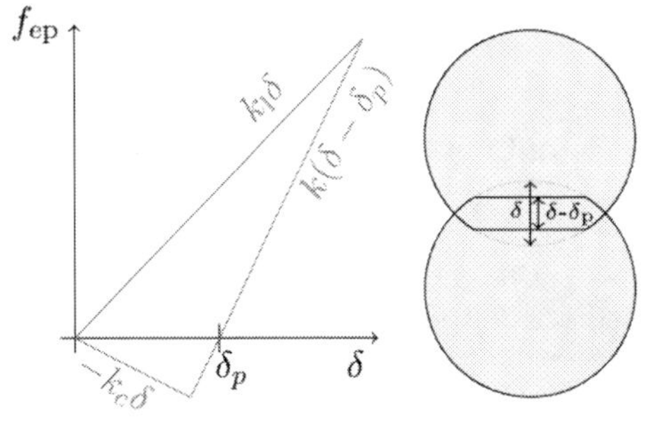

Figure 1. Elasto-plastic contact law.

All parameters of the contact model – and the particle density – may vary with temperature [6], but we neglect this effect here for simplicity, assuming constant values. At temperatures above the glass transition, viscous flow causes the particles to sinter, which creates an increase in plastic overlap, modelled here as

$$\dot{\delta}_p = \frac{f_n r_p}{n f_a \tau_s}\left(\frac{r_p}{x}\right)^{n-1}. \tag{4}$$

For two particles without compression ($f_n = f_a$) and of high stiffness ($\sqrt{m/k} \ll \tau_s$), we can assume that $\delta = \delta_p$, therefore the plastic overlap increases as $\delta_p/r_p = \sqrt[n]{t/\tau_s}$. Therefore, τ_s denotes the *sintering time scale* and n the *sintering power law exponent*. Such a law can be rigorously derived from the force laws derived in [7] for surface sintering, which yields an exponent $n = 3$. For $n = 1$, we obtain the well-known Frenkel law [8] for viscous sintering, for which $x/r_p \approx \sqrt{\delta/r_p} \propto \sqrt{t}$. Further, if the increase of the particle radius due to sintering would be taken into account, we obtain the modified Frenkel law derived in [9].

DOI: 10.1051/epjconf/201714013012

4 Simulation setup

We model the sintering and indentation process with discrete particle simulations, using a small cubic domain of 60 μm, with a flat base wall and periodic boundary conditions in both width and length. We introduce 2178 particles of radii 1.35 μm < r_p < 1.65 μm, producing a particle packing about 5 particle diameters thick (see Figure 2). To simulate sintering in a reasonable amount of time steps, the collision time is scaled up to t_c = 5 ms, which is still several orders of magnitude smaller than the time scales of gravity, sintering, and indentation, and thus has little effect on the results. The packing is sintered using a sintering time scale of τ_s = 666 s and n = 1 (thus assuming viscous sintering) to match the experimental results. For brevity, quantitative, more detailed results from numerical indentation tests will be presented in a later paper.

Figure 2. Vertical cut through centre of simulated sample during indentation.

5 Results and Discussion

5.1 Mechanical properties

After sintering of the PS particles samples under varying temperature–, size– and time–conditions, different densifications of the powder layer are obtained. It was found that the reduced elastic modulus (E_{red}) of each sintered film increased by a factor of max. 6 for sintering temperatures above T_g as the sintering time increased from 20 s to 180 s. E_{red} depends on the porosity of the film system.

According to Mazur et al. [10], particles smaller than a certain limiting radius are predicted to sinter to uniform density regardless of Newtonian viscosity. Consequently, the contact area initially grows much faster for particles with r_p < r_{crit} than predicted by the classical sintering models [8, 9], which neglect the contribution of surface forces as well as the resultant plastic and elastic contact deformation in the early stages of sintering. Assuming an initial packing fraction of 58%, a critical particle radius (r_{crit}) of 1.055 μm for sintering of PS particles is obtained. Figure 3 shows the E_{red} plotted against sintering temperatures for PS particle layers with varying radii from 0.25 μm to 2 μm. Particles with r_p < r_{crit} show a slightly higher E_{red} compared to larger particles even for temperature below T_g. This hints to an additional contribution of surface force and contact deformation, which leads to faster sintering in the initial stage of sintering. Moreover, Figure 3 shows a scaling of E_{red} with particle size. A transition from sintering dominated by contact deformation for r_p < r_{crit} due to surface forces and faster sintering to sintering dominated by viscous flow for r_p > r_{crit} is observed. Consequently, larger particles require higher sintering temperatures and times to show the same mechanical properties as smaller particles.

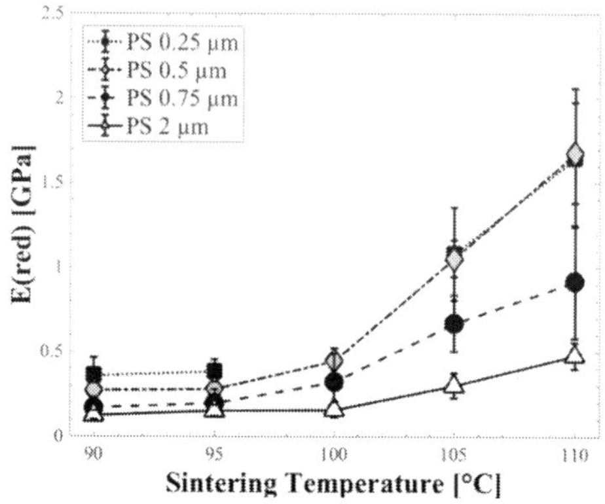

Figure 3. Mechanical properties of PS multilayers with particle radii of 0.25–2 μm sintered for 60 s. E_{red} increases with the densification of the porous solids ($T > T_g$).

Particles with radii close to r_{crit} (r_p = 0.75 μm) seem to show both mechanisms. The contribution of the sintering kinetics to the mechanical properties of the film system is therefore more complex and does not conform to a simple scaling with particle size.

In addition to the nanoindentation results, in-situ realtime topography deformation within the sintered particle layers during nanoindentation was studied with the help of confocal microscopy. Cross-sectional xz-plane LSCM images for (a) not sintered films and films sintered at (b) 90°C, (c) 100°C and (d) 110°C for 60 s at maximum indentation depth are shown in Figure 4.

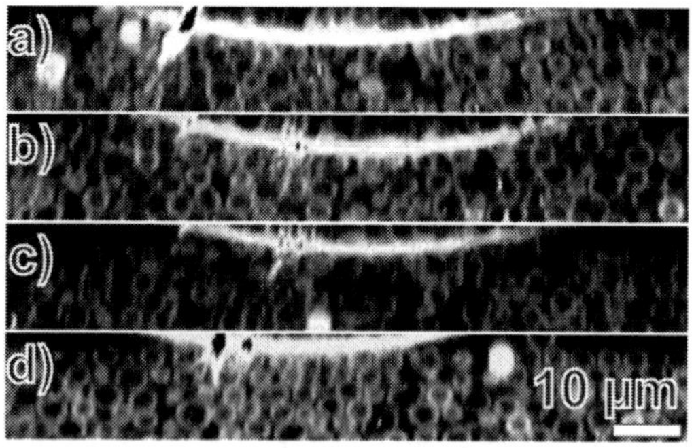

Figure 4. LSCM image (xz plane) during indentation with a spherical tip for a) not sintered PS multilayer film, b) sintered at 90°C for 60 s, c) sintered at 100°C for 60 s, and d) sintered at 110°C for 60 s. The maximum indentation decreases from a) to d) with increasing stiffness of the sample.

While the displacement of 5 μm is reached without exceeding the limited force of the nanoindenter in case of Figure 4a and 4b, the samples above T_g (Figure 4c and 4d) show a reduced indentation depth due to a higher E_{red} of the porous film system. A deformed region is observed after indentation for samples in Figure 4a-c. The sample sintered at a temperature of 110°C has a nearly fully elastic response. Such behavior can be attributed to a strong interconnection of the particles based on sintering necks.

5.2 Sinter kinetics

A representative plot of the time-dependent increase of the sintering neck radius x at 110°C for PS particles with radius of 1.5 μm is shown in Figure 5. The experimental results obtained with 3D reconstruction show a higher growth rate in the early stages of sintering compared to the prediction of the classical sinter models of Frenkel [8] and the modified Frenkel model [9], using a time scale of $\tau_s = 666$ s to match the experimental observations. Simulation of viscous sintering ($n = 1$, $\tau_s = 666$ s) agree well with the Frenkel model, up to the point where many overlaps approach the maximum plastic overlap ($t > 120$ s), while simulations using the surface sintering model ($n = 3$, $\tau_s = 110$ h) overpredict the early-stage sintering rate. This indicates that viscous sintering is aided by an additional contribution of surface forces to the sintering process.

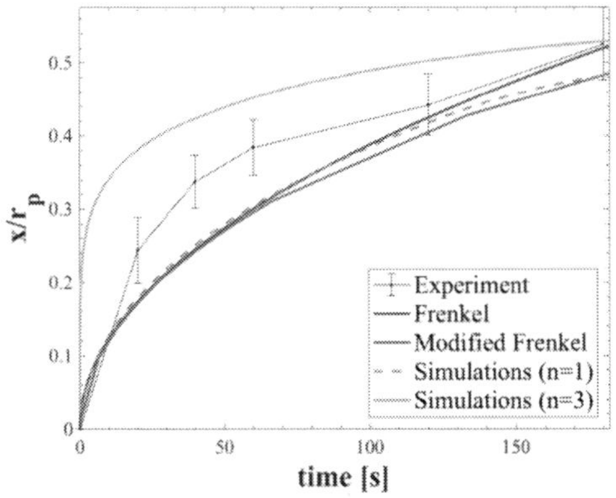

Figure 5. Sintering kinetic of PS particle multilayer ($r_p = 1.5\,\mu$m) at 110°C compared with Frenkel model (blue line), modified Frenkel model (red line), and simulations of viscous sintering ($n = 1$, green line) and surface sintering ($n = 3$, orange line).

6 Conclusion

Temperature-, time- and size-dependent sintering kinetics and mechanical properties of initial stage sintering of thin PS particle layers were investigated by utilizing 3D tomography (FIB/SEM), nanoindentation and confocal microscopy. Our experimental results show that the sintering kinetics and mechanical properties of particles with $r_p < 1\,\mu$m are dominated by contact deformation due to surface forces while sintering of larger particles is characterized by viscous flow as the dominant mechanism. Consequently, larger particles require higher sintering temperatures and times to reach the same extent of sintering. To calibrate a temperature- and pressure-dependent sintering model that includes the contribution of surface forces, we use the elasto-plastic model of Luding [6], and model sintering by introducing a rate of change for the permanent, plastic deformation at high temperatures. The contact model can simulate both contact sintering as well as compression, i.e., elastic repulsion, allowing the simulation of sintering and indentation tests in a single simulation framework. The preliminary simulation results shown here are for purely viscous and surface sintering, which agree well with the theoretical predictions, but under-, respectively overpredict, the experimental results. A model taking into account both surface forces and viscous sintering, as well as the increase in particle radius due to sintering may be able to fully explain the experimental observations.

Acknowledgments

The authors would like to thank the German Research Foundation (DFG) for financial support. This work is carried out within the framework of the Key Research Program (SPP 1486 PiKo "Particles in Contact") grants LU 450\10 and KA 1724\1. We thank Laurent Gilson for his help in confocal microscopy measurements and Nina Hoinkis for sample preparation. The numerical solutions of the contact models in this paper were carried out using the open-source code MercuryDPM (mercurydpm.org).

References

[1] E.N. Antonov, V.N. Bagratashvili, M.J. Whitaker, J.J.A. Barry, K.M. Shakesheff, A.N. Konovalov, V.K. Popov, S.M. Howdle, Advanced Mat. **17**, 327 (2005)

[2] R. Landers, U. Hubner, R. Schmelzeisen, R. Mulhaupt, Biomaterials **23**, 4437 (2002)

[3] L. Zhang, M. D'Acunzi, M. Kappl, G.K. Auernhammer, D. Vollmer, C.M. van Kats, A. van Blaaderen, Langmuir **25**, 2711 (2009)

[4] A.C. Fischer-Cripps, Vacuum **58**, 569 (2000)

[5] S. Luding, Granular Matter **10**, 235 (2008)

[6] S. Luding, K. Manetsberger, J. Müllers, Journal of the Mechanics and Physics of Solids **53**, 455 (2005)

[7] R. Besler, M. Rossetti da Silva, J.J. Rosario, M. Dosta, S. Heinrich, R. Janssen, Journal of the American Ceramic Society **98**, 3496 (2015)

[8] J. Frenkel, Journal of Physics **9**, 385 (1945)

[9] O. Pokluda, C.T. Bellehumeur, J. Vlachopoulos, AICHE Journal **43**, 3253 (1997)

[10] S. Mazur, R. Beckerbauer, J. Buckholz, Langmuir **13**, 4287 (1997)

An experimental investigation of the force network ensemble

Jonathan E. Kollmer[1,*] and *Karen E.* Daniels[1]

[1] *Department of Physics, North Carolina State University, Raleigh, North Carolina 27695, USA*

Abstract. We present an experiment in which a horizontal quasi-2D granular system with a fixed neighbor network is cyclically compressed and decompressed over 1000 cycles. We remove basal friction by floating the particles on a thin air cushion, so that particles only interact in-plane. As expected for a granular system, the applied load is not distributed uniformly, but is instead concentrated in force chains which form a network throughout the system. To visualize the structure of these networks, we use particles made from photoelastic material. The experimental setup and a new data-processing pipeline allow us to map out the evolution subject to the cyclic compressions. We characterize several statistical properties of the packing, including the probability density function of the contact force, and compare them with theoretical and numerical predictions from the force network ensemble theory.

1 Introduction

Although the positions of particles in a jammed granular system are fixed to a specific geometrical configuration, the particle positions alone are not sufficient to determine the force network that carries the load on that packing. As such, an underdetermined mechanical system, there are many ways in wich force and torque balance on each particle can be statistfied for any given packing geometry and boundary conditions [1, 2]. There are both stable configurations, and unstable configurations and initially stable granular systems can evolve into catastrophic failure. While two packings might have the same occupied volume or internal pressure they might have vastly different bulk material properties [3]. For frictional granular system it remains an open question to determine whether the statistics due to the stress state and the volume of the system can be decoupled from each other.

To make predictions for the physical behavior of granular systems, tools and concepts from statistical physics have become widely used [5], but what is the correct ensemble to describe jammed granular packings? To get more insight into these jammed packings one needs to look at the structure of force networks that form in loaded granular packings. When the packing is subjected to external load, not all particles share the load equally but the forces are highly localized into force chains. In this work, we present an experiment to look at the distribution of forces in a loaded granular system, while disentangling the effects of configuration from other influences. This is of interest to compare these distributions to predictions from the Force Network Ensemble theory [1].

*e-mail: jekollme@ncsu.edu

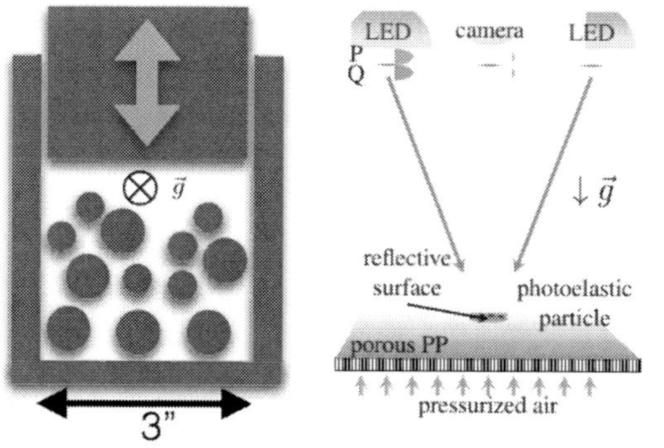

Figure 1. Experimental setup. Left: Particles are placed into a piston that cyclically compresses them. Right: The whole setup is horizontal and floated on an air cushion to eliminate the influence of gravity and basal friction. A camera and an unpolarized red light source are mounted overhead to image and track the particles. The particles are made from photoelastic material, providing visual access to their internal stress field when illuminated with polarized green light. The red and green color channels of the camera can later be seperated to process particle positions and force information from a single color image. Right subfigure adapted from [4].

The Force Network Ensemble

The Force Network Ensemble (FNE) is a concept introduced by Snoeijer et al. [1] in which they use an ensemble approach for examining the force distribution in static granular packings. Since forces in fixed granular packings are typically underdetermined, the ensemble averages over all microscopic variations of a packing, an approach that goes back to Edwards [6]. The FNE predicts, among other things [2], a fitnite value for $P(F)$ as $F \to 0$ and a faster than exponential decay of the probability density function (PDF) of the contact forces at large forces, depending on the dimensionality of the packing [1]. For a two dimensional system, $P(F)$ is predicted to have a gaussian tail. For a review see [2] (and references therein) where it is also discussed that the peak in the PDF should vanish for anisotropic stresses. Saithoh et al. [7], using molecular dynamics simulations to determine transition rates for contact changes, recently also found a master equation that describes the PDF of forces in soft particle packings.

2 An Experimental Approach

While a number of experimental works exist, e.g. [8, 9], PDFs of contact forces are most often produced from numerical simulations, including a recent pair of papers by Pugnaloni [10] and Kondic [11] where they study the structure of force networks in tapped particulate systems of disks superimposed by a gravity force. However, in most experimental studies the external load that probes the force network is not the only force applied to the system, there is additionally a load superimposed by gravity, or basal friction [12]. Further, there is only few experiments [13] probing granular ensembles.

In this manuscript, we present an experiment that is designed to enumerate how many force configurations of a single hyperstatic granular arrangement are practically accessible, while at the same time keeping the external load the only force that is beeing applied to the system. To achieve this we prepare a horizontal quasi 2D granular system that is floated on a gentle air cushion, thereby generating an effectively gravity free system without basal friction [4]. The particles are confined by a piston that can apply an uniaxial load to the packing. A schematic drawing of the experimental setup is detailed in Fig. 1. By cyclically loading and unloading the packing in a way that will not change the particle configuration (no neighbors changes), the system cycles trough many contact force configurations due to microscopic changes of the exact contact point. For the experiments done here, we compress the packing in steps of constant volume ($\Delta V = 0.002869 V_{initial}$) and the compression steps are applied quasistatically over 20 substeps of $\Delta x = 0.01$ mm. The initial volume $V_{initial}$ was chosen to be close to the onset of jamming, and the final volume so that the mean contact force rises by more than a factor of 3. We performe the experiment with several random configuration of 29 particles of two different radii ($r_1 = 5.5$mm and $r_2 = 7.6$mm) to prevent crystallization.

In order to extract force information from the experiment, the particles are made of photoelastic material (Vishay PSM-4), which will shift the polarization of light that is shined through it as a function of the applied load. A model of the force modulated light intensity can then be fitted to camera images of the particles [13, 14].

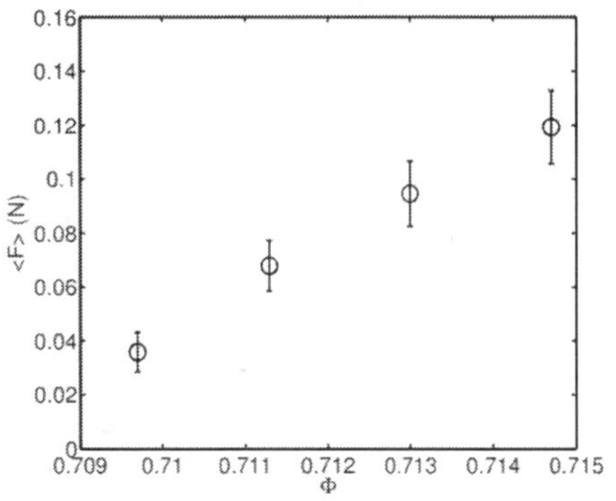

Figure 2. Average contact force F, averaged over all repetitions of a single experiment for each loading step (packing fraction). As can be expected F increases as compression increases, and the standard deviation (errorbars) increases with increasing load.

3 Results

When we start the load cycling we observe that even after an initial annealing period there are variations in the force network, (see Fig. 3), while the particle configuration is unchanged. This validates our experimental approach and allows us to probe the nature of the FNE.

We run the experiment for ≈ 1000 cycles and observe strong fluctuations in the contact number, determined by a minimum threshold force ($F_{th} > 0.01$ N) and in the number of load bearing particles (the number of particles with one ore more contacts above the threshold force). Figure 2 shows the average contact force

$$F = \left\langle \sqrt{F_N^2 + F_T^2} \right\rangle$$

and its standard deviation over all cycles, as a function of the applied compression. We see that the standard deviations in F grow with the applied compression. For higher packing fractions, the intervals given by the STD(F) around F begin to overlap for consecutive compression steps.

Figure 4 shows the PDFs of the contact forces for 5 compression steps over all cycles. The PDF exhibits a strong peak, decaying in the limit of both low and high forces. These results are qualitatively similar to the predictions by FNE theory: most notably a stronger than exponential decay in $P(F)$. We also see the peak in $P(F)$, along with the average contact force, move towards higher values as the system gets compressed stronger. In fact, the PDFs collapse onto a single curve when normalized by the average contact force at the corresponding compression step. Although we perform an experiment with anisotropic loading, we nonetheless identify a peak in the distribution, a feature that is suggested to vanish for anisotropic loads [2]. For small forces we find, $P(F)$ to rise exponentially, as approximately $F^{3/2}$. Wyart [15], showed that the exponent is determined by the pair distribution function g(r).

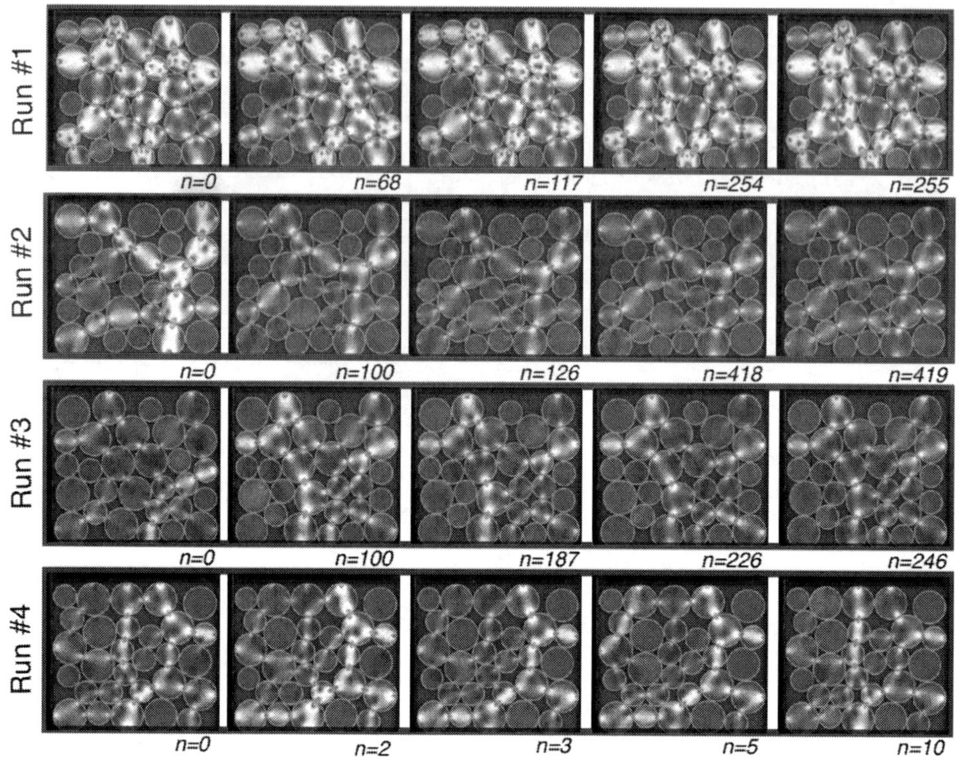

Figure 3. Snapshots after n loading cycles for four different initial configurations (Runs #1- 4). The force network is different for each configuration and for the same configuration fluctuates around several preferred sates.

However, in our small system with a fixed particle configuration, g(r) is undersampled, so the question arises what the exponent is set by here.

4 Conculsions and Outlook

We have designed an experiment that can explore the Force Network Ensemble of a two dimensional granular packing, while excluding forces other than the applied load. We find that we can qualitatively reproduce some of the features in contact force distribution predicted by the FNE theory.

Due to the limited number of cycles and initial packing configurations explored here, it is not immediately clear whether some features we observed in the contact force PDFs are due to sample size or the specific initial configuration. Repeating the experiment not only for more compression cycles but also many different packing configurations, would allow us to probe ergodic effects by contrasting the time average statistics to the ensemble average statistics.

Furthermore, future work should try to identify whether the are several subpopulations of networks which differ in their contact force PDFs. These subpopulations can be found by using tools from network theory such as community detection [16]. More generally, the data generated by this experiment will be useful in relating force network features to macroscopic packing properties [17, 18].

Acknowledgements

We gratefully acknowledge James Puckett for the design and construction of the air table on which the apparatus is based, and for the inspiration for the new parallelized version of the contact-force code. This research was suppored by the James S. McDonnell Foundation and the NSF through grants DMR-0644743 and DMR-1206808.

References

[1] J.H. Snoeijer, T.J.H. Vlugt, M. van Hecke, W. van Saarloos, Physical Review Letters **92**, 54302 (2004)
[2] B.P. Tighe, J.H. Snoeijer, T.J.H. Vlugt, M. van Hecke, Soft Matter **6**, 2908 (2010)
[3] S. Dagois-Bohy, B.P. Tighe, J. Simon, S. Henkes, M. van Hecke, Phys. Rev. Lett. **109**, 095703 (2012)
[4] J.G. Puckett, Phd thesis, North Carolina State University (2012), `http://www.lib.ncsu.edu/resolver/1840.16/7998`

 DOI: 10.1051/epjconf/201714002024

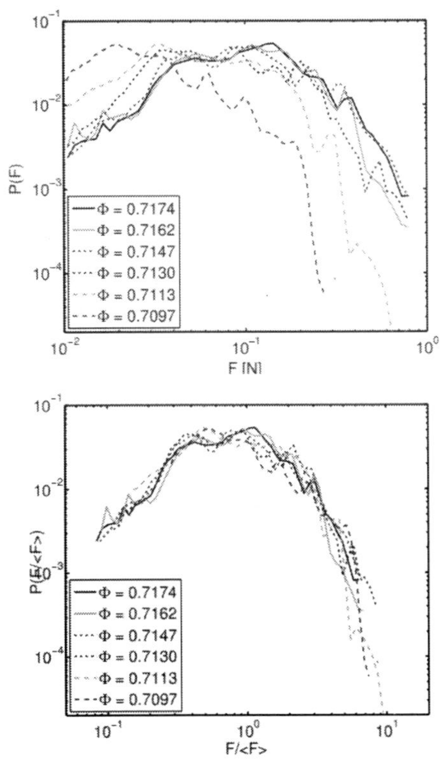

Figure 4. Left: Contact force probability distribution function for each loading step. The peak shifts to higher forces with increased load. for smaller loads the PDF seems to have a shoulder, this could be from a strong and weak force network or from two preferred configurations that the systems osciallated inbetween. Right: Contact force PDF for each loading step collapse when normalized by the average contact force of the corresponding loading step. In the limit of large forces the PDFs decay faster than exponential.

[5] D. Bi, S. Henkes, K. Daniels, B. Chakraborty, Annual Review of Condensed Matter Physics **6**, 63 (2015)
[6] S.F. Edwards, R.B.S. Oakeshott, Physica A **157**, 1080 (1989)
[7] K. Saitoh, V. Magnanimo, S. Luding, Soft Matter **11**, 1253 (2015)
[8] E.I. Corwin, H.M. Jaeger, S.R. Nagel, Nature **435**, 1075 (2005)
[9] D. Howell, R.P. Behringer, C. Veje, Physical Review Letters **82**, 5241 (1999)
[10] L.A. Pugnaloni, C.M. Carlevaro, M. Kramár, K. Mischaikow, L. Kondic, Phys. Rev. E **93**, 062902 (2016)
[11] L. Kondic, M. Kramár, L.A. Pugnaloni, C.M. Carlevaro, K. Mischaikow, Phys. Rev. E **93**, 062903 (2016)
[12] L. Kovalcinova, A. Goullet, L. Kondic, Physical Review E **93**, 042903 (2016)
[13] J.G. Puckett, K.E. Daniels, Physical Review Letters **110**, 058001 (2013)
[14] K.E. Daniels, J.E. Kollmer, J.G. Puckett, arXiv:1612.03525 (2016)
[15] M. Wyart, Phys. Rev. Lett. **109**, 125502 (2012)
[16] D.S. Bassett, E.T. Owens, M.A. Porter, M.L. Manning, K.E. Daniels, Soft Matter **11**, 2731 (2015)
[17] C. Giusti, L. Papadopoulos, E.T. Owens, K.E. Daniels, D.S. Bassett, Physical Review E **94**, 032909 (2016)
[18] L. Papadopoulos, J.G. Puckett, K.E. Daniels, D.S. Bassett, Physical Review E **94**, 032908 (2016)

Influence of dry cohesion on the micro- and macro-mechanical properties of dense polydisperse powders & grains

Robert Kievitsbosch[1,*], *Hendrik* Smit[1], *Vanessa* Magnanimo[1], *Stefan* Luding[1], and *Kianoosh* Taghizadeh[1]

[1] *Multi-Scale Mechanics, Faculty of Engineering Technology (ET), MESA+, University of Twente, Enschede, The Netherlands*

Abstract. Understanding how cohesive granular materials behave is of interest for many industrial applications, such as pharmaceutical or food and civil engineering. Models of the behaviour of granular materials on the microscopic scale can be used to obtain macroscopic continuum relations by a micro-macro transition approach. The Discrete Element Method (DEM) is used to inspect the influence of cohesion on the micro and macro behaviour of granular assemblies by using an elasto-plastic cohesive contact model. Interestingly, we observe that frictional samples prepared with different cohesion values show a significant difference in pressure and coordination number in the jammed regime; the differences become more pronounced when packings are closer to the jamming density, i.e. the lowest density where the system is mechanically stable. Furthermore, we observe that cohesion has an influence on the jamming density for frictional samples, but there is no influence on the jamming density for frictionless samples.

1 Introduction

Granular materials behave differently from usual solids and fluids. One way to understand the macroscopic particle behaviour is using the Discrete Element Method (DEM) [1–3]. Models based on this method rely on the contact force models used and solve a coupled system of equations of motion for all the interacting particles [4, 5]. Even though millions of particles can be simulated, this number (for fine powders) is generally too small to regard the systems modelled as truly macroscopic. Therefore, a transition from the micro to the macro level should be established. The microscopic properties can be used to derive macroscopic constitutive relations. These relations can be used to describe the particle behaviour on the large scale application/process level [6]. The testing and characterization of dry, non-sticky powders are well established. The main challenge comes when the powders are cohesive and less flowable like those relevant in food industry [7]. Research has already been done on cohesive granular materials (see refs [8, 9]), however the influence of cohesion on granular packings is still poorly understood. There are two cases where cohesion becomes important: When particles become very small the cohesive forces become larger than the other forces on each particle, as is the case for dry fine powders [5, 10]. Not only the size of the particles contributes to the influence of cohesive attractive forces, but also a liquid between the particles, as is the case for wet granular materials [11–13]. The research presented here will focus on dry cohesion and DEM is used to study granular packings made of polydisperse cohesive particles. The question arises how does the presence of attractive forces affect macroscopic properties of the packings? So far, only a few attempts have been made to answer this question. Gilabert et al. [14] focussed on a two-dimensional packing made of particles with short-range interactions (cohesive powders) under weak compaction. Yang et al. [15] studied the effect of cohesion on force structures in a static granular packing by changing particle size. Singh et al. [16] studied the effect of friction and cohesion on anisotropy in granular materials under quasi-static shear. The goal is to understand the influence of the microscopic parameters on the macroscopic properties of the packings. Knowing the influence of cohesion on particulate systems will advance of development of new constitutive models to predict the macroscopic material behaviour, to be used to model real life applications and to understand and optimize processes.

*e-mail: r.g.h.kievitsbosch@student.utwente.nl

2 Simulation details

We use the Discrete Element Method (DEM) to understand the behaviour of granular systems. In the model, we relate the force interacting between the particles to the overlap δ that the particles have with each other. DEM solves Newton's equations of motion for all forces $\mathbf{f}_i = f^n\mathbf{n} + f^t\mathbf{t}$ acting on particle i for the translational and rotational degrees of freedom.

2.1 Contact model

For the sake of simplicity, the linear visco-elastic normal contact force model can be used, but for cohesion the model in Sect. 2.2 is applied. It involves a linear repulsive and a linear dissipative force: $f^n = k\delta + \gamma_0\dot{\delta}$ with k as spring stiffness, $\delta = (a_i + a_j) - (\mathbf{r}_i - \mathbf{r}_j) \cdot \mathbf{n} > 0$ as particle overlap, $\mathbf{n} = \mathbf{n}_{ij} = (r_i - r_j)/|r_i - r_j|$ as normal unit vector, γ_0 as viscous damping coefficient and $\dot{\delta}$ the relative velocity in normal direction $v_n = -\mathbf{v}_{ij} \cdot \mathbf{n} = \dot{\delta}$. For simulations regarding frictional particles, we used the Coulomb friction law as explained in ref [5], with tangential force f^t and tangential direction vector(s) $\hat{\mathbf{t}}$.

An artificial damping force $\mathbf{f}_b$ is introduced to reduce dynamic effects and shorten relaxation times: $\mathbf{f}_b = -\gamma_b \mathbf{v}_i$. This force acts not on contacts but directly on particles, proportional to their velocity $\mathbf{v}_i$.

The integration time-step Δt_{MD} used for simulations needs to be much smaller than the contact duration t_c to make sure that the integration of the equations of motion is stable. Note that in extreme cases of an overdamped spring, t_c can become extremely, artificially large, i.e. dissipation γ should be neither too weak nor too strong.

2.2 Adhesive, elasto-plastic contact model

In this paper, a linear variant of the hysteric spring model is used (see refs [4, 5, 17, 18]). This model is a simpler version of nonlinear-hysteric force laws [17–20]. Different spring constants are assumed for loading and unloading. The (hysteric) force can be written as:

$$f^{\text{hys}} = \begin{cases} k_1\delta & \text{if } k_2(\delta - \delta_0) \geq k_1\delta \\ k_2(\delta - \delta_0) & \text{if } k_1\delta > k_2(\delta - \delta_0) > -k_c\delta \\ -k_c\delta & \text{if } -k_c\delta \geq k_2(\delta - \delta_0) \end{cases} \quad (1)$$

with $k_1 \leq k_2$. k_2 is variable, depending on the previous deformation amplitude, see refs. [5, 11]:

$$k_2(\delta_{max}) = \begin{cases} \hat{k}_2 & \text{if } \delta_{max} \geq \delta^*_{max} \\ k_1 + (\hat{k}_2 - k_1)\dfrac{\delta_{max}}{\delta^*_{max}} & \text{if } \delta_{max} < \delta^*_{max} \end{cases} \quad (2)$$

There are different phases during the particle contact (see figure 1). First there is an initial loading; The force increases linearly with the overlap δ. The unloading starts after the overlap reached its maximum value δ_{max}. This point defines the maximum force between two particles. During unloading, the force decreases linearly with a slope k_2, to zero, at overlap $\delta_0 = (1 - k_1/k_2)\delta_{\text{max}}$, which resembles the plastic contact deformation. Reloading will result in an increasing force with the same slope k_2. When unloading takes place below δ_0, then attractive forces will occur until the minimum cohesive force branch $-k_c\delta_{\text{min}}$ is reached. An overview of the parameters used in the DEM simulations can be seen in table 1.

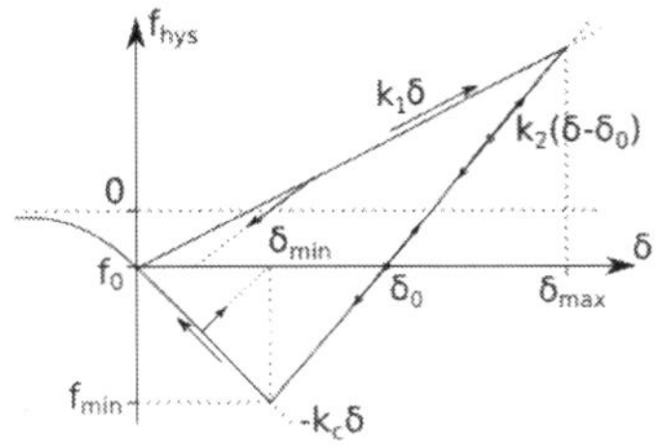

Figure 1. Schematic graph of the piece-wise linear, hysteric model. The adhesive force-displacement for normal collision. The non-contact forces (f_0) are kept equal to zero in this study and also the line for negative δ is neglected in this paper

Table 1. The microscopic contact model parameters values

Property	Symbol	Value	SI-units
Time unit	t	1	10^{-6} s
Length unit	x	1	10^{-3} m
Mass unit	m	1	10^{-9} kg
Particle radius	$\langle a \rangle$	1	10^{-3} m
Polydispersity	a_{max}/a_{min}	3	
Number of particles	N	5000	
Particle density	ρ	2000	2000 kg/m^3
Simulation time step	Δt_{MD}	0.0037	$3.7 \cdot 10^{-9}$ s
Elastic stiffness	$k = \hat{k}_2$	$15 \cdot 10^4$	$15 \cdot 10^7$ kg/s^2
Plastic stiffness	k_1/k	0.666	
Cohesive stiffness	k_c/k	0-20	
Friction stiffness	k_t/k	0.2866	
Static friction	μ_s	0.5	
Dynamic friction	μ_d	0.5	
Normal viscosity	$\gamma = \gamma_n$	1000	1 kg/s
Friction viscosity	γ_t/γ	0.2	
Background visc.	γ_b/γ	0.15	
Backgr. torque visc.	γ_{br}/γ	0.03	

3 Preparing samples

The sample preparation procedure consists of several steps. The first step of the preparation process is to compress isotropically the loose packing that was created randomly (with small overlaps) with a volume fraction of $\nu = 0.3$ up to a volume fraction of $\nu = 0.5$. The system is then relaxed at a constant volume fraction of $\nu = 0.5$ and the particles are allowed to dissipate their kinetic energy and reach a zero-pressure, relaxed configuration before jamming. Isotropic compression is then applied to reach a volume fraction of $\nu = 0.82$ and, in the last phase, the compressed packing is decompressed isotropically until a volume fraction of $\nu = 0.5$ is reached again [21]. The compression and decompression are performed by applying a constant strain rate to each particle. We use a simulation time of 4000 [μs] per phase for the whole procedure, which results in a strain rate of $\dot{\varepsilon} = 3.8 \cdot 10^{-5}$, for loading and unloading. By following the above procedure frictional and frictionless samples were studied for varying cohesion, from $0 \leq k_c/k \leq 20$. It is worth to mention that some particles have overlaps during the initial random generation. These overlaps form small clusters due to high cohesive force between particles which will last during sample evolution when $k_c/k > 1$. This causes inhomogeneity for samples with extremely high cohesion.

3.1 Loading

Here, we present the general definition of the investigated quantities. Pressure and coordination number were calculated during sample compression to observe the influence of cohesion. We use the dimensionless pressure which is calculated by using the average normal stress: $P^* = P/k^*$, where $P = (\sigma_{xx} + \sigma_{yy} + \sigma_{zz})/3$ and $k^* = k/(2\langle a\rangle)$. The coordination number is defined as the average number of contacts per particle ($C = M/N$, where M is the total number of contacts and N is the total number particles). Particles with zero number of contacts and particles having a too small number of contacts, so called rattlers, were excluded, because they do not contribute to the mechanical stability of the packing [22]. So the coordination number becomes: $C_4^* = M_4/N_4$, with M_4 = total number of contacts of particles with at least 4 contacts and N_4 = number of particles with at least 4 contacts. Note that C_4^* and C_3^* show similar behaviour for the results presented later, for that reason C_4^* is just shown here. The jamming density is the next quantity, it can be obtained from the performed simulations using the pressure and coordination number data. It is defined as the transition of the material from a fluid-like state to a solid-like state. When the constituent particles of such packings are so crowded that they are in mechanically stable contacts with one another, the whole packing undergoes a sudden dynamical arrest, where the static pressure becomes non-zero.

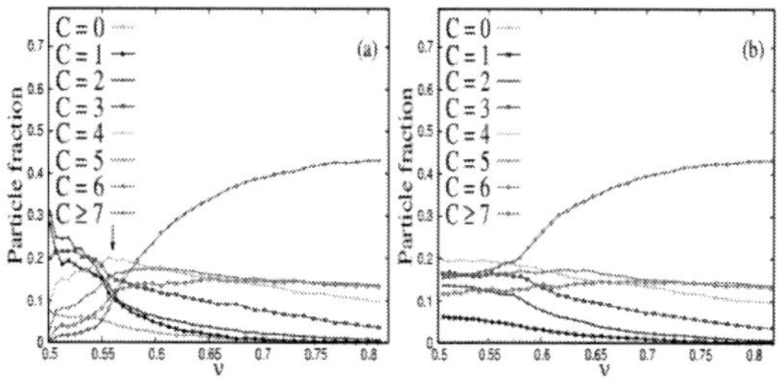

Figure 2. Fraction of particles possessing number of contacts from 0 to 6 or more, for the frictional sample with $k_c/k = 20$, during the sample (a) compression and (b) decompression. Where the arrow indicates the jamming density around $\nu = 0.56$

Figures 2(a) and (b) show the fractions of particles with a certain number of contacts during compression and decompression. Looking at figure 2(a), there is only a small number of particles having more than three contacts at low volume fraction. It is clear that by increasing the volume fraction, the number of particles with a high number of contacts increases ($C \geq 4$) and particles with a small number of contacts decreases ($C < 4$). Having a closer look, we can see that there is a regime of volume fractions, around the so-called jamming density, where the strongest changes occur in the number of contacts. This transition regime (around $\nu = 0.56$) displays in particular a change in slope for $C = 4$.

3.2 Unloading

Figure 2(b) shows the particle fractions during the sample decompression. As expected, the number of particles possessing a high number of contacts is greater at high volume fraction. By moving along the decompression path, it can be seen that there is not a big change in the number of particles with high number of contacts (especially for $C = 4, 5, 6$) and very few particles with contacts $C = 0$ and 1; this proves the forming of agglomerates during decompression due to the cohesive forces. Particles sticking together will stay attached to each other during decompression due to the high cohesive forces among them. Note that the compression path is used for further results, because decompression path is inhomogeneous, but this effect and the jamming transition of cohesive packings will be studied elsewhere.

4 Results and discussion

Figures 3 and 4 depict the evolution of pressure and coordination number, respectively, for frictionless (figure 3(a) and 4(a)) and frictional (figure 3(b) and 4(b)) samples. All data is taken during isotropic compression with a few cases during decompression.

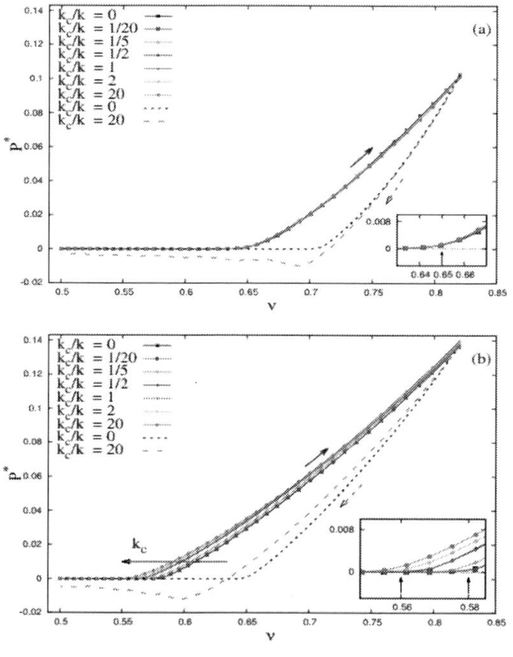

Figure 3. (a) Frictionless ($\mu = 0$) and (b) frictional ($\mu = 0.5$) simulation results of the dimensionless pressure P^* plotted against volume fraction ν for different cohesion values k_c, during compression (solid lines with points) and decompression (dashed lines), where the insert shows a zoom into the jamming regime. The arrows in the zoom indicate the range of $\phi_j \in [0.56, 0.58]$ for $\mu = 0.5$ and $\phi_j \approx 0.65$ for $\mu = 0$. The blue arrows in the plot indicate the compression and decompression paths for the largest and smallest k_c

In the case where the samples are prepared without friction, we see no significant changes in the pressure, throughout the whole regime (figure 3(a)) and in the coordination number at high volume fraction (figure 4(a)), for different cohesion values. For the lower densities, below jamming, a systematic increase in C_4^* can be seen with respect to k_c. On the contrary, we observe a change in

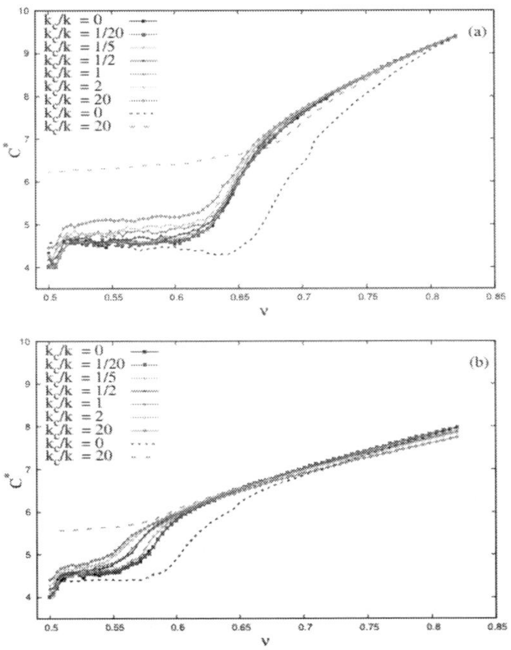

Figure 4. (a) Frictionless and (b) frictional coordination number C^* plotted against volume fraction ν for different cohesion values k_c during compression

pressure and coordination number with cohesion already at low volume fractions for frictional samples (see figures 3(b) and 4(b)). This is less pronounced for highly dense samples. Cohesive forces are relatively stronger at low volume fractions, so that the change of coordination number can be observed for both frictionless and frictional samples. Since, there is an extra force (tangential) in frictional samples, it causes particles to slide and rotate, causing rearrangements and a smaller C_4^*.

The effect of cohesive forces is thus already important at loose packings (close to the jamming density) since particles have more space to attract, repel and restructure. Far from the jamming density, there is not enough space for particles to establish new cohesive forces since they are strongly compressed and their normal force is dominated by elastic loading.

5 Conclusion

In this study the influence of cohesion on the macro- and microscopic properties, pressure, coordination number and jamming density, was observed. Frictionless and frictional samples were prepared with different interparticle cohesion. Sample preparation plays a key role to obtain a homogeneous medium; for more cohesive packings we observed stronger formation of clusters and agglomerates during sample preparation in particular during unloading. During compression, the influence of cohesion is more pronounced for samples with friction due to tangential forces that cause rotations and rearrangements. Moreover, cohesion plays a more significant role at low volume fractions, since samples have not been compacted yet and translational and rotational movement of particles can activate cohesive forces. The jamming density is a crucial state variable at which the transition from fluid to solid-like behaviour of configurations occurs. We observed that variation of cohesion does not change the jamming point of frictionless samples, but frictional samples get affected by cohesion. Increasing cohesion leads to a jamming transition at considerably lower densities. Future studies will focus on the effect of cohesive forces on the jamming density. Some packings with strong cohesive forces formed some tiny clusters from the beginning due to cohesion. We are currently quantifying how such clusters in the packings affect the results by creating cohesive packings without the early formation of clusters.

Acknowledgements

The financial support of the European-Union Marie Curie Initial Training Network, T-MAPPP, funded by FP7 (ITN 607453), is appreciated, see http://www.t-mappp.eu/ for more information.

References

[1] D.C. Rapaport, *The art of molecular dynamics simulation* (Cambridge university press, 2004)
[2] M. Gutiérrez, R. de Borst (Springer, 2001), pp. 1–16
[3] S. Luding, EJECE **12**, 785 (2008)
[4] S. Luding, in *Physics of dry granular media* (Springer, 1998), pp. 285–304
[5] S. Luding, Granular matter **10**, 235 (2008)
[6] S. Luding (Springer, 2009), pp. 453–492
[7] O.I. Imole, et al, Powder technology **287**, 108 (2016)
[8] S. Luding, Behavior of granular media **9**, 137 (2006)
[9] A. Singh, V. Magnanimo, K. Saitoh, S. Luding, Physical Review E **90**, 022202 (2014)
[10] R. Tykhoniuk, et al, Chemical Engineering Science **62**, 2843 (2007)
[11] N. Mitarai, F. Nori, Advances in Physics **55**, 1 (2006)
[12] Z. Fournier et al., Journal of Physics Condensed Matter **17**, S477 (2005)
[13] S. Roy, et al, CPM pp. 1–14 (2015)
[14] F. Gilabert, J.N. Roux, A. Castellanos, Physical review E **75**, 011303 (2007)
[15] R. Yang, et al, Physical Review E **78**, 031302 (2008)
[16] A. Singh, V. Magnanimo, S. Luding, AIP Conference Proceedings **1542**, 682 (2013)
[17] J. Tomas, KONA **18**, 157 (2000)
[18] O.R. Walton, R.L. Braun, Journal of Rheology **30**, 949 (1986)
[19] M.H. Sadd, Q. Tai, A. Shukla, International Journal of Non-Linear Mechanics **28**, 251 (1993)
[20] J. Tomas, Granular Matter **6**, 75 (2004)
[21] K. Taghizadeh, et al, IOP Conference Series: Earth and Environmental Science **26**, 012008 (2015)
[22] F. Göncü, O. Durán, S. Luding, Comptes Rendus Mécanique **338**, 570 (2010)

Experimental observations of root growth in a controlled photoelastic granular material

Jonathan Barés[1,★], *Serge* Mora[1,★★], *Jean-Yves* Delenne[2,★★★], and *Thierry* Fourcaud[3,★★★★]

[1] *Laboratoire de Mécanique et Génie Civil, Université de Montpellier, CNRS, Montpellier, France*
[2] *UMR IATE, CIRAD, INRA, Montpellier SupAgro, Université de Montpellier, Montpellier, France*
[3] *CIRAD, UMR AMAP, Montpellier, France*

Abstract.
We present a novel root observation apparatus capable of measuring the mechanical evolution of both the root network and the surrounding granular medium. The apparatus consists of 11 parallel growth frames, two of them being shearable, where the roots grow inside a photo-elastic or glass granular medium sandwiched between two pieces of glass. An automated system waters the plant and image each frame periodically in white light and between crossed polarisers. This makes it possible to follow (*i*) the root tips and (*ii*) the grain displacements as well as (*iii*) their inner pressure. We show how a root networks evolve in a granular medium and how it can mechanically stabilize it. This constitutes a model experiment to move forward in the understanding of the complex interaction between root growth and surrounding soil mechanical evolution.

1 Introduction

When growing, plants and trees adapt their morphology according to the mechanical and topological properties of the surrounding soil. Conversely, they change the mechanical and chemical properties of this environment [1]. Depending on the soil density or confining pressure, on the size and architecture of the voids between solid constituents of the soil, the roots will be blocked or grow farther [2–4], either by curving, changing their direction or by stressing the soil material [5, 6]. Therefore, the competing effects of soil mechanical impedance (mechanical property) [7] and penetrating capacities of the plant (biological property) lead to different strategies for the root development which can be optimized, and which in turn modifies differently the soil physical properties. Indeed, the root system architecture (RSA) impacts plant fitness and crop yield by facilitating efficient nutrient and water uptake from the soil. A better understanding of the effects of soil on RSA could improve crop productivity by matching roots to their soil environment [8]. Conversely, roots can change soil aeration, local packing fraction and constituent organization [9, 10] but also its repartition of water and cohesion [11, 12]. Hence, the mechanism of root growth in soil is a key issue to understand both how to improve plant development and how to improve grounds properties like stability for example. When a plant grows its roots have an effect on the stress development in the surrounding soil considered as a granular material. Conversely the induced pressure of the granular medium on the roots can affect the root network growth. We want to understand this complex root-grains, biological and mechanical, feedback loop between plant roots and soil structure and composition [13].

At the local scale, [5, 6] measured the force applied by a growing root tip on an obstacle. At the same scale, [1] have also studied the stress induced by a root when it goes through the gap between 2 bidimensional photoelastic grains. At the global scale [14] attempted to observe the root network in a packing of such particles. But root tip cannot initiate a path if there is no gap between grains, which is the case for a 2D dense packing of cylinders. From a numerical perspective, [11, 15] have modelled root network architectures in different soils while [16–18] have studied the tree anchorage and soil stabilization phenomenons.

Still, no experimental studies have been carried out to directly observe root network dynamics and surrounding stress modification while imposing specific grain configurations and density or specific mechanical loadings. It is therefore of primary importance to design a model experiment to understand progressively the intricate soil-root system from a mechanical point of view [1]. In this paper, we present a novel set-up which permits to observe the development of chickpea (*cicer arietinum L.*) root networks in a 2D granular material constituted of (*i*) home made bidisperse photoelastic discs or of (*ii*) glass beads. The photo-elastic particles (*i*) are molded with a groove on their edges so that the root tips are able to initiate a path between grains. Nevertheless the groove is small enough to observe a photoelastic signal when the root get thicker.

★e-mail: jb@jonathan-bares.eu
★★e-mail: serge.mora@umontpellier.fr
★★★e-mail: jean-yves.delenne@supagro.inra.fr
★★★★e-mail: thierry.fourcaud@cirad.fr

The position of the grains, the inter-granular spacing as well as the nature of the system confinement can be imposed in this set-up. The container of the granular material can be (*i*) an open cell, (*ii*) a confined cell and even a (*iii*) shearable cell. In this last case a slow shear loading can be imposed during the root growth or at the end of the development to test the stabilization effect on the granular medium. In the automated experimental apparatus several root development cells are treated in parallel to increase the statistical meaning of the observations. For each cell, each hour a picture is taken with and without polarized light. Then the evolution of the root network is followed as well as the position and pressure inside each particle by mean of classical image processing [19] and photo-elastic techniques [20, 21].

2 Experimental set-up

2.1 Growth frames

In the experimental set-up we observe the growth of chickpea (*cicer arietinum L.*) root networks in quasi-2D granular materials. Chickpea has been chosen because it is robust, the thickness of the roots (~ 1mm) is about the depth of the 2D medium (4mm), the root network is not too dense, it grows relatively rapidly (few weeks) and the seed is large enough to feed the plant during the first cycle of the growth. In a first configuration, these chickpeas are grown in simple 15cm^2 cells between 2 pieces of glass separated by 4mm, as presented in fig.1-A. In a second configuration, as presented in fig.1-B the left and write edges of the cell are movable so that the whole system (grains and roots) can be sheared by mean of a stepper motor once the root network is fully developed. In both case the plants grow in deionized water.

As shown in fig.1-A, to observe the growth, periodically (each hour) the cells are lit from the back with a flat circularly polarized light and imaged from the front with a 20Mp CCD camera[1] with and without a crossed polariser in front of the camera length. To observe the growth of several plants at a time, in several configurations, and to increase the statistical representativity of our results, 11 cells are observed at the same time (2 of them being shearable). As presented in fig.1-C the growth frames lay on a bar in parallel and the imaging part of the device, moved by a stepper motor, slides in front of each cell to take pictures. The whole system, including the light control is automated as well as a watering system which maintain the deionized water level constant in each cell (not shown here).

2.2 Granular material

These chickpeas are grown either in bidisperse photo-elastic discs (7 and 9mm in diameter) or in bidisperse glass beads (2 and 4mm in diameter). In the first case to avoid the problem with cylindrical particles pointed out by [14], as presented in fig.2-C, we used particles with a 0.5mm^2

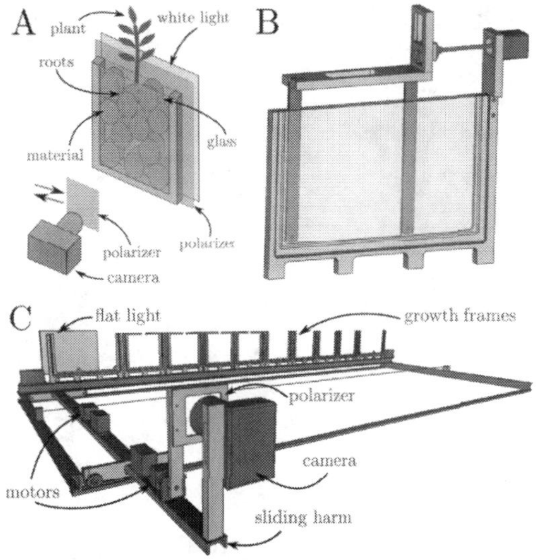

Figure 1. Experimental set-up. A: Imaging principle. The plant growths sandwiched between two pieces of glass in a 4mm thick flat frame with 15cm width and 15cm height. The roots growth in a *2D* granular material constituted of bidisperse glass beads or photo-elastic cylinders covered with deionized water. Periodically, the system is imaged by a front camera with a back-light with and without cross polarisers. B: Shearable growth frame. Some plants have been grown in frames equipped with a parallelogram mechanism to shear the granular medium when the growth is ended. The mechanism is actuated by a stepper motor. C: Complete set-up. The whole set-up is constituted of 11 growth frame (2 shearable, 9 simple) laying in parallel. Periodically a robotic harm slides in front of each frame to image it with and without polariser. Lighting is also controlled for pictures.

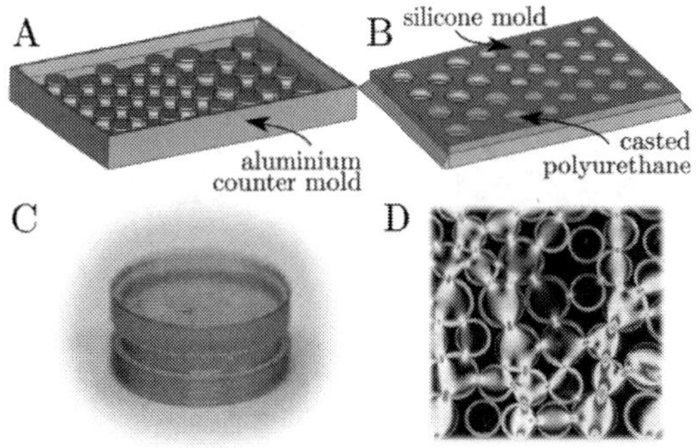

Figure 2. Photo-elastic grain molding. A: Aluminium counter mold of a set of 34 bidisperse particles. Particles are 4mm thick cylinders of diameter 7mm and 9mm with a 0.5mm deep and 1mm wide groove on the edge (see C). B: Silicone mold made from the aluminium counter mold. Colored polyurethane is molded in each alveolus. C: Photo-elastic polyurethane particle. D: Collection of compressed particles imaged between circular cross polarisers. Colored regions of the particles shows highly stressed areas.

[1] Canon® 70D

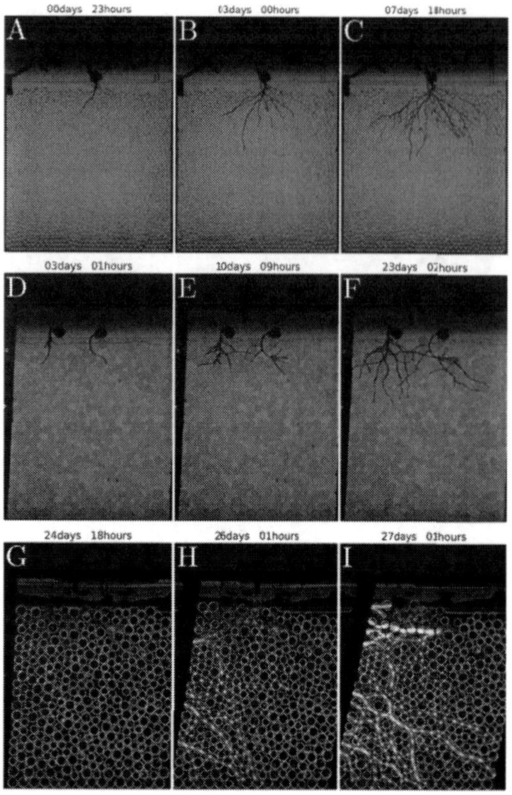

Figure 3. A, B, C: Development of a chickpea root network *cicer arietinum L.* in a bi-disperse quasi-monolayer of glass beads (2 and 4mm diameters). After a the main root network growth (A and B), a secondary one develops (C). Root oscillates betweens the beads. C, D, and E: Development of two chickpea root networks in a bi-disperse layer of photo-elastic particles. G, H and I: The system is then sheared quasistatically. The root networks avoid the top grains to rearrange and make the granular medium jammed. For A, B, C, D, E and F pictures are taken only with back-lighting while for pictures G, H and I the system is between crossed polarisers.

groove on the edge. Hence root tip can initiate a way between two adjacent grains and propagate deep inside the granular medium. These grooved photo-elastic particles are molded from polyurethane using the technique first introduced by [19]. As shown in fig.2-A a counter mold is machined out of aluminium. Then, a mold is made from silicon rubber[2] (see fig.2-B) and colored polyurethane[3] is molded in each hole. As shown in fig.2-D the photoelasticity of these particles is very good and classical techniques for photo-elastic granular materials pioneered by [20, 21] can by used.

3 Results and discussions

In fig.3-A, B and C we show the development of a chickpea root network in a bi-disperse quasi-monolayer of glass beads. After one day, the main root is already embedded in the granular material and secondary roots appear at the basis of the seed. Two days later, the full main root network is set and has moved the grains around the seed. After one week, the RSA is at its maximum of expansion without more nutriment. Secondary roots have grown from the main network and we clearly observe the root tortuosity imposed by the granular arrangement. Experiment is stopped and root dynamics can be studied.

In fig.3-D, E and F we show the development of two chickpea root networks in a bi-disperse layer of photo-elastic particles. Compared to the glass bead case the growth is much slower and plants need to be fed: deionized water is oxygenated and enriched. In this case it takes more then 3 weeks for the RSA to be fully developed. Tortuosity is also qualitatively lower since grains are larger than glass beads. After 23 days, the system is slowly and continuously sheared of 21% during 4 days. During this process even if the granular system is opened on its top and free to rearrange, a strong force network builds up. We see in fig.3-G, H and I that the root networks prevent the grains from rearranging and releasing stress. This create force chains forming from top left to the bottom right of the cell and from the left to the roots. This shows how the chickpea RSA can stabilize a granular system.

As presented in fig.3-A to F the back light pictures permit to observe the root network with a high contrast. This makes the detection of the RSA easy by thresholding the pictures. Hence, fig.4-A shows the position of the root network between photo-elastic grains once the RSA is fully developed. Just like already done in gels by [22], a homemade code as been written to detect and track the root tips during the development of the RSA. Blue and red lines in the figure are two examples of the path followed by two root tips. In both cases, we observe that the path does not follow the final root architecture. This means the root widely rearranges during its growing process either due to granular displacement or to the growing of secondary roots. Root-grain interaction is thus a dynamical process which needs to be followed on the long term. From the evolution of the root tip position, in fig.4-C, we plot the root speed for the two selected roots of the two plants. We see that the growth speed is not constant and not equivalent for both plants. However, it is difficult to analyse if this speed variation is due to biological or mechanical parameters. We also observed root tip circumnutation [23] in 2D (not shown here), how it is affected by the surrounding granular medium, how it changes the granular local organization and how it creates the root tortuosity.

Finally, from the back light pictures (see fig.3-F for instance), using image processing tools designed by [19], we detect the grain positions and sizes, and from the polarized pictures (see fig.3-I for instance) we compute the

[2] Smooth-On, MoldStar® 15 (slow).
[3] Smooth-On, ClearFlex® 50 with green SoStrong® ink.

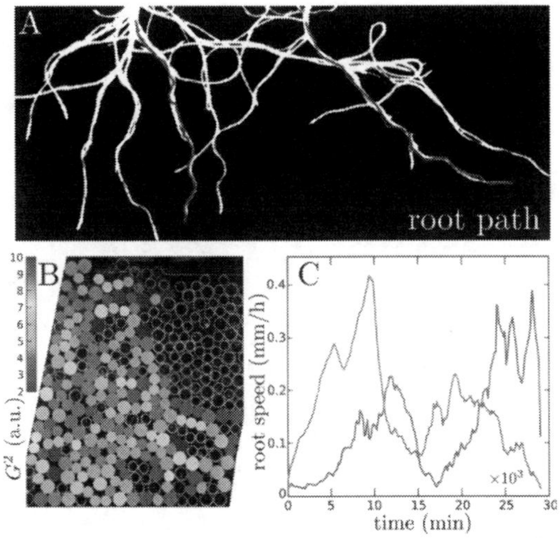

Figure 4. A: Root tip path. The black and white back ground shows the root networks of two chickpeas grown in photo-elastic discs just before the system is sheared. Blue and red line shows the path followed by two root tips (arbitrarily chosen). We see that due to the rearrangement of the root and grains the final root does not follow exactly the root tip path. B: Pressure measurement. For the maximum strain, when the root networks are fully developed, grain position has been tracked and pressure measured in each of them as G^2 [20]. By sake of clarity, grains with pressure lower than a certain value (2) are not presented. The force chains show the stabilization effect of the root networks C: Root tip speed. For root tips presented in blue en red in A, the evolution of their speed is shown during the plant growth before shearing the system. We see that the growth speed is not constant and not equivalent for both plants.

grain pressures in terms of G^2 as explained in [19–21]. For a system with a RSA fully developed and 21% sheared, fig.4-B shows the pressure in each grains. We observe force chains and measure the highest pressure near the root network. This is another observation of the fact that the RSA stabilizes the granular system.

4 Conclusion

We have described a novel experimental system that can observe the growth of chickpea root networks in different granular materials (namely grooved photo-elastic discs and glass beads) for different system confinements (namely open frames and sheared frames). The growth is observed both in white light and between crossed polarisers. This permits to measure the evolution of the RSA and more specifically to follow the root tips. The grain rearrangements and the evolution of their inner pressure is also recorded. This constitutes a model experiment to move forward in the understanding of the complex interaction between root growth and surrounding soil mechanical evolution.

References

[1] E. Kolb, C. Hartmann, P. Genet, Plant and Soil **360**, 19 (2012)
[2] A.M. Abdalla, D.R.P. Hettiaratchi, A.R. Reece, Journal of Agricultural Engineering Research **14**, 236 (1969)
[3] L.J. Clark, W.R. Whalley, P.B. Barraclough, in *Roots: The Dynamic Interface Between Plants and the Earth* (Springer, 2003), pp. 93–104
[4] S.R. Tracy, C.R. Black, J.A. Roberts, S.J. Mooney, Journal of the Science of Food and Agriculture **91**, 1528 (2011)
[5] R.K. Misra, A.R. Dexter, A.M. Alston, Plant and Soil **95**, 315 (1986)
[6] A.G. Bengough, C.J. Mackenzie, H.E. Elangwe, Plant and Soil **167**, 135 (1994)
[7] B.J. Atwell, Environmental and Experimental Botany **33**, 27 (1993)
[8] E.D. Rogers, D. Monaenkova, M. Mijar, A. Nori, D.I. Goldman, P.N. Benfey, Plant physiology pp. pp–00397 (2016)
[9] H.P. Cresswell, J.A. Kirkegaard, Soil Research **33**, 221 (1995)
[10] S.A. Materechera, A.R. Dexter, A. A. M. Alston, Plant and Soil **142**, 69 (1992)
[11] S. Baets, D. Torri, J. Poesen, M.P. Salvador, J. Meersmans, Earth Surface Processes and Landforms **33**, 1948 (2008)
[12] F. Danjon, D.H. Barker, M. Drexhage, A. Stokes, Annals of Botany **101**, 1281 (2008)
[13] A. Pierret, C. Doussan, Y. Capowiez, F. Bastardie, Vadose Zone Journal **6**, 269 (2007)
[14] D.M. Wendell, K. Luginbuhl, J. Guerrero, A.E. Hosoi, Experimental Mechanics **52**, 945 (2012)
[15] M. Heinen, A. Mollier, P.D. Willigen, Plant and Soil **252**, 251 (2003)
[16] L.X. Dupuy, T. Fourcaud, P. Lac, A. Stokes, American Journal of Botany **94**, 1506 (2007)
[17] A. Stokes, C. Atger, A.G. Bengough, T. Fourcaud, R.C. Sidle, Plant and Soil **324**, 1 (2009)
[18] S.B. Mickovski, A. Stokes, R. van Beek, M. Ghestem, T. Fourcaud, Ecological Engineering **37**, 1523 (2011)
[19] M. Cox, D. Wang, J. Barés, R.P. Behringer, Europhysics Letters **115**, 64003 (2016)
[20] D. Howell, R.P. Behringer, C. Veje, Physical Review Letters **82**, 5241 (1999)
[21] T.S. Majmudar, R.P. Behringer, Nature **435**, 1079 (2005)
[22] M.V. Thompson, N.M. Holbrook, Plant Physiology **135**, 1822 (2004)
[23] M. Stolarz, Plant Signaling & Behavior **4**, 380 (2009), pMID: 19816110, `http://dx.doi.org/10.4161/psb.4.5.8293`

Steady state rheology from homogeneous and locally averaged simple shear simulations

Hao Shi[1,⋆], *Stefan* Luding[1,⋆⋆], and *Vanessa* Magnanimo[1,⋆⋆⋆]

[1] *Multi Scale Mechanics, CTW, MESA+, University of Twente, 7500 AE Enschede, The Netherlands*

Abstract. Granular materials and particulate matter are ubiquitous in our daily life and they display interesting bulk behaviors from static to dynamic, solid to fluid or gas like states, or even all these states together. To understand how the micro structure and inter-particle forces influence the macroscopic bulk behavior is still a great challenge today. This short paper presents stress controlled homogeneous simple shear results in a 3D cuboidal box using MercuryDPM software. An improved rheological model is proposed for macroscopic friction, volume fraction and coordination number as a function of inertial number and pressure. In addition, the results are compared with the locally averaged data from steady state shear bands in a split bottom ring shear cell and very good agreement is observed in low to intermediate inertia regime at various confining pressure but not for high inertia collisional granular flow.

1 Introduction

Granular media are envisaged as a collection of microscopic particles which interact through dissipative contact forces; their natural discontinuity poses many challenges for both academia and industry in understanding their bulk behavior [1]. A continuum description of the flow behavior of granular media is highly desirable, for application in both natural phenomena studies and industrial applications. Despite numerous efforts, the constitutive relations describing the granular flow behavior is still a matter of debate. Solid mechanics and kinetic theory have been successful in predicting the solid and gas like behavior, respectively [2, 3]. At one extreme particles interact via enduring contacts, while in the gaseous regime, binary collisions are the mode of momentum exchange. The recently proposed inertial number framework has been successful in describing the flow behavior in the dense flow liquid like regime when the particles not only undergo collisions but also frictional interactions with other particles [4, 5]. Though it very well predicts the flow behavior in case of steady shear of rigid particles, it fails in cases of inhomogeneous shear and transient shear, where we need better constitutive models to describe dense granular flow [6]. Also a non-local constitutive models have been proposed and tested to improve the constitutive relations in inhomogeneous flow [7]. With the development of computational power nowadays, one can obtain simulation data with sufficient statistics that allow to derive constitutive relations that describe the local rheology and flow behavior [8–12].

⋆e-mail: h.shi-1@utwente.nl
⋆⋆e-mail: s.luding@utwente.nl
⋆⋆⋆e-mail: v.magnanimo@utwente.nl

A pressure density relation has been proposed in [13] in additive forms; the present paper proposes an extended local constitutive model in a multiplicative form and then tests the validity of this model for both homogeneous and inhomogeneous steady state soft granular flow.

2 Simulation Setups

2.1 Stress controlled simple shear box

We use MercuryDPM [14, 15], an open-source implementation of the Discrete Particle Method, to simulate a cuboid shear volume with Lees-Edwards periodic boundaries in x and y directions and normal periodic boundaries in z direction. The particles are sheared along x direction with a constant velocity V_x and the stress is kept constant along y direction as σ_{yy}. We vary the shear rate $\dot{\gamma} = \partial V_x/\partial y$ to understand the shear flow from quasi-static to dynamic regimes. The purpose of using Lees-Edwards periodic boundaries with constant stress is to represent dense granular flow in reality, e.g, sand or/and powders sheared in different shear cells or in an avalanche [10]. The piecewise linear elasto-plastic frictional contact model is used between particle contacts and the number of particles used in the simulation is 4096. The other input parameters are summarized in Table 1.

2.2 Split-Bottom ring shear cell

The same MercuryDPM software was used to simulate a shear cell with annular geometry and a split bottom, as explained detailed in [16–18]. The geometry of the system consists of an outer cylinder (radius $R_o = 110$ mm) rotating around a fixed inner cylinder (radius $R_i = 14.7$ mm) with

Table 1. Summary and numerical values of particle parameters used in the DEM simulations.

Parameter	Symbol	Simple Shear	Split-Bottom
Average Diameter	$< r >$	2.2 and 2 mm	2.2 mm
Polydispersity	$w = r_{max}/r_{min}$	2 and 3	2
Particle Density	ρ_p	2000 kg/m^3	2000 kg/m^3
Normal stiffness	k	10^8 kg/s^2	100 kg/s^2
Tangential stiffness	k_t/k_n	0.286	0.286
Rolling/torsion stiffness	k_r/k_n	0.286	0.286
Coulomb friction static	μ_s	0.01	0.01
Coulomb friction dynamic	μ_d	0.01	0.01
Rolling friction	μ_r	0	0
Restitution coefficient	e	0.804	0.469
Normal Viscosity	γ_n	1 kg/s	0.002 kg/s
Friction Viscosity	γ_t/γ_n	0.286	0.286
Rolling Viscosity	γ_r/γ_n	0.286	0.286
Torsion Viscosity	γ_r/γ_n	0.286	0.286

a rotation frequencies of $\Omega = 0.01$ to 0.75 revolutions per second. The granular material is confined by gravity between the two concentric cylinders, the bottom plate, and a free top surface. The bottom plate is split at radius R_s = 85 mm. Due to the split at the bottom, a narrow shear band is formed. It moves inwards and widens towards the free surface. This set-up thus features a wide shear band away from the bottom and the walls, which is thus free from boundary effects. The filling height ($H \approx 40$ mm) is chosen such that the shear band does not reach the inner wall.

In earlier studies [11, 17], a quarter of this system (0° $\leq \phi \leq$ 90°) was simulated using periodic boundary conditions. In order to save computation time, here we simulate only a smaller section of the system (0° $\leq \phi \leq$ 30°) with appropriate periodic boundary conditions in the angular coordinate, unless specified otherwise. No noticeable effect on the macroscopic flow behavior was observed in comparison between simulations done with a smaller (30°) and a larger (90°) angle.

3 Dimensionless numbers

The effects of varying shear rate, $\dot{\gamma}$, and confining pressure, p, can be modeled using two dimensionless numbers, expressed as a ratio of time-scales, as given in Table 2, where ρ_p is particle density and k is the contact loading stiffness. Note that the subscripts in time scale ratio, $\dot{\gamma}$, p and k denote shear rate, pressure and contact duration respectively.

Table 2. Dimensionless numbers for the model

Dimensionless number	Definition	Time scale ratio
Inertial number I	$\frac{\dot{\gamma} d_p}{\sqrt{p/\rho_p}}$	$t_p/t_{\dot{\gamma}}$
Softness p^*	$\frac{p d_p}{k}$	$(t_k/t_p)^2$

4 Rheological model

Here, a short summary of recent results on formulating an improved/accurate the local granular rheology is presented.

4.1 Classical $\mu(I)$-rheology

When formulating a granular rheology, the starting point is the successful, simple, and elegant so-called $\mu(I)$-rheology [5], which relates the so-called macroscopic (bulk) friction, i.e., the shear-stress to pressure ratio $\mu = \tau/p$, in a sheared particulate system, to the inertial number, I:

$$\mu(I) = \mu_0 + (\mu_\infty - \mu_0)\frac{1}{1 + I_0/I} \quad (1)$$

This relation describes well a wide variety of flows [13] of hard, cohesionless particles, where $\mu_0 = 0.15$ and $\mu_\infty = 0.42$ represent the zero and infinite strain-rate limits, respectively, and the characteristic dimensionless strain-rate where the inertial effect kicks in, is $I_0 = 0.06$, .

4.2 $\mu(I)$-rheology for soft particles

The first improvement is based on particle softness, p^*, as introduced in [11, 18, 19]:

$$\mu(I, p^*) = \mu_0 f_\mu(I) g_\mu(p^*) = \mu_0 \left(1 + \frac{\mu_\infty/\mu_0 - 1}{1 + I_\mu/I}\right)\left(1 - \left(\frac{p^*}{p^*_\mu}\right)^{0.5}\right) \quad (2)$$

This two multiplicative corrections account for a range of strain rates and particle softness (or stiffness), but also for a system with a confined stress condition, as in the most shear rheometers, and even for different magnitudes of gravity, as in a centrifuge or on the moon [11]. Additional corrections for cohesive particles involve the so-called Bond-number (Bo), as studied elsewhere [17] and ignored in the following. Apart from macroscopic friction, a rheological model has also been proposed for the bulk density (volume fraction) as a function of softness and shear rate:

$$\phi(I, p^*) = \phi_0 f_\phi(I) g_\phi(p^*) = \phi_0 \left(1 - \frac{I}{I_\phi}\right)\left(1 + \frac{p^*}{p^*_\phi}\right) \quad (3)$$

with the critical or steady state density under shear, in the limit of vanishing pressure and inertial number, ϕ_0. The inertial number for which dilation turns to fluidization, I_ϕ, and the typical pressure level for which softness leads to huge densities, p^*_ϕ [19]. Note that both correction terms are valid only for sufficiently small arguments. Too large inertial numbers would fully fluidize the system so that the rheology should be that of a granular fluid, for which kinetic theory applies, while too large pressure would lead to enormous overlaps, for which the contact model and the particle simulation become questionable. In the following, the considered inertial numbers are $I <$ 0.5, while the softness or pressure are $p^* < 0.1$.

In order to complete the rheology for soft, compressible particles, a new relation for the coordination number, C, as function of pressure and shear rate is proposed here, analogous to Eqs. (2) and (3) (but data are not shown):

$$C(I, p^*) = C_0 f_C(I) g_C(p^*) = C_0 \left(1 - \frac{I}{I_C}\right)\left(1 + \left(\frac{p^*}{p^*_C}\right)^{\alpha}\right) \quad (4)$$

with the critical or steady state coordination number under slow shear and low pressure, C_0, the inertial number order of magnitude for which dilation would turn to fluidization, I_C, the typical pressure level for which softness leads to a considerable increase in coordination number, p^*_C, and the power factor α. Note that all multiplicative terms $(1 \pm ...)$ are first order Taylor expansions of possibly more realistic functions only, and μ_0, ϕ_0 and C_0 are constants but varying with different materials, i.e. when changing the inter-particle friction coefficient μ_p.

5 Results

5.1 Globally averaged rheology for simple shear

For a system with very low inter-particle friction, $\mu_p = 0.01$, we simulated the stress controlled simple shear using the input parameters as shown in Table. 1, similar to Refs. [11, 12, 17], except for the loading stiffness and restitution coefficient. We vary both shear rate ($\dot{\gamma}$ from 0.0001 to 1 s^{-1}) and confining stress (p^* from 0.0025 to 0.1) to investigate the rate and softness dependencies. Unless specified, all the fitted functions shown in the following are using the data from homogeneous simple shear.

In Fig. 1, we plot the steady state bulk density (volume fraction) against the inertial number from our stress controlled homogeneous simple shear. Using the proposed $\phi(I, p^*)$ rheological function in Eq. (3), all our data are collapsing with the fitted curves; even for the highly inertial regime ($I > 0.1$) the bulk density is captured. However, if we check these curves focusing on only softness dependence as shown in Fig. 2, the lowest bulk density data (magenta) are actually deviating more from the fitting curve, and thus the low bulk density is not well captured.

In Fig. 2, we look at the confining stress dependence on the bulk density. When we keep the shear rate constant, the bulk density is increasing linearly with confining stress/pressure. Our data also fits well using the function from [11], which means our particle flow is also having a pressure dependence.

To check the quality of our fitting, we normalized our simple shear data using the fitted rheological model Eq. (3), $\phi(I, p^*)$. Up to I around 0.1, all the data points stay within 2% deviation range with no clear dependence on the inertial number, and this deviation increases slightly (still within 3%) when the inertial number goes above 0.1. However, the overall fitting quality is satisfactory (all within ± 3 %), and our rheological model is capturing well the bulk density of dense granular flow.

As for the macroscopic friction (Fig. 3), our improved rheological model is capturing well the friction behavior with deviation within ±5%. However, for coordination number (data not shown here), the deviation goes up to ±25%, which indicates a further improvement for Eq. (4) is still needed.

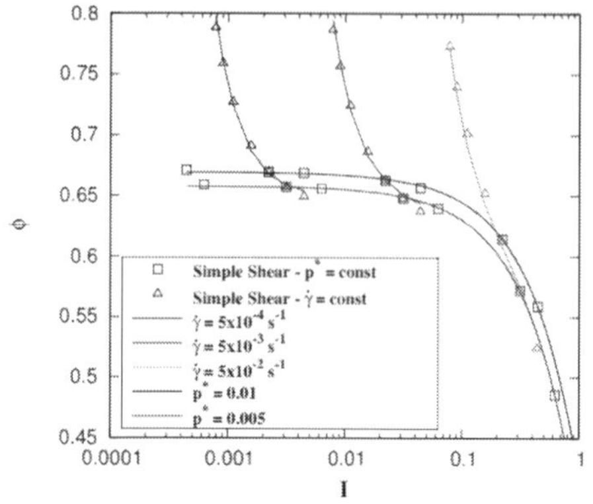

Figure 1. Bulk density, ϕ, plotted against inertial number, I, using stress controlled simple shear. Two paths are chosen here: (i) softness $p^* =$ const., vary shear rate $\dot{\gamma}$; (ii) $\dot{\gamma} =$ const., vary softness p^*, as shown in the legend, different colors refer to different shear rates or confining stresses. The homogeneous data are fitted using $\phi(I, p^*)$, with $\phi_0 = 0.65$, $I_\phi = 2.44$ and $p^*_\phi = 0.38$.

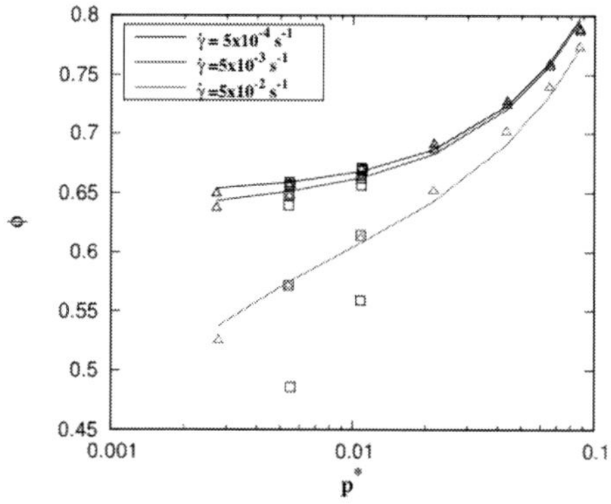

Figure 2. Bulk density, ϕ, plotted against softness, p^*, for the same simulations as in Fig. 1. Note that here we neglect the fitting lines from path (i), which are vertical straight lines with constant p^*.

5.2 Local rheology from split bottom ring shear cell

The steady state bulk density from the center of shear bands in the split bottom ring shear cell are plotted on top of simple shear data as shown in Fig. 4. The split bottom shear cell data in low softness regime ($p^* < 0.01$) are collapsing almost perfectly with the simple shear data with only a slightly larger scattering except for high inertia regime ($I > 0.01$). For high softness regime ($p^* > 0.01$), the split bottom shear cell data are consistently deviating up to +5% in all the investigated inertial numbers. And for the softness dependence as shown in Fig. 4(b), the deviation of split bottom ring shear cell data show the same increasing trend with increasing softness as simple shear data. The agreement between the two simulation setups is really good.

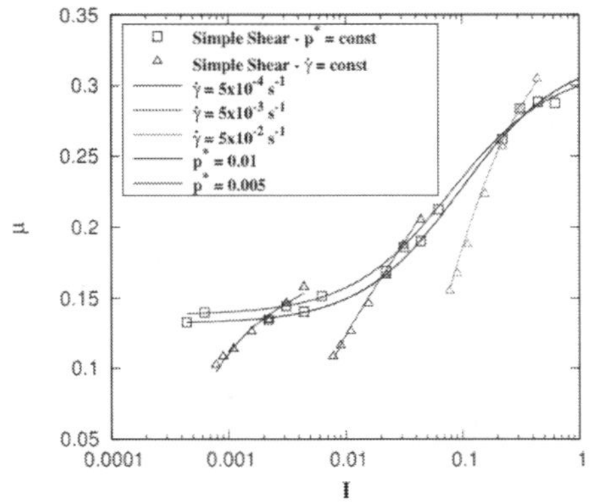

Figure 3. Macroscopic friction, $\mu = \tau^*/p^*$, plotted against inertial number, I, from the same simulations as in Fig. 1. The data are fitted using $\mu(I, p^*)$ as Eq. (2), with $\mu_0 = 0.15$, $\mu_\infty = 0.36$, $I_\mu = 0.09$ and $p^*_\mu = 0.60$.

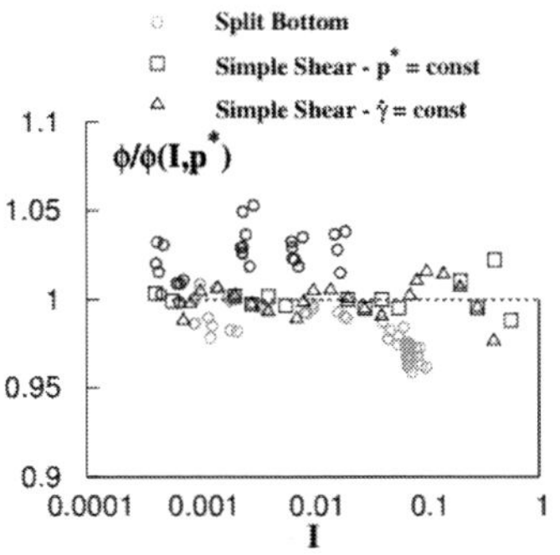

Figure 4. Bulk density comparison between simple shear and split bottom shear. The simulation data are normalized by the proposed rheology prediction $\phi(I, p^*)$ in Eq. (3). Blue and purple points are the split bottom shear cell data with softness $p^* > 0.01$ and $p^* < 0.01$ respectively.

6 Conclusion

The proposed rheological model for dense granular flow are checked using stress controlled simple shear simulations with looking at both inertial number and softness dependence. With increasing inertial number inside the system, the flow becomes more dilute, going towards collisional flow, where our model could not capture the behavior. Nevertheless, for granular flow with bulk density (volume fraction) above 0.5 and inertial number below 0.1, our model gives very good bulk density and macroscopic friction descriptions of the influences both from inertia and softness in a sheared granular flow. Good agreement between the globally averaged simple shear box data and locally averaged center of shear band data inside a split bottom ring shear cell is achieved and further verifies the universality/applicability of the improved local rheological model. Future work will include the development of extending our model towards non-local inhomogeneous rheology as well as for cohesive systems.

Acknowledgement

We would like to thank for the financial support through the European-Union-Funded Marie Curie Initial Training Network FP7 "T-MAPPP" project (ITN607453).

References

[1] H.M. Jaeger, S.R. Nagel, R.P. Behringer, Rev. Mod. Phys. **68**, 1259 (1996)
[2] I. Goldhirsch, Annu. Rev. Fluid Mech. **35**, 267 (2003)
[3] R.M. Nedderman, *Statics and kinematics of granular materials* (Cambr. Univ. Press, Cambridge, 1992)
[4] B. Andreotti, Y. Forterre, O. Pouliquen, *Granular media: between fluid and solid* (Cambridge University Press, 2013)
[5] P. Jop, Y. Forterre, O. Pouliquen, Nature **441**, 727 (2006)
[6] A. Ries, L. Brendel, D.E. Wolf, Comp. Part. Mech. pp. 1–8 (2015)
[7] M. Bouzid, M. Trulsson, P. Claudin, E. Clément, B. Andreotti, Phys. Rev. Lett. **111**, 238301 (2013)
[8] S. Luding, Particulate Science and Technology **26**, 33 (2007)
[9] S. Luding, F. Alonso-Marroquín, Granul. Matter. **13**, 109 (2011)
[10] T. Weinhart, R. Hartkamp, A.R. Thornton, S. Luding, Physics of Fluids **25**, 070605 (2013)
[11] A. Singh, V. Magnanimo, K. Saitoh, S. Luding, New J. Phys. **17**, 043028 (2015)
[12] S. Roy, S. Luding, T. Weinhart, Procedia Engineering **102**, 1531 (2015)
[13] S. Chialvo, J. Sun, S. Sundaresan, Phys. Rev. E **85**, 021305 (2012)
[14] A. Thornton, T. Weinhart, S. Luding, O. Bokhove, Int. J. Mod. Phys. C **23**, 1240014 (2012)
[15] T. Weinhart, A.R. Thornton, S. Luding, O. Bokhove, Granul. Matter. **14**, 289 (2012)
[16] J.A. Dijksman, M. van Hecke, Soft Matter **6**, 2901 (2010)
[17] A. Singh, V. Magnanimo, K. Saitoh, S. Luding, Phys. Rev. E **90**, 022202+ (2014)
[18] S. Roy, S. Luding, T. Weinhart, arXiv preprint arXiv:1609.03098 (2016)
[19] S. Luding, A. Singh, S. Roy, D. Vescovi, T. Weinhart, V. Magnanimo, The 7th International Conference on Discrete Element Methods. (2016)

Collapse of granular media subjected to wetting

Fatima Zahra El Korchi[1,*], *Frédéric* Jamin[1,2], and *Moulay Saïd* El Youssoufi[1,2]

[1] LMGC, Univ. Montpellier, CNRS, Montpellier, France.
[2] MIST Lab., IRSN, CNRS, Univ. Montpellier, France.

Abstract. This paper focuses on the collapse of granular materials subjected to wetting action. For soils, the collapse potential depends on several parameters such as liquid limit, matric suction, compactness, initial water content and the amount of fine particles. The effect of grain size, which plays a key role in the rearrangement of grains, remains little studied and poorly understood. To investigate the capillary origin of the collapse phenomenon, we present an experimental study on macroscopic and local scales. Our results show the effect of grain size and water content on collapse.

1 Introduction

In the pendular regime, the presence of isolated capillary bridges in a granular medium generates a cohesive force by capillary action between grains. These forces are responsible for the stability of a sand castle for example. The capillary forces depend on the quantity of liquid contained in the medium [11].

During wetting, the capillary bridges merge because the local geometry of the grains no longer allows the formation of isolated bridges. Thereby, big changes in the morphology liquid must take place in the granular assemblies, which must necessarily change its behavior. This phenomenon is still little studied experimentally considering the complexity of the phenomenon of coalescence at local scale. Deposited solutes and/or capillary bridges can bond the solid particles in soils. When these soils are subjected to wetting action, they can undergo a collapse phenomenon.

The collapse potential of soils depends on several parameters such as liquid limit [8, 2], suction [12], compactness [12, 4], initial water content [12, 4] and the amount of fine particles [3, 1]. The effect of grain size, which plays a key role in the rearrangement of grains, remains little studied and poorly understood.

This communication will focus on the collapse of granular media subjected to wetting action. To investigate the origin of capillary collapse phenomenon, we present an experimental study in macroscopic and local scales. Glass beads form the used samples. The wicking tests are performed on the macroscopic scale of a *VER* sample and at the local scale of 3 and 4 grains.

2 Macroscopic scale study

2.1 Materials and experimental protocol Materials

2.1.1 Materials

The used samples are made by glass beads of grain size Φ: < 50 μm, 60 – 100 μm, 80 – 200 μm and 300 – 500 μm. Some of the used material characteristics are presented in Table 1.

Table 1. Physical properties of the different samples.

Materials	Grain sizes Φ (μm)	Specific density ρ_s (kg/m³)	Liquid limit w_L (%)
Glass beads	< 50	2450	23
	60 - 100	2470	25
	80 - 200	2480	24
	300 - 500	2500	-

Fig. 1 shows the retention curves for all grain sizes obtained by the filter paper technique (*Whatman* “42” paper). A modelling of the retention curves is also plotted in Fig. 1 based on *Fredlund and Xing* [6].

2.1.2 Experimental protocol

The samples are cylindrical in shape with an initial height H_0 = 94 mm and an initial diameter D_0 = 50 mm. The samples were prepared with an initial compactness c_0 = 0.59 and initial water contents w_0 = 5%. The initial water content w_0 allows the creation of the capillary bridges in the sample that are at the origin of their mechanical strength by cohesion. The principle involves applying low water pressure (u_w = 5 kPa) at the bottom of the sample by the Pressure - Volume Controller while maintaining atmospheric air pressure at the top of the sample (u_a = 0 kPa) as explained in *El Korchi et al.* [5]. During the test, the water content w within the sample was determined by the volume of injected water V_w,

* Corresponding author: fatima-zahra.el-korchi@umontpellier.fr

directly reported by the Pressure - Volume Controller at a given time t of the wetting path [5].

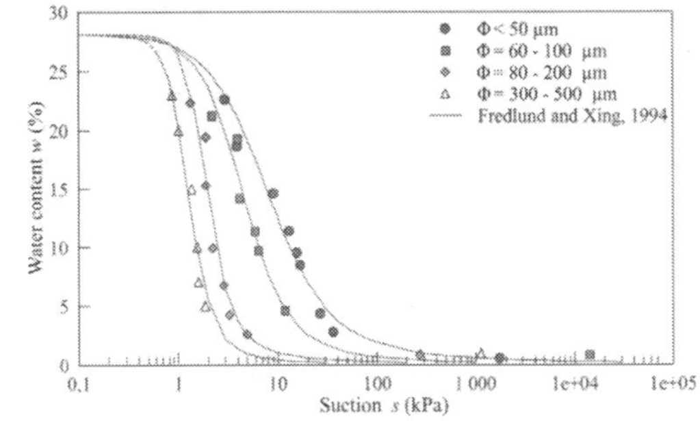

Fig. 1. Retention curves for different glass beads (c = 0.59).

2.2 Experimental results

2.2.1 Highlighting of collapse

The wetting test results obtained on glass beads of grain size distributions Φ < 50 µm, with compactness c_0 = 0.59 and initial water content w_0 = 5%, are shown in Fig. 2.

During the ascent of water in the granular sample, axial and radial deformations ε_a and ε_r evolve. A positive axial deformation corresponds to a decrease of the height of the sample and a negative radial deformation corresponds to an increase of its diameter. For a critical water content, deformations undergo a sudden variation assigned to the collapse phenomenon. This critical water content is subsequently called collapsing water content and noted w_{coll} in Fig. 2.

In what follows, we define the final axial deformation ε_a^f as the maximum axial deformation due to the wetting while the amplitude of collapse A is defined as the axial deformation due to the sudden collapse of the sample as shown in Fig. 2.

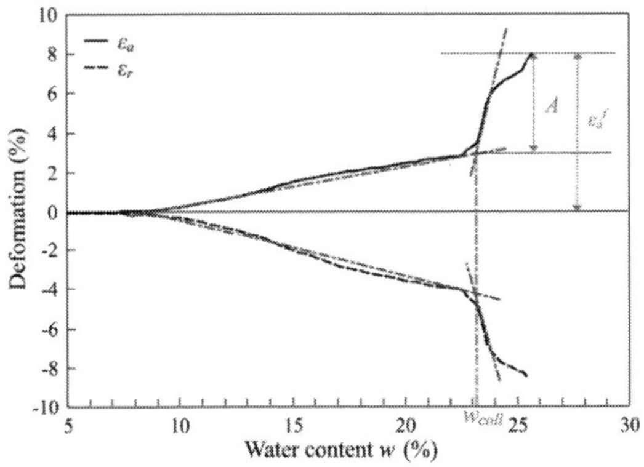

Fig. 2. Evolution of axial ε_a and radial ε_r deformations during wetting for glass beads Φ < 50 µm (w_0 = 5% and c_0 = 0.59).

2.2.2 Effect of grain size

The effect of grain size on deformations A and ε^f_a and collapsing water content w_{coll} is shown in Fig. 3. These figures are plotted with a horizontal uncertainty bar showing the spreading of the grain size of the glass beads.

Axial deformations ε^f_a and A decreased nonlinearly with the grain size (Fig. 3.a). This result shows the high sensitivity of small granularity samples to wetting. This sensitivity can be explained by the different microstructural morphology of the samples, having the same volume fraction. The finer the grain size, the more dominant are the capillary forces relative to the gravity [9].

The collapsing water content appears to depend on the grain size. A slight increase in the water content with the grain size shows the sensitivity of the granular materials for the wetting (Fig. 3.b).

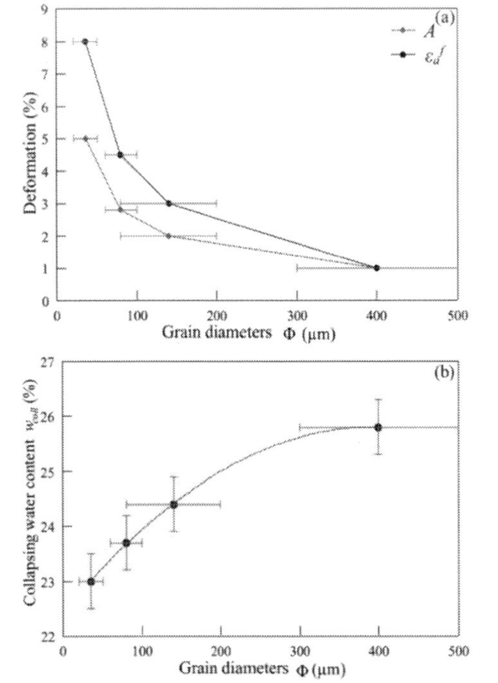

Fig. 3. a. axial deformation (ε^f_a and A) and **b.** collapsing water content w_{coll} as a function of grain size.

3 Local scale study

3.1 Materials and experimental protocol

3.1.1 Materials

At local scale, wetting tests were conducted on three and four borosilicate glass beads grade 25 and 8 mm in diameter. The capillary force is measured by the apparatus shown in Fig. 4. The used experimental device is composed of a precision balance 1/1000, a micrometric table on which is fixed an upper glass bead (Fig. 4.b) as well as supports of two or three spheres

(Fig. 5.a and Fig. 5.c) respectively to form 3 and 4 spheres configurations. Distilled water is used to form the capillary bridges between the glass beads. The provision of water in bridges is done with micro-syringes with the accuracy of 0.1 µl and the volume of 5 and 10 µl.

3.1.2 Experimental protocol

For the three sphere configuration, the centers of the three glass beads form a triangle in the vertical plane as shown in Fig. 5.a. and for the four sphere configuration, the centers of spheres form a tetrahedron (Fig. 5.b).

The upper sphere is fixed on the micrometer table which allows applying distances, denoted D_1 between the centers of the upper and the lower spheres. The lower spheres are fixed on a support with a distance, denoted D_2, separating their centers.

The test consists in forming two or three bridges between the upper sphere and the two or three lower spheres. Then, water is injected in each bridge with the micro-syringe (no bridges are formed between the lower spheres). For each water volume injected V_w into the bridges, the axial capillary force F_{cap} is obtained with the expression 1 [7]:

$$F_{cap} = (m_s + m_w - m_a).g \qquad (1)$$

where m_s is the mass of the support with the fixed spheres, $m_w = \rho_w V_w$ is the water mass cumulated, m_a is the apparent mass measured for each water volume injected, and g the acceleration of gravity.

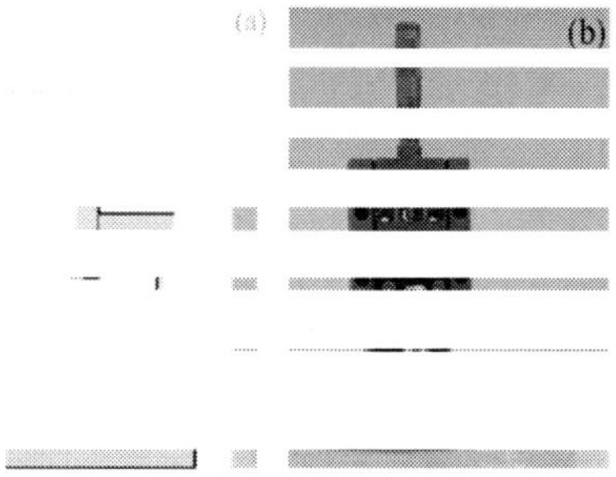

Fig. 4. a. Schematic of the used apparatus at a local scale, b. the micrometer table.

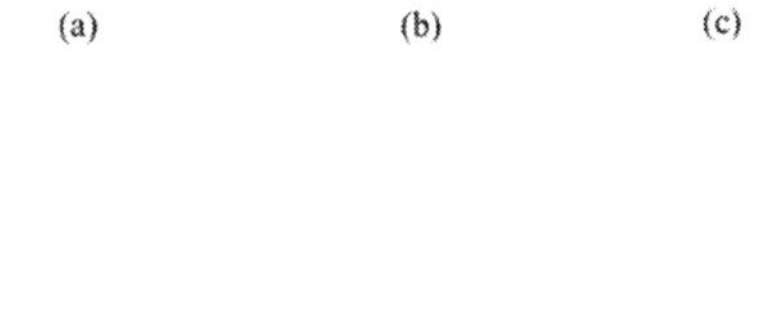

Fig. 5. a. Image of the three spheres configuration a. and the four spheres configuration b. with distances D_1 and D_2, and c. image of the three spheres with the intergranular distances D_2.

3.2 Experimental results

3.2.1 Three spheres configuration

Fig. 6 shows images of the three sphere configuration for each cumulative water volume V_w (D_1 = 8.7 mm and D_2 = 8.3 mm). After forming the two capillary bridges 2 µl each (V_w = 4 µl), adding 2 µl of water in each bridge increases their radii. When the two bridges touch, the coalescence phenomenon occurs to form a single bridge between the 3 spheres (V_w = 20 µl). By increasing the volume of water, a fused bridge lies between the 3 spheres (V_w = 24, 28, 32, 36, 40 and 44 µl). We begin to see the effect of gravity on the coalesced bridge, by exceeding a volume of water V_w = 36 µl.

Fig. 6. Photos of three spheres configuration with different cumulative water volume V_w (D_1 = 8,7 mm et D_2 = 8,3 mm).

For the same intergranular distances, Fig. 7.a presents the capillary force depending on the volume of water injected V_w. This figure shows four different domains. In the first, corresponding to the water volume between 0 and 4 µl, the capillary force greatly increases from 0 to 6.10^{-4} N following the creation of bridges between spheres. The second, corresponding to the volume of water of between 4 to 16 µl, where there is a stabilization of the capillary force with the volume of water injected into the two separate bridges. In the third, during the merger, the capillary force increases sharply (Fig. 6 (V_w = 20 µl)). The peak of the capillary force F_{cap} corresponds to the merger of two bridges and the development of the bridge between the 3 grains (Fig. 6 (V_w = 24 µl)). And the fourth, following the merger of the two bridges corresponds to water volumes of 24 µl to 36 µl, where the capillary force decreases slightly with the increasing volume of the coalesced bridge.

3.2.2 Four spheres configuration

For intergranular distances D_1 = 8.4 mm and D_2 = 8.3 mm, Fig. 7.b shows the capillary force as a function of volume of water V_w. As in the case of the three sphere configuration, changes in capillary force has(a)four domains in which the trend is different. The merger of the bridges is done in 3 steps, the first corresponds to the volume V_w of 30 µl where the two adjacent first bridges merge. When adding a water volume of 4 µl, the merger is made between the first coalesced bridge and the third bridge. And finally, with the addition of 4 µl, fusion is done on the 3 faces of the tetrahedron. At this stage, the capillary force is maximum 18.10^{-4} N (V_w = 38 µl).

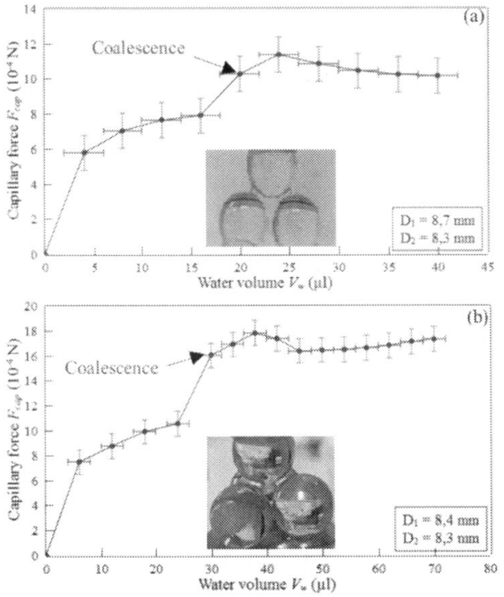

Fig. 7. Evolution of capillary force as function of water volume for **a.** 3 spheres and **b.** 4 spheres.

4 Discussion and conclusion

Wetting in a granular material induced volume changes of a cylindrical sample (Fig. 2). The deformations due to wetting took an abrupt form for a critical water content w_{coll}, due to which the sample instantly lost its mechanical strength. These abrupt deformations correspond to a homogeneous water content in the sample, contrary to the progressive deformations [5]. The observed collapse seems to result from coalescence of the capillary bridges between glass beads.

The increase of the water content, during wetting, impacts the finer grain sample more. This effect is illustrated on Fig. 3.a. We can notice that the fine grains present higher axial deformation (A and ε^f_a). Based on the retention curve (Fig. 1), changes in the suction Δs, between the initial state ($w = w_0$) and collapse $w = w_{coll}$, are shown as function of grain size in Fig. 8. It shows the same trend as the axial deformations (A and ε^f_a), which explains the effect of suction on the collapse.

At local scale, the capillary force increases due to the formation of capillary bridges (Fig. 7). This domain corresponds to low water contents when the capillary cohesion increases with the water content as shown by *Soulié et al.* [11] and *Richefeu et al.* [10]. During wetting, the capillary bridges merge and the liquid phase becomes continuous in the sample. In this state, the capillary force suddenly increases by scoring a jump depending on the volume of water. After bridge fusion, the capillary force decreases slightly (Fig. 7) with higher water volume and seems to be the origin of the observed collapse.

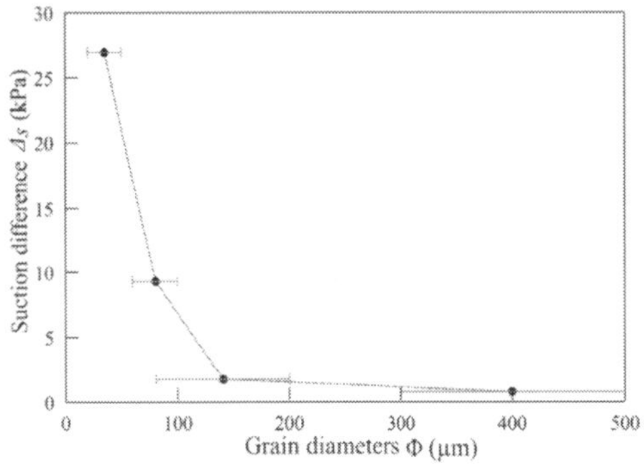

Fig. 8. Evolution of suction difference as a function of grain size.

References

1. T. Ayadat, B. Belouahri, R. Ait Ammar, RFG, **83**, 73-83 (1998)
2. T. Ayadat, S. Ouali, RFG, **86**, 53-56 (1999)
3. L. Barden, A. McGow, K. Collins, Engineering Geology, **7**, 49-60 (1973)
4. P. Delage, Y. J. Cui, P. Antoine, *Proceedings of International Conference on Problematic Soils*, 517-540 (2005)
5. F. Z. El Korchi, F. Jamin, M. El Omari, M. S. El Youssoufi, EJECE, DOI: 10.1080/19648189.2016.1177602 (2016)
6. D. G. Fredlund, A. Xing, Rev. Can. Géotech., **31**, 521-532 (1994)
7. J. P. Gras, Thèse de doctorat, Montpellier (2011)
8. H. Gibbs, J. Bara, Special Technical Publication N° **322**, ASTM, 277-283 (1962)
9. J. K. Mitchell, Wiley Inter Science, Second edition (1993)
10. V. Richefeu, M. S. El Youssoufi, F. Radjaï, Phys. Rev., **73**: 05130 (2006)
11. F. Soulié, M.S. El Youssoufi, F. Cherblanc, C. Saix, EPJ, **21**, 349-357 (2006)
12. R. Tadepalli, H. Rahardjo, D. G. Fredlund, GTJ, **15**, 115-122 (1992)

Electrostatic forces on grains near asteroids and comets

Christine Hartzell[1,*] and *Dylan* Carter[1]

[1]*Department of Aerospace Engineering, University of Maryland, 3178 Martin Hall, College Park, MD USA 20904*

Abstract. Dust on and near the surface of small planetary bodies (e.g. asteroids, the Moon, Mars' moons) is subject to gravity, cohesion and electrostatic forces. Due to the very low gravity on small bodies, the behavior of small dust grains is driven by non-gravitational forces. Recent work by Scheeres et al. has shown that cohesion, specifically van der Waals force, is significant for grains on asteroids. In addition to van der Waals cohesion, dust grains also experience electrostatic forces, arising from their interaction with each other (through tribocharging) and the solar wind plasma (which produces both grain charging and an external electric field). Electrostatic forces influence both the interactions of grains on the surface of small bodies as well as the dynamics of grains in the plasma sheath above the surface. While tribocharging between identical dielectric grains remains poorly understood, we have recently expanded an existing charge transfer model to consider continuous size distributions of grains and are planning an experiment to test the charge predictions produced. Additionally, we will present predictions of the size of dust grains that are capable of detaching from the surface of small bodies.

1 Introduction

Small planetary bodies, such as asteroids and comets, have increasingly become attractive targets for space exploration missions because of their scientific value, potential hazard to Earth, and as a possible human exploration destination (as an intermediate step before Mars). The dust on the surface of small bodies (called regolith) is of particular interests as this is the material with which landing vehicles will interact. Due to the small mass and resulting weak gravitational acceleration of small bodies, it has recently been shown by Scheeres *et al.* that nongravitational forces, specifically cohesion, will drive the behavior of the regolith [1]. Essentially, grains as large as centimeters in size on 500-m scale asteroids will behave like flour does on Earth [1].

In addition to cohesion, grains on small bodies will be influenced by electrostatic forces. Asteroids typically do not have magnetic fields. Thus, the surface of an asteroid interacts directly with the solar wind plasma and the solar UV radiation. The solar wind plasma tends to charge surfaces negatively. However, the solar UV radiation causes the regolith to photoemit, producing a net positive charge on the day-lit surfaces of the asteroid. The positively charged surface attracts electrons, resulting in a region of increased electron density near the surface called the plasma sheath. The plasma sheath serves to shield the charged asteroid surface from the free-stream plasma that has an electric potential of zero. Plasma sheaths are frequently studied experimentally in a variety of plasma physics applications. The charge separation in the plasma sheath causes an electric field, which points away from the surface on the day-side [2, 3]. On the night side, there is a plasma wake due to the higher mass of the ions than the electrons, which also causes an electric field [4]. The plasma environment has been measured at the Moon by the ARTEMIS mission and at comet Churyumov-Gerasimenko by the ROSETTA mission, however, these measurements have been confined to regions outside of the near-surface plasma sheath. Given our undestanding of the plasma environment, regolith grains on the surface of a small body will be charged and exposed to an electric field. The product of the charge and the electric field is an electrostatic force. It has been hypothesized that if the electrostatic force is greater than cohesion and gravity, it may cause dust to detach from the surface [5, 6]. Recent work by Zimmerman *et al.* and Wang *et al.* have independently shown (computationally and experimentally, respectively) shown that grain-scale interactions with the plasma electrons may cause the development of significantly larger electric fields than predicted considering the interaction of the plasma with a flat plate [7, 8]. Additionally, if grains are released from the surface due to other causes (e.g. micrometeoroid bombardment or spacecraft activities), their trajectories will be influenced by the electrostatic force while the grain is within the plasma sheath (i.e. within about 10m-15m of the surface). Electrostatic dust lofting has been observed experimentally in the lab [6, 9], but there have been no conclusive observations of lofted dust *in situ* at small planetary bodies.

In addition to grain charging due to interactions of the surface of an airless body with the solar wind plasma, grains may also charge during spacecraft activities via tri-

*e-mail: hartzell@umd.edu

bocharging; the exchange of charge when grains collide or slide against one another. Planetary regolith grains are typically dielectric. While it is known that tribocharging between dielectric grains of the same material occurs, there is currently no prevailing theory of the physical underpinnings of this charge exchange [10]. Currently, the leading theory of same-material dielectric tribocharging is that the charge exchange is due to electrons transitioning from high energy states on one grain to low energy states on the second grain [11]. Given this physical mechanism of charge exchange, it is possible to create semi-analytical predictions of the level of charging in a mixture of grains. By improving our understanding of the dielectric same-material charging, we will be able to predict the level of grain charging induced by spacecraft activity and the subsequent interaction of grains with the local plasma environment.

2 Grain Lofting

Grains will detach from the surface of an airless body if the forces holding the grains to the surface are less than the forces acting away from the surface. For small bodies, the forces working to hold grains on the surface are gravity and cohesion, although the gravitational acceleration is not likely to be directly normal to the surface given the highly aspherical shape of most asteroids. Acting to detach the grain from the surface is the centripetal acceleration (which can be large for fast spinning bodies), electrostatic forces, and, under certain circumstances, gas drag. It has recently been found that some asteroids experience outgassing, in addition to the known outgassing of comets [12]. Figure 1 shows the relative magnitude of these forces as a function of grain size, considering a location at the far end of the large lobe on the comet 67P/Churyumov-Gerasimenko.

In Figure 1, the gravitational acceleration is calculated using a bi-ellipsoid shape model. The bi-ellipsoid approximation consists of a 2.5x2.5x2 km 'head' perched on a 4.1x3.2x1.3 'body'. The comet is assumed to rotate with a period of 12.4hrs about a rotation axis through the comet's neck. The gas drag force on grains is modeled as:

$$F_d = 0.5\pi r_d^2 \rho C_d V_{rel}^2 \quad (1)$$

where a is the grain radius, ρ is the gas density, C_d is the drag coefficient (assumed to be 2, as in [13])), and V_{rel} is the relative velocity of the gas. For these calculations, the gas density is calculated assuming water molecules with a number density of 10^{17} m^{-3}. V_{rel} is assumed to be 300 m/s. The density and velocity numbers used are conservative when compared to the coma in order to allow us to explore other regions about the body. The drag force is assumed to act normal to the surface. The drag force may be neglected for most asteroids.

The electrostatic force is the product of the grain charge and the local electric field strength. We assume that the grain potential is 5 V and the local electric field is 5 N/C. These values are representative of the environment that we expect on a sunlit airless body, not considering any

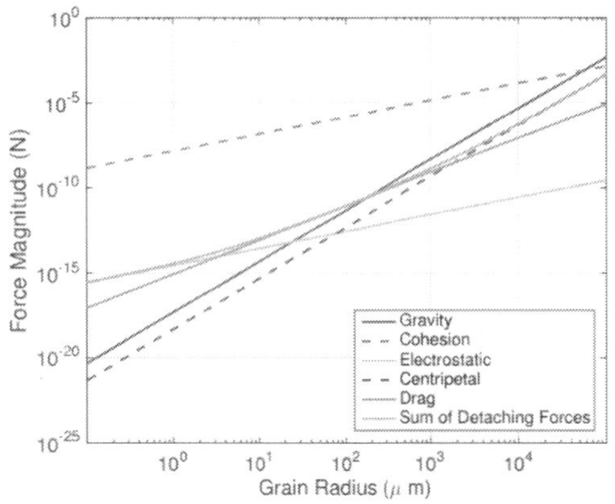

Figure 1. Relative magnitude of forces as a function of grain size on the surface of comet 67P/Churyumov-Gerasimenko. The gravitational acceleration is calculated on the large lobe at end farthest from the neck of the comet. Note that gravity and cohesion act to hold the grain on the surface, but centripetal acceleration, the electrostatic force and gas drag act to detach the grain.

tribocharging [14]. Local shadowing or the presence of a bow shock will change the local plasma environment.

The cohesion is calculated as:

$$F_{co} = \frac{DS^2}{48B^2}\left(\frac{r_d}{2}\right) \quad (2)$$

based on the work by [1]. D is the Hamaker constant (assumed to be 4.3×10^{-20} J, as predicted for lunar regolith [15]), B is the diameter of an oxygen ion (1.32×10^{-10} m), and S is a nondimensional measure of the cleanliness of the regolith (estimated to be 0.75 in this work [15]).

Figure 1 shows that, in the absence of cohesion, grains smaller than 10s of centimeters would be likely to spontaneously detach from the surface of the comet. We note that if the grain charge is increased beyond 5V through tribocharging, then it may be easier to detach grains due to electrostatic forces, called electrostatic lofting.

3 Tribocharging of Dielectric Same-Material Grains

Lacks and Levandovsky previously proposed a theory for the triboelectric charging of dielectric granular mixtures [11]. In their theory, dielectric charging between grains of the same material occurs due to the transfer of electrons from high energy states on one grain to low energy states on a second grain. Additionally, only one electron was transferred between grains during collision. However, the number of electrons transferred should be dependent on the surface area of the contact between the grains. We have modified the theory of Lacks and Levandovsky to include this area-dependent charge transfer term and have also generalized the theory to arbitrary size distributions (rather than purely bidisperse mixtures) [16]. With these modifications to the theory and considering a bidisperse mixture, we find that average charge on the larger grains is given by:

$$Q* = \frac{(1 - sr^2)msr^2(m - d)}{(msr^2 + d)(dsr^2 + m)} \tag{3}$$

Note that $Q*$ is non-dimensionalized by the charge of the high energy electrons at the initial state. m is the ratio of the total mass of the small grains divided by the total mass of the large grains ($m = m_2/m_1$), d is the ratio of the diameter of the small grains to the diameter of the large grains (always less than 1), $r = \left(2/(s(1 + d^3))\right)^{0.2}$ and $s = 1/4(1 + d)^2$.

A comparison of the resulting charge as a function of the grain size ratio d and the mixture mass ratio m is given in Figure 2. Figure 2 also shows the charging predicted by the Lacks and Levandovsky model for the same granular mixtures. Note that when the area-dependent charging is included, the predicted magnitude of charging decreases. Additionally, we note that for certain mixtures, the average charge on the large grains becomes negative. Prior experimentation has shown that large grains tend to charge positively, as is predicted by our model in most mixtures. However, there has not been a comprehensive experimental investigation of grain charging as a function of mixture properties, thus observation bias may have prevented detection of this phenomenon. We are currently developing an experiment to test the predictions shown in Figure 2 and described in [16].

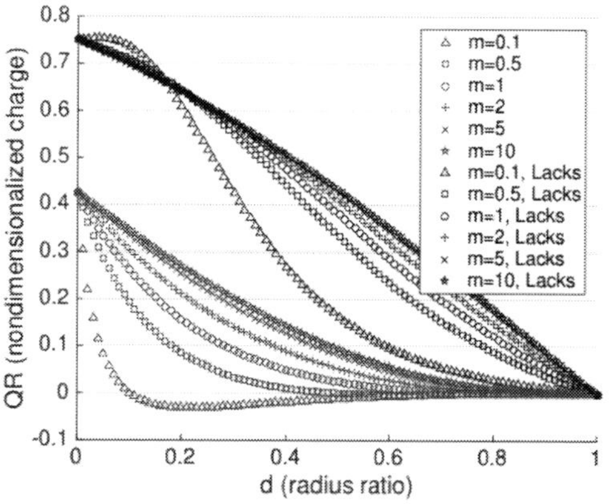

Figure 2. Nondimensionalized average charge on large grains of a bidisperse granular mixture as a function of the mixture mass ratio (total mass of small grains to total mass of large grains) and size ratio (small to large) predicted by the model by Carter and Hartzell [16] and compared to Lacks and Levandovsky [11]. Charge is nondimensionalized by the number of electrons initially in the high energy state on a single large grain.

The derivation of the average grain charge described here (and discussed in detail in [16]) is based on the theory that tribocharging between identical dielectric materials occurs when electrons trapped in high energy states on a grain are transferred to lower energy states on another grain. However, recent work by Waitukaitis *et al.*[17] measured high energy electrons through thermoluminescence and detected fewer high energy electrons than necessary to explain the tribocharging observed, weakening this hypothesized mechanism of triboelectric charging. Our model adds area-dependent charge transfer to the high energy electron transfer tribocharging model. The addition of area-dependent charge transfer changes predictions of the average polarity of tribocharged grains (specifically, it predicts that, in certain mixtures, large grains will charge negatively). It is significantly easier to experimentally measure the polarity of the net charge on a grain, as opposed to accurately measuring the exact charge on a grain. Thus, the model presented here provides a relatively simple test of the high energy electron transfer mechanism of tribocharging. If the predicted polarity is not observed experimentally, it will cast additional doubt on the hypothesis of high energy electron transfer mechanism of tribocharging. We are currently building an experiment to test this model.

Baytekin *et al.* have experimentally demonstrated that charge exchange between two polymer plates brought into contact occurs in patches (i.e. there are patches of positive and negative charge at scales of hundreds of nanometers across the plates) [18]. We expect that, on nanometer scales, there will be patches of positive and negative charge at the intersection surfaces of tribocharging dielectric regolith grains. Baytekin *et al* also show that, while the net charge on the polymer plates is relatively low, the charge density is higher than expected (1 elementary charge per $10nm^2$) [18]. Note that the surface charge density in our model is currently a variable, although the model aims to predict the net charge on grains. Finally, Baytekin *et al* observed that material was transferred between the plates during tribocharging. While significant for many applications, we plan to experimentally test our tribocharging predictions with grains of the same material. Although there may be some small chemical differences between grains, we do not expect that material transfer will be a significant source of any observed tribocharging in our planned experiment.

4 Conclusion

We have presented a brief review of the forces active on the surface of an airless body, with specific emphasis on electrostatic forces. Figure 1 shows the relative magnitude of these forces for an example location on comet 67P/Churyumov-Gerasimenko. There is significant uncertainty about the electric potential of grains on the surface of an airless body. Additionally, the physics of charge exchange through tribocharging of dielectric grains of the same material is poorly understood. We present results of a theory for dielectric grain tribocharging (Figure 2) which we plan to experimentally test in the coming year. By coupling an improved model of granular tribocharging with our survey of forces on grains, we will be able to predict the possibility of electrostatic dust lofting as a result of spacecraft activities.

Acknowledgements

Thanks to Jessica Sunshine and Tony Farnham for their helpful discussion of the drag forces.

References

[1] D.J. Scheeres, C.M. Hartzell, P. Sánchez, M. Swift, Icarus **210**, 968 (2010)

[2] C. Hartzell, D. Scheeres, Journal of Geophysical Research: Planets **118**, 116 (2013)

[3] T. Nitter, O. Havnes, F. Melandsø, Journal of Geophysical Research **103**, 6605 (1998)

[4] M. Zimmerman, W. Farrell, T. Stubbs, Icarus **226**, 992 (2013)

[5] C. Hartzell, D. Scheeres, Planetary and Space Sciences **59**, 1758 (2011)

[6] C. Hartzell, X. Wang, D. Scheeres, M. Horányi, Geophysical Research Letters **40**, 1038 (2013)

[7] M. Zimmerman, W. Farrell, C. Hartzell, X. Wang, M. Horányi, D. Hurley, K. Hibbitts, Journal of Geophysical Research: Planets **121** (2016)

[8] X. Wang, J. Schwan, H.W. Hsu, E. Grün, M. Horányi, Geophysical Research Letters **43**, 6103 (2016)

[9] X. Wang, M. Horányi, S. Robertson, Journal of Geophysical Research **114**, A05103 (2009)

[10] T. Matsuyama, E. Šupuk, H. Ahmadian, A. Hassanpour, M. Ghadiri, *Analysis of Tribo-Electric Charging of Spherical Beads Using Distinct Element Method*, in *Powders and Grains 2009, Proceedings of the 6th International Conference on Micromechanics of Granular Media*, edited by M. Nakagawa, S. Luding (American Institute of Physics, 2009), pp. 127–130

[11] D.J. Lacks, A. Levandovsky, Journal of Electrostatics **65**, 107 (2007)

[12] D. Jewitt, The Astronomical Journal **144** (2012)

[13] V. Tenishev, M. R.Combi, M. Rubin, The Astrophysical Journal **734** (2011)

[14] C. Hartzell, Ph.D. thesis, University of Colorado at Boulder (2012)

[15] H. Perko, J. Nelson, W. Sadeh, Journal of Geotechnical and Geoenvironmental Engineering **127**, 371 (2001)

[16] D. Carter, C. Hartzell, Physical Review E **95**, 012901 (2017)

[17] S.R. Waitukaitis, V. Lee, J.M. Pierson, S.L. Forman, H.M. Jaeger, Physical Review Letters **112**, 218001 (2014)

[18] H. Baytekin, A. Patashinski, M. Branicki, B. Baytekin, S. Soh, A. Grzybowski, Science **333**, 308 (2011)

Nano-granular texture of cement hydrates

Katerina Ioannidou[1,2,*], *Franz-Josef* Ulm[2], *Pierre* Levitz[3], *Emanuela* Del Gado[4], and *Roland J.-M.* Pellenq[1,2,5,**]

[1] *Multiscale Material Science for Energy and Environment, MIT - CNRS Joint Laboratory at Massachusetts Institute of Technology, Cambridge, Massachusetts 02139, USA*

[2] *Department of Civil and Environmental Engineering, Massachusetts Institute of Technology, Cambridge, Massachusetts 02139, USA*

[3] *PHENIX Laboratory, CNRS and University Pierre et Marie Curie, 75252 PARIS cedex 5, France*

[4] *Department of Physics and Institute for Soft Matter Synthesis and Metrology, Georgetown University, Washington, DC 20057, USA*

[5] *CINaM, CNRS and Aix-Marseille University, Campus de Luminy, 13288 Marseille Cedex 09, France*

Abstract. Mechanical behavior of concrete crucially depends on cement hydrates, the "glue" of cement. The design of high performance and more environmentally friendly cements demands a deeper understanding of the formation of the multiscale structure of cement hydrates, when they precipitate and densify. We investigate the precipitation and setting of nano-grains of cement hydrates using a combination of Monte Carlo and Molecular Dynamics numerical simulations and study their texture from nano up to the micron scale. We characterize the texture of cement hydrates using the local volume fraction distribution, the pore size distribution, the scattering intensity and the chord length distribution and we compare them with experiments. Our nano-granular model provides cement structure with realistic texture and mechanics and can be further used to investigate degradation mechanisms.

1 Introduction

Concrete is the most used manufactured material as it is the cheapest and durable, and has outstanding mechanical performances. Cement is the binder of concrete and calcium-silicate-hydrates (C-S-H) is the main hydration product of cement which acts as the glue of the material. Understanding the formation of the multiscale structure of C-S-H is important for controlling the macroscopic properties of cement upon chemical modifications.

When water is mixed with cement powder, cement begins to dissolve. Upon saturation of the ionic solution, C-S-H precipitates in the form of nano-grains. C-S-H is a nonstoichiometric hydration product with calcium to silicon (C/S) ratio ranging from 1 to 2.2 depending on the composition of the cement powder [1]. The inner structure of C–S–H nano-grains is complex, resembling that of Tobermorite minerals at low C/S and that of disordered glasses at high C/S. Recent advances of atomistic modeling of C-S-H have revealed characteristic of the structure and mechanical properties of the nano-grains [2]. Moreover they have provided a first description of the interaction potentials between such grains. However cement is a multiscale material and its macroscopic mechanical response relies not only on the chemistry of the nano-grains but also on the mesoscale structure that they form upon aggregation.

The setting, during which cement is transformed from a liquid paste to a soft and finally hard solid, is intimately connected to the hydration process, when C–S–H grains precipitate [3]. The complicated evolution with time indicates that C–S–H can have very different cohesion and space filling properties, not only due to initially different chemical composition but also due to the interplay between aggregation and arrested kinetics through the hydration and setting process [4, 5].

Here, we present a recent model for C–S–H grains precipitation and aggregation that is based on the potentials of mean field extracted at the atomistic scale and accounts for the structure and the mechanical response of the material all the way up to the micron scale. We characterize the nano-granular structure in terms of local volume fractions, pore size distribution, scattering intensity and chord length distributions.

2 Methods

2.1 Nano-granular model for cement hydrates

A cement hydrate is described as a spherical nano-grain of polydisperse size between 3.78 to 9.2 nm. These units (grains) have been reported in Small Angle Neutron Scattering (SANS) experiments [6, 7]. The nano-grains interact with short range attractive potential, prototypical of cement hydrates [8]. Recently, a group of such interaction potential for cement hydration has been investigated [5].

*e-mail: hekate@mit.edu
**e-mail: pellenq@mit.edu

Here, we investigate cement at late stages of cement hydration when the effective interaction is purely attractive and described with a generalized Lennard-Jones potential

$$V(r) = 6\epsilon\left[\left(\frac{\sigma}{r}\right)^{2\gamma} - \left(\frac{\sigma}{r}\right)^{\gamma}\right], \quad (1)$$

where r is the inter-particle distance, ϵ is the well depth between two particles with diameter σ and we have fixed the exponent to $\gamma = 12$ [4].

The precipitation of nano-grains during cement hydration is modelled by a combination of Molecular Dynamics with Grand Canonical Monte Carlo (GCMC) simulations. GCMC accounts for the interplay between the chemistry, in specific the free energy gain associated to the production of one new cement hydrate, and the attractive interactions between the nanograins. This mechanism drives the aggregation of grains and densification of the system. This non-equilibrium scheme can capture several features of the kinetics of cement hydration [4, 9].

A GCMC cycle consists of N_{MC} attempts of particle insertion or deletion followed by $N_{MD} = 100$ MD steps in NVT ensemble. $R = N_{MC}/(N_{MD} \cdot L^3 \cdot \delta t)$ where L is the length of the simulation box and $\delta t = 0.0025\sqrt{m\sigma^2/\epsilon}$, is the rate of hydrate production and in this work was $R = 25 \cdot 10^{-9}\delta t^{-1}nm^{-3}$. Simulations with higher values of R bring the system further out of equilibrium [4].Time was measured by MD unit time $\sqrt{m\sigma^2/\epsilon}$, the reduced temperature was $T = 0.15$ and the chemical potential $\mu = -1$ in Lennard-Jones units. The chemical potential is chosen to promote densification [5]. The simulations have been performed in box size 585.54 nm with particles up to 610000.

Averages have been performed over at least 3 independently generated samples and for the error bars of the results shown we used the standard deviation of the sample-to-sample fluctuations. The error bars are equal to the symbol size.

2.2 Characterization of the texture

The mesoscale structure obtained from the precipitation simulations is characterized using local volume fractions, the pore size distribution, scattering intensity and the chord length distribution.

The local volume fraction η_{local} is the volume occupied in the neighborhood of one cement hydrate and is computed in a spherical region of radius 35 nm around each spherical grain. The radius of a grain is defined by the equilibrium distance of two grains $r_i = \sqrt[12]{2}\sigma_i$ from Eq.1.

The pore size distribution is computed using the technique of Ref. [10] by the insertion of a probe sphere in a point k of the pore. The size of the pore at point k is the diameter R of the largest sphere that contains k and does not overlap with the solid grains. We construct a finite grid and compute the diameter of the largest sphere at each grid point. The pore size distribution is evaluated via the cumulative histogram of the probability that a point in the pore volume belongs in a sphere with diameter R greater than or equal of a pore width s.

The small angle scattering intensity $I(q)$ was computed from a digitized 3D image of the structure with a voxel size of 0.571 nm, much smaller than the minimum particle diameter 3.78 nm and well below the box size $L = 585.54$ nm. The two point fluctuation autocorrelation function $\eta^2(r)$ was first evaluated and then $I(q)$ is given by

$$I(q) = -\frac{2\pi(\Delta\rho)^2}{q}\frac{d(Re(\tilde{\eta}^2(q)))}{dq} \quad (2)$$

where q is the wave vector, $\Delta\rho$ is the scattering length density contrast and $\tilde{\eta}^2(q)$, the 1D Fourier transform of $\eta^2(r)$ [11].

The pore chord length distribution is obtained by propagating lines with random direction from random origin points in order to accumulate the histogram of the lengths in the pore [12]. Chord length distributions are sensitive to the surface roughness and have been proven effective in characterising porous media [13].

3 Results

The out-of-equilibrium aggregation of the nano-grains results in heterogeneous morphology of the solid at the mesoscale. Capillary pores separate the strands of the dense network of polydisperse cement hydrates. Here, we characterise the structure of the solid and the pore of a nano-granular C-S-H configuration at hardened cement paste.

The total volume fraction η of the system represents the fraction of the simulation box volume filled by spherical nano-grains. Figure 1 shows the local volume fraction distribution η_{local} of cement hydrates of a configuration with $\eta = 0.52$. The distribution displays a distorted Gaussian shape with maximum value at $\eta_{local} = 0.68$. This maxima can be used to estimate the volume fraction of the capillary pores as η/η_{max}. This is in agreement with X-ray tomographic experiments [14]. According to Powers and Brownyard model [15] that links the volume of the capillary pores to the initial water (w/c) to cement ratio our configuration corresponds to a cement paste with $w/c \simeq 0.45$.

Figure 2 shows the distribution of the pore sizes in the same cement configuration. There are nanopores of 0-3 nm, called gel pores in the cement literature, and mesopores of 3-30 nm, called capillary pores. The abundant amount of nanopores is the result of the local packing of grains. The larger pores have a broad distribution in sizes and result from the out-of-equilibrium aggregation of the nano-grains. The mesopores form a connected network which is crucial for the transport properties in cement. The pore size distribution of our configuration has all the characteristic reported by experiments of N_2 adsorption/desorption and NMR [1, 15–17].

SANS is a commonly used experimental technique to characterize the texture of hardened cement pastes [7, 18–20]. Figure 3 shows the SANS data from [21] compared with the scattering intensity computed for our configuration. The two curves have a good agreement especially for most of the range of q values. The plateau at low

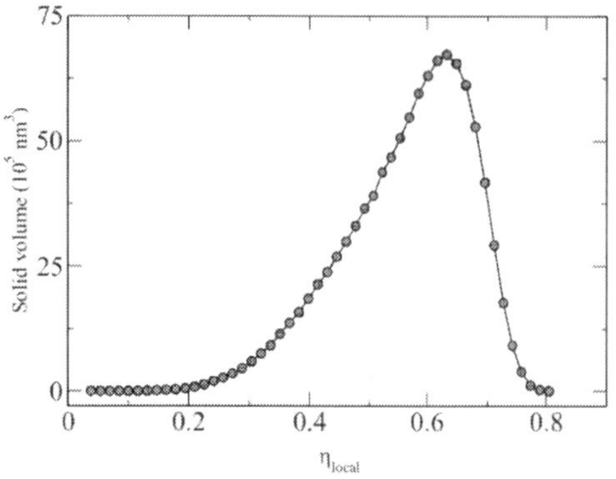

Figure 1: Local volume fraction distribution of hardened cement paste of total volume fraction $\eta = 0.52$.

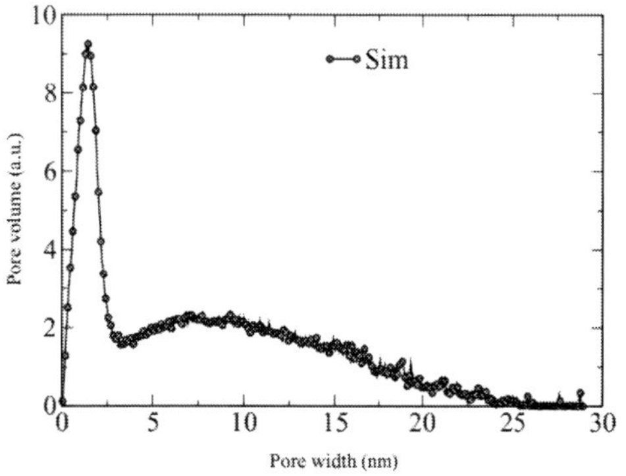

Figure 2: Pore size distribution of cement configurations of total volume fraction $\eta = 0.52$.

wave vector q, where the simulation data are restricted by the system size, is associated to a correlation length of about $\pi/q = 40$ nm, ten times smaller than the box size (585.54 nm). The shift in the intensity axis arises from the fact that we compare simulation data of pure C-S-H with experimental data of Ordinary Portland Cement (OPC) where other hydration products such as ettringite, portlandite etc. are present. At large q values we find a dependence $I(q) \sim q^{-4}$ typical of a Porod regime [22] indicating that the nano-grain surfaces have a sub-nano metric roughness [23]. At smaller q, $I(q)$ follows q^{-3} dependence for more than one order of magnitude in length as reported and discussed in the literature [6, 20, 24]. The low q regime indicates long range spatial correlations and heterogeneities over lengths of 3 – 40 nm pointing on the crucial role of the mesopores at the same length range.

The mesoscale structure of cement hydrates was further characterized by chord lengths that are sensitive to surface roughness. The comparison of the pore chord length distribution for our sample with that obtained from SEM images of a neat OPC paste in [13] is shown in Fig. 4). The pixel dimension in the experiments is assumed to be of $\simeq 0.56$ nm. The comparison shows good agreement

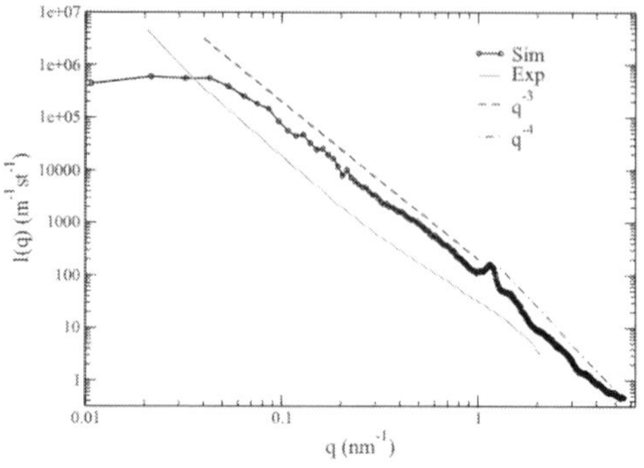

Figure 3: Scattering intensity computed from configurations of total volume fraction $\eta = 0.52$ (Sim), compared with SANS data from [21]. The Porod regime q^{-4} and q^{-3} dependence are depicted by dashed lines.

until lengths of few tenths of nm and the limitation of the simulation system size.

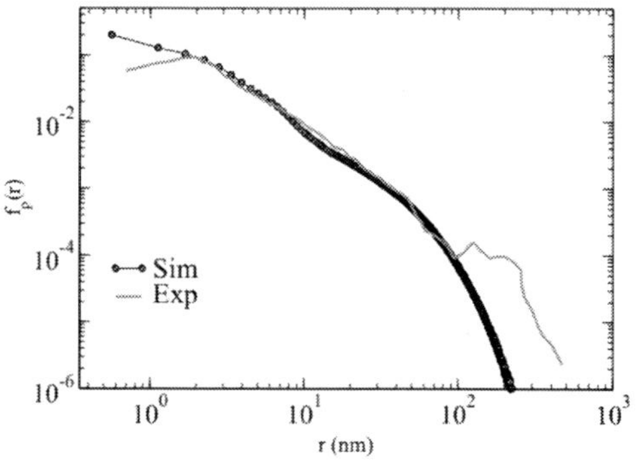

Figure 4: Pore chord length distribution of configurations of total volume fraction $\eta = 0.52$ compared with the pore chord length distribution computed from SEM images of Ref. [13].

4 Conclusions

In this work, we present a nanogranular model for precipitation of cement hydrates that is able to capture the complex structure of solid and pore of hardened cement paste. Cement hydrates precipitate, aggregate and densify to form a solid network of a continuous distribution of packing fractions with a connected mesopore network. An extensive comparison of our simulated structure with experiments reveals the crucial role of structural heterogeneities that arises during precipitation of the nanograins.

Moreover, variations of the local density in hardened cement pastes have been reported in nanoindentation experiments [25] with most of the experiments reporting values > 0.55 for local packing as nanoindentation cannot detect the softer areas. Here we find that areas of even lower packing fraction coexist with the denser ones. However in agreement with nanoindentation experiments the percolated dense areas carry the mechanical load [26].

Finally we note that our model is also able to reproduce the mechanical response of cement as reported by nanoindentation experiments [26]. The realistic description of both structure and mechanics of cement hydrates in our nano-granular model provides a model for investigating degradation mechanisms in cement such as fracture, creep, freeze-thawing etc.

Acknowledgements

This work is supported by Schlumberger under the XCEM project and by the CSHub at MIT. EDG acknowledges support from SNSF (Grant No. PP00P2_150738) and the National Science Foundation (Grant No. NSF PHY11-25915). This work has been carried out within the framework of the ICoME2 Labex (ANR-11-LABX-0053) and the A*MIDEX projects (ANR-11-IDEX-0001-02) co-funded by the French program "Investissements d'Avenir" which is managed by the ANR, the French National Research Agency.

References

[1] H.F. Taylor, Cement chemistry (Thomas Telford, 1997)
[2] R.J.M. Pellenq, A. Kushima, R. Shahsavari, K.J. Van Vliet, M.J. Buehler, S. Yip, F.J. Ulm, Proc Natl Acad Sci USA. **106**, 16102 (2009)
[3] D. Lootens, P. Hébraud, E. Lécolier, H. Van Damme, Oil Gas Sci. Technol. **59**, 31 (2004)
[4] K. Ioannidou, R.J.M. Pellenq, E. Del Gado, Soft Matter **10**, 1121 (2014)
[5] K. Ioannidou, M. Kanduc, L. Li, D. Frenkel, J. Dobnikar, E. Del Gado, Nature Communications **7**, 12106 (2016)
[6] A.J. Allen, J.J. Thomas, H.M. Jennings, Nat Mater **6**, 311 (2007)
[7] W.S. Chiang, E. Fratini, P. Baglioni, D. Liu, S.H. Chen, J. Phys. Chem. C **116**, 5055 (2012)
[8] R.J.M. Pellenq, H. Van Damme, MRS Bull **29**, 319 (2004)
[9] E. Del Gado, K. Ioannidou, E. Masoero, A. Baronnet, R.M. Pellenq, F.J. Ulm, S. Yip, Eur Phys J-Spec Top **223**, 2285 (2014)
[10] S. Bhattacharya, K.E. Gubbins, Langmuir **22**, 7726 (2006)
[11] P. Levitz, D. Tchoubar, J Physique I **2**, 771 (1992)
[12] J. Mering, D. Tchoubar, J Appl Crystallogr **1**, 153 (1968)
[13] P. Levitz, Adv Colloid Interface Sci **76**, 71 (1998)
[14] J.C. da Silva, P. Trtik, A. Diaz, M. Holler, M. Guizar-Sicairos, J.r. Raabe, O. Bunk, A. Menzel, Langmuir **31**, 3779 (2015)
[15] T. Powers, T. Brownyard, PCA Bulletin **22** (1948)
[16] R.S. Mikhail, L.E. Copeland, S. Brunauer, Can J Chem **42**, 426 (1964)
[17] R. Valckenborg, L. Pel, K. Kopinga, J Phys D: Appl Phys **35**, 249 (2002)
[18] A.J. Allen, J.C. McLaughlin, D.A. Neumann, R.A. Livingston, J Mater Res **19**, 3242 (2004)
[19] E. Fratini, A. Faraone, F. Ridi, S.H. Chen, P. Baglioni, J Phys Chem C **117**, 7358 (2013)
[20] S. Bae, R. Taylor, D. Shapiro, P. Denes, J. Joseph, R. Celestre, S. Marchesini, H. Padmore, T. Tyliszczak, T. Warwick et al., J Am Ceram Soc **98**, 4090– (2015)
[21] K.J. Krakowiak, J.J. Thomas, S. Musso, S. James, A.T. Akono, F.J. Ulm, Cem Concr Res **67**, 103 (2015)
[22] A. Guinier, G. Fournet, Small-Angle Scattering of X-rays (Wiley, 1955)
[23] S. Brisard, P. Levitz, Phys Rev E **87**, 013305 (2013)
[24] S. Brisard, R.S. Chae, I. Bihannic, L. Michot, P. Guttmann, J. Thieme, G. Schneider, P.J. Monteiro, P. Levitz, Am Mineral **97**, 480 (2012)
[25] G. Constantinides, F.J. Ulm, J Mech Phys Solids **55**, 64 (2007)
[26] K. Ioannidou, K.J. Krakowiak, M. Bauchy, C.G. Hoover, E. Masoero, S. Yip, F.J. Ulm, P. Levitz, R.J.M. Pellenq, E. Del Gado, Proceedings of the National Academy of Sciences **113**, 2029 (2016)

Structure of hexapod 3D packings: understanding the global stability from the local organization

Jonathan Barés[1,2,⋆], *Yuchen* Zhao[2], *Mathieu* Renouf[1], *Karola* Dierichs[3], and *Robert* Behringer[2]

[1] *Laboratoire de Mécanique et Génie Civil, Université de Montpellier, CNRS, Montpellier, France*
[2] *Department of Physics & Center for Nonlinear and Complex Systems, Duke University, Durham, North Carolina 27708, USA*
[3] *Institute for Computational Design, University of Stuttgart, Stuttgart, Germany*

Abstract.
Aggregates of non-convex particles have shown to be particularly stable which makes them good candidates to design new lightweight and reversible structures. However, few is known about the fundamental reason of their stability. In this paper we presents a novel experimental method to investigate the local structure of piles made of hexapod particles. This method is based on X-ray scanning and on an accurate homemade particle detection code. It permits to get the position and orientation of each particle as well as to detect their contact points. Measurement of the coordination numbers, statistics of the contact positions and local density evaluation for different packing configurations show a good agreement with the previous studies carried out at the global scale and permits to explain the main local mechanisms leading to stable structures.

1 Introduction

The use of aggregates made from non-convex particles or particles with anisotropic contact laws is an emerging area for both the research in physics [1–7] and the design of functionalized materials by tuning local properties of the grains [8–11]. This new field in granular matter science has already shown to be extremely promising for future lightweight and reversible architecture [12]. It permits to design easy-to-shape and remove strong structures. Indeed the shape of the grains [2, 13, 14] and the contact anisotropy [8] have been recently used to help crossing the jamming transition and to increase the stability/rigidity of granular systems. Different particle shapes, such as long and thin rods [15], staples [4], Z-shaped [6] and star-shaped [2, 5] particles have been shown to bring additional rigidity to 3D packings, compared with spherical shapes [16].

Two of the most striking characteristics of non-convex designed granulates, in terms of the granular system and possible design applications are their ability to form vertical structures with a 90° angle of repose, and to sustain small tilting or loading perturbations [2]. Even if this tremendous stability illustrated in the fig.1-A has already been evidenced in several experiments [2, 5, 17, 18] and if packing of such particles has been widely studied [19–21], few is know about the fundamental reasons of why non convex particle piles are much more stable than bead columns [16, 22, 23].

In order to understand the origin of this rigidity, in this paper, we investigate the local organization of columns made of star-shape or hexapod particles (see fig.1-A) using X-ray CT-scan. We have performed experiments by randomly pouring identical star-shape particles (2.5 to 10cm wide) into hollow cylinders, and removing the cylinders to get stable star columns, for different preparation conditions. Then, by mean of a X-ray CT-scan and a home made post-processing code, we have reconstructed the full structure of the packing: (*i*) position, (*ii*) orientation and (*iii*) contact points of each particles with a very good accuracy (0.1mm). We have analysed the coordination number and the distributions of the contact point positions.

In a first part we present the experimental set-up to prepare and scan hexapod piles. In a second part the method to measure the position, orientation and contact points for each particle is explained. Then, analysis tools are detailed before presenting the results.

2 Experimental method

The cylindrical columns have been prepared in two different ways and we used three kinds of hexapods particles as shown in fig.1-A. These last consisted of six orthogonal beams with square cross section, that tapered from a thickness of 2mm at the center of the particle to 1mm at the tips. The end-to-end size, L, varied between 2.5cm, 5cm and 10cm. They are made of laser-sintered white nylon PA2200, which has a friction coefficient 1.0 ± 0.3. In this paper we will only talk about 2.5cm, 5cm star packings because for 10cm stars the number of particles in the structure is not large enough to give statistically meaningful results.

⋆e-mail: jb@jonathan-bares.eu

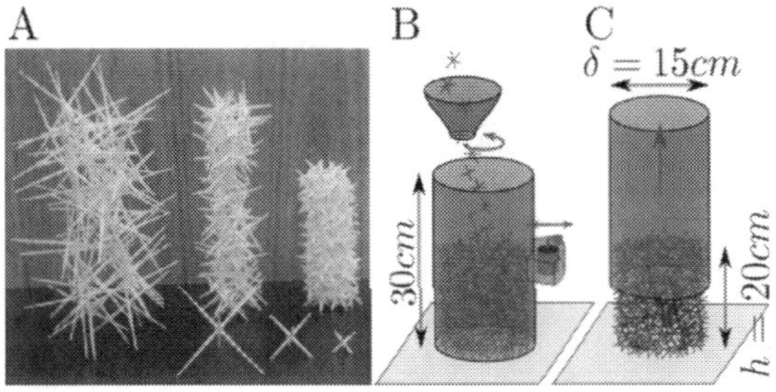

Figure 1. A: Columns of 2.5cm, 5cm and 10cm wide hexapods made of laser sintered nylon PA2200. Hexapods can form very stable structures capable of sustaining their own weight, tilt and external loading [2]. B: Cartoon where the non-convex particles are dropped one-by-one from an overhead hooper in a cylinder of diameter 15cm. The system can be vibrated by an eccentric, driven by a DC motor which is attached to the cylinder. C: The cylinder is then carefully and slowly removed by sliding lifting it vertically. Hexapods form a stable pile of height ~ 20cm.

The protocol to make particle piles consisted in filling thin-walled PVC tube with particles by dropping them one-by-one at a steady rate from an overhead hopper as shown in fig.1-B. The inner diameter of the PVC tube was 15cm. In some cases, during the filling process, the system could be vibrated by an eccentric that was attached to the outside of the tube as presented in the fig.1-B. This has been shown to increase the pile stability [2]. We tuned the speed of the motor to create a ~ 1700m/s^2 cyclic acceleration of the 16g eccentric. As shown in fig.1-C, the tube was then carefully and slowly removed by lifting it vertically. Friction between the particles and the cylinder was low enough not to significantly perturb particles inside the cylinder. This protocol is similar to what has been used by [2].

The hexapod column was then set inside a X-ray microCT scanner Nikon XT-H225 and scan with a X-ray source of 185kV and 235μA. Raw data were then post-processed using Nikon's Feldkamp [24] cone based CT algorithm to get 2000 density slices of size 2000×2000, with a spacial resolution of 0.1mm and a 16bits digital resolution. A slice in the middle of the column is shown in fig.2-A.

3 Data post-processing

Collecting the density slices we then formed a full 3D density matrix M_0 (see cartoon for two particles in fig.2-B) which we post-processed with a homemade Python algorithm to extract the position x, y, z and the orientation θ_1, θ_2, θ_3 of each particle. Because of the large dimension of the density matrix we first resize it to M a $400 \times 400 \times 400$ binarized matrix choosing a 60% density threshold (100% is the highest density, 0% the lowest). A virtual smaller binary matrix $m(\theta_1, \theta_2, \theta_3)$ is then formed. As presented in fig.2-C this matrix is null everywhere except for the pattern of an hexapod centered in the middle of the matrix and oriented along three axis: θ_1, θ_2, θ_3.

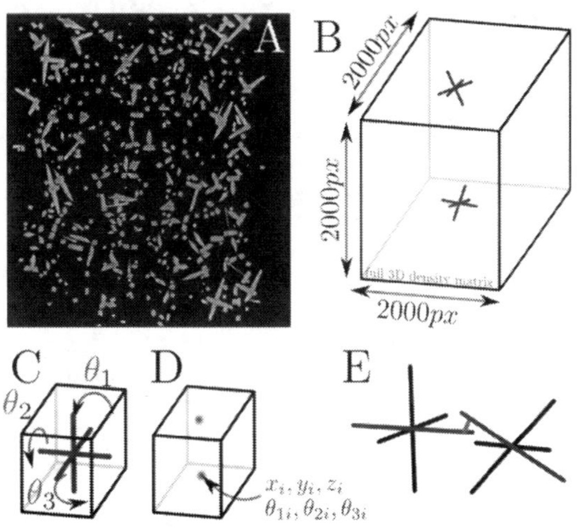

Figure 2. A: Slice of a density matrix in the middle of the hexapod pile. Blue is for low density (surrounding air) while red is for high density (particle nylon). B: Cartoon of a density matrix with only 2 hexapods. Matrices are 16bits $2000 \times 2000 \times 2000$ voxels. C: Cartoon of the convolution density matrix. This matrix is used for hexapod detection by convolving it with the density matrix (B) for particle with a certain orientation: θ_1, θ_2, θ_3. D: Synthesized convolved matrix. Blue high value islands show position and give orientation of the particles in the density matrix (B). See text for details. E: Cartoon of two neighbour particles. The red segment show the shortest distance between the two blue hexapod arms.

Taking advantages of the particle symmetry, for all the possible families of direction with θ_i going from 0° to 90° with 2° steps (91125 families), we convoluted the binarized density matrix M with $m(\theta_1, \theta_2, \theta_3)$ to get a convolution matrix $C(\theta_1, \theta_2, \theta_3)$ of the same size as M. This computation took 80h parallelized on a 60 nodes cluster. Then, a maxima matrix D of the same size as M is formed. For each coordinate i, j, k it took the maximum convolution value $C_{i,j,k}(\theta_1, \theta_2, \theta_3)$ over the 91125 possible orientations and keep the record of the corresponding $(\theta_1, \theta_2, \theta_3)$ family. As presented in the cartoon fig.2-D, this matrix shows high convolution values 3D islands at the position of the hexapod center. Using this property, a 3D local maxima research algorithm applied on the matrix D gave the position of each hexapod x_i, y_i, z_i and the record of their orientation gave θ_{1i}, θ_{2i}, θ_{3i}.

However, this measurement is not accurate enough since it has been done on a density matrix whose size has been divided by 5 (0.5mm accuracy on the particle center position) and orientation is only 2° accurate. To improve the measurement, for each detected particle center a sub-matrix of the full density matrix M_0 is extracted and binarized. A larger virtual binary matrix is made and a Powell optimization algorithm was implemented to improve the match between both matrices and give more accurate position (0.1mm accuracy) and orientation (0.5° accuracy) values. This part of the algorithm is reminiscent of the method used by [25].

DOI: 10.1051/epjconf/201714006021

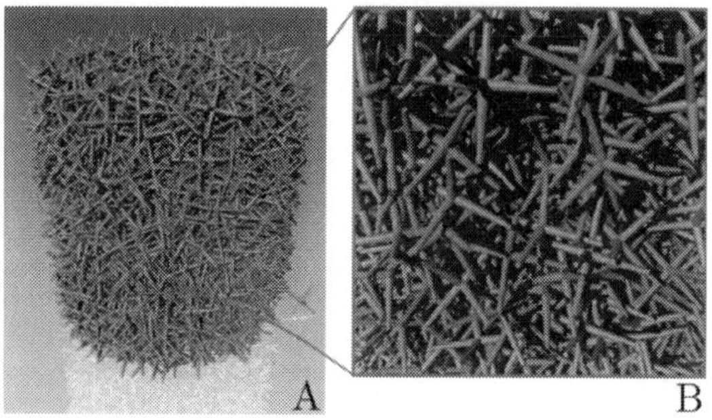

Figure 3. A: Full reconstruction of an hexapod packing from the X-ray scan of a non vibrated column of 2.5cm stars. The drawing has been generated using Vapory, the Pov-Ray wrapper in Python, knowing the hexapod geometry, and the measured position and orientation. Red spheres show the position of the contacts. B: Zoom in the specified area presented of A. A complete movie of the reconstruction for each particle size is given here: https://www.youtube.com/watch?v=2HUUAPet2rs.

Contact between the hexapods arms were finally detected from the particle position and orientation, and the density matrix M. As presented in fig.2-E, for each arm of each particle, the distance between the segment formed by the arm and the the segment formed by each arm of the neighbour particles was computed. If this distance were lower than 4.5mm, a $10\times10\times10$ submatrix (corresponding to 1mm^3) centred around the hypothetical point of contact was extracted from the density matrix M. This submatrix was binarized as previously and the number of connected domains computed. If 2 distinct domains were observed there was no contact. If there was only one large domain the two arms were in contact. This method gives contacts with a 0.1mm threshold accuracy. Further investigations are currently carried out to shows that this threshold is small enough not to detect too much fake contacts. A reconstruction of the 3D packing as well as the position of the contact is presented in fig.3.

4 Results

A coordination number (average number of contact per grain) of 6.58 have been measured for non vibrated 2.5cm particles. This value is larger than the theoretical coordination 6 at the limit of stability, so this explain the stability of such a system. When particles are vibrated, as emphasized by [2], we expect to have a more stable structure. Indeed, we measure a coordination number of 7.13 for such a configuration. For particle with longer arms (5cm) we expect something even more stable [2] which corresponds with the coordination number we measured: 9.18. Hence, this explain the increase is stability for vibrated system and large particle observed by [2] but also quantify this stability by giving the hyperstaticity of the different configurations.

In fig.4-A, we present the probability density function (PDF) of the contact position along a particle arm. For each system configuration, 2.5cm, 5cm particles, vibrated

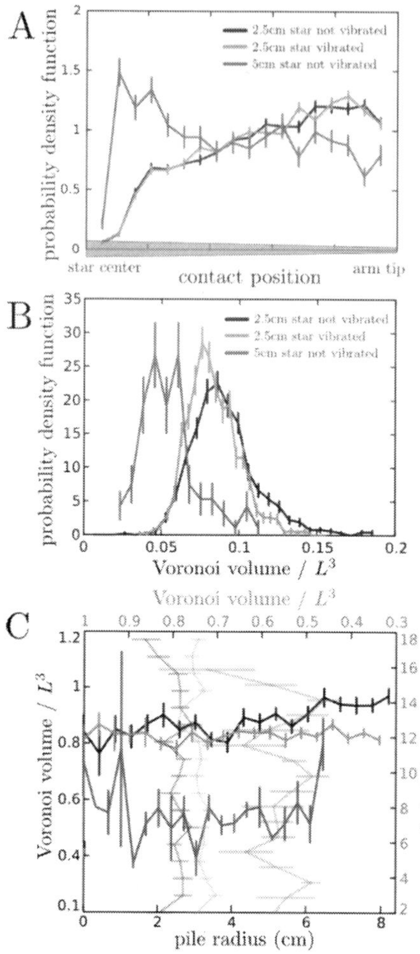

Figure 4. A: Probability density function of the position of a contact along a particle arm for the different system configurations: 2.5cm, 5cm particles, vibrated and non-vibrated. x-axis is rescaled to compare 2.5cm and 5cm stars. Vibration has no effect on the average contact position but the longer the arms, the closer the contact points are from the particle center. B: Probability density function of the particle local density. For each configuration the probability density function of the rescaled Voronoi volume around each star is plotted. The larger the particles or the more vibrated the system, the wider the statistical repartition. C: Averaged scaled Voronoi volume along the radial (horizontal plot) and height (vertical plot) directions for the 3 different packings. The particle density just slightly increases radially. Vertical error-bars give the 95% confidence intervals of the measurements.

and non-vibrated, for each arm of each particle, the distance of the contacts to the particle center has been measured. Then the PDF of this distance rescaled with the arm length L has been computed. We see that in the case of large particles, contacts are homogeneously spread along the whole arm, with an over-representation close to the particle center. On the contrary, for shorter particles the contacts are more concentrated at the arm tip. This explain the more important stability of large particle piles since hexapods are more intricated so more difficult to destabilize.

Fig.4-B present the PDF of the local packing fractions. For each system configuration, from the position coordinate of the particles, we have computed the 3D Voronoi tesselation. The volume of each tile is measured and rescaled with the an effective hexapod volume L^3. The PDF of this quantity is then plotted. For each case it vanishes before 10%, then presents a sharp peak and vanishes more progressively for larger volumes. We remarks that the vibration does not seems to have any significant effect on the statistical repartition of the local particle densities.However, the PDF peaks correspond to a larger scaled volume for 2.5cm particles than for 5cm ones. Indeed, for larger hexapods, the harm aspect ratio is thinner so particle can entangle deeper. It is remarkable to notice that this property corresponds with a more stable system configuration.

Fig.4-A shows the evolution of the average local density measured as the volume of the Voronoi tiles scaled with an effective hexapod volume L^3. This quantity has been averaged vertically and circularly to give the radial evolution of the local particle density for each packing configuration. We see that the pile is slightly denser in the inside and for larger particles but constant for 5cm ones. Averaging along the vertical dimension, we get the vertical evolution of the density. Surprisingly, within the error-bar, there is no variation of the density as a function of the pile height.

5 Conclusions

We presented an experimental method to investigate the local structure of aggregates made from non-convex particles. This method is based on X-ray scanning and on an efficient and 0.1mm accurate homemade particle detection code using parallelized matrix convolutions. This permits to measure the position and orientation of different sized hexapods forming large piles in vibrated and non-vibrated configurations. This local scale information is in good agreement with the macroscopic stability measurement made by [2]. Moreover, contact position statistics and local packing fraction variability provided interesting information about the effect of the pile preparation. However, these experimental results do not permit to detect the exact nature of the contact (point, line or plane) nor to measure the force. Hence, to test the accuracy of our experimental measurement as well as to improve the contact analysis, numerical simulations will be carried out.

Acknowledgements

This work was supported by the Labex NumEv (anr-10-labx-20), the William M. Keck Foundation, by NSF grant DMR1206351 and NASA grant NNX15AD38G and ITASCA Educational Partnership Program.

References

[1] S. Keller, H.M. Jaeger, Granular Matter **18**, 29 (2016)

[2] Y. Zhao, K. Liu, M. Zheng, J. Barés, K. Dierichs, A. Menges, R.P. Behringer, Granular Matter **18**, 24 (2016)

[3] M.Z. Miskin, H.M. Jaeger, Soft Matter **10**, 3708 (2014)

[4] S. V. Franklin, Europhysics Letters **106**, 58004 (2014)

[5] K. Dierichs, A. Menges, Granular Matter **18**, 1 (2016)

[6] K.A. Murphy, N. Reiser, D. Choksy, C.E. Singer, H.M. Jaeger, Granular Matter **18**, 1 (2016)

[7] P. Aejmelaeus-Lindström, J. Willmann, S. Tibbits, F. Gramazio, M. Kohler, Granular Matter **18**, 1 (2016)

[8] M. Cox, D. Wang, J. Barés, R.P. Berhinger, Europhysics Letters (2016)

[9] M. Z. Miskin, H. M. Jaeger, Nature Materials **12**, 326 (2013)

[10] B. Saint-Cyr, J.-Y. Delenne, C. Voivret, F. Radjai, P. Sornay, Physical Review E **84**, 041302 (2011)

[11] E. Brown, A. Nasto, A. G. Athanassiadis, H. M. Jaeger, Physical Review Letters **108**, 108302 (2012)

[12] K. Dierichs, A. Menges, Architectural Design **85**, 86 (2015)

[13] A. G. Athanassiadis, M. Z. Miskin, P. Kaplan, N. Rodenberg, S. H. Lee, J. Merritt, E. Brown, J. Amend, H. Lipson, H. M. Jaeger, Soft Matter **10**, 48 (2014)

[14] K. C. Smith, M. Alam, T. S. Fisher, Physical Review E **82**, 051304 (2010)

[15] M. Trepanier, S. Franklin, Physical Review E **82**, 011308 (2010)

[16] E. Lajeunesse, A. Mangeney-Castelnau, J.P. Vilotte, Physics of Fluids **16**, 2371 (2004)

[17] N. Gravish, S. V. Franklin, D. L. Hu, D. I. Goldman, Physical Review Letters **108**, 208001 (2012)

[18] S. V. Franklin, Physics Today **65**, 70 (2012)

[19] K. Miszta, J. de Graaf, G. Bertoni, D. Dorfs, R. Brescia, S. Marras, L. Ceseracciu, R. Cingolani, R. van Roij, M. Dijkstra, Nature Materials **10**, 872 (2011)

[20] J. de Graaf, R. van Roij, M. Dijkstra, Physical Review Letters **107**, 155501 (2011)

[21] Y. Jiao, F. H. Stillinger, S. Torquato, Physical Review E **79**, 041309 (2009)

[22] G. Lube, H. Huppert, R. Sparks, A. Freundt, Physical Review E **72**, 041301 (2005)

[23] G. Lube, H.E. Huppert, R. S. J. Sparks, M. A. Hallworth, Journal of Fluid Mechanics **508**, 175 (2004)

[24] L.A. Feldkamp, L.C. Davis, J.W. Kress, Journal of the Optical Society of America A **1**, 612 (1984)

[25] M. Neudecker, S. Ulrich, S. Herminghaus, M. Schröter, Physical review letters **111**, 028001 (2013)

DEM GPU studies of industrial scale particle simulations for granular flow civil engineering applications

Patrick Pizette[1,*], *Nicolin* Govender[1,2], *Daniel N.* Wilke[3], and *Nor-Edine* Abriak[1]

[1] *IMT Lille Douai, Univ. Lille, EA 4515 - LGCgE – Laboratoire de Génie Civil et géoEnvironnement, départment Génie Civil & Environnemental, F-59000 Lille, France*
[2] *Research Center Pharmaceutical Engineering, GmbH, Graz , Austria*
[3] *Centre for Asset and Integrity Management, University of Pretoria, Pretoria, 0086, South Africa*

Abstract. The use of the Discrete Element Method (DEM) for industrial civil engineering industrial applications is currently limited due to the computational demands when large numbers of particles are considered. The graphics processing unit (GPU) with its highly parallelized hardware architecture shows potential to enable solution of civil engineering problems using discrete granular approaches. We demonstrate in this study the pratical utility of a validated GPU-enabled DEM modeling environment to simulate industrial scale granular problems. As illustration, the flow discharge of storage silos using 8 and 17 million particles is considered. DEM simulations have been performed to investigate the influence of particle size (equivalent size for the 20/40-mesh gravel) and induced shear stress for two hopper shapes. The preliminary results indicate that the shape of the hopper significantly influences the discharge rates for the same material. Specifically, this work shows that GPU-enabled DEM modeling environments can model industrial scale problems on a single portable computer within a day for 30 seconds of process time.

1 Introduction

Granular materials are one of the most widely used materials in civil engineering. The geometrical properties that include particle shape, particle size distribution, surface roughness (texture) have a significant effect on the packing density of the particles. More importantly these properties significantly affect the macroscopic flow behavior i.e. they influence the mechanical properties that include dissipative and restoring forces between contact pairs.

The Discrete Element Method (DEM) proposed by Cundall and Stack [1] for geotechnical applications, is starting to mature to the degree that numerous and diverse industrial civil engineering applications are being considered [11]. The limiting factor has been the limited number of particles that can be viably simulated for industrial applications. Even though numerous assumptions such as limiting particle shapes to spheres have been employed, the current processing ability of the CPU has limited the advances for industrially relevant simulations.

Recent developments in simulating Discrete Element Models on Graphical Processor Unit (GPU) hardware bring new opportunities to tackle industrial granular applications. In particular, Govender et al [2, 3] have simulated industrial processes that include ball mill simulations and hopper discharge using 16 million particles on a single portable computer allowing new computational performance levels to be achieved. In recent collaborative studies between Mines-Douai (France), CSIR (South Africa) and University of Pretoria (South Africa) the GPU based DEM simulations have been extensively validated for silo discharge and milling simulations using both spherical and polyhedral shaped particles [4, 5, 7]. These studies clearly demonstrated that GPU based DEM can model complex particle shapes in a reasonable computational time when the number of particles are sufficiently large to be practically valid. These studies in particular proposed a systematic comparison between DEM–GPU simulations and experiments for idealized particles (spherical and convex polyhedra) [5, 9]. This allows for systematic validation of developed GPU codes as the particle complexity increases.

As an extension to the previous studies, this study considers the discharge of an industrial storage silo in an industrial concrete facility in Paris, France. This paper demonstrates the pratical utility of the GPU using a validated GPU-enabled DEM simulation environment to solve an industrial scale granular problem within a day for which purposes we selected 8 million particles.

2 Discrete Element Method

2.1 DEM GPU Approach: BlazeDEM3D-GPU Framework

The computational time requirement is the primary limitation on the application of DEM to industrial problems in particular in civil engineering [11].

*e-mail: patrick.pizette@imt-lille-douai.fr

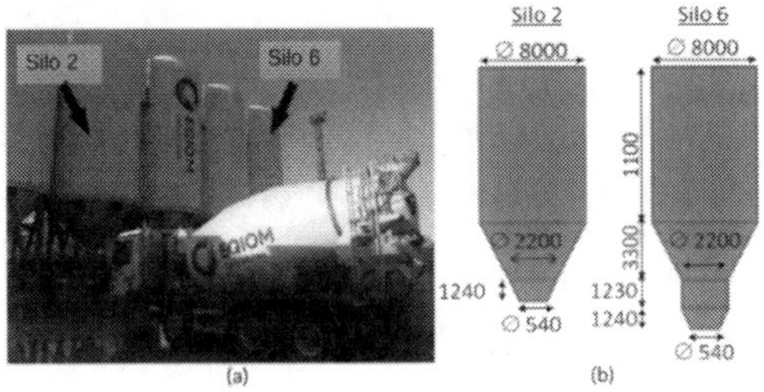

Figure 1. (a) Overview of the concrete central and the schematics of design of two of the storage silos (size in mm)

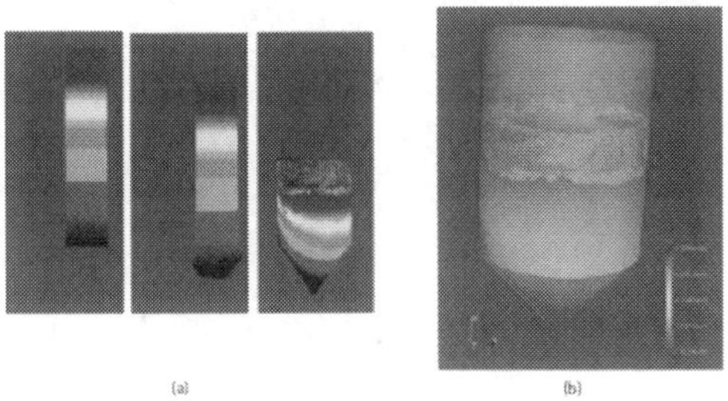

Figure 2. (a) Illustration of filling (color by ID) and (b) silo 2 at the initial packing (Range by ID)

The highly parallelized hardware architecture of GPU (Graphic Processor Unit) can offer new opportunities for industrial applications. In this study, we use the BlazeDEM3D-GPU framework developed by Govender et al. [2, 3, 6]. This numerical framework has been used and experimentally validated with success to simulate with sphere or more complex shape as for example regular polyhedron silo discharge [4, 5, 7] and milling operations [5, 6]. In this paper a laptop with a GTX 980 M GPU and 8GB of DDR5 RAM is used to conduct DEM simulations within a day for 30 seconds of process time using 8 millions of particles.

2.2 Contact Law Model

In this study, a linear spring dashpot model is used. The normal force between two particles in contact is calculated by :

$$F_n = K_n \delta \mathbf{n_{ij}} + C_n \dot{\delta} \mathbf{n_{ij}} \tag{1}$$

where δ , $\dot{\delta}$ and $\mathbf{n_{ij}}$ are respectively the interpenetration distance, the normal relative velocity and the normal contact vector between two particles in contact. K_n is the contact stiffness and C_n is the dissipative term. These two coefficients are determined from the coefficient of restitution (COR) and time duration [6, 10].

The tangential force is computed by a viscous damping model :

$$F_t = -C_t ||V_T|| \frac{\mathbf{V}_T}{||V_T||} \tag{2}$$

where C_t is the tangenial viscous coefficient and $\mathbf{V}_T$ the relative tangential velocity computed at the contact point. This tangential force is limited by the Coulomb criterion. Thus, effective friction coefficients μ_{pp} and μ_{pw} are defined respectively for particle-particle and particle-wall.

Note that a generated rolling resistance torque is added to counteract the relative rotational motion between two contacting particles involving a rolling resistance coefficient μ_{roll}.

3 Industrial scale simulations

3.1 Industrial problematics

To demonstrate the power of a GPU based simulation, we simulate a real-life concrete hopper using around 17 million particles. In particular, we focused on a DEM simulation relating to the new concrete facility of EQIOM located in Paris intra muros in the ecodistrict of Clichy-Batignolles. As shown in Figure 1(a), this concrete facility is composed of several elevated storage silos used to store sand or coarse aggregates (gravel, crushed stone, etc). Concrete is manufactured by first feeding the constituents into a central mixing drum. The required amount by mass of the constituents are fed by gravity.

The hoppers of the silos are composed of two coaxial conical shapes with two angles and a cylindrical extension between the two cones. The cylindrical silos have the same design with a diameter of 8 m and a height of 17 m. This study specifically focusses on silo 2 and 6. The two conical hoppers differ though, as the hopper for silo 6 has a cylindrical extension as depicted in Figure 1(b).

Table 1. Summary of DEM packing characteristics

Case	Particle diameter (mm)	Number of particles
A	20	16 777 216
B	40	8 388 608

3.2 Industrial storage silo modeling

The granular material was modeled as 20 mm or 40 mm diameter spheres for the lower or upper limits of an equivalent size for the 20/40-mesh gravel. As illustrated in Figure 2(a), the initial packing was prepared by two stages: firstly, the particles were generated in square box inside the silo. Secondly the particles fell freely in order to obtain a dense packing of particles. The offset between the box of particles with respect to the silo axis controls the asymmetry of the packing. It was ensured that the kinetic energy of the system was close to zero before extracting the final packing. As shown in the Figure 2(b), due to a slight offset in the initial particle packing position with respect to the silo axis, we obtained a non-symmetrical dense packing.

In this study, two packings A and B were generated with just over 8 million 40 mm diameter particles or just over 17 million 20 mm diameter particles (as tabulated in table 1).

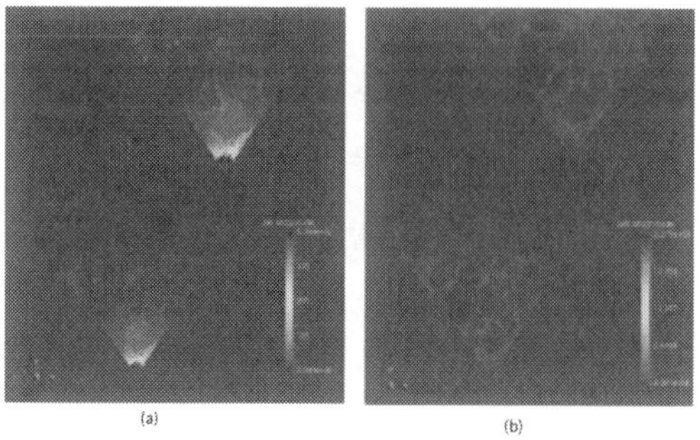

Figure 3. Silo 2 after 15s of discharge time for the particle case B depicting a cross-sectional view of (a) the magnitude of linear velocity (b) the magnitude of angular (magnitude)

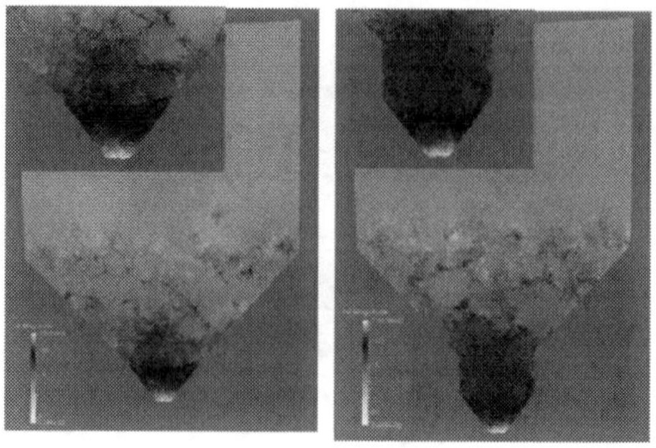

Figure 4. Cross-sectional views of linear velocity fields between the two shaped hoppers for silos 2 and 6 for case B

4 Silo Discharge

4.1 Discharge flow

The discharge is simulated by opening the hatch located at the bottom of last conical hopper. The number of particles amount passing through the orifice may be quantified in terms of number, mass or volume. The particles are systematically removed from the simulation after they have fallen a specified distance of 10 cm. The simulation model parameters are : $COR = 0.45$, $\mu_{pp} = \mu_{pw} = 0.3$, $\mu_{roll} = 0.01$, density = $2.4 g/cm^3$ for a time-step of $1.10^{-4} s$.

The linear and angular magnitude velocity fields after 15s depicted in Figure 3. The maximum velocity and velocity gradients correspond to the zone near the orifice as depicted in Figure 3(a). Above the conical hopper in the silo the flow is slower and more homogeneous than the hopper zone where the flow is dominated by shear behaviour. In particular the cross-sectional linear velocity in the hopper along the radius indicates velocity gradients that result in shear dominated flow between the granular particles. Higher in the hopper form blocks of uniform velocity that indicate domains of uniform flow in the granular material as paths of force chains develop in the material. These uniform domains are also highlighted by the angular velocity field in Figure 3(b)) where domains with higher rotation neighbour domains with lower rotation to develop local shear layers between these domains.

4.2 Hopper flow comparisons between silo 2 and 6

The height of the cylindrical extension is the main difference between the two storage silos considered. The resulting flow patterns is important as the silos are used to store several types of material with different particles sizes. Figure 5 shows the influence of this extension on the flow profile in the hopper for the 40 mm diameter spherical particles for the two silos for the first 30 seconds.

In particular, the flow pattern was followed by discriminating the particle color as a function of the position in sucessive layers (each layer is composed of around 500000 particles). The observed flow pattern in silo 6 highlights mass flow that are typical of hoppers with

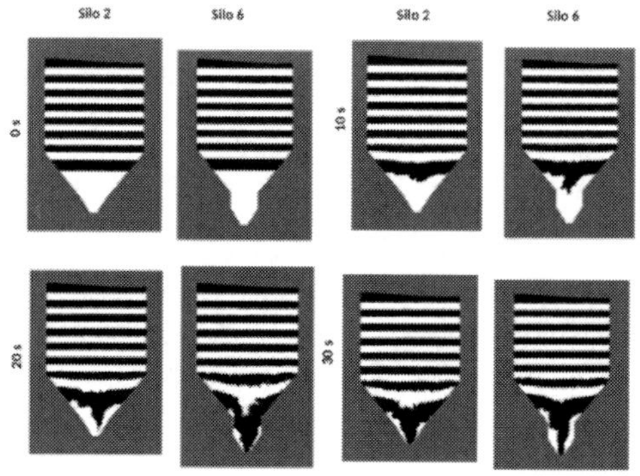

Figure 5. Cross-sectionnal views of flow patterns at several simulation times (0, 10, 20 and 30s) for silo 2 and silo 6 (case B)

smaller discharge angles. In addition, the flow pattern evolution in the hopper zone clearly show the effect of the nonsymetrical initial packing.

The mass flow rate for the two hoppers and for the two particle sizes are shown in Figure 6. This comparison confirms that the cylinder extension enhance the granular flow with a flow rate for silo number 6. This is in accordance with the flow patterns at 20s and 30s.

5 Applied mechanical stress on the silo

The understanding of the mechanical loading and the flow behaviour is an important key in the silo designs [12]. Practioners usually used empirical and continuum models to design hoppers and silos. Applying DEM at an industrial scale may allow for better designs with more fundamental understanding of the structural and particle loading in the various flow regimes. Towards this aim a mechanical loading study is conducted in which we compute the stress applied by the particles on the silos.

The resulting stress tensor of the particles on the silo boundaries is calculated from Love's formulation [8, 11] :

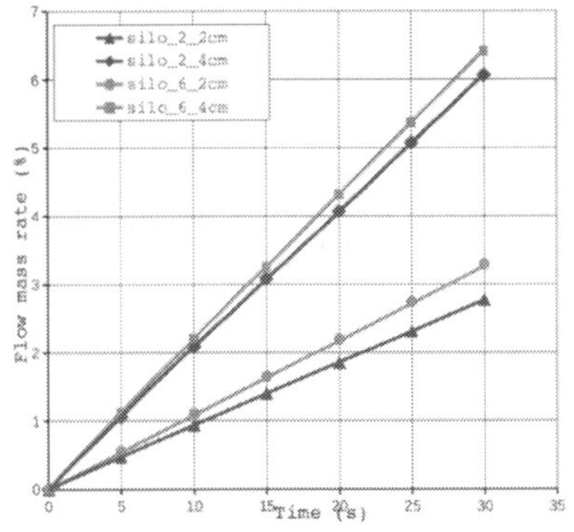

Figure 6. Flow rates (%) as a function of time (second) for the silo 2 and for the silo 6 and for the two case of packing

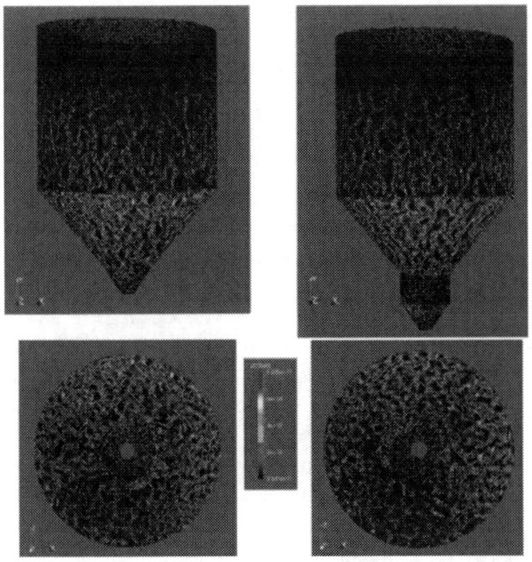

Figure 7. Von Mises stress patterns at 10 seconds for the silo 2 and the silo 6 for case B for the front (top row) and bottom view (bottom row)

$$\sigma_{ij} = \frac{1}{V}\sum_{N_{BF}} XC_j F_i, \qquad (3)$$

where the summation is made on all contacts N_{BF} act along the boundaries. V is the particle volume, F_i is the ith component of the total contact force and XC_j is the jth component of the contact location.

The Von Mises stress is computed from the computed stress tensor and plotted in Figure 7. The results show that the maximal shear zones are located near the junction surfaces of the silo and the hopper. The cylindrical extension reduces the shear stress near the orifice. Thus, the stress field exhibits some localizations of shear stress around the cohesive block zones that it is in agreement with the linear and angular velocity observations of the previous section.

6 Conclusion

A DEM GPU approach based with the recently developed BlazeDEM3D-GPU framework has been used to study the discharging flow of an industrial storage silos. The DEM simulation abilities are considerably increased by the GPU architecture allowing for typical industrial facility. In particular, the flow rate and loading have been studied for an industrial scale problem.

Acknowledgements

The exchanges with EQIOM Granulats and EQIOM Bétons about the concrete facility used in this study are greatly appreciated.

References

[1] Cundall P.A. and Strack O.D.L, A discrete numerical model for granular assemblies, Geotechnique **29**, 47-65 (1979)
[2] Govender N., Wilke D.N., Kok S., Blaze-DEMGPU: Modular high performance DEM framework for the GPU architecture, SoftwareX **5**, 62-66 (2016)
[3] Govender N., Rajamani R.K, Kok S., Wilke D.N, Discrete element simulation of mill charge in 3D using the BLAZE-DEM GPU framework, Miner. Eng. **79**, 152-168 (2015)
[4] Govender N., Pizette P., Wilke D.N, Abriak N.E., validation of the GPU based BLAZE-DEM framework for hopper discharge, PARTICLES 2015, Barcelona
[5] Pizette P, Govender N, Abriak N-E, Wilke D.N, GPU DEM simulations and Experimental studies of ball milling process for various particle shapes, Springer Proceedings Phys. **188**, 1345-1352 (2017)
[6] Govender N., Wilke D. and Kok S., Collision detection of convex polyhedra on the NVIDIA GPU architecture for the discrete element method, Appl. Math. Comput. **15**, 810-829 (2014)
[7] Govender N., Wilke D.N., Pizette P., Abriak N-E, A study of shape non-uniformity and poly-dispersity in hopper discharge of spherical and polyhedral particle systems using the BlazeDEM3D-GPU code, Appl. Math. Comput. (accepted)
[8] Weber J., Recherches concernant les contraintes inter granulaires dans les milieux pulverulénts, Bulletin de liaison des Ponts et Chaussées **20**, 1–20 (1966)
[9] Wilke D.N., N. Govender N. , Pizette P., Abriak N-E, Computing with non-convex Polyhedra on the GPU, Springer Proceedings Phys. **188**, 1371-1378 (2017)
[10] Pöschel T. and Schwager T., *Computational Granular Dynamics* (Springer, Berlin, Heidelberg, 2005)
[11] O'Sullivan C., *Particulate Discrete Element Modeling : A Geomechanics Perspectives* (Spon Press, New York, 2011) 561
[12] Jenike A. W.,Gravity Flow of Bulk Solids, Bulletin **108**, University of Utah Engineering Experiment Station (1961)

Highly deformed grain: from the Hertz contact limitation to a new strain field description in 2D

Serge Mora[1,*], *Thi-Lo* Vu[1,**], *Jonathan* Barés[1,***], and *Saeid* Nezamabadi[1,****]

[1] *Laboratoire de Mécanique et Génie Civil, Université de Montpellier, CNRS, Montpellier, France*

Abstract.
Most of the 2D studies on granular materials consider an assembly of cylindrical particles in the small deformation regime. Investigating the range of validity of the small deformation hypothesis is therefore of crucial importance in order to determine the quantitative relevance of the predictions emerging from these 2D studies. We introduce here a novel experimental set-up based on high resolution imaging and image correlation, and capable of measuring the strain field inside the particles with a cutting edge accuracy. We measure the strain field inside a unique circular cylindrical elastic particle undergoing an uniaxial compression. The significant deviations arising from the comparison with the predictions of the small deformation limit predicted by the Hertz's theory, and numerical simulations, give insight to improve the modeling of 2D granular materials.

1 Introduction

When local forces are measured or computed in granular matter physics, whether it is for numerical [1–3] or experimental studies [4–6], collections of circular grains with low polydispersity and in small deformation regime are very commonly considered. This permits to conform with the Hertz contact law [7] which, along with the superposition principle, provides the most common mechanical description of a grain contact. Unfortunately in real life most of the granular problems involve unevenly shaped grains, on a very broad range of sizes, in highly stressed situations [8–11]. The lack of a net demarcation line separating the regime of small deformations with a regime where it is necessary to take into account large deformations within the grains is a source of confusion. For example in the case of jammed granular systems some grains (the one involved in force chains [1]) can be far above the small deformation hypothesis which ruins the macroscopic description of the whole system.

Hence, in order to observe, model and simulate properly realistic granular media it is necessary to get a complete description of a grain behaviour beyond the linear elastic limit. This would permit to check already existing numerical models [12–14] and to provide inputs for improvements.

By means of novel 2D experimental approaches, we investigate when the Hertz strain field [7] is not valid any more and how it has to be modified beyond linear elasticity. To do so we introduce a new experimental measurement method suitable for granular material. We set-up a step by step single disk radial compression experiment attached to a very high resolution imaging apparatus (5μm accuracy over a 6cm diameter disc particle). The disk is one base surface of a thin cylinder made in an hyperelastic material. The displacement field in the plane of the disk is then precisely measured both spatially and in intensity via Digital Image Correlation (DIC). Strain is directly deduced and compared with theoretical results. The limitations arising from the linear elastic theory are then highlighted, for this common mechanical situation. These results are also compared to what is obtained by finite element (FE) simulations.

In the first part of this article, we present the experimental set-up, how we compress a particle and how we image it. Then, in a second part we give some image processing details to measure the displacement field inside the particle. Strain field is computed and compared to what is obtained with FE. Finally, in a last part, we present some results, conclusions and perspectives.

2 Experimental set-up and material

A manufactured in-house black silicone cylindrical particle [15] (Young modulus 2.7MPa) coated on one side with shiny very thin glitter [16] (average size 30μm) is led on the top glass of a flatbed scanner as presented in fig. 1-A. The initial diameter of the particle is 5.9cm and its thickness is 9.5mm. As presented in fig. 1-B the glitter coating forms an homogeneous thin random pattern on the bottom face of the particle with a high black/white contrast. Step by step, the flat cylinder is squeezed in its radial direction in a quasistatic manner by a manufactured in-house uniaxial compression apparatus (see fig. 1-A). This device

*e-mail: serge.mora@umontpellier.fr
**e-mail: thi-lo.vu@umontpellier.fr
***e-mail: jb@jonathan-bares.eu
****e-mail: saeid.nezamabadi@umontpellier.fr

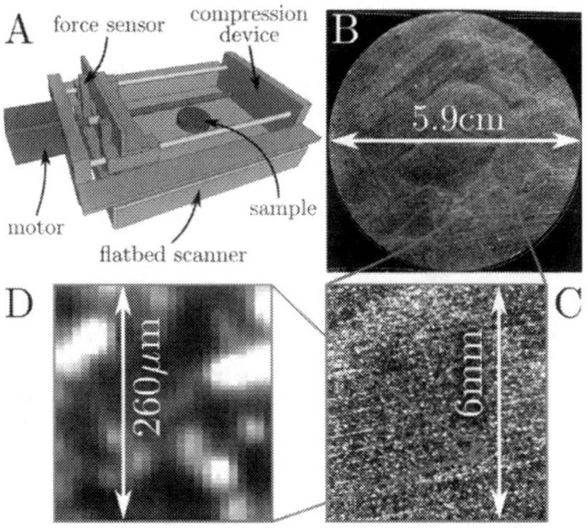

Figure 1. A: Sketch of the experimental set-up. A cylindrical particle (5.9 cm diameter and 9.5 mm thickness) lays on a flatbed scanner. A manufactured in-house compression device squeezes the particle step by step while the particle is imaged from below and the compression force is measured. B: Black and white scanner bottom view of the particle. A thin pattern has been deposited on the particle. C: Zoomed view of the particle pattern. D: Zoomed view to the size of a correlation length. The pixel size is 5.29 μm.

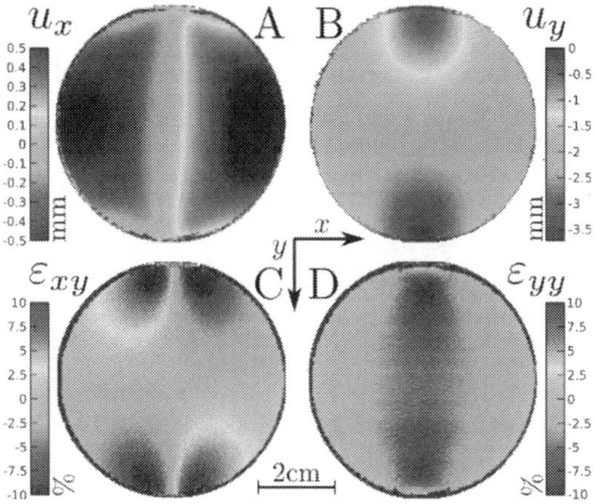

Figure 2. A,B: Displacement field measured from the local image correlation along the x and y direction respectively. C, D: ε_{xy} and ε_{yy} components (respectively) of the strain field. Fields are plotted on the undeformed geometry for a 8% total deformation of the particle. Measurement are noisy at the boundary because the correlation cells overlap patterned and unpatterned domains. The different fields are plotted on the undeformed coordinates (Lagrangian).

is accurately actuated by means of a stepper motor and a threaded rod with a very low distance/step ratio. The force is measured with two 1kg force cells. Vegetable oil with a very low viscous coefficient (60 mPa.s) is coated on the scanner glass to almost vanish static basal friction and improve optic transmission.

After each 0.5mm compression step, the system is imaged with the flatbed scanner. As emphasized by the results presented in this paper, it constitutes a very stable and accurate measurement apparatus both in space (tunable from 70dpi to 9600dpi) and numerical depth (tunable from 8bits to 16bits for each color) on a large area (maximum 21 × 29.7cm^2) without any optical distortion on a flat area. For the results presented in this paper, we used a scanner CanoScan 9000F Mark *II* with resolution 4800 × 4800dpi^2 on a 6 × 6cm^2 area on 8bits for black and white picture. As presented in fig. 1-C & D, we image the particle bottom on a wide area with an extreme accuracy.

3 Image post-processing

Image correlation for granular material has been widely used both in 2D and 3D at the continuum scale [17–21] to measure the grain flow but to our knowledge it has never been used at the scale of each grain to measure their deformation. However, as emphasized earlier in this paper it is of high interest to study stress networks in the most common granular media: unevenly shaped, highly stressed or deformed grains with any rheological behaviour. Until now, the only way to measure force between grains was to use photoelasticity [4, 22], thermoelasticity [5] or boundary deformation [6]. However for each of these techniques, particle materials and shapes are very restricted and it can be performed only in the small deformation regime. Here we propose a method capable of measuring the full stress field for any kind of grain and any loading in 2D.

By means of digital image correlation (DIC) technique [23, 24], adapted to high image size and large deformations we measure the evolution of the displacement field inside a compressed silicon particle. A regular mesh of 260 × 260μm^2 (50 × 50px^2) is first built on the undeformed initial picture. For each cell of the mesh, also called correlation cell, the displacement along x and y directions is measured incrementally between the picture of each compression step and the undeformed picture. To do so, a first measurement is done, with a 1px accuracy, using a Fourier transformed based algorithm [24]. Then, the measurement is improved to subpixel accuracy by means of computationally heavier algorithm based on an optimization algorithm [25] maximizing the correlation between deformed pictures and the undeformed one. To take the large deformations into account both algorithmic steps are performed sequentially for correlation cells going from 3 times bigger to 50 × 50px^2. Parallelized on 12, 3GHz nodes, the computation take around ~ 6 hours for 30 compression steps.

4 Results and discussions

As shown in fig. 2-A & B for a global compression of 8% strain, the x and y components of the displacement, u_x and u_y (y is for the direction of the global compression, x is for the in-plane orthogonal direction) are measured with

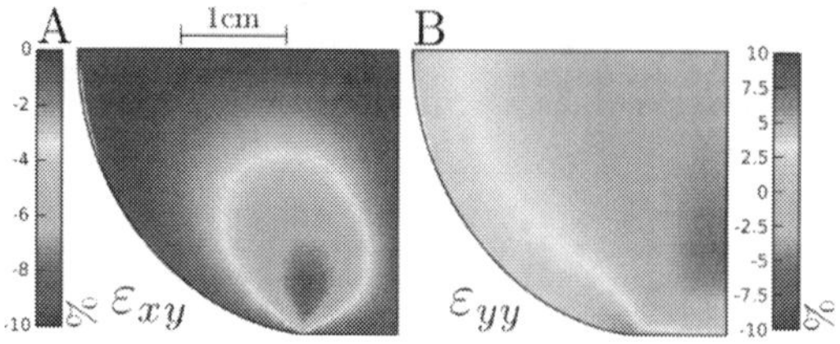

Figure 3. Computed strain. A, B: ε_{xy} and ε_{yy} components (respectively) of the strain field computed by FE using Cast3M [26]. Using symmetries only 1/4 of the the particle is simulated. The material is supposed to be linear, elastic and isotropic for a global 8% compression. Fields are plotted on the deformed system (Eulerian). Results are qualitatively similar to what is found experimentally. Difference is explained because the system is at the limit of the small deformation hypothesis.

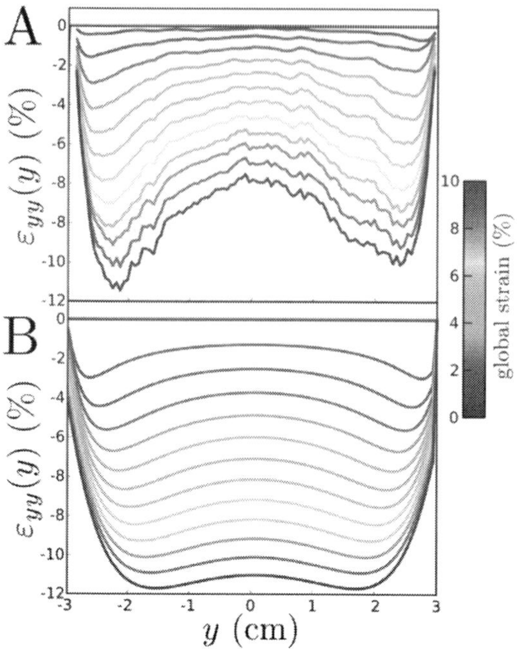

Figure 4. Strain evolution during the compression process along the y-direction. A,B: Evolution of the strain component $\varepsilon_{yy}(y)$ as function y, along the diameter parallel to y, for different compression levels going from 0% to 10% in the experimental case (A) and computed from Hertz's contact law (B). $y = 0$ corresponds to the center of the specimen. For the experiment, data close to the contact point have been cropped because they are noisy. The curve shape is sensibly different for large deformations.

an excellent smoothness and accuracy compared to what is classically done in the small deformation domain for disc compression [23]. Nevertheless around the narrow particle boundary, data are noisy since correlation cells overlap both particle and unpatterned surrounding space: correlation information in the particle is mixed with surrounding black area where no correlation is possible. Hence for these correlation cells displacement information is less accurate than displacement for cells in the middle of the particle. Deriving these raw displacement data u_x and u_y, as shown in fig. 2-C & D we then compute the strain inside the particle, $\varepsilon_{xy} = \frac{\partial u_x}{\partial y} + \frac{\partial u_y}{\partial x}$ and $\varepsilon_{yy} = \frac{\partial u_y}{\partial y}$ respectively. It is important to note that these data are obtained without any smoothing nor filtering at a scale higher than the cell scale. As expected from Hertz's theory [7], we find (from ε_{yy}) that the strain parallel to the compression direction is symmetric and maximum in a thin ellipse joining the two contact points. In fig. 2-D we also observed slight horizontal changes in the blue shade. They correspond with small instability of the scanner motor speed. This is the only artefact we have noticed coming from the scanner and it can be put in perspective to the fact that no noise filtering has been applied to the pictures nor to the post-processed data. Similarly, for ε_{xy}, results qualitatively in agreement with Hertz's theory are observed since the shear strain is anti-symmetric with two opposite lobes going from the contact areas. Then, assuming a linear rheology, the stress field is computed from the measured strain field with a Young modulus of 2.7MPa and a Poisson ratio of 0.49 (not shown).

In parallel to this, numerical simulations have been carried out, by means of the FE code Cast3M [26]. Taking advantage of the symmetries, 1/4 of the particle is simulated in the same mechanical condition as the experiments except the material is considered as linear, elastic and isotropic for the full deformation range. As presented in fig. 3-A & B the results are in qualitative agreement with what is measured experimentally but strain intensity do not seem to be the same.

In fig. 4-A, we show $\varepsilon_{yy}(y)$ the compression strain value along the loading direction. This has been plotted for different compression steps. We see that this value vanishes at the contact point, rapidly increases toward the center of the particle and then decreases before reaching a plateau in the middle of the sample. As presented in fig. 4-B, this is consistent with what is predicted by Hertz contact law for small deformations. Also, even for large deformations the minimum value of the strain near the contact points is well predicted. However when the global loading is larger than few percent, fig. 4-AB shows that the strain is quantitatively different in the center of the disc: the strain is largely overestimated. This can be explained by the fact that small deformation hypothesis is not valid any more or because out of plan displacement cannot be neglected any more in the experiment.

5 Conclusion and perspectives

In conclusion, in this paper we have presented an original experimental apparatus capable of compressing cylindrical granular particles while imaging them with an excellent accuracy. Along with a manufactured in-house fast DIC algorithm adapted to take large deformation into account, the evolution of the displacement field inside the particle is measured with an edge cutting accuracy. Strain field has been deduced and compared to what is expected

DOI: 10.1051/epjconf/201714005011

from Hertz's contact law. A fairly good qualitative agreement have been observed as long as the system stays in the small deformation regime.

This paper validates an experimental method for the measurement of the full strain and stress field at the grain scale for any loading and opens a new front of experimental exploration for a wide range of granular material: soft, brittle, mixed, *etc*. Hence in future works, the experimental measurement will be extended to different model materials: plastically deformable material, brittle material, zero Poisson ratio material,*etc*. The results will be compared to the ones obtained with different numerical models (finite element method for example) for different boundary conditions. The goal will be to show the limitation of these numerical models either by the fact they do not take the full complex rheology into account or by the fact they are limited in terms of spacial discretisation of the particles (for the bulk or for the contact model). This method can also be extended to 3D measurements in granular matter using tomography imaging.

Acknowledgements

This work is supported by the Labex NumEv (anr-10-labx-20).

References

[1] F. Radjai, M. Jean, J.J. Moreau, S. Roux, Physical Review Letters **77**, 274 (1996)
[2] F. da Cruz, S. Emam, M. Prochnow, J.N. Roux, F. Chevoir, Physical Review E **72**, 021309 (2005)
[3] T. Bertrand, R.P. Behringer, B. Chakraborty, C.S. O'Hern, M.D. Shattuck, Physical Review E **93**, 012901 (2016)
[4] T.S. Majmudar, R.P. Behringer, Nature **435**, 1079 (2005)
[5] P. Jongchansitto, X. Balandraud, M. Grediac, C. Beitone, I. Preechawuttipong, Soft Matter **10**, 8603 (2014)
[6] N. Brodu, J.A. Dijksman, R.P. Behringer, Nature communications **6** (2015)
[7] K.L. Johnson, *Contact mechanics* (Cambridge university press, 1987)
[8] G. Romeo, M.P. Ciamarra, Soft Matter **9**, 5401 (2013)
[9] N. Ouhbi, C. Voivret, G. Perrin, J.N. Roux, Procedia Engineering **143**, 1120 (2016), advances in Transportation Geotechnics {III}
[10] S.A. Galindo-Torres, D.M. Pedroso, D.J. Williams, L. Li, Computer Physics Communications **183**, 266 (2012)
[11] P. Menut, S. Seiffert, J. Sprakel, D.A. Weitz, Soft Matter **8**, 156 (2012)
[12] É. Azéma, F. Radjaï, R. Peyroux, G. Saussine, Physical Review E **76**, 011301 (2007)
[13] V. Topin, F. Radjai, J.Y. Delenne, F. Mabille, Powder Technology **190**, 215 (2009), selection of Papers from the Symposium Powder Science and Technology - Powders and Sintered Material STP-PMF 2007 Symposium on Powder Science and Technology - Powders and Sintered Material STP-PMF 2007
[14] S. Nezamabadi, F. Radjai, J. Averseng, J.Y. Delenne, Journal of the Mechanics and Physics of Solids **83**, 72 (2015)
[15] Smooth-On®, Mold Star® 15 SLOW colored with few drop of black Silc Pig®
[16] Smooth-On®, Cast Magic® (silver bullet)
[17] S.A. Hall, D.M. Wood, E. Ibraim, G. Viggiani, Granular matter **12**, 1 (2010)
[18] S.A. Hall, M. Bornert, J. Desrues, Y. Pannier, N. Lenoir, G. Viggiani, P. Bésuelle, Geotechnique **60**, 315 (2010)
[19] A.L. Rechenmacher, S. Abedi, O. Chupin, A.D. Orlando, Acta Geotechnica **6**, 205 (2011)
[20] E. Ando, S.A. Hall, G. Viggiani, J. Desrues, P. Bésuelle, Acta Geotechnica **7**, 1 (2012)
[21] C. Slominski, M. Niedostatkiewicz, J. Tejchman, Powder Technology **173**, 1 (2007)
[22] M. Cox, D. Wang, J. Barés, R.P. Behringer, Europhysics Letters (2016)
[23] F. Hild, S. Roux, Strain **42**, 69 (2006)
[24] B. Pan, K. Qian, H. Xie, A. Asundi, Measurement science and technology **20**, 062001 (2009)
[25] M. Bornert, F. Brémand, P. Doumalin, J.C. Dupré, M. Fazzini, M. Grédiac, F. Hild, S. Mistou, J. Molimard, J.J. Orteu et al., Experimental Mechanics **49**, 353 (2009)
[26] Cast3M, CEA, http://www-cast3m.cea.fr

Deformation profiles of elastic cylindrical tubes filled with granular media under an overload

V. Salomón Álvarez Salazar[1,*], *Abraham* Medina[2,**], and *Jaime* Klapp[3,4,***]

[1] *Instituto Politécnico Nacional, ESIME Zacatenco, México City, México, 07738*
[2] *Instituto Politécnico Nacional, ESIME Azcapotzalco, México City, México, 02250*
[3] *Instituto Nacional de Investigaciones Nucleares (ININ), Salazar, México, 52750*
[4] *CINVESTAV-ABACUS, Salazar, México, 52740*

Abstract. The deformation of a thin-walled vertical tube, filled with a liquid or a cohesionless granular material is investigated theoretically and experimentally. Experiments with an overload and without it were made with latex tubes filled with water or spherical glass beads and the results were compared with the theoretical profile derived from the Janssen model. The results suggest that the soft elastic tubes could provide a simple and convenient means to investigate the forces that arise in different materials.

1 Introduction

In 1895, H. A. Janssen, a German engineer concerned with the structural design of silos for corn storage, found the laws giving the pressures of dry, non-cohesive granular media on the vertical walls and the base of a vertical bin [1]. These pressures are not the same, contrary to the case a normal liquid. The model put forward by Janssen, and extensions thereof, is widely used today for the design of silos containing granular materials [2]-[5].
Pressures in quiescent liquids and in granular materials at rest are quite different. Consider a vertical cylindrical container filled with a granular material. The vertical pressure on an horizontal plane at a certain depth below the surface of the material does not increase linearly with depth, as it would do for a normal liquid. Instead, the Janssen's model predicts that this pressure tends to a constant value independent of depth. The origin of this behavior is in the static friction between the grains and the walls of the container. Due to this friction, the container walls can support part of the weight of the material.
The validity and possible shortcomings of the simple theory underlying the original Janssen model have been extensively tested experimentally and numerically, but, essentially, in the context of the pressure and force measurements [2], [4]-[6].
The purpose of this work is to explore the possibility of using the elastic deformation of the wall of a container to probe the distribution of pressure in the material filling the container. Specifically, we analyze how an elastic, latex-walled vertical tube is deformed by the pressure of a column of liquid or by the pressure of a column of granular material that fills the tube up to a certain height H.
Latex is well suited for our purposes because it is a soft elastic material with a very wide regime of linear elastic deformation [7]. This property allows visualizing, in a non destructive testing, the deformations originated by the hydrostatic pressure of a liquid or the stresses of a granular material. Comparison between both deformations clarifies the way the pressure distributes in each material.
The determination of deformations is also a preliminary step toward an analysis of the dynamic pressures and wall deformations arising during the flow of a granular material under the action of gravity. This is a very difficult problem due to the strong interaction between the silo structure and the moving granular material [8] and it will not be tackled here.
The paper is organized as follows. In next section the Janssen's original description of the pressures in a vertical tube is presented. Section 3 is devoted to the analysis of the elastic deformation of a thin-walled tube made of an elastic material due to a hydrostatic or granular load. Section 4 reports a series of experiments where the deformation of elastic tubes filled with water, with spherical glass beads with and without a solid overload is measured and compared with the theoretical results. Finally, the main conclusions of the work are summarized in Section 5.

2 Pressures of a granular medium in a vertical container

Consider a cylindrical vertical tube of radius r_0 filled with a dry granular material to a certain height H (Fig. 1). Take a horizontal section of the tube at a depth z in the granular material. According to Janssen's model, the force exerted

*e-mail: catalina-s@hotmail.com
**e-mail: amedinao@ipn.mx
***e-mail: jaime.klapp@hotmail.com

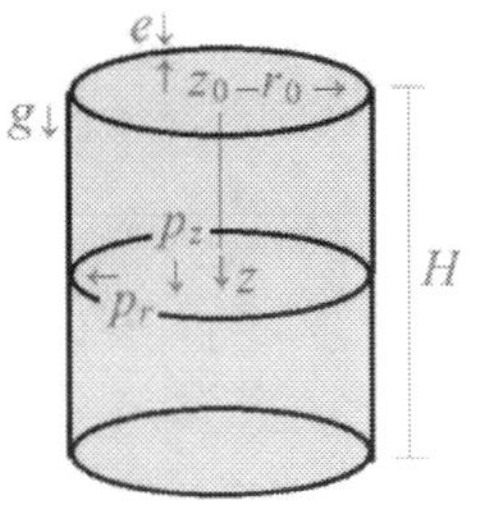

Figure 1. Schematic of the elastic tube filled with a granular material. The dimensions of the pipe and the vertical ($p_z(z)$) and horizontal ($p_r(z)$) pressures are indicated.

per unit area of this section by the material above it on the material below is a vertical pressure $p_z(z)$. This pressure is not equal to the horizontal pressure $p_r(z)$ of the material on the wall of the tube at the same depth, but the two pressures are linearly related [1]-[5]:

$$p_r(z) = Kp_z(z), \tag{1}$$

where K, named the Janssen parameter, is a constant independent of z that characterizes the conversion of vertical stress into horizontal stress within the granulate.

The horizontal pressure acting on the wall of the tube causes a friction stress τ between the wall and the granular material. A simple balance of the forces acting on a slab of the granular material reads

$$\pi r_0^2 \frac{dp_z}{dz} = \pi r_0^2 \rho g - 2\pi r_0 \tau, \tag{2}$$

where ρ is the bulk density of the granular material, assumed to be independent of z, and g is the acceleration due to gravity.

In the Janssen's model, the vertical friction stress is assumed to be given by Coulombs law, $\tau = \mu_w p_r$, with a constant static friction coefficient μ_w. The solution of Eq. (2) with the condition $p_z(0) = 0$ is then [1]-[4]

$$p_z(z) = P_J(z), \tag{3}$$

with

$$P_J(z) = \rho g \lambda [1 - exp(-\frac{z}{\lambda})], \tag{4}$$

and $\lambda = r_0/2\mu_w K$. Equation (4) shows that the vertical pressure first increases linearly with z, as $p_z(z) \approx \rho g z$ for $z/\lambda \ll 1$, and tends to the limiting value $\lambda \rho g$ for $z/\lambda \gg 1$. The length λ is the characteristic size of the region where the pressure undergoes this transition. In laboratory columns $\lambda \cong 1$ cm [9], so that the wall of the container supports most of the weight of the grains when H $\gg$ 1cm.

3 Deformation of the elastic wall

The deformations caused by the horizontal pressure $p_z(z)$ and the vertical shear stress τ depend on the mechanical properties of the wall of the tube. These deformations are computed here for a thin-walled open tube in the framework of the linear theory of elasticity. Let $e \ll r_0$ be the thickness of the wall. The radial pressure p_r produces an approximately uniaxial state of stress in which the only nonzero stress component is the circumferential traction $p_r r_0/e$; see, e.g., Ref. [10]. Hence, by Hooke's law, $E\delta r/r_{0]} = p_r r_0/e$ where E is the elasticity modulus of the material of the wall and $\delta r = r_z - r_0$ is the variation of the tube radius. Thus,

$$r(z) = r_0(1 + \frac{p_r(z)}{eE/r_0}). \tag{5}$$

When the tube is filled with a liquid, the hydrostatic pressure referred to the pressure of the gas outside the tube is $p_z(z) = \rho g z$ [11], which leads to a linear variation of the dimensionless radius, of the tube with depth,

$$\frac{r(z)}{r_0} = (1 + \frac{\rho g r_0}{eE}). \tag{6}$$

The volume of the deformed tube filled with liquid is

$$V = \int_0^H \pi r^2 dz = \frac{\pi r_0 eE}{3\rho g}[(1 + \frac{\rho g r_0 H}{eE})^3 - 1], \tag{7}$$

which is a relation between the volume of liquid and its height. When the tube is filled with a granular medium, the Janssen model [1] and [3]predicts the exponentially varying dimensionless radius

$$\frac{r(z)}{r_0} = [1 + \frac{Kr_0}{eE}P_J(z)] = (1 + \frac{Kr_0\rho g\lambda}{eE}[1 - exp(-z/\lambda)]). \tag{8}$$

The volume of the deformed tube can be easily computed but its analytical expression is rather cumbersome and will not be reproduced here.

3.1 Deformation with an overload

When is superimposed an overload on the top of the granular material, a different profile is obtained. To estimate such a deformation, it is necessary to solve Eq. (2) with the initial condition $p_z(0) = WA$ where W is the weight which originates the overload and A in the area of the upper granular free surface. Thus the resulting profile is given by

$$\frac{r(z)}{r_0} = 1 + \frac{Kr_0}{eE}\lambda g\rho[1 + e^{-\frac{z}{\lambda}}(\frac{W\pi r_0^2}{\rho g\lambda} - 1)]. \tag{9}$$

The respective plot for this last test will be given afterwards.

4 Experiments

In this section we describe and analyze some experiments performed to characterize the deformation of soft elastic tubes filled with water or with spherical glass beads, and to measure the compression force acting on the wall in

DOI: 10.1051/epjconf/201714002025

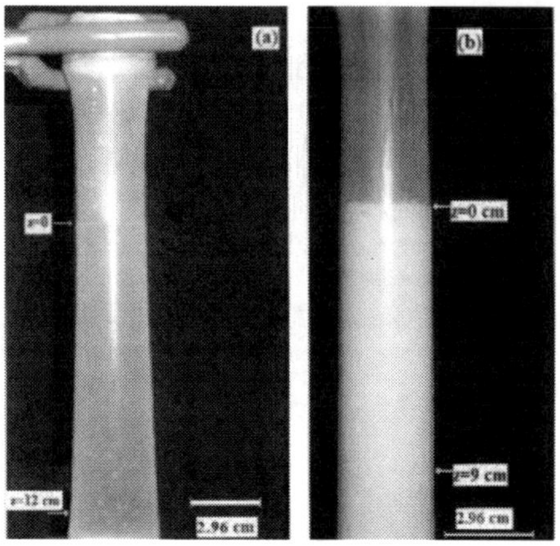

Figure 2. Pictures of deformed latex tubes filled with (a) water (b) spherical glass beads. Red-coloured vertical lines were depicted to show the approximately symmetric shape of the pipe and its relatively small deformation.

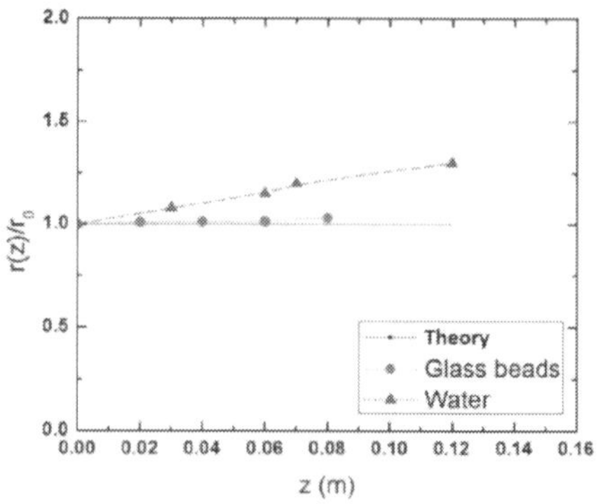

Figure 3. Theoretical deformation profiles, $r(z)/r_0$ as a function of z, for a latex tube filled with water (dotted straight line) obtained from Eq. (6). and glass beads (dashed curve) obtained from Eq.(8). Symbols ○ correspond to the mean values of measurements of the shape of tubes filled with water and symbols × correspond to mead values of measurements of the shape of tubes filled with glass beads. The line $(r(z)/r_0) = 1$ corresponds to the undeformed tube. Errors are of the size symbols.

the second case. The experimental results are compared with the theoretical results of the previous sections. Latex tubes were used because of the very wide range of linear deformation displayed by this material [7]. The tubes used in the experiments are 0.20 m long with wall thickness $e = 7 \times 10^{-5}$ m and internal radius $r_0 = 1.48 \times 10^{-2}$ m. The elasticity modulus is $E = 874.5$ kPa [7]. The tubes are closed at the bottom and open at the top. They were filled with water or with spherical glass beads of mean diameter $d_g = 1$ mm by gently pouring these materials from top. Solid toroidal disks of radius slightly larger than r_0 were fitted at the two ends of the tubes to allow clamping them to external supports and stretching them (see Fig. 2). These disks cause deformations near the ends of the tubes that are not accounted for in the previous section. The region where the experimental results can be meaningfully compared to the theoretical predictions extents from the upper level of the material $z = 0$ to about $z = 0.12$ m in the case of water and $z = 0.09$ m in the case of glass beads. These points are marked by arrows in Fig. 2.

Typical pictures of tubes filled with water and with glass beads are shown in Figs. 2(a) and 2(b), where clear differences between the two cases are apparent. Water causes a radial deformation that increases approximately linearly with depth, while the profile of a tube filled with glass beads is nearly uniform over most of its length. The explanation of this difference lies in the different manners in which the two material distribute the pressures.

Figure 3 shows the experimental profiles of the deformed tubes extracted from Figs. 2 and compares them with the theoretical predictions [6] and [8]. In the case of water ($\rho = 1000$ kg/m^3); circles in Fig.(3), the agreement between the theoretical Eq. (6) and the experimental results Fig. 2(a) is excellent we consider the fit as valid up to $z = 12.0$ cm because no far from this point the deformation induced by the clamp may affect the pipes shape, showing that the linear theory of elasticity gives a good approximation to the deformation induced by the hydrostatic pressure.

In the case of glass beads, some properties of the granular material must be determined or be specified before Eq.(8) can be used. The beads are made of glass of density $\rho_g = 2500$ kg/m^3. The bulk density is $\rho = \eta\rho_g$, where η is a packing factor which for a granular column has been reported to be about $\eta = 0.64$ for the case of dense packing of monodisperse spherical grains poured from top [12]. Consequently, the bulk density used in the calculation is $\rho = 1600$ kg/m^3. The friction coefficient between the glass beads and the latex wall was measured with the tilting plate method, i.e., by laid a thin layer of the glass beads onto a horizontal plate covered with latex and after the plate is slowly tilted and the angle recorded at which the layer of grains slides off [13]. It was found that $\mu_w = 1.27$. Finally, the value of the Janssen parameter is determined from the correlation $K = 1.1(1 - \sin\phi)$, where ϕ is the angle of internal friction [5]. Here, the angle of internal friction ϕ is a synonymous of the angle of repose, as is commonly accepted, see, for instance [14]. In our experiment we found that $\phi = 32°$, which was measured with the heap poured on flat surface method [13], therefore $K = 0.51$ and thus $\lambda = 1.14$ cm.

Figure 3 shows that the experimental deformation measured (by using a commercial software of digital pictures to compute the shape) on Fig. 2(b) compares reasonably with the theoretical prediction, when the aforementioned values of the properties of the medium are used. In Fig. 2(b) the red-coloured vertical lines were depicted to show visually the approximately symmetric shape and the small deformation of the elastic pipe. In this latest case a slight pixelation (individual pixels are clearly discernible along the vertical direction) is observed on the latex pipe image, however, on average, mean values of measurements

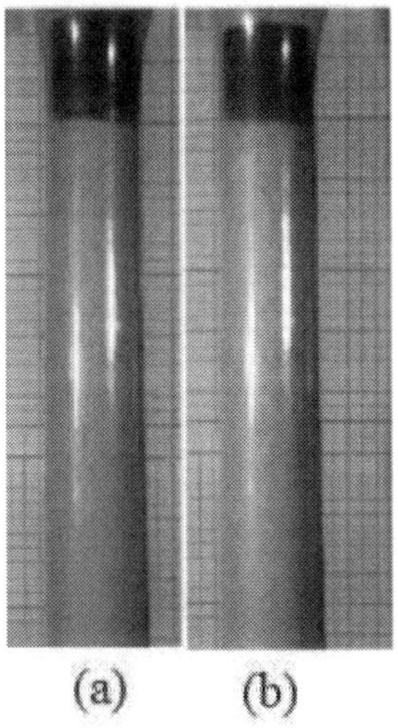

Figure 4. Sand-filled tube profile (a) without an overload and (b)with an overload.

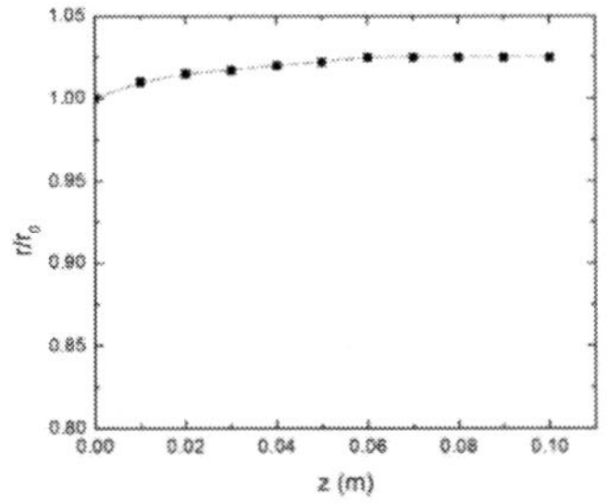

Figure 5. Theoretical profile of the elastic tube with the overload W.

fit reasonably the theoretical profile (dashed line), which we consider as valid up to z = 9.00 cm due to presence of the clamp no far from this point. Other possible origin of the apparent non uniform deformation of the tube could be due to the inhomogeneity of the thickness along its length, which is inherent to the manufacture process of this type of commercial material (latex condom).

Finally, in Fig. 4 is shown the profile of the elastic tube without overload Fig. 4(a) and with an cylindrical solid overload W = 1.24 N weight Fig. 4(b), which was gently located on the top region of the grains. In Fig. 5 the plot of the deformation derived from Eq. (9) is given and symbols are the corresponding experimental data.

5 Summary

An experimental and theoretical study has been presented of the deformation of a vertical tube made of a soft elastic material and filled with a liquid or a dry granular material. When the tube is filled with a liquid, the measured deformation increases linearly with depth in the liquid, in good agreement with the prediction based on Hooke's law. When the tube is filled with a granular material with and without an overload, the deformations attain different profiles among them. As a conclusion all these results are in line with the prediction of Janssen's model.

Acknowledgements.- AM acknowledges to the SIP-IPN project 20172107 for partial support of this study. V.S.A.S. acknowledges to CONACyT for a Ph. D. studentship.

References

[1] H. A. Janssen, Zeitschr. d. Vereines deutcher Ingenieure **39**, 1045 (1985).

[2] R. L. Brown and J. C. Richards, *Principles of Powder Mechanics* (Pergamon, New York, 1970).

[3] E. B. Pitmanr. Am. Math. Mtly **101**, 28-35 (1994).

[4] D. Schulze, *Powders and bulk solids* (Springer, Berlin, 2008).

[5] D. McGlinchey(Ed.), *Bulk Solids Handling: Equipment Selection and Operation* (Blackwell Publishing Ltd., Oxford, 2008).

[6] H. Pacheco-Martínez, H. Jan van Gerner, J.C. Ruiz-Suárez, Phys. Rev. E **77**, 021303(2008).

[7] J. A. Rayas, R. Rodríguez-Vera, A. Martínez, Rev. Mex. Fís **49**, 555-564 (2003).

[8] K. Wilde, M. Rucka, J. Tejchman, Powder Tech **186**, 113 (2008).

[9] P. G. de Gennes,Rev. Mod. Phys **71**, S374 (1999).

[10] A. P. Boresi, R. J. Schmidt, O. M. Sidebottom, *Advanced Mechanics of Materials* (John Wiley, New York, NY, 1993).

[11] F. M. White, *Fluid Mechanics* (McGraw-Hill, New York, NY, 1994).

[12] G. Ovarlez, C. Fond, E. Clément, Phys. Rev. E **67**, 060302(R) (2003).

[13] C.R. Woodcock, J.S. Mason, *Bulk Solids Handling: An Introduction to the Practice and Technology* (Blackie and Son Ltd, London, 1987).

[14] A. M. Deganutti, P.R. Tecca, R. Genovois, *Characterization of friction angles for stability and deposition of granular material, Italian Jour. Eng. Geol. Envnt.- Book: 5th International Conference on Debris-Flow Hazards "Mitigation, Mechanics, Prediction and Assessment"* (Casa Editrice, Università La Sapienza, Roma, 2011) (313-318).

Non-local rheology of dense granular flows

Mehdi Bouzid[1,2], *Martin* Trulsson[1,3], *Adrien* Izzet[1], *Adeline* Favier de Coulomb[1], *Philippe* Claudin[2,⋆], *Eric* Clément[1], and *Bruno* Andreotti[1]

[1] *Physique et Mécanique des Milieux Hétérogènes, PMMH UMR 7636 ESPCI – CNRS – Univ. Paris-Diderot – Univ. P.M. Curie, 10 rue Vauquelin, 75005 Paris, France.*
[2] *Department of Physics, Institute for Soft Matter Synthesis and Metrology, Georgetown University, 37th and O Streets, N.W., Washington, D.C. 20057, USA.*
[3] *Department of Chemistry, Lund University, PO BOX 124, SE-221 00 Lund, Sweden.*

Abstract. The rheology of dense granular flows is studied numerically in a shear cell controlled at constant pressure and shear stress, confined between two granular shear flows. We show that a liquid state can be achieved even far below the yield stress, whose flow can be described with the same rheology as above the yield stress. A non-local constitutive relation is derived from dimensional analysis through a gradient expansion and calibrated using the spatial relaxation of velocity profiles observed under homogeneous stresses. Both for frictional and frictionless grains, the relaxation length is found to diverge as the inverse square root of the distance to the yield point, on both sides of that point. We also make use of a micro-rheometer to determine the influence of a distant shear band on the local rheological behaviour. Finally, we compare various approaches based on different non-local constitutive relations and choices for the fluidity parameter. We emphasise that, to discriminate between the different approaches proposed in the literature, one has to go beyond the predictions derived from linearisation around a uniform stress profile, such as that obtained in a simple shear cell. We argue that future tests can be based on the nature of the chosen fluidity parameter, and the related boundary conditions, as well as the hypothesis made to derive the models and the dynamical mechanisms underlying their dynamics.

1 Introduction

The physics of granular material has reached today a good level of general understanding [1]. In particular, it is now well established that the rheology of dense granular flows takes a simple form when working at controlled pressure P [2]. It this case, the constitutive laws have been formalised in terms of the inertial number $I = |\dot{\gamma}|d/\sqrt{P/\rho}$, where $\dot{\gamma}$ is the shear rate, d and ρ are the grain size and density [3–6]. This dimensionless quantity compares the macroscopic shearing time scale $1/\dot{\gamma}$ to the microscopic time scale of local plastic rearrangements $d/\sqrt{P/\rho}$. In the limit of rigid grains, the effective friction coefficient μ, which compares the shear stress τ to the pressure P, as well as the volume fraction ϕ of the system, must be functions of I. The shape of the functions $\mu(I)$ and $\phi(I)$ have been determined by means of numerical simulations [4–6, 9], as well as by experimental measurements [7, 8]. This approach has been successfully applied in different geometries [2, 3], as well as to describe dynamic compressibility effects associated with spontaneous oscillatory motion [11, 12], and this constitutive law is now used as an input in hydrodynamics equations [10, 13].

Several experiments have however provided evidence for deviations from the $\mu(I)$-rheology, in the case of heterogeneous systems or when investigating the existence of a 'solid-liquid' interface [14–17]. These effects have been qualified as 'non-local', in the sense that flowing at a given location may be enhanced by grain motion in the neighbourhood of that location, so that the $\mu(I)$-rheology

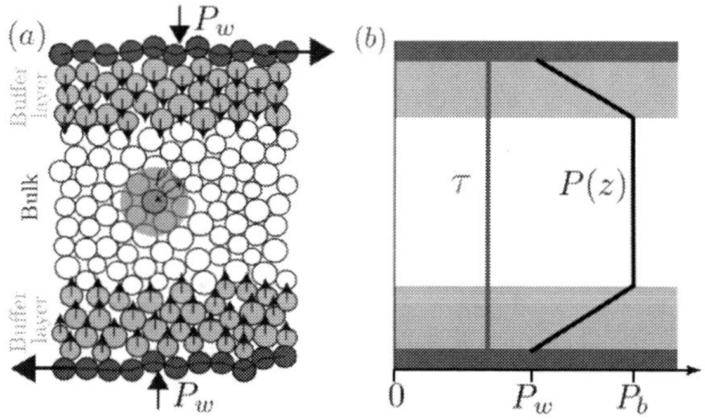

Figure 1. (a) Set-up of the numerical rheometer to probe non-local effects. x is the shear direction. The height of the bulk of the cell is $H \simeq 40d$. The buffer layers are $\Delta \simeq 5d$ thick. Green zoom: schematic of the influence of grain motion in a neighborhood of size ℓ on the local shear rate. (b) Schematic profiles of the pressure P (black line) and of the shear stress τ (red line) across the cell. In the bulk of the shear cell (white), the pressure is homogeneous: $P = P_b$.

⋆e-mail: philippe.claudin@espci.fr

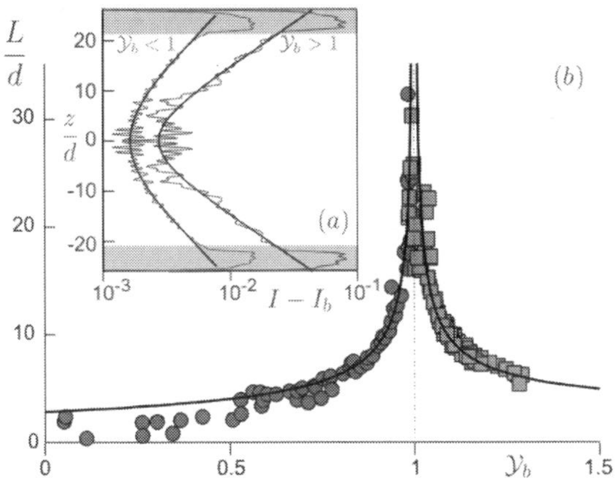

Figure 2. (a) I profiles above (blue line) and below (red line) yield conditions $\mathcal{Y}_b = 1$. The local rheology predics $I = I_b$. Solid black lines: fit of the function $I(z) - I_b = \frac{I(H/2)-I_b}{\cosh(H/(2L))} \cosh(z/L)$, where the relaxation length L is adjusted. (b) L as a function of $\mathcal{Y}_b$. These data are for frictional grains. A similar behavior is obtained with frictionless grains [22].

with the local values of the parameters is not valid. Non-local effects have also been reported in soft glass flows [18–20]. In this case, non-locality has been associated with long range interactions between Eschelby-like elasto-plastic events. These events are well defined in the over-damped regime. For inertial flows of rigid grains, however, the picture is different, and the identification of the mechanisms responsible for non-locality is an interesting yet open issue. In this paper, we summarise our recent results on the subject (see also [21]).

2 A rheometer to probe non-local effects

We have investigated non-local effects in granular flows by means of DEM numerical simulations [22]. The system is two-dimensional and constituted of $N \simeq 2000$ circular particles of mean diameter d, with a $\pm 20\%$ polydispersity. The shear cell is composed of two rough walls moving along the x-direction with opposite velocities (Fig. 1a). These walls are made of the same grains, but glued together. Their position is controlled to ensure a constant normal stress P_w. Periodic boundary conditions are applied along the x-direction. The particle and wall dynamics are integrated using the Verlet algorithm. Contact forces between particles are modeled as linear viscoelastic forces, with a normal contact stiffness large enough ($k_n/P_w > 10^3$) so that the rigid limit is reached. The tangential spring constant is $k_t = 0.5 k_n$. We have considered both frictionless and frictional grains, with a contact friction coefficient $\mu_p = 0.4$. The pressure and shear stress in the system are noted P and τ. We define the dimensionless Coulomb yield parameter $\mathcal{Y} = \tau/(P\mu_c)$, , where μ_c is the friction coefficient in the zero shear-rate limit, at the jamming volume fraction ϕ_c.

What makes the set-up original is the possibility to impose the profile of the $\mathcal{Y}(z)$ by means of gravity-like forces applied to the grains located in two buffer zones located close to the walls. These forces are oriented downward at the top of the cell, and upward at the bottom. In the bulk of the cell, the pressure is constant $P = P_b$. The local rheology predicts in the bulk a linear velocity profile $u_x = \dot{\gamma} z$ when $\mathcal{Y}_b > 1$ corresponding to a finite and constant I_b, and static grains ($u_x = 0$, corresponding to $I_b = 0$) when $\mathcal{Y}_b < 1$. What we find is that all grains of the system actually move, whatever the value of $\mathcal{Y}_b$. The stress ratio τ/P therefore does not control the liquid-solid transition. Rather than a constant value of the inertial number, we measure I profiles that exponentially relax towards I_b over a length L (Fig. 2a). Corresponding velocity profiles follow $u_x(z) = \dot{\gamma} z + \frac{u_x(H/2) - \dot{\gamma} H/2}{\sinh(H/(2L))} \sinh(z/L)$. The relaxation length L extracted from these profiles is displayed Fig. 2b as a function of $\mathcal{Y}$, showing a divergence on both sides of the local rheology yield point $\mathcal{Y}_b = 1$ with an exponent $1/2$.

3 A non-local model

In the usual accessible range of I (between 10^{-4} and 10^{-1}), the constitutive relations of the local rheology typically follow the empirical laws

$$\mu(I) = \mu_c + a I^\alpha, \quad (1)$$

$$\phi(I) = \phi_c - b I^\alpha, \quad (2)$$

where the exponent α is 0.5 in the frictionless case and 1 in the frictional case. In order to account for non-local effects, we perform a gradient expansion of the functional $\mathcal{Y}[I]$. Assuming that non-locality results from a statistically isotropic short-range interaction between shear zones (Fig. 1b), the lowest order operator is the Laplacian $\nabla^2 I$. As a direct consequence, I and its gradient must be continuous. Furthermore, we assume that the non-local correction remains finite as $I \to 0$, so that the expansion must be expressed in terms of $\kappa \equiv \ell^2 (\nabla^2 I)/I$, where ℓ is a length on the order of few grain diameters (Fig. 1a). At the linear order in κ, the constitutive relation is then :

$$\mathcal{Y} = \frac{\mu(I)}{\mu_c} (1 - \kappa) . \quad (3)$$

κ is positive when the point considered is surrounded by a more liquid region (higher I). This region flows more easily than expected from the local value of I, so that the corresponding shear stress is lower.

Linearising the rheology around I_b above yielding conditions, we find a second order differential equation, whose solutions are exponentials (hence sinh shapes for the velocity profiles), with a relaxation length $L_> = \ell \left[\mathcal{Y}_b/\alpha(\mathcal{Y}_b - 1)\right]^{1/2}$. Below yielding conditions, the non-local correction is of already of zeroth order, similarly leading to a relaxation length $L_< = \ell \left[1/(\mathcal{Y}_b - 1)\right]^{1/2}$. These expressions quantitatively fit the data shown in Fig. 2b, with $\ell \simeq 3d$. Further evidence for the square-root divergence, with log-log plots for both frictional and frictionless cases, can be found in [22]. Importantly, the

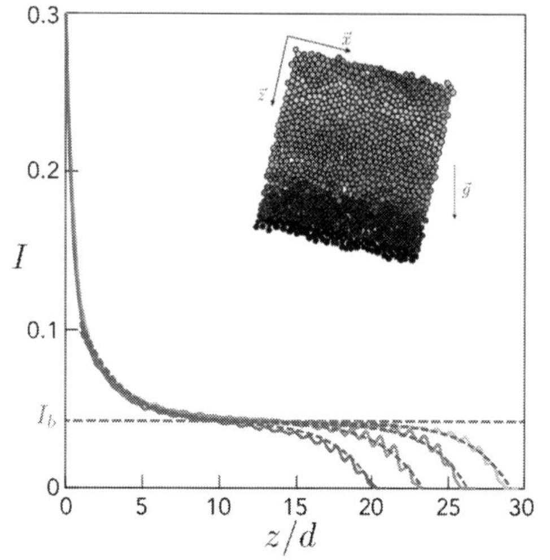

Figure 3. Profiles of the inertial number $I(z)$ for gravity flows of different thickness (dark blue: $H \simeq 20d$, light blue: $H \simeq 23d$, green: $H \simeq 26d$, yellow: $H \simeq 28d$,) at a given inclination $\theta \simeq 18°$. The free surface is at $z = 0$. Red dashed line: prediction from the local rheology. Black dashed lines: fits of the non-local model with the function $I = I_b + C_1 \exp(z/L) + C_2 \exp(-z/L)$. These data are for frictional grains ($\mu_p = 0.5$). A similar behavior is obtained with frictionless grains. Inset: schematic of the incline plane. Color code for the velocity of the grains along the $\vec{x}$ axis.

divergence of L is not the signature of a dynamic phase transition controlled by $\mathcal{Y}$. Both above and below yielding conditions, the system is constituted of a single liquid phase, with permanent fluctuations, in contrast to a solid that would flow by means of elasto-plastic events. The situation would of course be different looking at the initial shearing of a granular packing prepared in a non-critical solid state, where 'hot spots' corresponding to localised plastic events are visualised [23]. Here the stationary shearing maintains the liquid state in the whole system.

Importantly, one should realise that the validity of non-local constitutive relation is not restricted to the description of a creeping flow dynamics, and suited to go beyond the flow blockade as predicted by a local $\mu(I)$ rheology. But also, it is able to reproduce quantitatively flow profiles in inhomogeneous situations, which is actually of a much more general interest. This is illustrated in Fig. 3, in a gravity flows situation where the local rheology would predicts a constant inertia number $I_b = 0.4$. The actual result of the numerical simulation shows that the $I(z)$ profile is indeed reproduced quantitatively by the non-local rheology.

4 Discussion

Other approaches have been proposed to model non-locality in granular flows (see [21] and references therein). For example, Pouliquen and Forterre have proposed a formulation based on an analogy with Eyring's transition state theory for the viscosity of liquids, where mechanical fluctuations would play the role of temperature in thermal systems [24]. Another example is that of Rognon et al.[25]. Closer to our framework, Kamrin et al. have proposed a model [26–28] directly adapted from the Kinetic Elasto-Plastic theory introduced by Bocquet et al. [29] for soft matter.

A crucial issue is the choice of the quantity, called the fluidity f, for which the non-local constitutive relation is written. The fluidity must vanish in the solid phase and increase with the ability to flow. Here we have made the simplest choice in the context of the $\mu(I)$-rheology and take it as the inertial number itself $f = I$. As for comparison, Kamrin et al. rather took $f = \dot{\gamma}/\tau$ for the fluidity parameter. The choice for f is a central issue, which has strong physical consequences: boundary and continuity conditions will apply on f and its gradient.

Interestingly, Kamrin et al.'s approach and our model lead to the same predictions for the velocity profile in a situation where the stress is homogeneous, such that the shear cell described above. The differences between the approaches must be tested in situations that are strongly heterogeneous in space. In this spirit, we have performed numerical simulations where a secondary micro-rheometer is placed in the bulk of a shear cell as presented above [30]. Shearing within the micro-rheometer is obtained by means of localised bulk forces along two lines which induce a discontinuity of the shear stress. The pressure remains constant and the direction of shearing is reversed ($\dot{\gamma}$ changes sign at the discontinuity). We find that the local shear rate is proportional to that in the bulk and decreases roughly exponentially with the distance to the yield conditions. We finally measured numerically the ratio of the absolute value of the shear rate on one side and on the other side of the stress discontinuity, and shown that $|\dot{\gamma}|$ is indeed continuous, unlike $\dot{\gamma}/\tau$.

In conclusion, in spite of some success in the framework of simple rheological situations, a better clarification of the microscopic mechanisms responsible for the non-locality is needed to move forwards and provide a fully consistent picture of the constitutive relation hence proposed. We have for example explored the role of particle softness in the strain mode [21, 31]. Furthermore, to make such a theory fully practical, one should provide a meaningful determination of the boundary conditions, for example with rough-walls, using parameters derived on physical grounds and not from an a-posteriori fitting of the flow curves as it is currently done. This question is under current investigation.

Acknowledgements

BA is supported by Institut Universitaire de France. This work was funded by the ANR JamVibe and by a CNES grant.

References

[1] B. Andreotti, Y. Forterre and O. Pouliquen, *Granular Media, Between Fluid and Solid*, Cambridge University

Press (2013).
[2] Y. Forterre and O. Pouliquen, Ann. Rev. Fluid Mech. **40**, 1-24 (2008).
[3] GDR MiDi, Eur. Phys. J. E **14**, 341-365 (2004).
[4] F. da Cruz , S. Emam, M. Prochnow , J.-N. Roux and F. Chevoir, Phys. Rev. E **72**, 021309 (2005).
[5] P. Jop, Y. Forterre and 0. Pouliquen, Nature **441**, 727-730 (2006).
[6] T. Hatano, Phys. Rev. E **75**, 060301(R) (2007).
[7] P. Jop, Y. Forterre and O. Pouliquen, J. Fluid Mech. **541**, 167-192 (2005).
[8] A. Fall, G. Ovarlez, D. Hautemayou, C. Mézière, J.-N. Roux and F. Chevoir, J. Rheology **59**, 1065-1080 (2015).
[9] E. Azéma and F. Radjaï, Phys. Rev. Lett. **112**, 078001 (2014).
[10] T. Barker, D.G. Schaeffer, P. Bohorquez and J.M.N.T. Gray, J. Fluid Mech. **779**, 794-818 (2015).
[11] L. Staron, Phys. Rev. E **86**, 041307 (2012).
[12] M. Trulsson, M. Bouzid, P. Claudin and B. Andreotti, Europhys. Lett. **103**, 38002 (2013).
[13] P.-Y. Lagrée, L. Staron and S. Popinet, J. Fluid. Mech. **686**, 378-408 (2011).
[14] T.S. Komatsu, S. Inagaki, N. Nakagawa and S. Nasuno, Phys. Rev. Lett. **86**, 1757 (2001).
[15] I.S. Aranson, L.S. Tsimring, F. Malloggi and E. Clément, Phys.Rev.E **78**, 031303 (2008).
[16] K.A. Reddy, Y. Forterre and O. Pouliquen, Phys. Rev. Lett. **106**, 108301 (2011).
[17] E. Wandersman, M. Van Hecke, Europhys. Lett. **105**, 24002 (2014).
[18] J. Goyon, A. Colin, G. Ovarlez, A. Ajdari and L. Bocquet, Nature. **454**, 84 (2008).
[19] J.-L. Barrat and A. Lemaître, In 'Dynamical heterogeneities in glasses, colloids and granular media', **150**, 264 (2011).
[20] V. Mansard and A. Colin, Soft Matter **8**, 4025 (2012).
[21] M. Bouzid, A. Izzet, M. Trulsson, E. Clément, P. Claudin and B. Andreotti, Eur. Phys. J. E **38**, 125 (2015).
[22] M. Bouzid, M. Trulsson, P. Claudin, E. Clément and B. Andreotti, Phys. Rev. Lett. **111**, 238301 (2013).
[23] A. Amon, V.B. Nguyen, A. Bruand, J. Crassous, E. Clément, Phys. Rev. Lett. **108**, 135502 (2012).
[24] O. Pouliquen and Y. Forterre, Phil. Trans. R. Soc. A. **367**, 5091 (2009).
[25] P.G. Rognon, T. Miller, B. Metzger and I. Einav, J. Fluid Mech. **764**, 171 (2015).
[26] K. Kamrin and G. Koval, Phys. Rev. Lett. **108**, 178301 (2012).
[27] D.L. Henann and K. Kamrin, Proc. Natl. Acad. Sci. USA **110**, 6730 (2013).
[28] D.L. Henann and K. Kamrin, Phys. Rev. Lett. **113**, 178001 (2014).
[29] L. Bocquet, A. Colin and A. Ajdari, Phys. Rev. Lett. **103**, 036001 (2009).
[30] M. Bouzid, M. Trulsson, P. Claudin, E. Clément and B. Andreotti, Europhys. Lett. **109**, 24002 (2015).
[31] A. Favier de Coulomb, M. Bouzid, P. Claudin, E. Clément and B. Andreotti, Submitted.

DOI: 10.1051/epjconf/201714003072

Assessing the failure of continuum formula for solid-solid drag force using discrete element method in large size ratios

Payman Jalali[1,2,*], *Timo* Hyppänen[1]

[1]School of Energy systems, Lappeenranta University of Technology, 53851 Lappeenranta, Finland
[2]Department of Physics & Center for Nonlinear and Complex Systems, Duke University, Durham, North Carolina 27708, USA

Abstract. In loose or moderately-dense particle mixtures, the contact forces between particles due to successive collisions create average volumetric solid-solid drag force between different granular phases (of different particle sizes). The derivation of the mathematical formula for this drag force is based on the homogeneity of mixture within the calculational control volume. This assumption especially fails when the size ratio of particles grows to a large value of 10 or greater. The size-driven inhomogeneity is responsible to the deviation of intergranular force from the continuum formula. In this paper, we have implemented discrete element method (DEM) simulations to obtain the volumetric mean force exchanged between the granular phases with the size ratios greater than 10. First, the force is calculated directly from DEM averaged over a proper time window. Second, the continuum formula is applied to calculate the drag forces using the DEM quantities. We have shown the two volumetric forces are in good agreement as long as the homogeneity condition is maintained. However, the relative motion of larger particles in a cloud of finer particles imposes the inhomogeneous distribution of finer particles around the larger ones. We have presented correction factors to the volumetric force from continuum formula.

1 Introduction

The solid-solid drag force plays a key role in the hydrodynamics of gas-solid systems such as fluidized beds [1]. This force along with the gas-solid drag force determine how the particle clusters form, merge or break at any instant. The clusters are responsible in hydrodynamics as well as heat and mass transfer processes. In computer simulations of gas-solid systems using Eulerian approach, the solid-solid drag force is represented by a continuum formula known as the Syamlal equation [2]. This formula is derived based on the assumption of homogeneous mixtures of granular phases. In other words, it is assumed that the distribution of particles within any local volume remains uniform. This assumption could be precise enough when there is monodisperse packing, however, any large size ratio between the granular phases can violate the homogeneity assumption used in the Syamlal equation. Obviously, the large size ratio is associated with the presence of big particles in a cloud of small particles, which could normally result in the inhomogeneous distribution of cloud particles around any big particle. Then the distribution of solid volume fraction (of small particles) and the collision forces on the big particle would be dependent of the relative motion of the big particle with respect to the particle cloud. Such anisotropy in the distribution of collisional forces causes the deviation of mean collisional forces from the predicted values by the Syamlal equation. The focus of this work is to present a correction method of the relevant factors such as the solid volume fraction in terms of the direction of the relative motion of the two granular phases, which consequently lead to a corrected value of the solid-solid drag force.

2 Methods

In this study, the Lagrangian approach of discrete element method (DEM) is employed. The details of this method can be found in Refs. [3-4]. In DEM, the instantaneous positions, velocities and contact forces are calculated for each particle at successive time steps. This helps to calculate the averaged values including the interaction forces. The particle contact is modelled by the Hertzian contact theory where the contact forces are derived based on the spring, dashpot and frictional slider assembly between two overlapping particles. The normal and tangential stiffness coefficients are expressed as functions of particles diameters, their Young's moduli (or shear moduli) and Poisson's ratios [3]. The forces at any contact point include the forces in normal and tangential directions. A particle can have multiple contacts at the same time, therefore, the net contact force and its torque are the summation of all contact forces and torques. The net force and torque on any particle determine the translational and rotational accelerations within the corresponding time step. Then the velocities (translational and rotational) are obtained by integrating

* Corresponding author: pjalali@lut.fi

the accelerations. Similarly, the positions are calculated by the integration of the velocities within the time step. These calculations are performed in each time step for all particles in the system and the trajectory information of particles are stored for further calculations and averaging. The averaged contact force within a volume will result in volumetric drag force between the two granular phases, which can be calculated from DEM simulations as,

$$\mathbf{F}_{\mathrm{DEM}} = \frac{\sum_{i=1}^{N_{ml}} m_i(\mathbf{v'}_i - \mathbf{v}_i)}{\Delta V.\delta t}, \tag{1}$$

where the mass of particle involved in collision i is m_i with the post-collisional and pre-collisional velocities of $\mathbf{v'}_i$ and $\mathbf{v}_i$, where the total collisions of N_{ml} occurs within the volume ΔV and time δt.

On the other hand, Syamlal [2] derived the following equation to calculate the volumetric force by knowing the solid volume fraction, velocity and particle diameter of each granular phase along with the material properties including the coefficient of restitution and friction coefficient,

$$\mathbf{F}_{\mathrm{Syam}} = -S_{bs}(\mathbf{u}_s - \mathbf{u}_b), \tag{2}$$

in which $\mathbf{u}$ is the Eulerian velocity within the given volume with the subscripts s and b standing for the small and big particles. The drag coefficient is expressed by,

$$S_{bs} = \frac{3(1+e)(\pi/2 + \mu_f \pi^2/8)\varepsilon_b \varepsilon_s \rho_b \rho_s (d_b + d_s)^2 g_0 |\mathbf{u}_s - \mathbf{u}_b|}{2\pi(\rho_b d_b^3 + \rho_s d_s^3)}, \tag{3}$$

where e and μ_f represent the coefficient of restitution and the coefficient of friction for the particles. Other parameters include solid volume fraction ε, material density ρ, and particle diameter d with the same subscripts for granular phases described above. The radial distribution function at contact g_0 is expressed by,

$$g_0 = \frac{1}{\varepsilon_g} + \frac{3 d_b d_s}{\varepsilon_g^2 (d_b + d_s)}\left(\frac{\varepsilon_b}{d_b} + \frac{\varepsilon_s}{d_s}\right). \tag{4}$$

Here, ε_g is the volume fraction of the void space, which is $1-\varepsilon_b-\varepsilon_s$.

The simulation setup includes a rectangular box within which certain number of particles are randomly positioned to provide the desired initial volume fraction of the small particle size. The second granular phase includes particles with a bigger size. There is either a single big particle or a cluster of them up to 20 particles. Since the size ratio of particles is over 10, the number of particles in each size considerably differs. Figure 1a demonstrates a typical configuration initially in which the volume fraction of small particles (grey) is 0.08. The red particle is the big particle that is initially located at the top of the system. Its initial velocity is given downward along z-direction (here, it is 0.1m/s), as Fig. 1b shows its position after a short time. Only a slice of small particles is represented with the thickness of 0.014m, where the entire width is 0.04m. Here, the diameter of big particle (red) is 12mm, where it is 1mm for the small particles (grey).

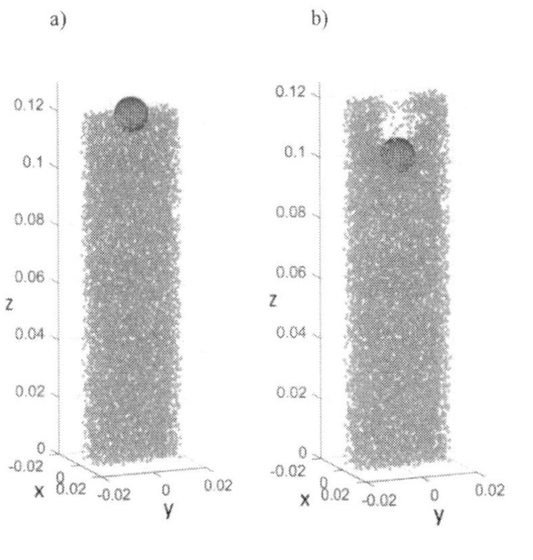

Figure 1. a) Initial configuration b) Intermediate configuration.

3 Results and Discussion

3.1 Force from DEM

The mean force exerted on the big particle is calculated via Eq. (1). The instantaneous force is displayed in Fig. 2a. As revealed in Fig. 2b, the mean force exhibits reduced spikes as the averaging given by Eq. (1) smooths out the ones observed in Fig. 2a.

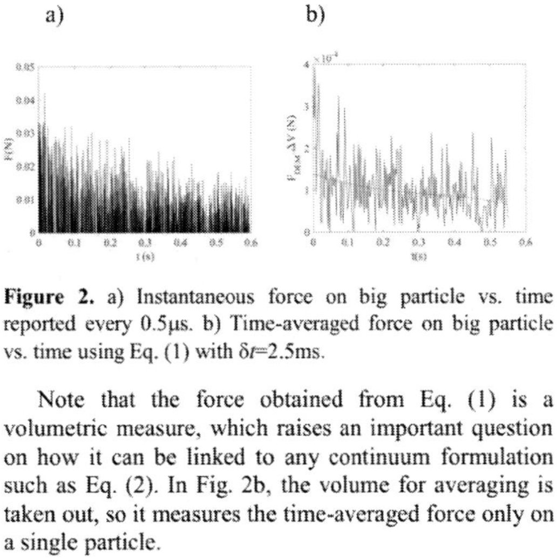

Figure 2. a) Instantaneous force on big particle vs. time reported every 0.5μs. b) Time-averaged force on big particle vs. time using Eq. (1) with δt=2.5ms.

Note that the force obtained from Eq. (1) is a volumetric measure, which raises an important question on how it can be linked to any continuum formulation such as Eq. (2). In Fig. 2b, the volume for averaging is taken out, so it measures the time-averaged force only on a single particle.

3.2 Continuum force

In case of single big particle, the interpretation of DEM force (Eq. 1) in a control volume is fuzzy. So, we implement Eqs. (2-4) to estimate the force on a single big particle as,

$$\mathbf{F}_b = \mathbf{F}_{Syam} \frac{V_b}{\varepsilon_b}, \tag{5}$$

where V_b is the volume of the big particle as $\pi\ d_b^3/6$. Also, ε_b is set to a very small value in Eq. (4). Figure 3 depicts the variation of the force on big particle F_b versus the relative velocity between the granular phases. The thin solid line stands for the time-averaged force from DEM ($F_{DEM}.\Delta V$) with the averaging time window of 2.5 ms, which fluctuates and the fitted line is shown by the dash line represented by the exponential function described in the caption of the figure. The thick solid line represents the continuum formula as described by Eqs. (2-4) with the parameters shown in Table 1. Note that $\Delta u_{sb}=|u_s-u_b|$.

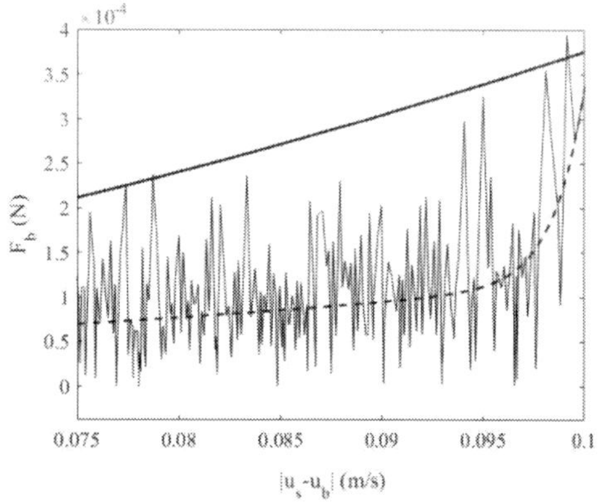

Figure 3. Force on big particle versus relative velocity between granular phases. The thick solid line represents the continuum formula whereas the thin solid line displays the time-averaged force from DEM with δt=2.5ms. The dash line is the exponential fit as: $F_b = a e^{b\,\Delta u_{sb}} + c e^{d\,\Delta u_{sb}}$, where a=8.6x10^{-5}, b=0.1465, c=1.3x10^{-8}, d=4.8.

There is a considerable discrepancy between the trends of the deterioration of drag force F_b with the relative velocity of granular phases corresponding to the DEM and continuum approaches. The continuum formula demonstrates a smooth decay of F_b with the relative velocity, while the DEM shows a sharp decay upon the start of the particle movement inside the cloud of small particles. Then the attenuation of F_b takes a gentle slope. The two approaches show the same starting points, which indicates the initial homogeneity in DEM setup yields the same force predicted by the continuum formula. However, the inhomogeneity driven by the size difference shows its effect in steepening the slope of force-velocity curve immediately.

Table 1. Parameters set in calculations

d_b (mm)	12	e	0.7
d_s (mm)	1	μ_f	0.3
ε_s	0.08	ρ_b (kg.m^{-3})	500
ε_b	0.001	ρ_s (kg.m^{-3})	2500

3.3 Correction to continuum prediction

In order to create any correction scheme to the continuum prediction of solid-solid drag force, depicted in Fig. 3, one must first quantify the inhomogeneity due to the size ratio. Figure 4 shows the vertical profile of the solid volume fraction of the granular phase of small size particles at the moment the big particle has penetrated to a middle position.

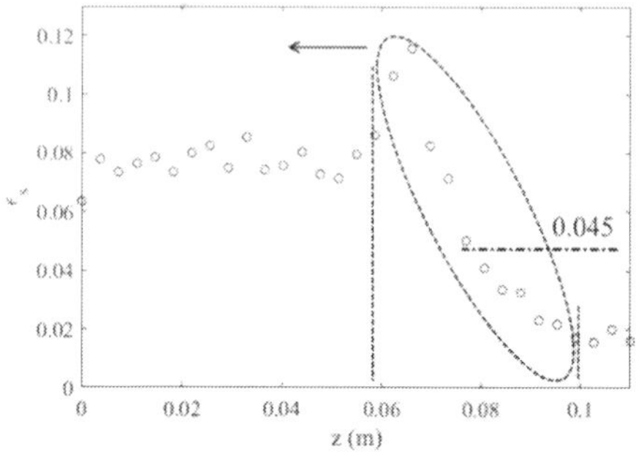

Figure 4. Profile of solid volume fraction of small grains when big particle is at z=0.070m. The area corresponding to each slice is 0.02x0.02m. The big particle is seen at the center of this slice. The affected region is marked by dash lines that gives an average solid volume fraction of 0.045.

The affected region marked by the dashed ellipse in Fig. 4 is persistent which has advection in the direction of the arrow. Therefore, the simplest way to correct the performance of the continuum formula is to modify the solid volume fraction within the affected region. The modified value of ε_s can be used in Eqs. (2-4) to obtain new prediction of F_b. Figure 5 illustrates the force-relative velocity diagram after correcting ε_s to the effective value of 0.045. The dash line represents the continuum formula prediction after correction.

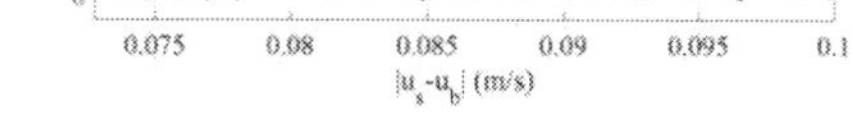

Figure 5. Modified continuum prediction of force on big particle shown by the dash line. The dots stand for the time-averaged values from DEM results and solid line is the exponential fit also shown in Fig. 3.

Note that the force variation is identical for both predictions by the DEM and the continuum formula, however, the initial sharp drop of the force is a transitional event in which the granular pattern develops around the big particle as started from a homogeneous state.

4 Summary and conclusions

Continuum formulation for solid-solid drag force is vital in Eulerian simulations of particulate flows, which can be found in gas-solid multiphase systems. The key assumption in the derivation of the continuum formula is the homogeneity of the mixture, however, this assumption fails when the size ratio between granular phases grows beyond 10. In this study, we employed DEM simulations to computationally measure the solid-solid drag force and test the performance of the continuum formula. It was observed that, 1) if the system is homogeneous, the predictions of drag are fairly close, 2) the size difference imposes a certain region within which the solid volume fraction obtains a profile indicating the inhomogeneous surrounding of the particle. The region has advection within the medium while the volume fraction profile remains fixed around the intruder. By obtaining the effective value of solid volume fraction, we found out that it can be used in the continuum formula to result the same trend of drag force variation with the penetration velocity. However, the correcting solid volume fraction requires a thorough investigation on various size ratios in a variety of solid volume fraction of the bulk. This is currently pursued as future study in continuation of the current study. Ultimately, a correction table will be created in which the solid volume fraction (e.g., the homogeneous value in a computational cell) is regulated depending its bulk value and the size ratio of particles. Single big particle is studied in the first step though clusters of big particles will be investigated in next steps.

Acknowledgement

P.J. would like to thank Walter-Ahlström foundation in Finland for providing the financial support for his research visit to Duke University, where this work was partially performed during this visit. Also, he acknowledges Prof. Robert P. Behringer's support for this visit.

References

1. D. Kunii, O. Levenspiel, *Fluidization Engineering* (R.E. Krieger Publishing Co. Huntington, New York, 1991)
2. M. Syamlal, Topical Report, National Technical InformationService: Springfield, VA, 1987; DOE/MC/21353-2373, NTIS/DE87006500.
3. P. Jalali, T. Hyppänen, Ind. Eng. Chem. Res. **49**, 5270 (2010).
4. Y. Tsuji, T. Tanaka, T. Ishida, Powder Technol. **71**, 239 (1992).

 DOI: 10.1051/epjconf/201714009036

Discrete modelling of front propagation in backward piping erosion

Duc-Kien Tran[1], *Noémie* Prime[2], *Francesco* Froiio[1,⋆], *Carlo* Callari[3], and *Eric* Vincens[1]

[1] *University of Lyon, LTDS - CNRS UMR 5513, Ecole Centrale de Lyon, 36 avenue Guy de Collongue, Ecully 69134 Cedex, France*
[2] *University Savoie Mont - Blanc, LOCIE - CNRS UMR 5271, Campus Universitaire - Savoie Technolac, 73376 Le Bourget du Lac Cedex, France*
[3] *University of Molise/DiBT, Via Francesco De Sanctis, 1, 86100 Campobasso, Italy*

Abstract. A preliminary discrete numerical model of a REV at the front region of an erosion pipe in a cohesive granular soil is briefly presented. The results reported herein refer to a simulation carried out by coupling the Discrete Element Method (DEM) with the Lattice Boltzmann Method (LBM) for the representation of the granular and fluid phases, respectively. The numerical specimen, consisiting of bonded grains, is tested under fully-saturated conditions and increasing pressure difference between the inlet (confined) and the outlet (unconfined) flow regions. The key role of compression arches of force chains that transversely cross the sample and carry most part of the hydrodynamic actions is pointed out. These arches partition the REV into an upstream region that remains almost intact and a downstream region that gradually degrades and is subsequently eroded in the form of a cluster. Eventually, the collapse of the compression arches causes the upstream region to be also eroded, abruptly, as a whole. A complete presentation of the numerical model and of the results of the simulation can be found in [12].

1 Introduction

Piping erosion within embankment dams and dykes or in their foundations is a frequent hazard that may lead these structures to failure [1, 2]. The process is generally described through four ordered phases of evolution: initiation, continuation, progression and breaching [3]. Throughout these different stages, two different erosion mechanisms can be singled out: the enlargement of the pipe is driven by tangential erosion caused by the turbulent flow, while backward erosion induced by Darcy flow prevails at the upstream-propagating pipe tip. Several experimental, analytical and numerical studies on the tangential erosion mechanism have led to substantial advances in the understanding and modelling of the pipe enlargement kinetics but few models for localised backward erosion processes in porous media have been developed [4]. This paper provides a brief account of a first, discrete numerical investigation of the backward erosion process at the pipe tip. The Reader is referred to [12] for a more extended description of this study. Modelling of both the cohesive granular soil and the pore fluid phase in the front region of the pipe is obtained herein by coupling the Discrete Element Method (DEM) with the Lattice Bolzmann Method (LBM), respectively. Recently, a similar DEM-LBM coupling scheme has been used to study the tangential erosion process at the walls of the erosion conduit [5, 6].

⋆e-mail: francesco.froiio@ec-lyon.fr

2 DEM-LBM coupling

An in-house 2D DEM code was used to investigate the micromechanical (grain-scale) processes underwent by the granular skeleton in the pipe tip region. The code was developed following a standard molecular dynamics approach [7]. Grains are represented by circular discs and their interactions are modelled by enabling normal (F_n) and tangential (F_t) forces at the contact points. They can result from a bilateral (bond between grains) or a unilateral contact (no bond). The failure envelope for the bonds is characterised by:

$$F_n > -A \quad \text{and} \quad |F_t| < \mu F_n + C \qquad (1)$$

where the strength parameters $A > 0$ (bond adhesion), $C > 0$ (bond cohesion) and $\mu > 0$ (friction coefficient) are chosen such that $C \geq \mu A$. As the bond breaks, the normal interaction becomes unilateral and the tangential force limit is computed on a purely frictional basis. Gravity is neglected. The same model is employed for the interaction between a grain and a rigid confining wall.

As for the fluid phase, LBM is a kinetic-theory-based numerical approach to fluid dynamics problems. The method is hinged upon a discrete form of the Boltzmann equation and therefore on the determination of the distribution function. The latter is defined as the probability density for the presence of a fluid particle with velocity **c** at position **x** (as random variables) and time t (as a parameter). In the in-house code developed for this study, the Boltzmann equation is discretised according to the nine-velocity square lattice model D2Q9 [13]. The Multiple Relaxation Times (MRT) approximation [10] is used for the collision operator of the Boltzmann equation, to overcome some known drawbacks of the standard Bhatnagar-Gross-Krook (BGK) approximation [8] and improve numerical stability [14]. A detailed description of the in-house LBM code and its validation can be found in [12].

No-slip conditions at stationary as well as moving solid boundaries, straight or curved, are imposed through an interpolated "bounce-back" scheme [9]. The actions exchanged between the fluid and the grains are computed by the LBM algorithm (in terms of momentum transfer from the fluid nodes) and summed, in the DEM algorithm, to the resultant forces and force moments acting on each grain due to contact interactions.

Fluid transport is not possible through a 2D granular packing, due to the pore network not being connected. To overcome this limitation and enable, in the 2D model, typical permeabilities and drag forces of 3D assemblies, the hydraulic radius of a grain in the LBM model is set (reduced) to 0.8 times the reference radius of the same grain in the DEM model [11].

3 Backward erosion test

3.1 Preparation of the granular specimen

The arrangement in Figure 1a represents the numerical specimen employed to model a granular soil REV located on the upstream side of the pipe face (see Figure 2). It consists of 800 circular grains with a core mass density of $2.65 \cdot 10^3$ kg m^{-2} and radii randomly dispersed in the range from 0.75 mm to 0.95 mm.

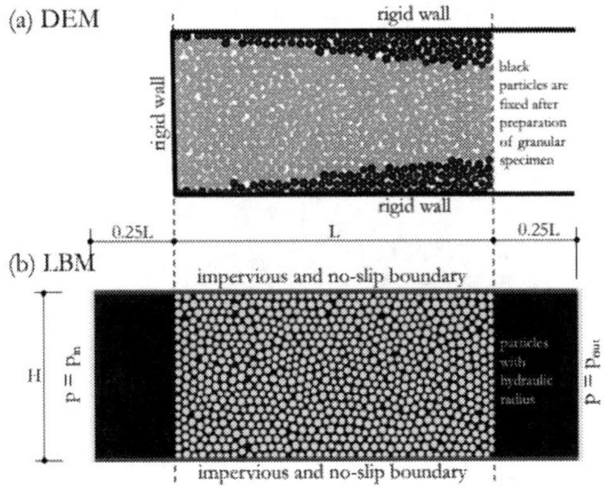

Figure 1. Simulation of backward erosion: geometric configurations (L = 66.8 mm, H = 33 mm), boundary conditions and initial particle arrangements for the coupled models developed with DEM (a) and LBM (b).

The specimen was obtained by a "dry" preparation procedure. A preliminary "lubricated" (*i.e.*, with null contact friction) isotropic compaction was performed, starting from an initial randomly-dispersed configuration, up to a confining pressure of 30 kN m^{-1}. The normal and tangential contact stiffnesses were both set to $5.4 \cdot 10^7$ N m^{-1}. The normal and tangential contact damping coefficients were set at about 80% of the critical value for an harmonic oscillator characterised by the same mass as the average grain. The contact friction coefficient μ (which is actually the same parameter appearing in the bond failure criterion) was chosen equal to 0.5. The contact adhesion was set as A = 1 N, which is consistent with an extremely weak macroscopic tensile traction resistance (of order of magnitude 10^2 N m^{-1}) and the contact cohesion was set as a multiple of the contact adhesion (C = 4 N). New contacts created after the initialisation of the bonds were considered by default as unilateral frictional-visco-elastic. The time step of the DEM algorithm was set to $2.5 \cdot 10^{-7}$ s.

3.2 Flow-induced erosion of the granular specimen

During the erosion test, the specimen resulting from the preparation procedure was confined by a rigid wall on the left (upstream) wall. The latter was maintained at the same position as at the end of the preparation procedure. The right wall used for confining the specimen during compaction had been carefully removed by the end of the preparation procedure. The right (downstream) side of the specimen is therefore unconfined. The degrees of freedom of the black-coloured grains in Figure 1a were locked in order to form two oblique rough boundaries at the top and the bottom, respectively. The resulting trapezoidal shape of this REV region, where the gray-coloured grains are mobile (erodible grains), is reminiscent of an angular sector at the supposedly curved soil-pipe interface in the front region. Figure 1b represents the hydraulic boundary conditions implemented in the simulation, as well as the "hydraulic configuration" of the specimen at the beginning of the test. An incompressible fluid flow, confined by the impervious walls located at the top and the bottom of the sample, was enforced through the REV by the pressure difference $\Delta p = p_{\text{out}} - p_{\text{in}} < 0$ between the right (outlet) and the left (inlet) boundaries of the fluid domain. The fluid is characterised by a mass density $\rho_{\text{f}} = 10^3$ kg m^{-2} and a dynamic viscosity of 10^{-3} N s m^{-1}. The evolution of the normalised pressure difference $-\Delta p/(\rho_{\text{f}} g L)$, where g is the standard gravity acceleration, is plotted in Figure 3i for the whole duration of the simulated test.

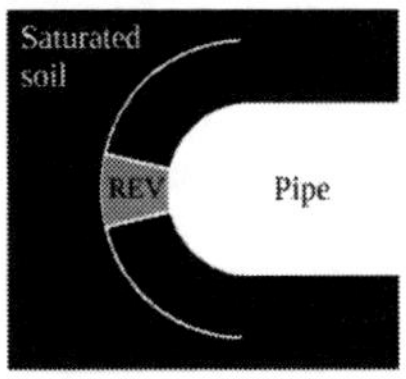

Figure 2. Sketch of the granular REV at the soil-pipe interface in the front region

The set of parameters controlling directly or indirectly the numerical implementation of the LBM model was chosen consistently with a number of requirements. In particular, the spacing between neighbouring lattice nodes was

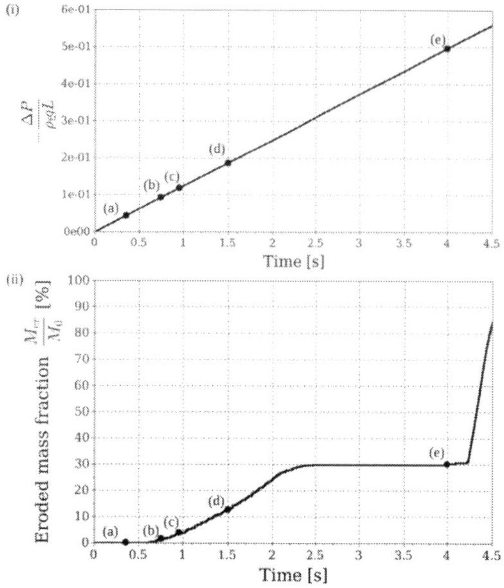

Figure 3. Evolution of the test: normalised pressure difference (i) and eroded mass fraction (ii). Labels refer to time $t = 0.35$ s (a), 0.75 s (b), 0.95 s (c), 1.5 s (d) and 4.0 s (e).

set to $5 \cdot 10^{-5}$ m (*i.e.*, 1/12 of the smallest hydraulic radius) and the lattice time step to 10^{-5} s, which enables an accurate estimate of the hydrodynamic actions. The complete set of LBM parameter and the relevant criteria can be found in [12].

The conservation of the fluid mass is not an implicit feature of the implemented no-slip condition for moving boundaries. Therefore, the total fluid mass was recorded during the first part of the erosion test, up to time $t = 1.55$ s, *i.e.*, before the first grain started exiting the fluid domain through the channel outlet. An increase of 0.63% was measured, which is considered consistent with the required accuracy.

3.3 Results

The evolution of the erosion process during the test is quantified in Figure 3ii in terms of the eroded mass fraction M_{er}/M_0, where M_{er} and M_0 are the eroded and erodible masses, respectively. At a given time instant, the eroded mass is defined as the cumulative mass of the grains having crossed the dashed line marking the right (downstream) limit of the configuration of the granular specimen in Figure1.

At the beginning of the test, for low-enough values of the pressure difference, no erosion was observed. Visual information on the process at this stage is provided in Figure 4a, corresponding to the marks with label (a) in Figure 3. Even at very low values, the hydraulic load affects the stress transmission through force chains within the specimen: a concentration of compressive and tensile force chains can be observed roughly at the upstream and downstream sides of the specimen, respectively. It can be inferred, already from Figure 4b and Figure 4c, that the main resistance mechanism developed in response to the action of the hydrodynamic forces consists of a few main arches of compressive force chains. These arches divide the sample into two parts: the compressive upstream region, which is essentially preserved during the process up to Figure 4(e), and the downstream region that is degrading due to bond breakage as the pressure difference increases. Backward erosion was observed for larger values of the driving pressure difference, roughly in the branch from (b) to (d) in Figure 3 (given the considered definition for eroded grains). The eroded mass is detached and transported by the flow in the form of a large, deformable cluster of grains. The increasing branch after point (d) of the eroded mass diagram in Figure 3ii is to be referred to previously mobilised grains progressively exiting the right limit of the measure volume: no significant further detachment of granular material was observed up to point (e). This was due to the resistance provided by the persisting arches of force chains. The erosion kinetics, by taking the form of released clusters of particles, is compatible with what is observed on site. After point (e), the final increasing branch in Figure 3ii corresponds to the final collapse of the persisting arches, resulting in all the remaining material being washed away simultaneously.

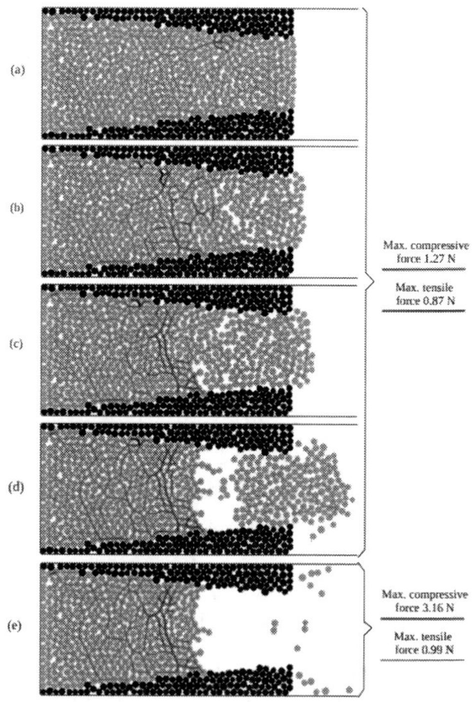

Figure 4. Evolution of the test: compressive and tensile force chains (blue-coloured and red-coloured lines, resp.) and configurations of the specimen at $t = 0.35$ s (a), 0.75 s (b), 0.95 s (c), 1.5 s (d) and 4.0 s (e).

 DOI: 10.1051/epjconf/201714009036

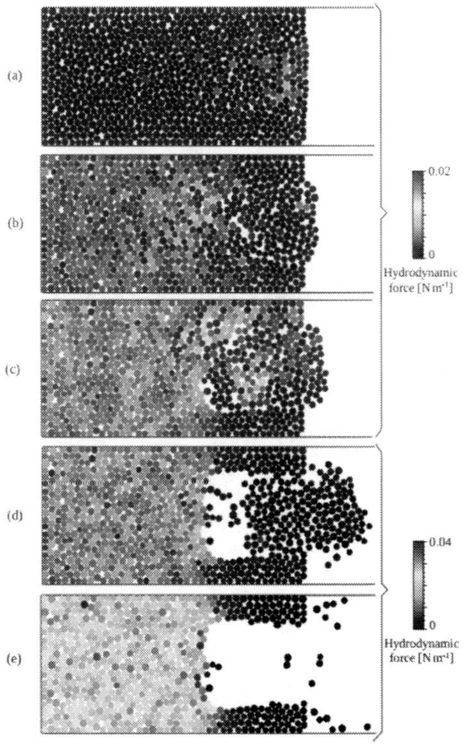

Figure 5. Evolution of the test: norm of the hydrodynamic force on each grain for the configurations of the specimen at t = 0.35 s (a), 0.75 s (b), 0.95 s (c), 1.5 s (d) and 4.0 s (e).

Finally, the plots in Figure 5(a-e) illustrate the evolution of the norm of the hydrodynamic forces on the grains, during the erosion test, and refer to the same configurations identified in Figure 3. Spatial variability of the hydrodynamic forces is observed at the scale of the specimen, with a tendency for lower values towards the downstream side of the granular region. Some randomness at the scale of a few grain diameters can also be observed, especially in Figures 5d and 5e, *i.e.*, for larger values of the imposed pressure difference. Figures 5b to 5d also illustrate the complexity of the interaction between the fluid and solid phases during the backward erosion process.

4 Conclusion

A numerical simulation of the backward erosion process for a REV of cohesive granular soil, at the front region of an erosion pipe, has been performed using a coupled DEM-LBM approach. Throughout the test, the most distinct effect that was observed was the marked arching through force chains, as a self-organised response of the contact/bond network to drag forces and couples on the grains. The main resisting arches that transversely crosses the sample partitioned the REV into a compressive upstream region, that remained almost intact during the largest part of the test, and a downstream region that gradually degraded as the pressure difference increased. Eventually, the downstream material was eroded and clusters of particles were released. The collapse of the compressive arches of force chains, by the end of the test, caused all the remaining material to be washed away as a whole.

Acknowledgment

Research supported by GIS VOR 2012, LTDS 2012 and PRIN 2010-2011 (2010BFXRHS_004) projects as well as by one PhD fellowship funded by the French Ministry of National Education, Higher Education and Research. The research also benefited from several invitations of C. Callari at École Centrale de Lyon – LTDS.

References

[1] da Silveira, A. F., *Safety of dams* (Balkema, Rotterdam, The Netherlands, 1984), p. 55–60
[2] Zhang, L.M and Xu, Y. and Jia, J.S., *Georisk*, **3** (3), p. 184–189 (2011)
[3] Fell, R. and Fry, J.J., *Internal Erosion of Dams and Their Foundations*, (Taylor and Francis/Balkema, Leiden, The Netherlands, 2007) p. 1–23
[4] Wang,D. and Fu,X. and Jie, Y. and Dong,W. and Hu, D., *Soils Found.*, **54** (5), p. 974–984 (2014)
[5] Lominé, F. and Scholtès, L. and Sibille, L. and Poullain, P., *Int. J. Numer. Anal. Methods Geomech.*, **37** (6), p. 577–596 (2013)
[6] Sibille, L. and Lominé, F. and Poullain, P. and Sail, Y. and Marot, D., *Hydro. Process.*, **29** (9), p. 2149–2163 (2015)
[7] Cundall, P. A. and Strack, O. D. L., *Géotechnique*, **29** (1), p. 47–65 (1979)
[8] Bhatnagar, P. L. and Gross, E. P. and Krook, M., *Phys. Rev.*, **94** (3), p. 551–525 (1954)
[9] Bouzidi, M. and Firdaouss, M. and Lallemand, P., *Phys. Fluids*, **13** (11), p. 3452–3459 (2001)
[10] d'Humières, D., *Rarefied Gas Dynamics: Theory and Simulations*, (American Institute of Aeronautics and Astronautics, 1992) p. 450–458
[11] Boutt, David F. and Cook, Benjamin K. and McPherson, Brian J. O. L. and Williams, J. R., *J. Geophys. Ressol. Ea.* , **112** (B10), (2007)
[12] Tran, D. K and Prime, N. and Froiio, F. and Callari, C. and Vincens, E., *Eur. J. Environ. Civ. En.*, (2016), http://dx.doi.org/10.1080/19648189.2016.1248794
[13] Qian, Y. H. and D'Humières, D. and Lallemand, P., , *EPL.*, **17** (6), p. 479–484 (1992)
[14] Lallemand, P. and Luo, L. S., *Phys. Rev. E*, **61** (6), p. 6546–6562 (2000)

Normal stress differences and beyond-Navier-Stokes hydrodynamics

Meheboob Alam[1,*] and *Saikat* Saha[1]

[1] *Jawaharlal Nehru Centre for Advanced Scientific Research, Jakkur PO, Bangalore 560064, INDIA*

Abstract. A recently proposed beyond-Navier-Stokes order hydrodynamic theory for dry granular fluids is revisited by focussing on the behaviour of the stress tensor and the scaling of related transport coefficients in the dense limit. For the homogeneous shear flow, it is shown that the eigen-directions of the second-moment tensor and those of the shear tensor become co-axial, thus making the first normal stress difference ($\mathcal{N}_1$) to zero in the same limit. In contrast, the origin of the second normal stress difference ($\mathcal{N}_2$) is tied to the 'excess' temperature along the mean-vorticity direction and the imposed shear field, respectively, in the dilute and dense flows. The scaling relations for transport coefficients are suggested based on the present theory.

1 Introduction

As in colloidal suspensions and polymeric fluids, the dry granular fluid possesses prominent non-Newtonian properties, like the normal stress differences, shear-thickening and shear-thinning; for example, the normal stress differences can be of the order of its isotropic pressure in a sheared dilute granular gas in contrast to its infinitesimal magnitude in its molecular counterpart [1, 2], see Fig. 1; both the first ($\mathcal{N}_1$) and second ($\mathcal{N}_2$) normal stress differences can change sign depending on the mean density of the granular fluid [3–5]. Such order-one effects must be incorporated in the theoretical modelling of granular fluids so that the theory remains valid from the dilute to the dense limit. Recently we outlined a Grad-like [6] extended hydrodynamic theory for granular fluids [7–9], and derived analytical expressions for normal stress differences and other transport coefficients that are valid for (a) a large range of density (encompassing both the dilute and dense limits) and (b) small restitution coefficient (far away from the limit of elastic collisions). After a brief overview of this theory in Sec. 2, we analyse the stress tensor in the uniform shear flow of granular materials with a focus on the dense-limit behaviour of normal stress differences and their scaling relations as detailed in Secs. 3 and 4.

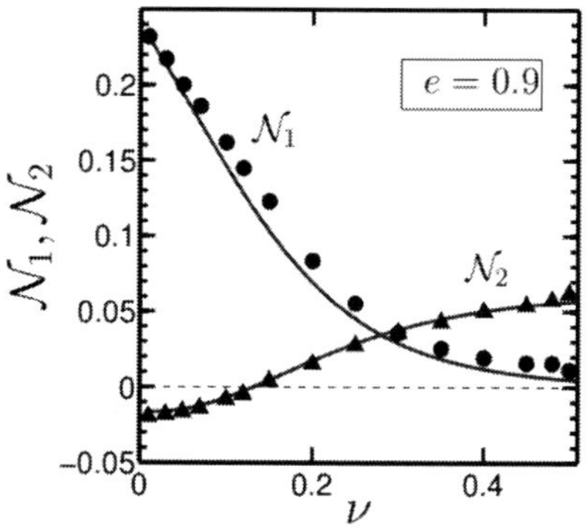

Figure 1. Variations of first ($\mathcal{N}_1$) and second ($\mathcal{N}_2$) normal stress differences with density for a restitution coefficient of $e = 0.9$; the symbols refer to data from event-driven simulation and the lines refer to present theory. Adapted from Fig. 1 of Ref. [8].

2 Overview of 10-Moment Theory

Going beyond the Navier-Stokes-order hydrodynamics of five field variables, here we consider an 'extended' set of 10 hydrodynamic fields: (i) the mass density $\rho(\boldsymbol{x},t) \equiv mn(\boldsymbol{x},t) = m\int f(\boldsymbol{c},\boldsymbol{x},t)\mathrm{d}\boldsymbol{c}$, (ii) the coarse-grained velocity $\boldsymbol{u}(\boldsymbol{x},t) \equiv \langle \boldsymbol{c}\rangle = \frac{1}{n(\boldsymbol{x},t)}\int \boldsymbol{c}f(\boldsymbol{c},\boldsymbol{x},t)\mathrm{d}\boldsymbol{c}$, and (iii) the second moment tensor $\boldsymbol{M}(\boldsymbol{x},t) \equiv \langle \boldsymbol{CC}\rangle = \frac{1}{n(\boldsymbol{x},t)}\int \boldsymbol{CC}f(\boldsymbol{c},\boldsymbol{x},t)d\boldsymbol{c}$, where $n(\boldsymbol{x},t)$ is the number density, $\boldsymbol{C} \equiv \boldsymbol{c} - \boldsymbol{u}$ is the peculiar velocity of particles and $f(\boldsymbol{c},\boldsymbol{x},t)$ is the single-particle distribution function. The last hydrodynamic field $\boldsymbol{M}(\boldsymbol{x},t)$ is required to account for normal stress differences [7, 8]. The balance equations for the mass, momentum and second-moment, respectively, can be obtained by taking appropriate moment of the Enskog-Boltzmann equation:

$$\frac{D\rho}{Dt} = -\rho\nabla\cdot\boldsymbol{u}, \tag{1}$$

$$\rho\frac{D\boldsymbol{u}}{Dt} = -\nabla\cdot\boldsymbol{P}, \tag{2}$$

$$\rho\frac{D\boldsymbol{M}}{Dt} = -\nabla\cdot\boldsymbol{Q} - \boldsymbol{P}\cdot\nabla\boldsymbol{u} - (\boldsymbol{P}\cdot\nabla\boldsymbol{u})^T + \boldsymbol{\aleph}, \tag{3}$$

where $D/Dt = (\partial/\partial t + \boldsymbol{u}\cdot\nabla)$ is the material derivative, $\boldsymbol{P}$ is the stress tensor, given by

$$\boldsymbol{P} \equiv \rho\boldsymbol{M} + \boldsymbol{\Theta}(m\boldsymbol{C}), \tag{4}$$

*e-mail: meheboob@jncasr.ac.in

Q is the flux of the second moment, a third-rank tensor, given by

$$Q \equiv \rho\langle CCC\rangle + \Theta(mCC), \tag{5}$$

and $\aleph$ is the collisional source of second moment, a second-rank tensor, given by

$$\aleph \equiv \aleph(mCC). \tag{6}$$

In Eqns. (4-5), the first and second terms refer to the kinetic and collisional contributions, respectively.

It is straightforward to show that the balance equation for the granular energy follows from the trace of (3),

$$\frac{3}{2}\rho\frac{DT}{Dt} = -\nabla\cdot\mathbf{q} - P : \nabla u - \mathcal{D}, \tag{7}$$

where $T \equiv M_{\alpha\alpha}/3$ is the granular temperature,

$$q_\alpha \equiv \frac{1}{2}Q_{\alpha\beta\beta} = \frac{1}{2}\rho M_{\alpha\beta\beta} + \frac{1}{2}\Theta_{\alpha\beta\beta} \tag{8}$$

is the heat flux vector, and

$$\mathcal{D} \equiv -\frac{1}{2}\aleph_{\beta\beta} = -\frac{1}{2}\aleph\left(mC^2\right) \tag{9}$$

is the rate of dissipation of kinetic energy per unit volume. The balance equations (1-2, 7), along with closure relations for (4), (8) and (9), constitute the Navier-Stokes-order hydrodynamics for which the equation for the deviatoric part of the second moment is satisfied identically.

For an extended hydrodynamic theory in terms of ten moments, the balance equation (3) for full second moment tensor along with mass and momentum balances (1-2) are needed. For a closure of (3), the deviatoric part of the third-order $Q_{\gamma\alpha\beta}$,

$$\widehat{Q}_{\gamma\alpha\beta} = Q_{\gamma\alpha\beta} - \frac{1}{5}\left(Q_{\gamma\xi\xi}\delta_{\alpha\beta} + Q_{\alpha\xi\xi}\delta_{\gamma\beta} + Q_{\beta\xi\xi}\delta_{\alpha\gamma}\right), \tag{10}$$

is assumed to be zero, leaving only its isotropic part, the heat flux vector (8), to be evaluated as a constitutive relation. Recently we showed [7] that the heat-flux follows a generalized Fourier law of the form

$$\mathbf{q} = -\boldsymbol{\kappa}\cdot\nabla T - \boldsymbol{\kappa}_\rho\cdot\nabla\rho - \boldsymbol{\Psi}\dot{:}\nabla\widehat{\boldsymbol{M}}, \tag{11}$$

where $\boldsymbol{\kappa}$ is the thermal conductivity tensor (of second rank) and $\boldsymbol{\kappa}_\rho$ is the Dufour tensor; the fourth-rank tensor $\boldsymbol{\Psi}$ is tied to non-equilibrium heat-flux due to the stress-gradient (note that $\widehat{\boldsymbol{M}}$ is the deviatoric part of the second moment which is proportional to kinetic shear stress and normal stress differences). Collectively, the anisotropic thermal conductivity $\boldsymbol{\kappa}$ and the stress-gradient conductivity $\boldsymbol{\Psi}$ are responsible for the "parallel" heat-flux (along the flow direction in the absence of temperature gradient); the latter is a well-known phenomenon in rarefied gases [6]. Closed-form expressions of $\boldsymbol{\kappa}$, $\boldsymbol{\kappa}_\rho$ and $\boldsymbol{\Psi}$ for a dilute granular gas undergoing shear flow can be found in Ref. [7, 9].

The balance equations (1-3), along with closure relations for the stress tensor (4), the source of second moment (6) and the heat-flux vector (11), constitute the 10-moment theory for a granular fluid – this represents the minimal continuum model that incorporates (i) normal stress differences as well as (ii) anisotropic heat flux [7, 8]. In the remainder of this paper, we will analyse the dense-limit behaviour of the stress tensor and related transport coefficients for uniform shear flow for which the anisotropic Gaussian [2, 7, 8] serves as the reference distribution function. The latter follows from the maximum entropy principle.

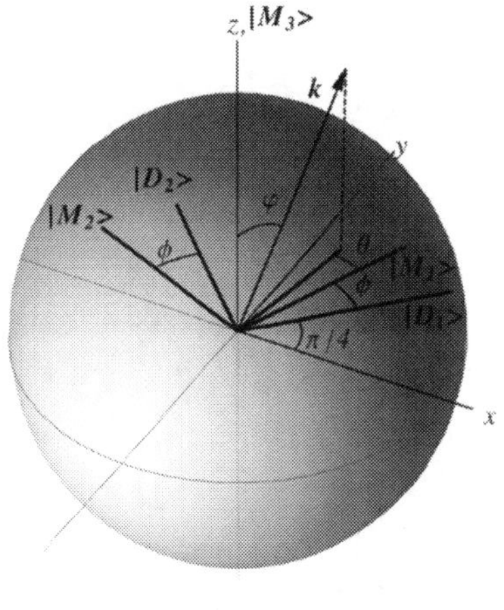

Figure 2. Schematic of the coordinate system displaying the eigen-basis of the shear tensor $\boldsymbol{D}$ and the second-moment tensor $\boldsymbol{M}$. Note that the contact vector $\boldsymbol{k}$ makes an angle φ with $|M_3\rangle$, and θ is the angle between $|M_1\rangle$ and $\boldsymbol{k}-(\boldsymbol{k}\cdot\boldsymbol{z})\boldsymbol{z}$, the projection of $\boldsymbol{k}$ on the shear plane. The uniform shear flow, $\boldsymbol{u} = (2\dot{\gamma}y, 0, 0)$, is directed along the x-direction, with the velocity gradient along the y-direction and the mean-vorticity along the z-direction. Adapted from Fig. 2 of Ref. [8].

3 Uniform Shear Flow (USF) and the Second-moment Balance

The coordinate system is depicted in Fig. 2, with the uniform shear flow (USF) occurring in the (x, y)-plane; the x and y denote flow and gradient directions, respectively, and the z direction is perpendicular to x-y plane. The velocity gradient tensor,

$$\nabla\boldsymbol{u} = \begin{bmatrix} 0 & 2\dot{\gamma} & 0 \\ 0 & 0 & 0 \\ 0 & 0 & 0 \end{bmatrix} \equiv \boldsymbol{D} + \boldsymbol{W}, \tag{12}$$

completely characterizes the uniform shear flow ($2\dot{\gamma}$ is the overall shear rate), with $\boldsymbol{D}$ and $\boldsymbol{W}$ denoting the shear and spin tensors, respectively. It is straightforward to verify that $\dot{\gamma}$, $-\dot{\gamma}$ and 0 are the eigenvalues of $\boldsymbol{D}$ and the corresponding orthonormal eigenvectors are, respectively,

$$|D_1\rangle = \begin{bmatrix} \cos\frac{\pi}{4} \\ \sin\frac{\pi}{4} \\ 0 \end{bmatrix}, \; |D_2\rangle = \begin{bmatrix} -\sin\frac{\pi}{4} \\ \cos\frac{\pi}{4} \\ 0 \end{bmatrix}, \; |D_3\rangle = \begin{bmatrix} 0 \\ 0 \\ 1 \end{bmatrix}. \tag{13}$$

While $|D_3\rangle$ is directed along the z-axis, the shear-plane eigenvectors $|D_1\rangle$ and $|D_2\rangle$ are rotated by 45° anticlockwise from the xy-axes.

3.1 Construction of second moment tensor

We can decompose the second moment tensor $\boldsymbol{M}$ into an isotropic tensor and a traceless deviatoric part $\boldsymbol{M} = T[\delta_{\alpha\beta}] + \widehat{\boldsymbol{M}}$, where $\widehat{\boldsymbol{M}}/T$ is the dimensionless counterpart of its deviatoric tensor whose eigenvalues ξ, ς and ζ satisfy $\xi + \varsigma + \zeta = 0$. Therefore the eigenvalues of $\boldsymbol{M}$ are $T(1+\xi)$, $T(1+\varsigma)$ and $T(1+\zeta)$, with the corresponding orthonormal set of eigen-directions being denoted by $|M_1\rangle$, $|M_2\rangle$ and $|M_3\rangle$, respectively. The second-moment tensor can be expressed as

$$\boldsymbol{M} = T(1+\xi)|M_1\rangle\langle M_1| + T(1+\varsigma)|M_2\rangle\langle M_2| + T(1+\zeta)|M_3\rangle\langle M_3|. \tag{14}$$

Referring to Fig. 2, we assume that the shear-plane eigenvectors $|M_1\rangle$ and $|M_2\rangle$ can be obtained by rotating the system of axes at an angle $(\pi/4 + \phi) \equiv \overline{\phi}$, with the angle ϕ being unknown, in the anti-clockwise sense about the z-axis which coincides with $|M_3\rangle$:

$$|M_1\rangle = \begin{bmatrix} \cos\overline{\phi} \\ \sin\overline{\phi} \\ 0 \end{bmatrix}, \quad |M_2\rangle = \begin{bmatrix} -\sin\overline{\phi} \\ \cos\overline{\phi} \\ 0 \end{bmatrix}, \quad |M_3\rangle = \begin{bmatrix} 0 \\ 0 \\ 1 \end{bmatrix}. \tag{15}$$

Inserting (15) into (14), we obtain the following expression for deviator of the second moment tensor

$$\widehat{\boldsymbol{M}} = T \begin{bmatrix} \lambda^2 + \eta\sin 2\phi & -\eta\cos 2\phi & 0 \\ -\eta\cos 2\phi & \lambda^2 - \eta\sin 2\phi & 0 \\ 0 & 0 & -2\lambda^2 \end{bmatrix}, \tag{16}$$

where we have introduced the following notations

$$\eta \equiv \frac{1}{2}(\varsigma - \xi) \geq 0 \quad \text{and} \quad \lambda^2 \equiv \frac{1}{2}(\varsigma + \xi) = -\frac{\zeta}{2} \geq 0. \tag{17}$$

Note that while $\eta \sim (T_x - T_y)$ is a measure of the *anisotropy* of the second-moment tensor in the shear plane, λ^2 is a measure of the *excess temperature* [8] [$T^{ex} = (T - T_z)$, where T_i is the temperature along the i-th direction] along the mean vorticity direction. It is straightforward to verify that the eigenvalues in the shear-plane can be expressed in terms of η and λ via

$$\xi = \lambda^2 - \eta \quad \text{and} \quad \varsigma = \lambda^2 + \eta > \xi, \tag{18}$$

with the eigenvalue, ζ, along the vorticity direction (z), being given by Eq. (17).

3.2 The balance of second moment in USF

In the steady uniform shear flow, the number density n, the velocity gradient $\nabla\boldsymbol{u}$ and the components of $\boldsymbol{M}$ are constants and the contracted third-moment vanishes. Therefore, the mass and momentum balance equations, (1) and (2), are identically satisfied. The remaining balance equation (3), $P_{\delta\beta}u_{\alpha,\delta} + P_{\delta\alpha}u_{\beta,\delta} = \aleph_{\alpha\beta}$, for the second-moment simplifies to

$$\rho M_{\delta\beta}(D_{\alpha\delta} + W_{\alpha\delta}) + \rho M_{\delta\alpha}(D_{\beta\delta} + W_{\beta\delta}) + \Theta_{\delta\beta}D_{\alpha\delta} + \Theta_{\delta\alpha}D_{\beta\delta} = A_{\alpha\beta} + \widehat{E}_{\alpha\beta} + \widehat{G}_{\alpha\beta}, \tag{19}$$

where we made use of $u_{\alpha,\delta} = (D_{\alpha\delta} + W_{\alpha\delta})$, $P_{\alpha\beta} = \rho M_{\alpha\beta} + \Theta_{\alpha\beta}$ and the following decomposition for the collisional source of second moment:

$$\aleph_{\alpha\beta} = A_{\alpha\beta} + \widehat{E}_{\alpha\beta} + \widehat{G}_{\alpha\beta} + \Theta_{\alpha\delta}W_{\beta\delta} + \Theta_{\beta\delta}W_{\alpha\delta}. \tag{20}$$

The Savage-Jeffrey number [10], or, the dimensionless shear-rate is defined as

$$R \equiv \frac{\dot\gamma}{4\sqrt{T/\sigma^2}} = \left(\frac{\sqrt{T}}{\sigma\dot\gamma/4}\right)^{-1} \equiv \frac{v_{sh}}{v_{th}}. \tag{21}$$

This parameter, which can also be interpreted as the inverse of the square root of dimensionless temperature, is a measure of the mean shear velocity ($v_{sh} = \sigma\dot\gamma/2$ over a particle diameter) relative to the thermal velocity ($v_{th} \propto \sqrt{T}$) associated with the random motion of particles. The second-moment tensor (14,16) in USF, constructed from its eigen-basis, is completely determined when R, η, ϕ and λ^2 are specified. Therefore the second-moment balance (19) can now be solved in terms of above four parameters for specified values of (i) the mean density ν and (ii) the restitution coefficient e.

4 Dense Limit: Solution and Discussion

The perturbation solution of equation (19), along with the resulting analytical expressions for the normal stress differences and other transport coefficients (shear viscosity, pressure, cooling rate, etc), have been detailed in Ref. [8]. Figure 1 illustrates the accuracy of theoretical predictions for two normal stress differences,

$$\mathcal{N}_1 = (P_{xx} - P_{yy})/p, \quad \text{and} \quad \mathcal{N}_2 = (P_{yy} - P_{zz})/p, \tag{22}$$

where $p = P_{\alpha\alpha}/3$ is the mean pressure, via a comparison with particle simulation data.

Here we determine an approximate solution of the second-moment balance equation in the dense limit ($\nu \to \nu_{\max}$). Since the collisional mechanism of momentum transfer dominates over its kinetic contribution as $\nu \to \nu_{\max}$, the stress tensor can be approximated by its collisional contribution, i.e. $P_{\alpha\beta} = \Theta_{\alpha\beta}$ which has an integral expression [8]. The balance equations (19) simplify to

$$\left.\begin{aligned} A_{x'x'} + A_{y'y'} + A_{z'z'} &= -3\widehat{\Gamma}_{z'z'} \\ 2\dot\gamma[(\Theta_{x'x'} - \Theta_{y'y'})\cos 2\phi - 2\Theta_{x'y'}\sin 2\phi] &= -3\widehat{\Gamma}_{z'z'} \\ 2\dot\gamma(\Theta_{x'x'} + \Theta_{y'y'})\cos 2\phi &= (\Gamma_{x'x'} - \Gamma_{y'y'}) \\ -\dot\gamma(\Theta_{x'x'} + \Theta_{y'y'})\sin 2\phi &= \Gamma_{x'y'} \end{aligned}\right\} \tag{23}$$

where $\Gamma_{\alpha\beta} = A_{\alpha\beta} + \widehat{E}_{\alpha\beta} + \widehat{G}_{\alpha\beta}$, with each tensorial quantity having an analytic integral expression which is evaluated as infinite series [7, 8] as functions of η, λ, ϕ and R^2.

4.1 Perturbation solution in the dense limit

Substituting the expressions [8] for the integrals $(\Theta_{x'x'} + \Theta_{y'y'})$ and $\Gamma_{x'y'}$ into the last equation of (23) yields $\phi = 0$; the remaining three equations can be solved perturbatively. The leading-order solution of (23) is given by

$$\eta = 0 = \phi, \lambda^2 = \frac{10(1-e)}{21(3-e)}, \quad R^2 = \frac{15(1-e)}{16(1+3e)}. \tag{24}$$

Since $\eta,\phi \to 0$, the first normal stress difference ($\mathcal{N}_1 \propto \eta \sin 2\phi$) vanishes as $\nu \to \nu_{max}$, but the second normal stress difference remains finite ($\mathcal{N}_2 \propto \lambda^2 \neq 0$) in the dense limit. The solution (24) indicates that $R, \lambda \sim \sqrt{(1-e)}$ and therefore both the shear-rate and the excess temperature scale with inelasticity as $\sqrt{(1-e)}$ in the dense limit.

The next-order corrections for η, R^2, λ^2 and ϕ have also been calculated and all transport coefficients (p, μ, $\mathcal{N}_1$ and $\mathcal{N}_2$) can be expressed as functions of ν and e (in dimensionless form):

$$p^* \approx \frac{(1+e)\nu^2 g_0}{110880R^2}\left(3465 + 7392R^2 + \eta R\chi_1 \cos 2\phi\right) \tag{25}$$

$$\begin{aligned}\mu^* \approx \frac{(1+e)\nu^2 g_0}{18480\sqrt{\pi}R}\Big[3\Big(616 + 77\sqrt{\pi}\frac{\eta\cos 2\phi}{R} \\ + 4\lambda^2(44 + 8\eta^2 - 11\lambda^2 + 4\lambda^4)\Big) + 128(11 - 4\lambda^2)R^2 \\ - 4\eta^2\left(22 + (11 - 12\lambda^2)(1 + 2\cos^2 2\phi)\right)\Big],\end{aligned} \tag{26}$$

$$\mathcal{N}_1 \approx \frac{12\left(231 - \frac{8}{\sqrt{\pi}}\eta R(11 - 12\lambda^2)\cos 2\phi\right)\eta \sin 2\phi}{(231(15 + 32R^2) + \eta R\chi_1 \cos 2\phi)}, \tag{27}$$

$$\mathcal{N}_2 \approx \frac{6\left(33(32R^2 - 7\eta\sin 2\phi + 21\lambda^2) + \eta R\chi_2 \cos 2\phi\right)}{(231(15 + 32R^2) + \eta R\chi_1 \cos 2\phi)}, \tag{28}$$

where

$$\chi_1 = \frac{8}{\sqrt{\pi}}\left(462 + 33\eta^2 - 352R^2 - 132\lambda^2 + 99\lambda^4\right)$$

$$\chi_2 = \frac{8}{\sqrt{\pi}}\left(66 + 6\eta^2 - 64R^2 - 33\lambda^2 + 18\lambda^4 + \eta(11 - 12\lambda^2)\sin 2\phi\right)$$

and g_0 is the contact value of the pair-correlation function [11]. A graphical comparison (not shown) of the above two dense-limit solutions for the first and second normal stress differences with the respective exact numerical-solution of second moment equation confirms the accuracy of present analytical solutions. The mechanistic origin of $\mathcal{N}_1$ and $\mathcal{N}_2$ is evident from (27-28).

4.2 Discussion

Although the Enskog-Boltzmann equation is unlikely to hold close to the maximum packing ($\nu \to \nu_{max}$) limit, it is interesting to know the predictions of the present theory on two specific issues in the dense limit: (i) the co-linearity between the eigen-directions of the stress and shear tensors and (ii) the scaling of the transport coefficients.

4.2.1 Are the stress and shear tensors co-linear?

Employing the analytical solution for η, λ, R and ϕ given in (24), the closed-form expressions for the non-zero elements of the collisional stress tensor, $\boldsymbol{P}^c = \boldsymbol{\Theta}(mC)$, can be obtained whose eigenvalues have been calculated (not written here). The corresponding eigenvectors are

$$|P_1^c\rangle = \begin{bmatrix} 1 \\ 1 \\ 0 \end{bmatrix}, \quad |P_2^c\rangle = \begin{bmatrix} -1 \\ 1 \\ 0 \end{bmatrix}, \quad |P_3^c\rangle = \begin{bmatrix} 0 \\ 0 \\ 1 \end{bmatrix}, \tag{29}$$

which coincide with the eigen-directions of the shear tensor, $\boldsymbol{D} = (\nabla \boldsymbol{u} + (\nabla \boldsymbol{u})^T)/2$, given in (13). Since (29) holds exactly at $\nu = \nu_{max}$, we conclude that the eigen-directions of $\boldsymbol{P}^c$ and $\boldsymbol{D}$ are co-linear at $\nu = \nu_{max}$, but they would be non-colinear when the density effects are taken into account for $\nu < \nu_{max}$. The latter has been checked via the numerical solution of second moment equation [7, 8].

4.2.2 Scaling relations

We see from (25) and (26) that both pressure and viscosity scale with g_0 in the dense limit. Using Torquato's form [11] for 'equilibrium' hard-sphere system, we have

$$\mu, p \sim g_0 \sim (\nu_{max} - \nu)^{-1}. \tag{30}$$

This exponent is much smaller than 2, found empirically from experiments and simulations on non-Brownian suspensions. (Leaving aside subtle many-body correlations, accurate data on how g_0 diverges near jamming is therefore important for any meaningful conclusion.) Among two normal stress differences, $\tilde{\mathcal{N}}_1 \to 0$ as $\nu \to \nu_{max}$ but the second normal stress difference diverges like

$$\tilde{\mathcal{N}}_2 = (P_{yy} - P_{zz}) \sim p\mathcal{N}_2 \sim (\nu_{max} - \nu)^{-1}, \tag{31}$$

(similar to the scaling of pressure) since its scaled version $\mathcal{N}_2$ remains finite at $\nu = \nu_{max}$ (see Fig. 1).

It is reasonable to argue that the onset of jamming is associated with a diverging length scale, $L_J \sim (\nu_{max} - \nu)^{-\alpha}$, with $\alpha \sim 0.73$ [12]. This can be tied with a lesser amount of collisional dissipation ($\mathcal{D} \sim L_J^{-1}$) as $\nu \to \nu_{max}$ and hence the temperature is likely to diverge as $T \sim (\nu_{max} - \nu)^{-2\alpha}$ which follows from the energy balance equation. Incorporating these informations into the expression of shear viscosity ($\mu \sim g_0 L_J \sqrt{T}$) yields an exponent for the viscosity-divergence [$\mu \sim (\nu_{max} - \nu)^{-\delta}$] as $\delta = (1 + 2\alpha) \approx 2.5$ which is very close to simulation results [13] of $\delta = 2.58 \pm 0.1$ for frictionless systems. On the other hand, for $\alpha = 1/2$, $L_J \sim (\nu_{max} - \nu)^{-1/2}$ and $T \sim (\nu_{max} - \nu)^{-1}$ which recovers the empirical scaling relation, $\mu \sim (\nu_{max} - \nu)^{-2}$, for shear-viscosity near jamming. These exponents can be verified in future simulations.

References

[1] N. Sela, I. Goldhirsch, J. Fluid Mech. **361**, 41 (1998)
[2] J. T. Jenkins, M. W. Richman, J. Fluid Mech. **192**, 313-328 (1988)
[3] M. Alam, S. Luding, J. Fluid Mech. **476**, 69 (2003)
[4] M. Alam, S. Luding, Phys. Fluids **15**, 2298 (2003)
[5] M. Alam, S. Luding, In *Powders and Grains* (Eds: R. Garcia-Rojo, H.J. Herrmann & S. McNamara; A. A. Balkema, Netherlands, 2005), pp. 1141-1144.
[6] H. Grad, Commun. Pure Appl. Math. **2**, 331-407 (1949)
[7] S. Saha, M. Alam, J. Fluid Mech. **757**, 251-296 (2014)
[8] S. Saha, M. Alam, J. Fluid Mech. **795**, 549-580 (2016)
[9] S. Saha, M. Alam, Preprint (2017, to be submitted)
[10] S. Savage, D. Jeffrey, J. Fluid Mech. **110**, 255 (1981)
[11] S. Torquato, Phys. Rev. E **51**, 3170-3184 (1995)
[12] T. Hatano, Phys. Rev. E **79**, 050301 (2009)
[13] P. Olsson, S. Teitel, Phys. Rev. Lett **109**, 108001 (2012)

Stopping dynamics of a steady uniform granular flow over a rough incline

Stéphanie Deboeuf[1,⋆], *Guillaume* Saingier[1,2], *Nitharshini* Thiruvalluvar[1], *Pierre-Yves* Lagrée[1], *Stéphane* Popinet[1], and *Lydie* Staron[1]

[1]*Sorbonne Universités, UPMC Université Paris 6, CNRS UMR 7190, Institut Jean le Rond d'Alembert, F-75005 Paris, France*
[2]*Surface du Verre et Interface, UMR 125, CNRS/Saint-Gobain, 93303 Aubervilliers, France*

Abstract. Granular material flowing on complex topographies are ubiquitous in industrial and geophysical situations. Even model granular flows are difficult to understand and predict. Recently, the frictional rheology $\mu(I)$ -describing the ratio of the shear stress to the normal stress as a function of the inertial number I, that compares inertial and confinement effects- allows unifying different configurations of granular flows. However it does not succeed in describing some phenomenologies, such as creep flow, deposit height, ... Is it attributable to the rheology, to non-local effects, ...? Here, we consider a thin layer of grains flowing steadily and uniformly on a rough incline, when the input mass flow rate is suddenly stopped. We focus on the arrest dynamics by using both experimental and numerical approaches. We measure the height and surface velocities of the granular layer during the long-time stopping dynamics and we compare our experimental results with computations of depth-averaged equations for a fluid of rheology $\mu(I)$.

1 Introduction

In order to be able to predict flows of divided matter encountered in the nature or in the industry, a first step is to gain in the understanding of small-scale controlled flows of model grains [1]. We study here a model gravitational flow of grains over an inclined plane, that is quasi one dimensional at first order [2]. We compare experimental results obtained with macroscopic hard frictional spheres and numerical results of depth-averaged equations in 1D of a thin continuous flow of frictional rheology $\mu(I)$ [2–4]. This relates the friction coefficient μ, equal to the ratio of the shear stress τ and the normal stress P, to the inertial number I, depending on the shear rate $\dot{\gamma}$ and the pressure P:

$$\mu(I) = \frac{\tau}{P}, \quad \text{with } I = \frac{\dot{\gamma} d}{\sqrt{P/\rho}}, \qquad (1)$$

d and ρ being the diameter and the density of grains. This friction coefficient allows to describe the dissipation occurring in the granular flow. Another way to write it, introducing a non Newtonian viscosity η, is [5]: $\tau = \eta\dot{\gamma}$, with $\eta = \mu(I)P/\dot{\gamma}$. Predictions from the rheology $\mu(I)$ were already compared to some steady flows and some unsteady flows, such as spreading of a small volume of grains over an incline [6, 7], initiation of granular surface flows in a narrow channel [8], or granular column collapses [5, 9–11]. The arrest of flowing grains was also studied in the case of a dam break [11] or flow over a slope [12, 13]. Here we focus on the stopping dynamics of a steady uniform flow over an incline when the mass flow rate vanishes suddenly.

⋆e-mail: sdeboeuf@dalembert.upmc.fr

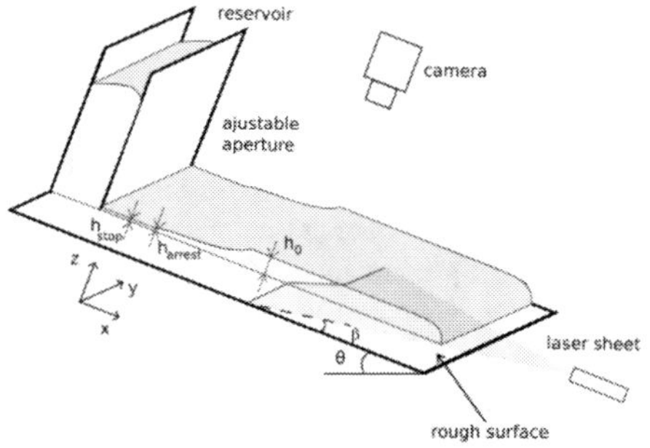

Figure 1. The reservoir aperture of an inclined plane can be suddenly closed to let the steady granular flow relax towards a static deposit

2 Experiments

The slope of a plane (Figure 1), 2m long and 40cm wide, with polyethylene plates for side walls, can be chosen at a fixed angle θ. The aperture of a reservoir of grains at the top of the plane can be opened or closed quickly, allowing for the release or the arrest of a steady uniform granular flow. A gate allows to adjust the aperture thickness h_{gate} in order to control the mass flow rate. The bottom surface of the plane is made rough by covering it with sand-paper of roughness about 200μm. This boundary condition should allow for a zero slip velocity at the bottom. Grains are spherical glass beads of diameter $d = 200 \pm 50\mu m$. This set-up allows for steady flows propagating at a constant velocity for some range of slope angles ($25° \leq \theta \leq 35°$)

and aperture thicknesses (5mm$\leq h_{gate} \leq$ 30mm). When opening the reservoir aperture, a granular front flows down the slope at constant velocity u_0 with a steady shape $h(x - u_0 t)$ [14, 15]. Behind the front, the layer has a uniform height h_0. How this steady uniform flow stops when the aperture is suddenly closed?

Height profiles are measured thanks to laser profilometry: one laser sheet illuminates transversally the surface at a low incident angle, giving the temporal evolution of height $h(t)$ at a given position. About 1% of grains are dyed, allowing for the measurement of the surface velocity $\mathbf{u}_s(t)$ using a particle imaging velocimetry algorithm. To this aim, movies of size 1280x720pixels with a spatial resolution of about 4pixels/mm at a frequency of 30Hz are recorded. It is possible to infer the mean behaviour of the granular flow from surface kinematics h and $\mathbf{u}_s(t)$, thanks to the zero slip velocity at the bottom, and assuming that the whole layer flows and is sheared. However this might be not true during the stopping dynamics.

3 Numerical simulations

In order to get predictions on the stopping of a flow for a frictional rheology $\mu(I)$, we solve numerically 1D depth-averaged equations [6]:

$$\frac{\partial h}{\partial t} + \frac{\partial}{\partial x}(h\overline{u}) = 0, \tag{2}$$

$$\frac{\partial}{\partial t}(h\overline{u}) + \alpha\frac{\partial}{\partial x}(h\overline{u}^2) = hg\cos\theta\,(\tan\theta - \mu(\overline{I}) - \frac{\partial h}{\partial x}), \tag{3}$$

using the open source Navier-Stokes code initiated by S. Popinet (Basilisk) [16]. The streamwise and vertical coordinates are x and z, h is the thickness of the layer, $\overline{u}$ denotes the depth-averaged streamwise velocity and g is the gravity acceleration. The granular layer is assumed to be thin enough, transversally uniform, and at a constant solid fraction $\phi \approx 0.60$ (incompressible). The shape factor $\alpha = \int_0^h u^2(z)dz/(h\overline{u}^2)$ is taken equal to $\alpha = 1$, despite recent results that demonstrate the relevance to make other assumptions [15]. The right hand side terms of Eq. (3) are the gravity along the slope, the friction and the pressure gradient. The friction term can be interpreted either as a basal friction [6] or as the averaged internal friction over the thickness [17]. The friction $\mu(I)$ [4] is expressed as:

$$\mu(I) = \mu_0 + \frac{\Delta\mu}{I_0/I + 1}, \tag{4}$$

where $\mu_0 = 0.41$, $\Delta\mu = 0.35$ and $I_0 = 0.38$ were measured from experiments of uniform and steady flows and successfully compared with data of non uniform flow fronts for the same set-up [15]. The inertial number $\overline{I}$ in the depth-averaged approach is computed for $\alpha = 1$ as:

$$\overline{I} = \frac{\overline{u}d}{h\sqrt{\phi g h\cos\theta}}. \tag{5}$$

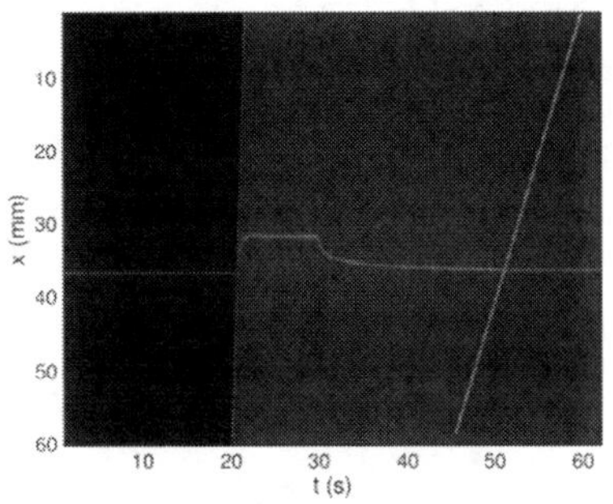

Figure 2. Space-time diagram showing the time evolution of one line of pixels taken along the streamwise direction x. The red and green lines indicate the velocity of the front and of the arrest wave respectively

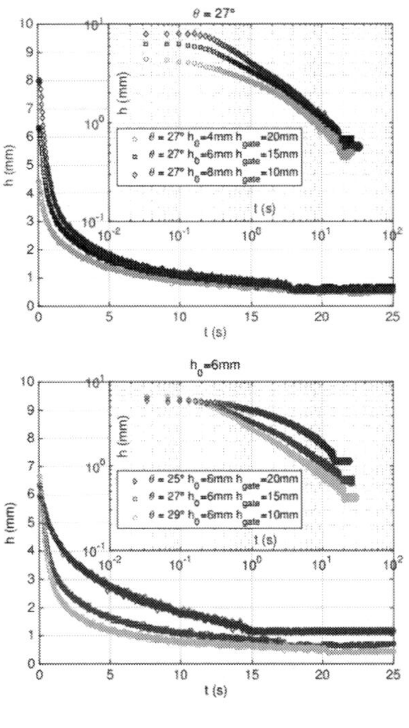

Figure 3. Time evolution of height $h(t)$ during the arrest of a flow when closing the reservoir gate from experiments at a constant slope angle $\theta = 27°$ for different thicknesses h_0 = 4mm, 6mm and 8mm (a) and at a constant initial height h_0 = 6mm at different slope angles θ = 25°, 27°, 29° (b) in linear and logarithmic scales

4 Transient stopping dynamics

4.1 Typical phenomenology

The typical phenomenology of the transient stopping dynamics of a steady uniform granular flow is depicted in Figure 1 and shown in Figures 2 and 6 for experiments and simulations respectively. First a steady uniform flow over the incline is generated. Second the aperture gate is

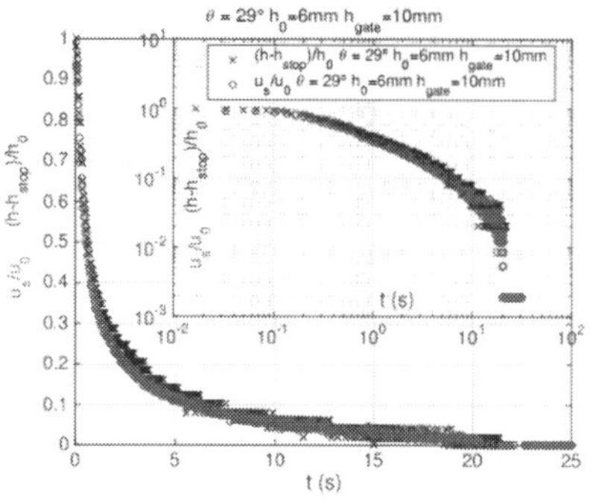

Figure 4. Time evolution of normalized height $(h(t) - h_{stop})/h_0$ and normalized surface velocity $u_s(t)/u_0$ for one experiment at $\theta = 29°$ and $h_0 = 6$mm in linear and logarithmic scales

suddenly closed, inducing a boundary condition $h = 0$ at the gate. This sudden variation of boundary condition creates a shock wave travelling at a velocity u_{shock} larger than the front velocity u_0, then the height and the velocity decrease slowly. Finally, a deposit lays on the rough plane of uniform height h_{stop} [18, 19]. It is known that $h_{stop}(\theta)$ is a function of the slope angle θ and depends on the properties of grains and of the bottom surface of the plane.

Another difference lays between experiments and simulations in the discontinuity of height observed experimentally at the end of the stopping dynamics: just before stopping, the still flowing height is equal to $h_{arrest} > h_{stop}$, that may be the signature of compressibility of the granular flow and its compression when stopping. By contrast, the height in the incompressible simulations continuously vanishes.

The stopping of the flow can be seen as the propagation of an arrest wave, defined as the transition between $u \neq 0$ and $u = 0$. Figure 2 shows a space-time diagram: the shift of the projected laser sheet is proportional to the thickness $h(t)$. The granular flow front crosses the image at about $t \simeq 20$s with a velocity u_0 shown by the red line. The height is uniform after a few seconds. At about $t \simeq 30$s, the shock wave generated by the closure of the gate crosses the image. Then the height decreases slowly for about 20s, when the arrest wave propagates at a velocity u_{arrest}, shown by the green line, larger than the front velocity u_0.

4.2 Relaxation dynamics

The height and the surface velocity decrease slowly after the passage of the shock wave (Figures 3 and ??). Figure 3 shows the height evolution during the stopping for different experiments realized at the same slope angle $\theta = 27°$ but different thicknesses $h_0 = 4$mm, 6mm and 8mm (a) and for the same uniform height $h_0 = 6$mm but at different slope angles $\theta = 25°, 27°, 29°$ (b). The typical time for the whole stopping of the granular layer is between 10s and 100s. It seems that this time depends on the slope angle θ only, but not on the height h_0 or velocity u_0 of the steady uniform flow. For now, we can not infer any obvious temporal evolution for $h(t)$ (exponential or power-law or logarithmic), but deeper studies are needed. While the height decreases until h_{stop} –the finite thickness of the deposit–, the surface velocity $u_s(t)$ continuously vanishes (Figure 4). The temporal evolution of $h(t)$ and $u_s(t)$ are exactly the same, as shown by Figure 4, where normalized surface velocity $u_s(t)/u_0$ and normalized height $(h(t) - h_{stop})/h_0$ have been plotted for one experiment.

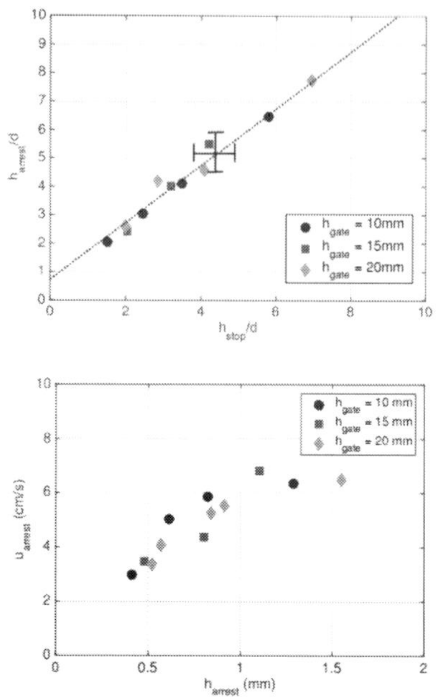

Figure 5. Experimental measurements of the arrest wave: height of the wave h_{arrest} as a function of the deposit height h_{stop} and velocity of the wave u_{arrest} as a function of h_{arrest}

We measured systematically the experimental values of the height h_{arrest} and the velocity u_{arrest} of the arrest wave, shown in Figure 5: it seems that they depend only on the slope angle θ. Just before stopping, the still flowing height h_{arrest} is not equal to the deposit thickness h_{stop}. Figure 5a shows that $h_{arrest} = h_{stop} + \delta h$ with $\delta h \simeq 1d$. The fact that the compression δh does not change with the thickness of the flow but is constant, may indicate that the arrest of the flow may not occur as a whole but may happen through the decrease of the thickness of flowing grains. Also, h_{arrest} and u_{arrest} are correlated.

One advantage of numerical simulations is to get the whole spatial and temporal evolution of the granular flow thickness. Figure 6 shows the spatial profile of the granular layer at different times t in the incline framework $h(x)$ (a), in the front framework $h(x - u_0t)$ (b) and in the shock wave framework $h(x - u_{shock}t)$ (c). For the present simulation, the shock wave propagates at a velocity $u_{shock} = 2.3u_0$, approximately twice larger than the front velocity

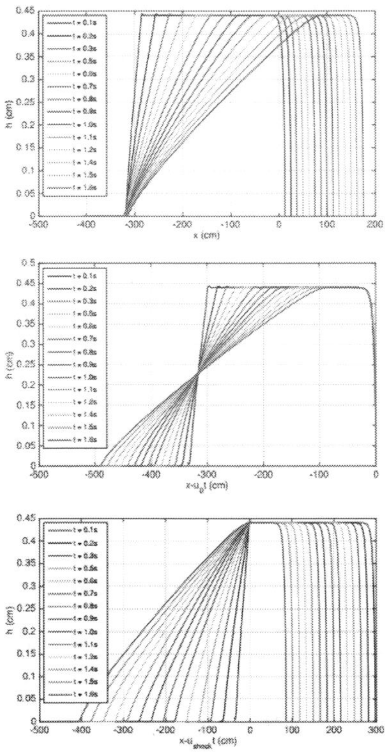

Figure 6. Streamwise height profiles from one numerical simulation at different times: $h(x)$ in the incline framework, $h(x-u_0t)$ in the front framework and $h(x-u_{shock}t)$ in the shock wave framework.

u_0. The arrest wave and the shock wave have the same velocity: $u_{shock} = u_{arrest}$ but propagate in opposite directions. As for a Newtonian fluid and for a fluid at constant friction μ in the case of a dam-break [11], all profiles in the front framework $h(x-u_0t)$ cross at the same point at first order (Figure 6b). Figure 6c suggests to use the same tools as for describing the spreading of a dam-break by taking into account the velocity propagation of the shock wave.

5 Conclusion

We study the long-time stopping dynamics of a steady uniform granular flow over a rough incline plane by using experimental and numerical tools. The stopping dynamics exhibit the propagation of a shock wave and of an arrest wave. More investigations are still needed to know more about the internal dynamics of the flow during the stopping. A possibility would be to measure velocity profiles through transparent lateral walls, to design a two dimensional flow or to use discrete element methods.

As the local frictional rheology $\mu(I)$ does not predict a static deposit or a height threshold, that is one major limitation of $\mu(I)$ until today, alternatives would be to consider a non-local rheology [20, 21], or to write the friction law as a function of I and h: $\mu(I, h)$ [7, 22]. We made some experiments with a smooth bottom plane. In this case, there is almost no deposit ($h_{stop} \simeq 0$), whatever the slope angle is. However, the flowing and stopping phenomenology are the same as for a rough bottom surface, suggesting that there is actually shear inside the granular layer in the steady uniform regime even in the case of a smooth bottom, that would allow for closer comparisons with predictions of $\mu(I)$ in the stopping regime.

References

[1] B. Andreotti, Y. Forterre, and O. Pouliquen, *Granular Media: Between Fluid and Solid* (Cambridge University Press, Cambridge, 2013)
[2] GDR MiDi, Eur. Phys. J. E, **14** 341-365 (2004)
[3] F. da Cruz, S. Emam, M. Prochnow, J.-N. Roux, and F. Chevoir, Phys. Rev. E **72** 021309 (2005)
[4] P. Jop, Y. Forterre, and O. Pouliquen, Nature **44** 727–730 (2006)
[5] P.-Y. Lagrée, L. Staron, and S. Popinet, J. Fluid Mech., **686** 378–408 (2011)
[6] S. B. Savage and K. Hutter, J. Fluid Mech., **199** 177–215 (1989)
[7] O. Pouliquen and Y. Forterre, J. Fluid Mech., **453** 133–151 (2002)
[8] P. Jop, Y. Forterre, O. Pouliquen, Phys. Fluids **19** 088102 (2007)
[9] E. Lajeunesse, A. Mangeney-Castelnau, and J. P. Vilotte, Phys. Fluids, **16** 2371 (2004)
[10] G. Lube, H.E. Huppert, R.S.J. Sparks, and M.A. Hallworth, J. Fluid Mech., **508** 175-199 (2004)
[11] R. R. Kerswell, Phys. Fluids, **17** 057101 (2005)
[12] S. Douady, B. Andreotti, P. Cladé, and A. Daerr, Adv. Compl. Syst. **4** 509–522 (2001)
[13] T. Boutreux and P.-G. de Gennes, C. R. Acad. Sci. Paris, Série II b **326** 257–262 (1998)
[14] O. Pouliquen, Phys. Fluids **11** 1956 (1999)
[15] G. Saingier, S. Deboeuf, and P.-Y. Lagrée, Phys. Fluids, **28** 053302 (2016)
[16] S. Popinet, J. Comput. Phys., **190** 572-600 (2003)
[17] J. M. N. T. Gray, and A. N. Edwards, J. Fluid Mech. **755** 503–734 (2014)
[18] O. Pouliquen, and N. Renaut, Journal de Physique II, EDP Sciences, **6** 923-935 (1996)
[19] A. Daerr and S. Douady, Nature, **399** 241-243 (199)
[20] M. Bouzid, M. Trulsson, P. Claudin, E. Clément, and B. Andreottti, Phys. Rev. Lett., **111** 238301 (2013)
[21] K. Kamrin and D.L. Henann, Soft Matter, **11** 179 (2015)
[22] A. Mangeney, F. Bouchut, N. Thomas, J.-P. Vilotte, and M.-O. Bristeau, J. Geophys. Res. **112** 02017 (2007)

Granular flow down a flexible inclined plane

Prasad Sonar[1,*], *Ishan* Sharma[1,**], and *Jayant* Singh[2,***]

[1] *Department of Mechanical Engineering.*
[2] *Department of Chemical Engineering.*
Mechanics and Applied Mathematics Group, Indian Institute of Technology Kanpur, Kanpur, UP 208016, India.

Abstract. Discrete and continuous systems are commonly studied individually, but seldom together. Indeed, granular flows are typically studied through flows over a rigid base. Here, we investigate the behaviour of granular flows over an inclined, flexible base. The flexible base is modeled as a rigid platform mounted on springs and has one degree of freedom. The base vibrations are introduced by the flow. We simulate such flows through a discrete element method and compare with experiments. We find that a flexible base increased the upper limit of the inclination up to which a steady flow is possible by at least 3 degrees. This stabilized zone may have important implications in applications such as conveyor belts and chutes.

1 Introduction

In most literature on granular flows over bumpy or smooth bases, the base is considered rigid and fixed for convenience (e.g[1–4]). The rigid base may be actively moved or vibrated to introduce energy into the system. However, in real applications like chute flow, and transport and conveying systems, the base's motion is driven by the granular flow. A vibrating base may couple with the flow in a manner which increase collisional energy dissipation. This can affect the flow considerably.

Here, as a first step, we introduce flexibility in the base by supporting it upon springs. We investigate the flow over such a base for various spring stiffnesses through discrete element (DE) simulations and experiments. We define a stabilized regime in parameter space for which flexibility of the base increases the dissipation of kinetic energy by collisions among the grains and, therefore, slows down the material making the flow steady, whereas it would be accelerated for rigid bases. At lower inclinations, we find that the flexible base may be employed to control the manner in which the granular flow is initiated. These features may have important applications in the handling and transport of grains. Finally, we attempt to characterize the onset of unsteady granular flow in our study by making contact with stability analysis performed for viscous fluid flow over soft solids [8] and for dense granular flow over rigid inclined plane [3, 4].

2 Methodology

Figure 1 shows our experimental setup. It consists of two 40 mm wide acrylic channels. We use steel balls with

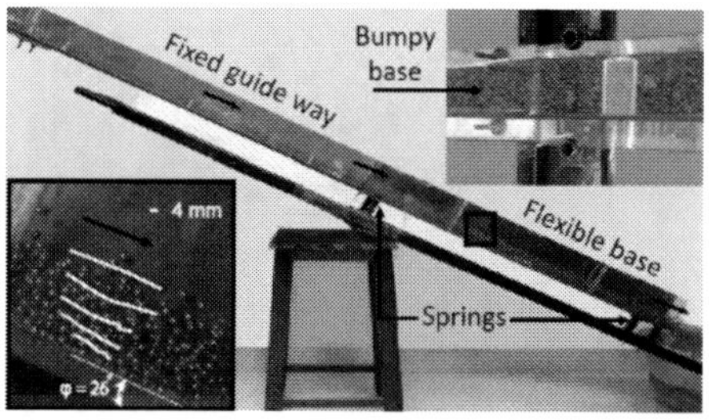

Figure 1. The experimental setup consists of a fixed guideway and the spring supported flexible base.

4 mm diameter as flowing granular material. The same balls are employed to make the base bumpy. First, the 2 m long guideway is fixed to the iron frame, and it guides the flowing particles into the second 1 m long 'active'channel whose base is made flexible by supporting it with springs on either ends. During experiments, the active channel oscillates up and down in its first mode of vibration primarily. When the flow has matured, the pitching motion of the base, corresponding to second mode of vibration, is observed to be small and, hence, ignored in this study.

During flow, we maintain a flow thickness of 40 mm, which corresponds to approximately 10 steel ball diameters, by controlling the outlet knob of the hopper. We perform experiments on angles of inclination φ varying between 22° and 30°. We also vary the stiffnesses of the supporting springs. This compressibility ζ is the inverse of the spring's stiffness k_s scaled by the stiffness k_n of the steel balls. We employ springs with stiffness values varying from 2 to 15 N/mm that corresponds to ζ varying from 70 to 260. The contact stiffness of steel balls k_n = 400 kN/m as found by [6]. We employ particle tracking and image processing techniques to obtain quantitative data from the flow.

*e-mail: prasads@iitk.ac.in
**e-mail: ishans@iitk.ac.in
***e-mail: jayantks@iitk.ac.in

We also simulate a three-dimensional granular flow with periodic boundary conditions (PBC) along the direction of the flow, as shown in Fig. 2. We employ DE simulations, whose details are available in Bhateja et al. [2]. The flow consist of spherical, cohesion-less, mono-dispersed, inelastic (coefficient of restitution $e = 0.88$), soft (non-dimensional normal spring stiffness $k_n = 2\times10^5$) and rough (friction coefficient $\mu = 0.5$) grains. The base is made rough and bumpy, and is supported on a linear spring in a direction normal to the flow. The spring stiffness k_s is kept lower than that of the grain's. The compressibility ζ in simulations, equivalent to experiments, is the ratio of the grain stiffness k_n to the spring stiffness k_s. We have not introduced any damper with the supporting springs. Thus, energy only dissipates through collisions among particles and with the base. Our simulations follow the same flow initiation process as in Silbert et al. [1].

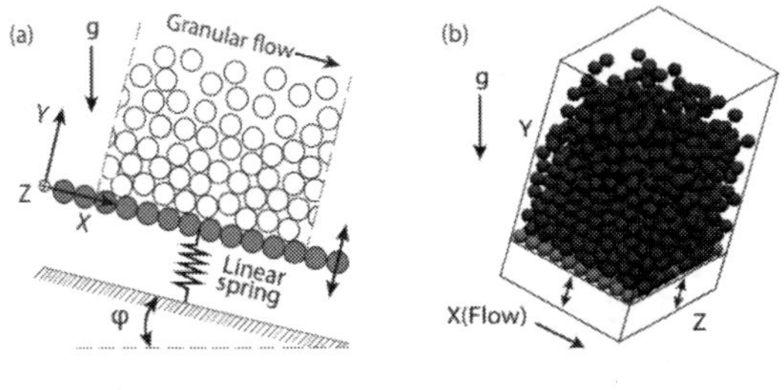

Figure 2. Granular flow over flexible base with one degree of freedom in direction normal to the inclined channel. Periodic boundary condition are imposed along the inclined plane (x-direction). (*a*) A two-dimensional schematic representation, and (*b*) a three-dimensional DE simulations of the flow.

In our simulations we have employed depth distribution of momentum as a criterion to differentiate between conditions of flow and no-flow. Figure 3(a) shows the distribution of momentum through the depth of the granular flow at various base inclinations. At lower inclinations, momentum is carried only by the top few layers, and the remaining grains are nearly static. However, at higher inclinations the momentum is well distributed through the flow depth. We define flow to have initiated when the minimum momentum per unit depth over half the depth at a cross-section is at least 10 % of the total momentum at that cross-section. A flow is said to be steady when the total kinetic energy (E_K^T) along the inclined plane is the same for all time. This happens when the power input by gravitational force is balanced by that lost to collisional dissipation and base vibration.

3 Results: Simulations

A typical result as shown in Fig. 3(b). We observe that on a rigid base flow accelerates when φ exceeds 29°, but on a flexible base with $\zeta = 80$, the flow remain steady even at $\varphi = 32°$. In passing we note that our result for steady granular flow on a rigid base with flow depth equaling 10 grain diameters are in agreement with Silbert et al. [1].

Our simulation results are summarized in Fig. 4 in terms of the parameters φ and ζ. The state of the flow is

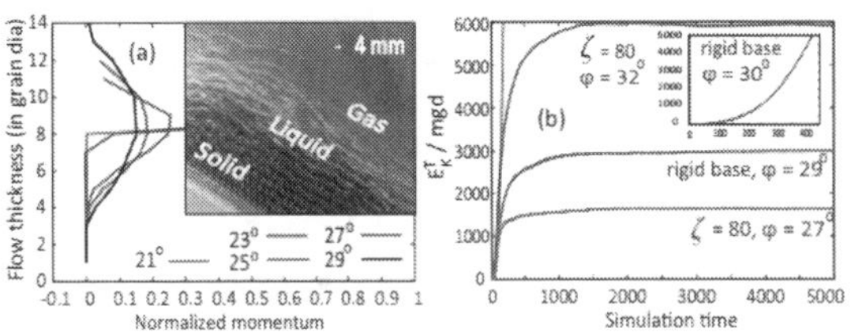

Figure 3. Flow criterion. (*a*) Momentum distribution through the depth at different angle of inclination φ. Inset shows states of flowing material, and is indicative of momenta carried by grains. (*b*) Kinetic energy of the granular flow on rigid and flexible bases. Inset shows increase in E_K^T on rigid base at $\varphi = 30°$. Here, energy is normalized using grain's mass m, diameter d and the constant of gravitational acceleration g.

represented in terms of the angle of inclination φ and compressibility ζ. The inclination angle φ is varied between 21° to 32° and ζ ranges between 0 and 100, with 0 corresponding to a rigid base, and 100 representing the softest base. The stabilized regime is the zone in parameter space

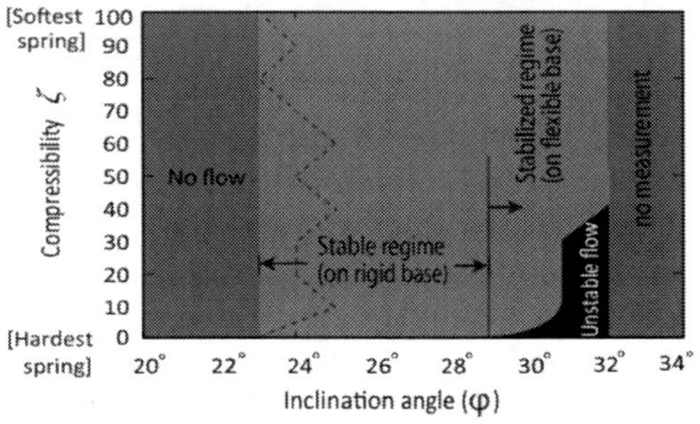

Figure 4. Behavior of granular flow down a channel with inclination φ supported by springs with compressibility ζ. Dashed line represents the inclination angle at which the flow initiates.

where the flow remains steady on the flexible bases. It is computationally expensive to track very rapid flows and, hence, inclinations above 32° are not simulated. It is very likely that some of these rapid flows, perhaps at greater ζ, are steady. Thus, the right boundary of the stabilized regime should not be understood as vertical.

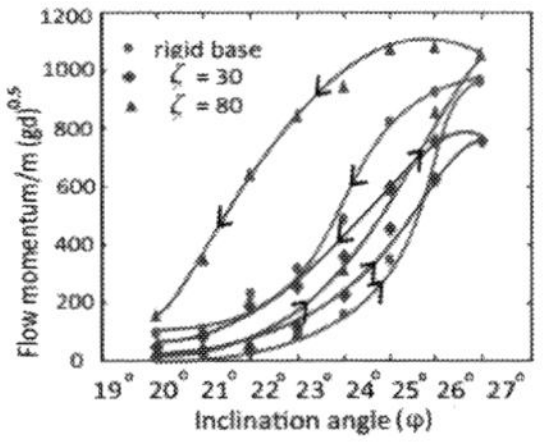

Figure 5. Effect of spring compressibility ζ on flow initiation and cessation. Y axis is scaled using grain's mass m, diameter d and the constant of gravitational acceleration g. Markers indicate simulation points.

As shown in Fig. 5, flow starts and stops at different φ when the inclination is increased and then decreased at the same constant rate. This happens due to the inertia of the grains. As the granular material is known to be hysteritic in nature, such behavior is not surprising. The width of hysteritic loop is affected by the base conditions which, as Fig. 4 suggests, may be used to reduce hysteresis. This will allow better control over the flow.

Figure 6 shows various flow properties at $\varphi = 26°$ for different compressibilities ζ. We find from Fig. 6(a) that on a rigid base, grains at the bottom do not contribute to the momentum of the flow. However, on flexible bases, these grains get energized due to base vibrations. Hence, even if the variation of number density with depth is comparable for various ζ, as shown in Fig. 6(b), grains at the bottom are more agitated when the base is flexible, indicating more collisions. The velocity profile in Fig. 6(c) further confirms that grains near the bottom are in motion where the base is flexible.

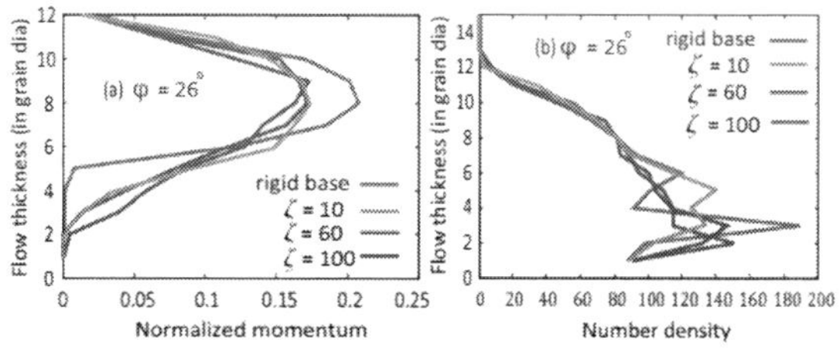

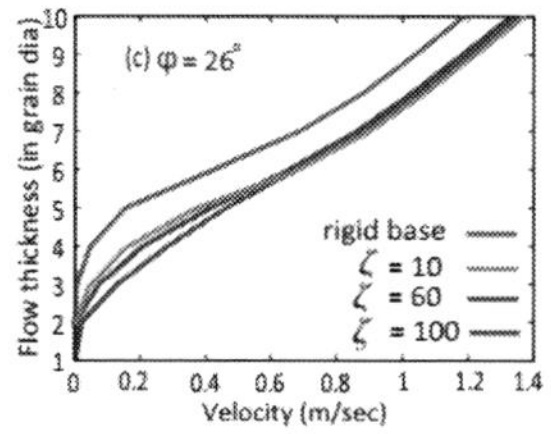

Figure 6. Granular flow properties at $\varphi = 26°$ for various ζ. (*a*) Momentum distribution, (*b*) number density, and (*c*) velocity profile. Flow thickness is measure in grain diameters.

To understand the source of the stabilized regime in Fig. 4, we investigate the effect of base flexibility on the granular flow. Consider first the simple one-dimensional computational experiment shown in Fig. 7. A grain is dropped on a stack of other grains resting on a spring mounted base. As shown, the impacting grain bounces many more times on a rigid base than on a flexible base. This suggests that energy is lost faster on a flexible base in contrast to rigid base. Together with increase in numbers of collisions (see Sect. 5) for flows over flexible bases this explains stabilization of rapid flows at high inclination by flexible bases.

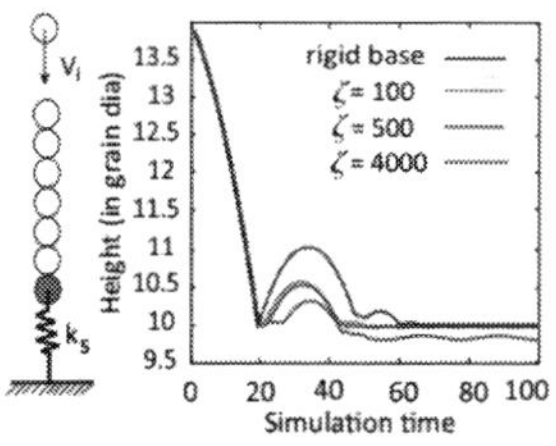

Figure 7. A grain bouncing on a stack of grains resting on a spring. Various compressibilities are considered.

4 Results: Experiments

Experimental and computational velocity profiles for various inclinations on rigid base are compared in Fig. 8. We find a qualitative agreement. Figure 9 shows experimentally measured total kinetic energy E_K^T at five different location along the inclined plane. Rigid and flexible bases inclined at various φ are considered. We have tracked E_K^T at five different locations along the flow channel. Flow enters the 1m long active base after traveling trough the 2m

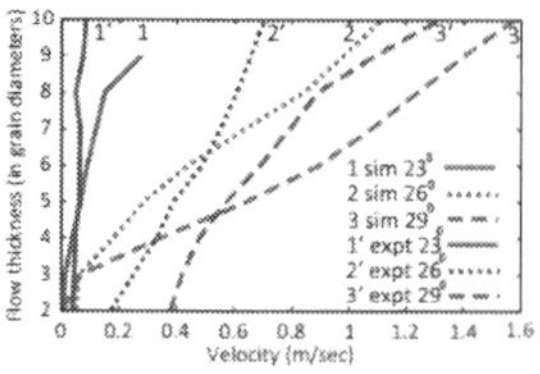

Figure 8. Comparison between experiments and simulations of velocity profiles of flows on a rigid base.

long guideway. We observed steady state regime between inclinations $\varphi = 25°$ and $28°$ on rigid base. At higher inclinations ($\varphi = 29°$), E_K^T of the flow increases on a rigid base, but remains constant for flows over flexible base. This difference is in line with the predictions of simulations about the stabilizing effect of a flexible base on a granular flow.

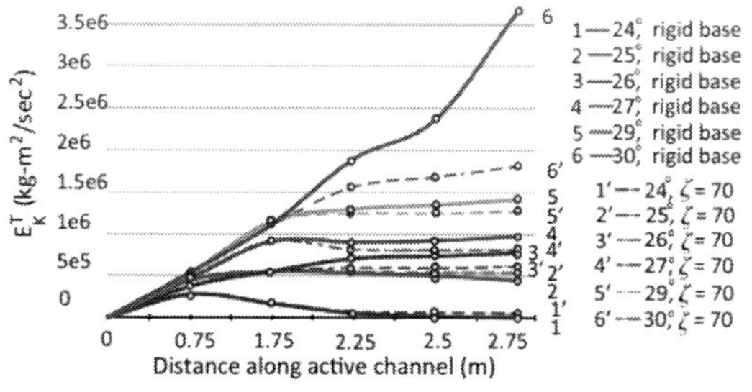

Figure 9. Kinetic energy of the flow E_K^T at various locations along the length for flows over rigid and flexible bases at various inclinations φ.

Figure 10 shows the variation of layer-wise kinetic energy E_K^L through the depth for flow over a flexible base inclined at $\varphi = 29°$ when compressibility $\zeta = 70$. It is seen that top grains accelerate while those at the bottom decelerate. Predictably, grains at the top appear to be unaffected by base conditions. However, grains at the bottom reflect the effect of a flexible base in terms of a reduced kinetic energy.

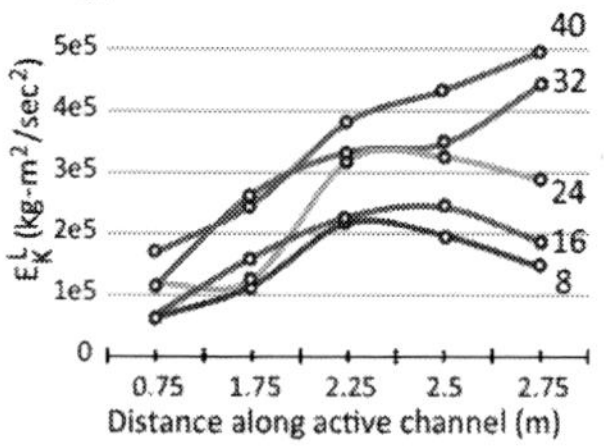

Figure 10. Depth-wise distribution of E_K^L. Labels 8 - 40 indicate depth (in mm) at which measurement is taken.

5 Discussion

We now investigate energy dissipation in such flows through simulations by comparing E_K^T of the flow at $\varphi = 29°$ on a rigid and flexible bases. We have seen that the kinetic energy, E_K^T of the flow is lower on a flexible base. We next trace the reason for the difference ΔE_K^T between the total kinetic energy of flows over rigid and flexible bases.

We find that the total number of grain contacts (N_C^T) per unit time step are higher in flows on a flexible base. The energy stored in the spring is very small compared to E_K^T and, so, may be ignored. Hence, increased N_C^T is the reason for the greater dissipation in flows over flexible bases. Independent of base conditions, approximately 85% of contacts occur among particles (N_C^P) and 15% are due to collisions with the base (N_C^B). As shown in Fig. 11, comparing N_C^T for flows at $\varphi = 29°$ on a rigid base with that over a flexible bases, we find that N_C^P increases from 330 to 414 and N_C^B from 52 to 74 per time step, i.e., increment of 25% and 42%, respectively. At $\varphi = 26°$, as the flow is denser, N_C^P increases by 50%, but N_C^B remains about the same. It is important to note that collisions at $\varphi = 29°$ are much more dissipative than those of flows at 26°, due to higher grain velocities. The energy stored in the spring does not vary much with inclination and changes only by 10% for compressibilities (ζ) varying between 10 and 100. Thus, we believe that base flexibility increases the number of inter-grain collisions which, in turn, cause more dissipation in the flow, thereby stabilizing it.

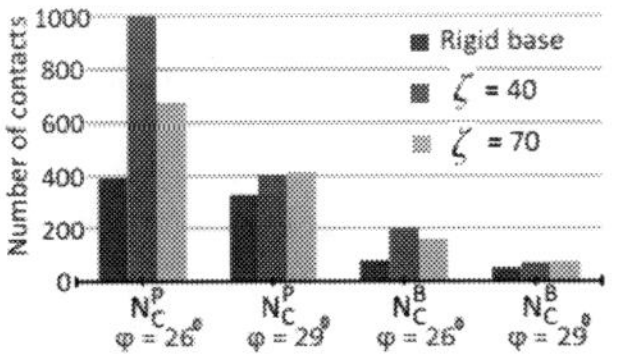

Figure 11. Comparison of number of inter-grain collisions (N_C^P) and number of collisions with base (N_C^B). See the text for details.

While we have observed increased energy dissipation in flows over flexible bases in both experimentally and in simulations, a steady flow in experiments is seen for narrow region. We believe that this happens because (a) channel used in experiments is not long enough, and (b) limitations on various grain and spring parameters available for experiments. Steel balls used in experiments are much harder compared to the grains simulated. We are forced to use grains with higher density, like steel balls, to allow us to use a variety of soft springs. Lighter grains like glass beads or plastic balls would force us to use softer springs that are hard to find and tend to fail easily in shear. At the same time, simulations with high stiffness coefficients k_n are computationally prohibitive.

6 Stability analysis

Here we make an initial attempt to connect onset of unsteady flow in our system with the stability of Newtonian fluids down deformable inclined surfaces [8]. The first correction to the wave-speed c_1, was found by [8] to be

$$c_1 = i[((6/5)Re - \cot\varphi) - 9\Gamma H], \quad (1)$$

where Re is the Reynolds number, and Γ is the solid elasticity parameter, which is a ratio of the viscous stress in flowing material to the elastic stress in the supporting flexible structure. The flow is unstable if the imaginary part of wave-speed, c_1 is positive. On a rigid base, where $\Gamma = 0$, the free surface undergoes long wave instability when $(6/5)Re > \cot\varphi$. On a deformable solid, the free surface instability is completely stabilized if $(9\Gamma H) > ((6/5)Re - \cot\varphi)$. This defines the stability criterion. In our simulations, $\Gamma = \mu(du/dy)(1/E)$, with $E = (k_n H/v^2 l^2)$, being the equivalent modulus for our spring supported base. Parameters μ, v, H and l are, respectively, bulk viscosity, volume fraction, equivalent thickness of elastic solid, and the length of the base. Here, du/dy and v are found from simulations and l is taken to be 10 grain diameters, and μ is obtained from kinetic theory [7] (eq. 7.161). The thickness H drops out in calculation and so need not be estimated.

The stability conditions of [8], when applied to our granular flow, predict that flow on a rigid base becomes unstable for $\varphi \geq 26°$. On flexible base with $\zeta = 40$ and 70, flow is predicted to be stable. Thus, the theoretical predictions of [8] also note an increase in stability in inclined flows in the presence of a deformable base. However, a quantitative match is not seen, as is to be expected given the difference in flowing material and base motion.

The spatial stability of steady uniform flow over a rigid inclined base was investigated by [3, 4]. Employing depth averaging and $\mu(I)$-rheology predicted that steady-uniform flows become unstable when $F > F_c = 2/3$. Where F is Froude number ($F_r = \underline{v}/(gH\cos\varphi)^{0.5}$) and F_c is critical Froude number. Parameters $g, \underline{v}$ and H are constant of gravitational acceleration, depth-average flow velocity and flow thickness respectively. In our simulations, flow on rigid base becomes unstable for $\varphi \geq 26°$, and remain stable for $\varphi \leq 28°$ on flexible base with $\zeta = 40$. The extension of these studies to granular flow over deformable surfaces remains a goal for further study.

7 Conclusion

We have shown that a flexible base stabilizes granular flow, which otherwise would accelerate on a rigid base. We found that a flexible base energizes the lower grains in the flow which results in more grain collisions. Increased collisions among the particles and with the base dissipate kinetic energy, stabilizing flows at higher inclination angles. Finally, we made a first step towards relating the stabilization of granular flows over flexible bases to stability of Newtonian fluids flowing over soft solids. A proper understanding of how base flexibility affects granular flows will allow us to better control them in engineering applications.

References

[1] Silbert L. E., Ertas D., Grest G. S., Halsey T. C., Levine D., and Plimpton S. J., Phys. Rev. E. 65, 051307 (2002).
[2] Bhateja A., Sharma I., and Singh J. K., Phys. Fluids. 28, 043301 (2016).
[3] Forterre Y. and Pouliquen O., J. Fluid Mech. 486, 21-50 (2003).
[4] Gray J. M. N. T. and Edwards A. N., J. Fluid Mech. 755, 503-534 (2014).
[5] Cundall P. A. and Strack O. D. L., Geotechnique 29, No. 1, 47-65 (1979).
[6] Mishra B. K. and Rajamani R. K., Appl. Math. Modelling. 16, 598-604. (1992).
[7] Rao K. K. and Nott P. R., *An Introduction to Granular Flow* (Cambridge University Press, UK, 2008) 490.
[8] Shankar V. and Sahu A. K., Phys. Rev. E. 73(1), 016301 (2006).

Understanding fluid transport through the multiscale pore network of a natural shale

Catherine A. Davy[1,*], *Thang* Nguyen Kim[2], *Yang* Song[3], *David* Troadec[4], *Anne-Marie* Blanchenet[5], and *Pierre M.* Adler[2]

[1] *Laboratoire de Mécanique de Lille (LML), FRE CNRS 3723 and Ecole Centrale de Lille, CS 20048, F-59651 Villeneuve d'Ascq Cedex, France*
[2] *Metis UMR CNRS 7619, UPMC, Paris, France*
[3] *Changzhou Institute of Technology, Chang Zhou Gong Xue Yuan, Changzhou Shi, Jiangsu Sheng, China, 213000*
[4] *Institut d'Electronique, de MicroElectronique et de Nanotechnologie (IEMN), UMR CNRS 8520, BP60069, 59652 Villeneuve d'Ascq, France*
[5] *Unité Matériaux et Transformations (UMET), UMR CNRS 8207, Université Lille 1, Bâtiment C6, 59655 Villeneuve d'Ascq, France*

Abstract. The pore structure of a natural shale is obtained by three imaging means. Micro-tomography results are extended to provide the spatial arrangement of the minerals and pores present at a voxel size of 700 nm (the macroscopic scale). FIB/SEM provides a 3D representation of the porous clay matrix on the so-called mesoscopic scale (10-20 nm); a connected pore network, devoid of cracks, is obtained for two samples out of five, while the pore network is connected through cracks for two other samples out of five. Transmission Electron Microscopy (TEM) is used to visualize the pore space with a typical pixel size of less than 1 nm and a porosity ranging from 0.12 to 0.25. On this scale, in the absence of 3D images, the pore structure is reconstructed by using a classical technique, which is based on truncated Gaussian fields. Permeability calculations are performed with the Lattice Boltzmann Method on the nanoscale, on the mesoscale, and on the combination of the two. Upscaling is finally done (by a finite volume approach) on the bigger macroscopic scale. Calculations show that, in the absence of cracks, the contribution of the nanoscale pore structure on the overall permeability is similar to that of the mesoscale. Complementarily, the macroscopic permeability is measured on a centimetric sample with a neutral fluid (ethanol). The upscaled permeability on the macroscopic scale is in good agreement with the experimental results.

1 Introduction

1.1 Scientific context

Claystones and shales are fine grained sedimentary rocks. Their pores are mainly located within their clay matrix or at its interface with non porous minerals [3–6, 10, 14]. The clay pores, of 1-100 nm size [1], most likely dominate the flow and transport properties [5, 11, 12]. Moreover, during sedimentation and compaction, clay minerals are deposited along preferred orientations, i.e. natural shales are generally anisotropic [7]. It is expected that the pore space geometry is also anisotropic, as are the transport properties of the shale [15].

The pore structure of claystones is still partly understood, both in terms of geometry and connectivity, though significant progress has been made over the last decade [3–5, 10–12, 14]. The inherent heterogeneity of these natural materials (which comprise swelling clay, soft and hard minerals), and the broad range of pore sizes, makes them complex to characterize. At least for the natural shale studied herein, at present, no technique alone provides a full overview of its porosity for bulk matter at the centimetric scale and above, due to its very wide pore range. A combination of techniques is usually preferred. They are either direct, i.e. imaging techniques, or indirect, i.e. they require a model to derive pore characteristics. Indirect techniques often rely either on gas adsorption (for the smallest nanoscopic pores) or on capillary intrusion (for pluri-nanometric to millimetric pores). The sample preparation technique may also limit the applicability to bulk matter.

The 3D characterization of the pore structure by direct imaging provides all necessary geometrical parameters, but this largely depends on the pore and sample size considered. Currently, X-Ray micro-tomography (micro-CT) provides data of the 3D pore space down to about 700 nm. This is insufficient for clay rocks, which display no connected porosity on this scale, although the analyzed sample sizes are representative of a "macroscopic" struc-

*e-mail: catherine.davy@centralelille.fr

[1] In all this contribution, unless otherwise stated, pore size refers to pore diameter and not radius.

ture (up to about 180 microns) [10]. Higher resolutions are available with Focused Ion Beam/Scanning Electron Microscopy (FIB/SEM) down to 10-20 nm voxel size, and Transmission Electron Microscopy (TEM) or Scanning Transmission Electron Microscopy (STEM) techniques, down to below 1 nm pixel/voxel size. However, analyzing one single sample with these techniques does not provide macroscopic representativeness due to the limited sample size. In practice, the representativeness issue is solved by (1): characterizing the sole clay matrix with FIB/SEM and (S)TEM on *several* samples to achieve *representativeness in a statistical sense* [2, 5], and by (2): coupling these with micro-CT using scale change methods [8]. Where no single characterization method provides adequate representativeness for such wide pore scales as in clay rocks, the combination of micro-CT, FIB/SEM and TEM is able to provide a full representative 3D pore space [2].

In this contribution, we benefit from synthetic images derived by training image techniques from micro-CT data for the natural shale from [10]. We also combine both FIB/SEM and TEM imaging of the mesoscopic clay matrix to describe the smallest 1-100nm pores. Four percolating FIB/SEM samples (27.8-146.7μm^3), and up to eight TEM images (18.5x10^3 - 41.0x10^3 nm^2) are used, which are fully analyzed in previous research [11, 12].

1.2 Aims and scopes

The images are used as a starting point for numerical simulations on all three scales. At the nanoscopic scale given by 2D TEM, in the absence of 3D images, the pore structure is reconstructed by using a technique, which is based on truncated Gaussian fields. Permeability calculations are performed with the Lattice Boltzmann Method (LBM) on the mesoscale given by 3D FIB/SEM pore structures, and also on the nanoscale, and on the combination of the two. Upscaling is finally done (by a finite volume approach) at the macroscopic scale by using synthetic images obtained by direct sampling from actual micro-tomography images of the same shale [13]. Complementarily, the macroscopic permeability is measured on a centimetric sample with a neutral fluid (ethanol). The upscaled permeability at the macroscopic scale is compared with the experimental results.

2 Methods

2.1 Experimental

For FIB/SEM or TEM imaging, two shale samples EST26095 and EST25679 are oven-dried at 80°C for at least 15 days and impregnated with non radioactive Methyl MethAcrylate resin (MMA) under partial vaccuum. Following impregnation, MMA is polymerized at a temperature of 55°C for 16h. Prior to polymerization, MMA has a very low viscosity (of 0.6cP at 20°C) smaller than that of water (1.0020 cP at 20°C). Therefore, it has excellent pore filling ability. Moreover, for further polishing, polymerized MMA (PMMA) limits the hardness difference between the paste and the aggregates, and proper flatness is achieved more smoothly than without any impregnation. For FIB/SEM observations, a third core origin EST27405 is used after 60°C oven-drying for at least 15 days, but without PMMA impregnation. For TEM imaging, the EST25679 sample is thinned to several tens of microns, prior to being ion milled using a Gatan Duomill^{TM} apparatus, with the use of a liquid nitrogen-cooled specimen holder, until sample edges show electron transparency. This corresponds to thicknesses below 100 nm.

Table 1. Overview of the 3D FIB/SEM samples (1)

Sample	EST27405-1	EST27405-2
Sample orientation wrt. bedding planes	perpendicular	perpendicular
Voxel size (nm^3)	8.49×10.78×10	5.94×7.54×10
Volume (μm^3)	6.55×7.97×1.03	3.99×5.95×1.17
Porosity (%)	2.83	4.89
Surface connected orientation	**z**	**z**

Table 2. Overview of the 3D FIB/SEM samples (2)

Sample	EST26095-1	EST26095-2	EST26095-3
Orientation wrt. bedding planes	parallel	parallel	parallel
Voxel size (nm^3)	8.48×10.77 ×20	8.48×10.77 ×20	8.49×10.78 ×20
Volume (μm^3)	6.55×5.21 ×4.30	6.30×6.99 ×2.02	4.50×6.22 ×3.98
Porosity (%)	1.71	5.9	5.21
Percolating direction	**y**	**z**	none

Table 3. Characteristics of the TEM images used for the numerical simulations of transport.

Sample	Pixel size (nm^2)	Sample size (nm^2)	Crop size (nm^2)	
T10$_1$	0.36×0.36	658×799	176×193	0.1
T08$_2$	0.85×0.85	1540×1886	140×132	0.142
T09$_1$	1.04×1.04	1903×2329	160×240	0.25

All grayscale images acquired by FIB/SEM and TEM are filtered, thresholded and binarized, to distinguish between the pores and the solids. For FIB/SEM, the direction perpendicular to imaging is **z** (it is the direction for slicing the sample at 10 to 20 nm thickness), whereas each image is in a (**x**, **y**) plane. Full details on the procedures are given in [11, 12]. The main properties of the samples are gathered in Tables 1 and 2.

2.2 Numerical

On the nanoscopic scale, in the absence of 3D images, the pore structure is reconstructed from the 2D TEM images, with the same porosity and the same correlation function, by using a technique based on truncated Gaussian fields

 DOI: 10.1051/epjconf/201714012016

Figure 1. Example of a 3D reconstructed pore structure from the 2D TEM image T08$_2$ (porosity 0.142, l_x=6, l_y=l_z=50), for a volume of 451x419x484 pixels^3 and pixel size a=2.12 nm.

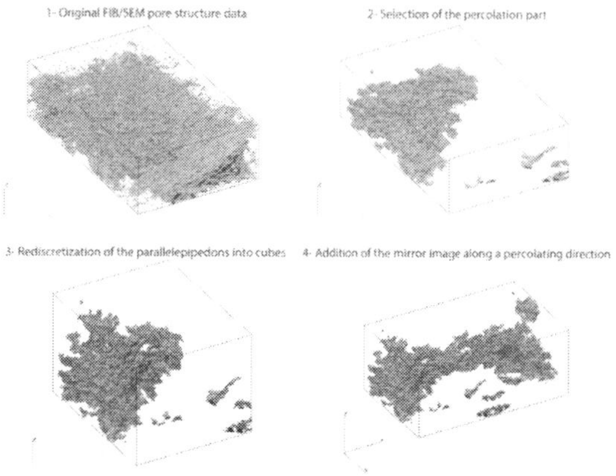

Figure 2. Numerical method providing a cubic and periodic pore volume from 1) the FIB/SEM image of EST26095-1. 2) The percolating part of the pore structure is retained. 3) The parallelepipedic voxels are rediscretized into cubes. 4) The mirror image of the percolating pore structure is added along a percolating direction.

[1]. A correlation function of the following form:

$$R_Y(\mathbf{u}) = R_Y(u_x, u_y, u_z) = \exp\{-(u_x/l_x)^2-(u_y/l_y)^2-(u_z/l_z)^2\} \quad (1)$$

where $l_x < l_y = l_z$ is used to generate flat-shaped nanopores, i.e. oblates (Fig. 1). It is assumed that directions **y** and **z** are equivalent.

On both the nanoscopic and mesoscopic scales, the permeability is calculated by solving the Stokes equation on the 3D pore structures. The same single phase Lattice Boltzmann code is used. According to the standard terminology, it is a D3Q19 code with two relaxation times. The classical bounce-back condition is used at the solid interface. For FIB/SEM data, each sample is completed by its mirror image along a percolating direction, which provides a periodic volume, in order to avoid overall boundary effects (Fig. 2). By construction, the 3D pore volume at the nanoscopic scale is periodic. More details can be found in [9].

3 Results and discussion

3.1 Permeability at the nanoscale

The dimensionless permeability K_n/a^2 (where a is the cubic voxel size) is determined along each spatial direction **x**, **y** and **z** (these are the K_{nxx}, K_{nyy} and K_{nzz} parameters) as a function of porosity ε (Table 4). In absolute values, the permeability at the nanoscale varies between 0.7×10^{-20} m^2 and 1.8×10^{-19} m^2. As expected, an anisotropy exists between the axes **x** and **y** (i.e. **z** also), as K_{nzz} is consistently greater than K_{nxx}.

Table 4. Dimensionless permeability on the nanoscale. Cubic voxel size a: 2.12 nm.

Porosity ε	Percolation axis	K_{nxx}/a^2	K_{nyy}/a^2	K_{nzz}/a^2
0.102	none	-	-	-
0.130	y only	0	0.113×10^{-2}	0.65×10^{-3}
0.142	x, y, z	0.487×10^{-2}	0.147×10^{-2}	$=K_{nyy}/a^2$
0.200	x, y, z	0.686×10^{-2}	0.138×10^{-1}	$= K_{nyy}/a^2$
0.252	x, y, z	0.802×10^{-2}	0.404×10^{-1}	$=K_{nyy}/a^2$

Table 5. Dimensionless permeability on the mesoscale.

Sample n.	Cubic voxel size a (nm)	K_{myy}/a^2 ($\times 10^{-2}$)	K_{mzz}/a^2 ($\times 10^{-2}$)	K_m (10^{-20} m^2)
EST26095-1	8.48	0.112	0	8.05
EST26095-2	8.48	0	0.0421	3.03
EST27405-1	8.48	0	0.379	27.3
EST27405-2	5.94	0	3.40	120

3.2 Permeability at the mesoscale

On the mesoscale, percolation of the FIB/SEM pore volumes occurs only along one direction, either **z** (for EST27405-1, EST27405-2 and EST26095-2) or **y** (for EST26095-1). The mesoscopic permeability K_m varies between 3.3×10^{-20} m^2 and 1.20×10^{-18} m^2, depending on the nature of the percolating volume (Table 5). The biggest K_m values, obtained with EST27405-1 and EST27405-2, correspond to percolating volumes identified as cracks [12].

3.3 Permeability by combining the nanoscale and the mesoscale pore structures

First, it is assumed that the permeability K_{nm} of the superposed nanoscale and mesoscale porosities is on the order of K_n, i.e. $K_{nm} \approx K_n$. Due to current limitations of the calculation means, the mesoscopic pore volume is subsampled and superposed to the nanoscopic pore volume.

For instance, for the FIB/SEM volume EST27405-1, a sub-sample provides an dimensionless permeability K_{mzz}/a^2 of 3.77, whereas the same volume superposed with the nanoscopic pore volume of porosity 0.142 provides a K_{nmzz}/a^2=3.92. For nanoscale porosity values ranging between 0.102 and 0.252, K_{nmzz}/a^2 ranges between 3.87 and 4.14. In this case of a cracked mesoscopic sample, the difference between K_{nm} and K_m is very limited.

A second sub-sample of the same mesoscopic pore volume provides K_{mzz}/a^2=0.133 × 10^{-1} (alone), whereas the nanoscopic pore volume of porosity 0.2 (and correlation lengths of 1.5, 12.5 and 12.5 nm) provides K_{nzz}/a^2=0.489 × 10^{-2} (alone). The combination of the two provides K_{nmzz}/a^2 =0.236 × 10^{-1}, which is significantly greater than K_{mzz}/a^2 alone. This means that the nanoscopic pore structure cannot be neglected to estimate the shale overall permeability.

3.4 Permeability at the macroscopic scale

On this scale, a finite volume approach is preferred. It is performed on synthetic 3D images, obtained by the direct sampling method [13] on actual micro-tomography images of the same shale as in [11]. The medium is composed of three phases. Phase 1 corresponds to the macropores (volume percentage 0.5), with an infinite permeability $K_{pore} = \infty$; for practical reasons, K_{pore} is at given values of 0, 1, 10, 10^2, 10^3, 10^4 and 10^5; Phase 2 corresponds to the porous clay matrix, with a volume percentage of 60, and a permeability K_{nm} given by the previous simulations. Phase 3 is the non porous minerals, with a volume percentage of 39.5 and zero permeability $K_{nonporous}$=0. The overall permeability K_{macro} is obtained by solving the Darcy's equation with a variable local permeability with spatially periodic boundary conditions. The calculation yields $K_{macro} \approx 0.4 K_{nm}$, and the medium is relatively isotropic ($K_{macroyy} \leq K_{macroxx} \approx K_{macrozz}$).

4 Conclusion

Compared to the mesoscopic permeability K_{nm} of the combined nanoscopic and mesoscopic pore structures, the macroscopic permeability represents only 40%. This means that these small scale pore structures actually drive the fluid transport through the shale. Our calculations show that the contribution of the nanoscopic pore structure cannot be neglected when assessing the transport of undisturbed matter. However, whenever the shale is cracked at the mesoscale, the permeability K_{nm} is of the same order as K_m, i.e. the nanoscopic pore structure can be negelected for assessing fluid transport. It is expected that such conclusion cannot be drawn for other phenomena relating to these shale, i.e. water sorption/desorption, ion diffusion, etc.

Acknowledgements

We acknowledge funding by NEEDS MIPOR (Grand Défi Inter-Disciplinaire Nucléaire: Energie, Environnement, Déchets, Société - MIlieux POReux, funded by CNRS, CEA, EDF, ANDRA, AREVA, IRSN, BRGM, France) in 2016, under the PERMEARGILE project label.

The SEM and TEM national facilities available at the Centre Commun de Microscopie de Lille (France) are supported by the Conseil Regional du Nord-Pas de Calais, and by the European Regional Development Fund (ERDF).

References

[1] Adler P.M., Jacquin C.G., Quiblier J.A., International Journal of Multiphase Flow **16** 691 (1990)

[2] Desbois, G. and Urai, J.L. and Pérez-Willard, F. and Radi, Z. and van Offern, S. and Burkart, I. and Kukla, P.A. and Wollenberg, U., Journal of Microscopy **249(3)**, 215-235 (2013)

[3] Keller, Lukas M and Holzer, Lorenz and Wepf, Roger and Gasser, Philippe, Applied Clay Science **52**, 85-95 (2011)

[4] Keller, L. M and Holzer, L. and Wepf, R. and Gasser, P. and Munch, B. and Marschall P., Physics and Chemistry of the Earth **36**, 1539-1544 (2011)

[5] Keller, Lukas M. and Schuetz, Philipp and Erni, Rolf and Rossell, Marta D. and Lucas, Falk and Lucas, Miriam and Gasser, Philippe and Holzer, Lorenz, Microporous and Mesoporous Materials **170**, 83-94 (2013)

[6] Marschall, P. and Horseman, S. and Gimmi, Th., Oil and Gas Science and Technology **60(1)** 121-139 (2005)

[7] Mitchell, J.K., *Fundamentals of Soil Behavior* (2nd Edition. John Wiley and Sons, New York, 1993)

[8] Moctezuma-Berthier, A. and Vizika, O. and Thovert, J.-F. and Adler, P.M., Transport in Porous Media, **56** 225-244 (2004)

[9] Pazdniakou A., Adler P.M., Advances in Water Resources, **62** 292 (2013)

[10] Robinet, J.C. and Sardini, P. and Coelho, D. and Parneix, J.C. and Prêt, D. and Sammartino, S. and Boller, E. and Altmann, S., Water Resources Research **48** (2012)

[11] Song Y., Davy C. A., Troadec D., Blanchenet A.-M., Skoczylas F., Talandier J., Robinet J.-C., Marine and Petroleum Geology **65**, 63-82 (2015)

[12] Song Y., Davy C. A., Bertier P., Troadec D., Microporous and Mesoporous Materials **228**, 64-85 (2016)

[13] Nguyen Kim T., *Direct sampling applied to porous media* (Université Pierre et Marie Curie, Paris, France, 2013) 1-189

[14] Yven, Béatrice and Sammartino, Stéphane and Geraud, Yves and Homand, Francoise and Villieras, Frédéric, Mémoires de la Société Géologique de France **178** 73-90 (2007)

[15] Zhang, C. L. and Rothfuchs, T., Applied Clay Science **26** 325-336 (2004)

DEM study of granular flow around blocks attached to inclined walls

Joel Samsu[1,3*], *Zongyan* Zhou[1,3], *David* Pinson[2,3], and *Sheng* Chew[2,3]

[1]Laboratory for Simulation and Modeling of Particulate Systems, Department of Chemical Engineering, Monash University, Clayton, VIC 3800, Australia
[2]Iron and Steelmaking Technology, BlueScope Ltd., P.O. Box 202, Port Kembla, NSW 2505, Australia
[3]ARC Research Hub for Australian Steel Manufacturing

Abstract. Damage due to intense particle-wall contact in industrial applications can cause severe problems in industries such as mineral processing, mining and metallurgy. Studying the flow dynamics and forces on containing walls can provide valuable feedback for equipment design and optimising operations to prolong the equipment lifetime. Therefore, solids flow-wall interaction phenomena, i.e. induced wall stress and particle flow patterns should be well understood. In this work, discrete element method (DEM) is used to study steady state granular flow in a gravity-fed hopper like geometry with blocks attached to an inclined wall. The effects of different geometries, e.g. different wall angles and spacing between blocks are studied by means of a 3D DEM slot model with periodic boundary conditions. The findings of this work include (i) flow analysis in terms of flow patterns and particle velocities, (ii) force distributions within the model geometry, and (iii) wall stress vs. model height diagrams. The model enables easy transfer of the key findings to other industrial applications handling granular materials.

1 Introduction

Dense granular flow is widely encountered in bulk material handling industries. In this particular flow regime, the flow dynamics are controlled by momentum transfer during particle collisions, and friction between grains or particle and containing walls, respectively [1]. In industrial applications, forces acting on containing walls, originating from particle flow dynamics, can cause serious problems to plant parts and equipment, resulting in interrupted operations and high costs for the plant operator [2].

The relationship between depth inside a silo structure and the resulting wall pressure had already been established by Janssen at the end of the 18th century [3]. Over the years the theory has been further developed for dynamic conditions and to account for geometrical discontinuities like the transition from vertical silo wall to inclined hoppers in the lower sections [4–6]. The onset of particle movement in the silo, as well as geometrical discontinuities along the wall, are associated with a sharp increased wall pressure, a so-called switch [7]. A great part of the available literature concentrates on performance aspects and the design of silos and hoppers (for example see [8,9]). This work, however, aims to describe the effect of complex wall structures, present for structural or functional reasons, inside an expanding geometry under continuous operations which are usually not considered when studying silo discharge. This particular setup has been chosen, as the technical motivation for the presented research is to better understand the flow at the near wall region of the ironmaking blast furnace shaft. The resulting outcomes, however are presented in a general way in order to be able to apply and extend them to other applications as well.

This paper presents the results of a DEM simulation study of the flow behaviour of steady-state dense granular flow around blocks attached to a wall in a hopper-like geometry. The effect of block spacing and different wall inclination angles, with respect to flow structure, particle velocities, and resulting forces and wall stresses have been studied.

2 Model description

In this work, the discrete particle simulation (DPS) technique based on the discrete element method (DEM) originally developed by Cundall and Strack [10] is used to model the granular flow system. The analyses have been carried out with the DEM code LIGGGHTS® [11].

Several model geometries were constructed featuring different wall angles (α) and number of blocks, i.e. spacing (s) between blocks (Fig. 1). All geometries exhibit the same model height, width at the top, model

* Corresponding author: joel.samsu@monash.edu

thickness, block size, and position of the first (top) and last (bottom) block. Periodic boundary conditions have been applied to the front and back of the model geometry to limit the total number of particles. A summary of the geometric dimensions of the considered cases is presented in Table 1.

Table 1. Geometric dimensions of the simulation cases.

Parameter	Value(s)
wall angle, α	90°, 85°, 80°
block spacing (mm), s	900, 600, 400
no. of blocks	5, 7, 10
model thickness (mm)	100
block size (mm)	150 x 150

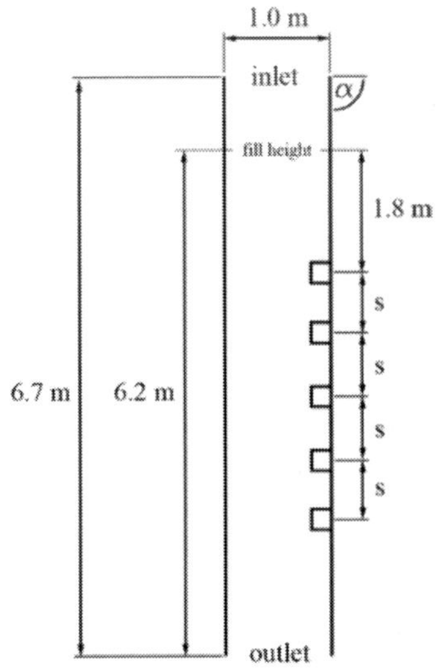

Fig. 1. Set-up and dimensions of the model geometry.

At the beginning of each simulation, the geometry is filled with monosized spherical particles exhibiting a diameter of 25 mm. After filling, particle settling takes place for 3 seconds, after which all particles above a height of 6.2 m are removed from the simulation domain. Subsequently, particles are inserted at the top with an insertion rate of 800 particles per seconds, which at a particle density of 1081 kg/m^3 corresponds to 6.15 kg/s. With the onset of particle insertion, particles at the bottom of the geometry are removed from the computational region at the same rate as particles are inserted at the top. This ensures that the number of particles inside the domain stays constant throughout the simulation. Only particles residing within a distance of two particle diameters from the bottom of the geometry are candidates for removal by spatially random selection. After a macroscopic steady-state flow is reached, particle positions, velocities and forces are recorded for further analysis. Material parameters and simulation conditions are given in Table 2.

3 Results and discussion

The results presented are representative simulation outcomes that have been chosen to demonstrate some of the key findings of this study.

Table 2. Material parameters and simulation conditions.

Parameter*	Value(s)
particle shape	spherical
particle diameter (mm), d_p	25
particle density (kg·m^{-3}), ρ_p	1081
Young's modulus (Pa), E	10^7
Poisson's ratio, ν	0.3
sliding friction coefficient, μ_s	0.4
rolling friction coefficient, μ_r	0.1d_p
coefficient of restitution, e	0.1
time step (s), Δt	0.00005

*uniform values for all material types/pairs

3.1 Flow structure

Different colours have been assigned to the particles in the simulation, which allows the identification of different flow zones inside the geometry. Fig. 2 illustrates the flow structure of the cases with 90° and 85° wall inclination angle and a block spacing of 600 mm, after the same number of particles have been inserted into and discharged from the domain. Blue particles originating from the initial filling process can be observed between and on top of blocks. These areas can be identified as stagnant zones. Each green or red particle layer consists of 1000 particles. In both cases the layered structure is preserved during descent, although in the case with the inclined wall, the layers become progressively thinner and layers misalign towards the bottom of the geometry. Zones of mixed particles occur in the transitioning region from stagnant to moving particles.

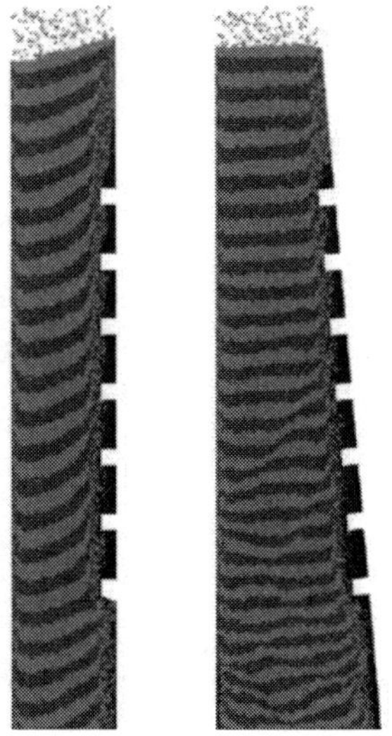

Fig. 2. Flow structure for the cases with 90° (the left) and 85° (the right) wall inclination angle, and 600mm block spacing.

3.2 Particle velocities

The same two cases are chosen to examine the influence of wall inclination angle on the particle velocity distribution inside the simulation domain (Fig. 3). The particle velocities of 100 different time steps have been averaged over a period of 200,000 simulation time steps, which corresponds to 10 s. In the case of 90° walls (left), the velocity is relatively constant along the geometry height, except for the regions near the blocks. In the case with an 85° wall angle (right), the velocity decreases as the cross-sectional area of the geometry increases. The effect of blocks on the velocity distribution is greater in the case of the inclined wall, as can be seen by the width of the velocity transition region between the block faces and the bulk flow region in Fig. 3. Phenomena observed in analysis of the flow structure in Fig. 2 can be explained by observations made from the velocity distribution. For example, with an 85° inclined wall, the diagonal misalignment of the layers in the middle region of the geometry coincides with the velocity contour line between 0.8 and 0.9 m/s. Similar observations can be made at other transition areas.

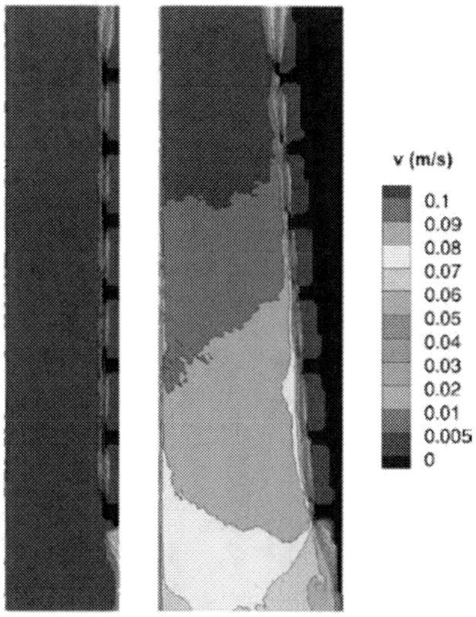

Fig. 3. Particle velocity distribution for the cases with 90° and 85° wall inclination angle, and 600mm block spacing.

3.3 Forces

Like the particle velocities, the magnitudes of the individual particle force vectors, i.e. total force on each particle, resulting from particle-particle and particle-wall interactions, respectively, as well as from gravity, have been averaged over their spatial location within the domain. The force distribution for different wall angles and a block spacing of 600 mm is shown in Fig. 4. Overall, the forces inside the domain are smaller at a decreasing wall angle (from left to right). As expected, the force within the domain increases with distance from the top due to the increasing weight exerted by overlying particles. Large forces in the regions near the blocks are most dominant in the case with vertical walls (left) and as wall angle decreases, the forces at the block edges also decrease.

The influence of block spacing on the force distribution is illustrated in Fig. 5. Large forces occur at the block edges and front faces of the protruding blocks. Forces at the particle-wall interface cannot be attained from these plots. To identify the influence of block spacing on the stress in the regions between blocks a wall stress analysis of these areas is presented in the next section.

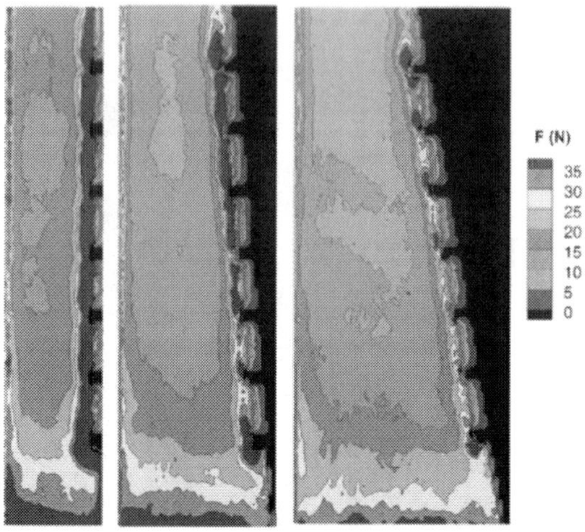

Fig. 4. Force distribution inside the geometry for cases with 90°, 85° and 80° wall inclination angle, and 600mm block spacing.

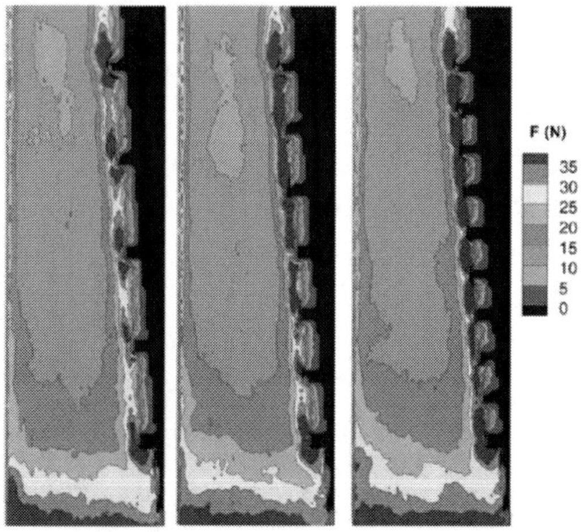

Fig. 5. Force distribution inside the geometry for cases with 900mm, 600mm and 400mm block spacing, and 85° wall inclination angle.

3.3.1 Wall stress analysis

Normal and shear stress acting on each mesh element of the model geometry have been recorded and averaged over time for the period of macroscopic steady-state particle flow. Each of the three diagrams in Fig. 6 compares the normal stress along the height of the wall for the cases with the same block spacing, but different wall angles. For easier comparison, only the stress acting on the wall area between blocks is shown. In all cases the overall stress increases from the top to the bottom of

DOI: 10.1051/epjconf/201714003075

the domain. With inclined walls (α<90°) wall stress decreases significantly compared to the cases with straight walls (α=90°), especially when block spacing is wide (see Fig. 6a). For the cases with 85° and 80° wall angle, the difference is less significant.

For each area between two neighbouring blocks, the stress increases vertically in the downward direction, i.e. the stress below a block is at a minimum and reaches a maximum in the area above the next block. The magnitude of these minimum and maximum values is dependent on the vertical location along the geometry, resulting in a steep stress gradient in the cases of 400 mm block spacing and a more gradual increasing wall stress for wider spacing.

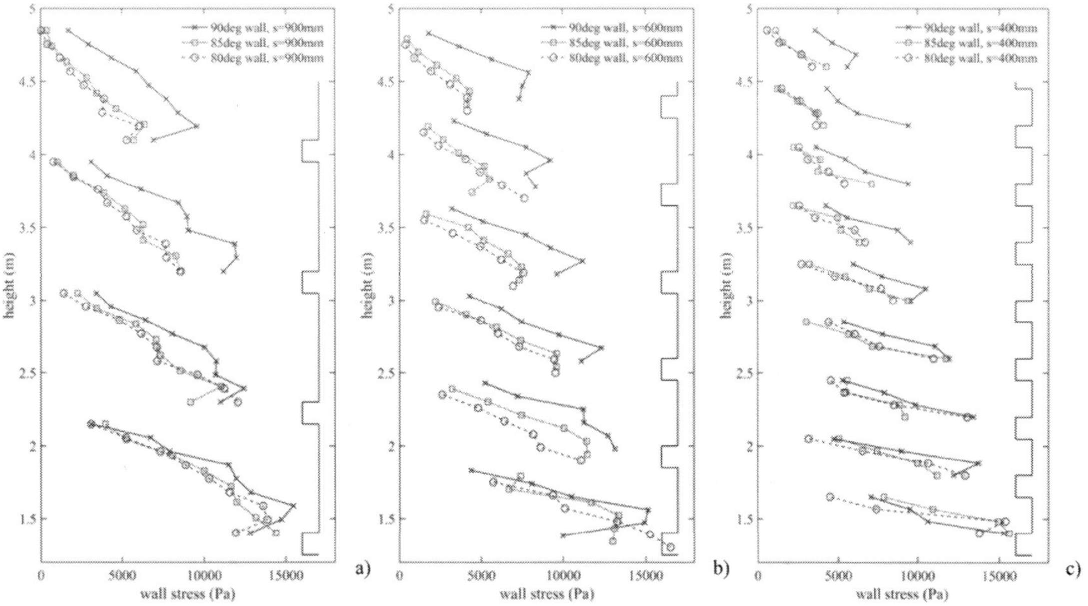

Fig. 6. Normal stress acting on the wall in the areas between attached blocks. The effect of wall angle is demonstrated for block spacing of a) 900mm, b) 600mm and c) 400mm.

4 Conclusion

These results demonstrate the influence of wall angle and block spacing on different flow characteristics and the resulting wall stress. The following conclusions can be made:

In the case of vertical walls the particle flow showed relatively uniform behaviour along the domain height with respect to flow structure and particle velocity. At inclined walls, however, particle layers became progressively non-uniform and the particle velocities were observed to decrease towards the bottom of the geometry.

Observations of flow patterns in cases with inclined walls were consistent with the results obtained from velocity analysis and can be explained by these.

The differences between the results of different wall inclination angles, e.g. between 85° and 80° were less significant than those between vertical and inclined walls. The trends observed in these results may need further study to establish their validity.

Large forces could be observed in the areas of the block edges protruding into the bulk flow area. Further analysis of this region, focusing on the interaction of particles and walls may be revealing, as this is where the transition from bulk continuum-like flow to discrete interaction dominates.

Funding from the Australian Research Council Industrial Transformation Research Hub Scheme (Project Number IH130100017) is gratefully acknowledged. This research was undertaken with the assistance of resources from the National Computational Infrastructure (NCI), which is supported by the Australian Government.

References

1. GDR Midi, Eur. Phys. J. E **14**, 341 (2004).
2. A. Dogangun, Z. Karaca, A. Durmus, and H. Sezen, J. Perform. Constr. Facil. **23**, 65 (2009).
3. H. A. Janssen, Zeitschr. D. Vereines Dtsch. Ingenieure **39**, 1045 (1895).
4. D. M. Walker, Chem. Eng. Sci. **21**, 975 (1966).
5. J. K. Walters, Chem. Eng. Sci. **28**, 13 (1973).
6. R. Artoni, A. Santomaso, and P. Canu, Chem. Eng. Sci. **64**, 4040 (2009).
7. J. K. Walters, Chem. Eng. Sci. **28**, 779 (1973).
8. P. W. Cleary and M. L. Sawley, Appl. Math. Model. **26**, 89 (2002).
9. S. Masson and J. Martinez, Powder Technol. **109**, 164 (2000).
10. P. A. Cundall and O. D. L. Strack, Géotechnique **29**, 47 (1979).
11. C. Kloss, C. Goniva, A. Hager, S. Amberger, and S. Pirker, Prog. Comput. Fluid Dyn. An Int. J. **12**, 140 (2012).

EPJ Web of Conferences **140**, 02026 (2017)
Powders & Grains 2017

DOI: 10.1051/epjconf/201714002026

Sample similarity analysis of angles of repose based on experimental results for DEM calibration

Yuan Tan[1,*], *Willibald A.* Günthner[1], *Stephan* Kessler[1] and *Lu* Zhang[2]

[1]Institute for Materials Handling, Material Flow and Logistics
Technical University of Munich, Boltzmannstraße 15, 85748 Garching, Germany
[2]Institute for Sensing Control Measurement and Precision Machining Technology
Dalian University of Technology, Linggong Road 2, Ganjingzi District, Dalian, Liaoning

Abstract. As a fundamental material property, particle-particle friction coefficient is usually calculated based on angle of repose which can be obtained experimentally. In the present study, the bottomless cylinder test was carried out to investigate this friction coefficient of a kind of biomass material, i.e. willow chips. Because of its irregular shape and varying particle size distribution, calculation of the angle becomes less applicable and decisive. In the previous studies only one section of those uneven slopes is chosen in most cases, although standard methods in definition of a representable section are barely found. Hence, we presented an efficient and reliable method from the new technology, 3D scan, which was used to digitize the surface of heaps and generate its point cloud. Then, two tangential lines of any selected section were calculated through the linear least-squares regression (LLSR), such that the left and right angle of repose of a pile could be derived. As the next step, a certain sum of sections were stochastic selected, and calculations were repeated correspondingly in order to achieve sample of angles, which was plotted in Cartesian coordinates as spots diagram. Subsequently, different samples were acquired through various selections of sections. By applying similarities and difference analysis of these samples, the reliability of this proposed method was verified. Phased results provides a realistic criterion to reduce the deviation between experiment and simulation as a result of random selection of a single angle, which will be compared with the simulation results in the future.

1 Introduction

According to the criterion of physical properties, granular material can be considered in a state between fluid and solid [1] and is able to maintain a static and stable condition with steepest angle of a classical conical sand-pile. The so-called angle of repose (AoR) is used to describe the flow behaviour and properties of granular material. At which, the material on the slope face is on the verge of sliding. As one of the most primary macroscopic parameters, AoR is widely studied and employed to calculate other parameters of granular material or indirectly describe solids motion, for example, calibration of material parameter of discrete element method (DEM) simulation.

To measure AoR, many methods are available and at least four basics of them are frequently applied: loading simulated measuring box, bottomless cylinder, rotary drum with frictional walls and free falling in a measuring device. All of which were summarized in [1]. In the first method, AoR is measured under in-situ static pressure, while it is measured after a dynamic flow process in the other three methods.

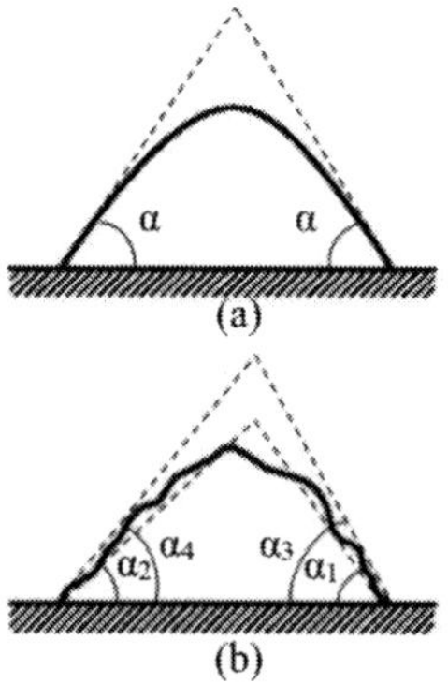

Fig. 1. Deviations of the formed willow chips heap from geometric cone solid: (a) cone with arc-shaped slope and linear generating lines; (b) cone with uneven slope and convex/concave generating line

Even though the measurement of AoR is predefined according to FEM [2] over several decades, just like the angle α pictured in Fig. 1 (a). In reality, the granular material forms an ideal cone shape like Fig. 1 (a) only in few cases. The results could widely range because of

*Corresponding author: tan@fml.mw.fml.de

particle characteristics (different shapes, wide size distribution) and material properties (e.g. sliding/rolling friction coefficient). In a previous study by Fraczek [3], three types of truncations (truncation of the top, cone with convex generating line and cone with concave generating line) were discussed for undesirable cones. Even though to measure extremely asymmetrical slope, such as Fig. 1 (b), either convex or concave generating line could also result in various AoR. The Angle α_2 was greater than α_1 in the case with concave generating line and the third case, and α_3 smaller than α_4 in the second case. Since it is generally reported that AoR increases with increasing sliding and rolling friction coefficients and deviation from spheres [4, 5], various angles would eventually lead to different calibrated material model for DEM simulation.

In light of the introduction above, predictable deviation between experiment and simulation by using traditional method could be even large in some extreme situations. As prescribed in [3], the difference of AoR between experiment and simulation should be within ±5%. We expect an initiate assessment process to minimize this deviation.

2 Material and method

Willow chips were chosen as the object of interest. Among all common bulk materials, it can build smooth top of heap and even arc-shaped surface because of the inhomogeneous particle size and low rolling as well as low sliding friction. As mentioned before, this kind of heap results in lower accuracy of calibration of material model. The experimental set-up, bottomless cylinder, consists of two essential parts (a cylindrical container and lifting equipment) and an extra rotatable plate.

2.1 Experiments

In the present study, measurements of AoR were conducted using a motor-driven bottomless cylinder with two dimensions, the inner diameter was 425 mm and 250 mm and the height was 595 mm and 350 mm, respectively. Biomass materials were lifted in two different velocities (142 mm/s and 33 mm/s) with three repetitions, respectively. For the purpose of 3D reconstruction of the formed heap, the new Xbox Kinect V2 sensor was utilized. Although the Kinect V2 was not designed for rehabilitation specifically. However, it allows free interaction between controller and full 3D-recognition [6, 7]. In total 12 independent experiments were completed. Since the form of heaps didn't reveal significant difference among them, the following content will be continued with only one experiment as example.

Two different views of the heap during one experiment are shown above (Fig. 2). As has been described earlier, the measurement of AoR is characterized by a considerable inaccuracy as a result of deviations of the shape of the cone generatrix from a straight line. Therefore, divergences of AoR of this same sand-pile suggested the poor quality of traditional methods.

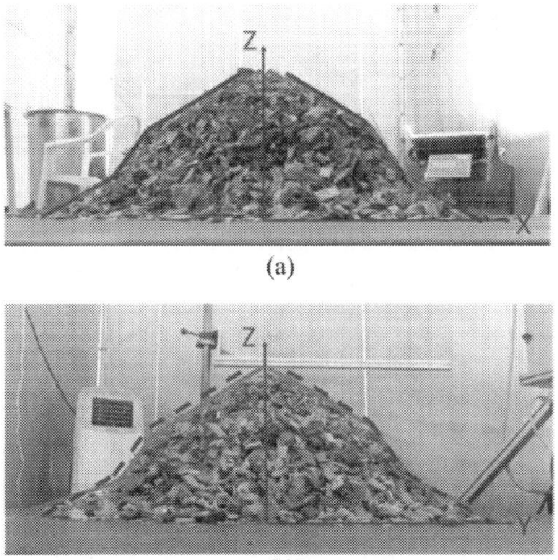

Fig. 2. Photos of the heap: (a) purple solid line-contour of view XZ; (b) blue solid line-contour of view YZ.

2.2 Data processing

Prior to data processing, the generated data need to be pre-processed in order to achieve desired accuracy within acceptable computation time. Specifically, overlapped and excessive points were removed by using MeshLab® [8] so that the file size was 100 times minimized.

The data processing of the surface of repose consists of six basic processes:

1. The scanned surface is obtained in STL format, thereby generating point clouds in Matlab R2015b® [9] (see Fig. 3 (a)).
2. The approach proceeds with analysis of each single slice, which consists of two parallel, 5 mm apart, planes (see Fig. 3 (a)). Since 5 mm is negligible compared to the dimension of the heap, it was hypothesized that all points are located approximately at the same tangential plane of the conical surface. The position of slice may not be precisely determined but two planes must symmetric to a same vertical plane (P_m), which passes through the origin. Thus, the points in each slice are successfully isolated (see Fig. 3 (b)).
3. These selected points are projected onto plane P_m, as shown in Fig. 3 (c).
4. An inclined line, which forms AoR with the horizontal line, need to be generated based on these scatter-plots. Then a LLSR technique was applied to the linearized regression (see Fig. 3 (d)).
5. Two angles of repose are computed for each slice.
6. Two planes of the first slice are rotated simultaneously around the origin of coordinate system to obtain new slices of the heap (see Fig. 3 (e)). Then, steps 3-5 are repeated.

These steps are schematically represented below. Following the steps, a great number of samples of AoR were studied. Here in this work, 17 slices, namely 34 angles were calculated. Based on statistic

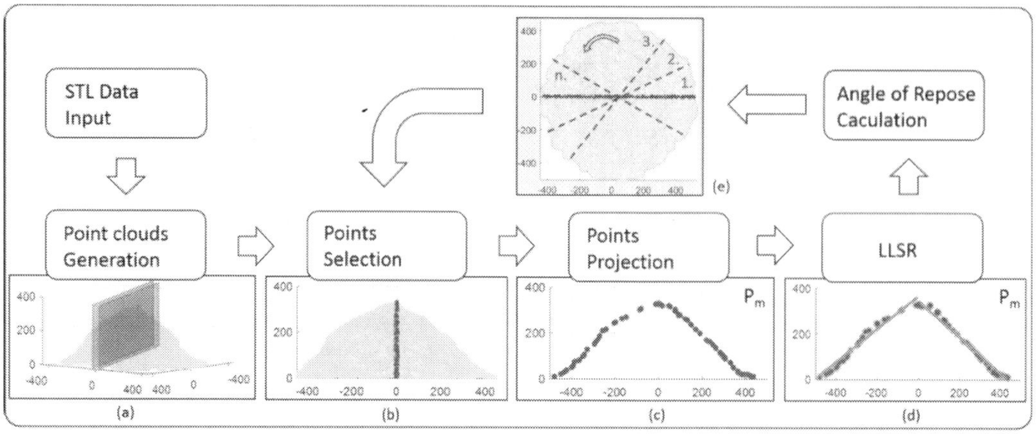

Fig. 3. Date processing of point cloud.

theory, the more samples and angles in each sample are, the more accurate analysis is in practice. Otherwise, the proposed algorithm was started with a low-density point clouds, in other words, coarse STL file. For sure it will be also feasible with intensive point clouds for the sake of more accurate analysis in actual applications, yet at the expense of large memory and computational effort.

3 Results and discussion

3.1 Correlation coefficient of linear regression

According to the trend of the selected points, the surface was considered as a linear regression in the whole calibration range, as shown in Fig. 4. To determine whether the regression equation truly represent the set of data, the correlation coefficient (r) was examined using the following mathematical formula [10]:

$$r = \frac{n\sum xy - (\sum x)(\sum y)}{\sqrt{n(\sum x^2) - (\sum x)^2}\sqrt{n(\sum y^2) - (\sum y)^2}}$$

Herein, the linear regression of only one sample is presented. Fig. 5 indicates that x and y of all samples have a strong linear correlation, with the absolute values of r extremely close to 1. The results of other samples were similar. In addition, the fitting lines could be more accurate using obtained overlapping plots from uncompressed STL file.

The two dashed lines in two diagrams of Fig. 4 represent different supposable measurements. Therefore, compared to angle (β_1 and β_2) of the linear regression line, these two lines are clearly deviated from each other in two cases: (1) the line was drawn tangentially to the points of lower part of the slope (angle β_{11} and β_{21}); (2) the line was drawn tangentially to the upper part (angle β_{12} and β_{22}). The difference between each two angles of dashed lines is over 5%, approaching the highest allowable limit for calibration. As a result, it could lead to distinct material models of DEM after calibration, if any of them is regarded as reference.

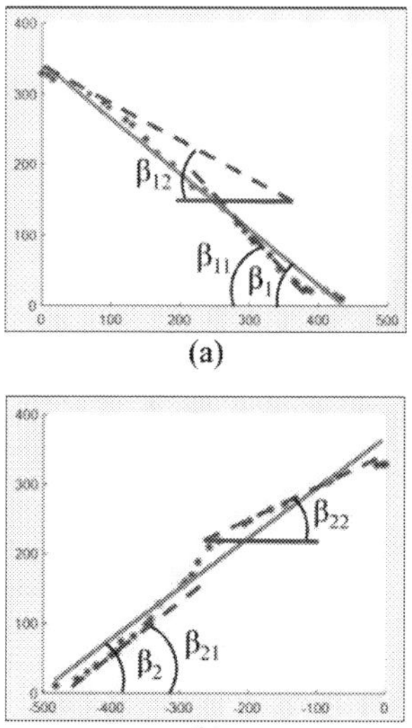

Fig. 4. Fitting lines according to LLSR and convex/concave generating lines of the selected section (a: right slope; b: left slope)

3.2 Sample similarity analysis

The proposed computational approach provides a way of obtaining a sample of AoR from the whole surface of heap, instead of one single AoR. In the next step, as mentioned in our objective, the similarity of different samples was investigated. Therefore, both the mean value and the standard deviation of each sample, were calculated to quantify the amount of variation or dispersion of a set of data values.

The mean values and standard deviations of 10 samples, each including 34 approximatively computed angles, are presented in Fig. 6. The values of them range from 37.4 to 37.75 and from 1.89 to 2.46 respectively.

Considering both of them could be taken as criterion to assess the similarity of heap, with closed mean values and low standard deviation it could be concluded, that the results show the efficiency of the proposed method.

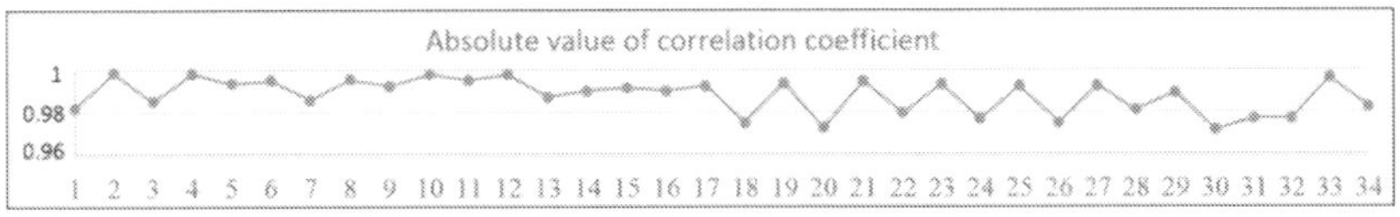

Fig. 5. Absolut value of correlation coefficient of LLSR (one sample).

As mentioned in section 2.2, every slice was located randomly and independently from each other. The result of similarity analysis demonstrated that any sample of AoRs of the same heap is able to characterize the slope, which is directly related to material parameters.

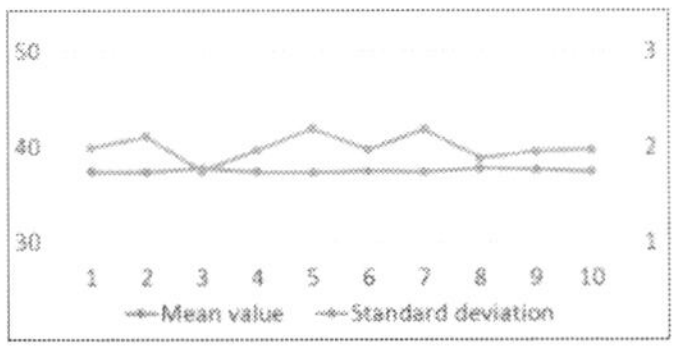

Fig. 6. Mean value and standard deviation of AoRs.

4 Conclusion

We believe that the proposed method, based on experimental evaluation, has a number of potential applications in the representation and expression of sand-pile of granular material with entire three-dimensional data. This study demonstrate that the Xbox Kinect V2 is a useful and cost-effective devise, which featured simple operation, time saving, and high performance.

Furthermore, this data processing procedure may be used to analyze the sand-pile form of simulation as well. In this way the DEM calibration could be accomplished by analyzing two random selected set of angles of experiment and simulation, instead of comparing two single angle of both processes.

4.1 Major advantages

Based on the result of this study, the proposed method is applicable to general problems and especially for cases with arc-shaped top or uneven slope.

As stated at the beginning of section 2.1, 12 tests were carried out. The similarity analysis of samples of AoRs for each test suggests a high repeatability regardless of objective factors, such as cylinder size, and lifting velocity, or subjective factors, like choice of view and point to measure as well as visual estimation about the tangent line of slope. As the physical performance of material is assessed based on the entire form of heap, namely, a sample of enough angles, instead of two-dimensional aspect.

It can be predicted that the propose method could also be applied in prospective analysis of simulation with bottomless cylinder. Hence, calibration of material parameter will be accomplished based on comparison between two samples of angles random selected from experiment and simulation. Furthermore, errors induced by objective and subjective factors could be considerably reduced.

4.2 Comparison with conventional methods

Unlike conventional methods, this method requires 3D-recognition instrument, which can obtain assessable data format of the surface of sand-pile at least. However, for some types of materials, the shape is somewhat angular and the particles have large size distribution range, which may build bridge during the lifting process, so that the correlation coefficient may not meet the requirements of linear regression. Therefore, further work is expected to enhance the applicability of the method for calibration of material parameters.

5 Outlook

In comparison with other traditional methods the proposed method, especially data processing and afterwards, is much more complex than conventional methods. The corresponding graphical user interface (GUI) is on-going, which enhances the efficiency and ease of use for the underlying logical design of stored program. Instead of text-based programming, such as typed command labels and text navigation, the automatic measurement analysis only requires data input from Kinect and simple icons. To facilitate calibration of material parameter, the ultimate goal is to find out a practicable assessment method. So the analysis of the similarity as well as difference between experiment and simulation should be performed in future studies.

References

[1] A. Wu, Y. Sun, *Granular dynamic theory and its applications*. 17 (2008).
[2] The European Federation of Materials Handling “FEM”, FEM 2.582, *General properties of bulk materials and their symbolization*, (1991)
[3] J. Frączek, A. Złobecki, J. Zemanek, Journal of food engineering, **83**, 17. (2007)
[4] Y. C. Zhou, B. H. Xu, A. B. Yu, Zulli, P. Powder technology, **125**, 45 (2002)
[5] W. Wang, J. Zhang, S. Yang, H. Zhang, H. Yang, G. Yue, Particuology, **8**, 482 (2010)
[6] ST. Smith, D. Schoene, *Aging Health 2012*, **8**, 243–52 (2012)
[7] MJD. Talyor, D. McCormick, T. Shawis, R. Impson, *Res Dev 2011*, **48**, 1171–1186 (2011).
[8] Meshlab, version V1.3.3 ISTI (CNR research center, Pisa, Italien, 2015)
[9] Matlab, version R2014b (The MathWorks Inc, Natick, Massachusetts, 2014)
[10] Bland, J.M., Altman, D.G, *Statistics notes: measurement error*. BMJ. **312 (7047)**, 1654 (1996)

Stress analysis of an agitated particle bed with different particle aspect ratios by the discrete element method

Wei Pin Goh[1], *Mojtaba* Ghadiri[1,*], *Frans* Muller[1], *Kushal* Sinha[2], *Nandkishor* Nere[2], *Raimundo* Ho[2], *Shailendra* Bordawekar[2] and *Ahmad* Sheikh[2]

[1]School of Chemical and Process Engineering, University of Leeds, Leeds, UK
[2]Process Research and Development, AbbVie Inc., 1 North Waukegan Road, North Chicago, Illinois 60064, USA

Abstract. The size distribution, shape and aspect ratio of particles are the common factors that affect their packing in a particle bed. Agitated powder beds are commonly used in the process industry for various applications. The stresses arising as a result of shearing the bed could result in undesirable particle breakage with adverse impact on manufacturability. We report on our work on analysing the stress distribution within an agitated particle bed with several particle aspect ratios by the Discrete Element Method (DEM). Rounded cylinders with different aspect ratios are generated and incorporated into the DEM simulation. The void fraction of the packing of the static and agitated beds with different particle aspect ratios is analysed. Principal and deviatoric stresses are quantified in the regions of interest along the agitating impeller blade for different cases of particle aspect ratios. The relationship between the particle aspect ratio and the stress distribution of the bed over the regions of interest is then established and will be presented.

1 Introduction

Agitated powder beds are found in many process industries for mixing and drying. The process performance is affected by the physical properties of the particles such as size and shape [1, 2]. Different packings of the bed give rise to different extents of resistance to flow and this may have adverse effects on particle breakage and transformations [3]. DEM is used to study the stress distribution within the region of interest as a function of particle aspect ratio. Rounded cylinders are used to represent particles with different aspect ratios. Polyhedra geometry [4] is used to make the particles to simulate particle shape. When a particle bed is agitated, the stresses arising within the bed change over time because of the impeller agitation. The interactions between the particles and the impeller as well as the evolution of stresses are of particular interest in this study. The bed packing void fraction is analysed and the stresses are characterised for various particle aspect ratios.

2 Materials and Methods

2.1. Simulation Setup

DEM simulations were carried out using the ROCKY software (ESSS, Florianópolis, Brazil) by incorporating particle shape as polyhedra. Particles with aspect ratios in the range 1, 1.25, 1.5, 1.75, 2, 3 and 5 were investigated. As can be seen in Figure 1, spheres are used to represent particles with unity aspect ratio while particles with other aspect ratios are represented by rounded cylinders with different lengths and diameters. The volume of the particles is kept constant at an equivalent-volume diameter of 3 mm regardless of the change of particle aspect ratio throughout the simulations. The geometry of the impeller and vessel used in the simulation are reproduced as shown in Figure 2. The details of the geometry and simulation conditions are shown in Table 1. The vessel is initially fed with particles until the bed height (0.06 m) is approximately three times the blade height (0.018 m). The bed is then agitated at a speed of 1 rev/s for 5 s.

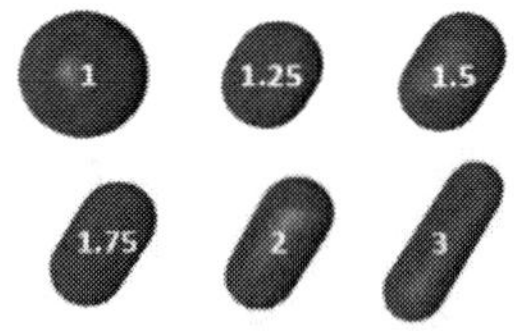

Figure 1. Sphere and rounded cylinders with different aspect ratios.

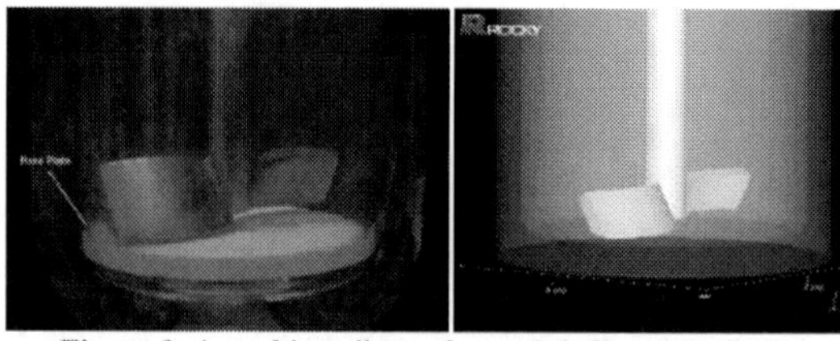

Figure 2. Actual impeller and vessel (left) and its CAD representation (right).

[*] Corresponding author: m.ghadiri@leeds.ac.uk

Table 1. Geometry and simulation details.

Operating Conditions	
Vessel Inner Diameter (mm)	110
Number of Blades (-)	2
Blade Height (mm)	18
Blade Inclination (°)	60
Blade Diameter (mm)	98
Clearance Height (mm)	2
Angular Velocity (rev/s)	1
Particle Feed Rate (kg/s)	3
Particle Density (kg/m^3)	1.6
Young's Modulus (Pa)	10^7
Coefficient of Sliding Friction (-)	0.3
Coefficient of Restitution (-)	0.3
Simulation Time (s)	5

2.2. Particle Bed Void Fraction

Particles with different aspect ratios pack differently to form a bed and this in turn affects the prevailing shear stress[5]. In this study, a cylindrical measurement cell with a height of 0.02 m is generated to measure the void fraction of the particle bed (Figure 3). The measurement cylinder is placed right above the impeller blades and below the free surface of the bed to eliminate the situation where unwanted entities (such as impeller blades and area above free surface of the bed) are taken into the calculation of the void fraction.

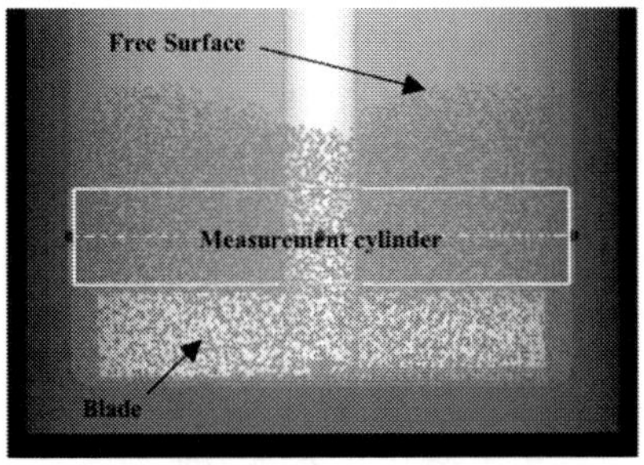

Figure 3. Position of measurement cylinder.

2.3. Stress Analysis

In an agitated bed, as the impeller rotates, particles in front of the blades experience shear and get pushed over and fall behind the blade to a region where the bed is dilated. The prevailing stresses of the region of particle bed that comes into direct contact with the rotating bed are analysed. Experimentally, the stress distribution within a particle bed is too difficult to be determined, but this is made possible in DEM simulation by introducing measurement bins in the regions of interest [6]. Five measurement bins created along the impeller blade that capture the particle interactions during each time step within their own domain are shown in Figure 4. The stress tensor of each bin is calculated using the equation below developed by Bagi [7],

$$\sigma_{ii} = \frac{1}{V}\sum_{1}^{N} F_{ii} \cdot r_p \tag{1}$$

where σ is the stress, V is the measurement bin volume, N is the number of particle within the bin and F is the force acting on a particle in direction ii with radius r_p. The average principal and deviatoric stresses are then given by equations 2 and 3:

$$\sigma_{ave} = \frac{P_1 + P_2 + P_3}{3} \tag{2}$$

where P_1, P_2 and P_3 are the three principal stresses, respectively. The deviatoric stress, σ_{dev} [8] can be calculated from the principal stresses and is defined as

$$\sigma_{dev} = \frac{\sqrt{(P_1 - P_2)^2 + (P_1 - P_3)^2 + (P_2 - P_3)^2}}{\sqrt{6}} \tag{3}$$

The calculation for both principal and deviatoric stresses is performed at each time step to capture the stresses experienced along the impeller blade. The size of the bin is kept in such a way that it has approximately more than 100 particles in each cell at each time step so that the dependency of individual particles is minimised while improving the statistical reliability for averaging the stress tensor in a measurement bin.

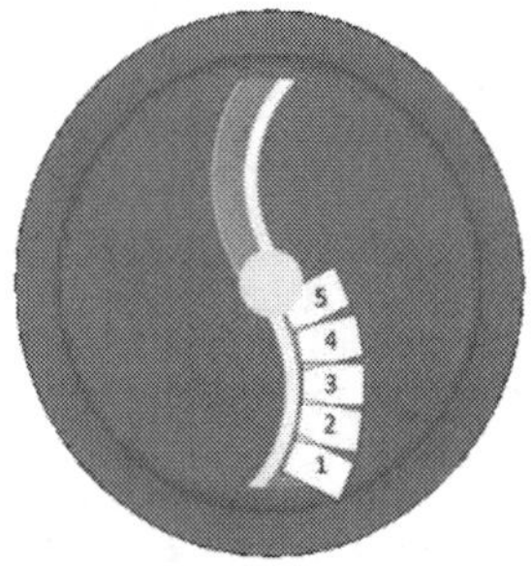

Figure 4. Measurement bins along the impeller blade.

3 Results and Discussions

3.1 Bed Packing

The void fraction of the static and agitated beds for different particle aspect ratios is shown in Figure 5. A bed made with spherical particles (aspect ratio 1) appears to have a void fraction of 0.5 during static state. As the particle aspect ratio increases, the void fraction of the bed decreases initially. A minimum void fraction value is observed when the particle aspect ratio is 1.5. A further increase in the particle aspect ratio increases the void fraction. While the bed is agitated, higher void fractions of the particle bed are observed, suggesting that the bed is dilated when it is in motion. The arrangement of particles at the middle of the bed (cross sectional view at XY plane) when it is static (0 s) and agitated (5 s) is shown in Figure 6. At low aspect ratio from 1 to 1.75, the particle arrangement appears random for both static and dynamic cases. As the particle aspect ratio approaches 2, the tendency of the particles to reorientate themselves and align with the neighbouring particles becomes prevalent. As the particle aspect ratio increases, the number of particles that cluster together increases as well. The effect is most significant and can be seen clearly for the case of particles with aspect ratio of 5. Even when the bed is agitated, the phenomenon of particles clustering as groups still prevails.

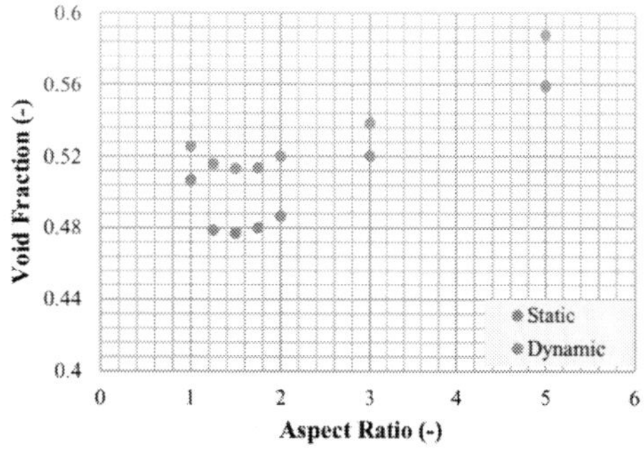

Figure 5. Void fraction for particle beds with different particle aspect ratios.

3.2 Impeller Torque

The impeller torque data obtained for the mono-sized particles are shown in Figure 7. A high torque is observed at the initial period when the impeller starts rotating. This is due a number of reasons: (i) rapid acceleration of the impeller that immediately achieves the required velocity, (ii) particle bed dilation and (iii) ordering of crystalline-like structure that occurs when the particles are mono-sized. As the particle bed starts to move, the torque reduces and begins to plateau. Increasing the particle aspect ratio increases the impeller torque. However, for particle aspect ratios more than 1.5, the torque is independent of the aspect ratio. A plot of the average impeller torque versus the aspect ratio shows a better visualisation of the trend (Figure 8). The impeller torque increases initially and starts to stabilise at a particle aspect ratio of around 2. A further increase in the particle aspect ratio only increases the impeller torque marginally, suggesting that there is a limiting particle aspect ratio where further any increase yields no significant effect on the impeller torque. Visual observation from Figure 6 suggested that the particles tend to align themselves and form groups of clusters that have preferred orientation when the aspect ratio is more than 1.75, which could potentially explain the trend observed. However, higher fluctuations of the impeller torque are observed as the particle aspect ratio increases and this could be attributed to the fact that particles with higher aspect ratio have a higher tendency of jamming.

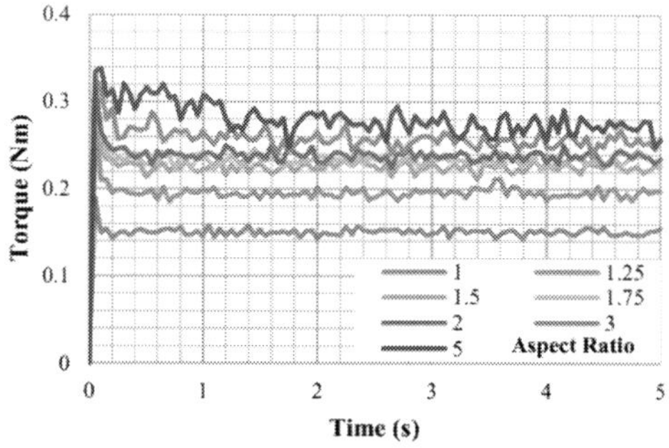

Figure 7. Impeller torque as a function of time for particle beds with different particle aspect ratios.

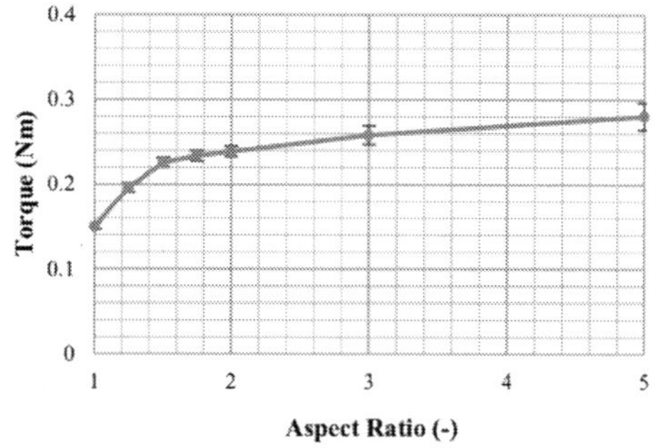

Figure 8. Average impeller torque versus particle aspect ratio.

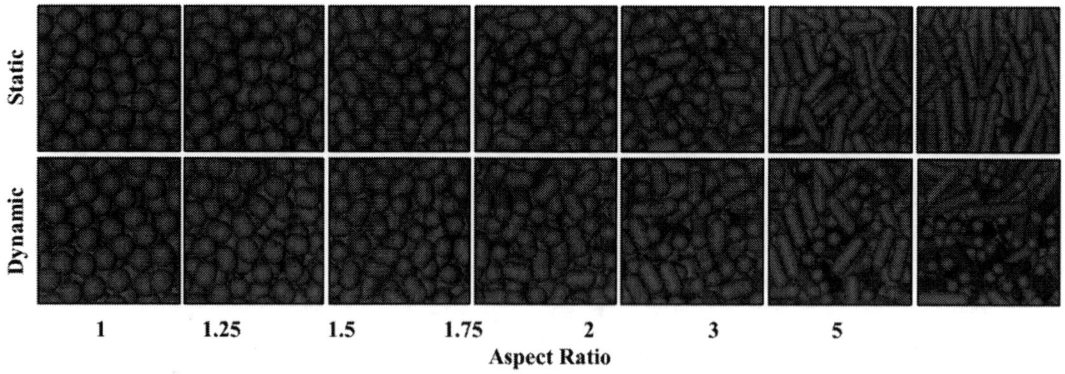

Figure 6. Particle arrangement in static and dynamic beds (cross sectional view at XY plane).

3.3 Principal and Deviatoric Stresses

When the impeller is rotating, for every 0.5 s, a blade comes into direct contact with the measurement bins. The region in front of the blade is constantly being compressed, sheared and pushed over the blade to a region behind it where the particle bed is relaxed. The deviatoric stresses experienced at the region nearest to the wall (bin 1) over time are shown in Figure 9. As the blade approaches the bins, an increase in the deviatoric stresses is observed. Upon reaching a maximum value, the deviatoric stresses decrease as the blade moves past the measurement bins. This cyclical event is observed for every measurement bin for both principal and deviatoric stresses but the results are not shown here. The average principal and deviatoric stresses experienced at the regions near the vessel wall (bin 1) and near the impeller shaft (bin 5) are shown in Figures 10 and 11. The principal stresses increase with the increase of particle aspect ratio for both the regions near the vessel wall and near the impeller shaft but the rate of increase is different. Similar trend is observed for deviatoric stresses. Compared to bin 5, a substantial increase of stresses (6 times) is observed at bin 1 as the particle aspect ratio is increased. This suggests that the particles are subjected to different extents of stresses at different radial positions. Both principal and deviatoric stresses show fluctuations as the particle aspect ratio is increased.

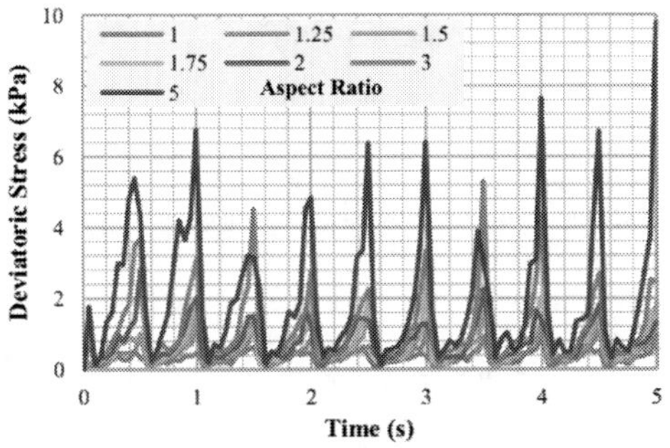

Figure 9. Deviatoric stress experienced at the region near the wall over time.

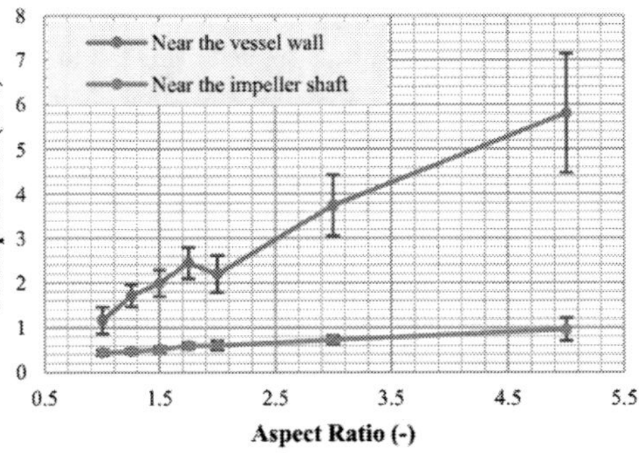

Figure 10. Average principal stress at the regions near the shaft and near the vessel wall.

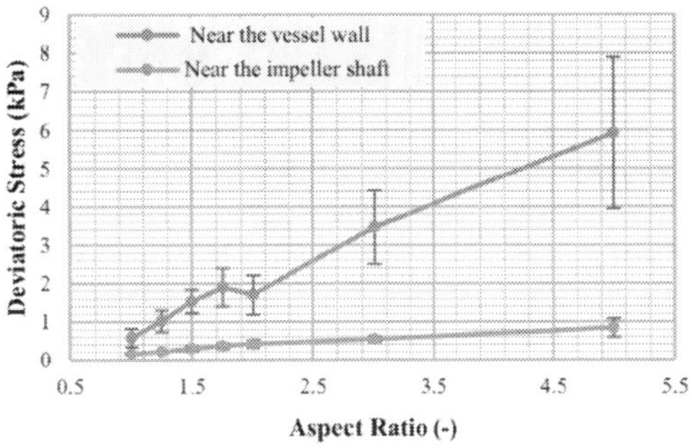

Figure 11. Average deviatoric stress at the regions near the shaft and near the vessel wall.

4 Conclusions

The packing of particle beds under shear deformation by an impeller is assessed using DEM for different particle aspect ratios of rounded cylinders from 1 to 5 and also for spherical particles for comparison. The void fraction of both static and agitated beds is calculated as compared. The beds undergo compression and dilation when being agitated. The impeller torque which is a measure of the resistance of the bed to flow is found to increase with the particle aspect ratio initially and it then reaches a plateau. Both principal and deviatoric stresses are evaluated at the regions of interest and are found to increase along the radial position.

Acknowledgment

The financial support of AbbVie Inc., Chicago, USA, for a studentship for the first author is gratefully acknowledged. Kushal Sinha, Nandkishor Nere, Raimundo Ho, Shailendra Bordawekar and Ahmad Sheikh are present employees of AbbVie Inc.

References

1. G. Cho, J. Dodds, and J. C. Santamarina, J. Geotech. Geoenviron. Eng., **132**, 591–602 (2006)
2. R. Guises, J. Xiang, and J. L. A. Munjiza, Granul Matter, **11**, 281–292 (2009)
3. A. Potapov and C. Campbell, Int. J. Mod. Phys. C., **7**, 717-729 (1996)
4. C. Hare, M. Ghadiri, and R. Dennehy, Chem. Eng. Sci., **66**, 4757–4770 (2011)
5. Y. Guo, C. Wassgren, W. Ketterhagen, B. Hancock, and B. James, J. Fluid Mech, **713**, 1–26 (2012)
6. K. Bagi, Mech. Mater., **22**, 165–177 (1996)
7. S. Luding, Particuology, **6**, 501–505 (2008)

Arching in three-dimensional clogging

János Török[1,*], *Sára* Lévay[1], *Balázs* Szabó[2], *Ellák* Somfai[2], *Sandra* Wegner[3], *Ralf* Stannarius[3], and *Tamás* Börzsönyi[2]

[1] *Department of Theoretical Physics, Budapest University of Technology and Economics, H-1111 Budapest, Hungary*
[2] *Institute for Solid State Physics and Optics, Wigner Research Center for Physics, Hungarian Academy of Sciences, PO Box 49, H-1525 Budapest, Hungary*
[3] *Institute of Experimental Physics, Otto-von-Guericke-University, Universitatsplatz 2, D-39106 Magdeburg, Germany*

Abstract. Arching in dry granular material is a long established concept, however it remains still an open question how three-dimensional orifices clog. We investigate by means of numerical simulations and experimental data how the outflow creates a blocked configuration of particles. We define the concave surface of the clogged dome by two independent methods (geometric and density based). The average shape of the cupola for spheres is almost a hemisphere but individual samples have large holes in the structure indicating a blocked state composed of two-dimensional force chains rather than three-dimensional objects. The force chain structure justifies this assumption. For long particles the clogged configurations display large variations, and in certain cases the empty region reaches a height of 5 hole diameters. These structures involve vertical walls consisting of horizontally placed stable stacking of particles.

1 Motivation

Discharge of silos and the flow of granular material through orifices are important questions of everyday life and technological processes. Several experimental and numerical studies were carried out to understand the processes of arch formation, force chains and arch breaking [1–4]. From the technological point of view one is interested in the prevention of silo clogging which blocks the outflow. The jamming probability as the function of the diameter of the orifice (D) and the aspect ratio (Q) of the granulate is also an unsolved question [5–7].

The accurate analysis of the microscopic details of the clogged configuration in three-dimensional (3D) hoppers is a challenge. In 2D it is straightforward to define the blocking arch by finding a continuous chain of particles from one side of the orifice to the other. Here, one can visualize the force network and the structure of the arch experimentally by using photoelastic particles [3, 4]. With the help of discrete element (DEM) simulations one can quantitatively compare the morphology of clogging arches with experimental data, and accurately map the force distribution inside the silo [1, 2]. In the present work we identify the blocking structure in three-dimensional hoppers with DEM simulations focusing on particle configurations and forces and compare our finding to our measurements using X-ray computed tomography. We compare the case of spherical particles with elongated grains.

*e-mail: torok@phy.bme.hu

2 Methods

2.1 Discrete element simulation

Discrete Element Method (*DEM*) simulations were performed to study three-dimensional clogging. We used a modified version [8] of the molecular dynamics code *LAMMPS* [9] to generate a cylindrical hopper with an orifice of given size on the bottom wall. We used elongated particles with given aspect ratios ($Q = 2, 3.3, 5$) and spheres ($Q = 1$). Elongated particles were created by gluing spheres together with overlaps between them. We created polydisperse samples with the variation of particle diameters by $\pm 20\%$, the length of particles by $\pm 10\%$ and the overlap of the spheres in elongated particles between $30 - 50\%$ in mean diameter units. The number of spheres required to build up a long particle was also varied to prevent linear gear effect. The initial state of simulations, the position of the particles was created using the *random sequential deposition* model in order to avoid the overlap of distinct particles. Particles were always created with average volume equivalent to a sphere with unit diameter.

The particles were put in a cylindrical container of 10 units radius, with a hole of diameter D in the bottom wall. The middle of the hole will be referred to as *center*. The height of the system was much larger than the height of the resulting static granular bed which was chosen to be about 25 particle diameters. After applying gravity, the system starts to flow out of the the container. A vertical cross-section of the simulated system is shown in Fig. 1. Periodic boundary conditions were used in the vertical direction in order to have the outflowing particles back in the

Figure 1. Sample cut of a clogged three-dimensional hopper with cylindrical orifice.

bulk again. The clogging states were defined by the lack of outflowing particles for a certain amount of time to allow for relaxation of the sample. In order to have several clogged states during simulations, we unclogged the system by raising every particle by 2 units. We measured the position and contact forces acting between particles. In the following we only deal with clogged states.

2.2 Experimental method

The experimental tests were realized using an axisymmetric nearly cylindrical container with the diameter of 19 cm and height of 21.4 cm [10]. The container was filled with peas which have nearly spherical shape with a mean diameter of 7.6 ± 0.23 mm. For the experiments presented here the orifice size was 23 mm, corresponding to $D = 3$. In order to determine the 3D configuration of the grains an X-ray tomogram (CT) was recorded using the robot-based flat panel X-ray C-arm system Siemens Artis zeego of the STIMULATE-lab, Otto von Guericke University, Magdeburg. The spatial resolution was 2.03 pixel/mm, with recorded volumes of 25.2 cm × 25.2 cm × 19 cm. A central vertical cross section of a sample tomogram is shown in Figure 2.

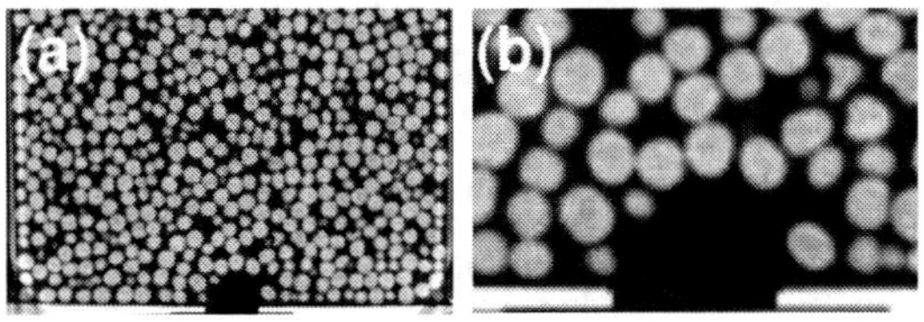

Figure 2. Vertical cross section of a tomogram covering (a) 188× 125 mm, and (b) 60 × 40 mm of the experimental system. The diameter of the hopper outlet was 24 mm.

2.3 Analysis methods

For characterizing the clogged state we used two methods: (i) *surface triangulation* of the clogged system and (ii) analysis of the average *packing density* (ϕ).

The clogged configuration is characterized by a cupola shaped cavity above the discharge hole. We need to identify the first layer of grains at the boundary of this cavity. Taking the center of mass of the particles, in principle this surface can be defined by a Delaunay triangulation, but this method unfortunately does not work well for such concave surfaces. Thus a novel method was developed in order to quantify the surface.

In our method triangles were identified by the edges between center of the particles, where the distance was not larger than 1.5 particle diameter. This ensures that all and only neighboring particles are connected. Using these triangles we determined the surface by projecting a grid from the orifice plane onto the triangles. We define the minimal $r_{\min}$ and maximal $r_{\max}$ radius from the points closest and farthest from the center.

The packing density was calculated by averaging several clogged states see Table 1. The spatial distribution in the $r - z$ plane is shown in Fig. 3 (a) for $Q = 1$ after averaging in the axial direction [10]. In these runs the the diameter of the outlet was $D = 3.4$. Approaching the orifice from the top along the vertical axis we find a decreasing packing density. Near the outlet there is an empty region, which is the region below the dome. The upper border of this region corresponds to the dome which holds the material above this clogged configuration.

3 The dome for spherical particles

The primary question we would like to answer is whether the dome is a complete three-dimensional structure, or it is rather an assembly of two-dimensional arches, similar to those observed in 2D hoppers?

The average density profile indicates a nice three-dimensional self supporting dome like structure which is taller than a hemisphere (Fig. 3 (a)). A comparison of the average density fields obtained in experiments and simulation are shown in Fig. 4 (a)-(b) for spheres, while numerically obtained average density fields for elongated grains are presented in Fig. 4 (c)-(d). Individual samples show similar features but often have large empty or sparser regions. The example shown in Fig. 3 (b) indicates empty regions above and sideway above the cupola. This structure with low density regions can be better seen in Fig. 3 (c), where only those particles are shown which are closer to the center than 2.9 particle diameter. We see a large hole in the cupola which is visible also on the force network Fig. 3 (d), where the red lines indicate the forces with line widths proportional to the normal force. One can observe a nice two-dimensional arch composed of 6 particles, which form a bridge above the outlet. Most probably the hole in Fig. 3 (c) is closed by an other smaller cupola placed on the hole of the first one.

We used the triangulation surface detection method to find these details. In Fig. 3 (e) a cut of the triangulation result is shown from the side with the hole facing the observer. The surface triangles indicate small auxiliary cupolas on the hole. The irregularity of the primary dome can be demonstrated by plotting the surface generated by the triangulation method, see Fig. 3 (f).

DOI: 10.1051/epjconf/201714003076

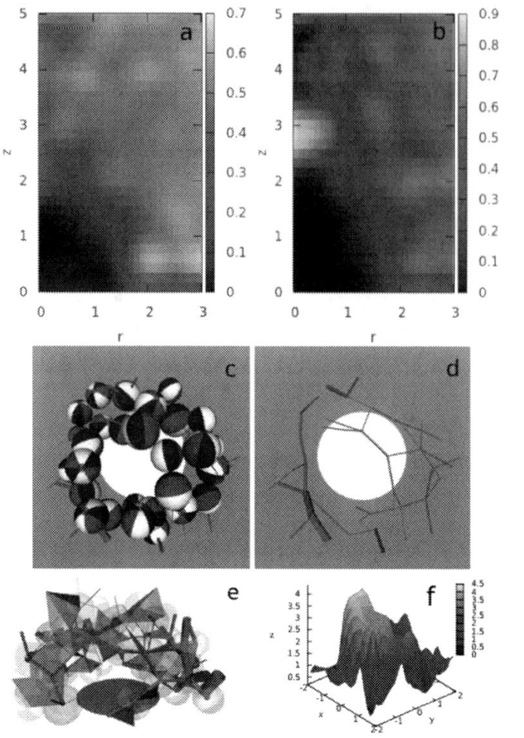

Figure 3. (a) Average density field for spheres. All the other figures are for the same single clogged configurations with spheres (b) two-dimensional average density field, (c) particles closer than 2.9 particle diameter to the center, (d) forces to latter particles, (e) middle 2 particle diameter cut of the surface triangulation, (f) detected surface from the triangulation.

The above example provides a strong hint that clogging in three-dimensional hoppers is actually a result of the formation of two-dimensional arches which eventually block the outlet. For a more general characterization we analyzed the statistics of $r_{max} - r_{min}$ the distance of the surface point closest and farthest from the outlet center. As can be seen in Table 1 the value of $(r_{max} - r_{min})/r_{min} \simeq 2.5$ indicating the presence of large holes in the cupola. Data for different values of D are very consistent, there is only little variation of both the minimal and maximal radii which affirms that different examples of clogged cupolas have essentially the same structure.

We may conclude that it is very likely that the clogged state of a three-dimensional hopper is basically composed of a complex structure of two-dimensional arches. For a three-dimensional system these arches consist of 5-6 particles, as in two-dimensional systems [11]. The three-dimensional system is thus composed of these small two-dimensional units. In order to better quantify the above observations we are currently working on the analysis of the force network in our three-dimensional system.

Table 1. Minimal and maximal distances of the surface from the center, their variances σ_{min} and σ_{max}, N is the number of configurations.

Q	$D/2$	r_{min}	r_{max}	σ_{min}	σ_{max}	N
1	1.5	1.52	3.53	0.20	0.12	21
1	1.6	1.71	3.62	0.10	0.12	8
1	1.7	1.80	5.02	0.11	0.43	5
2	1.5	1.21	4.74	0.34	0.55	8
2	1.6	1.36	4.47	0.18	0.35	14
2	1.7	1.54	6.77	0.23	0.3	22
2	1.9	1.69	6.45	0.19	1.71	13
2	2.1	1.8	5.61	0.22	1.11	19
3.3	2.0	1.01	4.39	0.23	0.37	40
3.3	2.3	1.23	5.05	0.28	0.87	44
3.3	2.6	1.49	5.96	0.28	1.65	31

4 The dome for elongated particles

It was shown [12] that elongated particles align themselves in shear flow. Our recent observations on hopper flows with elongated grains show that the particles approach the outlet with approximately their smallest cross section (they are aligned nearly parallel to the flow lines) [10]. Focusing on the position of a given particle near the outlet, clogging may happen at any instant, thus we expect that the fluctuations in the resulting cupola surface will be larger for longer particles. On the other hand, particles close the space with their circular surface, so we expect to see similar structures as in the case of the spheres.

Table 1 summarizes the numerical findings for different systems. The surface starts at around the orifice perimeter for spheres, but r_{min}/D decreases with elongation, indicating that trapped particles hanging inside reduce the effective orifice size. This indicates that the force chain arch is at about the same distance from the center but the inner end of the particles is closer.

In order to be able to unclog the system the outlet must be made larger with increasing particle elongation in spite of the fact that the particle diameter decreases as the volume is kept constant. Unfortunately our data are not good enough to determine any quantitative correspondence between equivalent D as a function of Q.

The other striking feature of the long particles is the strongly increased value of σ_{max} for larger D which indicates a large variation of the possible clogged configurations. The average density plots of these systems yield a large oval-shaped low density region which is compatible only with a fluctuating cupola height. Sometimes we get hemisphere like cupolas but other times very elongated ones. An example for an elongated cupola is shown in Fig. 5, where we observe a vertical wall composed of horizontally placed elongated particles. It seems that if the outlet is large and if it takes a long time to form a clogged state, particles can be placed upon each other to form a stable packing around the hole. This packing can take very important sizes and can reach five times D, even at moderate aspect ratios.

DOI: 10.1051/epjconf/201714003076

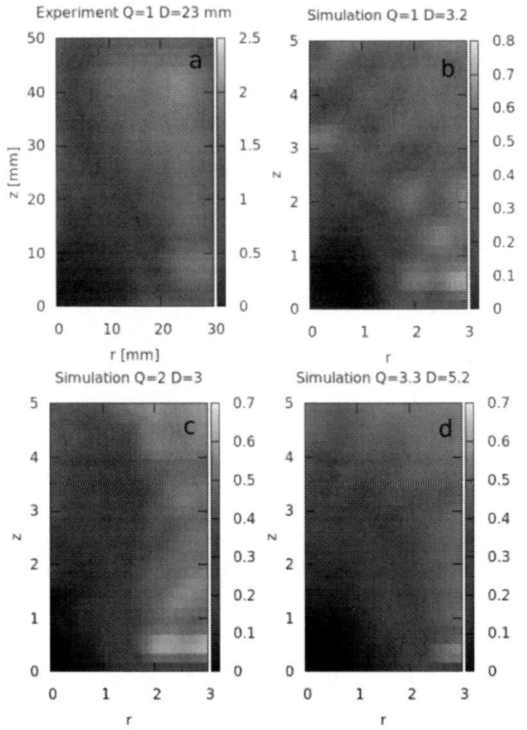

Figure 4. Average density profile for different particle aspect ratios. (a) spheres (experiment), (b) spheres simulation, (c) Q=2, (d) Q=3.3

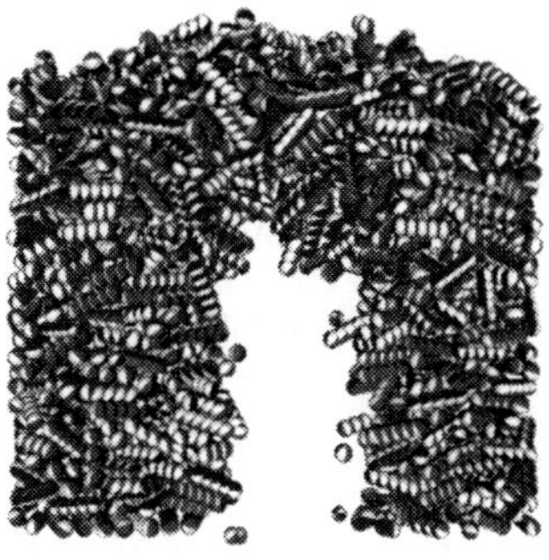

Figure 5. A sample snapshot cut for elongated clogged state. $Q = 3.3$, $D = 2.6$.

Systems consisting of long grains are characterized by large fluctuations, i.e. after shaking the above described structure collapses and the subsequent clogged state can be a small, hemisphere-like object.

5 Summary

Discrete Element Simulations and experiments were performed to study three-dimensional clogging of granular materials in a flat bottom hopper. The clogged state is characterized by a cupola-shaped configuration of particles around the outlet. Two methods were used to study the structure of the clogged material: We have developed a triangulation method which works for concave objects.

The inner shape of this object is close to a hemisphere but with large holes in it, indicating that it is essentially composed of two-dimensional arches which eventually fill the space. The above statement is justified by the fluctuations seen both on the individual density plots and on the variation of the maximal distance of the surface from the center of the outlet. Further studies are planned to characterize the clogging structures using the force network.

The clogging of elongated particles was also studied. In some cases, similar sphere-shaped cupolas were found, but in many cases, especially for long particles and large orifice, the cupola was standing on the top of a stable structure of horizontally placed long particles. These structures can grow up to five times the orifice diameter, even at the moderate aspect ratios considered here.

Acknowledgements

We thank G. Rose and the STIMULATE-lab of Otto von Guericke University Magdeburg for the opportunity to use the X-ray CT facilities (funded by BMBF, reference No. 03IP710) and support. Financial support from the DAAD/MÖB researcher exchange program (Grants No. 57141022/64975), the Hungarian Scientific Research Fund (Grant No. OTKA NN 107737) and the János Bolyai Research Scholarship of the Hungarian Academy of Sciences are acknowledged.

References

[1] A. Garcimartín, I. Zuriguel, L.A. Pugnaloni, A. Janda, Physical Review E **82**, 031306 (2010)
[2] R.C. Hidalgo, C. Lozano, I. Zuriguel, A. Garcimartín, Granular Matter **15**, 841 (2013)
[3] J. Tang, R. Behringer, Chaos **114**, 34002 (2016)
[4] J. Tang, R. Behringer, Chaos **21**, 041107 (2011)
[5] R. Arévalo, I. Zuriguel, Soft Matter **12**, 123 (2016)
[6] C.C. Thomas, D.J. Durian, Phys. Rev. Lett. **114**, 178001 (2015)
[7] I. Zuriguel, A. Garcimartín, D. Maza, L.A. Pugnaloni, J.M. Pastor, Phys. Rev. E **71**, 051303 (2005)
[8] L. Brendel, J. Török, R. Kirsch, U. Bröckel, Granular Matter **13**, 777 (2011)
[9] S. Plimpton, P. Crozier, A. Thompson, Sandia National Laboratories **18** (2007)
[10] T. Börzsönyi, E. Somfai, B. Szabó, S. Wegner, P. Mier, G. Rose, R. Stannarius, New J. Phys. **18**, 093017 (2016)
[11] K. To, P.Y. Lai, Physical Review E **66**, 011308 (2002)
[12] T. Börzsönyi, B. Szabó, G. Törös, S. Wegner, J. Török, E. Somfai, T. Bien, R. Stannarius, Phys. Rev. Lett. **108**, 228302 (2012)

Identification of tissular origin of particles based on autofluorescence multispectral image analysis at the macroscopic scale

Mathias Corcel[1,2], *Marie-Françoise* Devaux[2], *Fabienne* Guillon[2] and *Cécile* Barron[1*]

[1]UMR 1208 IATE, INRA-Supagro-UM-CIRAD, 2 Place P. Viala, 34 000 Montpellier, France
[2]UR 1268 BIA, INRA, Rue de la Géraudière, 44 316 Nantes, France

Abstract. Powders produced from plant materials are heterogeneous in relation to native plant heterogeneity, and during grinding, dissociation often occurred at the tissue scale. The tissue composition of powdery samples could be modified through dry fractionation diagrams and impact their end-uses properties. If tissue identification is often made on native plant structure, this characterization is not straightforward in destructured samples such powders. Taking advantage of the autofluorescence properties of cell wall components, multispectral image acquisition is envisioned to identify the tissular origin of particles. Images were acquired on maize stem sections and ground tissues isolated from the same stem by hand dissection. The variability in fluorescence intensity profiles was analysed using principal component analysis. The correspondence between fluorescence profiles and the different tissues observed in maize sections was assessed based on histology or known compositional heterogeneity. Similar variability was encountered in fluorescence profiles extracted from powder leading to the potential ability to predict tissular origin based on this autofluorescence multispectral signal.

1 Introduction

Plants are heterogeneous materials made up of organs (stem, grains, leaves…) which are themselves made up of various tissues including different cell types. The composition, structure and relative proportions of the different tissues in the organ are under biological control and determine human end-uses. A grinding step is often a prerequisite to any plant processing and the dissociation of the organs during milling is often observed at the tissue scale. As an example, the cereal milling industry aims at separating wheat grains and recovering the starchy endosperm into flour whereas the most external tissues are collecting in bran fractions. The proportions of each tissue in the resulting fractions impact their end-use properties and the distribution of tissue within a single particle is required to understand, control and optimize the fractionation processes. Therefore the tissular origin of powders must be identified at the particle level. If the identification of tissues in structured organ is commonly performed through morphological information together with chemical staining, it remains challenging for powders. Tissue identification in powder is mainly based on their specific biochemical composition, which is theoretically the same after grinding. To observe tissue dissociation within particles, identification must be carried out at the micrometric scale and lead to envision spectroscopic analysis coupled to imaging. Considering the overall composition of plant tissues and their quite small differences in between, molecular spectroscopies such as Raman, infrared or UV-visible fluorescence are classically used to analyse plant tissues in microscopic sections [1,2]. However, to be applicable to plant particles analysis, the spectral imaging must deal with: low sample preparation (eg avoid sample section), a fine resolution (<3µm per pixel), and at the same time, a large field of view to observe a large number of particles of variable size (from 50 µm to 2 mm). This implies a high speed of acquisition while keeping enough spectral information. Fluorescence macroimagers meet such requirements . ,In addition, many plant cell walls are known to be autofluorescent, with spectral signature specific to their phenolic composition. The objective of this work was to identify the tissue origin of plant particles based on their autofluorescence properties. Maize stem internode served as an example of plant material. It is made of two main regions: i) the inner one, the pith, which is constituted of vascular bundles (including both conducting tissues xylem and phloem surrounded by sclerenchyma cells) spread into parenchyma cells, and ii) the external one, the rind, which is rich in vascular bundles and sclerenchyma cells and surrounded by an epidermis on the most external part (Fig. 1) [3].

* Corresponding author: cecile.barron@supagro.inra.fr

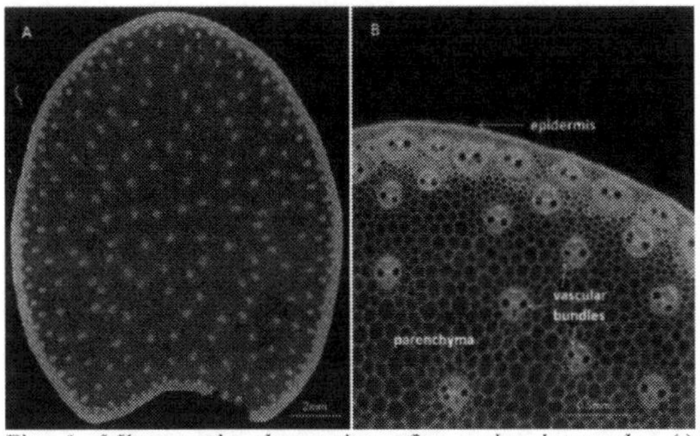

Fig. 1. Microscopic observation of a maize internode. A) Whole cross section issued of combined images B) one image with some maize tissues/structures underlined.

The method was developed on maize stem taken as a model plant, using a multispectral fluorescence imaging system. We propose to calibrate multivariate maps using data coming from the multispectral images of tissue sections and to validate the maps by projecting reference powders data issued from pure plant tissues.

2 Material and Method

2.1. Material

Cross sections and pure tissues were obtained from the same maize stem internode (Maxxis variety) in order to keep down the eventual compositional variability [4]. 150 μm thick cross sections were cut with a vibratome (Microm, Microtech France) and stored in 70% ethanol. Whole maize internode was also ground to obtain a powder including all the tissular variability (median particle size 154μm, with values between 25μm and 700μm). Pure tissues were obtained after meticulous hand dissection performed under a stereomicroscope. Considering the easiness of separation, parenchyma vascular bundle and epidermis were separated (Fig. 1), and then ground before image acquisition.

2.2. Multispectral image acquisition

A multi-zoom macroscope AZ100M (Nikon, Japan) has been equipped with 4 filter blocks to acquire images under UV, blue and green excitation light. Emission light was recovered through long pass filters and, taking advantage of the RGB channels of the colour camera (DSRiI, Nikon, Japan), each image were split in three channels and stack all together to obtain a 12 channels multispectral image (Fig. 2). The exposure times were fixed at U1 : 270 ms, U2 : 522 ms, Blue : 540 ms and Green : 990 ms. Channels names were: UA-B, UA-G, UA-R, UB-B, UB-G, UB-R, BL-B, BL-G, BL-R, GR-B, GR-G, and GR-R. The channels BL-B, GR-B and GR-G gave no autofluorescence signal considering the emission filters used but were kept to help to visualize each image in so-called multispectral images (Fig. 2).
The total magnification was set to 4x by combining the lens Plan Fluor 2x (NA: 0.2/WD: 45 mm) and setting the optical zoom to 2. The image size was 1280 x 1024 pixels, leading to a pixel size of 2.74 μm and a field of view of 3.5 x 2.8 mm^2. In order to image the whole internode cross section (Fig. 1A), mosaic image were obtained by acquiring multiple images across the section using automated mode and an overlapping of 15% and stitching of these images into a single large image.

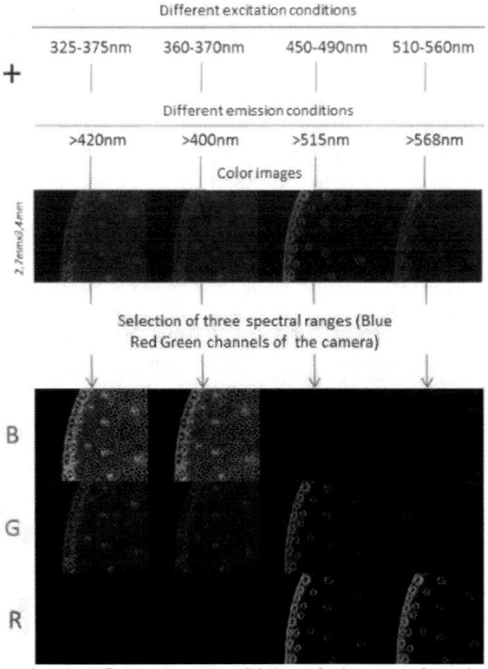

Fig. 2. Autofluorescence multispectral image of maize internode. Colour images and RGB channels.

2.3. Data treatment

Some pixels profiles from epidermis, parenchyma and sclerenchyma around vascular bundle were extracted using an interactive homemade function developed using the Matlab 2014a and the Image analysis toolbox (Mathworks, USA).

Principal component analysis was carried out, using Matlab 2014a., on the whole image section after normalisation and elimination of pixels out of the samples or saturated. Profiles of pixels acquired on powders were further projected on this principal component analysis.

3 Results

3.1. Multispectral images of maize stem sections and particles.

Autofluorescence multispectral images were acquired both on maize stem sections and maize powders (Figure 3). To enable comparison, exposure time ratios were set identical for all samples. The colour image is a representation of the 12 channels, obtained by averaging the four individual colour images obtained in each excitation/emission conditions.

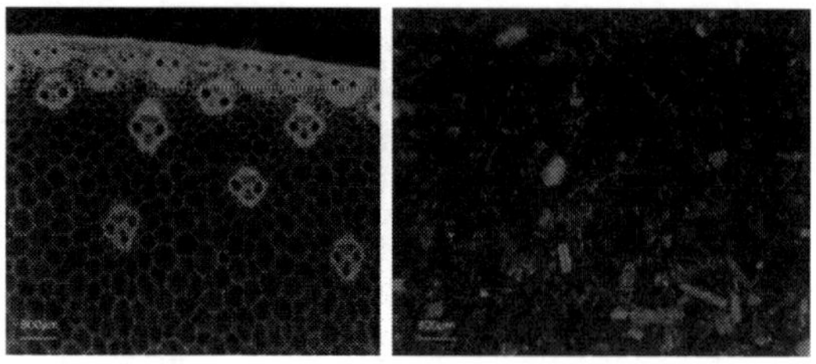

Fig. 3. Autofluorescence multispectral images of maize stem internode. A) cross section B) powder.

Maize stem structure could be easily differentiated in images of maize sections through their morphology (Fig.1) but also by their colour in the average representation of the multispectral image: parenchyma cell walls displayed a blue colour whereas epidermis or the sclerenchyma cells around vascular bundles were orange (Fig. 3A). Similar colour palette was observed in ground maize stem, with particles either orange or blue (Fig. 3B), reflecting the whole heterogeneity observed in a maize internode.

3.2. Fluorescence profiles variability encountered in maize section

In such multispectral images, each pixel is characterized by an intensity profile that can be considered as a spectral signature (Fig. 4).

Parenchyma showed intense fluorescence under UV excitation that could be related to the blue color observed in the averaged multispectral image (Fig 3A). The presence of phenolic compounds such as hydroxycinammic acid (ferulic or para-coumaric acid) could explain this strong blue fluorescence under UV excitation [5]. Profiles extracted from epidermis were different, with low fluorescence intensity recovered under UV excitation but intense under blue or green excitation. Similar profile was observed for sclerenchyma cells but with higher fluorescence intensity emitted under UV excitation. This could be related to the orange colour observed in multispectral image and to the high lignin content encountered in sclerenchyma tissue [6].

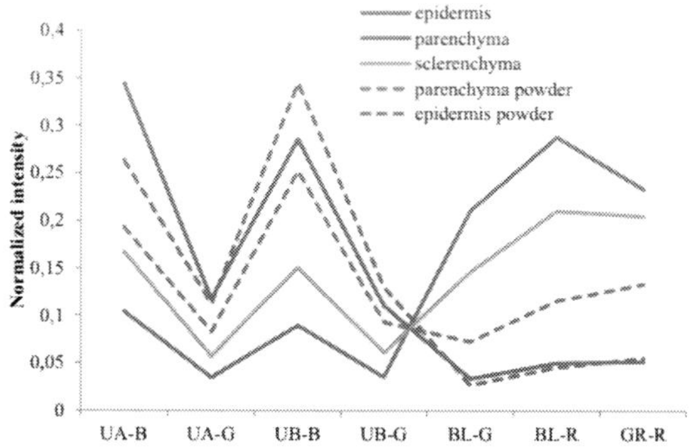

Fig. 4. Autofluorescence profiles of selected maize stem tissues either in cross-section or in powder.

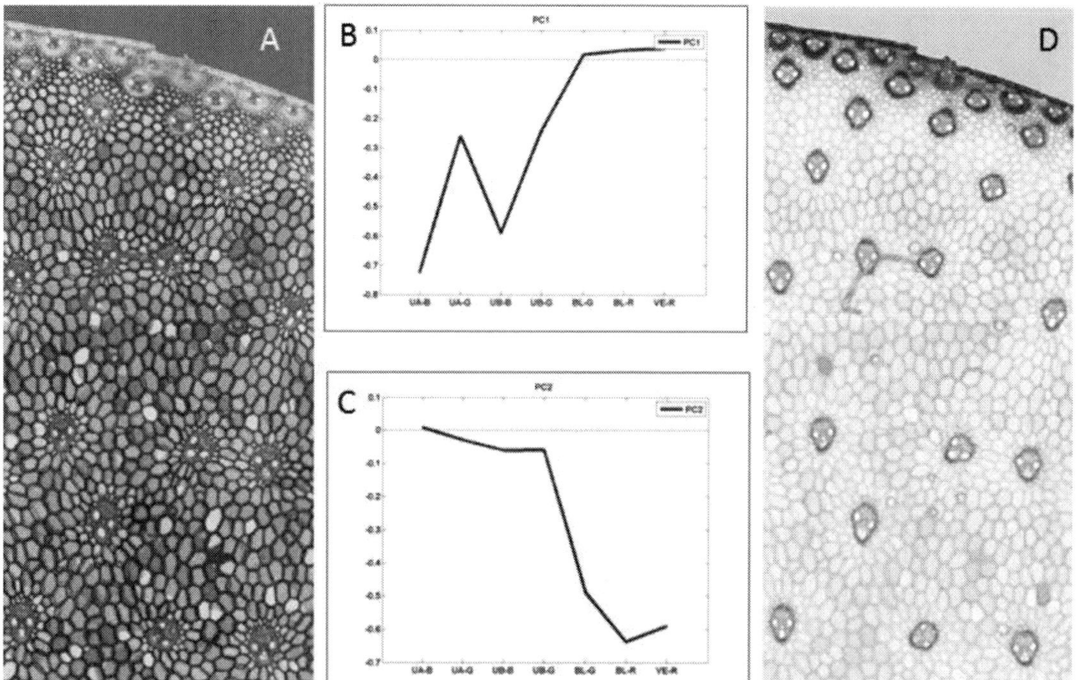

Fig. 5. Principal component analysis applied to autofluorescence images of maize stem cross section. A) Image of component 1, B) First loading, C) Second loading, D) Image component 2.

Principal component analysis was applied to the whole pixels from the maize stem cross section to analyze their similarities and taking into account the whole variability. The variance explained by the two first components was 67% and 29% respectively. The first loading (Fig. 5B) showed the autofluorescence intensities in UV filters in negative. In the component 1 image (Fig. 5A) a strong spatial heterogeneity is observed according to component 1 scores. The parenchyma pixels are observed in black indicating high fluorescence intensity under UV (Fig 5B) as observed in a typical fluorescence profile (Fig 4). On the contrary epidermis pixels were white indicating low fluorescence in these conditions, also in agreement with mean fluorescence profiles (Fig.4). Sclerenchyma pixels were in between, differently according to their location either in the pith or in the rind. The second loading showed the autofluorescence intensity in visible excitation (Fig. 5C). In the component 2 image (Fig. 5D) the parenchyma pixels are white indicating low fluorescence intensity under blue or green excitation, whereas epidermis or sclerenchyma pixels were black indicating strong fluorescence in these conditions.

3.3. Fluorescence profiles variability encountered in maize powders compared to maize section

Reference powders from parenchyma or epidermis were imaged with the same conditions as those applied to image internode sections. Pixels were extracted from pure tissue particles and their fluorescence profiles projected in the principal component analysis obtained from section (Fig. 6). Pixels of both tissues were largely found as expected from those obtained for tissues encountered in maize section. Parenchyma pixels showed slightly higher scores for component 2 that can be related to the overall higher fluorescence intensity observed under UV excitation in the fluorescence profiles (Fig. 4). Epidermis cells were found also with lower component 1 scores and higher component 2 scores. These both tissues projected in PC1/PC2 similarity map showed therefore a lower difference than the one expected from section analysis. This was in accordance with the average fluorescence profile in Fig. 4. If the structural state (powder vs section) was shown to have no impact on tissues extracted from wheat grain [7], this must be further investigated on maize stem in order to understand this discrepancy. The impact of particle size and thickness on fluorescence recovery should be integrated in the data exploration.

4 Conclusions

Autofluorescence multispectral imaging can be carried out either on plant sections or plant powders. The assignment of fluorescence signal was carried out specifically on maize stem sections based on histology or known compositional heterogeneity. Such methodology allows differentiating plant tissues according to their autofluorescence properties.

The variability in fluorescence signatures encountered in maize stem powder was similar to the variability observed in maize stem sections. According to the plant tissue, the spectral signature were kept or slightly modified. Confusions between plant tissues in powder seem higher than in plant section. This must be further investigated considering the specificity of each tissue and their powder characteristics.

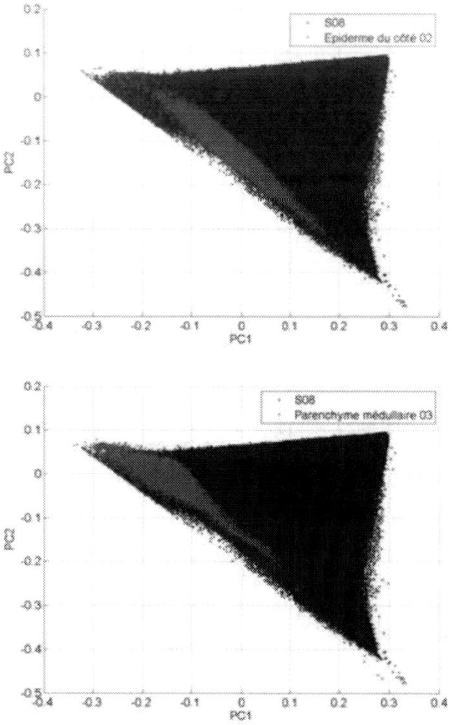

Fig. 6. Pixels from powder (in red) projected on principal component axis obtained from analysis carried out on pixels from maize section (in blue). A) Epidermis particles, B) Parenchyma particles

We would like to thank Aurélie Putois and Amélie Breysse for their help in sample preparation and image acquisition.

References

1. D.E. Akin, N. Ames-Gottfred, R.D. Hartley, R.G. Fulcher, L.L. Rigsby, 1990. Crop Sci., 30, 396-401, (1990)
2. L. Sun, BA Simmons, S. Singh. Biotechnol Bioeng, 108, (2), 286-295, (2011).
3. K. Esau. *Anatomy of Seed Plants*. 2'ed. *New York, (1977).*
4. T.A. Morrison, H.G. Jung H. G., D.R. Buxton, R.D. Hatfield, Crop Sci., **38**, 455-460, (1998).
5. P.J. Harris, R.D. Hartley, Nature, 259, 508-510, (1976).
6. A. Chesson, G.L. Provan, W. Russel, L. Scobbie. J. Sci. Food Agri., **73**,10-16, (1997).
7. M. Corcel, M.F. Devaux, F. Guillon, C. Barron Comp. Electron. Agri., 127, 281-288, (2016).

Agglomeration of powders with a new-coupled vibration-compaction device

Eric Serris[1,*], *Olivier* Desplat[1], *Olivier* Valfort[1] and *Phillipe* Grosseau[1]

[1]Ecole des Mines de Saint-Etienne, SPIN, CNRS 5307, LGF, F-42023 Saint-Etienne, France

Abstract. Inorganic powder recycling should be a crucial process for the "smart factories" in the future. A complex three-phase system (bauxite mixed with ordinary Portland cement and water) with a new-coupled vibration-compaction device is studied. The compressive stress of compacts seems to be improved by using this device at low compaction pressure leaving the other characteristics unchanged. The tomographic study of macroscopic porosities shows differences in the pores repartitions inside vibrated and untreated compacts. Classic porosity repartition is shown in the classic compacted bauxite compacts whereas in the vibrated-compacted bauxite exhibits inhomogeneities. Despite this, we find these results quite promising for further investigations.

1 Introduction

« Smart Factory» should have production systems ever increasingly efficient with useless by-products. This "eco-efficient plant" should be energy efficient, so recycling fine inorganic particles can be an essential part of this endeavor.

Our aim is to study a new coupled vibration-compaction device in order to produce inorganic compacts made from used or waste powder and reintroducing these objects into current production. Moreover, developing a successful hybrid machine: by combining concrete vibration stirring with pharmaceutical use of high compaction pressures. Densification by compaction is widely used with pharmaceutical (1) ceramic (2) and metallic powders (3). The overall process is quite realistically described and some interesting models are developed (4-6). Vibration is also used for densifying or packing granular media (7,8) to achieve compacities up to 0,7. Recent work on vibrated granular media by X –ray tomography (9) shows interesting measurements of internal local characteristics of packing parameters. Just a few studies are made on recycling by compaction and vibration-compaction (10,11) of inorganic particles.

2 Material and methods

Initially, the characteristics of the powder used to simulate the recycling of inorganic powders will be described. Then all the methods to produce and characterize compacts are detailed.

2.1 Materials

The inorganic powder used is manufactured bauxite referred to simply as bauxite in this paper. We added normal hydraulic binder from Calcia, ordinary Portland cement (OPC).

The helium pycnometry of the two powders obtained with the Accupyc 1330 from Micromeritics gave the density in table 1.

Table 1. Powder densities.

powder	Density : g.cm^{-3}
bauxite	3,74
OPC (cement)	3,19

The bauxite powder density is not so far from pure alumina corundum (3,97 g.cm-3). There is only little other oxide inside this powder (Al_2TiO_5). The composition of the OPC is the standard cement used worldwide.

The granulometry of both powders is plotted in figure 2. The bauxite powder is multimodal with a bigger mode at 200 μm and a lower mode at nearly 20μm. The cement is also multimodal with a main mode at 20μm.

* Corresponding author: serris@emse.fr

DOI: 10.1051/epjconf/201714013015

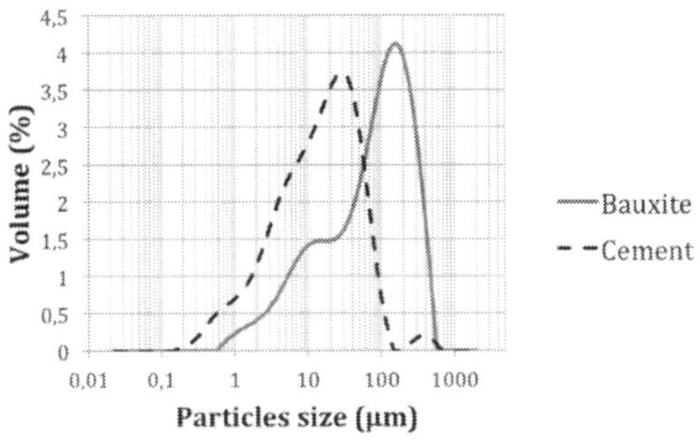

Fig. 1. PSD of the two powders

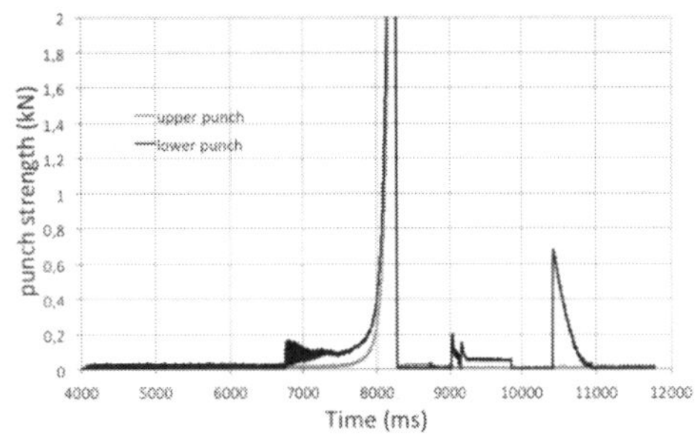

Fig. 2. Vibration just before compaction

2.2 Methods

The dry mixture is composed of 85% of bauxite and 15% of cement. Added to this dry mixture is 10% in weight of water. Everything is mixed inside a turbula for 20 minutes and then the wet mixture is inserted into the die.

The Medelpharm' Styl'One Evolution tablet press was employed. This R&D "tableting tool" can develop a 50 kN force. The die dimensions are 23x31x29.5mm. The maximal pressure applied on the powders is 67 MPa. The vibration stage is set to have two frequencies of 40 hz and 60 hz and acceleration from 2 g to 16 g. The duration of this stage is 3 seconds.

After compaction, the cement is hydrated by storing for 48 hours at 5°C with relative humidity at 100% and then maturing for 28 days at 20°C with a lower humidity of 60%.

After this phase, the compacts characterizations: compressive stress, porosity and density are determined.
* Compressive stress is provided by a normalized test called NF EN 196-1 on a uniaxial press LR 50 K from Lloyd Instruments. The punch velocity is 10 $mm.min^{-1}$, the value recorded is subsequently converted into stress using the horizontal area of the compact.
* Porosity is obtained by mercury intrusion porosimetry with the Micromeritics autopore IV 9500.
* Densities are calculated by direct measurement of the weight and volume of the compacts.

3 Results and discussion

The first result is the in-situ vibration obtained by special device made by Medelpharm company. In the next figure the vibration is shown in the variation of within the lower punch strength is zero during this time period.

3.1. Compacts properties

We only show results for compaction of non vibrated bauxite / OPC mixture, and compaction with 60 Hz vibration and accelerations of 11 and 16 g. The other experiments with lower frequency and acceleration exhibit no real differences from the non-vibrated compaction.

The first characteristic property of compacts is the compressive stress plotted in the Figure 3 versus compaction pressure (Pc).

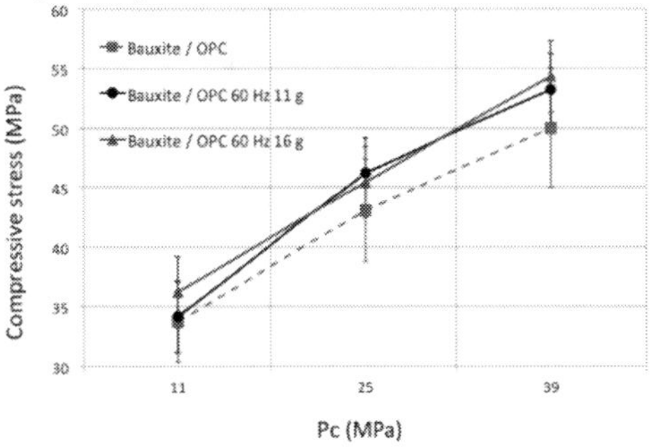

Fig. 3. Compressive stress versus compaction pressure of compacts.

This first result exhibits a small increase in the compressive stress with the vibration stage. For the three compaction pressures a 10% increase is obtained. This is a small but real impact of vibration on mechanical characteristic of compacts.

The densities are shown in the table below.

Table 2. Densities of compacts at different pressures and with and without vibration

Compaction pressure (MPa)	11	25	39
bauxite / OPC	2,29 $g.cm^{-3}$	2,41 $g.cm^{-3}$	2,44 $g.cm^{-3}$
bauxite / OPC 60 Hz 11g	2,28 $g.cm^{-3}$	2,38 $g.cm^{-3}$	2,43 $g.cm^{-3}$
bauxite / OPC 60 Hz 16g	2,26 $g.cm^{-3}$	2,37 $g.cm^{-3}$	2,45 $g.cm^{-3}$

We only can tell that the more acceleration the lower density we have (but not for 39 MPa). After hydration, we have a mixture of portlandite and C-S-H, the density is about 2,3 g .cm^{-3}. For a fullly hydrated system we should have 3,4 g .cm^{-3}. So in compacts we have from 33 to 28 % of porosity, this is quite large. The lower density can be explained by both an increase of porosity or a better hydration of the OPC.

Then the repartition of pores diameter of compacts realized with 11 MPa of compaction pressure is shown in the figure 4 below.

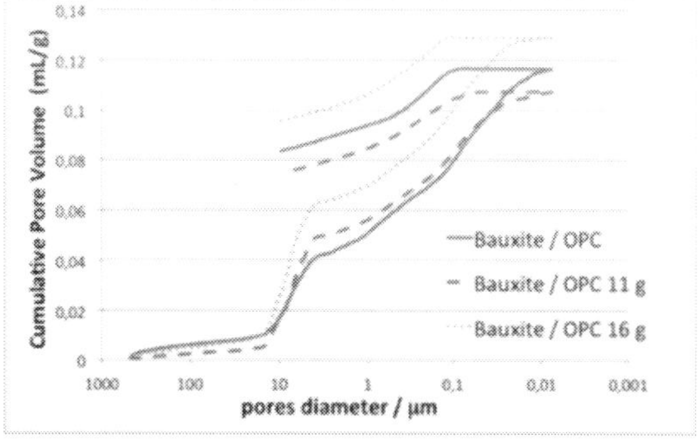

Fig. 4. Mercury intrusion porosimetry of compacts fragments

The porosity is quite the same for the three compacts. We only have a slight difference in the pores volume of the larger pores around 10 µm.

To explain the differences between those compacts we choose to study in detail the macroporosity inside the compact by using X-ray tomography.

3.2 X-ray tomography

The microtomography experiments were performed in a General Electric Phoenix Nanotom providing an X-ray beam with high-energy voltage on the range between 0 and 180 kV. In the following experiments, 90 kV were used, enabling X-ray to pass through the compacts with a good contrast for the final images. A thin layer of copper was used as a filter of the X-ray beam to reduce the beam hardening effects and neither streaking, shading and ring artefacts were observed in the reconstructed 3D images. 2800 radiographies were taken at different angles all around the 360° and the distance between the sample and the detector or X-ray source was adjusted to get a voxel size of 17 µm. The resolution in the images was chosen regarding the size of the whole sample and it was enough to get quantitatively information about the grains and porosity from a macroscopically point of view.

3D images were reconstructed (absorption coefficient) using the algorithm included in the Phoenix Nanotom software package. Then the whole 3D image volumes (800×1600×2000) were filtered with the median filters were used to enhance the contrast between grains and pores by using the AVIZO software from FEI.

A thresholding procedure was then applied to separate the solid phase from the porous one. We have this porosity first for a bauxite / OPC compact obtained with 11 MPa of compaction pressure. We only have a half compact in this figure 5 to have a better viewing (the top of the shortest thickness side).

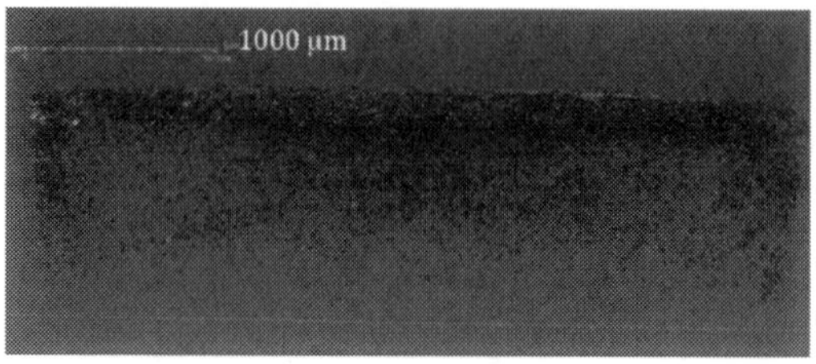

Fig. 5. X-ray tomography picture of porosity of the compact

We have a classic repartition of porosity in the upper part of the compact and in the corners. This tendency is also shown in the models in bibliography (4). We now want to calculate the homogeneity of the porosity. We choose only the bigger pores by labelling the porosity and separate pores by erosion. We only choose the 1000 bigger pores representing 50% of the overall porosity. In the figure 6 we plot the barycentre of all pores in the X axis. The same result is obtained in Y and Z axis.

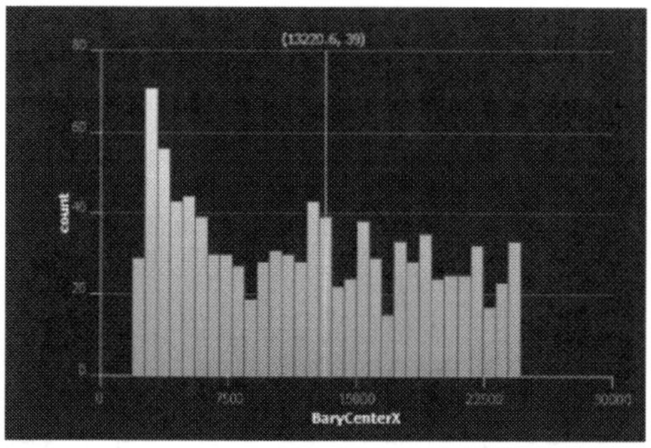

Fig. 6. Global porosity homogeneity inside reference compact

We have in this bar graph the same tendency observed in the picture in figure 5. We have high porosity near the edges of the compacts but the other pores are quite homogeneously positioned inside the compact.

The same procedure of image analysis is done for compact obtained with 60 Hz frequency with the acceleration of 11g and the compaction pressure of 11MPa.

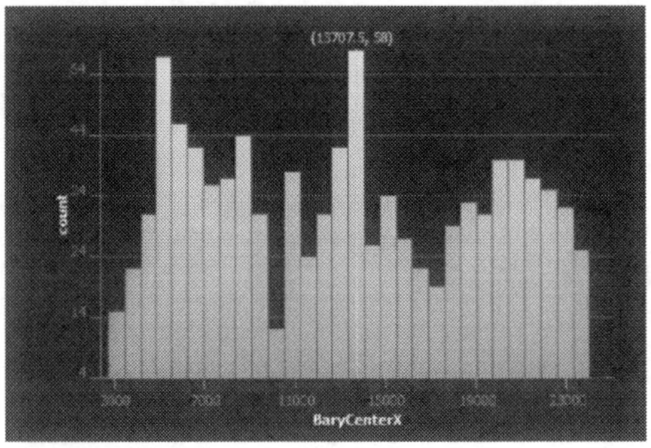

Fig. 7. Global porosity homogeneity inside vibrated compact (11 MPa and 11 g)

For the vibrated compacts it seems to have inhomogeneities. We will confirm this fact by viewing these pores inside the compact (the pores are labelled in colour and the solid phase in grey transparency)

Fig. 8 Global porosity (coloured dots) inside vibrated compact at 11 MPa and 11 g

We do not have small coloured spots that represent pores inside the compact equally positioned but only in some areas. If we only show the pores inside the upper right corner of the compact we have the following pores (figure 9).

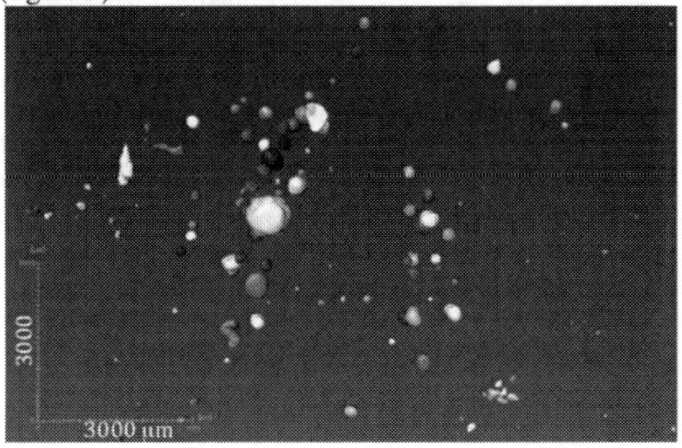

Fig. 9 Localised porosity at the right corner of vibrated compacts at 11 MPa and 11g

We have quite the same size and shape for these pores. In the compaction process, we have 10 % of water used to hydrate the cement. Those pores seem to be located where the water droplets were. The high acceleration during vibration may generate pellets with free water inside that does not react with OPC. During maturation stage, water evaporates and leaves round shape porosity inside the compact as we can see in figure 8.

So despite inhomogeneities and water that do not react with cement we increase the mechanical strength of the compact with the vibration stage, we found this result quite promising.

4 Conclusions

This work on the compaction of complex mixture of inorganic powder with hydraulic binder and water by an in-situ new coupled vibration compaction device show encouraging results. We have a small increase in the mechanical characteristics of compacts with quite the same porosity and density. Macroscopic X-ray tomography study shows inhomogeneities inside the vibrated compacts.

Former studies on other mixture with natural bauxites and different binder may show better results without inhomogeinies and better mechanical properties.

Acknowledgement

The authors wish to thank Chris YUKNA for his help with proofreading.

References

1. M.S. Kadiri, A.Michrafy, Powder Technol. **453**, 2 (2013)
2. T.J. Vogler, M.Y. Lee, D.E.Grady, Int. J. Solids Struct. **44**, (2007)
3. T. Sapanathan, S.Khoddam, S.H.Zahiri, A. Zarei-Hanzaki, R.Ibrahim, J Mater Sci, **51**, (2016)
4. V.Mazel, V.Busignies, S.Duca, B.Leclerc, P.Tchoreloff, Int J Pharm, **410**, (2011)
5. B.Harthong, J.F.Jérier, V.Richefeu, B.Chareyre, P.Doremus, D.Imbault, F.V.Donzé, Int J Mech Sci, **61**, (2012)
6. P. Lamy, G. Thomas, L. Brunet, Propellants Explos. Pyrotech., **30** (2005)
7. X.Z.An, C.X.Li, R.Y.Yang, R.P.Zou, A.B.Yu, Powder technol, **196**, 1, (2009)
8. P.J.Woodhead, M.Newton, J Pharm Pharmacol **36,** (1984)
9. L.Olmos, J.M.Chaix, S.Nadler, O.Bonnefoy, J.L.Gelet, G.Thomas, Granular Mat, **18**, 3, (2016)
10. X.An, Z.Xing, C.Jia, Metall Mater Trans *A*, **45 A**, (2014)
11. D.McGlinchey, A.J.Matchett, J.M.Coulthard, Trans IChemE, **75,** (1997),

Transformation around intruders in granular media

Pierre Jop[1,*], *Aymeric* Merceron[1], and *Alban* Sauret[1]

[1] *Surface du Verre et Interfaces, UMR 125 CNRS/Saint-Gobain, 33 quai Lucien Lefranc, 93303 Aubervilliers, France*

Abstract. Sintering, glass melting and other industrially relevant processes turn batches of grains into homogeneous products. Such processes involve coupled chemical and physical transformations of the granular packing. For sake of simplicity, we study how local evolutions on grains (volume decreases for example) entail mechanical rearrangements in the overall pile. Inert bidisperse metallic disks are mixed and confined in a vertical 2D cell. At the bottom of this set-up, initial intruders (one or two) have been previously set and mechanically linked to a linear motorized jack. While we quasi-statically pull the intruder(s) downward out of the cell at constant speed and constant liberated surface, we tracked the surrounding granular packing. Events largely distributed both spatially and temporally occur around intruders. We focus on the influence of the distance between the intruders on the local dynamic of the packing. We compare the distribution of the size of the event and their frequency as function of the relative position of the intruder, as a function of their radius. We show that their influence decreases rapidly. At short distance the mechanical perturbation induced by one intruder can destabilize the packing around the other.

1 Introduction

Sintering, glass melting and other industrially relevant processes turn batches of grains into homogeneous end-products. Such processes lead to complex and mostly misunderstood phenomena due to chemical and physical transformations occurring inside the granular packing. Volume changes due to dilatation coefficients [1] or phase transitions may obviously affect the particles [2]. Consequently, the particle's shapes of particles evolves [3] as well as their contact properties through lubrication [4] or formation of capillary bridges [5]. In addition, chemical reactions also modify the surface and the nature of each individual grain [6]. These different phenomena occur at the same time and lead to the final homogeneous product through complex paths. Modeling the behavior of such systems generally involve different fields of research. Typically, mechanics and statistical physics are used to predict the behavior of granular assemblies under flow, rearrangement or perturbation [7]. The rich behavior of granular media has led to extensive studies considering their rheology [8], the criteria of mechanical stability [9] or the properties of wet granular assemblies [10]. Other fields of research has been considered to tackle such problems: chemical engineering where the behavior of reactive powders is modeled [11] and geology where rock transformations have been characterized by focusing on thermodynamics of reactions [12].

Some recent studies have characterized the rearrangement of a reactive granular packing, often icy particles than can melt [2, 13], but experimental investigations remain rare, likely because of the difficulties in monitoring the location of the individual grains as well as their properties. Therefore, most studies have focused on granular flows around an unique intruder using a mechanical approach [14–16], rheological approach [17] or a purely kinematic point of view [18, 19]. Besides, much attention has been paid to the understanding of flows in silos and hoppers [20] through the arching problem [21] or for mining applications for which the grains are quasi-statically extracted [22].

In this paper we present experimental results on the response of bidimensional granular medium when two initial intruders are pulled out downward. These intruders are analog of melting grains that will disappear progressively. The 2D situation allows us to accurately characterize the position of each grain during the rearrangement process under gravity. We show that the behavior around an intruder is influenced by the other depending on their separation.

2 A bidimensional reactive pile

The reorganization of the granular medium around two intruders has been studied using a two-dimensional system (Fig. 1). The vertical cell is made of two glass plates (30 cm height and 60 cm width), separated by a gap of 3.1 mm. The cell is filled with 5000 bi-disperse metallic disks of 4 and 5 mm in diameter (proportion 10:7) to avoid crystallization. In the following of this paper, each displacement, each spatial coordinate and each length is expressed in terms of the diameter of smallest grains d_g. At the beginning of the experiment, both metallic intruders (black

*e-mail: pierre.jop@saint-gobain.com

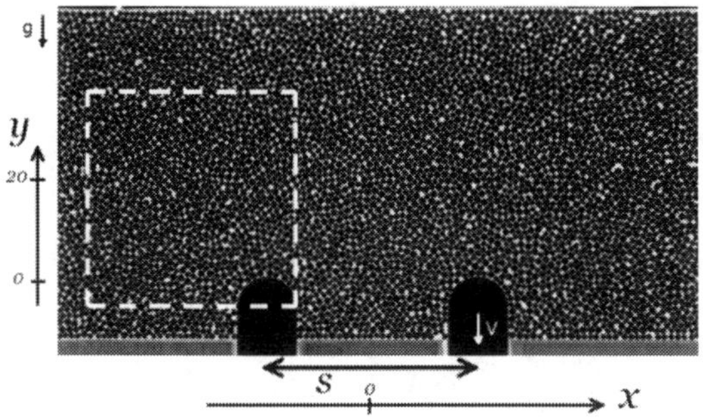

Figure 1. Picture of the experimental setup showing the two mechanical intruders in the 2D packing. The white dashed square shows the typical area studied to measure the influence of the distance s, which is 35 d_g on the picture.

round plates in Fig. 1) start to be extracted downward at a constant velocity V of 0.05 mm.s^{-1} involving a quasi-static evolution of the surrounding media due to the weight of the grains. We keep the same low velocity for all experiments.

During the experiment, we record the evolution of the cell recording images each 5 s. We then measure precisely the position of each grains, and identify their displacement between successive images using a tracking algorithm. We will change the distance s between the intruders, as well as their width w, to measure their impact on the reorganization of the media.

3 Results

3.1 Mean displacement field

We first maintain fixed the position of one intruder while the other is moving down (toward negative values of y). During this movement, we measure the average vertical displacement of the grains on the horizontal line $y = 1$. While the y coordinates of the grain is pointing upward, the vertical displacement are generally in the negative direction, so we always report their opposite values γ. The resulting curve is plotted twice in the figure 2a, each time at the position of one intruder. Then we compare the experimental vertical displacement observed when the two intruders are moving together at the same speed with the superposition of the previous curves as if there were no correlation (Fig. 2b), for two different initial positions of the pair of intruders. In both cases, the agreement between the black and the colored lines is fairly good, meaning that for such separation (s = 17.5 d_g and 35 d_g) the grains around one intruder are weakly influenced by the other intruder. On the contrary, for close intruders (Fig. 3a), the mean displacement in the middle of the cell is larger than the one expected from the simple superposition of the displacements from single intruders (fig. 3b). This result shows that at short distance, cooperation must be expected between intruders: the destabilization above one apex must propagate toward the grains around the other intruder.

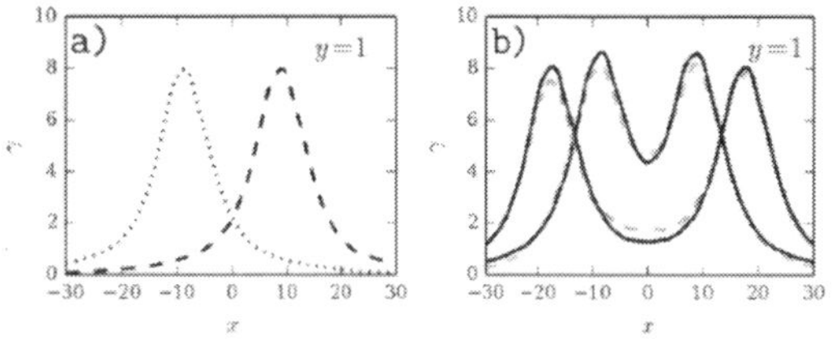

Figure 2. (a) Superposition of two vertical displacements of the grains along the axe $y = 1$, obtained from experiments with one fixed intruder in the cell. Its width is $w = 10\,d_g$ and we duplicated the profiles at the place of the twin intruder for $s = 17.5$. (b) Vertical displacements of the grains along the axe $y = 1$, when pairs of intruders are moving for $s = 17.5\,d_g$ (yellow dashed line) and $s = 35\,d_g$ (blue dashed line). The solid black lines are the sum of the displacement obtained from a single intruder plotted in a.

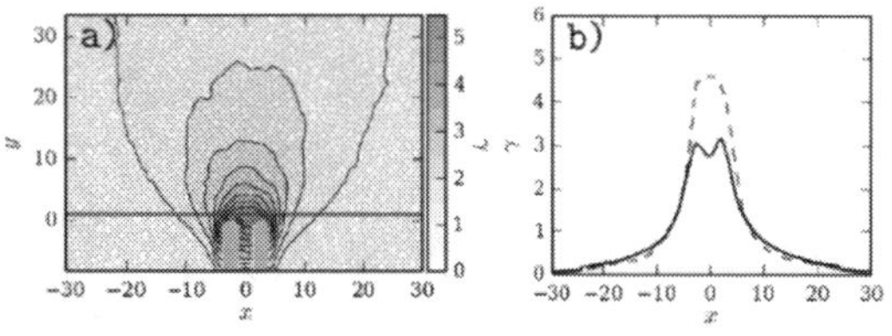

Figure 3. Vertical displacement γ between the first and the last image of the grains. (a) Displacement field for two close and narrow intruders of width $w = 2.5\,d_g$ separated by $s = 5\,d_g$. (b) Vertical displacements of the grains along the axe $y = 1$ for the close narrow intruders (green dashed line). The solid black line is the sum of the shifted displacements obtained from a single narrow intruder.

3.2 Avalanche sizes

To quantify this influence, we first define the size of an avalanche based on the displacement of the grains. Between two consecutive images, we count the number of grains that move at least as much as the intruder. To focus on a relevant area, we restrict this analysis to grains inside the white square of the figure 1, whose right border is aligned with the border of the left intruder. This number N_A defines the avalanche size. In the figure 4, we plot the distribution of the avalanche size for different distance s between the two intruders. First, we observe that the distribution for a single intruder is the asymptotic distribution when the distance increases between the pair of intruders. Second, when the distance is shorter than 15 grain-diameters, the distributions start to deviate and the maximum is shifting toward larger amplitudes, while the probability to observe a size of 0 decreases strongly. This means that the closer are the intruders, the less jammed is the system. Finally, the probability of the large events decreases as a power law with an exponent of −2.6.

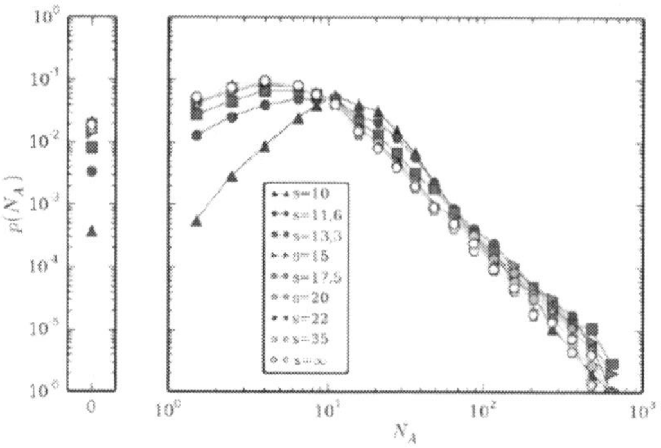

Figure 4. Avalanche size distributions in the white square of the figure 1 for a pair of intruders of width 10 d_g. From the lightest curve to the darkest curve, the distance s decreases. The case of a single intruder is denoted by $s = \infty$.

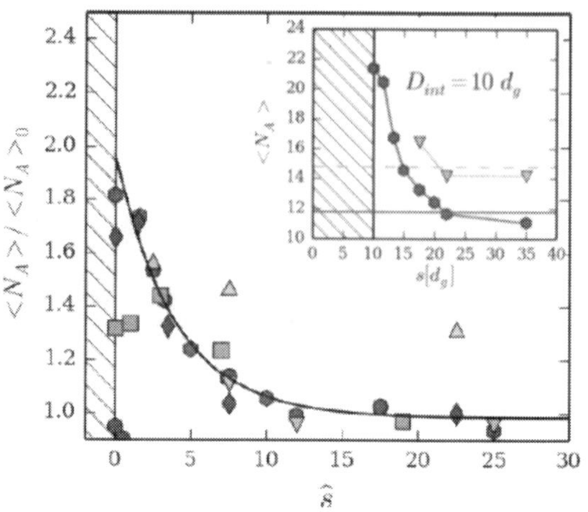

Figure 5. Normalized mean avalanche size as a function of the shifted distance ($\hat{s} = s - w$) for different widths of intruders w: 10 d_g (*hexagon*), 7.5 d_g (◇), 5 d_g (□)and 2.5 d_g (○) for the simulations and 10 d_g (▽) and 2.5 d_g (△) for the experiments. Inset: same data with the horizontal lines representing the measured reference value $< N_A >_0$ in each cases.

3.3 Characteristic lengths

The data show that a critical length exists to observe a notable influence of a near intruder on the distribution of avalanche size. We measure this effect for different widths of intruder. To compare the results, we plot only the mean avalanche size of the distributions $< N_A >$ as a function of the shifted distance ($\hat{s} = s - w$) for all intruders in figure 5. The sizes for far intruders converges toward the mean size obtained for a single intruder $< N_A >_0$, as shown by the horizontal lines in the inset of the figure 5, where the typical evolution of the mean size with s is shown for a simulation and an experiment. We then use this value to normalize our data. The data collapse on a single master curve except the experimental data for the narrower intruders. We can fit this curve by an exponential decrease of characteristic length 4 d_g. This results show that the destabilization near an intruder propagates toward the other site and may trigger an event if they are close enough. This situation is similar to the case of the silo with multiple orifices, where the flow through one outlet may unjam the flow of the second outlet [23].

We also note that the very first points at small separation depart from the master curve. This effect is due to the lack of space between the intruders. They behave like a single larger intruder. The increase near the axis $\hat{s} = 0$ is then attributed to the increase of the effective size of the intruder. Finally, the green triangle do not align with other points. These points correspond to the narrowest experimental intruder of 2.5 d_g in width. We did not identify the origin of this difference since the simulation with the same width collapses on the master curve. One explanation may be that for such narrow intruder, the formation of stable arches above the intruder is widely observed. These structures may be more stable in experiments due to friction against the glass walls and lead to a different dynamic.

4 Conclusion

We measured the behavior of a granular packing around two intruders to mimic two reactive sites, both numerically and experimentally. We have shown the influence of one intruder on the other. The effect increases exponentially when the distance between them decreases. The characteristic length does not depend on the width of the intruder but on the size of the grains in the system. We expect to observe a larger influence if the two sites are not on the same horizontal line, as it would be the case in a real reactive packing. Further works will investigate the role of a vertical gap between intruders and the role of the spatial concentration of reactive sites.

References

[1] F. Ludewig, N. Vandewalle, S. Dorbolo, M. Pakpour, G. Lumay, Physical Review E **92**, 010202 (2015)
[2] B. Turnbull, Physical Review Letters **107**, 258001 (2011)
[3] D.M. Kintea, T. Hauk, I.V. Roisman, C. Tropea, Physical Review E **92**, 033012 (2015)
[4] Q. Xu, A.V. Orpe, A. Kudrolli, Physical Review E **76**, 031302 (2007)
[5] S. Herminghaus, Advances in Physics **54**, 221 (2005)
[6] E. Gouillart, M.J. Toplis, J. Grynberg, M.H. Chopinet, E. Sondergard, L. Salvo, M. Suéry, M. Di Michiel, G. Varoquaux, Journal of the American Ceramic Society **95**, 1504 (2012)
[7] B. Andreotti, Y. Forterre, O. Pouliquen, *Granular media: between fluid and solid* (Cambridge University Press, 2013)

[8] P. Jop, Y. Forterre, O. Pouliquen, Nature **441**, 727 (2006)
[9] M. Wyart, Physical Review Letters **109** (2012), arXiv: 1202.0259
[10] N. Mitarai, F. Nori, Advances in Physics **55**, 1 (2006)
[11] O. Lame, D. Bellet, M. Di Michiel, D. Bouvard, Nuclear Instruments and Methods in Physics Research Section B: Beam Interactions with Materials and Atoms **200**, 287 (2003)
[12] P.z. Wong, J. Koplik, J.P. Tomanic, Physical Review B **30**, 6606 (1984)
[13] S. Dorbolo, F. Ludewig, N. Vandewalle, C. Laroche, Physical Review E **85**, 051310 (2012)
[14] E. Kolb, P. Cixous, N. Gaudouen, T. Darnige, Physical Review E **87**, 032207 (2013)
[15] C. Coulais, A. Seguin, O. Dauchot, Physical Review Letters **113**, 198001 (2014)
[16] E. Hamm, F. Melo, Europhysics Letters (EPL) **73**, 356 (2006)
[17] A. Seguin, C. Coulais, F. Martinez, Y. Bertho, P. Gondret, Physical Review E **93**, 012904 (2016)
[18] A. Seguin, Y. Bertho, P. Gondret, J. Crassous, Physical Review Letters **107**, 048001 (2011)
[19] E. Kolb, P. Cixous, J.C. Charmet, Granular Matter **16**, 223 (2014)
[20] J. Choi, A. Kudrolli, M.Z. Bazant, Journal of Physics: Condensed Matter **17**, S2533 (2005)
[21] A. Janda, I. Zuriguel, A. Garcimartín, L.A. Pugnaloni, D. Maza, EPL (Europhysics Letters) **84**, 44002 (2008)
[22] F. Melo, F. Vivanco, C. Fuentes, V. Apablaza, International Journal of Rock Mechanics and Mining Sciences **45**, 248 (2008)
[23] A. Kunte, P. Doshi, A.V. Orpe, Physical Review E **90**, 020201 (2014)

Stress distribution in conical sandpiles formed with ellipsoidal particles

Sida Liu, *Zongyan* Zhou*, *Ruiping* Zou and *Aibing* Yu

Laboratory for Simulation and Modelling of Particulate Systems, Department of Chemical Engineering, Monash University, Melbourne, VIC 3800, Australia

Abstract. The properties of a sandpile such as angle of repose and stress distribution are affected by many variables, among which particle shape is one of the most important ones. In this work, ellipsoids which can represent a large range of shapes varying from disk- to cylinder-type are used. The discrete element method (DEM) is employed in order to conduct controlled numerical experiments. The results show that 3D conical sandpiles have similar properties as 2D ones reported in the literature. It demonstrates that particle shape affects the magnitude of the contact force network significantly, with spheres being the smallest. As expected, the pressure distribution underneath sandpiles is featured with a relatively small pressure in the centre, and ellipsoids have a more significant stress dip region than spherical particles.

1 Introduction

A sandpile is one of the most typical particulate systems which are widely encountered in nature and many industries. It has two important properties including angle of repose and stress distribution, which have been investigated extensively in the literature. For example, angle of repose, as one of the most important parameters in characterizing the behavior of granular materials, is closely related to many phenomena, including packing, avalanching, stratification and segregation. Stress distribution is another interesting feature of granular materials which can affect many practical operations from silos and building foundations to road and dam construction, hence it is important to understand the physics underlying piling processes.

In the past decades, the two properties have been studied experimentally or mathematically. In particular, the so called discrete element method (DEM) [1], as a dynamic simulation technique, can take into account all dynamic factors related to either geometry or forces, hence it has been widely used to study sandpiles, as summarized by Zhu et al. [2]. Generally speaking, the shape and size of sandpiles are a macroscopic result from collected effects of various internal mechanisms involving many physical variables. For example, the magnitude of the angle of repose depends on pile formation method, and a range of other variables related to material properties such as sliding friction coefficient, rolling friction coefficient, density of particles, and particle characteristics such as size and shape. The stress distribution in a sandpile, particularly the pressure dip phenomenon, is often examined. It is known that the pressure distribution beneath a sandpile is affected by various factors, including formation scheme, particle properties, flow rate, pouring height, and bottom geometry. Among these factors, particle shape is an important parameter affecting the two properties, as illustrated by some investigators [3-7].

Ellipsoid is an ideal particle shape that can represent a wide range of shapes varying from disk like to cylinder like particles. Its shape for spheroids can be described by one single parameter, i.e. aspect ratio. Also, the surface of ellipsoids can be accurately described by the quadratic equations, and can be handled readily by DEM. Thus, ellipsoids have been increasingly used to examine the effect of particle shape on different behaviour of granular materials, for example, as demonstrated in recent studies of packing [8], fluidization [7] and hopper flow [9]. In the previous study, the effect of particle shape on sandpile properties was investigated in a 2D slot model [10]. It showed that ellipsoids have more pronounced stress dip than spheres, and the pressure distribution underneath sandpiles is featured a relatively constant normal pressure in the central region rather than a dip. It is not clear how the model geometry affects the two sandpile properties. In the present study, particle shape effect is carried out in a full 3D conical sandpile, and the results are compared with 2D slot models.

2 DEM for ellipsoids

DEM for ellipsoidal particles have been reported in the literature [8]. The governing equations can be written as:

$$m_i \frac{d\mathbf{v}_i}{dt} = \sum_{j=1}^{k_c} (\mathbf{f}_{c,ij} + \mathbf{f}_{d,ij} + \mathbf{f}_{v,ij}) + m_i \mathbf{g} \qquad (1)$$

* Corresponding author: zongyan.zhou@monash.edu

and

$$I_i \frac{d\omega_i}{dt} = \sum_{j=1}^{k_c} (\mathbf{M}_{t,ij} + \mathbf{M}_{r,ij} + \mathbf{M}_{n,ij}) \quad (2)$$

where v_i and ω_i are the translational and angular velocities of the particle, respectively, and k_c is the number of particles in interaction with the particle. The forces involved are: the gravitational force $m_i g$, and inter-particle forces between particles, which include elastic force $f_{c,ij}$, and viscous damping force $f_{d,ij}$. For fine particles, the van der Waals force $f_{v,ij}$ needs to be considered. These inter-particle forces can be resolved into the normal and tangential components at a contact point. The torque acting on particle i by particle j includes two components: $M_{t,ij}$ which is generated by the tangential force and causes particle i to rotate, and $M_{r,ij}$, commonly known as the rolling friction torque, is generated by asymmetric normal forces and slows down the relative rotation between particles. For ellipsoids, addition torque $M_{n,ij}$ should be added because the normal contact force and van der Waals force do not necessarily pass through the particle centre. Other critical features for ellipsoids are introduced briefly below:

- Determining the interaction forces by calculating the radius R*, which is strongly related to the radii of the curvature at a contact point;
- Using geometric potential algorithm to detect the contacts between ellipsoids; and
- Particle orientation is defined by three Euler angles (ϕ, Θ, Ψ) based on the quaternion method.

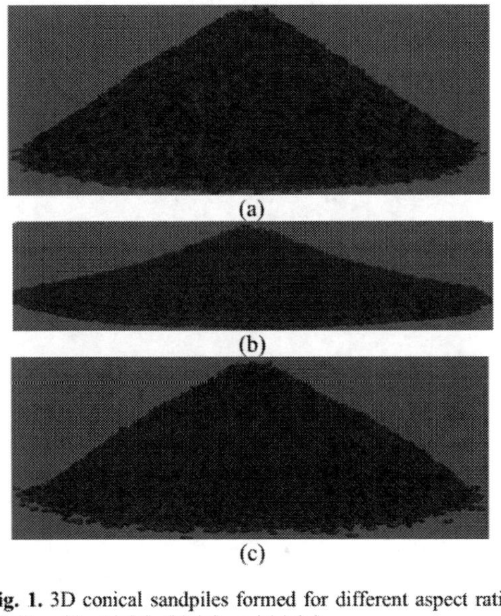

Fig. 1. 3D conical sandpiles formed for different aspect ratios of ellipsoids: (a) 0.3, (b) 1.0, and (c) 3.0.

3 Simulation conditions

A packing is formed first in a cylindrical hopper above a horizon plate, and an orifice is located in the central flat bottom. 80,000 mono-sized particles are poured into the hopper and allowed to settle under gravity. The orifice is then opened thus let particles discharge from the hopper. The instantaneous opening of the outlet starts a discharging process in which particles drop onto the bottom surface under gravity, and form a sandpile beneath the hopper. The distance between the orifice and the bottom plate is set to 0.5 m, and there are no side wall boundaries for the bottom surface. The formation of full 3D conical sandpiles are shown in Fig. 1.

Note that each ellipsoid has three diameters in three principal directions, where *a* is the principal radius in the polar direction, b and c are principal radii in the equatorial plane. For simplification, b is equal to c, resulting in all particles being in the class of spheroids. Hence, the aspect ratio is defined as the ratio of *a* to *b* (=*c*). Obviously, for a prolate spheroid, aspect ratio > *1*; for a sphere, aspect ratio = *1*; and for an oblate spheroid, aspect ratio < *1*. Spheroids with aspect ratio varying from *0.3* to *3.0* are used in this work to produce different shapes. Detailed parameters used in the DEM simulation are listed in Table 1.

Table 1. Parameters used in DEM simulation.

Quantity	Values
Number of particles	80,000
Dropping height	0.5 m
Sliding friction coefficient (p-p)	0.4
Sliding friction coefficient (p-w)	0.4
Young's modulus	1 x 10^7 Pa
Damping coefficient	0.3
Time step	3.2 x 10^{-5} s
Aspect ratio	0.3~3.0
Particle size (2a, 2b=2c,d_p)	d_p=2(abc)$^{1/3}$=0.01 m

4 Results and discussion

4.1 Angle of repose

Fig. 2 shows the variation of angle of repose with aspect ratio as well as orifice size. Note that a 3D stockpile could have slightly different angle of repose in different directions, thus the angle of repose measured in the present study is an average value of 8 directions. Spherical particles have the lowest angle of repose as observed in the literature [10]. For oblate particles, the angle of repose increases with aspect ratio deviating from sphere and reach a maximum value at aspect ratio 0.5, then decrease. For prolate particles, the angle of repose keeps increasing with aspect ratio but the rate of slope is decreased for aspect ratio larger than 1.5. This curve is similar to the variation of coordination number with aspect ratio for particle packing [8]. The orifice size of hopper flow in the upper part can make effect on angle of repose as well. Small orifice size usually leads to relatively large angle of repose.

4.2 Force structure

The contact forces between particles distribute non-uniformly and form a complex force network in a granular system. This has been confirmed by previous experimental studies in two dimensions and DEM simulation in a 2D slot model [10]. Fig. 3 shows the force networks for different aspect ratios. The forces between particles are demonstrated by the thickness and colors that are proportional to the magnitude of the normal contact force (red represents large forces and blue represents small forces). It can be observed that the large contact forces mainly propagate from the base toward the top of the sandpile for all the cases. Spheres have the smallest angle of repose thus the force network is weak correspondingly. With aspect ratio deviating from 1.0, the height of sandpiles increases and forms a relatively strong force network. The force network observed is similar to the slot models in reference [10].

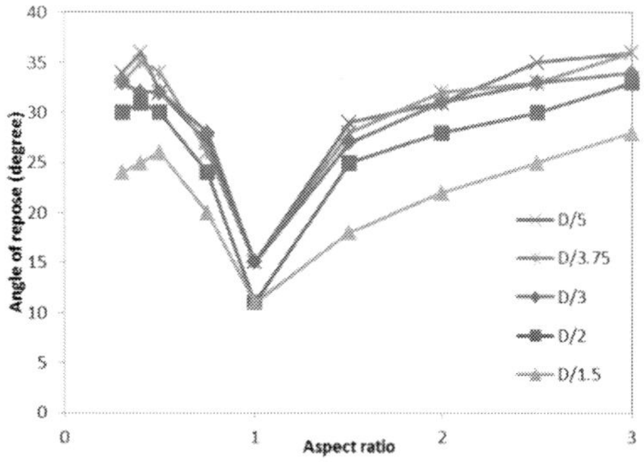

Fig. 2. Average angle of repose of sandpiles formed for different aspect ratios and discharge rate.

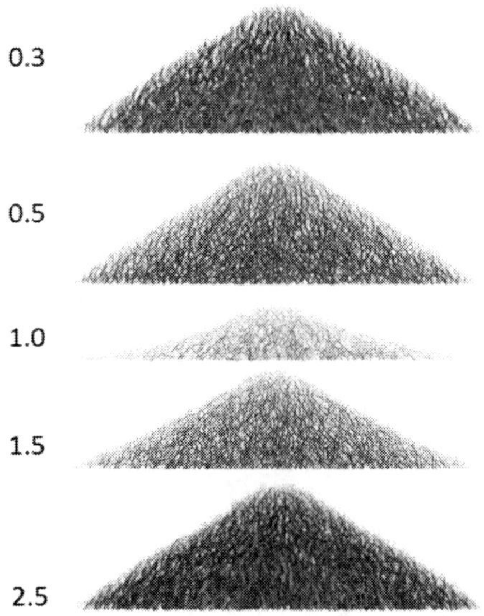

Fig. 3. Force structure in the central slice of sandpiles formed by particles of different aspect ratios.

One particular feature in the force network is the contact force orientation. Fig. 4 shows the normal contact force orientation of ellipsoids for different aspect ratios. For oblate spheroids, a strong peak can be observed at low angle. This is mainly due to particle orientation distribution that most particles are facing upward or downward, results in large number of vertical forces. As aspect ratio approaches 1.0 (spheres), the forces directions become more uniform. A hump is observed instead of peaks. The highest point of the hump shows up at about 45 degrees and another high point is at 90 degrees. This indicates a large number of contact forces are in random directions for spheres. For prolate particles, the distribution is similar to oblate particles, but the peak is not significant. This is because most prolate particles are lying horizontally.

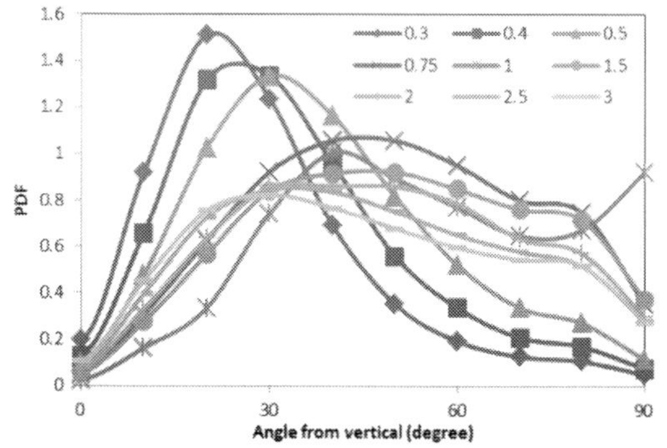

Fig. 4. Static of normal contact force directions of a sandpile for different aspect ratios.

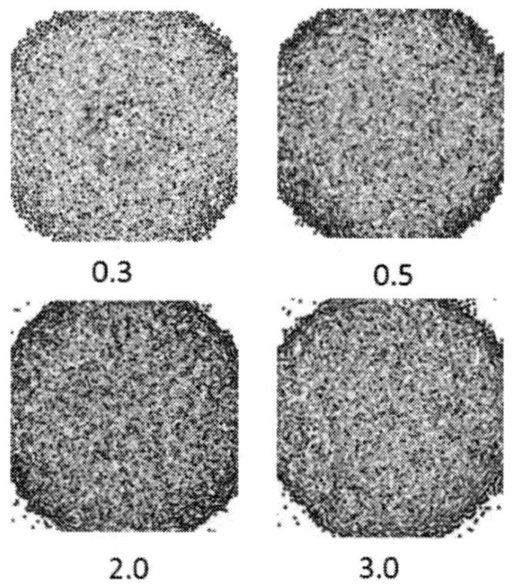

Fig. 5. Normal force distribution on bottom plate for different aspect ratios.

4.3 Bottom stress

Fig. 5 shows the normal force distribution on the bottom plate for different aspect ratios. Each point in the figure represents a contact point between particle and bottom plate, and its color reflects the magnitude of the force (red represents large forces and blue represents small forces). For some cases (e.g., aspect ratio of 0.3), relatively small forces can be observed in the central region and most large forces are distributed around the centre. For other cases, forces with various magnitudes are spread in a wide range, and it is hard to describe force distribution on the basis of these patterns.

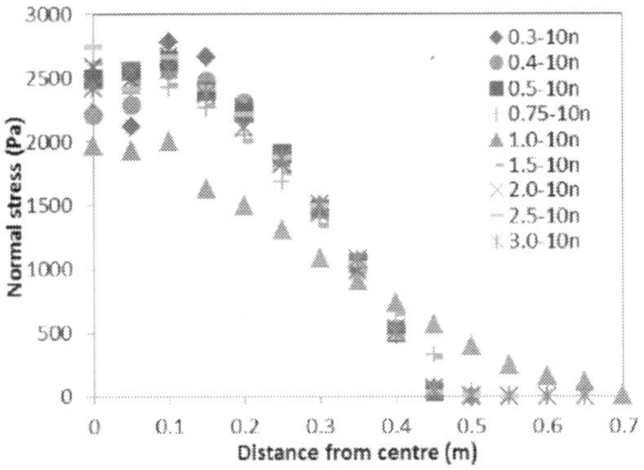

(a)

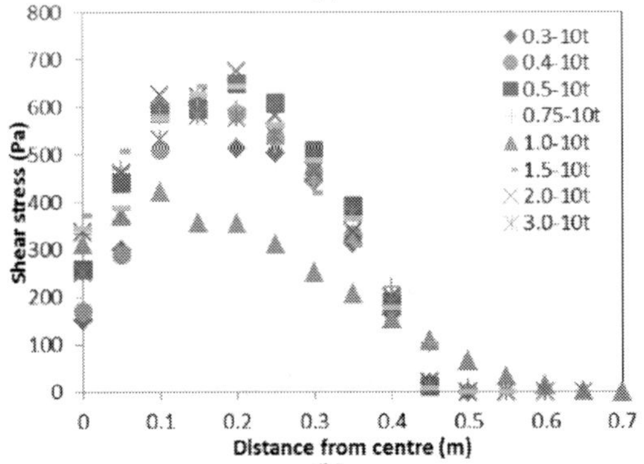

(b)

Fig. 6. Average bottom stress distribution for (a) normal stress and (b) shear stress.

It is often observed that a stress dip occurs beneath a sandpile. However, there is no consistency on this phenomenon as reported in the literature. Generally, it is known that the pressure distribution beneath a sandpile is affected by many factors, such as construction history, base deflection, particle size distribution and particle friction. Fig. 6 shows the normal and tangential stress beneath the sandpiles for different aspect ratios when orifice is 0.1m. To illustrate the stress dip distribution, the pressure is calculated using a ring that is 0.05m in width to obtain a relatively reliable value. For tangential stress, the stress dip can be observed for all the cases considered. But for normal stress, the dip is not so clear, consistent with the results as observed in the 2D slot model [10]. These results of stress distribution can generally be divided into three groups. The first group is stress dip group, for example, aspect ratio at 0.3 and 0.4, the stress dip phenomenon is very clear.The second group is terrace-like group, such as aspect ratio at 0.75 and 1.0. The stress in the middle section is tending to be uniform and form a terrace-like curve. The third group is stress rise group, such as aspect ratio at 1.5 and 2.5. For these cases, instead of a dip, the stress in the centre is higher than other areas, and the dip is observed at a small distance from the centre.

5 Further discussions and conclusion

The angle of repose obtained from sandpiles formed by ellipsoidal particles is comparable with previous studies. Spherical particles can have the smallest angle of repose and the trend for oblate and prolate spheroids differs. For oblate spheroids with aspect ratio deviating from sphere, the angle of repose firstly increases and the peak value is obtained when aspect ratio is 0.5, then the angle of repose will drop. For prolate spheroids, the angle of repose is larger than spheres but the increase rate is low for large aspect ratios. The trend is similar to average coordination number of packing of ellipsoids. The larger orifice size results in the smaller angle of repose.

Stress distribution is analysed to investigate the effect of particle shape and orifice size. Force distribution inside a sandpile is significantly related to aspect ratio. Vertical forces are more likely to be observed for ellipsoids, whilst large number of horizontal forces is obtained for spheres. Bottom force distribution also differs with aspect ratio. Stress dip can always be obtained for tangential stress. But for normal stress, certain cases shows a flat terrace in the centre (sphere), certain cases show stress dip (oblate spheroids with aspect ratio at 0.4), and certain cases show stress rise (prolate spheroids with aspect ratio of 2.5). There is no clear relationship between aspect ratio and stress dip phenomenon. But compare with previous 2D slot model studies, stress dip is more likely to obtain in the 3D DEM simulation as shown in the present study. Orifice is another factor can influence bottom stress. Most particle shapes can have stress dip when orifice is small, and enlarged orifice size leads to terrace-like or stress rise distribution.

References

1. Cundall, P.A. and O.D.L. Strack, *Geotechnique,* **29**, 1, (1979).
2. Zhu, H.P., et al., *Chemical Engineering Science,* **63**, 23, (2008).
3. Zhou, C. and J.Y. Ooi, *Mechanics of Materials,* **41**, 6, (2009).
4. Zuriguel, I. and T. Mullin, *Proceedings of The Royal Society A,* **464**, 2089, (2008).
5. Ai, J., et al., *Granular Matter,* **13**, 2, (2011).
6. Liu, J.G., Q.C. Sun, and F. Jin, *Powder Technology,* **212**, 1, (2011).
7. Zhu, J.Y., Y.Y. Liang, and Y.H. Zhou, *Powder Technology,* **234**, 0, (2013).
8. Zhou, Z.Y., et al., *Industrial & Engineering Chemistry Research,* **50**, 16, (2011).
9. Liu, S.D., et al., *Powder Technology,* **253**, 0, (2014).
10. Zhou, Z.Y., et al., *Granular Matter,* **16**, 5, (2014).

On the submerging of a spherical intruder into granular beds

Chuan-Yu Wu[1]*, *Ling* Zhang[1], and *Lan* Chen[2]

[1]*Department of Chemical and Process Engineering, University of Surrey, Guildford, GU2 7XH, UK*
[2]*School of Medical Instrument and Food Engineering, University of Shanghai for Science and Technology, Shanghai 200093, China*

Abstract. Granular materials are complex systems and their mechanical behaviours are determined by the material properties of individual particles, the interaction between particles and the surrounding media, which are still incompletely understood. Using an advanced discrete element method (DEM), we simulate the submerging process of a spherical projectile (an intruder) into granular materials of various properties with a zero penetration velocity (i.e. the intruder is touching the top surface of the granular bed and released from stationary) and examine its settling behaviour. By systematically changing the density and size of the intruder and the particle density (i.e. the density of the particles in the granular bed), we find that the intruder can sink deep into the granular bed even with a zero penetration velocity. Furthermore, we confirm that under certain conditions the granular bed can behave like a Newtonian liquid and the submerging intruder can reach a constant velocity, i.e. the terminal velocity, identical to the settling of a sphere in a liquid, as observed experimentally. A mathematical model is also developed to predict the maximum penetration depth of the intruder. The model predictions are compared with experimental data reported in the literature,good agreement was obtained, demonstrating the model can accurately predict the submerging behaviour of the intruder in the granular media.

1 Introduction

Penetration of a projectile into granular media is a ubiquitous process in nature, astrophysical, geophysical and technological applications, ranging from the penetration of crab's legs and human feet into beach sand, landing of planetary exploratory vehicles, asteroid impact on the moon and earth surface, to the high speed impact of military projectile to defensive materials. Two key questions associated with this process are how deep the projectile can penetrate and how large the impact crater is. Because of its technological relevance, penetration dynamics has been intensively explored over last three centuries. However, due to the complexity of the mechanical behaviour of granular materials and diversity of granular material properties, our understanding of the penetration processes is far from complete. For instance, majority of previous studies focused on the penetration of projectile at high impact speeds and explored the dependence of penetration depth and crater size on impact speed and materials properties. An interesting phenomenon reported in the literature published in the last decade is that a projectile can sink into granular materials even with zero initial impact velocity ([1], [2], [3], [4]). Recently, using a superlight granular material (expanded polystylene beads with a density of 14 kg/m3), Pacheo-Vazquez et al. ([4]) revealed that a projectile could penetrate indefinitely in the granular material with a terminal velocity, which is identical to the settling of a sphere in a Newtonian liquid. This indicates that during the penetration of a projectile into granular media, the granular material acts as a liquid, similar to conventional Newtonian liquids. In other words, for granular materials, there is a potential liquid phase.

*e-mail: c.y.wu@surrey.ac.uk

Here, with the aim of investigating whether there is a liquid phase for granular materials through which a projectile can reach a terminal velocity and penetrate indefinitely, we employ an advanced numerical method, the discrete element method (see the Methods section), and analyse the penetration of a spherical projectile of different sizes into granular media with zero initial speed, i.e. the projectile is placed on the top surface of the granular bed and released from rest. In contrast to the experimental study of Pacheo-Vazquez et al. ([4]) who used one particular granular material but varied the mass of the projectile, we systematically change the density of the particles in the granular materials for a given projectile, while keeping the initial packing structure of the granular bed identical, and examine how deep the projectile can sink into the granular materials and the kinematics of the projectile during the penetration process.

2 Submerging kinematics

We modeled a 2D system with the discrete element method (DEM), as detailed in Goey et al. [5]. A granular bed was first prepared by sedimentation of a particle system, which was randomly generated, under the action of the

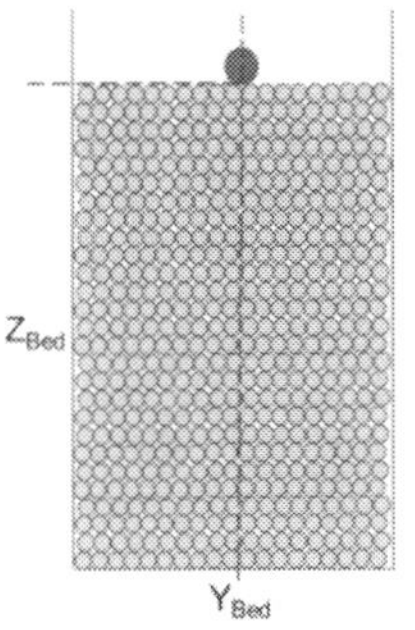

Figure 1. Illustration of the DEM model set up [5]

Table 1. Simulation parameters of granular beds and intruder

Parameter	Granular Bed	Intruder
Diameter	0.005 m	0.01 m
Density	14 to $490 kg/m^3$	$4,900 kg/m^3$
Young's modulus	3.37 GPa	210 GPa
Poisson's ratio	0.15	0.3

gravity. This bed was modelled with a random packing of 24,000 spheres, with a bed height of 0.8 m. This granular bed was contained in a rectangular container with rigid walls, as shown in Figure 1. The container was wide enough ($> 8d_p$, where d_p is the diameter of particles in the granular bed) to avoid any boundary effects from the side walls. During the sedimentation process, the motions of the spheres were traced until a final stable state was reached (when all the spheres in the bed had settled with a negligible velocity of $< 1x10^{-5} m/s$). A spherical intruder with a diameter of d_i was released with zero impact velocity, directly from the centre of the surface of granular beds. To examine the effect of the density ratio of the system (ranging from 1 to 350) on the subsiding behaviour, the density of the particles in the granular bed (ρ_p) was varied from $14 kg/m^3$ to $490 kg/m^3$, while keeping the density of the intruder constant at $\rho_i = 4900 kg/m^3$. The properties of the granular beds and intruder are presented in Table 1.

We first use a projectile of a diameter d_i that is twice as that of the particles, d_p, in the granular bed, i.e. the projectile/particle diameter ratio $\zeta = d_i/d_p = 2$. The density of the projectile is fixed at $\rho_i = 4900 kg/m^3$, but the density of the particles in the granular bed ρ_p is systematically varied to simulate various materials with a projectile/particle density ratio ($\psi = \rho_i/\rho_p$) in the range of 1 and 350, and to probe the effect of the density ratio ψ on the penetration dynamics. We observe that for $\psi > 1$ the projectile can sink into the granular bed to a certain depth before it reaches the bottom of the container and the penetration depth increases as the density ratio ψ increases (Fig. 2a). Once the projectile is released from stationary, it is accelerated under gravity to reach a maximum penetration velocity, thereafter it is decelerated to a halt at a finite penetration depth (Fig. 2b). These results are qualitatively in agreement with the experimental observations of Lohse et al. [2], Hou et al. [3] and Pacheco-Vazquez & Ruiz-Suarez [4].

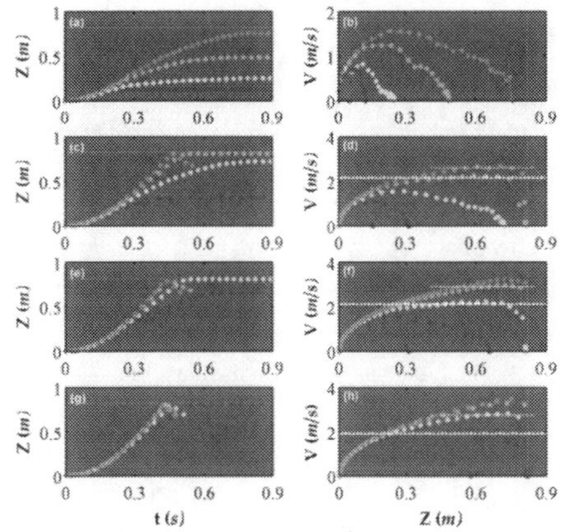

Figure 2. Kinematics of the intruder during submerging

Using the same granular beds, we then systematically increase the size of the projectile to $\zeta = 3\ 5$ (see Figs 2c- . 2h). We notice that, for these larger projectiles, the penetration behaviour can be classified into three distinctive regimes depending on the density ratio ψ: 1) At low density ratios, the projectile is accelerated initially but decelerated to a halt before it reaches the bottom of the container as observed for the cases with a diameter ratio $\zeta = 2$. And the maximum penetration depth increases as the density ratio increases and as the diameter ratio increases (Figs . 2d, . 2f and . 2h). 2) At high density ratios, the projectile is accelerated all the way as it sinks until it hits the bottom of the container and bounces back. 3) At the intermediate density ratios, the projectile is initially accelerated until a certain velocity is reached, thereafter it travels at this velocity until it reaches to the bottom of the container, instead of being decelerated, which is similar to the experimental observation of Pacheco-Vazquez et al. [4].

The penetration behaviour in the third regime is interesting as it implies that the projectile reaches a constant velocity, identical to the terminal velocity of a sphere settling in a Newtonian fluid, can penetrate indefinitely in the granular bed if there was no bottom boundary. This also indicates that a liquid phase exists for the penetration of a projectile in granular materials, in which the projectile can penetrate indefinitely with a terminal velocity. On examining the penetration behaviour shown in Fig. 2, we find that the transition from frictional granular solids to the liquid phase depends on not only the density ratio ψ but also the diameter ratio ζ.

3 Penetration model

To understand the phase transition behaviour, it is necessary to analyse the dynamics of the projectile during its penetration. As the projectile sinks in the granular bed, it is primarily subjected to two forces: the gravitational force

DOI: 10.1051/epjconf/201714002027

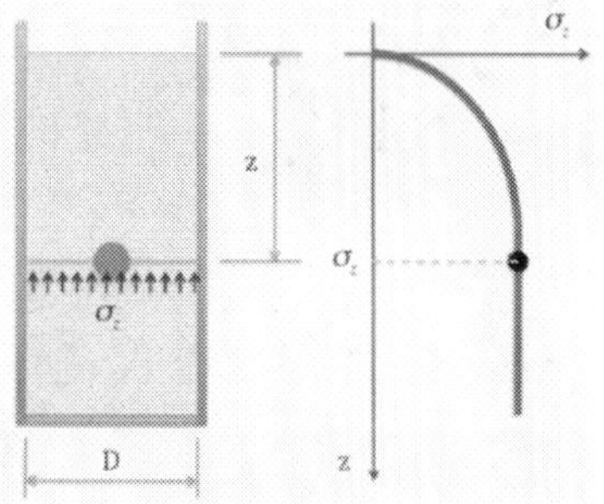

Figure 3. The resistance stress on the intruder

$m_i g$ (m_i – the mass of the projectile; and g – the gravitational acceleration) acting downwards and a upwards resistance force F_r induced by the contact with the particles in the granular bed. Based upon the Janson-Walker analysis [6] and assuming that the granular bed is homogeneous during submerging of intruder, the compressive stress σ_z in the vertical direction (Fig.3) at a distance z from the free surface of a packed granular bed is given as

$$\sigma_z = \frac{\rho_b g D}{4\mu_w K}\left[1 - e^{\left(\frac{-4\mu_w K}{D} z\right)}\right] \quad (1)$$

where $\rho_b = (1-\epsilon)\rho_p$ is the bulk density of the granular bed with ϵ being its porosity. D is the diameter of the container, μ_w is the wall friction coefficient and K is the stress transmission ratio. The resistance force F_r acting on the intruder can hence be approximated by

$$F_r = c_D A \sigma_z = \frac{C_D \pi D_i^2 (1-\epsilon)\rho_b g D}{4\mu_w K}\left[1 - e^{\left(\frac{-4\mu_w K}{D} z\right)}\right] \quad (2)$$

where A is the surface area of the projectile and c_D is a parameter related to the dynamic nature of the penetration process , in which the rearrangement of particles leads to the densification of the granular bed and intermittent strong force chain between the projectile and the granular bed. This will result in a much higher pressure than the static pressure given in Eq.1. Thus it is expected that c_D generally has a value greater than unity. Equations (1) and (2) show that σ_z and F_r increase from zero at $z = 0$ to a limiting value at a great depth. According to Newton's second law of motion,

$$m_i \frac{d^2 z}{dt^2} = m_i g - F_r \quad (3)$$

If the projectile penetrates to a finite depth Z_{max}, applying energy balance,

$$m_i g Z_{max} - \int_0^{Z_{max}} F_r d_z = 0 \quad (4)$$

Substituting Eq.(2) into (4), we obtain

$$m_i g Z_{max} - \int_0^{Z_{max}} \frac{C_D \pi d_i^2 (1-\epsilon)\rho_b g D}{4\mu_w K}\left[1 - e^{\left(\frac{-4\mu_w K}{D} z\right)}\right] d_z = 0 \quad (5)$$

Equation (6) can be integrated to obtain

$$m_i g Z_{max} - \frac{C_D \pi d_i^2 (1-\epsilon)\rho_b g D}{4\mu_w K} x \left[Z_{max} + \frac{D}{4\mu_w K} e^{\left(\frac{-4\mu_w K}{D} z\right)} - \frac{D}{4\mu_w K}\right] d_z = 0 \quad (6)$$

Re-arranging (6), we have

$$\frac{Z_{max}}{D} - \frac{6 C_D (1-\epsilon) D \rho_p}{4\mu_w K d_i \rho_i} \left[\frac{Z_{max}}{D} + \frac{1}{4\mu_w K} e^{\left(\frac{-4\mu_w K}{D} z\right)} - \frac{1}{4\mu_w K}\right] = 0 \quad (7)$$

Introducing dimensionless parameters

$$\gamma = \frac{Z_{max}}{D},$$

$$\eta = \frac{D}{d_i},$$

$$\delta_b = 6(1-\epsilon),$$

$$\xi_b = 4\mu_w K,$$

Equation (7) can be rewritten as

$$\frac{\xi_b \gamma}{\xi_b \gamma + e^{-\xi_b \gamma} - 1} = C_D \frac{\delta_b}{\xi_b} \frac{\eta}{\psi} \quad (8)$$

For shallow penetration (say $\gamma << 1$), Equation (8) can be approximated as

$$\gamma = \frac{2}{C_D \delta_b} \frac{\psi}{\eta} \quad (9)$$

Equation (9) indicates that 1) the penetration depth is linearly proportional to the density ratio and the diameter of the intruder; 2) the higher the packing fraction of the bed, the smaller the penetration depth. Equation (9) can also be expressed as

$$Z_{max} = \frac{2 d_i}{C_D \delta_b} \frac{\rho_i}{\rho_p} \quad (10)$$

Equation (10) implies that the maximum penetration depth is independent of the size of the container, but linearly proportional to the diameter of the intruder and the density ratio.

4 Model validation

To validate the model developed in the previous section, comparison between the model prediction and the experimental data for shallow and deep penetration reported in Lohse et al. ([3]), Hou et al. ([2]), Ambroso et al. ([1]), and Pacheco-Vazquez & Ruiz-Suarez ([4]) were presented in Figs 4-7. It is clear that the predictions using the developed model are in excellent agreement with the experimental data obtained for both shallow penetration and deep penetration.

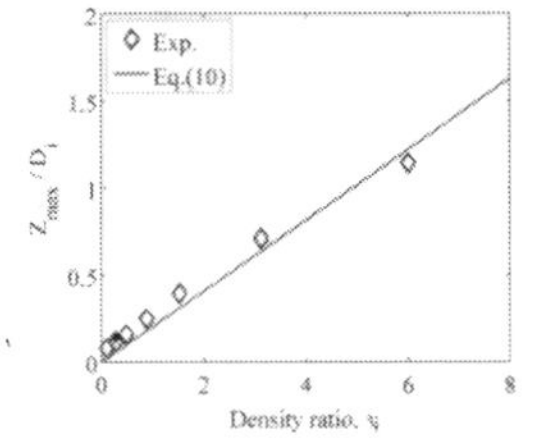

Figure 4. Maximum penetration depth: prediction vs. experiments (Ambrosso et al. [1])

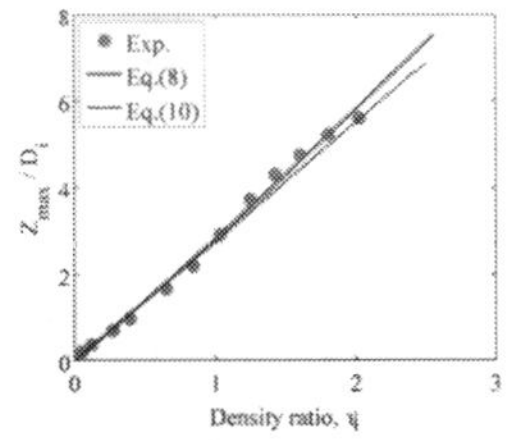

Figure 5. Maximum penetration depth: prediction vs. experiments (Lohse et al. [3])

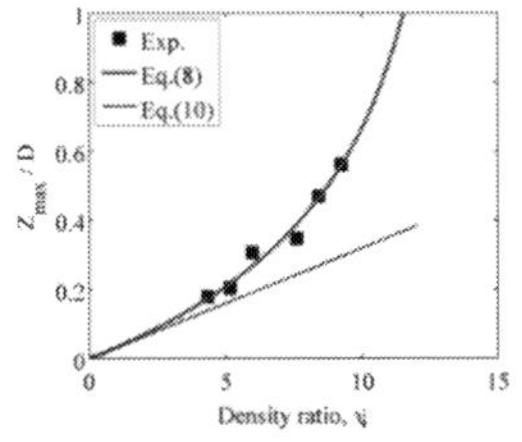

Figure 6. Maximum penetration depth: prediction vs. experiments (Hou et al. [2])

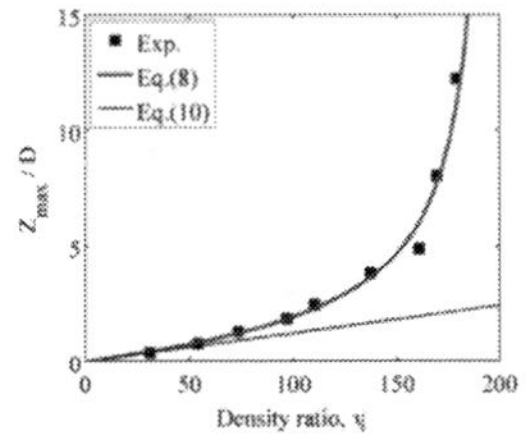

Figure 7. Maximum penetration depth: prediction vs. experiments (Pacheco-Vazquez et al. [4])

5 Conclusions

The submersing process of an intruder in granular materials was analysed numerically and analytically. An analytical model was developed and its accuracy was validated by comparison with experimental data reported in the literature. It was shown that the maximum penetration depth is a function of the diameter of the intruder, the packing density of the granular bed and the ratio of the intruder density to the density of particles in the granular bed.

Acknowledgements

CYW would like to acknowledge Chunlei Pei and Cher Hui Geoy for performing the DEM simulations discussed in Section 2.

References

[1] M.A. Ambroso, C. Santore, A. Abate and D. Durian, Phys. Rev. E **71**, 051305 (2005).
[2] M. Hou, Z. Peng, R. Liu, K. Lu and C. Chan, Phys. Rev. E **72**, 062301(2005).
[3] D. Lohse, R. Rauhe, R. Bergmann and D. van der Meer, Nature **432**, 689 (2004).
[4] F. Pacheco-Vazquez, G. Caballero-Robledo, J. Solano-Altamirano, E. Altshuler, A. Batista-Leyva and J. Ruiz-Suarez, Phys. Rev. Lett. **106**, 218001 (2011).
[5] C.H. Goey, C. Pei and C.Y. Wu, AIP Conference Proceedings **1542**, 437-440 (2013).
[6] J.P.K. Seville and C.Y. Wu, *Particle Technology and Engineering*, (Elsevier, London, 2016), 285.

The role of fluid viscosity in an immersed granular collapse

Geng Chao Yang[1], *Chung Yee* Kwok[1,*], and *Yuri Dumaresq* Sobral[2]

[1] *Department of Civil Engineering, The University of Hong Kong, Haking Wong Building, Pokfulam Road, Hong Kong*
[2] *Departamento de Matemática, Universidade de Brasília, Campus Universitário Darcy Ribeiro, 70910-900 Brasília, DF, Brazil*

Abstract. Instabilities of immersed slopes and cliffs can lead to catastrophic events that involve a sudden release of huge soil mass. The scaled deposit height and runout distance are found to follow simple power laws when a granular column collapses on a horizontal plane. However, if the granular column is submerged in a fluid, the mobility of the granular collapse due to high inertia effects will be reduced by fluid-particle interactions. In this study, the effects of fluid viscosity on granular collapse is investigated qualitatively by adopting a numerical approach based on the coupled lattice Boltzmann method (LBM) and discrete element method (DEM). It is found that the granular collapse can be dramatically slowed down due to the presence of viscous fluids. For the considered granular configuration, when the fluid viscosity increases, the runout distance decreases and the final deposition shows a larger deposit angle.

1 Introduction

Granular flows which are saturated by or immersed in fluids are ubiquitous phenomena in nature and industries, such as debris flow, submarine avalanche, food manufacturing and pharmacy. It is rather important to investigate these hydrogranular flows because of their crucial roles in sediment transport, shaping the landscape, risk assessment and industrial optimization. However, complex interactions between granular materials and fluids are yet to be fully understood.

Dry gravitational flows have been widely studied experimentally by a sudden release of granular column on a horizontal plane (see Fig. 1) over the last decades, focusing on flow kinematics and deposit morphologies [1, 2]. Scaling laws relating the deposit height and runout distance to the initial aspect ratio (h_0/x_0) of the granular column are proposed. As an important parameter in engineering and geophysics, the runout distance generally increases when the initial aspect ratio increases.

However, for granular flows in a viscous fluid, the runout distance also depends on other physical parameters, such as the initial volume fraction [3]. When the initial particle volume fraction is smaller than a critical value equal to 0.58, the scaled runout distance is longer compared to dry flows because of the generated excess pore pressure due to fluid expulsion [3]. In contrast, flows with high initial particle volume fraction travel similar distance as dry flows.

Apart from the significance of immersed granular collapse, the mechanisms and influence of fluids on flow kinematics and deposit morphologies are yet to be fully understood. The focus of this study is to investigate the role of fluid viscosity on granular collapse in a fluid using the coupled lattice Boltzmann method (LBM) and discrete element method (DEM).

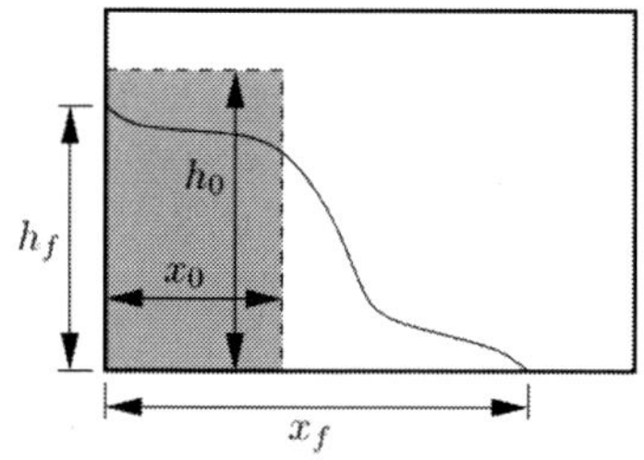

Figure 1. Granular Collapse with initial column (dashed line) of height h_0 and length x_0, and deposit of height h_f and length x_f

2 Methodology

LBM has been developed to become an alternative tool to traditional computational fluid dynamics (CFD) to couple with DEM, benefiting from the direct calculation of the drag force without ad hoc models [4–7]. LBM is a mesoscopic description of fluid flow based on the kinetic theory. Statistically, fluid behavior at macroscopic scale is nothing but the average micro-dynamics of fluid molecules. Meanwhile, macroscopic fluid behavior is also insensitive to the detailed dynamics of individual molecules. Therefore, LBM is able to describe the fluid system by a collection of molecules residing on a regular lattice. A set of density distribution functions are used to represent the collection of molecules at each lattice node, where *collision* between

*e-mail: fkwok8@hku.hk

"fluid" molecules takes place by following the conservation laws of mass, momentum and energy. At each time step, the post-collision density distribution functions are passed to the neighboring lattice nodes by a *streaming* process.

In LBM, the evolution equation for the density distribution function $f_i(\mathbf{x}, t)$ with a BGK approximation [8], can be written as:

$$f_i(\mathbf{x} + \mathbf{c}_i\delta_t, t + \delta_t) - f_i(\mathbf{x}, t) = -\frac{1}{\tau}\left[f_i(\mathbf{x}, t) - f_i^{eq}(\mathbf{x}, t)\right] \quad (1)$$

where the density distribution function f_i is related to the number of molecules at time t positioned at $\mathbf{x}$ moving with velocity c_i along the ith direction at each lattice node. The time step for evolution and the relaxation time are δ_t and τ, respectively. The equilibrium distribution function (EDF) is f_i^{eq}. Based on the conservation of mass and momentum, the fluid density ρ and velocity $\mathbf{u}$ can be easily reconstructed from the velocity moments of the discrete density distribution functions:

$$\rho = \sum_i f_i \quad (2)$$

$$\rho\mathbf{u} = \sum_i \mathbf{c}_i f_i \quad (3)$$

The pressure field p is related to the fluid density by the equation of state:

$$p = c_s^2\rho \quad (4)$$

The lattice speed of sound is c_s and equal to $1/\sqrt{3}$ for the D3Q19 lattice arrangement [9].

While the fluid is simulated using LBM, DEM is adopted to take care of the solid particles. DEM has been widely used in granular flow applications [10–12]. The particle-particle collisions are governed by the Hertzian contact laws [13]. By considering the gravity ($\mathbf{G}$), contact forces ($\mathbf{F}_c$ and $\mathbf{T}_c$) and hydrodynamic forces and torques ($\mathbf{F}_f$ and $\mathbf{T}_c$), the positions and velocities of particles can be updated according to the Newton's second law:

$$m\mathbf{a} = \mathbf{F}_c + \mathbf{F}_f + \mathbf{G} \quad (5)$$

$$\mathbf{I}\dot{\omega} = \mathbf{T}_c + \mathbf{T}_f \quad (6)$$

The mass and moment of inertia of particles are m and $\mathbf{I}$, respectively. The acceleration is $\mathbf{a}$ and the angular velocity is ω. The coupling between LBM and DEM is achieved by the immersed moving boundary method, initially proposed by Nobel and Torczynski [14]. For each cycle of LBM calculation, 100 sub-cycles of DEM calculation are conducted. During the sub-cycling process, the positions of DEM particles on the LBM lattice grid and the corresponding hydrodynamic forces and torques are not updated.

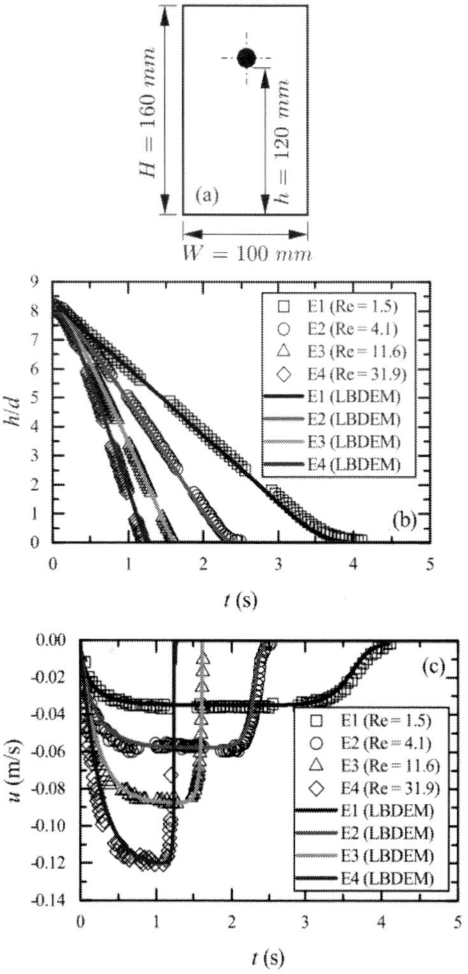

Figure 2. Single particle sedimentation test: (a) sketch of the experimental and numerical setup; (b) comparison between measured and simulated particle settling trajectory; (c) comparison between measured and simulated particle settling velocity

3 Results and discussions

3.1 Single particle sedimentation test

Fig. 2(a) shows the sketch of the experimental and numerical setups for the single particle sedimentation test conducted in [15]. A 15 mm diameter (d) particle is released at a height (h_0) equal to 120 mm with zero initial velocity within a container with dimension 100 × 100 × 160 mm. During sedimentation of the particle, the settling velocity first increases until a terminal velocity (u) is reached, at which the downward self-weight is balanced by the upward buoyancy force and drag force. We have repeated the same experiments referenced in [15] for four different values of Reynolds number (Re). The evolutions of particle trajectory and settling velocity from numerical simulations are compared to available experimental data in Fig. 2(b) and Fig. 2(c), respectively. It is shown that the LBM-DEM

DOI: 10.1051/epjconf/201714009037

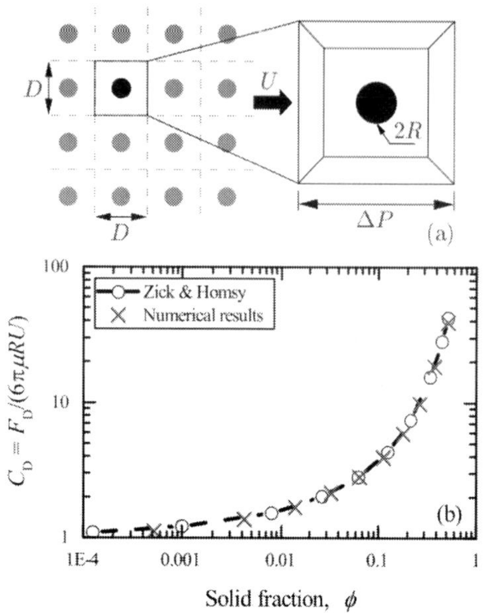

Figure 3. Flow through a three-dimensional periodic array of particles: (a) sketch of the numerical model; (b) comparison between numerical results and Zick & Homsy's solution for the drag coefficients at various solid fractions

coupling scheme presented in this study can capture well the fluid-particle interactions in the single particle sedimentation test.

3.2 Flow through a periodic array of particles

Fig. 3(a) shows the sketch of a viscous flow through three-dimensional periodic array of particles. In numerical simulations, only one cubic cell with size D equal to 0.01 m is considered with periodic boundary conditions defined in all faces. A small pressure gradient 0.001 Pa/m is applied to drive the flow from the left to the right. The simulations are run until the steady state is achieved at which point the flow velocity becomes steady. In order to get the drag forces at various solid fractions ϕ, the particle radius is changed from 0.0005 m to 0.005 m. Fig. 3(b) compares the normalized hydrodynamic forces from numerical simulations and Zick and Homsy's analysis [16], showing an excellent agreement over the entire range of solid fractions considered in this study.

3.3 Granular collapse

A sample of a granular column is first prepared using the gravitational deposition method. In this study, a rather dense initial packing is achieved by setting the particle friction coefficient to be zero, which is later adjusted to 0.4 before releasing the granular column. The other modelling parameters are summarized in Table 1.

In DEM, rigid walls are applied at all faces for the three-dimensional simulation domain. In LBM, non-slip

Table 1. Modelling parameters

	Parameters	Values
Particle	Diameter, d_p (m)	0.002
	Density, ρ_p (kg/m^3)	2500
	Young's modulus, E (Pa)	5E6
	Poisson's ratio, ν	0.35
	Coefficient of restitution, e	0.5
Fluid	Density, ρ_f (kg/m^3)	1000
	Viscosity, μ_f (Pa·s)	0.01 - 0.05
Granular column	Initial width, x_0 (m)	0.02
	Aspect ratio, AR	2.0
	Initial volume fraction, ϕ	0.62

boundary condition is defined at the walls following the bounce-back rule [4]. In this study, the influence of viscosity on the granular collapse in a fluid is considered. As the granular column collapses onto the horizontal plane (see Fig. 1), the particles will push the fluid forward at the lower part of the simulation domain. Because of mass conservation, the fluid will flow against the movement of particles at the upper part of the simulation domain, which is called as back flow here. As a result, a recirculation region is observed just above the free surface of the granular flow. In order to minimized the effect of back flows on the runout distance, a large enough domain size for the fluid field is set with length, height and width equal to 0.14 m, 0.06 m and 0.02 m, respectively.

Fig. 4 shows three cases of the granular flow, including one dry condition: (a) without surrounding fluid; and two wet conditions: (b) in a low-viscosity fluid (μ_f = 0.01 Pa·s) and (c) a high-viscosity fluid (μ_f = 0.05 Pa·s). Compared to the wet condition, it is found that the particles in the dry condition can travel much longer. Part of the particles in the dry condition keep moving forward after the collapse and eventually hit the right wall and bounce back.

In contrast, all the particles are stopped in the wet conditions, because a large part of the kinetic energy of particles is damped by the fluid via drag force. Some energy is turned into fluid kinetic energy and all of it is dissipated due to viscous effects in the end. Considering the granular collapse in a fluid, the drag force on the particles increases as the fluid viscosity increases, which works as a resistant force to prevent the granular column from collapse. Therefore, it takes a longer time for the granular column to collapse and reach its final deposition in the presence of a more viscous fluid. Furthermore, the runout distance is shorter. It is also worth mentioning that the deposit angle of the granular flow increases as the fluid viscosity increases.

The normalized runout distance from LBM-DEM simulations (colored cross symbols) is plotted against the initial aspect ratio in Fig. 5, together with the numerical results from [17] (hollow symbols). It is confirmed that when the aspect ratio and the initial volume fraction are equal to 2 and 0.62, respectively, both wet conditions in our study belong to the viscous regime [17], in which high-viscosity fluids will slow down the granular collapse and cause a reduction in runout distance.

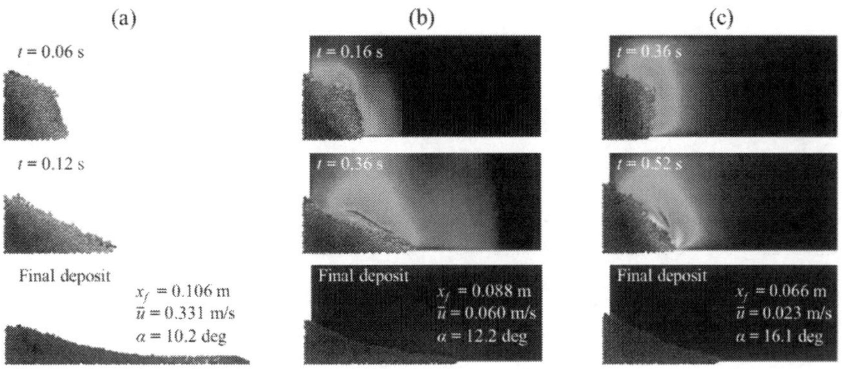

Figure 4. Three sequence of images corresponding to granular particles collapsing on a horizontal plane: (a) granular collapse without surrounding fluid; (b) granular collapse in a fluid with viscosity equal to 0.01 Pa·s; and (c) 0.05 Pa·s. The granular and fluid field are coloured by the corresponding velocity

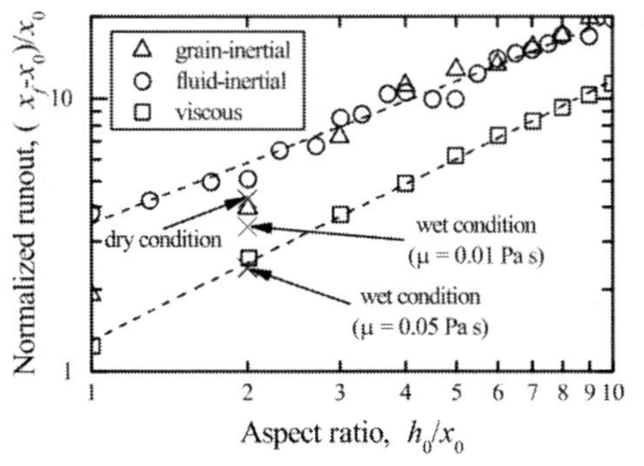

Figure 5. Normalized runout distance as a function of the initial aspect ratio in different regimes referenced in [17]

4 Conclusions

A qualitative investigation of the role of fluid viscosity in an immersed granular column is presented using the coupled LBM-DEM. The numerical scheme can successfully capture the hydrodynamic interactions between particle and fluid, which is verified against the available experimental data and analytical solutions. It is found that the presence of fluid slows down the granular flow dramatically. During the collapse of granular column, part of the particle kinetic energy is passed to the fluid through hydrodynamic interactions and finally dissipated inside the fluid due to viscous effects. As a result, when the fluid viscosity increases, the runout distance decreases and the final granular deposit shows a larger deposit angle.

In future works, we plan to fully examine the effect of viscosity and other parameters (for granular material, fluid and composition of granular packing) on the immersed granular collapse quantitatively and determine the appropriate scaling laws governing x_f and h_f. This can be an important risk assessment tool for particle-laden flows and provide optimization scheme for manufacturing industries.

Acknowledgements

This research is conducted in part using the research computing facilities and/or advisory services offered by Information Technology Services, the University of Hong Kong and under the support of FAP-DF, Brazil.

References

[1] Lajeunesse, E., J. Monnier, G. Homsy, Phys. Fluids **17**, 10 (2005)
[2] Lube, G., Huppert, H. E., Sparks, R. S. J., Phys. Rev. E **72**, 4 (2005)
[3] Rondon, L., O. Pouliquen, and P. Aussillous, Phys. Fluids **23**, 7 (2011)
[4] Chen, S., G.D. Doolen, Annu. Rev. Fluid Mech. **30**, 1 (1998)
[5] He, X., L.-S. Luo, Phys. Rev. E **56**, 6 (1997)
[6] Lallemand, P., L.-S. Luo, Phys. Rev. E **61**, 6 (2000)
[7] Aidun, C.K., J.R. Clausen, Annu. Rev. Fluid Mech. **42** (2010)
[8] Bhatnagar, P.L., E.P. Gross, M. Krook, Phys. Rev. **94**, 3 (1954)
[9] Mei, R., Shyy, W., Yu, D., Luo, L.-S., J. Comput. Phys. **161**, 2 (2000)
[10] Cundall, P.A., O.D. Strack, Geotechnique **29**, 1 (1979)
[11] Cleary, P.W., Miner. Eng. **11**, 11 (1998)
[12] Jing, L., Kwok, C., Leung, Y., Sobral, Y., Phys. Rev. E **94**, 5 (2016)
[13] Kloss, C., Goniva, C., Hager, A., Amberger, S., Pirker, S., Prog. Comput. Fluid Dy. **12**, 2-3 (2012)
[14] Noble, D., J. Torczynski, Int. J. Mod. Phys. C **9**, 8 (1998)
[15] Ten Cate, A., Nieuwstad, C., Derksen, J., Van den Akker, H., Phys. Fluids **14**, 11 (2002)
[16] Zick, A., G. Homsy, J. Fluid Mech. **115** (1982)
[17] Topin, V., Monerie, Y., Perales, F., Radjai, F., Phys. Rev. Lett. **109**, 18 (2012)

Effect of improved sol-gel method on composition and micro morphology of $Ca_3Co_4O_9$

Xuanye Hu[1,2], *Yanan* Li[1,2], *Ping* Wu[1,2,*], *Shiping* Zhang[1,2], and *Xiulan* Huai[3]

[1] University of Science and Technology Beijing, School of mathematics and Physics, Beijing 100083, China
[2] Beijing Key Laboratory for Magneto-Photoelectrical Composite and Interface Science, Beijing 100083, China
[3] Chinese Academy of Sciences, Institute of Engineering Thermophysics, Beijing 100190, China

Abstract. $Ca_3Co_4O_9$ is a potential high temperature thermoelectric material. We explore the hydrothermal treatment improved sol-gel method to prepare $Ca_3Co_4O_9$ thermoelectric materials. We find that this treatment can increase the thermoelectric performance of the samples compared with the conventional citric acid sol-gel method. We also find the samples prepared by the improved sol-gel method have more uniform grain distribution, better densification, and less voids. At about 1000K, compared with the conventional sol-gel method, the power factor of the samples prepared by improved sol-gel method improves about 168%.

1 Introduction

In recent years, due to the need of environmental protection and energy saving, thermoelectric materials have been paid more and more attention[1]. Thermoelectric oxides, which have the advantages of high stability under high temperature, no oxidation problem, no pollution, long service life and simple fabrication, have become an attractive research topic in the thermoelectric field. Especially, $Ca_3Co_4O_9$ becomes a strong candidate for high temperature applications. Its stability is better compared with the prior $NaCo_2O_4$ single crystal[2] for it can maintain its stable performance under more than 1100K in the air[3].

There are many methods to prepare $Ca_3Co_4O_9$, including solid state reaction method[4,5], sol-gel method[6-9], hydrothermal method and the like. In the solid state reaction, $Ca_3Co_4O_9$ is obtained directly at high temperature from the dry mixture of CaO and CoO in a stoichiometric ratio(3:4). This method is simple but the grain size is big and the reaction temperature has a great influence on the composition of final product. Sol-gel method is to treat the mixture of $Ca(NO_3)_2$ and $Co(NO_3)_2$ through solution, sol, gel, and then heat treatment to obtain $Ca_3Co_4O_9$. Hydrothermal method can obtain $Ca_3Co_4O_9$ by mixed solution of $Ca(NO_3)_2$ and $Co(NO_3)_2$ at certain temperature and pressure. The samples prepared by sol-gel method and hydrothermal method have a smaller grain size and more uniform distribution relative to the solid state reaction method, and these two factors are considered to be the key to improve the thermoelectric performance[10]. We speculate that the solution environment at high temperature and high pressure has an effect on the property of the samples. Therefore, we attempt to transfer the sol-gel process into the reactor, which is the difference between conventional sol-gel method and the improved sol-gel method.

In this work, the composition, structure and micro-morphology of the samples prepared by improved sol-gel method are investigated by comparing with those of the conventional method. Furthermore, the thermoelectric performance of the samples is investigated as well.

2 Experiment details

2.1 Preparation of $Ca_3Co_4O_9$ powders by conventional sol-gel method

$Ca_3Co_4O_9$ powders were prepared by conventional sol-gel method. A mixture of $Ca(NO_3)_2 \cdot 4H_2O$, $Co(NO_3)_2 \cdot 6H_2O$ in a stoichiometric ratio (3:4:7.7) was dissolved and thoroughly stirred in a citric acid solution. After that the solution was stirred by a magnetic mixer and heated at 80℃ until the wet gel was formed. Later the wet gel was dried at 120℃ in oven to get the dry gel. The dry gel was manually milled and then calcined at 750℃ for 2 hours in air. The powder were manually milled once more and uniaxially pressed at 30MPa to form pellets and sintered at 900℃ for 3 hours to get the $Ca_3Co_4O_9$ bulk samples.

* Corresponding author: pingwu@sas.ustb.edu.cn

DOI: 10.1051/epjconf/201714013016

2.2 Preparation of $Ca_3Co_4O_9$ powders by improved sol-gel method

$Ca_3Co_4O_9$ powders were prepared by improved sol-gel method. A mixture of $Ca(NO_3)_2 \cdot 4H_2O$, $Co(NO_3)_2 \cdot 6H_2O$ in a stoichiometric ratio (3:4:7.7) was dissolved and thoroughly stirred in a citric acid solution. After that the solution was stirred by a magnetic mixer and heated at 80℃until the sol was formed. Later the sol was transferred into a hydrothermal reactor and the reactor was placed into oven. After heated at a constant temperature for a period of time, the mixture in the reactor was filtered to get the sediment. Next the sediment was dried in drying baker at 120℃ for 5h to get rid of the deionized water and volatile organic. Then it was manually milled and heated at 900℃ for 2 hours to obtain the precursor powder of $Ca_3Co_4O_9$. Finally, the powder was pressed into planchet under 30Mpa and sintered at 900℃ for 3 hours to get the $Ca_3Co_4O_9$ bulk samples.

2.3 Characterization of samples

The phase was analyzed by X-ray diffraction instrument (Rigaku,Japan) with Cu Kα radiation, the range of scanning angle is 5° ~ 90° and the scan rate is 0.03° /s. The morphology of the samples was observed by scanning electron microscope (Zeiss Supra 55). The Seebeck coefficient and the electrical conductivity of the samples were measured by instrument SBA 458.

3 Results and discussion

3.1 XRD analysis

The X-ray powder diffraction patterns of all the samples are shown in Fig.1. The peak positions of all samples can be matched with the standard JCPDS card (23-0110) of $Ca_3Co_4O_9$ and (21-0139) of $Ca_9Co_{12}O_{28}$. The XRD patterns show the compositions of the samples prepared by the conventional sol-gel method and improved sol-gel method are almost the same. They are both composed of $Ca_3Co_4O_9$ and $Ca_9Co_{12}O_{28}$, with no other impurity phase.

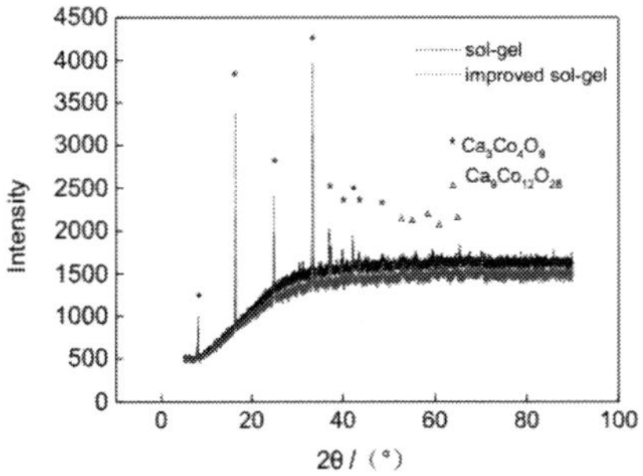

Fig.1. XRD patterns of two powders.

Furthermore, the relationship between $Ca_3Co_4O_9$ and $Ca_9Co_{12}O_{28}$ equals to the relationship between single crystal and polycrystalline and $Ca_9Co_{12}O_{28}$ can usually be expressed as $Ca_3Co_4O_{9+\delta}$. In conclusion, improved sol-gel method can be used to prepare the $Ca_3Co_4O_9$ samples.

3.2 SEM analysis

Fig.2(a) is the SEM image of the sample prepared by the conventional sol-gel method, and Fig.2(b) is the SEM image of the sample prepared by the improved sol-gel method. Fig.2. shows that the grain size are about the same size, near 1 μm. In contrast, the grain size of the samples prepared by improved sol-gel is more uniform, which will improve the thermoelectric performance of materials. As is shown in Fig.2., the places whose color is deeper than the surrounding are voids. And the voids of Fig.2(a) is more than that of the Fig.2(b), which leads to the decrease in carrier mobility, and then decreases the electrical conductivity of the sample.

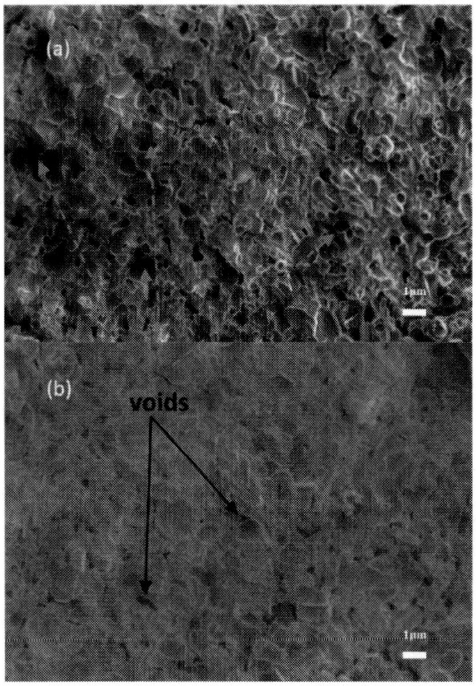

Fig.2. SEM spectra of two powders.[(a) is prepared by the conventional sol-gel method, (b) is prepared by improved sol-gel method.]

3.3 Thermoelectric performance analysis

We analyzed the thermoelectric performance of three different samples. The first sample, which is name after "ISM1", was prepared by improved sol-gel method and the oven was set to 200℃ and 48h. Compared with the first one, the second sample, which is name after "ISM2", was prepared by improved sol-gel method as well, but the oven was set to 150℃ and 12h. The third sample was prepared by conventional sol-gel method and named for "CSM".

3.3.1 Conductivity analysis

Fig.3. shows the curves of electric conductivity of different samples varies with temperature, the curves "ISM1" and "ISM2" are higher than the curve "CSM" at the same temperature. For the same temperature, the factors that influence the conductivity of the material are the carrier concentration and the carrier mobility. Hydrothermal treatment is to give a higher temperature and pressure, which will increase the carrier concentration of the samples. Combined with SEM image, the smaller void ratio makes the sample more dense, thereby increasing the carrier mobility. Therefore, improved sol-gel method does improve the conductivity of the sample.

It's surprising that different conditions of oven change the performance of the samples from semiconductor to metal. For the curve of "ISM1", the electrical conductivity decreases with the increase of temperature, which presents a metallic property. On the contrary, the curve of "ISM2" binding upward with the increase of temperature, presents a semiconductor property. This shows that the temperature has impacts on the conductivity of the sample, and can even changes in its performance irreversibly.

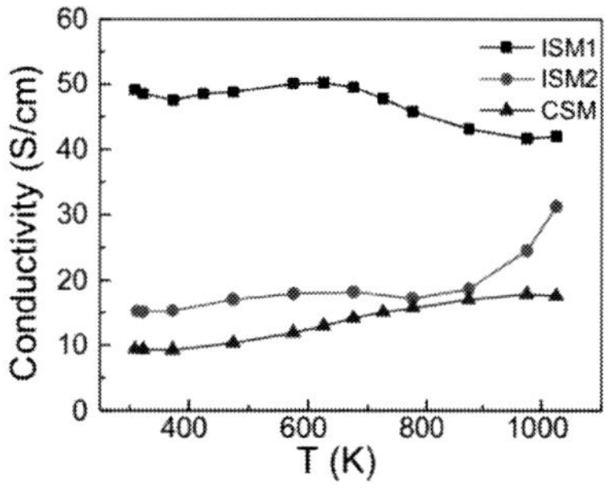

Fig.3. Temperature dependence of conductivity of all samples.

3.3.2 Seebeck coefficient analysis

The Seebeck coefficient of a material is a measure of the magnitude of an induced thermoelectric voltage in response to a temperature difference across that material, as induced by the Seebeck effect. Physically, the magnitude and sign of the Seebeck coefficient can be approximately understood as being given by the entropy per unit charge carried by electrical currents in the material. It may be positive or negative. In conductors that can be understood in terms of independently moving, nearly-free charge carriers, the Seebeck coefficient is negative for negatively charged carriers (such as electrons), and positive for positively charged carriers (such as electron holes). The maximum characteristics of thermoelectric materials is the transformation between heat and electric, so the measurement of Seebeck coefficient which can reflect the thermoelectric conversion is necessary.

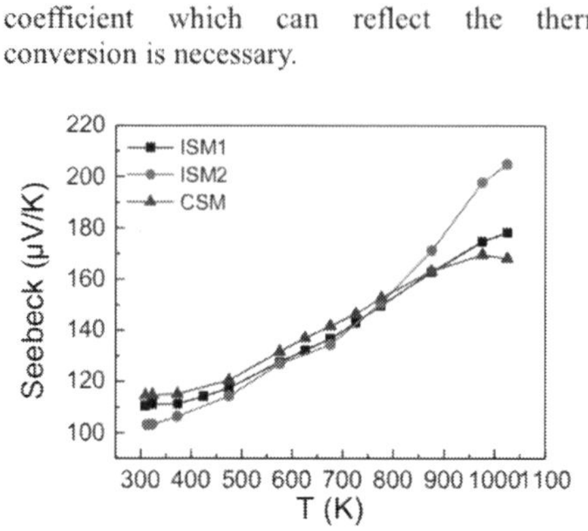

Fig.4. Temperature dependence of Seebeck coefficient of all samples.

Fig.4. shows the curves of Seebeck coefficient of different samples varies with temperature. All samples exhibit positive Seebeck coefficient indicating their hole carrier conduction , and the $Ca_3Co_4O_9$ sample is p-type semiconductor. The Seebeck coefficient of three samples increases with the increasing temperature. In contrast, the Seebeck coefficient of the samples prepared by improved sol-gel method are slightly higher.

3.3.3 Power factor analysis

Power factors, equals the product of the square of the Seebeck coefficient and the electrical conductivity. Fig.5. shows the curves of power factor of different samples varies with temperature.

The power factors of "ISM1" and "ISM2" reach 1.335×10^{-4}W/m·K^2 at 1000K, but that of "CSM" reaches 4.970×10^{-5}W/m·K^2 at the same temperature. By contrast, the power factor of the sample prepared by improved method increased about 168% at 1050K.

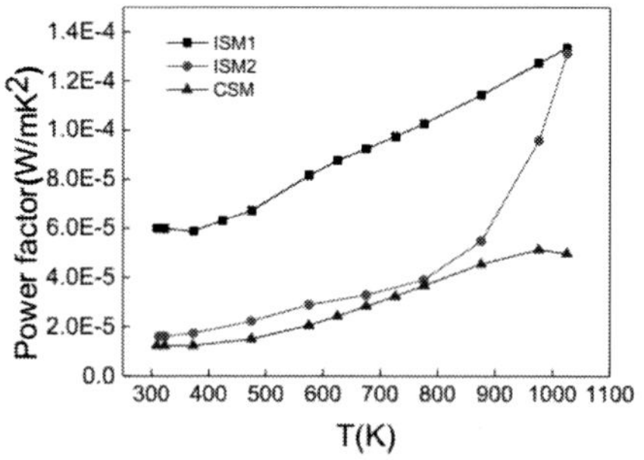

Fig.5. Temperature dependence of power factor of all samples.

4 Conclusion

We prepared the powders of $Ca_3Co_4O_9$ by the

hydrothermal treatment improved sol-gel method and the conventional citric acid sol-gel method. The following conclusions are obtained by the characterization of the phase composition, microstructure and thermoelectric performance of the material: (1) The samples prepared by the two methods are both composed of $Ca_3Co_4O_9$. (2) Compared with the sample prepared by the conventional sol-gel method, $Ca_3Co_4O_9$ thermoelectric material prepared by improved sol-gel method has more uniform grain distribution, better densification, and less voids, which lead to the increase of carrier mobility rate and conductivity; Seebeck coefficient has little change; The power factor of each temperature is increased. At about 1000K, compared with the conventional sol-gel method, the samples prepared by improved sol-gel method were improved by 168%.

We acknowledge the National Natural Science Foundation of China for financial support (No. 51476173).

References

1. F.J. Disalvo, Science **285**, 703(1999).
2. Terasaki I, Sasago Y, Uchinokura K, Phys Rev B **56**, 12685–12687(1997).
3. Woermann E, Muan A, J Inorg Nucl Chem **32**, 1455–1459(1970).
4. Kohei Obata, Yasunori Chonan, Takao Komiyama, et al, Journal of Electronic Materials **42**, 2221–2226(2013).
5. Bhaskar A, Huang Y.C, Liu C.J, et al. Journal of Electronic Materials **43**, 535-540(2013)
6. S. Pinitsoontorn, N. Keawprak, N. Lerssongkram, et al. J Mater Sci: Mater Electron **23**,1050-1056 (2012).
7. Song Chen, Yun Chen, Xueyan Song, et al. J Sol-Gel Sci Technol **64**, 627–636(2012).
8. F.P. Zhang, Q.M. Lu, J.X. Zhang. Physica B **404**, 2142–2145(2009).
9. Jie Xu, Changping Wei, Kun Jia. Journal of Alloys and Compounds **500**,227–230(2010).
10. S. Pinitsoontorn, N. Lerssongkram, A. Harnwunggmoung, et al, Journal of Alloys and Compounds **503**, 431–435 (2010).

Numerical analysis of the effect of particle shape and adhesion on the segregation of powder mixtures

Mohammadreza Alizadeh Behjani[1], *Ali* Hassanpour*[1], *Mojtaba* Ghadiri[1], and *Andrew* Bayly[1]

[1]School of Chemical and Process Engineering, University of Leeds, Leeds, LS2 9JT, UK

Abstract. Segregation of granules is an undesired phenomenon in which particles in a mixture separate from each other based on the differences in their physical and chemical properties. It is, therefore, crucial to control the homogeneity of the system by applying appropriate techniques. This requires a fundamental understanding of the underlying mechanisms. In this study, the effect of particle shape and cohesion has been analysed. As a model system prone to segregation, a ternary mixture of particles representing the common ingredients of home washing powders, namely, spray dried detergent powders, tetraacetylethylenediamine, and enzyme placebo (as the minor ingredient) during heap formation is modelled numerically by the Discrete Element Method (DEM) with an aim to investigate the effect of cohesion/adhesion of the minor components on segregation quality. Non-spherical particle shapes are created in DEM using the clumped-sphere method based on their X-ray tomograms. Experimentally, inter particle adhesion is generated by coating the minor ingredient (enzyme placebo) with Polyethylene Glycol 400 (PEG 400). The JKR theory is used to model the cohesion/adhesion of coated enzyme placebo particles in the simulation. Tests are carried out experimentally and simulated numerically by mixing the placebo particles (uncoated and coated) with the other ingredients and pouring them in a test box. The simulation and experimental results are compared qualitatively and quantitatively. It is found that coating the minor ingredient in the mixture reduces segregation significantly while the change in flowability of the system is negligible.

1 Introduction

Segregation of particulate solids is a phenomenon through which the homogeneity of a mixture is deteriorated by the separation of components due to different properties of the particles within the mixture. For instance, particles size, density, shape, and surface condition influence the segregation; therefore, controlling these factors can reduce the segregation effectively [1, 2]. To do so requires a good understanding of their role. In this study we address the effect of interparticle adhesion on the segregation tendency by coating a selected component with a thin layer of liquid to make it sticky
Particles surface condition can have a significant impact on inducing or reducing segregation mainly by influencing the flowability of the particles. Particles flowability is directly linked with the particles surface condition and surface texture [3-6]. The effects of surface condition is manifested in surface properties namely the coefficients of sliding friction and rolling friction and adhesion/cohesion of the particles.

Increasing the cohesivity of the particles by coating with a thin liquid layer is a typical method to reduce the segregation; as cohesive materials have a2 lower flowability [7, 8]. Particles adhesive tendency is expressed by the ratio of surface forces to gravitational force, known as Bond number [9]. Due to their lower momentum and higher Bond number, finer particles are more affected by coating compared to the coarse ones which may lead to their poor flowability or even caking and agglomeration. Nevertheless, making the minor component sticky in a mixture could be a remedy for reducing their segregation; while the flowability of the whole mixture is not significantly affected [10].

In the present study the segregation of a minor component in a ternary mixture of particles during heap formation is simulated using DEM. The particles properties including size, density, shape, and surface properties are measured experimentally and calibrated to be used in DEM. The minor component ingredient is a round enzyme placebo particle which is coated by PEG 400 to manipulate its stickiness. Segregation of the minor ingredients as affected by the coating in the mixture is simulated and analysed.

2 Methodology

DEM simulation of segregation of particles during the heap formation is carried out and compared with the experiment. Fig. 1 depicts the geometry of the test box in

* Corresponding author: a.hassanpour@leeds.ac.uk

experiment in which the heap formation is simulated. It is highly important in DEM modelling to set appropriate physical and mechanical properties in the model. While this part has been normally simplified or ignored in many previous studies, in this study the real physical parameters are measured and used where possible.

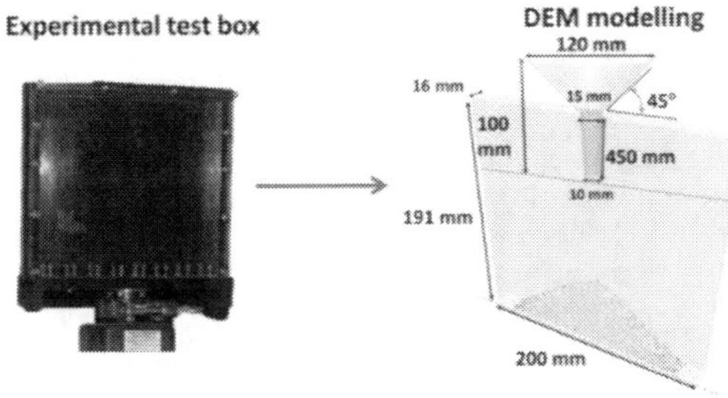

Fig. 1 : Image of the geometry of the test box used in experiment and modelling.

2.1 Particles physical and mechanical properties

Three different particles representing the common ingredients of home washing powders, namely, spray dried detergent powders (commonly referred to as Blown Powder or BP), tetraacetylethylenediamine (TAED), and enzyme granules (placebo used here) are utilized to form a ternary mixture. Particles are in different sieve-cut size ranges based on the mode of their size distribution. Table 1 shows the particles properties such as size, density, shear modulus, coefficients of restitution (COR) and sliding friction (COF) measured experimentally as described in [11, 12].

Table 1. Specifications of the modelling and the materials properties.

Material type	BP	TAED	Enzyme placebo	Perspex
Size (μm)	500-450	850-1000	600-700	----
Particles number	633597	4667	1401	-----
Total mass (g)	28.53	1.71	0.61	-----
Weight Percentage	92	6	2	-----
Particle shape	5-sphere	5-sphere	spherical	Wall
Repose angle	40	36	31	-----
Shear modulus (MPa)	100	100	100	1000
Density (kg/m^3)	780	850	2320	1180
Coefficient of rolling friction	0.10	0.01	0.05	0.01
Poisson's ratio	0.25	0.25	0.25	0.25
COF (BP)	0.62	0.69	0.70	0.42
COF (TAED)	0.69	0.75	0.75	0.36
COF (Placebo)	0.70	0.75	0.75	0.75
COR (BP)	0.20	0.30	0.20	0.28
COR (TAED)	0.30	0.32	0.20	0.32
COR (Placebo)	0.20	0.20	0.10	0.20

2.2 Particles shape

Particles shapes are generated using the clumped-sphere technique and based on the geometries obtained by X-ray tomograms (XRT). ASG software [13] is utilized to generate particle shape by clumping an optimum number of spheres of different sizes [11] as shown in Fig. 2. Generated particles shapes are then used in DEM accompanied with the physical properties of the particles given in section 2.1.

Fig. 2: The clumped spheres representing the real shapes of TAED and placebo particles.

2.3 Numerical modelling and contact models

EDEM 2.7.1 software, provided by DEM Solutions, Edinburgh, UK, is used to model the segregation process. The models used for particles contacts are Hertz-Mindlin [14-16] and JKR [14] by which the effects of collisions and cohesion/adhesion are taken into consideration. Hertz-Mindlin theory is used for calculating the normal and tangential forces during the collision between two particles; where Hertz model [15] calculates the normal impact and Mindlin theory [16] accounts for the tangential force. The details of these models are available elsewhere [17, 18].

Bonding forces in coated materials roots in both surface tension and viscous damping forces. By using the particles surface energy in the JKR model [14] the surface tension forces are taken into consideration and to account for viscous dissipation, low restitution coefficients are selected. To predict a valid surface energy for the particles in JKR model, a dimensionless number, i.e. Cohesion number [19], is used by which the surface energy is predicted for the current system.

$$Coh = \frac{1}{\rho g}\left(\frac{\Gamma^5}{E^{*2} R^{*8}}\right)^{1/3} \qquad (1)$$

Lower modulus of elasticity is selected in DEM modelling with the aim of having less computational effort but the cohesion number is kept constant by changing the surface energy. The rest of the parameters in the cohesion number are the same as those of the experiment. Enzyme placebo particles are coated by different mass ratios of PEG 400 and then the repose angle for each coating level is measured experimentally. The equivalent surface energy to obtain the same angle of repose in the modelling is then found by back-simulations (Fig. 3).

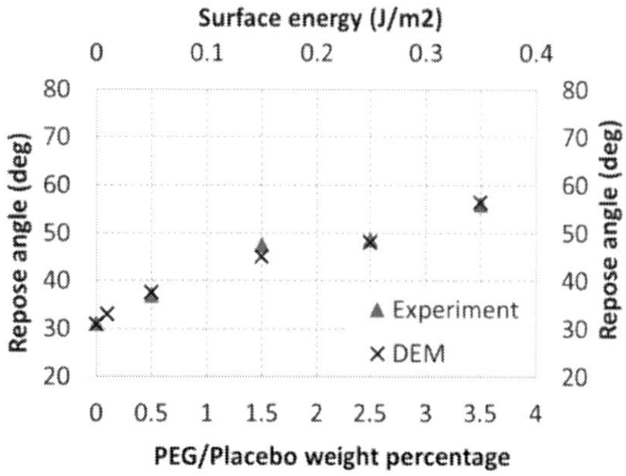

Fig. 3 : Calibration of particles surface energy versus their levels of coating using the angle of repose.

3. Results

Once all the particles are characterised and the surface energy of the coated placebo granules is calibrated, particles are mixed (Table 1) and introduced into the test box (Fig. 1) to form the heap. In the first test, particles are uncoated, i.e. they are free flowing, so their surface energy is set equal to zero. In this situation, although enzyme placebo granules are rounded and relatively larger than BP, they accumulate in the central zone of the heap due to their higher density, which is observable from the front view of both experiment and DEM simulation (Fig. 4).

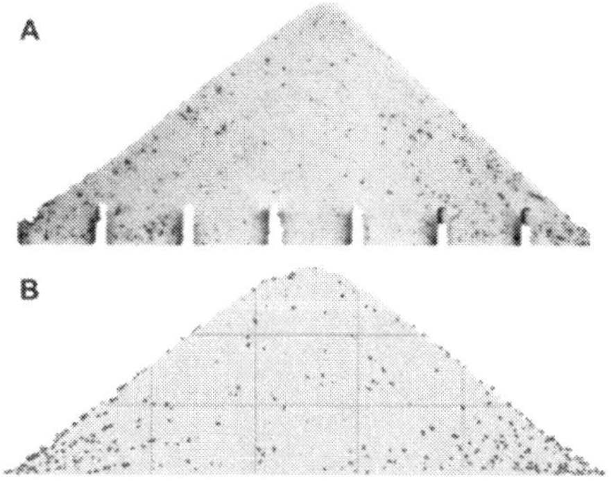

Fig. 4 : Heap formation of ternary mixture of BP (white), TAED (blue), and uncoated enzyme placebo (red) particles in A) experiment, B) DEM, and C) DEM-front thin layer when BP is removed.

In Fig. 4 (C) a thin layer from the front side of the heap (from DEM simulation) is displayed and the BP particles are removed to further analyse the mixture. This figure indicates the segregation of TAED particles in the corners of the heap and accumulation of the placebo granules in the centre which is in agreement with the experiment. Nevertheless, the experiment does not give a full picture of the mixture as the segregation pattern through the depth of the heap is not visible.

In the second test, the enzyme placebo granules are coated by 2.5 wt.% of PEG 400, mixed with BP and TAED and introduced into the test box to form the heap. Using the calibration study, shown in Fig. 3, and the cohesion number the equivalent interfacial surface energy values for placebo-BP, placebo-TAED, and placebo-placebo are predicted to be 0.037, 0.06, and 0.25 Jm^{-2}, respectively.

It is observed from Fig. 5 that the coated enzyme placebo particles do not accumulate in the centre anymore and the particles are distributed all over the heap leading to less segregation. Nonetheless, the distribution pattern of TAED particles is still the same as before meaning that the coated minor ingredients in the mixture has not affected the general behaviour and flowability of the mixture. This is also understood from the very similar repose angles of the uncoated (36°) and coated (37.6°) systems showing a negligible reduction in the mixture flowability after the coating.

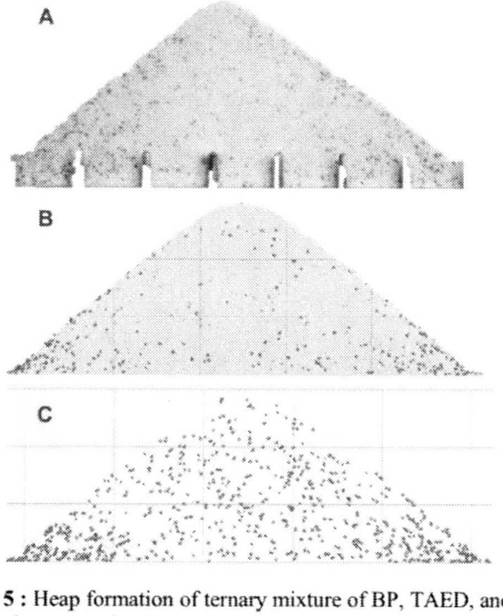

Fig. 5 : Heap formation of ternary mixture of BP, TAED, and coated enzyme placebo particles in A) experiment, B) DEM, and C) DEM-front thin layer when BP is removed.

To analyse the segregation quantitatively, the whole heap is divided into 21 bins including 7, 3, and 1 sections in width, height, and depth of the heap, respectively, as shown in Fig. 6. The Coefficient of Variations (COV) of the TAED and enzyme placebo particles are calculated based on their mass concentrations in each bin. It is observed that the COV of TAED particles is 0.76 for uncoated and 0.74 for coated systems which are very close. This result was expected as the coating has been done only on the enzyme placebo granules and the mixing pattern of the entire system is not affected by coating the minor components, i.e. enzyme placebo granules. On the other hand, a significant enhancement in distribution pattern of the enzyme placebo granules is observed after coating. The COV of the placebo granules before coating was 0.46, while it is reduced to 0.29 after coating, i.e. nearly 37% reduction in segregation index.

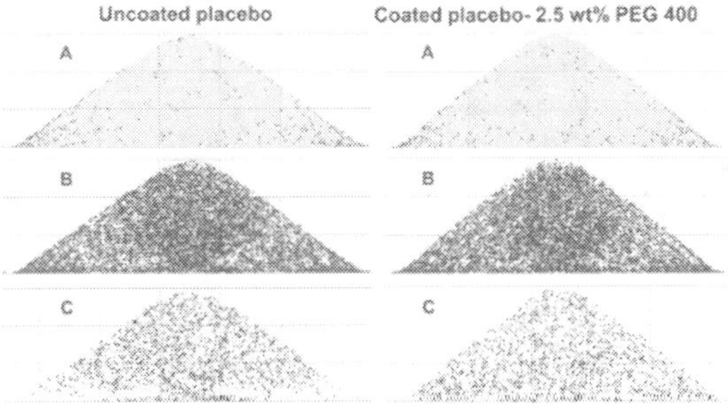

Fig. 6. Particles distribution before and after coating. Figure (A) the full image of the heap from the front, (B) the heap displaying TEAD and enzyme placebo particles, and (C) the heap displaying the distribution of enzyme placebo.

4. Conclusion

The segregation tendency of a ternary mixture of particles differing in shape, size, and density has been analysed for the case in which one component, being present as a minor content, readily segregates. To mitigate the extent of segregation this component has been made sticky by coating with a thin liquid layer before mixing with the other components. As a model system, the main ingredients of home washing powder (i.e. BP, TAED, and enzyme granules, the latter in placebo form) have been used and the segregation during the heap formation has been measured experimentally and analysed by numerical simulation using DEM. It is observed that uncoated enzyme placebo particles (as the minor component) segregate in the central area of the heap due to their high density; whereas, the coated particles are well distributed through the heap with 37% reduction in COV. At the same time, TAED particles segregate in the corners of the heap mainly because of their large size compared to the BP particles. It is observed that the distribution pattern of TAED does not change after coating of the enzyme placebos. Furthermore, comparing the repose angles of the mixture before and after coating shows a negligible reduction in flowability of the mixture after coating. These trends can be readily predicted by numerical simulation by DEM taking account of measured particle properties such as shape, density, size, and adhesion.

The authors would like to thank the Advanced Manufacturing Supply Chain Initiative (AMSCI) for their financial support. In addition, we would like to thank Claire Duckitt (Procter and Gamble) for project coordination.

References

1. J. C. Williams, Powder Technol **15** (2), 245-251 (1976)
2. J. Ottino and D. Khakhar, Annu Rev Fluid Mech **32**, 55-91 (2000)
3. E. H. J. Kim, X. D. Chen and D. Pearce, Colloids and Surfaces B: Biointerfaces **46** (3), 182-187 (2005)
4. A.-S. Persson, G. Alderborn and G. Frenning, Eur J Pharm Sci **42** (3), 199-209 (2011)
5. A. Spillmann, A. Sonnenfeld and P. Rudolf von Rohr, Plasma Processes and Polymers **5** (8), 753-758 (2008)
6. J. J. Nijdam and T. A. G. Langrish, J Food Eng 77 (4), 919-925 (2006)
7. H. Li and J. J. McCarthy, Phys Rev Lett **90** (18), 184301 (2003)
8. P. Begat, D. A. V. Morton, J. N. Staniforth and R. Price, Pharmaceutical Research **21** (9), 1591-1597 (2004)
9. W. H. Hager, J Hydraul Res **50** (1), 3-9 (2012)
10. A. Hassanpour, M. Eggert and M. Ghadiri, presented at the Powders and Grains 2009, Golden Colorado, USA, 2009 (unpublished)
11. M. A. Behjani, A. Hassanpour, M. Ghadiri and A. Bayly, Powder Technol (To be published)
12. M. Alizadeh Behjani, A. Hassanpour, M. Ghadiri and A. Bayly, presented at the International Congress on Particle Technology-PARTEC2016, Nuremberg, Germany, 2016 (unpublished)
13. M. Price, V. Murariu and G. Morrison, Proceedings of Discrete Element Modelling 2007 (2007)
14. K. L. Johnson, K. Kendall and A. D. Roberts, Proceedings of the Royal Society of London A: Mathematical, Physical and Engineering Sciences **324** (1558), 301-313 (1971)
15. H. Hertz, in Journal für die reine und angewandte Mathematik (Crelle's Journal) (1882), Vol. 1882, pp. 156–171
16. H. Deresiewicz, R. D. Mindlin, U. Columbia and E. Department of Civil, Elastic spheres in contact under varying oblique forces. (1952)
17. A. Di Renzo and F. P. Di Maio, Chemical Engineering Science **59** (3), 525-541 (2004)
18. M. M. Martín, *Introduction to software for chemical engineers* (CRC Press, 2014)
19. M. A. Behjani, N. Rahmanian and A. Hassanpour, Adv Powder Technol (2017)

Observation of microstructure of silty sand obtained from gel-push sampler and reconstituted sample

Muhamad Yusa[1,*], E.T *Bowman*[2], and Misko *Cubrinovski*[3]

[1]Civil Engineering Department, University of Riau, Indonesia
[2]Civil and Structural Engineering, University of Sheffield, United Kingdom
[3]Civil and Natural Resources Engineering, University of Canterbury, New Zealand

Abstract. Observation of microstructure study of natural sand i.e. clean sand with fines (particles adjudged to be smaller than 75μm) content < 5% (Gel Push A) and silty sand with 35% fine content (Gel Push B) obtained by gel-push sampling was described. In addition, some observations from reconstituted samples prepared by dry pluviation and moist tamping were presented. Microstructures were investigated statistically by measuring particle orientation. It was evidence that natural sand (either gel push A and B) have a preferred orientation i.e. horizontally oriented. Similar particle orientation trend were observed by dry pluviated sample. Undisturbed and dry pluviated samples shows that they are anisotropic in terms of particles orientation. Moist tamped sample on the other hand, results in fairly random orientation with a slight bias towards vertical, thus does not replicate natural sand fabric.

1 Introduction

Mechanical behaviour of granular material is well known to be dependend on its microstructure.However, obtaining natural undisturbed granular soils is difficult, due to its non-cohesive nature, which results in limited studies on it. Furthermore, most granular materials occurring in nature are composed of a broad range of particle sizes [1], whereas the few studies on soil microstructure have focused on clean, usually uniform soils, such as Bowman & Soga[2]. Indeed, some insight was obtained from numerical model using discrete element method

This research describes observations of the microstructure of silty sands in an 'undisturbed state'. The 'undisturbed samples' were obtained via gel push sampler. In addition this study also investigates microstructural of reconstituted samples prepared by dry funnel pluviation [3] and moist tamping[4].

2 Gel Push Sampling

Gel push sampling was first developed in 1999 in Japan to obtain samples of sandy, gravelly soil and fractured rock. The method then progressively has been developed and modified by Kiso-Jiban Consultants and the Taiwan Construction Research Institute under the guidance of Prof. Ishihara [5].

The method utilizes high density polymer solution at very low velocity without circulation (preventing loss of fine particles) to reduce the friction between the cut sample and the sample tube. Before sampling, the space for containing the sample is filled with high density polymer solution .When the sample comes into the space, the high density polymer solution is pushed out and coats the Gel Push and minimizes the sample disturbance caused by the friction and the sample tube. Figure 1 illustrates the sampling process. The polymer coated on the sample is gelated and semi hardened the surface to protect the sample. The method was brought to New Zealand as part of University of Canterbury Research Project on liquefaction assessment of Christchurch [6, 7].

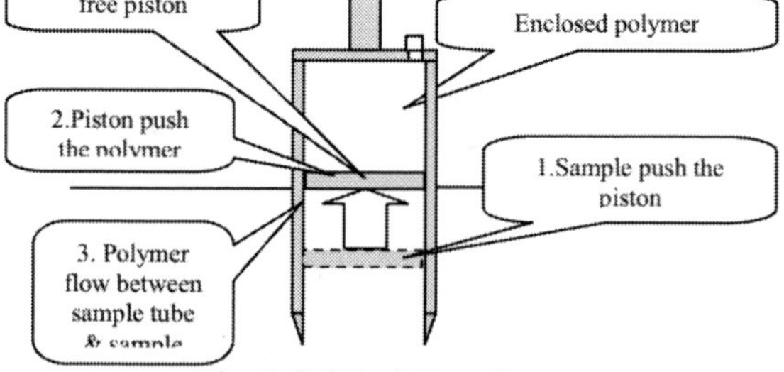

Fig. 1 Gel Push Sampler

3 Microstructure Measurement

In general there are two measurements used to quantify the microstructure changes i.e. particle orientation and

* Corresponding author: m.yusa@eng.unri.ac.id

distribution of spatial distance between particles. However due to limited space only the first parts is discussed herein. In addition, it is realized that two dimensional image analysis results in loss information of the pore structure and other attributes of three dimensional microstructure. However, this study is concentrated on retrieving some basic trends of microstructure of 'undisturbed' and reconstituted silty sands thus two dimensional microstructure measurements was deemed adequate.

The orientation of a non-spherical particle can be represented by the orientation of the long axis of the particle to a reference axis. The true three dimensional long axis of the particle is difficult to determine. Thus in this study particle orientation is defined by the angle of the apparent long axis of the projected particle in two dimensions (vertical plane in this study). The measure is represented by a rose diagram. Statistical parameters are then determined to summarize the distribution of particle orientation using Fisher distribution analysis [8].

Mean angle (α_m) can be determined as follows:

$$\alpha_m = \frac{1}{2}\tan^{-1}\left(\left\{\sum_{i=1}^{N}[l_i \sin(2\alpha_i)]\right\} / \left\{\sum_{i=1}^{N}[l_i \cos(2\alpha_i)]\right\}\right) \quad (1)$$

where l_i is the length of long axes of the particle, N is the number of measurements, α is the angle between the unit vector and the true direction. In addition, κ, was used to indicate degree of concentration and can be estimated from the equation below [9]

$$\kappa = \frac{N-1}{N-R} \quad (2)$$

where

$$R^2 = \sqrt{\left\{\sum_{i=1}^{N}[l_i \sin(2\alpha)]\right\}^2 + \left\{\sum_{i=1}^{N}[l_i \cos(2\alpha)]\right\}^2} \quad (3)$$

An increase in κ suggests a relative increase of isotropy in term of particle orientation. Rose diagrams and Fischer statistical analysis were drawn and calculated by the aid of Geo-Orient ver. 9.5.0 developed by Dr. Rod Holcombe [10].

4 Experimental Methodology

4.1 Material

Natural sands used in this study were ejected material from two sites around Christchurch City Centre i.e. Kilmore Street (Gel Push A) and Madras Street (Gel Push B) following the Canterbury Earthquake [11]. The particle size distributions (PSD) are shown by Figure 2. Gel Push A was relatively clean sand with 4 % fine particles (smaller than 75 micron), whereas Gel Push B was silty sand with 35% fines. The mean diameter of Gel Push A and B are 235 micron and 95 micron respectively.. Lab C is non-plastic silty sand with mean diameter of 150 micron and 15% fines. Lab C was prepared in order to make a comparison between the microstructure of natural undisturbed soil and reconstituted soil. Lab C was reconstituted by dry funnel pluviation (Lab C-DD) and mois tamping (Lab C-MT) to a relative density of 40% (loose).

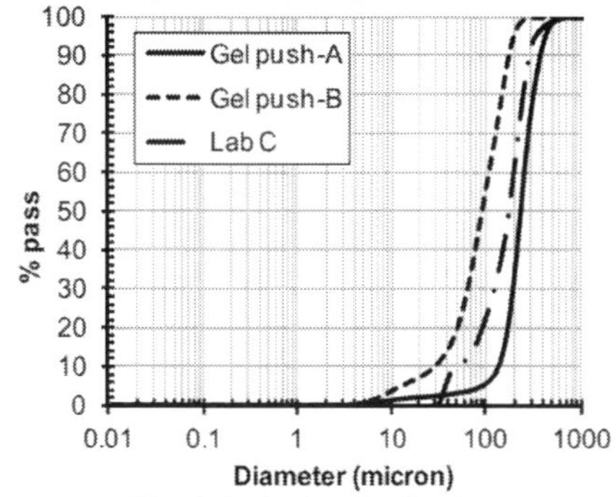

Fig. 2 Grain size distributions

4.2 Experimental methodology

The methodology used in this study is briefly described here. Detail of the methodology was explained in another paper [12].

4.2.1 Sample preservation

Considering that the natural soils used in this study involve irregular shape, wide particle distribution and fine particles, it was decided to adopt a direct method of preservation i.e. injection a low viscosity epoxy resin [13].

4.2.2 Sample preparation and loading

Gel Push-A and B samples were loaded one dimensionally (K_0 condition) to 80 kPa and 50 kPa vertical stress, respectively, which correspond to estimated field stress. For Lab C, the samples were loaded under a vertical stress of 125kPa, which was in accordance to the stress during triaxial K_0 consolidation with 60 kPa confining pressure.

4.2.3 Capturing image

Images were captured (at 45x magnification) using a scanning electron microscope via back scattered detection (Llyod, 1987). Material with higher atomic weight appears brighter compared to those with lower ones. This gives clear contrast between the grains and the epoxy filled voids.

4.2.4 Image processing

Captured images were processed and analyzed using image analysis software *ImagePro Plus v.7.0.* Steps involved in this process include contrast enhancement, correction for noise and thresholding. The resulting image is displayed as a binary image using black and white colors to distinguish foreground and background regions. Subsequent steps were filling the grain holes and watershed splitting to separate the grains.

5 Results

Examples of captured and processed images from gel push A, gel push B, lab C-DD and lab C-MT are presented in Figure 3 (a) to Figure 3(d) respectively. It is interesting to be noted from the figures, particularly for gel B and lab C, visible band of fines particles. For gel push B the silt band is visible running from the left edge to the top. This suggests that segregation between fines and larger particle is something natural in the field. Reconstituted lab C-DD shows band of silt running from the bottom to the right edge which indicates that segregation is encountered in dry pluviated sample. While lab C-MT shows no clear segregation. The statistical analysis involves approximately 1800, 18000 and 9000 particles for Gel Push A, gel Push B and lab C, respectively.

One factor that influences the final orientation of particles settling under gravity is the particle aspect ratio [14] .The particle aspect ratio exhibits quantitative influence on particle orientation, contact normal and sliding contact normal[15-17]. Figure 4 shows normalized frequency (divided frequency in each bin by total number of particles) distribution histograms of aspect ratio are shown by The calculations did not include particles touching the perimeter of the image because that would introduce s into the results. It can be seen that most particles (approximately 70%) for both Gel Push A and B have aspect ratio between 1.5 and 2, showing that many particles are elongated. A similar aspect ratio distribution was found for Lab C.

Table 1 shows result of circular statistical analysis. Separate analysis for 'sand size' particles and 'fines'-sized particles were also performed in order to see anisotropy between large and small grains. It is realized that it is difficult to determine which particles truly are fines (less than 75 micron in diameter in a sieve analysis) due to the 2D slice through each of the particles, which may not occur diametrically. Effort was trialled to differentiate fines from sand (for Lab C) based on the material composition utilizing tools on SEM via back scattered detector. However, the fines (99.8% silica dioxide) and some of the sand were found to have a very similar atomic weight. Thus it is not possible to differentiate whether fines sized particles were truly fines or corners of sand particles. Regardless it can be determined that particles having a feret minimum equal to or larger than 75 micron are sand. Feret minimum is the smallest the distance between two parallel tangents of the particle at an arbitrary angle, and is often used as the diameter equivalent to a sieve analysis [18]. In this study, the rest of the particles are assumed to be fines. This results in higher fines percentages than those from particle size distribution e.g. 5.2% (gel push-A) and 26% (lab C) compared to 4%, and 15 % of those from grain size distribution.

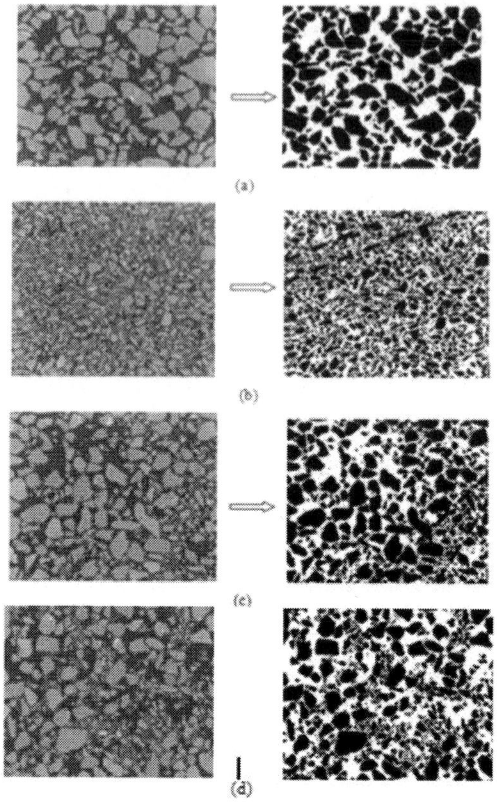

Fig. 3 Example of captured and processed images

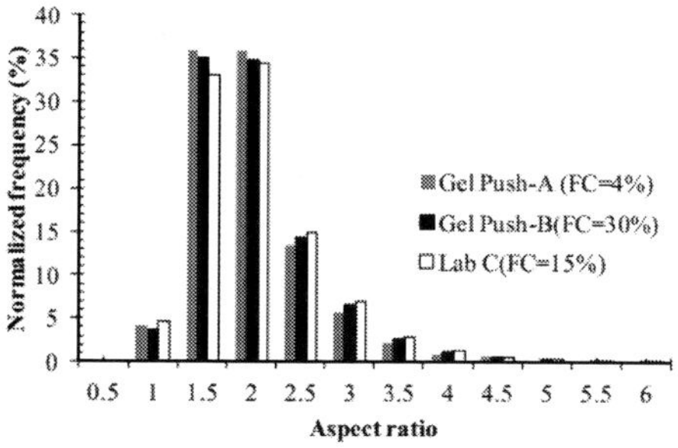

Fig. 4 Frequency distribution of aspect ratio

It can be seen from Table 1, Figure 5 and Figure 6 that the fabric of natural undisturbed sand (Gel Push A and B) has a preferred apparent orientation to the horizontal with mean angles of 2.2° and 0.9° to the horizontal axis. A similar trend was found for Lab C-DD, with a slightly larger mean angle i.e. 3.5° but still horizontally oriented. In addition, the Table also reveals that generally both sand size particles and fines from gel push and dry sample tend to be oriented in the same direction. Moreover, both gel push and reconstituted samples, in general, reveal that fines are less orientated than sand particles, as indicated by smaller κ values. On the other hand, figure 6 shows that moist tamped sample is more isotropic than dry pluviated samples. This is in accordance with κ values of moist tamped samples which is less than those of dry pluviated samples, indicating higher isotropy of moist tamped samples. Similar to gel push and dry pluviated sample, sand particles are more orientated than 'fines'. Furthermore, 'fines' of moist tamped samples are much less orientated than those of gel push and dry pluviated sample as suggested by lower values of κ values.

Table 1 Fischer statistical analysis results

Sample	α_m(°)*	κ
Gel Push A - All	87.8±1.2	0.53
Sand size	88.9±0.9	0.60
'Fines size'	81.9±3.0	0.33
Gel Push B - All	89.1±0.7	0.52
Sand size	87.3±0.9	0.80
'Fines size'	89.1±0.6	0.51
Lab C-DD - All	86.5±1.1	0.42
Sand size	86.5±0.8	0.65
'Fines size'	87.1±1.3	0.29
Lab C-MT - All	177.9±2.3	0.14
Sand size	0.1±1.8	0.30
'Fines size'	169.5±8.9	0.05

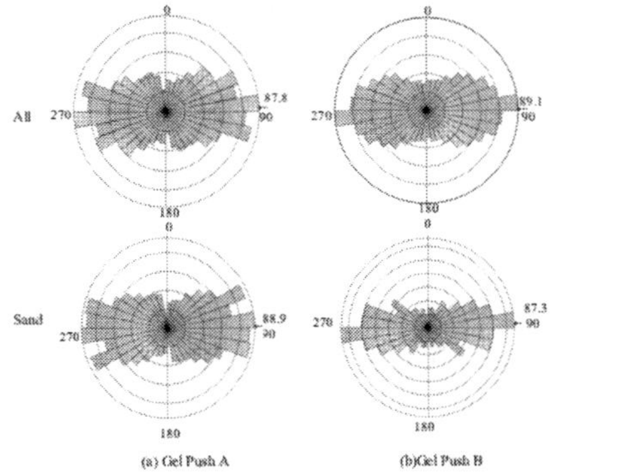

Fig. 5 Rose diagram based on gel-push sample

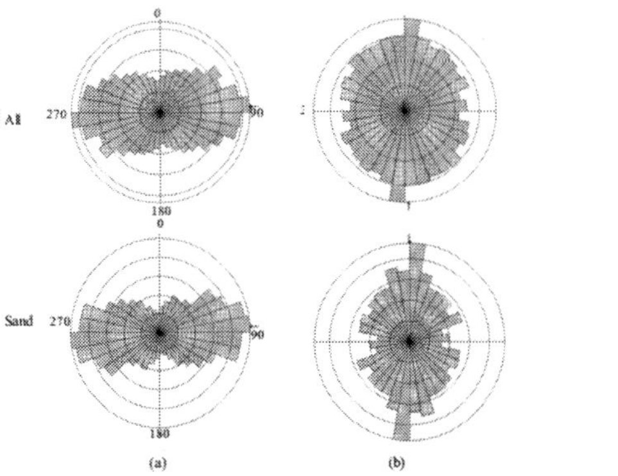

Fig. 6 Rose diagram based on reconstituted sample (a) dry pluviation (b) moist tamping

6 Conclusions

Method and observation of microstructure of sand containing fines from gel push samples and reconstituted sample have been described. Based on the circular statistical analysis, microstructure of natural sand and dry pluviated samples have similar preferred orientation which is horizontally oriented in this study. In addition both gel push sample sand and dry pluviated samples tend to be anisotropic in term of particle orientation. In contrast, moist tamping sample tend to be isotropic.This indicates dry funnel pluviation could be used to mimic microstructure of natural sand.

References

1. Mitchell, J.K. and K. Soga, *Fundamentals of soil behavior*. 3rd ed. (2005), New York: Wiley.
2. Bowman, E.T. and K. Soga, SOILS FOUND, (2003). **43**(4): p. 107-117.
3. Ishihara, K., GEOTECHNIQUE, (1993). **63**: p. 351 - 415.
4. Ladd, R., GEOTECH TEST J, (1978). **1**(1): p. 16-23.
5. Huang, A.B., et al. *3rd International Conference on Site Characterization*. (2008). Taipei: Taylor &Francis.
6. Taylor, M.L., M. Cubrinovski, and I. Haycock. (2012). New Zealand Society of Earthquake Engineering:2012 Annual Technical Conference.
7. Stringer, M.E., M. Cubrinovski, and I. Haycock. *ISC 5*. (2016). Gold Coast Australia.
8. Fisher, N.I., *Statistical Analysis of Circular Data*. (1993): Cambridge University Press.
9. Butler, R.F., *Magnetic Domains to Geologic Terranes.*, 1992.
10. Holcombe, R. *Structural Geology - Mapping/GIS Software*. n.d. [cited 2013 3 January]; Available from: http://www.holcombe.net.au/software/rodh_software_georient.htm.
11. Cubrinovski, M. *7th International Conference on Case Histories in Geotechnical Engineering*. (2013). Chicago, Illnois-USA.
12. Yusa, M. and E.T. Bowman. *Powder and Grain 2013*. (2013). Sydney.
13. Palmer, S.N. and M.E. Barton, J. SEDIMENT RES, (1986). **56**: p. 556-557.
14. Willmarth, W.W., N.E. Hawk, and R.L. Harvey, The Physic of Fluid, (1964). **7**(2).
15. Jiang, M.J., J.D. Liu, and M. Arroyo, International Journal for Numerical and Analytical Methods in Geomechanics, (2016). **40**(18): p. 2468-2488.
16. Jiang, M.J., T. Li, and Z.F. Shen, Granular Matter, (2016). **18**(3): p. 1-15.
17. Jiang, M.J., T. Li, and Z.F. Shen, Granular Matter, (2016). **18**(3): p. 1-14.
18. Sympatech GmBH. 2013 [cited 2013 February 12]; Available from: http://www.sympatec.com/EN/Science/Characterisation/05_ParticleShape.html.

DOI: 10.1051/epjconf/201714006025

BlazeDEM3D-GPU A Large Scale DEM simulation code for GPUs

Nicolin Govender[1,3,4,5,⋆], *Daniel* Wilke[2,⋆⋆], *Patrick* Pizette[3,4,⋆⋆⋆], and *Johannes* Khinast[1]

[1] *Research Center Pharmaceutical Engineering, GmbH, Graz , Austria*
[2] *Department of Mechanical and Aeronautical Engineering, University of Pretoria, South Africa*
[3] *Mines-Douai, LGCgE GCE, F-59508 Douai, France*
[4] *Université de Lille, 59650 Lille, France*
[5] *Department of Mechanical Engineering , University of Johannesburg, South Africa*

Abstract. Accurately predicting the dynamics of particulate materials is of importance to numerous scientific and industrial areas with applications ranging across particle scales from powder flow to ore crushing. Computational discrete element simulations is a viable option to aid in the understanding of particulate dynamics and design of devices such as mixers, silos and ball mills, as laboratory scale tests comes at a significant cost. However, the computational time required to simulate an industrial scale simulation which consists of tens of millions of particles can take months to complete on large CPU clusters, making the Discrete Element Method (DEM) unfeasible for industrial applications. Simulations are therefore typically restricted to tens of thousands of particles with highly detailed particle shapes or a few million of particles with often oversimplified particle shapes. However, a number of applications require accurate representation of the particle shape to capture the macroscopic behaviour of the particulate system. In this paper we give an overview of the recent extensions to the open source GPU based DEM code, BlazeDEM3D-GPU, that can simulate millions of polyhedra and tens of millions of spheres on a desktop computer with a single or multiple GPUs.

1 Introduction

The discrete element method (DEM) has become the *de facto* standard to simulate particulate materials for which the discrete nature of particles and importance of quantifying particle-scale effects cannot be neglected. Consequently, the need for large scale discrete element simulations that can quantify the effect of particle-scale changes on the macro-scale response is becoming increasingly more important in industrial bulk material handling applications, where computational efficient continuum models fail to give reliable predictions [1].

Established discrete element software frameworks rely mostly on multi-core central processing unit (CPUs) computing platforms [2]. This is exemplified by the number of commercial and open-source discrete element codes that include PFC by Itasca , EDEM by DEM Solutions, Star-CCM+ by CDAdapco, MecuryDPM by the Univeristy of Twente, BLOCKS3D by the University of Illinois, Y-DEM by Queen Mary University, Millsoft [3] by the University of Utah , LIGGGHTS by Johannes Kepler University in Linz, ELLIPSE3H by New Mexico University, YADE by Grenoble University and SAMADII/DEM by MetaRiver Technology.

Although the GPU (Graphics Processor Unit) is an ideal match for DEM simulations due to the highly threaded nature of the GPU, the technology is fairly new compared to the traditional CPU platform and the associated learning curve for GPU development high. Furthermore only an efficiently implemented GPU DEM solution will yield significant performance benefits over the CPU. Rocky by Granular Dynamics, XPS by RCPE Austria [4], and BLAZE-DEM3D-GPU (opensource) [5] are the only packages capable of efficiently modelling large scale simulations on the GPU. The parallelization benefits of the graphical processing unit (GPU) [6–10] is becoming increasingly more important as an alternative computational platform for discrete element simulations [11]. The GPU seems to be an essential computing platform to scale simulations to millions of particles within in a realistic computing time frame and financial budget. Although the GPU has numerous advantages it also presents various challenges to implementations of the discrete element method. The complexity of resolving contact for non-trivial particle shapes like polyhedra poses particular challenges on the GPU due to the divergent nature of polyhedral contacts. This is a direct result of the limited computational ability and memory restrictions on the GPU and consequently the majority of the effort has been directed to spherical particles on the GPU [2, 9, 10, 12].

⋆e-mail: govender.nicolin@gmail.com
⋆⋆e-mail: wilkedn@gmail.com
⋆⋆⋆e-mail: patrick.pizette@mines-douai.fr

2 BlazeDEM3D-GPU

BlazeDEM3D-GPU that has been demonstrated to simulate (i) tens of millions of mono-disperse spherical particles and (ii) millions of mono-disperse polyhedral particles within a realistic time frame on a desktop computer using a single or multiple GPUs [6][13][14]. BlazeDEM3D-GPU is developed for NVIDIA graphics cards using the available CUDA programming framework [15], allowing for a wide variety of GPU cards to be considered. BlazeDEM3D-GPU [13] has been extensively validated in the simulation of tumbling mills [14] and hoppers [16]. BlazeDEM3D-GPU was designed for the GPU hence the code design is similar to a dynamics engine, so the user is free to build a complex simulation using the various objects provided. This is in contrast to the fixed simulation types offered by commercial codes like Rocky and EDEM, where you can only do a "single" simulation eg ball mills, chutes, hoppers etc as standalone simulations. With BlazeDEM3D-GPU you can simulate a complete process e.g. a belt feeder that feeds into a silo that discharges through a cute into a ball mill. This allows a typical engineer to perform highly specialized simulations.

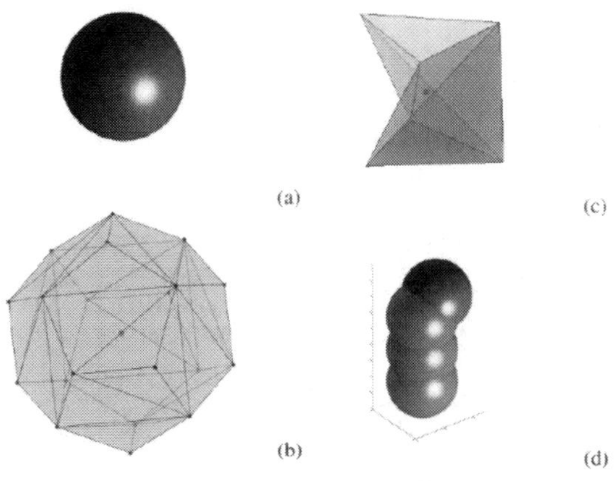

Figure 1. (a) Spherical and (b) native polyhedral shaped particles with angularity (c) native non-convex polyhedral shaped particles are supported by BlazeDEM3D-GPU, as opposed to (d) composite spherical particle approaches that neglect the angularity of particles.

2.1 Supported Particle Types

The BlazeDEM3D-GPU framework allows for efficient representation and computation for poly-disperse systems for both spherical and polyhedral shaped particles. Numerous particles are currently supported by BlazeDEM3D-GPU that include spheres, convex polyhedra and non-convex polyhedra as depicted in Figure 1(a)-(c). The native supported polyhedral shaped particles include angularity as opposed to popular clumped sphere approaches, as shown in Figure 1(d), that neglect particle angularity. The importance of particle angularity for industrial applications has been investigated a number of times [1, 17].

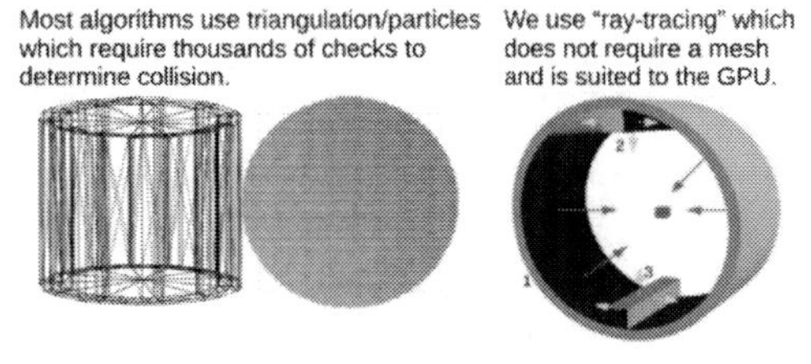

Figure 2. Two world geometry representation strategies.

2.2 World Object Representation

BlazeDEM3D-GPU efficiently represents world objects e.g. explicit representations of planar world objects as depicted in Figure 2.1(c) instead of conventional triangulation or particle representations of world objects as depicted in Figure 2(a) and Figure2(b) [10, 12, 18]. This minimalist representation allows for particle-world contact detection and resolution that is memory and computationally efficient and scalable with particles on the GPU.

While these conventional representations allow for generality in that any surface can be decomposed into particles or triangles it is computationally expensive and requires significant memory. Since we aim to utilize primarily the 48 KB of constant memory to get the most efficient algorithm on the GPU we use a minimalistic surface representation for first order surfaces which requires just the storage of the vertices of the surface. Note that any number of vertices can make up a surface provided they are coplanar. While a single surface can still be represented by a collection of smaller surfaces/triangles we use geometric primitives which we term "macro objects" for higher order surfaces such as cylinders which results in significantly reduced memory storage and computational cost. This approach of minimalistic world geometry representation is now also used in the CPU based open-source code MecuryDPM for spherical particles.

The BlazeDEM3D-GPU framework is well documented in [6, 13, 14] and is freely available under the modified BSD-3 license [5].

2.3 Particle-World Contact Detection

The framework considers particle objects that interact with each other and their surroundings described by world objects. The world objects are usually decomposed as volume or surface objects. The method used to determine the contact points between a particle and surface is termed a "ray-tracing approach" [6, 13]. Since convex polyhedra are represented as just a collection of surfaces in BlazeDEM3D-GPU, the same approach is used [13] to determine contact between polyhedra. This approach requires basic algebraic operations and is not iterative like the common-plane approach and hence well suited to the GPU. The details on the particle-world algorithms can be found in [13, 14], here we just provide a broad overview of the detailed particle-particle interaction.

2.4 Polyhedron Contact Detection

The initial BlazeDEM3D-GPU architecture resolved contact purely based on penetration depth [6, 13]. The contact strain energy therefore varied linearly with penetration depth independent of the volume of material involved during contact. This assumption proved to be not overly restrictive for loosely packed particulate systems but proved unstable for closely packed systems. This simplistic contact model also exhibited inconsistent and sometimes unstable contact resolution when contact was resolved close to the edges and vertices. This sparked the inclusion of a volume based contact model. In 2016, BlazeDEM3D-GPU included a volume-based contact model that exactly resolves the contact volume and contact normals as depicted in Figure 3(a)-(b). The contact volume between two convex particles in contact is also convex. Firstly, all the contact points are established. Thereafter an efficient in-house developed GPU-based convex-hull is used to resolve the vertices and faces of the contact volume. Finally, the contact volume is established and computed, while the contact normals are resolved by analytically integrating the contact surface associated with a particle.

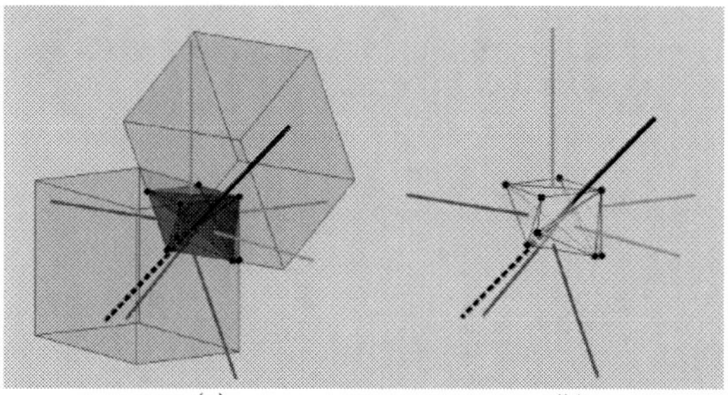

Figure 3. BlazeDEM3D-GPU initially resolved contact based on a penetration depth but recently included a (a) volume-based contact resolution in which the (b) contact volume and normals are exactly resolved enhancing the contact stability of polyhedral shaped particles. Red and green lines indicate face normals and the black solid and dashed lines indicate contact normals.

2.5 Multi-thread GPU collision detection design

Collision detection is computationally the most expensive part of a DEM simulation accounting for as much as 90% of the total simulation time for polyhedral shaped particles. Thus importance is placed on implementing our algorithms so that the take advantage of the parallel nature of the GPU. However regardless of how computational efficiently we implement an algorithm, if we do not utilize memory correctly and minimize transactions then the actual number of threads executing in parallel will be significantly reduced due to register pressure [19]. This is termed the occupancy of a particular method (kernel) executing on the GPU.

To increase the parallel efficiency of collision detection we split the detailed phase into three independent tasks which can be executed concurrently for spheres. Spheres only require a Neighbour Search while polyhedra require the Neighbour Search followed by a detailed particle contact search. Note that typically the particle-particle contact takes up most of the time. We also split the particle-world contact between surfaces and macro objects as different kernels to minimize divergence and reduce the number of registers used by the kernel.

Currently all tasks are run simultaneously on different streams on each GPU.

3 Multi-GPU Simulations

BlazeDEM3D-GPU is able to use multiple GPUs on the same node with an efficiency of 95 % for 2 GPUS and 90 % for 4 GPUS for polyhedra. This scaling is achieved due to our algorithms hiding most of the data transfer cost for polyhedra by overlapping it with compute steps. For spherical particles there a significantly less computational demand thus hiding the data-transfer as efficiently is not possible. While we get an efficiency of 90% for two GPUs the scaling to 3 is poor for less than 10 million particles. It should be noted that the scaling results is for an arbitrary particle distribution, for silos or chutes for example the domain can be decomposed, thus requiring only particles at the interface" to be transferred. In this case scaling is almost linear with the number of GPUs provided the problem size increases to ensure the compute loads are balanced.

4 Particle Shape Simulation Results

The spatial and velocity flow profiles is depicted in Figure 4 for polyhedra and spherical particles for a 0° hopper configuration. Consider the velocity field at 2.0s, 2.5s and 3.0s for the polyhedra. It is noteworthy that at 2.0 s the effective funnel discharge width is larger than after 3.0 s when the velocity profiles at the orifice are considered. In addition, the flow extends further to the walls up until 2.0s. After 3.0s the flow is more localized. This clearly demonstrates that particle shape needs to be taken into account.

To demonstrate the importance of large scale simulations we simulated a SAG mill with 35% fill using three models. Firstly, as a 2D model using N=6744 particles. The remaining two models were 3D models, a 10% slice of the length of the mill using N=385 534 particles and the full mill using $N=4 \times 10^6$ particles. The difference in response between the 2D and 3D models is evident resulting in an underestimation of the power draw for the simplified 2D model. We obtained a good match with an error of only 5 % for the power draw for the two 3D models further demonstrating the predictive capability of GPU DEM.

The simulations in this paper were done using the BlazeDEM3D-GPU code developed by the authors on a

 DOI: 10.1051/epjconf/201714006025

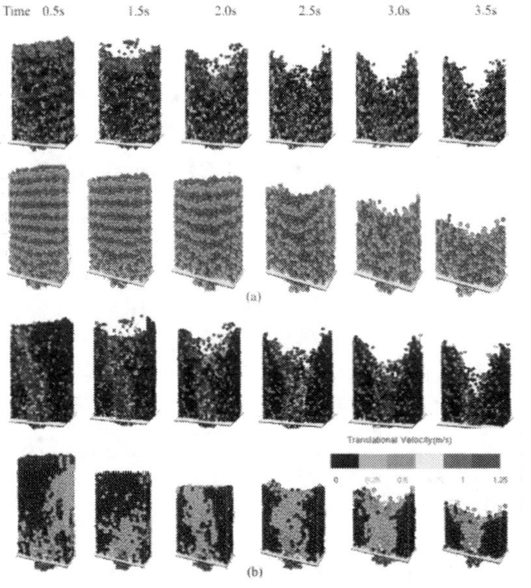

Figure 4. (a) Flow patterns and (b) velocity field for mono-disperse cubic shaped and spherical particles discharging from the 0° hopper configuration at six different time steps.

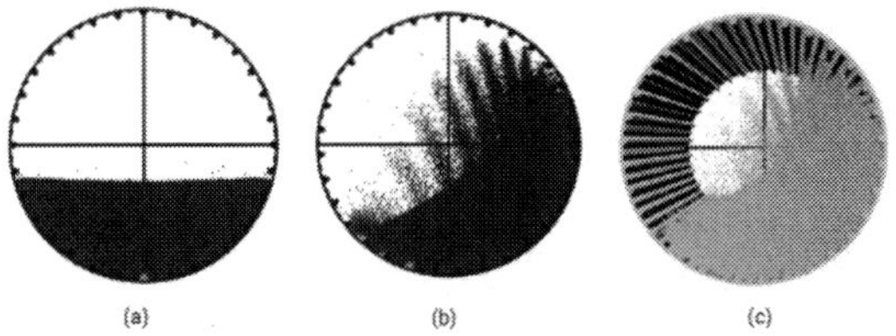

Figure 5. Ball mill steady state simulation with 35% filling modelled as a (a) 2D simulation using N=6744 particles, (b) 10% slice of the length using N=385 534 particles and (c) the full mill using N=4 × 10^6 particles, all views are orthogonal.

commodity NVIDIA GTX 980Ti graphics card. We required seven hours of wall-time and (900 MB of GPU memory) for a one second simulation time of over a million polyhedral shaped particles in a tightly packed state (before discharge) compared to 18 months using BLOCKS3D and 8.5 days using an impulse-based DEM approach [20] while using a more accurate contact detection resolution. A simulation of over a million spherical particles required just 3 minutes of wall time (300 MB of GPU memory) for one second simulation time.

5 Conclusion

In this paper the effect of particle shape and number on the simulation response was investigated. The need for large scale simulations and the value of our GPU DEM implementation to perform large scale simulations, within realistic time frames, was demonstrated with a 4 million particle simulation.

Acknowledgements

The support of NVIDIA with the donation of GPUs used in this paper is greatly appreciated.

References

[1] Y. Wang, J.Y. Ooi, Procedia Engineering **102**, 765 (2015)
[2] Shigeto, Sakai, Particuology **9**, 389 (2011)
[3] R. Venugopal, R. Rajamani, Powder Technology **115**, 157 (2001)
[4] C. Radeke, B. Glasser, J. Khinast, Chemical Engineering Science **65**, 6435 (2010)
[5] N. Govender, D.N. Wilke, S. Kok, SoftwareX (2016)
[6] N. Govender, D. Wilke, S. Kok, R. Els, JCAM **270**, 63 (2014)
[7] Zhang, et.al., Advances in Engineering Software **60**, 70 (2013)
[8] R. Rajamani, S. Callahan, J. Schreiner, *DEM Simulation of mill charge in 3D via GPU computing* (Proceeding of the SAG conference, Vancouver, 2011)
[9] J. Xu, et.al., Particuology **9**, 446 (2011)
[10] M. Hromnik, *Masters Thesis: A GPGPU implementation of the discrete element method applied to modeling the dynamic particulate environment inside a tumbling mill* (University of Cape Town, www.uct.ac.za, 2013)
[11] J. Gan, Z. Zhou, A. Yu, Powder Technology **301**, 1172 (2016)
[12] J. Longmore, P. Marais, M. Kuttel, Powder Technology **235**, 983 (2013)
[13] N. Govender, D. Wilke, S. Kok, Journal of Applied Mathematics and Computation **http://dx.doi.org/10.1016/j.amc.2014.10.013** (2014)
[14] N. Govender, R. Rajamani, D. Wilke, S. Kok, Journal of minerals engineering **http://dx.doi.org/10.1016/j.mineng.2015.05.010** (2015)
[15] J. Sanders, E. Kandrot, *CUDA by example*, Vol. 12 (2010)
[16] N. Govender, P. Pizette, D.N. Wilke, N. edine Abriak, *Validation of the GPU based BLAZE-DEM framework for hopper discharge*, in *Proceedings of IV International Conference on Particle-Based Methods* (2015)
[17] P.A. Langston, A.J. Matchett, F.Y. Fraige, J. Dodds, Granular Matter **11**, 99 (2009)
[18] H. Abou-Chakra, J. Baxter, U. Tuzun, Advanced Powder Technology **15**, 63 (2004)
[19] NVIDIA, *Cuda 6* (2014), `http://www.nvidia.com/cuda`
[20] S. Jaelee, *Developments in large scale discrete element with polyhedral particles simulations*

Systematic description of the effect of particle shape on the strength properties of granular media

Emilien Azéma[1,⋆], *Nicolas* Estrada[3,⋆⋆], *Itthichai* Preechawuttipong[2,⋆⋆⋆], *Jean-Yves* Delenne[4,⋆⋆⋆⋆], and *Farhang* Radjai[5,†]

[1] *LMGC, Université de Montpellier, Montpellier, France*
[2] *Departamento de Ingeniería Civil y Ambiental, Universidad de Los Andes, Bogota, Colombia*
[3] *Department of Mechanical Engineering, Faculty of Engineering, Chiang Mai University, 239 Huay Kaew Rd., Chiang Mai 50200, Thailand*
[4] *INRA, UMR IATE Montpellier*
[5] *LMGC, Université de Montpellier, Montpellier, France / $\langle MSE \rangle^2$, UMI 3466 CNRS-MIT, MIT Energy Initiative, 77 Massachusetts Avenue, Cambridge 02139, USA*

Abstract. In this paper, we explore numerically the effect of particle shape on the mechanical behavior of sheared granular packings. In the framework of the Contact Dynamic (CD) Method, we model angular shape as irregular polyhedral particles, non-convex shape as regular aggregates of four overlapping spheres, elongated shape as rounded cap rectangles and platy shape as square-plates. Binary granular mixture consisting of disks and elongated particles are also considered. For each above situations, the number of face of polyhedral particles, the overlap of spheres, the aspect ratio of elongated and platy particles, are systematically varied from spheres to very angular, non-convex, elongated and platy shapes. The level of homogeneity of binary mixture varies from homogenous packing to fully segregated packings. Our numerical results suggest that the effects of shape parameters are nonlinear and counterintuitive. We show that the shear strength increases as shape deviate from spherical shape. But, for angular shapes it first increases up to a maximum value and then saturates to a constant value as the particles become more angular. For mixture of two shapes, the strength increases with respect of the increase of the proportion of elongated particles, but surprisingly it is independent with the level of homogeneity of the mixture. A detailed analysis of the contact network topology, evidence that various contact types contribute differently to stress transmission at the micro-scale.

1 Introduction

In many particulate materials found in nature and industry, scientists and engineers need to quantify the effect of complex particle shapes. This is of major importance in the context of civil engineering and powder technology, where most processes need to be optimized or revised following the dramatic degradation of natural resources. Non-convex particles can be found in metallurgical and sintered powders, angular shaped particles are common in rocks and soils, whereas elongated and platy particles may occur in pharmaceutical and food products. Mixture of shapes is also used in order to reinforce soils or in civil engineering structures.

The first works dealing with particles shape parameters are not new. Indeed authors such Wadell in 1932 [1] or Krumbein Pettijohn 1938 [2] are the first to have developed methods for characterizing the shape of the particles.

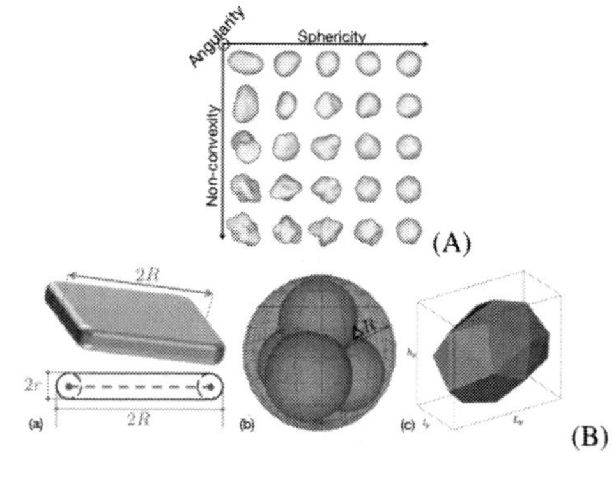

Figure 1. (A) 3D-diagram of particles shapes following Santamarina [3]. (B-a) Platy (elongated in 2D), (B-b) Non-convex and (B-c) Angular shapes .

These methods were primarily dedicated to the characterization of particles in the field of Sedimentology. Without going into details of the classification techniques, the particles can be grouped into three categories according to their angular, non-convex or elongation/platyness (also called sphericity) nature; see Figure 1(A).

⋆e-mail: emilien.azema@umontpellier.fr
⋆⋆e-mail: n.estrada22@uniandes.edu.co
⋆⋆⋆e-mail: itthichai.p@cmu.ac.th
⋆⋆⋆⋆e-mail: jean-yves.delenne@umontpellier.fr
†e-mail: franck.radjai@umontpellier.fr

Nevertheless, even if an increasing number of experimental and numerical studies have shown that the grain shape considerably affects the quasistatic behavior of granular materials [4, 6–9], systematic and quantitative investigations of shape or mixture of shape effects are still elusive. One of the difficulties is that the shape parameters mentioned above need to be defined conveniently in order to be able to generate particle shapes with continuously-variable shape parameters. In addition, particle shape is difficult to control experimentally.

A possible way for studying this problem is to analyze simplified systems with "idealized non-spherical" particles. With this respect, Discrete Element Methods are well suited for this type of simulations even if introducing particle shape in numerical simulations give raise to various technical difficulties such as contact detection and force calculation between particles of arbitrary shape.

Typically, angular shape represents a property of polygonal particles in 2D and polyhedral particles in 3D. Elongation and platyness can be analyzed throughout the aspect ratio parameter. Along the same line, interlocking is a basic feature of non-convex particles and can be represented by aggregate composed of spheres. Thus, the main objective of this work was to analyze numerically the effect of particle shape on the strength properties by varying systematically the shape from spheres to angular, non-convex and elongated/platy shape, as well as by considering mixture of shape.

2 Numerical procedures

2.1 Particle shape parameters

The level of sphericity can be defined by the three ratios $\lambda_1 = L_p/h_p$, $\lambda_2 = l_p/h_p$ and $\lambda_3 = L_p/l_p$, where L_p, l_p and h_p are the principal lengths axis ; see Fig.1(B-c). Consequently, spherical particles are such $\lambda_1 = \lambda_2 = \lambda_3 = 1$, In 2D, elongated particles are modeled as a juxtaposition of two half-disks of radius r and one rectangle of length L and width $2r$ as shown in Fig. 1(B-a). The aspect ratio $\lambda_3 = (L + 2r)/(2r)$ and $\lambda_1 = \lambda_2 = 0$. This particles are named "Rounded-Cap-Rectangle (RCR)". In the same way, in 3D platy particles are modeled as square plates with rounded edges, built as a sphero-polyhedra resulting from sweeping a sphere around a polyhedron. In this case the diagonal of the plate is equal to R and the aspect ratio is given by λ_3.

Non-convex particles are regular aggregates of 4-fold rotational symmetry composed of four overlapping spheres of the same radius r as shown in Fig. 1(B-b). In this case $\lambda_1 = \lambda_2 = \lambda_3 = 1$. This shape can be easily characterized by the ratio $\lambda^* = l/2r$, where l is the distance between the centers of spheres. This parameter varies from 0, corresponding to a sphere, to $\sqrt{3}/2$ corresponding to an aggregate where three coplanar spheres intersect at a single point, so that the radius R of the circumscribing sphere is given by $R = r(1 + \lambda^* \sqrt{3/2})$. Elongated, non-convex and platy shape may also be characterized by their degree of distortion η from a perfectly spherical shape, defined as

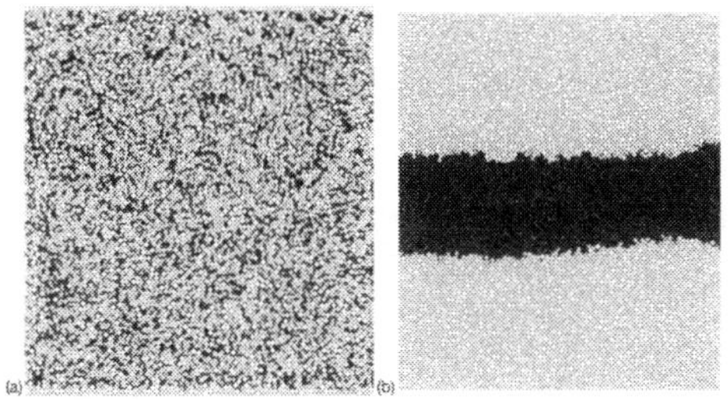

Figure 2. Examples of the generated packings at the initial state, here for binary mixture with $\alpha = 0.5$ and $M^* = 0$ (a) and $M^* = 1$(b). The aspect ratio is fixed at $\eta = 0.7$.

$\eta = \Delta R/R$, with $\Delta R = R - R'$ and R' the radius of the inscribed circle [10].

We define the angularity $\langle\alpha\rangle$ of a polyhedral particle as the mean exterior angle between its touching faces. In order to eliminate the effect of particle eccentricity, a strict procedure is implemented in order to isolate and control precisely the shape. First, a set of n_v vertices is randomly generated on a unit sphere. The convex hull of these points is created by associating three vertices for each face. Secondly, we impose that all ratios $\{\lambda_i\}_{i\in 1,3}$ to be greater than 0.9 ; see Fig.1(B-c). This means that, numerically, the sets of vertices are generated until this condition is satisfied. In this simple way, for a given aspect ratio, we can control the angularity of the particles with a single continuously-variable shape parameter depending only to the number of faces n_f.

2.2 Packing construction

We prepared different packings composed of large number of angular or non-convex or platy particles with $n_f \in [596, ..., 8]$ or $\eta \in [0, 1[$, respectively. In order to avoid long-range ordering, we introduce a size polydispersity by varying the circumradius R of the particles in the range $[r_{min}, 2r_{min}]$, where r_{min} is the minimum radius with a uniform distribution by volume fractions.

In order to analyse the effect of mixture by particle shape, we built also various packings composed with both discs and rcr particles with $\eta = 0.7$ in 2D. The mixture ratio parameter α is the proportion of elongated particles. We varied α from 0 for a packing composed of only disks to 1 for packing composed of only elongated particles. Mixture homogeneity is measured throughout Lacey mixture indice, which is directly proportional to the standard deviation from a given mean property of one of the constituents of a mixture. Knowing the fully segregated state where all elongated particles belongs in a band of thickness ϵ located at the center of the packing, and homogenous state, the Lacey parameter M^* is given by $M^* = \frac{S_r^2 - S^2}{S_r^2 - S_0^2}$, where S_r and S_0 are the values of the variance of the position of elongated particles calculated from the homogenous and fully segregated systems, respectively. Hence, numerically we constructed various packings with

M^* varies from 0, for fully segregated system, to 1, for fully homogeneous mixture. Figures 2 displays snapshots of the packings for several values of M^* and α at the end of isotropic compaction (see below).

By means of contact dynamic simulations (CD) [11] the particles are first compacted by isotropic compression inside a box. The gravity is set to 0 as well as friction coefficient μ and μ_w between particles and with the walls. A the end of isotropic compression, the connectivity number, defined as the mean number of contacts per particle (see Sec. 4) is nearly 12 for all packings which is in full agreement with the isostatic nature of our initial packings. These samples are then used as initial configuration for biaxial, respectively triaxial, quasi-static compression tests, for which friction between particles is set to 0.4.

3 Strength properties

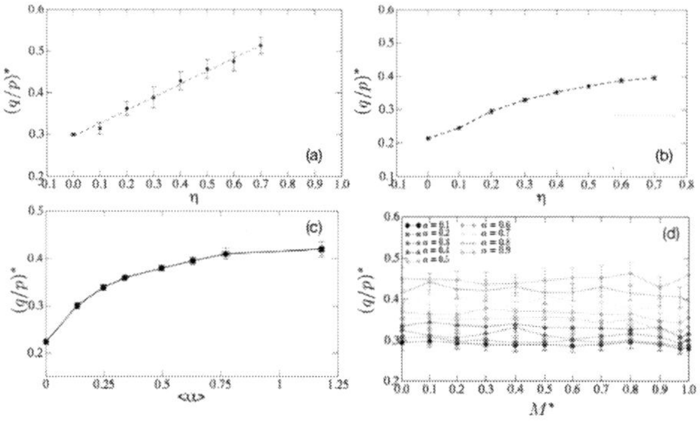

Figure 3. Normalized shear strength averaged in the steady state as a function of particle platyness/elongation (a), non-convexity (b), angularity (c), and as function of packing homogeneity for all values of mixture ratio (d).

In granular systems, the stress tensor σ is given by $\sigma_{\alpha\beta} = n_c \langle f_\alpha^c \ell_\beta^c \rangle_c$, where n_c is the number density of contacts c, $\mathbf{f^c}$ and ℓ^c refer to force and branch vector at the contact c, and the average $\langle ... \rangle_c$ runs over all contacts in a control volume [5]. In 3D with axial symmetry we define the stress deviator $q = (\sigma_1 - \sigma_3)/3$ and the mean stress $p = (\sigma_1 + \sigma_2 + \sigma_3)/3$, where $\sigma_{1,2,3}$ are the principal stress values. In 2D, we set $q = (\sigma_1 - \sigma_2)/2$ and $p = (\sigma_1 + \sigma_2)/2$. For our system of perfectly rigid particles, the stress state is characterized by the mean stress p and the normalized shear stress q/p [7]. During shear, the shear stress jumps initially to a high value before decreasing to a nearly constant value in the steady state. The steady-state shear stress $(q/p)^*$ characterizes the shear strength of the material.

Figure 3 shows the evolution of $(q/p)^*$ averaged in the steady state as a function of particle platyness/elongation (a), non-convexity (b), angularity (c), and as function of packing homogeneity for all values of mixture ratio (d). A general observation is that the shear strength increases as shape deviate from spherical shape and also as the proportion of non-spherical particle increases too. But surprisingly, the shear strength saturate at larger values of particles angularity and he remains independent with packing homogeneity. In this last case, this means that homogeneous mixtures have the same strength properties as segregated packings of the two particle shapes.

4 Particle connectivity and origins of strength

The main effect of particle shape is to allow for multiple contacts between particles as shown, for instance, in Fig. 4 for mixture of two shapes and non-convex aggregate. For non-convex particles as those used in this study we can identify seven different types of contacts : 1) "simple" contacts (s), 2) "double-simple" contacts (ds) defined as two simple contacts between two pairs of spheres, 3) " double contacts" (d) defined as two contacts between one sphere of one aggregate and two spheres belonging on the other aggregate, 4) "triple" contacts (t) defined as a combination of simple and double contacts or one sphere of one aggregate and three spheres of another aggregate or three simple contacts, 5) "quadruple" contacts (q) defined as a combination of two double contacts, and 6) five or six contacts with a negligible proportion (below 1%) compared to other contact types. With polyhedral particles, contacts can be grouped in terms of face-face(ff), face-vertex (fv), face-edge (fe) and edge-edge (ee) contacts. With elongated particles we have cap-cap (cc), cap-side (cs) and side-side (ss) contacts. With mixture of two shapes we can consider contact between disks (dd), rcr (rr)and disk-rcr (dr) particles.

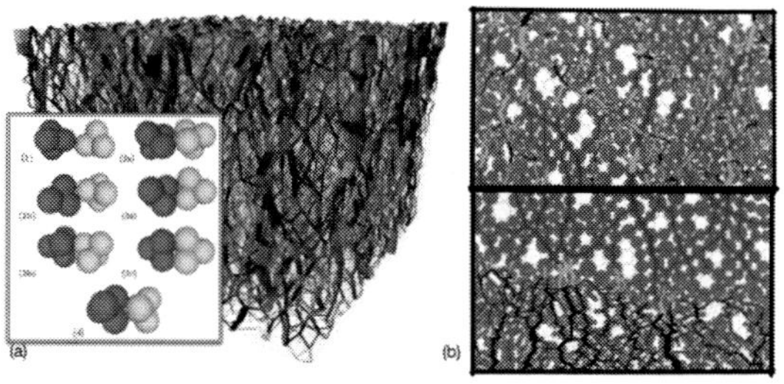

Figure 4. Snapshot of normal forces in (a) packing composed of non-convex particles (here η = 0.6), and (b) packings composed of mixture of two shape for homogeneous (up) and segregated (down) packings. The forces are plotted in different colors depending on contact types. The inset shows contacts type for non-convex aggregate whereas in mixture we can distinguish disks-disks, discs-elongated and elongated-elongated contacts.

To identify the impact of each contact type on the shear strength, we proceed by additive decomposition of the stress tensor according to contact type by $\boldsymbol{\sigma} = \sum_\gamma \boldsymbol{\sigma}_\gamma$, where γ stands alternatively for $\{ff, fv, fe, ee\}$ for polyhedral particles, $\{s, d, sd, t, q\}$ for non-convex particles, $\{cs, cc, cs\}$ for elongated particles and for $\{dd, dr, rr\}$ contacts for mixture of two shapes. The corresponding stress deviators q_γ, are then calculated and normalized by the mean pressure p such $(q/p)^* = \sum_\gamma q_\gamma/p$.

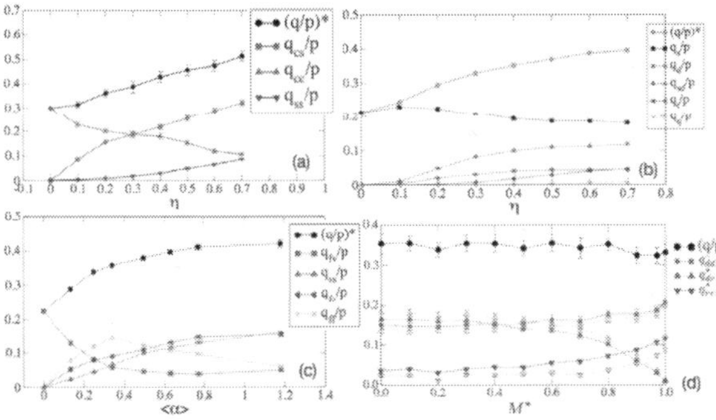

Figure 5. Normalized shear strength as a function of particle shape parameter or Mixing index, for (a) cs, cc and cs contacts in packing composed of elongated particles, (b) s, d, sd, t and q contacts in packing composed of non-convex particles, (c) ff, fv, fe and ee in packing composed of non-convex particles, and (d), dd, dr and rr contacts for mixture of two shapes composed by the same proportion of disk and rcr.

The corresponding normalized stress deviators q_γ/p averaged in the residual state are shown in Fig. 5 as a function of shape parameters, for packings of (a) elongated, (b) non-convex and (c) angular particles, as well as for mixture of two shapes as a function of mixing index M^* for packing composed by the same proportion of disk and rcr. We see that contact types contribute differently on strength properties as function of shapes. For instance, the rapid increase of the shear strength with angularity is due to a rapid increase of strength support by ff contact at low angularity, but the plateau observed for langer angularity is due to a rapid fall off of q_{ff}/p compensated by a rapid increase of q_{fe}/p and q_{fe}/p. For elongated particles $(q/p)^*$ follow q_{cs}/p. And, for mixture of two shapes, we see that q_{dd}/p is nearly constant, whereas remarkably q_{dr}/p follows a trend opposite to that of q_{rr}/p so that the decrease of q_{dr}/p with M^* is exactly compensated by the increase of q_{rr}/p, which explains the independence of $(q/p)^*$ with respect to M^*.

5 Conclusions

In this paper, a systematic numerical analysis of the effect of particle shape properties on the quasistatic rheology of sheared granular media, as well as packings composed of mixtures of two shapes was presented in the framework of Contact Dynamics simulations. The shape of the particles is systematically varied from spherical shape (as referential shape) to angular, non-convex and elongated or platy shape. We considered also binary mixture of disks and elongated particles characterized by the mixture ratio which varies from 0 for a packing composed of only disks to 1 for a packing composed of only rcr particles, and with an homogeneity parameter M^* which varies from 0 for an homogeneous packing to 1 for a fully segregated packing.

A major result of this work is that the particle shape effects are nonlinear and counterintuitive. We show that the shear strength increases as shape deviate from spherical shape. But, for angular shapes it first increases up to a maximum value and then saturates to a constant value as the particles become more angular. For mixture of two shapes, the strength increases with respect of the increase of the proportion of elongated particles, but surprisingly it is independent with the level of homogeneity of the mixture. At the micro-scale we show that various contact types, depending on the specificities of each shape, contribute differently to stress transmission. A general observation is that multiple contact are at the origins of the increase of strength.

In this investigation, the shape was systematically varied along "one" direction. Indeed, we used convex and smooth particles when varying elongation, convex and spherical particles for studying angularity, and spherical and smooth particles for analyzing the effect of non-convexity. However, it should be interesting to vary the shape considering combined effect of elongation and angularity, or elongation and non-convexity, or angularity and non-convexity. In the same way, by analogy with polydisperse packing by particles size, it will be very instructive to consider mixture of various shape, which imply the definition of higher order parameters such particle shape distribution function in order to systematically investigate such effects.

References

[1] H. Waddel, J. Geol. **40**, 443 (1932)

[2] W. Krumbrein, F. Petttijohn, *Manual of sedimentary petrography*, sepm class edn. (Cambridge University Press 1991, 1938), ISBN 0918985781

[3] J.C. Santamarina, G. Tech, G.C. Cho, *Soil behaviour : The role of particle shape*, in *Skempton Conference* (London U.K., 2004), pp. 1–14

[4] K. Mair, K.M. Frye, C. Marone, Journal of Geophysical Research **107** (2002)

[5] K. Bagi, Mechanics of Materials **22,3** (1996)

[6] W. Man, A. Donev, F.H. Stillinger, M. Sullivan, W. Russel, D. Heeger, S. Inati, S. Torquato, P. Chaikin, Physical Review Letters **94**, 198001 (2005)

[7] J. Mitchell, K. Soga, *Fundamentals of Soil behavior* (John Wiley, 2005), ISBN 978-0-471-46302-3

[8] S. Sacanna, L. Rossi, A. Wouterse, A.P. Philipse, Journal of Physics: Condensed Matter **19**, 376108 (2007)

[9] T. Börzsönyi, R. Stannarius, Soft Matter **9**, 7401 (2013)

[10] CEGEO, B. Saint-Cyr, K. Szarf, C. Voivret, E. Azéma, V. Richefeu, J.Y. Delenne, G. Combe, C. Nouguier-Lehon, P. Villard et al., European physical Letters **98** (2012)

[11] J.J. Moreau, European Journal of Mechanics - A/Solids **13**, 93 (1994)

Scaling of wet granular flows in a rotating drum

Ahmed Jarray[1,*], *Vanessa Magnanimo*[1], *Marco Ramaioli*[2] and *Stefan Luding*[1]

[1]Multi Scale Mechanics, CTW, MESA+, University of Twente, Enschede, Netherlands
[2]Department of Chemical and Process Engineering, University of Surrey, Guildford, United Kingdom

Abstract. In this work, we investigate the effect of capillary forces and particle size on wet granular flows and we propose a scaling methodology that ensures the conservation of the bed flow. We validate the scaling law experimentally by using different size glass beads with tunable capillary forces. The latter is obtained using mixtures of ethanol-water as interstitial liquid and by increasing the hydrophobicity of glass beads with an ad-hoc silanization procedure. The scaling methodology in the flow regimes considered (slipping, slumping and rolling) yields similar bed flow for different particle sizes including the angle of repose that normally increases when decreasing the particle size.

1 Introduction

A correct understanding of granular flow is critical to the success of a wide range of pharmaceutical processes including mixing, coating and tableting [1, 2].

Flow parameters are inherently scale-dependent: if the particle size is changed, characteristic properties such as the dynamic angle of repose (i.e. the angle near the midpoint of the flowing layer in a drum) and the depth of the free flowing layer will change. This makes the scaling of granular flows a complex endeavor.

A simple and practical geometry to study the flow of granular materials is the rotating drum. This has been extensively studied by many researchers including Taberlet et al. [3] who derived an equation to describe the S shape of the granular pile in a rotating drum. Rajchenbach [4] correlated the dynamic angle of repose to the rotation speed of the drum. Elperin and Vikhansky [5] proposed a model for describing the bed flow using a Mohr-Coulomb failure criterion.

While, most of these studies focused on the flow of dry particles, mechanisms governing particle flow in wet systems remain poorly understood. In fact, adding a small amount of liquid to a particulate system can significantly alter its flow behaviour. The work of Tegzes et al. [6] for instance showed that the capillary force determines the flow motion of particles. Brewster et al. [7] showed that the presence of interparticle cohesion reduces the concavity of the free flowing surface in the drum. Therefore, the effects of capillary force on wet granular flow cannot be ignored.

Dimensional analysis can be used to simplify the flow description by reducing the number of system parameters. Henein et al. [8] correlated the flow of particles in a drum to the magnitude of the Froude number and categorized it into six flow regimes; slipping, slumping, rolling, cascading, cataracting, and centrifuging. Orpe et al. [9] found that the flowing layer thickness is symmetric at low Froude number (Fr) and large particle size to drum radius (r/R). However, so far, scaling relationships for wet particulate systems in a rotating drum are unavailable, and it is not clear how particle size affects the flow of particles in the presence of liquid-induced cohesion.

In this work, we study the flow of wet particles in a rotating drum in the slipping, slumping and rolling regimes by using a silanization method to alter the surface properties of the glass beads, and hence, to indirectly control the bed flow in the drum. We begin by investigating the influence of the particle size, rotation speed and the capillary force on the bed flow motion. Then, we propose an approach that ensures particle flow similarity for different particle sizes.

2 Materials and Methods

2.1 Capillary force and silanization procedure

The capillary force between interacting particles is known to control the flow behaviour of wet granular materials; here we focus on the pendular regime [10]. For a given particle radius r, the capillary force depends linearly on the contact angle θ and the surface tension γ.

$$F^{c} \sim \gamma\cos\theta. \quad (1)$$

For a proper understanding of the problem, we isolate the roles of particle size, contact angle, and surface tension. Similarly to the work of Raux et al. [11], we treat the surface of the glass beads via silanization to increase their hydrophobicity, that is the contact angle θ. Using mixtures of ethanol-water as interstitial liquid and glass beads with different hydrophobicity, the capillary force between two adjacent beads can be manipulated over a wide range of values.

Chemical compounds used for silanization are: Silanization solution I~5% (v/v) (5% (v/v) dimethyldichlorosilane in Heptane, Selectophore), Hydrochloric acid (HCL, 0.1 M), Acetone and Ethanol.

The procedure for increasing the beads hydrophobicity is as follows: first, glass beads are cleaned for at least one hour by immersion into freshly prepared HCL solution under agitation using a rotor-stator homogenizer. Then, they are rinsed thoroughly with deionized water and oven dried for 3 hours at 60°C. Afterward, the freshly cleaned glass beads are immersed in 5 % (v/v) dimethyldichlorosilane in heptane (silanisation solution) under agitation for an hour. Finally, the glass beads are rinsed with acetone and allowed to air-dry under a fume hood for 24 hours.

* Corresponding author: a.jarray@utwente.nl

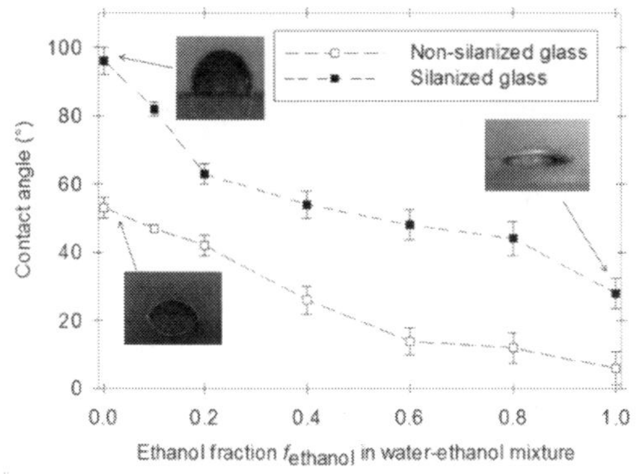

Fig. 1. Contact angles of water-ethanol mixtures as a function of the ethanol fraction on silanized and non-silanized glass.

To measure the contact angle θ, liquid droplets are placed on a glass surface. Then, pictures of the droplet are taken using a MotionBLITZ EoSens camera with close-up lenses. The contact angle is then deduced by image analysis using LBADSA plugin in the open source imageJ software. The LBADSA plugin is based on the fitting of the Young-Laplace equation to the image data [12]. As shown in Fig. 1, water–ethanol mixtures with different ethanol fraction allow to obtain intermediate contact angles, which continuously decrease from 96° to 30° for silanized beads as the ethanol fraction in the liquid increases from 0 to 1. The higher the contact angle, the lower the wettability of the glass.

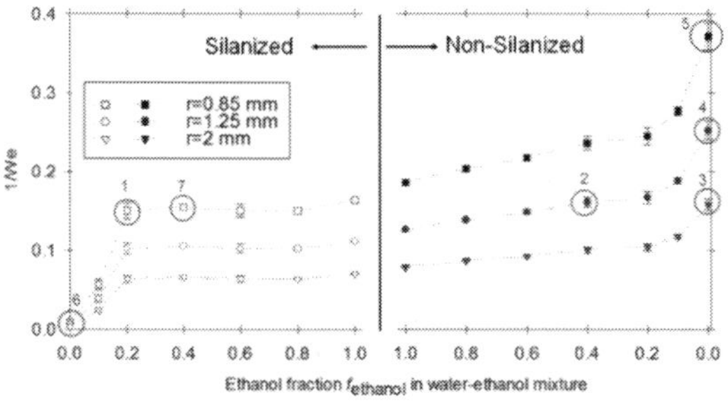

Fig. 2. Inverse of the Weber number *We* as a function of ethanol fraction for silanized and non-silanized glass for different particle radii *r*. The red circles mark to the configurations studied in section 3.3. Surface tension values are taken from Ref. [13].

To characterize the motion of wet particles, the particle based Weber number is often used [14, 15], given by the ratio of kinetic to capillary energy of the particle.

$$We = \frac{\rho_p r V^2}{\gamma \cos\theta} \qquad (2)$$

where r the particle radius, ρ_p the density of glass bead and V the average velocity of the flowing particles. This velocity is taken constant, $V = 0.11$ m/s. The Weber number in equation (2) also takes into consideration the effect of the particle-liquid contact angle θ.

In Fig. 2, we plot the inverse of the particle based Weber number *We*, defined in equation (2), against the volume fraction of ethanol in the water-ethanol mixture. Surface tension values are taken from Ref. [13]. Here, differently from e.g. [14, 15], the variability of the contact angle θ is taken into account. We vary *We* by varying the capillary force $F^c \sim \gamma cos\theta$, for different particle sizes. The Weber number increases (F^c decreases) with increasing ethanol concentration for particles with unmodified glass surface, while *We* decreases (F^c increases) in the case of silanized glass beads, in agreement with the literature [16]. This allows to control the strength of the capillary force and *We* number in wet granular flows.

2.2 The Drum apparatus

Samples of silanized and non-silanized glass beads with different ethanol-water mixtures as described above, are then placed in a rotating drum to study the granular flow. The drum is made by a cylinder of 121 mm inner diameter, 22 mm width (frontal view in Fig. 3 a), and held between two circular plexiglass (PMMA) plates of 5 mm thickness to allow optical access. The drum is placed vertically on a rotating axle driven by a variable-speed motor. The PMMA walls of the drum are coated with FEB coating to prevent wet glass beads from sticking on the wall. Images of the rotating drum are recorded using a MotionBLITZ EoSens high speed camera working at a speed of 460 frame.s^{-1}. Images are then post-processed using a particle tracking package in Matlab [17].

After each experiment, the drum is dried in an oven for one hour to let the remaining liquid from the previous experiment evaporate. In all experiments, the drum filling ratio is around 35% (w/w) (i.e. 125 g of glass beads) and the amount of added liquid is 4 ml, where the pendular state is expected.

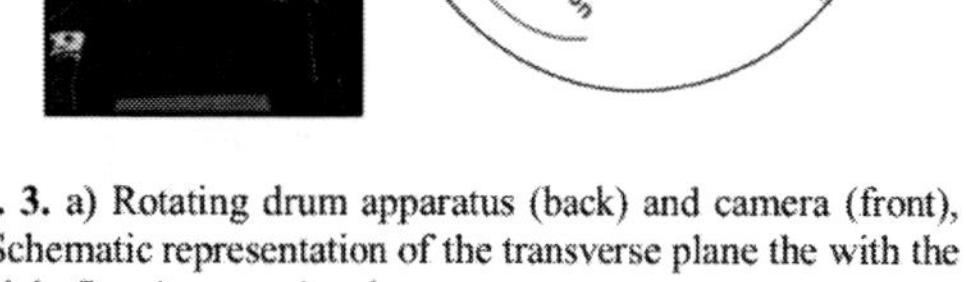

Fig. 3. a) Rotating drum apparatus (back) and camera (front), b) Schematic representation of the transverse plane the with the particle flow in a rotating drum.

We perform flow experiments in the drum for a selected set of particle types, with different hydrophobicity and sizes, namely 0.85 mm, 1.25 mm and 2 mm.

Parameters and characteristics of the drum and the glass beads are summarized in Table 1.

Table 1. Properties of the drum and the glass beads.

Properties	Value
Drum, R×L (mm)	60.5×22
Glass beads radii r (mm)	0.85, 1.25 and 2
Rotation speed (rpm)	3 to 57
Particle density ρ_p, (kg.m^{-3})	2500
Filling level β	35% (125 g)
Volumetric liquid content v_{liq} (ml)	4

3 Results and discussion

3.1 Effect of capillary force on the bed flow

In order to characterize the flow in the drum, we focus on the dynamic angle of repose, i.e. the angle between the top surface of the rolling bed and the horizontal plane (Fig. 3 b). This angle is numerically computed by linear regression of the positions of the particles on the surface of the flow, detected by a concave-hull algorithm [18].

First, we investigate the flow in the simplest case of non-salinized particles with pure-water as interstitial fluid. This will help us to gain knowledge on the interplay between particle size and capillary force.

Fig. 4 shows the dynamic angle of repose versus rotation speed for three particle sizes in wet system with pure water as interstitial liquid (configuration 3, 4 and 5 in Fig. 2). As the rotation speed increases, the dynamic angle of repose increases. Smaller glass beads have a higher dynamic angle of repose. This is because a decrease of the particle size decreases the Weber number We (see Fig. 2), i.e. capillary forces become stronger.

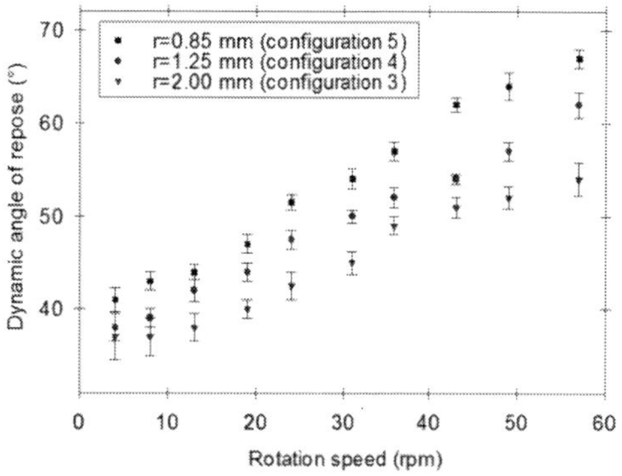

Fig. 4. Dynamic angle of repose as a function the rotation speed for different particle sizes in wet system.

In Fig. 5, we show the velocity gradient of the granular flow for particle size 1.25 mm at 25 rpm, with 0% ethanol concentration ($f_{ethanol} = 0$) in the case of wet glass beads (configuration 4 in Fig. 2) and silanized wet glass beads (configuration 6 in Fig. 2). For the sake of completeness, we also show the flow in the case of dry particles. The velocity profile of the particles in the drum is obtained using a matlab particle tracking method [17].

Clearly, the interparticle cohesion has a significant effect on the bed flow motion. Whereas the flow pattern has a convex shape for dry particles, an almost flat shape is observed for wet particles. The dynamic angle of repose for the wet particles shown in Fig. 5 (b) is about 46°, larger than that of dry particles (around 37° in Fig. 5 a). By visual inspection of Fig. 5, we also notice that the depth of the high velocity zone becomes larger, but slower, in the wet case, suggesting a gradual transition from shear flow to plug flow. This transition is also marked by lower flow velocity, indicating a reduction of the mobility of particles.

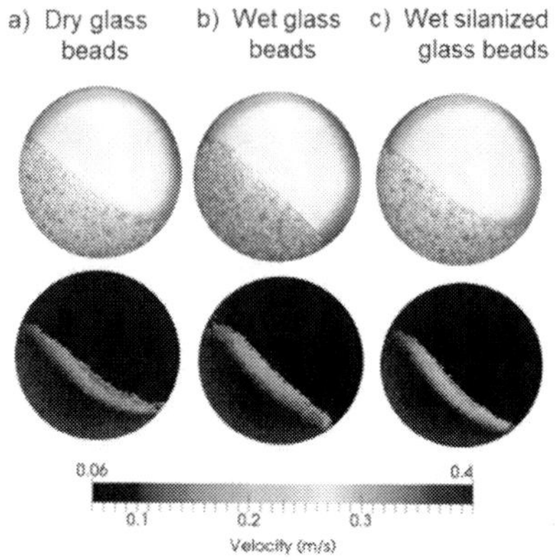

Fig. 5. Snapshots (top) and velocity (bottom) of the bed in the rotating drum for a) dry, b) wet and c) silanized glass beads.

On the other hand, silanized glass beads mixed with water show a lower dynamic angle of repose and a thinner flowing layer (see Fig. 5 c), close to that of the dry case. This is because the capillary force becomes weaker in the case of silanized surface due to the higher contact angle. Thus, by means of silanization, it is possible to alter the bed flow via the contact angle and even approach the flow motion of a dry system.

3.2 Scaling methodology

The concept of similarity based on dimensional analysis states that two processes might be considered similar if all involved length scales are proportional, and all dimensionless numbers needed to describe them have the same value [19].

Five dimensionless groups are presented in equation (3) including Froude number F_r (ratio of inertial force to gravity force), Weber number We, liquid to particle volume ratio, capillary number C_a, and drum fill ratio β.

$$F_r = \frac{2(R-r)\omega^2}{2g}\ ,\ We = \frac{\rho_p r V^2}{\gamma\cos\theta}\ ,\ \frac{N_p r^3}{v_{liq}} \qquad (3)$$
$$C_a = \frac{\mu V_r}{\gamma}\ ,\ \beta$$

where N_p is the total number of particles, μ is the dynamic viscosity, and V_r is the relative velocity between two particles in contact. In our study, we have kept the Froude number, liquid-particle volume ratio, and drum fill percentage constant among all experiments (see Table 1). In this study, we focus on the effect of the Weber number and neglect here the role of the Capillary number ($C_a << 1$).

3.3 Particle size scaling

Using the information from the previous sections, we propose a procedure to maintain dynamic similarity after scaling the particle size. In the case of granular flows, the relevant properties to compare are the dynamic angle of repose and the depth of the flowing layer. Here, we focus on the former and we test if we can obtain the same flow behavior by keeping the Weber number We constant. That is, for decreasing particle size, we must increase the capillary force accordingly.

We compare the flow motion of the smaller particle sizes (1.25 mm and 0.85mm) to the flow of the 2 mm particles and we check whether they have the same dynamic angle of repose. For that, we refer to Fig. 2 and we keep the inverse of the Weber number $1/We \sim 0.16$ for the three cases, starting from configuration 3 with pure water. For particles with $r = 0.85$ mm, rather than using water, we use a mixture of 20% ethanol-80% water (v/v) with silanized 0.85 mm glass beads, indicated in Fig. 2 as configuration 2. On the other hand, for particles of 1.25 mm, we use 40% ethanol-60% (v/v) water with non-silanized beads, marked in Fig. 2 as configuration 1.

Fig. 6 presents the dynamic angle of repose of wet samples for different rotation speeds. While the open red symbols represent the configurations with almost the same Weber number, solid symbols represent configs. 4 and 5 with 100% (v/v) water and lower Weber number.

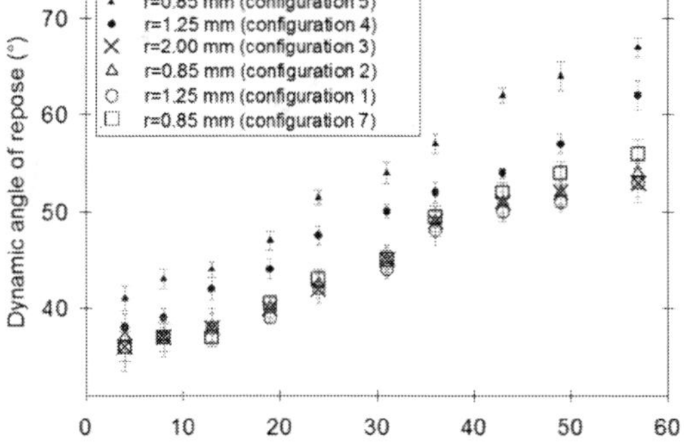

Fig. 6. Dynamic angle of repose as a function of the rotation speed for different particle sizes in wet systems.

The dynamic angle of repose of configurations 1 and 2 collapses with that of the 2 mm size case (config. 3), indicating that the latter can be rescaled to 0.85 mm or 1.25 mm sizes if the Weber number is kept constant, via the contact angle and the surface tension. This means that, experimentally, the scaling approach works in the considered flow regimes (slipping, slumping and rolling).

4 Conclusion

The effect of capillary force and particle size on the bed flow in a rotating drum was investigated at different rotation speeds, from 3 to 57 rpm. Specifically, we have shown that the presence of liquid increases the depth of the flowing layer and reduces the mobility of the particles. We established a procedure to alter the strength of capillary forces by using mixtures of ethanol-water as interstitial liquid with silanized and non-silanized glass beads. Then, we proposed a scaling methodology that ensures similarity of the dynamic angle of repose by keeping the Weber number constant after scaling the particle radius. We were able to obtain similar bed flow for different particle sizes, confirming that the proposed scaling methodology works for the few cases studied so far. A next step would be to test the scaling methodology on other quantities like the similarity of the flowing layer depth, and to include the effect of the liquid viscosity on the rather dynamic bed flow.

We thank H. Polman and B. Scheper for their help. Financial support through the "T-MAPPP" project of the European-Union-Funded Marie Curie Initial Training Network FP7 (ITN607453) is also acknowledged.

References

1. J. Osorio., F. Muzzio, Drug Dev. Ind. Pharm. **39**, 1464–1475 (2013).
2. J. Prescott, R. Barnum, Pharm. Technol. **24**, 60–85 (2000).
3. N. Taberlet, P. Richard, E.J. Hinch, Physical Review E **73**, 050301 (2006).
4. J. Rajchenbach, Phys. Rev. Lett. **65**, 2221 (1990).
5. T. Elperin and A. Vikhansky, Europhys. Lett. **42**, 619 (1998).
6. P. Tegzes, T. Vicsek, P. Schiffer, Phys. Rev. E **67**, 051303 (2003).
7. R. Brewster, G. S. Grest, and A. J. Levine, Physical Review E **79**, 011305 (2009).
8. H. Henein, J. K. Brimacombe, and A.P. Watkinson Metall. Trans. A B **14**, 191–205 (1983).
9. Orpe, A.V., Khakhar, D.V., Physical Review E **64**, 031302 (2001).
10. Z. Fournier, D. Geromichalos, S. Herminghaus et al. J. Phys. Condens. Matter. **17,** S477 (2005).
11. P. S. Raux, H. Cockenpot, M. Ramaioli, D. Quéré and C. Clanet, Langmuir **29**, 11 (2013).
12. A.F. Stalder, T. Melchior, M. Müller, D. Sage, T. Blu, Colloids Surf., A **364**, 72-81 (2010).
13. G. Vazquez, E. Alvarez, J. M. Navaza, J. Chem. Eng. Data , **40** (3), 611–614 (1995).
14. M. H. Klein Schaarsberg, I.R. Peters, M. Stern, K. Dodge et al., Phys. Rev. E **93**, 062609 (2016).
15. Q. Xu, I. Peters, S. Wilken, E. Brown, and H. Jaeger. J. Vis. Exp. **85**, 51249 (2014).
16. M. Tagawa, K. J. Gotoh, Adhes. Sci. Technol. **12,** 1341 (1998).
17. Matlab Particle Tracking Code Repository, (2016), http://site.physics.georgetown.edu/matlab/index.htm
18. A. L. Chaua, X. Lia, W. Yu, Neurocomputing **122**, 198–209 (2013).
19. F.M. Henderson, *Open channel flow. Macmillan Publishing* Co., Inc., (New York, 1966).

Barchan dunes on Pluto?

Eric J. R. Parteli[1,*] and *Thorsten* Pöschel[2,**]

[1] *University of Cologne, Cologne, Germany*
[2] *University of Erlangen-Nuremberg, Erlangen, Germany*

Abstract. We show that the orientation and morphology of bedforms occurring on top of Pluto's smooth ice coats are consistent with an aeolian origin under conditions of unidirectional flow. From scaling relations for dune size as a function of attributes of atmosphere and sediments, we find that the average diameter of the granular particles constituting such bedforms — assuming an aeolian origin — lies within the range $600\,\mu\text{m} < d < 750\,\mu\text{m}$. Our findings show that, owing to the effect of hysteresis in the minimal threshold wind velocity for saltation, dune migration on Pluto can occur under wind speeds that are common to Earth and Mars.

1 Introduction

Since New Horizon's encounter with Pluto in July 2015 we know that Pluto has a geologically active surface consisting of dark, cratered terrains alternating with huge icy mountains and flat, bright soils free of impact craters [1]. Several processes have been suggested to explain the complex topography, in particular chemical reactions triggered by ultraviolet radiation from the sun, heterogeneous rates of ice sublimation and deposition induced by gradients of surface temperature, and resurfacing due to radioactive heating from within the planetary body [2]. In this paper, we consider the role of wind as a geological process on Pluto. Based on images sent by New Horizons, we identify surface features on the surface of Pluto that resemble very much aeolian dunes on Earth and Mars. In particular, the referred bedforms display a dependence of shape on the size which is inherent of aeolian dunes. As we will show here, assuming an aeolian origin for these bedforms, theory that successfully describes the process of dune formation on Earth and Mars can be applied to constrain formative wind speeds and average grain sizes on Pluto [3–5].

2 The barchans

Bedforms with morphology similar to aeolian dunes of Earth and Mars occur on top of Pluto's bright, ice-coated crater-free surface, between Norgay and Hillary Montes (Fig. 1). Some of the bedforms display a dome-like shape while others resemble crescent-shaped barchan dunes – which form under unidirectional winds, in areas of limited sand supply, and have two limbs pointing in the wind direction [3–9]. The orientation of the barchan limbs indicates, thus, the prevailing sediment-moving wind trend, denoted by the dashed arrow in Fig. 1c (see Fig. 3 for reference).

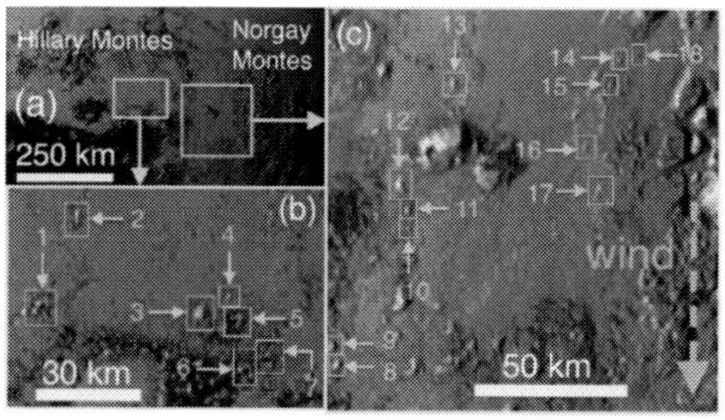

Figure 1. Dunes on the thin ice sheet between Hillary Montes and Norgay Montes on Pluto. The dunes occur within the areas enclosed by the rectangles in (a). These areas correspond to images (b) and (c), where each dune, enclosed by a box, is identified by a number for reference in the plot of Fig. 2. The dashed arrow in (c) denotes the mean formative wind direction, which is obtained from the orientation of the limbs of the barchan dunes (see Fig. 3). Images credit: NASA/JHUAPL/SWRI.

Figure 2 shows that the along-wind width L and cross-wind width W of the bedforms correlate linearly, which is the same behavior of barchan dune fields on Earth and Mars [3–5]. Some barchans display an asymmetric shape with one limb elongated downwind, which is common feature of dunes on Earth and Mars caused by wind trend fluctuations, local topography and inhomogeneous influx conditions due to the presence of upwind dunes [8, 9].

3 Model for aeolian transport on Pluto

The presence of dunes implies that winds on Pluto have occasionally exceeded the minimal threshold to transport granular materials – most probably constituted of nitrogen

*e-mail: eric.parteli@uni-koeln.de
**e-mail: thorsten.poeschel@fau.de

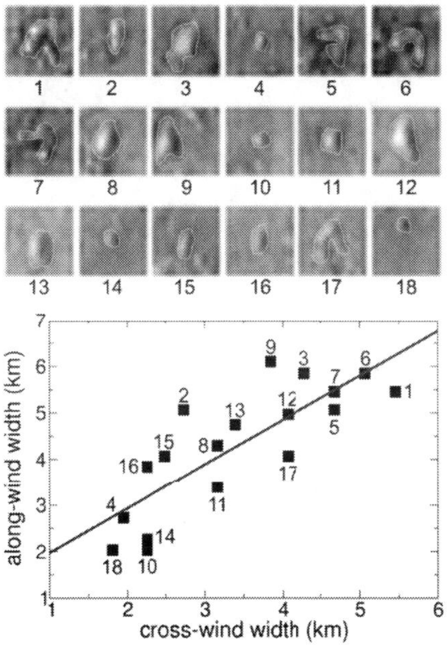

Figure 2. Bedform's morphometry. Left: dunes 1–18 from Fig. 1. The contour line was added in each image in order the highlight the dune shape. Each image is ≈9 km across. Right: along-wind and cross-wind widths (L and W, respectively) of the imaged dunes (squares). The best fit using the equation $L = a+b\cdot W$ yields $a \approx 1.02$ km and $b \approx 0.96$ (continuous line).

ice particles [1, 2, 10, 11] – along the surface. As shown in Ref. [5], natural grains occurring in dune fields have a broad range of sizes, and the transport mode depends on the particle size. While gravels, that is particles larger than 1 mm, are too heavy to enter the aeolian transport layer and thus roll or slide (creep) on the surface, dust particles, which are smaller than about 60 μm, are transported by the wind through suspension. However, particles in the intermediate range move mainly through *saltation*, which is the particle transport mode leading to dune formation and consists of grains moving in nearly ballistic trajectories and ejecting new particles upon collision with the soil [6].

Wind speeds 100 times larger than those shaping terrestrial dune fields are needed to initiate sediment transport on Pluto (whereas this difference is mainly due to the 10,000 times smaller atmospheric density of Pluto [12], compared to the Earth's). However, recent theory of aeolian sediment transport shows that the effect of hysteresis in the threshold wind speed for saltation transport [13–16] – which means that, once initiated, saltation can be sustained at much smaller wind speeds – is much stronger under rarified atmospheres. On Earth, the threshold for saltation maintenance (u_t) is about 80% the threshold for saltation initiation (u_{ft}), while on Pluto u_t is 10 – 20 times smaller than u_{ft} (see Fig. 4). In other words, dunes might form and migrate on today's Pluto as long as surface wind speeds on this planetary body occasionally exceed values

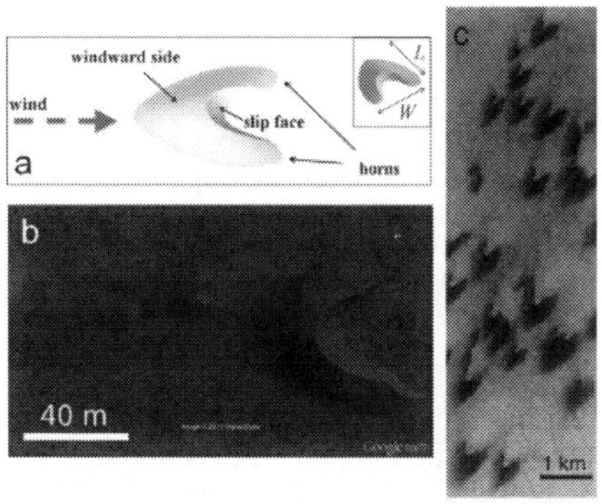

Figure 3. The barchan dune shape. **a**, Schematic diagram of a barchan dune. The dashed arrow indicates the prevailing wind direction. The definitions of along-wind width L and cross-wind width W of a barchan are provided in the inset. Sediments are transported downwind along the windward side of the barchan through saltation, thereby accumulating on the dune crest. As the amount of sediments on the crest increases, the dune's leeward side becomes increasingly steeper. When the slope of this side exceeds the angle of repose of the granular material (which for terrestrial and martian sand is about 34°), the downwind surface of the dune relaxes through avalanches in the direction of the steepest descent – a slip face forms. Furthermore, since the downwind migration velocity of aeolian bedforms scales with the inverse of the bedform size [6, 7, 18], the dune flanks – which are the lowest parts of the dune – appear advanced downwind relative to the central part of the dune, thus giving rise to the limbs that are characteristic of the barchan shape. **b**, barchan dunes in Morocco, near 26°5′N, 13°2′E (image credit: Google Earth); **c**, barchan dunes in Arkhangelsky crater, Mars, near 41°4′S, 24°8′W (image credit: NASA/JPL/MSSS). It can be seen that limbs are absent from barchans smaller than a minimal threshold size, which on Earth and Mars is about 10 m and 200 m, respectively. The same behavior is observed on Pluto (see Figs. 1 and 2).

that are not uncommon to dune fields of our planet. Mechanisms for transport initiation on Pluto are still uncertain. Candidates might be sporadic gusts of much stronger aeolian activity or particle lifting from the soil due to insolation-induced convective gas flows within the granular bed — which, as demonstrated recently [17], is a relevant direct particle entrainment process under conditions of low atmospheric pressure.

The fact that barchan limbs are absent from all bedforms of cross-wind width smaller than a threshold size W_{min} (which, from Fig. 1, is of typical size 3 km) is, indeed, a feature characteristic of dunes that migrate on top of bedrock under constant wind direction [5, 18–20]. This minimal dune size scales with the characteristic length needed by the mass flux of particles in saltation to adapt to a change in flow conditions — the flux saturation length, L_{sat} (see Refs. [13, 18, 21]). In particular, the saturation length dictates the transient distance needed for the flux to adjust to a topography-induced change in the average turbulent surface wind shear stress. Since the downwind position of the maximal wind shear stress over sand

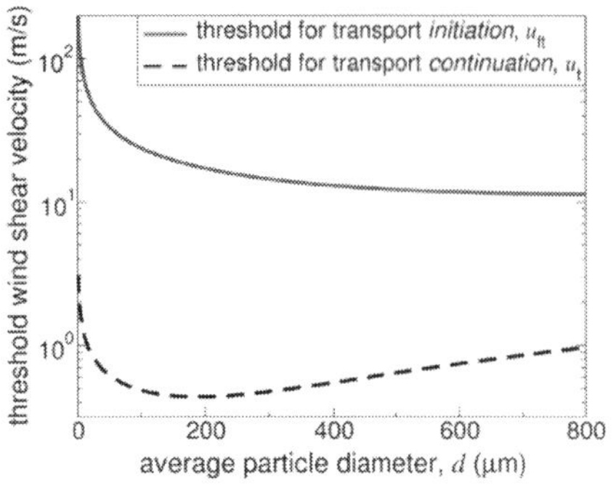

Figure 4. Minimal threshold wind shear velocity for saltation on Pluto, as a function of average particle diameter. The threshold for transport initiation, u_{ft}, is computed using the model of Ref. [24], while the threshold for transport continuation, u_t, is obtained using the model of Ref. [16]. The results correspond to average surface pressure of 20 μbar and temperature 44 K, assuming atmosphere composed mainly of N_2.

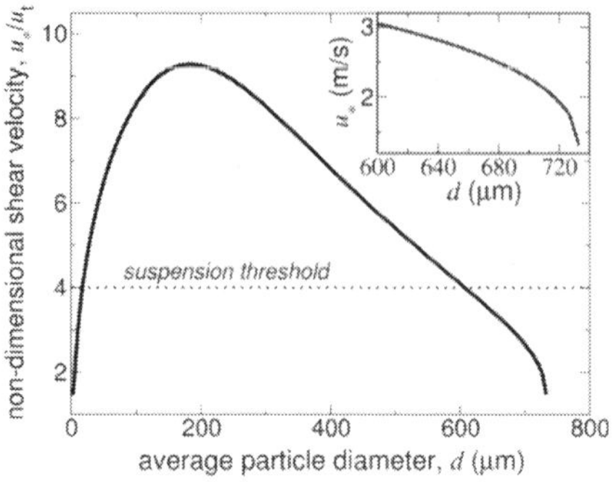

Figure 5. Range of formative wind shear-velocities and grain sizes from the scale of Pluto's dunes. The continuous line in the main plot denotes the solution for u^*/u_t and d obtained from the equation $12\,L_{\rm sat}\,(u^*/u_t, d) = W_{\rm min}$, where the saturation length of sediment transport, $L_{\rm sat}$, is computed with the model of Ref. [21] and $W_{\rm min} \approx 3$ km is the minimal barchan dune size on Pluto. Moreover, u_t is obtained as a function of the attributes of sediments and atmosphere with the model of Ref. [16]. The suspension threshold $u^*/u_t = 4$ is indicated by the dotted horizontal line [13, 16]. Based on the main plot, the most plausible range of shear velocity u^* and grain size d leading to dune formation on Pluto is shown in the inset.

hills and domes is shifted upwind of the bedform's crest [18, 20, 22], the development of a sand heap into a barchan occurs only if $L_{\rm sat}$ is smaller than this upwind shift, thus causing net deposition on the crest and bedform growth. It was shown [5] that this condition is fulfilled for bedforms that are larger than about $12 \times L_{\rm sat}$. Because $L_{\rm sat}$ is a function of the attributes of sediment, fluid and wind power [13, 18, 21], the value of $W_{\rm min}$ of the bedforms imaged by New Horizons poses one constraint to the physical conditions causing sediment transport and dune formation on Pluto.

Specifically, the scale of the minimal dune on Pluto provides a proxy for the range of average grain size d and wind shear velocity u^* — which is proportional to the mean flow velocity gradient in turbulent boundary layer flow [23] — leading to dune formation under atmospheric conditions valid for today's Pluto. This range is obtained from the equation $12\,L_{\rm sat}\,(u^*/u_t, d) = W_{\rm min}$, where $L_{\rm sat}$ is computed with the model of Ref. [13] and the minimal threshold for sustained transport, u_t, is a function of the atmospheric density, particle diameter and density, air viscosity and gravity [13, 16]. The main plot of Fig. 5 shows the values of d and u^*/u_t which satisfy the equation above (continuous line).

We can now constrain approximate value of wind speeds and grain sizes for the barchan-like bedforms as explained in the following. Since dunes-forming u^* values are upper-bounded by the minimal threshold for suspension — which is about $4\,u_t$ (see Ref. [21]) and is indicated by the dotted horizontal line in Fig. 5 — and given the prohibitively large values of the minimal threshold shear velocity for direct entrainment of powder-sized particles of diameter smaller than 20 μm (for which cohesive forces are large compared to particle weight and aerodynamic lift [24–26]), we estimate that the most likely diameter range of the sediments composing Pluto's bedforms is, based on the main plot of Fig. 5, 600 μm $\leq d \leq$ 750 μm. This range corresponds to particle sizes of coarse sand [6]. The corresponding range of shear velocities leading to the dune scale on Pluto is 1.5 m/s $\leq u^* \leq$ 3.0 m/s (inset of Fig. 5), which is comparable to the wind speeds required to initiate sand transport on Mars [27] and is about one order of magnitude larger than values of wind shear velocity of desert dune fields on Earth [28].

We have thus applied theory of sediment transport, which has proven to successfully describe dune formation in a broad range of physical environments (including Venus, Earth and Mars) [3, 13, 15, 16, 18, 21, 25], to estimate possible values of average grain size and wind speed formative of crescent-shaped bedforms occurring between Norgay and Hillary Montes on Pluto. In particular, the theory accounts for the stronger hysteresis in the threshold for saltation under rarified atmospheres (compared to the Earth). From the scale of the bedforms in Fig. 1, the theory predicts they are constituted by (coarse) sand-sized particles and were formed by wind speeds that are typical of Martian storms. Moreover, it is clear that better-quality images would be necessary to clarify the details of these bedforms, in spite of the resemblance of their crescent shape to barchan dunes on Earth and Mars. However, there is one further and stronger argument in favor of aeolian origin of the referred bedforms, that is the dependence of their shape (morphometry) on the size follows the same behavior of barchans on Earth and Mars (in particular the existence of a minimal size). Nevertheless, it remains uncertain how these bedforms could form, that is what is the origin of the sediments?

Possible sources for dune sediments, which have been identified in a previous discussion [10], include impact ejecta, as well as fine component of regolith impact; eruptive plume ("cryoclastic") fallout -- on Mars, the particle diameter of dunes sand, which is of volcanic origin, is most probably within the range between 100 – 600 μm [25, 29], thus smaller than the value obtained here for Pluto's dunes; atmospheric precipitation (snow), and abrasion of bed-ice surfaces. For the barchan dunes investigated here, one particular aspect in favor of last factor is the proximity of the dunes to the icy mountains at the dune's upwind area. Clarifying the origin of the sediments might help constrain the expected range of grain sizes which has been predicted from our calculations.

Whatever the origin the sediments, the present manuscript has adopted the assumption of an aeolian origin for the bedforms in Fig. 1, based on the observations described previously. However, despite the above discussion, it is pertinent question whether other processes but aeolian transport may have contributed to generate the sediments or even to shape these bedforms. While previous works already recognized the possible role of aeolian transport for shaping Pluto's surface, here we have shown how theory can be applied to obtain information on grain size and wind speed from the scale of dunes on Pluto. The method presented here should be now applied to other candidate dune fields at other places on Pluto. Our calculation results remain to be confirmed by data obtained by another approach, such as wind tunnel experiments of aeolian transport under conditions of pressure and temperature valid for Pluto.

While the present estimate of particle diameter is based on well-established theoretical expressions, which have proven to successfully predict the minimal dune size over 5 orders of magnitude in grain to fluid density ratio [18], it relies on the assumption that Pluto's bedforms have been formed under the average atmospheric density of *today's* Pluto. We note that global circulation models of Pluto have brought useful insights about wind behavior at heights larger than $\sim$ 200 m [30], and it would be interesting to compare our predicted range of average surface wind speeds with results from a mesoscale atmospheric model that resolves local topography. As shown recently [11], Pluto's surface displays a broad range of landform types and a diversity of terrain ages, and thus the quantitative investigation performed here shall provide one useful means for constraining possible origins of different types of aeolian bedforms to be imaged by New Horizons in the months and years ahead.

Acknowledgements

We acknowledge the German Research Foundation (DFG) for funding through its Cluster of Excellence "Engineering of Advanced Materials", ZISC, FPS, at the University of Erlangen-Nuremberg. E.J.R.P. acknowledges financial support from DFG through Grant RI 2497/3-1.

References

[1] https://www.nasa.gov/mission_pages/newhorizons/main/
[2] A. Witze, Nature, doi:10.1038/nature.2015.17986 (2015).
[3] E. J. R. Parteli, O. Durán and H. J. Herrmann, Phys. Rev. E **75**, 011301 (2007).
[4] R. A. Bagnold, *The Physics of Blown Sand and Desert Dunes* (Methuen, London, 1941).
[5] K. Pye and H. Tsoar, *Aeolian Sand and Sand Dunes* (Springer, Berlin, 2009).
[6] H. J. Finkel, J. Geol. **67**, 614 (1959).
[7] S. L. Hastenrath, Z. Geomorphol. **11**, 300 (1967).
[8] M. C. Bourke, Icarus **205**, 183 (2010).
[9] E. J. R. Parteli *et al.*, Aeolian Research **12**, 121 (2014).
[10] J. M. Moore, *et al*, Icarus **246**, 65 (2015).
[11] S. A. Stern, *et al.*, Science **350**, 6258 (2015).
[12] C. B. Olkin, *et al.*, Icarus **246**, 220 (2015).
[13] T. Pähtz, J. F. Kok, E. J. R. Parteli and H. J. Herrmann, Phys. Rev. Lett. **111**, 218002 (2013).
[14] M. P. Almeida, E. J. R. Parteli, J. S. Andrade and H. J. Herrmann, PNAS **105**, 6222 (2008).
[15] J. F. Kok, Phys. Rev. Lett. **104**, 074502 (2010).
[16] T. Pähtz, J. F. Kok and H. J. Herrmann, New J. Phys. **14**, 043035 (2012).
[17] C. de Beule, *et al.*, Nature Physics **10**, 17 (2014).
[18] K. Kroy, G. Sauermann and H. J. Herrmann, Phys. Rev. Lett. **88**, 054301 (2002).
[19] B. Andreotti, P. Claudin and S. Douady, Eur. Phys. J. B **28**, 321 (2002).
[20] E. J. R. Parteli, K. Kroy, H. Tsoar, J. S. Andrade Jr. and T. Pöschel, EPJST **223**, 2269 (2014).
[21] P. Claudin and B. Andreotti, Earth and Planetary Science Letters **252**, 30 (2006).
[22] P. Claudin, G. F. S. Wiggs and B. Andreotti, Boundary-Layer Meteorology **148**, 195 (2013).
[23] E. R. van Driest, Journal of the Aeronautical Sciences **23**, 1007 (1956).
[24] Y. Shao and H. Lu, J. Geophys. Res. **105**, 22437 (2000).
[25] J. F. Kok, E. J. R. Parteli, T. I. Michaels and D. Bou Karam, Rep. Prog. Phys. **75**, 106901 (2012).
[26] D. M. Burr, *et al.*, Nature **517**, 60 (2015).
[27] R. Greeley and J. D. Iversen, *Wind as a Geological Process on Earth, Mars, Venus and Titan* (Cambridge University Press, New York, 1985).
[28] S. G. Fryberger and G. Dean, *Dune forms and wind regime*. In: E. D. McKee (ed.), A Study of Global Sand Seas, no. 1052, Geol. Surv. Prof. Pap. 137 (United States Government Printing Office, Washington, 1979).
[29] K. S. Edgett and P. R. Christensen, J. Geophys. Res. **96**, 22765 (1991).
[30] A. M. Zalucha, Monthly Notices of the Royal Astronomical Society **459**, 902 (2016).

Balancing size and density segregation in bidisperse dense granular flows

Deepak R. Tunuguntla[1,*] and *Anthony R.* Thornton[1,**]

[1] *Multi-Scale Mechanics Group, Dept. of Thermal and Fluid Engineering, University of Twente, Enschede, The Netherlands.*

Abstract. Several experimental studies have illustrated a balance between the segregation forces arising due to size- and density-differences. However, no detailed studies have been carried out to quantify this balance. In 2014, by utilising discrete particle simulations, we presented a simple relationship between the particle size- and density-ratio, $\hat{s}^a = \hat{\rho}$, where 'a' determines whether the partial pressure scales with the diameter, surface area or volume of the particle. For a 50:50 mix (in volume) of bidisperse granular mixtures, we found the partial pressure to scale with the volume of the particle, i.e. $a = 3$. Moreover, there also exists a range of size- and density-ratios that satisfy the relation $\hat{s}^3 = \hat{\rho}$, where the bidisperse mixture remains homogeneously mixed. However, in this proceeding, we deviate from the conventional 50:50 mixes and consider a slightly extreme case of mixes, such as the 10:90 (in volume) mixes, which are often found in nature and industries. By doing so we observe that the partial pressure does not scale with the particle volume and, more importantly, the zero-segregation relation is not as simple as $\hat{s}^a = \hat{\rho}$. However, there does exist a range of size- and density-ratios for which the mixture weakly segregates.

1 Introduction

Granular mixtures often tend to arrange themselves in certain patterns when stirred, shaken or sheared [1]. These mixtures often comprise of constituents varying in size, density, inelasticity, shape, surface roughness etc. When such polydisperse mixtures are subjected to external forces, individual studies [2–10] confirm the influence of these constituent properties in forming patterns, e.g., segregation or de-mixing. However, in free-surface flows over inclined channels, differences in size and density are the primary factors for segregation.

Among several mechanisms responsible for segregation, *kinetic sieving* [11] is the dominant one in dense granular flows. *Kinetic sieving* is a trivial mechanism. As the mixture flows down the inclined channel, fluctuations in the local pore space cause smaller particles to fall into space/gaps created beneath them. The fines, i.e. small sized particles, easily fit into these pores leading to gradual percolation of them towards the base of the flow. Simultaneously, force imbalances lever/squeeze the coarser particles towards the surface. This simple mechanism results in stratified layers which one terms as segregation. Opposing kinetic sieving is diffusive remixing which causes random motions of particles as they collide and shear over each other [12]. Based on the relative strength of either mechanisms, the mixture strongly or weakly segregates. Apart from kinetic sieving, which is a purely size-based effect, buoyancy effects due to differences in particle density also play a major role in particle segregation [13]. For bi-disperse mixtures, varying in particle size and density, experiments [14] and simulations [6] indicate a balance between the two driving mechanisms, i.e between kinetic sieving and buoyancy, which in turn keeps the mixture homogeneously mixed.

Although experiments do provide authentic information, current state-of-the-art experimental techniques are still hampered due to some inherent constraints. Thereby, as an alternative to experiments, we utilise information-rich discrete particle simulations (DPMs) [15]. Given $\hat{s}$ and $\hat{\rho}$ is denoted as size- and density-ratio for a bidisperse mixture, in 2014, *Tunuguntla et al.* [6] proposed a DPMs validated theoretical relation

$$\hat{s}^a = \hat{\rho}, \tag{1}$$

where 'a' here determines whether the partial pressure scales with the diameter, surface area or volume of the particle[1]. More importantly, the relation further indicates that there exists a possible range of size- and density-ratios for which the bidisperse mixture very weakly or does not segregate. Although, the above relation has been validated using DPMs, the simulation setup in [6] considered a conventional shallow 50:50 bidisperse mixture alone. By 50:50 we imply that each particle type occupies 50% of the total solid volume. Nevertheless, for a given range of size- and density-ratios, they found the exponent $a = 3$ implying that the partial pressure scales with the volume of the particle.

In this proceeding, by utilising the discrete particle simulations as setup in Sec. 2, we aim to scrutinise the

* e-mail: d.r.tunuguntla@utwente.nl
** e-mail: a.r.thornton@utwente.nl

[1] For more details concerning the definition of partial pressures, see *Tunuguntla et al.* [6].

DOI: 10.1051/epjconf/201714003079

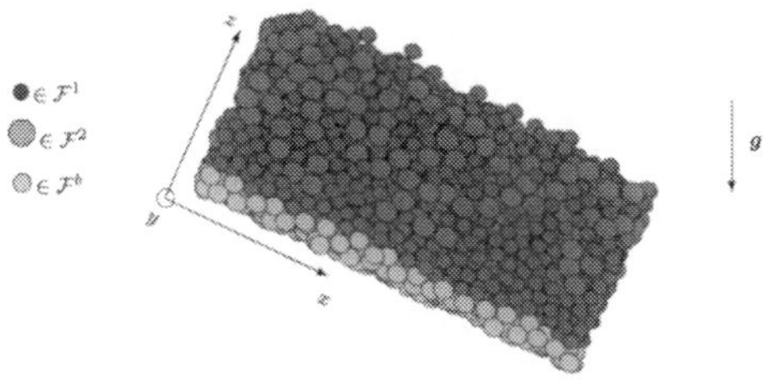

Figure 1. A illustration of a bidisperse mixture flowing in a periodic box inclined at 26° to the horizontal (discrete particle simulation), adapted from *Tunuguntla et al.* [16]. Colours/shades indicate the base/boundary (yellowish green, $\mathcal{F}^b$), species type-1 and type-2 (blue, $\mathcal{F}^1$ and red, $\mathcal{F}^2$). We define the *bulk* as $\mathcal{F}^1 \cup \mathcal{F}^2$.

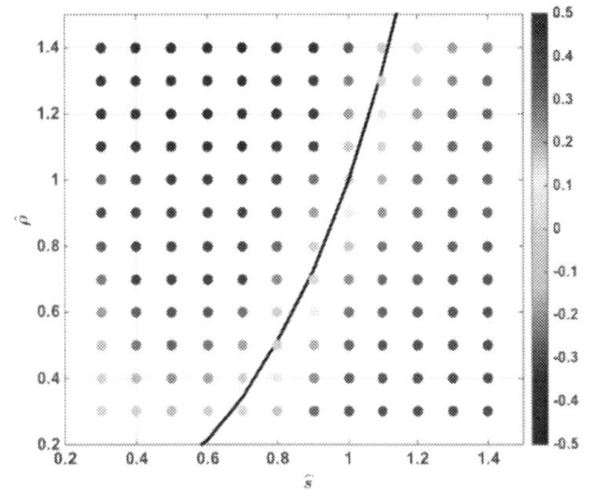

Figure 2. For a 50:50 initially homogeneously mixed bidisperse mixture, we plot $\widehat{D}_{com} = (\hat{z}^2_{com} - \hat{z}^{bulk}_{com})/\hat{z}^{bulk}_{com}$ for different values of $\hat{s}$ and $\hat{\rho}$. The black solid line represents the weak segregation line, which satisfies the simple relation in Eqn. 1 for $a = 3$. Note that the above plot is adapted from *Tunuguntla et al.* [6].

generality of the above relation (1). In order to do so, we sway away from the conventional 50:50 bidisperse mixtures and consider a 10:90 bidisperse mixtures, i.e. one of the particle type occupies 10% of the total solid volume in comparison to the other particle type. Moreover, it is the 10:90 mixes that are more common, in many natural and industrial scenarios, than the conventional 50:50 mixes. For example, in a pharmaceutical industry the final chemical composition of a drug often contains larger amounts of excipient than the active ingredient, and practically all marketed drugs contains excipients[2]. Thereby resulting in dilute granular mixtures such as the 10:90 bidisperse mixtures. When these dilute mixtures are subjected to external shear while transporting and processing, phenomena such as particle segregation often lead to unwanted inhomogeneities in the mixtures, which in turn hamper the final drug composition. Hence, demanding for thorough investigations concerning particle segregation in dilute mixtures. Here, we consider dilute bidisperse granular flows in inclined channels, which is a commonly utilised method of transport in industry.

2 Simulation setup

For a range of size- and density-ratios, $\hat{s} \in [0.3, 1.4]$ and $\hat{\rho} \in [0.3, 1.4]$, fully three-dimensional (3D) discrete particle simulations (DPM) are set up in our in-house open-source particle solver, MercuryDPM [17].

A cuboidal box, periodic in x- and y-direction, is inclined to the horizontal at 26°, with the dimensions $L \times W \times H = 20d_m \times 10d_m \times 10d_m$. Here, d_m represents the mean particle diameter defined as $d_m = \phi^1 d_1 + \phi^2 d_2$, where ϕ^i and d_i denotes the solid volume fraction and diameter of type-i particle species. In this proceeding, we define the particle size- and density-ratio as $\hat{s} := d_2/d_1$ and $\hat{\rho} := \rho_p^2/\rho_p^1$, where ρ_p^i is defined as the particle density of particle type-i.

In order to add a rough base (bottom) to the cuboidal box, we fill the periodic box with a randomly distributed set of particles with uniform diameter d_m. Once filled, they are simulated until a static layer of about 12 particles thickness is produced. To form a rough base, a slice of particles with centres between $z \in [9.3, 11]d_m$ is taken and translated 11 mean particle diameters downwards, see *Weinhart et al.* [18]. Additionally, to make sure no particles fall through the created rough base, a solid flat wall is placed underneath this static layer. After the creation of a rough base, the box is then filled with a homogeneously mixed bidisperse mixture with a specific size- and density-ratio, as seen in Fig. 1. For more details regarding the procedure for determining the number of particles and the contact model, please see *Tunuguntla et al.* [16, 19].

Given the particle size- and density-ratio, $\hat{s} \in [0.3, 1.4]$ and $\hat{\rho} \in [0.3, 1.4]$, total number of particles, N_1 and N_2, and contact model parameters, the particles are randomly positioned into the box with dimensions $\widehat{V}_{box} = 20 \times 10 \times H$ where H is defined as $(N_1 + N_2)/300$. However, when an inserted particle at any position overlaps with another particle, the insertion is rejected and the insertion domain is enlarged by increasing H to $H + 0.01H$ to ensure that there is enough volume in the box to accomodate all the particles. Thus, leading to an initially loosely packed mixture. As the simulation commences, the particles compact enough to reach a flow depth of $12d_m$, see Fig. 1, giving the particles enough energy to initialise flow. For more details see [16, 18, 19].

With the particle simulations setup in the above elucidated manner, the following section presents our observations.

3 Results

Homogeneously mixed bidisperse mixtures are simulated for a range of particle size- and density-ratios, $\hat{s} \in [0.3, 1.4]$ and $\hat{\rho} \in [0.3, 1.4]$, till a steady state is reached.

[2]By excipients, we mean bulking agents or fillers or diluents which facilitate drug solubility and easy dosage whereas the active ingredient is the one responsible for killing the virus for example.

DOI: 10.1051/epjconf/201714003079

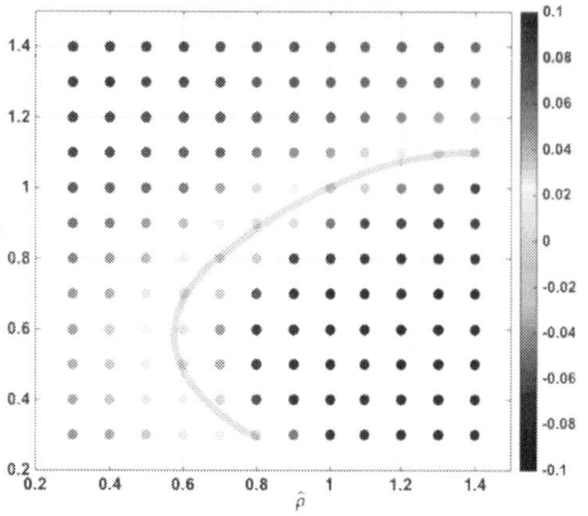

Figure 3. For a 10:90 initially homogeneously mixed bidisperse mixture, we plot $\widehat{D}_{com} = (\hat{z}^2_{com} - \hat{z}^{bulk}_{com})/\hat{z}^{bulk}_{com}$ for different values of size- and density-ratios, $\hat{s}$ and $\hat{\rho}$. Similar to the black solid line in Figure 2, the greyish transluscent line in the above plot approximately represents the weak segregation line in 10:90 dilute mixtures. However, this segregation line is no longer captured by the simple relation stated in Eqn. 1. Hence, implying that the value of a is dependent on the mixture volume fraction. Also note that the $x-$ and $y-$axis labels are flipped when compared to the ones in Fig. 2.

To define a steady mixture state, we track the vertical center of mass ($\hat{z}_{com}$) of both the type-1 and -2 species as a function of time, similar to the plots illustrated in Fig. 3 of *Tunuguntla et al.* [6]. As $\hat{z}_{com}$ reaches a steady value in time, we define our mixture to have reached a steady state. With this at hand, we first compute the relative difference between the vertical center of mass of the type-2 species ($\hat{z}^2_{com}$) and the bulk ($\hat{z}^{bulk}_{com}$) and then normalise it with the bulk's vertical center of mass, $\hat{z}^{bulk}_{com}$. By this, we arrive at a normalised relative measure $\widehat{D}_{com} = (\hat{z}^2_{com} - \hat{z}^{bulk}_{com})/\hat{z}^{bulk}_{com}$ as a function of $\hat{s}$ and $\hat{\rho}$.

As a brief recap, for 50:50 mixes *Tunuguntla et al.* [6] illustrated the normalised relative measure, $\widehat{D}_{com}$, for type-2 particle species, which is also shown in Fig. 2. Given the flows are in a steady state, for $\widehat{D}_{com} > 0$, type-2 particle species appear to be near the free surface of the flow; whereas for $\widehat{D}_{com} < 0$, type-2 particle species appear to be near the base of the flow. Moreover, Fig. 2 also illustrates the important aspect of very weak/zero segregation, which mainly occurs due to a balance between the segregation forces arising due to size- and density-differences. Additionally, the black solid line denotes the weak/zero segregation line which is represented by the relation $\hat{s}^{a=3} = \hat{\rho}$. Thus, also implying that the partial pressure is scaled by the volume of the particle, as $a = 3$.

On the other hand, to verify the generality of the relation $\hat{s}^a = \hat{\rho}$, we further simulated 10:90 mix of homogeneously mixed bidisperse mixtures, for the same earlier mentioned range of size- and density-ratios, $\hat{s} \in [0.3, 1.4]$ and $\hat{\rho} \in [0.3, 1.4]$. Once again, in Fig. 3, we plot the normalised relative measure, $\widehat{D}_{com}$, for type-2 particle species.

As illustrated in Fig. 3, similar to 50:50 mixes, there also exists a range of $\hat{s}$ and $\hat{\rho}$ for which the mixture exhibits weak or zero-segregation. However, the zero-segregation line is no longer a simple relation between the size- and density-ratio, as shown in Eqn. 1. There definitely exists a deviation from the relation stated in (1), implying that the relation $\hat{s}^{a=3} = \hat{\rho}$ cannot be generalised to dilute mixtures. Additionally, given the relation in (1) is not satisfied for $a \neq 3$, the partial pressure does not scale with the volume of the particle, as observed for mixtures with 50:50 volume ratio. Nevertheless, on closer observation of Fig. 3, we do also see that for $\hat{\rho} \in [0.6, 0.8]$ there exists a set of size-ratios for which the mixture remains homogeneously mixed. For example, for the same density-ratio, $\hat{\rho} = 0.7$, there are two size-ratios $\hat{s} = 0.4$ and $\hat{s} = 0.8$ at which the mixture weakly segregates. Thus, asking for a detailed future investigation.

4 Conclusions

By utilising discrete particle simulations, for both 50:50 and 10:90 mixes, we show that there exists a range of particle size- and density-ratios for which a homogeneously bidisperse, in size and density, mixture demonstrates weak/zero segregation. However, the simple relation (1) that captures the zero-segregation line for 50:50 mixes, does not apply for 10:90 mixes, implying that the relationship cannot be generalised to general mixes. Moreover, the exponent a in the relation (1) could be considered as a material parameter which is a function of particle volume fraction or material properties. However, further detailed study is required for investigating the segregation dynamics due to size- and density-differences.

4.1 Future outlook

Understanding particle segregation due to size- and density-differences is vital due to its wide range of applications in both nature and industry. Although, discrete particle simulations give detailed insight into the segregation dynamics, modelling realistic industrial flows is computationally expensive. Thereby, continuum models appear to be an efficient alternative. Similar to the several models developed to model size-based particle segregation models [19], developing efficient segregation models to account for, both, size- and density-differences would be useful. Although there exist a few models that account for these two effects [6, 20, 21], a detailed study using experiments or simulations would still be beneficial for better understanding and development of efficient continuum models.

Acknowledgements

The authors would like to thank the Dutch *Technology Foundation STW* for its financial support and user committee meetings of STW-Vidi project 13472, *Shaping Segregation: Advanced Modelling of Segregation and its Application to Industrial Processes*.

References

[1] J. Duran, *Sands, Powders, and Grains* (Springer, 2000)
[2] S. Wiederseiner, N. Andreini, G. Épely-Chauvin, G. Moser, M. Monnereau, J.M.N.T. Gray, C. Ancey, Phys. Fluids **23**, 013301 (2011)
[3] K. van der Vaart, P. Gajjar, G. Epely-Chauvin, N. Andreini, J.M.N.T. Gray, C. Ancey, Phys. Rev. Lett. **114**, 238001 (2015)
[4] F. Guillard, Y. Forterre, O. Pouliquen, J. Fluid Mech. **807** (2016)
[5] A. Tripathi, D.V. Khakhar, J. Fluid Mech. **717**, 643 (2013)
[6] D.R. Tunuguntla, O. Bokhove, A.R. Thornton, J. Fluid Mech. **749**, 99 (2014)
[7] H. Xiao, P.B. Umbanhowar, J.M. Ottino, R.M. Lueptow, *Modelling density segregation in flowing bidisperse granular materials*, in *Proc. R. Soc. A* (The Royal Society, 2016), Vol. 472, p. 20150856
[8] R. Brito, R. Soto, Eur. Phys. J. Special Topics **179**, 207 (2009)
[9] B.L. Pollard, H. Henein, Can. Metall. Q. **28**, 29 (1989)
[10] S. Ulrich, M. Schröter, H.L. Swinney, Phys. Rev. E **76**, 042301 (2007)
[11] J.A. Drahun, J. Bridgwater, Powder Technol. **36**, 39 (1983)
[12] S.S. Hsiau, M.L. Hunt, J. Fluid Mech. **251**, 299 (1993)
[13] D.V. Khakhar, J.J. McCarthy, J.M. Ottino, Chaos **9**, 594 (1999)
[14] G. Felix, N. Thomas, Phys. Rev. E **70**, 051307 (2004)
[15] S. Luding, European Journal of Environmental and Civil Engineering **12**, 785 (2008)
[16] D.R. Tunuguntla, A.R. Thornton, T. Weinhart, Comput. Part. Mech. **3**, 349 (2016)
[17] A.R. Thornton, D. Krijgsman, R. Fransen, S. Gonzalez, D.R. Tunuguntla, A. te Voortwis, S. Luding, O. Bokhove, T. Weinhart, Newsletter Enginesoft **1**, 48 (2013)
[18] T. Weinhart, A.R. Thornton, S. Luding, O. Bokhove, Granul. Matt. **14**, 531 (2012)
[19] D.R. Tunuguntla, T. Weinhart, A.R. Thornton, Comput. Part. Mech., DOI: 10.1007/s40571-016-0136-1 (2016)
[20] B. Marks, P. Rognon, I. Einav, J. Fluid Mech. **690**, 499 (2012)
[21] J. Gray, C. Ancey, Journal of Fluid Mechanics **779**, 622 (2015)

Effects of hydrodynamic interaction on random adhesive loose packings of micron-sized particles

Wenwei Liu[1], *Ran* Tao[1], *Sheng* Chen[1], *Huang* Zhang[1], and *Shuiqing* Li[1,*]

[1]*Key Laboratory for Thermal Science and Power Engineering of Ministry of Education, Department of Thermal Engineering, Tsinghua University, Beijing 100084, China*

Abstract. Random loose packings of monodisperse spherical micron-sized particles under a uniform flow field are investigated via an adhesive discrete-element method with the two-way coupling between the particles and the fluid. Characterized by a dimensionless adhesion parameter, the packing fraction follows the similar law to that without fluid, but results in larger values due to the hydrodynamic compression. The total pressure drop through the packed bed shows a critical behaviour at the packing fraction of $\phi \approx 0.22$ in the present study. The normalized permeability of the packed bed for different parameters increases with the increase of porosities and is also in consistent with the Kozeny-Carman equation.

1 Introduction

Understanding the properties of random packings of uniform spherical particles has drawing a lot of scientific attractions, since random packings are classical models of simple liquids, colloidal systems as well as granular matter [1–3] and also occur in a variety of engineering applications including fibrous filtration, industrial manufacturing, agriculture and physics [4, 5]. In packings of large grains where elastic and frictional forces dominate, the packing fraction ϕ expands between two well-known limits, the *random close packing* (RCP) limit at $\phi_{\mathrm{RCP}} \approx 0.64$ and the *random loose packing* (RLP) limit at $\phi_{\mathrm{RLP}} \approx 0.55$ [1, 2, 6, 7]. However, most of the real particles in the nature are micron-sized, which are subject to not only elastic and frictional forces, but also adhesive forces, electrostatic forces and hydrodynamic forces [4, 5]. In this case, both the structural and mechanical properties of micro-particle packings could be intrinsically different in presence of such interactions.

Previous simulations have found that both the volume fraction and the coordination number of micro-particle packings can go far below the RLP limit when considering the van der Waals adhesive forces [8, 9]. The effect of adhesion can be characterized by a dimensionless adhesion parameter, which uniquely determines the packing fraction ϕ and coordination number Z [8–10] with fixed friction coefficient. The introduction of electrostatic repulsion reduces both ϕ and Z due to the deceleration of the incoming particles, resulting in weaker inertial impaction and looser packing structure [11]. As for experimental investigations, a relatively low packing fraction of $\phi = 0.15$ was achieved by random ballistic deposition of $1.5\mu m$ particles in vacuum [12]. Similar results were obtained with fluidization technique that jamming of $7.8 - 19.1\mu m$ particles takes place in the range $\phi = 0.23 - 0.33$, which is related to the attractive energy [13]. Beyond jamming, the increase of hydrodynamic pressure results in the compression and densification of the micro-particle packing, and the applied stress is found to follow a power-law scaling $\sigma \sim (\phi - \phi_J)^{\beta}$ near jamming.

*e-mail: lishuiqing@tsinghua.edu.cn

In this work, we study the random packings of micron-sized spherical particles under a uniform flow field by means of a CFD-DEM coupled approach. The influence of the particle clusters on the fluid flow is well considered, through a two-way coupling of the fluid motion and the particle motion. The packing fraction with different material properties is examined and regressed with respect to the adhesion parameter. The pressure drop through the packing as well as the permeability are also analysed and discussed.

2 Models and method

2.1 Two-way coupled CFD-DEM approach

We perform numerical simulations of micro-particle packings via a discrete element method (DEM), which is coupled into an framework of the open-source CFD codes MFIX. The governing equations of the fluid-phase involve the mass and momentum conservation, which can be expressed as

$$\begin{aligned} &\frac{\partial \epsilon_f}{\partial t} + \nabla \cdot (\epsilon_f \mathbf{u}) = 0, \\ &\frac{D}{Dt}(\epsilon_f \rho_f \mathbf{u}) = \nabla \cdot \overline{S}_f - \mathbf{I}_{fs}. \end{aligned} \tag{1}$$

Here, ϵ_f is the fluid-phase volume fraction, ρ_f is the mass density of the fluid, $\mathbf{u}$ is the volume-averaged fluid velocity, $\overline{S}_f$ is the stress tensor of the fluid and $\mathbf{I}_{fs}$ is the momentum transfer term between the fluid and particle phases.

The computations of the flow field and the particles' information are two-way coupled through the the momentum transfer, which means that in every time step the flow field is first updated in the presence of the particle phase and then the forces acting on the particles are recalculated with the newest flow field. The DEM framework used in this work is specifically developed for micro-particles. The multi-coupling of adhesive force, friction and other interactions in the short-range particle-particle interaction zone is well tested and validated by particle/surface impact experiments [4, 5]. Both the translational and rotational motions of each particle follow the Newton's second law,

$$\begin{aligned} m\frac{d\mathbf{v}}{dt} &= \mathbf{F}_f + \mathbf{F}_a, \\ I\frac{d\mathbf{\Omega}}{dt} &= \mathbf{M}_f + \mathbf{M}_a, \end{aligned} \quad (2)$$

where $\mathbf{v}$ and $\mathbf{\Omega}$ are, respectively, velocity and rotation rate of an individual particle, m is the particle mass, and I is the moment of inertia. Gravity is not considered here. $\mathbf{F}_f$ and $\mathbf{M}_f$ are the fluid forces and torques exerting on a particle. $\mathbf{F}_a$ and $\mathbf{M}_a$ denote the van der Waals contact forces and torques acting on the particle. They include

$$\begin{aligned} \mathbf{F}_a &= F_n\mathbf{n} + F_s\mathbf{t}_s, \\ \mathbf{M}_a &= r_p F_s(\mathbf{n}\times\mathbf{t}_s) + M_r(\mathbf{t}_s\times\mathbf{n}) + M_t\mathbf{n}, \end{aligned} \quad (3)$$

where F_n is the normal force including adhesively elastic contact force F_{ne} and damping force F_{nd}, F_s is the tangential force due to the sliding friction, M_r is the rolling resistance and M_t is the twisting resistance. r_p is the particle radius. $\mathbf{n}$, $\mathbf{t}_s$ and $\mathbf{t}_r$ are the normal, tangential and rolling direction unit vectors, respectively. The Johnson-Kendall-Roberts (JKR) model is employed to account for F_{ne} between relatively compliant micro-particles [14]. The sliding, twisting and rolling friction terms in the presence of adhesion, are all approximated by a spring–dashpot–slider model with model parameters given in Ref. [15]. The critical values $F_{s,\mathrm{crit}}$, $M_{t,\mathrm{crit}}$ and $M_{r,\mathrm{crit}}$ are given by

$$\begin{aligned} F_{s,\mathrm{crit}} &= \mu_f|F_{ne} + 2F_C|, \\ M_{t,\mathrm{crit}} &= 3\pi a F_{s,\mathrm{crit}}/16, \\ M_{r,\mathrm{crit}} &= -4F_C(a/a_0)^{3/2}\Theta_{\mathrm{crit}}R, \end{aligned} \quad (4)$$

meaning that once these limits are reached, the particles start to slide, spin or roll and the resistances stay constant. Here Θ_{crit} is the critical angle for the relative rolling between two particles, F_C is the critical pull-off force expressed by the surface energy γ: $F_C = 3\pi\gamma R$, R is the effective radius between two contacting particles ($1/R = 1/r_{p1} + 1/r_{p2}$), and a is the radius of the contact area with a_0 at equilibrium in the JKR model.

2.2 Simulation conditions

As shown in Fig. 1, the simulation domain sizes with $H\times L\times L$, where $L = 20r_p$ and $H = 10L$. The domain width L is selected as the characteristic length scale, and the dimensionless computational domain is set as $-5 \le X^* \le 5$, $-0.5 \le Y^* \le 0.5$ and $-0.5 \le Z^* \le 0.5$, with x-axis set along the direction of height. Periodic boundary conditions are set along the Y^* and Z^* directions. Our simulation starts from successive random injection of 2,000 uniform spheres at $X^* = -5$ with an initial velocity U_0. Note that the gravitational effect is eliminated, so that all the particles are transported by the fluid flow and deposit onto a plane at $X^* = 5$. The initial flow field is set to be uniform, with $U_x = U_0, U_y = U_z = 0$, and updated with the information of particles during every computational time step. Hence when the particles enter the domain, the flow field could be changed to be nonuniform. The physical parameters used in the simulations, including the particle and fluid properties, are all listed in Table 1. The friction coefficient is kept constant at $\mu_f = 0.3$. Two series of simulations were performed to investigate the effects of the physical parameters, where only one parameter was tuned while the others remained unchanged in each series. The tuned parameters are velocity and surface energy, respectively, of which the ranges can also be found in Table 1.

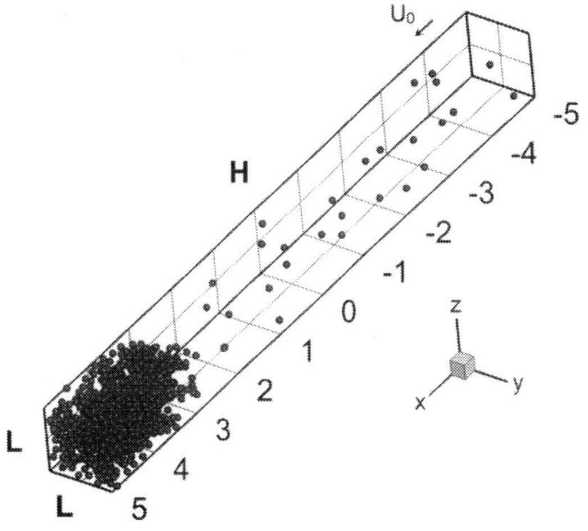

Figure 1. The schematic of the computational domain.

Table 1. Parameters used in the CFD-DEM simulations

Parameters	Value	Unit
Particle		
Particle radius (r_p)	1	μm
Mass density (ρ_p)	2500	kg/m^3
Elastic modulus (E_p)	2×10^7	Pa
Surface energy (γ)	0.03-125	mJ/m^2
Friction coefficient (μ_f)	0.3	
Fluid (Air)		
Mass density (ρ_f)	1.25	kg/m^3
Viscosity (μ)	1.79×10^{-5}	$Pa\cdot s$
Velocity (U_0)	0.01-5	m/s

3 Results and discussions

Figure 2 shows the packing fraction as a function of a dimensionless adhesion parameter [8, 11],

$$\mathrm{Ad} = \frac{\gamma}{\rho_p U_0^2 R}, \quad (5)$$

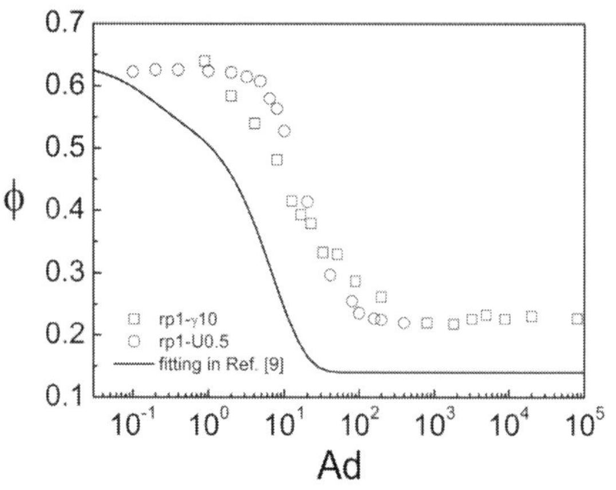

Figure 2. Packing fraction as a function of Ad. "rp1-γ10" means that the particle radius and surface energy are fixed at $r_p = 1\mu m$ and $\gamma = 10mJ/m^2$. The solid line is a fitting in Ref. [9] for micro-particle packing in vacuum.

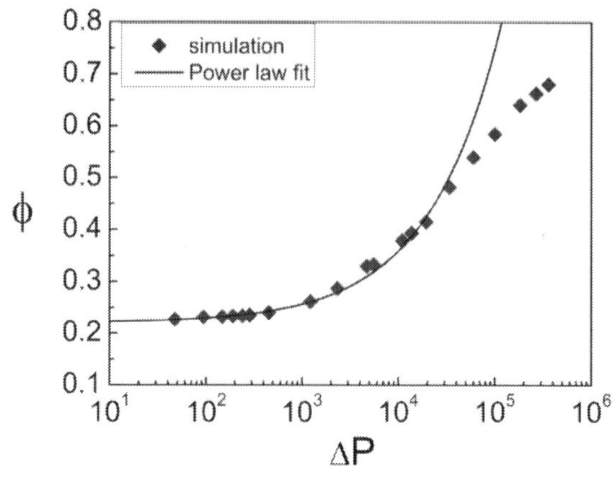

Figure 3. Variation of the applied stress with packing fraction. The solid line is a power-law fitting.

which represents the ratio of adhesion to particle inertia. It can be seen that with the increasing of Ad, packing fraction decreases exponentially from close to ϕ_{RCP} to a lower limit. A good regression can be observed for the two series simulations, as they still seem to fall on a single curve. However, the regression is not in consistent with that reported in Refs. [9] (the solid line), showing a shift to larger values of Ad. Note that the packings in [9] were generated in a vacuum environment. The introduction of the fluid flow exerts continuous compression on the packed aggregates and results in denser structure, which gives rise to a larger packing fraction at the same Ad. Since the flow velocity was fixed in each simulation case, a stable packing state will always come into being eventually due to the balance between fluid force and interparticle adhesive forces. The lowest packing fraction obtained in our simulations is around 0.22, which is higher than the conjectured ALP limit [8] due to the fluid compression. The highest packing fraction achieved is still around ϕ_{RCP}. This is because at this packing fraction, particles jam and geometrical constrains prevent the packing from reaching higher density through fluid compression.

In the study of the mechanical properties of the particle assembly, the consolidation stress plays an important role. Meanwhile, the pressure drop through packed bed is also an important parameter, since it serves as a fundamental property in fibrous filtration. Considering that gravity is not taken into account in our system, the normal consolidation stress applied on the packings equals the fluid pressure drop through the packed bed. It is well known that the pressure drop ΔP through a porous packed bed can be predicted by the Ergun function, or namely the Kozeny-Carman equation for a laminar flow field,

$$\frac{-\Delta P}{h} = \frac{180\mu U(1-\epsilon)^2}{d_p^2\epsilon^3}. \tag{6}$$

Here h is the height of the packed bed, U is the superficial fluid velocity and $\epsilon = 1 - \phi$ is the porosity of the packed bed. Figure 3 shows the variation of packing fraction with pressure drop. We see that the pressure drop derived from the simulations agrees well with the Kozeny-Carman equation, which testifies the validity of our numerical approach. Furthermore, when the fluid velocity is further decreased there comes a critical point at which the pressure drop gradually vanishes but the packing fraction stays. At this stage the clusters start to develop a yield stress, or in other words they are jammed [13]. It is also reported that the jamming point (ϕ_J) for adhesive particles depends on the attractive energy, i.e. the surface energy in our study, instead of the particle size [13]. Here the jamming point found in the present work is $\phi_J \approx 0.22$ for $\gamma = 10mJ/m^2$. More data is needed to further explore the relationship between ϕ_J and γ. Moreover, the stress near jamming is found to follow a power-law $\Delta P \sim (\phi - \phi_J)^\beta$ [13]. The power-law fitting is plotted as the solid line in Fig. 3. The critical exponent is $\beta = 1.788$ in our work, which is higher than $\beta = 1.5$ for Hertzian-type interactions [13].

Apart from the pressure drop, permeability is also critically important in industries, especially in the composite manufacturing processes. Permeability describes the ability of porous materials to allow for fluid to flow through. Based on the Darcy's law, permeability is defined as $K = U\mu h/\Delta P$. Combining the Kozeny-Carman equation (Eq. 6), permeability can also be calculated as $K = d_p^2\epsilon^3/(180(1-\epsilon)^2)$. From this definition, we can see that permeability depends on the particle size and the porosity. Thus, we can define a normalized permeability as

$$K^* = \frac{K}{d_p^2} = \frac{\epsilon^3}{(180(1-\epsilon)^2)}. \tag{7}$$

In this case, permeability is only a function of packing porosity. Figure 4 shows the normalized permeability for different series simulations along with the theoretical prediction of Eq. 7. Generally, we can see that the permeabil-

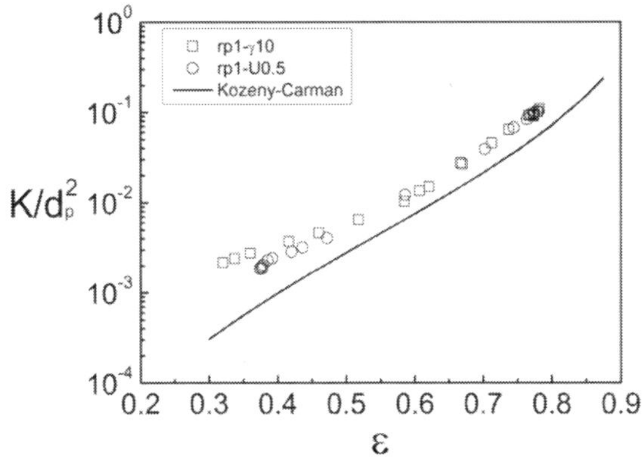

Figure 4. The normalized permeability as a function of the packing porosity. The solid line is the theoretical prediction derived from the Kozeny-Carman equation.

ity calculated from the simulation agrees with the theory, where a relatively large deviation occurs for small porosity.

4 Conclusions

In this work, a mesoscopic adhesive DEM incorporated with CFD approach was applied to investigate the random packings of monodisperse spherical micro-particles under a uniform flow field. The packing fraction with different material properties, the consolidation stress on the particle assembly and the permeability through the packed bed are examined and discussed. Conclusions are summarized as below.

(1) The packing fraction decreases with the increasing of a dimensionless adhesion parameter *Ad* but is higher than the reported value at the same *Ad*, which is due to the hydrodynamic compression.

(2) The pressure drop, or the consolidation stress on the packed bed agrees well with the Kozeny-Carman equation and follows the power-law scaling $\Delta P \sim (\phi - \phi_J)^\beta$ near the jamming point, which is found to be $\phi_J \approx 0.22$ for $\gamma = 10mJ/m^2$ under the present protocol.

(3) The normalized permeability agrees with the theoretical prediction derived from the Kozeny-Carman equation, except for a little deviation at low packing porosity.

Acknowledgements

This work has been funded by the National Natural Science Funds of China (Nos. 51390491) and the National Key Basic Research Program of China (No. 2013CB228506).

References

[1] J. D. Bernal, Nature **183**, 141-147 (1959).
[2] G. Parisi, F. Zamponi, Rev. Mod. Phys. **82**, 789-845 (2010).
[3] B. Andreotti, Y. Forterre, O. Pouliquen, *Granular Media between fluid and solid* (Cambridge University Press, Cambridge, 2013) p59-70.
[4] S. Q. Li, J. S. Marshall, G. Liu, Q. Yao, Prog. Energy Combust. Sci. **37**, 633-668 (2011).
[5] J. S. Marshall and S.Q. Li, *Adhesive Particle Flow: A Discrete-Element Approach* (Cambridge University Press, Cambridge, 2014) p2,p81,p98.
[6] C. Song, P. Wang, H. A. Makse, Nature **453**, 629-632 (2008).
[7] G. Y. Onoda, E. G. Liniger, Phys. Rev. Lett. **64**, 2727 (1990).
[8] W. W. Liu, S. Q. Li, A. Baule, H. A. Makse, Soft Matter **11**, 6492-6498 (2015).
[9] W. W. Liu, S. Q. Li, S. Chen, Powder Technol. **302**, 414-422 (2016).
[10] S. Q. Li, J. S. Marshall, J. Aerosol Sci. **38**, 1031-1046 (2007).
[11] S. Chen, S. Q. Li, W. W. Liu, H. A. Makse, Soft Matter **12**, 1836-1846 (2016).
[12] J. Blum, R. Schräpler, Phys. Rev. Lett. **93**, 115503 (2004).
[13] J. M. Valverde, M. A. S. Quintanilla, A. Castellanos, Phys. Rev. Lett. **92**, 258303 (2004).
[14] K. L. Johnson, K. Kendall, A. D. Roberts, Proc. R. Soc. London A **324**, 301-313 (1971).
[15] M. M. Yang, S. Q. Li, Q. Yao, Powder Technol. **248**, 44-53 (2013).

 DOI: 10.1051/epjconf/201714009038

Erosion of cohesive soil layers above underground conduits

Li-Hua Luu[1,*], *Pierre* Philippe[1], *Gildas* Noury[2], *Jérôme* Perrin[3], and *Olivier* Brivois[2]

[1] *IRSTEA, UR RECOVER, 3275 route de Cézanne, 13182 Aix-en-Provence, France.*
[2] *BRGM, DRP (Direction Risques et Prévention), 3 Av. Claude Guillemin, BP 36009, 45060 Orléans Cedex 2*
[3] *BRGM, D3E (Direction Eau, Environnement, Ecotechnologies), 3 Av. Claude Guillemin, BP 36009, 45060 Orléans Cedex 2*

Abstract. Using a recently developed 2D numerical modelling that combines Discrete Element (DEM) and Lattice Boltzmann methods (LBM), we simulate the destabilisation by an hydraulic gradient of a cohesive granular soil clogging the top of an underground conduit. We aim to perform a multi-scale study that relates the grain scale behavior to the macroscopic erosion process. In particular, we study the influence of the flow conditions and the inter-particle contact forces intensity on the erosion kinetic.

1 Introduction

Location of buried cavities, mostly in urban zones, represents a highly vulnerable area with regard to geotechnical aspects. Their collapse, that leads to sinkholes, is indeed a major cause of large soil subsidences. Our research focuses more specifically on the destabilization, after floods, of weak layers above existing conduits due to underground water flows, as depicted in Figure 1a. Up to now, the understanding of the resistance against erosion of cohesive soils remains poor, being mainly limited to empirical laws based on experimental tests performed in laboratory or *in-situ* [1, 2]. In this study, we propose to implement a numerical model, increasingly used in geomechanics [3, 4], that combines the Lattice Boltzman Method (LBM) and the Discrete Element Method (DEM) to efficiently describe the two-phase flow. As demonstrated in recent works, the proposed numerical model has the advantage to provide access to the local erosion behavior, at the grain scale [5–7].

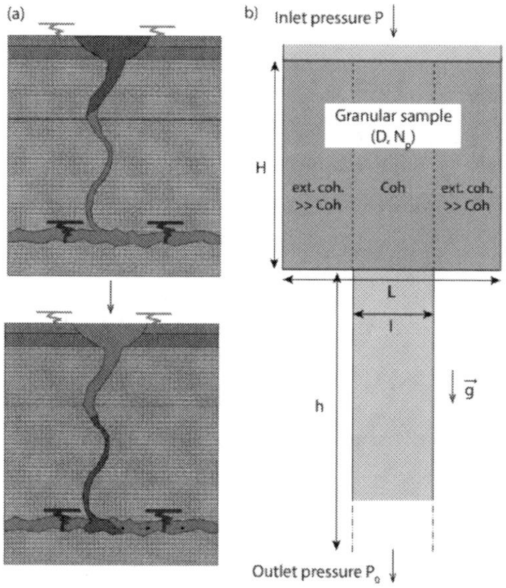

Figure 1. (a) Field situation of soil collapse within a flooded underground conduit. (b) Sketch of our numerical model whose parameters are indicated in Table 1.

2 Numerical model

On the one hand, the fluid phase is simulated by LBM, which is an explicit 2D finite difference scheme over a discrete lattice mesh of the continuous Boltzmann equation that involves the collision and propagation of fluid particles. We employ the classical 2D lattice with 9 particle velocity vectors, the so-called D2Q9 scheme. The collision process is described by the multiple relaxation time model (MRT) to overcome the shortcomings of the basic Bhatnaggar-Gross-Krook scheme (BGK) [8, 9].

On the other hand, our numerical model intends to model loose cemented granular materials using a 2D Discrete Element Method that includes cohesion *via* solid bonds between particles [10]. To this end, we use a simple elastoplastic model characterized by the normal and shear bond stiffnesses, $k_{n,b}$ and $k_{s,b}$, that includes a parabolic yield surface in the space of contact forces, as sketched in Figure 2. When a contact force reaches the yield surface, the bond is broken and the contact becomes purely frictional, *via* a Kelvin-Voigt relationship to model the normal force and a viscous-regularized Coulomb law for the shear force (see stiffness, damping and friction coefficients in

*e-mail: li-hua.luu@irstea.fr

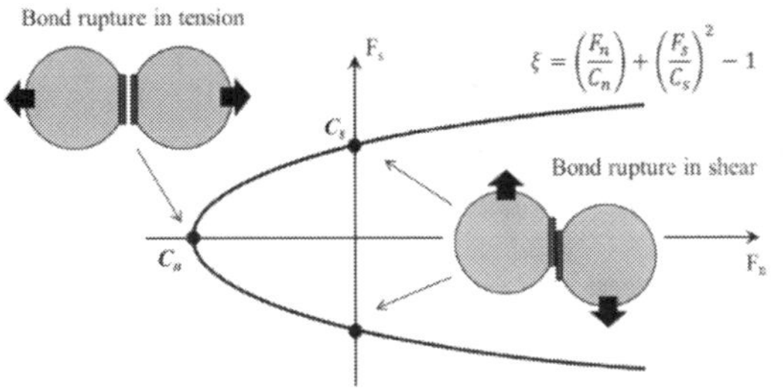

Figure 2. Yield surface of cohesive bonds in the space of interaction normal and shear forces (F_n, F_s), from [6]

Table 1). In the present numerical model, the normal and shear yield thresholds C_n and C_s are assumed to depend only on a single parameter $C=C_n= 2{\cdot}C_s$, which represents the inter-particle bond strength.

To deal with the solid-fluid coupling, we use a relationship based on a momentum-exchange method [11]. Finally, the LBM calculation is implemented with a hydraulic radius smaller than the real particule radius to recover a non-zero permeability in our 2D description.

Using this combined DEM-LBM method, we thus intend to model a specific soil erosion situation. The simulation setup consists in a conduit initially clogged by an immersed granular assembly as sketched in Figure 1b. In order to impose a preferential path for the flow within the granular sample, two soil cohesions are settled. In the external layers, the bond strength is fixed at C=1000 N, whereas in the central layer we set a much weaker cohesion magnitudes from 5 to 20 N. The erosion of the soil layer is triggered by a pressure gradient. As listed in Table 1, we simulate two samples of D=6 mm diameter grains, with a same height but two different lengths L, corresponding to particle numbers N_p equal to 11816 and to 3042.

In the following, the soil cohesion is characterized by a dimensionless cohesion number defined as the ratio of bond strength to the particle's own buoyant weight, $Coh = C/(\rho_s - \rho_f)gV$, where $\rho_s - \rho_f$ is the submerged apparent density, g is the gravitational acceleration and V is the particle's volume. The initiation of motion in sediment transport is classically analysed by defining the Shields number (the ratio of fluid force on the particle to the weight of the particle). Notice that for the cemented soil we model here, this criterion should be extended by introducing the cohesion number Coh. Work is under way on this issue.

3 Erosion regimes

The behavior of the granular layer submitted to a gradient pressure is parametrically studied. Table 1 reports the parameters implemented. By varying the conduit size Δ defined by $\Delta = l/D$, the granular cohesion given by the cohesion number Coh and the inlet fluid pressure P, we observe a "blocking" stable case where no erosion occurs and

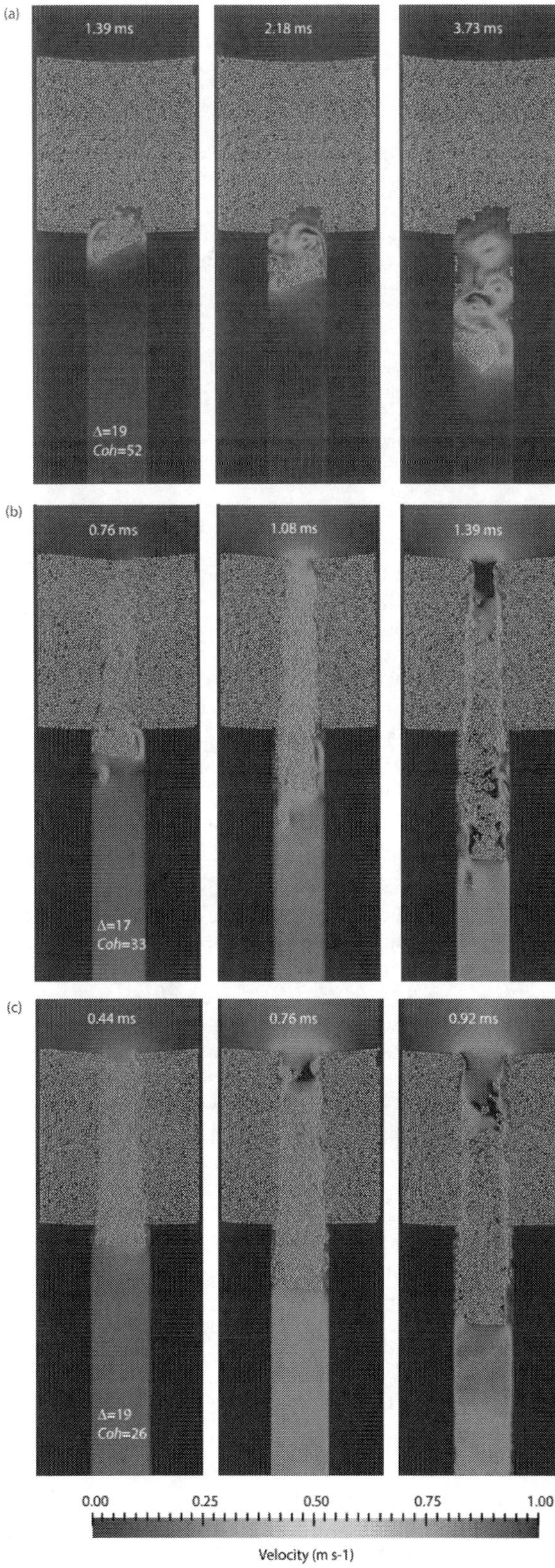

Figure 3. Different erosion regimes for the sample N_p=3042 and at P=10 Pa: (a) cavity formation, (b) backward extrusion and (c) plug extrusion

Table 1. Geometrical, solid and fluid model parameters implemented in the simulations.

Solid phase		Fluid phase	
Particle mean diameter, D	$6\cdot10^{-3}$ m	Conduit length, l	[3-27]·D
Polydispersity, D_{max}/D_{min}	1.5	Conduit height, h	100·D
Granular sample length, L	33·D, 67·D	Inlet pressure, P	[0-2000] Pa
Granular sample height, H	70·D	Outlet pressure, P_0	0 Pa
Particle density, ρ_s	$2.2\cdot10^3$ kg/m^3	Fluid kinematic viscosity, ν	$4\cdot10^{-5}$ m^2/s
Normal, shear and rolling stiffnesses $k_n = k_s = 10\cdot k_w$	$1.1\cdot10^5$ N/m	Fluid density, ρ_f	$8.5\cdot10^2$ kg/m^3
Friction coefficients, $\mu = 3\cdot\mu_w$	0.3	Lattice grid size, Δx	$6.3\cdot10^{-4}$ m
Damping coefficient, η	68.3 kg/s	Lattice speed, c_s	10 m/s
Bond stiffnesses, $k_{n,b} = 0.4\cdot k_{s,b}$	$0.5\cdot k_n$	Hydraulic radius factor, R_h	0.8
Bond strength, $C=C_n = 2C_s$	[5-20] N		

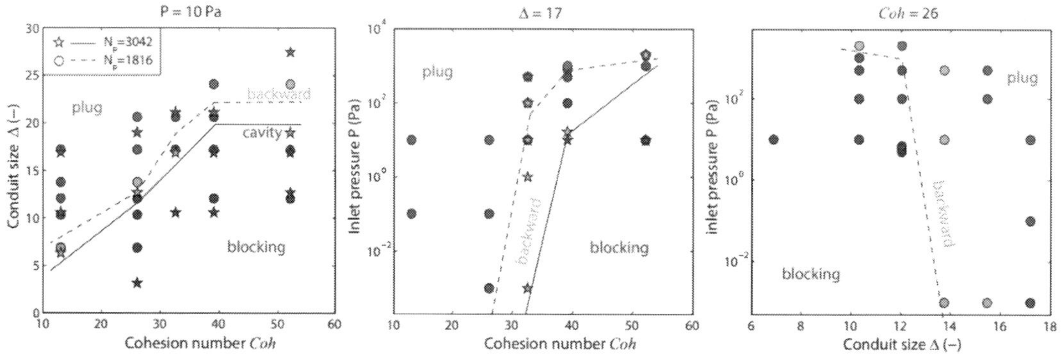

Figure 4. Phase diagrams: influence of the conduit size Δ, the cohesion number *Coh* and the inlet pressure on the different features, namely the blocking (blue), the cavity (grey), the backward extrusion (pink) and the collapse (red).

an intermittent evacuation of the granular sample through the predefined internal conduit. We here propose to separate this phenomenon into three different erosion regimes, as presented in Figure 3. At the lowest magnitude of erosion, a stable cavity can marginally be formed. If increasing eroding parameters, a backward erosion mechanism arises, where the soil destabilisation front direction is opposite to the flow direction. Once the erosion front reaches the top of the sample, the detached matter is evacuated by the flow. Finally, we call collapse, the regime where the whole granular column falls all at once, but not under its own weight. A bond breakage along the conduit, due to an actual erosion process triggered by the hydraulic gradient, precedes this collapse. It is important to note that one of the LBM limitation is that the fluid velocity has to be much smaller than the lattice speed c_s, so as to remain in the incompressible Navier-Stockes behavior. In this study, calculations stop when the fluid reaches 1 m/s. Consequently, we often cannot simulate the complete evacuation from the conduit, but as examinated later, its kinetic seems to reach a constant velocity that we are able to measure.

In Figure 4, we plot the three phase diagrams related to Δ, *Coh* and P for the two samples. We notably indicate the erosion threshold by the lines. Globally, erosion process can be triggered by a high enough conduit size, a high enough inlet pressure and a small enough cohesion. The first diagram shows that the higher *Coh* is, the higher Δ is required to erode soil. As regards the eroding effect of increasing P, the two other diagrams display very sharp threshold. This observation, that we are currently studying more thoroughly, could be related to Janssen effects.

Our simulations also point out the influence of the granular sample length L on the erosion threshold. This latter is slighthy "delayed" for the smaller sample, in the sense that a higher critical Δ, a smaller critical *Coh* and a higher critical P are involved in the erosion process. To capture the physics underlying this result, work under way is focusing on the intensity of inter-particle cohesion, in order to monitor how this microscale behaviour evolves towards bond rupture.

4 Erosion kinetic

From simulation pictures, we propose to plot spatio-temporal diagrams to analyse the erosion kinetic. Figure 5a displays a typical analysis following 3 steps (from left to right): (i) selection of the central part of the granular assembly, where particle motion occurs ; (ii) proceeding on spatio-temporal diagrams within this area along each vertical line with a spatial resolution of 1 mm ; (iii) calculation of the standard deviation over the set of diagrams. In the first case of backward extrusion, the time evolution of the grain's collective motion clearly presents two different kinetics (Fig. 5a). It first shows that the intiation of motion

Figure 5. (a) Backward extrusion: spatio-temporal analysis for the sample with N_p=1816, Δ=12, *Coh*=26 and P=10 Pa. (b) Plug extrusion: spatio-temporal diagram for the sample with N_p=1816, Δ=17, *Coh*=26 and P=10 Pa. More details in the text.

starts at the bottom of the granular sample while the top part stay static. This so-called backward mechanism is depicted by the green lines, horizontal at the top and tilted at the bottom. Second, particles move almost together with a constant velocity, represented by the two red lines having the same slope. Now let us slightly increase the conduit size to get the transition towards the purely plug extrusion. Figure 5b shows a typical spatio-temporal diagram of this regime. The red line indicates that we could consider, in a first approximation, a roughly constant evacuation of the whole block with the same velocity than during the final plug-like motion of the backward extrusion. This result therefore demonstrates a continuous transition between these two erosion processes.

5 Conclusion

This paper presents first results of a parametric study on numerical simulations of the destabilisation of a soil laying above an underground conduit due to a pressure gradient. We mainly discussed the influence of the inlet fluid pressure, the conduit size and the inter-particle cohesion on the different erosion regimes. One of the important findings is the role of the granular sample length on the erosion threshold. Complementary ongoing studies will consider effects of the particle size and the granular sample height, in the aim of capturing the relevant parameters for predicting the different erosion situations. Exploiting the grain scale behavior, we also analysed both the internal cohesion state of the soil and the erosion kinetic. More generally, this numerical study confirmed the relevance of employing a DEM-LBM coupling to get insight into micromechanical features of soil erosion realted to complex natural situations such as blocked karst conduits.

Acknowledgements

We thank Zeyd Benseghier, Pablo Cuéllar and Jean-Yves Delenne for their very helpful suggestions.

References

[1] G.J. Hanson, K.R. Cook, Engineering Agriculture, ASAE, **20(4)** 455-462 (2004)

[2] N. Benahmed, S. Bonelli, European Journal of environmental and Civil Engineering, **16(1)** 43-58 (2012)

[3] P. Mutabaruka *et al.*, Phys. Rev. E **89(5)** 052203(2014)

[4] L. Sibille *et al.*, Hydrol. Process. **29(9)** 2149-2163 (2015)

[5] J.-Y. Delenne, M.S. El Youssoufi, F. Cherblanc, J.-C. Bénet, Int. J. Numer. Anal. Meth. Geomech, **28** 1577–1594 (2004)

[6] P. Cuéllar *et al.*, Proc. 4th Int. Conf. on Particle-based Methods, PARTICLES (2015)

[7] P. Cuéllar *et al.*, Proc. 8th Int. Conf. on Scour and Erosion, ICSE (2016)

[8] S. Succi, the Lattice-Boltzmann Equation for fluid dynamics and beyond.Oxford university press (2001)

[9] P. Lallemand, L.-S. Luo, J. Comp. Phys. **184** 406-421 (2003)

[10] M. Jiang *et al.*, Granular Matter. **15(1)** 65-84 (2013)

[11] M. Bouzidi *et al.*, Phys. Fluids **13** 1111-1134 (2007)

 DOI: 10.1051/epjconf/201714013017

Flowability of lignocellusic biomass powders: influence of torrefaction intensity

John Pachón-Morales*, *Julien* Colin, *Floran* Pierre, *Thibaut* Champavert, *François* Puel and *Patrick* Perré

LGPM, Centrale Supélec, Université Paris-Saclay, Grande Voie des Vignes, 92295 Châtenay-Malabry, France

Abstract. The poor flowability of powders produced from raw lignocellulosic biomass may be an economically issue for the production of second-generation biofuels. Torrefaction is a pre-treatment step of the gasification process that improves the physical characteristics of biomass by making it more coal-like. Particularly, the loss of resilience allows a reduction of the grinding energy consumption and is likely to improve the flow behaviour of woody powders. In this study, we investigated the effect of particle size and shape distribution on flow properties (unconfined yield stress and flowability factor) of powder from raw and torrefied biomass *(Picea abies)*. Several intensities of torrefaction were tested, and its extent was quantified by the global mass loss, chosen as synthetic indicator of torrefaction intensity (its accounts for both the temperature level and the residence time). The intensity of torrefaction shifts the particle size distribution towards smaller sizes. An effect on the circularity and aspect ratio was also observed. A strong, positive correlation was obtained between the measured flowability of biomass powders at different consolidation stresses and the intensity of heat treatment. These results confirm the interest of torrefaction as a pre-treatment step and aim to provide new knowledge on rheological properties of biomass powders.

1 Introduction

In spite of the recent drop of oil price, the development of fuel from renewable resources is absolutely required to face global warming. After the first generation of biofuels that consumes edible products, a second generation using the lignocellulosic part of plants – such as wood –, is yet to be developed at the industrial scale. One possible *BtL* (biomass to liquid) route includes a gasification process followed by a Fischer-Tropsch synthesis. Most of gasification processes require a supply of ground biomass in the form of powder for three main reasons: to ease the conveying of the raw material along the processing chain, to optimize its injection into the gasifier and to increase the chemical reactivity of the biomass. Currently, biomass transport is a key obstacle for the cost-competitive production of second-generation biofuels. Indeed, due to its fibrous structure, the grinding of native biomass is an energy intensive process and the powder obtained does not flow well [1]. Thus, a pre-treatment step, such as torrefaction, is needed to improve the process.

Torrefaction is a thermal treatment operating at moderate temperature (200 to 300°C) under atmospheric pressure in absence of oxygen. The process is traditionally characterised by low particle heating rates (<50 $°C.min^{-1}$) and by a relatively long reactor residence time (typically higher than 30 minutes) [2]. The resulting product is intermediate between wood and charcoal and exhibits several advantages when compared to the original material such as a decreased hygroscopicity, a dimensional stabilization, an increased resistance to biological decay, a loss of mechanical resilience, etc. [3]

Several researchers have studied the question of estimating biomass flowability. Nevertheless, despite the frequent occurrence of handling problems, there is still a relatively poor knowledge of the flow properties of biomass materials [1]. More research is therefore needed in this field to ensure, the sustainability of the supply chain [4]. This includes a better understanding of the flowability of lignocellulosic biomass powders, namely the combined effect of the torrefaction and grinding parameters. This is needed for engineers to develop cost-effective storing, conveying and processing strategies.

In this study, we investigate the effect of different torrefaction intensities on the flow properties of ground biomass, and assess the relations between size and shape of the particles and the flow behaviour of these powders.

2 Materials and methods

2.1 Material preparation

2.1.1 Wood Chip preparation

A spruce (*Picea abies)* tree from a plantation in the forest of Le Châtaignier in Riotord (France) was used in this work.

All samples came from the same board. The size of the samples was chosen to fulfil several constraints such as the amount required for flowability tests (width), the grinding capacity of the grinder (maximum length) and the homogeneity of the heat treatment (small thickness). The size of each individual wood chip was thus fixed to

* Corresponding author: john.pachon@centralesupelec.fr

20x20x3 mm^3. Each sample number of Table 1 is composed of 32 chips of this size.

As it is well-known that the wood properties may vary in the tangential and radial directions, the 32 wood chips of each sample are selected following a recurrent stepped scheme along the board in order to average this variability and to obtain twin samples (Fig. 1).

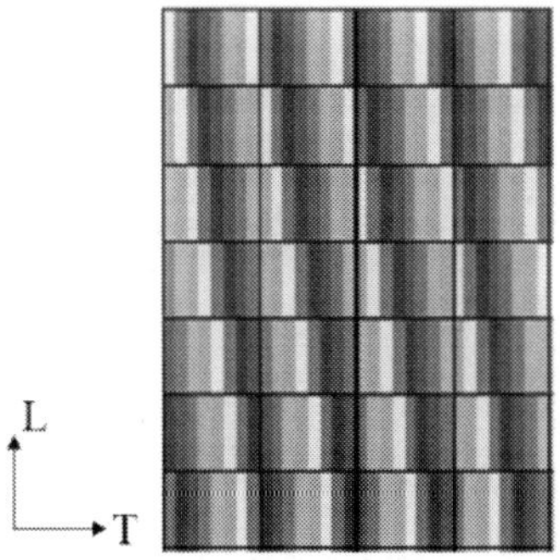

Fig. 1. Stepped cutting plan of the spruce board (Each colour corresponds to one individual sample of 32 chips). L: longitudinal direction, T: tangential direction.

2.1.2 Torrefaction procedure

A batch torrefaction device specifically developed in our laboratory was used to treat our sample [5]. The atmosphere of a hermetically closed *Memmert UFP400* chamber was controlled by a nitrogen flow rate (5 L.min^{-1}) to reduce the oxygen content, thereby avoiding oxidation and ignition. In all cases, the oxygen level remained below 1.5%. A powerful fan inside the chamber insures an efficient heat transfer either to heat-up the sample or to limit thermal overshot due to exothermic reactions. The gas temperature was measured nearby the samples using a calibrated thermocouple.

A total of 6 treatments were conducted at 220, 250 and 280 °C for 1 or 5 hours according to the following protocol: a. heating from room temperature (about 20 °C) at a rate of 5 °C min^{-1} up to 100 °C. b. plateau at 100°C for 30 minutes in order to evaporate the residual bound water before the heat treatment step. c. heating at a rate of 5°C min^{-1} until the treatment temperature. d. plateau at the treatment temperature for the desired duration and e. cooling down thanks to thermal losses and the nitrogen flow entering the reactor.

After treatment, samples were weighed to determine the mass loss. Mass loss (*ML*) due to heat treatment was calculated as follows:

$$ML\,(\%) = \frac{m_0 - m_t}{m_0} \cdot 100 \tag{1}$$

Where, m_0 and m_t are the oven-dry mass before and after torrefaction, respectively. Mass loss is known to be a good indicator of the treatment intensity [2].

Table 1. Heat treatment intensities.

Sample number	Treatment temperature (°C)	Duration of treatment (h)	Mass loss (%)
1	220	1	11,4
2	220	5	14,7
3	250	1	17,4
4	250	5	25,4
5	280	1	30,9
6	280	5	46,5
7	Non-treated		0

2.1.3 Grinding procedure

Heat-treated samples were ground in a *M20 IKA* semiautomatic laboratory 550W knife miller. The rotational speed of the knives was 20 000 rpm. The same grinding protocol was applied to all of the samples: a 95 s grinding made in several cycles (13) with intermediate waiting times of 15 s in order to avoid any additive heat treatment due to thermal dissipation. A circulation of cold water was applied to reduce heating during grinding.

2.2 Morphological granulometry

A *Sympatec-QICPIC* morphological granulometer with a M7 lens (measuring range from 10 to 10240 µm) with a *GRADIS* dispersion and *VIBRI* feeding units was used in this work [7]. To conduct the test, a sample of about 5 g of wood powder was loaded onto the hopper of the instrument. The size and shape of powders were recorded and analysed by the manufacturer software.

The Feret mean diameter, the mean value of the distance of two tangents to the particle contour over several orientations was measured: The mean diameter of the sample is then calculated as:

$$d_p(1{,}0) = \left(\sum_{i=1}^{n} n_i d_i\right) \Big/ \left(\sum_{i=1}^{n} n_i\right) \tag{2}$$

Where n_i is the number of particles with a Feret mean diameter d_i. Shape of the particles is characterised through the circularity and the aspect ratio. For the current study, circularity, ψ_S (from 0 to 1), is defined as the ratio of the perimeter of a circle having the same area as the projected area of a particle to its actual perimeter. A perfect sphere would give ψ_S=1 and the smaller the value of ψ_S, the more irregular is the shape of the particle. The aspect ratio, the ratio of the minimal to the maximal Feret diameter, gives an indication of the elongation of the particle.

2.3 Flow properties testing

A *RST-XS* Schulze ring shear tester was used to assess the flow properties of biomass powders. The ring shear tester is a widely used device to measure flow properties of powders, including angle of internal friction, wall friction and bulk density. The standard procedure leads to results with low variability [6].

A set of three pre-consolidation stresses (σ_{pre}), considered as representative of the stress range for industrial applications, were tested: 2, 5 and 10 kPa [12]. Three shear points at 25, 50 and 75 % of σ_{pre} for each consolidation stress were used to determine the yield locus of each sample. Using Mohr circle analysis, the yield loci obtained allowed to evaluate the consolidation stress σ_1 and the unconfined yield strength σ_c for the different normal stresses at preshear applied. The ratio *FFC* (flow function coefficient) of consolidation stress to unconfined stress (σ_1/σ_c) is used to characterize flowability numerically.

The larger *FFC* is, i.e., the smaller the ratio of the unconfined stress to the consolidation stress, the better a bulk solid flows.

3 Results and discussion

3.1 Particle Size distribution

In Fig. 2 the cumulative size distributions of the powder samples are represented. The intensity of heat treatment, represented by the mass loss of the samples, has an obvious impact on the size distribution of the biomass powders: the loss of resilience allows to obtain smaller particles by grinding.

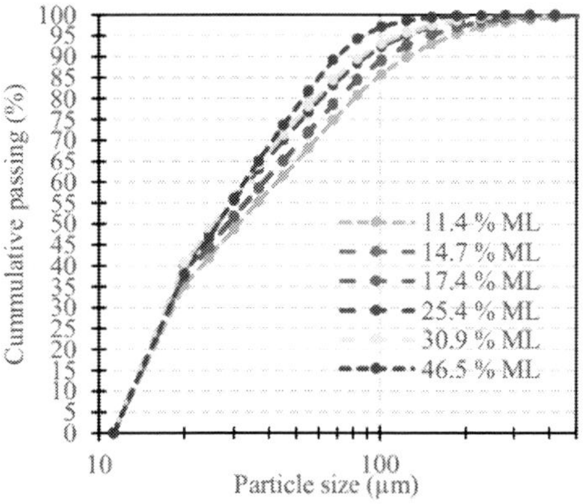

Fig. 2. Cumulative passing at several torrefaction intensities.

The reduction of the resilience at increasing intensity of heat treatment makes the samples brittle, thus the particle size distribution shifts towards smaller particle sizes and its dispersity is reduced. Similar conclusions have been also observed from other studies, for different kinds of biomass powders and torrefaction conditions [10], [11], [8].

Fig. 3 shows the value of the mean diameter and the cumulative percentage of particles with a Feret mean diameter below 101.2 µm at various torrefaction intensities. The cumulative number represents the fraction of particles having a diameter of 101.2 µm or less.

The mean diameter decreases dramatically whereas $d_{101.2}$ increases with the increase of torrefaction intensity, the grinding conditions remaining the same for all the samples. This shows that there is an improvement in the grindability characteristics of treated biomass. This result is in good agreement with those presented in previous works [8], [9].

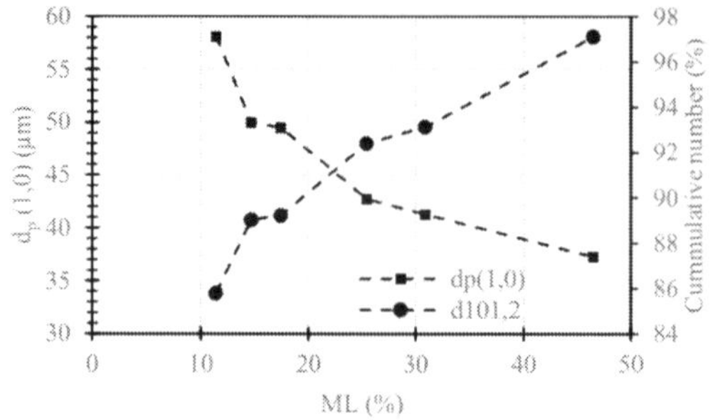

Fig. 3. Effect of intensity of heat treatment on the mean diameter and cumulative percentage of particles with a size below 101.2µm.

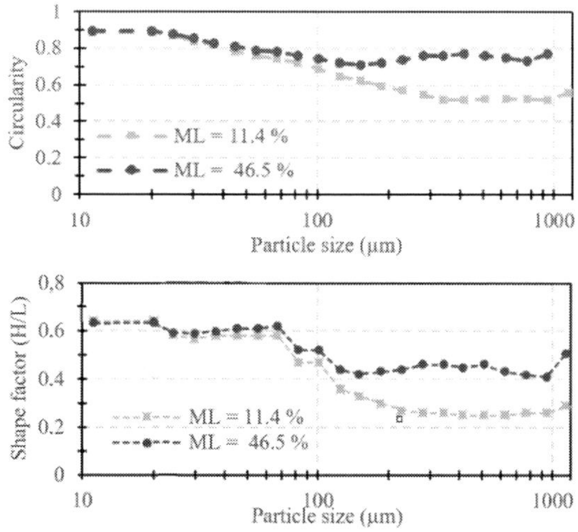

Fig. 4. Shape of the treated spruce powders at two different torrefaction intensities.

Fig. 4 represents the morphological parameters of the biomass powders for two torrefaction intensities. On the whole, circularity of spruce powders was improved as the intensity of treatment increased from *ML*=11.4 % to *ML*=46.5 % for particles size above 50 µm. More spherical particles are thus globally obtained when a severe treatment is applied (280°C, 5h). This denotes a loss of the fibrous behaviour. However, no significant difference in circularity of ground biomass was observed for torrefaction intensities from 11.4 % to 30.9 %. It seems that the finer particles tend to be more spherical. Nevertheless, a more sensitive morphological analysis is necessary in order to better characterize these small particles (d_p<100 µm). Similar conclusions can be drawn regarding the shape factor: more uniform particles are obtained when intensity of torrefaction is increased.

3.2 Flowability

Fig. 5 presents the flow functions for different torrefaction intensities along with the flowability regimes from not flowing to free-flowing behaviour. A clear

 DOI: 10.1051/epjconf/201714013017

dependence between the extent of the heat treatment and the flowability of spruce powders is depicted. The native ground biomass has a cohesive behaviour that decreases with the heat treatment until a nearly free flowing feature for the most intensively treated sample. In almost all cases, the unconfined yield stress increases with the consolidation stress to the same extent, so the regime of flowability for the powders is independent of the consolidation stress applied (except for the most intensely treated sample for which a free-flowing behaviour is reached at elevated consolidation stresses).

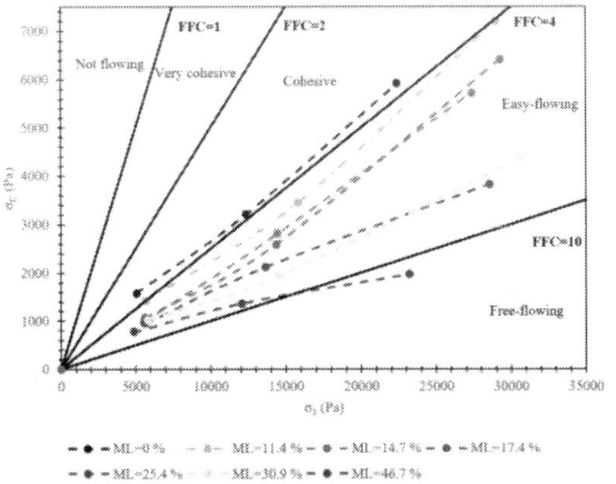

Fig. 5. Instantaneous flow functions for several intensities of torrefaction and lines of constant flowability FFC (solid lines).

For all samples, better flowability is obtained at a greater consolidation stress but there is no influence of the stress level on the qualitative ranking of the set of powders. This last feature is in good agreement with the general behaviour for most of the powders reported in literature [12].

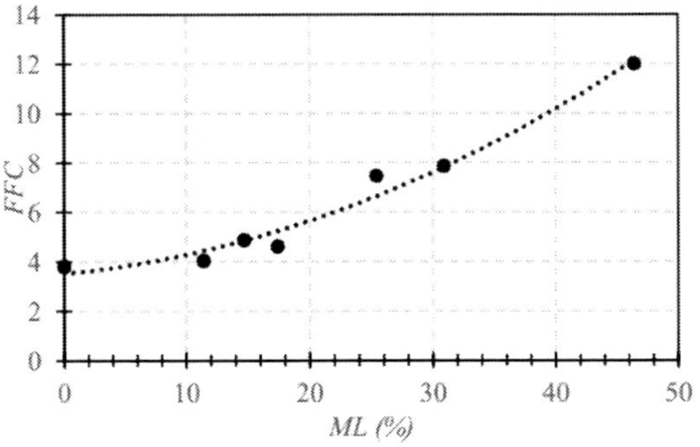

Fig. 6. Effect of torrefaction intensity on the flowability factor of the biomass powders. σ_{pre} = 10 kPa.

A direct relationship between the intensity of torrefaction and the flowability factor was found (Fig. 6). For sake of clarity, only the values at σ_{pre}= 10 kPa are presented, but similar tendencies were verified for the others values. The improvement of flow behaviour with the extent of heat treatment could be seen as a combined effect of several factors: the modifications on particle size and width of the distribution, the improvement of shape characteristics and the modification of the surface interactions. Further stage of our study will consist of isolating each factor in order to discriminate their effects on flowability. The lower values of circularity and larger aspect ratios lead to more frequent interparticle entanglements, which degrades flowability. These findings are in agreement with literature reports [13], [14].

4 Conclusions

This work evaluated the impact of six operating conditions of torrefaction (from 11.4 % to 46.5 % of mass loss) on the flowability of a biomass powder measured in a ring shear tester, as well as on particles size distribution and morphology. Several tendencies are clearly drawn from this work:

- the particle size distribution shifts towards smaller particle sizes and its dispersity is reduced with increase in torrefaction intensity,
- due to the loss of the fibrous behaviour, the circularity of spruce powders is improved at high-treatment intensities,
- an improvement of the flowability of spruce powders by heat treatment is clearly exhibited. This behaviour gradually shifts from a cohesive behaviour for the native biomass to a nearly free flowing behaviour for a heat treatment with 46.5 % of mass loss.

References

[1] Miccio, F., Barletta, D., & Poletto, M. (2013). Pow. Tech, 312-321.
[2] Bergman, P., Boersma, A., Zwart, R., & Kiel, J. (2005). Petten, The Netherlands: ECN.
[3] Almeida, G., Brito, J. O., & Perré, P. (2010). Bio. tech. 9778-9784.
[4] Ramirez-Gomez, A. (2016). Part. sci. and tech..
[5] Colin, J. (2007). U. H. Poincaré and AgroParisTech.
[6] ASTM D6773. (2002). ASTM International.
[7] Sympatec GmbH. (2016). Laser Diffraction Sensors. Retrieved from http://www.sympatec.com
[8] Arias, B., Pevida, C., Fermoso, J., Plaza, M., Rubiera, F., & Pis, J. (2008). Fuel pro. tech., 169-75.
[9] Phanphanich, S., Mani, S. (2011). Bior. tech., 1246-1243.
[10] Himmel, M., Tucker, M., Baker, J., Rivard, C., & Grohmann, K. (1985). *Biotech. and bioeng. symp.*, 39-58.
[11] Mani, S., Tabil, L. G., & Sokhansanj, S. (2006). Biom. and bioen. 30, 648-654.
[12] Schulze, D. (2007). *Powders and bulk solids.* Braunschweig: Springer.
[13] Crawford, N., Ray, A., Yancey, N., & Nagle, N. (2015). Fuel pro. tech., 46-56.
[14] Almendros, M., Bonnefoy, O., Govin, A., Nastoll, W., & Sanz,. *19th European Biomass Conference 2011.* Berlin.

A two-phase solid-fluid model for dense granular flows including dilatancy effects: comparison with submarine granular collapse experiments

F. Bouchut[1], *E.D.* Fernández-Nieto[2], *E.H.* Koné[3], *A.* Mangeney[3,*], and *G.* Narbona-Reina[2]

[1]*Université Paris-Est, Laboratoire d'Analyse et de Mathématiques Appliquées (UMR 8050), CNRS, UPEM, UPEC, F-77454, Marne-la-Vallée, France*
[2]*Universidad de Sevilla, Dpto. Matemática Aplicada I & IMUS, Universidad de Sevilla. E.T.S. Arquitectura, 41012 Sevilla, Spain*
[3]*Université Paris Diderot, Sorbone Paris Cité, Institut de Physique du Globe de Paris, 75005 Paris, France*

Abstract. We simulate here the collapse of granular columns immersed in a viscous fluid based on a simplified version of the model developed by [2]. The simulation quite well reproduces the dynamics and deposit of the granular mass as well as the excess pore fluid pressure measured in the laboratory experiments of [10] owing that dilatancy effects and pore pressure feedback are accounted for. In particular, the difference in the behaviour of initially loose and dense columns is reproduced numerically.

1 Introduction

Describing grain/fluid interaction in debris flows models is still an open and challenging issue with important impact on hazard assessment [3, 7]. A key issue in modeling grain/fluid flows is to describe the compression/dilation of the granular phase and its interaction with the pore fluid pressure. This is particularly challenging when trying to include these effects in models which are classically used for real applications, that assume thin-layer flows [1, 2, 9].

The depth-averaged thin-layer model used here is essentially an extension to non-uniform flows of the model proposed by [9]. We consider a two-phase layer that models the grain/fluid mixture with a fluid layer on top of it to account for fluid transfer into and/or out of the mixture [2]. The model describes explicitly the velocity of the solid and fluid phases, the compression/dilation of the granular media and its interaction with the pore fluid pressure. The system of mass and momentum conservation equations from Jackson's model is closed by a weak compressibility relation following [11]. This relation implies that the occurrence of dilation or contraction of the granular material in the model depends on whether the solid volume fraction is respectively higher or lower than a critical value. This description of dilatancy has been used in [9] to develop a thin-layer depth-averaged two-phase model for immersed granular flows (see also [6]). Moreover, following [9], the critical value is assumed to depend on the shear rate and the solid pressure.

Two models have been proposed in [2], depending on the order of magnitude of the friction coefficient between the two phases within the mixture. In both cases, the model reproduces the expected behaviour (see e. g. [9]): when dilation occurs, the fluid is sucked into the granular material, the pore pressure decreases and the friction force on the granular phase increases. On the contrary, in the case of contraction, the fluid is expelled from the mixture, the pore pressure increases and the friction force diminishes. The extra upper fluid layer considered in the model is essential to be consistent with the compression/dilation effects since it allows the fluid to be expelled or sucked from/in the mixture at its upper boundary. Additionally, this configuration also allows to use the model for the case of immersed flows, which corresponds to the experiments investigated in [10] and simulated here.

In this paper we present the first simulations obtained by solving numerically the equations proposed in [2] and compare the results to the laboratory experiments of granular column collapse immersed in a viscous fluid performed by [10]. These experiments have been chosen because they clearly show and quantify the strong impact of the initial compaction on the flow dynamics and deposit of the granular mass. Indeed, the initially loose column travels much further than the initially dense column due to the dilatancy processes described above. These simulations show that the model is able to capture at a first order the main effect of compression/dilation and its coupling with the pore fluid pressure.

In Section 2 the model and parameter settings are described and in Section 3 a quantitative comparison between simulation and experiments is presented and discussed.

2 Grain/fluid Model with dilatancy

2.1 Equations

The two-layer model proposed in [2] is made of a fluid layer on top of a two-phase mixture layer (Figure 1). It is described by three mass and three momemtum conservation equations and a closure relation. Moreover, it includes

*e-mail: mangeney@ipgp.jussieu.fr

mass transference between the upper fluid layer and the fluid phase within the mixture layer. Let us denote by h_f the thickness of the upper only-fluid layer and $\mathbf{u}_f$ its velocity, by h_m the thickness of the mixture layer, φ the solid volume fraction, $\mathbf{v}$ the solid phase velocity, and $\mathbf{u}$ the fluid phase velocity in the mixture layer. Then, the three continuity equations defining the model are:

$$\partial_t(\varphi h_m) + \nabla \cdot (\varphi h_m \mathbf{v}) = 0, \tag{1}$$

$$\partial_t((1-\varphi)h_m) + \nabla \cdot ((1-\varphi)h_m \mathbf{u}) = -\mathcal{V}_f, \tag{2}$$

$$\partial_t h_f + \nabla \cdot (h_f \mathbf{u}_f) = \mathcal{V}_f, \tag{3}$$

with $\mathcal{V}_f$ the fluid mass transference. The evolution of solid volume fraction is described by

$$\partial_t \varphi + \mathbf{v} \cdot \nabla \varphi = -\varphi \Phi, \tag{4}$$

where Φ is the closure function that takes into account dilatancy effects. From (1), (2), (3) and (4), one can indeed deduce the value of the fluid mass transference

$$\mathcal{V}_f = -h_m \Phi - \nabla \cdot ((1-\varphi)h_m(\mathbf{u}-\mathbf{v})). \tag{5}$$

Following the closure proposed by Roux and Radjai [11] we set $\Phi = K\dot{\gamma}(\varphi - \varphi_c^{eq})$, where the critical-state compacity is $\varphi_c^{eq} = \varphi_c^{stat} - K_2 \frac{\eta_f \dot{\gamma}}{p_{s|_b}}$ (see [9]). K and K_2 are two constants, φ_c^{stat} is a constant volume fraction corresponding to the static equilibrium, η_f is the fluid viscosity, $\dot{\gamma}$ the strain rate and $p_{s|_b}$, the solid pressure at the base of the flow (see [2, 9] for more details).

The momentum equations involve the bottom solid pressure $p_{s|_b} = \left((\rho_s - \rho_f) g_c \varphi h_m - (p^e_{f_m})_{|_b}\right)_+$ through the effective bottom solid friction

$$\widetilde{\tau}_b = p_{s|_b} \tan \delta_{\text{eff}} + K_1 \eta_f \dot{\gamma}, \tag{6}$$

where $\tan \delta_{\text{eff}} = \tan \delta + K(\varphi - \varphi_c^{eq})$, ρ_f and ρ_s are the density of the fluid and the solid, respectively, $g_c = g \cos \theta$, with g the gravity acceleration constant and θ, the slope of the inclined plane. K_1 is a constant and the excess pore pressure $p^e_{f_m}$ is such that the total fluid and solid pressures are

$$\begin{aligned} p_f(z) &= \rho_f\, g_c\, (b + h_m + h_f - z) + p^e_{f_m}(z), \\ p_s(z) &= (\rho_s - \rho_f)\, g_c\, \varphi\, (b + h_m - z) - p^e_{f_m}(z) \end{aligned} \tag{7}$$

where b denotes the bottom elevation with respect to the inclined plane, and where $p^e_{f_m}$ is the average with respect to z of $p^e_{f_m}(z)$. The main difference between a hydrostatic model and the one derived here is the contribution of $p^e_{f_m}$ in the solid friction (6) and its averaged gradient, that is defined as follows:

$$\overline{\nabla p^e_{f_m}} = \frac{1}{h_m}\left(\nabla(h_m p^e_{f_m}) + (p^e_{f_m})_{|b} \nabla b\right). \tag{8}$$

Two definitions are obtained in [2] for $(p^e_{f_m})_{|b}$ and $p^e_{f_m}$. In this work we consider the simplest case, associated to a large friction $\beta = 150\eta_f \varphi^2 / d^2(1-\varphi) = O(\epsilon^{-1})$ between the phases. ϵ is the aspect ratio assumed to be small according to the thin-layer framework and d is the solid particles mean diameter (see [2] for details). Hence

$$(p^e_{f_m})_{|b} = -\frac{\beta}{(1-\varphi)^2}\frac{h_m^2}{2}\Phi, \quad p^e_{f_m} = -\frac{\beta}{(1-\varphi)^2}\frac{h_m^2}{3}\Phi. \tag{9}$$

To simulate underwater granular flows, we consider the following simplifications: (i) we rewrite the hydrostatic pressure terms in the momentum equations by using the hypothesis of horizontal free surface: $h_m(t,x) + h_f(t,x) + b(x) + x\tan\theta = cst$ (Figure 1) and (ii) we neglect the friction between the upper fluid layer and the mixture layer. Moreover, we consider that the only-fluid layer has no tangential movement (i. e. $\mathbf{u}_f \equiv 0$). Let the fluid and solid contributions in the mixture mass be denoted by $h_1 = (1-\varphi)h_m$ and $h_2 = \varphi h_m$ and the fluid and solid discharges by $\mathbf{q}_1 = h_1 \mathbf{u}$ and $\mathbf{q}_2 = h_2 \mathbf{v}$, respectively. The system summarizes then in conservative form as

$$\begin{cases} \partial_t h_1 + \nabla \cdot \mathbf{q}_1 = -\mathcal{V}_f, \\ \partial_t \mathbf{q}_1 + \nabla \cdot (\mathbf{u} \otimes \mathbf{q}_1) = -\dfrac{1}{2}\mathcal{V}_f \mathbf{u} - \mathbf{f}, \\ \partial_t h_2 + \nabla \cdot \mathbf{q}_2 = 0, \\ \partial_t \mathbf{q}_2 + \nabla \cdot (\mathbf{v} \otimes \mathbf{q}_2) = -\dfrac{1-r}{2} g_c \nabla(\varphi h_m^2) + r\mathbf{f} \\ \qquad - (1-r) g_c h_2 \nabla \hat{b} - \dfrac{\widetilde{\tau}_b}{\rho_s}\text{sgn}(\mathbf{v}), \end{cases} \tag{10}$$

where $r = \frac{\rho_f}{\rho_s}$, $\hat{b} = b + x\tan\theta$, $\mathcal{V}_f$ is given by (5) and

$$\mathbf{f} = \frac{1}{\rho_f}\left(h_1 \overline{\nabla p^e_{f_m}} + \beta h_m(\mathbf{u} - \mathbf{v})\right).$$

2.2 Numerical method

The model is discretized with a combination of finite volume and finite difference schemes. The finite volume scheme is a Roe-type non-conservative method that handles the advection-convection part of the system, while a standard finite difference scheme deals with the source terms of orders 0 and 2.

As we will simulate the spreading of a granular column immersed in a fluid (see section 3.1), the model should be able to deal with the transition between the cases $h_m \neq 0$ and $h_m = 0$ (so-called dry bottom surface in numerical methods). This is a key and generally difficult issue in the design of numerical schemes solving hyperbolic equations. It is noteworthy that the current stage of the solver blows up when we use dry bottom surface to simulate granular column collapse. This is likely a numerical issue that we are investigating now. As a first step to simulate the experiments, we added here a thin artificial layer of 1 mm to the bottom surface. This is enough to prevent the code to blow up even though we may notice very small oscillations on the simulated profiles on the top of the mass for the dense initial packing and ahead of the front for the two packing configurations (Figure 2).

3 Simulation of granular collapses

3.1 Laboratory experiments

We simulate here the collapse of a granular column initially released from rest in a viscous liquid that has been

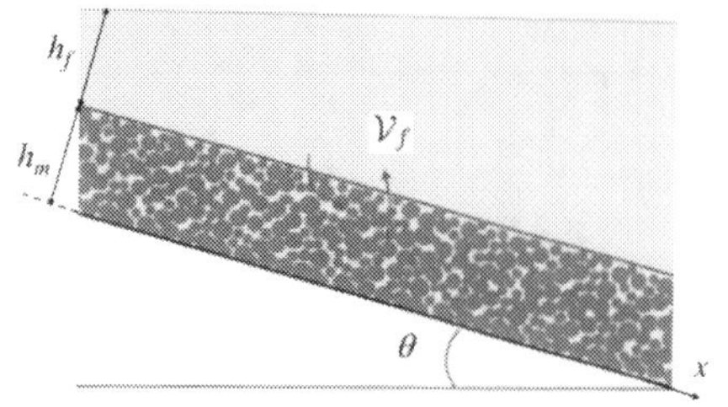

Figure 1. Immersed configuration.

experimentally investigated by [10]. The experimental setup consists of a box 70 cm long, 15 cm wide and 15 cm high, full of a viscous liquid. The horizontal bottom of the box is made rough by gluing a layer of particles on it. This experiment corresponds to a one-dimensional dam-break with an initial rectangular mixture mass of height H_i and width L_i that is at rest with a volume fraction φ_i. The release of an initially loose and initially dense granular column is tested experimentally with the data ($\varphi_i = 0.55, L_i = 6$ cm, $H_i = 4.8$ cm) and ($\varphi_i = 0.6, L_i = 6$ cm, $H_i = 4.2$ cm), respectively. In both cases, the particles are glass beads of density $\rho_s = 2500$ kg.m^{-3} and mean diameter $d = 225\,\mu$m, whereas the fluid has a density $\rho_f = 1000$ kg.m^{-3} and a viscosity $\eta_f = 12\times10^{-3}$ Pa.s. For these experiments, we have $\varphi_c^{stat} = 0.585$, $\tan\delta = 0.32$ and we set $K = 4.09$, $K_1 = 90.5$ and $K_2 = 25$ in the model as in [9].

3.2 Simulations

Overall, the quantitative agreement between simulation and observation is quite good (Figures 2, 3). The key point is that the spreading of the granular mass is, as stated in [10], mainly controlled by its initial volume fraction, with two different regimes corresponding to dense and loose packings. Loose packing columns have fast dynamics and lead to thin and long deposits, while dense ones produce slow flows and short runout distances (Figures 2 and 3(a)). Furthermore the maximum thickness of the deposit is much smaller for the initially loose case. While the deposit is quite well reproduced, the simulated mass spreads more rapidly than in the experiments. This is a common feature obtained when comparing thin-layer depth-averaged models with granular collapse experiments [4, 8]. Furthemore the effect of the gate that releases the granular column is not taken into account in the model. It has however a non-negligible effect on the flow dynamics while it almost does not change the deposit (see Figures 14 and 15 of [5]).

In addition, a positive (resp. negative) pore pressure is simulated below the column for loose (resp. dense) packings as in the experiments (Figure 3(b)). In case of dilation, the fluid is sucked into the mixture decreasing the pore fluid pressure and increasing thereby the effective friction on the granular phase. On the other hand, contraction induces an expulsion of the fluid from the mixture increasing the pore fluid pressure and decreasing the effective friction. This so-called pore pressure feedback claimed in [10] and references therein is well reproduced in the simulation. In the experiments, pressures are measured at the position $x = 2$ cm from the wall ($x = 0$). We represent here the simulated pressure at $x = 2$ cm (black), $x = 3$ cm (blue), $x = 4$ cm (red).

For the loose case, the simulated value of the pore pressure at $x = 2$ cm and its time change is in quite good agreement with the pressure measured at $x = 2$ cm. The model simulates a higher pick fluid pressure at the very beginning of the collapse which is not measured experimentally, at least at the acquisition rate of these measurements. For the dense case, even though the simulated pore pressure is of the good order of magnitude, it is quite different from the measured pressure. In particular, the pressure decrease at $x = 2$ cm, starts much later than in the experiments. When going closer to the initial gate position (i. e. $x = 3, 4$ cm), the pressure change starts earlier, as observed at $x = 2$ cm in the experiments. This time shift is only seen in the dense case and may correspond to the strong effect of the gate in that case or to more fundamental modeling issue.

Let us now compare the results obtained with the non-hydrostatic pressure (i. e. the whole model involving the excess pore fluid pressure and its gradient) with those obtained when only hydrostatic pressure is assumed (i. e. with $p_{fm}^e = 0$). Figure 2 clearly shows that the whole model results (middle plots) are much closer to the experiments (top plots) than the hydrostatic model (bottom plots), in particular in the loose case. The same observations holds for Figure 3(a), where we plot the time evolution of the front position. These results point out the relevance of including the excess pore fluid pressure. However, the numerical discretization of the excess pore fluid pressure and of its gradient is a very challenging process that is still under investigation. Improving the discretization may resolve the inability of the solver to simulate the granular collapse over a dry bottom surface and/or reduce the observed discrepancy in the triggering time of the pressure decrease.

4 Conclusion

The two-layer two-phase solid/fluid model introduced in [2] is solved here numerically in the case of immersed granular flows. This thin-layer depth-averaged model makes it possible for the fluid to be either expelled from or sucked into the mixture at its upper boundary depending on the compression or dilation of the granular phase. Simulations with the current stage of the solver are compared to the experiment of immersed granular collapse performed by [10]. We show that the quantitative behaviour is quite well reproduced by the model in terms of front dynamics, deposit and pore fluid pressure change with time. In particular, the simulation reproduces the strong change in behaviour of initially dense and loose columns. Discrepancies between the simulation and experiments are however observed. They are due to the depth-averaged thin-layer approximation, to the absence of the gate in the simulation but also probably to numerical issues. Indeed, the discretization of the pressure and

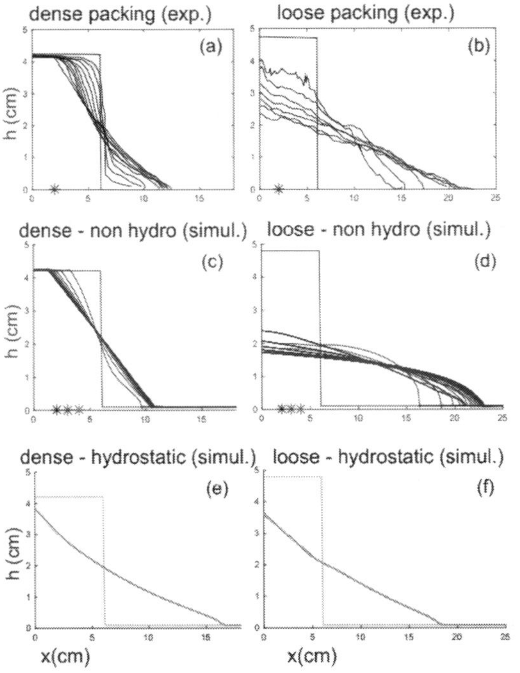

Figure 2. Granular mass profiles from experiments [top] and simulations (non hydrostatic [middle] and hydrostatic [bottom] pressures). Left: dense initial packing ($\varphi_i = 0.6, L_i = 6$ cm, $H_i = 4.2$ cm) every 3s. Right: loose initial packing ($\varphi_i = 0.55, L_i = 6$ cm, $H_i = 4.8$ cm) every 0.66s. The marker * is the position where the time series of the excess pore fluid pressure is measured in experiments.

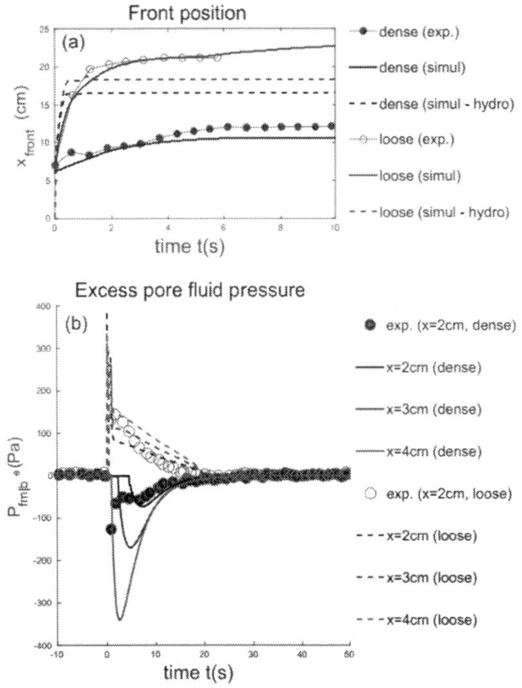

Figure 3. Time evolution of (a) the front position for both initial packing from experiments (dense -•- and loose -∘-) and simulations (non hydrostatic [dense –, loose –] and hydrostatic [dense – –, loose – –] pressures) and (b) the pore pressure below the column from experiments (dense • and loose ∘) at $x = 2$ cm and simulations at $x = 2$ cm (black), $x = 3$ cm (blue), $x = 4$ cm (red).

of its gradients as well as the numerical treatment of the front should be improved to go further in the quantitative comparison with the experiments. This is under investigation. We also show that when cancelling the excess pore fluid pressure by assuming hydrostatic pressure, very strong difference is observed with the experiments in terms of dynamics and deposit of the granular mass.

Acknowledgements

This work has been funded by the ANR-11-BS01-0016 LANDQUAKES, the ERC-CG-2013-PE10-617472 SLIDEQUAKES and FEDER and the Spanish Government through the research project MTM2015-70490-C2-2-R.

References

[1] Bouchut, F., Fernandez-Nieto, E. D., Mangeney, A., and Narbona-Reina, G., A two-phase shallow debris flow model with energy balance, Math. Model. Num. Anal., **49**(1), 101-140 (2015).

[2] F. Bouchut, E.D. Fernández-Nieto, A. Mangeney, G. Narbona-Reina, *A two-phase two-layer model for fluidized granular flows with dilatancy effects*. J. Fluid Mech, **801**, 166-221 (2016).

[3] Delannay, R., Valance, A., Mangeney, A., Roche, O., and Richard, P., *Granular and particle-laden flows: from laboratory experiments to field observations*, J. Phys. D: Appl. Phys., in press (2016).

[4] Fernandez-Nieto, E. D., Garres-Diaz, J., Mangeney, A., and Narbona-Reina, G., *A multilayer shallow model for dry granular flows with the mu(I) rheology: Application to granular collapse on erodible beds*, J. Fluid. Mech., **798**, 643-681 (2016).

[5] Ionescu, I., Mangeney, A., Bouchut, F., and Roche, O., *Viscoplastic modeling of granular column collapse with pressure-dependent rheology*, J. Non-Newtonian Fluid. Mech., **219**, 1-18 (2015).

[6] R. M. Iverson and D. L. George, *A depth-averaged debris-flow model that includes the effects of evolving dilatancy. I. Physical basis.* Proc. R. Soc. A, **470:0** 20130819 (2014).

[7] R. M. Iverson, M. Logan, R. G. LaHusen, and M. Berti, *The perfect debris flow? Aggregated results from 28 large-scale experiments.* J. Geophys. Res., **115:0** F03005 (2010).

[8] Mangeney-Castelnau, A., Bouchut, F., Vilotte, J.P., Lajeunesse, E., Aubertin, A., and Pirulli, M., *On the use of Saint Venant equations to simulate the spreading of a granular mass*, J. Geophys. Res., **110**, B09103 (2005).

[9] M. Pailha and O. Pouliquen, *A two-phase flow description of the initiation of underwater granular avalanches*. J. Fluid Mech., **633:0**, 115–135 (2009).

[10] L. Rondon, O. Pouliquen and P. Aussillous, *Granular collapse in a fluid: Role of the initial volume fraction*, Physics of Fluids, **23**, 073301 (2011)

[11] S. Roux and F. Radjai, *Texture-dependent rigid plastic behaviour*. In H. J. Herrmann et al., editor, *Physics of Dry Granular Media*, 305–311. Kluwer, (1998).

Fluid-particle energy transfer in spiral jet milling

Selasi Dogbe[1], *Mojtaba* Ghadiri[1,*], *Ali* Hassanpour[1], *Colin* Hare[1,+], *David* Wilson[2], *Richard* Storey[2] and *Iain* Crosley[3]

[1]School of Chemical and Process Engineering, University of Leeds, Leeds LS2 9JT, UK
[2]AstraZeneca, Macclesfield SK10 2NA, UK
[3]Hosokawa Micron, Runcorn WA7 3DS, UK

Abstract. Spiral jet milling is a size reduction process driven by the fluid energy of high velocity gas jets. Inter-particle and particle-wall interactions are responsible for size reduction. The process is energy intensive, but inefficient. The underlying mechanisms for size reduction in the mill are also not very well understood. The optimum grinding conditions are still currently found by trial and error experimentation.
In this work, the Discrete Element Method coupled with Computational Fluid Dynamics is used to investigate the effects of different parameters on the particle collisional behaviour in a spiral jet mill. These include the particle concentration in the grinding chamber, the particle size, and the fluid power input. We report on our work analysing the efficiency of energy transfer and how it can be improved by changing the milling conditions and particle properties.

1 Introduction

The spiral jet mill is a size reduction equipment capable of reducing particulate solids to micron sizes with narrow size distributions. This makes it particularly suitable for industries such as paint, food, and pharmaceuticals, where such product specifications are commonly required. For example, in the pharmaceutical industry the dissolution of the active pharmaceutical ingredient (API) in the target being, content uniformity of the API in a tablet dosage form, and drug delivery by dry powder inhalers all require control of the particle size.
The spiral jet mill functions using solely high-pressure gases to induce particle breakage and classification of the milled material within the grinding chamber. The spiral jet mill contains two gas inlets: the injector gas and grinding gas inlets. The injector gas is used to propel the feed material into the grinding chamber of the mill. The grinding gas creates a circular motion of fluid in the grinding chamber which entrains the entering feed material. At this point there are two competing forces acting on the particles: the centrifugal and drag forces. When the particle is large, the centrifugal force outweighs the drag force, and particles are pushed towards the mill wall, where they experience inter-particle and particle-wall interactions that lead to size reduction. When the particles are reduced to a certain size, known as the cut size, the drag force becomes the dominating force. This causes particles to be pulled towards the central outlet and escape from the grinding chamber.
Spiral jet milling is a highly inefficient process, with only as little as 2 % of the supplied energy being utilised in particle breakage [1]. Moreover the performance is strongly dependent on the mill design, affecting the relative contributions of fragmentation and surface chipping to the size reduction. The use of numerical simulations allows the role of important parameters to be identified and enables the cost efficient design and optimisation of the milling process. Previous researchers have simulated the spiral jet mill to investigate aspects of the process ranging from the velocity field to particle classification [2-5].
In this work the Discrete Element Method (DEM) is used coupled with Computational Fluid Dynamics (CFD) to analyse particle dynamics. The Hosokawa Alpine Aeroplex Spiral Jet Mill 50AS was used as the simulation geometry.

2 Method

Using CAD, the grinding chamber of the Hosokawa Alpine Aeroplex Spiral Jet Mill 50AS was drawn with the feed inlet and injector nozzle attached. Only this section of the mill was used in the simulation as this is the primary location of all milling activity. The chamber has a nominal diameter of 50 mm, and four evenly spaced grinding nozzles tangentially mounted to the wall.
DEM calculations were carried out using EDEM 2.7 (DEM Solutions, UK). ANSYS Fluent (ANSYS, USA) was used for CFD calculations. Particle properties common to all conducted simulation cases are listed in Table 1. The k-epsilon (k-ε) turbulence model was used for all simulations.

* Corresponding author: M.Ghadiri@leeds.ac.uk
+ Current address: Department of Chemical and Process Engineering, University of Surrey, Guildford GU2 7XH, UK

Table 1. Particle material properties

Poisson's ratio	0.35
Density (kg/m^3)	1525
Coefficient of restitution	0.5
Coefficient of sliding friction	0.5
Coefficient of rolling friction	0.01

Table 2. Number % of each particle size in the simulation

Particle size (μm)		10	20	50	75	100
Number %	**Case 1**	62.9	31.4	5.03	0.60	0.07
	Case 2	20	20	20	20	20

2.1 Collisional Energy

The collisional energy in the spiral jet mill was investigated for two cases containing 10,000 particles: one with a milled product size distribution of particles in the chamber, and one with an equal number of different sized particles. The size and number of each particle in the simulations are given in Table 2.

Both simulation cases were run using the Hertz-Mindlin contact model.

2.2 Effect of Particle Concentration

The spiral jet mill was simulated at grinding and injector pressures of 4.0 bar, with three different numbers of particles: 10,000, 100,000, and 1,000,000. The particle size distribution followed Case 1 shown in Table 2. The number of collisions and the dissipated energy of inter-particle and particle-wall collisions were investigated in each case. These simulations were carried out using the Hertz-Mindlin contact model.
Under the same grinding conditions, the relative velocity distribution of particles was investigated in cases containing 2,500, 5,000 and 10,000 monosized (425 μm) particles. The elasto-plastic-adhesive contact model developed by Pasha et al. [6] was used for these simulations, but adhesion was switched off.

2.3 Effect of Injector and Grinding Nozzles Gas Pressures

With a number of 2,500, 5,000 and 10,000 monosized (425 μm) particles, the spiral jet mill was simulated with various grinding and injector pressures (see Table 3). The Pasha et al. model [6] was used for these simulations.

Table 3. Grinding and injector pressure combinations

Pressure (barg)	
Grinding	**Injector**
1.0	1.0
2.0	2.0
3.0	3.0
4.0	4.0
5.0	5.0
6.0	6.0

3 Results

3.1 Collisional Energy

It is found that the dissipated energy between normal inter-particle collisions is greater than that of particle-wall collisions, indicating that the former contributes more towards size reduction than the latter (Fig. 1). This was found to be the case for all collisions except a 10 μm particle colliding with a 50 μm, 75 μm and 100 μm particle. This was due to the fact that the contact area between the 10 μm particle and the aforementioned particle sizes is smaller than that of the contact area between the 50 μm, 75 μm and 100 μm particles and the wall. The larger contact area means there will be a greater amount of deformation, and consequently dissipated energy. Another observed trend is the amount of dissipated energy increases as the particle size increases.

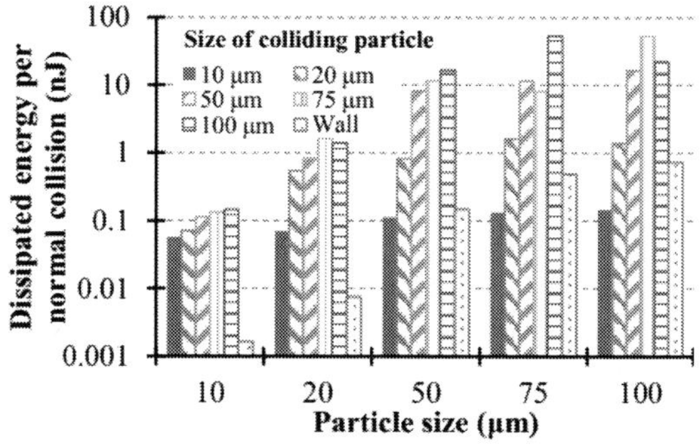

Fig. 1. Dissipated energy per normal inter-particle and particle-wall collisions

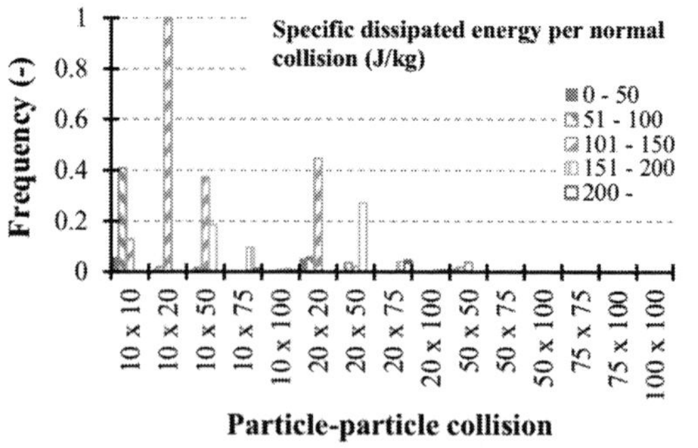

Fig. 2. Specific dissipated energy per collision and its frequency for every possible inter-particle collision (Case 1)

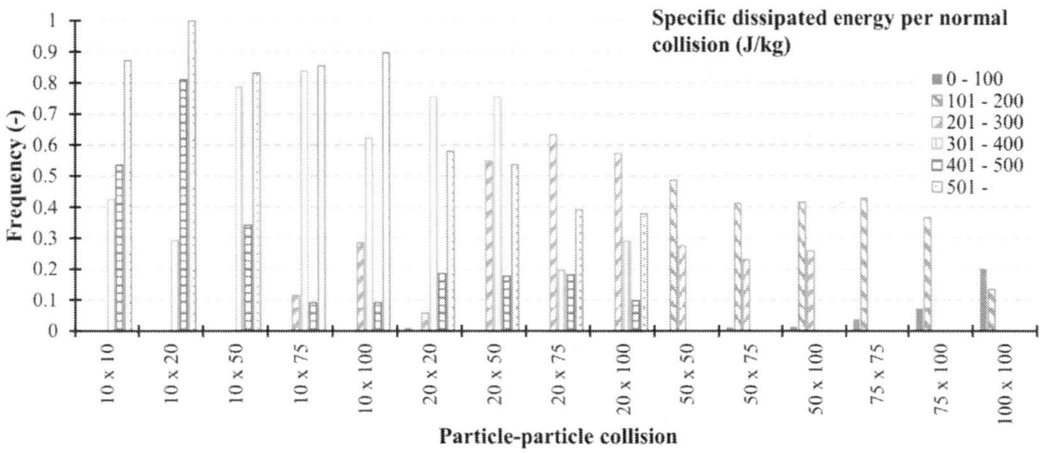

Fig. 3. Specific dissipated energy per collision and its frequency for every possible inter-particle collision (Case 2)

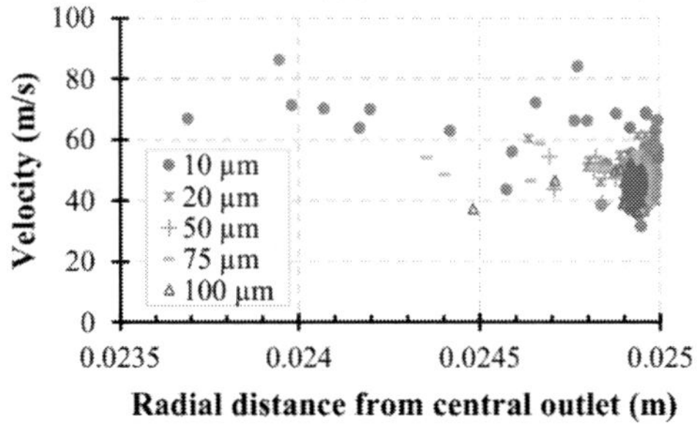

Fig. 4. Velocity magnitude and distance from central outlet of the different particle sizes (2000 of each particle size)

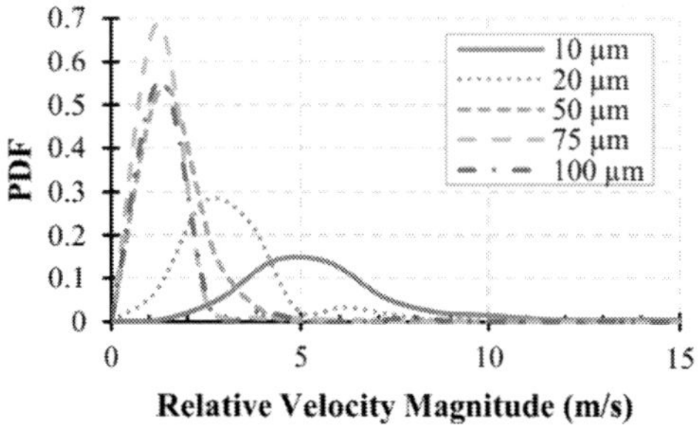

Fig. 5. Velocity distribution of different particle sizes

Analysis of the various possible inter-particle collisions in terms of their frequency and specific dissipated energy per collision shows that collisions between 10 and 20 μm particles have a much higher frequency compared to all other possible combinations, with a dissipated energy between 101 – 150 J/kg (Fig. 2). Therefore collisions with 10 and 20 μm particles have a higher rate of breakage than other possible collisions. This also suggests there exists an optimum size ratio for the rate of breakage in terms of inter-particle collisions.

When there is an equal number of each particle size (Case 2), it is again seen that collisions between 10 and 20 μm particles have a higher frequency and dissipated energy per collision than the other possible collisions (Fig. 3), indicating that this behaviour is not solely due to the size distribution.

Looking at the radial position of the different sized particles, there exists a size segregation of the particles as they circulate around the outer wall of the spiral jet mill (Fig. 4).

From the relative collisional velocity distribution of the different particle sizes in the mill with other particles (Fig. 5), it can be seen that 10 and 20 μm particles have a wider velocity distribution with generally higher relative velocity magnitudes. This combined with the relative positions of the particles indicates why there may exist an optimum size ratio. The design and operating fluid dynamics undoubtedly influence this size ratio.

3.2 Effect of Particle Number

Increasing the number of particles in the spiral jet mill influences the proportion of the total dissipated energy due to inter-particle and particle-wall collisions (Table 4). With 10^4 particles the majority of collisions and dissipated energy are due to particle-wall collisions. When the number of particles is increased to 10^5, inter-particle collisions become the dominant source of dissipated energy despite there still being a greater proportion of particle-wall collisions. With 10^6 particles, inter-particle collisions are now dominant in terms of number and energy loss.

Table 4. Change in dominant collision type with number of particles

No of particles	10^4	10^5	10^6
Total no of collisions	11×10^6	48×10^6	642×10^6
Inter-particle (%)	3.5	40.8	89.5
Particle-wall (%)	96.5	59.2	10.5
P-P energy (%)	4.1	71.2	97
P-W energy (%)	95.9	28.8	3

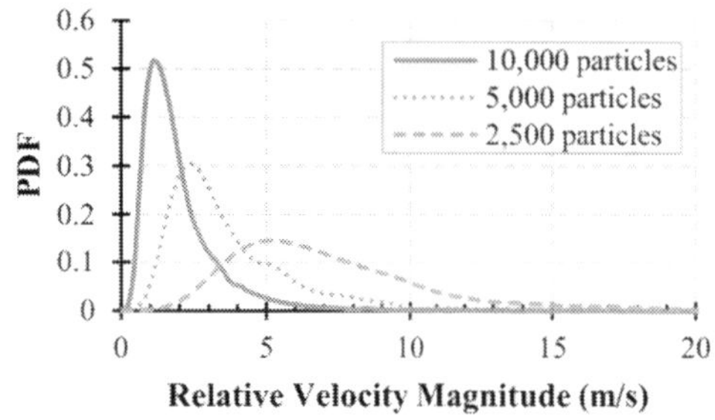

Fig. 6. Relative velocity distribution at [4.0,4.0] for different number of particles

Increasing the number of particles in the spiral jet mill leads to a decrease in the relative velocity between particles (Fig. 6). It can also be seen that the relative velocity distribution widens as the particle number in the spiral jet mill reduces. The higher relative velocity leads to a greater degree of dissipated energy and subsequent particle breakage.

3.3 Effect of Gas Pressure

With 10^4 particles, there is an increase in the particle velocity magnitude from 2.0 – 3.5 m/s as the grinding pressure increases (Fig. 7). From gas pressure of 4.0 bar, increasing the pressure has an insignificant effect on the relative velocity magnitude suggesting that the particle concentration is too dense to have a significant effect on breakage. The relative velocity distribution also widens as the gas pressure increases.

Reducing the number of particles to 5,000 leads to a greater increase in relative velocity magnitude with gas pressure, from 2.5 – 5.5 m/s (Fig. 8). This is due to particles having a greater amount of space in the chamber to accelerate and reach higher velocities before impact.

4 Conclusions

The spiral jet mill was simulated using DEM coupled with CFD to investigate the effect of different parameters on the energy transfer within the mill. Inter-particle collisions are found to have a higher dissipated collisional energy compared to particle-wall collisions, and therefore

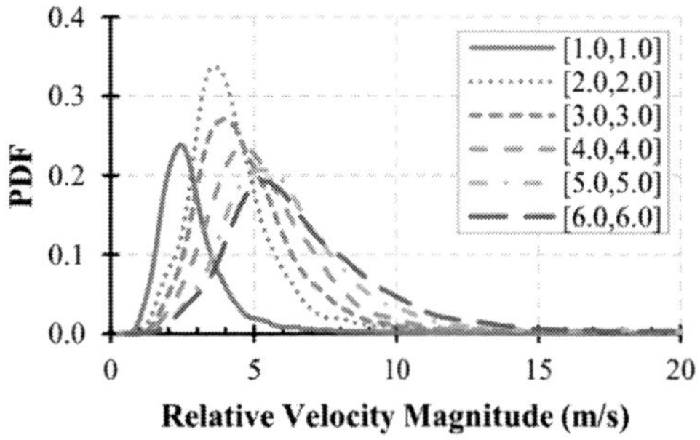

Fig. 7. Velocity distribution of 10^4 particles

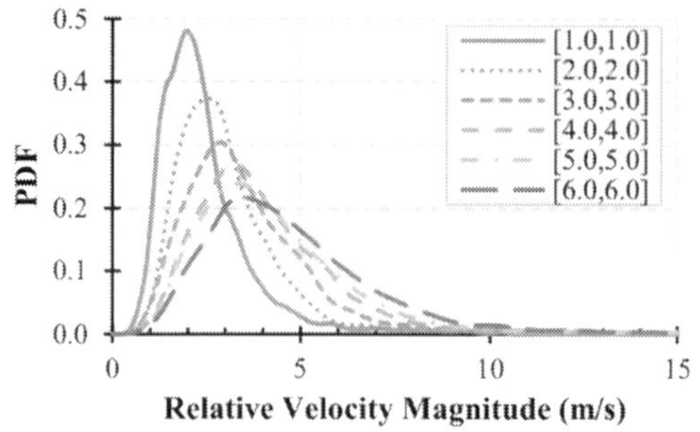

Fig. 8. Velocity distribution of 5,000 particles

expected to contribute more towards particle breakage. Collisions between 10 and 20 μm particles are the most frequent ones and are expected to contribute most to particle breakage compared to other possible collisions, also influenced by particle size segregation within the grinding chamber. Particle concentration influences the degree of inter-particle and particle-wall collisions, with the former becoming more dominant than the latter in denser systems. Increasing the particle concentration also leads to narrower relative velocity distributions between colliding particles, as well as generally smaller relative velocity magnitudes. For dense systems, increasing the grinding pressure has some effect on the relative velocity distribution, but to a lesser extent on the general velocity magnitude of the collisions.

Acknowledgements

The financial support in the form of an Industrial CASE Award by the EPSRC and AstraZeneca (Macclesfield, UK) for PhD studentship for the first author is gratefully acknowledged.

References

1. M. Metboul, J.F. Large, P. Guigon, Int. J. Miner. Process. **44-45** (1996)
2. T. Han, H. Kalman, A. Levy, Part. Sci. Technol. **20** (2002)
3. A. Levy, H. Kalman, Part. Sci. Technol. **25** (2007)
4. T. Brosh, Y. Batat, H. Kalman, A. Levy, A.B. Brown, Bulk Solids Powder – Sci. Technol. **3** (2008)
5. S. Teng, P. Wang, Q. Zhang, C. Gogos, Powder Technol. **208** (2011)
6. M. Pasha, S. Dogbe, C. Hare, A. Hassanpour, M. Ghadiri, Granul. Matter **16** (2014)

Granular compaction and the topology of pore deformation

Mohammad Saadatfar[1,⋆], *Hiroshi* Takeuchi[2], *Maryam* Hanifpour[3], *Vanessa* Robins[1], *Nicolas* Francois[1], and *Yasuaki* Hiraoka[2]

[1] *Department of Applied Mathematics, Research School of Physics and Engineering, The Australian National University, Canberra ACT 2601, Australia*
[2] *Graduate School of Science, Tohoku University. 6-3, Aramaki Aza-Aoba, Aoba-ku, Sendai, 980-8578, Japan*
[3] *Department of Applied Physics, Aalto University, Helsinki*

Abstract. The mechanism of crystallisation in highly dissipative materials such as foams or granular materials is still widely unknown. In macroscopic granular materials high levels of energy need to be injected to overcome the natural propensity of these dissipative materials to form amorphous structures [1, 2]. The transition from disordered to ordered packings in such systems triggers a wide range of geometrical, topological and mechanical changes at multi length scales [3]. Formation of cavities and patterns by aggregates of grains and their evolution during this transition requires a complete topological description of the system. Here, crystallisation of three-dimensional packings of frictional spheres is studied at the grain scale with x-ray tomography. Using a novel and powerful topological tool, Persistent Homology, we describe the complete formation process of perfect tetrahedral and octahedral patterns: the two building blocks of FCC and HCP crystalline arrangements. Additionally we present possible and allowable deformations of these components that accurately reproduce the main topological features of the system. These results give new insights into the crystallisation of these highly dissipative materials.

1 Introduction

A granular assembly is comprised of numerous particles therefore amenable to statistical mechanics approach. However, such an approach has proven to be extremely challenging due to the fact that the physics of granular systems are inherently different at micro and macro scales.
On one end of the granular matter spectrum is the thermal microscopic systems, including foams, colloids, and molecular glass formers, where glassy behaviour is observed. Crystallisation in these systems are described by equilibrium statistical mechanics, where their phase space can be fully described via entropic interactions.
At the other end of the spectrum are the macroscopic (mesoscopic) granular systems, which are athermal and frictional and therefor far from thermodynamic equilibrium. In these out-of-equilibrium systems the granular particles are so heavy that their thermal motion is negligible. Due to the presence of friction at this length scale, the inter-particle interaction is dissipative and the kinetic energy of the particles is irrecoverably lost to microscopic degrees of freedom. However, if a macroscopic hard-sphere packing is mechanically stable (e.g at rest under gravity), an analogy with equilibrium statistical mechanics can be drawn, where one assumes that all microstates (mechanically stable states) occur with the same probability . In Edwards' proposal to describe a macroscopic granular assembly using thermodynamics that is analogous to thermal systems, the volume of granular packing replaces the energy. In this approach, the configurational space (grain positions and their orientational arrangements in the packing) and the local free volume available to each grain play a crucial role [4–8].
In this paper we demonstrate that new ideas associated with computational topology provide an efficient, robust and faithful approach to implementing tractable models to decipher the complexity of the spatial structures of the configurational space of granular systems. More specifically, we present a novel topological characterisation tool, Persistent Homology (PH), to study dense granular systems. We show that PH is able to explore and characterise the configurational phase space of disordered and partially ordered macroscopic granular systems by identifying key features specific to granular systems. We first detail the experimental systems followed by the mathematical description of the persistent homology, and in particular its application for granular systems.

2 Experimental procedure

We analyse two sets of experimental granular packings each containing over $150,000$ monosized acrylic beads (diameter d = 1 mm, polydispersity = 0.025 mm): partially ordered cylindrical (see Fig. 1.a) and spherical (see Fig. 1.b) packing produced using a vibrational protocol with similar packing densities of $\phi \approx 0.68$ Details

⋆e-mail: mos110@physics.anu.edu.au

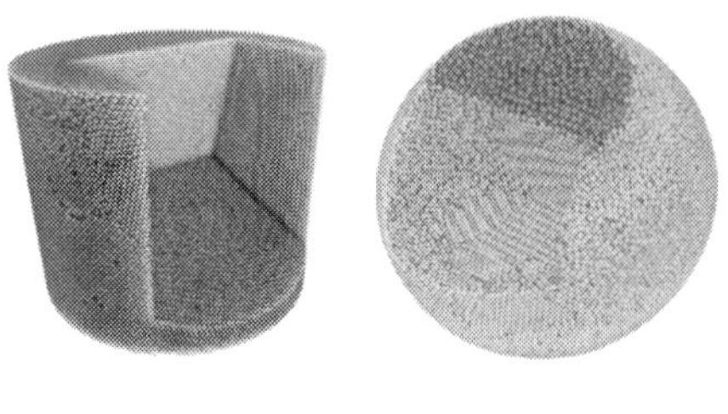

Figure 1. 3D visualisation of the internal structure of partially crystallised packing with a global packing density of $\phi = 0.685$ in (a) a cylindrical container, (b) a spherical container.

of the experimental procedure can be found elsewhere [2, 9]. Our experiments harness X-ray Computed Tomography (XCT) and three-dimensional (3D) image analysis to accurately determine grain centres with the precision of ($< 10^{-3}\mu m$) and grain's diameter with precisions greater than $\approx 10^{-2}\mu m$ [3]. Figure 1(a&b) show 3D renderings of such partially crystallised structures.

2.1 Persistant Homology

Homology is an algebraic tool for studying topological structures. Persistent Homology (PH) is a technique that facilitates the quantification of topological structures in data [10]. Persistent homology extends traditional homology by tracking how the homology groups change as an object grows. Each homology class has two values, which are calculated by varying a filtration parameter associated with it: a birth value and a death value. It is common practice to represent this information in a *Persistence Diagram* (PD) for each dimension of homology. PD_k contains all pairs (b, d), $b \leq d$, associated with a persistent homology class in dimension k.

For a perfectly mono-disperse bead pack with no "rattlers", all points in PD_0 have birth = 0 and death = r. Each point in PD_2 represents a kind of *pore* or *cavity* in the interstices between the beads. Perfectly crystalline sphere packings consist of a periodic arrangement of tetrahedral and octahedral cavities. This close packing configuration has a maximal packing density of $\phi = 0.74$. Below the RCP limit ($\phi = 0.64$), a great number of cavities are confined within a curved strip, one end of which is pointing to the the location of the tetrahedral coordinate in PD2. We note the presence of a less populated region located at higher death values ($d > \sqrt{2} \times D$) with a peak around $(b, d) = (0.04, 0.07)$ as seen in Figure 2.a. This region is a characteristic of packings containing amorphous structures with cavities that are irregular and larger than octahedral pores. Each point, (b, d), in this region of PD2 represents a connected chain of many pores, with the death value being the radius of the largest pore in the chain, and the birth value being the maximal throat (3-bead contact radius) that separates the extended pore from the others.

3 Cavity formation/deformation scenarios

A: Tetrahedral deformations

We consider two types of tetrahedral deformations, which are permitted in a close bead packing. The first tetrahedral deformation, D1, consists of letting one bead of a tetrahedron roll across the saddle formed by two neighbours, while the other beads stay fixed. A top view of this deformation (D1) is demonstrated in Figure2.b and the PD_2 associated curve produced by this deformation can be seen in Figure2.a. The second tetrahedral deformation, D2, is a symmetric process that transforms a regular tetrahedron into a flat square configuration. A top view of this deformation in shown in Figure2.b. The PD2 curves of this deformation can be seen in Figure2.a.

D1&D2 curves run along the lower and upper boundaries of a densely populated region in PD_2, where they are joined at the location of tetrahedral cavity (Figure2.a).

The curves D1 and D2 clearly delineate the highly populated strip which we identify as a region made of weakly distorted tetrahedral cavities. The extent of this domain confirms the variety and ubiquity of weakly distorted tetrahedra in partially crystallised packings It has recently been shown that the formation of polytetrahedral aggregates composed of weakly distorted tetrahedra is a geometrical principle of densification for amorphous frictional packings and a resilient feature in partially crystallised structures up to $\phi = 0.73$ [2, 11].

B: Octahedral deformation

Here we consider two relevant octahedral deformations each closely matching one of the densely populated paths connecting the octahedral hotspot to the diagonal line in PD_2. These dense regions are especially pronounced in the PD_2 of the partially crystallised packing ($\phi \approx 0.68\%$) as seen in Figure 2.c.

The first one is simply lengthening two opposite edges of the octahedron (symmetrically), while keeping all other edges equal to $2r$. This deformation scenario, D3, transforms two edge-adjacent tetrahedra into a regular octahedron as shown in the top cartoon in Figure 2.b. The second octahedral deformation, D4, is similar to the first tetrahedral deformation (D1), where only one of the beads roll along the saddle. This scenario is essentially the repeat of D1 deformation for an octahedral grain configuration as demonstrated in Figure 2.b. The birth-death curves of the above four deformation scenarios are superimposed on the PD_2 of our partially crystallised packing with the density of $\phi \approx 0.68$ (Figure 2.a). Moreover, D1-D4 curves and the evolution of (b,d) coordinates of these regular cavity deformations can be derived analytically using a combination of dihedral angles and grain radius [12].

4 Formation scenarios of regular cavities and mechanical features of packing crystallisation.

The above four formation mechanisms of regular cavities allow us to describe most of the evolution of PD_2 over a

DOI: 10.1051/epjconf/201714016009

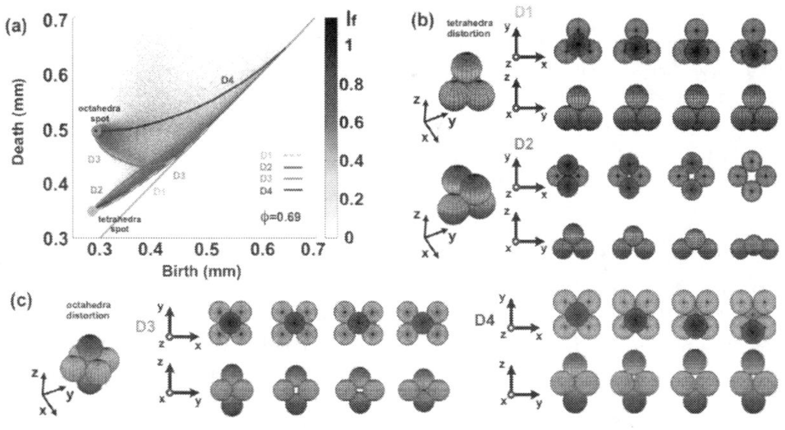

Figure 2. (a) PD_2 of a partially crystallised sphere packing with density $\phi = 0.685$. The superimposed curves correspond to analytically computed birth-death curves of the deformation scenarios shown in (b) and (c). (b) Top and side views of D1 and D2 deformations scenarios of a tetrahedral cavity. (c) Top and side views of D3 and D4 deformation scenarios of an octahedral cavity. The color code indicates the relative height of sections of the grain with respect to horizontal median plane.

broad density range $0.64 < \phi < 0.73$. This predictive power is quite remarkable given that these mechanisms are based on idealised deformation of regular cavities made of identical grains.

To gain more insight into this matter, let us now consider the four mechanisms depicted in Figs. 2(b,c) from a mechanical viewpoint. More precisely, we are interested in the mechanical contacts between those grains: i.e. the *internal* contacts within a specific motif made of N grains. In the following, we assume that the number of mechanical contacts surrounding the motif (i.e. the *external* contacts) remain unchanged. Table 1 shows the total number of *internal* mechanical contacts ΔC_{int} created during the formation of regular motifs according to the four scenarios. For each mechanism the number of contacts created per grain, $\Delta C_{int} \cdot 2/N$, is between $1/2$ and 1. This is an important feature of the formation mechanisms chosen: a small number of contacts is modified. To support this statement, we have tested other deformation mechanisms that allow more *internal* contacts to be modified and none of them describe accurately the evolution of PD_2.

The mechanical stability of packings depends on both mechanical and geometrical characteristics of the neighbourhood of each grain [1, 13–15]. This feature of packing stability has a huge influence on the crystallisation process [3]. One of the most remarkable consequences is the existence of a plateau for the mechanical contact number $< Z_m >$ (i.e. the average number of contact per grain) over the density range $0.64 < \phi < 0.68$. This feature has been observed in numerical simulations of frictionless spheres and experimentally in the case of frictional packings [3, 16, 17]. Quantitatively, $< Z_m >$ is constant at $< Z_m > \approx 5.6$ over the density range $0.64 < \phi < 0.68$, beyond $< Z_m >$ increases again and reaches $< Z_m > \approx 6.8$ at $\phi = 0.73$, a value well below $< Z_m > = 12$ as expected for perfect FCC or HCP crystals [3]. Therefore almost-crystalline packings made of weakly polydisperse beads still possesses a highly random mechanical backbone with properties close to that of amorphous packings [3, 18–20]. Quantitatively, there is a relatively good match between the values of $\Delta C_{int} \cdot 2/N$ and the fact that actual measurements of $< Z_m >$ show an increase of 1.2 over $0.64 < \phi < 0.73$ [3, 17].

There is no doubt that the deformation scenarios presented here are idealised (no grain polydispersity, no influence of the second shell of grains), however, they highlight essential features of crystallisation in granular packings:i) a small number of mechanical contacts per grain is created during the crystallisation. ii) since the mechanical structure of highly crystalline packing made of weakly polydisperse beads shares some similarities with that of amorphous packings, grain-scale structural evolution in the first shell of neighbours (rather than global periodic conditions [16, 21]) is expected to remain of prime importance to characterise packing stability [3, 13]. iii) the distorted cavities at the boundaries of the existence domain in PD_2 have to be dense configuration of grains, a natural consequence of crystallisation as a densification process based on the creation of tetrahedral and octahedral cavities.

Table 1. Number of *internal* mechanical contacts ΔC_{int} created during the formation of tetrahedral and octahedral cavities according to the scenarios D1, D2, D3, D4. N is the number of beads involved in the pattern. $C_{intfinal}$ is the total number of contacts in the final regular pattern. To define the number of contact created per grain, we divide ΔC_{int} by $N/2$ to account for the fact that a contact links two grains.

Scenario	N	$C_{intfinal}$	ΔC_{int}	$\Delta C_{int} \cdot 2/N$
D1	4	6	1	1/2
D2	4	6	2	1
D3	6	12	2	2/3
D4	6	12	2	2/3

5 Conclusions

Although persistent homology is a geometrical characterisation, we emphasise that the experimental nature of our packings ensures that the diagrams presented here actually describe the space of cavities that not only *exist* but are also *mechanically* stable. To illustrate this, we have identified four grain-scale deformation mechanisms that recover prominent features of the diagram evolution and highlight basic grain-scale rearrangements, pattern selection, and mechanical constraints underpinning packing crystallisation. We note that connections between topological features of packings and their mechanical stability have recently been uncovered [22]. The results reported here represent a basis on which to interpret at the grain-scale why granular crystallisation, an out-equilibrium phenomenon in a complex system can be mapped onto an established framework of statistical mechanics [16]. Our findings have also practical applications in domains such as pore description in soil- and geo-sciences, which are crucial for understanding natural systems mechanical stability, vibration induced compaction and flow permeability [23–25].

References

[1] Wyart, M. Marginal stability constrains force and pair distributions at random close packing. *Phys. Rev. Lett.* **109,** 125502 (2012).

[2] Francois, N., Saadatfar, M., Cruikshank, R. & Sheppard, A. Geometrical frustration in amorphous and partially crystallized packings of spheres. *Phys. Rev. Lett.* **111,** 148001 (2013).

[3] Hanifpour, M., Francois, N., Vaez Allaei, S. M., Senden, T. & Saadatfar, M. Mechanical characterization of partially crystallized sphere packings. *Phys. Rev. Lett.* **113,** 148001 (2014).

[4] Song, C., Wang, P. & Makse, H. A. A phase diagram for jammed matter. *Nature* **453,** 629-632 (2008).

[5] Blumenfeld, R., Jordan, J. F. & Edwards, S. F. Interdependence of the volume and stress ensembles and equipartition in statistical mechanics of granular systems. *Phys. Rev. Lett.* **109,** 238001 (2012).

[6] McNamara, S., Richard, P., Kiesgen de Richter, S., Le Caër G. & Delannay, R. Measurement of granular entropy. *Phys. Rev. E* **80,** 031301 (2009).

[7] Puckett, J. G. & Daniels, K. E. Equilibrating temperaturelike variables in jammed granular subsystems. *Phys. Rev. Lett.* **110,** 058001 (2013).

[8] Ashwin, S. S., Blawzdziewicz, J., O'Hern, C. S. & Shattuck, M. D. Calculations of the structure of basin volumes for mechanically stable packings. *Phys. Rev. E* **85,** 061307 (2012).

[9] Aste, T., Saadatfar, M. & Senden, T. J. Geometrical structure of disordered sphere packings. *Phys. Rev. E* **71,** 061302 (2005).

[10] Robins, V., Wood, P. J. & Sheppard, A. P. Theory and algorithms for constructing discrete Morse complexes from grayscale digital images. *Pattern Analysis and Machine Intelligence, IEEE Transactions*, **33,** 1646-1658 (2011).

[11] Anikeenko, A. V., Medvedev, N. N. & Aste, T. Structural and entropic insights into the nature of the random-close-packing limit. *Phys. Rev. E* **77,** 031101 (2008).

[12] Saadatfar, M., Takeuchi, H., Robins, V., Francois, N., & Yasuaki, H. Topological configuration landscape and pore deformation mechanisms during the crystallisation of sphere packing. *Nature Communications*, (2016.)

[13] van Hecke, M. Jamming of soft particles: geometry, mechanics, scaling and isostaticity. *J. of Physics: Condensed Matter* **22,** 033101 (2010).

[14] Charbonneau, P., Corwin, E. I., Parisi, G. & Zamponi, F. Universal microstructure and mechanical stability of jammed packings. *Phys. Rev. Lett.* **109,** 205501 (2012).

[15] O'Hern, C. S., Langer, S. A., Liu, A. J. & Nagel, S. R. Force distributions near jamming and glass transitions. *Phys. Rev. Lett.* **86,** 111 (2001).

[16] Jin, Y. & Makse, H. A. A first-order phase transition defines the random close packing of hard spheres. *Physica A* **389,** 5362 (2010).

[17] Hanifpour, M., Francois, N., Allaei, S. V., Senden, T. & Saadatfar, M. Structural and mechanical features of the order-disorder transition in experimental hard-sphere packings. *Phys. Rev. E* **91,** 062202 (2015).

[18] Mari, R., Krzakala, F. & Kurchan, J. Jamming versus glass transitions. *Phys. Rev. Lett.* **103,** 025701 (2009).

[19] Goodrich, C. P., Liu, A. J. & Nagel, S. R. Solids between the mechanical extremes of order and disorder. *Nature Physics* **10,** 578 (2014).

[20] Tong, H., Tan, P. & Xu, N. From Crystals to Disordered Crystals: A Hidden Order-Disorder Transition. *Scientific Reports* **5,** doi:10.1038/srep15378 (2015).

[21] Moukarzel, C. F. Isostatic phase transition and instability in stiff granular materials. *Phys. Rev. Lett.* **81,** 1634 (1998).

[22] Carlsson, G., Gorham, J., Kahle, M. & Mason, J. Computational topology for configuration spaces of hard disks. *Phys. Rev. E* **85,** 011303 (2012).

[23] Saadatfar M. et al., 3D mapping of deformation in an unconsolidated sand: A micro mechanical study. *SEG Tech. Prog. Expan. Abst.* **2012,** 1 (2012).

[24] Robins, V., Saadatfar, M., Delgado-Friedrichs, O. & Sheppard, A. P. Percolating length scales from topological persistence analysis of micro-CT images of porous materials. *Water Resources Research*, DOI: 10.1002/2015WR017937 (2016).

[25] Sufian, A., Russell, A. R., Whittle, A. J. & Saadatfar, M., Pore shapes, volume distribution and orientations in monodisperse granular assemblies. *Granular Matter*, **17(6),** 727-742 (2015).

 DOI: 10.1051/epjconf/201714003080

Acoustic monitoring of a ball sinking in vibrated granular sediments

Siet van den Wildenberg*, *Julien* Léopoldès, *Arnaud* Tourin, and *Xiaoping* Jia†

Institut Langevin, ESPCI Paris, PSL Research University, CNRS, 1 rue Jussieu, 75005 Paris, France

Abstract. We develop an ultrasound probing to investigate the dynamics of a high density ball sinking in 3D opaque dense granular suspensions under horizontal weak vibrations. We show that the motion of the ball in these horizontally vibrated glass bead packings saturated by water is consistent with the frictional rheology. The extracted stress-strain relation evidences an evolution of flow behaviour from frictional creep to inertial regimes. Our main finding is that weak external vibration primarily affects the yield stress and controls the depth of sinking via vibration-induced sliding at the grain contact. Also, we observe that the extracted rheological parameters depend on the size of the probing ball, suggesting thus a non-local rheology.

1 Introduction

A convenient method to determine the viscosity of a simple fluid is to drop a higher density ball in it. The ball accelerates under gravity until reaching a terminal velocity, which is inversely related to the viscosity in the laminar flow regime via the Stokes' law. In a yield stress fluid such as foams, emulsions and granular suspensions, the ball will arrest sinking at a certain depth due to the solid-like friction between the particles. The jamming phase diagram provides a general framework to explain such a transition from a liquid-like flowing state to a solid-like jammed state as a function of the packing density of particles, the confining pressure and the applied shear. However, studies of an intruder sinking in dense granular media are mostly limited to 2D packings [1] and investigations inside realistic 3D opaque dense granular media [2] remain a conceptual and practical challenge.

Here we report a new experiment investigating the dynamics of a sinking ball (steel) in horizontally vibrated granular sediments, i.e., glass bead packings saturated by water. Using a non-intrusive acoustic probing, we monitor precisely the position $z(t)$ of the steel ball in the 3D opaque granular suspension during sinking. The instantaneous settling velocity v or acceleration a is obtained by numerically differentiating z or v over a time interval. These experimental results are then analyzed with the granular rheology law $\mu(I)$ revealed in dry packings [3], being extended to dense granular suspensions [4-7]. We find that external shaking primarily affects the yield stress [8-10] and consequently controls the depth of sinking. This acoustic monitoring should improve the understanding of the mechanical behavior of dense granular suspensions, quicksands and landsliding.

2 Experiments

2.1 Set-up and procedure

Our setup consists of a plexiglass container which is coupled via a linker to a shaker and filled with a suspension of glass beads $d \sim 100$ μm in water. In our experiments, the ball sinking is generated by vibrating the container in the horizontal direction (Fig. 1a). The intruder is a steel ball (we experimented with intruders of different size) which is carefully placed on the surface of the suspension. In the absence of vibrations, the intruder remains on the surface indicating that the granular suspension is in a solid state with a packing density of about 60%.

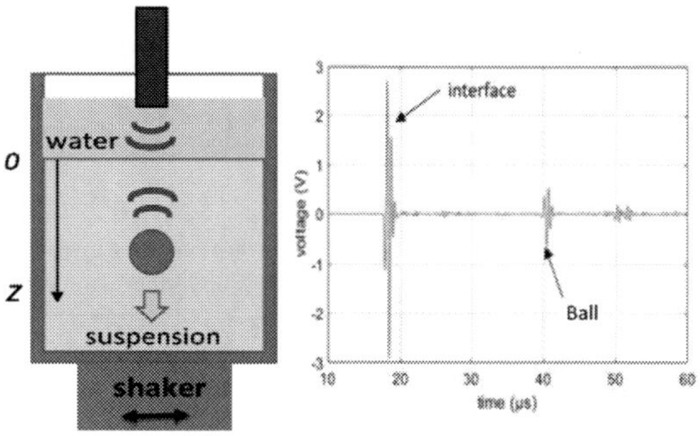

(a) (b)

Fig. 1. (a) Sketch of the experimental set-up. (b) Ultrasonic echos from the interface and the intruder (ball).

* Currently at Laboratoire Magmas et Volcans, Université B. Pascal, 63178 Aubière, France
† Corresponding author: xiaoping.jia@espci.fr

In contrast, when the suspension is vibrated the intruder starts to slowly sink. To follow the sinking of the intruder in the 3D-opaque suspension we employ acoustic monitoring. For this purpose, a transducer T is placed exactly above the intruder. The transducer is used in reflection mode and it emits short broadband pulses and detects the echos. In this way, a waveform of the echos is acquired at a sampling rate of ~ 5 Hz. In Fig. 1b we show a typical waveform which depicts two echos: one from the interface of the suspension and another one from the intruder. The time of arrival of the echo from the interface remains constant. Instead, the arrival time of the echo from the intruder increases as the distance between the transducer and the sinking intruder increases. From the stack of such waveforms (Fig. 2a), we may determine the position of the intruder z during sinking in the course of time t (Fig. 2b), by multiplying the arrival time of the echo by the sound velocity in water (c_w = 1500 m/s). As shown in Fig. 2b, we observe that the motion of the intruder is slow but continuous and we do not observe short periods of stalling as was reported for sinking in 2D dry granular media [1].

Notice that in certain case we observe not only a small compaction effect at short times but a small dilation of the granular suspension at larger times shown in Fig. 2b (lower trace). For following experiments we have developped a preparation protocol to yields reproducible and stable packings, suppressing this dilation effect.

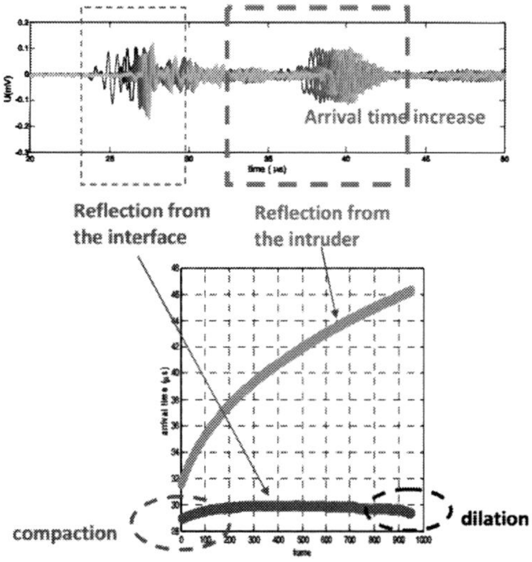

Fig. 2. (a) Stack of waveforms. (b) Position of the intruder during sinking (upper trace) and the position of the suspension-water interface (lower trace).

2.2 Results

To characterize in details the sinking behavior of the intruder, we examine how $z(t)$ depends on the acceleration of vibration Γ and the diameter of the intruder $2R$.

We start by investigating the influence of vibration intensity on the sinking dynamics of the intruder of R = 5 mm by varying Γ a decade between 0.27 to 2.7 m/s^2 and measuring $z(t)$. Generally, we find that the sinking of the intruder, at each Γ, can be divided in two regimes (Fig. 3): i) first, a regime in which the intruder decelerates fast (grey parts in the short time), and ii) a quasi steady flow regime in which the intruder moves with an almost constant velocity (in the long time range).

In order to separate these two regimes we calculated, for each Γ, the instantaneous velocity dz/dt and the acceleration d^2z/dt^2 of the intruder. The dz/dt was obtained, by numerically differentiating $z(t)$ via $[z(t+\Delta t) - z(t)]/\Delta t$, where the time interval Δt was set to 20 seconds so to obtain low noise while remaining much shorter than the duration of the data set. The acceleration was calculated from the velocity in a similar way. A cross-plot of the acceleration as a function of the velocity reveals a clear kink separating the fast decelerating and the quasi steady regime (data not shown). We determine the velocity at this kink $(dz/dt)^*$ and the associated depth z^* which is roughly equal to the diameter of the intruder, suggesting that the fast decelerating regime is likely due to partial submergence of the intruder and boundary effects such as crater formation on the surface just above the intruder. In the following, we will focus on the quasi steady flow regime. For low Γ we observe that the intruder decelerates until it finally arrests at a final depth. When increasing Γ the intruder reaches larger depths and it keeps decelerating but did not stop completely within our experimental time (Fig. 3).

Next, the influence of the intruder size is investigated by repeating the experiments with intruders of different size R = 4 mm and 7 mm. The phenomenology of sinking remains similar for all different intruders and we recover the fast decelerating regime and a quasi steady flow regime (data not shown). Interestingly, we find that in the steady flow regime, smaller intruders sink to larger depths; the motion of the intruder appears, therefore, not only to depend on Γ but also on its own size R.

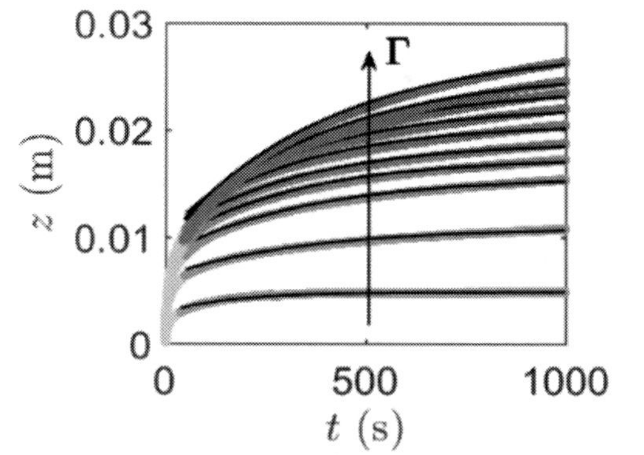

Fig. 3. Dynamics of the ball sinking as a function of time under different vibration intensity Γ.

3 Modelling and discussions

3.1. Frictional rheology

We consider the equation of motion for the intruder sinking along the vertical direction. There are three forces acting on the intruder during its trajectory through the suspension, a force due to gravity, a buoyancy force and a frictional force F_s:

$$m(d^2z/dt^2) = (4/3)\pi R^3 \Delta\rho g - F_s \tag{1}$$

where $\Delta\rho = \rho_s - \rho_{sus}$ with the density of the steel ball $\rho_s \sim 8000$ kg/m^3 and the density of the suspension $\rho_{sus} \sim 1960$ kg/m^3.

In order to specify the frictional rheology for our gravitational suspensions, we may estimate the viscous force and inertial forces acting on glass beads saturated by water [4, 5]: $F_v = 3\pi\eta_f d(dz/dt) \sim 10^{-11}$ N and $F_i = \rho_f \pi d^2 (dz/dt)^2/6 \sim 10^{-15}$ N, with the viscosity $\eta_f = 8.9\ 10^{-4}$ Pa.s and the density of water $\rho_f = 1000$ kg/m^3 and the flow velocity $(dz/dt) \sim 10^{-5}$ m/s. The Reynolds number $Re = F_i/F_v \sim 10^{-4}$ is very small and also the Stokes number $St = \rho_p d^2 \eta_f (d\gamma/dt) \ll 1$ with the density of glass grains $\rho_p = 2500$ kg/m^3 and the strain rate $d\gamma/dt \sim (dz/dt)/R$, indicating that viscous forces dominate at the particle scale and the internal time is given by a viscous scaling. In this case, the constitutive equation for the frictional rheology can be written as [4, 6, 7]:

$$\tau = \mu_0 P + \mu_1 \eta_f (d\gamma/dt) \tag{2}$$

where τ is the shear stress, $P = \Delta\rho g z$ the granular pressure due to gravity, μ_0 and μ_1 are the friction coefficient and viscous parameter, respectively. Eq. 2 shows that for small shear rate, the local stress in a gravitational suspension is a linear combination of a frictional stress and a viscous stress. Consequently, for slow driving the local shear stresse τ is proportional to P, while for fast driving the local shear stresses are proportional to the driving velocity [7, 8].

Assuming the frictional force $F_s = \tau\ (4\pi R^2)$ and considering the quasi-steady regime $d^2z/dt^2 \to 0$, we derive the corresponding solution of Eq. 1 using the initial condition $z = z^*$ at $t = t^*$:

$$z(t) = (A_0 - z^*)[(A_0/(A_0 - z^*) - e^{-k(t-t^*)}] \tag{3}$$

with $A_0 = \Delta\rho R/(3\mu_0 \rho_{sus} g)$ and $k = \mu_0 \rho_{sus} g/(\mu_1 \eta_f)$. This equation is used to fit the traces of $z(t)$ in the quasi steady flow regimes with the free fitting parameters μ_0 and μ_1 (black lines in Fig. 3).

3.2 Vibration effects on rheological parameters

The values for μ_0 and μ_1 obtained from fit are plotted as a function of the vibration acceleration Γ in Fig. 4 for different diameters of the sinking ball. We find that for the lowest Γ the static coefficient of friction is $\mu_0 \sim 0.15$, which is slightly lower than the values reported in literature for dense granular suspensions without vibrations $\mu_0 \sim 0.3$ [6]. For increasing vibration intensity, and, for $\Gamma > 2$ m/s^2, they approach a constant value.

Moreover, at low Γ the values for μ_0 and μ_1 depend on the size of the sinking ball R, suggesting that the local rheology that describes a macroscopic fields trough a constitutive coefficient (Eq. 2) is incomplete and there appear to be other important parameters such as correlation length of force chains in granular packings [11, 12].

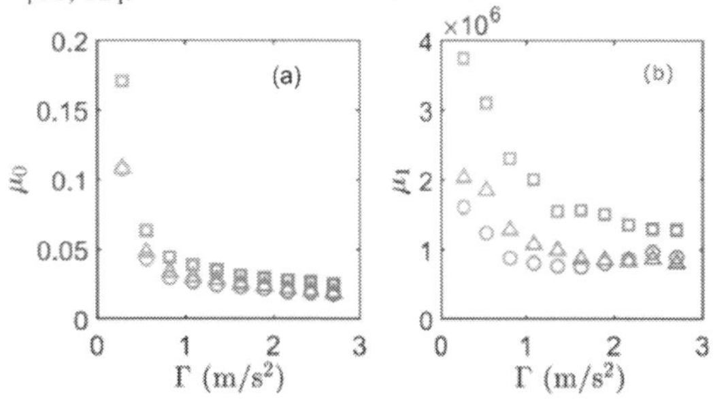

Fig. 4. Rheological parameters versus shaking intensity: (a) static friction coefficient μ_0 and (b) viscous coefficient μ_1, determined from fitting the $z(t)$ traces like in Fig. 3 by equation (3). The different symbols correspond to intruder size R= 4 mm (circles), 5 mm (triangles) and 7 mm (squares).

To compare our data with rheological measurements performed on the sample scale, we seek to plot the stress versus the strain rate. Specifically, we calculate the shear stress from Eq. 2 using the μ_0 and μ_1 obtained from the exponential fit and plot the friction coefficient $\mu = \tau/P$ as a function of shear rate $d\gamma/dt$ in Fig. 5. These plots show a plateau at low $d\gamma/dt$ pointing to the friction-like creep regime [13], while for larger shear rate the μ increases with increasing $d\gamma/dt$ corresponding to the inertial regime. These results are consistent with other rheological measurements in dense granular suspensions [6]. As shown in Figs. 4 and 5, the presence of external vibration strongly affects the yield stress, i.e., the coefficient of friction μ_0 as in dry granular packings [8]. Note that for moderate vibration applied here $\Gamma < 3$ m/s^2, the dilatancy effect caused by vibration would not be significant; the decrease of the yield stress should be primarily due to the reduction of friction coefficient between solid grains caused by acoustic lubrication [10].

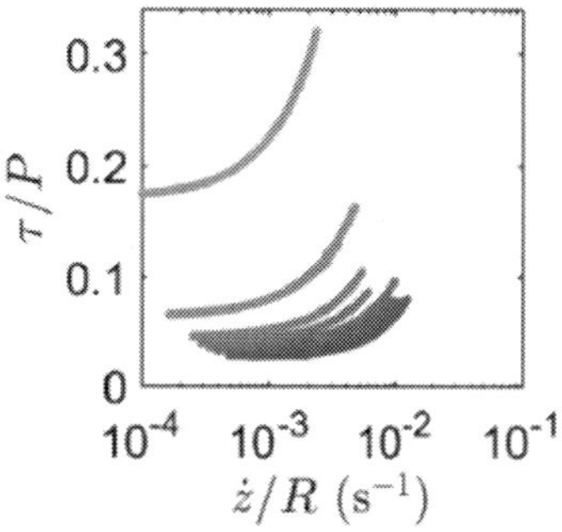

Fig. 5. Effective friction coefficients $\mu = \tau/P$ as a function of normalized sinking velocity (shear rate) of a steel ball $2R$ = 10mm, for different vibration intensity Γ.

DOI: 10.1051/epjconf/201714003080

4 Conclusions

We developed a novel non-intrusive tool by acoustic probing to follow the dynamics of an intruder emerged in an opaque dense granular suspension. This technique is based on the detection of echos of ultrasonic pulses that propagate through the suspension and are reflected by an intruder. The wavelength (~ 0.7 mm) is here much larger than the size of glass particles (~ 0.1 mm); nevertheless, we anticipate that this technique can also be adapted to investigate a variety of processes at a shallow depth in more scattering media.

Our main finding is that the sinking of an intruder in a gravitational suspension under vibration crosses over from an inertial viscous regime to a frictional creep regime and this cross-over depends on the vibration intensity. Moreover, the static and dynamic (viscous) friction coefficients decrease with increasing vibration intensity and they increase with increasing intruder size. The dependence of the constitutive friction coefficients on the size of the intruder suggests a non-local rheology.

We thank J.-L. Gennisson and A. Trabattoni for helpful assistance. This work was supported by French LABEX WIFI under references ANR-10-LABX-24 and ANR-10- IDEX-0001-02 PSL.

References

1. R. Harisch, T. Darnige, E. Kolb, E. Clément, Europhys. Lett. **96**, 54003 (2011)
2. K. Nichol, A. Zanin, R. Bastien, E. Wandersman, and M. van Hecke, Phys. Rev. Lett. **104**, 48 (2010)
3. Da Cruz et al, Phys. Rev. E **72**, 021309 (2005); GDR MiDi, Euro. Phys. J.E. **14**, 341 (2004)
4. C. Cassar, M. Nicolas, and O. Pouliquen, Phys. Fluids **17**, 103301 (2005)
5. S. Courrech du Ponts, P. Gondret, B. Perrin, and M. Rabaud, Phys. Rev. Lett. **90**, 044301 (2003)
6. F. Boyer, E. Guazzelli, and O. Pouliquen, Phys. Rev. Lett. **107**, 188301 (2011)
7. A. Fall, H. de Cagny, D. Bonn, G. Ovarlez, E. Wandersman et al, J. Rheol. **57**, 1237 (2013)
8. J. Dijksman, G. Wortel, L. van Dellen, O. Dauchot, M. van Hecke, Phys. Rev. Lett. **107**, 108303 (2011)
9. C. Hanotin, S. Kiesgen de Richter, P. Marchal, L. J. Michot, and C. Baravian, Phys. Rev. Lett. **108**, 198301 (2012)
10. J. Léopoldès, G. Conrad, X. Jia, Phys. Rev. Lett. **110**, 248301 (2013)
11. K. Kamrin and G. Koval, Phys. Rev. Lett. **108**, 178301 (2012)
12. M. Bouzid, A. Izzet, M. Trulsson, E. Clément, Ph. Claudin, B. Andreotti, Eur. Phys. J. E 38, 125 (2015)
13. T. Baumberger and C. Caroli, Adv. Phys. **55**, 279 (2006)

Investigation into macroscopic and microscopic behaviors of wet granular soils using discrete element method and X-ray computed tomography

Vinh-Du Than[1,2,*], *Anh-Minh* Tang[1], *Jean-Noël* Roux[1], *Jean-Michel* Pereira[1], *Patrick* Aimedieu[1], and *Michel* Bornert[1]

[1] Laboratoire Navier, UMR 8205, École des Ponts ParisTech, IFSTTAR, CNRS, Université Paris-Est, France
[2] University of Danang, College of Technology, Department of Civil Engineering, Danang, Vietnam

Abstract. We present an investigation into macroscopic and microscopic behaviors of wet granular soils using the discrete element method (DEM) and the X-ray Computed Tomography (XRCT) observations. The specimens are first prepared in very loose states, with frictional spherical grains in the presence of a small amount of an interstitial liquid. Experimental oedometric tests are carried out with small glass beads, while DEM simulations implement a model of spherical grains joined by menisci. Both in experiments and in simulations, loose configurations with solid fraction as low as 0.30 are prepared under low stress, and undergo a gradual collapse in compression, until the solid fraction of cohesionless bead packs (0.58 to 0.6) is obtained. In the XRCT tests, four 3D tomography images corresponding to different typical stages of the compression curve are used to characterize the microstructure.

1 Introduction

Wet granular materials are present in Nature (e.g., partially saturated soils) and industrial processes (such as granulation). For low saturations, the appearance of water menisci creates adhesive forces which strongly affect the macroscopic material behavior. Such cohesive materials, as opposed to cohesionless ones, may form stable structures at very low density; they are sensitive to stress intensity, not only to stress direction. Characterizations of the mechanical behavior of wet granular materials or unsaturated soils are most often carried out at the macroscopic scale [1], at which the irreversible compression and decompression curves, for a material subject to varying external load intensity, play a central role. To investigate contact networks and interstitial liquid morphology, several investigation techniques suitable to unsaturated granular assemblies have been recently used: fluorescence microscopy (FM) [2] and X-ray computed tomography (XRCT) [3]. As opposed to FM techniques, which require delicate choices of both granular materials and suitable (index-matched) liquids, XRCT, a high-resolution non-destructive observation technique, does not need any sample pre-treatment.

The present work reports on results of experiments on glass beads with capillary bonds in the pendular state, subjected to oedometric compression tests and observed at small scale by XRCT. Parallel to these experiments, as the material lends itself to numerical modeling [4], DEM simulations are carried out for the same test, providing access to detailed microstructural a micromechanical information. Numerical and experimental results are presented and compared in oedometric compression cycles, both for the plastic compression curve and the evolution of coordination numbers, as well as basic geometric features.

2 Experiments

The experimental procedure consists in an oedometric compression, supplemented with the acquisition of a XRCT scan of the central part of the sample at different steps.

2.1 Material and methods

2.1.1 Material and specimen preparation

The studied material is an assembly of spherical glass beads (density 2,460 kg.m^{-3}) with an average diameter of about 100 microns. To prepare the specimens the glass beads are first mixed with water. The low water content (3.5 % by mass) ensures that the material remains in at the pendular regime. The wet glass beads are then placed on a 200 μm opening sieve installed 80 mm above an oedometric cell of diameter 20 mm. This size allows passing a maximum of two grains at the same time through the grating. The sieve is vertically vibrated, so that the clusters of glass beads sequentially pass through and fall into the cell.

This procedure yields loose samples with solid fraction Φ_0 = 0.30 (corresponding to void ratio e_0 = 2.33), with an initial height of 10 mm.

* Corresponding author: vinh-du.than@enpc.fr

2.1.2 X-ray computed tomography tests

After a first compression reduces its void ratio from the initial state (e_0 = 2.33) to the first position S1 on Fig. 3 (e_1 = 2.21), an XRCT scan is carried out to visualize only the central zone of the sample (10 mm in diameter and 8 mm in height). Fig. 1 shows an example of reconstructed 3D image obtained from such a scan (here the scan time was 240 min and the voxel size 6.5 µm).

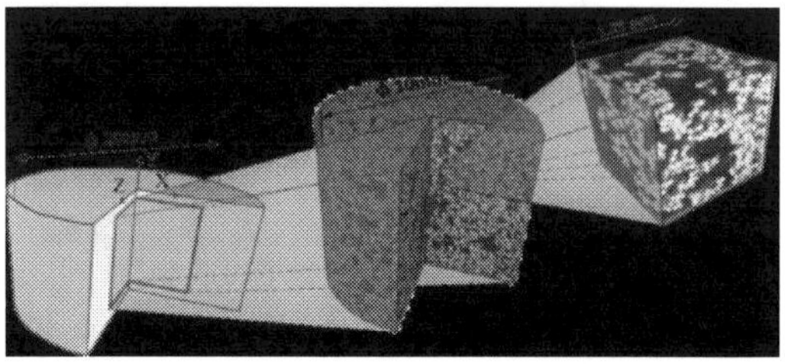

Fig. 1. Example of the position of the XRCT scan inside a sample and an example of investigated cube.

Fig. 1 clearly illustrates the ability of the XRCT technique to distinguish the grains and the pores - lighter grays characterizing larger X-ray absorption.

The same procedure is then repeated three times to obtain the XRCT images corresponding to positions marked S2 to S4 on the compression curve of Fig. 3.

2.1.3 Image processing

Image processing is carried out using software Fiji and the authors' codes, exploiting and adapting, for the detection of spherical objects, the algorithms developed by Xie *et al.* [6] (details are provided in [7].

From the 3D tomography images of the XRCT scans, cubical subregions (edge length =1.95 mm) are extracted in order to analyze the kinematics of the sample in compression. One such subdomain is shown in Fig. 1. Ten such cubes are randomly extracted at different positions within the scan, to take into account the sample microstructure heterogeneity.

Fig. 2 shows an example for one slice in a 3D detected image. The blue disks are the detected spheres, and the black points represent their centers. Incomplete white disks at the border are not used for center detection.

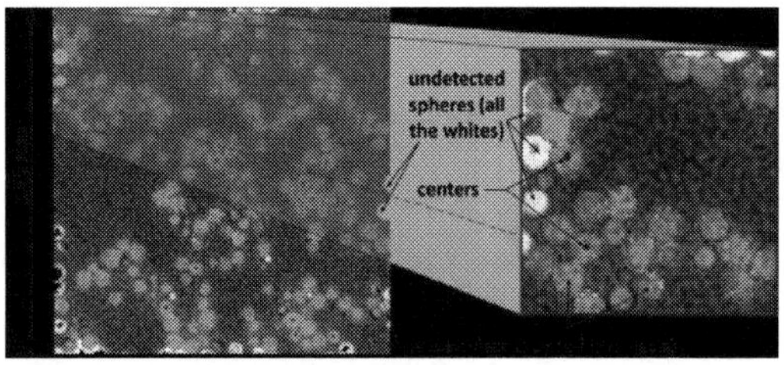

Fig. 2. A slice in a 3D detected image (not to scale).

2.2 Oedometric compression behavior

The specimen is carefully placed on the loading platform of the oedometric compression cell. The cell is then moved upward at a speed of 0.5 mm/min, towards a fixed piston where the vertical force is measured. Tests are duplicated to assess reproducibility.

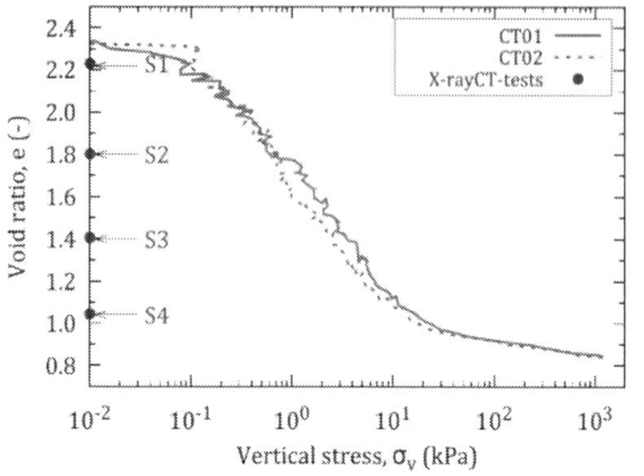

Fig. 3. Void ratio versus vertical stress during oedometric compression tests (CT01 and CT02). The circle points show the positions of XRCT scans.

Fig. 3 illustrates the compression curves for two different specimens (CT01 and CT02). Based on the shape of these curves and on the range of vertical stress, three stages, with the void ratio decreasing slowly in the first one, under low stress (< 0.1 kPa), then faster in the second one, and slowly again, in the third stage, at high stress (> 10 kPa). Blue dotes marked S1, S2, S3, and S4 show the *e* values of the XRCT scans.

2.3 Grain-scale analysis

2.3.1 Heterogeneity

The number of particles detected within the ten cubes at each of the four steps Si (i = 1 to 4) are shown in Fig. 4. These numbers increase with density, from approximately 4000 on average in the loosest state S1 (e = 1.04) to about 7500 in the densest one, S4 (e = 1.04).

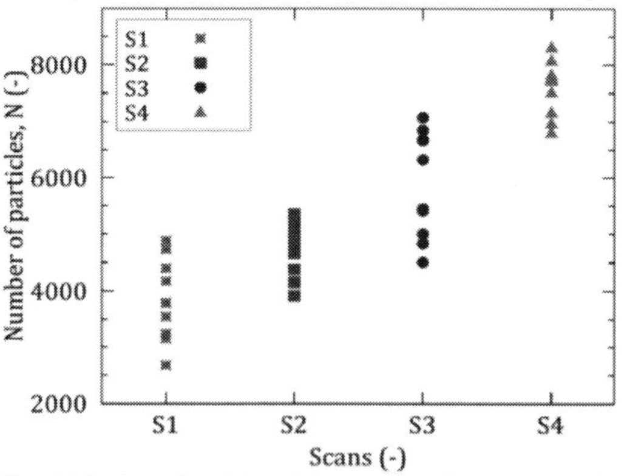

Fig. 4. Number of particles of the ten cubes for four steps after detection process.

The resulting solid fraction (averaged over all ten cubes) is compared in Fig. 5 to the macroscopic value (corresponding to the void ratio in Fig. 3). The average values are directly calculated from the tomography images, based on the IsoData method [8], applied to the segmentation process. In addition, after the detection process, the solid fraction is easily computed by dividing the volume of all the particles by the volume of the cube. The resulting differences between different values characterizes the heterogeneity of the microstructure of the sample. The obtained solid fraction is smaller after the detection process than before, decreasing respectively by 0.05, 0.06, 0.07, and 0.1 for scans S1, S2, S3, and S4.

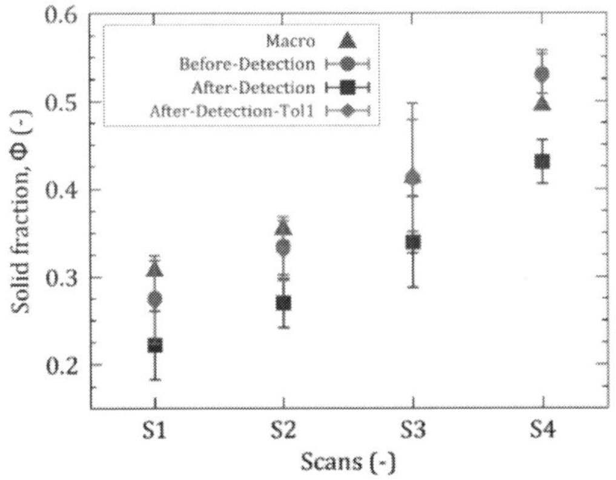

Fig. 5. Macroscopic solid fraction (triangle) and the average solid fraction of 4 steps before and after the detection process with tolerance of 0 and 1 voxel.

However, if it is recalculated with a tolerance of 1 voxel (6.5 μm) added to the diameter of the spheres the results come very close to the values obtained before detection. A tolerance of 1 voxel in all post-detection analysis thus appears quite suitable.

2.3.2 Total coordination number

We denote the coordination number by z. While it should represent the average number of contacts per grain, the finite resolution of the XRCT method, in the present case for which the particle diameter is only about 12 to 14 times as large as the voxel size, forbids a clear distinction between contacting and close neighbors.

Consequently, we compute z with a finite tolerance in terms of apparent interparticle distance.

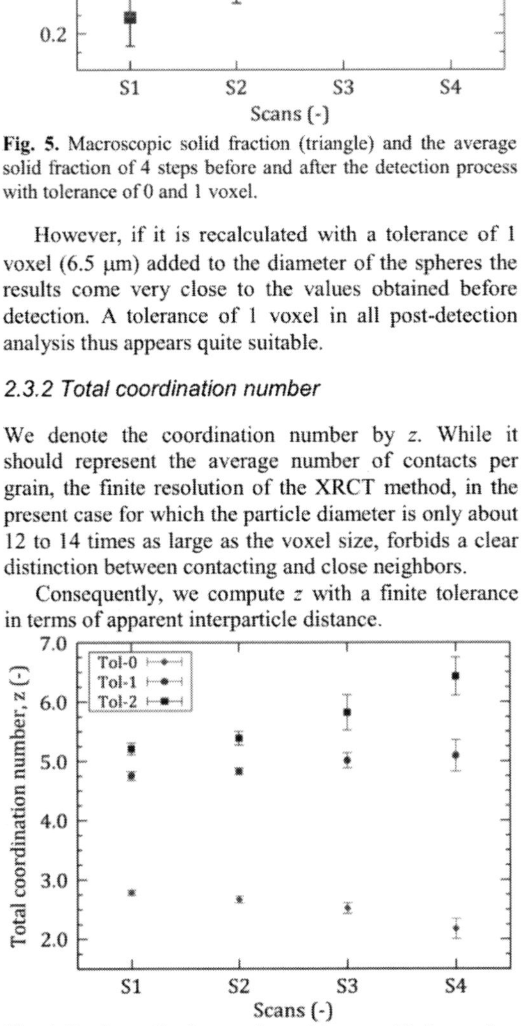

Fig. 6. Total coordination numbers of 4 steps with three values of tolerance (0, 1 and 2 voxels).

Fig. 6 shows the average coordination number at the four compression steps for three values of this tolerance (0, 1, and 2 voxels). The apparent value of z decreases under growing stress for tolerance zero, but measured z values are much larger (and realistic) as the tolerance is increased. The error bars are wider at a denser state or with a higher tolerance. The tolerance of 1 voxel is chosen for all calculations, this value is close to the one obtained in [9].

3 DEM simulations

3.1 Model material

The simulated systems are assemblies of equal-sized spherical beads of diameter a, interacting in their contacts by Hertz-Mindlin elastic-frictional forces, with the elastic properties of glass and friction coefficient 0.3.

As in Refs. [5,10], liquid menisci, with a fixed volume V_m, are assumed to form in contacting grains, with a rupture distance $V_m^{1/3}$ frictional. The capillary force is computed using the Maugis approximation [5,10]. Most calculations are carried out with the dimensionless meniscus volume $V_m/(a^3) = 10^{-3}$ [5].

3.2 Specimen preparation and compression process

A hard sphere event-driven method is first used to prepare disordered, low density configurations of 4000 grains. All particles are then launched with Gaussian-distributed random velocities with quadratic mean V_0. They collide and stick to one another within a cell of constant size, forming larger and larger aggregates. Finally, all grains are connected to one another by cohesive contacts and reach an equilibrium position.

The final packing structures are oedometrically compressed within periodic cells under growing axial stress. A stepwise pressure loading path is applied. In each compression step, vertical pressure σ_v is multiplied by a constant factor $10^{1/4}$, and one waits until the new equilibrium configuration is reached. The compression program is pursued until σ^{max}_v (maximum value $2x10^3$ kPa), well beyond complete plastic collapse is obtained. Upon decompression (not shown here), the system approximately keeps the highest density achieved under growing stress [4,5].

3.3 Compression curve and coordination number

The result of numerical oedometric compression is shown in Fig. 7, for an initial solid fraction Φ_0=0.30, as the conventional soil mechanics plot of e versus log(σ_v).

Three regimes can be identified in the compression curve, depending on the range of vertical stress σ_v [5]. Under low vertical stress σ_v, the initial structure can sustain the increasing stress without rearrangement and void ratio e remains nearly constant. The second regime starts as the void ratio starts to sharply decreases, due to the gradual collapse of the initial structure. Finally, void ratio e, in a third regime, approaches a minimum value e_{min}, which characteristics a structure stable without capillary cohesion. As in other models of cohesive granular systems [4], the collapse is controlled by the ratio of the applied stress to the characteristic stress built with the tensile strength of menisci and the particle diameter [5].

The coordination number z, is the sum of the contact coordination number z_c and the coordination number of distant interactions, through menisci joining non-contacting grains, z_d.

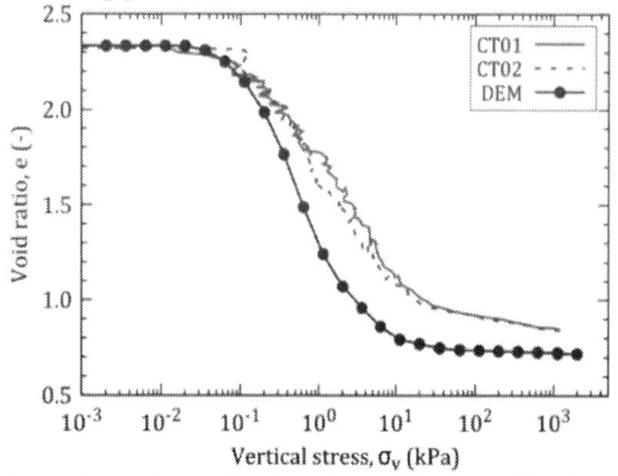

Fig. 7. Void ratio as a function of vertical stress σ_v. The circle-blue curve denotes the DEM's result. The continuous and dashed lines (CT01, 02) denote the results of oedometric tests.

Fig. 8 plots z, z_c, z_d versus vertical stress σ_v in the compression cycle and the total coordination number z in XRCT tests versus positions of scans. These positions are directly interpolated from the compaction curve of the DEM in Fig. 3. Initially, one has $z_d = 0$, because the gentle aggregation process, in that case, does not allow contact opening. z_c and z_d remain unchanged in the first regime. Both coordination numbers start to increase as the structure collapses and reorganizes in the second regime. They exhibit little change in the third regime, showing that the increase of the number of contacts is mainly due to the closing of narrow gaps between pairs of grains joined by a meniscus, as the structure is further compressed - a moderate effect.

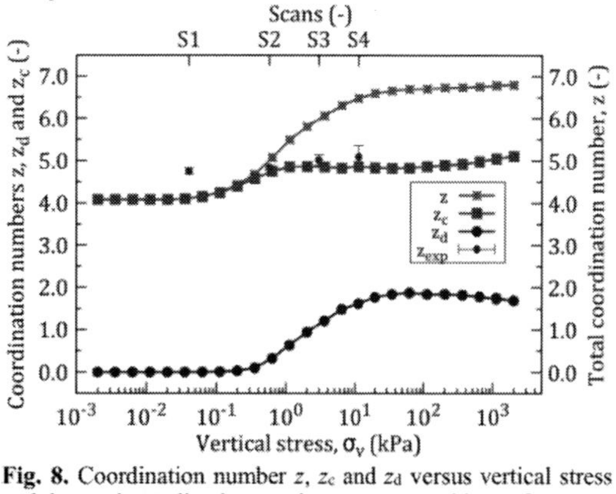

Fig. 8. Coordination number z, z_c and z_d versus vertical stress and the total coordination number z versus positions of scans.

A comparison with the total coordination number z measured in the experiment is also shown in Fig. 8, for the values of stress corresponding the scan positions. In general, the order of magnitude and the general trend of z_{exp} are well captured. We obtained a fairly good agreement of z_{exp} for the number of contacts z_c.

4 Conclusions

These results show that very loose, yet stable, states of wet granular assemblies can be prepared and subjected to mechanical tests, using very small water content. Oedometric tests reveal very similar behaviors in DEM simulations and in experiments. The same three stages of compression curve of cohesive soils are observed. Encouraging similarities are also observed in coordination number values as obtained through XRCT investigations of the microstructure and in DEM simulations. Microtomographic measurements enable assessments of sample homogeneity.

Acknowledgments

The first author is supported by the Scholarship of Ministry of Education and Training – Vietnam (the Project No 911) to do this work.

References

[1] D. M. Wood, *Soil Behaviour and Critical State Soil Mechanics* (Cambridge University Press, 1990).
[2] T. G. Mason, A. J. Levine, D. Ertaş, and T. C. Halsey, Phys. Rev. E **60**, R5044 (1999).
[3] J.-F. Bruchon, J.-M. Pereira, M. Vandamme, N. Lenoir, P. Delage, and M. Bornert, Granul. Matter **15**, 783 (2013).
[4] F. Gilabert, J.-N. Roux, and A. Castellanos, Phys. Rev. E **78**, 031305 (2008).
[5] V.-D. Than, S. Khamseh, A.-M. Tang, J.-M. Pereira, F. Chevoir, and J.-N. Roux, J. Eng. Mech. C4016001 (2016).
[6] L. Xie, R. E. Cianciolo, B. Hulette, H. W. Lee, Y. Qi, G. Cofer, and G. A. Johnson, Toxicol. Pathol. **40**, 764 (2012).
[7] V.-D. Than, P. Aimedieu, M. Bornert, J.-M. Pereira, J.-N. Roux, and A.-M. Tang, Eur. Phys. J. E (to be published).
[8] T. W. Ridler and S. Calvard, IEEE Trans. Syst. Man Cybern. **8**, 630 (1978).
[9] T. Aste, M. Saadatfar, and T. J. Senden, J. Stat. Mech. Theory Exp. **2006**, P07010 (2006).
[10] S. Khamseh, J.-N. Roux, and F. Chevoir, Phys. Rev. E **92**, 22201 (2015).

DOI: 10.1051/epjconf/201714008019

Rheology of wet granular materials under continuous shear: experiments and simulations

Michel Badetti[1,⋆], *Abdoulaye* Fall[1,⋆⋆], and *Jean-Noël* Roux[1,⋆⋆⋆]

[1] *Laboratoire Navier, UMR8205, CNRS-ENPC-IFSTTAR,2 allée Kepler, 77420, Champs-sur-Marne, France*

Abstract. The behaviour of wet granular media in shear flow is characterized by the dependence of apparent friction μ^* and solid fraction Φ_S on the reduced pressure P^* and the inertia number I. Reduced pressure, $P^* = \sigma_{22}a^2/F_0$, compares the applied normal stress σ_{22} on grains of diameter a to the tensile strength of contact F_0 (proportional to the surface tension Γ of the liquid and the beads diameter). A specifically modified rotational rheometer is used to characterize the response of model wet granular material to applied shear rate $\dot{\gamma}$ under controlled normal stress σ_{22}. Discrete Element Method (DEM) simulations in 3D are carried out in parallel and numerical results are compared with experimental ones. Cohesive, inertia, saturation and viscous effects on macroscopic coefficient of friction μ^* and solid fraction Φ_S are discussed.

1 Introduction

In the steady state, shear flows of granular materials under controlled normal stress σ_{22} and shear rate $\dot{\gamma}$ (see Fig. 1) are well described by the solid fraction Φ_S and the macroscopic friction coefficient $\mu^* = \sigma_{12}/\sigma_{22}$. Characterizing flow inertia of such materials with the inertial number $I = \dot{\gamma} a \sqrt{m/a\sigma_{22}}$ (m being the mass of a grain and a its diameter) is now a classical approach [1, 2]. However, recent studies showed that in the case of wet granular materials there is also a strong dependency of both μ^* and Φ_S on the cohesive effect induced by liquid bonding between particles [3–5]. For a pack of beads of diameter a with bonds strength f_0, the reduced pressure $P^* = a^2\sigma_{22}/f_0$ can characterize this cohesive effect.

Previous DEM studies proposed constitutive laws $\mu^*(I, P^*)$ and $\Phi_S(I, P^*)$ covering broad ranges of I and P^* [3]. The aim of the present commuication is to compare these DEM simulations of wet grain assemblies with new experimental measurements using a normal stress controlled shear cell, both in the quasistatic limit and in dense inertial flows.

Sec. 2 describes the model material and the shearing methods. Sec. 3.1 compares flow curves of our materials obtained numerically and experimentally, in Sec. 3.4 and 3.5 respectively the influence of liquid volume Φ_L and viscosity η in the quasistatic regime is discussed.

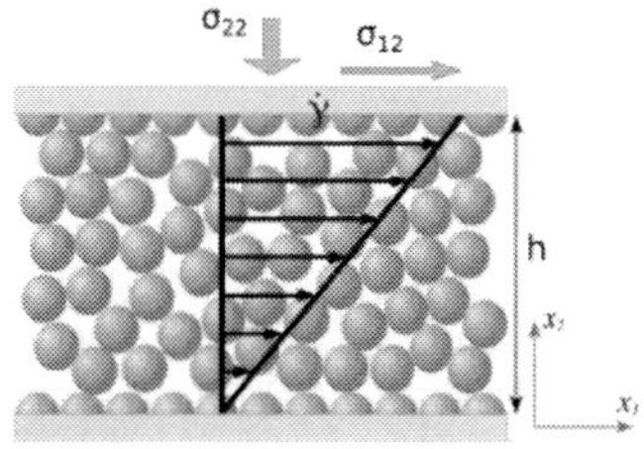

Figure 1. Ideal plane shear with confinement. shear rate is defined as $\dot{\gamma} = \partial \dot{x}_1/\partial x_2$.

2 Methods

2.1 Model material

In simulations and experiments, the model material proposed is a simple pack of monodisperse spherical polystyrene beads of diameter $a = 5 \times 10^{-4}$m mixed together with a newtonian silicon oil of surface tension $\Gamma = 47$mN.m^{-1} and variable viscosity η ranging from 20mPa.s up to 1000mPa.s. Liquid quantity is described by the ratio Φ_L/Φ_S which ranges from 0.003 up to 0.075 [6–8], thus ensuring a pendular state. In such state, for a perfecting wetting fluid bond strength at contact f_0 is well approximated by $\pi a\Gamma = 7.38 \times 10^{-5}$N [9].

⋆e-mail: michel.badetti@ifsttar.fr
⋆⋆e-mail: abdoulaye.fall@ifsttar.fr
⋆⋆⋆e-mail: jean-noel.roux@ifsttar.fr

2.2 Experimental setup

Materials are inserted in a specifically devised rheological cell (see Fig. 2). Constant force F_N and rotation rate w are applied on a volume V_S of material, Torque T and material width h (see Fig. 1) are measured once plateau values of the steady state are reached. Assuming plane shear with truncated center, σ_{12}, σ_{22}, $\dot{\gamma}$ and Φ_S are calculated as :

$$\sigma_{12} = \frac{3T}{2\pi R_e^3 (1 - (R_i/R_e)^3)}$$

$$\sigma_{22} = \frac{F_N}{\pi R_e^2 (1 - (R_i/R_e)^2)}$$

$$\dot{\gamma} = w\frac{R_e + R_i}{2h}$$

$$\Phi_S = \frac{V_S}{\pi h R_e^2 (1 - (R_i/R_e)^2)}$$

Materials are inserted to ensure a width $h > 20a$, the gap between inner R_i and outer radii R_e is equal to $20a$. Experimental measures with different values of h (see figure 3 showed no gap effect ruling out possibilities of localization effects. Similar devices have already been used to describe succesfully the rheology of dense suspensions and dry materials [1, 2, 10]. Means and error bars are obtained for materials tested at least 3 times.

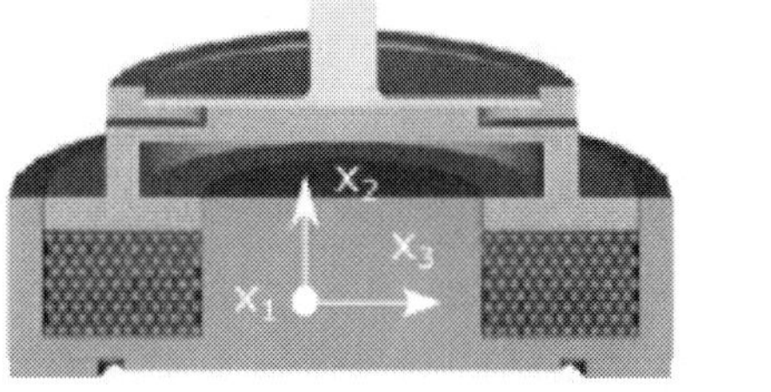

Figure 2. Rheological cell : gains (red) are trapped between the stator (grey) and the rotor (blue+yellow). Shear happens in the $\mathbf{x}_1$,$\mathbf{x}_2$ plane. Outer and inner radii are respectively R_e and R_i

2.3 Numerical simulations

Materials are placed in a 3D periodical cell sheared using Lees-Edwards boundary conditions. Shear rate $\dot{\gamma}$ and σ_{22} are applied, σ_{12} and Φ_S are measured. Contact interaction are implemented using an Hertz-Mindlin model with viscous dissipation, friction in contact satisfies Coulomb's law with a friction coefficient $\mu = 0.09$ chosen in order to fit values of $\mu^* = \mu_d^* = 0.257$ and $\Phi_S = \Phi_{S,d} = 0.616$ for $I \to 0$ and $P^* \to \infty$ found experimentally (see figure 4). Distant interactions involve meniscii assumed to form only when grain come into contact with a fixed volume V_m, they disappear at a certain rupture distance $D_0 = V_m^{1/3}$. Error bars are obtained using the technique of block transform estimates [11].

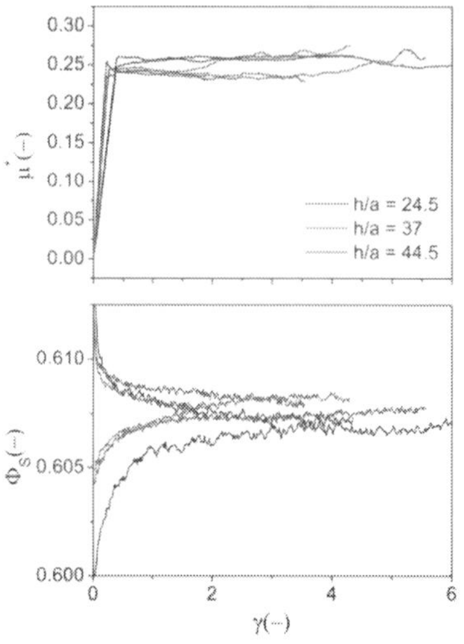

Figure 3. μ^* and Φ_S measured at different sample width. Absence of gap effect is highlighted. Measurements made with η = 20mPa.s, $\Phi_L/\Phi_S = 0.075$, $I = 10^{-4}$ and $P^* = 3$

3 Results

3.1 Inertia and cohesive effects

Figure 4 presents μ^* and Φ_S as functions of I for different values of P^* in experiments and simulations. For low values of I ($< 10^{-3}$), ie in the quasistatic regime, μ^* and Φ_S reach plateau values. For higher values of I, in the inertial regime, μ^* increases and Φ_S decreases upon increasing I.

Table 1 compares experimental and numerical data when $I = 10^{-3}$, values recorded with the two methods show very good agreements with less than 6% relative difference for μ^* and less than 3% relative difference for Φ_S.

The agreement between simulations and experiments remains good when comparing the solid fractions for inertial flows, the maximal relative difference between experiments and simulations is at most 0.5% for $P^* = 2$ and 0.8% for $P^* = 3$. There is, however, a discrepancy when looking at the values of μ^* in the inertial regime. When $I > 3 \cdot 10^{-3}$, the relative difference rises up to 30% for P*=3 and up to 17% for P*=2. Clearly, μ^* increases much faster with I away from its quasistatic value μ_0^* for $I \to 0$ in the experiments than in the numerical simulations. This discrepancy could emerge from questions of liquid repartition in the experimental samples.

Table 1. Values of μ_0^* and Φ_S obtained for $I < 10^{-3}$, $\Phi_L/\Phi_S = 0.03$ and η = 550mPa.s. exp and num subscripts respectively stand for experimental and numerical

P^*	μ^*_{exp}	μ^*_{num}	$\Phi_{S,exp}$	$\Phi_{S,num}$
∞	0.255	0.257	0.616	0.616
3	0.302	0.327	0.606	0.605
2	0.340	0.364	0.595	0.597

DOI: 10.1051/epjconf/201714008019

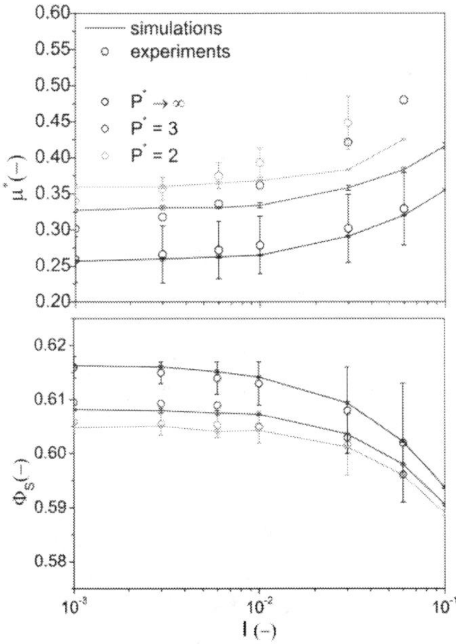

Figure 4. Flow curves $\mu^*(I, P^*)$ and $\Phi_S(I, P^*)$. filled circles represent experimental data while lines representnumerical data. Dry samples in black, $P^* = 3$ in blue and $P^* = 2$ in green. For wet samples viscosity $\eta = 550$mPa.s and $\Phi_L/\Phi_S = 0.015$.

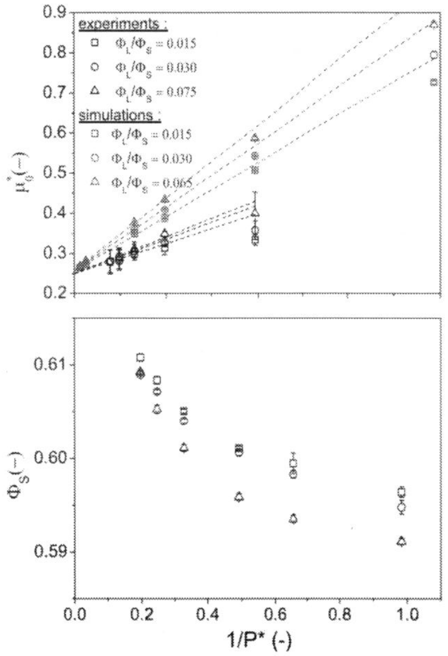

Figure 5. Numerical (open symbols) and experimental (filled symbols) values of μ_0^* and Φ_S versus $1/P^*$ for different values of Φ_L/Φ_S, $\eta = 550$mPa.s. Linear approximation of μ_0^* from equation 2 given in dashed lines.

3.2 Quasistatic regime : Saturation effect

Fig. 5 shows numerical and experimental μ_0^* versus $1/P^*$ for different ratio Φ_L/Φ_S. For poorly cohesive materials ($1/P^* = 0.033$), numerically increasing Φ_L/Φ_S from 0.015 up to 0.06, μ_0^* grows by 1%. For highly cohesive ones ($1/P^* = 1$), increasing Φ_L/Φ_S from 0.015 up to 0.075 experimentally and from 0.015 up to 0.06, μ_0^* grows by 16%. On the whole range of $1/P^*$, Fig. 5 also shows that Φ_S barely depends on Φ_L/Φ_S.

Assuming the quasistatic behavior of wet granular materials can be expressed through a Mohr-Coulomb criterion, e.g materials flow with :

$$\sigma_{12} = \mu_d^* \sigma_{22} + C \tag{1}$$

where cohesion C is defined as a rupture tangential stress in unconfined conditions. Ensuring $\sigma_{22} = 0$Pa is not possible with the methods at our disposal. However, knowing the constant macroscopic coefficient of friction μ_d^* of dry granular material, C can be estimated at various σ_{22}. Using equation 1, numerical estimation are averaged for $\sigma_{22} = 9000$Pa and 4500Pa, experimental ones are averaged for $\sigma_{22} = 1500$Pa and 1200Pa, the values obtained are reported in table 2. Estimates of C increase up to 39% within the range of Φ_L/Φ_S. Noting that Eq. 1 is equivalent to :

$$\mu_0^* = \mu_d^* + \frac{a^2 C}{f_0} \frac{1}{P^*} \tag{2}$$

one expects a linear evolution of μ_0^* with $1/P^*$, with a slope proportional to C and the constant ratio a^2/f_0. Fig. 5 exhibit linear predictions of μ_0^* with the values of cohesion given in Table 2, those predictions remain good with less than 5% relative difference with actual measurements when $1/P^* < 1$.

Assuming capillary forces introduce negligible contribution to shear stress, cohesion in granular media can also be approximated theoretically [3, 4] as :

$$C = \mu_d^* \frac{f_0}{a^2} \frac{\Phi_S z}{\pi} \tag{3}$$

In the whole range of $1/P^*$, the simulations show that the variations of the product $\Phi_S z$ are small (10% at most), while the cohesion estimated from equation 1 can vary up to 30% in simulations and up to 40% in experiments . It is likely that neglecting the contribution of capillary forces to shear stress, when the cohesive effect is high $1/P^* > 1$, is too strong an assumption.

Table 2. Values of cohesion C in Pa. As deduced from experiments (C_{exp}), simulations (C_{num}) and approximation (eq. 3, C_{app}).

Φ_L/Φ_S	C_{exp}	C_{num}	C_{app}
0.003		76.7	91.2
0.007		77.2	93
0.015	42	77.4	95.2
0.030	49	94.7	97.5
0.060		108.2	100
0.075	50	126.2	103.8

DOI: 10.1051/epjconf/201714008019

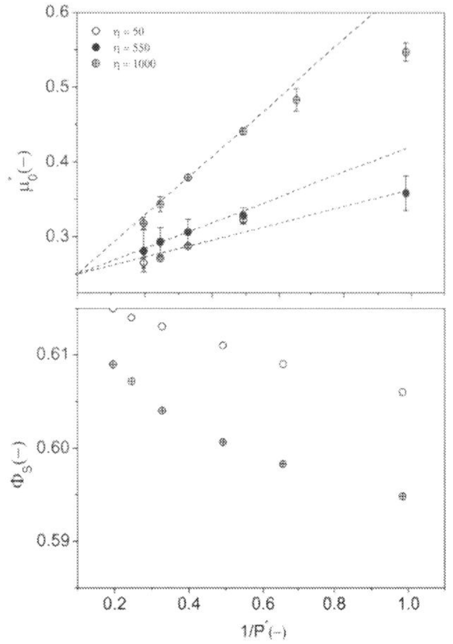

Figure 6. experimental values of μ_0^* and Φ_S versus $1/P^*$ for different values of η, $\Phi_L/\Phi_S = 0.03$. Linear approximations of μ_0^* from equation 2 are given by the dasehd lines.

3.3 Quasistatic regime : Viscosity effect

Fig. 6 shows the experimental value of μ_0^* and Φ_S versus $1/P^*$ for different values of viscosity η. As remarked above (Sec. 3.2), the linear variation of μ_0^* with $1/P^*$ shows the applicability of a Mohr-Coulomb criterion of plastic flow for $1/P^* < 1$.

Identified cohesion values appear to vary with the viscosity of the liquid, ranging from 33Pa for η = 50mPa.s up to 116Pa for η = 1000mPa.s. A strong influence of the viscosity on friction coefficient μ_0^* is also observed in the cohesion dominated regime, when $1/P^* > 1$. Solid fraction Φ_S, on the other hand, is not sensitive to η variations. While viscous forces through liquid bridges are negligible at low velocities [9], one may speculate that a higher viscosity affects the rate with which liquid is transported between different meniscii in the contact network, leading to more heterogeneous liquid distributions. In turn, such heterogeneities, possibly involving complex structures merging independent meniscii together (as apparent in Fig. 7, obtained by X-ray microtomography on one of the experimental samples) could explain the unexpected increase of μ_0^* with η.

Conclusion

While the effects of I,P^* and Φ_L/Φ_S, which are observed in the simulations are satisfactorily understood, the unexpected influence of the liquid viscosity could be attributed to inhomogeneous distributions of the liquid content in the sheared sample. Such inhomogeneities might affect the rheological properties (and possibly explain the stronger I dependence of experimental data, compared to numerical ones). More systematic microstructural studies using X-ray microtomography are planned to investigate these effects.

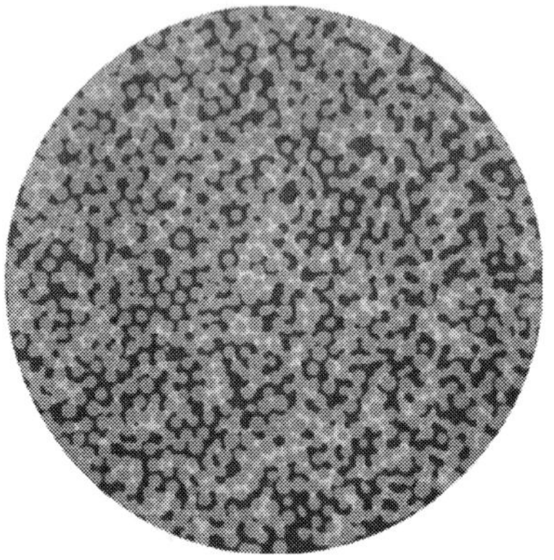

Figure 7. Microtomography image : slice of a sample, Φ_L/Φ_S = 0.075 and η = 1000mPa.s.

References

[1] A. Fall, G. Ovarlez, D. Hautemayou, C. Mézière, J.-N Roux, F. Chevoir, J. of Rheol. **59**, 2015

[2] A. Fall, M. Badetti, G. Ovarlez, F. Chevoir, J.-N Roux, *these proceedings*, 2016

[3] S. Khamseh, J.-N. Roux, F. Chevoir, Phys. Rev. E **92**, 2015

[4] V. Richefeu, M.S. El Youssoufi, F. Radjaï, Phys. Rev. E **73**, 2006

[5] F.A. Gilabert, J.-N. Roux, A. Castellanos, Phys. Rev. E**75**

[6] M. Scheel, R. Seemann, M. Brinkmann, M. Di Michiel, A. Sheppard, S. Herminghaus, J.of Phys.:Cond.Matt **20**, 2008

[7] N. Mitarai, F. Nori, J.of Phys.:Cond.Matt, 2006

[8] P.C.F. Moller, D. Bonn, EuroPhys. Letters **80**, 2008

[9] O. Pitois, P. Moucheront, C; Chateau, J. Coll. Interf. Science **231**, 2000

[10] F. Boyer, E. Guazzelli, O. Pouliquen, Phys. Rev. Letter **107**, 2011

[11] H. Flyvbjerg, H. G. Petersen, J. Chem. Phys. **91**, 1989

Influence of 3D particle shape on the mechanical behaviour through a novel characterization method

Noura Ouhbi[1,3], *Charles* Voivret[1], *Guillaume* Perrin[2], and *Jean-Noël* Roux[3]

[1]Department of Lines, Tracks & Environment, Engineering & Projects, SNCF-Réseau, Saint-Denis, France
[2]CEA/DAM/DIF, Arpajon, France
[3]Université Paris-Est, Laboratoire Navier (UMR 8205), CNRS, ENPC, IFSTTAR, Marne-la-Vallée, France

Abstract. The sensitivity of the mechanical behaviour of railway ballast to particle shape variation is studied through Discrete Element Method (DEM) numerical simulations, focusing on some basic parameters such as solid fraction, coordination number, or force distribution. We present an innovative method to characterize 3D particle shape using Proper Orthogonal Decomposition (POD) of scanned ballast grains with a high accuracy. The method enables not only shape characterization but also the generation of 3D distinct and angular shapes. Algorithms are designed for face and edge recognition.

1 Introduction

In the last years, particle shape analysis has become a topic of interest in several fields. In railway engineering, the importance of grain size and shape for the behaviour of ballast (the granular layer made of centimetre sized, angular grains, supporting the tracks) has been highlighted in numerous experimental and numerical studies [1-4]. Connections between shape parameters and mechanical performance are however not well known. This is the issue addressed in the present study.
Such a research objective requires, given its complexity, some modelling of ballast shape, with various levels of accuracy. Powerful methods have been suggested to characterize shape (e.g spherical harmonics) [5, 6]. These methods, on attempting to reconstruct such shapes, fail to reproduce the "facetted" and "angular" aspect of the particles.
To overcome these limitations, we propose (Sec. 2.1) a method based on Proper Orthogonal Decomposition (POD) [7] apt to provide optimal descriptions of complex shapes. Particle generation is carried out (Sec.2.2) by a stochastic method called multivariate Kernel Density Estimation (KDE) [8], which proved successful in taking into account the high level of correlation between data on ballast grains. Finally, DEM simulations are used to assess the sensitivity of mechanical properties to particle shape.

2 Model

2.1 Shape characterization

Our study is based on 3D scans of 1000 ballast grains, with millimetre accuracy. The point clouds, cleaned from possible outliers, first go through pre-processing steps which involve projecting onto a suitable common basis and aligning all grains according to their principal directions. A model reduction is then carried out through Proper Orthogonal Decomposition (POD) (commonly known as Principal Component Analysis in data analysis), a powerful tool to emphasize variation and bring out strong patterns in a dataset, thus easing data exploration and visualization. The POD method, as applied to ballast grain represented by magnitudes of radii in a chosen set of basis directions (as described in [9] in detail), does not assume any parametric description of grain shape but, rather, identifies appropriate basis functions by order of decreasing statistical relevance, thus capturing the dominant morphology features. Shape precision is controlled by choosing the size of the new reduced space representing data (Fig.1). A satisfactory convergence rate is obtained, in terms of morphology features, as the precision level increases [9]. One limitation, however, is the strong dependency of its efficiency on the sample size [10].

2.2 Complex shape generation

In order to increase the variability of shape in our numerical studies, we need a virtual grain generator. Basic parametric stochastic methods of generation applied to our grains [11] yield the right volume and surface area distributions, but fail to reproduce more advanced features such as angularity, faces, etc. Such features are statistically represented by the high-order correlations between points in space, which multivariate KDE, a non-parametric method proves apt to satisfactorily account for. An example of generated grains while keeping only 117 modes (99% of information) is presented in Fig. 2 and elongation

* Corresponding author: noura.ouhbi@sncf.fr

distribution is shown in Fig. 3. We define elongation E using the three principal lengths of the grains $S < I < L$: $E = \frac{I}{L}$. We also compared surface area, volume, average radii, flatness ($F = \frac{S}{I}$) and aspect ratio ($AR = \frac{S}{L}$). The results we got in terms of likelihood values according to the Kolmogorov-Smirnov test were all above 0.8, which is a very satisfactory result.

After verifying these shape properties, we now present a new indicator which consists of faces/edges detection. The procedure for face detection is based on a clustering algorithm. After computing neighbours for each vertex, we use clustering algorithms to identify groups of neighbouring points with nearly parallel normal directions. Based on this clustering, we obtain groups that represent faces or edges. Using these clusters, we define curvature as the deviation of the vertices normal directions and thus we are able to distinct faces from edges in each grain. The distribution of the number of large faces and edge curvatures and an illustration are presented in Figs. 4 to 6. We chose to represent large faces as those that contain more than 100 points with parallel normal. This number was calibrated according to

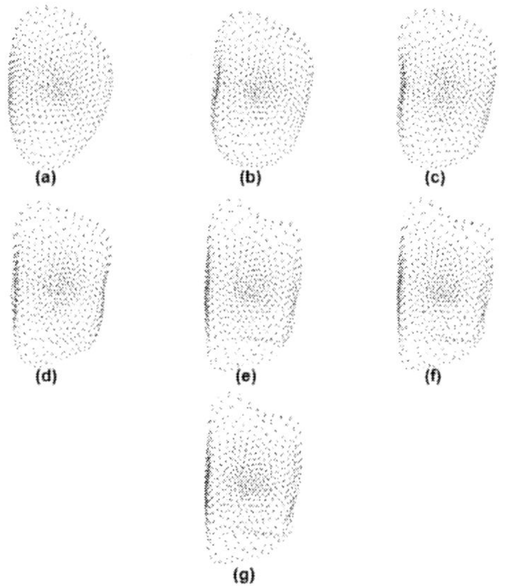

the noticed number of large faces obtained on some examples of grains.

Fig. 1. Reconstructed grain with different precision levels (from (a) to (g): 80%; 85%; 90%; 95%; 99%; 99,9%; 100% of information on shape)

Fig. 2. Two generated ballast grains

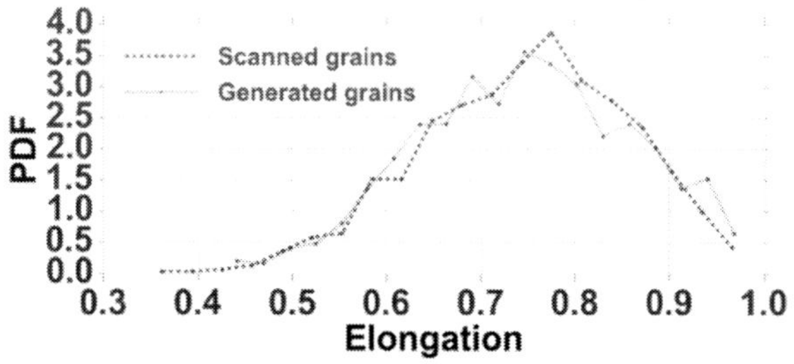

Fig. 3. Elongation Probability Function (PDF)

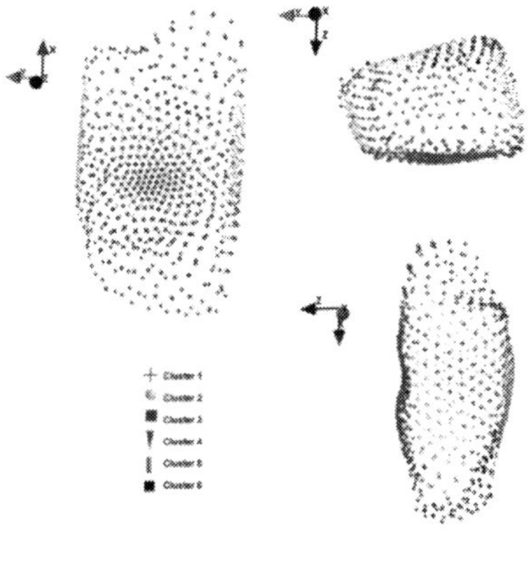

Fig. 4. Detection of faces for a generated ballast grain

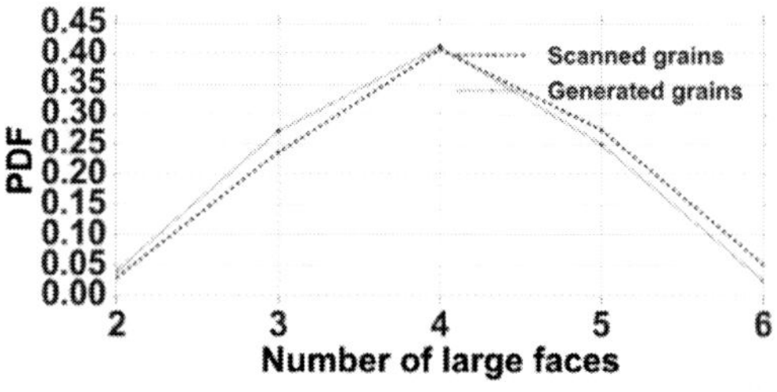

Fig. 5. Large faces (faces with number of points>100 points) (PDF)

 DOI: 10.1051/epjconf/201714006027

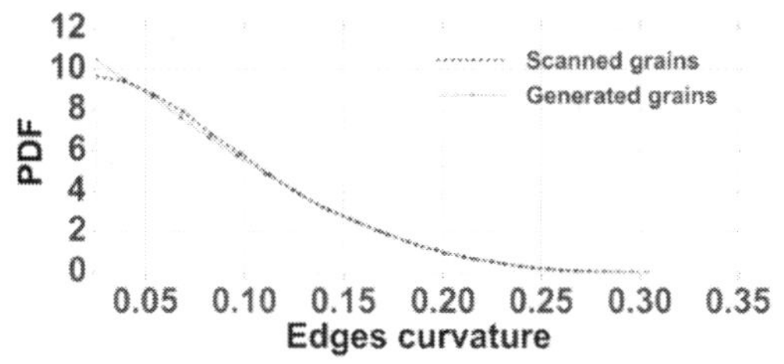

Fig.6. Edge curvature (Curvature = variation of normal directions; faces =low curvature, edges=High curvature) (PDF)

3 Numerical simulations

DEM simulations are carried out with LMGC90 [12], a software apt to deal with rigid polyhedral grains. After a suitable meshing, grains are modeled as polyhedra. In a first stage, the shapes are further simplified as convex polyhedra, which enables much faster computations. This entails a loss of precision on shape (in the sense of Fig.1) of about 20% (which would be reduced to less than 5% with non-convex polyhedra). This restricts the relevance of simulations with large numbers of modes. Here we simulate an assembling procedure via one-dimensional compression between parallel planes with periodic lateral boundary conditions, without gravity. It is carried out using 300 grains, with 100 faces and 3 different shape precision levels after convex meshing: 30% (1 mode), 68% (11 modes) and 80% of information (800 modes). The sample is first generated by depositing the spheres circumscribed to the polyhedral (Fig. 7) in a cell with a constant square section ($S = 30\,cm \times 30\,cm$). Three values of the intergranular friction coefficient μ are compared: 0, 0.3 and 0.8. As a reference, a similar simulation is carried out with meshed spheres with 100 faces (and μ=0.3). Contacts with planes are frictionless. A constant force of $F_0 = 500\,N$ is applied to the upper mobile plane until an equilibrium state is approached, in which the contacts resist further compression (Fig.7). Figs. 8 to 11 show how the solid fraction ρ and the coordination numbers corresponding to simple (vertex-face or edge-edge), double (edge-face) and triple (face-face) contacts respectively denoted as z_1, z_2, z_3 evolve in time towards their final equilibrium values. We note that the force supported by the bottom fixed plane converges to F_0after 3.8s. The ratio $K = \frac{Ec}{\sigma D^3}$ of the kinetic energy per grain,E_c, to the characteristic energy scale σD^3 (σ denoting stress $\frac{F_0}{S}$, and D the average grain diameter) gradually decreases to 4.10^{-6} in 10 s. Such time scales are related to the initial state density and to grain and upper plane inertia.

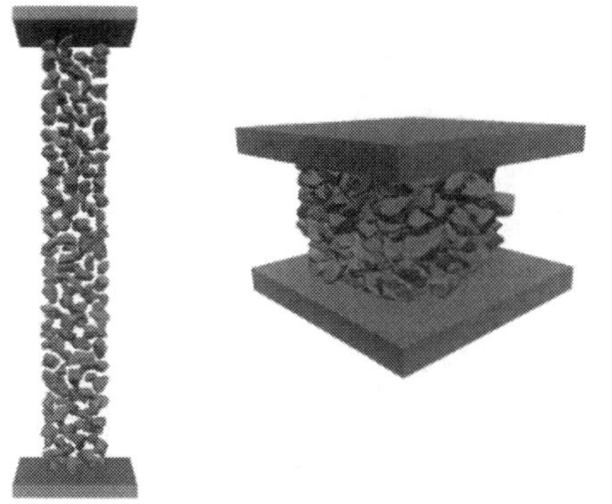

Fig.7. Assembling procedure: initial and final states.

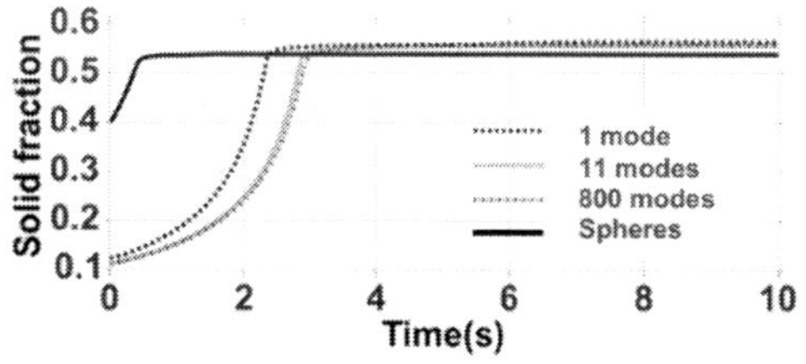

Fig. 8. Solid fraction vs time (μ=0.3)

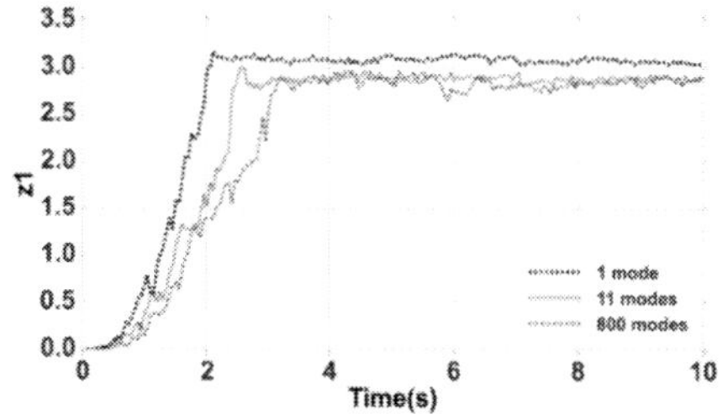

Fig.9. z_1 vs time (μ=0.3)

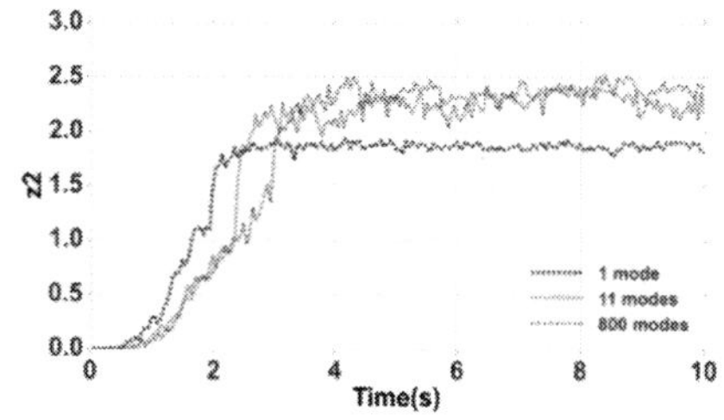

Fig.10. z_2 vs time (μ=0.3)

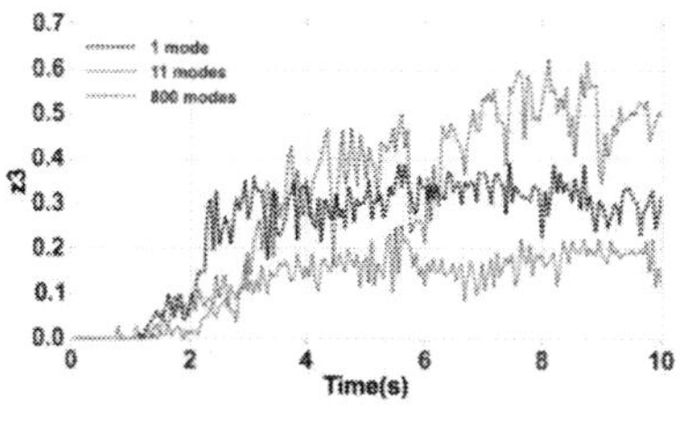

Fig.11. z_3 vs time (μ=0.3)

DOI: 10.1051/epjconf/201714006027

Final, equilibrium values of ρ, z_1, z_2, z_3 are given in Table 1, along with those of x_0, the proportion of "rattlers" (grains carrying no force) for different numbers of modes and values of μ.

Table.1. Evolution of some mechanical parameters

Parameter / N-μ: N	μ	ρ	z_1	z_2	z_3	x_0
11	0	0.690	4.360	3.766	0.320	0%
800	0	0.697	3.877	3.993	0.427	0%
1	0.3	0.590	3.046	1.206	0.213	7.6%
11	0.3	0.594	2.893	1.440	0.280	7.3%
800	0.3	0.602	2.874	1.593	0.332	5.08%
1	0.8	0.566	1.893	0.860	0.080	26.6%
11	0.8	0.567	1.886	0.980	0.080	26.33%
800	0.8	0.567	1.884	0.981	0.094	22.37%

We observe that the solid fraction increases with the number of modes, while z_1 decreases and z_2 and z_3 increase. Grains with more accurate shape description tend to form slightly denser assemblies, with an increase of double and triple contacts, at the expense of simple ones. This shape effect is stronger for low friction coefficients. A strong increase of rattler fraction x_0 should also be noted for growing μ, as in spherical bead assemblies.

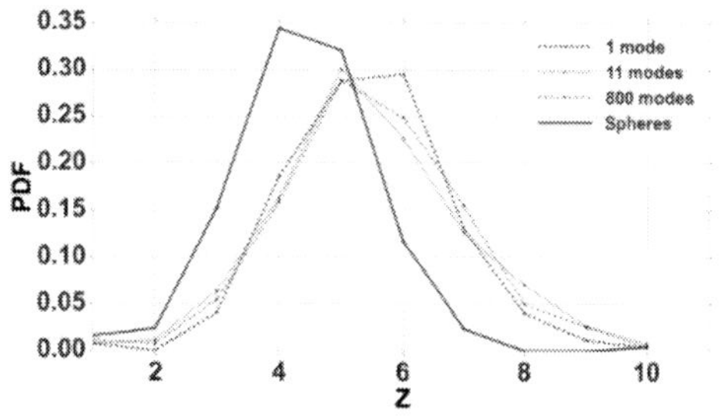

Fig. 12. Connectivity distribution (μ=0.3).

Fig.12 plots the distribution of connectivity Z (the fractions of grains having exactly Z contacting neighbours). A significant difference with spheres can be noted, while different shape models yield very similar results: Increasing the precision on shape modelling changes the type of contacts without affecting contact numbers.

4 Conclusions

The proposed new method to characterize the shape of ballast grains enables its description with controlled accuracy, and the generation of new sets of grains, similar to the real ones. Preliminary numerical simulations are carried out to quantify the impact of shape on mechanical properties, revealing that a very significantly reduced number of parameters (11 shape functions) may be sufficient for our model. Beyond the simulated assembling stage for a solid granular packing, we expect the shear response to exhibit a higher sensitivity to shape, as observed in [13] for polyhedra with varying numbers of faces inscribed within the same sphere.

Acknowledgements

The work has been supported by SNCF-Réseau Engineering and Projects, Lines Track & Environment Department.

References

1. E. Azéma, Y. Descantes, N. Roquet, J. N. Roux, F. Chevoir. Phys.Rev. E **86**(3), (2012).
2. E. Azéma. Doctoral dissertation, Montpellier. (2007).
3. E. Azéma, F. Radjai, G. Saussine. Mech. of Mat. **41**(6), 729-741. (2009).
4. J. C. Quezada, P. Breul, G. Saussine, F. Radjai. Comp. and Geotech. **55**, 248-253. (2014).
5. G. Mollon, J. Zhao. *Gran. Mat.* **15**(1), 95-108. (2013).
6. M. Grigoriu, E. Garboczi, C. Kafali. Powd. Tech. ***166***(3), 123-138. (2006).
7. Y. C. Liang, H. P. Lee, S. P. Lim, W. Z. Lin, K. H. Lee, C. G. Wu. Jour. of Sou. and vibr. **252**(3), 527-544. (2002).
8. D. W. Scott, S. R. Sain. *Handbook of Statistics*. V**24**. 229-261. (2005).
9. N. Ouhbi, C. Voivret, G. Perrin, J. N. Roux. *Procedia Engineering*, **143**, 1120-1127. (2016).
10. N. Ouhbi, C. Voivret, G. Perrin, J.N. Roux. *Proceedings of the 3rd International Conference on Particle-based Methods, held in Barcelona, Spain*, pp. 545–557. (2015).
11. G. Perrin, C. Soize, N. Ouhbi. Submitted to Journal of Comp. Stat. & Data An. (2016).
12. F. Dubois, F., M. Jean, M. Renouf, et al. LMGC90. *10e colloque national en calcul des structures. Giens, France*. (2011)
13. E. Azéma, F. Radjai, F. Dubois. Phy. Rev. E**87**. (2013).

High speed confined granular flows down inclined: numerical simulations

Velotiana Jean-Luc RALAIARISOA[1], *Alexandre* VALANCE[1], *Nicolas* BRODU[2], and *Renaud* DELANNAY[1,*]

[1] *Institut de Physique de Rennes, Université de Rennes 1, UMR CNRS 6251, 35042 Rennes Cedex, France*
[2] *GEOSTAT, Inria Bordeaux, 33405 Talence, France*

Abstract. New regimes in high-speed confined granular flows down inclined have recently been obtained in numerical simulations [1]. Increasing the angle of inclination reveals the destabilization of the well known unidirectional flows. Longitudinal rolls first appear. Upon further increase of the angle, a new regime, called supported, is observed. It is characterized by a dense core surrounded by granular gas. These numerical simulations have been performed for a fixed confinement width, $W = 68D$, where D is the size of the grains. Here, we perform numerical simulations with a smaller value of the confinement width: $W = 34D$. In spite of this strong confinement, we observe the transitions to the same regimes (rolls and then supported) by increasing the inclination angle. We characterize these transitions and highlight the robustness of the mass flow rate scaling law discovered in [1].

1 Introduction

Recently, several works have been devoted to the effect of lateral confinement on the properties of granular flows [2–4]. Both experimental and numerical studies have pointed out that frictional lateral walls induce new flow properties. For example, steady and fully developed flows (SFD flows) have been observed up to large angles of inclination whereas accelerated ones are usually expected [1, 5]. These SFD regimes present non-trivial features, including secondary flows (rolls) and flows with dense core supported by a very agitated dilute layer. Despite the diversity of the features of these states the velocity of the flow obeys a simple scaling law with the flow height. In [1] the effect of the variation of the inclination angle and of the mass holdup was explored extensively and systematically for a given value of the gap W between side walls. The domains of existence of the different regimes were reported for this value of W. We investigate here a smaller value $W/2$ of the gap to study the effect of a variation of the gap on the domains of existence of the flows regime. We also examine if the velocity scaling law observed in [1] remains valid for this smaller value of the gap.

2 Numerical method and simulation setup

We have conducted [1, 5] simulations of granular flows down flat and steep inclines with frictional sidewalls using a discrete element model (DEM). The principle of DEM simulations is to treat each grain as a sphere (of diameter D) subject to gravity and contact forces with both the other grains and the basal and lateral walls. Interaction model is performed by using linear visco-elastic approach. Two grains i and j interact when overlapping, such as $\delta_{ij} = [(D_i + D_j)/2] - r_{ij} > 0$, with r_{ij} the center-to-center grains separation. The force applied by grain i on grain j is decomposed into normal $\mathbf{F}_n^{i\to j}$ and tangential $\mathbf{F}_t^{i\to j}$ components. The normal contact force is given by: $F_n^{i\to j} = (k_n\delta_{ij} + \gamma_n v_n)$ where k_n is a spring constant, γ_n a damping coefficient set by the normal coefficient of restitution e_n, and v_n the normal component of the relative translational grain velocity. Similar model is used for the tangential component enforced by the Coulomb friction $|\mathbf{F}_t| \le \mu |\mathbf{F}_n|$ where μ is the model friction coefficient. The torque acting on a grain is given by: $\mathbf{q} = -(D/2)(\mathbf{F}_t \times \mathbf{n})$, where $\mathbf{n}$ is the normal unit vector. Table 1 shows parameters values used for the contact forces between interacted grains and for grain-wall interactions.

Table 1: Simulation parameters used for grain/grain and grain (glass)/wall interaction model.

	Grain/grain	Grain/wall
Normal restitution e_n	0.972	0.8
Tangential restitution e_t	0.25	0.35
Friction coefficient μ	0.33	0.596
Spring stiffness k_n (mg/D)	2×10^5	2×10^5

The equations of motion -including rotation - of each particle are calculated by applying Newton's second law and a Velocity-Verlet scheme is used for the numerical integration with a time step $dt = 10^{-4}(\sqrt{D/g})$. We repeat the whole approach for grain-wall interactions. The simulation consists of confined granular flows between side walls on flat inclined base in the gravity field. The periodic

*e-mail: renaud.delannay@univ-rennes1.fr

boundary conditions (PBC) are used in the X–direction (figure 1).

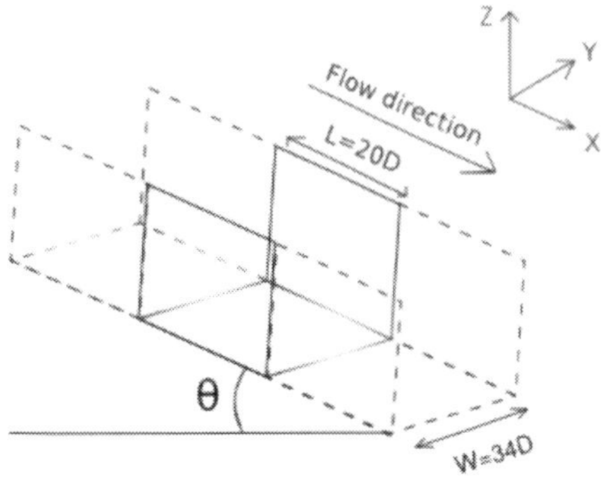

Figure 1: Sketch of the simulation configuration. The granular flow is confined by lateral walls with a gap W. Periodic boudary conditions are applied in the flow direction x with a period L. The bottom has an angle of inclination θ with the horizontal.

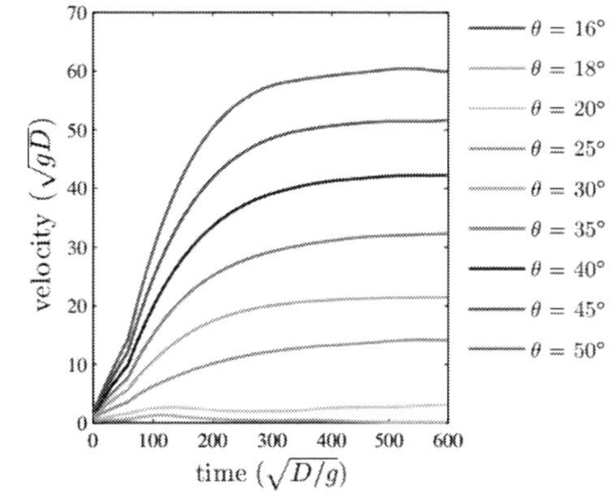

Figure 2: Temporal evolution of the averaged velocity for $W = 34D$, $\tilde{H} = 8D$. The steady-state is reached after an exponential saturation.

PBC are generally used to study steady and fully developed flows [6]. The periodic simulation cell has a length $L = 20D$ and a width $W = 34D$. In these numerical simulations, the control parameters are the angle of inclination θ and the mass hold up $\tilde{H}$. The mass hold up is an equivalent grain height corresponding to the mass per unit basal surface within the simulation box: $\tilde{H} = \frac{1}{\rho D}\sum_g \frac{m_g}{A}$ where ρ denotes the grain density, m_g is the mass of grain g and $A = L \times W$ is the basal area of the box. We run simulations for $\tilde{H} = 4, 6, 8D$ varying inclination angle between 16 and 50°. For each value of the control parameters, the simulations are run up to a stabilization of the kinetic energy of the set of flowing grains, which is always achieved after about 300 time units. Simulations lead to SFD flows. The flow shows a transient regime described by an exponential saturation (Figure 2) of the average grain velocity. The saturated velocity V_L is an increasing function of the mass holdup and of the inclination angle.

3 Transition to the supported regime

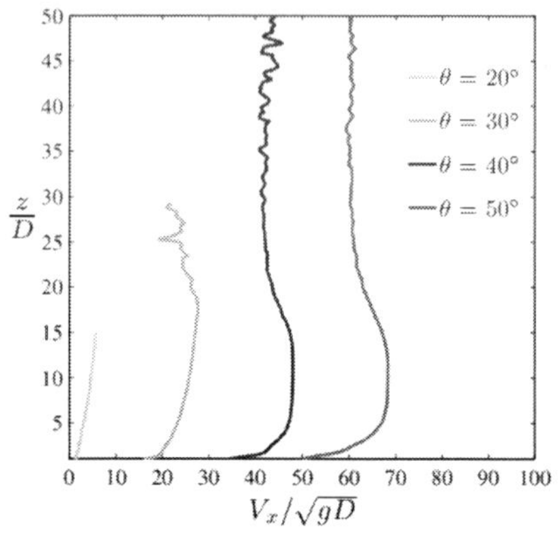

(a)

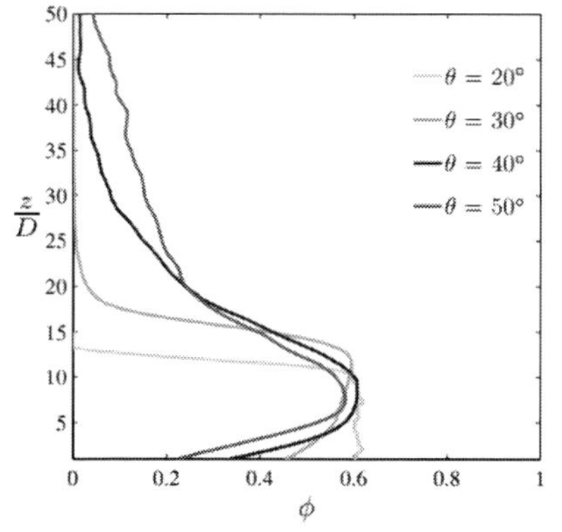

(b)

Figure 3: Height profiles of (a) the velocity in the main flow direction X and (b) the volume fraction, for $W = 34D$, $\tilde{H} = 8D$ and different angles of inclination θ. These quantities are measured at the center of the channel and averaged over $4D$ in the transverse direction.

Figure 3 shows the velocity and the volume fraction profiles for SFD flows at $\tilde{H} = 8D$ for $\theta = 20°$, $\theta = 30°$, $\theta = 40°$ and 50°. To get these profiles, data are averaged over the X-direction and over 300 time units and then averaged along Y-direction over +/- 2D from the channel center. At $\theta = 20°$ the SFD flow is unidirectional with a volume fraction which does not vary much in Z ($\phi \approx 0.6$) and with layering (figure 4a). The velocity profile is similar to what was obtained in [5] in the unidirectional regime: Bagnold profile with a slip velocity at the base. At $\theta = 30°$

 DOI: 10.1051/epjconf/201714003081

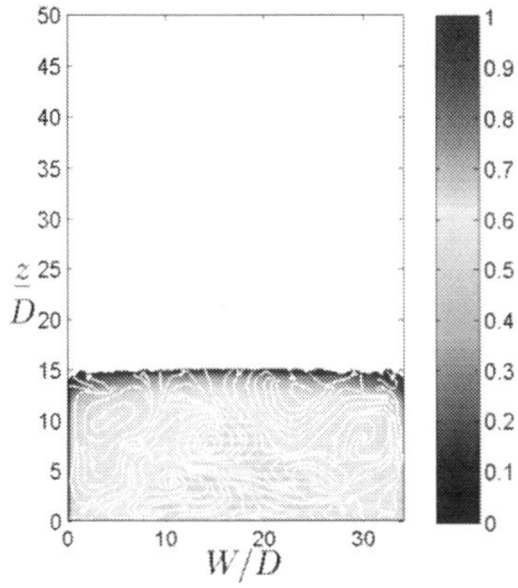

(a)

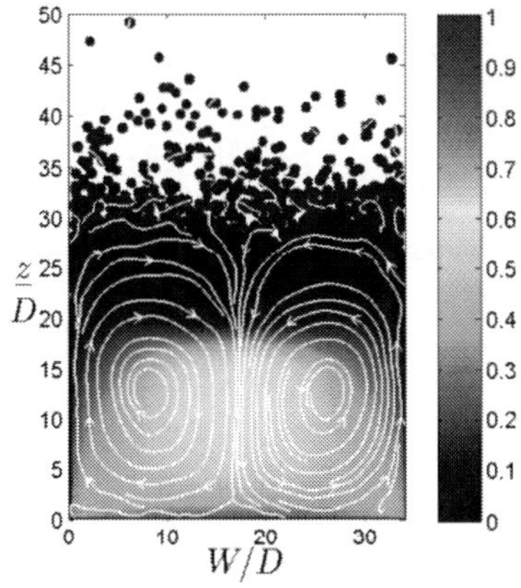

(b)

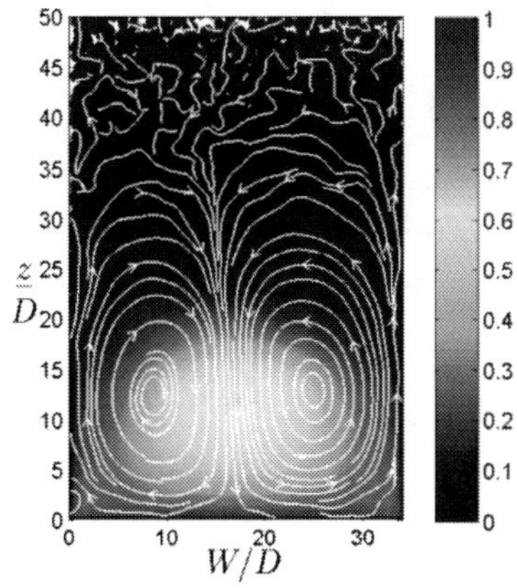

(c)

Figure 4: Two-dimensional maps representing the particle volume fraction and the velocity in the transverse direction, with the streamlines in white color, for $W = 34D$, $\tilde{H} = 8D$, (a) $\theta = 20°$, the particle volume fraction is higher at the base, (b) $\theta = 30°$, the streamline shows the presence of a convective regime (c) $\theta = 35°$, the dense core is clearly separated from the sidewalls and the base.

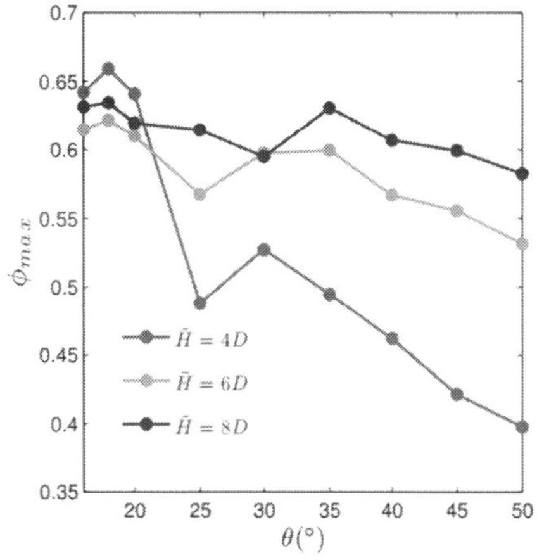

Figure 5: Maximum volume fraction ϕ_{max} on the height profile as a function of θ. The maximum volume fraction first decreases when the angle of inclination increases, then ϕ_{max} increases indicating the transition to supported regime.

the unidirectional flow is destabilized and exhibits longitudinal rolls. Figure 3b shows a lower volume fraction at the base than in the center of the flow. However, this depletion remains moderate and located at the flow base. The whole width of the channel is occupied by a pair of longitudinal vertices as illustrated by the streamlines in figure 4b. We find at higher angles of inclination ($\theta > 30°$) a dense core with a fast and quasi-uniform longitudinal velocity above a basal layer moving at a slower speed. The presence of the dense core is illustrated in Figure 3b by density profiles strongly inverted for $\theta > 30°$. The dense core floats above a very diluted and agitated granular layer. Furthermore, the dense core is also separated from sidewalls by strongly depleted regions (figure 4c). We tried to find a simple criterion to quantify the transition toward supported regime. The variation of the maximum value of the volume fraction ϕ_{max} within the flow with the inclination seems to give a good criterion, at least for small values of $\tilde{H}$. Figure 5 shows the variation of ϕ_{max} for $\tilde{H} = 4, 6, 8D$ as a function of the angle of inclination θ. At low inclination angles, ϕ_{max} decreases when θ increases. This decrease is interrupted by a sudden increase of ϕ_{max} corresponding to the transition toward supported regime. Then, in the supported regime, ϕ_{max} diminishes again with increasing θ but at a lower rate.

We report, in Figure 6 the domains of existence of the rolls and supported regimes for $W = 34D$ and $68D$ [1]. We indicate also the transition zone between the two regimes for both values of W. In [1], for $W = 68D$, the transition from rolls to the supported regime is identified between $\theta = 25°$ and $30°$ for $4D < \tilde{H} < 8D$. In our study, at $W = 34D$, this transition occurs between $\theta = 25°$ and $30°$ for $\tilde{H} = 4, 6D$. The same transition occurs between $\theta = 30°$ and $35°$ for $\tilde{H} = 8D$. We thus notice increase of the transition angle at $\tilde{H} = 8D$. This increase of the

 DOI: 10.1051/epjconf/201714003081

transition angle can be also clearly observed for $W = 68D$, in the phase diagram established by [1], but at higher mass hold up ($\tilde{H} > 10D$).

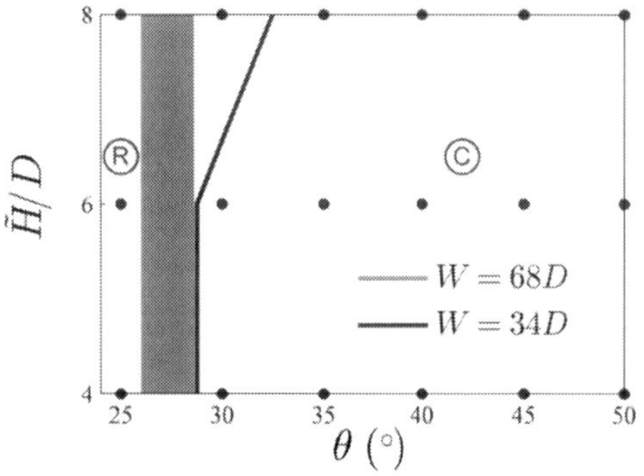

Figure 6: Phase diagram in mass hold up ($\tilde{H} = 4$ to $8D$) - angle of inclination ($\theta = 25°$ to $50°$) space. Blue dots are the sampling points where we performed simulations with a cell width $W = 34D$. Regimes: (R), convective flows with rolls [5, 7, 8], and (C), supported flows [1]. Grey band and blue line are the transition region from convective toward supported regime for $W = 68D$ and $34D$ respectively.

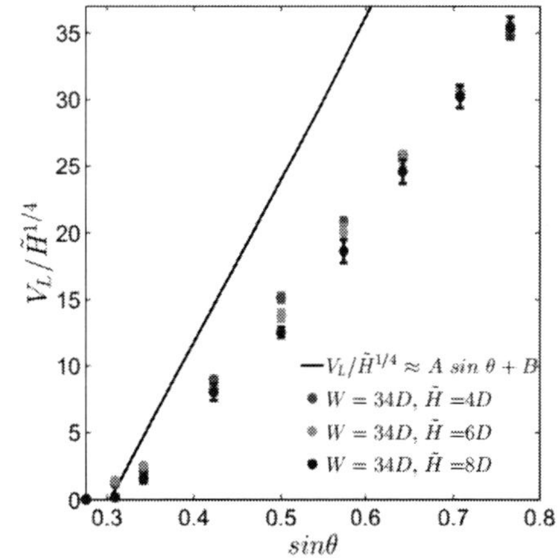

Figure 7: Rescaled limit velocity $V_L/\tilde{H}^{1/4}$ as a function of $sin\ \theta$, for $\tilde{H} = 4, 6, 8D$ respectively, with $W = 34D$. The black curve represents the scaling law established by [1] for $W = 68D$. Error bars represent one standard deviation of the fluctuations of V_L

4 Scaling law

As highlighted in [1], for $W = 68D$, the mass flow rate obeys a scaling law giving the dependence of the limit velocity V_L (figure 2) on the mass hold up and the inclination angle: $V_L/\tilde{H}^{1/4} \approx A\ sin\ \theta + B$ with $A \approx 122$ and $B \approx -37$. We thus calculated, for $W = 34D$, the limit velocity V_L - by averaging in time over 300 time units in the stationary regime the average grain velocity - rescaled by $\tilde{H}^{1/4}$, in function of the inclination angle θ for various mass hold up (figure 4). Our results follow the same scaling law but with a different slope. We can thus suspect that the scaling law of V_L in $\tilde{H}^{1/4}$ is universal, with coefficients A and B varying with W in a way which remains to be specified. This should be the object of further studies.

5 Conclusion

We performed numerical simulations of confined granular flow with a cell width $W = 34D$. Even for this relatively small value we observed the longitudinal rolls reported in [1, 7, 8]. We found that SFD supported regime appears for angles of inclination $\theta > 30°$. This is consistent with the results obtained with $W = 68D$ [1]. The transition toward the supported regime seems to be characterized by a sudden increase of the maximum volume fraction with the angle of inclination after a monotonous decrease. We recover, for $W = 34D$, the mass flow rate scaling law highlighted in [1]. This encourages systematic investigations for other values of W.

References

[1] N. Brodu, R. Delannay, A. Valance, P. Richard, Journal of Fluid Mechanics **769**, 218-228 (2015).

[2] P. Jop, Comptes Rendus Physique **16**, 62-72 (2015).

[3] E. Martínez, A. González-Lezcano, A. J. Batista-Leyva, E. Altshuler, Physical Review E **93**, 062906 (2016).

[4] G. G. Zhou, C. W. Ng, Q. C. Sun, Landslides, **11**, 369-384 (2014).

[5] N. Brodu, P. Richard, R. Delannay, Physical Review E **87(2)**, 022202 (2013).

[6] N. Taberlet, *Ecoulements gravitaires de matériaux granulaires* (PhD thesis, Université Rennes 1, 2005).

[7] Y. Forterre, O. Pouliquen, Journal of Fluid Mechanics, **467**, 361–387 (2002).

[8] T. Börzsönyi,R. E. Ecke, J.N. McElwaine, Physical review letters, **103**, 178302 (2009).

Experimental investigation of mode I fracture for brittle tube-shaped particles

Marta Stasiak[1,2,⋆], *Gaël* Combe[1,2], *Jacques* Desrues[1,2], *Vincent* Richefeu[1,2], *Pascal* Villard[1,2], *Gilles* Armand[3], and *Jad* Zghondi[3]

[1] *Univ. Grenoble Alpes, 3SR, F-38000 Grenoble, France*
[2] *CNRS, 3SR, F-38000 Grenoble, France*
[3] *Andra, R&D Division, Meuse/Haute-Marne Underground Research Laboratory, 55290 Bure, France*

Abstract. We focus herein on the mechanical behavior of highly crushable grains. The object of our interest, named shell, is a hollow cylinder grain with ring cross-section, made of baked clay. The objective is to model the fragmentation of such shells, by means of discrete element (DE) approach. To this end, fracture modes *I* (opening fracture) and *II* (in-plane shear fracture) have to be investigated experimentally. This paper is essentially dedicated to mode *I* fracture. Therefore, a campaign of Brazilian-like compression tests, that result in crack opening, has been performed. The distribution of the occurrence of tensile strength is shown to obey a Weibull distribution for the studied shells, and Weibull's modulus was quantified. Finally, an estimate of the numerical/physical parameters required in a DE model (local strength), is proposed on the basis of the energy required to fracture through a given surface in mode *I* or *II*.

1 Introduction

The discrete element method (DEM) is commonly used to model the response of rigid grains assemblies [1–3], but it has been also employed by various researchers in studying the behavior of crushable particles packings [4, 5].

Among all existing approaches capable of modeling particle breakage, two are frequently used. The first one takes into consideration the rigid particles that will be replaced by smaller ones, when the breakage must occur (according to the grain loading criterion) [6]. This approach has the advantage of the simplicity but may have the disadvantage of having material losses under certain conditions. In the second approach the grain is created by assembling smaller particles together [7, 8] – then the assembly bonds correspond to the pre-cracks of the grain. It is this second approach that we will choose in this article.

In this study, we focus on the peculiar 3D shape of grain which is a cylinder with ring cross-section, referred to as *shell*. Whereas we aim to study the micro-macro mechanical behavior of large, dense assemblies of crushable shells, we propose herein a strategy to model 3D grains, shaped as hollow cylinders, by means of DEM.

Since the packing of such shells is a new granular material, this research starts, at the grain scale, with the investigation of a single shell fracturing. Firstly, an experimental campaign has been performed, in order to extract the actual strength of the shells. Since these shells are made of baked clay, it is worth noting that the strength of brittle material may obey Weibull statistics, based on the theory of the weakest link. The distribution of the tensile strength, that prevents mode I fracture, is then analyzed assuming that it is related to the shell geometry. With the objective to model the shell fracture with a discrete element approach, an estimate of the DEM strength-parameters is proposed for the cluster of sphero-polyhedra (the shell is subdivided into parts bonded together by means of a force-law).

2 Mechanical properties of the shells

The particles, made of baked clay, are tube-shaped shells (Figure 1(a)) and result from an industrial process. In fact, two grain sizes are manufactured. Before been baked at high temperature, the particle diameter is either 18 or 20 mm. Once baked, the grain dimensions, sketched in Figure 1(a), show small geometrical variations as reported in Table 1.

A number of radial compression tests (Brazilian-like tests) have been performed (see inset in Figure 2(a)). This test has been chosen because it allows indirect measurement of a tensile strength and involves crack opening, that corresponds to mode *I* fracture. These experiments have

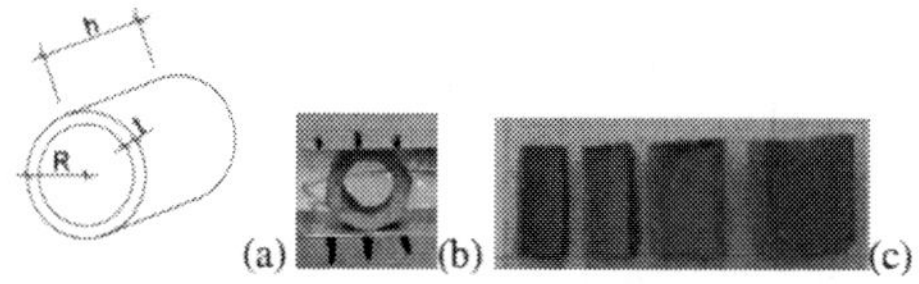

Figure 1: Studied shell: (a) its dimensions, (b) position in the experimental setup, and (c) four pieces after breakage.

⋆e-mail: marta.stasiak@grenoble-3sr.fr

Table 1: Name of the size, number of tests and geometrical description of tested tube-shaped shells. See Figure 1(a).

	No.	$D = 2R$ [mm]	h [mm]	t [mm]
D18	33	16.5 ± 0.4	17.4 ± 1.3	2.4 ± 0.1
D20	50	18.2 ± 0.4	18.4 ± 1.5	2.8 ± 0.1

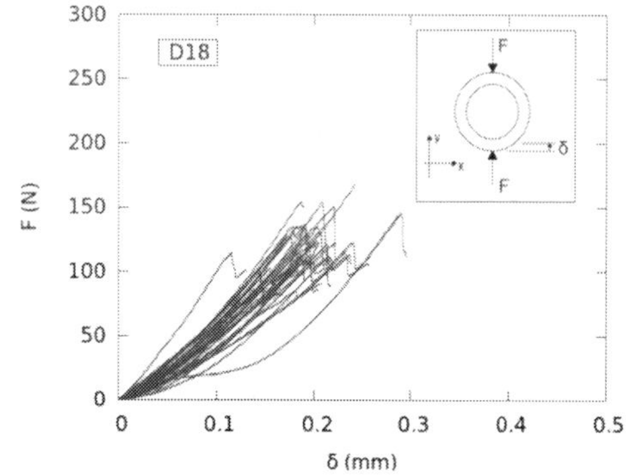

(a)

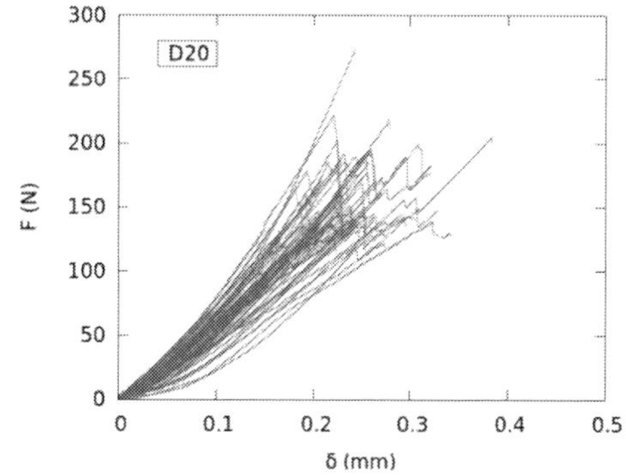

(b)

Figure 2: Force F *vs.* vertical displacement δ curves resulting from the radial compression tests, performed on two sizes of tube-shaped grains with a constant imposed displacement rate ($\dot{\delta} = 0.02$mm/s): (a) D18 and (b) D20. Inset: a sketch of the loading condition.

been performed using a standard loading frame. A vertical force F was measured on the outside wall of the shell; Figure 1(b). The total number of performed tests is given in Table 1. Each test ended when the shell broke most often into four pieces as shown in Figure 1(c). The radial shell shortening δ and the vertical compressive force F were recorded all along each test. Figures 2(a) and 2(b) show all force-displacement curves for tested shells D18 and D20, respectively.

From the compression tests, two quantities can be extracted: the maximum force at failure F_{max} and the corresponding vertical shortening δ_{max} of the shell. Observed variations in the mechanical response of the shells, as shown in Figure 2, may result from the geometrical heterogeneity of the particles (see Table 1), but also from the material heterogeneity. In particular, some internal flaws such as air bubbles can be trapped in the crude clay when it is formed.

When failure arises, the maximum tensile stress is localized at the vertical of the compression force on the inner cylindrical wall of the shell, as illustrated in Figure 3. For this configuration, it can be shown that σ_{max} is a function of the external radius R, the thickness t and the height h of the shell [11]. Assuming that the thickness t does not vary significantly from one shell to another, it is shown that

$$\sigma_{max} \propto \frac{F_{max}}{R\,h}\,\mathcal{K}(t)\,, \qquad (1)$$

where $\mathcal{K}(t)$ is a stress concentration factor that depends on t. Thanks to its very small variation from a shell to another, t is hereafter (and consequently $\mathcal{K}(t)$) assumed to be constant. Figure 4 reports the forces F_{max} as a function of the vertical shortening δ_{max} at breakage. In the inset of the same figure, F_{max} has been divided by Rh to scale with the maximum tensile stress at breakage, and δ_{max} has been divided by R to scale with the radial strain. That way, it is expected to cancel the scattering due to geometrical differences between the shells. However, no collapse is observed even if the cloud of points seems to be less spread. The variations in strength is thus a consequence of the heterogeneity of the clay matter itself.

3 Statistical analysis of tensile strength

The Weibull theory is supposed to be adequate for describing the statistical distribution of strength for quasi-brittle material. If $P(x)$ represents the cumulative distribution function of the occurrence of a strength x (this strength can be expressed in terms of force, stress or energy for example), then the survival probability $P_S(x) = 1 - P(x)$ is expressed as follows:

$$P_S(x) = \exp\left\{\left(-\frac{x}{x_0}\right)^m\right\} \qquad (2)$$

where m is a shape parameter (also called Weibull's modulus) and x_0 a scale parameter. Note that the higher is the shape parameter m, the smaller is the variation of strength x. It is convenient to use double logarithmic scale, that results from a linearization of Equation (2), so that the shape

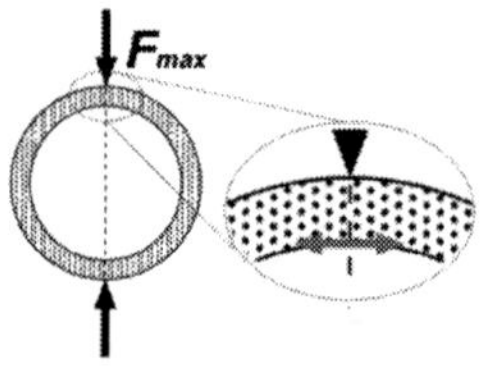

Figure 3: The maximum tensile stress is localized in the direction of the force on the inner wall of the shell. It is here shown by the double red arrows.

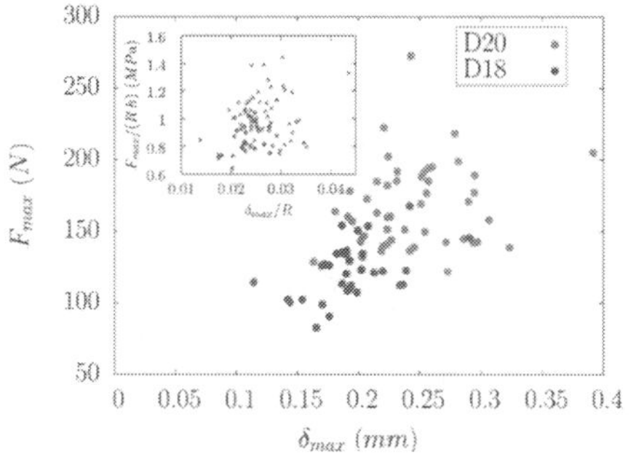

Figure 4: $F_{\max}$ at failure versus the corresponding displacement $\delta_{\max}$ for two grain sizes before normalization. Inset: Maximum force at failure $F_{\max}$ normalized by the radius and the length of the grain versus $\delta_{\max}$ normalized by the radius of the particles.

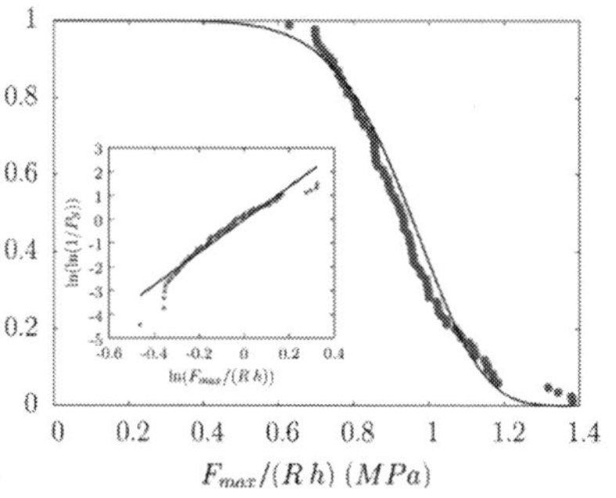

Figure 5: Weibull's distribution of the shell strength. Symbols are experimental data, and continuous line is related to Equation (2). Inset: the same plot in linearized scale – Equation (3) – used to identify Weibull's modulus $m = 6.7$. For three point bend test, in which the material fails also in tension, various types of ceramics have wide distributions described by Weibull's modulus ranging from 6 to 12.

parameter can be determined from the slope of this relation:

$$\ln \circ \ln\left(\frac{1}{P_S(x)}\right) = m\Big(\ln x - \ln x_0\Big) \tag{3}$$

This analysis can be applied with the normal strength $F_{\max}$ measured for the shells, normalizing it with Rh to limit the scattering caused by geometric variations. Figure 5(inset) clearly shows that most of the data follow a straight line according to Equation (3), and it obeys, as expected for quasi-brittle materials, the weakest link theory. Hence, Weibull distribution provides here a reliable first order statistical model for setting the shells strength (Figure 5) when submitted to vertical compression.

4 Towards a DE model for shell crushing

One final objective of this study is to explore the mechanical response of an assembly of breakable shells by using the DEM. We hope to determine the most relevant geometrical parameters that may influence significantly the shear resistance of such assemblies. Thus, a model of shell needs to be designed for this purpose.

Within an assembly, the loading mechanism of a single shell can be complex, and both mode I and mode II fractures can be activated. In case of mode I, it has been experimentally observed that the fracture surfaces are always radial (Figure 1(c)). Mode II characterization also requires breakage trough radial planes. Therefore, it was chosen to pre-define these surfaces by "sticking" some rigid, polyhedral elements in such a manner that they form the global shape of a shell (Figure 6 (a)). From the DE technical point of view, the adjacent elements are joined *via* four points, *i.e.*, spheres in contact. To define the contact forces, two directions are considered: the direction I normal to the predefined plan of fracture (Figure 6(b)) and the sliding direction II (Figure 6(c)). The corresponding elastic forces f_I and f_{II} are functions of the relative displacements (d_I and d_{II}) and the respective stiffnesses k_I and k_{II}. The behavior of the shell is in the elastic phase, until the following failure surface is reached:

$$\frac{f_I}{f_I^\star} + \left(\frac{f_{II}}{f_{II}^\star}\right)^q - 1 = 0 \tag{4}$$

where $f_I^\star$ and $f_{II}^\star$ are the yield cohesive forces for pure tensile and pure shear loading, respectively, and q is a parameter that shape the yield surface.

To complete the particle model, the threshold values $f_I^\star$ and $f_{II}^\star$ as well as the stiffnesses k_I and k_{II} have to be determined. To this end, one can use the *energy storage capacity* $\mathcal{W}^\star$, that for the selected mode of loading can be defined from the experimental force-displacement curve:

$$\mathcal{W}_{\text{exp}}(\delta) = \int_0^\delta F \mathrm{d}\delta \tag{5}$$

so that $\mathcal{W}^\star = \mathcal{W}_{\text{exp}}(\delta_{\max})$. Note that $\mathcal{W}_{\text{exp}}(\delta)$ includes an energy contribution, related to shell-device contact, however, it is negligibly small and omitted in modeling concept. From the DEM side, this quantity can be assessed

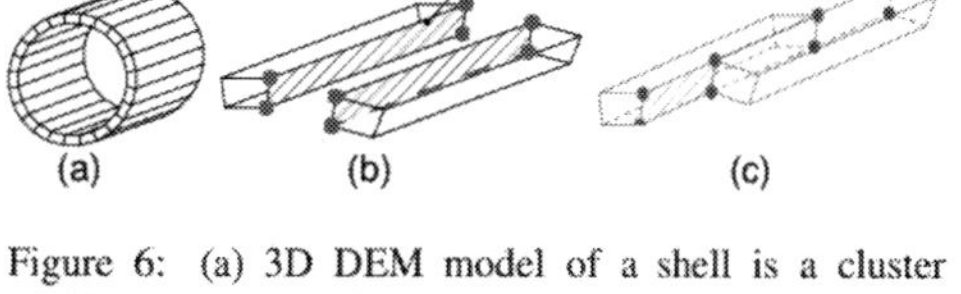

Figure 6: (a) 3D DEM model of a shell is a cluster of 24 non-deformable sphero-polyhedra. One sphero-polyhedron consists of 6 cylinders as edges, 8 spheres as corners and 6 polygons as faces. The adjacent parts are joined *via* four contact points between the spheres. (b) contact opening in fracture mode I, (c) contact opening in fracture mode II

 DOI: 10.1051/epjconf/201714007015

by means of the local force-displacement evolution of the bonded links. Assuming $k_{II} = \frac{2-2\nu}{2-\nu} k_I = \beta k_I$ [12], where ν is the Poisson coefficient , the elastic energy storage reads:

$$\mathcal{W}_{\text{DEM}}(\delta;\, k_I) = \frac{k_I}{2} \sum_{\text{contact } c} \left(\left(d_I^{(c)}\right)^2 + \beta \left(d_{II}^{(c)}\right)^2 \right) \quad (6)$$

The value of k_I can easily be determined using a single DEM simulation for an arbitrary stiffness $\tilde{k}_I$, so that following relation is satisfied:

$$\frac{k_I}{\tilde{k}_I} = \frac{\mathcal{W}_{\text{exp}}(\delta)}{\mathcal{W}_{\text{DEM}}(\delta;\, \tilde{k}_I)} \quad (7)$$

Depending on the main fracture mode involved in the experiment, a first-order estimate of the yield force can be proposed by considering the *energy storage capacity per unit surface* $\mathcal{W}^\star/S$, and by noting that each bond withstands a quarter of predefined surface of rupture $S' = th$:

$$\left\langle \frac{\mathcal{W}^\star_\bullet}{S} \right\rangle \asymp \frac{4\langle f^\star_\bullet \rangle^2}{k_\bullet\, t h} \quad \text{thus} \quad \langle f^\star_\bullet \rangle \asymp \sqrt{\left\langle \frac{\mathcal{W}^\star_\bullet}{S} \right\rangle \frac{k_\bullet\, t h}{4}} \quad (8)$$

where S is the total fracture surface; $\langle x \rangle$ is the mean of x; the symbol $\bullet$ stands for I or II depending on the main fracture mode; and the symbol $\asymp$ means that the left and right hand of the expression should evolve similarly and have a comparable order of magnitude.

To perform a simulation of an assembly, the variation of shell strength can be estimated from the distribution of the yield force. Such distribution can be assessed from the distribution of $\mathcal{W}^\star/S$, and Equation (8) suggests that:

$$P_S(f_I^\star) \propto P_S\left(\sqrt{\mathcal{W}_I^\star/S}\right) \quad (9)$$

For the shells, that are made of quasi-brittle material, a Weibull distribution is still expected, and Figure 7 shows such occurrence. It allows us to access the Weibull's modulus for the distribution of $\sqrt{\mathcal{W}_I^\star/S}$ and thus the distribution of $f_I^\star$ can be quantified.

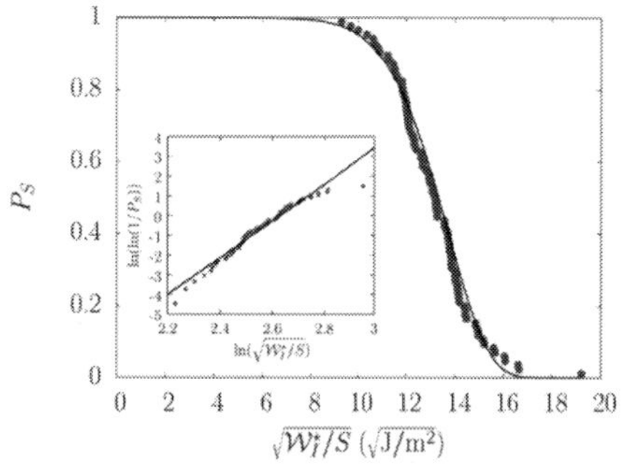

Figure 7: Weibull distribution of the square root of energy storage capacity per unit surface from radial compression tests, that reflects the distribution of the local yielding force for mode I rupture. Symbols are experimental data, and continuous line is related to Equation (2). Inset: the same plot with linearized scale - according to Equation (3) - used to identify Weibull's modulus $m = 9.26$

5 Conclusions and perspectives

An experimental campaign involving a great number of radial compression tests has been performed. The objective was to characterize the material resistance of tube-shaped grains made of baked clay, called shells. The variability of the macroscopic strength was studied within a statistical framework and it was shown to obey the Weibull theory. To model breakable shells by 3D-DEM, predefined fracture plans are bonded by cohesive elastic forces that can break according to a local yield criterion. A solution to assess experimentally the mean local strengths relies on the macroscopic energy storage capacity for each fracturing mode of the shell. Furthermore, the dispersion of these strengths can be characterized from macroscopic tests, and for the breakage in mode I it is shown to also obey a Weibull distribution.

The experimental campaign needs to be completed with shear tests on the shells to investigate their fracture in mode II. Once a single shell will be satisfactorily modeled, the next step will focus on the mechanical responses of an assembly of breakable shells subjected to triaxial and odometric loadings.

Acknowledgments

The Laboratoire 3SR is part of the LabEx Tec 21 (Investissements d'Avenir, Grant Agreement No. ANR-11-LABX-0030)

References

[1] CEGEO, B. Saint-Cyr, K. Szarf, C. Voivret, E. Azéma, V. Richefeu, J.-Y. Delenne, G. Combe, C. Nouguier-Lehon, P. Villard, *EPL* **98**(4), 44008 (2012)

[2] Roux J.-N. and Combe G., *IUTAM-ISIMM Symposium on Mathematical Modeling and Physical Instances of Granular Flows*, **1227**, 260-270 (2010)

[3] G. Mollon, V. Richefeu, P. Villard, D. Daudon, *Journal of Geophysical Research* **117**, F02036 (2012)

[4] G. R. McDowell & O. Harireche, *Geotechnique* **52**, 131-135 (2002)

[5] Y. P. Cheng, Y. Nakata & M. D. Bolton, *Geotechnique* **53**, 633-641 (2003)

[6] D. Cantor, N. Estrada & E. Azéma, *Geomechanics from Micro to Macro* (Taylor & Francis Group, London 2015) 257-262

[7] M. Renouf, B. Gezahengn, M. Abbas & F. Bourgeois, *Powders and Grains 2013* **1542**, 907 (2013)

[8] A. Aminzadeh & E. S. Hosseininia, *Powders and Grains 2013* **1542**, 915 (2013)

[9] W. Weibull, *ASME Journal of Applied Mechanics* **8**, 293-297 (1951)

[10] M. Mellor, I. Hawkes, *Eng. Geol.* **5**, 173-225 (1971)

[11] De-Li Ma, *Int.J.Pres.Ves.& Piping* **42**, 185-191 (1990)

[12] I. Agnolin and J.-N. Roux, *Phys. Rev. E* 76, 061304

Size Segregation of Binary Granular Mixtures

Towards more appropriate scaling of percolation velocity

Anurag Tripathi[1,]* and *Mohit* Nema[1]

[1] *Indian Institute of Technology Kanpur, 208016, India*

Abstract. We study size segregation of binary granular mixtures flowing over an inclined plane using DEM simulations. We critically examine the recently proposed scaling of the species percolation velocity with the local shear rate for different inclination angles and size ratios for an equal volume mixture of large and small spherical particles. Our DEM simulations explore a much wider range of inclination angles and shear rates for three different size ratios of large and small particles. Our results suggest that while the scaling of percolation velocity with the shear rate seems to work well for any given inclination angle, this scaling is unable to capture the influence of inclination angle on the segregation. We overcome this limitation and propose a more appropriate way of scaling the percolation velocity. This scaling, when used along with the convection-diffusion equation, is able to predict the steady state segregation of binary mixtures flowing at different inclination angles for different size ratios.

1 Introduction

Understanding and quantitatively predicting granular segregation due to size difference have been focus of researchers for decades [1, 2]. One highly promising approach toward quantitative prediction of size segregation is proposed recently by [3, 4] where the authors have studied segregation of binary granular mixtures in heap flow and in rotating cylinders. The authors report that percolation velocity of the species in the segregation direction, when scaled by the local shear rate [5, 6], shows a linear relation with the concentration of the other species with a constant of proportionality that depends logarithmically on the size ratio of the grains. In these studies, the expression for the segregation flux is symmetric in the species concentration. Recent studies [7–9], however, indicate that an asymmetric segregation flux function may be more appropriate to predict the size segregation in binary granular mixtures. In this study, we investigate the shear rate based scaling of the percolation velocity for a binary granular mixture consisting of particles of two different sizes flowing over an inclined plane using DEM simulations. In this configuration, we are able to explore a much wider range of shear rates compared to those used in [3, 4]. Our results establish that the shear rate scaling of the percolation velocity is not appropriate for chute flows. Instead, the percolation velocity scaled by the square root of the local shear rate seems to be more appropriate and can be used to predict the size segregation of binary mixtures for a range of inclination angles and size ratios.

*e-mail: anuragt@iitk.ac.in

2 Simulation methodology

Gravity driven flow of spherical, slightly inelastic, frictional particles over a rough bumpy surface is studied by means of discrete element method (DEM) simulations. Simulation box of length $L_x = 20d$ and width $L_z = 20d$ is used with periodic boundaries in x (flow) and z (neutral) direction (see fig.1a). The simulation box height in y direction is kept large enough so that the particles can not detect the presence of the top surface. The particles are initially placed in a simple cubic lattice arrangement with a small gap between the surface of the particles. Total number of particles is chosen to ensure that the layer thickness is around $25d$. The rough base is made by taking a $1.2d$ thick slice of particles from a random closed packed bed of particles of diameter d settled on a flat surface. The diameter (d) and mass (m) of small particles is fixed in all the simulations and three different ratios of large to small particle size ($r = d_L/d$), namely 1.5, 2 and 2.5 are considered. The large particles are placed near the base while the small particles are placed near the surface in the beginning of the simulations. The particles are settled on a horizontal surface ($\theta = 0°$). The inclination angle is then increased to desired angle (θ) and all the properties are observed as the flow becomes steady, fully developed. The contact forces between the particles are modelled using a linear spring (stiffness $k_n = 2 \times 10^5 mg/d$) with a viscous dashpot in the radial direction. The damping coefficient of the dashpot is chosen to obtain normal restitution coefficient ($e_n = 0.88$). A linear spring (stiffness $k_t = 2k_n/7$) is used to account for the elastic interaction in the tangential direction. The value of the friction coefficient between the grains is $\mu = 0.5$ and the spring deformation in the tangential direction is truncated so that the ratio of tangential to normal force for any contact is less than or equal to μ [10, 11]. Velocity-Verlet algorithm is used to integrate the equations of motion for all the particles and the time step of integration in the simulations is chosen to be small enough to resolve head-on collision of two small particles in 50 time steps. All the properties reported are non-dimensionalized using m, d and g where g is acceleration due to gravity. Properties of interest (velocity, packing fraction etc.) are averaged in a bin of thickness d_L spanning the entire simulation box in x and z direction. Fractional volume of the particles in a bin is accounted to accurately measure the properties in each bin. The solids volume fraction (ϕ_i) of species i is obtained as the ratio of the total volume of particles of species i in a bin to the volume of the bin. The local shear rate is obtained by numerically differentiating the velocity profile.

3 Results and discussion

We report results for an equal volume mixture of large and small particles with size ratio $r = 2$ in fig. 1b-d for four different inclination angles. As shown in fig. 1b and fig. 1c, the flow becomes faster (i.e., velocity in the flow direction v_x increases) and less dense (the packing fraction ϕ decreases) with increase in the inclination angle θ. The segregation due to the size difference is evident since the volume concentration of large grains (ϕ_L/ϕ) is close to unity (indicating mostly large grains are present) near free surface (fig. 1d). The large particle concentration reduces as the distance from the free surface increases, indicating that small grains concentrate near the base. Complete segregation will be indicated by a horizontal line in fig. 1d. As the inclination angle increases, the concentration profile moves away from the horizontal due to increased diffusive mixing at higher shear rates.

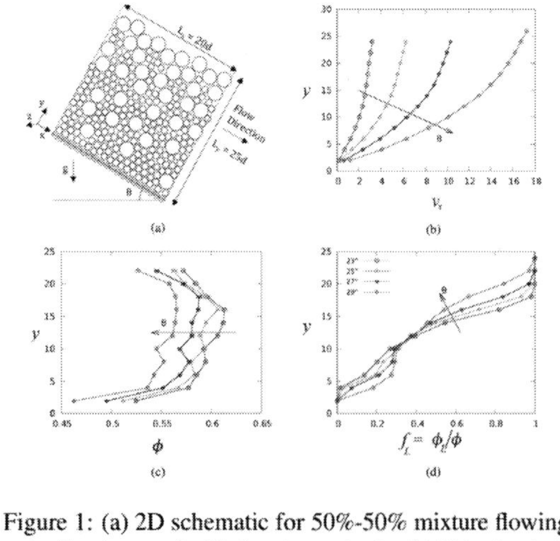

Figure 1: (a) 2D schematic for 50%-50% mixture flowing at inclination angle (θ) for size ratio 2. (b) Velocity (v_x), (c) total solids fraction (ϕ) and (d) concentration (f_L) of large particles.

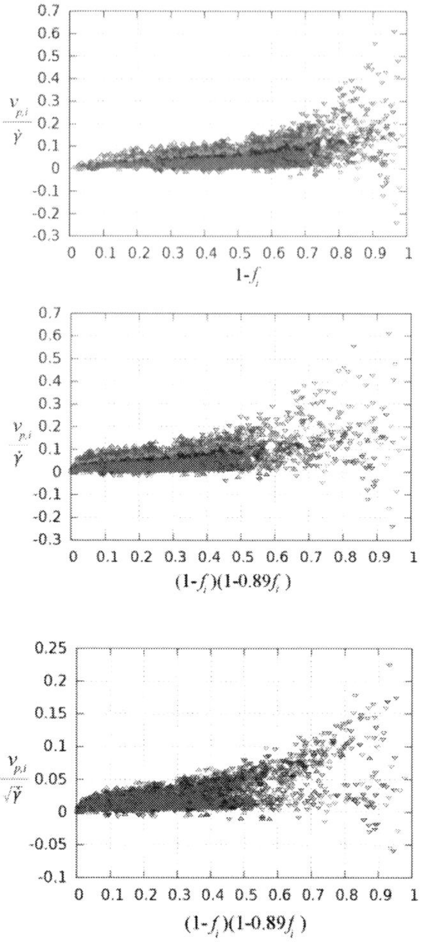

Figure 2: Top panel shows the variation of shear rate scaled percolation velocity with $1 - f_i$ and the middle panel shows the variation of the same with $(1 - f_i)(1 - 0.89f_i)$ for four different values of θ for size ratio $r = 2$ (colour scheme same as in fig.1). Bottom panel shows the variation of percolation velocity scaled by square root of shear rate with $(1 - f_i)(1 - 0.89f_i)$.

3.1 Percolation velocity scaling

The percolation velocity of a species in the segregation (y) direction is obtained by subtracting the volume average y-velocity (v_y) of all the grains in a bin from the volume average y-velocity of the species in that bin, i.e., $v_{p,i} = v_i - v_y$. In fig. 2, we plot the magnitude of the percolation velocity $v_{p,i}$ of the large (shown as upward triangles) and small (shown as downward triangles) grains scaled by shear rate or square root of shear rate for size ratio $r = 2$ at four different inclination angles. In fig. 2 (top), we use the symmetric segregation flux expression suggested by [3, 4] and plot the percolation velocity scaled by the local shear rate $\dot{\gamma}$ [5, 6] with the concentration of the other species $(1 - f_i)$. While the linear variation of the shear rate scaled percola-

DOI: 10.1051/epjconf/201714003082

tion velocity seems reasonably good (for most of the range of $1 - f_i$) for a particular inclination angle, the data for different inclinations do not follow the same linear variation, i.e., data for different inclinations do not fall on top of each other. The slope of the fitted line to this data is referred to as the 'percolation length' in [3] and it is reported that this percolation length varies logarithmically with the size ratio (r). Our results indicate that the slopes of the fitted lines (not shown here) differ substantially for different inclination angles, i.e., the 'percolation length' in [3] depends upon the inclination angle as well.

Next, we explore whether the asymmetric segregation flux function [7, 8] can be used to appropriately describe the variation of the percolation velocity with species concentration. We plot the magnitude of the percolation velocity scaled by the shear rate ($v_{p,i}/\dot{\gamma}$) with $(1 - f_i)(1 - \kappa f_i)$ in fig. 2(middle). While the value of κ might depend upon the size ratio and inclination angle [7, 9], following ref. [8], we have used a constant value of $\kappa = 0.89$ in present study. Again, while the linear variation of the shear rate scaled percolation velocity seems reasonable for a particular inclination angle, data for different inclinations do not follow the same trend and differ significantly from each other. By fitting a straight line through the data, we obtain the slope S_s so that $v_{p,i}/\dot{\gamma} = S_s(1 - f_i)(1 - 0.89 f_i)$. Fig.3 shows the variation of the fitted parameter S_s for three different size ratios $r = 1.5, 2$ and 2.5 at four different inclination angles ($\theta = 23°, 25°, 27°$ and $29°$). It is evident that the value of S_s for a given size varies substantially as the inclination angle is changed. This indicates that the effect of the inclination angle on the percolation velocity is very significant and can not be neglected. We note that in ref. [3, 4], the granular flow over erodible bases were considered and hence a limited range of shear rates were explored since the flow occurred close to the angle of the repose. However, if we scale the percolation velocity ($v_{p,i}$) by the square root of the shear rate ($\sqrt{\dot{\gamma}}$) and plot the variation of $v_{p,i}/\sqrt{\dot{\gamma}}$ with $(1 - f_i)(1 - 0.89 f_i)$, data for different inclinations fall on top of each other as shown in fig. 2(bottom). By fitting a straight line through this data, we obtain the slope S_{ss} so that $v_{p,i}/\sqrt{\dot{\gamma}} = S_{ss}(1 - f_i)(1 - 0.89 f_i)$. Fig.3 shows the variation of the fitted parameter S_{ss} for three different size ratios and four different inclination angles as mentioned above. For a given size, the values of S_{ss} do not differ much from each other suggesting that square root of the shear rate may be a more appropriate quantity to scale the percolation velocity. Further, the linear variation in S_{ss} with size ratio suggest that $S_{ss} = A(r - 1)$ can be used to describe the data for different inclination angles. Thus, the percolation speed for both large and small grains can be related to the other species concentration as

$$v_{p,i} = A\sqrt{\dot{\gamma}}(r - 1)(1 - f_i)(1 - 0.89 f_i) \quad (1)$$

where A is a constant determined from the simulations. We emphasise that the above relation accounts for the dependence of the percolation velocity on the local shear rate ($\dot{\gamma}$) for a wide range of inclination angles (and hence shear rates), species concentration (f_i) for different size ratio of the grains (upto size ratio 2.5) and hence, in principle, might be appropriate for other flow configurations as well.

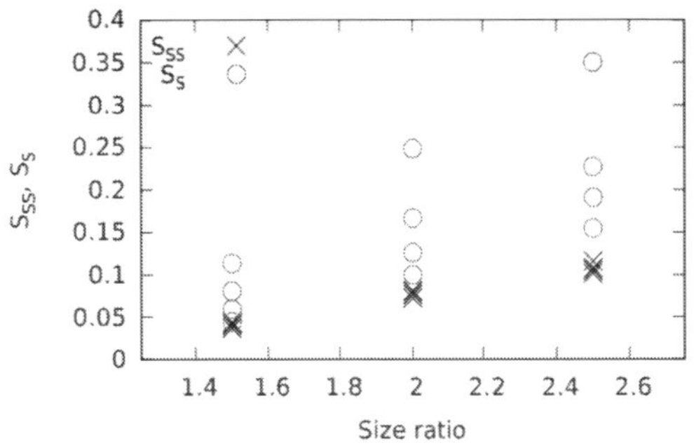

Figure 3: Variation of S_s and S_{ss} with size ratio for four different inclination angles.

3.2 Concentration profile prediction from convection-diffusion equation

The convection-diffusion equation for fully developed, steady chute flow of binary mixture of species reduces to the balance of segregation flux and diffusion flux in the segregation direction (y direction in fig. 1a) and is given as

$$v_{p,i} f_i = D\frac{df_i}{dy} \quad (2)$$

where D is the mass diffusivity of the binary mixture. The diffusivity D is calculated as half the slope of the mean square displacement of the particles in the y direction. Our simulation results indicate that the diffusivity is proportional to the shear rate (as in ref. [3]) and varies quadratically with the volume average diameter d_{mix}, i.e., $D = \beta\dot{\gamma}d^2_{mix}$. Local volume average diameter is given as $d_{mix} = (\phi_S d + \phi_L d_L)/\phi = (1 - f_L + r f_L)d$. The proportionality constant β depends upon the size ratio and we use the value of β obtained from simulations. Using the percolation velocity relation of eq. 1 and the above mentioned relation for the diffusivity in eq. 2, we obtain the following ordinary differential equation for large particle concentation (f_L)

$$\frac{df_L}{dy} = \frac{A(r - 1)f_L(1 - f_L)(1 - 0.89 f_L)}{\beta\sqrt{\dot{\gamma}}d^2(1 + (r - 1)f_L)^2} \quad (3)$$

which is solved as an initial value problem using Matlab ODE15s solver to obtain the concentration profile of large particles $f_L(y)$ at steady state.

Since the initial value of $f_L(y = 0)$ is not known, we use the fact that total concentration of the large particles in the layer of height h is known; for a binary mixture having equal volume of large and small grains $\frac{\int_0^h f_L(y)dy}{h} = 0.5$. Starting from an initial guess of $f_L(y = 0)$, we explore the solution $f_L(y)$ of eq. 3 that satisfies the condition that the total concentration of the large particles in the layer is 50%

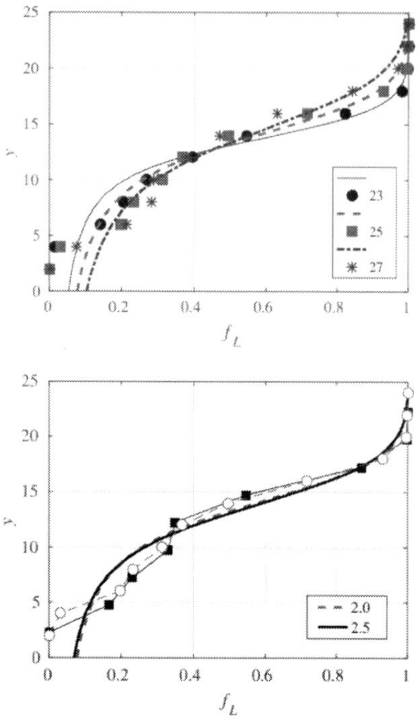

Figure 4: Comparison of the predicted concentration profile (lines) of large particles with simulation results (symbols) for three different values of θ for $r = 2$ (top) and for two different size ratios ($r = 2$ & 2.5) for $\theta = 25°$ (bottom).

within a tolerance limit of 10^{-3}, i.e., $|\frac{\int_0^h f_L(y)dy}{h} - 0.5| < 10^{-3}$. We use bisection method in the interval $0 < f_L < 1$ and iterate for the solutions of eq. 3 so that the tolerance criterion is satisfied. The values of A and β are obtained from the simulations. The shear rate profile used in eq. 3 is obtained by differentiating the function used to fit the steady state velocity profile at any particular inclination angle. Fig. 4(top) shows that the predictions using the percolation velocity scaling in eq. 1 is able to capture the dependence on the inclination angle θ. We emphasize that in the case of shear rate scaling, eq. 3 becomes independent of the shear rate and predicts that the concentration profile remains unaffected by the inclination angle. This is a crucial limitation of the existing segregation flux models which assume the percolation velocity to be proportional to the local shear rate. Fig. 4(bottom) shows that the concentration profile for $r = 2$ and 2.5 from the DEM simulations are nearly identical [10] and the same behaviour is observed in the predicted concentration profiles as well. We note that while the complete segregation in the upper part of the layer is well predicted by eq. 3, it fails to predict the near complete segregation of small particles close to the base observed in simulations.

4 Conclusion

Size segregation of binary granular mixtures, having an equal volume of large and small particles, is studied in chute flow configuration using DEM simulations for different inclination angles and size ratios. Suitability of the shear rate scaling of the percolation velocity, used by various researchers [3–9], has been investigated for both symmetric as well as asymmetric segregation flux functions. While the shear rate based scaling of the percolation velocity seems to be reasonable for a given inclination angle, it fails to describe the data when a wide range of inclination angles (and hence shear rates) are considered, i.e., the 'percolation length' used in ref. [3, 4] is found to depend on the inclination angle. This indicates that the shear rate based scaling of the percolation velocity is not appropriate for quantitative description of size segregation for chute flow. Our results establish that square root of the shear rate based scaling of the percolation velocity is more appropriate and is able to predict the large particle concentration for different size ratios and inclination angles in good quantitative agreement with simulation results.

Acknowledgements

AT gratefully acknowledges the financial support from I.I.T. Kanpur through the initiation grant for this study.

References

[1] J. Bridgwater, W. Foo, and D. Stephens, Powder Technology, **41**, 147-158 (1985)

[2] S. Savage and C. Lun, J. Fluid Mech., **189**, 311-335 (1988)

[3] Y. Fan, C. P. Schlick, P. B. Umbanhowar, J. M. Ottino, and R. M. Lueptow, J. Fluid Mech., **741**, 252-279 (2014)

[4] C.P. Schlick, Y. Fan, A.B. Isner, P.B. Umbanhowar, J.M. Ottino, and R.M. Lueptow, AIChE Journal, **61**, 1524-1534 (2015)

[5] B. Marks, P. Rognon, and I. Einav, J. Fluid Mech., **690**, 499-511 (2012)

[6] D.R. Tunuguntla, O. Bokhove, and A.R. Thornton, J. Fluid Mech., **749**, 99-112 (2014)

[7] P. Gajjar and J.M.N.T. Gray, J. Fluid Mech., **757**, 297-329 (2014)

[8] K. van der Vaart, P. Gajjar, G. Epely-Chauvin, N. Andreini, J.M.N.T. Gray, and C. Ancey, Phys. Rev. Lett., **114**, 238001 (2015)

[9] D.R. Tunuguntla, T. Weinhart, and A.R. Thornton, Comp. Part. Mech. (2016) doi:10.1007/s40571-016-0136-1

[10] A. Tripathi and D.V. Khakhar, Phys. Fluids, **23**, 113302 (2011)

[11] A. Tripathi and D.V. Khakhar, J. Fluid Mech., **717**, 643-669 (2013)

Element soil behaviour during pile installation simulated by 2D-DEM

Xiaohui Ji[1,*], *Yi Pik* Cheng[2], and *Junwei* Liu[3]

[1] MSc student, Department of Civil, Environment & Geomatic Engineering, University College London, London, UK
[2] Senior Lecture, Department of Civil, Environment & Geomatic Engineering, University College London, London, UK
[3] Academic visitor, Department of Civil, Environment & Geomatic Engineering, University College London, London, UK
Associate Professor, Qingdao Technological University, Qindao, China

Abstract. The estimation of the skin friction of onshore or offshore piles in sand is still a difficult problem for geotechnical engineers. It has been accepted by many researchers that the mechanism of driving piles in the soil has shared some similarities with that of an element shear test under the constant normal stiffness (CNS) condition. This paper describes the behaviour of an element of soil next to a pile during the process of pile penetration into dense fine sand using the 2D-DEM numerical simulation software. A new CNS servo was added to the horizontal boundary while maintaining the vertical stress constant. This should simulate the soil in a similar manner to that of a CNS pile-soil interface shear test, but allowing the vertical stress to remain constant which is more realistic to the field situation. Shear behaviours observed in these simulations were very similar to the results from previous researchers' lab shearing tests. With the normal stress and shear stress obtained from the virtual models, the friction angle and the shaft friction factor β mentioned in the API-2007 offshore pile design guideline were calculated and compared with the API recommended values.

1 Introduction

The study of the shaft resistance of offshore piles has become a hot spot in the geotechnical engineering. It has been widely accepted that the constant normal stiffness (CNS) shear test can provide insights into the mechanism of the skin friction developed along the pile shafts. [1- 5] Similar with the CNS shear tests, the frictional behaviours along the pile shaft is also significantly influenced by the behaviours of the soil-solid interface.

However, most of the CNS shear tests were conducted in the laboratory condition. These CNS tests often have high requirements on the experimental apparatus. [3] Also, due to the large scale of the pile-sand interface shear test, it is not easy to conduct the test in a laboratory. Therefore, the numerical simulation will be a better way to understand the behaviour of soil during the pile installation.

In this paper, the numeric analysis software based on the 2D discrete element method was used to conduct the CNS shear test simulation. A virtual model of the shear test with the CNS servo was established to investigate the soil behaviour. Also, the test results were compared with the data from the laboratory test. And the shaft resistance was also compared with the value recommended in the design code (API-2007) [6].

2 CNS pile-soil interface shear test

2.1 Material preparation

In the 2D-DEM numerical software, the input parameters of soil particles have huge impact on the macro properties of soil sample. Table 1 shows the particle parameters inputted into the virtual model.

Table 1. Input parameters in the model

Properties	Value
Soil element dimension (width x height),(mm)	10 x 20
Max. diameter of soil particle (mm)	0.3
Min. diameter of soil particle (mm)	0.15
Density of particles (kg/m3)	2650
Stiffness of particles k=ks=kn (N/m)	1e7
Stiffness of wall, kn (N/m)	1e9
Friction of particles μ (--)	0.7
Friction of wall μwall (--)	0

* Xiaohui Ji: xiaohuiji211@163.com

Diameter of particles composing pile surface (mm)	0.18
Normalized roughness of pile surface (--)	0.1

2.2 Virtual model creation

The virtual model for the CNS pile-soil interface direct shear test was made up of four parts: soil element, pile surface, four walls and a CNS servo.

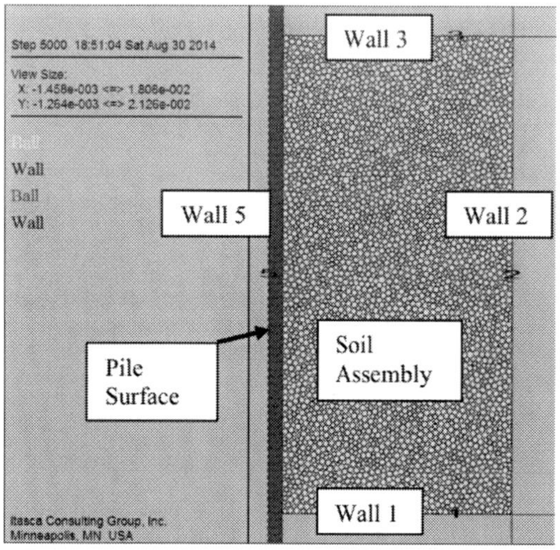

Figure 1. Model for the CNS pile-soil interface shear test

As it is shown in Figure 1, the tested soil element in the model is created as an assembly of small round balls, which stand for the sand particles. On the left of soil element, the vertical-arranged bigger balls, firmly connected with each other, composed the pile surface. Additionally, four frictionless walls around the soil element can prevent the soil particles from escaping during the simulation.

The constant normal stiffness condition during the shear test was obtained by a servo-controlled stress feedback system. This servo can automatically adjust the velocity of Wall 2 (the wall on the right side in Figure 1) to change the normal load (horizontal direction) on the soil according to the following equation with the measured stress on Wall 2.

$$\Delta\sigma_n = K * \Delta\mu \qquad (1)$$

Where,

$\Delta\sigma_n = \sigma_n - \sigma_{n0}$ is the increment of normal stress.

$\Delta\mu = \mu - \mu_0$ is the increment of normal displacement.

K is the constant normal stiffness.

It should be clearfied that the top and bottom walls (Wall 1 and Wall 3) control the vertical stress level, which is governed by how deep the element is located. These two walls are controlled by servo-mechanisms that ensures the vertical stress remains constant, they can move allow a change in volume of the soil element. Wall 5 is frictionless and fixed during the simulation. It can prevent the pile surface from moving towards the left without disturbing the measured vertical stress (shear stress) on the pile when dense soil dilates in the shear test.

2.3 Constant normal stiffness servo

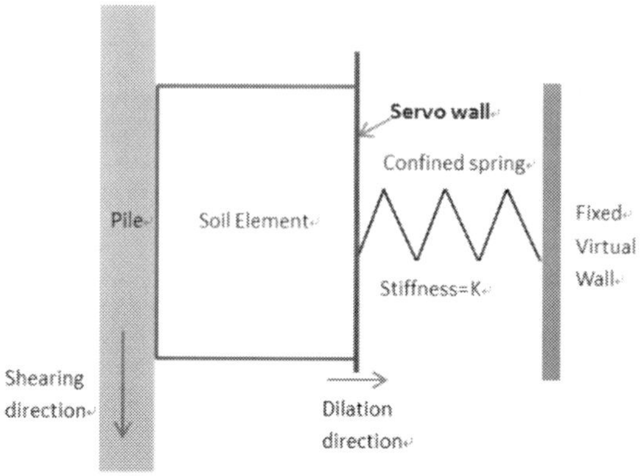

Figure 2. Schematic diagram of CNS pile-soil interface shear test

As it is shown in Figure 2, the soil sample (dense sand) will be sheared and dilated during the pile moving downward. It will result in an increase of the stress ($\Delta\sigma_n$) on the servo wall (Wall 2 in the model). However, due to the confinement of the spring, the servo wall will move to the right side, and compress the spring to get an equilibrium state. During this process, the displacement of the servo wall can be noted as $\Delta\mu$, and it can be determined by $\Delta\mu = \Delta\sigma_n / K$.

This equation is the base of designing the CNS servo wall. The specific algorithm is as follows:

(1) After the sample reach the initial stress condition, get the initial normal stress on the servo wall (σ_{n_old}). Then, fix Wall 5 and give a constant shearing rate (V_{pile}) to the pile. The shearing rate should not be too big, otherwise the unbalanced force in the soil cannot be fully dissipated.

(2) Allow the pile to move a certain time first (e.g. 10 cycles), and then stop the pile.

(3) After that, the new stress on the servo wall (σ_{n_new}) can be obtained by using "get_ss" FISH function in the PFC2D manual [7]. Then the change of the normal stress on the servo wall can be determined, which is $\Delta\sigma_n = \sigma_{n\_new} - \sigma_{n\_old}$. Hence, the required displacement of the servo wall under CNS condition should be calculated as $\Delta\mu = \Delta\sigma_n / K = (\sigma_{n\_new} - \sigma_{n\_old}) / K$.

(4) In order to reach this required displacement, the servo wall was given a constant velocity (V_{servo}). This velocity also should be small enough to allow the system to reach equilibrium.

(5) After the servo wall move $\Delta\mu$ distance, the wall will stop, and solve the unbalanced force ratio (muf/mcf) in the soil sample to reach equilibrium state. It should be noticed that during this step, the pile should also not move.

(6) These five steps above are included in a looping statement, which can be iterated thousands of times to get the expected shear displacement.

The most important point in this CNS algorithm is that the equilibrium state should always be achieved before any movement of the pile. It can reduce the unbalanced force in the system and increase the reliability of the test. The old stress on the servo wall (σ_{n_old}) will always be automatically updated by the new stress (σ_{n_new}) in that last loop.

2.4 Initial stress condition of soil element

In this paper, two soil samples in different depths (10m and 25m) were selected to simulate the shaft resistance of pile under different normal stress conditions. The normal stress σ_n was determined by the one-dimensional consolidation test with 2D-DEM simulation software.

Table 2. Initial stress condition of tested samples

Soil condition	Soil element 1	Soil element 2
Overburden stress σ_v (kpa)	220	550
Normal stress σ_n (kpa)	95	236
Input constant stiffness (pa/m)	5.69e8	6.57e8

3 Results

3.1 Normal stiffness in the test

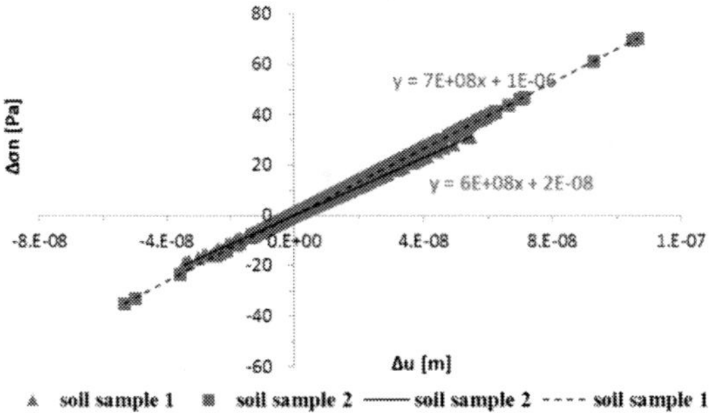

Figure 3. Typical trend of normal stiffness in the test

Figure 3 illustrated variations of normal stress on the servo wall ($\Delta\sigma_n$) against the variations of the wall displacement ($\Delta\mu$) in the CNS test. It is obvious that $\Delta\sigma_n$ shows a quite good linear relationship with $\Delta\mu$. This means the servo wall controlled the constant normal stiffness (CNS) boundary condition very well. The slopes of the lines are the value of constant normal stiffness, which is 6e8 Pa/m for soil sample 1 and 7e8 Pa/m for soil sample 2 respectively. These values are quite close to the input constant stiffness (5.69e8 for sample 1 & 6.57e8 for sample 2) in Table 2.

3.2 Shear behaviour compared with lab test

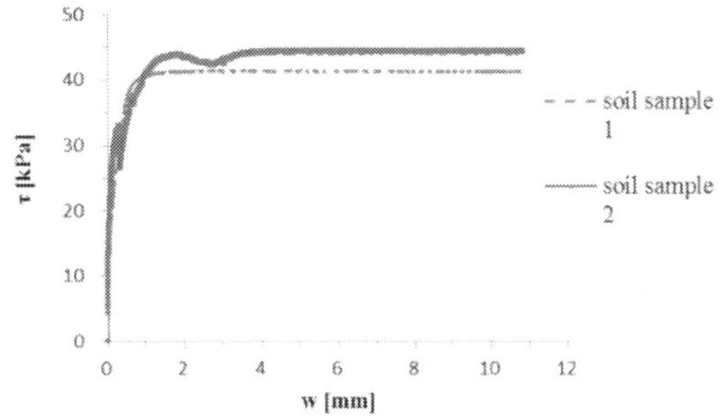

Figure 4. Shear stress from the CNS pile-soil shear test

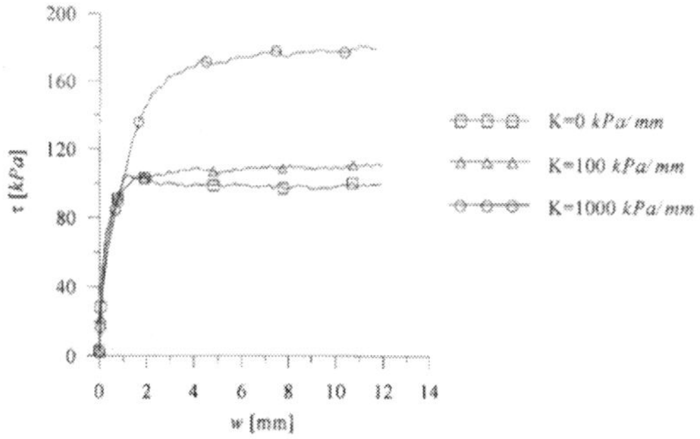

Figure 5. Shear stress from CNS lab test [1]

The shear stress obtained from the numerical simulation shown in Figure 4 shared a very similar trend with the results from the lab tests.

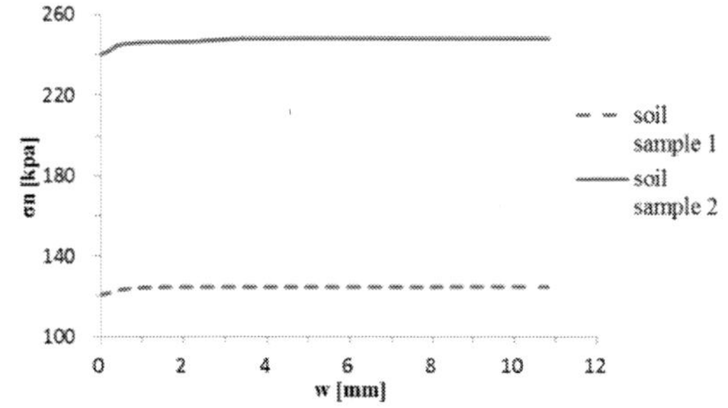

Figure 6. Variation of normal stress in CNS test

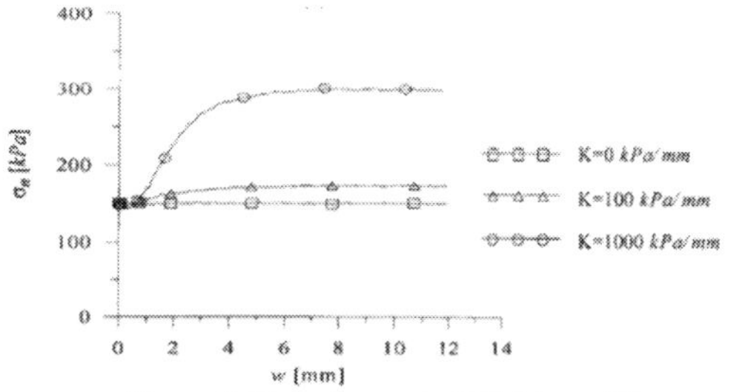

Figure 7. Normal stress in CNS lab test [1]

Figure 6 illustrated the normal stress against shear displacement (pile displacement) in the CNS shear test, while Figure 7 shown the corresponding results from the laboratory tests (D. Porcino et. al. ,2003).
It can be noticed that the normal stress in both tests were increased slightly. In the rough surface, the dense sand tends to dilate during the direct shear. Therefore the current normal stress on the servo wall was increased, as it is shown in Figure 7. And a higher stiffness will result in a bigger increase of normal stress.
However, it should be point out that the all the soil samples from Porcino's tests have the same initial normal stress, that's why the three samples in Figure 7 concentrated at the same point at the beginning of the test. In this paper, soil sample 1 and 2 have two different normal stresses due to their different depth, so they start to increase from two different points.

3.3 Link to the pile shaft resistance design

The lateral earth pressure method in API-2007 noted that the skin friction of the offshore piles is determined as

$$\tau_f = K\sigma_v' \tan\delta_f = \beta\sigma_v' \qquad (2)$$

Where,

K is the coefficient of lateral earth pressure ($k = \frac{\sigma_h'}{\sigma_v'}$), normally equals 0.8 or 1.0.
σ_v' is the vertical effective stress at any depth.
δ_f is the friction angle between the soil and pile shaft.
β is the shaft friction factor, which is equivalent to the "$K\tan\delta_f$" term.

According to the above numerical simulation results, the mobilized friction coefficient of the pile skin friction can be obtained as:

$$\tan\delta_f = \tau / \sigma_n \qquad (3)$$

Figure 8 illustrated the variations of the mobilized friction coefficient of two soil samples against shear displacement. The τ/σ_n ratio for two soil samples is 0.33 and 0.18 respectively. Therefore, the maximum mobilized friction angle for sample 1 is 18.3°, while for sample 2, friction angle is 10.2°. The recommended friction angle in API-2007 for medium or dense sand is around 25° to 30°, which is relatively higher than the results from the numerical simulations.

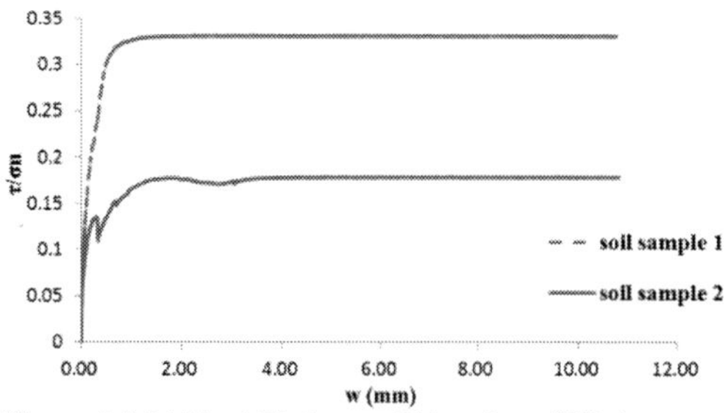

Figure 8. Mobilized friction coefficient from CNS shear test

Also, in Equation (2), if K equals to 0.8, the shaft friction factor β for sample 1 can be calculated as 0.27 and 0.14 for sample 2, which are also relatively smaller compared with recommended value (0.46 for dense sand) in API-2007.
The main reason of the shear friction angle and shaft friction factor β are both quite small, is likely because of the simulation of the sand particles is too ideal. In this paper, all the soil particles were simulated as round disks, which may reduce the roughness of the particles. Also, the influence of water could also be a possible reason, because the simulation of the soil sample is in a dry condition, but the design code is for the sand which is saturated.

4 Conclusions

This paper studied shaft resistance of offshore piles with discrete element method. The shaft resistance was obtained from a numeric simulation of the constant normal stiffness pile-soil interface direct shear test. The results from the tests were also analysed in details, and compared with offshore pile design codes. The following conclusions have been obtained from the tests.

1) A new method of simulating the constant normal stiffness shear test with 2D-DEM was proposed in this paper. Based on this new method, a virtual model of CNS interface shear test was created.
2) The soil sample interacted with the rough pile surface tend to dilate, which can increase the normal stress on the boundary (servo wall).
3) The initial stress condition has a big impact on the shear behaviour. The higher initial vertical stress, the higher shear stress and lower mobilized shear friction angle can be observed.
4) Although the shaft friction factor β obtained from the 2D-DEM model is relatively smaller than the recommended values from API-2007 codes, numeric simulations of CNS interface test with DEM can still be a feasible method to study the shaft friction resistance of offshore piles if some improvements of the model were made.

References

1. D. Porcino, V. Fioravante, V.N. Ghionna, S. Pedroni Geotechnical Testing Journal. **26**, 289-301 (2003)
2. L. H. Ooi, J. P. Carter Geotechnical Testing Journal, 10(1):10 (1987)
3. B. M. Lehane, D. J. White, Canadian Geotechnical Journal. **42**, 1039-1052 (2005)
4. V. Fioravante, Rivista Italiana Di Geotecnica. **3**, 7-22. (1999)
5. S. Liu, J. Wang, Geomechanics from Micro to Macro – Soga et al. (Eds), 455-458 (2015).
6. Recommended practice for planning, designing, and constructing fixed offshore platforms: working stress design. American Petroleum Institute. (2007)
7. Particle flow code 2D online manual. Itasca Consulting Group, Inc. Third Edition (2006).

Vertical slow drag of an intruder in a laterally confined granular medium

Agnès Duri*, *Sandra* Mandato, *Frédéric* Mabille, *Bernard* Cuq, and *Thierry* Ruiz

UMR IATE 1208 CIRAD/INRA/Montpellier SupAgro/Université Montpellier, 2 Place Pierre Viala, 34060 Montpellier Cedex 5, France

Abstract. The study of force transmission and typologies of motions in a powder bed under low shear solicitation constitutes a challenging issue to achieve a monitoring of most of processes. This work relies on the characterisation of particle mobility in the neighbourhood of a moving intruder in ascendant vertical motion. Vertical force measurements and Particle Image Velocimetry (PIV) technique are used to identify different typologies of particle motions (compaction, loading and rupture of the chain forces, lateral collapsing, avalanches...) of a granular bed ensiled in a semi-confined cell and with a free surface on its top. Force fluctuations during the blade displacement are highlighted and correlated with the dimensionless blade speed but also with the force profile in the static state. The velocity field analysis allows quantifying the shear-banding characteristics corresponding to the blade influence on grain displacements and the impact of the intruder in the motion direction. This signal contains periodical fluctuations which are correlated with the vertical force fluctuations in a specific zone of the granular bed.

1 Introduction

The description of stress and velocity fields induced by an object (or intruder) moving in a granular medium is of great interest in a wide range of industrial applications. For example, mechanisms involved in wet agglomeration are highly dependent on the knowledge of granular flows under the mechanical solicitation induced by the blade of the mixer. The ability of a mixer to agglomerate hydrated particles depends on its aptitude to induce a stress field in the powder bed in order to enhance the contact and adhesion between hydrated particles through capillary bridges. The study of stress transmission and typologies of motions in a powder bed under mechanical solicitation constitutes then a challenging issue to achieve a monitoring of process like agglomeration or mixing.

Experimentally, Particle Image Velocimetry (PIV) gives access to the powder behaviour by the analysis of velocity fields and particle trajectories. Different configurations have been studied: the vertical penetration of an intruder into a granular medium [1-3] or the horizontal motion of an intruder at a fixed depth in the medium, either entirely burried [4-6] or not [7-8]. A relevant result obtained by Albert et al. [7] is that the behaviour of the drag force in the low velocity regime is dominated by the inhomogeneous distribution of the stress in the static granular media.

This work relies on the analysis of the granular flow induced by an intruder in a semi-confined with a free-surface configuration. A rheophysical device and PIV technique are used to qualify the behaviour of granular materials. A cell with glass walls allows identifying particle motions in two dimensions. A rectangular blade is linked to a texture analyzer which allows an adjustable blade speed and the recording of the vertical force. Using a high-speed camera placed in front of the transparent glass of the cell, velocity fields are obtained by a PIV technique conducted during the blade motion. Different values of a characteristic speeds are tested by varying the blade speed from 1 to 40 mm.s^{-1}.

2 Materials and methods

2.1 Granular material

Experiments are conducted using durum wheat semolina of industrial quality as granular material. The main characteristics of semolina are: its dry basis water content of 16.8 ± 0.4 % measured by drying at 105°C for 24 h [9], its median diameter d_{50} of 287 ± 8 μm measured with a laser granulometer (Coulter TMLS 230, Malvern, England) and its true density of ρ_s = 1.478 ± 0.005 g.cm^{-3} determined with a nitrogen pycnometer (Ultrapycnometer 1000, Quantachrome Instruments).

2.2 Rheophysical device

The rheolphysical device (see Fig. 1), developed by Mandato *et al.* [10], is composed of an open glass cell container (160 mm height, 100 mm length, 51 mm width), a rectangular steel blade (0.97 mm thickness, 44.8 mm length, 19.7 mm width), screwed on a rod linked to a load cell of a texture analyzer (TA.XT plus, Table Micro System), which allows an ascendant vertical

* Corresponding author: agnes.duri@inra.fr

motion of the blade with an adjustable speed, and a recording of the vertical force in the powder bed.

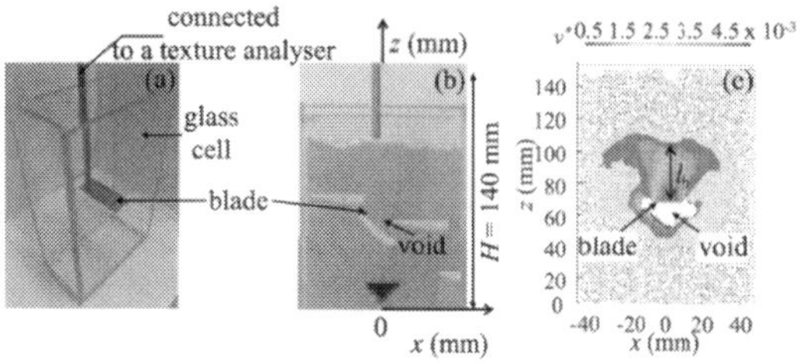

Fig. 1. (a) Picture of the experimental glass cell and the blade, (b) Picture of the granular bed, (c) Instantaneous velocity field image for a specific z position.

Force measurements, F_z, under dynamic conditions are recorded during the vertical rise of the blade in the cell. Before each test, the blade is first positioned at the bottom of the empty cell. Semolina is sprinkled with a funnel (outlet diameter of 12.3 mm) in the middle of the cell until it reaches a height H of 140 mm which corresponds to a compactness ϕ of 0.446. The blade is moved up for seven different vertical speeds v_b (1, 2.5, 5, 10, 20, 30, and 40 mm.s^{-1}) with three replications each. The blade speed is rescaled by the characteristic speed of grains in gravity, $v^* = v_b(gd_{50})^{-0.5}$ (g is the acceleration due to gravity). Thus, v^* ranges between 0 and 0.75. The lower v^*, the longer is the rearrangement time of the grains in the neighbourhood of the moving blade. The experiment at $v^* = 0$ corresponds to the static vertical force profile of the granular medium in the center (x=0 mm) of the bed.

The visualization of the semolina grain motion in the neighbourhood of the blade is carried out using a high-speed camera of type CMOS monochrome, with a recording frequency of 100 frames per second and image resolution of 1392 x 1024 pixels. In the case of granular flows, because the number of tracers is high, each 8-bit frame is divided into interest areas of 32 x 32 pixels, each with an overlap of 60. Velocity fields are obtained by PIV calculations performed by Direct PIV software (R&D Vision, France). From them, we have determined the blade vertical influence length, L_b, which is the vertical height of the zone above the blade, at x = 0 mm (*i.e.* in the middle of the cell) where the vertical velocity gradients are localized (Fig. 1c). Values of L_b are calculated as a function of the blade height during its vertical rise in the powder bed for each dimensionless speed.

3 Results and discussion

First of all, the static profile (v^* = 0) is different from the shape of the well-known Janssen profile [11-12]. From the surface, the local weight increases linearly with the depth until a maximum and then decreases until the bottom. In this experiment, as the height of the granular bed is less than the Janssen length (~ 260 mm), the level of the local vertical force does not attempt its saturation. However such a shape, with a decrease of the profile near the bottom, has already been highlighted experimentally by Mandato et al. [10] and numerically by Ovarlez and Clément [13]. This side effect is explained by the deflection of the vertical forces which takes place during the pouring of the cell and is similar with the well-known 'hole heap' [14].
When the blade rises from the bottom to the free surface, disturbances of the local vertical force profile are induced by its motion and the dissipative ability of the network. Transmission of the forces depends on the persistence of the subnetwork of large forces [15] during the blade motion, and this point is linked with the jamming state or the relative mobility of the particles in the bed. The non-uniformity of the initial force distribution in the granular packing and the difference in four different limit conditions between particles and blade, side walls, bottom and free surface, lead to a succession of phenomena. These latter ones can be identified from the measurements of the vertical force during the blade displacement for the different velocities (Fig. 2), and from the velocity fields obtained by PIV (Fig. 1c). With this experimental configuration, lateral shear banding is located in a thin zone at the vicinity of the blade. Considering that the lateral pressure of the shear zone is equal to the static pressure at the same depth and the shear rate approximated by the ratio of the blade speed and the thickness of the shear banding, the inertial number ranges between 10^{-4} and 10^{-3} [16]. The flowing regime in the sheared band laterally close to the blade is then found to be mostly quasi-static whatever the blade speed.

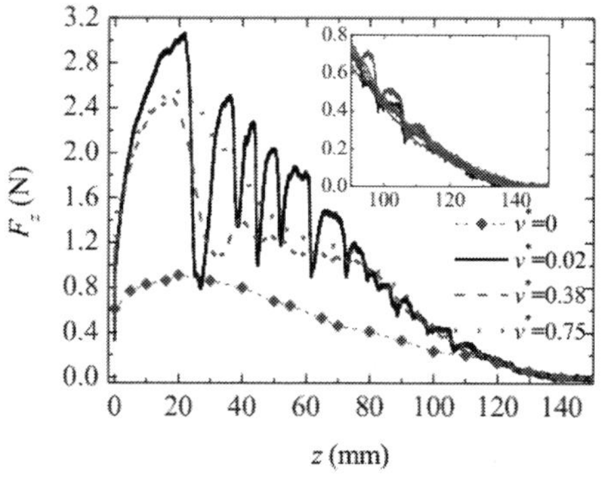

Fig. 2. Examples of F_z profiles obtained during ascendant blade displacement for v^*=0, 0.02, 0.38, 0.75. Inset: F_z profiles for all v^* in the relaxation phase.

For each solicitation, the shape of the force curves suggests that the dynamics of the granular medium can be subdivided from the bottom to the surface into the three following phases: (i) loading, (ii) fluctuations (loading/rupture), and (iii) final relaxation (Fig. 2).

Loading phase – At the bottom, the initial vertical force corresponds to the static value. During the rise of the blade, a first sharp increase of the force until a first maximum is observed (Fig. 2). At the first peak, the maximum vertical force value (mean value ≈ 2.59 ± 0.26 N) and its position (mean = 18.68 ± 2.41 mm) are quite uncorrelated with the dimensionless speed. However, the position of this peak corresponds to the position of the highest force of the static profile. During this loading phase, a rearrangement of the inter-granular voids above the blade is observed and corresponds to the creation of a void volume taking place just below the blade (Fig. 1b). The void volume comes from the compaction of the granular layers upon the blade. As the initial compactness of the powder bed (0.45) is lower than the critical compactness above which the granular medium dilates to allow particle rearrangements, it is relevant to assume that a compacted zone is formed just above the blade. During the formation of this compacted zone and all along the blade displacement, particles above the blade are evacuated by micro-avalanches in the free volume and form a concave heap (crater-like) on both sides of the blade. Whatever v^*, the height of the concave heap depends on the blade width (1.97 cm) and the angle of repose of semolina (43.8°). The formation of this hole takes place before the force reaches its first peak and corresponds to a small inflexion in the force profile.

Fluctuation phase – After the initial load followed by the break of the force network (decrease of the force just after the first peak), the level of force becomes equal to its static value (first minimum of the force profile), whatever is the blade speed. This result is interpreted as the weak extension of the zone in front of the blade which can be induced by the influence of the first intrusion in the granular media. Then periodic fluctuations are observed (Fig. 2). They correspond to successive phases of loading, compaction, breakage, and relaxation of the force network mobilized by the transmission of the shear stress. These fluctuations are different from the first load, but possess similar characteristics between them. Analysis of the intensities and the frequencies, developed at each cycle, highlights their periodicity which depends on the blade speed.

Loadings are linked with the force intensity of inter-granular vaults which act against the blade upward motion and deflect laterally the forces, and to the compaction of the bed zone located just at the top of the blade and up to the front of the disturbance zone vertically generated (Fig. 1c). When the applied force reaches the failure threshold of the force network, particles located in the rupture zones collapse and are then evacuated by avalanches in the void zone under the blade. In the same time, the force exerted on the blade is rapidly relaxed. After this discharge, the measured force increases progressively by the emergence of a new vertical contact network activated by the vertical motion (Fig. 2). This mechanism is periodic until it weakens entirely. Gravish et al. [8] also reported successions of compaction and breakage phenomena on experimental curves of force/displacement relative to a blade-like sensor moving horizontally at a constant speed in a bed of sand or glass beads. However, in this particular case, the constant weight of the displaced granular media ensures a regularity of the frequency.

Relaxation phase – When the blade enters in the highest part of the powder bed, vertical force progressively decreases until to reach the zero value (inset of Fig. 2). This relaxation is independent of the blade speed and can be explained by both a lack of global stiffness of the granular medium and the decrease of the mass weighing on the blade.

Vertical transmission of the force – The blade vertical influence length, L_b, is defined from the velocity field and represents the distance between the blade at $x = 0$ mm (*i.e.* in the middle of the cell) and the front of zone where particles are moving (Fig. 1c). Vertical velocity gradients are localized in this zone. Values of L_b are determined as a function of the blade height during its vertical rise in the powder bed and for each dimensionless speed (Fig. 3).

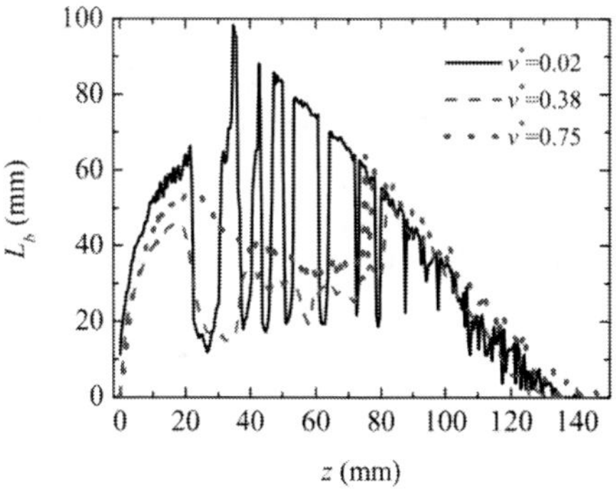

Fig. 3. Examples of L_b profiles obtained during ascendant blade displacement for v^*=0.02, 0.38, 0.75.

The vertical extension of the velocity gradient seems to be limited only by phenomenon of vaults rupture, which sets against the propagation of the vertical transmission of force. For each v^* and more particularly for $v^* < 0.2$, the blade vertical influence length fluctuates between two extrema (min and max). The persistence of a zone of grains in motion above the blade even after the rupture of the force network makes L_b always different from zero. A dihedral-shape assembly of grains constitutes a permanent disturbance zone induced by the blade motion. In this coherent zone, the horizontal components of velocity vectors are almost equal to zero and grains are in a jamming state The extension of this zone increases with v^*. It can be explained by a higher volume of particles dragged by the blade when v^* increases. On the contrary, the extension of the fluctuating zone above the permanent one decreases with v^*. At higher v^*, the characteristic time required for particle rearrangement is not sufficiently high relatively to the solicitation time. The force networks are alternatively recreated and removed and do not allow enough time to establish loading and rupture cycles of the vaults above the blade. The force transmission from the blade to the surface occurs gradually at the grain scale through a vertical contact network established above the blade. Thus, when the network is not persistent in a sufficiently long time comparatively to the characteristic time of solicitation, the transmission is limited in space. On the contrary, the lower is v^*, the higher is the distance over which the vertical contact network spreads between grains. The extension of the vertical transmission of force is then higher for lower v^*. Grains of the fluctuating zone are not anymore correlated with the blade which use the benefit of mobility to reach another arrangement.

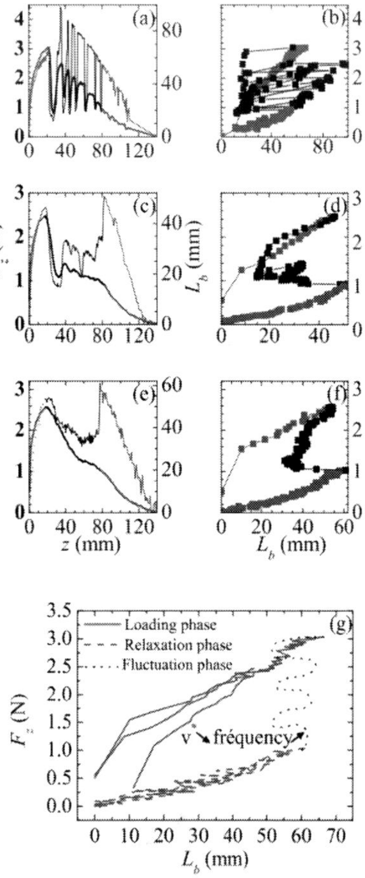

Fig. 4. (in color) F_z and L_b profiles obtained during ascendant blade displacement for v^*=0.02 (a), 0.38 (c), 0.75 (e). F_z as a function of L_b for v^*=0.02 (b), 0.38 (d), 0.75 (f). Focus on loading and relaxation phases for v^*=0.02, 0.38, 0.75 (g).

The coupled analysis of the dynamic and kinematic curves shows a marked correlation between the vertical force applied on the blade and the length influence of the blade with a small phase shift (Fig. 4). Loading and relaxation phases are independent of v^* (Fig. 4g). Fluctuation phase highlights a greater dependence with v^*, and it is amplified by slow solicitations as described just above.

In any case the force follows the decrease of L_b, which can be explained by the disjunction from the vertical contact network of a volume of grains which is weighing on the blade until then.

4 Conclusion

In this work, a better understanding of the different phenomena occurring in the neighbourhood of an intruder (*i.e.* flat blade) vertically moving in a laterally confined granular bed with a free-surface at its top is achieved. The PIV analysis allows identifying different types of particles motions (compaction, loading and rupture of the chain forces, lateral collapsing, avalanches, etc.). A detailed comparison between the length of the velocity gradient vertically generated above the blade (L_b) and the vertical force (F_z) perceived by the blade during its rise is carried out. It highlights periodical fluctuations of L_b which are correlated with F_z fluctuations in a specific zone between the bottom and a characteristic height. Such results contribute to give relevant data which will help in the understanding of granular mobility in a low shear mixer.

Acknowledgements

This research was financially supported by AgroFlow ANS project from CEPIA department of INRA.

References

1. Y.C. Zhou, A.B. Yu, R.L. Stewart, J. Bridgwater, Chem. Eng. Sci. **59**, 1343-1364 (2004)
2. F. Pacheco-Vázquez, J.C. Ruiz-Suárez, Nature Communications **1**, 123 (2010)
3. A. Seguin, Y. Bertho, P. Gondret, J. Crassous, Phys. Rev. Lett. **107**, 048001 (2011)
4. D. Chehata, R. Zenit, C.R. Wassgren, Phys. Fluids **15**, 1622 (2003)
5. Y. Ding, N. Gravish, D.I. Goldman, Phys. Rev. Lett. **106**, 028001 (2011)
6. F. Guillard, Y. Forterre, O. Pouliquen, Phys. Rev. Lett. **110**, 138303 (2013)
7. R. Albert, M.A. Pfeifer, A.L. Barabasi, P. Schiffer, Phys. Rev. Lett. **82**, 1, 205-208 (1999)
8. N. Gravish, P.B. Umbanhowar, D.I. Goldman, Phys. Rev. Lett. **105**, 128301 (2010)
9. American Association of Cereal Chemists, *Official Methods of the AACC*, Method 76-13, 10th edition, The Association, St. Paul, MN, USA, (2000)
10. S. Mandato, B. Cuq, T. Ruiz, Eur. Phys. J. E **35**, 56 (2012)
11. H.A. Janssen, Zeifschrift verein deutscher ingenieur **39**, 1045-1049 (1895).
12. M. Sperl, Granular Matter **8**, 59-65 (2006)
13. G. Ovarlez, E. Clément, Eur. Phys. J. E **16**, 421-438 (2005)
14. L. Vanel, D. Howell, D. Clark, R.P. Behringer, E. Clément, Phys. Rev. E **60**, 5, R5040 (1999)
15. F. Radjai, M. Jean, J.-J. Moreau, S. Roux, Phys. Rev. Lett. **77**, 2, 274-277 (1996)
16. G.D.R. MIDI, Eur. Phys. J. E **14**, 341-365, (2004)

 DOI: 10.1051/epjconf/201714006028

Towards reproducible experimental studies for non-convex polyhedral shaped particles

Daniel N. Wilke[1,*], *Patrick* Pizette[2,3,**], *Nicolin* Govender[2,3,4,***], and *Nor-Edine* Abriak[2,3,****]

[1] *Centre for Asset and Integrity Management, University of Pretoria, Pretoria, 0086, South Africa*
[2] *Mines-Douai, LGCgE GCE, F-59508 Douai, France*
[3] *Université de Lille, 59650 Lille, France*
[4] *Research Center Pharmaceutical Engineering, GmbH, Graz, Austria*

Abstract. The packing density and flat bottomed hopper discharge of non-convex polyhedral particles are investigated in a systematic experimental study. The motivation for this study is two-fold. Firstly, to establish an approach to deliver quality experimental particle packing data for non-convex polyhedral particles that can be used for characterization and validation purposes of discrete element codes. Secondly, to make the reproducibility of experimental setups as convenient and readily available as possible using affordable and accessible technology. The primary technology for this study is fused deposition modeling used to 3D print polylactic acid (PLA) particles using readily available 3D printer technology. A total of 8000 biodegradable particles were printed, 1000 white particles and 1000 black particles for each of the four particle types considered in this study. Reproducibility is one benefit of using fused deposition modeling to print particles, but an extremely important additional benefit is that specific particle properties can be explicitly controlled. As an example in this study the volume fraction of each particle can be controlled i.e. the effective particle density can be adjusted. In this study the particle volumes reduces drastically as the non-convexity is increased, however all printed white particles in this study have the same mass within 2% of each other.

1 Introduction

The two pillars on which science has been build is theory and experimentation [1]. Mathematical modeling initially aims to describe experimentally observed phenomena to ultimately predict the physical response of a system without requiring experimental observation, once it has been properly characterized and validated [2]. Mathematical models are often required to be characterized by conducting inverse analysis. Characterization aims to identify the essential information required to appropriately identify the respective mechanisms to be characterized in a mathematical model, although this is not always fully understood or appreciated. Having achieved this would allow for a predictive model over the domain of responses for which the respective mechanisms have been properly identified.

The availability of experimental data is critical for modelers to develop new models and also to characterize existing models. Numerous empirical studies have been conducted over the years [3], both at the lab scale and industrial scale. Experimental discrete element studies that have proven to be valuable include angle of repose investigations [4], particle shear interactions [5], hopper discharge [6], ball mill [7, 8] and dynamic impulse applications [9]. Numerous industrial applications have been studied in detail, but a common thread runs through all these studies, regarding the diversity of particle shapes that are considered. In these studies, particle shapes are almost exclusively convex with angularity, density, aspect ratio, moisture content etc. mainly investigated, however, non-convexity of particles remains largely elusive in experimental studies.

In this study we consider the random packing of non-convex polyhedral shaped particles as well as the hopper discharge of non-convex polyhedral shaped particles. However, to conduct a study in a laboratory is merely one part of the challenge. To allow for experimental studies that are reproducible when the non-convexity of particles are considered is an additional challenge but an important one for our scientific community. Hence, the goal of this study is to investigate the effect of the degree of non-convexity of particle shapes on the random packings in rectangular containers as well as the hopper discharge for flat bottomed hoppers such that it is completely reproducible by someone else, somewhere else, in the world.

Towards our first goal we consider particles that scale naturally between convex and varying degrees of non-convex shapes. We rely on carefully selected particles that can scale between convex and non-convex shapes using compact parameterizations.

* e-mail: nico.wilke@up.ac.za
** e-mail: patrick.pizette@mines-douai.fr
*** e-mail: govender.nicolin@gmail.com
**** e-mail: nor-edine.abriak@mines-douai.fr

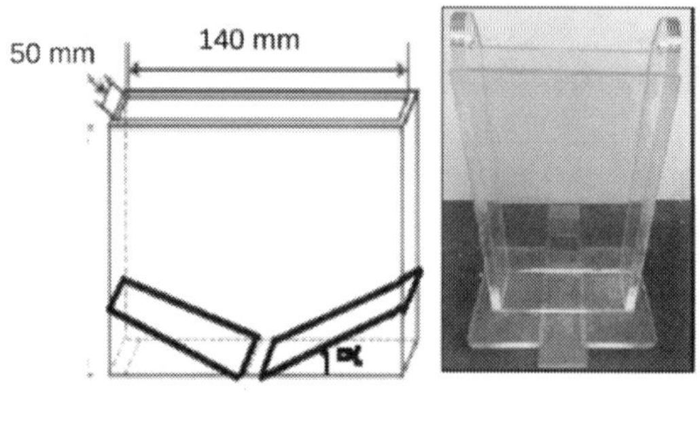

Figure 1. Schematic of the designed hoppers, with a width of 140mm and a depth of 50mm, while the hopper opening is 50mm. From left to right are three manufactured hoppers, flat bottomed hopper ($\alpha = 0$) used in this study as well as, $\alpha = 30^\circ$ and $\alpha = 60^\circ$ degree hoppers.

Figure 2. Makerbot Replicator utilized to print 153 (9x17) particles per batch using 100 micron layer resolution.

2 Reproducible Experiments

The plans and materials used for the container are given in detail and limited to standard components and materials that can be sourced internationally. The raw experimental data for each experiment is made available to allow for the extraction and processing of additional data. More importantly, this allows for comparison to other researchers' experimental results. Ideally, other researchers would make their experimental results available to improve the statistical significance of the experimental data on a large number of experiments for a variety of investigations. In addition, this may inspire numerous educational opportunities to make granular flow available to wider audiences at younger ages.

The hopper considered in this study is depicted by a schematic drawing and the actual manufactured hopper from Plexiglas[1] in Figure 1. The discharge width is not indicated in the drawing but is 50 mm. The interior of the hopper coincides with the smooth finish of the Plexiglass. The components were CNC manufactured and manually assembled at the Fabrication Laboratory (FabLab) at École des Mines de Douai, France.

3 Convex to Non-convex Particles

Towards our second goal of simplifying the replication of our experimental setup we base all particles on the ubiquitous fused deposition modeling (FDM) manufacturing approach [10]. All particles are 3D printed using standard white and black polylactic acid (PLA) [11]. In addition, all printed particles are made available as STereoLithography (STL) files for researchers to download and print on their own 3D printers. In addition the 3D printer settings are also made available to enhance reproducibility after a standard calibration has been performed. The particles are printed in a batch that is dependent on the size of the printer as depicted in Figure 2. The four printed particle types are depicted in Figure 3 in addition to sample STL objects of the particles.

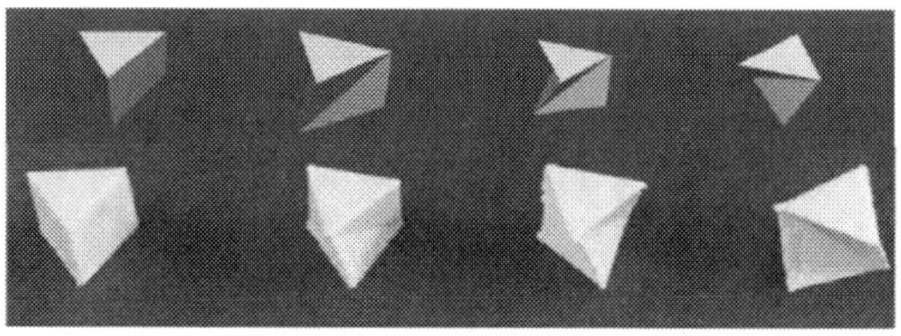

Figure 3. Four polyhedra types used in this study, starting with the triangular prism (0° twist) at the most left of the image. Three Schönhardt polyhedra are formed, from left to right, by applying 15°, 30° and 45° twist respectively to the triangular prism. The top row indicates the associated rendered particles while the bottom row depicts the actual printed particles using white PLA.

Before we proceed, we need to establish terminology of what we mean by varying between convex and degrees of non-convex shapes. Hence, we need a way to quantify the degree of non-convexity of a particle to allow us to quantify this concept appropriately. An example of a particle that scales naturally between convex and non-convex shapes is the Schönhardt polyhedron [12] and the variations thereof considered in this study. The Schön-

Table 1. Particle properties for the 0o,15o,30o and 45o twisted triangular prism and the mass of the white and black particles.

Particle Type	0o	15o	30o	45o
Print VF	20%	28%	36%	100%
Number Shells	1	1	1	1
Mass W kg/k (Min)	0.3353	0.3352	0.3351	0.3351
Mass W kg/k (Max)	0.3354	0.3354	0.3353	0.3351
Mass B kg/k (Min)	0.3630	0.3646	0.3648	0.3643
Mass B kg/k (Max)	0.3654	0.3652	0.3654	0.3655

1

- Density: 1.18 g/cm^3 at the standard state of 25 °C and 100 kPa,
- CAS Number: 9011-14-7
- IUPAC ID: Poly(methyl 2-methylpropenoate).

Figure 4. Schönhardt polyhedron is generated by twisting a triangular prism (green), while the edges of the triangles on opposite sides of the prism are connected to form edges that form non-convex particles. Indicated are 15° (purple), 30° (orange) and 45° (red) twist of the one triangular base relative to the other.

hardt polyhedron is well known in geometrical topology as the simplest polyhedron that cannot be tessellated without adding an additional vertex. In this study, we propose the twist angle as a single scalar measure for the degree of non-convexity of a Schönhardt polyhedron particle. The twist angle scales between 0 (convex) and 45 (most extreme non-convex particle). Needless to say, a single scalar measure to quantify the extent of non-convexity of a single particle is inherently limited. However, when considering the convexity to non-convexity measure for the same particle that has the same size and topology, that includes texture, sub-structure and global structure, it suffice for the purposes of our study. In actual fact it works surprisingly well as a quantitative measure.

The Schönhardt polyhedron consists of six vertices that are connected to form eight surfaces that encapsulates a non-convex volume. The initial shape when no twist is applied is a convex triangular prism as depicted, by the green particle, in Figure 4. As the twist of three vertices that form a triangle is applied relative to the other triangular surface, a non-convex shape results. Hence, the twist angle is the mechanism to scale the same topological particle between a convex particle and various non-convex particles, while keeping the aspect ratio the same. In Figure 4, three twist angles of 15°, 30° and 45° are depicted in purple, orange and red respectively. To ease the depiction of the particles the height is reduced, although the height of the printed particles remained the same.

The volume of the particles changes significantly as the twist angle is adjusted implying significant variations in particles mass between particle types. Fortunately, FDM allows for the density of the material to be controlled by changing the volume fraction of the printed material such that the particle mass only has a 2% variation between particle types for white and black particles respectively. This is a significant benefit of FDM in that the effective density of the particle can be changed to suit the investigation at hand. The properties for the four particle types are tabulated in table 1, and include the print volume fraction (VF), number of shells as well as the maximum and minimum masses measured for the white (W) and black (B) particles for five independent measurements. Note, the mass difference of around 10% by merely changing the color of the PLA filament. Although, not done in this study the VF for the white and black particles could be independently determined to ensure that the particle mass remains the same between particle colors. All particles were printed with a layer resolution of 100 micron, although the surface roughness is expected to remain unaffected, experimental validation of this is still required. Surface metrology of the printed particles will form part of a future investigation. Quantification of the effect of volume fraction on the experimental inertial properties against idealized inertial properties in under investigation and will be included in a future study.

Table 2. Maximum height (cm) of the random packing of the 0°, 15°, 30° and 45° twisted Schönhardt polyhedron.

Packing	0o	15o	30o	45o
1	12.4	12.2	12.7	12.9
2	12.3	12.2	12.6	12.9
3	12.5	12.3	12.7	13.0
4	12.5	12.4	12.7	12.8
5	12.5	12.3	12.6	12.9
Mean (cm)	12.4	12.3	12.7	12.9
Pack density (g/cm^3)	0.39	0.39	0.38	0.37

Documentation and standardization of the processes is being developed to allow for collaborative extension and reproduction of experimental investigations considering FDM within the DEM community. An online repository is under construction and will be communicated at the conference.

4 Experimental Results

A 1000 particles were randomly packed inside the hopper. The packing was repeated 10 times and the estimated height measured tabulated in table 2 using 5 samples. The effective density is computed for each random packing and it is clear that the 45o has the densest packing while the 30o particle has the least dense packing. In addition, the hopper discharge for the four particle types were conducted 15 times, while the mass as a function of time was tracked. The discharge rate graphs were then computed as well as the probability of stable arches forming for the various particle systems. The discharge at 0s, 0.8s, 1.6s and 2.4s are depicted in Figure 5 for the four particle types using a 1000 particles that are layered into 200 particles per layer.

5 Conclusion

Towards reproducible granular flow experiments we demonstrated that fused deposition modeling (FDM) is a meaningful technology, in particular when complex particle shapes are considered as with non-convex particles. Fused deposition modeling also allows for particle properties to be varied in isolation without implying an implicit

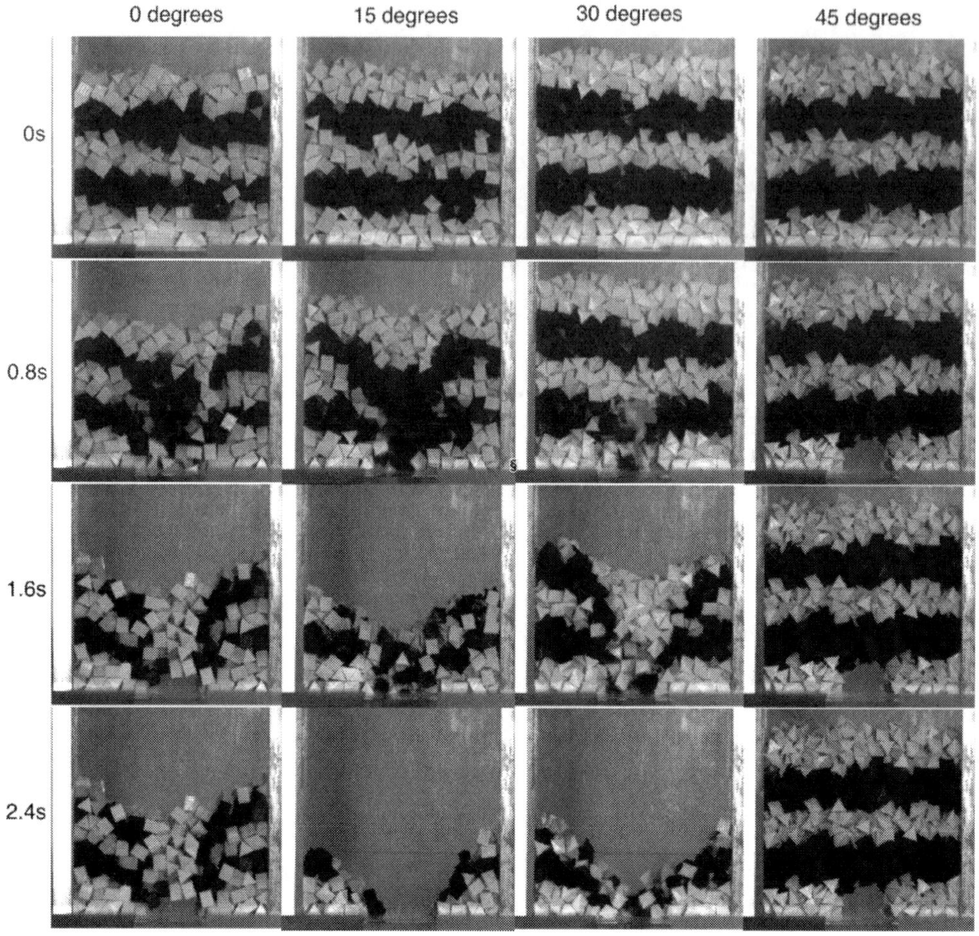

Figure 5. Flat bottomed hopper discharge for triangular prisms with 0°,15°,30° and 45° twist, depicted along the columns of the picture array, at 0s, 0.8s, 1.6s and 2.4s, depicted along the rows of the array of pictures.

variation in other properties. As a tangible and simple example we considered the discharge of a hopper. FDM allows for constant mass particles to be manufactured even when the volume changes significantly between particle types. Therefore FDM allows for changes in particle geometry to be isolated without implying an implicit variation in other particle properties, as for example particle mass, in our study.

Acknowledgments

The due diligence of the staff at the FabLab in École des mines de Douai, France is sincerely acknowledged. Dr. Pizette and LGCgE GCE, Mines Douai for providing all experimental data documented in this paper.

References

[1] M. Vardi, Communications of the ACM **53**, 5 (2010)
[2] ASME/NAFEMS pp. 1 – 4 (2006)
[3] K. Wieghardt, Annual Review of Fluid Mechanics **7**, 89 (1975)
[4] Y. Li, Y. Xu, C. Thornton, Powder Technology **160**, 219 (2005)
[5] C. Coetzee, Powder Technology **297**, 50 (2016)
[6] N. Standish, Powder Technology **45**, 43 (1985)
[7] P.W. Cleary, M.L. Sawley, Applied Mathematical Modelling **26**, 89 (2002)
[8] M. Kwapinska, G. Saage, E. Tsotsas, Powder Technology **161**, 69 (2006)
[9] E. Rougier, E.E. Knight, S.T. Broome, A.J. Sussman, A. Munjiza, International Journal of Rock Mechanics and Mining Sciences **70**, 101 (2014)
[10] S.S. Crump, *Apparatus and Method for Creating Three-Dimensional Objects* (1992)
[11] B. Wittbrodt, J. Pearce, Additive Manufacturing **8**, 110 (2015)
[12] E. Schönhardt, Math. Annalen **98**, 309 (1928)

EPJ Web of Conferences 140, 15029 (2017) DOI: 10.1051/epjconf/201714015029
Powders & Grains 2017

An experimental, theoretical and event-driven computational study of narrow vibrofluidised granular materials

Anthony Thornton[1,*], *Kit* Windows-Yule[1,2], *David* Parker[2], and *Stefan* Luding[1]

[1]*Multiscale Mechanics, Department of Fluid and Thermal Engineering, University of Twente, The Netherlands*
[2]*School of physics and astronomy, University of Birmingham, UK*

Abstract.
We review simulations, experiments and a theoretical treatment of vertically vibrated granular media. The systems considered are confined in narrow quasi-two-dimensional and quasi-one-dimensional (column) geometries, where the vertical extension of the container is much larger than one or both horizontal lengths. The additional geometric constraint present in the column setup frustrates the convection state that is normally observed in wider geometries.
We start by showing that the Event Driven (ED) simulation method is able to accurately reproduce the previously experimentally determined phase-diagram for vibrofludised granular materials. We then review two papers that used ED simulations to study narrow quasi-one-dimensional systems revealing a new phenomenon: collective oscillations of the grains with a characteristic frequency that is much lower than the frequency of energy injection. Theoretical work was then undertaken that is able to accurately predict the frequency of such an oscillation and Positron Emission Particle Tracking (PEPT) experiments were undertaken to provide the first experimental evidence of this new phenomenon.
Finally, we briefly discuss ongoing work to create an open-source version of this ED via its integration in the existing open-source package *MercuryDPM* (http://MercuryDPM.org); which has many advanced features that are not found in other codes.

1 Introduction

Granular materials, large collections of discrete, macroscopic particles, are very common in industry, the natural environment and our everyday lives [1]. Here we focus on vibrofluidised granular materials, where the material is 'fluidised' by a rapidly moving, normally high frequency low amplitude, base-plate that inputs energy into the system. The vibrofluidised system presents a wide range of behaviours, many of which are analogs of behaviours in molecular systems: phase separation [2, 3], Faraday-like pattern formation instabilities [4, 5], heap formation and convection [6, 7], segregation [8, 9], clustering [10] and periodic cluster expansions [11], etc.

The geometry of the system is known to have a large influence on the dynamics that can be observed within this vibrofluidised system. Here we investigate the effect of changing the system's lateral size and the transition from quasi-two-dimensional to quasi-one-dimensional systems, see Figure 1.

The rest of the paper is laid out as follows: in section 2 we start by introducing the Event Driven numerical methods and, in section 3, show it can be used to accurate reproduce all the known states of a vibrofluidised system. In section 4 we consider the quasi-one-dimensional system, introducing the new state: the low frequency oscillation (LFO) and (section 5) present both a theory that actually predicts the frequency and positron emission experiments that showed the first experimental evidence of such a state. Finally, in section 6 we draw conclusions and discuss on-

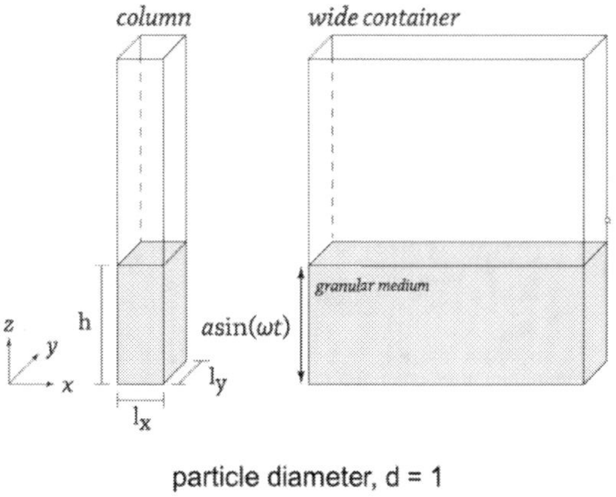

Figure 1. Setup of the geometries that are considered

[*]e-mail: a.r.thornton@utwente.nl

 DOI: 10.1051/epjconf/201714015029

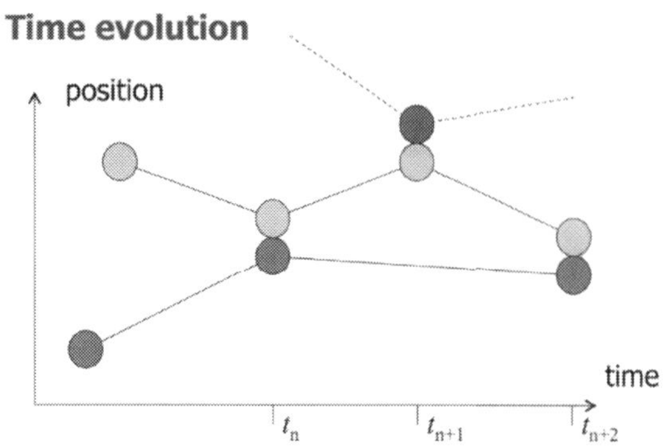

Figure 2. Schematic of how time is advanced in an event driven simulation. From one event n to the next, $n + 1$, and so on...

going work to integrate the ED method with the flexible open-source particle solver *MercuryDPM*.

2 Event driven simulation of vibrofluidised systems

In this paper, particle simulations are undertaken using the event driven method [12, 13]. In this approach, particles move freely under the effect of gravity until an event takes place, i.e. a collision with another particle or a wall. The motion of the particles in between successive events does not have to be simulated, the time is advanced in variable steps, per particle, from one event to the other, see Figure 2. This makes event-driven simulations faster than usual time-stepping simulations, where time is advanced in constant steps. However, the need for an analytical expression for the particle motion is a strong condition that limits the possible interactions between particles. In the following, we consider the most common approach: perfect hard-spheres, which imply binary collisions and no overlap or long-range forces between particles. In brief the ED algorithm works as follows:

1. For each particle compute its trajectory via
$$\boldsymbol{x}_i(t) = \boldsymbol{x}(t_0) + \boldsymbol{v}_i(t - t_0) + 1/2\boldsymbol{g}(t - t_0)^2$$
and find smallest $t = t_c$ for which $\exists i, j : |\vec{x}_i(t) - \vec{x}_j(t)| = r_i + r_j$.
2. Advance particle i and j to t_c
3. Apply collision rule to update $\boldsymbol{v}_i$ and $\boldsymbol{v}_j$.
4. Go back to step 1.

For the considered vibrofluidised system, the whole box (both the bottom and the side walls) is vertically vibrated in a bi-parabolic, quasi-sinusoidal way with a given frequency, f and amplitude a. The use of a quadratic instead of a sine function gives a considerable speed advantage in simulations, as the prediction of collision times with the moving walls becomes analytical and thus substantially faster. Full details of the simulations properties used for this paper can be found in [14].

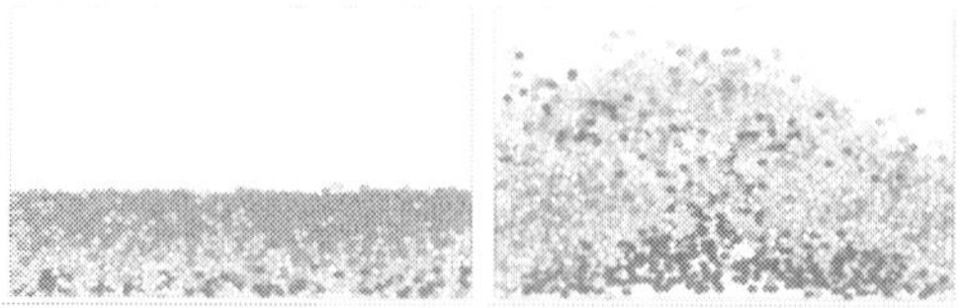

Figure 3. Snapshots of (left) Leidenfrost and (right) convection states from ED simulations. The colour indicates the fluctuating kinetic energy (or granular temperature) of the particles, with red high and blue low.

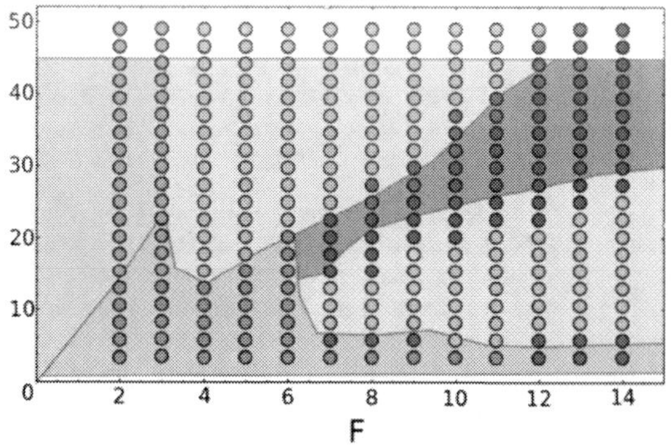

Figure 4. Determination of the states in the vibrofluidised system. States are in blue bouncing bed, yellow undulations. red Leidenfrost, green convection. Dots are from the ED simulations and the background is coloured with data from experiments. The grey circles are transition states.

3 Phase space from ED

Vibrofluidised systems are known to exhibit many different states, these include: bouncing bed, bursts, undulations, Leidenfrost, convection and gases [17]. Two of these states as produced in our ED simulations are shown in Figure 3. A phase-space can be constructed demarking the location of these states in terms of the number of filling layers, i.e. the number of particles per unit cross section (parallel to the vibrated base) and either the dimensionless acceleration $\Gamma = A\omega^2/g$ or the dimensionless shaking strength $S = a^2\omega^2/gd$, where a is the amplitude of vibrating base, ω the angular frequency, g the acceleration due to gravity and d is the dimeter of the particles.

Figure 3 shows a snapshot of a simulation in the Leidenfrost state. This state was first observed in simulations by [15] and takes its name from analogy with the fluid phenomena of the same name in which liquid is suspended over vapour [16]. As indicated by the colours in Figure 3 the granular equivalent is the same with high temperature particles supporting lower temperature particles. As the frequency of the vibrating wall is increased this state transits in to a convection state. The right panel of Figure 3 shows a snapshot of this state where hot particles are rising in the centre and cooling/sinking near the walls.

Figure 4 shows the full phase-space of states as determined by the current ED code compared with the experi-

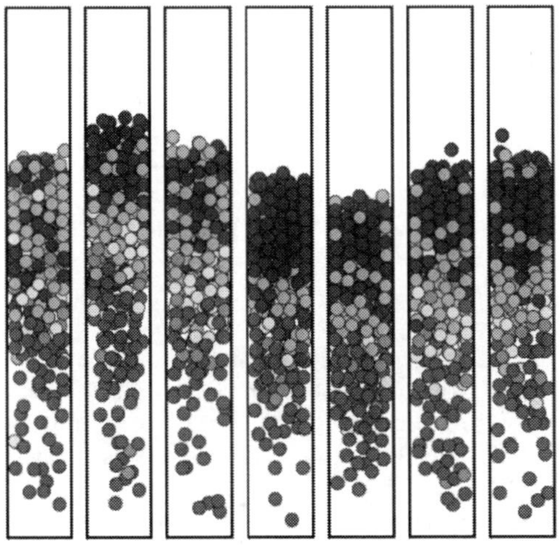

Figure 5. Snapshots of ED simulations of the narrow column. Time progresses from left to right. The time evolution clearly shows an oscillation of the upper surface.

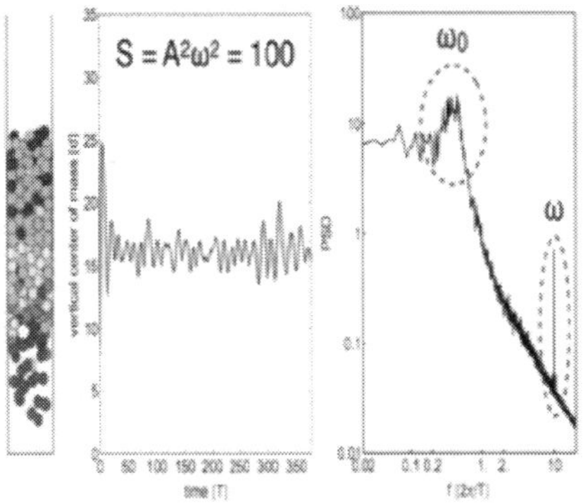

Figure 6. Left: Snapshots of ED simulations of the narrow column. Middle: Centre of mass of the column against time. Right: Fourier transform of the centre of mass. ω_0 is the characteristic low frequency oscillation and ω is the driving frequency on the vibrating plate.

mental and soft particle simulations of [17]. There is excellent quantitative agreement and this validates the use of the event driven method for studying vibrofluidised granular materials. It should be noted that there are a few differences; however, the error is small and could also be in the original data as some of these transitions are hard to exactly determine and experiments can have side effects that can be avoided in simulations. The ED method is several orders of magnitude faster than the soft particles method, in this regime, and gives a lot more information, e.g. granular temperature, than experiments. Therefore this adds a valuable new tool for investigating vibrofluidised granular materials. To emphasis how fast this ED tool is, it should be noted that **all** the ED data points for Figure 4 were generated on a normal desktop computer within a few hours.

4 Low frequency oscillations (LFO)

Using the new ED tool we now turn our attention to narrow column geometry as illustrated by the left schematic in Figure 1. As shown in [14] below a container size of around 10 particle diameters the convection state, illustrated by the right panel of Figure 3, is no longer possible. Therefore in this narrow column geometry it is possible to generate Leindenfrost states at very high accelerations (high Γ values) that in wider geometries would have broken into convection, as illustrated in Figure 3.

We next discuss the results of Event Driven (ED) simulations of this frustrated system, that where first reported in [14]. The confined geometry revealed a previous unreported collective oscillations of the grains with a characteristic frequency that is much lower than the frequency of energy injection. A time-series from these frustrated systems is shown in Figure 5 and clearly shows that the top of the gaseous region is not constant but is oscillating in time.

Careful analysis of the centre-of-mass of the narrow column system revealed a characteristic frequency in these oscillations that is much lower than the driving frequency: the LFO. Figure 6 shows a typical time-trace of the centre-of-mass of the corresponding Fourier transform, you can clearly see a peak at a considerably lower frequency than the driving frequency of the plate: the low-frequency oscillations (LFOs). In [14] the frequency and amplitude of these oscillations was studied as a function of the energy input parameters and the size of the container. It was later observed that, in the quasi-two-dimensional setup, LFOs are present even in the convective regime. This suggests that they may play a significant role in the transition from a density inverted state to convection. This transition and the role LFOs play in it was further studied by Rivas *et. al.* [18] and the fluctuations due to the finite number of particles.

5 Comparison with theory and experiments

In [14] two theoretical models are also presented; the first one, based on Cauchy's equations, is able to predict with high accuracy the frequency of the particles collective motion. The model shows that a sufficient condition for the existence of the low-frequency mode is an inverted density profile with distinct low and high density regions, a condition that may apply to other systems too. The second, simpler model just assumes an harmonic oscillator like behaviour and, using thermodynamic arguments, is also able to reproduce the observed frequencies with high accuracy. For the details of both models please see [14].

In [20] Positron Emission Particle Tracking (PEPT) was used to experimentally observe and study the novel LFOs. PEPT is a noninvasive technique whereby a single particle, physically identical to the others in the system, is labeled with a beta-plus-emitting isotope. The back-to-back pairs of gamma-rays emitted due to the rapid annihi-

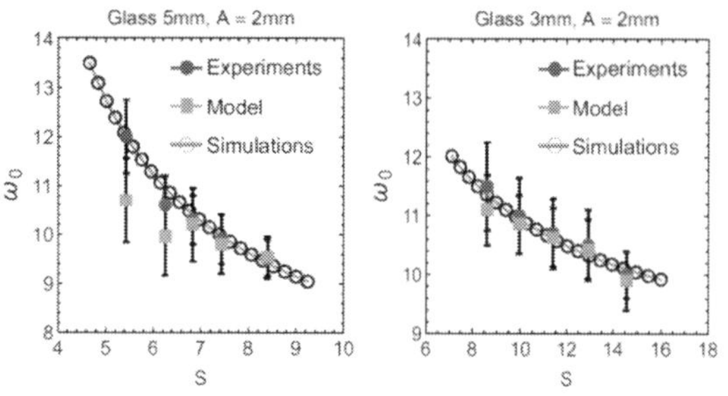

Figure 7. Experimental and theoretical values of ω_o as a function of the shaking strength $S = a^2\omega^2/gd$. Data for d = 3 mm and d = 5 mm glass particles, for A = 2 mm. Originally published in [19] and reproduced wth permission.

lation of positrons with electrons within this tracer particle are detected using a dual-headed gamma-camera, and can be used to triangulate its position with millimetre precision and millisecond time resolution. Figure 7 shows the comparison between the model, experiments and the simulations across a range of shaking strengths S for two differently sized glass particles. Where possible the known particle values were used in the simulations and the theoretical work. The experimentally acquired results show a close qualitative and quantitative agreement with both theory and simulations across the range of parameters tested. The agreement between all three (experiments, simulations, and theory) is always within 10% and improves as the value of the shaking strength S is increased. This makes sense as the model is only formally valid in the high shaking strength, S, limit and in this limit the agreement is very good. This was also tested for other materials and parameters and the agreement was always found to be good, illustrating that the LFOs are a robust, reproducible phenomena.

6 Conclusions

We have shown that event driven simulations are a valid tool for investigating vibrofluidised systems. The ED tool is so quick it is able to simulate thousands of particles in real time meaning you are able to interactively try new ideas. This is how the phenomenon of the low frequency oscillations was discovered in the first place. This highlights how new developments in computational tools can lead to new physical understanding that would have been hard without the rapid interactive feedback. The tool described in this paper has been used so far in various publications; for example[14, 18–20]. Work is underway to integrate this rapid ED tool with the open-source particle solver MercuryDPM (http://MercuryDPM.org) [21, 22]; which has many advanced features that are not found in other codes that could be integrated with this rapid event-driven code.

Acknowledgements

This work was financially supported by the NWO-STW Vidi Grant No. 13472 and NWO-STW Vici Grant No. 10828.

References

[1] H.M. Jaeger, S.R. Nagel, R.P. Behringer, Physics Today **49** (1996)

[2] J.S. Olafsen, J.S. Urbach, Phys. Rev. Lett. **81**, 4369 (1998)

[3] P. Melby, F.V. Reyes, A. Prevost, R. Robertson, P. Kumar, D.A. Egolf, J.S. Urbach, Journal of Physics: Condensed Matter **17**, S2689 (2005)

[4] F. Melo, P. Umbanhowar, H.L. Swinney, Phys. Rev. Lett. **72**, 172 (1994)

[5] S. Luding, E. Clement, J. Rajchenbach, J. Duran, EPL (Europhysics Letters) **36**, 247 (1996)

[6] S.G.K. Tennakoon, R.P. Behringer, Phys. Rev. Lett. **81**, 794 (1998)

[7] M. Medved, Phys. Rev. E **65**, 021305 (2002)

[8] K. Ahmad, I.J. Smalley, Powder Technol. p. 69-75 (1973)

[9] A. Kudrolli, Reports on Progress in Physics **67**, 209 (2004)

[10] S. Luding, H.J. Herrmann, Chaos **9** (1999)

[11] N. Rivas, S. Ponce, B. Gallet, D. Risso, R. Soto, P. Cordero, N. Mujica, Phys. Rev. Lett. **106**, 088001 (2011)

[12] B.D. Lubachevsky, Journal of Comp. Phys. **94** (1991)

[13] S. Luding, S. McNamara, Granular Matter **1** (1998)

[14] N. Rivas, S. Luding, A.R. Thornton, New J. Phys. **15**, 113043 (2014)

[15] B. Meerson, T. Pöschel, Y. Bromberg, Phys. Rev. Lett. **91** (2003)

[16] J.G. Leidenfrost, International Journal of Heat and Mass Transfer **9**, 1153 (1966)

[17] P. Eshuis, K. Weele, van der, D. Meer, van der, R. Bos, D. Lohse, Phys. Fluids **19** (2007)

[18] N. Rivas, A.R. Thornton, S. Luding, D. van der Meer, Phys. Rev. E **91**, 042202 (2015)

[19] C.R.K. Windows-Yule, N. Rivas, D.J. Parker, A.R. Thornton, Phys. Rev E **90(6)** (2014)

[20] C.R.K. Windows-Yule, N. Rivas, D.J. Parker, Phys. Rev. Lett. **111**, 038001 (2013)

[21] A.R. Thornton, D. Krijgsman, A. te Voortwis, V. Ogarko, S. Luding, R. Fransen, S. Gonzalez, O. Bokhove, O. Imole, T. Weinhart, *A review of recent work on the Discrete Particle Method at the University of Twente : An introduction to the open-source package MercuryDPM*, in *DEM 6: Conf. Proc.* (2013), pp. 393–399

[22] A.R. Thornton, D. Krijgsman, R. Fransen, S. Gonzalez, D. Tunuguntla, A. te Voortwis, S. Luding, O. Bokhove, T. Weinhart, Newsletter EnginSoft **10**, 48 (2013)

Numerical simulation of the compaction of crushable grains in 3D

David Cantor[1,2,⋆], *Emilien* Azéma[1], *Philippe* Sornay[2], and *Farhang* Radjai[1,3]

[1] *Laboratoire de Mécanique et Génie Civil (LMGC), Université de Montpellier, CNRS, Montpellier, France*
[2] *CEA, DEN, DEC, SFER, LCU, 13108, Saint Paul lez Durance, France*
[3] *Multi-Scale Materials Science for Energy and Environment <MSE>^{2}, UMI 3466 CNRS-MIT Energy Initiative, Massachusetts Institute of Technology, 77 Massachusetts Avenue, Cambrid02139, USA*

Abstract. Grain fragmentation is simulated by means of a three-dimensional discrete element approach called bonded-cell method (BCM). In this method, grains and potential fragments may have any polyhedral shape and size, capturing the geometrical complexity of brittle grain failure. As an application of this method, we present the uniaxial compaction of samples composed of several grains and we analyse the load-density relations, the grain size evolution, and the failure mechanism within the grains. This numerical approach permitted us to analyse the effect of the grains internal strength on the macroscopic compaction behaviour and to study the evolution of the grain size distribution towards a power-law distribution as several experiments have shown in literature. Finally, we present a brief micro-mechanical analysis on the failure modes within the grains, letting us know the kind of stresses that prompts grain fragmentation.

1 Introduction

Fragmentation of brittle grains has major effects on the compressive and shear strength of granular materials [1, 2], packing fraction [10], the yielding surface [11], grain size distribution [8], etc. Most of these observations have been conducted or analysed through experimental or empirical approaches and, together with the concepts of fracture mechanics, they provide a remarkable insight into the behaviour of breakable grains.

Nevertheless, the simulation of grain fragmentation remains still a challenge. Several models have been proposed to deal with fragmentation in the framework of continuum or discrete element methods (some examples can be found in Ref. [4] and Refs. [5, 15], respectively). These numerical methods contain, the most of the times, two main ingredients: the failure criterion and the geometry evolution. They define the stress level to split a grain apart and the shape of fragments. Due to the rapid evolution of grain geometry and grain size distribution during any fragmentation event, the computational cost of these models may be very important. Then, very simplified models have been, up to now, the way to tackle this phenomenon. For instance, the clustering of discs or spheres, in 2D and 3D simulations respectively, are the most common techniques to deal with grain failure. However, the shape complexity of grains and fragments is needed to properly map the macroscopic response from the grain scale interactions. For this reason, and inspired by the two-dimensional strategy of Nguyen et al. [12], we have introduced a three-dimensional discrete element approach to simulate fragmentation [3]. In this model, each grain is composed of a set of potential fragments, also known as cells, bonded with a couple of mechanical constrains or *bonding strengths*. When the intercell forces reach such strengths, the grains are allowed to fail in arbitrary shapes as their interactions and external loads may determine it.

This article is organized as follows: In Section 2, we briefly present the numerical model for grain fragmentation and the numerical procedure to perform uniaxial compressions on assemblies of grains. In Section 3, we present some results on the behaviour of such assemblies concerning the load-density relations, the grain size evolution, and the failure mechanism within the grains. Finally, we discuss the current results and potential perspectives of this work.

2 Numerical experiments

The discrete element approach that we use to simulate grain fragmentation is known as bonded-cell method (BCM) [3, 12]. As an application of the BCM, we performed uniaxial compression tests on assemblies of crushable grains. This kind of test was chosen given its simplicity and its well-known macroscopic response. In the two following Subsections, we briefly present the BCM-3D and the numerical procedures to perform the compression tests.

2.1 Bonded-Cell Method in 3D

This numerical method is based on the subdivision of grains in a set of potential fragments or cells. These frag-

⋆e-mail: david.cantor@umontpellier.fr

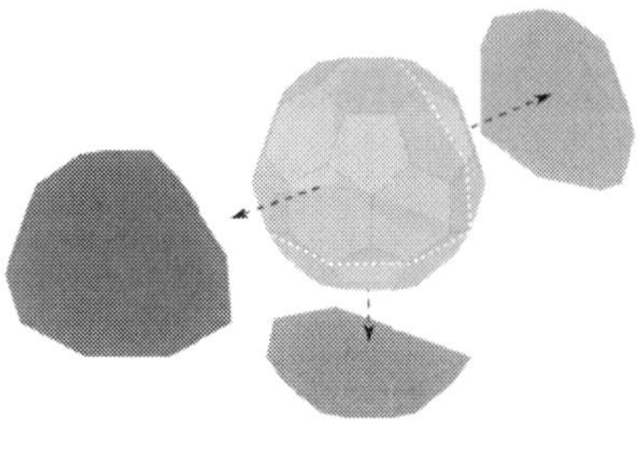

Figure 1. Exploded view of a convex grain composed of 3 cells.

ments are obtained after a Voronoi tessellation of the original grain. By changing the number of generators in the tessellation, it is possible to control the number of potential fragments (n_{cl}). Also, it is possible to modify the geometry of fragments by controlling the dislocation of the centroid of the cells with respect to the position of its corresponding generator [6]. A weighting of such dislocation for all the cells within a grain let us define a geometrical degree of disorder (λ) that may vary between 0 and 1. When λ tends to 0, the cells have almost-regular shapes and they are well distributed within the grain. As λ increases, the cells may have elongated shapes and varying sizes. Figure 1 shows an example of a grain with 3 cells highlighting the way it is composed in an exploded view. The intercell bond is determined by two parameters: a tensile strength (C_t) acting in the normal contact direction, and a shear strength (C_s) present in the tangential contact direction. Once the forces between two cells reach either $C_t \cdot S$ or $C_s \cdot S$, with S equals to the contact area, the bond is permanently lost (i.e., a fissure is created along the intercell contact surface). These broken surfaces will behave, in case of a new interaction with other cells, as dry surfaces with a coefficient of friction μ. It has been shown that parameters n_{cl} and λ can modify the strength of individual grains under diametral compression. However, it is the intercell strength that mainly controls the compressive behaviour of individual grains [3]. For the following numerical experiments, and for the sake of simplicity, we fixed $n_{cl} = 10$, $\lambda = 0.25$ and $\mu = 0.4$.

We used the discrete element method known as contact dynamics for our simulations, in which the equation of motion is integrated in an implicit time stepping scheme and the cells are considered as rigid bodies [13]. In effect, overlapping between bodies do not represent the contact deformations and only are taken into account during the contact detection. In addition, the cells velocities and contact forces are computed by means of a Gauss-Seidel iterative algorithm as function of the external loads and the rigid bodies interactions.

2.2 Compression tests

Assemblies of crushable grains were built by depositing polyhedral grains into cubic boxes. The assemblies were composed of 1000 grains, and each grain had 10 cells. For each sample, the intercell bonding strength varied between 0.1 and 10 MPa. Then, with the aim of having dense samples, isotropic compressions were carried out keeping in mind that the imposed stresses were light enough to prevent fragmentation during this test stage. Finally, uniaxial compression tests were undertaken on each assembly.

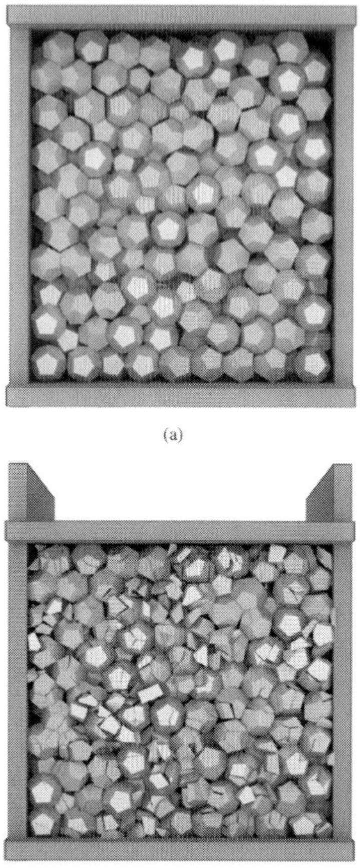

(a)

(b)

Figure 2. Front view of a sample composed of 1000 grains as the applied vertical stress increases from (a) $\sigma_h = 0$, to (b) $\sigma_h = 2C_t$. The frontal wall was removed for the visualization.

Figure 2 shows screenshots of a sample as the vertical load σ_h is increased. In this figure, it is possible to identify grains that are being fragmented as well as others that remain intact throughout the compaction. In general, grains transmitting an important part of the vertical load are more susceptible to break while grains outside such force networks are barely affected.

3 Results

In this section, we present some results concerning the relation between the external load and the sample density, the grain size evolution, and the fragmentation mode within the grains. The first two parameters mentioned above are measures at the sample scale, and they are usu-

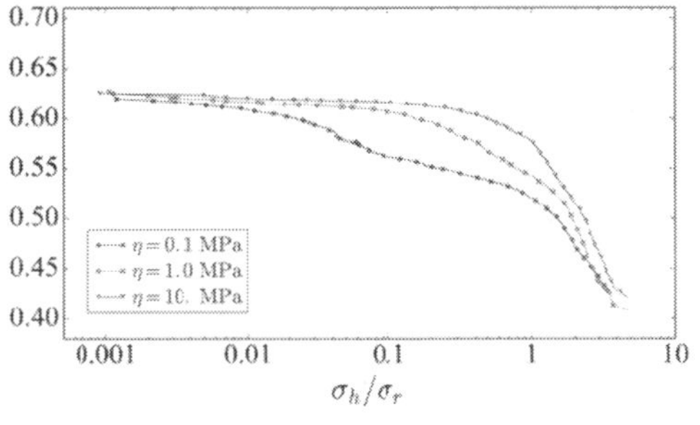

Figure 3. Evolution of the void ratio as function of the normalized vertical load for samples presenting different intercell bonding strength η.

ally found in experimental testing. The fragmentation mode is, instead, a measure at the grain scale and it describes the type of stresses that are causing the loss of bonds between cells.

3.1 Load vs density

In our experiments, the sample density is analysed through the void ratio (e) that is the parameter relating the volume of voids in the sample (V_v) and the volume of grains (V_s) ($e = V_v/V_s$). As the vertical stress increases, the volume of the container is reduced and the void ratio should decrease as well. Figure 3 shows the evolution of e as the ratio σ_h/σ_r increases (i.e., the imposed stress normalized by a stress of reference $\sigma_r = 1$ MPa). For the sake of simplicity, and given that C_t and C_n were increased simultaneously in different tests, let us simplify the notation as follows: $C_t = C_n = \eta$. Two different regimes during the compaction can be differentiated. The first one corresponds to that where the volume is slowly reduced as the load starts to increase. During the second stage, the void ratio decreases more rapidly. The transition between these two regimes is affected in particular by the intercell strength. Even if this effect seems to be relatively small, when compared with the general behaviour during the compaction, it evidences the increase of energy per unit of volume that is being stored in the intact bonds. We could also observe that fragmentation occurs since the beginning of the loading and it does not stop till the end of the second regime. At that point, not all the bonds are broken but the dense arrangement of grains prevents more fragmentation events.

3.2 Grain size evolution

Figure 4 shows the evolution of the grain size distribution (i.e., the cumulative probability h of having a grain of diameter d), for the sample with $\eta = 1$ MPa, as function of the measured vertical strain. As it can be observed, the sample starts as a mono-disperse assembly of grains. Different grain size distributions are presented for 4 values of vertical strain (ε_h) during the compaction. As indicated in the Figure, the dashed line corresponds to a power-law

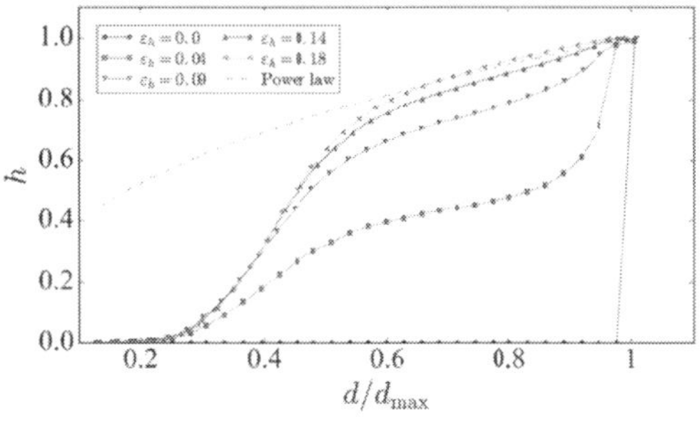

Figure 4. Evolution of the grain size distribution for 5 different instants during the compaction. The dashed line corresponds to the ultimate grain size distribution repetitively found in literature. Such distribution has a power-law shape of the form $h^*(d) = (d/d_{\max})^{3-\alpha}$, with $\alpha = 2.6$.

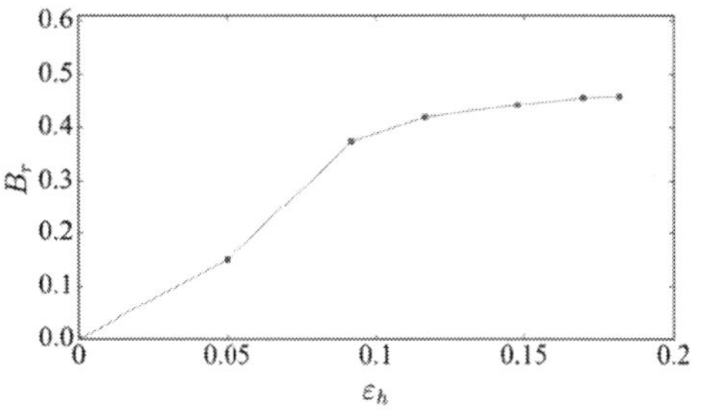

Figure 5. Evolution of the fragmentation index as function of the vertical strain for one of the samples tested.

distribution, being $d_{\max}$ the maximum particle size in the sample, and $\alpha = 2.6$. This exponent α has been repetitively found in experiments where granular samples undergo large shear deformations [14]. Regardless of the fact that in our model fragmentation events are constrained by the size of the smallest cell and the available space between grains, the grain size distribution tends partially to such power distribution. These grain size distributions may be compared by means of a fragmentation index B_r [7, 8], that can be calculated as follows:

$$B_r = \frac{\int_{d_{\min}}^{d_{\max}} (h_0(d) - h(d))\mathrm{d}d}{\int_{d_{\min}}^{d_{\max}} (h_0(d) - h^*(d))\mathrm{d}d}, \quad (1)$$

where h_0 is the grain size distribution at the beginning of the simulation. Figure 5 shows the evolution of B_r as function of the vertical strain. It indicates that, in the case of uni-axial compression tests, the fragmentation reaches at most 50% of the potential fragmentation $h^*(d)$ for this material . Nevertheless, we can observe the grain size distribution will hardly evolve further than such distribution shown in Fig. 4 for $\varepsilon_h = 0.18$. Then, the potential grain size distribution should be indeed revised to have a better evaluation of grain fragmentation during uni-axial compression tests.

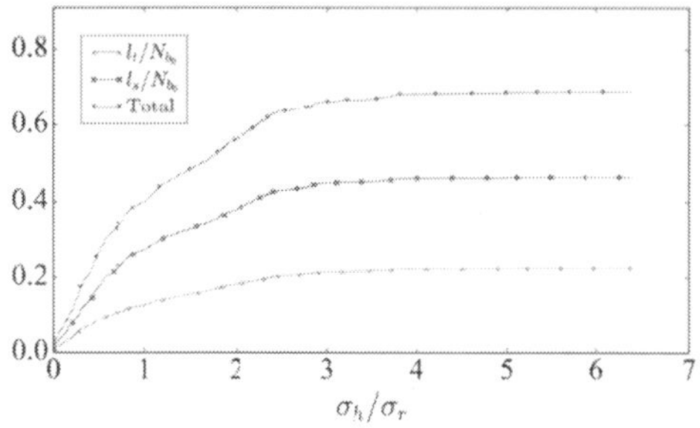

Figure 6. Intercell failure mode classification in contacts lost by tensile (l_t) or shear (l_s) stresses normalized by the initial number of bonds N_{b_0}.

3.3 Fragmentation mode

By means of our discrete element approach, we were able to identify the bonds that are lost by tensile or shear stresses. Figure 6 shows the number of bonds lost either in traction l_t or shear l_s modes normalized by the initial number of bonds (N_{b_0}), as function of the external load σ_h, for the samples with $\eta = 1$ MPa. In this figure, we can remark that most of the bonds are lost by shear stresses. We could determine that tensile failures are rare as the load increases, and they are usually consequence of previous shear failures that push apart other cells by induced indirect traction. Also, the increase of grain coordination (average number of contacts per grains), reduces the probability of having cells that fail in the tensile mode.

4 Discussion and perspectives

We have implemented a three-dimensional model to grain fragmentation known as bonded-cell method (BCM), based on the bonding of a grain potential fragments or cells. With this approach, the fragmentation events are function of the external loads and interaction between the grains, and it allows fragments to evolve into diverse shapes and sizes. We analysed the behaviour of assemblies of particles under uni-axial compression. In particular, we studied the load-density relations, the evolution of the grain size distribution, and the type of stresses within the grains that are causing failure. Our results are in very good agreement with experimental testing performed in soil and rock mechanics. For instance, this work presented the evolution of density as the vertical load increases. It let us observe how the grain size of fragments evolves towards a well-known power-law distribution, as well as we could identify the roll of the intercell strength on the energy storage within intact the bonds. Finally, we showed that shear failure is the principal mechanism of fragmentation in uniaxial compression tests. This discrete element approach permits to clearly map the macroscopic behaviour from the grain and contacts scale, making our 3D bonded-cell method an outstanding framework to study crushable materials. Future works with the BCM-3D will include the study of the shape variability of fragments, the evolution of number of contacts per fragment, and analysis on the consumption of energy as fragmentation occur.

Acknowledgement

This work was financially supported by the French Alternative Energies and Atomic Energy Commission (CEA).

References

[1] Aleai, E., Mahboudi, A.: Discrete model for simulating shear strength and deformation behaviour of rockfill material, considering the particle breakage phenomenon. Granular Matter **14**(6), 707–717, (2012)

[2] Arslan, H., Baykal, G., Sture, S.: Analysis of the influence of crushing on the behavior of granular materials under shear. Granular Matter **11**(2), 87–97, (2009)

[3] Cantor, D., Azéma, E., Sornay, P., Radjai, F.: Three-dimensional bonded-cell model for grain fragmentation. Computational Particle Mechanics, 1–10, (2016)

[4] Cecconi, M., DeSimone, A., Tamagnini, C., M.B. Viggiani, G.: A constitutive model for granular materials with grain crushing and its application to a pyroclastic soil. Int J Numer Anal Met **26**(15), 1531–1560, (2002)

[5] Cheng, Y. P., Bolton, M. D., Nakata, Y.: Discrete element simulation of crushable soil. Géotechnique **53**(7), 633–641, (2003)

[6] Du, Q., Faber, V., Gunzburger, M.: Centroidal Voronoi Tessellations: Applications and Algorithms. SIAM Review **41**(4), 637–676, (1999)

[7] Einav, Itai.: Breakage mechanics - Part I: Theory. J Mechan Phys Solids **55**(6), 1274–1297, (2007)

[8] Hardin, B. O.: Crushing of Soil Particles. Journal of Geotechnical Engineering **111**(10), 1177–1192, (1985)

[9] Hégron, L., Sornay, P., Favretto-Cristini, N. Compaction of a bed of fragmentable UO2 particles and associated acoustic emission. IEEE Transactions on Nuclear Science, **61**(4), 2175–2181, (2014)

[10] Kong, C. M., Lannutti, J. J.: Localized densification during the compaction of alumina granules: The Stage I-II transition. Journal of the American Ceramic Society **83**, 685–690, (2000)

[11] Miura, N., Murata, H., Yasufuku, N.: Stress-strain characteristics of sand in a particle-crushing region. Soils and Foundations **24**(1), 77–89, (1984)

[12] Nguyen, D-H., Azéma E., Sornay, P., Radjai, F.: Bonded-cell model for particle fracture. Physical Review E **91**(2), 022203, (2015)

[13] Radjai, F., Richefeu, V.: Contact dynamics as a nonsmooth discrete element method. Mechanics of Materials, **41**(6), 715–728, (2009)

[14] Sammis, C., King, G., Biegel, R.: The kinematics of gouge deformation. Pure and Applied Geophysics, **125**(5), 777–812, (1987)

[15] Tsoungui, O., Vallet, D., Charmet, J. C.: Numerical model of crushing of grains inside two-dimensional granular materials. Powder Technology **105** (1-3), 190–198, (1999)

Experimental and DEM analysis of the dissipation involved in the collision of a boulder with a substratum

Bruna Garcia[1,2,*], *Pascal* Villard[1], *Vincent* Richefeu[1], and *Dominique* Daudon[1]

[1] *Univ. Grenoble Alpes, 3SR, F-38000, Grenoble, France*
[2] *IMSRN, Ingénierie des Mouvements de Sol et Risques Naturels, 680 Rue Aristides Bergès, 38330 Montbonnot, France*

Abstract. Experiments and numerical analysis were both carried out to evaluate the amount of energy dissipated during the rebound between single boulder (concrete made blocks with many varied shapes) and a soft or stiff substratum. A high speed camera made it possible to analyze the collision kinematics and to determine the role played by different features such as the incident axial/angular velocities, the boulder shapes, the impact configuration and the substratum type, on the amount of dissipated energy. A three-dimensional discrete element code was used to simulate the experiments by accounting for the actual shape of the boulder and some dissipation mechanisms. The comparison between the experimental and numerical trajectories allowed for an physical interpretation of the dissipation mechanisms and an identification of a set of parameters to be introduced in the simulations. That way, the predictive ability of the discrete model of rock avalanche can be improved by including in-situ measurements.

1 Introduction

Rock falls or rock avalanches are becoming increasingly frequent due to climatic changes. Economic issues resulting from the closure of roads or the discontinuation of economic activities can be significant and their consequences on people and constructions are sometimes dramatic. The prediction of the propagation way and the stop position of a rock is highly uncertain given the complexity of the involved mechanisms. For a better understanding of such mechanisms,a numerical model was developed to take into account the quantities and modes of dissipated energies during the rebound of a rock on a smooth or a rigid substratum. The numerical model used is based on the DEM approach and takes account of real shape of blocks and energy dissipation modes due to collisions, friction and rolling resistance [1, 4]. First validated with experimental data resulting from small scale laboratory experiments involving the rebound of small parallelepiped blocks on a rigid substratum [2, 3], the ability of the numerical model to predict the rebound of a rock on a natural soil was evaluated using large scale laboratory experiments involving different block shapes. After validation, the numerical model could be used to conduct a statistic analysis of the influence of the block shape that was not possible in an experimental way due to the large number of tests that were needed. In this paper we focus on the methodology used to obtain the optimized set of parameters from experimental tests.

*e-mail: bruna.dasilvagarcia@3sr-grenoble.fr

2 Experimental tests

The device shown in Figure 1 is basically a metallic ramp in which a released boulder (1) rolls or slides, (2) flies freely, (3) collides with a substratum, and (4) bounces. The metal ramp is 2.7 meters long and its inclination can vary from 30 to 60 degrees, reaching a maximum elevation height of 2.30 m.

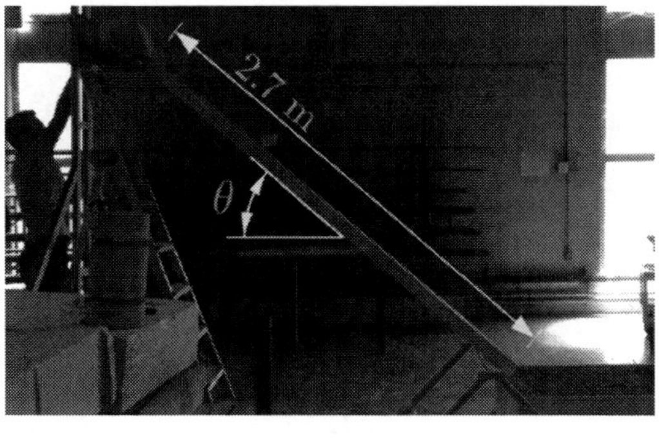

Figure 1. Experimental device. The inclination angle θ can vary from 30° to 60°.

The boulder kinematics is restricted to a vertical plan so that the stages 2 to 4 can be filmed with a single high-speed camera. Photographs were shot each 0.1 second to determine the boulder trajectories before and after impact. Both axial and angular positions were then tracked on the photographs thanks to two reference points marked in the visible face of the boulder.

 DOI: 10.1051/epjconf/201714016010

All the lengths were scaled by using the pixel sizes, which depends on the position and settings of the camera. The framework origin was arbitrarily set at the first collision point on the substratum plan; Figure 2. As a convenient consequence, (*i*) the incident trajectory has negative x-coordinates while the reflected trajectory has a positive x-coordinates; (*ii*) a positive y-coordinate corresponds to free flight motions, and a negative y-coordinate plays the same role as an overlap in discrete element simulations, and should be associated to the "contact phase".

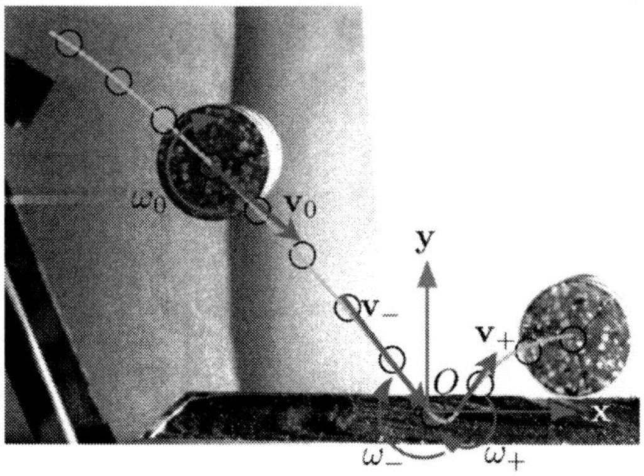

Figure 2. Sketch of the assessment of the trajectories before and after a collision. The circle symbols are the positions extracted from the picture shots; the green curve is the trajectory that can be obtained by means of statistical regression; $\mathbf{V}_0$ and ω_0 are initial velocities that can be used to perform a DEM simulation; $\mathbf{V}_-$ and ω_- are the incident velocities, and $\mathbf{V}_+$ and ω_+ are the reflected velocities.

Many geometries have been designed (Figure 3); the boulders have the same height (0.25 m) and they are made of ultra-high performance fiber-reinforced concrete (UHPC) in order to avoid their breakage. Luckily, the density of UHPC (~2400 kg/m^3) is close to the average density of natural rocks (~2700 kg/m^3).Two grooves intended to guide the boulder on the rail were designed so that the rotation was locked or free. The boulder weight m, axial inertia J and rotation hindrance in the ramp are listed in Table 1.

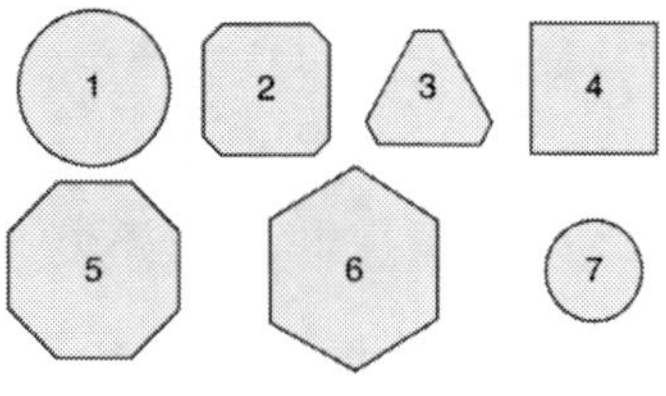

Figure 3. Boulder shapes used for the tests: (1) cylinder, 16 cm in diameter; (2) square with trimmed-corners, 150 mm in sides; (3) triangle with trimmed-corners, 155 mm in sides; (4) square, 160 mm in sides; (5) octagon, 400 mm in outer-diameter; (6) hexagon, 185 mm in outer-diameter; (7) cylinder, 110 mm in diameter.

Table 1. Mass properties of the boulders and ability to roll when released on the ramp.

Shape number	**Weight** m (kg)	**Axial inertia** J (kg·m·s^2)	**Rotation locked**
1	10.89	0.021820	No
2	11.82	0.040750	No
3	11.66	0.038314	No
4	13.80	0.060325	Yes
5	33.52	0.345403	No
6	16.92	0.087076	No
7	5.38	0.008177	No

To quantify the global dissipation as a function of the boulder shape and of the substratum, the restitution rate of the total kinetic energy has been determined from the axial velocities **v** and angular velocities ω, before (-) and after (+) the collision (Figure 2):

$$R = \frac{m\mathbf{v}_+^2 + J\omega_+^2}{m\mathbf{v}_-^2 + J\omega_-^2} \tag{1}$$

In practice, the time "just before" or "just after" the rebound is a matter of choice because the collision has a finite duration, which cannot be uniquely defined from the photographs. We however checked that a small variation in the delays before and after the rebound for the assessment of velocities has a very limited influence on the value of the restitution rate R. Thus, the last/first known velocity belonging to a parabolic trajectory of the mass center (free flight) were used for the -/+ values (respectively) in Equation 1.

Table 2 presents the restitution rates R for the 7 different shapes. All releases were performed from the same high and with a ramp inclination $\theta = 45°$. Except for shape 4 that slid on the ramp, the other shapes rolled. The impacted horizontal substratum was either a dry and loose sand (LS), a stabilized soil (SS), a bituminous concrete (BC) or a vegetated ground (VG).

From the data in Table 2, a number of facts can be drawn. The reproducibility for each drop is not so good excepted for the collision with the loose sand (LS), which is the softest substratum. This lack of repeatability may be due to geometrical or mechanical heterogeneities, however, the same order of R was obtained – at least two times over the 3 drops – for nearly all couples of shape-substratum. It is clear that both the shape of the boulder and the nature of the substratum affect the amount of kinetic energy restored after the collision. However, we stress here that the restitution rate R makes no distinction on how the kinetic energy is dissipated. For example, it does not evaluate what is the part of incident kinetic energy that has been transformed in angular velocity. As a consequence, an identification of the dissipation parameters involved in a model of collision may help to understand the mechanisms that take place as a function of the boulder shape and of the substratum.

Table 2. Values of the restitution rate R (see Equation 1) for 3 drops of different shapes on a ramp inclined at 45°. Four types of substratum were tested: Loose and dry Sand (LS), Bituminous Concrete (BC), Vegetated Ground (VG) and Stabilized Soil (SS).

Shape	Substratum	Drop 1	Drop 2	Drop 3
1	LS	0.45	0.36	0.49
	SS	0.50	0.74	0.58
	BC	0.62	0.94	0.65
	VG	0.58	0.35	0.95
2	LS	0.48	0.58	0.67
	BC	0.40	0.76	0.46
3	LS	0.39	0.56	–
	SS	0.58	0.26	–
	VG	0.34	0.10	0.34
4	LS	0.62	0.63	0.62
	SS	0.41	0.35	0.45
	BC	0.49	0.49	0.39
	VG	0.38	0.23	0.29
5	LS	0.41	0.45	(broken)
	BC	0.88	0.54	0.45
	VG	0.68	0.82	–
6	LS	0.28	0.28	–
	BC	0.50	0.47	0.52
	VG	0.82	0.73	0.64
7	BC	0.65	0.41	0.47

3 Collision parameters

A boulder trajectory can be computed by means of the discrete element method (DEM) and then compared with the 2D experimental trajectory. When the initial conditions (axial/angular velocities and positions) are correctly set, the computed trajectory matches the experimental one within a given tolerance. When the boulder leaves the horizontal plan that represents the substratum surface, the computed trajectory relies on the collision parameters. Here, the objective is to determine the DEM parameters that imply the reflected velocities having the closer match with the measured ones.

These DEM parameters are those of a local dissipative contact model developed for rock avalanches (see e.g. [1] or [2]), implemented in the code `DEMbox`. They are (1) the normal, tangential and rolling stiffnesses k_n, k_t and k_r, respectively, and (2) the associated dissipation coefficients, namely:

- The *kinetic energy restitution* e_n^2 in the normal direction.
- The *tangential dissipation coefficient* μ that relates to the friction and/or to the plasticity of the soil. Notice that μ is not named "friction coefficient" and it may be greater than 1.
- The *rolling dissipation coefficient* μ_r that acts like μ but for the rolling movements. Notice that this latter parameter is dimensionless.

Since the simulation results should only differ from the experiments after the collision, the relative-error functions are based on the reflected velocities, that are the axial and angular velocities of the boulder when it leaves the horizontal plan. For each component x of reflected velocity (x stands for either v_x, v_y, ω or R), it expresses as follows:

$$E_x(\{p\}) = \left(\frac{x_+^{\text{sim}} - x_+^{\text{exp}}}{x_+^{\text{exp}}}\right)^2 \quad (2)$$

where $\{p\} = (k_n, k_t, k_r, e_n^2, \mu$ and $\mu_r)$ is the set of collision parameters.

4 Selected results

Finding the best parameters $\{p\}$ can be done by optimizing them with a multivariate minimization technique like, *e.g.*, the conjugate directions descent or Powell's method. We, however, chose to use a *brute-force* optimization because it can inform us about what are the effects of varying each parameter independently. Two examples of such analyses are presented here for two different materials of the substratum: dry-loose sand (LS) and stabilized soil (SS). Both examples make use of the shape 1, that is a cylinder with 16 cm in diameter. The incident and reflected velocities, measured from the photographs of the experiments, are given in Table 3.

Table 3. Incident and reflected velocities extracted from the photographs, for the two cases selected. The values into brackets are suspicious since the horizontal velocity should not be increased with cylindrical boulders; in this case it was chosen in the optimization procedure $v_{x+} = v_{x-} = (3.17 + 3.32)/2$.

	v_{x-} m/s	v_{y-} m/s	ω_- s^{-1}	v_{x+} m/s	v_{y+} m/s	ω_+ s^{-1}
Soft	3.69	-3.29	-50.52	3.22	0.49	-37.09
Stiff	(3.17)	-3.51	-37.74	(3.32)	0.44	-40.80

For the brute-force optimization of the collision parameters between the cylinder and the loose sand (LS), a number of preliminary simulations have been carried out. The resulting trajectories have been compared with the experimental one, and it was found that mainly 3 parameters had a major influence on the vertical, horizontal and angular velocities; they are respectively e_n^2, μ and μ_r. The tangential and angular stiffnesses have been set to high values so that the corresponding efforts are rapidly mobilized. The normal contact stiffness k_n was adapted to the type of the substratum in order to reproduce similar ground settlements than those obtained experimentally ($k_n \simeq 10^6$ kN/m and $k_n \simeq 10^7$ kN/m for the soft and stiff soils, respectively).

Noticing that e_n^2 is the only parameter that affects the vertical velocity of the boulder, it was chosen to identify first this parameter. Figure 4 shows the relative-error function E_{v_y} as a function of e_n^2. We observe that the reflected vertical velocity is unambiguously optimized for $e_n^2 \simeq 0.04$ with a zero error (whatever the values of μ and μ_r).

The influence of μ and μ_r on the horizontal and angular velocities can be examined in a two-dimensional space as shown in Figure 5, where e_n^2 has been set to its optimum value. The horizontal velocity is shown to be controlled

DOI: 10.1051/epjconf/201714016010

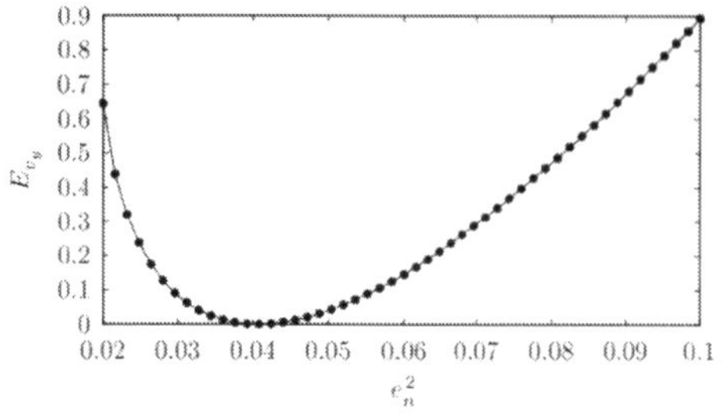

Figure 4. Relative-error function E_{v_y} as a function of e_n^2 in the case of a collision with a soft substratum.

only by μ but unaffected by μ_r. This is indeed expected since the horizontal component of the resultant force is the friction force, which dissipates by the only means of μ since the normal force is independently monitored. In contrast, the angular velocity seems affected by both μ and μ_r for the reason that the resultant moment of the cylindrical boulder is thresholded by the sum of two terms: (1) the resistant moment caused by friction $\ell(-\mu f_n)$, where ℓ is the lever arm (radius of the cylinder) and f_n the vertical force of contact, and (2) the moment caused by rolling resistance $\mu_r \ell f_n$. It results that the resistant moment, and thus the changes in angular acceleration, is controlled by the value of $(\mu_r - \mu)$. This observation is reflected in Figure 5 by the zero-level isocurve of E_ω which is a straight line given by $\mu_r - \mu = 0.08$.

Finally, for the studied collision of the cylindrical boulder on the loose sand, the optimum parameters are obtained at the intersection between the zero-level isocurves of E_{v_x} and E_ω, where $\mu \simeq 0.115$, $\mu_r \simeq 0.195$, and $e_n^2 \simeq 0.04$.

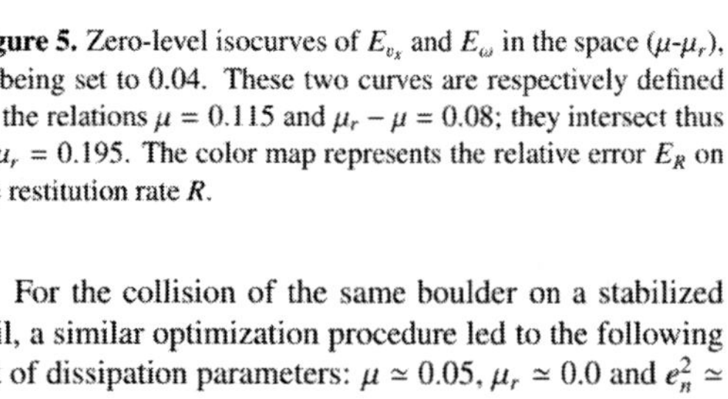

Figure 5. Zero-level isocurves of E_{v_x} and E_ω in the space (μ-μ_r), e_n^2 being set to 0.04. These two curves are respectively defined by the relations $\mu = 0.115$ and $\mu_r - \mu = 0.08$; they intersect thus at $\mu_r = 0.195$. The color map represents the relative error E_R on the restitution rate R.

For the collision of the same boulder on a stabilized soil, a similar optimization procedure led to the following set of dissipation parameters: $\mu \simeq 0.05$, $\mu_r \simeq 0.0$ and $e_n^2 \simeq 0.04$. In this case that involves a stiffer substratum, very low values of μ and μ_r were found because the horizontal and angular velocities were nearly unaffected. In fact, we noticed that an increase in the horizontal velocity was not obtainable because a very small value of μ caused only a slight increase in the angular velocity, but it also induced a sensible decrease in the horizontal velocity.

5 Concluding remarks

In this work, we intended to detect which of the 6 parameters implied in the proposed collision model were the most influential and caused the greater variations in reflected velocities after a collision. These parameters were found to be different depending on the nature of the substratum (soft or stiff) , and some conclusions are now beginning to emerge. For example, the rolling resistance was found to be inoperative in case of stiff substratum, while it is absolutely required with soft substratum. Another noticeable observation is the very low restitution of energy in the normal direction whatever the stiffness of studied substratum. The boulder shape plays also a crucial role in the dissipation mechanisms but it was not investigated in the present short paper. For the two types of soils taken into account for the numerical analysis, it was found that most of the vertical kinetic energy was dissipated during the rebound while the horizontal and rotational kinematics energy were only affected in the case of the soft soil.

The study is not yet completed and further analysis are still in progress or scheduled. In particular, more couples shape-substratum will be analyzed, and each analysis will be repeated to assess the variability of the identified parameters.

The authors thank T. Verchier, C. Grardel, T. Boudry and A. Colson who assisted the realization of the experimental tests in the framework of their final year study project. They also thank G. Vian for the conception and realization of the ramp. This study is funded by the company IMSRN (www.imsrn.com). The Laboratoire 3SR is part of the LabEx Tec 21 (Investissements d'Avenir, Grant Agreement No. ANR-11-LABX-0030).

References

[1] V. Richefeu, G. Mollon, D. Daudon, P. Villard, Engineering Geology **149**, pp. 78-92 (2012)

[2] V. Richefeu, P. Villard, *Modeling Gravity Hazards from Rockfalls to Landslides* (ISTE Press Ltd, Elvesier Ltd, UK, 2016) 161 pp.

[3] G. Mollon, V. Richefeu, P. Villard, D. Daudon, Journal of Geophysical Research – Earth Surface **117**, F02036 (2012)

[4] D. Daudon, P. Villard, V. Richefeu, G. Mollon, C. R. Mécanique, Mechanics of granular and polycrystalline solids **343**(2), pp. 166-177 (2015)

A discrete element model for damage and fracture of geomaterials under fatigue loading

Xiaofeng Gao[1,*], *Georg* Koval[1,**], and *Cyrille* Chazallon[1,***]

[1] *ICUBE Laboratory, UMR 7357, CNRS, National Institute of Applied Sciences of Strasbourg, 24 Boulevard de la Victoire, 67084, Strasbourg, France*

Abstract. Failure processes in geomaterials (concrete, asphalt concrete, masonry, etc.) under fatigue loading (repeated moving loads, cycles of temperature, etc.) are responsible for most of the dysfunctions in pavements, brick structures, etc. In the beginning of the lifetime of a structure, the material presents only inner defects (micro cracks, voids, etc.). Due to the effect of the cyclic loading, these small defects tend to grow in size and quantity which damage the material, reducing its stiffness. With a relatively high number of cycles, these growing micro cracks become large cracks, which characterizes the fracture behavior. From a theoretical point of view, both mechanisms are treated differently. Fracture is usually described locally, with the propagation of cracks defined by the energy release rate at the crack tip; damage is usually associated to non-local approaches. In the present work, damage and fracture mechanics are combined in a local discrete element approach.

1 Introduction

Fatigue is the weakening of a material caused by repeatedly applied loads of low intensity. The failure processes of geomaterials subjected to fatigue loading can be described by four stages, including crack nucleation (Stage I), short crack growth (Stage II), large crack growth (Stage III), and ultimate failure (Stage IV) [1]. In the beginning of the lifetime, the material presents only inner defects (micro cracks, voids, etc.). Due to the effect of the cyclic loading, these small defects tend to grow in size and quantity which damage the material, reducing its stiffness. When these defects become short cracks, the failure process turns into its second stage of short crack growth. With a relatively high number of cycles, these growing short cracks become large cracks, which characterizes the fracture behavior.

It is still difficult today to capture all the four stages with one single fatigue model. In the present work, a discrete element approach is proposed, based on a local description of damage and fracture which allows a smooth interaction between both formulations. The basic concepts of damage and crack growth models are presented in Sec. 2. The discrete element modeling in elasticity is discussed in Sec. 3, while the fatigue extension is analyzed in Sec. 4. Finally, the numerical results are compared to theoretical predictions in Sec. 5, followed by the conclusions.

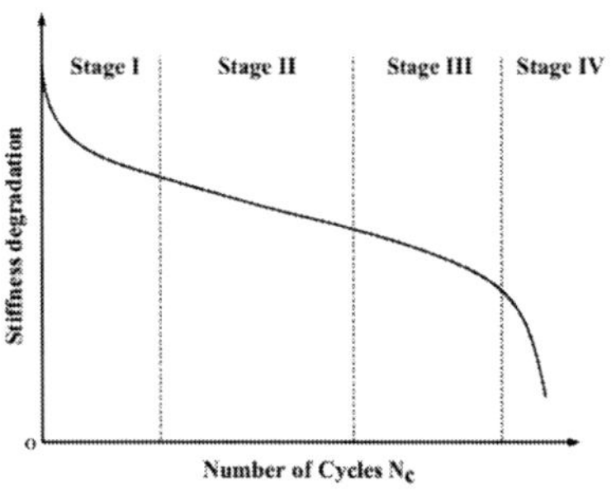

Figure 1. The fours stages of the fatigue failure process.

2 Continuum damage and fatigue crack growth models

2.1 Continuum damage model

The elastic isotropic continuum damage model characterizes the stiffness decrease with cyclic loading. As the material is deformed, the initiation, growth and coalescence of micro defects decrease the stiffness (degradation of material properties), which is represented by the growth of the damage variable D ($0 \leq D \leq 1$). For undamaged material, D is 0. While $D = 1$ corresponds to a completely damaged with zero stiffness. The stress strain relation can be written as [1]:

$$\sigma_{ij} = C_{ijkl}\varepsilon_{kl} = (1 - D)C^0_{ijkl}\varepsilon_{kl}, \quad (1)$$

where σ_{ij} and ε_{kl} are the stress and strain tensor components, C_{ijkl} and C^0_{ijkl} are damaged and initial (elastic) se-

*e-mail: xiaofeng.gao@insa-strasbourg.fr
**e-mail: georg.koval@insa-strasbourg.fr
***e-mail: cyrille.chazallon@insa-strasbourg.fr

cant stiffness of the material respectively. C^0_{ijkl} is a function of Young's modulus E and Poisson ratio ν.

The evolution of the damage proposed by [2] is controlled by the strain state of the material, defined by the scalar equivalent strain

$$\tilde{\varepsilon} = \sqrt{\sum_{i=1}^{3}\left(\frac{\langle\sigma_I\rangle_+ + \langle\sigma_{II}\rangle_+}{E}\right)^2}, \quad (2)$$

where $\langle\sigma_i\rangle_+ = \sigma_i$ if the principal stresses $\sigma_i \geq 0$ or $\langle\sigma_i\rangle_+ = 0$ otherwise. The Young's modulus of the damaged material E is a function of its initial value and the damage variable:

$$E = E_0(1 - D). \quad (3)$$

The rate of damage growth is defined as a function of local equivalent strain rate (local model):

$$\dot{D} = f(D)\tilde{\varepsilon}^{\beta}\left\langle\dot{\tilde{\varepsilon}}\right\rangle, \quad (4)$$

where β is a constant parameter and $f(D)$ is function of the damage. The expression given by Paas [1] is adopted in the following developments:

$$f(D) = \gamma D^{\alpha}. \quad (5)$$

where γ and α are constant parameters. In practice, the values of α, β and γ are calibrated according to the experimental results.

2.2 Fatigue crack growth model

The presence of a crack can significantly reduce the fatigue life of a component or structure. However, the stress singularity at the crack tip induce unreal estimation of the propagation of cracks in damage models.

Based on fracture mechanics, Paris' law [3] is the most popular fatigue crack growth model used in materials science, which relates the stress intensity factor range ΔK to the crack growth rate under a fatigue stress regime:

$$\frac{da}{dN_c} = B\Delta K^M, \quad (6)$$

where N_c is the number of loading cycles, da/dN_c is the crack growth rate, a is the crack length, B and M are material constants. Eq. 6 is well adapted for relatively long cracks, nevertheless it produces unrealistic small crack growth rates da/dN_c, if $a \to 0$.

3 Discrete element elastic model

3.1 Numerical scheme and contact model

The simulations are based on the discrete element method (DEM), which uses an explicit numerical scheme to evaluate the contact interactions of the particles and the their motion [4]. Each contact force presents normal and tangential components, respectively

$$\begin{cases} N = k_n\delta_n + c_n\dot{\delta}_n, \\ T = k_t\delta_n + c_t\dot{\delta}_t. \end{cases} \quad (7)$$

where δ_n and δ_t are the normal and tangential relative displacements at the contact, k_n and k_t are the normal and tangential stiffness, c_n and c_t are the normal and tangential viscous damping, respectively (see Fig. 2).

The state of the model is updated at each time step Δt after the following operations [5]: (1) position and velocity of each particle is calculated according to Newton's laws of motion; (2) contact detection; (3) contact forces, calculated by Eq. 7.

3.2 Continuum media description

An isotropic elastic behavior can be reproduced by the periodic compact arrangement of particles of diameter d shown on Fig. 2b [6, 7].

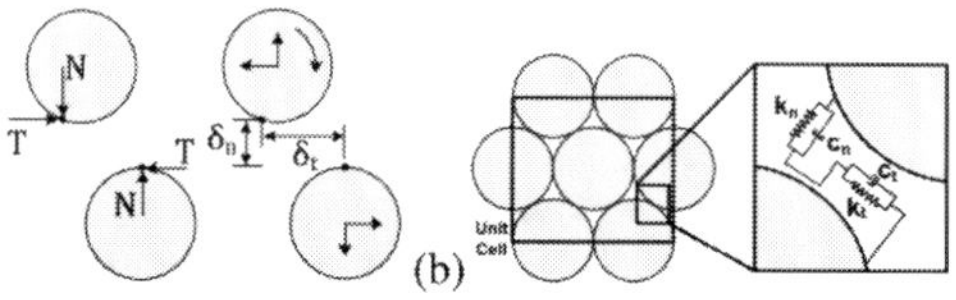

Figure 2. (a) Contact forces and associated displacements. (b) Periodic compact arrangement of particles and the contact law.

The relation between continuum elastic parameters (Young's modulus E and Poisson ratio ν) and discrete elastic parameters (normal and tangential stiffness, k_n and k_t, respectively) reads in plane stress [6]:

$$\begin{cases} k_n = \dfrac{E}{\sqrt{3}(1-\nu)}, \\ k_t = \dfrac{1-3\nu}{1+\nu}k_n = \dfrac{1-3\nu}{\sqrt{3}(1-\nu^2)}E. \end{cases} \quad (8)$$

The proposed approach can be generalized to non-periodic particle packings, however a calibration procedure would be necessary in order to identify the material parameters.

3.3 Principal stresses

The mean values of the components of the tensors of stress and strain are based on the behavior of one pair of contacts (ki and kj) associated to three particles (i, j and k) [8]. A local coordinate system (n; t) is defined; t virtually connects both contacts and n is an orthogonal axis (Fig. 3a).

The stress tensor (in two dimensions) can be defined by the value of the principal stresses $\overline{\sigma}_I$ and $\overline{\sigma}_{II}$, and their orientation. β is the angle between (n; t) and the coordinate system associated to the principal stresses. Consequently, the principal stresses may be written as:

$$\begin{cases} \overline{\sigma}_I = \overline{\sigma}_n + \overline{\sigma}_t \tan(\beta), \\ \overline{\sigma}_{II} = \overline{\sigma}_n - \overline{\sigma}_t / \tan(\beta). \end{cases} \quad (9)$$

where $\overline{\sigma}_n = \left(N_{ik}\sqrt{3} + T_{ik} + N_{jk}\sqrt{3} - T_{jk}\right)/(2d)$ is the normal stress and $\overline{\sigma}_t = \left(-N_{ik} + T_{ik}\sqrt{3} + N_{jk} + T_{jk}\sqrt{3}\right)/(2d)$ is the tangential stress, which are calculated based on the normal and tangential components of each contact (respectively N_{ik} and T_{ik}, N_{jk} and T_{jk}, as indicated in Fig. 3b).

The normal and the orthonormal strains, respectively $\bar{\varepsilon}_{nn} = \left(\delta_{nik}\sqrt{3} + \delta_{njk}\sqrt{3} + \delta_{tik} - \delta_{tjk}\right)/(d\sqrt{3})$ and $\bar{\varepsilon}_{tt} = \left(\delta_{nik} + \delta_{njk} - \delta_{tik}\sqrt{3} + \delta_{tjk}\sqrt{3}\right)/d$ are obtained by means of the normal and tangential (relative) displacements associated to the contacts ki and kj (respectively δ_{nik} and δ_{tik}, δ_{njk} and δ_{tjk}, as shown in Fig. 3a). Hence, the angle β is determined as $\beta = -(1/2)\arctan(2\bar{\sigma}_t/A)$, for $A = E(\bar{\varepsilon}_{nn} + \bar{\varepsilon}_{tt})/(1-\nu) - 2\bar{\sigma}_n$.

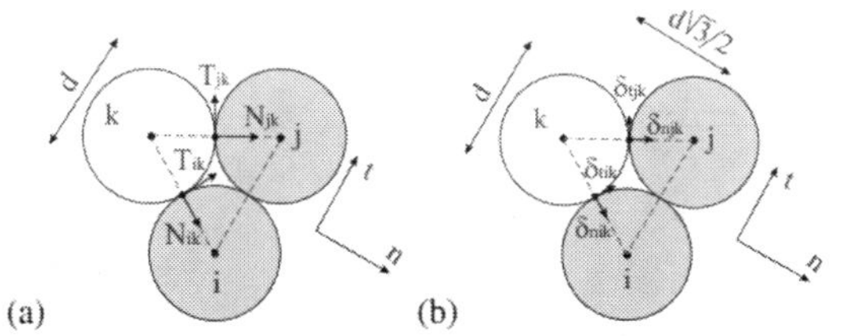

Figure 3. (a) Adjacent particles, respective contact displacements and (b) contact forces [8].

4 Discrete element fatigue model

In the present work, a damage approach (see Sec. 2.1) is adopted to describe the behavior before contact rupture. The rupture of a contact, giving rise to a crack propagation, is limited by a crack growth criterion (see Sec. 2.2), which depends on the stress intensity range ΔK.

4.1 Damage evaluation

The evaluation of the damage per contact can be summarized by the following operations: (1) DEM elastic analysis (as in Sec. 3.1); (2) evaluation of the principal stresses (Eq. 9); (3) calculation of the equivalent strain (Eq. 2) and strain rate; (4) evaluation of the damage growth (Eqs. 4 and 5); (5) evaluation of the damage and update of the value of the Young's modulus (Eq. 3) and the normal and tangential stiffness (Eq. 8).

4.2 Evaluation of the stress intensity range ΔK

In order to obtain consistent results with fatigue crack growth models for crack propagation, when a value of $D = 1$ is indicated by the fatigue model at a crack tip, the value of the energy release rate is evaluated and a potential crack extension is identified.

The normal components of the contact forces N and T, and contact displacements δ_n and δ_t, respectively write

$$\begin{cases} F = N\sin(\pi/3-\beta) + T\cos(\pi/3-\beta) \\ \delta = \delta_n \sin(\pi/3-\beta) + \delta_t\cos(\pi/3-\beta). \end{cases} \tag{10}$$

During a fatigue test, F and δ oscillates between a minimum and a maximum level, which depends on the shape of the cyclic loading. For a sinusoidal loading centered at zero stress, the positive values of F and δ naturally vary from 0 to max F and max δ, respectively. In this case, the damage of a contact induces a maximum energy release rate [9] G_{max}. The minimal energy release rate G_{min} is equal to zero; consequently, the variation of the energy release rate is simply defined as

$$\Delta G = G_{max} - G_{min} = \frac{\sum_1^{N_{cD}} g_i}{d^2/2}, \tag{11}$$

where N_{cD} is the number of cycles to reach $D = 1$ (total release of the contact energy), g_i is the surface of the triangle formed by the points $(0,0)$, $(\max\delta_{i-1}, \max F_{i-1})$ and $(\max\delta_i, \max F_i)$ as shown in Fig. 4.

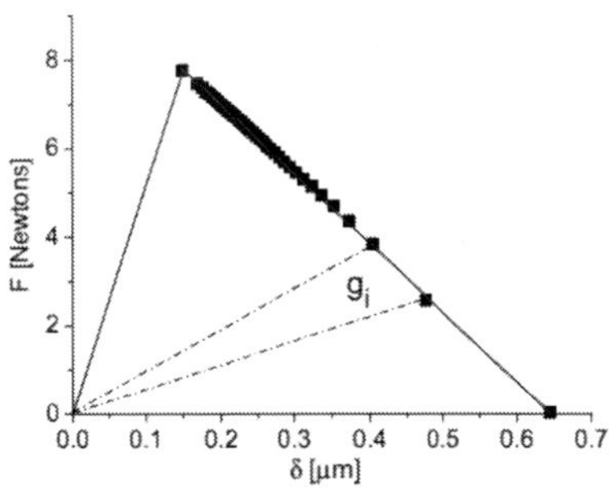

Figure 4. Evaluation of the energy release at the crack tip during a fatigue test.

Based on the relation between the energy release rate and the stress intensity factor in plane stress [9], the stress intensity range is simply defined as $\Delta K = E\sqrt{\Delta G}$.

4.3 Crack initiation and propagation

The evolution of the damage variable D characterizes the weakening of a contact before rupture. The rupture, associated with the propagation of a crack is defined by the value of da/dN_c (Eq.6). A contact which presents $D = 1$, is definitely broken (and cease to exist) only if

$$\frac{da}{dN_c} N_{cD} \geq \frac{d}{2}, \tag{12}$$

which represents a crack growth da equivalent to a particle radius $d/2$. Otherwise, D is set to 0, and the fatigue loading continues, until the number of cycles N_{cD} is sufficiently incremented to fulfill the rupture condition. This rupture criterion prevent deviations from the crack growth criterion induced by the damage model (Sec. 2.1) in conditions of singular stress at a crack tip.

5 Numerical results

A center cracked plate with dimensions $w = 80\,mm$ and $L = 120\,mm$ (as indicated in Fig. 5) is tested under sinusoidal fatigue loading with amplitude $\sigma_0 = 1.25\,MPa$ and frequency $f = 25\,Hz$. The material presents Young's modulus $E_0 = 30\,GPa$ and Poisson ratio $\nu = 1/3$. A viscous damping $c_n = c_t = 0.295\,Ns/m$ and a time step $\Delta t = 5 \times 10^{-7}\,s$ are adopted in the simulations. The damage model parameters are $\alpha = -2.25$, $\beta = 4.0$ and $\gamma = 5.0 \times 10^{16}$; the fatigue crack growth model parameters are $B = 3.0 \times 10^{-10}\,m/cycle[(Pa\sqrt{m})^M]$ and $M = 1.03$. Neglectful effects of the frequency indicate very low dynamic effects on the results (quasi-static regime).

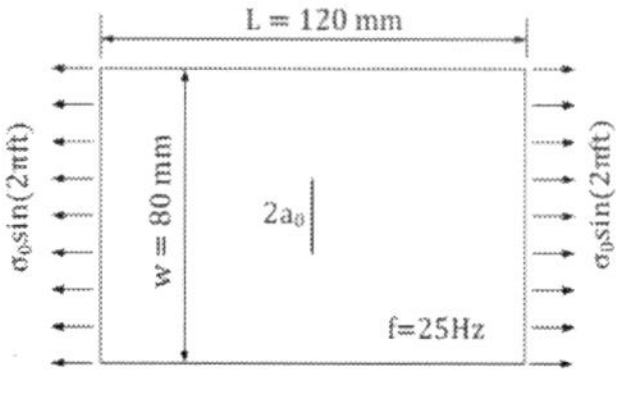

Figure 5. Center cracked plate subjected to fatigue loading.

In Fig. 6 the numerical evaluation of the stress intensity range ΔK is compared to the theoretical result for different diameters d. Deviations are observed only for very large cracks ($2a > 50\,mm$) and tend to decrease for smaller values of d, which indicates a convergent result.

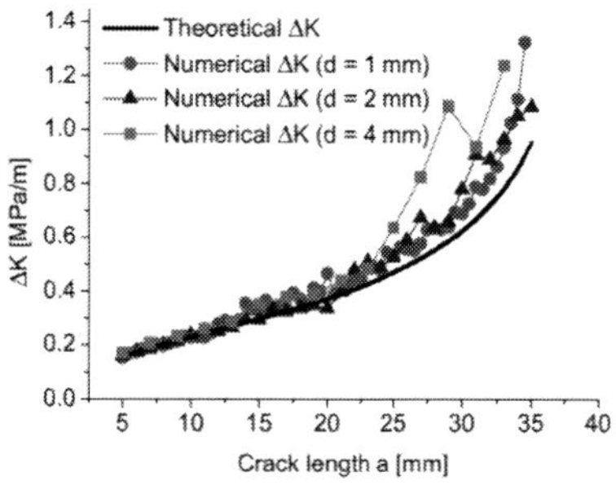

Figure 6. Comparison between the estimation of the stress intensity range ΔK and the theory for different crack lengths a ($a_0 = 5\,mm$).

The stiffness degradation can be quantified by the ratio between the initial amplitude of the displacements and its value during the test (which is a function of the number of cycle N_c). In Fig. 7 the fatigue stiffness degradation obtained for plates with different initial crack sizes a_0 are compared to theoretical predictions. For the uncracked plate, the numerical solution follows naturally the damage theoretical prediction. If the plate present a long crack, such as $a_0 \geq 10\,mm$ the numerical results tend to prediction of Paris' law. However, a smaller crack ($a_0 = 2\,mm$) deviates from Paris' law because of the supplementary contribution of the material damage on the stiffness degradation.

6 Conclusions

A numerical scheme coupling damage and fracture mechanics in a discrete element environment is presented. The association of these different mechanical formulations allows the reproduction of experimental evidences: before material rupture by damage models; during crack propagation by crack growth models. In parallel, important drawbacks of each approach are avoided, such as discretization effects, as shown by the convergent behavior of the results; and the unrealistic results of crack growth models for very short or simply no cracks.

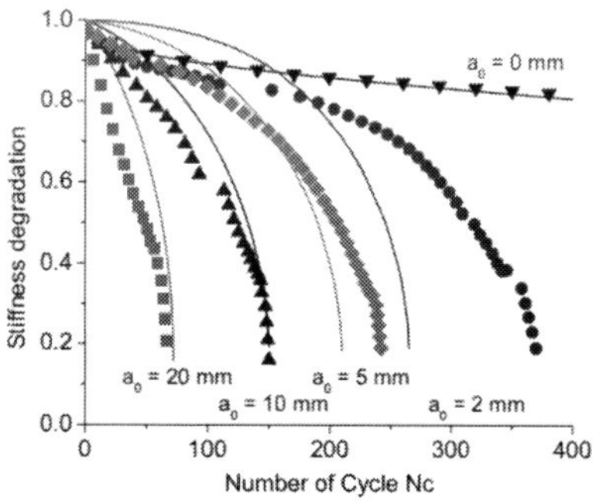

Figure 7. Stiffness degradation as a function of number of cycles N_c for different initial crack sizes. The lines indicate the theoretical solutions, while the symbols, the numerical predictions.

References

[1] M. Paas, Ph.D. thesis, Technishe Universiteit Eindhoven (1990)
[2] M. Jirásek, B. Patzák, Computers & structures **80**, 1279 (2002)
[3] P.C. Paris, M.P. Gomez, W.E. Anderson, The trend in engineering **13**, 9 (1961)
[4] P. Cundall, O. Strack, Géotechnique **29**, 47 (1979)
[5] *Particle Flow Code 5.0 documentation*, Itasca Consulting Group Inc. (2015)
[6] F.A. Tavarez, M.E. Plesha, International Journal for Numerical Methods in Engineering **70**, 379 (2007)
[7] B. Le, G. Koval, C. Chazallon, Engineering Computations **30**, 263 (2013)
[8] B. Le, G. Koval, C. Chazallon, International Journal for Numerical and Analytical Methods in Geomechanics **40**, 583 (2016)
[9] C. Sun, Z. Jin, *Fracture Mechanics*, Academic Press (Academic Press, 2012), ISBN 9780123850010

Multiple shear bands in granular materials

Sára Lévay[1,*] and *János* Török[1]

[1] *Department of Theoretical Physics, Budapest University of Technology and Economics, H-1111 Budapest, Hungary*

Abstract. The transition regime from quasi-static to dynamic shear regime is studied by discrete element method *(DEM)* and by means of a mesoscopic model. We show that at moderate shear rate multiple shear bands appear which eventually appear as a continuous shear profile for large shear rates. The model and the DEM simulations both show a minimum in the shear stress and a maximum for the number of simultaneous shear bands, however the position of the two do not coincide. The scaling analysis of the kinetic energy indicates that the peak in the latter is the relevant parameter. We also show how the long range order present in the quasi-static regime is gradually destroyed by the presence of multiple shear bands.

1 Motivation

Granular shear is a very complex process [1–3] with different characteristics according to the shear velocity [4] or the properties of the sheared material itself, e.g. wet or dry [5]. Here we only focus on cohesionless materials where still there is a qualitative difference in the behavior of the sample depending on the imposed strain rate. The slowly sheared granular material show intermittent shear characterized by shear bands, while fast shear results in a continuous shear velocity profile [6]. In this paper we will focus on the crossover of these two processes.

Slow or quasi-static shear is characterized by a slow increase of the stress which eventually released in a plastic event which is in general a shear band compatible with the boundary conditions. The plastic process restructures the material and if large strains are used the above process is repeated indefinitely [7, 8]: The stress increases again and gets released in a new plastic event. In general the position of the plastic event is determined by inhomogeneities of the material, however the structure of the material is modified by the plastic event. This creates a self-organized process which may result in long range correlations [7, 9].

In this paper we study the crossover from slow to fast shear of granular materials by performing discrete element simulation and presenting a mesoscopic model describing this. We show how long range correlations built up by the slow shear are destroyed by fast shear.

2 Discrete Element Method

2.1 Simulation

Discrete Element Method (DEM) simulations were performed to study the shear of granular materials. We used the molecular dynamics code LAMMPS [10] to determine

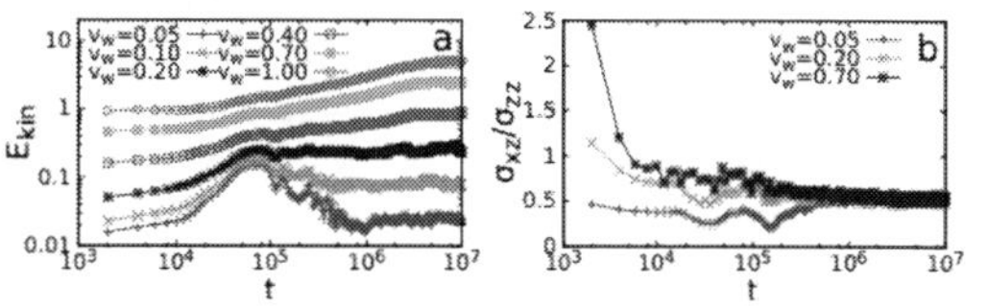

Figure 1. The time evolution of (a) the total kinetic energy and (b) the relative shear stress.

the crossover from slow to fast shear. Two-dimensional plane shear geometry was created where the granular material is placed between two rough walls in the xz plane. The walls are held together with confining stress σ_{zz}. The walls in our simulations were build up from randomly placed spherical particles with diameters 1.4 times larger than the average particle size to create as large wall friction as one can get [11]. We created polydisperse samples with the variation of particle diameters by ±20%. The mean particle diameter was chosen to be the distance unit. The coefficient of friction between particles was $\mu = 0.3$. The initial configurations were created by compressing samples of random sequential deposition by the confining walls. After the system is relaxed, walls start to move in direction x with constant velocities (v_w) equal in magnitude but opposite in direction. We apply periodic boundary conditions in the x direction. For each wall velocity 15 independent configurations were sheared for 10^5 DEM time units and the results show ensemble and time averages of these simulations. The natural units of the DEM simulation are $d = 1$, $\rho = 1$ and $\sigma_{zz} = 1$, particle stiffness was $k = 10^4$.

* e-mail: levay.sara93@gmail.com

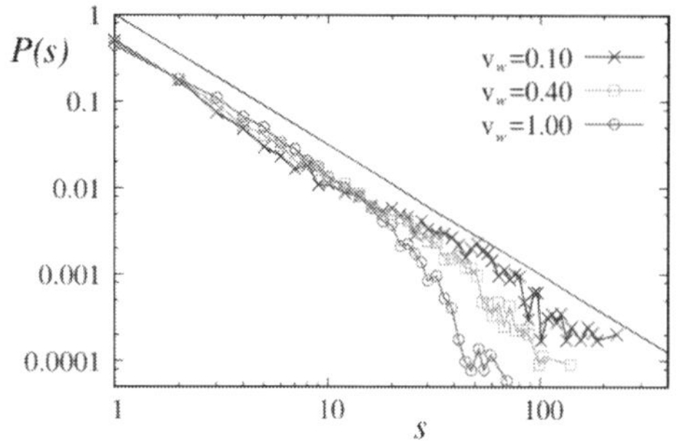

Figure 2. Avalanche size distribution for different values of wall velocity. The solid line is only to drive the eye.

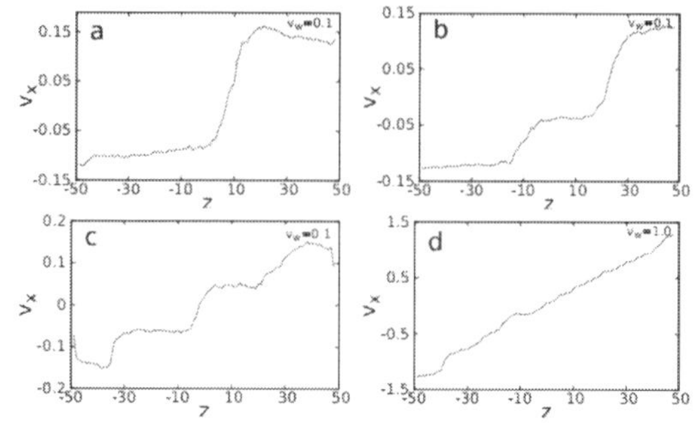

Figure 3. Sample velocity profiles at different time positions for the discrete element method with (a)-(c): $v_w = 0.1$, and (d) $v_w = 1.0$.

2.2 Results

Due to self-organized criticality quasi-static granular shear can build up long range correlations that span the entire system. There are many signs in the system which account for long range order: The non-monotonic behavior of the system in the transient [12] or the presence of critical exponents [13]. Therefore we start our investigation by testing this assumption. In the DEM simulations we measure the shear stress and the total kinetic energy as function of the shear strain. Fig. 1 shows the time evolution of the latter two quantities for different wall velocities. It is obvious that the transient behavior of both σ_{xz} and E_{kin} is non-monotonic for wall speeds up to $v_{\text{wall}} \simeq 0.3$ but above this threshold, simple relaxation can be seen.

Many critical exponents can be defined for sheared granular material [14], here we restrict ourselves to the study of avalanche size distribution $P(s)$ presented in Fig. 2 and Fig. 5. An avalanche is characterized by a series of correlated plastic events and can measured by correlations of the shear stress fluctuations. We define the length of an avalanche by the time interval between successive passes of the shear stress though given level [15]. The resulting avalanche size distribution is a power law for slow shear. As the wall velocity increases an exponential cutoff appears at earlier and earlier stages. Similar behavior was already found in [13].

The question we would like to answer is what causes the disappearance of the long range correlations for quicker shear. To answer this question we measured instantaneous velocity profiles. Four samples are shown in Fig. 3 for inertial number is $I = \dot{\gamma}d/\sqrt{P/\rho} = 2.45 \cdot 10^{-2}$ and $I = 2.45 \cdot 10^{-3}$ ($\dot{\gamma}$ is the shear rate, $d = 1$ is the average particle diameter, $P = 1$ is the pressure and $\rho = 1$ is the density). In the case of large inertial number the shear profile is linear [Fig. 3 (d)], but if the inertial number is slightly above the quasi-static limit of $I = 2.45 \cdot 10^{-3} > 10^{-3}$ we see a varying number of shear bands in the sample [Figures 3 (a)-(c)]. We believe that the presence of multiple shear bands is responsible for the appearance of a finite correlation length and the collapse of the scaling quasi-static regime.

3 Mesoscopic model

3.1 Description

In order to have a better understanding of the system we introduce a simple mesoscopic model motivated by the above observations which can describe simple shear for different shear rates. Dry granular materials are best described by Coulomb-friction. The material is stable for shear stress up to $\sigma_{xz} < \mu\sigma_{zz}$, where σ_{xz} is the shear stress, σ_{zz} is the confining stress and μ is the effective mesoscopic friction coefficient. In this publication we suppose that the shear rate is still relatively low and the change of volume fraction and friction coefficient with shear rate can be neglected.

We discretize the system on a mesoscopic scale which is taken to be the average particle diameter $\langle d \rangle$. Due to symmetry reasons the simple shear setup can be described by a one-dimensional system in the z direction. Thus the whole system is represented by a one-dimensional lattice where each site is characterized by a single static parameter the effective friction coefficient.

Shear stress, which increases due to wall displacement, will be released by localized plastic events which are the shear bands. Shear bands will appear at positions where the shear stress surpasses the local Coulomb limit. The observed macroscopic shear stress is determined from these local plastic events. The shear band, which is a dynamic event, has a timescale τ_0, and an effect on the local material properties which we consider here completely random. Thus the material will be modeled by local randomization of the effective friction coefficient at the shear band and in the neighboring sites. This local randomization is the essence of our mesoscopic model. In this publication for simplicity we always use uniform random numbers between 0 and 1.

Wall velocity in this mesoscopic picture will be represented by Δ which is the shear stress increase in a unit time which is the characteristic timescale of a shear band τ_0. We further assume that the stress redistribution is fast compared to τ_0 which is in line with our previous assumptions that the shear is not yet in the fully dynamic regime.

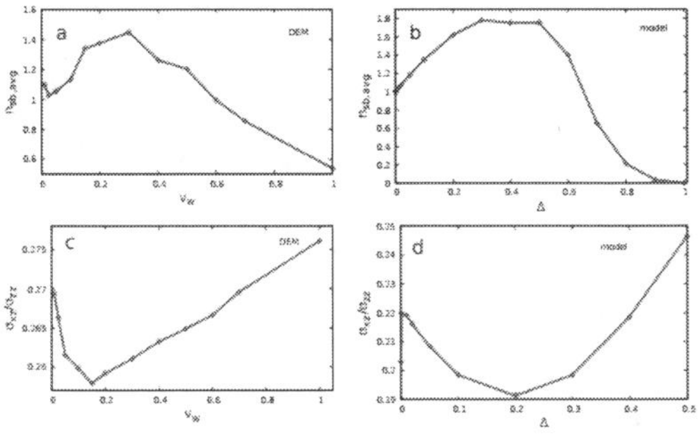

Figure 4. Average number of shear bands in (a) DEM simulation as function of the wall velocity and in (b) mesoscopic model as function of Δ. Average shear stress in (c) DEM simulation as function of the wall velocity and in (d) mesoscopic model as function of Δ.

In summary our mesoscopic model is defined as follows: The system is characterized by a one-dimensional mesoscopic system due to symmetries of the setup. The material is described by a single scalar: the local effective friction coefficient. The shear stress in the material is homogeneous and is increased by Δ during the timescale τ_0. All positions for which the shear stress exceeds the local Coulomb threshold, the material fails and a local shear band develops. The local shear band reorganizes the material under and in the immediate vicinity of the shear band modeled by random change of the local effective friction coefficient. With the plastic events the stress is released from the system but again it will be increased in steps of Δ.

3.2 Results

In Fig. 4 (a)-(b) we measured n_{sb} the average number of simultaneous shear bands both in the DEM simulation and the mesoscopic model. In the simulation a shear band was defined as follows: The strain rate at a given point was defined by the local velocity difference which was measured at 5 particle diameter distance to reduce error from fluctuations. The the average γ_a and the maximal γ_m local strain rate was determined. The value of γ_m was limited to $3\gamma_a$, to disregard excess fluctuations. A threshold was set to be $\gamma_t = (\gamma_m + \gamma_a)/2$, halfway between the maximum and the average. The number of shear bands S is defined by the number of intervals where the local shear rate exceeds the threshold γ_t. Only configurations with plastic event were considered.

In the mesoscopic model it is not enough to calculate the number of failed sites to have the number of shear bands if we want to compare it to DEM simulation. Failed sites next to each other appear as one single shear band in DEM data. Therefore in the mesoscopic model we imitated this and we counted clustered shear bands as one.

Clustering adjacent shear bands has the important characteristics that one should observe a peak at the transition, since on either limits of global shear rate the value of n_{sb} is small: For low shear rates there is only one shear band, for very large shear rate there is shear band everywhere so it is just a continuous homogeneous shear for which our method gives n_{sb}=0. In between there are cases with higher values of n_{sb}>1.

The resulting curves in Fig. 4 (a)-(b) show the predicted feature: the curve start from $n_{sb} = 1$ there is a maximum at v_w=0.3 and at Δ=0.4 and for large shear n_{sb} drops well below one. This result affirms the appearance of multiple shear bands in the transitional regime.

Let us note that in a completely different field of opinion dynamics in a collaborative environment similar transition was found. The crossover is between a step-wise function and an overall increasing one with maximum density of steps at the transition [16].

The average shear stress σ_{xz}/σ_{zz} in Fig. 4 (c)-(d) shows also interesting features. It has a minimum at lower values of v_w and Δ than the position of the maximum of the number of shear bands. The presence of a minimum in the macroscopic shear stress is counterintuitive as in general the shear stress increases with the shear rate [1]. It is thus surprising that we see an opposing trend for low wall velocities. The model can give an understanding of this feature.

In our model a shear band restructures three sites, the one under the shear band and the two neighboring ones. Since this is done uncorrelatedly of the previous state these local friction coefficients will be random. So in the next step the weakest one has a value which is the minimum out of 3 independent random numbers. In our case it is 1/4. In Fig. 4 (d) the obtained value for $\Delta \simeq 0$ is somewhat smaller which is due to the fact that the minimal site may jump to previous positions of the current avalanche.

The number of refreshed sites increases with the number of shear bands so a smaller minimum can be chosen. There is however a counter effect that not only the minimum but all sites in a Δ interval fail thus increasing the shear stress which effect eventually wins and increases the stress rate. At very high shear rates no minimization takes place as almost every site is a shear band.

In DEM simulations the same effect may happen, namely that if sheared moderately fast, then the multiple shear band weaken the system which can be sheared easier. On the other hand faster shear requires too many shear bands which does not allow for any optimization. Of course for even faster shear dynamic effects become more important and the shear stress will increase further.

Furthermore the correspondence between the model and the simulation indicates that the duration of a shear band is $\tau_0 \simeq 1$ which is supported by our preliminary results in DEM simulations.

The above observation helps us understand the difference between the position of the minimum in the macroscopic stress and the number of shear bands as the previous one incorporates two effects in a nontrivial way: the number of shear bands and the stress increase in unit time.

The kinetic energy in the steady state is only available in the DEM simulation. Figure 6 shows E_{kin}/v_w^2 as function v_w. For large wall velocities the velocity profile of the

DOI: 10.1051/epjconf/201714003084

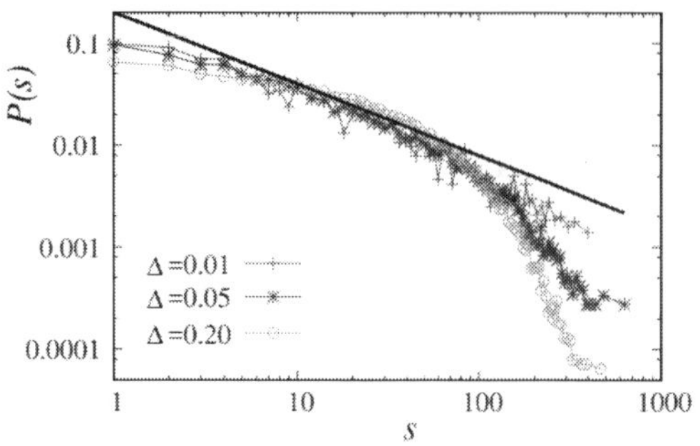

Figure 5. Avalanche size distribution for different values of Δ for the mesoscopic model. The solid line is only to drive the eye.

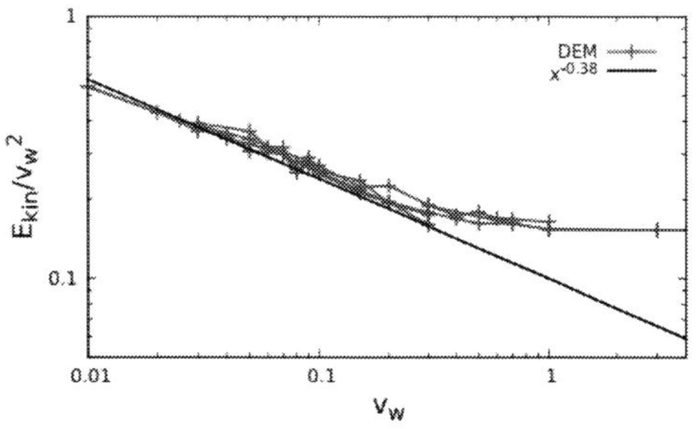

Figure 6. The kinetic energy rescaled with the square of the wall velocity as function of the wall velocity.

material is linear so the total kinetic energy stored in the system should be proportional to the square of the wall velocity. This is shown by the constant value for large v_w. On the other hand, for small wall velocities we have seen a power law increase which indicates super linear velocity profiles [17] which were found in quasi-static shear. There is a nice crossover between the two regimes at around $v_w \simeq 0.3$ which coincides with the maximum of the number of shear bands rather than with the minimum of the shear stress.

This latter finding indicates that the transition from quasi-static to dynamic regime is driven by multiple shear bands. The minimum of the shear stress is most probably the result of the crossover of two effects: one which decreases the shear stress (parallel shear bands in the avalanche), and an other one which is the mobilization of more stable regimes.

4 Summary

In this paper we have introduced a mesoscopic model for relatively slow granular shear. We studied the transition from quasi-static to dynamic shear. The model shows surprisingly similar features to the DEM simulations, namely there is a minimum in the shear stress and a maximum at the number of simultaneous shear bands at larger wall velocity. The long range order in quasi-static regime manifested in the power law distribution of avalanches is gradually destroyed by the presence of multiple shear bands in both cases. This indicates that real quasi-static system in infinite samples can be observed only by infinitely slow shear. The crossover in the kinetic energy suggests that the number of shear bands is the most important indicator of the transition to the fully dynamic regime.

It is still an open question how the presence of multiple shear bands is able to lower the average shear stress of the material, and what is the effect of the system size for the above quantities.

Acknowledgements

Support from the ÚNKP-16-2-I. New National Excellence Program of the Ministry of Human Capacities and from the Hungarian National Research, Development and Innovation Office NKFIH under grant OTKA K116036 is acknowledged.

References

[1] G. MiDi, The European Physical Journal E **14**, 341 (2004)
[2] P.C. Johnson, R. Jackson, Journal of Fluid Mechanics **176**, 67 (1987)
[3] C.S. Campbell, Annual Review of Fluid Mechanics **22**, 57 (1990)
[4] T. Börzsönyi, R.E. Ecke, J.N. McElwaine, Physical Review Letters **103**, 178302 (2009)
[5] V. Richefeu, M.S. El Youssoufi, F. Radjai, Physical Review E **73**, 051304 (2006)
[6] Z. Shojaaee, J.N. Roux, F. Chevoir, D.E. Wolf, Physical Review E **86**, 011301 (2012)
[7] L. Bocquet, A. Colin, A. Ajdari, Physical Review Letters **103**, 036001 (2009)
[8] M. Manning, J. Langer, J. Carlson, Physical Review E **76**, 056106 (2007)
[9] M. Talamali, V. Petäjä, D. Vandembroucq, S. Roux, Comptes Rendus Mecanique **340**, 275 (2012)
[10] S. Plimpton, P. Crozier, A. Thompson, Sandia National Laboratories **18** (2007)
[11] Z. Shojaaee, L. Brendel, J. Török, D.E. Wolf, Physical Review E **86**, 011302 (2012)
[12] B. Szabó, J. Török, E. Somfai, S. Wegner, R. Stannarius, A. Böse, G. Rose, F. Angenstein, T. Börzsönyi, Physical Review E **90**, 032205 (2014)
[13] K.A. Dahmen, Y. Ben-Zion, J.T. Uhl, Nature Physics **7**, 554 (2011)
[14] K.M. Salerno, C.E. Maloney, M.O. Robbins, Physical Review Letters **109**, 105703 (2012)
[15] P. Bak, K. Sneppen, Physical Review Letters **71**, 4083 (1993)
[16] J. Török, G. Iñiguez, T. Yasseri, M. San Miguel, K. Kaski, J. Kertész, Physical Review Letters **110**, 088701 (2013)
[17] F. Da Cruz, F. Chevoir, J. Roux, I. Iordanoff, Tribology Series **43**, 53 (2003)

 DOI: 10.1051/epjconf/201714003085

Flowability assessment of weakly consolidated powders

Alexandros Georgios Stavrou[1,*], *Colin* Hare[1], and *Ali* Hassanpour[2], and *Chuan-Yu* Wu[1]

[1]Department of Chemical and Process Engineering, University of Surrey, Guildford, GU2 7XH, UK
[2]School of Chemical and Process Engineering, University of Leeds, Leeds, LS2 9JT, UK

Abstract. The inability of cohesive powders to flow consistently and reliably is a major cause of process downtime and reduced efficiency across a wide range of powder processing industries. Most methods to assess powder flowability fail at low consolidation pressures (<1 kPa). In this paper, the ball indentation technique is used to assess the flow behaviour of two powders at low stresses by determining the bed hardness. In parallel, the powders are subjected to shear testing in a range of high stresses, with the derived unconfined yield strength used, along with the indentation hardness to define the constraint factor (*C*). By using the latter, which is considered independent of the preconsolidation stress applied, the unconfined yield strength of the powders at low stresses are inferred from the penetration hardness measurements.

1 Introduction

Nowadays, a large proportion of the materials processed and the products manufactured in industry are in the form of particulate solids. Several industries, such as pharmaceuticals, petrochemicals, cosmetics, food, materials, detergents and metallurgy comprise of processes that involve granular materials, such as fluidization, pneumatic conveying, blending, and storage in bins or hoppers. One of the major concerns regarding bulk solids handling is the flowability of powders in order to properly design the process and the equipment needed.

There have been various techniques developed for assessing powder flowability over the years, based on a variety of different principles. Some of them are quick and easily operable, such as the angle of repose, which are suitable only for a rough classification of flowability for quality control, while others, such as the uniaxial compression test and most notably shear testers are well established, providing quantitative results and offering stress control to the operator [1]. When it comes to evaluating the flow behaviour of cohesive powders at low consolidation pressures, which are involved in processes of great industrial interest e.g. in filling and dosing of powders in capsules, in feeding powders for packing and tableting machines, and dispersion in dry powder inhalers (DPI), all of the aforementioned methods fail. Furthermore, they are incapable of assessing flow at strain rates above the quasi-static regime as defined by Tardos *et al.* [2]. To address these issues, Hassanpour and Ghadiri introduced the ball indentation technique [3], which is based on penetration tests, with its operational window established by Zafar *et al.* and Pasha *et al.* [4, 5]. It involves a powder bed that is prepared inside a cylindrical die and is pre-consolidated by a piston. Then, the compressed bed is penetrated by a spherical indenter at a specified strain rate, until a certain load is reached, and then unloaded (Fig.1).

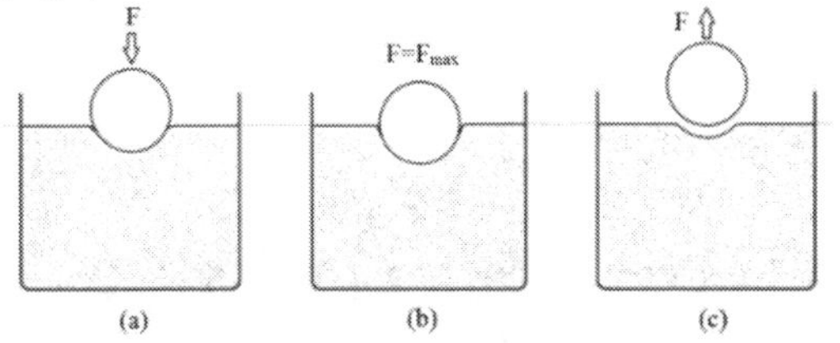

Fig. 1. Indentation step of the ball indentation technique [5].

Flowability is evaluated by calculating hardness, *H*, representing flow stress, from the force-displacement response of the bed. Hardness can be linked to the unconfined yield strength, *Y*, which is derived from shear testing or uniaxial compression tests, via a constraint factor, *C*, which is dependent on particle properties, although cannot yet be determined *a priori*. Shear cells are not consistent and reliable, and are often unable to generate steady-state shear in the region of low stresses, below 1 kPa, therefore a common practice is to extrapolate the yield locus to zero major principal stress, which often leads to an overestimation of unconfined yield strength [4]. On the contrary, ball indentation enables unconfined yield strength at low stresses to be inferred from hardness measurements via *C* at low stresses, providing invaluable data for hopper and silo design.

In the light of the above, this work has been centred on the evaluation of flowability of two different powders at low stresses by using the ball indentation technique to infer their unconfined yield strength.

* Corresponding author: a.stavrou@surrey.ac.uk

2 Materials and methods

2.1 Materials used

For this experimental investigation two different powders were chosen. The first powder was a model material, soda lime glass beads (Sigmund Lindner GmbH), which were sieved to obtain four consecutive single sieve cuts: 53-63, 63-75, 75-90 and 90-106 μm. The beads were made cohesive by applying a commercially available silane coating, known as Sigmacote (Sigma-Aldrich), as conventional glass beads have very low interface energy. The second material that was studied was titanium dioxide (TiO_2), also known as titania. The specific powder sample used was Tiona AT-1 (98.5% Anatase, Cristal Global), a technically pure, dry milled anatase pigment. It is made of smooth, rounded primary particles with a size of roughly 100 nm. The appearance of the powder is white and cohesive, with loose clusters easily forming during powder handling and storage. Titania has the ability to auto-agglomerate, due to possible agitation of the powder bed, such as mechanical vibration [6].

2.2 Methods used

In this work two methods were used to assess powder flowability, ball indentation and shear testing. Shear testing was performed using an automated Schulze Ring Shear Tester. A detailed description of the principles of shear testing can be found elsewhere [1]. From shear testing, the unconfined yield strength, Y, of a powder can be determined, which is the stress required to fail a consolidated mass of material to initiate flow. All the shear cell measurements are repeated three times at the same conditions and error bars indicate standard deviation of the measured values.

Ball indentation was investigated using an Instron 5566 mechanical testing machine (Instron Corp., USA). The samples were pre-consolidated to different levels by uniaxial loading in a die by a stainless steel piston using a 10 N load cell, which has a resolution of 0.25 mN. The cylindrical die used here is made of stainless steel and has a height of 20 mm and an internal diameter of 20.5 mm. The loading rate was kept constant at 1 mm/s, therefore testing at quasi-static conditions. The pre-consolidated samples were then subjected to indentation using a high precision glass spherical ball indenter of 3.969 mm diameter, supplied by Sigmund Lindner GmbH. The indenter was fixed to a small loading rod using super glue. Indentations were carried out only at the central zone on each specimen in order to avoid wall friction effect and increase the reliability of the data for different compression levels. The criteria for choosing bed diameter, bed height and indenter's radius are chosen based on the standardisation of the method by Zafar *et al.* and Pasha *et al.* [4, 5].

The flowability is determined by the force-displacement response of the powder bed. The applied load (F) and the displacement of the piston/indenter were continuously recorded throughout the process. Hardness, H, is given by the ratio of the maximum indentation load, F, to the projected area of the impression, A:

$$H = \frac{F_{max}}{A} \quad (1)$$

where A is the projected area of the impression of the indenter, which can be expressed in terms of the size of the indenter, d, and final depth of impression, h:

$$A = \pi(dh - h^2) \quad (2)$$

Tabor established a relationship between indentation hardness and yield stress for continuum materials that is represented by the following equation [7]:

$$H = CY \quad (3)$$

where Y is the yield stress and C is the constraint factor.

The value of C for a given powder is not known *a priori*. Zafar reported values of the constraint factor for various materials, ranging from 1.5-3 for silanised glass beads, to 6-8 for different lactohale particle sizes [4].

In this work all the hardness measurements are repeated five times at the same conditions and error bars indicate standard deviation of the measured values. Also, throughout the experiments of the current research, relative humidity and temperature were recorded and were in the range 15-25°C and 40-65% RH, respectively.

3 Results and discussion

3.1. Shear Testing

The two powders under investigation were subjected to shearing for a range of pre-consolidation stresses. Unconfined yield strength is investigated as a function of major principal stress as presented in Figures 2 and 3 for the two materials.

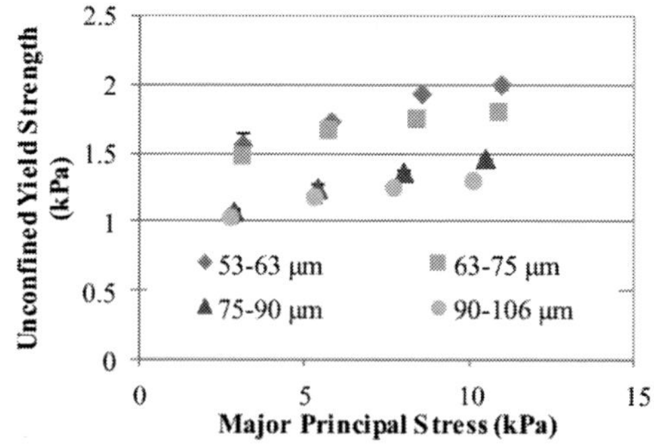

Fig. 2. Shear cell results for silanised glass beads.

 DOI: 10.1051/epjconf/201714003085

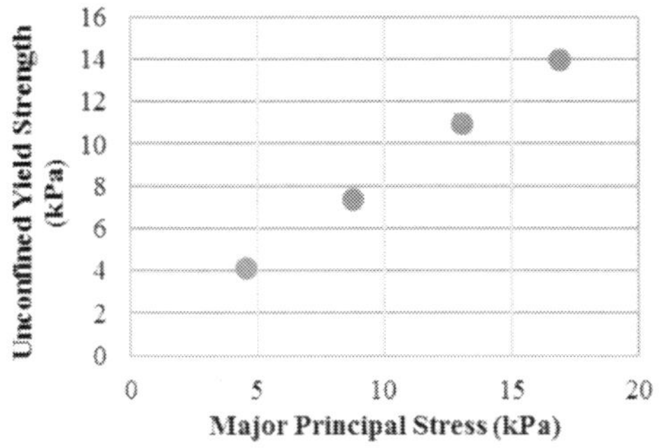

Fig. 3. Shear cell results for titania.

It can be seen from the data obtained from shear testing that the unconfined yield strength increases linearly with the increase in major principal stress due to an increased packing fraction, most notably for titania.

3.2 Ball Indentation

Hardness must be independent of indentation load in order to represent the plastic yield stress, therefore a range of penetrations depths that provide reliable hardness measurements has to be established first for both materials. The depth is presented as dimensionless penetration depth, which can be calculated from the following equation:

$$h_d = \frac{h}{R_i} \tag{4}$$

where R_i is the radius of the indenter.

For silanised glass beads, dimensionless penetration depths of 0.3-0.7 are found to give reliable hardness measurements (Fig. 4). At very low indentation loads/penetration depths, the force on the indenter is due to its interactions with only a few particles, and hence the calculations of the projected area of the impression, and therefore hardness, are inaccurate, since plastic deformation is not initiated.

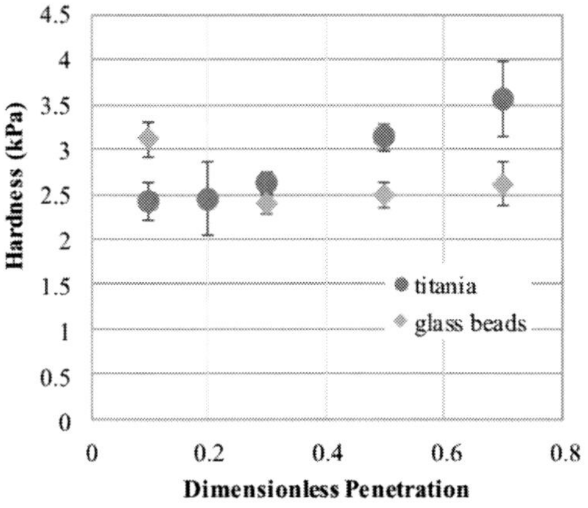

Fig. 4. Hardness vs dimensionless penetration at 1 kPa.

In contrast, for titania, it is observed that dimensionless penetration depths of 0.1-0.3 provide reliable hardness measurements. Titania's particle size is significantly smaller than the glass bead sieve cuts, therefore the indenter contacts a sufficient number of particles to cause plastic deformation and to obtain a reliable hardness measurement a short distance after penetrating the bed.

Now that a range of penetration depths, within which hardness remains stable has been established for both materials, indentation tests at dimensionless penetration depths of 0.2 for titania and 0.5 for silanised glass beads are conducted in a range of stresses with the results presented in Figures 5 and 6, respectively.

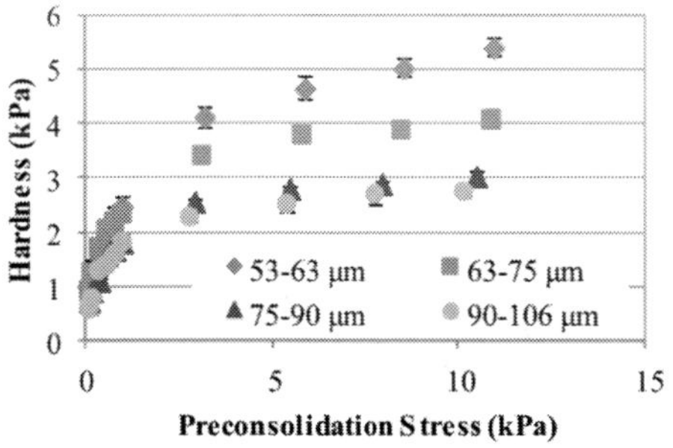

Fig. 5. Ball indentation results for silanised glass beads.

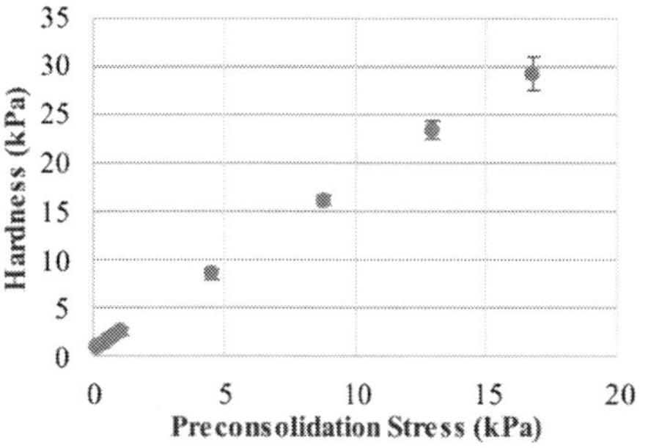

Fig. 6. Ball indentation results for titania.

Hardness is observed to increase with pre-consolidation stress, for both materials tested. Powders which have been pre-consolidated to lower pressure levels have a weaker resistance to plastic deformation at the same penetration depth, due to a decreased packing fraction. For titania, hardness increases linearly with pre-consolidation stress throughout the range of stresses applied. In contrast, for all the single sieve cuts of coated glass beads the increase is much more sharp in the range of low stresses (<1 kPa). It can also be seen that the smaller the particle size, the larger the exhibited hardness. This trend is clear among consecutive sieve cuts at high stresses, whereas at low stresses the difference is clear among alternating single sieve cuts.

3.3 Combined methods

The constraint factor is calculated using Eq. 3, and is found to be almost constant for both materials (Fig. 7). Also, in the case of silanised glass beads, the smallest particles were found to have the greatest value.

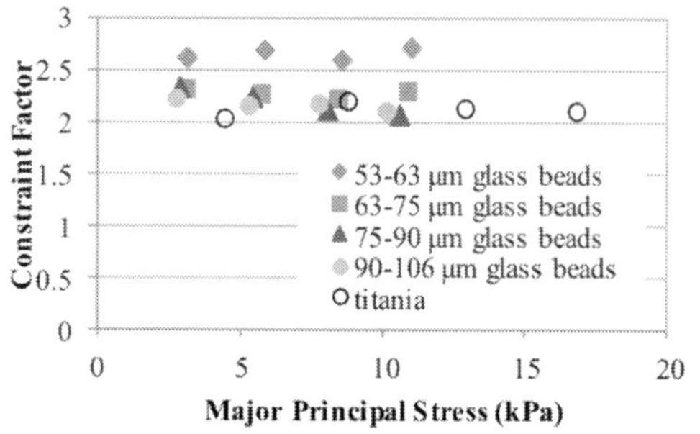

Fig. 7. Constraint factor for silanised glass beads and titania.

In order to calculate the unconfined yield strength at low consolidation pressures, it is hypothesized that the constraint factor remains constant throughout the whole range of stresses. Using the ball indentation hardness values, the unconfined yield strength is calculated at low consolidation levels using Eq. 3. (Figs. 8-9).

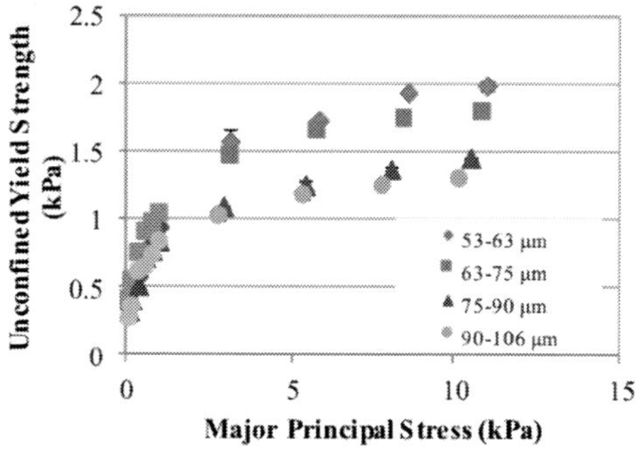

Fig. 8. Unconfined yield strength for silanised glass beads.

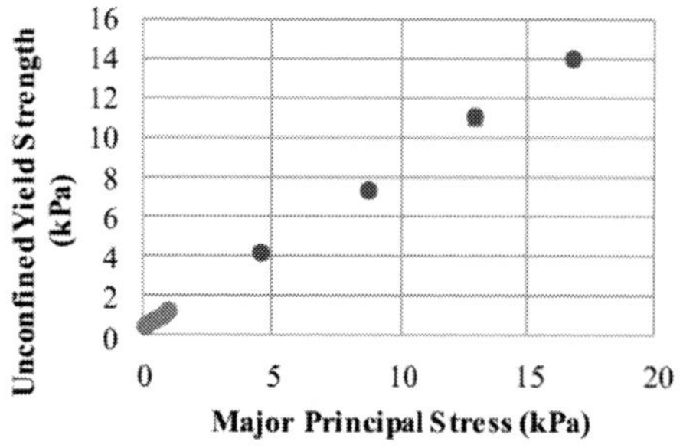

Fig. 9. Unconfined yield strength for titania.

In the case of silanised glass beads, shear tests predicted greater values of unconfined yield strength at low stresses than ball indentation, whereas for titania the extrapolation of the yield locus from shear cell measurements correlates well with the ball indentation findings.

4 Conclusions

Flowablity of titania and four consecutive silanised glass bead sieve cuts was assessed using the ball indentation technique and shear testing. Ball indentation results exhibited a very good reliability of the technique at both high and low stresses. Hardness was found to increase with pre-consolidation stress for both materials, but much more sharply for coated glass beads at low compaction levels. Also, ball indentation provided a very good sensitivity of hardness to particle size; the smaller the particle size, the greater the hardness. The constraint factor was found to be constant throughout the range of stresses for which it was calculated, having different values for different materials. Finally, by hypothesizing that the constraint factor remains constant at low stresses, for which it cannot be determined experimentally, the results of the yield stress inferred from ball indentation measurements showed that in the case of silanised glass beads shear cell extrapolation would result in an overestimation of yield stress. On the contrary, titania's inferred yield stress follows the same trend as the derived from yield locus extrapolation.

Future work will broaden the range of tested materials, as well as utilising the Distinct Element Method (DEM) to investigate the influence of particle properties on constraint factor and examine its variation at low stresses.

The authors greatly appreciate the financial support from the International Fine Particle Research Institute (IFPRI), without which this research would not have been possible.

References

1. J. Schwedes, J., Gran. Matter **5(1)**, 1-43 (2003)
2. G. I. Tardos, S. McNamara, I. Talu, Powder Tech. **131(1)**, 23–39 (2003)
3. A. Hassanpour, M. Ghadiri, Part. Part. Syst. Charact. **24(2)**, 117–123 (2007)
4. U. Zafar, C. Hare, A. Hassanpour, M. Ghadiri, Powder Tech. **310**, 300-306 (2017)
5. M. Pasha, C. Hare, A. Hassanpour, M. Ghadiri, Powder Tech. **233**, 80-90 (2013)
6. N. Ku, C. Hare, M. Ghadiri, M. Murtagh, R. A. Haber, Powder Tech. **286**, 223-229 (2015)
7. D. Tabor, *The Hardness of Metals* (1951)

Collapse of tall granular columns in fluid

Krishna Kumar[1,*], *Kenichi* Soga[2,**], and *Jean-Yves* Delenne[3,***]

[1] *Department of Engineering, University of Cambridge, UK*
[2] *University of California, Berkeley, USA*
[3] *INRA, CIRAD, University Montpellier 2 and SupAgro, France*

Abstract. Avalanches, landslides, and debris flows are geophysical hazards, which involve rapid mass movement of granular solids, water, and air as a multi-phase system. In order to describe the mechanism of immersed granular flows, it is important to consider both the dynamics of the solid phase and the role of the ambient fluid. In the present study, the collapse of a granular column in fluid is studied using 2D LBM - DEM. The flow kinematics are compared with the dry and buoyant granular collapse to understand the influence of hydrodynamic forces and lubrication on the run-out. In the case of tall columns, the amount of material destabilised above the failure plane is larger than that of short columns. Therefore, the surface area of the mobilised mass that interacts with the surrounding fluid in tall columns is significantly higher than the short columns. This increase in the area of soil - fluid interaction results in an increase in the formation of turbulent vortices thereby altering the deposit morphology. It is observed that the vortices result in the formation of heaps that significantly affects the distribution of mass in the flow. In order to understand the behaviour of tall columns, the run-out behaviour of a dense granular column with an initial aspect ratio of 6 is studied. The collapse behaviour is analysed for different slope angles: 0°, 2.5°, 5° and 7.5°.

1 Collapse of granular columns

The collapse of a granular column on a horizontal surface is a simple case of granular flow, however, a proper model that describes the flow dynamics is still lacking. Experimental investigations have shown that the flow duration, the spreading velocity, the final extent of the deposit, and the energy dissipation can be scaled in a quantitative way independent of substrate properties, grain size, packing density, and shape of the granular material and released mass [1–4].

Collapse of granular columns on a horizontal plane exhibit two distinct flow regimes: (a) for columns with an initial aspect ratio 'a' < 1.7, a linear relation between the spread and aspect ratio can be observed, and (b) for 'a' > 1.7, a power-law relationship exists. A simple frictional dissipation model is able to capture the flow dynamics for columns with small aspect ratios [5]. However, for tall columns the flow dynamics is controlled by the initial collisional regime, resulting in a power-law relationship with the initial aspect ratio of the columns. Simple mathematical models based on conservation of horizontal momentum capture the scaling laws of the final deposit, but fail to describe the initial transition regime [6]. Granular flow is modelled as a frictional dissipation process in continuum mechanics but the lack of influence of inter-particle collision and friction on the energy dissipation and spreading dynamics is surprising.

The amount of material destabilised above the failure plane, in tall columns, is larger than that of short columns. This increase in the mobilised mass results in a significant increase in the surface area of granular mass that interacts with the surrounding fluid. The increase in the area of soil - fluid interaction results in an increase in the formation of turbulent vortices, which alter the deposit morphology. It is observed that the vortices result in formation of heaps thereby redistributing the granular mass in the flow. The distribution of mass in a granular flow plays a crucial role in the flow kinematics [2]. In order to understand the behaviour of tall columns, the run-out behaviour of a dense granular column with an initial aspect ratio of 6 is studied using Lattice Boltzmann - Discrete Element Method coupling.

2 LBM-DEM coupling

The Lattice Boltzmann Method (LBM) is an alternative approach to the classical Navier-Stokes solvers for fluid flow and works on an equidistant grid of cells, called lattice cells, which interact only with their direct neighbours [7]. The fluid domain is divided into a rectangular grid or lattice, with the same spacing 'h' in both the x- and the y-directions. The present study focuses on 2-D problems; hence, the D2Q9 momentum discretization is adopted. The fluid particles at each node are allowed to

*e-mail: kks32@cam.ac.uk
**e-mail: soga@berkeley.edu
***e-mail: delenne@supagro.inra.fr

move to their eight immediate neighbours with eight different velocities $e_i(i = 1, \ldots, 8)$. The primary variables in LB formulation are called the *fluid density distribution functions*, f_i, each representing the probable amount of fluid particles moving with the velocity e_i along the direction i at each node. In the present study, Lattice-Boltzmann - Multiple Relaxation Time (LBM-MRT) is used for numerical stability. The nine eigen values of **S** are all between 0 and 2 so as to maintain linear stability and the separation of scales. In this study, the Large Eddy Simulation (LES) is adopted to solve for turbulent flow problems. The separation of scales is achieved by filtering of the Navier-Stokes equations, from which the resolved scales are directly obtained and unresolved scales are modelled by a one-parameter Smagorinski sub-grid methodology [8].

Lattice Boltzmann approach can accommodate large grain sizes and the interaction between the fluid and the moving grains can be modelled through relatively simple fluid – grain interface treatments. Further, employing the Discrete Element Method (DEM) to account for the grain – grain interaction naturally leads to a combined LB – DEM procedure [6]. The Eulerian nature of the LBM formulation, together with the common explicit time step scheme of both LBM and DEM makes this coupling strategy an efficient numerical procedure for the simulation of grain – fluid systems. A no-slip boundary condition is used to simulate grain – fluid and grain – grain interactions. The solid grains inside the fluid are represented by lattice nodes. The discrete nature of lattice, results in a stepwise representation of the surfaces, which are circular, hence sufficiently small lattice spacing is adopted. In order to model flow through interconnected pore space, a reduction in radius (The hydrodynamic radius r) is assumed during the LBM computation [5].

3 Numerical set-up

In this study, a 2D poly-disperse system ($d_{max}/d_{min} = 1.8$) of circular discs in fluid was used to understand the behaviour of granular flows on slopes (see fig. 1). The soil column was modelled using 2500 discs of density 2650 kg m^{-3} and a contact friction angle of 26° and a mean packing fraction of ≈ 0.82. The collapse of the column was simulated inside a fluid with a density of 1000 kg m^{-3} and a kinematic viscosity of 1×10^{-6} m^2 s^{-1}. The fluid domain was modelled with 5 to 9 million LB nodes. GPU computing was used to simulate the problem. The choice of a 2D geometry has the advantage of cheaper computational effort than a 3D case, making it feasible to simulate very large systems. A granular column of aspect ratio 'a' of 6 was used. A hydrodynamic radius of $r = 0.85\ R$ is adopted. Dry analyses were also performed to study the effect of hydrodynamic forces on the run-out distance. The collapse of a tall granular column on slopes of 0°, 2.5°, 5° and 7.5° are studied.

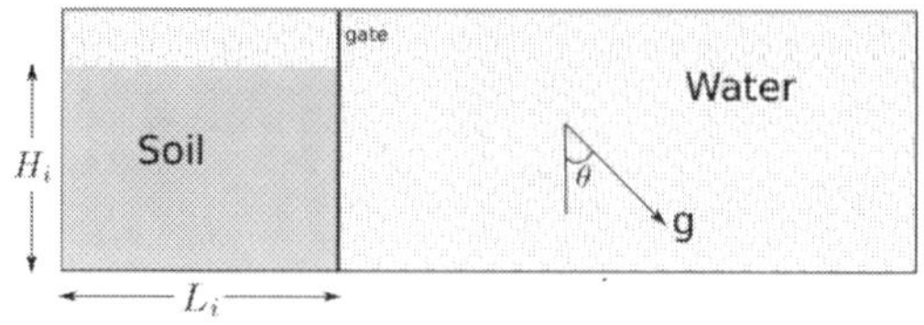

Figure 1: Underwater granular collapse set-up. The slope angle is represented by θ.

4 Collapse on a horizontal plane

Snapshots of the collapse of an aspect ratio 6 column in fluid on a horizontal surface are shown in fig. 2. The initial stage of collapse is characterised by the free-fall of grains above the failure surface. The height of the static region, which is below the shear-failure surface, is shorter than the total height of the column. As the grains experience free-fall they interact with the surrounding fluid. Unlike the dry condition, as the grains undergo free-fall due to gravity, they interact with the surrounding fluid experiencing drag forces. This results in a significant drop in the kinetic energy available for the flow. However, no vortices are observed during the initial stage of collapse.

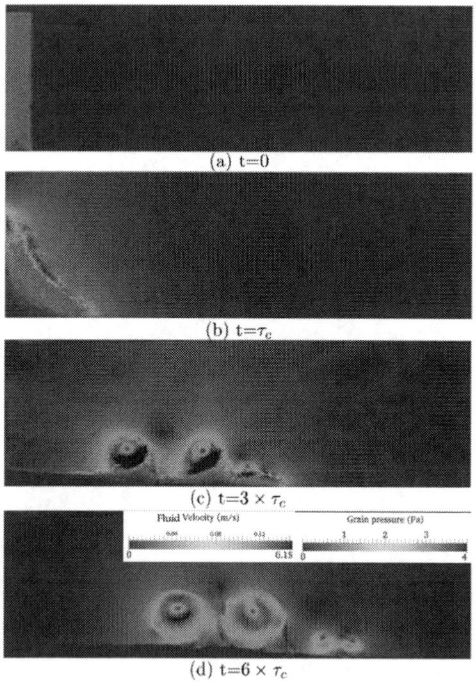

(a) t=0
(b) t=τ_c
(c) t=3 × τ_c
(d) t=6 × τ_c

Figure 2: Flow evolution of a granular column collapse in fluid (a = 6) on a horizontal surface.

As the grains reach the static region, they interact with the neighbouring grains and the kinetic energy gained during the free fall is converted into horizontal acceleration. Uniquely during this stage ($t = 3\tau_c$) the interactions between the soil grains on the surface with the surrounding fluid result in the formation of eddies. The number of eddies formed during the flow is proportional to the surface area of the granular mass interacting with the fluid. Hydroplaning can be observed at the flow front ($t = 3\tau_c$). Two large vortices with almost the same size can be observed at the final stage of collapse. The soil grains on the surface experience suction due to formation of eddies and this results in formation of heaps of granular mass in front of each vortex. Unlike short columns, these vortices have a significant influence on the mass distribution along the run-out. Heaps of granular material can be observed in front of each vortex. The number of vortices formed during a collapse is found to be proportional to the amount of material destabilised, i.e., the length of free-surface interacting with the fluid influences the number of vortices generated during the collapse. The reappearance of force chains at $t = 6\tau_c$ indicates the granular mass is consolidating resulting in an increase in the shear strength. The formation of heaps, although in evidence, doesn't significantly affect the distribution of mass in the case of collapse on a horizontal plane.

5 Collapse on slopes

Snapshots of the collapse of a granular column ($a = 6$) on a inclined plane at angle of 5° are shown in fig. 3. The collapse on a slope of 5° show flow evolution behaviour similar to the case of collapse on a horizontal plane. The vortices are formed only during the horizontal spreading stage $t = 3\tau_c$, but the number of vortices formed during the collapse is more than the collapse on a horizontal plane. However, as the flow progresses a single large vortex engulfs other smaller vortices, thus having a significant influence on the mass distribution.

Figure 4 shows the distribution of mass and the packing density at $t = 6\tau_c$ and $t = 8\tau_c$. A heap can be observed in front of the large vortex almost at the middle of the flow. The height of the heap formed in the middle of the granular flow is higher than the collapse height next to the wall. However, when the flow comes to rest and the vortex moves away from the flowing surface, the mass present in the heap gets redistributed (as seen at $t = 8\tau_c$). This behaviour is significantly different from that observed in the case of short columns. More denser packing can be observed in front of the eddies, resulting in spectrum of dense - loose regions.

5.1 Effect of slope angle

In order to understand the influence of slope angles on the run-out behaviour, the collapse of a granular column with an initial aspect ratio of 6 is performed on slopes of 0°, 2.5°, 5° and 7.5°. The run-out evolution with time for different slope angles are presented in fig. 5. The run-out distance increases with increase in the slope angle, however the run-out distance in the fluid is significantly shorter than the dry condition. The slow evolution of run-out in the submerged condition is due to the delay in the dissipation of large negative pore-pressures developed during the initial stage of the collapse. The formation of eddies during the flow indicates that most of the potential energy gained

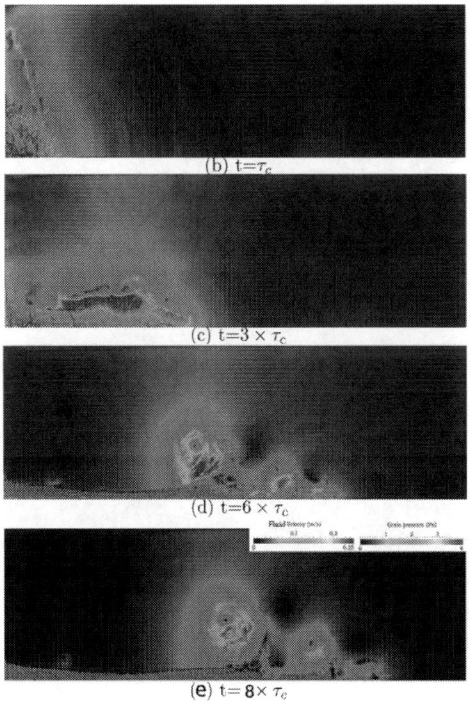

Figure 3: Flow evolution of a granular column collapse in fluid (a = 6) on a slope of 5°. Shows the velocity profile of fluid due to interaction with the grains (red - higher velocity).

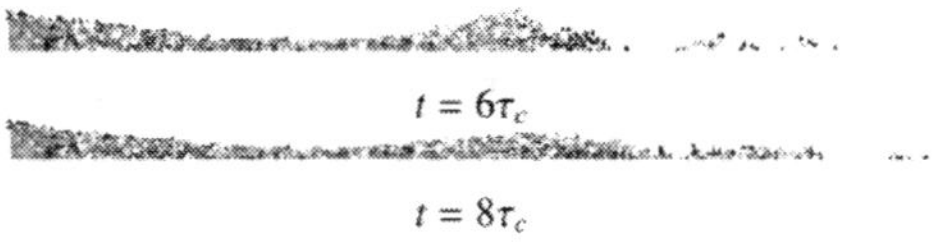

Figure 4: Packing density (black - maximum density) of a granular column collapse in fluid (a = 6) on a slope of 5°.

during the free-fall is dissipated through viscous drag and turbulence. This effect predominates over the hydroplaning that is observed during the flow resulting in a shorter run-out distance in the case of fluid. The evolution of the normalised height with time (fig. 6) for collapse on different slope angles indicates that the amount of material destabilised in fluid is less than the dry conditions. This is due to the drag forces experienced by the grains, which retards the quantity and the rate of collapse.

Figure 7 shows that in the dry condition, the kinetic energy available for the flow increases proportionally with the slope angle. The amount of kinetic energy available for the flow in the submerged condition is almost half that of the dry condition. However, in the case of submerged flow, slopes 0° and 2°, and 5° and 7.5° have similar kinetic energy evolution. The longer tail observed in the submerged granular collapse can be attributed to the formation and migration of eddies and subsequent mass redistribution into heaps.

 DOI: 10.1051/epjconf/201714009041

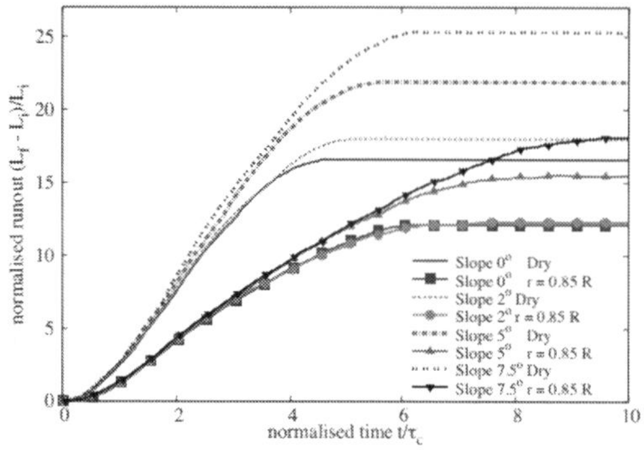

Figure 5: Evolution of run-out for a column collapse in fluid (a = 6).

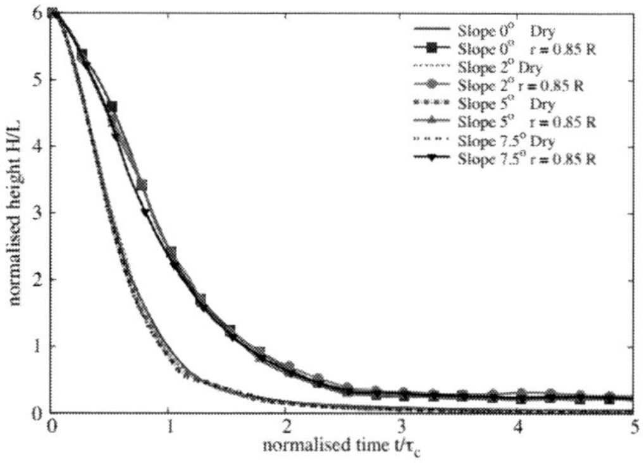

Figure 6: Evolution of height with time for a column collapse in fluid (a = 6).

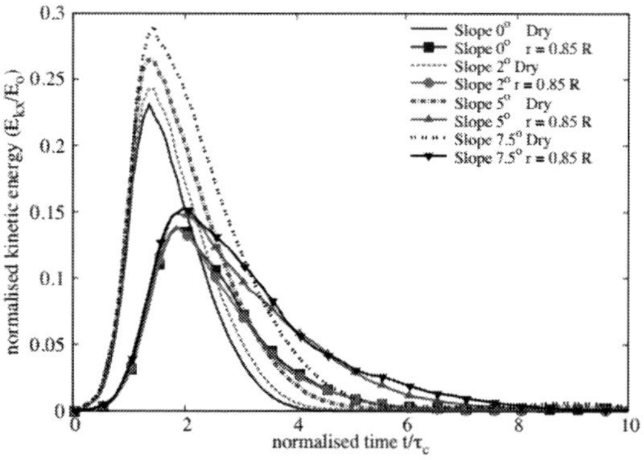

Figure 7: Evolution of horizontal kinetic energy with time for a granular column collapse in fluid (a = 6).

Figure 8 shows that the vertical kinetic energy in the fluid condition dissipates slower, in comparison to the free-fall release observed in the dry condition. The slower dissipation is attributed to the viscous drag force experienced by the grains. However, the slope angle has negligble effect on the evolution of vertical kinetic energy in submerged condition. Where as, dry conditions show ≈ 15% increase in the peak vertical kinetic energy when the slope angle changes from 0° to 7.5°.

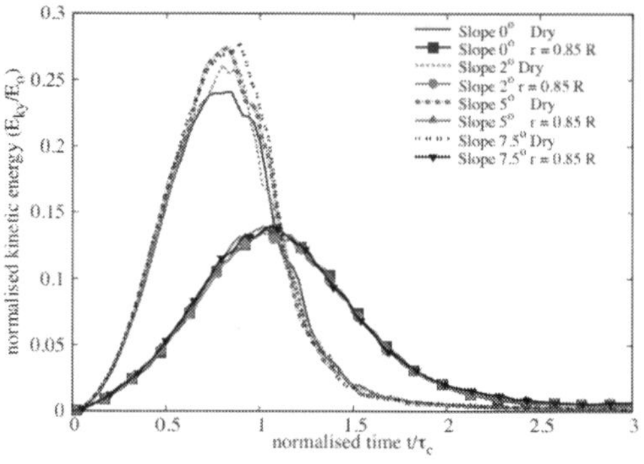

Figure 8: Evolution of vertical kinetic energy with time for a granular column collapse in fluid (a = 6).

6 Summary

The behaviour of tall columns is significantly different from that observed in short columns. The slope angle has a strong influence on the number and size of eddies during the flow. The eddies interact with the surface of the granular flow and forms heaps in front of each vortex. This significantly affects the mass distribution and in turn the run-out evolution. Although tall cliffs are quite rare in submarine condition, further research is required to understand the influence of permeability and packing density on the run-out behaviour.

Acknowledgements

The author would like to thank the Cambridge Commonwealth and Overseas Trust for the financial support to pursue this research.

References

[1] V. Topin, F. Dubois, Y. Monerie, F. Perales, A. Wachs, Journal of Non-Newtonian Fluid Mechanics **166**, 63 (2011)

[2] L. Staron, E.J. Hinch, Granular Matter **9**, 205 (2007)

[3] E. Lajeunesse, J.B. Monnier, G.M. Homsy, Physics of Fluids **17** (2005)

[4] G. Lube, H.E. Huppert, R.S.J. Sparks, A. Freundt, Physical Review E - Statistical, Nonlinear, and Soft Matter Physics **72**, 1 (2005)

[5] K.K. Soundararajan, Ph.D. thesis, University of Cambridge (2015)

[6] K. Kumar, K. Soga, J.Y. Delenne, *Granular flows in fluid*, in *Discrete Element Modelling of Particulate Media*, edited by C.Y. Wu (Royal Society of Chemistry, Cambridge, 2012), ISBN 978-1-84973-360-1

[7] X. He, Q. Zou, L.S. Luo, M. Dembo, Journal of Statistical Physics **87**, 115 (1997)

[8] J. Smagorinsky, Monthly weather review **91**, 99 (1963)

Dust generation in powders: Effect of particle size distribution

Somik Chakravarty[1,*], *Olivier* Le Bihan[2], *Marc* Fischer[1], and *Martin* Morgeneyer[1]

[1]Université de Technologie de Compiègne (UTC) – Sorbonne Universités, Laboratoire Transformations intégrées de la matière renouvelable (TIMR), Rue Roger Couttolenc, CS 60319, 60203 Compiegne Cedex, France
[2]Institut National de l'EnviRonnement Industriel et des RisqueS (INERIS), NOVA/CARA/DRC/INERIS, Parc Technologique Alata, BP2, F-60550 Verneuil-En-Halatte, France

Abstract. This study explores the relationship between the bulk and grain-scale properties of powders and dust generation. A vortex shaker dustiness tester was used to evaluate 8 calcium carbonate test powders with median particle sizes ranging from 2μm to 136μm. Respirable aerosols released from the powder samples were characterised by their particle number and mass concentrations. All the powder samples were found to release respirable fractions of dust particles which end up decreasing with time. The variation of powder dustiness as a function of the particle size distribution was analysed for the powders, which were classified into three groups based on the fraction of particles within the respirable range. The trends we observe might be due to the interplay of several mechanisms like de-agglomeration and attrition and their relative importance.

1 Introduction

Granular matter or bulk solids makes up for roughly 50% of products and 75% of the raw material used in industrial applications [1]. Applications and processes involving handling or transportation of bulk solids generate dust, referred to as small solid particles which remain suspended in the air for a prolonged period of time [2]. The propensity of a material to generate dust upon handling is known as its dustiness [3]. The risks of dust emission in a contained area such as in an occupational environment can involve inhalation of dust particles by industrial workers [4,5] or explosion of volatile dust cloud capable of creating substantial financial and human loss [6]. Regulatory measures such as the 2008/50/EC directive [7] or ATEX [8] in the EU underpins the need for assessment and containment of dust concentration in ambient air with an emphasis on the generation and exposure of fine particles such as PM2.5 and respirable fraction, responsible for significant negative impacts on human health.

Dustiness of a material and thus the risk of exposure while handling a material depends on its physical properties and the type of process at work [9]. Testing of dust generation is often practical when developing new products in industries before producing and distributing them in bulk scale. Lab scale testing of dustiness of granular material requires a low-cost tester capable of testing a wide range of material with relatively simple operations. Standardized testers such as the continuous drop and the rotating drum method according to the EN 15051'*Workplace atmospheres—Measurement of the dustiness of bulk materials—Requirements and reference test methods*' requires large amounts of powder (35cm^3 or 500g) and can give disparate results for a range of industrial minerals [4] due to the difference in stressing/agitation energy and the timescale of agitation, pointed out by several authors [4,10].

New testers such as the vortex shaker (VS) enable testing dustiness of powders using a small fraction of the powder quantity required for the standardized testers [3]. They are especially suitable for testing micro- and nano-scale powders typically used in catalysts and pharmaceutical industries where the powder test quantity is low and costs are high. Furthermore, they are capable of testing powders for different energy levels by varying vortex speed and time. Morgeneyer et al. [3] and [11] used the VS method to study the effects of tester parameters on dust generation of micron-sized alumina particles and carbon nano-tubes (CNTs), respectively. In [3], the effect of the VS speed and sample mass on the dust generated from alumina were studied whereas [11] dealt with the effect of change in tube diameter in addition to vortex speed on dustiness of CNTs.

Measuring dustiness and the risk of exposure in different environmental conditions requires testing and characterizing powder properties by their effect on dustiness as it can enable understanding the role of different powder properties on the dust generation mechanisms. Since dustiness of a powder depends on several parameters, studying the effect of each physical parameter on dust generation requires testing the same material while changing one parameter at a time keeping the other powder parameters and environmental conditions constant.

* Corresponding author: somik.chakravarty@utc.fr

In this study, respirable dustiness for eight calcium carbonate powders with similar physico-chemical properties were tested. The results were used to analyze the effect of particle size distribution (PSD) on the evolution of aerosol concentrations and size distributions. The powder samples were divided into three groups based on their PSD (and especially respirable fractions of particles already present in the powder) and attempted to identify *'potential scenarios'* or generation mechanisms prevalent in each group.

2 Material and experimental methods

2.1 Sample material

The Eskal series of calcium carbonate powders (KSL Staubtechnik GmbH, Germany) were used as the test material. They are standard test powders used in various industries and also as a reference powder for calibration of equipment [12] due to their high roundness (close to 0.9), and insensitivity to moisture and temperature changes. All powder samples were manufactured with the same process/technology with the same particle density (2,710kg/m^3), as reported by the manufacturer.

The volumetric size distribution of the powders (Table 1) were measured 'as received' using laser diffraction size analyser (Malvern Mastersizer 2000, Scirocco 2000M, UK). The samples were measured using dry dispersion of the powders with nozzle air pressure of 3 bars and the obscuration rate ranging from 1% to 5%, depending on grain size of the samples.

Table 1. Volumetric size distribution of the tests samples.

	Sample	X_{10}, µm (COV, %)	X_{50}, µm (COV, %)	X_{90}, µm (COV, %)
A1	Eskal 300	0.96 (3.3)	2.2 (3.1)	4.6 (3.4)
A2	Eskal 500	1.6 (0.57)	4.1 (0.12)	8.2 (0.34)
A3	Eskal 1000	1.7 (4.8)	4.6 (1.4)	10 (3.7)
B1	Eskal 10	5.7 (1.2)	10 (1.1)	16 (1.2)
B2	Eskal 14	8.3 (0.1)	14 (0.3)	23 (0.7)
B3	Eskal 15	8.8 (0.3)	16 (0.13)	25 (0.11)
B4	Eskal 20	11 (1.6)	20 (0.2)	33 (2.1)
C1	Eskal 150	99 (0.5)	136 (0.1)	187 (0.5)

The primary selection criteria for the Eskal powders were their median particle size (X_{50}), their PSD and the respirable fraction of particles already present in the powder. The powder samples were classified in three groups with group A (A1, A2 and A3) consisting mainly of particles with size smaller than 10µm, i.e., the maximum particle size sampled by a respirable cyclone. Group B (B1, B2, B3 and B4) consists of bi-modal powders, with modes at (1.1µm, 11µm) B1, (1.9µm, 15µm) B2, B3 (2.2µm, 17µm), and B4 (2.9µm, 23µm). Group C (C1) powder did not consist of particles in the respirable size range. Test samples from each group were designated by their group name followed by the sample number arranged in ascending order of their X_{50}. For example, A1, A2 and A3 are the three samples from group A arranged in ascending order of their X_{50}.

2.2 The vortex shaker dustiness tester

The experimental setup was similar to the one used by Morgeneyer et.al. [3] except that the released aerosol was sampled using a respirable cyclone (BGI GK2.69). The setup consists of 4 sections: generation, sampling, dilution, and measurement (Fig. 1). Aerosol is generated through the turbulent agitation of a powder-filled glass test-tube mounted on a digital vortex shaker (VWR Signature Digital Vortex Mixer).

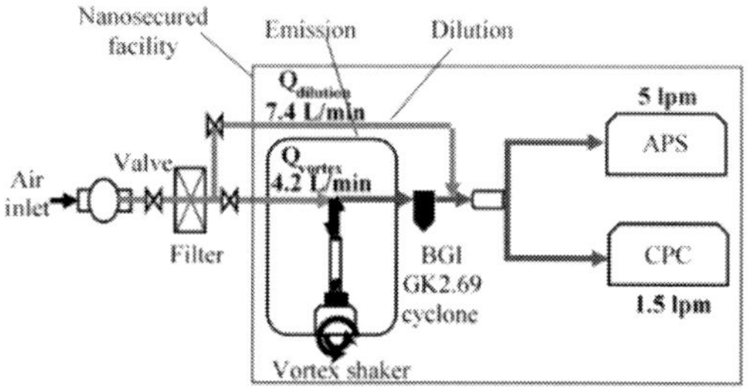

Fig. 1. Schematic of the vortex shaker experimental setup.

Airborne aerosol particles from the test-tube is carried to the respirable cyclone (50% cut point of 4µm) by the inlet flow (Q_v) of 4.2L/min (7e-05m^3/s). While the cyclone separates the larger particles, the respirable aerosol particles permeates through the cyclone and is diluted with (Q_D) of 7.4L/min (1.2e-04m^3/s) of filtered air before splitting into 2 channels for measurement and characterization [13]. The aerosol number concentration N_{CPC} over the 4nm to 3µm size range was measured using a condensation particle counter (CPC TSI 3775, TSI Inc.), whereas the aerosol number concentration N_{APS} and mass M_{APS} over the size range of 0.54µm to 20µm were measured by an aerodynamic particle sizer (APS TSI 3321, TSI Inc.). The APS records the particle numbers by their aerodynamic size which is based on times of flight of individual aerosol particles [3].

Three replicates were used for each test sample. Each test used 2g of powder weighed with an accuracy of ±0.001g using an analytical balance (MS1003S, Mettler-Toledo Inc.), manually filled in a centrifuge glass tube (diameter 0.025m, height 0.15m). The filled tube was sealed using a rubber stopper and carried to the isolator system. The powders were weighed within 1 hour of performing the experiments to limit the number of variables affecting the powder condition. The VS operated at 1500rpm and was run for 10 minutes (T). The background reference concentrations were measured for two minutes before the beginning and two minutes after the end of the vortex-shaker operation.

The total respirable aerosol number concentration measured by the CPC (4nm to 3µm) and the APS (3µm to 19.5μm) were combined to calculate the total number of generated particles S^{Number}_{Total} such as $S^{Number}_{Total} = S^{Number}_{CPC} + S^{Number}_{APS>3\mu m}$ adapted from Jensen [13], whereby

$$S^{Number}_{CPC} = [Q_V + Q_D].\Delta t_{CPC}.\Sigma_{i=0}^{T/\Delta t_{CPC}} N_{CPC}(t_0 + i.\Delta t_{CPC}) \quad (1)$$

$$S^{Number}_{APS>3\mu m} = [Q_V + Q_D].\Delta t_{APS}.\Sigma_{i=0}^{T/\Delta t_{APS}} N_{APS>3\mu m}(t_0 + i.\Delta t_{APS}) \quad (2)$$

where, Δt_{CPC} (1s) and Δt_{APS} (5s) are the time-step set for the CPC and the APS, respectively. $dN_{CPC}(t_0 + i.\Delta t_{CPC})$ and $dN_{APS}(t_0 + i.\Delta t_{APS})$ are the aerosol number concentration (#/cm^3) for the ith time interval measured by the CPC and the APS, respectively.

3 Results and discussion

In general, CPC and APS results reveal that all test samples emit respirable fractions of aerosol. The number concentration (N_{CPC}) curves from CPC (Fig. 2) are the average values over 3 repetitions. Also, S^{Number}_{Total} calculated for the test samples are shown (Fig. 2, top-right corner). All samples show an initial peak of N_{CPC} at the onset of the VS activity (around 120s to 135s from the start of the measurement). It is followed by a decline.

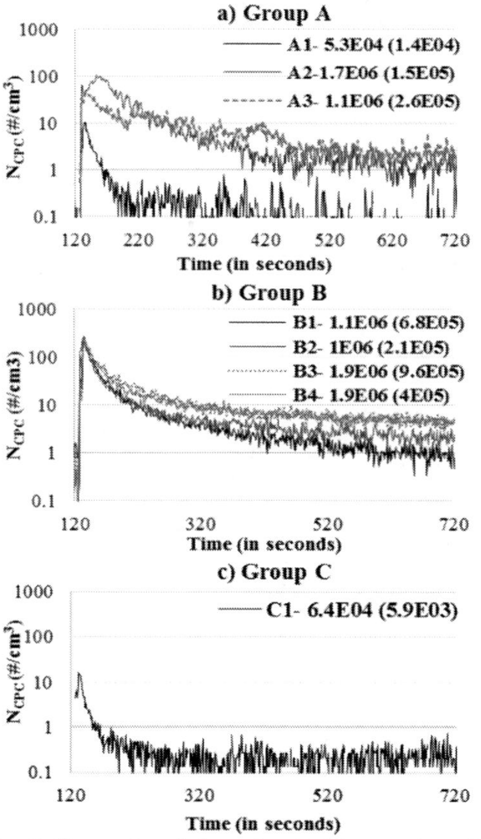

Fig. 2. Temporal evolution of respirable aerosol number concentration from the CPC for a) Group A; b) Group B; c) Group C. The total number of aerosol particles released S^{Number}_{Total} and their SD were calculated using (1) and (2) (see top-right corner).

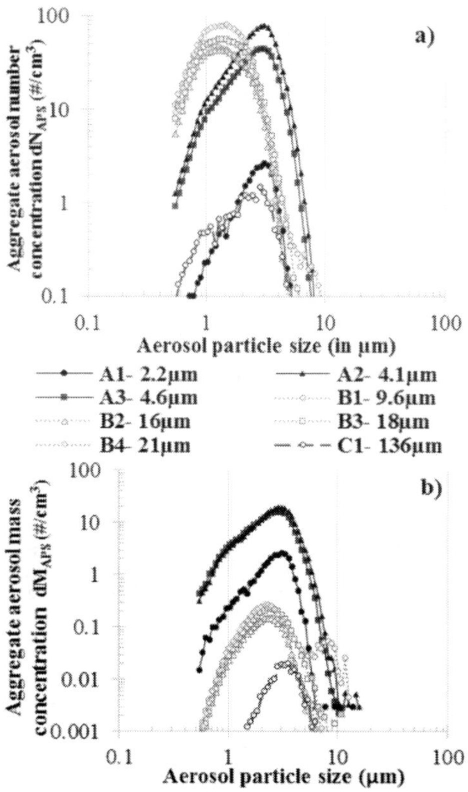

Fig. 3. APS measurements for aggregate aerosol concentration by particle size by a) number concentration (dN); b) mass concentration (dM).

3.1 Group A

Samples A2 and A3 show qualitatively similar dust generation behaviour, whereby they initially emit aerosols with concentrations up to 104#/cm^3 (A2) and 47#/cm^3 (A3) before a gradual decline. This stands in contrast to A1 which attains its maximum concentration of 10#/cm^3 before swiftly descending into emission of aerosols comparable to the background reference values. This relies on the assumption that the overwhelming majority of aerosol particles are in the size range measured by the CPC.

Measurements from the CPC (Fig. 2a) and APS (Fig. 3) shows the cumulative aerosol concentrations generated by A2 and A3 are greater than one order of magnitude higher than A1. Furthermore, the modal aerosol sizes of the three samples measured by the APS are similar and amount to approximately 3μm (Fig. 3).

Group A powders are cohesive in nature [14] and exist in the form of agglomerates of sizes greater than their primary particles. The dust generated from such powders could stem from the breaking and dispersion of the agglomerates into primary particles due to agglomerate-agglomerate collision or impacts against the wall. In the case of the VS, the centrifugal forces acting upon the agglomerates lead to collisions and impacts in the shaker which could result in the breakage and disintegration of agglomerates into particles. Such small disintegrated particles are more prone to be elutriated within a turbulent air flow developed in the VS. They follow a Stokesian flow regime [15].

A1 with a relatively small particle sizes (Table 1) forms agglomerates with high cohesive binding energies [14] such that the vortex agitation at 1,500 rpm might not be enough to separate and aerosolize high concentrations of primary particles. Contrarily to A1, samples A2 and A3 consist of particles relatively larger in size and displaying broader PSD. Thus, based on the theoretical relationship between particle sizes and Van der Waals cohesive forces [16], the cohesivities of A2 and A3 must be inferior to that of A1.

For A2 and A3 (similar in size) the temporal evolution of the total number of released aerosols and their PSD are similar (Fig. 2 and Fig. 3). At the onset of vortex agitation, both samples emit relatively elevated concentrations of aerosols with small particle sizes. This might be due to the fact that such particles are more prone to be elutriated within the turbulent air flow [17].

3.2 Group B

The four bi-modal powders in group B display similar aerosol emissions during the test, whereby a sharp initial peak is followed by a gradual decline of the aerosol concentration into values between 1 and 10 #/cm^3. B1 and B2 with the lowest X_{50} in this group emit relatively less respirable aerosols compared to B3 and B4. The modal aerosol size for all the samples measured from the APS is approximately 1µm (Fig. 3), which is in close proximity to the first mode particle size of B1 (1.1µm), B2 (1.9µm), B3 (2.2µm), and B4 (2.9µm) (not shown). Thus the respirable fraction of aerosols from group B could originate from small particles with sizes inferior to the respirable size fraction.

Furthermore, laser diffraction measurements of volume fractions of particles with diameters smaller than 4µm present in group B shows that the respirable dustiness of a sample increases with an increase in the volume of particles in the first population. The volume fraction of the group B samples are measured as 3.7% (B1), 4.9% (B2), 5.2% (B3), and 6.5% (B4), with their mode particle size close 1µm.

3.3 Group C

Group C represents powders with almost no particles within the respirable range. C1 initially releases an aerosol concentration peak at the onset of the VS followed by a rapid descent to aerosol concentration of 0 to 0.2#/cm^3, similar to background aerosol concentrations used as reference. The initial emission can be due to the presence of small impurities present in the powder or due to the generation of small fragments of powder due to the attrition of larger sized particles.

Several mechanisms can be responsible for the attrition of particles which depends on the particle mechanical properties, shape and mode of loading [18]. Large brittle particles are prone to generate dust by attrition as they contain more faults in the form of micro cracks or imperfections which can lead to fracture or breakage when compared to smaller particles.

An aerosol particle sampler, the mini-particle-sampler (MPS®) [13] could be used to capture and deposit aerosol particles on copper grids for off-site transmission electron microscope (TEM) analyses. This might provide evidence of the role of attrition in the generation of respirable aerosols.

4 Conclusion

The theoretical relationship between the properties of a powder and its dustiness remains poorly understood so that further experimental works are required [3,11].

We use a vortex shaker to test respirable dust generation from eight powders with similar physical properties except their particle size distributions which differ from each other. The powders were divided into three groups based on the fraction of particles within the respirable range. The interplay of several mechanisms like de-agglomeration and attrition and their relative importance might account for our observations.

Further studies combining experimental (atomic force microscopy in micro-scale and shear test in meso-scale) and numerical (discrete element method) techniques are required in order to confirm this.

We acknowledge EU FP7 MCA ITN T-MAPPP and Région Picardie/ Hauts de France and by the Programme 190 (French Ministry of Environment) for their support.

References

1. R. Nedderman, *Annals of Physics*, **54** (2005)
2. Int. Org. Standardization, ISO 7708 (1995)
3. M. Morgeneyer, O. Le Bihan, A. Ustache, O. Aguerre-Chariol, Powder Technol. **246**, 583 (2013)
4. I. Pensis, J. Mareels, D. Dahmann, D. Mark, Ann. Occup. Hyg. **54**, 204 (2010)
5. C. Pujara, D. Kildsig, In J.P. Wood, editor., Drugs and the pharmaceutical sciences, **108** (2001)
6. A. Klippel, M. Schmidt, U. Krause, J. Loss Prev. Process Ind. **34**, 22 (2015)
7. EU, Off. J. Eur. Communities., **152** (2008)
8. EU, Eur. Comm. Edition, **4** (2012)
9. M.A.E. Plinke, Am. Ind. Hyg. Assoc. J. **56**, 251 (1995)
10. S. Bach, E. Schmidt, Ann. Occup. Hyg. **52**, 717 (2008)
11. O.L.C. Le Bihan, A. Ustache, D. Bernard, O. Aguerre-Chariol, M. Morgeneyer, J. Nanomater., 7 (2014)
12. H. Zetzener, J. Schwedes, Part. Part. Syst. Charact. **19**, 144 (2002)
13. K.A. Jensen, D4.6: EU NANOGENOTOX (2012)
14. H. Shi, S. Luding, V. Magnanimo, *PGBSIA* (2016)
15. S.C. Liang, T. Hong, L.S. Fan, Int. J. Multiph. Flow. **22**, 285 (1996)
16. H. Rumpf, K. Sommer, K. Steier, Chemie Ing. Tech. **48**, 300 (1976)
17. K. Saleh, M. Jaoude, M. Morgeneyer, E. Lefrancois, Powder Technol. **255**, 141 (2014)

 DOI: 10.1051/epjconf/201714015030

Investigating the effect of unloading on artificial sandstone behaviour using the Discrete Element Method

Yueqin Huang[1,*], *Yi Pik* Cheng[2], and *Matthew* Coop[3]

[1] MSc student, Department of Civil, Environment & Geomatic Engineering, University College London, London, UK
[2] Senior Lecturer, Department of Civil, Environment & Geomatic Engineering, University College London, London, UK
[3] Professor, Department of Civil, Environment & Geomatic Engineering, University College London, London, UK

Abstract. The Discrete Element Method (DEM) was used to simulate the mechanical behaviour of a reservoir sandstone. Triaxial tests were carried out using 3D-DEM to simulate the stress-strain behaviour of a sandstone with comparisons made between the numerical tests and the laboratory tests. The influence of isotropic unloading was investigated, which was found to have an impact on bond breakage and was successfully captured in the 3D shearing processes. It was found that bond breakages correlated strongly with the stress-strain behaviour of the sandstone affecting the peak strength. It was also found that unloading affected the bond breakages, which then changed the mechanical behaviour of the sandstone. The tangent stiffnesses of simulated virgin and cored samples under different confining stresses were compared. From the tangent stiffnesses, gross yield envelopes and the yielding surfaces for unloaded samples and virgin samples were plotted and analysed in detail.

1 Introduction

Many triaxial laboratory tests have been carried out to study the mechanical behaviour of sandstones. In particular, Alvarado et al. [1] developed a series of unloading tests and found that the cementation was damaged by isotropic unloading, representing coring. The difference between the cored and the virgin samples was revealed by the difference in the positions of yielding surfaces. Samples that have never been unloaded to zero confining stress are called 'virgin samples' while samples that have been unloaded to zero confining stress are called 'cored samples'. Using the Discrete Element Method (DEM), a parametric study on Castlegate and Saltwash sandstone was carried out by Cheung et al. [2] calibrating the stiffness and strength of the parallel bonds (Table 2), where in DEM parallel bonds act as a finite size bond, transmitting both a force and a moment over a circular cross-section.

In this study, parallel bonds were applied to the DEM sandstone model with cementation and the data were calibrated with the stress-strain behaviour of the Castlegate sandstone under triaxial shearing. Unloading tests were carried out based on the laboratory tests performed by Alvarado et al. [1]. The gross yield envelopes of both cored and virgin samples were investigated and compared to the envelopes obtained from the laboratory tests. The results of bond breakage caused by unloading are described quantitatively.

2 Modelling procedures

In the 3D-DEM model, the particles of sandstone were assumed to be spherical. The preparation of the cylindrical artificial sandstone samples was based on the laboratory triaxial tests with a height to diameter ratio of 2. During the shear test, vertical forces were generated by applying a velocity of 0.1m/s to the top and bottom platens, simulating the movement of the platens in a real triaxial test. The average unbalanced force over average contact force ratio was 1.22 x 10^{-4}, which was below 10^{-3}. Therefore, the system was in "quasi-static" equilibrium during shearing.

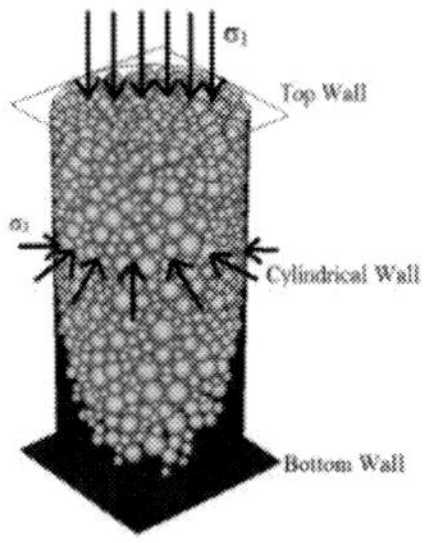

Fig. 1. Sketch of a DEM triaxial sample.

Parallel bonds were applied at 29MPa for both the cored and virgin samples; this was then followed by the loading and unloading tests (Table 1). Virgin samples were all then unloaded or loaded to 10, 29 and 50 MPa.

* Yueqin Huang: htxhyq2007@126.com

Table 1. Tested virgin and cored samples. (V= DEM virgin samples, C = DEM cored samples)

Sample	Unloading stress levels (MPa)	Shearing/Reloading stress levels (MPa)
V10	10	10
V29	29	29
V50	50	50
C10	0	10
C29	0	29
C50	0	50

Table 2. DEM input parameters.

3D input parameters		
Properties	Calibrated Parameters [2]	Parameters in this paper
Number of particles	34007	2812
Particle size distribution (mm)	0.525-1.5	75-100
Sample dimensions (widthxheight), (m)	0.05x0.1	2x4
Density, (kg/m^3)	265000	265000
Wall stiffness, (N/m)	-	8.8x10^{10}
Confining stress, (MPa)	29	5, 10, 29,50
Friction parameter, μ	0.5	0.7
Initial void ratio	0.623	0.372
Initial porosity	0.384	0.271
Poisson's ratio	0.17	0.17
Parallel bond size, α_{bond}	0.01-1.0	0.8
Normal and shear contact stiffness, Kc (N/m)	8.8x10^{10}	1.92x10^{10}
Parallel bond normal and shear stiffness, $K_{pb,}$ (Pa)	7.92x10^{11}	1.92 x10^{10}
Parallel bond normal strength, $S_{pb,S,normal}$, (Pa)	1.25 x10^9	1.25 x10^8
Parallel bond shear strength, $S_{pb,S,shear}$, (Pa)	1.5 x10^9	1.5 x10^8

The particle Young's modulus, E_c, was set to 20GPa and the damping coefficient was set to zero. A friction parameter μ of 0.7 was used; the input parameters were different from those of Cheung et al. [2] because the sample size and particle sizes were different in this study (Table 2).

3 Results

3.1 Friction angles

The shear stresses and the confining stresses at the peak states have been obtained from the 3D simulations and are plotted in Figure 2. The figure shows that the peak friction angle is about 53.7° which is comparable to the laboratory result. Regardless of this good match of peak strength, real sandstone particles are far more irregular in shape. Due to the fact that all the uncemented DEM particles were spherical and that they will interact only through forces without moment, the internal friction angle of the uncemented DEM samples could be far lower than that of a real soil unless a much larger value ofμ or elongated-shaped particles are used.

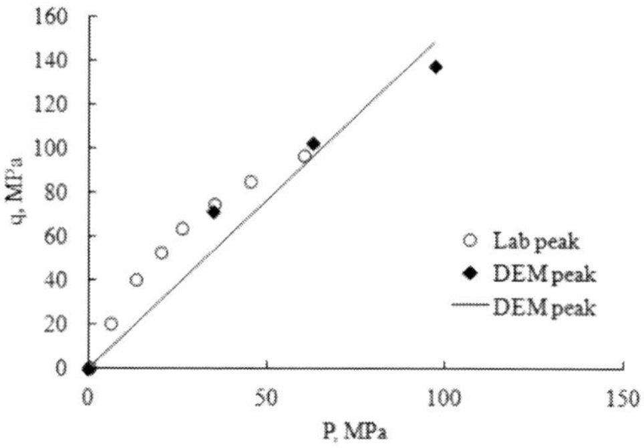

Fig. 2. Peak envelopes obtained from DEM simulations, and comparison with the data of Castlegate sandstone [1, 2].

3.2 Loading and unloading tests

3.2.1 Stress-strain behaviour and bond breakage

Figure 3 describes the stress-strain behaviour of both cored and virgin samples. The DEM samples shearing at higher confining stresses gave higher peak stresses. Samples shearing at the same confining stress showed similar behaviour, such as V10 and C10, regardless of the different unloading. As the confining stress became larger, C50 reached a slightly higher peak than V50.

The stress-strain behaviour shown in Figure 3 has a strong relationship with bond breakage. In Figure 4, the initial bond number of the unloaded (cored) samples was found to be lower than that of the virgin samples, showing that unloading damages the bonds between particles. Moreover, the bond numbers of all the cored samples remained the same until an axial strain of 1%

was reached, indicating that the initial increasing deviator stress was controlled by the strength of the cementing materials. Sample V50 had the highest bond numbers which might due to the higher confining pressure giving more bonded surfaces around the particles. C10 had the most remaining bonds when compared to the other samples at the strain-softening stage. The bond number of V29 reduced faster than C29 at the second stage and eventually reached the same state after 5% axial strain. With the bonds gradually breaking during shearing, the remaining bond numbers were similar for both the cored and virgin samples under the same confining pressure.

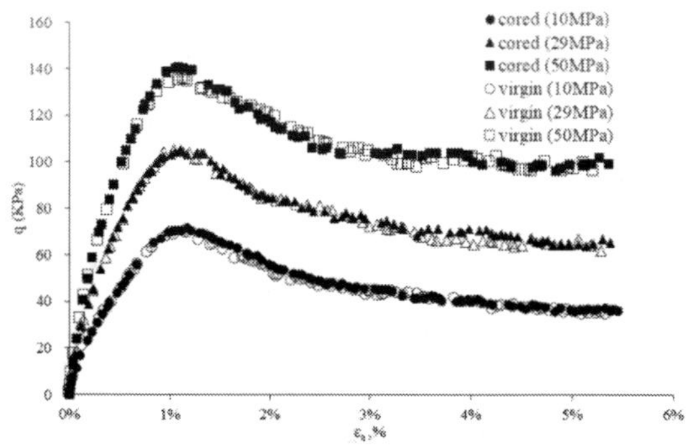

Fig. 3. Stress-strain behaviour of cored and virgin samples.

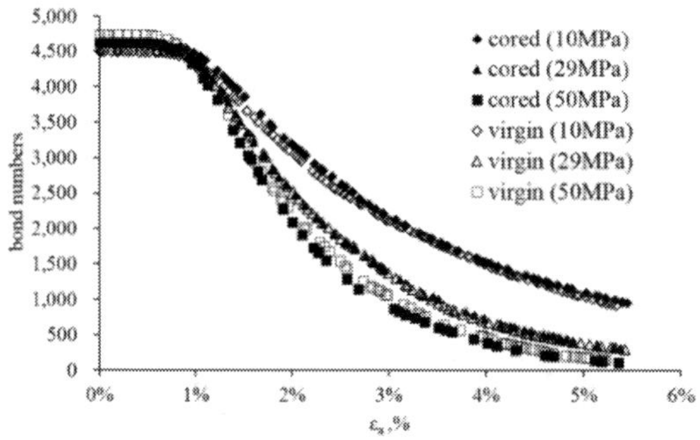

Fig. 4. Bond breakages for different confining stresses.

3.2.2 Tangent stiffness

The tangent stiffnesses of both cored and virgin samples are shown in Figures 5 and 6. Unlike Alvarado et al. [1] who used a Y2 method to identify the yield point, the gross yield points in this paper were obtained from the tangent stiffness plots at the end of the plateaus, where there was a sudden drop at around 0.6% axial strain, indicating the start of the major plastic deformation. Figure 4 indicates that these gross yield points happen when significant bond breakage starts. In the laboratory, the gross yield points were also obtained from the stiffness degradation curves. The gross yield point was different from the true yield point as pointed by Alvarado et al. [1], since the true yield points occurred at very small strains below which the behaviour was both linear and elastic. It can be seen from Figures 5 and 6 that below 0.10% axial strain, the tangent stiffnesses of the virgin samples were slightly lower than those of the cored samples and the plateaus were more obvious in the virgin samples than in the cored samples.

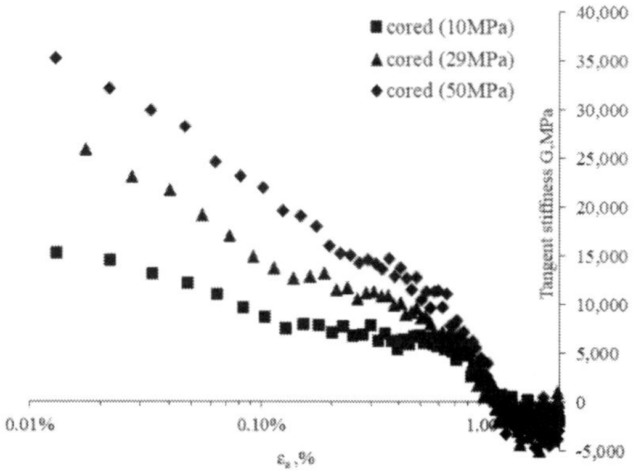

Fig. 5. Tangent stiffnesses of cored samples.

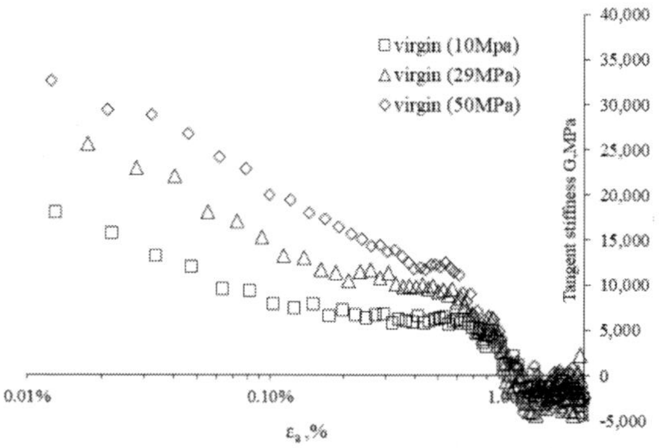

Fig. 6. Tangent stiffnesses of virgin samples.

3.2.3 Yielding surface

In the laboratory test, Alvarado et al. [1] used the Y2 method, which was defined by Jardine [5], and the yield points were identified when the direction of the strain increment changed. According to these yield points, it was found that the gross yielding envelopes and the yielding surfaces were similar to the laboratory results under relatively higher confining stresses. The stress paths of the three cored and virgin samples tested under 10, 29 and 50MPa confining pressures are shown in Figures 7 and 8 respectively. The critical state points were obtained from the end points of the stress-strain curves in Figure 3. Peak points at the top of the stress paths, form a peak envelope (dotted line) that curves at medium stress levels. The shape of the gross yield envelope for the cored samples is close to elliptical and is similar to the results of the real sandstone [3]. The gross yield envelope for the virgin samples, however, is much larger than that of the cored samples, which is similar to the results in the laboratory. The difference between the two yield envelopes was due to the unloading process, indicating that there were cementation bond breakages during the unloading of the cored samples.

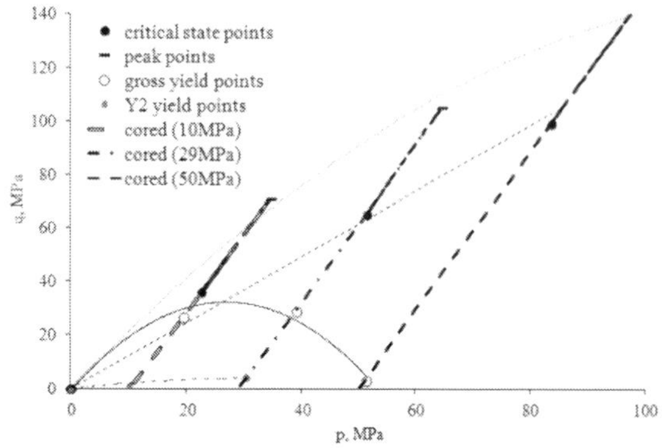

Fig. 7. DEM Y2 and gross yield surfaces of cored samples.

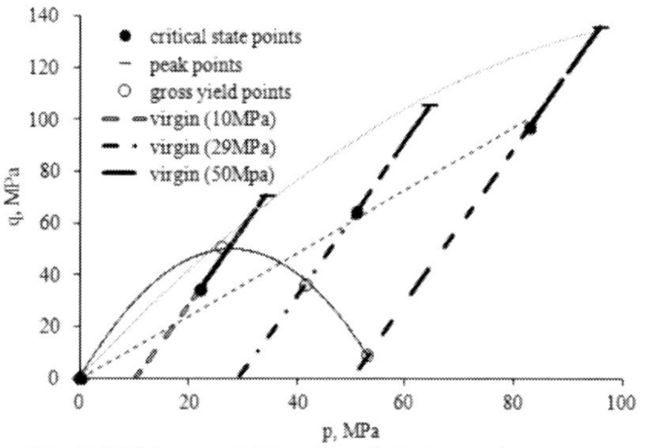

Fig. 8. DEM gross yield surface of virgin samples.

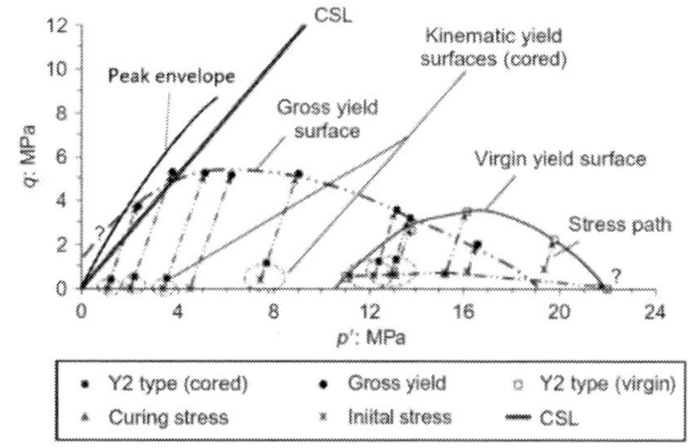

Fig. 9. Laboratory yielding surfaces of cored and virgin samples [1].

It should be noted that bond breakage is essential to reach the CSL, which was also studied by Cheng et al. [4] who pointed out that particle breakage also had a significant impact on the shearing behaviour of soil. The small amount of bond breakages revealed in the DEM simulations were difficult to identify in the laboratory tests. The laboratory tests also had difficulty in determining the yield points at high confining pressures due to the limitations of the apparatus.

4 Conclusions

This paper studied the mechanical behaviour of a reservoir sandstone based on the discrete element method. Qualitative comparisons were made between the numerical results and laboratory results mainly referring to the studies by Alvarado et al. and Cheung et al. [1, 2]. The results could potentially help engineers to have a better understanding of the shearing behaviour of the reservoir sandstone from a micromechanical perspective. The main conclusions of this study are:

1) The stress-strain behaviour of sandstones under triaxial shearing was confining pressure dependent. It was found that the microstructural parameters, in this case, the cementation bond breakage during unloading, had a great impact on the shearing behaviour of a sandstone.

2) There was a relationship between bond breakage and the stress-strain behaviour of a sandstone. The parallel bonds started to break when the deviator stresses reached the peak states. Isotropic unloading has been found to affect the behaviour of the sandstone, especially bond breakage. The breakage of bonds also reduced the tangent stiffnesses. However for real sandstone, there might be particle breakages (on top of cementation bond breakage) that further disturb the original structure of the sandstone, leading to a different tangent stiffness, when compared to the DEM data, due to increased locked stresses and frictional forces between the particles.

3) The gross yield envelopes between cored and virgin samples were compared. It was found that the gross yield envelope of the virgin samples was larger than that of the cored samples, both of them enclosed the curing stress (29MPa). It is difficult to tell the difference between the Y2 surface and the gross yield surface of the virgin samples.

References

1. G. Alvarado, M.R. Coop, S. Willson, Géotechnique, **62**, 303-316 (2012a)
2. L. Cheung, C. O Sullivan, M. Coop, International journal of rock mechanics and mining sciences. **63**, 93-103 (2014).
3. M.R. Coop, S.M. Willson, Journal of Geotechnical and Geoenvironmental Engineering, **129**, 1010-1019 (2003)
4. Y.P. Cheng, M.D. Bolton, Y. Nakata, P&G05, **2**, 1393-1397 (2005)
5. R.J. Jardine, Proceedings of the international symposium on pre-failure deformation characteristics of geomaterials, **2**, 855-886 (1995)

The role of force networks in granular materials

Karen E. Daniels[1,*]

[1] *Department of Physics, North Carolina State University, Raleigh, NC 27695, USA*

Abstract. One of the most visually striking features of granular materials is their heterogeneous pattern of force transmission, commonly known as force chains. In this paper and its associated talk, we will review several experiments on two-dimensional photoelastic granular materials which highlight methods for making quantitative interparticle forces measurements, theoretical frameworks for interpreting such data, and the specific findings which result from these methods.

1 Introduction

Granular materials are inherently heterogeneous, and continuum models of properties such as the shear modulus and sound speed often fail to quantitatively capture their dynamics. This lack of understanding has serious consequences for the engineering of both geotechnical applications, and the industrial storage and handling of granular materials. One likely reason for these difficulties is that boundary stresses are transmitted via a chain-like network of strong forces, as shown in Fig. 1.

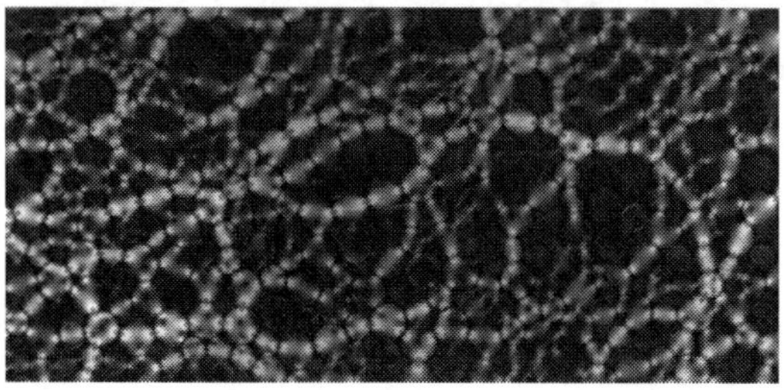

Figure 1. Sample image of force chains in a quasi-2D granular material composed of photoelastic disks. Dark particles carry little force, and bright particles carry more.

This paper presents several complementary methods of characterizing these systems: acoustic transmission at the particle- and network-scale [1, 2], the force-moment ensemble [3], and network science [4, 5]. For each case, we provide a description of both the method itself, and the type of information that can be gained through that approach. Collectively, these measurements highlight the importance of understanding the role of the force chain network in controlling the bulk properties of the material. In this paper and associated talk, we will primarily focus on 2D experiments on photoelastic materials, since these readily allow for the quantitative determination of the particle-scale forces [6]. The framework is general enough to be applied to 3D materials as similar data becomes available.

2 Photoelastic force measurements

The use of photoelastic materials, as shown in Fig. 1, allows for the quantification of the forces in a granular system by placing it between a pair of polarizing filters. In our experiments, we use darkfield photoelasticity: polarized light will only be be transmitted through the second polarizer if the stress on a particular grain has rotated the polarization of the light by an integer multiple of $\pi/2$. In the sample images shown here, dark particles are under low stress, and bright particles are under higher stress with an increasing pattern of dark/bright fringes as the stress increases. By using circular, rather than linear, polarized light, the resulting fringe pattern is independent of the relative angle between the applied stresses and the polarizer orientation.

These methods can be used either semi-quantitatively to measure the average pressure, or quantitatively to determine the vector force at each interparticle contact. Our experiments utilize either polyurethane or Vishay Photo-Stress particles, cut into centimeter-scale disks/ellipses using either a CNC mill or a water-jet.

By using a combination of unpolarized and polarized light (see Fig. 2), it is possible to obtain a snapshot of the entire state of the system. In the results presented below, we determine the positions of all particles from images taken in unpolarized light, and the pressure or contact forces from images taken in polarized, monochromatic light. In the most recent work described, this was done via LED lighting. In a single image, we obtain position-data

* e-mail: kdaniel@ncsu.edu

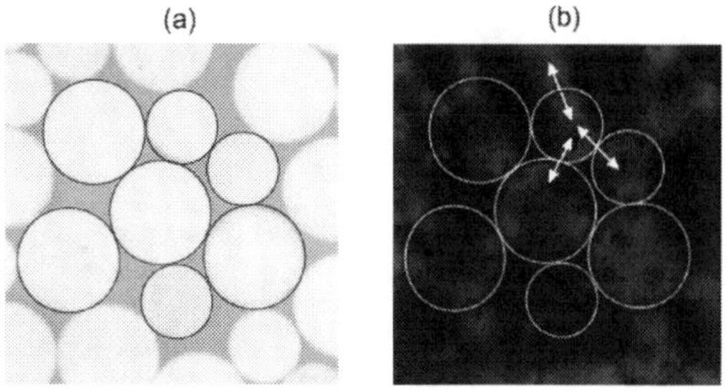

Figure 2. The use of multiple light fields — (a) unpolarized light and (b) polarized, monochromatic light — allows for the determination of all particle locations and all vector contact forces. Figure adapted from [3].

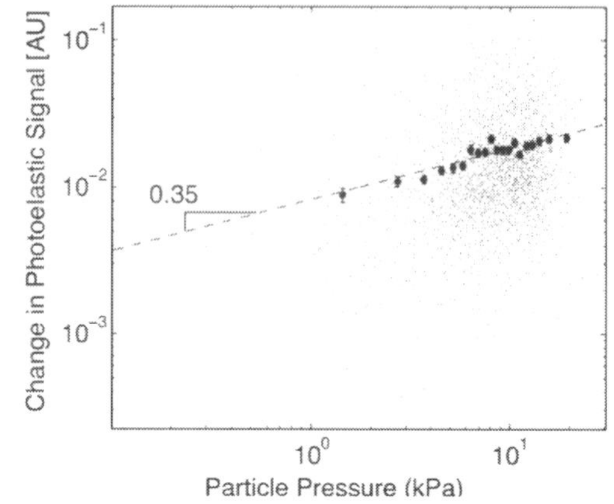

Figure 3. Data collected at the particle-scale (small gray symbols), averaged to reveal a relationship between the pressure on a particle and the amplitude of sound propagation through that particle. The dashed line is $\propto P^{0.35}$, from a fit to all points. The large, dark circles are the average and standard error of this dataset, calculated for equally-populated bins. Figure adapted from [1]

in the red channel of a color camera, and the force-data in the green channel. An open-source code for making vector contact force measurements, along with a review of the underlying theory and practical advice for constructing a photoelastic setup, is described in an upcoming review article [7].

3 Acoustics

By using a combination of high-resolution, low-speed imaging (to measure the stress on each particle) and low-resolution, high-speed imaging (to measure the amplitude of sound waves), we can examine the relationship between the force chain network and the spatial pattern of sound propagation. To measure the local amplitude of the propagating sound wave, we calibrated changes in the photoelastic signal against measurements from a small number of particle-scale piezoelectric sensors known to have a linear response. To measure the pressure on individual particles, we located all particles within a high-resolution image and used a combination of the average intensity I, the squared average intensity gradient $|\nabla I|^2$ [8], and the position of photoelastic fringes in the vicinity of each contact to provide an empirical estimate of the normal component of the contact force.

As shown in Fig. 3, we observe that the amplitude A of the acoustic signal is on average largest within particles experiencing the largest forces, with an empirical relationship that is approximately a power law $A \propto P^{0.35}$.

For Hertzian contacts, the interparticle contact force is given by $F \propto \delta^{\beta}$, where δ is the distance (measured from the edge) that each particle is compressed, and the exponent β depends on the particle geometry. The contact area between cylindrical particles is given by $a \propto F^{\frac{1}{2\beta}}$, for a rectangular contact area. For our particles, we observe that the interparticle force obeys a Hertzian-like contact force law $F \propto \delta^{5/4}$, a weak nonlinearity likely arising due to out-of-plane and asperity effects. For a Hertzian exponent of $\beta = 5/4$, the contact area between cylinders therefore grows as $P^{2/5}$, consistent with the observed pressure dependence in Fig. 3. We therefore conclude that increases

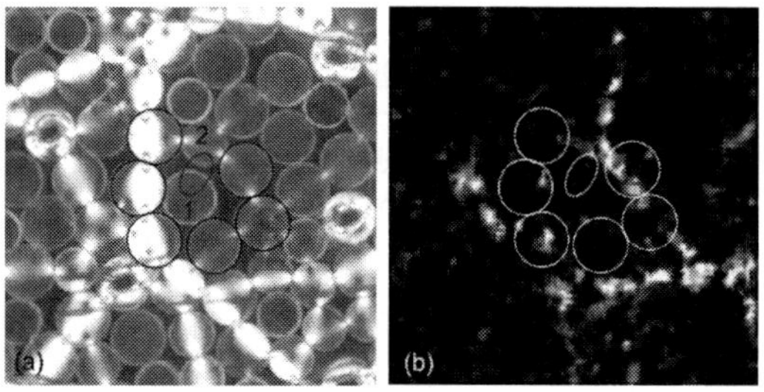

Figure 4. (a) Image of force chains, subject to acoustic driving, and (b) magnitude of difference in two images, taken during the transmission of an acoustic pulse. The small ellipse marks the formation of a new contact, which repeatedly opens and closes during each wave cycle while a transiently strong force chain forms to its right. Figure reprinted from [1].

in interparticle contact areas along force chains provides the main conduit for sound propagation.

Importantly, the force chain network does not remain a static structure during the propagation of sound. As shown in Fig. 4, the deformation of individual particles, for instance due to expansion via the Poisson ratio, can open and close individual contacts. This is exemplified by the particles labeled 1 and 2, which are clearly not touching in the static system (panel a, see elliptical region), yet transiently form a force chain during sound propagation (panel b, same elliptical region marked). For $\nu \approx 1/2$, the lateral expansion will be of the same order of magnitude as the compression δ; this is consistent with closing the gap shown.

Traditional discrete element methods, in which all particles in a simulation remain circular and "overlap" each other rather than deforming, are unlikely to be able to capture such effects. In fact, no preference for the sound waves to follow the force chains was seen in simulations that investigated the question [9], even though the effect would have been larger in that case (spherical particles with $\beta = 3/2$) than here. Future work will be necessary to resolve this discrepancy.

4 Athermal Statistical Ensembles

A key technique of statistical physics is to represent a system by an ensemble of allowed microstates, each occurring with a probability that depends on a small number of macroscopic state variables such as temperature, pressure, volume, and energy. Because granular materials exhibit reproducible ensembles of microscopic quantities, well-predicted by a few macroscopic parameters, these methods are a promising route to describing the state of a granular system. Importantly, these methods seem to be successful even though (1) the thermodynamic temperature is too small (by a factor of 10^{12} or so) to play a role in the state of a granular material, and (2) granular materials are both frictional and inelastic, ensuring that they are inherently non-equilibrium.

The Edwards ensemble and its later variations, including the force-moment ensemble, have gained significant attention as a promising approach to writing an equation of state for static granular materials [3]. A common feature of these ensembles is that they introduce a temperature-like variable of the form $\frac{1}{T} \equiv \frac{\partial S}{\partial E}$, where the entropy S is replaced by a configurational entropy and the temperature and energy (T, E) are replaced by a different pair of intensive (equilibrating) and extensive (additive) state variables. There has been considerable interest in the statistics of the locally-defined force moment tensor $\hat{\Sigma}$, examining the extent to which it provides an extensive variable analogous to energy in equilibrium statistical mechanics. It is calculated by summing the outer products of the moment arm $\vec{d}$ and the force $\vec{f}$ between the particles (m) and their contacting neighbors (n):

$$\hat{\Sigma} = \sum_{m,n} \vec{d}_{mn} \vec{f}_{mn} \,. \tag{1}$$

In two dimensions, it is convenient to decompose $\hat{\Sigma}$ into $\sigma_p = (\sigma_1 + \sigma_2)/2$ and the $\sigma_\tau = (\sigma_1 - \sigma_2)/2$, where $\sigma_{1,2}$ are the principal stresses. The average hydrostatic pressure in the system is given by $\Gamma_N = \mathrm{Tr}\,\hat{\Sigma}/N$, taken over clusters of N particles. Note that this formulation in terms of the force-moment tensor is related to the more familiar Cauchy stress tensor $\hat{\sigma}$ by the relation $\hat{\Sigma} \equiv V\hat{\sigma}$, where V is the total volume occupied by the particles [3].

In the force-moment ensemble, the corresponding temperature-like variable is known as the angoricity [10]

$$\alpha_i = \frac{\partial S(\sigma_i)}{\partial \sigma_i} \tag{2}$$

where $i = (p, \tau)$ for either the pressure or shear components. The equivalent of the Boltzmann distribution would be given by

$$P(\sigma_i) \propto e^{-\alpha_i \sigma_i} \tag{3}$$

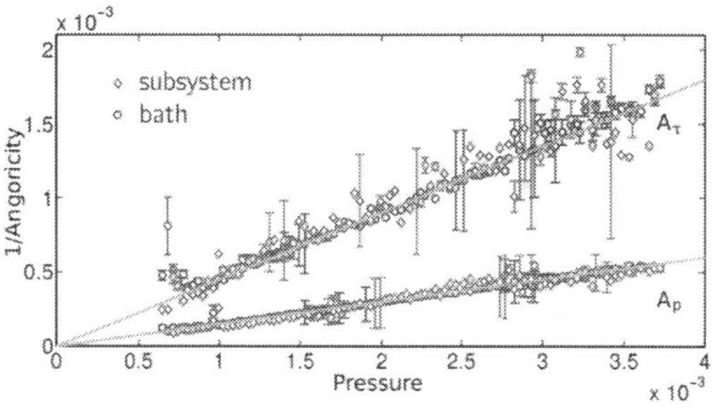

Figure 5. Equilibration of the two components of the temperature-like variable angoricity, in experiments on a low-friction subsystem (red ⋄) placed within a high-friction bath (black ∘). The low-friction particles are monitored for location via markings visible in blacklight, and are returned to the subsystem if they stray out of it; no mechanical confinement is present. Values are measured as averages over clusters of $N = 8$ particles, to reduce statistical fluctuations [11]. Pressure Γ is measured in units of [N/m]. Figure adapted from [12].

where an exponential distribution is expected [11] instead of σ_i^2. We make measurements of the relative-angoricity of the system using the method of overlapping histograms. For a stress state σ (either pressure or shear), consider a measurement of $P(\sigma)$ for the same system prepared at two different values of α, enumerated (1,2). The ratio of these histograms is conjectured to be given by [3]

$$\frac{P_1(\sigma)}{P_2(\sigma)} = \frac{Z(\alpha_2)}{Z(\alpha_1)} e^{(\alpha_1 - \alpha_2)\sigma} \tag{4}$$

where (α, σ) can be either the pressure (p) or shear (τ) components. By taking logarithm of the ratio, and fitting the data to determine the slope, it is possible to measure changes in the angoricity. Since $1/\alpha = 0$ at $\sigma = 0$, this can also be used as an absolute measurement.

We have conducted laboratory tests of the validity of angoricity, examining the extent to which these ensemble-based formalisms provide well-defined state variables. For these experiments, we use a quasi 2D granular material levitated on an air cushion to eliminate basal friction. By analogy with the canonical ensemble in conventional statistical mechanics, we consider whether state variables equilibrate between a smaller subsystem and a larger bath. For a low-friction subsystem within a high-friction bath, subject to biaxial compression, we observed that the temperature-like variable compactivity (associated with the system volume in the original Edwards ensemble [13]) fails to equilibrate [12]. This is a failure of the zeroth law of thermodynamics for this ensemble. However, for the force-moment ensemble, both α_p and α_τ are observed to equilibrate [12], as shown in Figure 5.

For a well-defined state variable, we would expect that the equation of state ($1/\alpha \propto \Gamma$) would be independent of loading history. To test this hypothesis, we performed experiments under three loading conditions — biaxial compression, uniaxial compression, and simple shear — as shown in Fig. 6. In the bottom half of the figure, we

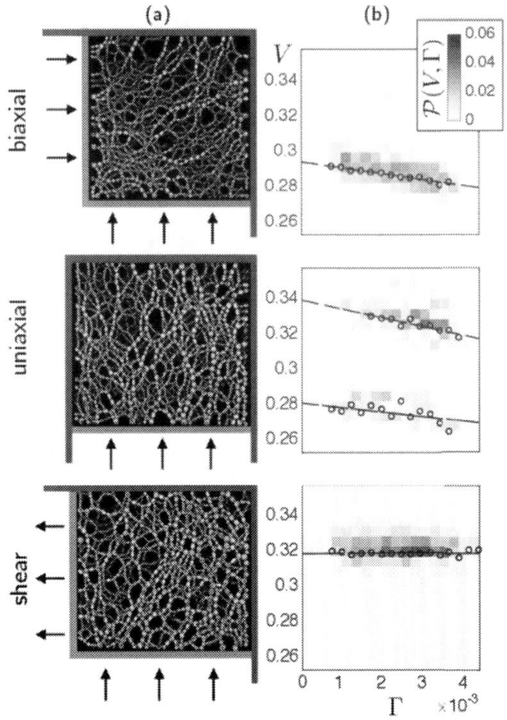

Figure 6. Left: Sample images of force chains until three loading conditions: biaxial, uniaxial, and shear. Right: Corresponding two-dimensional histograms of the volume V [AU] and pressure Γ [N/m], again measured over 8-particle clusters. Figure credit: [14]

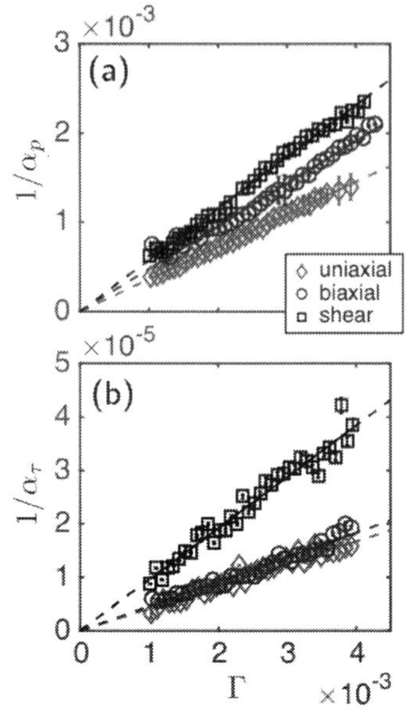

Figure 7. Equations of state for the (a) pressure-angoricity and (b) shear-angoricity, under all three loading conditions. Pressure Γ has units of [N/m]. Figure credit: [14]

report the volume and pressure measured over 8-particle clusters; these already show a strong history-dependence. In calculating the angoricities according to Eq. 2 and 3, we find that while σ_p is exponentially-distributed, σ_τ is Boltzmann-distributed (Eq. 3 and 4 have a σ^2 dependence instead of linear).

Subject to this modification, we observe (as expected) that both $1/\alpha_p$ and $1/\alpha_\tau$ are linearly related to the internal pressure Γ. However, the proportionality constant for this equation of state is both volume-dependent (not shown) and history-dependent (see Fig. 7). Therefore, we find the stress-based ensemble to be volume-dependent and angoricity to be a variable of process. This complicates its use as a state variable, but highlights the importance of tracking both normal and tangential components of interparticle contact forces.

5 Network Science

The two experiments described above approach granular problems from the standpoint that the particle-scale statistics control the bulk behaviors. However, the image in Fig. 1 clearly shows that the force chains create a secondary, mesoscale structure within the bulk. It remains an open question how best to extract length and force scales in a consistent, quantitative framework.

One approach to quantifying these structures is to consider a network representation in which the nodes (all particles) are connected by weighted edges with values corresponding to the measured contact forces. We construct the weighted force network **W** from a list of all inter-particle forces, obtained from photoelastic analysis. If particle m and n are in contact, then $W_{mn} = f_{mn}/\bar{f}$, where f_{mn} is the normal force (or, alternatively, the tangential force) between them and $\bar{f}$ is the average force in the whole network.

Such network representations — commonly used throughout the social, biological, and physical sciences — provide a mathematical framework which allows for the comparison of features at different spatial scales. A large catalog of statistical measures have been developed by the network science community [5], allowing for the quantification of particles, force-chains, or the whole system, as desired by the particular application. Here, we focus on developing measures which correspond to the force network.

We have observed [2] that domains of force-chains, located via automated community-detection [4], are more highly associated with sound propagation than are other smaller/larger scale measures. It is important, however, to have a way to determine the optimum length scale on which to define the communities. Community-detection algorithms work by optimizing the partition of the network into sub-units (communities) based on whether a node is more-strongly connected to members of its community

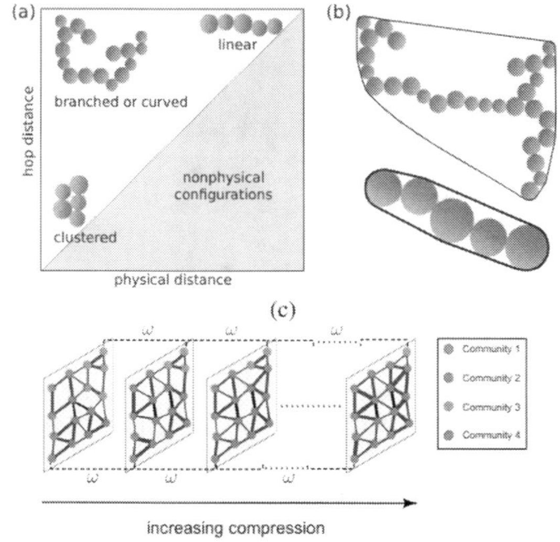

Figure 8. (a) Schematic plot of force chain communities with different degrees of correlation between hop distance (along the edges of a network) and physical distance (as the crow flies). (b) Illustration of force chain communities with a low or high ratio between their convex hull area and the physical area of the particles. (c) Schematic image of a multilayer network comprising multiple strain steps. Adapted from [20, 21].

than to others (the null model) [4, 15, 16]. To identify communities in a network, we use an open-source code [17] to maximize the modularity Q, defined as

$$Q = \sum_{m,n} [W_{mn} - \gamma P_{mn}] \delta(g_m, g_n) \quad (5)$$

where node (particle) m is assigned to community g_m, node (particle) n is assigned to community g_n, the Kronecker delta $\delta(g_m, g_n) = 1$ if and only if $g_m = g_m$, the quantity γ is a resolution parameter (discussed below), and P_{mn} (the null model, discussed below) is the expected weight of the edge that connects node m and node n. These methods require a well-justified choice of both the null model, and the resolution of comparison to that null model (for both the spatial and temporal scales).

Force networks in granular materials are an unusual class of network in that they are spatially-embedded: a grain can only be in contact with its geographical neighbors. Therefore, we have modified existing community-detection tools to include this constraint directly, and better-develop existing techniques so that they provide physically-meaningful results. The first modification we make is to improve the null model used in the community-optimization process (Eq. 5). Instead of using the more common Newman-Girvan null model, in which any particle could be connected to any other particle, we instead define a geographical null model [18] in which only a particle's geographical neighbors are possible edges (contacts).

Second, the choice of resolution parameter γ sets whether the optimization process probes small vs. large scales. For a single network (e.g Fig. 1), Eq. 5 provides the optimization. In order to follow the dynamics of such networks over time or strain, it is necessary to use a multilayer network which is a composite of many such networks. In this case, there are two resolution parameters: one for the spatial scale, and the other for the temporal scale. We have developed two methods to chose a spatial resolution parameter which best-captures the force chain network, illustrated in Fig. 8ab. The first method seeks a value for γ for which there is the lowest correlation between the physical (Euclidean) distance between two nodes and the 'hop' distance (measured only along network edges); this is illustrated in Fig. 8a [18]. The second method seeks a value for γ which maximizes the ratio between the volume of the convex hull (see Fig. 8b) and the volume of the particles within each network [19, 20]. In either case, the optimum value corresponds to choosing $\gamma \approx 1$: particles with the mean force are equally likely to be included/excluded from a network. Importantly, the optimization process does not impose a hard threshold.

Community detection techniques also work on on multilayer networks; we have adapted these methods to be used to understand granular dynamics by representing each layer as a step in time [20]. This is illustrated schematically in Fig. 8c, with the interlayer coupling constant ω controlling whether two adjacent layers are considered independent ($\omega = 0$) or coupled ($\omega > 0$). For $\omega = 1$, the interlayer coupling strength is matched to the mean edge weight (interparticle force). In that limit, we find that there is no flexibility in the community detection: a single set of communities is frozen throughout the dynamics. For $\omega = 0$, the flexibility is maximized: there is no correlation in community assignments for adjacent steps. We find that choosing a value of $\omega \approx 0.1$ to 0.2 gives approximately half the maximum flexibility and is a reasonable choice [20].

For each community detected, we can calculate the average size, strength (mean interparticle force), and sparsity (via the complex hull). This provides a rich set of characterizations through which we can quantify how the force chain network depends on particle properties, strain history, etc. These analyses can be performed on either the normal force network or tangential force network. One such comparison is shown in Fig. 9, in which we have analyzed the same dataset shown in Fig. 5 (from [12]).

As a test-case, we examine whether community detection is able to distinguish the subsystem of low-friction particles from the bath of higher-friction particles (see §4) when tangential forces are taken into account. We observe that the increasing community strength, measured from normal forces, is not significantly different between the subsystem and bath. However, the tangential network displays a clear distinction after a few steps, suggesting that the multilayer network model and community detection are sensitive to differences in friction. Importantly, the direction of this sensitivity agrees with the expectation from local force information alone (the modular structures found in the low friction bath are characterized by lower tangential intracommunity force), but the difference is more pronounced.

 DOI: 10.1051/epjconf/201714001006

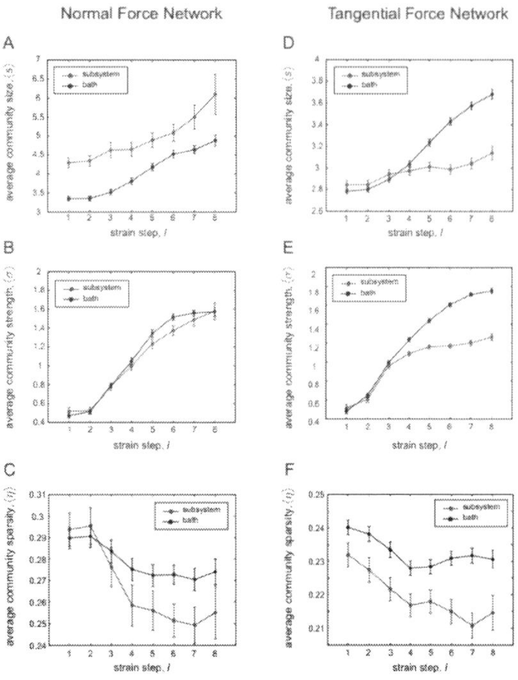

Figure 9. Curves showing the average community (a,d) size, (b,e) strength, and (c,e) sparsity of the subsystem (red curves) *vs.* bath (black curves) for both the normal (a,b,c) and tangential (d,e,f) force networks. Each strain step corresponds to a change in packing fraction of 0.009. Figure reprinted from [20].

6 Conclusions

We have used photoelastic contact force measurements to examine the network of forces within granular materials. We observe that the resulting force networks control the patterns of sound propagation, via the increase in contact area at the strongest contacts. Temperature-like variables associated with the force-moment ensemble obey some key thermodynamic-like features (e.g. obey the zeroth law) but their equations of state are invalidated through the presence of history-dependence. This suggests the presence of internal degrees of freedom — likely related to friction — which store the memory of how a granular packing was assembled. The application of network science tools within a granular context opens new horizons for investigating mesoscale phenomena within a granular material. In future work, the development of network techniques which allow for vector-weighed edges (to simultaneously account for both normal and tangential forces) will provide additional insight. Finally, while this paper focuses on measurements within 2D systems, due to the relative ease of making particle-scale force measurements, recent advances in 3D systems will soon make it possible to extend the techniques.

Acknowledgements

This work has been supported by the National Science Foundation under grant number DMR-0644743 and DMR-1206808, as well as by the James S. McDonnell Foundation. In addition, I am grateful to all of my coauthors on the works cited below for being instrumental in all the research described here.

References

[1] E.T. Owens, K.E. Daniels, Europhysics Letters **94**, 54005 (2011)
[2] D.S. Bassett, E.T. Owens, K.E. Daniels, M.A. Porter, Physical Review E **86**, 041306 (2012), `1110.1858`
[3] D. Bi, S. Henkes, K. Daniels, B. Chakraborty, Annual Review of Condensed Matter Physics **6**, 63 (2015)
[4] S. Fortunato, Physics Reports **486**, 75 (2010)
[5] M. Newman, *Networks: An Introduction* (Oxford University Press, 2010)
[6] T.S. Majmudar, R.P. Behringer, Nature **435**, 1079 (2005)
[7] K.E. Daniels, J.E. Kollmer, J.G. Puckett, (submitted to Review of Scientific Insruments) `arXiv:1612.03525`
[8] D. Howell, R.P. Behringer, C. Veje, Physical Review Letters **82**, 5241 (1999)
[9] E. Somfai, J.N. Roux, J.H. Snoeijer, M. van Hecke, W. van Saarloos, Physical Review E **72**, 21301 (2005)
[10] S.F. Edwards, Powders and Grains 2005 **353**, 114 (2005)
[11] S. Henkes, B. Chakraborty, Physical Review Letters **95**, 198002 (2005)
[12] J.G. Puckett, K.E. Daniels, Physical Review Letters **110**, 058001 (2013)
[13] S.F. Edwards, R.B.S. Oakeshott, Physica A **157**, 1080 (1989)
[14] E.S. Bililign, J.E. Kollmer, K.E. Daniels, (in preparation)
[15] M.A. Porter, J.P. Onnela, P.J. Mucha, Not Amer Math Soc **56**, 1082 (2009)
[16] M.E.J. Newman, Nat Phys **8**, 25 (2012)
[17] I.S. Jutla, L.G.S. Jeub, P.J. Mucha, *A generalized Louvain method for community detection implemented in MATLAB* (2011–2016), `http://netwiki.amath.unc.edu/GenLouvain`
[18] D.S. Bassett, E.T. Owens, M.A. Porter, M.L. Manning, K.E. Daniels, Soft Matter **11**, 2731 (2015)
[19] Y. Huang, K.E. Daniels, Granular Matter **18**, 85 (2016)
[20] L. Papadopoulos, J.G. Puckett, K.E. Daniels, D.S. Bassett, Physical Review E **94**, 032908 (2016)
[21] C. Giusti, L. Papadopoulos, E.T. Owens, K.E. Daniels, D.S. Bassett, Physical Review E **94**, 032909 (2016)

Deconvolution of grading curves during milling: example of wheat straw

Nicolas Blanc[1,*], *Vincent* Richefeu[2], *Claire* Mayer[1], and *Jean-Yves* Delenne[1]

[1] *IATE, UMR 1208 INRA – CIRAD – Montpellier Supagro – Université de Montpellier, 2 place Viala, 34060 Cedex, Montpellier, France*
[2] *Université Grenoble Alpes, 3SR, F-38000 Grenoble, France*

Abstract. The evolution of grading-curves during powder milling or agglomeration processes includes a wealth of information about the mechanisms involved at the scale of particles. However, such information can hardly be retrieved from the particle size distribution (PSD). Based on a minimization technique we developed a methodology for the decomposition of grading curves as sub-PSDs. In this paper we follow their evolution with time in the specific case of the comminution of wheat straw.

1 Introduction

Milling is a crucial operation for many industrial applications in pharmaceutical industry, cement production, food manufacturing... The main objective of the milling is to reduce the size of the raw material and to produce a target particle size distribution (PSD). The process involves many complex mechanisms at the scale of particles and is largely influenced by the microstructure of the material, the process parameters, and the environmental conditions. Milling is particularly important for the refinery of biomass in dry conditions [1] especially for the production of building block molecules, fibers, biocomposite fillers... In this case, several types of size reduction are generally distinguished: coarse milling (cm to mm) which disorganize the plant material at the scale of the organs, intermediate micronization at tissue scale (1 mm to 100 μm), fine grinding at cells scale (<100 μm) and ultrafine grinding (<20 μm) which implies the fracture of plant cell walls [2]. In a grinder, all these characteristic sizes can coexist as different populations of particles and their proportion evolves with time.

In this paper, we investigate the evolution of particle size reduction of a residue from agriculture industry: wheat straw. Wheat straw has a very heterogeneous microstructure (Figure 1) involving cells embedded in different tissues. During the milling process several *populations* of particles are generated leading to complex multi-modal PSDs. In this context we developed a tool to decompose the PSD as the sum of sub-distribution functions. The evolution of the characteristics of these sub-distributions (which are supposed to have specific properties in terms of grindability, chemical composition, shape...) is discussed in relation to the structure of wheat straw.

*e-mail: nicolas.blanc@supagro.inra.fr

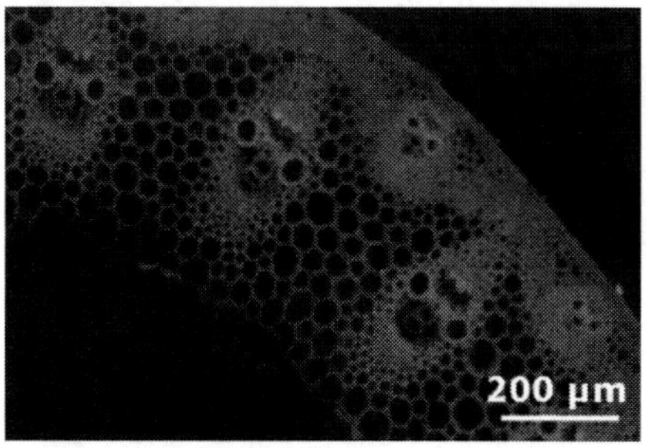

Figure 1. Cross section of an internode of wheat straw [3]

2 Comminution of wheat straw

We used the straw from a soft-wheat variety harvested in 2015. It was first milled in the Cutting Mill SM100 (Retsch®) with a selection grid of 1 mm aperture and sieved between 0.25 and 0.5 mm leading to a final median volume size $d50 = 511$μm. This raw powder was then milled in a jar rolling mill of intern diameter 175 mm and volume 2 L. The jar was filled at 33% with 3 kg of steel balls. The mass of the balls follow a uniform distribution per mass fraction with 3 diameters: 15, 20 and 25 mm. The mass of raw powder (63.1 g) was calculated in order to fill the space between the balls and to limit useless impacts between balls. The jar is rotating at 66 rpm which corresponds to 65% of the speed of centrifugation. Samples are withdrawn all along the grinding process until a d50 of 7 μm (milling time: 1342 min) was reached.

The particle size are measured by laser diffraction using a granulometer Malvern Mastersizer 2000. Ethanol was used as dispersant and three minutes of ultrasound was applied before the measurements to maximize the dispersion of the particles. Particle sizes were determined, by using the Mie method accurate for small particles at low concentration [4]. The size range (0.05 to 2000 µm) is divided in $N_s = 100$ classes i ($i \in [1, 100]$) following the relation:

$$x_i = x_{i-1}\left(\frac{x_{N_s+1}}{x_1}\right)^{\frac{1}{N_s+1}} \tag{1}$$

with x_i the ith lower-bound of the class i. The PSD is built by determining the volume fraction of particle in each class. Figure 2 shows all PSDs determined during the full grinding process. The presence of the different peaks reveals the existence of several populations.

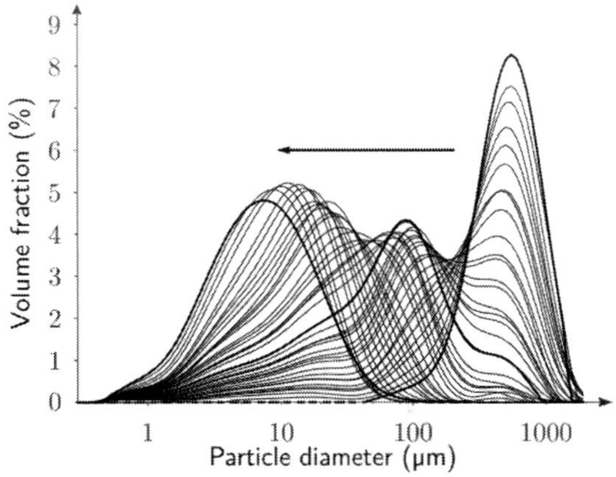

Figure 2. Evolution of PSD of wheat straw during grinding. The arrow indicates the time progression

3 Deconvolution of grading curves

The purpose of the methodology described below is to decompose the PSD into a sum of N_p sub-distributions (sub-PSDs). We denote N_s the number of discrete classes and the particle sizes s^i. From the shape of PSDs obtained in grinding processes it is clear that sub-PSDs have some level of asymmetry and should be bounded. For these reasons, following the works of Voivret *et al.* [5], we use the Beta law for each sub-distribution (*i.e.* for each population):

$$\mathcal{V}_p(s^i; \mathcal{X}_p) = \begin{cases} w_p \dfrac{(s_p^i)^{(\alpha_p-1)}(s_p^i - 1)^{\beta_p}}{B(\alpha_p, \beta_p)} & \text{if } s^i \in [a_p, b_p] \\ 0 & \text{otherwise} \end{cases} \tag{2}$$

where p is the population of particles and $\mathcal{X}_p$ is a set of parameters: w_p control the peak magnitude, α_p and β_p are parameters of the normalized Beta function $B(\alpha_p, \beta_p)$, and a_p and b_p define the distribution range scaled to [0, 1] using $s_p^i = (s^i - a_p)/(b_p - a_p)$.

The grading curve, which is the master distribution, is the sum of the sub-PSD:

$$\mathcal{V}(s^i; \mathcal{X}) = \sum_{p=1}^{N_p} \mathcal{V}_p(s^i; \mathcal{X}_p) \quad \text{with } \mathcal{X} = \bigcup_{p=1}^{N_p} \mathcal{X}_p \tag{3}$$

with

$$\sum_{p=1}^{N_p}\sum_{i=1}^{N_s} \mathcal{V}_p(s^i; \mathcal{X}_p) = 100\% \tag{4}$$

The core principle of the methodology is to compute sub-PSD parameters so that the master distribution best matches the experimental PSD. The latter is defined as a set of N_s pairs $\{(s^i, v^i)\}$, where s^i is the mean size of the class and v^i its volume fractions.

The set $\mathcal{X}$ contains thus $N = 5 \times N_p$ values x_n to be optimized. To do so, a classical minimization of the mean squared of $\sum_i r^2(s^i)$ is achieved, where

$$r(s^i) = v^i - \mathcal{V}(s^i; \mathcal{X}) \tag{5}$$

This minimization was performed thanks to the Levenberg-Marquardt algorithm [6, 7] implemented in the minpack library [8]. This iterative algorithm is known for its robustness as it is accurate even if its initial conditions are chosen far from the final solution. Note that, it is also possible to constrain the possible values of $\mathcal{X}$ by weighting the Jacobian terms $\partial r/\partial x_n$. For experimental reasons it is reasonable to assume that $w_p > 0$, and $\alpha_p > 1$ and $\beta_p > 1$ ensuring that $\mathcal{V}_p$ curve is bell-shaped.

Figure 2 shows the experimental PSDs measured by laser diffraction for 41 milling times of the straw. For each PSD, we observe a maximum number of four peaks. The deconvolution of the PSD at initial time is performed by positioning approximately the sub-PSDs on the 4 peaks. The algorithm is then run to determine the best fit. For the other PSDs we use the previous sub-PSDs as initial condition. It is worth noting that the initial guess may have an influence as several solutions may exist. This procedure requires an important number of experimental PSDs in order to capture a continuous evolution of the sub-PSDs parameters. Figure 3 shows an example of deconvolution of a grading curve into 4 sub-PSDs after 5 hours of milling. Some sub-PSDs, like the one of the population 4, may have very low weight. It is worth noting that this could leads to fluctuations, for example in the determination of the position of the pic as a function of time.

4 Analysis

Figure 4(a) shows the evolution of the volume fraction as a function of the mode of the distribution, which corresponds to diameter at the peak. It is interesting to note that the peak diameters can be related to the characteristic scales generally observed in the grinding of plants: tissue scale (peaks 1 between 450 and 550 μm and 2 around 100μm), cell scale (peak 3 between 10 and 40μm) and cell wall scale (peak 4 around 1μm) (see section 1).

During the grinding process, the largest particles (population 1) are broken into smaller sizes. The population 2 seems to act as a transition size. The volume fraction of the population 4 remains low even for a long grinding time (more than 20 hours) showing that this size is hardly reachable with our grinding process. The right-shifting in diameter of population 3 at the beginning of the milling may be

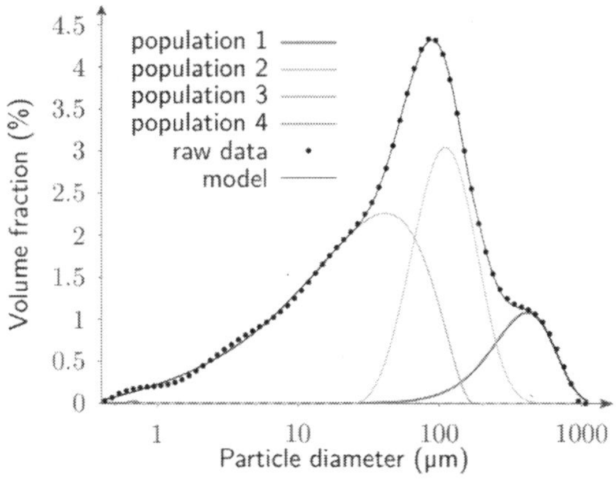

Figure 3. Example of deconvolution of the experimental PSD at a milling time of 4.78 hours.

attributed to deconvolution artifacts for weakly weighted sub-PSDs.

From Figure 3, one might think that the sub-PSD modes should decrease continuously with time as for the experimental PSD. Interestingly, the modes of population 1, 4 and partially 2 are nearly constant in the course of time; Figure 4(b). This can be due to a "shattering" process in which the particles are broken into multiple fragments feeding a limited number of smaller population with given sizes. On the contrary, the slow decrease of the mode of populations 2 and 3 probably denotes an erosion process [9] leading to a slow decrease in size of particles.

Figure 5 shows the evolution of the volume fraction with time. Four phases can be distinguished: *I*) (up to 4 hours) the volume fraction of populations 1, 2 and 3 evolve linearly. Population 1 decreases whereas both populations 2 and 3 increase with the same rate. *II*) From 4 to 7 hours, the population 2 decreases while populations 1 and 3 keep the same trend. *III*) Population 1 totally disappeared after 7 hours of grinding. Slopes of populations 2 and 3 change, corresponding to the time when their modes start to decrease. This indicates a possible erosion process and seems to be consistent with the increase in volume of finest particles (population 4). At the beginning of the process the stress is probably transmitted though the biggest particles while the smallest ones, into the pores network, are globally protected [10]. During this phase, populations 2 and 3 weakly evolve. Possible reasons are that erosion is less efficient than shattering, the remaining particles are the strongest or the increase in specific surface leads to a greater dissipation in friction and collisions. During this phase, the population 3 represents nearly the totality of the product (90% at 7 hours). Finally, in phase *IV*), after 12 hours, only populations 3 and 4 remain.

The mechanical energy transmitted to particles is directly proportional to milling time. Thus the variation rate of volume fraction depends on the milling ability of a given population.

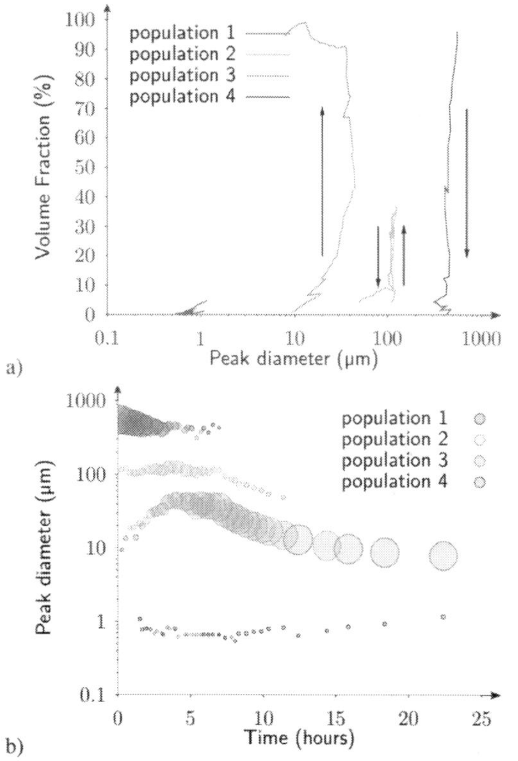

Figure 4. (a) Volume fraction as a function of the peak diameter (the arrows indicate the direction of time) (b) Peaks diameter as a function of time (the size of circles indicate the volume fraction)

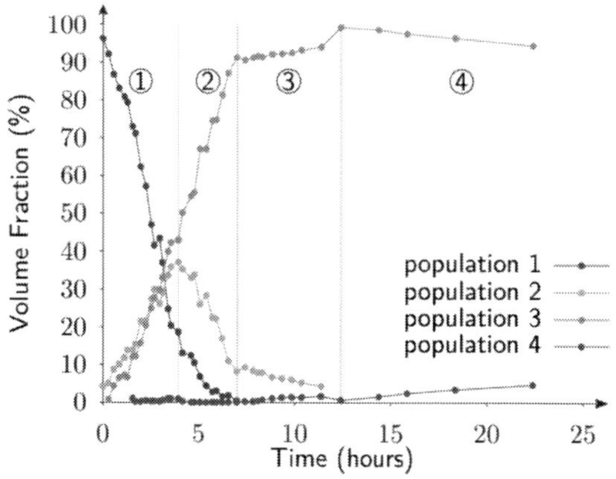

Figure 5. PSDs volume fraction as a function of time. Four different phases are identified.

However, since the population modes are not constant, it is difficult to compute an energy balance based on the variation of the volume fractions. In order to study the evolution of the specific surface, which reflects the energy used to create new surfaces, we assume that the particles are spherical and that their diameters corresponds to the size of the classes.

The Specific Surface Density (SSD) in the class i and for the population p is thus:

$$S_p^i = \frac{6\mathcal{V}_p(s^i; \mathcal{X}_p)}{s^i} \quad (6)$$

and the total specific surface per unit volume for population p is:

$$S_p = \sum_i S_p^i \quad (7)$$

In Figure 6 the total SSD of the powder is increasing at constant rate during the first two phases. Then, this rate seems to saturate denoting a drop in the ability of grinding of the product. The energy supplied by the mill can be insufficient or the modification of the particles' microstructure with the decrease of the size [11] may change the response to impact or shearing solicitations.

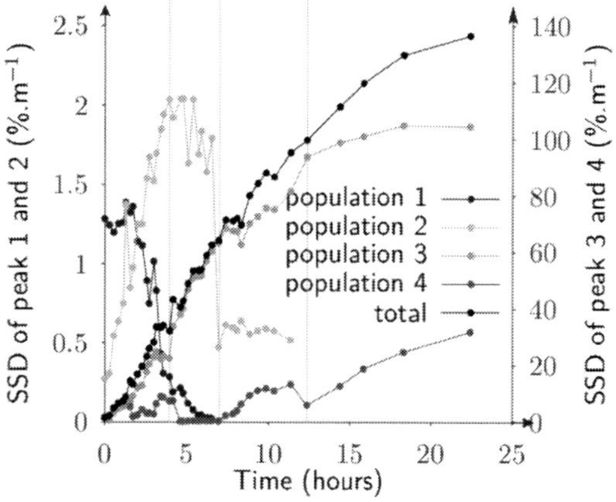

Figure 6. Specific surface density of the PSD and the different population as a function of time

It is interesting to note that the SSD curves do not reflect exactly the same phases as depicted in Figure 5.

The SSD of the population 2 exhibits instabilities during the second phase as its surface doesn't decrease regularly. This is due to instabilities in the values of a_p and b_p which can take very small values leading to a drastic increase of the SSD. Up to 1.7 hours, population 1 tends to stay constant showing that the increase in surface balances the loss of mass. The same behavior is visible for population 3 after 12 hours as its SSD is still increasing while the volume fraction declines. We note also that, because populations 1 and 2 represent large sizes, their contribution to the total SSD is small compared to the ones of populations 4 and 3.

5 Concluding remarks and perspectives

In this paper we proposed a methodology for the analysis of the grading curves of powders obtained by milling. The method is based on a deconvolution procedure which helps to identify characteristic sizes, corresponding to sub-PSDs. One important result was to be able to follow the evolution of the modes of the different populations in the course of grinding. This method can help to determine energy balance between the different populations, and to better understand how the grinding energy is employed at the particle scale for the creation of surface.

Regarding the jar mill process used in this work, we measured in details the evolution of the particle size distribution of wheat straw by taking 41 samples at different milling time. This process seems well adapted until the size of cellular scale corresponding to 20 μm (population 3). This size allows the separation of different chemical components interesting for applications in biofuel or biomaterials. Note however that a finer milling is of great interest for many applications as it increases the reactivity and the accessibility of specific molecules.

In future work we would like to study the effect of the fracture mechanisms at the scale of particles on the evolution of the populations. One possibility is to associate to the grading curves some chemical information obtained by X-ray photoelectron spectroscopy or Laser-induced breakdown spectroscopy [12] as a function of the particle size. Finally, an industrial outlook is to compare the efficiency of different milling technologies for several types of biomass.

Acknowledgments

We thanks Céline Contesto and Cécile Barron for helpful discussions.

References

[1] A. Barakat, H. de Vries, X. Rouau, Bioresour. Technol., **134**, 362-373 (2013)
[2] G. Ghizzi D. Silva, S. Guilbert, X. Rouau, Powder Technol, **208**, 2 266-270 (2011)
[3] G. Ghizzi D. Silva, *Fractionnement par voie sèche de la biomasse lignocellulosique : broyage poussé de la paille de blé et effets sur ses bioconversions*, (2011)
[4] D.L. Black, M.Q. McQuay, M.P. Bonin, Prog. Energ. Combust. **22**(3), 267-306 (1996)
[5] C. Voivret, F. Radjai, J.-Y. Delenne, M. S. El Youssoufi, Physical Review E **76**, (2007)
[6] K. Levenberg, Q. Appl. Math. **2**, 164–168 (1944)
[7] D. Marquardt, SIAM J. Appl. Math. **11**(2), 431–441 (1963)
[8] F. Devernay, C/C++ Minpack,http://devernay.free.fr/hacks/cminpack/ (2007)
[9] L. M. Popplewell, M. Peleg, Powder Technology, 21-30 (1992)
[10] C. Voivret, F. Radjai, J.-Y. Delenne, M. S. El Youssoufi, Phys. Rev. Lett. **102**, 178001 (2009)
[11] Yang, Y., Ji, G., Xiao, W. et al. Cellulose, **21**, 5, 3257-3268 (2014)
[12] Milena R. Martelli, F. Brygo, A. Sadoudi, P. Delaporte, and C. Barron, J. Agr. Food Chem., **58**,12, 7126-7134 (2010)

Structure Generation and Analysis from the Loosest to the Densest Random Packings

Gary W. Delaney[1,*]
[1]*CSIRO Data61, Private Bag 10, Clayton South, VIC 3168, Australia*

Abstract. We explore sets of jammed packings of frictional spheres and ellipsoids generated using the Discrete Element Method. A novel preparation method is presented and analysed, where the internal structure of the system is modified by iteratively removing grains to generate a continuous variation in packing fraction from the densest to the loosest mechanically stable systems. We present a number of measures quantifying the variation in the structure of the system as grains are removed, including the variation in packing fraction, distribution of grain contact numbers, degree of orientational ordering and the degree of displacements of the grains due to rearrangements to achieve mechanical stability after a grain is removed.

1 Introduction

Random packings of granular particles occur in abundance in natural and industrial systems [1–4]. One of the simplest ways to form a granular packing is to pour a set of particles into a container. This technique can lead to a broad range of densities being obtained depending on the details of the particle shape, the inter-particle friction, the viscosity of the fluid/gas of the containing medium and the fine details of the pouring setup.

In this paper, we will explore the transition between the loose and dense limits of the possible mechanically stable random granular packings. We will do this by employing a novel technique in which we iteratively remove grains and measure the response of the system in terms of the structural rearrangements and the resulting variation in density. We will consider two systems, the first composed of perfectly spherical grains and the second composed of prolate ellipsoids with an aspect ratio of 0.7.

Figure 1. DEM simulation of a packing of (left) 5000 spheres and (right) 4500 prolate ellipsoids with aspect ratio of 0.7.

2 Computational Model

We employ the Discrete Element Method (DEM) and in our implementation represent particles as superellipsoids [5, 6]. The normal force between two contacting particles is given by:

$$F_n = -k_n\xi_n + C_n v_n, \tag{1}$$

where k_n is a spring constant determining the stiffness of the particles, ξ_n is the linear overlap of the particles, v_n is the relative normal velocity and C_n is a constant related to the coefficient of restitution. The tangential force is given by:

$$F_t = min\left\{\mu F_n, k_t \sum v_t dt + C_t v_t\right\} \tag{2}$$

where the force vector F_t and velocity v_t are defined in the plane tangent to the surface at the contact point, and the summation is performed over the duration of the contact [6–8]. The total tangential force, F_t, is limited by the maximum Coulomb friction μF_n, at which point the surface contact shears and the particles begin to slide over one another. The grains are sedimented in a viscous fluid, where the fluid is modeled using the Stokes equations as described in [9].

Figure 2 shows the packing fraction variation with aspect ratio for systems of mono-disperse ellipsoids with two of their axes equal in length. The parameters used for the simulation are $k_n = 10^5$ N/m, a restitution coefficient of $e = 0.5$, a particle density $D = 2.7 \times 10^3$ kg/m^3 and values

*e-mail: gary.delaney@csiro.au

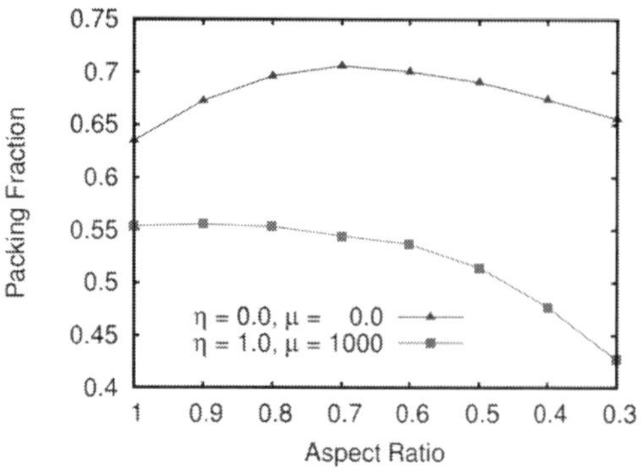

Figure 2. Packing fraction variation with aspect ratio for the densest (η = 0.0, μ = 0.0) and loosest (η = 1.0, μ = 1000) ellipsoid packings formed via DEM simulations of a sedimentation process.

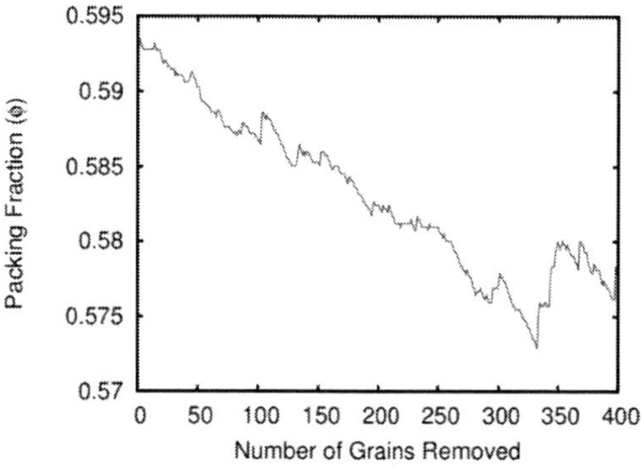

Figure 3. Evolution of the packing fraction as grains are iteratively removed from packing of spheres.

of inter-particle friction ranging from μ = 0 to 1000. Values of viscosity ranging from η = 0 Pa s to η = 1 Pa s are examined. The fluid has a density $\rho = 10^3$ kg/m^3, giving an effective gravity for the particles in the fluid of 0.69g. We show data for two cases, chosen to demonstrate the extremes of density that are obtainable using this preparation technique. The first case corresponds to zero inter-grain friction (μ = 0) without a background fluid (η = 0) which produces the densest packings and the second is in the high friction limit with an inter-grain friction (μ = 1000) and a high liquid viscosity (η = 1.0 Pa s). For dense packings created with zero inter-grain friction, the packing fraction for spheres (α = 1.0) is close to the random close packing limit (RCP) value of ϕ = 0.64. As the aspect ratio is varied from unity, we obtain a peak in packing fraction at an aspect ratio of around 0.7 followed by a near linear decline. For the loose packings prepared with high inter-particle friction (μ =1000) in a high viscosity fluid (η=1.0 Pa s), we obtain packing fractions similar to those measured for the random loose packing limit for spheres at $\phi \simeq 0.55$. As the aspect ratio is varied from unity, we see an initial plateau and then a steady decline between 0.8 and 0.6 followed by a faster rate of decrease. These results are similar to those seen for prolate ellipsoids packings prepared by the same technique [9] and also for random close packings of ellipsoids prepared by a particle expansion technique [1].

Having established the density limits that are obtainable by utilising a pouring (or sedimentation) technique in the previous section, we will now consider a more novel way of transitioning among states in between the loosest and densest mechanically stable packings (the area between the blue and red lines in Figure 3). Here we will constrain our consideration to two systems. The first is composed of 5000 spherical particles in a cylindrical container. Spheres have a packing fraction range for mechanically stable random packings between approximately ϕ = 0.55 and ϕ = 0.64 as shown in Figure 2. The second is a packing of 4500 prolate ellipsoids with aspect ratio 0.7 contained within a periodic box in the x and y direction (the 2 directions normal to gravity). These ellipsoids have a packing fraction range for mechanically stable random packings between approximately ϕ = 0.545 and ϕ = 0.71.

To study the transition between different mechanically stable structures, we will employ a simple technique of iteratively removing a single grain chosen at random from the packing and allowing the system to relax to a new mechanically stable state. The inter-particle friction is set to μ = 1000 to put us in the infinite friction limit, which allows for the potential to obtain any state between the densest and loosest mechanically stable limits, and the simulation has no background fluid present (η = 0). Depending on the grain chosen, its removal may lead to anything from no change in the positions of the remaining grains (as the system is still mechanically stable without the removed grain), up to very large scale global rearrangements of the packing as it finds a new mechanically stable state.

Figure 3 shows the evolution of the packing fraction ϕ with the number of grains removed for the packing of spherical grains. In this case, we initially start at a packing fraction of ϕ = 0.593, near to the middle of the range of mechanically stable packing fractions that can be obtained for random packings of spheres. As we iteratively remove grains from the system, the packing fraction shows a clear downward trend, with occasional avalanche like events that cause a temporary increase in the packing fraction. These events are caused by removing grains that were playing a particularly important role in the mechanical stability of the system, where their removal causes a large scale rearrangement of the grains and densification of the system under gravity. The increases in packing fraction at these avalanche events is generally only up of a maximum increase of $\Delta\phi$ = 0.002 until we get to looser packings below ϕ = 0.575 after approximately 300 grains have been removed. At these low densities the system has become much more susceptible to the further removal of grains and we see a more rapid increase in density over successive grain removals from the 330th to the 350th grain removals, where an increase of $\Delta\phi$ = 0.007 is observed.

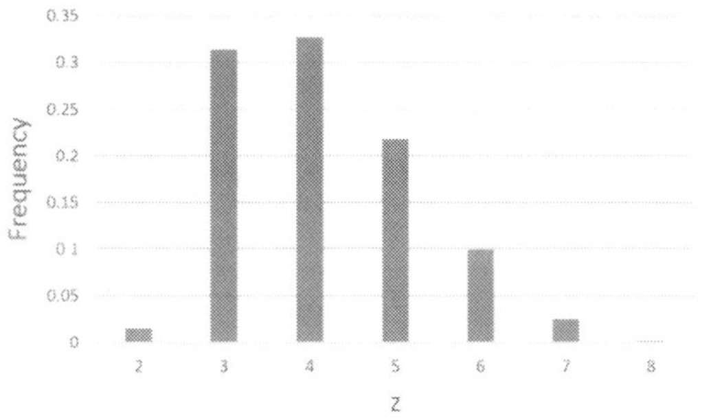

Figure 4. Distribution of the number of contacts z of the removed grains.

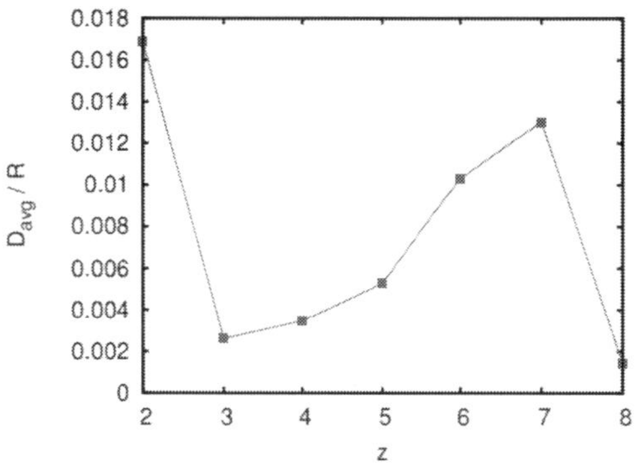

Figure 5. Variation in the average movement of a grain in the sample when a grain with a contact number of z is removed. Note very few grains have a z = 2, 7 or 8, meaning that considerable random variation in the value at these points would be expected.

Figure 4 shows the distribution of the number of contacts z of the removed grains. We see that a majority of the removed grains have a contact number between $z = 3$ and $z = 5$, consistent with the average contact number for the entire system of $Z \simeq 4.3$. Only a very small number of the removed grains have a value of $z = 2$, 7 or 8. This graph shows that by randomly removing grains from the system we are not uniformly sampling the possible contact numbers, but are instead sampling based on the distribution of contact numbers already present in the system.

To determine the effect of the contact number of the grain that is removed on the rearrangement of the system, in Figure 5 we plot the average movement of all grains within the packing (D_{avg}) normalized by the radius of the grains (R) against the contact number of the grain that was removed. The largest average movement of the grains in the sample occurs when a grain with $z = 2$ is removed, with on average all grains moving by a distance equal to 1.7% of the grain radius. Note however, that as the data in this plot is an average over all of the grain removal events and as we have very few events with $z = 2$, 7 and 8, there may be considerable random variation in the average values at these three points. Removing grains with $z = 3$ leads to the smallest average movement of the sample (0.25% of the grain radius) with a consistent increase in the average movement in the sample seen for larger contact numbers up to $z = 7$. It is unclear for the decrease observed for $z = 8$ if this is just due to being a single measurement point and hence highly susceptible to random variation.

Figure 6 shows the evolution of the packing fraction with the number of grains removed for the packing of ellipsoids with aspect ratio of 0.7. In this case we start with a slightly lower packing fraction of $\phi \simeq 0.575$ and perform 800 iterations of removal of grains. We see a similar variation in the packing fraction as grains are removed to the spherical case (Figure 3), with fewer increases in packing fraction due to avalanche events. The packing fraction levels off at $\phi \simeq 0.548$ after 700 grains have been removed, slightly above the loosest packings that are obtained by sedimentation of $\phi \simeq 0.545$.

The packing fraction variation with number of grains removed shows the ability of this technique to transition between a broad range of packing fractions right down to close to the loosest mechanically stable packings obtained by sedimentation processes. However, one key concern is the degree of ordering that is being induced as a result of this technique, and whether we are obtaining packings with very different structural properties to those obtained via sedimentation. In particular here we will investigate the degree of orientational ordering of the ellipsoids and how this varies as we remove grains. We quantify the degree of orientational alignment using an order parameter

$$\chi_\theta = \frac{3}{2}\left\{\frac{1}{N}\sum_{i=1}^{N}\cos 2\left(\theta_i - \frac{\pi}{2}\right) - \frac{1}{3}\right\} \tag{3}$$

where θ_i is the angle between the semi-major axis of the i-th particle and the vertical axis [10]. If all grains have random orientations, $\chi_\theta = 0$, while if the grains all lie flat $\chi_\theta = 1$. We have also determined the more commonly used S_2 order parameter [11], quantifying the degree of orientational ordering of the major axes of the grains about their average nematic director, and have found that this parameter follows the same trends observed for χ_θ.

Figure 7 shows the variation of the angular alignment of the grains with the number of grains removed. The ellipsoids initially have a small degree of angular alignment with $\chi_\theta \simeq 0.145$. As more grains are removed and the packing fraction reduces, the degree of alignment only varies very slightly, with an increase to $\chi_\theta \simeq 0.16$, before plateauing. There is a steady increase in the degree of alignment after 450 grains have been removed, followed by a very sharp drop that exactly coincides with a sharp increase in packing fraction of $\Delta\phi = 0.005$ when 500 grains have been removed. The degree of alignment then shows a consistent decrease to $\chi_\theta \simeq 0.13$ when 800 grains have been removed. The results clearly show that the reduction in packing fraction over the 800 grain removals is not due to any increase in ordering within the system, with the final orientational ordering being slightly lower than for the initial configuration.

DOI: 10.1051/epjconf/201714006029

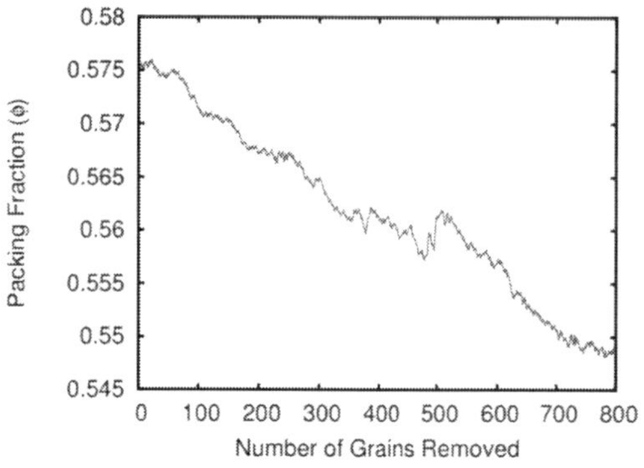

Figure 6. Evolution of the packing fraction as grains are iteratively removed from a packing of ellipsoids with an aspect ratio of 0.7.

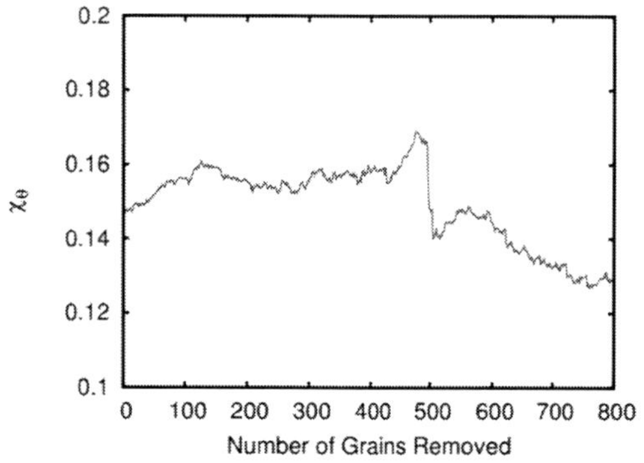

Figure 7. Evolution of the orientational order parameter χ_θ as grains are iteratively removed from a packing of prolate ellipsoids with an aspect ratio of 0.7.

3 Conclusions

We have presented results for a novel method for exploring the structure of random packings of granular matter at packing fractions between the loosest and densest mechanically stable random packings. We find that our technique of random removal of grains and relaxation of the sample allows us to study density variations for spheres down to packing fractions of $\phi \simeq 0.57$ and for ellipsoids down to $\phi \simeq 0.548$, close to the loosest mechanically stable packings obtained by sedimentation protocols. The technique also allows us to generate a continuous sequence of packings that ranges up to the densest packings. Our analysis of the effect on the structure of the sample of spheres from removing grains with different contact numbers shows the clear relationship between the contact number and the degree of rearrangement of the sample due to the removal of a grain. For the sphere system, we see an increase in the average displacement of the particles in the sample with the contact number of the removed grain for particle contacts from $z = 3$ to $z = 7$, representing the vast majority of the grains in the system. While our investigation of the orientational ordering in the sample of ellipsoids demonstrates that large variations in density can be achieved with this protocol without significant variations in the degree of orientational ordering within the system

The results presented here may be of particular interest in exploring more exotic granular structures and the very loose limits of mechanical stability. Understanding the degree of structural rearrangement from removing grains from the system may be especially useful in determining strategies for reducing the density of the system below the limits observed when generating granular packings using traditional techniques such as pouring.

References

[1] G.W. Delaney, P.W. Cleary, EPL (Europhysics Letters) **89**, 34002 (2010)
[2] F.M. Schaller, M. Neudecker, M. Saadatfar, G.W. Delaney, G.E. Schröder-Turk, M. Schröter, Phys. Rev. Lett. **114**, 158001 (2015)
[3] A. Baule, H.A. Makse, Soft Matter pp. 4423–4429 (2014)
[4] T. Aste, *The Pursuit of Perfect Packing* (Bristol and Philadelphia: IOP Publishing Ltd, 2000)
[5] P. Cundall, O. Strack, Geotechnique **29**, 65, 47 (1979)
[6] P.W. Cleary, Engineering Computations **21**, 169 (2004)
[7] P.W. Cleary, Powder Technology **179**, 144 (2008)
[8] G.G. Pereira, P.M. Dupuy, P.W. Cleary, G.W. Delaney, Progress in Computational Fluid Dynamics **12**, 176 (2012)
[9] G.W. Delaney, J.E. Hilton, P.W. Cleary, Physical Review E **83**, 051305 (2011)
[10] B.J. Buchalter, R.M. Bradley, EPL (Europhysics Letters) **26**, 159 (1994)
[11] A. Donev, S. Torquato, F.H. Stillinger, R. Connelly, Journal of Applied Physics **95**, 989 (2004)

EPJ Web of Conferences 140, 09042 (2017) DOI: 10.1051/epjconf/201714009042
Powders & Grains 2017

Properties of pendular liquid bridges determined on Delaunay's roulettes

Boleslaw Mielniczuk[1,2,*], *Olivier* Millet[1], *Gérard* Gagneux[1], and *Moulay Said* El Youssoufi[2,3]

[1]LaSIE, UMR-CNRS 7356, University of La Rochelle, 17042 La Rochelle Cedex 1, France.
[2]LMGC UMR CNRS 5508, University of Montpellier, 34095 Montpellier Cedex 5, France.
[3]Laboratory MIST, IRSN-CNRS-University of Montpellier, France.

Abstract. This work addresses the study of capillary bridge properties between two grains, with use of recent analytical model, based on solutions of Young-Laplace equation from an inverse problem. A simple explicit criterion allows to classify the profile of capillary bridge as a surface of revolution with constant mean curvature (Delaunay roulette) using its measured geometrical parameters (gorge radius, contact angle, half-filling angle). Necessary data are obtained from experimental tests, realized on liquid bridges between two equal spherical grains. Sequences of images are recorded at several (fixed) volumes of liquid and different separations distances between the spheres (from contact to rupture), in laboratory and in micro-gravity conditions. For each configuration, an exact parametric representation of the meridian is revealed. Mean bridge curvature, internal pressure and intergranular capillary force are also determined.

1 Introduction

Capillary bridge is a liquid link between solid particles (grains). It gives origin of several phenomena observed in unsaturated granular materials, in particular of interparticle capillary force. This force contributes to the formation, deformation and flow of granular materials (soils, powders, photonic crystal production, etc.), with significant influence on mechanical and physical properties of such materials. It imparts an apparent macroscopic strength (sand castle effect) to moist granular materials, even in the absence of intrinsic cohesion or confining stress (see e.g. [1, 2]). The knowledge on the behaviour of intergranular liquid at a local scale, in different conditions, seems to be indispensable in complete description of phenomena observed in humid granular materials and in modelling of such materials.

The most elementary capillary bridge is a pendular bridge: liquid bridge between two solid grains. Shape and evolution of pendular capillary bridges have been analysed for over hundred years from theoretical and experimental point of view (see e.g. [3-7]). However, the results of numerical modelling and physical experiments are not always in good accordance, because of several simplifications and approximations introduced often into the mathematical models.

This work addresses the experimental study of capillary bridge properties with use of the recent analytical calculation of bridge profile, based on solution of Young-Laplace equation [8]. In this method, the parameters of pendular bridge and the shape of its meridian may be estimated using appropriate theoretical solutions, based on an inverse problem, as the internal capillary pressure is unknown [8]. However, this pressure may be recovered with experimental measurements of geometrical parameters of liquid bridge: gorge radius y^*, contact angle θ, half-filling angle δ (see Fig. 1).

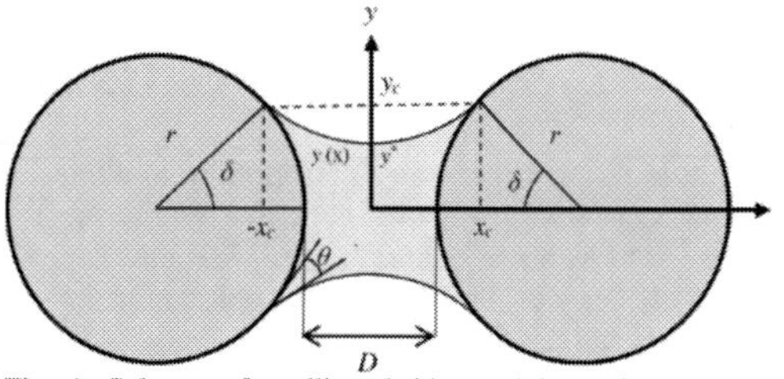

Fig. 1. Schema of capillary bridge and its main geometrical parameters.

2 Experimental setup

Experimental setup consists of photo camera, system of support which allows positioning of both grains, and background light (see Fig. 2). It is similar to configuration described in [9-11]. To examine capillary bridges in zero-gravity conditions, it is put into the rack, adapted for parabolic flight.

* Corresponding author: boleslaw.mielniczuk@umontpellier.fr

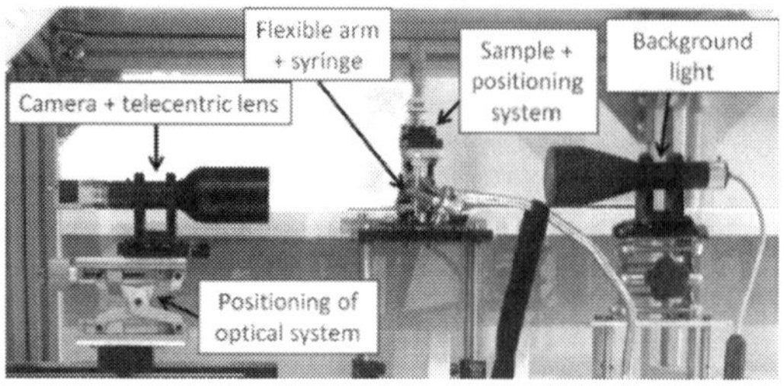

Fig. 2. Global view of experimental apparatus.

Capillary bridges between two borosilicate-glass spheres with diameter of 8 mm or 10 mm and with a fixed volume of water (from 1 to 10 μl) are examined. Initially, bridge is created for a separation distance between spheres D close to zero. Then, D is steeply increased up to the rupture of the bridge (step of 0.1mm). For each D the photo of the capillary bridge is recorded and the geometric parameters (gorge radius y^*, half-filling angle δ, contact angle θ) are measured by image processing. Then the shape of the capillary bridge, and its exact analytical parameterization, are deduced, according to the criterion given in [8].

3 Results of laboratory tests

Experimental data are analysed for each examined configuration (D, V). Measured geometrical parameters are used to determine the type of curve, to reconstruct and trace the bridge profile and to calculate bridge characteristics. Bridge profiles are approximated as surfaces of revolution with constant mean curvature (Delaunay roulettes). Evolution of their main geometrical parameters is briefly presented below.

Gorge radius y^* decreases almost linearly with increasing separation distance D for all examined volumes, with similar evolution for both sphere diameters. For higher bridge volumes V, higher final separation distance D_r (distance at rupture) is observed, from about D_r=1.2 mm for V=1 μl, to D_r=2.2 mm for V=10 μl.

Angle δ decreases from the beginning to about 50% of final separation D_r, than it remains constant up to the rupture. The value of contact angle θ initially decreases with increasing D, but it becomes constant very soon, while half-filling angle δ is still decreasing: pinning of contact line is observed (see details in i.e. [10]).

Evolution of mean curvature of the profile H is presented in Fig. 3. It depends on volume of water V and on separation distance D and it has slightly higher values for 10mm spheres diameter. In general, H is initially positive and it decreases almost linearly with increase of separation distance. At about 70-90% of the critical separation D_r, H passes through zero, becomes negative and stays negative up to the rupture (Fig. 3).

The sign of H determines also the sign of internal pressure $\Delta p=p_l-p_g$ (Laplace pressure, difference of pressure between the interior and the exterior of the capillary bridge) and the type of Delaunay's roulette used to approximate the bridge profile (see [8, 11]).

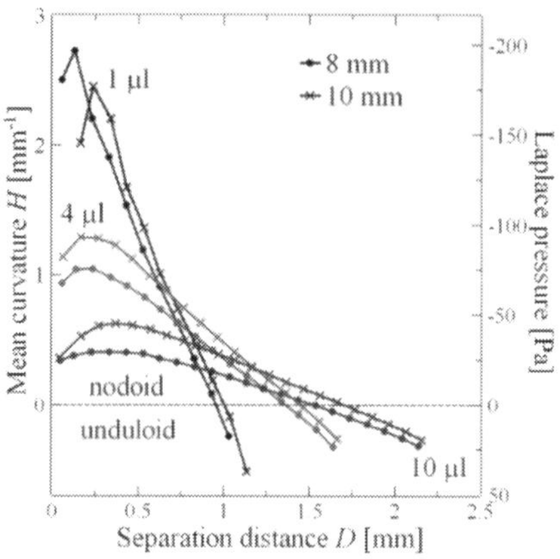

Fig. 3. Mean curvature H and internal pressure Δp of capillary bridge between spheres with diameter of 8 mm and 10 mm, as a function of separation distance D. H >0 corresponds to the nodoid case and H <0 to the unduloid one.

Two types of convex curves were observed in our experiments: a portion of nodoid (Delaunay hyperbolic roulette, from the creation of the capillary bridge to some distance D, H>0 and Δp<0) and a portion of unduloid (Delaunay elliptic roulette, before rupture, H<0, Δp>0).

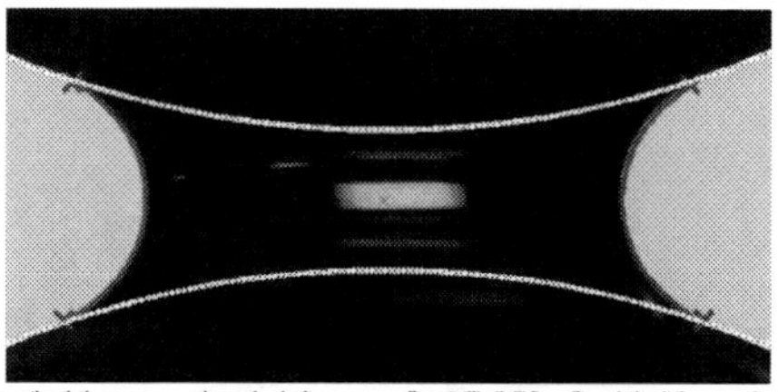

D= 0.66 mm, y^*= 1.16 mm, δ= 17.85°, θ= 10.5°, nodoid

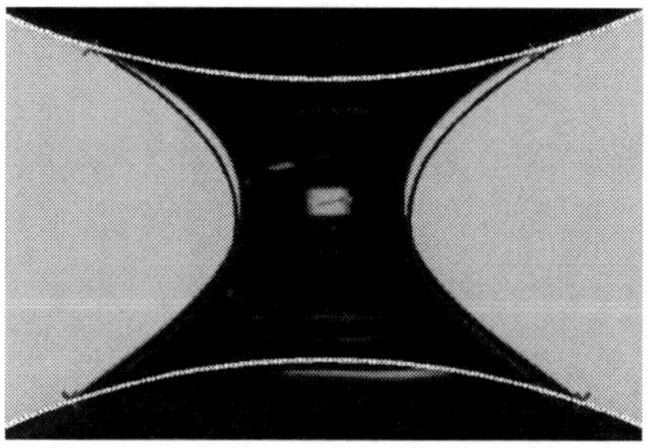

D= 1.66 mm, y^*= 0.52 mm, δ= 16.9°, θ= 14.8°, unduloid

Fig. 4. Resulting bridge profiles superposed on the original images for glass spheres of 10mm. The data are for fixed volume V= 4 μl and various distances D between grains (red: nodoid shape, blue: unduloid shape, calculated from theory).

Calculated bridge profiles are then superposed on the original photos to validate the approach, used in theoretical model. Example of sequence, obtained for capillary bridge between 10 mm spheres diameter, with V=4μl and different D are presented in Fig. 4. Calculated contact points and reconstructed bridge meridians are traced on recorded images.

Intergranular capillary force can be evaluated as $F_{cap}= F_{\Delta p}+ F_{ST}$, including two contributing forces: pressure resulting force $F_{\Delta p}=\pi\gamma H y^{*2}$ and surface tension resulting force $F_{ST}= 2\pi\gamma y^*$, where surface tension coefficient of water $\gamma= 0.072$ N/m. In general, the decrease of capillary force F_{cap} is observed with increasing D (Fig. 5). Only for 10mm spheres at V=10 µl, initial increase of F_{cap} is observed. At the moment of rupture, capillary force is higher for larger volumes of water (see Fig. 5).

For all realized tests, initially negative Δp becomes positive (with repulsive $F_{\Delta p}$) with increasing separation D. However, capillary force F_{cap} remains always attractive, because of dominating influence of attractive surface tension resulting force F_{ST} (see also [10]).

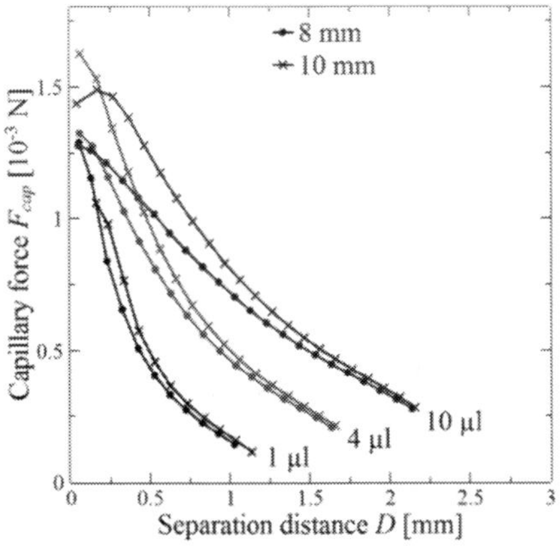

Fig. 5. Calculated capillary force F_{cap} as a function of interparticle distance D for capillary bridge between 8 and 10mm spheres diameter.

In Fig. 6, calculated F_{cap} is compared with capillary force measured in earlier experiments on extended capillary bridges, with use of laboratory balance [9, 10].

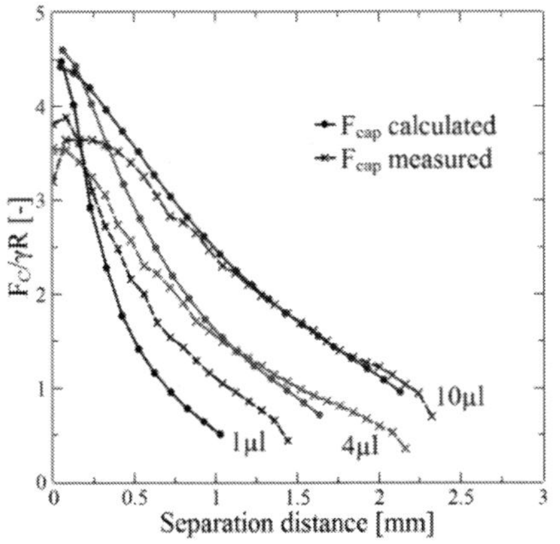

Fig. 6. Calculated and measured capillary forces F_{cap} in function of interparticle distance D for the spheres of 8 mm.

The liquids with different surface tension coefficients were used; both results are thus normalized with respect to γ, to make possible the direct comparison of measured and calculated values. The results are generally in good accordance. Although, it is seen that for small D, calculated force is higher than measured one. With increasing D, this difference becomes smaller, while close to the rupture, the measured force becomes slightly higher than the calculated one, especially for small V. Differences in both types of results may origin from different types of tests (continuous extension or quasi-static capillary bridges) and different surface tension coefficient of used liquids.

4 Micro-gravity tests

For larger liquid volumes and/or relatively high separation distance D, water is driven towards the lower part of the bridge due to the effects of gravity and capillary bridge loses its symmetry. The importance of the influence of the gravity on the bridge shape may be estimated with use of the Bond number $Bo=\Delta\rho g r^2/\gamma$, where $\Delta\rho$ is the difference in densities between water and air, and g is the acceleration of gravity.

The influence of gravity is negligible if $Bo/|2\xi r|<<1$ (where ξ is the bridge curvature) [12].

This is the case for capillary bridges containing low water volume and at low separation distance. One can see, than for such bridges, calculated profiles fits well with the ones recorded with use of photo camera.

When V becomes higher, the influence of gravity can no longer be neglected: geometry of the bridge is distorted and calculated variables are no more correct.

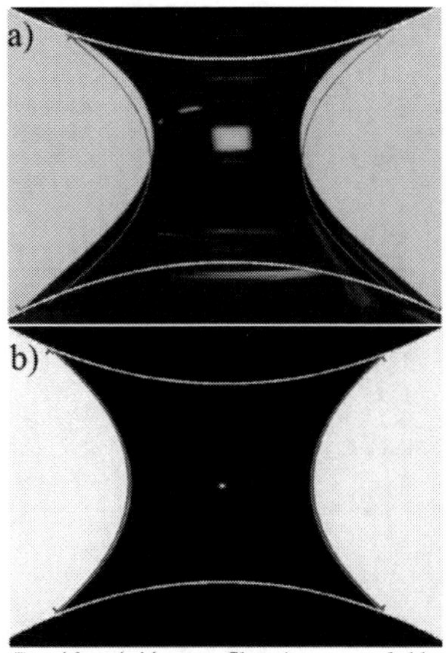

Fig. 7. Resulting bridge profiles (magenta: bridge shape, calculated from theory) superposed on the original images for capillary bridge between glass spheres of 10mm, with liquid volume V = 10µl and separation distance D= 2.06 mm, recorded in laboratory (a) and during parabolic flight (b).

An example showing the influence of gravity on a capillary bridge profile is shown in Fig. 7a). In this example, Bond number is close to 0.3 and theoretical profile is quite far from the recorded profile of the capillary bridge.

DOI: 10.1051/epjconf/201714009042

For now, taking into account the gravity in the general theoretical modelling do not allow to obtain simple analytical solution. It is therefore necessary to get rid of the effects of gravity in order to validate the modelling, carried out on a wide range of parameters (volume of the liquid bridge, grain size, separation distance, ...).

To minimize the influence of gravity, experimental tests were realized also in micro-gravity conditions, during parabolic flight (joint project between LMGC, LaSIE and CNES), with use of the same experimental protocol. Example of results obtained in micro-gravity conditions is presented in Fig. 7. Distortion of the bridge is observed in Fig. 7a) (laboratory test), while identical bridge examined during parabolic flight is not affected by the gravity (Fig. 7b).
Comparison of calculated mean curvature H and capillary force F_{cap} for capillary bridges between 8 mm spheres tested in laboratory and during parabolic flight is presented in Fig. 8.
It is seen, that calculated parameters are different for both tests, especially for higher separations D. Calculated F_{cap} (and also Δp) close to the rupture are larger for bridges tested in microgravity conditions. In addition, stability of capillary bridges is increased in such conditions and rupture of the bridge occurs at larger separations D. However, the passage by H=0, between two types of Delaunay's roulettes, occurs at the same separation distance for both experimental conditions.

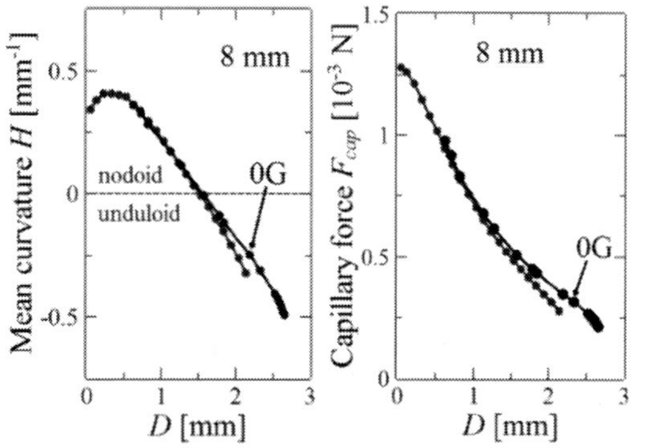

Fig. 8. Comparison of calculated mean curvature H (a) and capillary force F_{cap} (b) as a function of interparticle distance D for capillary bridges between 8mm spheres and V=10µl, in laboratory and micro-gravity tests.

5 Conclusions

In this paper we present some results of studies on capillary bridges between two equal spherical grains. Recent analytical model was used to determine variables and parameters associated with these capillary bridges. The accuracy and validity of the model was demonstrated by comparison with the experimental results. The results from theoretical modeling match very accurately experimental results for small liquid volumes and small separations between particles. For larger ones, the capillary bridge loses its symmetry due the effects of gravity.
In order to remove the distortion effect due to gravity, experiments in micro-gravity conditions, during a parabolic flight have been performed. For these tests, theoretical results are in good agreement with experimental ones, independently of liquid volumes and/or separations distances between grains.
We found that with increasing separation distance between grains the profile of capillary bridge profile evolves from nodoid to unduloid. This passage is accompanied by the increase of internal pressure, from initially negative values to positive pressure before the rupture, while capillary force continuously decreases. Experimental validation of proposed model justifies its further development, its adaptation to other configurations of capillary bridges, including dynamic processes and analysis of stability/rupture of liquid bridges, envisaged in the nearest future.

The part of the work concerning the experiments in micro-gravity was supported by the CNES (joint project PARABOLE 2015, VP118, between LaSIE-LMGC-CNES).

References

[1] H. Peron, J. Y. Delenne, L. Laloui, M. S. El Youssoufi, Comput. *Geotech.* **36**, 61 (2009)

[2] H. Rahardjo, D. G. Fredlund, *Can. Geotech. J.* **32**, 749 (1995)

[3] T. Young, *Phil. Trans.* **95**, 64 (1850)

[4] G. Mason, W. C. Clark, *Chem. Eng. Sci.* **20**, 859 (1965)

[5] G. Lian, C. Thornton, M. J. Adams, *J. Colloid Interf. Sci.* **161**, 138 (1993)

[6] Ch. D. Willett, M. J. Adams, S. A. Johnson, J. P. K. Seville, *Langmuir* **16**, 9396 (2000)

[7] F. Soulié, F. Cherblanc, M. S. El Youssoufi, C. Saix, *Int. J. Numer. Anal. Meth. Geomech.* **30**, 213 (2006)

[8] G. Gagneux, O. Millet, *Transport Porous Med.* **105**, 117 (2014)

[9] B. Mielniczuk, T. Hueckel, M. S. El Youssoufi, *Gran. Matt.* **16**, 815 (2014)

[10] B. Mielniczuk, T. Hueckel, M. S. El Youssoufi, *Powder Technol.* **283**, 131 (2015)

[11] G. Gagneux, O. Millet, B. Mielniczuk, M. S. El Youssoufi, *Eur J Environ Civ En.*, 10.1080/19648189.2016.1167782 (2016)

[12] M. J. Adams, S. A. Johnson, J. P. K. Seville, and C. Willett, *Langmuir* **18**, 6180 (2002)

Electrostatic separation of mineral and vegetal powders with a custom built corona separator: application to biorefinery of rice husk

Rova Karine Rajaonarivony[1], *Xavier* Rouau[1], *Lucien* Dascalescu[2], *Claire* Mayer-Laigle[1*]

[1]IATE UMR 1208 INRA-CIRAD-SupAgro-UM, 2 place Pierre Viala - 34060 Montpellier, France
[2] Institut PPRIME, UPR 3346, 4 Av. de Varsovie, 16021 Angoulême, France

Abstract. In a dry biorefinery scheme, the separation of plant materials into powders rich in the constituents of interest is a crucial step. In recent years, electrostatic separation of agri-resources has sparked a growing interest for its potentialities, but optimization efforts remain to be done especially in case of fine powders. In this study a custom-designed corona-electrostatic separator has been built and its usage for the separation of mineral (ash from rice husk) and vegetal powders (cellulose fibers) with different particle size distributions has been studied. First, the powders have been characterized by their electric charge decay curves and their behaviour on the separator was studied. At a second time, separation tests have been carried out with blends of these two powders, and with native, finely ground, rice husk powder (constituted of both mineral and vegetal particles). In each case, the efficiency of the process was evaluated.

1 Introduction

Plant materials, structurally organized in a multilayer composite containing different tissues [1] are excellent sources of raw materials for many applications as biofuels, green chemicals (surfactant, resin…) or biomaterials (fiber). In a plant bio-refinery scheme, the first step consists in the separation of agri-resources materials into their major constituents (carbohydrates, proteins, lignin, lipids, polyphenols, minerals…). Dry fractionation processes, could be seen as an interesting alternative to expensive hydo-chemical processes which induce a degradation of the native functionalities of the constituents and are known to have a detrimental environmental impact. These processes usually combine different unit operations of pre-treatment, grinding and physical separation to gradually de-structure and fractionate the plant materials at tissue (100 µm) or sub-cellular scale (10 µm) [2].

Among all the separation processes, the electrostatic separation emerged as an eco-friendly technology since it allows the preparation of fractions enriched in the constituents of interest from biomass particles ranging between 10 and 500 µm. In an electrostatic separator, particles are first charged by triboelectric charging, by conductive induction or by ion bombardment and then sorted in a high intensity electric field. Existing electrostatic separators are not directly transposable to agri-resources processing since the particles obtained in a dry bio-refinery scheme are smaller and/or lighter than those encountered in polymer recycling or mineral beneficiation industries. In the scientific literature, two devices, based on two different principles, are mainly used for the electrostatic separation of plant materials.

The first one, inspired by the industrial free fall electrostatic separator demonstrated its potentiality for processing many plant raw materials [3-5], but presents a weak flexibility. The second (belt-type corona-electrostatic separator), akin to a corona drum separator, have been used for removing fibers [6-7], in relatively coarse biomass powder (median size, d50 >1 mm). The present study is aimed at the separation of mineral (silica) and vegetal parts of a raw plant material: rice husk, using a custom-designed corona-electrostatic separator.

Rice husk is an agricultural by-product composed of roughly 85 % lignocellulosic material and 15 % amorphous silica. The high mineral content in the rice husk limited its utilization mainly to the energy production by combustion, with incorporation of the ash in the production of Portland cement or steel. In addition, amorphous silica is used in several other applications, such as ore metal smelting, water purification, rubber and glue additives, etc. but it is currently produced by wet or thermic processes, which are high energy consumers. The separation of the amorphous silica of rice husk by dry processes would allow to expand the field of potential usage and create added value for this by-product.

To dissociate the effects related to the powder properties and those of device parameters, a preliminary study was conducted on model powders simulating the vegetal part (cellulose fiber) and the mineral part (ash form burnt rice husk) of rice husk, with different particle size distributions. Powders have been characterized by their charge decay curves. Their behaviours on the separator were studied and separation tests were carried out with blends of mineral and vegetal powders. The different

* Corresponding author: claire.mayer@supagro.inra.fr

assays were compared using a separation index to define the best process parameters, which were then applied to the separation of ground rice husk.

2 Materials and Methods

In the device, a vibrating hopper is used to deposit the powder on the conveyor as isolate particles in order to limit the interparticle interactions during the separation process. In contact with the grounded conveyor, the conducting particles quickly loose the charge acquired by corona discharge ; when they enter in the electric field generated by the rotating roll electrode they get charged by electrostatic induction, detach from the conveyor, and may be recovered in bin B. Many of them change again the sign of their charge after the impact with the rotating roll electrode, and will be collected neither in bin B, nor in bin A. The poor- and non-conducting particles that loose their charge more slowly can be attracted to the rotating roll electrode by the electric Coulomb force. The corona discharge and the roll electrode in the custom built separator are powered by a high voltage generator between -60kV and + 60 kV. The velocity of the conveyor belt and the roll electrode, the distance between the conveyor and the roll electrode (d_e) are adjustable. Thus the charge of the particles, the contact time between the particles and the conveyor, and the intensity of the electrical field could be managed.

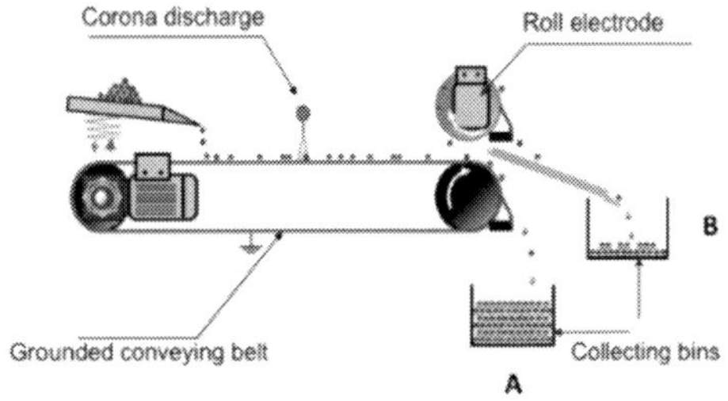

Fig 1: Custom built corona belt separator

From a theoretical point of view and in a first approximation, it can be assumed that a particle of mass m and carrying a charge q_a can be attracted by the roll electrode connected at a high voltage V, if the electric attraction force F_e surpasses the weight P, that is:

$$F_e = q_e * E > P = m * g \quad (1)$$

$$E = \frac{V}{d_e} \quad (2)$$

where g is the gravitational acceleration and E is the electric field strength. By increasing the ratio q_a/m the ability of a powder to be attracted by the roll electrode is enhanced. As a consequence, the efficiency of the separation is related to charge of the particle, its mass, its shape and its composition.
Powders of ash from rice husk, cellulose fibers and ground rice husk were studied in this work. The ash rice husk were recovered at the output of an industrial furnace at 900 °C. The ash was sieved with a screen of 0.5 mm to eliminate coarser particles. The powder obtained was called ARH cl2. A part of this fraction was ground with a ball mill to obtain a finer powder (d_{50}=30 µm) called ARH cl1. The cellulose fibers ARBOCEL came from the company JRS France. Two types of fibers were selected: Cel cl2 and Cel cl1for their differences in fiber length. To obtain a finer powder of rice husk (RH cl1), the native rice husk was ground with a ball mill during 5 hours until it reached a d_{50} = 30 µm. The main size characteristics of the different powders are summarized in table 1. Size dimensions of cellulose fibers are data from supplier and the size characteristics of rice husk and ash powder were determined by laser diffraction using a laser granulometer (Malvern, France).

Table 1: Main powder size characteristics

Powder name and description	Main size characteristics (d10, d50, d90)
ARH cl1 : Ash from rice husk sieved and ground	7 µm, 30 µm, 72 µm
ARH cl2 : Ash from rice husk sieved	81 µm, 232 µm, 565 µm
RH cl1 : Ground rice husk	3 µm, 33 µm, 126 µm
Cel cl1 : Cellulose fiber	Fiber diameter = 20 µm Fiber length : 60 µm
Cel cl 2 : Cellulose fiber	Fiber diameter = 20 µm Fiber length : 300 µm

All the tests described above were carried out at a relative humidity between 30 and 40%. In these conditions the dielectric breakdown field of the atmospheric air is roughly 30 kV.cm^{-1}.

3 Result and discussion

3.1 Charge decay curves

The electrical properties (conductivity, permittivity) are not known for these materials in powder form. To estimate their electrical behaviours, charge decay curves were recorded for each class of materials after a positive (+15 kV) or a negative (-15kV) corona discharge. The distance between the corona electrode and the conveyor was set to 4 cm, the velocity of the conveyor to 2,6 m/min and the position of the corona electrode was modified in order to adjust the contact time between the particles and the ground conveyor (residence time between 4 and 30 s). The charge of the particle was measured in a Faraday cage at the conveyor outlet, in two replicate experiments with 2 g of powders. The charge is expressed as mC/g since the attraction of the powder by the roll electrode is a function q_a/m (figure 2). The dotted lines in red and blue represent the residual charge of the finest and the coarsest powders, respectively. The residual charge is measured when no corona discharge is applied to the powder. It corresponds to the value towards the charge tends when the powder is in contact to the ground during an infinite time. For all powders, the charges do not evolve significantly with the range of time used. Thus, the residence time of the powder on the ground conveyor will not be a significant parameter in the optimization of the separation. The value of the charge of ground rice husk is almost twice the one of ash from rice husk and cellulose reflecting the fact that ARH and Cel are more conductive powders. Interestingly, it is well noticed that the value of the charge of the powder is less important and exhibits a higher error bar related to a weak repeatability of the different assays with a negative corona discharge. In addition, when a negative corona discharge is applied to the ARH, the measured charge of the powder is positive. These phenomena could be partially explained by an induction effect on highly conductive powders as described by Tilmatine and al [8].

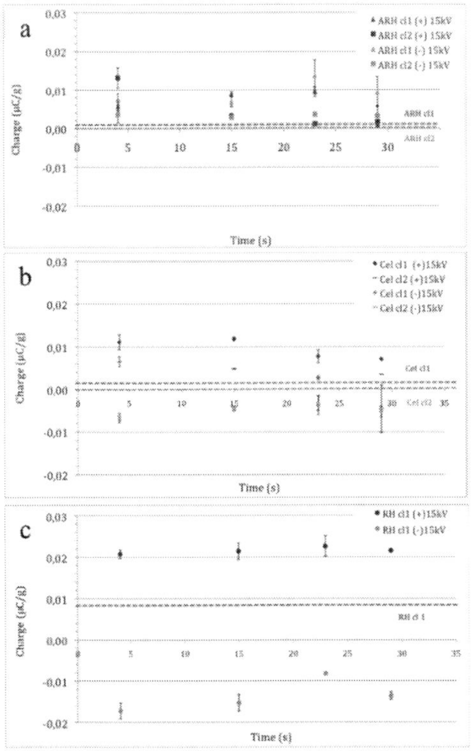

Fig. 2: Charge decay of ash from rice husk (a), cellulose (b) and ground rice husk (c), in contact with the grounded metallic conveyor belt

3.2 Behaviour of powders on the separator

The different model powders were tested on the separator in two different configurations (table 2). In configuration 1, the corona discharge was not activated and only the induction effects generated by the roll electrode was responsible for the charge and the attraction of the particle. In the configuration 2, the corona electrode was set up to -15 kV and the roll electrode to +15 kV. Assays were conducted with 5 g of powder. The products recovered in bin A (powder not attracted by the roll electrode) and B (powder attracted by the roll electrode) were weighted and the mass fraction in each bin was determined as the ratio of mass in the bin to the total mass (Figure 3). The losses correspond to particles that are first attracted by the roll electrode and then pushed after contact with it, so that they fell outside of the collecting bins. In both configurations, the mass of the powder attracted by the roll electrode was always low (less than 10%). However, the losses, which may constitute until 95% of the total mass for ARH cl2 correspond to particle reacting to the electrical field created by the roll electrode

Table 2: Configurations used in the separation tests

	Config 1	Config 2
Voltage of the corona electrode (V_c)	0 kV	-15 kV
Position of the corona electrode (d_c)	/	90 mm
Voltage of the roll electrode (V_r)	+15 kV	
Distance roll electrode/ conveyor (d_e)	25 mm	

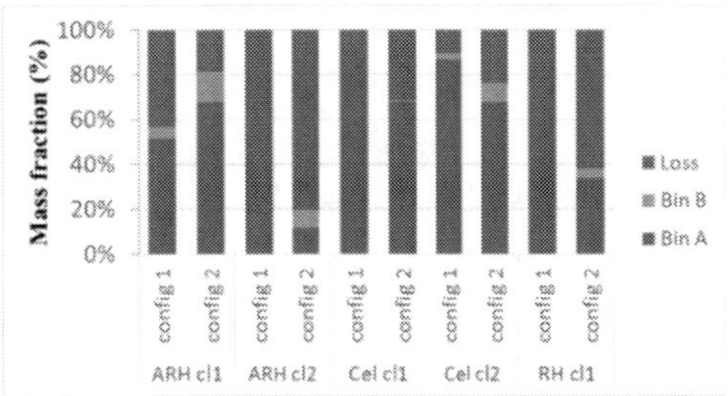

Fig. 3: Mass fractions recovered in each bin for configurations 1 and 2.

In figure 3, the mass fractions recovered in bin A are more important for Cel cl 1, although its charge and size are similar to those of ARH cl 1. This can be explained by the elongated shape of cellulose particles. Indeed, the electric image force that pins the particles to the surface of the earthed conveyor belt is stronger on the flat ones (i.e cellulose particles).

The different assays highlights the fact that the electrical field generated by the roll electrode has a significant effect on the different powders and it is slightly enhanced by the corona discharge. However it is important to optimize the different parameters of the process to increase its efficiency. This has been achieved by means of experimental design methodology, using a model mixture of ARH1 and Cel Cl1, as described in the following section.

3.3 Separation of a model mixture and of milled rice husk

A Blend composed of 2.5 g of ARH cl1 and 2.5 g of Cel cl1 was separated on the device. These powders were chosen because their size characteristics are closer to those of ground rice husk. An additional bin (C) was placed on the side of the conveyor to recover a significant part of the losses due to the rebound of the particles after impact with the roll electrode. The mass of each material in each collecting bins (A, B and C) was determined by combustion of the vegetal part in a furnace at 900°C during 2 hours. The efficiency of the separation was determined based on two separations indexes (equation 3), one for the cellulose and one for the ash from rice husk, to integrate both the yield of the separation and the purity of the fractions. These indexes vary from 0 to 100%, where 0 corresponds to the case where no separation is achieved (i.e. the composition of each fraction is equal to the composition of the initial blend) and 100% corresponding to the best separation (i.e. the case where each fraction is totally pure).

$$I_i = \Sigma^j (m_{i\_j} * (c_{i\_j} - c_i)) / (m_j * (1 - c_i)) \quad (3)$$

where m_{i_j} and m_j are the mass of product i recovered in the bin i and the mass of the powder in the bin j, respectively; c_{i_j} and c_i are the content of product i in the bin j and the initial content of the product i in the blend. Based on preliminary works, in different assays the corona discharge was not activated. The velocity of the roll electrode (v_r), and of the ground conveyor (v_c), as well as the voltage of the roll electrode (V_r) were varied. The distance roll electrode/conveyor was kept to 25 mm (minimum distance possible on the device) to maximize the electrical field effects. The separation indexes I_{ARH} and I_{Cel} were calculated and reported in table 3.

Table 3: Separation indexes for Ash from rice husk (I_{ARH})and Cellulose (I_{cel}) obtained in the different conditions.

Case	V_r (kV)	v_c (m/min)	v_r (rpm)	I_{ARH} (%)	I_{Cel} (%)
1.0	-15	2.6	12	7	10
2.0	-20	2.6	12	8	10
3.0	-25	2.6	12	11	10
2.1	-20	1.3	24	22	18
3.1	-25	1.3	24	27	24
2.2	-20	5.2	6	6	5

An increase of the voltage and the velocity of the roll electrode together with a decrease of the conveyor velocity increase the efficiency of the separation. The best separation was achieved in case 3.1 for the highest voltage and velocity of the roll electrode and the slowest velocity of the conveyor. In this configuration, the centrifugal force is at is maximum and probably helps the particle to detach from the belt. In addition in case of a high roll electrode speed, attracted particles are less time in contact to the surface of the roll and stay pinned to it until removed by the scrapper. In case 3.1, $m_A = 0.15$ g with a $c_{ARH\_A} = 53\%$, $m_B = 3.2$ g with a $c_{ARH\_B} = 37\%$ in the bin B and $m_C = 0.62$ g with $c_{ARH\_C} = 85.4\%$.

The conditions 2.1 and 3.1 were applied to the separation of ground rice husk, a composite of organic and mineral parts, and the corresponding separation indexes (I_m and I_v) are reported in table 4. The indexes obtained were very low and the best conditions for ground rice husk differed from those of ARS/Cel blend. Indeed, the mineral and vegetal powders chosen in these works are not completely representative of the ground rice husk. In addition, the efficiency of the dissociation of the mineral and vegetal part during the milling step has not been assessed in this study.

Table 4: Separation indexes obtained for ground rice husk

Case	V_r (kV)	V_c (m/min)	V_r (rpm)	I_{ARH} (%)	I_{Cel} (%)
2.1	-20	1.3	24	1	7
3.1	-25	1.3	24	0,3	2

4. Conclusion

The aim of this work was to evaluate the potential of a corona-electrostatic separator to separate mineral and vegetal fractions of plant material. The behaviour of pure mineral and vegetal powders was studied on a custom-designed separator and has shown that mineral particles are strongly affected by induction effects due to the roll electrode. Based on an experimental design, a blend of vegetal and mineral powders (50/50) was separated on the device with a significant enrichment in mineral in one fraction and vegetal in the other fraction. The conditions were applied to ground rice husk powder but the results obtained were not fully satisfactory. A new experimental bench, specifically adapted to ground rice husk powder will be set up in the future.

References

1. G. Ghizzi, D. Silva, S. Guilbert, X. Rouau, Powder Technol, 208, 2 pp. 266-270, (2011).
2. A. Barakat, C. Mayer-laigle, A. Solhy, R. A. D. Arancon, H. de Vries, R., Rsc Advances, 4(89): p. 48109-48127 (2014).
3. Y. Hemery, U. Holopainen, A-M. Lampi, P. Lehtinen, T. Nurmi, V. Piironen, M. Edelmann, X. Rouau, J Cereal Sci 53(1): p. 9-18, (2011)
4. J. Wang, M. de Wit, R. M. Boom, M.A.I. Schutyser, Separation and Purification Technology, 152(25): p. 164-171 (2015),
5. A. Barakat, and X. Rouau, Method for the fractionation of an oilseed cake, and applications of said method, Patent WO2015097290 A1. (2015).
6. T. Pandya, R. Srinivasan, and C. Thompson, Cereal Chem, 90(6): p. 535-539. (2013).
7. R. Delrue, and D.W.C.G. Van, Reduction of fibre content in fibre-containing oilseeds. Patent EP 1908355 A1 (2008)
8. A. Tilmatine, S. Flazi, K. Medles, Y. Ramdani, L. Dascalescu, J Electrostat, 61, p 21-30 (2004).

A DEM study of oedometric compression of model granular materials

Initial state influence, stress ratio, elasticity, irreversibility.

Mohamed Hassan Khalili[1,⋆], *Jean-Noël* Roux[1,⋆⋆], *Sébastien* Brisard[1,⋆⋆⋆], *Jean-Michel* Pereira[1,⋆⋆⋆⋆], and *Michel* Bornert[1,†]

[1] *Université Paris-Est, Laboratoire Navier, UMR 8205, École des Ponts ParisTech, IFSTTAR, CNRS, France*

Abstract. A DEM simulation study of spherical beads with elastic-frictional contacts in *oedometric* compression is carried out for a wide variety of initial states, differing in solid fraction Φ, coordination number z (independent of Φ in dense systems) and inherent anisotropy. Stress ratio $K_0 = \sigma_2/\sigma_1$, along with z, Φ and force and fabric anisotropies are monitored in compressions in which axial stress σ_1 varies by more than 3 orders of magnitude. K_0 tends to remain constant if the material was already one-dimensionally compressed in the assembling stage. Otherwise, it decreases steadily over the investigated stress range. K_0 relates to force and fabric anisotropy parameters by a simple formula. Elastic moduli may express the response to very small stress increments about the transversely isotropic equilibrated states, although oedometric compression proves an essentially anelastic process, mainly due to friction mobilization. Despite apparent nearly reversible increases of axial strain ϵ_1 (or density Φ), especially in dense samples, internal state evolutions are strongly irreversible, as evidenced by changes in z and K_0. Fabric changes are reflected by anisotropic elastic moduli.

1 Introduction

Oedometric compression, an axially symmetric process in which one principal strain component (ε_1) is increased, the others being maintained at zero ($\varepsilon_2 = \varepsilon_3 = 0$), is one of the simplest anisotropic loading processes, representative of natural materials under gravity (e.g. sediments consolidating under their weight). Oedometric compression leads to *transversely isotropic* structures, with the symmetry of revolution about axis 1. It is classically characterized by the ratio K_0 (termed coefficient of earth pressure at rest) of lateral stress $\sigma_2 = \sigma_3$ to axial stress σ_1[1, 2], which is often regarded as a constant. Oedometric compression is usually described as irreversible [2, 3], with a net plastic increase of solid fraction Φ (or a decrease of void ratio $e = -1 + 1/\Phi$) as σ_1 grows, while density changes occurring below the maximum past value of σ_1 (the preconsolidation pressure) are reversible.

The present contribution reports on a systematic DEM study of oedometric compression in a simple model material for a wide variety of initial states, with the aim of exploring, for such a simple case of an anisotropic material state under varying load intensity, stress ratio K_0, elasticity and irreversibility, with reference to microstructural and micromechanical aspects. After a brief description of the model and initial states (Sec. 2), Sec. 3 discusses stress anisotropy and its connections to force networks, and Sec 4 gives a brief account of elastic properties. Sec. 5 discusses irreversiblity, before a few words of conclusion (Sec. 6).

2 Material properties and sample preparation

Slightly polydisperse spherical beads (with diameters between D_1 and $D_2 = 1.2 \times D_1$, uniformly distributed by volume) with elastic-frictional contacts, using the simplified Hertz-Mindlin model of [4] and the elastic properties of glass (plane strain modulus $\tilde{E} = E/(1-\nu^2) = 77$ GPa), and friction coefficient $\mu = 0.3$, are enclosed in a fully periodic cuboidal box. They are first assembled into a solid pack under low stress, $\sigma_0 = 10$ kPa (or, in dimensionless form, $\kappa = (\tilde{E}/\sigma_1)^{2/3} = 39000$), by subjecting a disordered granular gas to either an isotropic ($\sigma_1 = \sigma_2 = \sigma_3 = \sigma_0$) or an oedometric ($\sigma_1 = \sigma_0$, $\varepsilon_2 = \varepsilon_3 = 0$) compression until contact forces accurately balance the applied load. One may regard the beads as frictionless in this assembling procedure, and possibly implement a "mixing" or "vibration" stage, as described in Ref. [4]. As a result, one obtains the 6 different initial states of Tab. 1, characterized by solid fraction Φ, fraction of *rattlers* (grains that carry no load) x_0, and coordination number z (z si the average contact number per grain, and $z^* = z/(1-x_0)$ denotes the coordination number of the force-carrying network). Those states are labelled with a three-letter code: the first letter, D or L, corresponds to dense or loose states, the second one is L for low or H for high coordination number, while the third one is i if the initial confinement

⋆e-mail: mohamed-hassan.khalili@enpc.fr
⋆⋆e-mail: jean-noel.roux@ifsttar.fr
⋆⋆⋆e-mail: sebastien.brisard@ifsttar.fr
⋆⋆⋆⋆e-mail: jean-michel.pereira@enpc.fr
†e-mail: michel.bornert@enpc.fr

is isotropic, o if it is oedometric. Note the peculiarity of DLi and DLo states which, as previously reported [4], are nearly as dense as DHo-DHi ones (with Φ values close to random close packing), but as poorly coordinated as the looser ones (LLo-LLi). Initial stresses, isotropic ($K_0 = 1$)

Table 1. Solid fraction Φ, coordination numbers z and z^*, rattler fraction x_0, and stress ratio K_0 for the different initial states. Values averaged over 3 configurations of $N = 4000$ grains.

	LLo	LLi	DHo	DHi	DLo	DLi
Φ	0.584	0.589	0.639	0.638	0.634	0.637
z	4.22	4.14	5.98	5.99	4.06	4.17
z^*	4.63	4.63	6.07	6.07	4.54	4.65
$x_0(\%)$	8.8	10.3	1.5	1.3	10.4	10.4
K_0	0.72	1	0.94	1	0.51	1

by construction in "XYi" states, might exhibit significant levels of anisotropy in oedometrically prepared ones.

3 Compression and stress ratio

All six initial states are subjected to an oedometric compression with σ_1 growing to 31.2 MPa ($\kappa \simeq 180$). In this process, while Φ increases by 0.01 to 0.02, coordination number z varies as in Fig. 1. In general, a gradual increase

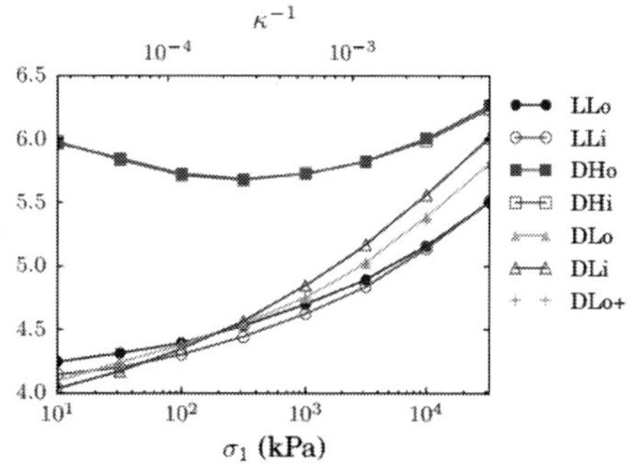

Figure 1. Coordination number z versus axial stress σ_1 for different initial states (DLo+ = DLo state with larger sample size).

of z under compression is observed, except for an initial stage in systems prepared with large coordination (DHo, DHi), in which z first decreases to a minimum. This shows that oedometric compression is a complex process that differs from a homogeneous uniaxial shrinking of the contact structure. The contact network, beyond its density, should also be characterized by fabric anisotropy, which is encoded, to first order, in parameter $\tilde{c}_2$ ($\tilde{c}_2 = 0$ in isotropic states):

$$\tilde{c}_2 = \langle n_1^2 \rangle - \frac{1}{3}. \tag{1}$$

$\tilde{c}_2$, initially zero in "XYi" systems by construction, gradually grows with σ_1, reaching typical values of 0.02–0.04 under high presure. It stays constant, near 0.04, in LLo systems, and achieves its highest values in the DLo states under low stress, slightly decreasing above 1 MPa (down to $\simeq 0.06$ for the highest σ_1 value).

Fig. 2 plots stress ratio K_0 versus σ_1 in the oedometric compression of all six initial states up to a large axial stress value. Initially equal to 1 for "XYi" systems, and

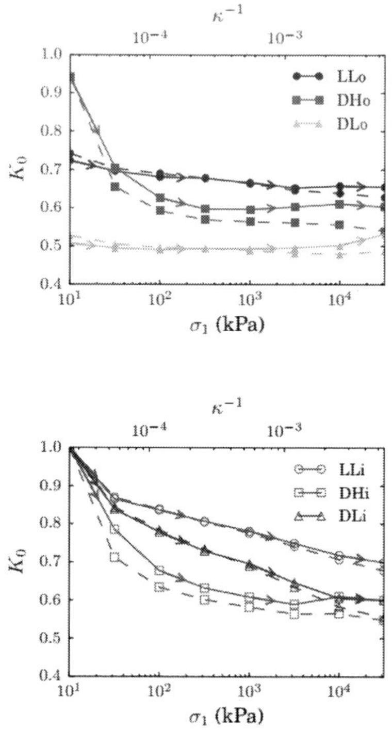

Figure 2. K_0 versus axial stress σ_1 in oedometric compression (dots and solid lines), in oedometrically (top) and isotropically (bottom) assembled systems. Dots joined by dashed lines: predictions of Eq. 3.

close to 1 in DHo ones, K_0 decreases, fast in DHo, much more gradually in isotropically assembled samples. It only stays roughly constant (near 0.5 in DLo, 0.7 in LLo, and somewhat above 0.6 in DHo for $\sigma_1 \geq 200$ kPa) in oedometrically assembled systems. K_0, using the classical formula relating stresses to contact forces in static granular packs [4], readily related to anisotropy parameters. A force anisotropy parameter, $\tilde{f}_2$ is defined from the angular distribution of average normal force, $\mathcal{F}(|n_1|)$, normalized by the average normal force in all contacts. By axial symmetry $\mathcal{F}$ is a function of the sole coordinate n_1 of the unit contact normal vector $\mathbf{n}$. One writes

$$\tilde{f}_2 = \frac{1}{4\pi} \int_{\Sigma} \mathcal{F}(|n_1|) n_1^2 d^2\mathbf{n} - \frac{1}{3}. \tag{2}$$

Using the standard formula for stresses [4], noting that the contributions of tangential forces to stresses *sigma*$_1$ *sigma*$_2$ are negligible, and dealing with anisotropies to leading order, one readily obtains an approximation for principal stress ratio $K_0 = \sigma_2/\sigma_1$:

$$K_0 = \frac{\sigma_{22}}{\sigma_{11}} \simeq \frac{2 - 3(\tilde{c}_2 + \tilde{f}_2)}{2 + 3(\tilde{c}_2 + \tilde{f}_2)}. \quad (3)$$

As apparent in Fig. 2, relation (3) provides a good approximation of K_0, which is thus simply related to anisotropy parameters $\tilde{c}_2$ and $\tilde{f}_2$ (as remarked in [5, 6]). $\tilde{f}_2$ steadily increases with σ_1, from zero in initially isotropic assemblies, the faster in highly coordinate systems, up to about 0.11 in the MPa range. $\tilde{f}_2$ and $\tilde{c}_2$ are of the same order in the most anisotropic systems, DLo and LLo.

4 Elastic properties

An elastic response is probed on subjecting well-equilibrated states along the compression curve to small stress increments that do not cause rearrangements and for which the effects of friction are negligible [7]. Elastic moduli are evaluated on building the stiffness matrix of the contact network, involving normal and tangential stiffnesses associated with small displacement and rotation increments [7]. Given the transversely isotropic symmetry in oedometric compression, there are 5 independent moduli [8, 9]. Defining a quasielastic range for incre-

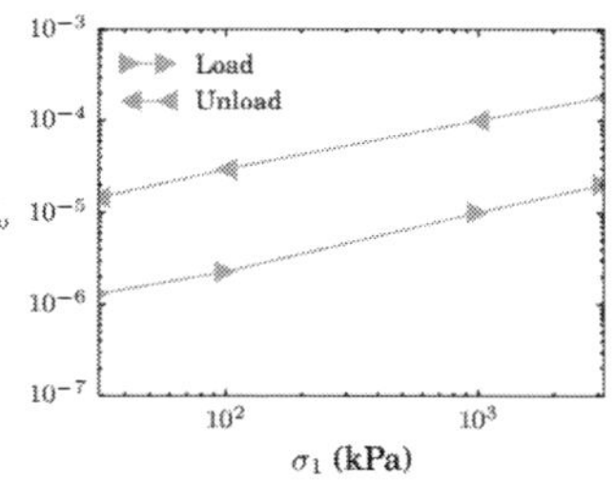

Figure 3. Quasielastic range versus σ_1 in one DLo system, for positive (loading) and negative (unloading) axial strain increments.

ments of axial strains, $\Delta\varepsilon_1$, by the requirement that axial strain increment $\Delta\sigma_1$ differs from the elastic response $C_{11}\Delta\varepsilon_1$ by less than 5%, its typical variation with σ_1, for both signs of $\Delta\varepsilon_1$, is shown in Fig. 3. A fully elastic response for the whole compression curve would imply stress ratios $\Delta\sigma_2/\Delta\sigma_1$ equal to the corresponding ratio of elastic constants, C_{12}/C_{11}. However, those two quantities strongly differ (typically one has $C_{12}/C_{11} < 0.4$, while $\Delta\sigma_2/\Delta\sigma_1$ values are similar to K_0). The oedometric compression entails friction mobilization and sliding (most frequent, somewhat surprisingly, in contacts with normal direction parallel to the transverse plane) and the quasielastic range is only associated to arrested, stabilized configurations (obtained after tiny creep intervals) in which full friction mobilization is suppressed. A full study of elastic moduli [10] reveals, on the one hand, similar trends as in isotropic systems [7], as to their dependence

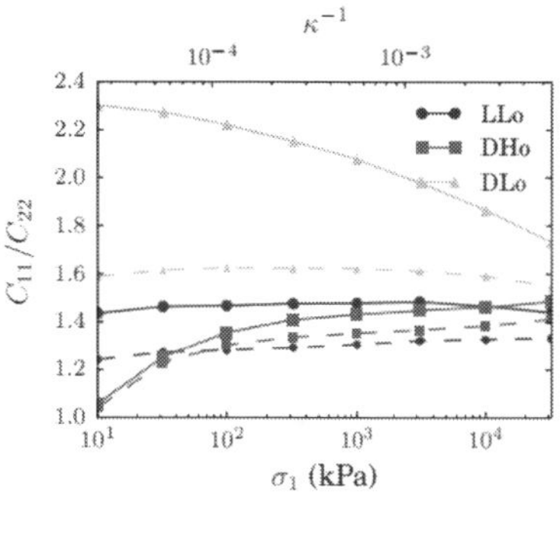

Figure 4. C_{11}/C_{22} versus σ_1 or κ^{-1} in initially anisotropic sample series. Dashed lines: prediction of the Voigt approximation.

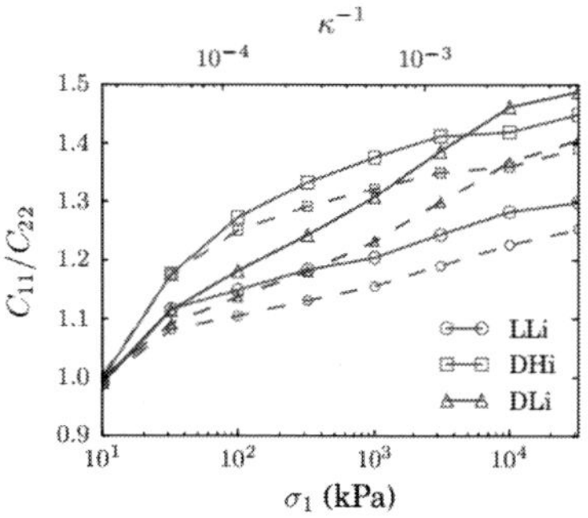

Figure 5. Analog of Fig. 4 for initially isotropic systems.

on stresses and coordination number (which is far mode influential than solid fraction), and, on the other hand, that moduli reflect the geometric and mechanical anisotropy of oedometrically compressed systems. Thus, Figs. 4-5 shows how the ratio of moduli C_{11} and C_{22} (associated to longitudinal waves propagating in the axial and transverse directions) achieves large values in the most anisotropic systems (DLo, and LLo), while lower ratios characterize the gradual gains of fabric (a moderate effect) and force anisotropy (a faster phenomenon) as initially isotropic packs are oedometrically loaded. Those figures also show the predictions of the Voigt ("effective medium theory") prediction, based on an assumption of homogeneous strains doan to the grain scale, for such ratios of moduli. While the approximation itself fails to correctly predict the values of the moduli [7, 10] its estimate of relative difference $C_{11}/C_{22} - 1$ is correct to within 20% for small levels of anisotropy (as $C_{11}/C_{22} < 1.2$). The variety of numerical results on elastic moduli for the different assembling procedures compares well [10] to that of published laboratory results on spherical bead assemblies [8, 9].

DOI: 10.1051/epjconf/201714002028

5 Unloading and irreversibilities

A compression cycle in which σ_1, after its maximum, is reduced back to its initial value, results in the variations of void index e and coordination number z shown in Fig. 6 The amount of irreversible compression (decrease pf e) af-

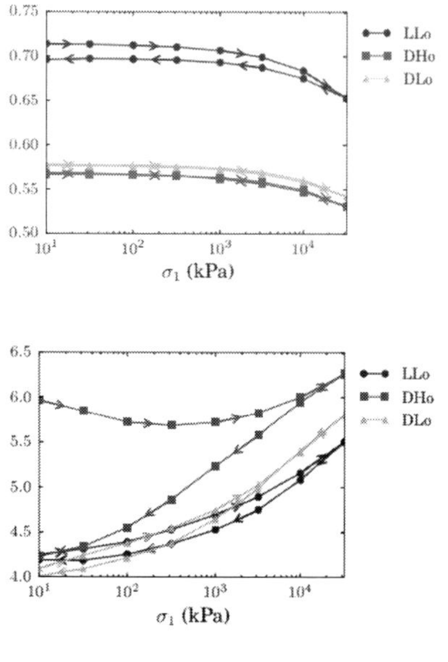

Figure 6. Void ratio e (top) and coordination number z (bottom) versus σ_1 in oedometric compression cycle.

ter one cycle is remarkably small, compared to typical experimental results on sands [11], a difference we attribute to the absence of plasticity or damage at the contact leve in the numerical model. The irreversibility is quite conspicuous, however, as regards z, which, as observed with isotropic bead packs [12], decreases to low values after one cycle in initially highly coordinated systems (this effect exists, to a smaller extent, in smaller stress cycles).

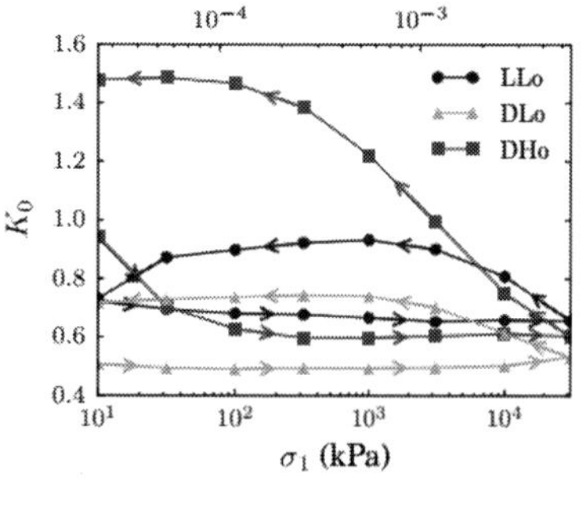

Figure 7. K_0 versus σ_1 in oedometric compression cycle.

Stress ratio K_0 also exhibits strongly irreversible behavior, with values larger than 1 obtained after a cycle (Fig. 7), as transverse stress σ_2 decreases slower than controlled, axial stress σ_1 along the unloading path.

6 Conclusions

This communication reports on some salient aspects of a systematic numerical study of oedometric compression of spherical grain assemblies [10], which investigates a wide variety of different initial states, subjected to a large increase of axial stress, and substantially extends the domains explored in the previous DEM literature [5, 6].

The coefficient of earth pressure, K_0, can be regarded as constant in systems prepared under similar initial oedometric load as the studied test. Otherwise, it appears to vary, reflecting a growing load-induced anisotropy. K_0 relates to both fabric and force anisotropy in a simple way. Despite the smallness of irreversible density changes, oedometric compression is a complex process involving friction-induced irreversible internal evolutions in the contact network, with possible decreases of initially large coordination numbers and hysteretic stress states. Elastic properties reflect material anisotropies and, given encouraging comparisons with experimental results, could provide access to fabric variables.

References

[1] I. Cavaretta, M. Coop, C. O'Sullivan, Géotechnique **60**, 413 (2010)
[2] J. Lee, S.Y. Tae, D. Lee, J. Lee, Soils and Foundations **53**, 584 (2013)
[3] J. Biarez, P.-Y. Hicher, *Elementary Mechanics of Soil Behaviour* (A. A. Balkema, Rotterdam, 1993)
[4] I. Agnolin, J.N. Roux, Phys. Rev. E **76**, 061302 (2007)
[5] J.C. Lopera Perez, C.Y. Kwok, C. O'Sullivan, X. Huang, K.J. Hanley, Géotech. Lett. **5**, 96 (2015)
[6] X. Gu, J. Hu, M. Huang, Granular Matter **17**, 703 (2015)
[7] I. Agnolin, J.-N. Roux, Phys. Rev. E **76**, 061304 (2007)
[8] R. Kuwano, R. J. Jardine, Géotechnique **52**, 727 (2002)
[9] Y. Khidas, X. Jia, Phys. Rev. E **81**, 021303 (2010)
[10] M.H. Khalili, J.-N. Roux, J.M. Pereira, S. Brisard, M. Bornert (2016), submitted to Phys. Rev. E
[11] A. Sanzeni, A.J. Whittle, J.T. Germaine, F. Colleselli, J. Geotech. Geoenviron. Eng. **138**, 1266 (2012)
[12] I. Agnolin, J.-N. Roux, Phys. Rev. E **76**, 061303 (2007)

Elastic waves in particulate glass-rubber mixture: experimental and numerical investigations/studies

Kianoosh Taghizadeh[1,*], *Holger* Steeb[2], *Vanessa* Magnanimo[1], and *Stefan* Luding[1]

[1] *Multi-Scale Mechanics, Faculty of Engineering Technology (ET), MESA+, University of Twente, Enschede, The Netherlands*
[2] *Institute of Mechanics, University of Stuttgart, Stuttgart, Germany*

Abstract. In this paper we study by wave propagation the elastic response of granular mixtures made of soft and stiff particles subjected under hydrostatic pressure/stress. This allows inferring fundamental properties of granular materials such as elastic moduli and dissipation mechanisms. We compare physical experiments in a triaxial cell equipped with piezoelectric wave transducers and Discrete Element Method simulations (DEM). In the experimental part, dense, static packings made of monodisperse glass and rubber beads are prepared at various levels of hydrostatic stress and species fractions. Small perturbations are generated on one side and the time of flight through the glass-rubber mixtures are measured to quantify the effect of the mixture composition on the elastic moduli. Interestingly, the experiments show that the behavior is non-linear and non-monotonic with increasing percentage of rubber particles. Wave velocity and modulus remain fairly constant when increasing the fraction of rubber to 30%, while they experience a sudden drop between 30% and 60%, to become again constant between 60% to 100%. DEM simulations offer deeper insights into the micromechanics in and at the transition between the glass- and rubber-dominated regimes. The simplest analysis with Hertzian spherical particles of different stiffness is performed as a preliminary step. The behavior of mixtures with high glass content is very well captured by the simulations, without need of any additional calibration, whereas the complex interaction between rubber and glass leave open questions for further study.

1 Introduction

In our daily life, we are surrounded by granular materials like soil, coffee, sand, nuts, etc. They constitute over 75% of raw materials feedstock to industry, including pharmaceutical, mining, agriculture, chemical, biotechnological, textile, etc. Despite its ubiquity, the behavior of granular materials is far from being fully understood, which leads to the loss of the world's energy consumption in processing and transport. Due to the wide range of application, particulate mixtures have received a lot of attention in the last decades. For example in geotechnical engineering recycled materials (e.g. shredded or granulated rubber, crushed glass) are often used into conventional designs and soil improvement projects [1–3]. Moreover, mixtures of asphalt and concrete are widely used to construct roads; thus exploring the effect of granular composition on the effective physical properties of mixtures can help optimizing industrial processes, engineering structures and make the pavement more robust and enduring [4–7].
The aim of this study is to explore the role of soft-stiff compositions on the bulk elastic response of the mixture. Particular attention is devoted to the dependence of sound velocity on the stress state, since this is an important controllable experimental parameter.

*e-mail: k.taghizadehbajgirani@utwente.nl

2 Experiment

2.1 Experimental setup

Uniform glass and rubber particles with similar size (d_r = d_g = 4 mm) are used in this study to prepare cylindrical specimens with different volume fractions of glass and rubber. Material properties are reported in table 1. Let ν be the ratio between the volume of rubber particles and the total volume of solids in the mixture. Glass-rubber samples were prepared with variable rubber content, ν = 0, 0.05, 0.1, ..., 0.9, 1.0, where ν = 0 composed of glass particles only and ν = 1.0 of rubber particles only. Care is taken to create homogeneous mixtures during manual inserting. Further segregation is avoided. All specimen are tested in our custom-made triaxial cell with sample diameter and height equal to 100 mm. Fig.1 shows a schematic drawing of the set up and peripheral electronics. Granular samples are compressed in the axial direction via the top piston in subsequent stress increments. At each instant step the radial stress is corrected to match the axial stress. Air is used as confining fluid for the samples enclosed by a rubber membrane. Seven pressure values are analyzed, namely $P = 50, 100, 150, 200, 250, 300$ and 350 kPa. At each pressure level, a high voltage burst signal is excited, to measure the time of flight. The top cap (the sound source) and the bottom plate (the detector) of the cell are instrumented with piezoelectric transducers with a diameter adjusted via a PMMT contact plate to the sample size (100mm), able to excite longitudinal waves and measure

Table 1. Properties of glass and rubber particles

Used material properties	Glass	Rubber
Diameter (mm)	4	4
Mass density (kg/m^3)	1540	860
Shear modulus (MPa)	29000	200
Poisson's ratio	0.3	0.5

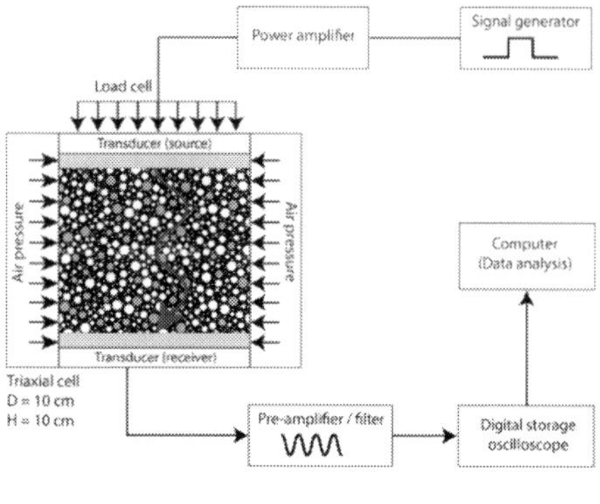

Figure 1. Schematic drawing of the experimental setup.

the arrival time, respectively. The transmitted signal is a ±400V step signal via the top cap. The signals transmitted and received are pre-amplified, filtered and recorded with a digital oscilloscope (LeCroy WaveSurfer 1GHz). Then, the wave velocity can be calculated from the travel time, given the height of the sample in the actual configuration and using the peak as reference. The signal-to-noise ratio is improved by repetitive averaging of 100 detected signals using the digital oscilloscope and then sent to a computer for further processing. We repeat the experiment five times for each rubber content and pressure level in order to avoid configuration-dependent results.

2.2 Experimental observations

In this section, we report results on the bulk stiffness of granular mixtures with diverse rubber content. In the long-wavelength limit, the longitudinal, P-wave modulus M is related to the velocity V_p in the medium by

$$M = \rho V_p^2, \tag{1}$$

where ρ is the mixture's density of the sample and is given by $\rho = (1-v)\rho_g + v\rho_r$, with ρ_g and ρ_r true mass densities of glass and rubber beads, respectively, and v rubber content. By means of Eq. (2) we can then calculate the P-wave modulus of the granular mixtures tested in the triaxial cell. Fig. 2 shows the evolution of M with rubber content, v [%], for all mixtures at different pressure levels. The figure shows that the compressional modulus remains fairly constant by increasing the volume of rubber content to 30%. In the case of high pressure, adding a small amount of soft particles surprisingly enhances the effective stiffness of the medium and the highest modulus is observed at $v \approx$ 20%. Hence, granular mixtures can be manipulated to obtain aggregates with even higher stiffness, but lighter and more dissipative thanks to rubber, when appropriate external conditions are matched (in this case the pressure) [8]. Between 30% and 60%, there is a considerable drop in the wave velocity. The increased rubber-to-rubber interaction gradually reduces the effective stiffness as the rubber content increases. The modulus is again relatively stable between 60% to 100%.

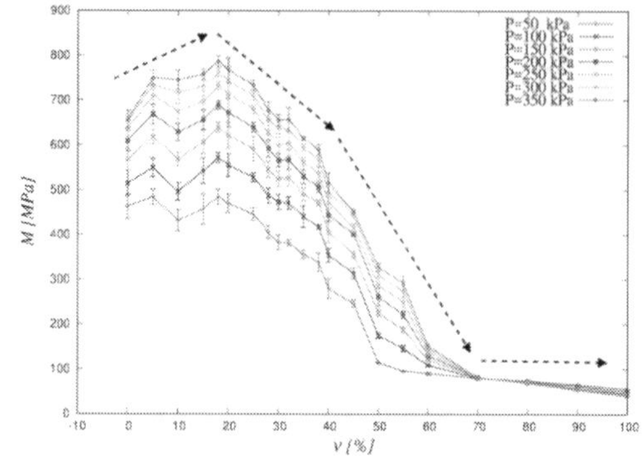

Figure 2. P-wave modulus versus fraction of rubber.

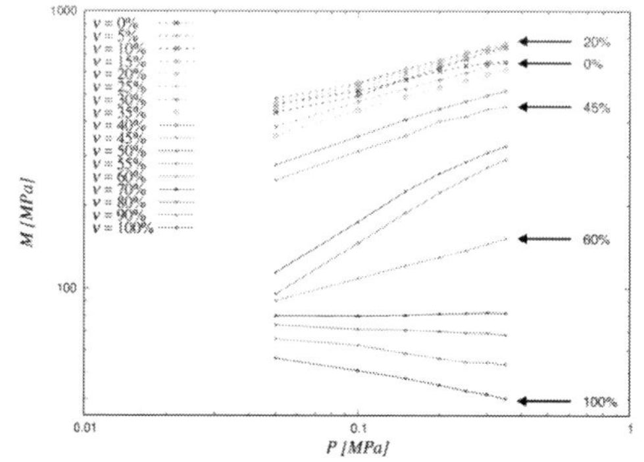

Figure 3. P-wave modulus versus applied vertical stress.

In Fig. 3, the same moduli are plotted against pressure P in log-scale to evaluate their functional behaviour with P. The slopes of the M-lines with P are almost constant for $v < 50\%$, while the behavior suddenly changes between $v = 50\%$ and 60% and the moduli become almost independent of pressure for higher rubber content. As already observed in Fig. 2, 70% to 100%-rubber packings show little dependence on pressure and it is a surprise to see a little decrease in the moduli with increasing pressure.

3 DEM study

3.1 Numerical setup

To understand the experimental observations we perform numerical simulations and try to reproduce the behavior glass-rubber mixtures tested in the experimental part [9]. The Discrete Element Method (DEM) allows to simulate large number of interacting particles that either move following Newton's second law or just in static mechanical equilibrium (before the wave-tests) [10]. Given two elastic spheres i and j of radii R_i and R_j at position $\vec{x}_i$ and $\vec{x}_j$, the contact force in the normal direction is given by the Hertz contact law, by

$$F_n = \frac{4}{3}E^* R^{1/2}\delta^{3/2} + F_d\ , \tag{2}$$

where $R = R_iR_j/\left(R_i + R_j\right)$ is the effective radius, $\delta = \left[R_i + R_j - |\vec{x}_i - \vec{x}_j|\right]$ is the normal overlap and E^* is the effective modulus

$$\frac{1}{E^*} = \frac{1-\nu_i^2}{E_i} + \frac{1-\nu_j^2}{E_j} , \qquad (3)$$

with E_i, E_j and ν_i, ν_j Young's moduli and Poisson's ratio, respectively. F_d is the damping force between two particles and given by

$$F_d = \gamma\sqrt{d\delta}\dot{\delta} , \qquad (4)$$

where γ is the Hertzian viscosity parameter and $\dot{\delta}$ is the relative velocity of two particles. Note that F_d contributes very small value (almost zero) in case of quasi-static simulations since $\dot{\delta} \approx 0$ (particles do not move).

In a tangential direction a linear-dashpot contact model with Coulumb threshold is used with stiffness k_t and damping γ_t as defined in [11, 12] and with coefficient of particle friction $\mu = \mu_s = \mu_d$. To find the inter-species parameters (numerical value of parameter between a glass and a rubber particle), the reciprocals of parameters are added and the reciprocal of the sum is taken (product over sum) as the inter-species numerical value. For example, inter-species density between glass and rubber particles is given by $\rho_{g,r=\frac{\rho_g\rho_r}{\rho_g+\rho_r}}$, note that other contact properties follow the same rule. For glass and rubber particles the material characteristics in Table 1 and contact properties in Table 2 are used.

After defining the microscopic interaction between grains, we describe the protocol to generate packings and measure the elastic moduli. We start our simulations from a set of non-overlapping particles randomly generated in a periodical cubic box at an initial volume fraction $\phi = 0.3$. The initial configuration is compressed isotropically by constant compression strain-rate until a given volume fraction $\phi = 0.5$, below the jamming point. The system is then allowed to relax at constant volume fraction until it reaches a stable state, which means that the particles dissipate kinetic energy and achieve a zero-pressure unjammed relaxed configuration. This is followed by an isotropic compression until the desired maximum volume fraction, $\phi = 0.82$ [13]. The same protocol is used for all glass-rubber mixtures from $\nu = 0.0$ to 1.0. In this way, samples resembling the set of experiments to some extent in Sec.2 are created. In simulations the preparation is isotropic, periodic without walls all the time, whereas in experiments, only hydrostatic stress conditions have been applied, thus isotropy of packing could not be guaranteed.

Once packings are created and compressed, various configurations are picked up at different pressure states above the jamming volume fraction. Those samples are allowed to relax with constant volume until a stable state is reached, which means the pressure remains unchanged over a period of time. Then a small strain perturbation is applied to these relaxed samples, either pure volumetric or pure deviatoric [11, 14]. The bulk and shear moduli B and G are calculated as the ratio between the measured increment in stress and the applied strain:

Table 2. Summary and numerical values of particle parameters used in the DEM simulations [12]

Property	Symbol	Value	SI-units
Time unit	t	1	10^{-6} s
Length unit	x	1	10^{-3} m
Mass unit	m	1	10^{-9} kg
Particle radius	$\langle a \rangle$	2	10^{-3} m
Number of particles	N	5000	
Particle density, g	ρ_g	2540	2540 kg/m^3
Particle density, r	ρ_r	1270	1270 kg/m^3
Simulation time step	Δt_{MD}	0.01	10^{-8} s
Tangential stiffness, g	k_t^g	10^4	10^7kg/s
Tangential stiffness, r	k_t^r	150	$150{\cdot}10^3$kg/s
Viscosity, g	γ_g	100	100 kg/s
Viscosity, r	γ_r	100	100 kg/s
Friction coefficient	μ	0.5	

$$\begin{aligned} B &= \frac{\delta P}{3\delta\varepsilon_v} = \frac{\delta(\sigma_{xx}+\sigma_{yy}+\sigma_{zz})}{3\delta(\varepsilon_{xx}+\varepsilon_{yy}+\varepsilon_{zz})} , \\ G_{xy} &= \frac{\delta(\sigma_{xx}-\sigma_{yy})}{\delta(\varepsilon_{xx}-\varepsilon_{yy})} , \ with \ \ \delta\varepsilon_{zz} = 0 , \end{aligned} \qquad (5)$$

where σ_{ij} and ε_{ij} are components of stress and strain tensors respectively. Since the aggregates are isotropic, the P-wave modulus is given by:

$$M = B + \frac{4}{3}G, \qquad (6)$$

as also measured directly by $M \simeq \frac{\delta\sigma_{xx}}{\delta\varepsilon_{xx}} \simeq \frac{\delta\sigma_{yy}}{\delta\varepsilon_{yy}} \simeq \frac{\delta\sigma_{zz}}{\delta\varepsilon_{zz}}$ when only one strain is activated.

3.2 Numerical results

In this section we show the results of the numerical moduli for different rubber content ν and pressure states. Note that a pressure range in simulation is wider than in experiments to gain more insight into the mechanical response of mixtures. In Fig.4 we plot the P-wave moduli versus pressure, decreasing monotonically with the rubber content from $\nu = 0.0$ to $\nu = 1.0$. In this respect, simulations are not able to reproduce the macroscopic behavior observed in Fig.3 in the experiment the maximum M at $\nu \neq 0$. We associate the mismatch to the adopted contact model not appropriate to describe rubber-rubber and rubber-glass interactions. Finding a better contact model is subject to ongoing study. However, when experiments and simulations are directly compared in Fig.5, interesting information can be inferred. For the sake of clarity, only three cases are shown, namely, $\nu = 0.05$, $\nu = 0.5$ and $\nu = 1.0$. Simulations with $\nu = 0.05$ capture the experimental data quantitatively, noticeably without any calibration. On the other hand, when looking at the the packing $\nu = 0.5$, the qualitative trend is well captured by simulations even if actual experimental values are higher than the simulated ones. Finally, for the case $\nu = 1.0$, where the moduli are pressure independent, simulations are far from experiments in qualitative trend.
By summarizing the previous observations, a three regime

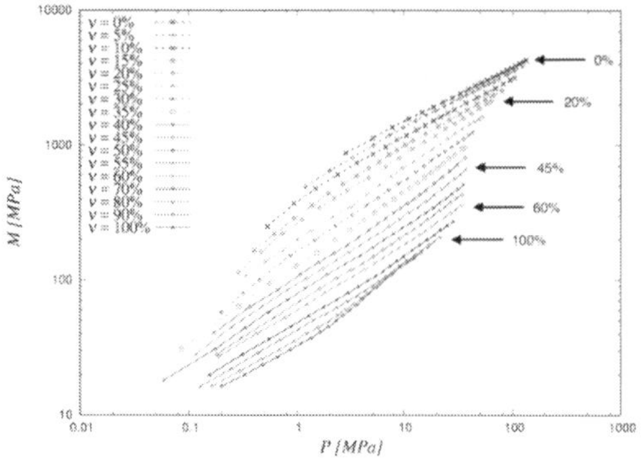

Figure 4. P-wave modulus versus applied vertical stress obtained by DEM simulations.

scenario shows up. In the glass-dominated regime G, waves do transmits via a glass beads network, where simulations based on Hertzian interactions are able to reproduce the macroscopic behavior. In the intermediate regime I, waves still have a preferential path via glass bead chains. Here two mechanisms concur to shape the bulk behavior: i) the density of glass beads in the sample reduces with respect to case G and the actual values of the moduli get lower; ii) the number of contacts increases with pressure faster than in the G-regime due to easy rearrangement of the rubber particles, that is the slope $M(P)$ gets higher. Finally, in the third regime R, the behavior of the mixture is dominated by the rubber beads, and the present simple DEM contact model can not offer an accurate representation of the system.

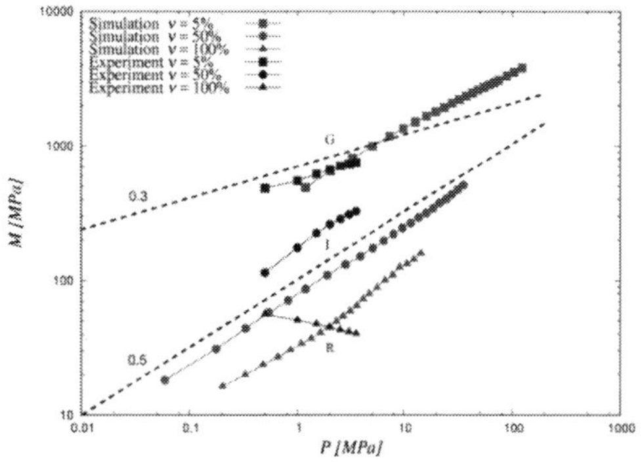

Figure 5. Experimental and numerical P-wave modulus plotted against pressure; comparison of DEM and experimental glass-rubber mixtures for $\nu = 0.05, 0.5$ and 1.0.

4 Conclusion

Experiments and numerical simulations were performed to examine the behavior of selected mixtures composed of glass and rubber particles. The experimental data indicate that the glass skeleton controls the behavior for $\nu < 0.3$, while the rubber skeleton prevails at $\nu > 0.6$. There is a considerable drop in modulus M only at intermediate mixtures ($0.3 < \nu < 0.6$) where the transition from stiff to soft occurs. Experimental data of high rubber content reveal that the modulus does not have any pressure dependency since there is not a strong stiff (glass) force chain for the wave to propagate through. Interestingly, we found that waves propagate faster in $\nu \approx 0.2$ than in for pure glass at all stress levels which might be of interest for many industrial applications to optimize their processes and materials.

Further, we studied numerically the evolution of granular mixtures with the volume fraction of rubber particles at different pressure levels to look more into the microstructure of the composites. Simulations data showed a systematical decrease of the P-wave modulus with increasing rubber fraction, unlike experimental observations where we could see a small peak of M at $\nu \approx 0.2$. Comparing results between experiments and simulations, the non-calibrated numerical model captures well the experimental results for low rubber content samples and to some extent (for few cases of mixtures) also for really high ν. However, simulations could not explain qualitatively the intermediate ν and the qualitative behavior of soft (rubber) dominated samples, because the Hertzian pair contact model is insufficient to depict the behavior of strongly deformable particles such as rubber. The classical contact laws predict the microstructure evolution during compression based on the assumption that contacts between particles are formulated locally as independent pair-interactions. In future, we thus will focus on the modification of contact models for strongly compressible particles [15].

Acknowledgement

The financial support of the European-Union Marie Curie Initial Training Network, T-MAPPP, funded by FP7 (ITN 607453), is appreciated, see http://www.t-mappp.eu/ for more information.

References

[1] P.J. Bosscher, et al, JGGE **123**, 295 (1997)
[2] V.K. Garga, V. O'Shaughnessy, Canadian Geotechnical Journal **37**, 75 (2000)
[3] J. Lee, et al, JGGE **125**, 132 (1999)
[4] T.C. Heimdahl, A. Drescher, JGGE **125**, 383 (1999)
[5] S. Pamukcu, S. Akbulut, JGGE **132**, 501 (2006)
[6] J.G. Zornberg, et al, Canadian Geotechnical Journal **41**, 227 (2004)
[7] B.H. Kjartanson, R.A. Lohnes, S. Yang (2002)
[8] H.K. Kim, J. Santamarina, Canadian Geotechnical Journal **45**, 1457 (2008)
[9] T.M. Evans, J.R. Valdes, Granular Matter **13**, 657 (2011)
[10] S. Luding, European Journal of Environmental and Civil Engineering **12**, 785 (2008)
[11] K. Taghizadeh, et al, in *IOP Conference Series: Earth and Environmental Science* (IOP Publishing, 2015), Vol. 26, p. 012008
[12] S. Luding, Granular Matter **10**, 235 (2008)
[13] N. Kumar, S. Luding, V. Magnanimo, Acta Mechanica **225**, 2319 (2014)
[14] V. Magnanimo, et al, EPL (Europhysics Letters) **81**, 34006 (2008)
[15] N. Brodu, J.A. Dijksman, R.P. Behringer, Physical Review E **91**, 032201 (2015)

CFD-DEM Analysis of Particle Attrition in a Jet in a Fluidised Bed

F.Fulchini[1], W. Nan[1,2], M. Ghadiri[1], M. Yazdan Panah[3], S. Bertholin[4], B. Amblard[4], A. Cloupet[4], T.Gauthier[4]*

[1]Institute of Particle Science and Engineering, University of Leeds, Leeds, UK
[2]State Key Labortary of Multiphase Flow in Power Engineering, Xi'an Jiaotong University, Xi'an 710049, China
[3]TOTAL Research & Technology Gonfreville, Le Havre, France
[4]IFP Energies Nouvelles, Solaize, France

Abstract. In fluidised bed processes, the solids are in vigorous motion and thus inevitably subjected to mechanical stresses due to inter-particle and particle-wall impacts. These stresses lead to a gradual degradation of the particles by surface wear, abrasion and body fragmentation commonly termed attrition. One significant contribution of attrition comes from the air jets of the fluidised bed distributor. Particles are entrained into the air jet, where they get accelerated and impacted onto the fluidised bed particles. The jet induced attrition only affects the part of the bed which is limited by the jet length, where the mode of attrition is largely collisional. The overall jet attrition rate is therefore the result of the combination of the single particle damage and the flux of particles entering into that region. The attrition behaviour of particles in the jet region is analysed by evaluating their propensity of breakage experimentally and by simulating an air-jet in a bed of particles by CFD-DEM. The frequency of collisions and impact velocities are estimated from which the attrition due to a single air-jet is predicted.

1. Introduction

Circulating fluidised beds are widely used for carrying out heterogeneous reactions in which particulate solids are needed to be regenerated in order to be reprocessed. The circulation of the solids provides plenty of chances for attrition, especially where the solids meet the most stressful conditions, i.e. the gas distributors jets, bubbling bed and cyclones [1]. The jetting region is one of the most significant contributors to particle breakage. According to Zhang et al. [2] and Boerefijn et al. [3], particles can undergo different types of stresses ranging from rapid shear to impact causing surface abrasion, chipping and fragmentation depending on the orifice to particle size ratio. The orifice velocity is the influential factor for accelerating the particles and projecting them upwards, where they collide with the bed particles leading to their attrition [4]. For orifice to particle size ratios greater than about 7.5 the main mechanism of jet attrition is collisional [2, 3]. Massimilla [5] developed a hydrojet dynamic model of the flow pattern in a fluidised bed jet from which particle velocity and flux can be calculated. Coupling of such information with a single particle breakage model can give an estimation of the overall jet induced attrition as proposed by Ghadiri et al. [6] using FCC catalyst. They first evaluated the attrition propensity by single particle impact testing and then coupled it with the hydrodynamic model of the jet of Donsi et al. [7] in order to analyse the dependence of the attrition rate on the orifice velocity and size. Two fundamental aspects are recognised to contribute to the jet attrition: (i) the jet hydrodynamics which dictate the particle velocity in the jet and the rate of entrainment of the particles into it, and (ii) the particle physical properties which influence the dependence of attrition on the operating conditions. This work aims to elucidate this phenomenon by evaluating the single particle attrition propensity of Chemical Looping Combustion (CLC) oxygen carrier particles coupled with an analysis of a single submerged jet by CFD-DEM simulation.

CLC is a process developed on the circulating fluidised bed design. The particulate solids are looped between two fluidised beds, referred to as risers, in which the solids alternate chemical oxidation to chemical reduction. CLC attrition propensity is evaluated by the single particle impact test [8] and by Scirocco disperser impact test [9] for a wide range of particle impact velocities. The outcomes of the two tests are joined and a single particle attrition model is presented based on the chipping model of Ghadiri & Zhang [10]. The CFD-DEM simulation is used to calculate process and their entrainment rate in a jet. Their coupling enables the overall attrition rate to be calculated.

2. Materials and Method

The material, provided by IFPEN, Solaize, France consisted of a batch of fresh natural ores manganese based CLC oxygen carriers. The particle size distribution (PSD)

* Corresponding author: M.Ghadiri@leeds.ac.uk

DOI: 10.1051/epjconf/201714007017

is very wide and the particle shape very irregular as shown in Figure 1. Particle density is 3300 kg/m^3.

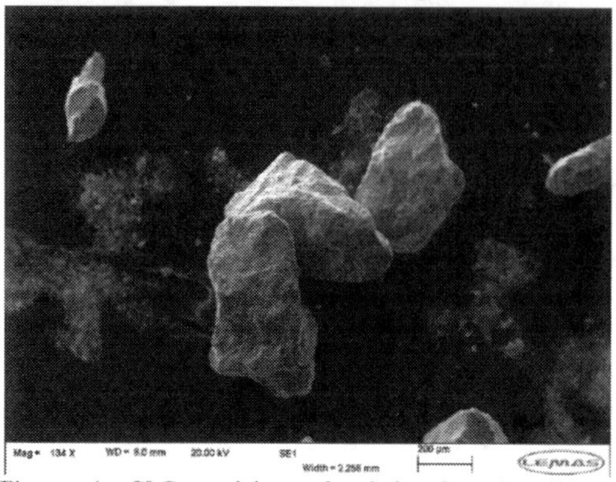

Figure 1. CLC particle analysed by Scanning Electron Microscopy (SEM)

was evaluated gravimetrically by sieving using BS sieves (as given in Table 1), so that the most representative sieve cuts of the distribution could be chosen and used to carry on single particle impact test and Scirocco disperser impact test. Therefore, the effect of particle size on breakage could be taken into account in the breakage model. The two impact tests are able to correlate the impact velocity with the extent of breakage. The extent of breakage is evaluated by gravimetric analysis: the material is collected post impact and sieved using a sieve which is two sizes below the lower limit of the feed particle cut size in order to separate the mother particles from the fines, as described below. The extent of breakage is calculated as the ratio of the mass of fines to mass of feed particles. Ghadiri and Zhang model for chipping for semi-brittle materials [10] is applied to the extent of breakage in order to determine the lumped parameter representing material properties. In parallel, the CFD-DEM simulation of a single jet by EDEM™ and ANSYS-Fluent is carried out to evaluate the fluid-particle interactions in the jetting region. The flow of entrained particles into the jet, the number of collisions of each particle and corresponding velocity are calculated and used to estimate the overall extent of breakage.

2.1. Particle Size Distribution

The particle size distribution was evaluated by mechanical sieving, and reported in Figure 2. The sample was first split to get a small representative sample, suitable for sieving, and avoiding errors due to segregation of the initial large batch [11].

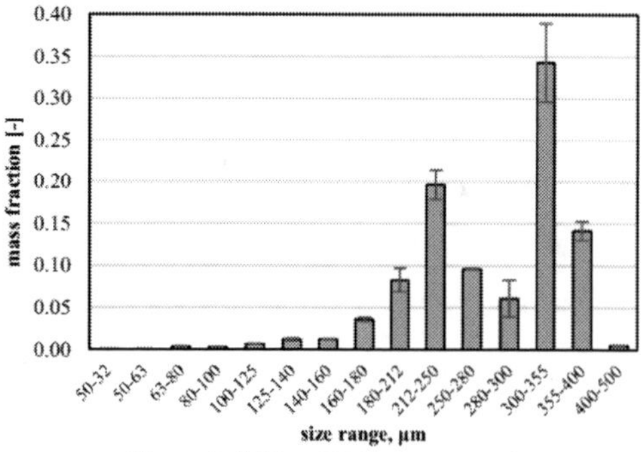

Figure 2. PSD of CLC oxygen carriers

Table 1. Representative size cuts used and sieve size for separating mother particles/fines for the breakage analysis

Representative size cuts (µm)	180-212	212-250	250-280	300-355	355-400
Cut off sieve size (µm)	125	150	180	212	250

2.2. Single Particle Impact test

The single particle impact test is according to the procedure described in [8]. Particle impact velocity is monitored and the extent of breakage is evaluated as:

$$R^* = \frac{m_D}{m_D + m_m} \times 100\% \qquad (1)$$

where m_D is the mass of fines and m_m is the mass of mother particles. The velocity range achievable for this test, for CLC, goes from free fall, about 1.8 m/s, up to 26 m/s depending on particle size.

2.3. Scirocco Disperser Impact Test

Scirocco is the dry disperser unit of Malvern Mastersizer 2000. Particles are fed and dispersed into an educator by a high pressure nozzle operating in the range from 0.5 to 4 barg. Given the L shaped bend of the educator, particles impact and experience attrition. Following passage through the disperser the particles are presented to the laser diffraction for the size measurement. They are recovered downstream the laser diffraction by a cyclone and analysed following the same procedure of the single particle impact test for consistency. A more detailed description is given by Ali et al. [9]. Varying the nozzle pressure from 0.5 to 4 barg, it is possible to achieve a range of impact velocity from 18 to 62 m/s depending on the particle size.

2.4. CFD-DEM Simulation of a Single Jet

The single jet simulation was carried out on a very simple geometry such as a cuboid of 100 mm×10 mm×600 mm mm size. The orifice is a 5 mm×10 mm rectangular slot and placed at centre bottom of the bed. Air, at normal condition, is issued from it at 40 m/s. The particles are considered spherical, having the diameter of 925 µm and the density of 3300 kg/m^3. The other DEM parameters such as coefficient of restitution, sliding friction and rolling friction, shear modulus and Poisson's ratio were specified as, 0.5, 0.5, 0.01, 0.1 GPa and 0.25, respectively, as given by default by the softwere. The total number of particles was set as 20000 for a total bed mass of 31.97 g. The system was simulated for 3.4 s.

3. Results and Discussion

3.1. Single Particle Impact Test and Scirocco Disperser Impact Test

Results from the impact tests, for each size cuts, are reported below in Figure 3. The extent of breakage is plotted as a function of particle impact velocity. The extent of breakage is corrected to take account of the loss of debris and mother particles. A very small amount of debris is lost in the Scirocco because the collecting efficiency of the cyclone is not 100%, so the extent of breakage is referred to as R^+. Whilst in case of the single particle impact test, the operation of manual sieving causes some losses of mother particles. In this latter case

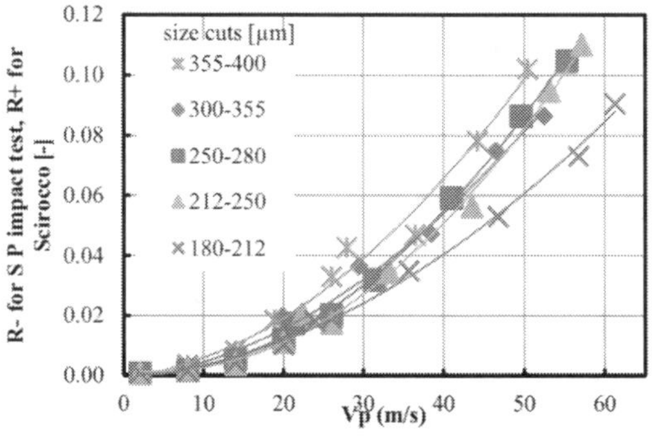

Figure 3. Gravimetric extent of breakage of CLC particles for the full range of velocities

the extent of breakage is R^-. Given that the same gravimetric analysis performed for the two tests, the results are combined for the full range of particle velocities, as shown in Figure 3.

The trend is parabolic, indicating that breakage is dependent on the incident impact energy. Besides, larger particles break more than the smaller ones which is consistent with the model of Ghadiri and Zhang [10], addressed below. Applying this model to the experimental data points a unification is achieved.

$$R = \alpha\eta = \alpha \frac{H}{K_c^2} \rho d_p V_p^2 \tag{2}$$

where ρ is the particle density, d_p is the particle size and V_p the particle velocity. The logarithm of the extent of breakage, log(R), is plotted as a function of $\log(\rho d_p V_p^2)$ in Figure 4.

The slope of the straight line is a lumped parameter, representing the mechanical properties of the particle, $\alpha H/K_c^2$, where α is a proportionality constant, H is the hardness and K_c is the fracture toughness. The chipping model describes very well the pattern of impact breakage. Interestingly, considering the intercept of the fitted line with abscissa, there is a minimum impact velocity for a

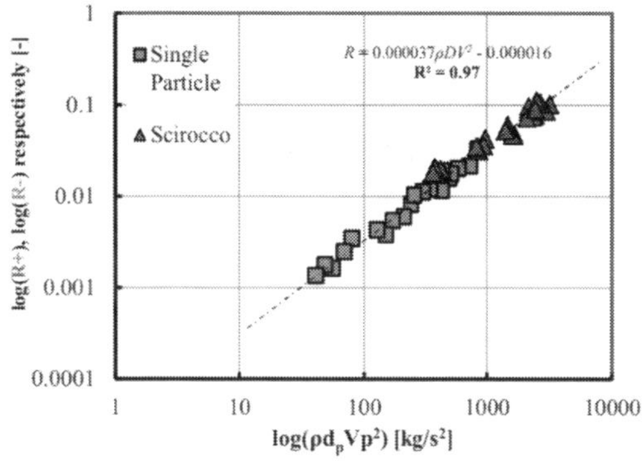

Figure 4. Extent of breakage: $\log(R^+)$ for Scirocco disperser impact test and $\log(R^-)$ for Single Particle Impact test, as a function of $\log(\rho d_p V_p^2)$ showing data unification

given particle size below which there is little/no breakage. The intercept can be used to estimate the transition velocity from no breakage to chipping for any particle size. In our case, considering particles of 925 μm, the transient velocity is 0.4 m/s.

3.2. CFD-DEM Simulation of a Single Jet in a Fluidised Bed

The dynamics of particle motion in the jet region was analysed in terms of particle velocity, number of collisions per particle and flux of particles into the acceleration zone. The breakage model developed in the experimental part is applied in order to estimate the overall attrition in the jet region. Particles enter the channel, created by the air jet, from the sides of the bed, Figure 5.

According to the review of Boerefijn et al. [3], in the case of large orifice to particle size ratios (5.4 in our case), particles entering the high velocity channel accelerate till they collide with the particles above the jet, experiencing attrition. The acceleration zone is limited to the channel height (about z=20 mm) as it is clear from Figure 6. Therefore, the whole analysis of attrition is limited to the jet core, where particles experience attrition by collision after they are accelerated. The breakage model of Ghadiri and Zhang was applied taking account of the flux of

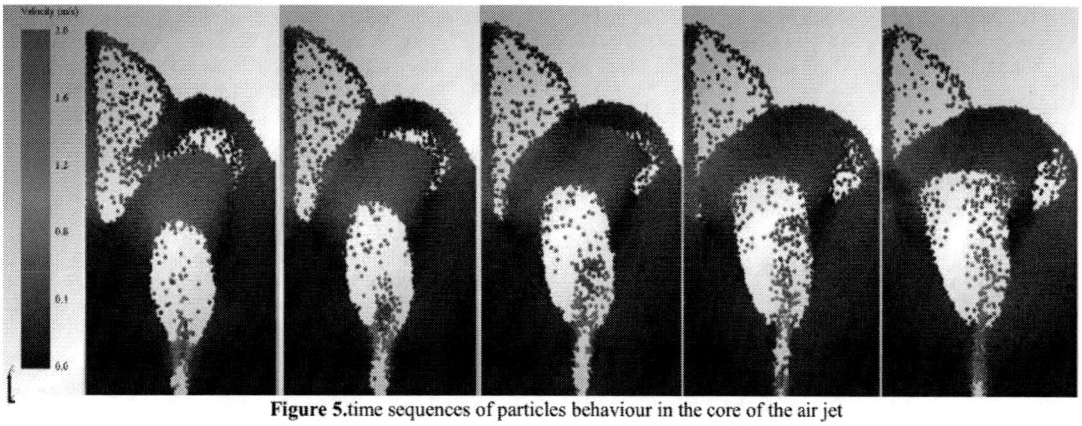

Figure 5.time sequences of particles behaviour in the core of the air jet

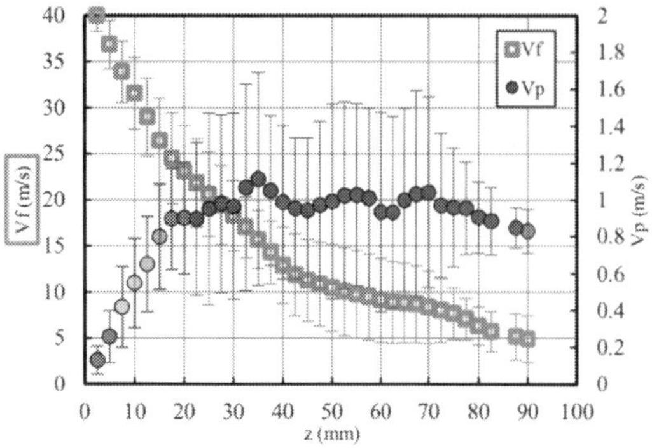

Figure 6. Time averaged air and particle velocity magnitudes at the centre bottom of the bed along z direction

particles entering the jet, Q, and the number of collisions per particle, f_c, thus obtaining the breakage rate B_{jet}, defined as the mass of fines formed due to attrition per unit time.

$$B_{jet} = R \cdot Q \cdot f_c \quad (3)$$

where R, f_c and Q are calculated every 0.01 s within the jet core, using the average impact velocity (V_p), the number of particles (N_p), the number of collisions (N_c), the mass of single particle (m_p) and Δt equal to 0.01 s.

$$Q = \frac{N_p \cdot m_p}{\Delta t} \quad (4) \qquad f_c = \frac{N_c}{N_p} \quad (4)$$

Table 2. Time averaged values for Q, f_c and V_p

Q [g/s]	f_c [-]	V_p [m/s]
35.39	5.30	0.83

In this condition, the breakage rate, i.e. the mass of debris formed by the jet per unit time, is calculated to be 0.025 g/s, which results in 0.084 g of mass of attrition product for the complete simulation time. At the end of the simulation, the bed inventory has therefore lost 0.26% of its mass due to the attrition of a single jet.

From the mass of fines produced, the debris size distribution of fines is calculated as follows, assuming that each impact generates only one fine particle.

$$d_D = \sqrt[1/3]{\frac{B_{jet} \cdot d_p^{\,3}}{Q \cdot f_c}} \quad (5)$$

The calculated particle size distribution after attrition is thus shown in Figure 7.

4. Conclusions

The attrition propensity of CLC particles is evaluated experimentally using two impact tests: the single particle impact test and Scirocco disperser impact test. Using the model of chipping for semi brittle material of Ghadiri and Zhang, a relationship between the extent of breakage and impact velocity is established, taking into account particle size and mechanical properties. The extent of breakage follows the square of the impact velocity and varies linearly with particle size. Particle mechanical properties such as hardness and fracture toughness were determined as a lumped parameter by plotting the extent of breakage as a function of particle density, size and velocity squared. A CFD-DEM simulation of a single jet in a fluidised bed is then performed and used to evaluate the overall attrition using the single particle breakage model and coupling it with the number of collisions per particle within the jet region and the flow of particles entering the jet core. The analysis reveals that, under these conditions, 0.025 g/s of fines is produced by attrition and that the average debris size is about 60 μm. This analysis is a starting point for a more detailed and realistic approach for predicting the attrition rate, since the single particle attrition propensity can be established for any system at any conditions and coupled with CFD/DEM to estimate the attrition rate. Validation of the simulation is planned next in order to finalise the model for a single jet.

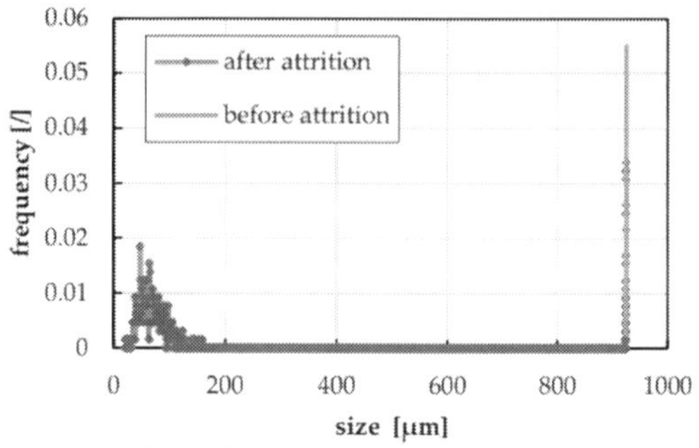

Figure 7. PSD before and after attrition

Acknowledgements

The financial support from Total, France, is gratefully acknowledged.

References

[1] K.G. Zenz FA, Powder Bulk Technol., 4th ed., 1980: pp. 13–20.
[2] S.-H. Zhang, R. Boerefijn, M. Ghadiri, Proc.World Congr. Part. Technol. 3, Brighton, 1998.
[3] R. Boerefijn, N.J. Gudde, M. Ghadiri, Adv. Powder Technol. 11 (2000) 145–174.
[4] T.A. Kutyavina, A.P. Baskakov, Chem. Technol. Fuels Oils. 8 (1972) 210–213.
[5] L. Massimilla, Fluid. 2, 2nd ed., Academic press, New York, 1985: pp. 133–175.
[6] M. Ghadiri, J.A.S. Cleaver, V.G. Tuponogov, J. Werther, Powder Technol. 80 (1994) 175–178.
[7] G. Donsi, L. Massimilla, L. Colantuoni, Fluidization, Matsen, New York, 1980.
[8] Z. Zhang, M. Ghadiri, Chem. Eng. Sci. 57 (2002) 3671–3686.
[9] M. Ali, T. Bonakdar, M. Ghadiri, A. Tinke, Powder Technol. 285 (2015) 138–145.
[10] M. Ghadiri, Z. Zhang, , Chem. Eng. Sci. 57 (2002) 3659–3669.
[11] J.T. Germaine, A. V. Germaine, Grain Size Analysis, in: Geotech. Lab. Meas. Eng., John Wiley & Sons, Inc., Hoboken, NJ, USA, n.d.: pp. 84–116.

Analysis of dense packing of highly deformed grains

Thi Lo Vu[1,*], *Saeid* Nezamabadi[1,**], *Jonathan* Barés[1,***], and *Serge* Mora[1,****]

[1] *Laboratoire de Mécanique et Génie Civil (LMGC), Université de Montpellier, CNRS, Montpellier, France*

Abstract. This paper concerns modeling of soft granular materials in which the grains are highly deformable. In order to simulate these materials, an approach based on an implicit formulation of the Material Point Method in the context of the finite strain theory, allowing for large deformations of grains, coupled with the Contact Dynamics method for the treatment of unilateral frictional contacts between grains, is proposed. In this context, the Mooney-Rivlin constitutive relationship is applied with two different set of elastic parameters. Considering these two material behaviors, a uniaxial compression of 2D soft granular packings is analyzed. The stress-strain relation and the evolution of the packing fraction as well as of the connectivity of the grains are discussed.

1 Introduction

Granular materials are an important class of matters found in the nature and in a large number of industrial fields like civil engineering, food and pharmaceutical industries, powder technologies (chemistry, cosmetics...), etc. In many situations, these materials can be highly stressed and their constitutive grains undergo relatively high elastic or inelastic deformations without rupture. It means that the behavior of soft granular systems is governed by both interactions between grains (contact, adhesion...) and their individual behavior (elasticity, plasticity...) [1, 2]. However, most of experiments and numerical simulations stay in the small deformation regime for theoretical convenience and are not able to catch the behavior of granular materials with highly deformable grains [3–6].

The discrete element methods (DEM) are known as powerful and robust numerical approches in order to model the granular materials [7–9]. However, these techniques are limited in the case of hard or weakly deformable grains and take into account essentially inter-grain interactions. To allow for high deformation of grains, the continuum approaches can be applied to model grain deformability. One of these numerical models is known as Material Point Method (MPM) [10]. It is based on the discretization of each grain by a collection of material points. The information carried by the material points is projected onto a background mesh, where equations of motion are solved. The mesh solution is then used to update the material points. In order to deal with contact interactions, in previous works [11, 12], it is proposed to compute the normal forces and, eventually, tangential ones by taking into account the unilateral stresses and the Coulomb frictional law by means of a DEM technique, Contact Dynamics (CD) method [8].

In this paper, numerical simulation of compression of a deformable granular system is carried out by mean of Material Point Method (MPM) extended to the finite strain theory. This numerical procedure allows the soft granular materials to reach the packing fraction beyond random close packing (RCP) state. We analyze the evolution of the packing and the effect of grain shape change. As we shall see, the material behavior of the grains affects the stress level and its evolution during compaction.

2 Numerical methodology

In a previous work [11], an implicit formulation of the material point method has been described to model the soft granular system. It was coupled with the contact dynamics method to deal with the frictional contact between deformable grains. Based on this approach which was developed in the context of the infinitesimal strain hypothesis, we will describe briefly below its extension to the finite strain theory.

Let us consider a continuum body occupying a domain Ω_0 in $\mathbb{R}^D$ (D being the domain dimension) in its reference configuration. If Ω is the actual occupied domain of this body, its conservations of mass and of linear momentum can be described by the following relations:

$$\frac{\partial \rho(\mathbf{x},t)}{\partial t} + \nabla \cdot (\rho(\mathbf{x},t)\,\mathbf{v}(\mathbf{x},t)) = 0 \quad \text{in } \Omega\,, \qquad (1)$$

$$\nabla \cdot \sigma(\mathbf{x},t) + \mathbf{b}(\mathbf{x},t) = \rho(\mathbf{x},t)\,\mathbf{a}(\mathbf{x},t) \quad \text{in } \Omega\,, \qquad (2)$$

where $\rho(\mathbf{x},t)$ is the material density, $\sigma(\mathbf{x},t)$ denotes the Cauchy stress tensor, $\mathbf{b}(\mathbf{x},t)$ represents the body force and, $\mathbf{v}(\mathbf{x},t)$ and $\mathbf{a}(\mathbf{x},t)$ are the velocity and the acceleration, respectively, at position $\mathbf{x}$ in the actual configuration and

[*] e-mail: thi-lo.vu@umontpellier.fr
[**] e-mail: saeid.nezamabadi@umontpellier.fr
[***] e-mail: jb@jonathan-bares.eu
[****] e-mail: serge.mora@umontpellier.fr

time t. The constitutive law in the framework of the finite strain theory is given by:

$$\mathbf{P}(\mathbf{X}, t) = \mathcal{F}(\mathbf{F}(\mathbf{X}, t)), \tag{3}$$

where $\mathbf{P}(\mathbf{X}, t)$ is the first Piola-Kirchhoff stress tensor and $\mathbf{F}(\mathbf{X}, t)$ denotes the deformation gradient tensor at position $\mathbf{X}$ in the reference configuration and time t. $\boldsymbol{\sigma}$ is related to $\mathbf{P}$ through:

$$\boldsymbol{\sigma}(\mathbf{x}, t) = \frac{1}{J}\, \mathbf{P}(\mathbf{X}, t)\, \mathbf{F}^T(\mathbf{X}, t) \tag{4}$$

with $J = \det(\mathbf{F})$.

In the MPM, the continuum body is divided into N_p material points with constant masses. This last assumption allows satisfying automatically the mass conservation relation (1). The material points represent the integration points to compute the integrals in the context of the finite element method. By supposing contact interactions between bodies, the discretized form of the momentum equation can be written as:

$$\mathbf{M}\,\mathbf{a}_{\text{node}}(t) = \mathbf{f}^{\text{int}}(t) + \mathbf{f}^{\text{ext}}(t) + \mathbf{f}^{C}(t), \tag{5}$$

where $\mathbf{a}_{\text{node}}$ is the nodal acceleration, $\mathbf{f}^C$ is the contact forces, $\mathbf{f}^{\text{ext}}$ represents the external applied force, and

$$\mathbf{M} = \sum_{p=1}^{N_p} m_p\, \mathbf{N}_p, \qquad \mathbf{f}^{\text{int}} = -\sum_{p=1}^{N_p} \mathbf{G}_p\, \boldsymbol{\sigma}_p\, V_p. \tag{6}$$

In the above equations, $\mathbf{M}$ is the lumped mass matrix and $\mathbf{f}^{\text{int}}$ gives the internal force vector resulting from the stress divergence. m_p and V_p are material point mass and volume, $\mathbf{N}_p$ denotes interpolation matrix or shape function matrix and $\mathbf{G}_p$ represents the gradient of the shape function $\mathbf{N}_p$ at $\mathbf{X}_p$.

The contact forces $\mathbf{f}^C$ between grains are computed using a contact algorithm accounting for the condition of non-interpenetration of matter as well as the Coulomb friction law. This algorithm combined MPM and CD method was presented in detail in a previous paper [11]. Moreover, a similar MPM implicit approach as [11] is implemented to solve the problem (5). The nodal solutions are then projected onto the material points, allowing for updating the information carried by these points.

For the applications that we target in the present work, two-dimensional simulations in plane strain conditions were performed. We also consider two hyperelastic Mooney-Rivlin constitutive laws to describe the material behavior in the finite strain framework [13]. The first one is the special case of the Neo-Hookean solid (NH) with the following constitutive relation:

$$\mathbf{S} = (\lambda \ln(J) - \mu)\, \mathbf{C}^{-1} + \mu\, \mathbf{I}. \tag{7}$$

The second one, simply called below the Mooney-Rivlin (MR) model is:

$$\mathbf{S} = 2(c_1 + c_2 \operatorname{Tr}(\mathbf{C}))\mathbf{I} - 2c_2\mathbf{C} + [2cJ(J-1) - d]\mathbf{C}^{-1}, \tag{8}$$

where $\mathbf{S}$ is the second Piola-Kirchhoff stress tensor related to $\mathbf{P}$ through $\mathbf{P} = \mathbf{FS}$. $\mathbf{C}$ is the right symmetric Cauchy-Green tensor ($\mathbf{C} = \mathbf{F}^T\mathbf{F}$), and $\boldsymbol{\gamma}$ denotes the Green-Lagrange strain tensor ($\boldsymbol{\gamma} = \frac{1}{2}(\mathbf{C} - \mathbf{I})$). λ, μ, c_1, c_2, c and d ($d = 2(c_1 + 2c_2)$) are the material constants.

These choices are motivated by wide range of materials that may be described by the considered constitutive laws. For instance, gels and hydrogels with deformable particles, which are present in foodstuffs, dictate to a large extent their rheological behavior [14]. Another exemple is vesicles being closed membranes which are highly deformable particles due to their internal viscous fluid and the flexibility and the quasi-inextensibility of their lipid viscous membrane [15]. On the other hand, for all applications, it is not clear to know if NH model is sufficient for describing the mechanical behavior of particules. Hence, in this paper, we present a comparison of obtained results between theses two constitutive laws in order to see the sensibility of granular behavior according to the choice of these laws. Note that if c_2 is set to zero and $c_1 = \frac{1}{2}\mu$ in equation (8) without considering volumetric part of stress, we can recover the Neo-Hookean law (7). So, one can consider that MR model is an extension of NH one.

3 Compaction of a deformable granular system

Let us now study the uni-axial compression of a packing of 300 deformable circular grains modeling by means of the proposed MPM-CD procedure. This packing is confined inside a rectangular box of width L and of height h_0. In the initial configuration, the grains don't touch each other and a small size polydispersity (diameters $\in [1.4, 2.4]$ mm) is introduced in order to avoid long-range ordering. To avoid stress gradients, the gravity effect is neglected. There is also no friction between the grains and, the grains and the walls. Two cases are investigated below by considering two above constitutive models for the deformable grains: (i) the Neo-Hookean model expressed by Eq. (7) with $\lambda = 61$ MPa and $\mu = 1.26$ MPa and, (ii) the Mooney-Rivlin model described by Eq. (8) with the parameters $c_1 = 0.63$ MPa, $c_2 = -0.0012$ MPa and $c = 6$ MPa. Note that we chose the high value of λ and of c in order to have quasi-incompressible grains during the simulations. The compaction process is performed by moving the top wall of the box downwards at a constant velocity of 1 m/s and a time step of $\delta t = 0.1\ \mu$s.

We are particularly interested here in the grain shape change due to the individual grains behavior, which may allow the packing fraction Φ to exceed the RCP state. Indeed, Φ being the ratio of the particles volume V_s to the total volume V: $\Phi = V_s/V$, varies due to both the volumetric particle deformation $\Delta V_s/V_s$ as a result of the compressibility of grains and the variation of V due to grain rearrangements and shape change:

$$\ln\left(\frac{V_s}{V_{s_0}}\right) = \ln\left(\frac{\Phi}{\Phi_0}\right) + \varepsilon, \tag{9}$$

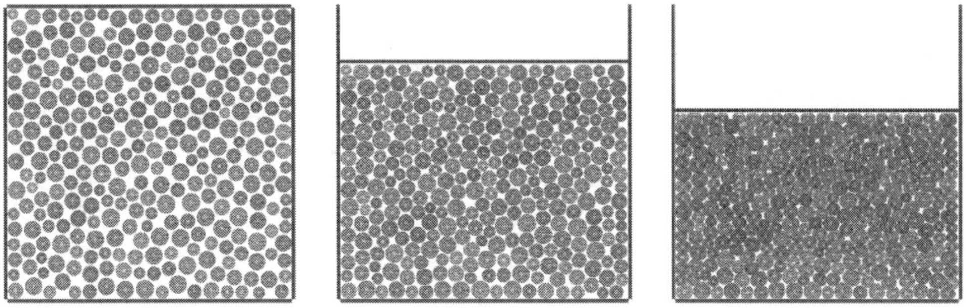

(a) $\varepsilon = 0$ and $\Phi = 0.65$ (b) $\varepsilon = 0.22$ and $\Phi = 0.8$ (c) $\varepsilon = 0.45$ and $\Phi = 0.99$

Figure 1: Three snapshots of the compaction of a packing of deformable grains. Filled Circles are material points.

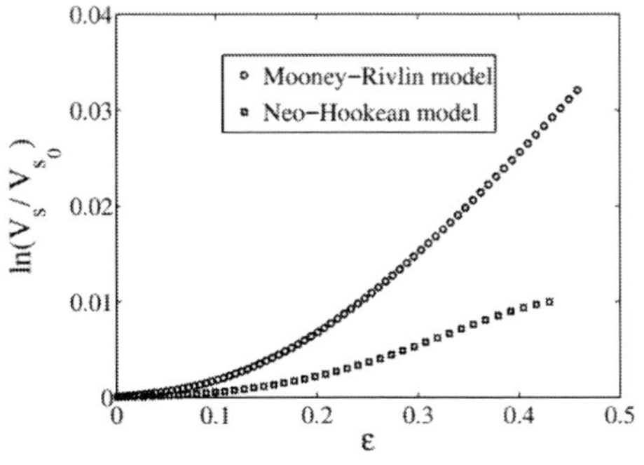

Figure 2: Total particles volume change $\ln(V_s/V_{s_0})$ as a function of the cumulative vertical strain ε.

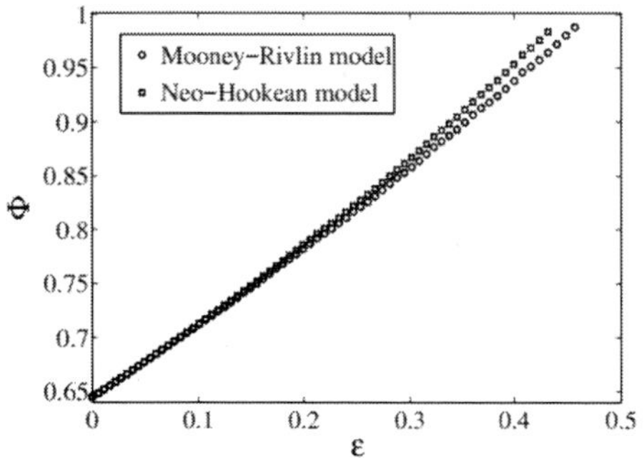

Figure 3: Packing fraction Φ as a function of cumulative vertical strain ε.

where $\varepsilon = \ln(V/V_0) \equiv \ln(h/h_0)$ is the true cumulative vertical strain. The evolution of $\ln(V_s/V_{s_0})$ with ε is represented in Fig. 2. Since, at the beginning, the compaction is mainly governed by the grain rearrangements, $\ln(V_s/V_{s_0})$ is quasi-null. The cumulative volume deformation varies approximately linearly with ε when $\varepsilon \in [0.1, 0.22]$. In this range, the grain strains are due to contact deflections, as in the Hertz regime. The 'jamming' transition occurs for $\varepsilon \simeq 0.22$ and beyond this value of the strain, the individual grain deformations are mainly dictated by the packing deformation. So, the volumetric deformation of the grains increases at a higher rate. We also observe that the MR grains semble more compressible than the NH ones. It may be explained by the fact that the ratio of λ/μ is larger than c/c_1. Indeed, these ratios are controlled the compressibility of the grains. However, the cumulative volume change of the grains remains small for both cases ($\ln(V_s/V_{s_0}) < 4\%$) and one can claim that the MR and NH grains are quasi-incompressible.

Fig. 3 shows the packing fraction Φ as a function of cumulative strain ε. Φ is a nearly linear function of ε for both cases since the grains volume approximately does not vary: $\Phi = \Phi_0' e^{\varepsilon} \simeq \Phi_0'(1+\varepsilon)$ ($\Phi_0' = 0.65$). Nevertheless, we observe a little difference between the two results for $\varepsilon > 0.22$, i.e. the regime of the individual grains deformation. It is due to the difference in the compressibility of the MR and NH grains.

The macroscopic evolution of the packing can be tracked through the vertical stress σ from the jamming point at $\varepsilon \simeq 0.22$ where $\Phi \simeq 0.8$. Fig. 4 represents σ, computed from the forces acting on the bottom wall and normalized by a consolidation stress σ_0, as a function of the cumulative vertical strain ε' from the reference state defined by $\sigma = \sigma_0$. The consolidation stress σ_0 corresponds to the jamming point. Similar behaviors are observed for both MR and NH grains. Nevertheless, we note a divergence of the results for $\varepsilon' > 0.05$. It can be explained by the difference behavior of the grains in two cases.

In order to analysis the microscopic evolution of the soft granular packings, we consider the variation of the mean coordination number Z as a function of the packing fraction Φ; see Fig. 5. We observe that Z increases with Φ. It is remarkable to note that the results for two cases are superimposed by normalizing $Z - Z_0$ per $Z_1 - Z_0$ and $\Phi - \Phi_0$ per $\Phi_1 - \Phi_0$, where Z_0 is the coordination number corresponding to $\Phi_0 \simeq 0.8$ (jamming transition) and Z_1

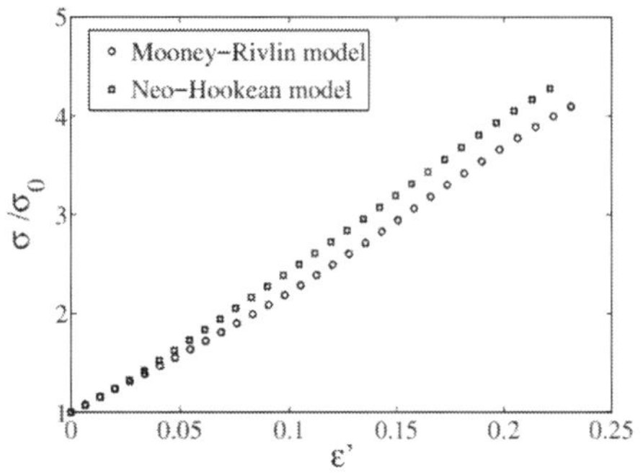

Figure 4: Normalized axial stress σ/σ_0 as a function of cumulative strain ε' from the reference state defined by $\sigma = \sigma_0$.

for $\Phi_1 \simeq 1$. One can proposed the following power law relation to fit these data points as shown in Fig. 5:

$$\frac{Z - Z_0}{Z_1 - Z_0} = \left(\frac{\Phi - \Phi_0}{\Phi_1 - \Phi_0}\right)^{0.5}. \qquad (10)$$

This power-law dependance of the coordination number with the packing fraction is almost independent of dimension, interaction potential or polydispersity [2, 6, 16]. It is worth noting that in the above relation, there is no adjusting parameter contrary to the previous works (see for example [16]) and it fits quite well the results of the MR and NH grains.

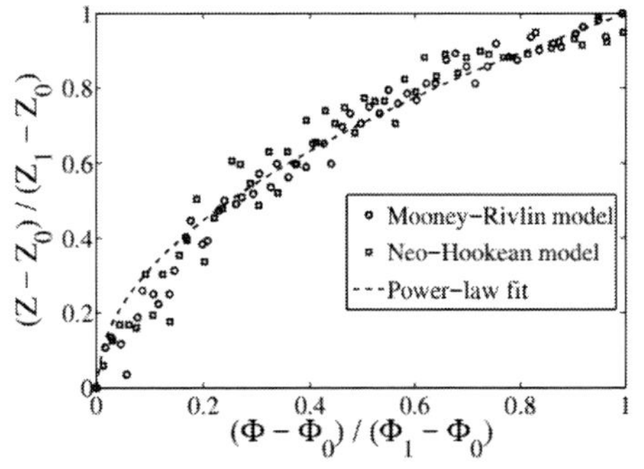

Figure 5: Evolution of the mean coordination number Z as a function of packing fraction Φ. The line is a power law fit $(\Phi - \Phi_0)^{0.5}$.

4 Conclusion

In this paper, the uniaxial compression of a deformable granular packing is studied using an extended approach of the MPM-CD procedure [11] in the context of the finite strain theory. This procedure combined the material point method (MPM) with the contact dynamics (CD) method. Two study cases are investigated by considering two hyperelastic constitutive relations: Neo-Hookean (NH) and Mooney-Rivilin (MR) models. Because of the material parameters choices, the MR grains are more compressible than NH ones although one can assume that both types of grains are quasi-incompressible. However, this little difference in the compressibility affects the behavior of the assemblies of soft grains. So, it may be interesting to study more the compressibility effects in a future work. Another important point concerns a simple power law which relates the coordination number directly to the packing fraction without any correction pre-factor. This point also merits further analysis by considering other types of material behavior.

Acknowledgment

This work has been carried out thanks to the support of the LabEx NUMEV project (no ANR-10-LABX-20).

References

[1] K. Liu, Journal of Physics D: Applied Physics **39**, 189 (2006)
[2] M. van Hecke, Journal of Physics: Condensed Matter **22**, 033101 (2010)
[3] F. Radjai, M. Jean, J.J. Moreau, S. Roux, Physical Review Letters **77**, 274 (1996)
[4] F. da Cruz, S. Emam, M. Prochnow, J.N. Roux, F. Chevoir, Physical Review E **72** (2005)
[5] T.S. Majmudar, R.P. Behringer, Nature **435**, 1079 (2005)
[6] J. Zhang, T.S. Majmudar, M. Sperl, R. Behringer, Soft Matter **6**, 2982 (2010)
[7] P.A. Cundall, O.D.L. Strack, Géotechnique **29**, 47 (1979)
[8] J.J. Moreau, European Journal of Mechanics A - Solids **13**, 93 (1994)
[9] F. Radjai, F. Dubois, eds., *Discrete-element Modeling of Granular Materials* (Iste-Wiley, London, 2011)
[10] S. Bardenhagen, J. Brackbill, D. Sulsky, Computer Methods in Applied Mechanics and Engineering **187**, 529 (2000)
[11] S. Nezamabadi, F. Radjai, J. Averseng, J.Y. Delenne, Journal of the Mechanics and Physics of Solids **83**, 72 (2015)
[12] S. Nezamabadi, T. Nguyen, J.Y. Delenne, F. Radjai, Granular Matter **19**, 8 (2017)
[13] S. Nezamabadi, H. Zahrouni, J. Yvonnet, Computational Mechanics **47**, 77 (2011)
[14] G. Lorenzo, N. Zartizky, A. Califano, Food Hydrocolloids **30**, 672 (2013)
[15] M. Abkarian, A. Viallat, Soft Matter **4**, 653 (2008)
[16] C. O'Hern, L. Silbert, A. Liu, S. Nagel, Physical Review E **68**, 011306 (2003)

Consideration of reinforcement mechanism in the short fiber mixing granular materials by granular element simulations

Kentaro Mori[1,*], *Kenji* Kaneko[1,**], and *Yutaka* Hashizume[1,***]

[1]*Hachinohe Institute of Technology, 88-1 Obiraki, Myo, Hachinohe City, Japan*

Abstract. The short fiber mixing method is well known as one of the method to improve the strength of granular soils in geotechnical engineering. Mechanical properties of the short fiber mixing granular materials are influenced by many factors, such as the mixture ratio of the short fiber, the material of short fiber, the length, and the orientation. In particular, the mixture ratio of the short fibers is very important in mixture design. In the past study, we understood that the strength is reduced by too much short fiber mixing by a series of tri-axial compression experiments. Namely, there is "optimum mixture ratio" in the short fiber mixing granular soils. In this study, to consider the mechanism of occurrence of the optimum mixture ratio, we carried out the numerical experiments by granular element method. As the results, we can understand that the strength decrease when too much grain-fiber contact points exist, because a friction coefficient is smaller than the grain-grain contact points.

1 Introduction

There is short fiber mixing reinforced soil which improves material strength by mixing short fibers with granular soil material whose performance is insufficient. In this method, short fibers made of polyethylene, polyester or natural palm having a length of about several cm are used. By mixing short fibers with granular soils, strength and toughness is improved. It is a basic concept of improvement that the short fiber mixing reinforced soil adds macroscopic cohesion by resisting tensile force of fibers. Many studies are carried out by a macroscopic viewpoint [1]. However, since various causes such as length, diameter, rigidity, mixture ratio and orientation of short fibers are complicatedly involved in the short fiber mixing reinforced soil, it is very difficult to elucidate its mechanical characteristics and reinforcement mechanism only by macroscopic viewpoint. There are several studies using DEM based on a microscopic viewpoint [2]. However, especially, relationship between the mixture ratio and the mechanical properties is not identified clearly.

In the theme of relationship between the mixture ratio and the mechanical properties of short fiber mixing reinforced soil, our research group carried out a series of laboratory tri-axial compression experiments using two kinds of silica sands and polyethylene short fibers [3]. As the results of these experiments, we can say that the macroscopic strength of short fiber mixing granular soils decreases when the mixture ratio is increased too much. In other words, the maximum strength exists for the short fiber mixture ratio and "the optimum mixture ratio" of the short fibers exists.

*e-mail: g139044@hi-tech.ac.jp
**e-mail: kaneko@hi-tech.ac.jp
***e-mail: hashizume@hi-tech.ac.jp

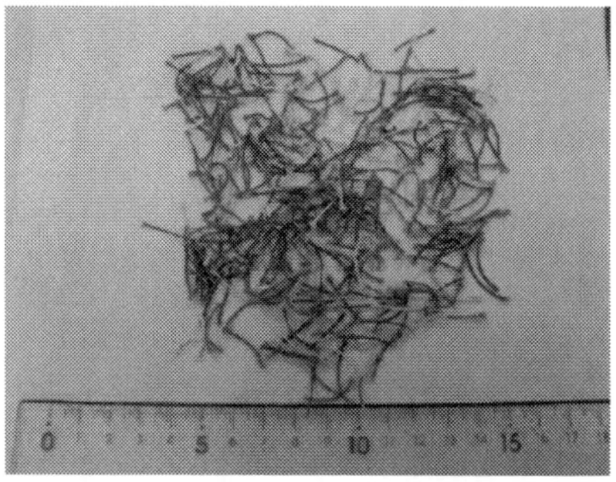

Figure 1. Short fibers used in our experiments

In this study, by microscopic numerical experiments, we examine the mechanism that the optimum mixture ratio exists. In our numerical experiments, we employ the Granular Element Method (GEM) proposed by Kishino [4, 5], as one of the discrete element method. Although GEM is similar to Distinct Element Method (DEM), GEM solves the quasi-static equilibrium equation of discrete system.

2 Outline of results of tri-axial compression tests

Focusing on the relationship between the mixture ratio of the short fiber and shear strength in the short fiber mixing reinforced soil, we carried out a series of the tri-axial compression test. Figure 1 shows short fibers, which are fish-

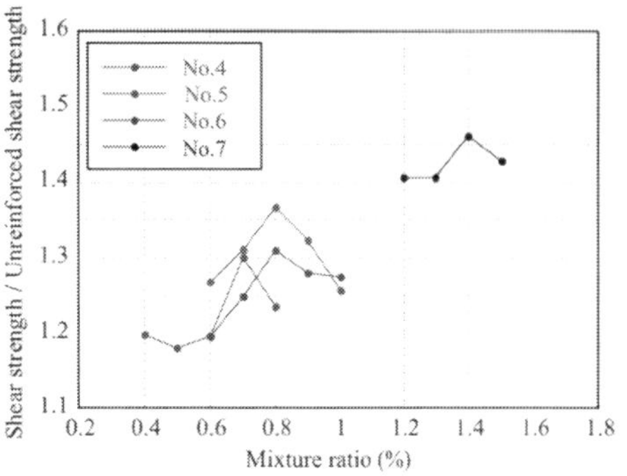

Figure 2. Shear strength and mixture ratio

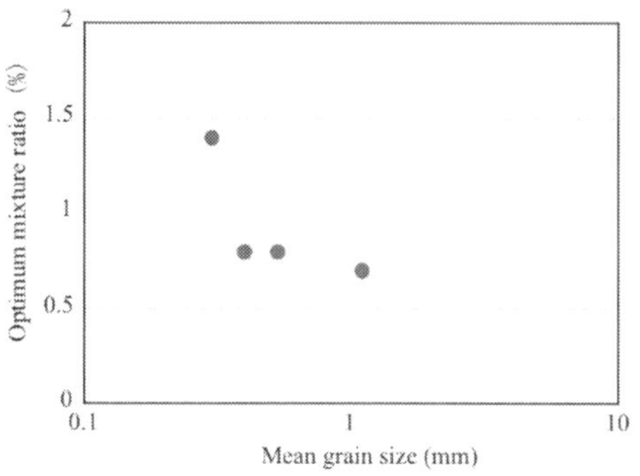

Figure 3. Optimum mixture ratio and mean grain size

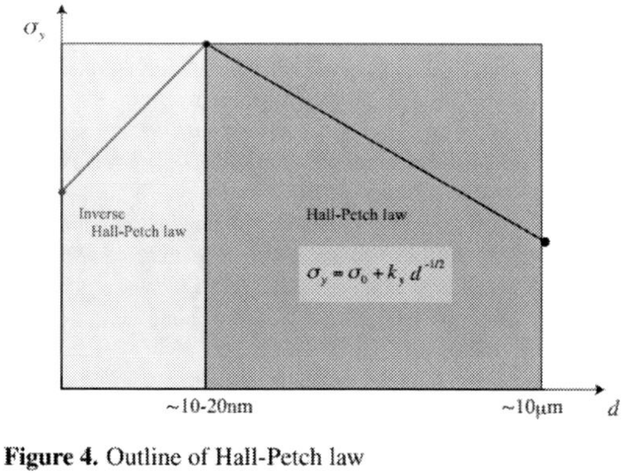

Figure 4. Outline of Hall-Petch law

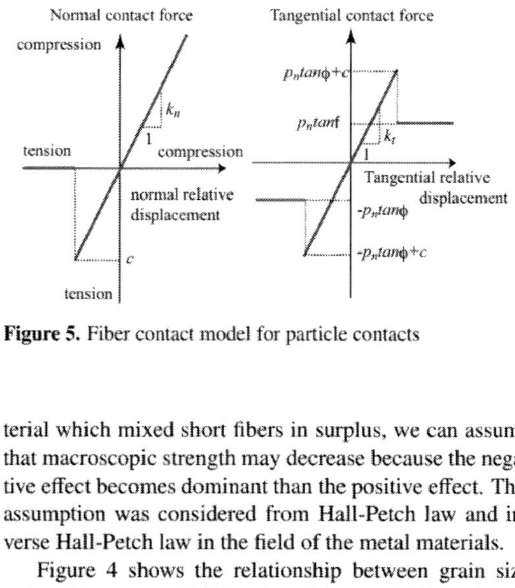

Figure 5. Fiber contact model for particle contacts

ing line made of polyethylene, used in our experiments as the reinforcement material. The diameter of the short fiber is about 0.52mm and length is about 20mm. We used four silica sands called No.4, No.5, No.6, and No.7. The average grain size of silica sands is different while the shape of grain size distribution is very similar. The mixture ratio of fiber used in the reinforced specimens is defined as the ratio of the weight of fibers (W_F) and the dry weight of sand (W_S):

$$w_f = \frac{W_F}{W_S} \times 100(\%) \quad (1)$$

The test specimen was made by sufficiently compacting so that the density becomes 90% of maximum dry density of sand. The confining pressure is 200kN/m^2 in all of the experimental cases.

Figure 2 shows the relationship between maximum principal stress difference and mixture ratio of short fibers obtained from the results in each case of tri-axial compression tests. In Fig. 2, normalized maximum principal stress difference of the vertical axis normalizes maximum principal stress difference in each case in that of the sands only. In all cases, shear strength defined by maximum principal stress difference is increased by mixing short fibers. However, shear strength decreases when exceeding a specific mixture ratio, and the reinforcement effect by mixing short fibers tends to become small. Consequently, it became clear that optimum mixture ratio related to shear strength in the short fiber mixing granular soils exists. Figure 3 shows the relationship between optimum mixture ratio of the short fiber and average grain size. As the sand has smaller particle size, necessary quantity of the fiber to get the maximum reinforcement effect increases.

From the results of the tri-axial compression tests, it is considered that when short fibers are mixed in granular media, two effects are mainly performed internally. One of them is a positive effect of restraining motion of grains by the tension of short fibers. However the granular medium without the cohesion is weak to tensile stress generally, the macroscopic cohesive strength is obtained by mixing fiber. The other one is a negative effect that the frictional force between the short fibers and the soil grains is lower than the friction between the soil particles, so that the macroscopic frictional force decreases. In the particulate material which mixed short fibers in surplus, we can assume that macroscopic strength may decrease because the negative effect becomes dominant than the positive effect. This assumption was considered from Hall-Petch law and inverse Hall-Petch law in the field of the metal materials.

Figure 4 shows the relationship between grain size and yield stress in Hall-Petch and inverse Hall-Petch law schematically. Generally, strength of polycrystalline metal increases when crystal particle size becomes small. This is called Hall-Petch equation, which is expressed as follows [6],

$$\sigma_Y = \sigma_0 + k_Y d^{-1/2}. \quad (2)$$

Where σ_Y is the yield stress, σ_0 and k_Y are the material constants, and d is the crystal grain size. Namely the yield

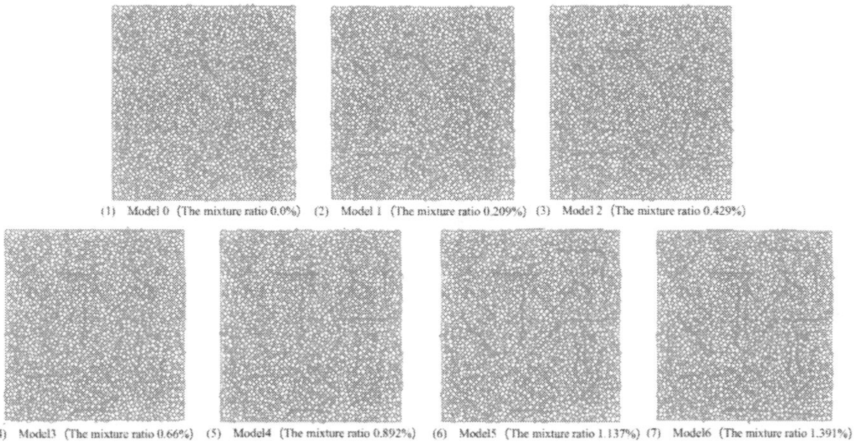

Figure 6. Analysis models

stress of a polycrystalline metal is in proportion to $d^{-1/2}$. The strength of the each crystal grain becomes strong as the size of the crystal grain becomes small, because the dislocation motion is more obstructed. However when the grain size is too small, the yield stress tends to decrease conversely. Because the strength of grain boundary, which is weaker than that of crystal grain, becomes more dominant for macroscopic strength, the macroscopic strength decreases when a crystal grain becomes small too much. This is called inverse Hall-Petch law. From such a law, it may be assumed that similar phenomena may also occur in short fiber mixing reinforced soil.

3 Numerical experiments by GEM

In this section, to explain the hypothesis in section 2, we carry out a series of numerical experiments by using 2-dimensional Granular Element Method with periodic boundary conditions [5].

3.1 Outline

At first, we explain the outline of numerical experiments of the short fiber mixing reinforced granular materials. This is aiming to explore the relationship between the shear strength and the mixture ratio.

As a model, the fiber connecting the small particles, is to be used to provide the tensile resistance between them. The schematic diagram of the contact model between the particles is shown in Figure 5. In the conventional GEM for analyzing the aggregate of only dry grains, if the normal relative displacement between the particles was tensile, the contact force between the particles would be determined as the non-contact. In the case of using fiber, the contact state would be maintained until the tensile normal contact force arrives at the tensile strength (c in Figure 5). In the tangential direction, the simple Coulomb's friction law is adopted as shown in

$$p_t \leq p_n \tan\phi \tag{3}$$

where p_n is the normal contact force, p_t is the tangential one and ϕ is the friction angle between particles. For the fiber particles–fiber particles contact, the tangential constraint model is taken as shown in

$$p_t \leq p_n \tan\phi + c. \tag{4}$$

To represent the small bending stiffness of short-fibers the small tangential spring constant is used compared to the normal direction.

In general, it is considered that the friction coefficient of short fibers is smaller than that of soil particles. The friction coefficient of short fibers has a large influence on the negative effect. Therefore, in this study, we carry out the numerical experiments of two cases which changed a coefficient of friction of the short fiber. In the first case, the friction coefficient of short fibers is equal to that of soil particles. It is assumed that there is no negative effect in this case. In the second case, we use a very small friction coefficient of the fiber to assume the case which has a very large negative effect. For each parameter on contact points between the soil particle and the short fiber particle, the average value of each parameter was used. Regarding c representing the tensile strength of short fibers, a huge value is used because short fibers never break during the tri-axial compression test.

Figure 6 shows numerical models prepared with different mixture ratios. In this figure, the initial state was created with an isotropic stress of 0.2MPa. Model 0, which has 5223 disk particles with 0.2mm-1mm diameters, is a model that does not mix short fibers. Models 1 to 6 were prepared by increasing short fibers particles and decreasing soil particles on the basis of model 0, but the void ratio was assumed to be almost constant in each model. On the assumption of a short fiber having a diameter of 0.5mm and a length of 20mm, 41 short fiber particles having a particle diameter of 0.52mm were connected and modeled. Numerical models are analyzed by stress control with constant lateral pressure.

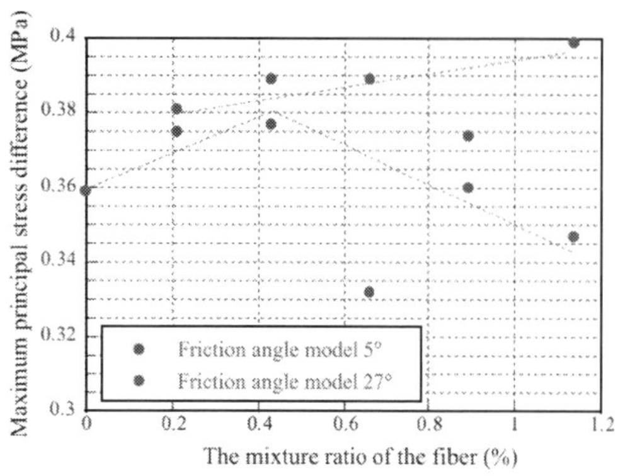

Figure 7. Mixture ratio and maximum principal stress difference

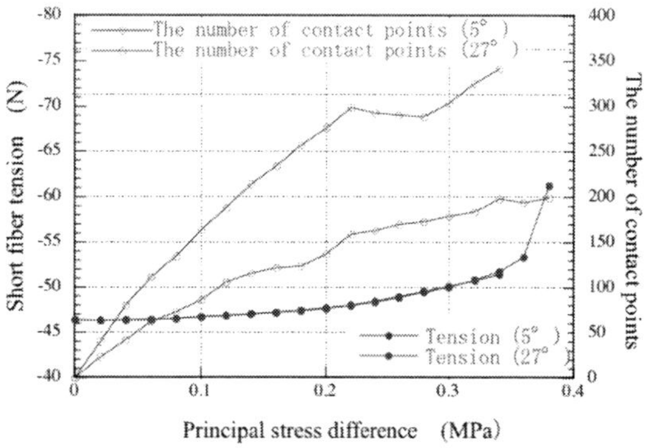

Figure 8. Fiber tension and slip occurrence (model 5)

3.2 Results

Figure 7 shows the relationship between the short fiber mixture ratio and the maximum principal stress difference obtained by changing the inter-particle friction angle of each model to 5° and 27°. When the friction angle between the particles of the short fibers is set to 27° which is the same as the soil particles, the maximum principal stress difference is larger at any mixture ratio compared to the case of the short fibers are not mixed. However, even if mixed rates increase, the maximum principal stress difference changes little. On the limited conditions, it seems that when the friction angle is 5°, even when the result of model 3 is regarded as an error, the maximum principal stress difference decreases with an increase in the mixture ratio of short fibers. It is considered that the shear strength of short fiber mixing reinforced soil is determined mainly by the balance between the positive effect due to the tension of the short fibers and the negative effect due to the decrease in the coefficient of friction between the soil particles. When the coefficient of friction between the short fiber particles is made small, it is considered that this is an increase in the contact point of a small coefficient of friction due to an increase in short fibers.

Figure 8 shows the relationship between the change in short fiber tension accompanying shearing loading in model 5 (the mixture ratio is 1.137%) and the change in the number of contact points where slip between short fibers and soil particles occurs. This is a model which has the largest influence difference of friction angle between particles . As can be seen from this figure, the tension is not so different even if the friction angle between the short fiber particles is changed, and almost the same tension is generated. On the other hand, as for the number of contact points between the short fibers and the soil particles where the slip is occurring, it can be seen that there are very many contact points where slipping occurs from the initial stage of loading. From the above, it is considered that the optimum mixture ratio with respect to the shear strength of the particulate material mixed with the short fibers appears because the negative action exceeds the positive action.

4 Conclusions

We clarified the appearance mechanism of optimum mixture ratio in reinforced soil by mixing short fiber from the results of the tri-axial compression tests and the numerical experiments. It may be possible that the optimum mixture ratio appears because the negative effect by mixing short fibers in surplus exceeds than the positive effect. In future, we plan to demonstrate a clear trend by conducting more numerical experiments. It is very important that there is an optimal solution for the mixture ratio of short fiber mixing reinforced soil. And it can contribute to the elucidation of the improvement mechanism of the mechanical properties of the granular material by mixing of short fibers.

Acknowledgements

This work was supported by JSPS KAKENHI Grant Number JP16K06493.

References

[1] A. Diambra, et al., Experiments and modeling, Geotextiles and Geomembranes, **28**, pp. 238–250 (2010).
[2] E. Ibraim & K. Maeda, New Horizons in Earth Reinforcement, Otani et al. eds., pp. 387–393 (2008).
[3] S. Nozoe, et al.,Powders & Grains 2013, pp. 305–308 (2013).
[4] Kishino, Y.,Powders and Grains, pp. 323–330, (1989).
[5] Kaneko, K., et al. Int. J. Solids Struct., **40**, pp.4043–4069, (2003).
[6] W. D. Callister, Jr.John Wiley & Sons, Inc, (2007).

Peridynamics simulation of the comminution of particles containing micro-craks

Nicolas Blanc[1,⋆], *Xavier* Frank[1], *Claire* Mayer-Laigle[1], *Farhang* Radjaï[2,3], and *Jean-Yves* Delenne[1]

[1] *IATE, CIRAD, INRA, Montpellier SupAgro, Université de Montpellier, F-34060, Montpellier, France*
[2] *Laboratoire de Mécanique et Génie Civil (LMGC), Université de Montpellier, CNRS, Montpellier, France*
[3] *<MSE2> UMI 3466 CNRS-MIT, DCEE, Massachusetts Institute of Technology, 77 Massachusetts Avenue, Cambridge CA 02139, USA*

Abstract. In this study, we rely on a 'bond-based' peridynamic approach to investigate the strength and failure of 2D particles containing a collection of 1D microcracks. The mechanical tests were performed on disks under diametral compression. In an extensive parametric study, the distribution of microcracks was varied for different particle sizes. The evolution of yield stress with diameter and the probability of failure in terms of Weibull distributions are investigated in detail. Finally, by means of a floodfill algorithm, we analyze the variation of the mean fragment size as a function of the density of defects.

1 Introduction

Comminution consists in reducing the average size of particles by different processes such as crushing, grinding or cutting. It is employed in many industrial applications and consumes large amounts of energy in applications such as cement manufacturing [1] and cereal milling [2]. Several models, mostly based on empirical parameters, have been developed to estimate the energy consumption of such processes. Population balance models [3, 4] involve the probability of failure and distribution of fragments obtained for a single particle of a given size. Brazilian tests (diametral compression) have been used for the determination of particle strength [5, 6] and in studies of the influence of the microstructure on the fracture and fragmentation [7, 8]. Numerical simulations based on the Discrete Element Method (DEM) have also been used for the investigation of particle crushing by modeling each grain as a cohesive agglomerate of smaller (primary) particles [9–12]. Such simulations can now reproduce the evolution of cracks and complex geometries of fragments in 2D [13, 14] and 3D [15–17]. But they fundamentally replace a continuum by a discrete medium with pre-existing potential fragments.

In this paper, we use a peridynamic approach applied to particles which can break into fragments of arbitrary sizes and shapes (up to the discretization limit) in the presence of microcracks. We investigate the crack patterns of 2D rounded particles containing defects of given density and distribution loaded between two horizontal plates in compression. We are interested in their yield stress and distribution of fragments as a function of the density of defects for a range of particle sizes.

⋆e-mail: nicolas.blanc@supagro.inra.fr

2 Bond-based peridynamics

The peridynamics method was introduced by Silling [18] as an alternative to the description of classical continuum mechanics by differential equations. Peridynamics has been successfully used for modeling the fracture of materials such as concrete [19], nickel nanowires [20], wood [21].... We use a simple implementation of peridynamics called bond-based approach which describes the mechanics of deformable bodies as a mass-spring system. A major interest of this model is its weak dependence on the geometry of the mesh contrary to other bond-based methods such as Fuse Models [22] and Lattice Element Method (LEM) [23]. The reason is that the computation of mechanical properties relies on a multinodal interactions within a distance h called the horizon. As other bond-based models, this method is allowing inclusion of discontinuities such as pores, cracks or stiffness and damage gradients without requiring special case-sensitive treatment of each problem [18].

The bonds are 1D elements characterized by a bond stiffness $k = c(\delta x)^4$, where c is the elastic micromodulus and δx the space step. The damage is introduced using a yield elongation s_0 above which the bond is broken. A mass m_i is associated to each node i. The stress tensor is upscaled from the forces connecting i to neighboring nodes j using the virial expansion $\sigma_i = \frac{1}{2V_i}\sum_j F_{ij} \otimes r_{ij}$, where V_i is a volume attributed to the node i and F_{ij} and the r_{ij} are the force and length of the bond ij, respectively. The Young modulus $E = \alpha c$ of the material is proportional to the micromodulus with $\alpha = \pi h^3(1-\nu)/6$ [24], where ν is Poisson's ratio. For isotropic linear elastic materials with central interactions, the value of Poisson's ratio is 1/3 in 2D. The fracture energy G is easily deduced from the critical elongation s_0 as $G = (9Ehs_0^2)/4\pi$ [24]. The computation of the evolution of the system is similar to Molecular Dynamics or Discrete Element Method simulations. We use the velocity-Verlet algorithm [25] to integrate the equations of motion with a time step δt.

3 Sample building

The particles are 2D disks meshed by a regular grid. We consider defects in the form of 1D linear segments. For all bonds crossing these zones we set the elastic modulus to a value below E. The defects are randomly distributed within the particle with number density $\bar{n}$. Their lengths and Young moduli follow Gaussian distributions: $\bar{\ell} \sim \mathcal{N}(\mu_{\bar{\ell}}, \sigma_{\bar{\ell}})$, and $\bar{E} \sim \mathcal{N}(\mu_{\bar{E}}, \sigma_{\bar{E}})$.

The particle is then subjected to a quasistatic diametral compression test. The bottom plate is kept fixed while the upper plate moves downward. The contacts between the particle and the plates are modeled by adding a normal repulsive force to the particle's nodes touching the plates. A regularized Coulomb friction law with a friction coefficient of 0.5 and a viscosity of 0.8 was used. The compression continues until the primary cracks fully propagate and before secondary fragments appear. As we rely on an explicit resolution of the equations of dynamics, we use a global viscous damping in order to dissipate low-frequency elastic waves.

The fracture of particle leads to the failure of bonds. From these failures, we determine at each node a damage parameter d_i defined as the ratio of the number of broken bonds to the initial number of bonds attached to this node. Fig.1a shows this damage field after fracture. The color level indicates the magnitude of the local damage, which has its maximum value in the middle of the cracks. As the peridynamics is a non-local approach, the crack-path dependence on the grid orientation is limited, compared with purely local lattice-based frameworks, like LEM.

To identify the fragments, we use a Floodfill algorithm (see Fig.1b, c and d) [26]. This algorithm paints connected zones made of the pixels which have damage values lower than a given threshold T. This threshold affects the width of the cracks and thus the number of potential fragments produced. Fig.2 shows the evolution of the number of fragments as a function of T. In the following we set T to 30%, which corresponds to a value for which the number of fragments is roughly constant.

4 Parametric study

The main objective of this parametric study is to investigate the influence of the particle size on the fracture process. It depends on the ratio of particle diameter D to the characteristic size $\mu_{\bar{\ell}}$ of the defects. For a reference diameter D_0, we introduce $\bar{n_0} = 100$ defects uniformly distributed in a square area (see Fig.3). The particle extracted from this area has thus an average number $\bar{n_0} \times \frac{4}{\pi} \simeq 78$ of defects. The average size of the defects was set to $\mu_{\bar{\ell}0} \simeq D_0/30$ with a standard deviation of $\sigma_{\bar{\ell}0} \simeq D_0/200$. The particle was then meshed with a spatial resolution of

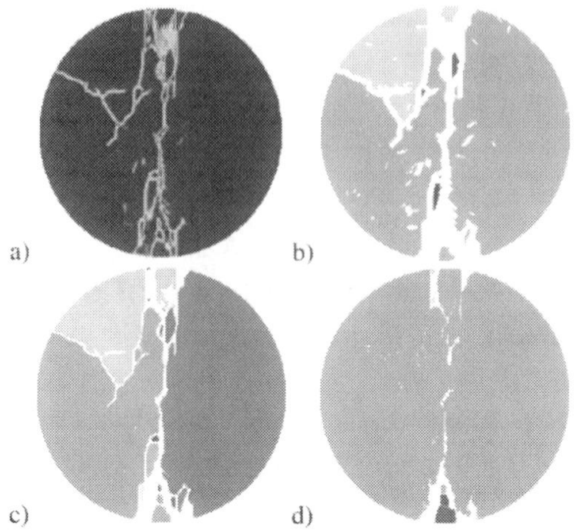

Figure 1. a): Damage field after particle fracture. The color level represents the magnitude of the local damage. b), c) and d): Results of the Floodfill algorithm with T=1%, T=33% and T=66% respectively.

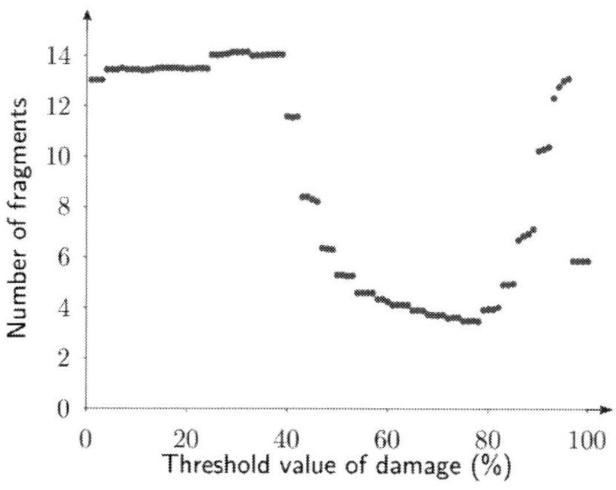

Figure 2. Average number of fragments of 100 particles as a function of damage threshold.

256 nodes per particle diameter. To reduce computational costs, we did not directly vary the particle size but the values of parameters related to the defects:

$$\mu_{\bar{\ell}} = \frac{D_0}{D}\mu_{\bar{\ell}0} \quad (1)$$

$$\sigma_{\bar{\ell}} = \frac{D_0}{D}\sigma_{\bar{\ell}0} \quad (2)$$

$$\bar{n} = \left(\frac{D}{D_0}\right)^2 \bar{n_0} \quad (3)$$

In all cases, the Young moduli of the defects are chosen to be narrowly distributed with an average $\mu_{\bar{E}}/E = 0.2$ and a standard deviation $\sigma_{\bar{E}} = 0.005$. Hence, the variability of particle strength mainly depends on the distribution of the lengths and the random positions of the defects. Table 1 shows the seven values of D/D_0 used in our study and the

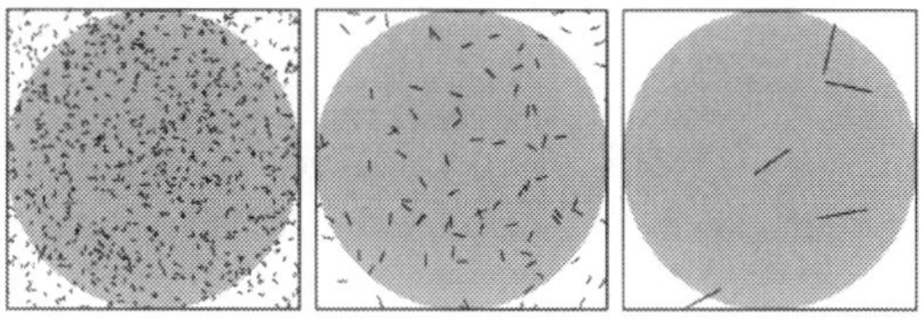

Figure 3. Population of defects in the particle for different diameters. From left to right: D/D_0=2.85, D/D_0=1 and D/D_0=0.2.

corresponding defect parameters. In order to obtain meaningful statistics on particle fracture, we broke 100 particles for each diameter.

Table 1. Parameters of defect distribution as a function of dimensionless particle diameter.

D/D_0	$\bar{n}$	$\mu_{\bar{l}}$	$\sigma_{\bar{l}}$
0.05	0.25	0.7	0.1
0.2	4	0.175	0.025
1	100	0.035	0.005
2	400	0.0175	0.0025
2.85	816	0.01225	0.00175
4	1600	0.00875	0.00125
10	10000	0.0035	0.0005

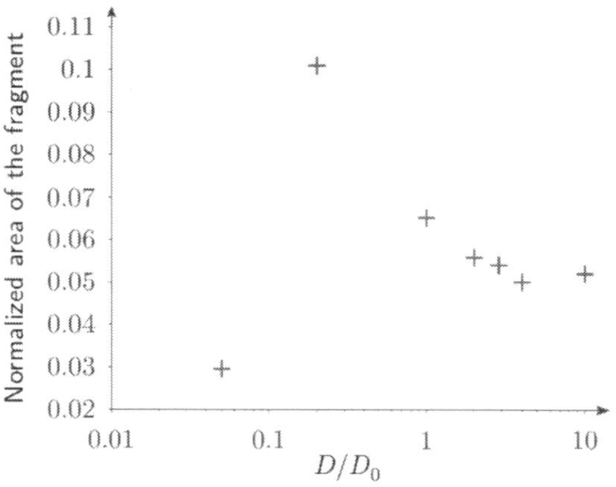

Figure 4. Mean fragment size normalized by particle area as a function of dimensionless diameter.

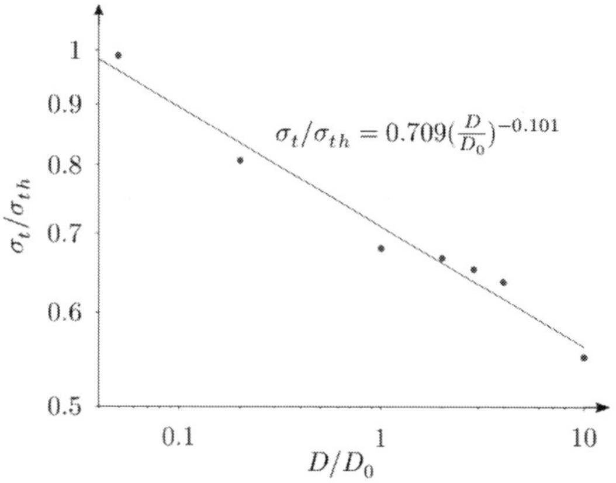

Figure 5. Average yield stress in tension normalized by the yield stress of the material without defects.

5 Results and discussion

Figure 4 shows the average size of the fragments as a function of dimensionless particle diameter. For a diameter above 0.2, the mean size of the fragments decreases with particle size. The increase in the number of defects with particle size leads to an increase of potential crack paths and thus to a larger number of fragments. For the smallest size (diameter of 0.005), the density of defects is too small so that the particle breaks homogeneously into numerous small pieces.

Assuming that the fracture occurs for the maximum value of tensile stress in the middle of the particle [27], we get:

$$\sigma_t = \frac{2F}{\pi D} \tag{4}$$

where F is the yield force at failure. In figure 5 we plot as a function of D/D_0, the average value of σ_t/σ_{th} where σ_{th} is the yield stress in tension of the material. We see that the strength of the particle decreases as the diameter increases, and follows a power law with an exponent of -0.101. This behavior is in agreement with experimental results [27].

In order to study the dispersion of the values of the yield stress, we consider the probability for a particle to survive under a given loading. For N particles, we assign the probability $P_S(i) = 1 - \frac{i}{N}$ to the ith smallest yield stress value. This distribution is then fitted by the Weibull function [28]:

$$P_S = exp\left[-\left(\frac{\sigma_t}{\sigma_w}\right)^m\right] \tag{5}$$

where σ_w corresponds to the stress for which 63% of particles are broken and m is the Weibull modulus characterizing the width of the distribution. Figure 6 shows $ln(ln(1/P_S))$ as a function of $ln(\sigma_t/\sigma_{th})$. If the data collapse on a straight line, the slope equals the Weibull modulus of the distribution. Note that a value of 0 on the x axis corresponds to a yield stress equal to σ_{th} and thus to a material without defect. Except for smallest particle diameters (0.05 and 0.2), the distributions are well fitted by the Weibull law.

The Weibull modulus globally increases with particle size (m = 4.4 to m = 41). As the number of defects is increasing with diameter, the probability of finding similar critical defects between particles increases. As a consequence, the distributions get narrower. On the contrary, the dispersion is smaller for 0.05 compared to 0.2. As the cracks are initiated in tension, the presence of a defect in the vicinity of the particle center has a dramatic impact. Because of the very low density of defects, many particles have no defect at their center. Thus, the probability distribution becomes narrow as the particles yield stress reaches the theoretical value of the material.

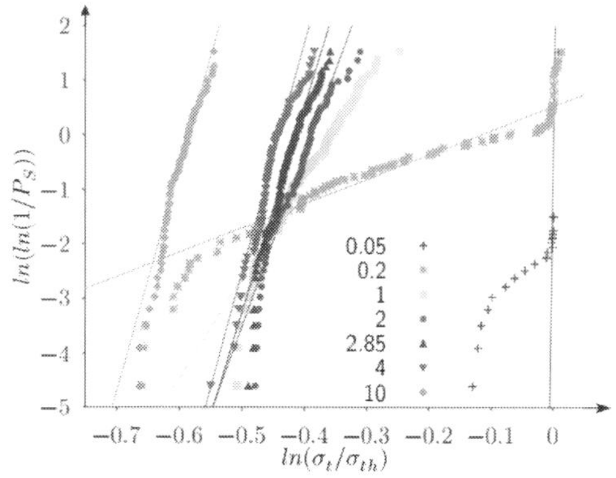

Figure 6. Surviaval probability of particles as a function of the applied stress for different values of relative particle size D/D_0. The straight lines represent Weibull fits to the data.

6 Conclusion

In this paper, we developed a methodology to investigate the fracture of 2D particles containing defects. More than 700 compression tests were performed with a peridynamic bond-based approach to characterize the evolution of yield stress and average fragment size as a function of particle diameter. We found that the Weibull modulus evolves with particle size and the average yield stress decreases as a power law with diameter.

In future work, we would like to analyze in detail the size distributions of the fragments. We would also like to analyze crack patterns. The simulations allow us also to evaluate the distributions of crack lengths and the interactions between the growing crack and defects. As the cracks tend to cross the defects, this leads to a lower number of defects in the fragments. We also plan to investigate the influence of defect lengths (Gaussian or power law distribution) on the probability distributions of the yield stress.

Acknowledgements

We thank BPI France for their financial support and Kerneos group for helpful discussions. We are also grateful to the genotoul bioinformatics platform Toulouse Midi-Pyrenees (Bioinfo Genotoul) for providing computing resources.

References

[1] D. Touil, S. Belaadi, C. Frances, International Journal of Mineral Processing **87**, 141 (2008)
[2] L. Wang, *Energy Efficiency and Management in Food Processing Facilities* (CRC Press, 2008)
[3] J. Herbst, D. Fuerstenau, International Journal of Mineral Processing **7**, 1 (1980)
[4] L. Austin, R. Klimpel, P. Luckie, *Process Engineering of Size Reduction: Ball Milling* (Society of Mining Engineers of the AIME, 1984)
[5] M. Aliha, M. Ayatollahi, Theoretical and Applied Fracture Mechanics **69**, 17 (2014)
[6] H. Guo, N. Aziz, L. Schmidt, Engineering Geology **33**, 177 (1993)
[7] D.Q. Dan, H. Konietzky, M. Herbst, International Journal of Rock Mechanics and Mining Sciences **58**, 1 (2013)
[8] A. Tavallali, A. Vervoort, Journal of Rock Mechanics and Geotechnical Engineering **5**, 366 (2013)
[9] A. Zubelewicz, Z.P. Bažant, *Interface element modeling of fracture in aggregate composites* (1987)
[10] F. Kun, H. Herrmann, Computer Methods in Applied Mechanics and engineering **7825** (1996)
[11] J. Åström, H. Herrmann, The European Physical Journal B **5**, 551 (1998)
[12] O. Tsoungui, D. Vallet, J.C. Charmet, Powder Technology **105**, 190 (1999)
[13] D.H. Nguyen, E. Azéma, F. Radjai, Geomechanics from Micro to Macro pp. 275–280 (2015)
[14] D.H. Nguyen, E. Azéma, P. Sornay, F. Radjai, Physical Review E **91**, 022203 (2015)
[15] M. Ciantia, M. Arroyo, F. Calvetti, A. Gens, Geotechnique **65**, 91 (2015)
[16] W. Zhou, L. Yang, G. Ma, X. Chang, Y. Cheng, D. Li, Granular Matter **17**, 497 (2015)
[17] S. Galindo-Torres, D. Pedroso, D. Williams, L. Li, Computer Physics Communications **183**, 266 (2012)
[18] S. Silling, Journal of the Mechanics and Physics of Solids **48**, 175 (2000)
[19] W. Gerstle, N. Sau, S. Silling, Nuclear Engineering and Design **237**, 1250 (2007)
[20] E. Celik, I. Guven, E. Madenci, Theoretical and Applied Fracture Mechanics **55**, 185 (2011)
[21] P. Perré, G. Almeida, M. Ayouz, X. Frank, Annals of Forest Science **73**, 147 (2016)
[22] J. Kertész, in *Statistical Models for the Fracture of Disordered Media*, edited by H.J. Herrmann, S. Roux (North-Holland, Amsterdam, 1990), Random Materials and Processes, pp. 261 – 290
[23] V. Topin, J.Y. Delenne, F. Radjai, L. Brendel, F. Mabille, The European Physical Journal E **23**, 413 (2007)
[24] Y.D. Ha, F. Bobaru, Engineering Fracture Mechanics **78**, 1156 (2011)
[25] M. Allen, D. Tildesley, *Computer Simulation of Liquids*, Oxford Science Publ (Clarendon Press, 1989)
[26] J.Y. Delenne, V. Richefeu, F. Radjai, Journal of Fluid Mechanics **762** (2015)
[27] G.R. McDowell, M.D. Bolton, Géotechnique **48**, 667 (1998)
[28] G. Mcdowell, A. Amon, Soils and Foundations **40**, 133 (2000)

Numerical study of the failure of materials embedding soft to hard particles

Xavier Frank[1,⋆], *Jean-Yves* Delenne[1], and *Farhang* Radjai[2,3]

[1] *IATE, CIRAD, INRA, Montpellier SupAgro, Université de Montpellier, F-34060, Montpellier, France*
[2] *Laboratoire de Mécanique et Génie Civil (LMGC), Université de Montpellier, CNRS, Montpellier, France*
[3] *<MSE>^{2}, UMI 3466 CNRS-MIT, CEE, MIT, 77 Massachusetts Avenue, Cambridge CA 02139, USA*

Abstract. In this study, we use a bond-based peridynamic approach to investigate the mechanical strength and cracking of composite materials with spherical inclusions. The total volume fraction of particles and the particle-matrix toughness ratio were varied to cover a range of soft to hard inclusions. The mean particle damage was characterized together with crack patterns at a sub-particle scale. Three types of crack patterns are identified depending on the toughness ratio.

1 Introduction

Damage and failure of materials including a granular phase is crucial in numerous natural processes and industrial applications. The case of tough inclusions was studied in detail by several authors for various applications such as the failure of concrete [1], the milling of wheat [2] (whose internal micro-structure can be seen as a cemented granular material composed of starch granules glued by a protein matrix), the strength of particle-enriched composites [3]... However, in many examples the particles can have a lower toughness than the matrix. This is the case of some biocomposites used in packaging [4] and in food products such as granola bars or rice chocolate in which the taste largely depends on the textural properties [5].

In this paper, we use a bond-based peridynamic approach to investigate this transition from fragile to tough inclusions in particle-laden composites. A major advantage of this approach is its versatile nature in application to discontinuities such as pores, cracks and jumps in mechanical properties... allowing one to simulate crack patterns [6, 7]. The bond-based peridynamic approach has been successfully applied for modeling elastic brittle materials [7] including the dynamic effects and fracture (for example, crack patterns in concrete [8], dynamic cracks branching [9], fracture of polycrystals [10], nanoscale mechanics [11, 12]...). In this paper, we investigate the crack patterns in 2D composites including a suspension of disk particles. We performed an extensive parametric study by varying the density of particles in the matrix and the yield stress of particles in order to simulate a range of very soft to very hard inclusions.

⋆e-mail: xavier.frank@inra.fr

2 Bond-based peridynamics

In recent years, various numerical approaches have been designed for the computation of crack patterns such as cohesive zone models [13], the cohesive discrete elements method [14, 15], the contact dynamics-based bonded-cell models [16] and the lattice element method [17, 18]. The discrete peridynamic approach was introduced by Silling [6]. In this approach the mechanical behavior is based on integro-differential equations rather than partial differential equations. The main advantage is to simplify the treatment of the discontinuities in the material. We use this method for its versatile and simple implementation to investigate the cracking of composites with soft and hard inclusions. In the following, we describe the basics of the method in two dimensions.

Lets consider a deformable body B_t defined at the time t as a subset of $\mathbb{R}^2$. For each point in this domain, the strain, stress, damage... can be directly upscaled from the behaviour of bonds defined at the microscale. Figure 1 shows, for the reference configuration B_0 (at time $t = 0$) a bond $\vec{\xi} = \vec{x}' - \vec{x}$ where $\vec{x}$ and $\vec{x}' \in B_0$. At each point $\vec{x}$ a set of bonds $H(\vec{x}) = \left\{\vec{\xi} \in \mathbb{R}^2 | \vec{x} + \vec{\xi} \in B_0 \wedge \left\|\vec{\xi}\right\| < h\right\}$ is defined, where h is the cutoff distance usually called horizon in peridynamics.

Under a deformation $B_0 \to B_t$ the material points $\vec{x}$ undergo a displacement $\vec{u}(\vec{x}, t)$. Thus, a bond $\vec{\xi} = \vec{x}' - \vec{x}$ becomes after deformation $\vec{\xi} + \vec{\eta}$ where $\vec{\eta} = \vec{u}\left(\vec{x}', t\right) - \vec{u}(\vec{x}, t)$ is the relative displacement (Figure 1).

The force exerted on a material point $\vec{x}$ results from the contribution of all bonds in $H(\vec{x})$. Here, we use the so-called bond-based approach in which we only consider radial pairwise forces $\vec{f}$ that depend on both $\vec{\xi}$ and $\vec{\eta}$ [19]. The value of the Poisson ratio of a homogeneous sample

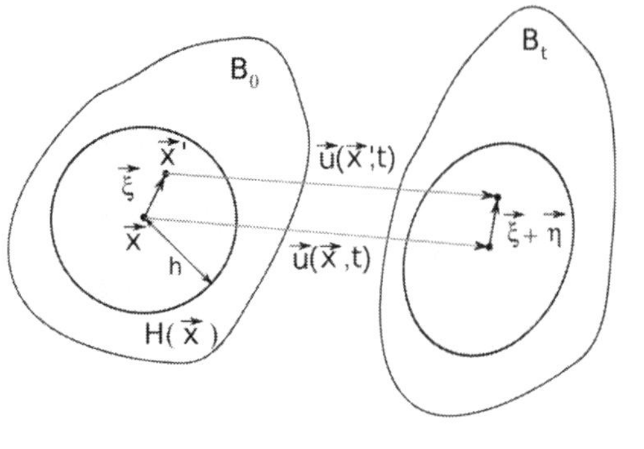

Figure 1. Initial configuration B_0: schematic representation of a bond $\vec{\xi}$ and the bond family $H(\vec{x})$ of a point $\vec{x}$ in a peridynamic body, definition of the horizon h. Schematic representation of the deformation of a peridynamic body from initial configuration B_0 to B_t and definition of both displacement $\vec{u}$ and relative displacement $\vec{\eta}$.

is fixed to $\nu = 1/3$ [8]. The forces can be expressed as

$$\vec{f}(\vec{\xi},\vec{\eta}) = f(\vec{\xi},\vec{\eta})\frac{\vec{\xi}+\vec{\eta}}{\left\|\vec{\xi}+\vec{\eta}\right\|} \tag{1}$$

where $f(\vec{\xi},\vec{\eta})$ is the bond force magnitude.

In this work, we use harmonic pairwise forces defined as:

$$s(\vec{\xi},\vec{\eta}) = \frac{\left\|\vec{\xi}+\vec{\eta}\right\| - \left\|\vec{\xi}\right\|}{\left\|\vec{\xi}\right\|} \tag{2}$$

$$f(\vec{\xi},\vec{\eta}) = cs(\vec{\xi},\vec{\eta}) \tag{3}$$

where s is the bond elongation and c is the elastic micromodulus. The value of the effective Young modulus of the material is given by [9]:

$$E = c(\pi h^3(1-\nu)/6) \tag{4}$$

which linearly depends on the micromodulus c. The damage of the material is introduced using a history-dependent scalar function μ as a prefactor modifying c to $\mu(\vec{\xi},t)c$, and defined as

$$\mu(\vec{\xi},t) = \begin{cases} 1 & s(\vec{\xi},\vec{\eta}(t')) < s_0 \ \forall t' < t \\ 0 & \exists t' < t \left| s(\vec{\xi},\vec{\eta}(t')) \geq s_0 \right. \end{cases} \tag{5}$$

where s_0 is the critical elongation of the bond. In other words, when at a time t' the elongation s exceeds s_0, the bond is definitely broken and $f(\vec{\xi},\vec{\eta}) = 0$. The fracture energy G is given by [9]

$$G = (9Ehs_0{}^2)/4\pi \tag{6}$$

For a heterogeneous material, each phase α is characterized by its elastic modulus E_α and its toughness $K_\alpha = \sqrt{E_\alpha G_\alpha}$, where G_α is the fracture energy of the phase α. If a bond connects two nodes i and j in the same phase α the mechanical properties of the bond are directly deduced from Equations 4 and 6. If the bond connects two different phases α and β, we choose $E_{\alpha\beta} = \min(E_\alpha, E_\beta)$ and we set $G_{\alpha\beta}$ so that $K_{\alpha\beta} = \frac{1}{2}(K_\alpha + K_\beta)$. More generally, the interphase properties can be independent of bulk properties.

3 Implementation

The discretization of the material is performed in 2D according to a regular grid of nodes $\vec{x}_i$ with a spatial resolution δx. A mass $m_i = (\delta x)^2\rho(\vec{x}_i)$ is attributed to each node. The mechanical system is similar to a mass-spring lattice in which each bond in $H(\vec{x}_i)$ represents a spring. The evolution of the system is computed by the discretized equations of motion:

$$m_i\ddot{\vec{u}}_i(t) = \sum_{\vec{x}_j \in H(\vec{x}_i)} ks(\vec{\eta}_{ij},\vec{\xi}_{ij})\frac{\vec{\xi}_{ij}+\vec{\eta}_{ij}}{\left\|\vec{\xi}_{ij}+\vec{\eta}_{ij}\right\|} + \vec{b}_i(t) \tag{7}$$

where $\vec{u}_i(t) = \vec{u}(\vec{x}_i,t)$, $\vec{\xi}_{ij} = \vec{x}_j - \vec{x}_i$, $\vec{\eta}_{ij} = \vec{u}_j - \vec{u}_i$, $\vec{b}_i(t) = (\delta x)^2\vec{b}(\vec{x}_i,t)$ and $k = c(\delta x)^4$.

External forces or displacements are applied at the boundaries of the samples. The stress-strain evolution of the system can be computed either in dynamics or quasi-static conditions. For the dynamic part we use the classical velocity-Verlet algorithm [20] with a time step δt. In order to efficiently damp low frequency modes, we use a global viscous term at all nodes. To ensure the quasi-static evolution of the system, we use a snap-back algorithm which consists in relaxing the loading strain periodically when the cumulative energy due to crack growth reaches a given value κ. For a sufficiently small value, the process is quasi-static but the computation time increases dramatically. We fixed the value of κ as the energy required to propagate a crack of distance equal to 1% of the sample width.

4 Sample building & mechanical tests

We prepared a series of cemented granular samples. These samples are composed of a collection of disks introduced into a matrix. For the granular phase, six samples were first prepared using a DEM code [21] and packed under a weak confining pressure. The particle radii are slightly dispersed in order to avoid local crystallization. These samples are then meshed using a rectilinear grid with a spatial resolution of approximately 50 nodes in the grain's mean diameter and the left boundary of each sample was notched to initiate crack propagation. To study the influence of particle density, we apply a geometrical shrinkage factor to each grain. Three particle solid fractions φ were used (0.3, 0.5 and 0.7) with six values of grain-matrix toughness ratio $\psi = K_g/K_m$ (0.25, 0.375, 0.5, 2.0, 3.0 and 4.0). Finally, 108 samples were meshed. As the computation effort is relatively high (with approximately 500 000 nodes and 7 000 000 bonds for each sample), we parallezied the code using Message Passing Interface (MPI).

All the samples were subjected to quasi-static tensile tests until full failure occurs. Figure 2 shows an example of the snap-back procedure and the stress-strain envelope. For each sample we consider the damage after full failure. We consider a scalar field $n(\vec{x})$ associated with the nodes. When a bond connecting two nodes $\vec{x}_i$ and $\vec{x}_j$ breaks, the values of $n(\vec{x}_i)$ and $n(\vec{x}_j)$ are incremented. The damage at a node can then evolve between 0 and the maximum number of bonds connected to a node which depends on the horizon. In the following, we consider only the total damage at the end of the simulation when the failure is fully developped and we compute the grain damage parameter

$$d = N_g/N_{tot} \tag{8}$$

where $N_g = \sum_{x_i \in \mathcal{G}} n(\vec{x}_i)$, $\mathcal{G}$ is the subset of $\mathcal{B}_0$ corresponding to the grain phase and $N_{tot} = \sum_{x_i \in \mathcal{B}_0} n(\vec{x}_i)$.

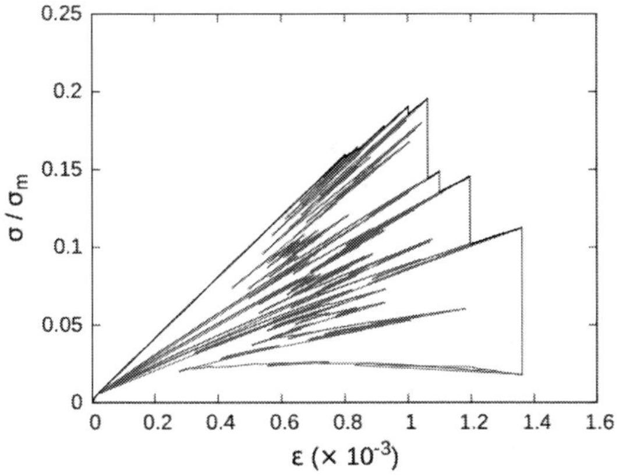

Figure 2. (color online) Evolution of the stress normalized by the yield stress of the matrix σ_m as a function of the strain ϵ for a sample with grain volume fraction $\varphi = 0.7$ and a toughness ratio $\psi = 3.00$.

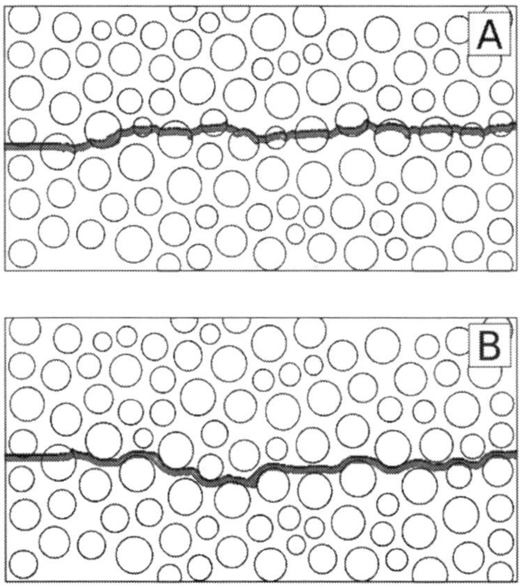

Figure 3. (color online) Crack path for grain volume fraction $\varphi = 0.5$. A: $\psi = 0.25$, B: $\psi = 4.0$.

Figure 3 shows two typical examples of crack patterns for $\varphi = 0.5$ and a toughness ratio ψ of 0.25 and 4.00, corresponding respectively to soft and hard inclusions. In these figures, we clearly observe different failure behaviors. For $\psi = 0.25$ the failure crosses the particles whereas for $\psi = 4$ the crack remains inside the matrix and bypasses the particles. This typical behavior was already observed for an single inclusion in a cement matrix [22].

5 Parametric study

In this section we discuss the results of the parametric study. For a quantitative description, we computed the total number of broken bonds in each phase cumulated on all samples for each parameter set φ,ψ. Figure 4 shows for three values of the grain volume fraction φ, the average grain damage parameter d (Equation 8) as a function of toughness ratio ψ. The amount of broken bonds in the grain phase increases with φ. It is interesting to note that

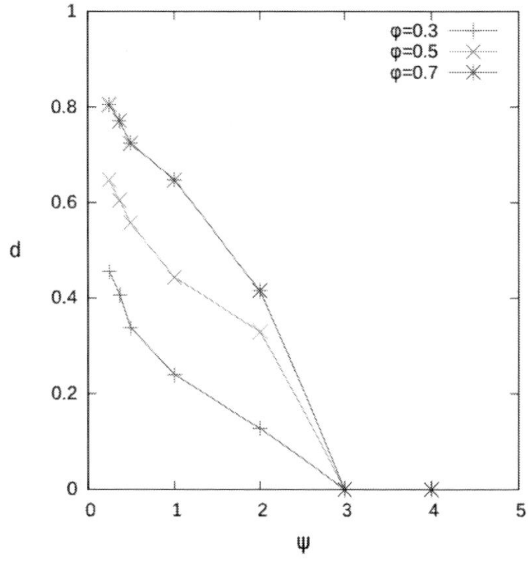

Figure 4. (color online) Grain damage parameter d corresponding to the average fraction of broken bonds inside grains (cumulated on all samples) as a function of toughness ratio ψ for three values of the grain volume fraction φ.

the value of ψ below which no failure occurs in grains is the same ($\psi \simeq 3$) whatever the value of φ.

For the same samples, Figure 5a and Figure 5b show the superposition of all crack paths in the vicinity and inside the damaged grains for a toughness ratio $\psi = 0.5$, and $\psi = 2.0$, respectively. As the grains do not have the same size, each crack trajectory was normalized by the corresponding grain radii. Note also that in the simulations, as the failure is initiated by a notch on the left hand side of the sample, the propagation occurs from left to right.

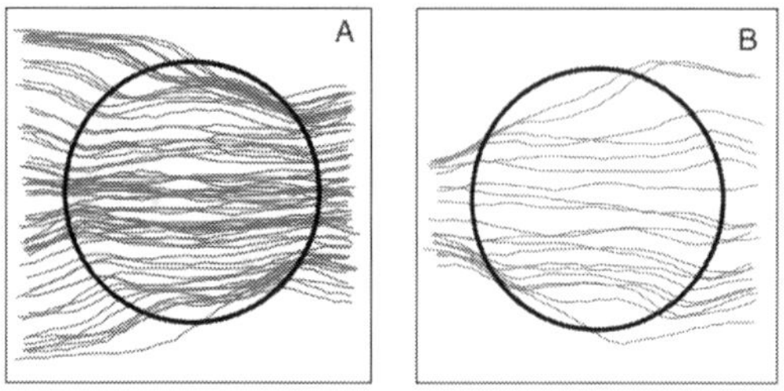

Figure 5. (color online) Superposition of all crack patterns (crack propagation from the left to the right) in the neighborhood of broken grains for grain volume fraction 0.4. Toughness ratio is A: 0.5, and B:2.0.

We see that, for a low value of the grain-matrix toughness ratio ψ, the grains tend to focus the cracks like a convergent optical lens. On the contrary, for $\psi > 1$, the grains deflect the cracks. This convergent-divergent transition occurs for a value of $\psi = 1$ which corresponds to a regime of straight failure.

6 Conclusion

In this paper we introduced a bond-based quasi-static peridynamic approach for the simulation of granular composites. This method is quite well suited to the computation of damage and crack patterns in highly heterogeneous media. An extensive parametric study was performed to investigate the effects of toughness ratio and grain volume fraction on the damage and fracture behavior across the grain-matrix interfaces.

Acknowledgment

We are grateful to the genotoul bioinformatics platform Toulouse Midi-Pyrenees (Bioinfo Genotoul) for providing computing resources.

References

[1] M. Issa, M. Issa, M. Islam, A. Chudnovsky, Eng. Fract. Mech. **70**, 125 (2003)
[2] V. Topin, F. Radjaï, J.Y. Delenne, A. Sadoudi, F. Mabille, J. Cereal Sci. **47**, 347 (2008)
[3] R. Eckner, M. Krampf, C. Segel, L. Krüger, Mechanics of Composite Materials **51**, 707 (2016)
[4] M.A. Berthet, H. Angellier-Coussy, V. Guillard, N. Gontard, Journal of Applied Polymer Science **133**, 42528 (2016)
[5] T. van Vliet, *Rheology and fracture mechanics of foods* (CRC Press, 2013)
[6] S. Silling, J. Mech. Phys. Solids **48**, 175 (2000)
[7] P. Perré, G. Almeida, M. Ayouz, X. Frank, Annals of Forest Science **73**, 147 (2016)
[8] W. Gerstle, N. Sau, S. Silling, Nuclear Eng. & Design **237**, 1250 (2007)
[9] Y. Ha, F. Bobaru, Eng. Fract. Mech. **78**, 1156 (2011)
[10] E. Askari, F. Bobaru, R. Lehoucq, M. Parks, S. Silling, O. Weckner, J. of Phys.: Conf. Series **125**, 012078 (2008)
[11] F. Bobaru, Modelling Simul. Mater. Sci. Eng. **15**, 397 (2007)
[12] E. Celik, I. Guven, E. Madenci, Theor. Appl. Fract. Mech. **55**, 185 (2011)
[13] X. Ren, J. Li, Phys. Rev. E **85**, 055102 (2012)
[14] J.Y. Delenne, M.S. El Youssoufi, F. Cherblanc, J.C. Bénet, International Journal for Numerical and Analytical Methods in Geomechanics **28**, 1577 (2004)
[15] H.A. Carmona, A. Guimarães, J. Andrade, I.N. Jr., F. Wittel, H. Herrmann, Phys. Rev. E **91**, 012402 (2015)
[16] D.H. Nguyen, E. Azéma, P. Sornay, F. Radjaï, Phys. Rev. E **91**, 022203 (2015)
[17] V. Topin, J.Y. Delenne, F. Radjaï, L. Brendel, F. Mabille, Eur. Phys. J. E **23**, 413 (2007)
[18] R. Affes, J.Y. Delenne, Y. Monerie, F. Radjaï, V. Topin, Eur. Phys. J. E **35**, 117 (2012)
[19] S. Silling, M. Epton, O. Weckner, J. Xu, E. Askari, J. Elasticity **88**, 151 (2007)
[20] M. Allen, D. Tildesley, *Computer Simulation of Liquids* (Oxford University Press, Oxford, 1986)
[21] F. Radjaï, F. Dubois, *Discrete Numerical Modeling of Granular Materials* (Wiley ISTE, 2011)
[22] O. Buyukozturk, B. Hearing, Int. J. Solids Struct. **35**, 4055 (1998)

Analysis of compaction of railway ballast by different maintenance methods using DEM

Jean-Francois Ferellec[1,*], *Robert* Perales[2], *Viet-Hung* Nhu[3], *Michel* Wone[1] and *Gilles* Saussine[1]

[1]SNCF Reseau, I&P-LVE, 6 av. F. Mitterrand, 93574 La Plaine St Denis cedex, France
[2]AD'missions, 20 rue Brunel, 75017 Paris, France
[3]Vibrateam, 28 chemin du Petit Bois, 69134 Ecully cedex, France

Abstract. Railway traffic continuously increasing, ballasted tracks need more efficient maintenance processes. Lines with long welded rails which are prone to buckling during heat waves require stabilisation before being fully operational. Stabilisation is performed either naturally using regular traffic at penalising lower speeds, dynamic stabilisation of sleepers or alternatively crib compaction. The objective of this paper is to apply the NSCD approach of DEM to simulate the processes of dynamic stabilisation and crib compaction as they are realised on site and compare their performance in terms of ballast compaction and lateral resistance. The results showed that NSCD is perfectly appropriate to simulate these maintenance processes and estimate their performance.

1 Introduction

As railway traffic of passengers and freight is continuously increasing, maintenance of ballasted tracks becomes more and more demanding particularly in the case of tracks using long welded rails prone to buckling during heat waves. To avoid this kind of incident, it is crucial to keep the lateral resistance of the track above a critical level. After their geometrical realignment by the process of tamping which loosens the ballast, tracks require stabilisation to achieve a high level of lateral resistance.

The simplest stabilisation method, called natural stabilisation, consists in successive passages of regular traffic trains at lower speed which is equivalent to applying cyclic vertical loads to the sleepers for a given time. This situation is equivalent to a dynamic uniaxial loading test on a sample. This natural stabilisation is progressively replaced by faster operations which do not disturb the regular train traffic. The most popular alternative method is the dynamic stabilisation where the rails are vibrated laterally while applying a vertical load on them using dedicated equipment. These conditions are typical of a biaxial loading test with varying confining pressure. The second option, called crib compaction and relatively new in railway maintenance, is the direct vertical packing of the ballast located between the sleepers and in the shoulders of the track. This approach is performed by a set of plates of different sizes and locations vibrating vertically while applying a vertical load. This case is similar to the compaction of a free surface granular medium.

In order to avoid in situ campaigns costly in terms of time and budget, both alternative stabilisation processes were analysed numerically by using the discrete element method (DEM) which models granular materials, like ballast, as objects interacting through contacts laws [1]. The code LMGC90 based on the non-smooth contact dynamics or NSCD approach of DEM was used in this study [2]. The aim of this study is to compare their performances in terms of ballast compaction and lateral resistance of sleepers.

The first section briefly presents the NSCD approach of DEM. The second one describes the simulations conditions. The third part shows a comparison of ballast compaction and lateral resistance obtained after using both maintenance processes.

2 NSCD approach of DEM

In conventional DEM, a granular material is considered as an assembly of objects interacting through a contact law [1], implementing explicitly the mechanical interaction using an elastic force depending on the overlap between these objects, a sliding criterion limiting the tangential force at contact and potentially a damping force function of the overlapping speed. The calculation cycle in this model consists of: first the detection of the contacts between the objects; second the calculation of the contact forces from the overlaps between them; third the calculation of their displacements by a double integration of their acceleration on a time interval called time step using Newton's Law. The displacements will lead to the update of the positions of the objects ready for another cycle. In this conventional DEM, the most common object is the sphere because it makes the contact detection step easier. However spheres reproduce

* Corresponding author: jean-francois.ferellec@reseau.sncf.fr

the behaviour of a granular material only in a qualitative way. Different approaches were developed to achieve results more quantitatively reliable but they were not really effective in capturing the irregular shape of real particles [3-7]. It is only recently that some models were more successful in reproducing real particles shape by using arrangement of spheres and particularly ballast stones [8-13].

The Non-Smooth Contact Dynamic (NSCD) is a particular approach of DEM, based on implicit time integration and non-smooth formulation of mutual exclusion and dry friction between objects [2, 14-16]. The contact law is defined by a non-smooth relation between the normal force and the normal relative velocity and Signorini conditions. A restitution coefficient relating the relative velocities before and after contact can be introduced in the contact law. The normal and tangential coefficients of restitution can be set equal to zero to reflect the total dissipation of kinetic energy of a particle into the dense packing of particles of ballast. The NSCD code called LMGC90 which is capable of modelling a collection of rigid particles of polyhedral shape has been used in the present study.

3 Conditions of simulations

Both dynamic stabilisation and crib compaction were simulated using a four-sleeper track section (fig.1) with ballast particles represented by 90,000 rigid polyhedrons constructed from scans of real ballast stones following French railway ballast requirements. It presents a slope on one side and a vertical wall on the other side as if a contiguous line was present. The friction coefficient between particles and between particles and sleepers has been determined equal to 0.8, the ballast stone density to 2700kg/m^3 and the time step to 2x10^{-4}s [17]. The sample is prepared by generating the ballast particles above the surface and letting them settle under gravity using a reduced friction to achieve the desired apparent density. Before stabilisation, the track goes through a process called tamping [18] which consists of lifting the sleepers and then squeezing the ballast underneath them using the tines which are represented in figure 1.

Fig. 1. Four-sleeper configuration model with tamping tines

Figure 2 shows how the six compacting plates are positioned during the crib compaction process: four between the sleepers and two at their ends. They compact the ballast by vibrating vertically during 2s at 50Hz with a load varying between 5 and 45kN, values which are typical for this process. The six-plate compactor is sequentially moved between the sleepers. In order to keep similar conditions for comparison, the analysis focused on the inside sleepers and cribs and hence only the inside sleepers were subjected to dynamic stabilisation. The dynamic stabilisation is processed in the simulations by exciting the inside sleepers with a signal based on the specifications of the stabilising equipment. The lateral velocity of the sleepers is oscillated using an amplitude varying in a sinusoidal way which corresponds to the displacement of the stabilising equipment along the track. Figure 3.a shows two vibration waves representing the passage of two successive stabilising equipments as seen by each sleeper. Simultaneously two sinusoidal shaped vertical loads are applied (Figure 3.b). In this study, the vibration signal has a frequency of 25Hz, a maximum amplitude of 2.5mm and a rolling speed of 2000m/s, values which are typical for this process. Finally the lateral resistance of the track is determined by applying a progressive load laterally on both inside sleepers and measuring simultaneously their lateral displacement.

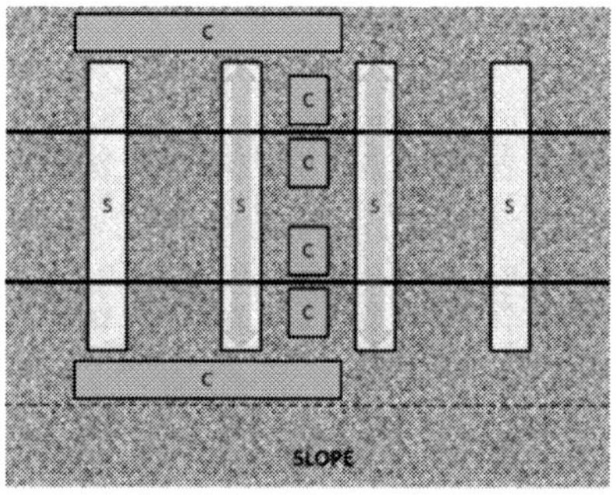

Fig. 2. Sample with sleepers (S) and compactor plates (C)

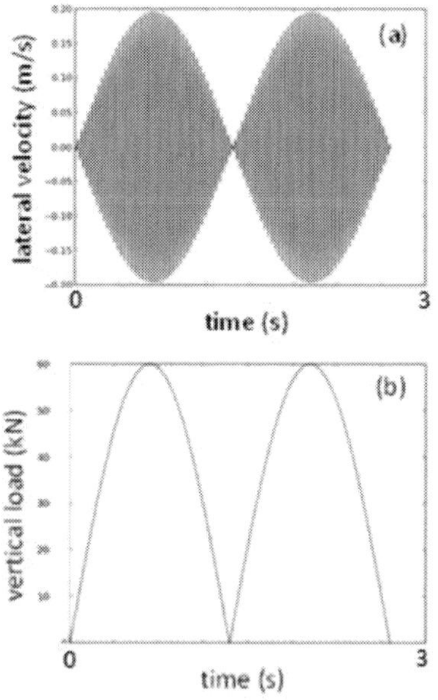

Fig. 3. Lateral velocity (a) and load (b) of dynamic stabilisation

4 Comparison of methods

Both processes of dynamic stabilisation and crib compaction were compared in terms of solid fraction gain under the sleepers and in the cribs (between sleepers). Figure 4 shows the evolution of the solid fraction for the ballast under the inside sleepers for both processes. The dynamic stabilisation lasts a little bit less than 3s while the crib compaction lasts 8s for the whole sample. Regarding the dynamic stabilisation, the effect of two waves is clearly observed, the second bringing less compaction. The ballast under the inside sleepers reach almost the same final solid fraction of 61.3%, even if they start with different initial values, homogenising the ballast density among the sleepers. The crib compaction seems to be linear with time. The black bottom line of figure 4 shows that the solid fraction under the first sleeper starts to increase during the 2s crib compaction on one side of the sleeper and the other one on the other side offering a solid fraction gain of more than 1%. The second sleeper obtains the same gain thus leading to a different final value. It does not seem to homogenise the solid fraction under the sleepers as the dynamic stabilisation which acts directly on the sleepers. This however could potentially be achieved by tuning the vibration frequency or vertical load of the crib compaction process.

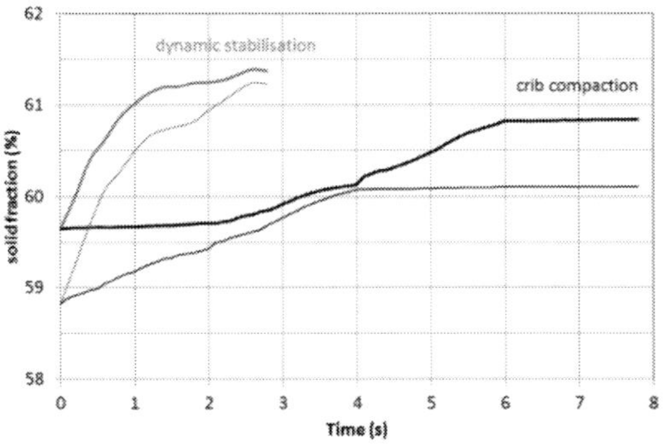

Fig. 4. Comparison of ballast solid fraction under sleepers for dynamic stabilisation (grey) and crib compaction (black)

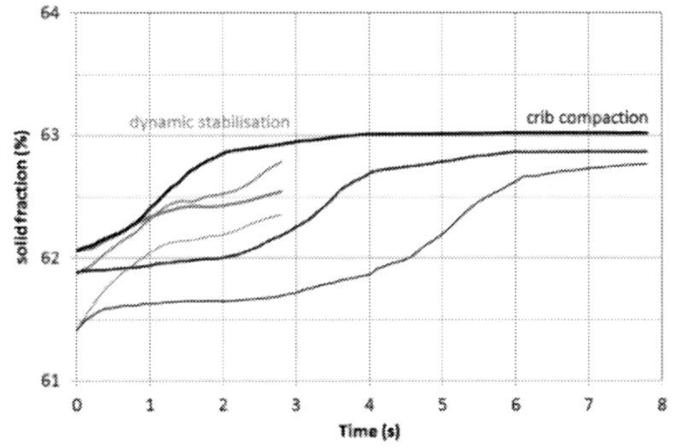

Fig. 5. Comparison of ballast solid fraction in cribs for dynamic stabilisation (grey) and crib compaction (black)

Figure 5 shows the evolution of ballast solid fraction in all three cribs along the inside sleepers under sleeper bottom level for both processes. The effect of the crib compaction process is clearly observable. The solid fraction increases during 2s for all cribs sequentially as expected reaching in the end almost the same final value. The dynamic stabilisation also increases the solid fraction in cribs but in a lesser way which is understandable as cribs are not directly vibrated.

Figure 6 shows the initial and final ballast solid fractions obtained along the track components for both processes. Both processes significantly compact the ballast under the sleepers although dynamic stabilisation seems more efficient. In cribs, the opposite effect is observed. Initially quite heterogeneous along the track, the ballast density presents a more homogeneous profile after dynamic stabilisation. The crib compaction seems to give the same gain to all areas as the ‘W’ shaped density profile along the track is almost unchanged.

Globally, in terms of ballast compaction, dynamic stabilisation seems more efficient than crib compaction using the parameters stated above which are typical of actual operations. Different sets of crib compaction parameters could lead to different final compaction levels inside the ballast.

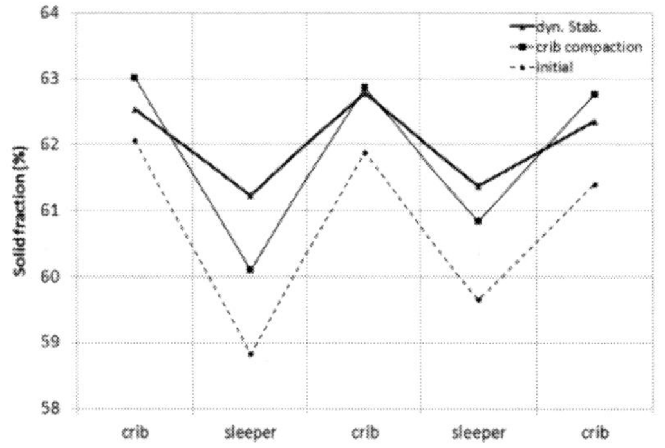

Fig. 6. Comparison of final solid fraction under sleepers and cribs

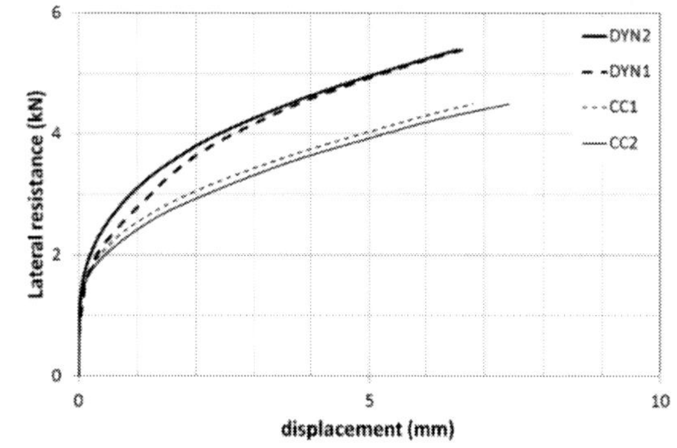

Fig. 7. Lateral resistance as a function of sleeper displacement for inside sleepers (1, 2) for dynamic stabilisation (DYN) and crib compaction (CC)

In terms of lateral resistance, dynamic stabilisation seems to give slightly better results. Figure 7 shows the lateral resistance obtained for the inside sleepers for both processes. For each process, both sleepers (identified as 1 and 2) present very similar lateral resistance. The difference of resistance between both processes seems to be around 1kN at 5mm displacement meaning that lateral resistance obtained after crib compaction is just 20% lower than the one after dynamic stabilisation. As an engineering point of view, this makes crib compaction a suitable alternative to dynamic stabilisation. By tuning its operating parameters, it could also be improved. The side effects of this tuning on the global ballast behaviour must also be identified. Dynamic stabilisation can also be improved by tuning its parameters. Previous studies have shown that decreasing the speed of dynamic stabilisation increased compaction [19].

5 Conclusions

The NSCD approach of DEM has been used to analyse dynamic stabilisation and crib compaction, two different ballasted tracks maintenance processes, in terms of compaction capabilities and track lateral resistance.

The results showed that NSCD is perfectly adequate for this type of analysis. Its application showed that both stabilisation methods fulfil their intended purpose of densifying the ballast under sleepers and increasing the lateral resistance of the track although dynamic stabilisation achieves it in a better way.

On the point of view of railway engineering, it shows that crib compaction is an adequate potential alternative to dynamic stabilisation even if it is slightly less efficient. This limitation in performance could be addressed by tuning its parameters, particularly the vibration frequency or vertical load.

NSCD is presently used to realise an impact study regarding parameters of both maintenance operations in order to optimise them. A deeper analysis of the effect of the parameters on various aspects of the phenomena occurring at the scale of the ballast stones will be realised.

Another study identifying the contribution of the different components of ballasted tracks to lateral resistance has been started. It includes the type of sleeper and the shoulder size among others.

References

1. P.A. Cundall, and O.D.L. Strack, Geotechnique, **29**, 1, 47-65 (1979)
2. J.J. Moreau, European J. Mech. A Solids, **13**, 4, suppl., 93-114 (1994)
3. J.R. Williams, and A.P. Pentland, Eng. Comput., **9**, 115-127 (1992)
4. X. Lin, and T.T. Ng, Geotechnique, **47**, 2, 319-329, 1997.
5. G.G.W. Mustoe, and M. Miyata, J. Eng. Mech., **127**, 10, 1017-1026 (2001)
6. P.W. Cleary, Engineering Computations, **21**, 2-4, 169-204 (2004)
7. L. Pournin, Granular Matter, **7**, 2-3, 119-126 (2005)
8. R. Hart, P.A. Cundall, and J. Lemos, Int. J. Rock Mech. Min. Sci. Geomech. Abstr., **25**, 3, 117-125 (1988)
9. L.B. Wang, J.Y. Park, and Y.R. Fu, Construction and Building Materials, **21**, 2, 338-346 (2007)
10. M. Price, V. Murariu, and G. Morrison, Proceedings of Discrete Element Methods Conference, Brisbane, Australia (2007)
11. Y. Lee, Granular Matter, **11**, 5, 307-315 (2009)
12. J.-F. Ferellec and G.R. McDowell, Granular Matter, 12, 5, 459-467 (2010)
13. J.-F. Ferellec, and G.R. McDowell, Geotechnique, **60**, 3, 227-232 (2010)
14. J.J. Moreau, European J. Mech. A Solids, **13**, 4, suppl., 93-114 (1994)
15. G. Saussine, C. Cholet, P.E. Gautier, F. Dubois, C. Bohatier, J.J. Moreau, Comput. Methods Appl. Mech. Eng., **195**, 2841-2859 (2006)
16. F. Dubois and M. Renouf, Proceedings of Multibody 2007, Politecnico di Milano, Milano, Italy (2007)
17. G. Saussine, Contribution to the modelling of tri-dimensional granular materials: application to ballast, PhD thesis (2004)
18. J.-F. Ferellec, R. Perales, P.-E. Laurens, M. Woné, J. Plu and G. Saussine, Proceedings of IV International conference on Particle-based Methods – Fundamental and Applications, Particles 2015, Barcelona, Spain (2015)
19. J.-F. Ferellec, M. Woné, R. Perales and G. Saussine, Proceedings of III International conference on Railway Technology: Research, Development and Maintenance, Cagliari, Italy (2016)

Small solar system bodies as granular systems

Daniel Hestroffer[1,*], *Adriano* Campo Bagatín[2,15], *Wolfgang* Losert[3], *Eric* Opsomer[4], *Paul* Sánchez[5], *Daniel J.* Scheeres[6], *Lydie* Staron[7], *Nicolas* Taberlet[8], *Hajime* Yano[9], *Siegfried* Eggl[1], *Charles-Edouard* Lecomte[7], *Naomi* Murdoch[10], *Fahrang* Radjai[11], *Derek C.* Richardson[12], *Marcos* Salazar[13], *Stephen R.* Schwartz[14], and *Paolo* Tanga[14]

[1] IMCCE, Paris Observatory, PSL Research University, CNRS, Sorbonne Univ., UPMC Paris 06, Univ. Lille, F-75014 Paris, France
[2] Departamento de Física, Ingeniería de Sistemas y Teoría de la Señal, Universidad de Alicante, E-03080 Alicante, Spain
[3] Institute for Physical Science and Technology, and Department of Physics, University of Maryland, College Park, MD 20742, USA
[4] GRASP, Research Unit CESAM, University of Liege, Belgium
[5] CCAR, University of Colorado Boulder, Boulder, CO 80309, USA
[6] Aerospace Engineering Sciences, University of Colorado Boulder, Boulder, CO 80309, USA
[7] Inst. Jean Le Rond d'Alembert, Sorbonne Univ, UPMC Univ Paris 06, CNRS, F-75005 Paris, France
[8] Université de Lyon, Laboratoire de Physique, Ecole Normale Supérieure de Lyon, CNRS, F-69364 Lyon, France
[9] JAXA/Institute of Space and Astronautical Science, 3-1-1 Yoshinodai, Chuo-ku, Sagamihara, Kanagawa 252-5210, Japan
[10] DEOS/SSPA, ISAE-SUPAERO, Université de Toulouse, F-31055 Toulouse, France
[11] LMGC, Université de Montpellier, CNRS, Montpellier, France
[12] Department of Astronomy, University of Maryland, College Park, MD 20742-2421, USA
[13] LICB, Univ. de Bourgogne, CNRS, F21078 Dijon, France
[14] Université Côte d'Azur, OCA, CNRS, IRD, Laboratoire Lagrange, F-06304 Nice, France
[15] Instituto Universitario de Física Aplicada a las Ciencias y la Tecnología, Universidad de Alicante, E-03080 Alicante, Spain

Abstract. Asteroids and other Small Solar System Bodies (SSSBs) are currently of great scientific and even industrial interest. Asteroids exist as the permanent record of the formation of the Solar System and therefore hold many clues to its understanding as a whole, as well as insights into the formation of planetary bodies. Additionally, SSSBs are being investigated in the context of impact risks for the Earth, space situational awareness and their possible industrial exploitation (asteroid mining). In all these aspects, the knowledge of the geophysical characteristics of SSSB surface and internal structure are of great importance. Given their size, constitution, and the evidence that many SSSBs are not simple monoliths, these bodies should be studied and modelled as self-gravitating granular systems in general, or as granular systems in micro-gravity environments in particular contexts. As such, the study of the geophysical characteristics of SSSBs is a multi-disciplinary effort that lies at the crossroads between Granular Mechanics, Celestial Mechanics, Soil Mechanics, Aerospace Engineering and Computer Sciences.

1 Introduction

Long seen as a point-like source in the sky, asteroids – first discovered in the 19th Century – are Solar System bodies named after their star-like appearance. More recent ground-based and space probe observations have revealed that asteroids have to be considered as small worlds on their own, some even forming systems with satellites. Those small Solar System bodies (SSSBs; comets, asteroids, meteoroids) orbiting around the Sun are found at different heliocentric distances ranging from close-by Near Earth Objects, between Mars and Jupiter, and beyond Neptune. The distinction between asteroids and comets, when analysing their physical properties is not very sharp. The transition, referred to as the "asteroid-comet continuum" shows that both have similar constituents, with likely larger proportion of volatiles for icy comets (those that are sublimated during passage at perihelion) than for rocky asteroids with higher concentration of carbon, silicates, and metals (that sometimes show some cometary-like activity). Due to their small sizes, the SSSBs are the pristine remnants of the formation of the Solar System, and hence important tracers of the history of planet formation, the evolution of the planetary system, and possibly for the origin of life on Earth. It is then of prime importance to better understand their constitution, as individual objects, and on a global basis as a system of objects. To this effect different techniques are used to characterise the physical properties of their surface and interior, and how they evolve on astronomical time scales due to dynamical perturbations, collisions, physical process, etc. One of the main features that has been revealed in the last decades is that small bodies can have low bulk densities compatible with "rubble pile" structures or gravitational aggregates [1], with low cohesion, and sustained plastic deformation. These bodies also show a large variety of irregular shape morphology, macro-porosity, surface properties (from smooth regolith covered terrain, to very rough collection of boulders), and topography (cratering,

* Corresponding author: hestro@imcce.fr

DOI: 10.1051/epjconf/201714014011

depressions, ridges) – see Fig. 1. All this has generated many studies and modelling to better understand the macro- and micro-scale structure and behaviour of these bodies, their formation and evolution, etc. In particular, as shown here, some key problems of mechanics and nonlinear dynamics of self-gravitating celestial bodies can best be approached by granular media modelling. A comprehensive review of asteroids and asteroid science was recently completed for the book Asteroids IV [2].

Fig. 1. Collage of images from various space missions to asteroids, comets, and satellites (see Table 1). The largest (21) Lutetia around 100 km diameter dominates its neighbours (top); the smallest (25143) Itokawa around 0.5 km size is enlarged (bottom left); the two Martian moons are added (bottom right). Credits: E. Lakdawalla (the Planetary Society), JAXA, ESA, NASA.

2 Mission and observations

The fundamental idea for why granular mechanics is relevant for the study of SSSBs emerged during space missions to visit these bodies. Theories developed in the seventies and eighties for the properties of SSSBs generally imagined asteroidal bodies to be comprised of large monolithic blocks with little or no surface covering of finer material (regolith). This changed drastically given the initial observations of asteroids Gaspra and Ida by the Galileo spacecraft in the early 1990's, the first direct images of such bodies. Snapshot images taken of these asteroids showed them to be covered with surface regolith, to be heavily cratered, and in the case of Ida to have a binary asteroid companion which itself had a regolith covering. Since these initial observations the space-based and ground-based observations of asteroids have grown, along with the realization that these bodies are fundamentally granular in nature and hence that their physical evolution must be described using principles of granular mechanics applied to these extreme environments.

There are a few key missions and ground-based observations that have shaped our view of these bodies as having a granular nature. Table 1 below presents a comprehensive list of asteroids visited by spacecraft.

Table 1. List of achieved space missions to small bodies.

Target	Mission	Date
Asteroids		
(4179) Toutatis	Chang'E 2	Dec. 2012
(21) Lutetia	Rosetta	July 2010
(2867) Steins	Rosetta	Sep. 2008
(25143) Itokawa	Hayabusa	June 2005
(5535) Annefrank	Stardust	Nov. 2002
(433) Eros	NEAR Shoemaker	Feb. 2000
(9969) Braille	Deep Space 1	Jan. 1999
(253) Mathilde	NEAR Shoemaker	June 1997
(243) Ida	Galileo	Aug. 1993
(951) Gaspra	Galileo	Oct. 1991
Comets		
67P/Churyumov–Gerasimenko.	Rosetta	Aug. 2014
103P/Hartley2	EPOXI	Nov. 2010
Tempel 1	Deep Impact, Stardust	Jul. 2005, 2011
Wild 2	Stardust	Jan. 2004
19P/Borrelly	DS 1	Sep. 2001
Halley	Giotto, Vega 1&2, Sakigake, Suisei, ICE	March 1986,
Giacobini-Zinner	ICE	Sep. 1985
Moons		
M1 Phobos	MEX, ODY, MGS, Phobos2, Viking1	2004, 2001, 1997, 1989, 1976
M2 Deimos	MEX, ODY, MGS, Viking2	2004, 2001, 1997, 1976

NEAR Mission: NASA's NEAR Shoemaker mission visited two asteroids, with a fast flyby of the large asteroid (253) Mathilde en route to a year-long orbital mission about the asteroid (433) Eros [3,4]. The flyby of Mathilde was notable as it was the first direct mass measurement of an asteroid and, when included with an estimate of the total volume of that asteroid, yielded a bulk density of 1300kg/m^3 that was surprisingly low for a C-type asteroid, and consistent with a highly porous structure. This was the first direct evidence of a high porosity within asteroids, consistent with a rubble pile structure. Asteroid Eros was observed at high resolution for an extended period of time, with the mission ending with a controlled landing to the asteroid surface, providing high-resolution imaging of the body. These observations showed the surface to be covered with a thick layering of regolith, and with the crater density of the surface virtually disappearing at small size scales, seemingly obliterated by resurfacing processes that moved the regolith across the surface. This has since been attributed to seismic shaking of the body due to cratering impacts that can resurface the body. Another feature of the surface was the presence of many areas of "ponded" material, consisting of finer regolith being preferentially trapped at the bottom of craters, in some cases partially embedding boulders. The transport of this material is not fully understood, but could be due to electrostatic levitation of dust fines which are then transported laterally over the surface to local geopotential lows. The presence of the fine regolith is attributed to the ability of the large (≈11 km) NEO Eros to retain more of the fine regolith after an impact.

Hayabusa Misson: The Japanese space agency's (JAXA's) Hayabusa mission visited asteroid Itokawa in 2005. The mission mostly focused on a 3-month period of characterization and sampling attempts. Due to its much smaller size, (≈0.5 km), not as much fine regolith is retained on this body, providing a clear view of its fundamentally rocky and block-covered surface. Despite this, there are specific regions where finer regolith has been seen to flow across the surface into the geopotential lows of the body. The observed boulders on the surface of Itokawa have been documented and determined to follow a roughly $1/d^3$ size distribution, which corresponds to an equal mass distribution for every size, with the largest boulder being on the order of a few decameters. Also observed were larger boulders of indeterminate length thrusting out of the interior of the asteroid. The asteroid probe also collected a small sample, which was seen to consist of a particle size distribution $1/d^x$ with x between 2 and 2.7. The body was also determined to have a high porosity of approximately 40% [5,6].

Ground-Based Observations: Evidence for the granular nature of small asteroids has also come from ground-based telescopic observations of these bodies. A key, and early, indication came from the statistics of asteroid spin rates vs. size, which showed that at small sizes the spin rates of asteroids is higher, indicating that these may be components of larger asteroids that could not spin as fast without falling apart. Also, binary asteroids which provide insight into the masses of these systems generally confirm the high porosity seen from space missions. Another source of insight comes from the shape models developed using range-Doppler imaging of asteroids. The accuracy of these models is sufficient to develop high resolution, non-convex shape models that provide insight into the surface slope distributions of these bodies. It has generically been found that these bodies all have the vast majority of their surface slopes below 35°, the usual angle of repose for granular materials, indicating surfaces that are likely covered with some sort of granular material. In at least one case however — for the asteroid 1950 DA — this is not true, the mass determination of this body indicates that it is highly porous and fast spinning, implying that this body has some level of cohesive strength. Other lines of evidence discovered recently include asteroid pairs and their link to rotational fission models and active asteroids which are seen to separate into multiple components.

Current Questions and Hypotheses: These recent observations have strengthened the concept of granular mechanics being important for the physical evolution and properties of these asteroids. Most asteroids smaller than 100 km are likely gravitational aggregates [7], which implies that they can be considered as large-scale granular structures, where grains are components of tens to hundreds of meters in size. The same applies to surfaces with grains of smaller sizes. Based on these ideas there have been a number of hypotheses and models for the behaviour of asteroids motivated by granular mechanics. First and foremost, the "Brazil nut effect" is invoked to hypothesise the percolation of larger boulders and grains to the asteroid surface. Other theories include a cohesion theory that posits the presence of cohesive strength in rubble pile asteroids arising from weak van der Waals forces between the finest components of asteroid regolith. Predictions from that theory have been seen to be consistent with constraints on cohesive strength within asteroids based on ground observations. These theories have, in turn, been tied to a range of possible models of internal porosity distribution within asteroids that lead to morphologies of different strength: depending on whether the interior of a gravitational aggregate has additional strength or has less strength (due to lower, or higher porosity), different failure patterns are expected to occur when the body is spun to high spin rates.

3 Numerical simulations

There are three main numerical methods that have been applied to the study of asteroid surfaces and interiors [8, and ref. therein]: Soft-sphere Discrete Element Method (SSDEM) [e.g. 9,10], Hard-sphere Discrete Element Method (HSDEM) [e.g. 11], and Finite Element Method (FEM) [e.g. 12].

FEM simulation codes, together with Soil Mechanics theory, model asteroids as a continuum in which the sizes of the constituent particles are much smaller than the characteristic length of the system. Here the geophysical characteristics are an input to the codes and system behaviour is most affected by the choice of yield criterion (Mohr-Coulomb, Drucker-Prager, etc.). Discrete methods model asteroids as a collection of self-gravitating particles (generally spherical) that conserve some of the geophysical characteristics of the bodies to be studied (angle of friction, cohesive strength, bulk density, etc.). These characteristics are the result of the interaction of individual grains and are, therefore, an output of the simulations that correspond to the material parameters of the grain-grain interaction models. Soft-sphere DEM codes integrate the equations of motion of every grain and allow them to overlap; it is this overlap that provides a repulsive force during collisions and sustained contact. Hard-sphere HSDEM avoids this overlap by transferring momentum in every collision and assuming it to be instantaneous. This makes it much faster than SSDEM, applicable to very kinetically active systems but not applicable to systems where sustained contacts are necessary. On the other hand, FEM codes are less computationally intensive, and can easily deal with realistic asteroid shapes and mass, but they can only show a static picture of the body, where the behaviour of the aggregate is assumed to follow a yield criterion. DEM simulations are far more computationally intensive, but have the advantage that they show a dynamical evolution and the material parameters are generally known and implemented.

In addition to topics common to granular mechanics research such as particle size segregation, stress states, and penetrometry, the granular physics of asteroids raises several unique challenges including: rotation induced deformation and disruption, collisional evolution, interior heterogeneities, gravitational tide induced deformation, interior heterogeneities, cratering, and sampling mechanisms, among the most important [12,13].

4 Laboratory experiments

To address the specific context of planetary science, the classical laboratory experiments of granular mechanics need to be adapted to micro-gravity or low-gravity and vacuum space. Thus, tests are generally performed in drop towers, in parabolic or ballistic flights, or on the international space station (see chart below). Unlike drop-towers or sounding rockets, parabolic flights (PF) enable the presence of scientists operating the experiment. The latter is of particular interest for studies requiring human interactions and manipulations. PF are adapted to short-duration (~20s) scientific and technological investigations in low-gravity. An example for PF experimentation in the field of granular matter is given by ESA's VIP-Gran instrument, which is designed for the study of the dynamics and the statistical mechanics of an ensemble of particles that interact through dissipative collisions. During ESA's PFC65 flight campaign, the segregation of driven binary granular media through vibrations was investigated within the frame of VIP-Gran. For specific filling conditions, different dynamical regimes as well as segregation of both species were reported [14, 15]. In another set of PF experiments, it was demonstrated that convective-like flows in a granular material are close to zero in microgravity [16]. As it has been observed, one of the main issues with the previous experimental methodologies is their duration. In order to avoid this, the Strata-1 experiment is being carried out for 1 year in the International Space Station [17]. This experiment consists of 4 tubes that contain a different granular material of a known size distribution in each one. The objective is to observe how the grains mix and segregate in an extremely low-gravity environment under gentle vibration during an extended period of time.

5 Future prospect and opportunities

Currently on the horizon are two space missions to small, primitive asteroids believed to have a granular "rubble pile" structure. These are the Japanese Hayabusa-2 mission to asteroid Ryugu and NASA's OSIRIS-REx mission to asteroid Bennu. Both of these missions have been launched and are en route to their targets. Hayabusa2 will arrive in the summer of 2018 and OSIRIS-REx in the late fall of 2018. Both will spend at least one year at their respective target bodies before coming back to Earth with surface samples. Data from both missions, numerical simulations, and laboratory experiments can be used to further probe the granular nature of their target bodies, and to further deepen the ties between granular mechanics and asteroid geophysics.

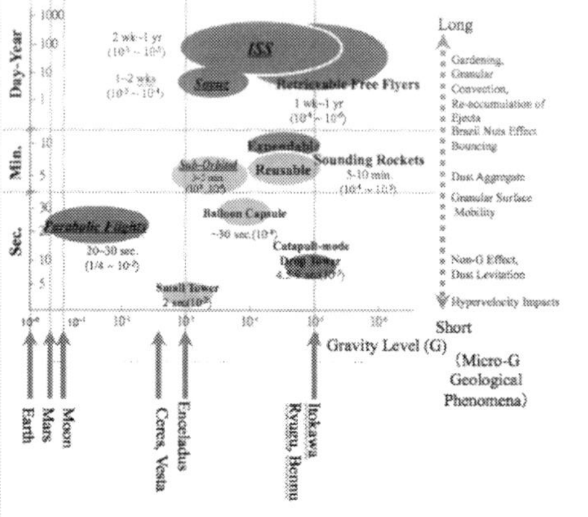

Fig. 2. Summary of the micro-gravity experiment facilities.

The authors acknowledge support by the International Space Science Institute (ISSI, Bern, Switzerland) and Paris observatory (France) to the international team « Asteroids & Self-Gravitating Bodies as Granular Systems[a] ». EO acknowledges European Space Agency program SpaceGrains and Prodex (Belspo, Brussels, Belgium) for financial support.

References

1. D.C. Richardson et al. In *Asteroids III*, p. 501 (2002)
2. *Asteroids IV*. Eds. P. Michel, F.E. DeMeo, W.F. Bottke. 785 pp. The Univ. of Arizona Press (2015)
3. A.F. Cheng. AdSR **33**, 1558 (2004)
4. J. Veverka et al. Science **278**, 2109 (1999)
5. A. Fujiwara et al. Science **312**, 1330 (2006)
6. H. Yano et al. Science **312**, 1350 (2006)
7. A. Campo Bagatin et al. Icarus **149**, 198 (2001)
8. N. Murdoch et al. In *Asteroids IV*, p. 767 (2015)
9. C. Maurel et al. MNRAS **464**, 2866 (2017)
10. P. Sánchez & D.J. Scheeres. ApJ **727**, 120 (2011)
11. D.C. Richardson et al. Icarus **143**, 45 (2000)
12. M. Hirabayashi & D.J Scheeres. ApJ **798**, L8 (2015)
13. D.J. Scheeres et al. In *Asteroids IV*, p. 745 (2015)
14. E. Opsomer, et al. Eur. Phys. J. E **37**, 115 (2014)
15. More information on http://spacegrains.org
16. N. Murdoch et al. 2013. PRL **110**, 018307 (2013)
17. M. Fries et al. LPIContrib. **1921**, 6547 (2016)

[a] ISSI team http://www.issibern.ch/teams/granularsystems/

EPJ Web of Conferences **140**, 03086 (2017) DOI: 10.1051/epjconf/201714003086
Powders & Grains 2017

Jamming/flowing transition of non-Brownian particles suspended in a iso-density fluid flowing in a 2D rectangular duct

Maxym Burel[1], *Sylvain* Martin[1], and *Olivier* Bonnefoy[1,*]

[1] *Ecole des Mines, SPIN, UMR CNRS LGF-5703, Saint-Etienne, F-42023 France*

Abstract. We present the results of an experimental study on the jamming/flowing transition. A suspension of neutrally buoyant large particles flows in an horizontal rectangular duct, where an artificial restriction triggers jamming. We show that the avalanche distribution size is exponential, that is memoryless. We further demonstrate that the avalanche size diverges when the restriction size approaches a critical value and that this divergence is well described by a power law. The parameters (critical opening size and divergence velocity) are compared to literature values and show a strong similarity with others systems. Another result of this paper is the study of the influence of the particle morphology. We show that, for a moderate restriction size, the dead-zone formed right upstream of the restriction is larger for angular particles but, paradoxically, that the avalanche size is larger for polyhedra compared to spheres by at least one order of magnitude.

1 Introduction

1.1 Context and objectives

Suspensions of solid particles in a fluid are of interest in many practical situations encountered in mining, polymer and oil/gas industries, just to name a few. The different models take into account a variety of interesting and challenging physical phenomena such as sedimentation/flotation, agglomeration, shear-thickening or shear-thinning. Most of these models rely on the assumption that the suspension is homogeneous (particles size very small compared to the flow length scales). In some contexts, however, solid particles may have a large diameter compared to the flow scales. For example, in the petroleum extraction industry, rock fragments suspended in oil may be, even individually, larger than the sediment pore throat. In the chemical industry, sticky polymer particles suspended in a liquid can form large agglomerates that are only one or two orders of magnitude smaller than the diameter of the transfer pipe. In this kind of situations, one may observe a clogging leading to a permeability decrease, a pressure loss increase or a pipe blockage. To reduce or mitigate the subsequent technical damages (pipeline erosion or failure), safety issues (fluid leakage) and economic losses (non-production costs), it is necessary to improve the understanding of such physical systems.

1.2 State-of-the-Art

In this section, we present a short survey of the literature dedicated to the transport of dense suspensions with a special care on the jamming event, which is the transition between the flowing and jammed states. More specifically, we will focus on flows involving 'large' particles, that is (1) the confinement effect is important: a discrete description prevails over a continuous description because the particle size is not much smaller than the system scale, and (2) the suspension is non-Brownian: the influence of the thermal fluctuations on the particles trajectory is negligible. As a consequence, particles are essentially driven by gravity, viscous drag and contact forces (particle-particle and particle-wall).

The jamming transition has been studied in several experimental conditions. The granular flows can be driven by viscous drag or gravity, in a cylindrical or rectangular pipe. Usually a restriction is located at the end (horizontal layout) or the bottom (vertical layout). Most authors count the number of grains flowing through the orifice between two jammed states (an avalanche) and perform a statistical analysis.

For gravity-driven flows of dry granular media, To [1] used a random walk model to predict the jamming probability. The model showed a good agreement with his experimental data. Zuriguel [2] and Garcimartin [3] results showed the independence between consecutive runs and found a geometric distribution for the jamming probability. For the same system, Janda [4] found that there is no critical size of restriction/orifice, which leads to the absence of jamming or, in other words, that every system can jam, whatever the restriction size. Gravity-driven flows of granular material immerged in a liquid have been tested by Wilson [5]. In contrast with dry granular material, he showed that the flowrate of the particles flowing through

*e-mail: olivier.bonnefoy@emse.fr

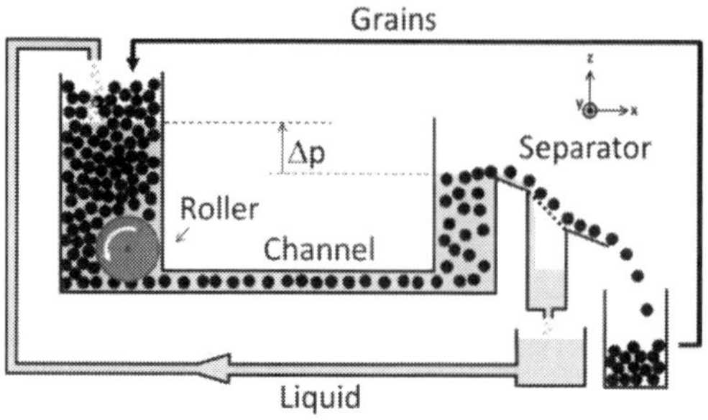

Figure 1. Side view of the experimental device.

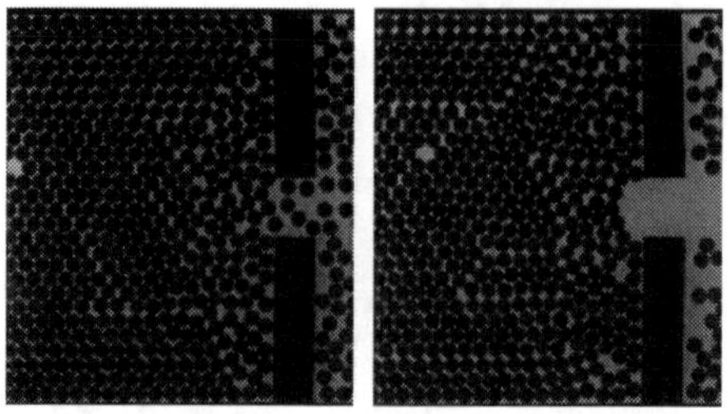

Figure 2. Snapshots of flowing vs. jammed suspension.

the orifice increases with the granular packing height. The underlying mechanism is not completely elucidated yet.

Viscous drag-driven particles flows have been investigated by Guariguata [6] and Lafond [7]. Guariguata studied the stationary flow of floating disks in an open channel and showed that the Beverloo law [8] is observed, and that the fluid velocity does not affect significantly the jamming probability. Lafond [7] used immerged polydisperse particles in an open channel and established a correlation between the jamming rate and the liquid flow rate.

For non-Brownian coarse particles suspensions, Vitthal and Sharma [9] identified two jamming mechanisms: the sequential bridging and the multi-bridging.

In this study, we explore the behavior of a suspension of completely immerged particles subjected to a liquid viscous drag, without buoyancy effects, in a horizontal quasi-2D configuration. To our knowledge, this layout has not been studied before. More precisely, we want to address the following questions: (i) what is the influence of the liquid flowrate on the avalanche size ? (ii) is there a critical restriction size above which the system never jams ? and (iii) how is the jamming probability affected by non-spherical (non-cylindrical) particle morphology ?

2 Material and Method

2.1 Experimental set-up

The experimental setup is mainly composed of a feeder, a rectangular duct and a separator (figures 1 and 2). The **feeder** is a vertical recipient made of PMMA that contains particles. The porous space is filled with the suspending liquid in the lower half and with air in the upper half. A soft roller located at the bottom of the feeder entrains the particles into the horizontal rectangular duct, at a delivery rate controlled by its rotational speed. It is designed in order to distribute the particles evenly along the transverse y direction. The **rectangular duct** (length L = 1200 mm and width W = 140 mm is also made of transparent PMMA. The height H is chosen equal to 1.25 times the particle diameter such that only one layer of particles is allowed in the duct (quasi-2D). The channel cross-section area is $\Sigma_c \equiv W.H$ = 10.5 cm^2. The suspension is forced to flow through a restriction of width w. The suspension flow is recorded by a camera at 200 fps with a resolution of 15 pixels by particle diameter. The **separator** is essentially composed of a sieve that removes the particles from the suspending liquid. The system operates in stationary conditions thanks to a pump.

2.2 Particles and liquid

The experiments are performed with a suspension of polyethylene particles. We use two sets of particles: spheres of 5.92 ± 0.02 mm diameter and polyhedral particles. The morphology is quantified by the departure to sphericity, as measured by the aspect ratio, equal to 1 for spheres and 0.75 ± 0.05 for polyhedra. In order to cancel buoyancy, we use a mixture of water and glycerol with the same density as solid particles (1050 ± 1 kg.m^{-3}). The fluid mixture is Newtonian, with a dynamic viscosity of 1.68 mPa.s. For all our experimental conditions, the suspension is non-Brownian (Peclet > 10^{12}) and discontinuous (Knudsen between 0.05 and 0.5).

2.3 Operating protocol

Initially, the feeder is filled with particles and liquid in its lower half, and the horizontal channel is filled with liquid at rest. A typical experimental run is composed of the following steps: (1) The roller is turned on at 200 rpm. This creates a dense 2D granular packing that is pushed towards the restriction until the system jams (arch formation). Then, the roller is stopped. (2) The peristaltic pump is turned on, with a rotational speed chosen to impose a given flowrate. Around 5 seconds are needed for the system to be stationary. (3) The horizontal duct is gently tapped to break the arch and start the discharge flow. (4) After some time, the system naturally jams again by forming a stable arch of particles. (5) The two previous steps are repeated 350 times to give robust statistics.

3 Results

Fig. 2 shows two typical top views. On the left, the particles are flowing through the restriction entrained by the viscous fluid. On the right, the system is jammed and a stable arch is clearly visible at the restriction level. It is worth noting that the liquid flowing through the restriction exerts

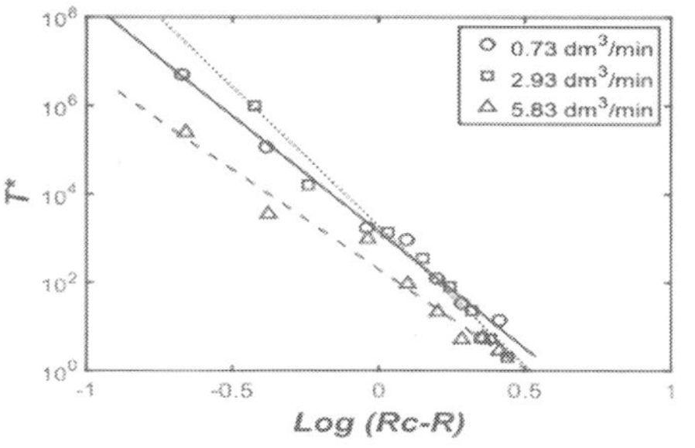

Figure 3. Power law divergence of avalanche size (spheres).

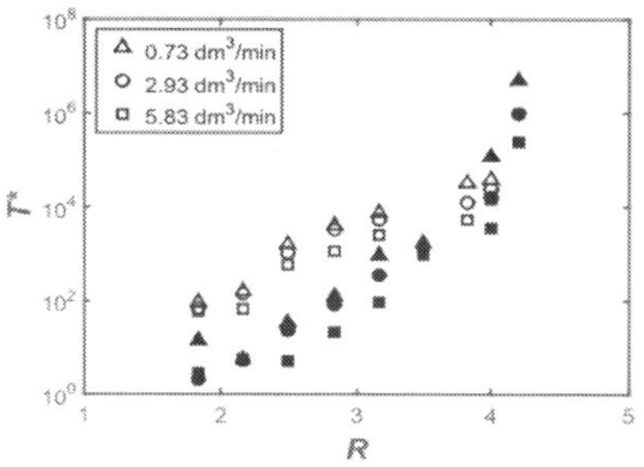

Figure 4. Avalanche size for spheres (filled) and polyhedra (empty).

a compressive stress on the granular assembly. Because of the high confinement ($W/d_P \approx 23$) and the monodispersity of the PSD, this leads to a strong ordering, very much like crystallization. Although it makes the interpretation of the results more complicated, this feature would have more degrees of freedom and the granular phase would experience the jamming later, on a statistically smoother basis. In this section, we study the influence of the restriction size, the liquid flow rate and the particle morphology on the avalanche size. We define the dimensionless opening ratio R as the restriction width w divided by the particle diameter d_P. It ranges between 1.83 and 4.2. The restriction cross-section area is labeled $\Sigma_r \equiv w.H = R.d_P.H$. Three flowrates Q_f are investigated: 0.73, 2.93 and 5.86 $dm^3.min^{-1}$. The particle morphology is characterized by the aspect ratio, which is 1 for spheres and 0.75 ± 0.05 for polyhedra. When the system is jammed, it forms a porous medium and the flow regime can be characterized by the Reynolds number $Re \equiv \rho_f.(Q_f/\Sigma).d_P/\mu_f$. In the channel upstream of the restriction, we have $\Sigma = \Sigma_c$ and the Reynolds number is equal to 70, 279 and 558 for the three different flow rates respectively. When approaching the arch, the fluid velocity increases and thus the Reynolds number. As an example, for the average values of $Q_f = 2.93$ $dm^3.min^{-1}$ and $R = 3$, its value is 2170 at the arch level ($\Sigma = \Sigma_r$). During an avalanche, the particles are entrained by the fluid and the slip velocity is reduced, presumably by a factor of the order of 1 to 10. All considered, our experimental conditions are characterized by a Reynolds number ranging between a few tens and a few hundred. According to Clift and Weber's terminology [10], it seems therefore reasonable to depicts the flow as weakly turbulent.

Experimental results show that the density probability decays exponentially with the avalanche size *i.e.*, the probability to have an avalanche of size T is proportional to $exp(-T/T^*)$, where T^* is a characteristic avalanche size. This is the signature of a Poisson process, characterized by the absence of memory effect. In our case, according to [9], the prevailing mechanism is the multi-bridging: when the local concentration is high enough right upstream the orifice, particles may try to simultaneously flow and instantaneously form a bridge. A strong analogy appears with the self-filtration phenomenon reported by Haw [11] for colloidal particles. Indeed, since the granular phase experiences more difficulties (collision, friction) to flow through the restriction than the liquid phase (which flowrate is imposed), the solid phase concentration is reduced upon crossing the restriction. This effect is obviously more pronounced in our experiment, where the dynamic arches are composed of a maximum of 5 to 10 grains. It would certainly deserve a more quantitative analysis. Whatever the liquid flowrate and the particle shape, experiments show that the characteristic avalanche size T^* gets very large when R is approaching a critical dimensionless opening ratio R_c. This divergence is also observed in classical thermodynamic phase transitions and thus suggests that it can be represented by a power law:

Table 1. Divergence parameters.

	Spheres			Polyhedra		
Q_f	A	R_c	γ	A	R_c	γ
0.73	1.4×10^3	4.4	5.3	1.4×10^7	5.9	7.9
2.93	1.9×10^3	4.6	6.3	1.9×10^5	5.6	6.3
5.86	2.0×10^2	4.4	4.5	2.0×10^6	5.9	6.8

$$T^* = A.(R_c - R)^{-\gamma} \quad (1)$$

where A is a constant and γ an exponent characterizing the divergence speed. To determine these parameters, the least squares method is used. Fig. 3 confirms that this law interpolates the experimental results with a very good agreement for spheres. The agreement is good for polyhedra as well, though a little bit more scattered. The results are summarized in Tab. 1, which gives the values of A, R_c and γ for the three flow rates Q_f investigated in our work.

4 Discussion

Tab. 2 compares the divergence velocity γ and the critical ratio R_c obtained in our work to the values given by Garcimartin [12], Thomas [13] and Lafond [14]. Whatever the system dimension (2D or 3D), the particle morphology (spheres or polyhedra), the driving force (viscous drag or gravity), the suspending fluid (gas or liquid), and the flow

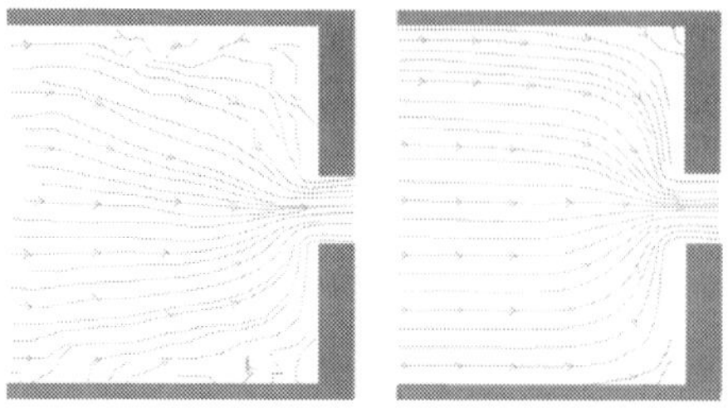

Figure 5. Particles streamlines for polyhedra (left) and spheres (right). Q_f =2.93 $dm^3.min^{-1}$ and R =3.83.

Table 2. Comparison with the literature.

Authors	Particles	Fluid	Geometry	R_c	γ
This work	PE spheres 5.92 mm	Water Glycerol	≈ 2D Hor Visc. driven	4.5	5.35
This work	PE polyhedra 5-7 mm	Water Glycerol	≈ 2D Hor Visc. driven	5.8	7
Lafond	PS spheres 10-16 mm	Water Nacl	3D Horiz Visc. driven	3.6	
Garci--martin	Glass spheres 0.42–3 mm	Air	2D Vertic Grav. driven	4.9	6.90
Thomas	Glass spheres 2 mm	Air	3D Vertic Grav. driven	4.7	5

direction (vertical or horizontal), we observe that the critical opening ratio R_c is 4.7±1.2 and the divergence velocity γ is 6.0±1.0. The relatively small dispersion of these parameters suggests that the mechanisms underlying the flowing/jamming transition share some important properties, that still need to be resolved. Despite the similarities, small but significant differences can be highlighted. Here, we restrain our discussion to the influence of the particle morphology for moderate opening ratios. The velocity fields of the solid phase exhibit two regions: a central flow region (particle velocity up to 18 $cm.s^{-1}$) and a lateral dead zone, where particles are at rest (velocity below 2 $mm.s^{-1}$). For spheres, the dead zone is limited to a small region upstream of the restriction. In contrast, for polyhedra, the dead zone extends further upstream and laterally. Fig. 4 compares the characteristic avalanche size for spheres and polyhedra. Experimental results clearly demonstrate that, for opening ratios R below 3.5, avalanches of polyhedra are larger than avalanches of spheres by at least one order of magnitude !

We propose the following explanation for this apparent paradox. At the particle scale, the angular shape of polyhedra prevents them from rolling around each other. This strong reduction of the rotational degree of freedom leads polyhedra to form larger dead zones than spheres. At the system scale, the size and shape of the dead zones dictates the size and shape of the zone, where unjammed particles keep flowing. More precisely, as suggested by Fig.5, a large dead zone is associated with a gently converging flow, whereas a vanishingly small dead zone is associated with an abrupt change of the streamlines direction when approaching the restriction. In the second configuration, the probability of head-on collisions is thought to be significantly increased and explain the greater jamming probability. To test this hypothesis, we calculate the fraction of particles converging to the restriction with an angle θ to the main flow direction. As expected, Fig.6 shows that the frequency of a head-on collision is larger for spheres (abruptly converging flow) than for polyhedra (smoothly converging flow). As a perspective, it would be interesting to use liquids with different viscosities in order to investigate the complex role of the fluid, which acts as a driving force but also as a lubricant through the film drainage.

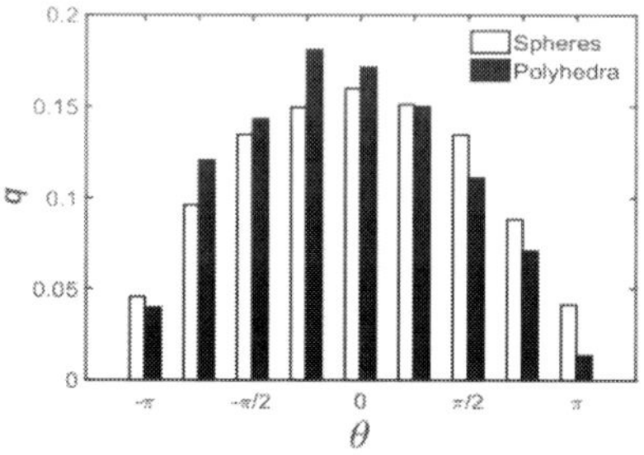

Figure 6. Normalized particle flow rate converging with angle θ towards the restriction center.

References

[1] K. To, Modern Physics Letters B **19**, 1751 (2005)
[2] I. Zuriguel, A. Garcimartín, D. Maza, L.A. Pugnaloni, J. Pastor, Phys Rev E **71**, 051303 (2005)
[3] A. Garcimartín, I. Zuriguel, D. Maza, J. Pastor, L. Pugnaloni, Experimental Chaos **742**, 279 (2004)
[4] A. Janda, I. Zuriguel, L. Pugnaloni, D. Maza et al., EPL **84**, 44002 (2008)
[5] T. Wilson, C. Pfeifer, N. Mesyngier, D. Durian, Papers in physics **6**, 060009 (2014)
[6] A. Guariguata, M.A. Pascall, M.W. Gilmer, A.K. Sum, E.D. Sloan, C.A. Koh, D.T. Wu, Phys Rev E **86**, 061311 (2012)
[7] P.G. Lafond, M.W. Gilmer, C.A. Koh, E.D. Sloan, D.T. Wu, A.K. Sum, Phys Rev E **87**, 042204 (2013)
[8] W. Beverloo, H. Leniger, J. Van de Velde, Chemical engineering science **15**, 260 (1961)
[9] M.M. Sharma, H. Chamoun, D.S.R. Sarma, R.S. Schechter, J. of colloid and interface science **149**, 121 (1992)
[10] R. Clift, J. Grace, M. Weber, Academic, New York pp. 171–202 (1978)
[11] M. Haw, Physical review letters **92**, 185506 (2004)
[12] A. Garcimartín, C. Mankoc, A. Janda, R. Arévalo, J.M. Pastor, I. Zuriguel, D. Maza, in *Traffic and Granular Flow'07* (Springer, 2009), pp. 471–486
[13] C. Thomas, D.J. Durian, PhysRev E **87**, 052201 (2013)
[14] P.G. Lafond, Ph.D. thesis, Colorado School of Mines. Arthur Lakes Library (2007)

DOI: 10.1051/epjconf/201714011015

A non-local plasticity theory for slow granular flows

Prabhu R. Nott[*]

Department of Chemical Engineering, Indian Institute of Science, Bangalore 560012, India

Abstract. Motivated by experimental observations of apparent non-locality in the mechanical response of granular materials, a few recent studies have proposed non-local rheological models. Though their predictions agree with some aspects of experimental observations, the tenets of the models are not experimentally verified, nor do they have an experimentally verifiable or plausible mechanical basis. Here we present a non-local model that is based on the simple idea of volume averaging, and show that it overcomes many of the shortcomings of the previously proposed models, and is superior in its predictive capabilities.

1 Introduction

A few recent studies on the mechanics of dense granular materials have shown evidence of non-locality in the mechanical response, wherein the motion of an intruder is seemingly aided by shearing the granular material far from it [1,2]. This behaviour is not explained by classical rate-independent plasticity theories; indeed, these theories have the more serious failings of being unable to predict the kinematics of simple shear flows, and being unable to account for the universally observed feature of deformation-induced dilation at constant pressure. Non-local theories have been proposed in the past, but they are of largely phenomenological origin [3,4], or based on the introduction of an additional field variable obeying a conservation equation whose mechanical origin is debatable [5]. In this work, we present a non-local, rate-independent plasticity theory whose mechanical origin is easy to comprehend, and involves no additional field variables – it captures rather simply the physical picture of plastic events in a spatial point influencing its neighbourhood. Moreover, it is based on the widely accepted method of coarse-graining that is used in multi-phase flows, and for complex fluids with internal structure. The theory predicts the kinematics of simple shear flows, such as the exponentially decaying (or saturating) velocity profile in simple shear, and shear-induced dilatancy. Finally, we show that our non-local theory plasticity theory is Hadamard well posed, a significant improvement over the previous theories.

2 Local plasticity theory

Before describing non-local plasticity models, it is useful to very briefly outline local critical state plasticity theory. The theory provides a relation for the stress when the material is in a state of plastic deformation. The two elements of the theory are the yield condition,

$$F(\boldsymbol{\sigma},v) \equiv \tau(\boldsymbol{\sigma}') - p_c(v)\sin\phi\, Y(\alpha,v) = 0, \qquad (1)$$

where $\alpha \equiv p/p_c(v)$, and the flow rule,

$$D_{ij} = \lambda\, \partial G/\partial\sigma_{ji}. \qquad (2)$$

In the above, the yield function F and the plastic potential G are scalar functions of the Cauchy stress $\boldsymbol{\sigma}$, Y is a function of the pressure p and the solids fraction v, τ is a scalar function of the deviator stress $\boldsymbol{\sigma}' \equiv \boldsymbol{\sigma} - p\boldsymbol{I}$, $\boldsymbol{D}$ is the rate of deformation tensor, and λ is fluidity variable. The quantity $p_c(v)$ represents the pressure at the critical state, a state of isochoric deformation.

A simple example of classical plasticity is the extended von Mises yield condition [6,7]

$$\tau = (\sigma_{ij}\,\sigma_{ji})^{1/2}, \quad Y = (n\alpha - (n-1)\alpha^{n/(n-1)}), \qquad (3)$$

and the associated flow rule $G = F$, for which (1) and (2) may be solved to yield the following form for the stress [7],

$$\boldsymbol{\sigma} \equiv p_c(v)(1 + \mu_b \boldsymbol{\nabla}.\mathbf{u}/\lambda)^{n-1} + 2\mu Y(\alpha,v) D_{ij}/\lambda, \qquad (4)$$

where $\lambda = -\gamma \equiv (D'_{ij} D'_{ji})^{1/2}$ is a scalar measure of the rate of deformation, n is a constant such that $n > 1$, $n-1 \ll 1$, and $\mu_b \equiv 1/(n\sin\phi)$, $\mu \equiv \sin\phi$ are the bulk and shear plastic moduli.

As already stated, classical plasticity has three important deficiencies [4,8]: (i) it does not predict the velocity field in simple shear flows, (ii) it does not capture dilation in in simple shear flows, and (iii) it is Hadamard ill-posed.

[*] Corresponding author: prnott@chemeng.iisc.ernet.in

3 Previous non-local plasticity models

Mohan et al. [4] proposed the Cosserat plasticity model to address some of the deficiencies of classical plasticity. The tenets of a Cosserat continuum are that the Cauchy stress is not necessarily symmetric, and the balance of angular momentum must be enforced. The theory introduces the additional kinematic variable of the mean particle spin $\boldsymbol{\omega}$ (which in a classical continuum is equal to the vorticity $\mathbf{w}$), and the additional dynamic variable $\mathbf{M}$, called the couple stress tensor, which is simply the angular momentum flux. With the following simple modifications of the extended von Mises yield condition (3) and the flow rule (2)

$$\tau = (a_1\sigma_{ij}\,\sigma_{ji} + a_2\sigma_{ij}\,\sigma_{i} + l^2 M_{ij}\,M_{ji})^{1/2}, \tag{5}$$

$$D_{ij} + \epsilon_{ijk}(\omega_k - w_k) = \lambda\,\partial F/\partial\sigma_{ji},\;\; \partial\omega_i/\partial x_j = \lambda\,\partial F/\partial M_{ji}, \tag{6}$$

they showed that the model captures several observed features of the kinematics within shear layers, including the exponential decay (or saturation) of the velocity within the shear layer in simple shear, the weak dependence of the shear layer thickness on the system size, and the deviation between $\boldsymbol{\omega}$ and $\mathbf{w}$. These features are a consequence of the mesoscopic length scale l that is present in the model. The drawback of the model is that experiments have thus far not been able to measure the couple stress, or confirm the asymmetry of the Cauchy stress, which are important elements of the model.

More recently, another non-local theory has been proposed [5], wherein the central tenet is that the constitutive relation (4) must be supplemented by a Helmholtz equation for the fluidity,

$$l^2\nabla^2\lambda = (\lambda - \lambda_{loc}), \tag{7}$$

where $\lambda_{loc} \equiv -\gamma$ is the "local" fluidity. This model too incorporates a mesoscopic length scale l, and has been shown to capture the kinematics in simple shear, and its predictions are in consonance with experimental observations of more complex flows. The drawbacks of this model are: (i) the basis for (7) is tenuous, and uses a boundary condition for λ that is unconvincing, (ii) the model does not capture the important phenomenon of dilatancy, and (iii) it does not escape the Hadamard ill-posedness of classical pasticity.

4 A new non-local model

We propose a model that repairs the aforementioned drawbacks of the previously proposed non-local models. The basis of the model lies in volume averaging the flow rule, an idea that follows from the intuitive expectation that plastic deformation at a point must be influenced by the stress state not simply at that point alone, but also in its neighbourhood (Fig. 1). This intuitive idea is in agreement with the now widely recognized fact that stress in granular materials is transmitted along quasi-linear clusters of grains called force chains – hence plastic deformation at one point can induce deformation at a distance from it due to reformation of the chains. With this in mind, we propose that the flow rule (2) (with $G = F$), be replaced by

$$D_{ij} = \int_y g(\boldsymbol{x}\text{-}\boldsymbol{y})\,[\lambda\,\partial F/\partial\sigma_{ji}]\,\mathrm{d}\boldsymbol{y}, \tag{8}$$

where $g(\boldsymbol{x}\text{-}\boldsymbol{y})$ is a weight function satisfying the properties

$$\int_y g(\boldsymbol{x}\text{-}\boldsymbol{y})\,\mathrm{d}\mathbf{y} = 1,\;\; \int_y g(\boldsymbol{x}\text{-}\boldsymbol{y})\,|\boldsymbol{x}\text{-}\boldsymbol{y}|^2\,\mathrm{d}\mathbf{y} = l^2, \tag{9}$$

where l is the radius of the averaging volume. Using the extended von Mises yield condition (3) for F in (8), expanding the term within square brackets in a Taylor series about $\boldsymbol{y}=\boldsymbol{x}$, and solving for $\boldsymbol{\sigma}$ as a perturbation series expansion in l^2, we get

$$\sigma'_{ij} = p_c(\nu)\sin\phi/\lambda(D_{ij} - l^2\nabla^2 D_{ij}). \tag{10}$$

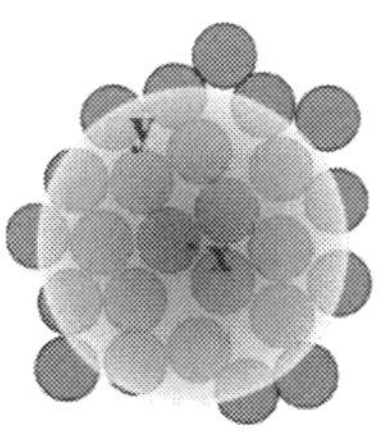

Fig. 1: Schematic diagram of the volume averaging of the flow rule about the point $\boldsymbol{x}$.

Substituting (10) in the yield condition, we get the relation for the fluidity

$$\lambda = -\gamma - l^2\nabla^2\gamma \tag{11}$$

Note the similarity between (11) and (7), and also the important difference that (11) is explicit in λ – it does not require a boundary condition. Moreover, the volume averaging process provides a clearer and physically acceptable process by which the mesoscopic length scale l is introduced into the constitutive equation for the stress. Most importantly, non-locality is reflected not just in λ, but also in the higher gradients in $\boldsymbol{D}$ in the relation for $\boldsymbol{\sigma}$. We note that the local flow rule is based on maximization of the plastic work – our proposed non-local flow rule is a simple extension, in the sense that the plastic work is maximized over a volume of dimension l, but not necessarily at every point.

It is straightforward to show that the model constitutive (10)-(11) captures all the kinematic features of slow granular flows that are captured by the previous non-local models, described in section 2. However, it has the following advantages over the previous models: (i) it does not require the solution of a separate differential equation for the fluidity λ, therefore circumventing the requirement for boundary conditions, (ii) it brings in non-locality in an intuitively acceptable manner, that is in consonance with experimental observations [1], and (iii) the model is unconditionally well posed in the sense of Hadamard.

Funding from the Science and Engineering Research Board, India is gratefully acknowledged.

References

1. K. A. Reddy, Y. Forterre, O. Pouliquen, Phys. Rev. Lett. **106**, 108301 (2011).
2. K. Nichol, A. Zanin, R. Bastien, E. Wandersman, M. van Hecke, Phys. Rev. Lett. **104**, 078302 (2010).
3. I. Vardoulakis, E. C. Aifantis, Acta Mech. **87**, 197–217 (1991).
4. L. S. Mohan, K. K. Rao, and P. R. Nott, J. Fluid Mech. **457**, 377–409 (2002).
5. K. Kamrin, G. Koval, Phys. Rev. Lett. **108**, 178301 (2012).
6. K. K. Rao, P. R. Nott, *An Introduction to Granular Flow*n (Cambridge University Press, New York, 2008).
7. P. R. Nott, Acta Mech. **205**, 151–160 (2009).
8. T. Barker, D. G. Schaeffer, P. Bohorquez, J. M. N. T. Gray, J. Fluid Mech. **779**, 794–818 (2015).

Gravity-driven granular flow in a silo: Characterizing local forces and rearrangements

Emma Thackray[1,*] and *Kerstin* Nordstrom[1,**]

[1] *Mount Holyoke College*

Abstract. While the gravity-driven flow of a granular material in a silo geometry can be modeled by the Beverloo equation, the mesoscale-level particle rearrangements and interactions that drive this flow are not well-understood. We have constructed a quasi-two-dimensional system of bidisperse, millimeter-scale disks with photoelastic properties that make force networks within the material visible. The system is contained in an acrylic box with an adjustable bottom opening. We can approach the clogging transition by adjusting this opening. By placing the system between cross-polarizers, we can obtain high-speed video of this system during flow, and extract intensity signals that can be used to identify and quantify localized, otherwise indeterminate forces. We can simultaneously track individual particle motions, which can be used to identify shear transformation zones in the system. In this paper, we present our results thus far.

1 Introduction

The gravity-driven flow and clogging of grains through a silo has been studied for many years, and many empirical laws have been developed with some theoretical ideas offered [1–4], however the picture is not complete. In particular, we aim to understand the abrupt transition from flowing to clogging behavior. It has been thought that the clogging transition might be similar to the jamming transition, where some critical parameters (i.e. opening size) tell whether a clog will occur. With this picture, one also expects the requisite fluctuations (such as velocity fluctuations or dynamical heterogeneities) to appear in the system as the opening size approaches its critical value. However, recent work [2] on glass spheres has shown that while fluctuations are present, there is no variation in these fluctuations on approach to clogging, suggesting it is not a true transition. While this may be true for glass spheres, does this also hold for different particles? And more importantly, regardless of whether or not a true transition exists, how do fluctuations and intermittent structures (force chains, contact networks, shear zones) conspire to create a clog? To answer this question, we use photoelastic disks so that we may visualize the force chains in addition to tracking the particles' translations and contacts.

2 Experimental Setup

Our experimental setup consists of a quasi 2D silo filled with disks, a schematic is in Fig. 1A. The silo is 20 cm wide by 40 cm tall by 3.2mm deep, made of clear plates screwed together, with the gap set by PTFE sheeting on the left/right sides. The gap in the silo is such that we can fit a single layer of disks in the 20 cm x 40 cm area. Generally, the disks will not touch the front/back clear walls, the average disk will only touch other disks. We treat these walls with a (very) fine dusting of Teflon powder to ensure any incidental contact is low-friction. On the other hand, the disks on the side will touch the PTFE sidewalls on the left/right side of the apparatus.

The disks themselves are made of polyurethane. They are bidisperse (50:50 ratio) to suppress crystallization, with sizes of 3.9 mm and 6 mm. The size of the large disks will be abbreviated with the symbol D for the rest of this manuscript. The silo is filled with disks to a depth h. In all flow and clogging trials, we start from a fixed depth of 33 cm. (A discussion of the height is included in Section 3). The Youngs modulus of the disks is $E \approx 3.5$ MPa,

The silo bottom is made of PTFE inserts that can slide in and out to set opening sizes. The opening can be up to about $8D$ in our current setup. For a fixed opening size, we insert a 3D printed plug into the gap. When we are ready to start a flow/clogging trial, we quickly remove this plug. The silo is attached to an external aluminum frame via an axle as depicted in Fig. 1B. The angle of the silo can be locked into place and is generally completely vertical. However, the silo can be removed or rotated easily to allow for quick resetting of the sample.

The silo is placed between crossed polarizers [4] and is backlit. This setup allows us to measure the photoelastic signal from the disks. Disks under stress will appear brighter than disks not under stress. Individual particle images can, in principle, be used to deduce the exact forces acting on each particle, but we have not set up sufficient spatial resolution to do this presently. We image the

*e-mail: thack22e@mtholyoke.edu
**e-mail: knordstr@mtholyoke.edu

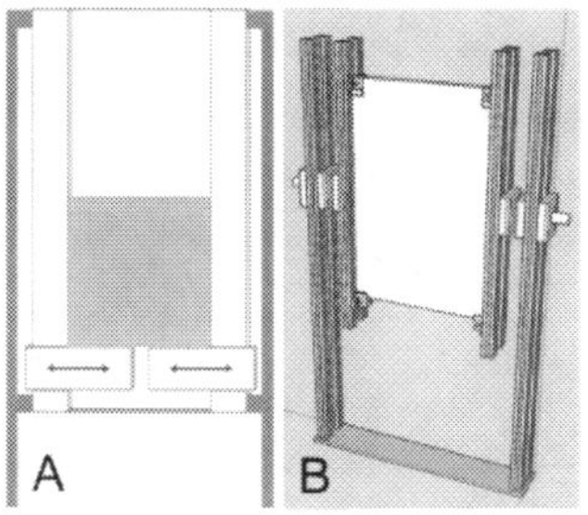

Figure 1. A) A schematic of the silo setup. PTFE inserts (double arrows) may be moved in or out to create openings of various sizes. B) The silo itself is coupled to an external frame with an axle. This allows the system to be rotated so the system may be reset, or completely inverted.

sample during flow/clogging at frame rates up to 10,000 fps using a Phantom CMOS camera. We do not need this high frame rate to capture the particle motion itself, but the force transmission (which appears as filaments of traveling brightness) between particles happens at much higher speeds. This points out another advantage to using these particles. Not only are they photoelastic, but since they are somewhat soft, the sound speed is much lower than an often-studied glass system, so we do have enough temporal resolution to capture this force transmission.

3 System Characterization

As this particular system has not been well-characterized before, we have started our investigation by seeing how this system behaves for two standard conditions. The first condition is static: do we see a characteristic pressure vs depth relationship? The second condition is in the free-flowing state: do we see flow curves that resemble other systems?

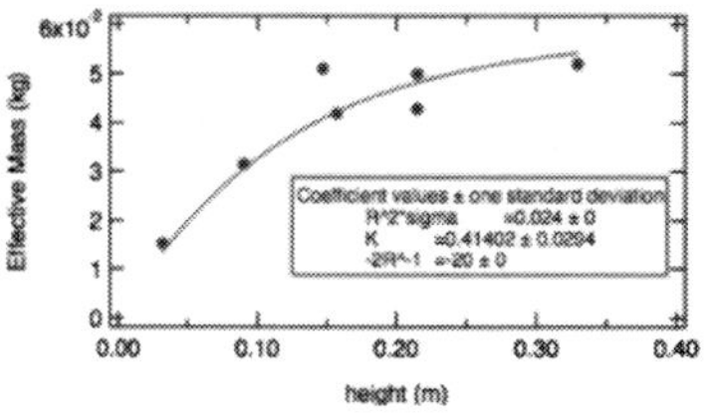

Figure 2. The effective mass of a system of 2D polyurethane disks as a function of height. We see that the Janssen effect is present in this system, and the data is consistent with Eq. 1.

To measure pressure vs depth in the sample, we remove the inserts that adjust the opening size and "plug" the bottom with a 3D printed rectangular insert that extends the width of the silo. This insert is then coupled to a force transducer. We fill the silo to various heights and measure the force at the bottom of the sample. We report this as "effective mass" in Fig. 2. The Janssen effect has widely been reported in granular systems, and is a pressure saturation vs depth, as the sidewalls can support some the weight of the pile. We present below a modified 2D expression based on an equation for a 3D system [5].

$$M_{eff}(h) = \frac{R^2\sigma}{K}(1 - e^{-2Kh/R}) \tag{1}$$

where R is the half-width of the silo, the h is the depth, σ is the effective areal mass density. The parameter $K = \mu P_h/P_v$, where μ is the coefficient of static friction and P_h/P_v is the ratio of horizontal to vertical pressure, which should be of order 1. Our data agrees with the functional form for Eq. 1. We find the value of K to be about 0.41, which coincides with our measurement of $\mu \approx 0.43$.

The second characterization of this system is in the free-flowing state. Generally, for these systems, free-flowing behavior is seen when the opening size is greater than about 5 times the particle size [1]. Calculation of the critical opening size involves experiments we will describe in a later section. However, even without this precise knowledge, it is simple to tell whether a particular opening size gives free-flow or not. We have looked at the flow rates vs time for several systems clearly in the free-flowing regime; we present example results from one trial in Fig. 3. To make this measurement, we fix an opening size and block the opening with a plug. We remove the plug and allow the particles to flow, and catch them in a pan. This pan is connected to a force transducer, which gives us a readout in real time. [It should be noted we have also made these same measurements using our video data, and we see consistent results.] We see some fluctuations and some transient behavior in the startup of flow, but overall the mass flow rate seems to be essentially constant. This is consistent with the Beverloo equation, an empirical equation developed for these systems, describing a constant flow rate. A 2D version is as follows [3]:

$$W(D) = C\rho_b\sqrt{g}(D - kd)^{3/2} \tag{2}$$

were C is an empirically-determined discharge coefficient, generally between about 0.55 and 0.65, ρ_b is effective density of the material, g is the acceleration due to gravity, and k is the empirically-determined shape coefficient with a value generally between 1 and 3. We see behavior consistent with this equation (Fig. 3).

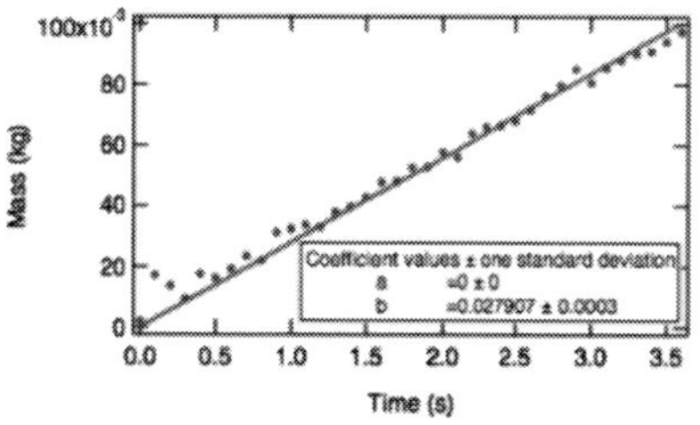

Figure 3. Mass discharged as a function of time for a free-flowing system. The slope obtained (the mass flow rate) is consistent with Eq. 2.

4 Bulk Avalanche Behavior

We are interested in how clogs develop and intermittency in the bulk flow. In this section, we focus on opening sizes that are below the critical opening size. In this regime, a clog will form after some amount of time passes. We characterize these clogs in two ways: by looking at the size of the mass discharge before a clog, and by looking at the photoelastic signal of the entire sample.

We call the event preceding a clog an "avalanche." We characterize each avalanche by the total amount of mass discharged (m). We start with the system filled, with a fixed opening size, and a plug blocking the opening. We remove the plug, and allow the sample to flow freely until a clog forms. To gather clogging statistics, we need tens to hundreds of trials at each opening size. We pour in particles lost during the last event and rotate the system on its side and shake it to reset the contacts. As a final measure to ensure a fresh run, we invert the entire sample; we have the same structure on the top and bottom to accommodate inserts and plugs. We create probability distributions of the discharged mass for each opening. We find that the data fit well to an exponential probability distribution: $P(m) = P_0 e^{-m/m_c}$, where P_0 is a normalization constant. and m_c is a fitting parameter with units of mass, which well call the "characteristic mass" of the opening size. We are able to this collapse the data in Fig. 4A. The fit to an exponential is consistent with the results in a 3D system of glass beads [2], and indicates that clogging can be considered a random Poisson process. By plotting characteristic mass vs opening size in Fig. 4B, we are able to see that m_c looks to diverge as we approach a critical opening size, i.e. $m_c \sim (D - D_c)^{-\gamma}$. Further trials are currently needed to complete this piece of the story. With enough trials, we will get two pieces of information from this plot: the critical opening size (D_c) and a critical exponent (γ). We already know approximately what the critical opening size is so this will be a matter of increasing our precision. However, we are interested if our critical exponent γ differs or agrees with those measured in [2].

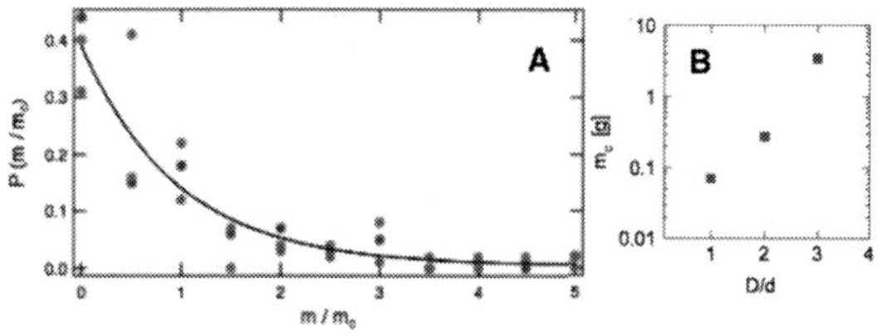

Figure 4. A) Probability distribution of mass discharged. Data are collapsed by rescaling m by the characteristic mass m_c, obtained by fitting to a decaying exponential. B) m_c obtained from fits in A) vs dimensionless opening size (D/d).

By taking high speed video of clogging events, we can characterize the bulk flow in temporal detail. An example is shown in Fig. 5. In this video we have captured the photoelastic signal from the sample, and can also see individual particles clearly to confirm what is happening in

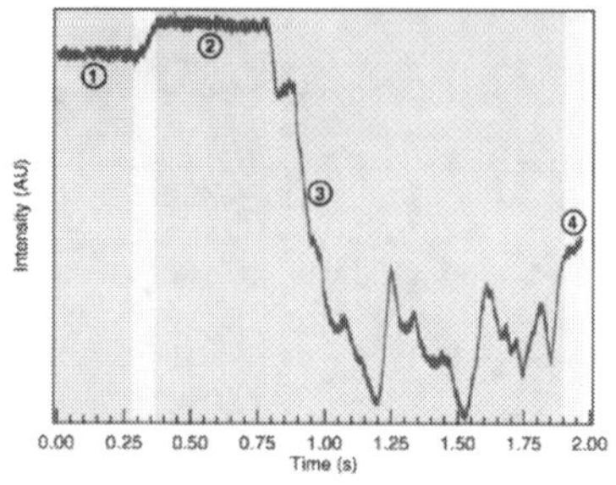

Figure 5. The intensity vs time of the photoelastic signal of the entire sample around a clogging event. 1) Before the plug is removed (pink, flat), then a brief period of avalanche (yellow, sloped) resulting in 2) a clog forming. The clog spontaneously breaks and a period of 3) free flow occurs, followed by 4) the formation of a new, permanent clog.

the sample. For this event, we have plotted the total intensity of a frame vs time. We have marked some key times with numbers in Fig. 5, and describe them here. 1) Before the plug is removed (pink, flat), then a brief period of avalanche (yellow, sloped) resulting in 2) a clog forming 3) free flow 4) A new, permanent clog forming. There are a few features of this event (and this method) to underscore. First, we see clogs as static intensity signals, and flow events as temporarily varying. This suggests a robust method to demarcate events in time as we have very high time resolution in the video signal. Secondly, there are three "static" states represented in this figure, all with different total intensities (stresses). This suggests a potential line of inquiry – are higher stress clogs more or less likely to spontaneously break? Third, we see intermittency in the flowing state, which echoes what we have seeing in the free-flowing state. Lastly, and most interestingly, we see a short-lived clog which breaks spontaneously. These temporary clogs are not captured in 3D (or 2D and low time resolution) studies, as the clog would be defined as the permanent clog (formed around 1.9 s for this data). Nonetheless these temporary clogs are very real and exist for finite amounts of time.

5 Local Behavior

We are interested in capturing local dynamics as clogs form. By taking photoelastic images at high frame rates, we are able to see local stresses within the sample at any given instant, and can also see forces being transmitted between particles. These force transmissions often occur much faster than the movement of the particles themselves. Of course, we are also able to track the particle positions. Briefly we use a circular Hough transform to find the particles, and MATLAB-based tracking algorithms to obtain their trajectories [6]. With the particle positions (and radii), we are also able to detect which particles are in contact with one another. In Fig. 6, we see these three complementary pieces of information (particle positions, force network, and contact network) plotted together.

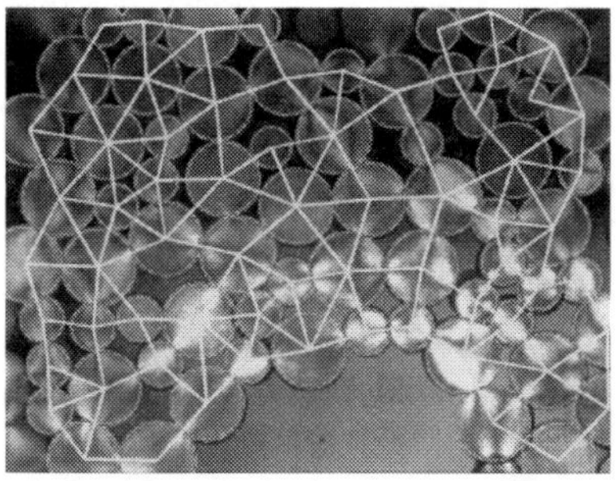

Figure 6. A zoomed-in image where an arch has formed over the opening, resulting in a clog. This composite image displays the photoelastic signal (grayscale intensity) which yields local force information, the particle positions (particles outlined in blue), and the contact network (yellow).

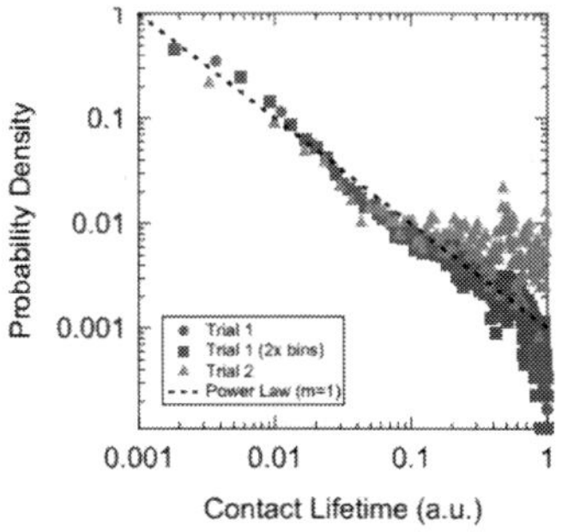

Figure 7. Probability density of contact lifetimes for two trials that clog $D/d = 4$. Two datasets are plotted for trial 1 (red circles and purple squares) this is to show the shape distribution does not depend on bin size. Trial 2 (blue triangles) clogs faster than 1.

Using the contact network data, we have made progress in characterizing the lifetime of contacts within the system. In Fig. 7, the probability density of contact lifetimes is plotted for two different trials which ultimately clog (D/d=4). A contact lifetime is defined simply as the time over which a single contact between two particles exists. In this figure, contact lifetimes are displayed for the whole sample (~1000 particles), not for just the zoomed-in area near the arch. The contact lifetime statistics are gathered from the start of flow until the formation of a permanent clog. Both trials display a power law decay with an exponent close to 1, as shown by the fit on the graph. In trial 2, however, the clog forms sooner. We also see a larger number of longer-lived contacts. Speculatively, this might be the reason why trial 2 clogs sooner. We are actively pursuing this matter.

Lastly, we are interested in local avalanches within the system, that is, where and are major rearrangement events occurring? To do this, we quantify nonaffine (NA) motion within the system. NA motion can be the result of a variety of phenomena, including irreversible rearrangements or force chain breakage. Nonaffine motions point to deviations in the mechanical behavior of the granular material from a continuous medium. To measure NA motion, we use the quantity D^2_{min} as we have in previous work [6]: $D^2_{min,i} = \min\{\sum_j[\Delta\bar{d}_{ij}(t) - E_i\bar{d}_{ij}]\}^2$. $D^2_{min,i}$ quantifies the NA motion of j particles in the neighborhood around a given particle i after removing the averaged linear response to the strain, given by tensor E_i; a larger D^2_{min} indicates more NA motion. The vector $\bar{d}_{ij}$ is the relative position of i and j, $\Delta\bar{d}_{ij}$ is the relative displacement. Fig. 8 shows a map of D^2_{min} for a particular time in the free-flowing system. We see that there are several regions of high D^2_{min}, but these do not always correspond to high-shear regions near the opening. By connecting these measurements with the force and contact networks we hope to elucidate why this is. Are these rearrangements far from the opening important? Indeed, we generally see that while the overall flow rate can be fit well to the (smooth) Beverloo equation, we can see by eye (and with the photoelastic signal) that the bulk motion is quite jerky. Presumably this influences clog formation when the system is in this regime.

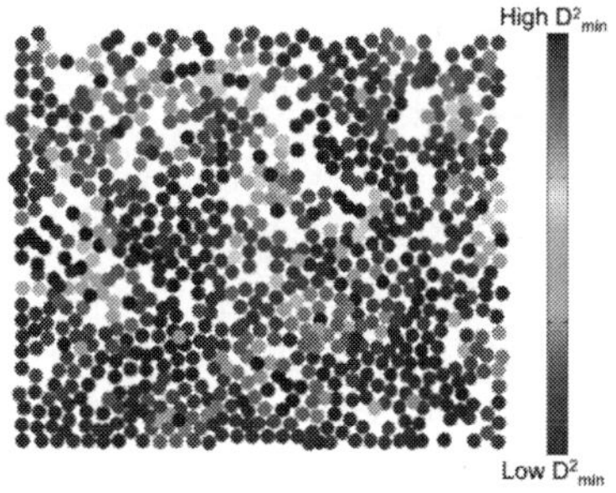

Figure 8. A plot of $D^2_{min,i}$ in a freely-flowing system partway through the outflow. The opening is about $7D$ and is bottom center in this plot. Warm colors denote "hotspots" of rearrangements and cool colors denote areas where the particles move according to the average flow.

Acknowledgements

The authors thank Kathryn Morrison, Len McEachern and Thomas Liimatainen. This work is partially funded by ACS award PRF 56888-UN19.

References

[1] K. To. *Physical Review E*, **71**, 060301 (2005).
[2] C. C. Thomas and D. J. Durian, *Physical Review E*, **87**, 052201 (2013).
[3] C. Mankoc, A. Janda, R. Arevalo, J. M. Pastor, I. Zuriguel, A. Garcimartin, and D. Maza, *Granular Matter*, **9**, 407 (2007).
[4] J. Tang and R. P. Behringer, *Chaos*, **21**, 041107 (2011).
[5] R. Back. *Granular Matter*, **13**, 723 (2011).
[6] K. N. Nordstrom, E. Lim, M. Harrington, and W. Losert *Physical Review Letters*, **112**, 228002 (2014).

Relaxation Times in Simple Shear and the Role of Walls

Lothar Brendel[1,⋆], *János* Török[2], *Alexander* Ries[1] and *Dietrich E.* Wolf[1]

[1]*Faculty of Physics and CENIDE, University of Duisburg-Essen, D-47048 Duisburg, Germany*
[2]*Department of Theoretical Physics, Budapest University of Technology and Economics, H-1111 Budapest, Hungary*

Abstract. We study the relaxation time of granular media in simple shear by means of DEM simulations (the methods being Molecular Dynamics as well as Contact Dynamics) in two and three dimensions with rough and with smooth frictional walls. While the system with rough walls behaves according to its steady state constitutive laws, the systems with smooth walls show a much stronger increase of the relaxation time with driving strength than to be expected. Employing a dynamic non-local rheology model[1] allows for a stronger increase but not sufficiently so.

1 Introduction

The long known non-Newtonian character of granular media[2] manifests itself in observations like a quadratic relation between stress and shear rate (*Bagnold scaling*[3]) or normal-stress differences[4, 5]. Such quantities can readily be measured in the steady state, but of course deviations from a constant viscosity influence transients like relaxation towards the steady state, too. In this work, we're studying the relaxation times of a simple shear situation (in two and three dimensions) for different boundary conditions (rough and smooth walls) and different simulation methods: Molecular Dynamics (MD) with soft particles[6] and Contact Dynamics (CD) with rigid particles[7]. After explaining the theoretical predictions for the asymptotic relaxation time by virtue of established constitutive relations in section 1.1, we present setup and results for the two-dimensional system with rough walls in section 2 and for the systems with smooth walls in section 3. For the latter we discuss the ostensible necessity of taking into account non-local rheology in section 4 and conclude our findings in section 5.

1.1 Diffusive Evolution

Assuming a homogeneous normal stress σ_{zz} in simple shear (driven by walls with normals in z direction), the evolution of the velocity field $v(z,t)$ is given by

$$\partial_t v = \frac{\sigma_{zz}}{\nu\rho}\partial_z\mu \quad , \tag{1}$$

with ρ being a particle's mass density. The volume fraction ν and the macroscopic friction $\mu = \sigma_{xz}/\sigma_{zz}$ are functions of the *inertial number*[8]

$$I = \tau_{\rm i}\partial_z v \quad \text{with} \quad \tau_{\rm i} = d\sqrt{\rho/\sigma_{zz}} \quad , \tag{2}$$

where d is the particle diameter. If ν is constant and $\mu(I)$ a linear function, (1) turns into a simple diffusion equation; otherwise it gets non-linear. But for walls with velocities $\pm v_{\rm W}$ having driven the system into the asymptotic regime, where $\delta v = v - 2z\,v_{\rm W}/L_z \equiv v - zI_\infty/\tau_{\rm i}$ is small, we can linearize (1) in δv to obtain

$$\partial_t\delta v = \frac{d^2}{\tau_{\rm i}}\frac{\mu'(I_\infty)}{\nu(I_\infty)}\partial_z^2\delta v \quad , \tag{3}$$

i.e. the "diffusion constant" $d^2/\tau_{\rm i}$ just gets scaled by a function of the *final* shear rate. The slowest relaxation mode has wavelength $2L(I_\infty)$, where the I_∞-dependence is due to dilatancy, which also implies $L(I_\infty)\nu(I_\infty)$ to be constant. With that we get

$$\tau(I_\infty) = \tau_{\rm i}\frac{L^2(I_\infty)\nu(I_\infty)}{\pi^2 d^2\mu'(I_\infty)} = \tau(0)\frac{\mu'(0)\nu(0)}{\mu'(I_\infty)\nu(I_\infty)} \quad . \tag{4}$$

That means, a decreasing μ' and a decreasing ν both give rise to an increase of the relaxation time. For local constitutive relations[9]

$$\mu(I) = \mu_{\rm s} + \frac{\alpha}{1/I + 1/I_{\rm c}} \quad \text{and} \quad \nu(I) = \nu_0 - \beta I \quad , \tag{5}$$

this factor is $(1 + I_\infty/I_{\rm c})^2/(1 - I_\infty\beta/\nu_0)$.

Considering directly the evolution of I instead of v, the counterpart to (1) is a bit more complicated, but its linearization corresponds exactly to (3), i.e. we can treat relaxation of v and of I on the same footing.

2 Rough Walls

2.1 Simulation Setup

Two dimensional simulations (xz plane) in the simple shear geometry were performed using the LAMMPS[10] simulation package. As natural units we chose average

⋆e-mail: lothar.brendel@uni-due.de

particle diameter d, confining pressure σ_{zz} and particle mass density ρ. Rough walls were made up from randomly placed disks of size 1.2 so that the effective friction coefficient of the wall was maximal[11]. The walls were kept together with constant stress in the z direction and were given shear velocities $\pm v_W$. Particles had diameters $d \in [0.8, 1.2]$ with a uniform distribution, a two dimensional density of 2/3, and a microscopic friction coefficient of μ_{mic}=0.3. The total kinetic energy in particle translation (E_{kin}) and rotation (E_{rot}) was recorded together with the shear stress σ_{xz}. The bulk shear rate was defined to be the average over the middle third of the system.

2.2 Results

After a transient, the system reaches a steady state characterized by a linear velocity profile. The stationary shear rate is thus $\dot{\gamma}_\infty$=$2v_W/L_z$, the inertial number I_∞= $\sqrt{2/3}\dot{\gamma}_\infty$. We measured constitutive relations of $\mu(I_\infty)$ and $\nu(I_\infty)$ which are in good agreement with the empirical predictions [9] as expressed in Eq. (5), but as shown in Fig. 1, there are alternatives like $\mu = \mu_s + \alpha I_c \arctan(I_\infty/I_c)$, which produces a factor $\mu'(0)/\mu'(I_\infty) = 1 + (I_\infty/I_c)^2$ in Eq. (4).

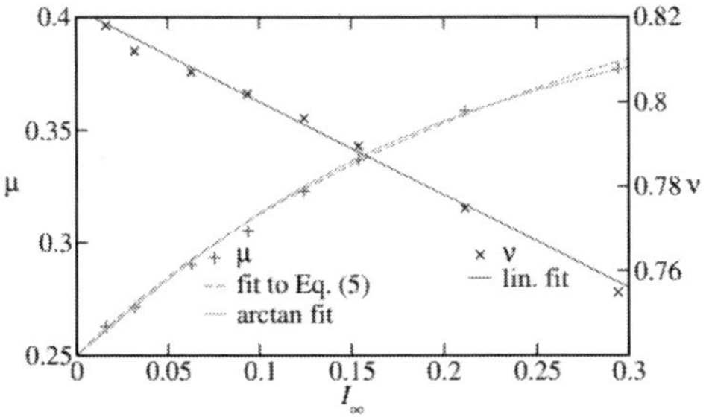

Figure 1. The friction coefficient μ (left axis) and the packing fraction ν (right axis) as function of the steady state inertial number I_∞ as points. Both data are fitted according to Eqs. (5) ($\Rightarrow \mu_s = 0.25, \alpha = 0.788, I_c = 0.377, \nu_0 = 0.822, \beta = 0.22$). The friction is (slightly better) fitted with $\mu = \mu_s + \alpha I_c \arctan(I_\infty/I_c)$, as explained in the text ($\Rightarrow \mu_s = 0.25, \alpha = 0.689, I_c = 0.124$).

Analogous to [12] we find that all quantities reach their stationary values in an exponential way, which enables us to define characteristic times $\tau_{kin}(I_\infty)$ for the kinetic energy, $\tau_{rot}(I_\infty)$ for the rotational energy, and $\tau_I(I_\infty)$ for the bulk shear rate. Figure 2 shows the dependence of the relaxation times as function of the inertial number. The kinetic energy and the bulk inertial number behave in essentially the same way, and their relaxation time can be fairly well described by Eq. (4) and the empirical arctan-law for μ, better than according to Eq.(5). While the difference in Fig. 1 seems minute, the curvature in μ matters.

Even though the diffusion model Eq. (3) describes the system quite well, it is interesting to note that the characteristic time for the rotational energy is different from the kinetic one indicating a different underlying mechanism. In Fig. 3 $E_{kin}(t)$ and $E_{rot}(t)$ are shown for three different positions. We find that initially the rotational energy is

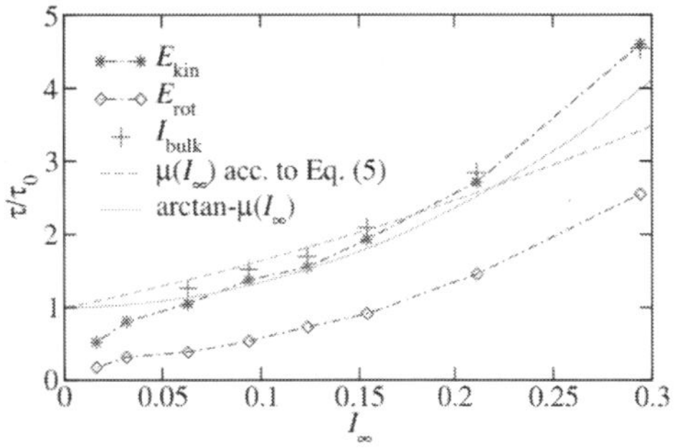

Figure 2. The relaxation time of E_{kin}, E_{rot} (dot-dashed) and I_{bulk} (points) as function of the steady state inertial number, showing a very similar behavior of E_{kin} and I, while E_{rot} relaxes quicker. The dashed and solid curves show the prediction of Eq. (4) when using Eq. (5) and when using the arctan-description of μ (cf. Fig. 1), respectively. τ_0 is the theoretical value $\tau(0)$ according to Eq. (4).

high at the walls and low in the middle but they approach the same value. The kinetic energy is always increasing at all positions, but the stationary values are different according to the velocity profile. It seems that the rotation is responsible for the sub-linear growth of $\mu(I)$ by switching to a new shear mode with rotating particles and less friction for larger shear rates. This is further manifested in the increasing relaxation time. This rotating layer of particles close to the wall decrease the effective friction coefficient and thus the stress in the bulk making the diffusion process slower as shown in Fig. 2.

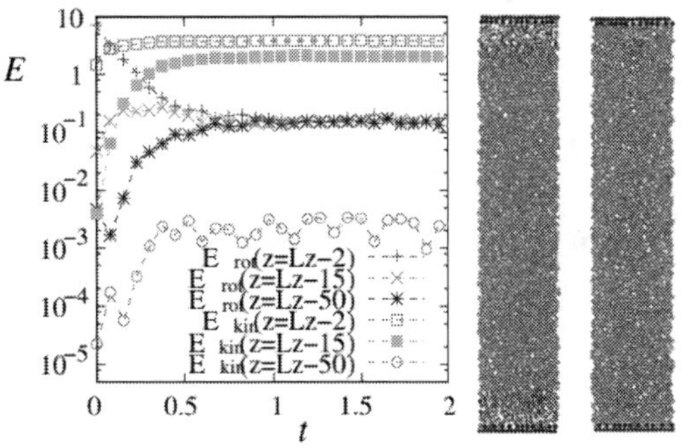

Figure 3. (a) Time evolution of E_{kin} and E_{rot} at positions z = 2, 15, 50. (b) Snapshots at time t = 0.07 and t = 5.5, amount of red is proportional to rotational energy.

3 Smooth Walls

3.1 Simulation Setup

Smooth walls represent a stronger perturbation of local packing geometry, and they can transmit force and torque only via contact friction: Sliding friction is set to μ_{mic} = 0.5 everywhere, rolling and torsion friction are absent.

In the three-dimensional Contact Dynamics simulations, $N = 15300$ particles with diameters $\in [0.8, 1.0]$ and zero coefficient of restitution are placed at rest in a cuboidal box with a base of size $L_x \times L_y = 10 \times 10$ and periodic boundary conditions. In z-direction, walls apply the unit pressure σ_{zz} by means of a constant force and have a fluctuating distance of $L_z \approx 100$ and a fixed velocity $\pm v_W$ in x-direction; cf. [12] for further details.

We also carried out 2D MD simulations with smooth walls, the parameters of which correspond essentially to the ones in the previous section, except for $\mu_{mic} = 0.5$ at contacts between particles and to the walls.

3.2 Results

In contrast to walls made out of particles, the smooth walls allow for a finite slip velocity, which itself is a dynamical quantity and is a non-trivial response to the driving velocity. Nevertheless, well defined boundary conditions for the bulk evolution are established in the sense that about 9 particle diameters away from each wall (defining an effective system size $L_z' < L_z$, which we will denote L in the following), the shear rate $\dot{\gamma}$ and thus the inertial number takes on a constant value I_0 (apart from fluctuations, of course) after less than 100 natural time units (not shown). The same time scale actually applies to the total rotational energy. During the subsequent relaxation of the bulk towards this shear rate, the corresponding velocity difference grows just as the slip velocity decreases. Due to the wall friction being Coulombian, this decrease has no influence on the driving force . Brief, the finite slip velocity provides a "buffer" as to enable I_0 as boundary condition (and thus Neumann BCs for the velocity).

In the steady state, the constitutive relations are measured, as shown in Fig. 4. Again, we measured the relaxation time of the bulk inertial number, the results of which is displayed in Fig. 5, for the 3D CD simulations as well as for the 2D MD simulations.

As already mentioned in section 1.1, a linear $\mu(I)$ does not contribute to an I_∞-dependency of the relaxation time, leaving only the very weak influence of $\nu(I)$, which, as shown in Fig. 5, is essentially negligible. A logarithmic law $\mu(I) = \mu_s + \ln(1 + \alpha I)$, on the other hand, actually leads to a factor $\mu'(0)/\mu'(I_\infty) = 1 + \alpha I_\infty$, which has the correct functional form seen in the simulation data in Fig. 5, but the slope is too small by a factor of six.

We omitted the comparison for the two-dimensional MD-simulations, because the discrepancy is the same. But we include the data to demonstrate that the strong increase of τ with I_∞ is quite generic and no artifact due to rigid particles or perfect collisional dissipation.

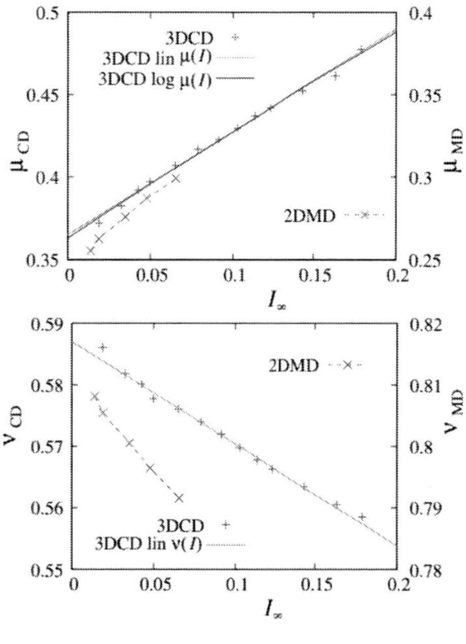

Figure 4. Macroscopic friction (top) and volume fraction (bottom) for the systems with smooth walls. Since there is no significant curvature to extract a value for I_c in Eq. (5), a linear function (corresponding to $I_c \to \infty$) is fitted to the 3D data ($\Rightarrow \mu_s = 0.365, \alpha = 0.624$). The reason for the logarithmic fit $\mu_s + \ln(1 + \alpha I_\infty)$, which is of similar quality (and yields $\mu_s = 0.363, \alpha = 0.665$) is explained in the text.

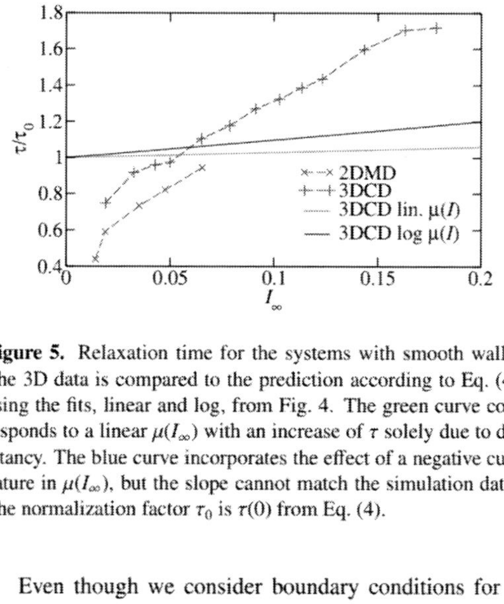

Figure 5. Relaxation time for the systems with smooth walls. The 3D data is compared to the prediction according to Eq. (4) using the fits, linear and log, from Fig. 4. The green curve corresponds to a linear $\mu(I_\infty)$ with an increase of τ solely due to dilatancy. The blue curve incorporates the effect of a negative curvature in $\mu(I_\infty)$, but the slope cannot match the simulation data. The normalization factor τ_0 is $\tau(0)$ from Eq. (4).

Even though we consider boundary conditions for I given within the system, about 9 particle diameters away from the actual walls, we cannot exclude any additional influence of the walls. They cause inhomogeneities, which have been shown to be captured successfully by non-local rheology[13].

4 Non-local Rheology

In the time dependent non-local rheology according to [1], an additional field g, the *granular fluidity*, is introduced to describe the shear response $I = \tau_i \dot{\gamma} = g\mu$ [1] and obeys

$$t_0 \partial_t g = A^2 \nabla^2 g - (\mu_s - \mu) g - \alpha \mu g^2 \quad , \tag{6}$$

[1]Here, g and A differ from the ones in [1] by factors of τ_i and d^2, respectively.

where the parameters t_0 and A set the intrinsic time and length scale of the evolution of g. For steady, homogeneous flow (constant $g > 0$) it yields $g = (\mu - \mu_s)/(\alpha\mu)$ which corresponds to $I_c \to \infty$ in (5). We can eliminate the field μ from the equations to keep only I and g:

$$\tau_i\,\partial_t I = d^2\partial_z\left(\frac{1}{\nu}\partial_z\frac{I}{g}\right) \tag{7}$$

$$t_0\,\partial_t g = A^2\partial_z^2 g - \mu_s g + I - \alpha I g \tag{8}$$

Then, linearizing around I_∞ and $g_\infty = I_\infty/(\mu_s + \alpha I_\infty) \equiv I_\infty/\mu_\infty$ leaves us with

$$\tau_i\,\partial_t\delta I = \frac{d^2}{\nu_\infty}\frac{g_\infty\partial_z^2\delta I - I_\infty\partial_z^2\delta g}{g_\infty^2} \tag{9}$$

$$t_0\,\partial_t\delta g = A^2\,\partial_z^2\delta g - \mu_\infty\delta g + \frac{\mu_s}{\mu_\infty}\delta I \quad , \tag{10}$$

where $\nu_\infty = \nu(I_\infty)$. A solution with vanishing δI and δg at the boundaries (i.e. Dirichlet BCs for I and g) is

$$\delta g = \exp(-t/\tau_{\rm nl})\cos(\pi z/L) \quad , \quad \delta I = \mu_*\delta g \quad , \tag{11}$$

and yields (besides the irrelevant μ_*)

$$\frac{2}{\tau_{\rm nl}} = \frac{1}{\tau'} + \frac{1}{\tau''} - \sqrt{\left(\frac{1}{\tau'} - \frac{1}{\tau''}\right)^2 + 4\frac{\mu_s}{t_0\tau''}} \tag{12}$$

$$= 2\frac{\pi^2A^2}{t_0\,L^2(I_\infty)} + O(I_\infty) \quad , \quad \text{where} \tag{13}$$

$$\tau' = \frac{\tau_i g_\infty\nu_\infty L^2(I_\infty)}{\pi^2 d^2} \quad \text{and} \quad \tau'' = \frac{t_0}{\pi^2A^2/L^2(I_\infty) + \mu_\infty} \quad .$$

The static limit $\tau_{\rm nl} = t_0L^2/(\pi^2A^2)$ is peculiar, because it does not contain parameters of the constitutive relations $\mu(I)$ and $\nu(I)$, but it is non-zero as opposed to the case considered in [12], using the model from [14]. On the other hand, $\tau_{\rm nl}$ is bounded by τ', which is clearly exceeded by the simulation data, as shown in Fig. 6. Hence, taking into account non-local rheology allows for a larger variability of the relaxation time, but is not enough.

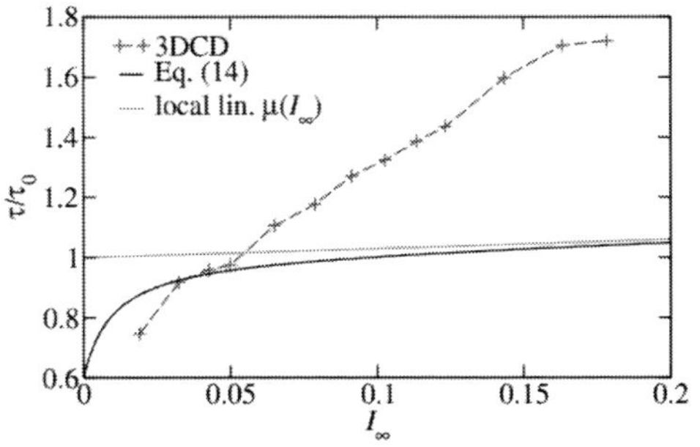

Figure 6. The relaxation for the 3D system with smooth walls and as predicted by the non-local rheology for $L_0 = 81.5$, $A = 4$ and $t_0 = 0.6A^2/(\nu_0\alpha)$, exhibiting a non-zero $\tau_{\rm nl}(I_\infty \to 0)$ and a pronounced, A-dependent slope for small I_∞. But it is bounded by the result for a locally valid, linear $\mu(I)$ (cf. Fig. 5), which prevents it from agreeing with the simulation data.

5 Conclusion

We have studied the very simple concept of the time scale needed by a granular system in simple shear to relax into its steady state of homogeneous bulk shear rate (and thus inertial number I_∞). The theory, linearized momentum conservation closed by constitutive relations for macroscopic friction and volume fraction, predicts the somewhat counter-intuitive increase of the relaxation time τ with increasing driving velocity. These predictions are pretty accurate for rough walls, but are much smaller than what we measured in the simulations with smooth walls.

Employing a non-local rheology model[1] provided a stronger variation of the relaxation time compared to the local rheology, but only in one direction (lower times). The role of the boundary conditions for the granular fluidity remains to be clarified, though (Dirichlet BCs have been chosen).

Acknowledgments

We gratefully acknowledge funding by the Mercator Research Center Ruhr (project 2015-0021) and support by the DFG via the grant WO 577/8 within the Priority Program SPP 1486 "Particles in Contact" (PiKo). JT acknowledges the support of the Hungarian Scientific Research Fund (Grant No. OTKA NN 107737).

References

[1] D.L. Henann, K. Kamrin, International Journal of Plasticity **60**, 145 (2014)
[2] S. Elaskar, L. Godoy, International Journal of Mechanical Sciences **40**, 1001 (1998)
[3] R.A. Bagnold, Proc. Royal Soc. London **225**, 49 (1954)
[4] C.S. Campbell, Ann. Rev. Fluid Mech. **22**, 57 (1990)
[5] M. Alam, S. Luding, Phys. Fluids **15**, 2298 (2003)
[6] S. Luding, *Molecular Dynamics Simulations of Granular Materials*, in *The Physics of Granular Media*, edited by H. Hinrichsen, D.E. Wolf (Wiley VCH, Weinheim, Germany, 2004), pp. 299–324
[7] M. Jean, Computer Methods Appl. Mech. Engrg. **177**, 235 (1999)
[8] GDR_MiDi, Eur. Phys. J. E **14**, 341 (2004)
[9] Y. Forterre, O. Pouliquen, Annu. Rev. Fluid Mech. **40**, 1 (2008)
[10] S. Plimpton, Journal of computational physics **117**, 1 (1995)
[11] Z. Shojaaee, L. Brendel, J. Török, D.E. Wolf, Physical Review E **86**, 011302 (2012)
[12] A. Ries, L. Brendel, D.E. Wolf, Computational Particle Mechanics pp. 1–8 (2015)
[13] K. Kamrin, G. Koval, Phys. Rev. Lett. **108**, 178301 (2012)
[14] D.L. Henann, K. Kamrin, PNAS **110**, 6730 (2013)

Statistical analysis of cone penetration resistance of railway ballast

Gilles Saussine, *Amine* Dhemaied, *Quentin* Delforge and *Selim* Benfeddoul

SNCF Réseau – I&P – LVE - CIR

Abstract. Dynamic penetrometer tests are widely used in geotechnical studies for soils characterization but their implementation tends to be difficult. The light penetrometer test is able to give information about a cone resistance useful in the field of geotechnics and recently validated as a parameter for the case of coarse granular materials. In order to characterize directly the railway ballast on track and sublayers of ballast, a huge test campaign has been carried out for more than 5 years in order to build up a database composed of 19,000 penetration tests including endoscopic video record on the French railway network. The main objective of this work is to give a first statistical analysis of cone resistance in the coarse granular layer which represents a major component of railway track: the ballast. The results show that the cone resistance (q_d) increases with depth and presents strong variations corresponding to layers of different natures identified using the endoscopic records. In the first zone corresponding to the top 30cm, (q_d) increases linearly with a slope of around 1MPa/cm for fresh ballast and fouled ballast. In the second zone below 30cm deep, (q_d) increases more slowly with a slope of around 0,3MPa/cm and decreases below 50cm. These results show that there is no clear difference between fresh and fouled ballast. Hence, the (q_d) sensitivity is important and increases with depth. The (q_d) distribution for a set of tests does not follow a normal distribution. In the upper 30cm layer of ballast of track, data statistical treatment shows that train load and speed do not have any significant impact on the (q_d) distribution for clean ballast; they increase by 50% the average value of (q_d) for fouled ballast and increase the thickness as well. Below the 30cm upper layer, train load and speed have a clear impact on the (q_d) distribution.

1 Introduction

The ballast layer is composed of coarse granular material and supports the loading of railway traffic. The design or track renewal are the most important operation in terms of cost. The characterization of the mechanical state is a fundamental work in order to optimize decision making [1,9,12]. The dynamic penetrometer test, known as Panda's test [2,3,11], offers the opportunity to study changes in the stiffness by measuring the dynamic cone resistance (q_d) at a given depth.

For more than 5 years, a huge campaign test has been realized on the French railway network using this technique. On the basis of strain energy measured at each hammer hit and the driving depth, the dynamic cone resistance is calculated. To this mechanical information is added visual information given by an endoscope camera. Thanks to the cone resistance values and the images, each layer can be identified accurately. This paper addresses some statistical analyses carried out on the database built from more than 19,000 penetration tests.

2 Analyzed data

The database has been constructed using Python programming language in order to examine statistical properties of cone resistance of railway layers based on different conditions.

2.1. The cone resistance data set

The database consists in several columns table where each line contains: cone resistance value at a given depth and several parameters related to this measure. The data set contains information on the characteristics of the railway where penetrometer test has been realized, which enables geographical, climatic factors, stratigraphy and thickness of identified layers by the endoscope, parameters related to railway type including speed and load capacity and identity of the survey. The Tab. 2.1 summarizes variables investigated in the study.

Tab. 2.1 Variables analyzed.

Value of the cone resistance q_d (MPa)
Depth of the measure (m)
Thickness of the layer (m)
Depth of the measure (m)
Name of the layer
UIC[a] class (2 to 9)

[a] It classifies railways in different groups based on their daily tonnage. The UIC class of the railway is an important factor which is discussed in more details in this article.

2.2. Filtering the data

The penetrometer test is validated when cone resistance and layers identification from the ballast to the platform are both known [8]. The database is composed of numerous tests which are a combination of mechanical investigation and visual information from endoscopic test which allow to identify the nature of material for each measurements. In some case the endoscopic investigation are not possible because the sounding conditions are not successful [5,6].

Therefore, the filtered data set used was reduced from 19,827 to 5,584 surveys, only 28% (Tab. 2.2) of the data is retained. The second step of our study will be to complete the missing data. A survey is defined in this paper as a sounding in railway track structure.

Tab. 2.2 Amount of raw data and cleaned data.

	Measures (value of cone resistance at a given depth)	Surveys
Initial database	1,836,659	19,827
Filtered database	489,538	5,557

Fig. 2.1 and Fig. 2.2 show the scatter of the cone resistance value (q_d) versus the depth of the measure in the interlayer using the raw data and the cleaned data.

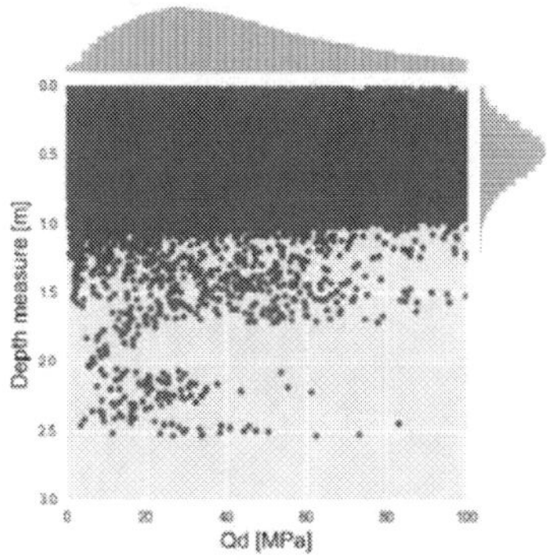

Fig. 2.1 Scatter plot of the depth versus the cone resistance in the interlayer using raw data.

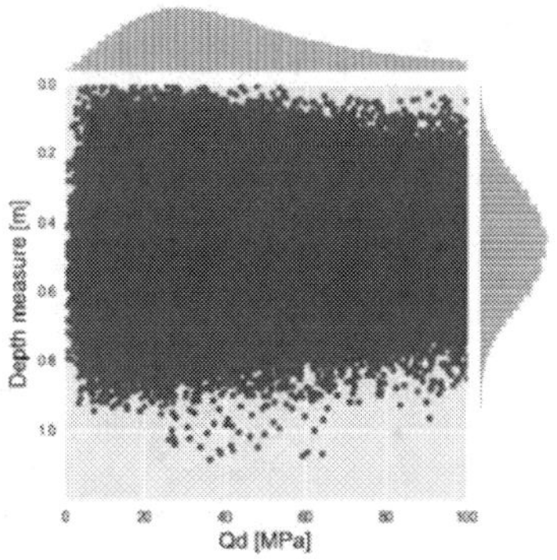

Fig. 2.2 Scatter plot of the depth versus the cone resistance in the interlayer using filtered data.

The interlayer [7,1] originally identified between 0 and 2.5m depth is now contained between 0.1 and 1.1m. On those figures, zero corresponds to the under sleeper level.

The distribution of the filtered data is represented in a pie chart (Fig. 2.3). This gives us a good indication about line traffic distribution which is commonly called UIC group.

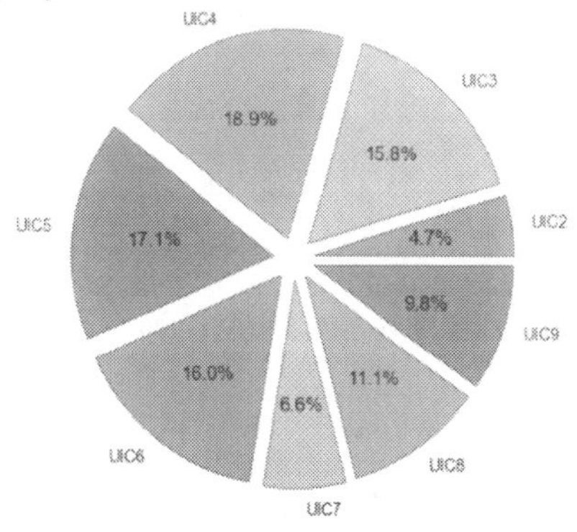

Fig. 2.3 Distribution of the filtered data by UIC groups.

The results obtained for the UIC groups 2, 7, 8 and 9 are less representative than those of the others groups (Fig. 2.3).

3 Observations

3.1. Cone resistance variation with depth

The first observation made on the filtered data concerns the variation of cone resistance with depth and the effect of the UIC group on this variation. Fig. 3.1 illustrates the variation of (q_d) with depth for each UIC group. Cone resistance values are generated by averaging the data contained in 2cm ranges.

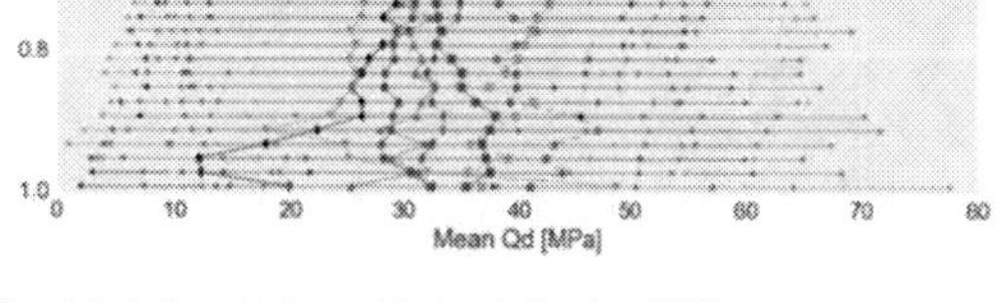

Fig. 3.1 (qd) variation with depth for four UIC groups.

We found that in the first zone corresponding to the first 30cm, (q_d) is increasing approximatively linearly with a slope of around 1MPa/cm [2] and seems to be independent of the UIC group. Bellow this zone, the cone resistance presents strong variations. We also found that UIC group has an impact since the highest (q_d) corresponds to the class of tonnage UIC3 (loading between 50,000 and 85,000 millions of tons) and lowest

(q_d) corresponds to UIC8 (loading to 1,500 millions of tons).

3.2. Effect of the UIC class on cone resistance

By plotting mean values of (q_d) for each UIC group and for each layer (Fig. 3.2), we validate the observation made in the previous section. The cone resistance tends to decrease with the reduction of load capacity railway, except for the clean ballast.

Besides, the statistical treatment of the data shows that train load and speed do not have any significant impact on the distribution of (q_d) for clean ballast. In the fouled ballast, average value of (q_d) increase by about 50% for most of UIC groups (this increase is less significant for the groups 7 and 9). The density of fouled ballast is about more than 1.5 more important of clean ballast.

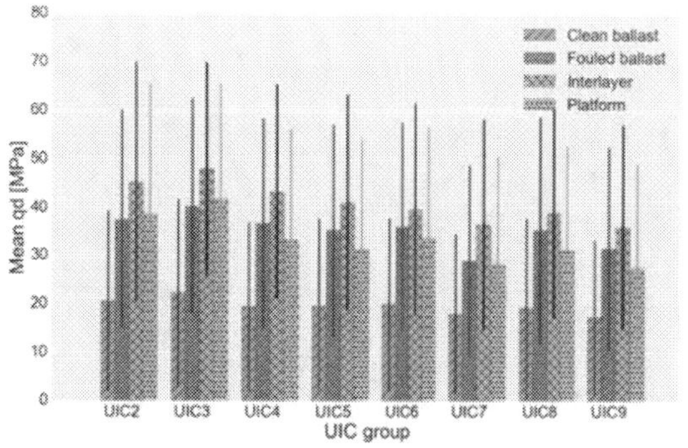

Fig. 3.2 Mean values of qd by UIC group and for each layer.

3.3. Effect of the UIC class on layers thickness

Fig. 3.3 illustrates that the thickness of ballast (clean and fouled) is linked with the UIC group in that a trend is discernible on the boxplots: for groups above UIC6, the thickness of ballast is less important which can be explained by the mean number of years from the last track renewal.

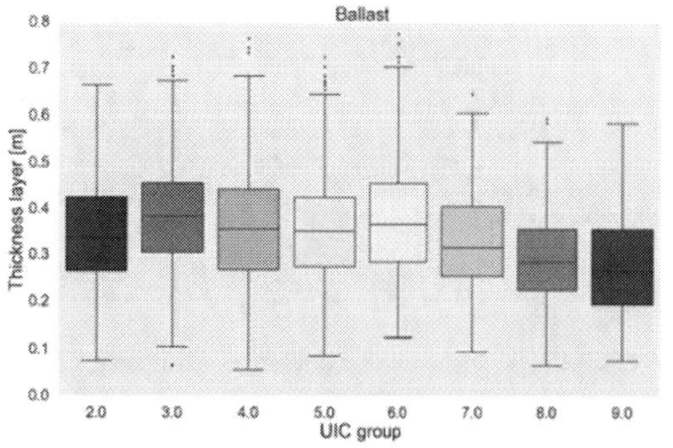

Fig. 3.3 Effect of the UIC group on the ballast thickness.

In the same way as the dynamic cone resistance, average value of the thickness layer is plotted for each UIC group (Fig. 3.4). The interlayer thickness [10,1] is independent of the UIC group. However, the clean ballast thickness declines from the UIC group 6.

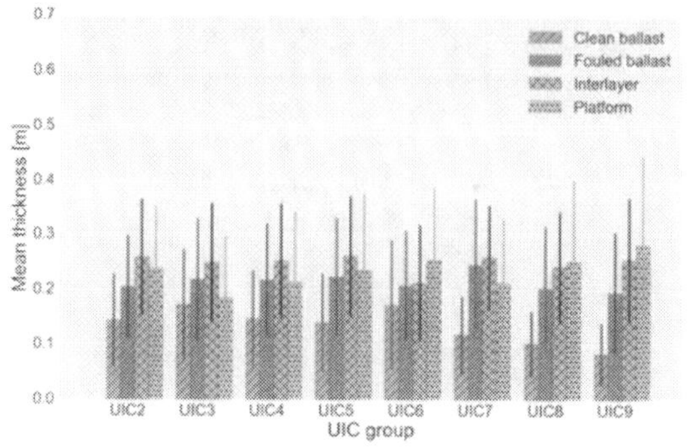

Fig. 3.4 Mean values of thickness by UIC group and for each layer.

3.4. Thickness layers effect on cone resistance

Fig. 3.5 illustrates the correlation between the fouled ballast and interlayer thicknesses and the cone resistance value for UIC groups we have the greater portion of data (e.g. groups 3, 4, 5 and 6). We found that no specific correlation can be observed.

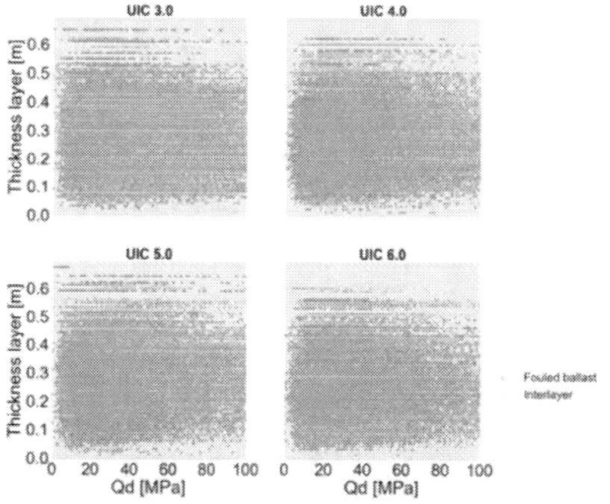

Fig. 3.5 Scatter plot of the fouled ballast and interlayer thickness versus the cone resistance value by UIC group

3.5. Influence of the penetrometer test location on the railway

We observe the impact of the location of the penetrometer test considering two areas: the first one located on the central position between the rails and the second located under the rail and called "outside" for the following of the study. We figure out the effect of the location test on the cone resistance and the thickness of clean and fouled ballast layers. Fig. 3.6 illustrates a classical feature: fouled ballast layer thickness is more important under the sleeper and rail which support the railway traffic.

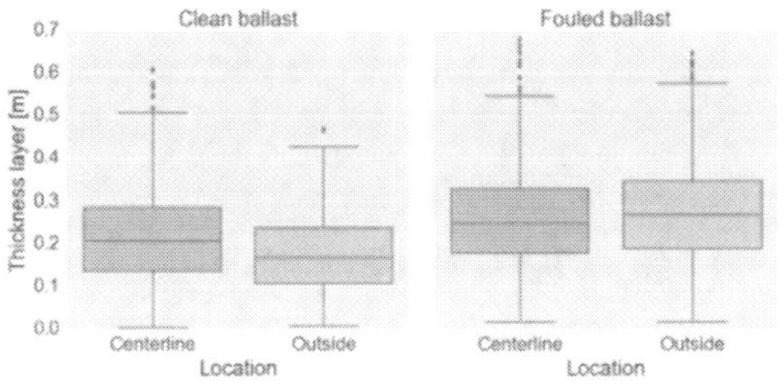

Fig. 3.6 Effect of the test location on the thicknesses of clean and fouled ballasts

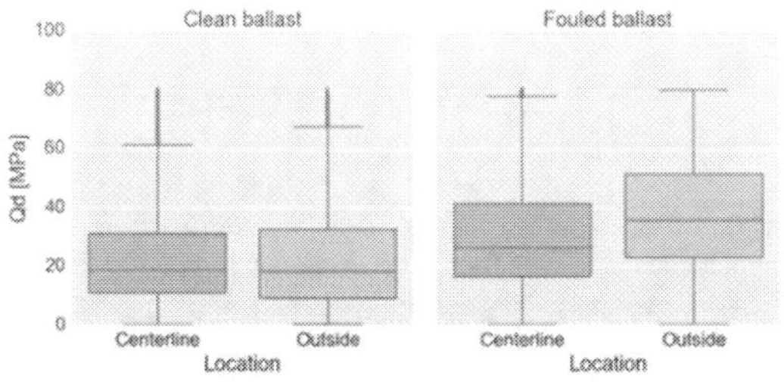

Fig. 3.7 Effect of the test location on the cone resistance for clean and fouled ballasts

On Fig. 3.7, we notice that statically the fouled ballast layer is important in centerline which is an important results that influence the track renewal operation in terms of fresh ballast add or effort for recycling existing material [7]

4 Conclusion

This paper has quantified the statistical properties of cone resistance of railway layers which is an important result for decision making of railway track renewal. This statistical study allows to evaluate the influence of thickness or traffic loading. The initial study carried on the filtered data reveals that cone resistance presents strong variations below 30cm deep and depends on the UIC class. Above this depth, in the clean ballast, (q_d) is independent of the UIC group. The most important dependence observed between the UIC group and the thickness layers concerns the clean ballast.

This preliminary analyses gives the opportunity to predict the track settlement by using a model based on the Chicago's density relaxation law [8] or the cone penetrometer improvements. The theoretical model was tested on experimental data and a prediction of settlement evolution with an error less than 10% was achieved.

References

1. SNCF – direction de l'ingénierie, "IN4103 : directive d'études voie, version N°3" du (13/07/2011)
2. R. Gourvès, "Le PANDA – pénétromètre dynamique léger à énergie variable", LERMES CUST, Université Blaise Pascal Clermont-Ferrand, (1991)
3. AFNOR, "NF P94-105 : Sols : reconnaissance et essais - Contrôle de la qualité du compactage - Méthode au pénétromètre dynamique à énergie variable - Principe et méthode d'étalonnage du pénétromètre - Exploitation des résultats – Interprétation", (Avril 2012)
4. P. Breul, R. Gourvès, Y. Haddani., "Géoendoscopie: caractérisation des géomatériaux par traitement et analyse d'images", revue française de génie civil, (2002)
5. Y. Haddani, "Caractérisation et classification des milieux granulaires par géoendoscopie", université Blaise Pascal, (2004)
6. Y. Haddani , G. Saussine, P. Breul, M.A Benz Navarrete, R. Gourvès, "Railway track stiffness and bearing capacity valuation thanks to panda and geoendoscopy tests", Georail, (2011)
7. P. Breul, R. Gourvès, A. Robinet, C. Bacconnet, "Assessment of the railway structures using an integrating testing and diagnosis methodology", Geotechnical and Geophysical Site Characterization, (2008)
8. G. Saussine, J.C. Quezada, P. Breul, F. Radjai, "Railway Ballast Settlement: A New Predictive Model", in J. Pombo, (Editor), "*Proceedings of the Second International Conference on Railway Technology: Research, Development and Maintenance*", Civil-Comp Press, Stirlingshire, UK, Paper 121, (2014). doi:10.4203/ccp.104.121
9. ASTM 6951, "Standard Test Method for Use of the Dynamic Cone Penetrometer in Shallow Pavement Applications", (2015)
10. N. Calon , M. Chanson., S. Nebieriedze, A. Robinet, "Auscultation des plates-formes ferroviaires par méthodes georadr et panda", Georail (2011) International symposium
11. Breul, P., Benz, M., Gourvès, R. & Saussine, G. Penetration test modeling in a coarse granular medium. Paper presented at Powders and grains 2009: *Proceedings of the 6th international conference on micromechanics of granular media*. Golden, Colorado. AIP Conference Proceedings. (2009)
12. M.A. Benz-Navarrete, E. Escobar, Y. Haddani, R. Gourves, S. Costa D'Aguiar, N. Calon, "Determination of Soil Dynamic Parameters by the Panda 3®: Railways Platform Case", in J. Pombo, (Editor), *"Proceedings of the Second International Conference on Railway Technology: Research, Development and Maintenance", Civil-Comp Press*, Stirlingshire, UK, Paper 56, (2014). doi:10.4203/ccp.104.56

3D scan line method for identifying void fabric of granular materials

Alexandros I. Theocharis[2,*], *Emmanouil* Vairaktaris[2], and *Yannis F.* Dafalias[1,2]

[1] Department of Civil and Environmental Engineering, University of California, Davis, CA, USA

[2] Department of Mechanics, School of Applied Mathematical and Physical Science, NTUA, Athens, Attica, Greece

Abstract. Among other processes measuring the void phase of porous or fractured media, scan line approach is a simplified "graphical" method, mainly used in image processing related procedures. In soil mechanics, the application of scan line method is related to the soil fabric, which is important in characterizing the anisotropic mechanical response of soils. Void fabric is of particular interest, since graphical approaches are well defined experimentally and most of them can also be easily used in numerical experiments, like the scan line method. This is in contrast to the definition of fabric based on contact normal vectors that are extremely difficult to determine, especially considering physical experiments. The scan line method has been proposed by Oda et al [1] and implemented again by Ghedia and O'Sullivan [2]. A modified method based on DEM analysis instead of image measurements of fabric has been previously proposed and implemented by the authors in a 2D scheme [3-4]. In this work, a 3D extension of the modified scan line definition is presented using PFC 3D®. The results show clearly similar trends with the 2D case and the same behaviour of fabric anisotropy is presented.

1 Introduction

Soil is a multiphase granular material consisting of solid particles and voids. Among other aspects of its multiphase character, the spatial distribution of the solids and the associated voids significantly characterizes the mechanical response of the material. The technical term fabric is used to define this spatial and geometric configuration of the soil structure elements that carries a strong flavour of orientation of these elements, related to the so-called fabric anisotropy in regards to mechanical response. As Oda et al [1] and Satake [5] suggested, the most important micromechanical elements whose orientations characterize fabric anisotropy are the solid particles, the interparticle contact normal directions, and void vectors related to corresponding voids.

Void fabric is related to the position, shape and orientation of the voids, the latter defined in various ways by means of appropriate unit void vectors. In order to analytically quantify these void fabric properties, a fabric tensor that incorporates all the important parts of this information is needed, and is defined by means of the void vectors. In this work, we present a 3D void fabric tensor based on the scan line method initially proposed by Oda et al [1] and modified in recent works by the authors in a 2-dimensional scheme [3-4].

The basis of the scan line method is the concept that by scanning a granular material sample with parallel lines with constant distance, voids can be very easily recognized and separated from solids (Fig. 1). Thus, by using parallel lines with inclination angle between -90° and 90°, the orientational distribution of voids is readily recognized, for every angle.

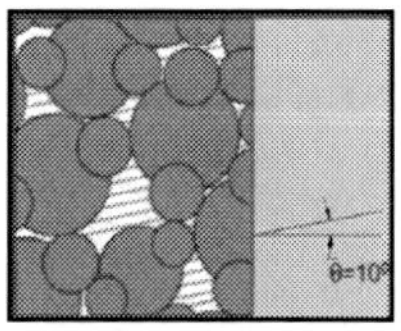

Fig. 1. Scan line 2D schematic application

Since the majority of constitutive models accounting for fabric are based on contact normal directions, and considering that contact normal fabric is more difficult to be experimentally determined than void fabric, our initial application of the modified scan line void fabric [3-4] aimed in finding a correlation between those two fabric types. This correlation has been strongly supported for 2 dimensions in [3-4]. The present work extends the applicability of that modified scan line method and explores the same correlation in 3 dimensions.

* Corresponding author: atheomec@mail.ntua.gr

2 3D fabric definitions

The scan line fabric tensor definition uses the mean length of the void segments in each direction as a weight to quantify the void segments distribution. Extending the original 2D definition by Oda et al [1] to 3D based on spherical coordinates, one has:

$$\mathbf{G}_{vs} = \frac{1}{\sum\limits_{\theta=-90}^{90}\sum\limits_{\varphi=0}^{90}\ell^{\varphi,\theta}} \sum_{\theta=-90}^{90}\sum_{\varphi=0}^{90}\ell^{\varphi,\theta}\mathbf{n}_{vs}^{\varphi,\theta}\otimes\mathbf{n}_{vs}^{\varphi,\theta} \quad (1)$$

where $n_{vs}^{\varphi,\theta}$ is the unit void vector for a point in spherical coordinates ([azimuth angle, polar angle, radial distance]=[θ,φ,1]) and is constant for a choice of these coordinates, ⊗ is the tensor product symbol, $\ell^{\varphi,\theta}$ is the mean length of scan lines (i.e. total length of line segments for angles (φ,θ) over the total number of scan line segments for the same angles), θ belongs to the interval (-90°, 90°] and φ belongs to the interval (0°, 90°]. Notice the trace of this tensor by definition is equal to 1.

Two important parameters apply on the calculation of this tensor: (a) the distance between the lines scanning each sample and (b) the minimum length (cut off) of the scan line void segments, above which they become significant. Both these and the reason for the cut off are analysed and explained in [3].

In this 3D application, Eq. 1 is implemented along with the modification of the cut off approach; that means that the void segments whose length is less than the cut off value, are ignored in the calculation of the fabric tensor. The cut off is defined similarly as in 2D, i.e. is equal to the minimum radius for a distribution of spheres in 3D and the distance between the parallel scanning lines is calibrated based on the particular sample.

The 3D contact normal fabric definition is straightforward and follows that of Satake [6]:

$$\mathbf{G}_c = \frac{1}{N_c}\sum_{N_c}\mathbf{n}_c\otimes\mathbf{n}_c \quad (2)$$

where $\mathbf{n}_c$ is the unit contact normal vector and N_c is the number of all the contact normal vectors.

For both cases of Eqs. (1) and (2), one can take the deviatoric part of the corresponding fabric tensors as characterizing fabric with no loss of generality, a process that can relate their evolution to deviatoric plastic strain rates in continuum theories motivated by DEM results.

3 DEM virtual experiments

The Discrete Element Method (DEM) has been used in order to calculate the scan line void fabric tensor for a 3D specimen with spheres as solid particles. Through the DEM it is easy to apply a loading path and follow the evolution of the fabric tensors, given that all the micromechanical details are known from the numerical analysis.

We follow a typical triaxial loading path with constant mean pressure (Fig. 2a). The particles' assemblage generation was made using the radius expansion method, since this method can produce initially isotropic fabric. Also, the final stress of the equilibrium state after the radius expansion was chosen to converge to the final mean stress; in that way, the homogeneity of the sample is improved relative to other methods of particles' assemblage creation, especially for loose samples and low confining pressures.

Notice also, that loose samples are preferred from dense samples in this analysis because the later enhance non-uniform deformation and strain localisation, which complicate fabric measurements and analysis. Finally, particles have been chosen to play the role of boundaries, and the shape of the sample was decided to be spherical (Fig. 2b). The spherical shape and the boundary particles decrease the possibility of arching, and thus, of non-uniform deformation during the triaxial loading. This specimen was used in the foregoing analysis.

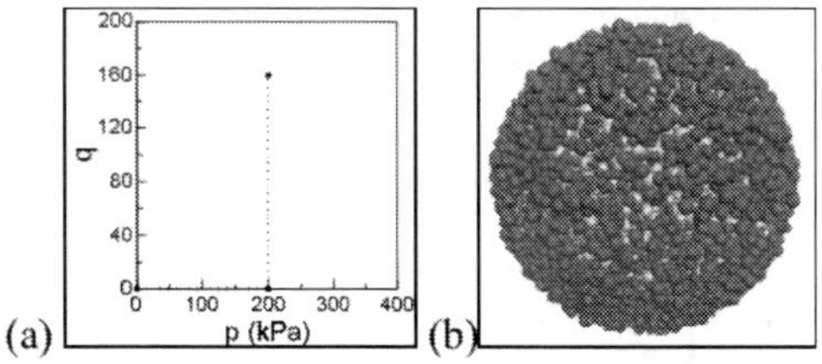

Fig. 2. (a) Loading path: isotropic compression and then triaxial compression with constant pressure p=200kPa and (b) sample's initial shape (dark grey are the boundary particles and light grey the sample inside)

The properties of the reference experiment are presented in Table 1. They were chosen based on the above considerations and also due to their relevance to those used in previous biaxial experiments in [3].

Table 1. Properties of the reference triaxial test.

Number of grains	13,723
Average grain radius (m)	0.001
Distribution; r_{min} ; r_{max} (m)	Uniform; 0.0005; 0.0015
Spherical shape initial radius (m)	0.0067
Particle Density (kg/m^3)	2600
Mean pressure (kPa)	200
k_n (N/m)	$2.2{\cdot}10^4$
k_s (N/m)	$2.2{\cdot}10^4$
Surface friction (ball-ball; ball-boundary ball)	0.5; 0.0

4 Fabric results

For the triaxial loading, by applying Eq. 1 and calculating the scan line fabric tensor in 3 dimensions, the final result is a 2nd order tensor having the non-diagonal terms very close to zero for our triaxial virtual experiment, i.e. fabric principal axes are aligned with the stress principal axes. The scan line fabric tensor

DOI: 10.1051/epjconf/201714012021

calculations also demonstrate the symmetry of the two horizontal axes, following the equality of the two horizontal stresses. Thus, the void fabric tensor actually has two independent principal values and subsequently the fabric anisotropy intensity, directly proportional to the fabric tensor norm, can be quantified by the difference of these two principal values denoted by α_{vs}.

The same observations apply for the contact normal fabric tensor and the respected fabric anisotropy intensity is quantified by the difference of the maximum and minimum principal values and denoted by α_c. The rest of the parameters related to the application of the 3-dimensional scan line method are defined and calibrated in the following paragraphs.

The distance between the scanning lines, after calibration, is set at $5{\cdot}10^{-5}$m=0.00005m; above this limit value the results of the scan line method are meaningfully different while below this value the results do not change significantly. This limit equals 1/10 of the size of the minimum grain radius.

Another important parameter of the implementation of the scan line method is the elimination of the inactive particles, i.e. particles that have less than two contacts. These particles (rattlers) in principle should be removed, as they do not contribute to the mechanical behaviour of the sample; they are thus treated as voids. On the other hand, in a real situation or physical experiment, the removal of these particles would be very difficult or impossible to account for. From that perspective two cases are examined: the first one where rattlers are treated as voids and the second one where they are treated as solids. The first case is completely compatible with the 2D case and is theoretically more precise, while the second one is more readily applicable.

Finally, another important issue is the volume of the scan line application. In our sample, the scan line method was applied both in the whole sample (excluding the boundaries) and inside a well-defined spherical Representative Volume (RV). The RV is representative for the macroscopic results of the sample (stress tensor, porosity); its radius is approximately 70% of the radius of the spherical sample and equal to 0.0045m and it contains approximately 5000 spheres.

In the first case, the total results of the scan line apply to the whole of the inside sample (boundary particles are not considered); in the second case the results are defined, in a uniform and representative area. Both results should not vary significantly, nevertheless, they are both considered and presented in the sequel to validate the results and prove uniformity of the samples.

The previous considerations provided 4 different types of application for the scan line method in the sample:

a: scan line applied to the whole sample and no elimination of grains

b: scan line applied to the whole sample and elimination of grains with less than 2 contacts

c: scan line applied to the RV and no elimination of grains.

d: scan line applied to the RV and elimination of grains with less than 2 contacts

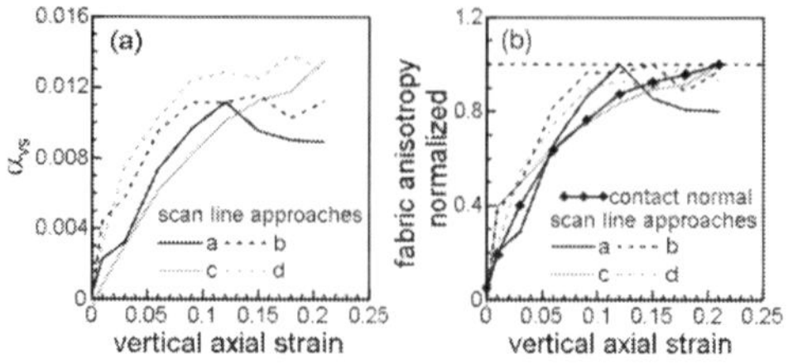

Fig. 3. Results of four variations of scan line void fabric anisotropy alongside the results for contact normal fabric anisotropy with respect to the vertical axial strain.

The evolution of the scan line anisotropy with respect to the vertical strain for all four scan line methods is presented in Fig. 3(a). The results are presented alongside the results for the contact normal fabric anisotropy in Fig. 3(b), normalized with respect to their maximum values; this kind of representation depicts the relation between the corresponding quantities with more clarity. Initially the fabric is isotropic and during the loading, anisotropy evolves gradually and monotonically - as typically for loose samples - towards its final steady state value.

The variations a and c, i.e. without elimination of rattlers for the RV and for the whole sample, evolve very similarly until 12% of axial strain; then the one for the whole sample, variation a, decreases, an unexpected response for this type of specimen and loading. This happens due to the measurement details of the boundaries of the whole sample (see Fig. 2(b)). While for the RV a solid sphere is defined as the boundary for the scan line measurement, on the whole sample no definite boundary shape has been considered; the parallel scanning lines stop at the final particle they intersect, so only "limit particles" exist. This produces the difference in the void anisotropy results, especially after significant deformation of the sample (12% of axial strain), when the sample's shape has evolved from spherical to an elliptical one.

Additionally, Fig. 3 demonstrates that the elimination of the rattlers slightly impacts the evolution of the anisotropy (variations b and d versus a and c). The elimination of the inactive particles produces an evolution smoother than without elimination; the intensity of the anisotropy reaches its final value in a lower axial strain and this critical state value is constant after 10% of axial strain. Notice also that variation b and d, because of these characteristics, are less qualitatively comparable to contact normal anisotropy intensity evolution, than variations a and c.

After Figure 3, despite the inconsistency presented, it is possible to use only the first variation, a, in our analysis, i.e. scan line to the whole sample with no elimination of grains with less than 2 contacts, because it is this kind of fabric tensor that can be most easily either calculated by DEM or experimentally measured by X-Ray techniques, for example. Thus, we focus on the direct comparison between the scan line void fabric anisotropy and the contact normal fabric anisotropy for our reference sample.

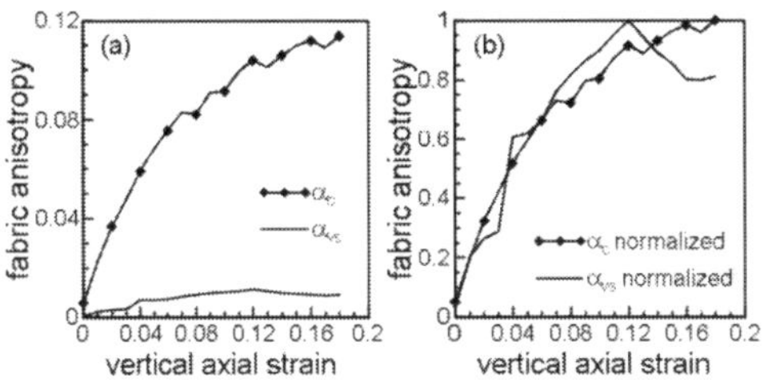

Fig. 4. Comparison between the scan line void fabric anisotropy and the contact normal fabric anisotropy (a) and their normalized values (b) with respect to the vertical axial strain.

The different scale in intensity, as presented in Fig. 4(a), is only quantitative, as the two anisotropies and the two fabric tensors are defined and calculated under a very different concept. Nevertheless, the normalized plots of anisotropies (Fig. 4(b)) show that the evolution of the two fabric anisotropies is very similar. The correlation between those two anisotropies becomes even clearer in Fig. 5. The final 5 points, that are below the correlation line, are due to the decrease of the void scan line anisotropy versus the contact normal that very slightly increases, and are due to the inconsistency discussed in the prequel.

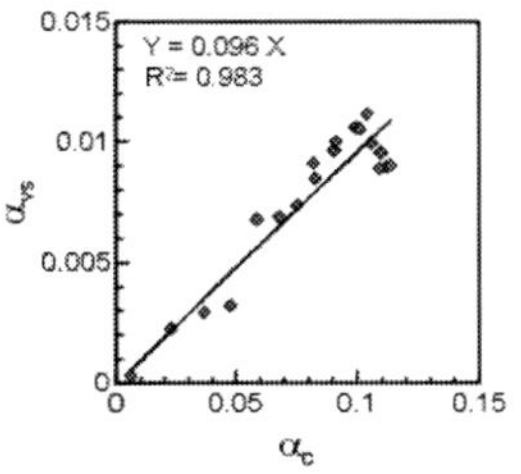

Fig. 5. Correlation between the scan line void fabric anisotropy and the contact normal fabric anisotropy

5 Conclusions

In this work a 3-dimensional application of the scan line method is proposed in order to quantify the void fabric of a granular medium. It is underlined that the scan line method as it has been already defined in 2D and now extended in 3D, seems to be able to adequately quantify the void fabric anisotropy and its evolution in 3 dimensions in granular materials. In addition, a similar correlation between scan line void fabric anisotropy and the contact normal fabric anisotropy seems to exist in 3 dimensions as in 2 dimensions. The linear slope of the 3D correlation is 0.096, significantly lower than the 2D analysis where the slope was 0.231. This happens since the contact normal anisotropy in terms of intensity is very similar for 2D and 3D, while the void scan line anisotropy intensity is in 3D one third of 2D.
The similarities between 2 and 3 dimensions are very advantageous; given that extensive research in 2 dimensions shows a clear correlation for various cases, similar behaviour can possibly be expected in 3 dimensions, towards which a more thorough investigation is in progress. Finally, a brief analysis on the importance of the inactive particles shows that the existence of those particles is not a significant barrier for the practical application of the scan line method in real cases.

The research leading to these results has received funding from the European Research Council under the European Union's Seventh Framework Program FP7-ERC-IDEAS Advanced Grant Agreement n° 290963 (SOMEF). Also, Itasca Education Partnership program is acknowledged for offering access to and mentorship on the software PFC v5.0.

References

1. M. Oda, S. Nemat-Nasser, J. Konishi, Soils and Foundations, **25**(3), 85–97 (1985)

2. R. Ghedia, C. O'Sullivan, Computers and Geotechnics, **41**, 1–12 (2012)

3. Theocharis A., Master's thesis, National Technical University of Athens, Athens, Greece, Retrieved from https://someferc.files.wordpress.com/2017/02/2015-thesis-theocharis_void_fabric_tensor.pdf (2015)

4. A. Theocharis, E. Vairaktaris, P. Fu, Y.F. Dafalias, *Geomechanics from Micro to Macro*, Cambridge U.K., K. Soga, K. Kumar, G. Biscontin, M. Kuo, CRC press 153 (2014)

5. M. Satake, Int. J. Eng. Sc., **30**(10), 1525–1533 (1992)

6. M. Satake, *Proceedings of the IUTAM conference on deformation and failure of granular materials*, Delft, Balkema press, H J Luger; P A Vermeer, 63 (1982)

Linear and nonlinear elastic properties of dense granular packings: a DEM exploration

Laure Lemrich[1,2,⋆], *Paul* Johnson[3], *Robert* Guyer[3,4], *Xiaoping* Jia[5], and *Jan* Carmeliet[1,2]

[1] *Chair of Building Physics, ETHZ, Wolfgang-Paulistrasse 15, CH-8093 Zurich, Switzerland*
[2] *Laboratory for Building Science and Technology, Überlandstrasse 129, CH-8600 Dübendorf, Switzerland*
[3] *Solid Earth Geophysics Group, Los Alamos National Laboratory, MS D443, Los Alamos, 87545 New Mexico, USA*
[4] *Department of Physics, University of Nevada, Reno (NV), USA*
[5] *Institut Langevin, ESPCI ParisTech, CNRS UMR 7587 - 1 rue Jussieu, 75005 Paris, France*

Abstract. Discrete Element Method modeling is used to study the frequency spectrum of particle motion in dense 3D packings of glass beads with Hertz-Mindlin contacts. Frequency sweeps show a dependency of the resonant frequencies on the drive amplitude and confining stress on the system, showing material changes in the system. The amplitude dependency of the second thickness mode $3\lambda/4$ as identified by the internal strain field scales as $f \propto \sigma^{1/6}$ while the confining stress dependency scales as $f \propto \sigma^{2/3}$, as predicted by Hertzian theory.

1 Introduction

Granular media exhibits complex behavior dependent on particle composition and contact network. Despite considerable efforts, it remains challenging to characterize a sample of material without disrupting or destroying it. Thus the development of nondestructive probing techniques to investigate the linear and nonlinear elastic behavior of granular media by wave propagation is an active field of research[1–3]. It remains however difficult to observe grain-scale mechanisms in a dense packing of particles, to understand and interpret probing results; DEM grain-scale modeling[4] allows us to study particle contacts and motion in granular media to understand how grain-scale mechanics affects wave transmission in the material. This work contributes to the research on nonlinear behavior in granular materials by studying resonant frequencies [5–8], and uses as reference the oedometric cell experiments of Jia et al.[9, 10] on acoustic waves propagation in dense packing of glass beads.

We build 3D dense packings of glass beads using Discrete Element Methods with Hertzian contacts (Fig. 1). On each packing equilibrated at a confining stress σ we perform a series of independent frequency sweeps with increasing drive strain ε_{drive} (Fig. 2). We report the local strain inside the material and extract one resonant mode of the system and study this mode as a function of the confining stress and drive strain (Fig. 3). We show that the results follow predictions from Hertzian theory. Finally, we present the dynamic vertical strain profile for a selection of frequency sweeps and formally identify the second longitudinal bulk mode $3\lambda/4$.

⋆e-mail: laure.lemrich@alumni.epfl.ch

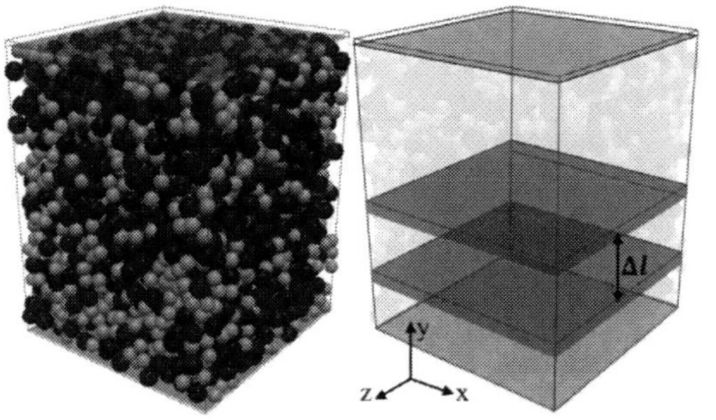

Figure 1. (left) Picture of a packing at σ = 10 kPa, the walls are in blue. Beads have radii $3 \cdot 10^{-6}$ m (yellow) and $4 \cdot 10^{-6}$ m (red). (right) scheme of the strain measurement protocol. The vertical position of two layers of particles (in red) are recorded over time, the differential distance Δl between each layer gives the local strain (see eq. 8).

2 Methodology

The simulations are implemented using the Discrete Element Method (DEM) code LIGGGHTS[11]. Particles are generated in a simulation box with periodic boundaries along the x- and z-axis, and enclosed by two walls along the y-axis. These walls are implemented as particles of infinite radius and mass.

Particles interact through Hertz-Mindlin contact mechanics with rolling resistance and viscous damping. Hertz-Mindlin contact law implements an accurate representation of normal elastic interaction between two rigid bodies, and a simplified version of tangential contacts which assumes that no micro-slip occurs at the contact interface between two particles; particles either 'stick' or 'slip', if they reach the Coulomb threshold. This nonlinear contact force allows us the study of nonlinear behavior in granular material, while limiting the cost of solving tangential contacts. The force $\boldsymbol{F}_i$ on each particle i is defined as the sum of its interactions with other particles j and viscosity factor:

$$\begin{aligned}\boldsymbol{F}_i &= \boldsymbol{F}_n + \boldsymbol{F}_t - \gamma_a \boldsymbol{v}_i \\ &= (k_n \delta n_{ij} \hat{\boldsymbol{n}}_{ij} - \gamma_n \boldsymbol{v}_{ij} \cdot \hat{\boldsymbol{n}}_{ij}) \\ &+ (k_t \delta t_{ij} \hat{\boldsymbol{t}}_{ij} - \gamma_t \boldsymbol{v}_{ij} \cdot \hat{\boldsymbol{t}}_{ij}) - \gamma_a \boldsymbol{v}_i\end{aligned} \quad (1)$$

where the tangential overlap $\boldsymbol{\delta t}_{ij}$ truncated to fulfil $\|\boldsymbol{F}_t\| < x_\mu \|\boldsymbol{F}_n\|$ (with the friction coefficient $x_\mu = 0.22$). $\hat{\boldsymbol{n}}_{ij}$ and $\hat{\boldsymbol{t}}_{ij}$ are the normal and tangential unit vectors of each contact ij. $\boldsymbol{v}_i$ is the particle velocity ($\boldsymbol{v}_{ij}$ the difference in particle velocity), δn_{ij} the normal overlap between particles i and j (Hertz theory) and δt_{ij} the tangential overlap (simplified Mindlin Theory).

A small viscous damping with coefficient $\gamma_a = 10^{-7}$ kg/s stabilises the algorithm and approximates the effect of dry air at room temperature.

For the grain material, the Young's modulus $Y = 65$ GPa, Poisson ratio $\nu = 0.25$ and loss coefficient $\beta = 0.0163$ (≥ 0) give a shear modulus $G = Y/(4(2-\nu)(1+\nu))$, with the elastic constants given by:

$$k_n = \frac{4}{3} Y \sqrt{R^* \delta n_{ij}} \quad (2)$$

$$k_t = 8G \sqrt{R^* \delta n_{ij}} \quad (3)$$

and the viscoelastic parameters given by:

$$\gamma_n = 2\beta (R^* \delta n_{ij})^{1/4} \sqrt{\frac{5}{3} m^* Y} \quad (4)$$

$$\gamma_t = 4\beta (R^* \delta n_{ij})^{1/4} \sqrt{\frac{10}{6} m^* G} \quad (5)$$

The effective radius R^* and mass m^* depend on the respective radii R and masses m of the two interacting particles: $1/R^* = 1/R_i + 1/R_j$ and $1/m^* = 1/m_i + 1/m_j$.

The rotation velocity of each particle i is updated as:

$$I_i \frac{d\omega_i}{dt} = \boldsymbol{r}_i \times \boldsymbol{F}_{t,i} + \boldsymbol{T}_{i,j} \quad (6)$$

with the moment of inertia I_i, the vector from the center of the particle i to the contact point $\boldsymbol{r}_i$, the tangential component of the force exerted on the particle i $\boldsymbol{F}_{t,i}$ and a torque at each contact:

$$\boldsymbol{T}_{i,j} = \mu_r k_n \delta n_{ij} R_i \frac{\boldsymbol{\Delta\omega}_{ij}}{\|\boldsymbol{\Delta\omega}_{ij}\|} \quad (7)$$

where $\mu_r = 0.01$ is the rolling resistance and $\Delta\omega_{ij} = \omega_i - \omega_j$ is the relative angular velocity[12].

The timestep for all simulations is $dt = 8 \cdot 10^{-8}$ s.The Discrete Element approach is a step-wise integration of Newton's equations.

Each sample is independently prepared (a sample packing is represented on the left of Fig. 1 (left)). A random packing of bidisperse particles is generated in the simulation box. The bottom wall (in $y = 0$) remains static, while the top wall is servo-controlled to apply a target confining stress σ. A gradient velocity field is applied to particles to smoothly compact them (maximum initial velocity $v_{max} = 6$ m/s). At each time-step, the algorithm will attempt to reach the confining stress specified for the top wall by testing several displacements and choosing the one closer to target stress (with a velocity limit of v_{max}). In the beginning of the simulations, particles and top wall move downward and particles compact against the bottom wall. After 0.2 s, a sinusoidal vertical movement $A\sin(\omega t)$ is applied to the bottom wall to perturb the system. The amplitude A of the perturbation is defined such that the applied dynamic strain equals $A/l_y = 5 \cdot 10^{-5}$. The frequency is set to $\omega = 2\pi \cdot 10$ kHz. This perturbation is stopped after 1 s and the sample is allowed to relax for 7.8 s. Energy is dissipated via viscous damping and the loss coefficient β.

Samples are prepared with simulation box dimensions of $l_x = l_z = 10^{-2}$ m and an initial height of $2.5 \cdot 10^{-2}$ m (the final sample height $l_y \approx 1.2 \cdot 10^{-2}$ m depending on the sample). Approximatively 4100 particles of radii $3 \cdot 10^{-4}$ m and $4 \cdot 10^{-4}$ m are created and compacted. This procedure creates independent packings identified by their confining stress $\sigma = 10$ kPa, 22 kPa, 55 kPa and 113 kPa.

Each frequency sweep is performed using the same initial sample preparation. A packing is loaded and the bottom wall is servo-controlled to maintain the target confining stress. The top wall is vertically displaced with a sinus $A\sin(\omega t)$, increasing $\omega = 2\pi f$ step by step. To produce the data presented in Fig. 2, a packing at 10 kPa was driven from $f = 1$ kHz to $f = 103$ kHz by steps of 300 Hz. Each frequency was maintained for 100 periods to let the system reach a dynamic quasi steady-state. The drive amplitude A ranges from $5 \cdot 10^{-8}$ to $6 \cdot 10^{-5}$ m (ε_{drive} from $4.13 \cdot 10^{-6}$ to $4.95 \cdot 10^{-3}$ m). The vertical position of two layers of particles is recorded 50 times per period over the whole frequency sweep. These layers include all particles with initial height $y_i \in 0.2 l_y \pm 2 \cdot 10^{-2}$ m and $y_i \in 0.4 l_y \pm 2 \cdot 10^{-2}$ m respectively (about 120 particles per layer) to eliminate boundary effects.

We define the strain $\varepsilon_{\omega_i}(t)$ with the height difference between the two encompassing layers $l1$ and $l2$ (shown on Fig. 1 (right)). The layer height is $l_y(t) = \boldsymbol{y}_{l2}(t) - \boldsymbol{y}_{l1}(t)$ and we define the local strain on the last 60 periods of oscillation at ω_i as:

$$\boldsymbol{\varepsilon}_{\omega_i}(t) = (\boldsymbol{\Delta l}(t) - < \boldsymbol{\Delta l}(t) >) \,/ < \boldsymbol{\Delta l}(t) > \quad (8)$$

The local strain amplitude $\varepsilon_{local}(\omega_i)$ is the variation amplitude of $\boldsymbol{\varepsilon}_{\omega_i}(t)$. We therefore detrend the strain $\boldsymbol{\varepsilon}_{\omega_i}(t)$ by applying a moving-average window of one-period length, and fit the sinus function $\varepsilon_{local}(\omega_i)\sin(\omega_i t + \phi_i)$.

The ratio of ε_{local} versus drive strain $\epsilon_{drive} = A/l_y$ is reported in Fig. 2 in function of the drive frequency f for different drive amplitudes.

The sample is excited by vertically moving the topwall and letting the bottom wall free to adjust to maintain con-

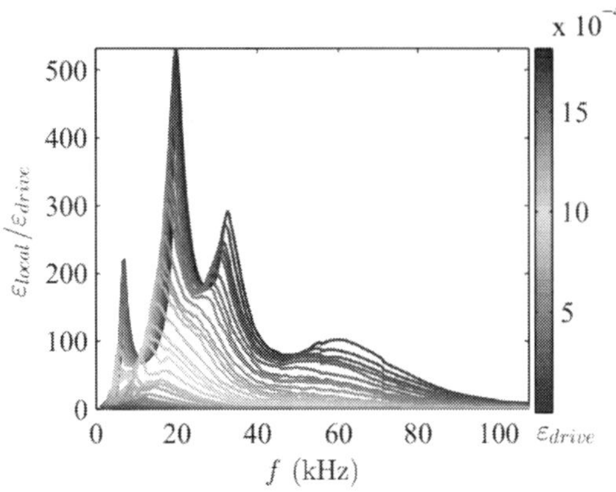

Figure 2. Frequency sweeps at progressively increasing amplitudes on a packing loaded at σ = 10 kPa. The top wall inputs a perturbation at frequency f and strain ε_{drive}. The strain amplitude ε_{local} is recorded close to the bottom of the box. The three first peaks correspond to the modes $1\lambda/4$, $3\lambda/4$ and $5\lambda/4$.

fining stress. As a consequence, a standing wave will have an antinode at the top wall, and a node at the bottom wall. The first mode to satisfy these conditions is 1/4 of a wavelength λ. The harmonics are then $3\lambda/4$ (outlined in black on Fig. 4 (left)) and $5\lambda/4$.

3 Results and discussion

Packings at 55 kPa and 113 kPa were probed with compression and shear tests; at σ = 55 kPa, the sample has a Young's modulus $E = 1.36 \cdot 10^8$ kPa and Poisson ratio ν = 0.39. The sample at σ = 113 kPa has $E = 1.66 \cdot 10^8$ kPa and ν = 0.40. These results are of the same order of magnitude as results from similar simulations[13] and from experiments[14].

Fig. 2 shows a series of frequency sweeps performed on a packing at σ = 10 kPa. Each sweep was independently performed on the same initial packing, varying only the drive amplitude. Three main peaks are visible at low drive strain ε_{drive}, which correspond to the modes $1\lambda/4$, $3\lambda/4$ and $5\lambda/4$. As the amplitude increases, peaks broaden (i.e dissipation increases in the system) and shift to lower frequencies. The dependency of the resonant modes on the drive amplitude is a nonlinear feature of the system, and shows a weakening of the material resulting from the increased drive amplitude. This result is in qualitative agreement with experimental observations [10]. Equivalent measurements were made for the confining stress 22 kPa, 55 kPa and 113 kPa.

For each packing and each sweep, the resonant frequency, drive strain and local strain of the mode $3\lambda/4$ are reported in Fig. 3 (top). Hertzian contacts define the contact force between two spheres as $F = Y\sqrt{R^*}\,\delta n^{3/2}$[14]. We therefore expect the applied confining stress σ and the

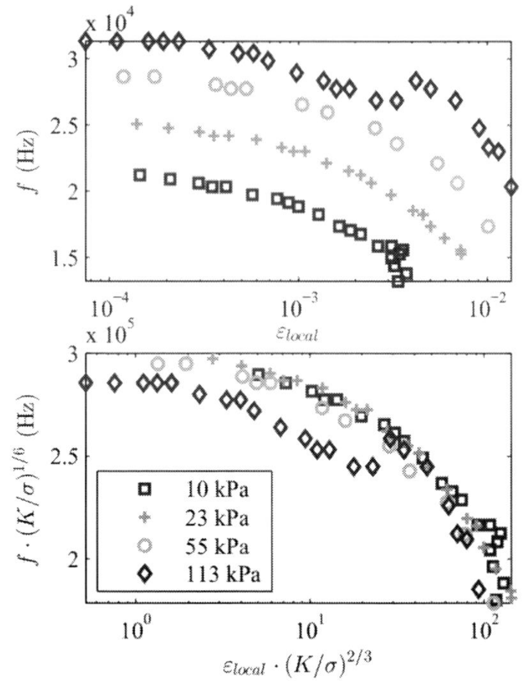

Figure 3. (top) The resonant frequency of the mode $3\lambda/4$ are extracted from frequency sweeps (see Fig. 2) of four packings at specific confining stresses σ, and reported here. (bottom) The frequency f and local strain ε_{local} of the mode are scaled according to predictions of Hertzian theory. The results overlap and show a weakening of the material at higher drive amplitude/local strain for all confining stresses.

quasi-static strain ε_0 due to σ to evolve as:

$$\sigma \approx \frac{F}{R^{*2}} = Y\left(\frac{\delta n}{R^*}\right)^{3/2} \tag{9}$$

$$\varepsilon_0 = \frac{\delta n}{R^*} = \left(\frac{\sigma}{Y}\right)^{2/3} \tag{10}$$

where the associated elastic constant $K_c = \partial\sigma/\partial\varepsilon_0 = Y(\sigma/Y)^{1/3}$. With the sample density ρ, the related frequency follows:

$$f = \frac{1}{l_y}\sqrt{\frac{K_c}{\rho}} \propto \left(\frac{\sigma}{Y}\right)^{1/6} \tag{11}$$

We therefore expect the resonant frequency to scale as $f \propto (\sigma/K)^{1/6}$ and $\varepsilon_{local} \propto (\sigma/K)^{2/3}$. In Fig. 3 (bottom) we present rescaled data from Fig. 3 (top). All four datasets collapse on the same line, indicating that our samples follow Hertzian behavior. Furthermore, we observe that for all confining stresses the material weakens as the local strain increases. This indicates a change in the material behavior due to particles rearranging.

In Fig. 4, we present the vertical strain profile over the packing for frequencies of the mode $3\lambda/4$ at three drive

 DOI: 10.1051/epjconf/201714015033

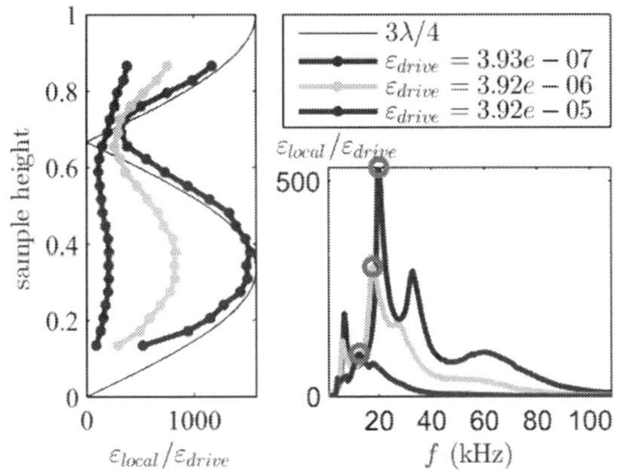

Figure 4. Formal identification of the mode studied in Fig. 3 for a packing at σ = 10 kPa. On the bottom right are the frequency sweeps of the three studied drive strain ε_{drive}. In red are the pairs (ε_{drive}, $f_{3\lambda/4}$). (left) The system is driven with each pair, and the local strain amplitude is reported along the height of the system. The black line shows the ideal result for the mode $3\lambda/4$.

strains ε_{drive}. On the right are shown the selected frequency sweeps with, circled in red, the peaks at $f_{3\lambda/4}$. The packing at σ = 10 kPa is vibrated for 200 periods at the three pairs (A, $f_{3\lambda/4}$). Each local strain is measured by two neighbouring layers of $2 \cdot 10^{-4}$ m of height, at a distance of $0.214l_y$ following the lock-in procedure presented in the methodology (figure on the right presents local strains computed over $0.2l_y$) and plotted in Fig. 4 (left). Outlined in black is the theoretical mode of $3\lambda/4$. We observe that all three local strains qualitatively follow the predicted resonant mode. The deviation from the theoretical outline (notably the node at $0.66l_y$) is likely due to heterogeneity in the material, leading to local differences in behaviour.

4 Conclusion

A 3D model of granular particles was built using Discrete Element Method with Hertz-Mindlin contact model and rolling resistance. Samples were packed at target confining stresses and probed across a range of frequencies. Model parameters and processing methods were carefully chosen to obtain a high precision and detect frequency shifts of less than 100 Hz. Clear resonant frequencies can be observed for drive strains ε_{drive} as low as $4.13 \cdot 10^{-6}$, which do not modify the sample and show linear behavior.

A shift in resonant frequency at an increasing drive strain, for several resonant modes in the system are observed. The mode $3\lambda/4$ is identified by vertical dynamic strain profiles through the samples, and its dependency on the confining stress shown to follow predictions of Hertzian theory. These results show a weakening of the sample at higher local strains, showing a change in the particle arrangement of the material. Research has shown that contacts have a significant impact on resonant frequency and damping, with humidity or ageing shifting resonance[15]. The relation of these factors, including confining stress, to an observed resonant frequency requires further research.

In future research we analyse the response of packings at confining stresses above 100 kPa. Since frequency sweeps are non-destructive probing techniques of material, we hope to contribute to knowledge regarding its application to the testing of granular samples.

Acknowledgements

Paul Johnson and Robert Guyer gratefully acknowledge the US Department of Energy Office of Science and support from EMPA.

References

[1] V. Tournat, V. Gusev, Acta Acustica united with Acustica **96**, 208 (2010)
[2] D. Johnson, L. Schwartz, D. Elata, J. Berryman, B. Hornby, A. Norris, Journal of Applied Mechanics **65**, 380 (1998)
[3] L. Ostrovsky, P. Johnson, Rivista del nuovo cimento **24**, 1 (2001)
[4] H. Zhu, Z. Zhou, R. Yang, A. Yu, Chemical Engineering Science **62**, 3378 (2007)
[5] K.A. Van Den Abeele, J. Carmeliet, J.A. Ten Cate, P.A. Johnson, Journal of Research in Nondestructive Evaluation **12**, 31 (2000)
[6] C.O. Reichhardt, L. Lopatina, X. Jia, P. Johnson, Physical Review E **92**, 022203 (2015)
[7] C. Wensrich, A. Collard, Powder technology **166**, 30 (2006)
[8] C. Windows-Yule, A. Rosato, A. Thornton, D. Parker, New journal of physics **17**, 023015 (2015)
[9] X. Jia, T. Brunet, J. Laurent, Physical Review E **84**, 020301 (2011)
[10] P.A. Johnson, X. Jia, Nature **437**, 871 (2005)
[11] C. Kloss, C. Goniva, Supplemental Proceedings: Materials Fabrication, Properties, Characterization, and Modeling, Volume 2 pp. 781–788 (2011)
[12] C. Goniva, C. Kloss, N.G. Deen, J.A. Kuipers, S. Pirker, Particuology **10**, 582 (2012)
[13] I. Agnolin, J.N. Roux, Physical Review E **76**, 061304 (2007)
[14] H.A. Makse, N. Gland, D.L. Johnson, L. Schwartz, Physical Review E **70**, 061302 (2004)
[15] C.J. Hsu, D.L. Johnson, R.A. Ingale, J.J. Valenza, N. Gland, H.A. Makse, Physical review letters **102**, 058001 (2009)

DOI: 10.1051/epjconf/201714006030

Towards minimal models for realistic granular materials: Tomographic analysis of bidispersed assemblies of ellipsoids

Fabian M. Schaller[1,⋆], *Horst* Punzmann[2], *Gerd E.* Schröder-Turk[3,1,⋆⋆], and *Mohammad* Saadatfar[4,⋆⋆⋆]

[1] *Institut für theoretische Physik, Friedrich-Alexander-Universität Erlangen-Nürnberg, Erlangen, Germany*
[2] *Centre for Plasmas and Fluids, Research School of Physics and Engineering, Australian National University, Canberra ACT 2601, Australia*
[3] *School of Engineering and Information Technology, Murdoch University, Murdoch WA 6150, Australia*
[4] *Department of Applied Mathematics, Research School of Physics and Engineering, Australian National University, Canberra ACT 2601, Australia*

Abstract. In this paper, we report experimental results on granular compaction in a model system made of mono- and bidisperse ellipsoidal packings as well as sand packings with grain size polydispersity. The packings are subject to vertical tapping of varying duration (number of taps) and their internal three-dimensional structure is obtained using x-ray computed tomography. Particles positions and orientations are reconstructed and the global packing densities are computed. The analysis of the vertical and horizontal local packing fraction profiles reveal a homogeneous densification in the ellipsoidal packings, however, sand packings exhibit radial density gradient, possibly linked to the onset of convection.

1 Introduction

Granular materials are large conglomerations of discrete macroscopic particles. Yet despite this seeming simplicity, a granular system exhibits complex behaviours, such as size segregation [1], dilatancy [2], creep [3, 4], glassy behaviour [5, 6], shear banding [7] and pattern formation [8, 9]. There are a number of factors that are responsible for this rich diversity; chief amongst them are: grain-grain friction, grain size distribution and particle shape[1]. Friction introduces a sharp nonlinearity in grain contact laws resulting in complex energy dissipation mechanisms. Particle shape and size distribution are two other primary parameters that determine granular properties at both micro- and macro-scale.

One particular physical phenomenon that has fundamental scientific value as well as outstanding industrial applications is the process of granular compaction. Understanding the connection between granular compaction and the slow relaxation of such out-of-equilibrium a-thermal systems [10], is one of the main themes of the field of Soft Matter [11, 12]. On the application side; due to the fact that granular compaction results in porosity reduction (increase in packing density) and permeability changes, it has become the research focus for pharmaceutical and transport industries. In geomechanics and carbon storage, granular compaction and deformation of soil and rocks are the key mechanisms in predicting landslides [13] and long term safety of carbon storage [14].

Characterisation of compaction process at the grain scale is complex due to the addition of the time dimension. As the compaction takes place in a granular medium, the structure settles down in a permanent but metastable static configurations. A large part of the problem arises from the strongly nonlinear contact law between continually rearranging rigid bodies coupled with the dynamical nature of the contact network. In this context, the spatial distribution and temporal evolution of the free volume available to individual particles is a salient quantity that expands from micro- (local Voronoi volume) to macro- (global packing density) scales.

Microscopic (grain-scale) models for granular materials need to incorporate physical parameters that are pertinent at the particle scale such as friction coefficients, compressibility, etc. They also need to take into account distribution properties such as the variability in grain size and shape. In this paper, we aim to reduce the model complexity to essential ingredients by identifying the key mechanisms at play that govern the *granular compaction*. The key question we aim to address is the extension of the 'zero-th order' model –the spherical case– to model with first- and second-order complexity. Recent years have seen many studies addressing the effects of the particle shape while maintaining uniform particles. Ellipsoids [15, 16], tetrahedra and polyhedra, pear-shaped particles, etc. have all been investigated. There has also been substantial in-

⋆e-mail: fabian.schaller@fau.de
⋆⋆e-mail: g.schroeder-turk@murdoch.edu.au
⋆⋆⋆e-mail: mohammad.saadatfar@anu.edu.au

[1] Assuming rigid particles.

DOI: 10.1051/epjconf/201714006030

Figure 1. Pharmaceutical placebo pills with two different aspect ratios

Figure 2. Loose packing preparation

vestigations into physical properties [15] that specifically change the compressibility [17].

Here, we focus on a different effect, namely the the compaction of bidisperse ellipsoidal & spherical particles via tapping. In addition, we study compaction in sand packings to better understand the effect of particle shape irregularity. We investigate the effect of mixtures of ellipsoids of identical aspect ratio but different size. Evidently, the question of particle size distribution has been addressed before in systems of spherical particles, specifically in the context of glass forming systems [18, 19], as well as in the Apollonian packings [20, 21]. Also, in studies of two-dimensional glass forming systems, bidispersity is a very commonly used tool to prevent crystallisation. However, the effect of size polydispersity of ellipsoidal packings is not extensively documented in the literature, even though it may have strong repercussions e.g. for the Voronoi cell distributions and hence local packing fractions.

Our results have practical applications in domains such as pore description in soil- and geosciences [22], which are crucial for understanding natural systems mechanical stability and flow permeability. Moreover, at a fundamental level, our results provide an opportunity to better our understanding of the non-equilibrium physics of compacting granular assemblies.

2 Experiment

To study the properties of bidisperse ellipsoid packings, we use two types of pharmaceutical placebo pills with an aspect ratio $\alpha \approx 0.57$ and large axis lengths of 8.9 mm and 10.2 mm, as shown in figure 1.

We start the preparation of the packing by poring the particles through a funnel into a cylindrical Perspex container of 144mm diameter, as shown in figure 1. A set of steel wire grids has been placed at the bottom of the container before the fill. The circular grids have a square mesh size of $\approx$ 20mm, spatially separated by $\approx$ 25mm in height and their mesh orientation is rotated by 45degrees with respect to each other. During the filling process, the container is placed on a motorised, slowly rotating platform ($\approx$ 0.2 Hz) to create a homogeneous and isotropic packing. The grid is slowly pulled out to loosen packing before the experiments. As shown in figure 3, this technique leads to a reproducible initial condition with a global packing fraction of 0.655$\pm$ 0.002. This preparation method is similar to the technique used for the study of monodisperse ellipsoidal particles [15].

The container is subsequently mounted atop an electro-dynamic shaker (TIRA TV51140). This 3kN shaker has a comparatively large diameter, 300mm armature which provides excellent lateral stability. The acceleration at the container base is measured using an accelerometer (B&K 4507, 1000mV/g) via the shaker systems motion controller (Vibration Research, VR9500). The packings are now compacted by applying sinusoidally shaped pulses to the shakers linear amplifier. The chosen peak acceleration of 2 g (where g = 9.81m/s^2) is higher than the earth's gravitational acceleration to allow the particles to move and rearrange [10]. The pulse width is 50 ms with a repetition rate of 3 Hz. By varying the number of taps from 0 up to 1,000,000, packings of different densities are created, see figure 3. Each experiment (dataset) starts with a new initial preparation. Packings of 0 taps (initial loose packings), 100 taps and 1,500 taps are repeated several times to assure consistency. The variation in packing fraction is less than 0.5% for the initial packings reduces to 0.05% for 1500 taps.

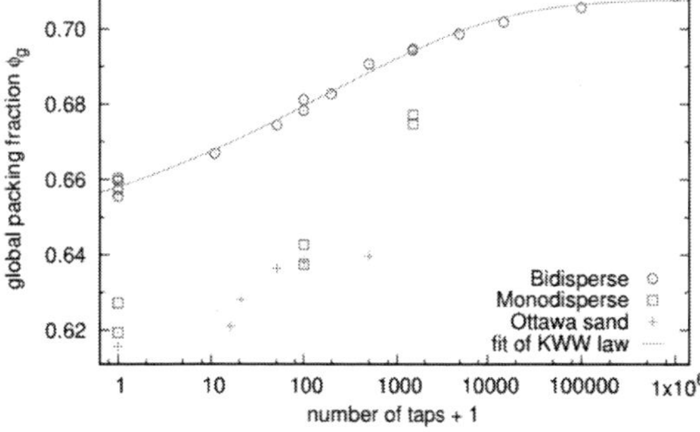

Figure 3. Number of taps vs. global packing fraction. Data of monodisperse and bidisperse ellipsoids is shown, as well as data of Ottawa sand packings. The orange curve is the best fit of the KWW law, see equation 1.

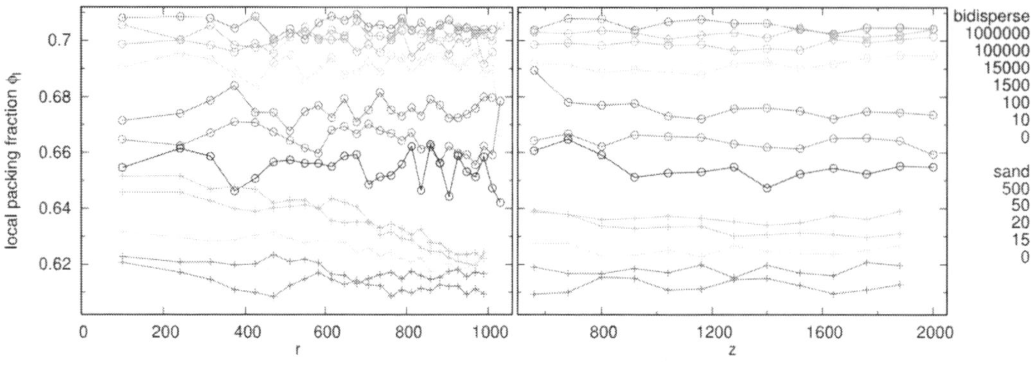

Figure 4. Radial and hight profile of the packing fraction for tapped packings of bidisperse ellipsoids and Ottawa sand. r is the radial distance from the rotation axis of the container and z the height from the bottom of the container in voxel units. Voxel size is 56μm for ellipsoids and 9μm for sand. The local packing fraction is averaged over all particles in each bin.

The resulting packings are imaged by helical x-ray tomography [23–25] and the particles positions and orientations are detected in the reconstructed image by a method based on a watershed algorithm [26, 27]. Particles close to the boundaries (cylinder walls, top and bottom) are removed from the analysis. The resulting packing is checked for spatial homogeneity in packing density. As a second dataset we show packings of Ottawa sand. The diameter of the sand grains is between 500 and 1000μm. Loose sand packings are prepared in the same way as the bidisperse ellipsoid packings, by pouring the particles in a cylindrical Perspex container of 24mm diameter and pulling a double grid through the packing to loosen it. The packings are again compacted by sinusoidal taps with peak acceleration of 2g. Figure 3 shows the global packing fraction vs. the number of applied taps for the bidisperse ellipsoid packing and the packings of Ottawa sand. As a reference, we include data for packings of monodisperse ellipsoids with approx. the same aspect ratio $\alpha = 0.59$ prepared with a similar preparation method, which was published elsewhere [15].

3 Results

Figure 3 indicates that both systems show densification and density saturation, however, the ellipsoidal packing shows a very slow density saturation with a gentle change of densification starting at $\approx$ 10 taps and another transition to density saturation at $\approx$1,000 taps. The sand packing, however, exhibits a relatively sharp transition from the onset of densification ($\approx$ 15 taps) to density saturation at $\approx$ 50 taps. The slow densification in the ellipsoidal packings is reminiscent of relaxation and compaction dynamics in glassy systems, which are commonly fitted by a stretched exponential law, the KWW (Kohlrausch, Williams, Watts) law [10, 12]

$$X_{\text{inf}} - (X_{\text{inf}} - X_0) * \exp\left[-(\#\text{taps}/\tau)^{\beta}\right] \quad (1)$$

Figure 3 shows a fit of the KWW law to our data of bidisperse ellipsoids, which is in very good agreement. In contrast, the global packing density in the sand packing, however, fully saturates at $\approx$ 50 taps.

To better understand this sharp contrast in the behaviour of global packing density, we compute radial (edge to centre) and vertical (direction of gravity) density profiles in these systems by using Set Voronoi diagrams [28]. The local packing fraction of a particle is defined as the ratio between particles volume and its Voronoi cell. Fig. 4 shows the radial and vertical density profile of selected packings. The plots show that unlike the ellipsoidal packing, the radial density profile of the sand packing is inhomogeneous and the density monotonically decreases towards the middle of sand packing. This is consistant with our observation of sand pile-up in the middle of sand packings after 15 taps. Both ellipsoidal and sand packings exhibit homogeneous vertical density profiles.

4 Conclusions and future work

We have shown that using a "double-grid" mesh system combined with pouring grains into a rotating container, we are able to generate reproducible loose (initial state of tapping) of bidisperse ellipsoid and sand packings. The packings are then subject to a fixed (2g) tapping with varying number of taps. We compute the local and global packing density after each tapping stage. The ellipsoidal packings densify very slowly with the number of taps without showing any sign of structural inhomogeneity. The sand packings, however, exhibit sharp density-increase only after 15 taps and the global packing density saturates at 50 taps. By analysing the radial and vertical density profile of the sand packings, we find strong structural inhomogeneity first appearing at 15 taps and increasing up to 50 taps. The radial density inhomogeneity combined with our observation of the formation of sandpile in the middle of the packings suggests that convection may be at play inside the sand packings.

In sand packings, compaction slows down to almost full saturation very quickly and we also observe convection. A possible reason for it is the frictional forces; between the particles and the container walls, which are more crucial because the sand particles are much smaller and lighter than the placebo ellipsoids. We are currently studying the effect that container roughness migh have on the onset of convection and the compaction of sand packings.

Another future direction is to study the effect of friction by using ellipsoids of the same size and shape but different materials. Furthermore, we aim to investigate particle size segregation by varying experimental parameters.

Acknowledgements

We thank Weimer Pharma GmbH for providing the placebo pills. FMS acknowledges funding by the German Science Foundation (DFG) through the research group "Geometry and Physics of Spatial Random Systems" under Grant No. SCHR-1148/3-2 and. FMS acknowledges funding by Marcelja Fellowship at ANU. MS thanks Tim Senden for his continued support.

References

[1] A. Kudrolli, Reports on progress in physics **67**, 209 (2004)
[2] O. Reynolds, **12** (1886)
[3] H.M. Jaeger, C.h. Liu, S.R. Nagel, Phys. Rev. Lett. **62**, 40 (1989)
[4] J.A. TenCate, E. Smith, R.A. Guyer, Phys. Rev. Lett. **85**, 1020 (2000)
[5] P. Charbonneau, J. Kurchan, G. Parisi, P. Urbani, F. Zamponi, Nature communications **5** (2014)
[6] A. Seguin, O. Dauchot, Phys. Rev. Lett. **117**, 228001 (2016)
[7] M. Saadatfar et al. (2012)
[8] G.H. Ristow, *Pattern formation in granular materials*, 164 (Springer Science & Business Media, 2000)
[9] N. Francois, M. Saadatfar, R. Cruikshank, A. Sheppard, Physical Review Letters **111**, 148001 (2013)
[10] P. Philippe, D. Bideau, EPL (Europhysics Letters) **60**, 677 (2002)
[11] M. van Hecke, Journal of Physics: Condensed Matter **22**, 033101 (2010)
[12] J.B. Knight, C.G. Fandrich, C.N. Lau, H.M. Jaeger, S.R. Nagel, Phys. Rev. E **51**, 3957 (1995)
[13] K. Hutter, *Avalanche dynamics* (Springer, 1996)
[14] S. Holloway, Energy Conversion and Management **38**, S241 (1997)
[15] F.M. Schaller, M. Neudecker, M. Saadatfar, G.W. Delaney, G.E. Schröder-Turk, M. Schröter, Phys. Rev. Lett. **114**, 158001 (2015)
[16] F.M. Schaller, S. Kapfer, J. Hilton, P. Cleary, C.D.M. K. Mecke, T. Schilling, M. Saadatfar, M. Schröter, G. Delaney, G. Schröder-Turk, EPL (Europhysics Letters) **111**, 24 (2015)
[17] M. Saadatfar, A.P. Sheppard, T.J. Senden, A.J. Kabla, Journal of the Mechanics and Physics of Solids **60**, 55 (2012)
[18] M. Maiti, S. Sastry, The Journal of chemical physics **141**, 044510 (2014)
[19] D. Fiocco, G. Foffi, S. Sastry, Physical Review E **88**, 020301 (2013)
[20] R.M. Baram, H. Herrmann, N. Rivier, Physical review letters **92**, 044301 (2004)
[21] D. Stäger, H. Herrmann, arXiv preprint arXiv:1607.08391 (2016)
[22] A. Sufian, A.R. Russell, A.J. Whittle, M. Saadatfar, Granular Matter **17**, 727 (2015)
[23] A. Katsevich, SIAM Journal on Applied Mathematics **62**, 2012 (2002)
[24] T. Varslot, A. Kingston, G. Myers, A. Sheppard, Medical Physics **38**, 5459 (2011)
[25] A. Sheppard et al., Nuclear Instruments and Methods in Physics Research Section B: Beam Interactions with Materials and Atoms **324**, 49 (2014)
[26] M. Saadatfar, A.N. Shepard, M.K. Knackstedt, Advances in X-ray Tomography for Geomaterials pp. 269–276 (2010)
[27] F.M. Schaller, M. Neudecker, M. Saadatfar, G. Delaney, K. Mecke, G.E. Schröder-Turk, M. Schröter, AIP Conference Proceedings **1542**, 377 (2013)
[28] F.M. Schaller, S.C. Kapfer, M.E. Evans, M.J. Hoffmann, T. Aste, M. Saadatfar, K. Mecke, G.W. Delaney, G.E. Schröder-Turk, Philos. Mag. **93**, 3993 (2013)

Study of pore fluid effect on the mobility of granular debris flows

Gordon G. D. Zhou[1,*], and *Q. C.* Sun[2]

[1]Key Laboratory of Mountain Hazards and Earth Surface Processes/Institute of Mountain Hazards and Environment, Chinese Academy of Sciences (CAS), Chengdu, China
[2]State Key Laboratory for Hydroscience and Engineering, Tsinghua University, Beijing, China

Abstract. Granular debris flows in nature are composed of a wide range of solids and viscous pore fluids, moving at high velocities down sloping channels. The pore fluids in a granular debris flow affect the interactions between the solid and fluid phases and thus govern the debris-flow mobility. Study of the pore fluid effect (i.e., excess pore water pressures correlated to solid structures, and the viscous shearing and dragging) is essential for understanding the high flow mobility of granular debris flows. This study critically reviews two dimensionless numbers with clear physical meanings, then demonstrates a new application of field monitoring data for identifying natural debris flows on large scales (i.e., surge and continuous debris-flows, respectively). This study illustrates that, the pore fluid viscous shearing stress dominates solid inertial stress due to solids collision. It is also found that different to continuous debris-flows, the high pore fluid pressures generated in surge debris-flow body dissipate quite slowly and mostly influence particle contact behaviour significantly. A new scientific criterion for identification of continuous and surge debris flow in nature can be given by this study.

1 Introduction

Debris flows occur when masses of poorly sorted sediment, agitated and saturated with water, surge down a slope in response to gravitational attraction [1]. Debris flows differ from rock avalanches and sediment-laden water floods in that both solid and fluid forces influence the motion of a debris flow and governs its rheological properties [1]. Stresses caused by solid particle collisions, contact friction, and solid-fluid interactions (e.g., buoyancy, drag, added mass, lift, and Basset, Faxen, and grain diffusion forces) generally govern the kinematic properties of debris flows [1-4].

When studying the rheological properties of natural debris flows, properly identifying the flow regimes is always the most important issue. Unfortunately, the classification of debris flows nowadays is generally based on empirical observations due to the complexity of the phenomena. For instance, the monitoring debris flows in China were classified into surge and continuous flows mostly based on the surge duration and empirical judgement, which is believed to be quite different for surge and continuous flows. However, sometimes the time difference is quite small and it is really difficult to identify whether the surge duration is relatively long or short. Pure empirical judgements can cause huge mistakes. However, Coussot and Meunier [5] demonstrated that the criterion used to distinguish phenomena of debris flows and other subaerial flows available vary and contradict one another, and they are incapable of covering all phenomena. Considering fine muddy debris flows (mudflows) are viscoplastic fluid flows, whose behaviour is dictated by the fine interstitial clay-water mixture, non-dimensional numbers for water flows are modified and extrapolated for muddy debris flows based on the concept of "global similarity" [6, 7]. It is noted that, different from mud flows, individual debris flows can contain grains that range from clay size to boulder size. Moreover, the sand, gravel, and larger grains compose most of the mass of debris flows and that silt and clay-sized grains which are the component of a slurry commonly constitute less than 10% of the mass [1,8,9]. The ability of debris-flow solids to exhibit dense, interlocked packings as well as loose, high-porosity packings has significant ramifications for mixture behavior, which varies markedly with subtle variations in solid volume fraction (concentration), shear rate, and grain size distribution [10]. Basic studies on the dynamics of solid-fluid two-phase flows are needed for the better understanding of mobility of debris flows.

Bagnold [11] reported that flows of solid-fluid mixtures can be divided into grain-inertial flows and macro-viscous flows. These flows depend on the relative significance of solid collisions and pore fluid viscosity, and a dimensionless number called the Bagnold number was developed for identifying flow regimes in this way. Particle collisions dominate in grain-inertia flows, which correspond to a large Bagnold number, and the momentum transfer and frequency of collisions are both proportional to the relative velocity difference between two neighbouring layers (shear rate). The corresponding stresses caused by collisions are proportional to the square of the shear rate. In macro-viscous flows, which correspond to a small Bagnold number, the effect of

* Corresponding author: gordon@imde.ac.cn

fluid viscosity is dominant. Obviously, then, dimensionless numbers are a good criterion for identifying flow regimes.

In this paper, we reinterpret the recorded field data of viscous debris flows in Dongchuan, China, and identify the different flowing regimes of natural granular debris flows using two dimensionless numbers based on the relative importance of solids collision, pore fluid viscous drag and pore fluid pressures. A uniform criterion which has clear physical meanings is developed to clarify different natural debris flows.

2 Field study at the Dongchuan Debris Flow Observation and Research Station

The Dongchuan Debris Flow Observation and Research Station (DDFORS), located in the transportation zone of the Jiangjia Ravine in the Dongchuan district, Yunnan Province of China (N26°14', E103°08'), is a facility of the Institute of Mountain Hazards and Environment, Chinese Academy of Sciences [12]. Since the 1960s, the station has conducted continual observations of field debris flows occurring in the Jiangjia Ravine due to intense rainfalls. The observations at DDFORS usually focus on a straight valley that is 270 m long on a 3.4° slope. The cross-section is rectangular, and instruments are installed on two banks for observing and recording flow properties (see Fig.1).

Fig. 1. Field observation area for natural debris flows in Jiangjia Ravine, China.

Two monitoring cross sections are marked along a straight channel to capture multiple debris-flow surges, with the predetermined distance ΔL used to measure the velocity of the surge front. The time for the surge front to pass through ΔL is recorded by a stopwatch and marked as Δt for each debris flow, and the mean front velocity is then calculated as $U=\Delta L/\Delta t$. The channel width W (m) and the ground level for each debris flow are determined by using cross section marks within the channel. A supersonic level meter installed above the front head measures the flow thickness h (m), allowing the front discharge Q to be calculated, where $Q=U\times W\times h$. The unit weight of each surge ρ_c (kg/m^3) is determined through direct sampling of the debris flow. Thus, the solid volume concentration C_S can be calculated from $C_S=(\rho_c-\rho_w)/(\rho_S-\rho_w)$, where ρ_w and ρ_S are the unit weights of the water and solid material in the debris flow, respectively, and ρ_S is generally taken to be 2750 kg/m^3. The solids discharge Q_S (kg/s) is then calculated as $Q_S=Q\times\rho_S\times C_S$. Furthermore, as the duration of each surge is recorded, the total (or sediment) runoff of each surge can be determined, allowing the flows to be classified into surge flows and continuous flows. By empirical estimation, either the flow time of a continuous flow is quite long or the discharge is quite large. If there is an obvious flow break between two continual surges, or the sustained time is relatively short, the flow is treated as a surge flow [13,14,15]. Field observation data for the 1961–2000 period has been collected and edited by Kang et al. [14], Zhang and Xiong [13], and Kang et al. [15]. It is from this dataset that we can analyze the dynamics of debris flows.

3 Dimensionless numbers with clear physical meanings for granular debris flows

3.1 Pore fluid viscous drag and solids collision: grain Reynolds number N_{Rey}

In fluid mechanics, the Reynolds number is well-known for identifying laminar and turbulent flows. For grain-fluid mixture flows, the grain Reynolds number N_{Rey} is applied to identify the effects of particle collisions and pore fluid viscosity. The definition of the grain Reynolds number is given in Eq. (1) (cf. Iverson 1997):

$$N_{Rey}=\frac{T_{s(i)}/T_{f(q)}}{\frac{C_s}{1-C_s}\frac{\rho_s}{\rho_f}}=\frac{\rho_f\dot{\gamma}d^2}{\mu_f} \quad (1)$$

Eq. (1) clarifies the physical meaning of the grain Reynolds number N_{Rey}: it is a ratio between the solid inertial stress due to particle collision, $T_{S(i)}$, and the fluid viscous shearing stress, $T_{f(q)}$,. The solid inertial stress may be estimated by $T_{s(i)}\sim C_S\rho_s\dot{\gamma}^2d^2$ (spherical grains of mean diameter d and density ρ_s, sheared under gravity at a shear rate of $\dot{\gamma}=dU/dh$) and the fluid viscous shearing stress derived from Newton's law of viscosity is $T_{f(q)}=C_f\dot{\gamma}\mu_f$ (C_f is the pore fluid volume concentration equal to 1-C_S and μ_f is the viscosity of the pore fluid).

3.2 Pore fluid pressure dissipation: N_P

To distinguish the different debris flows based on the field observation data, a new criterion correlating with the flow duration and pore fluid pressure dissipation is developed. Reconsider the dimensionless number N_P defined by Iverson et al. (2004)

$$N_p=\frac{\sqrt{l/g}}{\mu_f h^2/kE} \quad (2)$$

where l is the avalanche length and can be calculated by Ut (t is the surge duration), k is the hydraulic permeability. The definition shows that N_P is not only correlated to surge duration but also has clear physical meanings. Indeed, this number is the timescale ratio of avalanche motion $(l/g)^{0.5}$ and diffusion of disequilibrium pore fluid pressure normal to the flow direction ($\mu_f h^2/kE$). Small values of N_P indicate that high pore fluid pressure can be sustained quite long inside granular body and influence the avalanche motion significantly (Iverson et al. 2004). With a suggested hydraulic permeability (k=10^{-11}m^2) and a compressive stiffness (E=10^7 Pa) for typical loose granular soils (Iverson 1997; Iverson et al. 2004), the N_P values for surge and continuous debris flows can be calculated.

3.3 Identification of surge and continuous debris flows

We can calculate our dimensionless numbers (e.g., the grain Reynolds number and N_P number) of the natural debris flows at DDFORS by applying the material parameters listed in Table 1 and the field observation data.

TABLE 1. Material parameters of the debris flows at DDFORS

Parameter	Value
Unit weight of water, ρ_w	1000 kg/m^3
Unit weight of solids, ρ_S	2750 kg/m^3
Solids mean particle diameter, d	5 mm
Viscosity of pore fluid, μ_f	0.5 Pa·s
Hydraulic permeability, k	10^{-11} m^2
Compressive stiffness for typical loose granular soils, E	10^7 Pa

The dimensionless numbers values and the correlation between N_P and contact behaviour of solid particles in natural debris flows is shown in Figs. 2a and 2b. Obviously, most of the grain Reynolds numbers N_{Rey} are smaller than 1. This implies that the solid inertial stress is less dominant than the fluid viscous shearing stress. It also indicates that, for natural debris flows, fluid viscous shearing has a more significant effect on the mobility of the debris flows than do solid collisions. In contrast, if the pore fluid viscosity is small (e.g., 0.05 Pa·s suggested by Iverson (1997)), N_{Rey} is greatly larger than 1 and the particle inertial stress may be dominant. In addition, when the grain Reynolds number is reduced, the roles played by viscous drag forces on the solid particles are more and more important. Moreover, the increase trend of the grain Reynolds number N_{Rey} with N_P implies that when solid particles have strong collisions and the debris flows tend to be inertial flows, pore fluid pressure in granular body will dissipate quickly and has little influence on particle contact behaviour. However, what should be emphasized here is that debris-flow behaviour is also significantly governed by solids concentration and it appears solids contact can be negligible at solid fractions of 30% even at a grain Reynolds number about 10, perhaps higher [4]. Different stress mechanisms, including those due to surface tractions (viscous drag), acceleration, the velocity fluctuations (Reynolds stress), and differences in density of solids and liquid (effective stress), should be considered to explain the behaviour of inertial flow of particle-laden fluids. Otherwise, it may neglect significant effects of other particle-scale momentum transport mechanisms except solids collisions.

Generally, the N_P values of natural debris flows in DDFORS can be classified into three different zones. Most of the continuous flows possess high values of N_P and focus in zone 1 (N_P >0.03). In contrast, most of the surge flows have relatively small values of N_P and focus in zone 3 (N_P <0.01). Fig. 2b shows that surge flows usually possess relatively larger solid concentrations than continuous flows. Due to the high solid concentration and the large traveling velocity for surge flows, high pore pressures are more likely developed in these flows. Relatively small N_P values for surge flows indicate that the high pore fluid pressures can sustain quite long. Compared to surge flows, continuous flows with diluted solid particles possess high N_P values. This implies that high pore pressures developed inside the granular body may dissipate quickly, the effects of pore fluid pressure are likely negligible.

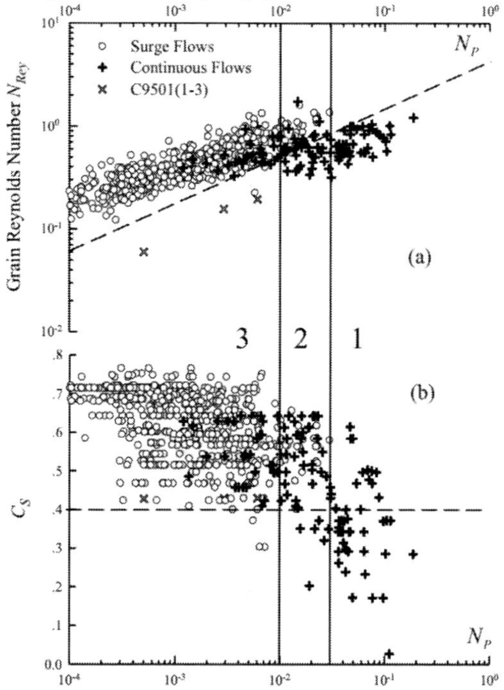

Fig. 2. Classification of natural debris flows as a function of dimensionless numbers: (a) the relationship between N_P and grain Reynolds number N_{Rey}; (b) the relationship between N_P and solid concentration C_S.

It is also obviously shown in the Fig. 2 that an overlapping zone exists: in zone 2 (0.01<N_P<0.03), continuous and surge debris flows have the same N_P values. Some flows in zone 2 are really difficult to identify as they possess both the continuous and surge flow properties. For this reason, the zone 2 can be treated as a transition zone. Based on the variation of N_P values with the dimensionless number N_{Rey}, in zone 1 and 3, the boundaries (dash lines) of distinguishing continuous and surge flows are approximately determined. Different debris flows in zone 2 can be classified by the dash lines (see Figs. 2a and 2b). For instance, by comparing the flows in zone 1 and 3 of Figs. 2a and 2b, the flows locating above the dash lines generally can be treated as surge flows. Comparatively, the flows locating below the dash lines are treated as continuous flows. By this method, the classification of natural debris flows can be obtained with little ambiguity as different flows focus in different zones.

Due to the limitations of observation and subjective judgment, some flows originally treated as "continuous flows" indeed locate in zone 3 above the dash lines (see Figs. 2a and 2b). In these flows, the solid contact behaviour, pore fluid viscous drag, and the relative significance of pore pressure developed in the granular body are quite similar to that of surge flows. In addition, Fig. 2b shows that the solid concentrations of these flows are sometimes relatively high (>0.6). These "continuous flows" should be considered as surge flows. It is also interesting to find that the first three "continuous flows" occurring in 1995 (C9501(1-3)) locate in zone 3 below the dash lines. Comparing to the surge flows, the surge durations of these flows are relatively long and the flowing velocity is relatively low. Similar to surge flows, the timescale of avalanche motion for these flows is relatively smaller than the timescale for diffusion of disequilibrium pore fluid pressure. The small N_P values imply that pore pressure inside the granular body may play a significant role. Moreover, the relative small N_{Rey} values (comparing to those surge flows above the dash line in Fig. 2a) further illustrate that the pore fluid viscous shearing stress dominates particles inertial stress due to solids collision. For these debris flows, both the pore fluid viscous drag and pressues play a key role on the dynamics.

4 Conclusions

By comparing different values of the dimensionless numbers N_{Rey} and N_P, we can predict the rheological properties of debris flows. Using this analysis on field data collected at the Dongchuan Debris Flow Observation and Research Station (DDFORS), we determined a number of properties of natural debris flows occurring on a large scale:

(1) Small values of N_{Rey} (< 1) illustrate that the natural debris flows at DDFORS are dense granular systems dominated by fluid viscous shearing stress rather than by particle collisions.

(2) Based on the dimensionless numbers of N_P and N_{Rey}, the continuous and surge debris flows can be identified. The surge flows with high solid concentrations have relatively small N_P values. It indicates that for surge flows, the high pore fluid pressures generated in granular body dissipate quite slowly and may influence particle contact behaviour significantly. When the natural debris flows tend to be solid collision dominated, pore fluid pressure will dissipate quickly and play a relatively negligible role to solid contact behaviour.

Acknowledgments

The authors acknowledge financial support from the National Natural Science Foundation of China (grant no. 11672318), the Key Research Program of Frontier Sciences, Chinese Academy of Sciences (grant no. QYZDB-SSW-DQC010), and the Youth Innovation Promotion Association Chinese Academy of Sciences.

References

1. R. M. Iverson, *Reviews of Geophysics* 35 (3), 245 (1997).
2. G. Johnson, K. R. Rajagopal, M. Massoudi, *USDOE Pittsburgh Energy Technology Center, PA (USA)*, (1990).
3. R. M. Iverson, Mathematical geology, 25(8): 1027-1048 (1993).
4. H. Haddadi, J. F. Morris, *Journal of Fluid Mechanics* 749, 431-459 (2014).
5. P. Coussot, M. Meunier, *Earth-Science Reviews*, 40(3): 209-227 (1996)
6. P. Coussot, *Journal of Hydraulic Research*, 32(4): 535-559 (1994).
7. P. Coussot, D. Laigle, Houille Blanche,3: 44-49 (1994). (in French)
8. J. E. Costa, Physical geomorphology of debris flows, in Developments and Applications of Geomorphology, edited by J. E. Costa and P. J. Fleisher, pp. 268–317, New York: Springer-Verlag (1984).
9. T. Takahashi, Debris Flow. IAHR Monograph Series. Rotterdam: Balkema (1991).
10. P. Coussot, J. M. Piau, *Can. Geotechn. J.*, 32, 263-270 (1995).
11. R. A. Bagnold, *Proceedings of the Royal Society of London. Series A, Mathematical and Physical Sciences*, Vol. 225 (No. 1160), pp. 49 (1954).
12. P. Cui, X. Q. Chen, Y. Y. Wang, K. H. Hu and Y. Li, "Jiangjia Ravine debris flows in south-western China", In *Debris-flow Hazards and Related Phenomena*,edited by M. Jakob and O. Hungr. Berlin: Springer, 2005.pp. 565-594.
13. J. Zhang and G. Xiong, Data collection of kinematic observation of debris flows in Jiangjia Ravine, Dongchuan, Yunnan (1987-1994), Beijing: Science Press, 1997.
14. Z. C. Kang, P. Cui, F. Q. Wei and S. F. He, Data collection of observation of debris flows in Jiangjia Ravine, Dongchuan Debris Flow Observation and Research Station (1961-1984), Beijing: Science Press, 2006.
15. Z. C. Kang, P. Cui, F. Q. Wei and S. F. He, Data collection of observation of debris flows in Jiangjia Ravine, Dongchuan Debris Flow Observation and Research Station (1995-2000), Beijing: Science Press, 2007.

Numerical modelling of suffusion by discrete element method: a new internal stability criterion based on mechanical behaviour of eroded soil

Nadjibou Abdoulaye Hama[1], *Tariq* Ouahbi[1*], *Said* Taibi[1], *Hanène* Souli[2], *Jean-Marie* Fleureau[3] *and Anne* Pantet[1]

[1]Normandie Univ, Unihavre, UMR 6294 CNRS, LOMC, 76600 le Havre, France
[2]UMR CNRS 8579, LTDS, Ecole Nationale d'Ingénieurs de Saint Etienne, France
[3]UMR CNRS, Laboratoire MSS-Mat, CentraleSupélec, France

Abstract. Non-cohesive soils subjected to a flow may have a behavior in which fine particles migrate through the interstices of the solid skeleton formed by the large particles. This phenomenon is termed internal instability, internal erosion or suffusion, and can occur both in natural soil deposits and also in geotechnical structures such as dams, dikes or barrages. Internal instability of a granular material is its inability to prevent the loss of its fine particles under flow effect. It is geometrically possible if the fine particles can migrate through the pores of the coarse soil matrix and results in a change in its mechanical properties. In this work, we uses the three-dimensional Particle Flow Code (PFC3D/DEM) to study the stability/instability of granular materials and their mechanical behavior. Kenney and Lau criterion sets a safe boundary for engineering design. However, it tends to identify stable soils as unstable ones. The effects of instability and erosion, simulated by clipping fine particles from the grading distribution, on the mechanical behaviour of glass ball samples were analysed. The mechanical properties of eroded samples, in which erosion is simulated and gives a new approach for internal stability. A proposal for a new internal stability criterion is established, it is deduced from the analysis of relations between the mechanical behaviour and internal stability, including material contractance.

1 Introduction

The construction of earth structures such as dams, levees and road embankments often requires the use of granular soils. These cohesionless granular media are often composed of two parts: a solid (coarse) skeleton that can support and transmit the external stresses, and fine particles that can move through the constrictions.

These granular soils can be internally stable or unstable, and more or less resistant to the phenomena of internal erosion and suffusion. These phenomena caused by seepage of water through the material affect its mechanical behaviour and may lead to its collapse.

In granular soils, suffusion may occur only when these three following conditions are satisfied: the fine particles are smaller than the size of constrictions formed by the coarse particles; fine particles do not fill the constrictions and are free to move; the hydraulic gradient is high enough. Kenney and Lau [1-2] proposed a criterion to assess the internal stability of granular media. This criterion is based on the assumption that each class of grain, characterized by its diameter d must be prevented to move by the grains whose diameter is comprised between d and 4d.

This criterion consists in plotting the PSD curve of a soil in a diagram with the mass fraction F of particles whose diameter is smaller than d on the X-axis and the mass fraction H of the particles whose diameter is comprised between d and 4d on the Y-axis. The soil is considered as stable if, for F < 20 or 30%, the curve obtained is located below the line H = F. Otherwise, if some portion of this curve passes above the line H = F, the soil is considered as unstable. According to Kenney and Lau [1-2], F values of 20% apply to the soils with widely graded primary fabric (Cu > 3) and the value of 30% to the soils with narrowly graded primary fabric (Cu < 3).

The mechanism of internal stability has also been widely studied by other authors [3-9], mainly in order to define criteria for soils likely to develop internal erosion.

Kézdi [4] proposed a criterion to assess the internal stability of soil based on the division of the grain size distribution curve into its coarser and finer components. Each component must satisfy the criterion D'15 / D''85 = 4 [5], where D'15 is the particle diameter of the filter corresponding to 15% of passing and D''85 is the particle diameter of the soil to be protected, corresponding to 85% of passing.

A unified approach combining the methods of Kézdi [4] and Kenney and Lau [1-2] was established by Li and Fannin [6]. The common feature of both methods is the examination of the slope of the gradation curve over a

* Corresponding author: tariq.ouahbi@univ-lehavre.fr

discrete interval of its length. This approach yielded a more reliable estimation of soil stability [7].
Internal instability has also been assessed by Andrianatrehina et al. [8]. The stability of nine samples with different grain size distribution curves, derived from direct grain size measurements carried out after triaxial and vibration tests was studied. The results were compared with the predictions of several methods [1, 4, 5, 9,10].
In the literature, many works have focused on the evaluation of the internal stability of materials, but few of them studied the change in the mechanical behaviour resulting from grain size distribution changes after flow.
The objective of this work is to establish a relationship between the internal stability parameters and the mechanical behaviour of materials. To achieve this, a Discrete Elements Method (DEM) is used to model soil sample submitted to a triaxial paths. A new criterion of internal stability taking into account the mechanical behaviour of soils is proposed.

2 Numerical modelling

This numerical study was performed with PFC 3D that uses the discrete element method (DEM) originally developed by Cundall and Strack [11, 12]. The DEM method is very suitable for numerical modelling of granular systems behaviour (granular soils; rocks...).
Materials are discretized in elements of simple shape (spheres) that interact with neighbouring contact elements.
A material is generated with a target granular distribution and porosity and then subjected to a triaxial test. Three stages were employed for the numerical specimen tests: (1) specimen preparation; (2) consolidation and (3) specimen shearing.

2.1 Numerical sample generation

The generation procedure of granular sample with a given grain size distribution curve and a given porosity developed by Abdoulaye Hama et al [13,14] is used.

2.2 Triaxial modelling

The triaxial test in PFC3D is performed by confining a cylindrical sample within three walls. The top and the bottom walls simulate loading platens and the lateral one simulates the confining pressure experienced by the sample sides. The void volume can be calculated using the total volume of the sample minus the total volume of the solid particles.
The axial strain can be calculated by the following equation (which can be obtained by a combination of the Eulerian and Lagrangian strains):

$$\varepsilon = \frac{H_1 - H_0}{\frac{1}{2}(H_1 + H_0)} \quad [1]$$

Where H0 is the initial height of the sample and H1 is the current height during the test.

The tridimensional particle flow determines the wall stress and adjusts the wall velocity to reduce the difference between the current wall stress and the target wall stress σ^w, σ^t. The wall velocity $\dot{u}^w$ must satisfy:

$$\dot{u}^w = G(\sigma^W - \sigma^t) = G\Delta\sigma \quad [2]$$

Where G is the "gain" parameter that is estimated by:

$$G = \frac{\alpha A}{k_n^w \Delta t} \quad [3]$$

3 Results and discussion

When it occurs in a soil sample the phenomenon of suffusion, its PSD changes and this affects its internal stability. To study the impacts of suffusion on the mechanical behaviour of granular soils, five PSD identified by S1; S2; S3; S4 and S5 were defined from an alluvial deposit from the Rhône River in France [14]. Table 1 summarises the Characteristics of these samples.

Table 1: Characteristics of different samples

Material	S1	S2	S3	S4	S5
D10 (mm)	0.56	1.01	1.26	1.29	7.61
D30 (mm)	1.62	2.31	3.52	2.50	6.5
D50 (mm)	2.70	8.00	5.87	7.61	8.00
D60 (mm)	3.25	6.15	5.85	6.20	6.3
CU= d60/d10	5.8	6.08	4.64	4.80	0.82
CC= d30/d60*d10	0.89	0.37	0.47	0.31	0.13
Porosity n (%)	40	40	40	40	40
Minimum ratio of stability $(H/F)_{min}$	1.85	2.47	3.87	3.38	0
Maximum coefficient of retention (D_{15}/ d85) max.	15.5	1.95	2.20	2.31	3.24

Table 1 shows that the only differences between samples are geometric. Samples PSD are all widely graded (Cu>3) with slight curvature. They have different minimum stability index and different maximum retention coefficient. We interpret this as a consequence of the phenomenon of suffusion. If it occurs, it causes loss or gain of particles. From the internal stability parameters ($(H/F)_{min;}$ $(D_{15}/d_{85})_{max}$.) the samples are classified according to the Kenney and Lau and Kezdi criteria. Samples internal stability state is presented on Table 2. One can note that the two criteria diverge on samples S1 and S5.

Table 2: Samples classification according to the Kenney and Lau and Kezdi criteria

Material	S1	S2	S3	S4	S5
Kenney & Lau (985-1986)	S	S	S	S	U
Kezdi (1979)	U	S	S	S	S
S stands for "Stable" and U for "Unstable"					

3.1. Triaxial results

Each sample was subjected to a consolidated and drained triaxial test simulation, at same initial void ratio of e=0.66 and constant confining stress σ_3=200 kPa. Results are presented in the deviatoric (ε_1; q/p) and the volume change (ε_1, ε_v) planes (Fig 1). They are used to calculate the stress ratio $(q/p)_{max}$ and the maximum contractancy (ε_{vmax}) corresponding at projection of the characteristic point in the (ε_1; q/p) plane.

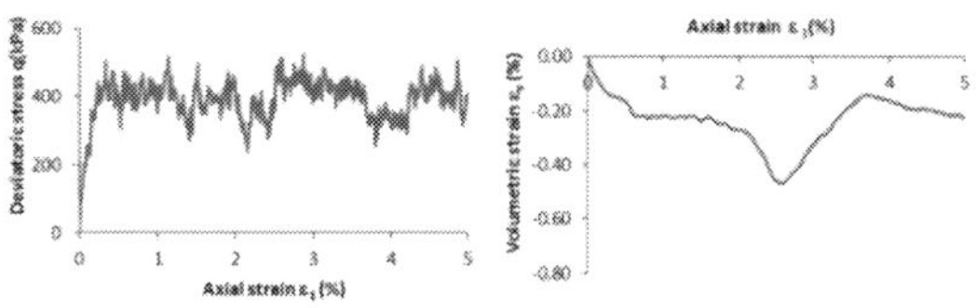

S1: $(H/F)_{min}$ = 1.85; $(D_{15}/d_{85})_{max}$= 15.5

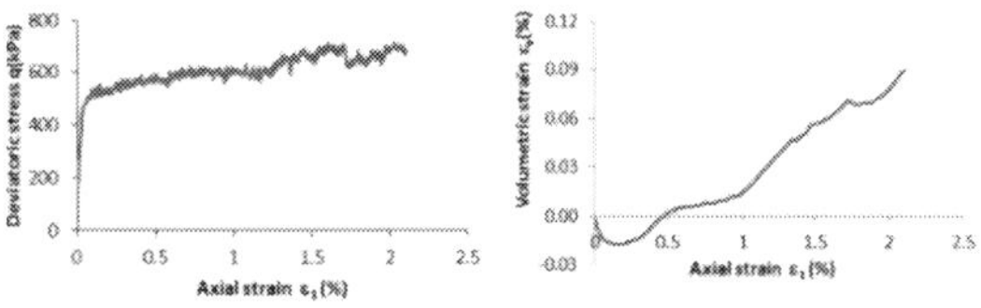

S2: $(H/F)_{min}$ =2.47; $(D_{15}/d_{85})_{max}$= 1.95

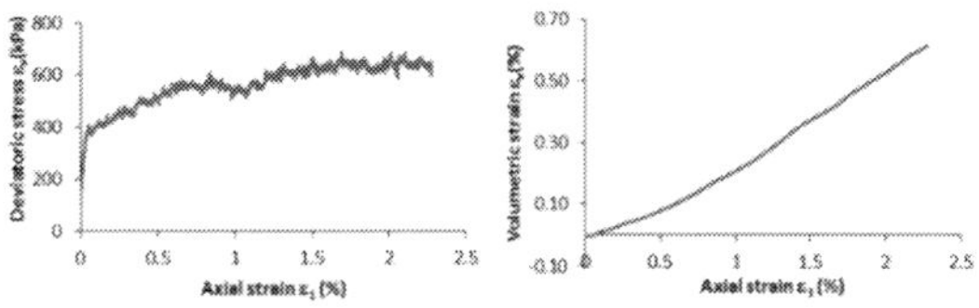

S3: $(H/F)_{min}$ = 3.87; $(D_{15}/d_{85})_{max}$= 2.20

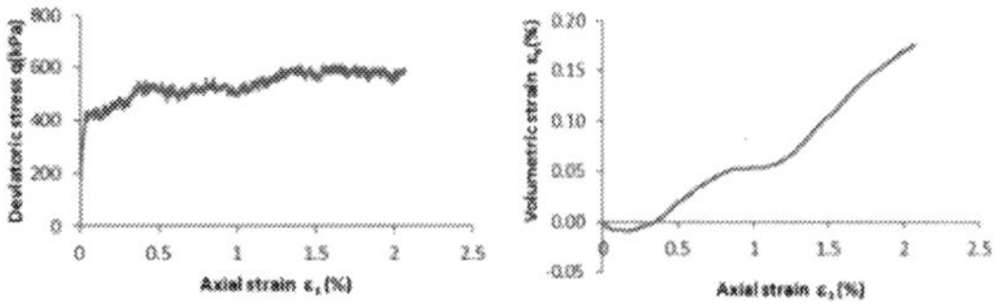

S4: $(H/F)_{min}$ = 3.38; $(D_{15}/d_{85})_{max}$= 2.31

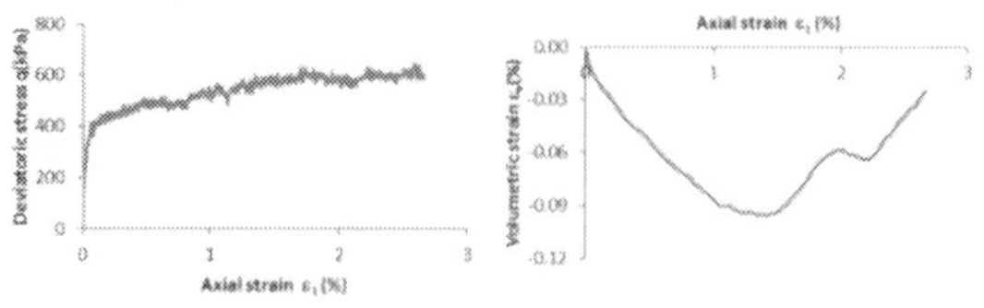

S5: $(H/F)_{min}$ = 0; $(D_{15}/d_{85})_{max}$= 3.24

Fig 1: Triaxial results

Despite the identical test conditions, samples exhibit different maximum shear strength and different maximum contractancy (Fig 1). Table 3 showed a comparison between mechanical parameters $(q/p)_{max}$ and ε_{vmax}) and internal stability ones (($(H/F)_{min}$; $(D_{15}/$ d85) max.).

Table 3: Comparison of mechanical and internal stability parameters

Material	S1	S2	S3	S4	S5
$(H/F)_{min}$	1.85	2.47	3.87	3.38	0
$(D_{15}/d_{85})_{max}$.	15.5	1.95	2.20	2.31	3.24
$(q/p)_{max}$	1.42	1.6	1.6	1.51	1.4
ε_{vmax}(%)	0.5	0.02	0	0.01	0.9

It is noted that the parameters of internal stability influence the mechanical behaviour of soil samples (Table 3). Samples with higher deviatoric stress and lower contractancy present a minimum ratio of stability higher than their maximum coefficient of retention: $(H/F)_{min} > (D_{15}/d_{85})_{max}$.

3.2 Proposition of a new stability criterion

Internal instability of a granular material is its inability to prevent the loss of its fine particles under flow effect. It is geometrically possible if the fine particles can migrate through the pores of the coarse soil matrix and results in a change in its mechanical properties.

From relationship $(H/F)_{min} > (D_{15}/d_{85})_{max}$, we propose a new criterion for assessing soil internal stability. It takes account mechanical behaviour and classifies stable sample with H/F>D15/d85. Fig 2 shows the classification of our samples according the proposed criterion. Samples S2, S3 and S4 are stable and S1 and S5 are unstable.

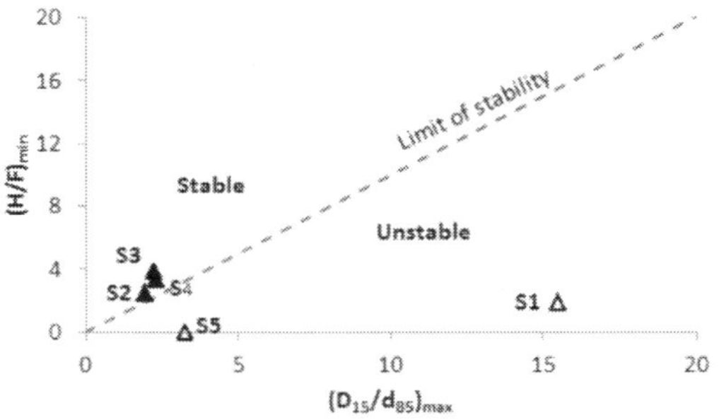

Fig 2: Proposed criterion

4 Validation of proposed criterion

To validate our criterion, 16 samples were taken from literature results: 4 from Kenney and Lau [3], 8 from Andrianatrehina et al [8], and 4 from Indraratna et al [10]. The choice of these samples is motivated by the diversity of their density and that of the analysis method of the authors. These 16 samples are classified using the proposed criterion. Fig 3 presents this classification and compares it with experimental result.

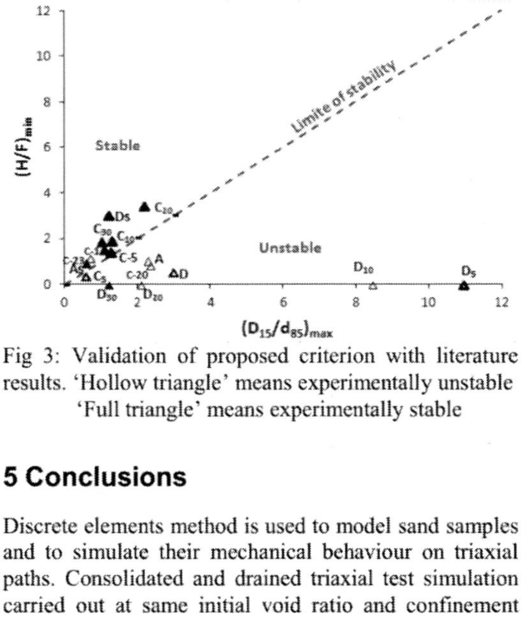

Fig 3: Validation of proposed criterion with literature results. 'Hollow triangle' means experimentally unstable 'Full triangle' means experimentally stable

5 Conclusions

Discrete elements method is used to model sand samples and to simulate their mechanical behaviour on triaxial paths. Consolidated and drained triaxial test simulation carried out at same initial void ratio and confinement stress. The maximum values of contractancy in (ε_1; ε_V) plane and the maximum deviatoric stress in the (ε_1; q/p) planee have been determined for each sample. Results showed that mechanical behaviour of granular material is influenced by the stability parameters $(H/F)_{min}$ and $(D_{15}/d_{85})_{max}$. Effects of suffusion on soil mechanical behaviour was analysed and a new internal stability criterion was proposed. Contrary to classical criteria, it takes into account the mechanical behaviour of the soil. It classifies as stable particle sizes distribution with $\frac{H}{F} \geq \frac{D_{15}}{d_{85}}$, corresponding to the better mechanical behaviour.

Acknowledgement

This work was supported by EOLIFT project financed by ADEME.

References

1. T.C. Kenney, D. Lau (1986) Internal stability of granular filters: reply. Canadian Geotechnical Journal; **23**: 420 – 423.
2. T.C. Kenney, D. Lau (1985) Internal stability of granular filters. Canadian Geotechnical Journal; **22** :215 – 225.
3. J. Lafleur, J. Mlynarek, A. Rolin (1989) Filtration of broadly graded cohesionless soils. Journal of Geotechnical Engineering 1989; **115** (12):1747 – 1768.
4. A. Kézdi (1979) *Soil Physics* — Selected Topics. Elsevier Scientific Publishing Co: Amsterdam.
5. T. Shire and C. O'Sullivan (2012). Micromechanical assessment of an internal stability criterion. Acta Geotechnica, **8**(1), 81-90.
6. M. Li, and R.J. Fannin (2008) Comparison of two criteria for internal stability of granular soil. Can. Geotech. J., **45**(9), 1303 – 1309.
7. V.F. De Mello (1975) Some lessons from unsuspected, real and fictitious problems in earth dam engineering in Brazil. In Proceedings 6th Regional Soil Mechanics and Foundation Engineering Conference for Afrika. p. 285-304.
8. L. Andrianatrehina, H. Souli, J.J. Fry , Q.T. Phan, J.M. Fleureau (2012) Internal stability of granular materials in triaxial tests. ICSE Paris
9. USACE (U.S. Army Corps of Engineers) (1953) Investigation of filter requirements for underdrains. Tech. Memo. No. 3-360, U.S. Waterways Experiment Station, Vicksburg, MI.
10. B. Indraratna, J. Israr and C. Rujikiatkamjorn (2015) Geometrical Method for Evaluating the Internal Instability of Granular Filters Based on Constriction Size Distribution. Journal of Geotechnical and Geoenvironmental Engineering,**141**(10), 04015045.
11. P.A. Cundall, O.D.L. Strack (1979) The Distinct Element Method as a Tool for Research in Granular Media, Part II. University of Minnesota: Minnesota.
12. P.A. Cundall, O.D.L. Strack (1979) Discrete numerical model for granular assemblies. Geotechnique 1979; **29** (1):47 – 65.
13. N. Abdoulaye Hama, T. Ouahbi, S. Taibi, S. Souli, J.M. Fleureau, A. Pantet. Analysis of mechanical behaviour and internal stability of granular materials using discrete element method. Int. Int. J. Numer. Anal. Meth. Geomech. (2016); **40**:1712–1729. DOI: 10.1002/nag.2510.
14. N. Abdoulaye Hama, T. Ouahbi, S. Taibi, J.M. Fleureau, A. Pantet, S. Souli, (2014). Numerical Analysis of Internal Stability of Granular Materials using Discrete Element Method. *Geomechanics from Micro to Macro*, CRC Press (2014). DOI : 10.1201/b17395-94.

A hierarchy of granular continuum models: Why flowing grains are both simple and complex

Ken Kamrin[1,*]

[1] *Massachusetts Institute of Technology, Department of Mechanical Engineering, Cambridge, MA 02139*

Abstract. Granular materials have a strange propensity to behave as either a complex media or a simple media depending on the precise question being asked. This review paper offers a summary of granular flow rheologies for well-developed or steady-state motion, and seeks to explain this dichotomy through the vast range of complexity intrinsic to these models. A key observation is that to achieve accuracy in predicting flow fields in general geometries, one requires a model that accounts for a number of subtleties, most notably a nonlocal effect to account for cooperativity in the flow as induced by the finite size of grains. On the other hand, forces and tractions that develop on macro-scale, submerged boundaries appear to be minimally affected by grain size and, barring very rapid motions, are well represented by simple rate-independent frictional plasticity models. A major simplification observed in experiments of granular intrusion, which we refer to as the 'resistive force hypothesis' of granular Resistive Force Theory, can be shown to arise directly from rate-independent plasticity. Because such plasticity models have so few parameters, and the major rheological parameter is a dimensionless internal friction coefficient, some of these simplifications can be seen as consequences of scaling.

1 Introduction

Granular materials have a well-deserved reputation as a complex rheological media [1]. Much debate still exists on what is the constitutive relation for granular flow. Dry granular systems display history- and preparation-dependent strengthening and dilation [2–4], flow anisotropy and normal stress differences [5–7], nonlinear rate-sensitive yielding [8–11], and nonlocality due to the finite size of grains [12–16]. All these phenomena depend sensitively on grain characteristics such as shape and size distribution, frictional properties, and stiffness [10, 17–20]. Continuum models have been proposed of varying complexity to represent these phenomena, with some taking a simpler form while representing fewer effects, and others more complicated forms with the trade-off of capturing more of the subtleties. Rather than choosing a particular model, the purpose of this review is to highlight the benefits of each model depending on the problem at hand. We will focus on well-developed flow phenomena.

Interestingly, the complexity of granular materials is somewhat problem specific. For motivation, let us compare granular phenomena to a relatively standard material model, say, a linear viscous fluid, whose shear stress τ and strain-rate $\dot{\gamma}$ are related by $\tau = \eta\dot{\gamma}$ for viscosity η. In inhomogeneous flow geometries, where the strain-rate field is not uniform, steady granular flow profiles display a characteristic size-effect whereby the size of shear features is influenced directly by the grain size rather than just the local homogenized fields, such as stress [10, 21, 22]. In this sense, granular phenomenology is more complex than a viscous fluid, in which the flow rule is fully local; i.e. stress at a position gives flow at that position. Indeed, to obtain predictive stress and flow fields in a granular material, one needs a nonlocal framework to account for flow cooperativity induced by the finite size of grains. However, if the question is reduced, the modeling complexity needed to obtain an adequate solution reduces enormously. For example, consider the problem of determining the resistive force on an arbitrary rigid solid object being dragged through a bed of grains. This problem has direct relevance in granular-solid interactions, such as locomotion in granular media. Here, the quantity being desired is a global net force, rather than spatial fields of flow and/or stress. By reducing the question in this way, it turns out that very simple models provide an accurate solution. In particular, reduced-order resistive force models from viscous flow theory can be translated to the granular problem, and often work better than in the original viscous fluid application [23–25]. It would seem that in these kinds of problems, interestingly, granular media may behave *simpler* than a viscous fluid.

Herein, we discuss the behaviors of granular media in terms of a nested family of continuum models. The basic features of granular flows can be predicted with simple frictional plastic models containing only a single (dimensionless) internal friction coefficient [26–28]. However, obtaining details at the next order of precision often incurs a rather large leap in modeling complexity, although certain specific observables are affected less than others by

*e-mail: kkamrin@mit.edu

which model is used. We discuss the Nonlocal Granular Fluidity (NGF) model as an example of a more complex model [12–14, 29], which augments the frictional plastic form with both rate-dependence and particle size dependence. We begin by describing the continuum models at hand, and then follow up with demonstrations focusing on weighing the need for simpler models, which are more amenable to analytical tools, with the need for more detailed forms that have greater field accuracy but at a larger computational cost.

2 Nested Family of Flow Models

The simplest continuum models for granular flow date back to the ideas of Coulomb in the 1700's. They are based on the concept of a frictional yield criterion, where the shear stress needed to achieve plastic flow grows linearly with the applied pressure. These notions have been extended to three-dimensions by Mohr and further simplified by Drucker and Prager [30] where they are common tools in geotechnical engineering [26, 31].

To write a steady flow model based on these principles, we first assume a standard momentum balance

$$\rho\dot{\mathbf{v}} = \nabla\cdot\boldsymbol{\sigma} + \mathbf{b} \tag{1}$$

where $\boldsymbol{\sigma}$ is the Cauchy stress, ρ is the mass density, $\mathbf{v}$ is the velocity, ∇ is the spatial gradient operator, and $\mathbf{b}$ is a body force (typically the gravitational body force).

For the sake of concreteness and simplicity, let us make a rigid-plastic assumption, neglecting the elastic response of the grains to approximate all flow as plastic flow. The constitutive response can now be expressed as constraints on the stress in terms of flow and any relevant state variables.

Let us define the (hydrostatic) pressure $p = -\sigma_{ii}/3$ and equivalent shear stress $\tau = \sqrt{\sigma'_{ij}\sigma'_{ij}/2}$ where $'$ denotes the deviator, i.e. $\boldsymbol{\sigma}' \equiv \boldsymbol{\sigma} + p\mathbf{I}$ for $\mathbf{I}$ the identity tensor. Further, we define the strain-rate tensor by $\mathbf{D} = (\nabla\mathbf{v} + \nabla\mathbf{v}^T)/2$ and the equivalent shear-rate by $\dot{\gamma} = \sqrt{2D'_{ij}D'_{ij}}$. It is reasonable to assume a roughly incompressible response since dilation/contraction should cease at the steady state, allowing us to write $\mathrm{tr}\mathbf{D} = 0$. We note that certain secondary flow phenomena may be missed by neglecting dilatation [32]. We can also make the assumption of codirectionality of the stress and strain-rate tensors, which when combined with incompressibility gives

$$\mathbf{D} = \lambda\boldsymbol{\sigma}' \tag{2}$$

where λ is a non-negative scalar multiplier. This assumption neglects the presence of anisotropy, which causes the eigenvectors of stress and strain-rate to misalign [6, 7, 9], as well as normal stress differences [5, 33], which cause the eigenvalues of strain-rate to differ by more than a single scalar factor from those of the deviatoric stress. Both these effects are known to exist; misalignment can be upwards of five degrees and normal stress differences in the range of ten percent are known to occur.

We now summarize three granular flow models, each adding more complexity to the one before it.

Frictional plasticity: In the well-developed limit, transient effects cease and we can express the Drucker-Prager failure criterion as

$$\mu \equiv \tau/p = \mu_s \ \text{ if } \dot{\gamma} > 0, \text{ and } \tau \le \mu_s p \ \text{ otherwise.} \tag{3}$$

Here, μ_s is an internal friction coefficient, deemed constant in this model. The above scalar relation combined with codirectionality and momentum balance provides a closed system of equations. Note this is a non-associated plasticity model. It should also be clarified that because we consider cohesionless grains, we must supply additional physics to ensure the pressure stays non-negative, i.e. no tension. We can achieve this by requiring

$$p = 0 \ \text{ if } \rho < \rho_c \tag{4}$$

where ρ_c is a critical packing fraction needed for the grains to just barely be in contact [27, 28]. If grains open up beyond this point, the system is modeled as disconnected and thus stress-free, as in a low-temperature granular gas. Note that this type of opening is distinct from plastic dilation in which grains dilate but still support a non-zero pressure. Numerically, the opening criterion can be implemented by reintroducing the elastic part of the deformation and requiring the elastic moduli to vanish when $\rho < \rho_c$. The constitutive model comprised of a Drucker-Prager failure criterion, codirectionality, incompressibility, and the opening constraint will be referred to as *frictional plasticity*.

Inertial rheology: When the inertial number, $I \equiv \dot{\gamma}\sqrt{\frac{\rho_s d^2}{P}}$, becomes large enough, typically $I > 10^{-1}$, the effect of rate on the plastic flow becomes noticeable. Here, d is the (mean) grain size and ρ_s is the solid density of the particle material. Rate effects in granular flow have been observed back to Bagnold in the 1950's [8] but were formalized more generally in the early to mid 2000's [9, 10]. The key notion is that the rate influences the plastic flow rule by causing the internal friction coefficient to depend on I. Most data (but not all, [34]) shows it grows with I. A common fit form [35] is

$$\mu = \hat{\mu}(I) = \mu_s + \Delta\mu/(bI/\Delta\mu + 1) \tag{5}$$

where μ_s is now the static friction coefficient, approached as $I \to 0$. Below μ_s no plastic flow can occur. $\Delta\mu$ and b are dimensionless material constants describing how μ increases with I.

It is useful to consider circumstances that induce non-negligible rate effects. Because I grows with strain-rate but decreases with pressure, inertial effects matter most where zones near free surfaces are flowing quickly. In fact, the clearest example of rate-sensitivity can be observed in flows of granular layers down inclined planes. A tall layer of grains on a plane tilted upward by θ experiences $\mu = \tan(\theta)$ everywhere. According to Eq 5, as verified in simulations [22, 33], steady flows down the incline occur for $\mu_s < \tan(\theta) < \mu_s + \Delta\mu$ with the flow rate increasing as θ increases. Without the dependence on I, one would predict that steady flow only occurs at $\tan\theta = \mu_s$ and above this angle, the grains accelerate without reaching a steady profile. The finite range of angles that admit steady flows down an incline exemplify the correction for rate presupposed by the inertial rheology.

Nonlocal Granular Fluidity: In a number of circumstances, it is found that the inertial rheology is still insufficient to describe observed flow profiles. Flows where the strain-rate field is inhomogeneous often have wider flow features than would be predicted by the inertial rheology. Counter to the μ_s yield criterion supposed by the inertial rheology, it is found that if part of the granular media is flowing, adjacent parts, even those with $\mu < \mu_s$, are also flowing even if just very slowly [21, 36, 37]. Moreover, regarding flows down inclines, it is observed that thinner layers of grains require higher θ to stop flow compared to thicker systems, even though both obey $\mu = \tan\theta$ [22, 38]. The explicit dependence of strength on system size goes beyond local mechanics. Other examples of behaviors that suggest an intrinsic length-scale must be introduced include the so-called secondary rheology of granular media [39–41], wherein flow induced at one point causes non-trivial changes to the rheology observed elsewhere. The effect decays away from the point of induced motion, scaled by the grain size.

What these phenomena have in common is that they suggest cooperativity of the grain motion over space must be represented within the rheology. The natural size-scale for this cooperativity is the size of the grains themselves, d. There are a number of general modeling frameworks in the literature that allow the inclusion of an intrinsic length-scale. These include micropolar or Cosserat models that introduce a couple stress field [15], gradient plasticity theories that permit power expenditure over strain gradients [42], thermal models that introduce a diffusing temperature variable (including specialized forms such as Granular Solid Hydrodynamics) [43, 44], and partial fluidization models that assume a diffusing order parameter governing the amount of fluid vs solid constitutive behavior at a point [16, 45]. Some of these models invoke assumptions that change the expression of momentum balance, Eq 1, while others leave momentum balance unchanged while introducing new field equations that must also be simultaneously solved. The model we will focus on, the NGF model, is of the latter type. It began as a granular extension to existing fluidity frameworks for emulsion flow [46, 47] but has recently been reconciled secondarily through alternative means [48].

The model is based on the introduction of a new scalar state field called the 'granular fluidity' denoted $g(\mathbf{x})$, which is assumed to exist throughout the body of material. Recent work has helped to establish the nature of the g field as a kinematically observable state variable [49]. Its dynamics are assumed to obey a reaction-diffusion form analogous to a non-standard Landau-type equation:

$$t_0\dot{g} = A^2d^2\nabla^2 g - \Delta\mu\left(\frac{\mu_s-\mu}{\mu_2-\mu}\right)g - b\sqrt{\frac{\rho_s d^2}{P}}\mu g^2 \qquad (6)$$

Here, t_0 is a constant time-scale and $\mu_2 \equiv \mu_s + \Delta\mu$. The constitutive relation then calls upon the g field to relate stress and strain-rate, i.e.

$$\mu = \dot{\gamma}/g. \qquad (7)$$

Because of the $A^2d^2\nabla^2 g$ term in the evolution of g, the spatial variation of the fluidity field is influenced directly by the mean particle size. This then causes the relation between stress and strain-rate at a point to depend on behavior not localized to that point, but influenced by nearby events. If the nonlocal amplitude A is set to 0, the steady solutions of the system reduce identically to the inertial rheology. A is the only new material parameter, beyond those of the inertial rheology. It is dimensionless and order one; $A \approx 0.5$ for glass beads and 0.75 for disks [12, 13, 17]. One could view Eq 6 like a Landau-type form where the last two terms behave like an 'energy derivative' of an energy function with a minimum at $g = 0$ if $\mu \le \mu_s$ and a non-zero minimum if $\mu > \mu_s$ [48]. As shall be seen in the next section, the model above is able to reconcile multiple manifestations of nonlocality in granular flows. Note that the difference between all three models can be seen in how each one relates μ to the other fields; compare Eqs 3, 5, and 7.

Equation 6 can be approximated to a steady-state-only form by performing linearization [48] to yield

$$g = g_{loc}(\mu, p) + \xi(\mu)^2\nabla^2 g \qquad (8)$$

where g_{loc} is what the fluidity would be if the inertial rheology were correct, and ξ is a dynamic correlation length. These take the form:

$$g_{loc}(\mu, p) = \max\left(0,\ \frac{1}{\mu}\sqrt{\frac{P}{\rho_s d^2}}\hat{\mu}^{-1}(\mu)\right)$$
$$\xi(\mu) = Ad\sqrt{\frac{\mu_2-\mu}{\Delta\mu|\mu-\mu_s|}}. \qquad (9)$$

where $\hat{\mu}$ comes from Eq 5. Equation 8 allows us to see that the fluidity is a sum of two influences. The term g_{loc} indicates the effect of local stress on the fluidity; it vanishes wherever $\mu < \mu_s$. The second term indicates that disturbances from neighboring media can also influence the fluidity. The function $\xi(\mu)$ diverges as $\mu \to \mu_s$ on both sides, which has been verified in numerical simulations [12] and comparisons with experimental data [13]. This bears similarities to diverging length-scales in phase transitions.

To provide intuition for how the NGF model works, suppose the stress is large enough to cause $g_{loc} > 0$ at some position. The above relation indicates that this region acts like a fluidity source, from which fluidity can then diffuse into neighboring zones, even those for which g_{loc} is zero, allowing material to flow in regions where local relations would predict no flow. Regarding boundary

DOI: 10.1051/epjconf/201714001007

conditions for g, two common choices used are $dg/dn = 0$ for n the outward normal to the surface, and $g = g_{loc}$. Both cause minimal 'disruption' at boundaries [12, 13]. As described mechanically in [48], the Neumann boundary condition can be seen as a sort of conduction-free condition that prevents the fluidity field from expending power at the boundary. The choice of which boundary condition to use matters less in large geometries than in thin ones; in the case of thin geometries, precise g boundary conditions can be inferred from existing data, as done in [14].

3 Nonlocal Granular Fluidity Model

The most obvious indication of nonlocal effects in steady flows is if the $\hat{\mu}(I)$ relation, obtained in planar shear tests, breaks down in flows that are inhomogeneous. Suppose for example an annular shear flow (Fig 1(a)). Locally, the flow resembles planar simple shearing. Each element of material, at steady-state, experiences a steady packing fraction and a steady stress state (modulo rotation). Absent a nonlocal effect, one would conclude these fields should obey $\hat{\mu}(I)$ since each material element experiences the same set of conditions that an equivalent planar shear flow experiences, and planar flows strongly confirm the inertial rheology at steady state. However, this is not the case.

In [21] a number of 2D simulations of annular shear were performed. At steady state, values of μ and I were coarse-grained at many locations and scatter-plotted. This was done for several values of inner wall speed v_{wall} and several values of R/d, see Fig 1(b-c). It was observed that, primarily for low I, the locus of μ and I was not one-to-one. Zones below μ_s were observed to flow. The inertial rheology, calculated from simple shear tests, can be seen to emerge in the limit $R/d \to \infty$ (stars in Fig 1(b)) and wherever I is sufficiently large. In Fig 1(c), zones flowing under μ_s display a curious rate-independence; the three data sets corresponding to the slowest wall speeds have essentially the same range of μ but have I ranges differing by orders of magnitude. The solid lines in the figure show predictions of the steady-state-only approximation of the NGF model. It is observed that all modes of branching of these different curves are correctly captured by the NGF model without any recalibration [12]. The NGF model captures the return to inertial behavior in the large R/d limit and large I limit because in these cases the nonlocal term $\xi^2\nabla^2 g$ is dwarfed by g_{loc} in Eq 8. The NGF model proved the ability to predict flow fields in these geometries over several orders of magnitude, well into the slow creeping zones, where inertial rheology alone fails.

The NGF model was then implemented in 3D [13]. This was done using a customized finite-element routine in Abaqus [50]. We first modeled the split-bottom cell geometry (Fig 2(a)), a flow environment that serves as a benchmark for flow rheologies, as no previous continuum model was able to correctly predict its flow fields. To compare against actual experiments, which used glass beads, the inertial law parameters, μ_s, $\Delta\mu$ and b, were chosen to match previously published values for beads, and A was chosen

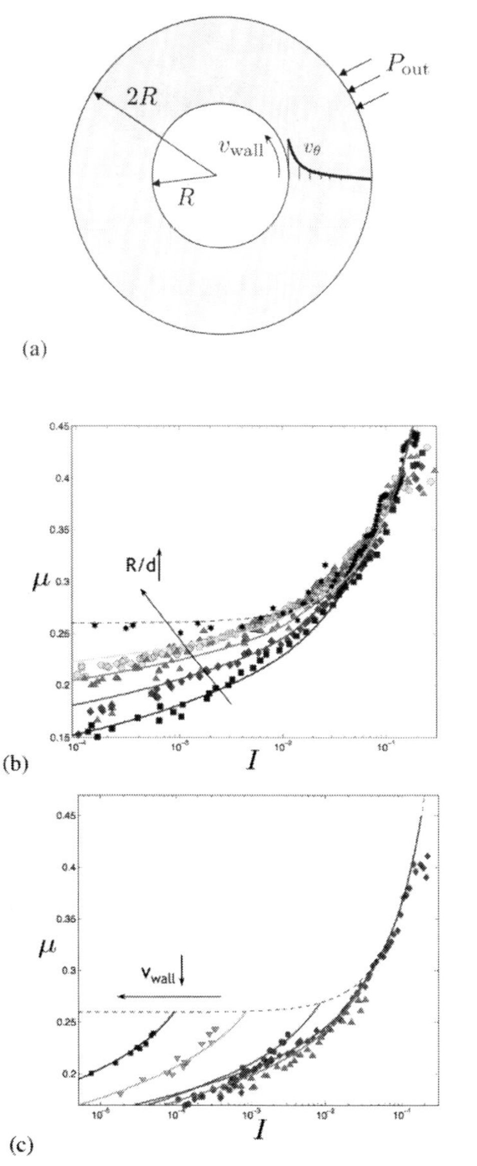

Figure 1: (a) Setup of the annular shear geometry. (b-c) Locus of μ vs I when geometry size relative to grain size is varied with wall speed held fixed (b), and when wall speed is varied with geometry fixed (c). Symbols are results from DEM simulations, dashed line is the prediction of the inertial rheology for all cases, and solid lines are predictions of the NGF model. Some images/data taken from [12].

at 0.48. Note that the inertial rheology on its own has been shown insufficient to describe these flows [51, 52].

When a split bottom cell achieves steady state the flow field at the top surface displays a wide shear band, as can be seen in plots of the normalized revolution rate $\omega(r) = v_\theta(r)/\Omega r$, where v_θ is the grain speed (which is all

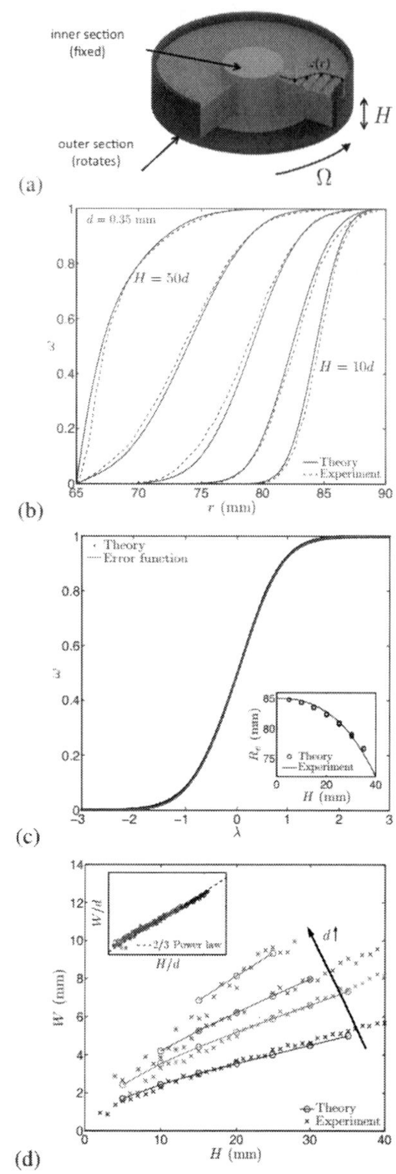

Figure 2: (a) Setup of the split-bottom cell. Grains fill an annular trough whose bottom is split along a circle so that the outer piece can rotate with respect to the inner. (b) Theory vs experiment for the flow field on the free surface as filling height increases. (c) In 22 NGF simulations of shallow flows for various H and d (black points), the top-surface flow profile collapses to an error function shape (green line). (d) Theory vs experiment for width of the shear-band on the top surface in terms of mean grain size and filling height. Some images/data taken from [13].

in the θ direction). The band is couched in the interior as long as H is small enough; for deeper layers, the shear band hits the inner wall. Predictions of the model vs the

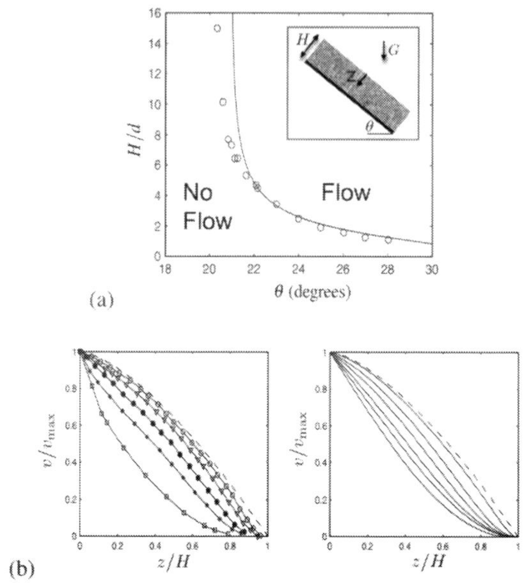

Figure 3: (a) Predicted (line) vs experimental (circles) H_{stop} values. Flow geometry shown in inset. G is gravitational acceleration. (b) [Left] Flow profiles normalized by $v_{\max} = v(z = 0)$ as given from discrete particle simulations. H/d increases from bottom to top with fixed inclination. [Right] Theoretical result using NGF. Dashed line is the Bagnold profile, approached at the largest H in both cases. Some data/images taken from [14].

experimental results for five values of H can be seen in Fig 2(b). The model correctly spreads the shear band and moves it inward as filling height increases.

One amazing fact about $\omega(r)$ is that it is observed experimentally, for various shallow H values, to have a universal shape. Letting R_c denote the center of the shear band and W the width, all such flows can be expressed with remarkable precision by an error function [36]: $\omega = \mathrm{erf}(\frac{r-R_c}{W})$. Interestingly, this collapse to an error function is equally present in solutions to the NGF model. Figure 2(c) shows the error function collapse for 22 combinations of d and H, each test contributing a set of black points in the figure. The model's prediction of where the shear-band center is, R_c, is also captured (see inset). Lastly, it should be noted that the split-bottom geometry forces a model to be able to disentangle the coupled effects of the micro-scale d and the geometric scale H, in determining the width of the shear-band at the top surface W. Figure 2(d) compares experiments to theory for many H and d values and it can be seen that the model consistently predicts the correct shear-band width.

As noted previously, there are multiple incarnations of grain size effects in granular media, beyond their influence on steady flow profiles. Another major size-effect is the so-called H_{stop} phenomenon. In flow down long inclined planes (see Fig 3(a) inset), μ is uniform throughout and

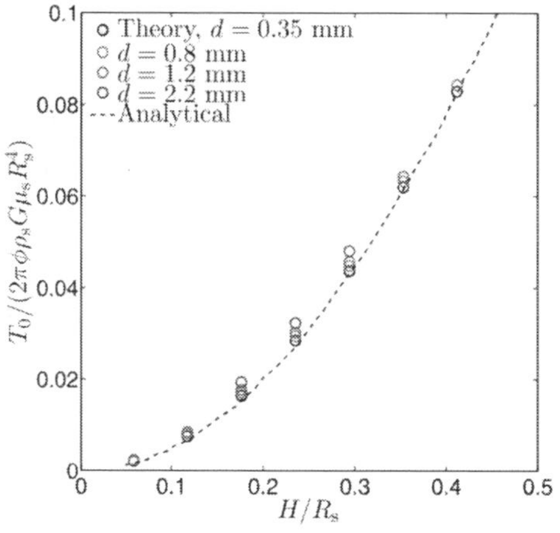

Figure 4: Total torque T_0 on the outer wall of a split-bottom cell. Symbols are solutions of NGF model for various particle sizes, line is an analytical form assuming an infinitely thin shear band [52]. R_s is the split radius, Φ is the packing fraction. Some data and/or image content taken from [13].

given by $\tan\theta$. The inertial rheology would predict, thus, that only θ determines if the layer will flow. However, experiments and particle simulations show that whether or not a flowing layer stops depends also on relative filling height, H/d [22, 38]. The function $H_{stop}(\theta)$ marks the phase boundary between flowing and static inclined layers, i.e. the locus of points where a flowing system comes to a stop.

Looking at Eq 6 in this geometry, using homogeneous boundary conditions for g implied by existing data [22], we can perform a stability analysis about the global no-flow solution, $g = 0$, to determine when a static state is the stable solution. The answer gives a prediction for the H_{stop} curve, which is shown against experimental data for glass beads in Fig 3(a) [14]. Note that the model uses all the same material parameters as the ones used to predict split-bottom flows, though this is answering a very different question — here we are asking whether flow should happen at all, whereas previously we were concerned with the spatial flow profile of grains in a driven process. That said, when flow does occur on the incline, the NGF model still succeeds in predicting its profile. In fact, when the filling height is not much larger than H_{stop}, the flows still show a marked difference from the 'Bagnold profile' predicted by the inertial rheology. As filling height increases, such that H/d is sufficiently large, the Bagnold flow profile re-emerges as the solution. This classical flow profile can also be obtained from the simple scaling of pressure with strain-rate squared when packing fraction is constant [53], which is largely the case in inclined plane flows. In view of Fig 3(b), it can be seen that the qualitative change in the shape of the flow profile as H grows above H_{stop} is well-captured by the NGF model.

4 Rate-Independent Frictional Plasticity

To understand circumstances where the machinery of the nonlocal model may not be necessary, consider the data in Fig 4. This shows the total normalized torque needed to rotate the outer wall of a split-bottom cell, comparing the NGF model for various grain sizes to an analytical result assuming an infinitely thin shear band [52]. Interestingly, the grain size has only a small effect on the torque. Although the size-effect within the NGF model can have a significant effect on the granular flow field, force-like quantities on boundaries appear to be minimally affected by the corrections due to particle size. Further, as noted previously, rate effects due to large I are most significant near free surfaces in fast flow. Taken together, we hypothesize that the problem of determining forces on macro-scale submerged boundaries may be modeled well with the simplest granular rheology, frictional plasticity.

This idea can be checked by comparing data on forces on buried plates moving in various ways in a granular bed. In [23], as shown schematically in Fig 5(a), a small plate buried in the grains (0.3mm beads) was angled in various directions γ and moved in various directions β. In each case the resistive force (F_x, F_z) of the grains acting on the plate is measured. It is convenient to scale the measured forces by plate depth times plate area, yielding resistive force coefficients, denoted by (α_x, α_z). Figure 5(b), left column, shows experimentally obtained plots of α_x and α_z. We have run finite-element simulations of rough plates moving through frictional plastic continuous media to attempt to model these experiments [28]. The parameters ρ_s and μ_s were selected using the density and angle of repose of the same grains as the experiments, as reported in [23]. The corresponding plots of resistive force coefficients are shown in the right column of Fig 5(b). A striking similarity, both in shape and scale can be seen between the experiment and model. It should be noted, that these demonstrations all utilize intruders that are sufficiently large compared to d and moving at a rate that makes I small enough to be neglected. Should either of these conditions be removed, it is unclear how strong the collapse would be.

In the study of resistive forces on generally-shaped intruding objects, a strange experimental correlation has been made, which we call the *resistive force hypothesis* [23, 24, 54]. It states that the stress acting on an arbitrary surface element of the intruding object can be computed by assuming it is an isolated plate and using the empirical relations from Fig 5(b) to approximate the local resistance. The experimental accuracy of this immense simplification is surprising since it presupposes that the stress acting on a small part of the intruding object is independent of the geometry and motion of the rest of the object; a complete lack of cross-correlation. Moreover, the accuracy of this hypothesis in grains is actually higher than the analogous local drag models developed for viscous fluids [55] — this is particularly ironic since the granular version was inspired by the fluid one, which dates back to the 1950's [56]. Since its first demonstration, it was unclear why the granular resistive force hypothesis works so well. However, as shown in Fig 6, this strange simplifi-

 DOI: 10.1051/epjconf/201714001007

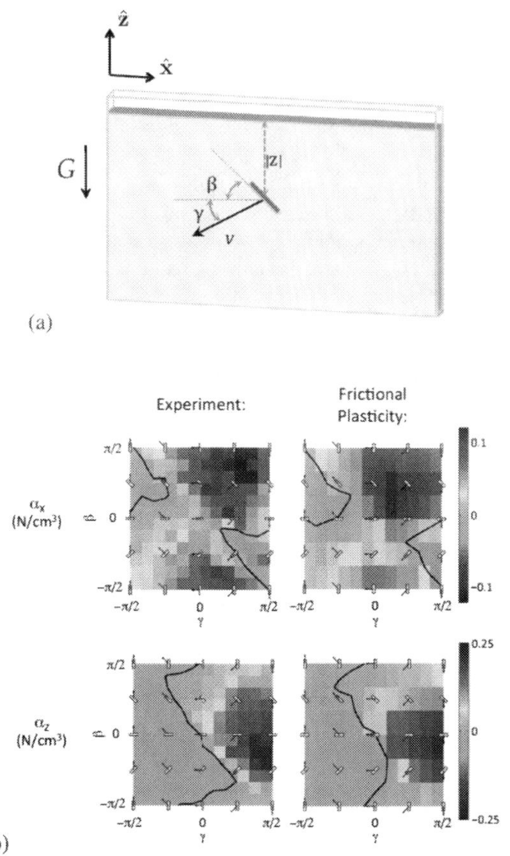

Figure 5: (a) Experimental schematic of submerged plate drag tests. At a prescribed depth $|z|$, the angle of intrusion γ given by the imposed velocity $\mathbf{v}$ is varied, as well as the tilt angle β. (b) The components of the resistive force per depth per area of the experiments compared against the corresponding results from frictional plasticity simulations. The experimentally determined repose angle and grain density were used to calibrate the frictional plasticity model. Images modified from [28].

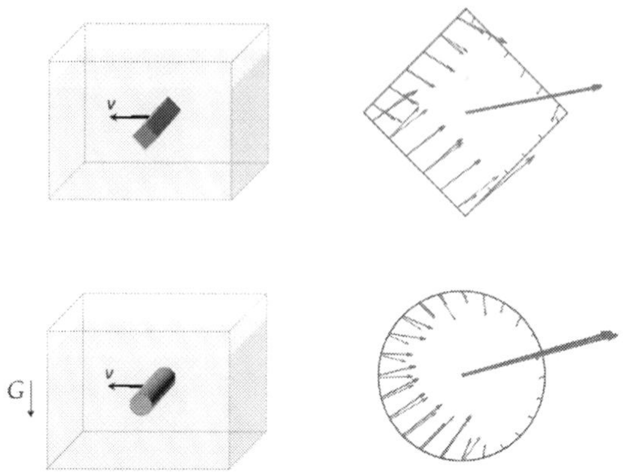

Figure 6: Circular and square cross-section intruders are simulated moving leftward in frictional plastic media, producing the force distribution shown with blue arrows. The red arrows are the force distribution arising from the resistive force hypothesis, obtained by assigning forces consistent with Fig 5, which were themselves obtained from dragging small flat plates. Some images/data taken from [28].

cation comes out in direct simulations of frictional plasticity. The figure shows, in blue, the force distributions arising from frictional plasticity simulations of a square cross-section intruder and a circular cross-section intruder. In red are the forces distributions implied by supposing Fig 5 can be applied on all surface elements of the leading surface of the objects. The distributions are very close, and, when summed to produce a resultant, the two agree almost exactly (resultant forces shown at the center of the objects). Further demonstrations of the resistive force hypothesis emerging in simulations of frictional plasticity can be found in [28].

A particular benefit of frictional plasticity is that it depends on very few input parameters, just ρ and μ_s. As a result, certain analytical tools are an accessible option that would otherwise be too difficult to use in more complex models. For example, dimensional analysis can be used to understand various behaviors of the frictional plastic flow rule. We have shown using dimensional analysis why, in certain cases, the resistive force hypothesis is so accurate. For example, dimensional analysis allows us to show, without solving any differential equations, that the total force on a large square 'garden hoe' with its top edge flush with the granular free surface must be exactly equal to the prediction made under the resistive force hypothesis [28]. We are currently studying the potential uses of scaling laws derivable from frictional plasticity in applications like granular locomotion [57].

With regard to the fact that the resistive force hypothesis emerges more strongly in frictional plasticity than in viscous fluid, consider the following. The shear behavior in viscous media is given by a dimensional parameter (viscosity) whereas it is given by a dimensionless parameter in frictional media (μ_s). Because viscosity has dimensions, the result is that the Reynolds number for fluids is a mixture of viscosity and system-defined length and velocity scales. On the other hand, for frictional plasticity, the equivalent dimensionless number is simply μ_s. Because no system-wide lengths or velocities are needed to build this characteristic dimensionless group, a number of simplifications arise in the scaling analysis of frictional systems compared to viscous ones. For example, from scaling alone, the same garden hoe test in a viscous fluid can be shown inconsistent with the resistive force hypothesis [28].

5 Conclusion

Granular flow models have a wide assortment of complexity. Here we have emphasized that the degree of complexity needed to capture certain phenomena varies dramatically and sensitively on the specific fields of interest and relative size of the geometry. The behavior of the granular velocity field shows subtle dependence on particle size, which influences the flow distribution in inhomogeneous flow geometries, and can cause jamming in smaller geometries even when local stresses would seem to be large enough to induce flow. However, global forces on macro-scale moving boundaries can be modeled well using basic frictional plasticity. It could be said then, that the simultaneous simplicity and complexity of different granular flow observables can be attributed to the immense simplification of going from those observables for which NGF is needed to those where frictional plasticity suffices.

References

[1] J.D. Goddard, Applied Mechanics Reviews **66**, 050801 (2014)
[2] A. Schofield, P. Wroth (1968)
[3] D.M. Wood, *Soil behaviour and critical state soil mechanics* (Cambridge university press, 1990)
[4] S. Roux, F. Radjai, in *Physics of dry granular media* (Springer, 1998), pp. 229–236
[5] M. Depken, J.B. Lechman, M. van Hecke, W. van Saarloos, G.S. Grest, Europhysical Letters **78**, 58001 (2007)
[6] T. Weinhart, R. Hartkamp, A.R. Thornton, S. Luding, Physics of Fluids (1994-present) **25**, 070605 (2013)
[7] L. S, Particulate Science and Technology **26**, 33 (2008)
[8] R.A. Bagnold, Proceedings of the Royal Society of London Series A **225** (1954)
[9] F. da Cruz, S. Emam, M. Prochnow, J.N. Roux, F. Chevoir, Physical Review E **72**, 021309 (2005)
[10] G. MiDi, The European Physical Journal E **14**, 341 (2004)
[11] P. Jop, Y. Forterre, O. Pouliquen, Nature **441**, 727 (2006)
[12] K. Kamrin, G. Koval, Physical Review Letters **108**, 178301 (2012)
[13] D.L. Henann, K. Kamrin, Proceedings of the National Academy of Sciences **110**, 6730 (2013)
[14] K. Kamrin, D.L. Henann, Soft matter **11**, 179 (2015)
[15] L. S. Mohan, K. K. Rao, P.R. Nott, Journal of Fluid Mechanics **457**, 377 (2002)
[16] I.S. Aranson, L.S. Tsimring, Physical Review E **65**, 061303 (2002)
[17] K. Kamrin, G. Koval, Computational Particle Mechanics **1**, 169 (2014)
[18] E. Azéma, F. Radjai, G. Saussine, Mechanics of Materials **41**, 729 (2009)
[19] E. Azéma, N. Estrada, F. Radjai, Physical Review E **86**, 041301 (2012)
[20] P.G. Rognon, J.N. Roux, M. Naaïm, F. Chevoir, Physics of Fluids (1994-present) **19**, 058101 (2007)
[21] G. Koval, J.N. Roux, A. Corfdir, F. Chevoir, Physical Review E **79**, 021306 (2009)
[22] L.E. Silbert, J.W. Landry, G.S. Grest, Physics of Fluids (1994-present) **15**, 1 (2003)
[23] C. Li, T. Zhang, D.I. Goldman, Science **339**, 1408 (2013)
[24] Y. Ding, N. Gravish, D.I. Goldman, Physical Review Letters **106**, 028001 (2011)
[25] R.D. Maladen, Y. Ding, P.B. Umbanhowar, A. Kamor, D.I. Goldman, Journal of The Royal Society Interface **8**, 1332 (2011)
[26] R.M. Nedderman, *Statics and kinematics of granular materials* (Cambridge University Press, 2005)
[27] S. Dunatunga, K. Kamrin, Journal of Fluid Mechanics **779**, 483 (2015)
[28] H. Askari, K. Kamrin, Nature Materials **15**, 1274 (2016)
[29] D.L. Henann, K. Kamrin, Physical Review Letters **113**, 178001 (2014)
[30] W. Prager, D.C. Drucker, Quarterly Journal of Applied Mathematics **10:2**, 157 (1952)
[31] J. Jaky, Journal of the Society of Hungarian Architects and Engineers **78**, 355 (1944)
[32] K. Krishnaraj, P.R. Nott, Nature Communications **7** (2016)
[33] L.E. Silbert, D. Ertas, G.S. Grest, T.C. Halsey, D. Levine, S.J. Plimpton, Phys. Rev. E. **64**, 051302 (2001)
[34] J.A. Dijksman, G.H. Wortel, L.T. van Dellen, O. Dauchot, M. van Hecke, Physical Review Letters **107**, 108303 (2011)
[35] P. Jop, Y. Forterre, O. Pouliquen, Journal of Fluid Mechanics **541**, 167 (2005)
[36] D. Fenistein, M. van Hecke, Nature **425**, 256 (2003)
[37] D.M. Mueth, Physical Review E **67**, 011304 (2003)
[38] O. Pouliquen, Physics of Fluids **11**, 542 (1999)
[39] K. Reddy, Y. Forterre, O. Pouliquen, Physical Review Letters **106**, 108301 (2011)
[40] E. Wandersman, M. van Hecke, EPL (Europhysics Letters) **105**, 24002 (2014)
[41] K. Nichol, A. Zanin, R. Bastien, E. Wandersman, M. Van Hecke, Physical Review Letters **104**, 078302 (2010)
[42] M.E. Gurtin, E. Fried, L. Anand, *The Mechanics and Thermodynamics of Continua* (Cambridge, 2010)
[43] C. Lun, S. Savage, D. Jeffrey, N. Chepurniy, Journal of fluid mechanics **140**, 223 (1984)
[44] Y. Jiang, M. Liu, Granular Matter **11**, 139 (2009)
[45] I.S. Aranson, L.S. Tsimring, Physical Review E **64**, 020301 (2001)
[46] J. Goyon, A. Colin, G. Ovarlez, A. Ajdari, L. Bocquet, Nature **454**, 84 (2008)
[47] L. Bocquet, A. Colin, A. Ajdari, Physical Review Letters **103**, 036001 (2009)

[48] D.L. Henann, K. Kamrin, International Journal of Plasticity **60**, 145 (2014)

[49] Q. Zhang, K. Kamrin, Physical Review Letters **118**, 058001 (2017)

[50] D.L. Henann, K. Kamrin, International Journal for Numerical Methods in Engineering (2016)

[51] P. Jop, Physical Review E **77**, 032301 (2008)

[52] T. Unger, J. Török, J. Kertéz, D.E. Wolf, Physical Review Letters **92**, 214301 (2004)

[53] S. Savage, M. Sayed, Journal of Fluid Mechanics **142**, 391 (1984)

[54] R.D. Maladen, Y. Ding, C. Li, D.I. Goldman, Science **325**, 314 (2009)

[55] B. Rodenborn, C.H. Chen, H.L. Swinney, B. Liu, H. Zhang, Proceedings of the National Academy of Sciences **110**, E338 (2013)

[56] J. Gray, G. Hancock, Journal of Experimental Biology **32**, 802 (1955)

[57] J. Slonaker, D.C. Motley, C. Senatore, K. Iagnemma, K. Kamrin, arXiv preprint arXiv:1604.02490 (2016)

Entrainment characteristics of fine particles under high speed airflow

Shaowu Yin[1, 2, *], *Li* Wang[1, 2], *Chuanping* Liu[1, 2], and *Lige* Tong[1, 2]

[1] School of energy and environmental engineering, University of Science and Technology Beijing, Beijing 100083, China
[2] Beijing Key Laboratory of Energy Saving and Emission Reduction for Metallurgical Industry, University of Science and Technology Beijing, Beijing 100083, China

Abstract. Fine silicon particles (mean size of 2.7 μm) are used as entrained materials, and the entrainment characteristics of fine particles are investigated in a cylindrical fluidized-bed (inner diameter of 28 mm and height of 1000 mm) under high speed airflow. The effects of the volume flow of gas (Q, 1 m^3/h to 2.5 m^3/h), the number of holes (N, 1 to 4), the size of holes (D, 1 mm to 3 mm), and the distance between holes and the upper surface of the material layer (H, -100 mm to 200 mm) on the entrainment characteristics (entrainment rate W and entrained powder-gas ratio R) are experimentally studied through orthogonal experiment. The experimental results show that an increase in Q and H constantly improves the entrainment characteristics; a decrease in D enhances such characteristics, whereas the number of holes N has no significant effect on the entrainment characteristics. An optimal operating condition can result in optimal entrainment characteristics (W, 3.1 g/min and R, 0.058 g/g), which can be achieved with a Q of 2.5 m^3/h, N of 1, D of 2 mm and H of 200 mm.

1 Introduction

Based on the fluidization technology and nitriding reaction of fine silicon powder, a fluidized bed reactor imposes strict requirements on raw silicon powder particle size for preparing fine silicon nitride powder, and requires the less agglomerations or the breaking of silicon aggregates in the entrainment process [1–3]. The dispersed cloud of powder feeding into the reactor must satisfy the following requirements: 1) low velocity to extend the residence time in the reactor; 2) small size to complete the conversion within a reasonable period; and 3) high powder–gas ratio to obtain high production efficiency and gas utilization [4]. Various feeders can be used for the feeding of powder, including mechanical feeders [5], pneumatic feeders [6], and fluidized bed feeders [7], but each of these feeders deals with specific types of materials.

The entrainment of fine particles presents a challenge to industry research. Geldart classified fine particles as group C particles [8]. A single fine particle cannot remain stable because of its extremely small size (less than 20 μm) and relatively large specific surface area. In this case, fine particles tend to agglomerate to achieve stability [9]. Group C particles also tend to agglomerate in the entrainment process, thereby reducing their entrainment quality. During the entrainment process, friction and collision continuously occur between group C particles, thereby reducing the distance among them. The van der Waals force emerges when such distance reaches the van der Waals distance scope. The friction between particles also produces static electricity [10]. The liquid bridging force is produced by the water in the fluidized gas, and all these forces will lead to the agglomeration of group C particles [11]. Therefore, many researchers have proposed various methods for preventing the agglomeration of group C particles or crushing the aggregates of these particles.

To determine the influence of coarse particles on the entrainment characteristics of fine particles, Yang et al. [12] investigated the entrainment characteristics of Geldart group C powder from solid mixtures (groups A and C) in a fluidized bed equipped with two vibrators and an agitator. Shin et al. [13] investigated the entrainment characteristics of fine powder from solid mixtures (Geldart groups A and C) in gas–solid fluidized beds with or without a tube. Yin et al. [14] studied the effects of the superficial velocity of nitrogen gas at room temperature, initial loading quantity of fine particles, vibration intensity, and stirring speed on the entrainment characteristics of fine silicon particles in fluidized bed with vibration, stirring, and coarse particles.

Silicon particles (mean size of 2.7 μm) are used as entrained materials, and the entrainment characteristics of fine particles are investigated in a cylindrical fluidized-bed (inner diameter of 28 mm and height of 1000 mm) under high speed airflow. The effects of the volume flow of gas, the number of holes, the size of holes, and the distance between holes and the upper surface of the material layer on the entrainment characteristics are experimentally studied through orthogonal experiment.

2 Experiment material and methods

* Corresponding author: yinsw@ustb.edu.cn

2.1 Experimental apparatus and procedure

The experimental apparatus, which comprises a air compressor, regulating valve, flow meter, conveying bed, copper tube with holes and bag filter, is schematically shown in Fig. 1. The conveying bed includes a PolyVinylChloride column with a 28 mm inner diameter and 1000 mm height. The lower surface of the copper tube with a 4 mm inner diameter has a number of small holes, which is buried in the material layer. Gas flow can be regulated using the regulating valve. The export of the conveying bed is connected to the bag filter, which has a favourable collection efficiency effect on the entrained powder in the experiment. Fine silicon particles (mean size of 2.7 μm) are used as the entrained materials. Air source uses compressed air generated by air compressor.

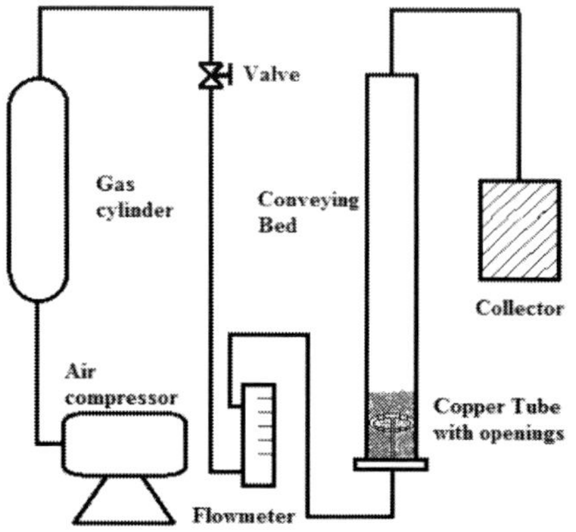

Fig. 1. Schematic diagram of fine silicon powder feeding device

Silicon particles are placed in an incubator at 105 °C for 24 h to remove the absorbed water. A weighed quantity of dried silicon particles are introduced into the bed. The copper tube is placed in a designated depth in the conveying bed. The bag filter is weighed using an electronic balance with a 0.001 g accuracy (denoted as m_1). The gas supply is activated. The airflow from the nozzle hole forms a high-speed jet airflow, which will impact and crush fine silicon powder and transport them. The entrained silicon powders from the conveying bed are captured by bag filter. The experiment begins after turning on the valve. After 1 min, the entire apparatus is switched off, and the bag filter is reweighed (denoted as m_2). These procedures are repeated five times in each experiment, and the average value is used as the final value for calculating the entrainment characteristics.

2.2 Experimental design

Five factors and four levels of orthogonal experiments are designed. The effects of the volume flow of gas Q, the number of holes N, the size of holes D, and the distance between holes and the upper surface of the material layer H on the entrainment characteristics (entrainment rate W and entrained powder-gas ratio R) are experimentally studied through orthogonal experiment.

The experimental factors are set as follows: Q is 1 m^3/h, 1.5 m^3/h, 2 m^3/h and 2.5 m^3/h; D is 1mm, 1.5mm, 2mm and 3mm; N is 1, 2, 3, and 4; H is 200mm, 100mm, 0mm and -100mm (i.e., 100mm above the surface of the material layer). In each experiment, the initial charge is 80g, and the loading height is 400 mm.

Entrainment rate W is defined as follows:

$$W = (m_2 - m_1)/t \tag{1}$$

where W(g/min) is the entrainment rate, m_1(g) and m_2(g) are the mass of the bag filter at the beginning and end of the experiment, respectively, and t (min) is the duration of the experiment.

The mass ratio of entrained powder to the needed nitrogen gas (i.e., entrained powder–gas ratio) R is defined as follows:

$$R = \frac{60 \times W}{1000 \times Q \times \rho} \tag{2}$$

where R(g/g) is the entrained powder–gas ratio, Q(m^3/h) is the volume flow of gas at room temperature, and ρ(kg/m^3) is the density of air at room temperature.

3 Experimental results

3.1 Orthogonal experimental analysis and variance analysis

3.1.1 Volume flow of gas Q

Effects of volume flow of gas Q on the entrainment characteristics are shown in Fig. 2. The increase of gas flow increases the apparent gas velocity in the bed. High gas velocity enhances the flow turbulence, and generates a great depth of action to reduce the sizes of agglomerates. W increases from 0.26 g/min to 1.56 g/min and R increases from 0.012 g/g to 0.029 g/g when Q increases from 1.0 m^3/h to 2.5 m^3/h. While the growth rate of R slows down after the Q exceeds 1.5m^3/h, indicating that the increase in W is less than the increase in Q.

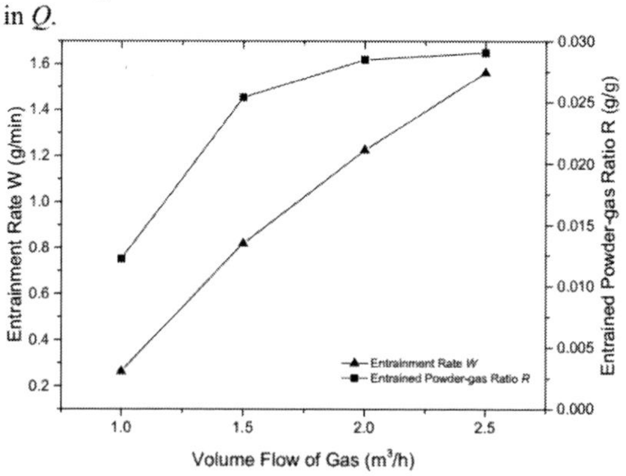

Fig. 2. Effects of volume flow of gas Q on the entrainment characteristics.

3.1.2 Distance between holes and the upper surface of the material layer H

The distance between holes and the upper surface of the material layer *H* determines the influence area of impact airflow, which influences the crushing and conveying effect of fine silicon powders in conveying bed. Effects of the distance between holes and the upper surface of the material layer *H* on the entrainment characteristics are shown in Fig. 3.

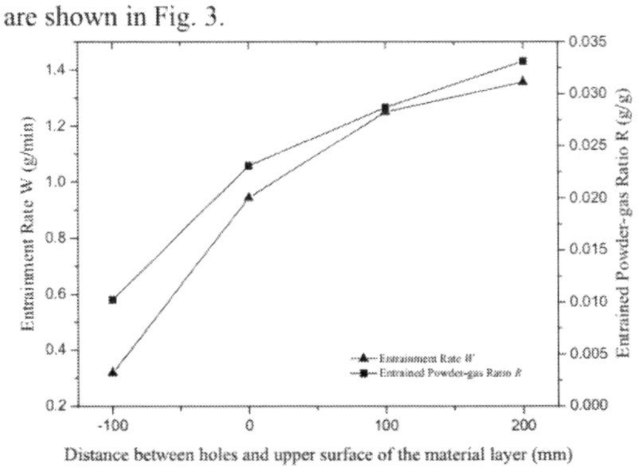

Fig. 3. Effects of the distance between holes and the upper surface of the material layer *H* on the entrainment characteristics.

When the holes are exposed above the material layer, the airflow from the holes releases into the cavity between the holes and material layer, and the airflow can't effectively play its impact. When the holes are buried in the material layer, the high-speed airflow released by the holes will penetrate a certain thickness of the material layer below, agitate the material layer and crush the agglomerates. Since the bottom of conveying bed is blocked, the airflow entrains fine silicon powders to form the upward motion of gas-solid two-phase flow and achieve the entrainment. When the buried depth of holes increases, the thickness of material layer above the holes increases. Therefore, the probability for fine silicon particles to be entrained out of the conveying bed increases along with *H*. However, the resistance and the load of the air compressor will increase, and it will lead to the slugging easily.

3.1.3 Number of holes N

The number and the size of the holes on copper tube determine the number, distribution and intensity of the impacting gas stream, thus affecting the effect of crushing the fine particles aggregates.

Effect of number of holes *N* on the entrainment characteristics are shown in Fig. 4. When the number of holes are small, the impact velocity is large, the airflow entrains fine silicon powders to form the upward motion of gas-solid two-phase flow and achieve higher entrainment rate and entrained powder–gas ratio. As the number of holes increases, the impact velocity decreases and the impact effect weakens. Upward movement of the two-phase flow entrained fine silicon powders encounters downward movement of the impact airflow, which will disperse the impact action of the high speed gas stream and reduce the mixing and crushing effect of airflow on the material layer. But the velocity of impact airflow decreases with the further increase in the number of holes, the hindering effect is also weakened, and the entrainment rate and entrained powder–gas ratio rise.

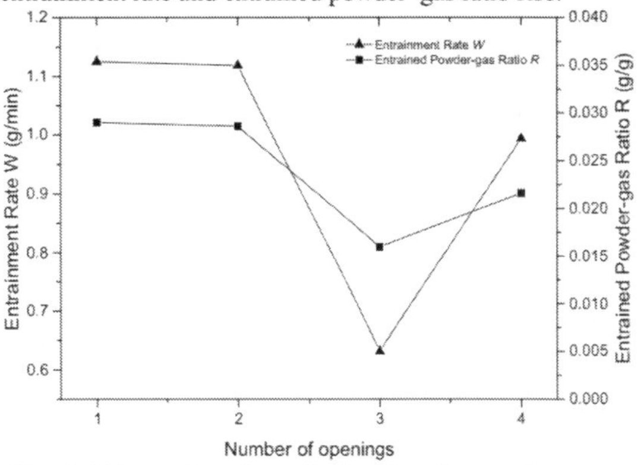

Fig. 4. Effects of number of holes *N* on the entrainment characteristics.

3.1.4 Size of holes D

Effects of holes size on the entrainment characteristics are shown in Fig. 5. When the size of holes are small, there are larger impact velocity, higher impact strength, greater the action depth, and higher entrainment rate and entrained powder–gas ratio. When the hole diameter D is 1mm, and the number of holes N is 1, the gas velocity at the outlet of the holes are about 884.2 m/s (supersonic speed) at the 2.5 m^3/h of the gas flow, so there is danger of burst at the pipeline interface.

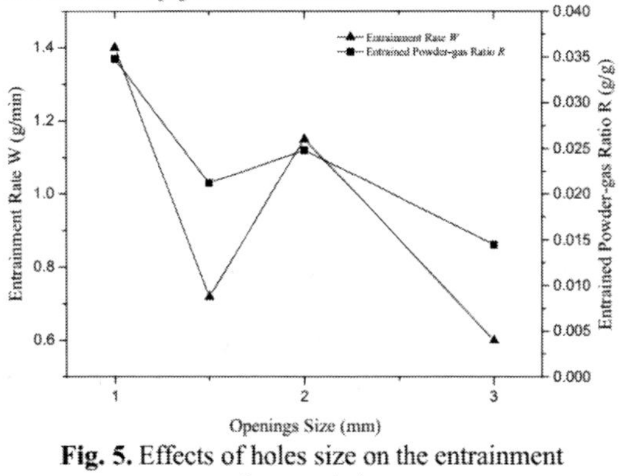

Fig. 5. Effects of holes size on the entrainment characteristics.

3.2 Analysis of variance

The variance analysis of *Q*, *N*, *D*, and *H* is performed according to the calculated orthogonal experimental results. Tables 1 and 2 show the variance analysis results for *W* and *R*, respectively.

Q and *H* both significantly affect *W*, but only *H* significantly influences *R*. Given that this study aims to obtain a high *R*, the optimal level combination conditions are set as follows: *Q* is 2.5 m^3/h, *D* is 1.0 mm, *N* is 1, and *H* is 200 mm.

In order to investigate the optimal combination of hole size and number under 2.5 m^3/h flow rate, some experiments are investigated through control variable experiments.The experimental results show that there are still airflow clogging under the conditions of hole diameter 1mm and hole number 3; Entrainment rate and entrained powder–gas ratio are generally low when the *D* is 1.5mm, which may be due to the large flow resistance at the square hole and the weakening of airflow impact. To sum up, an optimal operating condition can result in optimal entrainment characteristics (*W*, 3.1 g/min and *R*, 0.058 g/g), which can be achieved with a *Q* of 2.5 m^3/h, *D* of 2 mm, *N* of 1, and *H* of 200 mm.

Table 1. Variance analysis results of the four factors to investigate entrainment rate *W*.

Sources of variation	Sum of squares *SS*	Degree of freedom *df*	Mean square deviations *MS*	Value of F	Critical value Fa
D	1.67	3	0.56	6.30	
N	0.65	3	0.22	2.44	$F_{0.01}(3,3)$ =29.46
Q	3.76	3	1.25	14.20	$F_{0.05}(3,3)$
H	2.61	3	0.87	9.85	=9.28
Error	0.26	3	0.09		

Table 2. Variance analysis results of the four factors to investigate powder-gas ratio *R*.

Sources of variation	Sum of squares *SS*	Degree of freedom *df*	Mean square deviations *MS*	Value of F	Critical value Fa
D	0.00086	3	0.00029	5.78	
N	0.00047	3	0.00016	3.12	$F_{0.05}(3,3)$ =9.28
Q	0.00075	3	0.00025	5.01	$F_{0.1}(3,3)$
H	0.0012	3	0.00039	7.92	= 5.39
Error	0.00015	3	0.000050		

4 Conclusions

Using the self-designed high-speed airflow impact fine silicon powder feeding device, the effects of the volume flow of gas *Q*, the number of holes *N*, the size of holes *D*, and the distance between holes and the upper surface of the material layer *H* on the entrainment characteristics (entrainment rate *W* and entrained powder-gas ratio *R*) are experimentally studied through orthogonal experiment. The following conclusions were obtained:

(1) *W* and *R* increase obviously with the increase of *Q* from 1 m^3/h to 2.5 m^3/h; but when *Q* exceeds 1.5 m^3/h, the rising trend of *R* is slowed down.

(2) Under the premise of unobstructed gas flow, reducing *D* and *N* is conducive to enhance the mixing and crushing effect of impact airflow, which can get higher *W* and *R*, but the fluidization and transport of rising two-phase flow will be hindered if excessive jet holes are arranged in a small diameter conveying bed.

(3) When the holes are exposed above the material layer, the airflow can't effectively play its impact, and even obstruct the up-flow of the two-phase flow. When the holes are buried in the material layer, the high-speed airflow will agitate the material layer and crush the agglomerates. The deeper the buried depth, the better the transport effect, but the resistance is greater.

(4) The optimal operating conditions in this work are set as follows: *Q* is 2.5 m^3/h, *D* is 2 mm, *N* is 1, *H* is 200 mm.

This work was supported by the National Key Research and Development Program of China (No.2016YFB0601101), and the Beijing Higher Education Young Elite Teacher Project [No.YETP0360].

References

1. H.Z. Li, M.S. Guo, J. Chem. Ind. Eng. **01**, 52 (2013)
2. L. Wang, S.W. Yin, J.C. Dong, P. Chin. CN 1792774 06-28 (2006)
3. S.Y. Shan, Y.M. Wang, Q.M. Jia, J.F. Yang, J.Q. Gao, Z.H. Jin, Bull. Chin. Ceram. Soc. **01**, 138 (2009)
4. S.W. Yin, L. Wang, L.G. Tong, S.F Sun, J. Beijing University Chem. Technol. **35**, 70 (2008)
5. J.J. Dai, J.R. Grace, Powder Technol. **186**, 56 (2008)
6. C. Parker, L.T. Reist, Powder Technol. **107**, 36 (2000)
7. T.M. Francis, P.B. Kreider, P.R. Lichty, Powder Technol. **199**, 70 (2010)
8. D. Geldart, Powder Technol. **7**, 285 (1973)
9. D. Barletta, M. Poletto, Powder Technol. **225**, 93 (2012)
10. F. Babick, K. Schiebl, M. Stintz, Adv. Powder Technol. **22**, 220 (2011)
11. R.J. Cao, C.G. Lin, L. Sun, Y.L. Zhao, Z. Liu, C.C. Jia, P./M. Technol. **06**, 460 (2006)
12. F.M. Yang, L. Wang, S.W. Yin, Y.H. Li, C.P. Liu, L.G. Tong, Ind. Eng. Chem. Res. **52**, 1359 (2013)
13. M. K. Shin, E. M. Kim, B. S. Koo, G. Y. Han, K. J. Yoon, D. H Lee, Ind. Eng. Chem. Res. **46**, 1408 (2007)
14. S.W. Yin, W.Y. Tang, X.L. Zheng, L.G. Tong, L. Wang, C.P. Liu, Powder Technol. **299**, 150 (2016)

Effect of particle size distribution on the correlation between liquefaction resistance and shear wave velocity of granular soils

Changtao Zhou[1,*], *Xiao Min* Xu[2,†], and *Yi Pik* Cheng[1,‡]

[1] Department of Civil, Environmental and Geomatic Engineering University College London, London, UK
[2] Department of Engineering, University of Cambridge, Cambridge, UK

Abstract. Shear wave method has been increasingly popular in assessing the liquefaction potential of granular soils. Two particle-scale parameters, the inter-particle friction and the shear modulus of grains, play vital roles in correlation between Cyclic Resistance Ratio (*CRR*) and shear wave velocity corrected by overburden stress (V_{s1}). Series of drained one-dimensional compression tests were simulated on samples of different inter-particle friction angles assigned during preparation stage. Uniformity coefficients of these Particle Size Distribution (PSD) curves are 2 and 4 whose average particle size d_{50} are identical. The shearing results, as well as their assigned inter-particle friction angles form calibration curves for real sands. Dissimilar PSD curves result in different calibration outcomes. For Silica sand no.8, these curves give divergent inter-particle friction angles. This study calibrates particle shear modulus for Silica sand no.8 as well. Different PSD curves give divergent values of particle shear modulus. PSDs show impacts on calibrations of both vital parameters, which have converse effects on CRR-V_{s1} curves. This study suggests that the CRR-V_{s1} correlation should be independent of PSDs.

1 Introduction

Soil liquefaction is one of the hardest challenges that scientists and engineers are facing. Shear wave velocity corrected for overburden stress (V_{s1}, s for shear and 1 for principal direction of overburden stress) is widely considered as a promising alternative to assess the liquefaction resistance of granular soils. Previous researchers[1–3] performed numbers of cyclic tri-axial and dynamic centrifuge tests, whose work verified the reliability and suggested the CRR-V_{s1} correlation could be affected by soil types because the shear modulus (G_g) and the inter-particle friction parameter (μ_g) of soil particles play vital roles[4], as shown by equation (1).

$$CRR=\alpha_0/(G_g/G_{g0})^{m\beta}\cdot[\rho_g(V_{s1})^2/P_a]^{\beta/2} \quad (1a)$$

$$\beta=B\cdot\arctan(A\cdot\mu_g)+C \quad (1b)$$

where G_{g0}=1GPa, α_0=7.218 × 10^{-8}, m=0.32, A=4.890, B=0.971, and C=4.191. P_a is the atmospheric pressure. ρ_g is the density of grains.

It is difficult to directly measure the inter-particle friction parameter or the shear modulus of grains. By numerical tests, Xu *et al.* (2015) calibrated them for Silico Sand no.8. The value of φ_g obtained is 39.9°, and G_g is 6.649GPa[4, 5]. Using these two calibrated parameters, they produced a prediction curve which achieved 92% accuracy in separating liquefied and non-liquefied cases of centrifuge tests.

Note the calibration of micro mechanical properties is based on the DEM soil modelling. The PSD could have impacts on the calibration outcomes, which indirectly would affect prediction curves. DEM modelling enables us to modify micro parameters of particles, and see how they would affect the outcomes of macro tests. The increment of μ_g results in higher degree of induced structural anisotropy and a higher rates of dilation in shearing tests[6, 7]. The initial fabric of soil principally affects mechanical behaviours in numerical modelling, which can explain the experimental behaviour of sands[8]. Samples of different PSDs have dissimilar initial contacts fabric inside. Many novel laboratory tests show the PSDs have impacts on soil sample's peak friction angles[9, 10].

Particles' shear modulus could be calibrated by comparing numerical and laboratory shearing wave tests, based on the assumption that the real sand follows the same V_s-G_g power law[4]. PSD of samples could affect the outcome of G_g because a sample of lower void ratio tends to have a higher shear wave velocity inside[11], and a better graded assembly would achieve higher initial bulk density at any relative density level.

The objective of this paper is to show how and why PSDs affect calibrations for Silico sand no.8, in terms of the inter-particle friction and the shear modulus, and eventually the prediction curve for liquefaction resistance, in that the previous calibration[4] employed only one single uniform PSD curve. This paper conducted an integrated analysis on what PSD effects are.

* Changtao Zhou: changtao.zhou.14@alumni.ucl.ac.uk
† Xiaomin Xu: xx787@cam.ac.uk
‡ Yi Pik Cheng: yi.cheng@ucl.ac.uk

2 Methodology

2.1 Platform and Parallel Processing

The DEM simulations were performed on the PFC3D (Particle Flow Code in 3 dimensions) version 3.00 and version 3.10. The author developed codes based on PFC3Dv3.10, which can parallel-process assemblies and shearing tests.

Table.1 Particles numbers within samples

PSD curve	*Cu*	Particles
C2	2.00	43,764
C4	4.00	141,344
Xu *et al.* (2015)	1.17	28,335

Table 1 shows the particles numbers inside samples of PSD curves in this study (C2&C4), as well as previous study[4]. The limit of one single processor is of the order of 100,000 particles in three dimensions, in terms of the computation speed[12]. Performing particle generation and loading for PSD curve C4 could take extremely longer time. Parallel-supportive codes make advantage of powerful multiple-core processors. The assembly of particles is divided up spatially into several contiguous regions, and each processor takes charge of one region. In Fig. 2, particles are coloured according to their processors. Data of each region are stored locally in the corresponding processor's memory. Communication occurs merely for exchanges across interfaces between regions, in standard local network. Fig. 1 shows the time for the generation of increasing numbers of particles with different number of multi processors. Multi processors appear a significant efficiency in large-scale computation.

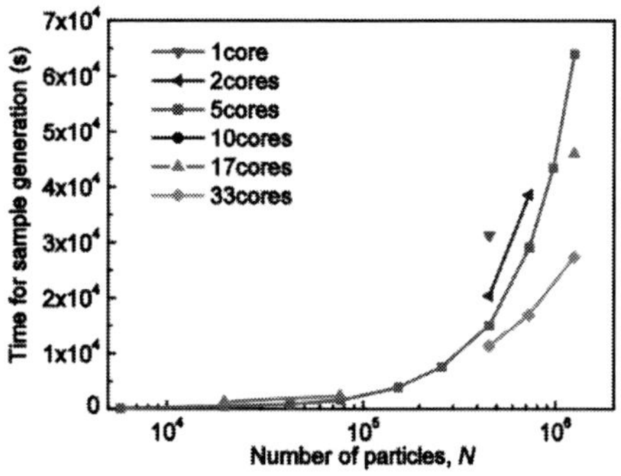

Fig. 1. Time needed for assemblies generation containing different numbers of particles

This paper imitates the practical tri-axial tests by making cylindrical assemblies. Samples were divided up along their long dimension, so that the exchanges of information between regions are kept to a minimum[12]. Walls were created on all processors. Tests' procedure remained the same as those single-processed ones. Test values were recorded locally by each processor, then accumulated by rejecting overlaps of shared balls at interfaces.

2.2 Sample preparation

Fig. 2 shows both PSD curves share the identical d_{50}, 0.175mm. Aspect ratio of height to diameter maintained 2.0. D/d_{50}=20 is the smallest size that a DEM specimen gives a converged and smooth results[5].

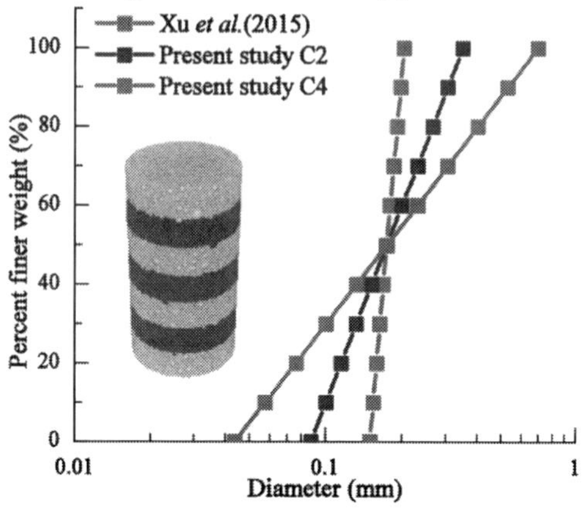

Fig. 2. PSD curves employed and a visualisation of the densest sample of C4

At beginning, particles were generated by 10 groups from bigger to smaller, randomly in a cylindrical region (4mm wide, 8mm high) constructed by rigid frictionless walls. Particles in one group, spheres of the same radius, were generated as smaller balls then expanded. The Hertz-Mindlin contact model was used due to its suitability for pressure-dependent behaviourr at small strain[4, 12]. Gravity was neglected. Each particle was prescribed with radius, shear stiffness (G_g=1GPa), Poisson ratio (υ_g=0.2), density (ρ_g=2630kg/m^3), and inter-particle friction (μ_g).

Then the assemblies were consolidated to an isotropic stress state. Note the initial void ratio (e_0) of this paper is defined after a specimen is iso-compressed. Loading was implanted by a servo-control mechanism written for the Hertz-Mindlin contact model. The e_0 of a specimen could be affected by the porosity input (void/bulk), ban or allowance of spinning, and the friction parameters in generation. Friction of particles was switched to the desired ones when isotropic stresses reached 90% of the required value and remained the same to the equilibrium state (the mean unbalanced force over the mean contact force$\bar{f}^u/\bar{f}^c \leq 10^{-3}$) at the desired stress.

Samples' generation follows a trial and error manner[4, 6, 13], to find the maximum and the minimum void ratios (e_{max}, e_{min}) at the required stress state. For each PSD curve, the e_{max} is reached by setting friction parameter higher and banning particles spinning at generation stage, while the e_{min} is achieved by freeing the spinning and setting particles frictionless. The trial of initial parameter falls to the porosity setting at the beginning of generation. The relative density (D_r) of the DEM specimens is defined as (e_{max}-e)/(e_{max}-e_{min}), where e is the void ratio of each one.

Samples of PSD curve C2 have void ratios ranging 0.481-0.694. These of C4 have void ratio range of 0.338-0.519.

2.3 Drained strain-controlled tri-axial tests

The densest sample (of e_{min}) of each PSD curve was used to perform tri-axial shearing tests.

Densest specimens were assigned with different inter-particle friction parameters (range from 0.1 to 0.5). Every time-step, the servo-control function adjusts the radial velocity of the cylindrical wall, according to differences between the actual stress acting on and the required pressure, to keep the confinement pressure at 100kPa during shearing. The strain rate was checked by using the inertial number defined as $I = \dot{\varepsilon} d\sqrt{\rho/p'}$, where $\dot{\varepsilon}$ is the axial-loading rate, ρ is the grain density and p' is the mean effective stress. It kept the inertia $I \leq 2.5e10^{-3}$ to ensure the quasi-steady conditions[14].

The axial and radical stresses are recorded by dividing the accumulated contact forces on walls by their area. The maximum internal friction angle (φ_{max}) was obtained by finding the highest ratio of the deviator stress (q) over the mean effective stress (p'). The correlation of φ_{max}-φ_g is established by aligning the maximum internal friction angles (φ_{max}) with the inter particle friction angles (φ_g). In this study we call this correlation φ_g calibration curves.

$$\sin(\varphi)=3(q/p')/(6+q/p')=3(\sigma_1-\sigma_3)/(5\sigma_1+7\sigma_3) \quad (2)$$

2.4 Shear wave velocity measurements

This study follows the method of previous research[4] to calibrate grains' shear modulus. G_g was calibrated based on the linear V_s–D_r relationship. For each PSD curve, samples of different relative density (D_r) were tested for wave propagation velocity (V_s) inside. It is well known that the value of small strain shear modulus (G_{max}) could be obtained from the shear wave velocity (V_s).

$$G_{max}=\rho_b V_s^2 \quad (3)$$

where ρ_b is the bulk density of tested samples.

This study uses equation (3) reversely to get V_s of tested samples. Assemblies were rested to a state that velocities of particles inside are lower than 10^{-3}m/s. Then specimens were subjected to axial loading to a very small strain ($<10^{-6}$), with stress state remaining 100kPa and isotropic. It gives the dynamic shear modulus (G_{max}), with which equation (3) could calculate an equivalent V_s.

3 Results and observations

3.1 Inter-particle friction calibration

All these φ_{max}-φ_g relationships for DEM assemblies used could be expressed as equation (4)

$$\varphi_{max}=a(\varphi_g)^b \quad (4)$$

Table.2 Calibration results of Inter-particle friction angle by using various PSD curves

PSD curve	a	b	R^2	φ_g (°)
C2	11.524	0.312	0.999	**33.3**
C4	10.812	0.356	0.999	**25.7**
Xu *et al.* (2015)	10.121	0.332	/	**39.9**

Fig. 3 presents these φ_{max}-φ_g relationships obtained from shearing tests based on PSD curves C2 and C4. Sets of a, b was obtained as presented in Table 2, which also gives their relative coefficients. The calibrated inter-particle friction angle (φ_g) for Silico sand no.8 could be determined by both φ_{max}-φ_g relationships. The right-hand column of Table 2 presents the calibration results based on both calibration curves, and the previous study[4] as well. It shows that the better gradation, the smaller calibrated μ_g value.

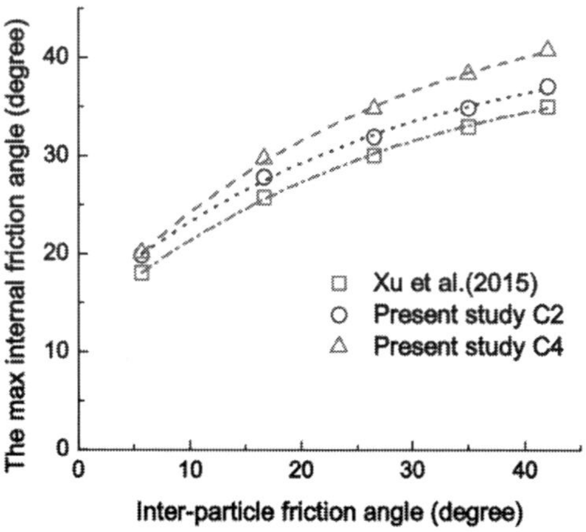

Fig. 3. φ_g calibration curves of PSD curves

3.2 Grains' shear modulus calibration

Specimens of different D_r were assigned with a same value of μ_g, 0.5. Based on the assumption that real sand follows the same V_s-G_g power law, the particle shear modulus (G_g) could be calibrated by comparing values of V_s obtained from both lab tests and the equivalent shear wave velocity tests at similar relative density range[4, 5].

$$V_s=\alpha'(G_g)^{\beta'} \quad (5)$$

where α'=94.9813, and β'=0.3204.

As shown in Fig. 4, shear wave velocity tests produced various V_s–D_r relationships for different PSD curves.

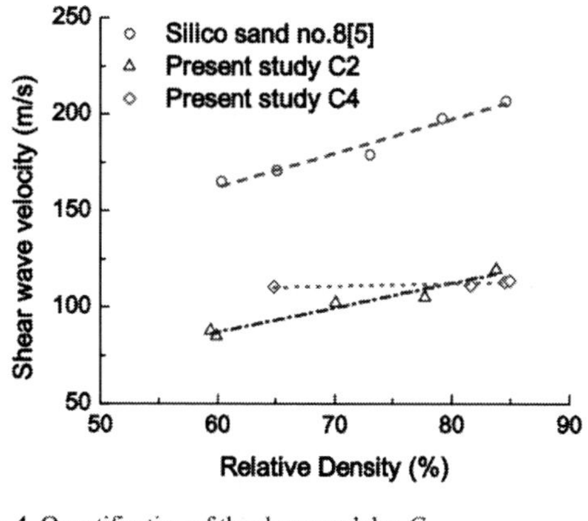

Fig. 4. Quantification of the shear modulus G_g

V_s–D_r fitting lines could be expressed as

$$V_s=m \cdot D_r+n \quad (6)$$

Numerical shear wave velocity tests assigned shear modulus of 1GPa to each particles. Doing ratio of the lab test over numerical results on both sides of equation (5) could calibrate the shear modulus of Silico sand no.8. PSD curves, C2 and C4, gave dissimilar values of G_g, as listed in Table 3. It indicates a better gradation would lead to a smaller calibrated value of G_g.

Table.3 Calibration results of grains' shear modulus by using various PSD curves

PSD curve	m	n	R^2	G_g(GPa)
C2	127.21	10.542	0.953	**6.090**
C4	13.677	101.22	0.727	**5.141**
Xu *et al.*(2015)	176.55	56.108	0.962	**6.649**

3.3 *CRR-V_{s1}* curves

The increase of C_u leads to the decreases of calibrated values of both μ_g and G_g. The influence of smaller inter-particle friction and Shear modulus is converse to the *CRR*-V_{s1} correlation. It shows in single case of V_{s1}, from equation (1), lower values of both calibrated parameters have little effect on *CRR*.

Fig. 5 presents the predicted *CRR*-V_{s1} curves with both calibration results of μ_g and G_g and 204 CSR data from the dynamic centrifuge tests[3]. It shows all newly produced curves are very close to the reference curve[4]. Effects of calibrated μ_g and G_g could eliminate each other. Although PSD of assemblies affects the calibration of μ_g and G_g, little on *CRR*-V_{s1} correlation has be observed.

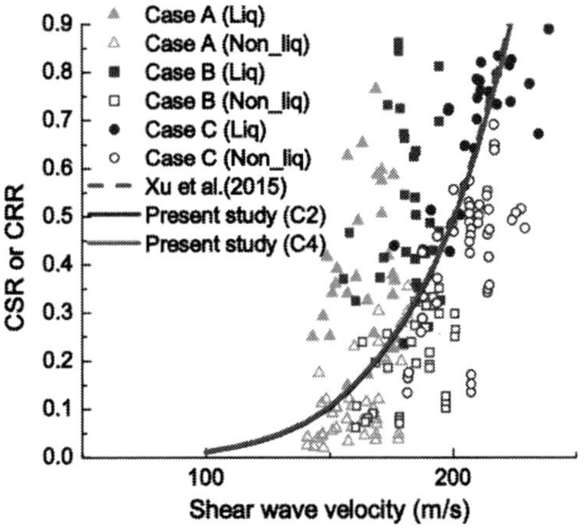

Fig. 5. Super-Influence of calibrated μ_g and G_g on *CRR*-V_{s1} correlation

4 Conclusions

By following the systematic parametric method pioneered by previous researches[3, 4, 11, 15], this study has explored the influence of PSDs on correlation between liquefaction resistance and shear wave velocity (*CRR*-V_{s1}) of granular soils by performing a series of drained strain-controlled tri-axial tests and shear wave velocity tests, on DEM specimens with various relative density and particle mechanical properties. The findings of this study is that the *CRR*-V_{s1} correlation is possibly independent of soils' gradation. PSD affects two key parameters' calibrations which determining this correlation. But changes that a new PSD brings to inter-particle friction parameter (μ_g,) and shear modulus (G_g) have converse effects on the eventual *CRR*-V_{s1} correlation, and eliminate each other. The *CRR*-V_{s1} curve remains still when coefficient of uniformity is increasing.

This study does not consider the breakage nor shape effects of particles. The coefficient of uniformity of 6.0 categorizes soil samples into "well-graded" or "poorly-graded" groups. Both PSD curves employed above have coefficient of uniformity lower than 6.0. Some researches[16–18] on particles crushing within soil samples have shown a tendency that more particles would crush or break in poorly graded soil than in well graded soil under shearing conditions. Spheres were generated as particles in this study, which was meant to isolate effects of contact friction. Spheres tends to undertake more rotation comparing to very angular particles. Assemblies of long or angular particles in DEM simulation, show behaviours closer to the practice while bearing shearing[15, 19].

References

1. Y. Zhou, Y. Chen, J Geotech Geoenvironmental Eng **133**, 959–972. (2007)
2. Y. Chen, Y. Zhou, H. Ke, 14th World Conf Earthq Eng, 1–8 (2008)
3. Y. Zhou, Y. Chen, Y. Shamoto, J Geotech Geoenvironmental Eng **136**,165–177 (2010)
4. X. Xu, D. Ling, Y. Cheng, Y. Chen, Geotechnique **65**, 337–348 (2015)
5. X. Xu, *Study on the micromechanism of sand liquefaction and its evaluation*, Zhejiang University (2012) (not published)
6. C. Thornton, Geotechnique, **50**, 43-53 (2000)
7. X. Huang, K.J. Hanley, C. O'Sullivan, F.C.Y. Kwok Particuology **15**, 107–115 (2014)
8. K. Soga, S. Yimsiri, Géotechnique **60**, 483–495 (2010)
9. E.M. Kara, Eng Technol Appl Sci Res **3**, 497–501 (2013)
10. A. Hamidi, E, Azini, B. Masoudi, Sci Iran **19**, 393–402 (2012)
11. X. Xu, Y. Cheng, D. Ling, AIP Conf. Proc., 201-204 (2013)
12. Itasca, *Manual of particle flow code in 3-dimension version 3.00* (2003)
13. C. Salot, P. Gotteland, P. Villard, Granul Matter **11**, 221–236 (2009)
14. J.C. Lopera Perez, C.Y. Kwok, C. O'Sullivan, X. Huang, K.J. Hanley, Géotechnique **66**, 725–740 (2016)
15. Y. Yu, Y. Cheng, X. Xu, K. Soga, *Geomechanics from Micro to Macro*, 207–212 (2015)
16. M.R. Coop, F.N. Altuhafi, Géotechnique **61**, 459–471. (2011)
17. F.N. Altuhafi, B. Baudet, P. Sammonds, Can Geotech J **790**, 775–790 (2010)
18. F. Altuhafi, B.A. Baudet, Eng Geol **120**, 1–9 (2011)
19. R.P. Jensen, P.J. Bosscher, M.E. Plesha, T.B. Edil, Int J Numer Anal Methods Geomech **23**, 531–547 (1999)

Discharge of repulsive grains from a silo: experiments and simulations

David Hernández-Enríquez[1], *Geoffroy* Lumay[2], and *Felipe* Pacheco-Vázquez[1,*]

[1] *Instituto de Física, Benemérita Universidad Autónoma de Puebla, Apartado Postal J-48, Puebla 72570, Mexico.*
[2] *GRASP, Université de Liège, B-4000 Liège, Belgium.*

Abstract. In granular matter, sliding friction and collisions among grains are fundamental mechanisms of energy dissipation that determine the particles dynamics. Here we consider an unconventional granular system composed of magnetic repelling grains confined in a two dimensional cell that interact only through their magnetic field. The repulsive interaction prevents contact among grains and therefore produces a different dynamics compared to the dynamics of classical granular systems. In particular, we present experiments and simulations on the discharge of this repulsive granular medium from a silo. The results reveal an inverted density profile and a plug-flow through the aperture that contrast with the dynamics displayed by contacting grains. Moreover, the simulations allow to estimate the friction coefficient generated by the lateral confinement.

1 Introduction

Because of its omnipresence in industry and agriculture, one of the most widely studied systems in granular matter is the discharge of grains from a silo [1-10]. In the pioneering work of Beverloo [2] describing the mass rate Q of grains of size d flowing through an orifice of diameter D under the influence of gravity g, it was found that $Q = C\rho_b \sqrt{g}(D - \kappa d)^{5/2}$, where ρ_b is the apparent density of the material and C and κ are constants related to the particle nature and silo geometry. The 5/2 power dependence has been verified for different kind of particles, independently of their packing fraction, density, surface properties or shape [3]. For small apertures, where clogging is possible ($D/d \sim 10$), an exponential corrective factor must be introduce to predict the flow due to a lower density of material near the exit [4]. Concerning the discharge of a two dimensional silo, an expression for the flow rate is given by $Q = \langle\rho_b\rangle \sqrt{2g\beta D}(D-kd)$ with $\beta \sim 0.5$. One of the most relevant aspects of these expressions is the independence of Q on the column height, in contrast to the hydrostatic dependence at the bottom of a water column. Recently, it has been also investigated the granular discharge of submerged hoppers, and a surge in the flow rate was measured at the end of the process[5].

A granular material is usually defined as a conglomerate of macroscopic particles characterized by a loose of energy due to elasticity and friction when particles collide or slide [6]. In a recent work [7], the author of this manuscript and colleagues presented experiments in a two dimensional cell filled with an assembly of magnets that do not interact through contact forces but through their repelling magnetic field. These contact-less particles are able to form stable piles with an average angle of repose, to flow through an aperture when it is large enough, and to form arches and produce jamming as it happens with conventional grains. Our previous experimental study was focused on determining the discharge flow rate for repelling grains in a two-dimensional silo. In contrast to classical granular flows, the results showed a plug flow velocity profile along the silo aperture. It was also found that the average density of grains $\langle\rho\rangle_{x,t}$ above the aperture decreases with D (another difference with classical granular materials) and it was well fitted by:

$$\langle\rho\rangle_{x,t} = \rho_\infty + (\rho_0 - \rho_\infty)e^{-D/\alpha} \quad (1)$$

where ρ_0 and ρ_∞ are the density of grains for the static column previous to discharge and the density when the bed is in free fall, respectively, and α is a constant related to the minimum value of D for a continuous flow. Then, the flow rate is given by:

$$Q = \left[\rho_\infty + (\rho_0 - \rho_\infty)e^{-D/\alpha}\right] \sqrt{2g\beta}D^{3/2} \quad (2)$$

This expression is independent of the column height, and therefore, Q was found to be constant as in the case of a discharge of contacting grains.

In the present paper, we show numerical results obtained from molecular dynamics(MD) simulations which are compared with the above experimental findings. These simulations allow us to better understand the effect of friction between the grains and the walls and to avoid finite size effects by keeping the grains number constant.

2 Numerical simulations

In our experiments only 565 magnets fit into the 1.8×0.5 m^2 compartment due to the grain-grain and grain-wall repulsive interactions. The small number of particles produced finite size effects so that the average density and

*e-mail: fpacheco@ifuap.buap.mx

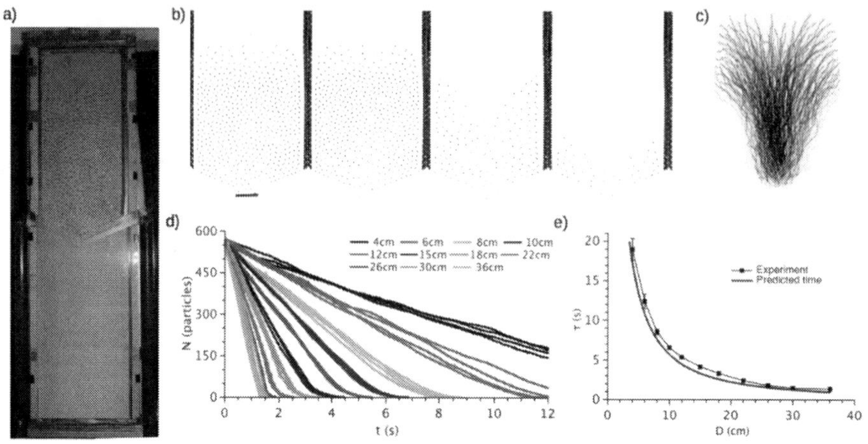

Figure 1. a) Experimental setup. b-c) Snapshots and grain paths for a silo discharge of aperture $D = 8$ cm. d) Number of particles N remaining in the upper compartment as a function of time t for different values of D. e) Total time of discharge τ vs D.

velocities were computed in a reduced time interval of the whole discharge. Since building a larger cell was unworkable, and aiming to validate the experimental results, we performed MD simulations considering disks of diameter d= 5 mm with a repulsive potential in two-dimensions. The non-linear repulsive force $\vec{F}(r,\theta)$ between two magnetic dipoles separated by a distance r with an angle θ between the direction of the magnetic field and the line joining the dipoles centres is given by: $\vec{F}(r,\theta) = (3\mu m^2/4\pi r^4)\left[\left(1-3\cos^2\theta\right)\hat{r} + \sin 2\theta\hat{\theta}\right]$, where μ_0 is the vacuum magnetic permittivity and m is the dipolar magnetic moment. The total repulsive force over each magnet $\vec{F}_i(x_i(t))$ was calculated considering its interaction with the rest of the disks at distances r_{ij}. In the two-dimensional configuration, $\theta = \pi/2$, and therefore the force over the i-th magnet at a given time t is given by:

$$\vec{F}_i(x_i(t)) = \sum_{j\neq i}^{n} \frac{3\mu_0 m^2}{4\pi r_{ij}^4}\widehat{r_{ij}} \tag{3}$$

This expression was introduced in the velocity Verlet algorithm[11] programmed in Matlab ® to integrate the equations of motion, where positions x_i, velocities v_i and accelerations a_i at time $t + \Delta t$ are obtained from the same quantities at time t according to:

$$\vec{x}_i(t+\Delta t) = \vec{x}_i(t) + \vec{v}_i(t)\Delta t + \frac{1}{2}\vec{a}_i(t)\Delta t^2 \tag{4}$$

$$\vec{v}_i(t+\Delta t) = \vec{v}_i(t) + \frac{\vec{a}_i(t) + \vec{a}_i(t+\Delta t)}{2}\Delta t \tag{5}$$

$$\vec{a}_i(t+\Delta t) = \frac{\vec{F}_i(x_i(t)) + \vec{\xi}_i}{m_i} \tag{6}$$

In eq.(6), the term $\vec{\xi}_i$ represents a dissipative force associated to friction due to the glass walls.

The simulation begins with 565 disks in a rectangular array with random velocities taken from a Gaussian distribution. The discs are dropped under gravity into a silo of appropriate dimensions and geometry according to the experiments. Once all the grains have dropped and the change in the average kinetic energy between two consecutive time steps is less than a certain value ϵ, the gate is open to start the discharge and the flow is measured by counting the number of particles that have fallen for each time step Δt. In order to increase the total discharge time and neglect finite size effects, the number of grains N was kept constant during the simulation by reinserting the outgoing grains in the upper part of the silo. The simulation ends when 565 magnets have passed through the aperture. The simulations were performed with and without the dissipative term to determine the effect of the confinement.

3 Results

Let us first describe the experimental set-up and some results that were necessary to calibrate our simulations. Figure 1a shows the cell filled with $N_{tot} = 565$ cylindrical Neodymium magnets oriented with their magnetic dipoles pointing in the same direction to induce a repulsive force among them; the lateral walls were also repulsive and the aperture size D was set prior to discharge. Because there is no contact among grains, they can be easily distinguished and tracked. Figure 1b shows snapshots of the discharge for a small aperture $D = 8$ cm; the paths followed by the grains in fig. 1c indicate a funnel flow pattern as it occurs with non-repelling grains. From videos taken at 250 fps, we counted the number of particles N remaining at the upper cell compartment as a function of time t for different values of D ranging from 4 cm to 36 cm; the results are shown in fig. 1d. There are two different behaviours: a linear decrease during most of the discharge that reflects a constant flow rate Q (constant slope) and a decrease of the flow at the end of the process. The total discharge time τ is plotted as a function of D in fig. 1e for the experiments (points) and compared with the expected value $\tau = N_{tot}/Q$ (blue line).

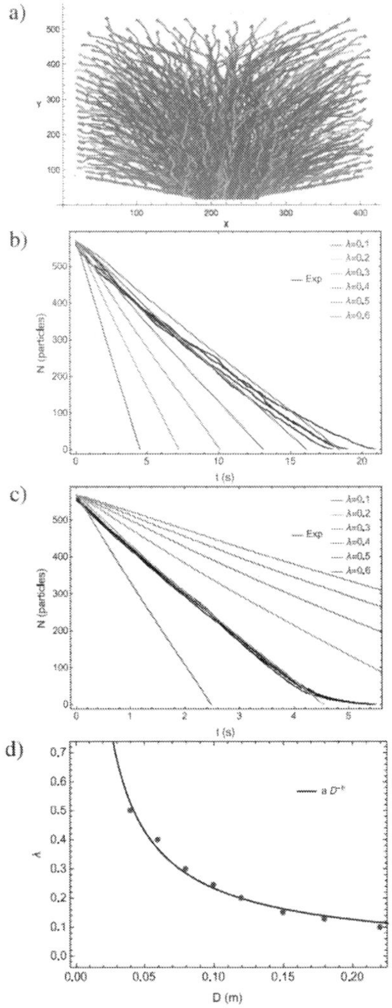

Figure 2. MD simulations: a) Particles paths for D = 8 cm, b-c) N vs t for D = 4 cm and 12 cm, respectively, for different values of λ. Black lines represent the experiments. d) λ vs D obtained from the best comparison with experimental results (red points); the black line corresponds to $\lambda = (0.03 \pm 0.002)D^{-0.89\pm0.07}$.

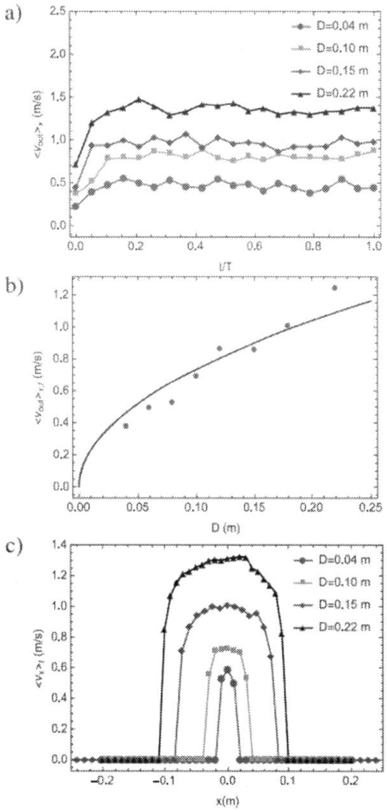

Figure 3. a) $\langle v_{out}\rangle_x$ vs t for four different apertures; t is normalized by the discharging time T of 565 magnets. b) $\langle v_{out}\rangle_{x,t}$ vs D. c) Velocity profile along the aperture.

Figure 2a shows the trajectories of 565 repulsive particles obtained from the simulations. Note that, as in the experimental case, the magnets move mainly along the central part of the cell developing a funnel flow. We performed simulations with different dependences for $\vec{\xi}_i$ but only a viscous term of the form $\vec{\xi}_i = \lambda\vec{v}_i$ was able to reproduce the linear behaviour of N vs t found in the experiments. The friction coefficient λ was varied until obtaining the best comparison with the experimental data for each value of D, see figs. 2b-c. Figure 2d shows that λ decreases with D and it is well fitted by $\lambda = (0.03 \pm 0.002)D^{-0.89\pm0.07}$. To explain this dependence, let us consider that the torque $\vec{\tau}$ on a magnetic dipole $\vec{\mu}$ placed in a magnetic field $\vec{B}$ is directly related to the strength of the field by $\vec{\tau} = \vec{\mu} \times \vec{B}$; if the dipole is free to move, it will rotate until $\vec{\mu}$ is parallel to $\vec{B}$. In the cell, the frontal walls prevent the magnet rotation but the torque over each magnet produces an increase of friction with the walls which is proportional to $\vec{\tau}$. In the experiments, the separation among particles is observed to increase with D, reducing the local density; therefore, the magnetic field over each particle $\vec{B}_i$ decreases and also $\vec{\tau}_i$, this causes the decreasing friction coefficient observed in fig. 2d.

Once λ was found for each value of D, we computed the average particle velocity at the output $\langle v_{out}\rangle_x$ as a function of time; the results for four different apertures are shown in fig. 3a (notation $\langle\rangle_x$ indicates the average over the aperture). Note that $\langle v_{out}\rangle_x$ practically has a constant value during the whole discharge as it is expected for a constant flow rate. By plotting $\langle v_{out}\rangle_x$ averaged on time as a function of D and using $\langle v_{out}\rangle_{x,t} = \sqrt{2g\beta D}$ to fit the data, see fig. 3b, one finds $\beta \approx 0.27 \pm 0.05$, in contrast with $\beta \approx 0.5$ found for contacting grains in which case a free-fall semicircular arch region of size proportional to the aperture is assumed. The different value obtained with repelling grains could be associated to the fact that the free-fall region does not really exist [8], since even in the aperture the outgoing magnets are interacting with the magnetic field of the neighbouring particles. Figure 3c

 DOI: 10.1051/epjconf/201714003089

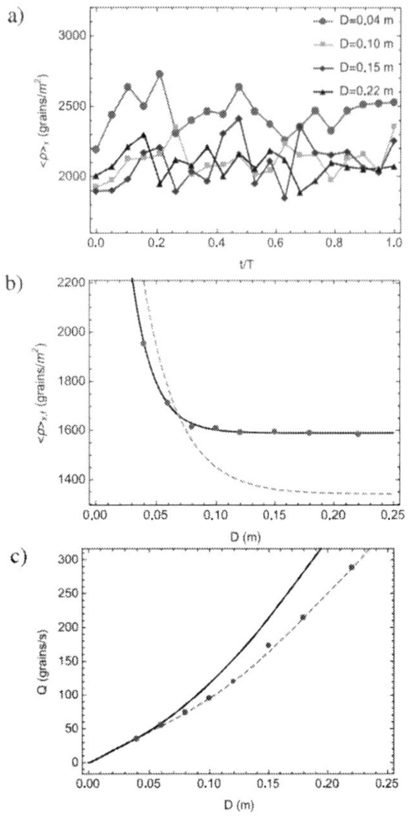

Figure 4. a) $\langle\rho\rangle_x$ vs t/T found in simulations. b) $\langle\rho\rangle_{x,t}$ vs D (•), data fitting using eq.1 (–), and experimental fitting reported in [7] (- -). c) $Q = N_{tot}/\tau$ vs D (•) and solution of eq. 2 with parameters found in simulations (–) and those found in experiments [7] (- -).

shows the velocity profile along the aperture obtained from the simulations, $\langle v_x\rangle_t$ is the average velocity in each coordinate x during the total time of discharge and $\langle v_x\rangle_t = 0$ outside the aperture corresponds to the static magnets on the gates. As in the experiments, an important velocity gradient is observed near the borders of the aperture with a plug flow in the central zone; this plug flow constitutes another strong differentiation with classical granular flows.

Finally, we computed the density of particles $\langle\rho\rangle_x$ inside a rectangular box of $10d \times D$ centred in the silo aperture as it was done with the experimental measurements in [7]. We found fluctuations around a mean value during the evolution of the discharge, see fig. 4a. In the experimental results, it was found a constant density only during an intermediate interval of the process due to the reduced number of grains; this effect is avoided in the simulations by reinserting the outgoing grains at the top of the silo, giving as a result a constant $\langle\rho\rangle_x$. The average density taken over time $\langle\rho\rangle_{x,t}$ decreases with D, see red points in fig. 4b, supporting the previous discussion related to the friction factor and in agreement with experiments. Using eq. (1) to fit the numerical results we obtain the values $\rho_\infty = 1589.26$ grains/m^2, $\rho_0 = 5049.47$ grains/m^2 and $\alpha = 0.0177$. These values were used to calculate Q as a function of D with eq. (2); the results are compared in fig. 4c with the fit of the experiments reported in [7]. The difference between simulations (black line) and the experimental data fit (dashed blue line) is again associated to finite size effects.

4 Conclusions

We studied numerically the discharge of repelling magnetic grains from a two-dimensional silo. Our results are in agreement with previous experimental findings[7] without the limiting effects produced by the small number of grains and finite size of the system. The magnets move towards the aperture developing a funnel flow pattern as it occurs in classical granular materials, but in contrast, a plug-flow profile appears at the exit. The density of grains measured just above the silo output is found to decrease exponentially with the aperture size in accordance with the experiments. The magnetic interaction produces a torque that generates friction with the walls, the simulations allowed us to evaluate this friction factor which diminishes with D as a consequence of the decreasing density. Of particular interest could be to compare our results with the dynamical scenario that typically occurs when cars pass through a bottleneck, where no surface-surface interactions also exist.

Acknowledgements

This research was supported by CONACyT Mexico through the project No. 242085 of the Sectoral Research Fund for Education, PROFOCIE-SEP 2015-2016 and VIEP-BUAP. The authors thank J. Schockmel, and S. Dorbolo for contributions and discussions. G.L thanks FNRS (Grant PDR T.0043.14) and ULG (Grant C-13/88).

References

[1] I Zuriguel, Papers in Physics **6**, 060014 (2014)

[2] W. A. Beverloo, H. A. Leniger, J. van de Velde, Chem. Eng. Sci. **15**, 260 (1961).

[3] D. Hirshfeld, Y. Radzyner and D. C. Rapaport, Phys. Rev. E **56**, 4404-4415 (1997).

[4] C. Mankoc, *et al.*, *Granular Matter* **9**, 407 (2007).

[5] T. Wilson, C. Pfeifer, N. Mesyngier, and D. Durian, Pap. Phys. **6**, 060009 (2014).

[6] J. Duran, *Sands, Powders, and Grains: An Introduction to the Physics of Granular Materials*, (Springer-Verlag, New York, 2000).

[7] G. Lumay, *et al.*, Pap. Phys. **7**, 070013 (2015).

[8] S. M. Rubio-Largo, A. Janda, D. Maza, I. Zuriguel, and R. C. Hidalgo, Phys. Rev. Lett. **114**, 238002 (2015).

[9] R. O. Uñaca, A. M. Vidales and L. A. Pugnaloni, J. Stat. Mech. **2012**, 04008 (2012)

[10] H. Ahn, Z. Başaranoğlu, M. Yilmaz, A. Buğutekin, and M. Zafer Gül, Powder Technol. **186**, 65 (2008).

[11] Q. Spreiter and M. Walter, J. Comput. Phys. **152**, 102–119 (1999).

Distinct element analysis of inter-particle coating variability in a batch seed coater

Mehrdad Pasha[1,*], *Mojtaba* Ghadiri[1], *Alfeno* Gunadi[2], and *Patrick M.* Piccione[3]

[1] School of Chemical and Process Engineering, University of Leeds, United Kingdom
[2] Process Studies Group, Technology and Engineering, Jealott's Hill International Research Centre, Syngenta, United Kingdom
[3] Process Studies Group, Technology and Engineering, Syngenta, Münchwilen, Switzerland

Abstract. Coating of particulate solids by a thin film layer is of interest in many industrial applications such as seed and tablet coating. In seed processing, seeds are commonly coated with a protective coating layer consisting of fertilisers and crop protection products. Rotary drum batch coaters are typically used for this purpose. The coater consists of a cylindrical vessel with a rotating spray disk in the centre, onto which the coating liquid is fed. The seeds are driven around the vessel by its rotating base, and are mixed by two baffles; one on either side of the vessel. In the present study, DEM simulations are used to analyse the seed coating process. Corn seed is used as a model material and its shape is captured using X-Ray micro-tomography (XRT). The shape is incorporated into the simulations by the clumping multiple spheres to form a particle assembly. The coating uniformity of the seeds is predicted by implementing a coating model, whereby the coating mechanism is represented in the DEM by considering that once a droplet contacts a corn seed, it is removed from the simulation and its mass is attributed to the coating mass of the corn seed. The distribution of mass of sprayed spheres on the corn seeds and the coefficient of variation are evaluated for a range of process conditions, such as spinning disk rotational speed, droplets size and baffle arrangement and designs. In addition to evaluation of coating uniformity of particles, the scale-up rules for rotary drum batch coaters are investigated and reported. The outcome provides guidelines on scale-up rules and improvement of coating uniformity for rotary batch seed coaters.

1 Introduction

Coating of particulate solids by a thin film layer is widely utilised in a variety of industrial applications including seed and tablet coating. In seed processing, rotary drum batch coaters are mainly used to coat the seeds with a protective layer consisting of fertilisers and crop protection materials. The effectiveness of the coating liquid formulation and the level of coverage of the coating on the seeds are important for optimisation of the seed coating process as these have a direct impact on the quality of the final product. Coating is influenced by motion, mixing of seeds, and liquid spreading over them, which are directly controlled by process operation.

The use of Distinct Element Method (DEM) [1] for the simulation of particulate systems has gained popularity in recent years since it provides a robust and realistic way of simulating the behavior of particulate matter during processing. More recently, it has been used to address the coating uniformity of pharmaceutical tablet coating in pan coaters [2-10]. There are two critical parameters for the analysis of the coating variability: inter- and intra- particle coating variability. The former is the variation in the average coating mass from a granule to another, whilst the latter is the distribution of coating liquid on the surfaces of individual granules, both of which can be quantified using the coefficient of variation [3].

The effect of seed coating process parameters (i.e. baffle angle and clearance gap to the wall, atomiser distance relative to the base and baffle geometry) on coating uniformity are investigated and their optimal values are identified. Moreover, the outcome of this study helps establishing scale-up rules for which little work has been reported in the literature.

2 Materials and methods

2.1 DEM simulations

A series of DEM simulations were performed for a rotary batch seed coater (0.3 m in diameter and 0.21 m in height), as shown in Fig. 1, using EDEM® software package (DEM-Solutions) to predict the impacts of various coating process conditions on the coating uniformity of corn seeds. Atomisation of a liquid stream introduced by a nozzle onto a spraying disc was simulated by generating spheres at a radial position (represented by a ring-shaped virtual geometry) corresponding to the tip of the rotating disc and at a rate conforming to the liquid flow rate. The droplet size and velocity distributions corresponded to the experimental

* Corresponding author: m.e.pasha@leeds.ac.uk

values. In order to represent the coating mechanism in the DEM simulation, a droplet is removed from the simulation after it contacts a corn seed and its mass is attributed to the coating mass of the corn seed. The base rotates to mobilise the seeds, whilst the vertical plates act as baffles, turning the bed over and ensuring adequate mixing of the seeds in order to increase the uniformity of coating.

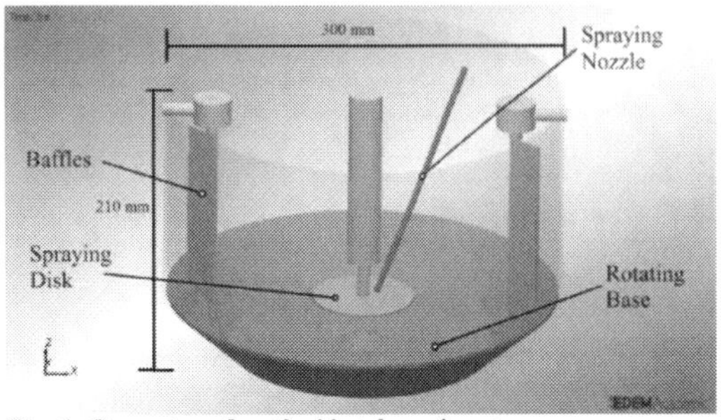

Fig. 1. Geometry of vertical batch seed coater.

2.2 Particle shape

A Phoenix Nanotom© CT scanner was used to obtain the three dimensional (3D) image of a single corn seed in the 7.1 - 8.0 mm sieve-cut, as shown in Figure 2a. The shape of corn seeds is then approximated with five clumped spheres (approximately 0.1 % deviation in volume compared to volume of the actual seed) with different sizes, as shown in Figure 2b. The validity of the approximation for predicting the motion of corn seeds in the coater has been previously demonstrated by Pasha *et al.* [11]. A bed consisting of approximately 4200 seeds (21,000 spheres), corresponding to 1.4 kg of corn seeds, is used.

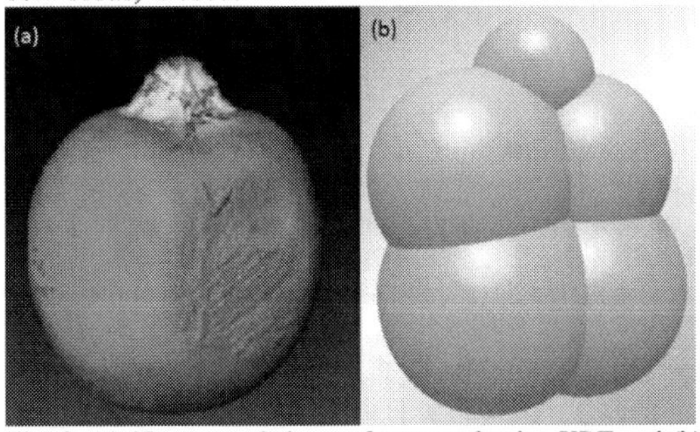

Fig. 2. (a) 3D captured shape of corn seed using XRT and (b) representation of corn seed shape by five overlapped spheres.

2.3 Coating model

The spray droplets are considered as non-interacting spheres until they contact the corn seeds. Video motion analysis is used to capture the motion of particles and determine their size and velocity distributions, which are then incorporated in the DEM simulations. In the simulation stage, a particle property termed 'coating mass' is introduced in addition to the existing properties. The model of Toschkoff *et al.* [10] is employed to model the coating process. According to this model, following a contact between droplets and corn seeds, the droplet mass is added to the coating mass of the corn seed and the droplet is removed from the simulation. The momentum of droplets is considered negligible compared to the corn seeds. As the impact of droplets on the corn seeds would not have a strong influence on the momentum of corn seeds, it is ignored in the force calculations. The Hertz-Mindlin contact model with rolling friction based on viscous dissipation [12] is used to simulate inter-particle interactions.

By taking into account the measured droplet size and velocity distributions and the direction of their movement, a custom code is implemented in EDEM software for generation and movement of droplets. A ring-shaped virtual geometry with 1.5 mm height and outer diameter corresponding to the disc atomiser diameter is considered as the particle factory for generation of the droplets. The droplets are randomly generated inside the particle factory and at the first time step of generation of each droplet; they are given an initial acceleration, which leads to the desired tangential velocity at the end of the subsequent time step. In the simulations, the rate of generation of droplet spheres is chosen such that their volumetric flow rate was equal to that used in the experiments, i.e. 2.67 ml/s. This produces about 36,300 spheres per second. The coating simulations are done for 25 seconds of real time operation and the droplets are generated after one second of process time where the motion of seeds had reached steady state. The particle and simulation properties are summarised in Tables 1 and 2 where particle size follows a normal distribution.

Table 1. Particle and wall properties used in the DEM.

Property	Seeds	Walls
Particle diameter (mm)	7.5 ± 3%	–
Shear modulus (GPa)	0.01	70
Density (kg/m^3)	1163	7800
Poisson's ratio	0.25	0.3

Table 2. Particle interaction properties used in the DEM.

Property	Seed –Seed	Seed – Wall
Sliding friction	0.3	0.3
Rolling friction	0.01	0.01
Coefficient of restitution	0.6	0.69

3 Results and discussions

3.1 Baffle angle and wall clearance

The effect of baffle angle and clearance to the walls on the coating uniformity of corn seeds are investigated by varying from 25° to 55° and 5 mm to 20 mm, respectively. The latter is considered such that it represents a clearance corresponding to approximately 1 to 4 corn seed diameters. The definitions of the angle, α, and clearance, *d*, are shown in Fig. 3. For all the investigated cases, the rotational speed of the coater base, atomiser disc distance to the base and flow rate of coating droplets are kept constant at 300 rpm, 30 mm and 2.67 ml/s, respectively. The coating simulations are carried out for 25 s of real time operation.

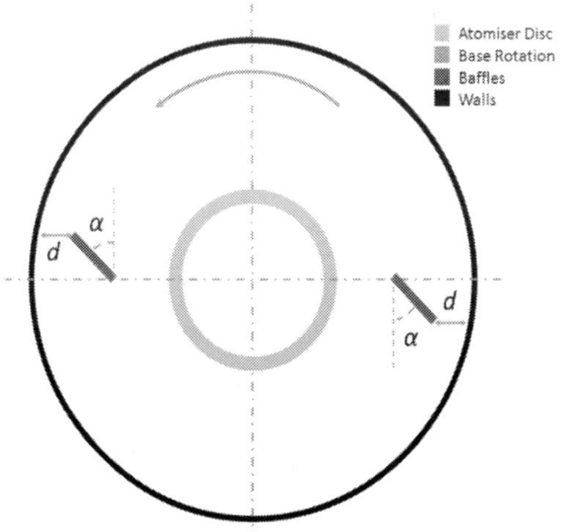

Fig. 3. Schematic representation of baffle angle and clearance gap to the wall.

The effect of baffle clearance to the wall on coating uniformity of corn seeds is shown in Fig. 4. For all investigated cases, the coating mass variability is reduced as the clearance gap is increased to 15 mm, after which it increases again. Therefore, the optimum baffle clearance is around 15 mm for the investigated system here for all the baffle angles. This clearance represents approximately two to three particle diameters. The difference between the two extreme cases of clearance gap (5 and 20 mm) investigated is considered significant, where 5 % improvement in terms of coating uniformity is achieved. For clearance of 10 to 20 mm, the coating uniformity is improved by increasing the baffle angle to 35 – 45°, beyond which the CV is slightly increased. However, in the case of a 5 mm clearance, the optimum baffle angle is found to be 25°.

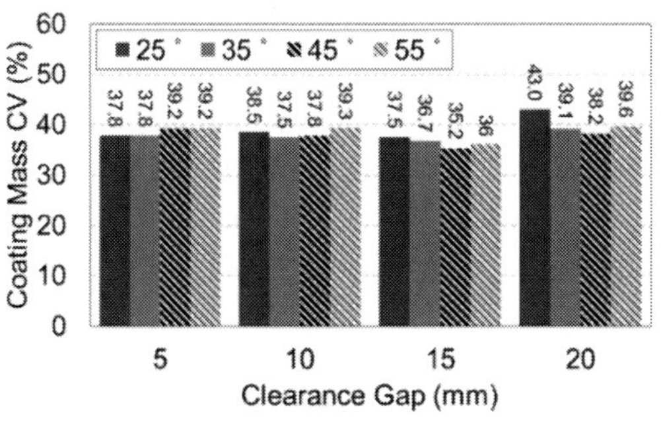

Fig. 4. Effect of baffle angle and clearance gap to the wall on coating uniformity of corn seeds in DEM simulations.

It can be concluded that optimum values of baffle angle and clearance to the wall of the coater are 45° and 15 mm, respectively, where the coating uniformity of corn seeds can be improved by 7 % between the two extremes of coating mass CV. The clearance gap to the wall controls the fraction of the surface that is renewed after each circulation of base, since this parameter controls the thickness of the particle layer being sliced off from the bed surface. Moreover, the baffle angle controls the direction and the position of the particles being sliced from the bed. In combination, these two parameters control the surface renewal rate of the bed, which directly influences the coating uniformity of corn seeds since only the particles at the surface receive coating liquid.

3.2 Atomiser vertical position

Since the coating droplets are generated using a rotating disc atomiser rather than a sprayer, a thin layer of droplets are generated. The position of the atomiser disc relative to the base of the coater affects the fraction of seeds being coated on the surface of bed. A series of DEM simulations were carried out by varying the vertical position of atomiser disc relative to the base of the coater from 30 to 70 mm, using the flat design baffle (30 mm wide) with clearance and angle of 15 mm and 45°, respectively. The particle coating mass CV values using the proposed vertical positions of the atomiser disc are shown in Fig. 5. The coating uniformity of the particles improves as the vertical position is increased to 50 mm from the base; however, beyond this point it starts to deteriorate (CV increases). Comparing the optimised vertical position of the spray disc (50 mm) with the initial position used in the earlier simulations (30 mm), the position of the disc relative to the base of coater has a notable influence on the coating variability and approximately 5 % improvement of coating variability is achieved as shown in Fig. 5. Hence, it can be concluded that this process parameter is one of the key influential parameters in the coating process using this type of coater.

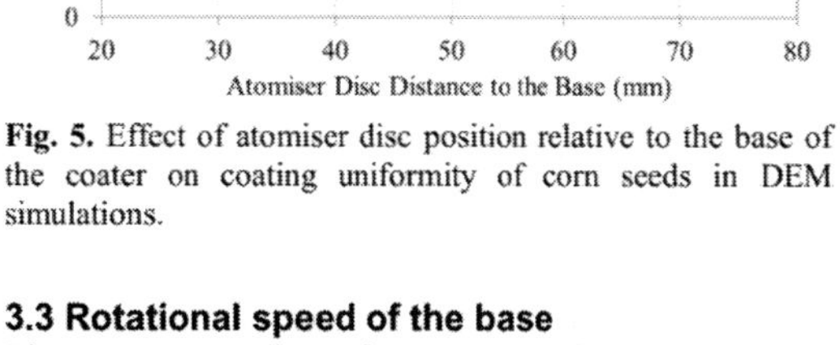

Fig. 5. Effect of atomiser disc position relative to the base of the coater on coating uniformity of corn seeds in DEM simulations.

3.3 Rotational speed of the base

There are a number of parameters that are likely to influence the coating variability of the particles in seed processing such as the dynamics of droplet generation, surface renewal rate, and number of coating events (e.g. number of bed turnovers). A larger number of rotations should lead to a higher probability of new and less coated particles being available for coating at the surface of the bed, hence leading to a higher probability that the particles will be more uniformly coated in the system. It is expected that a minimum number of bed turnovers is required to let all the particles become coated. However, an excessively high rotational speed would lead to higher impact velocities of the seeds to the baffles, which may damage both the particles and their coating.

In this study, DEM was applied to investigate the effects of various process parameters on coating variability of corn seeds. The focus here was on the coating variability of seeds rather than any damage on their coating. Since the potential damage to the coating of the seeds was to be avoided, only three base rotational speeds of 300, 400 and 500 rpm were tried to investigate their effect on the coating variability. It was found that increasing the base rotational speed improved the coating uniformity of the corn seeds, but only very slightly, as approximately 1.4 % improvement was achieved.

4 Conclusions

The effect of various process parameters on coating variability of corn seeds in a batch seed coater was investigated using DEM simulations. In the coating model, the coating liquid droplets were represented by small spheres. Once brought into contact with a corn seed surface, the mass of droplets was stored in the memory as the corn seed coating mass, and it was removed from the simulation. The uniformity of coating was then assessed based on coefficient of variation of the coating mass on each seed. It was shown that using this model the amount of the coating on each particle could be tracked; hence useful information such as distribution of coating mass among the particles and the coefficient of variation of coating mass of the particles could be assessed.

Among the investigated process parameters, it was concluded that the position of the atomiser disc relative to the base of coater and baffle clearance to the walls had a strong influence on the coating uniformity of the particles. In the case of baffle clearance to the wall, a clearance to the wall of 15 mm provided the lowest inter-particle coating variability when compared to small and large clearance gaps. An improvement of approximately 7 % was achieved between the best and worst case combinations of baffle angle and clearance gap investigated in this study. Additionally, it was found that the influence of other process parameters including baffles angle and base rotational speed on coating uniformity was insignificant since they only lead 1 to 2 % change in the coefficient of variation of the coating mass.

Acknowledgements

The financial support of Syngenta through its Technology & Engineering's Strategic & Enabling Technology programme is gratefully acknowledged. The authors would like to thank Dr Neil George, Syngenta Ltd, UK for his support and encouragement, and Phil Taylor of Syngenta Jealott's Hill Formulation Technology Group for providing the seed coater. The authors would also like to thank DEM Solutions for providing support and additional licences for the purpose of this work.

References

1. P.A. Cundall, O. Strack, *Gotechnique* **29** (2) (1979) 47-65.
2. H. Zhu, Z. Zhou, R. Yang, A.B. Yu, *Chem. Eng. Sci.* **63** (23) (2008) 5728-5770.
3. A. Kalbag, C.R. Wassgren, *Chem. Eng. Sci.* **64** (11) (2009) 2705-2717
4. A. Muliadi, P. Sojka, *Atomization and Sprays* **20** (7) (2010) 611-638
5. K. Yamane, T. Sato, T. Tanaka, Y. Tsuji, *Pharmaceutical Research* **12** (9) (1995) 1264-1268.
6. J. Park, C.R. Wassgren, *Part. Sci. & Tech.* **21** (2) (2003) 157-175.
7. D. Suzzi, G. Toschkoff, S. Radl, D. Machold, S.D. Fraser, B.J.Glasser, J.G. Khinast, *Chem. Eng. Sci.* **69** (1) (2012) 107-121.
8. J. Li, C. Wassgren, J.D. Lister, *Chem. Eng. Sci.* **95** (2013) 203-210.
9. S. Just, G. Toschkoff, A. Funke, D. Djuric, G. Scharrer, J. Khinast, K. Knop, P. Kleinebudde, *Int. Jour. of Phar.* **457** (1) (2013) 1-8.
10. G. Toschkoff, S. Just, A. Funke, D. Djuric, K. Knop, P. Kleinbudde, G. Scharrer, J.G. Khinast, *Chem. Eng. Sci.* **101** (2013) 603-614.
11. M. Pasha, C. Hare, M. Ghadiri, A. Gunadi, P.M. Piccione, *Powder Technology* **296** (2016) 29-36.
12. Y. Zhou, B. Wright, R. Yang, B. Xu, A. Yu, *Physica A* **269** (24) (1999) 536-553.

Cohesive strength of iron ore granules

Rafael Jaimes Contreras[1], *Nicolas* Berger[2,3], *Edouard* Izard[4,*], *Jean-François* Douce[4], *Alexey* Koltsov[4], *Jean-Yves* Delenne[5], *Emilien* Azema[3], *Saeid* Nezamabadi[3], *Frédéric* van Loo[1], *Roland* Pellenq[2] and *Farhang* Radjai[2,3]

[1] Centre for research in Metallurgy, 4000 Liège, Belgium
[2] UMI 3466 CNRS-MIT, Massachusetts Institute of Technology, MA 02139, USA
[3] Université de Montpellier, CNRS, LMGC, 34095 Montpellier, France
[4] ArcelorMittal R&D Maizières, Voie Romaine, F-57283, Maizières-Lès-Metz, France
[5] SupAgro, INRA, 34060 Montpellier, France

Abstract. We present an experimental and numerical investigation of the mechanical strength of crude iron ore (Hematite) granules in which capillary bonds between primary particles are the source of internal cohesion. The strength is measured by subjecting the granules to vertical compression between two plates. We show that the behavior of the granules is ductile with a well-defined plastic threshold which increases with the amount of water. It is found that the compressive strength scales with capillary cohesion with a pre-factor that is nearly independent of size polydispersity for the investigated range of parameters but increases with friction coefficient between primary particles. This weak dependence may be attributed to the class of fine particles which, due to their large number, behaves as a cohesive matrix that controls the strength of the granule.

1 Introduction

Since thirty years, iron ores dedicated to sinter plant show a downward trend of their quality (finer particles, broader size distributions, lower grades and larger fluctuations of properties [1]). This trend leads to a significant increase of the input rate of very fine particles in the sinter plants causing productivity (and sometimes sinter quality) drops. A direct and intuitive countermeasure to this evolution is to enhance the input materials in the so-called granulation step, where fine particles are agglomerated into coarse granules allowing for adequate sinter bed permeability on the strand. The granulation process is carried out in continuous drum granulators where water is used as binder [2]. Besides iron ores, this process involves multiple components such as return fines, solid fuel (coke breeze or anthracite), by-products and fluxes (limestone, olivine and dolomite [3]).

The present study aims at developing a better understanding of the physical mechanisms involved in the agglomeration process in view of quantifying and predicting the mechanical strength of the granules. To isolate those mechanisms, we study a laboratory model of granules. These model granules are produced in a laboratory-scale batch drum granulator with iron ore controlled in size and nature. Rather than impact tests ([4, 5]), we perform compression tests [6, 7] in order to measure the strength of the granules as a function of granulation parameters. In parallel, simulations of compression tests are performed by means of the Discrete Element Method [8, 9] (DEM) coupled to a model of capillary bonding. Our goal is to establish a clear relationship between grain-scale cohesion and texture, on one hand, and the granule macroscopic strength, on the other hand.

In the following, we first describe the procedures used to prepare the laboratory model of granules and the compression experimental tests. Then, discrete element simulations using a cohesive force are briefly introduced and simulations of compression tests are presented.

2 Experimental methods and results

The raw material used in the granulation experiments is a commercial iron ore identified as hematite ore from which a sieve size fraction of diameters below 0.25 mm was extracted. The chemical analysis of the raw material is presented in Table 1. The primary particles are mixed to achieve a homogeneous mixture (Figure 1A) that was used throughout this work.

Table 1. Chemical analysis of the raw material

Constituents	%
Fe total	64
CaO	0.06
SiO2	6.46
Loss of ignition	1.03

* Corresponding author: edouard.izard@arcelormittal.com

Let us first present the experimental procedure used to create laboratory model granules. A stainless steel batch granulation drum of 8.5 cm in diameter and 4.25 cm long, without scraper or lifter bar, was used to create granules which were then subjected to compression test. Tap water is injected in the drum and acts as binder to agglomerate the particles. A needle with an internal diameter of 0.4 mm is used to spray the binder with the same flow rate and injection time for all tests. The homogeneous mixture of hematite ore is introduced into the drum, as shown in Figure 1B. The drum starts to rotate up to the constant value of 17 rpm. Then, a pre-defined amount of water is poured by means of the needle. The drum rotates during three minutes while the granules are being formed. Finally, the granules are collected and their size distribution is measured by standard manual sieving.

Compression tests are performed on the granules between two parallel plates as shown in Figure 2. A force gauge allows for the measurement of the compressive force in combination with a motorised motion system which stops or reverses at a set load. The load increments and displacements of the plates are collected thanks to the EmperorTM software. As the load increases, the granule deforms and fails. The axial force increases, peaks and drops at failure to a lower value as shown in Figure 3. The deformation of the granule beyond the peak is plastic with a nearly constant force level F^* (up to fluctuations around the mean). F^* has nearly half the peak value. A sensitivity analysis on the platen speed was conducted, showing that a velocity of *1* mm/min is sufficiently low to reproduce peak force results. The statistical errors are evaluated by repeating the test. Figure 4 shows the range of the values of the peak force for two different compression speeds. Ten granules of identical diameters (with an error of 11.48%), collected from drum granulator experiments were used for each series of tests. We see that the variability is smaller at lower speed. The peak force depends also on the moisture content as observed in Figure 5. The peak force increases with the water content up to 10.1% in volume, then levels off at a value of 5 N for higher water content. At this level, the liquid is in the funicular state [10].

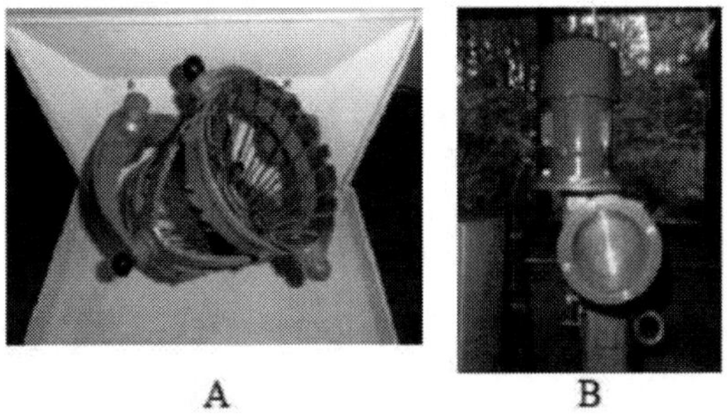

Fig. 1. Experimental set-up for iron ore granulation: picture (A) shows the turbula shaker-mixer used for homogenisation procedure while picture (B) illustrates the drum granulator used to create granules.

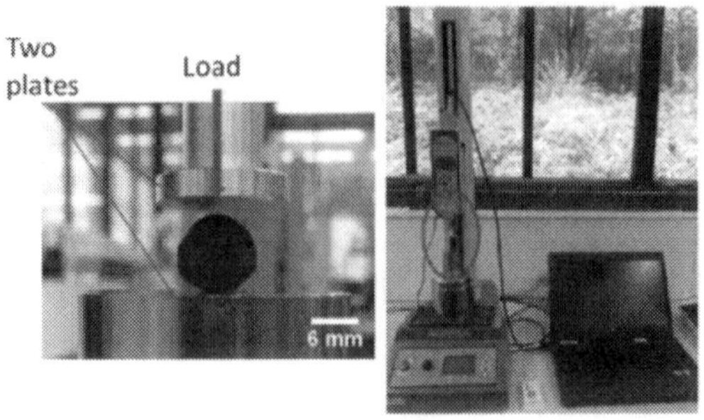

Fig. 2. The setup used for compression test on iron ore granules. The granule is compressed between two platens and the time series of the compressive force with the downward displacement of the top plate is recorded.

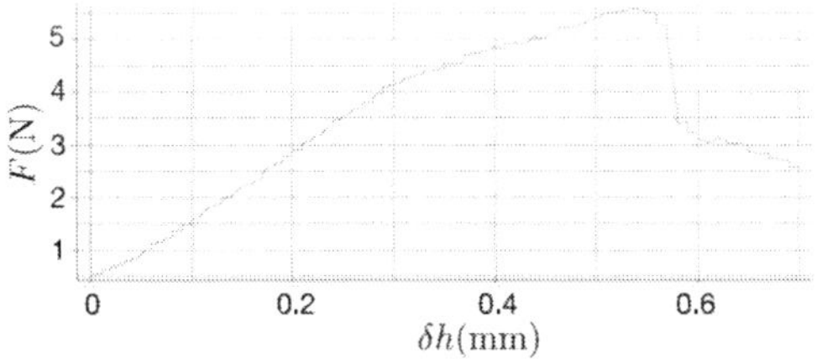

Fig. 3. Typical force-displacement curve during diametrical compression of a granule for water content $w = 11.1\%$.

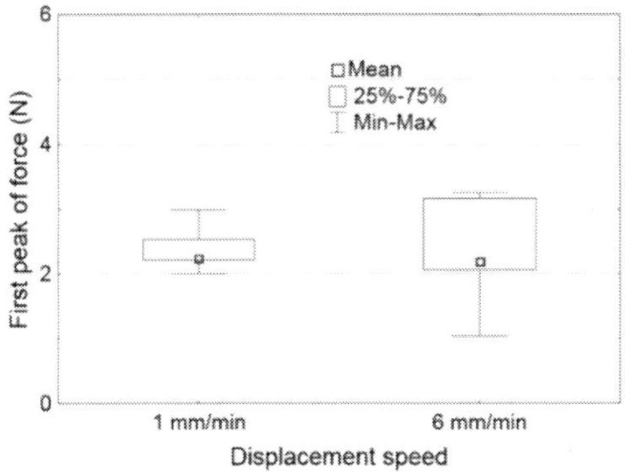

Fig. 4. Impact of load speed on the first peak force measured in the compression experimental test.

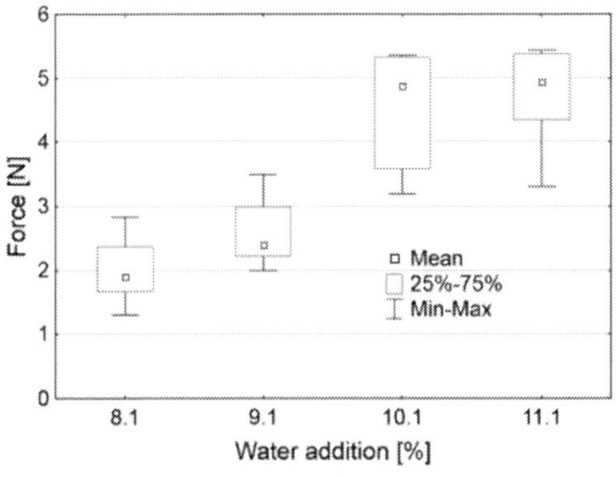

Fig. 5. First peak force as a function of water content in the hematite granule. The force increases with water content. The granule size varies slightly with the amount of water.

3 Discrete Element simulations

The compressive strength of iron ore granules, as measured from experiments, reflects the capillary cohesion of primary particles. For smallest amounts of water used in experiments, it may be assumed that the water is distributed as binary liquid bridges between particles [11, 12]. For the simulations, we also assume that the capillary forces act only at the contact points between particles and the contact angle is zero. The capillary force at the contact point between two particles of diameters d_i and d_j is given by $f_c = \pi\gamma_s\sqrt{d_i d_j}$ where γ_s is the liquid-vapor surface tension. These forces induce a cohesive strength proportional to $2\pi\gamma_s/\langle d\rangle$.

The strength σ_a of the granule can be defined from value of the peak force F divided by the average section πR^2 of the granule, e.g. $\sigma_a = F/(\pi R^2)$.
We expect that σ_a is proportional to the cohesive strength. Hence, we can define the following dimensionless strength:

$$\eta_c = \frac{\sigma_a}{2\pi\gamma_s/\langle d\rangle}$$

This ratio may depend on the size distribution of primary particles, friction coefficient between particles and the connectivity of the contact network. From the experimental value of σ_a in the low water content limit, the average size of the primary particles ($\langle d\rangle \approx 6.8$ μm) and using $F^* \approx 1$N, we get $\eta_c \approx 0.2$ for our experiments.

In order to investigate the effects of various parameters on the value of η_c, we used DEM simulations with varying size polydispersity and different values of the friction coefficient between particles. For the numerical modeling of granules, we used the Contact Dynamics (CD) method [8, 9] with spherical primary particles together with the expression of the capillary force f_c. The size span of the primary particles in experiments being quite broad, we are computationally limited by the number of particles. The number of particles required to account correctly for size polydispersity increases rapidly with the ratio $\lambda = d_{max}/d_{min}$ between the maximum diameter d_{max} and minimum diameter d_{min}. It is necessary to satisfy both the statistical representativeness of all sizes in their number (especially for the largest particles) and in their volume (especially for the classes of finest particles) [13]. For this reason, we consider four size classes with diameters d_i ($i \in [1,2,3,4]$) and 8 values of λ ranging from 1 (monodisperse) to 4.5 with increments of 0.5. For the shape of the size distribution, we used a uniform distribution by volume fractions of particles such that the particle volumes in all classes are equal. This choice was motivated by the fact that a uniform volume distribution of particles leads to a sufficient number of fine particles to allow the effect of size polydispersity to be manifest on the compressive strength of the granule. The granules are composed of nearly 5000 primary particles.

To build the granules, we first create a dense assembly by raining frictionless particles into a box and allowing them to relax to equilibrium. Then, we extract a spherical volume from this assembly. This method allows us to control the characteristics of the granules such as their size, compactness, connectivity and homogeneity. In the absence of friction, the granules are rather dense with the packing fraction increasing from 0.59 to 0.68 as λ varies from 1 to 4.5. As in experiments, the granules are subjected to diametrical compression between two walls. Figure 6 displays a granule at three instants of compression with force chains between primary particles.

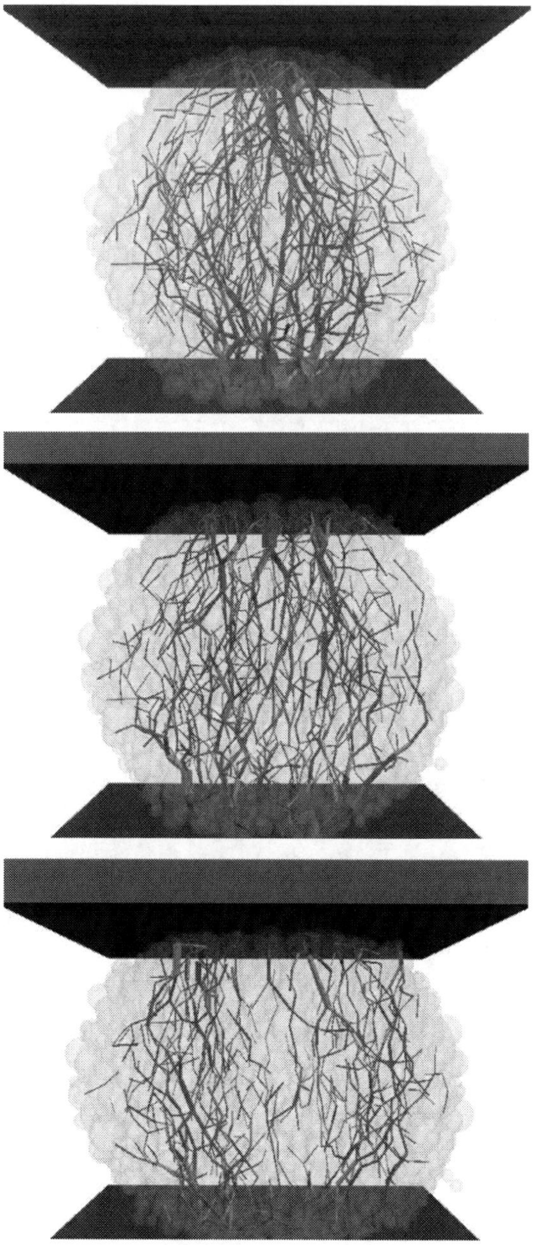

Fig. 6. A granule at three instants of diametral compression. Line thicknesses are proportional to normal forces between primary particles. The compressive forces are in green (black for black and with print) color whereas tensile forces are in orange color (gray for black and with print).

The stress-strain curves for all granules with friction coefficient μ_s=0.7 between primary particles are plotted in Figure 7. As a result of force fluctuations due to finite size effects, the first stress peak cannot be easily identified; thus, we consider the mean value F^* of the force in the plastic regime. Table 2 presents the values of the mean calculated force and dimensionless strength parameter η_c. We see that $\eta_c \approx 0.2$ independently of λ and in excellent agreement with experiments.

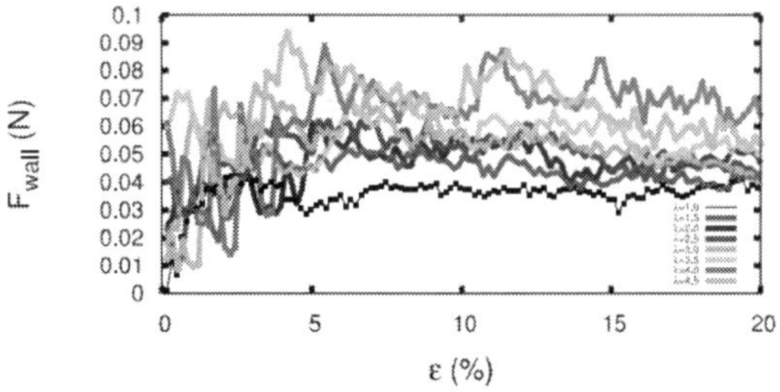

Fig. 7. A Compressive force on the top wall as a function of vertical deformation $\varepsilon = \delta h/2R$ for all value of size ratio λ.

Table 2. The plastic threshold F^*, initial radius R, mean diameter $\langle d \rangle$ of the primary particles and the value of dimensionless strength parameter η_c for different granules with different size ratios λ.

λ	$F^*(N)$	$R\ (mm)$	$\langle d \rangle\ (mm)$	η_c
1	0.035	16.5	2.00	0.20
1.5	0.048	19.8	2.34	0.22
2	0.048	21.9	2.46	0.19
2.5	0.055	22.8	2.49	0.21
3	0.057	23.3	2.48	0.21
3.5	0.056	23.6	2.45	0.19
4	0.065	24.3	2.41	0.21
4.5	0.072	24.5	2.40	0.23

Figure 8: A zoom on the organization of the class of finest particles in a granule with λ=4.

We checked that the same granules with $\mu_s = 0.4$ lead to the same strength. But for $\mu_s = 0.1$, the value of η_c declines by a factor 2. This shows that the friction coefficient can influence the granule strength in the range of 0 to 0.4. Microstructural analysis (not shown in this paper) indicates that, for a uniform distribution of particle volume fractions, the cohesion is mainly controlled by the class of finest primary particles and hence it weakly depends on the size distribution although it crucially depends on the average size. This class behaves as a cohesive matrix with a large interface with all size classes. Figure 8 shows the class of fine particles in the simulations with λ=4, where we see how they form a connected matrix filling the whole volume of the granule. Hence, as far as the class of fine particles is large in number, including larger primary particles in the granule does not affect much the overall cohesion of the granules. This is intuitively the case when nuclei of large size are added to the agglomeration process.

4 Conclusion

In this paper, we presented an experimental and numerical investigation of the strength of agglomerates composed of primary spherical particles as a function of water content (in experiments) and increasing size polydispersity (in simulations). The relevant parameter in the analysis of the strength is the plastic threshold normalized by the cohesive capillary stress induced by liquid bonds. We found nearly the same value both from experiments and DEM simulations using the Contact Dynamics method. The independence of normalized strength with respect to size polydispersity can be understood from the fact that the cohesion of the granule is mainly controlled by the class of fine particles. This class, because of its large number of particles and volume fraction, behave as a homogeneous matrix with a large interface with particles in all size classes.

References

1. R. Remus, M. A. Aguado Monsonet, S. Roudier, & L. Delgado Sancho, JRC reference report. *Best Available Techniques (BAT) Reference Document for Iron and Steel Production* (2013)
2. J. Litster, B. Ennis. *The science and engineering of granulation processes* (Vol. 15). (Springer Science & Business Media, 2013)
3. D. F. Ball, *Agglomeration of Irons Ores*, (New York : American Elsevier Pub. Co., 1973)
4. S. Antonyuk, M. Khanal, J. Tomas, S. Heinrich & L. Mörl, Chemical Engineering and Processing: Process Intensification, **45**(10), 838-856 (2006)
5. S. Antonyuk, J. Tomas, S. Heinrich, L. Mörl, Chem. Eng. Sci., **60**(14), 4031-4044 (2005)
6. S. Antonyuk, S. Heinrich, J. Tomas, N. G. Deen, M. S. van Buijtenen, J. A. M. Kuipers, Granular Matter, **12**(1), 15-47 (2010)
7. A. Russell, P. Müller, J. Tomas, Chem. Eng. Sci., 114, 70-84 (2014).
8. M. Jean, J. J. Moreau, Proceedings of Contact Mechanics International Symposium, pp. 31-48 (1992)
9. F. Radjai, V. Richefeu, Mech. Mat. **41**, 715-728 (2009)
10. J.-Y. Delenne, V. Richefeu, F. Radjai, J. Fluid Mech., **762** R5 (2014)
11. T. C. Halsey, A. J. Levine. Phys. Rev. Lett., 80(14), 3141 (1998)
12. A. Gladkyy, R. Schwarze, Granular Matter, **16**(6), 911-920 (2014)
13. C. Voivret, F. Radjaï, J.-Y. Delenne, M. S. El Youssoufi, Phys. Rev. E **76**, 021301 (2007)

A local view on the role of friction and shape

Matthias Schröter[1,*]

[1] *Institute for Multiscale Simulation, Friedrich-Alexander-University (FAU), 91052 Erlangen, Germany*

Abstract. Leibniz said "Naturam cognosci per analogiam": nature is understood by making analogies. This statement describes a seminal epistemological principle. But one has to be aware of its limitations: quantum mechanics for example at some point had to push Bohr's model of the atom aside to make progress. This article claims that the physics of granular packings has to move beyond the analogy of frictionless spheres, towards local models of contact formation.

On earth solid assemblies of granular particles are by far the most frequent phase of granular matter; we encounter granular packings everywhere from our kitchen cabinet to civil engineering textbooks. In order to make their handling, transport, and storage more efficient, we strive for a theory that predicts their mechanical properties, such as shear and bulk modulus or yield stress, starting from a few state variables only. Efforts to develop such a theory often start by modeling granular packings as an assembly of frictionless spheres. This is a rather unsuitable starting point, for a number of reasons:

1. All granular particles are frictional.
2. Frictional particles have lower isostatic numbers than frictionless particles.
3. Granular physics happens at volume fractions inaccessible to frictionless particles.
4. The volume fraction of soft particles can be changed by compression. The volume fraction of frictional particles is changed by changing their geometry.
5. Friction is one reason for history dependence in granular systems.
6. Real world granular media are rarely spherical. Shape adds complexity, e.g. to history dependence.

These six theses are also the outline for the following sections. They are intended to provoke discussions with a sizeable subgroup of the theoretically or numerically working scientists. Many experimentalists, applied scientist, and engineers might find them, at least in part, well-known. For simplicity, we will discuss in the following only monodisperse spheres; except for section 6.

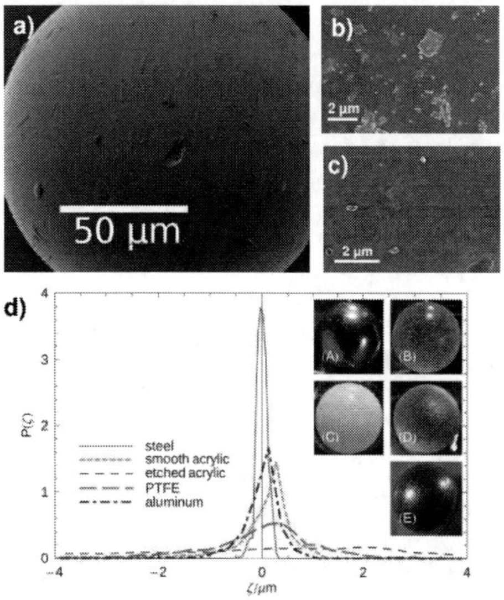

Figure 1. Granular particles are rough particles. **a)** Scanning Electron microscope (SEM) image of a factory-new soda-lime glass bead. Image by courtesy of Karina Sand. **b)** SEM image of the asperities on the surface of a new soda-lime glass bead. **c)** After 30900 flow pulses in a water fluidized bed, abrasion has removed many of the surface asperities, resulting in a measurable difference in packing properties. Images b) and c) are taken from [1]. **d)** Histograms of surface roughness ξ of spherical particles measured with a profilometer. The inset shows images of the corresponding particles: (A) steel, (B) smooth acrylic, (C) PTFE, (D) solvent-etched acrylic, (E) aluminum. Reproduced from [2].

1 All granular particles are frictional.

Contrary to other particulate systems, such as foams or emulsions, the constituents of granular media are solid particles. This implies that their surface is geometrically rough, cf. figure 1. If two particles get into contact, their surface asperities will interlock, allowing for the existence of tangential forces at the contact [3]. In the context of granular packings, friction is sufficiently well described by

* e-mail: matthias.schroeter@fau.de

 DOI: 10.1051/epjconf/201714001008

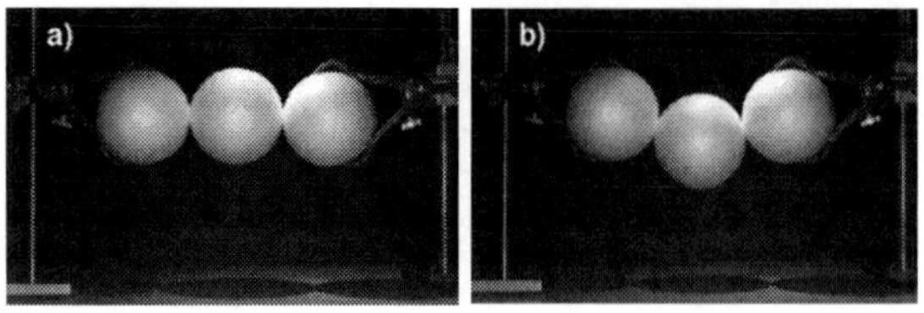

Figure 2. Friction increases the number of mechanically stable configurations. In both images the center Styrofoam sphere is only hold in place by the tangential forces at the contacts. Neither arrangement would be possible with frictionless spheres. In fact, under gravity the only mechanically stable configuration of three frictionless spheres is the perfect vertical alignment.

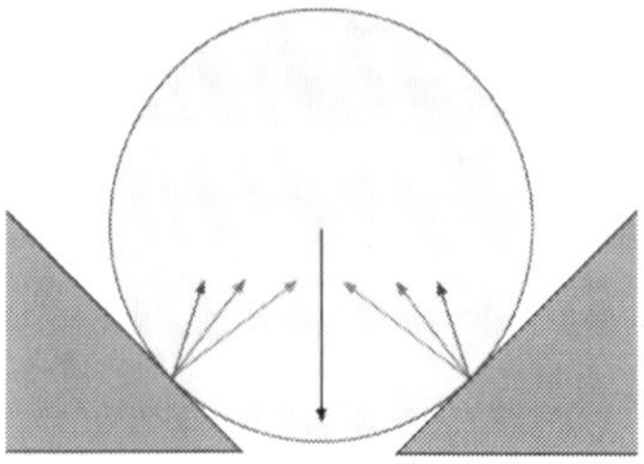

Figure 3. The contact forces of a sphere in a wedge depend on the preparation history. Each of the blue, red, or green pairs of contact forces can balance the weight of the sphere (black arrow). Which one is realized depends how the sphere was placed.

the Amontons-Coulomb law:

$$F_t \leq \mu F_n \tag{1}$$

F_n and F_t are the normal and tangential components of the contact force, and μ is the static coefficient of friction.

All granular media consist of frictional particles. Even hydrogel spheres, which consist of up to 99.5% water, have a friction coefficient of ≈ 0.01 [4, 5]. Moreover, while it is possible to relax all tangential forces in a packing by vibrating it at small amplitudes and high frequencies [6], this will also compactify the packing to values falling into the range of frictionless packings, thereby bypassing the interesting granular physics as discussed in section 3.

In loose granular systems, such as e.g. granular gases, the dynamics is more controlled by collisions than contacts. Friction changes the way particles exchange momentum during collisions, but this seems to be often only a higher order perturbation. Granular packings on the other hand consist of enduring contacts, here the existence of friction changes the physical picture completely:

First, the presence of tangential forces provides additional ways to satisfy the force and torque balance, which will be discussed more quantitatively in section 2. This leads to a massive increase of the number of mechanically stable states; figure 2 gives a simple example. Most granular packings studied in nature or experiment are looser than the loosest packing that can be created without friction. The consequences of this will be discussed in sections 3 and 4.

Secondly, equation 1 is an inequality. For a given normal force at a contact it allows for a whole range in tangential force, as shown in figure 3. The actual tangential force will be a consequence of how the contact was formed. This property is one of the reasons for the so called history-dependence of granular matter, which will be discussed further in section 5.

2 Frictional particles have lower isostatic numbers than frictionless particles.

For any granular packing to be solid, the average number of contacts a particle forms with its neighbors Z has to be at least so large that all degrees of freedom (DOF) of the particle can be fixed. This minimal value, the so called isostatic contact number Z_{iso}, does depend on the dimension, shape, and most importantly friction of the particles considered.

In the absence of friction, the rotational DOF of a perfect sphere are not relevant and only the three translational DOF have to be blocked by the contacts. At each contact there exists one normal force, which is however "shared" between the two particles which means that each contact blocks on average only 0.5 DOF per particle. A packing of frictionless spheres needs therefore to have at least $Z^0_{iso} = 6$ contacts to be mechanically stable.

If we assume $\mu = \infty$, then each contact has 3 independent force components (one normal and two tangential), which fix 1.5 constraints per particle. On the other hand, we now also have to consider the rotational DOF which results in 6 DOF per particle. The isostatic number in the presence of infinite friction Z^{μ}_{iso} is therefore 4.

The inequality $Z^0_{iso} > Z^{\mu}_{iso}$ holds for all particle shapes and in 2 and 3 dimensions [7]. One consequence is the massive increase in the number of mechanically stable packings of frictional particles, as discussed in the next section. Another consequence is that granular packings are typically *hyperstatic* i.e. their actual contact number is larger than the minimum number needed for stability: $Z > Z^{\mu}_{iso}$. From this follows that for a given spatial configuration of particles, there exists a multitude of possible force networks that will all satisfy the boundary conditions of the system [8]. This property is intimately connected to the history dependent behavior discussed in section 5.

There is a possible caveat regarding hyperstaticity. In real granular media μ is finite and the contacts might have tangential forces which are exactly at the Coulomb threshold, so called fully mobilized contacts. As this type of contact will block only 1 DOF, the constraint counting argument has to be modified and the effective Z^{μ}_{iso} becomes larger. However, a number of numerical studies [9–12] have shown that the number of fully mobilized contacts is not sufficient to regain isostaticity in any other situation than when preparing a pressure free packing very slowly, cf figure 4. Which is also the recipe to approach the limit of Random Loose Packing, the loosest packing possible (discussed in section 3.2).

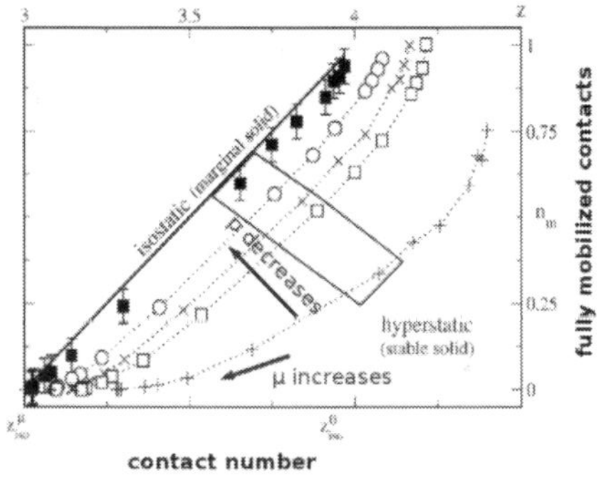

Figure 4. Most granular packing are also hyperstatic when fully mobilized contacts are taken into account. In disc packings both Z (x-axis) and the number of fully mobilized contacts (y-axis) changes with pressure and μ. However, only in the limit of vanishing pressure the system will approach isostaticity (solid line). From [9].

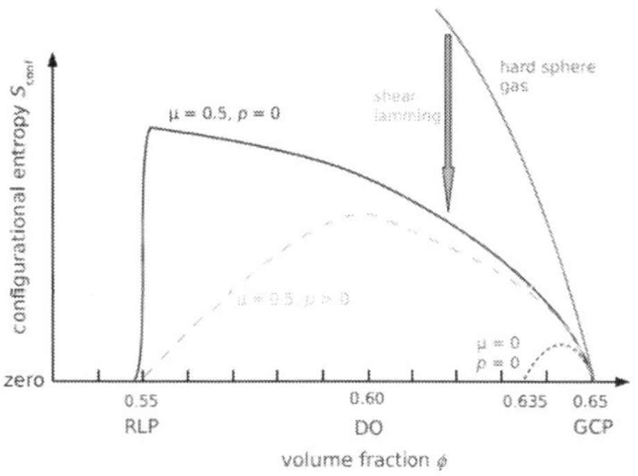

Figure 5. A schematic how the configurational entropy S_{conf} of a sphere packing depends on the friction coefficient μ and the confining pressure p. The solid magenta line indicates the configurational entropy of an amorphous hard sphere gas i.e. a "packing" with non overlapping particles but no requirements on the mechanical stability or number of contacts. The solid blue and dashed cyan line represent packings with an approximate real world value of μ and the dotted red line corresponds to frictionless particles. While the points with $S_{conf} = 0$ are well supported by experiments and simulations, the actual shape of the different curves is speculative.

3 Granular physics happens at volume fractions inaccessible to frictionless particles.

The consequences of the lower isostatic contact number of frictional particles are best discussed using the the concept of the configurational entropy S_{conf} of the packings. S_{conf} was first introduced by Sam Edwards [13, 14], it is proportional to the logarithm of the the number of mechanically stable packing configurations that fit in a given volume and support given boundary conditions. As we are interested in the thermodynamic limit, we will discuss here S_{conf} as a function of the global volume fraction ϕ_g. More specifically, we are interested in a comparison of the upper and lower bounds of ϕ_g between which S_{conf} becomes non-zero for both frictional and frictionless systems.

The main results of this discussion are summarized in figure 5. But a word of caution is necessary: The ϕ_g values of the upper and lower bounds are well supported by the numerical and experimental work discussed below. But the functional form of S_{conf} connecting these boundaries is speculative and only supported by the heuristic arguments given below.

Moreover, we are only interested in packings that exist in a thermodynamic sense. This excludes both packings crystallizing in FCC and HCP configurations at volume fractions above $\phi_g \approx 0.65$ [15–21] [1] and the "tunneled crystal" packings at $\phi_g = 0.49$ [22]. Neither of these two configurations are extensive, i.e. their number does not grow exponentially with the number of particles in the system. Which means that in the thermodynamic limit their entropy is zero.

[1] Readers familiar with equilibrated hard sphere systems might expect crystallization to occur in the range $\phi_g \approx 0.494 - 0.61$. However, these systems are driven by the entropy increase due to newly gained vibrational DOF. These DOF do not exist in athermal granular packings where all particles are permanently in contact.

An important upper bound on S_{conf} is the configurational entropy S_{HS} of an amorphous *hard sphere gas* where the only remaining condition for a valid configuration is that particles do not overlap. Mechanical stability and consequentially Z do not matter. In figure 5 this boundary is indicated by a solid magenta line. The pressure of a hard sphere gas diverges at the so called Glass Close Packing (GCP) point with $\phi_{GCP} \approx 0.65$ [20, 21, 23]. From which follows that the system runs out of non-overlapping configurations and S_{HS} goes to zero.

3.1 Frictionless sphere packings: $0.635 < \phi_g < 0.65$

As S_{HS} is an upper limit for any sphere packing, S_{conf} of a frictionless packing also needs to go to zero at $\phi_{GCP} \approx 0.65$. For any smaller value of ϕ_g, S_{conf} has to be smaller than S_{HS} because we now additionally require an isostatic number of contacts. In fact, the set of mechanically stable configurations should be of measure zero compared to hard sphere gas: there are infinite more possibilities of two spheres to be not in contact compared to the one configuration where they are. Luckily, we know that S_{conf} is still extensive. This was shown for soft frictionless disk and sphere systems of different sizes by dividing the total accessible phase space volume by that of an average basin of attraction [24, 25].

The total range of $S_{conf} \neq 0$ is shown as a red dotted line in figure 5. There is still some debate [7, 21, 26, 27] if the onset of mechanical stability happens for frictionless spheres at the so called Jamming point of $\phi_J \approx 0.64$ or slightly below at $\phi_g \approx 0.635$. However, it is known that the actual volume fraction of a packing of uncompressed frictionless spheres will depend on the preparation history. For an extended discussion and further references see [28].

3.2 Frictional sphere packings: $0.55 < \phi_g < 0.65$

All mechanically stable configuration of frictionless particles will stay valid if we allow for additional tangential forces. Therefore S_{conf} of frictional packings will always be larger than its frictionless counterpart. Because S_{HS} is also an upper bound to frictional systems, GCP will still be the upper limit for uncompressed packings.

The lower boundary, commonly referred to as Random Loose Packing (RLP), is however considerably lower than in frictionless systems: As mentioned in the last section, the inequality $Z^0_{iso} > Z^\mu_{iso}$ holds for all particle shapes and in 2 and 3 dimensions. Moreover, as Z can be generically expected to decrease monotonically with decreasing ϕ_g (i.e. larger average separation between particles), the onset of mechanical stability will happen at a lower volume fraction for frictional particles. The actual value of ϕ_{RLP} does depend on pressure [29, 30] and the friction coefficient[2, 30]: the higher μ the smaller is ϕ_{RLP}. For the experimentally common values of $\mu \approx 0.5$ and vanishing pressure, ϕ_{RLP} approaches 0.55 [2, 29–33].

In figure 5 the solid blue and the dashed cyan line represent S_{conf} of sphere packing with a finite value of μ and either zero or finite confining pressure p. Besides the points with $S_{conf} = 0$, the shape of the curves is speculative as there are few analytical or experimental results on how S_{conf} depends on ϕ_g. What we do understand is that for any given value of ϕ_g, S_{conf} will grow monotonically with μ because allowing larger tangential forces will never destabilize any existing packing, but allow for new, additional configurations [34]. Moreover, we can use the Widom insertion method in combination with experimental or numerical packings to obtain an upper bound on S_{conf} [35]. However, a lower bound would be more helpful. Finally, under certain additional assumptions, S_{conf} can be computed from the volume fluctuations of a repeatedly driven granular packing [36–38]. However, the results obtained this way do not agree with each other.

Another way of assessing the shape of $S_{conf}(\phi_g)$ is to consider experimental preparation protocols and to use the additional assumption that the state the system will end up in is the most likely one: the one with the highest value of S_{conf} under the given circumstances. For example, packing prepared by slow sedimentation in an almost density matched fluid (i.e. in the limit $p \to 0$) will always end up at ϕ_{RLP} [2, 29, 30]. Which indicates that S_{conf} has a maximum at RLP for $p = 0$, cf. the blue line in figure 5.

Without density matching (i.e. at a finite static pressure) and with an increased sedimentation speed (meaning that the settling particles will transfer more momentum on the already existing packing), the most likely packing fraction moves up to $\phi \approx 0.6$ [1]. The cyan dashed line in figure 5 represents the idea that this becomes the new maximum in S_{conf}. Which can be rationalized by assuming that configurations with less excess contacts compared to an isostatic packing are more likely to not posses a force network capable of supporting the increased stress at the boundaries. $\phi \approx 0.6$ is incidentally also the value for the onset of dilatancy, which is discussed in the next subsection.

Finally, getting the system to compactify to values of ϕ_g above 0.6 requires repeated driving under confining gravitational pressure, either by flow pulses [1] or mechanical taps [39, 40]. This indicates that these states become more and more unlikely which agrees with the idea that S_{conf} goes to zero for ϕ_g approaching GCP.

But the main point of this section is untouched by this discussion of the shape of S_{conf}: The range of volume fractions of mechanically stable packings is *seven times larger* for frictional particles than for frictionless particles.

3.3 Dilatancy in frictional packings

Most of the interesting physics of granular packings happens at intermediate values of ϕ_g (i.e. between RLP and GCP); these volume fractions are inaccessible to frictionless packings. A good example is dilatancy: If a dense granular packing is sheared (at a finite hydrostatic pressure), it will expand [41]. However, dilatancy does not occur in sufficiently loose samples; those will instead compactify.

Now if dense packings expand during shear and loose packings collapse, there will be an intermediate density, usually termed Dilatancy Onset (DO) or critical state, where the volume fraction ϕ_{DO} stays constant during shear. Due to effects like shear banding the exact determination of ϕ_{DO} is not straightforward, but most experiments point to $\phi_{DO} \approx 0.6$ for frictional spheres at comparatively small confining pressures [42–47]

Dilatancy is closely related to a phenomenon labeled shear-jamming [48–50]. For $\phi_g > \phi_{DO}$ one can start from a hard sphere gas configuration (experiments are done in horizontal two-dimensional system and therefore effectively in the absence of gravity), shear it at a constant volume and arrive at a mechanically stable configuration. This protocol is indicated as a blue arrow in figure 5. Below ϕ_{DO} this is not possible.

One way of interpreting ϕ_{DO} is to see it as a "natural attractor" for all sheared systems starting at other volume fractions. This interpretation agrees well with the idea of S_{conf} being maximal at $\phi_{DO} \approx 0.6$.

4 The volume fraction of frictional particles is controlled by their geometry.

The exact value of ϕ_g of a frictionless packing at zero pressure does depend on the preparation history [20, 21, 28, 51]. However, the scaling laws of these packings are normally studied by preparing pressure free packings with a given protocol and then increasing ϕ_g by compressing this packing [7, 26].

With respect to the contact number Z this amounts to a study of the pair correlation function, or more precisely the slope of the right shoulder of the first peak which describes the close-by particles which will form contacts when compressed. This slope leads to an equation for Z:

$$Z(\phi_g) = Z^0_{iso} + c(\phi_g - \phi_{iso})^{0.5} \quad (2)$$

Here ϕ_{iso} is the volume fraction of the uncompressed, isostatic packing and the constant c depends on the dimension and polydispersity of the system. For compressed frictionless packings such as emulsions and foams [52], equation 2 is indeed a good description.

However, real world granular particles are normally not very squishy; they change their volume fraction by isobarically changing their packing geometry not by compression. We all intuitively know this from kitchen physics: if we want to fill more grains into a storage container we do not compress them with a piston, but we tap the container a couple of times on the counter top to change its packing structure.

For a more quantitative example lets compare two glass spheres (Young's modulus = 70 GPa, diameter = 250 μm) which are either uncompressed at the upper surface or compressed below a 1 m high column of other glass spheres. Using Hertz law we can derive that this increase in pressure will deform the sphere by approximately 10 nm at each contact. This deformation is an order of magnitude smaller than the vertical surface roughness of typical glass spheres[53]. Assuming that the sphere is compressed symmetrically, this corresponds to a change in volume fraction of 7×10^{-5} compared to the uncompressed sphere. This illustrates that the large range of $0.55 < \phi_g < 0.65$ available to frictional sphere packings can not be explored by compression.

Please note that granular experiments can be performed in a way to test frictionless models. E.g. the compression of frictional but sufficiently soft photoelastic discs (with a Young's modulus of 4MPa [54]) can be used to verify equation 2 [54] or study glassy behavior [55]; provided that the tangential forces are relaxed by additional tapping or vibration. These experiments do however not prove that frictionless models describe generic frictional particles.

4.1 Friction with your neighbors? Think locally!

Because Z and ϕ_g are in frictional packings not simultaneously controlled by the globally defined parameter pressure, the idea expressed in equation 2 of a function $Z(\phi_g)$ runs into an epistemological problem. Contacts are formed at the scale of individual particles and their neighbors. At this scale the global ϕ_g is not only undefined; due to local volume correlations [56, 57] it would even be impossible for a particle scale demon to compute ϕ_g by averaging over the volume of the neighboring particles.

What is needed for the theoretical description of frictional particles is an ansatz which explains Z using only locally defined (i.e. on a particle level) parameters [58–61]. The most important [61] of these local parameters is the local volume fraction ϕ_l which describes by how much free volume an individual particle is surrounded. ϕ_l is computed by dividing the volume of the particle by the volume of its Voronoi cell (a tessellation method assigning all points in space to the closest particle). However, for a complete local description more parameters such as the shape of the Voronoi cells [62] or the fabric anisotropy [63] are needed.

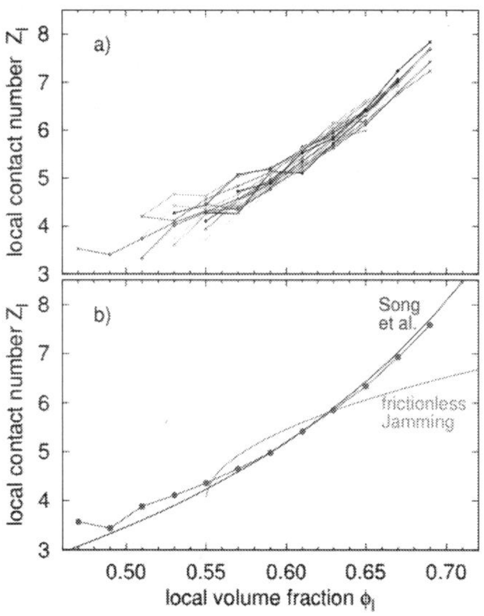

Figure 6. Understanding contact numbers in sphere packings requires a local approach. **a)** The average local contact number Z_l of individual spheres, measured by X-ray tomography and averaged in local volume fraction ϕ_l bins of size 0.02. Data corresponds to 15 different sphere packings with global volume fractions ϕ_g in the range from 0.56 to 0.625. Within experimental scatter, Z_l depends only on ϕ_l, not on ϕ_g. **b)** The red dots correspond to a bin-wise average of all data shown in panel a. The local mean field theory by Song *et al.* (eq. 3, no fit parameter) provides a fair description of the data. This can not be said about the local interpretation of the scaling law for frictionless, compressed spheres (eq. 2, one fit parameter). From [61].

Figure 6 substantiates this claim for the necessity of local theories. Panel a shows that the number of contacts Z_l an individual particle will form does only depend on its own ϕ_l, not the ϕ_g value of the packing it resides in. Figure 6 b) demonstrates that Z_l can be well explained by the local theory presented in [31], which predicts:

$$Z = \frac{2\sqrt{3}\,\phi_l}{1 - \phi_l} \quad (3)$$

A local reinterpretation of equation 2 for frictional systems (ϕ_g becomes ϕ_l, Z^0_{iso} becomes Z^{μ}_{iso}, ϕ_{iso} becomes ϕ_{RLP}) fails to describe the experimental data.

5 Friction is one reason for history dependence in granular systems.

A number of experiments demonstrate history dependent behavior in granular materials: Two seemingly identical packings, which only differ in their preparation histories,

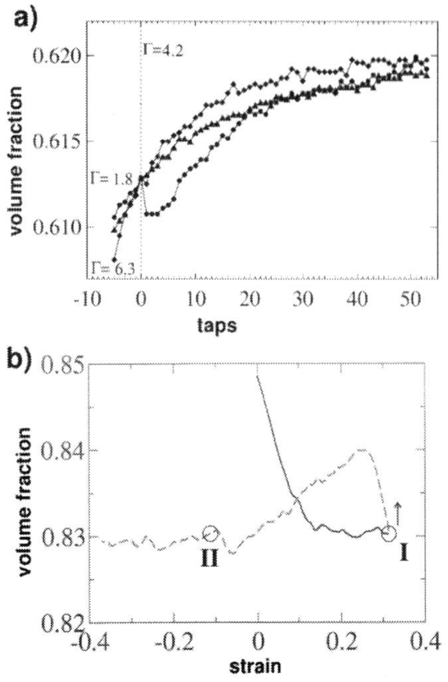

Figure 7. History dependence in granular systems. **a)** Three samples of glass beads are compactified to the same value of ϕ_g =0.613 using three different initial tapping strengths. At this point (vertical dotted line) the tapping strength is set to the same value of 4.2 g in all three experiments. The system does however respond differently depending on its preparation history. From [64]. **b)** Volume fraction alone is not sufficient to characterize dilatancy onset. In this contact dynamics simulations an initially dense system of discs has been sheared long enough to dilate to its critical state (solid line). When then the shear direction is reversed (dashed line), the system responds first by compaction before it dilates again. This implies that the packings at points I and II have identical volume fractions but respond differently to shear in the same direction. From [65].

respond either differently to an external excitation, or they differ in some of their not immediately obvious mechanical properties.

An example of the first type is shown in figure 7a: three samples are compactified to the same volume fraction ϕ_0 but using three different driving strengths Γ_i. When these samples at ϕ_0 are then driven with the same strength Γ_0, their response depends on the past Γ_i, not Γ_0 [64]. Similar results can be obtained for periodically shearing glass spheres in a parallelepiped shear cell [66] or going to large strains in a simple shear cell (fig. 7b) [65].

Examples of the second type of history dependence include how the pressure distribution below a sandpile depends on its preparation history: If the sand rained down from a large sieve, the maximum pressure at the bottom plate will be below the tip; which is the point with the largest column of sand on top. However, if the sand flowed out of a small funnel opening, which means that the pile grew from many downhill avalanches, the maximum pressure at the bottom plate will be at a ring with a diameter of roughly one third of the total pile diameter [67]. An similar example is the history dependence that exists in the so called Janssen effect: The pressure at the bottom of a cylindrical column filled with grains will be lower than the total weight of the grains because tangential forces at the sidewalls carry a part of the load. The amount of this reduction will again depend on the preparation history [68]. Finally it has also been shown numerically, that the number of contacts formed in a packing depends on the preparation history [69].

History dependence does also exist in frictionless packings (see [28, 51] for a novel approach how the jamming volume fraction can be used as a state variable to characterize the history). However, most of the examples listed above seem to require friction. Either because the extra degrees of freedom allow variability in the contact number or the geometric fabric formed by the contacts (pressure distribution at the sand pile bottom, shear response at critical state). Or because the memory of a previous state can be encoded as a particular configuration in the force phase space spanned by hyperstaticity (Janssen).

In all examples discussed here the apparent identity of the initial states has only been established in terms of global variables such as shape of the sample and ϕ_g. In fact, in all these cases history dependence can also be viewed as another name for: "we do not know all relevant parameters which characterize the system".

6 Granular matter is rarely spherical.

Most of the readers will have heard some variant of the "spherical cow in vacuum" joke [72]. But in fact there are not only good reasons for theorists to use spheres as a first approximation, also numerical scientists appreciate the easy collision detection algorithm coming with spheres. And experimentalist like spheres because they are the only monoschematic particles (all particles have the same shape) which are easily available in large quantities. Being monoschematic is a big advantage during image processing where the a priori knowledge of their shape helps to identify the individual particles [73].

Still, real world granular materials are basically always non-spherical in shape. This adds additional complexity which every theory suitable for practical purposes will need to take into account. Figure 8 discusses some of this complexity using packings of tetrahedra as an example. Contrary to spheres, tetrahedra can form four different types of contacts, cf. figure 8b. This has important consequences for the pair correlation function shown in figure 8a. Because perfectly aligned face-to-face contacts are less frequent than slightly shifted face-to-face or low angle face-to-edge contacts, the closest possible distance between two particles is *not* the most likely contact configuration [70]. The different shape of the first peak of $g(r)$ brings as a consequence that scaling laws developed for compressible sphere packings [7, 74], such as eq. 2, will not work for compressible tetrahedra. Moreover, figure 8 c) shows, that tetrahedra packings are even stronger hyperstatic and history-dependent than sphere packings.

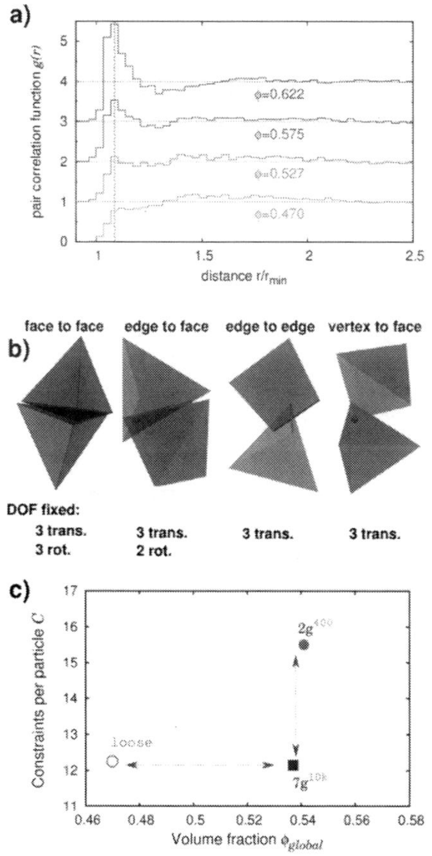

Figure 8. Packings of tetrahedra differ from sphere packings. **a)** The pair correlation functions of experimental tetrahedra packings shows that contrary to spheres the shortest possible distance $r_{\min}$ (approximately 0.408 times the side-length) is not the most likely distance between individual particles (indicated by a vertical dashed line) [70]. Offsets have been added for improved readability. **b)** Tetrahedra form four different types of contacts, which block different numbers of translational and rotational degrees of freedom. For the purpose of defining a distance to isostaticity we need to determine the total number of constraints per particle C blocked by the combination of all four contact types. **c)** Isostaticity corresponds to $C = 6$ (each tetrahedra has three rotational and three translational DOF). Even the loosest tetrahedra packings have twice the constraint number needed for isostaticity, which means that tetrahedra packings are much more hyperstatic than sphere packings. Moreover do tetrahedra packings exhibit a strong history dependence. Using different tapping protocols it is possible to create pairs of packing which differ strongly in only one of the two variables C and ϕ_g [71].

Experimentally, one way of moving away from the spherical cow paradigm is to take advantage of the quickly improving 3D printing technology to create large samples of monoschematic but non-spherical particles [75, 76]. Alternatively, we can directly use natural materials such as sand and improve our 3D image processing to obtain trustworthy segmentation results [77]. Additionally, there is also progress towards theories for non-spherical particles [59, 60].

Conclusion

Frictionless spheres are a great model for emulsions, foams, glasses, and colloids. They give reasonable results for granular gases and describe even glassy behavior in driven granular systems. But a *frictionless granular packing* is a a self-contradicting statement, describing a theoretical model that for the most part has outlived its usefulness.

Acknowledgments

The ideas in this article have taken shape in many invaluable discussions with colleagues: Vasili Baranau, Karen Daniels, Max Neudecker, Charles Radin, Fabian Schaller, Gerd Schröder-Turk, Harry Swinney, Brian Tighe, Song-Chuan Zhao.

References

[1] M. Schröter, D.I. Goldman, H.L. Swinney, Phys. Rev. E **71**, 030301 (2005)
[2] G.R. Farrell, K.M. Martini, N. Menon, Soft Matter **6**, 2925 (2010)
[3] F.P. Bowden, D. Tabor, *Friction* (Anchor Books, Garden City, 1973)
[4] J.A. Dijksman, H. Zheng, R.P. Behringer, in *Powders & Grains* (2013), Vol. 1542 of *AIP Conf. Proc.*, pp. 457–460
[5] J. Dijksman, *private communication* (2017)
[6] G.J. Gao, J. Blawzdziewicz, C.S. O'Hern, M. Shattuck, Phys. Rev. E **80**, 061304 (2009)
[7] M. van Hecke, J. Phys.: Condens. Matter **22**, 033101 (2010)
[8] B.P. Tighe, J.H. Snoeijer, T.J.H. Vlugt, M.v. Hecke, Soft Matter **6**, 2908 (2010)
[9] K. Shundyak, M. van Hecke, W. van Saarloos, Phys. Rev. E **75**, 010301 (2007)
[10] L.E. Silbert, D. Ertaş, G.S. Grest, T.C. Halsey, D. Levine, Phys. Rev. E **65**, 031304 (2002)
[11] H.P. Zhang, H.A. Makse, Phys. Rev. E **72**, 011301 (2005)
[12] S. Henkes, M. van Hecke, W. van Saarloos, Europhys. Lett. **90**, 14003 (2010)
[13] S.F. Edwards, R.B.S. Oakeshott, Physica A **157**, 1080 (1989)
[14] A. Mehta, S.F. Edwards, Physica A **157**, 1091 (1989)
[15] A.V. Anikeenko, N.N. Medvedev, Phys. Rev. Lett. **98**, 235504 (2007)
[16] C. Radin, J. Stat. Phys. **131**, 567 (2008)
[17] Y. Jin, H.A. Makse, Physica A **389**, 5362 (2010)
[18] S.C. Kapfer, W. Mickel, K. Mecke, G.E. Schröder-Turk, Phys. Rev. E **85**, 030301 (2012)

[19] N. Francois, M. Saadatfar, R. Cruikshank, A. Sheppard, Phys. Rev. Lett. **111**, 148001 (2013)
[20] V. Baranau, U. Tallarek, Soft Matter **10**, 3826 (2014)
[21] V. Baranau, U. Tallarek, Soft Matter **10**, 7838 (2014)
[22] S. Torquato, F.H. Stillinger, J. App. Phys. **102**, 093511 (2007)
[23] G. Parisi, F. Zamponi, Rev. Mod. Phys. **82**, 789 (2010)
[24] D. Asenjo, F. Paillusson, D. Frenkel, Phys. Rev. Lett. **112**, 098002 (2014)
[25] S. Martiniani, K.J. Schrenk, J.D. Stevenson, D.J. Wales, D. Frenkel, Phys. Rev. E **93**, 012906 (2016)
[26] C.S. O'Hern, L.E. Silbert, A.J. Liu, S.R. Nagel, Phys. Rev. E **68**, 011306 (2003)
[27] M. Pica Ciamarra, M. Nicodemi, A. Coniglio, Soft Matter **6**, 2871 (2010)
[28] N. Kumar, S. Luding, Granul. Matter **18**, 58 (2016)
[29] G.Y. Onoda, E.G. Liniger, Phys. Rev. Lett. **64**, 2727 (1990)
[30] M. Jerkins, M. Schröter, H.L. Swinney, T.J. Senden, M. Saadatfar, T. Aste, Phys. Rev. Lett. **101**, 018301 (2008)
[31] C. Song, P. Wang, H.A. Makse, Nature **453**, 629 (2008)
[32] L.E. Silbert, Soft Matter **6**, 2918 (2010)
[33] G.W. Delaney, J.E. Hilton, P.W. Cleary, Phys. Rev. E **83**, 051305 (2011)
[34] Y. Srebro, D. Levine, Phys. Rev. E **68**, 061301 (2003)
[35] V. Baranau, S.C. Zhao, M. Scheel, U. Tallarek, M. Schröter, Soft Matter **12**, 3991 (2016)
[36] C. Briscoe, C. Song, P. Wang, H.A. Makse, Phys. Rev. Lett. **101**, 188001 (2008)
[37] S. McNamara, P. Richard, S.K. de Richter, G. Le Caer, R. Delannay, Phys. Rev. E **80**, 031301 (2009)
[38] S.C. Zhao, M. Schröter, Soft Matter **10**, 4208 (2014)
[39] E.R. Nowak, J.B. Knight, E. Ben-Naim, H.M. Jaeger, S.R. Nagel, Phys. Rev. E **57**, 1971 (1998)
[40] P. Ribière, P. Richard, P. Philippe, D. Bideau, R. Delannay, Eur. Phys. J. E **22**, 249 (2007)
[41] B. Andreotti, Y. Forterre, O. Pouliquen, *Granular Media* (Cambridge University Press, 2013)
[42] M. Schröter, S. Nägle, C. Radin, H.L. Swinney, Europhys. Lett. **78**, 44004 (2007)
[43] A.J. Kabla, T.J. Senden, Phys. Rev. Lett. **102**, 228301 (2009)
[44] N. Gravish, P.B. Umbanhowar, D.I. Goldman, Phys. Rev. Lett. **105**, 128301 (2010)
[45] P. Umbanhowar, D.I. Goldman, Phys. Rev. E **82**, 010301 (2010)
[46] J. Métayer, D.J. Suntrup III, C. Radin, H.L. Swinney, M. Schröter, Europhys. Lett. **93**, 64003 (2011)
[47] N. Mueggenburg, Phys. Rev. E **85**, 041305 (2012)
[48] D. Bi, J. Zhang, B. Chakraborty, R.P. Behringer, Nature **480**, 355 (2011)
[49] J. Ren, J.A. Dijksman, R.P. Behringer, Phys. Rev. Lett. **110**, 018302 (2013)
[50] H.A. Vinutha, S. Sastry, Nature Phys. **12**, 578 (2016)
[51] S. Luding, Nature Phys. **12**, 531 (2016)
[52] G. Katgert, M. van Hecke, Europhys. Lett. **92**, 34002 (2010)
[53] S. Utermann, P. Aurin, M. Benderoth, C. Fischer, M. Schröter, Phys. Rev. E **84**, 031306 (2011)
[54] T.S. Majmudar, M. Sperl, S. Luding, R.P. Behringer, Phys. Rev. Lett. **98**, 058001 (2007)
[55] C. Coulais, R.P. Behringer, O. Dauchot, Soft Matter **10**, 1519 (2014)
[56] F. Lechenault, F.d. Cruz, O. Dauchot, E. Bertin, J. Stat. Mech. **2006**, P07009 (2006)
[57] S.C. Zhao, S. Sidle, H.L. Swinney, M. Schröter, Europhys. Lett. **97**, 34004 (2012)
[58] T. Aste, M. Saadatfar, T.J. Senden, J. Stat. Mech.: Theory and Exp. **2006**, P07010 (2006)
[59] A. Baule, R. Mari, L. Bo, L. Portal, H.A. Makse, Nat. Commun. **4**, 2194 (2013)
[60] A. Baule, H.A. Makse, Soft Matter **10**, 4423 (2014)
[61] F.M. Schaller, M. Neudecker, M. Saadatfar, G.W. Delaney, G.E. Schröder-Turk, M. Schröter, Phys. Rev. Lett. **114**, 158001 (2015)
[62] G.E. Schröder-Turk, W. Mickel, M. Schröter, G.W. Delaney, M. Saadatfar, T.J. Senden, K. Mecke, T. Aste, Europhys. Lett. **90**, 34001 (2010)
[63] F. Radjai, in *Powders & Grains* (2009), Vol. 1145 of *AIP Conf. Proc.*, pp. 35–42
[64] C. Josserand, A.V. Tkachenko, D.M. Mueth, H.M. Jaeger, Phys. Rev. Lett. **85**, 3632 (2000)
[65] F. Radjaï, S. Roux, in *The Physics of Granular Media*, edited by H. Hinrichsen, D.E. Wolf (Wiley-VCH, Weinheim, 2004), pp. 165–187
[66] M. Nicolas, P. Duru, O. Pouliquen, Eur. Phys. J. E **3**, 309 (2000)
[67] L. Vanel, D. Howell, D. Clark, R.P. Behringer, E. Clément, Phys. Rev. E **60**, R5040 (1999)
[68] L. Vanel, E. Clément, Eur. Phys. J. B **11**, 525 (1999)
[69] I. Agnolin, J.N. Roux, Phys. Rev. E **76**, 061302 (2007)
[70] M. Neudecker, S. Ulrich, S. Herminghaus, M. Schröter, Phys. Rev. Lett. **111**, 028001 (2013)
[71] N.N. Thyagu, F. Schaller, S. Weis, M. Neudecker, M. Schröter, arXiv:1501.04472 (2015)
[72] https://www.youtube.com/watch?v=oUwlEdz42xo (2013)
[73] S. Weis, M. Schröter, submitted to Rev. Sci. Instr. (2017), http://arxiv.org/abs/arXiv:1612.06639
[74] A.J. Liu, S.R. Nagel, Ann. Rev. Cond. Matt. Phys. **1**, 347 (2010)
[75] A.G. Athanassiadis, M.Z. Miskin, P. Kaplan, N. Rodenberg, S.H. Lee, J. Merritt, E. Brown, J. Amend, H. Lipson, H.M. Jaeger, Soft Matter **10**, 48 (2013)
[76] C. Scholz, S. D'Silva, T. Pöschel, New J. Phys. **18**, 123001 (2016)
[77] I. Vlahinić, E. Andò, G. Viggiani, J.E. Andrade, Granul. Matter **16**, 9 (2014)

Breaking of rod-shaped model material during compression

Kulaviak Lukas[1,*], *Penkavova* Vera[1], *Ruzicka* Marek[1], *Puncochar* Miroslav[1], *Zamostny* Petr[2], *Grof* Zdenek[3], *Stepanek* Frantisek[3], *Schongut* Marek[3], and *Havlica* Jaromir[1]

[1] Czech Academy of Science , Institute of Chemical Process Fundamentals, 16502 Rozvojova 2/135, Prague, Czech Republic
[2] University of Chemistry and Technology Prague, Department of Organic Technology, 16500 Technicka 5, Prague, Czech Republic
[3] University of Chemistry and Technology Prague, Department of Chemical Engineering, 16500 Technicka 3, Prague, Czech Republic

Abstract. The breakage of a model anisometric dry granular material caused by uniaxial compression was studied. The bed of uniform rod-like pasta particles (8 mm long, aspect ratio 1:8) was compressed (Gamlen Tablet Press) and their size distribution was measured after each run (Dynamic Image Analysing). The compression dynamics was recorded and the effect of several parameters was tested (rate of compression, volume of granular bed, pressure magnitude and mode of application). Besides the experiments, numerical modelling of the compressed breakable material was performed as well, employing the DEM approach (Discrete Element Method). The comparison between the data and the model looks promising.

1 Introduction

Anisometric granular particles commonly occur in many different materials and products (e.g. crystals, pharmaceutical ingredients, [1-2]). Many technological operations (filtration, washing, drying, transportation) can lead to uncontrolled changes in grain size distribution caused by attrition and breakage [3]. When a randomly packed bed of anisometric particles is compressed, either by its own weight or by an external force, the task is to find the relationship between the applied stress and the grain compressibility and breakage. The goal of this study is to experimentally measure the breakage dynamics, find some quantitative descriptive variables and estimate their values, as they depend on few important control parameters. Such an effort is of both scientific and application interest. The experimental data are used to validate numerical modelling of granular beds [4], as is also briefly shown in this study.

2 Experimental

2.1 Material properties

Based on the preliminary tests, the rod-shaped pasta was used as the model granular material. Cylindrical dry noodles Capellini Major (Italy) 26 cm long of diameter 1.06±0.02 mm were cut into 8 mm particles with multifunction tool DREMEL® 4200. The bundle of noodles, tightly constricted by a thermo-sensitive sleeve, was cut with a high-speed rotating (up to 35 000 rev. per min) thin (0.75 mm) diamond cutting disc (type SC545). The cut particles had a narrow size distribution and for the purpose of our experiments were considered as monodisperse (mean length 8 mm, standard deviations $1\sigma_{68\%}$ = 0.25 mm, $2\sigma_{95\%}$ = 0.65 mm). Their mechanical material properties were determined by the three point rigidity test using Brookfield CT3 Texture analyser with the following result: the elastic modulus is 4.9 GPa, stiffness 292 kN/m, bending modulus 5.67 mNm and bending stress 57.9 MPa.

2.2 Apparatus and methodology

2.1.1 Compression

The material was subjected to the uniaxial compression tests (force-displacement analysis) carried out in the apparatus GTP-1 (recently labelled as Gamlen R-Series, Gamlen Tableting Ltd.), which is capable to generate load up to 500 kg. We used 3 cylindrical stainless steel measuring cells with height 30 mm and of diameters D = 25, 33, 41 mm, referred to as: S (small), M (medium), L (large). The maximum reachable pressure in the small cell was 10 MPa. Several metal disk-shaped spacers were employed to adjust the initial height of the particles layer to three values: H = 10, 15, 20 mm. The initial volume V of particles was put in the cell manually, by pouring through a funnel. Piston speed v and applied load P were set and the measurement started. The piston moved downward at a constant speed and its position and the bed resistance force were measured at the sampling rate 200 Hz, till the value of P was reached. From the piston position, the actual bed volume was obtained and the compressibility calculated $C = \Delta V/V$, usually is shown in %. The typical reproducibility of the results from five measurements in terms of the standard deviation is 5%. The result was the dependence $C(P)$ shown in diagrams known as compressibility curves.

* Corresponding author: kulaviak@icpf.cas.cz

After reaching the terminal pre-set pressure P and the corresponding final compression, the piston returned to its initial position and the layer slightly relaxed due to the elastic relieve. Two modes of the applied pressure were tested: (i) single-run and (ii) repetitive. In the non-recurring mode, after the compression, the particles were taken out of the measuring cell for granulometry. In the recurring mode, after granulometry, the particles returned to the cell for the next run with the same pressure P.

2.1.2 Granulometry

The particle size distribution (PSD) of the grains before and after the compression tests was measured by the dry granulometry device (Solids particle size analyser, CANTY, Industrial SolidSizer). An ensemble of particles (typically 2000) passed through the vibratory conveyor forming a slowly moving mono-layer. It slid along the inclined plane to ensure 2D motion of the particles in front of the black&white camera (320 x 240 pxls, 5 fps). The particle images were analysed by the device software. Each 2D image of a particle was fitted inside enclosing rectangle whose major size was the particle length. The histograms of particle sizes were produced, using 200 regular bins. From this, 4 coarse bins were also made that are used here: B1 (1-3 mm), B2 (3-5 mm), B3 (5-7 mm), B4 (7-8 mm). The average length of particles L and distribution of particle length characterized by polydispersity Z were calculated from the individual moments $\mu_n = \sum_{j=1}^{P} l_j^n$ as:

$$L = \frac{\mu_1}{\mu_0}, \quad (1)$$

$$Z = \frac{\mu_0 \cdot \mu_2}{\mu_1^2}, \quad (2)$$

where P is the number of particles, j is the current particle index (j = 1, 2, ... P), l_j is the length of particle j, l_j^n is the n-th power of l_j, μ_n is the n-th moment.

2.1.3 Modelling

Computer simulation was employed, using the DEM approach (Discrete Element Method), to better understand the breakage mechanism whose results were observed in the experiments. The multi-element particle model was used [5]. The model details are present in [4, 6]. The model particle is a row of 21 partially overlapping rigid spherical elements connected by bonds that break when shear force or bending torque exceed a pre-set threshold. Selected modelling results regarding the pressure effects are shown in Sections 3.3 and 3.4, for demonstration.

3 Results and discussion

3.1 Rate of granular bed compression

The measuring cell M was used with the bed height H = 15 mm. The applied pressure was P = 5200 kPa. Six different compression rates were employed within the measuring range of the device: v = 0.1, 0.5, 1, 5, 10 and 60 mm/min. The result is shown in Fig. 1, where no apparent effect of the piston speed can be observed. Based on this, the medium piston speed v = 5 mm/min was used for the further tests.

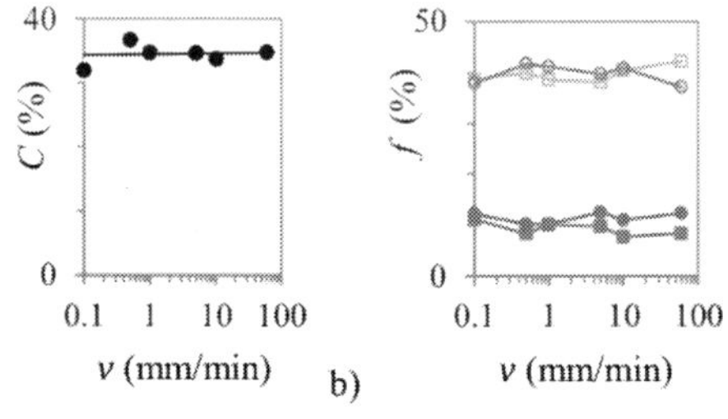

Fig. 1. Effect of compression rate v. (a) Compressibility C. (b) Particle fractions f in coarse bins: B4 - blue filled squares, B3 - red filled circles, B2 - green empty squares, B1 - purple empty circles. Applied pressure P = 5200 kPa.

3.2 Volume of granular bed

The initial volume of the compressed bed was adjusted by combining three cell sizes and three bed heights (9 combinations). Each measurement was repeated 5x and the results were averaged. The applied pressure was P = 3350 kPa. The result is shown in Fig. 2. The final compressibility in cells S and M is almost identical, and is slightly lower in cell L. The coarse bins' populations are not much influenced too. Thus the wall effects in our experiments were likely similar, in all cases. The medium cell size M and the middle layer height H = 15 mm were used for the further tests.

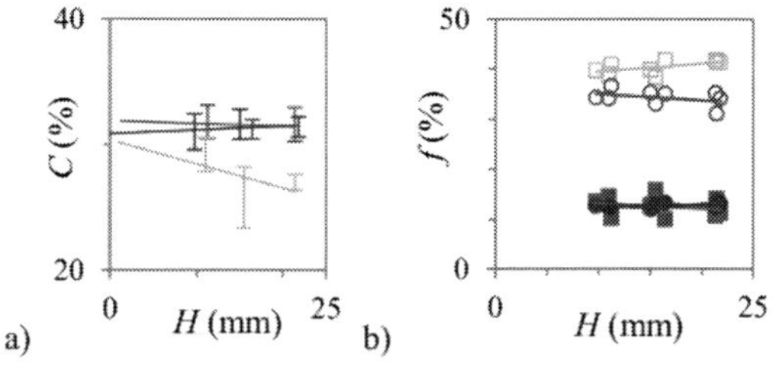

Fig. 2. Effect of bed volume. (a) Compressibility C; blue, red, green line for cells S, M, L. (b) Particle fractions f in coarse bins: B4 - blue filled squares, B3 - red filled circles, B2 - green empty squares, B1 - purple empty circles. Three marks at one H are 3 different cells. Applied pressure P = 3350 kPa.

3.3 Pressure magnitude

The effect of one-time applied pressure (non-repetitive) was tested, for eight values, P = 100, 500, 1000, 1400, 1900, 2600, 3350, 5200 kPa. The cell size was M and the layer height was H = 15 mm. The result is shown in Fig. 3. The shape of $C(P)$ graph obeys the typical saturation curve (3a), for both the data and model. At the load of 500 kPa, the particles break preferentially in half (green curve). At higher loads, the particles break into smaller pieces and progressively populate the size bins B1 and B2 (3b). The mean particle size is reduced by a high enough pressure, both in the data and model, Fig. 4. The wave-like course of the polydispersity is seen in Fig. 5.

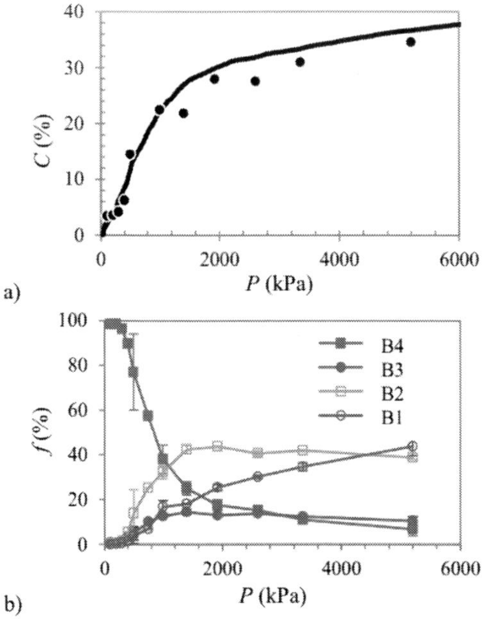

Fig. 3. Effect of applied pressure P (single compression). (a) Compressibility C; data - marks, model - line. (b) Particle fractions f in coarse bins B1-B4.

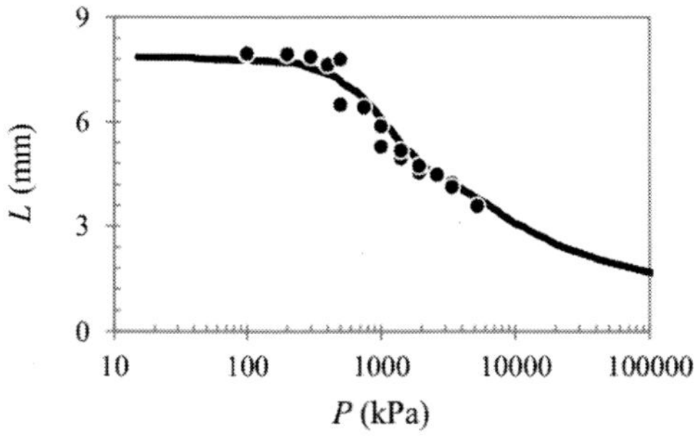

Fig. 4. Effect of applied pressure P (single compression) on particle average length, L; data – marks, model - line.

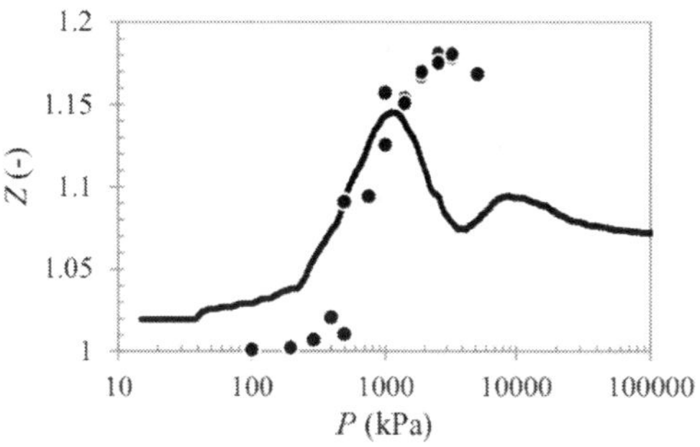

Fig. 5. Effect of applied pressure P (single compression) on particle polydispersity Z; data – marks, model - line.

3.4 Repeated compression

The relation between single and multiple compressions was studied too. The cell size was M and the layer height was 15 mm. The applied pressure P was set to 500 or 3350 kPa. The variable parameter was the number N = 1, 2, 3, 4, 5, of how many times the layer was consequently compressed, after emptying and re-filling the cell. The result is shown in Fig. 6. The $C(N)$ graphs are similar for both applied pressures (6a): weak almost linear rise over the limited range of N. The size distribution (6b) shows enhanced destruction of the particles, as compared with that in Fig. 3b. After two runs (N = 2), the initial length (bin B4) is missing and the population in the half-size bin (B2) stops growing. With more runs, the small particles (B1) grow in number at the expense of the others, mainly of B2.

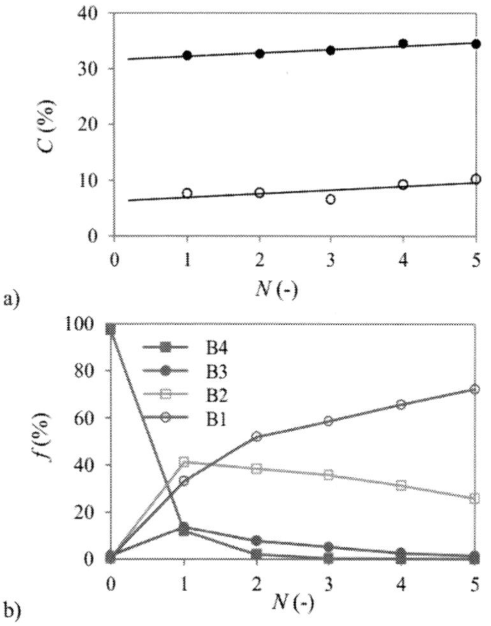

Fig. 6. Effect of pressure: multiple compressions. (a) Compressibility C; filled and empty circles with trendlines for P = 3350 and 500 kPa respectively. (b) Particle fractions f in coarse bins B1-B4. Pressure P = 3350 kPa was applied N-times.

The repeating breakage with a lower final pressure seems to be more effective than the single breakage with a larger pressure. It may be due to the particles re-arrangement at the sample handling (taking for granulometry). Particle arrangement in recompressed samples likely offers more breakable contacts and leads to higher bulk density.

The relation between the single/multiple compression is shown in the compound Fig. 7. The single compression data series (blue marks) starts at the point $[L,Z] = [8,1]$ where we have the initially monodispersed bed. Each point is the result of a single compression experiment. As we increase the pressure P (from 0 to 5200), the data go toward lower size L and higher polydispersity Z, due to the breakage. The multiple compression data at low fixed $P = 500$ kPa (green marks) follow the same trend and are parametrized by the number of repetitions N. The same holds for the recompression at high fixed $P = 3350$ kPa (red marks). All these data behave in a similar way and have a peak in polydispersity, in the about half-size region (4 mm). The repetitive compression can be, in certain sense, equivalent to the single compression. Therefore, the existence of a possible relation between (P and N) of the former and (P) of the latter can be anticipated. The model (line) agrees qualitatively and needs some parameter adjustment to agree in values too. The difference in Z between the model and experiment shows that at the comparable stress P (Fig. 5) and average particle length L (Fig. 7), the particle size distribution is wider in the experiment than in the model. Detailed analysis revealed that this is caused by longer (unbroken) particles that survive higher loads in the experiment. A possible explanation can be that the real particles can be cracked but not separated, as seen in Fig 8. This effect is not included in the model where the crack means separation. This phenomenon needs to be studied more closely in our future work.

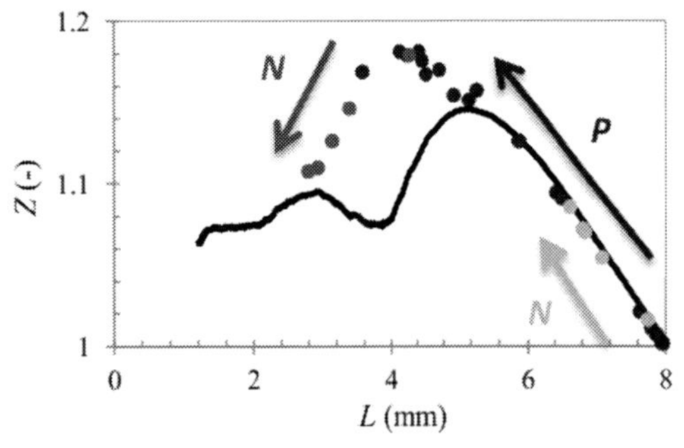

Fig. 7. Comparison of single and multiple compression. Polydispersity Z plotted vs particle average length L. Single compression with P growing from 0 to 5200 kPa (blue marks, arrow = increase). Recompression at low (500 kPa) and high (3350 kPa), fixed P with increasing N (green and red marks). Model - line.

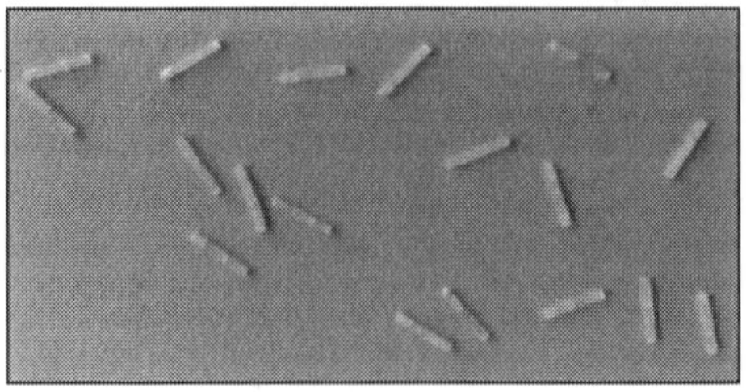

Fig. 8. Demonstration of cracks in experimental particles.

4 Conclusions

Under given conditions, the rate of compression was not the main factor of the breakage dynamics. The bed volume showed very weak effect with no obvious trend. The applied pressure determines the breakage result to a high degree. The relation between the repetitive and single compression is open for further investigation. The DEM modelling proves promising for simulating the granular media breakage.

The support by Czech Science Foundation GACR through contract No. 15-05534S is gratefully acknowledged.

5 Nomenclature

C : compressibility [%]
D : diameter of cell [m]
f : fraction of particles [%]
H : height of layer [m]
L : particles average length [m]
l : length of a particle [m]
N : number of repetition [-]
P : applied pressure [Pa]
v : piston speed [m/s]
V : initial volume of granular bed [m^3]
Z : polydispersity [-]
μ : moment

References

1. D. Schulze., *Powders and bulk solids* (Springer, Berlin, 2008)
2. H.G. Merkus., G.M.H. Meesters, (Eds.), *Production, handling and characterization of particulate materials* (Springer Cham, Switzerland, 2016)
3. H. Kalman, KONA Powder Part. J. **18** (2000)
4. Z. Grof, F. Stepanek, Phys. Rev. E **88** (2013)
5. J.F. Favier, M.H. Abbaspour-Fard, M. Kremmer, A.O. Raji, Eng. Comput. **16** (1999)
6. Z. Grof, F. Stepanek, Adv. Powder Technol. **27** (2016)

Compaction of granular materials composed of deformable particles

Thanh Hai Nguyen[1,2,*], *Saeid* Nezamabadi[2,**], *Jean-Yves* Delenne[3,***], and *Farhang* Radjai[2,4,****]

[1] *The University of Danang - University of Science and Technology, 54 Nguyen Luong Bang, Danang, Vietnam*
[2] *LMGC, UMR 5508 CNRS, Université de Montpellier, Montpellier, France*
[3] *IATE, UMR1208, INRA, Université de Montpellier, Cirad, SupAgro, Montpellier, France*
[4] *<MSE>[2], UMI 3466 CNRS-MIT, CEE, Massachusetts Institute of Technology (MIT), Cambridge, USA*

Abstract. In soft particle materials such as metallic powders the particles can undergo large deformations without rupture. The large elastic or plastic deformations of the particles are expected to strongly affect the mechanical properties of these materials compared to hard particle materials more often considered in research on granular materials. Herein, two numerical approaches are proposed for the simulation of soft granular systems: (i) an implicit formulation of the Material Point Method (MPM) combined with the Contact Dynamics (CD) method to deal with contact interactions, and (i) Bonded Particle Model (BPM), in which each deformable particle is modeled as an aggregate of rigid primary particles using the CD method. These two approaches allow us to simulate the compaction of an assembly of elastic or plastic particles. By analyzing the uniaxial compaction of 2D soft particle packings, we investigate the effects of particle shape change on the stress-strain relationship and volume change behavior as well as the evolution of the microstructure.

1 Introduction

Granular materials made of soft deformable particles represent a broad class of materials. Besides granular materials such as metallic powders, in which the particles can undergo large plastic deformations, a large number of cosmetic and food products, emulsions and biological cells may be included in this class. The deformation of the individual soft particles has been a subject of experimental and theoretical investigations at low level of deformation and more recently for large deformations [1, 2]. Although the behavior of individual particles has been studied experimentally with tools designed for this purpose or inside a shear flow, the behavior of a dense packing of soft particles under stress remains largely unexplored. Unlike weakly deformable particle materials, the mechanical behavior of an assembly of soft particles is governed by both rearrangements and shape change of the particles, and also partially by their compressibility. Particle shape change modifies drastically the collective behavior of the particles, allowing the packing to reach high packing fraction beyond Random Close Packing (RCP) state.

A major obstacle to the numerical study of soft particle materials has been the lack of a proper discrete modeling for the soft particles. Discrete Elements Methods (DEM) (like Contact Dynamics (CD) [3, 4]) often used for hard particles, are based on the approximation of small strains of contacts between particles. Thus, the only degrees of freedom considered in these techniques are those of particles (displacements and rotations). For soft particles, in contrast, it is necessary to take into account the internal degrees of freedom of each individual particle as well as the contact interactions.

In this paper, we consider the Material Point Method (MPM) and CD method, both recently implemented and applied to small assemblies of particles [5]. We also consider a Bonded Particle Model (BPM) in which each particle is represented by an aggregate of primary rigid particles with appropriate long-range interactions such that the aggregate can undergo plastic deformations without rupture of the aggregate and dispersion of primary particles [6]. The major advantage of this latter approach is that it uses the framework of the DEM. However, as we shall see below, the resulting behavior is plastic. In the following, we first briefly introduce the applied numerical approaches. Then, we present the behavior of a single deformable particle. Finally, we study the compaction of an assembly of soft circular particles. We conclude with a brief summary and scope of this work.

2 Numerical procedures

There are two general possible representations of ultra-soft particles: 1) as an aggregate of rigid primary particles with attractive interactions (BPM) and 2) as a set of material points (MPM). In the MPM, the information carried by material points is projected onto a background mesh, where equations of motion are solved. The mesh solution is then used to update the material points. The important

*e-mail: nthai@dut.udn.vn
**e-mail: saeid.nezamabadi@umontpellier.fr
***e-mail: jean-yves.delenne@supagro.inra.fr
****e-mail: franck.radjai@umontpellier.fr

point for the treatment of soft particles with MPM is the necessity of coupling it with the CD approach for a rigorous treatment of frictional contacts. More details of this procedure can be found in [5].

The BPM approach has been mainly used for the simulation of breakable particles [7]. Its application to soft particle materials involves the introduction of a long-range attraction force between primary particles. Hence, the interaction force between primary particles is composed of two forces: i) a long-range center-to-center attraction force F_a, as in colloidal systems, and ii) a short-range repulsion force F_r [8]. In this manner, each BPM particle is an aggregate of primary rigid particles, which can move and rearrange according to the external forces acting on its boundaries by other aggregates.

To define the interaction force between primary particles, let us consider two rigid circular particles of the radii a_1 et a_2 separated by a distance δ. We use a Lennard-Jones attractive force:

$$F_a = -F_0\left(1+\frac{\delta}{a_0}\right)^{-\gamma}, \tag{1}$$

where $a_0 = a_1 + a_2$, and F_0 is the maximum value of the attraction force at contact between particles ($\delta = 0$) [8, 9]. The choice of the exponent γ is not crucial for this study. It has been shown that this law with different values of this exponent may represent the potential of the mean force in complex materials such as cement and clay [9].

The repulsive part of the interaction between primary particles can be defined in different ways. But, as we use the CD method for the simulation of primary particles, we consider that the particles are perfectly rigid so that the repulsion force F_r is activated only when the two particles touch ($\delta = 0$). Therefore, the interaction force $F = F_a + F_r$ as a function of δ consists of only attraction at arbitrary distance $\delta > 0$ and takes a value $\in [-F_0, +\infty]$ at contact ($\delta = 0$). In the CD method, this force is calculated by using nonsmooth an implicite integration scheme [3, 4, 10]. To avoid strain shear localization and volume change within aggregates, the coefficient of friction between primary particles belonging to a particle is set to zero. Detailed descriptions of the BPM method can be found in [6].

3 Diametral compression of one particle

In this section, we study the accuracy and efficiency of the proposed models by considering the behavior of a single particle by using the MPM and BPM approaches. To do so, we compressed a soft particle of radius $R = 5$ mm between two rigid walls by fixing the bottom wall and moving downwards the top wall at a constant velocity of 0.2 m/s. The BPM particle is composed of 1750 rigid frictionless primary particles with a weak size polydispersity (diameters $\in [0.16, 0.26]$ mm). Their interactions are governed by equation (1) with $F_0 = 100$ N and $\gamma = 7$. For an efficient implementation of this law, a cut-off distance $\delta = a_0$ is introduced. Indeed, for $\gamma = 7$, the attraction force is negligibly small beyond $\delta \simeq a_0$, in which case only the first and second neighbors of each primary particle are involved in the interactions with other primary particle.

Two-dimensional MPM simulations in plane strain conditions were performed in the same conditions. The computation domain was meshed with four-node quadrangular elements, and an initial distribution of four material points per element was used. Young's modulus, Poisson's ratio and density of the particles were set to $E = 10$ MPa, $\nu = 0.45$ and $\rho = 990$ kg/m^3, respectively. We chose a high value of Poisson's ratio in order to get nearly constant volumes of the particles as in BPM particles. The spatial relative resolution is $\frac{\Delta r}{R} = 0.029$, where Δr is the mean distance between material points.

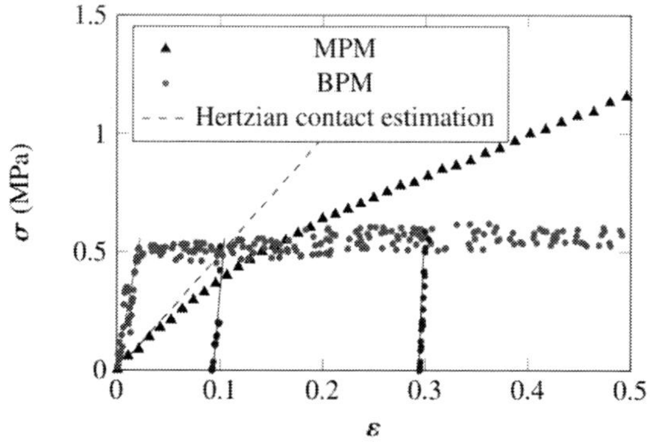

Figure 1. Vertical stress as a function of cumulative axial strain for a single particle subjected to diametrical compression by means of BPM and MPM simulations.

Figure 1 shows the vertical stress $\sigma = \frac{F}{L}$, where L is the horizontal diameter of the particle at any time step, as a function of the cumulative axial strain $\varepsilon = \ln(1 + d/R)$, with d being the displacement of the center of the particle. The MPM particle shows a linear elastic behavior at low strain as predicted by the Hertz analysis for a disk [5]:

$$F = \frac{\pi}{4}\frac{E}{1-\nu^2}R\varepsilon\,, \tag{2}$$

Deviation from linear behavior is observed for $\varepsilon > 0.05$. This range of the Hertz scaling (with an exponent 3/2 in 3D) is generally used in molecular dynamics simulations of granular materials composed of spherical particles by assuming small contact strains [11, 12]. A transition to a linear mode with a lower slope is observed beyond $\varepsilon =$ 0.15.

The BPM particle is characterized by a linear response at small deformations of the order of 2% and a plastic behavior beyond. The plastic strain threshold remains constant and is almost equal to a characteristic stress defined by $\sigma_l = \frac{F_0}{<a_0>} = 0.5$ MPa, where $< a_0 > \simeq 0.2$ mm, is the mean distance between primary particles. Moreover, to demonstrate the plastic behavior of BPM particle, we applied an unloading at two strain levels (10% and 30%); see Fig. 1. The stress decreases linearly during unloading with a slope that appears to increase slightly with strain.

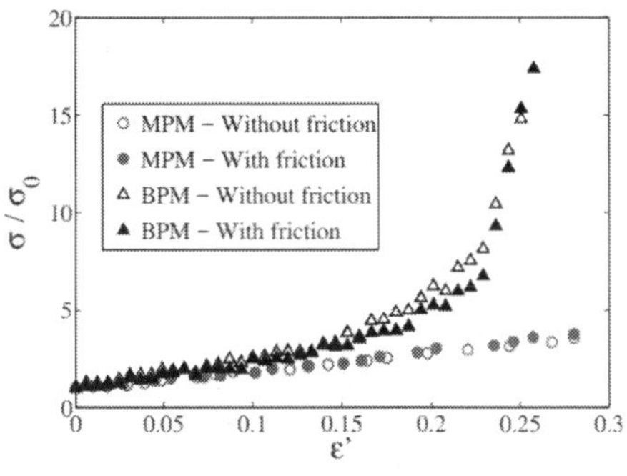

Figure 2. (a) Vertical stress σ normalized by the consolidation stress as a function of cumulative vertical strain ε' from the reference state σ_0 for frictionless and frictional particle packings by BPM and MPM simulations.

4 Uni-axial compression of a soft particle system

In this section, we use the BPM and MPM procedures to analyze the compaction of a packing of soft particles. It is composed of 300 soft circular particles confined inside a rectangular box of width L in which only the top wall is mobile and moves downwards at constant velocity of 5 m/s with a time step of $\delta t = 0.1$ μs. In the initial configuration, a small size polydispersity is introduced in order to avoid long-range ordering (diameters $\in [1.4\ ,\ 2.4]$ mm). The gravitational acceleration and friction between the particles and the walls are set to be zero in order to avoid stress gradients. We consider two cases below: 1) without friction and 2) with a coefficient of friction $\mu = 0.5$ between the particles. In the BPM simulations, each particle is composed of a minimum of 200 primary particles. The same material parameters are considered for the two methods as in previous section.

Figure 2 shows the vertical stress σ, calculated from the forces acting on the bottom wall and normalized by the consolidation stress σ_0, defined for the packing fraction Φ_0, as a function of the cumulative vertical strain $\varepsilon' = \ln(1 + \Delta h/h_0)$. The couple (σ_0, Φ_0) defines a reference state. We chose $\Phi_0 = 0.8$ which corresponds to the beginning of particle jamming. In the MPM simulations, the relationship between stress and the cumulative strain is linear whereas in the BPM simulations, a nonlinear increase of stress with ε' is observed. Three deformation mechanisms control this development: (1) the elastic and/or plastic volume change of particles, (2) the rearrangement of particles and (3) particles shape change. The latter dominates the behavior of the system and allows the packing fraction to exceed considerably the RCP state. It is also worth noting that in the BPM simulations, the vertical stress grows at an increasing rate with ε' as in the compaction of metallic powders [13]. Moreover, in both cases, the friction between particles has little effect on the deformation.

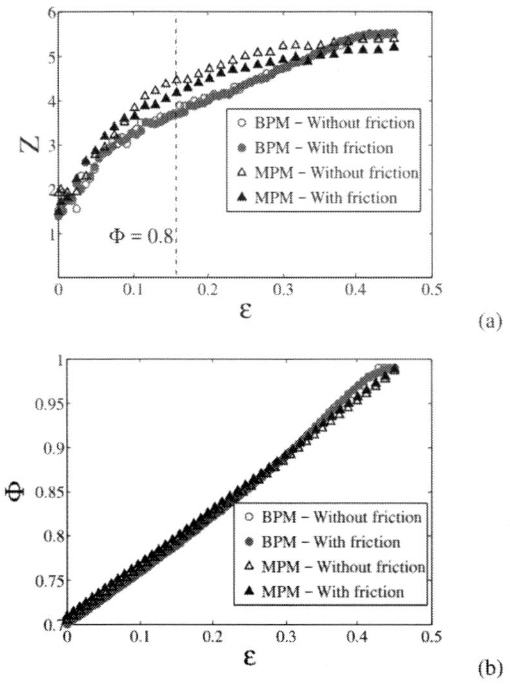

Figure 3. Mean coordination number Z (a) and packing fraction Φ (b) as a function of cumulative vertical strain ε for frictionless and frictional particle packings by BPM and MPM simulations.

The microstructure of an assembly of soft particles can be characterized in terms of the coordination number Z. Above 20% of deformation, the particles are so much deformed that the contacts can no more be modeled as point contacts. In this regime, the particle shapes are well approximated by polygons, and Z then represents the number of sides of these polygons.

Figure 3 shows the mean coordination number Z and the packing fraction Φ as a function of cumulative vertical strain ε defined from the beginning of the simulations. Φ is an almost linear function of ε. Indeed, as the volume of the particles does not vary, we have $\Phi = \Phi'_0 e^{\varepsilon} \simeq \Phi'_0(1+\varepsilon)$ ($\Phi'_0 = 0.7$). In the case of MPM particles, the behavior is linear elastic, and therefore the remaining small pores for high values of packing fraction can only be filled at much higher stresses (requiring the radii of curvature to be increasingly smaller). However, in the BPM simulations, the pores can be filled more efficiently by plastic deformation of the particles for a much lower stress level. The pore-filling process is similar in both approaches as the pore size between the BPM particles is greater than the size of the primary particles. The limit where the internal pore sizes of the particles are of the same order as the pores at the interface between two particles, corresponds to 40% of deformation and a packing fraction of 0.93 in the BPM simulations.

The mean coordination number Z increases with ε. In the BPM simulations, we observe no difference between frictional and frictionless particles, which is due to the absence of friction between the primary particles. Indeed, the friction force at the contact between two primary particles a t the boundary of two particles in contact, can not be mobilized because these primary particles can rotate freely without dissipation. In MPM simulations, we see slightly higher values of Z for frictionless particles while the packing fraction is the same in all cases with or without friction at all levels of deformation. The maximum value of Z is 5.5 in all simulations. This value reflects the low size polydispersity of particles.

The jamming transition occurs for $\varepsilon \simeq 0.08$ where $Z \simeq 3$ and $\Phi \simeq 0.75$. From this state, the vertical stress begins to increase. However, the assembly remains fragile (rearrangements of particles) until a critical packing fraction of $\Phi_0 \simeq 0.8$ for a stress σ_0, which was used earlier as the the consolidation stress.

Figure 4 shows the evolution of Z as a function of packing fraction for both approaches and both cases of with and without friction. We see that by normalizing $Z - Z_0$ per $Z_1 - Z_0$, where Z_0 is the coordination number corresponding to $\Phi \simeq 0.8$ and Z_1 for $\Phi \simeq 1$, all data points collapse on a function well-fitted by a power law. Beyond Φ_0, Z increases as $(\Phi - \Phi_0)^{0.5}$ as in the case of emulsions and foams [1, 14]; see Fig. 4. This dependence of the coordination number with the packing fraction is quite robust and almost independent of the interaction potential, polydispersity and friction [14].

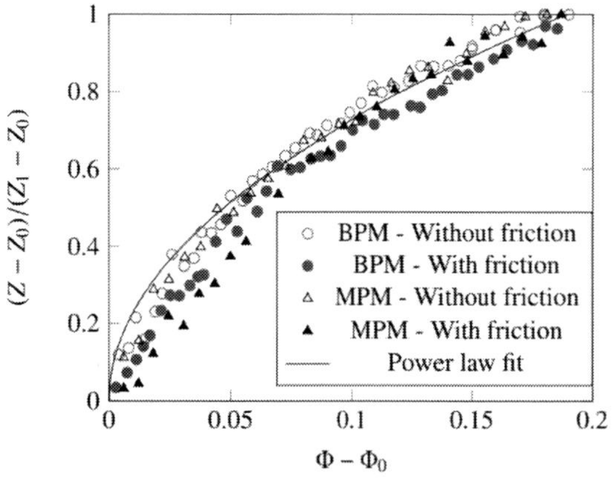

Figure 4. Average coordination number Z as a function of packing fraction Φ. The line is a power-law fit $(\Phi - \Phi_0)^{0.5}$.

5 Conclusion

In this paper, we analyzed the uniaxial compression of a soft-particle assembly beyond the RCP state using two numerical techniques: Material Point Method (MPM) and Bonded Particle Model (BPM). We characterized the macroscopic behavior, and simulations by two methods allowed us to highlight several significant differences between the assemblies of elastic and plastic particles. One interesting finding concerns the relationship between the packing fraction and the coordination number by a simple power law as also observed in emulsions.

Acknowledgment

Thanh Hai Nguyen would like to acknowledge the support of the University of Danang, University of Science and Technology, code number of project: T2017-02-02-91.

References

[1] J. Zhang, T.S. Majmudar, M. Sperl, R. Behringer, Soft Matter **6**, 2982 (2010)
[2] P. Menut, S. Seiffert, J. Sprakelae, D. Weitz, Soft Matter **8**, 156 (2012)
[3] J. Moreau, European Journal of Mechanics A - Solids **13**, 93 (1994)
[4] F. Radjai, V. Richefeu, Mechanics of Materials **41**, 715 (2009)
[5] S. Nezamabadi, F. Radjai, J. Averseng, J.Y. Delenne, Journal of the Mechanics and Physics of Solids **83**, 72 (2015)
[6] S. Nezamabadi, T.H. Nguyen, J.Y. Delenne, F. Radjai, Granular Matter **19**, 8 (2017)
[7] D.H. Nguyen, E. Azéma, P. Sornay, F. Radjai, Physical Review E **91**, 022203 (2015)
[8] J. Israelachvili, *Intermolecular and surface forces* (Academic Press, London, 1993)
[9] K. Ioannidou, R. Pellenq, E. Del Gado, Soft Matter **10**, 1121 (2014)
[10] F. Radjai, F. Dubois, eds., *Discrete-element Modeling of Granular Materials* (Iste-Wiley, London, 2011)
[11] C. Thornton, KONA Powder and Particle **15**, 81 (1997)
[12] H.A. Makse, D. Johnson, L. Schwartz, Physical Review Letters **84**, 4160 (2000)
[13] F. Toussaint, Ph.D. thesis, Institut National Polytechnique de Grenoble - INPG (2001)
[14] C. O'Hern, L. Silbert, A. Liu, S. Nagel, Physical Review E **68**, 011306 (2003)

Scaling behavior of immersed granular flows

L. Amarsid[1,2,3,★], *J.-Y. Delenne*[4], *P. Mutabaruka*[5], *Y. Monerie*[2,3], *F. Perales*[1,3], and *F. Radjai*[2,5,3]

[1] *IRSN, PSN, CE Cadarache, BP3-13115 St Paul-Lez-Durance Cedex, France*
[2] *LMGC, CNRS - University of Montpellier, 163 rue Auguste Broussonnet, 34090 Montpellier, France*
[3] *Laboratoire MIST, IRSN-CNRS-University of Montpellier, France*
[4] *IATE, UMR1208 INRA - CIRAD - University of Montpellier - SupAgro, 34060 Montpellier, France*
[5] *$(MSE)^2$, UMI CNRS-MIT, Massachusetts Institute of Technology, 77 Massachusetts Avenue, Cambridge 02139, USA*

Abstract. The shear behavior of granular materials immersed in a viscous fluid depends on fluid properties (viscosity, density), particle properties (size, density) and boundary conditions (shear rate, confining pressure). Using computational fluid dynamics simulations coupled with molecular dynamics for granular flow, and exploring a broad range of the values of parameters, we show that the parameter space can be reduced to a single parameter that controls the packing fraction and effective friction coefficient. This control parameter is a modified inertial number that incorporates viscous effects.

1 Introduction

Immersed granular materials occur in many natural processes such as sediment transport, landslides and submarine avalanches, as well as in many applications involving particle-laden fluids in powder technology and food and pharmaceutical industries [1–6]. The central issue in the field of dense suspensions of non-Brownian particles is how the suspended particles affect the rheology [7] whereas in the granular community the query is how the inertial, viscous, capillary and lubrication effects in the presence of an interstitial fluid modify the shear behavior and packing fraction in granular flows [8, 9].

Fluid-grain flows involve a large parameter space: fluid viscosity η_f, fluid density ρ_f, particle mean size d, particle density ρ_s, confining pressure (acting on the particle phase) σ_s and shear rate $\dot{\gamma}$. In the absence of the fluid, it is well-known that the parameter space can be reduced to the *inertial number* $I = \dot{\gamma} d(\rho_s/\sigma_s)^{1/2}$ [10], and the rheology is described by effective friction coefficient μ and packing fraction Φ as functions of I. On the other hand, it has been suggested that, when particle-inertial effects can be neglected, the parameter space is reduced to the *viscous number* $I_v = \eta_f \dot{\gamma}/\sigma_s$, and the rheology is described by μ and Φ as a function of I_v [8].

Recently, it was suggested that a general control parameter combining I and I_v may account for both inertial and viscous effects in a unified framework [11]. Trulsson et al. simulated sheared granular materials by means of Molecular Dynamics (MD) and the effect of fluid was accounted for through the application of a drag force on all particle centers. However, since the fluid is introduced through its drag force effect on particles, it is unclear whether the unified approach accounts for other major fluid parameters such as density and volume effects and lift forces induced by shear. In particular, the relative density $r = \rho_s/\rho_f$ may also influence the flow regime [12]. Moreover, the fluid in the pore space carries dynamic pore pressures, which may come into play in a dense suspension. For all these reasons, the visco-inertial flow regime needs to be investigated by using computational fluid dynamics coupled with granular dynamics.

In this paper, we use MD simulations for particle dynamics coupled with the Lattice Boltzmann Method (LBM) for the dynamics of the fluid phase to investigate the rheology of dense granular flows immersed in a viscous fluid. The simulations cover a broad range of parameter values. As we shall see, the effective flow properties are well described by a single modified inertial number, in agreement with the approach suggested by Trulsson et al. [11]. We first briefly present the numerical approach and system parameters. Then, we describe our main results and conclude with the scopes of this work.

2 NUMERICAL METHOD

We employed the MD method interfaced with the Lattice Boltzmann Method (LBM) for the simulations. The fluid is modeled by a time-dependent distribution function $f(\vec{r}, \vec{v}, t)$ of particle positions $\vec{r}$ and velocities $\vec{v}$. The spatio-temporal evolution of f is governed by the Boltzmann equation:

$$\left(\frac{\partial}{\partial t} + \vec{v} \cdot \frac{\partial}{\partial \vec{r}} + \frac{\vec{F}(\vec{r})}{m} \cdot \frac{\partial}{\partial \vec{v}}\right) f(\vec{r}, \vec{v}, t) = \Omega_{coll} \qquad (1)$$

where m is particle mass, $\vec{F}(\vec{r})$ represents external forces, and Ω_{coll} is the collision operator. The simplest collision

★e-mail: lhassan.amarsid@umontpellier.fr

model is the BGK operator [13]:

$$\Omega_{BGK} = -\frac{1}{\tau}(f - f^0) \tag{2}$$

where τ is a relaxation time and f^0 is the Maxwell-Boltzmann distribution. The Boltzmann equation together with the BGK collision operator yields the Navier-Stokes equations [14].

In LBM [15–17] the velocity vector space is discretized into a finite number of directions. We used the D2Q9 lattice (two dimensions with nine velocity directions), as shown in Fig. 1. An independent distribution function f_i corresponds to each velocity direction $\vec{e}_i$.

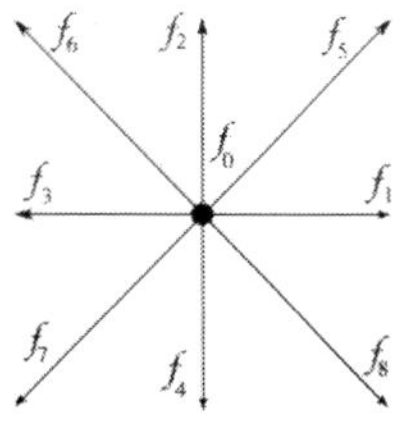

Figure 1. D2Q9 scheme.

The discretized equations for different directions are solved in two steps:

$$\begin{array}{ll} \text{Collision :} & f_i^{out}(\vec{r}, t) = f_i(\vec{r}, t) + \Omega_i \\ \text{Streaming :} & f_i(\vec{r} + \Delta t\vec{e}_i, t + \Delta t) = f_i^{out}(\vec{r}, t) \end{array} \tag{3}$$

where Δt is the time step, and the $f_i^{out}(\vec{r}, t)$ are the distribution functions after the collision step. The density $\rho(\vec{r}, t)$ and fluid velocity $\vec{u}(\vec{r}, t)$ are obtained as follows:

$$\begin{array}{l} \rho(\vec{r}, t) = \sum_i f_i(\vec{r}, t) \\ \rho(\vec{r}, t)\vec{u}(\vec{r}, t) = \sum_i f_i(\vec{r}, t)\vec{e}_i \end{array} \tag{4}$$

The BGK operator is simple but leads to fluctuating velocity fields. In our simulations, we used a Multi-Relaxation Time (MRT) collision approach [18, 19]. Nine moments are attributed to every fluid node, corresponding to the nine distribution functions, through a matrix **M** such that

$$\mathbf{m} = \mathbf{Mf} \tag{5}$$

where $\mathbf{m} = (m_0, m_1, \ldots, m_8)^T$ is the moment vector, $\mathbf{f} = (f_0, f_1, \ldots, f_8)^T$ and **M** is given by:

$$\mathbf{M} = \begin{pmatrix} 1 & 1 & 1 & 1 & 1 & 1 & 1 & 1 & 1 \\ -4 & -1 & 2 & -1 & 2 & -1 & 2 & -1 & 2 \\ 4 & -2 & 1 & -2 & 1 & -2 & 1 & -2 & 1 \\ 0 & 1 & 1 & 0 & -1 & -1 & -1 & 0 & 1 \\ 0 & -2 & 1 & 0 & -1 & 2 & -1 & 0 & 1 \\ 0 & 0 & 1 & 1 & 1 & 0 & -1 & -1 & -1 \\ 0 & 0 & 1 & -2 & 1 & 0 & -1 & 2 & -1 \\ 0 & 1 & 0 & -1 & 0 & 1 & 0 & -1 & 0 \\ 0 & 0 & 1 & 0 & -1 & 0 & 1 & 0 & -1 \end{pmatrix}$$

Hence, the collision step is applied in the moment space, each moment m_i being relaxed to its equilibrium state m_i^{eq} with a relaxation time s_i. The moments corresponding to the density $\rho(\vec{r}, t)$ and the flux $\vec{j}(\vec{r}, t) = \rho(\vec{r}, t)\vec{u}(\vec{r}, t)$ are conserved. The moment vector $\mathbf{m}^{out}$ resulting from collision can be written as

$$\mathbf{m}^{out} = \mathbf{m} - \mathbf{S}(\mathbf{m} - \mathbf{m}^{eq}) \tag{6}$$

where $\mathbf{S} = diag(0, s_1, s_2, 0, s_4, 0, s_6, s_7, s_8)$ is a diagonal 9×9 matrix.

All relaxation times are proportional to τ^{-1} [20]. The equilibrium moment vector $\mathbf{m}^{eq}$ is given by

$$\mathbf{m}^{eq} = \begin{bmatrix} \rho \\ -2\rho + 3\left(j_x^2 + j_y^2\right)/\rho \\ \rho - 3\left(j_x^2 + j_y^2\right)/\rho \\ j_x \\ -j_x \\ j_y \\ -j_y \\ \left(j_x^2 - j_y^2\right)/\rho \\ j_x j_y/\rho \end{bmatrix} \tag{7}$$

The distribution functions $f_i^{out}(\vec{r}, t)$ after the collision step are $\mathbf{f}^{out} = \mathbf{M}^{-1}\mathbf{m}^{out}$. Finally, the streaming step is applied in the velocity space.

The no-slip boundary conditions are implemented using the bounce-back rule, which consists in reflecting back the incoming distribution functions at a boundary node to their original fluid nodes in the opposite direction *iopp* ($\vec{e}_i + \vec{e}_{iopp} = \vec{0}$):

$$f_{iopp}^{in}(\vec{r}, t + \Delta t\vec{e}_i) = f_i^{out}(\vec{r}, t) \tag{8}$$

The equations of motion of the particles are integrated by means of the MD method [21, 22]. The normal force $\vec{F}_N$ between a pair (i, j) of touching particles is governed by a viscoelastic law:

$$\vec{F}_N = \begin{cases} \left(-k_n\delta_{ij} - \gamma_n\dot{\delta}_{ij}\right)\vec{n}_{ij} & \text{if } \delta_{ij} < 0 \\ \vec{0} & \text{otherwise} \end{cases} \tag{9}$$

where δ_{ij} is the gap or the overlap, $\dot{\delta}_{ij}$ its derivative with respect to time, k_n is the stiffness and γ_n is a viscous damping parameter that controls restitution coefficient.

The friction force $\vec{F}_T$ obeys the dry Coulomb friction law:

$$\vec{F}_T = -\min\left\{\gamma_t\|\vec{v}_t\|; \mu_s\|\vec{F}_N\|\right\}\vec{t}_{ij} \tag{10}$$

where $\vec{v}_t$ is the tangential velocity at contact, γ_t is the tangential viscosity parameter, and μ_s is the friction coefficient.

The particles are meshed on the lattice grid and represented by solid nodes. The interaction between particles and fluid occurs at their interface. The solid nodes are considered as moving boundaries over which the no-slip condition is imposed [23]. The hydrodynamic forces acting on particles are calculated by the momentum exchange method proposed in [24].

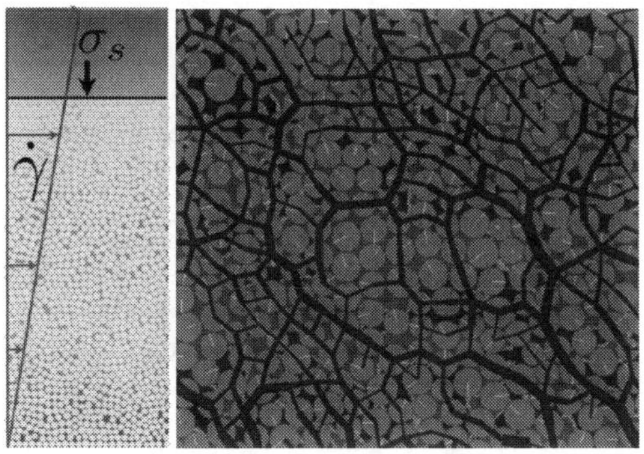

Figure 2. A snapshot of the suspension and its boundary conditions (left); a snapshot of the force network and negative and positive fluid pressures in blue and red, respectively.

3 Simulated system

The simulated system is shown in Fig. 2. The fluid fills a rectangular domain which is periodic in the horizontal direction. The particles are disks of mean diameter $d = 2.5 \times 10^{-3}$ m and distributed in the range $[d_{min}, d_{max}]$ with $d_{min} = 0.6d_{max}$. A confining pressure σ_s is applied on the top wall in contact with the granular column. This wall is permeable to the fluid filling a larger domain of constant volume. A velocity gradient $\dot{\gamma}$ is applied to the fluid nodes at the domain boundary to generate a plane Couette flow. The particles are initially distributed randomly in the box and assembled by downward motion of the top mobile wall under the action of the normal stress σ_s.

All simulations were performed with 1253 particles and a well-resolved fluid with lattice step $< 0.03d$. The friction coefficient μ_s is set to 0.4 between the particles and to 0 with the top wall. The bottom wall is made rough by sticking a layer of particles to the bottom of the box. The shear rate $\dot{\gamma}$ is varied in the range $[0.28, 5.6]$ s^{-1}, confining stress σ_p in the range $[20, 120]$ Pa, the relative density $r = \rho_s/\rho_f$ in the range $[0.5, 3]$ and the fluid viscosity η_f is varied from that of water η_w to $2500\eta_w$. More than 70 simulations with a total CPU time of about 10^5 hours were carried out.

4 Results and discussion

In order to reduce the parameter space, we consider the characteristic stresses which, in contrast to characteristic times, have the advantage of being additive. They are of three different origins: 1) the static stress σ_s 2) the viscous stress $\sigma_v \sim \eta_f\dot{\gamma}$ and 3) the inertial stress $\sigma_i \sim \rho_s(d\dot{\gamma})^2$. The corresponding characteristic times are $t_s = d(\rho_s/\sigma_s)^{1/2}$, $t_v = d(\rho_s/\eta_f\dot{\gamma})^{1/2}$ and $t_i = \dot{\gamma}^{-1}$, respectively. We also have the Stokes time $t_{St} = \eta_f/\sigma_s = t_i(t_s/t_v)^2$. From these three time scales two independent dimensionless numbers can be build: $I = t_s/t_i = (\sigma_i/\sigma_s)^{1/2} = \dot{\gamma}d(\rho_s/\sigma_s)^{1/2}$, the inertial number, and $J = t_s/t_v = (\sigma_v/\sigma_s)^{1/2} = (\eta_f\dot{\gamma}/\sigma_s)^{1/2}$, which is the square root of I_v. The ratio $St = (t_v/t_i)^2 = \sigma_i/\sigma_v = I^2/J^2 = \rho_s d^2\dot{\gamma}/\eta_f$ is the Stokes number.

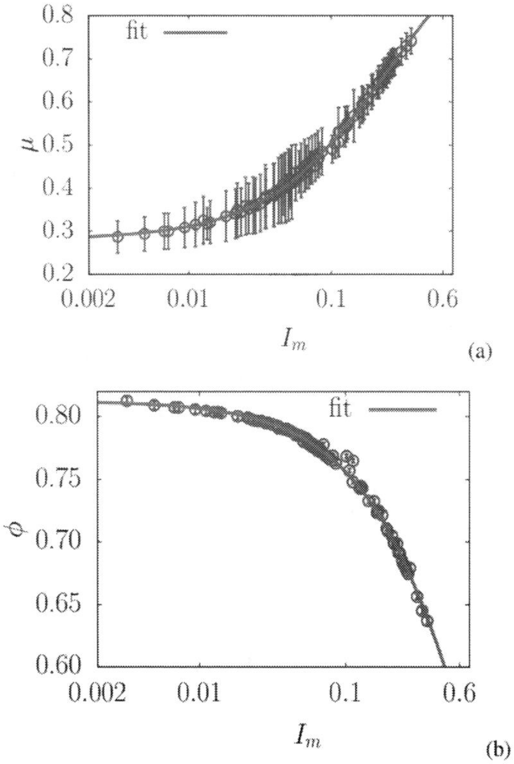

Figure 3. Effective friction coefficient μ (a) and packing fraction Φ (b) as a function of the modified inertial number I_m for all data points obtained with different values of system parameters. The solid lines are the fitting forms (12) and (13).

The total shear stress in steady flow is the sum of static, inertial and viscous stresses, and we expect that the rheology is governed by the additive effects of viscous and inertial stresses on particles compared to the static stress. Hence, owing to this additivity, we can build a unique dimensionless number $(\sigma_i + \alpha\sigma_v)/\sigma_s = I^2 + \alpha_v J^2 = I^2(1 + \alpha_v/St)$, where α_v is a constant to be determined. Taking the square root of this ratio, we get a *modified* inertial number

$$I_m = I\left(1 + \frac{\alpha}{St}\right)^{1/2} = \left\{\frac{\rho_s d^2}{\sigma_s}\dot{\gamma}^2 + \alpha\frac{\eta_f}{\sigma_s}\dot{\gamma}\right\}^{1/2}, \quad (11)$$

In this way, the parameter space can be reduced only if the packing Φ and effective friction coefficient μ are unique functions of I_m for a constant value of α.

Figure 3 shows μ and Φ as a function of I_m for $\alpha = 2.0$. We see that, all data points remarkably collapse on well-defined curves both excellently fit by the following functional forms:

$$\mu(I_m) = \mu_c + \frac{\delta\mu}{1 + b/I_m}, \quad (12)$$

$$\Phi(I_m) = \frac{\Phi_c}{1 + aI_m} \quad (13)$$

with $\mu_c = 0.280 \pm 0.002$ the quasi-static effective friction coefficient, $\Phi_c = 0.8123 \pm 0.0003$ the steady-state packing fraction, $b = 0.246 \pm 0.008$, $\delta\mu = 0.783 \pm 0.014$ and $a = 0.750 \pm 0.003$. Hence, all system parameters, including the relative fluid-particle density, affect the rheology only through I_m. The transition from viscous to inertial regime is governed by the Stokes number. According to the definition of I_m, this transition occurs for $St \simeq \alpha \simeq 2$. This result is in agreement and extends the scope of a unique framework introduced by Trulsson et al. [11] to a more general parameter space.

Note that, according to the defintion (11) of the modified inertial number, when the shear rate $\dot{\gamma}$ tends to zero, the leading term is the viscous stress (the term in $\dot{\gamma}$) irrespective of the values of other parameters. For this reason, at low shear rates, we generally expect a viscous behavior, where the control parameter is $I_m \simeq \sqrt{\alpha I_v}$. But the confining stress σ_s plays the same role with respect to viscous and inertial forces and its variation does not lead to a transition between inertial and viscous regimes.

5 Conclusion

Our extensive simulations using a sub-particle well-resolved computation of a viscous fluid coupled to rigid particles suggest that the visco-inertial flow regime of immersed granular materials can be described by a modified inertial number combining the inertial number with the Stokes number. This behavior can also be expressed by using a viscous description of the flow. In this description, both shear viscosity and bulk viscosity can be described by unique function of the packing fraction provided the fluid viscosity η_f is replaced by $\eta_f + \rho_s d^2 \dot{\gamma}/\alpha$. This remarkable scaling reflects the joint effects of particle inertia and fluid viscous forces on the granular microstructure. A detailed analysis of local evolutions of pores pressures and contact networks will be presented elsewhere.

Finally, it is important to note that the unified picture discussed in this work is with regard to the *functional* dependence of flow properties on those parameters that control particle dynamics. It is thus implicitly assumed that parameters such as particle size distribution and friction coefficient between particles enter the rheology only through their effect on the model parameters μ_c, Φ_c, b, $\delta\mu$ and a. This assumption needs to be checked by further simulations.

References

[1] R.M. Iverson, Reviews of Geophysics **35**, 245 (1997)
[2] F. Legros, Engineering Geology **63**, 301 (2002)
[3] K. Hewitt, American Scientist **98**, 410 (2010)
[4] K.D. Nguyen, S. Guillou, J. Chauchat, N. Barbry, Advances in Water Resources **32**, 1187 (2009)
[5] C. Lareo, P. Fryer, M. Barigou, Food and Bioproducts Processing **75**, 73 (1997)
[6] M.J. Rhodes, *Introduction to particle technology* (John Wiley & Sons, Chichester, 1998)
[7] J.J. Stickel, R.L. Powell, Annual Review of Fluid Mechanics **37**, 129 (2005)
[8] F. Boyer, E. Guazzelli, O. Pouliquen, Phys. Rev. Lett. **107**, 188301 (2011)
[9] F. Blanc, F. Peters, E. Lemaire, Journal of Rheology **55**, 835 (2011)
[10] GDR-Midi, The European Physical Journal E **14**, 341 (2004)
[11] M. Trulsson, B. Andreotti, P. Claudin, Phys. Rev. Lett. **109**, 118305 (2012)
[12] S. Courrech du Pont, P. Gondret, B. Perrin, M. Rabaud, Physical Review Letters **90**, 044301 (2003)
[13] P.L. Bhatnagar, E.P. Gross, M. Krook, Phys. Rev. **94**, 511 (1954)
[14] C. Bardos, F. Golse, D. Levermore, Journal of Statistical Physics **63**, 323 (1991)
[15] G.R. McNamara, G. Zanetti, Phys. Rev. Lett. **61**, 2332 (1988)
[16] F.J. Higuera, J. Jiménez, EPL (Europhysics Letters) **9**, 663 (1989)
[17] Y.H. Qian, D. D'Humières, P. Lallemand, EPL (Europhysics Letters) **17**, 479 (1992)
[18] D. d'Humières, Progress in Aeronautics and Astronautics **159**, 450+ (1992)
[19] P. Lallemand, L.S. Luo, Phys. Rev. E **61**, 6546 (2000)
[20] A. Mussa, P. Asinari, L.S. Luo, J. Comput. Phys. **228**, 983 (2009)
[21] S. Luding, *Collisions and Contacts between two particles*, in *Physics of dry granular media - NATO ASI Series E350*, edited by H.J. Herrmann, J.P. Hovi, S. Luding (Kluwer Academic Publishers, Dordrecht, 1998), p. 285
[22] P.A. Cundall, O.D.L. Strack, Géotechnique **29**, 47 (1979)
[23] M. Bouzidi, M. Firdaouss, P. Lallemand, Physics of Fluids **13**, 3452 (2001)
[24] A.J.C. Ladd, Journal of Fluid Mechanics **271**, 285 (1994)

Wall roughness and nonlinear velocity profiles in granular shear flows

Paul Schuhmacher[1,*], *Farhang* Radjai[1,2], and *Stéphane* Roux[3]

[1]*LMGC, UMR 5508 CNRS – Université de Montpellier, 163 rue Auguste Broussonnet, 34090 Montpellier, France*
[2]*(MSE)^2, UMI 3466 CNRS-MIT, CEE, Massachusetts Institute of Technology, 77 Massachusetts Avenue, Cambridge 02139, USA*
[3]*LMT, ENS Paris-Saclay / CNRS / Univ. Paris-Saclay, 61 Av. du Président Wilson, 94235 Cachan Cedex, France*

Abstract. Inhomogeneous velocity profiles in granular flows are well known from both experiments and simulations, and considered as a hallmark of nonlocal behavior. By means of extensive contact dynamics simulations, we show that the sigmoidal velocity profiles in 2D flows of rigid disks are controlled by the roughness of driving boundary walls. We find that the velocity profile becomes linear for a critical value of wall roughness up to an exponential decay close to the walls with a characteristic length that does not depend on the flow thickness and rate. We describe the velocity profiles by introducing a state parameter that carries wall perturbation. By assuming that the local shear rate is a linear function of the state parameter, we obtain an analytical expression that fits velocity profiles. In this model, the nonlinear velocity profiles are explained in terms of the effects of wall roughness as boundary condition for the state parameter.

1 Introduction

Many natural processes, such as debris flows [1, 2], and industrial operations in powder technology [3], involve granular flows with various particle properties, boundary geometries and driving mechanisms. Dissipative contact interactions and the absence of long-range cohesive forces lead to a complex dynamics, in which the energy input and stresses from the boundary are balanced by internal dissipation and momentum exchange between particles. In the classical Mohr-Coulomb model, only quasi-static deformations and limit states are considered and the behavior is modeled in terms of a yield surface characterized by an isotropic effective (or internal) friction coefficient $\mu = \sigma_t/\sigma_n$, where σ_t is the shear stress and σ_n is the normal stress [4]. This model was later supplemented in soil mechanics with a *flow rule* by considering dilatancy as a function of a state variable measuring the distance to the critical state (steady isochor flow) and a phenomenological *evolution rule* of the state variable [5].

More recently, the frictional behavior was extended to inertial states, in which both the effective friction coefficient μ and packing friction Φ are described as functions of the inertial number $I = \dot{\gamma} d_s(\rho_s/\sigma_n)^{1/2}$, where $\dot{\gamma}$ is the shear rate, d_s is the mean particle diameter and ρ_s is particle density [6, 7].

A key pending issue is the observation of *flow inhomogeneities*, which can not be predicted or analyzed within the afore-mentioned models of granular flow. The most common example is spontaneous strain localization in shear bands of small thickness [8]. In the Mohr-Coulomb framework, this phenomenon is described as a split of the material into two rigid blocks slipping past each other and explained as material instability [9]. In simple shear flows, it is observed that, depending on the wall roughness and shear rate, the flow may consist of a solidlike region that does not flow or creeps and a fluidlike region that flows, with an interface that may evolve in time [10].

Even when the flow occurs in the whole bulk of the material, the shear rate is often observed to be nonuniform despite uniform stress across the flow [11]. This feature is obviously in contrast with the $\mu(I)$ rheology, in which μ is assumed to be a unique function of I so that a constant value of μ across the flow implies a uniform value of I and thus a uniform shear rate. Similar inhomogeneous velocity profiles have been observed in other soft glassy materials such as emulsions and colloidal suspensions [12, 13]. Recently, several models based on a *nonlocal* formulation of the rheology have been introduced to account for these observations [14–16].

In this paper, we use contact dynamics simulations to investigate the velocity profiles in a systematic way in stress-homogeneous 2D shear flows driven by moving rough walls and confined by a constant pressure perpendicular to the flow. We are interested in the effects of wall roughness and flow thickness. As we shall see, independently of the inertial number, wall roughness controls the velocity profile as long as the flow thickness is below $130d$, where d is mean particle size. Above this thickness, the shear strain is uniform across the flow. We show that this observation can be rationalized in a model accounting for a state parameter as carrier of perturbation introduced by wall roughness.

* e-mail: paul.schuhmacher@umontpellier.fr

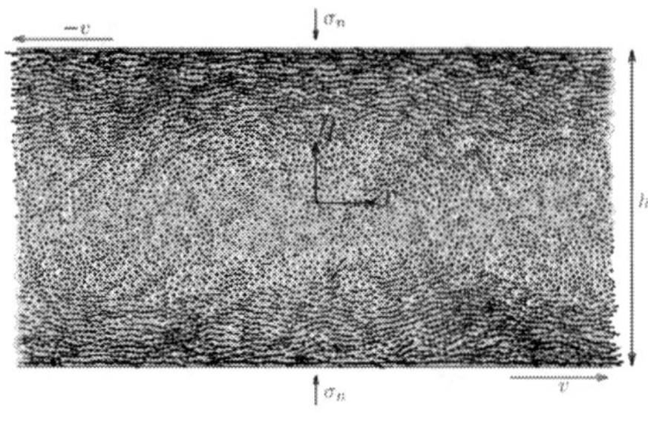

Figure 1. Boundary conditions for simple shear flow. The arrows represent the particle velocities. The boundary walls and particle velocities are periodic in the horizontal direction.

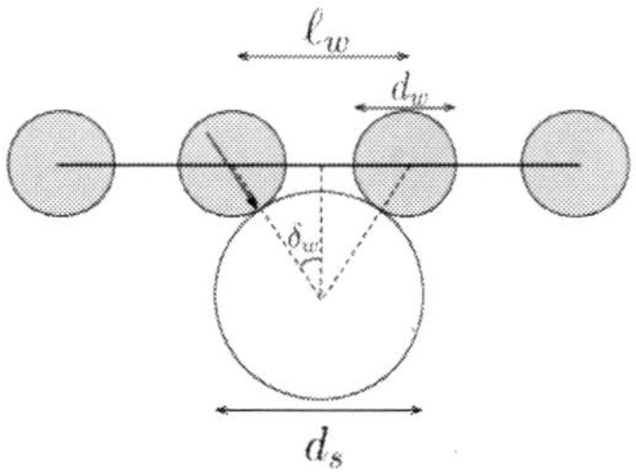

Figure 2. Illustration of wall roughness. The gray particles are attached to the wall while the white particle is mobile and belongs to the flow.

2 Numerical procedures

An assembly of 6000 disks is confined between two rough horizontal walls as shown in Fig. 1. The particle size distribution $P(d)$ is uniform in particle volumes ($P(d) \sim d^{-2}$) with diameters in the range $[d_m, d_M]$ where $d_M = 2d_m$. This particle size polydispersity is necessary in 2D to favor long-range geometrical disorder. The system is periodic along the flow. The walls are made rough by attaching a periodic array of particles ('wall particles') of diameter d_w which can not rotate. The wall roughness is controlled by the size ratio $R = d_w/d_s$ and the distance ℓ_w between wall particles; see Fig. 2. The top and bottom walls are moved horizontally at constant velocities v and $-v$, respectively, in the horizontal direction x. The bottom wall is immobile along the vertical direction y whereas the top wall is free to move vertically and subjected to a constant compressive stress $\sigma_n = \sigma_{yy}$. The nominal inertial number I_e defined from boundary conditions is set to 0.01. In these conditions, we are sure to avoid the intermittent regime at low inertial number in which averages and velocity profiles depend on the integration time [10].

We assume that the particles are very stiff with a Young modulus well above the confining stress σ_n. In this limit, the particles may be considered as fully undeformable and simulated by the contact dynamics method (CDM), which is a discrete element method based on *nonsmooth contact laws* [17, 18]. In this framework, the equations of dynamics for all degrees of freedom are integrated together with the kinematic constraints arising from unilateral contacts and Coulomb friction law. Hence, unlike molecular dynamics method, which employs particle overlaps as strain variable, in CDM the forces are calculated simultaneously with particle velocities via an iterative implicit scheme. In our simulations, we set the friction coefficient between particles to $\mu_s = 0.3$ and the normal and tangential restitution coefficients to zero. In dense granular flows, the flow behavior is insensitive to the values of restitution coefficients since energy is mostly dissipated by frictional slipping events, and the collisions have a multi-contact nature.

3 Velocity profiles

In order to extract average velocity profiles from the simulation data, we map the information carried by the particles to an Eulerian regular grid of 128 × 64 points. At each point $\boldsymbol{r}$ of the grid at a given time t, the local mass and momentum are respectively given by $\rho_m(\boldsymbol{r}, t) = \sum_i m_i \delta(\boldsymbol{r} - \boldsymbol{r_i})$ and $\boldsymbol{p_m}(\boldsymbol{r}, t) = \sum_i m_i \boldsymbol{v_i} \delta(\boldsymbol{r} - \boldsymbol{r_i})$, where $\boldsymbol{r_i}(t)$ is the position of particle i at time t, m_i is its mass and $\boldsymbol{v_i}$ is its velocity. Equivalent continuum fields are defined by convolution with a Gaussian distribution: $\widetilde{\rho}(\boldsymbol{r}) = G * \rho_m$ and $\widetilde{\boldsymbol{p}}_m(\boldsymbol{r}) = G * \boldsymbol{p}_m$, where $G(\boldsymbol{r}) = (1/\pi w^2) \exp(-(\boldsymbol{r} - \boldsymbol{r_i})^2/w^2)$ is a Gaussian function of width $w = 2d$, defining the spatial resolution. The velocity field is then given by $\boldsymbol{v}(\boldsymbol{r}, t) = \widetilde{\boldsymbol{p}}_m/\widetilde{\rho}$, which satisfies the mass and momentum conservation [19]. This convolution leads to smooth derivable fields.

At a given value of the nominal inertial number I_e defined from the nominal shear rate $\dot{\gamma}_e = 2v/h$, the average height h of the flow remains constant. We divided the flow into 50 horizontal layers of thickness $\delta h \simeq d$. The velocity profiles were obtained by averaging over time along each layer. The profiles shown below were calculated in the steady state over a cumulative shear strain higher than 6.

Figure 3 shows velocity profiles for $I_e \simeq 0.01$ and several values of the *roughness angle* δ_w defined by (see Fig. 2):

$$\sin \delta_w = \frac{\ell_w}{d_s(1+R)}. \quad (1)$$

The velocity profiles clearly depend on the roughness angle and they present a sigmoidal shape, in agreement with measurements reported by several authors [11, 20] . But we also see that for a critical value of this angle $\delta_w^c \simeq 0.75$ radians or nearly 45°, the profile is remarkably linear. We also have nearly the same profile for two very different values of δ_w above and below δ_w^c.

The use of roughness angle is motivated by the observation that the velocity profiles for different combinations of size ratio R and ℓ_w are controlled only by the roughness angle, which combines these parameters (not shown here). It is a physically plausible parameter as the effect of roughness depends on how momentum is transmitted from wall particles to the mobile particles of the sheared sample. The amplitude is clearly given by $m_s v$, but its effect depends on its direction. For given d_w, the lowest value of δ_w is $d_w/(d_w + d_s)$, which vanishes only when $d_w/d_s \to 0$.

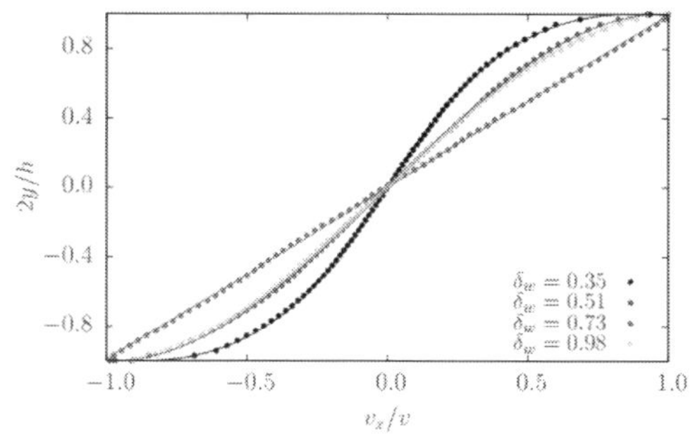

Figure 3. Velocity profiles for four different values of roughness angle δ_w and a constant nominal inertial number $I_e \simeq 0.01$ in the steady state, for a fixed average flow thickness $h/d \approx 50$. Solid lines are fits by equation (6).

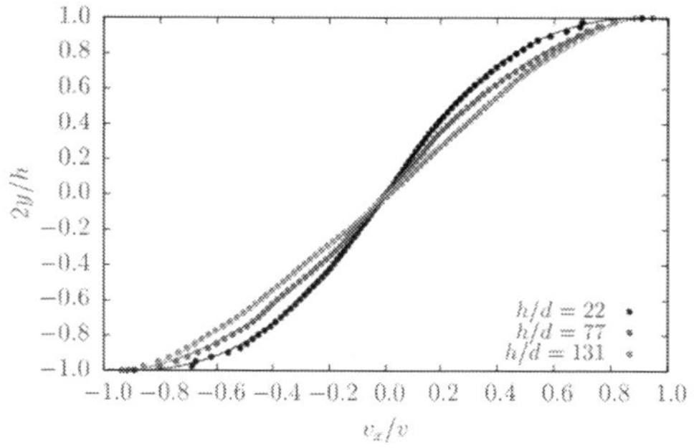

Figure 5. Velocity profiles for three different values of flow thickness in the steady state at constant nominal shear rate $\dot{\gamma}_e$ and wall roughness $\delta_w = 0.44$.

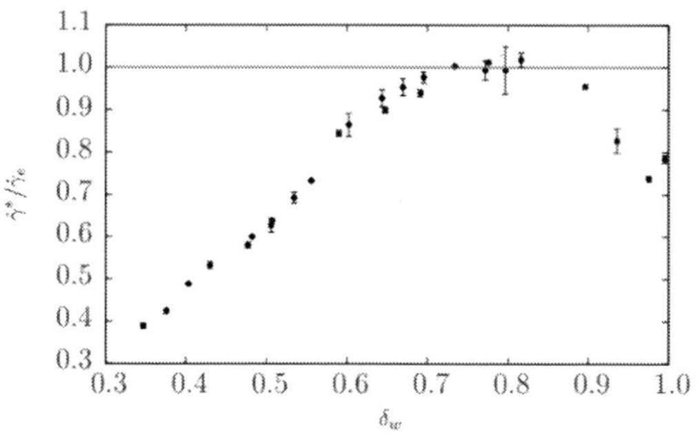

Figure 4. Shear rate at the center of flow normalized by the nominal shear rate as a function of roughness angle.

Figure 4 displays the shear rate $\dot{\gamma}^*$ at the center of the flow normalized by the nominal shear rate $\dot{\gamma}_e$ for a set of simulations in which both ℓ_w and d_w have been changed. At the minimum value of δ_w, the velocity profile has its largest deviation from the linear profile, corresponding to the lowest value of $\dot{\gamma}^*$. As δ_w increases from its minimum value, $\dot{\gamma}^*$ increases and reaches its highest value at δ_w^c. At this point, the velocity profile is linear and the shear rate at the center of flow coincides with the nominal shear rate $\dot{\gamma}_e$. When δ_w is further increased, $\dot{\gamma}^*$ again declines. This non monotonous dependence of the velocity profile on roughness angle reveals the nature of the wall-flow interaction. Indeed, the angle $\delta_w^c \simeq 45°$ corresponds to the major principal direction of the strain-rate tensor or the direction of maximal contraction. Hence, at $\delta_w = \delta_w^c$, the direction along which the wall momenta are on the average transmitted to the granular material is consistent with that of the flow. For larger and smaller values of δ_w, the mismatch between the two directions leads to lower transmission of wall momenta and hence lower shear rate at the center of the flow.

Note that in many experiments and numerical simulations, the walls are made rough by attaching contiguous particles of the same size as flowing particles to the wall. For such a 'natural' roughness, we have $R = 1$ and $\ell_w = d_w$, so that the roughness angle is 30°, which is below δ_w^c. According to our data, this angle leads to a nonlinear profile with $\dot{\gamma}^*/\dot{\gamma}_e \simeq 0.6$, thus the counterintuitive conclusion that 'natural' roughness is not the best way of setting a granular material in motion. Given the value of the critical roughness angle, for contiguous wall particles the best size ratio is $R \simeq 2.3$ in 2D, for which a linear velocity profile is expected.

4 Perturbation model

The picture briefly presented above is all the more plausible that the nonlinear velocity profiles find here their origin in the diffusion of momenta from the boundary layer. Put differently, the nonlinear profiles are a consequence of the perturbation of a linear profile by the wall and its roughness. Hence, for large enough values of flow thickness h, the velocity profile is expected to become linear. Fig. 5 displays the velocity profiles at constant values of wall roughness and inertial number but for three different values of h. We see that the profile is linear for $h/d = 131$. The distance over which the perturbation produced by wall roughness propagates (the exponential falloff of velocity near the walls) is essentially the same in these simulations (note that the distances are normalized by h). Most experimental or numerical studies in plane-strain geometry have been performed for a flow thickness below $40d$. Thick flows ($100d$) were simulated by da Cruz *et al.* but without focusing on boundary effects [7].

In order to capture the wall effect on velocity profiles, let us consider a scalar state parameter χ that carries the effect of wall roughness across the flow. This parameter can be a descriptor of the microstructure or dynamic heterogeneities of the flow. The flow behavior will depend on both the shear rate and this state parameter. This parameter reflects the effects of the two walls so that $\chi(y)$ at a position y is the sum two contributions $f(h/2+y)$ and $f(h/2-y)$ by the same function f but depending on the distances $h/2+y$ and $h/2-y$ from the two walls:

$$\chi(y) = f(h/2+y) + f(h/2-y) \qquad (2)$$

It is easy to see that all even derivatives of χ with respect to y have the same structure as in (2) so that all even derivatives are proportional to $\chi(y)$. The only regular function satisfying this condition is hyperbolic cosine:

$$\chi(y) = \chi^* \cosh(y/\xi) \tag{3}$$

where $\chi^* = \chi(0)$ and ξ is a characteristic length, which can be interpreted as the distance over which wall perturbation propagates. It should be intrinsic, proportional to d and independent of h.

The shear strain field $\dot{\gamma}(y)$ is assumed to depend on the state parameter. As $\dot{\gamma}(y)$ has the same symmetry as $\chi(y)$, it can be expanded to first order as $\dot{\gamma}(y) \simeq B + A\chi(y)$. The constant B can be determined as a function of shear rate $\dot{\gamma}^*$ at the center of flow:

$$\dot{\gamma}(y) \simeq \dot{\gamma}^* + A\chi^* \left(\cosh\frac{y}{\xi} - 1\right) \tag{4}$$

By integrating with respect to y and given the symmetry of the velocity field, we get

$$v_x(y) = \dot{\gamma}^* y + A\chi^* \left\{\xi \sinh\frac{y}{\xi} - y\right\} \tag{5}$$

This functional form fits excellently our data as shown in Fig. 3. The effect of flow thickness h can be analyzed by considering equation (5) at $y = h/2$ where $v_x = v$:

$$v = \dot{\gamma}^* h/2 + A\chi^* \left\{\xi \sinh\frac{h}{2\xi} - h/2\right\} \tag{6}$$

This relation suggests that, for a given driving velocity v, A declines as h increases. Hence wall perturbation declines everywhere in the flow; see Fig. 5. This result is independent of the precise mechanism of wall perturbation or the state variable that carries the perturbation. It only requires that ξ be independent of system size. By analyzing the simulation data for different values of the inertial number (up to $I_e = 0.08$), wall roughness δ_w and flow thickness h, we find that ξ declines only slightly from $7d$ to $6d$ as I_e increases. For a given flow thickness, on the other hand, the amplitude A declines as δ_w increases and nearly vanishes at δ_w^c, then increases again for larger values of δ_w. This is consistent with linear profiles observed at critical roughness.

5 Conclusion

In this paper, we used extensive contact dynamics simulations to analyze the effect of wall roughness on simple shear flows of granular materials in the steady state. It was shown that the velocity profile is linear for a flow thickness of the order of $130d$. For smaller thickness, the velocity profile has a sigmoidal shape that varies with wall roughness. 'Roughness angle' that encodes the geometry of wall roughness and sensitively describes the velocity profile. We showed that, independently of flow thickness, the velocity profile becomes linear at a critical value of roughness angle. Physically, we argued that this angle coincides with the most contractive direction in the flow, suggesting thus that for this critical angle momentum is optimally transmitted from the walls to mobile particles of the flow. We also introduced a simple model based on a state parameter that carries wall perturbation across the flow. Assuming linear perturbation, we found an expression for velocity profiles that accurately fits the data and explains some of the observed features. This framework suggests that a state variable can be used to enrich a purely local rheological law.

We are grateful to Jean-Noel Roux for useful discussions and references that he brought to our attention.

References

[1] R.M. Iverson, Reviews of Geophysics **35**, 245 (1997)
[2] F. Legros, Engineering Geology **63**, 301 (2002)
[3] M.J. Rhodes, *Introduction to particle technology* (John Wiley & Sons, Chichester, 1998)
[4] D. Wood, *Soil behaviour and critical state soil mechanics* (Cambridge University Press, Cambridge, England, 1990)
[5] X.S. Li, Y.F. Dafalias, Géotechnique **50**, 449 (2000)
[6] GDR-MiDi, Eur. Phys. J. E **14**, 341 (2004)
[7] F. da Cruz, S. Emam, M. Prochnow, J.N. Roux, F. Chevoir, Phys Rev E Stat Nonlin Soft Matter Phys **72**, 021309 (2005)
[8] Desrues, J. Lanier, P. Stutz, Eng. Fracture Mechanics **21**, p909 (1983)
[9] F. Darve, F. Laouafa, Mechanics of cohesive-frictional materials **5**, 627 (2000)
[10] Z. Shojaaee, J.N. Roux, F. Chevoir, D.E. Wolf, Phys. Rev. E **86**, 011301 (2012)
[11] T. Miller, *The kinematics of dense granular materials under indefinite plane-shear* (2014)
[12] L. Isa, R. Besseling, W.C.K. Poon, Phys. Rev. Lett. **98**, 198305 (2007)
[13] J. Goyon, A. Colin, G. Ovarlez, A. Ajdari, L. Bocquet, Nature **454**, 84 (2008)
[14] L. Bocquet, A. Colin, A. Ajdari, Phys. Rev. Lett. **103**, 036001 (2009)
[15] T. Miller, P. Rognon, B. Metzger, I. Einav, Phys. Rev. Lett. **111**, 058002 (2013)
[16] M. Bouzid, A. Izzet, M. Trulsson, E. Clément, P. Claudin, B. Andreotti, The European Physical Journal E **38**, 125 (2015)
[17] J.J. Moreau, European J. Mech. A Solids **13**, 93 (1994)
[18] F. Radjai, V. Richefeu, Mechanics of Materials **41**, 715 (2009)
[19] J. Zhang, R.P. Behringer, I. Goldhirsch, Progress of Theoretical Physics Supplement **184**, 16 (2010)
[20] K. Kamrin, G. Koval, Phys. Rev. Lett. **108**, 178301 (2012)

Effect of particle size distribution on 3D packings of spherical particles

Mahdi Taiebat[1,2,*], *Patrick* Mutabaruka[2,**], *Roland* Pellenq[2,***], and *Farhang* Radjai[2,3,****]

[1] *Department of Civil Engineering, University of British Columbia - 6250 Applied Science Lane, Vancouver, BC V6T 1Z4, Canada*
[2] *$\langle MSE\rangle^2$, MIT-CNRS, Massachusetts Institute of Technology, 77 Massachusetts Avenue, Cambridge, MA 02139, USA*
[3] *LMGC, CNRS-University of Montpellier, 163 rue Auguste Broussonnet, 34090 Montpellier, France*

Abstract. We use molecular dynamics simulations of frictionless spherical particles to investigate a class of polydisperse granular materials in which the particle size distribution is uniform in particle volumes. The particles are assembled in a box by uniaxial compaction under the action of a constant stress. Due to the absence of friction and the nature of size distribution, the generated packings have the highest packing fraction at a given size span, defined as the ratio α of the largest size to the smallest size. We find that, up to $\alpha = 5$, the packing fraction is a nearly linear function of α. While the coordination number is nearly constant due to the isostatic nature of the packings, we show that the connectivity of the particles evolves with α. In particular, the proportion of particles with 4 contacts represents the largest proportion of particles mostly of small size. We argue that this particular class of particles occurs as a result of the high stability of local configurations in which a small particle is stuck by four larger particles.

1 Introduction

Granular materials with broad particle size distributions are very common in nature and industrial applications. Broad size distributions often arise as a result of particle breakage [1–4] and strongly influence the space-filling and strength properties of granular materials [5–11]. Despite their importance, most studies of packing structure are devoted either to weakly polydisperse systems [12, 13] or to ideal apollonian or random apollonian constructions characterized by the assumption that there is no lower bound on particle size [7, 8, 14]. Real granular materials may, however, have a variety of size distribution functions and bounds on the particle size. Many geometrical algorithms have been proposed to generate jammed packings of particles [7, 9, 10, 15–18]. In contrast, dynamical simulations of polydisperse packings are quite rare as the generation of statistically representative samples of different size classes in a packing requires increasingly large numbers of particles as the ratio α between the largest and smallest sizes increases [18].

In this paper, we investigate polydisperse packings of spherical particles by means of molecular dynamics simulations up to a ratio $\alpha = 5$. Since we are interested in well-defined granular states that do not depend on the construction method, we consider only frictionless particles, that assemble into an isostatic packing with the largest possible packing fraction up to possible finite size effects. The absence of friction leads also to a nearly isotropic packing. We first describe the system characteristics and numerical procedures. Then, we analyze the geometrical properties of the packings such as packing fraction and connectivity of the particles.

*e-mail: mtaiebat@civil.ubc.ca
**e-mail: pmutabar@mit.edu
***e-mail: pellenq@mit.edu
****e-mail: fradjai@mit.edu

2 Method and system parameters

We consider size distributions with uniform volume fractions in a range $[d_{min}, d_{max}]$ of particle diameters. This means that the normalized distribution $P_d(d)$ of particle diameters varies as d^{-3} so that the volume V_d is the same and equal to the total volume V divided by the total number N_p of particles for each size d. Hence, the normalized distribution is

$$P(d) = \frac{2}{d_{min}^{-2} - d_{max}^{-2}} d^{-3} \quad (1)$$

Since only size ratios matter, we normalize all particle diameters by d_{min}:

$$d_r = d/d_{min} \quad (2)$$

$$\alpha = d_{max}/d_{min} \quad (3)$$

where d_r is the reduced size and α is the 'size ratio'. The distribution of reduced diameters is given by

$$P(d_r) = \frac{2}{1-\alpha^{-2}} d_r^{-3} \quad (4)$$

In practice, for a discrete system with a finite number of particles, this distribution must be discretized by dividing the range $d_r \in [1, \alpha]$ into N_c size classes. Each class i contains $N_{p/c}(i)$ particles. The total number

Table 1. Dispersity properties of the samples considered in this study.

ID	Dispersity†	α	$N_p^{\min\ddagger}$
S1	Monodisperse	1	–
S2	Polydisperse	2	1,581
S3	Polydisperse	3	3,467
S4	Polydisperse	4	6,376
S5	Polydisperse	5	10,548

†For polydisperse samples, uniform distributions in volume is used.
‡Min. number of particles needed for satisfying $N_{p/c}^{\min} = 50$ and $N_c = 10$.

of particles is $N_p = \sum_i N_{p/c}(i)$. Since the system is polydisperse, for a given number N_p of particles, the number N_c of size classes and the values of $N_{p/c}(i)$ must be such that in each class both particle volumes and their numbers are correctly represented. This requires a large enough number of largest particles $N_{p/c}^{min}$ and large enough volume of smallest particles. By choosing $N_{p/c}^{min} = 50$, and given that the volume is the same in all size classes, we get the minimum number of particles N_p^{min} required to construct the packing by summing up the values of $N_{p/c}^{min}$. The values are given in Table 1 where we see that the largest number of particles requires to ensure a minimum of 50 particles in each class is below 11,000 particles for $\alpha = 5$ and $N_c = 10$. Hence, for a representative size distribution, we used 25,000 particles in our simulations for 5 values of α varying from 1 (monodisperse) to 5.

The particles are smooth spheres (friction coefficient $\mu = 0$). They are initially placed randomly inside a rectangular cell and subjected to uniaxial compaction by applying a constant load σ_{zz} on the top wall, the bottom wall imposed to be immobile, and using periodic boundary conditions along x and y directions. The normal contact force f_n between particles is governed by a linear visco-elastic law. Hence, the energy is only dissipated by inelastic collisions between particles and with the top and bottom walls. The compaction is fairly fast at the beginning but considerably slows down in approaching the jamming point. We monitor the total kinetic energy, contact network and packing fraction Φ. When kinetic energy is low compared to $\sigma_{zz}V$, which represents the potential energy of the system, and the packing fraction and contact network are stable, we stop the simulation. The simulation parameters are given in Table 2. An important parameter is the maximum contact deflection δ_{max}. We set the applied pressure to $\sigma_{zz} = 2 \times 10^5$ Pa, so that $\delta_{max}/d_{max} \sim 10^{-3}$. This means that the particles can be considered as nearly undeformable and the calculated values of the packing fraction are not influenced by contact deflections. The 5 samples generated for increasing value of α are denoted as S1, S2, S3, S4 and S5. A snapshot of a portion of the packing S4 is displayed in Fig. 1. The force chains are displayed for a portion of the packings S1 and S4 in Fig. 2.

Table 2. Parameters used in the DEM simulations.

Description	Parameter	Value
Normal stiffness	k_n [N/m]	1×10^6
Normal damping coefficient	c_n [kg/ms]	1500
Density	ρ [kg/m^3]	2500

Figure 1. Snapshot of the S4 packing. Gray level is proportional to particle size.

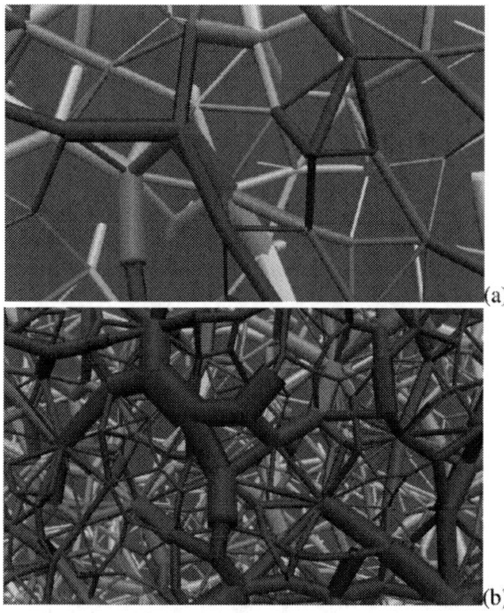

Figure 2. Force chains in packings S1 (a) and S4 (b). Line thickness is proportional to normal force.

3 Packing fraction

The packing fraction Φ is calculated by simply dividing the total volume of the particles by the total volume of the packing. Fig. 3 shows Φ as a function of α. We see that for considered range of size polydispersity, Φ is a nearly linear function of α. Consistently with the measured values of the random close packing fraction, we get $\Phi \simeq 0.643$ in the monodisperse case ($\alpha = 1$) [19–22]. For $\alpha = 5$, Φ is as large as 0.71. Obviously, the linear relationship can not be extended to much larger values of α as the packing

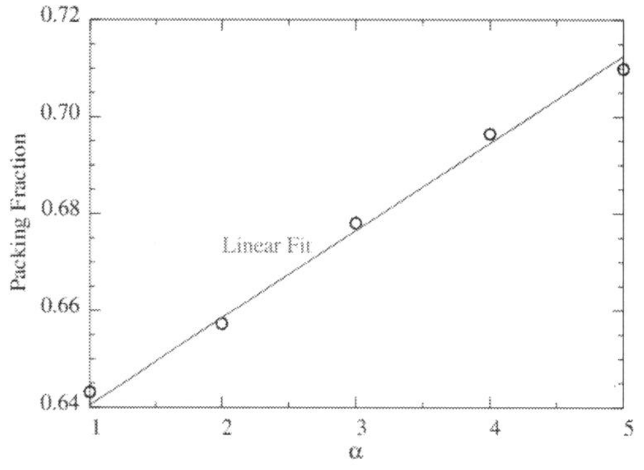

Figure 3. Packing fraction as a function of size ratio α for the packings.

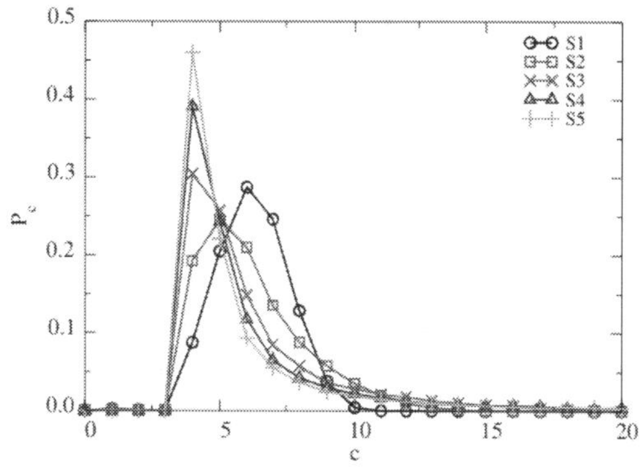

Figure 4. The connectivity $P(c)$ of the contact network for the packings S1–S5.

fraction can not exceed 1. It can thus be the beginning of an exponential relation, for example, which tends to 1 as $\alpha \to \infty$. Note also that for a size ratio $\alpha \geq 4$, a sphere of size $d_r = 1$ can be fit inside a pore between four spheres of size $d_r = \alpha$. But a close examination of the samples reveals that there is almost no such floating particles in samples S4 and S5. This is mainly because, given the much lower proportion of large particles, the probability of a configuration with four such touching particles is quite low. But, it is expected that for larger values of α, the number of small floating particles increases, as it was also shown in 2D packings of disks [18].

4 Contact networks

The contact network can be analyzed using various descriptors. We analyze only partially here the connectivity of the particles described by the proportion $P_{dc}(d_r, c)$ of particles of reduced size d_r and having exactly c contacts. The connectivity of the particles with no distinction between particle sizes, is $P_c(c) = \sum_{d_r} P_{dc}(d_r, c)$, which is simply the proportion of particle with c contacts. The coordination number is given by $Z = \sum_c P_c(c)$. In the absence of friction, our packings are isostatic and thus Z is nearly equal to 6 in all our packings. But the connectivity depends on α. Fig. 4 displays the distributions $P_c(c)$ for all values of α. In S1, we observe a peak on $c = 6$, but we have a large number of particles with $c = 4$, $c = 5$ and $c = 7$ contacts. We also observe that $P_c(3) \simeq 0$ for all values of α. Indeed, for $c = 3$, force balance is possible only if the three force vectors are on the same plane, which is quite improbable. In the same way, force balance for $c = 2$ implies equal forces on diametrically opposite points, that is a configuration of zero chance to occur.

As α increases, we observe an decreasing number of particles with $c > 6$. Interestingly, the peak value of P_c tends to $c = 4$ and $P_c(4)$ increases from $\alpha = 3$ to $\alpha = 5$. For $\alpha = 5$, the proportion of particles with $c = 4$ contacts is almost half the total number of contacts. This observation indicates that the configurations in which a particle is equilibrated by 4 neighboring contact particles

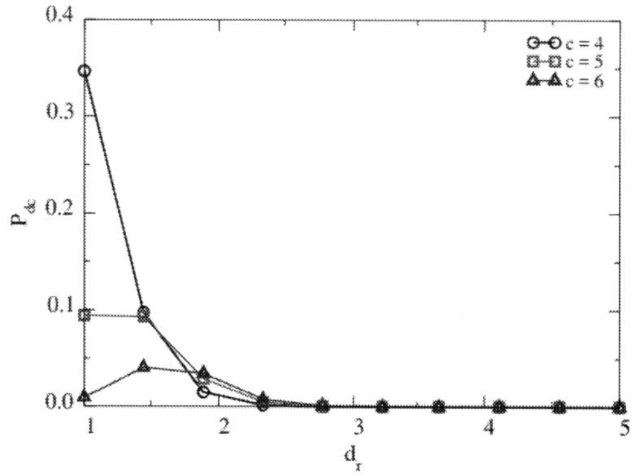

Figure 5. The proportions of particles with $c = 4$, 5 and 6 contacts as a function of their reduced size d_r for $\alpha = 5$.

are *stable* in the mechanical sense. But since the peak value increases with α, it is expected that all such particles are of small size. This is indeed the case, as we see in Fig. 5 where for $\alpha = 5$ we have plotted the distribution $P_{dc}(d_r, c = 4)$, representing the proportion of particles with 4, 5 and 6 contacts as a function of their reduced size d_r. The largest proportion (above 30%) of all particles is for $c = 4$ and for small particles ($d_r = 1$).

The presence of such a high number of small particles stuck by four larger neighboring particles can be understood by the following argument. When a small particle (the one that is stuck between four particles) is in contact with three larger particles, the orientations of the normals of the three contacts (that are also the orientations of the forces in the absence of friction) define a broader range than when the particle is of the same size as its neighbors; see Fig. 6. For force balance, the force vector of the forth contact should be in this range but in the opposite direction. In the limit where the diameter of the central particle is smaller than 1/4 the large particle diameters, it can perfectly fit into the pore. The evolutions of $P_c(4)$, $P_c(5)$ and $P_c(6)$ are displayed as a function of α in Fig. 7. We see that $P_c(5)$ is nearly constant and represents

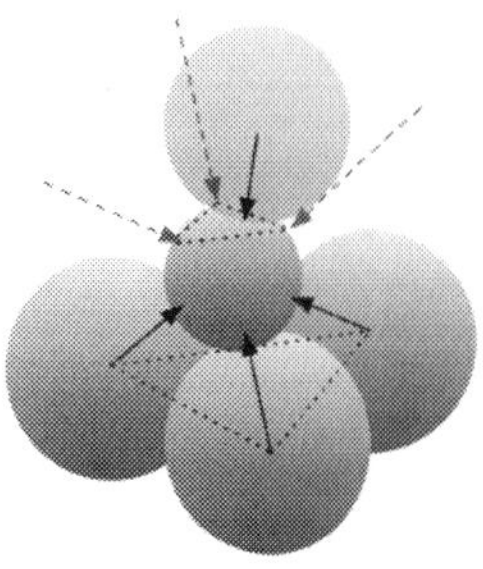

Figure 6. A small particle stuck by four larger particles. The range of possible positions of the top particle for the equilibrium of the central particles depends on the size ratio.

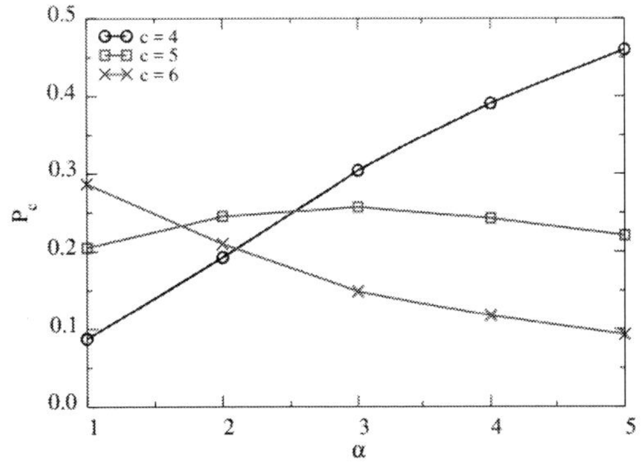

Figure 7. The connectivity of the contact network for the packings S1–S5, illustrating the variation of $P_c(4)$, $P_c(5)$, and $P_c(6)$.

20% of particles whereas $P_c(4)$ increases at the expense of $P_c(6)$ and particles of higher number of contacts.

5 Conclusions

In this paper, we studied polydisperse packings of frictionless spherical particles by means of molecular dynamics simulations. We showed that, for a uniform size distribution in particle volume fractions, the packing fraction is a nearly linear function of size ratio between largest and smallest particles up to a ratio 5. While the coordination number is nearly 6 due to the isostatic nature of the packings, the structure of the contact network evolves with size ratio. In particular, we evidenced the existence of a large proportion of particles with exactly four contacts for size ratios above 3. The force distributions and their correlations with the microstructure are other interesting features of polydisperse packings that we will investigated in a future paper.

Acknowledgements

M.T. warmly thanks the support and hospitality of the joint MIT-CNRS laboratory during his sabbatical stay in MIT.

References

[1] G. McDowell, M. Bolton, D. Robertson, Journal of the Mechanics and Physics of Solids **44**, 2079 (1996)
[2] T. Fukumoto, Soils and foundations **32**, 26 (1992)
[3] J.K. Mitchell, K. Soga, *Fundamentals of Soil Behavior, third edition* (Wiley, 2005)
[4] V. Bandini, M.R. Coop, Soils and Foundations **51**, 591 (2011)
[5] F. de Larrard, *Concrete Mixture Proportioning* (E & FN Spon, 1999)
[6] T. Aste, D. Weaire., *The Pursuit of Perfect Packing* (Institute of Physics Publishing, Bristol and Philadelphia, 2000)
[7] P.S. Dodds, J.S. Weitz, Phys. Rev. E **65**, 056108 (2002)
[8] H.J. Herrmann, R. Mahmoodi Baram, M. Wackenhut, Physica A: Statistical Mechanics and its Applications **330**, 77 (2003)
[9] O.U. Uche, F.H. Stillinger, S. Torquato, Physica A: Statistical Mechanics and its Applications **342**, 428 (2004)
[10] A. Donev, F.H. Stillinger, S. Torquato, Phys. Rev. Lett. **96**, 225502 (2006)
[11] C. Voivret, F. Radjai, J.Y. Delenne, M.S.E. Youssoufi, Phys. Rev. Lett. **102**, 178001 (2009)
[12] M. Ammi, D. Bideau, J.P. Troadec, J. Phys. D: Appl. Phys. **20**, 424 (1987)
[13] A. Gervois, D. Bideau, *Some geometrical properties of two-dimensional hard disk packings*, in *Disorder and Granular Media*, edited by D. Bideau (Elsevier, Amsterdam, 1993)
[14] T. Aste, Phys. Rev. E **53**, 2571 (1996)
[15] R. Jullien, P. Jund, D. Caprion, D. Quitmann, Phys. Rev. E **54**, 6035 (1996)
[16] R. Jullien, P. Meakin, Colloids and Surfaces A: Physicochemical and Engineering Aspects **165**, 405 (2000)
[17] A.R. Kansal, T.M. Truskett, S. Torquato, JOURNAL OF CHEMICAL PHYSICS **113,number 12**, 4844 (2000)
[18] C. Voivret, F. Radjai, J.Y. Delenne, M.S. El Youssoufi, Physical Review E **76**, 021301 (2007)
[19] W.S. Jodrey, E.M. Tory, Phys. Rev. A **32**, 2347 (1985)
[20] P.E. Peyneau, J.N. Roux, Physical Review E **78** (2008)
[21] P.E. Peyneau, J.N. Roux, Physical Review E **78** (2008)
[22] J.N. Roux, Phys. Rev. Lett. **61**, 6802 (2000)
[23] F. Radjai, Comptes Rendus Physique **16**, 3 (2015)

Landsliding generated by thermomechanical interactions between rock columns and wedging blocks: Study case from the Larzac Plateau (Southern France)

Alfredo Taboada[1*], *Hadrien* Ginouvez[1**], *Mathieu* Renouf[2], *Pierre* Azemard[3]

[1]Laboratoire Géosciences Montpellier, UMR 5243, Université de Montpellier, France
[2]Laboratoire de Mécanique et Génie Civil, UMR5508, Université de Montpellier, France
[3]CEREMA, Direction Territoriale Méditerranée, Laboratoire d'Aix-en-Provence, France

Abstract. The Larzac Plateau is delimited by vertical cliffs whose geometry is controlled by vertical joints. Cliff's erosion involves landslides initiated by incremental enlargement of joints that progressively detach rock columns at very low velocities (1.2 mm/yr). We find that enlargement of joints is linked to intraseasonal thermal cycles ranging between 2-15 days in relation with dilation/contraction of rock blocks trapped inside the joints. The mechanism involves two successive stages in which blocks create a wedging and a ratcheting effect on the rock column. Wedging is associated with compressional forces acting on the rock column, resulting from temperature increase and dilation of the shallow rocks. Ratcheting is associated with downward displacement of blocks by gravity to a new equilibrium position, resulting from temperature decrease and contraction of shallow rocks. The displacement vector in a thermal cycle is split into a plastic and a thermal component; plastic displacements range between 10 – 200 μm according to the seasons, and are absorbed along a shear plane dipping ~40° beneath the rock column: they are largest during autumn and winter, minor during spring and negligible in summer. This deformation mechanism is termed thermomechanical creep as permanent deformations are associated to mechanical forces induced by short-term thermal cycles.

1 Introduction

The role of temperature fluctuations on the initiation of landslides is a controversial subject. Although thermal effects are normally considered to be secondary, recent studies suggest that temperature variations may induce permanent plastic deformations within jointed rock slopes, which may trigger landslide processes [1-3]. In particular, some authors suggest that temperature variations during annual cycles may induce shearing beneath rock columns through a mechanism combining wedging and ratcheting [4-6]. The wedging-ratcheting mechanism (which will be termed thermomechanical creep) is associated to cycles of dilation and contraction of blocks that are trapped inside the major vertical joint located between the column and the upslope cliff.

We propose a simplified theoretical model for thermomechanical creep, based on displacement and temperature measurements on a rock column, and on many field observations in the Larzac Plateau (Southern France). In our model, short-term (and not annual) thermal cycles linked to varying climatic conditions are at the origin of incremental plastic deformations beneath the rock column. Moreover, thermal cycles differ according to the seasons, inducing different plastic deformation patterns that are analysed in detail.

The paper is organized as follows: section 2 presents the geological setting of our field observations, section 3 presents data form the instrumented rock column, section 4 explains the theoretical model based on measurements and observations, and section 5 discusses the results and perspectives of this research topic.

2 Geological and Morphological Setting

The Larzac Plateau is primarily composed of a thick sequence of Jurassic limestone and marl layers whose thickness ranges between 1500 – 2000 m; these rocks were uplifted without folding by the combined effect of large-scale mantellic processes (asthenospheric diapir) and two compressional tectonic phases (the Alpine and Pyrenean).

The plateau is dominated by thick and massive capping layers of karstified limestones and dolostones incised by a few rivers, forming river valleys and gorges whose morphologic profile is controlled by the lithology of the exposed layers: we observe rock cliffs of decametric scale along massive competent rocks (10-100 m), and low to intermediate slopes within marl-limestone weaker layers.

The geometry of rock cliffs is controlled by preexisting vertical joint sets. The erosion mechanisms observed along the valleys and gorges mainly involve gravitational processes such as rock falls, rock topples and rotational landslides of varying scales. The hillslopes are covered by talus deposits and soils overlying the bedrock.

3 Instrumented Rock Column (Southern Margin of the Larzac Plateau)

To study the influence of temperature on the stability of rock columns we have analyzed data from a rock column located in the La Lergue river valley along the cliffs that limit the upstream karstic cirque (Pas-de-l'Escalette,

*taboada@gm.univ-montp2.fr; **hadrien.ginouvez@sfr.fr

Figure 1) [7-8]. The rock column is composed of massive dolostones (Bathonian age) underlaid by marl-carbonate weak layers with low shear strength (Infra-Bathonian age), enhancing the formation of landslide processes. The unstable column has a parallelepipedal shape with an approximate volume of 40 000 m^3 (the height, width and length are 50 m, 10 m, and 80 m, respectively). The column is detached from the massif by a vertical open joint striking N130°E. The opening of the joint ranges from 1.5 m at the foot of the column to 2m at the top. The joint has a winding geometry with rock blocks that are blocked at different heights (Figure 1). The marl-limestone layers at the foot of the column are overlaid by a plurimetric deposit of colluvium.

Fig. 1. Instrumented rock column showing trapped blocks within the vertical joint (Pas-de-l'Escalette, La Lergue valley).

3.1 Kinematics of the Rock Column and Intraseasonal Temperature Cycles

The horizontal and vertical displacements of the rock column relative to the uphill cliff were measured during a 30 months period, using high-precision displacement sensors (0.01 mm). The ambient temperature near the sensors was also measured to analyse possible correlations with displacements. The rock column is composed of at least 3 parts delimited by fractures, with different kinematics. The measurements from a couple of sensors located at the bottom of the open joint indicates that the central (NW) part of the column exhibits non-reversible intraseasonal displacements relative to the uphill cliff (Figure 2). These displacements are mainly observed during the autumn and winter, and to a lesser degree during the spring: cumulated horizontal and vertical displacements during an annual cycle indicate that the column moves at a rate of ≈1,2 mm toward the free boundary and dips downward. The displacements during the summer are negligible.

The trajectory of the rock column relative to the uphill cliff indicates that there is a correlation between intraseasonal temperature oscillations and displacement cycles (Figure 3). Typical non-reversible displacement cycles indicate that the joint opens as the average temperature decreases, while it tends to close as the average temperature increases (colors represent moving averages of temperature for a 48 hours interval). Note that the displacement cycle is non-reversible: the opening phase is subhorizontal while the closing phase follows a linear/concave trajectory. The average net displacement of the column during a displacement cycle is oriented toward the free boundary with a dip angle that varies according to the seasons: during the winter the displacement vector dips between 30-50°, whereas during the autumn it dips between 0-10°. The duration of thermal cycles is between 2-15 days, while average temperature variations are roughly between 1-5°C. Thus, thermal cycles are consistent with climatic cycles observed in Southern France (e.g., 'Cévenol' storm episodes during the autumn).

4. Simplified Theoretical Model

We present a simplified model explaining the correlations between short-term intraseasonal displacement and thermal cycles, based on: a) the analysis of data from the instrumented column presented in the previous section ; and b) geological and morphological observations of many rock columns in the Larzac Plateau.

The model supposes that incremental enlargement of major joints is linked to intraseasonal temperature fluctuations in relation with movement of rock blocks that are blocked inside the joints. The mechanism involves two successive stages in which rock blocks create a wedging and a ratcheting effect on the rock column.

Wedging is associated with compressional forces applied on the rock column, resulting from temperature increase and dilation of shallow rock material in the uphill main cliff, the blocks and the emerged part of the rock column. On the contrary, the base of the rock column, which is often buried beneath a thick layer of talus deposits, remains at relatively constant temperature and exhibits negligible volume changes. Thermal dilation tends to close the joint inducing compressional forces in the blocks and an active lateral pressure on the rock column; if compressional forces are large enough the rock column moves outward an incremental (submillimetric) distance along a basal shear surface (e.g. a landslide) while the vertical joint is enlarged.
Ratcheting is associated with the downward displacement of blocks by gravity to a new equilibrium position, resulting from temperature decrease and contraction of the cliff – blocks – rock column system. Downward block movement sets the initial condition for the next wedging cycle as the temperature rises.

Three distinct types of thermal cycles with periods ranging between 2-15 days can be identified according to the seasons, each of which is characterized by well identified deformation cycles (Figure 4). These short-term cycles occur during three long-term thermal phases that are easily identified in the annual temperature curve (we suppose that the long-term temperature during the year has a first-order sinusoidal shape): a) An automn

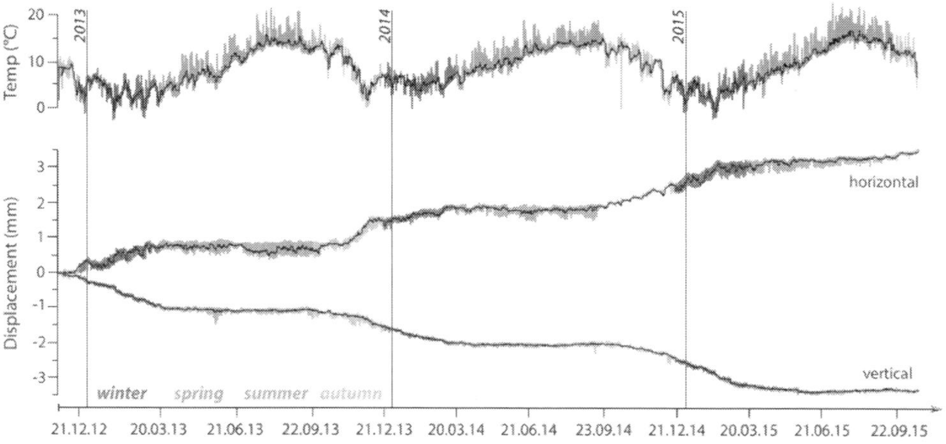

Fig. 2. Temperature and displacement data from the rock column (colored and black curves indicate continuous and moving averages over 48h) (positive horizontal displacement is toward the free boundary).

cooling phase extending from the end of summer to the end of autumn; b) A winter thermal plateau extending from the end of autumn to the end of winter; and c) A spring heating phase extending from the end of winter to the end of spring.

Note that wedging and ratcheting result from thermal fluctuations around the long-term temperatures observed during the year.

For each short-term thermal cycle the displacement vector δ is decomposed into two components, namely, a) a reversible thermal displacement δT linked to the net variation in temperature (Tc - Ta), without considering the wedging effect; and b) a permanent displacement δp linked to plastic deformation at constant temperature (Figure 4): ***Autumn thermal cycles (cooling phase) :*** During the autumn the long-term average temperature decreases and shallow rocks cool down progressively. Short-term thermal cycles display a high amplitude thermal decrease followed by a low amplitude thermal increase. The thermal displacement δT is oriented toward the right (i.e., the fracture opens as temperature decreases); δT is defined between points *a* and *a'* along the trajectory, such that $Ta' = Tc$. The plastic displacement δp joins points *a'* and *c* located along the trajectory and dips downward toward the free boundary.

Winter thermal cycles (thermal plateau) : During the winter the average long-term temperature remains constant, whereas short-term cooling and heating fluctuations of high amplitude are observed. Thus, the thermal displacement δT is negligible and the total displacement is equivalent to the plastic displacement ($\delta = \delta p$), which joins points *a* and *c* and dips at a moderate to steep angle toward the free boundary.

Spring thermal cycles (heating phase) : During the spring the long-term average temperature increases and shallow rocks heat up progressively. Short-term thermal cycles display a low amplitude thermal decrease followed by a high amplitude thermal increase. The thermal displacement δT is oriented toward the left

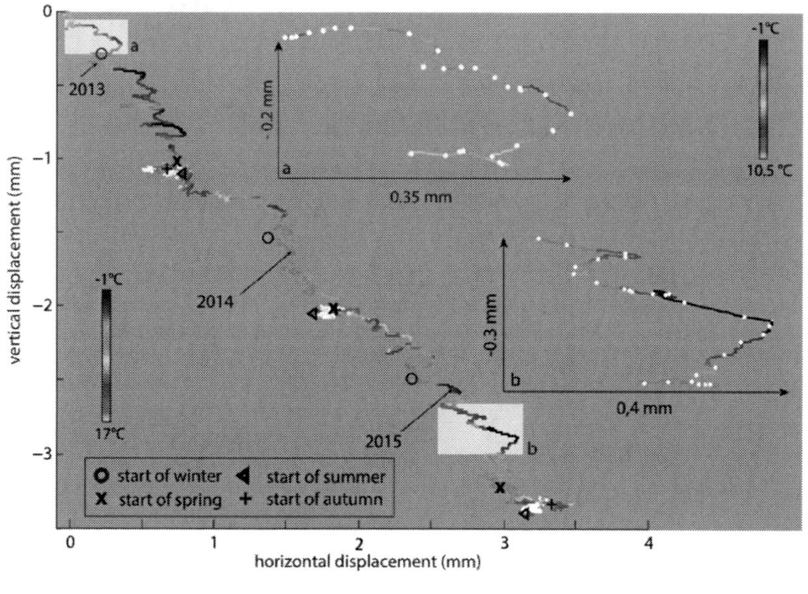

Fig. 3. Trajectory of rock column during a 30 months period (moving averages over 48h); colours indicate temperatures, (lower-left scale) (the fracture opens rightward). The two zooms (a, b) indicate intraseasonal displacement cycles (temperature = upper-right scale, white circles = daily intervals).

(i.e., the fracture closes as temperature increases). The plastic displacement δp dips downward toward the free boundary, yet its magnitude is much smaller than in the winter cycle. This results from the reduction of the ratcheting effect, which attenuates the wedging effect.

The average displacement vector over an annual period roughly dips 40° toward the free boundary and is equivalent to the plastic displacement (i.e., the cumulative thermal displacement during an annual period is negligible). This orientation can be interpreted as resulting from shearing along a basal plane located beneath the rock column.

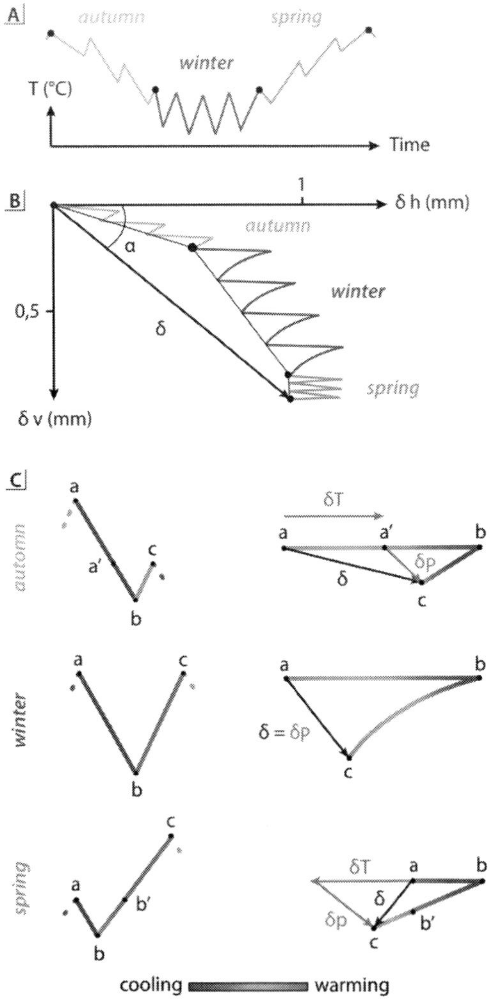

Fig. 4. Schemes illustrating temperature fluctuations and the trajectory of the rock column during an annual cycle (A, B). Displacement of the column toward the free boundary is rightward. Short-term thermal and displacement cycles of the rock column according to seasons (C).

5. Discussion

The analysis of displacement and temperature measurements shows that intraseasonal short-term deformation cycles of wedging and ratcheting induce incremental downslope movements of rock columns. These deformation cycles as a whole will be termed **thermomechanical creep** as permanent deformations are associated to mechanical forces induced by short-term thermal cycles. As for soil creep, the finite deformation results from the sum of incremental non-reversible deformation cycles in which materials expand and contract. The balance between applied shear forces (including the gravity component) and frictional strength of rocks beneath the column controls the rate of thermomechanical creep.

We are currently performing thermomechanical simulations in order to study the influence of parameters such as the duration of short-term climatic cycles, the size of the columns relative to the blocks, and the geometry of the upslope joint, on the rate of thermomechanical creep. Many questions remain open such as the coupling of successive deformation cycles over the seasons.

Acknowledgements

This research project was financed by the programme Tellus-ALEAS, INSU CNRS (2017).

References

1. Y. Gunzburger, V. Merrien-Soukatchoffa, Y. Guglielmi, *Int. Journal of Rock Mech. Mining Sc.* **42**, 331–349, (2005).
2. E. do A. Vargas Jr, R.Q. Velloso, L.E. Erik Chavez, L. Gusmao, C. Palmeiro do Amaral, *Rock Mech Rock Eng* **46** : 123–134, (2013).
3. B. @Collins, G. Stock, *Nature Geoscience*, **9**, 395-400, (2016).
4. D. Bakun-Mazor, Y.H. Hatzor, S.D. Glaser, C. Santamarina, *Int. J. Rock Mech. Mining Sci.* **61**, 196–211, (2013).
5. C .Pasten, M. Garcia, D.D. Cortes, *Géotech. Lett.*, **5**, 186-190 (2015).
6. H. Ginouvez, Géomorphologie et dynamique de chandelles rocheuses au Pas-de-l'Escalette, geoscienstist master report, Université Montpellier, 30pp, (2015).
7. P. Azemard et al., 2009. A75 Tunnel de l'Escalette : Inspection des colonnes rocheuses dominant la tête Sud. CETE méditerranée, laboratoire régional d'Aix-en-Provence, service Géotechnique et Mécanique des Sols.
8. N. Berthoz, R. Batteux, L. Valette., 2014. A75 Lodève – Chandelles de l'Escalette : Surveillance 2014. CETE méditerranée, laboratoire régional d'Aix-en-Provence, service Géotechnique et Mécanique des Sols.

Modeling root growth in granular soils: effects of root stiffness and packing fraction

Mahmoud Fakih[1,2,*], *Jean Yves* Delenne[3,**], *Farhang* Radjai[2,4,***], and *Thierry* Fourcaud[1,****]

[1] *CIRAD, UMR AMAP, TA A51/PS2, 34398 Montpellier, France*
[2] *LMGC, CNRS - University of Montpellier, 163 rue Auguste Broussonnet, 34090 Montpellier, France*
[3] *IATE, UMR1208 INRA - CIRAD - University of Montpellier - SupAgro, 34060 Montpellier, France*
[4] *$\langle MSE \rangle^2$, UMI CNRS-MIT, Massachusetts Institute of Technology, 77 Massachusetts Avenue, Cambridge 02139, USA*

Abstract. We use molecular dynamics simulations to investigate the effects of root bending stiffness and packing fraction on the path followed by a growing root in 2D packings of grains representing a soil. The root is modeled as a chain of elements that can grow in length and change their direction depending on the forces exerted by soil grains. We show that the root shape is mainly controlled by the bending stiffness of its apex. At low stiffness, the root randomly explores the pore space whereas at sufficiently high stiffness, of the order of soil hardness multiplied by mean grain size, the root follows a straight path across the soil. Between these two limits, the root shape can be characterized by the standard deviation of its re-directions at the scale of soil grains. We find that this shape parameter varies as a power-law function of the normalized bending stiffness.

1 Introduction

Besides many other factors such as developmental instability on the cellular level [1], the mechanical interactions between a growing root and its surrounding soil can have major impact on the growth variability and plant biomass production [2–6]. In particular, the inhomogeneous structure of soil at the particle scale and broad distributions of contact forces can significantly affect root growth trajectories [7–11]. Continuum representation of soils and roots has been used for describing soil stability [12–15]. In such models, the effect of soil is incorporated mainly through its average resistance to the root ingrowth and the interactions between soil grains and roots are not considered. There are, on the other hand, functional-structural plant models [16] that describe plant growth and incorporate biological and environmental factors but do not account for the soil strength. Many experiments and observations at the grain scale have shown that grain configuration and local cracks or obstacles have crucial effect on root growth [17–23].

In this paper, we report on discrete modeling of soil-root interactions and its application to a single root growing in a granular soil. Our model accounts for the elastic behavior of the elongation zone of the root and for stiffening at higher zones where root maturation occurs. We consider granular packings prepared by means of different procedures in order to evaluate the effects of packing fraction. We present here only the results regarding root trajectories. We first briefly describe the numerical model. Then, we analyze root trajectories and the effects of root and soil parameters. We conclude with a short summary of the results and scopes of this work.

*e-mail: mahmoud.fakih@umontpellier.fr
**e-mail: jean-yves.delenne@supagro.inra.fr
***e-mail: fradjai@mit.edu
****e-mail: thierry.fourcaud@cirad.fr

2 Root-soil model

The root is modeled as a chain of connected elongated elements; see Fig. 1. The first root element is fixed on the top of soil sample. Starting from the bottom end of this element, a new circular element is added and elongated by moving the circle along a direction at a given growth rate until a length ℓ_r is reached. Then, a new element is added by linking the top end of the second element to the bottom end point of the first element, and the same process is iteratively applied to each new element. Two successive elements are linked together via linear normal and angular springs defined by their stiffnesses. For angular stiffness, the reference angle α_0 is 180 degrees so that when no forces or torques are exerted on the root elements, they simply align themselves along the direction of the first element. All angles α are allowed and the torque is given by

$$M = K_b(\alpha - \alpha_0), \tag{1}$$

where K_b is the bending stiffness. The grain-grain and grain-root contacts are governed by elastic forces acting along normal and tangential directions, as well as by dry friction law with coefficients μ_{pp} and μ_{pr}, respectively. Viscous terms are also added to control energy restitution.

We use molecular dynamics method with velocity-Verlet time-stepping scheme to integrate the equations of

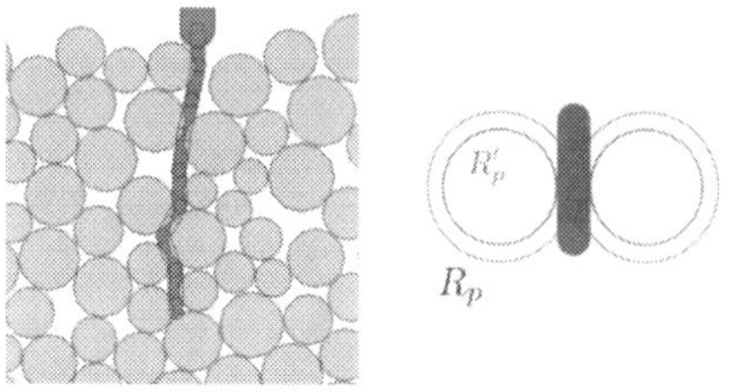

Figure 1. A growing root inside a 2D packing of disks (*left*) and the definition of the inner and outer grain radii (*right*); see text.

motion of both soil grains and root elements [24, 25]. The simulation box is partitioned into cells to keep track of the neighborhood lists of grains and root elements. Soil samples were prepared by gravitational deposit of grains in the simulation box. The grain diameter distribution is uniform in volume fractions between a minimum d_{min} and a maximum d_{max} (these are outer grain diameters; see below) [26].

An important feature of roots is that the bending stiffness is not uniform along the root. The cap and elongation zones are soft, allowing the root to explore the available pore space. Above elongation zone, the bending stiffness increases with time as a result of the accretion of new material, change in internal material properties and radial growth. We do not take radial growth into account but we attribute a high bending stiffness $10^3 K_b$ to the root elements that do not grow anymore, where K_b is the bending stiffness at the joint between the growing root element and the one to which it is attached. In other words, the root cap can more easily change its direction when it touches soil grains as compared to other root elements.

Another feature is the reference angle α_0, which is initially set to 180 degrees, but every time the bending stiffness at a joint is increased, the angle α_0 is changed to its current value at that joint. In this way, the root can deform only elastically and its shape reflects the path followed by its growing cap element, which keeps its value of $\alpha_0 = 180$ degrees as long as it continues to grow. It is frozen to its current value when it stops growing and a new root element is created. We see that this model combines two aspects that are important for a growing root: 1) The root growth is an outcome of both a constant growth rate and interactions with soil grains, and 2) The root shape fully reflects ('memorizes') the growth dynamics.

One issue in 2D modeling is that the pore space is not connected. In 3D, the pore space is connected throughout the soil and the roots can spread inside the pore space. To fix this problem in 2D, we use two different grain diameters: 1) An outer diameter $2R_p$ used for grain-grain contacts, and 2) An inner diameter $2R'_p$ for grain-root contacts, as illustrated in Fig. 1. Hence, the root elements can cross the gap, which has a thickness $d_g = 2(R_p - R'_p)$ when two grains touch. A crucial parameter for root growth is thus the ratio $\beta = 2R_r/d_g$ of root diameter $2R_r$ to gap thickness. If $\beta < 1$, the root can cross the gap without interacting with the two grains whereas for $\beta > 1$ the root will be unable to cross the gap without pushing the two grains apart.

3 Shape parameter

The main parameters of our system are the root stiffness parameters, growth rate, maximum length ℓ_r of root elements, root diameter R_r, outer and inner grain diameters (defining the gap thickness), limit outer grain diameters d_{min} and d_{max} (the distribution being uniform in grain volume fractions), mean grain diameter $\langle d \rangle$, friction coefficients μ_{pp} and μ_{pr}, packing fraction and time step. For the roots, the normal stiffness is set to high values. They are less important than bending stiffness K_b that closely controls the root shape. We set $\mu_{pp} = 0.4$ and $\mu_{pr} = 0.4$. These can change depending on the nature of the roots and grains. The time step and growth rate are small compared to the grain relaxation time under gravity. We performed extensive simulations for different values of the parameters. The packings are composed of 1500 grains and prepared by gravitational deposit inside a rectangular box. Here, we consider only the effects of β and packing fraction Φ.

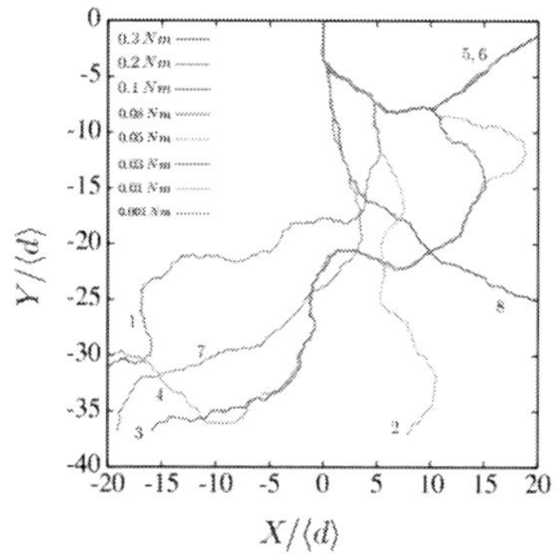

Figure 2. Examples of root trajectories for several values of bending stiffness K_b. The numbers are the values of K_b in increasing order.

Figure 2 displays several examples of root trajectories for different values of K_b and for $\beta = 1.54$, $d_{min} = 2$ mm, $d_{max} = 4$ mm. The positions are normalized by mean grain diameter $\langle d \rangle$. For high values of K_b ($K_b > 2$ Nm), the root grows downwards by dislodging all grains on its way and is only slightly deviated from the vertical due to elastic deformation. In this limit, the root behaves as a penetrometer pushed vertically downwards into the granular material. For low values of K_b, horizontal and vertical excursions are observed in the trajectories of the root, and the root becomes more and more "noisy" at increasingly lower values of K_b. The root cap, due to its lower bending stiffness, follows a rectilinear motion inside a pore. It can change its direction only if it meets the surface of a grain. Therefore, the length over which the root re-directions are significant is the mean grain diameter. We will therefore quantify

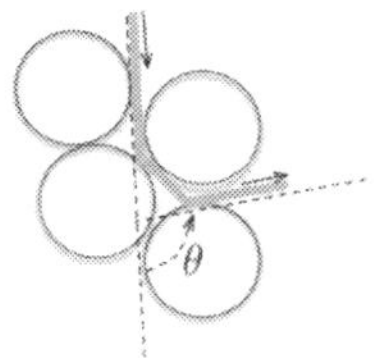

Figure 3. Definition of re-direction angle θ of a root.

these re-directions θ of the root by the total change of direction of root elements over a distance equal to $\langle d\rangle$; see Fig. 3.

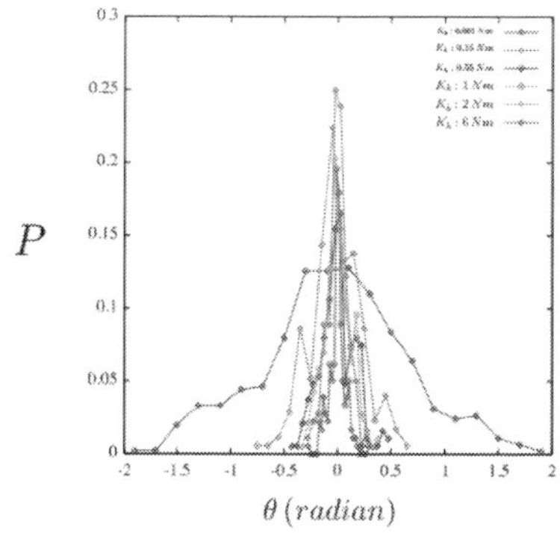

Figure 4. Distributions of the re-direction angles θ for different values of bending stiffness K_b.

Figure 4 shows the distributions of re-direction angles $P(\theta)$ for different values of K_b. All distributions are peaked on zero but their width increases as K_b declines. This means that the shape of the root may simply be characterized by the standard deviation S of the set $\{\theta_i\}$ of re-direction angles:

$$S = (\theta_i^2 - \langle\theta\rangle^2)^{1/2}, \quad (2)$$

to which we will refer as *shape parameter*.

4 Effect of root bending stiffness

Figure 5 shows the shape parameter S as a function of K_b for two different values of β. We distinguish two regimes separated by a characteristic value K_b^* of bending stiffness. For $K_b < K_b^*$, S decreases as a power law as K_b increases, i.e. the root becomes less and less noisy. For $K_b > K_b^*$, S continues to decrease even faster. This latter range corresponds actually to very small elastic deviations of the root as a whole whereas the range $K_b < K_b^*$ reflects the structure of the root. The power-law behavior allows to characterize the shape by an exponent:

$$S \propto K_b^{-\gamma}. \quad (3)$$

The exponent is $\gamma \simeq 0.16$ for $\beta = 0.86$ and $\gamma \simeq 0.21$ for $\beta = 1.54$. The higher value of γ for $\beta = 1.54$ is consistent with

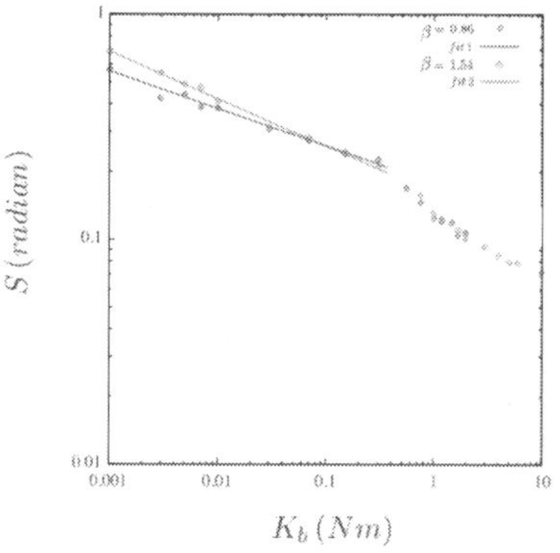

Figure 5. Shape parameter S as a function of root bending stiffness for two values of the parameter β. The lines are linear fits (corresponding to power laws).

the fact that in this case the root diameter is thicker than the gaps between grains, leading thus to a faster decrease of S.

5 Effect of packing fraction

Figure 6(a) displays S on linear scale as a function of K_b for three different values of the packing fraction Φ and for $\beta = 1.54$. S declines as K_b increases and its values are bigger for higher values of the packing fraction. But for these higher values, the resistance to penetration is also higher. The resistance to penetration is the force H or the penetration stress $H/(4\pi\langle d\rangle^2)$ required to push a rigid rod into the granular material. This force H multiplied by the average grain size $\langle d\rangle$ is homogeneous to a torque that can be compared to K_b. In other words, we expect that when the bending stiffness is normalized by $H\langle d\rangle$, then the data points collapse on a single curve. This is what we observe in Fig. 6(b). We also see that the transition point is $K_b^* \simeq H\langle d\rangle$, and we have $\gamma \simeq 0.13$. This scaling with K_b^* works also for other soil parameters such as particle size and cohesion. The variations of the exponent γ are generally small and, within the natural variability of soil configurations, γ varies between 0.12 and 0.2.

6 Conclusions

In this paper, we introduced a model for root growth inside granular materials. This model is characterized by a constant root growth and accounts for the mechanical interactions between the root cap and grains. The root bending stiffness appears as a key parameter for the root shape. We introduced a 'shape parameter' that accounts for the degree of noisiness of the root shape, and showed that it varies as a power law with bending stiffness. The data points for different values of the packing fraction can be collapsed by normalizing the bending stiffness by a characteristic torque defined from the resistance to penetration and mean grain size. For bending stiffness above this characteristic value, the root grows by dislodging grains so that

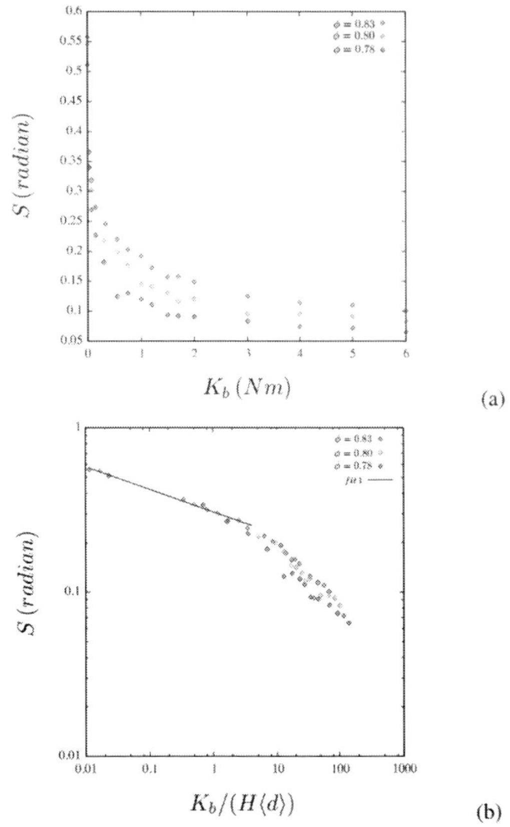

Figure 6. Shape index S as a function of root bending stiffness K_b for three different values of packing fraction Φ in linear and log-log scales.

the shape is nearly rectilinear and the shape parameter is negligibly small. Below the characteristic value, the root can not dislodge the grains but follows the pore space. We also investigated the forces exerted by soil grains on the growing root and their relation with root shape. Those results will be published elsewhere.

Acknowledgements

This work was carried out within the framework of the RoSoM project. This project is supported by Agropolis Fondation under the reference ID 1202-073 through the Investissements d'avenir program (LabexAgro: ANR-10-LABX- 001-01).

References

[1] B. Forde, Journal of Experimental Botany **60**, 3989 (2009)
[2] A. Bengough, C. Croser, J. Pritchard, Plant and Soil **189**, 155 (1997)
[3] A.G. Bengough, M.F. Bransby, J. Hans, S.J. McKenna, T.J. Roberts, T.A. Valentine, Journal of Experimental Botany **57**, 437 (2006)
[4] A.G. Bengough, B.M. McKenzie, P.D. Hallett, T.A. Valentine, Journal of Experimental Botany **62**, 59 (2011)
[5] I. De Smet, P.J. White, A.G. Bengough, L. Dupuy, B. Parizot, I. Casimiro, R. Heidstra, M. Laskowski, M. Lepetit, F. Hochholdinger et al., Plant Cell **24**, 15 (2012)
[6] C. Bécel, G. Vercambre, L. Pagès, Plant and Soil **353**, 169 (2012)
[7] D.M. Mueth, H.M. Jaeger, S.R. Nagel, Phys. Rev. E **57**, 3164 (1998)
[8] F. Radjai, D.E. Wolf, M. Jean, J. Moreau, Phys. Rev. Letter **80**, 61 (1998)
[9] E. Azéma, F. Radjaï, R. Peyroux, G. Saussine, Phys. Rev. E **76**, 011301 (2007)
[10] I. Agnolin, J.N. Roux, Phys. Rev. E **76**, 061302 (2007)
[11] F. Radjai, J.Y. Delenne, É. Azema, S. Roux, Granular Matter **14**, 259 (2012)
[12] A.S. L. Dupuy, T. Fourcaud, American Journal of Botany **94**, 1506 (2007)
[13] A.S. L. Dupuy, T. Fourcaud, European Journal of Soil Science **56**, 319 (2005)
[14] A.S. L. Dupuy, T. Fourcaud, Plant and Soil **278**, 119 (2005)
[15] A.S. T. Fourcaud, J. Ji, Annals of Botany **101**, 1267 (2008)
[16] T. Fourcaud, X. Zhang, A. Stokes, H. Lambers, C. Körner, Annals of Botany **101**, 1053 (2008)
[17] A. Dexter, J. Hewitt, Journal of Agricultural Engineering Research **23**, 17 (1978)
[18] G.M. Whiteley, A.R. Dexter, Plant and Soil **77**, 141 (1984)
[19] B. Atwell, Environmental and Experimental Botany **33**, 27 (1993)
[20] A. Bengough, B. McKenzie, Journal of Experimental Botany **48**, 885 (1997)
[21] E. Kolb, C. Hartmann, P. Genet, Plant and Soil **360**, 19 (2012)
[22] J.L. Silverberg, R.D. Noar, M.S. Packer, M.J. Harrison, C.L. Henley, I. Cohen, S.J. Gerbode, Proc Natl Acad Sci U S A **109**, 16794 (2012)
[23] N. Ahmadi, A. Audebert, M. Bennett, A. Bishopp, A. de Oliveira, B. Courtois, A. Diedhiou, A. Diévart, P. Gantet, A. Ghesquière et al., Rice **7**, 29 (2014)
[24] L. Verlet, Phys. Rev. **165**, 201 (1968)
[25] M. Allen, D. Tildesley, *Computer simulation of liquids*, Oxford science publications (Clarendon Press, 1987), ISBN 9780198553755
[26] C. Voivret, F. Radjaï, J.Y. Delenne, M.S. El Youssoufi, Phys. Rev. E **76**, 021301 (2007)

Strength of wet agglomerates of spherical particles: effects of friction and size distribution

Thanh-Trung Vo[1,2,⋆], *Patrick* Mutabaruka[3,⋆⋆], *Jean-Yves* Delenne[4,⋆⋆⋆], *Saeid* Nezamabadi[1,⋆⋆⋆⋆], and *Farhang* Radjai[1,3,†]

[1] *Laboratoire de Mécanique et Génie Civil (LMGC), Université de Montpellier, CNRS, Montpellier, France.*
[2] *Bridge and Road Department, Danang Architecture University, 566 Nui Thanh St, Hai Chau Dist, Da Nang city, Vietnam.*
[3] *<MSE>[2], UMI 3466 CNRS-MIT, CEE, MIT, 77 Massachusetts Avenue, Cambridge 02139, USA.*
[4] *IATE, UMR1208 INRA - CIRAD - Université de Montpellier - SupAgro, 34060 Montpellier, France.*

Abstract. We investigate the mechanical behavior of wet granular agglomerates composed of spherical particles by means of molecular dynamics simulations. The capillary cohesion force is modeled as an attraction force at the contact between two particles and expressed as an explicit function of the gap and volume of the liquid bridge. We are interested in the effect of the friction coefficient between primary particles. The agglomerates are subjected to diametrical compression tests. We find that the deformation is ductile involving particle rearrangements. However, a well-defined stress peak is observed and the peak stress is used as a measure of the compressive strength of the agglomerate. The strength increases with friction coefficient but levels off at friction coefficients above 0.4. Furthermore, the compressive strength is an increasing function of particle size span.

1 Introduction

Agglomeration of fine particles is used in many sectors including powder metallurgy [1], chemical engineering [2], iron-making industry [3] and pharmaceutical industry [4]. Fine granular materials are prepared and mixed in required proportions, compacted into a green granule or tablet and finally sintered to reach sufficient mechanical strength and toughness required to resist subsequent loading. The strength is generally measured by performing diametrical compression on the granules or tablets [5, 6]. It is crucially dependent on the agglomeration process but also on the nature of binding mechanism.

For a binding liquid, the important parameters are the amount, surface tension and viscosity of the liquid [7–9]. However, there are other important parameters that influence the strength of a green granule. The size distribution of primary particles controls the structure of the granule and the capillary force depends on particle size. Furthermore, although capillary attraction force is the main source of cohesion, the friction force between primary particles should affect the behavior of granules under loading. None of these parameters has been investigated in a systematic way in the past since particle size distribution and homogeneous blending of particles with the binding liquid are difficult to control in experiments. Numerically, broad size distributions require large numbers of particles in 3D.

In this paper, we present a numerical investigation of the effects of particle size distribution and friction on the strength of granules of spherical shape composed of primary spherical particles. For the simulations, we use the molecular dynamics (MD) method [10] and the capillary cohesion is modeled as an explicite function of the the distance between primary particles, liquid volume, surface tension and particle diameters [11–16]. We also assume that the liquid is homogeneously distributed between all eligible particle pairs. In the following, we first introduce numerical procedures. Then we present the a few results and conclude with a brief discussion.

2 Numerical model

The primary particles interact via normal repulsion force f_n^e, normal damping force f_n^d, Coulomb friction force f_t and capillary force f_n^c. The total normal force is given by

$$f_n = f_n^e + f_n^d + f_n^c. \qquad (1)$$

The normal repulsion force is modeled as a linear spring: $f_n^e = k_n \delta_n$, where δ_n is the normal elastic deflection approximated by the overlap between two particles. The normal damping force is a viscous contribution proportional to the relative normal displacement: $f_n^d = \gamma_n \dot{\delta}_n$, where γ_n is the damping coefficient.

The capillary force is due to the liquid bond and it depends on the gap δ_n, water volume V_b, surface tension γ_s,

⋆e-mail: thanh-trung.vo@umontpellier.fr
⋆⋆e-mail: pmutabar@mit.edu
⋆⋆⋆e-mail: jean-yves.delenne@inra.fr
⋆⋆⋆⋆e-mail: saeid.nezamabadi@umontpellier.fr
†e-mail: franck.radjai@umontpellier.fr

and the particle-liquid-gas contact angle θ. The capillary cohesion force can be determined by integrating Laplace-Young equations [17], but we use the following explicite expression [11]:

$$f_n^c = \begin{cases} -\kappa R, & \text{for } \delta_n < 0, \\ -\kappa R\, e^{-\delta_n/\lambda}, & \text{for } 0 \le \delta_n \le \delta_n^{max}, \\ 0, & \text{for } \delta_n > \delta_n^{max}, \end{cases} \quad (2)$$

where $R = \sqrt{R_i R_j}$ is the geometrical mean of particle radii, $\kappa = 2\pi\gamma_s \cos\theta$, and δ_n^{max} is the debonding distance, distance beyond which the bond breaks. It is related to liquid volume by [17]

$$\delta_n^{max} = \left(1 + \frac{\theta}{2}\right) V_b^{1/3}. \quad (3)$$

The length λ is is a function liquid volume V_b, the harmonic mean ($R' = 2R_iR_j/(R_i + R_j)$) and the size ratio $r = \max\{R_i/R_j; R_j/R_i\}$:

$$\lambda = \alpha\, h(r) \left(\frac{V_b}{R'}\right)^{1/2} \quad (4)$$

where α is a constant prefactor. This form fits the capillary force obtained from direct integration of the Laplace-Young equation by setting $h(r) = r^{-1/2}$ and $\alpha \simeq 0.9$ [18]

The total amount of liquid is evenly distributed among all pairs of particles that have a gap below δ_n^{max}. When a bond breaks, the liquid volume is redistributed among the contacts belonging to the same particles in proportion to particle sizes [11]. The largest value of the capillary force occurs when two particles are in contact ($\delta_n \le 0$) [18]. This is the case of most liquid bonds, and the capillary force at those contacts is independent of liquid volume.

For the tangential force f_t, we use a combination of elastic force and a frictional threshold [11, 19]:

$$f_t = -min\left\{(k_t\delta_t + \gamma_t\dot{\delta}_t), \mu(f_n - f_n^c)\right\} \quad (5)$$

where k_t is tangential stiffness, γ_t is the tangential damping parameter, δ_t is tangential displacement, μ is friction coefficient, $\dot{\delta}_t$ is tangential velocity.

To create the granules, we first prepare a large sample under isotropic compaction in a box. The distribution of particle diameters is either monodisperse or uniform in particle volume fractions between d_{min} and d_{max}. This distribution generally leads to high packing fraction. The size ratio $\alpha = d_{max}/d_{min}$ was 1 (monodisperse), 2 or 5. The largest particle size was kept to a constant value and the size of the smallest size was decreased. We extract spherical agglomerates from this sample comprising nearly 5000 particles. The cohesion forces are activated when the granule is subjected to diametrical compression between two platens. The friction coefficient was varied from 0 to 1. For each value of μ, several independent samples are simulated. Table 1 displays the values of all parameters used in the simulations.

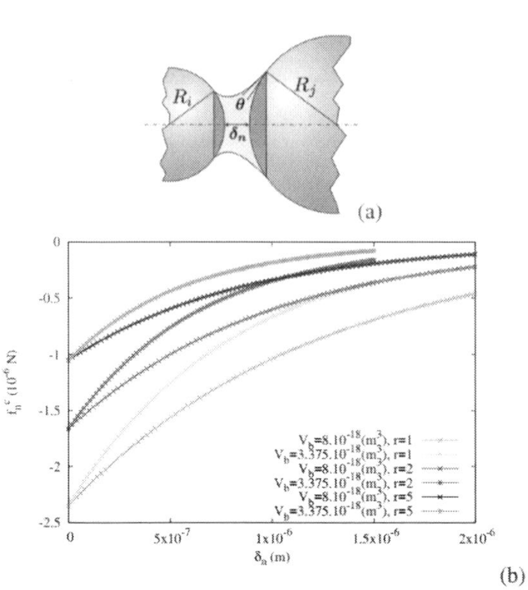

Figure 1. Geometry of the capillary bridge (a) and capillary force f_n^c for different values of parameters as a function of the gap δ_n, liquid volume V_b and size ratio (b).

Table 1. Parameters used in these simulations

Parameter	Symbol	Value
Maximum diameter	d_{max}	10.0 μm
Size ratios	α	[1; 2; 5]
No. of primary particles	N_p	5000
Friction coefficient	μ	[0,1]
Normal stiffness	k_n	1000 N/m
Tangential stiffness	k_t	800 N/m
Normal damping	γ_n	5.10^{-5} Ns/m
Tangential damping	γ_t	5.10^{-5} Ns/m
Capillary force prefactor	κ	0.45 N/m
Surface tension of water	γ_s	0.072 N/m
Density of primary particles	η	2600 kg.m^{-3}
Time step	dt	10^{-9} s

3 Compression of a granule

Figure 2 shows three snapshots of a granule during diametrical compression. During diametrical compression at slow constant downward velocity, we measure the vertical force F between the top mobile wall and the granule. By dividing this force by the sectional area of the granule, we get the mean vertical stress inside the granule, which can also be calculated by using the values of forces

$$\sigma_{zz} = \frac{4F}{\pi D^2} \approx \frac{1}{V} \sum_{k=1}^{N_c} f_z^k \ell_z^k \quad (6)$$

where D is granule diameter, V is the total volume of the granule, N_c is the number of contacts, f_z^k and ℓ_z^k are the z-components of the contact force vector and branch vector, respectively, at contact k. This stress can be normalized by the reference stress

$$\sigma_c = \frac{\gamma_s}{\langle R \rangle} \quad (7)$$

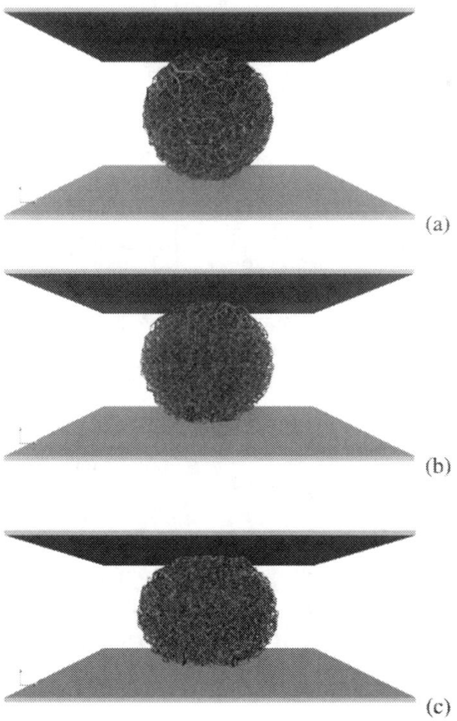

Figure 2. Three snapshots of a granule under diametrical compression. The lines represent force chains.

The vertical strain ε is calculated from the initial hight $h_0 = D$ of the granule and the total downward displacement Δh.

$$\varepsilon = ln\Big(1 + \frac{\Delta h}{h_0}\Big) \tag{8}$$

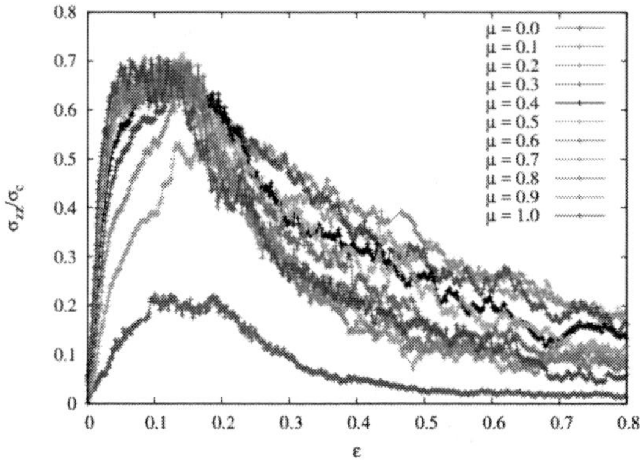

Figure 3. Vertical strength normalized by the reference stress σ_c for values of friction coefficient μ as a function of cumulative vertical strain.

Figure 3 shows the vertical stress as a function of ε for 11 different values of μ and $\alpha = 1$. We see that the stress increases to a peak value with strain all the more rapidly that the friction coefficient is higher. After the peak, the stress declines quite smoothly with cumulative strain towards zero. This ductile behavior is a consequence of particle rearrangements and dissipation due to friction and debonding. The debonding events between primary particles do not lead to spontaneous formation of a fracture surface. The particles are well-connected and the loss of one or two bonds of a particle does not lead to macroscopic rupture. Furthermore, the cohesion is reversible as the liquid is redistributed at each step homogeneously between particles within the debonding distance.

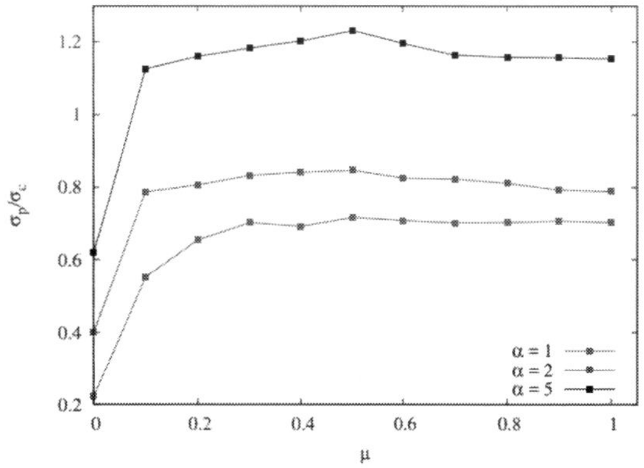

Figure 4. Normalized peak strength as a function of friction coefficient between primary particles for three different values of size ratio α.

The peak stress σ_p increases with friction coefficient. Fig. 4 shows σ_p/σ_c as a function of friction coefficient for three different values of size ratio. It is a rapidly increasing function of μ but levels off beyond $\mu = 0.4$. This means that the granule strength for most geomaterials, which generally have friction coefficients above 0.4, has practically no effect. But lubrication forces reducing friction coefficient may substantially reduce the granule strength. For example, during agglomeration in a rotating drum, the liquid lubrication forces activated by collisions between primary particles may affect the process [7, 8].

We also observe that the peak strength at high friction coefficient increases with size ratio for all values of friction coefficient. As shown in Fig. 5(a), it is nearly a linear function that may be a consequence of the uniform distribution by particle volume fractions. Its value is nearly doubled for a size ratio $\alpha = 5$ as compared to the monodisperse case. Both the packing fraction Φ and wet coordination number Z (including all capillary bonds) increase with size ratio α.

4 Conclusions

In this paper, we used molecular dynamics simulations to investigate the effects of friction coefficient and size polydispersity on the mechanical strength of agglomerates (granules) composed of spherical particles interacting

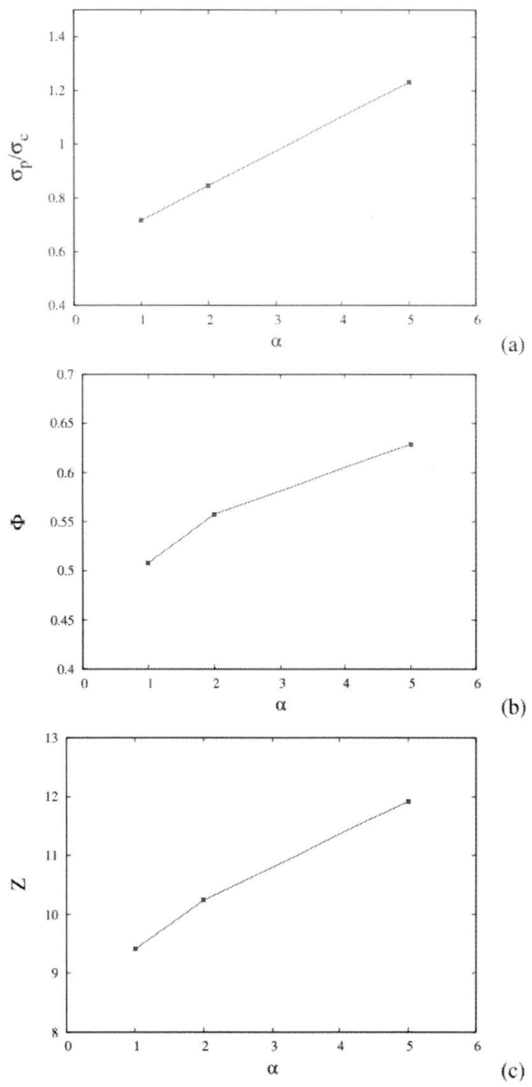

Figure 5. Normalized compressive strength (a), packing fraction Φ (b) and coordination number Z (c) of the granules as a function of size ratio α.

via capillary cohesion and Coulomb friction. The stress-strain behavior under diametrical compression is ductile and shows no abrupt rupture of the granule. The peak strength is of the order of the liquid-gas surface tension divided by the mean particle size, but increases considerably with both friction coefficient and size ratio. For a uniform size distribution in particle volume fractions, this increase of the mechanical strength as a function of size ratio can be attributed to the larger connectivity of the particles, which increases nearly in the same proportion. A more detailed analysis of the microstructure should later allow us to better understand the effect of different particle size classes inside the granule.

References

[1] A. Nosrati, J. Addai-Mensah, D.J. Robinson, Hydrometallurgy **125–126**, 90 (2012)

[2] S.H. Chien, G. Carmona, L.I. Prochnow, E.R. Austin, J Environ Qual **32**, 1911 (2003)

[3] R. Aguado, S. Roudier, L. Delagado, eds., *Best available techniques (BAT) reference document for iron and steel production.*, Joint Research Centre of the European Commission (Luxembourg: Publications Office of the European Union, 2013)

[4] C.Y. Wu, O. Ruddy, A. Bentham, B. Hancock, S. Best, J. Elliott, Powder Technology **152**, 107 (2005)

[5] J. Fell, J. Newton, Journal of Pharmaceutical Sciences **59**, 688 (1970)

[6] D. Bika, M. Gentzler, J. Michaels, Powder Technology **117**, 98 (2001)

[7] S. Iveson, J. Litster, B. Ennis, Powder Technology **88**, 15 (1996)

[8] S. Iveson, J. Litster, AIChE journal **44**, 1510 (1998)

[9] A.M. Bouwman, M.J. Henstra, D. Westerman, J.T. Chung, Z. Zhang, A. Ingram, J.P.K. Seville, H.W. Frijlink, Int J Pharm **290**, 129 (2005)

[10] F. Radjai, F. Dubois, *Discrete-element modeling of granular materials* (Wiley-Iste, 2011)

[11] V. Richefeu, F. Radjai, M.S.E. Youssoufi, Eur. Phys. J. E **21**, 359 (2007)

[12] F. Radjai, V. Topin, V. Richefeu, C. Voivret, J.Y. Delenne, E. Azéma, M.S. El Youssoufi, in *Mathematical Modeling and Physical Instances of granular Flows*, edited by J.D. Goddard, J.T. Jenkins, P. Giovine (AIP, 2010), pp. 240–260

[13] V. Richefeu, M.S. El Youssoufi, E. Azéma, F. Radjai, Powder Technology **190**, 258 (2009)

[14] V. Richefeu, M.S. El Youssoufi, R. Peyroux, F. Radjai, International Journal for Numerical and Analytical Methods in Geomechanics **32**, 1365 (2007)

[15] F. Radjai, V. Richefeu, Phil. Trans. R. Soc. A **367**, 5123 (2009)

[16] V. Richefeu, M.S. d El Youssoufi, F. Radjai, Phys. Rev. E **73**, 051304 (2006)

[17] G. Lian, C. Thornton, M. Adams, Journal of Colloid and Interface Science **161**, 138 (1993)

[18] F. Radjaï, V. Richefeu, Philosophical Transactions of the Royal Society A **367**, 5123 (2009)

[19] S. Luding, *Collisions and Contacts between two particles*, in *Physics of dry granular media - NATO ASI Series E350*, edited by H.J. Herrmann, J.P. Hovi, S. Luding (Kluwer Academic Publishers, Dordrecht, 1998), p. 285

DEM simulation of the granular Maxwell's Demon under zero gravity

Wenguang Wang[1,2], *Zhigang* Zhou[1,2], *Jin Zong*[1], *and Meiying* Hou[1,2,*]

[1]Key Laboratory of Soft Matter Physics, Beijing National Laboratory for Condense Matter Physics, Institute of Physics, Chinese Academy of Sciences, 100190 Beijing, China
[2]School of Physical Sciences, University of Chinese Academic of Sciences, 100049 Beijing, China

Abstract. In this work, granular segregation in a two-compartment cell (Maxwell's Demon) under zero gravity is studied numerically by DEM simulation for comparison with the experimental observation in satellite SJ-10. The effect of three parameters: the total number of particles N, the excitation strength Γ, and the position of the window coupling the two compartments, on the segregation ε and the waiting time τ are investigated. In the simulation, non-zero segregation under zero gravity is obtained, and the segregation ε is found independent of the excitation strength Γ. The waiting time τ, however, depends strongly on Γ. For higher acceleration Γ, $|\varepsilon_i|$ reaches steady state value ε faster.

1 Introduction

Granular materials' extremely rich dynamical behaviours have attracted attentions of physicists of different fields in recent years[1-3]. Examples are the heap formation of a granular bed[4-6] and the size segregation of a granular system with grains of various sizes under vertical vibration known as the Brazil Nut effects[7,8]. In these phenomena, energy is being injected continuously into the system by the oscillating boundaries and propagated into the bulk by the inelastic collisions of the grains. A steady state of the whole system is reached when the dissipation of the system is balanced by the input of the energy. Since both the energy input and dissipation depends crucially on the configurations of the system, many intriguing steady states, such as granular segregation in a two-compartment cell known as Maxwell's Demon phenomenon[9,10] and even oscillatory states can be observed. The segregation phenomenon relies on the existence of a non-monotonic flux, which determines the number of particles per unit time flows between the two compartments[11]. The flux function is derived from the equation of gas state and therefore depends on the gravity. If applicable, the phenomenon can be used to transport granular materials in space[12,13]. In this work, we therefore perform simulation study to find the condition for possible segregation in two-compartment granular gas system in an environment of zero gravity. We find that (1) density gradient exists in the cell along excitation direction that particles gather near the end away from the vibration wall; (2) segregation among the two compartments exists even for very small excitation acceleration Γ as long as number of particles N exceeds a critical number and the waiting time is long enough; (3) different from the situation in gravity, segregation ε is independent of Γ when in zero gravity; (4) the waiting time τ is shorter when the acceleration Γ is higher, but reaches a minimum when Γ is large enough.

2 Model

The simulation is based on discrete element method, in which each particle is treated as a discrete element[14]. The motion of each particle obeys the Newton's Second Law. The interaction between them is considered if and only if two particles collide. Taking the rotation into consideration, the particles' equations of motion are as follows:

$$m_i\ddot{\vec{r}}_i = \sum_{j,j\neq i} \vec{F}_{ij}^C + \vec{F}_i^O ,$$

$$I_i\dot{\vec{\omega}}_i = \sum_{j,j\neq i} \vec{R}_i \times \vec{F}_{ij}^C + \vec{M}_i^O ,$$

where m_i , I_i , $\vec{r}_i$ and $\vec{\omega}_i$ are the mass, the moment of inertia, the displacement and the angular velocity of particle i , respectively. $\vec{F}_{ij}^C$ is the contact force that the particle i is applied by the particle j . $\vec{R}_i$ is the radius vector of particle i . $\vec{F}_i^O$ and $\vec{M}_i^O$ are other forces and torques applied to the particle i , such as the gravity force. Generally, $\vec{F}_{ij}^C$ can be decomposed into two parts: a normal component and a tangential one,

$$\vec{F}_{ij}^C = \vec{F}_{ij}^n + \vec{F}_{ij}^t .$$

The normal and the tangential contact forces between two particles are handled separately.

* Corresponding author: mayhou@iphy.ac.cn

Two particles interact with each other only during contact. The overlap in the normal direction between particles i and j with radii R_i and R_j, respectively, is

$$\delta = (R_i + R_j) - (\vec{r}_i - \vec{r}_j)\cdot\vec{n},$$

Where δ is a positive number when the two particles interact. $\vec{n}$ is the unit vector of the relative displacement $\vec{n} = (\vec{r}_i - \vec{r}_j)/|\vec{r}_i - \vec{r}_j|$. Considering the dissipative effect, the equation of the normal contact force is

$$F_{ij}^n = k_n\delta + \gamma_n v_n ,$$

where k_n and γ_n are the spring stiffness and the dissipative coefficient, respectively, and v_n is the normal component of the relative velocity $\vec{v}_{ij} = \vec{v}_i - \vec{v}_j$. This is called Linear Spring Dashpot (LSD) model.

The tangential contact force is determined by the tangential component of the relative velocity and the normal contact force

$$F_{ij}^t = \min\left(\gamma_t v_{ij}^t, \mu F_{ij}^n\right),$$

Where γ_t is a numerical constant, μ is the friction coefficient and v_{ij}^t is the tangential component of $\vec{v}_{ij}$. Fig. 1 shows the mechanism of the contact model in this work.

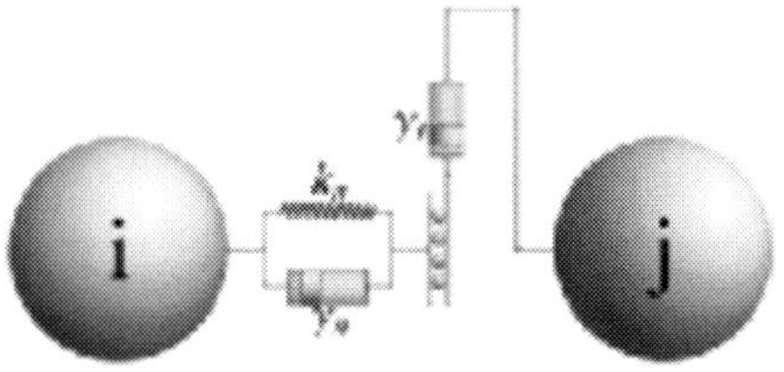

Fig. 1. The mechanism of the contact model.

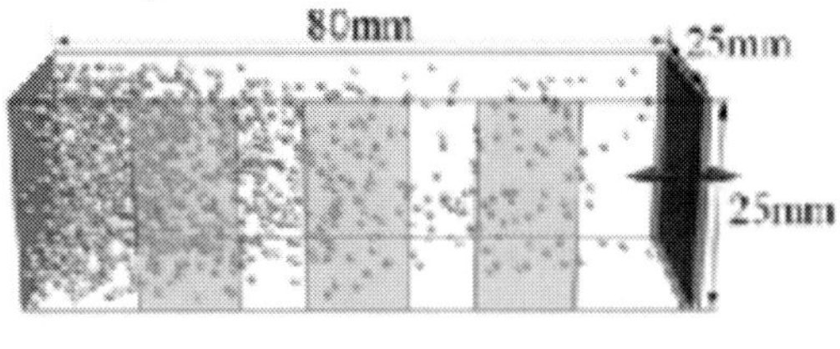

Fig. 2. Schematic of the virtual-window single cell. Particles are excited by the right-end moving wall. The virtual window at three different locations are indicated by rectangular shades at the side wall.

The numerical parameters of the granular cell system are taken same as that in the experiments. In the simulation, the particle radius R is 0.5mm. The density ρ of particles is 4500kg/m³. The coefficient of restitution e for grain-grain collision is taken as 0.8. The coefficient of frication μ is 0.3. The total number of particles in the connected two-compartment cell is taken as N. Let the number in cell one (say upper one) be N_1, the number in the other (lower) cell is then $N_2 = N - N_1$. To start with, we put particles evenly in each compartment, $N_1 = N_2 = N/2$. The initial velocity is arbitrarily set to be 0.1m/s with random direction.

In order to find a flux function for the segregation condition in microgravity, i.e., to know the number of particles flow instantaneously through the window per unit time, we count the number of particle-wall collisions at a virtual window of a single cell[9] shown in Fig. 3. The size of the single-cell is 80mm×25mm×25mm. Three rectangular shades along the side wall are shown in Fig. 3 to indicate the virtual windows at different positions. We record how many times particles collide with a given virtual window at a time interval for a sufficiently long time. The flux profile can then be found by the counts of collisions per unit time across a given virtual window.

3 Results

3.1 Flux function obtained in virtual-window single-cell simulation

Finding flux function is helpful in modelling the occurrence of the segregation. The segregation is governed by the following equation:

$$\frac{dN_i(t)}{dt} = -F(N_i(t)) + F(N - N_i(t)), \quad i = 1,2 ,$$

where the $N_1(t)$ and $N_2(t)$ are the counts of particles in the two compartments at time t, and $F(N_i)$ is the number of particles per unit time at time t that flow from a compartment containing N_i particles to another compartment. Once knowing the function $F(N_i)$, $N_i(t)$ can be obtained from Eq. (10).

In order to characterize the segregation of particles in the two cells, a dimensionless parameter ε_i is introduced:

$$\varepsilon_i = (N_i - N/2)/N, \quad i = 1,2 .$$

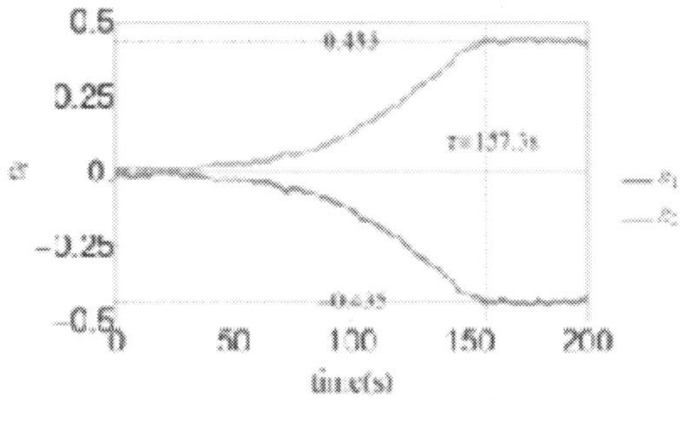

Fig. 3. One example of simulation result of $\varepsilon_i(t)$

Fig. 3 shows a plot of simulation result how ε_i changes with time. It takes some time for the population $N_i(t)$ to reach a steady state. We call this time as the waiting time τ . At any time $\varepsilon_1(t) = -\varepsilon_2(t)$, and the absolute value of ε_i at the steady state is indicated as ε . $\varepsilon = 0$ means particles are equally populated in the two cells, and $|\varepsilon| = 0.5$ means all the particles gather in one compartment, i.e., fully segregated.

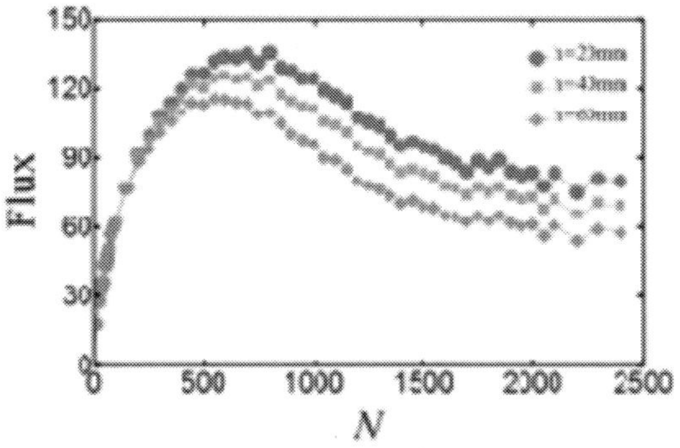

Fig. 4. Flux as a function of the number N .

The flux function obtained by the virtual-window single-cell simulation method[9] is shown in Fig. 4 for three locations of the window. A non-monotonic function $F(N_i)$ guaranteed the occurrence of the segregation. A weakly position dependence of the flux function shown in Fig. 4 tells us that flux value is greater when the window position is moving away from the oscillating wall.

3.2 The effect of total number N on the segregation ε

Initially particles are distributed equally in the two cells. Segregation appears after about 100 seconds, and is fully developed 200 seconds afterwards.

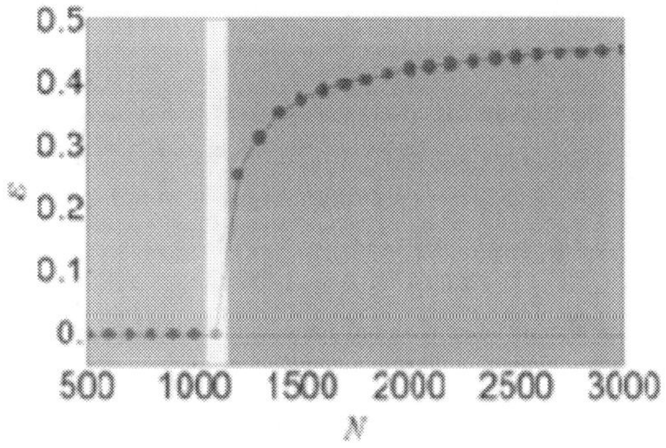

Fig. 5. Variation of the asymmetry parameter ε in the steady state as a function of N .

With the designed setup parameters, Fig. 5 gives us the simulation results how $\varepsilon_i(t)$ changes with the total number N from 800 to 1500. When the total number N is less than or equal to 1100, the asymmetry ε fluctuate around 0. Segregation does not occur. Only when N exceeds a critical value (here the critical value is about 1100) the segregation occurs. The asymmetry parameter ε grows from 0 and approaches 0.5 with increasing N .

3.3 Effect of excitation acceleration on the segregation

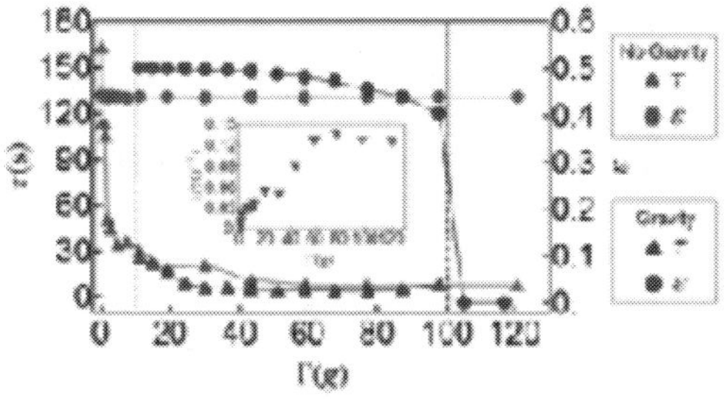

Fig. 6. Effects of the vibration intensity on the asymmetry parameter ε and the waiting time τ with and without gravity.

Fig. 6 gives the simulation results of the asymmetry parameter $|\varepsilon_i|$ changing with time under different Γ. For higher acceleration Γ, $|\varepsilon_i|$ reaches steady state value ε faster.

Since in microgravity experiments the operational time is normally limited, the waiting time τ is important to know in advance, which tells us how long to wait for seeing the effect. Shown in Fig. 6, the rate $1/\tau$ seems linearly proportional to Γ at low Γ. Only when Γ is high enough the rate $1/\tau$ reaches a maximum value. For the Γ value available in our microgravity experiment (normally less than 1g), Fig. 6 tells us the time duration τ shall be of the order of 100 seconds. This result helps us in determining the experimental design.

In Fig. 6, it shows that the waiting time τ does not go to zero, there is a minimum time required for particles to go to the steady segregation state no matter how strong the vibration intensity goes. In our case the value is of the order of 10 seconds.

The above numerical results are quite different with the experiments in gravity. With gravity particles act like gas molecules, and distribute in both cells. No segregation of the granular gas is observed at high Γ when particle random collisions dominate. When Γ is below a certain threshold, no segregation is observed either as the gravitational force dominates. With no gravity, the segregation ε becomes a constant and does not depend on Γ, as seen in Fig. 6, but the waiting time τ depends strongly on Γ with or without gravity.

4 Conclusion

Our simulation shows granular segregation is achievable at zero gravity, although the condition can be different from the case under gravity. With gravity, the segregation depends on Γ and can be divided into three regimes: one at low Γ when the gravitational force is dominant that no segregation appears and ε is zero; the second regime when Γ is at an intermediate value that segregation appears ε becomes non-zero; the third regime when Γ is high enough that random particle collisions dominate and ε becomes zero again. Under zero gravity, segregation ε does not depend on Γ. It is a constant and is determined by the geometry of the cell. A minimum excitation time is necessary to observe the phenomenon with or without gravity. This time τ, however, depends strongly on the value Γ. With the affordable acceleration Γ in satellite condition, the excitation time is on the order of a few minutes, which means no segregation can be seen in drop tower or parabolic experiments[1,15]. Both offer short time, 3 seconds to 22 seconds, microgravity condition.

Acknowledgments

Project supported by the "Strategic Priority Research Program-SJ-10" of the Chinese Academy of Sciences (Grant No.XDA04020200), and the National Natural Science Foundation of China (Grant Nos. 11474326, 11274354).

References

1. D. van der Meer, K. van der Weele and D. Lohse, Phys. Rev. E **63** 061304 (2001)
2. B. Meerson, T. Pöschel, P. V. Sasorov and T. Schwager, Phys. Rev. E **69** 021302 (2004)
3. G. Costantini, D. Paolotti, C. Cattubo and U.M.B. Marconi , Physica A **347** 411-428 (2005)
4. R. Lambiotte, J.M. Salazar and L. Brenig Phys. Lett. A **343** 224-230 (2005)
5. Liu R, Li Y and Hou M, Phys. Rev. E **75** 061304(2007)
6. Liu R, Li Y and Hou M, Phys. Rev. E **79** 052301(2009)
7. Hou M, Li Y, Liu R, Zhang Y and Lu K, Phys. Status Solidi A **207** 2739-2479 (2010)
8. Li Y, Hou M and P. Evesque, J. Phys. Conf. Ser. **327** 012034(2011)
9. Li Y, Liu R, M. Shinde and Hou M, Granular Matter **14** 137-143(2012)
10. Li Y, Liu R and Hou M, Phys. Rev. Lett. **109** 198001(2012)
11. J. Eggers, Phys. Rev. Lett. **83** 5322-5325(1999)
12. E. Opsomer, M. Noirhomme, N. Vandewalle and F. Ludewig, Phys. Rev. E **88** 012202(2013)
13. M. Noirhomme, E. Opsomer, N. Vandewalle and F. Ludewig, Eur. Phys. J. E **38**(2015)
14. S. Luding, Granular Matter **10** 235-246(2008)
15. Wang H, Chen Q, Wang W G and Hou M Y, Acta. Phys. Sin. **65** 014502(2016)

The Potential of Mean Force concept for bridging (length and time) scales in the modeling of complex porous materials

Katerina Ioannidou[1,2,3], *Benoit* Carrier[4], *Matthieu* Vandamme[4], and *Roland* Pellenq[1,2,3,*]

[1] *Department of Civil and Environmental Engineering, Massachusetts Institute of Technology (MIT), Cambridge, MA, USA*
[2] *Centre Interdisciplinaire de Nanosciences de Marseille, Aix-Marseille Université, CNRS, Campus de Luminy, Marseille, France*
[3] *<MSE>[2] The joint CNRS / MIT / Aix-Marseille Université laboratory, MIT, Cambridge, MA 02139, USA*
[4] *Laboratoire Navier, Université Paris-Est, CNRS, ENPC, IFSTTAR, Marne-la-Vallée, France*

Abstract. We introduce the concept of Potential of Mean Force, PMF, as a way to implement upscaling modeling from the nano-scale to micron-scale. A PMF is a free energy function representing in an effective way the interactions between objects (cement hydrates, clay platelets, etc.) at thermodynamics conditions. The PMF is therefore the key piece of information allowing to coarse-grained Physical-Chemistry information in a meso-scale model formulation. The use of PMF offers a huge computational advantage as it allows a straight up-scaling to the meso-scale while keeping essential interactions information that are the hallmark of Physical-Chemistry processes. Such a coarse-grained modeling integrates atomistic response into inter-particle potentials that fully propagate molecular scale information all the way to the meso-scale.

1 Introduction

In this paper, we describe the concept of Potential of Mean Force as the tool for modeling soft and hard amorphous porous materials across length and time scales. This Bottom-Up approach integrates different scales of knowledge into accurate models of relevant phenomena and complex systems across multiple length and time scales. We outline the originality of the method, its basic ingredient, and give two examples of applications, namely cement and clays minerals. We believe that this approach provides a unique opportunity for developing a full predictive modeling and simulations for amorphous complex porous materials from molecular to engineering scales. Besides a few examples such a zeolites, the vast majority of porous materials are multi-scale disordered complex systems. They are usually classified according to their characteristic pore size Ø: microporous solids have Ø lower than 2 nm, mesoporous 2 nm < Ø < 20 nm and macroporous Ø > 20 nm. Such materials are ubiquitous on Earth, be it in the Earth Crust (clays, porous carbons, zeolites, serpentines rocks, asphalt, wood etc.) or man-made (concrete, carbons, gas storing porous materials, nuclear fuels, etc.). While macroscopic porous media theory has a long tradition in all fields of engineering, ranging from Civil, Environmental, Chemical, Biochemical, many current and future challenges in the field of energy and environment heavily rely on breakthroughs in our understanding of the physics and mechanics of the intimate interplay between the two ingredients involved in the make up of such porous materials: their complex solid texture that results often from an initial soup chemistry and the role of liquids confined to their pore voids; all of which can be gathered under the term Poromechanics when it come to understanding the link between Physical-Chemistry and Mechanics.

Cement and clays minerals are salient examples of multi-scale porous materials. While clays is obvious the important natural materials in Soil Science and Agriculture, cement is widely produced more than any other synthetic material on Earth. The current worldwide cement production stands at three billion tons, enough to produce more than twenty billion tons or one cubic meter of concrete per capita per year. There is no other material that can replace cement in the foreseeable future to meet our societies' needs for housing, shelter and infrastructure by the unique property of cement transforming from liquid to stone at room temperature. Nevertheless cement faces an uncertain future, due to a non-negligible ecological footprint that amounts to 5-10% of the worldwide CO_2 production. On the other hand, thanks to breakthroughs in science and engineering, cement and concrete have a novel potential to contribute to a sustainable development encompassing economic growth, social progress while minimizing on the ecological footprint. Manufacturers are willing to produce high-performance concrete that is not only environmentally more efficient (consuming less energy and giving off less CO_2) but qualitatively transformed. The recent disasters of the Japanese nuclear plant of Fukushima (with the explosion of the reactor's concrete dome) and the oil spill in the Gulf of Mexico (due to an uncontrolled cement liner setting during well completion) highlight the lack of materials technology and the need of renewing engineering practices from science-based principles. A similar yet less dramatic lack of knowledge expands into many fields of environmental and energy related materials research, from the environmental compatibility of shale-gas or nuclear wastes storage (for which a thorough understanding of the interaction between cement and clay remains to be accomplished. Refreshingly, concrete starts to achieve its integrity through research and craftsmanship — much like any other material. This requires a holistic approach in which progress comes from the nano-scale and is featured into innovative structural engineering

* Corresponding author: pellenq@mit.edu

applications whose impact on sustainable development can be evaluated with advanced environmental-econometric impact studies.

Up to now, innovation in the field of cementitious and clay materials was a classical top-down approach where a problem at one scale was solved at the scale just below. This was somewhat successful to address mechanical properties but never addressed environmental issues up front. Today the advancement of fundamental knowledge leans toward developing a predictive Bottom-Up approach from the electronic and atomic level to micro-texture and macroscopic engineering properties. In the particular case of the cement paste, the first outcome of this approach was a realistic and consistent atomistic model of cement hydrates (see Fig. 1a), deciphering the "DNA of cement". Since then, progresses have been made on many fronts: in particular the realistic micron-scale model (see Fig. 1b) opens unprecedented possibilities and constitutes the starting point for modeling fracture, creep and aging of the cement paste in the context of durability that is the core of the present paper. A new multi-scale approach is needed to integrate different scales of knowledge into accurate models of relevant phenomena and complex systems across multiple length and time scales, ultimately allowing for the development of concept materials with improved or new properties. We propose a shift of paradigm based on a bottom-up approach that can be applied to and tested for most important technological, economical and environmental materials: cement, soils and sediments, ceramics, solid nuclear fuels, geo-materials, geo-polymers, rocks, organic minerals, organic-inorganic composites, biological tissue; all being important classes of multi-scale materials.

Figure 1: a) At the root of the bottom-up approach the first atomistic model of cement [1]. b) Up-scaling to the first meso-scale model of cement hydrates [2]. The color code of the meso-scale model indicates locally dense and loose regions of matter (the density increases from gray, blue, orange to red).

The bottom-up approach that we propose is amenable to bridge time and length scales based on the handshake between numerical simulations and experiments. Ultimately, this approach should be transformational, shifting existing technology from empirical practice to science-based knowledge in the context of sustainability, durability, energy and waste management with the emergence of a new interdisciplinary field of research at the frontier between Sciences and Engineering. In the proposed approach, novel simulation methods from the large toolbox of computational soft matter and glass physics can be combined to enable science-based engineering of new materials.

2 The Potential of Mean Force Concept at the core of the Bottom-Up Approach

2.1 Atomistic calculations

Our starting point is the atomic structure (of cement hydrates or clays, ceramics, etc.) from which, using state-of-the art inter-atomic potentials, one can derive the so-called "Potential of Mean Force" (PMF) between different phases using the perturbation approach from Statistical Physics for free energy calculations. A PMF is a free energy function representing in an effective way the interactions between objects (cement hydrates, clay platelets, etc.) at a given temperature. The PMF is therefore the key piece of information allowing to coarse-grained Physical-Chemistry information in a meso-scale model formulation. The use of PMF offers a huge computational advantage as it allows a straight up-scaling to the meso-scale while keeping essential interactions information that are the hallmark of Physical-Chemistry processes. Such a coarse-grained modeling integrates atomistic response into inter-particle potentials that fully describe the atomic interactions and enables to reach much larger time and length scales. We thus propose to use the PMF coarse-grained colloidal modeling approach to translate the understanding at atomistic scales to the scale of micro-texture for amorphous porous materials.

In the recent years, we have elaborated further the idea and developed new meso-scale colloidal model of cement hydrates (in the hardened cement paste) and clay minerals. In both cases, the complex pore network and the structural heterogeneities naturally emerge from the short-range cohesive interactions typical of nano-scale grains of cement hydrates and nano-platelets of clay minerals and from non-equilibrium conditions. The key point of this approach is that if the coarse-graining unit is well represented, the access to the physics depends only on the appropriate level of modeling the interactions. The simulation techniques at the colloidal scale are the same as the ones used at the atomistic scale but applied to grains of matter containing thousands of atoms or molecules. These are Molecular Dynamics (MD) and Monte Carlo (MC) in different statistical ensembles. New scheme combining MD and Grand Canonical Monte Carlo (GCMC) have been applied with success to simulate the early and late stages of cement hydration, imitating the precipitation of cement particles [2-5] and the sedimentation of laponite clay mineral [6]. The bottom-up approach therefore provides a rigorous link between the physical chemistry of materials and their overall functional and mechanical properties including elasticity, strength, fracture and permeability, etc...

It opens the route to the engineering scale by providing the relevant scale for analytical coarse-graining and homogenization of the properties. At the micron-scale, the texture has either the features of a granular material, with well-defined particles of different shapes and sizes and cohesive-frictional contact interactions embodied in a solid or liquid phase, or the features of a porous aggregate. The mechanical properties at this scale involve thus granular-type disorder and multi-phase coupling, which can be more efficiently addressed by using discrete-element modeling (DEM) or semi-continuum approaches such as Lattice Element Methods (LEM) or Material Point Models (MPM) possibly coupled with a liquid phase or a binding solid phase; see for example Ref. 6. The method starts with the atomistic calculations of the change in interaction free energy as a function of the separation distance between two nano-grains at the atomistic scale using a thermodynamics integration scheme based on a perturbation theory appraoch. Note that the integration constant is usually set to zero. Then the Gay-Berne potential type for oblates particles, including platelet-shaped object, can be fitted to the PMF at the meso-scale. Note that the Gay-Berne potential reduces into the well-known Lennard-Jones form for spherical grains). The GB potential is defined as:

$$U = 4\,\varepsilon \left\{ \left(\frac{\sigma}{h_{12}+\sigma}\right)^{12} - \left(\frac{\sigma}{h_{12}+\sigma}\right)^{6} \right\} \eta_{12}\chi_{12} \tag{1}$$

where $\varepsilon = 1$ determines the energy scale, σ is the atomic interaction radius and h_{12} is a function that approximates the anisotropic interparticle distance. η_{12} and χ_{12} characterize anisotropic interactions of particles due to their shapes and relative orientations. In the case of clay platelets, the PMF for face-to-face interactions between layers following the procedure of Ebrahimi et al. [7], is given in Figure 2. It represents the change in system free energy at 300K as a function of the distance between layer and is given as an energy per surface area hence allowing the scaling as a function of the particle size. This function shows 3 minima that indeed correspond to the known stability distances for sodium Montmorillonite. The PMF is then taken as an effective interaction between particle of given sizes and allows implementing an efficient mesoscale modeling scheme as the atomic details (water content, ions density etc...) are gathered in the PMF function. It worth noting that extracting a PMF from atomistic scale is computationally expensive and that it si remains a state function that is only valid for the thermodynamic conditions (P, T, ion concentration, water content...) that were used to establish it. This is clearly a drawback of the approach and other less expensive schemes have to be explored (see below). The fit of the PMF for sodium Montmorillonite extracted from atomistic scale requires going beyond the Gay-Berne potential that can only describe a single well PMF. In the case of cement hydrate (CSH), extracting the PMF is more difficult as CSH is not a well defined compound with a weill defined chemistry but rather a non stoechimetric coumpound with a range of possible compositions

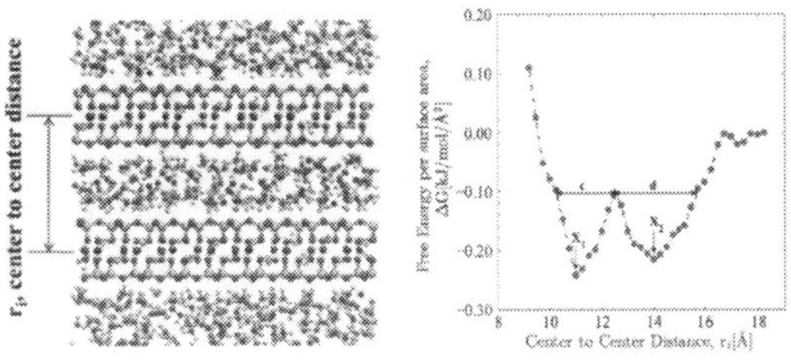

Figure 2: Face-to-face PMF for Sodium-Montmorillonite clay at 300K [7]

measured in terms of the Ca/Si ratio spanning the range from 1 to 2. CSH is hovever a (defective) layered material for Ca/Si ratio retainting the layered structure of a parent mineral called Tobermorite. In Figure 3, we give the PMF extracting from atomistic scales for the Ca/Si=1.65 (that is the most common/probable value of CSH in a real cement paste [8]. PMF was recently obtained for tobermorite [9] (Ca/Si=1) predicting essentially a single well potential behaviour. Again, as for clay Montmorillonite, CSH-PMF cannot be stricly fitted to GB potential. A better description can be obtained using a 12-24 Lennard Jones together with a Yukawa term to account for the repulsive shoulder that is observes at larger distances [3-5].

Validation of the PMF bottom-up approach came through the development of advanced analytical tools to compute quantities that are measured in experiments at the scale well below the micron, such as scattering intensity, pore size distribution, surface area, local volume fractions, indentation modulus and hardness. All these quantities that were "measured" in the simulations and compared successfully with experiments provided the first consistent characterization of the complex and elusive meso-scale structure of cement [2]. The experimental tools applied to measure such quantities in cement were electron-microscopy imaging, nano-indentation tests, small angle X-rays and neutron scattering (SAXS and SANS), nuclear magnetic resonance (NMR) spectroscopy and adsorption desorption experiment of N_2 and water. A remarkable agreement with experiment was achieved for both cement and clays [2,7].

2.2 Improving Primitive model predictions

The cohesion and stability of such charged layers separated by a solvent containing counterions has long been discussed in the framework of the Derjaguin-Landau-Verwey-Overbeek (DLVO) theory. The DLVO theory is the combination of the repulsive electrical double layer interaction with the attractive van der Waals interactions [10]. The repulsive electrostatic interaction is obtained by solving analytically the Poisson-Boltzmann equation of a system made of positive point

DOI: 10.1051/epjconf/201714001009

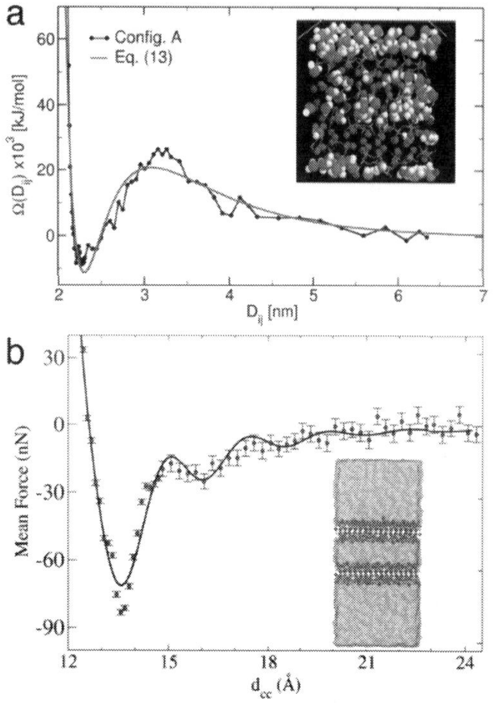

Figure 3: a) Face-to-face PMF for CSH for Ca/Si=1.65 at 300K Black line, atomistic simulation (Ref), green line the fit using a Yukawa LJ-(12-24) form [8]. b) Face-to-face PMF for CSH for Tobermorite, Ca/Si=1 at 300K [9]

charges between negatively charged surfaces. Making use of the Poisson-Boltzmann equation, the electrostatic pressure P between two charged surfaces in function of the surface separation D is given by:

$$P(D) = k_B T\left(\rho_0(D) - \rho_0(\infty)\right) \qquad (2)$$

where k_B is the Boltzmann constant, T the temperature and ρ_0 is the cation concentration at the mid-plane for a surface separation x. This pressure depends only on the increase on the ionic concentration at the center of the pore and therefore its origin is entropic. However, it is now known that a system of charged plates and counterions can generate strong attractive electrostatic interactions, the so called ionic correlation forces [10].

To study such systems numerically, the so-called Primitive Model wad introdiced in the 80's. In the Primitive Model, the charged lamellae and the hydrated cations are represented by uniformly charged infinite planar surfaces and by charged hard spheres, respectively, while the solvent is represented by a dielectric continuum. Monte Carlo simulations in the framework of the Primitive Model showed that fluctuations of the ionic concentration in the interlayer space result in an instantaneous polarization and give rise to an attractive ionic fluctuation interaction, the so-called ionic correlations forces . Thus, the behavior of the system is governed by a competition between repulsive entropic forces and attractive energetic forces. For divalent or trivalent cations, the attractive energetic contribution can exceed the repulsive entropic component of the pressure and results in one or several attractive wells of the pressure between the lamellae. Figure 5 shows the pressure between two montmorillonite layers in function of the distance between the layers, as predicted by the Primitive Model. Two cases are considered: a montmorillonite with sodium counterions and a montmorillonite with calcium counterions. As predicted by the Poisson-Boltzmann equation, in the case of sodium counterions, the pressure is always repulsive reflecting the fact that it is a mean-field solution ignoring thermally activated ion density

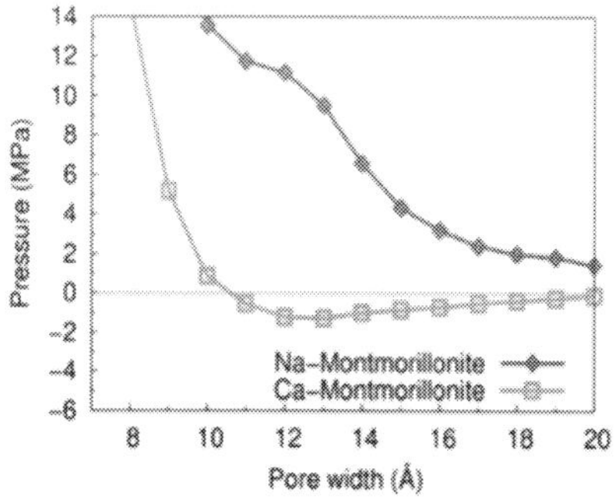

Figure 4: Interlayer pressure computed with the Primitive Model versus pore width for montmorillonite with sodium (red diamond symbols) or calcium (green square symbols) interlayer cations [11].

fluctuations. In contrast, in the case of calcium counterions, attractive pressures are observed for pore sizes larger than 10 Å and explains the experimental observation that exfoliating completely calcium montmorillonite layers is difficult. Note that the corresponding PMF is simply the integral form of the pressure curve along the inter particle direction. The Primitive Model has obviously its own limitations. The surfaces of the slit pore are assumed to be flat and uniformly charged, the ions are supposed to be fully hydrated and mobile and the solvent is considered to be a dielectric continuum the dielectric constant of which is the one of bulk water. These hypotheses make the Primitive Model questionable, in particular at low surface separation. In this case, the molecular structures of the solvent and the surfaces have to be taken into account. Therefore, numerous works adopted a full-atomistic description of the system to study for instance the intralamellar swelling of smectite clays [7]. However, as mentioned above, the computational cost of these atomistic simulations is very significant preampting coarse-grained simulations. Accordingly, it is necessary to determine more precisely the complexity which needs to be introduced into the models of lamellar systems and the key physical ingredients responsible for the cohesion of such materials. Moreover, the link between the Primitive Model and the full-atomistic model has never been made clear until the PhD work of Carrier in 2013 [11]..

It is crucial to determine what level of complexity is required in the modeling of the system in order to get meaningful results, depending on the charge of the surfaces, on the valency of the cations, and on the separation between the charged surfaces. In particular, we want to answer the following questions: is the Primitive Model sufficient to study the effect of confined electrolytes on the cohesion of the system, especially when the pore width is small? At which critical value of the so-called electrostatic coupling do the fluctuations of the ionic concentrations have to be considered? The electrostatic coupling is defined as the ratio between the interaction energy between a counterion and a charged surface and the thermal energy, and is therefore proportional to the product of the counterion valency and the surface charge. To address these questions, we present computer simulations of confined electrolytes between charged clay layers. In addition to the DLVO theory, we used various models of increasing complexity: the Primitive Model (PM), the Explicit Solvent Primitive Model (ESPM) and the full-atomistic model. The Explicit Solvent Primitive Model is a model we propose to take into account explicitly the solvent while keeping the simplicity of the Primitive Model. As in the Primitive Model, in the ESPM model the charged lamellae are represented by infinite uniformly charged planes. The counterions and the solvent particles are represented by point charges and point dipoles, respectively, embedded in soft spheres; solvent molecules being decored with a dipole moment whose norm is adjusted to reproduce bulk water dielectric constant at room temperature. A dielectric constant equal to 78.5 was obtained with a dipole moment equal to 1.7 D. This value was lower than the experimental dipole moment of water molecules in a liquid phase (i.e. about 2.6 D). Indeed, this simplistic model is not detailed enough to capture the dielectric response of water nor the real hydrogen bond governed structure of liquid water. Figure 6 shows the interaction pressure between layers of montmorillonite versus pore width either with Na interlayer cations or with Ca interlayer cations (calculated from the virial definition in the the direction perpendicular to the layers). It compares the results from the full-atomistic appraoch with the prediction of the ESPM model. The quatitative agreement between the full-atomistic model and the ESPM model suggests that the ESPM model is able to capture the relevant physical mechanisms governing the cohesion of charged layers in a solvent, for a much lower computational cost than full-atomistic models (i.e. the same accuracy could be obtained about one order of magnitude faster in terms of computational cost). The results of the ESPM model contrast with the ones of the Primitive Model. Indeed the repulsive pressure predicted by the Primitive Model at any pore size in the case of Na-montmorillonite is invalidated by the ESPM model. Therefore, an explicit solvent proves to be critical to explain the cohesion of charged lamellae. The cohesion of Na-montmorillonite can be explained neither with the mean-field DLVO theory nor with the Primitive Model. According to the Primitive Model, the entropic repulsion is always greater

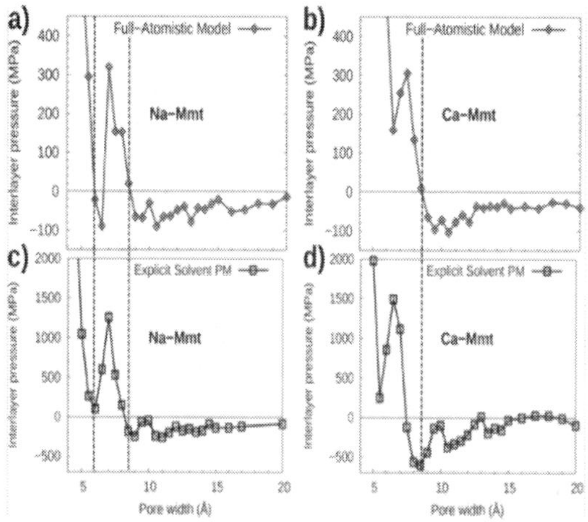

Figure 5: Interlayer pressure versus pore width for (a and b) Na-Montmorillonite and for (c and d) Ca-Montmorillonite computed with (a and c) the full-atomistic model or with (b and d) the Explicit Solvent Primitive Model. [11].

than the electrostatic attraction due to ionic fluctuations. However, the assumption that the dielectric constant of the confined fluid is equal to the dielectric constant of bulk water leads to a significant underestimation of the energetic attractive contribution. The ESPM model, as well as the full-atomistic model, showed that with sodium interlayer cations, stable states and attractive pressure wells exist no matter the surface charge, while the results of the Primitive Model showed that for the same systems the interaction pressure was always repulsive. Therefore, the solvent plays an important role in the cohesive behavior of the system. The effect of the solvent cannot be captured by a simple dielectric continuum as in the Primitive Model and the solvent has to be taken into account explicitly. Moreover, the results of the ESPM model suggest that ionic correlation forces play a significant role even at low electrostatic couplings.

3 Conclusion

We introduced the concept of Potential of Mean Force (PMF) as a tool to implement up-scaling simulations from the nano-scale to the micron-scale reaching out to so-called engineering scales in the cases of cement and clay minerals. There are different levels of calculations that can be performed to extract the PMF. The most computationally expensive and most accurate is certainly to use an atomistic description together with a thermodynamic integration route (perturbation theory approach).

The interaction between charged objects immersed in an electrolyte such as cement hydrates or clay particles has long been studied using colloidal theories such as DLVO, ionic correlation forces, etc... These are easier ways to extract PMFs between particles using a direct integration scheme at a much smaller computational expense. However, these colloidal approaches can be seen as over-simplistic as they might not give even a qualitative account of experiments in terms of the stability and cohesion of these charged objects mainly because of a poor description of the confined solvent. Here we showed that taking into account the solvent explicitly is necessary to describe the cohesion of charged lamellar systems. Indeed, the solvent exhibits a complex dielectric response that impacts strongly the electrostatic interactions in the system. In order to keep the simplicity of the Primitive Model while modeling water explicitly, we introduce the Explicit Solvent Primitive Model, in which each water molecule is modeled by a point dipole. The ESPM model gives results that are in good agreement with the results of the full-atomistic model, but at a lower computational cost (i.e. approximately one order of magnitude faster). We recommend to use the ESPM model to compute interactions between charged grains and extract PMFs at the colloidal scale.

This approach further gives the foundation of meso-scale investigations of fracture and creep of complex systems such as cement and clays. Long time scale is crucial for investigating creep, the slow drifting of the nano-scale cement or clay units when under stress. Large length scales are equally critical in order to understand long-range space correlations and to obtain microscopic features when investigating fracture.

This work was supported by the ICoME2 Labex (ANR-11-LABX-0053) and the Aix-Marseille University AMIDEX foundation.

References

1. Pellenq et al., Proc. Natl. Acad. Sci. USA **106**, 16102–1610 (2009)
2. K. Ioannidou, K. J. Krakowiak, M. Bauchy, C. G. Hoover, E. Masoero, S. Yip, F.-J. Ulm, P. Levitz, R. J.-M. Pellenq, and E. Del Gado, Proceedings of National Academy of Science USA, **113** (8), 2029-2034 (2016)
3. K. Ioannidou, R. J.-M. Pellenq and E. Del Gado, Soft Matter, **10**, 1121-1133 (2014)
4. K. Ioannidou, M. Kanduc, L. Li, J. Dobnikar, D. Frenkel and E. Del Gado, The crucial effect of early-stage gelation on the mechanical properties of cement hydrates, Nature Communications, 7, 12106 (2016)
5. E. Del Gado, K. Ioannidou, E. Masoero, A. Baronnet, R. J.-M. Pellenq, F.-J. Ulm, and S. Yip, Eur. Phys. J. – Spec. Top. **223**, 2285-2295 (2014)
6. F. Radjai and F. Dubois, (Wiley-Iste, Berlin), ISBN 978-1-84821-260-2 (2011)
7. D. Ebrahimi, A. Whittle, J. and R. J.-M. Pellenq, J. Chem. Phys., **140**, 154309 (2014)
8. P. A. Bonnaud, C. Labbez R. Miura et al, Nanoscale **8**, 4160-72 (2016)
9. S. Masoumi, H. Valipour, M. J. Abdolhosseini Qomi, J. Phys. Chem. C, **121**, 5565–5572 (2017)
10. J. Israelachvili, Intermolecular and Surface Forces (Third Ed.) ISBN: 978-0-12-375182-9 (2011)
11. B. Carrier, PhD thesis, Ecole Nationale de Ponts et Chaussées, Paris (in English)

Micromechanics of complex granular materials: a focus on small strain behavior

V. Magnanimo*

[1] *MultiScale Mechanics, ET, MESA+, University of Twente, Enschede, Netherlands*

Abstract. We study the bulk properties of isotropic and anisotropic granular assemblies using discrete element simulations and experiments. The focus is on the influence of the microstructure on the elastic properties of the aggregate. By studying three selected examples, namely assemblies of monodisperse glass beads, bidisperse mixtures in size and soft-stiff bimodal mixtures, we show that the effective moduli of a dense granular assembly can be manipulated by properly combining material characteristics and system properties.

1 Introduction

Granular materials behave differently from usual solids or fluids and show peculiar mechanical properties like dilatancy, history dependence or ratcheting due to their intrinsically discrete nature. To understand the behavior of granular materials it is first necessary to characterize their state by introducing an appropriate set of (independent) internal variables. In early studies, macroscopic variables measurable in laboratory experiments, i.e., stress and porosity, were thought to be sufficient to characterize the elastic stiffness of a granular assembly, see [4] and references therein. However, with further investigations it has been observed that other factors are also important [1, 8, 10, 14, 16]. Among those are the geometric arrangement of the grains, i.e., the geometric fabric, and the interparticle forces, probability distributions and correlations and in particular the concept of weak and strong force network. There are different mechanisms that lead to a change of the structure. When a granular assembly is deformed, neighboring particles may form new contacts and, at the same time, some existing contacts may disappear, i.e., a geometric fabric develops [15]. On the other hand, collective deformations of many grains can lead to re-arrangements on a larger scale, (non-affine), which also brings change to structure. While the former local contact creation or destruction is closely related to the compressive and tensile directions of strain, the latter phenomenon is at the origin of the interesting behavior of granular media.

In many industrial applications as well as in natural phenomena, granular assemblies are constitute not only by one type of particles but by of combination of them. Processes involving milling, segregation, agglomeration, filtration and sieving are common and often lead to the generation of granular systems with large size ratios. Dealing with mixtures systems is exceptionally challenging and often requires heuristic assumptions to be made, as prediction/control of the behavior is still an unsolved issue. A common and interesting sub-class are bimodal granular mixtures. When two types of particles are mixed (homogeneously), the granular material displays a peculiar macroscopic behavior due to the non-trivial arrangement of the two species. The behavior is non-linear and often non-monotonic, in some cases showing dramatic sudden changes. For example, it has been well established that deformation of powder compacts can be severely retarded by the presence of heterogeneities such as inclusions or reinforcements, where the role of the volume fraction and shape of reinforcements on the densification behavior needs to be further highlighted [9]. A well known example is the binary composite consisting of a conducting phase and an insulating phase used for electromagnetic shielding and anti-static devices. These composites show a sharp transition from low to high conductivity at a critical concentration of the conducting phase [2]. It is well known in the geomechanical community that the presence of fines strongly influences the mechanical behavior of granular soils. Scientific work on the topic is extensive and several studies have de- scribed the dependence of elastic bulk stiffness and stress-strain behavior of granular mixtures on the volume of fines (see [20] and references therein).

Despite the massive amount of work on granular mixtures, a deep understanding from a micromechanical perspective is still missing. which experimentally is difficult to achieve for very small particles. This is because the role of the different components can not be discerned by classical experiments. Here we combine discrete element simulations and experiments to highlight the influence of the microstructure on the effective elastic stiffness of monondisperse and bidisperse granular assemblies under isotropic and anisotropic conditions. We present a set of examples, namely monodisperse granular assemblies

*e-mail: v.magnanimo@utwente.nl

DOI: 10.1051/epjconf/201714001010

under difference stress paths, bidisperse mixtures with different size and bimodal mixtures made of soft-stiff particles We show that the bulk properties of the material can be controlled by indirectly tuning the microstructure, where increasing complexity offers more degree of freedom.

2 Single specie samples

In order to investigate the relation between elasticity and microstructure, we use the distinct element method [3] and prepare random assemblies of identical, frictional, elastic spheres. We deal with aggregates of 10,000 particles with radius $R = 0.1$ mm randomly generated in a periodic cubic cell. We employ material properties typical for glass spheres: a shear modulus $G_g = 29$GPa and Poisson's ratio, $\nu = 0.2$. The interaction between particles is a non-central contact force in which the normal component follows the non-linear Hertz's law. For the tangential component we incorporate a bilinear relationship with elastic displacement followed by Coulomb sliding with friction coefficient μ (details are given in [12]).

2.1 Isotropic samples

After random generation, particles are isotropically compressed in the absence of gravity from an initial gas to the desired solid volume fraction. Because we are interested in dense aggregates of frictional particles, we require that the solid volume fraction ϕ is close to the random close packing value $\phi_{RCP} \simeq 0.64$. It is well known that different packing structures are obtained experimentally according to the preparation protocol. Our approach is to generate numerical packings with different final structure. To do this we employ a protocol where ϕ and p are set in two different stages of the deformation (details are given in [14]). The result is that, for a given pressure ($p = 100$kPa) and identical solid volume fraction ($\phi \simeq 0.638$), we obtain isotropic packings with different coordination number, that is the average number of contacts per particle Z_0^* [19]. Here we consider three aggregates with initial coordination number $Z_0^* = 6.17, 5.36, 4.88$ respectively.

2.2 Axialsymmetric compression

The three samples described above are subjected to axisymmetric deformation at constant mean stress $p = p_0 = 100$ kPa [19]. During each loading step, the target pressure is maintained with a servo mechanism which continuously adjusts the applied strain rate according to the difference between the desired stress state and the stress measured at that time step [14]. In order to reproduce a quasi-static, triaxial loading we compress the sample very slowly ($\gamma_{\partial t}/\Delta_0$ is lower 5×10^{-3}, where $\gamma_{\partial t}$ is the accumulated strain in the time step ∂t). After each increment we let the system to relax until a new equilibrium state is reached.

2.3 Elastic moduli

For each one of the isotropic samples in Sec.2.1 as well as tt different steps along the axisymmetic loading paths in Sec.2.2, we estimate the elastic stiffness of the granular assemblies. We focus on two moduli, the bulk modulus K and the axial shear modulus G_{hv}. They are calculated by applying an incremental strain and measuring the corresponding incremental stress response [13, 14]. We apply a shear strain, $\Delta\epsilon_{hv}$ and measure the stress response σ_{hv} to obtain G_{hv}. We also apply an isotropic strain, $\Delta v = \Delta\epsilon_{11} + \Delta\epsilon_{22} + \Delta\epsilon_{33}$, to measure the bulk modulus K, where ϵ_{ij} are components of the strain tensor. The friction coefficient is set on a very high value to prevent sliding between grains because we are interested in the elastic resistance of the aggregate. For each calculation, we verify that the applied strain is small enough to be in the linear response regime.

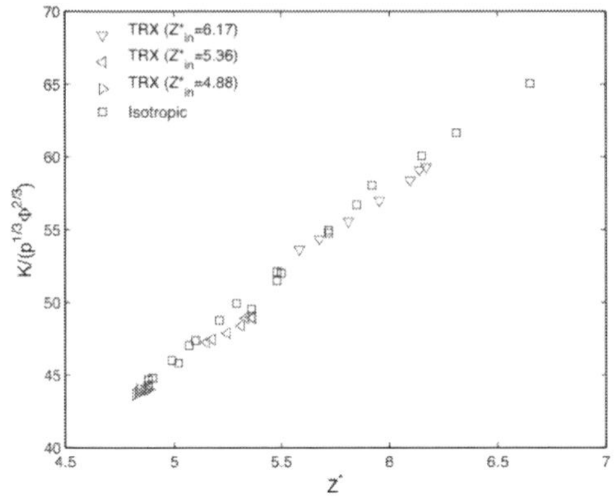

Figure 1. Evolution of the normalized bulk modulus versus coordination number, for axisymmetric compression (TRX) of the three samples with different initial microstructure Z_0^* and for isotropic samples.

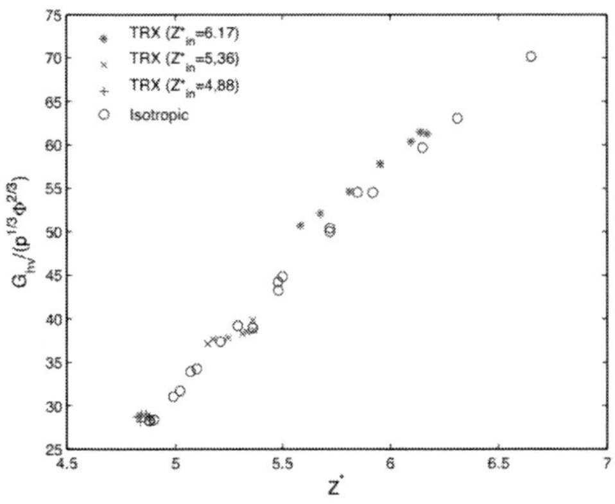

Figure 2. Evolution of the normalized shear modulus versus coordination number, for axisymmetric compression (TRX) of the three samples with different initial microstructure Z_0^* and for isotropic samples.

In Fig.1 and Fig.2 the bulk and shear moduli are collected for all numerical packing and plotted against the coordination number. The moduli are normalized with a power of pressure and volume fraction, $p^{1/3}$ and $\phi^{2/3}$, following the prediction of effective medium theory [21]. In this ratio, the dependence on pressure and volume fraction cancel out and all that remains is the relation with the actual value of coordination number Z^*. Surprisingly, the plots of $B(Z^*)$ and $G_{hv}(Z^*)$ result in universal curves, irrespective of induced anisotropy or (deviatoric) stress state, as predicted by classical theories [17]. That is, once the actual state has been achieved, the measurement of the bulk and shear stiffness can be associated with a unique coordination number, which acts as a state variable. This is able to characterize the microstrusture of any granular sample, along with pressure and void ratio, no additional specification of the stress states needed. It is noteworthy that the major role is taken by the scalar coordination number for both isotropic and anisotropic cases, while the orientation of contacts (fabric tensor) most likely plays second-order effects.

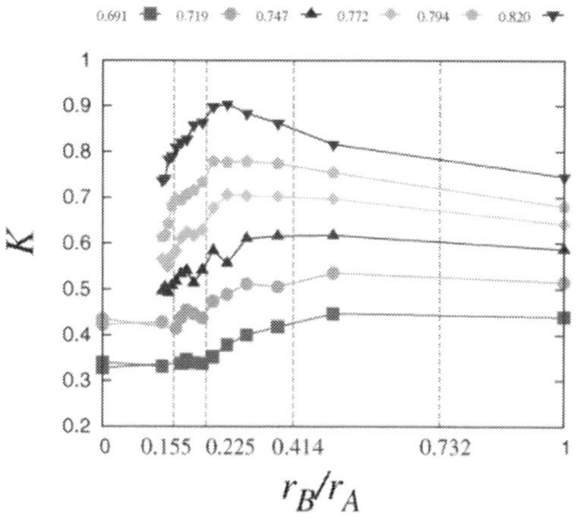

Figure 3. Evolution of the bulk modulus normalized by $2r_A/k$ versus size ratio for bidisperse mixtures at different values of volume fraction (from [7]).

3 Toward complexity

3.1 Bidisperse mixtures of particles with different sizes

As further step toward complexity, we look at how the elastic properties properties of a monodisperse, isotropic granular assembly vary in the case of a bidisperse mixture, with the main focus on the bulk modulus. In order to keep the discussion as simple as possible, we limit ourself to the case where only 5% of the volume is substituted with particles of different size (and same characteristics). We analyze the properties of the granular mixture on two phase spaces: (i) by varying the size ratio of fines to coarse and (ii) by spanning a wide range of volume fraction, and find the optimum size ratio that maximizes the bulk modulus at each volume fraction.

We use again a numeric approach via DEM to discuss the mixture behavior. The reference sample consists of N_A^0 monodisperse particles A of radius r_A. Starting from this base sample, many mixtures are created by substituting a given number of of $N_A^0 - N_A^T$, with particles of species B of different radius $r_B \leqslant r_A$, such that the same volume $(N_A^0 - N_A^T)(4\pi/3)r^3A = N_B^T(4\pi/3)r^3B = V_B^T$ of material $A$is replaced by B. Further details are given in [7]. For the sake of simplicity a linear contact model for frictionless particles is used in this part of the study. However details of the implementation do not affect the results.

After preparing isotropic samples at different values of volume fraction, $\phi = 0.69 - 0.82$ per each size ratio r_B/r_A, the bulk modulus is calculated in a similar fashion as in Sec.2.3. Fig.3 shows the evolution of the bulk modulus scaled by $2r_A/k$, with k contact stiffness, plotted against size ratio for different volume fractions. As expected, K increases systematically with density. For loose states ($phi = 0.69, 0.72$), K mostly increased with increasing r_B/r_A. The behavior associated with denser states is much more interesting, as we observe an increase in K to a maximum, followed by a decrease for larger r_B/r_A. Note that the value of r_B/r_A where K reaches its maximum varies with increasing densities and the maximum becomes stronger. From Fig.3, we extract very important insights: (i) The bulk modulus of a granular assembly can be manipulated by only substituting 5/105% of the base material with fines. (ii) We can control the direction of the change (enhanced or lowered bulk modulus) and the magnitude of change through the density and the size of the small particles. (iii) For dense material, for a given density, there is an ideal size of fines that leads to the maximum in the bulk modulus.

Interestingly, we see in [7] that the trend observed for K in Fig.3 is mainly related to the behavior of AB interactions between small and big particles, while the actual value depends on the contributions of AA main network. We associate the different trends observed for loose and dense systems with the ability of the fines (material B) to fill the voids formed by particles A. With increasing density, the void size gets smaller and compatible with the size of particles B, and thus contribute to the active contact network. For dense assemblies, K attains maxima near the size ratio corresponding to tetrahedron configurations, meaning that the dense state is more likely (possibly due to ongoing crystallization) to create such a configuration and this is the efficient mean arrangement. For the looser systems, the maximum of K moves towards higher size ratios r_B/r_A, corresponding to cubic-like configurations.

Hence the bulk stiffness of a granular material is strongly affected by the addition of a different component (even in a very small amount) that leads to modifications in the microstructure of the main component. The interest arises from geophysical hazards, like earthquakes, where the material volume remains constant, but the size of few particles can decrease quickly due to breakage. The change in the size distribution, even limited to a very small volume, has been shown to play an important role in soil stability. Important applications exist for example in pharmaceutical and food processes such as chromatographic separation or extraction, where the flow through a deformable granular material depends strongly on the effective bulk modulus and can be enhanced by by adding inclusions to the main particle consistent. In these industrial processes the focus is optimizing the performance of a given (granular) material with minimum modification, minimum costs. Other obvious applications of this study cover building mixtures like concrete or asphalt, and shock absorbing materials. The presence of small particles in a granular mass, which is often associated with weakness, can be an asset for functionality.

3.2 Bimodal mixtures of soft-stiff particles

Finally we want to highlight the influence of particle materials on the elastic behavior of a granular mixtures. To address this topic, we have performed wave-propagation experiments on isotropic mixtures particles of two different species, namely glass and rubber beads, but identical size. Details of the experiments are reported in [18]. Results are consistent with evidences by previous studies [6, 11] and show how granular mixtures can be engineered to exhibit exceptional properties.

When we look at the evolution of the elastic modulus with rubber content and pressure, we observe that, similarly to bidisperse mixtures in Sec.3.1, the bulk behavior can be tuned by a proper combination of material parameters and state variables, that eventually lead to modifications of the glass network microstructure.

The elastic stiffness decreases non-linearly with rubber content v. The modulus is fairly constant for $v \lesssim 30\%$, decreases sharply between $30\% \lesssim v \lesssim 70\%$ and is again independent of rubber content for $v \gtrsim 70\%$. In the case of high pressure, adding a small amount of soft particles surprisingly enhances the effective stiffness of the medium and the highest modulus is observed at $v \simeq 20\%$, indicating that a mixture of 20% rubber content may become stiffer than a pure glass aggregate. That is proper mixtures of soft and stiff particles exhibit enhanced stiffness along with low porosity, high compressibility and high attenuation. Eventually, the combination of these properties is highly desirable in engineering practice for many constructions applications, including highway embankments or backfill for retaining walls [6].

4 Conclusion

We have used DEM simulations and experiments to analyze the bulk properties of granular assemblies, in the cases of i) monodisperse particles under isotropic/anisotropic stress paths, ii) isotropic mixtures bidisperse in size and iii) soft-stiff isotropic mixtures. We have focused on the effective elastic moduli, measured by applying small perturbations to the system. We have shown that the microstructure controls the bulk stiffness and can be manipulated to enhance the material properties as useful to contrast natural hazards or for industrial applications. In particular, the scalar coordination number is the main parameter that characterize bulk and shear moduli for isotropic and anisotropic aggregates of monodisperse particles. In the case of bidisperse mixtures, while K decreases monotonically for loose systems, for dense systems, K first increases, reaches an optimum, and later decreases, the peak value depending on the density of the mixture, that is the optimal geometrical arrangement of the particles. Finally glass-rubber mixtures show an enhanced bulk stiffness in combination with low density and high damping for small amount of rubber. An interesting extension of this study would consider a *hierarchical granular medium* in which the particle size distribution, type and eventually shape [5] are engineered to obtain desired mechanical properties, e.g. the bulk or shear modulus.

References

[1] I. Agnolin and J. N. Roux, Phys. Rev. E **76**, 061304 (2007).
[2] C.-H. Kuo and P. K. Gupta Acta metall. mater. **43**(1), 397403 (1995).
[3] P. A. Cundall and O. D. L. Strack, Géotechnique **29**, 47 (1979).
[4] J. D. Goddard, Proc. R. Soc. Lond., Ser. A **430**, 105 (1990).
[5] M.Z. Miskin and H.M. Jaeger, Nature Materials **12**, 326 (2013).
[6] H.-K. Kim and J. C. Santamarina, Can. Geotech. J. **45**(10), 1457 (2008).
[7] N.Kumar, V. Magnanimo, M. ramaioli and S. Luding, Powder Tech. **293**, 94 (2016).
[8] R. Kuwano and R. J. Jardine, Géotechnique, **52**, 727 (2002).
[9] D. Bouvard and F. F. Lange Acta metall. mater. **39**(12), 3083 (1991).
[10] L. La Ragione and V. Magnanimo, Phys. Rev. E **85**, 031304 (2012).
[11] C. Lee, Q. H. Truong, W. Lee and J.-S. Lee, J. Mat. Civ. Eng. **22**(4), 323 (2010).
[12] H. A. Makse, N. Gland, D. L. Johnson, and L. Schwartz, Phys. Rev. Lett. **83**, 5070 (1999).
[13] H. A. Makse, N. Gland D. L. Johnson, and L. M. Schwartz, Phys. Rev. E. **70**, 061302 (2004).
[14] V. Magnanimo, L. La Ragione, J. T. Jenkins, P. Wang, and H. A. Makse, Europhys. Lett. **81**(3), 34006 (2008).
[15] M. Oda, Soils Found. **12**, 17 (1972).
[16] F. Radjai, D.E. Wolf, M. Jean, J. Moreau, Phys. Rev. Lett. **80** 61 (1998).
[17] S. Roesler, J. Geotech. Eng. Div. **105**, 871 (1979).
[18] K. Taghizadeh, H. Steeb, V. Magnanimo, and S. Luding, In this volume (2017).
[19] C. Thornton, and S. J. Antony, Roy. Soc. of London Phil. Tr. A **356**, 2763 (1998).
[20] L.E.Vallejo, Can. Geotech. J. **38**, 1097 (2001).
[21] K. Walton, J. Mech. Phys. Solids, **35**, 213 (1987).

AUTHOR INDEX

AUTHOR INDEX

AUTHOR INDEX

AUTHOR INDEX

AUTHOR INDEX

AUTHOR INDEX

AUTHOR INDEX

AUTHOR INDEX

AUTHOR INDEX

AUTHOR INDEX